AA004524

8th World Conference on Photovoltaic Energy Conversion (WCPEC-8)

Milan, Italy
26 – 30 September 2022

Volume 1 of 2

Editors:

A. Scognamiglio
R. Kenny
J. Bergmiller
Y. Ueda
S. Hayase
J. De Gregorio
S. Hubbard
A. Smets

ISBN: 978-1-7138-7180-4

Printed from e-media with permission by:

Curran Associates, Inc.
57 Morehouse Lane
Red Hook, NY 12571

WIP – Renewable Energies
Sylvensteinstr. 2
81369 Munchen
Germany

Phone: +49 89 72012735
Fax: +49 89 72012791

wip@wip-munich.de

Additional copies of this publication are available from:

Curran Associates, Inc.
57 Morehouse Lane
Red Hook, NY 12571 USA
Phone: 845-758-0400
Fax: 845-758-2633
Email: curran@proceedings.com
Web: www.proceedings.com

8th World Conference on Photovoltaic Energy Conversion

Proceedings of the International Conference

26 September – 30 September 2022

Edited by:

A. SCOGNAMIGLIO
ENEA - Italian National Agency for New Technologies, Energy and Sustainable Economy
Italy

Y. UEDA
Tokyo University of Science, Japan

S. HUBBARD
RIT- Rochester Institute of Technology
USA

R. KENNY
European Commission Joint Research Centre
Italy

S. HAYASE
Kyushu Institute of Technology
Japan

A. SMETS
TU Delft- Delft University of Technology
The Netherlands

J. BERGMILLER
WIP Renewable Energies
Germany

J. DE GREGORIO
WIP Renewable Energies
Germany

Coordination of the Technical Programme:

European Commission Joint Research Centre
Via E. Fermi 1
21020 Ispra (VA)
Italy

Institutional Support:

European Commission
WCRE – World Council for Renewable Energy

Supporting Organisations:

APVI – Australian PV Institute
ENEA – Italian National Agency for New Technologies, Energy and Sustainable Economic Development
EREF – European Renewable Energies Federation
ESA – European Space Agency
ESMC – European Solar Manufacturing Council
EUFORES –The European Forum for Renewable Energy Sources
EUREC – The Association of European Renewable Energy Research Centres
ETIP PV – European Technology & Innovation Platform PV
Guidehouse Insights
GÜNDER – The Turkish Solar Energy Society
IEA PVPS – International Energy Agency, Photovoltaic Power Systems Programme
INES – National Solar Energy Institute
NUS /SERIS – National University of Singapore / Solar Energy Research Institute of Singapore
ODTÜ-GÜNAM – Center for Solar Energy Research and Applications
REN21 – Renewable Energy Policy Network for the 21st Century

Institutional PV Industry Cooperation:

SolarPower Europe
SOLARUNITED

Supporting Associations:

ABENS – Associação Brasileira de Energia Solar
AREA – African Renewable Energy Alliance
CABA – Continental Automated Buildings Association

WCPEC-8 realised by:

WIP Renewable Energies
Sylvensteinstr. 2, 81369 München, Germany
Tel: +49 89 720 12 735, Fax: +49 89 720 12 791
Email: pv.conference@wip-munich.de
www.wcpec-8.com
www.wip-munich.de

Proceedings produced and published by:

WIP
Sylvensteinstr. 2, 81369 München, Germany
Tel: +49 89 720 12 735, Fax: +49 89 720 12 791
Email: pv.conference@wip-munich.de
www.wcpec-8.com
www.wip-munich.de

ISSN 2196-0992

8th WORLD CONFERENCE ON PHOTOVOLTAIC ENERGY CONVERSION
26 SEPTEMBER – 30 SEPTEMBER 2022

WCPEC-8 COMMITTEES

CONFERENCE ORGANISING COMMITTEE

Conference General Chairs

A. Scognamiglio, ENEA - Italian National Agency for New Technologies, Energy and Sustainable Economy, Italy
Y. Ueda, Tokyo University of Science, Japan
S. Hubbard, RIT, USA

Committee Members

M. Topič, University of Ljubljana, Slovenia
P. Helm, WIP Renewable Energies, Germany
J. Bergmiller, WIP Renewable Energies, Germany
J. de Gregorio, WIP Renewable Energies, Germany
F. Lambert, Genvia, France
N. Pearsall, Northumbria University, United Kingdom
J. Serra, Lisboa University, Portugal
R. Kenny, European Commission, Joint Research Centre, Italy
J. Carabe, CIEMAT, Spain
P. Delli Veneri, ENEA - Italian National Agency for New Technologies, Energy and Sustainable Economy, Italy
M. Kern, WIP Renewable Energies, Germany
A. Schöning, WIP Renewable Energies, Germany
P. Gerlinger, WIP Renewable Energies, Germany
T. Morrison, WIP Renewable Energies, Germany
P. Malbranche, International Solar Aliance, India
H. Ossenbrink, Band Gap, Germany
S. Nowak, NET Nowak Energie & Technologie, Switzerland
A. Grassi, ETA Florence Renewable Energies, Italy
J. Trube, VDMA, Germany
A. Michaelsen, WIP Renewable Energies, Germany
G. Masson, Becquerel Institute, Belgium
R. Drozdowski, IPVF, France
R. Kopecek, ISC Konstanz, Germany
D. Schultz, WIP Renewable Energies, Germany
D. Moser, EURAC, Italy
P. Fath, RCT Solutions, Germany
P Mints, Consultant, USA
M. Woodhouse, NREL, USA
Z. Holman, ASU, USA
P. Verlinden, Amrock, Australia
T. Jester, NEXT Energy Technologies, USA
M. Beck, Department of Energy, USA
K. Hinzer, University of Ottawa
S. Haouka, IPVF, France
P. Frankl, iea, France
I. Weiss, WIP Renewable Energies, Germany
D. Mugnier, Planair, France
L. Großhans, WIP Renewable Energies, Germany
D. Lecoque, ARE, Belgium
C. Pedretti, ARE, Belgium
S. De Iuliis, ENEA - Italian National Agency for New Technologies, Energy and Sustainable Economy, Italy
R. Janssen, WIP Renewable Energies, Germany
R. Blanchard, Loughborough University, United Kingdom
H. M. Albuflasa, University of Bahrain, Bahrain
S. Caneva, WIP Renewable Energies, Germany
G. Ruggieri, Insubria University, Italy
S. Leanza, WIP Renewable Energies, Germany
C. Lins, GWNET, Austria
D. Ngei, GWNET, Austria
W. van Sark, Utrecht University, The Netherlands
R. Egan, UNSW, Australia
W. Sinke, TNO, The Netherlands
C. Ballif, EPFL, Switzerland
J. Poortmans, imec, Belgium
F. Dimroth, Fraunhofer-Institut für Solare Energiesysteme ISE, Germany
P. Szymanski, European Commission, Joint Research Centre, The Netherlands
M. Konagai, Tokyo City University, Japan
Y. Zhao, Nankai University, China
C. Park, Yeungnam University, Korea
B. Hoex, UNSW, Australia
S. Hayase, Kyushu Insititute of Technology, Japan
J.-J. Lee, Dongguk University, Korea
Y. Kurokawa, Nagoya University, Japan
J. Vasi, IIT Bombay, India
A. Kabu, International Solar Aliance, India
J. H. Yun, Korea Institute of Energy Research, Korea
M. K. Tan, SERIS, Singapore
T. Minemoto, Ritsumeikan University, Japan
M. Yamaguchi, Toyota Technological Institute (TTI), Japan
S. Bremner, UNSW, Australia
A. Freundlich, University of Houston, USA
S. Kurtz, Merced University, USA
S. Marsillac, Old Dominion University, USA
A. Smets, TU Delft, The Netherlands
I. Sellers, University of Oklahoma, USA
T. Grassman, Ohio State University, USA
C. Zimmermann, Airbus Defence and Space, Germany
D. Wilt, US Department of Energy, USA
L. Kazmerski, retired from NREL, USA
B. Smith, NREL, USA
K. Hinzer, University of Ottawa, USA
A. Nayfeh, Khalifa University, United Arab Emirates
E. Warren, NREL, USA
P. Bermel, Purdue University, USA
J. McNatt, NASA, USA
S. Polly, RIT, USA
S. Tomasulo, Naval Research Lab., USA
S. Ringel, Ohio State University, USA

INTERNATIONAL ADVISORY COMMITTEE TO THE WCPEC (IAC)

Chairs

H. Ossenbrink, Band Gap, Germany
M. Yamaguchi, Toyota Technological Institute (TTI), Japan
S. Ringel, Ohio State University, USA

Committee Members

M. Topic, University of Ljubljana, Slovenia
W. Sinke, TNO, The Netherlands
J. Serra, University of Lisbon, Portugal
A. Jaeger-Waldau, European Commission, Joint Research Centre, Italy
P. Malbranche, International Solar Alliance, India
S. Nowak, NET Nowak Energy & Technology, Switzerland

J. Poortmans, imec, Belgium
P. Verlinden, Amrock, Australia
F. Ferrazza, ENI, Italy
W. Hemetsberger, SolarPowerEurope, Belgium
P. Helm, WIP Renewable Energies, Germany
A. Scognamiglio, ENEA - Italian National Agency for New Technologies, Energy and Sustainable Economy, Italy
R. Kenny, European Commission, Joint Research Centre, Italy
A. Smets, TU Delft, The Netherlands
M. Konagai, Tokyo City University, Japan
N. Usami, Nagoya University, Japan
B. Hoex, University of New South Wales, Australia
A. Aberle, SERI, Singapore
M. Green, University of New South Wales, Australia
D. Kim, Korea University, Korea
C. Lan, National Taiwan University, Taiwan
Y. Ueda, Tokyo University of Science, Japan
J. Vasi, India India Institute of Technology, India
J. Yi, Korea, SungKyonKwan University, Korea
D. Yang, China, Zheejiang University, China
Y. Zhao, Nankai University, China
S. Hayase, Kyushu Institute of Technology, Japan
A. Freundlich, University of Houston, USA
S. Marsillac, Old Dominion University, USA
A. Reinders, University of Twente, The Netherlands
T. Anderson, University of Massachusetts, USA
S. Bailey, retired from NASA Glenn Research Center, USA
A. Barnett, retired from University of Delaware, USA
R. King, Arizona State University, USA
R. Raffaelle, Rochester Institute of Technology, USA
B. Stanbery, HelioVolt, USA
D. Wilt, Air Force Research Laboratory, USA
L. Kazmerski, retired from NREL, USA
R. Todi, IEEE, USA
S. Kurtz, University of Merced, USA
S. Hubbard, RIT, USA
A. Smets, TU Delft, The Netherlands
Y. Matsumoto, SEES, Mexico
A. Nayfeh, Khalifa University, UAE
K. Hinzer, University of Ottawa, Canada

WCPEC-8 SCIENTIFIC COMMITTEE

Programme Technical Chairs
R. Kenny, European Commission, Joint Research Centre, Italy
S. Hayase, Kyushu Institute of Technology, Japan
A. Smets, TU Delft, The Netherlands

Topic Chairs

Topic 1: Silicon Materials and Cells
S. Glunz, Fraunhofer-Institut für Solare Energiesysteme ISE, Germany
Q. Wang, Jinko solar, China
K. Ding, Forschungszentrum Jülich, Germany

Topic 2: Evolving and Emerging Technologies: Tandems; Thin Film absorbers; III-V; New Materials and Concepts; Advanced Modelling
M. Edoff, Uppsala University, Sweden
U. Bach, Monash University, Australia
C. Hages, University of Florida, USA

Topic 3: Photovoltaic Modules and BoS Components
U. Jahn, VDE Renewables, Germany
A. Lennon, UNSW, Australia
T. Barnes, NREL, USA

Topic 4: PV Systems Engineering, Integrated/Applied PV
David Moser, EURAC, Italy
Thomas Reindl, National University of Singapore (NUS), Singapore
Clifford Hansen, Sandia National Laboratories, USA

Topic 5: Energy Transition – Integration, Storage, Sustainability, Policy, Economics, Energy Poverty, Society
Stefan Nowak, NET Nowak Energy & Technology, Switzerland
Keiichi Komoto, Mizuho Research and Technologies, Japan
Annick Anctil, Michigan State University, USA

Topic Organisers and Paper Review Experts

Topic 1: Silicon Materials and Cells
D. Bredemeier, ISFH Institut für Solarenergieforschung, Germany
S. Riepe, Fraunhofer-Institut für Solare Energiesysteme ISE, Germany
S. W. Glunz, Fraunhofer-Institut für Solare Energiesysteme ISE, Germany
G. Hahn, Universität Konstanz / Department of Photovoltaics, Germany
K. Ding, Forschungszentrum Jülich, Germany
D. Muñoz, CEA / INES, France
P. Roca i Cabarrocas, CNRS-LPICM LPICM Laboratoire de Physique des Interfaces et Couches Minces, France
A. W. Weeber, TNO Energy Transition, The Netherlands
J. Cárabe, CIEMAT, Spain
M. Topic, University of Ljubljana UL-FE, Slovenia
K. Bothe, ISFH Institut für Solarenergieforschung, Germany
F. Ferrazza, ENI S.p.A., Italy
S. Peters, Hanwha Q CELLS, Germany
Q. Wang, Zhejiang Jinko Solar Co., Ltd., China
M. P. Bellmann, STIFTELSEN SINTEF, Norway
M. Benmansour, CEA / INES, France
T. Bruton, Progress in Photovoltaics, United Kingdom
A. Ciesla, UNSW Australia, Australia
C. Hagendorf, Fraunhofer-Center für Silizium Photovoltaik CSP, Germany
K. Hesse, Wacker Chemie AG, Germany
X. Yu, Zhejiang University, China
R. Brendel, ISFH Institut für Solarenergieforschung, Germany
S. Dubois, CEA / INES, France
T. Dullweber, ISFH Institut für Solarenergieforschung, Germany
M. Ernst, ANU - The Australian National University, Australia
M. Hermle, Fraunhofer-Institut für Solare Energiesysteme ISE, Germany
R. G. Naber, Hanwha Q CELLS, The Netherlands
W. Nemeth, NREL - National Renewable Energy Laboratory, USA
R. Preu, Fraunhofer-Institut für Solare Energiesysteme ISE, Germany
B. Terheiden, Universität Konstanz / Department of Photovoltaics, Germany
R. Turan, METU - Middle East Technical University, Turkey
G. Yang, Delft University of Technology, The Netherlands
W. Dauksher, Arizona State University, USA

A. Descoeudres, CSEM - Centre Suisse d'Electronique et de Microtechnique SA, Switzerland
M. Despeisse, CSEM - Centre Suisse d'Electronique et de Microtechnique SA, Switzerland
W. Favre, CEA / INES, France
J. Horzel, Fraunhofer-Institut für Solare Energiesysteme ISE, Germany
Y. Ohshita, Toyota Technological Institute, Japan
J. Rentsch, Fraunhofer-Institut für Solare Energiesysteme ISE, Germany
J. Schmidt, ISFH Institut für Solarenergieforschung, Germany
J. Zhao, CSEM - Centre Suisse d'Electronique et de Microtechnique SA, Switzerland
P. Delli Veneri, ENEA - Italian National Agency for New Technologies, Energy and Sustainable Economy, Italy
S. Gall, Helmholtz-Zentrum Berlin für Materialien und Energie, Germany
T. Matsui, AIST - National Institute of Advanced Industrial Science and Technology, Japan
A. Meeder, Geo-En Energy Technologies, Germany
J. Meier, Meier Technologies, Switzerland
M. Vetter, IPHT Institut für Photonische Technologien Jena e.V., Germany
A. Augusto, Arizona State University, USA
N. Azkona, UPV/EHU - Instituto de Tecnologia Microelectronica (TiM), Spain
D. MacDonald, ANU - The Australian National University, Australia
A. Otaegi, UPV/EHU - Instituto de Tecnologia Microelectronica (TiM), Spain
M. C. Schubert, Fraunhofer-Institut für Solare Energiesysteme ISE, Germany
L. Serenelli, ENEA Casaccia Research Center (SP 064), Italy
K. Sporleder, Fraunhofer-Center für Silizium Photovoltaik CSP, Germany
A. Yamada, Tokyo Institute of Technology, Japan
D. L. Bätzner, Meyer Burger Research AG, Switzerland
A. Danel, CEA / INES, France
P. Fath, RCT-Solutions, Germany
A. Metz, , Germany
H. J. Nonnenmacher, Meyer Burger, Germany
P. J. Verlinden, AMROCK Pty Ltd., Australia
W. Wu, Tongwei Solar (Meishan), China
W. Zhang, Gonda Electronic Technology, China

Topic 2: Evolving and Emerging Technologies: Tandems; Thin Film absorbers; III-V; New Materials and Concepts; Advanced Modelling

J. C. Goldschmidt, Philipps-Universität Marburg, Germany
I. Gordon, imec vzw, Belgium
U. Bach, Monash University, Australia
W. Tress, Zurich University of Applied Sciences, Switzerland
S. Veenstra, TNO Energy Transition, The Netherlands
M. Edoff, Uppsala University, Sweden
C. Hages, University of Florida, USA
A. N. Tiwari, EMPA - Swiss Federal Laboratories For Materials Science and Technology, Switzerland
G. Flamand, imec vzw, Belgium
G. Siefer, Fraunhofer-Institut für Solare Energiesysteme ISE, Germany
A. Martí Vega, UPM - Universidad Politécnica de Madrid, Spain
J. J. Poortmans, imec vzw, Belgium
P. Pearce, UNSW Australia, Australia
I. Ramiro, Universidade Nova de Lisboa, Portugal
S. Albrecht, Helmholtz-Zentrum Berlin für Materialien und Energie G, Germany
G. Coletti, TNO Energy Transition, The Netherlands
F. Dimroth, Fraunhofer-Institut für Solare Energiesysteme ISE, Germany
U. W. Paetzold, KIT Karlsruher Institut für Technologie, Germany
A. Virtuani, EPFL-STI-IMT-PVLAB, Switzerland
A. Walter, CSEM - Centre Suisse d'Electronique et de Microtechnique SA, Switzerland
M. J. Brites, LNEG Laboratório Nacional de Energia e Geologia, Portugal
S. Cros, CEA / INES, France
C. Fell, CSIRO - The Commonwealth Scientific and Industrial Research Organisation, Australia
S. Huang, Macquarie University, Australia
N. Kyranaki, CEA / INES, France
C. F. Lin, National Taiwan University, Taiwan
B. Mihaylov, European Commission JRC, Italy
Y. Saygili, ODTÜ-GÜNAM , Turkey
N. Barreau, University of Nantes, France
V. Bermudez Benito, QEERI- Qatar Environment & Energy Research Institute, Qatar
T. Dalibor, Avancis, Germany
D. Dimova-Malinovska, Bulgarian Academy of Sciences, Bulgaria
S. C. Irvine, Swansea University, United Kingdom
S. Ishizuka, AIST - National Institute of Advanced Industrial Science and Technology, Japan
P. Kratzert, SOLARWATT, Germany
M. Lira-Cantú, CIN2, Spain
M. C. Lux-Steiner, Helmholtz-Zentrum Berlin für Materialien und Energie, Germany
M. Meuris, UHasselt - IMO IMOMEC, Belgium
S. Paetel, Zentrum für Sonnenenergie- und Wasserstoff-Forschung Baden-Württemberg (ZSW), Germany
R. Po, ENI S.p.A., Italy
A. Redinger, University of Luxembourg, Luxembourg
A. Romeo, University of Verona, Italy
V. Sittinger, Fraunhofer-Institut für Schicht- und Oberflächentechnik IST, Germany
M. Theelen, TNO/Solliance, The Netherlands
R. King, Arizona State University, USA
T. Minemoto, Ritsumeikan University, Japan
G. Timò, RSE S.p.A. (Ricerca sul Sistema Energetico), Italy
J. P. Connolly, GeePs-CentraleSupelec, France
J. Govaerts, imec vzw, Belgium
M. A. Hernandez-Fenollosa, UPV - Technical University of Valencia, Spain
K. Hummelen, University of Groningen, The Netherlands
J. P. Kleider, CNRS/GeePs, France
D. König, UNSW - Integrated Materials Design Centre (IMDC), Australia
I. Konovalov, University of Applied Sciences Jena, Germany
Y. Okada, University of Tokyo, Japan
F. Roca, ENEA - Italian National Agency for New Technologies, Energy and Sustainable Economy, Italy
M. Rusu, Helmholtz-Zentrum Berlin für Materialien und Energie, Germany
N. Ekins-Daukes, UNSW Australia, Australia
C. Lévy-Clément, CNRS - Institut de Chimie Matériaux Paris Est, France
B. Rech, Helmholtz-Zentrum Berlin für Materialien und Energie, Germany
R. I. Schropp, Utrecht University, The Netherlands
T. Tayagaki, AIST - National Institute of Advanced Industrial Science and Technology, Japan
W. Warta, Fraunhofer-Institut für Solare Energiesysteme ISE, Germany

Topic 3: Photovoltaic Modules and BoS Components
W. J. Gambogi, DuPont Company, USA
A. Lennon, UNSW Australia, Australia
T. Barnes, NREL, USA
R. Gottschalg, Fraunhofer-Center für Silizium Photovoltaik CSP, Germany
U. Jahn, VDE Renewables, Germany
G. Bardizza, European Commission JRC, Italy
W. Herrmann, TÜV Rheinland Energy, Germany
C. Buerhop-Lutz, Helmholtz-Institut Erlangen-Nürnberg für Erneuerbare Energien HI ERN, Germany
N. Pearsall, Northumbria University, United Kingdom
A. Faes, CSEM - Centre Suisse d'Electronique et de Microtechnique SA, Switzerland
M. Mittag, Fraunhofer-Institut für Solare Energiesysteme ISE, Germany
M. Á. Muñoz-García, Universidad Politécnica de Madrid. UPM-ETSIAAB., Spain
H. Nagel, Fraunhofer-Institut für Solare Energiesysteme ISE, Germany
I. Tsanakas, CEA / INES, France
C. Zhu, LONGi, China
G. Beaucarne, Dow Silicones, Belgium
N. Bosco, NREL - National Renewable Energy Laboratory, USA
C. Camus, LayTec AG, Germany
R. Einhaus, Apollon Solar, France
S. Oktik, Kadir Has University Central Campus, Turkey
G. Oreski, PCCL Polymer Competence Center Leoben, Austria
M. Pander, Fraunhofer-Center für Silizium Photovoltaik CSP, Germany
T. Sample, European Commission JRC, Italy
G. Friesen, SUPSI University of Applied Sciences and Arts of Southern Switzerland, Switzerland
Y. Hishikawa, AIST - National Institute of Advanced Industrial Science and Technology, Japan
J. Lopez-Garcia, QEERI, Qatar
C. Monokroussos, TÜV Rheinland (Shanghai) Co., Ltd., China
H. Müllejans, European Commission JRC, Italy
H. Nussbaumer, Zurich University of Applied Sciences, Switzerland
M. Paverttoni, SERIS, Singapore
J. Roy, IIT Indian Institute of Technology Kharagpur, India
R. Sinton, Sinton Instruments, USA
P. Bauwens, UGhent Ghent University, Belgium
F. Dross, STS, USA
G. Graditi, ENEA - Italian National Agency for New Technologies, Energy and Sustainable Economy, Italy
J. Moschner, KU Leuven Katholieke Universiteit Leuven, Belgium

Topic 4: PV Systems Engineering, Integrated/Applied PV
A. M. Gracia Amillo, Fraunhofer-Institut für Solare Energiesysteme ISE, Germany
W. M. van Sark, Utrecht University, The Netherlands
F. P. Baumgartner, Zurich University of Applied Sciences, Switzerland
C. Hansen, Sandia National Laboratories, USA
J. M. Almeida Serra, University of Lisbon - FCiencias.ID, Portugal
D. Moser, Eurac Research, Italy
G. Mütter, Alternative Energy Solutions, Austria
F. Frontini, SUPSI University of Applied Sciences and Arts of Southern Switzerland, Switzerland
A. Scognamiglio, ENEA - Italian National Agency for New Technologies, Energy and Sustainable Economy, Italy
T. Reindl, NUS National University of Singapore, Singapore
R. E. Valckenborg, TNO Energy Transition, The Netherlands
E. E. Voroshazi, CEA / INES, France
G. Adinolfi, ENEA - Italian National Agency for New Technologies, Energy and Sustainable Economy, Italy
U. Muntwyler, Dr. Schüpbach&Muntwyler, Switzerland
I. Antón Hernández, Universidad Politécnica de Madrid - IES, Spain
K. Araki, University of Miyazaki, Japan
R. Campesato, CESI S.p.A, Italy
S. Taylor, European Space Agency, The Netherlands
P. Blanc, MINES ParisTech, France
A. Louwen, Eurac Research, Italy
T. Oozeki, AIST - National Institute of Advanced Industrial Science and Technology, Japan
C. Protogeropoulos, EEPS, Greece
J. Remund, Meteotest, Switzerland
M. Sengupta, NREL - National Renewable Energy Laboratory, USA
C. Alonso-Tristán, UBU - University of Burgos - Escuela Politécnica Superior, Spain
M. Carbone, ENEL Green Power, Italy
P. Darez, 350 Renewables, Chile
K. Lappalainen, Tampere University, Finland
C. Nyman, Soleco Ltd, Finland
M. Grottke, WIP Renewable Energies, Germany
P. Hacke, NREL - National Renewable Energy Laboratory, USA
K. Kiefer, Fraunhofer-Institut für Solare Energiesysteme ISE, Germany
P. Lechner, Zentrum für Sonnenenergie- und Wasserstoff-Forschung Baden-Württemberg (ZSW), Germany
J. Lin, PV Guider, Taiwan
O. Mayer, Bayern Innovativ - Bayerische Gesellschaft für Innovation und Wissenstransfer, Germany
A. Migan-Dubois, GeePs, France
S. Rodrigues, Forschungszentrum Jülich, (HI-ERN) Helmholtz Institut Erlangen Nürnberg, Germany
J. S. Stein, Sandia National Laboratories, USA
D. Stellbogen, Zentrum für Sonnenenergie- und Wasserstoff-Forschung Baden-Württemberg (ZSW), Germany
P. Suratkar, PV Diagnostics, India
M. Theristis, Sandia National Laboratories, USA
P. Alonso, BWP Baumanagement, Germany
P. Bonomo, SUPSI University of Applied Sciences and Arts of Southern Switzerland, Switzerland
G. C. Eder, OFI, Österreichisches Forschungsinstitut für Chemie und Technik, Austria
I. Hagemann, Architekturbüro Hagemann, Germany
B. Kaempfen, Kämpfen Zinke + Partner, Switzerland
T. Reijenga, BEAR-iD, The Netherlands
E. Román Medina, Fundación Tecnalia Research & Innovation, Spain
K. Sinapis, Trespa International BV, The Netherlands
L. H. Slooff-Hoek, TNO Energy Transition, The Netherlands
L. Tautenhahn, Avancis, Germany
M. Dörenkämper, TNO Energy Transition, The Netherlands
M. Heinrich, Fraunhofer-Institut für Solare Energiesysteme ISE, Germany
R. Peibst, ISFH Institut für Solarenergieforschung, Germany
B. van Aken, TNO Energy Transition, The Netherlands
G. Barchi, Eurac Research, Italy
V. Efthymiou, University of Cyprus, Cyprus
K. Nishioka, University of Miyazaki, Japan
M. Victoria, Aarhus University, Denmark
R. J. Ellingson, Toledo University, USA
V. Khorenko, Azur Space Solar Power, Germany
M. Á. Vázquez Rodríguez, DHV Technology, Spain

Topic 5: Energy Transition – Integration, Storage, Sustainability, Policy, Economics, Energy Poverty, Society

K. P. Kairies, ACCURE Battery Intelligence, Germany
R. Pestana, R&D NESTER, Centro de Investigação em Energia REN - State Grid, Portugal
I. Weiss, WIP Renewable Energies, Germany
A. Anctil, Michigan State University, USA
K. Komoto, Mizuho Research & Technologies, Ltd., Japan
K. Wambach, Wambach-Consulting, Germany
M. Getsiou, European Commission DG RTD, Belgium
P. Malbranche, International Solar Aliance, India
S. Nowak, NET Nowak Energy & Technology, Switzerland
C. Breyer, Lappeenranta University of Technology, Finland
G. Masson, Becquerel Institute, Belgium
H. Ossenbrink, Band Gap, Germany
S. De Iuliis, ENEA Casaccia Research Center (SP 064), Italy
J. Vasi, IIT Bombay, India
N. della Valle, European Commission JRC, Italy
K. Yenneti, University of Wolverhampton, United Kingdom
P. J. Alet, CSEM - Centre Suisse d'Electronique et de Microtechnique SA, Switzerland
M. C. Brito, University of Lisbon - Faculty of Science, Portugal
R. Bründlinger, AIT Austrian Institute of Technology, Austria
F. Carigiet, Zurich University of Applied Sciences, Switzerland
J. S. da Costa Fernandes, Hochschule Offenburg, Germany
B. Gaiddon, HESPUL, France
J. M. Klein, CEA / INES, France
P. Lalanne, HydroClapeyron, France
A. C. Neves Foles, University of Évora, Portugal
E. Schüpbach, Bern University of Applied Sciences BFH, Switzerland
S. Arancón, WIP Renewable Energies, Germany
S. Capaccioli, ETA - Florence Renewable Energies, Italy
V. Fthenakis, Columbia University, USA
G. Heath, NREL - National Renewable Energy Laboratory, USA
W. Palitzsch, LuxChemtech, Germany
E. Parnham, DuPont, United Kingdom
G. G. Arrowsmith, EUREC, Belgium
D. Polverini, European Commission DG GROWTH, Belgium
N. Taylor, European Commission JRC, Italy
T. Haarberg, BNW-Energy AS, Germany
I. Kaizuka, RTS Corporation, Japan
E. Vartiainen, Fortum Growth Oy, Finland
R. Blanchard, Loughborough University, United Kingdom
J. McDonald, ITP Renewables, Australia
A. Nayfeh, Khalifa University, United Arab Emirates
C. Pedretti, ARE, Belgium
S. Caneva, WIP Renewable Energies, Germany
G. Ruggieri, Insubria University, Italy
S. Tay En Rong, National University of Singapore, Singapore
E. Warren, NREL - National Renewable Energy Laboratory, USA

Awards Coordinators

Student Awards Coordinator

A.H.M. Smets, Delft University of Technology, The Netherlands

Poster Awards Coordinator

J. Cárabe, CIEMAT, Madrid, Spain

SUBJECT INDEX

Silicon Materials and Cells

Sessions 1CP.2, 1AO.4, 1AO.5, 1AO.6, 1BO.1, 1BO.2, 1BO.3, 1BO4, 1BO.5, 1DO.11, 1DO.12, 1CV.2, 1CV.3, 1DV.4

Evolving and Emerging Technologies: Tandems; Thin Film absorbers; III-V; New Materials and Concepts; Advanced Modelling

Sessions 2CP.3, 2AO.1, 2AO.2, 2AO.3, 2BO.6, 2BO.7, 2BO.8, 2BO.9, 2BO.10, 2CO.1, 2CO.2, 2CO.3, 2DO.6, 2DO.7, 2DO.8, 2DO.9, 2DO.13, 2AV.1, 2AV.2, 2AV.3, 2BV.2

Photovoltaic Modules and BoS Components

Sessions 3AP.1, 3BO.11, 3BO.12, 3BO.14, 3BO.15, 3CO.4, 3CO.5, 3CO.6, 3DO.17, 3DO.18, 3DO.19, 3EO.1, 3BV.3, 3DV.1, 3DV.3

PV Systems Engineering, Integrated/Applied PV

Sessions 4CP.1, 4AO.7, 4AO.8, 4AO.9, 4BO.16, 4BO.17, 4CO.8, 4CO.9, 4DO.1, 4DO.2, 4DO.3, 4DO.4, 4DO.5, 4DO.16, 4EO.2, 4BV.1, 4BV.4, 4BV.5, 4CV.1

Energy Transition – Integration, Storage, Sustainability, Policy, Economics, Energy Poverty, Society

Sessions 5EP.1, 5CO.10, 5CO.11, 5DO.10, 5DO.14, 5DO.15, 5EO.3, 5DV.2

Topic Code

1 Silicon Materials and Cells
2 Evolving and Emerging Technologies: Tandems; Thin Film absorbers; III-V; New Materials and Concepts; Advanced Modelling
3 Photovoltaic Modules and BoS Components
4 PV Systems Engineering, Integrated/Applied PV
5 Energy Transition – Integration, Storage, Sustainability, Policy, Economics, Energy Poverty, Society

Session Type

P = Plenary Session
O = Oral Session
V = Visual Session

Day Codes

A = Monday, 26 September 2022
B = Tuesday, 27 September 2022
C = Wednesday, 28 September 2022
D = Thursday, 29 September 2022
E = Friday, 30 September 2022

e.g. 1AO.1 ⇒ 1= Silicon Materials and Cells, A=Monday, O=Oral session, 1=Session 1

TABLE OF CONTENTS OF WCPEC-8 PROCEEDINGS PAPERS

Visual PRESENTATIONS 3DV.1 Module Design and Manufacturing Posters

Visual PRESENTATIONS 4BV.4 Solar Resource and Forecasting

FOREWORD

The World Conference on Photovoltaic Energy Conversion (WCPEC) is the world's biggest platform for photovoltaic (PV) research and development, offering a unique possibility of international exchange by combining the three largest and most prominent international PV conferences: the European Photovoltaic Solar Energy Conference (EU PVSEC), the Photovoltaic Specialists Conference (IEEE PVSC) and the International PV Science and Engineering Conference (PVSEC). After being in place for more than a quarter of a century, the WCPEC, being held every four years, is firmly established among the global PV community as the event for the greatest minds and experts in various PV sectors to meet and discuss, to exchange and network.

The WCPEC-8 (combining the 39th EU PVSEC, the 49th IEEE PVSC and the 32nd PVSEC) took place from 26 – 30 September 2022 in Milan, Italy, was great success and highlighted the reality that a global and full-scale PV deployment needs collaboration. In the year of the achievement of the first Terawatt of installed PV capacity, the conference concluded with the vision that with its exponential growth, the multiTW scale PV deployment will be a reality in few years and that despite the many challenges that it presents (the PV research is more than ready to solve them), there are massive opportunities for the PV sector. Research communities around the world and their innovations across the sector, together with the fast deployment and numerous application options are making PV a key pillar of energy production. This year, more than 1500 participants from 60 countries attended this highly inspiring platform delivering over 1100 presentations. It provided a vital forum for information exchange in the field of PV solar energy research, demonstration and applications. In the exhibition 74 companies from all parts of the world welcomed visitors and presented their products and services.

The WCPEC-8 Proceedings give a comprehensive overview of the photovoltaic solar sector, its current status and future prospects in science, research, innovation, development and deployment on almost 3,700 pages. Selection for inclusion in the conference was made by paper review experts and topic organisers (see the listing on pages III-VII), to whom we express our sincere gratitude for their comprehensive review work and overall contribution to the success of the conference. In addition to the 354 submitted papers, the proceedings include 90 presentations (slides) shown during the plenary and oral presentations as well as 140 poster files of the visual presentations. In total this amounts to 584 publications.

The WCPEC-8 maintains its commitment to providing quick and open access to high quality scientific results. They constitute a powerful tool for targeted and quick information search and retrieval, enabling you to search by keyword, paper title and authors. We are sure that these features will help to simplify the use and exploit the full potential of this extensive source of information.

The Conference Proceedings are published as downloadable DVD file and are also fully accessible online. A DOI code (Digital Object Identifier) has been assigned to each paper. This ensures unequivocal and permanent identification and full citability. The WCPEC-8 papers can be viewed and downloaded in a full free open access from the Proceedings website www.eupvsec-proceedings.com.

We are confident that these Proceedings will play an important role in providing a comprehensive overview of the current actors and activities in the global PV sector and that they will disseminate information on the state-of-the-art of technologies and applications. This can generate further research, add momentum to innovation and promote interest in PV worldwide.

We would like to cordially thank all authors and participants of the WCPEC-8 for their contributions and look forward to welcoming you in Lisbon, Portugal from 18 – 22 September 2023 at the EU PVSEC 2023, the 40th European Photovoltaic Solar Energy Conference and Exhibition

The Editors

LESSONS LEARNED FROM 25 YEARS PRODUCTION TECHNOLOGY RESEARCH & DEVELOPMENT

R. Preu[1*], J. Rentsch[1], S. Rein[1], F. Clement[1], J.-F. Nekarda[1], P. Saint-Cast[1], S. Lohmüller[1], E. Lohmüller[1], J. Greulich[1], N. Wöhrle[1], S. Mack[1], A. Wolf[1], S. Pingel[1], A. Steinmetz[1], P. Baliozian[1], B. S. Goraya[1], S. Nold[1], S.W. Glunz[1,2], A.W. Bett[1,3],

[1]Fraunhofer Institute for Solar Energy Systems, Heidenhofstraße 2, 79110 Freiburg i. Br., Germany
[2] Institute of Physics, University of Freiburg, Hermann-Herder-Straße 3, 79104 Freiburg i. Br., Germany
[3] Laboratory for PV Energy Conversion, University of Freiburg, Georges-Köhler-Allee, 105, Freiburg i. Br., Germany

*Corresponding author, e-mail to: ralf.preu@ise.fraunhofer.de

ABSTRACT: In 1997, a visionary study addressed limitations of the cost reduction potential for high efficiency solar cell devices. The results of this study triggered increased efforts for the development of production technology at Fraunhofer ISE. Many new approaches with substantial impact have been developed to the demonstration level. Economic analysis and a technology readiness terminology have been established to aid in assessing technological approaches, while the set-up of the PV technology evaluation center (PV-TEC) allowed for testing equipment, processes and materials on a higher maturity level. Reference processes building the backbone of the technology have been developed consistently further and device efficiencies follow a roughly 0.5%/a progress gradient since 2005, which is in good agreement with reports for an industrial pilot line. Today, we can rely on reference processes for PERC, TOPCon and SHJ in the PV-TEC with total and large area efficiencies of up to 22.7%, 24.0% and 24.1%, respectively. Further learning includes that for production technology R&D for PV devices high average power conversion efficiencies must be targeted and that scaling should be combined with thorough testing of products along the value chain. All cost and value determining parameters like production efficiency distribution, yield, capital and operational expenditure as well as degradation and performance under field conditions have to be monitored.

1 INTRODUCTION

The last decade of the last century was decisive for the future development in photovoltaics (PV) and many milestones were achieved. The German Feed-In-Law from 1990 was extended to the renewable energy law with attractive fixed tariffs in 2000. In the late 1990s, for the first time, 100 MWp/a nominal PV module power were sold globally [1], and single junction silicon solar cells close to 25% efficiency were demonstrated, applying the passivated emitter and rear cell (PERC) concept with local diffusion under the rear contacts (PERL) in laboratory [2]. In 1997, PV production in Europe was based on approx. 20 MWp/a production of solar cells, mostly following the concepts of the laser grooved buried contact (LGBC), the screen-printed Al-BSF (Back-Surface Field) and metal-insulator-semiconductor (MIS) approach. Just then, a study by leading PV companies and institutions for setting up a visionary 500 MWp/a production line was published at the EU-PVSEC with seven different approaches being used [3]. The cell technologies under investigation were Al-BSF, LGBC, MIS and the LBSF (local BSF) approach from Fraunhofer ISE, with a design structure and process sequence similar to that of the PERL device. Most approaches were calculated to similar anticipated production costs on a module level. A substantial advantage of 20% vs. 16% conversion efficiency on cell level was assumed for the LBSF approach, but the approximately seven times more expensive cell processing could not be compensated, leading to substantially higher module cost. A main cost driver was the photo-lithographical structuring used to allow for the fine contact and selective diffusion structures in these cells. The increased cost could be attributed to both high material consumption but also low productivity of the used equipment. Based on this analysis, other techniques were to be developed at Fraunhofer ISE with the main objective to realize device features for efficiency increase, high productivity, and low material cost. Thus, a team was formed to focus on production technology development for highest efficiency solar cells, setting up a first lab. Moreover, an industry consortium was built to investigate cost saving potential for PV fabs. For the latter, techno-economic assessment based on bottom-up cost analysis was implemented at the institute. Our methodology has been then tested in our production research and development platform PV-TEC with high quality standards and allowing for in-depth characterization and the objective to support the industry targeting efficiency increase, productivity scaling and economical use of resources. Our methodology focusses on the reduction of the photovoltaic electricity generation cost, applying advanced techno-economic metrics.

This paper is to present the findings of this work on production technology over a period of 25 years and highlight important general results and their impact.

2 AIM AND APPROACH

The aim of the research and development work was guided by the relevance of efficiently bridging the gap between new ideas and concepts achieving outstanding laboratory results to the introduction of industrial pilot processing and mass production. Therefore, the efforts have been focussed on developing innovative equipment and new functional materials feasible for high throughput.

Compared to many state-of-the-art institutional R&D environments, a higher industrial maturity of the technologies was addressed, frequently assessed by its technology readiness level (TRL), which is commonly known to be introduced by NASA for flight missions. In 2016, we have released a first draft of a TRL-methodology specifically defined for devices like PV cells, including a review and more detailed discussion on the TRL-terminology [4]. We have further extended this draft to the version shown in Table I. Such a more specific TRL-methodology allows to be also used as quality gates for further budget decisions throughout technology development.

Table I: TRL stages for PV cell production. The definitions are an advanced version of the one published in [4].

TRL	Phase: Description
0	**Creation**: Idea of a new approach with beneficial features or potential, typically based on findings from modelling and/or experiment.
1	**Early proof-of-concept**: Laboratory process, idea works and is reproducible on small/reduced area.
2	**Advanced proof-of-concept**: Demonstrating potential in lab environment and reaching performance benchmarks using adapted or standardized measurement procedures.
3	**Early scaling**: Laboratory process, functional and reproducible on industrial standard large and complete areas / substrates. Demonstration of a product with essential features allowing for standardized measurement procedures.
4	**Early feasibility**: Laboratory process, efficiency potential demonstrated on industrial standard large and full areas. Successful first feasibility testing in next value chain step.
5	**Advanced scaling**: First high-throughput tools with still high parameter variability are tested and show successful throughput scaling of a process on large and full area. Products from the next value chain step are manufactured and first performance tests are successfully performed.
6	**Feasibility**: Industrial throughput level is reached for all equipment of the approach, but not set-up in a production-oriented mode. Cost drivers are identified, manufacturing can be performed at high throughput. Critical tests for certification have been performed, field tests are prepared.
7	**Pilot**: Process used in temporarily continuous industrial pilot, with operational throughput, yield and efficiency distribution targets met. Substantial quantities of next value chain step are manufactured, and fully certified, successful field tests are performed.
8	**Industrialization**: Process used in mass production with 24 hours/7 days on production tools. More equipment being sold, but still less output compared to benchmark approach. Still substantial potential in equipment and material productivity and resource efficiency. Field application becomes standardized and reliability parameters are monitored.
9	**Maturity**: Thorough production, a standard and dominant technology/product in its application field. Development is focused on further improvements of materials and equipment, where high output ensures continued learning.

3 RESULTS

3.1 Technology milestones

In the beginning, the production technology work was addressed to set up a solar cell process to produce Al-BSF cells and to be able to test technological approaches to achieve higher efficiencies on these cells. Based on this, the knowledge and technical infrastructure has been developed continuously to address the given targets. Today, many different technologies are tested based on reference processes for PERC, TOPCon and SHJ cell device features. If approved on lower technology readiness levels, technologies were typically developed until the pilot level. For complete devices this was obtained for example for screen-printed PERC and MWT (metal wrap through) solar cells. MWT was transferred to TRL 7 with an industrial partner. In 2010, the production technology know-how for screen-printed PERC devices was transferred to industrial partners and their pilot lines with TRL 7 and further matured to TRL 9, according to Table I. An overview of contributions including work of other groups in the development of PERC devices is given in [5]. We also developed an MWT-PERC hybrid and produced cells in a pilot at the PV-TEC, which have been manufactured to modules and installed at a building at the Fraunhofer ISE.

3.2 Findings

Most of the learning has naturally occurred for the respective production technologies, including device structures, processes, materials and respective equipment. Nevertheless, there are a few general learnings. First the establishment of reference processes for relevant device structures is very important. For the selection, we typically focus on device types which have a major share of production volume or show the highest prospects indicated by the gradient in production capacity or new installations. Such device structures allow for the most value for the development in the intermediate TRL levels 4-6. Another learning is the importance of high quality standards for a research and development line, as PV-TEC today includes more than 30 large process tools. Due to this high number of tools, the number of operators including engineers and PhD students also rises. Thus, advanced methods for quality management must be developed compared to lab type processing, including e.g., traffic light procedures, which define how to deal with sudden or drifting changes of control parameter values. This also implies to set parameter value boundaries, which ensure successful processing. In 2017, PV-TEC was re-built and extended to two parts for front-end and back-end processing, with adequate infrastructure for the respective technologies. Since then, we have focused to establish reference processes for the currently most important cell technologies PERC, TOPCon and SHJ, with representative structures of specific variants shown in Figure 1.

One general finding is visualized in Figure 2. For fast market introduction, it is essential to develop technologies to higher technology maturity while simultaneously pushing the development along the value chain. Benefits from an innovative approach must be validated on the system level. Thus, if the development of cell production technology on medium throughput level is conducted, it is essential to manufacture modules from the cells, too. The most relevant pull for introducing new technologies comes eventually from the motivation of reduced electricity generation price (€/kWh). To reflect this fundamental relevance, the initial development of in-depth cost assessment has been continuously extended. Today, we can calculate levelized cost of electricity for a new technological approach, based on thorough cost data and metrics. We also perform life cycle assessment and address criticality of resources to evaluate the sustainability of different technologies.

Table II: Milestones of solar cell production technology development at Fraunhofer ISE

Period	Milestones in production technology development at Fraunhofer ISE	Ref
1997-1998	Production technology group and expert round founded, set-up of in-depth cost calculation, first screen-printed solar cells manufactured.	[6]
1999-2001	Reference process for screen-printed solar cells set up, first attempts to include PERC features, including laser ablation and laser firing.	[3] [7–9]
2002-2004	First innovative alpha-tools set-up: e.g. galvo-scanner based laser tool, inline plasma etching tool, inline sputtering tool, low contamination inline diffusion furnace.	[10–12] [13]
2005-2007	Set-up of PV-TEC: Full pilot Al-BSF reference process (TRL 5-6), target >100 wafers/h, metal-wrap-through variant, 125-210 mm edge length.	[14] [15, 16]
2008-2010	PERC in PV-TEC: high-throughput oxidation, PECVD-AlOx, fully screen printed, PVD and Al-foil metallization.	[17, 18, 18, 19, 19, 20]
2011-2013	Transfer of PERC to industrial pilot line, MWT-PERC, PERC pilot process in PV-TEC, advanced quality management, modelling and assessing the technical and economical potential (TRL 6).	[21–23]
2014-2016	Set-up of n-type technology in the new PV-TEC Select, bifacial PERC/PERT, dispensing, ultrafast regeneration, machine-learning-enhanced inline PL characterization demonstrated.	[24–29]
2017-2019	PV-TEC front- and back-end, green manufacturing, development of PERC, TOPCon and SHJ reference processes, edge passivation, shingled cells, high-throughput, ultra-fineline metallization.	[30–32, 32, 33]
2020-2022	Highly stable PERC reference line, pilot processes for TOPCon/SHJ and shingled cells (TRL 5-6), first silicon tandem cells in PV-TEC (TRL 3-4 upscaling).	[5]
2023-2025	Pilot process for leading (production and/or growth rate) Si single junction cell technologies (TRL 6), production technologies for Si-tandem technologies (TRL 5-6).	

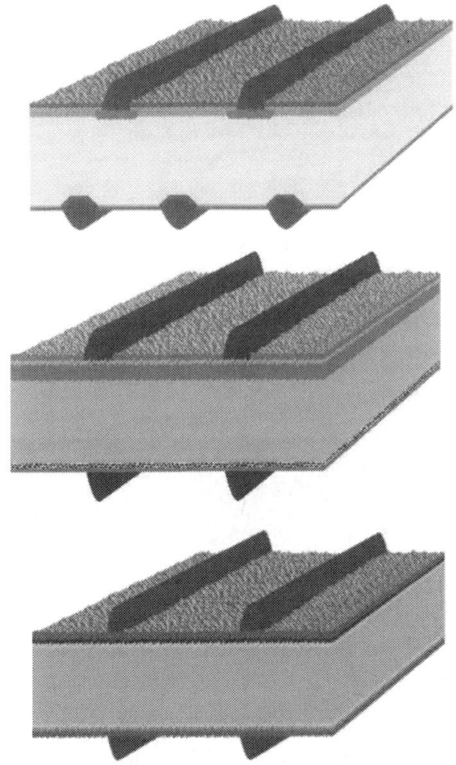

Figure 1: Representations of cell structures of the devices currently produced in PV-TEC in their bifacial versions (PERC with selective emitter, upper graph, TOPCon, center graph and SHJ, lower graph). For PERC devices most cells have been processed applying a lean process with homogeneous front emitter and full area aluminum paste coverage on the rear.

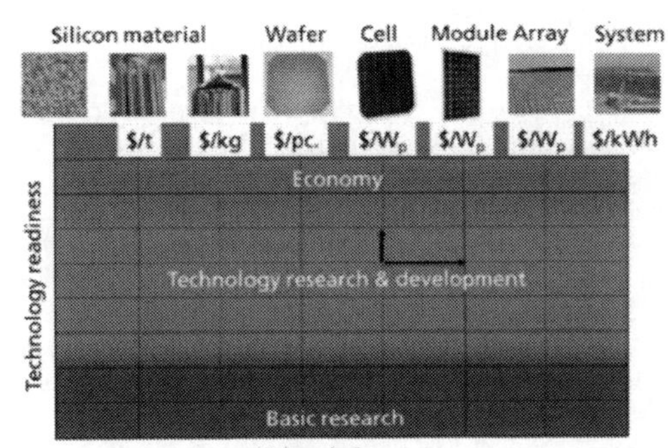

Figure 2: Technology readiness vs. value chain. The arrows indicate that technology development must consider both, TRL of the product but also the subsequent steps.

3.3 Reference process development for current devices

In the early years of the production technology development at Fraunhofer ISE, reference processes for screen-printed cells were developed but not sufficiently often repeated to work as a reference. With the set-up of the PV-TEC, we learned the importance of reliable reference processes based on well-established quality management features. The development of efficiencies is shown in Figure 3 for the first solar cells processed in PV-

TEC in December 2005 to the currently highest, independently confirmed (Callab PV Cells, Fraunhofer ISE) values of 22.7%, 24.0% (both with homogeneous emitter) and 24.1% for our reference processes of PERC, TOPCon and SHJ devices, respectively. Remarkably for PERC solar cells, the deviation of our best to the worst cell is less than $0.4\%_{abs}$ within a batch of 250 cells. Such high reliability with relevant efficiencies ensures valuable research and development for the currently dominating cell structures.

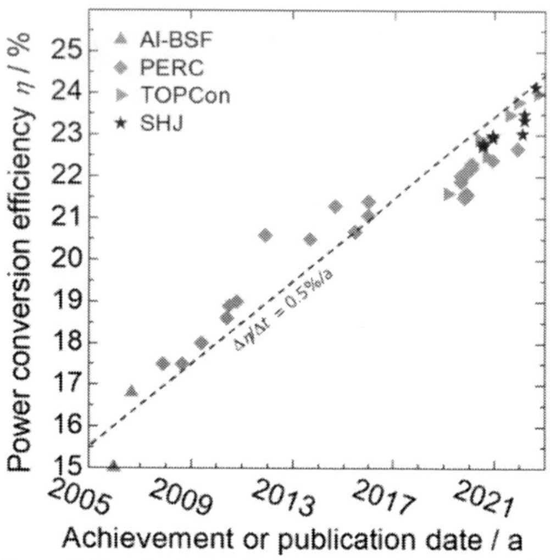

Figure 3 Development of power conversion efficiencies within the PV-TEC for different cell structures independently determined at the Callab PV cells (extended from [5]. The two cells which yield the highest efficiency values for SHJ and TOPCon feature a second MgF_2 anti-reflection layer. All TOPCon and most PERC cell results are prepared with a homogeneous emitter, only. The efficiency progress gradient of approx. 0.5%/a (dashed line) is similar to the report from industrial pilot lines [34].

4 TRAINING AND DISSEMINATION

One of the most important results is the training of scientifically, technologically but also economically qualified individuals who disseminate the knowledge. Within the 25 years, almost 100 PhD students and several 100 master and diploma students have performed their scientific work within our production technology environment. It was mandatory to install high standards for scientific quality, which have resulted in several hundred of conference and journal publications within this period, contributing to the knowledge, which set the unforeseeable pace in the industrial deployment of photovoltaics. For all this work, the general approach to set-up a respective technology environment and to combine this with in-depth techno-economic analysis has proven to be a very powerful and stringent approach.

5 SUMMARY

This paper gives a summary of findings of the work on production technology research and development performed at Fraunhofer ISE. It further introduces an advanced technology readiness scheme, which allows to evaluate maturity and support budget decisions.

ACKNOWLEDGEMENTS

The authors would like to sincerely thank the whole PV R&D team at Fraunhofer ISE for setting-up and maintaining such an excellent infrastructural and personal environment for applied research and development. Special thanks to all the persons involved in the progress of PV production technology for their contributions, Dilara Subasi and Ivo Rahlff for support in graph preparation. Trustful cooperation with numerous PV device, equipment, and material manufacturer and continuous support for PV production technology development by the German Federal Ministry of Economic Affairs and Climate Action, furthermore the German Federal Ministry of Research and Education, the EU with numerous programmes including Horizon 2020 as well as the Land Baden-Württemberg are gratefully acknowledged.

REFERENCES

[1] S. Philipps and W. Warmuth, *Photovoltaics Report.* [Online] Available: https://www.ise.fraunhofer.de/content/dam/ise/de/documents/publications/studies/Photovoltaics-Report.pdf.

[2] J. Zhao, A. Wang, and M. A. Green, "24·5% efficiency silicon PERT cells on MCZ substrates and 24·7% efficiency PERL cells on FZ substrates," *Prog. Photovolt: Res. Appl.*, vol. 7, no. 6, pp. 471–474, 1999.

[3] R. Preu, J. O. Schumacher, P. Hahne, H. Lautenschlager, I. Reis, S. W. Glunz, and W. Warta, "Screen printed and RT-processed emitters for crystalline silicon solar cells," in *2nd World Conference on Photovoltaic Solar Energy Conversion Vienna*, Vienna, 1998, pp. 1503–1506.

[4] P. Baliozian, S. Mourad, D. Morse, S. Kim, L. Friedrich, and R. Preu, "Photovoltaic Development Standardizing Based on Roadmaps and Technology Readiness Levels," in *32nd EU PVSEC*, Munich, 2016.

[5] R. Preu, E. Lohmüller, S. Lohmüller, P. Saint-Cast, and J. Greulich, "Passivated Emitter and Rear Cell – Devices, Technology and Modelling," *Applied Physics Reviews*, vol. 7, no. 041315, pp. 1–41, 2020.

[6] R. Preu, R. Lüdemann, G. Emanuel, W. Wettling, W. Eversheim, G. Güthenke, D. Untiedt, and G. Schweitzer, "Innovative production technologies for solar cells - SOLPRO," in *16th EU PVSEC*, Glasgow, 2000, pp. 1451–1454.

[7] I. Reis, D. Huljic, P. Hahne, R. Preu, D. Zickermann, B. Bucher, and P. V. Fleischer, "Fine-line screen printing for solar cells," in *11th International PVSEC Sapporo*, Hokkaido, 1999, pp. 335–336.

[8] R. Preu, S. W. Glunz, S. Schäfer, R. Lüdemann, W. Wettling, and W. Pfleging, "Laser ablation—a new low-cost approach for passivated rear contact formation in crystalline silicon solar cell technology," in *16th EU PVSEC*, Glasgow, 2000, pp. 1181–1184.

[9] E. Schneiderlöchner, R. Preu, R. Lüdemann, S. W. Glunz, and G. Willeke, "Laser-fired contacts

(LFC)," in *17th EU PVSEC*, Munich, 2001, pp. 1303–1306.
[10] R. Preu, D. Biro, G. Emanuel, A. Grohe, M. Hofmann, D. Huljic, I. E. Reis, J. Rentsch, E. Schneiderlochner, W. Sparber, W. Wolke, and G. Willeke, "Technology path to the industrial production of highly efficient and thin c-Si solar cells," in *3rd World Conference on Photovoltaic Energy Conversion: Joint conference of 13th PV Science & Engineering Conference, 30th IEEE PV Specialists Conference, 18th European PV Solar Energy Conference; Osaka International Congress Center "Grand Cube"*, Kurokawa, Kosuke, 2003.
[11] Schneiderlöchner, E. Grohe, A., S. W. Glunz, R. Preu, and G. Willeke, "Scanning Nd:YAG laser system for industrially applicable processing in silicon solar cell manufacturing," in *3rd World Conference on Photovoltaic Energy Conversion: Joint conference of 13th PV Science & Engineering Conference, 30th IEEE PV Specialists Conference, 18th European PV Solar Energy Conference; Osaka International Congress Center "Grand Cube"*, Kurokawa, Kosuke, 2003, pp. 1364–1367.
[12] J. Rentsch, D. M. Huljic, S. Reber, R. Preu, and R. Ludemann, "Progress in screen printed front side metallization schemes for CSiTF solar cells," in *29th IEEE Photovoltaic Specialists Conference New Orleans*, New Orleans, 2002, pp. 134–137.
[13] J. Rentsch, G. Emanuel, C. Schetter, T. Aumann, D. Theirich, J. Gentischer, K. Roth, M. Fritzsche, K.-H. Dittrich, and R. Preu, "Plasma etching for industrial in-line processing of c-si solar cells," in *3rd World Conference on Photovoltaic Energy Conversion: Joint conference of 13th PV Science & Engineering Conference, 30th IEEE PV Specialists Conference, 18th European PV Solar Energy Conference; Osaka International Congress Center "Grand Cube"*, Kurokawa, Kosuke, 2003, pp. 1376–1379.
[14] R. Preu, D. Biro, J. Rentsch, F. Clement, D. Erath, G. Emanuel, A. Grohe, M. Hofmann, A. Mette, C. Voyer, W. Wolke, S. W. Glunz, and G. P. Willeke, "The status of silicon solar cell production technology development at fraunhofer ISE," in *4th World Conference on Photovoltaic Energy Conversion Hawaii*, Hawaii, 2006, pp. 1040–1043.
[15] F. Clement, R. Preu, I. Reis, D. Biro, J. Rentsch, M. Zimmer, S. Rein, G. Emanuel, A. Krieg, A. Müller, and F. Martin, "Pilot-line processing of very large (21 x 21 cm^2) and thin mc-si solar cells," in *22nd EU PVSEC*, Milano, 2007, pp. 1411–1414.
[16] F. Clement, M. Lutsch, T. Kubera, M. Kasemann, W. Kwapil, C. Harmel, N. Mingirulli, D. Erath, H. Wirth, D. Biro, and R. Preu, "Processing and comprehensive characterisation of screen-printed mc-si metal wrap through (mwt) solar cells," in *22nd EU PVSEC*, Milano, 2007, pp. 1399–1402.
[17] M. Hofmann, S. Kambor, C. Schmidt, D. Grambole, J. Rentsch, S. W. Glunz, and R. Preu, "PECVD-ONO: A new deposited firing stable rear surface passivation layer system for crystalline silicon solar cells," *Advances in OptoElectronics*, vol. 2008, pp. 1–10, 2008.
[18] D. Biro, S. Mack, A. Wolf, A. Lemke, U. Belledin, D. Erath, B. Holzinger, E. A. Wotke, M. Hofmann, L. Gautero, S. Nold, J. Rentsch, and R. Preu, "Thermal oxidation as a key technology for high efficiency screen printed industrial silicon solar cells," in *34th IEEE Photovoltaic Specialists Conference Philadelphia*, Philadelphia, CA, 2009, pp. 1594–1599.
[19] J. Nekarda, S. Stumpp, L. Gautero, M. Hörteis, A. Grohe, D. Biro, and R. Preu, "LFC on Screen Printed Aluminium Rear Side Metallization," in *24th EU PVSEC*, Hamburg, 2009, pp. 1441–1445.
[20] P. Saint-Cast, D. Kania, M. Hofmann, J. Benick, J. Rentsch, and R. Preu, "Very low surface recombination velocity on p-type c-Si by high-rate plasma-deposited aluminum oxide," *Appl. Phys. Lett.*, vol. 95, no. 15, p. 151502, 2009.
[21] E. Lohmuller, B. Thaidigsmann, M. Pospischil, U. Jager, S. Mack, J. Specht, J. Nekarda, M. Retzlaff, A. Krieg, F. Clement, A. Wolf, D. Biro, and R. Preu, "20% efficient passivated large-area metal wrap through solar cells on boron-doped Cz silicon," *IEEE Electron Device Lett.*, vol. 32, no. 12, pp. 1719–1721, 2011.
[22] A. Mohr, P. Engelhart, C. Klenke, S. Wanka, A. A. Stekolnikov, M. Scherff, R. Seguin, S. Tardon, T. Rudolph, M. Hofmann, F. Stenzel, J. Y. Lee, S. Diez, J. Wendt, W. Brendle, S. Schmidt, J. W. Müller, P. Wawer, P. Saint-Cast, J. Nekarda, D. Erath, J. Rentsch, and R. Preu, "20%-efficient rear side passivated solar cells in pilot series designed for conventional module assembling," in *26th EU PVSEC*, Hamburg, 2011, pp. 2150–2153.
[23] U. Jäger, A. Wolf, B. Steinhauser, J. Benick, J. Nekarda, and R. Preu, "Laser doping for high-efficiency silicon solar cells," *Proceedings of SPIE*, vol. 8473, 2012.
[24] F. Fertig, K. Krauß, J. Greulich, F. Clement, D. Biro, R. Preu, and S. Rein, "The BOSCO solar cell - a both sides collecting and contacted structure," *phys. stat. sol. (RRL)*, vol. 8, no. 5, pp. 381–384, 2014.
[25] M. Pospischil, M. Kuchler, M. Klawitter, C. Rodriguez, A. Padilla, R. Efinger, H. Gentischer, M. König, M. Hörteis, L. Wende, O. Doll, F. Clement, and D. Biro, "Ultrafine front side metallization on silicon solar cells by industrial dispensing technology," in *29th EU PVSEC*, Amsterdam, The Netherlands, 2014, pp. 1304–1306.
[26] P. Rothhardt, S. Meier, K. Jiang, A. Wolf, and D. Biro, "19.9 % efficient bifacial n-type solar cell produced by co-diffusion-cobin," in *29th EU PVSEC*, Amsterdam, The Netherlands, 2014, pp. 653–655.
[27] K. Krauß, F. Fertig, J. Greulich, S. Rein, and R. Preu, "biPERC silicon solar cells enabling bifacial applications for industrial solar cells with passivated rear sides," *Phys. Stat. Sol. (a)*, vol. 213, no. 1, pp. 68–71, 2016.
[28] M. Demant, M. Glatthaar, J. Haunschild, and S. Rein, "Analysis of luminescence images applying pattern recognition techniques," in *25th EU PVSEC*, Valencia, 2010, pp. 1078–1082.
[29] M. Demant, J. Greulich, M. Glatthaar, J. Haunschild, and S. Rein, "Modelling of physically relevant features in photoluminescence images," *Energy Procedia*, vol. 27, pp. 247–252, 2012.

[30] J. Rentsch, M. Bivour, D. Erath, F. Feldmann, A. Fischer, S. Mack, A. Moldovan, J. Temmler, M. Hermle, S. W. Glunz, and R. Preu, "A New Pilot Research Facility for HJT and Selective Contact Solar Cells – PV-TEC SELECT," in *33rd EU PVSEC*, Amsterdam, The Netherlands, 2017, pp. 441–446.
[31] N. Wöhrle, T. Fellmeth, E. Lohmüller, P. Baliozian, A. Fell, and R. Preu, "The SPEER solar cell – simulation study of shingled PERC technology based stripe cells," in *33rd EU PVSEC*, Amsterdam, The Netherlands, 2017, pp. 844–848.
[32] F. Clement, A. Lorenz, M. Pospischil, D. Biro, H. Brocker, D. Bangel, R. Greutmann, M. Lehner, T. Ott, F. Hage, K. Oppelt, T. Honold, L. Wende, M. Ringel, A. Senne, J. Rohde, and R. Preu, "High Throughput Printing for Highly Efficient Cost-Effective Si Solar Cells," in *33rd EU PVSEC*, Amsterdam, The Netherlands, 2017, pp. 373–376.
[33] F. Clement, M. Linse, S. Tepner, N. Wengenmeyr, L. Ney, K. Krieg, A. Lorenz, M. Pospischil, S. Bechmann, K. Oehrle, S. Steckemetz, and R. Preu, "Project "FINALE" - Screen and Printing Process Development for Ultra-Fine-Line Contacts below 20µm Finger Width," in *36th EU PVSEC*, Marseille, 2019, pp. 255–258.
[34] F. Fertig, I. Höger, M. Schaper, R. Lantzsch, F. Kersten, M. Bartzsch, F. Frühauf, B. G. Lee, A. Mette, B. Klöter, and J. W. Müller, "Q.ANTUM on p-type Cz silicon: high-end performance and reliability," in *7th World Conference on Photovoltaic Energy Conversion*, Waikoloa, Hawaii, 2018, 0993 - 0995.

PROGRESS IN THE DEVELOPMENT OF UPGRADED METALLURGICAL SILICON SOLAR CELLS FOR INDUSTRIAL PRODUCTION

Carlos del Cañizo*[1], David Fuertes Marrón[1], Nerea Dasilva-Villanueva[1], Manuel Funes[1,5], Juan José Torres[1], Bo-Kyung Hong[1], Bülent Arikan[2], Hasan Hüseyin Canar[2], Rasit Turan[2,3], Hasan Asav[2], Ahmet Emin Keçeci[2], Sümeyye Koçak Bütüner[2], Guillermo Sánchez Plaza[4], Laura Méndez[5], Eduardo Forniés[5]
[1] Instituto de Energía Solar, ETSI Telecomunicación, Universidad Politécnica de Madrid, Spain
[2] ODTÜ-GÜNAM (Center for Solar Energy Research and Applications), Turkey
[3]Middle East Technical University, Department of Physics, Turkey
[4]Nanotechnology Center – Universidad Politécnica de Valencia, Valencia, Spain
[5]Aurinka PV Group, Rivas-Vaciamadrid, Spain
* Corresponding author: Instituto de Energía Solar. ETSI Telecomunicación, Avenida Complutense 30, 20840 Madrid, Spain, carlos.canizo@upm.es, phone: +34-910672521

ABSTRACT: This contribution reports on significant improvements achieved in the field of upgraded metallurgical silicon (UMG-Si) as an alternative route to Siemens-refined poly-Si for the manufacture of high-efficiency multicrystalline (mc) silicon solar cells and modules. Recent results in the whole cycle from feedstock production to module performance in the field are summarized. To further improve UMG-Si material quality, we have implemented a multilevel research program, using the charge carrier lifetime at different stages of wafer processing as an effective metric for the assessment of the strategies developed for the upgrading of the material electronic quality. Carrier lifetimes up to 722 μs have been demonstrated by the introduction of an optimized sequential Phosphorus Diffusion Gettering (PDG) which ensures, on the one hand, the non-limiting nature of bulk-related recombination mechanisms in our UMG-Si wafers, and, on the other hand, the suitability of adopting high-efficiency device architectures based on this material. Furthermore, the optimization of metal-assisted nano-texturing and of Al_2O_3/SiN_x-based rear side passivation coating have also been monitored by means of spectrophotometric and carrier lifetime measurements. Remarkable values for the passivation quality have been demonstrated and promising efficiencies in pilot production-line for multicrystalline PERC devices have been obtained. Advanced cell architectures on UMG-Si are currently under investigation, in particular a TOPCore structure, with preliminary tests indicating that efficiencies higher than 22% are within reach. Our results, obtained in the frame of the European SOLAR-ERA-NET CHEER-UP project, indicate the effectiveness of isolated processing actions, compatible with industrial up-scaling, in the upgrading of the electronic performance of mc-UMG-Si wafers; thereby improving the efficiency of related devices and the possibility of implementing advanced device architectures based on mc-UMG material. In this regard, the advantages of UMG-Si related to a lower economic and environmental impact in comparison to standard poly-Si can, in principle, be fully exploited.
Keywords: Feedstock. Upgraded Metallurgical Silicon. Multi-crystalline.

1 INTRODUCTION

The advancement of alternative routes to the Siemens purification process (that produces the conventional feedstock known as "polysilicon") has been identified to have the neat potential for reducing both the manufacturing costs and the energy consumption associated to the fabrication of solar silicon. Among them, the "metallurgical route" has been historically developed in waves associated to events such as the oil crisis of the 1970s and the silicon shortage of the 2000s, producing the so-called Upgraded Metallurgical Silicon (UMG-Si). While UMG production was mostly stopped throughout the second half of the last decade, the need of fast capacity expansions to limit the impact of polysilicon's increases in price, as well as a heightened awareness of environmental aspects, may allow the re-entry of UMG in the current market, provided these advantages are accomplished without penalty on the level of device performance of the current Photovoltaic (PV) technology.

In this work we report on significant advances achieved in the electronic quality of silicon solar cells based on the multicrystalline UMG-Si produced by FerroSolar, compatible with industrial scale and with minimal impact on the value chain of current technology. The upgrading strategy is focused at three stages, namely: (i) the improvement of the bulk quality of UMG-Si wafers by means of conditioning steps prior to emitter provision, using sequential phosphorous diffusion gettering steps optimized to the properties of the material; (ii) the implementation of metal-assisted etching (MAE) nano-texturing and surface passivation coating schemes, which have been proven effective for the minimization of reflective losses and surface recombination of photogenerated carriers, respectively; and (iii), the adoption, supported by the demonstrated upgrade of bulk and surface properties of mc-UMG-Si wafers following the previous points, of high-efficiency solar cell architectures for the full exploitation of the UMG-Si electronic, economic and environmental potential.

2 THE UMG ROUTE BY FERROSOLAR

Details of the whole purification process from metallurgical silicon to solar cell are given elsewhere [1]; they are summarized in the following paragraphs.

Having metallurgical silicon as starting material, the FerroSolar process concatenates several steps that include slagging to reduce the boron content, vacuum evaporation to reduce the phosphorus content, and directional solidification to remove metallic impurities. The resulting UMG-Si product is characterized by a boron concentration below 0.2 ppmw, phosphorus concentration in the range of 0.1–0.3 ppmw, and total metal concentration below 0.5 ppmw.

High Performance Multicrystalline ingots were grown at Pillar in Ukraine with this silicon feedstock. As UMG-Si is a compensated material (similar concentration of boron and phosphorous), certain quantities of Ga (from 5 to 10 ppmw) and P (0.12 ppmw) were added to control the resistivity variation along the ingots, as to keep it in the range of 1 to 3 Ωcm. Ingots were further cut into 25 bricks with a square area of 156.75 × 156.75 mm^2 and later sliced into wafers.

45.000 of these wafers were sent to an independent Tier 1 solar cell manufacturer and introduced in the manufacturing line to produce multicrystalline PERC solar cells. The UMG-Si-based cell performance was then compared with that obtained with poly-Si material. With an average cell efficiency of 20.4% for conventional poly-Si, the UMG-Si solar cells showed a wider distribution in the efficiency histogram towards the lower efficiencies, but in average there was only a slight reduction in cell efficiency of less than 1.4% relative.

In a parallel test, whose details can be found in [2], Al BSF solar cells were produced, with efficiencies over 18%, and modules were manufactured. 7 modules with UMG-Si solar cells were installed in Tudela (north of Spain), together with 7 modules made of poly-Si solar cells, and their module performance was tracked for a two-year period. The first year, the Performance Ratio corrected to 25°C (PR_{25}) decreased from 1 to 0.972 for UMG-Si and to 0.976 for poly-Si, which is in concordance with the initial light induced degradation (LID) of Al-BSF solar cells. After 2 years of degradation, the PR_{25} is even higher for UMG-Si modules (PR_{25}=0.980) than for poly-Si ones (PR_{25}=0.977).

Note that these relevant results have been obtained without any specific adaptation of the conventional solar cell processing to the characteristics of the UMG-Si. Hence, the interest in exploring the potential of the material when tailoring process conditions is being addressed in the frame of the European SOLAR-ERA-NET project "Low-cost high-efficient and reliable UMG PV cells" (CHEER-UP, Reference JTC-2_2019_030) [3], whose main results to date are reported in the following sections.

3 OPTIMIZING UMG SOLAR CELL PROCESSING STEPS

3.1 Improving the bulk

To effectively remove impurities from the bulk of the UMG-Si wafers, a phosphorous diffusion gettering (PDG) step is introduced prior to cell processing.

The wafers used in this analysis were randomly selected from different bricks coming from three ingots, as to account for positional effects and batch variability. They were cut into 52 x 52 mm^2 samples, and firstly characterized by Photoconductance Decay (PCD) measurements using a Sinton Instruments WCT-120 tool, with surfaces passivated with iodine-ethanol (IE). They were then chemically treated with CP4 (HNO_3/HF) etching, RCA1 (HN_3:$H2O_2$) surface cleaning and a treatment in 2% HF to ensure native oxide removal before the thermal process.

P-diffusion was performed in a tubular furnace, varying the processing time (30, 60 and 90 minutes) and temperature (780°C, 800°C, 820°C and 850°C). It was followed by a 10-minute drive-in. The P-source was liquid $POCl_3$ kept at 26°C, assuring super-saturation. After PDG, the emitter formed atop the surface was removed by etching and PCD measurements were conducted again.

The impact of sequential PDG processes carried out under different conditions (duration and temperatures) on the bulk carrier lifetime of UMG-Si at a carrier injection level Δn = 10^{15} cm^{-3} can be seen in Figure 1. Best results for the first gettering process are systematically obtained at a low temperature (780°C), resulting in lifetimes upgraded to an average level of 300 µs. Further improvements can be observed after a second PDG process, reaching an outstanding maximum lifetime of 722 µs. It is also interesting to note that this second P-diffusion provides an efficient gettering effect under a relatively wide range of temperatures, in particular and most conveniently, those compatible with P-diffusion processes for emitter provision and subsequent cell processing.

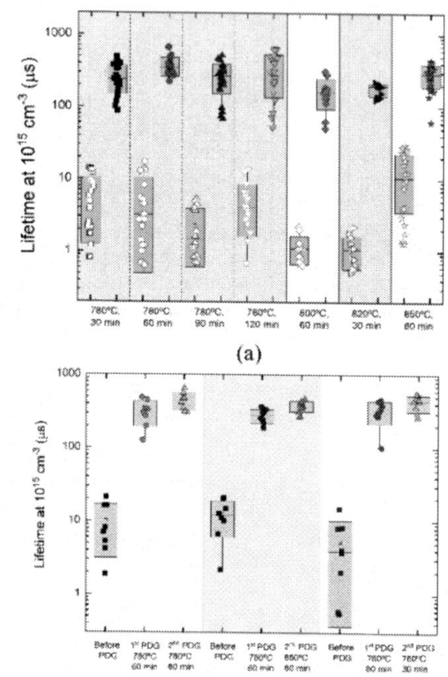

Figure 1: Effective lifetime variation after several PDG conditions for IE-passivated UMG-Si samples, after a single PDG (a) and with a subsequent one (b).

More details on the strategy deployed to improve wafer quality through PDG can be found in reference [4].

3.2 Improving the surfaces

One of the leading techniques compatible with industrial solar cell production for surface structuring is the Metal-assisted nano-texturing method. With this method, the surface reflection of mc wafers can be significantly reduced and even lower reflection values can be achieved than alkaline texturing applications used in monocrystalline wafers [5, 6]. A silver-assisted nano-texturing has been developed for UMG-Si wafers (see Figure 2), with which reflective losses were reduced to the 9% target level in the optical range of interest. Afterwards, the rear side of the nano-textured surface was etched and polished, continuing with the fabrication of monofacial UMG PERC solar cells.

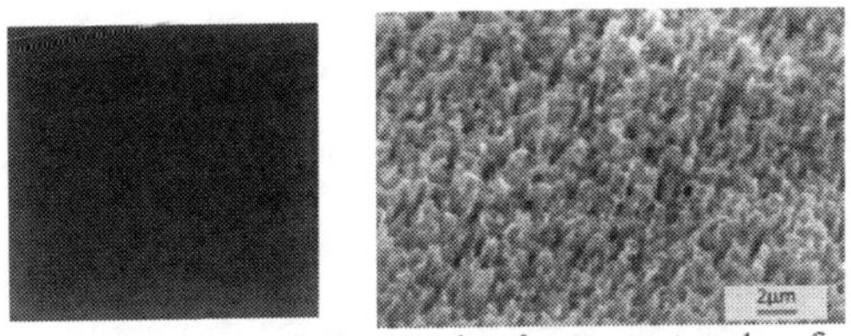

Figure 2: Images of silver-assisted nano-textured surface.

In parallel, in-depth studies were conducted with mc UMG wafers on the development of the rear side passivation strategy for PERC type solar cells. Al_2O_3/SiN_x stack layer configurations were deposited on both sides of polished UMG wafers by atomic layer deposition (ALD) and plasma-enhanced chemical vapor deposition (PECVD), respectively. Al_2O_3 thickness is used as an optimization parameter and effective lifetime measurements were performed on symmetrically passivated and gettered samples as an estimation of the passivation quality. An example of the results obtained, which are described in more detail in [4], is shown in Figure 3, confirming the excellent upgrading of mc-UMG-Si wafers achieved, both in bulk and surface properties.

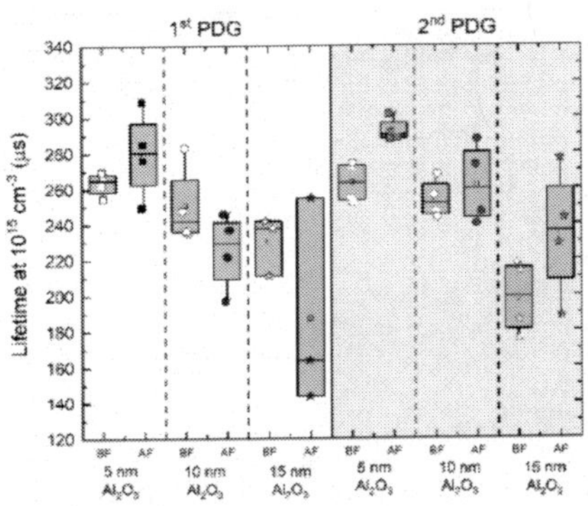

Figure 3: Carrier lifetimes obtained symmetrically Al_2O_3/SiN_x passivated samples, for different thicknesses of the Al_2O_3, before (BF) and after firing (AF). Left side corresponds to measurements after a PDG and right side to measurements for two subsequent PDGs.

3.3 Implementing PERC solar cells

The nanotexturing, gettering, doping and passivation strategies developed have fed a PERC-type cell production process in pilot line using industrial-scale equipment, on mc-UMG wafers with $156.75x156.75mm^2$ size and base resistivity of 1.1-1.4 Ωcm. Although partial positive effects of the second gettering were observed in the passivation studies, it was decided that a single gettering operation would be sufficient to reach the targeted efficiency values. Sheet resistance for homogeneous emitter created after phosphorus doping was optimized in the range of 95-105 Ω/□.

Front and rear passivation strategies were studied separately and $SiO_2+SiN_x+SiO_xN_y$ on the front and AlO_x+SiN_x stack layer configurations on the rear side were chosen as antireflection coating and passivation layers, respectively (Figure 4).

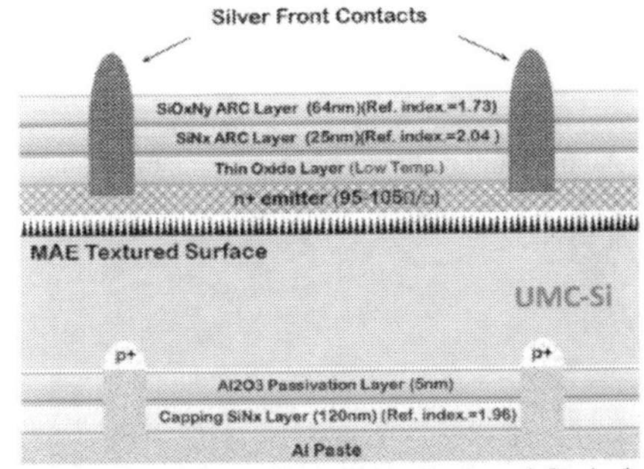

Figure 4. PERC solar cell structure designed for industrial scale production.

Metal contacts have been formed by screen-printing with a 5-busbar, 129 finger configuration in the front grid, following a careful and ongoing optimization of the firing process as a function of peak temperature and processing time. The observed tradeoff between fill factor and open-circuit voltage, linked to the two mentioned process parameters together with the hydrogenation processes, is currently serving as a guide for reaching maximum efficiency. With the help of the ramp-up processes and new approaches to be developed for the ongoing industrial scale recipe, there is no obstacle to the production of high efficiency mc-UMG PERC solar cells compatible with the high lifetime values achieved in the passivation tests in the near future.

4 EXPLORING ADVANCED CELL ARCHITECTURES ON UMG SILICON

To assess the potential of UMG to keep up and follow the rapid evolution of silicon technology, experiments are underway at Fraunhofer ISE to implement on the FerroSolar UMG-Si substrates a TOPCon rear emitter structure, called TOPCoRE [7].

As a first step, bulk lifetimes after PDG and TOPCon anneal steps have been measured by calibrated Photoluminescence (PL), confirming that values of carrier lifetime in the range of hundreds of microseconds are obtained in most of the area of the processed wafers, even locally surpassing the millisecond in some regions.

Also, samples have been satisfactorily plasma-textured, achieving a reflectance (without ARC) lower than 5.9% in the range 300 to 1000 nm. It has been confirmed the high passivation level that can be reached on these textured surfaces with an Al_2O_3/SiN_x stack, measuring implied V_{oc} >700 mV in good areas of the processed wafers.

Based on the injection-dependent PL results, an ELBA analysis is performed, i.e., device simulations have been implemened to give the spatial efficiency distribution that can be expected from the material quality [8]. As shown in Figure 5, a global efficiency higher than 22% is within reach with these multicrystalline UMG-Si substrates.

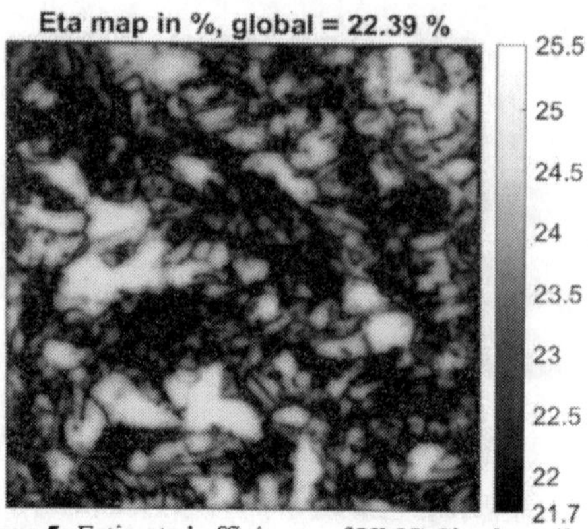

Figure 5: Estimated efficiency of UMG-Si solar cells according to an ELBA analysis for a TOPCore structure.

5 CONCLUSIONS

Recent progress on the enhancement of the electronic quality of UMG-Si has been presented. Optimized PDG treatments, surface texturing, passivation schemes, and metallization-related parameters, all of them compatible with industrial manufacturing, are shown capable of significantly reducing the gap observed in the electronic quality between conventional poly-Si and UMG-Si both at wafer and device levels, thereby opening the path to high-efficiency cell architectures on UMG material.

UMG-Si can be the basis of a low-CAPEX low-cost environmental-friendly supply of solar silicon, keeping up with the continuous progress that crystalline silicon technology is experimenting.

6 ACKNOWLEDGEMENTS

This work is part of the R&D SOLAR-ERA.NET Cofund 2 project CHEER-UP, funded by the Spanish MCIN/ AEI/10.13039/501100011033/ through projects PCI2019-111834-2 and PCI2019-111903-2, by the Spanish Centro para el Desarrollo Tecnológico Industrial (CDTI), and by the Turkish TÜBITAK through Project 219M029. It has also been partially funded by the Regional Government of Madrid with the support from FEDER Funds (Project MADRID-PV2, S2018/EMT-4308).

The involvement of Fraunhofer ISE's Group Innovative Cleanroom Technologies lead by Dr. Jan Benick is greatly acknowledged.

7 REFERENCES

[1] E. Forniés, B. Ceccaroli, L. Méndez, A. Souto, A, Pérez Vázquez, T. Vlasenko, J. Diéguez, Mass Production Test of Solar Cells and Modules Made of 100% UMG Silicon. 20.76% Record Efficiency, Energies. 12 (2019) 1495. doi:10.3390/en12081495.

[2] E. Forniés, C. del Cañizo, L. Méndez, A. Souto, A. Pérez Vázquez, D. Garrain, UMG silicon for solar PV: from defects engineering to PV module degradation, Solar Energy. 220 (2021) 354–362. doi:10.1016/j.solener.2021.03.076.

[3] https://www.era-learn.eu/network-information/networks/solar-cofund-2/solar-cofund-2-joint-call/low-cost-high-efficient-and-reliable-umg-pv-cells-cheer-up

[4] N. Dasilva-Villanueva *et al.*, submitted to Solar Energy Materials and Solar Cells.

[5] X. Li *et al.*,, High-efficiency multi-crystalline black silicon solar cells achieved by additive assisted Ag-MACE, Solar Energy 195 (2020) 176–184, doi:10.1016/j.solener.2019.11.045.

[6] J. Zhang *et al.*, Chemically microsurface-sphericizing on diamond wire-sawn multi-crystalline silicon wafers by metal-assisted chemical etching and rounding etching, Solar Energy 220 (2021) 1032–1039, doi:10.1016/j.solener.2021.01.064.

[7] A. Ritcher *et al.*, Design rules for high-efficiency both-sides contacted silicon solar cells with balanced charge carrier transport and recombination losses, Nature Energy 6 (2021), 429-438, doi:10.1038/s41560-021-00805-w

[8] B. Michl *et al.*, Efficiency limiting bulk recombination in multicrystalline silicon solar cells, Solar Ener. Mat. Solar Cells 98 (2012) 441–447, doi:10.1016/j.solmat.2011.11.047

DESIGN OF A GROWTH PROCESS OF HIGH-QUALITY QUASI-MONOCRYSTALLINE SILICON INGOT INTEGRATING EXPERIMENTAL, THEORETICAL, COMPUTATIONAL, AND DATA SCIENCES

Noritaka Usami[1], Xin Liu[1], Yusuke Fukuda[1], Hiroyuki Tanaka[1], Kentaro Kutsukake[2], Takuto Kojima[3]
[1]Graduate School of Engineering, Nagoya University, [2]Center for Advanced Intelligence Project, RIKEN,
[3]Graduate School of Informatics, Nagoya University
[1,3] Furo-cho, Chikusa-ku, Nagoya, 464-8603, Japan
[2] Nihonbashi, Chuo-ku, Tokyo 103-0027, Japan

ABSTRACT: We report on a methodology to design a growth process of high-quality quasi-monocrystalline silicon ingot integrating experimental, theoretical, computational, and data sciences. First, the model of crystal growth simulation was validated by utilizing experimental data, and the resulting model was used to collect training data showing the relationship between multi-dimensional process parameters and the dislocation density distribution after growth. A machine learning model of the simulation was created to instantly predict the dislocation density from the process parameters, allowing the application of an optimization algorithm. As a result, efficient design of the optimal growth process was possible. Furthermore, by using seed crystals with artificial grain boundaries and their spontaneous change during the growth, we have clarified the effect of the grain boundary structure on the direction of grain boundary growth and the carrier recombination velocity at the grain boundary. These permits to design of an appropriate seed crystal and process, which will result in the growth of high-quality quasi-monocrystalline silicon ingot.
Keywords: Crystallisation, c-Si, Defects, Dislocation, Machine Learning

1 INTRODUCTION

Materials informatics, which accelerates the discovery of novel high-performance materials, is advancing, and methodologies are being developed to answer the question, "What kind of materials should we make?" For example, candidates for solid-state lithium superionic conductors [1], lightweight high-entropy alloys with superior mechanical properties [2], and high-performance absorbers for thin film solar cells [3] can be efficiently obtained by the high-throughput computational method. On the other hand, there are great expectations for the advancement of process informatics to answer the question of "how to make candidate materials?". In general, controlling parameters for materials synthesis are diverse, and their impacts are not fully understood. Therefore, improving and optimizing individual processes have been the main focus in actual manufacturing, while a scientific search for the optimal process has rarely been taken.

In the growth of quasi-monocrystalline silicon ingot, seed crystals are placed at the bottom of a large crucible. The silicon raw material and the top part of the seed are carefully melted in the temperature gradient in the furnace. Then, the silicon melt is directionally solidified from the bottom to the top, thereby growing a single crystal that has inherited the crystal orientation of the seed crystal. In reality, however, grain boundaries are always present because multiple seed crystals must be aligned to cover the entire bottom surface of the crucible. In addition, the crucible wall acts as an inhomogeneous nucleation site. Therefore, the suppression of dislocation generation and polycrystallization from the seed joints is an issue. The use of functional grain boundaries is known to be effective in suppressing polycrystallization [4]. In general, grain boundaries are harmful defects that act as carrier recombination sites and reduce the conversion efficiency of solar cells. However, it is possible to intentionally exploit the interaction between grain boundaries and impurities or other defects to provide beneficial functions for the overall material properties. In the proposed method, functional grain boundaries are artificially formed near the crucible walls using artificially designed seed crystals, and these grain boundaries suppress the expansion of polycrystalline grains. The concept of functional grain boundaries was extended by proposing the seed manipulation for artificially controlled defects technique (SMART), in which a high-density defect layer intentionally formed by using seed crystals is used for stress relaxation and dislocation confinement [5]. Furthermore, TOPCon cells have been fabricated using SMART wafers, and an energy conversion efficiency of 23.3% has been reported [6]. Therefore, the design of seed crystals is extremely important for achieving high-quality quasi-monocrystalline silicon ingot.

In this study, we attempted to develop a methodology to optimize the process with multiple parameters by integrating experimental, theoretical, computational, and data sciences. The crystal growth process of a quasi-monocrystalline silicon ingot was employed as a model process, for which the fundamental equations to describe the growth process are relatively simple. We have utilized experimental results to improve crystal growth simulations and developed a machine learning model to predict the simulation results, enabling fast crystal quality prediction. As a result, we could apply an optimization algorithm to the crystal growth recipe and efficiently optimize the process. We also report on systematic crystal growth experiments and analysis on crystals with artificial grain boundaries to establish design guidelines for seed crystals.

2 GROWTH PROCESS OPTIMIZATION

Optimization of the growth process consists of refining the crystal simulation model using experimental data, creating a machine learning model of the simulation, and optimization algorithms [7]. Crystal growth experiments were performed using a vertical Bridgman crystal growth furnace with three-zone heaters. The seed crystal contains four artificial grain boundaries. Details of the preparation of seed crystals are reported elsewhere [8]. The power of each heater and temperatures at ten different points were measured as multiple data. The

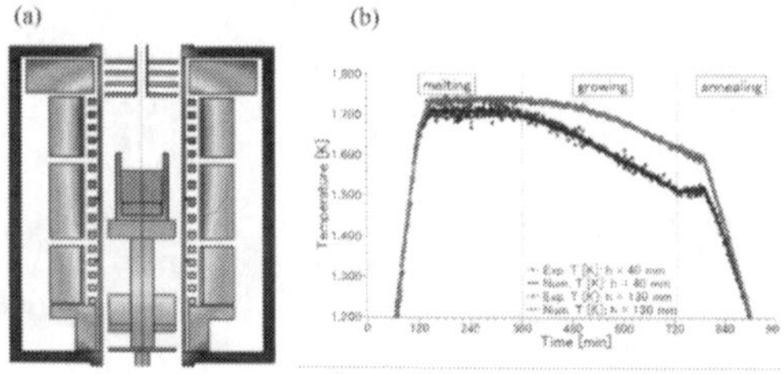

Figure 1: (a) A typical temperature distribution during the growth (b) A comparison of temperature profiles obtained by the experiment and simulation at two different heights.

solidification completion time was obtained by observing the surface condition of the silicon during the process. The shape of the solid-liquid interface and its temporal change was obtained from the electrical resistivity distribution of the grown crystal.

A commercial software, CGSim (STR), was used to simulate directional solidification of quasi-monocrystalline silicon ingot from melting to annealing. To save the computational cost, we employed a two-dimensional axisymmetric model, and the square crucible and other non-axisymmetric parts were converted to an axisymmetric shape according to the principle of the equivalent volume method. This approximation introduces uncertainties in the model, for example, relative positions of the heater and thermocouples, susceptor width, and so on. We optimized such parameters to minimize the error between experiments and simulations. Figure 1 shows (a) a typical snapshot of the temperature distribution and (b) a comparison of temperature profiles obtained by the experiment and simulation at two different positions. It is seen that the simulation after optimization of the model well reproduces the experimental results.

Our most important task is to minimize dislocation density by controlling various process parameters. The Haasen-Alexander-Sumino model in CGSim can be used to predict the spatial distribution of stress and dislocations. While simulation with an elaborated model is a useful tool, it is not realistic to exhaustively calculate every combination of process parameters, as the computation time for a single process can be as long as 6 hours. Therefore, we attempted to construct a machine learning model that can predict the results of simulations at high speed without having to run the simulations. Specifically, we created an artificial neural network (ANN) model that predicts stress and dislocation density after growth. The change rates in temperatures of multiple heaters and the pulling speed of the crucible and the position in the crystal are used as inputs of ANN. 150 training data sets were generated by simulation. Different ANN structures in terms of the number of hidden layers and neurons were trained, and the structure to show the lowest root-mean-square errors was selected. As a result, the machine learning model could instantly predict stress and dislocations, which permits the application of an optimization algorithm to require calculations with a vast set of input parameters.

We employed non-dominated sorting genetic algorithm (NSGA-II), which is a multi-objective optimization algorithm. NSGA-II generated 160 random individuals and then recommended the optimum parameter settings after 400 generations of evolution. The

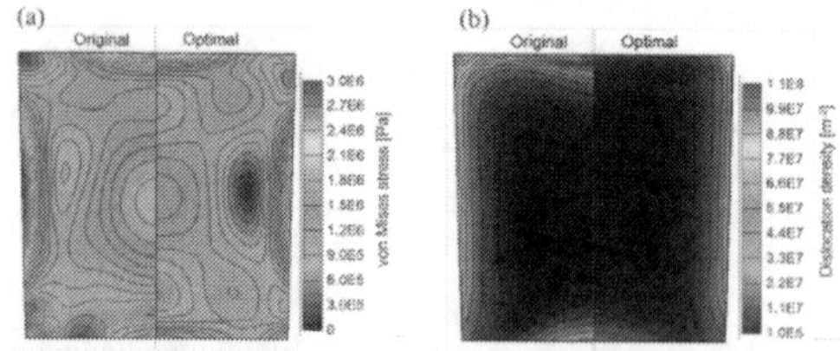

Figure 2: Comparison of distribution of (a) stress and (b) dislocation density in the crystal grown using the original and optimal recipe.

optimal recipe could be chosen from the Pareto front to decrease the stress and dislocation density while keeping a shorter growth time. The original recipe empirically determined by crystal growth experts consists of the constant temperatures for all three heaters and the constant pulling rate of the crucible. On the other hand, the optimized recipe suggested that we should gradually decrease the temperature of the top heater and increase that of the middle heater. In addition, it was recommended that we should slow down the pulling rate of the crucible during the growth.

To verify the usefulness of the optimal recipe, we performed the simulation using CGSim with the original and optimal recipe. Figure 2 compares (a) von Mises stress and (b) dislocation density. It is seen that the dislocation density in the crystal grown using the optimal recipe is much smaller than that grown using the original recipe especially at the top part of the ingot. The slower growth rate under a lower temperature gradient in the optimal recipe would be responsible for the reduction of dislocation density.

3 TOWARDS SEED OPTIMIZATION

Although the design of seed crystals is crucial to realize high-quality quasi-monocrystalline silicon ingot, the grain boundary structure is diverse and complex with five macroscopic degrees of freedom. Therefore, the atomic structure and physical properties of grain boundaries are known only for a limited variety of grain boundaries even for simple materials such as silicon. Specifically, we investigated the electrical properties at grain boundaries, which are essential for designing high-performance polycrystalline materials, and the effect of grain boundary structure on the grain boundary growth direction during the solidification process.

Seed crystal plates cut from <100> and <110> oriented Czochralski single crystal silicon ingots were aligned with the same crystal orientation in the z direction to design <100> and <110> tilt grain boundaries including asymmetric tilt grain boundaries. By controlling the solid-liquid interface shape during crystal growth using this designed grain boundary and the directional solidification, two of the five degrees of freedom are fixed, and the grain boundary can be expressed by three grain boundary structure factors (relative orientation difference α, asymmetry angle β, and deviation angle θ from the z direction). The electrical properties at the grain boundaries were quantitatively evaluated by the carrier recombination velocity (v_{GB}), which was estimated using a machine learning model from photoluminescence images near the grain

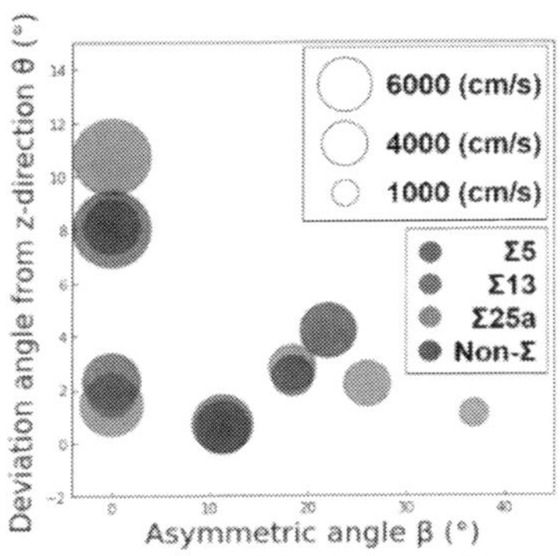

Figure 3: Comparison of distribution of (a) stress and (b) dislocation density in the crystal grown using the original and optimal recipe.

boundaries [9].

Most of the grain boundaries fabricated using <110> oriented seed crystals (<110> design grain boundaries) grew along the orientation direction. On the other hand, grain boundaries fabricated using <100> oriented seed crystals (<100> design grain boundaries) grew away from the orientation direction. Figure 3 shows the effect of <100> design grain boundary structure on the recombination rate. θ was found to have the largest effect on v_{GB} among the three factors. The originally designed grain boundary is a tilted grain boundary, and a change in θ corresponds to the introduction of a twist component. On the other hand, changes in α and β are changes in the tilt component. We believe that this result is due to the formation of a three-dimensional complex relaxation structure during reconstruction, as the designed grain boundary changes to a mixed grain boundary with the introduction of the torsional component [10].

The reason why the grain boundary growth direction differs from the seed crystal orientation direction is discussed in terms of energy using a simple linear model. Specifically, we believe that this is due to the generally higher grain boundary energy of <100> tilt grain boundaries compared to <110> tilt grain boundaries [8].

These findings are helpful in designing functional defects in SMART ingot production.

4 SUMMARY

We demonstrated the usefulness of a methodology that combines experimental, theoretical, computational, and data sciences to optimize the multi-dimensional controlling parameters of the growth process of quasi-monocrystalline silicon ingots. In addition, we attempted to elucidate the fundamental governing factors of multicrystalline material properties using an artificial grain boundary whose structure can be expressed by three factors and our original machine learning model. As a result, we obtained knowledge about the structures that have a significant influence on electrical properties and the direction of grain boundary growth, which will be helpful in designing seed crystals for quasi-monocrystalline silicon ingot.

This work was partly supported by JST/CREST under Grant No. JPMJCR17J1 (2017–2023).

[1] Y. Wang, W.D. Richards, S.P. Ong, L.J. Miara, J.C. Kim, Y. Mo, G. Ceder, Nat. Mater. 14 (2015) 1026–1031.
[2] R. Feng, C. Zhang, M.C. Gao, Z. Pei, F. Zhang, Y. Chen, D. Ma, K. An, J.D. Poplawsky, L. Ouyang, Y. Ren, J.A. Hawk, M. Widom, P.K. Liaw, Nat. Commun. 12 (2021) 6–15.
[3] K. Choudhary, M. Bercx, J. Jiang, R. Pachter, D. Lamoen, F. Tavazza, ArXiv. (2019) 1–30.
[4] K. Kutsukake, N. Usami, Y. Ohno, Y. Tokumoto, I. Yonenaga, Appl. Phys. Express. 6 (2013) 025505.
[5] I. Takahashi, S. Joonwichien, T. Iwata, N. Usami, Appl. Phys. Express. 8 (2015) 6–10.
[6] M.C. Schubert, F. Schindler, J. Benick, S. Riepe, P. Krenckel, A. Richter, R. Müller, B. Hammann, S. Nold, Sol. Energy Mater. Sol. Cells. 219 (2021) 110789.
[7] X. Liu, Y. Dang, H. Tanaka, Y. Fukuda, K. Kutsukake, T. Kojima, T. Ujihara, N. Usami, ACS Omega. 7 (2022) 6665–6673.
[8] Y. Fukuda, K. Kutsukake, T. Kojima, N. Usami, CrystEngComm. 24 (2022) 1948–1954.
[9] K. Kutsukake, K. Mitamura, N. Usami, T. Kojima, Appl. Phys. Lett. 032105 (2021) 1–6.
[10] Y. Fukuda, K. Kutsukake, T. Kojima, Y. Ohno, N. Usami, J. Appl. Phys. 132 (2022) 025102.

This is a short version of the paper submitted to the journal Solar RRL.

It has been selected by the Committee of the WCPEC-8 for submission to Solar RRL.

HIGH LIFETIME GA-DOPED CZ-SI FOR CARRIER SELECTIVE JUNCTION SOLAR CELLS

Jörg Horzel, Sebastian Mack, Ioan Voicu Vulcanean, Karin Zimmermann, Nico Jung, Sebastian Pingel, Wolfram Kwapil, Felix Maischner, Hannes Höffler, Sattar Bashardoust, Dirk Wagenmann, Johannes Greulich, Johannes Seif, Anamaria Steinmetz, Jochen Rentsch
Fraunhofer Institute for Solar Energy Systems ISE, Heidenhofstrasse 2, 79110 Freiburg, Germany
Tel: +49-761-4588-2180, e-mail: joerg.horzel(at)ise.fraunhofer.de

ABSTRACT: Industrial mass production of solar cells is moving towards carrier selective junction solar cells with passivating contacts such as TOPCon, POLO and HJT solar cells. At the same time many manufacturers consider switching from p-type Cz-Si to n-type Cz-Si wafers.
This contribution indicates that Ga-doped p-type Cz-Si material is still a viable option for the new type of devices while giving an opportunity to benefit from 8-10% lower wafer cost. In that respect a Ga-doped Cz-Si ingot has been specified and characterized. We report minority carrier diffusion lengths that are an order of magnitude larger than the thickness of the studied HJT and TOPCoRE devices. So far best p-type HJT solar cells perform 0.2 % absolute lower in efficiency than co-processed n-type reference HJT solar cells. Best TOPCoRE solar cells on Ga doped Cz-Si, however, show a 0.2% to 0.3% higher efficiency than their co-processed n-type TOPCon counterparts.
Keywords: Ga-doped Cz-Si, carrier selective junctions, minority carrier lifetime, light induced improvement

1 INTRODUCTION

For decades, p-type crystalline silicon wafers with resistivities of 0.5-3.0 Ω·cm have been the work horse in mass production of Al BSF and PERx-type solar cells. As the performance of such solar cells improved and diamond wire sawing of wafers enabled a reduction of wafering cost there was a clear shift from multicrystalline to Czochralski silicon (Cz-Si) monocrystalline wafers [1]. Now with a next transition to passivated contact device architectures such as tunnel oxide passivated contact (TOPCon), poly silicon on oxide (POLO) and heterojunction technology (HJT) many PV manufacturers prepare for switching to n-type Cz-Si. Outstanding solar cell efficiencies have been reported on such material [2-5].

However, moving from p-type to n-type Cz-Si comes at an additional cost for the wafer material. During the last years the cost as tracked for instance at PVInsights [6] for n-type compared to p-type Cz-Si wafers has been 8-10% higher. Leading wafer manufacturers see fundamental reasons why n-type Cz-Si wafers will remain more expensive than their p-type counterparts. This contribution indicates that switching the doping level towards higher resistivities and the doping species to Ga doping offers an alternative that might be economically advantageous compared to n-type material.

A complete Ga-doped Cz-Si ingot has been grown according to our specification. The base resistivity is in the range 5-12 Ω·cm after thermal donor annihilation. We investigate tunnel oxide passivated contact rear emitter (TOPCoRE) [7] and heterojunction technology (HJT) solar cells [8] and respective precursors for lifetime and implied open circuit voltage (i-V_{OC}) assessment. For this purpose, wafers are representatively selected along the ingot length. Thus, we look at material aspects with respect to high as well as to low temperature processing.

We demonstrate that such material allows for minority carrier diffusion lengths that are an order of magnitude larger than the solar cell thickness. Those diffusion lengths are larger than those typically obtained for 1 Ω·cm n-type Cz-Si material and should not limit the performance of the discussed solar cells.

Furthermore, we look at stability aspects of such devices with respect to typical conditions such as illumination at room and elevated temperatures as encountered in PV modules in the field. Those investigations are reported in a parallel publication [9]. Surprisingly, we observe a significant light or injection induced improvement in minority carrier lifetime (MCL), i-V_{OC}, V_{OC}, pseudo fill factor (pFF) and FF and a moderate improvement in short circuit current density (J_{SC}) within a very short illumination time (second to minute range) that is leading to efficiency improvement of up to 0.8% absolute on TOPCoRE solar cells. This improvement is clearly higher than improvement observed in parallel for TOPCon devices on 1 Ω·cm n-type Cz-Si that have been co-processed. The rear junction p-type Cz-Si TOPCoRE and front junction n-type Cz-Si TOPCon solar cells apply in this comparison the same processing sequence prior to studying the stability under illumination. The best TOPCoRE solar cells on p-type Cz-Si perform 0.2% absolute better in efficiency than co-processed TOPCon reference solar cells.

For the HJT low temperature processing route the same processing sequence has been applied as well to, both, 5-12 Ω·cm Ga-doped p-type and to 1 Ω·cm n-type Cz-Si wafers. In this case we compare front junction p-type Cz-Si solar cells to n-type Cz-Si rear junction solar cells. So far, we see a gap in average efficiency of 0.36 % in favor of the n-type reference process for which the processing sequence has been developed. The lower HJT performance on the p-type wafers suffers so far from a lower FF while J_{SC} and V_{OC} are slightly higher compared to the n-type reference HJT solar cells. It is not clear yet whether this gap is fundamental or can be solved by adapted processing. However, it can be stated that the material quality (diffusion length) of the high resistivity Ga-doped p-type Cz-Si is not limiting the p-type HJT cell performance. A high temperature pre-treatment step to de-activate thermal donors and to getter potential transition metal impurities does not seem to be required to obtain good solar cell performance along the complete ingot even though a small i-V_{OC} improvement from gettering is noticed. The advantage in i-V_{OC}, MCL and V_{OC} for gettered samples increases with increasing ingot length.

2 LOW TEMPERATURE PROCESSING

The heterojunction technology (HJT) offers a low

temperature processing route. In all processing steps, wafer temperatures do not exceed 250 °C. Cz-Si material contains oxygen related thermal donors (TDs) in the as received state [10]. For moderately doped material such as the investigated high resistivity Ga-doped p-type Cz-Si material the base resistivity is significantly increased by those TDs if no thermal processing at elevated temperatures is applied to de-activate the thermal donors. Material improvement from processing steps like external impurity gettering during diffusion processes may not be applied during HJT processing. For a part of the studied i-V_{OC} precursors and solar cells we apply prior to HJT processing a $POCl_3$ gettering step followed by a complete removal of the diffused and gettered zone by wet chemical etching. This pre-gettering step does not only getter impurities but also annihilates TDs. With that comparative investigation we intend to quantify potential improvement that could be expected by adding such additional processing steps.

2.1 Ingot material evaluation

For both, high and low temperature routes, wafers have been selected in equidistant steps along the complete ingot length of 10000 wafers from seed to tail end of the Ga doped Cz-Si ingot. Figure 1 shows the processing sequence for the low temperature HJT route including $POCl_3$ gettering prior to HJT processing for half of the selected wafers.

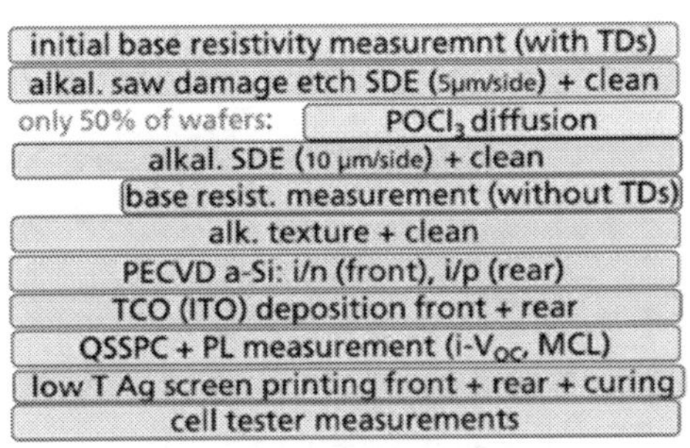

Figure 1: Evaluation process sequence for HJT precursors and solar cells

The high temperature diffusion step for the pre-gettered wafers is sufficient to completely de-activate oxygen related thermal donors in the material. As the etching steps are the same for gettered and non-gettered wafers and as the etching completely removes the diffused region (10 µm per side) the resistivity with and without thermal donors may be compared. Figure 2 shows this comparison along the complete ingot length. A significant difference is visible in the seed near region of the Ga-doped Cz-Si ingot. In this region there are the highest oxygen concentrations along the ingot.

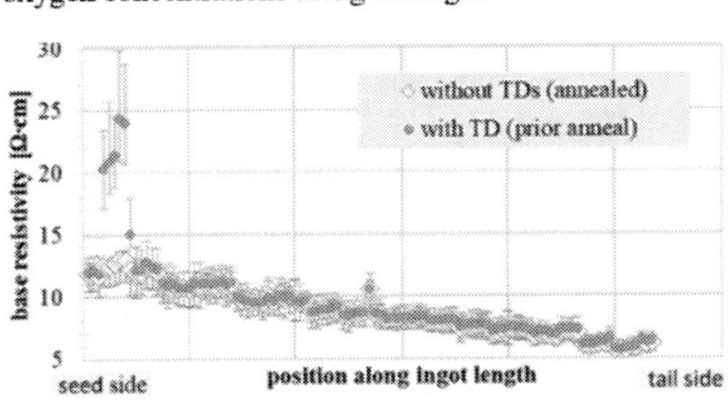

Figure 2: Base resistivity measurement results along the complete Ga doped Cz-Si ingot for wafers with and without TDs.

The oxygen-related thermal donor concentration in the seed near region compensates to a large extent the p-type Ga doping concentration in this region. After high temperature processing, however, the thermal donors are de-activated and the resistivity along the ingot follows the typical resistivity gradient caused by segregation behavior of the Ga doping atoms at the melt/solid interface during ingot growth. The Ga doping concentration increases from seed to tail end of the ingot.

Figure 3 shows the effective minority carrier lifetime (MCL) as assessed at an injection level of $5 \cdot 10^{15}$ cm^{-3} for the asymmetric solar cell precursors that had been representatively selected along the complete ingot length. This assessment is performed after TCO deposition (see Figure 1). The base resistivity shown in green (right y axis) indicates the position of the precursors in the ingot.

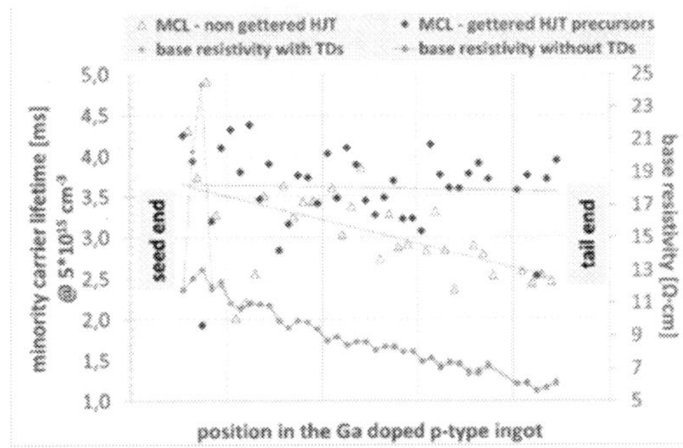

Figure 3: Evaluation of effective minority carrier lifetime along the Ga-doped ingot for HJT i-Voc precursors with and without pr-gettering prior to HJT processing.

The pre-gettering step that had been performed for half of the precursors improved the effective MCL compared to HJT precursors that have not seen any high temperature processing step. This improvement in effective MCL increases towards the tail end of the ingot where more getterable impurities (transition metals) can be expected due to segregation during ingot growth. However, with an effective MCL of more than 2.5 ms along the complete p-type ingot it is unlikely that the minority carrier diffusion length is limiting the solar cell performance. Due to the 2.6 times higher mobility of electrons (minority carrier in p-type Si) compared to holes (minority carrier in n-type Si) the diffusion length is clearly higher than for our reference process on 1 Ω·cm n-type Cz-Si and in all cases more than 10 times of the solar cell thickness. From that perspective the chosen Ga doped Cz-Si material is consequently not limiting the solar cell performance.

2.2 HJT solar cell results

Applying the processing sequence described in section 2.1 (see Figure 1) HJT solar cells have been co-processed on n-type reference Cz-Si wafers (1 Ω·cm) and those selected from the p-type Ga-doped Cz-Si ingot. The processing sequence has been applied as optimized for n-type Cz-Si. Thus, for the Ga-doped wafers the p-n junction is on the front side while for the n-type material it remains on the rear. Neither a-Si layers nor TCO have been adapted yet for the Ga-doped Cz-Si wafers. The resulting solar cell efficiency along the Ga-doped Cz-Si ingot is in average 0.36% absolute lower than for the reference process on n-type Cz-Si. The delta efficiency distribution is shown in Figure 4.

The HJT solar cells on high resistivity Ga-doped p-type Cz-Si show in average a 0.3 mA/cm² higher J_{SC}. This is likely a consequence from having the junction at the front. For the pre-gettered wafers the V_{OC} is in average 2.5 mV higher than for the n-type HJT solar cells. On the non-gettered p-type wafers V_{OC} on resulting HJT solar cells is in the same range as for (non-gettered) n-type wafers. The performance difference is so far dominated by a significantly (+1.5% absolute in average) higher fill factor for the n-type material. There is no significant difference in efficiency between pre-gettered and non-gettered p-type Cz-Si HJT solar cells. Neither do we see a change in performance for non-gettered p-type HJT solar cells in the region of the ingot that has significantly higher base resistivity due to the presence of thermal donors. The scattering in FF so far dominates the efficiency distribution.

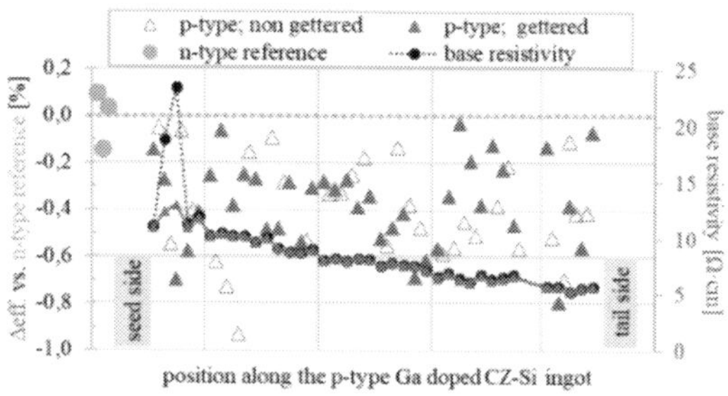

Figure 4: delta efficiency distribution along the Ga doped p-type ingot compared to the n-type reference case

The efficiency disadvantage (-0.36% absolute) results at this stage of evaluation (not yet optimized) for the Ga doped p-type ingot in a cost of ownership [11] penalty that is slightly higher than the cost advantage from the 8-10% lower p-type Cz-Si wafer prices. The top efficiencies in this first evaluation differ only by 0.2% absolute in favor of the n-type reference. Therefore, HJT process optimization for p-type wafers leading to higher FF values will be required to benefit from an economical advantage when using this alternative wafer source. Such improvement seems possible when adapting the processing sequence. In addition, there is still room for going to a slightly lower resistivity range.

3 HIGH TEMPERATURE PROCESSING

High temperature processing routes for carrier selective junction solar cells such as TOPCon are considered to have economic advantages [11] compared to HJT with respect to levelized cost of energy (LCOE) and capital expenditure (CAPEX). The TOPCoRE (on p-type Cz-Si) or TOPCon solar cells (on n-type Cz-Si) apply high-temperature processing steps including boron diffusion, doped poly-silicon deposition, annealing as well as metallization firing. Therefore, Cz-Si material should preferably be compatible with these high-temperature processing steps that might alter the material quality.

3.1 Precursors with symmetric TOPCon passivation

To assess the effective MCL that can be achieved when applying a $POCl_3$ pre-gettering step and TOPCoRE passivation on both, front and rear flat surfaces, we applied the processing sequence depicted in Figure 5 to wafers that had been selected at equidistant locations along the complete Ga-doped ingot length. In that way, both, the response of the material along the complete ingot of 10000 wafers to high T processing including gettering as well as the surface passivation by the TOPCoRE passivation stack can be assessed. Results are shown in Figure 6. Those effective minority carrier lifetimes of the Ga-doped Cz-Si correspond at all sections of the ingot to diffusion lengths that are at least an order of magnitude larger than the device thickness and should not limit the solar cell performance. This is a requirement for a rear junction high efficiency solar cell. Furthermore, excellent TOPCoRE passivation is demonstrated with effective median MCL values exceeding 4 ms along the complete ingot without a clear MCL trend along the ingot.

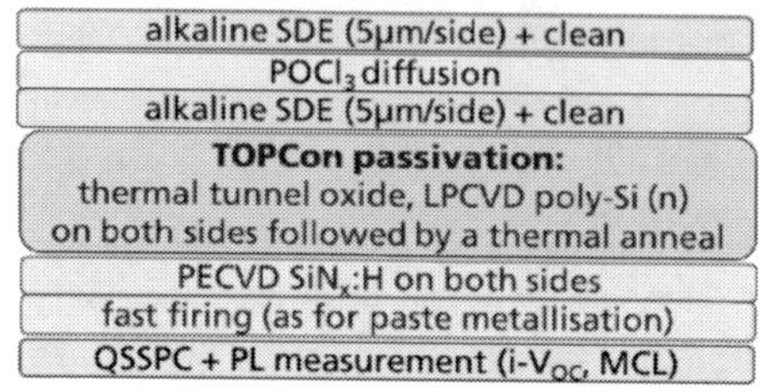

Figure 5: Process sequence for symmetric TOPCoRE effective MCL precursors with flat surfaces.

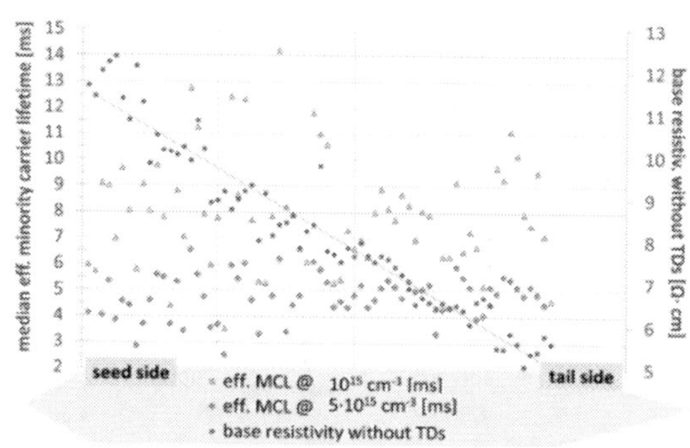

Figure 6: Median effective MCL at two injection levels for TOPCoRE passivation on wafers selected along the Ga-doped Cz-Si ingot.

3.2 TOPCoRE versus TOPCon solar cells

The processing sequence depicted in Figure 7 applied to both, n-type and p-type Cz-si wafers results in TOPCon front junction [12] and TOPCoRE rear junction solar cells, respectively.

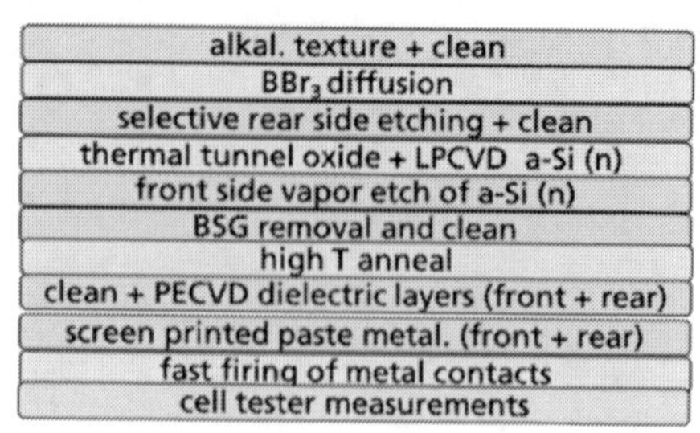

Figure 7: Process sequence for TOPCon (n-type Cz-Si) front junction and TOPCoRE (p-type Cz-Si) rear junction solar cells.

The processing sequence had been originally optimized for n-type wafers of 1 Ω·cm resistivity. Nevertheless, already a first comparison identified that similar solar cell performance can be achieved. Unfortunately, for the delta efficiency evaluation (co-processed n-type TOPCon cells serving as reference) shown in Figure 8 we noticed unwanted variation in one processing step, namely the single sided gas phase etching of poly-Si(n) that had been deposited on top of the front BSG layer and the wafer edges.

The TOPCoRE solar cells encircled in green towards the tail end of the ingot performed better than the reference TOPCon process. All these cells had been co-processed in vapor etching in one batch while other TOPCoRE cells of neighboring regions in the ingot and from the first half of the ingot had been processed in other vapor etching batches. Therefore, the scattering in efficiency along the ingot is not dominated by material properties but by processing during single sided vapor etch. However, this indicated to us a road how TOPCoRE processing can result in advantageous solar cell performance as compared to TOPCon processing.

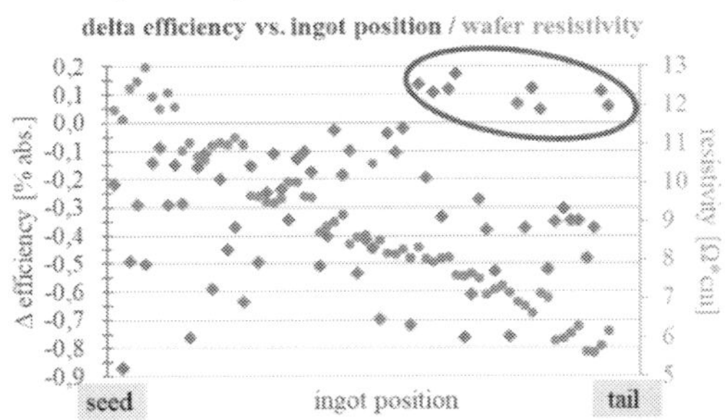

Figure 8: Delta efficiency distribution of TOPCoRE solar cells from Ga doped p-type wafers that been selected along the complete Cz-Si ingot. The efficiency is compared to the TOPCon reference process on n-type Cz-Si.

Furthermore, we identified to our surprise that for our processing sequence (identical processing steps for TOPCon and TOPCoRE), that finished solar cells show a very rapid light induced improvement (LII) behavior. Within repeated cell tester flashes (80 ms, 1 sun) there is a very significant improvement in efficiency within an accumulated (1 sun) illumination duration in the order of seconds to a few minutes. The improvement is more pronounced for TOPCoRE than for TOPCon solar cells that have been co-processed and leads to an improvement of upto 0.8% absolute in efficiency. That LII behavior results from MCL improvement, i-V_{OC} and V_{OC} improvement, pFF and FF improvement, and in case of TOPCoRE solar cells also from J_{SC} and series resistance improvement. More details on that finding are given in the accompanying journal publication to this conference contribution [9]. The observed LII is annoying for cell tester characterization but would not have negative impact on module performance in the field due to the very fast improvement as compared to a comparably slow relaxation to the original state when stored in the dark. Investigations to understand the cause(s) of this LII are on-going. In parallel also investigations with respect to LeTID behavior are ongoing and reported in [9].

Besides the significant improvement from LII, we applied an additional processing step to further improve the efficiency of TOPCoRE and TOPCon solar cells. After contact firing Laser Enhanced Contact Optimization (LECO) [13] has been applied to further increase the FF of our solar cells. With adapted LECO processing for both, TOPCoRE and TOPCon solar cells, we see by now for co-processed solar cells an advantage of 0.2% to 0.3% absolute in efficiency for TOPCoRE solar cells on the studied Ga-doped p-type Cz-Si material as compared to TOPCon on n-type Cz-Si of 1 Ω·cm resistivity. That finding needs to be confirmed with more statistics and on wafers selected from all regions of the ingot in future.

4 CONCLUSIONS

We specified, ordered and analyzed a p-type Ga-doped Cz-Si ingot with a comparably high resistivity range and compared it to n-type material that is typically used for HJT and TOPCon solar cells.

The p-type Cz-Si material has not only a significant cost advantage compared to commercially available n-type material but showed also along the complete ingot minority carrier diffusion lengths that do not limit the solar cell efficiency of carrier selective junction and passivated contact solar cells, such as HJT and TOPCoRE. The diffusion length for the studied Ga doped p-type material is even superior to that of the n-type reference Cz-Si material.

Both, high temperature and low temperature processing routes have been applied to manufacture i-V_{OC} and MCL precursors as well as industrially relevant high efficiency solar cells. In both cases, HJT processing and TOPCoRE processing the processing sequence was kept identical as for the reference process on n-type Cz-Si. By this the p-n junction moved for HJT from rear to front and for TOPCon/TOPCoRE from front to rear.

HJT precursors just prior to metallization show i-V_{OC} values above 745 mV [9] and MCL above 2.5 ms along the complete studied p-type ingot. If pre-gettering is applied prior to HJT processing the i-V_{OC} and MCL is slightly improved. Towards the tail end of the ingot improvement from pre-gettering is more pronounced. So far, the outstanding diffusion length and i-V_{OC} values >750 mV result in improved J_{SC} and in case of pre-gettered HJT cells also in higher V_{OC} values for p-type compared to n-type HJT solar cells. Unfortunately, the FF of the p-type Cz-Si HJT solar cells is inferior to the counterparts on n-type Cz-Si and suffers from large scattering in FF. Thus, in average the resulting efficiency is so far 0.36% lower for the p-type material. Respective best efficiencies of co-processed p-type HJT solar cells are 0.2% lower than for the n-type reference process. For HJT a FF improvement will be required to make p-type Cz-Si due to its lower wafer cost commercially advantageous.

For TOPCoRE application in high temperature processing the material proved also to result in excellent effective MCL when evaluated with symmetric lifetime samples. The minority carrier diffusion length showed to be along the complete ingot at least an order of magnitude higher than the device thickness of TOPCoRE solar cells.

In addition, we could demonstrate that the rear junction TOPCoRE solar cells have not only for lab cells [7] but also for industrial application the potential to result in better solar cell performance than their n-type counterparts. Thus, there are two commercial advantages, one coming from the significantly lower wafer cost and one from the better solar cell performance that make this route highly attractive. With the currently high wafer prices and the price difference between p-type and n-type

Cz-Si wafers those advantages for p-type TOPCoRe solar cells offer in a 10 GW/year TOPCon production scenario the potential to save > 100 million US$/year.

With respect to long term stability of the respective material we do not see disadvantages for the studied Ga doped p-type Cz-Si compared to n-type Cz-Si. To our surprise we see in both cases a light induced improvement LII behavior on a very short time scale that seems more pronounced for the back junction TOPCoRE solar cells. The cause of this LII is not yet fully understood and further under investigation. We do not find this behavior on low temperature processed HJT solar cells nor on the TOPCoRE lifetime precursors. With respect to LeTID we need to extend our investigations from lifetime samples to solar cells. So far, we do not see a disadvantage for the Ga-doped p-type compared to n-type Cz-Si.

5 ACKNOWLEDGMENTS

The authors acknowledge that part of this work was supported by funding from the German Federal Ministry for Economic Affairs and Climate Action within the research project "TOPConCluster" (contract no. 03EE1065A).

The authors wish to acknowledge the support of LONGI in supplying Cz-Si ingot material according to our requirements.

5 REFERENCES

[1] S. Riepe et al., Proc. 37th EUPVSEC, Lisbon 2020, pp 121-125, ISBN 3-936338-73-6

[2] A. Bhambhani, 25.5% efficiency for 210 mm i-TOPCon cell, https://taiyangnews.info/technology/25-5-efficiency-for-210mm-i-topcon-cell/ , accessed: March, 2022.

[3] A. Bhambhani, 25.7% efficiency record for n-type mono Si solar cell, https://taiyangnews.info/technology/25-7-efficiency-record-for-n-type-mono-si-solar-cell, accessed: April, 2022.

[4] A. Bhambhani, 26.07% efficiency record for HJT solar cell, https://taiyangnews.info/technology/26-07-efficiency-record-for-hjt-solar-cell/, accessed: March, 2022.

[5] A. Bhambhani, 26.5% efficiency for silicon HJT solar cell, https://taiyangnews.info/technology/26-5-efficiency-for-silicon-hjt-solar-cell/, accessed: June, 2022.

[6] Weekly snapshots and Price Downloads at http://www.pvinsights.com as assessed between 2019 and 2022.

[7] A. Richter et al., Nature Energy, 6, p 429, April 2021, https://doi.org/10.1038/s41560-021-00805-w

[8] A. Descoeudres et al., Prog Photovolt Res Appl., 28, p 569, 2020, https://doi.org/10.1002/pip.3227

[9] J. Horzel et al., accompanying journal publication to this conference contribution; Solar RRL 8/2022, https://doi.org/10.1002/solr.202200613

[10] J. Horzel et al., Proc. 27th EUPVSEC, 2012, p780, DOI: 10.4229/27thEUPVSEC2012-2CO.15.6

[11] B. Kafle et al., Solar Energy Material and Solar Cells, 227, 2021,111100, https://doi.org/10.1016/j.solmat.2021.111100

[12] S. Mack et al., Proc. of this conference

[13] T. Fellmeth et al., PIP, 8/2022, https://doi.org/10.1002/pip.3598

This is a short version of the paper submitted to the journal Solar RRL.

It has been selected by the Committee of the WCPEC-8 for submission to Solar RRL.

STUDY ON THE DISSOCIATION OF GALLIUM-HYDROGEN PAIRS IN CRYSTALLINE SILICON DURING ILLUMINATED ANNEALING

Jochen Simon, Axel Herguth, Fabian Geml, Giso Hahn
University of Konstanz, Department of Physics, 78457 Konstanz, Germany

ABSTRACT: In this study, fundamental investiations are performed on gallium-doped Cz-Si samples with regard to the hydrogen content introduced from PECVD SiN_x:H layers in a firing step. The change in gallium-hydrogen (GaH) pair concentration was monitored with highly sensitive resistivity measurements since pair formation counteracts doping. It is found that GaH pairs form during dark annealing at elevated temperatures. In contrast, they partially dissociate within a few seconds at 240°C when excess charge carriers are generated by illumination. This fast dissociation is reversible such that GaH pairs form again during a subsequent dark anneal at 180°C. The dependence on excess charge carrier concentration suggests an electron-driven dissociation reaction.
If illumination at elevated temperature is applied for days, even more GaH pairs dissociate as indicated by the change in resistance. Since it is not possible to form GaH pairs in a subsequent dark anneal, it is proposed that two different dissociation processes are taking place which might be independent or a cascade reaction.
Keywords: Hydrogen, Defects, Gallium-Hydrogen pairs

1 INTRODUCTION

The release of hydrogen from a dielectric layer is the most common method in solar cell production to use its ability to passivate dangling bonds and neutralize several defects and impurities in crystalline silicon. However, hydrogen has also a particular role in the generation of defects formed during light- and elevated temperature-induced degradation (LeTID), which can reduce solar cell efficiency significantly [1–2]. Since the underlying mechanisms of LeTID are still unclear and the direct microscopic impact of hydrogen is not tangible, quantifying hydrogen might be the key in the understanding of this defect. With only 10^{14} to 10^{15} cm^{-3}, the total hydrogen concentration is fairly low, making it difficult to be observed directly. However, H tends to form acceptor-hydrogen pairs influencing bulk resistivity giving access to the hydrogen content. Therefore, the better accessible reaction of hydrogen with acceptors is a promising route for further investigations.

Such studies got some attention recently [3–6] on boron-doped silicon. This contribution focuses entirely on the study of gallium-hydrogen pairs (GaH) in Cz-Si. Adopted from the case of boron [3], the predominant overall reaction during annealing in the dark at elevated temperatures is

$$2\,Ga^- + 2\,h^+ + H_{2A} \rightarrow 2\,GaH^0 \qquad (1)$$

where this process consumes holes h^+ changing hole concentration p. The change in pair concentration Δ[GaH] can be derived from the change in (negative) hole concentration $-\Delta p$, thus $\Delta[GaH] = -\Delta p$. However, as Δp is only a measure of the change, it does not carry any information on the pair concentration already existent before annealing. Therefore, it is the aim of this study to clarify whether and under what circumstances the back reaction takes place so that (already existing) GaH pairs may be deliberately dissociated and thus quantified.

2 EXPERIMENTAL

Mono-crystalline Cz-Si:Ga (0.7 Ωcm) wafers were saw damage etched (KOH for 8 min at 80°C, final thickness 160 µm) and cleaned (surface oxidation by ozone followed by an HF dip). Subsequently, SiN_x:H layers with thickness of 80 nm and index of refraction $n = 2.05$ were deposited on both sides using PECVD. The non-metallized samples were fired in a belt furnace with a measured sample peak temperature of 870°C to provide H-rich samples. Contacts were produced by thermal evaporation of aluminum and subsequent laser pulses to form laser fired contacts [7] allowing for direct contact resistance measurement. Resistivity measurements were performed in a four-terminal configuration on a temperature-controlled measurement setup as described in [8]. The negative change in hole concentration $-\Delta p = p_0 - p$ was calculated with respect to the hole concentration of a reference state p_0 from the resistance R of the sample. Hole mobilities for the respective hole concentration were taken from PVlighthouse mobility calculator [9].

To trigger formation of the pairs, the samples were dark annealed at 180°C on a hotplate (Praezitherm PZ-28, uncertainty ±0.5°C) until resistivity does not change any more. The dark anneal was repeatedly interrupted for ex-situ resistance measurements. To investigate the GaH pair dissociation, the samples were subjected to variable excess charge carrier concentrations Δn by illumination with a monochromatic, homogeneous light source ($\lambda = 808$ nm) at $T = 240$°C. In addition to the resistance, the effective lifetime τ_{eff} of excess charge carriers was determined via photoconductance decay measurements (WCT-120 lifetime tester, Sinton Instruments) regularly to allow for intensity adaptions in order to achieve constant

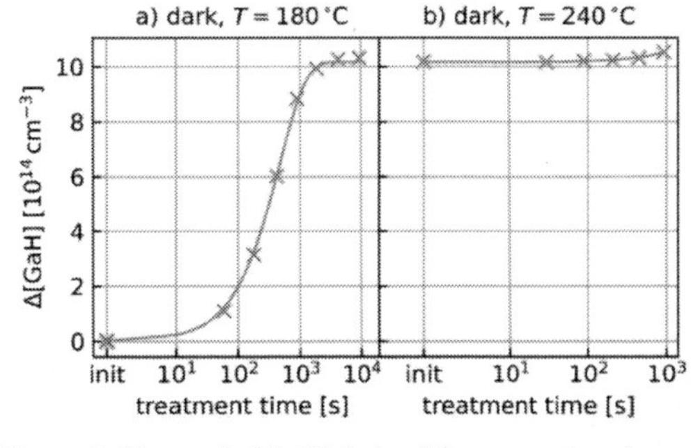

Figure 1: Change in [GaH] derived from negative change in hole concentration $-\Delta p$. a) Formation of GaH pairs during dark anneal at 180°C. b) Subsequent treatment at 240°C. Lines are guide to the eye only.

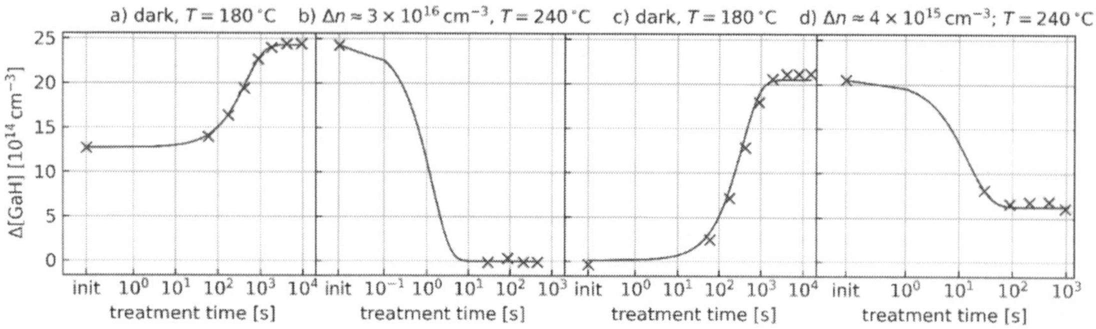

Figure 2: Change in [GaH] derived from negative change in hole concentration $-\Delta p$. a) Formation of GaH pairs during dark anneal at 180°C. b) Subsequent illuminated anneal at 240°C. c) Formation of pairs in the dark. d) Dissociation again with different excess charge carrier concentration Δn. Lines are guide to the eye only.

excess charge carrier concentration despite changes in τ_{eff}. It should be mentioned that samples were at 30°C during lifetime measurement. Due to the temperature dependence of τ_{eff}, an uncertainty in Δn is expected, since the illuminated anneal takes place at 240°C.

Three experiments were conducted: dark annealing only (Fig. 1), a sequence of dark and "short" illuminated annealing steps (Fig. 2), and a sequence of dark and "long" illuminated annealing steps (Fig. 3). Note that the first dark annealing step is the same for all experiments and is intended to form GaH pairs for later dissociation steps.

3 RESULTS

The relative change in pair concentration Δ[GaH] derived from resistance measurements during dark anneal at 180°C in three experiments is plotted in Fig. 1 a), 2 a) and 3 a). The scaling of Δ[GaH] is different in each figure, since the reference state p_0 was deliberately set to the minimum value for each sample in the course of the experiment.

The concentration of GaH pairs is rising until ~2 h, after which a saturation value seems to be reached.

In order to check whether annealing at 240°C even without illumination already leads to a dissociation of GaH pairs, the change in charge carrier concentration Δp during a subsequent anneal at $T = 240$°C and $\Delta n = 0$ was studied (Fig. 1 b). There is no significant change despite a small increase which is likely to be attributed to the formation of thermal donors [5]. Things change drastically if illumination is applied, as shown in Fig. 2 b). A fast change in resistance which translates into a steep decrease of Δ[GaH] is visible for treatment times below 1 min. Note that due to the lack of data in the first minute the shown line does not necessarily reflect the true progression. The change in resistance is partially reversible as shown in Fig. 2 c) and 2 d), where dark annealing followed by illuminated annealing reproduces the temporal behavior observed before. This gives strong evidence that the aforementioned reaction runs forward without excess charge carriers and backwards with charge carrier injection. Since the extent of the decay is higher with higher injection, it is reasonable to assume that the equilibrium of formation and dissociation can shift with excess charge carrier concentration.

It turns out that not only the equilibrium value as a result of formation and dissociation but also the dissociation rate constant is dependent on injection level Δn and temperature [10]. This is probably the reason behind the different limiting values in Figs. 2 b) and 2 d) obtained under different illumination conditions.

While most of the effect seems to occur in several seconds for the presented combinations of temperature and injection, this process can take up to an hour at temperatures of 180°C [10]. It is likely that the aforementioned dependence of the reaction rate on injection is caused by the fraction of different charge states of hydrogen. The fraction depends on Fermi level [11,12] which in turn is determined by excess charge carrier concentration. As described elsewhere in more detail [10], the fraction of negatively charged hydrogen can explain

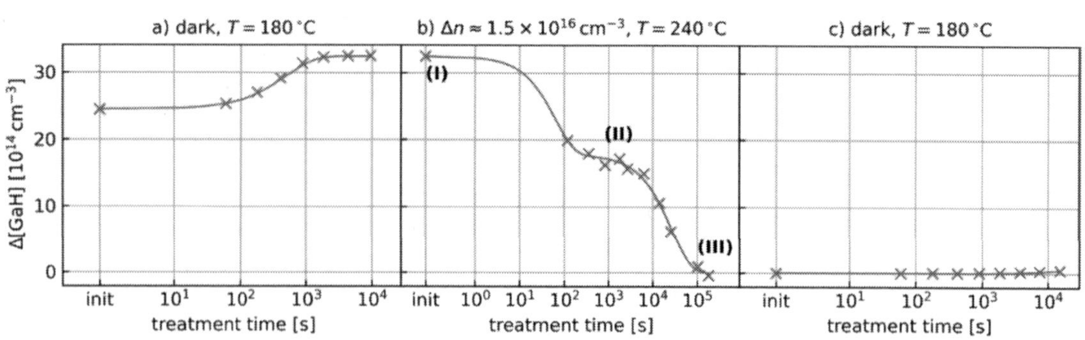

Figure 3: Change in [GaH] derived from negative change in hole concentration $-\Delta p$. a) Formation of GaH pairs during dark anneal at 180°C. b) Subsequent extended illuminated anneal at 240°C. c) No pair formation in the dark. Lines are guide to the eye only.

the dependence of the reaction rate on injection Δn.

However, cycling between formation and dissociation in Fig. 2 does not seem to be perfectly reversible as the limiting values in Figs. 2 a) and 2 c) are not the same. The reason behind this might be explained by the third experiment (Fig. 3). For illuminated annealing for more than 1 day, another increase in hole concentration indicates a second, slower decrease in GaH pair concentration. This behavior is depicted in Fig. 3 b) as transition between (II) and (III). After the state indicated by (III) is reached, pair concentration Δ[GaH] does not change significantly under dark annealing conditions (Fig. 3 c) making this an irreversible process. Hence, the difference in limiting values in Figs. 2 a) and 2 c) might indicate that some irreversible dissociation already occurs there as well.

Due to the difference in time constants of these two effects and their different behavior (reversible vs. irreversible), there are probably two processes ongoing. However, the presented data are not sufficient to tell whether a cascade reaction takes place or two independent reactions. However, it is plausible that the first, reversible dissociation of GaH under illumination pairs shifts reaction (1) towards its precursors H_{2A} and unpaired Ga^-. The second drop in pair concentration may be attributed to a transition of hydrogen into a more stable configuration. Effusion cannot be excluded as well.

3 CONCLUSIONS

The formation and dissociation of GaH pairs in Ga-doped Cz:Si has been investigated successfully via resistivity measurements.

The formation of these pairs can be triggered in hydrogen-rich samples by dark annealing, the dissociation only occurs in presence of excess charge carriers. If injection is applied, a balance between formation and dissociation is established. Since the dissociation only takes place in the presence of excess charge carriers, the assumption of an electron-driven process is obvious. GaH pairs form again during a subsequent dark anneal.

In contrast, illuminated annealing durations of days lead to the dissociation of additional GaH pairs. This dissociation is reversible in a sense that these pairs do not form again during a subsequent dark anneal.

The results presented here open the way for concise hydrogen determination in Ga-doped silicon and the insight into the behavior of GaH pairs can help in the understanding of the LeTID effect.

4 ACKNOWLEDGMENTS

The authors would like to thank S Sanz-Alonso for performing the wet chemistry work during sample preparation.

Part of this work was supported by the German Federal Ministry of Economic Affairs and Climate Action under Contract No. 03EE0152A. The content is the responsibility of the authors.

REFERENCES

[1] M.A. Jensen et al., Journal of Applied Physics 124(8) (2018) 085701.
[2] S. Jafari et al., IEEE Journal of Photovoltaics 11(6) (2021) 1363.
[3] V.V Voronkov, R. Falster, Physica Status Solidi B 254 (2017) 1600779.
[4] D.C. Walter et al., Solar Energy Materials and Solar Cells 200 (2019) 109970.
[5] C. Winter et al., Physica Status Solidi (A) 218 (2021) 2100220.
[6] D. C. Walter et al., Journal of Applied Physics 131(16) (2022) 165702.
[7] E. Schneiderlöchner et al., Progress in Photovoltaics: Research and Applications 10(1) (2002) 29.
[8] A. Herguth, C. Winter, IEEE Journal of Photovoltaics 11(4) (2021) 1059.
[9] PV Lighthouse, Mobility Calculator. Available online: https://www2.pvlighthouse.com.au/ calculators/ mobility%20calculator/mobility%20 calculator.aspx
[10]J. Simon et al., Physica Status Solidi RRL, accepted, DOI 10.1002/pssr.202200297
[11]C. Sun et al, Journal of Applied Physics 117(4) (2015) 045702.
[12]B. J. Hallam et al., Progress in Photovoltaics: Research and Applications 28(12) (2020) 1217.

STRESS-INDUCED DE-PASSIVATION OF POLY-SI/SINX LAYER STACKS

Jan Hoß*, Jan Lossen, Jonathan Linke, Radovan Kopecek, Florian Buchholz
[1]ISC Konstanz, Konstanz, Germany
Rudolf-Diesel-Str. 15, 78467 Konstanz, Germany
Tel: +49 (0) 7531 / 36 183 - 0
Fax: +49 (0) 7531 / 36 183 – 11
*Corresponding author: +49 7531 36 183 366, jan.hoss@isc-konstanz.de

ABSTRACT: The present study investigates the passivation of layer stacks consisting of an interfacial oxide, LPCVD deposited polycrystalline silicon (poly-Si) and PECVD deposited silicon nitride (SiN_x) layers when subjecting them to high temperature processes after the deposition of SiN_x. We show that the passivation of such poly-Si/SiN_x layer stacks can be severely degraded by high temperature processes. Further, we found that the passivation can be recovered after removing the annealed SiN_x layer and subsequent annealing of the c-Si/SiO_2/poly-Si interface. We conclude that thermal stress, induced by the densification of SiN_x contributes to the loss of passivation at the c-Si/SiO_2/poly-Si interface. Comparing symmetric and asymmetric sample structures, where the poly-Si/SiN_x stack is present on both and on only one side respectively, shows that stress induced de-passivation and the re-passivation of poly-Si/SiN_x stacks cannot be appropriately studied with symmetric test structures under certain circumstances. The importance of these findings is underlined when developing cell concepts that require the presence of such stacks in high temperature process steps. As an example, results for TOPCon cell precursors fabricated with an alternative process sequence with poly-Si deposition before BBr_3 emitter diffusion are presented. Although the poly-Si layer was capped with SiN_x during the emitter diffusion, cell precursors with iV_{oc} values of up to 709 mV could be produced with this process. We further propose to consider stress-induced de-passivation as potential root-cause for the degradation of the passivation of poly-Si/SiN_x stacks during the firing process of such samples.

Keywords: Annealing, Degradation, Hydrogen, Passivation, Polycrystalline

1 INTRODUCTION

Tunnel oxide passivated contact (TOPCon) solar cells with a rear side contact passivation by polycrystalline silicon (poly-Si), have been developed by many research institutes and companies in recent years [1-4] and are now increasingly manufactured in industrial pilot and mass production [5, 6]. Surface and contact passivation of the rear side of such solar cells are achieved with a layer stack consisting of a thin interfacial oxide (SiO_2), a highly n^+ doped poly-Si layer, and a capping layer of silicon nitride (SiN_x) for hydrogenation. In what has become a standard process route, the as-deposited silicon thin film is crystallized to poly-Si in an annealing step before the SiN_x deposition. After hydrogenation such layer stacks can achieve saturation current densities (J_0) of well below 5 fA/cm^2.

The present study investigates the passivation of such layer stacks when subjecting them to high temperature process steps after the deposition of SiN_x. Since the standard process route for the manufacturing of TOPCon solar cells usually does not include any high temperature process steps after SiN_x deposition other than a very short contact firing step, little attention has been given so far to the effects occurring when poly-Si/SiN_x stacks are subjected to diffusion or annealing processes.

However, with the increasing maturity of this passivation scheme, alternative process sequences and other cell concepts with poly-Si passivated contacts will receive growing attention from the scientific community. Examples for such processes include an alternative TOPCon process sequence with poly-Si deposition before the BBr_3 emitter diffusion ("poly-Si first process"), as depicted in Figure 1, as well as process sequences for manufacturing poly-Si passivated interdigitated back contact (IBC) solar cells [7]. Both of these cell concepts require that poly-Si layers that are capped with SiN_x for protection against diffusion of dopant species of opposite polarity are subjected to subsequent diffusion or annealing processes. Exemplary schematics of such cell structures in the state as during the subsequent high temperature process are shown in Figure 2.

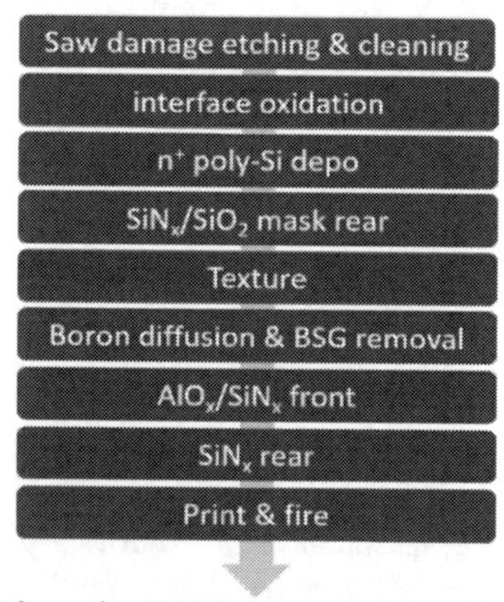

Figure 1: Alternative TOPCon process flow with poly-Si deposition before emitter diffusion.

Additionally, potentially cheaper processes towards TOPCon solar cells in which SiO_2, poly-Si and SiN_x layer are deposited without breaking the vacuum, e.g. via physical vapor deposition (PVD) deposition, would entail such structures. For the optimization of such processes, a sound understanding of the effects relevant for the passivation of annealed poly-Si/SiN_x stacks is needed.

Van de Loo et al. have presented an extensive study on the role of hydrogenation of similar layer stacks, showing that the passivation crucially depends on hydrogen management in the system and that the loss of hydrogen passivation is reversible [8]. A loss of surface passivation of similar layer stacks during firing has also been reported by Kang et al. [9], where it is found that the degradation can depend on the layer properties of the SiN_x layer. Furthermore, Hollemann et al. [10] find that the loss of surface passivation is correlated with an increase in defect state density at the c-Si/SiO_2/poly-Si interface when samples are fired at too high temperatures and hypothesize that excessive supply of hydrogen degrades the passivation.

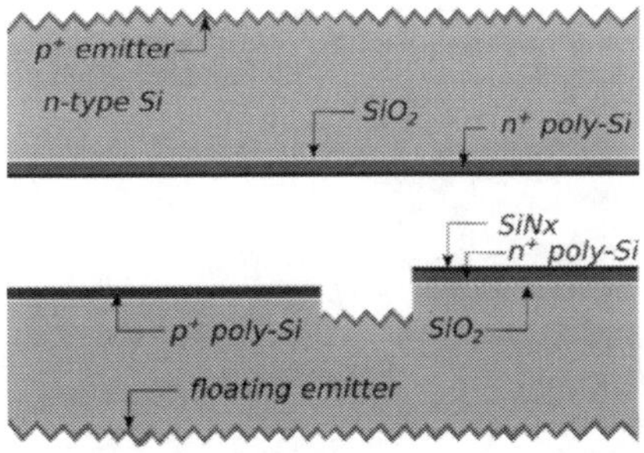

Figure 2: TOPCon cell with alternative process flow (top) and poly-Si passivated IBC cell (bottom) as during emitter diffusion and p^+ poly-Si annealing, respectively. Both cells feature poly-Si/SiN_x stacks that are subjected to a high temperature process.

While the aforementioned publications focus on firing of poly-Si/SiN_x layer stacks, our study contributes to the understanding of these layer systems by investigating effects occurring when subjecting poly-Si/SiN_x layer stacks to longer high temperature processes such as annealing and diffusion processes.

2 EXPERIMENTAL DETAILS

2.1 Sample preparation and poly-Si/SiN_x stack annealing

The study comprises symmetric and asymmetric lifetime samples from n-type M2 CZ wafers with a base resistivity of 14 Ωcm. As shown in the upper part of Figure 3 (left), the symmetric samples feature SiO_2/poly-Si/SiN_x layer stacks on both sides. The samples were prepared by saw damage etching in NaOH, cleaning, dry thermal interface oxidation ($d_{SiO2} \approx 2.2$ nm), LPCVD poly-Si deposition with in-situ phosphorous doping ($d_{poly} \approx 150$ nm), annealing in N_2 at 950°C for 30 min, and PECVD deposition of SiN_x ($d_{SiNx} \approx 40$ nm) using a centrotherm c.PLASMA direct plasma batch reactor. The SiN_x layer used in this study are standard SiN_x layers as used for coating the rear side of bifacial TOPCon cells and feature a refractive index of $n \approx 2.05$.

Asymmetric lifetime samples as shown in Figure 3 (right) feature the same layer stack on one side, while the poly-Si was stripped on the other side by etching in 22% NaOH solution. The poly-Si layer on one side was protected during this step by SiN_x.

Thereafter, the poly-Si/SiN_x layer stacks are annealed in N_2 at 950°C for 30 min for both symmetric and asymmetric samples. The thermal budget of the SiN_x annealing has been chosen to resemble a BBr_3 diffusion to mimic the effect that the thermal budget of such a diffusion has on poly-Si/SiN_x layer stacks, e.g., when forming the emitter of TOPCon solar cells manufactured with the alternative process sequence from Figure 1, or when forming a floating emitter for a poly-Si passivated IBC solar cell.

2.2 Retaining and removing the annealed SiN_x layers

As illustrated in Figure 3, the symmetric and asymmetric groups were split after the annealing step in different sub-groups. The annealed SiN_x layer on top of the poly-Si was retained for groups S1 (symmetric) and A1 (asymmetric). The samples from both groups were then coated with an additional layer of SiN_x (≈ 60 nm) on both sides on top of the annealed SiN_x layers and, in case of A1, on the non-passivated side of the wafer. The SiN_x deposition on the bare rear side of the asymmetric samples, serves to provide basic passivation as needed for meaningful measurements of the passivation quality. Finally, the samples were fired in an industrial conveyor belt furnace with a standard temperature profile for TOPCon metallization.

For groups S2 (symmetric), A2 and A3 (asymmetric) the annealed SiN_x layers were stripped in 8% HF after the annealing process. Subsequently, the samples from groups S2 and A2 were annealed again under the same conditions as before, while group A3 was not subjected to this last annealing step. Finally, the process was completed for these three groups by the deposition of SiN_x on both sides and firing of the samples together with groups S1 and A1.

After each process step, five samples were extracted to evaluate the level of surface passivation by quasi-steady-state photo conductance decay (QSSPC) measurements using a Sinton lifetime tester WCT-120. The implied open-circuit voltage (iV_{oc}) was evaluated at one sun illumination and the surface recombination current density J_0 was extracted with the method of Kane and Swanson [11] using the Auger recombination model recently introduced by Niewelt et al. [12]. Each five measurements per sample were performed on different positions across each sample.

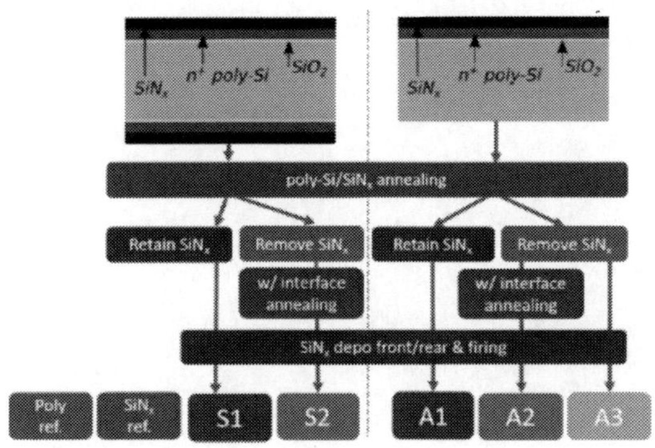

Figure 3: Schematic representations of symmetric (left) and asymmetric (right) lifetime samples as during the annealing of the poly-Si/SiN_x stacks (top). After poly-Si/SiN_x annealing, the samples are split in groups with retained and removed SiN_x layers and are partially subjected to a further annealing process. Finally, SiN_x is deposited on both sides and the samples are fired. The colors associated with the group names in the lower row match the colors in the Figures below.

2.3 Reference groups

Two additional groups of symmetric samples, depicted on the lower left part of Figure 3, were processed as reference groups. A poly-Si reference group was produced following the standard process for TOPCon layer stacks, consisting of interface oxidation, LPCVD n^+ poly-Si deposition, annealing, PECVD SiN_x deposition and firing.

A second reference group was used to assess the

passivation quality of the rear side of the asymmetric samples, where SiN_x is deposited directly on the c-Si wafer without poly-Si. The samples followed to same process as the asymmetric samples described above, however, here the poly-Si was removed on both sides prior the SiN_x passivation and firing

3 PASSIVATION OF SYMMETRIC LIFETIME SAMPLES

3.1 Reference groups

Figure 4 and Figure 5 show results for iV_{oc} and J_0 from QSSPC measurements for different groups of symmetric lifetime samples after various process steps. Reported are the mean values and the standard deviation from five samples with five measurement points each. The poly-Si reference group follows the expected behavior that was qualitatively also reported elsewhere [8]: directly after poly-Si deposition the median iV_{oc} is 689 ± 3.1 mV. After annealing the passivation is significantly improved by crystallization and dopant activation (730 ± 1.7 mV) and then further improved by SiN_x deposition (742 ± 1.4 mV) due to hydrogenation of the c-Si/SiO_2/poly-Si interface. A slight degradation of the passivation with iV_{oc} dropping to 733 ± 0.7 mV is obverted after firing, an effect that is also reported in literature [9, 10]. The corresponding J_0 values in Figure 5 follow the same trend, indicating that the observed iV_{oc} value changes are caused by altered surface passivation. After firing J_0 values of 6.7 ± 0.4 fA/cm^2 are achieved for the poly-Si reference group.

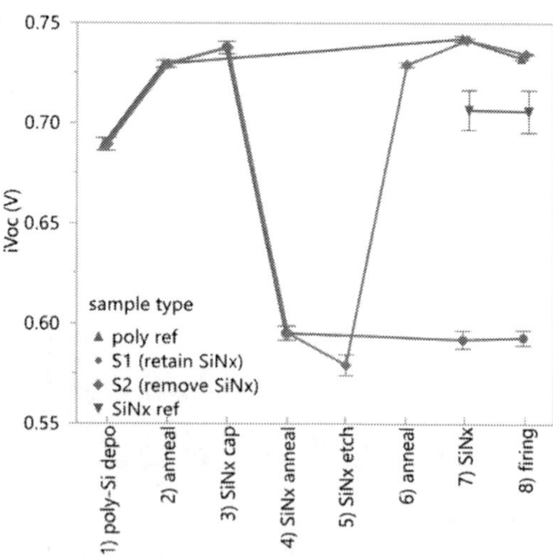

Figure 4: Implied V_{oc} values of symmetric lifetime samples after various process steps.

Additionally, Figures 4 and 5 show iV_{oc} and J_0 for the symmetric reference group with SiN_x passivation without underlying SiO_2/poly-Si. Here, iV_{oc} values of 707 ± 10 mV and J_0 values of 14.4 ± 4.2 fA/cm^2 are reached after SiN_x deposition. Firing these samples does not change these values significantly with iV_{oc} remaining at 706 ± 10 mV and J_0 at 12.2 ± 2.2 fA/cm^2. The passivation of SiN_x on c-Si will be needed for the interpretation of the asymmetric lifetime samples as discussed below.

3.2 Annealing of symmetric poly-Si/SiN_x samples

Groups S1 and S2 are processed identically as the poly-Si reference group until the first annealing process (step #2 in Figure 4 and Figure 5). After capping the poly-Si with a 40 nm thick SiN_x layer, iV_{oc} reaches 738 ± 3.1 mV (#3), which is in agreement with the iV_{oc} of the poly-Si reference group after SiN_x deposition. However, the passivation severely degrades to 596 ± 3.5 mV after annealing the poly-Si/SiN_x stack at 950°C for 30 min (#4). Notably, the passivation is significantly worse than directly after poly-Si deposition (#1) and after the initial poly-Si annealing (#2).

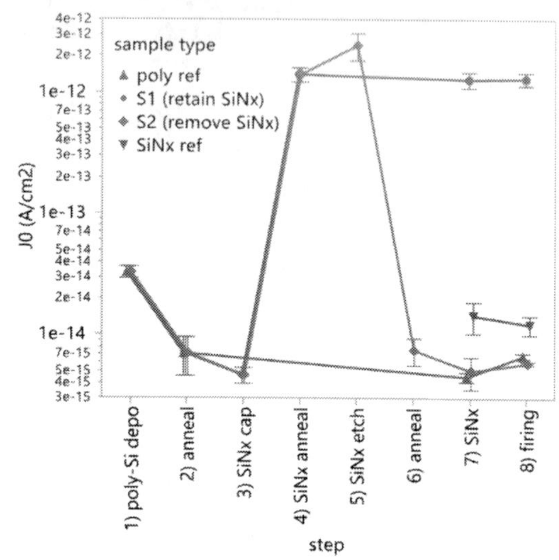

Figure 5: J_0 values of symmetric lifetime samples after various process steps. The J_0 values refers to the recombination current density on one side of the symmetric sample.

After annealing the poly-Si/SiN_x stack, groups S1 and S2 are split. As shown in Figure 4, iV_{oc} remains below 600 mV for the group for which the annealed SiN_x layer is retained (S1), even after an attempt to re-hydrogenate the interface with an additional SiN_x deposition on top of the annealed SiN_x layers (#7) and subsequent firing (#8). This means that the de-passivation introduced by annealing the poly-Si/SiN_x layer stacks can either not be passivated by hydrogen, or that the additional hydrogen cannot reach the c-Si/SiO_2/poly-Si interface because of the blocking effect of the annealed SiN_x layers.

In contrast, the passivation can be recovered when the annealed SiN_x layers are removed for group S2: etching the annealed SiN_x layers in HF (#5) does not improve the passivation (iV_{oc} = 579 ± 5.2 mV), but a further annealing step in N_2 (#6) at 950°C for 30 min improves iV_{oc} significantly to 729 ± 1.2 mV. After re-hydrogenating the interface with SiN_x deposition and subsequent firing, iV_{oc} reaches 735 ± 0.4 mV, rendering the recovered passivation equivalent to the poly-Si reference group without initial degradation of the passivation.

3.3 Discussion

Van de Loo et al. show that the passivation of c-Si/SiO_2/poly-Si passivation stacks can be degraded and recovered in a reversible process by depleting the interface of hydrogen and by re-hydrogenation, respectively [8]. However, the degradation of the poly-Si/SiN_x passivation found in the present study cannot be explained by this effect as set out by the following considerations.

First of all, the passivation of the annealed poly-Si/SiN_x stacks drop significantly below the values observed directly after poly-Si deposition (#1) and after subsequent annealing without SiN_x capping layer (#2). In these states no significant hydrogen passivation is to be expected, considering that as-deposited LPCVD silicon thin films already contain very little hydrogen when deposited at similar temperatures as in our experiment [13], let alone after annealing at 950°C. This suggests that the loss of passivation cannot only be due to a loss of hydrogen.

Alternatively, the loss of passivation after annealing the poly-Si/SiN_x stack could be due to an excess of hydrogen at the interface as proposed for passivation degradation after firing at too high temperatures [9, 10]. However, this cause also seems unlikely to be responsible here, given the long annealing time of 30 min at 950°C. Hydrogen effusion experiments show that nitrogen rich SiN_x layers, produced with comparable process parameters as the ones used in this study, release hydrogen at temperatures above 900°C within a few minutes of annealing time [14]. We therefore assume that the poly-Si/SiN_x stack system is depleted of hydrogen after the annealing step (#4).

Finally, the fact that the passivation can be recovered after SiN_x removal and subsequent annealing in N_2, i.e. without the introduction of hydrogen, shows that the loss and recovery of the passivation observed in the present study cannot be explained with hydrogen alone. We therefore propose stress-induced de-passivation as a contributing factor for the significant loss of passivation when annealing poly-Si/SiN_x layer stacks. The initially hydrogen-rich SiN_x layer densifies during the annealing process. This is underlined by the fact that the SiN_x etch rate in HF is reduced by a factor ≈ 7 after annealing. We hypothesize that this causes compressive stress that increases the defect state density at the c-Si/SiO_2/poly-Si interface. This would explain, why iV_{oc} drops below the values measured in the as-deposited and as-annealed state. Furthermore, this interpretation is compatible with the observed recovery of the passivation after the annealed SiN_x layer, i.e. the source of the compressive stress, is removed and the interface is annealed at a high temperature.

4 PASSIVATION OF ASYMMETRIC LIFETIME SAMPLES

Given these results, it could be concluded that for any cell concept that employs SiN_x masks on poly-Si during high temperature processes, removal of the SiN_x mask and subsequent annealing of the c-Si/SiO_2/poly-Si interface is needed to restore the passivation. This would render such processes relatively complex. To further investigate this de-passivation effect, we additionally examine asymmetric lifetime samples where the poly-Si/SiN_x layer stack is only present on one side of the sample.

As shown in Figure 6 (iV_{oc}) and Figure 7 (J_0), group A2, for which the annealed SiN_x was removed after annealing of the poly-Si/SiN_x stack, behaves similar as its symmetric counterpart S2 with removed SiN_x. The passivation of the poly-Si can be recovered after removing the annealed SiN_x layer in HF, annealing in N_2 (#6, iV_{oc} = 729 ± 1.5 mV) and re-applying a SiN_x passivation for hydrogenation (#7, iV_{oc} = 734 ± 0.9 mV). Note that the side of the sample without poly-Si was passivated after the respective process step by SiN_x deposition before the QSSPC measurement, since completely unpassivated wafer surfaces would render such measurements useless. The corresponding surface recombination current density of this side of the sample is 14.4 ± 4.1 fA/cm^2, as estimated with the symmetric SiN_x reference group. The absolute iV_{oc} values of symmetric and asymmetric samples may therefore be different.

One notable difference between the groups S2 and A2 is that the recovery for A2 apparently already starts after SiN_x etching (#5, iV_{oc} = 687 ± 1.8 mV), while this process step did not improve the passivation for the symmetric lifetime samples. This differences is likely due to partial interface annealing during the aforementioned SiN_x deposition on the non-passivated side of the sample (T ≈ 440°C).

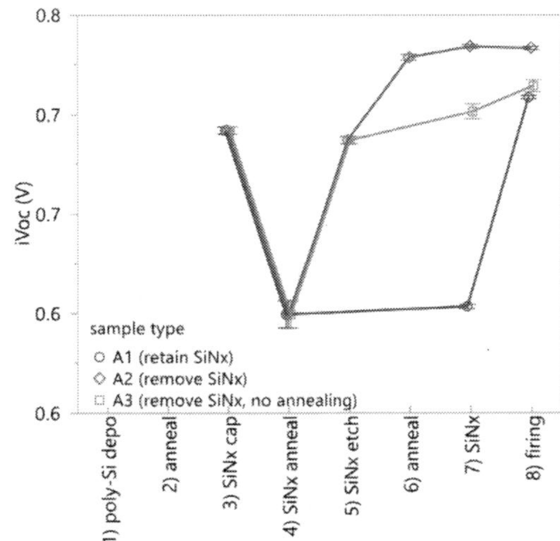

Figure 6: Implied V_{oc} of asymmetric lifetime samples after various process steps. The front side of the samples is passivated with SiO_2/poly-Si. The presence and the annealing state of an additional SiN_x layer on top of the poly-Si is indicated by the horizontal axis label. The rear side of the sample is passivated with SiN_x.

Given the observation that the passivation can also be improved by annealing of the c-Si/SiO_2/poly-Si interface during the PECVD deposition with a deposition temperature of ≈ 440°C, an additional group A3 has been processed similarly to A2, but without the dedicated high temperature annealing process at 950°C. As shown in Figure 6 and Figure 7 the passivation can be significantly improved even without the annealing step with iV_{oc} reaching 714 ± 3.1 mV after SiN_x deposition and firing. However, the passivation remains below the passivation of group A2 with the dedicated interface annealing step, suggesting that elevated temperatures 440°C < T < 950°C are necessary for complete interface healing.

The most important difference, however, is found for the asymmetric samples for which the annealed SiN_x layer is not removed (A1). The passivation remains poor with iV_{oc} values of 604 ± 1.0 mV after additional SiN_x depositions on the side with the annealed poly-Si/SiN_x stack and on the other non-passivated side with bare c-Si. This implies that the passivation of the annealed poly-

Si/SiN$_x$ stack remains poor after this process step, as expected based on the results of the symmetric samples. However, the passivation of the asymmetric samples significantly improves after firing with iV$_{oc}$ reaching 709 ± 0.9 mV. While this is below the value achieved for the group with removed SiN$_x$ (A2), it appears to be in stark contrast to the difference in iV$_{oc}$ after firing observed for the degraded (S1, iV$_{oc}$ = 593 ± 3.7 mV) and recovered (S2, iV$_{oc}$ = 735 ± 0.4 mV) symmetrical samples.

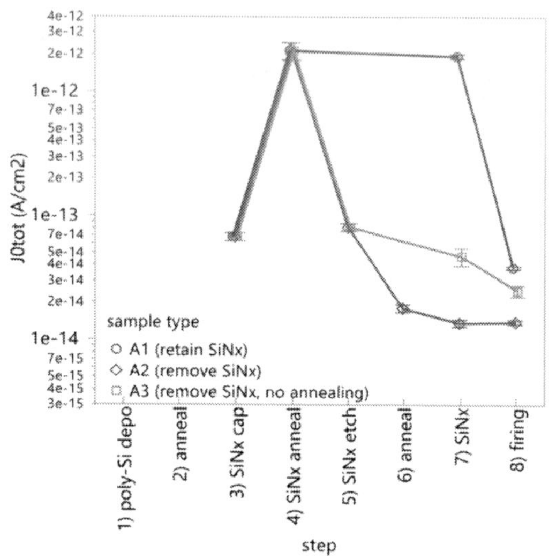

Figure 7: Total J$_0$ values of asymmetric lifetime samples after various process steps. The front side of the samples is passivated with SiO$_2$/poly-Si. The presence and the annealing state of an additional SiN$_x$ layer on top of the poly-Si is indicated by the horizontal axis label. The rear side of the samples is passivated with SiN$_x$. The reported values refer to the total J$_0$ from both sides of the samples.

5 COMPARISON OF SYMMETRIC AND ASYMMETRIC SAMPLE STURCTURES

Table I summarizes expected and actually measured J$_0$ contributions for the asymmetric samples of group A1 before and after firing. The first two lines of the table show the expected contributions for the two different sides based on QSSPC measurements of the respective symmetric samples, i.e. for poly-Si/SiN$_x$ with retained SiN$_x$ after annealing (group S1) and for the SiN$_x$ passivated reference samples without poly-Si. The reported values refer to the J$_0$ contribution from one side of the respective sample (c.f. Figure 5). Adding these contributions yields an expected value for the total J$_0$ for the asymmetric sample with the annealed poly-Si/SiN$_x$ stack on one side and SiN$_x$ passivation on the other side. The uncertainty of this value is calculated by adding the squared uncertainties of the summands. This expected value can be compared to the actually measured total J$_0$ value for this sample structure, which is shown in the last line of the table.

Before firing both expected and measured total J$_0$ values of the asymmetric samples are very high (>1000 fA/cm^2) and the measured value exceeds the expected value slightly. This is fundamentally different after firing: While the expected value remains high at $J_{0,exp}^{tot}$ = 1312 ± 160 fA/cm^2, the measured value drops significantly to $J_{0,meas}^{tot}$ = 38.4 ± 1.1 fA/cm^2. This discrepancy shows that the assessment of surface passivation by use of symmetrical test structures is not appropriate to infer the surface passivation of asymmetric sample structures for these types of samples. Moreover, the result has important implications for the evaluation of cell processes that include high temperatures processes after the deposition of poly-Si/SiN$_x$ layer stacks: While the results of the symmetric sample structures suggested that the surface passivation of such stacks cannot be recovered after annealing without removing the annealed SiN$_x$ layer, the asymmetric structures show that this can be achieved.

Table I: Expected J$_0$ contributions before and after firing for group A1 (asymmetric with retained SiN$_x$) based on QSSPC measurements of the corresponding symmetric lifetime samples from group S1 (poly-Si/SiNx with retained SiN$_x$) and the SiN$_x$ reference group. Additionally, the actually measured total J$_0$ of the asymmetric samples is shown.

group	item	J$_0$ before firing (fA/cm^2)	J$_0$ after firing (fA/cm^2)
S1	poly-Si/SiN$_x$	1280 ± 187	1300 ± 160
Ref.	SiN$_x$	14.4 ± 4.2	12.2 ± 2.2
A1	total expected	1295 ± 187	1312 ± 160
A1	total measured	1929 ± 75	38.4 ± 1.1

The poor passivation of the symmetric samples with annealed poly-Si/SiN$_x$ stacks after additional SiN$_x$ deposition on top of the annealed SiN$_x$ layer and subsequent firing (Figure 5, group S1 with retained SiN$_x$) shows that re-hydrogenation of the c-Si/SiO$_2$/poly-Si interface through the annealed SiN$_x$ layer is not possible. Therefore, the improvement of the passivation of the poly-Si/SiN$_x$ passivated side of the asymmetric sample after firing (Figure 7, group A1) is likely achieved by passivating the degraded c-Si/SiO$_2$/poly-Si interface with hydrogen provided from the rear side of the sample. Hints for the hydrogenation of interfaces across the wafer bulk has been reported for similar sample structures elsewhere [15]. Overall, the finding suggests that stress-induced de-passivation of annealed poly-Si/SiN$_x$ stacks can, at least partially, be recovered by providing hydrogen to the affected c-Si/SiO$_2$/poly-Si interface.

This can be applied to produce TOPCon cell precursors with the alternative process sequence with LPCVD poly-Si deposition before emitter diffusion as illustrated in Figure 1. In this experiment TOPCon cell precursors with median iV$_{oc}$ of 706 mV (champion 709 mV) were produced in our lab without removing the annealed SiN$_x$ capping layer on the poly-Si after BBr$_3$ diffusion. While this result is slightly below precursors from a conventional TOPCon process with BBr$_3$ diffusion before poly-Si deposition with iV$_{oc}$ ≈ 715 mV, it exemplifies that annealing poly-Si/SiN$_x$ layer stacks can be a viable option for advanced cell processes and that findings from symmetric lifetime samples are not necessarily suitable for estimating J$_0$ contributions for such samples. Adjusting the SiN$_x$ deposition parameters or the deposition method to adapt the as-deposited density of the layer, its hydrogen content and its thickness might help in further reducing the stress induced by this layer on the c-Si/SiO$_2$/poly-Si interface during high temperature process. This would allow to further improve the passivation and to make this lean process flow a viable alternative to the standard TOPCon process route. Similarly, experimental results for IBC solar cells that also feature poly-Si/SiN_x layer stacks that are subjected to high temperature processes show that such structures can be successfully applied in high efficiency cell concepts [7].

6 DISCUSSION

Accepting stress-induced de-passivation of poly-Si/SiN_x layer stacks as a result of high temperature processes begs the question if other observations of degraded passivation of such stacks, e.g. during firing at too high temperatures, as reported by other authors, are also related to the proposed phenomenon. Kang et al. [9] found that the loss of passivation upon firing is correlated with an increased hydrogen concentration at the c-Si/SiO_2/poly-Si interface and propose a causal relation between the two observations. In addition, Hollemann et al. [10] came to a similar conclusion by comparing the hydrogen concentration at the interface with estimated defect state densities for samples after annealing and after firing at different temperatures. The observation that the hydrogen concentration exceeds the defect state density after firing is interpreted as evidence that hydrogen not only populates dangling bonds but also introduces additional defect. However, the defect state density after firing was estimated using samples that were fired without hydrogen containing capping layers. This approach potentially underestimates the actual defect state density if additional stress-induced defects are created when firing samples with a hydrogen containing capping layer as suggested by the findings of the present study.

We therefore propose stress-induced de-passivation of poly-Si/SiN_x as a possible root-cause for the degradation of the passivation of poly-Si/SiN_x stacks, not only in high temperature annealing and diffusion processes, but also during firing at too high temperatures. With this interpretation the passivation does not degrade because of the excess hydrogen but despite of it. The increased hydrogen concentration at the interface found by other authors would be due to the increased number of dangling bonds as a result of thermal stress. If these additional defects are not sufficiently compensated by hydrogen passivation it would lead to the observed degradation of the passivation.

7 SUMMARY AND CONCLUSION

The study examines the passivation quality of symmetric and asymmetric lifetime samples featuring SiO_2/poly-Si/SiN_x layer stacks on both and on one side of the sample, respectively. It was found that the passivation of such layer stacks can be heavily degraded by subjecting them to high temperature processes such as annealing or diffusion processes. It was found that the passivation drops significantly below the level achieved with as-deposited poly-Si. Furthermore, the passivation can be restored by removing the annealed SiN_x layer and subsequent annealing of the c-Si/SiO_2/poly-Si interface in N_2 at high temperatures of ≈ 950°C. The absence of any hydrogen in this recovery process indicates that the defect state density at the interface is reduced by reorganization of the interface rather than by hydrogen passivation of persisting dangling bonds. We therefore conclude that the degradation of the passivation of annealed poly-Si/SiN_x stacks is at least partially induced by compressive stress exerted on the c-Si/SiO_2/poly-Si interface by the densification of the SiN_x layer on top of the poly-Si during high temperature processes.

Results for the asymmetric sample structures show that assessing J_0 contributions for asymmetric lifetime samples by measuring the passivation of corresponding symmetric samples is not necessarily a valid approach. Here it was found that the degraded passivation of annealed poly-Si/SiN_x stacks can be significantly improved by hydrogenating the stress-induced defects at the c-Si/SiO_2/poly-Si interface with hydrogen from the rear side of the sample in a firing process – a process that is impeded in the case of symmetric samples with annealed poly-Si/SiN_x stacks on both sides.

This observations makes cell processes that require poly-Si/SiN_x stacks in high temperature processes viable despite of the temporary stress-induced de-passivation of these stacks. Specific adaptations of the properties of the SiN_x capping layer on the poly-Si, e.g. adjusting the layer thickness, its hydrogen content and the as-deposited layer density might help to minimize thermal stress during the high temperature process and hence to reduce de-passivation.

Finally, we propose that stress-induced de-passivation of poly-Si/SiN_x layer stacks is not only relevant for long high temperature processes such as diffusions, but might also be a contributing factor to the degradation of the passivation quality of such layer stacks during a short firing process that has been reported by several authors.

7 ACKNOWLEDGMENTS

The authors thank the German Federal Ministry for Economic Affairs and Climate Action (BMWK) for funding this work under grant number FKZ: 03EE1068B (project SpuTOPiA) and all our colleagues at ISC Konstanz who were involved in conducting the experiments presented in this study.

8 REFERENCES

[1] S.W. Glunz, B. Steinhauser, J.-I. Polzin et al., Silicon-based passivating contacts: The TOPCon route, Prog Photovolt Res Appl. 2021; 1- 19. doi:10.1002/pip.3522

[2] D. Yan, A. Cuevas, J. I. Michel, et al., Polysilicon passivated junctions: The next technology for silicon solar cells?, Joule, Volume 5, Issue 4, 2021 https://doi.org/10.1016/j.joule.2021.02.013.

[3] A. Richter, R. Müller, J. Benick et al., Design rules for high-efficiency both-sides-contacted silicon solar cells with balanced charge carrier transport and recombination losses, Nat. Energy 6 (2021) 429–438, https://doi.org/10.1038/s41560-021-00805-w.

[4] F. Feldmann, B. Steinhauser, T. Pernau et al., Industrial TOPCon Solar Cells Realized by a PECVD Tube Process, Presented at the 37th European PV Solar Energy Conference and Exhibition, 7-11 September 2020

[5] J. Bao, C. Chen, M. Limin et al., Towards 24% efficiency for industrial-type bifacial passivating-contact solar cells with homogeneous emitter, Presented at the 37th European PV Solar Energy Conference and Exhibition, 7-11 September 2020

[6] Y. Chen et al., Mass production of industrial tunnel oxide passivated contacts (i-TOPCon) silicon solar cells with average efficiency over 23% and modules over 345 W, Prog. Photovolt: Res. Appl., pp. Vol. 27, pp. 827– 834, 2019
[7] F. Buchholz, J. Linke, C. Peter, J. Hoß, J. Lossen and Radovan Kopecek, Fully Passivating Contact IBC Solar Cells Using Laser Processing, WCPEC-8 Milan, to be published (2022)
[8] Bas W.H. van de Loo, B. Macco, M. Schnabel et al., On the hydrogenation of Poly-Si passivating contacts by Al2O3 and SiNx thin films, Solar Energy Materials and Solar Cells 215 (2020)
[9] D. Kang, H.C. Sio, D. Yan, J. Stuckelberger, R. Liu, D. Macdonald, Firing stability of doped polysilicon passivation layers. 37th European photovoltaic solar energy conference and exhibition, 2020
[10] C. Hollmann, N. Folchert, S. P. Harvey et al., Changes in hydrogen concentration and defect state density at the poly-Si/SiOx/c-Si interface due to firing, Solar Energy Materials and Solar Cells 231 (2021)
[11] D.E. Kane, R.M. Swanson, Measurement of the emitter saturation current by a contactless photoconductivity decay method, in: 18th IEEE Photovolt. Spec. Conf., Las Vegas, 1985, p. 578
[12] T. Niewelt, B. Steinhauser, A. Richter et al., Reassessment of the intrinsic bulk recombination in crystalline silicon, Solar Energy Materials and Solar Cells 235 (2022)
[13] Choi, S., Kwon, O., Min, K.H. et al., Formation and suppression of hydrogen blisters in tunnelling oxide passivating contact for crystalline silicon solar cells. Sci Rep 10, 9672 (2020) https://doi.org/10.1038/s41598-020-66801-4
[14] S. Jafari, J. Hirsch, D. Lausch, et al., Composition limited hydrogen effusion rate of a-SiNx:H passivation stack, AIP Conference Proceedings 2147, 050004 (2019); https://doi.org/10.1063/1.5123853
[15] R.C.G. Naber, B.W.H. van de Loo, J.R.M. Luchies, LPCVD In-Situ n-Type Doped Polysilicon Process Throughput Optimization and Implementation into an Industrial Solar Cell Process Flow, 36th European Photovoltaic Solar Energy Conference and Exhibition, 2019

STATUS AND PERSPECTIVE OF INDUSTRIAL TOPCON SOLAR CELL DEVELOPMENT AT FRAUNHOFER ISE

Sebastian Mack, Bishal Kafle, Christopher Tessmann, Katrin Krieg, Sattar Bashardoust, Elmar Lohmüller, Udo Belledin, Pierre Saint-Cast, Hannes Höffler, Daniel Ourinson, Tobias Fellmeth, Bernd Steinhauser, Sven Kluska, Andreas Wolf
Fraunhofer Institute for Solar Energy Systems ISE
Heidenhofstrasse 2, 79110 Freiburg, Germany
Tel.: +49 761 4588-5048, email: Sebastian.Mack@ise.fraunhofer.de

ABSTRACT: This work reports on the status of industrial TOPCon solar cell development at Fraunhofer ISE. The baseline process makes use of Cz-Si:P wafers of M2 size, a TOPCon stack consisting of a thermal silicon oxide layer in combination with an 80 nm thin in-situ phosphorous-doped low-pressure chemical vapor deposited polysilicon layer as well as screen printed metal contacts. A comparison between different technologies for deposition of aluminum oxide (AlO_x) layers reveals advantages for atomic layer deposition, which considerably reduces minority carrier surface recombination, and as a result leads to an overall increase in solar cell efficiency. Potential challenges in contacting by screen-printed and fired silver contacts, which might arise due to the presence of an AlO_x layer on the rear side, are overcome by implementation of a laser-enhanced contact formation step, which reduces the series resistance of such solar cells by more than 2 Ωcm^2. These modifications of our baseline process in combination with a new generation of metallization pastes allows for the fabrication of solar cells with a champion efficiency of 23.8%. Our stable baseline process allows for evaluation of other TOPCon-related technologies or plated contacts, as well as identifying challenges for so-called TOPCoRE solar cells.

Keywords: TOPCon, Silicon Solar Cell, Manufacturing and Processing, Passivation

1 INTRODUCTION

More than 30 years passed between the initial presentation of the passivated emitter and rear cell (PERC) [1] and the actual large-scale manufacturing of that solar cell type. While the producers keep striving to push solar cell efficiencies, PERC is approaching its practical peak efficiency, however, due to physical limits of the cell architecture.

To keep up the learning rate of an increase of the conversion efficiency of industrial solar cells of around 0.5% absolute per year [2], a reduced time from lab to fab is required. After less than 10 years that the tunnel oxide passivated contact (TOPCon) solar cell has been initially published [3], industrial TOPCon (i-TOPCon) solar cells have already entered large-scale manufacturing in several production lines [4,5].

TOPCon technology, here used as an abbreviation for high-temperature passivating contact technology in general, i.e., excluding heterojunction technology, has been subject to increased research by numerous companies and research bodies over the last years. TOPCon combines a very thin silicon oxide (SiO_x) layer with a doped silicon layer to allow for very low surface recombination and to minimize metal contact related recombination.

Several technologies for formation of the tunnel oxide layer have been investigated. The tunnel oxide layer is very thin, typically in the range of 1 nm to 2 nm, to allow an efficient majority carrier transport, whereas others use oxide layers over 2 nm thickness, which are then deliberately broken by annealing at higher temperatures [6]. In most cases, SiO_x layers are being used that are formed by e.g., thermal oxidation, wet chemical oxidation, gaseous oxidation, or deposited layers; but also other materials are being investigated.

On top of the tunnel oxide layer, polysilicon layers are deposited. For this purpose, several technologies are possible, among which are low pressure chemical vapor deposition (LPCVD) [5], atmospheric pressure CVD (APCVD) [7], plasma-enhanced CVD (PECVD) [3,8], or sputtering / physical vapor deposition (PVD) [9]. In addition, some approaches come along with in-situ doping of the layer, i.e. addition of a doping precursor during deposition, whereas other choose to dope the layers ex-situ in a separate process step, such as e.g. $POCl_3$ diffusion [5]. Regardless of the doping type, most TOPCon routes require some sort of longer high-temperature treatment for crystallization/dopant activation and dopant drive-in, whereas some approaches have been reported to only require a short contact firing step [10].

In addition to shallow dopant profiles, hydrogen management can play an important role for low recombination on the TOPCon side, and different approaches have been recommended, such as $Al_2O_3/SiN_x:H$ or $SiN_x:H$ layers, hydrogenation by illumination and heating [4] or water vapor annealing [11].

In this paper, we will show the actual status of our research on i-TOPCon solar cells at Fraunhofer ISE, which is conducted mainly in our pilot line laboratory PV-TEC [12], which in the mean-time consists of two separate labs for front-end and back-end processing. All research is being performed using M2 sized wafers; however, we are in the process of upgrading our line to keep up with industry needs and be able to process wafers up to G12 size. In addition, we will also give an update on TOPCon Rear Emitter (TOPCoRE) solar cell technology.

2 SAMPLE PREPARATION

2.1 Solar cells

For solar cell fabrication, phosphorous-doped M2-sized Cz-Si wafers are used. Saw damage removal and random pyramid formation in alkaline solution represents the first process. Conventional tube furnace diffusion at atmospheric pressure using a BBr_3 liquid precursor forms the boron emitter [13] with a sheet resistance of around 110 Ω/sq. The unwanted rear emitter is removed by a single side borosilicate glass (BSG) layer removal step in an inline tool and a silicon removal step, both of which either in an inline system in acidic ambient, or alternatively in a combination of inline system for BSG removal and batch system in alkaline ambient. The targeted Si removal is 4 to 6 µm, which ensures an efficient isolation of the wafer edge. In both cases, the front BSG needs to be kept intact. The TOPCon stack, which consists of a thermally grown SiO_2 tunnel oxide layer and an in-situ doped polysilicon layer, both formed subsequently in one process in a low-pressure chemical vapor deposition (LPCVD) tube. The addition of PH_3 during polysilicon deposition leads to an in-situ doping of the around 80 nm thick deposited silicon layer.

As the polysilicon layer is wanted solely on the rear side of the sample, the parasitically deposited layer on the front side needs to be removed. This is being accomplished by etching in fluorine gas (F_2) at atmospheric pressure (atmospheric dry etching, ADE) in an inline tool, which also removes the polysilicon at the edges of the wafers and ensures a very high shunt resistance and sufficient reverse bias stability [14]. Due to its high selectivity, the front BSG layers is not etched during ADE, it acts as an etch stop for F_2 gas. The BSG layer is removed however in the next process step in fluoric acid (HF) solution, followed by cleaning of the wafers and thermal annealing for dopant activation in the TOPCon layer. For front side passivation, the SiO_x layer, which is formed during annealing, is removed during wafer cleaning. Surface passivation is being realized by deposition of a AlO_x layer by atomic layer deposition (ALD) in single slot configuration in a tube furnace, a subsequent outgassing step in another tube furnace, and the deposition of SiN_x:H layers on front and rear by plasma-enhance chemical vapor deposition (PECVD). Alternatively, the AlO_x/SiN_y:H stack on the front side is deposited by PECVD. Contact formation is realized by single step screen printing of a commercial Ag paste on the rear side, a AgAl paste on the front side and contact firing in a conventional conveyor belt furnace, yielding bifacial solar cells. In our industrial cell tester, we include a laser-enhanced contact optimization (LECO) step [15]. Its effect on IV parameters will be described later in the paper.

The described baseline process features a high flexibility and enables the implementation of alternative processes, such as PECVD TOPCon deposition [16] or plated metallization [17].

2.2 Lifetime samples

As an extension to our solar cell experiments, we typically fabricate asymmetric lifetime samples, i.e., cells without metallization, i.e., $p^+/n/n^+$ structure, for determination of the open circuit voltage limit iV_{oc}. For assessing the recombination at the front side of the cells, symmetric lifetime samples with random pyramids as well as Boron emitter, i.e., $p^+/n/p^+$ structure, and AlO_x/SiN_y:H passivation layers are used. The latter allow for extracting the dark saturation current density j_{0e} in high injection from quasi steady-state photoconductance (QSSPC) measurements by the slope method. Similarly, symmetric $j_{0,rear}$ rear side samples, i.e., $n^+/n/n^+$ structure, enable a better understanding of the recombination at the rear side of the sample.

3 EXPERIMENTAL RESULTS

3.1 LPCVD TOPCon

As mentioned above, our TOPCon layer consists of a thermally grown tunnel oxide layer and an in-situ doped LPCVD polysilicon layer of only 80 nm thickness, both formed during one process. To evaluate the dopant profile of our process, both within the polysilicon layer as well as the tail in the crystalline Si wafer, electro-chemical capacitance voltage (ECV) profiling has been used. The result of such a measurement is depicted in Figure 1, performed after thermal activation of the layer. The polysilicon layer features a rather constant dopant concentration of $2*10^{20}$ cm^{-3}, which allows for a low-ohmic contacting of this layer by screen-printed Ag pastes. Beyond the tunnel oxide layer, the carrier concentration decreases rapidly, leading to a shallow phosphorous tail in the wafer itself. Please note that no correction of the dopant profile depth has been performed, thus the tunnel oxide layer seems to be located at an etch depth of around 60 nm, instead of the actual position at around 80 nm depth. The dark saturation current density $j_{0,rear,SDE} = 4$ fA/cm^2 at a planar alkaline etched surface and $j_{0,rear,txt} = 10$ fA/cm^2 at an alkaline textured surface reveal a decent surface passivation quality of this layer. However, by further tailoring of the dopant profile, lower recombination values seem realistic.

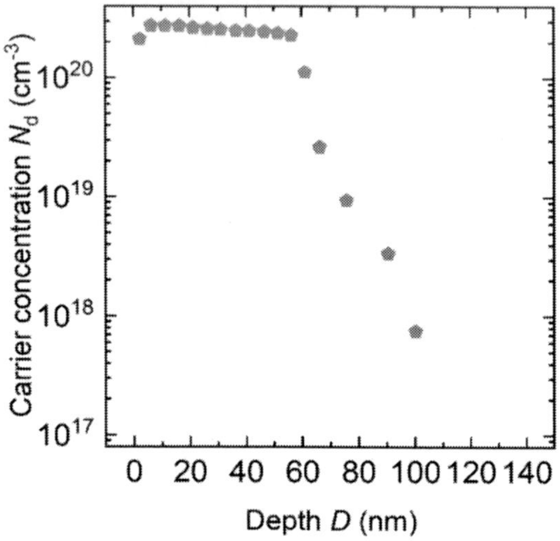

Figure 1: Active carrier dopant concentration N_d of a n-doped polysilicon layer, after thermal annealing, measured on a planar surface. Please note that no depth correction of the etched profile to a SEM thickness measurement has been performed.

One possible challenge for in-situ doped polysilicon layer deposition by LPCVD is the reduced deposition rate, when PH_3 gas is added for in-situ doping of the layer. To compensate this, often intrinsic layers are deposited, which are doped subsequently using $POCl_3$ diffusion furnaces (labelled "ex-situ doping"). Our results show that this approach yields similar recombination values $j_{0,rear}$ for planar and textured surfaces as the in-situ doping route.

As mentioned above, for the time being, LPCVD is our method of choice for i-TOPCon layer deposition. Two alternative methods for silicon layer deposition are PECVD and PVD. Both come with the potential advantages of strict single-sided deposition, which would eliminate the need for the one-sided polysilicon layer removal that is necessary for LPCVD layers [14]. While the PECVD process will be handled in the following section of this paper, there is a dedicated paper for the PVD route [18].

3.2 PECVD TOPCon

One significant challenge with direct-plasma PECVD of conductive layers can be the isolation of the electrodes. If the isolation is exposed to the plasma, as commonly the case for industrial direct-plasma PECVD tools, the isolation will be coated by deposition of conductive phosphorus doped a-Si. This leads to a gradual shunting of the isolation and thus reduced process efficiency. Consequently, the deposition rate decreases until at some point the plasma collapses. In our case, when using a 24-wafer horizontal carrier, after approx. 3 to 4 deposition runs the carrier must be dismantled to replace the isolators. Therefore, we developed new isolators that can keep up the isolation even after multiple deposition runs with highly doped n-TOPCon. Figure 2 shows the determined deposition rate for 19 n-TOPCon process runs with the same process boat. The deposition rate is mostly stable and only shows some slight upwards trend, also influenced by the uncertainty in the thickness measurement on Cz wafers due to the rougher surface. Combined with a few depositions (not shown here), a total layer of approx. 9 µm thickness was deposited so far using the modified isolators.

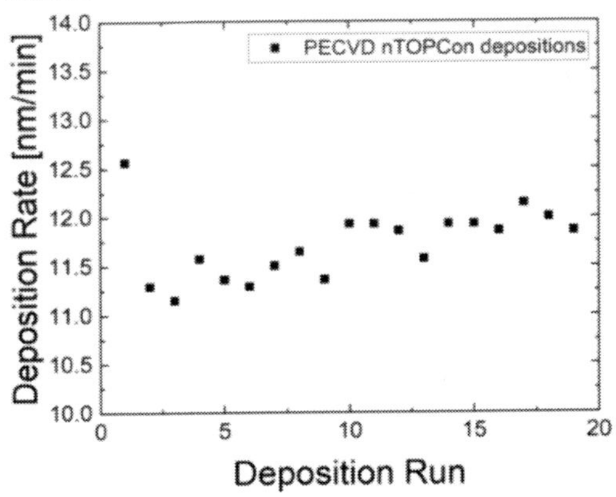

Figure 2: Deposition rate of PECVD n-TOPCon for 19 measured runs. Each deposition amounts to a thickness of 150 nm to 250 nm, depending on the target thickness. The total deposited thickness was approx. 9 µm.

3.3 Front side passivation

From around February 2020 until mid 2022, our boron emitter passivation was based on an AlO_x/SiN_y:H stack, which has been deposited by PECVD. However, in mid 2022, we were able to setup a process for high throughput thermal, temporal ALD in a tube furnace batch system, and we compared this ALD process to our PECVD reference process. By putting one wafer in each slot of the ALD process boat, we chose to deposit Al_2O_3 on both sides of the samples of this group, which is achieved easily due to nature of double-sided deposition during the process. The results reveal that the ALD process reduced j_{0e} to 15 fA/cm², compared to j_{0e} = 24 fA/cm² for the PECVD passivation approach. This is in agreement with values shown by others [4,19] for similar dopant profiles, which yields an increase of the median implied V_{oc} of asymmetric cell precursors of around 4 mV, see Figure 3. Nevertheless, although the results already make hope for a positive transfer towards solar cells, it is mandatory to also fabricate such solar cells, as the possible existence of an Al_2O_3 layer on the rear side might negatively affect the contact formation between Ag paste and poly-Si layer. However, the metallization process (see next section) showed to be tolerant to the presence of the Al_2O_3 layer and the solar cell conversion efficiency increased by 0.5% absolute for the ALD route, because of a 6 mV higher V_{oc} and an increase in fill factor *FF* by 0.6% absolute. The different increase of iV_{oc} and V_{oc} hints towards another positive effect of the Al_2O_3 deposition on the rear side, and in fact Al_2O_3/SiN_x:H stacks on top of polysilicon layers have been reported to allow for an improved hydrogenation compared to single SiN_y:H layers [20]. An alternative explanation might be that the existence of a thin Al_2O_3 layer on the rear side retards the contact formation and a thus lower $j_{0,met}$ would result in a higher V_{oc}. Further insight into that topic is necessary to determine the underlying effect and to be able to separate between the two explanations.

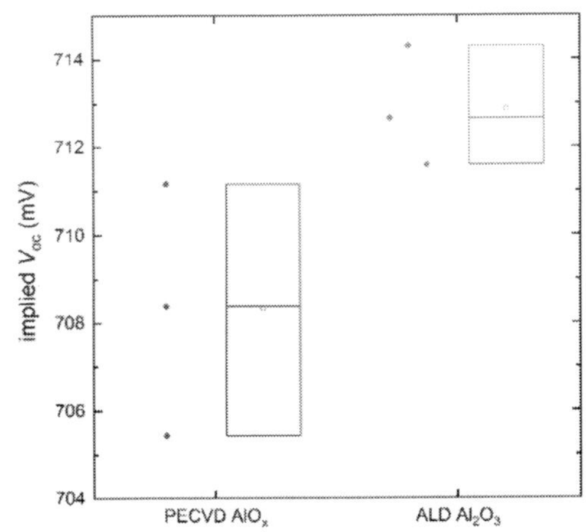

Figure 3: Implied open circuit voltage of asymmetric lifetime samples (non-metallized TOPCon cell precursor) for determination of the V_{oc} limit, measured after firing and illumination anneal.

3.4 Contact optimization

At Fraunhofer ISE, we implemented a laser enhanced contact formation (LECO) treatment into our cell tester. Explaining all the details of this LECO step is beyond the scope of this paper, but more information can be found in literature [21,22]. The LECO process allows for under-firing the contacts by reducing the required firing set temperature, which reduces the damage induced by the metal contact and thus e.g. $j_{0,met}$ at the front side [23]. In this case, LECO leads to a significant improvement in IV parameters. For our TOPCon solar cells, LECO leads to an increase of the conversion efficiency of up to 4% absolute (see Figure 4), by considerably reducing the average series resistance of the solar cells from 2.81 Ωcm² to 0.46 Ωcm², which results in an increased fill factor. This LECO treatment is especially important in case of Al_2O_3 layers on the rear side, since these layers hinder contact formation for non-adapted Ag pastes, resulting in extremely low fill factors for such cells. In such cases, LECO can demonstrate its full potential. The overall champion solar cell treated with LECO and measured at our industrial cell tester is listed in Table 1.

Table 1: IV parameters extracted from measurements in an industrial cell tester, after LECO, using a GridTouchTM unit with 30 wires for current and 5 wires for voltage, and measured versus a black and electrically non-conductive background.

Poly Thickness	η (%)	j_{sc} (mA/cm²)	V_{oc} (mV)	*FF* (%)
80 nm	23.5	40.7	706	81.7

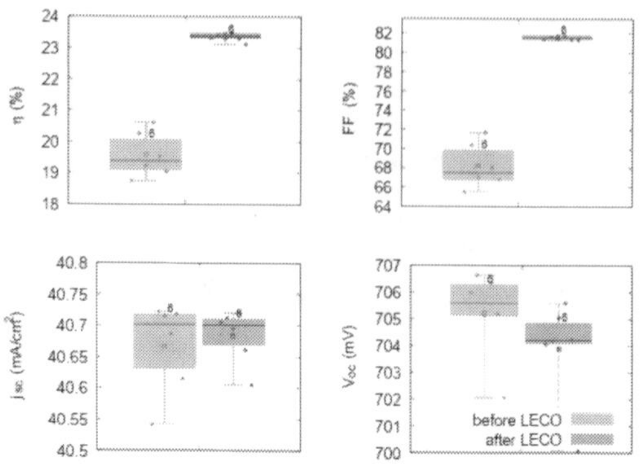

Figure 4: Impact of LECO processing on typical IV data, measured at an industrial cell tester.

Table 2 shows the IV parameters of another industrial M2 sized i-TOPCon solar cell from another run, which has been subject to a calibrated measurement at Fraunhofer ISE CalLab PV Cells, using a golden, reflective chuck and 30 current wires. The solar cell features V_{oc} = 706 mV, j_{sc} = 41.3 mA/cm², and FF = 81.6%, which results in a conversion efficiency η = 23.8 %. As described above, the polysilicon thickness on the rear side is only 80 nm, as confirmed by scanning electron microscopy. The front side passivation consists of an ALD deposited Al_2O_3 layer in combination with a PECVD SiNx layer. Single step screen printing and single step firing form the metal contacts, followed by LECO treatment.

Table 2: IV parameters from our champion M2 solar cell with area 244.5 cm² and a polysilicon thickness of 80 nm (n-doped, in-situ). The calibrated measurement has been performed at Fraunhofer ISE CalLab on a golden reflective chuck with full area rear side contacting, using 30 wires for current and 5 wires for voltage, neglecting the grid resistance.

Status	η (%)	j_{sc} (mA/cm²)	V_{oc} (mV)	FF (%)
As processed	23.8	41.3	706	81.6
After MgF_2	24.0	41.6	708	81.6

3.5 TOPCon vs. TOPCoRE

So far, in all experiments described above, n-type Si wafers have been used. However, for reasons of cost and dopant segregation over the ingot, p-type Cz-Si wafers are still of relevance. In addition, Ga-doped Si-wafers feature a high minority carrier lifetime and are favorable to boron-doped Si wafers with respect to potential iron contamination. This makes them interesting for use in so-called TOPCoRE solar cells, i.e., a TOPCon solar cell, where instead of a n-type Si wafer, a p-type Si wafer is used. This transition puts the emitter on the rear side of the cell, in this case then a polysilicon passivated contact emitter, and the boron diffused front side becomes a front surface field, which due to the high conductivity of the wafer itself does not need to contribute to lateral conductivity and which can be thus optimized with respect to passivation quality and contact formation only. On the one hand, this cell structure is very sensitive to front side recombination and carrier diffusion length, as photogenerated electrons need to reach the cells rear side, and on the other hand to carrier recombination at the rear side metal contacts due to presence of the pn junction. A possible too deep contact at the rear emitter side would lead to a direct shunting of the device.

So far, in few experiments, we used high quality, high-ohmic Ga-doped Cz-Si wafers for fabrication of such TOPCoRE solar cells. For reasons of stability, we increased the thickness of the polysilicon layer to 160 nm, however first tests with 80 nm of n-doped polysilicon were successful, too, and showed almost an identical conversion efficiency, with differences in the range of 0.1%. All other processes from the above-described process i-TOPCon sequence have been left unchanged. The measurements at Fraunhofer ISE CalLab PV Cells show an efficiency of 23.2% for such p-type TOPCoRE devices, which forms a promising starting point for further progress on such cells. More information on this cell structure can be found in the respective paper [24].

3.6 Plating

The combination of local laser ablation of passivation layers and a plated stack of nickel (≈0.5µm thickness) / copper (5 to 10 µm) / silver (<0.5 µm) [25] offers the possibility to reduce more than 90% in silver consumption for industrial TOPCon solar cells. Furthermore, recent publications from Fraunhofer ISE [26] could demonstrate that low-damage laser contact opening for plating enables to further reduce the poly-Si thickness of the TOPCon contact down to 60 nm and by that further reduce processing cost of TOPCon solar cells.

Figure 5 shows the V_{oc} of industrial TOPCon precursors metallized with i) either both sides plated Ni/Cu/Ag or ii) screen-printed silver (no LECO) contacts on the poly-Si rear side, and screen-printed AgAl paste on the front side. The plated solar cells feature an increased V_{oc} of about 5 mV compared to the screen-printed reference. Photoluminescence imaging (PL) of test fields on i-TOPCon solar cell precursors with and without laser contact openings (LCO) on a sample with 60 nm poly-Si thickness demonstrate that the application of UV picosecond laser ablation yields a marginal laser-induced damage with an average iV_{oc} decrease of only 1 mV. These results show the potential to decrease the poly-Si thickness by introducing laser defined plated contacts for i-TOPCon solar cells. Furthermore, recent developments at Fraunhofer ISE [6] show that the contact resistivity ρ_c below 1 mΩcm² allows to decrease the LCO width down to 5 µm, which leads to plated contact widths of about 12 µm only.

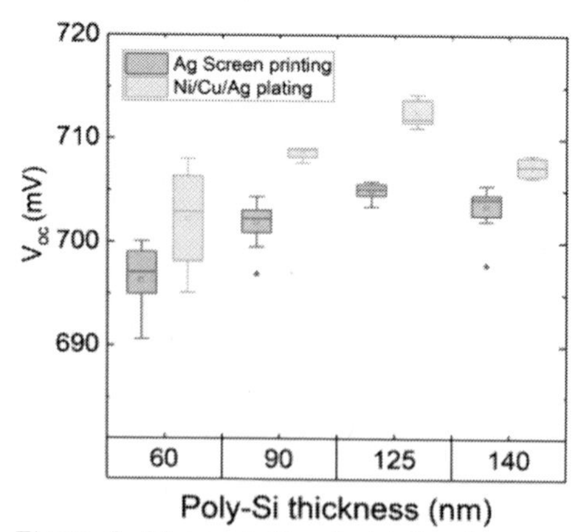

Figure 5: Measured V_{oc} of plated and screen printed TOPCon solar cells with variation of poly-Si thickness on the TOPCon rear side (from [26]).

The recent progress in reducing the impact of laser-induced damage in LCO improved contact resistivity and contact aspect ratio [17,27], which allowed to achieve champion solar cell efficiencies of 24% [26] (see Table 3) for industrial TOPCon solar cells with plated Ni/Cu/Ag contacts. The low-damage, narrow plating contacts allowed an improvement of 0.5% absolute compared to the screen-printed metallization at the R&D line of the cell supplier.

Table 3: Calibrated IV measurements of champion TOPCon solar cells (poly thickness: 125 nm) with plated or screen printed metallization measured at Fraunhofer ISE CalLab using industrial TOPCon precursors [26].

Metallization	η (%)	J_{sc} (mA/cm²)	V_{oc} (mV)	FF (%)
Ag/AgAl screen printed (supplier, no LECO)	23.5	40.7	705	81.9
Plated Ni/Cu/Ag (FhG-ISE)	24.0	41.0	715	82.0

4 SUMMARY

In this paper, we have shown an overview over our activities in the field of i-TOPCon solar cells. Within a short time of only a few years, TOPCon technology has matured from a proof of concept to an industrial cell concept, which has already been transferred into production by many cell manufacturers. Several technologies are available for both interface oxide formation and polysilicon layer deposition, and while each of them seems to allow for a high quality TOPCon layer, in the end it might come down to questions like uptime, stability, process flow integration, and of course cost.

We have developed a process flow for i-TOPCon solar cells, which makes use of an in-situ doped polysilicon layer formed by LPCVD, with screen-printed metallization. Champion results are currently in the range of 24% for M2-sized Cz-Si:P wafers, with a polysilicon thickness of only 80 nm. Notable efficiency improvements have been achieved recently by the implementation of ALD AlOx deposition and LECO treatment. In addition, we plan to upgrade all relevant tools until end of 2023, to be able to process wafers up to G12 size.

Acknowledgements
The authors want to thank all colleagues at Fraunhofer ISE for the excellent atmosphere and the support of TOPCon activities in general, by means of processing, simulation, discussions or financing.
This work was funded by the German Federal Ministry for Economic Affairs and Climate Action within the research projects "TOPConCluster" (contract no. 03EE1065A), "BUSSARD" (contract no. 03EE1071A), "TALER" (contract no. 03EE1021B), "SpuTOPiA" (contract no. 03EE1068C), and "StroKoFo" (contract no. 03EE1055B).

References

[1] A. W. Blakers, A. Wang, A. M. Milne, J. Zhao, and M. A. Green, Appl. Phys. Lett. 55 (1989), 1363.
[2] F. Fertig, R. Lantzsch, A. Mohr, M. Schaper, M. Bartzsch, D. Wissen, F. Kersten, A. Mette, S. Peters, A. Eidner, J. Cieslak, K. Duncker, M. Junghänel, E. Jarzembowski, M. Kauert, B. Faulwetter-Quandt, D. Meißner, B. Reiche, S. Geißler, S. Hörnlein, C. Klenke, L. Niebergall, A. Schönmann, A. Weihrauch, F. Stenzel, A. Hofmann, T. Rudolph, A. Schwabedissen, M. Gundermann, M. Fischer, J. W. Müller, and D. Jeong, Energy Proced. 124, 338 (2017).
[3] F. Feldmann, M. Bivour, C. Reichel, M. Hermle, and S. W. Glunz, Sol. Energy Mater. Sol. Cells 120 (2014), 270.
[4] R. Chen, M. Wright, D. Chen, J. Yang, P. Zheng, X. Zhang, S. Wenham, and A. Ciesla, Prog Photovolt Res Appl 29 (2021), 1213.
[5] Y. Chen, D. Chen, C. Liu, Z. Wang, Y. Zou, Y. He, Y. Wang, L. Yuan, J. Gong, W. Lin, X. Zhang, Y. Yang, H. Shen, Z. Feng, P. P. Altermatt, and P. J. Verlinden, Prog Photovolt Res Appl. 27 (2019), 827.
[6] D. Tetzlaff, J. Krugener, Y. Larionova, S. Reiter, M. Turcu, F. Haase, R. Brendel, R. Peibst, U. Hohne, J. D. Kahler, and T. F. Wietler, Sol. Energy Mater. Sol. Cells 173 (2017), 106.
[7] A. Merkle, S. Seren, H. Knauss, B. Min, J. Steffens, B. Terheiden, R. Brendel, and R. Peibst, Proceedings 35th EU PVSEC, (2018), 782.
[8] J. Temmler, J.-I. Polzin, F. Feldmann, L. Kraus, B. Kafle, S. Mack, A. Moldovan, M. Hermle, and J. Rentsch, Phys. Status Solidi A 215 (2018), 1800449.
[9] Di Yan, A. Cuevas, S. P. Phang, Y. Wan, and D. Macdonald, Appl. Phys. Lett. 113 (2018), 061603.
[10]A. Ingenito, G. Nogay, Q. Jeangros, E. Rucavado, C. Allebé, S. Eswara, N. Valle, T. Wirtz, J. Horzel, T. Koida, M. Morales-Masis, M. Despeisse, F.-J. Haug, P. Löper, and C. Ballif, Nat. Energy 3 (2018), 800.
[11]Z. Zhang, M. Liao, Y. Huang, X. Guo, Q. Yang, Z. Wang, T. Gao, C. Shou, Y. Zeng, B. Yan, and J. Ye, Sol. RRL 3 (2019), 1900105.
[12]J. Rentsch, M. Bivour, D. Erath, F. Feldmann, A. Fischer, S. Mack, A. Moldovan, J. Temmler, M. Hermle, S. W. Glunz, and R. Preu, Proceedings 33rd European Photovoltaic Solar Energy Conference and Exhibition (2017), p. 441.
[13]S. Werner, E. Lohmüller, U. Belledin, A. Vlooswijk, R. Naber, S. Mack, and A. Wolf, Proceedings 31st European Photovoltaic Solar Energy Conference and Exhibition (2015), p. 637.
[14]B. Kafle, S. Mack, C. Teßmann, S. Bashardoust, L. Clochard, E. Duffy, A. Wolf, M. Hofmann, and J. Rentsch, Sol. RRL 6 (2022), 2100481.
[15]E. Krassowski, S. Großer, M. Turek, and H. Höffler, Proceedings 37th European Photovoltaic Solar Energy Conference and Exhibition (2020).
[16]F. Feldmann, T. Fellmeth, B. Steinhauser, H. Nagel, D. Ourinson, S. Mack, E. Lohmüller, J. Polzin, J. Benick, A. Richter, A. Moldovan, M. Bivour, F. Clement, J. Rentsch, M. Hermle, and S. W. Glunz, Proceedings 36th European Photovoltaic Solar Energy Conference and Exhibition (2019), p. 304.
[17]S. Kluska, B. Grübel, G. Cimiotti, C. Schmiga, H. Berg, A. Beinert, I. Kubitza, P. Müller, and T. Voss, EPJ Photovolt. 12 (2021), 10.
[18]V. Linss, E. Schneiderlöchner, T. Dietsch, J. Baumann, U. Graupner, J. Hoß, J. Linke, J. Lossen, J.-I. Polzin, S. Mack, H. Nagel, and M. Bivour, this conference.
[19]B. Liao, X. Wu, W. Wu, C. Liu, S. Ma, S. Wang, T. Xie, Q. Wang, Z. Du, W. Shen, X. Li, W. Li, and B.

Hoex, Prog Photovolt Res Appl (2022), https://doi.org/10.1002/pip.3607.

[20]M. Stodolny, K. Tool, B. Geerligs, J. Loffler, A. Weeber, Y. Wu, J. Anker, X. Lu, J. Liu, P. Bronsveld, A. Mewe, G. Janssen, and G. Coletti, Proceedings 46th IEEE Photovoltaic Specialists Conference (PVSC), p. 1456.

[21]T. Fellmeth, H. Höffler, S. Mack, E. Krassowski, K. Krieg, B. Kafle, and J. Greulich, Prog Photovolt Res Appl (2022), https://doi.org/10.1002/pip.3598.

[22]E. Krassowski, B. Jaeckel, U. Zeller, M. Pander, P. Schenk, E. Hofmueller, and H. Hanifi, Sol. RRL 6 (2022), 2100537.

[23]E. Lohmüller, S. Werner, B. Thaidigsmann, N. Wöhrle, S. Mack, F. Clement, and D. Biro, Proceedings 29th European Photovoltaic Solar Energy Conference and Exhibition. Proceedings (2014), p. 635.

[24]J. Horzel, S. Mack, I. Voicu Vulcanean, K. Zimmermann, S. Pingel, W. Kwapil, F. Maischner, H. Höffler, S. Bashardoust, D. Wagenmann, J. Greulich, J. Seif, A. Steinmetz, and J. Rentsch, Sol. RRL (2022), 2200613.

[25]B. Grübel, G. Cimiotti, C. Schmiga, V. Arya, B. Steinhauser, N. Bay, M. Passig, D. Brunner, M. Glatthaar, and S. Kluska, IEEE J. Photovolt. 11 (2021), 584.

[26]S. Kluska, R. Haberstoh, B. Grübel, G. Cimiotti, C. Schmiga, A. A. Brand, A. Nägele, B. Steinhauser, M. Kamp, M. Passig, M. Sieber, D. Brunner, and S. Fox, Solar Energy Materials and Solar Cells 246 (2022), 111889.

[27]B. Grübel, G. Cimiotti, C. Schmiga, S. Schellinger, B. Steinhauser, A. A. Brand, M. Kamp, M. Sieber, D. Brunner, S. Fox, and S. Kluska, Prog Photovolt Res Appl. 30 (2022), 615.

TOWARDS COST-EFFECTIVE HIGH-EFFICIENCY POLO IBC SOLAR CELLS WITH MINIMAL CONVERSION INVEST FOR EXISTING PERC+ PRODUCTION LINES

Thorsten Dullweber*[1], Verena Mertens[1], Maximilian Stöhr[1], Jonathan Langlois[1], Larissa Mettner[1], Ulrike Baumann[1], Felix Haase[1], Rolf Brendel[1,2], Joris Libal[3], Angelika Hähnel[4], Alexander Müller[4], Volker Naumann[4], Aaron Vogt[5], Norbert Ambrosius[5], Thomas Pernau[6], Helge Haverkamp[6]

[1]Institute for Solar Energy Research Hamelin (ISFH), Am Ohrberg 1, 31860 Emmerthal, Germany
[2]Institute of Solid-State Physics, Leibniz Universität Hannover, Appelstrasse 2, 30167 Hannover, Germany
[3]ISC Konstanz, Rudolf-Diesel-Str. 15, 78467 Konstanz, Germany
[4]Fraunhofer CSP, Otto-Eißfeldt-Straße 12, 06120 Halle, Germany
[5]LPKF Laser & Electronics AG, Osteriede 7, 30827 Garbsen, Germany
[6]centrotherm international AG, Württemberger Str. 31, 89143 Blaubeuren, Germany

*corresponding author: Tel.: +49 (0)5151-642, dullweber@isfh.de

ABSTRACT: In this paper, we report a novel manufacturing process sequence for POLO IBC solar cells applying a local PECVD SiO_xN_y/n-a-Si deposition through a glass shadow mask to form the structured carrier selective n-poly-Si layer in one process step. Using a lab-type single-wafer PECVD tool, the POLO IBC precursors without metal contacts exhibit an excellent implied V_{oc} of 741 mV. A TEM analysis confirms that the N_2O in-situ plasma oxidation indeed incorporates N in the interfacial SiO_xN_y layer which could facilitate the excellent firing stable surface passivation. The laser contact openings and screen-printed Al and Ag rear contacts were successfully aligned to the shadow mask features within ± 20 µm precision. Fully processed POLO IBC solar cells on M2-sized Ga-doped Cz wafers exhibit conversion efficiencies of up to 23.0% with V_{oc}=708 mV, J_{sc} =41.2 mA/cm^2 and FF=78.7%. We attribute the 33 mV difference between implied V_{oc} and V_{oc} to additional J_0 contributions of the Ag and Al metal contacts which is subject to future improvements. As next step towards POLO IBC mass production readiness, we transfer the PECVD SiO_xN_y/n-a-Si process through shadow masks from the single-wafer lab-tool to a mass-production c.plasma PECVD tool from centrotherm, which is installed at the ISFH SolarTeC. Test wafers with double-sided in-situ PECVD SiO_xN_y/n-a-Si deposition reveal an excellent J_0 = 3 fA/cm^2 after annealing, SiN capping and firing. We apply the glass shadow mask in the industrial c.plasma PECVD tool in combination with the PECVD SiO_xN_y/n-a-Si deposition. The first POLO IBC solar cells processed with the c.plasma tool with shadow masks achieve conversion efficiencies up to 22.3% and average V_{oc} values of 710 mV. In the future, we aim at further increasing the conversion efficiency towards 25% and implementing an automated shadow mask loading into the industrial c.plasma tool.
Keywords: Silicon Solar Cell, Back Contact, PECVD, PERC, High-Efficiency

1 INTRODUCTION

Today, around 80% of the worldwide produced solar cells apply p-type Ga-doped silicon wafers and the monofacial PERC or bifacial PERC+ solar cell design [1]. In mass production, PERC and PERC+ cells obtain conversion efficiencies around 23% applying a best-in-class cost-effective manufacturing process [1]. Nevertheless, the carrier recombination in the phosphorus-doped emitter and at the Ag front contacts limits the V_{oc} and the efficiency potential of PERC+ cells to below 700 mV and 24%, respectively [2].
Based on fundamental carrier selectivity considerations [3], ISFH developed the POLO IBC solar cell [3,4] as next-generation cell technology. The POLO IBC cell design builds on today's industrial PERC+ solar cells by continuing to use Ga-doped Cz wafers, an AlO_x/SiN_y rear passivation and Al finger base contacts. However, it replaces the efficiency limiting phosphorus emitter by a carrier selective polysilicon on oxide (POLO) [5,6] contact on the rear side thereby drastically increasing the V_{oc} potential up to 733 mV with an efficiency potential of up to 25.5% as confirmed by Quokka simulations [7]. As process technology for POLO IBC cells, ISFH developed a PECVD SiO_xN_y/n-a-Si deposition process for the POLO stack using a lab-type tool with N_2O in-situ plasma oxidation resulting in an excellent passivation quality after firing with J_0 = 2 fA/cm^2 [8]. Glass-based shadow masks enable the confined local PECVD deposition of the SiON/n-a-Si layer stack onto the silicon wafer [9] thereby facilitating a very lean POLO IBC process sequence, as proposed in [10,11]. In this paper, we report for the first time fully-processed POLO IBC solar cells with local PECVD SiO_xN_y/n-a-Si deposition through a shadow mask applying a lab-type single-wafer PECVD tool as well as a mass-production-type PECVD tool. Our cost calculations demonstrate that POLO IBC cells manufactured with this process can be cost-competitive to today's mainstream PERC+ cells.

2 PECVD SiO_xN_y/n-a-Si process transfer and analysis

2.1 Process transfer from lab tool to industrial tool

As next step towards manufacturability of POLO IBC cells with shadow masks, we transfer the local PECVD process as published in [8] using a lab-type single wafer tool (Clustertool, Von Ardenne) to an industrial c.plasma tool from centrotherm at the ISFH SolarTeC. Both tools apply a capacity-coupled plasma source, an in-situ plasma oxidation using N_2O to grow an approx. 1.7 nm thin SiO_xN_y, and SiH_4, PH_3 and H_2 to deposit a ca. 120 nm thick in-situ-doped n-a-Si layer. To assess the passivation quality, symmetrical n-type Cz test wafers are processed with the PECVD SiO_xN_y/n-a-Si followed by annealing in N_2 at 850°C, PECVD deposition of a 100 nm thick SiN_x layer, and firing at 810°C.
As reference, Figure 1 a) shows the results obtained with the lab-tool as published in [8], where the SiO_xN_y interface exhibits an excellent firing stability in contrast

to a thermal SiO_x interface, which slightly degrades during firing. After adapting, implementing, and optimizing the PECVD SiO_xN_y/n-a-Si recipe, the industrial c-plasma tool also achieves very good J_0 values of 3 fA/cm² after firing as displayed in Fig. 1 b). For comparison, we also develop an in-situ plasma oxidation using O_2 at the industrial tool. As shown in Fig. 4b) the PECVD SiO_x/n-a-Si layer stack exhibits higher J_0 values of 6 fA/cm² post SiN deposition which degrade to $J_0 = 10$ fA/cm² after firing. Hence, for both tools the SiO_xN_y interface grown by in-situ N_2O plasma oxidation delivers the best J_0 values post firing.

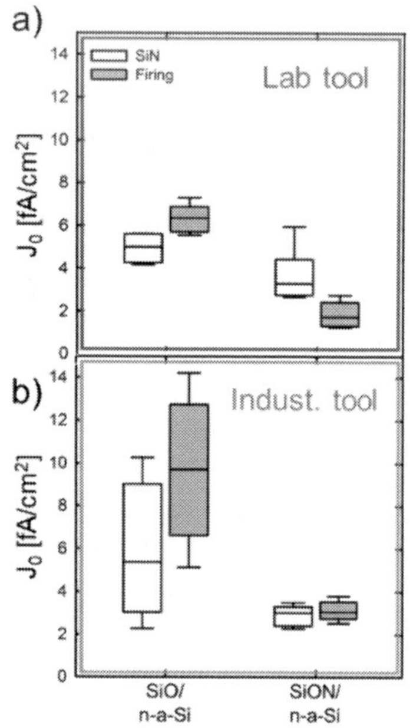

Figure 1: Saturation current density J_0 of symmetrical test wafers passivated with either SiO_x/n-a-Si or SiO_xN_y/n-a-Si. Both, the lab PECVD tool a) as well as the industrial PECVD tool b) achieve excellent J_0 values post firing of 2 and 3 fA/cm², respectively, when applying the interfacial SiO_xN_y. In contrast, when using an SiO_x either by thermal oxidation a) or by plasma oxidation b), the surface passivation degrades after firing to J_0 values post firing of 6 and 10 fA/cm², respectively

2.2 TEM/EDX analysis of the SiO_xN_y/n-a-Si layer stack

To analyze the root cause of the superior firing stability of the SiO_xN_y interface shown in Fig. 1, a silicon wafer deposited with a SiO_xN_y/n-a-Si layer stack from the lab-tool and annealed at 850°C was analysed by Transmision Electron Microscopy (TEM) at Fraunhofer CSP. For this sample, the SiO_xN_y layer was intentionally grown thicker to about 3 nm to enable a depth dependent elemental resolution by Energy Dispersive X-ray (EDX) analysis. The TEM image in Fig. 2 a) shows that annealing at 850°C converts the amorphous n-a-Si layer into a polycrystalline n-poly-Si layer. The EDX analysis in Fig. 2 b) shows the elemental distribution determined by standard-less Cliff Lorimer quantification of the EDX analysis which yields up to roughly 10% nitrogen in the SiO_xN_y layer close to the interface to the c-Si. Gatz et al. published in Ref. 12 that the incorporation of N in amorphous silicon increases the Si – H bond energy and thereby improves the firing stability of the SiN surface passivation compared to a-Si layers which degrade during firing. In Ref. 8 we published the hypothesis, that nitrogen in the SiO_xN_y layer may similarly improve the firing stability of SiO_xN_y/n-a-Si layer stacks by increasing the bond energy of Si-H at the c-Si wafer surface thereby suppressing the H out-diffusion and formation of Si dangling bonds during firing which act as recombination centers. The accumulation of N at the c-Si/SiO_xN_y interface shown in Fig. 2 b) supports this hypothesis.

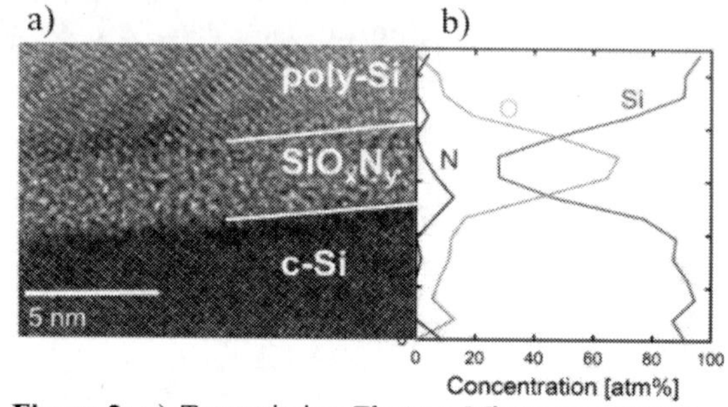

Figure 2: a) Transmission Electron Microscope (TEM) image of a PECVD SiO_xN_y/n-a-Si layer stack after annealing at 850°C converting the amorphous n-a-Si layer in a polycrystalline n-poly-Si layer. The SiO_xN_y layer was intentionally grown thicker to about 3 nm to enable the depth dependent elemental resolution by Energy Dispersive X-ray (EDX) analysis in b). The EDX analysis confirms that roughly 10% nitrogen is incorporated in the SiO_xN_y layer close to the interface to the c-Si wafer. The N may facilitate the firing stability of SiO_xN_y/n-a-Si layer stacks shown in Fig. 1 as explained in the text.

3 POLO IBC SOLAR CELLS WITH SHADOW MASKS

3.1 Process sequence with shadow masks

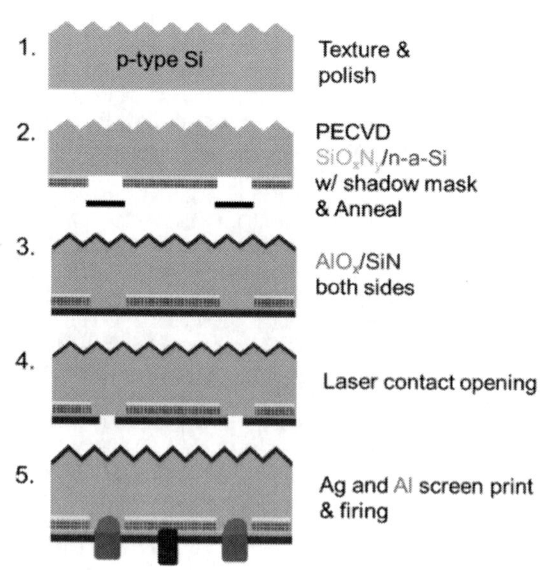

Figure 3: Schematic drawing of the lean POLO IBC process flow applying the local PECVD SiO_xN_y/n-a-Si deposition through a glass shadow mask.

Figure 3 shows schematic drawings of the POLO IBC process flow applying the local PECVD SiO_xN_y/n-a-Si deposition through a glass shadow mask. We use 1Ωcm Ga-doped M2-sized Cz wafers which are double-sided textured and subsequently polished at the rear side according to step 1 in Fig. 3. Afterwards, we locally deposit the SiO_xN_y/n-a-Si layer stack through a glass shadow mask provided by LPKF by using either the lab-type PECVD tool or the industrial PECVD tool. Step 2 in Fig. 3 is completed by annealing the wafers in nitrogen atmosphere at 850°C. Both wafer sides are passivated by a AlO_x/SiN layer stack (step 3) followed by laser contact opening (LCO) of the AlO_x/SiN at the rear side base region (step 4). Finally, the Al contacts are screen-printed on top of the LCOs and the Ag contacts are printed aligned to the n-poly-Si. The process sequence is completed by firing the wafers at around 810°C where the Al contacts locally alloy with the silicon wafer forming an Al-BSF and where the Ag paste dissolves the AlO_x/SiN layer contacting the n-poly-Si.

Compared to an industrial PERC+ process, this POLO IBC process uses the same process steps such as texturing, rear polishing, AlO_x/SiN deposition, LCO and screen-printing of Al and Ag pastes. Only the $POCl_3$ diffusion and laser doping is replaced by the local PECVD SiO_xN_y/n-a-Si deposition, see also the process flow comparison in the cost calculation displayed in Figure 7.

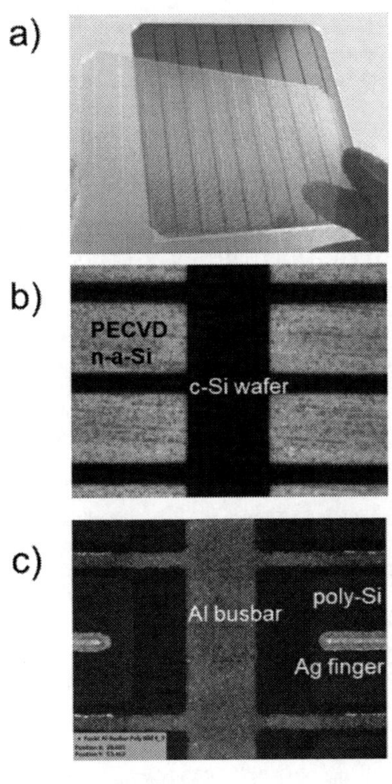

Figure 4: a) Photograph of a M2-sized glass shadow mask in front of a Cz wafer with local SiO_xN_y/n-a-Si layer stack. b) Light microscope image of local SiO_xN_y/n-a-Si layer stack deposited through the shadow mask in the industrial PECVD tool. c) Light microscope image of the rear side of a fully processed POLO IBC cell including screen-printed Al and Ag fingers which are well aligned to the local SiO_xN_y/n-a-Si layers.

Figure 4 a) shows a photograph of a M2-sized glass shadow mask in front of a Cz wafer which received the local SiO_xN_y/n-a-Si deposition. A close up image by light microscopy in Fig 4 b) reveals that the local SiO_xN_y/n-a-Si layer width matches the shadow mask layout within ± 10 µm accuracy. Figure 4 c) displays a light microscope image of the rear side of a fully processed POLO IBC cell including screen-printed Al and Ag fingers which are well aligned to the local SiO_xN_y/n-a-Si layers within an accuracy of ± 20 µm.

3.2 Implied V_{oc} results of POLO IBC precursors without metal contacts

In order to assess the V_{oc} potential of the POLO IBC cells, we process implied V_{oc} (iV_{oc}) precursors according to the process flow in Fig. 3 up to step 3 without LCO (step 4) and without screen printing (step 5) but including firing. When depositing the local PECVD SiO_xN_y/n-a-Si layer in the lab-type tool, we obtain excellent average iV_{oc} values of 741 mV determined by infrared lifetime mapping (ILM) as can be seen in Fig. 5. This high iV_{oc} corresponds to a total J_0 = 10 fA/cm^2 of the POLO IBC precursor revealing the excellent passivation qualities of the local SiO_xN_y/n-a-Si and of the AlO_x/SiN layers as well as a very low $J_{0,bulk}$ of the Ga wafer, which is in accordance with the simulations in Ref. 7.

When depositing the local PECVD SiO_xN_y/n-a-Si layer in the industrial tool, we obtain average iV_{oc} values of 721 mV as can be seen in Fig. 5. However, these wafers have not yet received the optimized PECVD SiO_xN_y/n-a-Si recipe in Fig. 1 b) but a previous recipe exhibiting higher J_0 values. When applying the improved recipe of Fig. 1 b) we expect that the iV_{oc} values of POLO IBC precursors processed in the industrial tool will approach the high iV_{oc} values obtained with the lab-type tool.

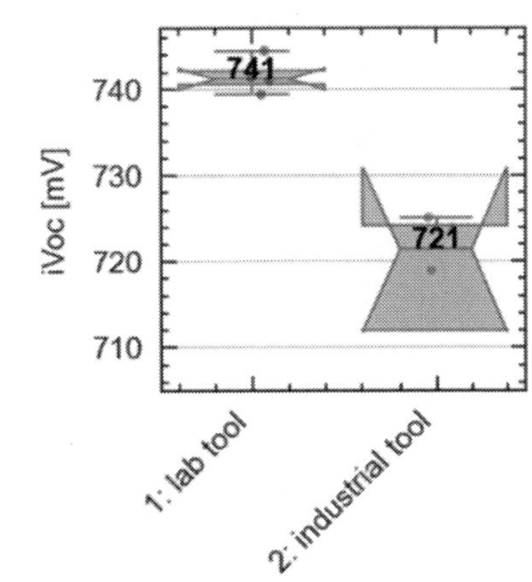

Figure 5: Implied V_{oc} (iV_{oc}) values of POLO IBC precursors without metal contacts where the local PECVD SiO_xN_y/n-a-Si deposition was performed in the lab tool (group 1) or the industrial tool (group 2).

3.3 POLO IBC solar cell results using shadow masks

Since the local PECVD SiO_xN_y/n-a-Si is quite new, so far we have processed only 2 POLO IBC cell batches using the lab-type PECVD tool and only one cell batch using the industrial PECVD tool. The cells are measured with an IV tester (LOANA from pv tools) at ISFH which uses contact bars to contact the busbars at full length at the cells rear side and which include sense pins to measure the voltage of the cell. The illumination is calibrated using the internal calibration of the IV tester since we do not yet have a POLO IBC cell measured at a certified calibration lab.

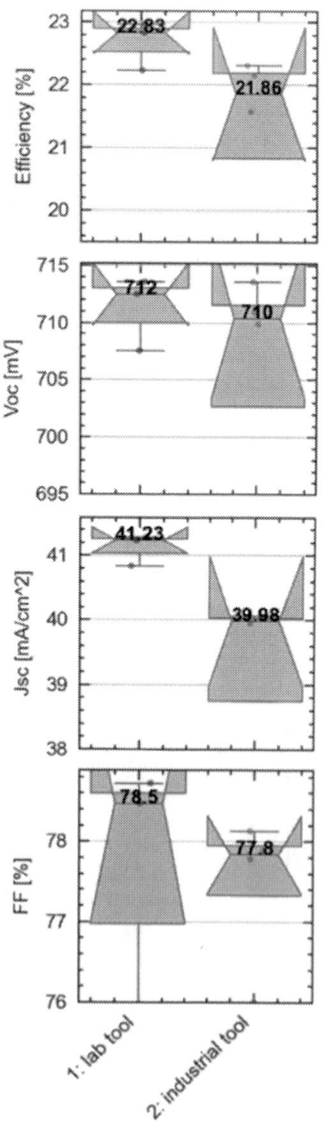

Figure 6: IV parameters of POLO IBC solar cells with local PECVD deposition through a shadow mask using either the lab-type tool or the industrial tool.

Figure 6 displays the IV parameters efficiency, open circuit voltage V_{oc}, short circuit current J_{sc}, and fill factor FF of POLO IBC solar cells from the second run using the lab-type tool for local PECVD SiO_xN_y/n-a-Si deposition through a shadow mask. In average, we obtain 22.8% efficiency and V_{oc} = 712 mV. The best POLO IBC cell of this run exhibits a conversion efficiency of 23.0% with V_{oc}=708 mV, J_{sc} =41.2 mA/cm^2 and FF=78.7%. We attribute the 29 mV difference between implied V_{oc} = 741 mV (see Fig. 5) and V_{oc} = 712 mV to the high Al and Ag contact area fractions of approx. 3% and 10%, respectively. When assuming a typical $J_{0,Al\text{-}BSF}$ = 500 fA/cm^2 [2] and a $J_{0,Ag}$ = 150 fA/cm^2, the full metallization would account for an area-weighted $J_{0,met}$ of 30 fA/cm^2, which explains the delta between the measured V_{oc} and iV_{oc}. By reducing the metal area coverage and by further improving the Ag contact properties, we expect to achieve higher V_{oc} values and higher conversion efficiencies in the near future.

In the first batch of POLO IBC cells using the industrial PECVD tool we processed two different split groups. Split group 1 received the full PECVD SiO_xN_y/n-a-Si deposition through the shadow mask. Split group 2 received a wet chemically grown SiO_x interface by O_3 dissolved in DI water, followed by depositing the PECVD n-a-Si layer through the shadow mask. Figure 6 shows the POLO IBC cell results obtained with split 2 using the wet chemical SiO_x interface and the local PECVD n-a-Si. We obtain average efficiencies of 21.8% and V_{oc} values of 710 mV with a best cell efficiency of 22.3%. The split with the full PECVD SiO_xN_y/n-a-Si deposition through the shadow mask is not shown in Fig. 6. The resulting V_{oc} and J_{sc} values are very similar but the FF is significantly lower which we attribute to a too thick PECVD SiO_xN_y thickness. We are currently optimizing the SiO_xN_y thickness for the industrial tool and will apply it in combination with the improved PECVD SiO_xN_y/n-a-Si recipe of Fig. 1 b) to the next POLO IBC cell run thereby expecting that the conversion efficiencies obtained with the industrial tool will approach the values obtained with the lab tool.

4 COST ASSESSMENT OF POLO IBC CELLS MANUFACTURED WITH SHADOW MASKS

To assess the future economic competitiveness of POLO IBC cells manufactured with shadow masks versus todays mainstream PERC+ cells, we compare the cost of ownership of each cell manufacturing step using a cost model of ISC Konstanz as shown in Fig. 7. Assuming M6 wafer size, 5 GWp production in EU, and Ag paste costs of 700 USD/kg, we calculate the POLO IBC cell processing costs to 4.6 US$cent/Wp applying the cost of ownership values per process step as shown in Fig. 7 and assuming a POLO IBC cell efficiency of 25%. As a benchmark, 23% efficient PERC+ cells are currently at very similar 4.5 US$cent/Wp. The marginally higher processing costs of POLO IBC originate from the PECVD SiO_xN_y/n-a-Si process, which is more expensive than the $POCl_3$ diffusion used for PERC+. Based on preliminary experimental results, we target that one shadow mask can be used for about 1000 subsequent PECVD depositions and hence only minimally contributes to the PECVD SiO_xN_y/n-a-Si processing costs.

Whereas POLO IBC cell processing costs are comparable to PERC+, using an ISC Konstanz cost model we calculate that the POLO IBC modules will produce electricity in utility scale (and residential) applications ca. 4% (and ca. 7%) cheaper since the 9%rel higher cell efficiency will substantially reduce the area-related balance of system costs per kWp. Hence, this POLO IBC technology is attractive for new cell production lines as well as for upgrading PERC+ fabs to POLO IBC with minimal conversion invest.

	PERC+	POLO-IBC
Cell efficiency	**23.00%**	**25.00%**
	USDct/Wp	USDct/Wp
KOH batch texture	0.52	0.48
P-Diffusion	0.27	
Edge isolation and PSG removal SSE	0.56	0.52
PECVD SiON/na-Si with shadow mask		0.40
Anneal /SiO2 or Anneal N2	0.15	0.20
PECVD SiNx/AlOx	0.49	0.45
PECVD SiNx or AlOx/SiNx	0.43	0.45
Laser SiNx ablation	0.09	0.09
Printing 4 Stage (w/o paste cost)	1.12	1.03
pure Ag cost USD/cell	0.88	1.01
Total cell processing cost (w/o wafer)	4.51	4.62

Figure 7: Cost-of ownership calculation of POLO IBC versus PERC+ cell process costs applying a cost model of ISC Konstanz. Whereas the cell processing costs are comparable, we expect that the levelized costs of electricity (LCOE) of POLO IBC modules in utility scale (residential) applications will be up to 4% (7%) cheaper due to the higher conversion efficiency potential.

5 SUMMARY

We have developed a novel manufacturing process sequence for POLO IBC solar cells applying a local PECVD SiO_xN_y/n-a-Si deposition through a glass shadow mask to form the structured carrier selective n-poly-Si layer in one process step. After developing the process using a lab-type tool, we transfer the PECVD SiO_xN_y/n-a-Si to an industrial c.plasma tool obtaining an excellent J_0 = 3 fA/cm^2 after firing. A TEM analysis confirms that the N_2O in-situ plasma oxidation indeed incorporates N in the interfacial SiO_xN_y layer which could facilitate the excellent firing stable surface passivation. Fully processed POLO IBC solar cells on M2-sized Ga-doped Cz wafers exhibit conversion efficiencies of up to 23.0% with V_{oc}=708 mV, J_{sc} =41.2 mA/cm^2 and FF=78.7% when processed with the lab PECVD tool and also demonstrate an excellent iVoc of 741 mV. We attribute the 33 mV difference between implied V_{oc} and V_{oc} to additional J_0 contributions of the Ag and Al metal contacts which is subject to future improvements. First POLO IBC solar cells processed with the c.plasma tool with shadow masks achieve conversion efficiencies up to 22.3% and average V_{oc} values of 710 mV. In the future, we aim at further increasing the conversion efficiency towards 25% and implementing an automated shadow mask loading into the industrial c.plasma tool.

6 REFERENCES

[1] ITRPV 2021

[2] T. Dullweber et al., Solar Energy Mat. Solar Cells **212** (2020) 110586.

[3] R. Brendel et al, Proc. 35th EUPVSEC, pp. 39 (2018).

[4] F. Haase et al., Proc. 46th IEEE PVSC, Chicago, IL, USA, 2200-2206 (2019).

[5] R. Brendel et al., Proc. 28th EUPVSEC, pp. 676 (2013).

[6] F. Feldmann et al. Proc. 28th EUPVSEC, pp. 988.(2013).

[7] C.N. Kruse et al., Scientific Reports, 11, 996 (2021).

[8] M. Stöhr et al., ACS Appl. Energy Mater., 4 (5), 4646–4653 (2021).

[9] M. Stoehr et al., EUPVSEC (2020), pp. 521 – 524.

[10] T. Dullweber et al., presented at SNEC (2021).

[11] V. Mertens et al., Proc. 38th EUPVSEC, 135-139 (2021).

[12] S. Gatz et al., Sol. Energy Mater. Sol. Cells 96 (2012), pp. 180–185.

WET-CHEMICAL OZONE-BASED SILICON OXIDE LAYERS AND ELECTRON-BEAM EVAPORATED SILICON LAYERS FOR PASSIVATING CONTACTS

Nathan Nicholson*[1], Johannes Frisch[1], Maurice Nuys[2], Uwe Breuer[2], Mohammed Jaber[2], Jannik Kleesiek[1], Daniel Amkreutz[1], Lars Korte[1], Steve Albrecht[1], Bernd Rech[1]
[1]HZB, Berlin, Germany, [2]Forschungszentrum Jülich, Jülich, Germany

We investigate TOPCon-like passivating contacts using silicon from electron-beam evaporation (E-Beam evaporation), plasma doping and wet-chemical ozone oxides. Results show iVoc values up to 722 mV for symmetrical lifetime samples and 708 mV for cells using the passivating contact on the rear side. SIMS measurements were performed on samples before and after annealing to investigate the distribution of phosphorous within the samples for different process conditions. Synchrotron-based XPS measurements were performed to investigate compositional changes both for the as-grown oxide and after poly-Si formation and etch-back of the poly-Si layer. A clear decrease in the intensity of the transitional area can be observed between non-annealed and annealed samples, which is assumed to be the reason for the non-existent surface passivation quality of freshly grown silicon oxide layers.
Keywords: Doping, Passivating, SiO2

1 Introduction

Solar cells using passivating silicon oxide in combination with poly-Si contacts have shown to be a promising candidate for the next generation of industrial solar cells. However, the methods used for the growth of each layer still remain a point of discussion. For the poly-Si layer, both LPCVD and PECVD have been used on an industrial scale reaching high efficiencies [1]. For LPCVD the main issue is the wrap around during layer deposition, which requires additional steps for removal. For PECVD the issue arises from the preference for thicker poly-Si layers, which are more easily combined with contact firing. Another option is electron-beam evaporation (E-beam evaporation) as shown by J. Lossen et. al. [2], who were able to reach iVoc values up to ~731 mV on symmetrical lifetime samples with the passivating contact on each side. E-Beam evaporation combines several properties that are of interest for the growth of silicon layers for passivating contacts such as high growth rates, one-sided depositions, depositions without hazardous gases and low hydrogen content. While J. Lossen et. al. used NaOH solutions for the oxide growth, we believe that the use of ozone containing solutions, which have been previously used for passivating contacts [3], is preferable. Wet-chemical ozone enables low oxidation without requiring chemicals for the process besides water. The process is also used for the cleaning of silicon wafers. Therefore, we investigated the combination of wet-chemical ozone oxidation with E-Beam evaporation for passivating contacts. Part of this investigation also focused on the composition of the oxide layer as well as the doping process that was employed.

2 Experiment details

Saw-damage was removed from the surface of 100 n-type CZ wafers with a base resistivity of 1-5 ohm*cm using KOH. The samples were then labelled and cut into 5x5 cm2 squares. A thick silicon oxide layer was then deposited on the rear side of some samples to protect the surface during the following texturing process. The frontside of those samples was then textured in a solution containing KOH. The silicon oxide capping layer was then removed using 5% HF. Subsequently, Silicon oxide was grown on top of the substrates using a solution containing H20 and 30 ppm O3 at room temperature for 30 mins. A VON ARDENNE CS400PS cluster was used for the deposition of amorphous silicon using E-Beam evaporation. Set temperature and electron-beam emission current were kept constant across all experiments at 450°C and 190 mA respectively. A quartz crystal microbalance was used to determine the deposition rate and adjust the deposition time accordingly. Deposition rates would typically vary between 6-18 nm/min depending on the condition of the silicon ingot. After depositing the initially intrinsic silicon layer with a thickness of ~320 nm the samples were moved in-system to an adjacent PECVD chamber. Samples were doped using a plasma doping process that was developed by the Helmholtz-Zentrum Berlin (HZB). During the process a plasma is ignited containing PH3 and H2 at a temperature of 450°C. The doping step requires a high-temperature annealing step, which is typically performed for passivating contacts, to activate the Phosphorous as a dopant. Samples were therefore annealed in a tube furnace at 860°C - 915°C for 30 min. A hydrogenation step was then performed in the same tube furnace using forming gas with a 5% hydrogen content at 450°C for 30 min. For the front side of the cell a typical silicon-heterojunction design was used (fig. 1). Details regarding deposition methods and conditions can be found in [4]. Rear metallization was realized using thermal evaporation using titanium (10 nm) and silver (1 μm). The front contact was printed using silver pastes via screen-printing. Finally, cells were characterized using Suns-Voc and JV measurements. Non-textured symmetrical lifetime samples using CZ wafer were also created for the passivation contact to evaluate its passivation quality. These samples were characterized using photoconductance decay ECV and SIMS measurements. ECV and SIMS were used to better understand the doping process. 111 n-type polished FZ wafers were used for XPS experiments. Here, the composition of the oxide before and after processing was investigated. For this purpose, passivated contacts were realized on the front side of the wafer. After the hydrogenation step, poly-Si was removed using 5% TMAH at 60°C. XPS measurements were performed using synchrotron radiation with an energy of 170 eV. Results were compared to samples that were just oxidized in wet-chemical ozone. Deconvolution of the signal was performed according to [5].

3 Results and discussion

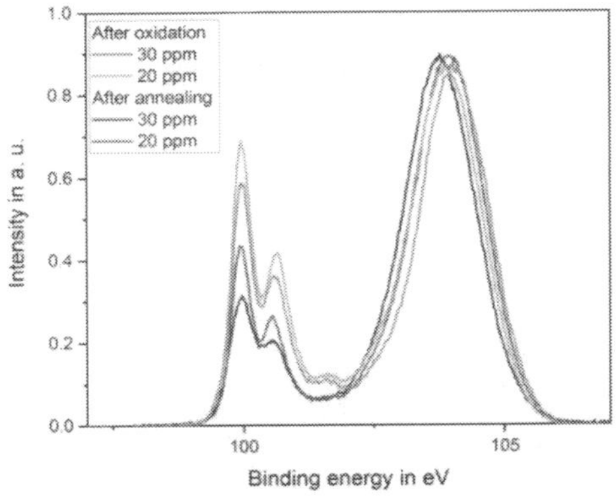

Figure 1: Silicon 2p spectrum of wet-chemical ozone-based silicon oxide layers after oxidation and after removal of poly-Si layer from passivating contacts annealed at 900°C.

Table I: FWMH values of each silicon oxidation state for annealed and freshly grown samples

		FWHM in EV			
Temp.	Si^0	Si^{1+}	Si^{2+}	Si^{3+}	Si^{4+}
900°C	0.46	0.72	0.51	0.99	1.39
800°C	0.44	0.66	0.42	0.91	1.29
RT	0.46	0.59	0.62	1.21	1.25

Fig. 1 shows XPS Si 2p spectra for silicon oxides grown using solutions with varying ozone concentrations that were freshly grown or annealed in passivating contacts. Annealing causes a decrease in Si 0 signal relative to the Si4+ peak relating to an increase in silicon oxide thickness. Samples with the freshly grown oxide on top show a more significant contribution from interfacial oxides leading to a noticeable peak in that area. For samples that have been annealed and etched the peak disappears indication a structural change during annealing. These changes can be partly related to structural changes for the Si^{3+} peak, which were quantified through the full width at half maximum (FWHM). A significant decrease in FHWM can be observed for the Si^{3+} peak state indicating a more ordered structure. These structural changes are also assumed to be partly responsible for the poor passivating that wet-chemical ozone oxides initially show. Here it is assumed that through the restructuring of the oxide stress and stress related defects will decrease leading to a better passivated wafer surface. For other states changes were not as significant when comparing freshly grown samples with those annealed at 800°C. However, an increase in FWHM can be observed when increasing the annealing temperature to 900°C. This broadening may be related to the degradation of the silicon oxide layer, which was previously reported by other groups for wet-chemical silicon oxides annealed at similar temperatures [6]. Values for the Si^{2+} state are assumed to be the result of error minimization as the contribution of the Si^{2+} state to the total model fit was minor (0% - ~0.4% of the total fitted area). For the other states similar trends were also observed when excluding the Si^{2+} state from the fitting method. A quantification and comparison between oxide growth conditions is still ongoing, although the differences between the investigated oxides might be not sufficient to determine a correlation between growth condition and shape/form of the transitional area. Further experiments may be required.

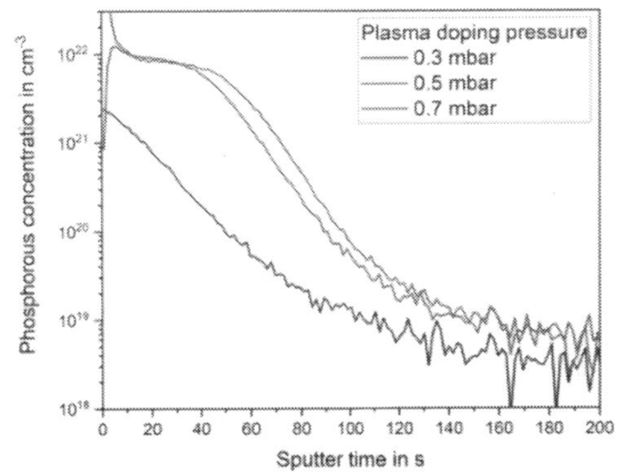

Figure 2: SIMS data for a-Si layers that were conditioned with plasma doping

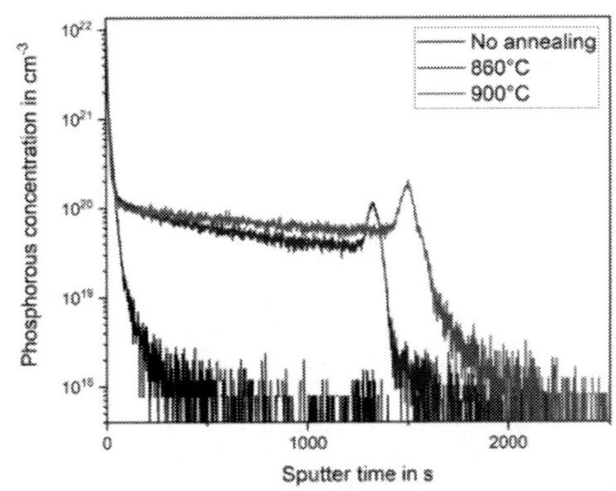

Figure 3: SIMS data for samples that received the plasma doping treatment after annealing.

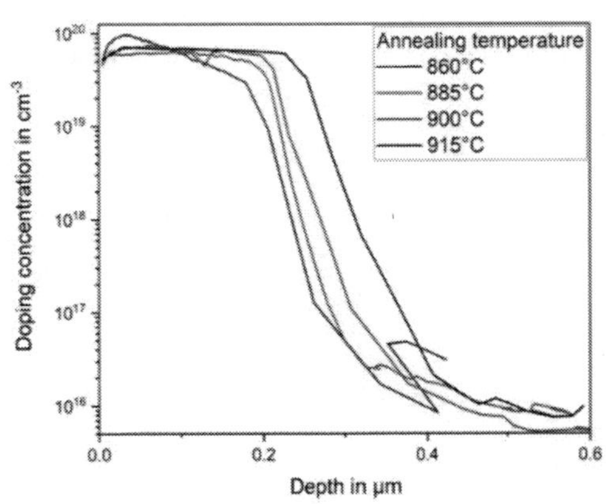

Figure 4: ECV data for samples that received the plasma doping treatment after annealing.

During the initial plasma doping process Phosphorous will diffuse into the sample. An increase in plasma doping pressure increases the total amount of Phosphorous in the sample as shown in **fig. 2**. However, the difference between 0.5 and 0.7 mbar seems minor. While more Phosphorous is able to diffuse deeper into the sample, the maximum concentration seems to be the same. This could indicate a saturation of the sample regarding the amount of Phosphorous it can contain within a volume. During the annealing step Phosphorous will diffuse deeper into the sample with a significant decline in the Phosphorous concentration once it reaches the silicon oxide layer, which limits the diffusion of Phosphorous. Annealing the samples at 900°C causes an even distribution of Phosphorous within the poly-Si layer. However, for 860°C a gradient can be seen, which is also visible in the ECV profile. An increase in annealing temperature causes a more even Phosphorous distribution across the poly-Si layer for the total amount of Phosphorous as well as the electrically active part, although the profiles of samples annealed at elevated temperatures also show more significant indiffusion.

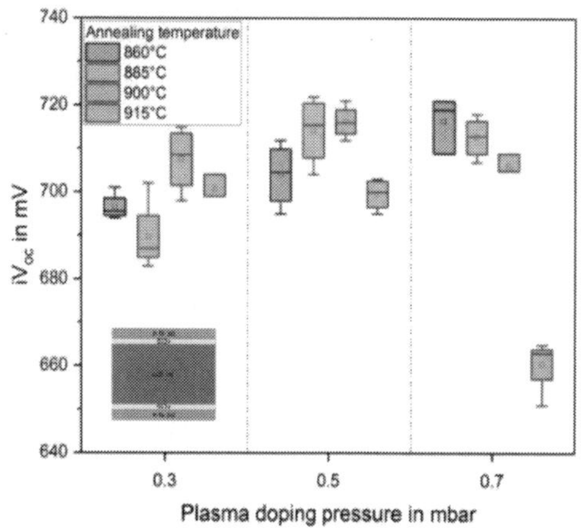

Figure 5: iVoc data for symmetrical lifetime samples with a passivating contact on each side. Plasma doping pressure and annealing temperature were varied. An optical constant of 0.7 was assumed.

Lifetime samples annealed at 860°C show a general increase of iVoc with plasma doping pressure. An increase in annealing temperature should lead to an increase in the concentration of electrically active phosphorous atoms within the sample. Similarly, an increase in plasma doping pressure should also lead to an increase in electrically active phosphorous concentration by introducing more phosphorous during the initial doping process. The increase in dopant concentration can improve the selectivity and therefore passivation quality of the contact as seen here. For samples annealed at higher temperatures a general decrease in iVoc can be observed when increasing the plasma doping pressure from 0.5 to 0.7 mbar. It is likely that this decline is caused by a more significant diffusion of phosphorous through the silicon oxide layer, which is known to limit the passivation quality of the passivating contact [7]. However, sample annealed at lower temperatures showed generally good results as long as the doping pressure was high enough. Therefore, a low temperature approach can still be considered here as long as it is compensated by diffusing more Phosphorous into the sample during the plasma doping step. Samples annealed at 915°C show generally poor results with a significant decline in passivation quality at higher doping pressures. This could be relating to the more severe indifussion that was visible in the ECV data in **fig. 4**. A degradation of the oxide can also be considered here, but has not been confirmed beyond the broadening in FWHM extracted from XPS data. Overall, the best result of ~ 722 mV implied or ~4ms was found for samples annealed at 885°C with a plasma doping pressure of 0.5 mbar.

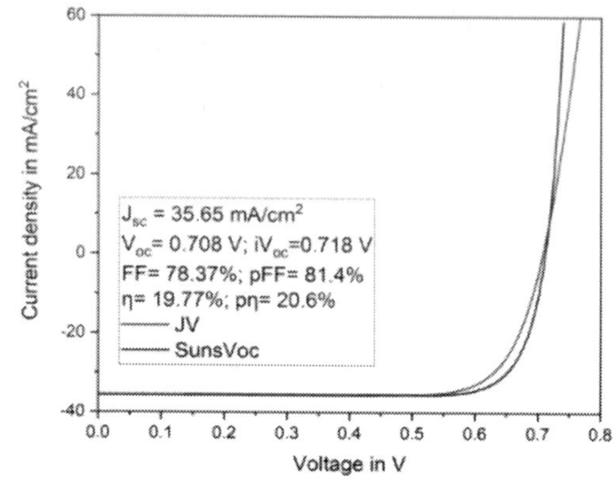

Figure 6: JV measurements performed on the best cell that used a passivating contact on the rear. A pseudo JV curve created from SunsVoc measurements is also shown.

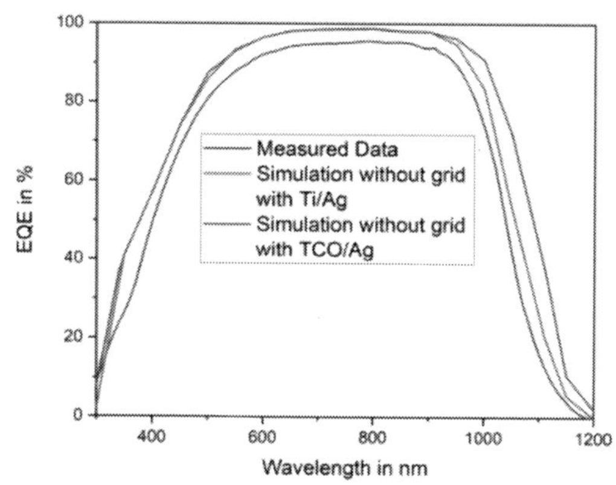

Figure 7: Comparison of measured and simulated EQE data.

The same conditions that enabled an iVoc of 722 mV were also used for the passivating contact that was used for the cell shown in **fig. 6**. It was possible to achieve an efficiency of 19.77% with a Voc of around 708 mV. It is assumed that the Voc is currently limited by the front side passivation which could explain the significant difference between Voc and iVoc from the symmetrical samples. However, investigations are still ongoing. Optical simulations showed that the titanium layer on the rear caused significant parasitic absorption. Other contacting schemes using TCO/Ag contacts instead of Ti/Ag were tested and showed similarly good contact resistivities of ~5 mOhmcm². Using these contacting schemes an increase of ~1 mA/cm² is expected for the cell according to optical simulations conducted.

4 Conclusions

We investigated the combination of wet-chemical ozone oxides with E-beam evaporated silicon layers. It was possible to achieve iVoc values of up to 722 mV (optical constant =0.7). Investigations performed allowed us to gain insight into the plasma doping process as well as structural changes that wet-chemical ozone-based silicon oxides undergo during the processing of passivating contacts. While further investigations are necessary to better understand these structural changes, it was possible to show some aspects using XPS measurements.

References

[1] B. Kafle et. al., TOPCon – Technology options for cost efficient industrial manufacturing, Solar Energy Materials and Solar cells Volume 227 (2021)

[2] J. Lossen et. al., Electron beam evaporation of silicon for poly-silicon/SiO2 passivated contacts, 35 EUPVSEC (2018)

[3] A. Moldovan et. al., Tunnel oxide passivated carrier-selective contacts based on ultra-thin SiO2 layers, Solar Energy Materials and Solar Cells Volume 142 (2015)

[4] P. Wagner et. al., Low-resistance hole contact stacks for interdigitated rear-contact silicon heterojunction solar cells, IEEE Volume 11 Issue 4 (2021)

[5] F. J. Himpsel, F. R. McFeely, A. Taleb-Ibrahimi, J. A. Varmoff and G.Hollinger, Microscopic structure of the SiO2/Si interface, Phy. Rev. B 38 (1988)

[6] J. Polzin et. al., Study on the interfacial oxide in passivating contacts, AIP (2019)

[7] H. Park et. al., Role of polysilicon in poly-Si/SiOx passivating contacts for high-efficiency silicon solar cells, RSC advances Issue 40 (2019)

TOWARDS HIGH EFFICIENCY INVERSION LAYER CELLS BASED ON ION-CHARGED DIELECTRICS

Mingzhe Yu[1], Matthew Wright[1], Jingyan Chen[1], Yifu Shi[1], Brett Hallam[2], En-Te Hwu[3], Nicholas E. Grant[4], John D. Murphy[4], Pietro P. Altermatt[5], Peter Wilshaw[1], Ruy Sebastian Bonilla[1]
[1]University of Oxford, Department of Materials, Oxford, United Kingdom
[2]University of New South Wales, School of Photovoltaic and Renewable Energy Engineering, Sydney, Australia
[3]Technical University of Denmark, Department of Micro- and Nanotechnology, Lyngby, Denmark
[4]School of Engineering, University of Warwick, Coventry, CV4 7AL, United Kingdom
[5]Trina Solar, State Key Laboratory for Photovoltaic Science and Technology (SKL PVST), Xinbei District, Changzhou, Jiangsu Province 213031, China

ABSTRACT: This work investigates the production and performance of p-type Inversion Layer (IL) Si solar cells, manufactured with an ion-injection technique that produces a highly charged dielectric nanolayer. Ions are applied to the front dielectric layer and then driven towards the c-Si/SiO_2 interface by an electric field before stabilisation with an anneal. As this process can be performed in minutes at temperatures below 500 °C, it potentially provides a fast, yet controllable way for IL cell manufacturing. We demonstrate by simulations using Sentaurus TCAD that for the 1 Ω·cm p-type Si/thermal oxide model defined in this work, the sheet resistance of the field-induced electron layer can reach 1.1 kΩ/sq in the dark by reducing band-tail interface state density to below 10^{14} $eV^{-1}cm^{-2}$ and increasing the dielectric charge density to above 2×10^{13} cm^{-2}. Additionally, we present a proof-of-concept p-type IL cell on a non-gettered, non-hydrogenated substrate with an efficiency of 10.8%, and an open-circuit voltage (V_{OC}) equivalent to that in a cell with a diffused phosphorous emitter. Lastly, we perform Sentaurus TCAD simulations to assess the efficiency potential of such IL cells. By incorporating optimal passivation, gettering, hydrogenation and metallisation, IL cells are predicted to reach an efficiency of 24.5% on 5 Ω·cm, and beyond 24.8% on 10 Ω·cm p-type substrates, provided the dielectric charge density reaches 2×10^{13} cm^{-2}, which has been experimentally demonstrated to be possible. IL cells are therefore, in principle, a potential competitive candidate in the photovoltaic industry.
Keywords: Inversion-layer, Dielectric Charge, Laser Processing, Interfaces, Silicon Solar Cell

1 INTRODUCTION

In p-type IL (inversion layer) cells, the pn junction is formed by introducing positive charges into a front dielectric layer, such that an n-type inversion layer is formed in the underlying p-type base. IL cells have been studied in detail two decades ago, yet they did not progress to successful commercialisation. In such IL cells intrinsic charge in the dielectric layer was used to induce the emitter [1]–[5]. Their development was limited by the high emitter resistivity due to the low intrinsic dielectric charge. The ion-injection technique in this work allows embedding of ionic charge into the dielectric, thus has the potential of reaching charge densities beyond the highest (>10^{13} cm^{-2}) obtained before [6]. Compared with a P-diffused emitter, the induced emitter would inherently lead to less Auger, or dopant-mediated recombination. The conditions required for the technique includes an application of ion precursor, a strong electric field across the dielectric, and a temperature below 500 °C. Considering the low cost and the high compatibility of the process to the mainstream technology, the induced emitter can be a viable replacement for the P-diffused emitter.

In a p-type IL cell, the p-n junction is formed by introducing positive charges into a front dielectric layer such that an n-type inversion layer is formed in the underlying p-type base. Previously, fixed charge in titanium oxide, silicon oxide, tantalum oxide, and aluminium oxide thin-films have been used to induce inversion layers in silicon [1], [2], [7]. Silicon nitride was first used in IL cells in the 1980s and has since been widely studied due to its high charge density and good charge stability [4]. Introducing cesium before nitride deposition was found to boost the charge density up to nearly 10^{13} cm^{-2} [8]. However, this concentration was not high enough to induce a sufficiently conductive emitter to compete with P-diffused emitters at the time. Therefore, it is necessary to produce high enough charge densities in the dielectric layer to reduce lateral resistive losses. In this study, we use a technique that embeds ionic charge into the dielectric layer, with the potential of reaching a charge density beyond 10^{13} cm^{-2} [9]. We aim to develop an IL cell with a charge-induced emitter that is comparable to, or beyond, a P-diffused emitter in performance while maintaining simpler and lower temperature (hence lower-cost) manufacturing processes.

This paper is composed of two main results sections. In section 2, various aspects of the inversion layer induced by ion injection are discussed. Modelling is employed to investigate the impact of surface charge on the characteristics of the emitter. Such structures are then demonstrated experimentally. In section 3, a proof-of-concept inversion layer solar cell is fabricated. Further modelling indicates the efficiency potential of this cell design.

2 FIELD-INDUCED INVERSION LAYER

2.1 Ion migration method

The ion migration technique involves delivery of ion precursors, application of a corona discharge to generate an electric field across the dielectric, and an anneal to migrate the ions into the dielectric and reach the semiconductor-dielectric interface. Fig. 1 shows the processing method and a schematic of the wafer with a field-induced electron layer. Our previous work [10] has demonstrated that the ion-injection technique can produce a field-induced electron accumulation layer with a dark sheet resistance of as low as 0.95 kΩ/sq on 1 Ω·cm n-type silicon substrates, which is the lowest reported in the literature.

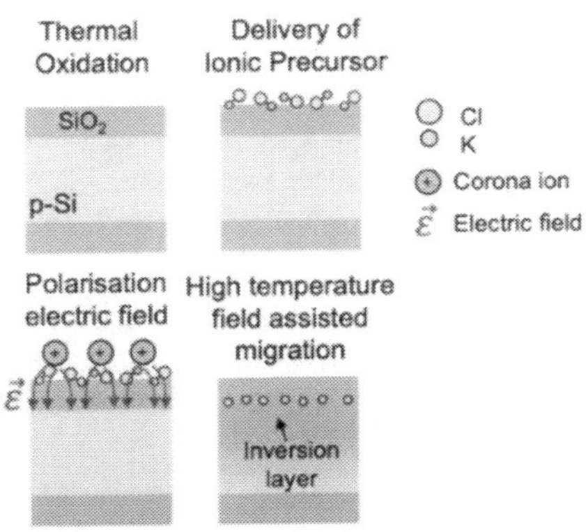

Figure 1: Processing of migration of ions into surface dielectrics: ion precursor delivery, applying electric field, and anneal.

2.2 Simulating conductivity of inversion layers

A model was developed in Sentaurus TCAD to simulate the interface carrier conditions of field induced layers. The model comprises a 200 µm thick, 1 Ω·cm p-type Si substrate, with an oxide passivating layer on both sides. Dielectric fixed charge was defined at the front Si/SiO_2 interface. The interface was defined following the common parametrisation as reported in [11], with interface state densities at midgap and band tails as primary parameters. The Lombardi model [12], [13] was used to model the mobility throughout the sample bulk and in the space charge layer. A modified local-density approximation (MLDA) quantum-mechanical model was used to account for the confined carrier distributions occurring near semiconductor–insulator interfaces [14]. It has been demonstrated that in the case of strong band bending, the band-tail interface states are able to store charged carriers and can therefore compensate the dielectric charge, leading to a reduced net charge density to induce the inversion layer [10], [15], [16]. In the model, the band-tail interface state density is described by Eq. 1 and Eq. 2:

$$D_{it}(acc) = D_{it}\mathrm{CB} \times e^{\frac{(E_g - E)}{E_{0,CB}}} \quad (1)$$

$$D_{it}(don) = D_{it}\mathrm{VB} \times e^{\frac{E}{E_{0,VB}}} \quad (2)$$

where D_{it} (acc) and D_{it} (don) represent the acceptor/donor band-tail interface state density. D_{it}-CB and D_{it}-VB are the maximum interface state density at conduction/valence band edge. E_g is the band gap of crystalline Si (1.12 eV). E is the energy from valence band edge in eV. $E_{0,CB}$ and $E_{0,VB}$ represent the slope of the tail at the conduction band and valence band, respectively. Here $E_{0,CB}$ and $E_{0,VB}$ were set to 0.028 eV and 0.024 eV, respectively, to reflect average values extracted from previous works [17]–[19]. In this model, D_{it}-midgap was set to 5×10^{10} $eV^{-1}cm^{-2}$. D_{it}-CB and D_{it}-VB were set to 10^{14} and 5×10^{13} $eV^{-1}cm^{-2}$, respectively [16]. Fig. 2 shows the schematic of the model and the band diagram of the model near the interface. To calculate the inversion layer sheet resistance from the model, the equilibrium state was simulated, and the carrier density and mobility in bulk silicon were extracted as a function of the distance from the Si/SiO_2 interface. The thickness of an inversion layer is defined by where the hole density exceeds that of electrons. The sheet resistance can then be calculated using the following equation:

$$R_{sh} = \frac{1}{\int_0^{thickness} (nq\mu_n + pq\mu_p)\, dt} \quad (3)$$

where n and p are electron/hole density, μ_n and μ_p are electron/hole mobility, and q is the elementary charge.

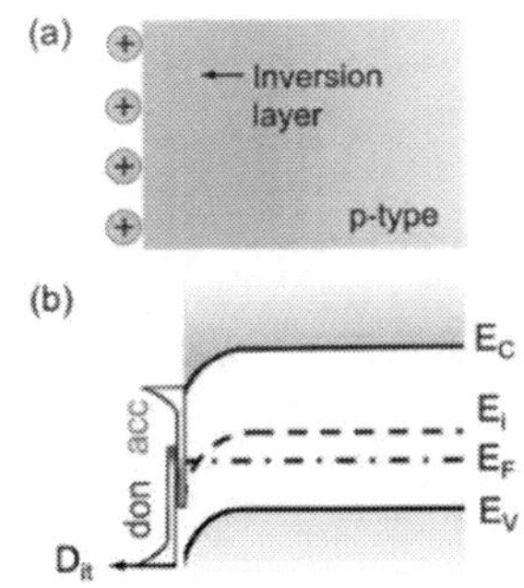

Figure 2: (a) Schematic of the model developed in Sentaurus TCAD. (b) Band diagram of p-type Si substrate near Si/SiO_2 interface in the presence of positive surface charge.

Fig. 3 shows the simulated inversion layer sheet resistance and its reciprocal, inversion layer sheet conductance, as a function of positive surface charge density. It is shown that the inversion layer sheet resistance decreases with increased charge density. The conductivity plateauing at high charge densities indicates that extra charge introduced in this range does not lead to an as significant increase in conductivity. Since conductivity is dependent on both the carrier mobility and the carrier density (as shown in Eq. 3), the plateauing effect can be attributed to two factors. The first is that the quasi-Fermi level of the trapped charges in the band-tail states will reach the band-tail near the conduction band edge at high enough charge densities. This will allow field-induced electrons to occupy band-tail interface states rather than remaining in the bulk and contribute to conductivity in the inversion layer, thus resulting in a smaller slope in the conductivity curve in the high charge density range. The other factor is the increased surface scattering, and therefore reduced carrier mobility, with increased dielectric charge density. The profiles of electron density, mobility, and the product of electron density and mobility near the interface with varied dielectric charge densities are plotted in Fig. 4. Fig. 4a shows that the electron density profile increases with charge density. Fig. 4b shows the reduced electron mobility near the interface with increased charge density. As a result, Fig. 4c shows the product of electron density and mobility. This demonstrated that the increase in conductivity induced by extra charge reduces with charge density. The simulation results demonstrate that the conductivity in the inversion layer is limited by the high density of interface states at the band-tail and the reduced carrier mobility due to increased surface scattering. It is noted that the lowest sheet resistance of the field-induced electron layer obtained (0.95 kΩ/sq) from our previous work [10] is close to the inversion layer sheet resistance expected at a charge density of $2-4 \times 10^{13}$ cm-

^{2}. This suggests that the charge density obtained with the ion migration method is sufficiently high for the sheet resistance to plateau.

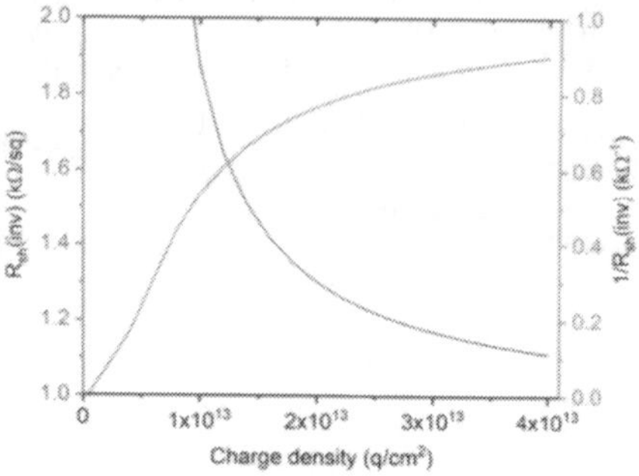

Figure 3: Simulated inversion layer sheet resistance (left axis) and conductivity (right axis) on a 1 Ω·cm p-type Si base as a function of positive interface charge density.

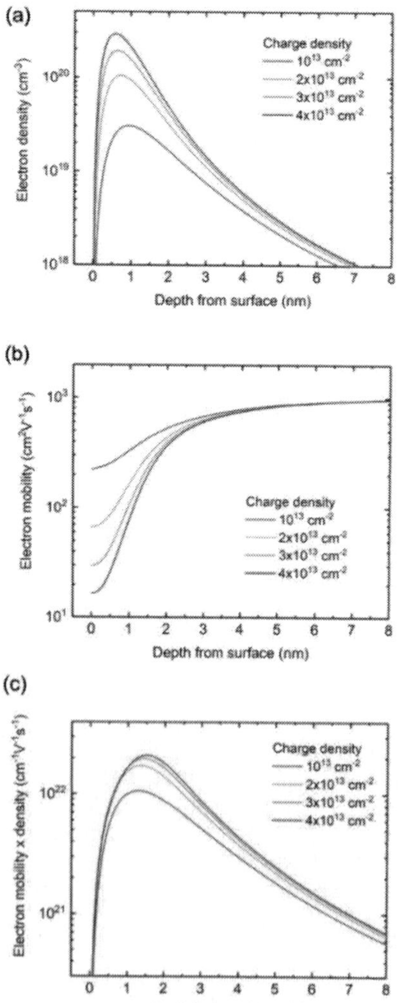

Figure 4: (a) Electron density, (b) electron mobility, and (c) electron density × mobility profiles near the interface on a 1 Ω·cm p-type base in the presence of varied positive interface charge density.

Since the surface scattering and therefore the reduced carrier mobility cannot be avoided, the impact of band-tail interface state density on the conductivity is explored. Fig. 5a shows the inversion layer conductivity as a function of charge density with varied D_{it}-CB. A significant increase in conductivity can be observed by reducing the D_{it}-CB from 10^{15} to 10^{14} eV^{-1}cm^{-2}, and further to 10^{13} eV^{-1}cm^{-2}, such that less interface states are available to accommodate the field-induced electrons. Fig. 5b demonstrates that passivating the band-tail interface states is an effective way to improve the inversion layer conductivity, yet it plateaus upon reducing the D_{it}-CB further. Since the band-tail interface states originate from termination of the periodicity of the bulk silicon, and that the energy slope has been demonstrated to be a function of overall bulk defect density [20], the band-tail interface states reflect the level of disorder. A characterisation technique has been developed in our previous work to detect D_{it}-CB [15], [16], which can be used to explore methods to passivate D_{it}-CB.

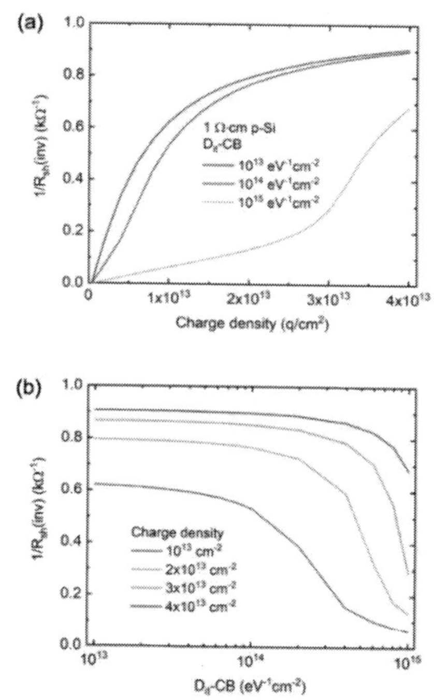

Figure 5: (a) Calculated inversion layer conductivity on a p-type base as a function of positive interface charge density with varied D_{it}-CB. (b) Calculated inversion layer conductivity on a p-type base as a function of D_{it}-CB with varied positive interface charge density.

Fig. 6 shows the simulated carrier density profile near the interface of both a field-induced inversion layer with a charge density of 2×10^{13} cm^{-2}, and a phosphorus diffused emitter in the dark and under illumination. The generation profile was obtained from ray tracing using Sunrays [21] and altered to mimic a pyramid surface with 0.1 μm height. A positive charge density of 10^{12} cm^{-2} is defined at the semiconductor-dielectric interface for the P-diffused emitter, which is the typical charge density of a SiN_x passivating layer [22]. D_{it}-midgap was set to 5×10^{10} eV^{-1}cm^{-2}. D_{it}-CB and D_{it}-VB were set to 10^{14} and 5×10^{13} eV^{-1}cm^{-2}, respectively [16]. The conductivity of the emitter is calculated by integrating the product of carrier density and mobility over the thickness of the pn junction. The depth of the pn junction is defined by where the hole density exceeds that of electrons. In the dark, the electron density in the inversion layer drops rapidly with the distance from the interface, whereas the phosphorus diffused emitter depends on the doping profile, which leads to a much deeper junction. Under illumination, both the majority and the minority carrier density will be increased by the light injection, increasing the local conductivity. As shown in Fig. 6, the pn junction depth will also be altered by light injection. The junction shifted from ~375 to 340 nm for the diffused emitter (reduction by 10%), and from ~100 to 60 nm for the field-induced emitter (reduction by 40%). As a result, the sheet resistance of the diffused emitter dropped from 196.6 to 195.1 Ω/sq upon illumination, corresponding to a 0.8% increase in conductivity. For the field-induced emitter, the sheet resistance dropped from 1.30 to 1.21 kΩ/sq, corresponding to a 7.4% increase in conductivity. This will lead to reduced resistive losses under operating conditions with illumination. Compared with the standard diffused emitter, the field-induced emitter appears to be thinner, more resistive with the sheet resistance more sensitive to light injection. For the induced emitter, the region with high carrier concentration that leads to strong Auger recombination is thinner, which leads to an overall reduced Auger recombination compared with the deep diffused emitter. As shown in Fig. 5, the sheet resistance can be reduced by increasing the charge density and further passivating the band-tail interface states. In the cell design, a smaller metal contact pitch can be used to compensate the resistive losses in the emitter. The efficiency potential of the inversion layer cell will be discussed in Section 3.

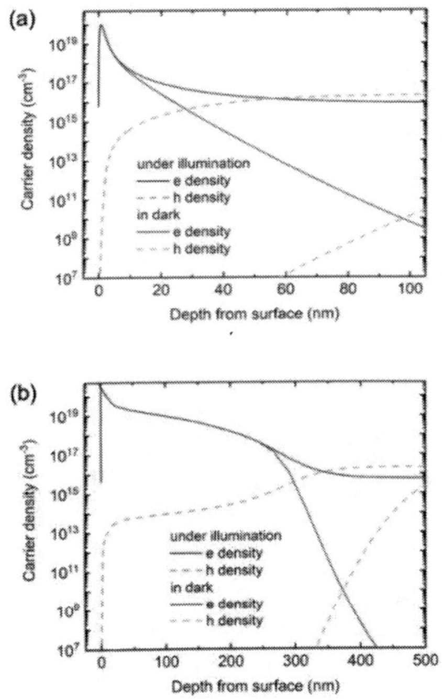

Figure 6: Simulated carrier density profiles of (a) a field-induced inversion layer with a positive charge density of 2×10^{13} cm^{-2}, and (b) a P-diffused emitter near the interface in dark and under illumination.

2.3 Field-induced electron accumulation layer

The Van der Pauw method is used to measure the sheet resistance of the induced electron layer [23], [24]. Fig. 7a shows the sample structure designed for the measurement. Direct electrical contact to the induced layer is required to measure the sheet resistance. For an inversion layer, as illustrated in Fig. 7b, the electrical contact to the inversion layer requires local doping underneath the meta contact. Due to the experimental difficulty in measuring the sheet resistance of an inversion layer, an electron accumulation layer was induced on an n-type Si substrate using the ion migration method. A 200 μm thick, 1 Ω·cm n-type Si substrate with 100 nm of thermal oxide on both sides was used for the experiment. Our previous work has shown an accumulation layer sheet resistance of as low as 950 Ω/sq on the n-type Si substrate but with enhanced chemical passivation by a forming gas anneal in 5 % hydrogen ambient at 425 °C for 30 min [10]. In this work, the ionic charge was introduced by spin coating of 0.05 M KCl solution (DI water:IPA 30:70) followed by a drive-in process consisting of 8 cycles of corona (30 kV, 30 s) – anneal (430 °C, 10 s). The KCl spin coating plus drive-in was repeated 4 times. Fig. 8 shows the average accumulation layer sheet resistance measured from four identical samples after drive-in of K^+ ions by each spin coating/drive-in step. The samples were then stored at room temperature in dark until the sheet resistance stabilised. In equilibrium, the sheet resistance is determined by the charge density and the D_{it} profile. However, the recorded sheet resistance shows a decay within the 25 days after processing. The hypothesis is that the high voltage and anneal involved in the ion migration process drives electrons in bulk Si into the interface states, or states in the dielectric material, compensating the net ionic charge. The gradual decay of such electrons leads to an increase in net dielectric charge and therefore a reduction in sheet resistance. The stability of the sheet resistance was then tested against elevated temperature and UV irradiation separately for a period of 2 months. During the period, no increase is observed for the samples kept at 120 °C nor the ones kept under UV irradiation, demonstrating a promising stability of conductivity in the field-induced electron layer.

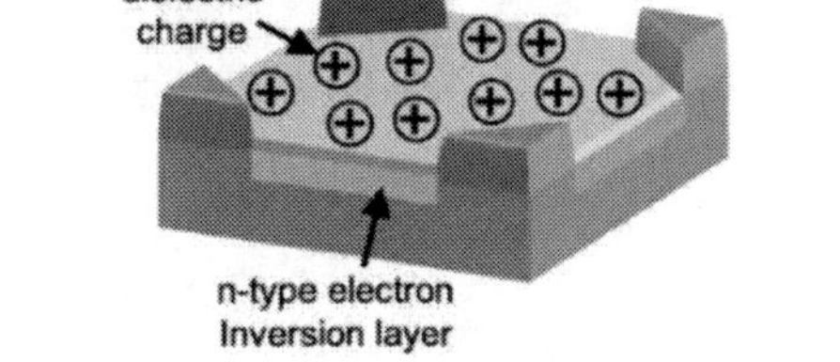

Figure 7: (a) Structure of sample designed for Van der Pauw measurement. (b) Schematic of electron inversion layer on a Van der Pauw sample structure with p-type Si substrate.

The impact of the K^+ ions on surface passivation is explored by characterising D_{it}-midgap prior to and after the ion migration. D_{it}-midgap was obtained by an analytical fitting of effective lifetime of the sample as a function of surface potential. A detailed description of this method can be found in [25]. Fig. 9a shows the schematic diagram of the experimental setup. A transparent PEDOT:PSS gate electrode was applied on both surfaces to monitor the surface potential, while effective lifetime of the sample was recorded as a function of surface potential using the lifetime tester. Fig. 9b shows the experimental data and the analytical fitting of one fresh sample and one with K^+ ions introduced on both sides by one spin coating and the subsequent driven-in. The interface parameters used in the fitting are listed in Table I. The shift in gate bias in Fig. 9b corresponds to an increase in charge density by 9.6×10^{12} cm^{-2}. Upon drive-in of the K^+ ions introduced by one spin coating, D_{it}-midgap increased from 2.8×10^{10} to 3.6×10^{11} $eV^{-1}cm^{-2}$, demonstrating the damage of the K^+ ions to the surface passivation. One hypothesis is that the accumulation of K^+ ions at the interface causes a strain that generates more defects and therefore leads to an increase in D_{it}-midgap. The Si substrates used in this experiment are passivated solely by thermal oxide without any hydrogenation steps. A natural subsequent experiment is the addition of hydrogenation introduced by depositing a SiN_x or AlO_x layer and conducting an activation anneal for improved surface passivation. Such methods will be explored in future work.

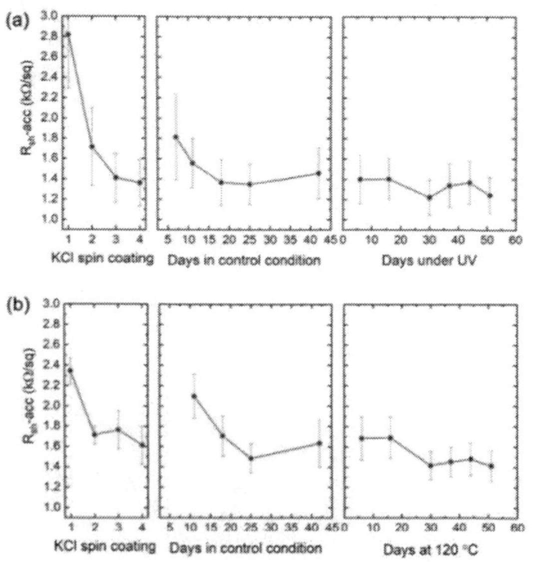

Figure 9: Calculated accumulation layer sheet resistance on the n-type samples upon introduction of ionic charge by spin coating of KCl solution by 4 times and kept (a) under UV irradiation, (b) at 120 °C for 2 months.

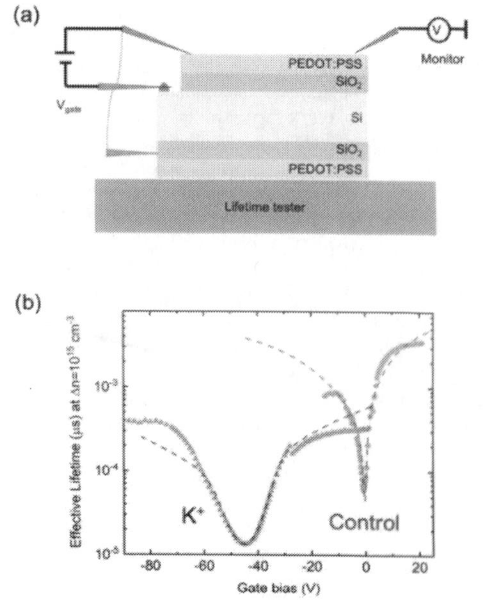

Figure 8: (a) Schematic diagram of experimental setup to obtain effective lifetime of a sample as a function of surface potential. (b) Experimental data (dotted line) and its analytical fitting (dashed line) of effective lifetime of one fresh sample and one sample with K^+ ions introduced by one spin coating and the subsequent drive-in.

Table I: Interface parameters used in the analytical model to fit the experimental lifetime data as a function of surface potential of one fresh sample and one sample with K^+ ions applied by one spin coating driven in.

	Control	K^+ ions introduced
D_{it}-midgap ($eV^{-1}cm^{-2}$)	2.8×10^{10}	3.6×10^{11}
D_{it}-CB ($eV^{-1}cm^{-2}$)	9×10^{13}	7.5×10^{14}
D_{it}-VB ($eV^{-1}cm^{-2}$)	5×10^{13}	3.5×10^{14}
Q_f (cm^{-2})	10^{11}	9.7×10^{12}

3 P-TYPE INVERSION LAYER CELLS

3.1 Proof-of-concept inversion layer cell

In IL cells the inversion layer is interrupted at regions beneath the metal contacts, hence local doping is necessary to ensure the continuity of the n-type layer. Laser doping is a fast and cost-effective way to achieve this. Fig. 10a shows a schematic of a prototype IL cell. The prototype cells start with 180 µm thick, 1 Ω·cm, p-type Cz Si wafers. These wafers were pyramid textured at the front surface and had 20 nm thermal oxide on both sides. The effective lifetime of this sample structure is ~50 µs since they were made in solar grade silicon without gettering. Rear openings were produced using a laser scriber prior to rear screen-printing of aluminium paste. The wafers were then fired at ~750 °C for 3 s in a belt furnace. A localised emitter was fabricated across the front surface by laser doping. An 85% H_3PO_4 solution was spin coated onto the sample surface at 12,000 rpm for 50 s as a phosphorus source. A 355 nm nanosecond pulsed laser was scanned across the sample surface according to the pattern shown in Fig. 10b to both ablate the oxide and melt the silicon beneath to allow in-diffusion of phosphorus atoms to dope the molten region. The fingers were 10 mm long and were spaced apart by 1 mm. Two busbars of 0.5 mm wide with a line spacing of 30 μm were designed at the two ends of the fingers. The samples were then rinsed in DI water to remove the phosphorous residues. An anneal at 560 °C for 1 hour was found necessary to allow regeneration of the laser-doped and damaged region as shown in Fig. 10a, so that optimal availability of K^+ ions is ensured. An inversion layer emitter was induced by spin coating of a KCl solution 4 times each followed by 8 Corona anneal cycles as described in Section 2. Before metallisation, the oxide in the laser doped grooves formed during laser doping and anneal was etched off by 1% HF solution. Fig. 10c shows a schematic of the process of light-assisted nickel electroplating. A voltage of 1.1 V was applied between a Ni wire and the rear metallisation of the sample for 4 min while the front is illuminated by a white LED lamp. A ready-for-use commercial nickel sulfamate solution (NB Semiplate Ni 100) was used as the electrolyte. Since prior to front metallisation, the inversion layer emitter has been formed with its electrical connection to the local doped region ensured, upon illumination, the photocurrent contributes to the electroplating current, accelerating the process. A 10 × 10 mm² proof-of-concept IL cell is fabricated in this manner including Ni electroplated metallisation.

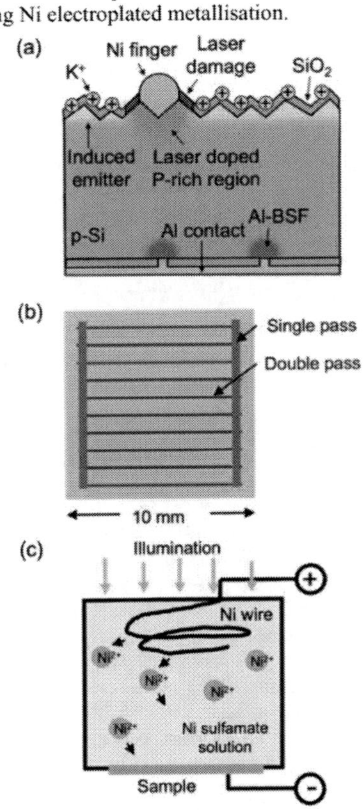

Figure 10: Schematic of a p-type cell precursor in (a) cross section and (b) plan view from top surface. (c) Process of light-assisted Ni electroplating.

The formation of the inversion layer emitter was monitored by Suns-V_{OC} and is shown in Fig. 11a. After the drive-in of K^+ ions introduced by the first spin coating, the V_{OC} increased by about 35 mV to 590 mV. No evident increase of V_{OC} was observed while further K^+ ions were embedded after the 2nd, 3rd and the 4th KCl spin coating, with the highest V_{OC} reaching 600 mV. Fig. 11b shows an LBIC image of 500 × 500 μm² area of the IL cell with the pixel size being 2 × 2 μm², as described in [26]. The contrast in the image indicates the photocurrent collected by the cell while a red laser is scanned across the sample surface. The blue low-photocurrent region (3-5 μA) in the middle represents a Ni finger. The region adjacent to the Ni finger with a photocurrent of 11-13 μA is affected by the recombination sites introduced by the laser-induced damage. The large area with high photocurrent (13-14 μA), together with the increase in V_{OC} upon introducing K+ ions, indicates the presence of a uniform inversion layer emitter and the continuity of the n-type layer beneath the dielectric film. Fig. 11c shows the IV curve of a device with both the busbars and the fingers doped by single laser pass. The IV curve gives a V_{OC} of 559 mV, a J_{SC} of 33.9 mA/cm², and an efficiency of 10.8%. These experiments demonstrate the potential of fabricating IL cell using the ion migration method. This cell represents a proof-of-concept for the ion injection inversion layer solar cell. However, there were several aspects of the cell design, shown schematically in Fig. 10a, which limited the performance. Better performance is expected if the following issues are addressed: (i) Start with Si substrates with lifetime in ms level (rather than 50 μs as in the proof-of-concept cell) and high resistivity (5/10 Ω·cm); (ii) Add SiN_x coating for hydrogenation and anti-reflection; and (iii) Anneal Ni-Si contact to form NiSi alloy for reduced contact resistance. Although an efficiency of only 10.8% is obtained by the proof-of-concept IL cell, the potential of IL cells is evaluated by simulations in Section 3.2.

The stability of V_{OC} of the precursor cells was monitored against elevated temperature and UV irradiation. Performance of the cells were monitored by taking Suns-V_{OC} measurements throughout the 2-month period. Fig. 12 shows the recorded V_{OC} of the cells during drive-in of the K^+ ions and its stability against elevated temperature, UV radiation, and a control sample set kept in dark at room temperature for comparison. In all 3 groups of samples, during the ion migration process, a tendency that V_{OC} decreases with the number of ions embedded is observed. This agrees with the finding observed in Section 2.3 that the interface is damaged during the ion injection process. There is hence a trade-off between improved emitter conductivity and limiting the damage to the interface. Methods to introduce atomic hydrogen are to be tested for improved interface passivation. One point to note is that the V_{OC} values measured after K^+ drive-in is 50-60 mV lower than that measured in Fig. 11a. This can be explained by the increased laser damage caused by the non-optimised laser parameters. Upon keeping the samples at 120 °C, under UV irradiation, and at the control condition, a reduction in V_{OC} ranging from 3 to 8 mV is observed for all 3 groups within 7 days. The variations afterwards are not significant enough to show any further degradation. Since similar trends are observed in all of the three figures in Fig. 12, the conclusion is that the degradation in V_{OC} caused, or accelerated at elevated temperature and under UV irradiation is minor in the 2-month period.

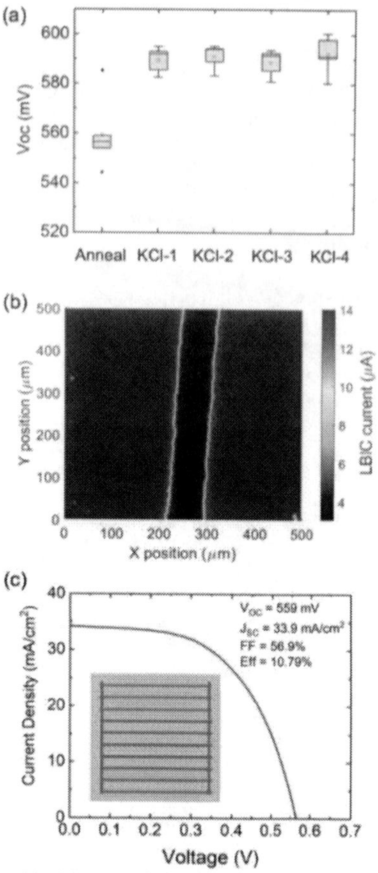

Figure 11: (a) Recorded V_{OC} of IL cell precursors after anneal and introducing ionic charge. (b) LBIC photocurrent map of an IL cell. (c) IV curve of an IL cell with a single laser-doped line as busbar and a single laser pass for laser doping.

3.2 Simulating the maximum potential of inversion layer cells

Next, we turn our attention to the efficiency potential of the proposed inversion layer cell design. A model was developed in Sentaurus TCAD to simulate the performance of p-type IL cells. The model comprises a 170 μm, boron-doped silicon substrate, with both surfaces passivated by dielectric layer. Base resistivity of 1, 5, and 10 Ω·cm was simulated to reflect the influence of doping density. The bulk lifetime was set to τ_p=10·τ_n=3.7 ms for 1 Ω·cm, as may be reached in Ga-doped wafers if Fe concentration is sufficiently small, and τ_p=10·τ_n=15 ms for 5 Ω·cm, τ_p=131ms, τ_n=1.14 ms for 10 Ω·cm wafers as referred from [27]–[29]. Dielectric charges were defined at the silicon-dielectric interface, with the positive charge at the front dielectric to induce the inversion layer, and the negative charge at the rear dielectric to limit the access of electrons to the rear surface for reduced carrier recombination. The front finger width was set to 25 μm since recent progress in metallisation technologies shows that such thin metal fingers are possible using screen printing, or a combination of laser doping and electroplating [30], [31]. Due to the absence of charge-induced inversion layer beneath the front metal contact, n-type local doping is required for the inversion layer to connect to the fingers via an n-type local contact. A phosphorous doping profile measured from an industrial phosphorus diffused emitter was defined beneath the contact. Similarly, an aluminium profile was defined beneath the rear contact to resemble Al-BSF for reduced contact resistivity. While a flat surface is defined in the model, the generation profile was obtained from ray tracing using Sunrays [21] and altered to mimic a pyramid surface with 0.1 μm height. Despite the potential of developing a bifacial cell, in this work, only light injection from the front surface is considered. A summary of parameters used to describe the model is listed in Table II.

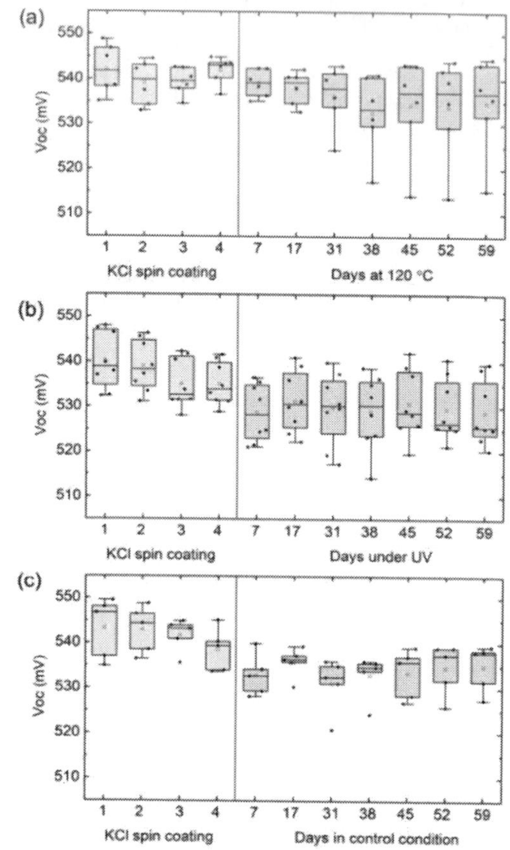

Figure 12: Recorded V_{OC} of IL cell precursors kept (a) under UV irradiation, (b) at 120 °C, and (c) at room temperature in dark for 2 months.

Fig. 13 shows the simulated IL cell efficiency with wafer resistivity being 1/5/10 Ω·cm at various charge densities and finger spacings. At the interface, D_{it}-midgap was set to 5 × 10^{10} $eV^{-1}cm^{-2}$. D_{it}-CB and D_{it}-VB were set to 10^{14} and 5 × 10^{13} $eV^{-1}cm^{-2}$, respectively. A negative charge density of 5 × 10^{12} cm^{-2} was defined at the rear interface, which is typical for a SiO_2/AlO_x passivating stack [32]. It is noted that higher efficiency is expected for substrates with higher resistivity. This is due to the higher bulk lifetime estimated for more resistive bulk. Another point to note is that the increase in efficiency with higher charge density is not evident, especially for 5/10 Ω·cm substrates. This can be explained by noting that with D_{it}-CB of 10^{14} $eV^{-1}cm^{-2}$ defined, the increase in emitter conductivity with charge density is not as effective for charge densities beyond 2 × 10^{13} cm^{-2}, which agrees with the conclusion drawn from Fig. 5a. Considering that the accumulation of K^+ ions at the interface damages the passivation as demonstrated in Fig. 9, it might be beneficial to maintain a relatively low charge density. Further studies on the interface and simulations are required to determine the optimum charge density. The peak in the efficiency vs. finger spacing curve occurs when the performance-gain due to reduced shading losses is outweighed by the increased resistive losses at large finger spacings. The efficiency peak shifting from < 0.8 mm for the 1 Ω·cm base to about 1.4-1.6 mm for the 5/10 Ω·cm base indicates a more conductive emitter, and therefore less resistive losses. This is because for a high resistivity substrate, the initial hole density is lower for the field-induced electrons to compensate and eventually turned into n-type. Low emitter conductivity has been considered as one of the major concerns for IL cells. For the 5/10 Ω·cm base, with the efficiency peak expected to appear beyond 1.4 mm in finger pitch on a well passivated cell, and that 0.7-1.6 mm [33] was used in industry for the mainstream PERC, emitter resistivity is no longer limiting the efficiency potential for IL cells. These simulation results suggest: 1) the necessity of moving to high resistivity substrates; 2) with well passivated surfaces, increasing the charge density to beyond 2×10^{13} cm^{-2} is less effective in improving the cell performance; and 3) and internal efficiency of 24.8% can be obtained with optimised processing.

Table II: Summary of cell parameters used for simulation of full inversion layer cells in Sentaurus TCAD.

Parameter	Value
Cell thickness	170 μm
Bulk base resistivity	1/5/10 Ω·cm
SRH bulk lifetime	1 Ω·cm: $\tau_n = 0.371$ ms, $\tau_p = 3.71$ ms 5 Ω·cm: $\tau_n = 1.5$ ms, $\tau_p = 15$ ms 10 Ω·cm: $\tau_n = 1.14$ ms, $\tau_p = 131$ ms
Front finger width	25 μm
Rear finger width	67 μm
Finger spacing	0.8-1.8 mm
SRV at front contacts	$S_n = 10^7$ cm/s, $S_p = 10^7$ cm/s
SRV at rear contacts	$S_n = 4 \times 10^4$ cm/s, $S_p = 10^7$ cm/s
Front contact resistivity	2 mΩ·cm²
Rear contact resistivity	5 mΩ·cm²
D_{it}-midgap	5×10^{10}-10^{11} $eV^{-1}cm^{-2}$
D_{it}-tail	CB: 10^{13}-10^{15} $eV^{-1}cm^{-2}$ VB: 10^{13}-5×10^{14} $eV^{-1}cm^{-2}$

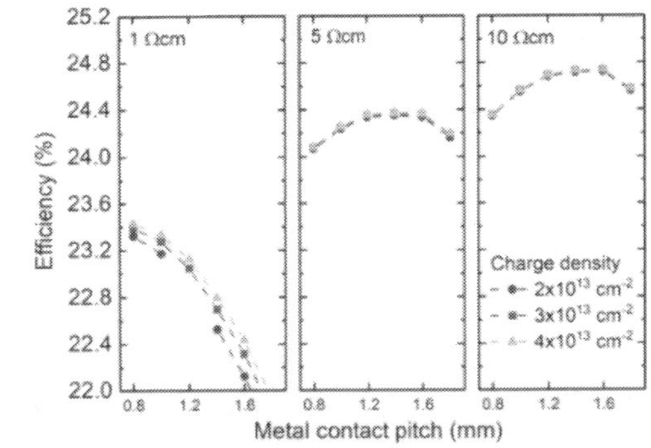

Figure 13: Simulated IL cell efficiency with wafer resistivity being 1/5/10 Ω·cm at various charge densities and finger spacings. D_{it}-midgap, D_{it}-CB, and D_{it}-VB were set to 5×10^{10}, 10^{14}, and 5×10^{13} $eV^{-1}cm^{-2}$, respectively.

D_{it}-CB was varied and efficiency plotted in Fig. 14. As concluded from Fig. 5b that the emitter conductivity reaches its maximum from D_{it}-CB of 10^{14} $eV^{-1}cm^{-2}$ and does not increase further at lower D_{it}-CB. In Fig. 14, an evident increase in efficiency is observed by reducing D_{it}-CB from 10^{15} to 10^{14} $eV^{-1}cm^{-2}$, while the improvement is minor by further reducing the D_{it}-CB to 10^{13}. One other point to note is that by reducing D_{it}-CB from 10^{15} to $10^{14}/10^{13}$ $eV^{-1}cm^{-2}$, the efficiency peak shifts from 1.2 mm to 1.4-1.6 mm in finger pitch, indicating an increase in emitter conductivity, which agrees with the conclusion drawn from Fig. 5b.

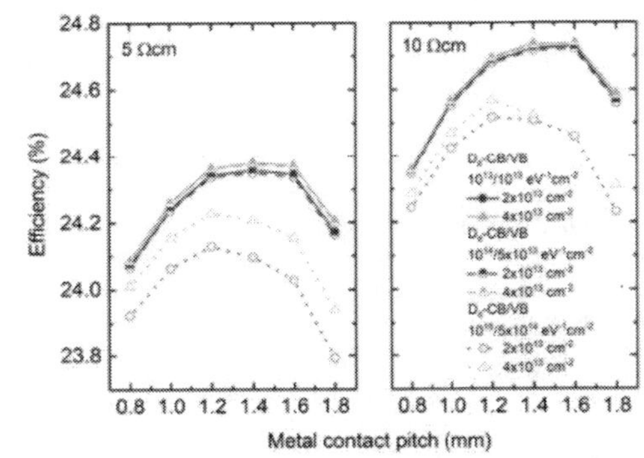

Figure 14: Simulated IL cell efficiency with wafer resistivity being 5/10 Ω·cm at various charge densities, finger spacings, and D_{it}-CB/VB.

In the model, negative charge is defined at the rear interface for field-effect passivation. A typical value of 5×10^{12} cm^{-2} was used in previous simulations. However, charge densities of above 10^{13} cm^{-2} have been demonstrated at a c-Si/AlO_x interface [34] and a c-Si/SiO_2/AlO_x stack [35]. In Fig. 15, simulations with rear negative charge densities as high as the front positive charge densities were plotted for comparison. No significant improvement in efficiency is observed by increasing the rear negative charge density from 5×10^{12} to 2×10^{13} cm^{-2} for a front positive charge density of 2×10^{13} cm^{-2}. This is due to the already well passivated surfaces and that the recombination at the rear surface is not the limiting factor. For the front positive charge density of 4×10^{13} cm^{-2}, more prominent improvement is observed upon increasing the rear negative charge density. This is due to the recombination at the front surface being further reduced, so that stronger field-effect passivation at the rear leads to an increase in efficiency. Since the resistive losses in the emitter and therefore the optimum finger spacing is determined by finger width, the advancement in metallisation technology needs to be considered. The finger width by screen printing in industry is expected to be narrowed down to 18 µm by 2032 [36]. Cu plating allows narrower fingers down to 12 um has been demonstrated [37]. For the last group of data with hollow symbols and dashed lines in Fig. 15, the finger spacing was set to 15 µm, which allows reduced shading losses and therefore more room for the finger spacing to be reduced for lower resistive losses. This leads to a shift in peak efficiency from 1.6 to 1.2 mm finger pitch, and an increase in efficiency to beyond 24.8%.

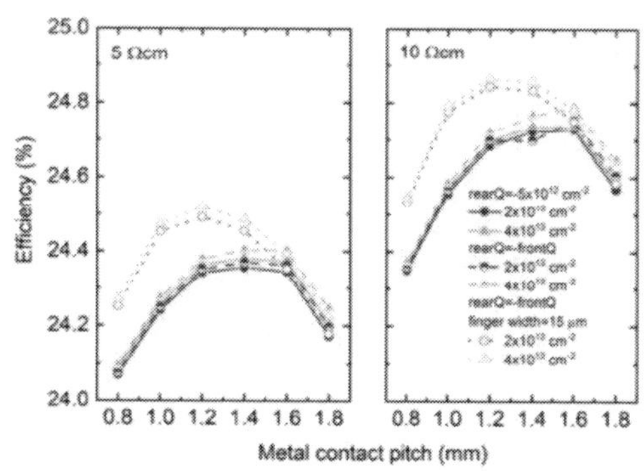

Figure 15: Simulated IL cell efficiency with wafer resistivity being 5/10 Ω·cm at various charge densities, finger spacings, and negative rear charge densities for field-effect passivation. Band-tail interface states are assumed to be well passivated, with Dit-CB/VB both set to 10^{13} eV^{-1}cm^{-2}.

The D_{it}-midgap of 5×10^{10} eV^{-1}cm^{-2} and Dit-CB/VB of $10^{14}/5 \times 10^{13}$ eV^{-1}cm^{-2} used in the above simulations were measured from a float zone Si substrate with a planar {1 0 0} plane passivated by thermal oxide. Since pyramid-textured surfaces are mostly used in commercial solar cells and that the interface passivation will be compromised because of the {1 1 1} facets [38]. In Fig. 16, D_{it}-midgap, D_{it}-CB, and D_{it}-VB were increased to values of 10^{11}, 10^{15}, and 5×10^{14} eV^{-1}cm^{-2}, respectively. The simulations show significant increase in efficiency by increasing the front charge density from 2×10^{13} to 4×10^{13} cm^{-2}, and by increasing the rear charge density for stronger field-effect passivation. This again suggests the importance in passivating the interface and maintaining high charge densities at both the front and the rear for reduced surface recombination.

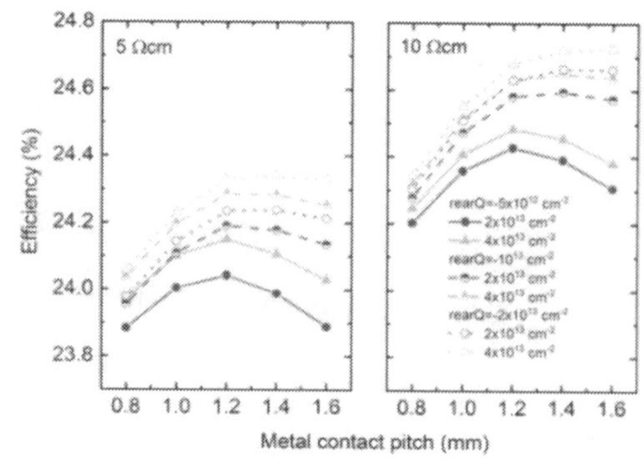

Figure 16: Simulated IL cell efficiency with wafer resistivity being 5/10 Ω·cm at various charge densities, finger spacings, and negative rear charge densities for field-effect passivation. D_{it}-midgap, D_{it}-CB, and D_{it}-VB were set to 10^{11}, 10^{15}, and 5×10^{14} eV^{-1}cm^{-2}, respectively.

4 CONCLUSIONS

In this work, the formation of field-induced electron surface layers and the potential of inversion layer cells is explored. For inversion layer cells, simulation results indicate that higher cell efficiencies are expected for substrates with higher resistivities due to improved bulk lifetime and lower doping density. This suggests the necessity of fabricating inversion layer cells on substrates with resistivity above 5 Ω·cm. Both high charge densities and well passivated surfaces are necessary to obtain high emitter conductivity and therefore cell efficiency, with the best efficiency reaching beyond 24.8%. However, further increasing the charge density to beyond 2×10^{13} cm^{-2} and reducing D_{it}-CB to below 10^{14} eV^{-1}cm^{-2} are not as effective in improving the emitter conductivity. Beyond the charge density of 2×10^{13} cm^{-2}, the effect of increased electron density on emitter conductivity is balanced by the reduced carrier mobility due to increased surface scattering. For D_{it}-CB below 10^{14} eV^{-1}cm^{-2}, the released electrons from the interface states are too little compared with the electron density in the emitter to lead to an observable improvement. It is noted that the accumulation of K^+ ions at the interface damages the passivation. Further research on the damaged interface is required to determine its effect on the cell performance. The optimum finger spacing of 1.2-1.6 mm for 5/10 Ω·cm substrates is equivalent to that of the mainstream PERC. This suggests that although the field-induced emitter is more resistive, the optimised finger spacing does not lead to significantly higher shading losses than that of the cells with a P-diffused emitter. A 10.8% efficient proof-of-concept inversion layer cell is fabricated. It is observed that the sheet resistance of a field-induced electron layer and V_{OC} of a precursor cell with K^+ ions embedded in the same method show good stability against elevated temperature (120 °C) and UV irradiation within 2 months. Although the efficiency of the prototype cell is low, an internal efficiency of beyond 24.8% is expected by simulations with optimised processing.

5 ACKNOWLEDGEMENTS

All the authors are thankful to Radka Chakalova for assistance in clean-room processing. M.Y would like to thank the China Scholarship Council for funding her doctoral studies. R.S.B was supported by the Royal

Academy of Engineering under the Research Fellowship scheme. This work was supported by the UK Engineering and Physical Sciences Research Council grant numbers EP/V038605/1 and EP/V037749/1 and by the Oxford University John Fell Fund. For the purpose of Open Access, the author has applied a CC BY public copyright licence to any Author Accepted Manuscript (AAM) version arising from this submission.

6 DATA AVAILABILITY

All data created during this research and published in this article is openly available from the Oxford University Research Archive and can be downloaded free of charge from http://ora.ox.ac.uk.

7 REFERENCES

[1] J. V. M. Paul Van Halen, Robert P. Mertens, Roger J. Van Overstraeten, Raye E. Thomas, "New TiOx-MIS and SiO2-MIS Silicon Solar Cells," *IEEE Trans. Electron Devices*, vol. 25, no. 5, pp. 507–511, 1978.

[2] R. E. Thomas, R. B. North, and C. E. Norman, "Low Cost - High Efficiency MIS/Inversion Layer Solar Cells," *IEEE Electron Device Lett.*, vol. 1, no. 5, pp. 79–80, 1980.

[3] F. Werner, Y. Larionova, D. Zielke, T. Ohrdes, and J. Schmidt, "Aluminum-oxide-based inversion layer solar cells on n -type crystalline silicon: Fundamental properties and efficiency potential," *J. Appl. Phys.*, vol. 115, no. 7, 2014, doi: 10.1063/1.4865962.

[4] R. Hezel, "Silicon nitride for the improvement of silicon inversion layer solar cells," *Solid State Electron.*, vol. 24, no. 9, pp. 863–868, 1981, doi: 10.1016/0038-1101(81)90103-9.

[5] K. Jager and R. Hezel, "Optical stability of silicon nitride MIS inversion layer solar cells," *IEEE Trans. Electron Devices*, vol. 32, no. 9, pp. 1824–1829, Sep. 1985, doi: 10.1109/T-ED.1985.22204.

[6] R. Hezel, K. Blumenstock, and R. Schorner, "Interface States and Fixed Charges in MNOS Structures With APCVD and PECVD Silicon Nitride," *J. Electrochem. Soc.*, vol. 131, p. 1679, 1984.

[7] F. Werner, Y. Larionova, D. Zielke, T. Ohrdes, and J. Schmidt, "Aluminum-oxide-based inversion layer solar cells on n -type crystalline silicon: Fundamental properties and efficiency potential," *J. Appl. Phys.*, vol. 115, no. 7, 2014, doi: 10.1063/1.4865962.

[8] K. Jager and R. Hezel, "Optical Stability of Silicon Nitride MIS Inversion Layer Solar Cells," *IEEE Trans. Electron Devices*, vol. 32, no. 9, pp. 1824–1829, 1985, doi: 10.1109/T-ED.1985.22204.

[9] R. S. Bonilla and P. R. Wilshaw, "A technique for field effect surface passivation for silicon solar cells," *Appl. Phys. Lett.*, vol. 104, no. 23, pp. 2–7, 2014, doi: 10.1063/1.4882161.

[10] M. Yu *et al.*, "Assessing the Potential of Inversion Layer Solar Cells Based on Highly Charged Dielectric Nanolayers," *Phys. Status Solidi - Rapid Res. Lett.*, vol. 15, no. 12, 2021, doi: 10.1002/pssr.202100129.

[11] S. W. Glunz, S. Rein, J. Knobloch, W. Wettling, and T. Abe, "Comparison of boron- and gallium-doped p-type Czochralski silicon for photovoltaic application," *Prog. Photovoltaics Res. Appl.*, vol. 7, no. 6, pp. 463–469, 1999, doi: 10.1002/(SICI)1099-159X(199911/12)7:6<463::AID-PIP293>3.0.CO;2-H.

[12] M. N. Darwish, J. L. Lentz, M. R. Pinto, P. M. Zeitzoff, T. J. Krutsick, and H. H. Vuong, "An improved electron and hole mobility model for general purpose device simulation," *IEEE Trans. Electron Devices*, vol. 44, no. 9, pp. 1529–1538, 1997, doi: 10.1109/16.622611.

[13] C. Lombardi, S. Manzini, A. Saporito, and M. Vanzi, "A physically based mobility model for numerical simulation of nonplanar devices," *IEEE Trans. Comput. Des. Integr. Circuits Syst.*, vol. 7, no. 11, pp. 1164–1171, 1988, doi: 10.1109/43.9186.

[14] G. Paasch and H. Übensee, "A Modified Local Density Approximation. Electron Density in Inversion Layers," *Phys. Status Solidi*, vol. 113, no. 1, pp. 165–178, 1982, doi: 10.1002/pssb.2221130116.

[15] M. Yu, S. McNab, I. Al-Dhahir, C. E. Patrick, P. P. Altermatt, and R. S. Bonilla, "Extracting band-tail interface state densities from measurements and modelling of space charge layer resistance," *Sol. Energy Mater. Sol. Cells*, vol. 231, no. May, p. 111307, 2021, doi: 10.1016/j.solmat.2021.111307.

[16] M. Yu, M. Wright, S. Mcnab, I. Al-dhahir, P. Pietro, and R. S. Bonilla, "Probing the Interface State Densities Near Band Edges from Inductively Coupled Measurements of Sheet Resistance," in *AIP Conference Proceedings, Silicon PV Conference*, 2022.

[17] H. Flietner, W. Füssel, N. D. Sinh, and H. Angermann, "Density of states and relaxation spectra of etched, H-terminated and naturally oxidized Si-surfaces and the accompanied defects," *Appl. Surf. Sci.*, vol. 104–105, pp. 342–348, 1996, doi: 10.1016/S0169-4332(96)00168-7.

[18] H. Flietner, "U-Shaped Distributions at Semiconductor Interfaces and the Nature of the Related Defect Centres," *Phys. Status Solidi*, vol. 91, no. 1, pp. 153–164, 1985, doi: 10.1002/pssa.2210910120.

[19] W. Füssel, M. Schmidt, H. Angermann, G. Mende, and H. Flietner, "Defects at the Si/SiO2 interface: Their nature and behaviour in technological processes and stress," *Nucl. Instruments Methods Phys. Res. Sect. A Accel. Spectrometers, Detect. Assoc. Equip.*, vol. 377, no. 2–3, pp. 177–183, 1996, doi: 10.1016/0168-9002(96)00205-7.

[20] M. Powell and S. Deane, "Defect-pool model and the hydrogen density of states in hydrogenated amorphous silicon," *Phys. Rev. B - Condens. Matter Mater. Phys.*, vol. 53, no. 15, pp. 10121–10132, 1996, doi: 10.1103/PhysRevB.53.10121.

[21] R. Brendel, "Coupling of Light into Mechanically Textured Silicon Solar Cells: a Ray Tracing Study," *Prog. Photovoltaics Res. Appl.*, vol. 3, pp. 25–38, 1995.

[22] G. Kaur *et al.*, "Ultra-thin LPCVD SiNx/n+poly-Si passivated contacts - A possibility?," *Conf.*

Rec. IEEE Photovolt. Spec. Conf., pp. 2679–2683, 2019, doi: 10.1109/PVSC40753.2019.8980624.

[23] L. J. van der Pauw, "A Method of Measuring Specific Resistivity and Hall Effect of Discs of Arbitrary Shape," *Philips Res. Reports*, vol. 13, no. 1, pp. 1–9, 1958.

[24] R. Chwang, B. J. Smith, and C. R. Crowell, "Contact size effects on the van der Pauw method for resistivity and Hall coefficient measurement," *Solid State Electron.*, vol. 17, no. 12, pp. 1217–1227, 1974, doi: 10.1016/0038-1101(74)90001-X.

[25] R. S. Bonilla, I. Al-dhahir, M. Yu, P. Hamer, and P. P. Altermatt, "Charge fluctuations at the Si–SiO 2 interface and its effect on surface recombination in solar cells," *Sol. Energy Mater. Sol. Cells*, vol. 215, no. May, p. 110649, 2020, doi: 10.1016/j.solmat.2020.110649.

[26] Y. Shi, M. Voss, M. Yu, E. T. Hwu, and R. S. Bonilla, "NanoLBIC Characterisation of Silicon Solar Cells Based on a Blue Laser Pick-up Unit," in *AIP Conference Proceedings, Silicon PV Conference*, 2021.

[27] P. P. Altermatt *et al.*, "Learning from the past to look beyond the roadmap of PERC Si solar cell mass production," *35th Eur. Photovolt. Sol. Energy Conf. Exhib.*, vol. September, pp. 215–221, 2018, doi: 10.4229/35THEUPVSEC20182018-2BP.1.1.

[28] B. Min *et al.*, "A Roadmap Toward 24% Efficient PERC Solar Cells in Industrial Mass Production," *IEEE J. Photovoltaics*, vol. 7, no. 6, pp. 1541–1550, Nov. 2017, doi: 10.1109/JPHOTOV.2017.2749007.

[29] N. E. Grant *et al.*, "Gallium-Doped Silicon for High-Efficiency Commercial Passivated Emitter and Rear Solar Cells," *Sol. RRL*, vol. 5, no. 4, pp. 1–8, 2021, doi: 10.1002/solr.202000754.

[30] G. Beaucarne, G. Schubert, L. Tous, and J. Hoornstra, "Summary of the 8th Workshop on Metallization and Interconnection for Crystalline Silicon Solar Cells," *Energy Procedia*, no. September 2019, p. 1, 2019, doi: 10.1016/j.egypro.2016.10.076.

[31] R. S. Davidsen *et al.*, "Black silicon laser-doped selective emitter solar cell with 18.1% efficiency," *Sol. Energy Mater. Sol. Cells*, vol. 144, pp. 740–747, 2016, doi: 10.1016/j.solmat.2015.10.018.

[32] G. Agostinelli *et al.*, "Very low surface recombination velocities on p-type silicon wafers passivated with a dielectric with fixed negative charge," *Sol. Energy Mater. Sol. Cells*, vol. 90, no. 18–19, pp. 3438–3443, 2006, doi: 10.1016/j.solmat.2006.04.014.

[33] Y. Zhang, L. Wang, D. Chen, M. Kim, and B. Hallam, "Pathway towards 24% efficiency for fully screen-printed passivated emitter and rear contact solar cells," *J. Phys. D. Appl. Phys.*, vol. 54, no. 21, 2021, doi: 10.1088/1361-6463/abe900.

[34] F. Werner and J. Schmidt, "Manipulating the negative fixed charge density at the c-Si/Al 2O3 interface," *Appl. Phys. Lett.*, vol. 104, no. 9, 2014, doi: 10.1063/1.4867652.

[35] H. Patel, C. Reichel, A. Richter, P. Masuch, J. Benick, and S. W. Glunz, "Effective charge dynamics in Al2O3/SiO2 multilayer stacks and their influence on silicon surface passivation," *Appl. Surf. Sci.*, vol. 579, no. November 2021, p. 152175, 2022, doi: 10.1016/j.apsusc.2021.152175.

[36] "International Technology Roadmap for Photovoltaic (ITRPV)," *Itrpv*, vol. 13th Editi, no. March, pp. 1–81, 2022.

[37] B. Grübel *et al.*, "Progress of plated metallization for industrial bifacial TOPCon silicon solar cells," *Prog. Photovoltaics Res. Appl.*, vol. 30, no. 6, pp. 615–621, 2022, doi: 10.1002/pip.3528.

[38] S. C. Baker-Finch and K. R. McIntosh, "The contribution of planes, vertices, and edges to recombination at pyramidally textured surfaces," *IEEE J. Photovoltaics*, vol. 1, no. 1, pp. 59–65, 2011, doi: 10.1109/JPHOTOV.2011.2165530.

SIMPLIFIED CELL CUTTING, EFFICIENT EDGE PASSIVATION, COPPER METALLIZATION: TACKLING THE LAST HURDLES FOR OPTIMIZED SHJ INTEGRATION IN SHINGLE MODULE CONFIGURATION

S. HARRISON[1], V. BARTH[1], C. CARRIERE[1], M.ALBARIC[1], B.MARTEL[1], M.GALIAZZO[2], E.SOVERNIGO[2], N.FRASSON[2]

[1]Univ Grenoble Alpes, CEA-LITEN, INES, 50 avenue du Lac Léman, F-73375 Le Bourget-du-Lac, France
[2]AMAT-BACCINI Systems; via postumia ovest, 31050, Olmi di S.Biagio di Callalta, Italy

ABSTRACT: In this abstract, we propose key improvements paths to specifically optimize the performance potential of SHJ devices in Shingle module configuration. But the different processes developed can also further be applied to any cut-based module technology, which are now the norm for any module supplier. We thus believe that the progress shown for example for edge cut-and-passivation can benefit to other technologies and module configurations as well. In particular, the cut-cell process described is highly innovative, as to our knowledge such wafer configuration has never been evaluated and integrated for PV applications before. This 45° rotated wafer based approach, is fully compatible with industrial constraints, and allows a significant simplification of the cell cut-step, opening new alternative technological solutions to the traditional laser-based processes usually used The development of a low-temperature and highly conformal passivation layer is also key for many silicon cut-cell based technologies. Furthermore, to our knowledge, the very high recovery levels demonstrated (up to 90%) are among the highest observed on PV cut-cells, whatever technology (cell or module) is considered. Such layers should thus be of interest for many PV applications, where edge defectivity is a clear performance limitation.

Keywords: Heterojunction, Shingle, passivation, wafer edge, metallization, Copper

1 INTRODUCTION

Shingle interconnection is gaining more and more interest as it combines high efficiency potential and improved aesthetics thanks to the denser silicon integration within the final module panel. However, latest high efficiency PV cell architectures, mostly relying on high level of passivation are not an obvious match for such module configuration. This is particularly the case for silicon heterojunction devices (SHJ), for which optimized integration needs to overcome two major intrinsic limitations:

1-High efficiency losses after cell cut in Shingle Stripes (up to 1% Abs losses can be observed on final cut-cells)

2-High metal resistance linked to the deported busbar metal pattern, and the high resistivity of low-temperature silver metal paste

So innovative solutions are needed to fully profit from the high efficiency potential brought by the use of SHJ. Proposed alternatives must remain simple, cheap, and compatible with industrial constraints. In this abstract, we will firstly focus on cell edge impact, describing the promising post-cut edge repassivation solutions developed. Indeed, with simple low-temperature AlOx layer, more than 90% performance recovery has been achieved. SHJ shingle cut-cells fabricated with such edge passivation process reached indeed efficiencies very close to the mother wafer initial properties before cut. To achieve such remarkable outcomes, as described in previous work [1,2], it is mandatory that the cut process leads to minimized edge degradation, for both the active layers and the silicon bulk. But very few cut techniques can currently fulfill these requirements. We thus developed an innovative and simplified cut approach, based on 45° rotated silicon wafers. The idea is to properly square the initial ingots, so that the natural silicon crystallographic cleavage plane gets perfectly aligned with the final desired cut within the wafer. With this simple idea, no laser based-cut or additional complex tool is needed for the cleavage, as a simple scribe on the edge of the wafer is now enough to initiate and propagate the cut.

Finally, we will also provide insights on latest developments done at metallisation and interconnection level to minimize (eventually suppress!) the amount of silver/and or ECA used. First promising shingle modules with full-cell copper metallization have for example been finalized. Moreover, drastic reduction of ECA deposition was successfully achieved with high reliability kept for shingle modules produced with less than 1mg of ECA deposited on each individual shingle cell.

In conclusion, with proper optimization of the different technological paths presented in this abstract, we demonstrate that all major hurdles left for larger scale SHJ shingle integration can be rapidly and efficiently addressed, opening promising future perspectives for such technology.

2 ROTATED INGOT/WAFER CUT APPROACH

As mentioned in previous sections, to consider high passivation levels on cut-edges, extremely smooth and uniform edge morphology is required after cell separation. As only very few cut techniques can meet these constraints (all based on thermo-mechanical approaches as proposed by companies such as 3D-Micromac [2] or Innolas), new approaches need to be explored as well. We thus propose an alternative and innovative cut-process, based on 45° rotated ingot. Such process allows a tremendous simplification of the overall cut procedure, as simple mechanical scribing on the wafer edge is now enough to insure proper cell separation, with propagation of the crack occurring without any heating of the cell or need of dedicated laser and/or cleavage tool. This mechanical propagation occurs along the (011) crystallographic silicon plane, so the idea is to rotate the silicon ingot during the squaring step, to align that specific plane with the pattern of the cell [4]. A 45° rotation is thus needed, as illustrated in Figure 1. All further wafering steps remain identical (sawing, cleaning, rectification…etc.).

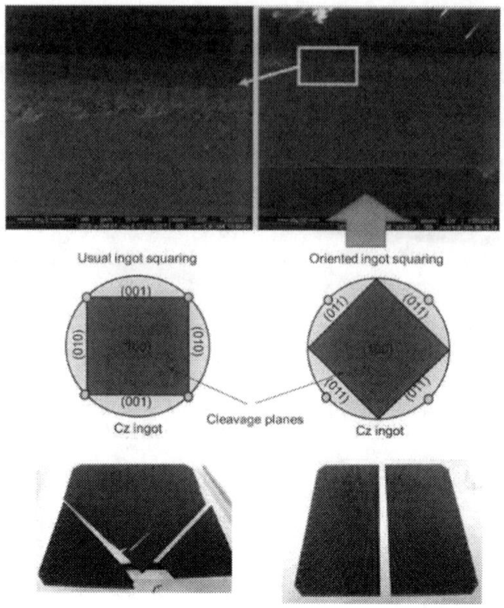

Figure 1: Experimental validation of new cut-cell approach based on 45° rotated ingots. Preferential cleavage plane is aligned with the desired cut direction, allowing simple and smooth cut of the cell as shown on the SEM pictures.

Such 45° rotated wafers were firstly generated on in-house pulled ingot, but process has also been successfully transferred recently on industrial ingots (M2 size). Several hundred of such 45° SHJ shingle cells were then fabricated on CEA-INES production pilot-line [3], with no need of adaptation or modification of any automation or process step. Identical breakage rate was even observed when compared to standard industrial CZ wafers, demonstrating thus the full compatibility of such "45° rotated" cells with industrial constraints. Representative electrical data of different batches produced at CEA are presented in figure 2. As shown, for equivalent bulk quality, similar final cell efficiencies are achieved, demonstrating as expected that there is no impact of the modified ingot orientation on the device performances.

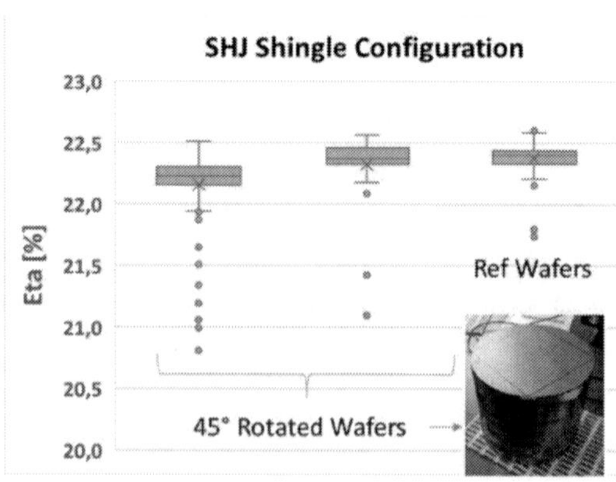

Figure 2: Typical electrical data achieved on 45° rotated wafers processed in shingle configuration on the CEA-INES industrial pilot-line. Efficiencies obtained are fully comparable with those measured on usual reference wafers integrated in the same shingle configuration.

3 DEVELOPMENT OF EDGE PASSIVATION LAYER

With such high cut-edge quality obtained, the challenge is to be able to develop a suitable passivation layer, compatible with the low-temperature SHJ constraints. Following the promising first results presented in [1], we continued the developments of dedicated low-temperature Aluminium Oxide (AlOx) layers. Still using industrial ALD equipment for the deposition, we managed to reach very high passivation levels thanks to specifically tuned deposition conditions, highlighting the importance for example of the precursor gaz sequence during deposition. But we also evidence the crucial role of final annealing conditions, as best results were obtained with an optimized light soaking sequence applied on the final cells. More process details can be found in [5]. 15nm thick AlOx layers are deposited on the sunny side of the cut cells, as illustrated on figure 3. Typical PL observations done on SHJ shingle wafers after cut, and after proper edge repassivation are presented in figure 4, demonstrating the clear local improvement of passivation properties allowed by the passivation process applied.

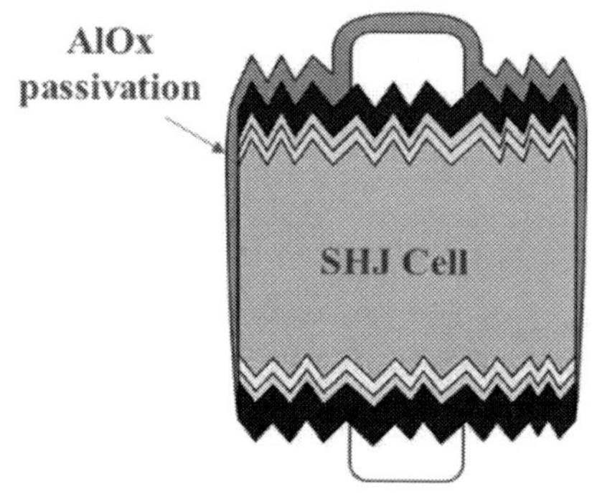

Figure 3: Schematic of the edge passivation protocol. Low-T° Alox layer is deposited on the full area of the cut-cell, covering as well the naked edges generated during cleavage.

PL signal improvement measured correlates well with the significant gain observed on final cell pFF/FF (figure 5), which clearly indicates the lower impact of the edge recombination mechanisms. However, as already presented in previous work [1], despite the tremendous progress made, perfect edge repassivation still does not seem achievable, with some intra-cell edge PL variability observed, reflecting as well on wafer to wafer reproducibility of the process. Passivation seems highly sensitive to local defectivity (mechanical, contamination) occurring during cut step and overall cell manipulation before passivation. Cut and subsequent AlOx deposition in controlled atmosphere should probably improve even further the recovery potential of the edge passivation process developed. It is also worth to note that the pFF/FF gain obtained with the improved edge passivation seems highly stable in time, with no degradation of performances observed after more than 40 days storage in ambient air conditions.

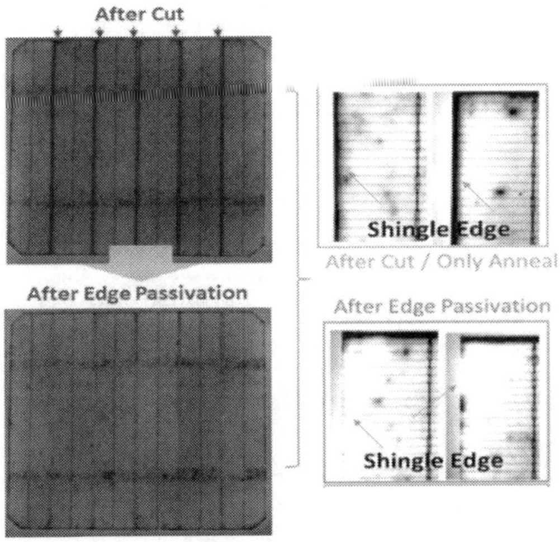

Figure 4: Typical PL observations made on SHJ shingle cells after cut (top pictures), and after applying the edge passivation process described in this paper (bottom pictures). Clear PL improvement is observed locally on the edge of the cut-cells with the passivation process.

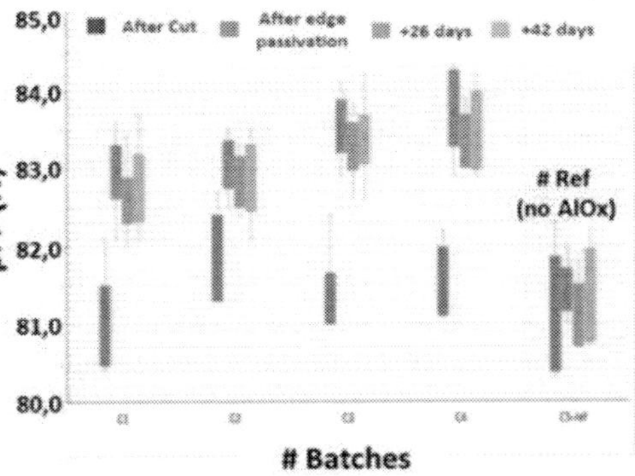

Figure 5: Evaluation of edge passivation impact on pFF (SHJ Shingle cell configuration). Clear improvement of cell performances is observed thanks to the additional passivation layer.

Both PL and pFF boost brought by the edge passivation process applied, directly translates in significant cell performance recovery, as illustrated in figure 6. Indeed, IV measurement are done at several steps of the process, demonstrating first a performance drop with the cut of the mother cell in 6 individual Shingle cells. After the AlOx layer deposition, followed by proper Light Soaking anneal, we can observe an impressive improvement of efficiencies for all the cells processed. More than 60% of performance recovery is achieved for all samples, up to 90% for the best cells, demonstrating the effectiveness of the additional passivation, and the potential of the process developed. Good stability and repeatability of the process described in this paper has also been proven on repeated batches, with higher volume of cells processed, proving the robustness of the passivation process proposed. Further developments are on-going to improve even further the AlOx layer quality and industry compatibility, trying in particular to reduce even further the layer thickness and passivation potential, and working on deposition transfer on PECVD tools, more suited to high volume wafer treatment, as required for the PV industry.

Best Cell Output (AlOx 15nm) ~90% Eff recovery

	Voc (mV)	Isc (A)	FF (%)	Eff (%)	Delta Eff (%)
Initial	740,6	9,18	81,2	22,59	
Post Cleavage	734,3	9,132	78,3	21,48	-1,1
Passivation	741,6	9,136	81,1	22,48	+1,0

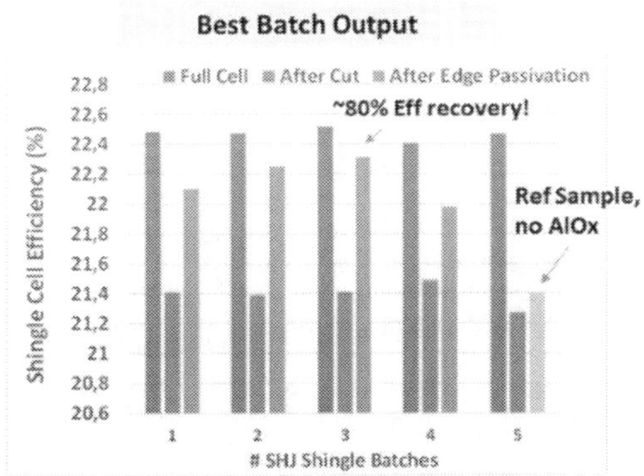

Figure 6: Typical IV measurements for a SHJ Shingle cell batch, demonstrating an effective efficiency recovery after proper edge passivation layer deposition and activation. Up to 90% recovery has been obtained for the best cells.

4 PASSIVATED CELL MODULE INTEGRATION

Following promising edge passivation results presented in previous section, it is important to asses if the progress obtained at cell level can be maintained after module integration. Furthermore, the presence of the AlOx layer over the cell surface may degrade the interconnection quality insured by the ECA dots deposited between two successive cells in the shingle module configuration.

Several mini-modules (6 tiles) were thus built, half including shingle cells without edge passivation, half including shingle cells with the AlOx passivation layer. Firstly, all modules built are functional and similar EL characterization is observed between both module configurations. This tends to indicate that the additional AlOx layer used for the passivation does not degrade the quality of the interconnection (good adhesion between tiles kept after stringing) and good electrical contact seems to be kept between two successive cells (apparent limited increase of resistance induced by the passivation layer).

Finally, electrical measurement on final modules clearly show an improved output power for all the modules with edge passivation, which is the first confirmation that performance improvements achieved at cell level can be maintained at module level. Power gain is linked to both improved FF and Voc on the passivated modules as expected. Negligible impact on Isc has been observed. Modules with passivated edge show however a greater performance dispersion than the reference ones, indicating that further integration optimization is probably needed to fully profit from the higher cell efficiency potential. This is an overall very promising first validation that the passivation process proposed in this paper should be suitable with all cell and module production constraints. New experiments are planned, and overall reliability assessment of the module produced will be done in the next months.

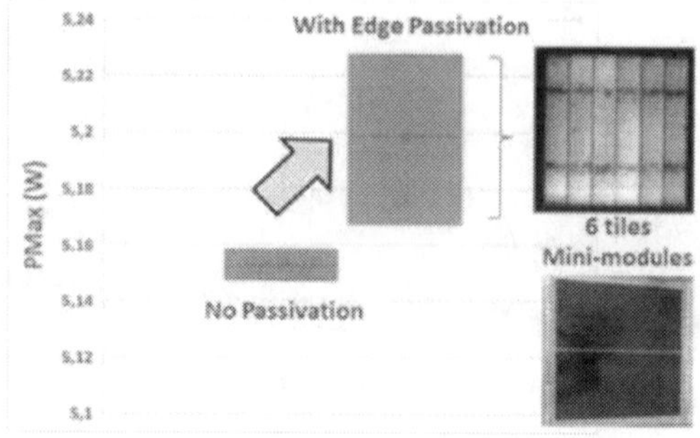

Figure 7: Power improvement is achieved as well for SHJ shingle modules integrating the edge passivation layer described in this paper.

5 METALLIZATION CHALLENGES

Even if this paper mainly focused on cell cut and edge passivation, it is important to highlight current development paths considered and progresses made to overcome the metallization limitations identified for the SHJ shingle configuration. Main objective is indeed to reduce drastically the amount of silver used (cost issues), without degradation of the device performances. Main approach considered is the replacement of silver by copper, either using copper pastes deposited by screen-printing, either using copper plating techniques. First modules integrating SHJ cells produced with copper paste metallization were thus successfully achieved. If this promising first result shows the compatibility of full copper metallization with shingle configuration, dedicated reliability tests conducted (not shown here) indicate a significant degradation of performances after only a few TC cycles, probably linked to uncontrolled oxidation of the copper. Further optimization is thus needed, and new tests are initiated with alternative Cu pastes and adapted ECA's. Most promising combination(s) should soon be identified, and shingle modules will also take profit of general progresses made for copper integration in standard module interconnection configuration.

Figure 8: First SHJ shingle modules successfully assembled with full Copper metallization

Finally, if we only consider the interconnection [6], it is also important to reduce the amount of ECA used, which cannot be neglected as well when cost issues and overall module sustainability are considered. Several module were thus built, with progressive reduction of ECA amount deposited on each Shingle cell before interconnection. As shown on figure 9, we managed to build shingle modules with extremely low ECA deposits (less than 0.7mg per cell, ~1/3 of the usual ECA deposit for reference modules). No apparent degradation of module output powers, and even more impressive, no degradation of overall final module reliability is observed on such low-ECA modules (less than 2% power losses after up to 600 TC cycles). This excellent result, allowed by the unique combination of SHJ (low internal stress) and shingle interconnection show that clear optimization path is achievable to improve further the overall competitivity of the SHJ shingle technology.

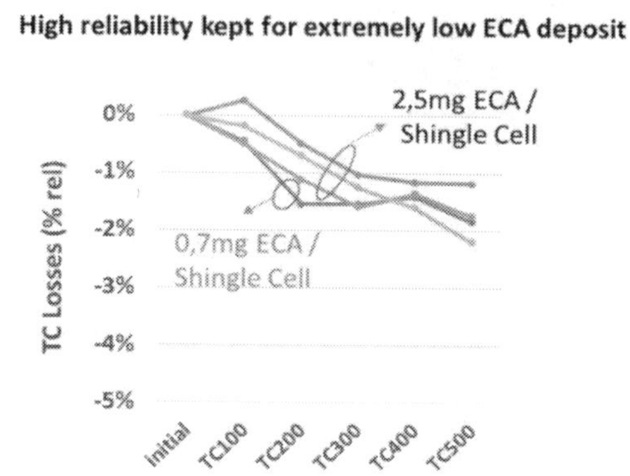

Figure 8: Extremely high reliability demonstrated on SHJ Shingle modules with very low-ECA deposits, down to less than 0.7mg per cell.

6 CONCLUSION

Clear solutions have been identified to overcome current identified limitations for SHJ shingle module configuration. In particular very promising performance recovery (up to 90%) has been demonstrated with the deposition of AlOx layers combined with innovative 45° rotated wafer cut-approach. At module level, drastic reduction in ECA usage is possible, and first promising outputs generated with full copper metallization, driving the technology closer to industrial requirements in terms of cost and sustainability constraints.

Acknowledgement

This work iss supported by the HighLite project that has received funding from the European Union's Horizon 2020 research and innovation programme under grant agreement No 857793.

REFERENCES

[1] S. Harrison et. al. EU PVSEC 2021 proceedings: 38th. 2021. doi: 10.4229/EUPVSEC20212021-2BO.15.3.
[2] P.Baliozian et al, "Postmetallization passivated edge technology for separated silicon solar cells", IEEE Journal of Photovoltaics, Vol 10, No 2, March 2020.
[3] A.Danel et al, Silicon Heterojunction cells with open circuit voltage over 750mV, 35th EUPVSEC, Belgium (2018).
[4] J. F. Lelièvre et al., 37th European Photovoltaic Solar Energy Conference and Exhibition (2020) 487–489
[5] B.Martel et al, "Addressing separation and edge passivation challenges for high efficiency Shingle Heterojunction solar cells",to be published.
[6] C.Carrière et al, "Shingle interconnection of HJT solar cells", proceeding of 38th EUPVSEC 2021

REDUCTION OF ITO BY A LOW-COST SPRAYED TiO_x CAPPING LAYER FOR SHJ SOLAR CELLS

Ulrike Heitmann, Leonard Tutsch, Angela De Rose, Dominik Dreja, Leonie Jakob, Radwa Elgazzar, Martin Bivour, Jonas Bartsch
Fraunhofer Institute for Solar Energy Systems ISE
Heidenhofstraße 2, 79110 Freiburg, Germany

ABSTRACT: We demonstrate a spray pyrolysis process for deposition of titanium oxide (TiO_x) on ITO on silicon heterojunction (SHJ) solar cells. Layer stacks of 20 nm ITO + 55 nm TiO_x, 50 nm ITO + 25 nm TiO_x and references with 75 nm ITO were deposited on M2+ (157 × 157 mm²) solar cells. All three layer stacks yield similar cell efficiencies of close to 21 %. Homogeneous sprayed coatings were achieved despite using a small laboratory spray coater. Single cell modules were manufactured from cells of all three groups. Only slight modifications in the soldering process were necessary, even though the screen-printed metallization is covered with TiO_x for two groups. Module efficiencies are also on par for the three cell types. A slightly increased series resistance is measured on both cell and module level for the cells with 20 nm ITO layer, due to higher resistance for lateral current transport in this layer. Consumables costs of <1ct/wafer are estimated based on the laboratory process.
Keywords: Heterojunction, Antireflection Coating, Spray Pyrolysis, Soldering

1 INTRODUCTION

The global photovoltaic production capacity is steadily rising, with currently >200 GW_P production per year and with an estimated capacity of >1TW_P around 2030 [1]. Demand for material resources needed for PV production will increase likewise, unless technical solutions are found to lower the material needed per W_P. For some materials, a five-fold demand as compared with today can be satisfied in principle (though the practical realization of this capacity growth will still be challenging). For other, more scarce materials, high shares would need to be taken for PV production, meaning increasing costs and competition with other uses. Silver, indium and bismuth have been identified as critical materials in this respect [2].

Indium consumption for PV production originates from thin film PV such as CdTe and CIGS, and from silicon heterojunction (SHJ) solar cells in the field of crystalline silicon PV. Indium is part of the transparent conductive oxide (TCO), which is needed for contacting the semiconductor material, for lateral conductance of the extracted charge carriers and for antireflection purposes. Tin-doped indium-oxide (ITO) is mostly used for this layer at present. Indium-based oxides offer the most attractive properties for solar cells, as they combine excellent transparency with high conductivity. They are typically deposited using physical vapor deposition (PVD), namely sputtering. While this process delivers high-quality layers, it is more costly than some alternative processes that do not require vacuum conditions.

Several approaches have been proposed to lower the Indium consumption of SHJ solar cells [3-7]. Using alternative, indium free TCOs such as Aluminum doped zinc oxide (AZO) can mitigate Indium altogether, but comes at the cost of reduced stability and slightly affected performance. AZO can also be used in TCO stacks, covered with thin layers of ITO, which reduces In consumption effectively and relies on known sputtering processes.

In other approaches, a thin ITO layer is capped with a dielectric layer. A thinner ITO layer would have less ideal antireflective properties, so the dielectric layer is selected to compensate or even overcompensate for this loss. In addition, the dielectric layer can improve the stability of SHJ solar cells encapsulated into modules in the field. Different techniques can be used to deposit the dielectric layer, such as APCVD, PECVD [4,7] or Spray Pyrolysis [8]. Some of these techniques are potentially cheaper than sputtering. The dielectric layer is typically applied after the metallization has been printed onto the SHJ solar cell. This brings on challenges for the module interconnection, but can also be exploited, e.g. for a subsequent copper plating process [7] which may additionally lower the consumption of another scarce material, silver.

In this work, we demonstrate the application of a dielectric layer deposited by spray pyrolysis, a potentially cheap and mass-production capable deposition technique. The work is based on our previous work on small-area solar cells [8], where we first demonstrated that this process can work effectively. Here, we show the transition to an industrial wafer format (M2+,157 × 157mm²) and the fabrication of single-cell-modules.

2 EXPERIMENTAL

SHJ solar cells were manufactured at Fraunhofer ISE from 157 × 157 mm² *n*-type Cz wafers, 1 Ωcm base resistivity. Wet-chemical saw damage etching and texturization were made prior to a cleaning step with ozone. After an HF dip, intrinsic and doped *a*-Si:H layers were deposited by PECVD, followed by sputtering of ITO.

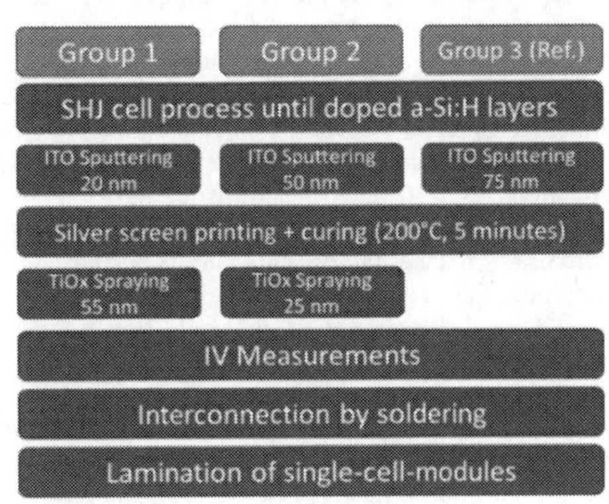

Figure 1: Overview of the experimental groups in the presented experiment.

The TCO deposition process was altered for some of the solar cells (all layer thicknesses refer to texturized surfaces). Besides a reference group with 75 nm ITO for ideal antireflective properties, two groups with thinner ITO (20 nm and 50 nm) were created. An overview of the experimental groups is given in Figure 1.

After screen-printing and curing (200 °C, 5 min) of silver metal contacts (5 busbar layout) on all cells, the latter two solar cell groups were coated with a 55 nm and a 25 nm titanium oxide (TiO_x) layer by spray pyrolysis. A precursor solution containing a titanium alkoxide dissolved in an alcoholic solvent was used. A spray coater from Equisonics was used, which is equipped with an ultrasonic nozzle (40 kHz) and a pneumatic nozzle. The nozzles can be moved by an X-Y-Z robot over the sample surface, which can be heated to up to 500°C. Substrate set temperature in the experiments to fabricate solar cells was 200 °C. Both nozzle types were tested. Better homogeneity of the TiO_x layer was obtained with the pneumatic nozzle, while material consumption is lower with the ultrasonic nozzle (0.8 ml of precursor solution per wafer). The used setup is laboratory scale and requires multiple passes of the nozzle over the surface. Spraying times of one minute (25 nm TiO_x layer thickness) and two minutes (55 nm TiO_x layer thickness) were required for the deposition of the layer. For better comparison, the cells from both groups were left on the heated substrate platform after spraying for a total duration of 5 minutes.

After performing IV measurements of the SHJ cells, single-cell-modules were fabricated. It has been shown, that infrared (IR) soldering is well suited for the interconnection of SHJ solar cells [9]. For our experiment, a semi-automatic soldering station with IR heating was used. The cells were pre-heated on a hotplate and soldered using Sn60Pb40-coated Cu ribbons and no-clean flux. After cross connection, small-format modules were laminated with glass, POE encapsulant and polymer-based black backsheet. The finished modules were characterized by EL imaging and IV measurements.

For optical investigations, borofloat glass with a thickness of ~0.7 mm was spray coated using different substrate set temperatures (200 °C, 300 °C and 400 °C). Other conditions in these spray processes were as discussed above.

3 RESULTS

3.1 Optical properties of the spayed TiO_x layers

The requirements for the application as a capping layer in SHJ solar cells are high transparency and a refractive index close to that of the underlying ITO. The transparency of the sprayed TiO_x was analyzed by coating of a 0.7 mm thin Borofloat glass slide with approx. 50 nm TiO_x. The transparency was evaluated by measuring the reflection (R[%]) and the transmission (T[%]) and calculating the absorption according to A=100-R-T. Different substrate temperatures during spraying were compared. The results are shown in Figure 2. It should be noted that the accuracy of the reflection-transmission measurement is ≈ 2 % absolute, which could explain the calculated negative absorption. However, when comparing the absorption of the TiO_x thin film deposited at 200 °C with the one deposited at 300 °C, it can be seen that the absorption between 400-650 nm decreases with reduced substrate temperature, resulting in a highly transparent thin film when depositing at a substrate temperature of 200 °C.

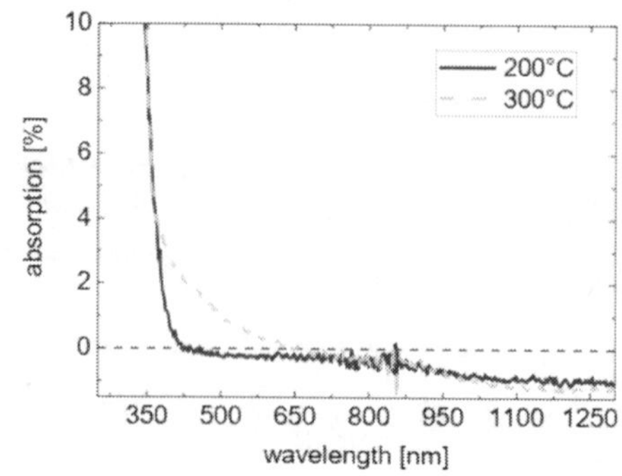

Figure 2: Calculated absorption of spray coated TiO_x at substrate temperatures of 200 °C and 300 °C from transmission/reflection measurements.

The refractive index of the TiO_x thin film was determined by spectroscopic ellipsometry (SE) measurements on TiO_x coated silicon substrates (TiO_x thickness ≈ 100nm). The results are shown in Figure 3. The refractive index was analyzed for different substrate set-temperatures during the deposition as well. It decreases with reduced substrate temperature. At 200 °C it is very close to the refractive index of the sputtered ITO.

In conclusion, the optical properties of the thin sprayed TiO_x film deposited at a substrate temperature of 200 °C are ideal for the use as a capping layer on ITO layers and compatible with the low-temperature regime (max. 200 °C) of SHJ solar cells.

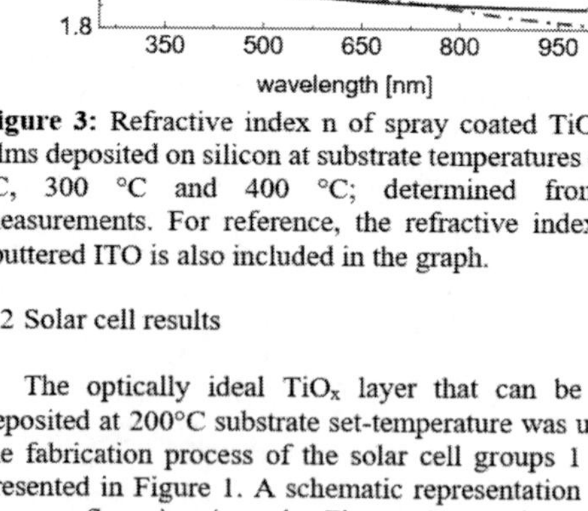

Figure 3: Refractive index n of spray coated TiO_x thin films deposited on silicon at substrate temperatures of 200 °C, 300 °C and 400 °C; determined from SE measurements. For reference, the refractive index of a sputtered ITO is also included in the graph.

3.2 Solar cell results

The optically ideal TiO_x layer that can be spray deposited at 200°C substrate set-temperature was used in the fabrication process of the solar cell groups 1 and 2 presented in Figure 1. A schematic representation of the process flow is given in Figure 4, together with a photograph of a finished M2+ wafer.

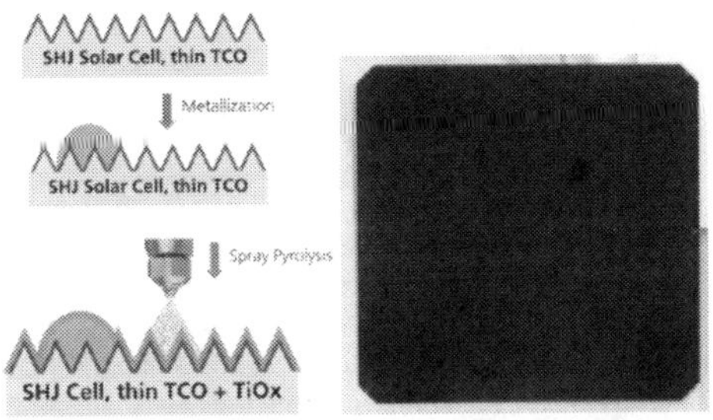

Figure 4: Top layers of the cells processed in this work (left), photograph of cell (without metallization, right)

Groups 1 and 2 were IV measured before and after spray pyrolysis of the TiO_x layer. In the post spray coating measurement, the reference group 3 was included. All IV results on cell level are given in Table I. The overall efficiency level is poor, as the fabrication process was only ramped up at the time these cells were made. The measurements still allow to assess the proposed process and to compare it to the reference process.

Table I: IV results of groups 1 – 3 on cell level, groups 1&2 before and after TiOx spray pyrolysis.

	V_{OC} [mV]	J_{SC} [mA/cm²]	FF [%]	η [%]	R_S [Ωcm²]
20 nm ITO (group 1)					
Before TiO_x	725	34.7	76.7	19.3	1.12
After TiO_x	725	37.0	77.1	20.7	**1.14**
50 nm ITO (group 2)					
Before TiO_x	723	36.3	77.0	20.3	0.87
After TiO_x	722	37.0	77.2	**20.6**	**0.87**
75 nm ITO (Ref., group 3)					
	727	**37.1**	77.2	**20.8**	**0.88**

Groups 1 and 2 show a clear disadvantage in short circuit current density before TiO_x spray pyrolysis (marked red in Table I). After the deposition of the TiO_x layer, this disadvantage is fully compensated (marked green and bold). Group 1 also shows a higher series resistance as compared to the other groups, which does not mirror in the cell efficiency. The thinner ITO layer limits the lateral transport of the extracted charge carriers towards the printed metal contacts. This can partly be compensated by the silicon bulk, but for the 20 nm thin ITO layer, a measurable effect results.

3.3 Single cell module results

After the cells were soldered and laminated, the single-cell-modules were measured for their IV parameters. Two modules were built for group 1 and 2 each, and four modules were built for group 3. The soldering process was similar to a standard soldering process as used on an industrial stringer. The IV results of the modules are presented in Table II. A photograph of one of the modules is shown in Figure 5.

Table II: IV results of single-cell-modules after lamination

	V_{OC} [mV]	J_{SC} [mA/cm²]	FF [%]	η [%]	R_S [Ωcm²]
20 nm ITO 55 nm TiO_x (group 1)	724	35.1	74.0	18.8	1.92
50 nm ITO 25 nm TiO_x (group 2)	719	34.9	74.5	18.7	1.69
75 nm ITO (Ref., group 3)	727	34.9	73.5	18.7	1.88

As for the cells, the efficiency level of all three module groups is on the same level. The fill factor drops as compared to the cells for all three groups, as additional resistances are added to the system. The increasing R_S confirms this interpretation. The measured values are in the expected range for cell-to-module losses. The J_{SC} drops as well, due to optical losses and the larger reference area.

Figure 5: Photograph of single-cell-module fabricated from a M2+ SHJ solar cell with ITO/TiO_x layer stack.

4 DISCUSSION

We present a simple and potentially cheap process to reduce the indium consumption in SHJ solar cells by as much as 70 %, without sacrificing efficiency. The TiO_x thin film can be sprayed on the full wafer, no masking or subsequent partial etching of the TiO_x is necessary. The spraying process is already quite quick in the lab setup (1-2 minutes), in an industrial multi-nozzle setup, process times should be significantly faster. The spraying process might even be combined with the curing of the screen-printed Ag grid, which is also carried out at a temperature of 200 °C. However, this was not done for the presented results. Standard soldering is feasible and small-scale modules were successfully fabricated..

At the current indium price level, it still remains uncertain if this process will be economically advantageous. Calculations presented at the EU PVSEC 2021 [10] show that additional 55 nm of ITO currently cost around 1.75 €ct/wafer. In the same publication, it has been discussed that replacing sputtering of ITO with PECVD of SiN_x does not lower the production cost. In the present work, the material costs for the chemical precursor are 2.5 €ct/wafer in our labs, where we buy in small volume and in lab quality. We can estimate that the material cost will be below 1 €ct/wafer for industrial throughput production. The cost for the machine is currently difficult to estimate, but the small cost advantage for avoiding ITO sputtering make it very improbable that an established production line will be altered.

Three factors might change this:

1. A demonstration of superior stability of modules with top layer stack of thin sputtered ITO and thick spray coated TiO_x.
2. A demonstration of a fully sprayed top layer stack
3. Increasing indium prices and/or reduced availability of indium.

5 CONCLUSION AND OUTLOOK

In this paper, a simple and effective spray pyrolysis process of TiO_x to coat a thin layer of ITO is demonstrated on M2+ SHJ solar cells. Small-scale modules are fabricated by IR soldering and laminating these cells. It is shown that the amount of indium per cell can be reduced by as much as 70 % using this approach without sacrificing efficiency. While the advantage in costs and performance may be insufficient to motivate cell manufacturers to change their current or planned production at present, this may change in the future. Besides the perspective of rising indium costs, we plan to demonstrate that this approach can yield superior long-term stability of modules in the field, and that sputtering can be avoided altogether by spraying both TCO and TiO_x layer. This may be an attractive option for newly installed SHJ production lines.

6 ACKNOWLEDGEMENT

This work has received funding from the Federal Ministry for Economic Affairs and Climate Action within the Projects NextTec (03EE1001A) and CUSTCO (03EE1032).

7 REFERENCES

[1] Goldschmidt, Jan Christoph; Wagner, Lukas; Pietzcker, Robert; Friedrich, Lorenz (2021): Technological learning for resource efficient terawatt scale photovoltaics. In: Energy Environ. Sci. 14 (10), S. 5147–5160. DOI: 10.1039/D1EE02497C.

[2] Zhang, Yuchao; Kim, Moonyong; Wang, Li; Verlinden, Pierre; Hallam, Brett (2021): Design considerations for multi-terawatt scale manufacturing of existing and future photovoltaic technologies: challenges and opportunities related to silver, indium and bismuth consumption. In: Energy Environ. Sci. 14 (11), S. 5587–5610. DOI: 10.1039/D1EE01814K.

[3] Han, Can; Santbergen, Rudi; van Duffelen, Max; Procel, Paul; Zhao, Yifeng; Yang, Guangtao et al. (2022): Towards bifacial silicon heterojunction solar cells with reduced TCO use. In: Progress in Photovoltaics 30 (7), S. 750–762. DOI: 10.1002/pip.3550.

[4] D. L. Bätzner, P. Papet, B. Legradic, D. Lachenal, R. Kramer, T. Kössler, L. Andreetta (2019): 'HJT 2.0' PERFORMANCE IMPROVEMENTS AND COST BENEFITS FOR SILICON HETEROJUNTION CELL PRODUCTION. In: Proceedings of the 36th European Photovoltaic Solar Energy Conference and Exhibition, Marseille, S. 300–303, zuletzt geprüft am 28.08.2020.

[5] Morales-Vilches, Anna B.; Cruz, Alexandros; Pingel, Sebastian; Neubert, Sebastian; Mazzarella, Luana; Meza, Daniel et al. (2019): ITO-Free Silicon Heterojunction Solar Cells With ZnO:Al/SiO 2 Front Electrodes Reaching a Conversion Efficiency of 23%. In: IEEE J. Photovoltaics 9 (1), S. 34–39. DOI: 10.1109/JPHOTOV.2018.2873307.

[6] Wang, Jianqiang; Meng, Chuncai; Liu, Huan; Hu, Yunfei; Zhao, Lei; Wang, Wenjing et al. (2021): Application of Indium Tin Oxide/Aluminum-Doped Zinc Oxide Transparent Conductive Oxide Stack Films in Silicon Heterojunction Solar Cells. In: ACS Appl. Energy Mater. 4 (12), S. 13586–13592. DOI: 10.1021/acsaem.1c02209.

[7] Adachi, Daisuke; Terashita, Toru; Uto, Toshihiko; Hernández, José Luis; Yamamoto, Kenji (2017): Effects of SiOx barrier layer prepared by plasma-enhanced chemical vapor deposition on improvement of long-term reliability and production cost for Cu-plated amorphous Si/crystalline Si heterojunction solar cells. In: Solar Energy Materials and Solar Cells 163, S. 204–209. DOI: 10.1016/j.solmat.2016.12.029.

[8] Bartsch, J.; Heitmann, U.; Jakob, L.; Mahmoud Algazzar, R.; Tutsch, L.; Hermann, R. et al. (2020): Spray Pyrolysis – A Versatile Technique for Thin Film Deposition in PV. 4 pages / 37th European Photovoltaic Solar Energy Conference and Exhibition; 44-47. DOI: 10.4229/EUPVSEC20202020-1AO.3.6.

[9] A. De Rose, T. Geipel, D. Eberlein, A. Kraft, and M. Nowottnick, "Interconnection of Silicon Heterojunction Solar Cells by Infrared Soldering - Solder Joint Analysis and Temperature Study," in *Proceedings of the 36th European Photovoltaic Solar Energy Conference and Exhibition (EU PVSEC)*, Marseille, France, 2019, pp. 229–234.

[10] L. Tutsch, L. Jakob, C. Luderer, T. Hatt, B.S. Goraya, U. Heitmann, J. Bartsch, B. Bläsi, S. Pingel, A. Moldovan, S. Nold, M. Hermle, M. Bivour (2021), Strategies for Minimizing ITO Consumption in SHJ Modules: Combined Consideration of Electro-Optical Performance and Costs, In: Proceedings of the 38th European Photovoltaic Solar Energy Conference and Exhibition, Lisbon

INSIGHTS INTO CUT-EDGES OF SHJ SOLAR CELLS BY EL AND LBIC CHARACTERIZATION

Matevž Bokalič[1], Miha Kikelj[1], Benjamin Lipovšek[1], Samuel Harrison[2], Marko Topič[1]
[1] University of Ljubljana, Faculty of Electrical Engineering
Tržaška 25, SI-1000 Ljubljana, Slovenia
[2] 2CEA, INES,
50 Av. du Lac Léman, 73370 Le Bourget-du-Lac, France
E-mail: Matevz.Bokalic@fe.uni-lj.si

ABSTRACT: Further improvement of shingled cut-cell PV modules requires precise characterization of cut cell edges in order to evaluate, understand and improve cutting process and subsequent re-passivation quality. We use light beam induced current (LBIC) and electroluminescence (EL) measurements to characterize different wafer-based solar cell cutting techniques applied to bifacial silicon heterojunction (SHJ) solar cells: laser ablation and cleaving from the front (L&Cf) and the back (L&Cb) side, thermal laser separation (TLS), and cleaving only. We characterize cut edges from both, the front and the back side of the cell and compare them to the natural (uncut) edge. We present results in terms of normalized response profiles over different edge types and correlate LBIC and EL profile steepness with edge-zone recombination rate.

Measured LBIC profiles are much steeper than EL profiles due to forced charge carrier extraction and lower carrier concentration. By decreasing current injection, EL profiles become steeper and approach LBIC profiles in the limit case. LBIC profiles are also steeper when illuminated from the back, because the minority charge carriers are thereby mostly generated near the main junction and immediately extracted from the bulk. Front and back EL profiles do not show this distinction, because here, in both cases, there are much more charge carriers present in the bulk and radiative recombinations are distributed more homogeneously across the whole depth of the bulk.

TLS, cleave only and L&Cf provide overall the steepest profiles, but since TLS and cleave only offers better re-passivation potential, they are preferred options. L&Cb shows the worst result as laser ablation introduces recombination centres very close to and also in the main junction region. Cleaved only separation provides the best results observed by EL, which is better correlated with normal operation conditions than LBIC due to higher charge carrier density; however 45° rotated ingot is required for this technique. The natural edge profiles exhibit the least steep profiles measured with front and back LBIC, however this is not due to recombination processes but rather due to TCO masking near the edge.

Keywords: Electroluminescence (EL), Light beam induced current (LBIC), Silicon heterojunction (SHJ), Silicon solar cells, edge recombination

1 INTRODUCTION

The rapid growth of photovoltaic market is based on silicon solar cells. As silicon solar cells are approaching their practical efficiency limit of around 29% [1] with new cell technologies such as interdigitated back contact (IBC), Tunnel oxide passivated contact (TOPCon) and silicon heterojunction (SHJ), a gap between the cell and the module efficiency is being minimized by using cut cells and innovative interconnection techniques. While these approaches minimize series resistance and maximize module area utilization on one hand, on the other hand cell cutting introduces edge recombination losses, which reduce the efficiency. In order to minimize edge losses and fully exploit new technologies, we must understand the mechanisms behind edge recombination losses.

Herein we focus on how different cut-edge types influence the performance of cut SHJ cells around cut edges. We obtain a comprehensive set of characterization results by injection dependent electroluminescence (EL) imaging and wavelength dependent light beam induced current (LBIC) measurements from front and backside. We present the results in a form of normalized profiles and discuss the results.

The paper is organized as follows. In the experimental section we describe the cells, two measurement techniques and edge types that we have characterized. In the results section we first focus on the results obtained by LBIC measurements, before focusing on EL results and comparing EL and LBIC results. Finally, we summarize the paper and draw some conclusions.

2 EXPERIMENTAL

2.1 Cells and edge types

We have investigated the properties of various edges of 24% efficient state-of-the-art silicon heterojunction cells produced by CEA-INES [2]. The cells are based on an n-type wafer and passivated with an intrinsic amorphous silicon (a Si) on both sides. An n+ a Si layer at the front and a p+ a Si layer at the back form carrier selective junctions. The latter is the main junction. We use injection-dependent (0.05 – 1.0 J_{SC}) EL imaging and wavelength-dependent (520 – 1060 nm) LBIC measurements. They are both performed on in-house developed setups.

The cells were cut to small pieces about 2 by 3 cm^2. Different cutting techniques were used, enabling us to characterize the following edge types: natural (non-cut, as manufactured), laser ablated and cleaved from the front (L&Cf) and the back (L&Cb) side (in-house process), thermally laser separated (TLS, by 3D Micromac) [3] and cleaved only (in-house, similar to CEA-INES technique [4]).

The cells were mounted in a custom-built holder that allowed bifacial illumination and EL emission acquisition.

2.2 Wavelength dependent LBIC

Wavelength dependent LBIC measurements were acquired using in-house developed LBIC system based on single-mode laser diodes with 520, 642, 830, 940 and 1060 nm wavelengths (Figure 1 a). Measurement step for edge profile acquisition was 10 µm and spot size was selected appropriately to cover the whole scanned area. The cell's response was measured as a voltage on a resistor with a resistance low enough to establish circumstances similar to short circuit. A lock-in technique was used to provide a reliable reading of very low voltage on the resistor.

2.3 Injection dependent EL

Injection dependent EL was measured using the in-house developed setup built around a 1.4 megapixel cooled silicon CCD camera and a programmable power supply via a Matlab based automation system (Figure 1 b). Images were acquired at five different injection current densities, namely 0.05, 0.1, 0.25, 0.5 and 1.0 times J_{SC}. Image processing include background value subtraction and vignetting correction. The resolution of final images and acquired profiles is 4 μm.

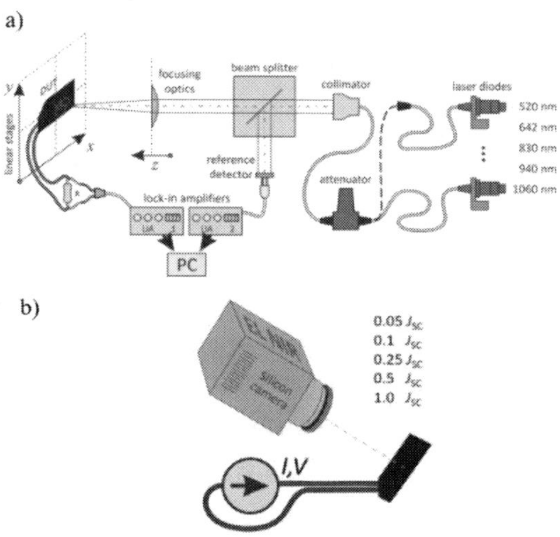

Figure 1: Schematic of the measurement systems. a) LBIC and b) EL system.

2.4 Results preparation

Front and back LBIC scans and front and back EL images of a cut SHJ cell are presented in Figure 2 a and b, respectively. LBIC images acquired by excitation wavelengths of 940 nm and 642 nm are chosen for the front and the back side, respectively. EL images were acquired at one quarter of short circuit current. The locations of edge profiles are labelled in the figures. TLS and L&C cut edge profiles were acquired parallel to contact fingers, while natural edge profiles were acquired perpendicular to the fingers, and cleaved edge profiles were acquired under 45° angle against the fingers, because cleaved edge forms in the orientation of the crystal structure. The latter two profiles (cleaved and natural) therefore include sharp drops, which correspond to electrical fingers either on the front or the back side. The L&Cb profiles were acquired on the other cell (images not shown here).

All profiles in Figure 3 and Figure 4 are presented in normalized values to better compare the edge phenomena. All profiles are also oriented in the same direction and aligned, so that zero coordinate is set at the assumed cell edge. For LBIC, the cell edge was determined as a point where the linear fit to the steepest part of the 642 nm profile crosses the y-axis, while for EL, the cell edge was determined as a point where the linear fit to the steepest part of the 0.25 J_{SC} profile crosses the linear fit of the background values.

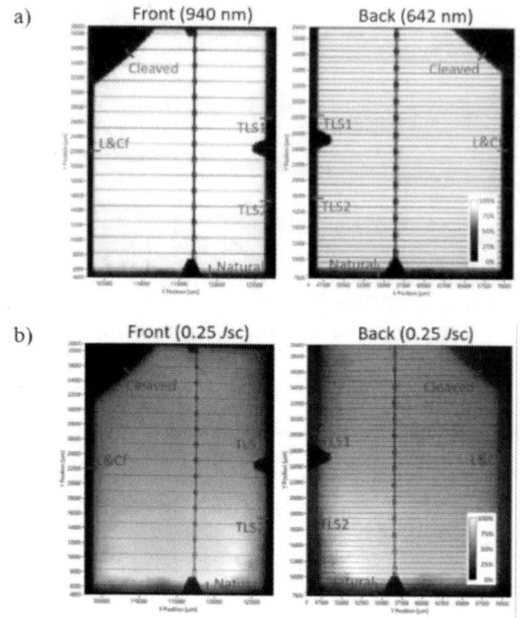

Figure 2: EL and LBIC images with labelled profile locations. a) LBIC images acquired from the front at 940 nm and from the back at 642 nm excitation wavelength. b) EL images acquired at one quarter of short-circuit current injection from the front and from the back.

3 RESULTS AND DISCUSSION

3.1 Wavelength dependent LBIC

Figure 3 presents selected normalized LBIC profiles of cell edges. Subfigures a) and b) enable comparison between different edge types, acquired from a) front and b) back side with 642 nm beam. The third figure shows how L&Cf cut influences front LBIC profiles at different excitation wavelengths.

All profiles acquired from the back (Figure 3 b) are steeper than corresponding profiles acquired from the front (Figure 3 a), because, when illuminated from the back, the minority charge carriers are generated closer to the main junction, while they need to traverse the whole bulk when illuminated from the front and thus have a higher probability of recombining at the (cut) edge.

TLS and L&Cf cuts perform the best and identically for all wavelengths (except 1060 nm) when illuminated from the front, while there are some minor discrepancies when illuminated from the back.

Cleave only edge profile is slightly less steep when acquired from the front (Figure 3 a), however when acquired from the back (Figure 3 b) the steepness is on par with TLS and L&Cf. This discrepancy is attributed to unknown cleave angle.

Natural edge results in profiles with a very gradual rise and they also behave very similarly for illuminations from front or back (not the case for other edges), which is a consequence of the TCO masking near the edge, due to both, optical and electrical influences.

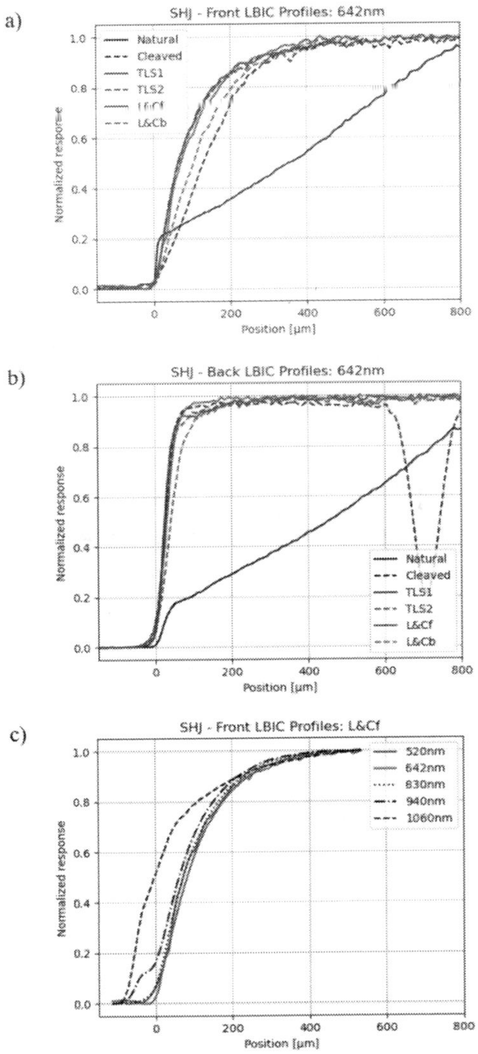

Figure 3: Normalized LBIC edge profiles of SHJ cells: - of different cuts at 642 nm from a) front side and b) back side, - at different wavelengths from the front for c) L&Cf cut.

Comparing front LBIC profiles of the same cut at different wavelengths (L&Cf shown in Figure 3 c) reveals that the profiles are in general insensitive to wavelength for wavelengths with shallower penetration depths between 520 and 940 nm for the scans acquired from the front side. However, this is not the case for the profiles acquired from the back (not shown), where some wavelength dependence is visible and is correlated to different penetration depths of light of different wavelengths.

3.2 Injection dependent EL

Figure 4 presents selected normalized EL profiles of cell edges. Subfigures a) and b) show comparison between different edge types, acquired from the front and the back side, respectively. They were acquired at a short-circuit injection and enable comparison between different edge types. The subfigure c) shows how the front EL profiles of L&Cf cut are influenced by different injection current densities (Other cut types show very similar dependence).

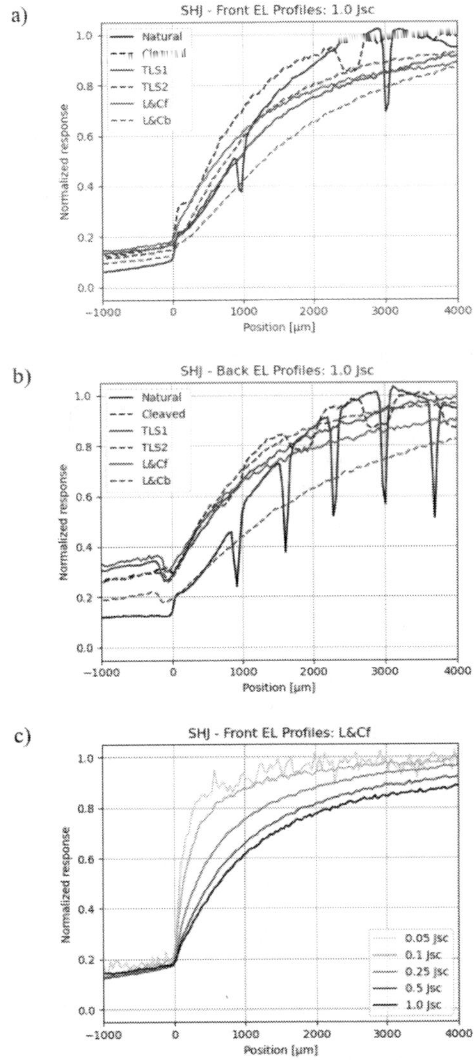

Figure 4: EL edge profiles of SHJ cells: - of different cuts at 1 J_{SC} from a) front side and b) back side, - at different currents from front side for c) L&Cf cut.

Cleaved edge has the sharpest cut off for all EL measurements (Figure 4 a and b). TLS and L&Cf cuts show very similar edge profiles from the front side (Figure 4 a), whereas from the back side L&Cf performs slightly worse (Figure 4 b). The L&Cb cut shows the worst performance (Figure 4 a and b) because the laser ablation damages the main junction at the back of the cells.

The normalized EL profiles are less sharp at higher injection current for all edge types (L&Cf shown in Figure 4 c).

3.3 LBIC vs. EL

Comparison between LBIC(λ) (Figure 3 c) and EL(J) (Figure 4 c) profiles of SHJ cells shows that LBIC profiles are much steeper (mind the x-axis scale) and that the lower the injection, the higher the steepness of EL profiles. Since LBIC profiles are acquired at overall very low injection, we can conclude that LBIC represents a limit for EL profiles at zero injection current, which we also observed for IBC cells [5].

4 SUMMARY AND CONCLUSION

We have used injection dependent EL and wavelength dependent LBIC to investigate cut edge recombination loss of SHJ solar cells. While LBIC is proportionate to EQE and provides a straightforward evaluation of the results, it only reveals one part of the story – it is acquired at very low illumination and close to thermal equilibrium. EL on the other hand shows the second part of the story. Here, the concentration of free charge carriers is significantly higher and the applied voltage is on the level of open circuit voltage, thus the circumstances are much closer to the actual operation point of solar cells under illumination. Therefore, the EL results are more relevant when trying to understand the edge recombination effects under real operating conditions.

Regarding the different cutting techniques, the measurements show that TLS and L&Cf perform similarly, but since TLS offers better potential for subsequent repassivation it is the preferred technique. Cleave only technique also provides very good results, especially at EL, but it requires that wafers' crystal orientation must be rotated for 45° prior to finalizing the cell manufacturing. Also, for the simplest L&C cutting technique the laser ablation must not be performed on the side of the active junction because it introduces edge recombination in the most sensitive region of the device, which leads to the highest edge recombination losses.

We have shown that injection-dependent EL and wavelength-dependent LBIC measurements provide additional information about efficiency losses due to cut-edge recombination. It allows more precise characterization and deeper understanding of physics behind edge recombinations. Wavelength dependent LBIC enables determination of quantum efficiency versus light penetration depth, whereas injection dependent EL takes into account the voltage distribution as well, and illustrates the influence of cut edge recombination at normal operating point of solar cells.

5 ACKNOWLEDGMENTS

The authors acknowledge 3D Micromac for cutting the cells using the TLS technique. This work was supported by the HighLite project that has received funding from the European Union's Horizon 2020 research and innovation programme under grant agreement No 857793 and by the Slovenian Research Agency under the programme P2-0415.

6 REFERENCES

[1] W. Long, S. Yin, F. Peng, M. Yang, L. Fang, X. Ru, M. Qu, H. Lin, X. Xu, *Solar Energy Materials and Solar Cells* **2021**, *231*, 111291.

[2] L. Tous, J. Govaerts, S. Harrison, F. Buchholz, A. Halm, A. Faes, G. Nogay, A. Ingenito, F.-J. Haug, F. Feldmann, D. Raine, T. Fellmeth, M. Heinrich, M. Mittag, D. Reinwand, F. Haase, A. Morlier, M. Bokalič, K. Brecl, M. Topic, J. C. P. Kester, S. Wendlandt, M. Galiazzo, A. Voltan, G. Galbiati, L. Theunissen, F. Torregrosa, M. Grimm, J. Denafas, T. Radavicčius, P. Lukinskas, J. Kaakkunen, T. Savisalo, T. Regrettier, C. Carrière, *38th European Photovoltaic Solar Energy Conference and Exhibition* **2021**, 533.

[3] S. Eiternick, F. Kaule, H.-U. Zühlke, T. Kießling, M. Grimm, S. Schoenfelder, M. Turek, *Energy Procedia* **2015**, *77*, 340.

[4] J. F. Lelièvre, S. Harrison, M. Albaric, L. Carton, B. Portaluppi, V. Barth, *37th European Photovoltaic Solar Energy Conference and Exhibition* **2020**, 487.

[5] M. Bokalic, M. Kikelj, K. Brecl, M. Jankovec, F. Buchholz, V. Mihailetchi, M. Topic, *37th European Photovoltaic Solar Energy Conference and Exhibition* **2020**, 308.

ACCURATE LINEARITY MEASUREMENTS OF THE SHORT-CIRCUIT CURRENT USING A SPECTRAL SHAPING SETUP

Matthias Mühleis, J. Kasimir Reichmuth, Michael Rauer, Jochen Hohl-Ebinger
Fraunhofer Institute for Solar Energy Systems ISE
Heidenhofstraße 2, 79110 Freiburg
Phone +49 761 4588-5359, matthias.muehleis@ise.fraunhofer.de

ABSTRACT: An accurate quantification of the linearity of the short-circuit current with respect to the irradiance is an integral part in the electrical characterisation of solar cells. It is especially important for the calibration of irradiance sensors and reference devices. We demonstrate that an in-house developed spectral shaping setup can reproduce the global AM1.5 solar reference spectrum with a high match. The generated irradiance of < 1 W/m² is perfectly adjusted to conduct a dual beam configuration with a differential measurement technique to measure short-circuit currents in the irradiance range from 0 W/m² to over 1000 W/m². We also investigate the short-circuit current spectrally dependent owing to the freely programmable spectral shaping setup with approximately 1088 accessible spectral subsections in the wavelength range from 355 nm to 1200 nm. Deviations from linearity below 0.1 % are detected on a WPVS solar reference cell demonstrating the high sensitivity of the presented setup. A rear side of a PERC solar cell was investigated to demonstrate the measurement of a strong spectrally dependent nonlinearity of 40 %.
Keywords: Characterisation, Calibration, Solar Radiation

1 INTRODUCTION

Reference solar cells are widely used in the photovoltaics community to determine irradiance levels of natural and artificially generated sunlight [1]. In this context, an accurate characterisation of the short-circuit current's linearity with respect to the irradiance is essential. Additionally, deviations from linearity are an important aspect for the prediction of the energy yield of solar cells because the yearly naturally occurring irradiance levels are widely distributed.

This work presents results of using a spectral shaping setup for accurate short-circuit current measurements at various irradiance levels of up to 1000 W/m². The spectral shaping setup can reproduce a variety of spectra with distinct spectral features, including the global AM1.5 solar reference spectrum [2] with a high spectral match in the wavelength range from 355 nm to 1200 nm [3]. The generated irradiance of < 1 W/m² of this spectral shaping setup is perfectly suited to apply a differential, dual beam method [4, 5] at which additionally installed bias lamps irradiate the solar cell with a tunable irradiance to over 1000 W/m². In this White Light Response (WLR) method [5], the first beam, generated by the spectrally shaped radiation, determines the slope of the short-circuit current versus irradiance curve. The purpose of the second beam, generated by the bias lamps, is to determine this slope at different irradiance levels. The linearity determination is thus an integral part of the WLR method.

In comparison to linearity measurements with a sun simulator [6], the N-lamp method [1] or a WLR setup without the spectral shaping ability [5], the WLR setup, presented in this study, benefits from the very well-matching spectrum, as deviations to the reference spectrum can cause errors or time-consuming corrections. Whilst the accuracy of the new WLR setup is seen as comparable to a Differential Spectral Responsivity (DSR) setup in linearity measurements [7], the WLR setup is approximately 100 times faster than the DSR setup by omitting highly spectrally resolved measurements.

However, we can gain an insight into a spectrally dependent nonlinearity by reproducing subsections of the AM1.5 spectrum and investigating the corresponding short-circuit currents.

In this paper, we present a brief introduction to the theory, the experimental setup with the spectral shaping setup and the application of this setup at two test devices. The ability to quickly measure a very high linearity is demonstrated at an exemplarily chosen WPVS solar reference cell which is presented in this work for the first time. Additionally, we demonstrate a new method which provides insights into the spectrally dependent nonlinearity at a nonlinear test solar cell.

2 METHODS

2.1 Theory

In this work, we use the WLR method [5] combined with a new optical spectral shaping setup to determine the short-circuit current of a solar cell at various irradiance levels. The WLR method uses a similar approach as the DSR method [4]. The WLR method utilizes a dual beam configuration at which one beam has a low irradiance but a well-matched spectrum in comparison to the target reference AM1.5 spectrum. This beam has a chopped irradiance with a known frequency generating a small alternating current signal which is precisely detected by a lock-in amplifier. The second beam of the dual beam configuration is generated by additional bias lamps which produce a steady irradiance from 0 W/m² to over 1000 W/m² leading to the measurable steady bias short-circuit currents I_{b}.

The measured alternating current signal ΔI, which has been previously calibrated with a reference device, is used to determine the AM1.5-weighted differential responsivity $\tilde{s}_{\mathrm{AM1.5}}$ which describes the slope of the short-circuit current versus irradiance $I(E)$ curve. The $\tilde{s}_{\mathrm{AM1.5}}$ value is determined at various bias irradiance levels and thus, bias currents I_{b}:

$$\tilde{s}_{\mathrm{AM1.5}} = \left.\frac{\Delta I(E)}{\Delta E}\right|_{E_{\mathrm{b}}=E_{\mathrm{b,0}}(I_{\mathrm{b,0}})} . \quad (1)$$

A perfect linear solar cell has irradiance-independent $\tilde{s}_{\mathrm{AM1.5}}$ values.

The short-circuit current $I(E)$ at a specific irradiance E level is determined by integration [4, 8]:

$$E = \int_0^{I(E)} \frac{1}{\tilde{s}_{\mathrm{AM1.5}}} dI_{\mathrm{b}} . \quad (2)$$

The linearity of the short-circuit currents with respect

to the irradiance is quantified with the parameter percentage deviation from linearity NL [8]:

$$NL = \left(\frac{I(E)}{E}\frac{E_{STC}}{I_{STC}} - 1\right) \cdot 100 \tag{3}$$

The parameter NL describes the deviation of the measured short-circuit currents from a linear relation with the same short-circuit current I_{STC} at $E_{STC} = 1000$ W/m². NL is zero at 1000 W/m² by definition. For a linear device, NL is zero for other irradiance levels as well. For a nonlinear device, NL differs from zero.

The propagation of the uncertainties of the measured AM1.5-weighted differential responsivity $\tilde{s}_{AM1.5}$ values to the uncertainties of the short-circuit currents and the parameter NL is conducted with the Monte Carlo method [9] which allows to consider manifold correlations.

2.2 Experimental setup

The experimental setup involves a dual beam setup at which the chopped light beam is generated by a spectral shaping setup and the other steady light beam is generated by steady bias lamps as illustrated in Fig. 1.

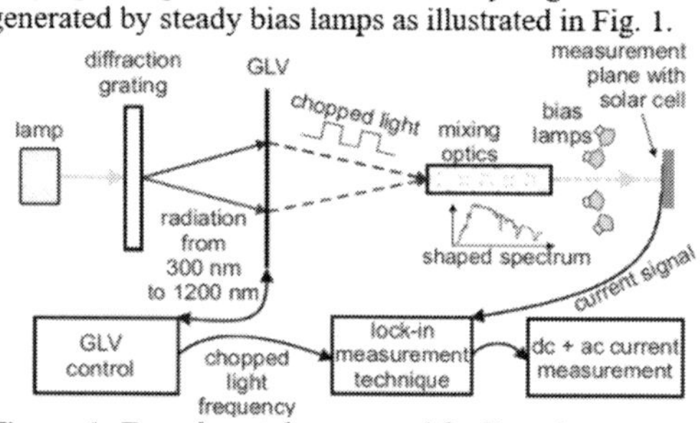

Figure 1: Experimental setup used for linearity measurements of the short-circuit current with respect to the irradiance. Reused from [7].

Cold mirror halogen lamps with less requirements on the spectrum [10] are installed as bias lamps.

The spectral shaping setup consists of a laser-driven lamp with a high radiance, a diffraction grating and a grating light valve (GLV) from Silicon Light Machines [11]. The GLV is a programmable phase diffraction grating and is used as a spectral shaping tool. Details of this setup are given in [3]. The optical setup can reproduce spectra in the wavelength range from 355 nm to 1200 nm with a spectral resolution of 7 nm to 15 nm. The large number of 1088 independently addressable segments of the GLV and the wavelength-dependent irradiation of these segments enables us to reproduce many different spectra with sharp spectral features.

Further details about the measurement conditions, uncertainties and equipment can be found in [7].

2.3 Reproduced spectral irradiance distributions

As seen in Fig. 2, the global AM1.5 solar reference spectrum can be reproduced with a high match in the wavelength range from 355 nm to 1200 nm. The irradiance between 355 nm and 400 nm is slightly boosted so that a typical broadband silicon solar cell would generate the same current under the AM1.5 reference spectrum as under the shaped spectrum in the wavelength range from 300 nm to 400 nm. The spectral mismatch factor [12] of this generated spectrum to the AM1.5 spectrum is equal to unity within its uncertainty of approximately 0.1 % for a wide range of spectral responsivity curves [13]. Thus, a spectral mismatch correction is not necessary.

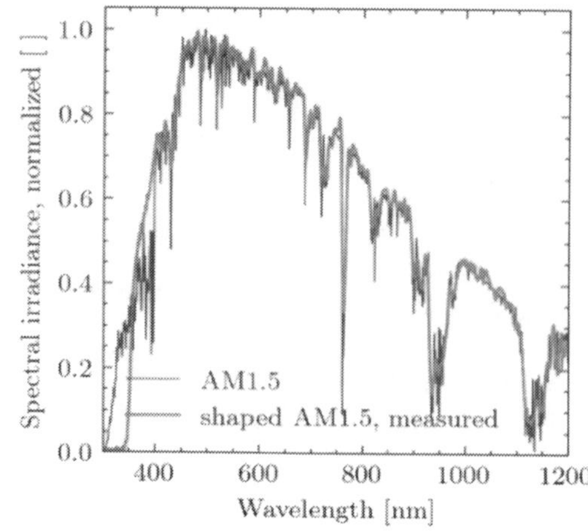

Figure 2: Global AM1.5 solar reference spectrum and the measured, shaped spectrum generated with our optical spectral shaping setup installed at the CalLab PV Cells at Fraunhofer ISE. The high optical resolution of the setup enables a high spectral match.

Additionally, we can use the spectral shaping ability with a high resolution of our setup to generate subsections of the AM1.5 spectrum with abrupt transitions as shown in Fig. 3. The number of subsections

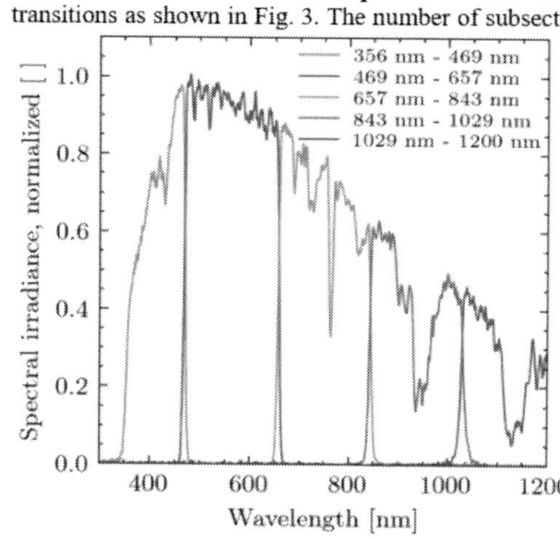

Figure 3: Spectral subsections of the global AM1.5 solar reference spectrum. The subsections can be generated separately and are applied to a solar cell to characterise a possible spectrally dependent nonlinearity.

and the wavelength boundaries of these subsections can be freely chosen in a wide range.

The presented subsections are generated separately and applied to a test solar cell consecutively. This can be used to detect spectrally dependent nonlinearities in a short time. The measurement time of one data point, generated from the shaped AM1.5 spectrum or of a subsection of the AM1.5 spectrum, was approximately 2 - 10 seconds in this work.

2.4 Test devices

Two different test devices were used in this work. The ability of the presented setup to quantify a high linearity at many adjustable irradiance values from 0 W/m² to over 1000 W/m² in a short time is demonstrated at a WPVS solar reference cell [14]. The ability of the setup to characterise a spectrally dependent nonlinearity is exemplarily shown at a rear side of a piece of a PERC solar cell. In [7, 13], it has been published that this cell is spectrally nonlinear which was characterised by using the Differential Spectral Responsivity (DSR) setup at the Physikalisch-Technische Bundesanstalt (PTB) [10] and at the CalLab PV Cells at Fraunhofer ISE. Additionally, it was reported that the DSR results were consistent with the results of the WLR setup. In this work, we use the generated spectral subsections of the AM1.5 spectrum (Fig. 3) to quantify the spectral nonlinearity.

3 RESULTS & DISCUSSION

3.1 Measuring the high linearity of a WPVS solar reference cell

In Fig. 4, the AM1.5-weighted differential responsivity $\tilde{s}_{\mathrm{AM1.5}}$ values are shown as a function of the bias current. The relative difference of the $\tilde{s}_{\mathrm{AM1.5}}$ values

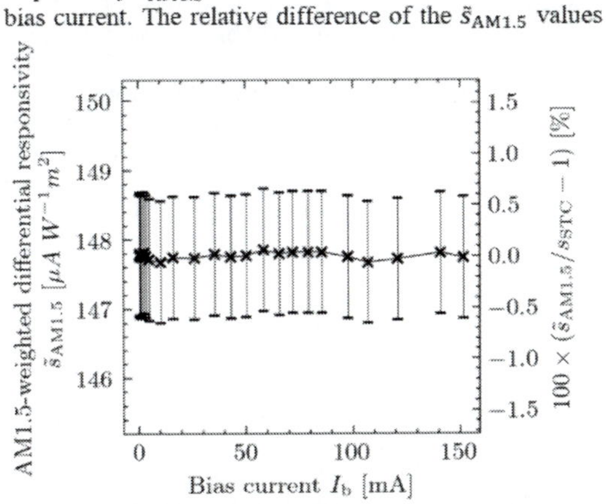

Figure 4: Measured AM1.5-weighted differential responsivities $\tilde{s}_{AM1.5}$ of a WPVS reference solar cell. The relative differences (right axis) of these values to the derived s_{STC} value remain below 0.1 % indicating a very linear solar cell. Uncertainty values refer to the left axis.

to the s_{STC} value is also shown (right axis). The s_{STC} value is the short-circuit current I_{STC} at 1000 W/m² divided by the irradiance of 1000 W/m². The assigned uncertainty values (coverage factor k=2) refer to the left axis with absolute values. The uncertainties of the relative differences would be significantly smaller. The reason is that the uncertainty components which include correlations between the $\tilde{s}_{\mathrm{AM1.5}}$ values cancel out in calculating the relative differences between the $\tilde{s}_{\mathrm{AM1.5}}$ values and the s_{STC} value. The correlated uncertainty components propagate to the s_{STC} value.

The $\tilde{s}_{\mathrm{AM1.5}}$ values remain within a range of approximately 0.1 % around the s_{STC} value. Consequently, it can be stated that this solar cell is highly linear.

The short-circuit current versus irradiance curve is calculated with Eq. (2) using the measured $\tilde{s}_{\mathrm{AM1.5}}$ values and is shown in Fig. 5. Due to the short measurement time of approximately 2 - 10 seconds for each data point, it was possible to determine 29 data points on this short-circuit current versus irradiance curve in a reasonable time.

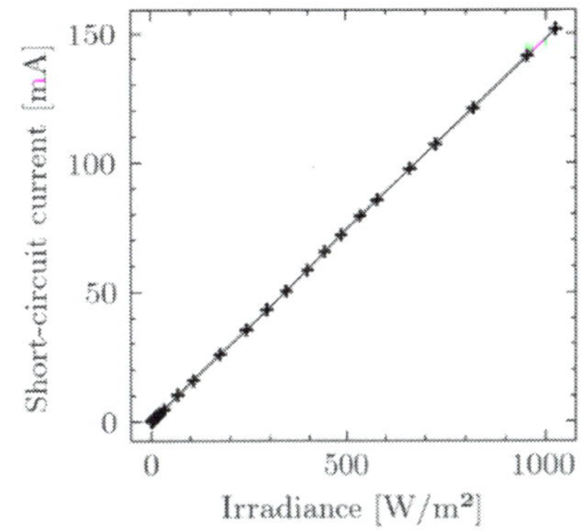

Figure 5: Short-circuit currents as a function of the irradiance. Test device: WPVS reference solar cell. Many measurement points (29) could be determined owing to the fast WLR method.

To take a further look into the linearity of this short-circuit current versus irradiance curve, we calculate the nonlinearity parameter NL with Eq. (3). As seen in Fig. 6, the NL parameter is precisely determined and remains below 0.1 % in the whole irradiance range for the tested WPVS solar reference cell. The assigned uncertainties

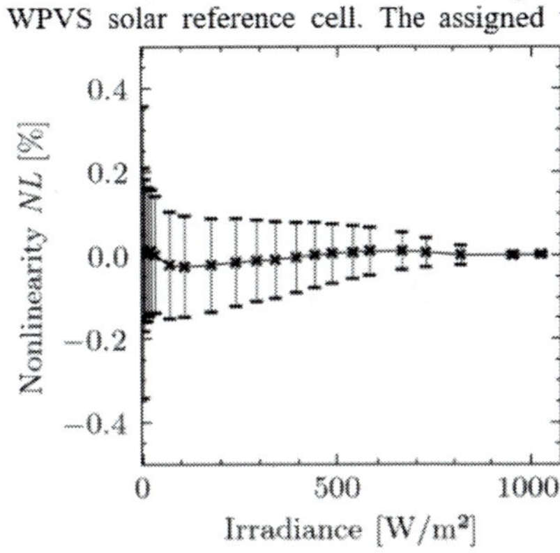

Figure 6: Nonlinearity values for a WPVS reference solar cell which describe the deviations of the short-circuit current versus irradiance curve in Fig. 5 from a straight line. The values are close to zero indicating a very linear solar cell.

(k=2) decrease with rising irradiance values due to the involved correlations between the $\tilde{s}_{\mathrm{AM1.5}}$ values and the used Eq. (2) and (3) which are considered during the conducted Monte Carlo method. These findings confirm that a high linearity could be measured with a low measurement uncertainty.

3.2 Reproducing subsections of the AM1.5 spectrum to detect spectral nonlinearities

The short-circuit current versus irradiance curve of the rear side of a PERC solar cell is shown in Fig. 7 (black curve). Additionally, the fractions of the short-circuit currents corresponding to the spectral subsections of the AM1.5 spectrum (Fig. 3) are plotted (colored curves). The sum of these colored curves would match with the black curve. The fractions of the short-circuit currents are assigned to equivalent irradiance levels which consider the whole AM1.5 spectrum.

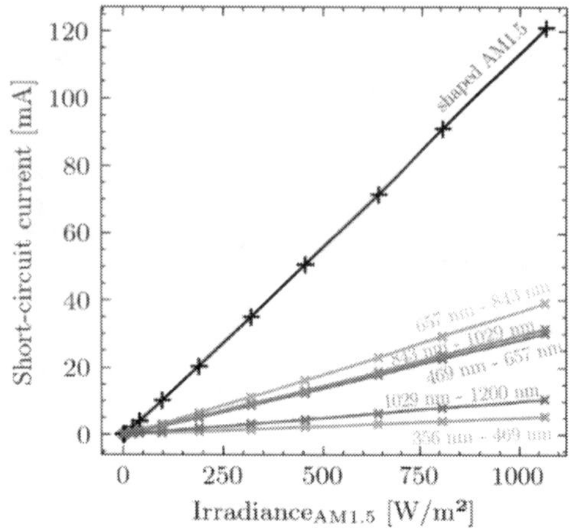

Figure 7: Short-circuit current as a function of the AM1.5 spectrum equivalent irradiance. Test device: Rear side of a PERC solar cell piece. The short-circuit currents generated by the AM1.5 spectrum at different irradiance levels and the fractions of the short-circuit currents corresponding to the generated subsections of the AM1.5 spectrum (Fig. 3) are plotted.

On the first sight, the short-circuit current versus irradiance curve seems to be linear. However, a closer analysis reveals that this solar cell is strongly nonlinear with respect to the irradiance as illustrated in Fig. 8.

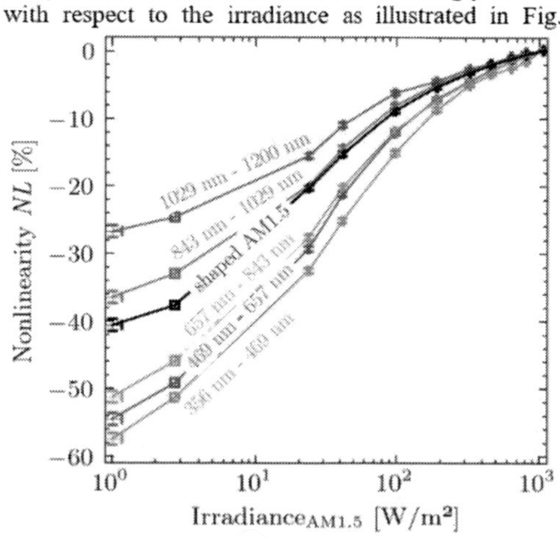

Figure 8: Nonlinearity values produced through the AM1.5 solar reference spectrum (black curve) and spectral subsections of it. Test device: Rear side of a PERC solar cell. The nonlinearity value depends on the irradiance and on the chosen spectral subsection indicating a spectral nonlinearity.

Deviations from linearity of up to 40 % are observed considering the whole AM1.5 spectrum.

It is evident that the nonlinearity is also dependent on the generated spectral subsection of the AM1.5 spectrum. A stronger nonlinearity is observed towards shorter wavelengths. As a result, we can state that the spectral nonlinearity of this solar cell can be detected and quantified with the presented setup in a short measurement time.

4 CONCLUSIONS

Short-circuit current measurements at different irradiance levels from 0 W/m² to over 1000 W/m² were conducted on two different solar cells. The presented spectral shaping setup can reproduce the global AM1.5 solar reference spectrum with a high spectral match. The generated irradiance of < 1 W/m² for this spectrum is perfectly suited to apply a dual beam approach with a differential measurement technique.

An exemplarily chosen WPVS solar reference cell was characterised to be highly linear. Deviations from linearity were < 0.1 %. The short measurement time for one data point of approximately 2 - 10 seconds enabled us to determine 29 data points on the short-circuit current versus irradiance curve in a reasonable measurement time. The strong nonlinearity of up to 40 % of a rear side of a PERC solar cell was also measured considering the AM1.5 spectrum. By reproducing several subsections of the AM1.5 spectrum and applying these to the solar cell consecutively, we observed that the nonlinearity of the short-circuit current depends on the chosen spectral subsections.

We conclude that the large number of possible data points, owing to the fast measurement technique, allows to have a detailed insight into irradiance-dependent short-circuit current curves. The results of this study demonstrate that our presented setup is highly sensitive in quantifying linearities.

5 REFERENCES

References
[1] H. Müllejans and E. Salis, "Linearity of photovoltaic devices: quantitative assessment with N-lamp method," *Measurement Science and Technology*, vol. 30, no. 6, p. 65008, 2019.
[2] *Photovoltaic devices – Part 3: Measurement principles for terrestrial photovoltaic (PV) solar devices with reference spectral irradiance data*, IEC 60904-3, IEC 60904-3, Feb. 2019.
[3] M. Mühleis and J. Hohl-Ebinger, "Programmable spectral shaping demonstrated at the solar spectral irradiance distribution," *Opt. Express*, vol. 29, no. 6, pp. 8223–8234, 2021, doi: 10.1364/OE.417790.
[4] J. Metzdorf, "Calibration of solar cells. 1: The differential spectral responsivity method," *Applied Optics*, vol. 26, no. 9, pp. 1701–1708, 1987.
[5] J. Hohl-Ebinger, Siefer Gerald, and W. Warta, "Non-linearity of solar cells in spectral response measurements," in *22nd EU PVSEC*, Milano, 2007, pp. 422–424.
[6] M. Bliss *et al.*, "Interlaboratory comparison of short-circuit current versus irradiance linearity measurements of photovoltaic devices," *Solar Energy*, vol. 182, pp. 256–263, 2019.
[7] M. Mühleis, I. Kröger, and J. Hohl-Ebinger, "Fast and accurate short-circuit current versus irradiance determination of a spectrally nonlinear solar cell using a spectral shaping setup," *Solar Energy Materials and Solar Cells*, vol. 230, p. 111208, 2021, doi: 10.1016/j.solmat.2021.111208.
[8] *Photovoltaic devices – Part 10: Methods of linear dependence and linearity measurement*, IEC 60904-10, IEC 60904-10, Sep. 2020.
[9] *Evaluation of measurement data — Supplement 1 to the "Guide to the expression of uncertainty in*

measurement" — Propagation of distributions using a Monte Carlo method, JCGM 101:2008, BIPM, IEC, IFCC, ILAC, ISO, IUPAC, IUPAP and OIML, 2010.
[10] I. Kröger, D. Friedrich, and S. Winter, "Calibration of solar cells beyond STC using the DSR method," in *IEEE 7th World Conference on Photovoltaic Energy Conversion (WCPEC) (A Joint Conference of 45th IEEE PVSC, 28th PVSEC & 34th EU PVSEC)*, 2018, pp. 3492–3497.
[11] D. T. Amm and R. W. Corrigan, "Optical performance of the Grating Light Valve technology," in *Projection Displays V*, 1999, pp. 71–78.
[12] *Photovoltaic devices – {P}art 7: Computation of the spectral mismatch correction for measurements of photovoltaic devices*, IEC 60904-7, IEC 60904-7, Aug. 2019.
[13] M. Mühleis, I. Kröger, and J. Hohl-Ebinger, "Highly resolved spectral shaping for fast and accurate solar cell characterization," in *2021 IEEE 48th Photovoltaic Specialists Conference (PVSC)*, 2021, pp. 48–52.
[14] C. R. Osterwald *et al.*, "The world photovoltaic scale: an international reference cell calibration program," *Progress in Photovoltaics: Research and Applications*, vol. 7, no. 4, pp. 287–297, 1999.

ACCURATE HIGH THROUGHPUT CURRENT-VOLTAGE CLASSIFICATION OF HIGH EFFICIENCY SILICON SOLAR CELLS WITH RAPIDWAVE®

B. Mitchell, B. Klöter, D. Clugston, B. Mette, H. Wagner-Mohnsen, S. Esefelder
WAVELABS Solar Metrology Systems GmbH
Spinnereistr. 7, 04179 Leipzig, Germany

ABSTRACT: Modern manufacturing of advanced PERC, TOPCon or HJT silicon solar cells results in high excess minority carrier lifetimes in the finished device. Such cells are inherently slower to respond to transients in terminal voltages, leaving steady-state current-voltage sweeps incompatible with today's industrial throughput requirements. This work discusses the use of the RapidWAVE® method implemented in WAVELABS SINUS LED cell testers to address the requirements for throughput without sacrificing accuracy, repeatability, and reproducibility of the reported STC performance and sorting parameter. Using the capabilities of a controlled pulsed or steady-state LED light source, we verify millisecond RapidWAVE® with second long steady-state current-voltage sweeps on the same tool. This allows for a unique discussion of the accuracy of the RapidWAVE® method in correcting charging and discharging currents during rapid transients. A 20ms RapidWAVE® efficiency classification of the most challenging, i.e., highest voltage, heterojunction solar cells with an open circuit voltage of 745mV is demonstrated with negligible deviation from steady-state direct current-voltage sweep. RapidWAVE® enables comprehensive state-of-the-art mass production testing (IV front & rear, R_s, R_{sh}, I_{rev}, EL, EQE, SunsVoc, IR, others) realizing up to 5000 full cell equivalent pieces per hour on the most modern cell production lines. RapidWAVE® is also compatible with half or third cell measurements realizing up to 13500 cut cell measurements per hour.

1 INTRODUCTION AND AIM

Modern manufacturing of advanced silicon solar cells leads to high excess minority carrier lifetimes in the finished device due to advanced surface and contact passivation. Such cells are inherently slower to respond to transients in terminal voltages, placing steady-state current-voltage sweeps incompatible with today's industrial throughput requirements outlined by the ITRPV [1].

Fast transient current-voltage sweeps of such high minority-excess-carrier lifetime solar cells are strongly influenced by charging and discharging currents from the diffusion capacitance build-up in the device, when not in equilibrium with its terminals, as shown in Figure 1 for a reverse sweep in a PC1D simulation.

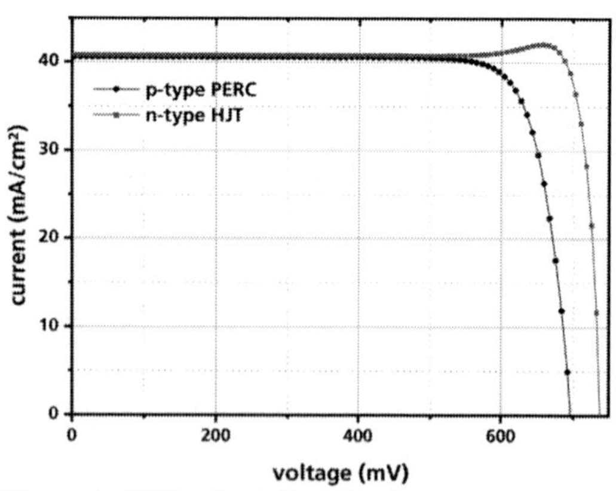

Figure 1: PC1D simulation of a fast reverse current-voltage sweep of a p-PERC and n-HJT solar cell

The aim of this paper is to present and discuss a new method for the ultra-fast IV measurement of high efficiency silicon solar cells affected by diffusion charge storage during fast terminal voltage transients [2] The RapidWAVE® method is implemented on the WAVELABS SINUS-300 a triple A+ solar simulator to the fulfilment of IEC 60904 [3]

RapidWAVE® uses the theoretical framework generalized by Sinton et al. [4], [5], [5]. Two transient sweeps of opposite directions are used to reconstruct the steady-state curve. It provides a quasi-steady-state curve not using any error prone diode model current-voltage curve fitting [6], [7] or any of the other published approaches including applying voltage pulses [8], slow dark-IV sweeps [9] or multiple flashes [10].

2 EXPERIMENT AND RESULTS

We use the steady-state illumination functionality of the SINUS-300 LED solar simulator and cell tester to verify the accuracy of RapidWAVE® sweeps. A true steady-state sweep is achieved when the distribution of excess carrier profiles within the cell arrived in the equilibrium state regarding the respective terminal voltage applied. Only in steady-state influences from charging or discharging the cell on terminal values are cleared.

For cells with high lifetime, i.e., high open circuit voltage, each current-voltage sample requires several tens of milliseconds to enter the equilibrium state, resulting in a total steady-state sweep time of up to several seconds in case of heterojunction cells.

Both illumination intensity and cell temperature of the cell must be well controlled or corrected across the experiment to obtain an accurate and reproducible reference. This was achieved by using a temperature stabilized measurement chuck, pre-measurement illumination, and monitoring with an IR spot pyrometer.

Here, we provide the data extracted from the most challenging device, i.e., a 745mV heterojunction silicon solar cell, using the SINUS-300 without recontacting the cell as illustrated in Figure 2. A sequence of direct reverse sweep of various sweep rates was followed by RapidWAVE® sweeps of 20 to 80ms. The latter were repeated 30 times to also include the repeatability of the measurement.

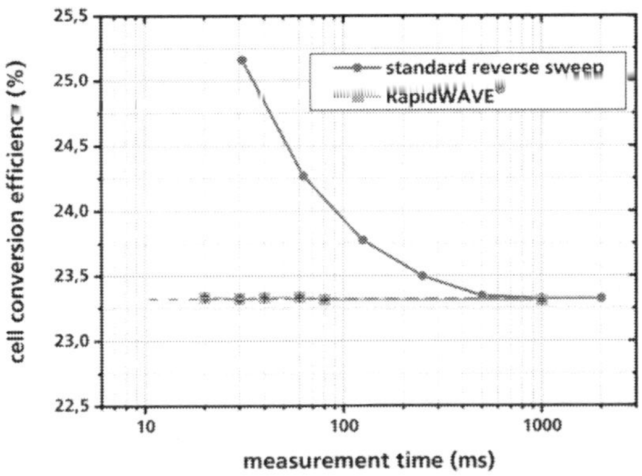

Figure 2: Measured efficiency of a high voltage heterojunction solar cell as a function of IV classification (log-scale) time for direct and RapidWAVE® sweeps. 30 repeats of the RapidWAVE® sweeps are included (see error bars). All data was extracted on the same SINUS-300 cell tester without recontacting.

RapidWAVE® accelerates such highly demanding current-voltage sweeps to 20ms for a complete sweep as shown in Figure 3. However, it not just maintains the accuracy and precision on the main IV parameter, i.e., short circuit current, open circuit voltage and fill factor, but also precisely reproduces the full steady-state IV curve.

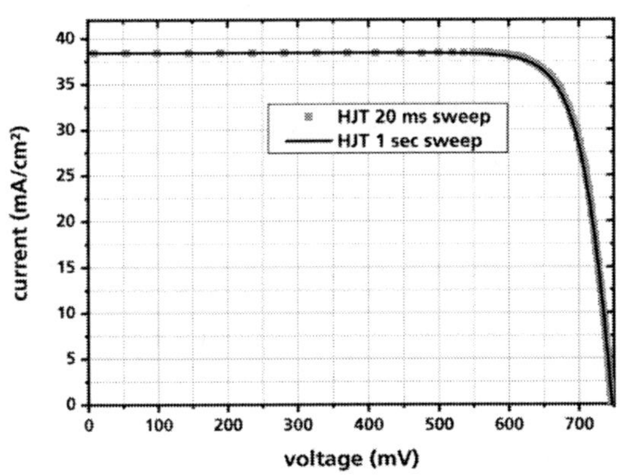

Figure 3: 20ms RapidWAVE® current-voltage curve of a high voltage heterojunction solar cell coinciding with a 1 second steady-state direct sweep.

3 CONCLUSION

RapidWAVE® introduces a new accurate, precise and ultra-fast method for the current-voltage measurement of modern highest efficiency silicon solar cells. It is available for all WAVELABS silicon cell testing equipment.

In this work, we demonstrated the methods accuracy and precision on the most demanding silicon heterojunction solar cell. RapidWAVE® maintains the tight repeatability specification for high lifetime cells such as advanced PERC, TOPCon and HJT in mass-manufacturing.

20ms current-voltage sweeps allow for a throughput of 5000 cells per hour even when including some additional measurements such as EL, IR, SunsVoc, EQE during end-of-line testing using the SINUS-300 LED cell tester.

RapidWAVE® is compatible with half or third cell measurements available with SINUS-300, thus realizing up to 13500 cut cell measurements per hour.

RapidWAVE® is in use at GW scale. In addition, ultra-fast measurement results in minimal temperature change on the cells and can therefore also provide an easier path to high-precision measurements in the laboratory when evaluating long-life solar cells.

8 REFERENCES

[1] VDMA, "International Technology Roadmap for Photovoltaic (ITRPV) - 2021 Results," 2022.

[2] D. L. King, J. M. Gee, and B. R. Hansen, "Measurement precautions for high-resistivity silicon solar cells," *Conference Record of the IEEE Photovoltaic Specialists Conference*, vol. 1, pp. 555–559, 1988, doi: 10.1109/pvsc.1988.105763.

[3] IEC, *IEC 60904-9-Ed3 Classification of solar simulator characteristics*. 2020.

[4] R. A. Sinton, H. W. Wilterdink, and A. L. Blum, "Assessing Transient Measurement Errors for High-Efficiency Silicon Solar Cells and Modules," *IEEE J Photovolt*, vol. 7, no. 6, pp. 1591–1595, Nov. 2017, doi: 10.1109/JPHOTOV.2017.2753200.

[5] H. Vahlman, J. Lipping, J. Hyvärinen, A. Tolvanen, and S. Hyvärinen, "Capacitive effects in high-efficiency solar cells during I-V curve measurement: Considerations on the error of correction," 2018. doi: 10.4229/35thEUPVSEC20182018-2AO.4.4.

[6] K. Ramspeck, S. Schenk, L. Komp, A. Metz, and M. Meixner, "Accurate Efficiency Measurements on Very High Efficiency Silicon Solar Cells," in *Proceedings of the 29th European Photovoltaic Conference*, 2014, pp. 1253–1256.

[7] S. Winter, "Analyse und Verbesserung der rückführbaren Kalibrierung von Solarzellen," 2003.

[8] A. Virtuani, G. Rigamonti, G. Friesen, D. Chianese, and P. Beljean, "Fast and accurate methods for the performance testing of highly-efficient c-Si photovoltaic modules using a 10ms single-pulse solar simulator and customized voltage profiles," *Meas Sci Technol*, vol. 23, no. 11, 2012, doi: 10.1088/0957-0233/23/11/115604.

[9] Y. Hishikawa, "ACCURATE I-V CURVE MEASUREMENTS OF HIGH-CAPACITY C-SI SOLAR CELLS: EFFECTS OF I-V TESTERS AND NEW TECHNOLOGY FOR RAPID MEASUREMENT," in *Proceedings of the 28th European Photovoltaic Conference and Exhibition*, 2013, vol. 1, pp. 3159–3162.

[10] R. Sinton, "A solution to the problem of accurate flash-testing of high-efficiency modules," *21st European Photovoltaic Solar Energy Conference*, no. September, pp. 634–638, 2006.

ROLE OF THE ELECTRON DARK CURRENT AND EARLY EFFECT IN THE DEGRADATION OF THE OPEN CIRCUIT VOLTAGE AND S-SHAPE DISTORTION IN a-Si:H(p)/c-Si HETEROJUCNTION SOLAR CELLS WITH SCHOTTKY CONTACT

Moustafa Y. Ghannam, Yaser M. Abdulraheem. Ivan Gordon[1], Jef Poortmans[1]
Electrical Engineering department, Kuwait University, Kuwait
[1]Imec, Kapeldreef 75, 3000 Leuven, Belgium

ABSTRACT: A bipolar equivalent circuit model is developed for silicon hetero-junction silicon solar cells (HJT) with a Schottky contact between the p-type amorphous silicon and the transparent conducting oxide (TCO). The Schottky barrier is treated as a bipolar collector junction through which holes are transported by thermionic emission, while electrons reaching the barrier are swiftly collected. The electron current is shown to exhibit an Early effect especially in cells with low a-Si:H(p) doping level. The results of the equivalent circuit analysis in the dark and under illumination perfectly match the *J-V* characteristics resulting from AFORS-HET device simulations of cells with low and with high doping levels in a-Si:H(p). The model provides a clear interpretation for the role of the Schottky diode in the degradation of the cell open circuit voltage and in the S-shape distortion of the *J-V* characteristics under illumination.

1 INTRODUCTION

In many cases, the TCO/a-Si:H(p) contact encountered in silicon HJT solar cells [1,2] does not ensure an ohmic behavior especially when the work function of the TCO material (WF_{TCO}) is relatively small, and/or when the a-Si:H(p) doping level is not high enough. In such cases, a Schottky barrier is formed at TCO/a-Si:H contact, which affects the cell series resistance, cell current in the dark and under illumination and causes the degradation of V_{OC}, of the *FF* and leads to a distorted S-shaped *J-V* characteristics under illumination [3-9]. The origin of the S-shape is not firmly established but has been previously attributed in several reports to different effects including high defect density in the a-Si:H bulk and interfaces with c-Si and with TCO, to values of a-Si:H energy bandgap and electron affinity not matching those of c-Si which causes conduction and valence band-offsets at the a-Si:H/c-Si hetero-interface, and more [e.g., 10-14]. On the other hand, circuit modeling has rarely been used to directly interpret the degradation and distortion of the cell performance although it groups different device properties into a limited number of measurable electrical parameters.

In the present work, an equivalent circuit model is proposed for the of HJT cell in which the Schottky contact is treated as a bipolar junction collector [15]. So far, such a specific attribute and its impact on the current voltage dependence have never been considered in different equivalent circuit models proposed for the solar cells. Fitting the current-voltage characteristics of the HJT cell obtained from AFORS-HET device simulations [16] in the dark and under illumination with the current-voltage characteristics obtained from the proposed equivalent circuit analysis is carried out and used to confirm the validity of the model and to evaluate its accuracy.

2 CELL STRUCTURE BAND DIAGRAM AND CURRENTS

2.1 Cell structure

The HJT cell structure with Schottky front contact under study is sketched in Fig.1. The device physical and geometrical properties are listed in Table 1. Two a-Si:H(p) doping levels are considered: 1) low doping with $N_{a\text{-}Si} = 10^{19}$ cm^{-3} and 2) heavy doping with $N_{a\text{-}Si} = 10^{20}$ cm^{-3}. The TCO work fucntion equals 4.9 eV but ideal flat band front contact is occasionally considered for comparison.

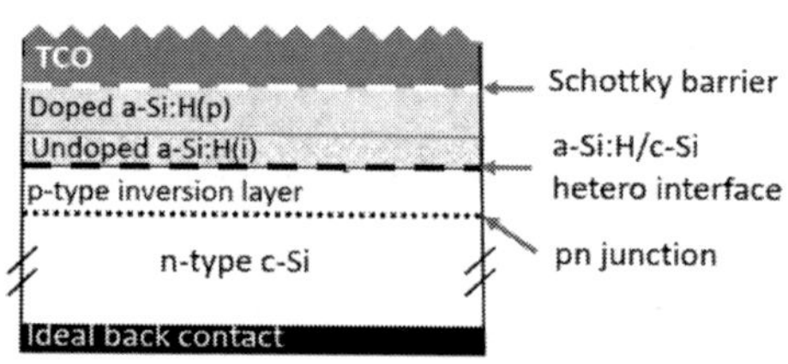

Fig.1a: Silicon HJT cell under study

Table I: Physical and geometrical properties of the cell under study

Layer	Description of property
SUBSTRATE	
Thickness	200 μm
Doping type, concentration	n-type, 1.5x10^{16} cm^{-3}
Carrier lifetime	1 ms
Back contact	Ideal ohmic, S=1cm/s
a-Si:H/c-Si interface	
Dangling bond shape, peak	Gaussian, 10^{10} cm^{-2}eV^{-1}
a-Si:H	
Energy band gap $E_{G\,a\text{-}Si}$	1.72 eV
Electron affinity	3.9 eV
UNDOPED a-Si:H(i)	
Thickness	5 nm
Dangling bond shape, peak	Gaussian, 1.6x10^{16} cm^{-3}eV^{-1}
Doped a-Si:H(p)	
Thickness	10 nm
Doping type, concentration	p-type, $N_{a\text{-}Si}$ cm^{-3}
Dangling bond shape, peak	Gaussian, $N_{a\text{-}Si}$ cm^{-3} eV^{-1}
TCO	
Work function	4.9 eV

2.2. Band Diagram and hole current

The simulated band diagram of the cell under study at

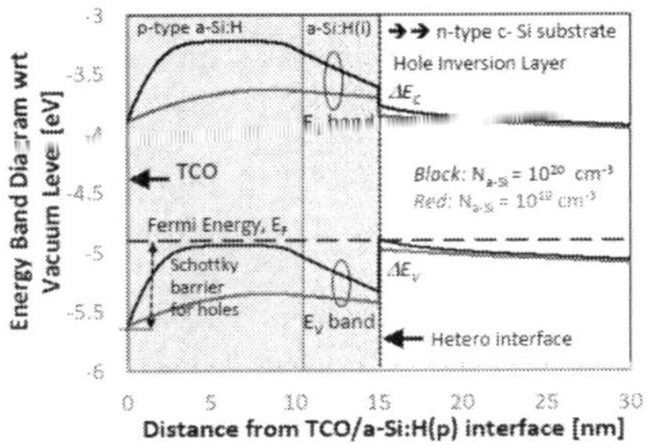

Fig,1b. Simulated energy band diagram of lowly and heavily doped cell at equilibrium.

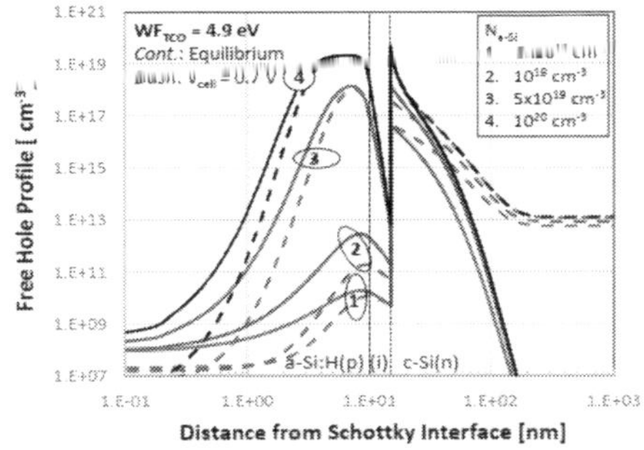

(b)

Fig.2. Free hole concentration profile in HJT cells with different values of N_{a-Si} simulated by AFORS-HET: (a) in the dark at equilibrium and (b) at V=0.7 V. Simulated profiles in cells with flat band front contact are plotted in (a) for comparison.

equilibrium is plotted in Fig.2. As previously reported the cell pn junction occurs inside the c-Si substrate between the p-type inversion layer and the c-Si substrate a few tens of nanometers away from the a-Si:H/c-Si hetero interface [17]. Due to band discontinuities, carriers have to cross such an interface by thermionic emission which may cause serious increase in the cell series resistance under illumination [18]. What is more relevant to the present study is the Schottky barrier created at the TCO/a-Si:H(p) interface. Such a barrier acts on the majority holes but is transparent for electrons. At forward biased in the dark, the cell is biased such that the positive polarity is applied to the TCO which forces the Schottky barrier to be reverse biased. Hence, the maximum hole current flowing through the Schottky barrier when the cell is forward biased is limited to the reverse Schottky saturation current, $J_{0,Sch}$. Under illumination, however, the photogenerated holes flows accross the Schottky barrier from its p-type a-Si:H(p) side hence following the forward bias Schottky current equation. Hole tunneling through the Schottky barrier may occur in the dark and under illumination if the doping level in a-Si:H(p) is degenerate or very high.

2.3 Electron current, hole depletion and Early effect.

As mentioned above, the Schottky barrier is transparent for electrons. In principle the electron dark current in the silicon HJT cell is negligible which is the reason for the enhanced cell voltage. Band bendings created by the Schottky barrier create a surface depletion region that extends deeply in the a-Si:H(p) layer especially when the doping level in a-Si:H(p) is low, as confirmed in the hole profiles simulated at equilibrium and displayed in Fig.2 (a). Such depletion of free holes occupies a large part of the neutral a-Si:H(p) layer which tends to enhance the electron dark current. In addition, the depletion region widens and extends deeper in a-Si:H(p) as the reverse bias

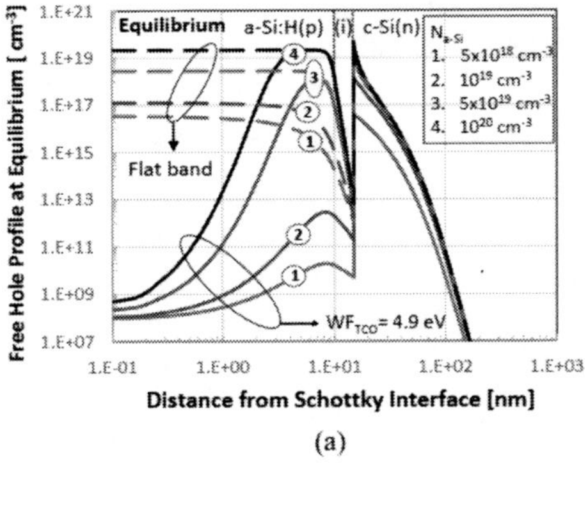

(a)

on the Schottky barrier increases, that is when the cell forward bias increases, as depicted in Fig.2 (b). Thus, it is expected that the electron dark current would exhibit an Early effect [19] due to "a-Si:H(p) width modulation" that is more apparent when the doping level in a-Si:H(p) is lower and when the a-Si:H(p) layer is thinner. Indeed such an effect is confirmed in the simulated dark I-V characteristics displayed in Fig.3. As the cell forward terminal voltage increases the hole current increases

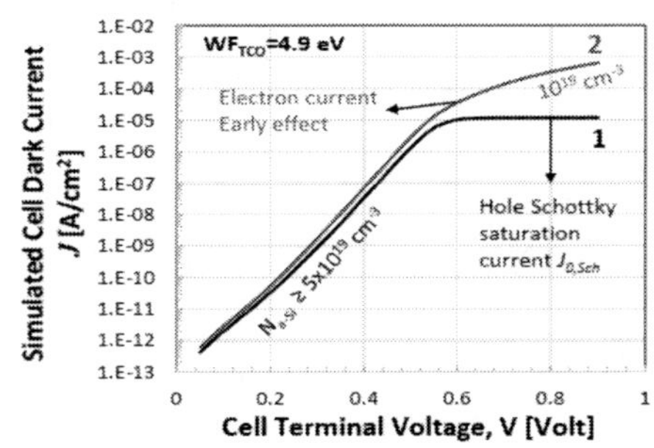

Fig.3. Simulated dark I-V characteristics of silicon HJT cells with $N_{a-Si} = 10^{19}$ cm^{-3} and 10^{20} cm^{-3} at 300 K.

exponentially then saturates once $J_{0,sch}$ is reached Such saturation of the hole current crossing the pn junction is accompanied with a pinning of the juncion voltage. Despite such voltage pining the electron current does not saturate but continues to increase with voltage pulled by the Early effect when $N_{a-Si} = 10^{19}$ cm^{-3}. Such an effect is not observed when $N_{a-Si} = 10^{20}$ cm^{-3} due to the neligible electron current and negligible depletion taking place at such a high doping level. In addition, the Early effect disappears from the lowly doped cell when the a-Si:H(p) layer gets thicker than 14 nm, as depcicted in Fig.4.

3. EQUIVALENT CIRCUIT MODEL AND ANALYSIS IN THE DARK

Based on the knowledge acquired from the results of cell simulations, and by analogy to the Ebers Moll model of bipolar junction transistor [20] the back-to-back diode structure of the HJT could be modelled by two sub-circuits D and S in connected in series, with 1) sub-circuit D representing the pn emitter junction, 2) sub-circuit S representing the Schottky bipolar collector junction, and 3) R_S representing the cell series resistance. Sub-circuit D is nothing but the standard two diode pn junction solar cell

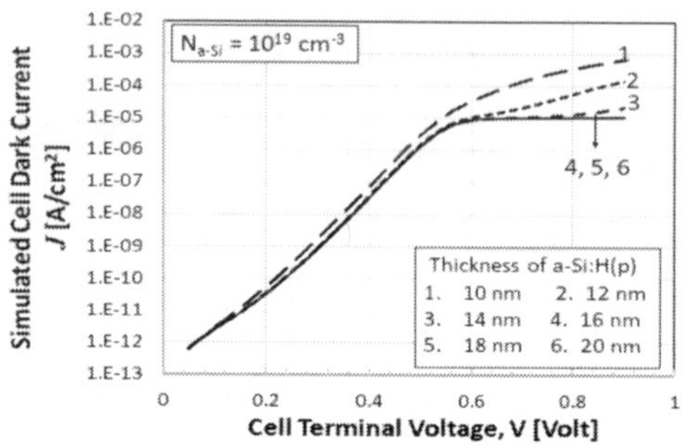

Fig.4. Simulated dark I-V characteristics of silicon HJT cells with $N_{a\text{-}Si} = 10^{19}$ cm^{-3} for different a-Si:H(p) layer thickness at 300 K.

model with ideal pn junction diode D1 in parallel with a non-ideal pn junction diode D2 and with a shunt leakage conductance G_D. Sub-circuit S has the TCO Schottky diode DS connected back-to-back with D1 and D2. Only the hole current through S obeys the Schottky diode equation. The electron current, however, is injected through the cell pn junction in sub-circuit D and hence depends exponentially on its bias. Once it reaches the front contact the electron current is swept through the TCO Schottky barrier and hence is represented by a voltage dependent electron current source J_{Sn}. An Early effect [19] acts on the electron current due to a-Si:H(p) width modulation which is represented by the conductance G_{Sch} connected in parallel with DS and with J_{Sn}. Such a supplementary electron current component continues to flow in sub-circuit D through the voltage dependent current source $G_{Sch}V_S$, as depicted in Fig.5. Finally, the conductance G_T represents probable hole tunneling through the TCO/a-Si:H(p$^+$) Schottky contact. When the a-Si:H(p) layer is very heavily doped like when $N_{a\text{-}Si} = 10^{20}$ cm^{-3} the current source J_{Sn}, the conductance G_{Sch} and the current source $G_{Sch}V_S$ are negligible and can be removed from the model. In such a situation, tunneling is highly probable and the Schottky barrier will be by-passed by the large conductance G_T. In the present analysis, however, G_T is omitted since tunneling is not considered in the simulations.

The circuit analysis proceeds by solving the voltage loop equation given by

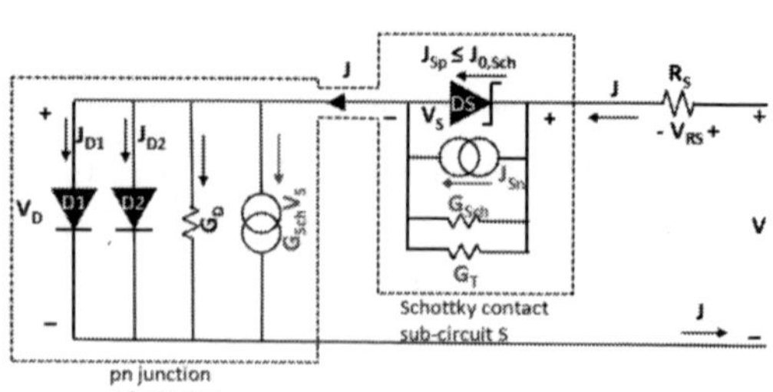

Fig.5. Proposed HJT equivalent circuit model in the dark including Schottky contact at forward biased cell terminals with proper voltage polarities and current directions.

$$V = V_{RS} + V_S + V_D \tag{1}$$

with V the cell terminal volage, V_D, V_S and V_{RS} are the voltage drops on sub-circuits D and S, and on R_S, respectively. Then the continuity of the hole and electron current equations in the equivalent circuit sketched in Fig.5 or any of its simplified versions must be ensured. The optimum values of the parameters involved in equation (1) and in the J-V relationships of all devices included in the circuit model that lead to perfect fitting of the simulated dark J-V characteristics are determined using an in-house developed MATLAB optimization code that formulates a constrained least-squares optimization problem, as detailed in [21]. The procedure is repeated at each voltage step and reiterated to finally extract a single optimum value for each parameter valid for all voltage steps. Such an optimization procedure results in the perfect fitting depicted in Fig.6 between the simulated dark I-V characteristics and the currents obtained from the equivalent circuit analysis for cells with $N_{a\text{-}Si} = 10^{19}$ cm^{-3} and 10^{20}cm^{-3}.

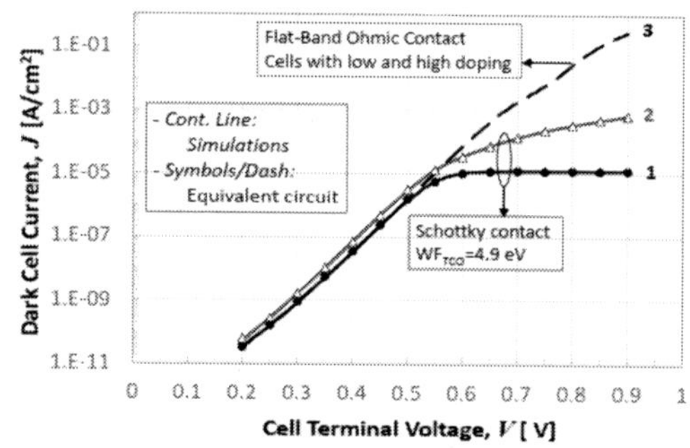

Fig.6. Dark current of cells with Na-Si = 10^{19} cm^{-3}and 10^{20} cm^{-3} having a Schottky front contact with WF_{TCO}= 4.9 eV. The current of a similar cell with flat-band ohmic contact is plotted for comparison *lines*: results of AFORS-HET simulations, *Markers:* results of equivalent circuit analysis.

As depicted in Fig.4, at low voltage and current below $J_{0,Sch}$ the cell current is governed by the pn junction diffusion equations and increases exponentially with V_D hence V. The dark current is larger when $N_{a\text{-}Si}=10^{19}$ cm^{-3} due to a larger electron current that exhibits an Early effect and continues to increase with voltage while the hole current saturates at $J_{0,Sch}$. As expected, with a flat band ohmic contact, the dark current does not saturate and continues to increase with voltage.

4. Equivalent Circuit Model Under Illumination

4.1 Up to Open circuit

Under illumination and below V_{OC} the hole current flows through the Schottky contact in the forward direction. The Schottky depletion region is shallow and the Early effect is much less significant. Hence, the conductance G_{Sch} and the consequent current source $G_{Sch}V_S$ are removed from the equivalent circuit of both cells. The conductance G_D can also removed from the equivalent circuit since its value is very small to affect the operation under illumination. Consequently, the equivalent circuit reduces to the circuit sketched in Fig.7 in which the photo-current sources J_{ph} and J_{phS} are included by superposition in sub-circuits D and S, respectively, and amounting to 40.2 mA/cm^2 and 2.2 mA/cm^2, respectively, as suggested by the simulations. Straight forward analysis of this equivalent circuit is carried out using the optimum model parameters extracted from the perfect fitting of the dark current displayed in Fig.6. The fitting procedure under illumination is carried out by determining the terminal voltage $V(J)$ from the main loop equation (1) at each current level J. It is worth mentioning

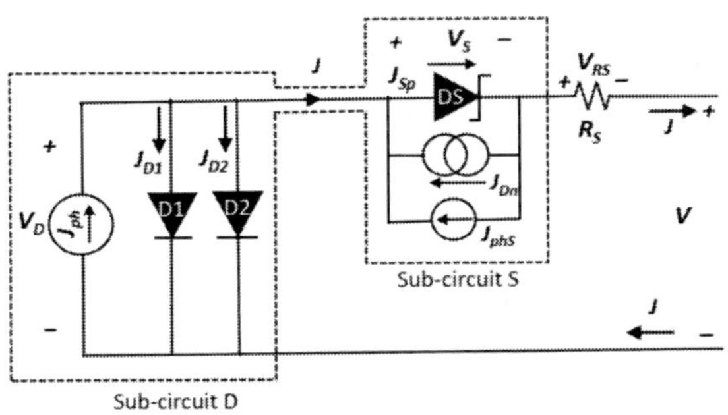

Fig.7. Proposed complete HJT equivalent circuit model under illumination for the voltage range from short circuit up to open circuit terminals.

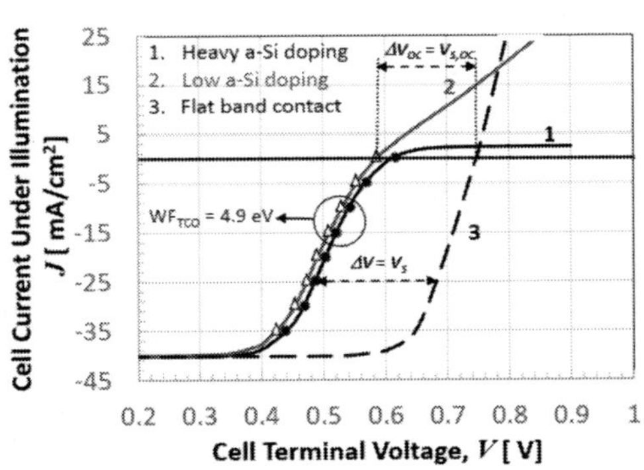

Fig.8. J-V characteristics under illumination resulting from AFORS-HET simulations (lines) and equivalent circuit fitting (markers up to V_{OC}) for cells with $N_{a\text{-}Si} = 10^{19}$ cm^{-3} and 10^{20} cm^{-3} and with WF_{TCO}=4.9 eV. The identical J-V characteristics at flat-band are used as reference.

that these results are obtained assuming that the cell series resistance under illumination is constant which imposes drift-diffusion transport across the a-Si:H/c-Si hetero-interface to be adopted in the simulations [18].

At open circuit terminals, the total current through sub-circuit S is zero such that J_{Sp}, $J_{ph,S}$, and J_{Dn} are looping and a voltage drop $V_{S,OC}$ is established on the terminals of DS. In that case equation (1) becomes

$$V_{OC} = V_{D,OC} - V_{S,OC} \tag{2}$$

The degradation in V_{OC} is therefore attributed to $V_{S,OC}$ and hence to the Schottky barrier reverse saturation current $J_{0,Sch}$. The latter lumps several effects associated to the material and device properties and hence can be used effectively as a specification criteria for the quality of the Schottky contact and it impact on the device performance.

4.2 Beyond open circuit: The S-shape

Beyond V_{OC} the cell current reverses its direction and the forward bias V_S on DS decreases then reverses its polarity. The reverse hole current J_{Sp} through DS increases up to $J_{0,Sch}$ then saturates causing V_D pinning. Any further increase in V is then supported by V_S and V_{RS}. In such a configuration, G_{sch} and the current source $G_{Sch}V_S$ must be maintained, as depicted in the equivalent circuit model sketched in Fig.9. In the heavily doped cell, however, the dark electron current J_{Dn} as well as the Early conductance G_{Sch} are negligible and can both be removed from the equivalent circuit. In such a case, the cell current J saturates at $(J_{0,Sch}+J_{phS})$ forming an S-shaped J-V characteristics with a flat edge, as depicted in Fig.8 (curve 1). In the lowly doped cell, however, the electron current through the current sources J_{Dn} is not negligible and exhibits an Early behavior through G_{Sch}. Consequently, the cell current continues to increase beyond V_{OC} and forms an S-shaped voltage dependence with a rising edge, as depicted in Fig.8 (curve 2).

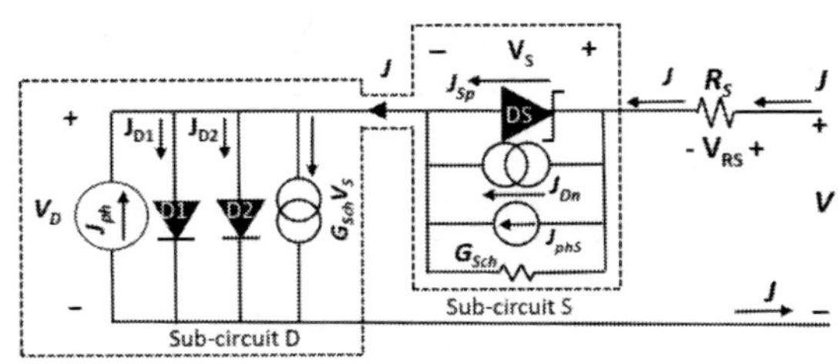

Fig.9. Proposed HJT equivalent circuit model under illumination beyond open circuit.

5. Summary and Conclusions

An accurate equivalent circuit model is developed for HJT pn junction solar cells with a Schottky barrier at the TCO/a-Si:H(p) contact. The model considers the bipolar nature of the Schottky barrier in such a cell and takes into account the collection of the electrons injected through the pn junction and reaching the TCO contact. The proposed model is based on the Ebers-Moll model of bipolar junction transistor including bias dependent a-Si:H(p) layer width modulation. The J-V characteristics simulated by means of AFORS-HET simulations could be perfectly reproduced by the equivalent circuit analysis in the dark and under illumination including the S-shape. The V_{OC} degradation in such cells is attributed to Schottky photo-generated current and cell electron current looping through the Schottky diode at open circuit. In cells with heavily doped a-Si:H(p) the Schottky contact results in moderate V_{OC} degradation and weak FF degradation and in an S-shape distortion beyond V_{OC} that saturates at a level given by the sum of the Schottky photo-generated and reverse saturation currents. On the other hand, in cells with lowly doped a-Si:H(p) the V_{OC} and FF degradation are more severe, and the S-shape has a rising edge beyond V_{OC} due to an effective Early effect in a-Si:H(p).

ACKNOWLEDGEMENT
This work is supported by Kuwait University/IMEC KFAS funded bilateral project CN18-15EE-01.

6. REFERENCES

[1] K. Yoshikawa, W. Yoshida, T. Irie, H. Kawasaki, K. Konishi, H. Ishibashi, T. Asatani, D. Adachi, M. Kanematsu, H. Uzu, and K. Yamamoto, Sol. Energy Mater. Sol. Cells, vol. 173 (2017) 37.
[2] M. Tanaka, M. Taguchi, T. Matsuyama, T. Sawada, S. Tsuda, S. Nakano, H. Hanafusa, and Y. Kuwano, Jpn. J. Appl. Phys., vol. 31-Part 1 (1992) 3518.
[3] N. Corpus-Mendoza, M.M. DeSouza, and F.U. Hamelmann, IEEE J. Photovolt., vol. 5 (2014) 22.
[4] L. Zhao, C. Zhou, H. Li, H. Diao, and W. Wang, Phys. Status Solidi A, vol. 205 (2008) 1215.
[5] Rossler, C. Leendetz, L. Korte, N. Mingirulli, and B. Rech, J. Appl. Phys., vol. 113 (2013) 144513.
[6] M. Ghannam, Y. Abdulraheem, and G. Shehadah; Sol. Energy Mater. Sol. Cells, vol. 145 (2015) 423.
[7] C. Luderer, C. Messmer, M. Hermle, and M. Bivour, IEEE J. Photovol., vol. 10 (2020) 952.
[8] M. Leilaeioun, W. Weigand, M. Boccard. Z. J. Yu, K. Fisher, and Z.C. Holman, IEEE J. Photovolt., vol. 10 (2020) 54.
[9] R. Gogolin, M. Turcu, R. Ferré, J. Clemens, N-P Harder, R. Brendel, and J. Scmidth, IEEE J. Photovolt., vol. 4 (2014) 1169.
[10] R.V. K Chavali, S. De Wolf, and M.A. Alam, Prog. Photovolt. Res. Appl., vol. 26 (2018) pp241.
[11] E. Centurioni and D. Inencinella, IEEE Electron Dev. Lett., vol.24 (2003) 177.
[12] J. Allen, B. Shu, L. Zhang, U. Das, and S. Hegedus, Proceedings of the. 37th IEEE PVSC, Seattle, USA (2011) 2545.
[13] G. Nogay, J. P. Seif, Y. Riesen, A. Tomasi, Q. Jeangros, N. Wyrsch, F-J Haug, S. Dewolf, and C. Ballif, IEEE J. Photovol., vol. 6 (2016) 1654.
[14] R. Saive, IEEE J. Photovolt., vol. 9 (2019) 1477.
[15] G.A.May, Solid St. Electron., vol.11 (1968) 613.
[16] R. Stangl, C. Leendertz, and J. Haschke, in R.D. Rugescu (Ed.), Solar Energy, Intech, Vukovar, Croatia, ISBN:978-953-307-052-0 (2010) 319.
[17] M. Ghannam, G. Shehadah, Y. Abdulraheem, and J. Poortmans, Sol. Energy Mater. Sol. Cells, vol. 132 (2015) 320.
[18] M. Ghannam and Y. Abdulraheem, Appl. Sci. (MDPI), vol. 8 (2018) 1846 (1-19).
[19] J.M. Early, Proceedings of the IRE, vol. 40 (1952) 1401.
[20] J.J. Ebers, and J.L. Moll, Proceedinsg of the IRE, vol. 40 (1954) 1761.
[21] M.Y Ghannam, Y.M. Abdulraheem, H.S. Radhakrishnan, I. Gordon and J. Poortmans, IEEE Trans. Electron Devices, vol. 69 (2021) 613.

This is a short version of the paper submitted to the journal Solar RRL.

It has been selected by the Committee of the WCPEC-8 for submission to Solar RRL.

APCVD (N) POLY-SI PASSIVATING CONTACTS ON TEXTURED AND PLANAR SURFACES

Raphael Glatthaar[1], Tobias Okker[1], Frank Huster[1], Beatriz Cela Greven[2], Sven Seren[3], Giso Hahn[1], Barbara Terheiden[1]
[1] University of Konstanz, Department of Physics, 78467 Konstanz, Germany
[2] Fenzi AGT, Fregatweg 38, 6222 NZ Maastricht, Netherlands
[3] Schmid Group, Robert-Bosch-Str. 32-36, D-72250 Freudenstadt, Germany

ABSTRACT: Planar and textured APCVD (n) poly-Si/SiO_x passivating contacts with a ~60 nm poly-Si layer were metallized by screen-printing. Higher contact resistivity on textured samples $\rho_{C,textured}$ (~6-7 mΩcm² compared to $\rho_{C,planar}$ ~3 mΩcm²) can be attributed to a thinner poly-Si layer and to a more aggressive etching behaviour of the paste around the pyramid tips. Since this behaviour also occurs in classical contact formation on monocrystalline Si, we conclude that the geometrical structure of the pyramids affects the contact formation more than the structure of the used layers. Passivation quality was measured in the non-metallized region between 2-4 fA/cm² for planar samples and to a minimum of 17 fA/cm² for textured samples. In the metallized regions, $J_{0,Met}$ increases towards higher firing temperatures from ~20 fA/cm² to ~140 fA/cm² for planar samples and from 140 fA/cm² to 830 fA/cm² for textured samples.

1 INTRODUCTION

Passivating contacts, which are based on polycrystalline silicon (poly-Si) and a thin SiO_x layer, enable high efficiencies and are therefore more and more used in the global PV market [1,2]. The very good passivation quality, especially under the metal contacts, is the distinguishing factor compared to previous generations of cell structures, which is caused by a high selectivity [3,4]. In particular, a kind of spatial separation is created between metal contacts and minority charge carriers, which leads to low recombination often expressed as saturation current density ($J_{0,Met}$). For application in high performance passivating contacts with a very thin poly-Si layer, the metallization must satisfy both, a low $J_{0,Met}$ and a low contact resistivity (ρ_C) between the metal finger and the poly-Si layer. Various groups have already achieved very good results in their investigations on the contact formation of (n) poly-Si passivating contacts. Thereby, the required amorphous-Si layer (a-Si) is mostly deposited by low pressure chemical vapor deposition (LPCVD) (e.g., [5,6]) and plasma enhanced chemical vapour deposition (PECVD) (e.g., [7]), which both need a vacuum system and are thus less cost-effective.

In contrast, for the experiments presented here, we deposit a single sided (n) a-Si layer by atmospheric pressure chemical vapour deposition (APCVD). The inline deposition tool from SCHMID GROUP stands out above other deposition techniques by its high throughput since no vacuum system is required and thus enables a cost-effective alternative to the beforementioned techniques [8]. For these APCVD (n) poly-Si/SiO_x passivating contacts, we are investigating the contact formation with non-commercial fire-through Ag pastes on planar and textured surfaces through screen-printing.

2 EXPERIMENTAL

For this work, n-type (4.5 Ωcm), n-type (35 Ωcm), and p-type (1.3 Ωcm), Czochralski grown crystalline silicon (c-Si) substrates with a size of 156x156 mm² and a thickness of 180 µm were saw damage etched by potassium hydroxide. n-type 35 Ωcm samples and parts of the p-type c-Si substrates were treated with an alkaline texture before all samples were subsequently exposed to a wet chemical ozone cleaning. Afterwards, a nominal 1.5 nm thick thermal oxide was grown in a classical tube diffusion furnace. For p-type samples, a 64(15) nm thick (n) amorphous-silicon (a-Si) layer was deposited on both sides using the APCVD tool from SCHMID GROUP, which corresponds to 55(15) nm for textured samples. On planar n-type base material samples ~65(5) nm and on textured samples ~60(5) nm (n) a-Si were deposited. The junction formation was performed subsequently in a tube furnace at a temperature of 920°C for 30 minutes. During this step, the a-Si layer was crystallized and in-diffusion of phosphorous through the thermal SiO_x occurred. For hydrogenation, a 75 nm thick SiN_x layer was deposited on the samples using a PECVD system, whereby the deposition conditions for the textured samples were adjusted so that the SiN_x thickness for planar and textured samples were equal. The metallization was carried out by screen-printing with a non-commercial fire-through Ag paste. While for the samples with the p-type substrate transfer length method (TLM) patterns with finger width of 40 µm were printed, on the passivation samples (n-type base substrate) 45 µm fingers with varying finger pitches were printed leading to six 4x4 cm² areas with different metallization fraction between 0%-21%.

The subsequent contact formation was achieved by fast firing in a standard belt furnace at five different set peak temperatures. Therefore, the TLM samples were laser cut to a size of 5x5 cm², resulting in five samples per group for each set peak firing temperature. Samples for evaluation of passivation were fired in 156x156 cm² size, but set peak firing temperatures were lowered to 765°C-865°C.

The contact resistivity (ρ_C) was evaluated by TLM measurement with a pv-tools setup. For a further investigation of the contact formation between the silver paste and the (n) poly-Si layer, the metal finger was removed. This was achieved through etching the bulk Ag finger in a nitric acid solution and etching the glass layer in diluted hydrofluoric acid. The now visible microstructure was then examined by scanning electron microscopy (SEM). In order to characterize the samples with respect to their passivation quality, photoconductance decay (PCD) was measured on non-metallized reference samples using a Sinton lifetime tester. In addition, the saturation current density at Ag metal contacts was determined using a time-resolved photo-luminescence setup. The measurement provides a lifetime map of the single-sided metallized sample for different minority charge carrier densities. In high injection, a J_0 can be extracted for each pixel by a linear fit according to the method of Kane and Swanson [9]. By extrapolating the J_0 for the areas with different metallization fractions to 100%, a $J_{0,Met}$ value can be determined. A detailed explanation and the limitations of this evaluation can be found in Glatthaar et al. [10].

3 RESULTS

In the following, passivation values for APCVD (n) poly-Si passivating contacts are shown first, with a distinction between metallized and non-metallized samples. A contact formation study shows in addition to TLM measurements the contact area between the metal paste and the poly-Si/SiO_x surface after chemical removal of the metal finger.

The results presented below are mostly and also in more detail presented in another publication [10].

3.1 Passivation quality

In Figure 1, the passivation for metallized and non-metallized (n) poly-Si/SiO_x stacks on n-type c-Si substrates is shown in terms of the saturation current density (J_0). For planar samples, very low $J_{0,Pass}$ 2-4 fA/cm² can be measured. These are corresponding to an implied open circuit voltage (iV_{OC}) of ~740 mV and demonstrate the high potential for passivation of the APCVD (n) poly-Si/SiO_x junction. While for planar samples the firing temperature does not have a strong impact on the passivation, for textured samples $J_{0,Pass,textured}$ increases from ~20 fA/cm² to ~70 fA/cm². At this point, it should be noted that for no particular reason the silicon base material has different resistivities for both groups of samples, as well as the (n) poly-Si thickness is not exactly the same (compare therefore Section Experimental).

The passivation under the metallized region of the planar samples shows a $J_{0,Met,planar}$ of ~20 fA/cm² at 765°C, increasing to ~140 fA/cm² for a set peak firing temperature of 865°C. An increasing firing temperature affects the passivation of textured samples much stronger than planar samples, resulting in an enhancement of $J_{0,Met,textured}$ from ~140 fA/cm² to ~830 fA/cm².

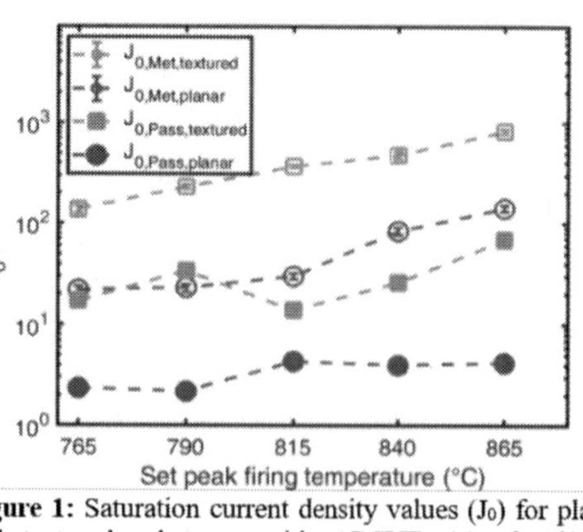

Figure 1: Saturation current density values (J_0) for planar and textured substrates with APCVD (n) poly-Si/SiO_x structures. $J_{0,Pass}$ values were obtained by PCD measurement on symmetrical samples. $J_{0,Met}$ values were obtained by injection-dependent, time-resolved photo-luminescence evaluation on single-sided screen printed samples. Resulting J_0 values for six different metallization fractions allows an extrapolation to the $J_{0,Met}$ at 100% metallization. While each shown data point corresponds to a measurement on one sample, uncertainty is a result of the used analysis. Dashed lines are a guide to the eye. Note the actual sample temperature is in part significantly below the here indicated set peak firing temperature.

3.2 Contact resistivity measurements

In a second experiment on samples with p-type c-Si substrate, the differences in contact resistivity between planar and textured samples will elaborated. Note that, in addition to a different base material, the sample size in the following is 5x5 cm², which means that, especially at the fast firing process, the actual sample temperature behaviour can differ significantly to the samples prepared for passivation measurements. Figure 2 shows ρ_C for planar and textured (n) poly-Si/SiO_x structures for set peak firing temperatures of 825-900°C. For samples with planar surfaces, contact resistivity slightly decreases towards higher firing temperatures with a minimum mean value of ρ_C ~3 mΩcm² at 850°C. The textured samples show higher ρ_C of ~6-7 mΩcm², whereby an influence of the firing temperature is not clearly visible. In literature, it is often reported that contact formation is significantly better for textured samples than for planar ones [11]. A part of the observed difference in ρ_C can be explained by the thinner (n) poly-Si layer for textured samples, since the layer thickness has also a large impact on contact formation [5]. However, we expect that this difference in thickness cannot explain that ρ_C for planar and textured samples shows the opposite trend compared to diffused wafer surface without poly-Si.

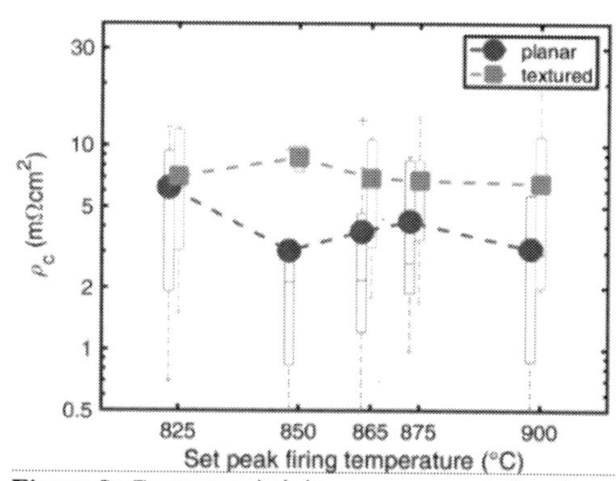

Figure 2: Contact resistivity measurements for planar and textured APCVD (n) poly-Si/SiO_x structures. Measurement points are taken from five different 5x5 cm² samples. TLM measurements with 40 µm finger widths are shown. Dashed lines between the mean values are a guide to the eye. Note the actual sample temperature is in part significantly below the here indicated set peak firing temperature.

3.3 Surface analysis

To understand the differences between planar and textured samples in contact formation microscopically, SEM images of the surfaces were taken after the chemical removal of the metallization. Figure 3a shows the surface of a planar (n) poly-Si sample fired at a set peak temperature of 825°C. Although some areas of the sample are still covered with SiN_x (blue colored), most of the sample surface is covered by the poly-Si layer (yellow colored), which is at the surface slightly etched by the Ag paste (not shown here). However, at some places the Ag paste consumed the whole poly-Si layer. After firing these imprints are filled with Ag crystallites. Interestingly, the interface to the c-Si bulk substrate is not damaged (a detailed analysis on that can be found in [10]). Furthermore, at a set peak firing temperature of 900°C, the amount of contact imprints in the poly-Si is increased as can be seen in Figure 3b. However, a clear reduction of the SiN_x covered area is not detected.

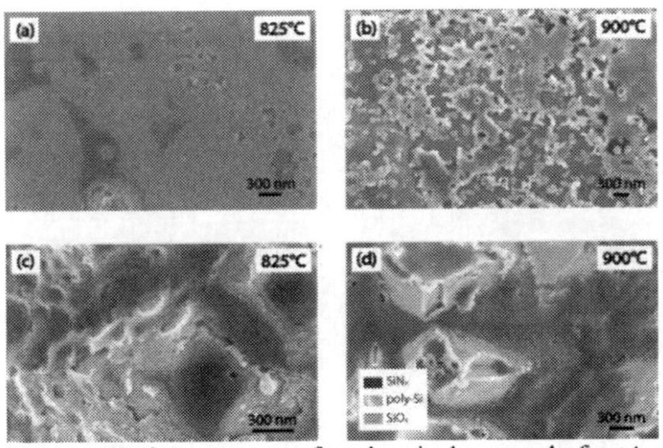

Figure 3: Microstructure after chemical removal of an Ag finger on planar (a,b) and textured (c,d) (n) poly-Si/SiO_x structures by an etching cascade of HNO_3-HF-HNO_3. Samples in (a) and (c) are fired at a set peak firing temperature of 825°C and (b) and (d) at a set peak firing temperature of 900°C. Blue colored areas highlight the remaining SiN_x layer, yellow colored area the (n) poly-Si layer and red colored area the non-penetrated base substrate and thus presumably the SiO_x layer. Non colored areas (here grey) correspond to imprints in the base substrate caused by the massive etching of the Ag-paste.

For textured samples it is, firstly, noticeable that the etching through the SiN_x layer by the Ag paste takes only place around the tips of the pyramids, whereby in the pyramid valleys the SiN_x layer persists. This can be seen in Figure 3c for a firing temperature of 825°C. Secondly, besides the consumption of the poly-Si layer also the c-Si base substrate is etched by the Ag paste. These imprints are predominantly at the pyramid tip and are increased in size significantly towards a set peak firing temperature of 900°C (Figure 3d). At this firing temperature, the entire highly doped poly-Si layer in the SiN_x free areas is consumed by the Ag paste.

As a result, the grown Ag crystallites are in direct contact with the bulk c-Si leading to a high recombination. Furthermore, since the diffusion of dopants from the poly-Si into the c-Si wafer forms only a shallow emitter, locally high contact resistivities are expected for the interface between Ag crystallite and the lowly doped c-Si substrate. Since similar observations of deep imprints where not made for planar samples, this effect might explain that unexpectedly the contact resistivity for textured samples is not lower than for planar ones.

Interestingly, these massive contact imprints at the pyramid tips also exist in the classical contact formation of textured c-Si emitters [12]. In contrast, the stopping of the Ag paste at the thermal SiO_x interface is not known in this respect. We therefore assume that the geometric structure and the exposure of the pyramid tip to the Ag paste have a high impact on the etching behavior of the Ag paste.

4 CONCLUSION

In this work, very good passivation values for unmetallized ~60 nm APCVD (n) poly-Si/SiO_x passivating contacts on planar substrates with 2-4 fA/cm^2 and also good passivation values for unmetallized textured samples with a minimum of 17 fA/cm^2 could be reached. The passivation quality after contact formation of a fire-through Ag paste and the APCVD (n) poly-Si/SiO_x passivating contacts is further quantified by a $J_{0,Met}$ evaluation. Increasing fast firing temperatures enhances $J_{0,Met}$ from ~20 fA/cm^2 to ~140 fA/cm^2 for planar samples and from 140 fA/cm^2 to 830 fA/cm^2 for textured samples.

In a second experiment, contact resistivity was measured for planar sample structures with mean values of ρ_C ~3 mΩcm^2. For textured samples ρ_C is slightly higher. This can be attributed to a thinner poly-Si layer and to the etching behavior of the paste around the pyramid tips. The contact imprints for textured samples are comparable to standard Ag contact formation in monocrystalline silicon emitters, in contrast to planar samples. Therefore, we assume that the geometrical influence of the sample's topography on contact formation is stronger than that of the stopping mechanism at the poly-Si/SiO_x interface.

5 ACKNOWLEDGEMENTS

This work was partially funded by the German Federal Ministry for Economic Affairs and Climate Action contract number 03EE1022C (project FUN) within SOLAR-ERA.NET cofund 2.

REFERENCES

[1] F. Haase et al., Laser contact openings for local poly-Si-metal contacts enabling 26.1%-efficient POLO-IBC solar cells, Solar Energy Materials and Solar Cells 186 (2018) 184.

[2] D. Yan et al., Polysilicon passivated junctions: The next technology for silicon solar cells?, Joule 5 (2021) 811.

[3] M. Hermle et al., Passivating contacts and tandem concepts: Approaches for the highest silicon-based solar cell efficiencies, Applied Physics Reviews 7 (2020) 021305.

[4] B. Nemeth et al., Polycrystalline silicon passivated tunneling contacts for high efficiency silicon solar cells, Journal of Materials Research 31 (2016) 671.

[5] P. Padhamnath et al., Development of thin polysilicon layers for application in monoPoly™ cells with screen-printed and fired metallization, Solar Energy Materials and Solar Cells 207 (2020) 110358.

[6] A. Chaudhary et al., Influence of polysilicon thickness on properties of screen-printed silver paste metallized silicon oxide/polysilicon passivated contacts, physica status solidi (a) 218 (2021) 2100243.

[7] F. Meyer et al., Localisation of front side passivating contacts for direct metallisation of high-efficiency c-Si solar cells, Solar Energy Materials and Solar Cells 235 (2022) 111455.

[8] A. Merkle et al., Atmospheric pressure chemical vapor deposition of in-situ doped amorphous silicon layers for passivating contacts, Proc. 35th EU PVSEC (2018) 785.

[9] D.E. Kane, R.M. Swanson, Measurement of the

emitter saturation current by a contactless photoconductivity decay method, Proc. 18th IEEE PVSC (1985) 578.
[10]R. Glatthaar et al., Contact formation of silver paste and APCVD (n) poly-silicon passivating contacts on planar and textured surfaces, submitted (2022).
[11]P. Padhamnath et al., Progress with passivation and screen-printed metallization of boron-doped monoPoly™ layers, Solar Energy 231 (2022) 8.
[12]E. Cabrera et al., Experimental evidence of direct contact formation for the current transport in silver thick film metallized silicon emitters, Journal of Applied Physics 110 (2011) 114511.

POTENTIAL OF SPUTTERED AZO LAYERS FOR THE INDUSTRIAL MANUFACTURING OF HETERO JUNCTION SOLAR CELLS

Martin Dimer[1a)], Alexandros Cruz[2)], Stefan Janke[2)], Stefan Wendlandt[3)], Uwe Graupner[1)], Darja Erfurt[2)] Winfried Wolke[4)], Martin Bivour[4)], Bernd Stannowski[2)], Eric Schneiderlöchner[1)]
[1)] VON ARDENNE GmbH, Am Hahnweg 8, 01328 Dresden, Germany
[2)] Helmholtz-Zentrum Berlin, PVComB, Schwarzschildstr. 3, 12489 Berlin, Germany
[3)] PI Photovoltaik-Institut Berlin AG, Wrangelstr. 100, 10997 Berlin, Germany
[4)] Fraunhofer Institute for Solar Energy Systems ISE, Heidenhofstr. 2, 79110 Freiburg, Germany
[a)] dimer.martin@vonardenne.biz

ABSTRACT: Aluminium doped zinc oxide (AZO) is an interesting transparent conductive oxide (TCO) for silicon heterojunction (SHJ) solar cells to replace indium tin oxide (ITO) to enable cost reduction and omit the use of the limited global indium reserve. However, AZO layers have a much higher sheet resistance and lower damp heat stability compared to ITO. This work shows that the AZO stability can be improved by using an ITO capping layer, which is also decreasing the sheet resistance of the AZO/ITO layer stack compared to a pure AZO layer. Here, high rate rotatable magnetron sputtering was used to deposit AZO and ITO films. Under these conditions ITO capped AZO layers deposited on a-Si/p can result in lower fill factors. It will be shown that heating before sputtering and introduction of a thin ITO layer between AZO and a-Si/P avoid the fill factor decrease and lead to fill factors above 80%. The properties of the solar cells with the ITO/AZO/ITO layer stack are comparable to those solar cells with the pure ITO reference layer. Similar solar cell properties with the same ITO/AZO/ITO layer stack on the a-Si/n of the front side have been achieved. Furthermore, the TCO materials costs of ITO, AZO/ITO and ITO/AZO/ITO layers stacks for industrial manufacturing are discussed.
Keywords: SHJ solar cells, TCO, ITO, AZO, Damp Heat Test

1 INTRODUCTION

The industrial mass production of silicon heterojunction (SHJ) solar cells has gained significant interest over the last few years as the technology offers high efficiencies of 26.7% for an interdigitated back contacted (IBC) solar cell [1] and of 26.4% for two-side contacted solar cells on industrial relevant processes [2]. Today a major focus in industrial SHJ manufacturing is to reduce the operative expenditures (OPEX). One well known approach to cut OPEX is to replace the expensive indium tin oxide material (ITO) used as TCO layer on both sides of the solar cell with a cost-effective material as AZO. AZO as an alternative TCO material has been investigated in different material compositions already for decades and lower electrical conductivity of the films and an insufficient stability in the module have been reported [3]–[6].

This work focuses on the development of a sputtering process technology that allows a significant TCO material cost reduction of up to 60% by using $ZnO:Al_2O_3$ tube targets with 1% Al_2O_3. This has been tested by other authors showing the possibility of strongly ITO reduction without mayor efficiency penalty[7]–[9].

Beyond the presentation of the optoelectrical properties of the produced ITO single layers, AZO/ITO as well as ITO/AZO/ITO stack layers, it is also investigated how the produced TCO films perform on solar cell level. Furthermore, produced SHJ solar cells (Cz, M2) with AZO/ITO have been assembled into single cell mini modules for damp heat testing of up to 3000 hours in order to assess the durability of the AZO films and AZO/ITO layer stacks against moisture and heat.

2 EXPERIMENTAL AND RESULTS

2.1 AZO and ITO deposition and layer properties

Design on experiment investigations to assess optimum single layer properties were performed in a horizontal pilot line sputtering tool installed at one of the contributing partner´s site. The process pressure, total gas flow and hydrogen content have been varied in order to further understand the most important deposition parameters to tune in optimal electrical and optical film properties.

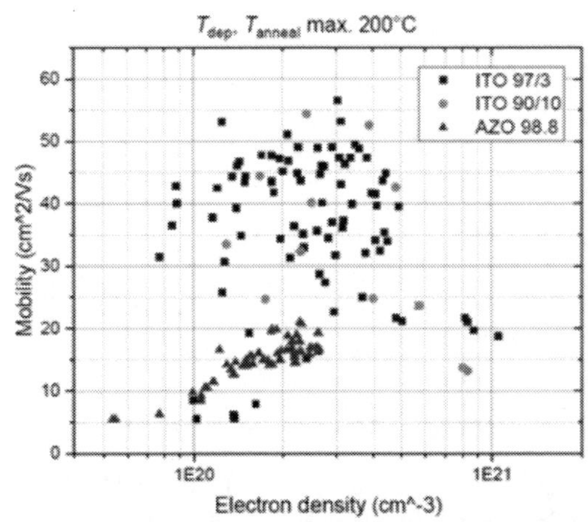

Figure 1: Mobility μ_e vs carrier concentration n_e plot extracted from Hall measurements of 100±10nm ITO and AZO layers deposited on 1 mm thick glass.

Figure 1 compares the Hall mobilities (μ_e) and the charge carrier densities (n_e) of layers which were sputtered from three different rotatable tube targets based on AZO 98.8/1.2, ITO 90/10 and ITO 97/3. The mobility of the AZO films is between 10 and 20 cm^2/Vs and thus three times lower compared to the mobilities of the ITO films. This can be explained by an increased grain boundary scattering of the AZO which is especially affected by the low temperature deposition and post annealing and by the relatively low material thicknesses used for SHJ solar cell application[3], [10]–[12].

For the evaluation of the TCO layers on solar cells an in-line sputtering system was used, which is equipped with industrial rotatable magnetron sputtering sources. In order to be capable to transfer the DC sputtering processes to production the line speed of the carrier with the substrates and the magnetron power were chosen as needed in sputtering systems for manufacturing.

ITO, AZO/ITO double and ITO/AZO/ITO triple layers were deposited on a-Si coated Cz-wafers. The length of the industrial type targets was 965 mm and the power levels were between 3 and 7 kW. For the deposition of the AZO and ITO films the rotatable magnetrons were equipped with ceramic tube targets (ZnO:Al_2O_3 with 1 wt% Al_2O_3 and In_2O_3:SnO_2 with 3 wt% SnO_2). Firstly, the films were deposited on glass at different O_2 partial pressures to find best layer properties. In contrast to the ITO and ITO/AZO/ITO layer stack the AZO/ITO layer stack was not heated prior the film deposition. The spectral absorbance of selected layer stacks that showed highest short current densities (J_{SC}) of solar cells is shown in Figure 2.

To replicate the curing procedure during silver screen printing on the TCO layers the TCO layer stacks on glass were annealed at 200°C for 30 min in ambient. The decrease of the absorbance of the AZO/ITO layer stack after annealing is related to the solid phase crystallization of the more amorphous AZO/ITO layer stack compared to the ITO and ITO/AZO/ITO films deposited on glasses with a higher temperature. This change of the AZO/ITO film properties after annealing can be recognized in the increased mobility (table 1). On the other hand, no or less change of the absorbance, mobility and carrier density is observed for the ITO and the ITO/AZO/ITO layer stack.

It is interesting to observe that there is a higher absorbance for the AZO containing layer stacks below 700 nm compared to pure ITO.

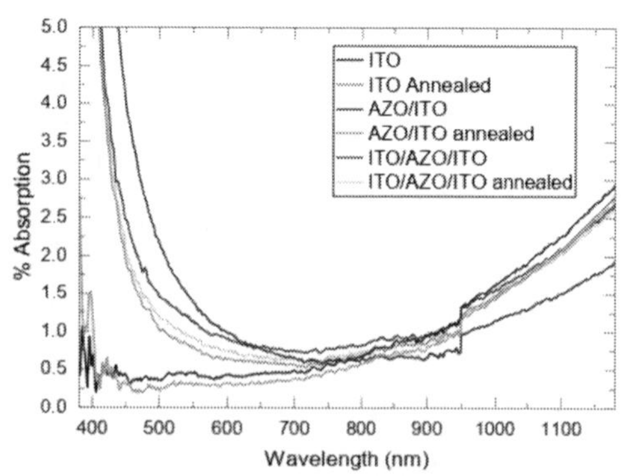

Figure 2: Absorption spectra of ITO, AZO/ITO and ITO/AZO/ITO films on glass.

Table I: Properties of ITO reference, AZO/ITO and ITO/AZO/ITO stack on 1 mm thick B270 glass. Left as-deposited and right after 200°C annealing for 30 min in ambient.

Stack	Thick (nm)	R_{Sh} (ΩSQ)	Mobility (cm2/Vs)	Carrier density (1E20cm^{-3})
ITO	90	38 / 40	60 / 60	3.2 / 3.2
AZO/ITO	105	284 /121	19 / 24	1.4 / 2.1
ITO/AZO/ITO	109	109 / 120	27 / 31	1.7 / 1.6

Remarkable is the high sheet resistance of 284 Ohms of the AZO/ITO and only the low sheet resistance of 38 Ohms of the ITO reference film. The AZO/ITO sheet resistance decreases to 121 Ohms after annealing. The reason of the relative low mobility of the film stacks is in the low mobility of the AZO of around 15 cm^2/Vs.

2.2 Manufacturing and SHJ solar cell properties

Figure 3 shows a schematic illustration of the SHJ solar cell, which were used to evaluate the different TCO layer stacks.

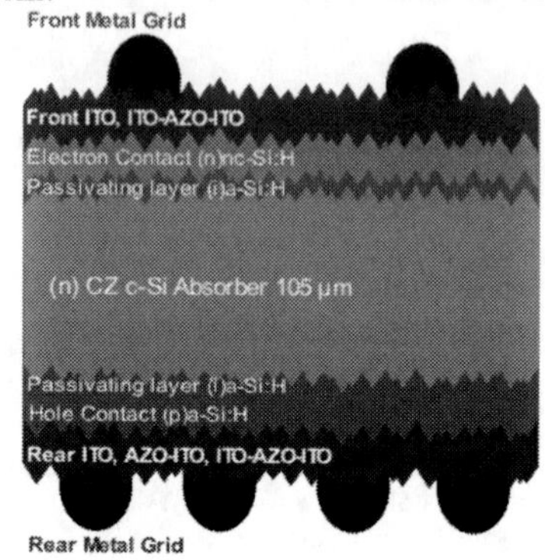

Figure 3: Schematic cross-section of a standard 4 cm^2 bifacial rear-junction SHJ solar cell as the ones utilized for experiments.

To investigate the performance of the developed AZO/ITO layer stacks on rear-junction SHJ solar cells, four experimental groups were prepared for which the *J-V* parameters are shown in Figure 4. Please note that the TCO thicknesses are nominal values measured on glass and these are reduced on a textured wafer approximately by a factor of 0.68.

The first group is the reference with front and back 110 nm ITO layers, the second group has front 110 nm ITO and back 90 nm AZO + 20 nm ITO, the third group has 110 nm ITO front and 20 nm ITO + 70 nm AZO + 20 nm ITO back layers, and the fourth group has same layers as group three but with inverted position for back and front. Figure 4 x-axis labels show the front TCO on the left and the rear TCO configuration on the right for each group. For each of the three non-reference experimental groups, an oxygen variation was carried out and these are listed with an oxygen reduction from left to right in Fig. 4.

Considering the short circuit current density (J_{SC}) presented in Figure 4, we can observe that adding AZO to the rear layer stack slightly reduces J_{SC} compared to the ITO reference. Furthermore, reducing the O_2 for the AZO/ITO and ITO/AZO/ITO rear stacks reduces J_{SC}. Bringing the ITO/AZO/ITO layer stack to the front increases the impact on J_{SC} reduction. The low oxygen samples result in an efficiency (η) reduction to 22.8% and 22.6% for the back and front positioned ITO/AZO/ITO stacks compared to the 22.9% median η of the ITO reference cells. This is driven by the aforementioned J_{SC} losses.

The open circuit voltage (V_{OC}) is very stable independently of the TCO variations applied. It shows a small variation from 742.5 for the ITO reference to 745 mV median values for the highest ITO/AZO/ITO group.

Observing now the fill-factor (*FF*) it is very clear that the cells contacted with AZO on the rear show a reduced *FF* due to increased series resistance (R_S). This hints to a bad contact formation at this interface [13], [14]. This behavior is reproducible, and its nature needs further investigation. The decreased *FF* for the AZO rear contacted cells derives in a lower median efficiency (η) of 22.6% compared to the 22.9% ITO reference. On the other side, the ITO/AZO/ITO front and rear stacks with higher oxygen content result in a good optoelectrical compromise. These show very similar median η compared to the ITO reference when positioned at the front with 22.9% and a slightly increased median η of 23.1% when positioned at the back. This means that implementing the ITO/AZO/ITO layer stack at both the electron and hole contact of the solar cell can be realized without an η penalty compared to the ITO reference. This enables a reduction of ITO by roughly 60%. The best efficiencies are of 23.7% for the reference ITO cell compared to 23.5% for the ITO-AZO-ITO stack positioned at the back and 23.4% when it's implemented at the front.

During the preliminary investigations of AZO as a back contact layer on the hole selective contact improved FF were observed, if the wafer was as good as nothing preheated. Hence the AZO/ITO deposition in this work was carried out without preheating. Surprising were the low FF results with the not preheated AZO/ITO. We assume that the high power of the industrial sputtering process used in this work could be the cause of the low FF of the solar cells with the AZO/ITO layer stack. As shown in Figure 4 an ITO layer between the AZO and the hole selective layer avoids the decrease of the FF.

Further investigation is needed to obtain consistent good hole rear-side contacting for the AZO to be able to omit the ITO initial seed layer improving the ITO reduction to 80%. Furthermore, a remaining open aspect and discussed in section 2.3 is the long-term stability of the AZO based layer stacks as AZO has been known to show poor damp-heat stability.

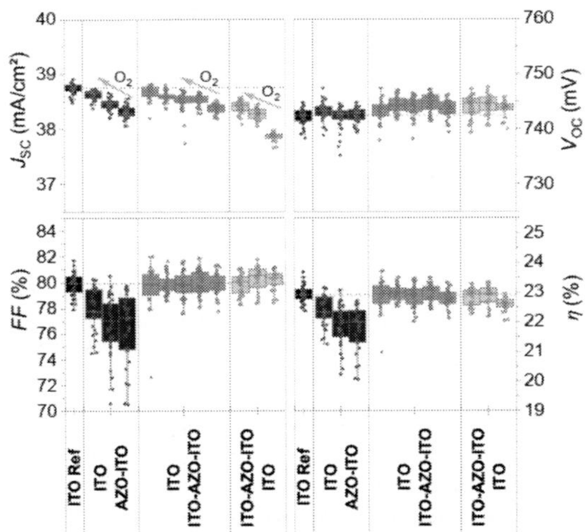

Figure 4: *J-V* Parameters of SHJ solar cells with varied front and rear ITO, AZO/ITO or ITO/AZO/ITO layer stacks. In the x-axis labels, the left is for the front and the right for the rear TCO layers, respectively. The ITO reference median values are shown with a dotted line in each graph.

2.3 Module preparation

M2 size SHJ solar cells featuring 5 different TCO layer stacks symmetrically at the front and back were prepared on 1-3 Ohmcm Cz-wafers to study the impact of this stress test on the device performance at a module level: 1. ITO (110nm) as a reference, 2. pure AZO 1 (110nm) 3. pure AZO 2 with increased oxygen (110nm) 4/5. AZO 1/2 & ITO stack (90nm/20nm). An UV-open PolyOlefin Elastomer (POE) with a transmission of ~90% was selected as the encapsulation material. The selected glass-glass design was used due to the high recycle rate of glass and a glass on the rear has a high transmission for the bifacial application as well as glass has high moisture barrier properties.

The interconnection is based on Electrically Conductive Adhesives (ECA) and ribbon. The focus of the ECA selection was to ensure that there was no damage to the TCO layer and that an industrial process could be used, such as a short curing time.

2.4 Damp heat test and module degradation

After the initial characterization (IV-Curve at STC and EL@1.0x Isc) six damp heat test cycles of 500 hours (DH500) were executed. Figure 5 shows the relative power degradation to the initial data of each sample.

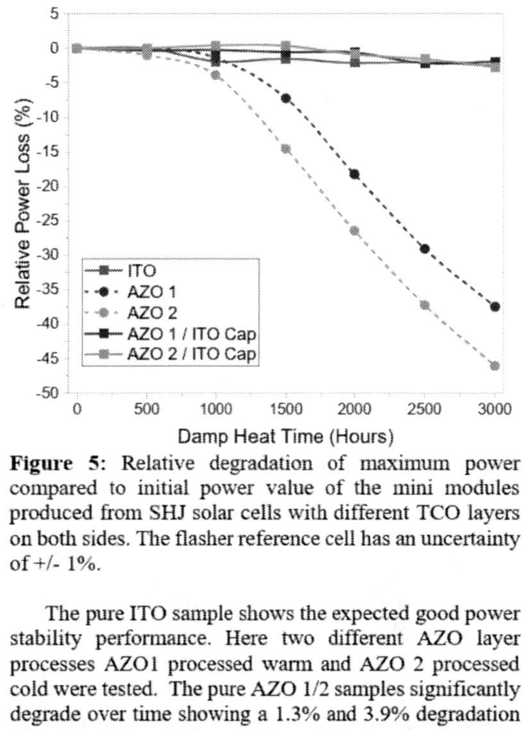

Figure 5: Relative degradation of maximum power compared to initial power value of the mini modules produced from SHJ solar cells with different TCO layers on both sides. The flasher reference cell has an uncertainty of +/- 1%.

The pure ITO sample shows the expected good power stability performance. Here two different AZO layer processes AZO1 processed warm and AZO 2 processed cold were tested. The pure AZO 1/2 samples significantly degrade over time showing a 1.3% and 3.9% degradation after DH1000 as well as 37.4% and 43.6% after DH3000. On the contrary both AZO/ITO stacks show very similar degradation behavior compared to pure ITO throughout the whole damp heat test period with 2% and 2.8% after 3000 hours of damp heat testing, respectively. This shows the effectiveness of the ITO capping on top of AZO enabling the substitution of ITO by 60% considering the front and the rear side of the solar cell.

Corresponding Electroluminescence measurements (EL) images are shown below in Figure 6.

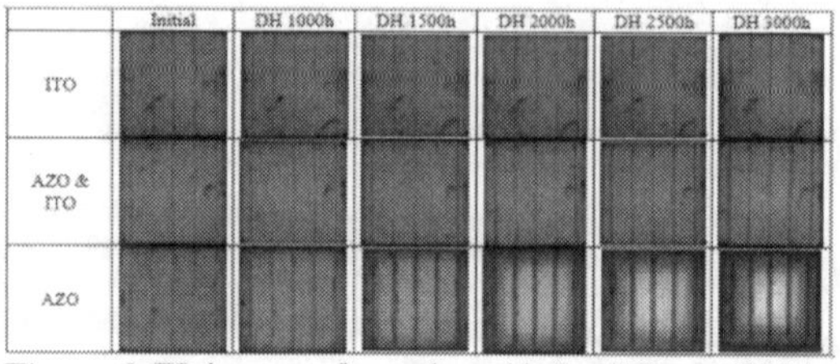

Figure 6: EL images of samples with the different TCO's.

Over the duration of 3000 hours of damp heat testing there are no major changes noticeable in the EL images for the pure ITO as well as for the AZO with ITO cap samples. At the AZO sample, the moisture penetrating from the edges and at the ribbons becomes clearly visible.

2.5 Target costs

A significant reduction of the target costs for the sputtering of the front and rear contact can be achieved by at least partly replacing ITO with AZO. The main cause for that is the cheap ZnO powder compared to the cost-intensive indium oxide powder, which is needed for the manufacturing of the ceramic tube targets. In contrast of ITO the manufacturing costs of AZO tube targets are much more determined by the manufacturing and not by the material costs. Therefore, the price surcharge for increasing the target thickness is low compared to ITO. This helps to further reduce the AZO target costs. Figure 1 shows a comparison of the costs of ITO and AZO tube targets which are needed for the deposition of ITO or AZO films or film stacks of both materials. The assumed target thicknesses of ITO were 9 mm and of AZO 18 mm. The indicated film thicknesses are related to thicknesses, which are achieved on flat substrate surfaces like glass. In case of a textured wafer the thicknesses are around 75 nm. The calculation is based on target prices of well established target manufacturer.

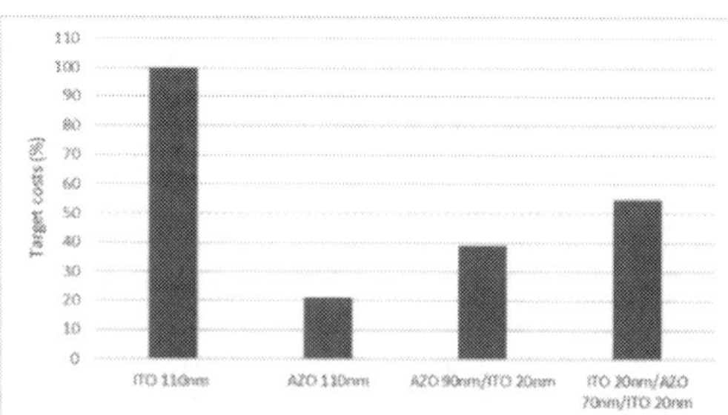

Figure 7: Impact of target material and film stack design on the target costs, film thicknesses on glass

The figure 7 shows that a complete replacement of ITO by AZO will reduce the target cost by 80%. The application of film stack of AZO 90 nm with a capping ITO layer of 20 nm can still reduce the costs by 60%.

3 CONCLUSION AND OUTLOOK

Commonly the scarce and cost intensive ITO is used in SHJ solar cells as contact layer. In this work the ITO content of the contact layers was partially replaced by AZO. ITO, AZO/ITO double as well as ITO/AZO/ITO triple layers were deposited by high power industrial magnetron sputtering as TCO front and back contact.

In case of coating the hole selective layer on the back side of the solar cell with AZO/ITO the FF is 1.5% lower compared to the ITO reference. This is attributed to a bad contact formation between the AZO and the hole selective layer. Introducing a thin ITO interlayer between AZO and the hole selective layer avoids the decrease of the FF.

The combination of a thin ITO interlayer, a thick AZO layer and a thin ITO capping layer leads to the best efficiencies. 23.5 and 23.4% are the best efficiencies, which were achieved with the preheated ITO/AZO/ITO on the back and front side respectively. This is nearby the best efficiency of 23.7% of the ITO coated reference solar cell.

In addition, there were damp heat test of 3000 h performed on mini modules with AZO/ITO coated SHJ solar cells. The power loss of AZO/ITO coated SHJ solar cells in the damp heat test is less than 3% and on the same level as the ITO reference. AZO is stable with ITO capping. That is an additional prerequisite to use AZO in commercial solar cells.

Finally, the sputtering target costs for the different layer stacks for mass production are calculated. 45% of the target cost and 64% ITO can be saved by using the layer stack ITO/AZO/ITO instead of pure ITO. Further reduction of costs and ITO could be possible by decreasing the ITO thicknesses of the layer stack

Still, further investigations are needed to increase the maximum J_{sc} of 38.7 mA/cm^2 and FF of 81% of ITO/AZO/ITO coated SHJ solar cells. A better understanding and improvement of the contact between AZO and the hole selective layer could make the interlayer ITO superfluous.

ACKNOWLEDGEMENT

This work was funded by the German Federal Ministry for Economic Affairs and Climate Action.

REFERENCES

[1] K. Yoshikawa *et al.*, "Silicon heterojunction solar cell with interdigitated back contacts for a photoconversion efficiency over 26%," *Nat. Energy*, vol. 2, no. 5, p. 17032, May 2017, doi: 10.1038/nenergy.2017.32.
[2] "SunDrive achieves 26.41% efficiency with copper-based solar cell tech," *pv magazine International*. https://www.pv-magazine.com/2022/09/05/sundrive-achieves-26-41-efficiency-with-copper-based-solar-cell-tech/ (accessed Sep. 26, 2022).
[3] K. Ellmer and R. Mientus, "Carrier transport in polycrystalline transparent conductive oxides: A comparative study of zinc oxide and indium oxide," *Thin Solid Films*, vol. 516, no. 14, pp. 4620–4627, Mai 2008, doi: 10.1016/j.tsf.2007.05.084.
[4] D. Greiner, S. E. Gledhill, Ch. Köble, J. Krammer, and R. Klenk, "Damp heat stability of Al-doped zinc oxide films on smooth and rough substrates," *Thin Solid Films*, vol. 520, no. 4, pp. 1285–1290, Dezember 2011, doi: 10.1016/j.tsf.2011.04.190.
[5] T. Tohsophon *et al.*, "Damp heat stability and annealing behavior of aluminum doped zinc oxide films prepared by magnetron sputtering," *Thin Solid Films*, vol. 511–512, pp. 673–677, Jul. 2006, doi: 10.1016/j.tsf.2005.12.130.
[6] A. Cruz *et al.*, "Effect of front TCO on the performance of rear-junction silicon heterojunction solar

cells: Insights from simulations and experiments," *Sol. Energy Mater. Sol. Cells*, vol. 195, pp. 339–345, Jun. 2019, doi: 10.1016/j.solmat.2019.01.047.
[7] O. Madani Ghahfarokhi, K. Chakanga, S. Geissendoerfer, O. Sergeev, K. von Maydell, and C. Agert, "DC-sputtered ZnO:Al as transparent conductive oxide for silicon heterojunction solar cells with μc-Si:H emitter," *Prog. Photovolt. Res. Appl.*, vol. 23, no. 10, pp. 1340–1352, 2015, doi: 10.1002/pip.2570.
[8] H. Li *et al.*, "Application of Room Temperature Sputtered Al-doped Zinc Oxide in Silicon Heterojunction Solar Cells," in *2018 IEEE 7th World Conference on Photovoltaic Energy Conversion (WCPEC) (A Joint Conference of 45th IEEE PVSC, 28th PVSEC & 34th EU PVSEC)*, Jun. 2018, pp. 2151–2154. doi: 10.1109/PVSC.2018.8547891.
[9] A. B. Morales-Vilches *et al.*, "ITO-Free Silicon Heterojunction Solar Cells With ZnO:Al/SiO2 Front Electrodes Reaching a Conversion Efficiency of 23%," *IEEE J. Photovolt.*, pp. 1–6, 2018, doi: 10.1109/JPHOTOV.2018.2873307.
[10] K. Ellmer, "Resistivity of polycrystalline zinc oxide films: current status and physical limit," *J. Phys. Appl. Phys.*, vol. 34, no. 21, pp. 3097–3108, Oct. 2001, doi: 10.1088/0022-3727/34/21/301.
[11] K. Ellmer and R. Mientus, "Carrier transport in polycrystalline ITO and ZnO:Al II: The influence of grain barriers and boundaries," *Thin Solid Films*, vol. 516, no. 17, pp. 5829–5835, Jul. 2008, doi: 10.1016/j.tsf.2007.10.082.
[12] A. Cruz *et al.*, "Influence of Silicon Layers on the Growth of ITO and AZO in Silicon Heterojunction Solar Cells," *IEEE J. Photovolt.*, pp. 1–7, 2019, doi: 10.1109/JPHOTOV.2019.2957665.
[13] M. Bivour, C. Reichel, M. Hermle, and S. W. Glunz, "Improving the a-Si:H(p) rear emitter contact of n-type silicon solar cells," *Sol. Energy Mater. Sol. Cells*, vol. 106, pp. 11–16, Nov. 2012, doi: 10.1016/j.solmat.2012.06.036.
[14] S. Kirner *et al.*, "The Influence of ITO Dopant Density on J-V Characteristics of Silicon Heterojunction Solar Cells: Experiments and Simulations," *Energy Procedia*, vol. 77, pp. 725–732, Aug. 2015, doi: 10.1016/j.egypro.2015.07.103.

P-TYPE TOPCON SOLAR CELLS WITH N-POLY REAR EMITTER

Shih-Peng Hsu, Chao-Ping Huang, Chai-Hung Tsai, Sung-Yu Chen

Green Energy & Environment Research Laboratory, Industrial Technology Research Institute
No. 360, Gaofa 2nd Rd., Guiren Dist., Tainan City 711010, Taiwan

ABSTRACT: A p-TOPCon rear emitter solar cell was implemented in this study. The front surface field is formed by low pressure BCl_3 diffusion. The recombination current density of the front surface is reduced by adjusting the in-situ oxidation parameters in the drive-in step of the diffusion process. The rear emitter is an n-poly silicon layer made by in-situ doping of low pressure chemical vapor deposition process, and there is no peeling issue in thicker n-poly layer. The recombination current density on the rear side can be reduced through the modification of the PH_3 gas flow rate, tunnel oxide thickness and annealing temperature. The implied open circuit voltage of these cells before screen printing can reach 712mV, and conversion efficiency is 23.1% for the best cell with 12-busbar electrode pattern.

Keywords: Tunnel oxide passivated contacts, rear emitter, Ga-doped

1 Introduction

Solar cells have always played an important role in renewable energy applications, and silicon solar cells account for more than 90% of the market share of the entire solar cell today. In the history of silicon solar cells, Al-BSF (aluminum back surface field) solar cells are the mainstream of silicon solar cells for a long time before 2017 due to the simple processes and low manufacturing cost. PERC (passivated emitter and rear contact) solar cells with modified rear surface passivation and compatible with the processes of the Al-BSF cells, replaced the Al-BSF cells later due to superior efficiency. However, due to the backside laser opening process and the direct contact of the metal electrode with the silicon substrate, they will lead to high recombination losses and the efficiency of the PERC solar cells is limited to around 23~24%. To mitigate such deleterious effects, the passivated contacts are being developed via the insertion of the very thin tunnel oxide and hydrogenated poly-silicon layers between the silicon wafer and the overlying metal electrodes. Fraunhofer ISE implemented an n-type solar cell featuring a rear-side passivated contacts with an efficiency of 25.1%, which is called tunnel oxide passivated contact (TOPCon) solar cell [1]. A p-type rear emitter TOPCon cell with 26% efficiency were slso demonstrated recently [2]. The high efficiency potential via the carrier selective contacts and high temperature process compatible features significantly attract the attention of te industry.

In our approach we use Ga-doped p-type CZ wafers with thin thermal oxide and low pressure chemical vapor deposition (LPCVD) n-poly silicon layers as rear emitter and uniform boron-diffused layer as front surface field (FSF), as shown in Figure 1. The manufacturing processes are very similar to those of n-TOPCon solar cells.

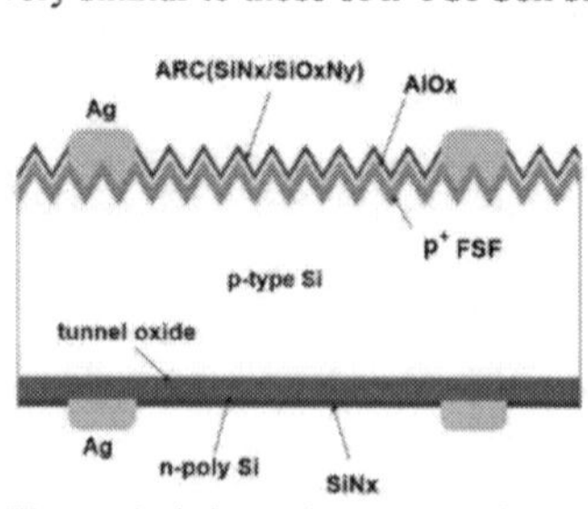

Figure 1: Schematic cross section of p-TOPCon solar cell with full-area passivated contact rear emitter.

2 Experimental

The p-TOPCon solar cells fabricated in this study have an area of 158.75x158.75 cm². The wafers are (100)-oriented Ga-doped p-type CZ silicon wafers with a thickness of 180μm. The wafers were polished in alkaline solution to remove saw damage layer followed by ultrathin thermal oxide formation in tube furnace. The n-poly Si layer was deposited on both sides by LPCVD method. The flow rate of SiH_4 was fixed at 300sccm, in-situ doping concentration was increased by increasing the PH_3 flow rate. The deposition temperature and pressure were kept at 580°C and 150mtorr, respectively. The wafers were annealed at 890°C and then a thickness of 80nm plasma enhanced chemical vapor deposition (PECVD) SiN_x capping layer was deposited on both sides. The lifetime and implied V_{oc} were measured by WCT-120.

The front side of the solar cells were textured in alkaline solution to form a random-pyramids surface. The boron-doped p^+ FSF was formed by low pressure BCl_3 furnace tube pre-deposition at 850°C followed by drive-in and in-situ oxidation at 950°C. There are three different types of boron diffusion profile for FSF preparation according to the different times of drive-in and in-situ oxidation. Additional (100)-oriented n-type polished wafers are diffused in the same boron diffusion conditions for boron doping profile measurement by the Electrochemical Capacitance-Voltage (ECV) method. After a SiN_x protection layer deposited on front surface, the full-area passivated contact emitter on the rear side was formed by an ultrathin thermal oxide layer covered by a 170nm thick phosphorus-doped polycrystalline silicon layer deposited with LPCVD. Removal of wrap-around poly silicon from front side was followed by annealing of phosphorus-doped polycrystalline silicon layer on rear side at 890°C. The rear side was additionally coated with a 50nm thick SiN_x layer grown by PECVD. The FSF was passivated with a 7nm thick AlO_x layer deposited with atomic layer deposition. The AlO_x layer was covered with a triple-layer antireflection coatings of $SiN_x/SiN_y/SiO_xN_y$ grown by PECVD. Both sides metallization are realized by screen printing of silver pastes and then co-firing in IR belt furnace. The process flow chart is shown in Figure 2.

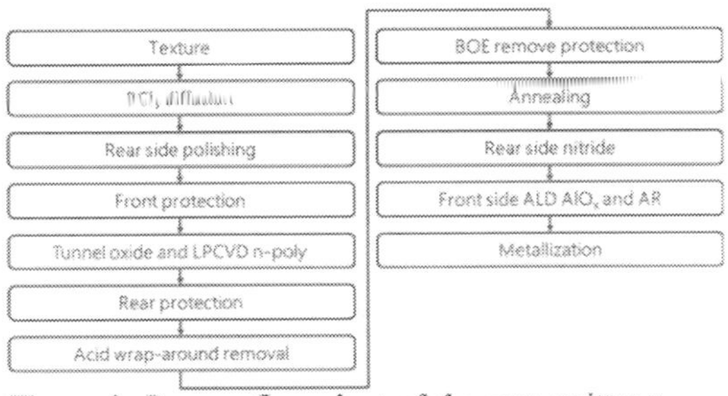

Figure 2: Process flow chart of the rear emitter p-TOPCon solar cells.

The 1-sun I-V characteristics of the solar cells were measured under standard testing conditions (STC: AM1.5g, 100mW/cm^2, 25°C). The active boron concentration profiles are measured with Nanometrics ECVPro system. The eternal quantum efficiency (EQE) values are measured with PVmeasurement QX-7 system.

3 Results and discussion

The sheet resistance, lifetime and implied V_{oc} are shown in Table I. The lifetime and implied V_{oc} increase with increasing PH_3 flow rate until PH_3 flow rate reaches 13sccm. When the PH_3 flow rate is 15sccm, the lifetime is slight lower than that of 13sccm. In view of this, the PH_3 flow rate of the LPCVD step was kept at 13sccm in the following experiments of rear emitter TOPCon solar cells.

Table I: Sheet resistance, lifetime and implied V_{oc} of the wafers with the same oxide and different n-poly Si prepared with various PH_3 flow rates

PH_3 flow rate (sccm)	Resistance (ohm/sq)	lifetime (μsec)	iV_{oc} (mV)
3	N/A	2.7	540
5	N/A	2.7	545
7	N/A	3.1	550
9	150	167.2	687
11	96	488.1	723
13	57	720.8	735
15	65	713.8	735

The results of electrochemical capacitance-voltage (ECV) measurement are shown in Figure 3. The surface concentrations of shallow FSF wafer and middle FSF wafer are similar, the boron concentration in deep FSF wafer is much lower than that of the other two wafers due to the 3-hour in-situ oxidation. The corresponding emitter sheet resistance values of these three different diffusion conditions in n-type wafers are 160, 105 and 125 ohm/sq, respectively.

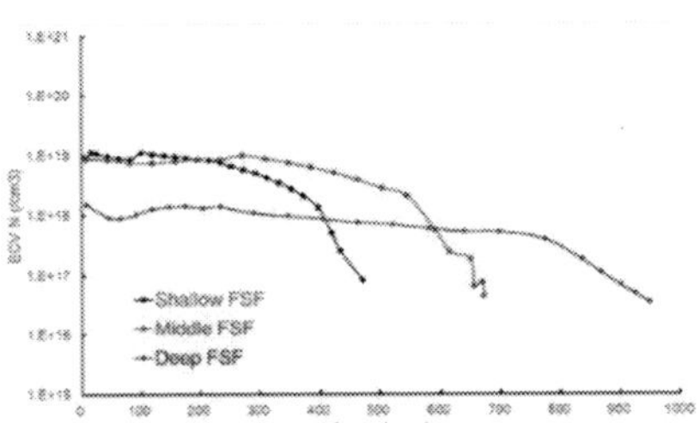

Figure 3: Diffusion profiles of boron concentration measured by ECV method.

The measured I-V characteristics of the p-TOPCon solar cells are shown in Table II. The J_{sc} is much lower than that of n-TOPCon solar cells due to the lower lifetime of Ga-doped p-type wafers. The I-V characteristics of shallow FSF cell and middle FSF cell are similar. The V_{oc}, J_{sc} and FF of the deep FSF cell are higher than those of the other two types of cells. The implied V_{oc} (iV_{oc}) and recombination current density of the deep FSF cells are 712mV and 24fA/cm^2, which are better than those of shallow FSF cells. The results of the EQE measurement are shown in Figure 4, EQE curve of the deep FSF cell is higher than that of the other two types of cells in all wavelength range. In order to improve the FF of deep FSF cell, 5-busbar finger pattern was replaced by 12-busbar finger pattern, the FF increases to 81.93%. J_{sc} also increases to 40.40mA/cm^2 due to the probe shading effect correction in I-V measurement. The efficiency of the best cell is 23.1%.

The rear emitter p-TOPCon cells without any FSF were also implemented in this study, the V_{oc}, J_{sc}, FF and efficiency of the best cell are 668mV, 36.26mA/cm^2, 75.59% and 18.31%, respectively. Although the efficiency of the cell is much lower, the implied V_{oc} before metallization can reach 720mV. Since there are no complicated boron diffusion and wrap-around removal steps, the processes can be simplified significantly, as shown in Figure 5. The manufacturing cost will be decreased dramatically if the suitable silver paste for such FSF-free rear emitter p-TOPCon cell is developed in the future.

Table II: I-V characteristics of the cells (w/ 5-busbar finger pattern) made with three types of FSF

	V_{oc} (mV)	J_{sc} (mA/cm^2)	FF (%)	Efficiency (%)
Shallow FSF	682	39.01	79.17	21.0
Middle FSF	683	38.86	79.37	21.1
Deep FSF	698	39.35	81.18	22.3

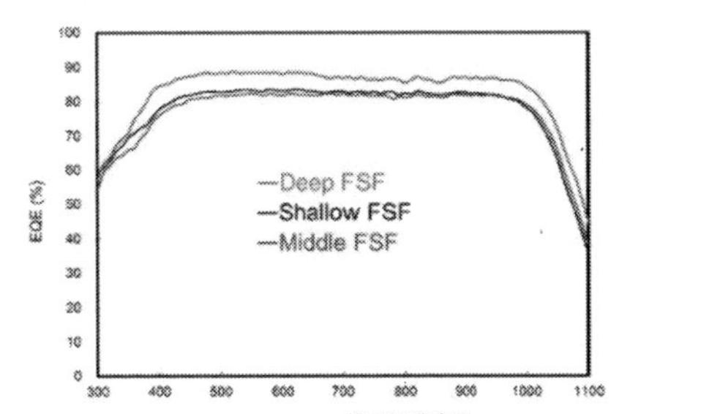

Figure 4: EQE curves of p-TOPCon solar cells made with three different types of FSF

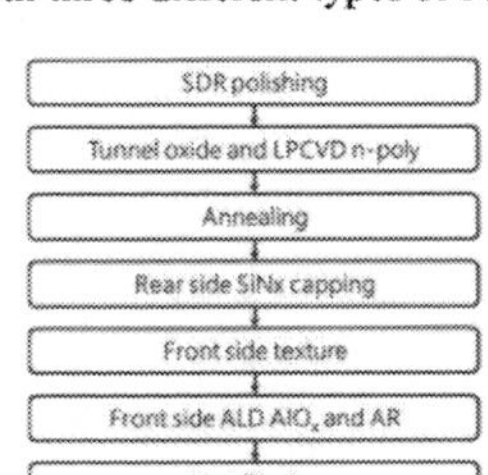

Figure 5: Process flow chart of the FSF-free rear emitter p-TOPCon solar cells

4 Conclusion

The rear emitter p-TOPCon solar cells were implemented in this study. The cells with deeper and lower boron concentration profile FSF can have better performance. The efficiency of the best cells with deep FSF and 12-BB is 23.1%.

Acknowledgements

The financial support provided by Bureau of Energy, Ministry of Economic Affairs, Taiwan (R.O.C.) (Grant No. 111-S0102) is gratefully acknowledged.

References

[1] A. Richter, J. Benick, R. Mueller, F. Feldmann, C. Reichel, M. Hermle, S. W. Glunz, Progress in Photovoltaics Research and Applications, 2018, 26, 579–586.

[2] A. Richter, R. Mueller, J. Benick, F. Feldmann, B. Steinhauser, C. Reichel, A. Fell, M. Bivour, M. Hermle, S. W. Glunz, Nature Energy 6 (2021) 429–438.

FABRICATION OF IBC SOLAR CELLS WITH EFFICIENCY HIGHER THAN 24% USING TOPCON TECHNOLOGY

Han-Chen Chang[1], Chien-Chi Huang[2], Bo-Cheng Kung[1], Sung-Yu Chen[1], Ming-Tsun Kuo[1], Peichen Yu[2] and Chorng-Jye Huang[1]

[1]Photovoltaic Technology Division, Green Energy and Environment Research Laboratories, Industrial Technology Research Institute, Tainan, Taiwan

[2]Department of Photonics, National Yang Ming Chiao Tung University, Hsinchu, Taiwan

ABSTRACT: In this article, an interdigitated back passivated contact (IBPC) solar cell, a new type of heterojunction solar cell with a rear side tunnel oxide passivated contact (TOPCon) structure is presented. Numerical simulations of cell structure have been investigated, including the thickness of anti-reflection coating (ARC), the thickness of p^+ poly-Si layer, anode finger width and p^+-n^+ distance. The photovoltaic properties of the prepared solar cell are Voc = 701.6 mV, Jsc = 41.5 mA/cm^2, FF = 82.53%, and η = 24.03%.

Keywords: Interdigitated back contact solar cell, tunnel oxide passivated contact, interdigitated back passivated contact solar cells

1 INTRODUCTION

Since the concept of back contacted (BC) solar cells was proposed, it has become one of the popular research topics of solar cell technology due to its advantages bringing the excellent architecture properties. The earliest developed BC solar cells are called interdigitated back contact (IBC) cells. They refer to PN junction and metal electrodes are arranged on the rear side of cell in the shape of interdigitation, and the front of cell is completely free of the shielding of the metal electrodes.

According to ITRPV annual report as shown in Figure 1 [1], the market share of IBC solar cells will gradually increase to 8% by 2032. There are two reasons for the phenomenon: (1) the cost of n-type wafers is gradually lower than that of p-type wafers (2) the gradual maturity of n-type solar cell technologies.

The commericalized high-efficiency solar cells (e.g. IBC, HIT and TOPCon) based on n-type silicon wafers are fabricated with a conversion efficiency of more than 23% [2-10]. Compared to HIT solar cells, IBC solar cells are more compatible with the current industrial manufacturing process, and have potentials for low-cost production. However, the manufacturing processes of traditional IBC solar cells are complicated, and the industry is difficult to accept. Therefore, we developed a simple, new type, low cost and high efficiency IBPC solar cell using a tunnel oxide passivated contact structure.

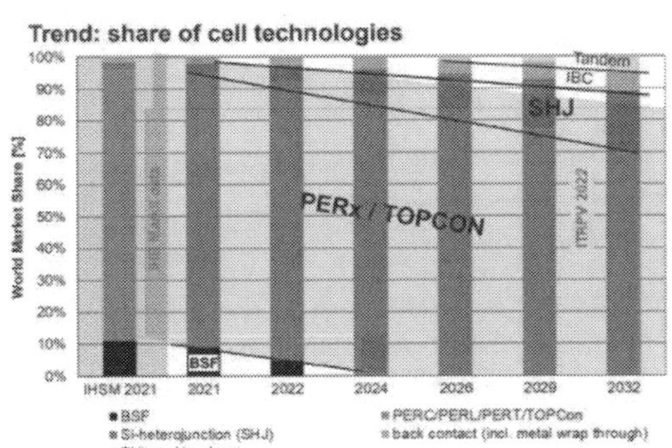

Figure 1 : Market share of c-Si PV cell technologies [10].

2 EXPERIMENTAL METHODS

In this work, we focused on the process integration of interdigitated back passivated contact solar cells with the recently matured TOPCon technology. Electrical contacts were fabricated using laser scribing techniques and screen printing with high precision alignment stages to get rid of cost-intensive lithography process typically used by IBC. We employed three main approaches to develop IBPC solar cells including (1) Numerical simulation of device structures to predict device characteristics and optimize designs, (2) Deposition of doped polycrystalline silicon replacing furnace diffusion or ion implantation to improve uniformity and cost, and (3) Development of high accuracy alignment technology to reduce series resistance and increase fill factor (FF).

In the conventional heterojunction solar cell architecture, the cell has high open circuit voltage (Voc) and short circuit current density (Jsc) due to the excellent surface passivation of amorphous silicon (a-Si) layer; however, it always leads to a low FF and a "S" curve in the IV plot due to the influence of potential barriers on charge transport across heterojunction [11]. These barriers were either created through a mismatch in energy band alignment with the selective contact or with the passivation layer. In order to achieve high efficiency, we use the 2D simulation software that can simulate the interface defect and junction energy band to conduct the simulation of IBPC solar cell and determine the most suitable experimental parameters to achieve high conversion efficiency.

The process of preparing IBPC solar cell is shown in Figure 2. First, the as-cut wafer was polished by an alkali etching to remove saw damage, and then the front side surface of silicon wafer was textured to reduce reflection of the incident light. After cleaning the wafer by hydrochloric acid-hydrogen peroxide-water mixtures (HPM) and then dipped by diluted hydrofluoric acid (HF) to removal native oxide from wafer surfaces. Tunnel silicon oxide was formed on both sides with an oxidation furnace. Boron-doped polycrystalline silicon (poly-Si) film was deposited by plasma enhanced chemical vapor deposition (PECVD) on the rear tunnel silicon oxide.

The boron-doped p^+ emitter and back surface field (BSF) regions were defined by laser ablation of boron-doped polycrystalline silicon, followed by HF etching to remove residual surface contamination and laser damage. Prior to laser ablation, SiN_x was deposited as a mask to protect boron-doped p^+ emitter from etching during laser damage removal. After laser damage removal, both side SiN_x and front side oxide were removed by HF.

n^+ BSF and n^+ front surface field (FSF) were simultaneously fabricated by $POCl_3$ diffusion. Prior to $POCl_3$ diffusion, SiN_x was deposited at rear side as a diffusion mask to protect boron-doped p^+ poly-Si emitter from phosphorus doping, and the SiN_x was removed by HF after $POCl_3$ diffusion. Then, the anti-reflection coating (ARC) layer of SiN_x by PECVD were deposited at the front side, and the tunnel silicon oxide on both BSF and emitter grown by oxidation furnace, and silicon oxynitride (SiON) growth on the front side during oxidation [12]. At the rear side of solar cell, BSF and emitter were passivated by a stack of Al_2O_3 and SiN_x successively deposited by atomic layer deposition (ALD) and PECVD, respectively. After the removal of front side Al_2O_3 and SiN_x, silver paste and aluminum paste were screen printed on the BSF and p^+ emitter, respectively. The co-firing process was carried out after screen printing processes. During co-firing, the silver paste penetrated $SiO_x/Al_2O_3/SiN_x$ resulting in ohmic contact with the p^+ emitter; in addition, an Al-Si alloy formed between the emitter and the aluminum electrode resulting in n^+ BSF.

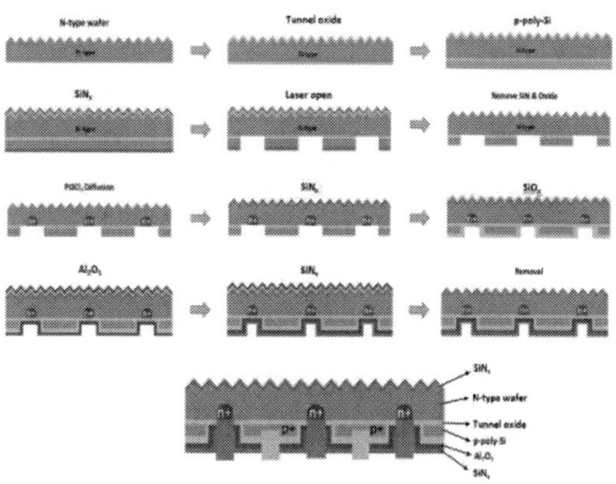

Figure 2 : Schematic diagram of the processes flow for fabricating IBPC cells.

3 RESULTS AND DISSUSION

In this work, the ARC layer is a bilayer of SiO_x and SiN_x. The refractive index and absorption coefficient of ARC layer obtained by optical measurement were used as the parameters for simulation. This simulation provided optimal combination of refractive index and thickness for a high-efficiency IBPC solar cells shown in Figure 3. An ARC layer with a refractive index of 1.96 and a thickness of 70nm is the best combination, compared to previous ARC layer with a refractive index of 2.03 and a thickness of 90nm, the Jsc and efficiency of the optimal IBPC solar cell estimated to increase by 0.55mA/cm^2 and 1.68%, respectively.

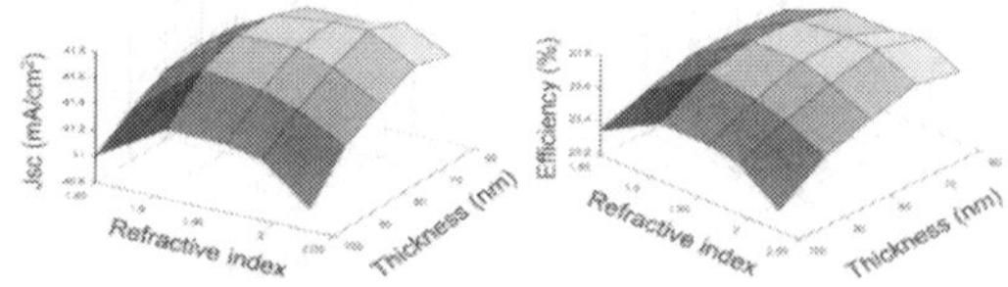

	Thickness (nm)	n	Voc (mV)	Jsc (A/cm^2)	FF (%)	Efficiency (%)
Reference	90	2.03	685	41.19	78.41	22.13
POLO IBPC	70	1.96	730	41.75	78.05	23.81

Figure 3 : IV characteristics simulations (a) Jsc (b) efficiency of IBPC solar cells with ARC layers of different refractive indices and thicknesses.

The IBPC solar cell structure is divided into 5 key areas, including (1) front electric field passivation area, (2) n^+ BSF $POCl_3$ diffusion area, (3) p^+ poly-Si emitter passivation area, (4) p^+-n^+ gap (i.e. the distance between p^+ and n^+) passivation area, (5) electrode and polycrystalline silicon contacting area. The n^+ $POCl_3$ diffusion and p^+ poly-Si deposited on the tunnel SiO_x areas were optimized by improving the passivation characteristics, mainly sheet resistance and contact resistance. Before optimizing the patterns of back contact electrodes, it is necessary to simulate the p^+ poly-Si thickness and doping concentration for a high-efficiency IBPC solar cell. To search for the best combination of p^+ poly-Si thickness and doping concentration via simulation, p^+ poly-Si thickness and doping concentration were varied from 50nm to 300nm and from 1e18 cm^{-3} to 5e19 cm^{-3}, respectively. Figure 4(a) shows that Jsc periodically fluctuates with increasing p^+ poly-Si film thickness, and the efficiency also has a consistent trend as shown in Figure 4(d). It also shows that different doping concentrations have almost no effect on Jsc. As shown in Figure 4(b) and 4(c), Voc and FF do not differ significantly with the film thickness; however, they significantly increase with increasing doping concentration. Figure 4(d) indicate that the efficiency increase with increasing doping concentration for various p^+ poly-Si thicknesses. It contributes to high Voc and FF because of larger band-bending and higher electronic conductivity.

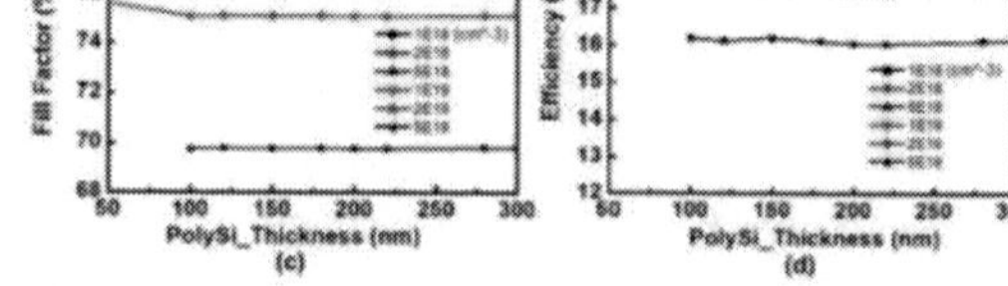

Figure 4 : Simulated IV characteristics (a) Jsc (b) Voc (c) FF (d) Efficiency as a function of p^+ poly-Si layer thicknesses with different doping concentrations.

In the IBPC solar cell, n^+ $POCl_3$ diffusion area and p^+ poly-Si area collect electrons and holes, respectively. The electrons drift to the cathode under the influence of the built-in electric field. Accordingly, increasing the distance between adjacent n^+ $POCl_3$ diffusion areas (i.e. the fingers of cathode) can increase the resistance and decrease the fill factor. On the other hand, p^+ poly-Si collects minority carriers (i.e. holes) which move to the electrodes (i.e. the fingers of anode) mainly by diffusion.

The design of geometric pattern of electrode fingers was optimized by simulation, and a total pitch of 1100 μm between the first cathode finger and the last anode finger were assumed. Simulations were performed at p^+-n^+ gaps of 50, 100 and 150 μm, with anode finger widths ranging 100 to 800 μm as the cathode finger width decreased accordingly. Figure 5(a) and 5(d) show that the efficiency increasing with anode finger width is corresponding with the trend of Jsc. Figure 5(b) surprisingly shows that Voc slight decreases with increasing anode finger width for all different p^+-n^+ gaps, though it doesn't significantly affect efficiency. Figure 5(b) also shows that at different anode finger widths, Voc increases with increasing p^+-n^+ gap. In Figure 5(c), for the p^+-n^+ gap of 150 μm, the fill factor increases as anode width increases upto 700 μm, a reverse drop seen at an anode finger width of 800 μm. The anode finger width of 800 μm represents an extremely narrow cathode finger width of 150 μm, which leads to poor cathode contact and solar cell performance.

It is worth noticing that the extremely narrow cathode finger width makes it difficult to prepare solar cells, such as a high risk of mistaks in forming contacts by laser processing. How to avoid the short circuit caused by the contact of the anode finger and cathode finger is one of the key techniques in the manufacturing process of IBPC solar cells.

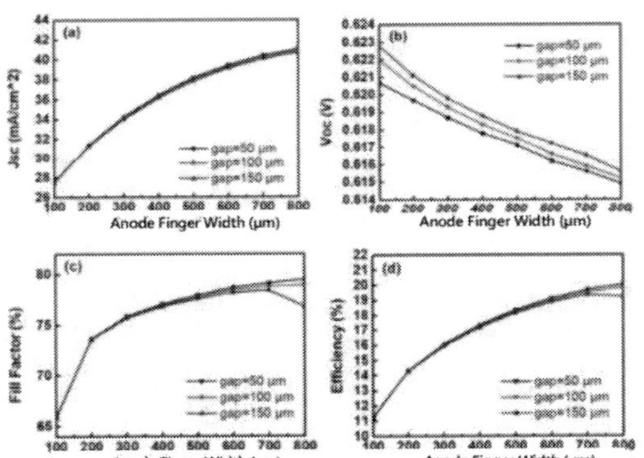

Figure 5 : Simulated IV characteristics (a) Jsc (b) Voc (c) FF (d) Efficiency as a function of anode finger width along with p^+-n^+ gap.

The passivation-related parameters, minority carrier lifetime, and reverse saturation current can be obtained through numerical calculation and the results can be used for practically preparing IBPC solar cells. The front side of the IBPC solar cells has a textured surface to reduce light reflection, and the rear side is polished to reduce carrier recombination at or near the surface. For surface passivation, the different materials prepared on textured and polished silicon surfaces are studied. In Figure 6, all surface passivation materials show lower recombination current densities (J_0) on polished silicon surfaces than on textured silicon surfaces. It is worth noting that the J_0 of Al_2O_3 deposited on the polished silicon surface is dramatically higher than that of the textured silicon surface. In contrast, the J_0 of SiN_x has little difference between textured and polished silicon surfaces, implying that SiN_x passivation is independent of surface morphology and crystallographic plane. All of passivation materials, the Al_2O_3+SiN_x stack has the best passivation on the polished silicon surfaces, with a lowest J_0 below 150 fA/cm^2.

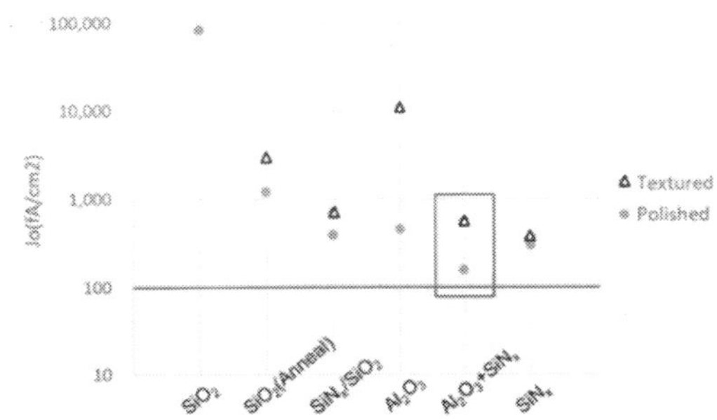

Figure 6 : The recombination current density (J_0) of various surface passivation materials on the textured (△) and polished (●) surfaces of wafer.

In the fabrication of electrodes, laser ablation and screen printing techniques were applied. The purpose of laser ablation is to remove the stacked dielectric layers, thereby bringing the BSF and emitter into direct contact with the metal-based conductive pastes for low series resistance. The anode fingers and cathode fingers are interdigitated on the rear side of the solar cell; therefore, reducing the alignment error of laser ablation can effectively avoid short circuiting. Wafer cleaning after laser ablation is another key process. Since the laser ablation was carried out in an atmospheric environment, and the surface of silicon wafer reacted with oxygen in the surrounding atmosphere to from a SiO_x layer identified by X-ray photoelectron spectroscopy (XPS). If this SiO_x layer is left in place, it will be covered by other layers deposited in subsequent processes and then act as a barrier to conduction inside the device. Figure 7 shows that the efficiency of the solar cell impregnated with diluted HF is significantly improved by a factor of 10 over that of un-impregnated. Furthermore, the efficiency of solar cell is also substantially affected by laser damages. Laser damages could be solved by surface etching, through laser parameter optimization and using short-pulse laser source [13].

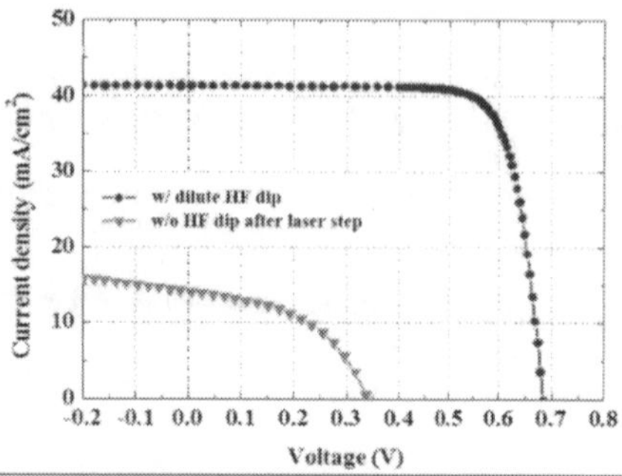

IV test	J_{SC} (mA/cm^2)	V_{OC} (mV)	FF (%)	η (%)
w/ dilute HF dip	41.21	685.25	78.48	22.16
w/o HF dip after laser step	14.16	340.68	47.60	2.29

Figure 7： J-V characteristics of IBPC solar cells. After laser ablation, dipped with (●) and without (▼) diluted HF.

Finally, in order to prevent P and N fingers from short circuiting, we selected less fine P:N ratio of 2:1 (i.e. 600μm : 300μm) and a p^+-n^+ gap of 100 μm for device fabrication. An Al_2O_3 layer with a film thickness of 7 nm was deposited by atomic layer deposition (ALD) on the front side of solar cell to improve passivation. The photovoltaic properties of the prepared solar cells were measured under AM 1.5G simulated light and the results are shown in Figure 8. The measured IV characteristics are Voc = 701.6 mV, Jsc = 41.5 mA/cm^2, FF = 82.5%, and η = 24.03%.

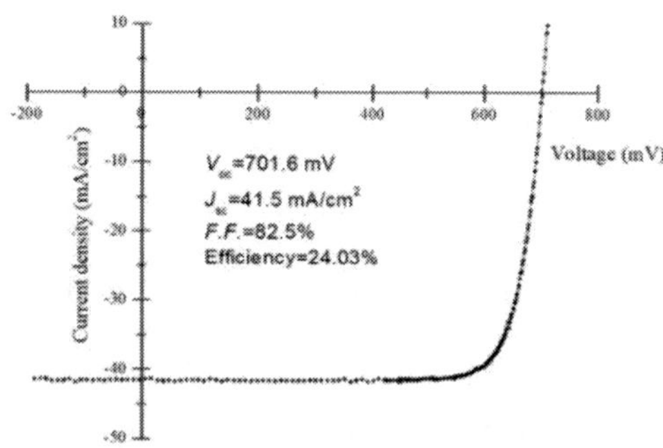

Figure 8. J-V curve and measured characteristics of an IBPC solar cell with an area of 4mm^2.

4 CONCLUSION

We have performed device design and computational simulations of back contact solar cell incorporating TOPCon technology. The cell efficiency is closely related to the doping concentration of the p^+ poly-Si. We also demonstrated the process integration of an IBPC solar cell with an efficiency of 24.03% prepared under optimized parameters and process conditions. Furthermore, it is optimistic to use laser contact openings and screen printing to replace time-consμming and cost-intensive lithography processes typically used by IBC.

5 ACKNOWLEDGEMENTS

The financial support provided by Bureau of Energy, Ministry of Economic Affairs, Taiwan (R.O.C.) (Grant No. 111-S0102) is gratefully acknowledged.

6 REFERENCE

[1] International Technology Roadmap for Photovoltaic (ITRPV) Results 2022 incl. maturity repot, 13th edition.
[2] S. Werner, 33rd Eur. PV Solar Energy Conf. Exhib., Amsterdam, The Netherlands, (2017) 406-412.
[3] K. Masuko et al., IEEE J. Photovolt., 4 (2014) 1433-1435.
[4] J. Nakamura et al., IEEE J. Photovolt., 4 (2014) 1491-1495.
[5] U. Römer et al., Solar Energy Mater. Solar Cells, 131 (2014) 85-91.
[6] A. D. Upadhyaya et al., IEEE J. Photovolt., 6 (2016) 153-158.
[7] A. Moldovan et al., Solar Energy Mater. Solar Cells, 142 (2015) 123-127.
[8] Y. Peng et al., IEEE Journal of Photovoltaics, 10 (2020) 383-389.
[9] F. Haase et al., Solar Energy Materials and Solar Cells, 186 (2018) 184-193.
[10] K. Gensowski et al., Prog Photovolt Res Appl, 28 (2020) 538-544.
[11] R. Saive, IEEE Journal of Photovoltaics, 10 (2019) 1477-1484.
[12] S. I. Raider et al., J. Electrochem. Soc., 123 (1976) 560-565.
[13] Z. Du et al., IEEE J. Photovolt., 6 (2016) 617-623.

BORON AND PHOSPHORIS DOPED FIREABLE PASSIVATING TUNNEL OXIDE CONTACTS FOR CRYSTALLINE SILICON SOLAR CELLS

Franz-Josef Haug, Mario Lehmann, Sofia Libraro, Ezgi Genç, Audrey Morisset, Christophe Ballif)
Ecole Polytechnique Fédérale de Lausanne (EPFL), Photovoltaics and Thin Film Electronics Laboratory (PV-Lab)
rue de la Maladière 71b, CH-2000 Neuchâtel, Switzerland

ABSTRACT: We present passivating contacts for silicon solar cells using an interfacial layer of SiO2 grown by immersion in hot HNO3 and in-situ doped layers of both polarities grown by plasma enhanced chemical vapor deposition (PECVD). The doped layers are crystallized and their dopants are activated by rapid thermal processing (RTP) with a temperature profile similar to the firing of screen-printed metallization pastes, hence we call them fired passivating contacts (FPCs). After RTP, we apply hydrogenation to passivate defects, using thermal release of hydrogen from a SiNx reservoir layer during hot plate annealing at 450°C. We find that the kinetics of the hydrogenation process is characterized by a fast process associated to the passivation of defect states at the Si/SiO2 interface and a slower process likely related to the hydrogenation of bulk defects that are formed during RTP in the floating zone wafers used in this experiment. Despite the issue with bulk defects, we test the suitability of the FPCs for solar cells by depositing n- and p-type layers on the textured front and on the flat rear of single-side textured wafers, respectively. After co-processing in a single RTP step, we apply the optimum hydrogenation conditions identified above. Using an ITO/Ag metallization made at low temperature to fabricate proof-of-concept solar cells, we obtain efficiencies up 20.5%.

Keywords: Silicon Materials and Cells, High-Temperature Route for Si Cells

1. INTRODUCTION

Silicon solar cells with passivating contacts on the basis of a tunnelling oxide and a doped layer of silicon have gathered significant attention as they yield efficiencies of 26% and above [1], [2], and because they are compatible with standard fire-through metallization using rapid thermal annealing processes often called firing [3], [4]. Most work on passivating contacts applies to n-type contacts where the silicon layer is grown either with in-situ doping or implanted after growth. In both cases, the layer is subjected to a thermal annealing process that crystallizes the layer, activates dopants in the layer, and in most cases diffuses some dopants across the interfacial oxide to yield a shallow doped region in the wafer below the interfacial oxide. In conjunction with a SiNx passivation and metallization with standard fire-through Ag pastes, the n-type contacts normally require film thicknesses in the order of 100 nm or more to avoid penetration of Ag through the passivating oxide [3], and therefore they are put at the rear to avoid parasitic absorption. Accordingly, the p-type contact is applied to the front to form the pn-junction, generally by diffusion of boron, application of an AlOx/SiNx front passivation and contacting with a Ag-Al paste.

Alternatively, passivating contacts can also be manufactured with rapid thermal processing (RTP) using conditions very similar to the firing process of the metallization. The thermal budget of RTP with a heating ramp of 50°C/s, dwell times between 3 and 30 s, and an equally fast cooling ramp was found sufficient to crystallize the doped layers, and to activate boron concentrations equal to the thermodynamic solubility limit [5]. However, the thermal budget of RTP is too low for the formation of a shallow in-diffused region [6]. In this case, surface passivation likely relies to a large extent on the passivation of defects at the Si/SiO_2 interface which can be achieved by the standard hydrogenation procedure used in solar cell manufacturing, i.e. thermal release of hydrogen from a SiNx:H reservoir layer [6]. The integration of a full-area p-type fired passivating contacts (FPC) at the rear of PERC solar cells could thus reduce recombination losses with minor changes to the manufacturing process.

In this contribution, we present FPCs of both polarities using in-situ doped layers grown by plasma enhanced chemical vapor deposition (PECVD). We investigate the impact of RTP on the bulk lifetime and on the effective lifetime of wafers covered with SiCx. Subsequently, we investigate the kinetics of defect passivation in the latter configuration by thermal release of hydrogen from a SiNx reservoir layer, using either hot plate annealing for several minutes at 450°C or firing for 3 s at 800°C. Finally, we test the suitability of the FPCs for solar cells by depositing n- and p-type layers on the textured front and on the flat rear of a single-side textured wafer, respectively. After co-processing in a single RTP step, we apply an ITO/Ag metallization made at low temperature to fabricate proof-of-concept solar cells, resulting in efficiencies up 20.5%.

2. RESILTS AND DISCUSSION

To assess the impact of RTP on the bulk lifetime of the p-type FZ wafers, we exposed them to different peak temperatures and subsequently covered them with a bilayer of intrinsic and n-type amorphous silicon. Owing to the excellent surface passivation of this layer stack [7], the lifetimes shown in Figure 1 are a good approximation for the bulk lifetimes. We find that the bulk lifetimes are little changed by RTP up to 450°C, but RTP at 600°C decays them to values between 10 and 20 μs. Eventually, RTP at higher temperatures yields a slight improvement up to values around 100 μs.

Our observation is similar to results of Hiller et al., but apparently our RTP process does not reproduce the almost complete recovery that they observed for their annealing processes [8]. The explained their observations in terms of nitrogen which is commonly added during pulling of FZ ingots; they argue that N interacts with silicon vacancy defects (V_{Si}) to form substitutional N_{Si} defects which in turn become passivated with H [8]. Loss of hydrogen from the passivated N_{Si}-H complex would thus express the N_{Si} defect state, explaining the low bulk lifetime after RTP at 600°C. Effusion of nitrogen during thermal treatment at higher temperature would explain the recovery of the lifetime for peak temperatures of 900°C and higher.

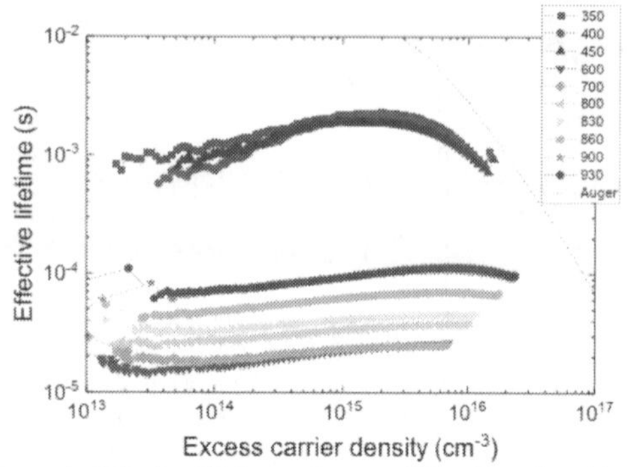

Figure 1: Effective lifetimes of p-type FZ wafers after firing at different temperatures and subsequent surface passivation with an amorphous i/n stack.

Figure 2 shows effective lifetimes measured on FZ wafers covered with a SiOx/SiCx layer stack. Looking first at the samples before hydrogenation (open symbols), it emerges that the effective lifetimes behave very similarly to the bulk lifetimes of Figure 1, i.e. they show a slight recovery as the peak temperature of the RTP in increased from 770 to 860°C (lower temperatures were not tested in this experiment). However, the trend reverses at a peak temperature of 900°C, indicating a deterioration of the interface passivation.

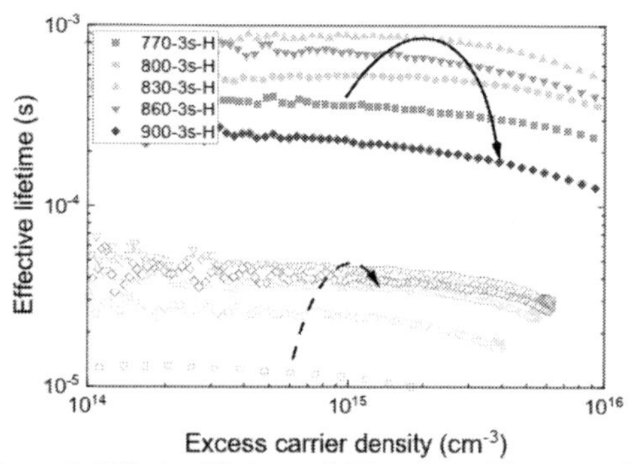

Figure 2. Effective lifetimes of FZ wafers with p-type SiCx passivating contacts on both sides after firing (open symbols) and after hydrogenation by firing (full symbols).

We can use the bulk lifetimes after firing from Figure 1 and the effective lifetimes before hydrogenation from Figure 2 to estimate the surface recombination velocity S according to the relation $\tau_{eff}^{-1} = \tau_{bulk}^{-1} + 2S/w$ [9].

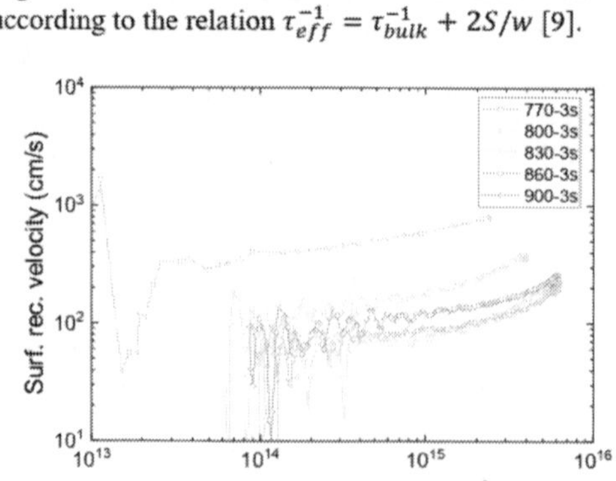

Figure 3: Surface recombination velocities of the SiOx/SiCx interfaces after firing at the shown temperatures.

The characteristics shown in Figure 3 suggest that the SiOx/SiCx interface is improved as the peak temperature of RTP is increased to from 770°C with an optimum between 830 and 860°C whereas higher peak temperatures increase the surface recombination velocity. This is likely due to damage of the interfacial oxide by the formation of pinholes which was reported for temperatures of 900°C and above [10], [11].

After hydrogenation, the full symbols Figure 2 show a substantial improvement of the effective lifetimes to values of ca. 10x higher than the bulk lifetimes shown in Figure 1. Thus, we conclude that the hydrogenation process passivates bulk as well as interfacial defects. As before, the trend of improving lifetime with increasing temperature of the RTP is inversed, in this case already for 860°C ad indicated by the arrow. Evaluation of the surface recombination velocity S for the samples after hydrogenation is ongoing.

To better understand the changes in bulk and surface passivation, we etched the SiOx/SiCx layers after each of the individual processing steps and passivated the surface again with a bilayer of intrinsic and n-type amorphous silicon. Thus, the photoluminescence images in Figure 4 and the overlaid values of the effective lifetime at $\Delta n = 10^{15}$ cm^{-3} are indicative of the bulk lifetimes after each processing step. We conclude that the deposition process of the SiCx at 200°C layer does not interfere with the passivation of bulk defects. After RTP for 3 s at 800°C, the PL image shows a ring-shape, indicating a high density of bulk defects in the centre of the wafer and a lower density in the peripheral region, similar to earlier reports on the spatial variation of nitrogen-related defects in FZ wafers [12]. After deposition of the SiNx layer, the peripheral region show a slight improvement, possibly due to passivation of interfacial defects by the hydrogen plasma. Finally, hydrogenation by thermal release of H from the SiNx:H layer improves the central region.

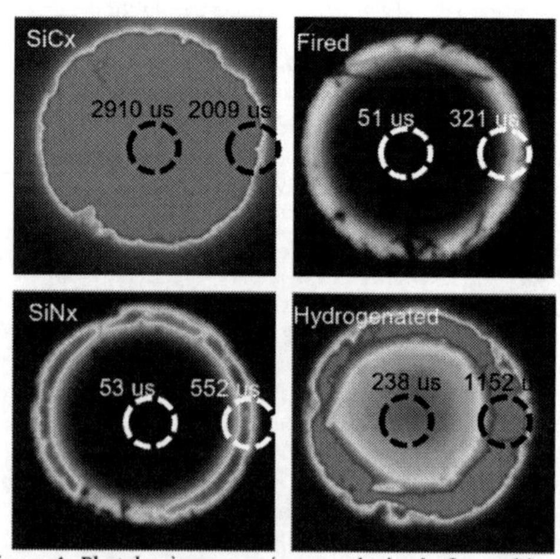

Figure 4: Photoluminescence images obtained after etching off the SiCx after each of the individual steps of the fabrication process and re-passivation with a-Si:H. Numbers indicate the effective lifetime at $\Delta n = 10^{15}$ cm^{-3}, overlaid circles illustrate the area of the lifetime measurement.

Figure 5 shows PL images recorded at different stages of the hydrogenation process. For the initial state at 0 min we find a uniformly low PL intensity. We attribute this to poor passivation by the un-hydrogenated SiOx/SiCx interface which yields a low effective lifetime also in those regions where high bulk lifetime is retained (c.f. the lower left image of Figure 4). After hydrogenation for 2 min, the PL image show an improvement of the peripheral region, indicating defects at the wafer surface are already passivated to a large extent. After prolonged hydrogenation, PL is also improved in the central region, indicating a slower hydrogenation process of the bulk defects.

We note that the hydrogenation kinetics proceeds very similarly for an n-type SiCx layer (not shown).

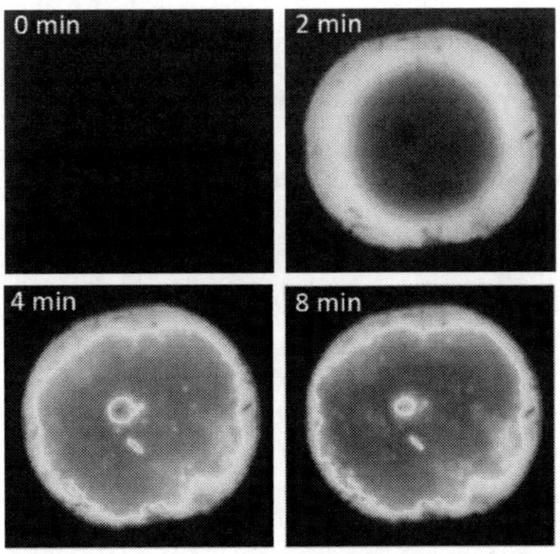

Figure 5: Photoluminescence images of SiOx/SiCx contacts of both polarities, recorded after different durations of hydrogenation on a hotplate at 450°C.

In order to test the passivation and the current extraction of the SiOx/SiCx passivating contacts, we manufactured proof-of-concept solar cells. To this end, we used single-side textured wafers, grew an interfacial oxide in ozone, and deposited n- and p-type SiCx layers on the textured front and on the flat rear, respectively. After co-processing in a single RTP step and hydrogenation on the hotplate, we removed the SiNx layer by HF dipping. Subsequently, we applied a low-temperature metallization. First, a bilayer of ITO/Ag was sputtered at the rear, followed by sputtering of a single layer of ITO at the front. Next, the front metallization was completed by screen-printing of a low-temperature Ag paste, followed by curing at 130°C.

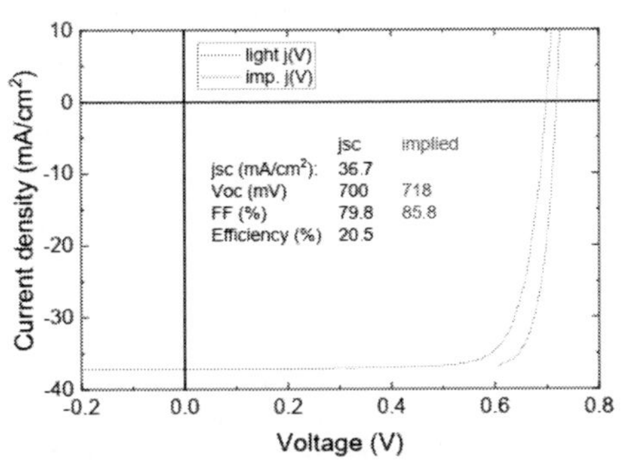

Figure 6. Current-voltage characteristics of a proof-of-concept solar cell for testing passivation and current extraction of FPCs at front and rear. Red and black curves represent the implied j(V) characteristics of the cell precursor and the measured characteristics, respectively.

The current voltage characteristics of the resulting device is shown in Figure 6. The Voc of the device is 700 mV which is ca. 30 mV less than the implied Voc before ITO deposition, presumably related to sputtering damage [10]. Overall, this approach for solar cell processing with the low thermal budget of RTP yields cells with promising efficiencies up to 20.5%. The FF and the current density are very similar to values of cells with annealed SiCx passivating contacts [10]. Nevertheless, the external quantum efficiency of the cell (not shown) indicates that parasitic absorption in the n-SiCx layer at the front can account for the moderate j_{sc}.

3. ACKNOWLEDGEMENT

We thankfully acknowledge funding by the EU project HighLite (contract No. 857793) and the Swiss Federal Office of Energy (SFOE).

4. REFERENCES

[1] F. Haase *et al.*, "Laser contact openings for local poly-Si-metal contacts enabling 26.1%-efficient POLO-IBC solar cells," *Solar Energy Materials and Solar Cells*, vol. 186, pp. 184–193, 2018.

[2] A. Richter *et al.*, "Design rules for high-efficiency both-sides-contacted silicon solar cells with balanced charge carrier transport and recombination losses," *Nature Energy*, vol. 6, no. 4, pp. 429–438, 2021.

[3] D. Chen *et al.*, "24.58% total area efficiency of screen-printed, large area industrial silicon solar cells with the tunnel oxide passivated contacts (i-TOPCon) design," *Solar Energy Materials and Solar Cells*, vol. 206, pp. 110258–110258, 2020, doi: 10.1016/j.solmat.2019.110258.

[4] J. Chen, "The development of n-type TOPCon cell and module in Jolywood," 2019.

[5] F.-J. Haug *et al.*, "Contributions to the contact resistivity in fired tunnel-oxide passivating contacts for crystalline silicon solar cells," *IEEE Journal of Photovoltaics*, vol. 9, no. 6, pp. 1548–1553, 2019.

[6] A. Ingenito *et al.*, "A passivating contact for silicon solar cells formed during a single firing thermal annealing," *Nature Energy*, vol. 3, no. 9, pp. 800–800, 2018.

[7] F.-J. Haug, S. Libraro, M. Lehmann, A. Morisset, A. Ingenito, and C. Ballif, "Impact of rapid thermal processing on bulk and surface recombination mechanisms in FZ silicon with fired passivating contacts," *Solar Energy Materials and Solar Cells*, vol. 238, p. 111647, 2022.

[8] D. Hiller *et al.*, "Kinetics of Bulk Lifetime Degradation in Float-Zone Silicon: Fast Activation and Annihilation of Grown-In Defects and the Role of Hydrogen versus Light," *physica status solidi (a)*, vol. 217, no. 17, p. 2000436, 2020.

[9] A. Sproul, "Dimensionless solution of the equation describing the effect of surface recombination on carrier decay in semiconductors," *Journal of Applied Physics*, vol. 76, no. 5, pp. 2851–2854, 1994.

[10] G. Nogay *et al.*, "Crystalline silicon solar cells with coannealed electron- and hole-selective SiCx passivating contacts," *IEEE Journal of Photovoltaics*, vol. 8, no. 6, 2018, doi: 10.1109/JPHOTOV.2018.2866189.

[11] D. Tetzlaff *et al.*, "A simple method for pinhole detection in carrier selective POLO-junctions for high efficiency silicon solar cells," *Solar Energy Materials and Solar Cells*, vol. 173, pp. 106–110, 2017.

[12] F. E. Rougieux, N. E. Grant, C. Barugkin, D. Macdonald, and J. D. Murphy, "Influence of annealing and bulk hydrogenation on lifetime-limiting defects in nitrogen-doped floating zone silicon," *IEEE Journal of Photovoltaics*, vol. 5, no. 2, pp. 495–498, 2014.

INVESTIGATION OF THE PASSIVATION QUALITY OF EX-SITU DOPED (P+) AND (N+) POLY-SI LAYERS FOR IBC SOLAR CELLS

Vaibhav V Kuruganti[1, 3*], Radovan Kopecek[1], Sven Seren[2], Olindo Isabella[3], Valentin D. Mihailetchi[1]
1. ISC Konstanz e.V., Rudolf-Diesel-Str. 15, 78467 Konstanz, Germany
2. SCHMID, Robert-Bosch str. 32-36, 72250 Freudenstadt, Germany
3. PVMD Group, Delft University of Technology, Mekelweg 4, 2628 CD, Delft, Netherlands

ABSTRACT: In recent years, crystalline silicon solar cell efficiency in mass production has increased by 0.5 to 0.6% per year. New cell technologies must keep this developing speed to enter mass production. This work presents a unique process route for fabricating complete ex-situ doped poly-IBC solar cells using a single high-temperature step and a single laser step. Patterning of the rear side of IBC solar cells was achieved using selective laser ablation of an (i) poly-Si capping layer. Excellent passivation quality was demonstrated with iV_{oc} and J_0 of 741 mV and 1.35 fA/cm^2 for (n+) poly-Si and 720 mV and 17.4 fA/cm^2 for (p+) poly-Si layers, respectively. However, upon fast firing in the industrial belt furnace, the passivation quality of the laser patterned n+ poly-Si regions degrades. The extent of passivation loss was found to depend on the laser ablation fluence values used for patterning. The observed passivation loss is hypothesized to be due to stress-induced blistering of the layer stack.

Keywords: Passivating contacts, Ex-situ doping, IBC, laser patterning.

1 INTRODUCTION

One of the main objectives of the photovoltaic community is to lower the levelized costs of electricity (LCOE) [1]. An effective way to achieve this target is to improve the efficiency of solar cells, thereby enhancing the photovoltaic energy yield [2]. Passivated contact based on highly doped polysilicon (poly-Si) and a thin interfacial oxide incorporated into an Interdigitated Back Contacted (IBC) architecture is an elegant solution to achieve high-efficiency solar cells [3]. Since the thin interfacial oxide separates the bulk c-Si layer and metal contacting poly-Si, we can benefit from both low contact resistance (ρ_c) as well as an excellent passivation quality, which majorly influences the Fill factor (FF) and open-circuit voltage (V_{oc}), respectively [4]. Furthermore, IBC architecture eliminates the optical shading losses caused due to metal fingers and bus bars on the front side. Moreover, the lack of the doped poly-Si layer on the front diminishes the parasitic absorption, thereby boosting the short circuit current density (J_{SC}) [5].

An extensive study has been carried out to understand and optimize the passivation quality of in-situ doped poly-Si layers. As a result, outstanding passivation quality of in-situ doped p+ poly-Si and n+ poly-Si layers have been reported with J_{0pass} values lower than 5 fA/cm^2 [6][7] and 2 fA/cm^2[8][9], respectively. However, in-situ doping of poly-Si may be cumbersome depending on the deposition technique used, since the introduction of dopant gases complicates the control of the layer thickness, dopant uniformity, and deposition rate. Also, adding dopants during deposition can affects the physical properties such as grain size and orientation. With all the mentioned shortcomings, ex-situ doping can be an interesting alternative for doping poly-Si layers.

Hence in this work, we have used complete ex-situ doped poly-Si layers for fabricating poly-IBC solar cells. As a result, we have demonstrated excellent passivation quality with iV_{oc} and J_0 of 741 mV and 1.35 fA/cm^2 for (n+) poly-Si layer and 720 mV and 17.4 fA/cm^2 (p+) poly-Si layer. However, upon fast-firing the samples in the industrial belt furnace, the passivation quality of the (n+) poly-Si regions formed using the laser ablation patterning degraded due to blistering of the layer stack. Also, it was observed that the extent of blistering depends on the laser ablation fluence used for ablating the (i) poly-Si capping layer.

2 EXPERIMENT

All test samples used in this work were fabricated using large area M2 format (156.75 x 156.75 cm^2) n-type (180 µm, 4 ± 0.8 Ω.cm) Czochralski-grown crystalline silicon (c-Si) wafers. A simplified schematic of the process sequence is depicted in Figure 1.

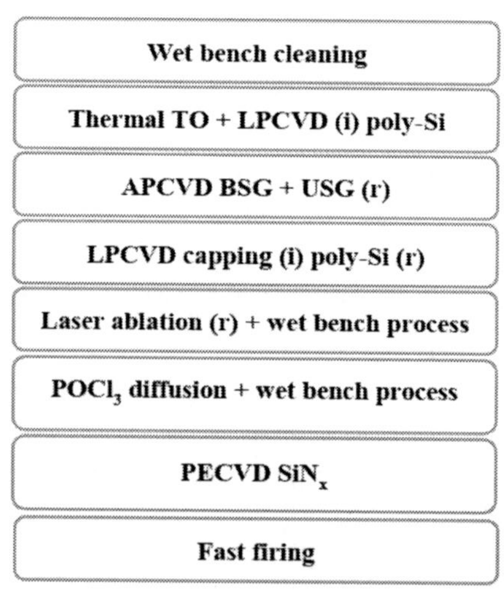

Figure 1: Process flow of the symmetrically processed lifetime samples. USG stands for undoped silicon glass.

The samples were saw-damage etched in an alkaline bath followed by cleaning in a Piranha bath to remove any organic and inorganic impurities. The interfacial oxide of 2.2 nm was grown in the thermal oxidation tube. Low-pressure chemical vapour deposition (LPCVD) was used to deposit a 250 nm intrinsic poly-Si layer at a temperature below 600°C in a quartz tube furnace. The samples were then transported to SCHMID for the Atmospheric pressure chemical vapour deposition (APCVD), Borosilicate glass (BSG) and undoped silicon glass (USG) depositions on both sides. The BSG layer acts as a dopant source, whereas the USG layer functions as a capping layer for the transport and handling of the samples. The deposition of the BSG layer is carried out by injecting N_2-diluted Silane (SiH_4), Diborane (B_2H_6) and oxygen (O_2) through a chemical vapour deposition (CVD) injector at ambient pressure. On top of the APCVD glass layer, we deposited a 50 nm capping (i) poly-Si layer using the same LPCVD technique. We performed a gentle laser ablation of the capping (i) poly-Si using frequency-doubled Nd: YVO_4 nanosecond pulse laser with VIS (532nm) wavelength and a pulse duration of 120 ns. Subsequently, the glass layers under the laser ablated regions were removed in the wet bench process, thereby patterning the rear side of the IBC solar cells. The ex-situ doping to form (n+) poly-Si areas was done using a high-temperature $POCl_3$ diffusion process, which also serves as an annealing step for the crystallization of the poly-Si layers. The temperature profiles were optimized in the high-temperature $POCl_3$ diffusion step to drive both dopant types and tailor the dopant profile to attain the best passivation quality. The wet bench process removed all the unwanted poly-Si capping and dopant glass layers. During the etching of the capping poly-Si layer on top of (p+) poly-Si, we also etch some part of (n+) poly-Si on the other side and hence the thickness of the (n+) poly-Si layer is lower than that of the (p+) poly-Si layer (as depicted in Figure 2). The symmetrically processed lifetime samples were passivated using Plasma Enhanced Chemical Vapor Deposition (PECVD) of a SiN_x layer with a refractive index of 2.03. Finally, the samples were fired in a belt-type fast-firing furnace for hydrogenation.

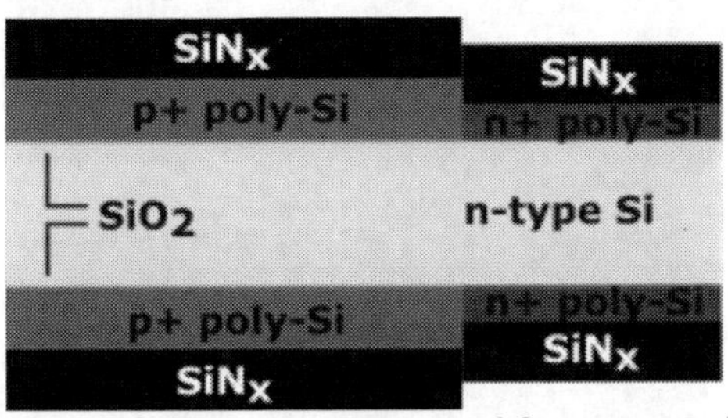

Figure 2: Schematic representation of the symmetrically processed lifetime samples

3 CHARACTERIZATION

The passivation quality of the poly-Si layers was measured using Quasi Steady State Photoconductance (QSSPC) measurements from Sinton instruments and Photoluminescence (PL) imaging tool. In addition, the thickness of the poly-Si layers was measured accurately using SE 800-PV ellipsometer from SENTECH. Finally, we used a laser scanning microscope (LSM) from Olympus to study the surface topography poly-Si layers.

4 RESULTS

4.1 Investigation of the laser ablation settings

One of the most critical processing steps that define the success of the patterning used in this study is the selective laser ablation of the capping (i) poly-Si layer and the subsequent wet bench process. Hence, extensive research was carried out in this work to investigate the suitable laser ablation fluence needed for gently ablating the capping (i) poly-Si layer without causing any damage to the underlying thick (i) poly-Si layer.

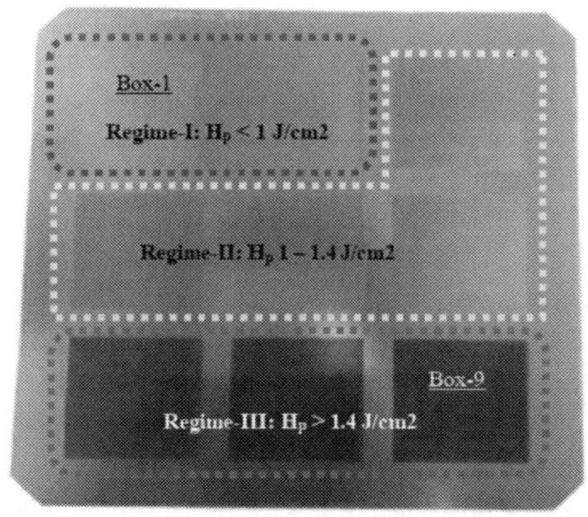

Figure 3: Photo depicting the different laser fluence.

Figure 3 shows the wide range of laser ablation fluence (H_p) tested in the study. After laser ablation, the sample was dipped in an alkaline bath for a short time to remove any residual capping (i) poly-Si. The laser ablation fluence were varied from 0.75 J/cm^2 in box-1 to 2.8 J/cm^2 in box-9. Laser ablation can be categorized into three regimes. In Regime-I, the laser fluence was lower than 1 J/cm^2, so the laser energies are insufficient to ablate the capping (i) poly-Si layer. In Regime-II, where the H_p varies from 1 to 1.4 J/cm^2, we observe partial ablation of the capping (i) poly-Si and visually observe vague light blue colour traces (which denotes the 100 nm APCVD glass layers). Once we enter regime-III, where H_p exceeds 1.4 J/cm^2, we observe the complete ablation of the capping (i) poly-Si, and for visual inspection, we can observe the 100 nm APCVD glass layers in a clear dark blue colour. A suitable ellipsometer model was used to measure the samples from regime-III, and with a perfect fit, 100nm of BSG was measured with no traces of (i) poly-Si on these samples. Hence, the H_p higher than 1.4 J/cm^2 was used for the symmetrical lifetime samples.

4.2 Passivation quality of doped poly-Si layers

The symmetrical lifetime samples were fabricated using the process flow shown in Figure 1. Using the learnings from previous sections, laser ablation H_p > 1.4 J/cm^2 was used for ablating the capping (i) poly-Si layers. Figure 4 shows the passivation quality before and after the fast firing of the non-lasered (p+) poly-Si (box-5 in Figure 4) layers and (n+) poly-Si layers formed by different laser ablation fluence settings (except box-5 in Figure 4). At first glance, excellent passivation quality has been achieved for both polarities of poly-Si layers before firing. For the best case iV_{oc} and J_0 of 741 mV and 1.35 fA/cm^2 (n+) poly-Si and 720mV and 17.4 fA/cm^2 (p+) poly-Si were achieved respectively. It is observed that the passivation quality improves gradually from 726 mV to 738 mV as the fluence increases from 1.5 to 2.3 J/cm^2. At very high fluence, i.e. 2.8 J/cm^2, the passivation quality drops drastically. From our understanding, at such high H_p instead of laser ablation of capping (i) poly-Si layer, we might perform laser doping of the underlying thick (i) poly-Si layers to form (p+) poly-Si. The above-explained theory was confirmed by ECV measurements which showed cross contamination (with boron) in the (n+) poly-Si layer after $POCl_3$ diffusion, and the tunneling oxide was destroyed due to high fluence, thereby lowering the passivation quality.

Upon firing, we observed that the non-lasered (p+) poly-Si passivation improves marginally. In contrast, there is a noticeable decrease in the passivation quality of all the laser patterned (n+) poly-Si (*Figure 5*). Also, interestingly the loss in the passivation quality after firing was found to be a function of laser fluence used for the ablation of the capping (i) poly-Si. Among the laser ablation fluence tested in the optimal laser ablation settings, we observe that at very low laser fluence i.e 1.5 J/cm² results in marginal loss in iV_{oc} (≈ 5 mV). As we further increase the laser fluence from 1.7 to 2.3 J/cm², we observed that the losses in the passivation quality flatten-up with iVoc losses of 25-30 mV. At very high H_p i.e 2.8 J/cm² due to significant laser damages induced during ablation iV_{oc} drops dramatically by 48 mV. To further investigate the source of the passivation loss in the lasered (n+) poly-Si layers, high-resolution images were taken using a laser scanning microscope (LSM), depicted in Figure 6.

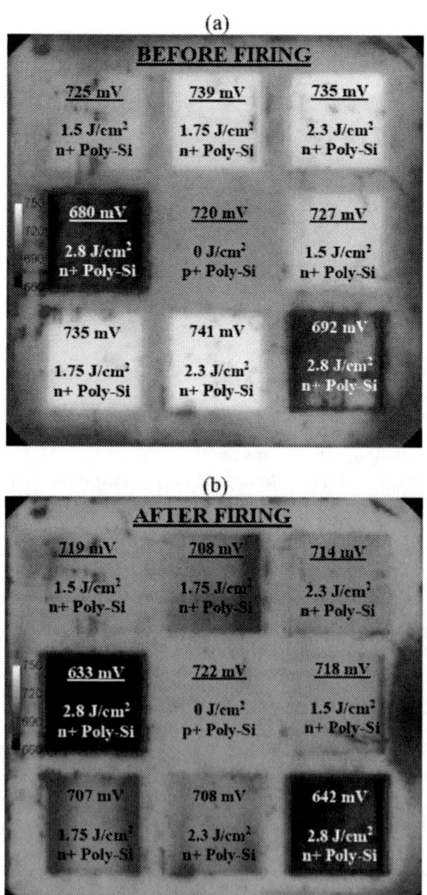

Figure 4: Voltage calibrated Photoluminescence images (iVoc) of the symmetrical lifetime with non-lasered box-5 form (p+) poly-Si and surrounding lasered 8 boxes with different laser ablation fluence forming (n+) poly-Si (a) before firing and (b) after firing.

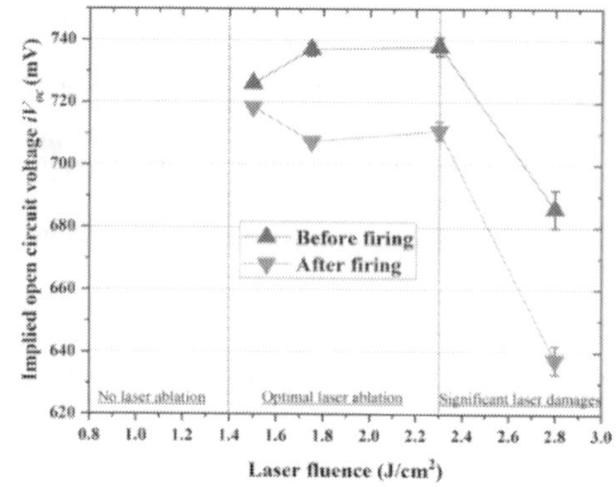

Figure 5: Influence of laser fluence on the passivation quality of laser damaged (n+) poly-Si layer before and after firing.

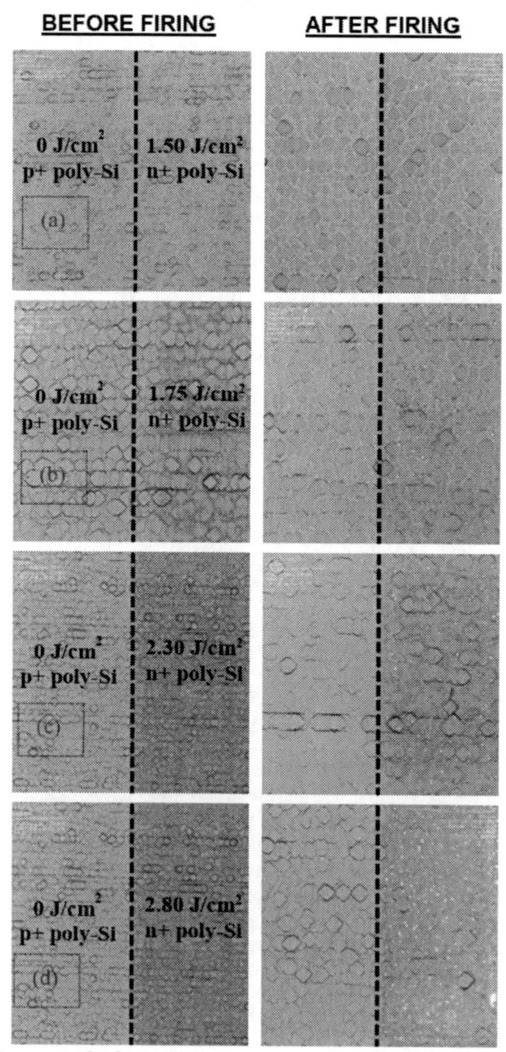

Figure 6: The laser scanning microscope observations of the wafer surface topography before (left) and after firing (right) with varying laser ablation fluence (H_p) a) 1.5 J/cm², b) 1.75 J/cm², c) 2.3 J/cm² and d) 2.8 J/cm².

Some interesting inferences include: firstly, for H_p > 1.5 J/cm^2, we observe that the surface morphology of the laser processed areas is altered compared to the non-laser regions even before firing. Secondly, with increasing laser fluence the blister density increases and, finally, increasing laser fluence also increases the size of the blisters formed. Our hypothesis that could explain the observed blistering phenomenon is that during the laser ablation of the capping (i) poly-Si, the underlying (i) poly-Si gets unintentional laser damage, thereby creating some mechanical stresses in the layer. Upon fast firing, due to the difference in thermal expansion of SiO_x as compared to c-Si and laser-damaged poly-Si, the whole stack undergoes stress-induced blistering, thereby lowering the passivation quality. Currently, work is underway to prove the hypothesis mentioned above. Once we understand the blistering phenomenon, the process sequence will be further optimized to develop blister-free high temperature stable and high passivation quality poly-Si layers using APCVD layers. These layers would be used to fabricate cost-effective and high-efficiency poly-Si IBC cells.

6 SUMMARY AND OUTLOOK

The process fabrication route for complete ex-situ doped Poly-IBC solar cells has been presented. A systematic study was carried out to investigate the suitable laser ablation fluence needed for patterning the rear side. The high-temperature step was optimized for ex-situ doping and crystallization of both polarities of poly-Si layers. As a result, excellent passivation quality was demonstrated for both polarities of poly-Si layers. However, the passivation quality of the laser patterned (n+) poly-Si layers was found to degrade upon fast–firing. The observed passivation loss was hypothesized to be due to stress-induced blistering. Currently, work is underway to develop blister-free poly-Si layers, which will be used for fabricating complete ex-situ doped poly-IBC solar cells.

6 ACKNOWLEDGEMENTS

This work was funded by the German Federal Ministry for Economic Affairs and Energy within the research project "HOBIT" (contract no. 03EE1121B).

7 REFERENCES

[1] International Energy Agency, "Snapshot of Global Photovoltaic Markets - 2018," Rep. IEA PVPS T1-332018, pp. 1–16, 2018.

[2] J. Melskens, B. W. H. Van De Loo, B. Macco, L. E. Black, S. Smit, and W. M. M. Kessels, "Passivating Contacts for Crystalline Silicon Solar Cells: From Concepts and Materials to Prospects," IEEE J. Photovoltaics, vol. 8, no. 2, pp. 373–388, 2018, doi: 10.1109/JPHOTOV.2018.2797106.

[3] C. Hollemann et al., "Separating the two polarities of the POLO contacts of an 26.1%-efficient IBC solar cell," Sci. Rep., vol. 10, no. 1, Dec. 2020, doi: 10.1038/s41598-019-57310-0.

[4] A. Chaudhary et al., "Influence of Polysilicon Thickness on Properties of Screen-Printed Silver Paste Metallized Silicon Oxide/Polysilicon Passivated Contacts," Phys. Status Solidi Appl. Mater. Sci., vol. 218, no. 18, pp. 1–9, 2021, doi: 10.1002/pssa.202100243.

[5] C. E. Chan, B. J. Hallam, and S. R. Wenham, "Simplified interdigitated back contact solar cells," Energy Procedia, vol. 27, pp. 543–548, 2012, doi: 10.1016/j.egypro.2012.07.107.

[6] P. Padhamnath, N. Nampalli, A. Khanna, B. Nagarajan, A. G. Aberle, and S. Duttagupta, "Progress with passivation and screen-printed metallization of Boron-doped monoPoly™ layers," Sol. Energy, vol. 231, no. November 2021, pp. 8–26, 2022, doi: 10.1016/j.solener.2021.11.015.

[7] A. H. Soeriyadi et al., "Impact of firing and capping layers on long-term stability of doped poly-Si passivating contact layers," SiliconPV 2021, 11th Int. Conf. Cryst. Silicon Photovoltaics, vol. 2487, no. August, p. 050006, 2022, doi: 10.1063/5.0089658.

[8] A. Chaudhary et al., "Influence of Polysilicon Thickness on Properties of Screen-Printed Silver Paste Metallized Silicon Oxide/Polysilicon Passivated Contacts," Phys. Status Solidi Appl. Mater. Sci., vol. 218, no. 18, 2021, doi: 10.1002/pssa.202100243.

[9] G. Yang, A. Ingenito, O. Isabella, and M. Zeman, "IBC c-Si solar cells based on ion-implanted poly-silicon passivating contacts," Sol. Energy Mater. Sol. Cells, vol. 158, pp. 84–90, Dec. 2016, doi: 10.1016/j.solmat.2016.05.041.

FULLY PASSIVATING CONTACT IBC SOLAR CELLS USING LASER PROCESSING

Jonathan Linke, Florian Buchholz*, Christoph Peter, Jan Hoß, Jan Lossen, Valentin Mihailetchi, Radovan Kopecek
International Solar Energy Research Center Konstanz e.V.
Rudolf-Diesel-Straße 15, D-78467 Konstanz, Germany
*Corresponding author: +49 7531 36 183 553, florian.buchholz@isc-konstanz.de

ABSTRACT: A novel approach for interdigitated back contacted (IBC) solar cell production featuring polycrystalline silicon on interfacial oxide (poly-Si/SiO_x) passivating contacts on both polarities is introduced, namely the polyZEBRA concept, which is based on the IBC solar cell process ZEBRA. The process sequence is compatible with standard industrial equipment, i.e. IBC patterning via laser processing and metallization for both polarities in one single screen-printing step. This approach allows for an upgrade of existing ZEBRA production lines with minimal additional costs. The champion polyZEBRA cell produced so far reached an in-house efficiency of 23.5% on full area wafer size. Besides the introduction of the polyZEBRA concept, an optimization of the gap/front side BBr_3-diffusion towards low recombination is presented, enabling 6 mV higher open-circuit voltage on cell level compared to the baseline.

Keywords: Interdigitated Back Contact, poly-Si, passivating contact, laser processing, c-Si, n-type, Silicon Solar Cell

1 INTRODUCTION

Interdigitated back contacted (IBC) solar cells feature metal contacts of both polarities on the rear side, enabling higher short current densities in absence of any metal on the front side. It is straight forward to combine this advantage with the excellent passivation of polycrystalline silicon on interfacial oxide (poly-Si/SiO_x) passivating contacts. *Haase et al.* designed a proof-of-principle IBC cell architecture with poly-Si/SiO_x contacts for both polarities, which lead to the current laboratory world record efficiency on p-type wafers [1]. Besides this, poly-Si/SiO_x IBC cells on n-type wafers are in production by high-end module producers such as Sunpower. However, these cell designs seem to rely on non-standard industrial equipment and thus are not suited for low price applications. Processes compatible with standard industrial technologies were proposed but did not reach the aimed high efficiencies on full wafer area [2,3] or used a poly-Si/SiO_x contact only for one polarity [4].

In this work, a novel IBC solar cell fabrication process is presented, which features poly-Si/SiO_x contacts on both polarities and relies on standard industrial equipment only. The IBC patterning is achieved by laser mask ablation and laser crystallization/doping, while the screen-printed metallization is compatible with the standard ZEBRA contacting pattern [5,6]. This implies in particular simultaneous screen-printing of both polarities (n^+ and p^+ poly-Si). In this way it is ensured that existing ZEBRA production lines can be upgraded to polyZEBRA production utilizing most of the existing tools. In contrast to other known IBC structures, the cell architecture features a p-diffused gap and front side, leading to low recombination, UV stability and a soft and early breakdown in reverse bias without shunting the cell, which could make bypass diodes obsolete [7].

Besides the introduction of the polyZEBRA concept, an optimization of the gap/front side BBr_3-diffusion towards is presented, in which only the recombination current density is minimized regardless of the sheet resistance. This approach may be counter-intuitive, since in classical homo-junctions a diffused emitter is optimized towards both quantities in order to enable lateral conductivity and at the same time low recombination currents. An optimization of such a diffusion process aims at the best compromise between both properties. In contrast, for the gap/front side diffusion within the polyZEBRA concept, the sheet resistance plays a minor role as the lateral conductivity is ensured by the p^+ poly-Si in the emitter region on the rear side of the cell. Thus, the diffusion can be optimized towards low recombination currents only.

2 EXPERIMENTAL

2.1 polyZEBRA process sequence

Figure 1 shows the processing sequence of the IBC polyZEBRA cells (in the following, reference to the steps therein in parenthesis). The process starts with cleaning of Czochralski-grown (Cz) base material (n-type, 1 - 7 Ωcm, M2 wafer size) (1). After thermal interfacial oxide growth (~1.7 nm) and a low-pressure chemical vapor deposition (LPCVD) of a 150 nm thick in-situ doped n^+ poly-Si layer (2), a plasma-enhanced chemical vapor deposited (PECVD) SiN_x mask with a thickness of 60 nm is deposited (3). Then, the mask is locally laser ablated in the emitter and gap regions (4). The remaining SiN_x protects the underlying n^+ poly-Si and defines the back surface field (BSF) pattern. After alkaline etching of the emitter and gap region (5), another thermal interfacial oxide is grown (~1.7 nm) and on top an intrinsic poly-Si layer with a thickness of 190 nm deposited using LPCVD (6). This intrinsic layer is then lightly doped by a soft BBr_3-diffusion (7). After removal of the boron silicate glass (BSG) in hydrofluoric acid (HF), the pattern of the emitter region is formed by local laser treatment for crystallization and doping of the p^+ poly-Si (8) as previously published in [8]. This process creates an etch selectivity in alkaline solution of 500-700 compared to the surrounding lightly doped p poly-Si without laser treatment [9]. During the subsequent alkaline etching step, the p poly-Si without laser treatment in the gap regions and on the front side is removed and the underlying c-Si wafer is textured, while the laser treated p^+ poly-Si in the emitter region is preserved. In order to achieve low recombination current densities and UV stability in the gap and on the front side, these regions are p-doped in an additional BBr_3-diffusion step (10), whereby the thermal budget of this diffusion forms both the n^+ and p^+ poly-Si/SiO_x contacts. Furthermore, the diffused gap leads to a soft and early breakdown in reverse bias without shunting the cell, which could make bypass diodes obsolete (patent pending). Eventually, the whole structure is passivated with an AlO_x/SiN_x stack on both sides (11) and metallized with the standard ZEBRA contact pattern (12) [5,6], which implies simultaneous screen-printing of both polarities (n^+ and p^+ poly-Si).

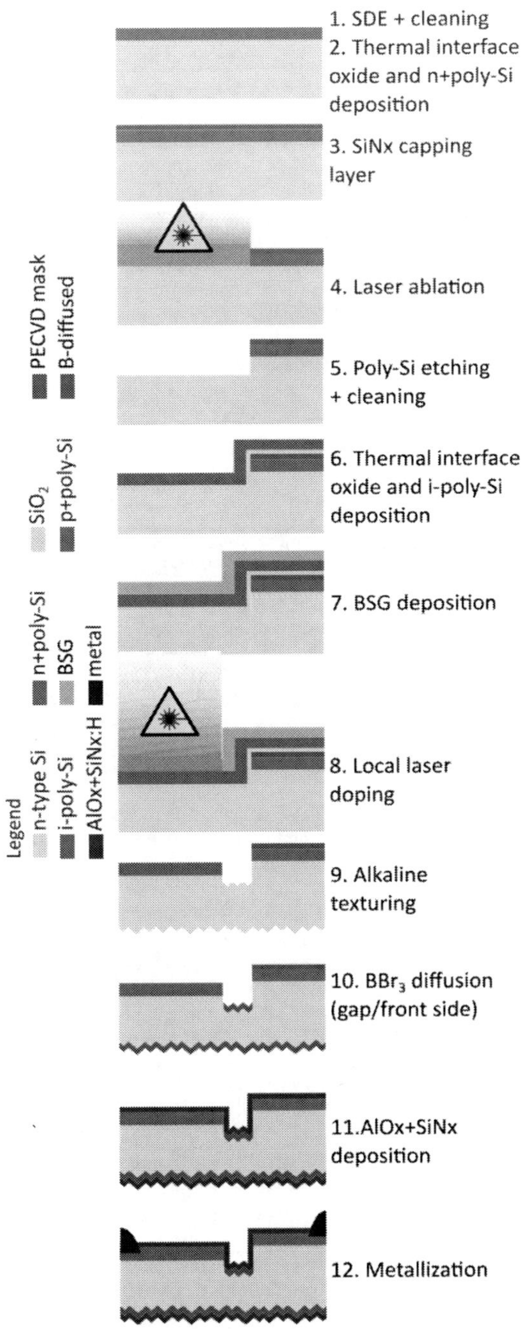

Figure 1: polyZEBRA process sequence and final device structure (12). Patent pending.

Figure 2 shows the learning curve of the polyZEBRA concept with respect to the best cell of each run. The process sequence described above was fixed around 05/2021 and initially limited to cell efficiencies of 17 – 19%. These cells were strongly limited by fill factor (FF), a problem that was finally overcome by using a different contacting paste (color-coded in Figure 2) and optimization of the screen pattern (triangles in Figure 2). To date, FF >80% and iV_{oc} to V_{oc} losses as low as 5 mV prove that simultaneous screen-printing of both polarities is possible also for poly-Si/SiO_x passivating contacts.

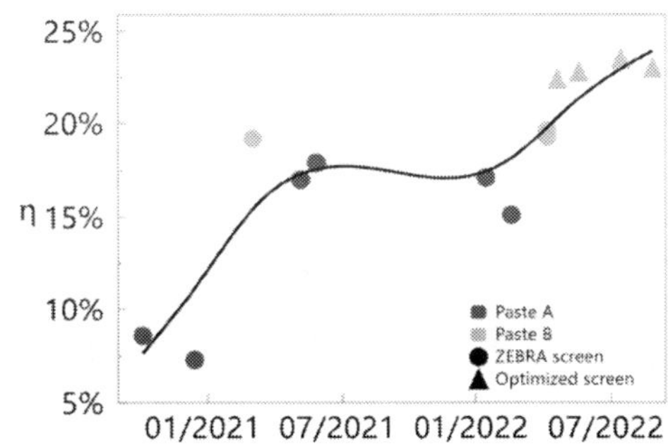

Figure 2: Learning curve of the polyZEBRA concept. Major improvements were achieved by changing the contacting paste (color-coded) and by an optimization the contacting screen pattern (triangles).

2.2 Optimization of gap/front side diffusion

For the optimization of the gap/front side diffusion towards low surface recombination several BBr_3-diffusion recipes were tested on symmetrical flat and textured Cz wafers. The temperatures and process step durations during BSG deposition and drive-in phase are summarized in Table I. After BSG removal in HF, the sheet resistance was measured on one wafer per group, while all other wafers were passivated by the same AlO_x/SiN_x stack as for the cells.

Table I: Tested gap/front side BBr_3-diffusion recipes A-D and the baseline recipe.

Recipe	Deposition	Drive-in
Baseline	840-900°C, 16 min	985°C, 20min
A	830°C, 16 min	960°C, 5 min
B	820°C, 16 min	950°C, 20 min
C	820°C, 16 min	925°C, 20 min
D	810°C, 16 min	925°C, 20 min

Afterwards, polyZEBRA cells were fabricated using the process sequence described in Section 2.1 with i) the baseline gap/front side diffusion and ii) the most promising gap/front side diffusion according to the pre-test described above. Additionally, cell precursors were taken out before metallization and dedicated test structures were produced for recombination current density measurements with squares symmetrically on both sides of the wafer. These squares were created following the cell process but applying different laser process patterns and reflect all relevant regions, namely n^+ poly-Si (BSF), p^+ poly-Si (emitter) and gap/front side.

2.3 Characterization

The sheet resistance was obtained by standard four point probe (4PP) measurements, while the recombination current density J_0 was measured by the quasi-steady-state photoconductance decay (QSSPC) method. The homogeneity of the passivation was assessed by photoluminescence (PL) imaging. Characterization of the solar cells was done using a Halm IV flasher.

3 RESULTS AND DISCUSSION

3.1 Results of the pre-test

The tested BBr_3-diffusions A-D (Table I) have lower deposition and drive-in temperatures compared to the baseline in order to achieve a lower doping level at the surface and a more shallow dopant profile. Both properties are expected to enable lower recombination current densities. Furthermore, the drive-in phase was designed to provide the necessary thermal budget for n^+ and p^+ poly-Si/SiO_x contact formation, if applied on cell level.

Figure 3 shows the recombination current density J_0 and sheet resistance for flat and textured wafers. Regarding the sheet resistance, a continuous trend is observed towards higher sheet resistance from recipe A to D, which holds for both flat and textured wafers, while the values for the textured wafers are consistently higher. This trend correlates to decreasing deposition and drive-in temperature. However, the difference between recipe C and D is very little, inferring that the influence of the deposition temperature is rather weak. Overall, the drive-in temperature dominates the sheet resistance.

Regarding the passivation, a similar trend is observed towards lower J_0 values from recipe A (mean flat: 24 fA/cm², mean textured: 20 fA/cm²) to D (mean flat: 18 fA/cm², mean textured: 15 fA/cm²), but the scattering of the data is quiet strong so that at least on flat wafers for recipe B-D no statistically relevant difference is observed. In contrast for the textured wafers, recipe C and D show slightly lower values than recipe B. Thus, the drive-in temperature, which is lowest for recipe C and D, dominates also J_0.

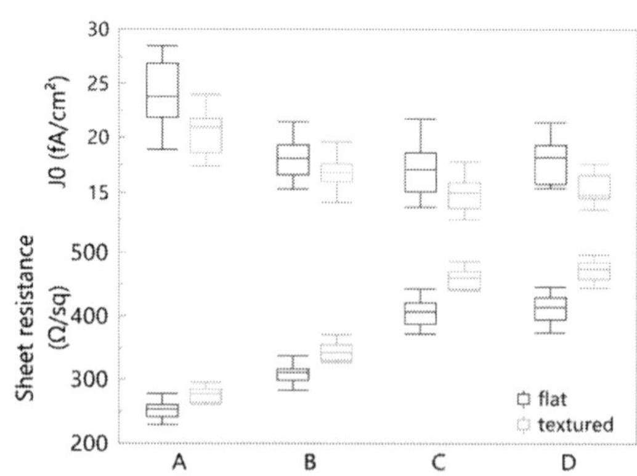

Figure 3: Recombination current density J_0 from QSSPC and sheet resistance from 4PP measurements for the tested gap/front side diffusion recipes according to Table I for flat and textured samples.

Considering the sheet resistance and passivation simultaneously, J_0 decreases as the sheet resistance increases. An explanation for this behavior is, that a lower surface doping and shallower doping profile increases the sheet resistance and decreases J_0, as desired. This does also explain the surprising observation, that J_0 is lower on the textured wafers, since the larger surface area of the textured wafers leads to even lower doping compared to the flat wafers. In conclusion, the desired reduction of the recombination was achieved and recipe C was chosen to be applied in polyZEBRA cells.

3.2 Lifetime wafer and polyZEBRA cell results

During the time between the pre-test and the gap/front side diffusion of the cell process it turned out that the drive-in phase of recipe C (925°C, 20 min) did not provide enough thermal budget for the activation of the n^+ and p^+ poly-Si passivation. Thus, the diffusion recipe was amended by a preceding thermal annealing step in N_2 atmosphere (950°C, 30 min) before starting the BBr_3-diffusion without breaking the vacuum. It is expected, that this extra step can be omitted in the future after fine-tuning the properties of the interfacial oxide below the n^+ and p^+ poly-Si.

Figure 4 exemplarily shows PL images of a precursor (left) and of the test structures (right) of the group diffused with recipe C. The precursor shows only minor handling traces but otherwise quite uniform appearance, indicating the suitability of the process for full wafer area solar cell production.

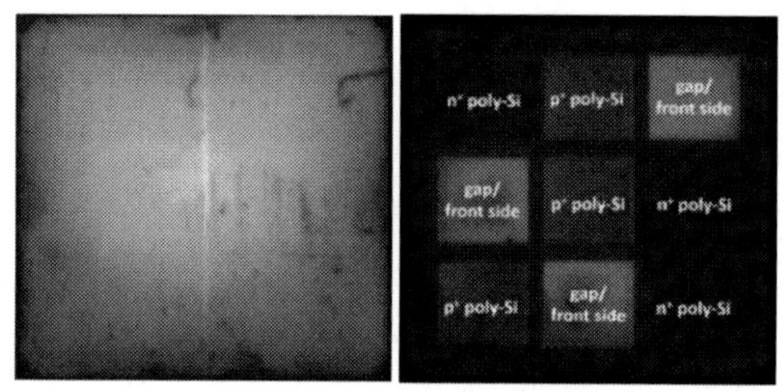

Figure 4: Exemplarily PL images of a precursor (left) and a symmetric lifetime wafer (right) of the group diffused with recipe C.

The test structure shows the brightest squares for the gap/front side region, and darkest squares for the n^+ poly-Si region. The corresponding J_0 values for both groups are shown in Figure 5 together with the values measured on cell precursors. First of all, for the n^+ poly-Si region no difference is observed between both recipes. However, the absolute value ~40 fA/cm² is much higher than expected. Since the PL image of the precursor (Figure 4, left) does not show any dark lines reflecting the poor n^+ poly-Si passivation as observed in the PL image of the test structures (Figure 4, right), it is concluded, that the specific test structure is the reason of the high recombination. According to results by *Hoß et al.* [10], a strong stress-induced depassivation can be observed if wafers with n^+ poly-Si/SiN_x stacks on both sides are annealed in conventional tube furnaces as it is the case for these specific test structures: The n^+ poly-Si region is protected by a SiN_x mask (Figure 1, 3) and this stack is annealed during the gap/front side diffusion (Figure 1, 10). It is noteworthy that *Hoß et al.* found that this depassivation can be reversed if such a stack is only on one side, a hydrogen-supplying dielectric is deposited on the other side and the wafer subsequently fired. This sequence is included in the polyZEBRA process flow (Figure 1, 11/12). Thus, the n^+ poly-Si region of the symmetric test structures is affected, while the depassivation of same region on the cell precursors is reversed during firing of the AlO_x/SiN_x stack on the front side.

For the p^+ poly-Si regions, higher J_0 values were observed for recipe C (~16 fA/cm²) as compared to the baseline (~8 fA/cm²). This indicates, that the extra thermal budget of the preceding annealing step before the recipe C diffusion (950°C, 30 min) was still not enough compared to the thermal budget of the baseline diffusion (985°C, 20 min). However, the preceding annealing step can easily be adapted in the future in order to achieve <10 fA/cm² in the p^+ poly-Si region also for recipe C.

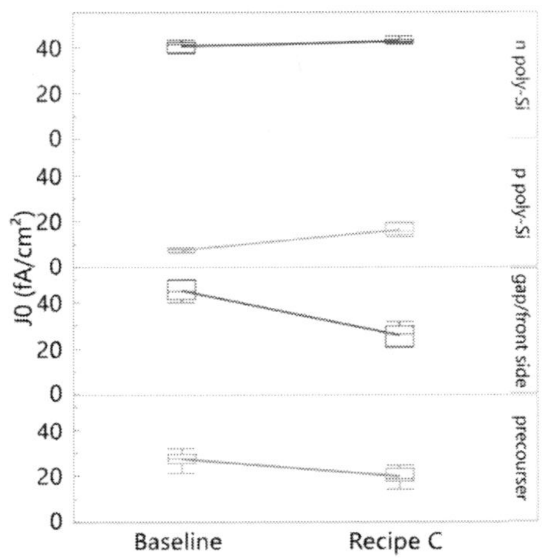

Figure 5: Recombination current density J_0 of the respective regions in the solar cells, measured by QSSPC on the squares of the test structures and on the cell precursors.

The gap/front side region shows the desired reduction from ~45 fA/cm² (baseline) to ~26 fA/cm² (recipe C). However, the reason why the 15 fA/cm² measured in the pre-test for recipe C could not be reproduced was not yet found due to the complexity of the cell process. Nevertheless, the reduction to roughly half of the baseline recombination is expected to significantly increase the open circuit voltage of the cell, since more than 50% of the cell is covered by the gap/font side structure. Indeed, the recombination on the precursors could be reduced from ~27 fA/cm² (baseline) to ~20 fA/cm² (recipe C), meaning that the gains on the larger area fraction of the gap/front side region overcompensated the losses in the p^+ poly-Si region.

On cell level, this translates to an improvement of the open-circuit voltage V_{oc} by 6 mV from 696 mV (baseline) to 702 mV (recipe C) as shown in Figure 6. However, the mean values of the measured cell efficiencies are ~22.9% for both recipes. This is attributed rather to the relatively larger loss in pseudo fill factor by ~0.8%$_{abs}$ (1%$_{rel}$) than to the relatively smaller loss in short-circuit current density J_{sc} by ~0.1 mA/cm² (0.2%$_{rel}$). In summary, the desired V_{oc} improvement was achieved by optimizing the gap/front side diffusion, but accompanied by a loss in pFF, which hindered a gain in cell efficiency.

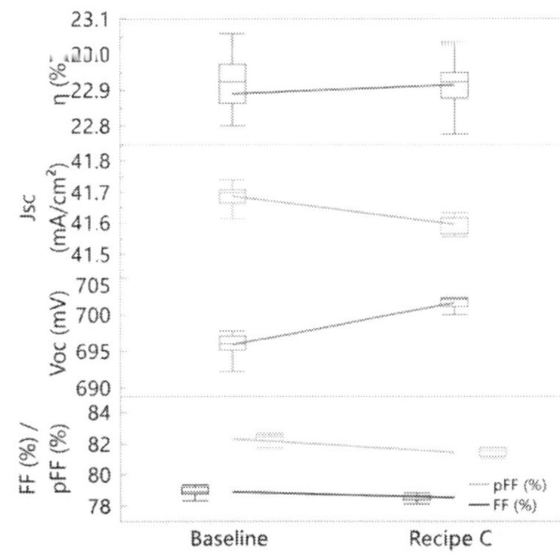

Figure 6: IV parameters of the polyZEBRA cells for the baseline and recipe C gap/front side diffusions.

4 SUMMARY AND OUTLOOK

In this paper the poly-Si/SiO_x passivating contact based IBC polyZEBRA concept was introduced, which relies on standard industrial equipment only. The gap and front side of the cell are p-doped by a BBr_3-diffusion, which was optimized towards low surface recombination in these regions. Applying the optimized diffusion, recombination current densities of the gap/front side region as low as 15 fA/cm² were achieved and successfully implemented in solar cells, resulting in V_{oc} values up to 702 mV, which was ~6 mV higher than achieved with the baseline diffusion..

The best measured polyZEBRA cell so far reached an efficiency of 23.5% on full wafer area (Table II). Both the high FF and low iV_{oc} to V_{oc} loss proves that simultaneous screen-printing of both polarities is possible also for poly-Si/SiO_x passivating contacts. However, the efficiency is limited by pFF and V_{oc}, which will be addressed in the next experiments. Given that not yet all individual process steps have been optimized and aligned to each other, we expect to reach 24% efficiency in the near future.

Table II: Solar cell parameters of best polyZEBRA cell measured so far (in-house, G1 wafer format, M2 metallization, aperture measurement).

η (%)	V_{oc} (mV)	J_{sc} (mA/cm²)	FF (%)
23.5	701	41.5	80.6
pFF (%)	**iV_{oc} (mV)**	**iV_{oc} - V_{oc} (mV)**	**pFF - FF (%)**
82.3	706	5	1.7

5 ACKNOWLEDGEMENTS

This work was supported in part by the project Highlite that has received funding from the European Union's Horizon 2020 research and innovation program under grant agreement No 857793.

6 REFERENCES

[1] F. Haase et al., "Laser contact openings for local poly-Si-metal contacts enabling 26.1%-efficient POLO-IBC solar cells", Solar Energy Materials and Solar Cells, 186 (2018), doi:10.1016/j.solmat.2018.06.020

[2] G. Yang et al., "High-efficiency black IBC c-Si solar cells with poly-Si as carrier-selective passivating contacts." Solar Energy Materials and Solar Cells, 186 (2018), doi:10.1016/j.solmat.2018.06.019

[3] P. Wang et al., "Development of TOPCon tunnel-IBC solar cells with screen-printed fire-through contacts by laser patterning", Solar Energy Materials and Solar Cells, 220 (2021), doi:10.1016/j.solmat.2020.110834

[4] F. Haase et al., "Design of large poly-Si on oxide interdigitated back contact silicon solar cells with local Al-p^+ contacts in the constraints of measurement and module integration", Solar RRL 2200583 (2022), doi:10.1002/solr.202200583

[5] S. de Vecchi et al., "New metallization scheme for interdigitated back contact silicon heterojunction solar cells", Energy Procedia 38 (2013), doi:10.1016/j.egypro.2013.07.335

[6] R. Kopecek et al., "ZEBRA technology: low cost bifacial IBC solar cells in mass production with efficiency exceeding 23.5%", 47th IEEE Photovoltaic Specialists Conference (2020), doi:10.1109/PVSC45281.2020.9300503

[7] J. Linke et al., "The role of masking layers during metallization of poly-Si/SiO_x contacts", To be published in: AIP Conference Proceedings (SiliconPV 2022)

[8] F. Buchholz et al., "Local passivating contacts from laser doped p^+ polysilicon", Proceedings of 38th EU-PVSEC Conference (2021), doi:10.4229/EUPVSEC20212021-2BO.11.3

[9] R. Charavel et al., "Advantages of p^{++} polysilicon etch stop layer versus p^{++} silicon", Proceedings of SPIE (2003), doi:10.1117/12.498107

[10] J. Hoß et al., "Stress-induced de-passivation of poly-Si/SiN_x layer stacks", to be published in: Proceedings of WCPEC 2022

23.1 %-EFFICIENT POLO BACK JUNCTION SOLAR CELL

B. Min[1], L. Nasebandt[1], C. Hollemann[1], D. Bredemeier[2], L. Thiemann[1], T. Brendemuehl[1], K. Bothe[1], R. Peibst[1] and R. Brendel[1, 2]
[1]Institute for Solar Energy Research Hamelin (ISFH), Am Ohrberg 1, 31860 Emmerthal, Germany
[2]Dep. Solar Energy, Inst. Solid-State Physics, Leibniz University of Hannover, Appelstr. 2, 30167 Hannover, Germany

ABSTRACT: This work presents the recent progress of all screen-printed p-type back junction (BJ) solar cells featuring Al-front contacts and passivating polysilicon on oxide (POLO) rear contacts. We achieve a new best cell efficiency of 23.1 % on full M2 Cz wafers as independently confirmed by ISFH CalTeC in Germany. We mainly improve the fill factor of our POLO BJ solar cells by increasing the number of Ag fingers at the rear side. This increase of Ag rear contacts does not affect the open-circuit voltage because we apply an Ag paste with less aggressive chemistry, hence the recombination losses at Ag contacts are negligibly small. In addition, we analyze the temperature-dependent cell performance of an in-house fabricated POLO BJ solar cell. We extract –0.2953 %/K as the efficiency temperature coefficient (TC_η) of the investigated POLO BJ solar cell, which is superior to –0.36 %/K of PERC solar cells in the literature [1], mainly due to the higher open-circuit voltage TC of the POLO BJ device, that amount to –0.2462 %/K. Our study indicates that particular high annual yields are to be expected from POLO BJ cells in the regions with high temperatures such as Middle East and North Africa.
Keywords: Polycrystalline, Passivating Contacts, Back Junction

1 INTRODUCTION

The POLO back junction (BJ) cell concept is an attractive option for introducing passivating polysilicon on oxide (POLO) contacts to industrial silicon solar cells. It utilizes major innovations of mainstream p-type PERC solar cells such as AlO_x/SiN_y passivation stack or screen-printed Al contacts as shown in Figure 1 [2].

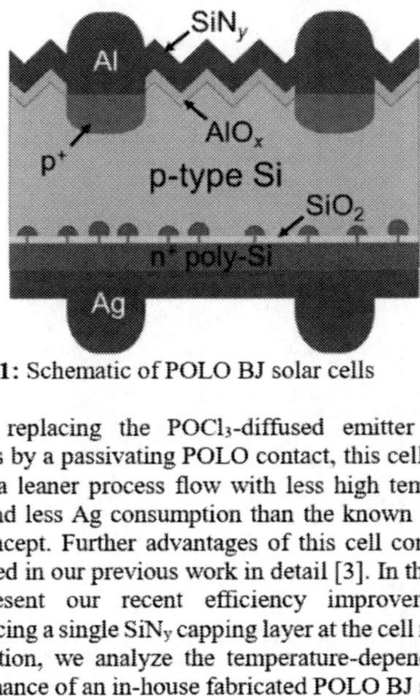

Figure 1: Schematic of POLO BJ solar cells

By replacing the $POCl_3$-diffused emitter and its contacts by a passivating POLO contact, this cell concept allows a leaner process flow with less high temperature steps and less Ag consumption than the known TOPCon cell concept. Further advantages of this cell concept are discussed in our previous work in detail [3]. In this paper, we present our recent efficiency improvement by introducing a single SiN_y capping layer at the cell rear side. In addition, we analyze the temperature-dependent cell performance of an in-house fabricated POLO BJ solar cell in a temperature range from 16 °C to 60 °C.

2 POLO BJ SOLAR CELL FABRICATION

We use gallium-doped Cz grown wafers with a resistivity of 0.8 Ωcm. After a saw damage etch and cleaning, an interfacial oxide layer is grown wet-chemically in a solution of deionized water and ozone. After the deposition of an in-situ phosphorus-doped poly-Si layer using LPCVD tool, the samples are oxidized in a tube furnace, remaining 120 to 130 nm thick poly-Si layers. Simultaneously, the interfacial oxide layer breaks up and forms the POLO contacts. After single-sided removal of the oxide layer at the front, the samples receive a KOH-based texturing step. After removal of the protective oxide layer at the rear and cleaning, we deposit an AlO_x/SiN_y stack at the front side and a SiN_y layer at the rear side. A picosecond laser with a wavelength of 532 nm opens the front dielectric stack locally. The contacts are realized by aluminum-paste screen printing at the front side and silver-paste screen printing at the rear side followed by a co-firing process.

3 RESULTS

3.1 SiN_y capping layer at the rear side

We replace the AlO_x/SiN_y capping layer stack used in our previous work [4, 5] with a SiN_y capping layer on the rear side as is shown in Figure 1. With a single SiN_y layer, the screen-printed Ag paste doesn't need to fire-through the AlO_x layer and thus less aggressive chemistry in the paste is sufficient. Therefore, we have less damage of poly-Si layers by spiking-through of Ag-crystallites.

Figure 2 shows the contact resistivities between screen-printed silver and n^+-type poly-Si measured with TLM-SCAN tool from pvtools.

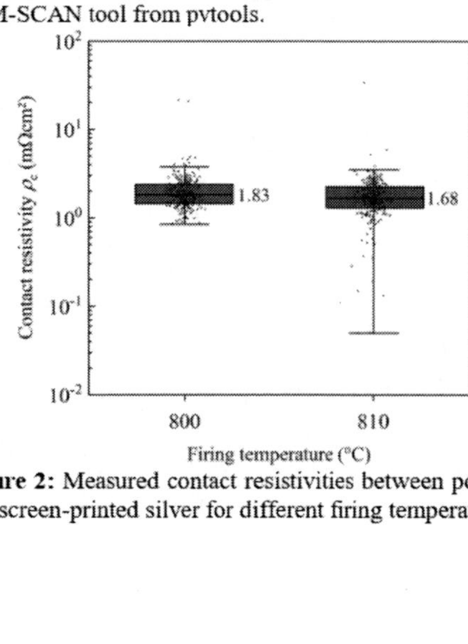

Figure 2: Measured contact resistivities between poly-Si and screen-printed silver for different firing temperatures

The median contact resistivities is 1.7 and 1.8 mΩcm² for the two firing temperatures, respectively. For cell fabrication, we decide to apply 810 °C because higher firing temperature improves the p^+-type layer formation beneath Al fingers at the front side (not shown here).

Figure 3 shows the measured *IV* parameters of POLO BJ solar cells with a single SiN_y layer at the rear side. Despite the increased number of Ag fingers from 112 to 156, the V_{oc} remains at same level, indicating that the recombination losses beneath screen-printed Ag fingers are negligibly small. The *FF* increases by about 0.9 $\%_{abs}$ because the resistive losses at the cell rear side are reduced by the increased number of Ag fingers. Therefore, we obtain a 0.4 $\%_{abs}$ gain in cell efficiency.

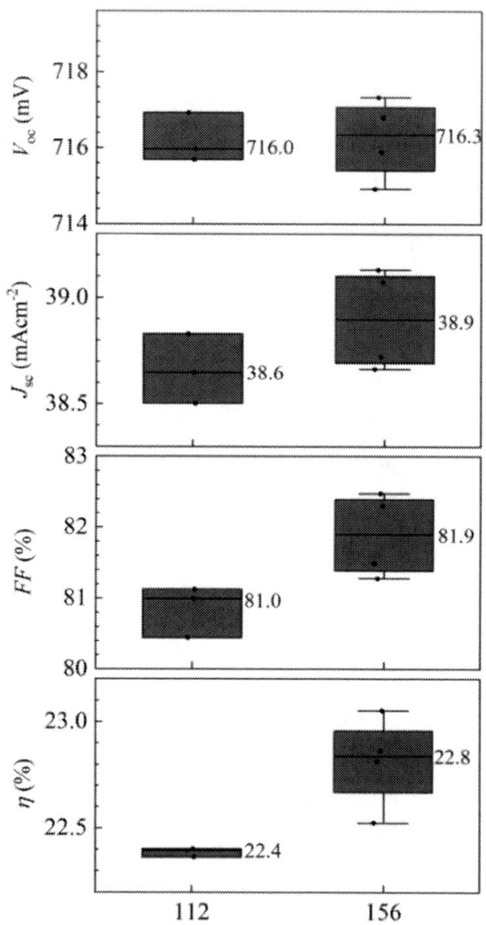

Figure 3: Measured *IV* parameters of POLO BJ solar cells for different number of Ag fingers at the rear side

As the best POLO BJ solar cell in this experiment has 23.1 % as was independently confirmed by ISFH CalTeC. Compared to our previous work [5], mainly the *FF* improves 0.8 $\%_{abs}$ due to the increased number of rear fingers as shown in Table I.

Table I: *IV* parameters of the best POLO BJ solar cell

	V_{oc} [mV]	J_{sc} [mA/cm²]	*FF* [%]	η [%]
Best cell*	717	39.5	81.7	23.1

*independently confirmed by ISFH CalTeC

3.2 Temperature-dependent performance of POLO BJ solar cells

The temperature-dependent *IV* measurements are performed with a LOANA tool from pvtools as presented in our previous work [6]. An in-house fabricated POLO BJ solar cell with Ga-doped M2 wafer is investigated. We either set the wafer temperature to 60 °C and decrease it from measurement to measurement, or start from 16 °C while increasing the temperature between the measurements. Both procedures show comparable results. The measurements are thus done under close to stationary conditions. All temperature values given in this study are measured by the sensor touching the solar cell from below. The TCs are extracted from the slopes of linear fits of the cell parameters as a function of temperature. All the TCs are normalized to the value at 25 °C and reported as relative TC values in this study.

Figure 4 shows the temperature-dependent *IV* parameters of the investigated POLO BJ solar cell.

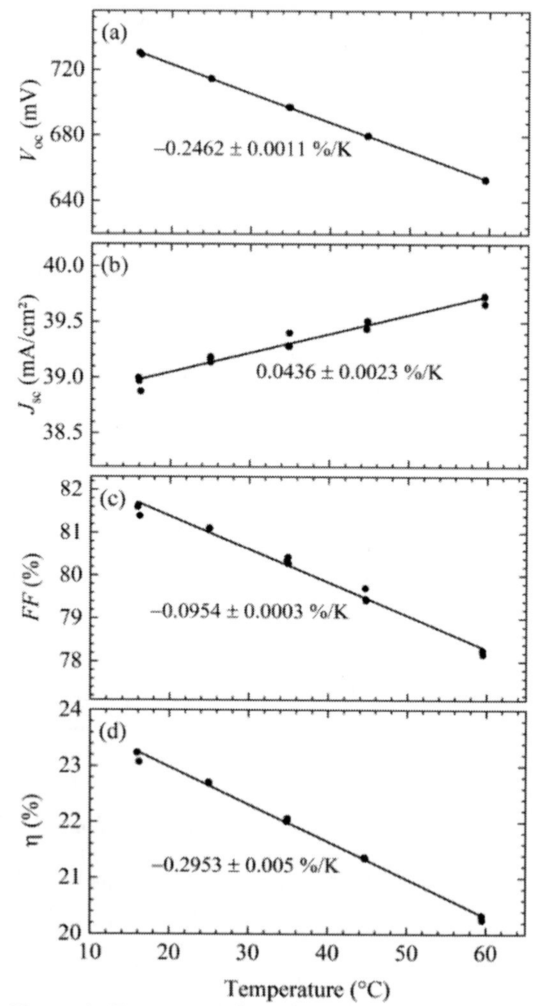

Figure 4: Temperature-dependent *IV* parameters of the investigated POLO BJ solar cell

Figure 4(a) shows that the POLO BJ solar cell exhibits a higher open-circuit voltage TC (TC_{Voc} = −0.2462 %/K) than PERC solar cells (TC_{Voc} = −0.29 %/K) in the literature [1]. This is due to the high V_{oc} value of POLO BJ solar cell, which is 714 mV at 25 °C. The extracted TC_{Voc} of −0.2462 %/K is even slightly better than values reported for SHJ solar cells (−0.25 %/K and −0.254 %/K [1, 7]), despite lower V_{oc} at 25 °C (714 mV for POLO BJ, 733 mV and 731 mV for SHJ [1, 7]). We found only a single SHJ solar cell in the literature with better TC_{Voc} (−0.238 %/K) than our POLO BJ cell. This SHJ cell has a V_{oc} of 741 mV at 25 °C [8]. Both the short-circuit current density TC (TC_{Jsc}) and the fill factor TC (TC_{FF}) of the POLO BJ solar cell are slightly better than that of PERC solar cells [1]. Finally, the efficiency TC of the investigated POLO BJ solar cell is $TC_{\eta} = -0.2953$ %/K as shown in Figure 4(d) and thus superior to that of PERC solar cells showing $TC_{\eta} = -0.36$ %/K.

4 SUMMARY

In the first part of this paper, we presented our best POLO BJ solar cell with an efficiency of 23.1 % as independently confirmed by ISFH CalTeC. We improve the *FF* by increasing the number of Ag fingers at the rear side by maintaining high V_{oc} of 717 mV, since the recombination losses beneath Ag fingers are negligible.

In the second part, we demonstrated the excellent temperature-dependent performance of POLO BJ solar cells by extracting an efficiency temperature coefficient of −0.2953 %/K which is superior to the reported value of PERC solar cells ($TC_{\eta} = -0.36$ %/K) [1]. Therefore, we expect particular high annual yields from POLO BJ cells in the regions with high temperatures which will be verified in future work.

5 ACKNOWLEDGEMENTS

The authors thank to M. Pollmann, B. Gehring, T. Neubert and D. Sylla for processing solar cells and M. Wolf and D. Hinken (all ISFH) for fruitful discussions about cell characterization. We also thank S. Huebner, T. Dippell and B. Wattenberg from Singulus Technologies AG for constructive discussions. This work was financially supported by the state of Lower Saxony and the German Federal Ministry for Economic Affairs and Climate Action (BMWK) under contact number 03EE1012A (NanoPERC).

6 REFERENCES

[1] J. Haschke, J.P. Seif, Y. Riesen, A. Tomasi, J. Cattin, L. Tous, P. Choulat, M. Aleman, E. Cornagliotti, A. Uruena. "The impact of silicon solar cell architecture and cell interconnection on energy yield in hot & sunny climates", Energy & Environmental Science 10, 1196-1206 (2017).

[2] R. Brendel, C. Kruse, A. Merkle, H. Schulte-Huxel, F. Haase, R. Peibst. "Screening Carrier Selective Contact Combinations for Novel Crystalline Si Cell Structures", Proc. of the 35th European Photovoltaic Solar Energy Conference and Exhibition, Brussels, Belgium, 2018, pp. 39 - 46.

[3] B. Min, N. Wehmeier, T. Brendemuehl, A. Merkle, F. Haase, Y. Larionova, L. David, H. Schulte-Huxel, R. Peibst, R. Brendel. "A 22.3% Efficient p-Type Back Junction Solar Cell with an Al-Printed Front-Side Grid and a Passivating n+-Type Polysilicon on Oxide Contact at the Rear Side", Solar RRL 2000435 (2020).

[4] B. Min, N. Wehmeier, T. Brendemuehl, F. Haase, Y. Larionova, L. Nasebandt, H. Schulte-Huxel, R. Peibst, R. Brendel. "716 mV Open-Circuit Voltage with Fully Screen-Printed p-Type Back Junction Solar Cells Featuring an Aluminum Front Grid and a Passivating Polysilicon on Oxide Contact at the Rear Side", Solar RRL 5, 2000703 (2021).

[5] B. Min, N. Wehmeier, H. Schulte-Huxel, R. Witteck, T. Brendemühl, T. Daschinger, F. Haase, Y. Larionova, L. Nasebandt, K. Tsuji, M. Dhamrin, R. Peibst, R. Brendel. "Approaching 23% with p-Type Back Junction Solar Cells Featuring Screen-Printed Al Front Grid and Passivating Rear Contacts", Proc. of the 38th European Photovoltaic Solar Energy Conference and Exhibition, Online, 2021, pp. 158 - 161.

[6] R. Peibst, F. Haase, B. Min, C. Hollemann, T. Brendemühl, K. Bothe, R. Brendel. "On the chances and challenges of combining electron-collecting nPOLO and hole-collecting Al-p+ contacts in highly efficient p-type c-Si solar cells", Prog. Photovoltaics n/a, (2022).

[7] A.H.T. Le, J. Dréon, J.I. Michel, M. Boccard, J. Bullock, N. Borojevic, Z. Hameiri. "Temperature-dependent performance of silicon heterojunction solar cells with transition-metal-oxide-based selective contacts", Prog. Photovoltaics n/a, 1-13 (2021).

[8] S.M.F. Zhang, J.P. Seif, T.G. Allen, R. Basnet, A.H.T. Le, I. Perez-Wurfl, Z. Hameiri. "Temperature- and Illumination-Dependent Characterization of Solar Cells Using Suns-Voc(T) and I-V(T)", Proc. of the 2021 IEEE 48th Photovoltaic Specialists Conference (PVSC), 2021, pp. 0737-0740.

INFLUENCE OF ALO_X DEPOSITION TEMPERATURE ON THE PASSIVATION MECHANISM OF ALO_X/SIN_Y:H STACKS

Fabian Geml, Sarah Sanz, Heiko Plagwitz, Giso Hahn
University of Konstanz, Department of Physics
78457 Konstanz, Germany

ABSTRACT: High efficiency solar cell concepts (e.g. passivated emitter, rear totally diffused (PERT) cells) rely on thin-film surface passivation layers, often deposited by chemical vapor deposition (CVD) due to the high parameter variability of the CVD. Many of these concepts use AlO_x/SiN_y:H stacks which are investigated within this work. The passivation mechanism of these stacks after firing is analyzed depending on the deposition temperature of APCVD AlO_x layer. It is shown that annealed single layers show an opposite behavior in passivation quality depending on deposition temperature compared to the stacks. With corona charging, capacitance voltage (CV) measurements and secondary ion mass spectrometry (SIMS) it can be explained that this behavior can be attributed to the H diffusion kinetics through the AlO_x barrier layer. A higher deposition temperature leads to a layer with a higher mass density which suggests a slower diffusion of H through this layer. Fixed charges are shown to be more pronounced in single layers deposited at higher temperatures. However, the resulting stronger field effect for these layers is shown to be the subordinated passivation mechanism as compared to chemical passivation. The understanding of passivation mechanism enables further potential of optimization of CVD AlO_x.
Keywords: CVD based deposition, passivation, aluminum oxide

1 INTRODUCTION

High efficiency solar cell concepts (e.g. passivated emitter, rear totally diffused (PERT) cells) rely on thin-film surface passivation layers, often deposited by chemical vapor deposition (CVD) due to the high parameter variability of the CVD. Especially CVD AlO_x layers have shown to provide excellent surface passivation [1]. In this contribution, we investigate the passivation mechanism of atmospheric pressure (AP) CVD deposited AlO_x films based on trimethyl-aluminum and oxygen as precursors [2], which are a relevant alternative to coatings produced by plasma-enhanced (PE) CVD or atomic layer deposition [3], [4]. We aim at separating the effect of chemical passivation from the field effect by fixed charges in these layers, which enables the possibility to optimize the layer properties in production-relevant stacks like AlO_x/SiN_y:H.

2 EXPERIMENTAL

During sample preparation, the mono-crystalline Czochralski (Cz) Si wafers (n-type, 2 Ωcm) are saw damage etched, cleaned and chemically polished using an HNO_3/HF solution. Float-Zone(FZ)-Si wafers (n-type, 200 Ωcm; p-type, 200 Ωcm) undergo a single HF etching step. AlO_x is deposited by an APCVD roller system at different set temperatures between 570-690°C with constant layer thickness. Part of the wafers are subsequently coated with SiN_y:H or SiN_y:H/D (using ND_3 instead of NH_3) using a semi-remote CVD tool and fired at a sample peak temperature of 800°C. Reference wafers with single AlO_x layers are divided into two groups. One is fired in the same way as above and the other one is annealed for 30 min at 400°C. Both single layer groups show the same qualitative behavior, however, the passivation quality of the annealed group is higher and more comparable to the AlO_x/SiN_y:H stacks and therefore shown in the following. The general process flow is schematically depicted in Fig. 1. The passivation quality is determined by applying photo conductance decay (PCD) measurements to determine the effective minority charge carrier lifetime τ_{eff}. For investigations of the passivation mechanism, samples are measured applying the corona charging (CC) and capacitance voltage (CV) techniques.

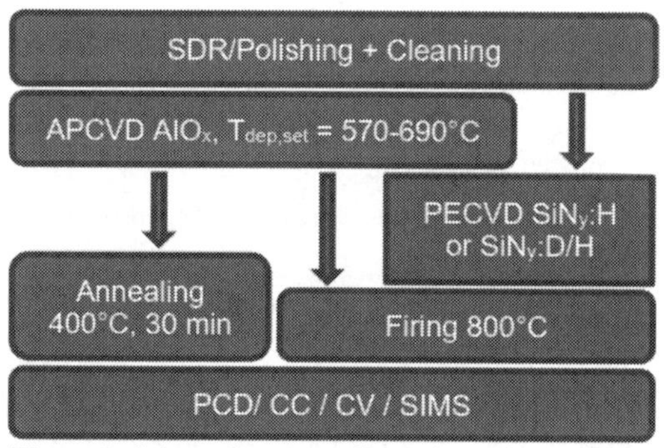

Figure 1: Schematic process flow of samples shown in the results section.

3 RESULTS AND DISCUSSION

Fig. 2 shows the effective surface recombination velocity S_{eff} of annealed AlO_x single layers and fired AlO_x/SiN_y:H stacks. The calculation of S_{eff} is based on [5], [6]. For the single layers, S_{eff} decreases with increasing deposition temperature, while for the stacked layers S_{eff} increases. Thus, the additional SiN_y:H layer seems to have a system-changing influence on the passivation quality, which can be attributed to a change in the passivation mechanism.

To separate chemical passivation and field effect passivation, CC is used. For this purpose, charges are gradually applied one-sidedly to the sample surface, which eliminates the field effect passivation on this side. Since the field effect of the other side may still be intact, this may not result in the pure chemical passivation, but an additional offset from field effect. The quality of chemical passivation for layers deposited at different temperatures can be compared via their respective maximum of S_{eff} or minimum of τ_{eff} reached when the amount of applied CC eliminates the field effect of the fixed charges.

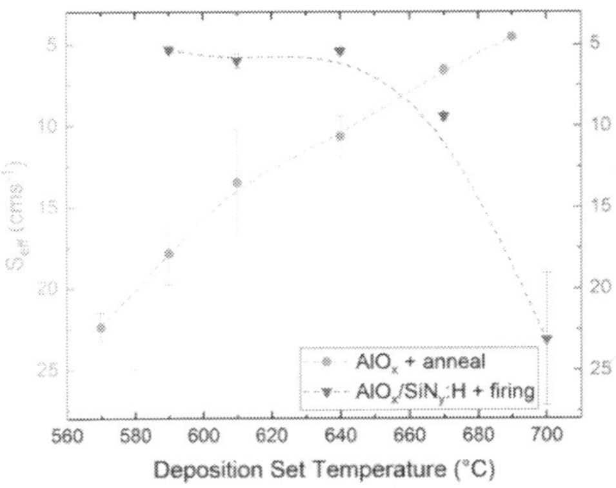

Figure 2: S_{eff} of annealed AlO_x (left axis) and of fired AlO_x/SiN_y:H stacks (right axis) dependent on the APCVD deposition set temperature. Lines serve as guide to the eye.

Fig. 3 shows τ_{eff} as a function of the corona charging duration, for annealed AlO_x single layers and fired AlO_x/SiN_y:H stacks. For the single layers, the height of the minimum of τ_{eff} increases with increasing deposition temperature. Accordingly, the chemical passivation of samples with a higher deposition temperature is more effective, meaning a likely higher density of passivated dangling bonds.

The fired stacks show in general a higher chemical passivation than the annealed single layers. This is likely to be attributed to a higher H content at the Si interface after firing with the additional H source on top. Nevertheless, the behavior of strength of chemical passivation for the stacks compared to the single layers is the other way around: the height of the minimum decreases slightly with increasing deposition temperature. This can be explained due to H diffusion through the AlO_x which is known to act as a barrier layer [7], and the barrier properties might be changed with changing deposition temperature.

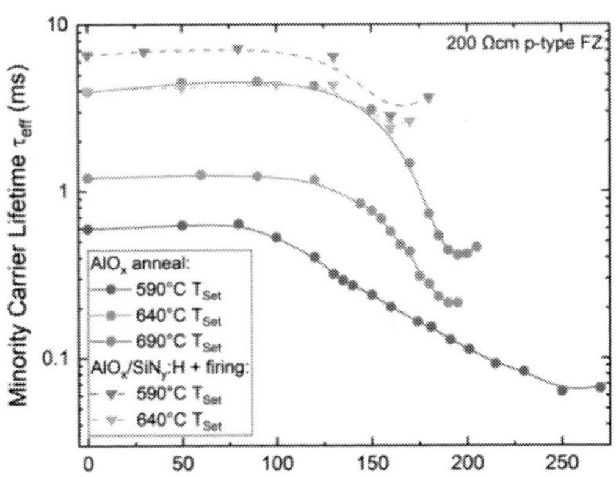

Figure 3: Measured τ_{eff} during application of Corona Charging dependent on the Corona Charging time for annealed AlO_x and fired AlO_x/SiN_y:H stacks depending on the deposition set temperature.

To determine the amount of H diffusing from a SiN_y:H layer through the AlO_x to the Si interface, a depth-resolved secondary ion mass spectrometry (SIMS) measurement is performed. AlO_x/SiN_y:H/D stacks at different extreme (one relatively cold, one hot) AlO_x deposition temperatures and a SiN_y:H/D single layer are deposited. The SiN_y:H/D is prepared using deuterated ammonia (ND_3) to distinguish the D from the SiN_y:H/D from the H that originates from the AlO_x layer. Afterwards the samples have been fired.

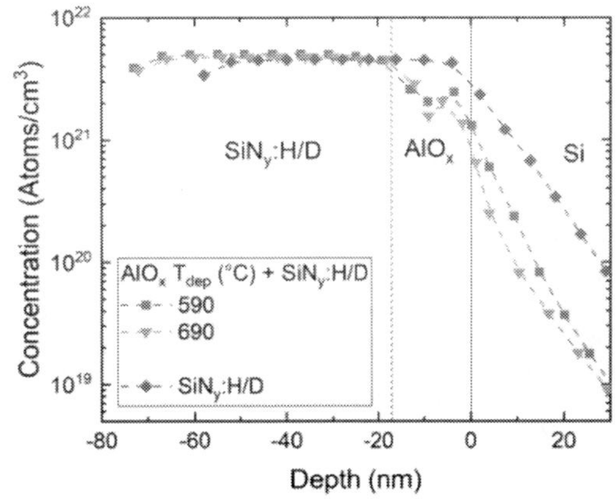

Figure 4: Concentration of D measured with SIMS in an AlO_x/SiN_y:H/D stack dependent on the depth for different deposition set temperatures and a SiN_y:H/D reference without AlO_x layer.

Fig. 4 shows the concentration of D with depth, as measured with SIMS. The signals are shifted so that the Si interface is located at zero depth in each case, calibrated with the Al signal (for samples with AlO_x) and by depth calibration (for the sample without AlO_x). The boundary of the AlO_x layer for the two samples with AlO_x layer is indicated by the dashed line. At the interface to Si, the D signal of the single SiN_y:H/D layer shows the highest concentration.

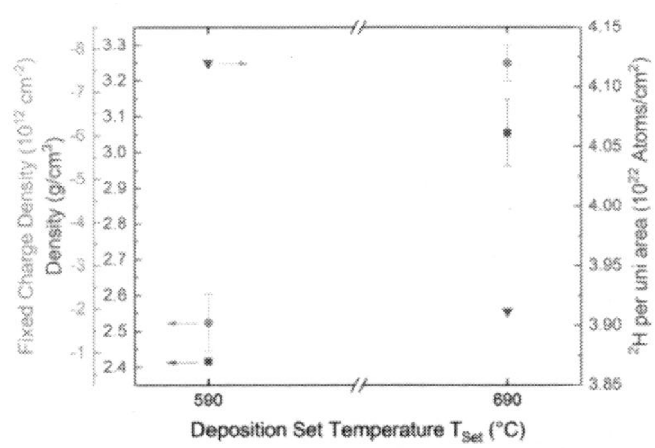

Figure 5: Density of AlO_x (left axis, grey), fixed charge density of annealed AlO_x (left axis, blue) and area concentration of D in AlO_x in an AlO_x/SiN_y:H/D stack (right axis, brown) dependent on the APCVD deposition set temperature.

Fig. 5 shows the integral of the D concentration in the AlO_x layer plotted against the deposition set temperature (brown, right axis). Numerical integration was performed over the area attributed to the AlO_x layer in Fig. 4. The concentration of deuterium per unit area in the AlO_x layer decreases significantly with increasing deposition temperature. This suggests that with increasing deposition temperature, less D diffuses from the SiN_y:H/D layer through the AlO_x layer to the interface and also into the Si.

Additionally, Fig. 5 shows the density of the AlO_x layer for the two different deposition temperatures (grey, left axis). The density increases with increasing deposition temperature. This can be an explanation for the fact that less H/D can diffuse from the SiN_y:D/H layer through the AlO_x to the Si. Thus, the chemical passivation achieved by H saturation of interface defects may be lower for higher AlO_x deposition temperatures in fired AlO_x/SiN_y:H stacks.

With a capacitance-voltage measurement, the fixed charge density Q_f in a layer can be determined and thus a prediction can be made about the field effect passivation. Fig. 5 shows Q_f for the two different deposition temperatures of annealed AlO_x single layers (blue, left axis). Q_f increases with increasing deposition temperature and thus the field effect passivation. Literature indicates [8] that this field effect can be weakened if fixed charges of the opposite polarization are added due to stacked layers, e.g. AlO_x/SiN_y:H. However, the negative charges at the interface seem to dominate the field effect compared to the additional positive fixed charges from the SiN_y:H, which means that the chemical passivation quality must be crucial to explain the behavior for the fired AlO_x/SiN_y:H stacks regarding the higher passivation quality at lower AlO_x deposition temperatures.

4 CONCLUSION

In conclusion, the better passivation quality of a fired AlO_x/SiN_y:H stack with AlO_x deposited at a lower temperature can be attributed to the lower density of the AlO_x layer. This allows more H to diffuse from the SiN_y:H layer through the AlO_x to the interface. Despite the stronger charges in AlO_x layers deposited at higher temperatures, the chemical passivation by H is essential for the observed better passivation quality of AlO_x/SiN_y:H stacks deposited at lower temperatures.

5 ACKNOWLEDGEMENTS

Part of this work was supported by the German BMWi under contract 0324226A. The content of the publication is the responsibility of the authors.

6 REFERENCES

[1] K.O. Davis et al., Phys. Stat. Sol. RRL 7 (2013) 942.
[2] L.E. Black and K.R. McIntosh, Appl. Phys. Lett. 100 (2012) 202107.
[3] F. Geml et al., Proc 37th EUPVSEC, 2020, 198.
[4] L.E.Black et al., Energy Procedia 92 (2016) 317.
[5] J. Brody et al., Solid-State Electronics 45(9) (2001) 1549.
[6] M.J. Kerr et al., J. Appl. Phys. 91 (2002) 399.
[7] A. Schmid et al., IEEE J. Photovolt. 11(4) (2021) 967.
[8] V. Naumann et al., Energy Procedia 27 (2012) 312.

HIDDEN TANDEM SOLAR CELLS

Z.T. Kuznicki
SEGTON Advanced Technology,
Address: 13 rue SAINT HONORE, 78000 VERSAILLES France 78000
Email: contact@segton-adt.com, Site internet: https://www.segton-adt.fr

ABSTRACT: According to Shockley-Queisser theory, the efficiency of single junction Si solar cells is unequivocally limited to about 30%. This limit can be overcome by localized nanostructuring of the crystal lattice inside the silicon wafer. The nanocomponents of the device taking the form of electronic walls give the converter fascinating optoelectronic properties that are particularly important in the photovoltaic field. The device, called a hidden tandem solar cell, is enriched by a secondary conversion, which adds without disturbance to the initial performance of the high-efficiency host converter. The conditioned electron wall is occupied by low-energy generation centers that are enough numerous to interact efficiently with hot electrons resulting from the primary conversion. The integration of the electronic walls has become possible thanks to specific methods of local energy deposition and subsequent technological treatment that only allows the desired physical effects to be achieved.
A selection of experimental results obtained on laboratory and commercial-sized industrial devices is presented in the paper. The different approaches providing proofs of concept take into account structural and optoelectronic aspects of the new type of silicon converters. The resulting conversion yield at this stage of R&D overpasses 27.0%.
Keywords: Silicon Solar Cell, High-Efficiency, Interfaces and Nanocomponents, Low-energy secondary generation and multiplication, Electron wall.

1. INTRODUCTION

The objectives of this work are to provide the conventional silicon solar cell with a booster powered by part of the solar spectrum energy usually lost. This postulate implies the use of an additional mechanism that allows going beyond the Shockley-Queisser theory, limiting the conversion efficiency to about 30%.

A second conversion mechanism completes the primary conversion due to physical phenomena that had not been exploited until now. A promising solution to the Shockley-Queisser "legacy" are the device nano-components embedded in the crystal lattice of a silicon wafer.

The solution is based on local nano-restructuring taking the form of electronic walls enabling the harmonious work of two mechanisms: one is the photon-electron interaction of the primary conversion and the other is the electron-electron interaction of the secondary conversion. Indeed, the secondary conversion is triggered by its primary counterpart; the population of electrons generated by the energetic photons is multiplied by the specific low-energy generation mechanism.

Scientific publications, cited in the references [1-3], concern the theoretical and experimental results that paved the way to new ideas guaranteeing an increase in the efficiency of light-to-electricity conversion based on modified silicon cells. The main part of the R&D work has already been patented.

2. LOCALIZED NANOSCALE TRANSFORMATIONS

2.1 Amorphizing ion implantation.

The method of local transformations of crystal lattice is based on the amorphizing ion implantation, the associated thermodynamic treatment and the physical conditioning of the embedded nanocomponents. The ion implantation destroys the Previous crystalline order –▶ then the Amorphized disorder –▶ is transformed again into New crystalline order by a Thermodynamic treatment. The Technology cycle has to respect Temperature profile & Thermal balance allowing required device Conditioning.

The fabrication method is illustrated in Fig. 1, where the paths of the ions entering the crystal lattice are visualized by Monte Carlo simulation (code SRIM). The resulting buried amorphization by P^{31} ions at 180 keV is visualized by RBS and TEM giving an insight into nano transformations.

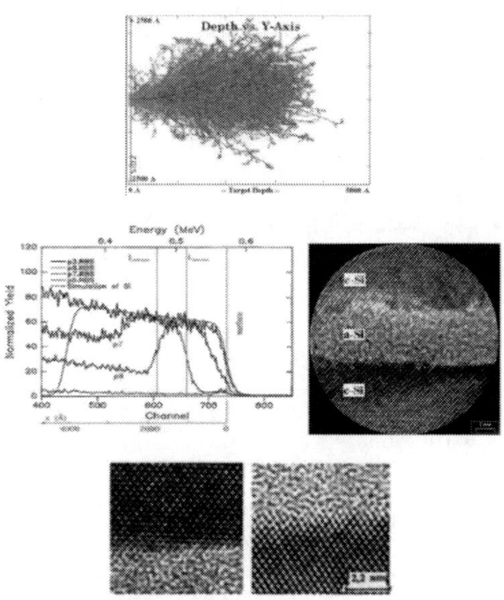

Figure 1: Upper: Amorphizing implantation by P^{31} ions at 180 keV (Monte Carlo simulation – SRIM code). Middle left: RBS characterization revealing a specific structural transformation. Middle right: TEM image of a buried amorphization inside the crystalline lattice. Lower: HREM images of intimate interfaces of two structural phases. The a-Si nanolayer is limited by the upper (left) and lower (right) heterointerfaces.

The integration of the secondary mechanism in the light-electricity converter can be expressed most simply as the integration of nanocomponents in a conventional high-efficiency cell. This leads to the design of a hidden tandem. The size of the nanocomponent must be very limited in the bulk of the wafer so that it can be integrated with the least possible structural and functional disturbances. Of course, it must be invisible from the viewpoint of the action/performance of the host cell, i.e. its performance must be perfectly additive to the initial high performance of the cell. Some microscopy images of embedded nanocomponents are shown in Fig. 2.

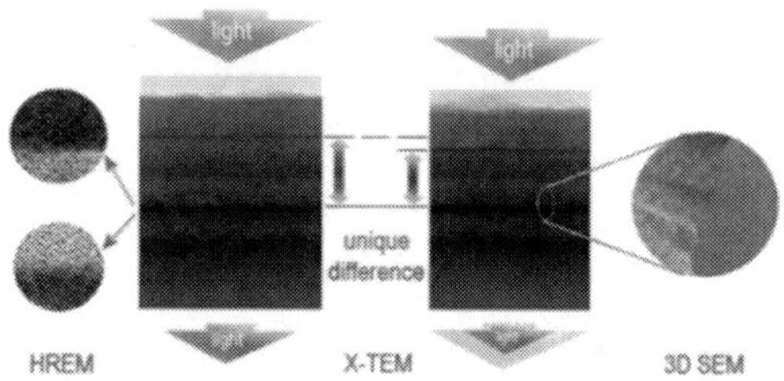

Figure 2: HREM, X-TEM and 3D SEM images of a continuous nanoscale Si-layered system buried within a silicon wafer. The difference in light transmission is due to the superficial absorbance at the surface area.

The first step consists of the localized insertion of a crystalline disorder in the form of an amorphized nanolayer. The next step uses this disorder to create a new crystalline order through solid-state epitaxy. Amorphization acts as a seed for the phase transformation. The third and final step is the conditioning of the nanoscale Si-layered system containing two nanolayers of a new crystalline phase of the silicon. This phase is metastable at room temperature and at not too high temperatures (< 600°C) due to strong physical conditioning, which is only available at the nanoscale. The shaped metamaterial can be usefully packaged as an electron wall with specific optoelectronic characteristics. In particular, it can:

- transform an untapped part of the UV photon energy into an additional population of free electrons;
- be conveniently pumped by incident light energy transmitted through the hot electrons and hopping transport.

2.2 Electron wall

The functioning of the electron wall is schematized in Fig. 3. Specific aspect concerns the energy space of the semiconductor, in which the electrons are exchanged in different ways.

The locally increased populations of weakly bound electrons, at electron walls, were visualized experimentally by positron annihilation as shown below in Fig. 5. Specific electron occupancy by weakly bound electrons appears in two metamaterial nanolayers of SEG-Matter. The metamaterial represents a specific crystalline phase of silicon wrought by physical forces available at the nanoscale.

For comparison, are also shown TEM and 3D SEM microscopy images of the buried nanoscale Si-layered system that appears after proper treatment and conditioning.

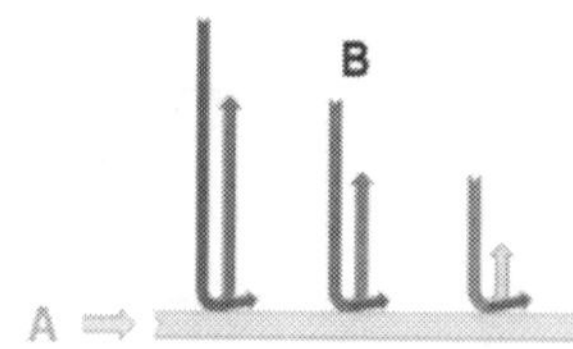

Figure 3: An artist's view of the "pumping" electron wall and the three-stage generating collisions of a single hot electron. A: The hoping pumping of low-energy generation centers located in the electron wall. B: Hot electron paths at the origin of collisional generations of secondary electrons with the energy of about 0.3 eV.

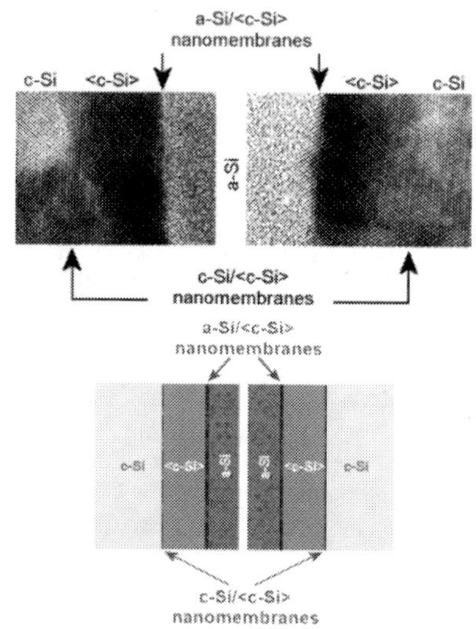

Figure 4: HREM images and associated diagram of electron walls (<c-Si>) bounded by nanomembranes: c-Si /a-Si interfaces and transitions between relaxed and strained c-Si lattice.

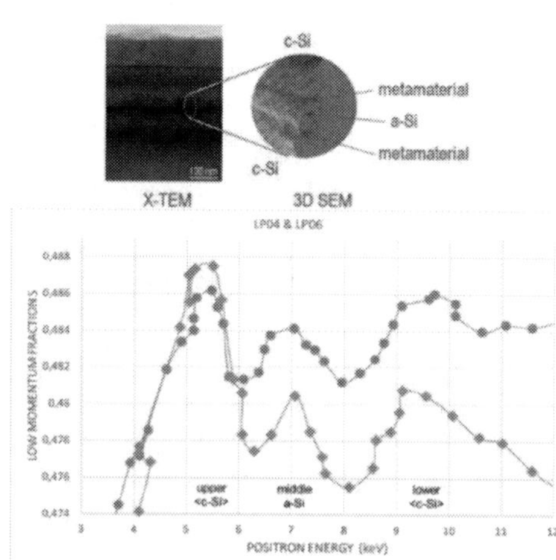

Figure 5: Locally increased populations of weakly bound electrons as visualized by positron annihilation. Experiments were conducted on test devices, which X-TEM and 3D SEM results are also shown.

As mentioned above, the technique allowing the embedding of an electron wall in the crystalline Si wafer bases on the amorphizing ion implantation, subsequent thermal treatment, and finishing conditioning [6, 7]. This is available due to local physical fields, crystal lattice arrangement, geometric properties and design of the device. Fig. 6 illustrates the composition details of a nanocomponent embedded in a silicon wafer by ion implantation and appropriate treatment.

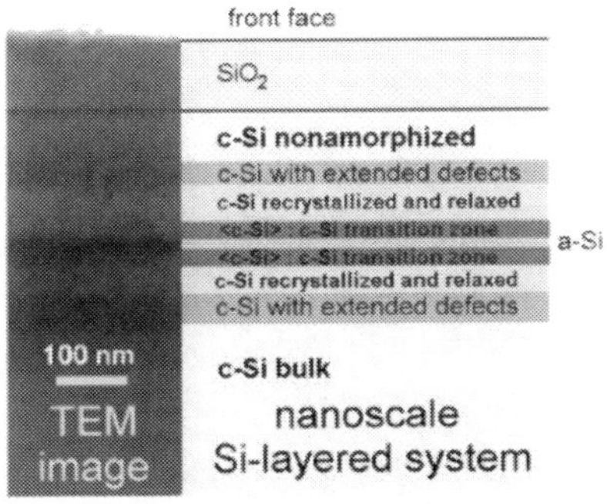

Figure 6: Detailed structure of an amorphized nanocomponent embedded in a silicon wafer.

2.3 Electron wall features

In short, to take advantage of an electron wall, the hidden tandem must meet the requirements, which can be grouped as follows.

Structure – a local nanostructuring using buried amorphization and post-implant thermal treatment.

Conditioning – the creation of secondary dense enough generation centers filled by weakly bond electrons and effectively electrically screened from recombination centers.

Operation – a near-permanent electric state assuming the great probability of collisional generations and the rapidity of the electron recharging.

Functioning – low-energy generation centers filled with weakly bonded electrons and plunged in the device area where there are numerous hot electrons.

Requirements – the adjusted level of n-type impurity doping, suitable geometrical factors and position in device bulk assuming a good electron transport, interactions time constants, and the high probability of collection of all photogenerated carriers.

The electron wall functioning properly as the hidden tandem nanocomponent:

- transforms an untapped part of the UV photon energy into an additional population of free electrons;
- is conveniently pumped by incident light energy transmitted through the hot electrons and hopping transport.

3. NANOSTRUCTURE CHARACTERIZATIONS

To improve the efficiency of light-to-electricity conversion, a number of specific physical and geometric conditions must be met. In this case, a specific role plays the buried semiconductor interfaces [4] that allow locally producing metastable states thanks, among other things, to their transition zones and the great physical forces available only inside the crystal lattice. For these reasons, the integrated nanostructures were previously tested by various methods to obtain detailed information about their structural and electronic properties when embedded inside the silicon wafer.

Selected properties of localized nanostructuring were examined from the first step of buried amorphization that permits the subsequent formation of a nanoscale Si-layered system. Examples of these characterizations are grouped below in Fig. 7.

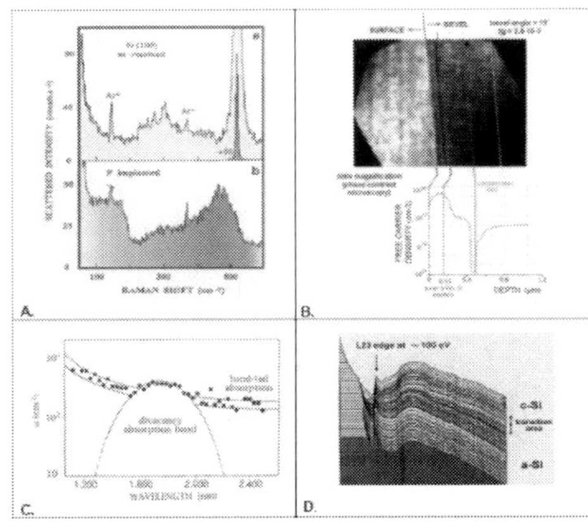

Figure 7: Results of nanostructure characterizations. **A**: Raman characterization of buried amorphizations reveling a specific structural transformation of the crystalline semiconductor. **B**: Spreading resistance showing a P^{31} (free carrier) profile compared with channeling profile. **C**: PDS characterization of buried amorphization relevant a specific structural transformation of the crystalline semiconductor. **D**: Series of EELS spectra obtained by e line scan through the a-Si/c-Si interface.

The results confirm the fact that the transformations leading to the insertion of electron walls can be carried out on large scale wafers with industrial tools. Moreover, the appropriate nanostructuring allowing to extend the useful properties of the semiconductors can be modified efficiently by using relatively very low energies.

Therefore, nanostructured components can be suitably positioned and distributed in the bulk of the wafer. Finally, new effects become available at the nanometric scale thanks to feedback: there are no mechanisms sought without a dedicated device, there is no appropriate device without specific mechanisms, the two, conditioning and mutually optimizing each other together.

4. EXTRINSIC ELECTRON LEVELS

Performance enhancement would not be possible without the incorporation into a silicon crystal of appropriate extrinsic electron levels. Electron walls solve the problem, where two levels occupy particular positions inside the forbidden band. What is important, they usefully communicate in a specific way with each other.

Insight into characteristic energies of electron walls has been given experimentally by photoluminescence (PL) at helium temperatures, as shown in Fig. 8. Such conditions make it possible to their emptying and thanks to it their clear visualization, because at room temperatures, they are completely filled with electrons.

These levels appear as a result of two additional atom bonds, in addition to the primary covalent bond, in the modified crystal silicon lattice of electron walls: shorter (0.77 eV level), and longer (0.90 eV level). This crystalline reorganization gives the origin of a new unit of Secondary

Electron Generation Tuned On Nanoscale (called SEGTON). SEGTONs are incrusted in conditioned material (metamaterial) for Secondary Electron Generation (SEG-Matter), which represents a metastable crystalline phase of modified silicon.

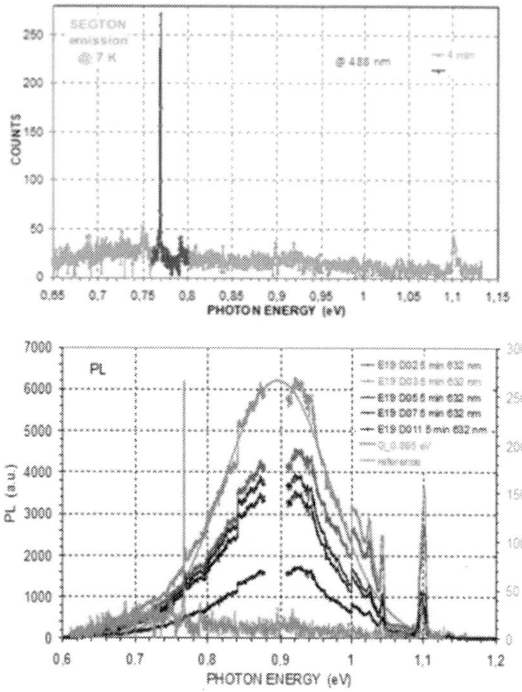

Figure 8: Electron energy levels revealed by PL at helium temperatures. **Upper**: A reference PL signature of SEGTONs. PL emission peak around 0.77 eV at 7 K from a Si wafer containing a buried SEG-Matter, under the laser excitation at 488 nm. **Lower**: PL signatures of SEGTONs around 0.9 eV at 7 K from a Si wafer containing a buried SEG-Matter, under the laser excitation at 632 nm.

The PL signature at 0.77 eV confirms an excellent quality of the crystal structure of the host high-efficiency cell despite embedded nanocomponents, while the 0.90 eV level, which is only about 0.3 eV below the bottom of the conduction band, plays a particularly important role in light-to-electricity conversion. This is at the origin of the low energy generation and multiplication of secondary electrons.

5. HIDDEN TANDEM - SPATIAL DISTRIBUTION OF PHOTOGENERATION

A new field in silicon photovoltaics based on the localized structuring of the crystal lattice inside a silicon wafer is an effective method to break the Shockley-Queisser thermodynamic barrier of 30% of the conversion efficiency of single-junction semiconductor devices. Such an improvement is obtained thanks to an undisturbed performance superposition of two complementary mechanisms, primary conventional conversion (photon-electron interactions) and secondary additional conversion (electron-electron interactions)

It became possible thanks to the feedback available at the nanoscale, there are no mechanisms without a dedicated device, we called it hidden tandem. The sum of the two undisturbed performances (primary and secondary) exceeds the efficiency of the best conventional silicon cells.

A numerical deconvolution reveals the individual contributions of the hidden tandem strata in the final collection efficiency as shown in Fig. 9. This occurs when the transmission of incident photons in the active areas of the device is correct and when almost all of the photogeneration can be correctly collected.

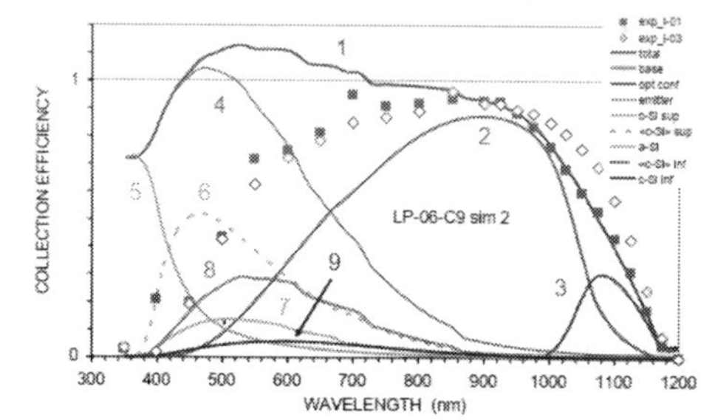

Figure 9: Individual contributions to the collection efficiency of hidden tandem components resulting from the numerical deconvolution of optimal photogeneration with the use of several experimental results, adjustments and analysis of test devices. Curves: 1) CE resulting from all the components 2) collection in the base, 3) collection enhancement due to optical confinement, 4) collection resulting from the whole emitter, 5) collection from the non-modified Si (c-Si) conventional part, 6) additional collection in the upper emitter <c-Si> due to secondary conversion, 7) collection in the buried amorphized (a-Si) layer, 8) additional collection in the lower emitter <c-Si> due to secondary conversion, 9) conventional collection of the lower c-Si emitter; <c-Si> describes nanolayer of conditioned metamaterial embedded the emitter..

The experimental points (squares and diamonds) were obtained on the same test sample with different stages of its thermal treatment. The curve with diamonds showing an important IR improvement after a 3 min additional annealing at 500°C.

The result reveals the importance of electronic walls in the performance of hidden tandems, sum of curves (6) and (8) in Fig. 9.

6. ACTIVITY OF NANO COMPONENTS LOW-ENERGY GENERATION

Low energy generation and multiplication [5] allows a significant increase in the population of free electrons as shown by the comparison of theoretical data presented in Fig. 10.

The experimental proof of the secondary generation mechanism with a high probability of interaction has been made on dedicated test devices. A stepped curve of low-energy secondary generation, reflecting the low-energy multiplication orders, appeared in the test devices is shown in Fig. 11.

The electronic activity of the electron wall of SEG-Matter has been experimentally proven based on the comparison of the EQE and IQE characteristics of commercial size test devices and equivalent reference cells, Fig.12. Results as differences in EQE and IQE are shown in Fig. 13.

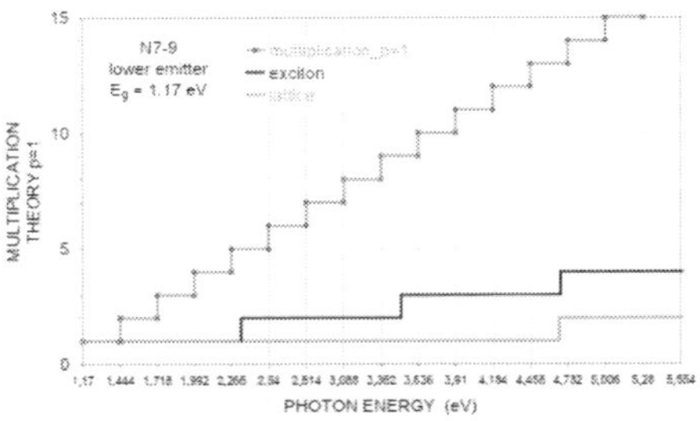

Figure 10: Comparison of currents from three multiplication effects in Si (collected with a maximal probability of p = 1.0): conventional impact ionization (green curve), exciton multiplicity in Si nanocrystals (dark blue curve) and low-energy generation in a Si metamaterial (clear blue curve). For example, for $h\nu$ = 3.365 eV (368.5 nm) the conventional impact ionization allows exclusively a primary generation, the excitons can double the density of the free-carrier population and the low-energy generation can lead to a nine-fold increase of the primary population. The clear blue curve corresponds to the new effect.

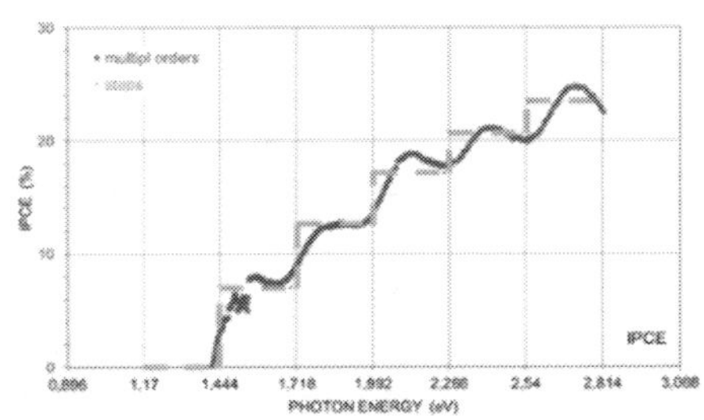

Figure 11: Induced Photon Collection Efficiency of an experimental test cell – steps of a secondary generation. On the ordinate axis, there are photon energy ranges, corresponding to successive orders of low-energy electron multiplication. This curve reveals the electronic activity of the SEG-Matter electron wall.

The spectral response was measured on bifacial devices with the SEG-Matter and on their conventional counterparts, both fabricated on a pilot line.

In fact, the electron wall of SEG-Matter has been integrated into the equivalent conventional host cell with 20% efficiency. All the technological processes were exactly the same for the reference and test cell, except for the SEG-Matter insertion sequence.

The results of Figs 13 and 14 testify to the good reproducibility of the Experimental Protocol used, which was initially designed as a method of technological quality control of large-format cells (commercial-sized) manufactured using SEGTON technology.

For this proof of concept, the SEG-Matter is on the opposite side to the illuminated side of measured bifacial cells.

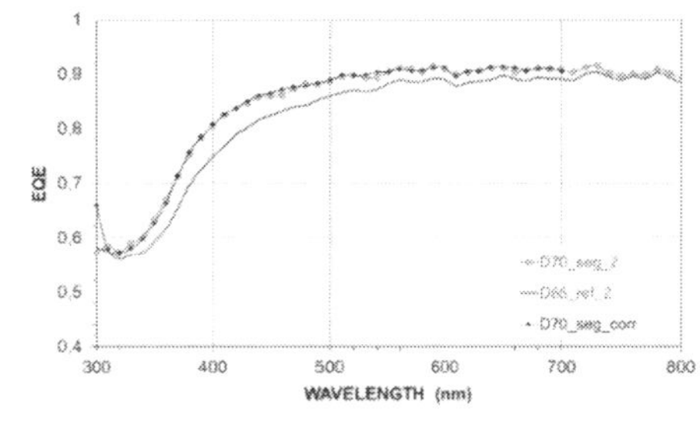

Figure 12: Spectral response (EQE) of the test cell in the range of the activity of the SEG-Matter in comparison with that of its conventional equivalent (η = 20%) as a function of the incident wavelength.

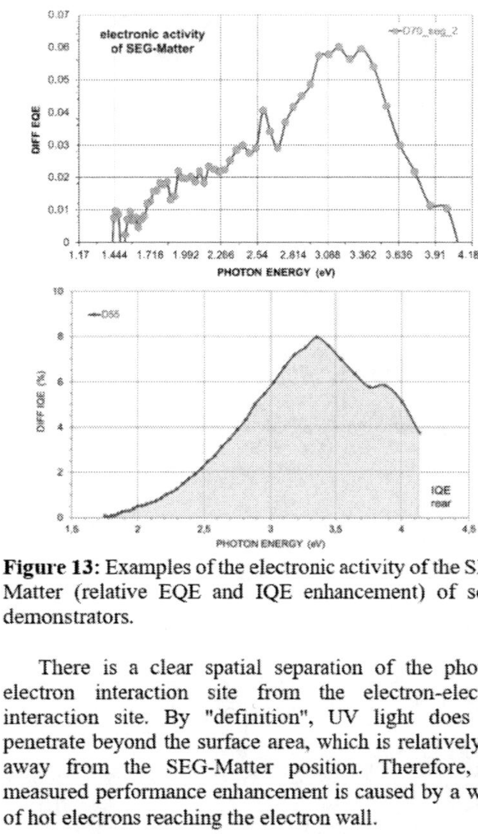

Figure 13: Examples of the electronic activity of the SEG-Matter (relative EQE and IQE enhancement) of some demonstrators.

There is a clear spatial separation of the photon-electron interaction site from the electron-electron interaction site. By "definition", UV light does not penetrate beyond the surface area, which is relatively far away from the SEG-Matter position. Therefore, the measured performance enhancement is caused by a wave of hot electrons reaching the electron wall.

7. ENHANCEMENT OF CONVERSION PERFORMANCE

The improvement in the conversion efficiency available by hidden tandem silicon cells equipped with the electron wall of SEG-Matter can be compared with the solar spectrum distribution, as shown in Fig. 14. At this stage, the result was obtained on commercial-size demonstrators using their spectral response and the final performance by numerical analysis with our SIM-SEG code [6].

The electron walls are active in the unused/inaccessible zone of the solar spectrum (blue areas in Fig. 14). Their operation complements the conversion efficiency of high-performance silicon cells as they are electrically screened against the main wafer volume in which the primary conversion takes place.

Fig.15 shows a photo of the device that contains continuous electronic walls of SEG-Matter suitably buried in a large wafer.

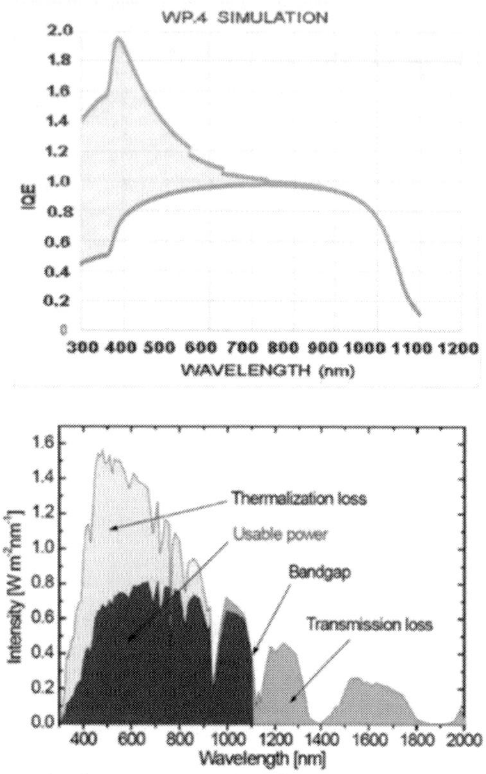

Figure 14: Improvement of the light-to-electricity conversion by the low-energy mechanism. **Left**: IQE enhancement by the SEG-Matter shown as a function of the wavelength: i) conventional reference cell $\eta = 22.14$ % (blue curve); ii) tandem demonstrator with a monoface architecture, SIM-SEG code [8], η = 26.90 % (orange curve). **Right**: Solar spectrum to compare.

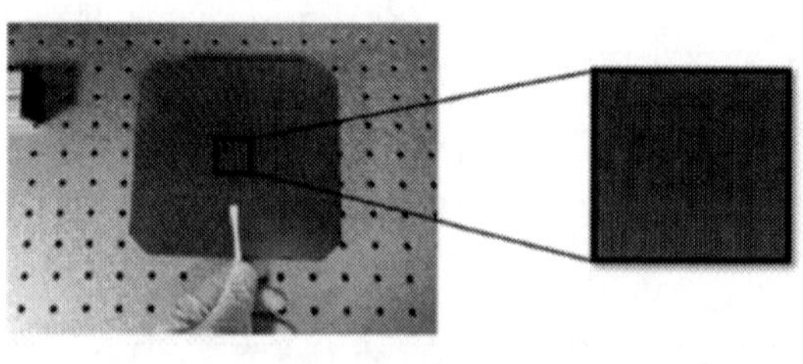

Figure 15: A commercial-sized demonstrator (156 mm x 156 mm) fabricated on a pilot line.

CONCLUSIONS

Discoveries and innovations focused on the field of silicon photovoltaics have reached the point of designing and manufacturing a new generation of solar cells with significantly improved efficiency, we have called them hidden tandems.

These single-junction converters match two complementary mechanisms: conventional conversion (photon-electron interactions) and secondary conversion (electron-electron interactions). The sum of these two undisturbed performances clearly exceeds that of the best conventional silicon cells.

References

[1] Multi-interface novel devices, model with a continuous substructure, Chapter 8 of the book "3rd Generation Photovoltaics for High Efficiency through Full Spectrum Utilization", Eds. A. Luque & A. Marti, Institute of Physics Publishing, Bristol, UK, 2003, Z.T. Kuznicki.

[2] Low energy free-carrier generation in nanoscale Si-layered systems: Experimental evidence, Z.T. Kuznicki, P. Meyrueis, WCPEC-4, May 7-12, 2006 Hilton Waikoloa Village, Hawaii, USA, Proceedings, pp. 107-110.

[3] Application of innovations grouped under the name giant photoconversion, Z.T. Kuznicki, P. Meyrueis, M. Hosatte; M. Basta, Proc. SPIE 12150 Photonics for Solar Energy Systems IX, 121500A (24 May 2022; doi:10.1117/12.2620963.

[4] Low-High homojunction in the stationary state, Z.T. Kuźnicki, Journal of Applied Physics, 69(1991)6526-6541.

[5] Low energy free-carrier generation in nanoscale Si-layered systems: Experimental evidence, Z.T. Kuznicki, P. Meyrueis, WCPEC-4, May 7-12, 2006 Hilton Waikoloa Village, Hawaii, USA, Proceedings, pp. 107-110.

[6] Multi-interface Solar Cells; I. Limits, Modelling and Design, Z.T. Kuznicki, E-MRS First Polish-Ukrainian Symposium "New Photovoltaic Materials for Solar Cells", Cracow-Przegorzaly, Poland, 21-22 October 1996, Proceedings, pp. 58-78.

[7] Application of innovations grouped under the name giant photoconversion, Z.T. Kuznicki, P. Meyrueis, M. Hosatte; M. Basta, Proc. SPIE 12150 Photonics for Solar Energy Systems IX, 121500A (24 May 2022; doi:10.1117/12.2620963.

[8] SIM-SEG code giving a complete insight into electronic and photovoltaic performances basing on nondestructive optical measurements, M. Basta, Z.T. Kuznicki, M. Hosatte, Proc. SPIE 12150, Photonics for Solar Energy Systems IX, 121500B (24 May 2022); doi: 10.1117/12.2622101.

HETEROJUNCTION GERMANIUM SOLAR CELLS WITH A RECORD HIGH EFFICIENCY OF 8.6%

°Makoto Konagai[1], Kimihiko Saito[2], Kazuyoshi Nakada[1], Yukimi Ichikawa[1], Naoki Suyama[1], Ryosuke Ishikawa[1]
Tokyo City Univ.[1], Fukushima Univ[2]
[1]Advanced Research Laboratories, Tokyo City University, 8-15-1 Todoroki, Setagaya-ku, Tokyo 158-0082, Japan
[2]Fukushima University, Faculty of Symbiotic Systems Science, Fukushima, Japan

ABSTRACT: Ge solar cells were prepared using three types of p-Ge wafers with carrier concentrations ranging from 1.0×10^{16} / cm^3 (0.28 Ωcm) to 3.0×10^{18} / cm^3 (0.0063 Ωcm). The surface treatment of Ge wafers and the passivation of the n-a-Si:H (n-μc-Si:H) / p-Ge interface have a great impact on the conversion efficiency. As for the substrate pretreatment, the HF treatment and the method of adding the HCl treatment to the HF treatment were also attempted. At the n-a-Si:H (n-μc-Si:H) / p-Ge emitter interface, i-a-Si:H with a thickness of 4 to 14 nm was inserted as a passivation layer. A BSF structure of p-Ge /i-a-Si:H /p-μc-Si was deposited on the back surface side. *Voc* was greatly improved by lowering the resistivity of the substrate. In a cell in which an i-a-Si:H layer with a thickness of 14 nm was introduced at the emitter interface on a 0.016 Ω cm substrate, a conversion efficiency 8.6 % (*Voc* = 291 mV, *Jsc* = 45.0 mA/cm^2, *FF* = 0.656, AM1.5, 100mW/cm^2, 25 ° C) was obtained. The interface was observed with a transmission electron microscope (TEM). As a result, It was found that the island-shaped heteroepitaxial growth suppressed the misfit dislocations that would otherwise occur due to lattice mismatch, and exhibited good interfacial characteristics.
Keywords: Germanium, Heterojunction, Interfaces

1 INTRODUCTION

Ge solar cells are attracting attention as bottom cells for multi-junction solar cells. Several research groups have reported conversion efficiencies of 5 to 7 % so far for Ge heterojunction(GHJ) solar cells[1]-[6], but further improvement in conversion efficiency exceeding 8% is required . Therefore, to clarify the limiting factors of the photovoltaic characteristics, we conducted a device simulation and found that the open-circuit voltage *Voc* greatly depends on the resistivity of the Ge wafer, unlike the Si heterojunction(SHJ) solar cell. Based on these theoretical analyses, we aimed to improve the conversion efficiency by using low resistivity Ge wafers.

The surface treatment of Ge wafers and the passivation of the n-a-Si:H (n-μc-Si:H) / p-Ge interface have a great impact on the conversion efficiency. As for the substrate pretreatment, the HF treatment and the method of adding the HCl treatment to the HF treatment were also attempted. *Voc* was greatly improved by lowering the resistivity of the substrate. In addition, by increasing the thickness of the i-a-Si:H layer on the low resistivity substrate, the solar cell characteristics tended to improve in all parameters.

As a result, a conversion rate of 8.6% (AM1.5) was obtained in a cell with an area of 1 cm^2. This value is the highest in the world among the Ge heterojunction solar cells reported so far.

To clarify the mechanism of conversion efficiency improvement, analysis was performed using cross-sectional TEM(Transmission Electron Microscope), SCM(Scanning Capacitance Microscope), and SIMS(Secondary Ion Mass Spectrometry). As a result, it was found that the key to improving the conversion efficiency is the island-like epitaxial growth of Si on the Ge wafer.

2 STRUCTURE AND PREPARATION OF GHJ SOLAR CELLS

2.1 Structure of GHJ solar cells

Fig.1 compares the structures of GHJ solar cells and SHJ solar cells. The preparation method of SHJ solar cells is described in detail in [7] to [10]. Basically, an n-Si wafer with a textured surface is used, and 2 nm of i-a-SiO_x:H and 4.4 nm of i-a-Si:H are inserted at the interface between the p-layer and n-Si. The role of i-a-SiO_x:H is to prevent crystallization of i-a-Si:H at the interface. In SHJ solar cells, when i-a-Si:H crystallizes at the interface, the solar cell characteristics, especially the open-circuit voltage *Voc*, are greatly reduced. Wide gap p-μc-SiO_x:H containing 5% to 10% oxygen is used for the p-layer. A similar n-Si/i-a-SiO_x:H/i-a-Si:H passivation structure is used on the rear surface of the n-Si wafer.

GHJ solar cells, on the other hand, are prepared on p-type Ge wafers. For Ge, an i-a-SiO_x:H buffer layer was inserted at the beginning of research and development, but a tendency for the fill factor (*FF*) to decrease was observed, so in this report, i-a-SiO_x :H buffer layer is not inserted. The biggest difference between SHJ and GHJ structures is the thickness of i-a-Si:H. For SHJ, the optimum thickness is 4.4 nm, but for Ge, a wide range of 4 nm – 17 nm was investigated.

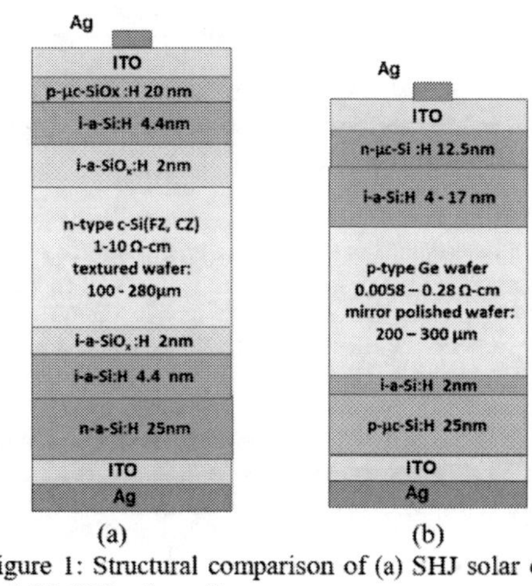

(a) (b)

Figure 1: Structural comparison of (a) SHJ solar cells and (b) GHJ solar cells.

2.2 Preparation of solar cells

In device simulations of SHJ solar cells, the photovoltaic (PV) characteristics do not depend much on the carrier concentration of the wafer. On the other hand, in the GHJ solar cells, the PV characteristics are very dependent on the acceptor concentration of the wafer.

For example, the results of simulating the relationship between the carrier concentration N_D of SHJ solar cells wafers and cell characteristics using AFORS-HET show that when N_D ranges from $1\times10^{14}/cm^3$ to $1\times10^{17}/cm^3$, the N_D dependence of Voc is not recognized [8]. In SHJ, Voc is limited by the diffusion potential determined by p-μc-SiO_X:H and n-a-Si:H. On the other hand, in the GHJ device simulation using SILVACO Atlas, we obtained the result that Voc greatly depends on the carrier concentration N_A of the wafer.

Fig.2 shows the theoretical relationship between N_A and Voc in GHJ solar cells. In the figure, Voc is calculated using the carrier lifetime of the wafer as a parameter. A monotonous increase in Voc is observed in the range from $1\times10^{16}/cm^3$ to $1\times10^{18}/cm^3$.

In this work, Ge wafers with three different resistivities were prepared for the fabrication of GHJ solar cells. Their resistivity is 0.28 Ω-cm ($1\times10^{16}/cm^3$) for high-R (resistivity) wafers, 0.0158 Ω-cm ($5\times10^{17}/cm^3$) for middle-R wafers, and 0.0058-0.0063 Ω-cm ($2\times10^{18}/cm^3$) for low-R wafers. After fabricating the cell structure shown in Fig. 1(b), the effective carrier lifetimes were measured by μ-PCD and found to be 50 – 125 μs, 3-9 μs, and < 1μs, respectively. It can be seen that the Voc of SHJ solar cells is greatly affected by the effective carrier lifetime, but that of GHJ solar cells is greatly dependent on the N_A of the wafer in addition to the effective carrier lifetime.

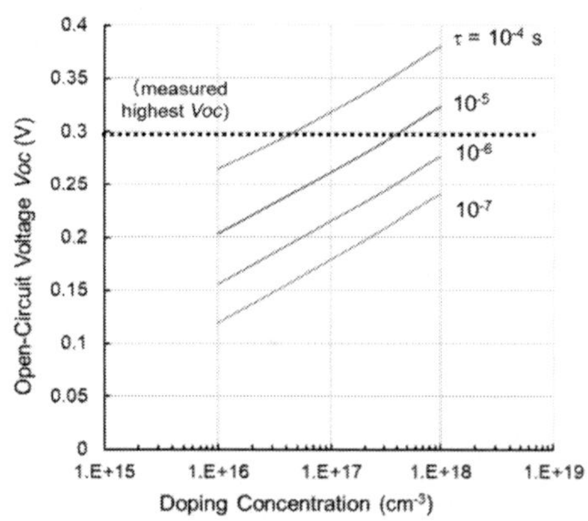

Figure 2: Theoretical relationship between N_A and Voc in GHJ solar cells using SILVACO Atlas.

2.3 Cleaning of Ge wafer and the passivation layer

In heterojunction Ge solar cells, the surface treatment of wafers and the passivation of the n-a-Si:H (n-μc-Si:H) / p-Ge interface are very important. As for the substrate pretreatment, the HF treatment and the method of adding the HCl treatment to the HF treatment were also attempted. At the n-a-Si:H (n-μc-Si:H) / p-Ge emitter interface, i-a-Si:H with a thickness of 4 to 14 nm was inserted as a passivation layer. A BSF structure of p-Ge /i-a-Si:H /p-μc-Si was deposited on the back surface side.

3 PHOTOVOLTAIC CHARACTERISITICS

3.1 Effect of a-Si:H buffer layer thickness

Fig. 3 shows the dependence of the characteristics of Ge heterojunction solar cells prepared by performing with HF + HCl pre-treatment on the emitter-side i-type a-Si:H layer thickness.

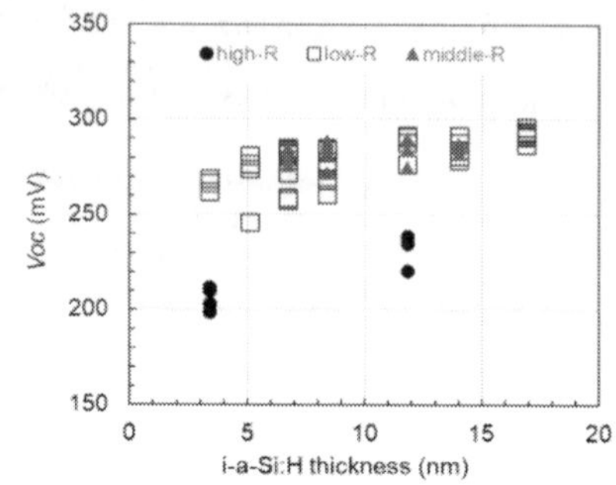

(a) Open circuit voltage, Voc

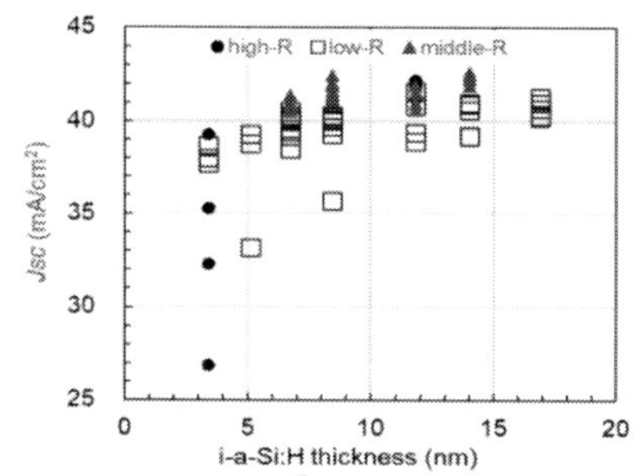

(b) Short circuit current density, Jsc

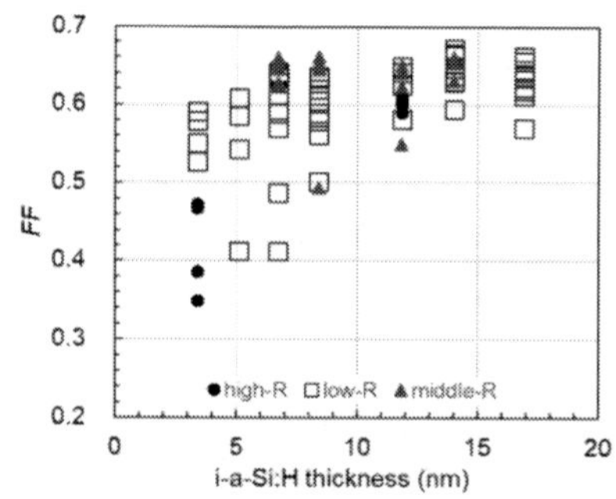

(c) Fill factor, FF

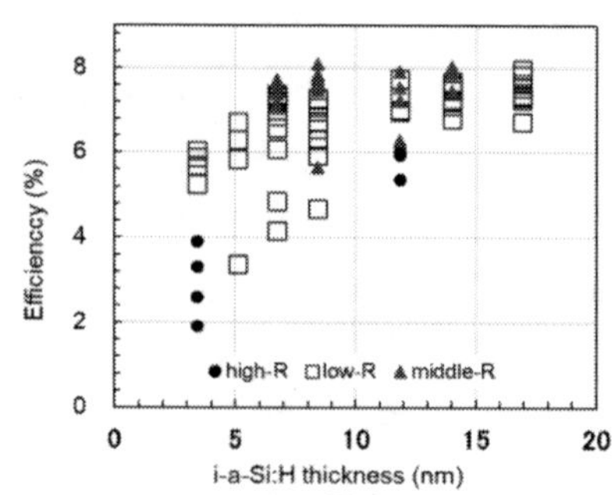

(d) Efficiency

Figure 3: Effect of a-Si:H buffer layer thickness on photovoltaic parameters. AM1.5, $100mW/cm^2$, 25°C.

At the beginning of the research, we repeated experiments using high-R wafers with an effective carrier lifetime after passivation of 50 – 125 μsec. In particular, we tried optimizing the stacked structure and thickness of the passivation layer, but the *Voc* was often as low as 200 mV or less, and the data varied greatly.

In the next step, when we repeated preparing using wafers with low resistivity (low-R wafer), a significant improvement in *Voc* was observed even with an i-a-Si:H passivation film thickness of 4 nm, and the conversion efficiency improved to 6%. Then, when the thickness of the i-a-Si:H passivation layer was increased from 4 nm to 17 nm, *Voc* increased to around 300 mV and the data dispersion decreased. In particular, an improvement in *Voc* and *FF* was observed. As a result, a stable conversion efficiency of 7% to 8% can be obtained.

Furthermore, when a wafer with a middle resistivity (middle-R wafer) is used, a performance close to that obtained when using a low-R wafer was obtained. The experimental results of *Voc* using wafers with three types of resistivity showed the same trend as the theoretical analysis results shown in Fig.2. On the other hand, for *Jsc*, the low-R wafer, which has a low Ge carrier lifetime, has a slightly lower *EQE* in all wavelength ranges than the middle-R wafer. Especially when using a low-R wafer, making the wafer thinner is effective in further improving the conversion efficiency.

3.2 Record high efficiency solar cell

Fig.4 shows the I-V characteristics under light irradiation of a Ge solar cell fabricated on a low-R Ge wafer (0.0158 Ω-cm, 317 μm thick wafer) that showed the highest efficiency so far. A conversion efficiency of 8.6% (*Voc*=291 mV, *Jsc*=45.0 mA/cm^2, *FF*=0.656) was obtained with an area of 1 cm^2. The irradiation conditions of the solar simulator are AM1.5, 100mW/cm^2, and 25°C. The cell structure is shown in Fig. 1(b). In this cell, ITO with a thickness of 80 nm also serves as an antireflection film. On the other hand, for Ge wafers, unlike Si, there is no known simple texture etching method to reduce the surface reflectance, so we tried to reduce the reflectance by two-layer coating using MgF_2.

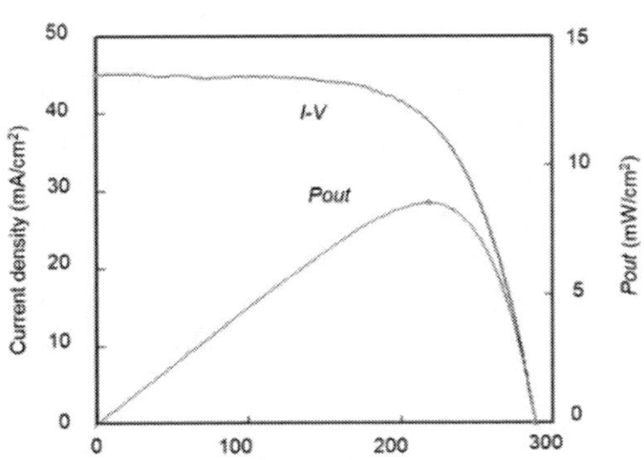

Figue 4: Illuminated I-V characteristics of a Ge solar cell with an efficiency of 8.6% (AM1.5, 1cm^2)

OPAL2 was used for the optimal design of the double-layer antireflection coating [11]. In this case, the ITO thickness was fixed at 80 nm, and the MgF_2 thickness was designed to maximize *Jsc*. The optimal MgF_2 thickness was 111 nm. Fig.5 shows the *EQE* and

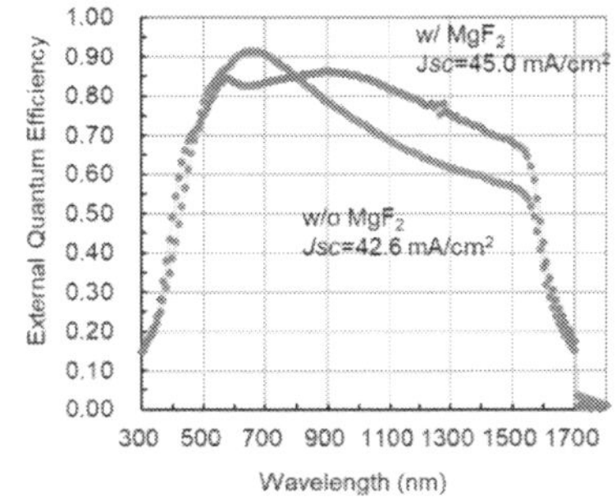

Figure 5: External quantum efficiency of a Ge solar cell with an efficiency of 8.6% (w/ MgF_2) and of 8.03% (w/o MgF_2).

Jsc values before and after MgF_2 deposition. The introduction of the double-layer coating film increased *Jsc* from 42.6 mA/cm^2 to 45.0 mA/cm^2.

4 DISCUSSIONS

4.1 TEM analysis

In this work, the world's highest conversion efficiency of 8.6% has been achieved. To clarify the mechanism of conversion efficiency improvement, analysis was performed using cross-sectional TEM, SCM, SIMS, etc.

Fig.6 shows a cross-sectional TEM image of GHJ cell in which an i-a-Si: H layer of 4 nm is deposited on a Ge substrate with a resistivity of 0.1-5 Ω cm. As for the substrate pretreatment, only the HF treatment was performed and the conversion efficiency of the cell is around 5%. As shown in the figure, the interface between the Ge wafer and i-a-Si:H is very clear and no crystalline phase is observed.

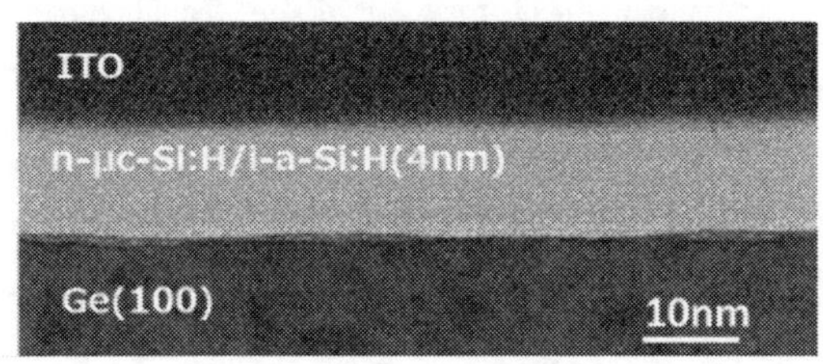

Figure 6: Cross-sectional TEM image of the n-μc-Si:H/i-a-Si:H/Ge interface. Substrate pretreatment: only HF.

On the other hand, Fig. 7 is the TEM observation result of a solar cell showing a conversion efficiency of 8.6%. In Fig. 7, it can be seen that Si epitaxially grows in island form over a region of 5 to 8 nm at the Si/Ge interface. In addition, no misfit dislocation is observed in Si crystalline phase. In this sample, the Ge wafer was treated with HF and HCl prior to buffer layer deposition.

Furthermore, in order to precisely analyze the crystal structure at the Si/Ge interface, the FFTM (Fast Fourier Transform Mapping) method, which maps the lattice spacing, was used to obtain a diffraction pattern by Fourier transform of a miro-region from the obtained high-resolution TEM image. As a result, in the (111) interplanar mapping image, it was confirmed that the crystallized parts of Si existed in a triangular or trapezoidal shape. Also, the (200) interplanar spacing in these regions is smaller than the interplanar spacing of Si (2.72 Å). This is because the lattice constant in the [100] direction perpendicular to the interface decreased, and it is considered that Si was epitaxially grown with the lattice spacing widened in the [011] direction of the Ge substrate. The island-like epi-growth suppresses misfit dislocations that would otherwise occur due to lattice mismatch, and exhibits good interfacial characteristics. In Fig. 3, it can be seen that the thicker the i-a-Si:H, the higher the *Voc* and the smaller the data variability. This indicates that a stable high efficiency can be obtained by making the buffer layer thicker than the height of the island-shaped Si layer.

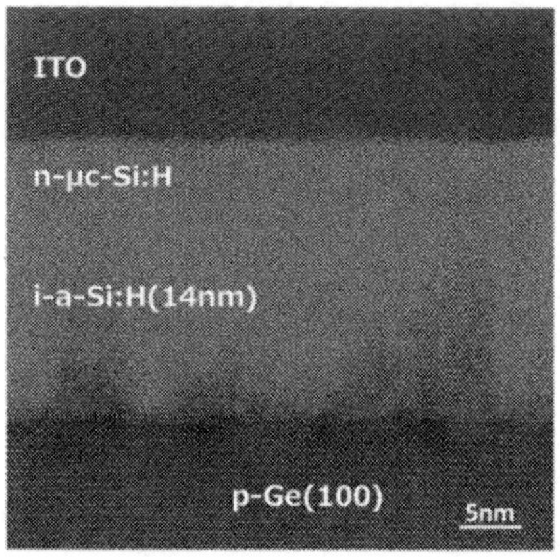

Figure 7: Cross-sectional TEM image of the n-μc-Si:H/i-a-Si:H/Ge interface of a cell with an efficiency of 8.6%. Substrate pretreatment: HF+HCl.

4.2 Effect of HCl treatment

In this work, it is found that the HCl treatment of Ge wafers before depositing the buffer layer greatly contributes to the improvement of *Voc*. Therefore, in order to clarify the effect of HCl, SIMS analysis near the interface was performed. It is generally said that the surface of Ge after HF treatment is covered with a sub-oxide layer such as GeO_2. Therefore, we measured depth profiles of oxygen and chlorine near the interface by SIMS and TEM-EDX. As a result, it was found that 2.9×10^{14} atoms/cm^2 of oxygen existed at the Si/Ge wafer interface in a cell with a conversion efficiency of 8.6%. This corresponds to about 46% of the surface density of (100) Ge atoms. In addition, it was found that the oxygen content ratio at the interface decreased to 0.65 when normalized by the oxygen content of the sample in which Si was not crystallized at the interface. This decrease in the amount of oxygen at the interface is the driving force behind the island growth of Si at the interface.

7 SUMMARY

Optimum deposition conditions for improving conversion efficiency differ greatly between Si heterojunction solar cells and Ge heterojunction solar cells. The new findings obtained in this work are as follows.

(1) In the GHJ system, the solar cell characteristics are very dependent on the acceptor concentration of the wafer. In this work, it was found that the conversion efficiency increased monotonically in the range from 1.0×10^{16}/cm^3 to 1.0×10^{18}/cm^3.
(2) Cleaning methods for Si wafers have been established, but cleaning Ge wafers is a very difficult technique. In this work, various methods were tried, but the most important point to improve the performance of solar cells was the HCl treatment.
(3) The optimum structure at the interface between the wafer and doping layer differs greatly between SHJ and GHJ solar cells. In the Ge system, the island-shapepd epitaxial growth at the interface is a key to reduce interface defects, and due to this island structure, the optimum thickness of the undoped-a-Si:H layer becomes extremely thick at 14 nm.

By using these technologies, *Voc* was greatly improved. In a cell in which an i-a-Si:H layer with a thickness of 14 nm was introduced at the emitter interface for a 2.5×10^{17} /cm^3 substrate, a conversion efficiency 8.6 % (*Voc* = 291 mV, *Jsc* = 45.0 mA/cm^2, *FF* = 0.656, AM1.5, 100mW/cm^2, 25 ° C) was obtained.

ACKNOWLEDGEMENT
This study was supported by JST/MIRAI/17943202.

References

[1] M. Konagai, R. Kondo, K. Sawano and Y. Ichikawa, Proceedings 27th Photovoltaic Science and Engineering Conference, 12-17 Nov. 2017, Otsu (2017) Page
[2] N. E. Posthuma, G. Flamand, W. Geens, J. Poortmans, Solar Energy Materials and Solar Cells, 88 (2005) 37
[3] T. Kaneko and M. Kondo, Jpn. J. Appl. Phys., 50 (2011) 120204
[4] S. Nakano and M. Shiratani, Thin Solid Films, 685 (2019) 225
[5] B. Hekmatshoar, D. Shahrjerdi, M. Hopstaken, K. Fogel, and D. K. Sadana, Appl. Phys. Lett., 101, 032102 (2012); https://doi.org/10.1063/1.4737166
[6] E.U. Onyegam et. al., Solar Energy Materials and Solar Cells, 111 (2013) 206
[7] H. Zhang, K. Nakada, M. Konagai, Thin Solid Films 628 (2017) 214
[8] M. Konagai, Proceedings 44th IEEE-PVSC (2017) DOI: 10.1109/PVSC.2017.8366298
[9] M. Konagai, R. Kondo, T. Otani, S. Takahashi, K. Sawano, Y. Ichikawa, K. Nakada and Y. Kurokawa, Proceedings 2918 IEEE 7th World Conf. on Photovoltaic Energy Conversion, IWWW Xplore (2018) DOI: 10.1109/PVSC.2018.8548235
[10] M. Konagai, T. Takamura, R. Iwata, R. Kondo, Y. Ichikawa and K. Saito, Proceedings 2020 47th IEEE-PVSC (2020) 88, **DOI:** 10.1109/PVSC45281.2020.9300432
[11] https://www2.pvlighthouse.com.au/calculators/OPAL%202/OPAL%202.aspx

COMPARISON OF AMORPHOUS SILICON DEPOSITION METHODS FOR HETEROJUNCTION SOLAR CELLS

Carsten Strobel*[1], Sebastian Leszczynski[1], Matthias Albert[1], Johann W. Bartha[1], Thomas Mikolajick[1], Frank Stahr[2], Sören Röhlecke[2], Olaff Steinke[2]
[1]Technische Universität Dresden, Institute of Semiconductors and Microsystems, Nöthnitzer Str. 64, 01187, Dresden, Germany,
[2]FAP GmbH, Gostritzer Str. 67B, 01217, Dresden, Germany

ABSTRACT: Various amorphous silicon (a-Si:H) deposition methods for heterojunction solar cells were investigated regarding their suitability for effective wafer passivation and high solar cell efficiencies. Among them, radio-frequency plasma-enhanced chemical vapor deposition (RF-PECVD) is the standard approach for low temperature a-Si:H fabrication, while very high frequency plasma-enhanced chemical vapor deposition (VHF-PECVD) and hot-wire chemical vapor deposition (HW-CVD) are emerging technologies to achieve softer deposition conditions. High effective carrier lifetimes of about 9 ms and high efficiencies of about 22 % for heterojunction solar cells were achieved with all deposition methods. A better passivation with soft VHF-PECVD and HW-CVD could not be verified. However, the latter deposition methods enable higher deposition rates but therefore require elaborate power coupling compared to state-of-the-art RF-PECVD. For reference reasons, a high temperature a-Si:H fabrication process, namely low-pressure chemical vapor deposition (LPCVD), was also applied and the layer properties were compared to the standard low-temperature approaches.

Keywords: RF-PECVD, VHF-PECVD, HW-CVD, LPCVD, a-Si:H, heterojunction solar cell

1 INTRODUCTION

Heterojunctions with intrinsic thin-layer solar cells (HIT [1]) are a very promising technology with highest efficiencies for single junction devices up to 26.7% [2]. This solar cell benefits from the full area passivation of the crystalline silicon wafer surface leading to very high open circuit voltages up to 750 mV. The passivation layer of HIT devices is made from hydrogenated amorphous silicon (a-Si:H) mostly deposited by radio frequency plasma enhanced chemical vapor deposition (RF-PECVD) [3]. Alternative deposition methods (very high frequency PECVD [4], hot wire CVD [5]) for a-Si:H have also been applied successfully to heterojunction solar cells. In general, a very soft deposition process is required to achieve the best electronic interface quality. Thereby, the softness of the deposition is related to the ion-bombardment energy in the plasma. For very high frequency plasma-enhanced chemical vapor deposition (VHF-PECVD) the ion-energies in the plasma are reduced due to the decreased plasma potential. For hot-wire chemical vapor deposition (HW-CVD) no ions are involved in the deposition process at all, so this method can be considered the softest. An additional benefit for HW-CVD and VHF-PECVD compared to RF-PECVD is, that much higher deposition rates can be obtained for high quality a-Si:H layers.

Until now, high efficiencies for heterojunction solar cells have been achieved with both RF-PECVD, VHF-PECVD and HW-CVD [1,2,3,4]. However, different groups used different deposition parameters (pressure, power, temperature, gas composition e.g.) to obtain these results. A uniform picture of the influence of the various deposition methods on heterojunction solar cell properties with single a-Si:H layers fabricated in the same deposition system with similar process parameters is still missing. Therefore, the present study aims for a direct comparison of RF-PECVD, VHF-PECVD and HW-CVD for heterojunction solar cells. In case of VHF-PECVD a plasma excitation frequency of 140 MHz is applied. This frequency is much higher than used before for heterojunction solar cells (< 60 MHz). As the ion bombardment energy decreases with increasing plasma excitation frequency [6] it can be assumed that the softness of PECVD with 140 MHz is increased compared to the lower state of the art frequencies used so far. The influence of such a high plasma excitation frequency on the passivation and solar cell properties will be discussed. First, intrinsic a-Si:H passivation layers are deposited with similar parameters by RF-, VHF- and HW-CVD on polished and textured wafers with different orientations (100), (111). The passivation quality of the various layers is evaluated by means of the effective carrier lifetime (τ_{eff}). Then, these passivation layers are incorporated into complete heterojunction solar cells. The solar cell parameters (e.g. η, FF, Jsc, Voc) as a function of the applied deposition method will be presented. For reference reasons a high-temperature approach, namely low-pressure chemical vapor deposition (LPCVD), is applied as well to deposit a-Si layers at a temperature of 580°C. Although this method is incompatible with the heterojunction solar cell technology due to epitaxial growth constraints it is frequently applied for the TOPCon solar cell technology. Up to now, only a very limited number of cross comparison studies between high-temperature LPCVD and low-temperature PECVD growth of a-Si:H layers exist [7]. Therefore, in this study we also present main differences in layer properties between LPCVD grown a-Si layers and PECVD grown a-Si:H layers.

2 EXPERIMENTAL

The inline deposition system used in this study to fabricate a-Si:H layers by PECVD is based on the linear plasma source concept developed by Dresden University of Technology and FAP GmbH Dresden [8]. A HW-CVD module is included in the same deposition system, as well. For the LPCVD studies another reconstructed single chamber CVD system was used. For the deposition of intrinsic layers, a gas mixture of silane and hydrogen is used. When producing p- or n-conducting layers, trimethylborane (TMB - $B(OCH_3)_3$) or phosphine (PH_3) are added. The PECVD electrode configuration consists of a powered electrode and a grounded carrier, which can be moved horizontally inside a tunnel with a velocity controlling the desired thickness of the layer. The linear electrode used in this study is equipped with a showerhead gas distribution system. The input signal for PECVD is generated using a signal generator and an amplifier with a frequency of 13.56 and 140 MHz.

The intrinsic thin silicon films were fabricated on 1.1 mm thick Corning Eagle XG glass substrates to investigate various layer properties such as e.g. conductivity, absorption, and layer thickness. The passivation studies were performed by depositing a-Si:H layers on both sides of 300 µm thick n-type double side polished float zone silicon (100) wafers and textured (111) wafers. Finally, silicon heterojunction (SHJ) solar cells were fabricated on textured (111) 300 µm n-type float zone silicon wafers (n-FZ).

The deposition rates respectively the thickness of the layers was measured by a surface profilometer from Veeco. The passivation properties of each amorphous silicon layer were investigated by effective minority carrier lifetime measurements using microwave detected photoconductivity (MDP [10]) at an injection level of larger than 10^{15} cm^{-3}. The optimized a-Si:H i-layers were incorporated into standard SHJ solar cells on planar (100) and texturized (111) substrates and measured under AM 1.5 illumination (25°C, 100 mW/cm²) to analyze the influence of deposition parameters on the device characteristics. For this purpose, a sun simulator developed in house at Technische Universität Dresden was used. A xenon arc lamp in combination with an infrared-filter ensures a good fit of the emitted light to the AM 1.5 spectrum. The solar simulator is calibrated against a class A simulator at the Helmholtz-Zentrum Berlin.

3 RESULTS

At first, the passivation of a-Si:H i-layers deposited by the different methods (RF-PECVD, VHF-PECVD, HW-CVD) was investigated. Therefore, a set of i-n-i samples was prepared at varied silane concentration ratios (see Fig. 1). After the i-layer preparation a hydrogen plasma was applied on both sides of the wafer to increase the lifetime. From Fig. 1 it can be seen that the effective carrier lifetime at 100 % silane concentration is almost equal for VHF- and RF-PECVD and only slightly lower for the HW-CVD process. For lower silane concentration (SC) ratios the lifetime remains high for VHF-PECVD and HW-CVD and drops below one millisecond only for very low SC values. Instead, the lifetime for RF-PECVD drops to below one millisecond even at medium silane concentration ratios of 50 %. Evaluation of Raman measurements indicate the onset of epitaxial growth for silane concentrations of 50 % and lower for all deposition methods. It is known, that epitaxial growth leads to very low carrier lifetimes. This is indeed the case for RF-PECVD at SC < 50 %. Contrary, for VHF-PECVD and HW-CVD the lifetime remains high (> 1ms) despite the onset of epitaxial growth at SC = 50 %.

The epitaxial growth is also substrate dependent. At (100) surfaces the tendency towards epitaxial growth is increased compared to (111) surfaces. Therefore, a second set of i-n-i samples was prepared at 50 % silane concentration on polished (100) and textured (111) wafers. Fig. 2 shows the effective carrier lifetime on different wafer surfaces for VHF- and RF-deposited samples as a function of the deposition rate. On (100) substrates the discrepancy of the lifetime between RF- and VHF-deposited samples remains high as already observed in Fig. 1 for SC = 50%. However, on (111) samples the lifetime is similar for RF- and VHF-PECVD as well as HW-CVD. Note that it is not easy to achieve high deposition rates with hydrogen dilution and RF-PECVD even at very high applied power levels. On the other hand, for VHF-PECVD high rates (60nm/min) and high carrier lifetimes (> 1ms) were obtained (see Fig. 2 - red triangles) at relatively low power applied to the system. This demonstrates a clear advantage for high volume VHF-PECVD mass production of heterojunction solar cells.

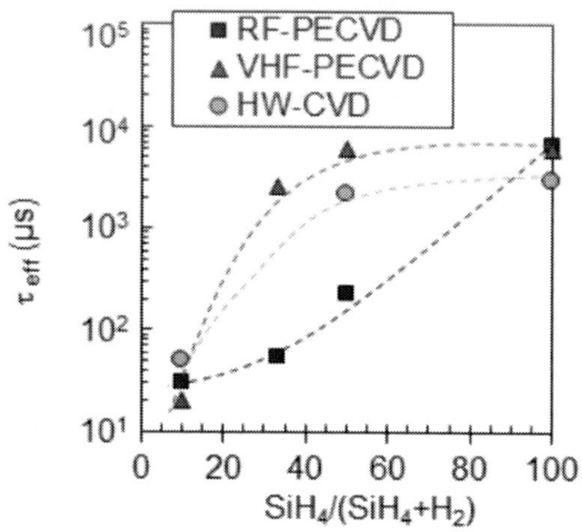

Figure 1: Effective carrier lifetime (τ_{eff}) as a function of the silane concentration ratio ($SiH_4/(SiH_4+H_2)$) for a-Si:H i-layers deposited on both sides of polished (100) n-type wafers by means of RF-PECVD, VHF-PECVD, and HW-CVD.

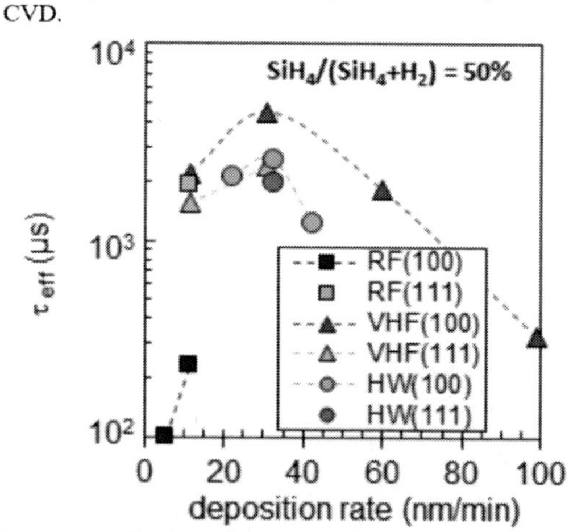

Figure 2: Effective carrier lifetime versus deposition rate for VHF-, HW- and RF-deposited i-layers at 50% silane concentration ratio on (100) and (111) substrates.

To further increase the carrier lifetime, a double layer passivation scheme is used (see Fig. 3). The first layer is a thin porous a-Si:H seed film deposited with a silane concentration ratio of 100 % in order to prevent epitaxial growth. The second thicker layer is a high-quality amorphous silicon film deposited with SC = 10 %. Between layer 1 and layer 2 a hydrogen plasma is applied. Tab. I shows the maximum effective carrier lifetimes obtained with the different passivation scheme and deposition methods. A strongly increased $\tau_{eff} \approx 9$ ms for the double layer passivation scheme can be observed compared to the single layer passivation ($\tau_{eff} \approx 6$ ms, see Fig. 1). This can be attributed to the lower defect density of the bulk layer with high hydrogen dilution in the double layer passivation scheme. On the other hand, for VHF-, RF-, and HW-deposited test structures with the double layer passivation scheme similar lifetimes were obtained.

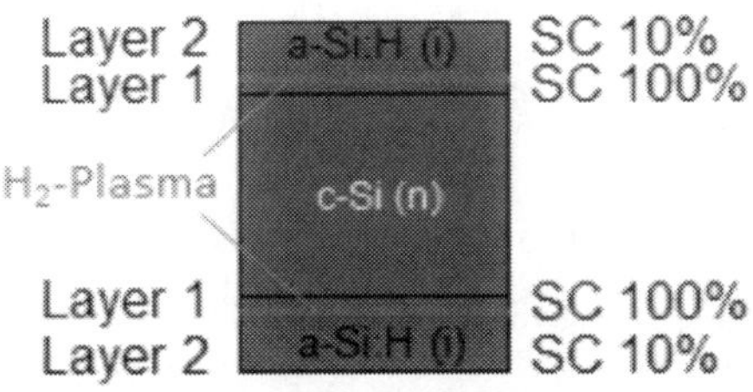

Figure 3: a-Si:H double layer passivation scheme with different silane concentration ratios of 100% / 10% (total a-Si:H layer thickness < 10 nm), substrate: polished Si (100).

Table I: Maximum effective carrier lifetime (τ_{eff}) of i-n-i test structures with the double layer passivation scheme according to Fig. 3 for RF-, VHF- and HW-deposition.

Deposition method	τ_{eff} (ms)
RF-PECVD	9.2 [0.9]
VHF-PECVD	8.8 [0.8]
HW-CVD	8.6 [0.7]

Further, amorphous silicon was deposited by LPCVD at 580°C for reference reasons. Fig. 4 shows the absorption coefficient alpha measured by Fourier transform infrared (FTIR) spectroscopy as a function of the wavenumber for a-Si layers deposited by LPCVD and VHF-PECVD. The most important difference is the absence of Si-H bonds for LPCVD a-Si layers. These bonds are visible as peaks at around 630 cm^{-1} and 2000-2100 cm^{-1} in the FTIR spectrum for VHF-PECVD layers. For LPCVD layers these peaks are absent due to the high temperature during LPCVD growth, which leads to hydrogen effusion.

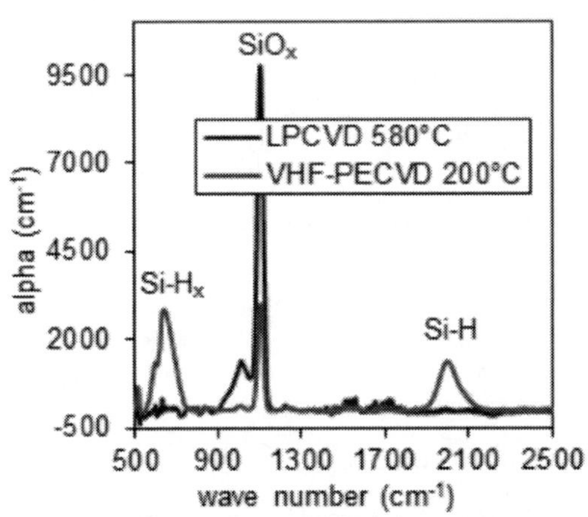

Figure 4: Absorption coefficient alpha as a function of the wavenumber for a-Si layers deposited by LP-CVD (blue curve) and VHF-PECVD (red curve).

The I-V characteristics of the a-Si layers deposited by LPCVD and PECVD was analyzed as well (see Fig. 5). For the PECVD material much lower dark currents and dark conductivities can be achieved compared to the LPCVD material. The hydrogen present in the PECVD material saturates the dangling bonds which is not the case in the LPCVD material. Thus, a completely different conduction mechanism is present in the two different materials. For LPCVD a hopping conduction via defects takes place leading to high dark currents. In the PECVD material a "mobility" bandgap exists which leads to band like transport with much lower currents.

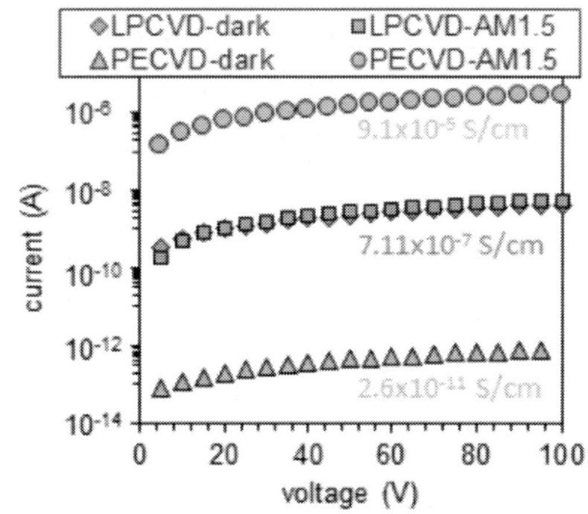

Figure 5: I-V characteristics of a-Si layers deposited by LPCVD (blue data points) and PECVD (orange data points) in the dark and under AM1.5 illumination. Conductivities are presented beyond the corresponding data points.

Note that the AM1.5 current is almost the same as the dark current for the LPCVD material which is a direct consequence of the high defect density which leads to strong recombination of excited carriers. On the other hand, the AM1.5 current for the PECVD material exceeds the dark current by about six orders of magnitude which is a sign of much lower recombination.

The absorption coefficient in the visible range is presented in Fig. 6 for LPCVD, VHF-PECVD, and RF-PECVD deposited a-Si:H layers. The absorption for the LPCVD a-Si layers starts at lower photon energies leading also to a lower optical bandgap of $W_{04} \approx 1.72$ eV. On the other hand, the optical bandgaps for the PECVD materials are increased, which is directly related to the hydrogen being present in these materials. The refractive index can be extracted from the optical data as well. For LPCVD a-Si the refractive index is $n_{\infty} \approx 3.3$ while for the RF-PECVD layers the refractive index is $n_{\infty} \approx 3.2$ and for VHF-PECVD a-Si:H the refractive index is $n_{\infty} \approx 3.1$. As the refractive index is an indicator for the porosity of the material it is obvious that the LPCVD material is the most compact one while the VHF-PECVD layer is the most porous one. Thereby, the differences in the material structure are again related to the different growth conditions. LPCVD layers are grown slowly at high temperatures which leads to few voids in the material. On the other hand, the PECVD materials are grown faster at lower temperatures of about 200°C and additional ion-bombardment takes place during growth. This leads to increased porosity of these PECVD layers.

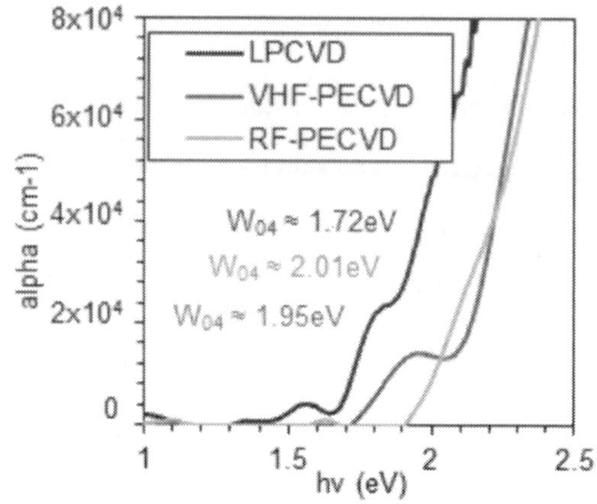

Figure 6: Absorption coefficient alpha as a function of the photon energy for LPCVD, VHF-PECVD, and RF-PECVD deposited amorphous silicon.

Next, the optimized a-Si:H i-layers deposited by VHF-PECVD, RF-PECVD and HW-CVD were incorporated into standard HIT solar cells on texturized (111) wafer substrates. The structure of the test devices is given in Fig. 7. To identify the influence of the various deposition methods on the n-Si / a Si:H interface quality, only the i-layers were deposited by the different techniques (RF-PECVD, VHF-PECVD and HW-CVD) while the p- and n-doped layers were deposited by RF-PECVD. In Tab. II the solar cell parameters of HIT devices fabricated by the different deposition methods are shown. High open circuit voltages of approximately 720 mV have been obtained with both RF- and VHF-PECVD. Note that these V_{oc} values correspond to a wafer thickness of 300 µm. By using wafers with lower thickness (100 µm), V_{oc}'s of more than 740 mV can be expected. In case of HW-deposited i-layers, slightly lower V_{oc}'s (~ 711 mV) have been obtained on (111) substrates.

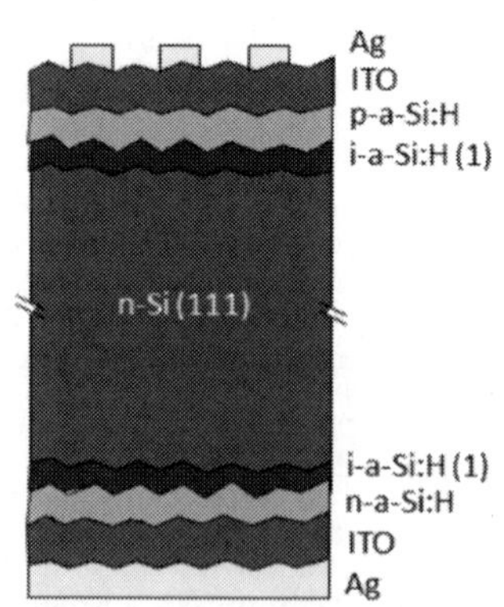

Figure 7: Stucture of HIT solar cells on texturized (111) n-type float zone wafers (300 µm).

Further, the device with i-layers fabricated by VHF-PECVD exhibit similar device performance as the solar cell with RF-deposited i-layers. However, the deposition rate of the VHF cell is much higher. This demonstrates a clear advantage of the alternative deposition technique VHF-PECVD. The SHJ solar cell with HW-CVD i-layers shows slightly fill factor values compared to the RF- and VHF-deposited cells regardless of the i-layer deposition rate. However, high efficiencies (> 21 %) were also obtained for HW-CVD i-layers.

Table II: Heterojunction solar cell properties with structure according to Fig. 7 and different i-layer deposition methods.

Method (i)	rate (nm/min)	η (%)	FF (%)	Jsc (mA/cm²)	Voc (mV)
RF-PECVD	11	22.1	78.0	39.36	722
VHF-PECVD	54	22.1	77.8	39.12	725
HW-CVD	16	21.3	76.9	39.02	711

5 CONCLUSION

During this study the compatibility of HW-CVD, RF-PECVD and VHF-PECVD for the fabrication process of highly efficient SHJ solar cells was demonstrated. The presented results show that, using all compared deposition techniques (RF-PECVD, VHF-PECVD and HW-CVD), high effective carrier lifetimes of about 9 ms and high quality SHJ solar cells with efficiencies of about 22% can be achieved. The results were reached with very high deposition rates, especially for VHF-PECVD fabricated cells. Although not compatible with the heterojunction technology, LPCVD was applied as well to fabricate a-Si layers and differences in the main material parameters were highlighted. The bonding configuration, conductivity and optical properties differ considerably for LPCVD a-Si layers compared to PECVD a-Si:H layers.

6 REFERENCES

[1] M. Taguchi, A. Yano, S. Tohoda, K. Matsuyama, Y. Nakamura, T. Nishiwaki, K. Fujita, and E. Maruyama, IEEE journal of photovoltaics, Vol. 4, No. 1, January (2014) 96-99.

[2] K. Yoshikawa, H. Kawasaki, W. Yoshida, T. Irie, K. Konishi, K. Nakano, T. Uto, D. Adachi, M. Kanematsu, H. Uzu and K. Yamamoto, Nature Energy Vol. 2, (2017) 17032.

[3] L. Korte , E. Conrad, H. Angermann, R. Stangl, M. Schmidt, Solar Energy Materials & Solar Cells 93 (2009) 905–910.

[4] A. Descoeudres, Z.C. Holman, L. Barraud, S. Morel, S.De Wolf, and C. Ballif, IEEE Journal of Photovoltaics, Vol. 3, No. 1, January (2013) 83-89.

[5] T.H. Wang, E. Iwaniczko, M.R. Page, D.H. Levi, Y. Yan, H.M. Branz, Q. Wang, Thin Solid Films 501 (2006) 284–287.

[6] M. Heintze and R. Zedlitz, J. Non-Cryst. Solids 198 (1996) 1038-1041.

[7] A. Stoffel, A. Kovacs, W. Kronast and B. Müller, J. Micromech. Microeng. 6 (1996) 1–13.

[8] C. Strobel, T. Zimmermann, M. Albert, J.W. Bartha, J. Kuske, Sol. Energy Mater. Sol. Cells 93, (2009) 1598–1607.

7 ACKNOWLEDGEMENT

This work is supported by the Bundesministerium für Bildung und Forschung (BMBF) under contract number 03SF0538. Financial support by the Federal Ministry for economic affairs and climate action (BMWK) in the framework of the DYNASTO initiative (0324293D) is gratefully acknowledged.

HIGH-MOBILITY FLOURINE DOPED INDIUM OXIDE FILMS FOR HIGH EFFICIENCY SILICON SOLAR CELLS

Thomas Tom[a,b,c], Leonard Tutsch[a], Martin Hermle[a], José Miguel Asensi[b,c], Julian López-Vidrier[b,c], Joan Bertomeu[b,c], Martin Bivour[a]
[a]Fraunhofer Institute for Solar Energy Systems ISE, Heidenhofstr. 2, 79110 Freiburg, Germany
[b]Departament de Física Aplicada, Universitat de Barcelona, Martí i Franquès 1, 08028 Barcelona, Spain
[c]Institute of Nanoscience and Nanotechnology (IN[2]UB), Universitat de Barcelona, Barcelona, Spain

ABSTRACT: Transparent conducting oxides with high infrared transparency and high mobility are crucial for further enhancing the efficiency of crystalline silicon solar cells. Numerous hydrogen-doped or cation-doped indium oxide films have been intensively studied, while anion doping in indium oxide films is still under development. Here we investigate the fluorine doped indium oxide films deposited by industrial DC sputtering by varying the oxygen and hydrogen percentages. To gain a holistic understanding on the role of hydrogen and oxygen we performed a very wide parameter variation of these gases. The films were processed at room temperature and were annealed at 220 °C on a hot plate in ambient conditions. Electrical, optical, and structural properties of the films were investigated and highest mobility of 87 cm^2/Vs at a carrier density of 2.5×10^{20} cm^{-3} was achieved when depositing under 2.3% O_2 and 2.0% H_2 flow. The strong correlation between the defined figure of merit and the electron mobility of the films is observed. The solid phase crystallization of the annealed films was identified by X-ray diffraction. The highest mobility films achieved an efficiency of 21.8% in silicon heterojunction solar cells.

Keywords: High mobility, Transparent conducting oxides, Silicon solar cell, Fluorine-doped Indium oxide, Industrial sputtering.

1 INTRODUCTION

Silicon heterojunctions (SHJ) are gaining prominence in the field of photovoltaics due to their ease of fabrication and processing, as well as their ability to achieve high efficiencies[1][2][3]. These SHJ solar cells use transparent conductive oxides (TCO) on the front side as an anti-reflection coating and to ohmic conduct charge laterally to the metal fingers. The sheet resistance of these layers must be low as possible to reduce the power lost due to Joule heating during conduction through the metallic grids[4][5]. In order to reduce the sheet resistance of these layers the thickness of the films could be adjusted, but for the best antireflection the thickness of these films must be fixed at 65–80 nm. The other way to reduce the sheet resistance is to increase the electron density in these films by doping. However, this will in turn increase near-IR free-carrier absorption (FCA), which affects the transparency of these films. Therefore, the most adequate way to reduce the sheet resistance is to increase the mobility of these TCO layers[6][7]. Additionally, they must have a high lateral conductivity and good contact with the doped hydrogenated amorphous Si (a-Si:H) layers.

Tin-doped indium oxide (ITO) is the predominantly used material in industry and in research labs as the TCO layer. Nonetheless, optoelectronic properties of ITO are limited due to its high carrier density and moderate electron mobility that can lead to significant FCA. Therefore, TCOs with high mobility and reduced carrier density are referred to combine both excellent optical and electrical properties[8].

Recently, several doping techniques have been explored to fabricate the high-mobility indium oxide-doped films. Hydrogen or cation-doped (H, Ce, Zn, Zr) In_2O_3 films have attained a mobility greater than 60 cm^2/Vs[9][10][11][12]. Nevertheless, anion doping in these films is much less explored. Among them, fluorine-doped indium oxide (IFO) is found to be a suitable candidate. Here, the substitutional replacement of F^- with O^{2-} takes place as a consequence of the ion radius of F^- being similar to that of O^{2-}, thus causing a minor distortion in the lattice. These films show enhanced mobility and were fabricated by RF magnetron sputtering from a planar target using water vapor as H_2 source, leading to an improved efficiency of SHJ cells[13][14].

In this work, the electrical, optical, and structural properties of IFO films were investigated in more detail using a more industrial approach, i.e., DC magnetron sputtering from a rotary target and H_2 diluted in Ar as hydrogen source. Highest mobility of 87 cm^2/Vs at a carrier density of 2.5×10^{20} cm^{-3} was achieved for the films deposited at 2.3% O_2 and 2.0% H_2. As a proof-of-concept, monofacial SHJ cells demonstrated increase in efficiencies and short-circuit current density (J_{sc}) values. The best films achieved the highest efficiency of 21.8%, demonstrating IFO's versatility as a candidate for SHJ cells and other optoelectronic devices.

2 EXPERIMENTAL DETAILS

An in-line DC magnetron sputtering system using rotary IFO targets was used. During the film depositions in Ar atmosphere, varying percentages of O_2 and H_2 gas were introduced, ranging from 0 to 3.5% and from 0 to 2.5%, respectively. Throughout all sets of depositions, the power (3000 W) and deposition pressure (4×10^{-3} mbar) were kept constant. Base pressure of 7×10^{-7} mbar was achieved before depositions. Planar test substrates (glass and Si) were used for the deposition. All the depositions were done at a room temperature. The thickness of the films was ~100 nm, which was measured by ellipsometry. In ambient air, post-deposition annealing was performed for 20 minutes at 220 °C. The van der Pauw's method was used to measure the Hall parameters. The optical properties of the films were determined using a UV-Vis spectrometer, and the weighted average absorbance was compared to the Hall mobility and carrier concentration. The crystalline nature of 100-nm-thick films was studied via X-ray diffraction (XRD) using the PANalytical X'Pert PRO MRD X-ray diffractometer. To test the IFO performance at the device level, monofacial SHJ cells using the rear emitter design were created. Active cell area of ~4 cm^2 was defined by using shadow-mask front TCO and rear TCO + Ag sputtering. The front side was finished by a front grid with screen printing. Finally, the current density-voltage (J–V) characteristics of the SHJ cells were assessed using a solar simulator under AM1.5 and 100 mW/cm² illumination.

3 RESULTS AND DISCUSSION

The Hall parameters of the IFO films after annealing are shown in Fig. 1. The evolution of electron mobility (μ) and charge carrier density (N_e) in response to the atmospheric species during deposition is depicted in Fig. 1(a, b). After solid phase crystallization, the electrical characteristics of annealed films are significantly impacted by the film structure, which can also be influenced by deposition parameters. To gain a holistic understanding on the role of H_2 and O_2 we performed a very broad parameter variation and will first discuss the role of either changing O_2 or H_2. By doing so the heat map can be categorized into two regions.

Region1 (R1): This region shows the influence of changing O_2 without H_2 being intentionally added during deposition. Here we could see a huge drop in N_e by adding O_2 in the films from 3.0×10^{20} cm^{-3} to 0.3×10^{20} cm^{-3}, and an increase in the mobility of the films from 33 cm²/Vs to 42 cm²/Vs (and correspondingly an increase in the sheet resistance from 59 Ω/sq to 799 Ω/sq, not shown). As there is an increase in the oxygen supply, N_e decreases due to the reduction in the amount of O vacancies, typically doubly-charged donors (Vo^{++})[15]. This, in turn, will slightly increase the mobility of the films due to reduced ionized impurity scattering by Vo^{++}.

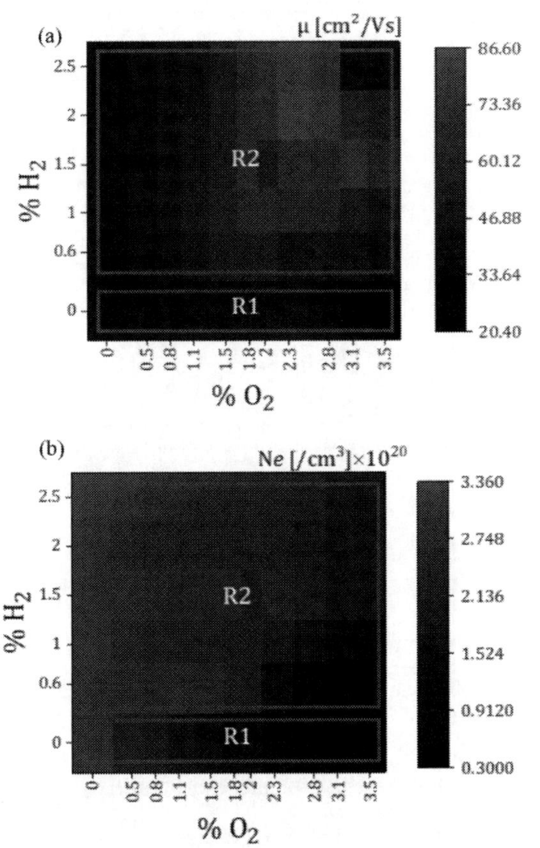

Figure 1: Development of the Hall parameters of 100-nm-thick IFO films deposited on glass substrates: a) electron mobility (μ), and b) charge carrier density (N_e).

Region 2 (R2): In this region, both O_2 and H_2 flows are introduced and varied independently. Here the N_e of films decreases, and a huge increase of mobility is observed. These regions have the superposition of impact on addition of H_2 and O_2. Increase of these two parameters (O_2, H_2) will have counter imparting impact on N_e and μ. With the increase in the percentage of H_2 supply during deposition, the amount of substitutional and interstitial (Ho^+ and H_i^+, respectively) hydrogen increases[9]. Highest mobility of 87 cm²/Vs is obtained for the films sputtered at 2.0% of H_2 and 2.3% of O_2 after annealing. This clearly explains the importance of O_2 and H_2 during growth of films (visible in Fig. 1a). Addition of H_2 till a particular concentration could positively impact on the structural properties of the films, such as formation of a larger grain size, and passivation of grain boundaries and defects. This also acts as efficient single-charged donor compared to the doubly-charged Vo^{++}.

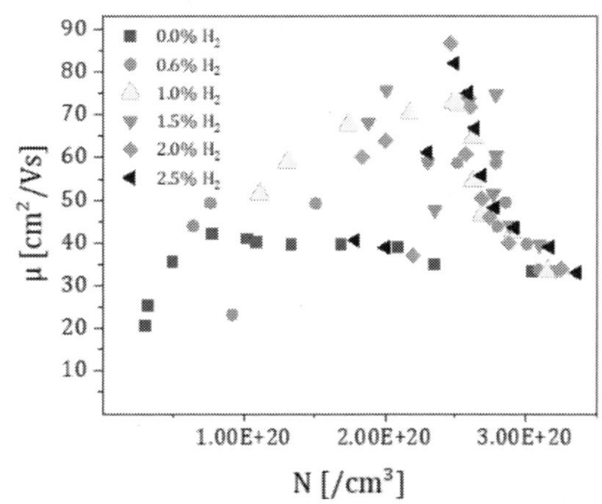

Figure 2: Mobility vs. charge carrier density graph corresponding to 100-nm-thick IFO films.

As a conclusion, the low N_e and higher μ after solid phase crystallization can be presumably attributed to the interplay between these three possible dopants: oxygen vacancies (Vo^{++}), fluorine dopants and atomic hydrogen (Ho^+ and H_i^+) [13][14][16]. This could be easily observable from the Fig. 2, which shows the influence of N_e on μ when increasing the percentage of H_2 during deposition.

Furthermore, the increase in the μ due to H_2 concentration increase is related to a) structural rearrangements causing the hydrogen atom filling the oxygen vacancies, b) effusion of hydrogen during annealing that reduces H_i^+ concentration and in turn lowers the doping, and c) passivation of the grain boundaries by hydrogen atoms.

Optoelectronic properties of the IFO films deposited in varying percentages of O_2 and H_2 atmospheres is shown in Fig. 3. All properties were measured for films deposited on flat glass. To highlight the high-performance of the IFO films, the optical absorption of the films is calculated and compared with the sheet resistance of the films. The absorptance of the IFO film with different percentages of O_2 and H_2 in films is shown in Fig. 3a. Then, we utilize the absorptance (A) value to compare the quantity of light parasitically absorbed by an electrode irrespective of substrate and surface roughness: $A = 1 - TT - TR$; where TT is the total transmittance and TR is the total reflectance.

Then, absorptance is spectrally weighted to the 1 Sun air mass 1.5 global (AM 1.5G) expressed as,

$$A_{\text{weighted}} = \frac{\int_{\lambda_1}^{\lambda_2} I_{light}(\lambda)\, A(\lambda)\, d\lambda}{\int_{\lambda_1}^{\lambda_2} I_{light}\, d\lambda} \qquad (1)$$

where $A(\lambda)$ is the absorptance of the transparent electrode as a function of wavelength, I_{light} is the intensity of AM 1.5G radiation, and λ_1 and λ_2 are 300 and 1200 nm, respectively[17].

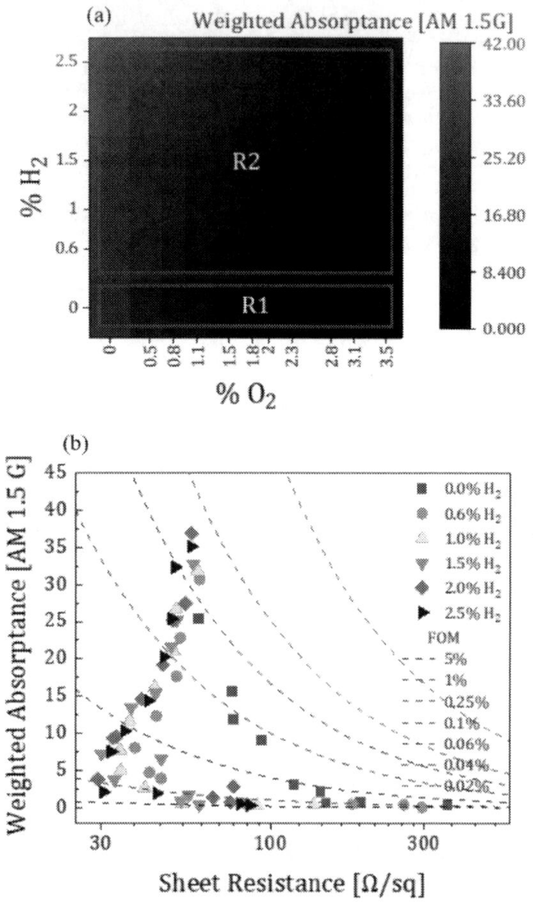

Figure 3: Optoelectronic properties of the IFO films after annealing: (a) weighted absorbance of the films deposited at varying O_2 and H_2%, and b) weighted absorbance as a function of the sheet resistance. The dashed lines indicate curves of constant figure of merit (FOM).

In Fig. 3b, the absorptance, which is weighted with the AM 1.5 solar spectrum according to equation (1), is plotted over as a function of the sheet resistance for films sputtered in varying atmospheres. A figure of merit (*FOM*), quantifying the electro-optical performance of the films simply by being the inverse product of A and R_{sh}, and isolines thereof, are indicated by dashed lines in the figure. The FOM is expressed as:

$$FOM = \frac{1}{A_{\text{weighted}} * R_{\text{sh}}} \qquad (2)$$

where A_{weighted} is the weighted absorptance as calculated in (1) and R_{sh} is the measured sheet resistance.

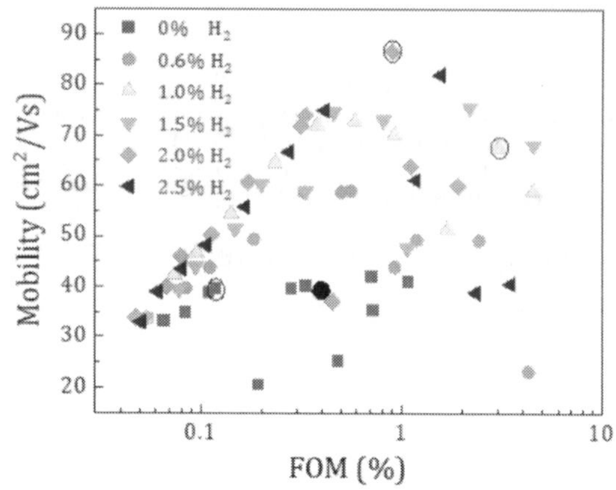

Figure 4: Mobility as a function of the figure of merit of the films.

Fig. 4 displays μ vs. FOM of IFO films for different H_2 atmospheres. This clearly shows the importance of hydrogen in achieving a good optical-electrical trade-off. This marks significance of IFO films, which exhibit a high FOM with great mobility, in comparison to typical ITO films of 0.5 FOM and μ of 40 cm²/Vs (marked as a black spot in Fig. 4), as M. Morales-Masis *et al.* reported [17][18].

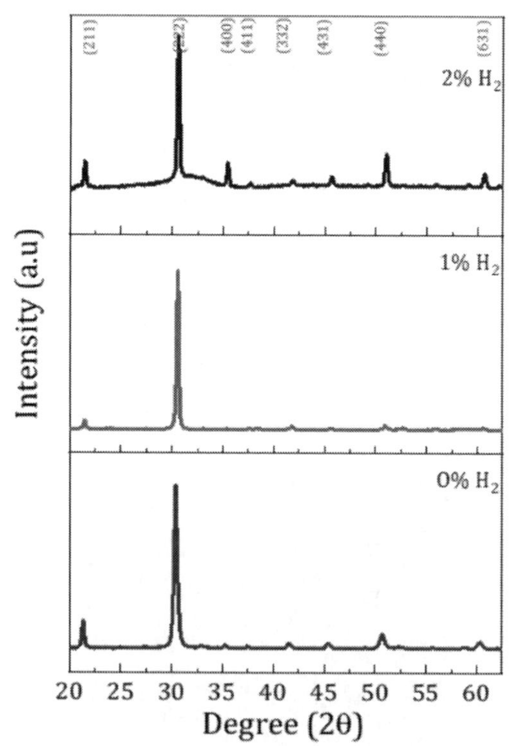

Figure 5: X-ray diffraction patterns corresponding to the optimized IFO sputtered at different H_2 percentages.

The crystal structure of the IFO films with three different H_2 concentration was studied using XRD analysis (Fig. 5). All the peaks match with the ICDD database which is assigned to the cubic bixbyite structure of In_2O_3[19][20]. The XRD patterns of the annealed films grown at different hydrogen percentages achieves polycrystalline cubic In_2O_3 structure. This also indicates the solid phase crystallization of the films after annealing at 220 °C[21]. The increase in the H_2 percentage in the films helps to suppress the polycrystalline nature and thus to increase the mobility of the films. The mean crystallite size of the films was calculated from the full-width at half-maximum (FWHM) of the X-ray peak of (222) following the Scherrer's formula,[13] With values that grow from 20 nm for films without hydrogen to 40 nm for films with higher percentage of H_2, which is a clear indication of the increase in the mobility of the films with addition of H_2.

To test the TCO films on device level, SHJ cells with the scheme shown in Fig. 6 were fabricated. On the rear side, 130-nm-thick layer and on front side 75-nm-thick film of IFO were sputtered. The cells were fabricated using three different H_2 percentages during deposition. Table 1 shows a clear correlation between J_{sc} and mobility: a significant decrease of 3.3 mA/cm^2 is observed for the cell with the lowest mobility IFO (40 cm^2/Vs) compared to cell with the highest mobility IFO (87 cm^2/Vs). Colored circles in Fig. 3c shows the μ and *FOM* values of different solar cells, which clearly indicates the necessity of H_2 to have better J_{sc}. The maximum power conversion efficiency of 21.8% is obtained for the cell with TCO of highest mobility, proving that IFO could be a versatile candidate for SHJ and for different optoelectronic devices.

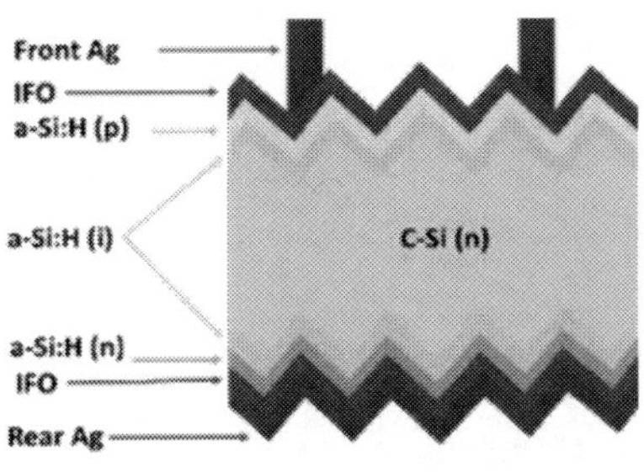

Figure 6: Schematics of the fabricated solar cell.

Table I: Solar cell parameters in comparison with the electro-optical properties.

H_2 (%)	V_{oc} (mV)	J_{sc} (mA/cm^2)	*FF* (%)	*PCE* (%)	μ (cm^2/Vs)
0	718.7	36.2	74.9	19.5	40
1	732.4	38.9	74.0	21.1	68
2	732.0	39.5	75.7	21.8	87

4 CONCLUSIONS

In this work, the opto-electrical properties of IFO films were investigated in more detail using an industrial approach, in particular DC-magnetron sputtering from a rotary target and H_2 diluted in Ar as hydrogen source. We were able to develop highly conductive and broadband transparent thin films, compatible for SHJ contacts through additional analysis and optimization. During the annealing the film quality increases leading to less defects and a more efficient incorporation of singly-charged donors (hydrogen, fluorine) and reduction of doubly-charged oxygen vacancies. Highest mobility of 87 cm^2/Vs at a carrier density of 2.5×10^{20} cm^{-3} was achieved for the films deposited at 2.3% O_2 and 2.0% H_2. The reduction of free-carrier absorption in the films is also explained by a significant increase in the mobility and a decrease in the carrier density after annealing. By using a figure of merit (*FOM*) that takes into account the weighted absorptance and the sheet resistance, a significant improvement with respect to standard ITO was demonstrated, and a strong correlation between *FOM* and electron mobility was evidenced. The solid phase crystallization of the high mobility films was proved by XRD studies. The high mobility films could then successfully be transferred to monofacial SHJ cells. This resulted in increased J_{sc} in the cells, with increase in mobility of the films. The high mobility film achieved the highest efficiency of 21.8%, demonstrating IFO's versatility as a candidate for SHJ cells and other optoelectronic devices.

5 ACKNOWLEDGEMENTS

This research has been supported by Spanish government through grant PID2019-109215RB-C43 funded by MCIN/AEI/10.13039/501100011033. One of the authors (T. Tom) acknowledges the support of the Secretaria d'Universitats i Recerca de la Generalitat de Catalunya and European Social Fund (2019 FI_B 00456).

6 REFERENCES

[1] S. DeWolf, A. Descoeudres, Z. C. Holman, C. Ballif, *Green* **2012**, *2*, 7.

[2] K. Yoshikawa, H. Kawasaki, W. Yoshida, T. Irie, K. Konishi, K. Nakano, T. Uto, D. Adachi, M. Kanematsu, H. Uzu, K. Yamamoto, *Nat. Energy* **2017**, *2*,17032.

[3] T. Tom, E. Ros, N. López-Pintó, J. Miguel Asensi, J. Andreu, J. Bertomeu, J. Puigdollers, C. Voz, *Materials (Basel)*. **2020**, *13*, 4905.

[4] C. Battaglia, A. Cuevas, S. De Wolf, *Energy Environ. Sci.* **2016**, *9*, 1552.

[5] H. T. Nguyen, E. Ros, T. Tom, J. Bertomeu, J. M. Asensi, J. Andreu, I. M. Garcia, P. Ortega, M. Garin, J. Puigdollers, C. Voz, R. Alcubilla, *IEEE J. Photovoltaics* **2019**, *9*, 72.

[6] L. Barraud, Z. C. Holman, N. Badel, P. Reiss, A. Descoeudres, C. Battaglia, S. De Wolf, C. Ballif, *Sol. Energy Mater. Sol. Cells* **2013**, *115*, 151.

[7] S. Calnan, A. N. Tiwari, *Thin Solid Films* **2010**, *518*, 1839.

[8] E. Aydin, M. De Bastiani, X. Yang, M. Sajjad, F. Aljamaan, Y. Smirnov, M. N. Hedhili, W. Liu, T. G. Allen, L. Xu, E. Van Kerschaver, M. Morales-Masis, U. Schwingenschlögl, S. De Wolf, *Adv. Funct. Mater.* **2019**, *29*.

[9] T. Koida, H. Fujiwara, M. Kondo, *Jpn J. Appl. Phys* **2007**, *46*, L685.

[10] L. Tutsch, H. Sai, T. Matsui, M. Bivour, M. Hermle, T. Koida, *Prog. Photovoltaics Res. Appl.* **2021**, 835.

[11] M. Morales-Masis, S. Martin De Nicolas, J. Holovsky, S. De Wolf, C. Ballif, *IEEE J. Photovoltaics* **2015**, *5*, 1340.

[12] M. Morales-Masis, E. Rucavado, R. Monnard, L.

Barraud, J. Holovsky, M. Despeisse, M. Boccard, C. Ballif, *IEEE J. Photovoltaics* **2018**, *8*, 1202.
[13] C. Han, L. Mazzarella, Y. Zhao, G. Yang, P. Procel, M. Tijssen, A. Montes, L. Spitaleri, A. Gulino, X. Zhang, O. Isabella, M. Zeman, *ACS Appl. Mater. Interfaces* **2019**, *11*, 45586.
[14] C. Han, G. Yang, A. Montes, P. Procel, L. Mazzarella, Y. Zhao, S. Eijt, H. Schut, X. Zhang, M. Zeman, O. Isabella, *ACS Appl. Energy Mater.* **2020**, *3*, 8606.
[15] D. Erfurt, M. D. Heinemann, S. Körner, B. Szyszka, R. Klenk, R. Schlatmann, *Mater. Sci. Semicond. Process.* **2019**, *89*, 170.
[16] T. Koida, Y. Ueno, *Phys. Status Solidi Appl. Mater. Sci.* **2021**, *218*, 27.
[17] M. Morales-Masis, S. De Wolf, R. Woods-Robinson, J. W. Ager, C. Ballif, *Adv. Electron. Mater.* **2017**, *3*,1600529.
[18] F. Qu, T. Zhang, H. W. Gu, Q. Q. Qiu, F. Z. Ding, X. Y. Peng, H. Y. Wang, *Rare Met.* **2015**, *34*, 173.
[19] Z. G. Yu, J. Sun, M. B. Sullivan, Y. W. Zhang, H. Gong, D. J. Singh, *Chem. Phys. Lett.* **2015**, *621*, 141.
[20] Q. Hou, J. Buckeridge, T. Lazauskas, D. Mora-Fonz, A. A. Sokol, S. M. Woodley, C. R. A. Catlow, *J. Mater. Chem. C* **2018**, *6*, 12386.
[21] B. Macco, M. A. Verheijen, L. E. Black, B. Barcones, J. Melskens, W. M. M. Kessels, *J. Appl. Phys.* **2016**, *120*, 085314.

LOW RESISTANCE WITH THE HIGH TRANSPARENCY OF Al-DOPED ZnO TCO THIN-FILMS FROM ALD PROCEDURE FOR c-Si BASED PASSIVATED CONTACT SOLAR CELL APPLICATIONS

Namitha Dsouza[1], Ashish K. Singh[2], Rajesh Maurya[1], Ramakrishna Madaka[2], Narendra Bandaru[2], Rajesh Kanakala[1], Md. Seraj Uddin[1,2], Jatindra K. Rath[1,2]
[1]Department of Physics, Indian Institute of Technology Madras, Chennai
[2]DST Solar Energy Harnessing Centre, Indian Institute of Technology Madras, India (600036)

ABSTRACT: The study aims to develop a suitable TCO via different modes of the ALD process for c-Si based passivating contact solar cells (PCS) and we have proposed a new type of deposition scheme. This TCO deposition, along with other layers (passivation and electron/hole transport layers) can be deposited in a single ALD run, which helps to fabricate a complete PCS device without exposing the device outside of the vacuum chamber. ZnO:Al (AZO) films on the glass/c-Si substrates from thermal (TH) ALD and plasma enhanced (PE) ALD processes were deposited at different temperatures (100-250 °C), and the growth per cycle (GPC) of these layers were calculated from spectroscopic ellipsometry and X-ray reflectometry. The GPC of films deposited by PE ALD at 200 °C is measured to be 1.45±0.1 Å for Al_2O_3 and 1.78±0.2 Å for ZnO films; similarly for the TH ALD deposited films, a GPC of 1.00±0.1 Å and 1.20±0.1 Å, respectively. The (1: n) (Al_2O_3: ZnO) pulse ratio was used for the film deposition and optimized GPC of Al_2O_3 and ZnO was used to deposit about 200 nm AZO films for different Al concentrations (0.5 to 5 at %). The (1:21) (Al - 2%) ratio provides an optimum film property for the PE ALD process, and the same ratio was tried with TH ALD, where the film properties were significantly enhanced. Further, a new mode of deposition named "DTH" (DEZ-TMA-H_2O) was explored, and an additional decrease in the film resistivity was observed. We have altered the sequence step of precursor dose and co-reactant in this mode. Compared to the conventional TH and PE ALD technique, the DTH mode of deposition proves to be the best fit for the TCO application here. A 200nm AZO film deposited by DTH ALD has a minimum resistivity in the range of 1.91×10^{-3} Ω-cm and high transparency of more than 85% in the visible region. DTH approach in the ALD proves to be an innovative deposition technique for developing films with enhanced quality, indicating the potential use in PCS device applications.

1 INTRODUCTION

The advancement in photovoltaics demands the replacement of the extensively used transparent conducting oxide, ITO which is toxic and contains scarce element Indium. The well-known alternative, zinc oxide is a wide bandgap (~3.30eV) material, with inexpensive and abundant raw materials, and offers tunability in its optoelectronic properties with doping [1]. Our previous comparative study between ITO and ZnO of varying thicknesses shows that ZnO is the most suitable TCO material for silicon heterojunction solar cell [2]. As many as 22 elements have so far been used in doping ZnO[3]. Among these, Aluminum is the most widely used dopant producing an n-type semiconductor. Al doped zinc oxide (AZO). films are conventionally deposited by sputtering with resistivity of thick films ~10^{-4} Ω-cm [4]. Also, pulsed laser deposition (PLD) [5] and chemical vapor deposition (CVD) have been employed to deposit ZnO films [6]. AZO with different Al concentration has also been studied as an efficient TCO helping in light confinement in thin film silicon solar cell [7].
Our study mainly aims at the development of a TCO for c-Si based passivating contact solar cells. The primary demand here is a lower deposition temperature (~200 °C) and fabrication of a complete device without breaking the vacuum. Atomic layer deposition (ALD) serves this purpose, in addition to excellent conformity, control over film thickness, and uniformity of the films [8]. ALD also is a very efficient technique to dope thin films with known atomic fractions of elements. It works on the principle of self-limiting chemisorption of precursors and oxygen source on the substrate heated at an appropriate temperature (considerably low) to help in depositing ultrathin films having superior quality [9]. The two primary deposition techniques in ALD differ by the source of co-reactant during deposition. The co-reactant in our work, oxygen, is provided for the deposition through O_2 plasma and H_2O vapour in Plasma-Enhanced (PE) ALD and Thermal (TH) ALD, respectively. Energy for the surface reaction is supplied by the temperature provided to the substrate in TH ALD and by plasma species in PE ALD [10]. Al-doped ZnO (AZO) deposited by Thermal ALD with superior quality was reported with the resistivity of 5×10^{-4} Ω-cm with carrier concentration of 2×10^{22} cm^{-3} and optical transparency 90% in the visible range for 30nm thick films [11]. However, a few reported PE ALD deposited AZO films, though transparent, tend to show higher resistivity. AZO films were deposited with both TH ALD and PE ALD in this study. A new sequence of deposition was also tried and a systematic study of structural, optical, and electrical characterization was done by depositing the films with various Al concentrations to understand the difference in the doping process by these techniques.

2 EXPERIMENTAL DETAILS

AZO films were deposited at the temperature range of 100 to 250 °C from the Oxford Instruments, UK (Model: FlexAL) ALD reactor. Before the deposition, a piece of glass (Corning Eagle XG Glass, USA) and <111> double side polished Si wafer (Longi Green, China) samples were cut into the required size with the diamond/laser cutter; then, these samples were cleaned and immediately loaded in the load lock chamber for processing the ZnO and Al_2O_3 and AZO layer deposition. Diethylzinc (DEZ) and trimethylaluminum (TMA) (EpiValance, UK) are the precursors used for ZnO and Al_2O_3 film fabrication. Deionized H_2O for TH ALD and O_2 plasma for PE ALD were used as co-reactant for oxygen sources. The (1, 1) pulse sequences were used for the film deposition processes

(ZnO and Al_2O_3). To incorporate a specific Al at. % doping into the ZnO lattice, a single Al_2O_3 cycle (TMA- H_2O or TMA- O_2) was inserted after a certain number of ZnO cycles (DEZ- H_2O or DEZ- O_2). This (1: n) constituted one "super-cycle" of AZO, composed of n cycles of ZnO followed by one cycle of Al_2O_3. Three different modes of depositions are tried for AZO films in this study [Figure 1]. The PEALD AZO was fabricated using a pulsing sequence of [n(DEZ| O_2 plasma)|TMA| O_2 Plasma] cycle. In THALD [n(DEZ| H_2O)|TMA| H_2O] sequence is employed. A slight variation in the THALD array is done to obtain the DTH recipe, [n-1(DEZ| H_2O)|DEZ|TMA| H_2O], where the H_2O dose prior to TMA is skipped.

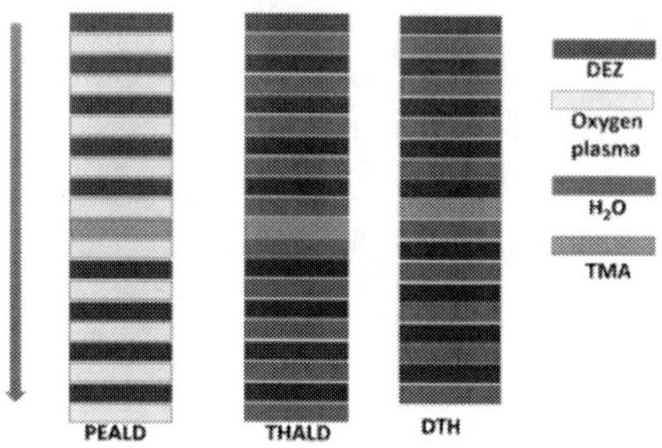

Figure 1: Schematic of pulsing sequences employed in different modes of deposition; PEALD, THALD and DTH recipes.

Rigaku SmartLab X-ray diffractometer, equipped with a Cu K_α source with a wavelength of 1.5418 Å, records X-ray diffraction either with grazing incidence (GI-XRD) or under Bragg's Brentano configuration was used for structural characterization of the thin films deposited on glass substrates. Film thickness was calculated from X-ray reflectometry (XRR), and J. A. Woollam M-2000 ellipsometer. JASCO UV-VIS spectrophotometer was used to study the transparency of the as-deposited films in the wavelength range of 300-1800 nm. Electrical properties of the deposited films were measured from the Lake Shore Hall Effect tool, and AZO sheet resistance was measured by four probe assembled with a Keithley source meter.

3 RESULTS AND DISCUSSION

The deposition temperature for our experiment was carefully selected to be at 200 °C because of the growth saturation observed around this temperature window (ALD window) for Al_2O_3 thin films and the reported data suggesting the growth of highly conducting AZO at this temperature[12]. The Al_2O_3 and ZnO films were grown on polished Si with various cycles and were characterized with XRR and spectroscopic ellipsometry. The thickness and number of cycles data were tabulated to get the growth per cycle (GPC) of the films grown by PEALD and THALD in Table I.

All the samples studied here are of 200nm thick. Initially, AZO samples were deposited with PE ALD where the concentrations of aluminum were varied from 0.5 to 4 at. % in ZnO, and the films were characterized with four probe measurements to obtain the sheet resistance. These films were found to be highly resistive and the AZO sample with 2 at. % Al showed a resistivity of 1.05 Ω-cm. It should be noted that these samples are as deposited and no post-deposition processing is done.

Table I Growth per cycle(GPC) of Al_2O_3 and ZnO films grown on polished Si(111) wafer at 200 °C by PEALD and THALD.

Deposition process at 200 °C	Deposited films on polished Si	GPC (Å) from XRR	GPC (Å) from spectroscopic ellipsometry
TH ALD	Al_2O_3	1.12±0.2	1.0±0.2
PE ALD	Al_2O_3	1.45±0.2	1.38±0.2
TH ALD	ZnO	1.78±0.2	1.70±0.2
PE ALD	ZnO	1.20±0.2	1.25±0.2

Similarly, AZO samples with varying aluminum concentrations were fabricated with TH ALD, and the films were found to be good conducting. The resistivity of these samples is as indicated in [
Figure *2* (a)]. The least resistivity of AZO was obtained for an aluminum doping of 2 at. % that is 2.78×10^{-3} Ω-cm. Following this trend, a comparative study of electrical properties of AZO with (1:21) cycles of Al, deposited by PEALD, THALD, and DTH recipes is done using the hall measurement data at room temperature [
Figure *2* (b)]. The AZO sample fabricated with DTH recipe shows a further decrease in the resistivity to 1.91×10^{-3} Ω-cm with a carrier concentration of 3.02×10^{20} cm^{-3} and mobility of 10.78 cm^2 / V s. This indicates the efficient incorporation of Al into the ZnO lattice, increasing the oxygen vacancies and Zn interstitials resulting in the improvement of the carrier concentration compared to Plasma and Thermal recipe.

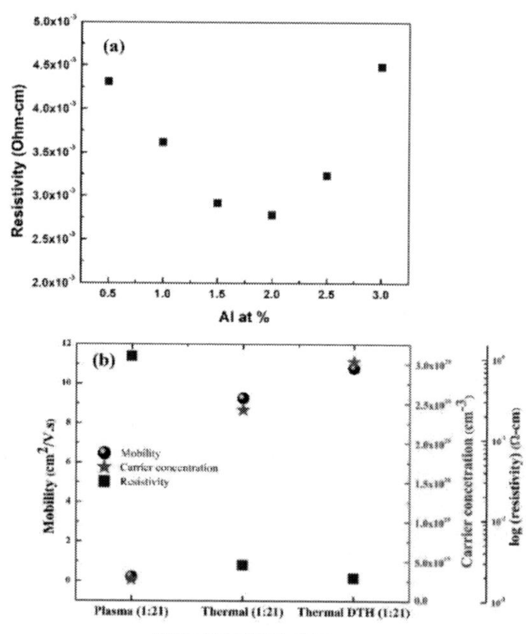

Figure 2 (a): Resisitivity of AZO samples (200nm) deposited by THALD at 200 °C with different Al concentration. **(b)**: Room temperature mobility (black), carrier concentration(red), and resistivity(blue) of AZO (1:21) samples deposited at 200 °C by PE, TH and DTH ALD recipes.

In the fabrication of Si solar cell, the essential requirement for a potential TCO material is the high optical transmittance in the visible wavelength region, which enables the light to reach the absorber without significant loss. The optical transmittance of the AZO thin films deposited using different ALD modes on glass with different Al concentrations is studied. The deposition is done for an optimized thickness of 200nm utilizing growth per cycle of Al_2O_3 and ZnO. The transmittance spectra were measured in the wavelength range of 300nm to 1800nm, the results shown in [Figure 3 (a)] are for the AZO films deposited by TH ALD. The optical transmittance of all AZO samples with various atomic concentrations of Al shows an average transmittance greater than 85% in the visible region. The doping efficiency can be clearly visualized in the near infrared (NIR) range, where the free carrier concentration increases with an increase in the Al concentration in the films. The free carrier absorption according to the Drude model [13] observed is consistent with the electrical data of the films.

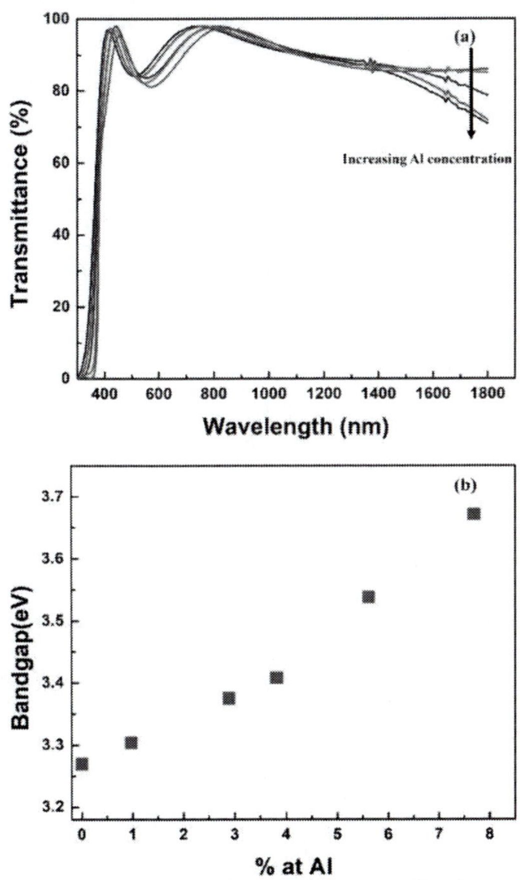

Figure 3 (a): Transmittance (%) as a function of wavelength of AZO samples (200 nm) deposited by THALD at 200 °C with different Al concentrations. **(b):** Bandgap variation of samples with different Al at. % doping in THALD AZO films.

The optical bandgap of these films is measured by fitting the absorption edges. It is determined using the $(\alpha.\ h\nu)^2$ vs. $(h\nu)$ plots, where α is the absorption coefficient and $h\nu$ is the light energy. A considerable change in the bandgap of AZO films can be observed with a change in Al concentration. The ideal bandgap of ZnO and Al_2O_3 is reported to be 3.27eV and 8.7eV, respectively [14] [15]. The bandgap changes from 3.28eV for undoped ZnO to 3.67eV for doping of 4 at % of Al. This corresponds to the Burstein-Moss effect [16] [17] , where the incorporation of Al cation into the ZnO lattice, enhances the charge carriers, which raises the fermi level beyond the conduction level giving rise to degeneracy in the films. This results in the shift of the bandgap energy towards the higher edge and the variation can be seen in [Figure 3 (b)], where the dependency is quite linear.

After confirming the doping efficiency, the transmittance spectra of the samples deposited by TH, PE, and DTH modes of ALD with a cycle of (1:21) were studied. The transmittance of all three samples is greater than 85% in the visible region as required [Figure 4 (a)]. But, in the NIR region, we see that the Al incorporation in ZnO lattice in the case of PE ALD is quite inefficient compared to TH ALD samples resulting in resistive films. However, the DTH sample shows enhanced free carrier absorption depicting the increase in the carrier concentration which is in accordance with electrical data. The bandgap variation of these three samples also represents the efficiency of AZO film fabrication for TCO application using different modes of ALD [Figure 4 (b)]. The AZO with (1:21) super-cycle sample deposited using the DTH recipe has a bandgap of 3.45eV which is greater than that of the TH ALD sample which is 3.40eV.

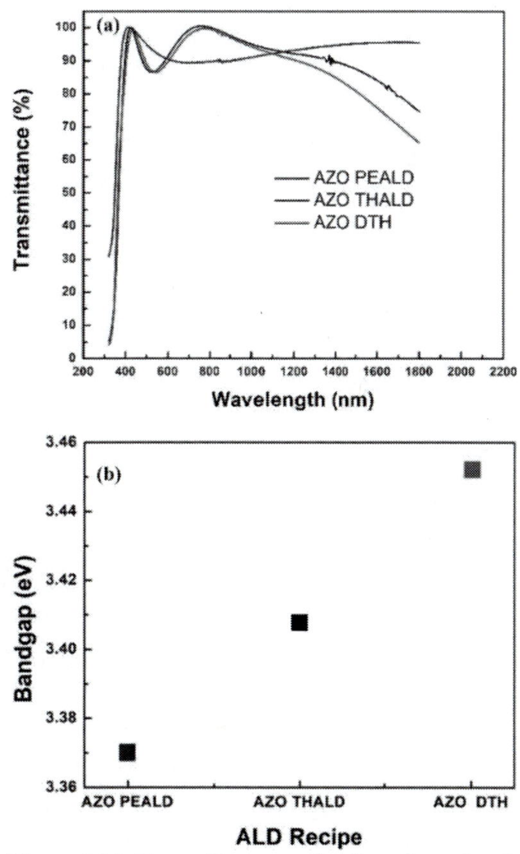

Figure 4 (a): Transmittance as a function of wavelength and **(b):** Bandgap variation of AZO (1:21) samples deposited by PE, TH, and DTH ALD recipes deposited at 200 °C.

The ZnO as-deposited film without any doping shows a polycrystalline behavior, but as the Al doping is increased, we see a preferential growth in (100) direction suppressing the other peaks. This (100) peak is occurring because of the breakdown of negatively charged hydrocarbon ligands in DEZ which disturb the charge distribution on (002) planes [18]. ALD ZnO is known to have textured growth. In our study, the AZO (1:21) grown by PEALD has a (100) peak of lower intensity, but this peak intensifies in THALD, and in DTH samples. We see two additional peaks at (002) and (101) orientation depicting the crystalline nature of the film deposited and thus aiding in the enhanced electrical conductivity.

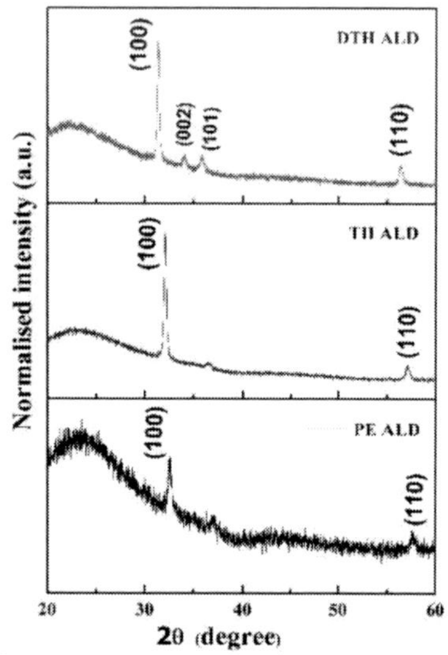

Figure 5: XRD patterns of AZO (1:21) films deposited at 200 C with different modes of ALD

4 CONCLUSION

ALD is a useful technology in fabricating a passivating contact solar cell as it allows soft deposition of pinhole free ultrathin films and also gives an opportunity to tune the properties of materials by changing various parameters. The AZO samples of 200nm were deposited at 200 °C using the different sequencing of precursors and co-reactants. The AZO samples deposited by PEALD technique is found to be resistive at all concentrations of Al doping. The THALD AZO samples, show enhanced crystallinity and conductivity. AZO with 2 at.% Al has the least resistivity of 2.78×10^{-3} Ω-cm and mobility of 9.27 cm^2 / V s. Also, these samples have an average transmittance of 85% in the visible region indicating the properties of an efficient device quality TCO material. A slight variation in the pulsing sequence of reactants during THALD deposition is made to get samples with the DTH recipe, which has better electrical properties than the conventional deposition (PE/TH) techniques in ALD. The reduced resistivity of 1.91×10^{-3} Ω-cm of only 200nm thin AZO DTH films with 2 at.% Al is attributed to the efficient doping of Al in ZnO and increased crystallinity.

These recipes affect the surface reactions during deposition and influence the film properties. The innovative pulsing sequences in ALD along with post-deposition processing can be utilized for further research and fabrication of AZO films with superior quality for TCO application.

5 REFERENCES

[1] K. Ellmer, A. Klein, and B. Rech, Eds., *Transparent Conductive Zinc Oxide: Basics and Applications in Thin Film Solar Cells*, vol. 104. Berlin, Germany,: Springer, 2004.

[2] J.-W. Schüttauf, J. K. Rath, F. Roca, and E. Portici, "State-of-the-Art Transparant Conductive Oxide Layers for silicon heterojunction solar cells"

[3] Z. Gao and P. Banerjee, "Review Article: Atomic layer deposition of doped ZnO films," *Journal of Vacuum Science & Technology A*, vol. 37, no. 5, p. 050802, Sep. 2019, doi: 10.1116/1.5112777.

[4] T. Minami, H. Nanto, and S. Takata, "Highly Conductive and Transparent Aluminum Doped Zinc Oxide Thin Films Prepared by RF Magnetron Sputtering", *Japanese Journal of Applied Physics*, vol. 23, no. 5, 1984

[5] F. K. Shan, G. X. Liu, W. J. Lee, and B. C. Shin, "The role of oxygen vacancies in epitaxial-deposited ZnO thin films," *J Appl Phys*, vol. 101, no. 5, 2007, doi: 10.1063/1.2437122.

[6] W. S. Lau and S. J. Fonash, "Highly Transparent and Conducting Zinc Oxide Films Deposited by Activated Reactive Evaporation," *Journal of Electronic Materials*, Vol. 16, No. 3, 1987. doi: 10.1007/BF02655478

[7] J. K. Rath *et al.*, "Transparent conducting oxide layers for thin film silicon solar cells," *Thin Solid Films*, vol. 518, no. 24 SUPPL., Oct. 2010, doi: 10.1016/j.tsf.2010.03.104.

[8] C. S. Hwang, "Erratum to: Atomic Layer Deposition for Semiconductors," Boston, MA: Springer US, 2014, pp. E1. doi: 10.1007/978-1-4614-8054-9_10.

[9] R. L. Puurunen, "Surface chemistry of atomic layer deposition: A case study for the trimethylaluminum/water process," *Journal of Applied Physics*, vol. 97, no. 12. 2005. doi: 10.1063/1.1940727.

[10] S. E. Potts, W. Keuning, E. Langereis, G. Dingemans, M. C. M. van de Sanden, and W. M. M. Kessels, "Low Temperature Plasma-Enhanced Atomic Layer Deposition of Metal Oxide Thin Films," *J Electrochem Soc*, vol. 157, no. 7, p. P66, 2010, doi: 10.1149/1.3428705.

[11] M. L. Lin, J. M. Huang, C. S. Ku, C. M. Lin, H. Y. Lee, and J. Y. Juang, "High mobility transparent conductive Al-doped ZnO thin films by atomic layer deposition," *J Alloys Compd*, vol. 727, pp. 565–571, Dec. 2017, doi: 10.1016/j.jallcom.2017.08.207.

[12] P. Banerjee, W. J. Lee, K. R. Bae, S. B. Lee, and G. W. Rubloff, "Structural, electrical, and optical properties of atomic layer deposition Al-doped ZnO films," in *Journal of Applied Physics*, Aug. 2010, vol. 108, no. 4. doi: 10.1063/1.3466987.

[13] J. I. Pankove, "Optical Processes in Semiconductors", New York, Dover Publications, Inc, 1971.

[14] Chennupati Jagadish and Stephen Pearton, "Zinc Oxide Bulk, Thin Films and Nanostructures," Elsevier Science, 1st Edition, 2006.

[15] G. D. Wilk, R. M. Wallace, and J. M. Anthony, "High-κ gate dielectrics: Current status and materials properties considerations," *J Appl Phys*, vol. 89, no. 10, pp. 5243–5275, May 2001, doi: 10.1063/1.1361065.

[16] E. Burstein, "Anomalous Optical Absorption Limit in InSb." *Physical Review Journal*, vol. 93(3) pp. 632-633. doi: 10.1103/PhysRev.93.632

[17] T. S. Moss, "The Interpretation of the Properties of Indium Antimonide," *Proceedings of the Physical Society. Section B*, vol. 67, no. 10, pp. 775–782, Oct. 1954, doi: 10.1088/0370-1301/67/10/306.

[18] S. Y. Pung, K. L. Choy, X. Hou, and K. Dinsdale, "In situ doping of ZnO nanowires using aerosol-assisted chemical vapour deposition," *Nanotechnology*, vol. 21, no. 34, Aug. 2010, doi: 10.1088/0957-4484/21/34/345602.

IMPACT OF A RENEWED SiN_x:H PASSIVATION LAYER AND/OR A RENEWED FIRING STEP ON LeTID AND REGENERATION

Andreas Schmid, Annika Zuschlag, Giso Hahn
University of Konstanz, Department of Physics, 78457 Konstanz, Germany
andreas.schmid@uni-konstanz.de, Tel: +49 7531 882082, Fax: +49 7531 883895

ABSTRACT: Light and elevated temperature induced degradation (LeTID) significantly affects the performance of crystalline Si solar cells. The underlying mechanisms of LeTID and following regeneration are still largely unknown, but the phenomenon seems to depend on the H content in the Si bulk. Therefore, in this work, we investigate the influence of a renewed H-rich SiNx:H passivation layer (with a subsequent firing step) as well as solely re-firing of the sample after certain points in time during LeTID and regeneration. By re-depositing a SiNx:H layer in combination with a new firing step, we show that the LeTID and regeneration cycle is run through again without any significant change in kinetics or strength being detected. Assuming that H enters the Si bulk by firing a H-rich passivation layer, a renewed passivation would represent a new reservoir of H and thus could lead to renewed degradation. Furthermore, comparable kinetics show up before and after renewed firing of the already fired SiNx:H layer, but the extent of degradation slightly decreases with each new firing step, which could be due to a depletion of the H-source.
Keywords: Degradation, Hydrogen, Diffusion, Silicon-Nitride, Firing, c-Si

1 INTRODUCTION

Multicrystalline (mc) Si PERC (passivated emitter and rear contact) solar cells, but also lifetime samples, show a strong degradation of excess minority charge carrier lifetimes τ_{eff} under illumination at elevated temperatures (*e.g.* [1-3]). The so-called light and elevated temperature induced degradation (LeTID) [4] cannot be explained by the already known degradation mechanisms like BO-correlated defects or the dissociation of FeB pairs [1, 5]. Even though the underlying mechanisms of LeTID are largely unknown, a wide range of influencing factors has already been analyzed. In addition to the impact of the degradation conditions (*e.g.* [4, 6]), the influence of individual steps in the sample production, such as a gettering step [7] or the peak firing temperature [8-12], could be shown. Moreover, Kersten *et al.* [13] demonstrated that LeTID was only observed on samples passivated with H-containing dielectric layers during firing and Vargas *et al.* [14] further correlated the severity of LeTID with the amount of H released from the dielectric layer during firing. These findings were further verified by studies of Bredemeier *et al.* [15] and Schmidt *et al.* [16].

Recently, many studies assume that H is directly involved in LeTID defect formation (*e.g.* [11, 15, 17-19]). To determine the influence of H on the LeTID formation dynamics, the manipulation of the H content in the c-Si bulk gains more and more importance. For example, various investigations manipulated the bulk H content by applying different stacks of passivation layers and thus were able to control the severe of LeTID and regeneration [20-22]. With regard to the regeneration, Bredemeier *et al.* [23] already considered H diffusing from the surface into the bulk and subsequently passivating bulk defects as a possible mechanism. A more recent study by Mehler *et al.* [24] was able to establish a delayed onset of regeneration through co-doping with B and Al, which they linked to the binding energy of H with the dopant.

Furthermore, Chen *et al.* [25] previously demonstrated that temperature treatment ($T < 275°C$) directly after firing can have an effect on LeTID kinetics. An extended study by Fung *et al.* [26] performed this treatment after reaching different points in time of degradation/regeneration and they were able to show that LeTID kinetics are altered regardless of the time of application.

In this study, the influence of high temperature steps ($T >> 275°C$) after certain points in time of LeTID and regeneration (in terms of a renewed SiN_x:H deposition or a renewed firing step) are being examined. The aim of this research is to investigate the H reservoir in more detail and to what extent a new firing step influences the amount of H released and thus LeTID and regeneration.

2 EXPERIMENTAL

In order to investigate the influence of a renewed firing step or a renewed passivation on the degradation and regeneration process, 1.1 Ωcm B-doped p-type mc-Si wafers were used. For comparison, sister samples with comparable grain and defect structure were used. A simplified scheme of the applied process sequence is given in Fig. 1. After different cleaning steps, all samples received double-sided 75 nm thick SiN_x:H layers by plasma-enhanced chemical vapor deposition (PECVD). Afterwards, they were fired in a belt furnace at a set temperature of 850°C. For degradation, the lifetime samples are held at a temperature of approx. 110°C on a hotplate under illumination with halogen lamps (1.0±0.1 suns). τ_{eff} is measured repetitively via the photoconductance decay (PCD) method at room temperature. For the evaluation, lifetimes are extracted at

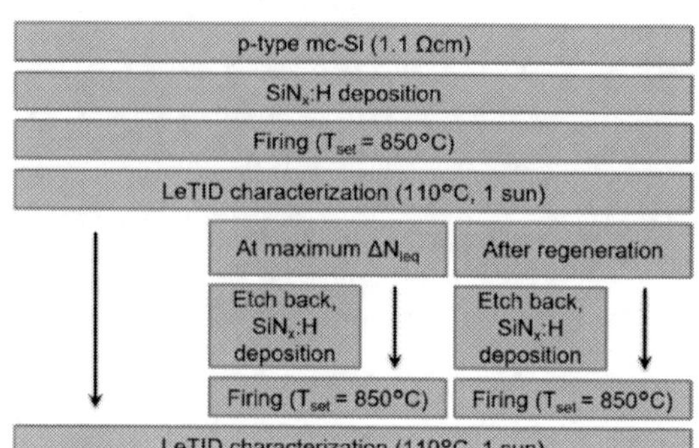

Figure 1: Process sequence of the investigated lifetime samples.

fixed charge carrier concentration of $\Delta n = 1.5 \times 10^{15}\ cm^3$.

One group of samples was removed from the hotplate at maximum defect density while another group of samples was removed only after complete regeneration. One wafer from each group was re-fired at a set temperature of 850°C. The SiN_x:H layer of another sample from each group was etched-back before the wafers received a renewed SiN_x:H passivation and were again fired at a set temperature of 850°C. In addition, a third group of samples was removed from the hotplate at the same times as the other sample groups, but no further processes were applied. This group was stored in the dark at room temperature for reference.

Furthermore, the SiN_x:H layer of five sister samples of each group was etched-back at relevant points in time, meaning after the initial firing step, after degradation (and subsequent re-firing) and after complete regeneration (and subsequent re-firing). To verify whether the observed behavior is a bulk or surface defect, a wet-chemical iodine-ethanol (IE) passivation was applied to these samples. As a reference measurement of an IE passivated 2 Ωcm FZ sample revealed lifetimes of up to 2 ms, surface recombination of the IE passivated mc-Si of lower material quality should be negligible. Thus, effective lifetime which was measured via the PCD method at room temperature should represent the bulk material quality of the IE passivated samples.

3 RESULTS AND DISCUSSION

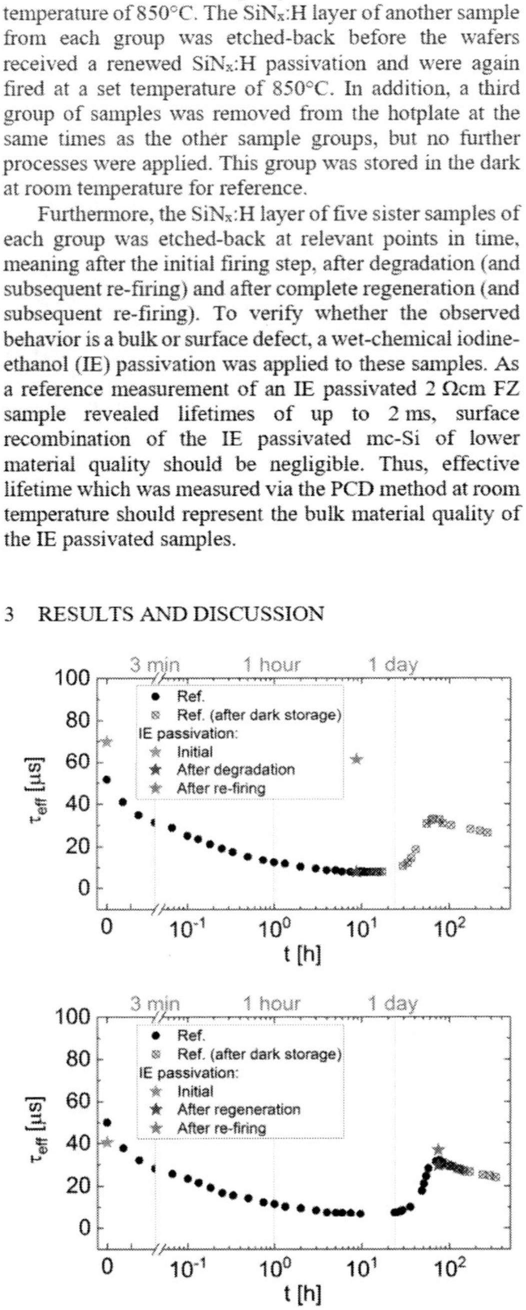

Figure 2: Degradation and regeneration of a mc-Si sample at 110°C and 1 sun. The stars represent sister samples which were wet-chemically (IE) passivated in certain LeTID and regeneration states or after re-firing.

As shown in Fig. 2, the mc-Si wafers show a pronounced degradation and subsequent regeneration behavior at 110°C and 1 sun illumination within approx. 100 h. Furthermore, the samples which were stored in the dark at room temperature while the other samples were further processed show that interruption has no effect on degradation and regeneration kinetics. The lifetimes of the IE passivated wafers after the initial firing step indicate no limitations of the SiN_x:H surface, as both, IE and SiN_x:H passivated samples show similar τ_{eff}. An IE re-passivated sample at maximum defect density (ΔN_{leq}) does not show any improvement in τ_{eff}, indicating a bulk defect that forms during degradation. Interestingly, after re-firing a sample at maximum ΔN_{leq} (and re-passivating with IE), lifetime increases again up to its initial value. This shows that the bulk defect can be extinguished with a new firing step. IE passivated samples after complete regeneration (and subsequent re-firing) do not show any significant improvement of the lifetime. This demonstrates an improvement of the bulk during regeneration, which cannot be significantly further enhanced by a renewed firing step.

In the left graph of Fig. 3 the degradation (and regeneration) at 110°C and one sun illumination of two samples is shown, before they were removed from the hotplate and a renewed SiN_x:H passivation with subsequent firing at a set temperature of 850°C was applied. The right graph in Fig. 3 shows the newly passivated samples treated under the same conditions as before (110°C, 1 sun).

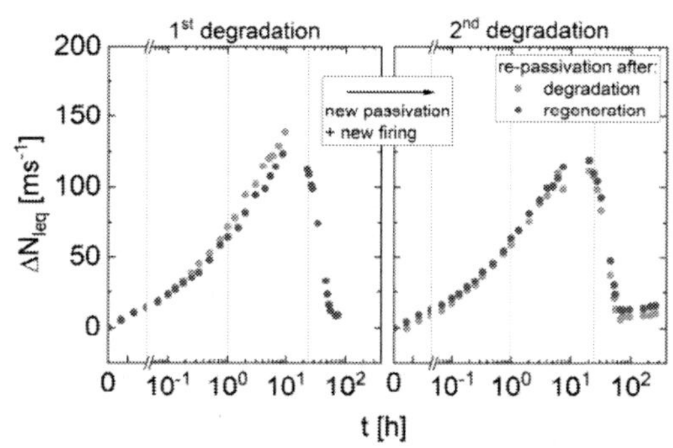

Figure 3: ΔN_{leq} of two mc-Si samples at 110°C and 1 sun. One sample was etched-back at maximum degradation, the other after complete regeneration. Both samples were then re-passivated with SiN_x:H (+subsequent firing) and treated again at 110°C and one sun.

First of all, it should be mentioned that both samples returned to their initial τ_{eff} value of approx. 50 µs after re-passivation ($\Delta\tau_{eff}$: 1 or 4.5 µs). Interestingly, both samples show a very similar degradation and regeneration behavior compared to each other and especially compared to the status before the new passivation. It does not seem to matter whether the sample was previously taken from the hotplate at maximum degradation or regeneration. Thus, by re-depositing a SiN_x:H layer in combination with a new firing step, the LeTID and regeneration cycle is run through again without any significant change in kinetics or strength being detected. This is in contrast to the findings of Fung *et al.* [26], who performed a temperature-only treatment (< 250°C) after comparable points in time of the LeTID cycle and demonstrated a decrease in LeTID strength. Recently, it could be shown that H plays an essential role in the LeTID formation process (*e.g.* [11, 15, 17-19]). Assuming that H enters the Si bulk by firing a H-rich passivation layer [13, 14, 27], a renewed passivation would represent a new source of H. Firing could replenish the H reservoir in the Si-bulk and thus leading to renewed degradation and regeneration. This would also explain the discrepancy with [26], who explained the decrease in LeTID strength with a limited supply of a defect precursor (in this case H reservoir). According to [26], each new temperature treatment in the dark redeposits H from other binding states and replenishes the H reservoir to some extent, which is gradually depleted. However, new H cannot diffuse from the passivation layer into the Si bulk at the low temperatures (< 250°C) and completely replenish the H reservoir.

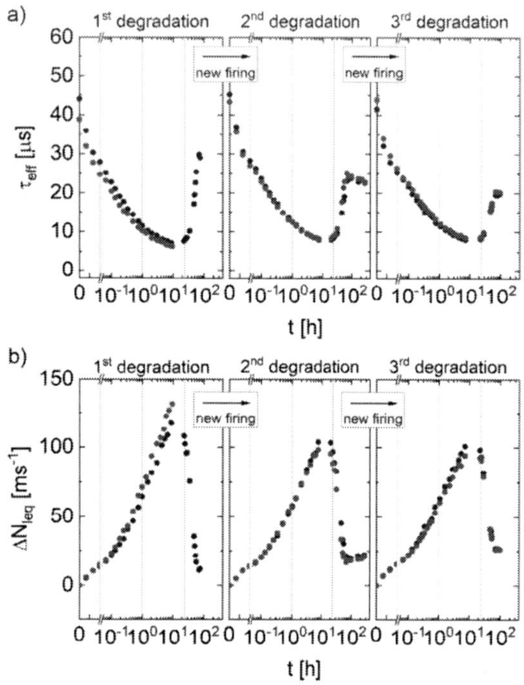

Figure 4: τ_{eff} (a) and ΔN_{leq} (b) of mc-Si samples at 110°C and 1 sun. The samples were re-fired twice. After both re-firing steps, the samples were treated again at 110°C and 1 sun.

Since these findings do not rule out the possibility that a renewed firing step alone is sufficient to trigger the LeTID and regeneration cycle once again, a further group of samples was only re-fired after maximum degradation or regeneration. Subsequently, the samples were treated again at 110°C and one sun illumination. Once both samples were fully regenerated, they were again re-fired and treated under the same conditions as before (110°C, one sun). Fig. 4 shows all degradation and regeneration cycles of the examined samples.

First, it can be seen that each new firing step causes the initial τ_{eff} values to increase again to comparable values. Again, it is not relevant whether the sample was re-fired after complete degradation or regeneration. As observed for the samples with renewed passivation, comparable LeTID and regeneration kinetics of the samples before and after renewed firing can be detected. However, there is a slight tendency towards a lower LeTID extent. This is evident on the one hand from somewhat higher τ_{eff} and on the other hand from slightly decreasing maximum ΔN_{leq} at the point of maximum degradation with each additional firing step. According to the theory mentioned above, the firing of an H-rich passivation layer leads to an in-diffusion of H which can refill the defect precursor reservoir in the Si bulk. Consequently, with each renewed firing step, the H source (passivation layer) would be slowly depleted, leading to a not completely refilled defect precursor reservoir and finally to a lower LeTID extent. Nevertheless, such a severe drop in LeTID strength as shown by Fung *et al.* [26] at lower temperatures (< 250°C) cannot be observed. Again, one explanation for this distinction could be that at such low temperatures there is only a rearrangement of the H-bonded states and no in-diffusion of new H from the passivation layer, leading to a stronger decrease in LeTID strength.

Furthermore, it can be seen that the extent of regeneration decreases slightly with each new firing step. For example, the highest lifetime reached after regeneration of the sample shown in blue (Fig. 4 a)), decreases from 29.8 µs after the first LeTID and regeneration cycle to 23.8 µs (2nd cycle) to 19.5 µs after the third cycle. Thus, this could also be an indication of a lower H content after each renewed firing step in the sample bulk, assuming that H plays a particularly important role during regeneration, as described by Mehler *et al.* [24]. Since less H is available with each new firing step, this could be an explanation for the decrease in regeneration strength. However, an accelerating onset of surface degradation could also be responsible for this observation. In order to verify this first rough model in more detail, however, further experiments on this subject are necessary.

4 CONCLUSION

In conclusion, wet-chemically (IE) passivated samples confirmed that the applied SiN_x:H passivation layer does not represent a limitation of τ_{eff} for the investigated mc-Si material. Furthermore, an IE re-passivated sample at maximum defect density does not show any improvement in lifetime, indicating a bulk defect that forms during degradation. After re-firing of the existing SiN_x:H layer, lifetime increases again up to its initial value, which indicates that the bulk defect can be extinguished with a new firing step. IE passivated samples after complete regeneration confirm an improvement of the Si bulk which cannot be further enhanced by a new firing step.

By re-depositing a new SiN_x:H layer in combination with a new firing step, the LeTID and regeneration cycle is run through again without any significant change in kinetics or strength being detected. Assuming that the H enters the Si bulk by firing a H-rich passivation layer, a renewed passivation would represent a new source of H, that could completely replenish the defect precursor reservoir, leading to a renewed degradation.

A renewed firing step alone, on the other hand, leads to a slight decrease in LeTID strength. This could be due to a depletion of the H-source (passivation layer) with each renewed firing, leading to a not completely refilled defect precursor reservoir and finally to a lower LeTID extent. In addition, the extent of regeneration decreases with each new firing step as well.

5 ACKNOWLEDGEMENTS

Part of this work was supported by the German Federal Ministry for Economic Affairs and Climate Action under contract 03EE1051C. The content of the publication is the responsibility of the authors. The authors would like to thank U. Wohlert and B. Rettenmaier for technical support and C. Fischer, D. Skorka and M. Mehler for supporting discussions.

6 REFERENCES

[1] K. Ramspeck, S. Zimmermann, H. Nagel, A. Metz, Y. Gassenbauer, B. Birkmann, A. Seidl, Light induced degradation of rear passivated mc-Si solar cells, *Proc. 27th EUPVSEC*, 2012, 861-865.

[2] F. Fertig, K. Krauss, S. Rein, Light-induced degradation of PECVD aluminium oxide passivated silicon solar cells, *Phys. Stat. Sol. RRL* 9 (2014) 41-46.

[3] A. Zuschlag, D. Skorka, G. Hahn, Degradation and regeneration analysis in mc-Si, *Proc. 43rd IEEE PVSC*, 2016, 1051-1054

[4] F. Kersten et al., Degradation of multicrystalline silicon solar cells and modules after illumination at elevated temperature, *Sol. Energy Mat. Sol. Cells* 142 (2015) 83-86.

[5] J. Lindroos, H. Savin, Review of light-induced degradation in crystalline silicon solar cells, *Sol. Energy Mater. Sol. Cells* 147 (82016) 115–126.

[6] J. M. Fritz, A. Zuschlag, D. Skorka, A. Schmid, G. Hahn, Temperature dependent degradation and regeneration of differently doped mc-Si materials, *Energy Procedia* 124 (2017) 718–725.

[7] A. Zuschlag, D. Skorka, G. Hahn, Degradation and regeneration in mc-Si after different gettering steps, *Prog. Photovolt. Res. Appl.* 25 (2017) 545–552.

[8] D. Bredemeier, D. Walter, S. Herlufsen, J. Schmidt, Lifetime degradation and regeneration in multicrystalline silicon under illumination at elevated temperature, *AIP Adv.* 6 (2016) 035119.

[9] D. Chen et al., Evidence of an identical firing-activated carrier-induced defect in monocrystalline and multicrystalline silicon, *Sol. Energy Mater. Sol. Cells* 172 (2017) 293–300.

[10] R. Eberle, W. Kwapil, F. Schindler, M. C. Schubert, S. W. Glunz, Impact of the firing temperature profile on light induced degradation of multicrystalline silicon, *Phys. Stat. Sol. RRL* 10 (2016) 861–865.

[11] T. Niewelt, F. Schindler, W. Kwapil, R. Eberle, J. Schön, M. C. Schubert, Understanding the light-induced degradation at elevated temperatures: Similarities between multicrystalline and floatzone p-type silicon, *Prog. Photovolt. Res. Appl.* 26 (2018) 533–542.

[12] C. E. Chan et al., Rapid stabilization of high-performance multicrystalline p-type silicon PERC cells, *IEEE J. Photovolt* 6(6) (2016) 1473–1479.

[13] F. Kersten, J. Heitmann, J. W. Müller, Influence of Al_2O_3 and SiN_x passivation layers on LeTID, *Energy Procedia* 92 (2016) 828–832.

[14] C. Vargas, K. Kim, G. Coletti, D. Payne, C. Chan, S. Wenham, Z. Hameiri, Carrier-induced degradation in multicrystalline silicon: Dependence on the silicon nitride passivation layer and hydrogen released during firing, *IEEE J. Photovolt.* 8(2) (2018) 413–420.

[15] D. Bredemeier, D. C. Walter, R. Heller, J. Schmidt, Impact of hydrogen-rich silicon nitride material properties on light-induced lifetime degradation in multicrystalline silicon, *Phys. Stat. Sol. RRL* 13 (2019) 1900201.

[16] J. Schmidt, D. Bredemeier, D. C. Walter, On the defect physics behind light and elevated temperature-induced degradation (LeTID) of multicrystalline silicon solar cells, *IEEE J. Photovolt.* 9(6) (2019) 1497–1503.

[17] T. H. Fung et al., A four-state kinetic model for the carrier-induced degradation in multicrystalline silicon: Introducing the reservoir state, *Sol. Energy Mater. Sol. Cells* 184 (2018) 48–56.

[18] D. Chen et al., Hydrogen induced degradation: A possible mechanism for light- and elevated temperature induced degradation in n-type silicon, *Sol. Energy Mater. Sol. Cells* 185 (2018) 174–182.

[19] A. Ciesla et al., Hydrogen-induced degradation, *Proc. WCPEC-7*, 2018, 0001–0008.

[20] A. Schmid, C. Fischer, D. Skorka, A. Herguth, C. Winter, A. Zuschlag, G. Hahn, On the role of AlO_x thickness in AlO_x/SiN_y:H layer stacks regarding light- and elevated temperature-induced degradation and hydrogen diffusion in c-Si, *IEEE J. Photovolt.* 11(4) (2021) 967-973.

[21] L. Helmich, D. C. Walter, D. Bredemeier, J. Schmidt, Atomic-layer-deposited Al_2O_3 as effective barrier against the diffusion of hydrogen from SiN_x:H layers into crystalline silicon during rapid thermal annealing, *Phys. Stat. Sol. RRL* 14 (2020) 2000367.

[22] U. Varshney et al., Controlling light- and elevated-temperature-induced degradation with thin film barrier layers, *IEEE J. Photovolt.* 10(1) (2020) 19–27.

[23] D. Bredemeier, D. C. Walter, J. Schmidt, Possible candidates for impurities in mc-Si wafers responsible for light-induced lifetime degradation and regeneration, *Sol. RRL* 2 (2018) 1700159.

[24] M. Mehler, A. Schmid, A. Zuschlag, M. Trempa, G. Hahn, Delay of regeneration by adding aluminum in boron-doped crystalline Si, *Phys. Stat. Sol. A* 218 (2021) 2100603.

[25] C. Chan et al., Modulation of carrier-induced defect kinetics in multicrystalline silicon PERC cells through dark annealing, *Sol. RRL* 1 (2017) 1600028.

[26] T. H. Fung, C. E. Chan, B. J. Hallam, D. N. R. Payne, M. D. Abbott, S. R. Wenham, Impact of annealing on the formation and mitigation of carrier-induced defects in multi-crystalline silicon, *Energy Procedia* 124 (2017) 726-733.

[27] J. Lindroos, A. Zuschlag, D. Skorka, G. Hahn, Silicon nitride deposition: Impact on lifetime and light-induced degradation at elevated temperature in multicrystalline silicon, *IEEE J. Photovolt.* 10(1) (2020) 8–18.

EFFECT OF DIFFERENT AlO_x PASSIVATION LAYERS DEPOSITED BY APCVD AND ALD ON LETID

Melanie Mehler, Fabian Geml, Andreas Schmid, Annika Zuschlag, Giso Hahn
University of Konstanz, Department of Physics, 78457 Konstanz, Germany

ABSTRACT: In this work the influence of AlO_x, acting as a barrier layer for in-diffusion of H from a H-rich dielectric layer, on the strength and kinetics of the LeTID (light- and elevated temperature-induced degradation) phenomenon was investigated. For this purpose, different AlO_x passivating layers deposited by different deposition methods, APCVD (atmospheric pressure chemical vapor deposition) and ALD (atomic layer deposition), are considered. Since the composition and material properties of the AlO_x layers, fabricated in the two different deposition systems, are not exactly the same, the LeTID defect kinetics may differ. In the first experiment, by varying the AlO_x layer thickness, it could be confirmed that on Cz material the AlO_x layers grown by ALD and APCVD both function as a barrier layer for H. In the second experiment, the comparison of the different AlO_x layers from APCVD and ALD was considered in more detail. The deposition temperature was varied for the respective methods, but the AlO_x layer thickness of 10 nm remains constant for all samples. To obtain conclusions on the LeTID behavior and the H content in the Si bulk, excess charge carrier lifetime and BH measurements have been performed on the samples. Besides the AlO_x single layers, AlO_x/SiN_y:H stacks were also investigated. The results show that the APCVD AlO_x layer is more permeable to H than the AlO_x layer grown by ALD. Alternatively, the obtained results could also be explained assuming that the APCVD layer contains generally more H in the layer than the ALD AlO_x layer. The findings can help to better understand the origin of LeTID and find new processing sequences that can avoid or minimize the occurrence of LeTID, leading to higher stabilized efficiencies for high efficiency cell concepts.
Keywords: Degradation, Hydrogen, CVD Based Deposition

1 INTRODUCTION

Many high efficiency solar cell concepts such as PERC (passivated emitter and rear cell) are susceptible to degradation issues, reducing the efficiency over time. One of these phenomena is the LeTID (light- and elevated temperature-induced degradation) effect. The mechanism and the root cause of LeTID are not sufficiently understood yet. However, it has been shown that H plays an important role in LeTID [1-2]. The occurrence of LeTID was shown after firing of H-rich dielectric passivation layers such as SiN_y:H [3]. Since AlO_x layers represent a diffusion barrier for atomic H, the H content in the Si bulk can be affected by them [4].

The aim in this work is to investigate different AlO_x barrier layers that can influence the diffusion of H into the Si bulk in order to achieve a positive effect on the LeTID degradation behavior. Besides the variation of the layer thickness and the composition of the passivation layer, the deposition method can influence the LeTID effect. For this, the two different deposition methods APCVD (atmospheric pressure chemical vapor deposition) and ALD (atomic layer deposition) are investigated. In order to draw conclusions about the H content in the Si bulk, the concentration of BH pairs was determined via resistivity measurements.

2 EXPERIMENTAL

2.1 Sample preparation for ALD AlO_x as diffusion barrier

B-doped Cz-Si wafers with a bulk resistivity of ~1 Ωcm serve as base material and were lasered to 5x5 cm^2. After saw damage removal and cleaning, the samples are gettered using a $POCl_3$ diffusion. The resulting emitter is removed in an etching step and the samples are cleaned again. Afterwards, the samples are deposited with AlO_x from ALD at 300°C and the thickness of the layer is varied from 10-25 nm. Since the deposition time is different for each layer thickness, the time was adjusted so that the samples obtained the same temperature profile. In addition, 75 nm SiN_y:H using PECVD (plasma-enhanced chemical vapor deposition) was deposited on the AlO_x layer. After firing at a temperature of $T_{sample,peak}$=813°C, samples were degraded at 80°C and at 0.9(1) suns. The effective lifetime τ_{eff} is determined using photoconductance decay (PCD) and evaluated at Δn=0.1 p_0, with p_0 being the doping density. To allow a better comparison of the degradation and regeneration behaviour, the lifetime equivalent defect density ΔN_{leq} is calculated from τ_{eff}.

2.2 Sample preparation for AlO_x from different deposition tools

Again, B-doped Cz-Si (~1 Ωcm) serve as base material. After saw damage removal and cleaning, AlO_x was deposited on the 5x5 cm^2 samples with two different deposition tools, APCVD and ALD. In addition to change of deposition methods, different deposition temperatures are also considered. With APCVD, AlO_x layers are deposited at 590°C and 690°C, and with ALD at 170°C and 300°C. However, the AlO_x layer thickness of 10 nm is the same for all samples. To investigate the influence of an additional SiN_y:H layer on LeTID and on the H content in the Si bulk, some of the samples were additionally coated on both sides with 75 nm SiN_y:H and fired at a temperature of $T_{sample,peak}$=813°C. The remaining samples, which are only coated with AlO_x, are also fired at the same temperature.

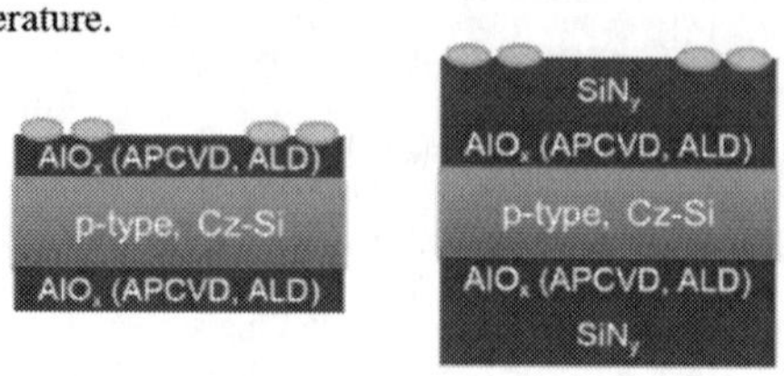

Figure 1: Schematic sample structure for BH pair and lifetime measurements on fired samples with AlO_x (left) and fired samples with AlO_x/SiN_y stacks (right). Al contacts shown in orange.

For the determination of the BH pairs via resistivity measurement [5], aluminum was deposited on one side of the samples followed by laser fired contacting (LFC), see Fig. 1. For better statistics, two samples were processed for each variation. Analogous to the previous procedure, τ_{eff} is determined using PCD and the samples were degraded at 80°C and at 0.9(1). For the BH measurements the samples were annealed at 220°C in the dark.

3 RESULTS

Fig. 2 shows the resulting ΔN_{leq} of the samples, which are deposited with 0-25 nm thick ALD AlO_x layers and an additional 75 nm thick SiN_y:H layer. As expected, the progression of ΔN_{leq} values shows a degradation and regeneration behavior. Within the first hour, ΔN_{leq} increases and reaches the maximum after about 0.5-2 h, followed by a decrease of ΔN_{leq}. It can be seen that the different AlO_x layer thicknesses lead to a decrease in ΔN_{leq}. Compared to the 0 nm thick AlO_x layer, the maximum value of the $\Delta N_{leq,max}$ decreases from 12 ms^{-1} to below 4 ms^{-1} for the 10 nm thick AlO_x layer. Increasing the layer thickness to 15-25 nm does not lead to a further decrease of ΔN_{leq}. This could be due to a saturation value of AlO_x with regard to H diffusion. These data confirm the results of [6].

The same experiment with variation of AlO_x layer thickness was performed with APCVD. It could be shown that the APCVD AlO_x layers also function as a barrier layer [7] (not shown here).

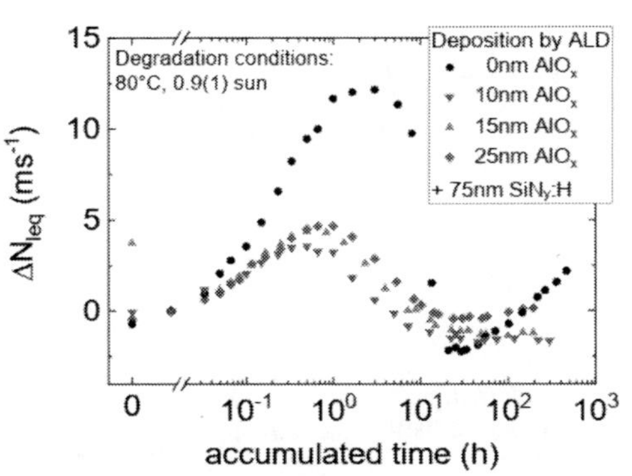

Figure 2: Lifetime equivalent defect density ΔN_{leq} over accumulated time (80°C, 0.9 suns) of AlO_x layers of different thickness with an additional 75 nm SiN_y:H layer after firing.

When comparing the different AlO_x layers from the two deposition methods, the layer thickness of 10 nm, which acts as a diffusion barrier (see Fig. 2), is the same for all samples. As mentioned earlier, the determination of the BH pairs is done by resistance measurement, but not every change in resistance can be attributed to the change in BH pairs. Depending on the treatment conditions, thermal donors could occur which can influence the change in resistance [8]. Since one hole is consumed for the formation of BH pairs from H-dimers in boron-doped silicon [9], the change in hole concentration -Δp is given. Thus, -Δp includes the change in BH pairs and the possible occurrence of thermal donors.

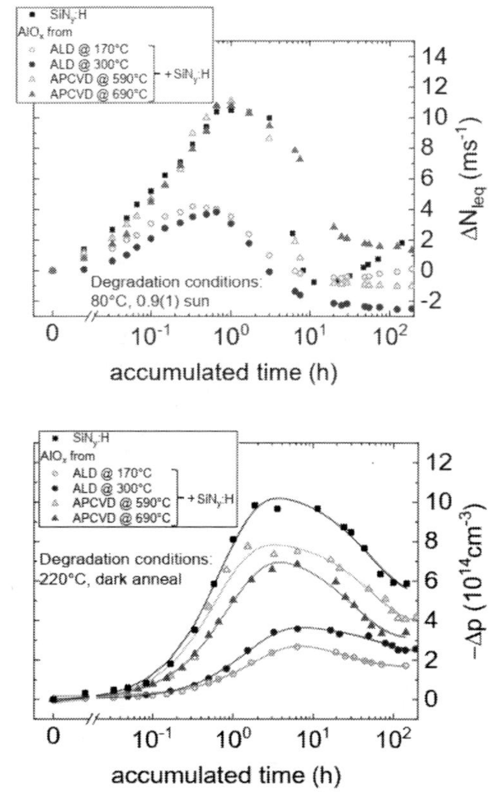

Figure 3: Lifetime equivalent defect density ΔN_{leq} (top) and change in hole concentration −Δp with fit (bottom) over accumulated time of AlO_x layers deposited by APCVD and ALD with an additional 75 nm SiN_y:H layer after firing.

Fig. 3 shows the defect density ΔN_{leq} and changing hole concentration -Δp over treatment time of the fired samples with AlO_x/SiN_y:H layers and SiN_y:H layer as reference. The ALD AlO_x layers were deposited at 170°C or 300°C, respectively, and the APCVD AlO_x layers at 590°C or 690°C, respectively. Also here, the progression of ΔN_{leq} shows a degradation and regeneration behaviour. The maximum of ΔN_{leq} lies for all samples at around 1 h and, as can be seen, the maximum of ΔN_{leq} for the APCVD AlO_x layers (~11 ms^{-1}) and for the SiN_y:H reference (~11 ms^{-1}) is higher than for the ALD layers (<4 ms^{-1}). During degradation the SiN_y:H reference and the APCVD AlO_x layer have comparable ΔN_{leq} values but the behavior differs during regeneration. APCVD AlO_x deposited at 690°C shows a flatter progression of regeneration compared to AlO_x deposited at 590°C and the SiN_y:H reference. Furthermore, except for the SiN_y:H reference and the ALD AlO_x layer deposited at 170°C, the samples do not reach the minimum of ΔN_{leq} within 200 hours.

In the lower graph of Fig. 3 the changing hole concentration −Δp is given. As mentioned in the experimental part, for the BH measurements two samples were processed for each variation, but since the data of the two samples is reproducible, only one sample per variation is shown in the graph. In the first hours -Δp increases and reaches the maximum after about 2-10 h, afterwards -Δp starts to decrease again before -Δp starts to rise again after 200 h. Since it can be assumed that this second increase is not caused by the formation of BH pairs but by other processes in the Cz-Si, e.g. formation of thermal donors, the data are normalized to t=0. As can be seen, the maximum of $-\Delta p$ for the APCVD AlO_x layers ($6\text{-}8\cdot10^{14}$ cm^{-3}) is higher than for the ALD layers ($<4\cdot10^{14}$ cm^{-3}). Together with the results of the defect densities ΔN_{leq} the APCVD layer appears to be more permeable to H diffusion from the SiN_x:H layer than the ALD layer. It could also suggest that the APCVD AlO_x layer generally already contains more H than the ALD layer.

To get more insight into the influence of the different AlO_x layers, AlO_x samples with or without additional SiN_x:H layer are investigated. Fig. 4 shows the resulting ΔN_{leq} and $-\Delta p$ of the fired samples passivated either with AlO_x only or with AlO_x/SiN_y:H. The AlO_x layers are deposited using ALD at 170°C and 300°C. Since the single APCVD AlO_x layers show very low effective lifetime, these samples were not further analysed. Also here, the normalization was performed at time t=0.

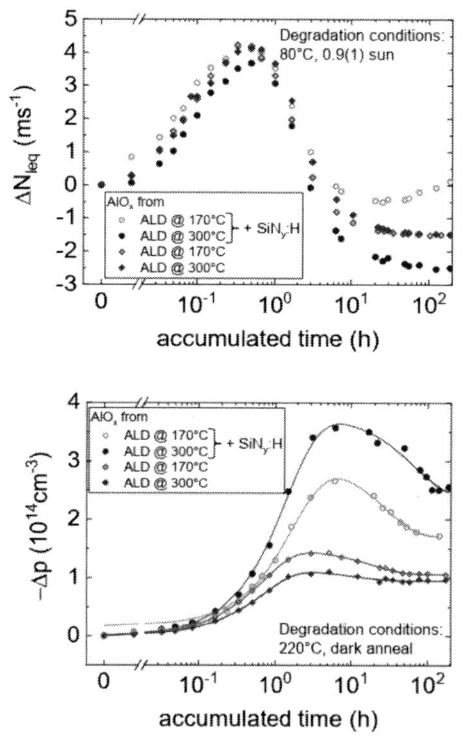

Figure 4: Lifetime equivalent defect density ΔN_{leq} (top) and change in hole concentration $-\Delta p$ (bottom) over accumulated time of single AlO_x and AlO_x/SiN_y:H layers. AlO_x layers deposited with ALD at 170°C and 300°C.

The ALD AlO_x samples with and without SiN_y:H show the same defect kinetics. During degradation the deposition temperature does not seem to have any influence on ΔN_{leq}. However, the presence of an additional SiN_y:H layer appears to affect the regeneration kinetics. A minimum of ΔN_{leq} is reached after less than 30 hours for AlO_x/SiN_y:H stack at an ALD deposition temperature of 170°C while the minimum is not reached within 200 hours for samples with single AlO_x layer and for the AlO_x/SiN_y:H stack at ALD deposition of 300°C. Thus, the deposition temperature seems to have an influence on the regeneration kinetics/behavior through the presence of a SiN_y:H layer. In the lower graph of Fig. 4 the changing hole concentration $-\Delta p$ is given. As can be seen, the maximum $-\Delta p$ for the AlO_x/SiN_y:H layer ($2\text{-}4\cdot10^{14}$ cm^{-3}) is higher than for the single AlO_x layer ($<1.5\cdot10^{14}$ cm^{-3}). Thus, as expected, an additional SiN_yH leads to more H in the Si bulk. Therefore, the additional SiN_y:H layer leads to a doubling of the hole concentration The deposition temperature does not appear to have a significant effect on $-\Delta p$.

4 CONCLUSION

Since the ALD AlO_x shows lower defect density ΔN_{leq} and changing hole concentration $-\Delta p$ compared to AlO_x deposited by APCVD, it seems that ALD AlO_x has a stronger barrier effect than APCVD AlO_x. Besides the barrier properties, a generally higher amount of H in the APCVD AlO_x layer compared to ALD AlO_x can also be an explanation. The influence of different deposition temperatures of ALD or APCVD show no significant influence on the degradation kinetic but on the regeneration behavior. Also, the samples with and without additional SiN_yH layer show comparable ΔN_{leq} and $-\Delta p$ values during degradation, but differ during regeneration. The influence of the temperature input during deposition is further described in [7]. Due to the possible influence of thermal donors, the BH measurements on FZ material seem to be more suitable.

5 ACKNOWLEDGEMENT

Part of this work was financially supported by the German Federal Ministry for Economic Affairs and Climate Action (FKZ 03EE1051C). The content is the responsibility of the authors.

6 REFERENCES

[1] M.A. Jensen et al., Journal of Applied Physics 124(8) (2018) 085701.
[2] J. Schmidt, D. Bredemeier, D. C. Walter, IEEE Journal of Photovoltaics 9(6) (2019) 1497.
[3] F. Kersten et al., Energy Procedia 92 (2016) 828.
[4] A.A. Dameron et al., Journal of Physical Chemistry C 112(12) (2008) 4573.
[5] A. Herguth, C. Winter, IEEE Journal of Photovoltaics 11(4) (2021) 1059.
[6] A. Schmid, et al., IEEE Journal of Photovoltaics 11(4) (2021) 967.
[7] F. Geml and M. Mehler et al., to be published, 2022.
[8] C. Winter et al., Physica Status Solidi A 218(23) (2021) 2100220.
[9] C. Sah et al., Applied Physics Letters 43 (1983) 204.

MODELLING OF EDDY-CURRENT MEASUREMENT FOR INHOMOGENEOUS CHARGE CARRIER DEPTH PROFILE

D. Krisztián[1,2,*], F. Korsós[1], M. Kovács[1], F. Ujhelyi[1], P. Tüttő[1]
[1]Semilab Co. Ltd. 4/A Prielle K. str., H-1117 Budapest, Hungary
[2]Budapest University of Technology and Economics, Hungary;
*Address: Prielle Kornélia str. 4/A. H-1117 Budapest, Hungary,
email: david.krisztian@semilab.hu, phone: +36 1 505 4690

ABSTRACT: We developed a simple one-dimensional model of the eddy-current measurement for samples with inhomogeneous carrier depth profile. The model is well suited for the evaluation of transients recorded from ingots or thick slices of silicon using radio frequency (RF) eddy-current based photoconductance decay (PCD) technique [1]. The three-dimensional distribution of the RF magnetic field, the induced eddy-currents and depth sensitivity of the eddy-current sensor was thoroughly analyzed using a finite element (FE) simulation. The experimental results confirmed the reliability of the 3D FE simulations to model the eddy-current measurement. Despite the three-dimensional nature of the problem, we found that in the case of laterally homogenous carrier profile, a simplified one-dimensional model can calculate the eddy-current signal with sufficient accuracy. The few calibration parameters of the model can be easily determined using homogenous resistivity etalon samples of different thicknesses. Such rapid calculation of the expected eddy-current signal enables to apply the model even for real-time RF-PCD measurement. To verify the reliability and applicability of the applied model, we injected a controlled density of excess carriers using a large spot laser illumination at various flux levels. The basic parameters to compute the excess carrier depth profiles was determined using an iterative method. The agreement between the measured, the calculated (1-D model) and simulated (3-D) eddy-current signals confirm the reliability of our model.

Keywords: Silicon, Recombination, Lifetime, Simulation, Ingot

1 INTRODUCTION

Eddy-current based photoconductance decay (e-PCD [1]) is the most commonly used measurement technique to characterize silicon ingots in the early phase of the solar cell manufacturing. Beyond single bulk lifetime τ_b value, its dependence on the excess carrier density Δn can provide further information about the quality of the final cells. Therefore, the final aim of our project is to determine the bulk lifetime in the function of injection level from a single e-PCD transient.

As it was recently published, a consistent carrier lifetime can be determined at $\Delta n = 10^{15}\ cm^{-3}$ using the e-PCD technique [2]. However, to resolve τ_b in the function Δn in the full accessible range, the proper determination of injection level is required. In this study, we introduce a model which is significantly enhances the accuracy of the calculated injection level.

In this approach, Δn is determined from the measured eddy current signal, and the time evolution of the signal is also considered. This calculation requires the detailed knowledge of the sensitivity of the radio frequency (RF) detector.

During a typical e-PCD measurement, a light source generates a strongly inhomogeneous $\Delta n(x)$ depth profile near the surface, which increases the eddy-current signal. A significant challenge is to calculate the average injection level Δn_{avg} from the signal, as the penetration depth of RF field and the depth of the $\Delta n(x)$ profile roughly coincide. The evaluation routine of commercially available RF-PCD systems applies a static depth sensitivity of eddy-current sensor. However, at high excitation levels, the Δn gets comparable to the doping density or even exceeds it. In this case, the skin depth and so the depth sensitivity varies significantly in the depth.

We aimed to develop a universal description of the depth sensitivity of the eddy-current signal, which enables to estimate Δn_{avg} during the lifetime measurement.

The approach is summarized in Fig.1. First, we developed a three-dimensional finite element (3D-FE) simulation of our eddy-current sensor tool. The instrumental parameters were determined using the results of eddy-current measurements from thick silicon slugs with known resistivity and without illumination.

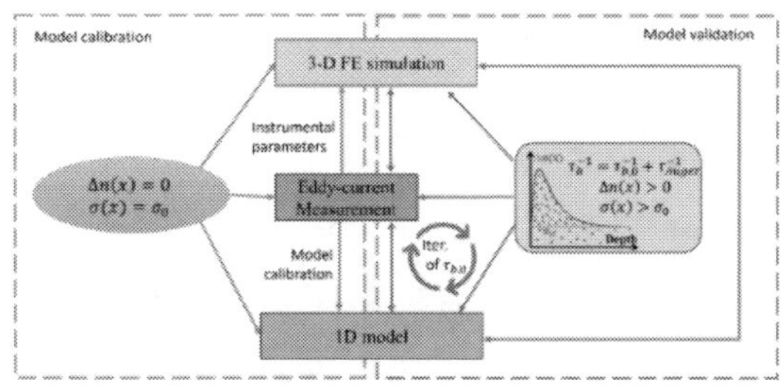

Figure 1: *The illustration of the approach to develop a simple 1D model to calculate the depth and injection level dependence of the eddy-current sensing. Results from resistivity etalons are used for the calibration, while samples illuminated by a 1064nm CW laser are used to validate the model.*

Then, we simulated steady state excess carrier profiles corresponding to a sample illuminated using a monochromatic steady-state light source. We applied a continuous wave (CW) laser for this purpose. The expected eddy-current signals were computed using the 3D-FE software. Although the results confirmed the accuracy of the simulation, this method is too time consuming to apply for the in-situ evaluation of the PCD transients. Therefore, we developed a simple 1-D approximate model based on the results of the 3-D simulations to overcome this limitation.

The sensitivity of eddy current used in e-PCD tools has

been already investigated in several studies. Black et al. showed that the thickness of silicon wafers must be considered as the sensitivity decreases within the sample [3]. He also proposed a model including a skin-effect based correction. However, it cannot be applied in case of thick samples. Swirhun et al. found a resistivity-dependent sensitivity depth of the RF coil in a commercial RF-PCD system [4]. They also noticed the problem of the decreased sensitivity depth of illuminated samples but made his simulations in low injection to avoid its distortion. We propose a more general depth sensitivity model, which considers the impact of the $\Delta n(x)$ depth profile, as well.

2 1-D AND 3-D SIMULATIONS

2.1 Calibration of the 3D-FE model

We modelled our eddy-current sensor in COMSOL finite element simulation software as a cylindrical symmetrical coil and the corresponding primary electrical components. The radius of the coil and its distance to the sample surface corresponds to the physical parameters of Semilab WT-1200IL e-PCD measurement system. We validated the reliability of the simulation by comparing the results to the measured values of some selected silicon slugs (with a width of 2-5 cm) from a wide resistivity range without illumination. For the simulation, the resistivity ρ is considered to be homogeneous with a ρ value provided by 4-point probe measurements. The simulated eddy-current and the measured signals correlate very well, the slope defines the instrumental factor (Fig. 2/a.).

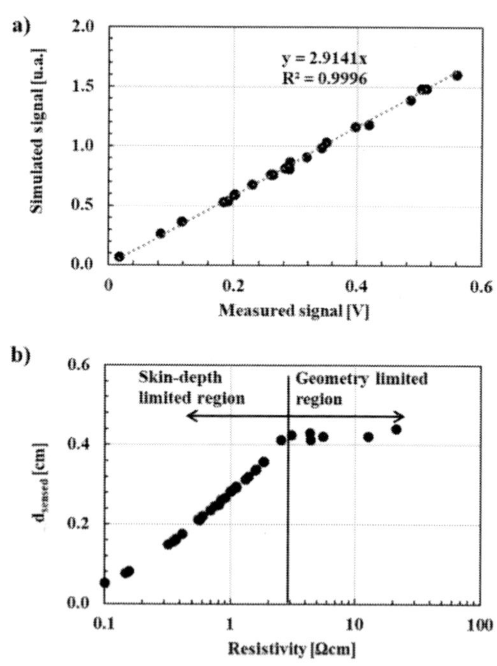

Figure 2: *The excellent correlation between the measured and the simulated (3D-FE COMSOL) signals validates the accuracy of the simulation (a). The effective depth of sensitivity can be either limited by the skin-depth or the geometry of the RF coil depending on the resistivity (b).*

In the homogeneous charge carrier density case, we can define the effective sensed depth, $d_{sense}(\sigma)$, as the ratio of sensed sheet conductance $\sigma_{\square,sensed}$ and the dark conductivity σ_0 of the sample:

$$d_{sense} = \frac{\sigma_{\square,sensed}}{\sigma_0},$$

where $\sigma_{\square,sensed}$ is calculated after calibration with lightly doped wafers, much thinner than the skin depth. Two distinct regions of d_{sense} can be observed in the function of resistivity in Fig. 2/b. Above $\rho = 3\ \Omega cm$ the finite size of the RF coil limits the penetration of the RF-field, for lower ρ values the skin effect becomes dominant.

2.2 Validation of the simple 1-D model

In the general form, we can define a 1-D depth sensitivity function, $\varphi(x,\sigma)$, as the ratio of sensed sheet conductance of very thin layers at each step in depth. To simplify the description, an "effective" value of d_{sense} is defined, which corresponds to the injected region with its average injection level.

$$\sigma_{\square,sensed} = \int_0^{\infty} \sigma(x)\cdot\varphi\big(x, d_{sense}(\sigma_0, \Delta n_{avg})\big)dx.$$

Our aim is to determine this $\varphi(x, d_{sense})$ function using the 3D-FE simulation, which calculates the eddy-current signal corresponding to the inhomogeneous $\Delta n(x)$ profile originating from the illumination during the e-PCD measurement. For this purpose, we created such depth profiles by solving the continuity equation of the excess carrier density using a 1-D simulation:

$$\frac{\partial \Delta n(x,t)}{\partial t} = G(x,t) - \frac{\Delta n(x,t)}{\tau_b} + D_n(\Delta n, x)\frac{\partial^2 \Delta n(x,t)}{\partial x^2}.$$

The $\Delta n(x)$ profiles (Fig. 3/a.) supposed to be laterally homogenous are used as input to the 3-D simulation and the 1-D calculation as well. The $\varphi(x,\sigma)$ can be acquired using the 3-D simulation by the analysis of the decay of the current densities in depth. In Fig. 3/b. a typical eddy-current density distribution corresponding to an illuminated sample is depicted.

These results are used for the finetuning of the 1-D model. The rather complex formula of $\varphi(x,\sigma)$ considers the change in the strength of the skin effect in the depth and the shift of the phase of the RF magnetic field, as well.

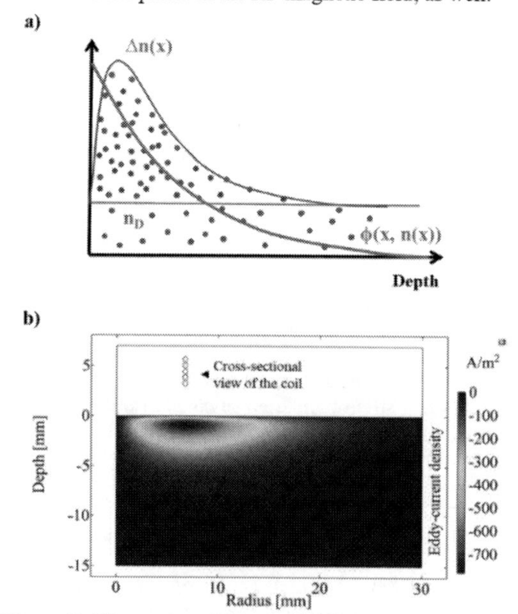

Figure 3: *Illustration of the simplified 1-D model of depth sensitivity. The penetration depth of RF field and the profile depth coincides. (a). Simulated eddy current densities in an illuminated sample (b).*

3 EXPERIMENTAL RESULTS

3.1 Comparison of the 1-D calculation and the 3-D simulation

To test the validity of the simplified 1-D model, we computed the σ(x) steady state profiles corresponding to a sample illuminated laterally homogenously by a CW laser. The bulk lifetime $\tau_{b,0}$ is varied in a broad range from 1 µs to 10 ms, and the Auger recombination was also considered, then the bulk lifetime:

$$\tau_b^{-1} = \tau_{b,0}^{-1} + \tau_{Auger}^{-1}.$$

As the excess sheet conductance increases for higher lifetimes the eddy-current signal flattens, since the sensed volume decreasing. It indicates the usefulness of a proper depth sensitivity function. The results of the very simple 1-D calculation agree very well to the 3-D simulation for moderate lifetimes, up to the *ms* range, but at the very high injection levels it shows a slight overestimation (Fig. 4/b.).

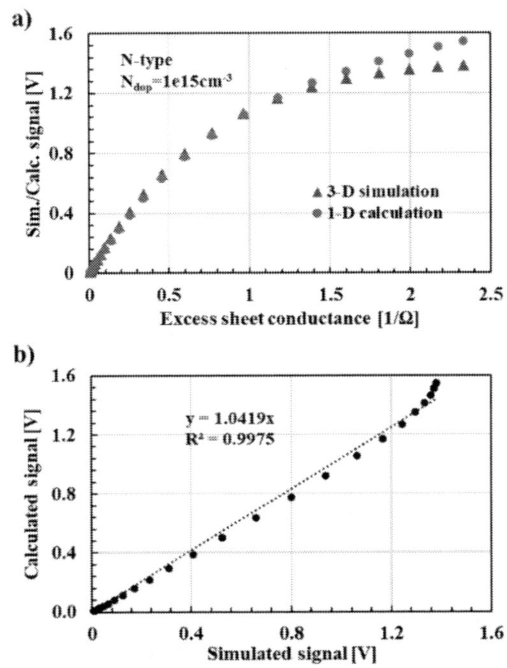

Figure 4: *The simulated and calculated signals show non-linear behavior with the excess sheet conductance due to dynamic skin-depth phenomenon (a). The 1-D calculated excess signals correlate well to the 3-D simulation results up to the ms carrier lifetime range (b).*

3.2 Comparison of the simulations and the measurement

We selected a Cz mono-Si slug suitable to form a deep and strongly varying σ(x) profile (n-type, $N_D = 1.1 \cdot 10^{15}\ cm^{-3}$, $\tau_{b,0} = 3\ ms$). We executed the simulations for different illumination intensities (generation rates). The resulting $\sigma(x)$ profiles are depicted on Fig. 4/a. At the high intensities, the peak σ value is over ten times larger compared to σ_0, so the penetration depth of the RF-field reduces significantly. In such cases, the involvement of $\sigma(x)$ profile is essential to calculate the depth sensitivity.

The corresponding excess signals were calculated with both the 3-D simulation and the based on the 1-D model. The $\sigma(x)$ simulation parameters were chosen to meet the properties of the actual sample.

The sample was measured using a calibrated WT-1200IL tool and adjusting the laser power according to the simulation. Fig. 4/b. depicts the excellent correlation of the 1-D model calculated, the 3-D simulated and the measured excess eddy-current signals. The slightly smaller measured signal level can be attributed to the minor inhomogeneity of the illumination.

This result confirms the capability of both the 1-D calculation and the 3D-FE simulation to describe the eddy-current sensing for PCD measurements.

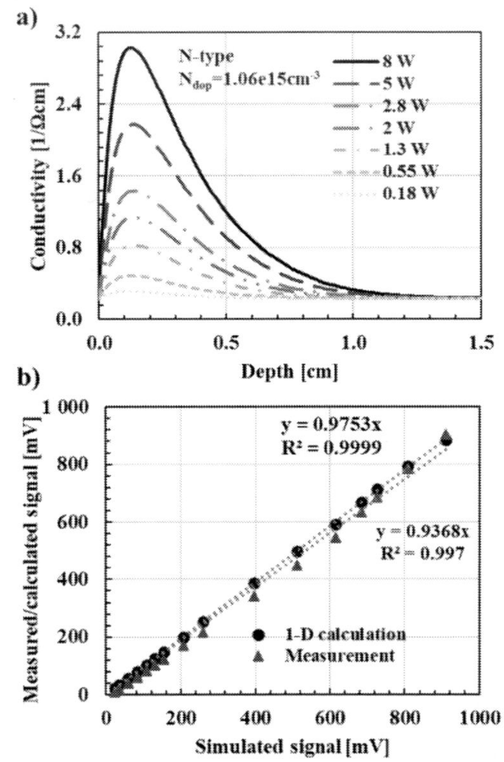

Figure 5: *Computed steady state σ(x) profiles with different generation rates are used to develop and validate the depth sensitivity model (a). The simulated excess signals are well correlating with both the calculated and measured results (b).*

4 CONCLUSIONS

We studied the eddy-current sensing for inhomogeneous carrier density depth profile experimentally and by computer simulations. We found that 3D finite element simulation is a very powerful method for the accurate computation of the eddy-current signal corresponding to any realistic $\Delta n(x)$ profiles during the PCD measurement. The simulated eddy-current density depth profiles helped us to develop a simple approximate 1-D model which describes the depth sensitivity sufficiently. Results of these 1-D calculations agreed very well with both the 3D-FE simulation and the measured eddy-current signals. Therefore, it is applicable for the more accurate evaluation of e-PCD measurements.

ACKNOWLEDGEMENT

We thank Ferenc Simon for useful discussions.

D. Krisztián acknowledge the support of the doctoral student scholarship program of the co-operative doctoral program of the Ministry for Innovation and Technology from the source of the National Research, Development and Innovation Fund.

4.3 References

[1] G. Paráda, et al., "Transient method for lifetime characterization of monocrystalline Si ingots", proceedings of the 29th EU-PVSEC, pp. 1264 - 1266 (2014).

[2] D. Krisztián et al., "Improved accuracy of eddy-current sensor based carrier lifetime measurement using laser excitation", EPJ Photovoltaics 13, 3 (2022).

[3] L. E. Black and D. H. Macdonald, "Accounting for the Dependence of Coil Sensitivity on Sample Thickness and Lift-Off in Inductively Coupled Photoconductance Measurements", IEEE Journal of Photovoltaics 9 (6), 1563 - 1574, (2019).

[4] J.S. Swirhun et al., "Contactless measurement of minority carrier lifetime in silicon ingots and bricks", Prog. Photovolt: Res. Appl., 19: 313-319. (2011)

EVALUATION OF THE EFFECT OF SURFACE STRUCTURE ON FLEXIBILITY OF SI WAFERS FOR FLEXIBLE C-SI SOLAR CELLS

Koki Ide[1], Yutaka Hara[1], Tappei Nishihara[1,6], Ryo Yokogawa[1,2], Kyotaro Nakamura[3], Yoshio Ohshita[3], Tomoyuki Kawatsu[4], Toshiki Nagai[4], Noboru Yamada[5], Yukio Miyashita[5], Yuma Aoki[5], Hayato Kobayashi[5], and Atsushi Ogura[1,2]
1. School of Science and Technology, Meiji University., 2.Meiji Renewable Energy Laboratory, Meiji University
3.Toyota Technological Institute., 4.Komatsu NTC Ltd., 5.Nagaoka University of Technology, 6.JSPS Research Fellow DC2
ce221011@meiji.ac.jp

Abstract – **The Ball-on-Ring test, a type of biaxial bending test that can evaluate the three dimensions flexibility using the maximum bending deflection at crystalline Si wafer fracture. The wafer flexibility was improved by slice damage removal etching process, significantly. And texture formation on the wafer surface decreased flexibility. Furthermore, we found that the flexibility was recovered by mixed acid etching. However, the recovery was not completed as good as before the texturing. SEM observation of the area near the fracture starting point of these samples showed that fracture proceeded along the valley of the pyramidal structure. Further magnification showed that the valleys were rounded off by mixed acid etching. The rounded valleys of the textured structure, mitigated the fracture. Meanwhile reflectance measurements showed that the mixed acid etching had no significant effect on reflectance.**

Keywords – c-Si, Silicon solar cells Flexible substrate, Module, Texturisation,wafering

1. INTRODUCTION

In recent years, the introduction of photovoltaic into new fields such as vehicle are attracting growing attention. In addition to light weight, flexibility is required from the viewpoint of design. In particular, since aerodynamic performance is directly related to fuel efficiency in automobiles, 3D curved surfaces are frequently used in the body of the vehicle, and it is essential to develop solar cells with high stability and conversion efficiency, CIGS solar cells and perovskite solar cells have attracted attention as flexible solar cells candidates [1,2]. On the other hand, crystalline Si (c-Si) solar cells have higher stability and conversion efficiency. We have reported on the improvement of flexibility by reducing the wafer thickness as shown in Fig. 1, indicating that c-Si solar cells are expected to be flexible solar cell [3]. It has been recognized that the texture structure of the Si wafer surface to improve the light absorption affects the mechanical properties while the effect of the wafer surface structure on the three-dimensional flexibility has not yet been clarified sufficiently [4].

In this study, in order to evaluate the effects of the wafer surface structure on the flexibility, we performed the Ball-on-Ring test, which is a kind of biaxial bending test that can evaluate three-dimensional flexibility, on Si wafers with various surface structures.

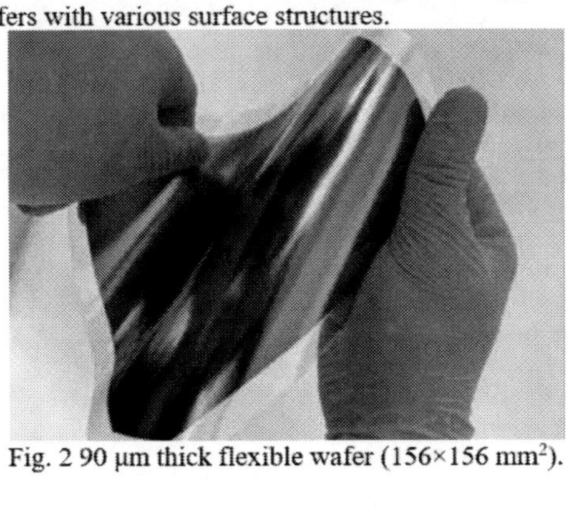

Fig. 2 90 μm thick flexible wafer (156×156 mm²).

2. EXPERIMENTS

2.1. Sample fabrication procedure

We prepared multi-wire sliced Si(100) wafer with the area of 156×156 mm² (here after "As-sliced"). Then, the damage removal etching with 20 wt% KOH at 70°C for 10 min was performed on "As-sliced" ("Etched"). To the texture formation, single side of "Etched" was protected by SiO_2 film deposited by atmospheric pressure CVD system (APCVD) to texturing etch on only single side, and texturing etching was performed on it with 3 wt% KOH solution including some additives at 80°C for 10 min ("Textured"). After texture formation on the single side, "Textured" were then etched by mixed acid etched with HNO_3:HF (50:1) at room temperature for 20 s and 120 s to round off the pyramidal structure (Rounded) in order to improve the wafer flexibility.

2.2. Wafer flexibility in 3D

To evaluate the flexibility of each sample, quantitatively a Ball-on-Ring test, a type of biaxial bending test that can evaluate bending in three dimensions, was conducted as shown in Fig. 1. The force was applied at 1 mm/min to 180 μm-thick sample center, the radius of the ring was 44 mm, the point where the force is applied is at the center of the wafer 78 mm from the wafer edge, and the radius of the loading ball was 4.76 mm. The flexibility was evaluated as the deflection to fracture. In addition, the results are shown as the radius of curvature. The following formulas (2.1) and (2.2) were used to calculate the radius of curvature.

$$r^2 = (r-d)^2 + 44^2 \ldots (2.1)$$

$$r = \frac{d}{2} + \frac{44^2}{2d} \ldots (2.2)$$

r : radius of curvature, d : deflection to fracture.

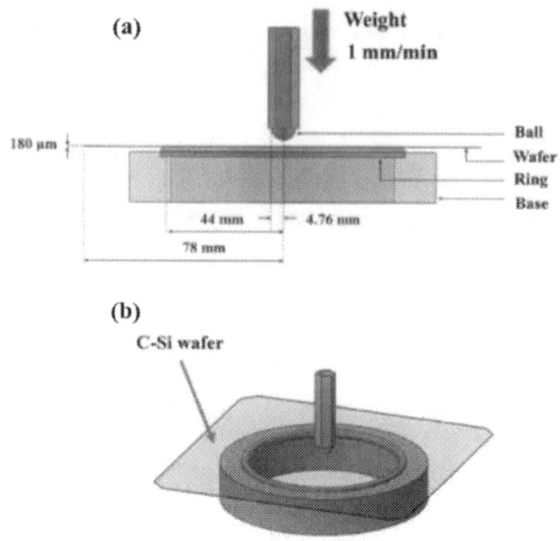

Fig. 1. Schematics of Ball-on-Ring test.
(a)Side view, (b)Top view

To clarify the mechanism of the change in flexibility, the area near the fracture starting point of each sample that was destroyed by the Ball-on-Ring test was observed by scanning electron microscope (SEM). In addition, the pyramid structures were observed before and after mixed acid etching in detail.

2.3. Reflectance measurement

We measured the reflectance of Textured, Rounded (20 s) and Rounded (120 s) by UV-visible and near-infrared spectrophotometer to evaluate the change in reflectance due to mixed acid etching on textured structures. Textured structures have a role in improving light absorption to the Si wafer. Therefore, we investigated the optical effects of the change in texture shape by the mixed acid etching and examine the application to the flexible c-Si solar cells. In addition, a comparison of light reflectance was compared among the different in surface structure.

3. RESULTS AND DISCUSSION

3.1. Flexibility evaluation using Ball-on-Ring test

As shown in Fig. 2, we confirmed the well-sliced and damage-etched 90 µm-thick Si is remarkably flexible. In order to evaluate the flexibility and the effect of the surface structure quantitatively, we carried out the Ball-on-Ring test for the approximately 150 µm-thick single side textured Si wafers with different surface structure.

Fig. 2 90 µm thick flexible wafer (156×156 mm^2).

3.2. Flexibility evaluation using Ball-on-Ring test

We applied force from a flat surface with no texture. Figure 3 shows the curvature radius calculated from the results of Ball-on-Ring test for samples with difference surface structure. And Fig. 4 shows the fractured Si wafer with different surface structure after the Ball-on-Ring test. From Fig. 3, compare to "As-sliced" and "Etched", it is apparent the "Etched" flexibility is improved significantly and achieving a curvature radius of 0.3 m. Because the damage removal etching on the as-sliced wafers removed cracks and other defects that could be the starting point of fracture. In this case, the wafer was broken into small pieces (Fig. 4(a)). On the other hand, the flexibility of "Textured" was less than the "As-sliced" level, and the wafer was broken into four large pieces (Fig. 4(b)) along the <110> crystalline orientation which is known to be cleavage direction. "Rounded" was improved in the flexibility compared to "Textured". The longer the etching time of the mixed acid etching, the more the flexibility was improved, indicating that the mixed acid etching can improve the flexibility. After 120 sec rounding etching, the flexibility was significantly recovered although there were still slight degradation from the "Etched" sample. At this time, the wafer was broken into more pieces than the "Textured" (Fig. 4(c)).

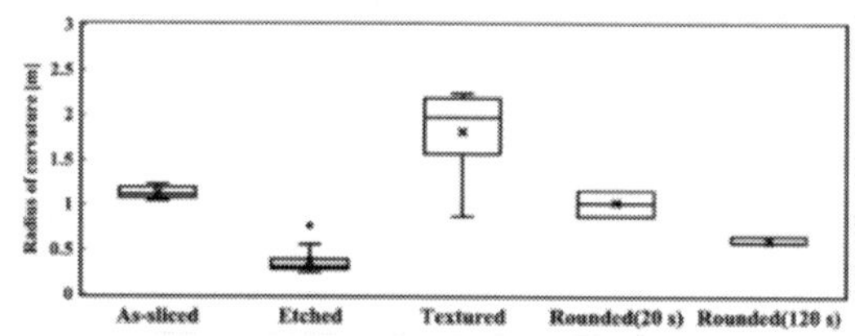

Fig. 3. Results of the Ball-on-Ring test.

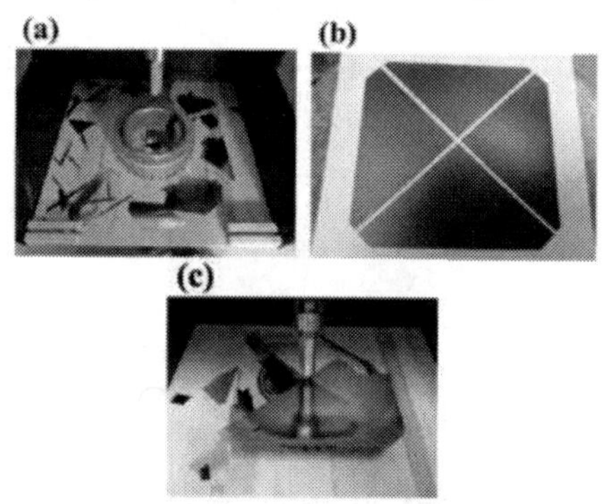

Fig. 4. The images of fractured Si wafers after Ball-on-Ring test.
(a)Etched, (b)Textured, and (c)Rounded (120 s)

3.3. Structures observation by SEM

To clarify the mechanism of flexibility improvement, the area near the fracture starting point of each sample was observed by SEM. Figure 5 shows that the destruction is proceeds along the valleys of the pyramidal structure. This is because the valleys of the pyramidal structure are along the <110> crystalline orientation which is known to be cleavage direction. In particular, "Textured" is particularly progressed to be fractured along the valley of the pyramidal structure.

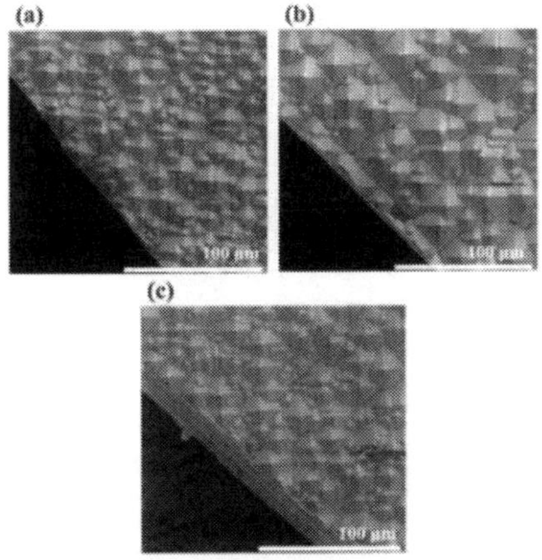

Fig. 5. SEM images near the fracture starting point of each sample.
(a)Textured, (b)Rounded (20 s), (c)Rounded (120 s)

The average sample thickness of each sample are 148.5 μm (Textured), 152.5 μm (Rounded (20 s)), and 145 μm (Rounded (120 s)), respectively. Thus, the process provided relatively small variation within approximately 3.5 μm between the "Textured" and the "Rounded (120 s)". As forhe texture size, we consider that mixed acid etching consumed the smaller size textures first and thus the average texture size might increase by mixture acid etching as well as the rounding effect at the both top and bottom corner of the texture.

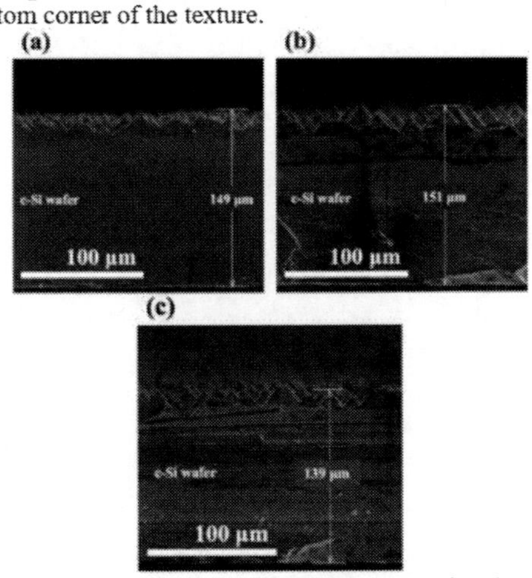

Fig. 6. Texture size observation by cross-sectional SEM. (a)Textured, (b)Rounded (20 s), and (c)Rounded (120 s)

Further magnified images of the valley are shown in Fig. 7. From Fig. 7, valleys of "Rounded (20 s, 120 s)" are slightly rounded. The reason for the improvement in flexibility by mixed acid etching is that the rounding of the valleys disperses the stresses applied to the valleys.

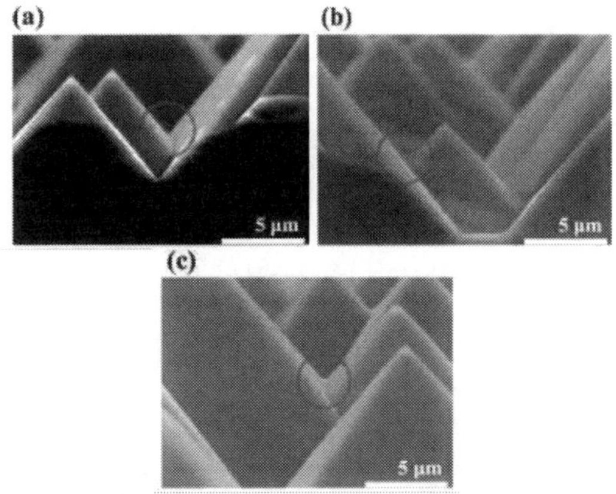

Fig. 7. Enlarged view of the pyramid structure of each sample. (a)Textured, (b)Rounded (20 s), (c)Rounded (120 s)

3.4. Reflectance measurement

Figure 8 shows the results of reflectance measurements. There was only slight change in reflectance between "Textured", "Rounded (20 s)" and "Rounded (120 s)". Therefore, there might be a few effect on reflectance by mixed acid etching.

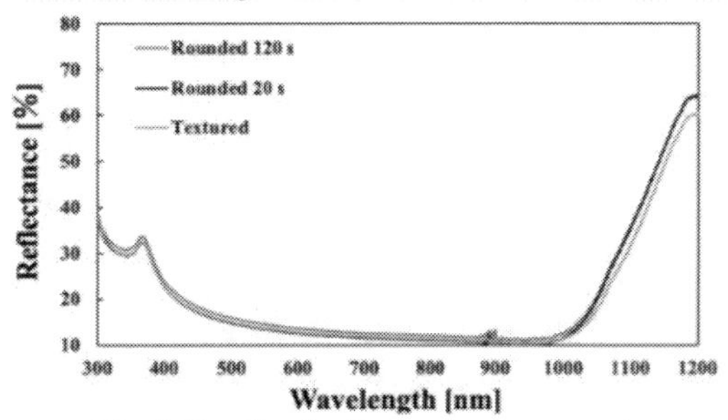

Fig. 8. Results of reflectance measurements.

4. CONCLUSIONS

In order to realize a c-Si solar cell that can be applied for the three-dimensional curved surfaces, we performed three-dimensional bending tests on Si wafers with different surface structures (As-sliced, Etched, Textured, and Rounded), to evaluate their flexibility, observed by SEM, and measured reflectance. We found that damage removal etching improved the flexibility of the wafers, but texture formation decreased the flexibility. However, flexibility was recovered by rounding off the valleys of pyramid structure by mixed acid etching. The minimum radius of curvature for each sample is 0.25 m (Etched), 0.90 m (Textured), and 0.60 m (Rounded), respectively. Furthermore, SEM observation of the area near the fracture starting point of each sample shows that the reason for the reduced flexibility of "Textured" is that the valleys are along the direction of the crystalline cleavage. And the detail observation of the textured structure shows the valleys of the pyramid structure are slightly rounded by the mixed acid etching. The rounding of the valleys disperses the stresses on the valleys and improves wafer flexibility. Moreover, reflectance was rarely affected by the mixed acid etching. From these results, the flexibility for three-dimensional curved surfaces was confirmed, indicating the possibility of realizing flexible c-Si solar cells that can be used for three-dimensional curved surfaces.

5. ACKNOWLEDGMENT

This study was supported by New Energy and Industrial Technology Development Organization (NEDO).

6. REFERENCES

[1] P. Jackson et aI., Progr. Photovoltaics, 19, 894-897 (2011).
[2] Cristina Rolbán-Carmona et al., Energy Environ. Sci., 7, 994-997 (2014).
[3] K. Onishi et al., Proc. of 36th EUPVSEC, 126-130 (2019).
[4] J.H. Woo et al., Scripta Materialia 140, 1-4 (2017).

A NEW TECHNOLOGICAL AND ENVIRONMENTALLY SUSTAINABLE LONG-TERM SUPPLY CHAIN SOLUTION FOR THE MASS PRODUCTION OF HIGH PURITY SILICA FROM CANADA

Weeden P.Eng., Laura, Bullen MBA, Brent, Needham P.Geo., Troy
Sio Silica Corporation
1930, 440 2nd Ave SW Calgary, Alberta T2P 5E9 Canada

ABSTRACT: A cost effective and sustainable ESG (Environmental, Social and Governance) compliant supply of high purity silica is required to advance photovoltaic technology moving to net zero 2050 targets. Although silica is one of the most abundant minerals on Earth, it is the rarer pure silica which is in the highest demand for photovoltaics and emission reduction efforts. Discovered in Manitoba, Canada, and approximately 55 m below surface exists an unconsolidated, freshwater aquifer formation of high purity silica sand. Once brought to surface by an environmentally friendly patent pending extraction process, and passing through a simple, chemical free process, the purity averages 99.9% with zero ppm to trace amounts of Boron and Phosphorous. Continued lab testing and processing studies have demonstrated successful beneficiation results of 99.99%. The Extraction Method utilizes a perforated Extraction Apparatus in a wellbore and air circulation to create a low-pressure zone allowing silica sand to flow into the apparatus and to surface. This extraction process eliminates many of the environmental impacts and excessive costs of open-pit mining. With over 13 billion tonnes of high purity silica in place, and a focus on environmental stewardship and sustainable mining, this unique silica sand mine will contribute to the worldwide effort to reduce emissions.
Keywords: PV Materials, Polycrystalline, Silicon, SiO_2, Feedstock.

1 INTRODUCTION

A new high purity silica sand mine has been proposed in southeastern Manitoba, Canada. Once constructed it will extract high purity silica sand from 55 m below ground in an unconsolidated sandstone, freshwater aquifer using a newly developed environmentally friendly patent pending mining method. A proposed sand processing facility will wash, dry and magnetically separate this sand bringing it to 99.9% purity levels and suitable for feedstock inputs into photovoltaics such as polysilicon, ingots, and solar grade glass, including transparent solar glass, among other industries such as, ceramics, medical glass, low iron glass, coatings and others. This method and facility design reduces the overall impact of mining and manufacturing raw materials for Photovoltaic manufacturing.

This paper provides an outline of the silica resource that has been discovered and its geological properties. It will explore the unique high purity case of such a deposit and the potential for feedstock into many different applications. It will also describe the historical work that has been undertaken to examine and prove purity capabilities with examples of intermediary products and where this feedstock can be used. The developed mining method (Extraction Method) is thoroughly explained and the potential environmental impact of this new method and processing design as it compares to traditional open-pit mining is discussed.

Overall, this paper seeks to share new innovation in silica sand mining and potential for feedstock in key PV manufacturing and technology development for energy conversion.

2 SILICA RESOURCE

2.1 Regional Geology

The unique high purity silica sand resource is found in southeastern Manitoba, Canada in a confined aquifer approximately 55 m below ground. This formation, known as the Winnipeg Sandstone, specifically the Carman Sand Member is of Ordovician age and covers a vast area trending east-west with a maximum recorded thickness of 63.4 m [1]. The Carman Sand Member is relatively continuous and uniform with well-rounded grains of high purity quartz silica and is largely unconsolidated. The sand body extends approximately 240 km from Sandilands Provincial Forest and pinches out to the west to Pelican Lake, MB. The width of the sand body increases easterly towards the Sandilands Provincial Forest from less than 25 km to greater than 95 km [1].

Overlying the Winnipeg Sandstone Formation are interbedded layers of Red River Shale. These sediments were deposited during the Middle Ordovician, in shallow marine seas. The shales are generally light olive-grey in colour, kaolinitic, with variable sand and silt content [1]. This shale layer, averages 2.4 m – 3 m thick, and acts as an aquitard between the Winnipeg Sandstone Formation below and the Red River Carbonate aquifer above.

The Red River Carbonate is of Ordovician age and ranges in thickness in the area up to 60 m. This Carbonate aquifer is the most common target for water wells in the area because of its high-quality water. Typically it does not require a well screen, and due to its shallower nature, has lower costs associated for construction. The Red River Carbonate is also significant for its structural properties and ability to act as a competent cap rock above the Winnipeg Sandstone formation when silica sand is extracted below it.

Above the Red River Formation is Quaternary diamicton, which ranges in composition from silty to rocky, and is calcareous [2].

A lithology overview is provided below in Table I for the local area.

Table I: Local Lithology

Eon	Era	Period	Geologic Unit	Member	Lithology
Phanerozoic	Cenozoic	Quaternary			Diamicton
	Paleozoic	Ordovician	Red River Formation	Selkirk, Cat Head, Dog Head members	Carbonate
					Shale
			Winnipeg Formation	Carman Sand Member	Sand
				Equiv. Ice Box Member	Shale
				Black Island Member	Sand
Archean					Granitoid

2.2 Resource Development

The Winnipeg Sandstone Formation is a vast formation containing high purity silica sand. The project has carried out ongoing drilling exploration programs over many years, drilling as many as 89 exploration wells within a 40 km by 70 km area of the resource area. Through the exploration activities it has been determined that the resource for all staked claims contains over 10 billion tonnes of sand and an estimated 7,051 Mm³ in volume. More specifically, based on recoverable sand volumes, the claimed area contains 66,000,000 tonnes of indicated resource of sand and 465,000,000 tonnes inferred resource of sand. This resource accounts for 406 mineral claims and covers an area of 93,324 hectares of land.

During the exploration drilling programs samples are collected at intervals and put through various levels of testing to determine silica purity. Raw unwashed purity ranges in the exploration area from 99.74% to 99.87% normalized (lab specific WDXRF) (40/70) and the majority (97%) of the particle sizes fall within 40/140 mesh size (420 μm to 105 μm) which is well within the range for PV feedstock requirements. See Figure 1 below for Average Sieve Distribution.

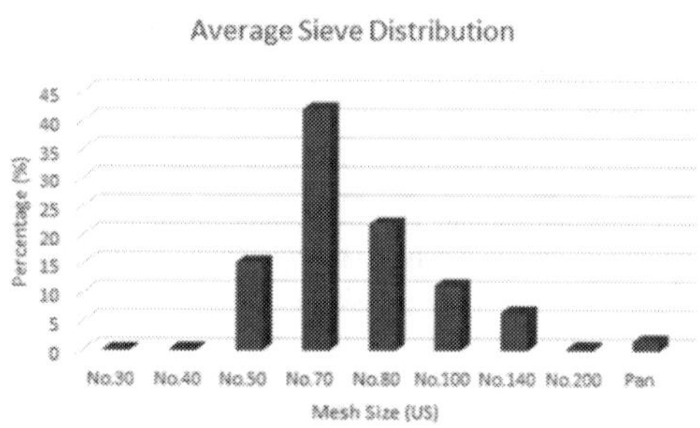

Figure 1: Average Sieve Distribution.

With a planned facility to process 1.36 Mt per year of this high purity silica sand, this resource has the capabilities to provide high purity silica as a feedstock for more than 350 years and make a significant contribution to the net zero targets of the planet in both the long term and short term.

3 SILICA SAND CHARACTERISTICS AND QUALITY

Traditionally, high purity silica is mined and prepared as feedstock into many different products by mining out large quartz blocks, crushing and beneficiation consisting of many steps to achieve a 99.9% purity or higher. What makes this silica sand resource unique is that the purity needed to reach required feedstock levels can be achieved by simple beneficiation steps. The silica sand found in this resource area in the Winnipeg Sandstone is extracted raw and then processed utilizing simple methods with no chemicals required to clean the sand bringing the purity to an average of 99.9% (ICP-OES purity testing).

3.1 Determining Purity

Since 2019 Sio Silica Corporation (Sio Silica) has been working with various labs to explore and prove the full extent of possibilities of purity for the silica sand found in the Winnipeg Sandstone. When the sand comes to surface via a patented Extraction Method, it comes to surface as raw sand grains with water from the aquifer. This raw sand, unwashed or treated in anyway, historically yields testing results utilizing a Wavelength Dispersive X-Ray Fluorescence spectroscopy (WDXRF) lab procedure for determining purity between 99.74% and 99.87%.

More in-depth testing has determined that the grains contain minor contaminations such as aluminum and iron. These minor contaminations are often surface contaminations and can be removed using a simple water wash across a dewatering screen using high pressure water. The sand can then be passed through a magnetic separator to further remove trace contaminants. This process has been found to bring the purity up to an average of 99.9% utilizing Inductively-Coupled Plasma Optical Emission Spectrometry (ICP - OES) Aqua Regia digestion normalized purity testing. Testing shows it also reduces iron concentrations below 100 ppm, which is often a minimum specification requirement for feedstock consideration. See Figure 2 for a sample after simple wash and magnetic separation and Figure 3 for an example of contaminant particles removed during magnetic separation.

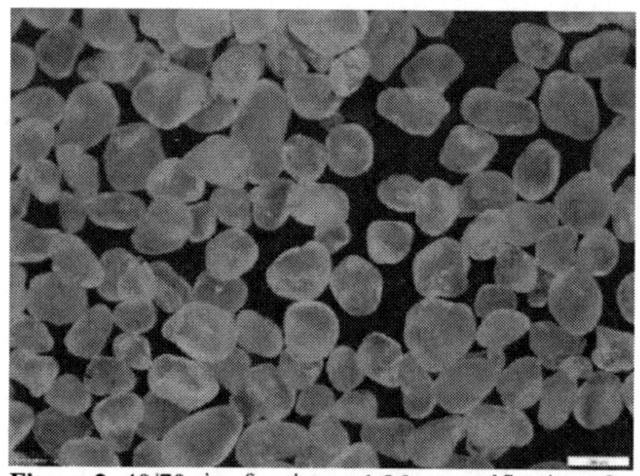

Figure 2: 40/70 size fraction at 1.26x magnification after wash and magnetic separation. Scale 500 μm.

Figure 3: Magnetically separated contaminants from waste stream. 1.89x magnification and scale 200 μm.

By utilizing an ICP – OES Aqua Regia digestion purity testing method, impurities can be determined at the most accurate level possible. Table II below demonstrates an example of ICP results following a wash and magnetic separation of a typical sample. Of note, reduction of Fe (iron), Al (aluminum) in particular and none detected of B (boron) or P (phosphorous).

Elemental Value (ppm)	Washed	Wash & Magnetically Separated
P (ppm)	ND	ND
S (ppm)	86.2673	79.3798
Mg (ppm)	38.1602	27.5699
B (ppm)	ND	ND
Fe (ppm)	241.7772	39.2854
Al (ppm)	467.4101	308.4654
Ca (ppm)	96.0379	90.3295
Ni (ppm)	0.4259	0.1589
Ti (ppm)	4.8859	2.7793
K (ppm)	17.5805	13.6377
Na (ppm)	11.2899	14.1718
SiO_2 (%)	99.9019	99.9413

Table II: Washed vs washed and magnetically separated ICP – OES Aqua Regia digestion normalized testing results.

This simple processing method utilizes no chemical treatments of any further beneficiation yet yields a 99.9% purity silica sand. This level of purity is considered High Purity Quartz (HPQ) and within spec for feedstock into solar grade glass including transparent solar glass, medical glass, smart glass, ceramics, coatings, paints, and silicon metal, which is an input product for the polysilicon and wafers needed for various types of solar panels.

3.2 Beneficiation to 99.99% - 4N purity

Once 99.9% was achieved using a simple process, further testing beneficiation methods have been explored to increase purity beyond 99.9% targeting 99.99%, otherwise known as 4N. To date, a 4N level purity has been reached utilizing different laboratory testing.

Testing was conducted to determine if utilizing an organic based descaler, can further reduce the trace elements found on the sand grains. This type of descaler is organic and much more environmentally friendly than other inorganic alternatives. It has been proven that this can increase the purity, although, often just as effective as a quality water wash and magnetic separation in many cases.

Therefore, to further beneficiate the silica sand it has been found that a common concentration combination of 12% HCl with 3% HF (Hydrofluoric) acid wash is effective in removing other trace elements to increase purity to a 4N classification level. ICP-OES normalized testing results yielded 99.9932% SiO_2 with Al <1.5 ppm and Fe <10ppm.

This 4N level of silica purity is required for epoxy molding compounds (EMCs), resins, optical glass, and silicon metal feedstock, polysilicon, or other intermediary product inputs.

It is currently being investigated if a 5N level purity could be achieved. This would involve a final beneficiation technique utilizing high temperature chlorination [3] and is still being investigated. Crucibles are the main output product utilizing 5N and significant for their usage in many applications, including smelting and production of silicon ingots [4]. Crucibles used to form silicon ingots must be replaced every six to eight ingot cycles. Therefore, as demand for ingots continues to increase, so will the demand for crucibles [5]. This is significant, as crucibles are used in the formation of ingots for monocrystalline solar panels.

3.3 Product Conversion

Silica sand, or silicon can be utilized in many different applications. For higher purity applications it must first be converted to an intermediary product such as silicon metal, cristobalite, or silicon carbide. These intermediary products are used as inputs into dental products, coatings, polysilicon, ingots and wafers.

Typical methods for forming silicon metals involve an alternating current electric arc furnace and a carbon source (often petroleum coke) combined with lump quartz (minimum 95% purity or more) [6]. To date, Sio Silica has been able to work with a lab to successfully create silicon metal by providing washed, dried, magnetically separated and descaled silica sand utilizing a proprietary direct current arc furnace design instead. See Figure 4 below for a photo of the silicon metal piece shortly after its formation.

Figure 4: Silicon metal formed using a DC electric arc furnace with extracted, washed, dried magnetically separated and descaled silica sand.

Cristobalite is also an intermediary feedstock that can be created by superheating (1500 °C) silica sand altering the structure of the grains, but not the chemistry (also known as a polymorph). By working with the University of Alberta, Sio Silica's high purity silica sand has been converted into cristobalite. Cristobalite is used in filling agents, paints, adhesives and compounds.

4 SILICA SAND EXTRACTION

With the targeted silica sand more then 55 m below the ground surface, traditional mining methods of open-pit mines are not suitable to mine the sand. A new mining method had to be developed to mine sand efficiently and safely from the Winnipeg Sandstone Formation without harming or impacting the aquifer that many rely on in the area for their water. By applying a combination of Darcy's Law, with traditional water well drilling methods and a movable production piping system, a new mining method was developed (Extraction Method) and is now patent pending.

4.1 Well Design

The Extraction Method begins with a vertical water well. The design is a modified version of a traditional water well which utilizes an outer PVC casing installed to the top of the carbonate aquifer utilizing a Dual Rotary drilling method for install. A second smaller PVC casing is then installed from surface to a depth of approximately 58 m and approximately 3 – 5 m into the unconsolidated sandstone. This casing is then cemented in place along the sandstone, shale, limestone interface with a 10 – 12 m annular cement plug (otherwise known as a seal). It is critical that a quality cement plug is placed to prevent inter-aquifer connection. Cement is conveyed using 25 mm PVC tremie line along the annulus in a mixture of cement, bentonite and calcium chloride to reduce cement cure time. Bentonite allows for some flexibility since cement is quite brittle. The casing is installed in this manner to allow for its removal during well abandonment and protect against any surface contamination possibilities.

Once the well is no longer active, the Extraction Apparatus is removed, and the well is abandoned. Well abandonment is conducted following strict environmental regulations and company policies and must ensure no potential for contamination of either aquifer. Materials such as cement, bentonite, pea gravel and drill cuttings are utilized in the process. Most Extraction sites are either previously disturbed areas, such as gravel quarries, or privately owned grazing and shrub lands of local residents. Therefore, once the well is abandoned, the site is reclaimed and seeded or re-planted to revegetate the disturbed site to the same or better state then it was prior to activities.

4.2 Extraction Apparatus

The Extraction Apparatus consists of a 200 mm steel production pipe suspended from a fit for purpose rig. The 200 mm production pipe has 25 – 50 mm in diameter perforations from the end of the casing for approximately 3m. This set of perforations stems from the well-known Darcy's Law. Thereby increasing the surface area proportionally to the pipe size to encourage a minimum flow rate into the pipe to move a grain of sand but not overwhelm the system which could result in plugging or restricting flow that would otherwise starve the well.

Inside the 200 mm production pipe a 50 mm steel airline with a drill bit is lowered to rest just above the perforations. This airline is fed by a dry screw compressor on surface and the pressure and CFMs (cubic feet per minute) are controlled on surface to not overwhelm the system or create the potential for plugging.

The combination of the outer steel production pipe and the inner airline is the Extraction Apparatus. This entire apparatus is capable of moving up and down in the wellbore to adjust the depth at which sand is mined. Additionally, both components are designed to be manipulated independently of each other for increased efficiency.

4.3 Extraction Method

Combining the use of the Extraction Apparatus and the addition of air through the 50 mm airline placed at the top of the perforations a circulation of air and water is created. As the airline is inside the casing, the air does not have anywhere to go except back up the wellbore in the annular space between the casing and the 50 mm airline. The air and water circulates up to surface and leaves the wellbore through a robust hose design. This action of circulating the air increases the ΔP between the natural formation pressure and the inside of the 200 mm production pipe by drawing down the water level in the well and thereby creating a lower pressure zone inside the casing compared to outside the production pipe. As the pressure is higher outside the wellbore, the water and unconsolidated sand begin to flow into the wellbore through the perforations and open end of the production pipe. The continuous circulation of air carries the sand and water to the surface.

Once on surface the water and sand are separated on site from each other using a dewatering screen. The sand is then transported to the facility for processing and transportation to market. The water remains at the well site where it is filtered, treated with ultraviolet (UV) light as a precautionary step to aid in the destruction of microbes, and then re-injected into an active extraction well via the annular space between the 200 mm production pipe and the cemented larger wellbore casing. No pressure is required for re-injection. This practice allows for a close to net zero impact on the drawdown effects of water removal in the aquifer and removes the need to discharge any water on surface. Figure 5 below illustrates a well cross section of the extraction method.

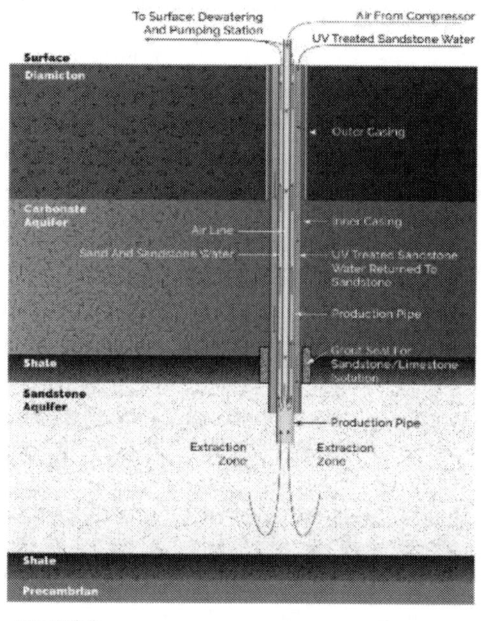

Figure 5: Cross sectional view of Extraction Method components and well.

5 ENVIRONMENTAL FOOTPRINT REDUCTION

5.1 Manufacturing Green Products

The target input for this final high purity silica sand are green products and electronics. With modern day society's push to net zero by 2050, the footprint of raw minerals mining and feedstocks into manufacturing solutions must be considered. This mining method and minimal processing will have an impact on reducing the footprint and energy required to mine and manufacture products needed now and for the future to get to net zero 2050. According to the IAE's May 2022 forecast report, renewable capacity is expected to increase 8% in 2022 and 60% of that increase is expected to come from solar PV. [7] As the demand continues to rise for PVs so does the demand for silica as an input product for manufacturing components, specifically polysilicon, ingots and solar grade glass including transparent solar glass, making up anywhere from 46 to 85% of a solar panel depending on the type of panel. [5]

5.2 Reducing Mining Emissions Footprint

Due to the depth of the target sand and the unique mining method, there are many opportunities for non-traditional mining activities to reduce the environmental footprint. Unlike traditional mining, there is no open pit mine, eliminating the need for explosives and truck traffic moving sand out of an open pit.

This type of mining method also allows for recycling of materials. The casing that is installed at the time of well drilling, is utilized until the well is abandoned, at which time, 65% or more of the PVC casing is removed to be reused in another well. This practice reduces overall waste of well materials during operations.

In addition, since the sand is extracted through a casing and the extracted sand comes to surface with water, the sand remains wet. The water that is extracted with the sand is separated from the sand, as described in Section 4.3, and returned to the aquifer resulting in a near net zero water drawdown except for a small percentage of water remaining on the sand grains prior to sand transportation to the facility.

Sand is transported to the facility for processing via a high-density polyethylene (HDPE) slurry line. Therefore, the usual sand mine and transportation dust risks are eliminated by keeping the sand in water throughout this entire process. This reduces the land footprint, dust from dry sand and truck traffic, and exposure risk to silica is reduced to virtually zero.

5.2 Electrification

In the province of Manitoba, 98% of the electricity generated is from clean and renewable sources including hydroelectric and wind [8]. By electrifying the vast majority of the equipment at the processing facility, the greenhouse gas (GHG) emissions for the processing of the high purity silica sand is drastically reduced when compared to traditional equipment of diesel source generation or other forms.

As the technology becomes available the field equipment such as pumps, extraction rigs and water treatment systems will be electrified to reduce the need for diesel generators. At this time, pumps and most of the surface handling equipment is already capable of utilizing an electrical source, however the extraction rigs that are utilized are still in development phase for electrification.

The availability of clean, low-cost electricity is also a consideration when looking at the footprint and efficiency requirements for PV components manufacturing. Particularly, polysilicon ingot and wafers are energy intensive. For wafers, electricity accounts for nearly 20% of production costs, and for polysilicon over 40% as is currently mostly supplied by coal. [5]

It should be noted that the rotary dryer and heating systems must still run-on natural gas or propane, as no alternative exists at this time.

5.3 Minimal Processing

As demonstrated in Section 3.1, the high purity silica sand processing methods required for a 99.9% average purity are simple and environmentally friendly in nature. The processing facility does not require the use of any chemicals or harmful products, nor crushing or grinding operations as are often seen in silica quartz mining and processing. The process sees the sand brought to the facility by slurry line and then includes a wash, dry and magnetic separation, prior to being stored in silos and shipped by rail to various markets national and internationally. Water is recycled through the facility in a continuous loop and supplied for steps requiring water. The water is treated after usage using a biodegradable food grade flocculant. This practice drastically reduces water consumption such that the facility consumes 757 – 1,136 L/day, which is the approximate daily usage of a household of four to six people. [9]

6 CONCLUSION

It is clear that high purity silica is in demand for feedstock into critical items such as medical glass, electronics and specifically photovoltaics. As the world moves towards net zero, the need for new resources of raw minerals to manufacture key components are increasing. The introduction of more environmentally friendly sustainable mining methods for these minerals becomes critical in the movement to net zero.

The proposed high purity silica sand mine in Manitoba, Canada marks the beginning of a new more environmentally friendly way of mining silica sand. This proposed mine will utilize a newly developed extraction method which eliminates truck traffic and open-pit mining, and a simple process without the use of chemicals to obtain 99.9% purity. Furthermore, by utilizing clean renewable energy, high purity silica sand mining suitable for PV manufacturing will have a significant impact on emissions reduction and feedstock availability.

7 REFERENCES

[1] Bezys, R.K. and Conley, G.G., 1998. Geology of the Ordovician Winnipeg Formation in Manitoba; Manitoba Energy and Mines, Stratigraphic Map Series, Ow-1, 1:2 000 000.

[2] Matile G.L.D., and Keller G.R., 2004. Surficial Geology Compilation Map Series. Manitoba Geological Survey, GIS Map Gallery. Geographic Information System. Accessed at: http://www.gov.mb.ca/iem/geo/gis/sgcms/legend.html

[3] Huiming,Tian. (2012). Method for producing high-purity quartz sand with high-temperature chlorination process and chlorination device (China Patent No. CN102701223B). Accessed at: https://patentscope.wipo.int/search/en/detail.jsf?docPN=CN102701223

[4] Fisher, Graham & Seacrist, Mike & Standley, Robert. (2012). Silicon Crystal Growth and Wafer Technologies. Proceedings of the IEEE. 100. 1454-1474. 10.1109/JPROC.2012.2189786.

[5] IEA (2022), Solar PV Global Supply Chains, IEA, Paris. Accessed at: https://www.iea.org/reports/solar-pv-global-supply-chains

[6] Abdellatif, Masud. (2011) DC arc smelting of silicon: Is it technically feasible? Mintek, Randburg, South Africa. Proceedings of the Southern African Institute of Mining and Metallurgy, Johannesburg. Accessed at: https://www.mintek.co.za/Pyromet/Files/2011Abdellatif1.pdf

[7] IEA (2022), Renewable Energy Market Update - May 2022, IEA, Paris. Accessed at: https://www.iea.org/reports/renewable-energy-market-update-may-2022

[8] Manitoba Environment, Climate and Parks. Hydroelectricity. n.d. Accessed at: https://www.gov.mb.ca/sd/environment_and_biodiversity/energy/initiatives/hydro.html

[9] AECOM. 2020. Vivian Sand Facility Project Environment Act Proposal. Prepared for CanWhite Sands Corp. July 2020.

IN-DEPTH EXPOSURE OF NEW PHASE IN POLY-SILICON OF TOPCON DEVICE: IS PHOSPHORUS BONDED OXYGEN PRESENT?

Z.Q.Ma[1, 2], Y.L.Wang[1] and Z.X.Lan[1]

[1] SHU-SOEN's R&D Lab, Department of Physics, Shanghai University, Shanghai 200444, China

[2] Instrumental Analysis & Research Center, Shanghai University, Shanghai 200444, China

ABSTRACT: For the TOPCon PV device, the heavy phosphorus (P) – doped polycrystalline silicon layer (n^+-poly-Si) with ultra-thin silicon oxide (a-SiO_x) is extremely beneficent for passivating contact and electron selectivity as its exceptional coupled microstructure (chemical components and physical phase) and optoelectric properties, compared with other passivation materials. However, an unknown configuration of oxygen (O) and phosphorus (P) in the n^+-poly Si film is still open for investigation. The findings of the P-O bonds (bridge and non-bridge linkages) in the whole scope of n^+-poly Si film have been implemented by X-ray photoelectron spectroscopy (XPS) in the mode of depth profile measurements of P 2p, O 1s, and Si 2p states. The symbolic binding energies (BEs) with a shoulder in the range of 126 – 142 eV for the P 2p state and the other two photon-emission peaks of the O 1s state centered at 532 and 535 eV, respectively, were obtained by an occasional survey of SiO_x composition. Furthermore, the Grazing incidence X-ray diffraction (GIXRD) measurements of the n^+-poly Si films showed that the typical nano-crystallinity with grain sizes of 38-45 nm is obtained. The derivatives of oxidized phosphorite localized at grain boundaries (GBs) were tentatively analyzed through thermodynamic functions. In addition, the segregation of P and O compositions was detected by energy dispersive X-ray (EDX) patterns for the surface-textured (pyramid) wafer, while there is no segregation of elements or compounds in the planar n^+-poly Si film.

Keywords: XPS profile, TOPCon device, poly-Si film, P-O bonds, thermodynamic analysis

1 INTRODUCTION

Currently, it has been widely recognized by the PV community that silicon-based tunneling oxide passivating contact (TOPCon) solar cell is the most promising route towards the next-generation passivating contact technologies [1, 2], demonstrating highly efficient solar cells as well as remarkable efficiency potentials. In contrast with passivated emitter and rear contact (PERC) solar cell [3], TOPCon employs indirect contact of metallic electrodes with a polycrystalline silicon material, resulting in significantly reduced recombination of carriers at electrical contacts and elevated open-circuit voltage (V_{OC}) of the solar devices. So far, p-TOPCon, n-TOPCon, and bi-facial TOPCon solar cells have been developed so fast. Not surprisingly, bifacial TOPCon demonstrated substantial efficiency potential due to superior surface passivation and complete elimination of metal/c-Si contact. The limiting efficiency for bi-facial TOPCon is 28.7 %. In other words, these technologies are capable of achieving higher conversion efficiencies, while the efficiency gap is insignificant [4]. Of course, this great progress is beneficial from heavy phosphorus (P)-doped polysilicon films, and its functions are more than just "passivation" and low-resistance contacts.

Due to the high concentration of phosphorus elements in the n^+-poly Si film, (10^{20}-10^{21}) / cm^3, and high temperature (~ 920 °C) recrystallization process, making the diffusion and redistribution of phosphorus and internal oxygen elements, will change its microstructure (configuration and phase), leading to the favorable variation of optical and electric properties for the TOPCon device. However, in addition to phosphorus substitutional doping, what we are concerned about is the production of new compounds resulted from phosphorus atoms, which has not been reported.

To explore the chemical components within the heavy phosphorus (P) doped polycrystalline silicon film of the TOPCon device, X-ray photoelectron spectroscopy (XPS) with depth-profile by Ar ion milling was applied to the test sample through the variation of binding energies (BEs) of P 2p, O 1s, and Si 2p core-levels. Where poly-Si films were fabricated by normal low-pressure chemical vapor deposition (LPCVD) and subsequently 860-920°C annealing. As a subsidiary method, a grazing incidence X-ray diffraction (GIXRD) was utilized for the determination of the crystallography of the n^+-poly Si film. Furthermore, energy dispersive X-ray spectroscopy (EDX) was used to detect the film's elemental distribution on the textured surface.

Based on the above-mentioned three X-ray correlated energetic or optical emission characterization methods, the P-Si and P-O bonds within the whole n^+-poly-Si film have been recognized and the derivatives of oxidized phosphorus have been determined to be present in the grain boundaries. The nano-crystal Si film with grain sizes of 38-45nm is obtained and the crystallite size is estimated by Scherrer's formula in terms of diffraction peaks at (111), (220) and (311) of GIXRD.

Significantly, the thermodynamic parameters, such as formation energies, molar Gibbs free energies and molar enthalpies of silica and phosphorus oxide (P_2O_5), are employed to analyze the P-O bonds and their derivatives.

2 EXPERIMENTAL IN DETAILS

2.1 Preparation of n^+-poly Si film

There are two types of samples prepared for the research. One is a double-side-textured surface for the PV device; another is a planar surface covered by ~ 9 nm SiO_2 for the test. In any case, *n*-type (100)-Cz Si wafers were cleaned with the standard RCA method. To form a front boron-doped emitter, boron diffusion at 860 °C with a BBr_3 source was applied to the n-Si substrate. Subsequently, a low-temperature silicon oxide (~1.5 nm) was grown on both sides of the wafer, followed by the growth of intrinsic Si layer of ~130 nm in a low-pressure chemical vapor deposition (LPCVD) on both sides of the wafer. To gain a TOPCon device with n⁺-doped polysilicon, phosphorus dopant was introduced into the intrinsic amorphous silicon layer by a phosphorus diffusion process, then thermal annealing in a tube furnace at 920 ℃ was applied to activate phosphorus elements accompanied by crystallization of amorphous to polycrystalline state.

For the fabrication of ultrathin SiO_x(s) layer (< 2.0 nm) on the textured c-Si, through LPALD (Low-Pressure Atomic Layer Deposition) thin film growth technology [5], silicone source (tri-dimethylamino silane, i.e., $SiH(N(CH_3)_2)_3$) and oxygen (O_2) were mixed to be absorbed and deposited on the surface of the n-Si substrate at a deposition temperature of 180°C, a deposition pressure of 120 Pa. A flow ratio between the silicon source and oxygen was 1:2. The reaction equation is as follows [6]

$$SiH(N(CH_3)_2)_3(s) + O_2(g) \rightarrow SiO_x(s) + H - N(CH_3)_2(g) \quad (1)$$

Subsequently, SiH_4, PH_3, and N_2 were in-situ introduced and, using LPCVD technology. A high concentration of phosphorus-doped amorphous silicon films (a-Si) was deposited on top of SiO_x/n-Si, with a thickness of 130 nm and a deposition temperature of 380° C, with a ratio of 1:2:2.5 for SiH_4, PH_3, and N_2。 Finally, the stacked film's system of n-Si/SiOx/n⁺-poly Si was annealed at a high temperature of 920°C for 45 minutes in an N_2 atmosphere, so that the a-Si film was transformed into an n⁺-poly Si polycrystalline state.

After removing n⁺-doped polysilicon stuck on the edge by the wet etching method, a rear SiN_x film was deposited via PECVD on the rear side to enhance the passivation effect and bifacial power generation. To reduce emitter recombination rate and minimize reflection loss, a stacked layer with 4-nm-thick ALD-Al_2O_3 and 75-nm-thick PECVD-SiN_x was coated on the textured surface.

The test samples in this work were an area of 158 x158 mm^2 (M6). After finishing the electrode on both sides, the open-circuit voltage, the short-circuit current density, and the filling factor of the photovoltaic device was measured to be 715.01 mV, 41.05 mA/cm^2, and 83.55%, respectively, under a standard sunlight exposure of AM 1.5G. So that the average photoelectric conversion efficiency of about 24.53% is reached. On the other side, the shunt and series resistances of the device are approximately 1715.8 Ω and 0.57 mΩ, respectively.

Based on this standard process, in order to accurately measure the chemical state in the n⁺-poly Si film, we specifically prepared the plane type of samples to analyze the depth distribution of the chemical states.

2.2 Characterization methods

To identify the chemical bond states within the heavily P-doped Si thin films, near normal angle X-ray photoemission spectroscopy (XPS) coupled with depth etching was carried out using (Thermo-fish Scientific) monochromated Al Kα X-ray (hν=1486.6 eV) source. The pass energy window of 200 eV and 40 eV for the wide-scan (survey) and core-level modes, respectively. No extra signals were detected within the sensitivity of the XPS measurements except those from chemical elements that make up the chemical composition of P-doped Si films and bulk Si. All the binding energies (BEs) were precisely determined within ± 0.3 eV using an XPS handbook. To precisely identify the chemical bond states of spin-orbit doublet peaks in P 2p and O 1s core-level spectra, peak fitting was performed using Shirley background subtraction and Gaussian-Lorentzian functions [7].

The grazing incidence X-ray diffraction (GIXRD) measurements were carried out by using an X-ray diffractometer (SmartLab-X) in the θ-2θ configuration with a CuKα source (λ =1.5418 Å). All of the measurements were performed with a small scanning step of 0.02°. The fixed grazing angle incident on the n⁺-poly Si surface was set to be 0.3°. The X-ray tube was operated at 40 kV and 40 mA. Additionally, a wide range of 20-70° for 2θ angle was scanned so that all the major diffraction peaks could be detected.

The lattice morphology, as well as elemental distribution of P and O in the n⁺-poly Si film, were examined by using high-resolution transmission electron microscopy (HR-TEM, JEOL-2100F) at 200 keV, together with energy dispersive X-ray (EDX) emission.

3 RESULTS

3.1 Characterization of n⁺-poly Si film

In the case of the depth profile of the chemical components of P 2s, O 1s, and Si 2p in n⁺-poly Si film, the relative atomic percent (%) was obtained through the integration of fitted peaks of XPS spectra, as shown in Fig.1. For the P 2p state, this integral is taken from the BEs of 126-142 eV, as illustrated in Fig.2a. In addition, the experimental collected spectra taken at different depth inside n⁺-poly Si film and three fitted P 2p peaks for each one were obtained at 127.5–131.9 (P1), 131.5–135.4 (P2) and 131.2-142.0 eV (P3), respectively, which were assigned to be the phosphide species ($P^{\delta-}$, δ < 5) and oxidized species of phosphorus (P^{5+}) [8]. We found that the fitted peaks centered at 129.0, 133.0, and 134.9 eV should be ascribed to be P-Si (or P-P), P-O, and PO_x bonds, respectively [9, 10]. Especially, there was no more chemical shift of BEs for the peaks as the phosphorus oxides were uniformly distributed in the n⁺-poly Si film. This depth profile data were gained through time (s)-dependent Ar ion etching point-by-point for the 130 nm film with a planar surface.

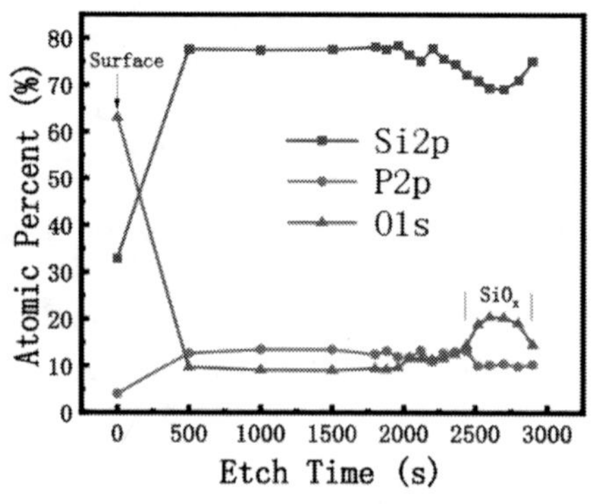

Fig.1 The relative atomic percent (%) of P 2p、O 1s and Si 2p optoelectrons in n⁺-poly Si film visa etching time (depth).

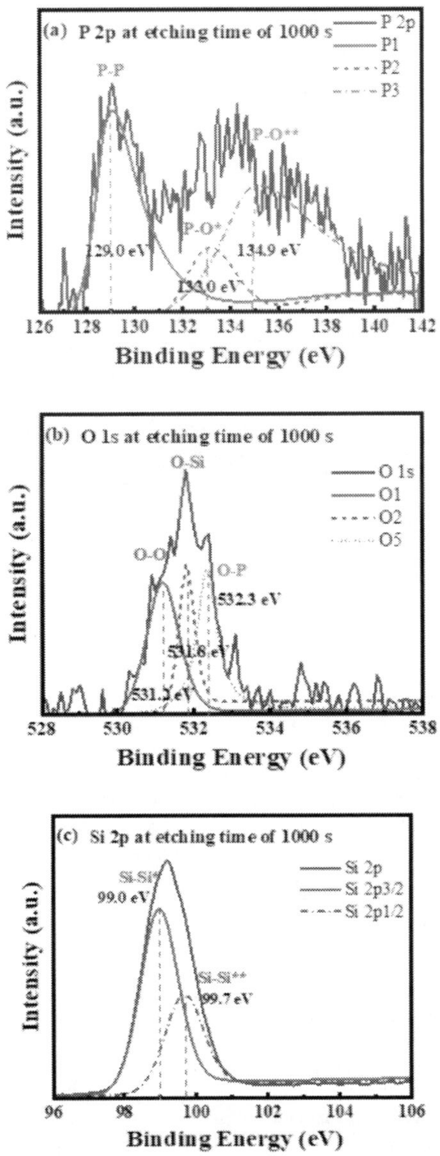

Fig.2 The fitted spectra of XPS - P 2p、O 1s、Si 2p states at 1000s, as an example. Also, the peak's shape and intensities at other depths were nearly the same.

Similarly, for the O 1s and Si 2p states, their experimental and fitting spectra were shown in Figs.2b and 2c, respectively. In Fig.2, the collected spectra of P 2p, O 1s, and Si 2p states were at 1000 seconds inside the n^+-poly Si film and, the shape of the collected spectra within the whole n^+-poly Si film was almost identical.

The phase and components in the n^+-poly Si film were verified by GIXRD patterns, as demonstrated in Fig.3. The typical diffraction peaks oriented at (111), (220), and (311) for the n^+-poly Si films were manifested, corresponding to the 2θ = 28.4, 47.3 and 56.2°, respectively. The measured full-width-at-half-maximum (FWHM) peak at (111) of about 0.28° (silicon substrate was 0.05°).

Grain sizes in the poly Si films can be calculated using Scherrer's formula [11]:

$$D = k\lambda / \beta \cos\theta \qquad (2)$$

where D is the average crystallite size, λ (1.54 Å) is the wavelength of the used X-ray, β is the FWHM of the measured peak in radians and θ is the angle of diffraction. k is a constant which is determined by the geometry of the crystallites and taken as 0.9. The average grain sizes of the films were in the range of 38-45 nm [12, 13].

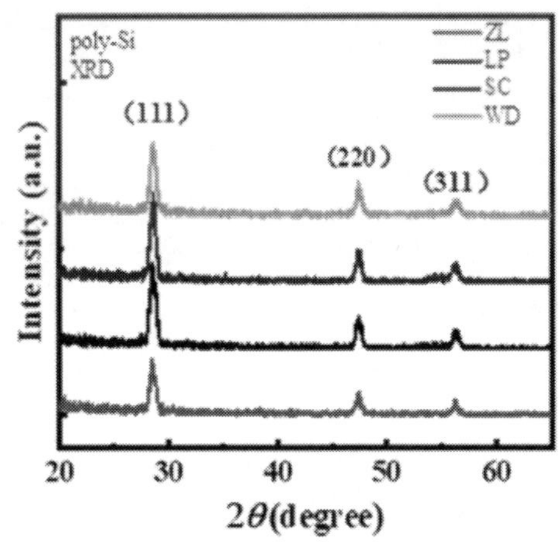

Fig.3 GIXRD patterns of n^+-poly Si films with four different deposition conditions: LPCVD (ZL, LP); PECVD (SC, WD). The grazing angle is 0.3° at the θ-2θ mode of 2θ: 25-65°.

The island-like segregation of O and P elemental distribution was exhibited in Fig. 4, which is disclosed through EDX for the surface-textured wafer. However, there are no segregation phenomena of O or P elements in the n^+-poly Si film for the planar surface of the c-Si wafer.

The crystallite quality of the n^+-poly Si film with orderly aligning and lattice array, as well as a few of the grain boundaries are displayed in Fig.5. It can be seen that the more clear-cut atomic arrays with a few grain boundaries are distributed in the n^+-poly Si film. Thus, it can be verified that a high quality of large crystallite has been obtained in the state-of-the-art processing of CVD methods [14]. The lattice morphology and orientations inside the n^+-poly Si film indicated that the merit of the figures indeed reflected the case of high-quality polycrystalline silicon thin films.

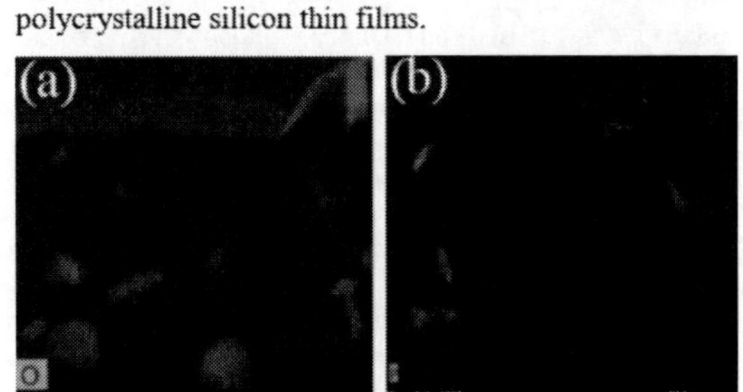

Fig.4 EDX spectroscopies of the segregation for oxygen (O) and phosphorus (P) in the n^+-poly Si film at rear side of the textured c-Si wafer.

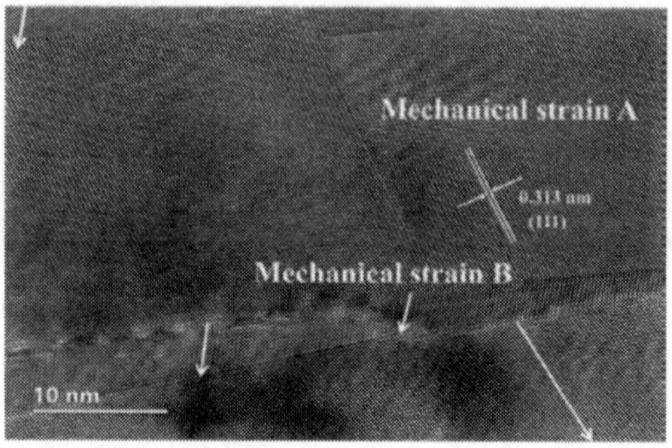

Fig.5 The cross-sectional image of HR-TEM for the different crystallite's face orientations and mechanical strains in n^+-poly Si film of n-TOPCon device.

3.2 Thermodynamic verification

The thermodynamic parameters, such as formation energies, molar Gibbs free energies, and molar enthalpy of silica (SiO_2) and phosphorus oxide (P_2O_5), are measured to be about -9.4 and -15.6 eV, -856.3, and -1348.9 kJ/mol, and -910.7 and -3010.4 kJ/mol under the standard conditions (one-atmosphere pressure, 298.15K), respectively [15-17]. Thus, according to the minimum principle of energy in the equilibrium state, at high temperature (860-920 °C) and high pressure (arisen from tensile stress of crystallites), the oxygen and phosphorus elements in the n^+-poly Si film are prone to chemical bonds and oxidized phosphorus and enriched on the grain boundaries. Nevertheless, the probability of forming silicon oxide is small. The tensile stress in the polysilicon film is induced by the effective doping of phosphorus instead of the silicon atom, and at the same time, compressive stress is generated in the grain boundaries.

3.3 Presence of GBs and P-O bonds in n^+-poly Si film

The grain boundary (GBs) in the polysilicon film has not only an important impact on the conductance performance (resistivity ρ_c) of the film but also the carrier recombination parameters (J_o) of TOPCon photovoltaic devices. Many results of correlated studies have been reported on the influence of thin film conductivity through modeling and calculation, which belongs to the category of basic knowledge. Impurities and dangling bonds in the GBs, which exert direct recombination effects on the transport of carriers in the device, are also well known. Therefore, the structure of the GBs, segregation of doped elements, and free carrier absorption (FCA) are keys to a deeper understanding of the properties of the n^+-poly Si film. To highlight the connotation of the results of this study, we focused on whether phosphorus elements, outside the activated state, can form P-O bonds, and if so, where they are in the film. Since it is very difficult to directly detect compounds with P within the scale of only 5-8 Å of GBs, direct evidence cannot be experimentally obtained. So, we focus on the tensile stress in the crystallite and the compressive stress within the GBs in this film. Based on the thermodynamic analysis and minimum principle of Gibbs energy under an equilibrium state, we inferred that most of the P-O bonds and their molecules are at the GBs and it is reasonable to passivate the GBs and reduce the resistance of the film. Also, the behaviors of GBs and stress during the growth and subsequent thermal annealing of the polycrystalline films were previously reported in the correlated literature [18-26].

Polycrystalline silicon is composed of a large number of monocrystalline silicon particles (called grains, or crystallites) and inter-granular boundaries (also known as grain boundaries). For the n^+-poly Si thin films in our case, the grain size was around several ten's nanometers, which has been obtained by GIXRD, and the grain boundary width was between 0.5–1.0 nm. Those defects of grain surface dangling bonds, vacancy, interstitial, and impurities were localized in the region. The surface dangling bonds of individual grains can be saturated by the segregated P or O elements within GBs, furthermore, the increase of P-O bonds in GBs can also reduce the density of other defects. The indirect evidence can be referred to the relevant literature [27-31].

To fully understand the above experimental facts and objective conditions for the existence of P, O, and P-O bonds in polysilicon thin films, comprehensive consideration of the following factors, including, but not limited to the characteristics of the GBs mechanical structure and, excess elements and their chemical bonds, GBs energy, and the minimum principle of thermodynamic ensemble energy in equilibrium, referring to the research results of others, as listed literature mentioned-above

After heavy doping, annealing, and thermal relaxation of the phosphorus element, those excess, deactivated P elements, and oxygen elements after diffusion, must gather in the compressively stressed crystal boundary, rather than elsewhere. This provides conditions for P-O bond formation and fits the variation of thermodynamic functions. Therefore, it is reasonable to have an electrical passivation effect on the crystal boundary. And the corresponding material structure and semiconductor device performance changes have been quoted in the literature [32-35].

As for the oxygen sources in GBs of n^+-poly Si film, there are three kinds of origin: The first is the raw material, including SiH_4 and carrier gas in the deposition chamber environment, therein, LPCVD at low temperature (550°C) is used in the initial process. The second is the oxygen (O) desorption from the ultrathin SiO_x layer during the thermal annealing process at the high temperature of 920°C, leading to the rapid diffusion of oxygen along the GBs in the poly Si (n^+) film from both sides, and this is the main source. The third is the subsequent thermal relaxation and the segregation of oxygen atoms in the GBs [36, 37].

Recently, M. Firat et al.[38] put TCAD (Technology computer-aided design) simulation together with SSRM (scanning spreading resistance microscopy) experiments, have obtained a pinhole profile of 1.2-1.5 nm within the tunneling SiO_x layer. From this, it can be determined that the desorbed oxygen element from the SiO_x layer certainly diffuse into the polysilicon film. Based on the former literature [39, 40], it is evident that O elements enriched in the GBs for the minimum energy state and were incorporated in the extended structural defect of SiO_x clusters [41, 42]. And then, high-temperature annealing at 920°C broke the Si-O bonds, and part of the oxygen rearranged and bonded with P elements in GBs for lower Gibbs energy of phosphorus oxide [43]. As for the diffusivity and depth distribution of oxygen composition in the n^+-poly Si film, a consistent depth profile of the XPS-O1s state has been obtained in our XPS experimental analysis, which means that the oxygen concentration should be changeless through the whole n^+-poly Si film.

4. CONCLUSION

In this work, we investigated the chemical bonding states in heavily P-doped poly Si thin films grown on textured and planar SiO_x/ n-Si (100) substrates using X-ray photoelectron spectroscopy (XPS). XPS-P 2p core-level spectra manifested that the oxidized phosphorus through bridging or non-bridging linking bonds within n^+-poly Si films, such as P-P, P-Si, P-O, and O-P-O bonds, corresponding to the indicative binding energies (BEs). The elemental segregation phenomena of phosphorus (P) and oxygen (O) in the n^+-poly Si film of textured sample were revealed from EDX spectroscopy. It is addressed that many excess oxygen and phosphorus elements are condensed on the crystal boundary, which may arise from diffusion at high temperature and compressive stress near the crystal boundary, which is also verified by the minimum energy principle. Furthermore, the appeared physical phase and chemical compositions correlative to phosphorus oxide were also reflected in GIXRD patterns by a solid reaction of P and O, or Si as well. These results indicate that more P atoms are incorporated into the substitutional sites of the Si lattice with the increase of P concentrations. We observed that the presence of three well-resolved diffraction peaks in GIXRD and three characteristic binding energies in XPS-P 2p spectra only for the heavily P-doped n^+-poly Si film samples, which means the new chemical states produced. Thus, experimental findings in this study using XPS, GIXRD and EDX open up new doors for evaluating the chemical states of P-doped Si materials in fundamental research as well as in photovoltaic industrial applications.

5. ACKNOWLEDGEMENT

This work is mainly supported by the National Natural Science Foundation of China (No. 61874070, 61674099 and 61274067) and the R&D Foundation of the SHU-SOENs PV Joint Laboratory (No. SS-E0700601).

6. REFERENCES

[1] T.G. Allen, J. Bullock, X. Yang, A. Javey, S. De Wolf, Passivating contacts for crystalline silicon solar cells, Nature Energy 4 (2019) 914.

[2] M. Hermle, F. Feldmann, M. Bivour, J.C. Goldschmidt, S.W. Glunz, Passivating contacts and tandem concepts: approaches for the highest silicon-based solar cell efficiencies, Appl. Phys. Rev. 7 (2020) 021305.

[3] R. Preu, E. Lohmüller, S. Lohmüller, P. Saint-Cast, J.M. Greulich, Passivated emitter and rear cell—devices, technology, and modeling, Appl. Phys. Rev. 7 (2020) 041315.

[4] J. Schmidt, R. Peibst, R. Brendel, Surface passivation of crystalline silicon solar cells: present and future, Sol. Energy Mater. Sol. Cells 187 (2018) 39.

[5] M. Lozac'h, S. Nunomura, K. Matsubara, Sol. Energy Mater. Sol. Cells **207** (2020) 110357.

[6] L. Han, Z. Chen ECS J. Solid State Sci. Technol. **2** (2013) N228.

[7] D. Briggs, M.P. Seah, Practical Surface Analysis: By Auger and X-ray Photoelectron Spectroscopy, Wiley (1990) 572–575.

[8] A. Thogersen, M.Syre, B. R. Laisen, S. Diplas, Studies of the oxidation states of phosphorus gettered silicon substrates using X-ray photoelectron spectroscopy and transmission electron microscopy, Journal of Applied Physics 113 (2013) 044307.

[9] W.B Ying, Y Mizokawa, Y Kamiura, K Kawamoto, W.Y Yang, The chemical composition changes of silicon and phosphorus in the process of native oxide formation of phosphorus-doped silicon, Appl. Surf. Sci. 181 (2001) 1.

[10] M. Lee, H.-Y. Ryu, E. Ko, D.-H. Ko, Effects of phosphorus doping and post growth laser annealing on the structural, electrical, and chemical properties of phosphorus-doped Si films, ACS Appl. Electron. Mater. 1 (2019) 288.

[11] A. Monshi, M. R. Foroughi, M. R. Monshi, Modified Scherrer equation to estimate more accurately nano-crystallite size using GIXRD World Journal of Nano Science and Engineering, 2 (2012) 154.

[12] J. Langford, A. Wilson, Scherrer after sixty years: a survey and some new results in the determination of crystallite size, J. of Applied Crystallography, 11 (1978) 102.

[13] K.J. Chen, et al., Measurement of poly-Si film thickness on textured surfaces by X-ray diffraction in poly-Si/SiO_x passivating contacts for monocrystalline Si solar cells, Solar Energy Materials and Solar Cells 236 (2022) 111510.

[14] A.S. Kale, et al., Effect of crystallographic orientation and nanoscale surface morphology on poly-Si/SiO_x contacts for silicon solar cells, ACS Appl. Mater. Interfaces 11 (45) (2019) 42021.

[15] W. Ying, Y. Mizokawa, Y. Kamiura, K. Kawamoto, and W. Yang, The chemical composition changes of silicon and phosphorus in the process of native oxide formation of heavily phosphorus-doped silicon, Appl. Surf. Sci. 181 (2001) 1.

[16] https://materialsproject.org/#apps/interfacereactions

[17] H. L. Lukas, S. G. Fries, B. Sundman, Computational Thermodynamics, The Calphad Method, Cambridge University Press, Cambridge, 2007.

[18] N. K.Upreti, S. Singh, Grain boundary effect on the electrical properties of boron-doped polysilicon films. Bulletin of Materials Science, 14 (1991) 1331.

[19] H. Watanabe, Statistics of grain boundaries in polysilicon, IEEE Transactions on Electron Devices, 54 (2007) 38.

[20] C. H. Seager, Grain boundaries in polycrystalline silicon. Annual Review of Materials Science, 15 (1985) 271.

[21] T. Kai, In grain boundary segregation of impurity in polycrystalline silicon, The 3rd International Workshop on Crystalline Si Solar Cells, Norway, Norway, 2009; pp 1-6.

[22] F. A. Trumbore, Solid solubilities of impurity elements in germanium and silicon. Bell System Technical Journal, 39 (1960) 205.

[23] J. M. Serra, R. Gamboa, A. M. Vallera, Optical absorption coefficient of polycrystalline silicon with very high oxygen content. Materials Science and Engineering: B, 36 (1996) 73.

[24] A. J. Mueller, R. D. White, In residual stress variation in polysilicon thin films, ASME 2006 International Mechanical Engineering Congress and Exposition, 2006; pp 1-7.

[25] T. W. Kim, B. Gogoi, K. G. Goldman, A. C. Mcneil, D. J. Koch, In substrate and annealing influences on

the residual stress of polysilicon, 1998 Solid-State, Actuators, and Microsystems Workshop, South Carolina, June 8-11; South Carolina, 1998.

[26] X. Zhang, T.-Y. Zhang, M. Wong, Y. Zohar, Residual-stress relaxation in polysilicon thin films by high-temperature rapid thermal annealing. Sensors and Actuators A: Physical, 64 (1998) 109.

[27] D. D. Zhao, Y. J. Li, Revealing the factors influencing grain boundary segregation of P, As in Si: Insights from first-principles, Acta Materialia, Volume 168, (2019) 52.

[28] H. Watanabe, Statistics of grain boundaries in polysilicon, IEEE Transactions on Electron Devices February, 54 (2007) 38.

[29] B. Kim, H. Jang, S.-W. Kim, D.-S. Byeon, S. Koo, Growth and electrical properties of in situ phosphorus-doped polycrystalline silicon films using Si_3H_8 and PH_3, Journal of Vacuum Science & Technology A 32 (2014) 031510.

[30] S. Duguay, A. Colin, D. Mathiot, P Morin, D. Blavette, Atomic-scale redistribution of dopants in polycrystalline silicon layers, Journal of Applied Physics, 108 (2010) 034911.

[31] D. Nobili A. Armigliato, M. Finnetti, S. Solmi, Precipitation as the phenomenon responsible for the electrically inactive phosphorus in silicon, 1982 Journal of Applied Physics, Vol. 53, (1982) 1484.

[32] K. Peter, S. Talid, A mechanistic study of impurity segregation at silicon grain boundaries. Journal of Applied Physics 118:9, (2015) 095301.

[33] M. Mohammad, K. Mandurah, C. Saraswat, C. Robert Helms, Dopant segregation in polycrystalline silicon, Journal of Applied Physics, 51 (1980) 5755.

[34] M. M. Mandurah, K. C. Saraswa, T. I. Kamins, A model for conduction in polycrystalline silicon—Part I: Theory, in IEEE Transactions on Electron Devices, vol. 28, no. 10, Oct. 1981, pp. 1163-1171.

[35] M. M. Mandurah, K. C. Saraswat, T. I. Kamins, A model for conduction in polycrystalline silicon—Part II: Comparison of theory and experiment, in IEEE Transactions on Electron Devices, vol. 28, no. 10, Oct. 1981, pp. 1172-1176.

[36] M. Fırat, M. R. Payo, F. Duerinckx, J.-M. Luchies, M. Lenes, J. Poortmans, Characterization of absorption losses in rear side n-type polycrystalline silicon passivating contacts. In 15th International Conference on Concentrator Photovoltaic Systems (CPV-15), 2019.

[37] R. A. B. Devine, D. Mathiot, J. B. Xu, I. H. Wilson, M. Gauneau, W. L. Warren, Grain boundary enhanced oxygen out-diffusion in annealed polycrystalline Si/SiO2/crystalline Si structures. Thin Solid Films, 286 1996) 317.

[38] M. Fırat, L. Wouters, P. Lagrain, F. Haase, J.-I. Polzin, A. Chaudhary, G. Nogay, T. Desrues, J. Krügener, R. Peibst, L. Tous, R. H. Sivaramakrishnan, J. Poortmans, Local enhancement of dopant diffusion from polycrystalline silicon passivating contacts. ACS Applied Materials & Interfaces, 14 (2022) 17975.

[39] A. K. Fedotov, A. V. Mazanik, A. G. Ulyashin, Charge relaxation at oxygen-enriched silicon grain boundaries. Solid State Phenomena, 82-84 (2001) 515.

[40] S. Pizzini, H. P. Strunk, J. H.Werner, Carrier transport at silicon and germanium grain boundaries enriched by impurities. Solid State Phenomena, 51-52(1996) 385.

[41] V. Borjanović, I. Kovačević, B. Šantić, B. Pivac, Oxygen-related deep levels in oxygen doped EFG poly-Si. Materials Science and Engineering: B, 71 (2000) 292.

[42] B. Pivac, A. Sassella, A. Borghesi, Infrared study of oxygen segregation at structural defects in polycrystalline silicon. Mikrochimica Acta, 14 (1997) 485.

[43] P. Gawobzewski, E. Hild, F. G. Kirscht, L. Vecsernyés, Infrared spectroscopical and TEM investigations of oxygen precipitation in silicon crystals with medium and high oxygen concentrations. Physica status solidi (a), 85 (1984) 133.

CUT OUT FOR EFFICIENCY: A DETAILED LBIC STUDY OF EDGE RECOMBINATION IN CUT-CELL SILICON PHOTOVOLTAICS

Miha Kikelj[1], Benjamin Lipovšek[1], Matevž Bokalič[1] and Marko Topič[1]
[1]University of Ljubljana, Faculty of Electrical Engineering
Tržaška cesta 25, 1000 Ljubljana, SI-Slovenia

ABSTRACT: In our work we aim to, with the help of numerical modelling on one side and subsequent correlation with spatially resolved measurements on the other, qualitatively evaluate and quantitatively determine the recombination parameters of cut edges of silicon solar cells, which govern the efficiency decreases of cut-cells obtained by different cell separation techniques (cleaving, laser ablation and cleaving, thermal laser separation). With the proposed approach, we were able to recreate an LBIC scan of a healthy part of a separated Zebra IBC cell in high detail.
Keywords: LBIC, simulation, cut cells, Zebra IBC

1 INTRODUCTION

Silicon solar cell cutting and fabrication of highly efficient cut-cell shingled modules proves a relatively simple yet cost effective solution to drive further module efficiency increase of silicon photovoltaics. As solar cell separation increases the edge-to-surface ratio, the effect of edge recombination becomes increasingly important to the solar cell's final efficiency [1], [2]. With the aim to characterize and model the recombination processes at cut-cell edges of interdigitated back contact silicon solar cells, namely the Zebra IBC [3], we have followed the route of spatially resolved imaging methods, namely the Laser Beam-Induced Current (LBIC) method. In comparison with the evaluation of the change in integral *J-V* parameters (V_{OC}, J_{SC}, FF, P_{MPP}) as pursued in [4], [5], the LBIC method provides an insight into the spatial distribution and influence of damaged areas. Further on, it also proves advantageous to other spatially resolved characterization methods (EL, PL) as employed by [6]–[8], because it allows for a step-by-step local calibration of shorter lifetimes of damaged areas instead of a single calibration parameter (e.g. surface recombination velocity on the cut edge) or a spatially resolved profile of lifetime decrease towards the edge. In addition, even though LBIC is an optically simpler characterization method compared to quicker EL or PL, it still suffers from some optical limitations which will be briefly described in later sections.

Regardless of the separation technique, our characterization approach of the cut edges is based on the combination of experimental and numerical procedures. First, we employ (LBIC), to measure the local response profiles of cut solar cells in the micrometer resolution range. Different wavelengths of the incident beam are used to extract the effects of cut edges. Next, we recreate the exact structure of the cut solar cell together with the LBIC illumination conditions within a detailed 3D optical modelling environment in order to extract the spatial distribution of optical generation within the bulk at multiple wavelengths. These results are then coupled to a 2D electrical model to calculate the response of the solar cell at the given illumination conditions. Finally, by calibrating the modelling results with the LBIC measurements, it is possible to extract the key electrical parameters associated with the cut-edge effects, including the mapping of the local minority carrier lifetimes.

The combined opto-electrical model presents us with the opportunity to explore in detail how different parameters of the LBIC setup – wavelength, beam size, beam position relative to the cell and cell's intrinsic parameters – influence the LBIC response of the studied device. The results reveal the true nature of light absorption within the bulk, which becomes increasingly delocalized as the wavelength increases and thus significantly influences the final results of the LBIC scan. In addition to obvious factors (wavelength, beam size, beam position), the position of defects (decreased lifetimes) relative to the device's period also elicits significantly different LBIC responses.

2 MODELLING APPROACH

For that purpose, we have established a combined, full-3D opto-electrical modelling approach, based on a previously validated combined optical model [9], [10], which in the first step allows us to obtain an accurate distribution of optical generation rate from external, multi-wavelength (520 – 1060 nm) light beam illumination at every spatial step of external illumination. It is important to note that at lower wavelengths only a single optical simulation was needed for all characterization steps, and the resulting generation rate was simply shifted laterally for each illumination step. At longer wavelengths of light which reach the rear interface of the device under test (DUT) and exhibit larger lateral spreading, a separate optical simulation was needed for each discrete step. Optical simulation results for a single illumination step at 640 nm, 940 nm and 1060 nm respectively are shown in Fig. 1. One can clearly observe the lateral spreading of light through the cell, especially at longer wavelengths. This raises a question as to what extent locally measured properties (LBIC intensities) actually correspond to underlying local electrical properties. This, however, will be the topic of future work, and within this work we will therefore focus on characterization at lower wavelengths, namely the 640 nm.

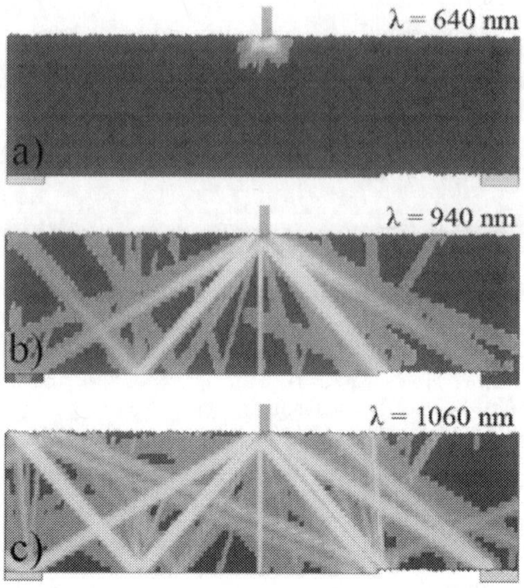

Figure 1. Wavelength dependent optical generation for a 10 µm wide, top side illumination beam incident halfway between the n+ and the p+ fingers of the Zebra IBC silicon solar cell for a) 640 nm, b) 940 nm, and c) 1060 nm.

After obtaining the optical generation profiles, we transformed the 3D optical generation distribution into its 2D representation, paying special attention to preserve the shape and the integral of the optical generation rate as best as possible, in order to use the distribution in further 2D drift-diffusion electrical simulations. After applying the optical generation rate to the 2D electrical mesh the cumulative generation rate was scaled one again in order to compensate for any possible errors due to interpolation between different finite-element meshes. This step proved crucial to obtaining smooth LBIC profiles in the simulation. The 2D electrical model is based on a set of measured parameters (doping profiles, bulk lifetimes) and a set of commonly accepted electrical models and parameters [11], previously validated for the exact same device structure [10].

We used the combined opto-electrical model in an iterative fashion where in each step we first calculated the exact distribution of the optical generation rate, then transformed it into its 2D counterpart, ran the electrical simulation and as a last step extracted the J_{SC}. In this manner we scanned the entire region-of-interest (ROI) of our DUT. The extracted positionally dependent short-circuit current densities were then compared to the measured LBIC response of the same ROI.

2.1 Modelling of Cut-Edges

In order to model laser-cut and cleaved edges solar cell edges, we formed a 20 µm wide region at the edge of the cut cell in order to accurately recreate the effects of laser ablation. The width of the region was approximated from the laser spot size as well as microscopic images of the damaged edges. Even though technically laser ablation removes material we modelled the damaged area as a bulk region with homogeneously reduced lifetime.

3 RESULTS AND DISCUSSION

3.1 Modelling of a Healthy Cell

As a first step, we validated our modeling approach on a healthy, non-damaged part of the IBC. The results of the LBIC scans of a laser-cut and cleaved DUT at 640 nm are shown in Fig. 2, with the marked healthy part of the cell used for the validation of the LBIC model. Fig 2. Also shows the extracted coarse (200 µm step) and detailed (20 µm step) LBIC scans of the Zebra IBC from the front side within the marked ROI. For the comparison the simulated LBIC profile using our combined opto-electrical approach is also shown in Fig. 2.

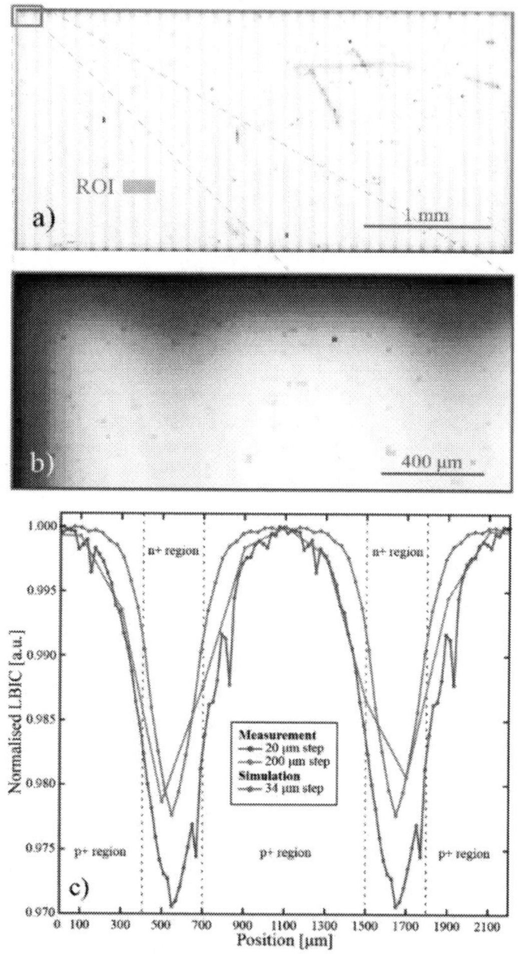

Figure 2. a) Coarse and b) detailed LBIC scans of the Zebra IBC solar cell at 640 nm, and c) comparison of normalized LBIC responses between two measurements with steps of 200 µm and 20 µm (period is extended for easier visual comparison) and a simulation with a step of 20 µm.

As a first approximation all three profiles seem to agree rather well, exhibiting the same periodicity as well as roughly the same magnitude of the normalized LBIC response decrease in the n+ region of the IBC. The only major discrepancy between the measured and the simulated profiles is the width of the dip in the n+ region which is wider in the measurements. This could indicate that the width of our chosen simulation illumination beam of 10 µm is actually narrower than the width of the measurement setup's beam. The validation of our model gives us confidence that we can accurately recreate LBIC profiles also at the edges of the cells with proper lifetime reduction.

3.2 Lifetime Reduction in the Cut-Edge Area

To model the effects of cut edge we included the minority carrier lifetime reduction region in the vicinity (20 µm) of the cut edge, spanning half the thickness of the bulk (75 µm). The effects of lifetime reduction are shown in Fig. 3. We found that a severe reduction to approx. 0.05‰ of the bulk lifetime is necessary in order to match the LBIC profile at the edge. It is important to note that all the recombination of the cut edge is attributed to the laser damaged part of the cut, while the cleaved edge is considered as perfectly passivated. This is obviously a simplification of the true electrical situation, but as the recombination on the cleaved edge is much lower than in the laser ablated part, considering only that part of the recombination should be sufficient.

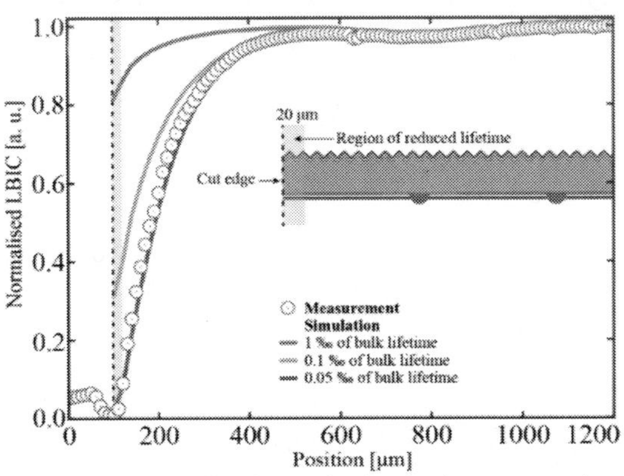

Figure 3. Effect of lifetime reduction in the 20 µm region at the cut cell edge.

3.3 Effect of Laser Cut Depth

Furthermore, we studied the influence of the laser cut depth on the recombination parameters on the edge. For that purpose, we laser scribed a Zebra IBC cell in six different locations, with a different number of passes between the cuts. The normalized LBIC profiles of such laser scribed edges are shown in Fig. 4. One can observe that with increasing number of laser passes the normalized profiles do not exhibit much change in their lowest point. One can observe, that a slight change in profile slope as well as a slight widening of the affected area is present. Therefore, one can conclude, that the number of laser passes, hence the cut depth does not play a major role in the shaping of edge profiles.

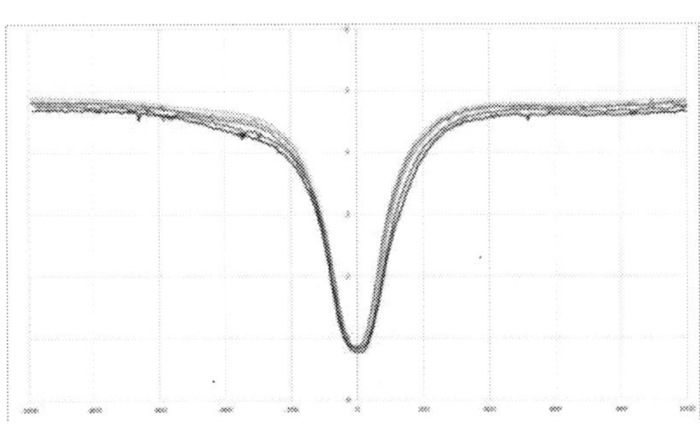

Figure 4. Effect of number of laser scribe passes / cut depth on the LBIC profiles around the damaged area.

This is further confirmed by changing the depth of the defect region in the simulations from 25 µm to the whole thickness of the bulk (150 µm). The results are shown in Fig. 5, and one can observe, that the depth of the lifetime reduction region does not significantly influence the LBIC profiles. As with the measurements in Fig. 3 the depth of the cut does not change the value of the lowest point at 0 µm, but rather affects the slope of the profile asnd slightly widens the affected area. Nevertheless, the best match between the simulation and the measurement is achieved with the lifetime reduction region taking up roughly half of the bulk thickness, which coincides well with the measured cut depth also shown in Fig. 5.

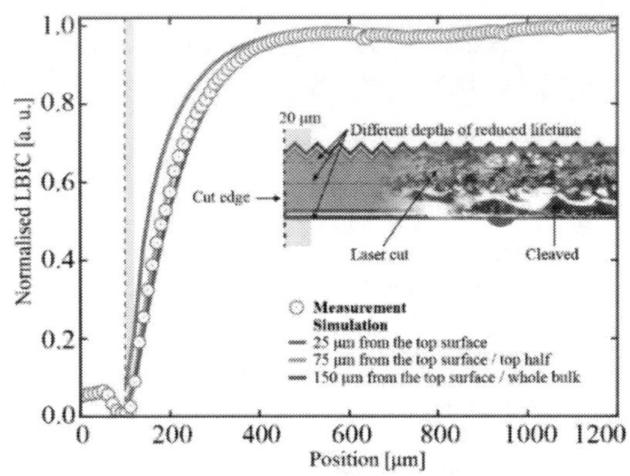

Figure 5. Effect of lifetime reduction in the 20 µm region at the cut cell edge.

4 CONCLUSION AND OUTLOOK

In this work we established a detailed opto-electrical workflow for accurate LBIC response simulation of healthy as well as cut solar cells. We showed that LBIC simulation can be used for extraction lifetime reduction attributed to cut edges. We found that a severe reduction of the bulk minority carrier lifetime is necessary (0.05‰) to correctly model the laser-cut and cleaved edges, while the depth of the damaged region only plays a secondary role. This same approach will be used in our future work to simulate the multi-wavelength LBIC responses at the cut edges in order to extract positionally dependent lifetime profiles (or rather lifetime decrease profiles) corresponding to the specific types of the aforementioned cell separation techniques. The extracted lifetime reduction profiles will then be used to thoroughly understand the processes that govern the performance reduction of cut solar cells. Based on the knowledge we will establish possible guidelines for edge separation procedures or edge re-passivation.

5 ACKNOWLEDGMENTS

7 REFERENCES

[1] M. Kessler, T. Ohrdes, P. P. Altermatt, and R. Brendel, "The effect of sample edge recombination on the averaged injection-dependent carrier lifetime

in silicon," *J. Appl. Phys.*, vol. 111, no. 5, Art. no. 5, Mar. 2012, doi: 10.1063/1.3691230.

[2] A. Fell, J. Schön, M. Müller, N. Wöhrle, M. C. Schubert, and S. W. Glunz, "Modeling Edge Recombination in Silicon Solar Cells," *IEEE J. Photovolt.*, vol. 8, no. 2, Art. no. 2, Mar. 2018, doi: 10.1109/JPHOTOV.2017.2787020.

[3] R. Kopecek *et al.*, "ZEBRA technology: low cost bifacial IBC solar cells in mass production with efficiency exceeding 23.5%," in *2020 47th IEEE Photovoltaic Specialists Conference (PVSC)*, Jun. 2020, pp. 1008–1012. doi: 10.1109/PVSC45281.2020.9300503.

[4] K. C. Fong *et al.*, "Perimeter Recombination Characterization by Luminescence Imaging," *IEEE J. Photovolt.*, vol. 6, no. 1, Art. no. 1, Jan. 2016, doi: 10.1109/JPHOTOV.2015.2480225.

[5] S. Schäfer *et al.*, "26%-efficient and 2 cm narrow interdigitated back contact silicon solar cells with passivated slits on two edges," *Sol. Energy Mater. Sol. Cells*, vol. 200, p. 110021, Sep. 2019, doi: 10.1016/j.solmat.2019.110021.

[6] M. D. Abbott, J. E. Cotter, T. Trupke, and R. A. Bardos, "Investigation of edge recombination effects in silicon solar cell structures using photoluminescence," *Appl. Phys. Lett.*, vol. 88, no. 11, Art. no. 11, Mar. 2006, doi: 10.1063/1.2186510.

[7] H. Stolzenburg *et al.*, "Edge recombination analysis of silicon solar cells using photoluminescence measurements," Fes, Morocco, 2019, p. 020017. doi: 10.1063/1.5123822.

[8] P. Saint-Cast, D. Herrmann, P. Baliozian, H. Stolzenburg, H. Hoeffler, and A. Fell, "Extracting Metal and Edge Recombination Parameters which are Compatible with Multi-Dimensional Cell Simulations," p. 5, 2019.

[9] M. Kikelj, B. Lipovšek, M. Bokalič, F. Buchholz, and M. Topič, "Detailed 3D Optical Modelling of Interdigitated Back Contact Solar Cells," in *2021 IEEE 48th Photovoltaic Specialists Conference (PVSC)*, Jun. 2021, pp. 0997–1000. doi: 10.1109/PVSC43889.2021.9518850.

[10] M. Kikelj, M. Bokalič, M. Topič, and B. Lipovšek, "Detailed luminescence modelling in high-efficiency solar cells for precise calibration of spatially resolved characterisation methods: A bottom-up opto-electrical approach," *Sol. Energy Mater. Sol. Cells*, vol. 248, p. 111990, Dec. 2022, doi: 10.1016/j.solmat.2022.111990.

[11] A. Fell *et al.*, "Input Parameters for the Simulation of Silicon Solar Cells in 2014," *IEEE J. Photovolt.*, vol. 5, no. 4, Art. no. 4, Jul. 2015, doi: 10.1109/JPHOTOV.2015.2430016.

CHARACTERISTICS OF PEDOT:PSS/Si SOLAR CELLS WITH A FLAT AND TEXTURED INTERFACES

S.V. Mamykin, I.B. Mamontova, T.S. Lunko, O.S. Kondratenko, N.V. Kotova, V.R. Romanyuk
V. Lashkaryov Institute of Semiconductor Physics National Academy of Sciences of Ukraine
41 Nauky ave., Kyiv 03680, Ukraine, Tel: +38 (044) 5256546; E-mail: mamykin@isp.kiev.ua

ABSTRACT: Solar cell (SC) heterostructures based on PEDOT:PSS/Si heterojunction with a flat and quasigrating type microrelief interface and two types of PEDOT:PSS films – pure and chemically doped have been fabricated and studied. It is shown that an increase in the SC efficiency occurs both due to texturing of the Si interface (increase in absorbed light due to decrease in optical losses), and improvement of electrical parameters of the polymer layer due to chemical doping (with DMSO) and post-fabrication treatment (washout of the PSS group). The maximum efficiency (4.82%) was obtained for SC heterostructures with a quasigrating Si interface and modified PEDOT film, which is ~2-6 times higher than that for the similar structure and for structure with a flat interface.
Keywords: solar cells, heterojunction, silicon, PEDOT:PSS, texturization

1 INTRODUCTION

Poly(3,4-ethylenedioxythiophene)-poly(styrene sulfonate) (PEDOT:PSS) is a *p*-type organic polymer mixture which has been widely used to create organic-inorganic solar cells based on crystalline silicon (c-Si). This is the vacuum-free spin coating technology at room temperature, which greatly reduces the cost of production. PEDOT:PSS thin films have high transparency, conductivity, passivating and anti-reflection properties [1]. In a solar cell, such films serve as a conductive front electrode and a charge-selective layer that reflects electrons and extracts holes, thereby significantly reducing recombination [2]. But the intrinsic conductivity of PEDOT:PSS is still not sufficient to obtain the maximum characteristics of the photoconverting heterostructures. Therefore, the search for the methods for increasing the conductivity of thin PEDOT:PSS films while maintaining its transparency continues in various directions: chemical doping, creating composites, and post-fabrication treatments.

In general case, a rather large (up to 30%) part of the incident radiation can be reflected back from the surface of Si-based solar cell and not contribute to photoconversion [3]. Layered anti-reflective coatings reduce reflectivity mainly in a limited spectral and angular range, because the interference pattern depends on thickness of the film and angle of light incidence. Patterning of semiconductor surface is widely used to reduce reflection losses in the solar cell structures as well [4]. Nano- and microstructuring of the SC surface can significantly suppress the reflection of front surface, in a much wider spectral range and without pronounced angular dependence.

To increase the absorption of incident radiation by solids, the different type of micro- and nanoreliefs of various morphologies can be formed by chemical etching of the semiconductor surface, for example, such as nanowires [5], nanoholes [6], silicon nanotubes [7], nanocones [8], nanopyramids [9], inverted pyramids [10], porous ("black") silicon [11], dendrite type and quasigrating [12].

In this work, the effect of interface patterning as well as organic film electrical properties modifications on photoelectrical characteristics of organic/inorganic solar cell heterostructures based on a PEDOT:PSS/Si heterojunction are investigated.

2 EXPERIMENTAL DETAILS

Polished silicon wafers of (100) orientation, *n*-type conductivity, doped with phosphorus, with a resistivity of 0.5-2.4 Ohm·cm, were used as a substrate for the fabrication of heterostructures with flat interface.

Microtextured Si surfaces of quasigrating type were obtained on *n*-type Si (110) doped with phosphorus (0.5-1.0 Ohm·cm) by anisotropic chemical etching in an aqueous solution of 20% KOH at 75°C, t = 35-45 minutes.

As-purchased, undiluted transparent conductive PEDOT:PSS (3.0-4.0% in H_2O, Sigma Aldrich) polymer formulation was used for the fabrication of reference (PEDOT:PSS/Si) heterostructures.

For the investigation of chemical doping on the properties of organic polymer film an aqueous PEDOT:PSS (2 wt% in H_2O, Sigma Aldrich) polymer formulation was used. Dimethyl sulfoxide (DMSO) (2.5 wt.%) was added to increase the electrical conductivity of the film and Triton X100 (2 wt.%) to ensure wettability of the hydrophobic Si surface. Besides, these films were subjected to rinsing in ethanol after annealing to improve their conductivity by washout of the non-conductive water-soluble PSS fraction [13].

Films with a target thickness of 100-200 nm on the surface of the flat and patterned substrates were obtained by multiple spin-coating procedure and posterior annealing at a temperature of 100-110°C for 1 minute [14].

The thickness of the films and their optical properties were extracted from the spectroscopic ellipsometric measurements (SE-2000, Semilab) in the spectral range of 0.6-5.0 eV.

The light current-voltage characteristics of solar cells were measured under simulated AM0 illumination with a radiation power of P=136 mW/cm^2.

3 RESULTS AND DISCUSSION

The optical micrograph and AFM images of the Si surface with a microrelief of the quasigrating type are shown in Fig. 1 and 2. AFM measurements showed that the quasigrating period is 2.2 μm, the average depth is 250-400 nm. The angle between the faces at the top or in the grooves is of the order of 145°, that means, the relief is rather shallow.

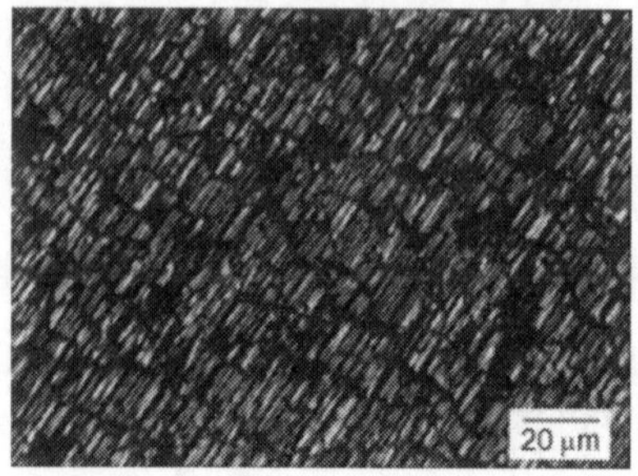

Figure 1: Optical micrograph of Si (110) surface with microrelief of quasigrating type fabricated by wet anisotropic chemical etching

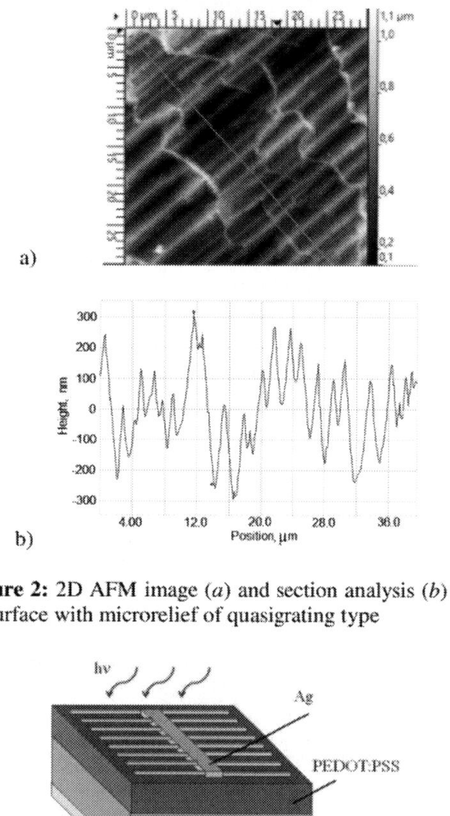

Figure 2: 2D AFM image (*a*) and section analysis (*b*) of Si surface with microrelief of quasigrating type

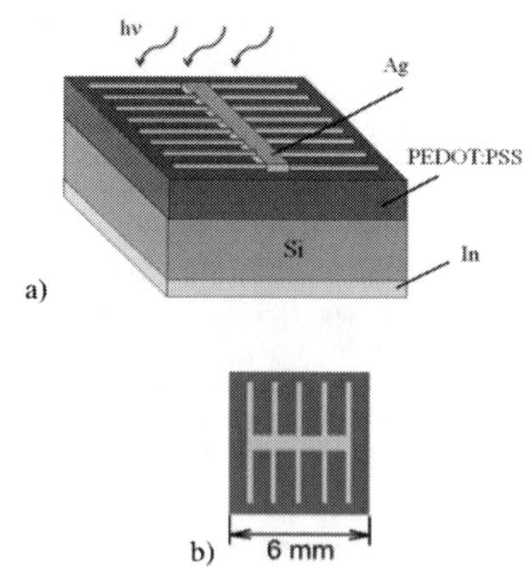

Figure 3: Schematic representation of the solar cell structure design based on PEDOT:PSS on the flat surface of Si including Ag gride as top contact and In backside ohmic contacts (*a*); top-side image of the structure used in the experiments (*b*)

Finally, to complete fabrication of solar cells heterostructures, the Ag contact grids have been deposited by thermal evaporation through the masks of corresponding geometry. The In ohmic contact was fabricated on the back side of Si plate. The general design of the solar cells is shown schematically in Fig. 3.

Current-voltage characteristics reflect all the processes occurring in solar cells and, in general, are described by formula derived from the Shockley diode equation [15]:

$$I(U) = I_{ph} - I_0\left(\exp\left(\frac{q(U - IR_s)}{nkT}\right) - 1\right) - \frac{U + IR_s}{R_{sh}}, \qquad (1)$$

where I_{ph} is the photocurrent, I is the total current, I_0 is the saturation current, U is the voltage, R_s and R_{sh} are the series and shunt resistances, respectively, n is the diode ideality factor, T is the temperature, k is the Boltzmann constant, q is the electron charge.

The efficiency of a solar cell is determined by the formula [15]:

$$\eta = \frac{P_m}{p} = \frac{fI_{sc}U_{oc}}{P}, \qquad (2)$$

where P_m is the maximum power of the solar cell, P is the power of incident radiation on the solar cell, f is the fill factor of the current-voltage characteristic, I_{sc} is the short-circuit current and U_{oc} is the open circuit voltage.

The photoelectrical parameters of solar cells (efficiency η, open circuit voltage, short-circuit current, fill factor, etc.) are calculated based on the measured current-voltage (I-V) characteristics (Fig. 4 and 5) and presented in Table I and II.

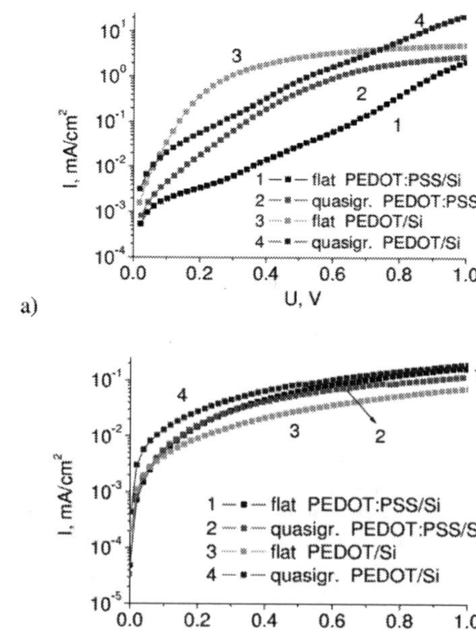

Figure 4: The dark direct (*a*) and reverse (*b*) current-voltage characteristics of solar cell structures fabricated on the flat (1, 3) and textured (quasigrating type) (2, 4) Si substrate with untreated PEDOT:PSS (1, 2) and modified (3, 4) PEDOT polymer layers

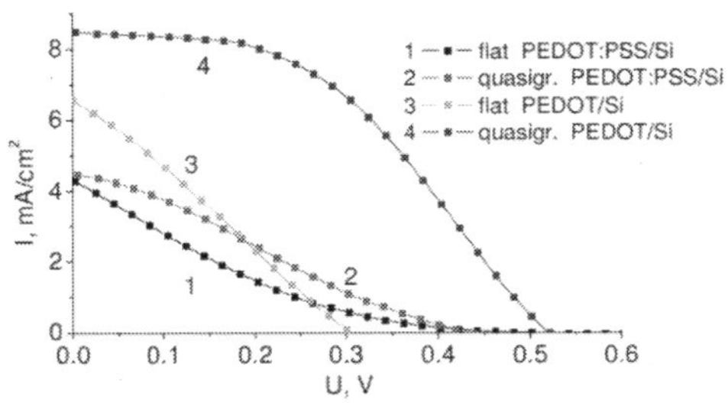

Figure 5: Light current-voltage characteristics of solar cell structures fabricated on flat (1, 3) and textured (quasigrating type) (2, 4) Si substrate with non-modified PEDOT:PSS (1, 2) and modified (3, 4) PEDOT polymer layers

Table I: Structure and electrical parameters of PEDOT:PSS/Si solar cells

Sample number	Structure of solar cells	R_s, Ohm	R_{sh}, Ohm
#1	PEDOT:PSS/Si (flat)	1535.9	69.3
#2	PEDOT:PSS/Si (quasigrating type)	156.8	252.9
#3	PEDOT/Si (flat)	50.8	69.4
#4	PEDOT/Si (quasigrating type)	41.5	513.6

Table II: Photoelectrical parameters of PEDOT:PSS/Si solar cell structures

Sample number	I_{sc}, mA/cm^2	U_{oc}, mV	f, %	η, %
#1	14.2	504.6	14.3	0.75
#2	26.5	431.1	25.3	2.13
#3	21.5	304.5	26.6	1.28
#4	27.8	520.0	45.3	4.82

It can be seen that with adding DMSO and Triton X100 to PEDOT:PSS formulation in certain proportions, as well as washout of the PSS fraction, the characteristics of solar cell structures (efficiency η, fill factor f, open circuit potential V_{oc}, and short circuit current I_{sc}) were significantly improved. The application of the interface patterning increases the efficiency of photoconverting about 3 times compared to a structure with a flat interface. From a physical point of view, texturing of the interface leads to a decrease in optical losses and an increase in absorbed light, which contributes to the increasing of the short-circuit current.

Obviously, not all of this photoconversion efficiency increase can be attributed to enhanced absorption of incident radiation. It is likely that interface texturing also changes the electronic properties of the semiconductor surface. As a consequence the maximum efficiency (4.82%) was obtained for solar cell structures manufactured on the microrelief surface of a semiconductor and using a modified organic polymer film, which is ~2-6 times higher than that for the similar structure and the structure with a flat interface.

The proposed microtexturing of the quasigrating type with subsequent technology of multiple PEDOT deposition not only leads to a decrease in light reflection, but also does not reduce the open circuit voltage, as usually happens after texturing due to a pinhole in the PEDOT layer.

4 CONCLUSIONS

Solar cell heterostructures based on PEDOT:PSS/Si heterojunction with a flat and microrelief (quasigrating type) interface and two types of PEDOT:PSS films – unmodified and chemically doped have been fabricated.

Heterostructures on microtextured surfaces show up to 6 times higher efficiency than on flat ones. Chemical doping of formulation and physical removal of the non-conductive component from the PEDOT:PSS film improves its electrophysical parameters and leads to a decrease in heterostructure series resistance and an increase in shunt resistance, which contributes to an increase in the efficiency of SC based on them.

The maximum efficiency (4.82%) was obtained for SC structures with a quasigrating interface and PEDOT film with removed PSS part, which is ~2-6 times higher than that for the similar structure and for the structure with a flat interface.

ACKNOWLEDGMENT

The authors express their gratitude for funding from the Ministry of Education and Science of Ukraine (project No. 22BF051-04, State No. 0122U001956 "Nanostructured metasurfaces for sensors and photovoltaics").

5 REFERENCES

[1] Z. Sun, Y. He, B. Xiong, S. Chen, M. Li, Y. Zhou, K. Sun, Y. Zheng, C. Yang, Angewandte Chemie International Edition 60(10) (2020) 5036.

[2] P. Gao, Z. Yang, J. He, J. Yu, P. Liu, J. Zhu, J. Ye, Advanced Science 5(3) (2017) 1700547.

[3] K. Forberich, G. Dennler, M.C. Scharber, K. Hingerl, T. Fromherz, C.J. Brabec, Thin Solid Films 516(20) (2008) 7167.

[4] P. Singh, S.K. Srivastava, B. Sivaiah, P. Prathap, C.M.S. Rauthan, Solar Energy 170 (2018) 221.

[5] H. Wang, J. Wang, L. Hong, Y.H. Tan, C.S. Tan & Rusli, Nanoscale Research Letters 11(1) (2016) 311.

[6] L. Hong, Rusli, X. Wang, H. Zheng, H. Wang, H. Yu, Nanotechnology, 24(35) (2013) 355301.

[7] H. Jeong, H. Song, Y. Pak, I.K. Kwon, K. Jo, H. Lee, G.Y. Jung, Advanced Materials 26(21) (2014). 3445.

[8] H. Wang, J. Wang, Rusli, Nanoscale Research Letters 10(1) (2015) 191.

[9] X. Wang, Z. Liu, Z. Yang, J. He, X. Yang, T. Yu, P. Gao, J. Ye, Small 14(15) (2018) 1704493.

[10] W. Chen, Y. Liu, L. Yang, J. Wu, Q. Chen, Y. Zhao, Y. Wang, X. Du, Scientific Reports 8(1) (2018) 3408.

[11] X. Liu, P.R. Coxon, M. Peters, B. Hoex, J.M. Cole, D.J. Fray, Energy Environmental Science 7(10) (2014) 3223.

[12] N.L. Dmitruk, O.Y. Borkovskaya, I.N. Dmitruk, I.B. Mamontova, Solar Energy Materials and Solar Cells 76(4) (2003) 625.
[13] O.S. Kondratenko, S.V. Mamykin, T.S. Lunko, I.B. Mamontova, V.R. Romanyuk, Molecular Crystals and Liquid Crystals 717(1) (2021) 92.
[14] S.V. Mamykin, I.B. Mamontova, T.S. Lunko, O.S. Kondratenko, V.R. Romanyuk, Semiconductor Physics, Quantum Electronics & Optoelectronics 24(2) (2021). 148.
[15] S.M. Sze, Physics of Semiconductor Devices. 2nd ed. New York: John Wiley and Sons; 1981.

IRON QUANTIFICATION IN CRYSTALLINE SILICON SOLAR CELLS
FROM OPEN-CIRCUIT VOLTAGE MEASUREMENTS

Axel Herguth
University of Konstanz
78457 Konstanz, Germany

ABSTRACT: Iron contamination in crystalline silicon can be quantified with high sensitivity thanks to its peculiarity to switch between interstitial and dopant-paired form, each exhibiting specific recombination activity. Typically, this is done by an iso-injective comparison of effective excess carrier lifetimes with iron either being interstitial or paired form. Within this contribution it is shown by means of PC1D simulations that the procedure works for solar cells as well, exploiting that effective lifetimes can be calculated from Voc data. Furthermore, it is shown that even a comparison of Voc's taken at a constant intensity, meaning differing injection, already allows for a good estimation of iron contamination. Using multiple intensities allows an even better quantification.
Keywords: characterization, iron, quantification, silicon solar cell, simulation

1 INTRODUCTION

Iron is a common but unfortunately quite harmful contamination in crystalline silicon solar cells, even though it is often present 'only' in concentrations in the $10^{10}\,cm^{-3}$ range even after phosphorous gettering. Iron may exist in interstitial form (Fe_i) or paired with acceptors like boron (FeB) {or gallium (FeGa)}. In experiments iron can be easily toggled between Fe_i and FeB form {or FeGa} by short illumination steps and storage in darkness [1, 2]. Both forms are strongly recombination active, each affecting injection-dependent lifetime and, in consequence, solar cell performance (esp. V_{oc}) in a specific way as shown in Fig. 1. In fact, this can be used for the quantification of iron in crystalline silicon [1, 2]. Typically, this is done by means of excess carrier lifetime spectroscopy on lifetime samples. Within this contribution, it is evaluated by means of PC1D simulations whether iron quantification is possible on cell level using open-circuit voltage (V_{oc}) measurements only.

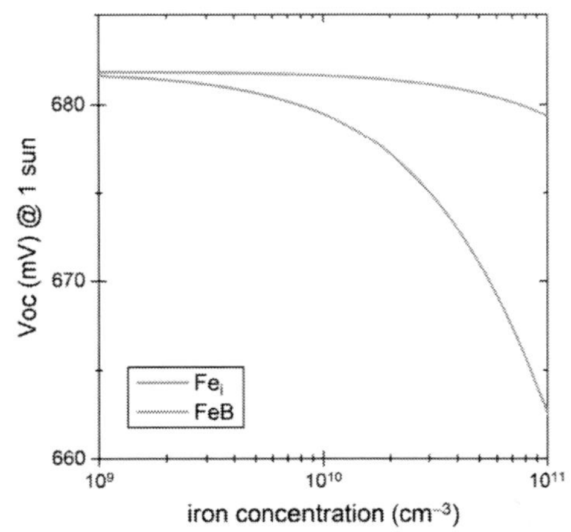

Figure 1: Impact of iron either in interstitial form or paired to boron on the open-circuit voltage of PERC-like solar cell.

2 SIMULATION APPROACH

PC1D simulations [3] with well-known input data were used to evaluate the possibility to quantify iron contamination from V_{oc} measurements. The simulated PERC-like cell features a base doping of $1\times10^{16}\,cm^{-3}$ and a total front and rear surface saturation current density of $110\,fA/cm^2$. Series resistance is not of importance as the presented method relies on V_{oc} only. Iron contamination either in form of FeB or Fe_i was taken into account as Shockley-Read-Hall (SRH) defect with the well-known SRH parameters summarized in Table I. No other defect was assumed to exist. For ease of computation of generation rate, monochromatic illumination at 800 nm was used with an intensity of $62\,mW/cm^2$ (photon-flux equivalent to 1 sun with AM 1.5g). As charge carriers fairly redistribute under open-circuit conditions, the deviating generation profile has a negligible impact on V_{oc}. The impact of iron contamination ranging from 10^9 to $10^{11}\,cm^{-3}$ on V_{oc} is shown in Fig. 1.

Table I: Defect-specific SRH parameters (from [2])

	σ_n (cm^{-2})	σ_p (cm^{-2})	E (eV)
FeB	5×10^{-15}	3×10^{-15}	$E_c - 0.26$
Fe_i	4×10^{-14}	7×10^{-17}	$E_v + 0.38$

3 DATA ANALYSIS APPROACH

As pointed out by Kerr et al. [4], mass action law allows to translate pn-junction voltage (or V_{oc}) into an excess charge carrier density Δn according to

$$n \cdot p = \Delta n^2 + \Delta n \cdot (n_0 + p_0) + n_0 p_0 = n_i^2 \cdot \exp\left(\frac{qV_{oc}}{kT}\right)$$

where n and p are total electron and hole densities while n_0 and p_0 denote electron and hole equilibrium densities, n_i the effective intrinsic carrier density and kT/q the mean thermal energy.

With known generation rate G, an effective lifetime τ_{eff} can then be calculated according to

$$G \cdot \tau_{eff} = \Delta n$$

Hence, V_{oc} may be converted to τ_{eff} valid at a certain injection Δn. In this context it is important to note that a difference in V_{oc} (Fig. 1) implies different injection Δn.

Effective lifetime can be approximated as the sum of inverse lifetimes related to iron (FeB or Fe_i) and a non-iron related background (τ_{bg}).

$$\frac{1}{\tau_{eff}} = \frac{1}{\tau_{bg}} + \frac{1}{\tau_{FeB}} + \frac{1}{\tau_{Fei}} = \frac{1}{\tau_{bg}} + f_{FeB}[FeB] + f_{Fei}[Fe_i]$$

where the different recombination activity of FeB and Fe_i

is captured by the known specific pre-factors f_{FeB} and f_{Fei} (following from SRH theory with parameters from Tab. I), and where the iron related lifetime components scale inversely with the respective concentrations [FeB] and [Fe_i]. Total iron concentration [Fe] resembles the sum [FeB] + [Fe_i]. In this context, it is important to note that FeB or Fe_i form may be toggled almost completely by storage in darkness (FeB forms) and illumination steps (FeB splits) so that either [FeB] = [Fe] or [Fe_i] = [Fe] while τ_{bg} remains the same. If some iron does not switch its form or is clustered, it becomes part of the unchanged background. Insofar, [Fe] relates only to the part of iron toggled between FeB and Fe_i.

Solving the above equation system with either [FeB] = [Fe] or [Fe_i] = [Fe] yields

$$[Fe] = \frac{\left(\frac{1}{\tau_{eff,Fei}} - \frac{1}{\tau_{eff,FeB}}\right) - \left(\frac{1}{\tau_{bg,Fei}} - \frac{1}{\tau_{bg,FeB}}\right)}{f_{Fei} - f_{FeB}}$$

where all quantities with index FeB refer to the system with iron in the paired form and all quantities with the index Fe_i refer to the system with iron being in interstitial form. However, as mentioned earlier, a difference in V_{oc} occurring in iso-generative conditions (Fig. 1) implies a difference in injection Δn so that the quantities in the above equation refer to different injections: $\Delta n_{FeB} \neq \Delta n_{Fei}$, and background lifetime (even though unchanged) is not the same: $\tau_{bg,FeB} \neq \tau_{bg,Fei}$.

The situation is exemplarily shown in Fig. 2 for an iron contamination of $5\times10^{10}\,cm^{-3}$ where the effective lifetimes at V_{oc} in the FeB and Fe_i state are highlighted. As depicted in more detail in the inset, background lifetime is not constant in this injection range, implying that the background difference term does not cancel out and that [Fe] cannot be calculated from the change in τ_{eff} due to a lack of information on the injection-dependence of τ_{bg}.

In the following, it is discussed what this lack of information in iso-generative conditions means for the quantification of iron contamination.

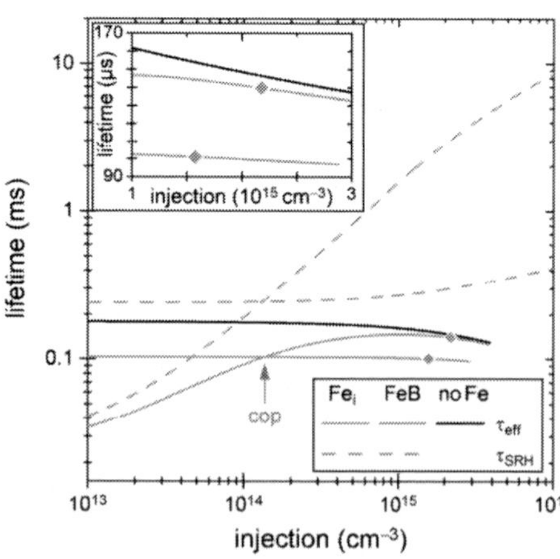

Figure 2: Injection-dependent effective lifetimes and iron-related components with $5\times10^{10}\,cm^{-3}$ iron in FeB or Fe_i form (or absent for reference). The lifetimes calculated from V_{oc} are highlighted. Inset in linear scale.

4 DISCUSSION

In principle, it would be better to adapt illumination intensity so that V_{oc} and thus Δn in the FeB and Fe_i state match and the background lifetime difference term in the above equation cancels out. In reality this would mean to measure V_{oc} not just at one intensity but rather in a certain intensity range. Or in other words, to use Suns-V_{oc} measurements and select iso-injective conditions for analysis. As can be seen in Fig. 3, the corresponding green line yields reasonable results close to unity. Note that the exact result of the Suns-V_{oc} analysis depends to some degree on the chosen injection level.

However, a Suns-V_{oc} measurement might not be available. Hence, what does the iso-generative change in V_{oc} tell about the iron contamination?

The advantage of simulation is that background is well-known from simulation without iron contamination. Hence, the above equation may be used with this knowledge to evaluate whether the iso-generative approach works or not. As can be seen in Fig. 3, the corresponding black line is fairly close to unity as well, demonstrating that the above equation describes the situation correctly. But without any available knowledge on background lifetime, this approach is not viable in reality.

One possible approach is to simply neglect the change in background lifetime so that the background difference term cancels out even though this is obviously not true in the shown example (inset of Fig. 2). As can be seen in Fig. 3, this results in a certain underestimation of up to 15% (for 1 sun). Shifting the analysis to lower intensities like 1/3 sun yields an even better estimation of iron contamination. The reason behind this is that an analysis at lower intensity shifts V_{oc} to lower injection where background lifetime (black line in Fig. 2) flattens out so that the background difference term loses importance. The same happens in Fig. 3 for higher iron contamination which imply lower effective lifetimes and thus lower injection as well.

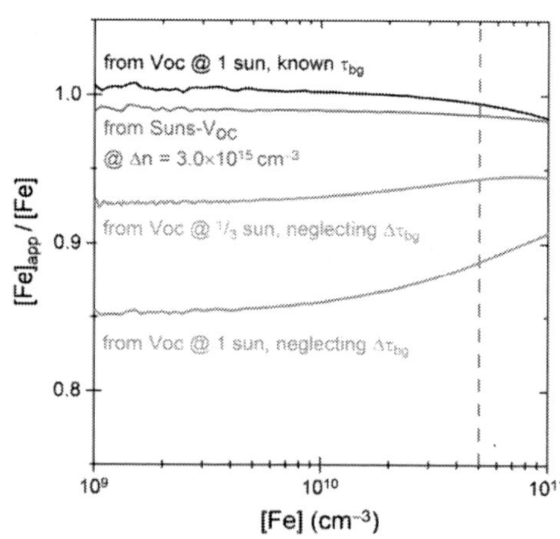

Figure 3: Ratio of apparent $[Fe]_{app}$ and known iron contamination [Fe] evaluated across the assumed iron contamination range for different approaches calculated from the V_{oc} values shown in Fig. 1. The gray dashed line marks the situation from Fig. 2.

5 CONCLUSION AND OUTLOOK

Considering the simplicity of the approach – just a V_{oc} measurement before and after splitting of FeB and some calculations – neglecting the change in background lifetime seems a viable approach for the estimation of iron contamination on cell level.

An in-depth analysis of the presented approach including the impact of doping and a thorough error analysis is published elsewhere [5]. It is furthermore elaborated how the combination of V_{oc} measurements taken at different intensities can eliminate the lack of information on background lifetime leading to higher accuracy in the quantification of iron contamination on cell level [5].

REFERENCES

[1] G. Zoth, W. Bergholz, Journal of Applied Physics 67 (1990) 6764.

[2] D. Macdonald, L. Geerligs, A. Azzizi, Journal of Applied Physics 95 (2004) 1021.

[3] H. Haug, J. Greulich, A. Kimmerle, E.S. Marstein, Solar Energy Materials & Solar Cells 142 (2015) 47.

[4] M. Kerr, A. Cuevas, R. Sinton, Journal of Applied Physics 91 (2002) 399.

[5] A. Herguth, IEEE Journal of Photovoltaics 12 (2022), 937

USING LOW-COST SCANNERS FOR OPTICAL INSPECTION OF SOLAR CELLS

Mattias Klaus Juhl, Zhuangyi Zhou, Binesh Puthen Veettil, and David Neil Roger Payne
Macquarie University, School of Engineering
Sydney, NSW, 2113, Australia

ABSTRACT: This paper investigates the possibility of using low-cost consumer flatbed scanners for optical inspection of solar cell precursors. Two types of scanners, with contact image sensor (CIS) and charge coupled device (CCD) technologies, are used and compared for monitoring the texturing quality between and across multi-crystalline and mono-crystalline silicon wafers. The performance of the low-cost scanner in monitoring wafer reflectance is evaluated by comparing the outputs with the results from a high-end industry scanning spectrophotometer LOANA. The study shows good qualitative agreement and statistical correlation of multi-crystalline silicon wafer images measured with a low-cost flatbed scanner and the LOANA.
Keywords: optical inspection, low cost, texturing

1 INTRODUCTION

Several consumer electronics devices have been reported to be useful in technologically complex scientific equipment designs. For example, office scanners have been used for radiation monitoring [1-4], biomedical imaging [5], physical dimension tracking [6] etc. A contact image sensor (CIS) and a charge coupled device (CCD) based flatbed scanner are examined in this work for the possibility to monitor the reflectance of silicon wafers. Notably, these types of scanners have also previously been used in holographic microscopy [7].

Spatial uniformity of a solar cell is desirable for high efficiency but hard to achieve because the state-of-the-art solar cells are large area (in some cases greater than 220 cm^2) devices. The reflection characteristics between and across solar cells is often monitored to quantify spatial uniformity. Stand-alone point-by-point characterisation equipment such as the laser-based reflectance mapping Semi-lab tool and the scanning spectrophotometer LOANA from pv-tools are commonly used in the photovoltaics (PV) industry for texturing quality inspection, however, the measurements can take as long as 20 mins per sample. Such a slow measurement speed is far behind the production speeds (over one sample per second per line) in the PV industry. Furthermore, the costs of these tools are typically in the range of tens to hundreds of thousand dollars, making them prohibitively expensive for many research groups with limited budgets.

If low-cost (< $100 USD), off the shelf flatbed scanners with high spatial resolution (40 μm pixel) can be applied for optical inspection of solar cells, they can substantially reduce the equipment costs and measurement time, resulting in optical characterisation of solar cells being more accessible for more people.

In this paper, we compare two flatbed scanners based on CIS and CCD technology. The analysis of their performance shows that the CIS flatbed scanner is preferred to monitor the variation of reflectance across solar cell precursors and can take fewer than 10s per sample for optical inspection of wafer texturing quality.

2 METHODOLOGY

A conventional flatbed scanner consists of a built-in white light source that enables measurements across three different colour channels (red, green and blue) and a line sensor that has its field of view sequentially moved across a region to build up a two-dimensional image. As the design intention of such scanners is to scan documents, a typical scan area is at least the size of an A4 page (210 mm x 297 mm), larger than the current and projected size of a solar cell.

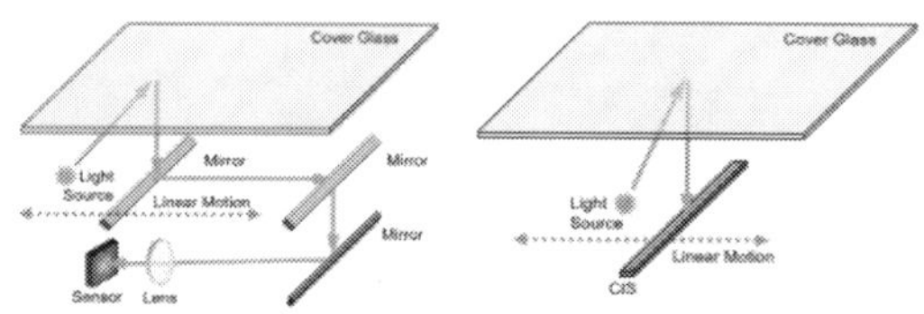

Figure 1: Comparison of the experimental configurations between (a) a standard CCD scanner and (b) a CIS scanner.

The experimental configurations of the two types of flatbed scanners investigated in this work are shown in Figure 1. To compare the reflectance measured with these two types of scanners, the same p-type random pyramid textured mono-crystalline solar cell precursor was scanned with a CIS scanner Cannon LiDE 300 and a CCD scanner Epson V800. The measurements were conducted using different colour channels, i.e., red, green and blue.

Then, a preferred scanner was selected for performance comparison with a LOANA (pv-tools) at an illumination wavelength of 405 nm laser. The samples used for this comparison included several p-type multi-crystalline and mono-crystalline silicon wafers. The multi-crystalline wafers were textured using $AgNO_3$ nano-pitting solution followed by an HF-HNO_3 etch [8] [9], and intentional variation of sub-micron inverted texture was created. The mono-crystalline wafers were random pyramid textured using an alkaline solution. The LOANA measures a total hemispherical reflectance of the wafer surface, however, the metal probes designed for quantum efficiency measurements for metallised solar cells create three dark regions in the produced reflectance images and they are unrelated to the optical properties of the samples, therefore, the data from these regions was excluded from the analysis.

Since the results from the scanners and the LOANA are in different metrics, reported in counts and hemispherical reflection, respectively. To fairly compare them, two methods are used to observe the similarities and differences between the scanner and the LOANA measured images. The first method is to compare pixel level data by quantifying the signals by setting the colour scale between 2nd and 98th percentile of the frequency distribution of the values for the images. The values are normalised by subtracting the mean and divided by the standard deviation. The second method is to evaluate the correlation of the raw data collected from the CIS scanner and the LOANA by plotting it on the same graph.

3 RESULTS

Significant differences in the performance comparison between CIS and CCD scanners were observed. The results show that the CIS based scanner could provide stronger signals of silicon wafer images than the CCD scanner, particularly from the blue channel in response to the SiN_x dielectric layer. Moreover, the image counts from the CIS scanner show less variation across the same distance across the scanner. Therefore, the CIS scanner is selected to monitor the texturing quality of multi- and mono-crystalline silicon wafers for later investigation. Then, through the comparison of the images produced by the CIS scanner and a high-end laboratory tool LOANA, a positive linear correlation was found for multi-crystalline silicon wafers.

3.1 CIS and CCD based scanners

A p-type random pyramid textured mono-crystalline solar cell precursor from a 2020 manufacturing line was used to compare the performance of the CIS and the CCD scanner, clear differences were observed as shown in Figure 2. The intensity of the signals varies in respect of the sample positioned across the scanner, with larger relative differences from the CCD scanner than the CIS scanner, irrespective of the sample rotation. Therefore, the region that the sample is positioned also affects the final image quality, particularly for the CCD based scanner. The image counts increase when the sample is positioned away from the centre. Here two distinct axes are used in this work to clarify the distance along the scanner head and the distance of the scanning movement.

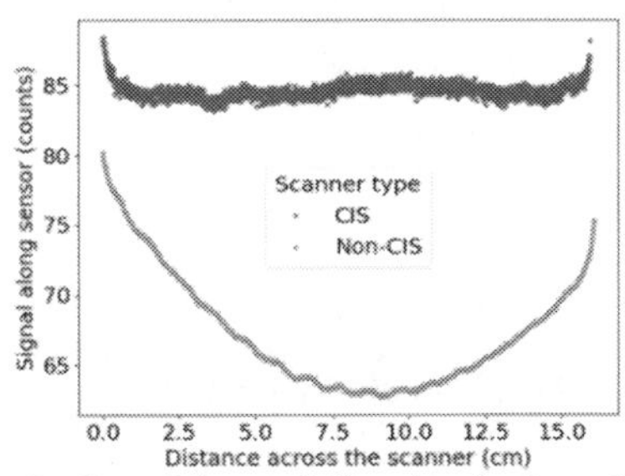

Figure 2: Comparison of the signals from the green channel between a (a) CIS and (b) CCD based scanner. Points are taken from the averages of the corresponding pixel across the solar cell precursor in the scanner direction of movement.

The image signals from the CIS scanner show overall higher average counts and less dependency on the sample positions across the same distance. Therefore, the CIS flatbed scanner was chosen over the CCD scanner for monitoring the spatial uniformity of multi- and mono-crystalline silicon wafers in this work, and the results are shown in the next section.

3.2 Multi- and mono-crystalline silicon texturing

One of the advantages of using the CIS scanner to monitor texturing of solar wafers is that this method takes less time per measurement and provides higher resolution than hemispherical reflection mapping technique using the LOANA. It takes the LOANA approximately 20 mins to produce a 155 x 155 pixel image of one multi-crystalline wafer or a 160 x 160 pixel image of one mono-crystalline wafer, but the scanner takes around 10 s to provide a 1800 x 1800 pixel image of one wafer. The comparison of silicon wafer images measured with the CIS scanner and the LOANA is shown in Figure 3. Two example multi-crystalline and two example mono-crystalline silicon wafers are shown. The similarities between the images taken with the CIS scanner and the LOANA indicates the possibility of using CIS flatbed scanners to monitor spatial uniformity of solar cells.

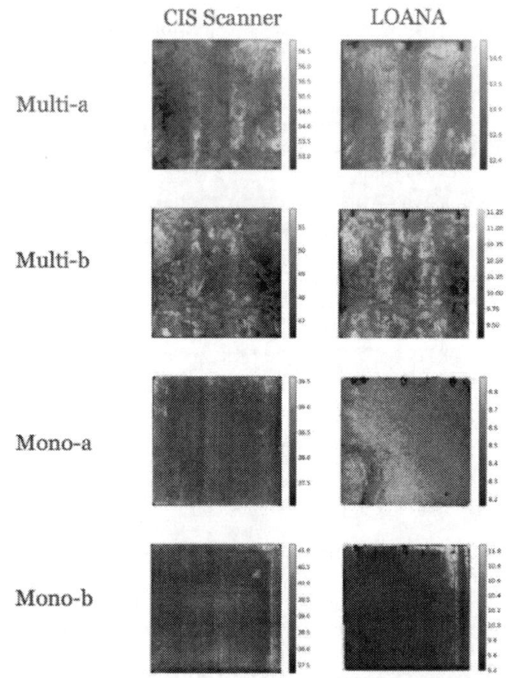

Figure 3: The left column shows the images measured with the CIS scanner in the blue channel and the right shows hemispherical reflection map measured with the LOANA. Each row represents a different sample.

To further validate that the CIS scanner can be used for texturing inspection of solar cell precursors, datasets collected from the CIS scanner and the LOANA are compared using two techniques, normalised frequency histogram comparison and statistical comparison.

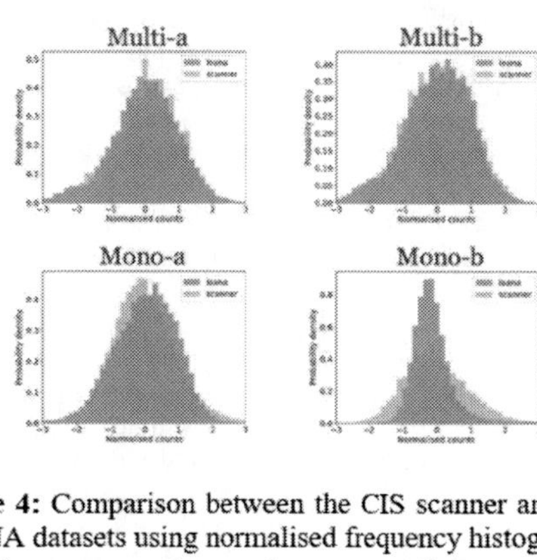

Figure 4: Comparison between the CIS scanner and the LOANA datasets using normalised frequency histograms. Each plot represents the sample corresponding to each row in Figure 3 in the same order.

Normalised frequency histogram comparison is used to quantify similarities of the signals received from the CIS scanner and the LOANA, the frequency distribution of normalised count rate is shown in Figure 4 for the same image data as shown in Figure 3. Similarities in the shapes of frequency histograms are observed between signals received from the CIS scanner and the LOANA for the two multi-crystalline wafers and one mono-crystalline wafer mono-a. However, there is clear disagreement in the histograms for the LOANA and scanner data for the mono-b sample.

To statistically analyse the correlation between the CIS scanner and the LOANA images, all the data points from each sample in Figure 3 are recorded in Figure 5 along with additional multi-crystalline and mono-crystalline samples (c and d). A positive correlation is observed between the CIS scanner and the LOANA results of multi-crystalline wafers, but not mono-crystalline wafers.

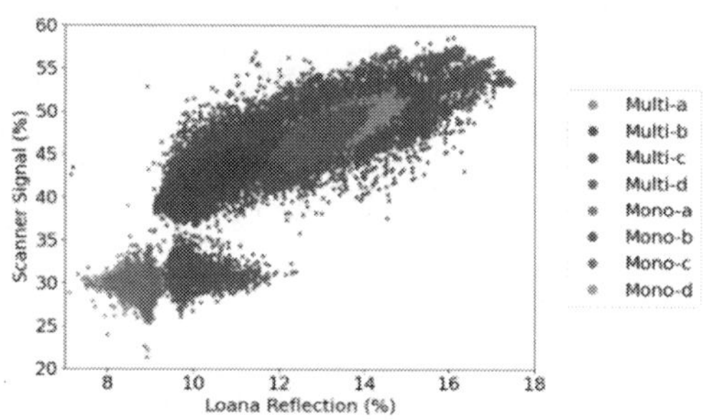

Figure 5: Comparison of pixel intensity of the same wafers.

4 DISCUSSION

The reason why the solar wafer images produced by the CIS scanner show smaller variation than the CCD scanner (see Figure 2) is likely due to their different collection optics. For every pixel collected by a CIS scanner, one lens is dedicated for x 1 magnification, which results in the same field of view across the sensor. Whereas multiple mirrors in a CCD scanner optics setup results in a changing field of view (see Figure 1), then the changes in the view are corrected in a process called flat-field correction. Such a correction is designed to correct the scattering profile of paper or photos for its original purpose. Indeed, no bow similar to what is shown in silicon wafer scans as shown in Figure 2 is observed when paper or photos are scanned by the CCD scanner, suggesting more scattered light from silicon wafers is detected by the CCD scanner.

Since the CIS scanner shows no requirement for flat-field correction, the results are more sample independent than the ones from the CCD scanner. Therefore, the comparison between images from the CIS scanner and the reflection maps from a LOANA tool is used to qualify if the low-cost scanner can be used to monitor spatial uniformity of the reflection of textured silicon wafers.

A positive linear correlation is found between and across all the multi-crystalline wafers of varying textures using statistical analysis, indicating it is possible and suitable to use the consumer scanner technology for monitoring texturing quality of multi-crystalline silicon wafers. However, no clear correlation between the CIS scanner data and the LOANA images is observed for mono-crystalline wafers, indicating the reflection of the random pyramid texture caused by alkaline is not well captured by the scanner.

This can be explained by how light scatters differently on different textured surfaces created by metal assisted chemical and alkaline texturing processes. The metal assisted chemical texture on the multi-crystalline silicon is created via a more statistical process, resulting in varied geometric shapes and generally isotropic reflection [10], whereas the alkaline texture on the mono-crystalline silicon results in more well-define pyramidal shapes. The reflection variation of the silicon surface is from the reflected light that comes off one pyramid and onto another, and therefore the geometry of the pyramids affects the light scattering mechanism from the silicon surface [11].

Additionally, the silicon wafer images from the CIS scanner (see Figure 3 left column) show multiple vertical and horizontal lines (in the mono-a and mono-b image) across the wafers. These fine features are likely to be remnants of wire sawing and can only be seen clearly in the high-resolution scanned images.

5 CONCLUSIONS

Two different types of consumer flatbed scanners using CIS and CCD technology are compared in this work to evaluate the possibility of monitoring texturing quality of multi- and mono-crystalline solar cell precursors. It is shown that CIS scanners, without flat-field correction, can produce as-acquired scanned images that are more accurate than CCD scanners. Moreover, through comparison between the images scanned with a CIS scanner and the results from a lab reflection mapping tool, the CIS scanners are found to be more suitable for monitoring spatial uniformity of multi-crystalline with faster measurement speed and higher-resolution results than the conventional laboratory LOANA tool. In an unmodified state, the CIS scanner technology is not able to accurately monitor the more subtle texture variations of the monocrystalline samples tested, however, this may be achievable with low-cost scanner modifications, to be investigated in future work.

6 REFERENCES

[1] van Battum, L. J., Huizenga, H., Verdaasdonk, R. M. & Heukelom, S., How flatbed scanners upset accurate film dosimetry. Physics in Medicine and Biology 61, 625–649 (2015).

[2] Schoenfeld, A. A., Wieker, S., Harder, D. & Poppe, B., The origin of the flatbed scanner artifacts in radiochromic film dosimetry—key experiments and theoretical descriptions. Physics in Medicine and Biology 61, 7704–7724 (2016).

[3] Lárraga-Gutiérrez, J. M., García-Garduño, O. A., Treviño-Palacios, C., & Herrera-González, J. A., Evaluation of a LED-based flatbed document scanner for radiochromic film dosimetry in transmission mode. Physica Medica 47, 86–91(2018).

[4] Shameem, T., Bennie, N., Butson, M. & Thwaites, D. A comparison between EPSON V700 and EPSON V800 scanners for film dosimetry. Physical and Engineering Sciences in Medicine 43, 205–212 (2020).

[5] Göröcs, Z., & Ozcan, A., Biomedical imaging and sensing using flatbed scanners. Lab Chip 14, 3248–3257 (2014).

[6] De Vicente, J., A. M. Sánchez-Pérez, P. Maresca, J. Caja, & E. Gómez., A model to transform a commercial

flatbed scanner into a two-coordinates measuring machine. Measurement 73, 304–312 (2015).
[7] Shimobaba, T., Yamanashi, H., Kakue, T., Oikawa, M., Okada, N., Endo, Y., Hirayama, R., Masuda, N. and Ito, T., In-line digital holographic microscopy using a consumer scanner. Scientific Reports 3, 2664 (2013).
[8] Zou, S., Ye, X., Wu, C., Cheng, K., Fang, L., Tang, R., Shen, M., Wang, X. & Su, X., Complementary etching behavior of alkali, metal-catalyzed chemical, and post-etching of multicrystalline silicon wafers. Progress in Photovoltaics: Research and Applications 27, 511–519 (2019).
[9] Chen, K., Zha, J., Hu, F., Ye, X., Zou, S., Vähänissi, V., Pearce, J.M., Savin, H. & Su, X., MACE nano-texture process applicable for both single-and multi-crystalline diamond-wire sawn Si solar cells. Solar Energy Materials and Solar Cells, 191, pp.1-8 (2019).
[10] Scheul, T. E., Khorani, E., Rahman, T., Charlton, M. D. & Boden, S. A. Light scattering from black silicon surfaces and its benefits for encapsulated solar cells. Solar Energy Materials and Solar Cells 235, 111448 (2022).
[11] Scheul, T. E., Khorani, E., Rahman, T., Charlton, M. D. B. & Boden, S. A. Wavelength and angle resolved reflectance measurements of pyramidal textures for crystalline silicon photovoltaics. Progress in Photovoltaics: Research and Applications 28, 1248–1257 (2020).

INFLUENCE OF BULK CARRIER LIFETIME ON EFFICIENCY RECOVERY IN GATED SOLAR CELLS

Anil Kumar[1,2], Nithin Chatterji[3], Aldrin Antony[4], and Pradeep R. Nair[1,2]
[1]Department of Electrical Engineering, Indian Institute of Technology Bombay, Mumbai-400076, Maharashtra, India.
[2]National Centre for Photovoltaic Research and Education, Indian Institute of Technology Bombay, Mumbai-400076, Maharashtra, India.
[3]Department of Electronics Engineering, Sardar Vallabhbhai National Institute of Technology Surat, Surat-395007, Gujarat, India
[4]Department of Physics, Cochin University of Science of Technology, Kochi-682022, Kerala, India.
Email: anilrawat@ee.iitb.ac.in, prnair@ee.iitb.ac.in.

ABSTRACT: Gated solar cell architectures promise to recover the efficiency loss due to interface degradation. Here, we address the impact of the material quality of the absorber layer and the efficiency recovery window of such gated solar cells. For a poorly passivated interface, the performance is limited by the interface recombination. With additional control gate architecture, the interface recombination at the rear interface can be reduced leading to the dependence of performance on the bulk lifetime of the solar cell. For surface recombination velocity (S_b) of 1000 cm/s, in comparison to the reference solar cell, the efficiency of the cell is improved by 1.28 % and 1.81 % for bulk lifetimes of 0.1 and 10 ms, respectively. Thickness-dependent gated solar cell results show that thicker cells require a higher bulk lifetime to achieve high efficiency, while thinner cells are rather independent of the bulk lifetime. Our results indicate that gated solar cells could be of interest to achieving high-efficiency solar cells with heterogeneous materials with low-cost fabrication.
Keywords: Field effect solar cell, light and elevated temperature-induced degradation (LeTID), photovoltaic cell

1 INTRODUCTION

Solar photovoltaics relies on efficient generation and collection of electron-hole pairs at distinct contacts[1]. To ensure the same, c-Si solar cells demand excellent material quality in the bulk and as well as at the interfaces [2]–[4]. In addition, doped junctions are necessary to ensure carrier separation and collection at separate contacts. This whole process involves significant material and manufacturing costs. However, interface degradation and the subsequent degradation in power output due to environmental conditions in the field [5]–[8] is a major concern.

Solar cell architectures with an additional control gate[9], [10] are known to mitigate some of these issues. The main features of the gated solar cell are: (i) material agnostic with respect to back surface passivation layers, (ii) processes with less precision can be used for fabricating the cells, (iii) it can partially recover the power loss due to the degradation in the field, and (iv) it can be implemented with existing processing used for the fabrication of IBC and PERC solar cells[11], [12]. In this work, we identify the conditions under which an additional gate results in performance improvement. Through detailed numerical simulations, we show that gated solar cells are well suited for solar cell architectures with excellent carrier lifetime in the bulk of Si absorber. This architecture is capable of efficiency recovery of the order of 1-2% for solar cells with bulk lifetimes of the order of 0.1-10ms.

2 MODEL SYSTEM

Figure 1 depicts the structure of the gated solar cell used for the simulations. The Poisson and drift-diffusion equations were solved self consistently to simulate the solar cells[1], [13]. The back surface passivation layer (BSPL) is used to passivate the rear surface of the silicon surface. The properties of the BSPL are like that of silicon oxide without considering any charges and it

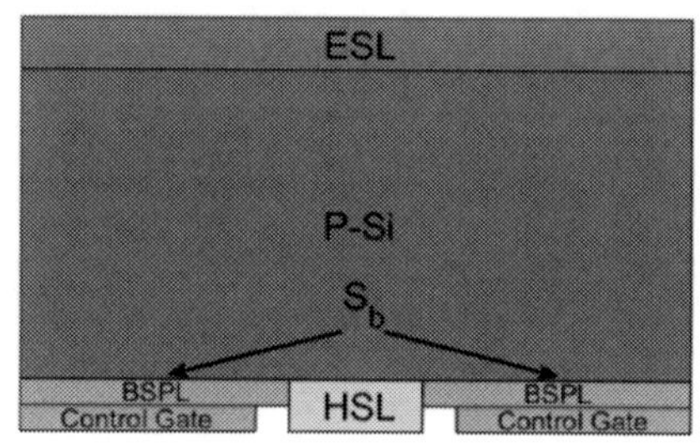

Figure 1: Structure of the gated solar cell used in this study

covers 90% of the cell's rear side area. The control gate is separated from the bulk silicon with BSPL and used for the field-effect passivation of the rear side. The surface recombination velocity at the interface between silicon and BSPL (S_b, for electrons) is assumed as 10 cm/s corresponding to a good passivated interface and 1000 cm/s for a poorly passivated interface. The surface recombination velocity between the electron selective layer (ESL) and silicon (for holes) is 100 cm/s. Similarly, the surface recombination velocity between the hole selective layer (HSL) and silicon (for electrons) is also assumed as 100 cm/s. The difference in the work function of the bulk silicon and the control gate is assumed to be -0.345 eV and the simulation results indeed depend on this work function difference. The generation rate used for the simulations was calculated by considering the AM1.5G spectrum and the wavelength-dependent absorption coefficient of the silicon[14], [15].

3 RESULTS AND DISCUSSIONS

The reference solar cell (i.e., without the additional

gate control) characteristics as a function of silicon bulk lifetime for different thicknesses of silicon absorber are shown in figure 2. The short circuit current density increases with the increase in the thickness of the silicon absorber due to increased absorption of light. The short circuit density is lower with an S_b of 1000 cm/s due to increased recombination at the interface. For both the surface recombination velocities, the bulk lifetime had no impact on the short circuit current density. For an S_b of 10 cm/s, with a low bulk lifetime, the open-circuit voltage decreases with an increase in the thickness of silicon bulk. This is because lower diffusion length leads to higher recombination in the bulk for higher thicknesses. For higher bulk lifetimes, open circuit voltage is almost similar for all the thicknesses. For an S_b of 1000 cm/s, the open-circuit voltage is decreased due to higher recombination at the interface and bulk lifetime had little impact on it. For higher wafer thicknesses, the bulk recombination dominates, and hence the open-circuit voltage is lower for 0.1ms and higher for 10ms bulk lifetime. In contrast, for lower thicknesses the interface recombination dominates, and bulk lifetime does not play a role in performance degradation. Overall, the efficiency of the cells increases with the increase in the thickness of the silicon absorber mainly due to increased short circuit current density. For an S_b of 10 cm/s, the efficiency is nearly the same for lower thicknesses for all bulk lifetimes. With the increase in the thickness of the absorber layer, the efficiency gap between lower and higher bulk lifetime increases. This is because of the increased recombination in bulk with a lower lifetime. For an S_b of 1000 cm/s, and for a given thickness, the efficiencies are the same for all the bulk lifetimes but are lower than that with an S_b of 10 cm/s because the interface recombination dominates in the cell.

Figures 3 (a) and (b) show the performance parameters of the gated solar cell with a thickness of 200 µm and S_b of 10 and 1000cm/s, respectively. For an S_b of 10 cm/s, the performance of the cells is limited by bulk recombination. As evident from figure 3(a), with the increase in the bulk lifetime the performance of the solar cell becomes better while the addition of the control gate does not aid in performance enhancement as the recombination in the cell is already very low. In the case of an S_b of 1000 cm/s, the performance of the cells is limited by the interface recombination. With our gated solar cell architecture, we can improve the performance of the cells. In comparison to the reference solar cell with the above-mentioned S_b, the short circuit current density, the open-circuit voltage, and the efficiency of the cells are improved by 1 mA/cm^2, 37 mV, and 1.81 % respectively with the application of a gate voltage of -0.7 V for a bulk lifetime of 10 ms. In the reference solar cell, the bulk lifetime had a negligible impact on the efficiency of the solar cell as the interface recombination dominates the cell performance. With the additional control gate, the interface recombination can be controlled leading to the dependence of the cell performance on the bulk lifetime.

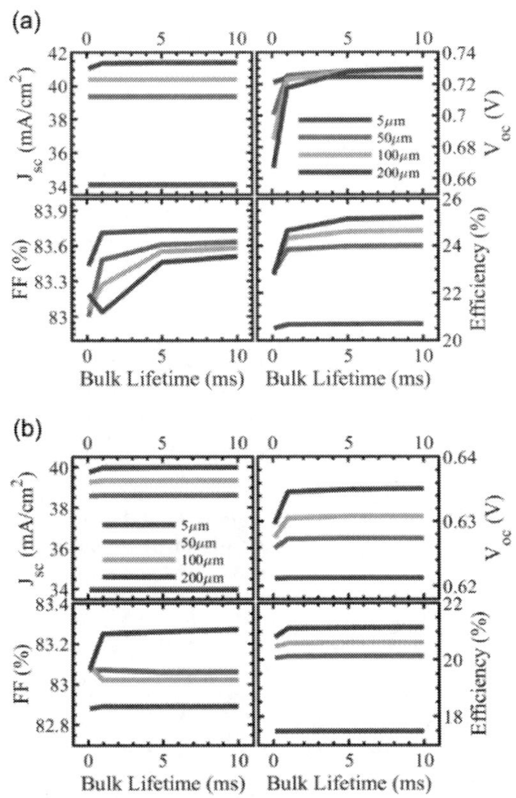

Figure 2: Performance parameters of the reference solar cell without any gate control as a function of silicon bulk lifetime for different thicknesses of silicon bulk for S_b of (a) 10 cm/s and (b) 1000 cm/s.

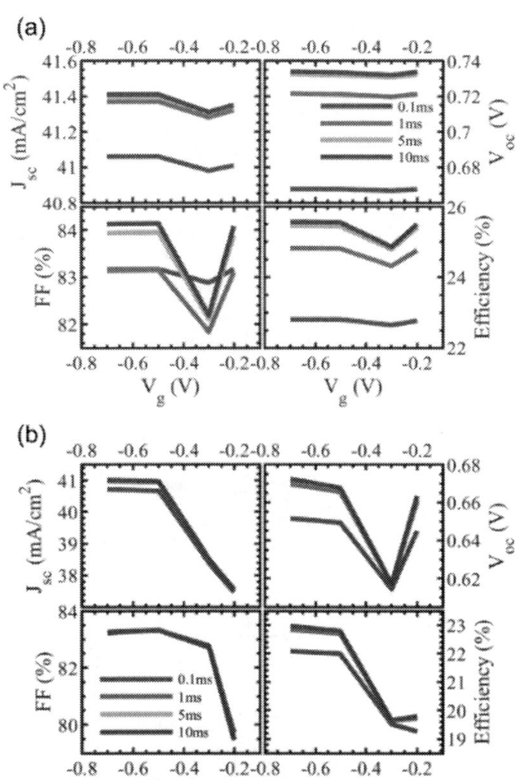

Figure 3: Short circuit current density (J_{SC}), open circuit voltage (V_{OC}), fill factor (FF) and efficiency of the gated solar cell with S_b of (a) 10 cm/s (b) 1000 cm/s.

With a bulk lifetime of 10 ms, the efficiency of the solar cell with an S_b of 1000 cm/s can reach nearly 23 % as compared to 22 % for the bulk lifetime of 0.1 ms.

We have also studied the impact of the additional control gate on the solar cell with different thicknesses of the silicon bulk with different schemes of rear passivation. Figure 4 shows the plot of efficiency versus thickness for different bulk lifetimes and surface recombination velocity of 10 and 1000 cm/s with a gate voltage of -0.7 V. For S_b of 10 cm/s and higher thicknesses of the silicon bulk, the performance is dominated by the bulk recombination. Therefore, the higher the bulk lifetime higher the efficiency of the gated solar cells. As the thickness of the silicon bulk decreases, the gap between the efficiency of the cells with a bulk lifetime of 0.1 and 10 ms decreases. This is because the interface recombination also starts to impact the performance. For lower thicknesses, the interface recombination dominates, and hence the efficiency is controlled by the interface recombination. Therefore, the efficiency is almost the same for the thickness of 5 µm for all bulk lifetimes. A similar explanation holds for a higher S_b of 1000 cm/s. The only difference is the variation in the efficiency is smaller between the bulk lifetimes of 0.1 to 10 ms. This is because the performance is now limited by both the bulk and interface recombination in the solar cell and the complete domination of interface starts at a higher thickness. Overall, the bulk lifetime is important when the interface passivation quality is very good.

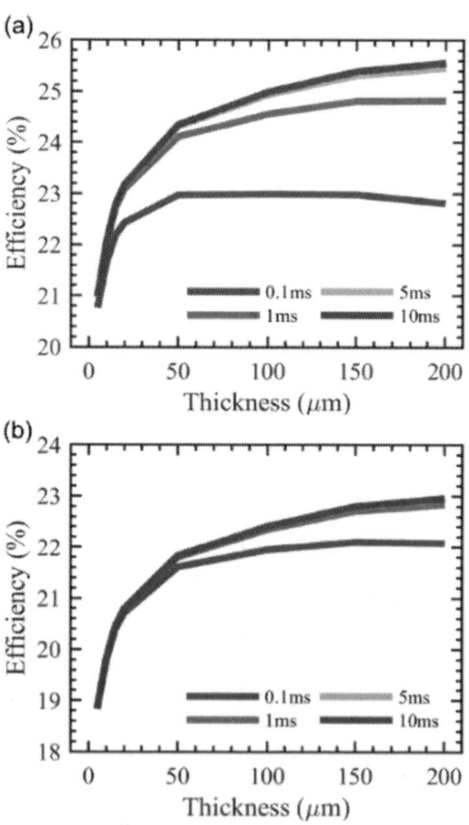

Figure 4: Efficiency as a function of silicon absorber thickness for different silicon bulk lifetimes at a gate voltage of -0.7 V for an S_b of (a) 10 cm/s and (b) 1000 cm/s.

4 CONCLUSION

In conclusion, we have studied the influence of the bulk carrier lifetime on the efficient recovery of gated solar cells. For the reference solar cells with the poorly passivated rear surface, the performance is limited by the interface recombination. On the other hand, for good passivation of the rear interface, the performance is limited by bulk recombination. With a surface recombination velocity of 10 cm/s, the gate voltage had little impact and the bulk lifetime plays the deciding factor in the efficiency of the cell. For surface recombination velocity of 1000 cm/s, appropriate control gate bias leads to reduced interface recombination and makes the performance of the cell bulk lifetime dependent. Although unable to completely reduce interface recombination, the control gate results in an efficiency of 23% for devices with a bulk lifetime of 10 ms and thickness of 200 µm.

ACKNOWLEDGMENT

This work was supported in part by Science and Engineering Research Board (SERB, Project Code CRG/2019/003163), Department of Science and Technology, India. The authors would like to acknowledge CEN and NCPRE, IIT Bombay, India for experimental facilities. Anil Kumar acknowledges the financial support of University Grants Commission (UGC), India. Pradeep R. Nair also acknowledges Visvesvaraya Young Faculty Fellowship.

Table I: Parameters used in the simulations

Parameter	p-Si	ESL	HSL
Thickness	200 µm	10 nm	10 nm
N_A, N_D (cm^{-3})	10^{16}, 0	0, 10^{16}	10^{18}, 0
E_g (eV)	1.12	3.4	2.12
χ (eV)	4.05	4.05	3.05
τ_{SRH}	0.1, 1, 5, 10ms	1µs	1µs
B (cm^3s^{-1})	1.1×10^{-14}	--	--
C_n (cm^6s^{-1})	1×10^{-31}	--	--
C_p (cm^6s^{-1})	0.79×10^{-31}	--	--

References

[1] J. Nelson, The Physics of Solar Cells, 1st ed. London, UK: Imperial College Press, 2003.

[2] T. Dullweber and J. Schmidt, "Industrial Silicon Solar Cells Applying the Passivated Emitter and Rear Cell (PERC) Concept—A Review," IEEE Journal of Photovoltaics, vol. 6, no. 5, pp. 1366–1381, 2016, doi: 10.1109/JPHOTOV.2016.2571627.

[3] S. de Wolf, A. Descoeudres, Z. C. Holman, and C. Ballif, "High-efficiency Silicon Heterojunction Solar Cells: A Review," Green, vol. 2, no. 1, pp. 7–24, 2012, doi: doi:10.1515/green-2011-0018.

[4] J. Melskens, B. W. H. van de Loo, B. Macco, L. E. Black, S. Smit, and W. M. M. Kessels, "Passivating Contacts for Crystalline Silicon Solar Cells: From Concepts and Materials to Prospects," IEEE Journal of Photovoltaics, vol. 8, no. 2, pp. 373–388, 2018, doi: 10.1109/JPHOTOV.2018.2797106.

[5] V. Sharma and S. S. Chandel, "Performance analysis of a 190 kWp grid interactive solar photovoltaic power plant in India," Energy, vol. 55, pp. 476–485, 2013, doi: https://doi.org/10.1016/j.energy.2013.03.075.

[6] C. R. Osterwald, A. Anderberg, S. Rummel, and L. Ottoson, "Degradation analysis of weathered crystalline-silicon PV modules," in Conference Record of the Twenty-Ninth IEEE Photovoltaic Specialists Conference, 2002., 2002, pp. 1392–1395. doi: 10.1109/PVSC.2002.1190869.
[7] M. Padmanabhan et al., "Light-induced degradation and regeneration of multicrystalline silicon Al-BSF and PERC solar cells," physica status solidi (RRL) – Rapid Research Letters, vol. 10, no. 12, pp. 874–881, 2016, doi: https://doi.org/10.1002/pssr.201600173.
[8] M. Selinger et al., "Spatially Resolved Analysis of Light Induced Degradation of Multicrystalline PERC Solar Cells," Energy Procedia, vol. 92, pp. 867–872, 2016, doi: https://doi.org/10.1016/j.egypro.2016.07.095.
[9] N. Chatterji, S. Bhatia, A. Kumar, A. Antony, and P. R. Nair, "Material- and Process-Tolerant High-Efficiency Solar Cells With Dynamic Recovery of Performance," IEEE Transactions on Electron Devices, vol. 68, no. 4, pp. 1676–1681, 2021, doi: 10.1109/TED.2021.3056954.
[10]A. G. Aber;e, S. Glunz, and W. Warta, "Field effect passivation of high efficiency silicon solar cells," Solar Energy Materials and Solar Cells, vol. 29, no. 2, pp. 175–182, 1993, doi: https://doi.org/10.1016/0927-0248(93)90075-E.
[11]R. Vasudevan et al., "Laser-induced BSF: A new approach to simplify IBC-SHJ solar cell fabrication," AIP Conference Proceedings, vol. 1999, no. 1, p. 40024, 2018, doi: 10.1063/1.5049287.
[12]"PERC".
[13]Synopsys, "Synopsys TCAD." Mountain View, CA, USA, 2016.
[14]M. A. Green and M. J. Keevers, "Optical properties of intrinsic silicon at 300 K," Progress in Photovoltaics: Research and Applications, vol. 3, no. 3, pp. 189–192, 1995, doi: https://doi.org/10.1002/pip.4670030303.
[15]M. A. Green, "Self-consistent optical parameters of intrinsic silicon at 300K including temperature coefficients," Solar Energy Materials and Solar Cells, vol. 92, no. 11, pp. 1305–1310, 2008, doi: https://doi.org/10.1016/j.solmat.2008.06.009.

ADVANTAGES OF TOPCON CELL WITH REAR FULL AREA HOLE SELECTIVE CONTACT ON P-TYPE SI WITH FRONT N+ EMITTER OVER PERC

Jayshree Bhajipale[1,2], Anil Kottantharayil[1,2], and K. P. Sreejith[1,2]
[1] National Centre for Photovoltaics Research and Education (NCPRE), Indian Institute of Technology, Bombay, Powai, Mumbai, 400076, India.
[2] Department of Electrical Engineering, Indian Institute of Technology, Bombay, Powai, Mumbai, 400076, India.

ABSTRACT: Tunnel Oxide Passivated Contact structure on p-type silicon (Si) with rear full area hole selective passivating contact and n^+ emitter (p-TOPCon) would be ideal successor for Passivated Emitter and Rear Contact (PERC). Hence, effort to predict the performance gain of p-TOPCon over the PERC architecture is needed. In this work, Sentaurus TCAD based simulation study is performed to investigate the advantages of p-TOPCon cell over PERC with respect to surface passivation. Front and rear surface passivation effects are examined by considering surface recombination velocities (S_{rear} and S_{front}) at non-metallized Si surfaces. Cell recombination characteristics are studied in terms of open circuit voltage (V_{oc}). p-TOPCon and PERC outperform Al-BSF for identical cell parameters and p-TOPCon performance is comparatively less sensitive to S_{rear} than PERC. V_{oc} and efficiency of p-TOPCon decreases for an order of magnitude higher S_{rear} (>10^3 cm s^{-1}) compared to PERC (>10^2 cm s^{-1}). V_{oc} and efficiency trends are identical for p-TOPCon and PERC with S_{front} variation for fixed emitter sheet resistance (R_{sheet}) of 95 Ω/sq. There is a scope for performance improvement with enhanced front surface passivation. Conclusively, p-TOPCon consistently shows better performance compared to PERC for identical bulk lifetime and front and rear passivation conditions.

Keywords: Passivation, PERC, p-TOPCon, Silicon Solar Cell, Simulation

1 INTRODUCTION

Recently, the photovoltaic industries have adopted the Passivated Emitter and Rear Contact (PERC) architecture over the conventional Aluminum Back Surface Field (Al-BSF) architecture for solar cell production due to its superior rear surface performance [1]. However, the performance of PERC solar cells is still limited by the recombination at the localized metal-silicon (Si) interface and higher series resistance due to the patterned contacts at the rear side. The Tunnel Oxide Passivated Contact (TOPCon) structures completely mitigate the above drawbacks as it uses a combination of ultrathin oxide and doped polysilicon (poly-Si) layer, which offers an excellent full area rear side passivation and low contact resistance [2, 3].

TOPCon architecture on p-type Si is not widely implemented due to its inferior bulk quality and non-availability of efficient metallization schemes over its counterpart [4, 5]. However, recent studies suggest that gallium doped Si wafers drastically improve the bulk lifetime and do not show light-induced degradation (LID) and light and elevated temperature-induced degradation (LeTID) [6]. In addition, it has also been demonstrated that the bulk lifetime of the boron doped Si wafers could be enhanced after deactivation of boron-oxygen complex [7]. Further, Mack et al. identified commercially available silver (Ag) paste that allows low contact resistivity of 4-5 mΩ cm^2 [8]. These open up the window for fabrication of TOPCon architecture over p-type Si. Switching from PERC to TOPCon architecture with rear full area hole selective passivating contact and n^+ emitter on p-type wafers (p-TOPCon) will be more feasible as most of the existing PERC line processes can be used without major alterations. Implementation of p-TOPCon would provide hole selectivity and improved passivation at the rear compared to PERC. Therefore, the influence of carrier selective structure on the cell performance in terms of passivation needs to be evaluated against the rear dielectric passivated PERC.

Numerical simulations of p-TOPCon and PERC would provide an insight into the performance potential of the device prior to its industrial fabrication challenges. There are few recent reports on the simulation study of p-TOPCon. However, these are mainly focused only on the effect of cell design parameters including bulk lifetime and resistivity, tunnel oxide thickness, and carrier selectivity of the passivating rear contact [9, 10].

In this work, numerical simulations are realized in Sentaurus Technology Computer Aided Design (TCAD) to compare the performance of PERC and p-TOPCon test structures with respect to surface passivation. Both front and rear surface passivation effects are studied by considering surface recombination velocity.

2 SIMULATION APPROACH

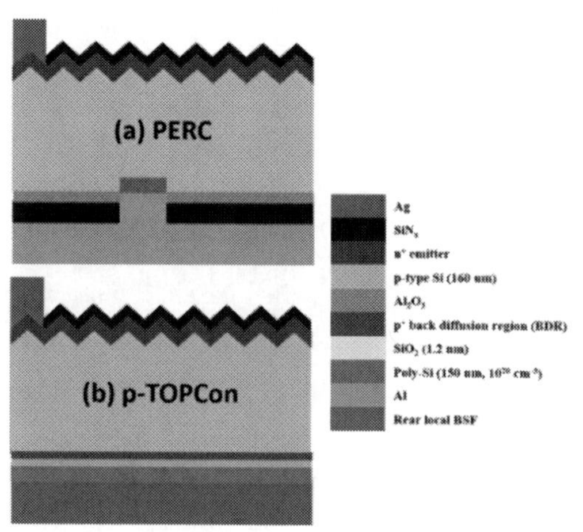

Fig. 1. 2D schematic of (a) PERC and (b) p-TOPCon unit cell.

The TCAD simulation environment is calibrated using the measured data of the Al-BSF cell fabricated at our lab with 19.4% efficiency. The methodology followed is that the generation profile obtained from 3D optical simulation is used in 2D electrical simulations. Identical models as in Al-BSF architecture are used for PERC simulations and a negative oxide charge of 10^{12} cm^{-2} is considered at the rear side aluminum oxide (Al_2O_3) dielectric layer and Si interface. In addition to these models, the tunneling models used for p-TOPCon simulations are calibrated with respect to the data published by Steinkemper et al. [11]. The 2D schematics of the PERC and p-TOPCon unit cells used for the simulation study are shown in Fig. 1.

In this simulation work, the p-type Si substrate with a resistivity of 2 Ω-cm, thickness of 160 μm, and bulk lifetime of 900 μs is considered [12]. An emitter with sheet resistance (R_{sheet}) of 95 Ω/sq is considered [13]. These values closely match the current process parameters used for PERC solar cell production in the industry. Surface recombination velocity at the Si/metal interface is fixed to 10^7 cm s^{-1}. The ultrathin tunnel oxide was modeled using a non-local tunneling model with effective electron and hole tunneling masses of $0.40m_0$ and $0.32m_0$, respectively [14].

3 RESULTS AND DISCUSSION

The surface recombination and therefore the surface passivation characteristics of the cell are evaluated in terms of open circuit voltage (V_{oc}). In all the simulations, passivation at the non-metalized interfaces is studied.

3.1 Rear side passivation

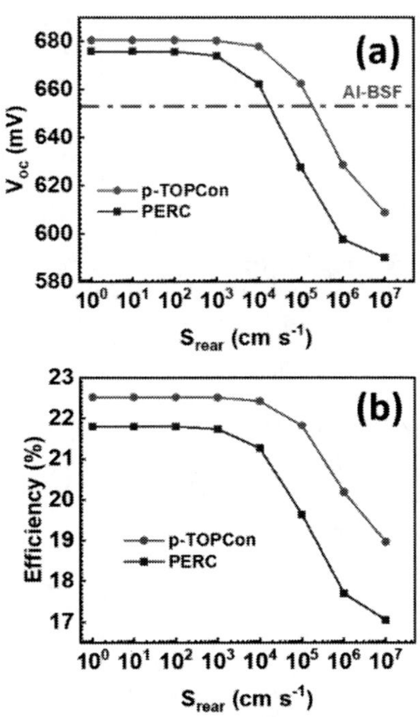

Fig. 2. (a) V_{oc} and (b) efficiency variation with S_{rear} for p-TOPCon and PERC architecture.

In Fig. 2 (a), V_{oc} is plotted against the rear surface recombination velocity (S_{rear}). A similar trend is observed for both cells. However, the PERC cell is more sensitive to the rear surface passivation in comparison to p-TOPCon. The latter has a higher V_{oc} of 680.6 mV, constant for low S_{rear} ($\leq 10^3$ cm s^{-1}) and starts to decrease for $S_{rear} > 10^3$ cm s^{-1}. In PERC cell, V_{oc} is constant at 675 mV and decreases for $S_{rear} > 10^2$ cm s^{-1}. Therefore, p-TOPCon has an order of magnitude broader range of S_{rear} before the rear surface recombination starts dominating over the bulk.

For identical cell parameters, V_{oc} of 653.1 mV is achieved for the Al-BSF cell. Therefore, $S_{rear} \leq 10^4$ cm s^{-1} and $S_{rear} \leq 10^5$ cm s^{-1} for PERC and p-TOPCon, respectively is required to ensure these cells outperform conventional Al-BSF from the passivation aspect. Therefore, PERC solar cell need comparatively stringent conditions on S_{rear} to surpass the V_{oc} obtained with Al-BSF cell. This is in contrast with the p-TOPCon cell where a broader range of S_{rear} could suffice for it to outperform over both Al-BSF and PERC cells. Further, it can be observed from Fig. 2(b) that the efficiency follows the V_{oc} trend implying that the V_{oc} dominates the overall cell performance.

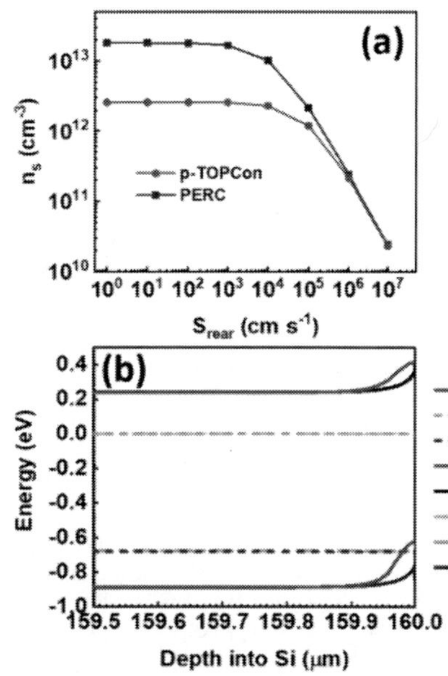

Fig. 3. (a) Surface minority carrier concentration (n_s) and (b) band diagram at Si/SiO_2 interface for p-TOPCon and at Si/Al_2O_3 interface for PERC for S_{rear} of 1 cm s^{-1} at open circuit condition.

V_{oc} trend can be verified from the minority carrier electron concentration (n_s) at the rear passivating interface; Si/Al_2O_3 for PERC and Si/SiO_2 for p-TOPCon, at open circuit condition. For the identical base doping and front and rear surface passivation, n_s is lower in p-TOPCon leading to low surface recombination and therefore higher V_{oc} in comparison to PERC (Fig. 3(a)). The low n_s at low S_{rear} is clear from the energy band diagram at the rear interface of both the cells considered at open circuit condition (Fig. 3(b)). Larger band bending at Si/SiO_2 interface in p-TOPCon ensures a higher electric field repelling minority carriers electrons away from the rear surface. This minority carrier repulsion leads to reduced rear surface recombination in p-TOPCon in comparison to PERC.

3.2 Front side passivation

With excellent rear surface passivation, the front surface limits the overall performance of advanced Si solar cells. In this work, a low emitter dark saturation current density $j_{oe,pass}$ ~ 56 fA cm^{-2}, which corresponds to front surface recombination velocity (S_{front}) of 37000 cm s^{-1} was considered for an emitter R_{sheet} of 95 Ω/sq [13]. Passivation potential is studied by varying the front surface recombination velocity (S_{front}) for different S_{rear} (10-1000 cm s^{-1}).

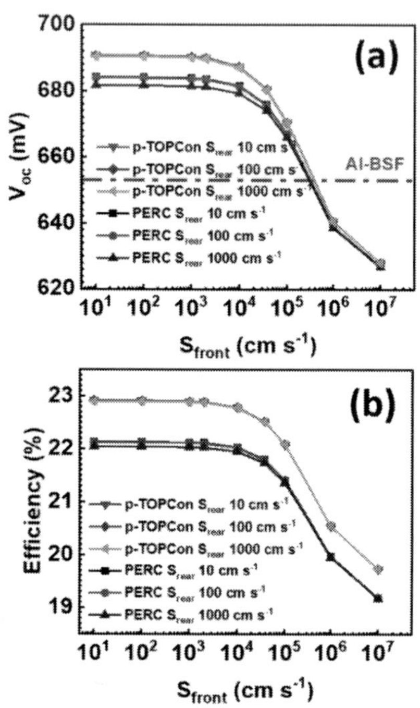

Fig. 4. (a) V_{oc} and (b) efficiency variation with S_{front} for S_{rear} of 10, 10^2, 10^3 cm s^{-1} for p-TOPCon and PERC architecture.

A similar V_{oc} and efficiency trend with S_{front} variation is observed for both PERC and p-TOPCon. Due to better rear passivation, PERC and p-TOPCon cell shows significant improvement in V_{oc} in comparison to Al-BSF (653.1 mV). At this S_{front}, PERC and TOPCon have V_{oc} of 680.6 mV and 675 mV, respectively considering S_{rear} of 10 cm s^{-1}. The sensitivity of PERC to S_{rear} can be verified from (Fig. 4 (a)). Increasing S_{rear} from 10^2 to 10^3 cm s^{-1}, there exists no change in V_{oc} and efficiency of p-TOPCon, however, their values decrease in the case of PERC. There is a scope to increase the V_{oc} and efficiency of both the cells by improving the front emitter passivation. The results clearly indicate that p-TOPCon outperforms PERC for identical S_{rear}.

4 CONCLUSION

In this work, Sentaurus based TCAD simulations were successfully performed to study the effect of front and rear passivation on PERC and p-TOPCon architecture with front n$^+$ emitter on p-type Si. The performance of p-TOPCon is less sensitive and has a broader range of operation to rear surface passivation than PERC. The V_{oc} and further, the efficiency decrement is at an order of magnitude higher S_{rear} in p-TOPCon than in PERC. Whereas, the V_{oc} and efficiency trend for S_{front} variation is similar for both the cells. However, it is observed that p-TOPCon outperforms PERC for identical bulk, front emitter, and rear and front surface passivation.

5 ACKNOWLEDGEMENT

This work was supported by the National Centre for Photovoltaic Research and Education (NCPRE) at IIT Bombay, funded by the Ministry of New and Renewable Energy of the Government of India through Project No. 31/09/2015-16/ PVSE-R&D dated 15th June 2016.

6 REFERNCES

[1] ITRPV 2021. https://itrpv.vdma.org. (accessed February 10, 2022).

[2] F. Feldmann, M. Bivour, C. Reichel, M. Hermle, S. W. Glunz, Proceedings 28th European Photovoltaic Solar Energy Conference, (2013) 988-992.

[3] F. Feldmann, M. Bivour, C. Reichel, M. Hermle, S. W. Glunz, Solar Energy Materials and Solar Cells 120 (2014) 270–274.

[4] M. K. Stodolny, M. Lenes, Y. Wu, G. J. M. Janssen, I. G. Romijn, J. R. Luchies, L. J. Geerligs, Solar Energy Materials and Solar Cells 158 (2016) 24–28.

[5] C. Liu, D. Chen, Y. Chen, Y. Ling, Y. Zou, Y. Wang, J. Gong, Z. Feng, P. P. Altermatt, P. J. Verlinden, Solar Energy Materials Solar Cells, 215 (2020) 110690.

[6] J. Horzel, S. Mack, I. V. Vulanean, K. Zimmermann, S. Pingel, W. Kwapil, F. Maischner, H. Hoffler, S. Bashardoust, D. Wagenmann, J. Greulich, J. Seif, A. Steinmetz, J. Rentsch, Solar RRL, (2022) 2200613.

[7] D. C. Walter, B. Lim, J. Schmidt, Progress in Photovoltaics: Research and Applications, 24 (2016) 920–928.

[8] S. Mack, D. Herrmann, M. Lenes, M. Renes, and A. Wolf, Solar RRL, 5 (2021) 2100152.

[9] T. Sugiura, S. Matsumoto, N. Nakano, "Japanese Journal of Applied Physics, 59 (2020) SGGF03.

[10] Y. Zeng, Q. Yang, Y. Wan, Z. Yang, M. Liao, Y. Huang, Z. Zhang, X. Guo, Z. Wang, P. Gao, C. Wu, B. Yan, J. Ye, Solar Energy, 178 (2019) 249–256.

[11] H. Steinkemper, F. Feldmann, M. Bivour, and M. Hermle, IEEE Journal of Photovoltaics, 5 (2015) 1348–1356.

[12] Y. Zhang, L. Wang, D. Chen, M. Kim, and B. Hallam, Journal of Physics D: Applied Physics, 54 (2021) 214003.

[13] E. Lohmuller, J. Greulich, P. Saint-Cast, S. Lohm€ uller, S. Schmidt, U. Belledin, T. Fellmeth, S. Mack, G. Emanuel, K. Krieg, M. Zimmer, R. Kunert, F. Zobel, M. Linse, J. Horzel, M. Meßmer, A. Wolf, R. Preu, Proceedings 37th European Photovoltaic Solar Energy Conference and Exhibition, (2020) p-11.

[14] A. Gehring, S. Selberherr, IEEE Transactions on Device and Materials Reliability, 4 (2004) 306–319.

CHARACTERIZATION OF THE DAMAGING POTENTIAL OF HIGH THROUGHPUT CELL MANUFACTURING PROCESSES BY SENSITIVE STRENGTH ANALYSIS

Ringo Koepge, Stephan Großer

Fraunhofer Center for Silicon Photovoltaics CSP
Otto-Eissfeldt-Strasse 12, 06120 Halle, Germany

Current motion and process sequences already lead to breakage in cell and module manufacturing under industry standard wafer cycle times. Increasing wafer throughput can lead to extreme handling challenges if the wafer needs to survive cell production undamaged. This investigation shows a method to analyze cell manufacturing processes regarding wafer damaging. A prototype handling process was developed, and damage was measured indirectly by the wafer strength. The statistical strength distributions of wafers are compared. The introduced procedure for wafer strength analysis was applied for the first time to evaluate high throughput manufacturing processes. Two handling operations modes, that means printing processes are compared to each other regarding the fracture stress distribution. The results show a decrease of wafer strength after handling by novel printing technologies. The rotary printing process and screen-printing was performed. Strength reduction was observed.

Keywords: Manufacturing, Reliability, Mechanical Strength, Defects, Automation

1 INTRODUCTION

These days the production capacity of the leading PV manufacturers is growing [1], high throughput processes are needed and the production technology must follow this development. Novel handling processes, like the rotary printing process, are strongly motivated by the number of wafers that must be moved per hour [2, 3, 4]. But wafer handling in general was observed since mass production in photovoltaic market grows in Europe. Wafer breakage during manufacturing needs to decrease down to zero to ensure economic efficiency. Several topics raised, the characterization of simple transport operations for instance [5] show different damaging potential of gripper technologies. Wafer damage characterization over the entire cell manufacturing processes is a useful method to figure out critical process steps like texture process or firing process [6]. In general, a variation of cell strength was observed that can be correlate to solar cell types or stages in cell production as well as certain handling devices or processes [7].

This paper presents a comparison of two different printing technologies within an automated handling and printing process regarding their damage potential and influence on the wafer's strength using a statistical approach. Based on the results of this analysis, the damage potential of the printing process especially for high throughput processes can be quantified.

2 METHODS

Determination of a breakage rate is common for the evaluation of handling processes. Damaging of wafer and cells do not lead to breakage in all cases. Mostly the wafer breakage occurs at the following processes or manufacturing steps. Thus, a more sensitive method is needed to characterize handling processes, in this case high throughput handling operations. Measuring the wafer strength will give an indirect value of wafer damaging and show the impact of handling or manufacturing. The following Table 1 shows an overview of the three main DoE's that were performed.

Table 1: Overview of three handling tests DoE's. The Wafers runs parallel$^{(p)}$ and cross$^{(c)}$ with respect to their saw mark trough the printing process. The reference wafers only run through without any printing procedure.

DoE A	Screen printing	
Name	Printing	Velocity
#1 Reference-Batch$^{(p)}$	No	1 Wafer/ s
#2 Reference-Batch$^{(c)}$	No	1 Wafer/ s
#3 Test-Batch$^{(p)}$	Yes	1 Wafer/ s
#4 Test-Batch$^{(c)}$	Yes	1 Wafer/ s

DoE B	Rotary printing	
Name	Printing	Velocity
#1 Reference-Batch$^{(p)}$	No	1.1 Wafer/ s
#2 Reference-Batch$^{(c)}$	No	1.1 Wafer/ s
#3 Test-Batch$^{(p)}$	Yes	1.1 Wafer/ s
#4 Test-Batch$^{(c)}$	Yes	1.1 Wafer/ s

DoE C	Rotary printing *optimized*	
Name	Printing	Velocity
#1 Reference-Batch$^{(p)}$	No	1.1 Wafer/ s
#2 Reference-Batch$^{(c)}$	No	1.1 Wafer/ s
#3 Test-Batch$^{(p)}$	Yes	1.1 Wafer/ s
#4 Test-Batch$^{(c)}$	Yes	1.1 Wafer/ s
#5 Test-Batch$^{(p)}$	Yes	1.8 Wafer/ s
#6 Test-Batch$^{(c)}$	Yes	1.8 Wafer/ s

Three DoE's (A, B, C) are shown in Table 1 that were perform on two handling devices. Screen printing (A) was done on a machine that was using a conventional screen-printing process. The rotary printing (B, C) was done on prototype machine, that was built for high throughput processes [2, 3, 4].

2.1 Sample Preparation

For this investigation p-type (M2-Size) monocrystalline silicon wafers were used. The wafers had a standard thickness of 175 µm with a total thickness variation of less than 30 µm. All wafers a taken out of the same diamond wire wafering process. The wafers in the main batch were all neighbored wafers. To get comparable results, the main batch wafers were sorted into test batches with 60 wafers for every mechanical characterization. First wafer of the main batch was assigned to test batch one, the second to test batch on and so on. After assignment of the wafer to the last test batch the next wafer of the main batch was assigned to the first test batch. See Figure 1, an example of 4 test batches is shown.

Figure 1 Principle of the 4-point bending test [5]

Thus, every batch consists of nearly identical wafers with similar grain structure and mechanical properties.

2.2 Strength Determination

To investigate the damage of the wafers after handling different methods can be applied. The resolution of automated micro-crack detection systems is too low to identify cracks in the micrometer range and the manual analysis by optical and infrared microscopy can be very time consuming. Thus, the damage was measured indirectly by the wafer strength. Since the strength of a brittle material is defined by the biggest defect in the stress field, changes in defect type and size can be measured by measuring the material strength. Mostly, the detailed defect type is unknown and thus the statistical strength distributions of the batches are compared.
The wafer strength was measured in a 4-point bending test, which is also known as 4-line bending test. The principal setup is shown in Figure 2. A parallel set of loading and support rollers are bending the wafer uniaxially until the wafer fractures. As experimental result the force-deflection charts and the fracture force or deflection is derived. In this investigation the wafer sawing direction is placed parallel to the rollers for every wafer. The contact surface, where the printers touched the wafer was on top in the 4-point bending. The printing process causes a tensile stress field to the bottom side that was also placed in the constant tensile stress field within the inner rollers of 4-point bending test. It was used a distance of $l = 80$ mm between the lower rollers und $b = 40$ mm between the upper rollers. The rollers had a diameter of 10 mm and were covered by a polymer tape to reduce contact pressure.

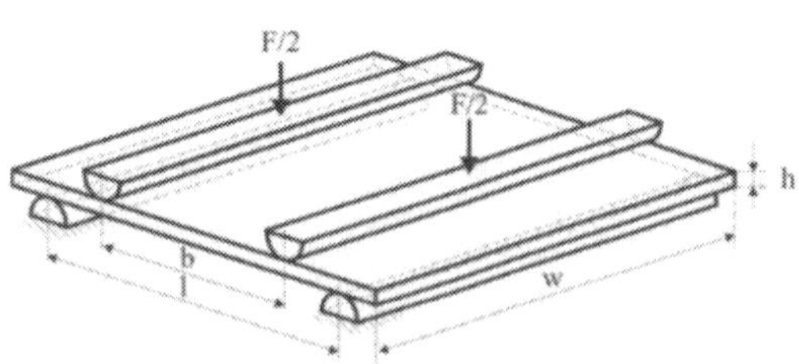

Figure 2 Principle of the 4-point bending test [5]

The 4-point bending tests were performed on universal testing machine ZWICK 005. The force was measured by 500 N load cell and the deflection was measured by the position of the machine truss. The mean thickness of every wafer was measured by the weight in order to consider thickness variations within the batch. The fracture stress was determined according to DIN SPEC 91351 [8] to consider the non-linear behavior of the thin wafers. The Weibull distribution [9] was used for statistical evaluation of the fracture stresses of every batch. This distribution function is based on the weakest link theory and is commonly used for brittle materials as silicon. The two parameters are the characteristic fracture stress σ_θ, at which 62.3% of all samples fail and the Weibull modulus m, which represents the scattering. The Weibull parameters, estimated by the Maximum-Likelihood-Estimation, as well as the confidence bounds were determined following [8].

3 RESULTS

The statistical evaluation of fracture stresses was done with the use of Weibull distribution. The distribution correlates to failure behavior of brittle materials like silicon. Table 2 shows an overview of the determined Weibull parameter for each DoE and wafer batch.

Table 2: Overview Weibull parameter, characteristic Fracture strength and the Weibull modulus for each DoE and Batch.

DoE [ ]	char. Fracture Strength [MPa] *Standard Error*	Weibull Modulus [ ] *Standard Error*
A#1	121.9 *1.3*	14.2 *1.4*
A#2	120.0 *1.1*	16.0 *1.7*
A#3	118.8 *1.0*	17.5 *1.9*
A#4	115.7 *1.2*	14.6 *1.6*
B#1	119.4 *1.0*	16.0 *1.5*
B#2	119.1 *0.8*	18.6 *1.9*
B#3	114.1 *1.2*	12.0 *1.3*
B#4	117.7 *1.0*	15.8 *1.5*
C#1	113.5 *0.9*	16.9 *1.7*
C#2	111.6 *1.0*	16.1 *1.7*
C#3	113.9 *0.9*	19.0 *2.0*
C#4	116.9 *0.9*	19.3 *1.8*
C#5	113.3 *0.9*	18.9 *2.1*
C#6	116.2 *0.9*	19.3 *2.0*

The range of characteristic fracture strength and Weibull modulus is typical for diamond wire sawn wafers that were tested parallel to their saw marks in 4-point bending tests. The drop of characteristic fracture strength after handling was measured for DoE A (screen-printing) and DoE B (not optimized rotary printing. The characteristic fracture strength of the optimized rotary printing tests shows no decrease of strength. But wafers that were moved cross to their saw marks through the printing process show a slight increase of the char. fracture strength that is not explainable yet. The Weibull modulus significantly decrease only for batch B#3. That means the damage type changes. That fits to a simultaneous reduction of characteristic fracture strength for that wafer batch.

3.1 Screen-Printing

In detail the Figure 3 shows the Weibull plot of DoE A, the conventional screen-printing process. Two reference batches and two wafer batches that run through the printing process can be seen. Both reference batches show no significant difference of the distribution that means the prepared wafer batches are equal regarding the existing mechanical damage that comes from the wafering process. A slight drift to the left of the entire distribution of both printing batches was observed. That means damaging of the wafers take place and a reduction of the characteristic fracture strength of approx. up to 6 MPa was observed. The Weibull modulus does not change significantly that mean no change of the dominating damage type takes place.

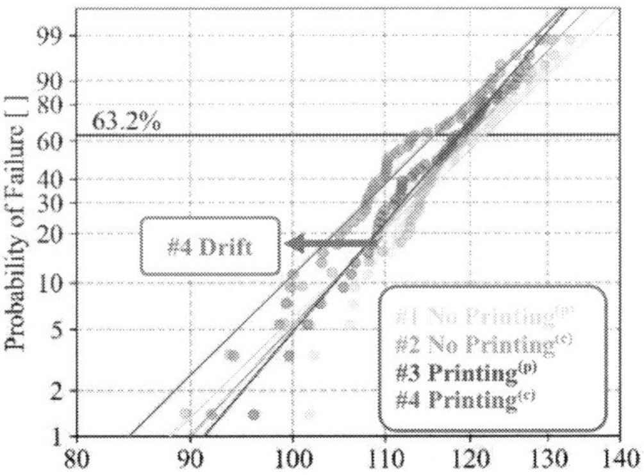

Figure 3 Weibull plot of DoE A (screen-printing process)

3.2 Rotary Printing

Figure 4 shows the Weibull plot of DoE B, the not optimized rotary printing process. Two reference batches and two wafer batches that run through the printing process can be seen. Both reference batches show no significant difference of the distribution as expected. A drift to the left of the entire distribution of batch #3 (parallel configuration for movement) was observed as well as a significant change of the slope equal to the Weibull modulus. That means damaging of the wafers take place and a reduction of the characteristic fracture strength of approx. 5 MPa was observed. The Weibull modulus change significantly that mean the dominating damage type change. Other damages were taken onto the wafers like new cracks or change of the crack formation / orientation.

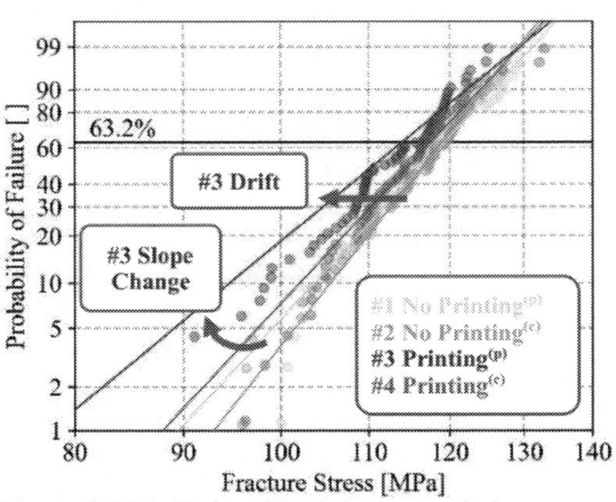

Figure 4 Weibull plot of DoE B (rotary printing process that is not optimized)

3.3 Rotary Printing *optimized*

Figure 5 shows the Weibull plot of DoE C, the optimized rotary printing process. Two reference batches and two wafer batches that run through the printing process can be seen. Both reference batches show no significant difference of the distribution as expected. The batches of the printing process at a speed of 1.1 wafer per second show not damaging of the wafers. No significant decrease of the fracture strength as well as the Weibull modulus was observed. Similar results were determined for wafers that were moved with a speed of 1.8 wafer per second. Again, the characteristic fracture strength and the Weibull modulus were not decreased by process operations. A slight increase of the wafer strength was observed for the wafer batches that runs cross with respect to their saw marks through the printing process. There is no mechanical explanation yet.

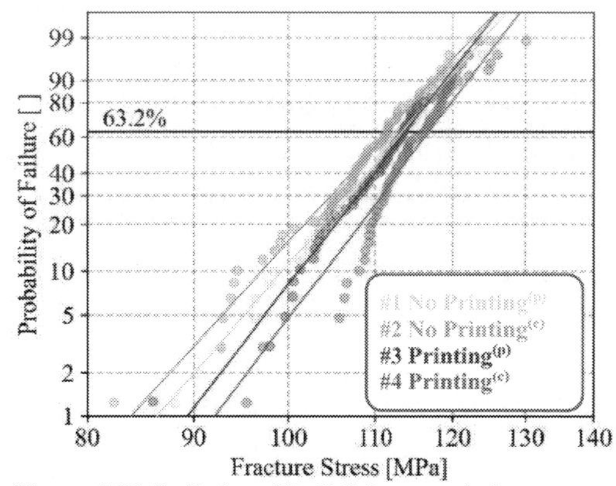

Figure 5 Weibull plot of DoE C (rotary printing process at 1.1 wafer/sec)

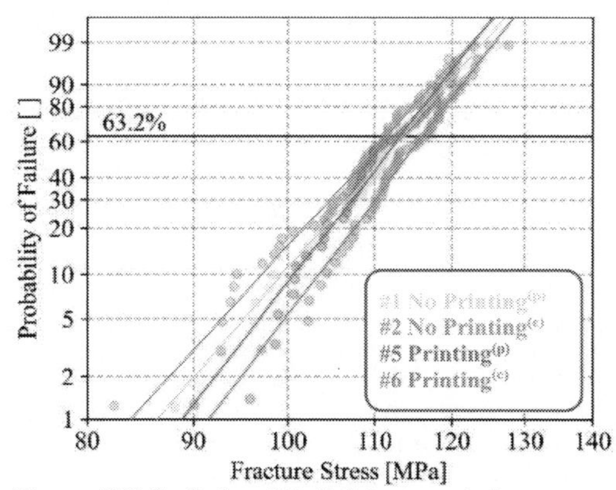

Figure 6 Weibull plot of DoE C (rotary printing process at 1.8 wafer/sec)

4 SUMMARY AND CONCLUSION

Characterization of process damage potential by strength analysis was performed successfully. Equal Weibull distributions of the reference batches indicate a well-done sample preparation. The screen-printing process and not optimized rotary printing process show a damaging potential. The drop of characteristic fracture strength and the Weibull modulus was detected. A mechanical optimized printing process show now wafer damaging as well as no change of the defect type equal to a change of Weibull modulus. Finally, a method was presented to evaluate any type of process regarding a mechanical damaging to ensure process quality and minimize product fails.

5 ACKNOWLEDGMENTS

Financial support by the Federal Ministry for Economic Affairs and Climate Action funded project "NextTec" (FKZ: 03EE1001A) is gratefully acknowledged as well as Fraunhofer ISE for performing the Design of Experiment handling operations.

6 REFERENCES

[1] VDMA, International Technology Roadmap for Photovoltaic (ITRPV), 2022, 13th Edition, March 2022

[2] A. Lorenz, M. Klawitter, M. Linse, L. Ney, S. Tepner, S. Pingel, M.S. Sabet, J. Reiner, K. Oehrle, R. Greutmann, J. Röth, M. Drews, K. Muramatsu, S. Ikarashi, F. Clement, Rotary Screen-Printed Metallization of Heterojunction Solar Cells: Toward High-Throughput Production with Very Low Silver Laydown, Energy Tech (2022). https://doi.org/10.1002/ente.202200377.

[3] A. Lorenz, M. Klawitter, M. Linse, S. Tepner, J. Röth, N. Wirth, R. Greutmann, M. Lehner, A. Senne, D. Reukauf, M. Drews, S. Gombert, H. Brocker, A. Mette, J. Rohde, E. Dörsam, F. Clement, The project »Rock-Star«: The evolution of rotary printing for solar cell metallization, in: AIP Conference Proceedings, Vol. 2367, No. 1. 5–6 October 2020, Genk, Belgium, AIP Publishing LLC, p. 20008 (2021).

[4] A. Lorenz, J. Röth, K. Zengerle, M. Linse, M. Klawitter, S. Tepner, N.Wirth, R. Greutmann, M. Lehner, A. Senne, D. Reukauf, A. Mette, F. Hage, M. Drews, S. Gombert, H. Brocker, J. Rohde, E. Dörsam, F. Clement, R. Preu, Project "Rock-Star" – High-Speed Rotary Printing for Solar Cell Metallization: From Vision to Reality, in: Proc. of the 37th European Photovoltaic Conference (EUPVSEC). 7 - 11 Sept., Virtual Conference, pp. 400–403 (2020).

[5] R. Koepge, S. Schoenfelder, T. Giesen, C. Fischmann, A. Verl, J. Bagdahn, „The Influence of Transport Operations on the Wafer Strength and Breakage Rate", 26th European Photovoltaic Solar Energy Conference and Exhibition, 2072 – 2077, 2011

[6] Ringo Koepge, Frank Wegert, Sven Thormann, Stephan Schoenfelder, "Characterization of damage and mechanical strength of wafers and cells during the cell manufacturing process", PV International, Vol.22, 2013

[7] Felix Kaule, Marcus Oswald, Ringo Koepge, Carola Klute, Stephan Schoenfelder, "Comprehensive analysis of strength and reliability of silicon wafers and solar cells regarding their manufacturing processes", PV International, Vol. 29, 2015

[8] DIN SPEC 91351:2017-04, "Strength Testing für Photovoltaic Wafers", April 2017

[9] W. Weibull, A Statistical Distribution Function of Wide Applicability, Journal of Applied Mechanics, Vol. 18, 293-297, 1951

INVESTIGATION OF THE FRONT SIDE ANTI-REFLECTION COATING LAYER ON M10 PERC SOLAR CELLS

Cheng-Wen Kuo, Ta-Ming Kuan, Yung-Chih Li, Wei-Lo Chueh, Li-Guo Wu, Shih-Chieh Lin and Cheng-Yeh Yu
TSEC Corporation, No.85, Gaungfu N. Rd., Hsin-Chu 30351, Taiwan.

ABSTRACT: In this work, we demonstrated that the p-type PERC solar cells can also be improved front side performance based on front side SiN/SiON anti-reflection coating (ARC) layer structure. Improved ARC layer was obtained by depositing the SiON layer after the SiN layer. By changing the n value of SiON layer through the process parameters, it can be found that the short wavelength can be improved, and the photocurrent can also be increased by about 56mA. About 0.1% gain in conversion efficiency was obtained compared to the result of reference solar cell. Finally, the potential induced degradation (PID) test of SiON PERC solar cell is also in compliance with IEC TS 62804-1 specifications.

Keywords: PERC, SiO_xN_y, PID

1 INTRODUCTION

In photovoltaics, the refractive index of the amorphous hydrogenated form of SiN (SiNx:H) has been found to be ideal because it leads to very low absorption in the solar spectral range and has excellent passivation properties on the silicon surface and within the silicon bulk in the case of mono and multi-crystalline material. As a result, over the past decade, silicon nitride has become the standard for implementing front-side anti-reflection (AR) coatings for industrial solar cells. Plasma-enhanced chemical vapor deposition (PECVD) is currently one of the preferred deposition methods for SiN thin films due to its lower thermal budget than other CVD processes and the high hydrogen content in the layers [1]. The amorphous hydrogenated form of SiON dielectric (SiOxNy:H) shares some similarities with SiN which can be deposited by PECVD at low temperature, it exhibits low absorbance and an adjustable refractive index from below 1.5 to as high as 3.0 by hanging the precursor gas flow ratio during the deposition process[2]. Therefore, by adjusting the optical index and thickness of the SiON films, these properties are suitable for antireflection multilayer coating structures for solar cells. Previous literature [3] also reported the anti-reflection coating ARC on crystalline silicon solar cell plays an important role not only on optical absorption but also on PID sensitivity. Refractive index of SiNx ARC is reported to influence the PID sensitivity.

In this work, we introduce novel SiOyNx/SiNx in the ARC layer to improve the optical properties of silicon solar cells. It can be found that the short wavelength can be improved, and the photocurrent can also be increased by about 56mA. Finally, the potential-induced degradation (PID) test of SiON PERC solar cells also complies with the IEC TS 62804-1 specification, which is also similar to SiN PERC solar cell degradation.

2 EXPERIMENTAL

Figure 1 showed the PERC+ structure in this work. The M10 size (182.0 mm x 182.0 mm) bifacial PERC solar cells were fabricated on 160 μm thick p-type boron doped Cz-silicon wafers with a resistivity of 0.4~1.1 Ω-cm.

The process sequence in this work was shown in Figure 2. Chemical wet etching methods were used for saw damage removal and surface texturing, which include both acidic etching for multi-crystalline silicon wafers and alkaline etching for mono-crystalline silicon wafers. The solution composition of acidic etching was nitric acid (HNO3) : hydrofluoric acid (HF) : deionized water (DI) = 1 : 2 : 1.5, and the solution composition of alkaline etching was potassium hydroxide (KOH) : deionized water (DI) = 1 : 50. Conventional phosphorous diffusion was employed by furnace to form the n+ emitters with a sheet resistance of 80 - 100 ohm/sq. The diffusion process also improved to uniformity for this work.

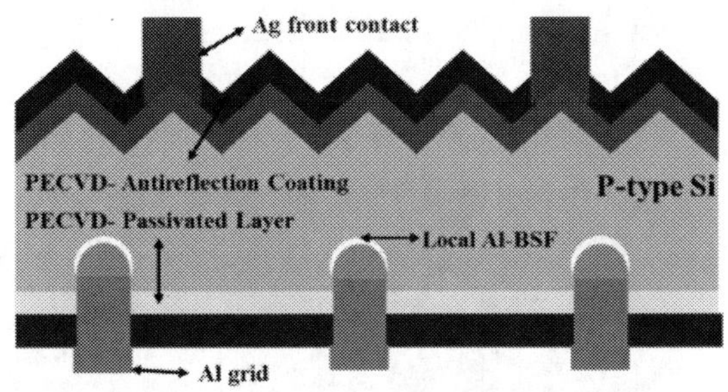

Figure 1: PERC+ structure as investigated in this work.

The phosphosilicate glass (PSG) was then removed by HF-contained solutions, formed edge isolation in the meantime. Then, the rear surface is polished which removes 5 μm silicon in this work, in order to reduce the roughness and its impact on the effective charge carrier lifetime [4]. In order to decrease PID effect and increase efficiency, we optimized the CVD process parameter to improve. An antireflection coating (ARC) layer of silicon nitride (SiN_x: H) and/or SiO_xN_y/SiN_x dual ARC was deposited by plasma enhanced chemical vapor deposition (PECVD) with thickness of 90 nm. The color of ARC layer is shown in Figure 3. The rear side passivation was carried out on rear side also using a PECVD. The thickness of rear side passivation thickness was 280 nm [5]. Afterward, the rear side laser opening ratio is also optimized. All PERC solar cells have the laser-ablated opening widths in the rear-side passivation stacks which were about 30 μm, as shown in Fig 3. The rear Al finger width is 150 μm in this work. The detail parameters were listed in Table 1. The front side silver and rear side aluminum metallization was formed by screen printing technique and followed by co-firing process, which resulted in formation of front silver grid contact and local back-surface-field. Finally, the current-voltage measurements of solar cells under AM 1.5 illumination were performed.

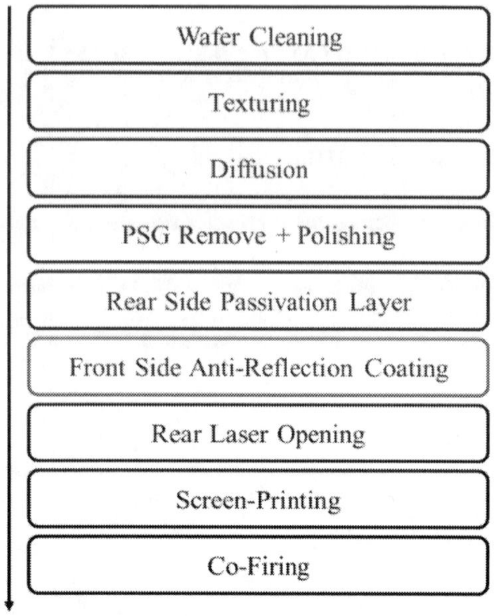

Figure 2: Process flow for the fabrication of PERC solar cell as developed in this work

Figure 3: The color of different ARC layer for this work. Left is M10 REF and right is SiON PERC.

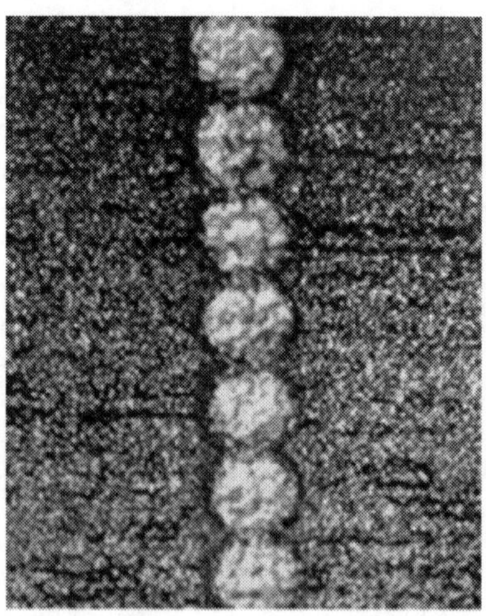

Figure 4 : OM image shown the laser ablated opening width in this work.

3 EXPERIMENTAL RESULTS AND DISCUSSION

We have performed an optical reflectance analysis for the different ARC layer, as shown in Figure 5. The distinct reduction of the reflectance at short wavelengths (300-500nm) is apparent for the optimized SiON stack layers compared with the baseline M10 REF, the front average reflection of short wavelengths is about 2.6% and 5.9%, respectively.

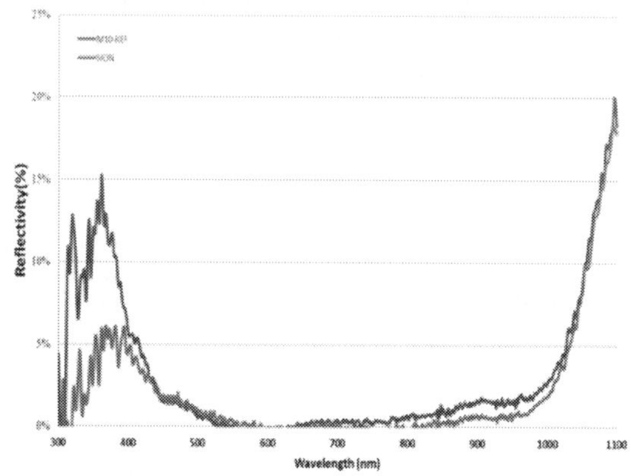

Figure 5: Measured reflectivity for the optimized M10 REF (red) PERC solar cell and SiON (green) PERC solar cell.

Table 1 shows the average electrical performance of M10 REF and SiON PERC PERC solar cells. The average efficiency of SiON solar cell was 22.9%, independent on the thickness of ARC layers of SiON PERC solar cells, the average efficiency of 22.8% was obtained for the M10 REF PERC solar cells. It shows that the photocurrent can also be increased by about 56mA 0.1% and the absolute gain in conversion efficiency can be obtained by introducing SiOyNx /SiNx as an ARC instead of multi SiNx layer. Figure 6 shows scanning electron microscope (SEM) images of the different front side ARC thickness for M10 REF and SiON PERC solar cell. It can be found that the ARC thickness is similar to the process setting. It means that the use of SiON stacked ARC layers can improve the optical characteristics, however, the appearance color need to be considered.

Table 1: The average electrical results of M10 REF and SiON PERC cells.

	EFF (%)	FF (%)	V_{OC} (V)	I_{SC} (A)
M10 REF PERC	22.8	79.77	0.6878	13.730
SiON PERC	22.9	79.77	0.6876	13.786

On the other hand, the ARC layer structure of PERC solar cells has an effect on PID [3], the potential-induced degradation (PID) test of the solar cell also conforms to the IEC TS 62804-1 specification and was used in this experiment. The M10 REF and SiON PERC solar cells also passed PID test which condition was indicated to be 85 degree Celsius and 85% RH in climatic chamber apply -1000 V for 96 hours as shown in Figure 7. Observed the power loss < 5% and no EL darkened area were found after PID test for the M10 REF and SiON PERC solar cells.

Figure 6: Scanning electron microscope images of rear side passivation thickness for the (a) M10 REF, (b) SiON PERC

	Pre-PID		Post-PID		EL		
Sample	Rsh	Pmpp	Rsh	Pmpp	**Pre-PID**	**Post-PID**	**ΔPmpp**
M10 REF	268.1	6.2504	148.3	6.1777			1.16%
M10 REF	326.5	6.1868	212.1	6.0822			1.69%
M10 REF	353.9	6.1437	221.9	6.0734			1.14%
SiON	331.1	6.2523	299.8	6.1372			1.84%
SiON	627.8	6.0378	264.2	5.8747			2.70%
SiON	274.7	6.2334	204.3	6.0561			2.84%

Figure 7: Summary of power loss and EL images of pre-PID and post-PID of M10 REF and SiON PERC solar cells. PID test condition is 85 degree Celsius / 85% RH biased -1000V for 96 hours.

4 CONCLUSIONS

In this work, we have demonstrated that the p-type PERC solar cells could be improved front side performance based on front side SiN/SiON anti-reflection coating (ARC) layer structure. Improved ARC layer was obtained by depositing the SiON layer after the SiN layer. The distinct reduction of the reflectance at short wavelengths (300-500nm) is apparent for the optimized SiON stack layers compared with the baseline M10 REF. It can be found that the short wavelength can be improved, and the photocurrent can also be increased by about 56mA. About 0.1% gain in conversion efficiency was obtained compared to the result of reference solar cell. Finally, PID test of M10 REF and SiON PERC solar cell is also in compliance with IEC TS 62804-1 specifications, the power loss of both is similar in this experiment.

5 REFERENCES

[1] J. Dupuis, E. Fourmond, J. Lelièvre, D. Ballutaud, M. Lemiti, "Impact of PECVD SiON stoichiometry and post-annealing on the silicon surface passivation." *Thin Solid Films*, vol. pp.6954, 2008.

[2] J. Dupuis, E. Fourmond, D. Ballutaud, N. Bererd, M. Lemiti, "Optical and structural properties of silicon oxynitride deposited by plasma enhanced chemical vapor deposition" *Thin Solid Films*, vol. 519, pp.1325, 2010.

[3] C. Zhou, J. Zhu, S.E. Foss, H. Haug, Ø. Nordseth, E.S. Marstein, W. Wang, "SiOyNx/SiNx Stack Anti-reflection Coating with PID-resistance for Crystalline Silicon Solar Cells," Energy Procedia, vol. 77 pp.434, 2015.

[4] E. Urrejola, R.Petres, J. Glatz-Reichenbach, K. Peter, E. Wefringhaus, H. Plagwitz, G. Schubert., "High Efficiency Industrial PERC Solar Cells with all PECVD-Based Rear Surface Passivation" 26th EUPVSEC, pp. 2233, 2011.

[5] C.-W. Kuo, T.-M. Kuan, W.-L. Chueh, L.-G. Wu, S.-C. Lin, C.-Y. Yu, "Investigation of the Rear Side Passivation Layer on Bifacial PERC Solar Cells" 38th EUPVSEC, pp. 288, 2021

LOW-AREA SCREEN-PRINTED CONTACTS FOR INDUSTRIAL PERC SOLAR CELLS

Yuchao Zhang, Li Wang, Ran Chen, Catherine Chan, Brett Hallam
School of Photovoltaic and Renewable Energy Engineering, UNSW Sydney
UNSW Sydney, Kensington, 2052, Australia

ABSTRACT: Screen-printing technology has already become the dominant metallization technique for silicon solar cells in the PV industry. Despite significant progress and improvements in the performance of screen-printed contacts during the past decades, one remaining challenge for screen-printed solar cells is the formation of highly recombination-active metal/Si interfaces, which can result in substantial carrier recombination losses in screen-printed PERC solar cells and the emerging high-efficiency TOPCon solar cell technology. With the bulk and surface passivation quality continuously improving in industrial silicon solar cells, the impact of such contact recombination losses will become more and more detrimental and eventually become the limiting factor to the efficiency of screen-printed solar cells. To minimize such losses, the metal/Si interface area needs to be greatly reduced for screen-printed contacts. In this work, preliminary results on the development of two novel metallization designs are presented. Both approaches aim to significantly reduce contact recombination of screen-printed contacts via reducing the metal/Si interface areas while maintaining compatibility with current screen-printing capabilities. The fabrication of proof-of-concept PERC solar cells with the proposed contact structures demonstrated a promising V_{OC} gain of up to 8 mV by reducing the metal/Si interface area from ~3.1% with industrial standard H-pattern grids to 0.5% with the dash metallization design. The key challenges encountered throughout the development are identified, and potential methods to overcome these challenges and achieve higher efficiencies in future work are also discussed.
Keywords: Solar Cell, Metallization, Screen Printing

1 Introduction

Since the 1970s [1], the adoption of screen-printing technology in the PV industry has provided a simple, robust, and cost-effective technique for the metallization of silicon solar cells. Currently, screen-printed contacts are being used in all major industrial silicon solar cell technologies, such as passivated emitter and rear cells (PERC), tunnel oxide passivated contact (TOPCon), and silicon heterojunction (SHJ) solar cells, with a market share of more than 98% in the PV industry [2]. Despite the greatly improved performance of screen-printed contacts after half a century's advancements and developments, overcoming the performance limitations of current screen-printed contacts still provides an effective pathway to improve the efficiency of industrial silicon solar cells.

One major loss mechanism associated with screen-printed contacts in silicon solar cells is the formation of highly recombination-active metal/Si interfaces [3], [4]. Such interfaces are essential to the power extraction in PERC solar cells and on the front side of TOPCon solar cells. However, to allow the direct contacts between screen-printed electrodes and the silicon surface, the surface passivation dielectrics need to be removed by either the use of fire-through screen-printing pastes or a dedicated laser contact opening process, resulting in the loss of surface passivation at such direct metal/Si interfaces. In addition, the surface recombination can be further increased by the etching of the silicon surface with screen-printing pastes during the high-temperature firing. This is particularly a concern for the use of Ag/Al pastes on the front side of TOPCon solar cells, which could lead to the formation of large-area Ag/Al spikes and excessive etching of silicon on the front side of TOPCon solar cells [5].

In current industrial screen-printed PERC solar cells, despite the implementation of laser-doped selective emitters (LDSEs) underneath the front silver contacts, a saturation current density (J_0) as high as 500-900 fA/cm^2 [6] can still be expected at metal/Si interfaces on the front. With a contact area of 3-5% of the total cell area, recombination at front metal/Si interfaces could contribute 15-45 fA/cm^2 to the total J_0 of such screen-printed industrial PERC solar cells, leading to up to 0.3%$_{abs}$ losses in the efficiency [4]. Notably, as the bulk and surface passivation quality continuously improve, the impact of such significant recombination losses on the efficiency will become even more detrimental and eventually become the limiting factor to the efficiency of industrial screen-printed PERC solar cells.

On the other hand, although tunnel oxide passivated contacts are well-known for their capabilities of mitigating contact recombination losses, such passivated contacts are only featured on the rear side of current industrial TOPCon solar cells [7], which leaves the front-side screen-printed contacts still susceptible to significant contact recombination losses. Due to intrinsically low bulk and surface recombination in TOPCon solar cells, the presence of highly recombination-active metal/Si interfaces on the front is even more problematic for TOPCon solar cells compared to PERC. The detailed loss analysis of a 24.8%-efficient screen-printed TOPCon solar cell carried out by Zheng *et al.* [8] shows that the front contact recombination has already become the single largest recombination mechanism in their solar cell. As such, the development of new technologies to greatly reduce recombination losses at metal/Si interfaces of screen-printed contacts will be not only critical for current mainstream PERC solar cells but also urgently needed for the emerging high-efficiency TOPCon solar cells.

In this work, novel metallization designs featuring a greatly reduced coverage area of metal/Si interfaces for the mitigation of contact recombination in screen-printed contacts are presented. Preliminary results on the implementation of these metallization designs on the front side of industrial PERC solar cells are presented, and key challenges that need to be overcome in future work are identified and discussed.

2 Low-Area Laser-Doped Contacts

2.1 The Concept of Low-Area Laser-Doped Contacts

The current implementation of screen-printed contacts uses fire-through (FT) silver pastes on the front side of industrial PERC solar cells. In this way, significant reductions in the metal/Si interface area will only be allowed if the printed width of fingers can be substantially reduced. Meanwhile, the reduced finger width will also likely trigger a reduction in the finger spacing on the front side of PERC solar cells to minimize the emitter lateral resistive losses, which will subsequently lead to undesirable increases in the metal/Si interface area again.

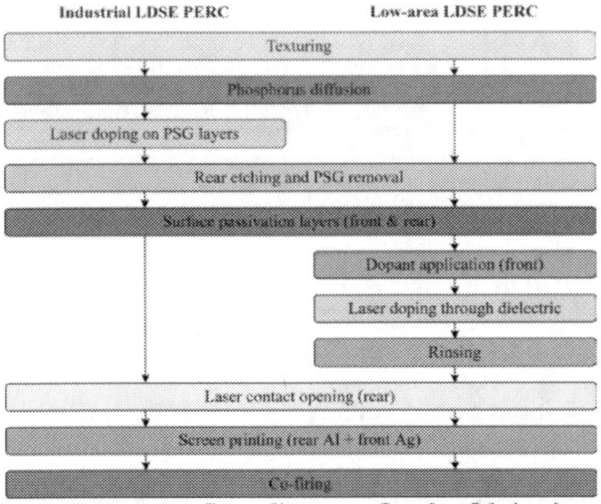

Figure 1: Process flow diagrams for the fabrication of screen-printed PERC solar cells with industrial LDSEs and the low-area laser-doped contacts.

In this work, a novel solar cell structure is developed based on the concept of low-area laser-doped contacts. The proposed contact scheme utilizes an alternative laser-doping technology in conjunction with the non-fire-through (NFT) floating silver contacts to greatly reduce the metal/Si interface area to less than 1% of the total cell area on the front side of PERC solar cells while maintaining compatibility with current screen-printing capabilities and alignment requirements. The process flow of the proposed low-area laser-doped contacts is shown in **Figure 1** in comparison with the industrial standard LDSE contacts. One key feature of the low-area laser-doped contacts is the formation of a series of laser-doped dashes or wavy lines spaced along the finger direction after the deposition of surface dielectrics using externally applied dopant sources. Subsequently, floating silver fingers using NFT Ag pastes are aligned with and printed on laser-doped regions, whereby the metal/Si interface is limited to the area of intersection between the laser-doped dashes/wavy lines and screen-printed floating silver fingers.

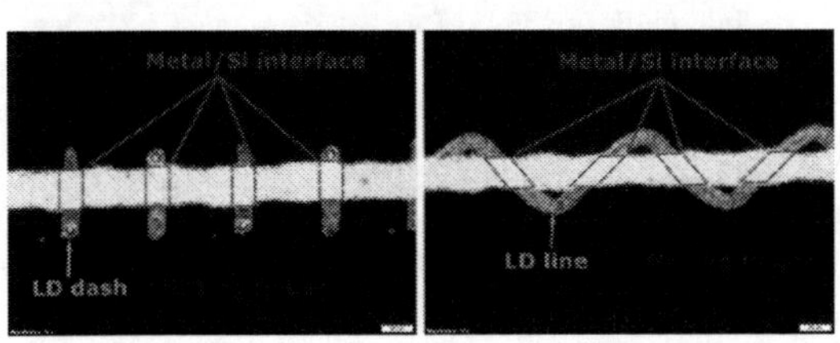

Figure 2: Microscope images of low-area laser-doped contacts with laser-doped (left) dashes and (right wavy lines).

With this design, the metal/Si interface area can be easily adjusted by manipulating the spacing between dashes or the period of wavy lines regardless of the printed width of fingers. In addition, the use of ~100 μm wide dashes or wavy lines with 100 μm peak-to-peak amplitude will ensure sufficient alignment tolerance (± 30 μm) to the screen-printing of fingers without risking the yield in mass production environments. Microscope images of the proposed low-area laser-doped contacts can be found in **Figure 2**.

2.2 Device fabrication

In this experiment, proof-of-concept PERC solar cells were fabricated using the proposed low-area laser-doped contact structure on 6-inch rear-side metallized cell precursors without front-side metallization grids sourced from an industrial supplier. Prior to the laser doping, spin-on dopant (85%$_{v/v}$ H_3PO_4) was applied on the front surface of cell precursors as dopant sources. Subsequently, all samples were laser-doped using a continuous-wave (CW) 532 nm Newport/Spectra Physics Millennia Prime Pump Laser with a Gaussian beam shaper. To achieve a 1% metal/Si interface area between laser-doped regions and NFT Ag fingers, a series of 100 μm long and 30 μm wide laser-doped dashes were formed along the finger direction with a spacing of 98 μm. Fingers and busbars were printed using an industrial screen printer with commercially available standard FT and NFT Ag pastes, followed by a metallization co-firing using a standard firing temperature profile with a peak temperature of 780 °C. Prior to IV measurements, small-area cells were laser cleaved from the full-area samples.

2.3 Results and Discussions

The benefits of mitigating contact recombination losses with reduced metal/Si interfaces can be clearly seen in **Figure 3**, where the solar cell fabricated with laser-doped dashes and NFT Ag fingers showed substantially improved photoluminescence (PL) response compared to the control cell with conventional FT Ag fingers. It should be noted that for cell regions that received laser doping, slightly lower PL counts have been observed compared to control regions without any laser doping. This indicates that the laser doping process, particularly with the formation of exposed laser-doped regions at each side of the dashes, could result in additional recombination losses on the front surface.

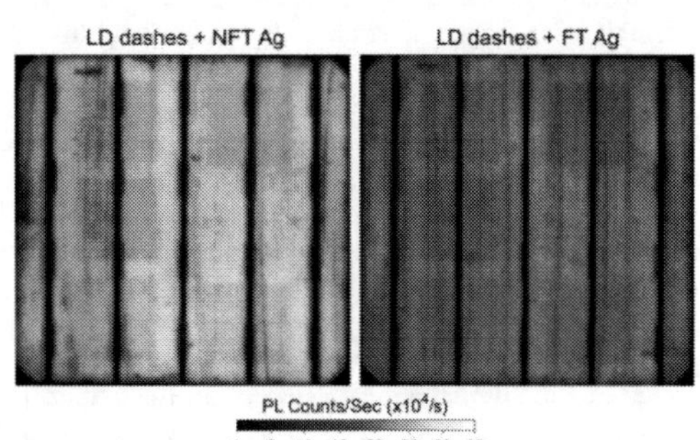

Figure 3: Photoluminescence images of PERC solar cells fabricated with (left) low-area laser-doped contacts and

(right) conventional FT Ag contacts. Note: each sample contains five identically laser-doped cell regions and one region without any laser doping.

To minimize such losses, a high dopant concentration within the laser-doped regions would be highly appreciated, which could lead to more effective field-effect passivation at such exposed silicon surface. On the other hand, reducing the length of dashes by improving the alignment accuracy of the industrial screen-printing process could provide another effective way to minimize the coverage area and surface recombination losses of exposed laser-doped regions. Nevertheless, in this experiment, higher PL counts have been seen in cell regions with laser-doped dashes and NFT Ag fingers compared to regions with FT Ag fingers and without laser doping. This suggests that slight increases in recombination due to the laser doping process could be well justified by the overwhelming benefits of reduced contact recombination.

Table I: I-V results of fabricated PERC solar cells.

	Eff (%)	V_{OC} (mV)	J_{SC} (mA/cm^2)	FF (%)
FT Ag fingers	18.6	645.1	39.93	72.19
NFT Ag fingers	1.8	641.8	17.07	16.43

The I-V data of the fabricated PERC solar cells can be found in **Table I**. Unfortunately, a very low efficiency has been obtained on PERC solar cells fabricated with low-area laser-doped contacts and NFT Ag fingers. The root cause of such poor efficiency can be largely attributed to the low FF as a result of extremely high contact resistivity (> 300 mΩ.cm^2) between NFT Ag fingers and laser-doped regions. Given that a low contact resistivity of 0.5 mΩ.cm^2 was measured between conventional FT Ag pastes and the laser-doped silicon surface, it is more likely that the contact formation issues of NFT Ag pastes are related to the properties of the pastes rather than the laser doping process.

Figure 4: Scanning electron microscope (SEM) images of metal/Si interfaces with (left) NFT Ag fingers and (right) FT Ag fingers after chemical etching of bulk silver and glass layers.

Figure 4 shows the Ag crystallites on the silicon surface with both NFT and FT Ag pastes used in this work. Since the presence of Ag crystallites plays an essential role in the contact formation between screen-printed Ag contacts and the silicon surface, the considerably lower density of Ag crystallites formed with NFT Ag pastes can be seen as the root cause of the contact formation issues experienced in this experiment. Notably, the lack of effective formation of Ag crystallites with NFT pastes could be closely related to the special glass frit system used in such pastes, where the relatively low reactivity of the glass frits is expected to have detrimental impacts on the direct reaction between Ag^+ ions dissolved in glass layers and the silicon surface. As a result, to allow the proper contact formation between NFT Ag pastes and the laser-doped silicon surface, modifications to the chemistry of existing NFT Ag pastes will likely be required, particularly in the properties and reactivity of the glass frit system used in NFT Ag pastes.

3 Alternative screen-printed dash metallization design

In the meantime, to overcome the contact formation issues identified for the low-area laser-doped contacts, an alternative screen-printed dash metallization design utilizing a dual-print process and a combination of both FT and NFT Ag pastes is developed as another pathway to reduce the metal/Si interface area and contact recombination losses on the front side of PERC solar cells. Despite the great similarity in the overall geometry of both approaches, the most important feature of the new dash metallization design is that it still uses the well-developed conventional FT Ag pastes for the formation of metal/Si interfaces, which eliminates the complications related to the laser doping process and contact formation with unconventional NFT Ag pastes. To allow significant reductions in the metal/Si interface area with this new approach, conventional FT Ag pastes are firstly printed as a series of 100-200 µm dashes perpendicular to the finger direction for the formation of metal/Si interfaces while also providing sufficient alignment tolerance. Subsequently, floating fingers and busbars are printed using NFT Ag pastes on top of existing FT Ag dashes for the purpose of electrical conduction only. The microscope images of such dash metallization design are shown in **Figure 5**.

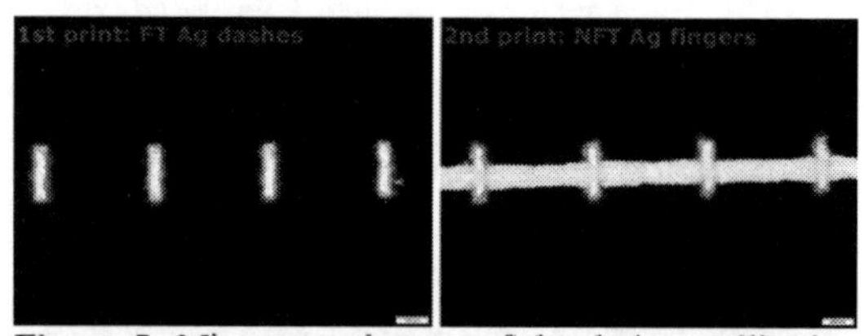

Figure 5: Microscope images of the dash metallization design after (left) the 1st print of FT Ag dashes and (right) the 2nd print of NFT Ag fingers.

3.1 Device fabrication

Finished PERC solar cells were fabricated using the proposed dash metallization design on rear metallized cell precursors in this experiment. FT Ag dashes (200 × 50 µm) were printed on the front surface of cell precursors using an industrial screen printer and commercially available FT Ag pastes. The pattern design consists of four small-area printed regions with the dash spacing of 1550, 775, 517 and 388 µm in each of them to achieve metal/Si interface areas of 0.5%, 1.0%, 1.5%, and 2%, respectively. After the printing of FT Ag dashes, all samples were fired using an industrial standard firing temperature profile with a peak temperature of 780 °C. Subsequently, floating fingers and busbars of H-pattern grids were printed using NFT Ag pastes on top of dashes, followed by another firing process. It should be noted that with properly optimized printing and drying processes, the firing step between the first and the second printing can be removed to avoid unnecessary thermal treatment in the cell fabrication sequence. In addition, another group of control cells were fabricated with only H-pattern grids printed on the front surface using conventional FT Ag pastes. Prior to I-V measurements, all small-area (6 × 6 cm²) were laser cleaved off from the full-area samples.

3.2 Results and Discussions

The PL images and R_s mapping of PERC solar cells fabricated with the dash metallization design and conventional H-pattern grids are shown in **Figure 6**. This result clearly demonstrates the trade-off between contact recombination losses and contact resistance losses for reduced metal/Si interface areas, where the use of smaller contact areas leads to lower contact recombination losses and higher V_{OC} of solar cells but at the cost of increased losses from contact series resistance due to reductions in the contact area. Preliminary simulation results suggest that, with a contact resistivity of ~ 1 mΩ.cm² in typical industrial PERC solar cells, the front metal/Si interface area can be reduced to a range of 1.0-1.3% of the total cell area before start seeing any efficiency losses due to increased contact resistance. As such, despite the slight disadvantages of increased contact resistance, reducing the metal/Si interface area should still be considered as a viable and effective way to mitigate contact recombination losses of screen-printed contacts in PERC solar cells.

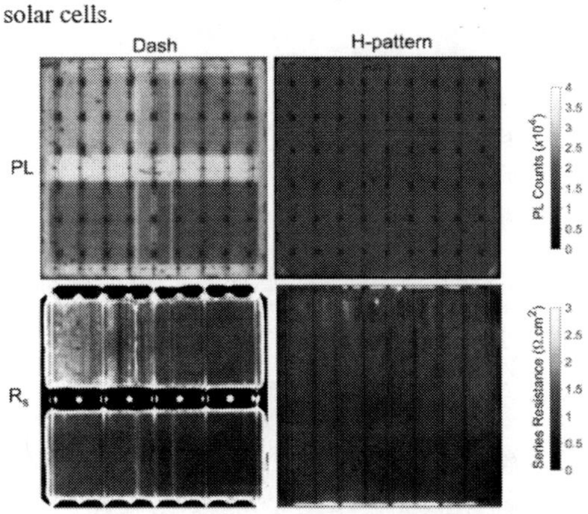

Figure 6: PL images and series resistance (R_s) mapping of solar cells fabricated with the dash front metallization design and conventional H-pattern design. Note: the dash coverage area from top left to bottom right regions is 0.5%, 1.0%, 1.5%, and 2%, respectively.

The I-V results of fabricated small-area PERC solar cells can be found in **Table II.** By reducing the metal/Si interface area from ~3.1% with conventional H-pattern to 0.5% with the dash metallization design, a remarkable V_{OC} improvement of up to 8 mV has been achieved in this experiment, which clearly shows the effective mitigation of contact recombination losses with the reduced metal/Si interface area. As the metal/Si interface area reduces from 2.0% to 0.5% for the dash metallization design, substantial reductions in FF from over 80% to less than 70% were triggered largely due to increased contact resistance. It should be noted that a high contact resistivity of 5-10 mΩ.cm² has been measured for the dash metallization design as a result of non-optimized firing profiles and the mismatch between the FT Ag pastes and emitter dopant profiles of samples used in this work. Such a high contact resistivity could significantly aggravate the negative impact of reducing the contact area on the contact resistance and FF. Moreover, despite a much larger contact area with the conventional H-pattern grids, the FF of control cells was found to be significantly lower than solar cells fabricated with the dash metallization design. The low FF of control cells was largely attributed to a higher-than-expected contact resistivity of ~ 20 mΩ.cm², which was most likely caused by fabrication issues encountered in this experiment. As a result, even though an efficiency gain of up to 0.2%$_{abs}$ has been achieved by the dash metallization design compared to the control cells with conventional H-pattern grids, such efficiency gain should not be considered as the true reflection of the performance benefits of the dash metallization design.

Table II: I-V results of fabricated PERC solar cells.

	Eff (%)	V_{OC} (mV)	J_{SC} (mA/cm²)	FF (%)
Control H-pattern	19.8	666.1	39.12	76.09
Dash – 2.0%	19.6	665.8	36.74	80.25
Dash – 1.5%	20.0	669.3	37.30	80.14
Dash – 1.0%	19.6	671.7	37.65	77.65
Dash – 0.5%	17.8	674.0	38.11	69.11

As for the short-circuit current density (J_{SC}), noticeable losses in J_{SC} have been observed for the dash metallization design compared to the conventional H-pattern grids. Although the use of 200 µm-long dashes could indeed contribute to additional optical shading on the front, the majority (> 80%) of losses in J_{SC} were found to be a result of higher optical shading from fingers due to a large width of ~ 100 µm of NFT Ag fingers printed on top of dashes. With slight modifications in the rheology of NFT Ag pastes and optimizations in printing parameters, such issues can be expected to be easily resolved in future work.

4 Conclusion and Outlook

In this work, preliminary results on the development of two novel metallization designs are presented. Both approaches aim to greatly reduce contact recombination of screen-printed contacts via reducing the metal/Si interface areas while maintaining compatibility with current screen-printing capabilities. For the proposed low-area laser-doped contact structure, one key challenge identified for it is the difficulty in forming high-quality ohmic contacts between NFT Ag pastes and laser-doped silicon surface, which resulted in poor efficiency of solar cells fabricated with this design. To overcome this challenge, modifications in the chemistry of such NFT Ag pastes will be required in future work, where the reactivity of the special glass frit system particularly needs to be improved for the effective formation of Ag crystallites on the silicon surface. In addition, understanding and further investigation in potential power losses associated with exposed laser-doped regions will also be necessary to ensure no significant losses will be introduced to solar cells fabricated with such a design.

The alternative dash metallization design provides another effective way to mitigate contact recombination losses without the complications related to the laser-doped process and contact formation with NFT Ag pastes as identified for the low-area laser-doped contact approach. The use of a dual-print process in the dash metallization design is highly relevant to the PV industry, where a continuous transition from single print to dual print has been projected by ITRPV. With the dash metallization design, a promising improvement of up to 8 mV in V_{OC} has been achieved in solar cells fabricated with this design. Although a high efficiency and significant efficiency gain have yet to be achieved with the dash metallization design in this work, most efficiency losses were found to be caused by fabrication issues, such as high contact resistivity and wide printed width of fingers, rather than fundamental issues with the design itself. By overcoming these fabrication issues in future work, much more significant performance benefits of the proposed dash metallization design can be expected. Moreover, although the current implementation of the dash metallization design features no LDSEs underneath FT Ag dashes, it is expected that industry-style LDSEs can be incorporated into this metallization design with the additional benefits of replacing the long LDSE stripes with localized LDSEs only under FT Ag dashes for reducing the coverage area of LDSE regions.

5 Acknowledgements

This work was supported by the Australian Government through the Australian Renewable Energy Agency (ARENA) (2020/RND005) and Australian Centre for Advanced Photovoltaics (ACAP). The views expressed herein are not necessarily the views of the Australian Government, and the Australian Government does not accept responsibility for any information or advice contained herein.

6 References

[1] E. Ralph, "Recent advancements in low-cost solar cell processing," in 11th Photovoltaic Specialists Conference, 1975, pp. 315–316.

[2] ITRPV, "International Technology Roadmap for Photovoltaic (ITRPV), 2020 Results," 2021.

[3] D. Inns, "Understanding Metal Induced Recombination Losses in Silicon Solar Cells with Screen Printed Silver Contacts," Energy Procedia, vol. 98, pp. 23–29, 2016, doi: 10.1016/j.egypro.2016.10.077.

[4] Y. Zhang, L. Wang, D. Chen, M. Kim, and B. Hallam, "Pathway towards 24% efficiency for fully screen-printed passivated emitter and rear contact solar cells," J. Phys. D. Appl. Phys., vol. 54, no. 21, p. 214003, May 2021, doi: 10.1088/1361-6463/abe900.

[5] S. Fritz, M. Konig, S. Riegel, A. Herguth, M. Horteis, and G. Hahn, "Formation of Ag/Al screen-printing contacts on B emitters," IEEE J. Photovoltaics, vol. 5, no. 1, pp. 145–151, 2015, doi: 10.1109/JPHOTOV.2014.2364117.

[6] T. Fellmeth, A. Born, A. Kimmerle, F. Clement, D. Biro, and R. Preu, "Recombination at metal-emitter interfaces of front contact technologies for highly efficient silicon solar cells," Energy Procedia, vol. 8, no. April, pp. 115–121, 2011, doi: 10.1016/j.egypro.2011.06.111.

[7] D. Chen et al., "24.58% total area efficiency of screen-printed, large area industrial silicon solar cells with the tunnel oxide passivated contacts (i-TOPCon) design," Sol. Energy Mater. Sol. Cells, vol. 206, no. September 2019, pp. 1–8, 2020, doi: 10.1016/j.solmat.2019.110258.

[8] P. Zheng et al., "Detailed loss analysis of 24.8% large-area screen-printed n-type solar cell with polysilicon passivating contact," Cell Reports Phys. Sci., vol. 2, no. 10, 2021, doi: 10.1016/j.xcrp.2021.100603.

TRANSITIONING TO HIGHER EMITTER SHEET RESISTANCE FOR PERC SOLAR CELL PRODUCTION

Mehul C. Raval[1]*, Mehmet Siddik Memetemin[2], Guru Prasad A[3], Mehmet Ender[2], Peter Fath[1]
1-RCT Solutions GmbH, Line - Eid - Str. 1, D-78467 Konstanz, Germany.
2- Kalyon Günes Teknolojileri Üretim A.,S., 06909 Ankara, Turkey.
3- centrotherm India Ltd., Bengaluru, India.

ABSTRACT: Increasing the sheet resistance of the phosphorus-doped PERC solar cell is a way to improve the solar cell efficiency. In current work, trials for increasing the sheet resistance for PERC solar cell from 140-150 Ω/□ to 180-185 Ω/□ by varying the $POCl_3$ flow and the pre-deposition & drive-in temperatures have been carried out. By stopping the $POCl_3$ flow during heat-up between pre-deposition and drive-in step, the sheet resistance increased to 230 Ω/□ with a higher standard deviation. By increasing the pre-deposition and drive-in temperature by 5°C, the I_{sc} and V_{oc} improved while the average FF decreased due to higher zone-wise sheet resistance variation. Altering the recipe with no $POCl_3$ flow during heat-up from pre-deposition to drive-in phase and decreasing the $POCl_3$ flow in the pre-deposition step provided a better way to increase the emitter sheet resistance. Reduction of $POCl_3$ flow during the pre-deposition phase, enabled to obtain average sheet resistance values in the range of 170 Ω/□ to 185 Ω/□. A sheet resistance value of 175 Ω/□ led to a gain of 0.05% absolute compared to the baseline production recipe. Better fine-tuning of zone-wise sheet resistance can lead to increased gain in the V_{oc} and I_{sc} without adversely affecting the FF.

Keywords: PERC, emitter, sheet resistance

1 INTRODUCTION

As per the ITRPV roadmap [1], the emitter sheet resistance of phosphorus-doped emitter for selective emitter PERC solar cells is expected to increase from 150 Ω/□ to 170 Ω/□ by 2026. Moving to higher sheet resistance emitter can help to improve the I_{sc} and V_{oc}, while good values of the FF can be maintained based on the laser-doped contacts.

In this work, trials for increasing the emitter sheet resistance from 140-150 Ω/□ to 180-190 Ω/□ have been carried out to check the influence on PERC solar cell I-V parameters. Such a study will help to understand the influence of emitter diffusion parameters on the emitter sheet resistance and finally on the solar cell I-V parameters.

2 EXPERIMENTAL DETAILS

2.1 Production of PERC solar cells

The study is carried out in the cell section of Kalyon PV, which is an integrated 1 GW bifacial p-type mono PERC PV factory. P-type (Boron doped) mono c-Si ingots are manufactured from poly-Si, which are then cropped, squared and chamfered into bricks. The bricks are then sliced into wafers using 50 µm diamond wire into wafers of 175 +20/-10 µm thickness. The wafers are fed to the cell section to make bifacial p-type mono PERC cells with G1 size wafers, which are in turn used to make glass-glass bifacial modules with half-cut cells. The phosphorus emitter diffusion is carried out by low pressure (LP) $POCl_3$ process with production emitter sheet resistance (R_{sh}) in the range of 140-150 Ω/□.

2.2 Variation of the emitter diffusion process

The emitter sheet resistance was varied in two ways. In the first method, the $POCl_3$ flow during heat-up and the pre-deposition & drive-in temperatures were varied to increase the sheet resistance from the current production value. The various trial 1 groups are shown in Table I. These trials were carried out in Phase I of the cell factory which produces G1 wafer size PERC solar cells.

Table I: Description of trial 1 groups evaluated for increasing the emitter sheet resistance.

Trial 1 groups	Pre-deposition temperature (°C)	Drive-in temperature (°C)	Variation
Baseline-1	780-790	840-855	Reference for production with $POCl_3$ flow during heat-up from pre-deposition to drive-in stage.
Group 1	780-790	840-855	No $POCl_3$ flow during heat-up from pre-deposition to drive-in.
Group 2	785-795	840-855	No $POCl_3$ flow during heat-up from pre-deposition to drive-in and increase pre-deposition temperature by 5°C.
Group 3	785-795	845-860	No $POCl_3$ flow during heat-up from pre-deposition to drive-in and increase pre-deposition temperature & drive-in temperature by 5°C.

In another variation of the trials, the emitter diffusion recipe was changed in which there was no $POCl_3$ flow during heat-up between the $POCl_3$ pre-deposition and drive-in phase and the $POCl_3$ flow during the pre-deposition was increased to obtain production R_{sh} in the range of 140-150 Ω/□. The trial 2 groups are summarized in Table II below. The trial 2 cells were produced in Phase 2 of the cell factory which makes M10 wafer size PERC solar cells.

For all trial groups, the batch size was 1,200 wafers which were processed in the same diffusion tube and then via similar equipment to ensure minimum influence of the process variation.

Table II: Description of trial 2 groups evaluated for increasing the emitter sheet resistance.

Trial 2 groups	Pre-deposition temperature (°C)	Drive-in temperature (°C)	Variation
Baseline-2	770-785	835-850	Reference for production without $POCl_3$ flow during heat-up from pre-deposition to drive-in stage and increase in $POCl_3$ flow during pre-deposition.
Group 1	770-785	835-850	Reduction in $POCl_3$ flow during pre-deposition to obtain 170-175 Ω/□.
Group 2	770-785	835-850	Reduction in $POCl_3$ flow during pre-deposition to obtain 180-185 Ω/□.

3 RESULTS AND DISCUSSIONS

The results for trial batches 1 and 2 are discussed in the sub-sections ahead.

3.1 Emitter sheet resistance variation for trail 1 groups

The emitter sheet resistance variation for the trial groups with the corresponding baseline groups are shown in Figure 1. For group 1 with no $POCl_3$ flow during the heat-up phase, the emitter sheet resistance increases to 225 Ω/□ from baseline average value of 140 Ω/□. For group 2, increasing the pre-deposition temperature by 5°C, did not decrease the sheet resistance significantly, but it became more uniform with less deviations compared to group 1. Increasing the drive-in temperature by 5°C for group 3 led to a further decrease in the sheet resistance by around 10 Ω/□ to 219 Ω/□.

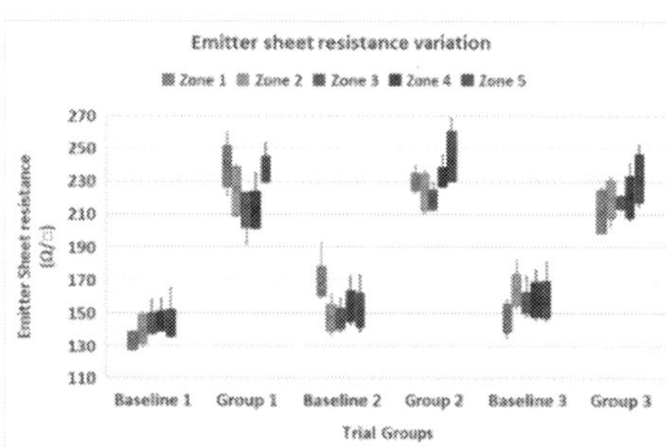

Figure 1: Variation of emitter sheet resistance for the different groups in trial 1 batch.

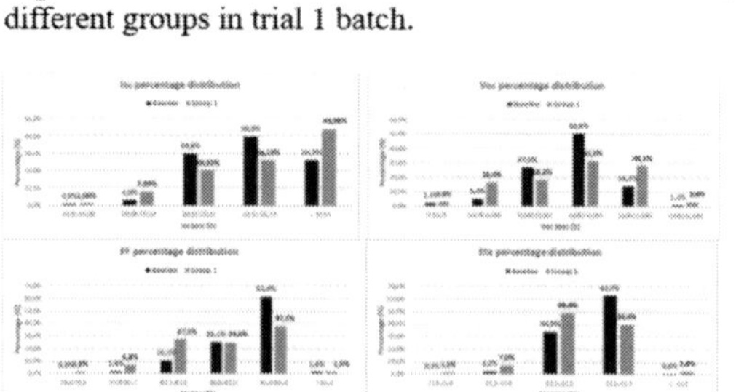

Figure 2: Comparison of I-V parameters for group 1 without $POCl_3$ flow during the heat-up phase for trial group 1.

The comparison of I-V parameters for the group 1 is shown in Figure 2. The percentage of cells with I_{sc} > 10.14 A is significantly higher compared to the baseline group. Similarly, the percentage of cells with V_{oc} > 684 mV is also higher for group 1. However, it can be observed that the cells in group 1 with FF < 80.6% is higher than the baseline group. This is attributed to the higher R_{sh} and the corresponding higher standard deviation. As can be observe from Table III, the average I_{sc} and V_{oc} values are same, while the FF decreased due to the higher sheet resistance.

Table III: Comparison of I-V parameters for the baseline-1 and group 1 batches in trial 1.

Group	I_{sc} (A)	V_{oc} (mV)	FF (%)	Eta (%)
Baseline-1	10.13	682.4	81.0	22.21
Group 1	10.13	682.4	80.7	22.15

From Figure 3, it can be observed that the percentage of cells with I_{sc} and V_{oc} values above 10.14 A and 684 mV, respectively are much higher for group 2 compared the baseline cells. This leads to I_{sc} and V_{oc} gain of around 30 mA and 2 mV, respectively as shown in Table IV. However, the percentage of cells with lower FF are still high which leads to a FF loss of around 0.8% compared to the baseline group. The results for group 3 were similar to that of group 2.

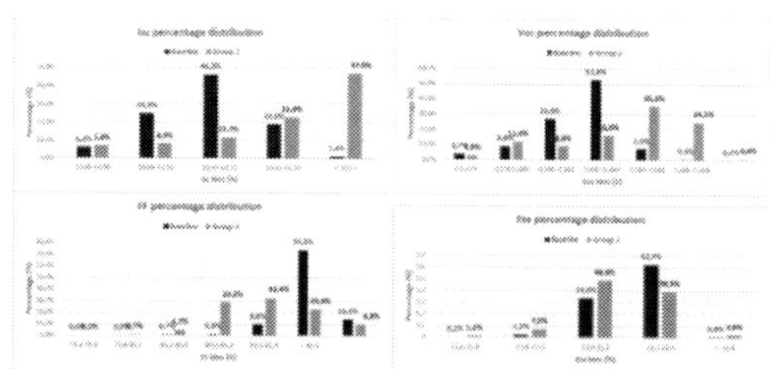

Figure 3: Comparison of I-V parameters for group 2 without $POCl_3$ flow during the heat-up phase and higher pre-deposition temperature for trial group 2.

Table IV: Comparison of I-V parameters for the baseline-1 and group 2 batches in trial 1.

Group	I_{sc} (A)	V_{oc} (mV)	FF (%)	Eta (%)
Baseline-1	10.10	682.0	81.2	22.21
Group 2	10.13	683.8	80.8	22.20

3.2 Emitter sheet resistance variation for trail 2 groups

The results of the trial 2 groups are shown in Figure 2 and summarized in Table V. It should be noted that the normal average production efficiency is in the range of 22.2%-22.3%. The trial groups were affected by some production delays during which time the over-all line efficiency was little lower. From the I_{sc} distribution graph, it can be observed that group 2 has higher share in higher current bins compared to the other two groups. It can be inferred that higher R_{sh} is beneficial to have shallower junction and hence increase the I_{sc}. Group 1 is better from V_{oc} point of view with a gain of around 1 mV compared to the baseline group. Group 2 is influenced by higher percentage of cells in lower voltage bins. The average FF of group 1 is similar to that of the baseline process, while group 2 has higher share in the lower value range. From the zone-wise sheet resistance data it was observed that for group 2, one zone's sheet resistance was in the range of 194 Ω/□ to 215 Ω/□. This could have led to non-uniform junction formation in the specific zone and lead to lower V_{oc} and FF values.

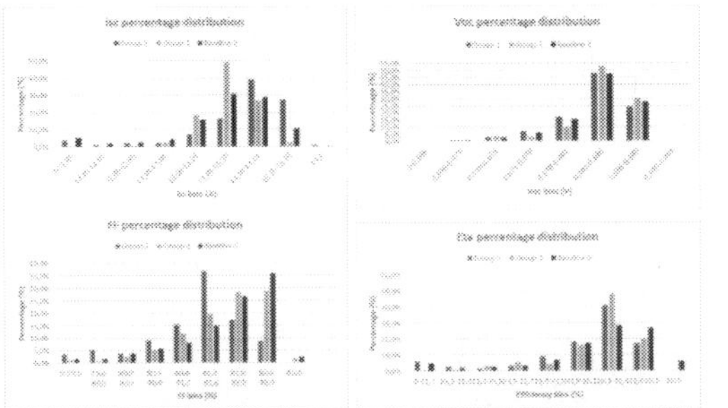

Figure 4: Comparison of I-V parameters for various groups in trial 2 batch.

Table V: Comparison of I-V of various groups in trial 2 batch.

Group	R_{sheet} (Ω/□)	I_{sc} (A)	V_{oc} (mV)	FF (%)	Eta (%)
Baseline-2	141	13.06	683.3	81.55	22.05
Group 1	175	13.08	683.9	81.59	22.10
Group-2	183	13.10	683.0	81.15	21.99

The results of both trial groups indicate that when the emitter sheet resistance is increased, maintaining the uniformity of the zone-wise sheet resistance value becomes critical which can otherwise adversely affect the V_{oc} and FF of solar cells.

4 SUMMARY

The baseline emitter diffusion recipe with $POCl_3$ flow during the heating phase from the pre-deposition to drive-in phase proved to be difficult to increase the R_{sh} in a controlled way with values of greater than 200 Ω/□. Even with temperature increase in the pre-deposition and drive-in steps after stopping $POCl_3$ flow between the steps, the average FF value was less compared to the baseline group though the I_{sc} and V_{oc} were improved. This led to the average efficiency of the trial group to be comparable to the baseline efficiency. This indicates that for emitter sheet resistance above 200 Ω/□, the increase in series resistance due to higher R_{sh} leads to a significant decrease in the FF. In addition, control of the zone-wise R_{sh} also becomes difficult.

With an altered baseline diffusion recipe with no $POCl_3$ flow between pre-deposition and drive-in steps, the control of sheet resistance by varying the $POCl_3$ flow during pre-deposition was more convenient. By reducing the $POCl_3$ flow up to 120 sccm, increased R_{sh} of up to 185 Ω/□ could be obtained. An average R_{sh} value of 175 Ω/□ lead to efficiency gain of 0.05% absolute compared to the baseline production efficiency. Further zone-wise fine-tuning can lead to more uniform R_{sh} values and possibility to increase R_{sh} upto 185 Ω/□ and increasing the average production efficiency by 0.05%-0.1% absolute.

ACKNOWLEDGEMENTS

Part of this work was supported by the German Federal Ministry for Economic Affairs and Climate Action under contract number 03EE1081A (FlexFab2). The content is the responsibility of the author.

We thank KalyonPV management & cell team for the support.

REFERENCES

[1] International Technology Roadmap for Photovoltaic (2021) 13th Edition. March, 2022.

SINGLE SIDED HIGH THROUGHPUT SPUTTER PROCESS TECHNOLOGY FOR IN-SITU DOPED N-TYPE AMORPHOUS SILICON LAYERS FOR HIGH EFFICIENCY TOPCON SOLAR CELLS

Volker Linss[1,a)], Tina Dietsch[1)], Jens Baumann[1)], Uwe Graupner[1)], Jan Hoß[2)], Jonathan Linke[2)], Jan Lossen[2)], Jana-Isabelle Polzin[3)], Sebastian Mack[3)], Henning Nagel[3)], Martin Bivour[3)], Eric Schneiderlöchner[1)]
[1)] VON ARDENNE GmbH, Am Hahnweg 8, D-01328 Dresden,
[2)] International Solar Energy Research Center Konstanz e.V., Rudolf-Diesel-Straße 15, D-78467 Konstanz,
[3)] Fraunhofer Institute for Solar Energy Systems ISE, Heidenhofstrasse 2, D-79110 Freiburg
[a)] Linss.Volker@VONARDENNE.biz

ABSTRACT: In this work we report on the development of an industrially relevant in-situ doped sputtering process for depositing 100-150 nm thick aSi(n) layers to passivate the rear contact of TOPCon solar cells. All relevant properties of the poly-Si layers have been optimized using symmetrical lifetime samples. The sputtered aSi(n) layers do not show blistering effects and can be doped with Phosphorus concentrations larger than 1E21 at/cm³ which has been evaluated via calibrated SIMS measurements. ECV measurements of optimized layers after a high temperature anneal at 850-980 °C for 10-30 min confirm the necessary electrically active concentration $c_P \sim$ 1-3E20 at/cm³. After tuning the interfacial oxide layer and optimizing the PVD process parameters as well as the annealing process implied open circuit voltages $iV_{oc} > 735$ mV could be measured on symmetrical lifetime structures at one sun injection density on 1-3 Ωcm n-type Cz-wafers. Corresponding sheet resistances of the 140 nm thick poly-Si layers range from 60-320 Ω/sq and sufficiently low specific contact resistivities below 10 mΩcm² for screen-printed contacts have been determined accompanied by low recombination of $J_{0,met} < 65$ fA/cm². The best TOPCon solar cell with sputtered poly-Si layers has an efficiency of 23.4 % measured by ISE CalLab PV Cells.
Keywords: TOPCon, Magnetron Sputtering, Amorphous Silicon, High Deposition Rate

1 INTRODUCTION

Over the span of the last five to ten years passivated contacts based on tunnel oxides and highly doped poly-silicon thin films, known as tunnel oxide passivated contacts - TOPCon [1] or poly-silicon on oxide - POLO [2], have been identified as the key technology milestone [3, 4] that will enable to overcome the efficiency limitations of industrially manufactured PERC solar cells. This is, although the underlying principles of this type of passivated contacts were investigated and published already in the 1980s [5, 6]. To use this technology in high volume mass production a low-cost high-quality deposition method has to be applied that allows for the reduction of process steps. Inline magnetron sputtering is such a scalable high-volume deposition process and it is also a single sided deposition process so that no etching of any residual aSi-deposition on the front side of the solar cell has to be performed. This feature allows for a leaner TOPCon process sequence compared to CVD based poly-Si deposition methods [7]. The sputtering approach to produce the doped poly silicon layer for TOPCon contacts has first been reported by Yan and Cuevas in 2018 [8, 9] and later by other groups [10, 11, 12, 13] as an interesting alternative preventing a wraparound deposition of amorphous silicon around the wafer edge. It is also a cost competitive, high volume manufacturing technology commonly used in mass production industries as thin film PV, silicon PV, architectural glass coating and others. An industrial application of the inline sputter process demands for in-situ doping of the deposited aSi layers e.g. as discussed in [14] and at the same time state of the art high quality properties of the resulting poly-Si layers. Furthermore, a high deposition rate allowing for a high manufacturing throughput as well as layers without problems due to blistering or film stress and the like are mandatory. Thus, the development of the sputter process to "production ready" consists of two main parts, (a) demonstrate that in-situ doped aSi can be grown at high deposition rates with a sufficiently high level of doping and (b) proof that the so grown films can be converted into high quality poly-silicon layers.

This work is a cooperation between the deposition tool supplier VON ARDENNE GmbH (VA) and the research institutions International Solar Energy Research Center Konstanz e.V. (ISC) and Fraunhofer Institute for Solar Energy Systems (ISE).

2 EXPERIMENTAL APPROACH

The development of the sputtered aSi(n) for TOPCon process technology was conducted in an iterative way. Both, the properties of the sputtered layer (VA) and other cell processes, in particular the interface oxidation, annealing, and metallization of the cells, have to be optimized (ISE, ISC) as a process bundle to get the best results. Both institutes, ISE and ISC, followed their own optimized process routes for pre-processing of wafers and post processing after sputter deposition giving an additional variation for the testing of the developed sputter process.

2.1 Sputter deposition at VA

The experiments of sputter deposition with in-situ doping were carried out in an industrially relevant vertical inline deposition tool equipped with one meter long rotary targets, gas manifolds for the adjustment of the sputter gas atmosphere, substrate heaters, power supply and a carrier transport system. Substrates were attached to a carrier und deposited inline by passing the target with ignited sputter plasma (dynamic deposition). Glass substrates and small silicon wafer pieces were used for the basic sputter process characterization like estimation of the deposition rate and determination of the chemical composition of the grown films by calibrated SIMS. Important sputter parameters are among others the applied power, the sputter atmosphere, and the substrate temperature. The film thickness was adapted between 100 nm and 150 nm. The next step after the basic process characterization was the deposition of symmetric lifetime samples provided by the partner institutions (1-3 Ωcm saw damage etched CZ wafers and 10 Ωcm shiny etched FZ wafers with interfacial oxide on both sides) and finally the deposition of solar cell precursors delivered by ISE and ISC as well. In each wafer deposition run different sputter parameter sets were applied in order to test the influence of the sputter parameters on the electro-optical properties of the films using lifetime samples as well as cell precursors. After sputter deposition, the wafer substrates were sent back to the according institution for further processing.

2.2 Sample preparation and characterization at ISE

Different types of sample wafers were prepared and then shipped to VA. After deposition of the sputtered doped silicon films the wafers were shipped back to ISE and processed further. Fig 1 summarizes the process sequence how i-TOPCon solar cells have been produced.

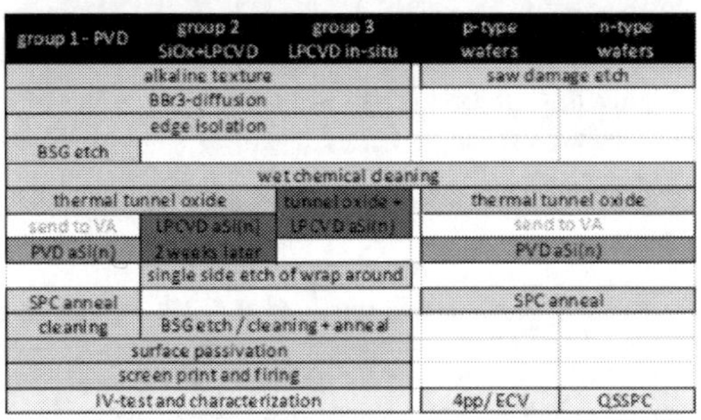

Figure 1: Process sequence of a solar cell batch (cells and symmetrical lifetime samples). PVD solar cells do not need a single side back etch to remove wrap around.

A complete experiment consists of lifetime samples for a first check of the interplay of PVD parameters and annealing temperature and of solar cell precursors to produce i-TOPCon solar cells. Lifetime samples were saw damage etched CZ n-type wafers of M2 format with thermal tunnel oxide on both sides. These wafers were sputter deposited with three different PVD parameter sets and annealed at different temperatures and durations to find an optimum annealing process. Solar cell precursors (M2 CZ) were then sputtered with the same PVD parameters but annealed at optimized temperatures, passivated, screen printed without busbars and fired followed by IV-test and other characterization. The benchmark for the sputter deposited solar cells was a standard LPCVD process developed at ISE. P-type CZ wafers were also coated by aSi(n) sputter deposition in order to measure the sheet resistance of the grown films after annealing with a 4-point probe (4PP) independently of the wafer substrate. After the first experiment, a second run was conducted with a similar process sequence as shown in Fig. 1 and optimized cell precursors (different emitter and tunnel oxide). The PVD parameters for the sputter deposition at VA of the second experiment were varied based on the results of the first experiment.

2.3 Sample preparation and characterization at ISC

The basic approach and the PVD process technology were the same as for the experiments with ISE substrates. Fig. 2 summarizes the process sequence for wafers processed at ISC and VA. Two different types of tunnel oxide, thermally (TO) and wet chemically grown (NAOS), were used. Wafers of both tunnel oxide types were sputter coated with the same PVD parameter set and an annealing temperature variation was conducted. The TO wafers have also been sputter coated with varying PVD parameter sets. The metallization of the cells was conducted with 9 busbars. In addition to the cell characterization, a number of substrates was used to measure the contact resistance by TLM measurements and to determine $J_{0,met}$ after firing at different peak temperatures.

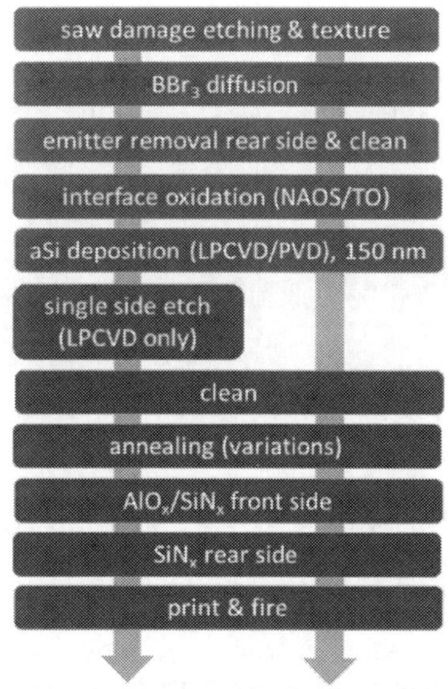

Figure 2: Process sequence of ISC solar cell batch with different tunnel oxide (thermal and wet) for the PVD sputtered groups as well as the LPCVD reference. PVD solar cells do not need a single side back etch to remove wrap around.

3 RESULTS AND DISCUSSION

3.1 Sputter deposition and film properties

Fig. 3 shows the dynamic deposition rate (ddr) in dependence on the fraction of the maximum applied power. In this experiment the maximum power was limited to about 60 % of the usually recommended maximum power. A variation in deposition atmosphere at constant power can lead to differences in the ddr of about 10...20 %. Using maximum recommended power, a ddr of about 100 nm*m/min can be reached.

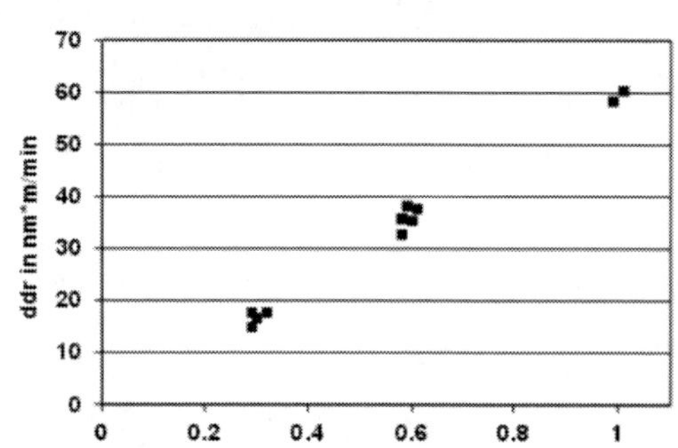

Figure 3: Dynamic deposition rate (ddr) of the aSi(n) sputter process in dependence on the fraction of the maximum applied power to the tube target

Fig. 4 compares the SIMS-depth profiles of phosphorus of samples deposited with two PVD processes (VA) and a PECVD process (ISE). All samples have a uniform distributed P profile after deposition. The P-density is even higher for the PVD-deposited films, and a variation in the sputter process leads to different phosphorus contents. In the shown sample a density of 2E21 at/cm³ has been reached using the developed PVD-process technology.

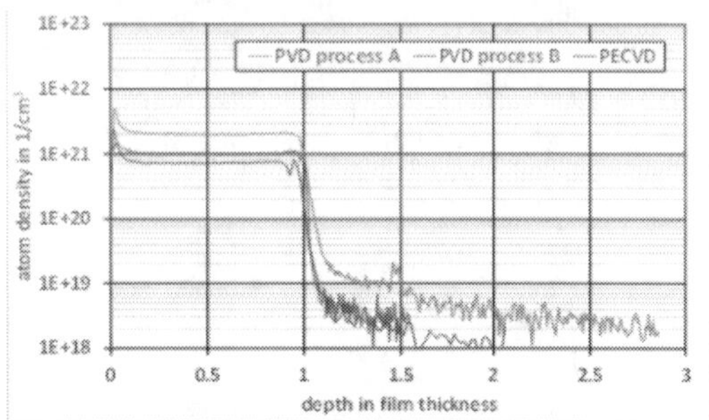

Figure 4: Phosphorus SIMS depth profiles of one PECVD- and two PVD deposited aSi(n) layers

During annealing, the deposited layer is crystallized to poly-Si and the phosphorus is activated. The level of activation depends on the annealing temperature but also on the PVD parameters. Tab. I shows the sheet resistance values of the annealed films grouped by PVD parameter sets for different annealing temperatures. PVD parameter set 3 produces the films with the lowest sheet resistance followed by parameter set 2 and then 1.

Table I: R_{sh} of annealed 140 nm thick films measured by 4PP in Ω/sq (annealing temperature in brackets, duration was 10min)

	Annealing 1	Annealing 2	Annealing 3
PVD1	323 (920 °C)	278 (935 °C)	200 (950 °C)
PVD2	274 (920 °C)	230 (935 °C)	177 (950 °C)
PVD3	100 (935 °C)	62 (960 °C)	71 (980 °C)

It should be mentioned that after annealing the sheet resistance distribution was very uniform over the whole substrate wafer as is confirmed by Fig. 5.

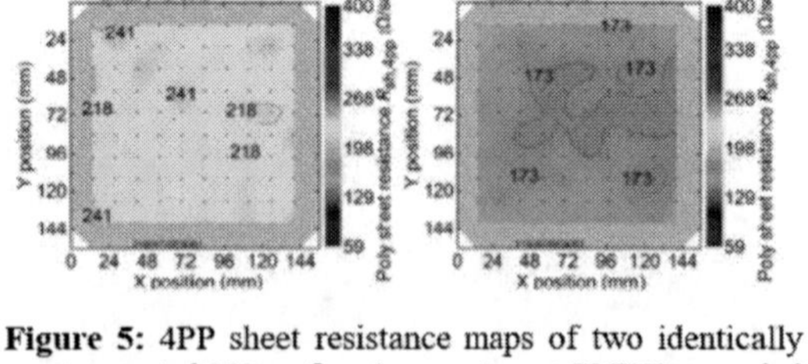

Figure 5: 4PP sheet resistance maps of two identically sputter coated M2 wafers (parameter set PVD2) annealed at two different temperatures (935 °C, 950 °C) showing a homogeneity of ±3 %

A high enough absolute density of incorporated phosphorus is the precondition for a suitable density of activated phosphorus after annealing. Fig. 6 compares the electrochemical capacitance–voltage (ECV) measurement profiles of a PVD sample annealed at three different temperatures with an ECV profile of a PECVD sample. The density of activated donors of the PECVD process can also be reached with PVD deposited samples. Increasing temperature leads to two effects. Firstly, the density of active phosphorus increases, and secondly there is increasing diffusion of phosphorus into the wafer bulk. The PVD sample annealed at 960°C has similar activation density and in-diffusion like the PECVD sample annealed at 950°C.

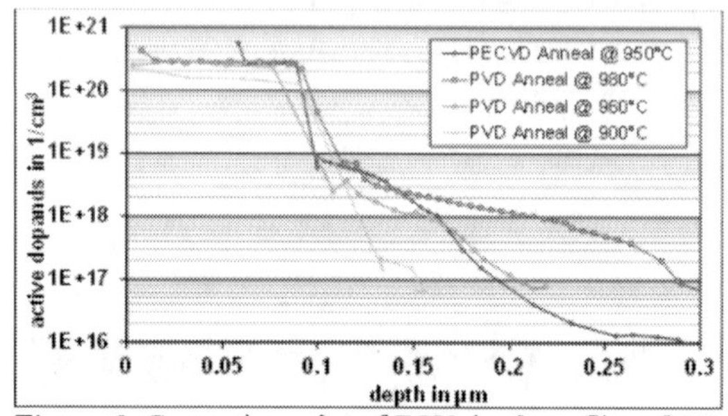

Figure 6: Comparison plot of ECV depth profiles of 100 nm thick PVD- and a 50 nm thick PECVD poly-Si layer annealed at different temperatures. PECVD profile is shifted in depth for better comparison.

3.2 Properties of lifetime samples first run

Fig. 7 gives an overview of the iV_{oc} values of the symmetric lifetime samples of the reported experiment with ISE grouped by PVD parameter sets and annealing conditions. The figure shows that the passivation quality of the sputtered aSi(n) films depends on the sputter parameters on the one hand and on the annealing conditions on the other. There are clear differences between the parameter sets which gives the possibility to optimize the process further. The parameter sets PVD1 and PVD2 produce films with similar iV_{oc} values with slightly higher iV_{oc} for PVD1 up to 740 mV. The best results are reached for the lowest investigated thermal budget (920 °C, 10 min) whereas for PVD3 the highest iV_{oc} values were found for annealing at 960 °C for 10 min. Based on these findings the annealing of the sputtered solar cell precursors was adapted in particular for PVD3.

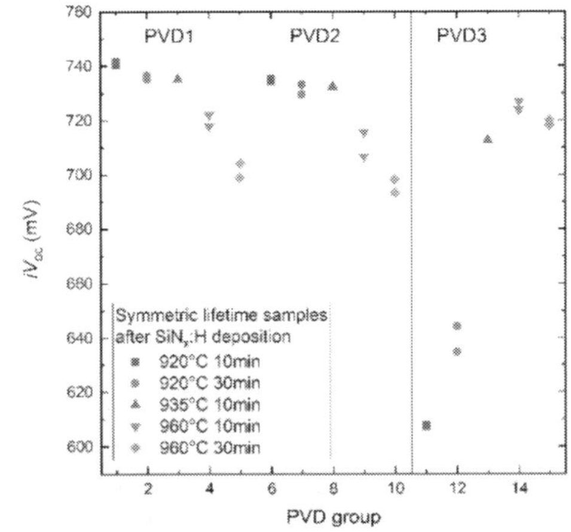

Figure 7: Overview of iV_{oc} values of symmetrical lifetime samples reached for different sputter process parameters (PVD1-3) and annealing conditions (temperature and duration)

Fig. 8 presents an example of the measurement of $J_{0,met}$ and ρ_c values as measured by ISC of samples sputtered with a specific PVD parameter set in dependence on the peak firing temperature. Values below 65 fA/cm² ($J_{0,met}$) and 10 mΩcm² (ρ_c) can be realized which are necessary to obtain good solar cell results.

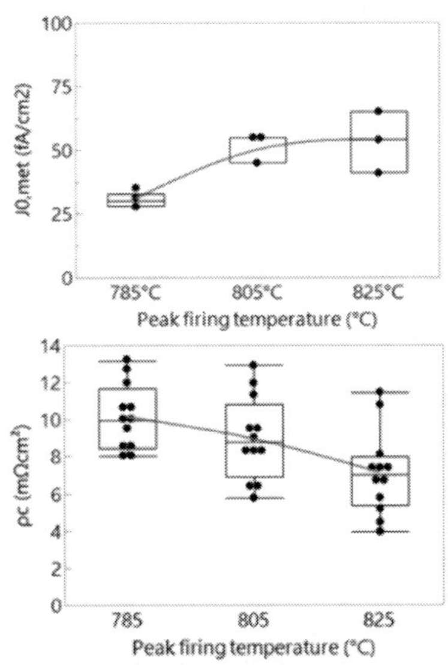

Figure 8: $J_{0,met}$ (upper diagram) and ρ_c values (lower diagram) versus peak firing temperature reached for a specific PVD parameter set used for optimization of the fast firing process for screen-print metallization

3.3 Properties first solar cell batch (ISE)

Fig. 9 summarizes the IV results of the first cell run. There are clear differences between the cells sputtered with different PVD parameter sets. Whereas the parameter set PVD1 was the best for high iVoc of the lifetime samples the parameter set PVD2 was the most successful to produce solar cells. PVD1 performed slightly lower and the samples deposited with PVD3 show even lower efficiencies. Samples deposited with PVD2 have nearly equal efficiency and fill factor compared to the LPCVD reference and have even higher short circuit current density j_{sc} but lower open circuit voltage V_{oc}.

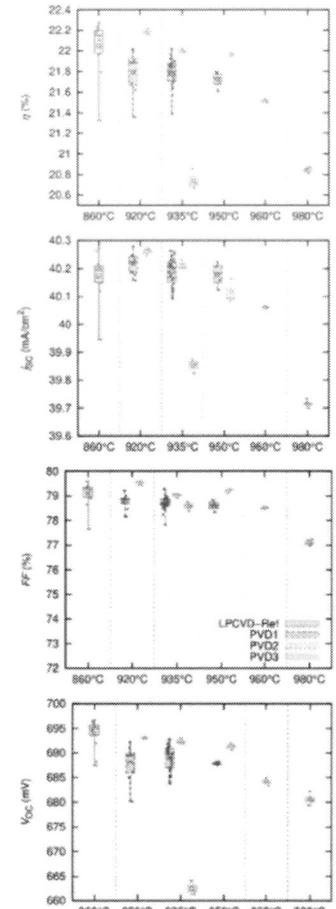

Figure 9: IV-parameters of screen-printed TOPCon solar cells (zero busbars) featuring sputtered, in-situ doped aSi(n) layers for the rear passivated contact in dependence on the annealing temperature

3.4 Properties second solar cell batch (ISE)

With one of the next solar cell optimization runs the efficiency could be increased for both, the PVD deposited cells and the LPCVD deposited cells because all steps, pre-processing, sputter deposition, and post processing were tuned. Fig. 10 compares the results for the LPCVD reference with four different PVD parameter sets. The variation in PVD parameters has impact on different properties of the final cell. For instance, the deposition with the two PVD parameter sets that lead to the highest j_{sc} result in lower efficiency and open circuit voltage. The variation of the sputter parameters causes different compositions and densities of the amorphous Si films that lead to varying specific contact resistances, dopant activation, crystallization behavior during annealing and hence impacting different solar cell properties. Therefore, the optimal adaptation of the growth parameters of the sputtered aSi-films is a multidimensional optimization task.

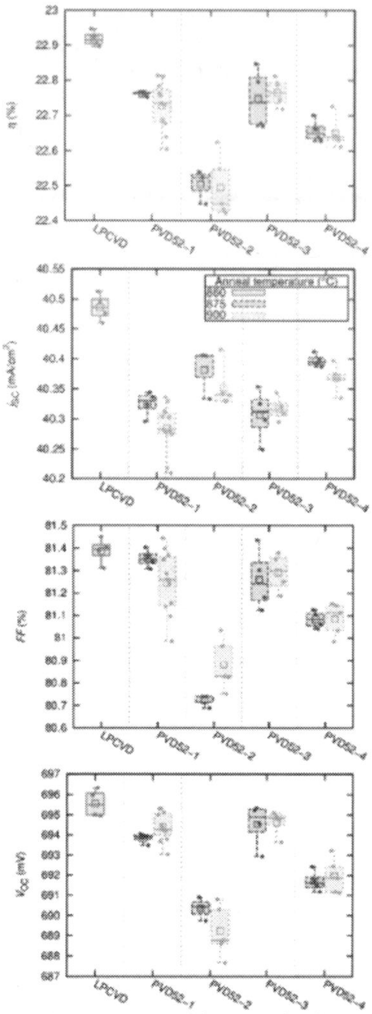

Figure 10: IV-parameters of the second cell run (ISE)

One of the best sputter deposited cells measured in the uncalibrated characterization was sent to ISE CalLab PV Cells and was measured under calibrated conditions. This cell had a confirmed efficiency of 23.4%. The parameters of the champion cell and the according I-V measurement are shown in Fig. 11.

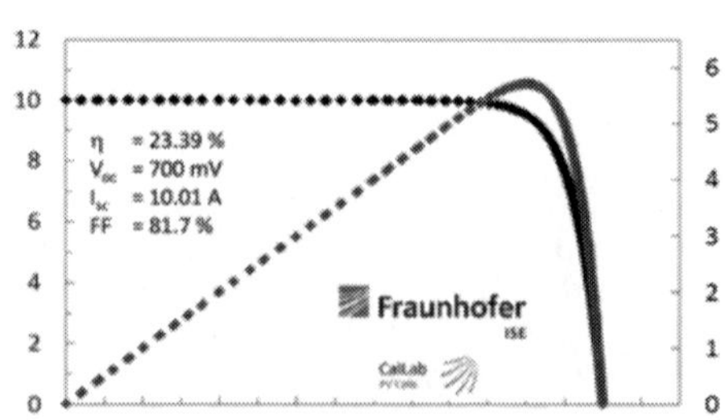

Figure 11: I-V measurement from ISE CalLab PV Cells for one of the champion cells in the second ISE cell run

3.5 Cell properties third run (ISC cells)

The two tested types of tunnel oxides (NAOS, TO) can be distinguished by the color in Fig. 12. Reference wafers deposited by LPCVD are the groups G1 (NAOS) and G2 (TO).The wafers processed with the same PVD parameter set for the annealing temperature variation are the groups G3-5 for NAOS and G6-9 for TO, wafers sputtered with other PVD parameters are the groups G10 and G12. It has to be mentioned that the variation between the PVD parameters for G6-9 and G10 and G12 was not very strong to see if slight differences in PVD parameters show a clear impact on the cell properties (in this case mainly the energy of the depositing particles was changed).

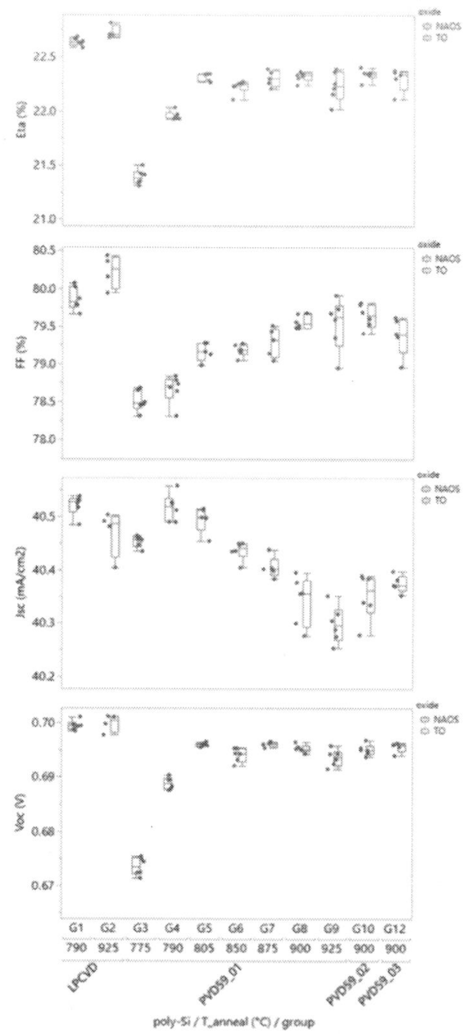

Figure 12: IV parameters of the ISC cell run grouped by Si-deposition, annealing temperature, and tunnel oxide

The highest efficiency is found for the wafers deposited with the LPCVD reference process with a slight advantage for the TO wafers. The PVD deposited samples with NAOS show a clear trend of increasing efficiency with increasing annealing temperature. It seems that the maximum possible efficiency was not reached, and a higher temperature would have further closed the gap between the best LPCVD results and the best PVD result. For the TO samples there is only a very small trend in annealing temperature, and it can be supposed that the best annealing temperature range was already chosen. Furthermore, the chosen variation in PVD process parameters does not have a clear impact on the efficiency of the TOPCon solar cells. The same applies to the other solar cell parameters displayed in Fig. 12. Whereas we see a trend for the FF (increasing) and J_{sc} (decreasing) in annealing temperature (G6-9) there is no clear difference for the samples deposited with different PVD parameter sets and annealed at 900 °C (groups G8, G10, G12). However, it cannot be concluded that the applied variation in PVD parameters would not have impact on the properties of the NAOS samples neither because the difference in energy of the depositing particles can lead to very different effects in dependence on the density and thickness of the tunnel oxide. The results of the ISC cell run confirm that parameters of the sputter deposition of the doped silicon layer together with the optimization of the tunnel oxide can be used to tune the properties of the solar cells towards the best results reachable with the standard LPCVD process. This will be part of future work.

4 CONCLUSIONS

It has been shown that the scalable and industrially relevant sputter deposition is a suitable method to produce in situ doped aSi(n) films that allow to produce i-TOPCon solar cells of high efficiency. Our future work will focus on the further optimization of the overall processing sequence in order to provide a high quality PVD technology solution for the manufacturing of TOPCon solar cells as an alternative to the deposition of the doped Si thin films by LPCVD or PECVD. Although CVD-based methods provide a high process quality they also come along with a residual deposition around the wafer edges that would create edge shunts and hence needs to be removed in an additional process step. In contrast, sputter deposition is a single-sided process preventing undesired wrap-around of the deposited film and hence no additional removal process, as e.g. single side etching, is necessary. We have shown that sputter parameters (for example sputter atmosphere, sputter process power, film thickness) as well as the selection of the type of the tunnel oxide have strong impact on the efficiency and electrical properties of PVD TOPCon solar cells. In this work, a maximum efficiency of 23.4 % has been demonstrated on M2 CZ-silicon wafers using an industrially applicable PVD-process technology that is compatible to be used in high volume manufacturing inline sputter tools.

ACKNOWLEDGEMENT

This work was funded by the German Federal Ministry for Economic Affairs and Climate Action.

REFERENCES

[1] F. Feldmann, M. Bivour, C. Reichel, M. Hermle, S.W. Glunz, Proceedings of 28th EU-PVSEC, (2013) 988–992.

[2] U. Römer, R. Peibst, T. Ohrdes, B. Lim, J. Krügener, E. Bugiel, T. Wietler, R. Brendel, Solar Energy Materials and Solar Cells 131 (2014) 85–91.

[3] J. Schmidt, R. Peibst, R. Brendel, Solar Energy Materials and Solar Cells, 187 (2018) 39-54.

[4] A. Richter, R. Müller, J. Benick, F. Feldmann, B. Steinhauser, C. Reichel, A. Fell, M. Bivour, M. Hermle, S.W. Glunz, Nature Energy 6 (2021) 429-438.

[5] E. Yablonovitch, T. Gmitter, R.M. Swanson, Y.H. Kwark, Applied Physics Letters 47 (1985) 1211–1213.

[6] J.Y. Gan, R.M. Swanson, Proceedings of the 21st IEEE Photovoltaic Specialists Conference Kissimmee, Kissimmee, 1990, pp. 245–250.

[7] J. Chen, "The Research Progress and Industrial Status of n-Type Bifacial TOPCon Technology in Jolywood", Proceedings of SiliconPV 2022, to be published.

[8] D. Yan, A. Cuevas, S.P. Phang, Y. Wan, D. Macdonald, Appl. Phys. Lett. 113 (2018) 061603.

[9] A. Cuevas, D. Yan, S.P. Phang, Y. Wan, D. Macdonald, Proceedings of 35th EU-PVSEC, 2018, pp. 233-237.

[10] J. Hoß, J. Baumann, M. Berendt, U. Graupner, R. Köhler, J. Lossen , M. Thumsch, E. Schneiderlöchner, AIP Conference Proceedings 2147 (2019) 040007.

[11] J. Baumann, E. Schneiderlöchner, J.-I. Polzin, F. Feldmann, M. Bivour, M. Hermle, "Deposition and Properties of Sputtered Phosphorus Doped Amorphous Silicon Films for Passivating Contacts", Proceedings of SiliconPV 2022, to be published.

[12] J. Linke, J. Rinder, G. Hahn, B. Terheiden, AIP Conference Proceedings 2147 (2019) 040018.

[13] L. Nasebandt,B. Min,C. Hollemann,S. Hübner,T. Dippell,R. Peibst,R. Brendel, Solar RRL 6 (9) (2022) 2200409.

[14] C. Allebé, J.J. Diaz Leon, A. Descoeudres, A. Ingenito, S. Hübner, T. Dippell, B. Paviet Salomon, C. Ballif, "Sputtered Poly Si for the Formation of Passivating Contacts at the Rear of Large Area n-PERT C-Si Solar Cells", Proceedings of SiliconPV 2022, to be published.

OPTIMIZATION OF ELECTRICAL LID REGENERATION FOR PERC SOLAR CELL PRODUCTION

Mehul C. Raval[1]*, Furkan Kaya[2], Gorkem Besik[2], Andreas Teppe[1], Mehmet Ender[2], Peter Fath[1]
1-RCT Solutions GmbH, Line - Eid - Str. 1, D-78467 Konstanz, Germany.
2- Kalyon Günes Teknolojileri Üretim A. ,S., 06909 Ankara, Turkey.

ABSTRACT: Mitigating LID for PERC solar cells has been extensively researched and severe degradation after field-level exposure has been minimized due to regeneration processes integrated in the solar cell production lines. It is observed that without the LID regeneration process, the relative degradation in PERC cell efficiency is more than 5.0%. With the production electrical injection based LID regeneration recipe, the relative efficiency improvement is more than 1.0%, whereas the trial recipes had a relative gain of 0.3%-0.9%. LID light-soaking show that the baseline production recipe also has a degradation of less than 0.5% compared to the trial recipes. The new trial recipes demonstrate lower degradation compared to baseline production regeneration recipe under LeTID light soaking. Module level light-soaking results show a degradation of 1.4% for trial recipes compared to the baseline regeneration recipe which had a degradation of 1.8%. The results indicate that for electrical injection based LID regeneration, a constant temperature of 160-180°C and constant current of 10 A is more suitable than the current production recipe.

Keywords: PERC, LID, regeneration, electrical injection

1 INTRODUCTION

Extensive study has been carried out on the LID of PERC solar cells which is related to the presence of B-O complexes in the boron-doped p-type bulk material. Evaluation of various LID regeneration methods have also been carried out and it seems that electrical injection based method is better suited for the regeneration process [1].

In this work, optimization trials for the electrical injection based regeneration has been carried out. Such a study will help to understand the influence of regeneration recipe on the cell parameters after the regeneration and light soaking processes.

2 EXPERIMENTAL DETAILS

2.1 Production of PERC solar cells

The study is carried out in the cell section of Kalyon PV, which is an integrated 1 GW bifacial p-type mono PERC PV factory. P-type (Boron doped) mono c-Si ingots are manufactured from poly-Si, which are then cropped, squared and chamfered into bricks. The bricks are then sliced into wafers using 50 µm diamond wire into wafers of 175 +20/-10 µm thickness. The wafers are fed to the cell section to make bifacial p-type mono PERC cells with G1 size wafers, which are in turn used to make glass-glass bifacial modules with half-cut cells. The LID regeneration of the production PERC solar cells is done using electrical injection method after the fast-firing process.

2.2 Variation of LID regeneration recipe

The aim of the work is to optimize the electrical injection based LID regeneration recipe in the PERC solar cell production line in the integrated Kalyon PV factory. The various groups for the LID regeneration recipes along with the details are outlined in Table I. The trial groups were selected based on the work report by Jiaxing Y, et. al [2]. The injection current for the baseline recipe was in the range of 7.0-9.0 A for the different stages, while for the trial groups it was maintained constant at 10 A [2]. For group 6, the injection current was varied from 12 A to 10 A to check the influence due to lower regeneration temperature.

The experimental plan for evaluation of different recipes is as shown in Table II. For each experimental group 25 cells were taken from the production line to measure the efficiency difference before and after the regeneration process. In addition to above groups, a reference group of cells which did not undergo any LID regeneration treatment was also monitored to serve as a reference. The production cell efficiency was in the range of 22.0%-22.2% for the trial batches. The LID light soaking was done under an illumination of 1,000 W/m^2 at 35°C for 24 hours, while the LeTID soaking was done subsequent to LID soaking at 1,000 W/m^2 at 75°C for 120 hours.

Table I: Temperature profile and time duration for different electrical injection LID regeneration recipes

Group	Temperature in °C (Stage 2 to Stage 8)	Total time (minutes)
Baseline	145-170	45.5
Group 1	180	45.5
Group 2	160	45.5
Group 3	180	30.3
Group 4	200	45.5
Group 5	180	40.8
Group 6	160	45.5

Table II: Procedure for evaluating the LID regeneration recipes

Step	Procedure
1	I-V measurement after fast-firing
2	Regeneration based on specific recipe
3	I-V measurement after regeneration
4	LID/LeTID light soaking
5	I-V measurement after light soaking

In addition to making the trials with production solar cells, cells from selected trail groups were used to fabricate bifacial modules which were light-soaked in outdoor conditions and then remeasured again to understand the degradation for solar cells which had undergone different LID regeneration recipes.

3 RESULTS AND DISCUSSIONS

The degradation of the PERC solar cells without any LID regeneration and with different regeneration recipes are discussed in the sub-sections ahead.

3.1 Degradation for cells without regeneration

The efficiency trend before and after light soaking of solar cells without any LID regeneration treatment is shown in Figure 1.

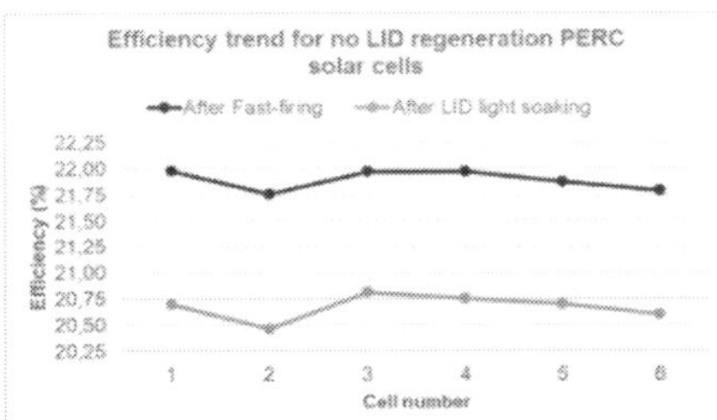

Figure 1: Efficiency trend for reference cells without LID regeneration treatment.

The average solar cell efficiency before light-soaking was 21.9% which reduced to 20.7% (Decrease by 5.6% relative) after the LID light-soaking. The decrease in V_{oc} was 2.5% relative, while I_{sc} and FF decreased by 1.6% relative, respectively. It can be inferred that the decrease in V_{oc} has a more significant contribution in the LID degradation of the PERC solar cells.

3.2 Comparison of different LID regeneration recipes

The comparison of different LID regeneration recipes are shown in Table III. The baseline recipe leads to a gain of 1.4% relative after the regeneration process, while the groups 1 to 4, lead to a gain of 0.3% relative to 0.9% relative. The group 5 and 6 show nearly no gain and even degradation, respectively.

Table III: Summary of the efficiency trend for different LID regeneration recipes.

Group	η after fast-firing (%)	η after regeneration (%)	Change in η after regeneration (%)
Baseline	21.80	22.10	1.4
Group 1	21.99	22.18	0.9
Group 2	21.80	21.96	0.7
Group 3	21.94	22.01	0.3
Group 4	21.90	22.02	0.5
Group 5	21.85	21.87	0.1
Group 6	21.89	21.00	-4.1

It implies that 160°C is the critical regeneration voltage below which regeneration is not expected to occur within a reasonable time duration. In addition, for regeneration temperature of 180°C, a time durations of less than 40 minutes don't seem to be effective for the regeneration process. From the results, group 1, 2 and 4 were selected for further trials.

The LID and LeTID soaking results for the selected groups are shown in Table IV and Table V below. All solar cells had an efficiency in the range of 22.2%-22.3% after the regeneration process. It can be observed from the LID results that the baseline recipe is better than the other trials recipes. However, the LeTID results indicate that group 1 and 2 have lower degradation than the baseline recipe. Group 4 has higher degradation for both results. Based on these results, group 1 and 2 were selected for light-soaking experiment on the module-level.

Table IV: Summary of the LID soaking results for selected groups with 4 cells in each group.

Group	Regenerated efficiency (%)	Efficiency after LID soaking (%)	Relative efficiency difference (%)
Baseline	22.25%	22.2%	-0.4%
Group 1	22.25%	21.7%	-2.7%
Group 2	22.25%	22.1%	-0.7%
Group 4	22.25%	20.6%	-7.2%

Table V: Summary of the LeTID soaking results for selected groups with 4 cells in each group.

Group	Regenerated efficiency (%)	Efficiency after LID soaking (%)	Relative efficiency difference (%)
Baseline	22.25%	21.9%	-1.4%
Group 1	22.25%	22.1%	-0.9%
Group 2	22.25%	22.0%	-1.0%
Group 4	22.25%	21.0%	-5.8%

3.3 Module-level light soaking results

Bifacial glass-glass modules were fabricated from the solar cells which had undergone different LID regeneration recipes. Two modules of each group were fabricated from cells in the efficiency bin of 22.0%-22.2%. The I-V parameters of module before and after light soaking for 15 days are shown in Table VI. It can be observed that the main degradation parameter of the modules is I_{sc} for all the groups. The module were remeasured again after five days and the results are shown in Table VII. It can be observed that the module have some recovery, which can be attributed to the annealing effect for modules stored in dark conditions [3]. However, it should be noted that the baseline LID regeneration recipe module still show a degradation of more than 1.0% compared to the trail groups.

Table VI: Summary of the module-level I-V results before and after light-soaking of 15 days.

Group	Light-soaking	Pmax (W)	V_{oc} (V)	I_{sc} (A)	FF(%)
Baseline	Before	403.8	49.32	10.29	79.2
Baseline	After	396.6	49.35	10.15	79.1
Difference		-1.8%	0.0%	-1.3%	-0.06%
Group 1	Before	399.4	49.18	10.29	78.9
Group 1	After	393.8	49.30	10.14	78.9
Difference		-1.4%	0.0%	-1.54%	0.04%
Group 2	Before	403.4	49.4	10.28	79.4
Group 2	After	397.7	49.4	10.15	79.3
Difference		-1.4%	0.0%	-1.33%	-0.1%

Table VII: Summary of the module-level I-V results before and after light-soaking of 15 days and measured again after 5 days of light-soaking.

Group	Light-soaking	Pmax (W)	V_{oc} (V)	I_{sc} (A)	FF(%)
Baseline	Before	403.8	49.32	10.29	79.2
Baseline	After	397.9	49.34	10.23	78.8
Difference		-1.46%	0.05%	-0.61%	-0.44%
Group 1	Before	399.4	49.18	10.29	78.9
Group 1	After	397.7	49.32	10.22	78.9
Difference		-0.43%	0.29%	-0.67%	-0.04%
Group 2	Before	403.4	49.4	10.28	79.4
Group 2	After	400.3	49.41	10.23	79.1

Difference	-0.78%	0.02%	-0.50%	-0.43%

4 SUMMARY

Our studies indicate that for PERC solar cells without any LID regeneration, the degradation is greater than 5.0%. The baseline recipe for electrical injection in the factory is already good with an efficiency gain of > 1.0% after the regeneration process. Regeneration in the range of 160-200°C had an efficiency gain in the range of 0.5%-1.0%, while a total time duration of less than 40 minutes is not effective for the regeneration process. Increasing the injection current above 10 A led to a degradation due to the regeneration process. The LID results indicate that the baseline recipe is better than the trial groups, but the LeTID results indicate that group 1 and 2 have lower degradation at increased cell temperature. Light soaking on the module level confirm these results. The baseline recipe had a degradation of 1.8%, while the trial recipes had a degradation of 1.4%. The results indicate that fine-tuning the LID regeneration recipe in the range of 160-180°C with an injection current of 10 A can lead to a reduced degradation of the modules installed in a PV plant.

ACKNOWLEDGEMENTS

Part of this work was supported by the German Federal Ministry for Economic Affairs and Climate Action under contract number 03EE1081A (FlexFab2). The content is the responsibility of the author.

We thank KalyonPV management & cell team for the support.

REFERENCES

[1] H Li, et. al, Proceedings 33rd European Photovoltaic Solar Energy Conference (2017), 983-985. DOI: 10.4229/EUPVSEC20172017-2CV.2.87

[2] Jiaxing Ye, et. al, International Journal of Photoenergy (2019), Article ID 5357370. DOI: 10.1155/2019/5357370

[3] Z. Y. Yeo, et. al, Renewable and Sustainable Energy Reviews, 159 (2022), 112223. DOI: 10.1016/j.rser.2022.112223

This is a short version of the paper submitted to the journal EPJ Photovoltaics.

It has been selected by the Committee of the WCPEC-8 for submission to EPJ Photovoltaics.

MODEL FOR CONTACT FORMATION OF NOVEL TeO_2 CONTAINING PB-FREE SILVER PASTE ON N^+ AND P^+ DOPED CRYSTALLINE SILICON

Fabian Geml[1,*], Benjamin Gapp[1], Simon Johnson[2,+], Patricia Sutton[2], Angela Goode[2], Jonathan Booth[2], Heiko Plagwitz[1], Giso Hahn[1]
[1]University of Konstanz, Department of Physics, 78457 Konstanz, Germany
*fabian.geml@uni-konstanz.de, Tel: +49 7531 884995
[2]Johnson Matthey Technology Centre, Sonning Common, Reading, RG4 9NH, UK
[+]Now with Fenzi AGT Netherlands BV, 6222 NZ Maastricht, The Netherlands

ABSTRACT: Silver (Ag) pastes are widely used in the global market for most solar cell architectures. In particular, lead (Pb) is no longer wanted in productions for environmental reasons. In this work, a model for the contact formation between Pb-free, tellurium (Te) oxide containing screen-printable Ag pastes and silicon is presented. It is shown that Te plays a key role in this model. Te is not only an important part in etching the surface passivation layers with TeO_2 dissolving the dielectric layer, but also for a formation of the contacts with Te forming a compound consisting of Ag_2Te. Using EDX mapping, precise contact regions can be examined and interpreted for contact formation. The used paste is a novel developed commercial paste which is on a par with other pastes used in industry concerning the resulting contact properties. This is also demonstrated in this work by the very low contact resistances of less than 1 mΩcm^2 over a wide range of fire peak temperatures. It is additionally shown that good resistances can be achieved on both n- and p-doped regions.
Keywords: Silver, Screen printing, Metallization

1 INTRODUCTION

Most Si solar cell concepts are relying on contacting n^+-doped Si surfaces with Ag [1, 2]. The Ag paste, which is mostly screen-printed, is then fired to etch through the passivation layers and form the contact [3, 4]. The global Ag paste market is moving from the former standard of PbO-based paste system towards lead-free formulations, e.g. Bi_2O_3/TeO_2 based systems [5].
Therefore, it is essential to understand how metals such as Bi and Te affect contact formation in order to optimize paste formulations. In this work, we present results leading to a new model. Therefore, instead of usual commercial pastes with glass frits, a novel developed Ag paste consisting of no glass frits but milled crystalline metal oxides such as TeO_2 is used.

2 EXPERIMENTAL

During sample preparation, the monocrystalline n/p-type Czochralski (Cz) Si substrate (4-5 Ωcm) is etched and cleaned. Samples are textured using a KOH-based alkaline solution with alcohol-based additives. Emitters are formed within a diffusion step in classic $POCl_3$ or BBr_3 tube furnace diffusions, respectively, aiming for an emitter sheet resistance of 75 Ω/sq. Afterwards, the samples are etched in diluted HF and then coated with PECVD SiN_x:H for surface passivation. In a subsequent step, all samples are screen-printed with the same Ag paste in TLM (transfer length method) and grid structures. All samples are (co-)fired in a belt furnace at various peak temperatures. The samples are first analyzed by applying TLM measurements and afterwards investigated by SEM (scanning electron microscopy) / TEM (transmission electron microscopy). For the latter, small lamellas are cut by FIB (focused ion beam) and also analyzed by mapped EDX (energy dispersive X-ray spectroscopy).

3 RESULTS AND DISCUSSION

The contact formation model is shown in Fig. 1. First, the Ag paste consisting of Ag powder and crystalline mixed oxides (Si oxides as well as metal oxides) is screen-printed on a SiN_x:H passivated n^+-emitter. The SiN_x:H is here denoted as a stochiometric version in order to give a suggestion for reaction equations. (A)

During firing the paste reaches a partly molten state where the mixed oxides including TeO_2 move to the interface with the passivation layer and start etching this layer. (B)

A Si-rich glass layer is formed in the liquid phase. The TeO_2, which is partly responsible for the etching process, forms atomic Te according to the reaction equation $3TeO_2 + Si_3N_4 \rightarrow 3Te + 3SiO_2 + 2N_2(g)$. (C)

While Ag alloys with the Si, the Te moves away from the interface and reduces the Ag atoms present in the liquid phase, mostly as ions. In the process, it forms Ag_2Te. (D)

In the final state, Ag nanoparticles are present in the liquid phase, which form precipitates in the glass during solidification and may contribute to the current flow via tunneling or when connected to the silver bulk [6]. (E)

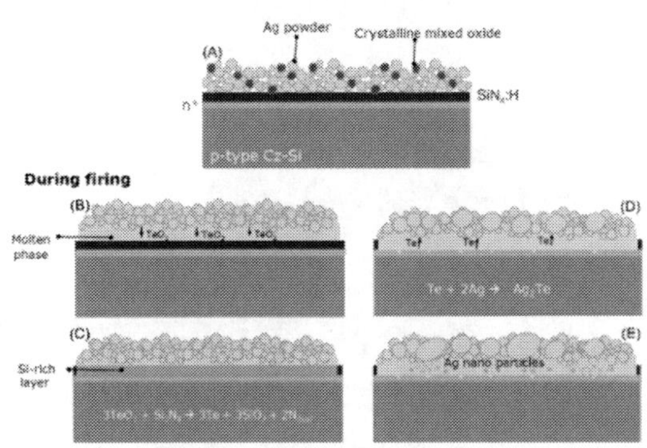

Figure 1: Schematic model of contact formation with a PbO-free paste based on TeO_2.

Fig. 2 shows an EDX cross section of a Si pyramid. A single Ag crystallite sticks into the Si bulk. We see that SiN_x:H is removed in the vicinity of this contact point, which suggests the etching of the passivation layer. The constituents of the glassy paste components can be found moving towards the "troughs" between pyramids during firing.

The glass, identifiable mainly by the high oxygen concentration, also shows a corresponding proportion of Bi. Te is also present in the glass but accumulates more strongly in some places.

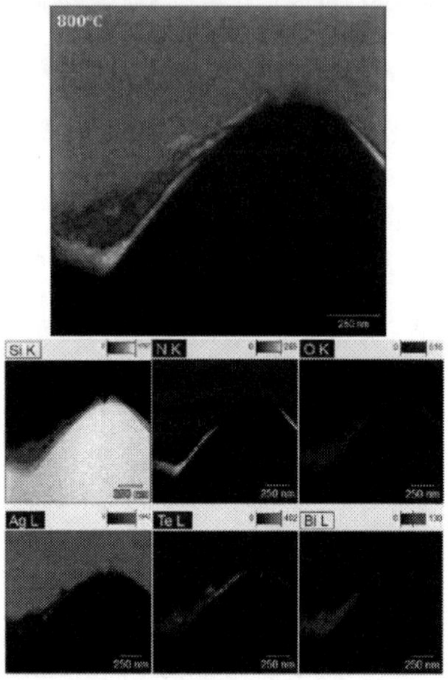

Figure 2: EDX mapped cross section of a Si pyramid with Ag crystallite at the top, combined on the top, for each element respectively on the bottom.

Fig. 3 shows an EDX cross-section of a Si pyramid trough with Ag crystallite at the top. Bi moves to the pyramid trough and to the wafer surface. Te moves upward and the alloying of Te with Ag is also visible. Within the glass layer, some thicker and thinner Ag spots are visible, which seem to intersperse the entire glass.

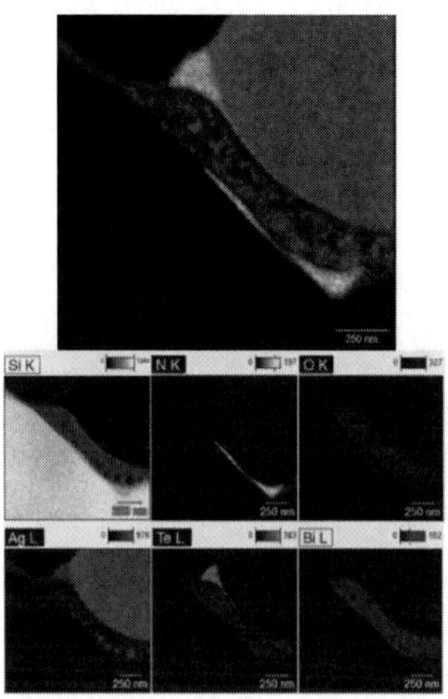

Figure 3: EDX mapped cross section of a Si pyramid trough with Ag crystallite at the top, combined on the top, for each element respectively on the bottom.

In addition, Fig. 4 and Fig. 5 show EDX cross-sections of a Si pyramid with Ag crystallite at the top which are connected to the Ag paste bulk via an Ag_2Te phase. The formation of an Ag- and Te-rich metallic region (green region in the center) is clearly seen, which suggests the formation of Ag_2Te alloy as described in our model. The nitrogen signal is negligible in this area, which supports the hypothesis that the TeO_2 in the paste plays an important role in etching the SiN_x:H passivation layer, reacting to Te and then formingAg_2Te.

Again, it is clear that Bi is incorporated in the glass regions in addition to O and Si. Te can also be detected here. Additionally, in Fig. 5 Bi-containing glass is detected in the region of the Ag_2Te. Occasionally, Ag and Te containing particles (Ag and Ag_2Te) can be detected in the glass which may explain a possible additional current path via tunneling/hopping apart from direct contact of Ag crystallizes with the Ag paste matrix.

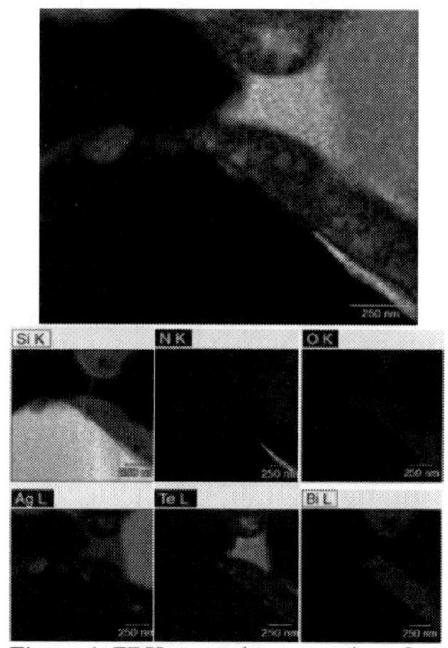

Figure 4: EDX mapped cross-section of a Si pyramid with Ag crystallite at the top, combined on the top, for each element respectively on the bottom.

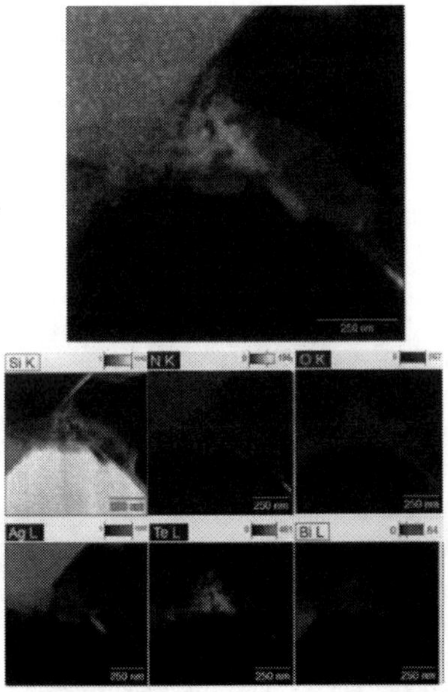

Figure 5: EDX mapped cross section of a Si pyramid trough with Ag crystallite at the top, combined on the top, for each element respectively on the bottom.

It can be shown that the Ag paste is capable of simultaneously contacting both polarities very well ($POCl_3$ formed n^+-emitter and BBr_3 formed p^+-emitter). Fig. 6 shows the specific contact resistance as a function of peak firing temperature. Values of <1 mΩcm^2 can be achieved for a wide temperature window. Efficiencies of >22.5% could be reached at an industry partner's production line with this paste, proving the industrial applicability.

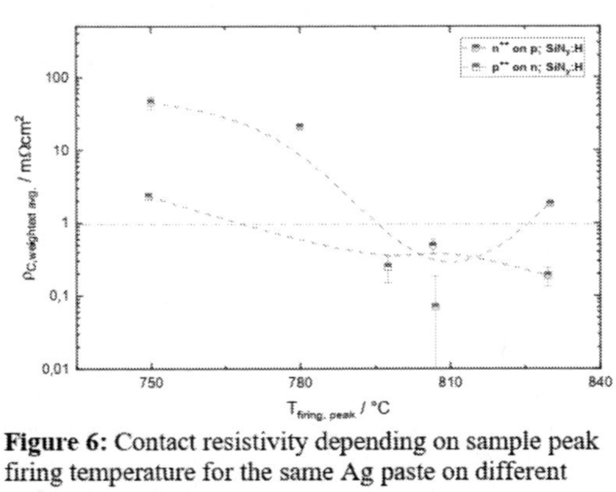

Figure 6: Contact resistivity depending on sample peak firing temperature for the same Ag paste on different emitter types.

4 CONCLUSION

In summary, Te seems to play a key role in the contacting of crystalline Si by Pb-free, Te-containing Ag pastes. It was shown that due to the presence of Si, Ag can react with Te via a redox reaction. Ag paste is capable of simultaneously contacting both n^+ and p^+ polarities with contact resistances below 1 mΩcm^2. For more details, see the invited journal publication [7].

5 ACKNOWLEDGEMENTS

Part of this work was supported by the German BMWK under contract 0324226A. The content of the publication is the responsibility of the authors.

6 REFERENCES

[1] S. Fritz, S. Riegel, S. Gloger, D. Kohler, M. König, M. Hörtheis, G. Hahn, Energy Procedia 38 (2013) 720.

[2] S. Fritz, M. König, S. Riegel, A. Herguth, M. Hörteis, G. Hahn, IEEE Journal of Photovoltaics 5(1) (2015) 145.

[3] S. Kontermann, M. Hörteis, M. Kasemann, A. Grohe, R. Preu, E. Pink, T. Trupke, Solar Energy Materials and Solar Cells 93(9) (2009) 1630.

[4] M.I. Jeong, S.E. Park, D.H. Kim, J.S. Lee, Y.C. Park, K.S. Ahn, C.J. Choi, Journal of the Electrochemical Society 157(10) (2010) H934.

[5] ITRPV, International Technology Roadmap for Photovoltaics (ITRPV): Results 2021, VDMA e.V., Frankfurt 2022.

[6] G. Schubert, F. Huster, P. Fath, Solar Energy Materials and Solar Cells 90(18-19) (2006) 3399.

[7] F. Geml, B. Gapp, S. Johnson, P. Sutton, A.E. Goode, J. Booth, H. Plagwitz, G. Hahn, EPJ Photovolt. (2022) (submitted).

PROJECT "INNOMET" - EVALUATION OF INNOVATIVE GLASS-BASED PRINTING FORMS FOR SOLAR CELL METALLIZATION

A. Lorenz[1]*, S. Tepner[1], K. Gensowski[1], T. Wenzel[1], K. Karimi[1], A. Vogt[2], M. Schulz-Ruhtenberg[2], S. Bechmann[3], M. Kuchler[4], M. Pospischil[4], R. Akyüz[1], M. Singler[1], S. Auerbach[1], A. Jimenez[1], L. Ney[1], F. Clement[1]

[1]Fraunhofer Institute for Solar Energy Systems ISE, Heidenhofstraße 2, 79110 Freiburg i.Br., Germany
[2]LPKF Laser & Electronics AG, Osteriede 7, 30827 Garbsen, Germany
[3]Christian Koenen GmbH, Otto-Hahn-Str. 24, 85521 Ottobrunn, Germany
[4]Highline Technology GmbH, Merzhauserstr. 76, 79100 Freiburg, Germany

*Corresponding author, e-mail to: andreas.lorenz@ise.fraunhofer.de

ABSTRACT: Within this work, we discuss the evaluation of innovative glass-based stencils starting with in-depth computational fluid dynamics (CFD) simulation of different channel geometries using COMSOL Multiphysics©. Based on these simulations, a special hour-glass shaped channel geometry has been identified which theoretically enabled the direct printing of contacts with a triangular cross-section using stencil printing. With the expertise of the involved partners LPKF Laser & Electronics AG and Christian Koenen GmbH, several sets of LIDE-structured glass foil stencils have been fabricated and tested on a semi-automatic screen printer at Fraunhofer ISE. Within these tests, the prediction of stencil-printed triangular contacts could be impressively verified on Cz-Si PERC solar cells. Due to the optimal reflection properties of such a contact shape, this could enable effectively "transparent contacts" in the future and thus represents a long-lasting dream of all solar cell metallizing specialists. Furthermore, glass foil nozzle plates were manufactured for the parallel dispensing technique by LPKF Laser & Electronics AG to demonstrate the proof of concept for an alternative, cost-effective material including production costs for nozzle plates compared to the standard material. LIDE-structured glass foils showed promising results in the multi-nozzle dispensing process to be used in the future within further optimization in the manufacturing process.

Keywords: silicon heterojunction (SHJ), metallization, fine line printing, stencil printing, parallel dispensing

1 INTRODUCTION

Optimizing the fine line metallization of silicon solar cells is one of the key factors for further increase of the cell performance. Impressive results have been demonstrated in the field of fine line screen printing recently [1, 2]. Besides, the performance aspects on the solar cell (reduction of shading, lateral conductivity of the grid), further aspects like reduction of silver laydown, process and material costs and throughput must be considered for the metallization step. Within the joint project "Innomet", a consortium of project partners including Fraunhofer ISE, LPKF Laser & Electronics AG and Christian Koenen GmbH has developed a highly innovative approach to fabricate very precise printing forms for solar cell fine line metallization based on thin glass foil structured with the so-called laser induced deep etching (LIDE) process [3]. This process enables the cost-effective and highly precise structuring of very fine openings in glasses with a thickness of up to 1000 µm. The LIDE process enables high precision and quality processing of glasses in a two-step process: In a first step, the glass (i.e. borosilicate glass) is modified and structured by a laser step. The modified areas are then etched with a high selectivity in a subsequent wet etch process (**Figure 1** and **Figure 2**).

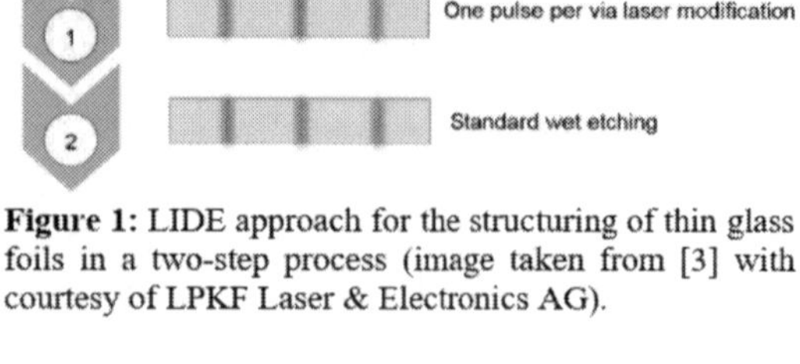

Figure 1: LIDE approach for the structuring of thin glass foils in a two-step process (image taken from [3] with courtesy of LPKF Laser & Electronics AG).

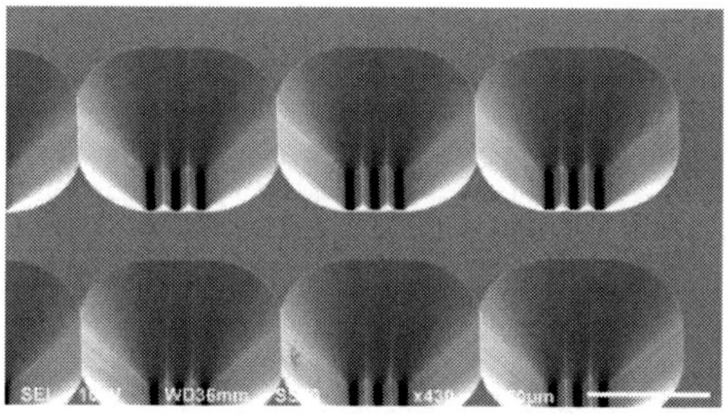

Figure 2: SEM images of LIDE-structured glass foil (image with courtesy of LPKF Laser & Electronics AG).

Within the project, two approaches are considered: On the one hand stencil printing [4] which has been applied for decades for printing of solder pads in the field of printed circuit board (PCB) production [5] as well as solar cell metallization [6, 7] using metal-foil stencils. On the other hand multi-nozzle parallel dispensing [8–11], a highly promising approach for the contactless application of homogeneous fine line contacts on silicon solar cells. The potential of parallel dispensing as a highly promising metallization technique for solar cell applications with strong benefits regarding throughput, silver reduction and effective silver usage has been demonstrated in multiple studies [12, 13]. Parallel dispensing has been successfully applied in silicon photovoltaics [14] as well as in thin-film photovoltaics [15].

Within this work, we evaluate the applicability of the LIDE technology to fabricate high-precision fine line stencils as well as nozzle plates for parallel dispensing based on thin glass foil. We show that the LIDE approach could enable a promising and cost-effective alternative to elaborately structured metal foils.

2 EXPERIMENTAL

2.1 Simulation of channel geometry

In order to evaluate the optimal channel geometry of glass-based stencils regarding paste flow and printing result, a comprehensive computational fluid dynamics (CFD) numerical simulation of different channel geometries using COMSOL Multiphysics© has been carried out which is described in detail in [16, 17]. Based on the results of various simulation runs, a special hour-glass shaped channel geometry (**Figure 3**) has been identified as a very promising channel design which theoretically enables printing of contacts with a triangular shape (cross-section).

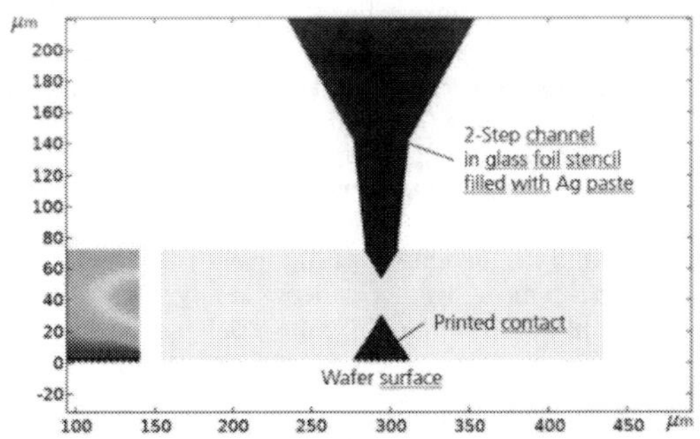

Figure 3: COMSOL Multiphysics© CFD simulation of channel geometry and stencil printing process with the result of triangular-shaped contacts [16, 17].

2.2 Fabrication and evaluation of glass foil stencils

With the expertise of the involved partners LPKF Laser & Electronics AG and Christian Koenen GmbH, several sets of glass foils with an initial thickness of $d_{glass} = 400$ µm are structured using the LIDE process. Based on the results of the previously conducted comprehensive CFD simulation, prototype glass stencils with a solar cell front side grid layout (busbarless layout, 100 fingers) using an optimal two-step channel geometry are realized. Microscopy measurements of the finger channels determine the width of the LIDE-structured channels with approx. 200 µm on the squeegee side and approx. 21 µm on the printing side of the stencil. The integration of structured glass foils into a stencil frame with an adequate stability for the printing process proved to be particularly challenging. This challenge could be solved by developing a new technique to mount the glass foil into the frame using the expertise of the project partner Christian Koenen GmbH. Using this method, stencils with a printing area of 300 mm x 300 mm are fabricated (**Figure 4**) for fine line printing tests using an EKRA XS-2 semi-automatic screen printing machine at Fraunhofer ISE and commercially available PERC silver (Ag) paste. In order to evaluate the optimal printing parameters, the printing test is carried out with varying printing speed ($v_{print} = 20$ mm/s to $v_{print} = 300$ mm/s) and printing pressure ($p = 0 - 20$ N).

Industrially pre-fabricated Cz-Si PERC solar cell precursors without metallization are used for the metallization process. The printing result and the three-dimensional geometry of the printing contacts is analyzed with confocal laser-scanning microscopy (CLSM) and subsequent image analysis using the Fraunhofer ISE tool "Dash" [1]. Furthermore, scanning electron microscopy (SEM) is applied to further analyse the finger geometry on selected contacts.

Figure 4: LIDE-structured glass foil stencil with busbarless fine line grid.

2.3 Fabrication of LIDE-structured glass foil nozzle plates and evaluation of parallel dispensing process

First pre-tests regarding the application of the LIDE-process for the fabrication of dispensing nozzle plates revealed several challenges:

- Reproducible geometry of the LIDE-structured nozzle openings within the nozzle plate
- Mounting of the nozzle plate on the dispensing head with sufficient sealing and mechanical stability
- Mechanical stability of the glass nozzle plate during the dispensing process
- Wetting behavior of the solvent on the bottom side of the nozzle plate

To evaluate and assess the optimal parameters of the glass material and the LIDE process for dispensing nozzle plates, 24 different variations of glass nozzle plates (each plate with 15 parallel nozzle openings) are fabricated by LPKF Laser & Electronics AG and analysed at Fraunhofer ISE. The following parameters have been varied:

- Initial thickness d of glass foil: $d_{glass} = 300$ µm and $d_{glass} = 400$ µm
- Nominal nozzle diameter D: D = 25 µm, D = 30 µm, D = 35 µm
- Nozzle shape (with/without trench)
- Coating: glass foil with and without Titan (Ti) coating (wetting behaviour)

A test series in five steps has been carried out for each glass nozzle plate design. In the first step, three individual nozzle plates of each variation were characterised by means of microscopy regarding the nozzle diameter in x- and y-directions of all nozzle openings. Based on the optical characterisation, the nozzle plate with the smallest deviation of the nozzle diameter over the 15 nozzles is selected and bonded to a nozzle plate carrier. This bonded nozzle plate was mounted to the R&D print head 'GECKO' [18] and is used for stability analysis using a commercially available low-temperature silver paste. The stability tests are carried out on a specific test setup using the R&D print head in combination with a rotating roll.

Within the stability analysis, it is investigated whether homogeneous, straight paste threads can be extruded from the nozzle openings and whether the glass nozzle plate is mechanically stable when applying the process pressure load. Subsequently, the glass nozzle plate is cleaned to visually characterise the nozzle plate regarding hairline cracks in the glass substrate and changes of the nozzle openings. If the glass nozzle plate shows no damage, dispensing tests are carried out on Cz-Si wafers with a size of 156.75 mm x 156.75 mm. The process velocity is varied between $v_{process}$ = 50 mm/s and $v_{process}$ = 500 mm/s to evaluate the optimal dispensing conditions for a homogeneous result on the whole wafer area. Finally, the printed contacts are characterised regarding finger height and width using a 3D confocal laser scanning microscope.

3 RESULTS AND DISCUSSION

3.1 Glass foil stencil printing

Glass foil stencils with an initial thickness of d_{glass} = 400 µm proved to be stable during the process and showed no traces of mechanical damage or micro-cracks after the process. The best printing result is achieved at a printing velocity of v_{print} = 150mm/s. At velocities below and above 150mm/s, the paste transfer was insufficient, resulting in very inhomogeneous contact fingers with many interruptions. CLSM analysis of selected homogeneous contact fingers results in a mean core finger width of w_f = 29 µm and a mean finger height of h_f = 24 µm. Due to remaining local deviations along the LIDE-structured fine line grid, it was not possible to print a completely homogeneous metallization grid with sufficient quality for solar cell metallization onto the Cz-Si precursors. Ongoing effort is done by LPKF to improve the homogeneity of the glass channels.

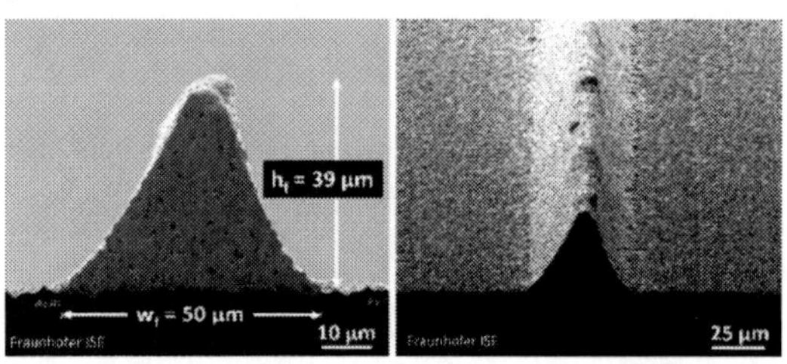

Figure 5: SEM images of contacts with triangular shape printed with LIDE-structured glass foil stencils.

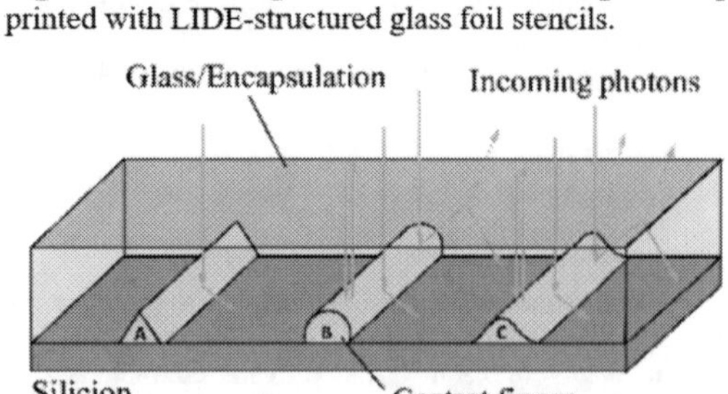

Figure 6: Schematic of internal reflection properties of solar cell contacts within the module. Triangular-shaped contacts are particularly beneficial due to internal back-reflection. Modified image based on [19, 20].

However, a particularly remarkable result could be achieved by printing contacts with almost perfectly shaped triangular cross-section which impressively confirms the prediction of the previously conducted CFD simulation [17]. SEM images of selected contacts confirm the triangular cross-section of the contacts (**Figure 5**). Fingers with a triangular shape are expected to have particular advantageous reflection properties within the module due to an effective internal reflection of incoming light into the solar cell [20] (**Figure 6**). In an optimal case, such fingers can even be considered as "effectively transparent" [21, 22]. Further improvement of the LIDE-structured glass foil stencils could enable a direct printing process of triangular fingers over the whole metallization grid.

3.2 Glass foil dispensing process

The microscopic analysis of the LIDE-structured nozzle plates revealed a sufficient, yet still expandable quality of the nozzle openings. The maximum deviation between 15 nozzle openings within one glass nozzle plate is up to 2 µm. The average diameter of nozzle openings is determined with $D_{n,25}$ = 24.8 µm ± 0.6 µm for the 25 µm-nozzles and $D_{n,35}$ = 36.7 µm ± 2.6 µm for the 35 µm nozzle plate. Simulation results by Pospischil et al. show that already a small variation in nozzle diameter leads to an inhomogeneous flow profile and consequently to an inhomogeneous mass flow and thus to an irregular printing image and local disruptions within the dispensing process [23]. Thus, further improvement will focus on reducing the deviation between the nozzle openings to a value < 1 µm.

Within the stability tests, the glass foil nozzle plates showed a comparable mechanical stability to the reference nozzle plates. The nozzle plate configuration with an initial glass thickness of d_{glass} = 300 µm, Titan coating and a nozzle opening of d_{nozzle} = 30 µm obtained the most promising results during the stability plate and thus is selected for a metallization test run on Cz-Si wafer material. Using this configuration, a full-area solar cell wafer (156.75 mm x 156.75 mm) is successfully metallized with a busbarless grid of 150 dispensed fine line contacts using low-temperature silver paste for silicon heterojunction solar cells at a dispensing velocity of $v_{process}$ = 240 mm/s (**Figure 7**).

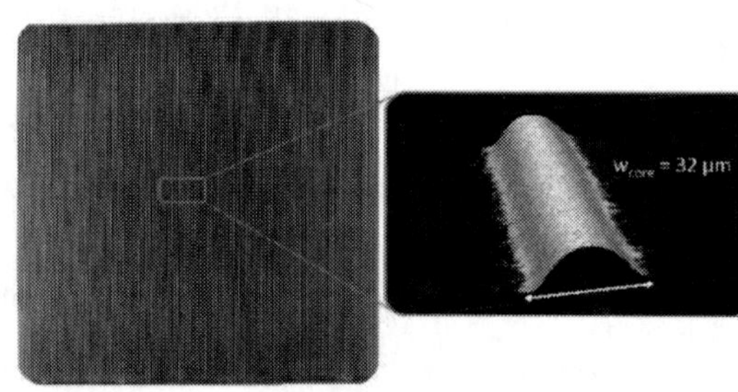

Figure 7: Dispensed full-area solar cell and CLSM image of a dispensed contact using glass foil dispensing process.

Subsequent CLSM analysis of the finger geometry revealed a mean core finger width of $w_{f,c}$ = 32 µm. Similar results are achieved with the reference dispensing process at the time of the experiment (2021). A wet paste laydown of m_{Ag} = 0.44 mg per contact finger is determined on R&D level using glass foil nozzle plates. Comparing these results to the reference dispensing process using standard metal nozzle plates, the results are encouraging. Yet, further optimization of the process will be necessary to obtain comparable results. Recently, a wet paste laydown of m_{Ag} = 0.30 mg per contact finger for low-temperature curing Ag pastes has been demonstrated by HighLine Technology GmbH on industrial level [10]. In 2022, the paste laydown could be even reduced to m_{Ag} = 0.27 mg per contact finger on R&D level for the dispensing process by using standard nozzle plates [13].

To obtain similar results, further R&D activities by LPKF will focus on improving the LIDE-process with respect to a reduced deviation of the nozzle openings. It is expected that comparable results to the reference can be achieved using optimized glass foil nozzle plates.

4 CONCLUSION AND OUTLOOK

In summary, the adressed solutions and results of the project "Innomet" provide two highly promising approaches for the next-generation fine line metallization of high-performance solar cells. Stencil printing based on LIDE-structured glass foils with funnel-shaped channel geometry offers the very promising opportunity to realize contacts with triangular-shaped cross-section. This offers the very interesting potential to realize effectively transparent contacts on silicon solar cells using a reliable stencil printing process. While the proof of principle has been successfully demonstrated within the project, further effort is required to enable an easy-to-handle and stable manufacturing procedure for glass foil stencil.

In the field of parallel dispensing, a successful proof of principle has been conducted regarding the fabrication and application of LIDE-structured nozzle plates for parallel dispensing of Si solar cells. Glass-based nozzle plates can be a promising approach to reduce costs and complexity of the nozzle plate manufacturing process. The feasibility of this approach has been successfully demonstrated with promising results regarding finger width and uniformity as well as process stability. Current R&D effort is focussed on improving the process reliability to reduce deviations of the nozzle opening size.

5 ACKNOWLEDGEMENTS

This work was partly supported by the German Federal Ministry of Economic Affairs (BMWi) within the funding program Photonics Research Germany under the contract number 0324308B (Innomet). The authors would further like to thank our associated project partner Wickon Hightech s.r.o. for supporting the project with advanced solutions for solar cell inspection.

6 REFERENCES

[1] T. Wenzel *et al.*, "Progress with screen printed metallization of silicon solar cells - Towards 20 μm line width and 20 mg silver laydown for PERC front side contacts," *Solar Energy Materials and Solar Cells*, vol. 244, p. 111804, 2022, doi: 10.1016/j.solmat.2022.111804.

[2] S. Tepner *et al.*, "Screen pattern simulation for an improved front-side Ag-electrode metallization of Si-solar cells," *Prog. Photovolt: Res. Appl.*, vol. 28, no. 10, pp. 1054–1062, 2020, doi: 10.1002/pip.3313.

[3] R. A. Krüger, M. Schulz-Ruhtenberg, B. Rösener, O. Ostermann, R. Ostholt, and N. Ambrosius, "LIDE: high aspect ratio glass processing technology for the mass production of microfluidic devices for biomedical applications," in *Microfluidics, biomems, and medical microsystems xvii*, San Francisco, United States, 2019, p. 5. [Online]. Available: https://www.spiedigitallibrary.org/conference-proceedings-of-spie/10875/2509899/LIDE--high-aspect-ratio-glass-processing-technology-for-the/10.1117/12.2509899.full

[4] T. Dullweber, Ed., *Silicon Solar Cell Metallization and Module Technology*. Stevenage: Institution of Engineering & Technology, 2021. [Online]. Available: https://ebookcentral.proquest.com/lib/kxp/detail.action?docID=6801142

[5] K. Petherbridge, P. Evans, and D. Harrison, "The origins and evolution of the PCB: A review," *Circuit World*, vol. 31, no. 1, pp. 41–45, 2005, doi: 10.1108/03056120510553211.

[6] H. Hannebauer, S. Schimanke, T. Falcon, P. P. Altermatt, and T. Dullweber, "Optimized Stencil Print for Low Ag Paste Consumption and High Conversion Efficiencies," *Energy Procedia*, vol. 67, pp. 108–115, 2015, doi: 10.1016/j.egypro.2015.03.294.

[7] J. Hoornstra and B. Heurtault, "Stencil print applications and progress for crystalline silicon solar cells," in *Proc. of the 24th Photovoltaic Solar Energy Conference (EUPVSEC)*, Hamburg, Germany, 2009.

[8] M. Pospischil, A. Lorenz, S. Kühnhold-Pospischil, and A. Adrian, "Alternative printing technologies," in *Energy Engineering Ser, Silicon Solar Cell Metallization and Module Technology*, T. Dullweber, Ed., Stevenage: Institution of Engineering & Technology, 2021, pp. 255–308.

[9] M. Pospischil *et al.*, "High Speed Dispensing with Novel 6" Print Head," *Energy Procedia*, vol. 98, pp. 61–65, 2016, doi: 10.1016/j.egypro.2016.10.081.

[10] K. Gensowski *et al.*, "Dispensing of low-temperature silver pastes," in *AIP Conference Proceedings, Vol. 2367, No. 1*, Genk, Belgium, 2021, p. 20007.

[11] M. Pospischil *et al.*, "Applications of parallel dispensing in PV metallization," in *AIP Conference Proceedings, Vol. 2156, No. 1*, 2019, p. 20005.

[12] K. Gensowski, M. Much, E. Bujnoch, S. Spahn, S. Tepner, and F. Clement, "Filament stretching during micro-extrusion of silver pastes enables an improved fine-line silicon solar cell metallization," *Scientific reports*, vol. 12, no. 1, p. 12318, 2022, doi: 10.1038/s41598-022-16249-5.

[13] K. Gensowski *et al.*, "Filament stretching during parallel dispensing – A way to reduce silver consumption in SHJ metallization," *Sol. Energy Mater. Sol. Cells*, vol. 245, p. 111871, 2022, doi: 10.1016/j.solmat.2022.111871.

[14] K. Gensowski *et al.*, "CIGS Mini-Modules with Dispensed Metallization on Transparent Conductive Oxide Layer," *Sol. RRL*, vol. 4, no. 12, p. 2000475, 2020, doi: 10.1002/solr.202000475.

[15] K. Gensowski, T. Freund, S. Tepner, and F. Clement, "Increasing the Effective Area of Small-Sized CIGS Modules by Printed Ag Front Contacts," *Phys. Status Solidi RRL*, vol. 16, no. 7, p. 2200040, 2022, doi: 10.1002/pssr.202200040.

[16] S. Tepner *et al.*, "Printed Triangle-Shaped Contacts for Metallization of Silicon Solar Cells," Constance, Germany, 28th March 2022.

[17] Printed Triangle-Shaped Contacts for Metallization of Silicon Solar Cells, "S. Tepner; K. Karimi; T. Wenzel; L. Ney; S. Auerbach; M. Linse; M.

Schulz-Ruthenberg; S. Bechmann; A. Lorenz; F. Clement," in *12th International Conference on Crystalline Silicon Photovoltaics (Silicon PV)*, Constance, Germany, 2022.

[18] M. Pospischil *et al.*, "Process Development for a High-throughput Fine Line Metallization Approach Based on Dispensing Technology," *Energy Procedia*, vol. 43, pp. 111–116, 2013, doi: 10.1016/j.egypro.2014.02.001.

[19] A. Lorenz, "Evaluierung von Rotationsdruckverfahren für die Metallisierung von Silicium-Solarzellen," PhD Thesis, Institut für Mikrosystemtechnik und Informatik, Albert-Ludwigs-Universität, Freiburg, 2017.

[20] A. Mette, "New concepts for front side metallization of industrial silicon solar cells," PhD Thesis, Albert-Ludwigs-Universität, Freiburg, 2007.

[21] R. Saive *et al.*, "Effectively Transparent Front Contacts for Optoelectronic Devices," *Advanced Optical Materials*, vol. 4, no. 10, pp. 1470–1474, 2016, doi: 10.1002/adom.201600252.

[22] R. Saive *et al.*, "Silicon heterojunction solar cells with effectively transparent front contacts," *Sustainable Energy Fuels*, vol. 1, no. 3, pp. 593–598, 2017, doi: 10.1039/C7SE00096K.

[23] M. Pospischil *et al.*, "Development of a high-throughput fine line metallization process using CFD-simulation," in *Conference Records of 39th IEEE Photovoltaic Specialists Conference (PVSC)*, Tampa, FL, USA, 2013, pp. 2250–2253.

This is a short version of the paper submitted to the journal Progress in Photovoltaics.

It has been selected by the Committee of the WCPEC-8 for submission to Progress in Photovoltaics.

NOVEL SEMI-TRANSPARENT SOLAR CELL BASED ON ULTRATHIN MULTIPLE SI/GE QUANTUM WELLS

Hosni Meddeb[1], Maximilian Götz-Köhler[1], Kai Gehrke[1], Martin Vehse[1]
1. German Aerospace Center (DLR), Institute of Networked Energy Systems, Urban and Residential Technologies, Carl-von-Ossietzky-Str. 15, 26129 Oldenburg, Germany

ABSTRACT: In this work, we demonstrate a novel semi-transparent solar cell based on ultrathin hydrogenated amorphous Si/Ge multiple quantum wells (MQW). This concept allows extra degree of freedom for both optical design and bandgap engineering compared to single QW structures. Mainly, the multiplication of the QWs number in a periodic configuration, taking advantage of effective synergy between electronic and photonic confinements, leads to an improvement of photocurrent, while preserving high voltage and fill factor and ensuring significant transparency. The MQW new concept yields a boost in power conversion efficiency up to 3.4% and a considerable average visible transmission of about 33%. The achieved performance points out the promising potential of the proposed semitransparent ultrathin QW solar cell for glass façades and solar windows applications.

Keywords: quantum well, optoelectronic modelling, thin-film solar cell, semi-transparent photovoltaics, integrated PV

1 INTRODUCTION

Ultrathin solar cells are one of the key approaches in transparent photovoltaics. These concepts rely on thin enough photoactive semiconductors that allow partial transmission of visible light [1,2]. Such PV technologies are promising and efficient providing considerable advantages in terms of low material consumption, fast fabrication processes and cost reduction [3]. Different inorganic and organic material systems have been employed with various degrees of optical transmission and leading to a direct trade-off between photoconversion efficiency and average transmission [2,4]. By employing low-dimensional nanoscale semiconductors, size-dependent changes in the optoelectronic properties are manifested, due to the quantum confinement (QC) effects. An outstanding representative technology from this category is ultrathin quantum well (QW) solar cell based on hydrogenated amorphous silicon (a-Si:H) as barrier and germanium (a-Ge:H) as QW nanostructures [5]. Integrated in an absorbing photonic resonator, this concept allows a significant thickness-dependent tuning of all the cell characteristics. An enhancement of open-circuit voltage (Voc) and fill factor (FF) is counterbalanced by a drop in the photocurrent (Jsc) when reducing the thickness of the QW in single QW (SQW) solar cells. The opposite effect can be observed when increasing the QW thickness [5]. In semi-transparent architecture, it is expected that thicker SQW are less advantageous for reaching high light transmission, while, ultrathin SQW yield lower power conversion efficiency (PCE). In this work, an innovative concept for semitransparent solar cells is explored by integrating ultrathin multiple QWs (MQW) as photoactive material to overcome the intrinsic limitations imposed by SQW configurations. The main output of this work is the demonstration of the effectiveness of amorphous Si/Ge MQW concept in boosting the power generation performance up to 3.4%, while preserving a considerable average visible transmission level of about 33%. Therefore, a superior tradeoff between efficiency and transparency is achieved by integrating MQW structures compared to SQW counterparts. The successful demonstration of ultrathin semitransparent MQW solar cells indicates the promising potential for integration in window applications as electrical power generation source and heat insulation approach [6].

2 EXPERIMENTAL METHODS

Different SQW and MQW semi-transparent solar cells were fabricated using cost-effective and industrial-compatible processes. A 1 μm-thick aluminium-doped Zinc Oxide ZnO:Al (AZO) transparent conductive oxide layer formed the front contact. The solar cell with n-i-p superstrate configuration consisted of (n-a-Si:H/i-a-Si:H/SQW or MQW/ i-a-Si:H/p-a-Si:H) multilayers with thicknesses (7 nm/3 nm/SQW or MQW/3 nm/7 nm). The intrinsic regions were composed of an alternation between i-a-Si:H (barrier)/i-a-Ge:H (QW) films. For comparison purposes, two different i-a-Ge:H SQW thicknesses were considered, with ultrathin L_{QW} = 2.5 nm and thick L_{QW} =20 nm, as well as, i-a-Si nanoabsorber of 20 nm composed of full barrier material. In the MQW region having the same total nanoabsorber thickness as the thick SQW, a periodic stratified medium was incorporated with 6 QWs of L_{QW} = 2.5 nm each, separated by 1 nm-thick quantum barrier (QB) regions. The front contact opening through the glass substrate was performed by laser scribing. Layer stacks of Ag (15 nm)/AZO (60 nm) were deposited as a back transparent electrode through a shadow mask to define 1 × 1 cm^2 cell areas. The front and back electrodes were deposited by DC magnetron sputtering technique. Whereas, the semiconductor functional layers were fabricated by low-temperature Plasma Enhanced Chemical Vapor Deposition (PECVD) method at 13.56 MHz in the same cluster tool system. Further details on the fabrication processes can be found elsewhere [5,7]. The optical transmission and reflection measurements of the cell devices at normal incidence were carried out using a spectrophotometer with an integration-sphere. Through optical modelling, the experimental optical characteristics of different semi-transparent QW solar cells were analyzed by means of software combination including CODE/Scout (by W. Theiss Hardware and Software) based on transfer matrix method [8]. Detailed inputs of the optoelectronic numerical simulation can be found in previous works [7].

3 RESULTS AND DISCUSSIONS

3.1 Structure of semi-transparent ultrathin Si/Ge MQW solar cell

Figure 1 illustrates the structure of a periodic MQW solar cell featuring a subwavelength optical nanocavity.

The photoactive region is composed of an alternation between i-a-Si:H (barrier) and i-a-Ge:H (QW) layers. The n-i-p region consist of deep-subwavelength lossy a-Si and a-Ge media with high and comparable refractive and extinction indices (n~k), sandwiched between partially reflective AZO front and ultrathin Ag/AZO back electrodes with low refractive indices. This creates a low finesse optical nanocavity where optical interference and photonic confinement occur [9]. Moreover, the quantum confinement effect in narrow bandgap ultrathin a-Ge:H QW embedded between two a-Si:H QB regions, leads to an upward shift of the conduction band edge to higher discrete level due to the energy quantization, and hence, a bandgap widening is manifested [5]. This affect both the photogeneration process and the carrier transport at a-Si:H/a-Ge:H heterojunctions.

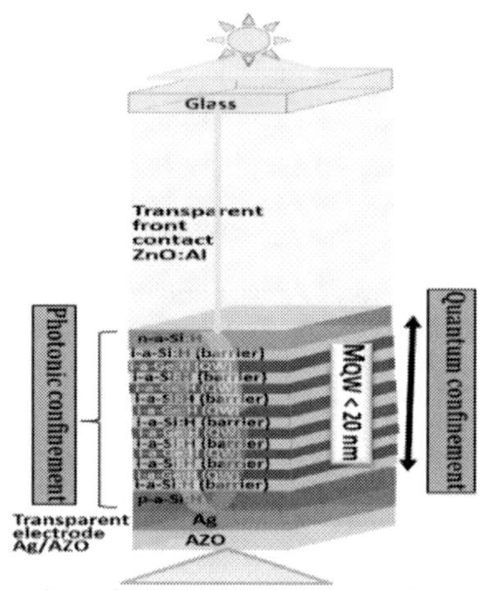

Figure 1: Schematic of semitransparent ultrathin solar cell based on multiple quantum wells MQW as a-Si/a-Ge nanostructures inside an absorbing resonant nanocavity

3.2 Optical modelling of semi-transparent ultrathin Si/Ge MQW solar cell

In Figure 2, the wavelength-dependence of the electric field profile through different semi-transparent QW solar cells i.e. thick SQW (20 nm), ultrathin SQW (2.5 nm), and MQW (6 × 2.5 nm) devices is shown. It is clearly observed that the increase of the photonic resonator length from ultrathin SQW (2.5 nm) to thick SQW (20 nm) or MQW (6×2.5 nm) results in a shift of the maximum amplitude of the electric field to longer wavelengths. In the MQW configuration, high optical field intensities are obtained over all the QWs sub-regions. Another observation is that regions of maximum electric field amplitude in the visible range are spatially restricted to the bottom zone of the thick SQW (20 nm), whereas they are extended over the total nanoabsorber depth in MQW (6×2.5 nm) structures. This implies a more effective contribution in the photocurrent generation.

Regarding the transparency aspect, it is noticed that the highest propagating optical field through the back AZO and escape from the semitransparent cell is obtained for ultrathin SQW (2.5 nm). However, the MQW configuration leads to comparable transmitted optical field with clearly higher amplitude than the thick SQW (20 nm). This is expected to results in a considerable optical transmission of the semitransparent MQW solar cell device.

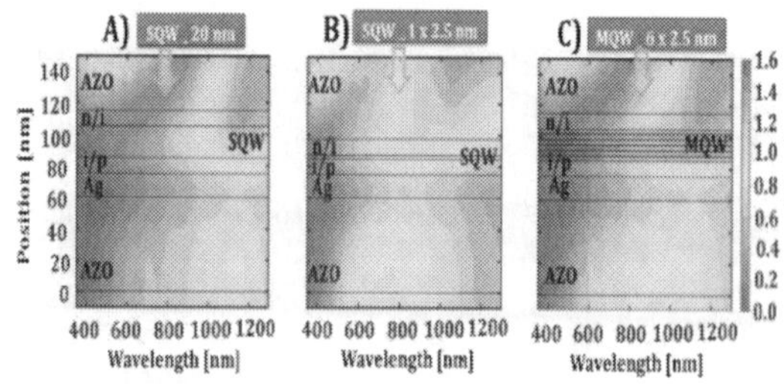

Figure 2: Optical modeling results of local absorption in semitransparent solar cells with ultrathin SQW with 2.5 nm (a), thick SQW of 20 nm (b) compared to ultrathin MQWs (6 × 2.5 nm) (c).

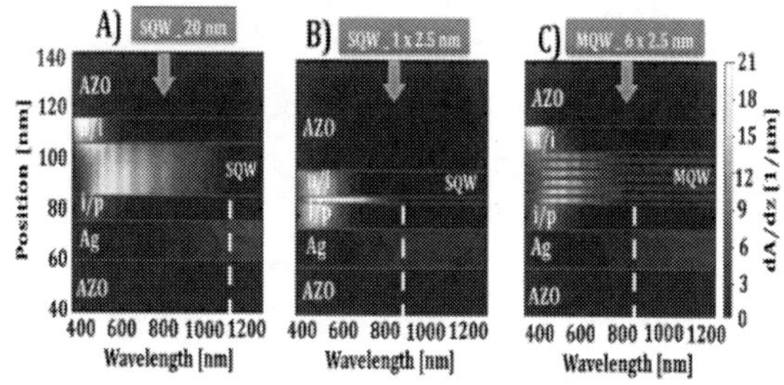

Figure 3: Optical modeling results of local absorption in semitransparent solar cells with ultrathin SQW with 2.5 nm (a), thick SQW of 20 nm (b) compared to ultrathin MQWs (6 × 2.5 nm) (c).

Figure 3 presents the variations of local absorption as a function of wavelengths in semi-transparent solar cells with different QW nanoabsorber configurations. Across the semi-transparent device, the optical absorption in different functional layers mainly depends on the amplitude of light electromagnetic field and on the optical refractive indices (n, k) of materials, given by [10]:

$$Absorption\ (z,\lambda) = \frac{2\pi c\ \varepsilon}{\lambda}\ n(z,\lambda) \times k(z,\lambda)\ |E(z,\lambda)|^2$$

where c is the speed of light, ε is the permittivity of free space and $|E(z,\lambda)|^2$ is the electric field amplitude as a function of depth z and wavelength λ. Accordingly, due to higher refractive index of a-Ge QW compared to a-Si barrier, higher local absorption is distinguished in all QW regions compared to the barrier zones. The regions of QW nanoabsorber near to the back electrode is characterized by stronger local absorber in all the QW configuration. It is important to note that the absorption edge shifts towards shorter wavelength for both SQW and MQW with reducing the corresponding QW thickness in contrast to thick SQW, as highlighted by the dashed white lines. This is expected to affect the profile of EQE spectra of the semitransparent solar cells especially at long wavelengths.

3.3 Experimental characteristics of semi-transparent Si/Ge MQW solar cell

Figure 4 shows the J-V curves under illumination for different semitransparent QW solar cells. The short-circuit current density (Jsc) values are in correlation with the optical modelling expectations, giving an intermediate photocurrent value for the MQW configuration (Jsc = 10.3 mA/cm^2) relative to SQW-2.5 nm (Jsc = 5.4 mA/cm^2) and

SQW-20 nm (Jsc = 10.3 mA/cm^2). Interestingly, the multiplication of QWs in a configuration of (6 × 2.5 nm) clearly enables to overcome the intrinsic limitations of low absorbance for ultrathin SQW (2.5 nm).

Another advantage of MQW configuration can be noticed on the open-circuit voltage (V_{oc}) parameter. Owing to the nanoabsorber bandgap E_g widening under QC effect, the implementation of MQW (6 × 2.5 nm) allows an enhancement of Voc up to 535 mV compared to only 320 mV for thick SQW (20 nm) with narrow bandgap. Whereas the MQW (6×2.5 nm) preserves a comparable V_{oc} level to SQW (2.5 nm) [11].

Moreover, the fill factor (FF) also presents a significant improvement for the device with MQW (6 × 2.5 nm) compared to thick SQW (20 nm) counterparts. This indicates the enhancement of the carrier transport and the lowering of resistive losses, due to the reduction of the conduction band discontinuity (**Δ**Ec) [11].

One can conclude that the MQW configuration allows the combination of the benefits from both thick and ultrathin SQW counterparts. This is attributed to the rise of the photocurrent due to the extension of the volume of photoactive medium, as well as the enhancement of both Voc and FF thanks to the band gap engineering. Importantly, the semitransparent MQW device yields a largely superior photoconversion efficiency (PCE) over to all other configuration counterparts reaching a PCE above 3.4%.

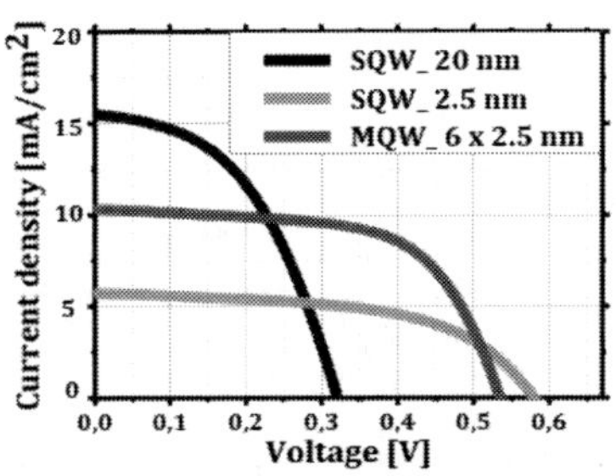

Figure 4: Illuminated J-V curves of different semi-transparent solar cells with MQWs (6 × 2.5 nm) compared to ultrathin SQW with 2.5 nm and thick SQW of 20 nm.

Table I: Characteristic outputs of semi-transparent solar cell devices with different nanoabsorbers.

	Jsc [mA/cm^2]	Voc [mV]	FF [%]	PCE [%]	AVT [%]
SQW (20 nm)	15.5	320	47	2.33	25
SQW (2.5 nm)	5.4	580	59	1.89	39
MQW (6×2.5 nm)	10.3	535	61	3.41	33

Regarding the optical visibility aspect, we quantify the average transmission (AVT) parameter which measures the transmission of AM1.5G irradiation weighted with the photopic response of the human eye [12].

The MQW leads to higher AVT than thick SQW (20 nm) with similar nanophotonic resonator length. This can be attributed to both the reduction of a-Ge absorbing material amount and the increase of QW bandgap. Surprisingly, despite the multiplication of QWs, a considerable light transmission is preserved in the case of MQW, which is comparable to ultrathin SQW counterpart.

Overall, the integration of MQW as nanoabsorber in novel semitransparent ultrathin Si/Ge solar cells leads to an optimum trade-off between efficiency and transparency by boosting the power generation performance while preserving a considerably high visible transmission.

4 CONCLUSIONS

In this work, we report the prime proof-of-concept for a novel semitransparent solar cell based on multiple quantum wells as ultrathin nanoabsorber. The here presented device concept employing amorphous Si/Ge quantum confined nanostructures integrated in a subwavelength nanophotonic resonator is unique of its kind. Therefore, such emerging semitransparent technology takes advantage of an effective synergy between quantum and photonic confinements to boost both the power generation and visible transmission levels. The aim of the QWs multiplication in a configuration of 6 × 2.5 nm is to overcome the intrinsic limitations imposed by single QW configurations, mainly, the low absorbance for ultrathin SQW (2.5 nm) and the inappropriate electronic structure for thick SQW (20 nm). Based on the conducted optoelectronic modelling, a considerable room of improvement through optical design and bandgap engineering is indicated. The successful demonstration of ultrathin semi-transparent Si/Ge MQW solar cells indicates the promising integration potential of this emerging photovoltaic technology for supplying systems in relevant applications such as in buildings, vehicles and greenhouses.

ACKNOWLEDGMENTS

The authors acknowledge funding from the Energy branch of the German Aerospace Center (DLR).

References:

[1] C. J. Traverse, R. Pandey, M. C. Barr, R. R. Lunt, Nat Energy 2017, 2, 849.

[2] K. Lee, H.-D. Um, D. Choi, J. Park, N. Kim, H. Kim, K. Seo, Cell Reports Physical Science 2020, 1, 100143.

[3] Massiot I, Cattoni A, Collin Nat. Energy 2020; 5(12): 959–72.

[4] O. Almora et al, Adv. Energy Mater. Adv. Energy Mater. 2021; 11(11): 2102526.

[5] H. Meddeb, N. Osterthun, M. Götz, O. Sergeev, K. Gehrke, M. Vehse, C. Agert, Nano Energy 2020, 76, 105048.

[6] Meddeb H, Götz-Köhler M, Neugebohrn N, Banik U, Osterthun N, Sergeev O, Berends D, Lattyak C, Gehrke K, Vehse M. Advanced Energy Materials 2022: 2200713.

[7] H.Meddeb, M. Götz-Köhler, N. Osterthun, O. Sergeev, K. Gehrke, M. Vehse, C. Agert, IEEE J. Photovoltaics 2022, 12, 3, 760-770.

[8] WTheiss Hardware and software- CODE/Scout https://www.wtheiss.com/ (Accessed June 2022).

[9] M.A Kats, F. Capasso, Laser & Photonics Rev. 2016; 10(5): 735–49.

[10] K-T. Lee, S. Seo, J-Y. Lee, L-J. Guo. Advanced materials 2014; 26(36): 6324–8.

[11] Meddeb H, Götz-Köhler M, Osterthun N, Sergeev O, Gehrke K, Vehse M, Agert C. In: 2021 IEEE 48th Photovoltaic Specialists, pp. 975–978.

[12] Yang C, Liu D, Bates M, Barr MC, Lunt RR. Joule 2019; 3(8): 1803–9.

IMPACT OF LATERAL EFFECTS ON EQE MEASUREMENTS OF LARGE SCALE TANDEM SOLAR CELLS

S. Kasimir Reichmuth[1,2], A. Fell[1,3], G. Siefer[1], M. Schachtner[1], D. Chojniak[1], O. Fischer[1,2], M. Mühleis[1], M. Rauer[1], J. Hohl-Ebinger[1], M. C. Schubert[1]
[1]Fraunhofer Institute for Solar Energy Systems ISE, Heidenhofstrasse 2, 79110 Freiburg, Germany
Email: kasimir.reichmuth@ise.fraunhofer.de,
[2]Albert-Ludwigs-University, INATECH, Emmy-Noether-Strasse 2, 79110 Freiburg, Germany
[3]AF simulations, Landstr. 33a, 79232 March, Germany

ABSTRACT: Laterally inhomogeneous cell parameters in large scale perovskite/silicon (PSC/Si) tandem solar cells may significantly influence the device performance. The lateral quality of the absorber can be analysed with electroluminescence (EL), photoluminescence (PL) and thermographic methods. In addition to a global impact on cell performance, such lateral effects are often not considered in *EQE* and *I-V*-characterization of tandem devices, yet may lead to erroneous measurement results. Therefore, we see the necessity to employ large-area 3D PSC/Si-tandem simulations to understand the impact of lateral inhomogeneities, and the interaction with non-ideal measurement conditions such as small-area or non-uniform illumination of the solar cell. We use a tandem addon of the 3D simulation software Quokka3 for our full-cell 3D tandem simulations which treats the perovskite top cell skin, and optionally also the Si bottom cell, with an "equivalent-circuit" model instead of solving the drift-diffusion model. We present simulations and experiments to quantify the impact of inhomogeneous cell properties such as low localized shunt resistances or non-uniformities of the cell absorber, in interaction with the illumination and biasing conditions during the *EQE* measurement. It is specifically interesting to gain insight into lateral effects by simulation, as the experimental investigation of such detailed effects in often metastable PSC/Si tandem cells is highly challenging.
Keywords: Multijunction Solar Cell, Calibration, Simulation, Perovskite, III-V Semiconductors

1 INTRODUCTION

Recently, perovskite/silicon tandem cells (PSC/Si) have shown efficiencies of 31.25% [1] for a lab sized sample and 6" wafer scale PSC/Si have been certified with an efficiency of 26.8 ±1.2 % [2]. At the same time, first commercialization has been announced for the current year, aiming for upscaling size and throughput [3]. In the industrial implementation, processes that have been established for lab-sized cells are in adaptation to high scale output. Compared to small laboratory cells, lateral effects may be more important for full wafer sized cells. This can explain the efficiency decrease seen for perovskite absorbers in the upscaling process [4].

The spatial inhomogeneity has an influence on both, cell performance and the characterization of these cells, e.g., if the methods rely on local illumination only and do not analyse the whole area of the device. This can be critical for *EQE* and *I-V*-characterization, which may lead to significant deviations from the true characteristics and thus to misinterpretations or even misdirected cell development.

To showcase the significance, we examine the impact of lateral effects by the example of local and full illuminating *EQE* measurements both by experiment and simulation. Besides optical lateral non-uniformities which can be expected *e.g.* from non-uniform film thicknesses, we also investigate the impact of a further/more complex electrical *EQE* measurement artifact. This artifact, which is common in two-terminal multi-junction devices, is caused by low shunt resistance (R_{shunt}) or reverse breakdown characteristics [5–7], and depends on the bias voltage and spectral irradiance of the bias illumination.

With the recently released tandem functionality of the 3D solar cell simulation tool Quokka3, we examine how lateral defects such as localized shunts influence this *EQE* artifact.

2 LOCAL *EQE* EFFECTS

2.1 Experimental: Differences between locally and fully illuminated *EQE* measurements

EQE measurements are often performed using monochromatic beam sizes smaller than the actual cell area, mainly due to setup limitations. However, when only one fraction of cell area is chosen for *EQE* characterisation, possible non-uniform optical or electrical cell absorber properties are emphasized or ignored and can possibly lead to results not representative for the whole device.

It is common to calculate the short-circuit current density j_{sc} from measured *EQE*s. Especially for sub-cells of a two-terminal multi-junction device this is not straightforward. j_{sc} determination from *EQE* when measured with a monochromatic light beam smaller than the cell area is influenced by various factors. Obviously, the illumination of a solar cell in a small area can lead to different amounts of shading due to metallization compared to illuminating the whole cell. The same is true for the reference solar cell used for the calibration of the setup.

On the other hand, locally measured EQE neglects possible non-uniform optical or electrical cell absorber properties.

For the investigation we use the *EQE* equipment described in [8], which provides a monochromatic beam of up to more than 6" square in the test plane with low non-uniformity of the bias light (~9.0 %) and, more importantly, of the monochromatic beam (~1.0 %) [9].

As both metastable effects as well as systematic deviations due to local illumination of laterally inhomogeneous samples affect the EQE measurement result, we need to separate both effects. Here, we chose a stable two terminal multi-junction device as a stable test device to exclude meta-stable behaviour which typically is observed for the perovskite top cell.

The sample for investigation is a silicon-based wafer-bonded monolithic triple-junction solar cell

(GaInP/$Al_{0.0635}GaAs$//Si) from a batch with record efficiency (35.9 %) described elsewhere [10, 11].

For the investigation we use a poor sample of the batch (efficiency ~30%) where we observed visible non-uniformity in the top cell's PL and EL signals (Figure 1).

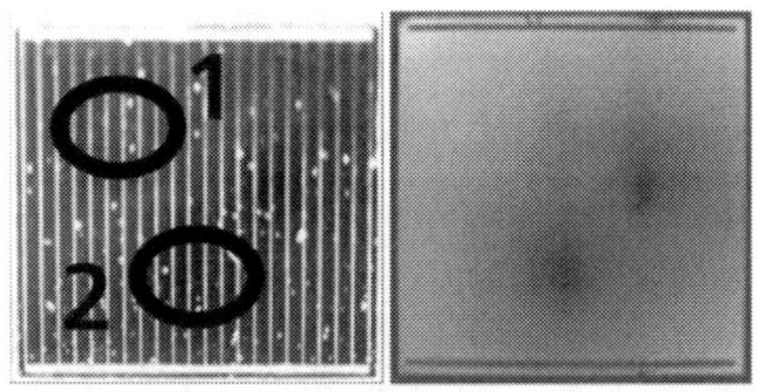

Figure 1: Left: Image of the visible top cell photoluminescence of the III-V//Si sample (2x2 cm²). Circles 1 and 2 indicate the positions of the monochromatic beam in the local-illuminated *EQE* experiments. Right: EL image of the top cell of the same sample.

The *EQE* of all three sub cells was measured with the above mentioned equipment by fully illuminating the whole device and is plotted in Figure 2. The spectral bias irradiance that was used for measuring the top cell is also shown in the graph as areal plots.

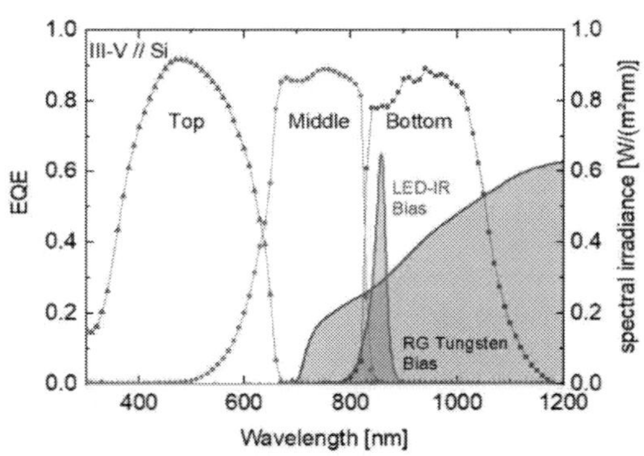

Figure 2: Measured *EQE* of the three sub cells of the investigated III-V//Si sample with full monochromatic illumination as dot line chart. The spectral irradiance of the LED and tungsten bias illumination for top cell limitation is shown as area plot referring to the right Y-axis.

To explore the impact of local *EQE* measurements we illuminated the sample with a monochromatic beam (∅ ~5.8 mm) in two positions as indicated in Figure 1. Position 1 is in a region with high PL signal and position 2 in a region with lower PL signal. The bias light however was illuminating the whole cell area uniformly for all measurements. For both positions *EQE* measurements of the top cell were performed for different bias voltages exploring the above-mentioned artifacts occurring in two terminal devices. The spectral irradiance of the bias illumination was thereby chosen so that the top cell limits the current of the cell (Figure 2).

The results are plotted in the top graph of Figure 3. It can be noted that for the same bias voltage, the absolute *EQE* values differ significantly in the spectral range of absorption of the top cell (~300…640 nm) for the positions 1 and 2.

For the bias voltages below 1.6 V the *EQE* is measured lower in the wavelength region of interest, whereas an unwanted signal is measured in the wavelength region with no actual response from the investigated sub cell (~640…830 nm), see Figure 3. Interestingly, the height of the artifact signal measured in the spectral range of the middle cell is very similar for positions 1 and 2 of the local *EQE* measurements of the top cell for same bias voltage values.

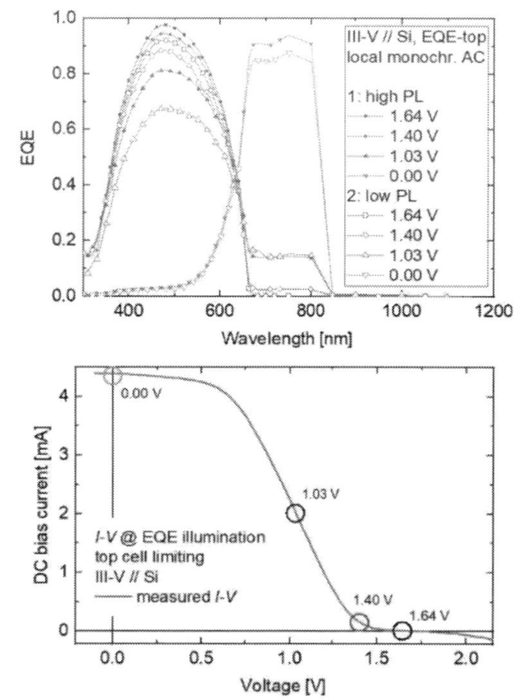

Figure 3: Top: *EQE* measurements with monochromatic light spot (∅ ~5.8 mm) on the 2x2 cm² III-V//Si-sample for different bias-voltages and top cell limiting bias-light at two locations with different PL signal (1 and 2, as in Figure 1). Bottom: *I-V*-sweep performed at *EQE* setup and top cell limiting bias-light. The circles indicate bias voltages used in the *EQE* measurements; their colour coding is used accordingly.

To gain more insight into the artifact behaviour, the lower graph in Figure 3 shows the results of an *I-V*-sweep performed at the *EQE* setup under the same bias illumination as used for the *EQE* measurement of the top cell (Figure 2, right Y-axis).

The different bias voltages used for the *EQE* artifact analysis are shown in the *I-V*-plot as coloured circles. When the bias light irradiance (Figure 2) is used to calculate the current generated in the sub cells, it becomes clear that the photo current of the quasi non-illuminated top cell can be found at ~1.64 V. An early reverse breakdown of the top cell is then observed, and a plateau corresponding to the middle cell current is present at around ~0.00 V. The *I-V*-characteristics can explain how the bias voltage leads to different operating conditions for the *EQE* measurement in dependence of the applied bias voltage: When measuring at 1.64 V, the top cell operates at a (low) current corresponding approximately to its photo current. The measured *EQEs* at this voltage show no artifact signal. For lower bias voltages the multi-junction device is operating in the reverse characteristics of the top cell. The steep slope leads to an artifact equivalent to a low R_{shunt} as described in [5]. Measurements at 0.00 V show the middle cell *EQE* characteristic which is expected as it is limiting the current at this region of the *I-V*-curve.

The results of the *EQE* measurement when the sample was fully illuminated with the monochromatic beam larger than the cell area are shown in Figure 2. When comparing the maximum *EQE* values of the top cell reached, the fully illuminated *EQE* is similar to the *EQE* values measured locally in the region with low PL signal.

Intuitively the top cell *EQE* measured with full area monochromatic illumination would be expected to show values between the two locally measured *EQE*. However, the influence of the edges on the *EQE* were not covered with the conducted experiment with local monochromatic illumination so that their impact for the 2x2 cm^2 cell could lead to the observed behaviour of similar *EQE* values for low PL region *EQE* and fully illuminated *EQE*. This underlines the importance of full-area illumination for the determination of accurate *EQE*s.

As a result, it can be stated that local *EQE* measurements can differ significantly depending on the location of the monochromatic light spot. This may lead to erroneous results, especially if *EQE* values are taken to calculate the sub cell j_{sc}s. If j_{sc} values need to be calculated from measured *EQE*s, the tandem device under test should be fully illuminated with monochromatic and spectrally tuned bias light to avoid lateral effects. This requires highly uniform monochromatic and bias light in the measurement plane and careful adjustment of the bias voltage conditions for accurate results.

As stated above, calculated j_{sc} values can be erroneous for different reasons. As a more reliable and proven way to derive absolute *EQE* values for two terminal devices, we advise to perform a spectrometric characterization at a calibrated solar simulator [12, 13].

2.2 Simulations: local R_{shunt} in *EQE* measurement

To analyse the effect of lateral shunts, we use the newly introduced tandem capability [14] of the solar cell simulation software Quokka3 to simulate PSC/Si tandem cells. Briefly, the lumped skin boundary condition to the 3D bulk drift-diffusion solver [15] supports to define *I-V* curves with current generation to represent the perovskite top cell. Two-diode model parameters are accepted for the parameterization of the *I-V* curve, see Figure 4. Thus, local R_{shunt} values can be directly set and varied.

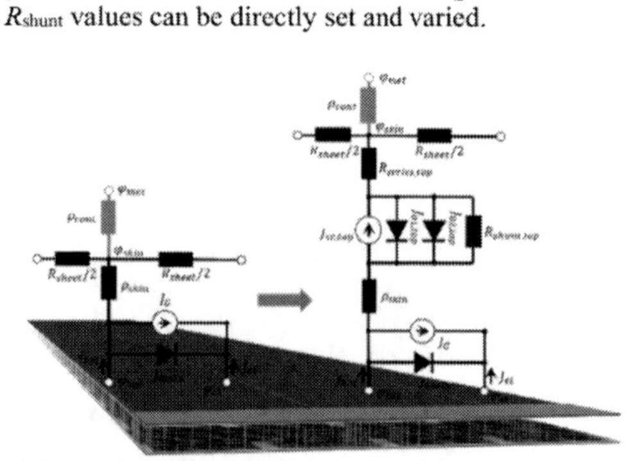

Figure 4: Sketch of Quokka3 tandem concept where the front skin boundary condition is extended to support a top cell *I-V* curve via a 2-diode model.

Optically, the top cell is spectrally defined by its *EQE* and the transmission to the Silicon bottom cell. Both characteristics need to be known from measurements or separate optical modelling. The quasi-analytical treatment of the top cell as a boundary condition results in a similar computational effort compared to single-junction Si cell simulations, which makes large-area 3D tandem cell simulations practically feasible with Quokka3. As several skins can be defined region-wise in Quokka3, it is possible to implement laterally inhomogeneous top cell characteristics, see Figure 5. Different skin regions are connected laterally via the bottom cell, as well as via a sheet resistance representing the front TCO conductive layer, thereby fully accounting for distributed lateral transport effects.

The impact of inhomogeneous cell properties on cell measurements can be understood via 3D modelling. We investigate the case of a significant shunt in the perovskite top cell. The *EQE* measurement procedure of an exemplary PSC/Si tandem cell is simulated in Quokka3: (i) full area bias illumination with a wavelength of 1000 nm is used to "flood" the Si bottom cell, (ii) a terminal voltage corresponding to the V_{oc} of the Si sub cell is applied, and (iii) full area monochromatic illumination of low irradiance and varying wavelength is applied. The resulting incremental current is used to calculate *EQE*.

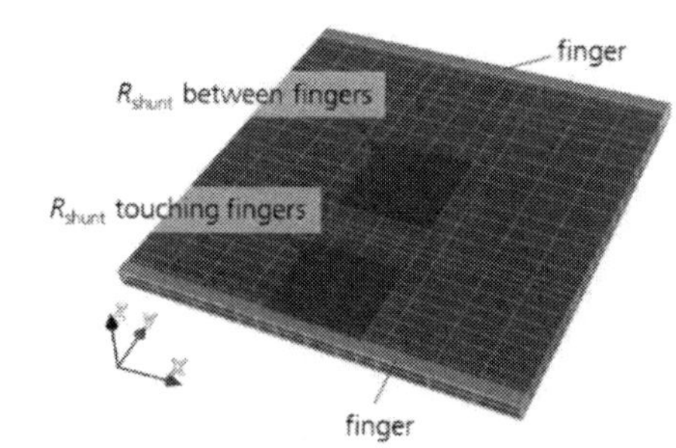

Figure 5: 3D simulation domain in Quokka3 tandem for the analysis of impact of localized shunts.

In an 1D simulation, which represents a uniform cell with homogeneously distributed shunt, we are able to reproduce the effects described by Meusel et al [5], see Figure 6. The main effect of such measurement artifacts is that the *EQE* of another sub cell is measured, while at the same time the signal of the limiting sub cell is too low. In this case the *EQE* of the underlying middle cell is observed.

It is however well known that shunts are rather of localized nature, *e.g.*, occurring at the cell edges or as local defects within the cell area. To represent a localized shunt, we create a 3D solution domain in Quokka3 and confine the shunt only to a small subregion, see Figure 5. We further vary this region by placing it either in the centre between the metal fingers, or touching one of the metal fingers. This is supposed to change lateral transport effects. The local and global R_{shunt} values are chosen so that in each case the shape of the *I-V*-curve under AM1.5g appears to show a similar shunt characteristic, which notably required to choose values somewhat lower than expected from area-averaging (20 Ωcm^2 for homogeneous case, ~1 Ωcm^2 for local cases).

The results in Figure 6 (bottom plot) reveal that the localized form of the shunt leads to a smaller *EQE* artifact compared to a homogenous shunt, although the impact on the *I-V* curves is identical. The strength of the effect also depends on the position of the shunt relative to the metallization. This means that the non-uniformity strongly impacts this *EQE* measurement artifact. Especially the ratio of reduction of measured EQE and the strength of the artifact signal is changed. This ratio is used in the classical correction procedure [6] based on a homogenous assumption. The procedure may therefore be not always correct, when strong influence from local R_{shunt} is present in the respective device under test.

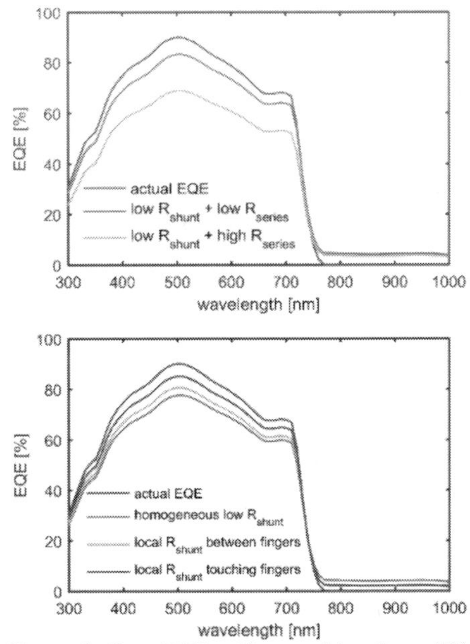

Figure 6: Top: *EQE* artifact resulting from different R_{shunt} and from a combination of R_{shunt} and R_{series}. Bottom: comparison of true *EQE* with 3D simulations for a homogeneous and two local R_{shunt} positions.

3 CONCLUSIONS

The influence on *EQE* measurement of a spatially non-uniform top cell in a two-terminal multi-junction device has been analysed by experiment and simulation. By choosing a III-V//Si-sample for the experiments, the typical metastable behaviour of PSC/Si tandem cells could be avoided, while the non-uniform top cell properties and their impact on *EQE*-characterization in a two-terminal device could be explored.

Two measurements with a monochromatic beam smaller than the cell area have revealed a significantly different *EQE* for the examined top cell. With the light spot in the region of a high PL signal, the *EQE* had higher values compared to the light spot in a region with lower PL signal. This outcome shows that locally measured *EQEs* as it is done by many laboratories can contribute to a non-realistic calculation of the sub cells' j_{sc}s. Instead *EQE* or I_{sc} should be measured using light with high uniformity fully illuminating the device under test (non-uniformity at CalLab PV Cells ~1.0 %, > 6").

During EQE measurement an artifact was observed detected in the spectral region of the middle cell. Apart from the anomalous setting of 0 V, a similar strength for both measurement locations for all applied bias voltages occurred. An *I-V*-sweep recorded under *EQE* bias light revealed an early breakdown of the limiting top cell. The observed artifact in the *EQE* measurement can therefore be explained with the corresponding reverse characteristic of the limiting top cell. In contrast to the artifact caused by a R_{shunt}, the artifact strength due to reverse breakdown can be reduced to zero when the bias voltage is set close to the photo-current of the limiting top cell (due to the chosen bias illumination here at tandem-V_{oc}). This emphasizes the relevance of bias voltage adjustments.

We further investigated electrical *EQE* measurement artifacts via Quokka3 simulations. For the case of a shunt, which is homogeneously distributed over the entire cell area, the *EQE* results show a behaviour as described with a 1D model from literature.

With 3D simulations the influence of a localized shunt, which is likely more realistic than homogenous shunts in experimental cells, artifact signals in the *EQE*-measurement have been quantified exemplarily. It is shown that both the locality and also the lateral position of the shunt defects have a significant influence on the artifact strength. Additionally, it is observed, that the artifacts caused by local R_{shunt} lead to a different ratio of reduced *EQE* and artifact signal strength as it is known from the artifact caused by homogeneous R_{shunt}. As a consequence, the typical correction procedure for *EQE* measurement artifacts based on the assumption of a uniform R_{shunt} might lead to deviations from the true *EQE* for real experimental tandem cells.

4 ACKNOWLEDGEMENTS

We want to thank the BMWK for financial support under contract number 03EE1087A (Katana). Part of the results have been achieved in the European Union's Horizon Europe funded project VIPERLAB (GA number 101006715).

We thank P. Schygulla for epitaxial growth, R. Koch and R. M. da Silva Freitas for solar cell processing and R. Müller for providing the III-V//Si sample.

5 REFERENCES

[1] CSEM SA & EPFL, EPFL and CSEM smash through the 30% efficiency barrier for perovskite-on-silicon-tandem solar cells—setting two certified world records: press release, 2022. Accessed: Sep. 21 2022.

[2] M. A. Green et al., "Solar cell efficiency tables (Version 60)," Progress in Photovoltaics: Research and Applications, vol. 30, no. 7, pp. 687–701, 2022, doi: 10.1002/pip.3595.

[3] Oxford PV: press release, 2021.

[4] F. Fu et al., "Monolithic Perovskite-silicon Tandem Solar cells: from the Lab to Fab?," Advanced materials (Deerfield Beach, Fla.), e2106540, 2022, doi: 10.1002/adma.202106540.

[5] M. Meusel, C. Baur, G. Létay, A. W. Bett, W. Warta, and E. Fernandez, "Spectral response measurements of monolithic GaInP/Ga(In)As/Ge triple-junction solar cells: Measurement artifacts and their explanation," Progress in Photovoltaics: Research and Applications, vol. 11, no. 8, pp. 499–514, 2003, doi: 10.1002/pip.514.

[6] G. Siefer, C. Baur, and A. W. Bett, "External quantum

efficiency measurements of Germanium bottom subcells: Measurement artifacts and correction procedures," in 35th IEEE Photovoltaic Specialists Conference (PVSC), 2010: 20-25 June 2010, Honolulu, Hawaii ; conference proceedings, Honolulu, HI, USA, 2010, pp. 704–707.

[7] M. A. Steiner et al., "Measuring IV Curves and Subcell Photocurrents in the Presence of Luminescent Coupling," IEEE J. Photovoltaics, vol. 3, no. 2, pp. 879–887, 2013, doi: 10.1109/JPHOTOV.2012.2228298.

[8] M. Mundus, "Ultrashort laser pulses for electrical characterization of solar cells," Dissertation, Fraunhofer-Institut für Solare Energiesysteme; Fraunhofer IRB-Verlag, 2016.

[9] M. Mühleis, Spectral Shaping for Accurate Solar Cell Characterization: accepted dissertation, 2022.

[10] R. Müller et al., "Silicon-based monolithic triple -junction solar cells with conversion efficiency > 34%," in Presented at the 37th European PV Solar Energy Conference and Exhibition, 2020, p. 11.

[11] P. Schygulla et al., "Two-terminal III–V//Si triple-junction solar cell with power conversion efficiency of 35.9 % at AM1.5g," Progress in Photovoltaics: Research and Applications, vol. 30, no. 8, pp. 869–879, 2022, doi: 10.1002/pip.3503.

[12] M. Meusel, R. Adelhelm, F. Dimroth, A. W. Bett and W. Warta, "Spectral mismatch correction and spectrometric characterization of monolithic III-V multi-junction solar cells," Progress in Photovoltaics, Vol.10 (2002), No.4, N-17060, 2002, doi: 10.1002/pip.407.

[13] A. J. Bett et al., "Understanding Sub-Cell Effects and their Influence on the Performance of Monolithic Perovskite/Silicon Tandem Solar Cells: submitted," Adv. Energy Mater., vol. 2022.

[14] A. Fell, O. Schultz-Wittmann, C. Messmer, M. C. Schubert, and S. W. Glunz, "Combining drift-diffusion and equivalent-circuit models for efficient 3D tandem solar cell simulations," IEEE Journal of Photovoltaics, vol. 2022.

[15] A. Fell, J. Schön, M. C. Schubert, and S. W. Glunz, "The concept of skins for silicon solar cell modeling," Solar Energy Materials and Solar Cells, vol. 173, pp. 128–133, 2017, doi: 10.1016/j.solmat.2017.05.012.

HARVESTING SOLAR ENERGY BELOW SI BANDGAP USING METAL-SILICON JUNCTION THROUGH HOT CARRIER EFFECT

Chung-Han Chang[1], Jia-Ci Jhou[1], Hsin-Ting Lin[2], Ching-Fuh Lin[1,3,4*]
1. Graduate Institute of Photonics and Optoelectronics, National Taiwan University, No. 1, Sec. 4, Roosevelt Road, Taipei 106319, Taiwan
2. Graduate Institute of Advance Technology, National Taiwan University, No. 1, Sec. 4, Roosevelt Road, Taipei 106319, Taiwan
3. Graduate Institute of Electronics Engineering, National Taiwan University, No. 1, Sec. 4, Roosevelt Road, Taipei 106319, Taiwan
4. Department of Electrical Engineering, National Taiwan University, No. 1, Sec. 4, Roosevelt Road, Taipei 106319, Taiwan
*lincf@ntu.edu.tw

ABSTRACT: Traditional silicon-based solar cells generally can only absorb the solar spectrum with wavelengths below 1.1 microns, and others cannot be absorbed but consumed in the form of heat. In this research, the metal/semiconductor junction is used to break through the limitation of silicon bandgap, and hot carrier effect is applied. The metal thin layer is used to absorb sunlight to generate hot carriers, quickly passing through the energy barrier to generate photocurrent, maximizing the use of solar energy. Compared with conventional silicon solar cells on the market, metal/silicon junction solar cells have better absorption and photovoltaic performance in the infrared regime, expanding the absorption range of silicon-based solar cells, and more effectively utilizing the entire solar spectrum.
Keywords: Metal-Semiconductor junction, hot carrier effect, Schottky energy barrier, mid-infrared light absorption, ultra-thin metal films

1 INTRODUCTION

The radiant energy of the entire AM 1.5 sunlight spectrum is mainly concentrated in the region of wavelength 0.15-4 μm, of which ultraviolet regime accounts for 5%, visible regime 52%, and infrared regime 43%.[1] According to the limitation, only photons with energy higher than the forbidden band width of the semiconductor material can be absorbed. This determines the minimum photon energy, that is, the maximum wavelength, At present, the absorption spectrum of commercially available Si solar cells are mostly between 300-1100 nm, accounting for 80.43% of the entire AM 1.5 solar spectrum radiant energy, while the remaining solar energy, that is, the near-infrared light to the mid-infrared light above 1100 nm, accounts for 19.57% and cannot be used effectively.

In order to effectively utilize infrared light with wavelengths above 1100 nm, several materials with an energy gap lower than 1.1 eV are applied to solar cells as absorption layers. In the past, Group III-V and Group II-VI compounds materials were mostly used, such as GaSb, InAs, CIS or InGaAsSb compounds, etc. Although these materials have mature epitaxial technology and are used in high-efficiency tandem cells, they have several disadvantages, such as the high material costs due to some rare sources on earth, and difficulties on epitaxial and manufacture process. These disadvantages make it unsuitable for broad use. Other materials that can absorb infrared light at wavelengths above 1100 nm are metals.[2] However, metal has no bandgap, causing additional difficulties. Here we report how we overcome such challenges.

2 EXPERIMENTS

First of all, silicon wafers are spliced into squares in size of 2.5 cm*2.5 cm, and then cleaned using standard procedure. After cleaning, silver in the thickness of 9.5 nm was deposited on the Si substrate as the absorption layer via thermal evaporation. It also forms metal/semiconductor junction structure.[3][4] Then silver fingers are evaporated on the metal layer as a transparent electrode. On the other side of the substrate, Al is deposited to form an ohmic contact with silicon as the back electrode. Metal-silicon junction solar cell is completed.

Current density and voltage curves were measured by solar simulator connected with Keithley 2400 under light source condition AM 1.5G, intensity 100 mW/cm^2. Besides, in order to investigate the photovoltaic performance of metal/silicon junction solar cell in the infrared regime, an optical filter is applied to the measurement system to block the wavelength below 1100 nm in the solar spectrum. The transmittance of the filter was measured via a spectrophotometer, and transmitted residue radiation intensity was checked by a power meter.

3 RESULTS AND DISCUSSION

Utilizing the broad wavelength spectrum absorption characteristic of metal film, the metal/silicon structure can absorb infrared regime with wavelength above 1100 nm. Utilizing the hot-carrier effect, when sunlight irradiates on the metal film, the electrons transit to high energy levels and then hot carriers are generated. Because metal has no energy band gap, these hot carriers will cool down in a few picoseconds if it is not exported in time. In order to quickly export hot carriers, the metal film should be thin enough to allow the hot carriers to pass through the metal layer before cooling, and then to cross the energy barrier caused by schottky contact between metal and silicon to generate photocurrent, as illustrated in Fig 1.

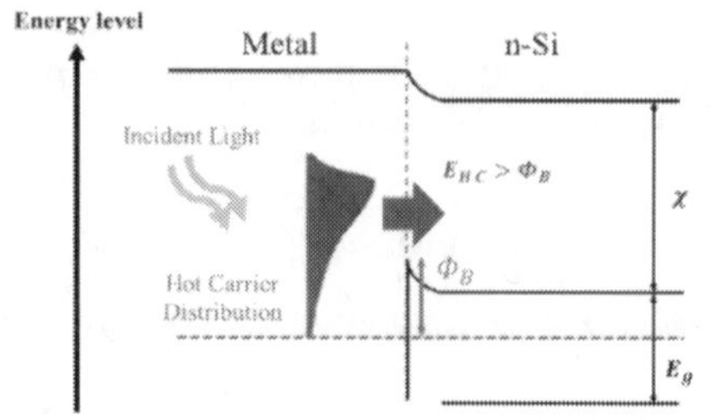

Figure 1: Hot carriers with energy(E_{HC}) higher than the barrier height(Φ_B) across the energy barrier to generate photocurrent

Finally, conversion efficiency of the metal-silicon junction solar cell to the infrared regime is measured. In order to ensure that the incident light received by the metal/silicon junction is above the wavelength of 1.1 um, we used a Si wafer as the filter to block the solar spectrum. Before the measurement, the transmittance of the filter was checked via spectrophotometer. The transmission spectrum of the filter confirms that the light below 1.1 micron is opaque, transmittance above 1.1 micron is 55 %, and the residual light intensity measured via power meter is 13.85 mW/cm^2. To confirm that the incident light which can be absorbed by silicon is effectively blocked by filter, we measured the current-voltage curve of a commercial available p-n junction silicon solar cell. The measurement results are shown by the black line in Fig 2(b), and the efficiency in the infrared regime above 1.1 μm is 0.254 %, very close to zero, which means that the filter has indeed filtered out the wavelength below 1.1 μm. If the filter is removed, the efficiency of this p-n junction silicon solar cell is 17.1 %, the measurement result is shown in Fig 2(a). The black line shows that its photon-to-electricity conversion performance is normal.

Then, the same measurement is performed on the metal-silicon junction solar cell. Under the incident light of AM 1.5 solar spectrum, cell efficiency is 2.992 %. The result is shown in the red line in Fig 2(a). The photovoltaic performance is worse than that of the p-n junction silicon solar cell. Next, filter was applied, under infrared regime, residue intensity is 13.85 mW/cm^2, and the measured cell efficiency is 3.296 %. The result is shown in the red line in Fig 2(b), which is significantly higher than usual for wavelengths above 1.1 microns, compared to the p-n junction silicon solar cell.

4 CONCLUSION

In this study, we achieved a silicon-based solar cell with higher infrared conversion performance. Metal/silicon junction solar cell allows an access to absorb infrared regime in a rather simple and low-cost process, In terms of the photon-to-electricity conversion efficiency, under AM 1.5 solar spectrum, cell efficiency is measured to be 2.992 %, for infrared regime above 1.1 μm. Cell efficiency is measured to be 3.296 %, which is several times higher than that of traditional p-n junction silicon solar cells. Moreover, the above result is the preliminary test. In the future, we would investigate detailed parameter of metal thin film to raise the ability of hot carrier extraction. Besides, anti-reflective design for metal-silicon structure is necessary to increase the light absorption capacity of the cells. Thus, higher infrared conversion efficiency is expected.

(a)

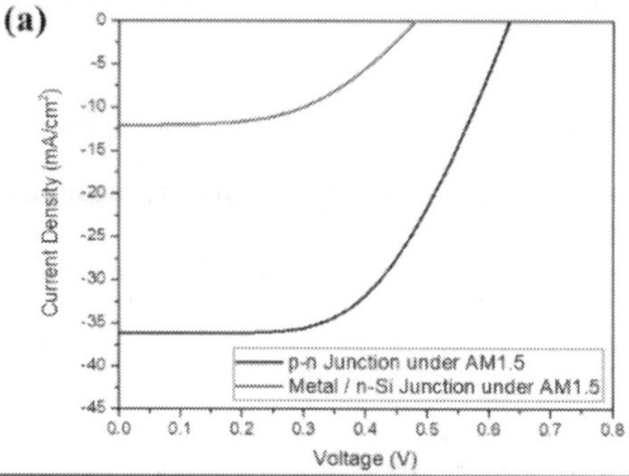

AM 1.5	Voc (V)	Jsc (mA/cm²)	Fill Factor (%)	Cell eff. (%)
p-n Junction	0.637	36.20	74.25	17.1
Metal / n-Si	0.480	12.1	51.49	2.992

(b)

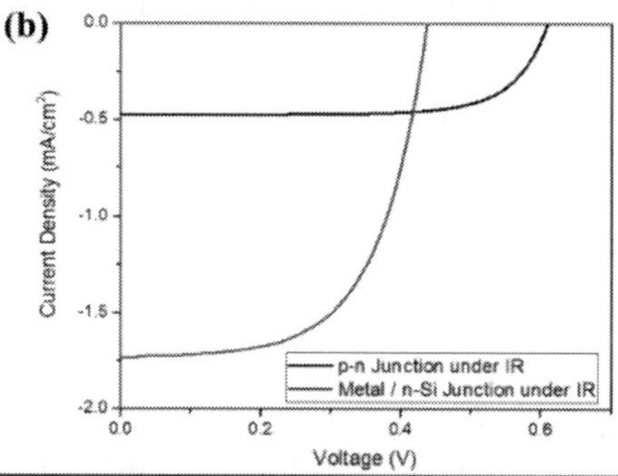

IR (>1.1 μm)	Voc (V)	Jsc (mA/cm²)	Fill Factor (%)	Cell eff. (%)
p-n Junction	0.615	0.47	60.27	0.254
Metal / n-Si	0.438	1.73	60.17	3.296

Figure 2: Photovoltaic characteristic curve and comparison of photovoltaic characteristics between metal-silicon junction and silicon-crystalline solar cells under (a) AM1.5 solar spectrum and (b) infrared regime (>1.1 μm)

5 ACKNOWLEDGMENT

The authors gratefully acknowledge Ministry of Science and Technology in Taiwan for supporting the research with the contract numbers: MOST 108-2221-E-002-032-MBK, MOST 108-2221-E-002-145-MY3, MOST 107-2221-E-002-172-MY3. Besides, we would like to thank Ms. Yuan-Tzu Lee of the Instrumentation Center at National Taiwan University for their assistance with the SEM operations.

6 REFERENCES

[1] https://en.wikipedia.org/wiki/Sunlight

[2] Hilsum, C. "Infrared absorption of thin metal films." JOSA 44.3 (1954): 188-191.

[3] St-Gelais, Raphael, et al. "Hot carrier-based near-field thermophotovoltaic energy conversion." ACS nano 11.3 (2017): 3001-3009.

[4] Anderson, W. A., A. E. Delahoy, and R. A. Milano. "Thin metal films as applied to Schottky solar cells:

optical studies." Applied optics 15.6 (1976): 1621-1625

FABRICATION OF SILICON HETEROJUNCTION SOLAR CELLS WITH LIGHT TRAPPING STRUCTURES SPECIALIZED FOR NEAR-INFRARED LIGHT BY NANOIMPRINTING

Yuto Kimata[1], Kazuhiro Gotoh[1], Satoru Miyamoto[1], Yasuyoshi Kurokawa[1], and Noritaka Usami[1]
[1] Department of Materials Process Engineering, Graduate School of Engineering, Nagoya University, Japan
Correspondence: Yasuyoshi Kurokawa
Furo-cho, Chikusa-ku, Nagoya, 464-8603 Japan
Tel: +81-52-789-3246, Fax: +81-52-789-4516,
E-mail: kurokawa.yasuyoshi@material.nagoya-u.ac.jp

ABSTRACT: Vehicle-integrated photovoltaics (VIPV) are gaining attention to realize a decarbonized society in the future, and the specifications for solar cells used in VIPV are predicated on a low cost, high efficiency, and the ability to be applied to curved surfaces. One way to meet these requirements is to make the silicon substrate thinner. However, thinner substrates result in lower near-infrared light absorption and lower efficiency. To increase light absorption, light trapping structures (LTSs) can be implemented. However, conventional alkali etched pyramid textures are not specialized for near-infrared light and are insufficient to improve near-infrared light absorption. Therefore, in this study, as an alternative to alkaline etching, we employed a nanoimprinting method that can easily fabricate submicron-sized LTSs on solar cells over a large area. In addition, as a master mold fabrication method, a method using silica particles as a mask was adopted. As a result, control of the LTSs structure was confirmed by silica particles. The sample with LTSs achieved a reduction of reflectivity in the near-infrared region (1000~1200 nm) and a theoretical maximum gain of J_{sc} of 1.65 mA/cm^2.

1 INTRODUCTION

In recent years, VIPVs have attracted much attention to achieve carbon neutrality [1]. One approach to achieve long distances to drive is an implementation of highly efficient tandem solar cells on VIPVs. Schygulla *et al.* reported a two-terminal wafer-bonded III–V//crystalline Si (c-Si) triple-junction solar cell with the efficiency of 35.9% [2]. To implement solar cells on curved surfaces like a roof and hood of vehicles, solar cells should be flexible and a silicon substrate must be thin. However, the challenge is that thin silicon substrates do not absorb near-infrared light well enough [3]. In general, alkaline etching [4] and photolithography [5,6] have been commonly used to fabricate optical confinement structures. However, the alkaline etching method is not sufficient to absorb near-infrared light and is difficult to fabricate on thin silicon substrates and in a large area. Photolithography also has the problem of a low throughput. There are some new approaches to enhance the near-infrared light absorption [7-10]. One of them is nanoimprinting technique, which has enable us to control the morphology of submicron light trapping structures (LTSs) and to fabricate the LTSs on a thin Si substrate in a large area [10-19]. Although conventional LTS fabricated by nanoimprinting technique has periodicity, a non-periodic LTS enhances light absorption in a wide wavelength such as a conventional random pyramid texture for c-Si solar cells or ZnO texture structure for Si-based or chalcopyrite thin film solar cells. In this study, to obtain a non-periodic LTS, the fabrication of master molds by colloidal lithography using silica particles [15] was combined with nanoimprinting. The colloidal lithography has the advantage of easily controlling the morphology of LTSs in a submicron range. The diameter and height of LTSs were controlled by varying the diameter of silica particles (D) and etching time (t), respectively. We implemented the LTSs on the backside of c-Si/indium tin oxide (ITO) substrates by nanoimprinting technique and evaluated the light confinement performance.

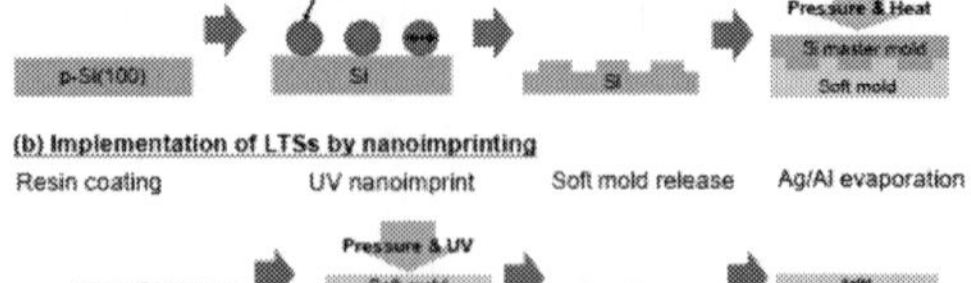

Fig. 1. Schematic diagram of (a) the fabrication process of a master mold of LTSs and (b) implementation process of LTSs on a back side of solar cells by nanoimprinting.

2 EXPERIMENTAL

2.1 Fabrication of a master mold

The fabrication process of a master mold of LTSs is shown in Fig. 1(a). A 2 × 2 cm^2 c-Si substrate was used. c-Si substrates were cleaned by the RCA process. Subsequently, they were immersed into 5% hydrofluoric (HF) acid for 1 minute to remove a native oxide. The substrates were immersed in ethylenediamine and irradiated with ultraviolet (UV) light for 3 hours. After that, ethylenediamine is removed with toluene, ethanol, and DI water. Silica particles terminated by carboxyl groups were stirred with DI water in a homogenizer for 30 minutes. The silica particles with the diameters of 500, 800, 1000, and 1500 nm were used. The solution was spin-coated on the c-Si substrates at 200 rpm for 20 seconds and dried at 6000 rpm for 5 seconds. It took 2 hours to make carboxyl groups on silica particles combined with amino groups on Si substrates adequately. The substrates were etched by reactive ion etching (RIE). During the RIE, the silica particles work as an etching mask. The flow rates of CF_4 and O_2 were 80 and 8 sccm, respectively. Plasma power was kept constant at 100 W. The substrate was immersed in 5% HF acid for 1 minute to remove remaining silica particles. The structure of fabricated molds were observed by scanning electron microscopy (SEM; JEOL JSM-7001F).

2.2 Implementation of LTSs by nanoimprinting

The implementation process of LTSs on a back side of c-Si/ITO substrates were shown in Fig. 1(b). A master mold is transferred to a soft mold by thermal nanoimprinting. The pressure, temperature, and duration were 6 bar, 130 °C, and 1 minute, respectively. An UV curable resin (mr-NIL 210, microresist technology GmbH) were applied by spin coating. Rotation speeds at the first and second steps were 500 rpm for 30 seconds and 3000 rpm for 25 seconds. The soft mold pattern was transferred to the resin by UV nanoimprinting. The pressure and duration were 2 bar and 1 minute, respectively. Silver and aluminum electrodes were deposited on the LTSs. For optical evaluation, light was incident from the front side of the c-Si/ITO/LTSs/electrode structure. Reflectance was measured by a spectrophotometer (Jasco V-570). The short-circuit current density (J_{sc}) was calculated from the reflectance spectra and AM1.5G spectra.

3 RESULT AND DISCUSSION

3.1. Structure of a Si master mold

Figure 2 shows the height of LTSs on the Si master molds fabricated by silica particles with different diameters. t was varied from 3 to 25 minutes. Regardless of D, the height of LTSs has a maximum with respect to t. As D increases, the etching time when the height of LTSs becomes maximum (t_{max}) increases. For D=500 nm, the maximum height was about 200 nm at t=5 minutes. On the other hand, at D=1500 nm, the maximum height is about 900 nm around t=25 minutes. Figure 3 shows SEM images of LTSs at D=800 nm and (a) t=10 minutes and (b) 15 minutes. It can be seen that the structure changes from a dome-shape to a cone-shape at 15 minutes, which corresponds to the t_{max}. Around $t=t_{max}$, silica particles started to disappear on the LTSs. By the absence of etching masks, the LTSs themselves were etched and the shape was changed and the height of the LTSs was decreased. Therefore, t_{max} depends on the etching selectivity between Si and SiO_2. Etching selectivity is decided by the flow rate of CF_4 and O_2 [20]. If etching selectivity was decreased, silica particles would disappear at shorter t. and t_{max} would be decreased and the height of LTSs would be also decreased.

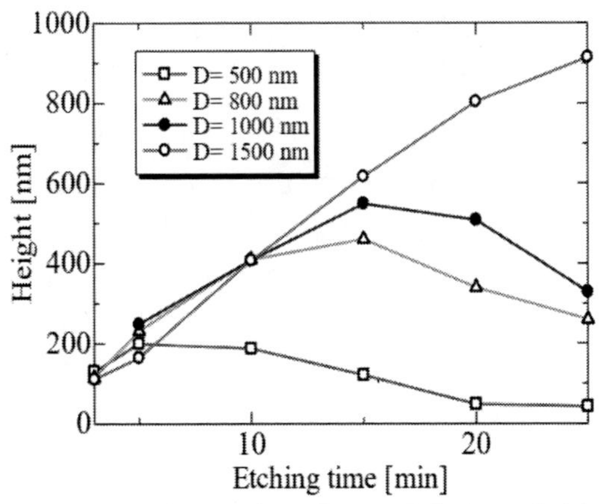

Fig. 2. Etching time dependence of the LTS height fabricated by silica particles with a different diameter.

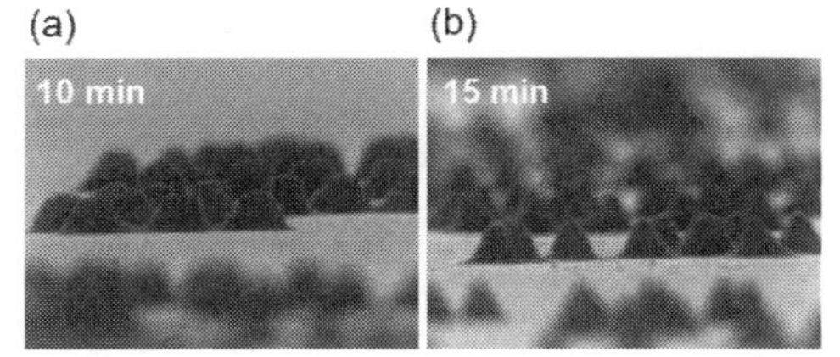

Fig. 3. SEM images of LTSs on a master mold fabricated by (a) D=800 nm, t=10 minutes and (b) D=800 nm, t=15 minutes.

3.2 Characteristics of LTSs

Figure 4(a) shows the reflectance spectra of LTSs fabricated by silica particles with different diameters at t=3 minutes. In the wavelength range from 1000 to 1200 nm, the reflectance is significantly influenced by the reflection on the back side, since the Ag/Al electrode was deposited on the back side. At t=3 minutes, the height of LTSs was about 100 nm for all the diameters. From Fig. 4(a), the LTSs reduced the reflectance in all the wavelengths compared to the flat sample. The reflectance became the lowest at D=500 and 800 nm. Fig. 4(b) shows the reflectance at the wavelength (λ) of 1100 nm. The lowest reflectance was obtained at D=800 nm. This suggests that there is an optimum aspect ratio of LTSs to reduce the reflectance.

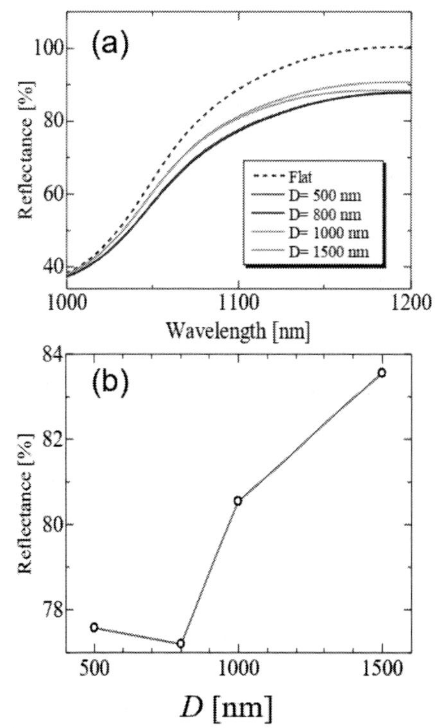

Fig. 4. (a) Reflectance spectra and (b) reflectance at λ=1100 nm of LTSs fabricated by silica particles with different D at t=3 minutes.

Figure 5 shows the reflectance spectra of LTSs fabricated by silica particles with D=1000 nm. The t was varied from 3 to 25 minutes. At all etching times, the reflectance was lower than that of the flat sample. The reflectance was the lowest at t=20 minutes. From a SEM image, the morphology was similar with Fig. 3(b). Therefore, not only the height but also the cone-shape enhanced the light scattering effectively. Figure 6 shows the projected J_{sc} (pJ_{sc}) calculated by the reflectance spectra shown in Fig. 5. The pJ_{sc} was calculated by the following equation

$$pJ_{sc} = \int_{1000\,\mathrm{nm}}^{1200\,\mathrm{nm}} qAN_{\mathrm{photon}}\, d\lambda \quad (1)$$

where q is the elementary charge, λ is the wavelength of incident light, A is the absorptance, and N_{photon} is the number of photons in AM1.5G spectrum. The pJ_{sc} was improved compared to the flat sample at all D and t. When t was short, pJ_{sc} was higher at smaller diameter. On the other hand, when t was long, pJ_{sc} was higher at larger diameter. This suggests pJ_{sc} was increased when the height of LTSs was lower regardless of the diameter. Therefore, lower aspect ratio leads to higher pJ_{sc}. A maximum gain of 1.65 mA/cm^2 was observed at D=800 nm and t=5 minutes, suggesting that D=800 nm may be the optimal width for the near-infrared light in the range from 1000 to 1200 nm.

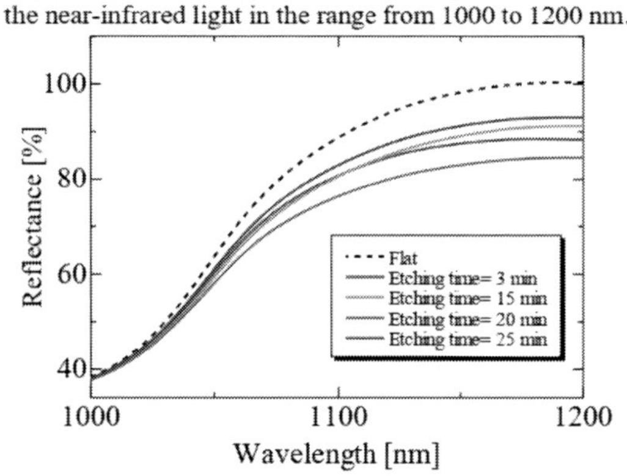

Fig. 5. Reflectance spectra of LTSs fabricated by silica particles with D=1000 nm.

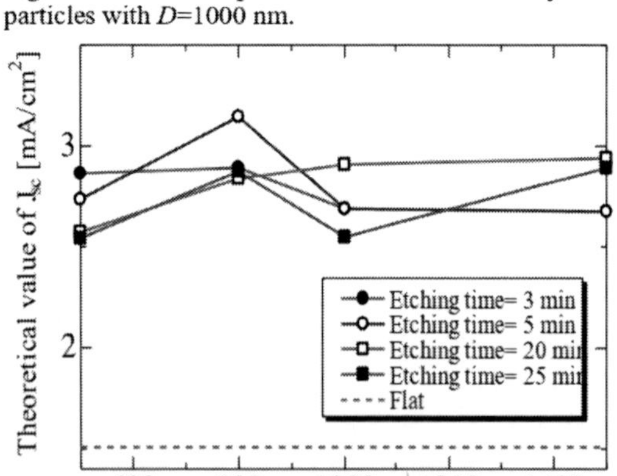

Fig. 6. pJ_{sc} calculated by reflectance spectra shown in Fig. 5 and Eq. (1).

4 CONCLUSIONS

Si master molds were fabricated by colloidal lithography using silica particles. By controlling t and D, the LTSs could be controlled. Regardless of D, the height of LTSs has a maximum with respect to t. As D increases, the etching time when the height of LTSs becomes maximum (t_{max}) increases. The SEM image revealed that the LTSs were changed from a dome-shape to a cone-shape around t=t_{max}. Around t=t_{max}, silica particles started to disappear on the LTSs. By the absence of etching masks, the LTSs themselves were etched and the shape was changed and the height of the LTSs was decreased. The reflectance of c-Si/ITO/LTSs/electrode structure was evaluated by spectrophotometer. Since light was incident from the front side of the c-Si/ITO/LTSs/electrode structure, in the wavelength range from 1000 to 1200 nm, the reflectance is significantly influenced by the reflection on the back side. The reflectance measurement showed that the LTSs reduced the reflectance in all the wavelengths compared to the flat sample. This indicates that the fabricated LTSs exhibits optical confinement performance. The pJ_{sc} was calculated from the measured reflectance. Lower aspect ratio leads to higher pJ_{sc}. A maximum gain of 1.65 mA/cm^2 was observed at D=800 nm and t=5 minutes, suggesting that D=800 nm may be the optimal width for the near-infrared light in the range from 1000 to 1200 nm.

5 ACKNOWLEDGEMENT

This work was supported in part by the New Energy and Industrial Technology Development Organization (NEDO) under the Ministry of Economy Trade and Industry of Japan.

6 REFERENCES

[1] M. Yamaguchi, K. Araki, K.-H. Lee, N. Kojima, T. Masuda, K. Kimura, A. Satou, and H. Yamada, Proc. the 5th IEEE Int. Conf. on Smart Energy Grid Engineering, 2017, pp. 394.

[2] P. Schygulla, R. Müller, D. Lackner, O. Höhn, H. Hauser, B. Bläsi, F. Predan, J. Benick, M. Hermle, S. W. Glunz, and F. Dimroth, Progress in Photovoltaics: Research and Applications, 30 (2022) 869.

[3] H. Sai, T. Oku, Y. Sato, M. Tanabe, T. Matsui, and K. Matsubara, Progress in Photovoltaics, 27 (2019) 1061.

[4] W. L. Bailey, M. G. Coleman, C. B. Harris, and I. A. Lesk, US Patent 4137123 (1979).

[5] A. Wang, J. Zhao, and M. A. Green, Applied Physics Letter, 57 (1990) 602.

[6] J. Zhao, A. Wang, X. Dai, M. A. Green, and S. R. Wenham, Proc. the 22nd IEEE Photovoltaic Specialists Conference, 1991, pp. 399.

[7] Y. Kurokawa, O. Aonuma, T. Tayagaki, I. Takahashi, and N. Usami, Japanese Journal of Applied Physics, 56 (2017) 08MA02.

[8] A. Hombe, Y. Kurokawa, K. Gotoh, S. Akagi, Y. Yamamoto, D. Yurasov, A. Novikov, and N. Usami, Japanese Journal of Applied Physics, 57 (2018) 08RF09.

[9] M. Sei, Y. Kurokawa, S. Kato, and N. Usami, Japanese Journal of Applied Physics, 57 (2018) 08RB21.

[10] Y. Ota, A. Hombe, R. Nezasa, D. Yurasov, A. Novikov, M. Shaleev, N. Baidakova, E. Morozova, Y. Kurokawa, and N. Usami, Japanese Journal of Applied Physics, 57 (2018) 08RB04.

[11] H. Hauser, A. Mellor, A. Guttowski, C. Wellens, J. Benick, C. Müller, M. Hermle, and B. Bläsi, Energy Procedia, 27 (2012) 337.

[12] C. Trompoukis, O. E. Daif, V. Depauw, I. Gordon, and J. Poortmans, Applied Physics Letter, 101 (2012) 103901.

[13] A. Mellor, H. Hauser, C. Wellens, J. Benick, J. Eisenlohr, M. Peters, A. Guttowski, I. Tobías, A. Martí, A. Luque, and B. Bläsi, Optics Express 21 (2013) A295.
[14] E.-C. Wang, S. Mokkapati, T. P. White, T. Soderstrom, S. Varlamov, and K. R. Catchpole, Progress in Photovoltaics: Research and Applications 22 (2014) 587.
[15] N. Tucher, J. Eisenlohr, H. Hauser, J. Benick, M. Graf, C. Müller, M. Hermle, J. C. Goldschmidt, and B. Bläsi, Energy Procedia, 77 (2015) 253.
[16] A. Peter Amalathas and M. M. Alkaisi, Materials Science in Semiconductor Processing 57 (2017) 54.
[17] D. Liu, Q. Wang, and Q. Wang, Applied Surface Science 439 (2018) 168.
[18] H. Hauser, K. Mühlbach, O. Höhn, R. Müller, S. Seitz, J. Rühe, S. W. Glunz, and B. Bläsi, Optics Express, 28 (2020) 10909.
[19] S. Kato, Y. Watanabe, Y. Kurokawa, A. Yamada, Y. Ohta, Y. Niwa, and M. Hirota, Japanese Journal of Applied Physics, 51 (2012) 02BP09.
[20] See-Eun Cheon, Hyeon-seung Lee, Jihye Choi, Ah Reum Jeong, Taek Sung Lee, Doo Seok Jeong, Kyeong-Seok Lee, Wook-Seong Lee, Won Mok Kim, Heon Lee and Inho Kim. Scientific reports 7 (2017) 7336.

This is a short version of the paper submitted to the journal EPJ Photovoltaics.

It has been selected by the Committee of the WCPEC-8 for submission to EPJ Photovoltaics.

OXIDE FERROELECTRIC PEROVSKITE NANOPARTICLES DISPERSED IN BIOPOLYMER-MATRIX FOR HIGH-PERFORMANCE SOLAR CELL (FERRO-OPV) FOR INDOOR APPLICATIONS

Rémi Ndioukane[1*], Fanta Baldé[1], Diouma Kobor[1], Jeanne solard[2] and Laurence Motte[3]
[1]Laboratoire de Chimie et de Physique des Matériaux (LCPM), University Assane Seck of Ziguinchor, Senegal
[2]Centrale de Proximité en Nanotechnologies de Paris Nord, Institut Universitaire de Technologie de Villetaneuse, France
[3]Laboratory for Vascular Translational Science (LVTS), Université Paris 13, France
Address: Assane Seck University of Ziguinchor, PoB 523 Ziguinchor, Senegal, Email: remindioukane@gmail.com

ABSTRACT: The present work proposes another way to obtain solar cells based on PZN-PT ferroelectric oxides with very high efficiency beyond that of halogenated perovskites and requiring only ferroelectric nanoparticles dispersed in biopolymer matrix thus allowing the reduction of their environmental impact. In our recent studies, we successfully fabricate thin films on ITO substrate by dispersing these nanoparticles in a biopolymer. We demonstrated that they conserve the ferroelectric properties of the PZN-PT bulk crystal. The idea to use these ferroelectric materials (PZN-PT) came from the fact that the ferroelectric nature of these layers could facilitate electric charges accumulation on the interfaces of the solar cell. Thus, it would increase the open circuit voltage Voc which could reach more than 10 V. This would directly impact the efficiency which is proportional to Voc, thus hoping to obtain solar efficiency never equaled by the halide perovskites which are less stable and less resistant in aggressive environments. The cell produced gave an exceptional record efficiency with an open circuit voltage (Voc) of 3.50 V, a short-circuit current density (Jsc) of 0.118 mA/cm^2 and an FF of 0.72 measured in the positive polarization direction under 3825 lux (5.6 W/m^2) lightning.
Keywords: Perovskite, Nanoparticles, Thin Film, Ferrophotovoltaic

1 INTRODUCTION

From conservation efforts and cleantech to tracking of environmental conditions and reductions in energy usage, every imaginable facet in the quest to reduce our carbon footprint is being explored through the Internet of Things (IoT). Thus, for indoor applications, the power supply of billions of independent electronic devices and equipment, with the advent of the IoT, is a huge energy requirement that risks compromising reduction decisions of the impact of energy consumption on the climate [1-5]. One of the solutions would be the use of photovoltaic cells capable of efficiently converting low intensity light in the indoor environment into megawatt to microwatt [6-10] class electrical power. These types of materials would be excellent candidates for powering low power equipment and appliances. In general, conventional solar cells use the standard solar spectrum with the best yields unlike these new cells which operate optimally with emission spectra from light sources at low intensities often less than 1 mWcm^{-2} [11]. Indeed, crystalline silicon cells exhibit low power conversion efficiencies (PCEs) under indoor lighting [12, 13]. On the contrary, although some of the emerging photovoltaic cells, such as organic photovoltaic (OPV) cells and dye solar cells, are not as efficient as crystalline silicon cells under standard solar lighting, the absorption properties of the highly adjustable light of their materials make them promising candidates for interior applications [14-18]. Compared with dye solar cells, OPV cells have better business prospects due to the advantages of low cost, large area module through solution printing and coating techniques. In addition, they could be used as the rear face of bifacial solar cell to exploit more efficiently the low and diffuse irradiance of the sun permitting to improve its energy conversion.

Here, we present ferroelectric perovskite nanoparticles dispersed in a biopolymer thin film on ITO glass for highly efficient indoor energy harvesting. First, we successfully fabricated and demonstrated large improvement of photovoltaic performances due to the presence of the PeryDA+EB layer. For this purpose, we realized Ag/np+CC/TiO2/ITO, Ag/EB/np+CC/TiO2/ITO, Ag/PeryDA/np+CC/TiO2/ITO and Ag/PeryDA+EB/np+CC/TiO2/ITO solar cells via spin coating and droplet procedures. Measurements were achieved under low lightning levels and at high lightning for comparison. PCEs obtained are respectively 0.14 %, 0.010 %, 9 % and 39.32 % at 4781 lux (7 W/m^2) illuminance. This is a record of power conversion of solar cells for indoor application with an open voltage (Voc) of 3.50 V, a short-circuit current density (Jsc) of 0.118 mA/cm^2 and an FF of 0.72.

2 EXPERIMENTAL PROCEDURE

2.1 Solar cells fabrication

To fabricate the solar cell, the ITO used has a square resistance of 9 - 15 Ω/□, a thickness of 1200 to 1600 Å. A 30 nm TiO_2 charge transport layer was deposited on pre-cleaned ITO conductive glass. The mesoporous TiO_2 was deposited on the substrate by spin coating at a speed of 3500 rpm for 10 min with an acceleration ramp of 4 s, from a paste of titanium oxide powder diluted in acetic acid, the weight ratio of TiO_2 and acetic acid was 6:1, then the substrates were sintered at 450 °C for 30 min [19].

The PZN-PT single crystals were ground in a silicon carbide mortar to obtain perovskite nanopowders [20]. The biopolymer (CC) solution, which is the organic matrix in which the nanoparticles are dispersed, is obtained by organic extraction procedure. The nanoparticles are then dispersed in this CC solution and placed under magnetic stirring for 30 min to one hour. The result is a homogeneous solution containing ferroelectric oxide nanoparticles. The film of perovskite nanoparticles was deposited by spin-coating on the TiO_2 layer. The spin coating deposition procedure was carried out at room temperature with a rotational speed of 3500 tr/ min for 5 min with an acceleration of 4s. Subsequently, the sample was placed in an oven for 30 min at 105 °C, then it was annealed at 450 °C for 30 min.

After cooling to room temperature, a charge carrier material Perylene Diimide Aniline (PeryDA) was deposited on the the perovskite layer. The droplet deposition procedure was carried out in an ambient room. The PeryDA solution was synthesized by a chemical reaction between perylene and aniline in the presence of propanoic acid at 80 °C for 20 h thus giving a reddish powder. This powder will be dissolved in distilled water or ethanol to deposit a few drops on the perovskite layer. After drying at room temperature, a natural biomolecule of (EB) was deposited on the PeryDA layer. The droplet deposition procedure was carried out in ambient conditions. The EB solution was extracted directly from a EB plant by making an incision in the bark to extract the white latex.

2.2 Electrical characterization

Finally, a layer of a few nm of silver paste was deposited by screen printing using a mask on a non-conductive glass and annealed at 100 °C was produced to act as an electrode for electrical measurement. The current-voltage characteristics are measured using Keithley model 2612B source meter driven by a computer under TSP Express program. The light source was a 100 W xenon lamp. The incident power is measured using a luxmeter and desired value can be obtained by varying the distance between the sample and the incident source.

3 RESULTS AND DISCUSSION

The architecture of the device used in this study is given in fig.1c and Fig. 1d. The as fabricated cell gave an exceptional record efficiency of 39.32 % with an open voltage circuit (Voc) of 3.50 V, a short-circuit current density (Jsc) of 0.118 mA/cm^2 and an FF of 0.72 with an incident power of 3825 lux (5.6 W/m^2) measured in positive direction (Fig. 2a). The power - voltage dependence curve of this cell is illustrated in Fig. 2b. A maximum power of 220 μW/cm^2 is obtained. This is explained by the low short-circuit current density. However, this value is large compared to the majority of generated power from indoor solar cells in literature which record is 301.6 μW/cm^2 [21].

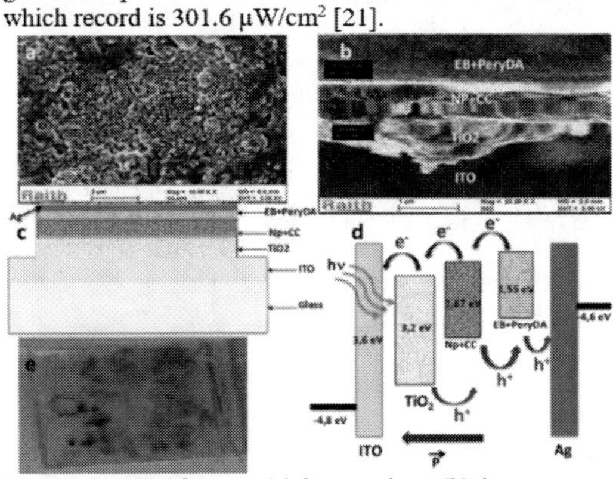

Figure 1: SEM images (a) in top view, (b) in transverse view, (c) illustration in transverse mode of the different layers deposited on ITO substrate, (d) diagram of the energy bands of the different layers and (e) photo of a as fabricated Ferro-OPV solar cell

For the J - V measurement in negative direction with a power of 4781 lux (7 W/m^2), we obtained a current density of 2 mA/cm^2, an open circuit voltage of 2.30 V and a fill factor of 0.35. The behavior of this J - V curve is typical of ferrophotovoltaic materials which have a characteristic of being either in the quadrant (V > 0 and J < 0) for the direction opposite to the direction of polarization, or in quadrant (V < 0 and J > 0) in the same direction. This is what we observe in Fig. 2c and Fig. 2d. We notice that the current density is higher than that obtained in the first case (2 mA/cm^2 instead of 0.118 mA/cm^2) with a small open circuit voltage Voc of 2.30 V instead 3.5 V for the first one. The interesting result is the improving of the power delivered by the device from 220 μW/cm^2 to a record of 1,700 μW/cm^2 for indoor application solar cells knowing the record is 301.6 μW/cm^2 as published in [21] in 2021. These high values of Voc and generated power would open large potential application for many electronic devices empowering.

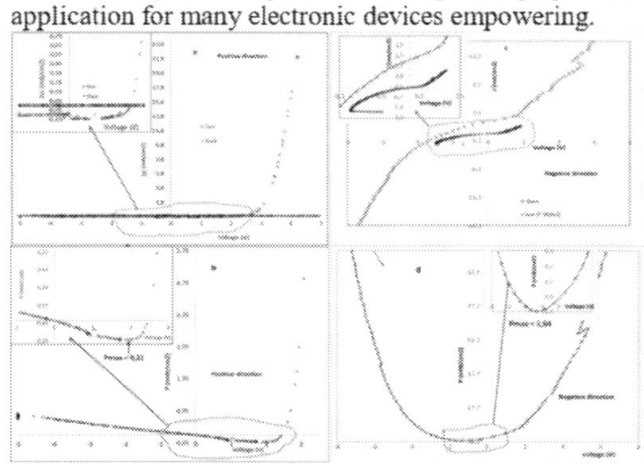

Figure 2: Characteristics (a) current - voltage (J-V) and (b) power - voltage (P–V) of the Ferro-OPV solar cell under dark and under an incident lightning of 5.6 W/m^2 in the direction of polarization of the ferroelectric layer (+ on ITO), (c) current - voltage (J-V) and (d) power - voltage (P–V) of the Ferro-OPV solar cell under dark and under lightning of 7 W/m^2 in the opposite direction to the direction of the polarization of the ferroelectric layer (+ on Ag)

For investigation on the effects of different components and layers of the solar cells we realized I-V characteristics of different solar cells composed by ITO/TiO2/np-CC (Fig.3, blue star curve), ITO/TiO2/np-CC/EB Fig. 3, black x curve), ITO/TiO2/np-CC/PeryDA (Fig. 3, purple triangle curve) and ITO/TiO2/np-CC/EB-PeryDA (Fig. 3, red +) under 4781 lux incident light.

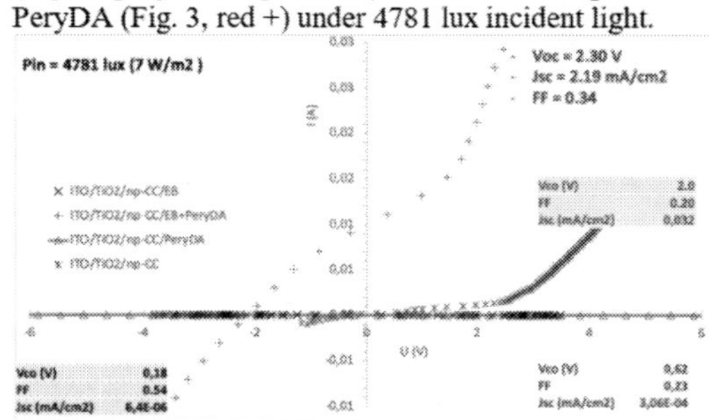

Figure 3: I-V characteristics of different solar cells composed by (blue star) ITO/TiO2/np-CC, (black x) ITO/TiO2/np-CC/EB, (purple triangle) ITO/TiO2/np-CC/PeryDA and (red +) ITO/TiO2/np-CC/EB-PeryDA

From this figure, the values of Voc, Jsc and FF were 2V, 0.032 mA/cm^2 and 0.20 respectively for the PZN-4.5PT nanoparticles in CC solar cell. The value of Voc decreased from 2 to 0.19 V and the short current density from 0.032 to 0.001 mA/cm^2, showing that the high value of Voc is coming from the ferroelectric nanoparticles presence in the device. In the other hand the fall down of the Jsc value revealed that the EB is like more insulating material than conductive one and can inhibit the ferroelectric effect. Adding a PeryDA on the nanoparticles in CC layer, the Voc increased from 0.18 to 0.60 V while the Jsc remains very low. This means that the excellent values from the cell are not from the PeryDA only also. The combination of these different components in the detailed architecture permitted to reach very high values for Voc of 2.30 V, for Jsc of 2.19 mA/cm2 and a FF of 0.34. These results indicates that the presence of any of these different materials layers is fundamental to get such important electrical characteristics.

4 CONCLUSION

We, successfully, fabricated a ferroelectric PZN-4.5PT nanoparticles thin film using a biopolymer as matrix. More than 4 cm^2 hybrid ferro-OPV solar cell, using low cost technique with very high efficiency (η = 39.2 %) and a record Voc of 3.5 V for indoor application, was performed. A record of 1.64 mW/cm^2 generated power was reached under 4781 lux (7 W/m^2) boosting the possible application of this type of solar cell for low power devices energy autonomy.

References

[1] L. Atzori, A. Iera & G. Morabito, The internet of things: a survey, Comput. Netw. 54, 2787–2805 (2010).

[2] J. Gubbi, R. Buyya, S. Marusic & M. Palaniswami, Internet of things (IoT): a vision, architectural elements, and future directions, Future Gener. Comp. Syst. 29, 1645–1660 (2013).

[3] A. Al-Fuqaha, M. Guizani, M. Mohammadi, M. Aledhari & M. Ayyash, Internet of things: a survey on enabling technologies, protocols, and applications, IEEE Commun. Surv. Tut. 17, 2347–2376 (2015).

[4] J. A. Khan, H. K. Qureshi & A. Iqbal, Energy management in wireless sensor networks: a survey, Comput. Electr. Eng. 41, 159–176 (2015).

[5] H. Yin et al. Designing a ternary photovoltaic cell for indoor light harvesting with a power conversion efficiency exceeding 20%, J. Mater. Chem. A 6, 8579–8585 (2018).

[6] H. K. H. Lee, Z. Li, J. R. Durrant, & W. C. Tsoi, Is organic photovoltaics promising for indoor applications? Appl. Phys. Lett. 108, 253301 (2016).

[7] B. Minnaert & P. Veelaert, A proposal for typical artificial light sources for the characterization of indoor photovoltaic applications. Energies 7, 1500–1516 (2014).

[8] M. Freitag et al., Dye-sensitized solar cells for efficient power generation under ambient lighting, Nat. Photon. 11, 372–378 (2017).

[9] A. S. Teran et al. Energy harvesting for GaAs photovoltaics under low-flux indoor lighting conditions, IEEE Trans. Electron. Dev. 63, 2820–2825 (2016).

[10] B. Li, B. Hou and G. A. J. Amaratunga, Indoor photovoltaics, The Next Big Trend in solution-processed solar cells, InfoMat., 3, 445–459. (2021).

[11] S. Mori et al., Investigation of the organic solar cell characteristics for indoor LED light applications, Jpn J. Appl. Phys. 54, 071602 (2015).

[12] F. De Rossi, T. Pontecorvo & T. M. Brown, Characterization of photovoltaic devices for indoor light harvesting and customization of flexible dye solar cells to deliver superior efficiency under artificial lighting, Appl. Energy 156, 413–422 (2015).

[13] C. L. Cutting, M. Bag & D. Venkataraman, Indoor light recycling: a new home for organic photovoltaics, J. Mater. Chem. C 4, 10367–10370 (2016).

[14] Y. Cao, Y. Liu, S. M. Zakeeruddin, A. Hagfeldt & M. Grätzel, Direct contact of selective charge extraction layers enables high-efficiency molecular photovoltaics, Joule 2, 1108–1117 (2018).

[15] M. Li et al. Interface modification by ionic liquid: a promising candidate for indoor light harvesting and stability improvement of planar perovskite solar cells, Adv. Energy Mater. 8, 1801509 (2018).

[16] H. K. H. Lee et al. Organic photovoltaic cells promising indoor light harvesters for self-sustainable electronics, J. Mater. Chem. A 6, 5618–5626 (2018).

[17] X. Liu, S. Huettner, Z. Rong, Sommer, M. & Friend, R. H. Solvent additive control of morphology and crystallization in semiconducting polymer blends, Adv. Mater. 24, 669–674 (2012).

[18] Y. Z. Lin & X. W. Zhan, Non-fullerene acceptors for organic photovoltaics: an emerging horizon, Mater. Horiz. 1, 470–488 (2014).

[19] R. Ndioukane et al., Optical properties of PZN-PT nanoparticles thin layer on ITO glass for photovoltaic application, OAJ Materials and Devices, 4, 1502 (2019).

[20] R. Ndioukane et al, Ferromagnetism evidence and size dependence in ferroelectric PZN-4.5PT nanoparticles, EPL, 125, 47004 (2019).

[21] X. He et al, 40.1% Record Low-Light Solar-Cell Efficiency by Holistic Trap-Passivation using Micrometer-Thick Perovskite Film, Adv. Mater., 2100770 (2021).

This is a short version of the paper submitted to the journal Progress in Photovoltaics.

It has been selected by the Committee of the WCPEC-8 for submission to Progress in Photovoltaics.

INTERFACIAL LIMITATIONS OF ALD-GROWN TIN OXIDE ELECTRON SELECTIVE LAYER IN PEROVSKITE-BASED SOLAR CELLS

Félix Gayot[1], Elise Bruhat[1], Matthieu Manceau[1], Eric De Vito[2], Denis Mariolle[3], Stéphane Cros[1]
[1]Univ. Grenoble Alpes, INES — CEA Liten, Departement of Solar Technologies, , F-73375 Le Bourget du Lac, France
[2]Univ. Grenoble Alpes, CEA Liten, Departement of Nanomaterial Technologies, 38054 Grenoble Cedex France
[3]Univ. Grenoble Alpes, CEA Leti, F-38000 Grenoble, France
felix.gayot@cea.fr

ABSTRACT: Lower performances are usually recorded for perovskite and silicon/perovskite tandem solar cells that use tin oxide (SnO_2) grown by Atomic Layer Deposition as an electron selective layer instead of solution-processed SnO_2. This work attempts to determine possible causes for such limitations. A comparative study on chemical, electrical, optical and topographical properties of ALD-grown SnO_2 and solution-processed SnO_2 thin films and on their integration in perovskite-based solar cells is presented. This study points out the SnO_2/perovskite interface as the main performance-limiting element. Specifically, a larger workfunction for ALD-grown SnO_2 may create a potential barrier for electron extraction at PK interface.
Keywords: perovskite, ALD, electron transport layer, SnO_2

1 INTRODUCTION

Tin(IV) oxide (SnO_2) is one of the main candidates to make highly efficient ESL for perovskite (PK) and perovskite/silicon (PK/c-Si) tandem solar cells[1]. Even though solvent-based processes are mainly reported for SnO_2 ESL deposition in state-of-the art devices[2]–[4] such techniques might not be appropriate on top of large and textured substrate[5], [6] [4], [5]. With many advantages ALD is a promising alternative for ESL[7]. Indeed, its cyclic nature of segregated injections of precursors allows ultra-conformal, dense and non-aggressive thin film growth, with a fine control over thicknesses[8].

Although effective ESL were fabricated using ALD-grown SnO_2 (SnO_2^{ALD}) for PK and PK/Si tandem solar cells[9]–[11], these studies are few and outline interfacial defects between SnO_2^{ALD} and the PK absorber[12]. Adding another thin film such as C_{60} at the PK/ SnO_2^{ALD} has proven to passivate such defects[12], [13]. Nevertheless, reported SnO_2^{ALD}-based PK cells performances remain lower than what can be achieved when solution-processed SnO_2 (SnO_2^{SP}) is employed[2]–[4].

This work aims at better understanding SnO_2^{ALD} limitations when employed as single-layer ESL. A systematic comparison between a SnO_2^{SP} and SnO_2^{ALD} has been conducted. Chemical, electrical, optical and topographical properties of each type of film were investigated. Similarly, some interfacial properties of the ESL/PK junction were compared for each kind of SnO_2. Whenever possible, the investigated SnO_2 thin films were integrated in single-junction PK solar cells in an attempt to link material properties and device performances.

3 EXERIMENTAL DETAILS

30 nm-thick SnO_2^{ALD} layers were grown on top of glass/ITO substrates at 150°C from Tetrakisdimethylamino-tin(IV) (TDMASn) and hydrogen peroxide (H_2O_2). 30 to 40 nm-thick SnO_2^{SP} layers were prepared by spin-coating of a water-based SnO_2 nanoparticles dispersion on UV-O_3 treated substrates. Spin-coated samples underwent annealing at 80°C for 1 minute. Bi-layered ESL were fabricated by evaporating 10nm of C_{60} on top of SnO_2. The PK precursor solution was subsequently spin-coated under N_2 atmosphere on top of SnO_2 layer to form $Cs_{0.05}FA_{0.95}Pb(I_{0.83}Br_{0.17})_3$ double-cation α-phase PK. In the case of SnO_2^{ALD}, substrates underwent a 30 minutes UV-O_3 treatment before PK coating to improve surface wettability. 150μL of chlorobenzene was dropped 5s before the spinning of the PK precursors solution ends. Afterwards, films were annealed at 100°C for one hour. A hole selective layer was made of a spin-coated doped Poly(triarylamine) thin film. Electrodes were made of evaporated gold.

SnO_2 chemical environment was examined through Hard X-ray Photoelectron Spectroscopy (HAXPES) with a Cr-Kα X-ray source. Steady-state photoluminescence (PL) spectra were recorded using a Xenon lamp. Atomic Force Microscopy (AFM) and Kelvin Probe Force Microscopy (KPFM) measurements were performed in tapping mode in amplitude modulation mode. Device cross-section images were recorded acceleration voltage of 200 keV and elemental detection was performed by Energy Dispersive X-ray (EDX).

3 RESULTS

3.1 SnO_2^{ALD}-induced limitations of PK cell performances

A comparison of typical current density-voltage

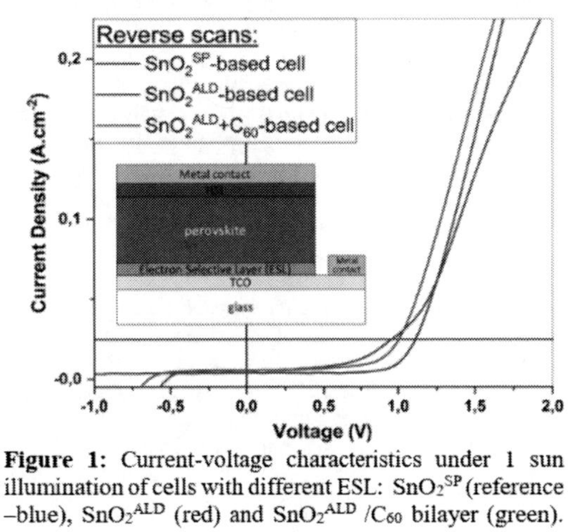

Figure 1: Current-voltage characteristics under 1 sun illumination of cells with different ESL: SnO_2^{SP} (reference –blue), SnO_2^{ALD} (red) and SnO_2^{ALD} /C_{60} bilayer (green). Inset: Schematic of a NIP PK solar cell.

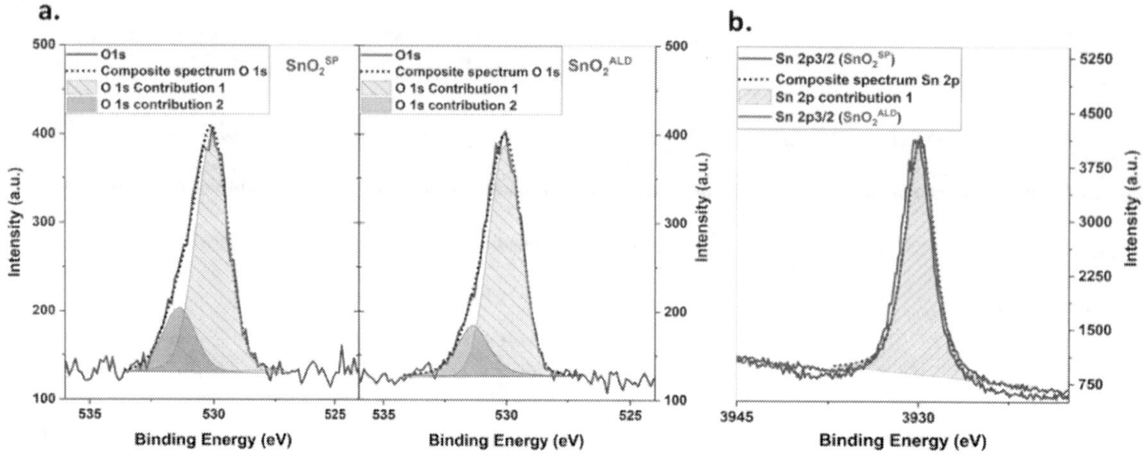

Figure 2: Fitted HAXPES spectra centred on the O 1s binding energies region for SnO_2^{SP} and SnO_2^{ALD} (a.) and HAXPES spectra centred on $Sn2p_{3/2}$ binding energies region of SnO_2^{SP} and SnO_2^{ALD} (b.), where only the fitting curve for SnO_2^{SP} sample is shown.

curves for cells fabricated with SnO_2^{ALD} or SnO_2^{SP} can be seen in Fig. 1. In agreement with literature, SnO_2^{ALD}-based cells perform less than reference ones, and all photovoltaics parameters are affected. Average short-current density (J_{SC}), open-circuit voltage (V_{OC}) and fill factor (FF) values are respectively 18.5 mA.cm^{-2}, 0.88 V and 46 % whereas reference cells feature on average 21.1 mA.cm^{-2}, 1.08 V and 69 % for J_{SC}, V_{OC} and FF respectively. A so-called "s-shape" that strongly degrades the maximum power output of SnO_2^{ALD}-based cells can be observed. It may be the signature of an inadequate energy bands alignment which could involve an electron barrier in the system[14], [15]. These cells also show a very weak reverse breakdown voltage, which can be attributed to poor hole blocking ability of the ALD-grown ESL. These two features might illustrate a lack of charge selectivity. It can be seen from Fig. 1 how adding a C_{60} thin film in between SnO_2^{ALD} and PK improves the device characteristics. Although V_{OC}, J_{SC} and reverse breakdown voltage remain similar due to a weak shunt resistance, the elimination of the s-shape improves the FF to an averaged value of 59 %. Series resistance seems also reduced with the adjunction of C_{60}, owing for a better charge transport throughout the structure.

In a previous work[16], a comparison of PK films grown either on SnO_2^{SP} or on SnO_2^{ALD} did not give rise to any significant differences between analyzed properties. This study also showed that SnO_2^{ALD} features a conductivity sufficiently high for a use as an ESL while its optical properties cannot be responsible of the J_{SC} loss of observed in PK cells. Based on these results and in regard of the degraded behavior highlighted by the current density-voltage curves, the current study focuses on intrinsic properties of SnO_2^{ALD} and SnO_2^{SP} thin films and on their interface with PK.

3.2 SnO_2 thin film properties

HAXPES measurements were performed on 30 nm-thick ALD-grown and solution-processed films. The large X-rays energy provided by the Cr lamp can probe the films over a depth of about 20nm. Such in-depth X-ray penetration enables to ignore the contribution of surface chemical environment in photoelectrons signal. Still, in Fig. 2.a, a clear shoulder at high binding energies is seen for the O 1s peak of both SnO_2^{SP} and SnO_2^{ALD}, which is attributed to -OH species. In Fig. 2.b, Sn 2p3/2 peak of each SnO2 shows a single contribution. Quantitative analysis after peaks fitting allowed to check SnO_2^{ALD} stoichiometry with an O to Sn ratio close to 2. One can further note in Fig. 2.b that peaks rise at same binding energies for both SnO_2 films, which ensures a similar chemical environment for both types of SnO_2. In addition, the absence of signal in the N 1s and C 1s binding energies regions testified that no precursor residuals remain in ALD-grown films.

SnO_2^{ALD} thin film exhibits also a weaker wettability with respect to its solution-processed counterpart, indicating a different surface energy due possibly to a different surface area (roughness) and/or a different surface chemistry. AFM characterization performed on glass/ITO, glass/ITO/SnO_2^{ALD} and glass/ITO/SnO_2^{NP} presented an averaged surface roughness of 4 nm for the two former and of 2 nm for the latter. This outlines that ALD-grown films are totally conformal to the ITO substrate whereas the SnO_2 nanoparticles of the solution-processed film tend to fill ITO roughness, thus unveiling a larger surface roughness of glass/ITO/SnO_2^{ALD} surface.

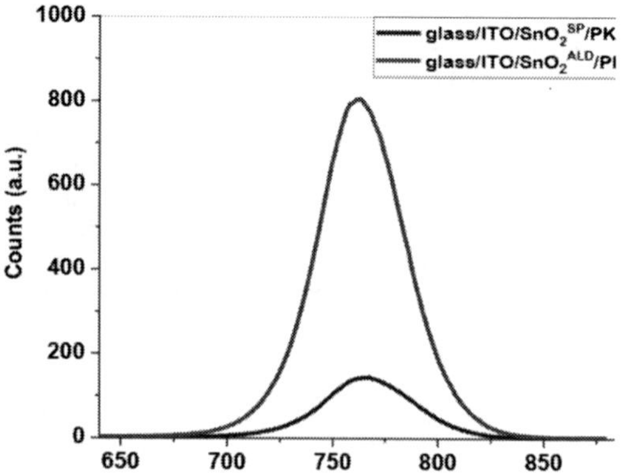

Figure 3: Absolute intensity recorded from steady-state photoluminescence measurement on glass/ITO/SnO_2^{SP}/PK and glass/ITO/SnO_2^{ALD}/PK samples.

3.3 SnO_2-PK interface characterization

Charge selective layers/PK interfaces are known to be a critical issues in PK-based photovoltaics [16]. It is of interest to focus on SnO2ALD/PK and SnO2SP interface with PK in view of observed device limitations.

As one can see in figure 3, PL measurements performed on glass/ITO/SnO_2^{ALD} /PK samples yielded about 4 times higher PL intensity than on glass/ITO/ SnO_2^{SP}/PK. Since SnO_2^{ALD}-based solar cells give lower V_{OC} than reference ones, this increase of PL intensity supports the assumption of a poor electron extraction at the SnO_2^{ALD}/PK interface rather than being caused by a reduced non-radiative recombination rate at this interface[17]. According to AFM characterizations results, this limited electron extraction seems not to be caused by a lack of contact area with PK.

On another hand, an energy barrier for electron transport at SnO_2^{ALD}/PK interface could be involved. On this matter, KPFM was performed to probe SnO_2 workfunction. SnO_2^{ALD} exhibits a 120 meV larger workfunction than SnO_2^{SP}. In principle, a low workfunction is preferable for efficient electron extraction[6], [18], hence such workfunction difference may create a potential step hindering electron extraction out of the PK. However, in order to conclude properly, this analysis must be done in view of respective valence band maxima and conduction band minima for both ESL and PK layers[19]. Further characterization by KPFM on PK and by Ultra-Violet Photoelectron Spectroscopy (UPS) on both SnO_2 and PK should give a clearer picture.

Additionally, TEM was performed on Si/ITO/ESL/PK stacks for both types of SnO2 layers. As one can see from Fig. 4, SnO_2 layers are well covering ITO substrates and do not show any visible pinhole. However, EDX profile scans revealed the presence of I, Br and Pb in the SnO_2^{SP} film, whereas the SnO_2^{ALD} dense nature prevent any penetration of these PK elements. This nanoporous nature for SnO2SP could explain its much higher wettability with respect to the ALD-grown layer. From the cell characteristics, it is clear that this intermixing of PK-related elements in the SnO_2^{SP} is not performance limiting.

3 CONCLUSION

This work seeks to clarify the causes of PK-based solar cells performance limitations when SnO_2^{ALD} is employed as the ESL instead of SnO_2^{SP}. SnO_2^{ALD} and SnO_2^{SP} thin film properties and their interface with PK were comparatively characterized. The systematic comparison of SnO_2^{ALD} thin films properties to reference SnO_2^{SP} ones ensured the relatively pure SnO2 growth by ALD, and highlighted their slightly rougher surface (when deposited onto ITO). Steady-state PL showed a more intense emission for PK films deposited on top of SnO_2^{ALD}, owing for a reduced electron extraction ability compared to SnO_2^{SP}. The larger workfunction exhibited by SnO_2^{ALD} may cause a potential barrier that limits electron extraction. Ongoing characterization by KPFM and UPS shall give a more refined picture of the bands alignment at ESL/PK interface. Besides, TEM-EDX profiling showed the presence of PK-related elements within SnO_2^{SP} films only. This intermixing could hypothetically a positive impact (passivation effect, enhanced electron extraction…). Nevertheless, at this time, a better insight in the chemical structure of this PK-based phase is still required to conclude on this last assumption.

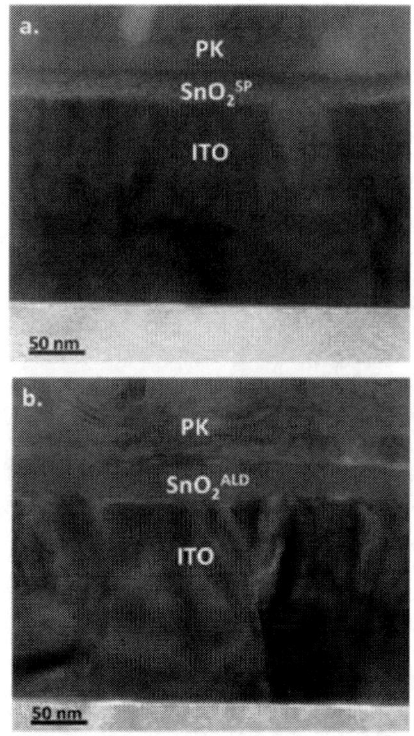

Figure 3: TEM cross-section images of a Si/ITO/ SnO_2^{SP}/PK (a.) and Si/ITO/SnO_2^{ALD}/PK (b.) stacks.

ACKNOWLEDGEMENT

Part of this work, carried out on the Platform for Nanocharacterisation (PFNC), was supported by the "Recherches Technologiques de Base" program of the French National Research Agency (ANR).

REFERENCES

[1] C. Altinkaya *et al.*, "Tin Oxide Electron-Selective Layers for Efficient, Stable, and Scalable Perovskite Solar Cells," *Advanced Materials*, vol. 33, no. 15, p. 2005504, 2021, doi: https://doi.org/10.1002/adma.202005504.

[2] Q. Jiang *et al.*, "Surface passivation of perovskite film for efficient solar cells," *Nature Photonics*, vol. 13, no. 7, Art. no. 7, Jul. 2019, doi: 10.1038/s41566-019-0398-2.

[3] J. J. Yoo *et al.*, "Efficient perovskite solar cells via improved carrier management," *Nature*, vol. 590, no. 7847, Art. no. 7847, Feb. 2021, doi: 10.1038/s41586-021-03285-w.

[4] H. Min *et al.*, "Perovskite solar cells with atomically coherent interlayers on SnO2 electrodes," *Nature*, vol. 598, no. 7881, pp. 444–450, Oct. 2021, doi: 10.1038/s41586-021-03964-8.

[5] W. Ke *et al.*, "Effects of annealing temperature of tin oxide electron selective layers on the performance of perovskite solar cells," *J. Mater. Chem. A*, vol. 3, no. 47, pp. 24163–24168, Nov. 2015, doi: 10.1039/C5TA06574G.

[6] T. Hu *et al.*, "Indium-Free Perovskite Solar Cells Enabled by Impermeable Tin-Oxide Electron Extraction Layers," *Advanced Materials*, vol. 29,

no. 27, p. 1606656, 2017, doi: 10.1002/adma.201606656.

[7] V. Zardetto *et al.*, "Atomic layer deposition for perovskite solar cells: research status, opportunities and challenges," *Sustainable Energy Fuels*, vol. 1, no. 1, pp. 30–55, 2017, doi: 10.1039/C6SE00076B.

[8] S. M. George, "Atomic Layer Deposition: An Overview," *Chem. Rev.*, vol. 110, no. 1, pp. 111–131, Jan. 2010, doi: 10.1021/cr900056b.

[9] J. P. Correa Baena *et al.*, "Highly efficient planar perovskite solar cells through band alignment engineering," *Energy Environ. Sci.*, vol. 8, no. 10, pp. 2928–2934, 2015, doi: 10.1039/C5EE02608C.

[10] Y. Lee *et al.*, "Efficient Planar Perovskite Solar Cells Using Passivated Tin Oxide as an Electron Transport Layer," *Adv. Sci.*, vol. 5, no. 6, p. 1800130, Jun. 2018, doi: 10.1002/advs.201800130.

[11] Y. Kuang *et al.*, "Low-Temperature Plasma-Assisted Atomic-Layer-Deposited SnO2 as an Electron Transport Layer in Planar Perovskite Solar Cells," *ACS Appl. Mater. Interfaces*, vol. 10, no. 36, pp. 30367–30378, Sep. 2018, doi: 10.1021/acsami.8b09515.

[12] A. F. Palmstrom *et al.*, "Interfacial Effects of Tin Oxide Atomic Layer Deposition in Metal Halide Perovskite Photovoltaics," *Advanced Energy Materials*, vol. 8, no. 23, p. 1800591, 2018, doi: https://doi.org/10.1002/aenm.201800591.

[13] C. Wang *et al.*, "Low-temperature plasma-enhanced atomic layer deposition of tin oxide electron selective layers for highly efficient planar perovskite solar cells," *Journal of Materials Chemistry A*, vol. 4, no. 31, pp. 12080–12087, 2016, doi: 10.1039/C6TA04503K.

[14] R. Scheer and H.-W. Schok, "Appendix A: Frequently Observed Anomalies," in *Chalcogenide Photovoltaics*, John Wiley & Sons, Ltd, 2011, pp. 305–314. doi: 10.1002/9783527633708.ch7.

[15] M. Neukom, S. Züfle, S. Jenatsch, and B. Ruhstaller, "Opto-electronic characterization of third-generation solar cells," *Science and Technology of Advanced Materials*, vol. 19, no. 1, pp. 291–316, Dec. 2018, doi: 10.1080/14686996.2018.1442091.

[16] F. Gayot, E. Bruhat, M. Manceau, E. D. Vito, and S. Cros, "Study of ALD-Grown SnO2 as an Electron Selective Layer for NIP Perovskite-Based Solar Cells," presented at the EU PVSEC 2021, Sep. 2021. doi: 10.4229/EUPVSEC20212021-3BV.1.5.

[17] T. Kirchartz, J. A. Márquez, M. Stolterfoht, and T. Unold, "Photoluminescence-Based Characterization of Halide Perovskites for Photovoltaics," *Advanced Energy Materials*, vol. n/a, no. n/a, p. 1904134, doi: 10.1002/aenm.201904134.

[18] J. Zhang *et al.*, "3,5-Difluorophenylboronic acid-modified SnO2 as ETLs for perovskite solar cells: PCE > 22.3%, T82 > 3000 h," *Chemical Engineering Journal*, p. 133744, Nov. 2021, doi: 10.1016/j.cej.2021.133744.

[19] M. A. Martínez-Puente *et al.*, "Unintentional Hydrogen Incorporation into the SnO2Electron Transport Layer by ALD and Its Effect on the Electronic Band Structure," *ACS Applied Energy Materials*, 2021, doi: 10.1021/acsaem.1c01836.

ELECTRO-THERMAL MODEL FOR LOCK-IN INFRARED IMAGING OF DEFECTS IN PEROVSKITE SOLAR CELLS

Ennio Luigi Comi[1*], Evelyne Knapp[1], Mattia Battaglia[1], Stefano Weidmann[2], Kirsch Christoph[1], Sandra Jenatsch[2], Hiestand Roman[2], Mathias Bonmarin[1], Beat Ruhstaller[1, 2]
[1]Institute of Computational Physics, Zurich University of Applied Sciences, Technikumstr. 9, 8401 Winterthur, Switzerland
[2]Fluxim AG, Katharina-Sulzer-Platz 2, 8406 Winterthur, Switzerland
*Ennio Luigi Comi, ZHAW, Winterthur, Switzerland, ennio.comi@zhaw.ch

ABSTRACT: The production of uniform layers without defects is crucial for the efficient upscaling of perovskite solar cells. To understand the origin of defects and their impact on efficiency, we compare steady-state (DC) and alternating current (AC) measurements with simulation results obtained by an electro-thermal 2D+1D finite element method (FEM) implemented in the software Laoss. The software supports the upscaling process from small- to large-area devices by solving for the potential and temperature distribution in 2D top and bottom electrode domains, which are coupled by a vertical 1D coupling law. Recently, we extended this FEM model to the frequency domain in order to study both DC and AC characteristics of solar cells. Here, we report on the extension of this frequency-dependent FEM model to the thermal domain, allowing us to calculate amplitude and phase images of solar cells that are voltage-modulated in the dark. We measured and modelled a screen-printed carbon-based hole-transporter-free perovskite solar cell with a defect, appearing as a hotspot in temperature measurements. In contrast to the traditional DLIT method using a large voltage modulation, we introduce a small-signal DLIT (SS-DLIT) imaging method which our model is capable to reproduce. Fitting thermal AC simulations to measured data, allowed to quantify the defect and examine its behaviour and origin.

Keywords: Perovskite, Simulation, Modelling, Experimental Methods, Defects

1 INTRODUCTION

Thin film solar cells such as single-junction perovskite and monolithic perovskite/silicon tandem photovoltaic cells have attracted researchers' attention across the globe in recent years due to rapidly increasing efficiencies. On laboratory scale, single-junction perovskite solar cells (PSCs) currently achieve a power conversion efficiency (PCE) of 25.7 %, while perovskite/silicon tandem cells reached a certified PCE of 31.3 % [1]. Most PSCs use an organic hole transport layer (often spiro-OMeTAD) to facilitate the transport of holes from the perovskite layer to the counter electrode [2]. However, organic hole transport materials (HTM) have been reported to inhibit long-term stability [3]–[5]. In 2012, Etgar et al. [6] presented the first hole transporter-free perovskite solar cell using the methylammonium lead iodide ($MAPbI_3$) perovskite layer as a photon absorber and hole transporter simultaneously. This type of perovskite solar cell is associated with high stability as recently reported by Mei et al. [7].

When upscaling PSCs, a loss in efficiency is to be expected, since with larger areas, the series resistance increases due to the sheet resistance of the electrodes and the need for current collecting metal grids leads to shadow losses. Furthermore, the probability of defect-occurrence due to the inhomogeneity of the films [8]–[10] increases as well. Jaysankar et al. [11] shows in a recent study that the inhomogeneity of the deposited layers is a key loss factor when scaling up perovskite solar cells.

In this study, we use a commercial software tool called Laoss [12]–[17] developed in-house to better understand and quantify the effects of inhomogeneities and non-ideal electrodes in semiconductor devices through simulations and direct comparison with measurements. Laoss contains a FEM solver that solves Ohm's law for charge transport and Fourier's law for heat transport on 2D top and bottom electrodes. The 2D electrodes are connected by a 1D coupling law for vertical current and heat transport, neglecting oblique flows of current due to the large aspect ratio of the cell cross-section. This 1D coupling law represents a quasi-ideal device that is not influenced by lateral effects like the sheet resistance and leads to a 2D+1D approach which saves computation time compared to a full 3D simulation.

The software tool has recently been extended by an electrical [15] and thermal small-signal mode to study the influence of defects and non-ideal electrodes in the frequency domain. It calculates and displays potential, source current, and temperature distributions, which for instance help to analyse defect types, predict the performance of an upscaled device or optimise metal grids of semiconductor surfaces, as recently shown by Burwell et al. on an organic solar cell [16].

Since the simulation tool includes heat generation and transport in a semiconductor device caused by a small-signal voltage variation around a certain bias/DC voltage, we have therefore developed a dark lock-in thermography (DLIT) measurement setup with an infrared (IR) camera. In our presented case, a DC voltage is applied to the device and a small rectangular voltage signal with a chosen lock-in frequency is superimposed. A similar method called MPP-SLIT was presented by Siegloch et al. in 2014 [18], where a small-signal voltage modulation was applied around the bias of an illuminated cell's MPP. In the following, we refer to this simulation and measurement method as small-signal dark lock-in thermography (SS-DLIT) as this approach differs from the widely used and established DLIT method in which a voltage variation oscillating with a larger amplitude at the lock-in frequency is applied at zero offset voltage [19], [20].

Here, we present Laoss' thermal AC mode and compare SS-DLIT measurements of a screen-printed carbon-based hole transporter-free PSC at different bias voltages with the corresponding electro-thermal SS-DLIT simulations.

The device has revealed a hotspot which we model and quantify with our software by manually fitting the simulation to SS-DLIT data. In addition, we analyse an electro-luminescence (EL) image of the screen-printed cell qualitatively to draw further conclusions about the analysed defect and other fabrication related artefacts.

2 METHODS

2.1 Device Fabrication

The HTM-free perovskite solar cell was deposited in an n-i-p configuration onto a 400 nm fluorine-doped tin oxide (FTO) coated 2 mm thin glass substrate. A compact 50 nm titanium dioxide layer ($cTiO_2$) was applied by spray pyrolysis to the FTO-coated glass plate. Mesoporous titanium dioxide ($mTiO_2$, 500 nm), mesoporous zirconium dioxide ($mZrO_2$, 1000 nm) and the 10-12 μm carbon layer were screen-printed and the $MAPbI_3$ precursor was inkjet infiltrated through the carbon top layer and then crystallized at ambient temperature. It is assumed that most of the porous layer stack is filled with perovskite crystals.

In this cell structure, the $cTiO_2$ functions as a hole blocking and electron transport layer (ETL) and the $mTiO_2$ and $mZrO_2$ layers act as a scaffold for the perovskite. The $mZrO_2$ also serves as a spacer layer between the TiO_2 anode and the carbon cathode (hole collector and electrode at the same time).

The cell was produced by the Swiss Federal Laboratories for Materials Science and Technology (Empa) and the Swiss company Solaronix SA. For this study, we selected a particular cell that appeared to have a hotspot in IR measurements. A schematic cross-section including the defect and an image of the device with an active area of 1.44 cm^2 is shown in Figure 1.

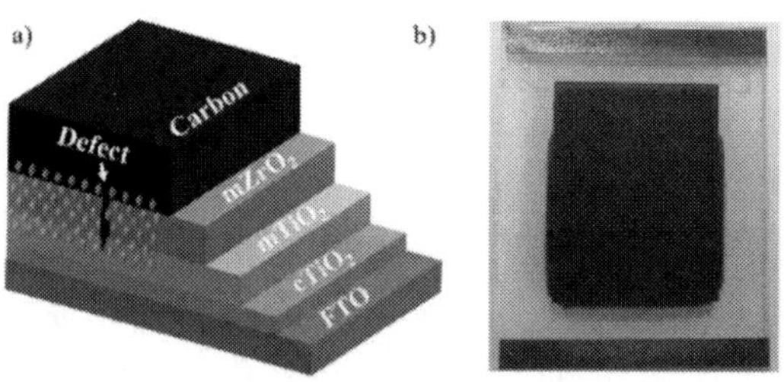

Figure 1: Cross-section of the HTM-free perovskite solar cell with an included pinhole defect (a) and an image of the device facing the carbon top electrode (b).

2.2 Experimental Characterization

For the current-voltage characteristic, the all-in-one measurement platform Paios 4.3.1 [21] from Fluxim AG was used. The cells were measured with 320 sequential steps with 0.2 s measuring time and 1.2 s settling time at each step to reduce hysteresis [22]. The device was measured in the dark with a voltage range from -0.2 to 1.5V in forward direction, resulting in a slow scanning rate of 3.8 mV/s.

The electro-luminescence image presented here, was taken in a dark-chamber, equipped with a Nikon D800 digital camera with a removed infrared radiation filter and an installed longpass filter, blocking visible light. The image was taken after a DC potential of 1.2V was applied to the PSC for approximately 150 s, within which a steady-state current has been reached.

In a DLIT measurement of solar cells, a periodically pulsed voltage signal, with a specific frequency $f_{lock-in}$ is applied to a device to generate a periodic temperature signal induced by Joule heating [19]. With the LIT method, local heat sources can be better investigated compared to steady-state IR images, due to an increased spatial resolution by several orders of magnitude.

The time-dependent thermal images acquired by the IR camera are demodulated according to the synchronous two-channel correlation with two sets of weighting factors, which approximate a sine and a cosine function. This algorithm allows to compute the amplitude A and the phase Φ of the temperature signal from the "in-phase" $S^{0°}$ and the "in quadrature" $S^{-90°}$ components. For this study, we have solely examined the temperature amplitude image, which shows the dissipated power and is calculated as

$$A = \sqrt{(S^{0°})^2 + (S^{-90°})^2}. \quad (1)$$

In this study, we investigate thin-film solar cells with a small-signal dark lock-in thermography measurement. The most pronounced difference between DLIT and our SS-DLIT method is the applied voltage oscillation. To compare our thermal AC simulations with measurements, we had to apply a small harmonic voltage signal around a DC bias of a device (s. Figure 2). Using the lock-in algorithm with IR images, it is possible to analyse the temperature increase caused by the small-signal.

To acquire the IR images, we used a PI450i camera (Optris GmbH, Germany) with an image resolution of 382 x 288 pixels and a frame rate of 80 Hz. We use a lock-in frequency $f_{lock-in}$ of 1 Hz resulting in a thermal diffusion length of approximately 1 mm for the carbon electrode. IR images could not be taken through the glass substrate, so they were acquired with the camera facing the carbon electrode. We assume that this will not distort the measured temperature, since the cells are extremely thin compared to their lateral dimension. During the SS-DLIT measurement a DC voltage signal of 0.9, 1.1, 1.3 and 1.6 V and a modulation amplitude of 0.1 V was applied.

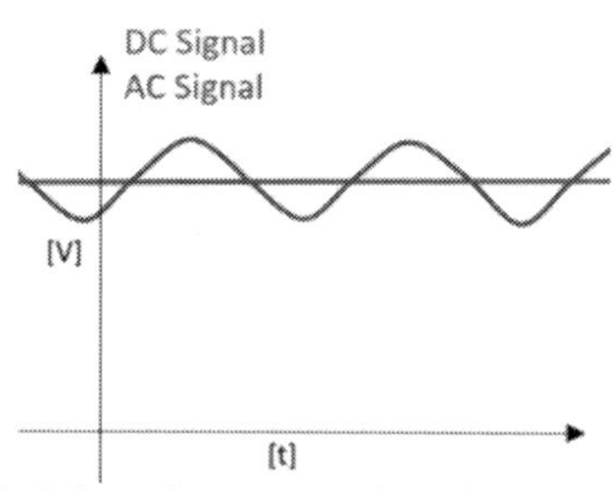

Figure 2: Schematic representation of the superimposed sinusoidal voltage signal applied to the device for the SS-DLIT measurements.

2.3 Modelling

In a large-area semiconductor device, charge carriers travel laterally across the top and bottom electrodes to the charge collecting contacts. The potential distribution on the electrodes can be influenced by various types of defects and ohmic losses due to the sheet resistance R_{sh}, which is a significant part of the resulting series resistance R_s in current-voltage characteristics [23]. The software Laoss accounts for these losses by solving Ohm's law on the 2D electrodes with the finite element method. In the steady-state (DC, direct current) case, the electrodes are coupled by a 1D current-voltage law (e.g. diode model) [13] and in the time-dependent small-signal (AC) case, the coupling requires an additional frequency and bias voltage φ_{DC} dependent admittance $Y(\omega, \varphi_{DC})$ with the angular frequency $\omega = 2\pi f$. The coupling law represents a quasi-ideal device, which is not influenced by surface effects or defects (e.g. shunts). This 2D+1D concept is illustrated in Figure 3 with the 1D coupling laws symbolized by diodes for the DC case and additional RC elements for the AC case.

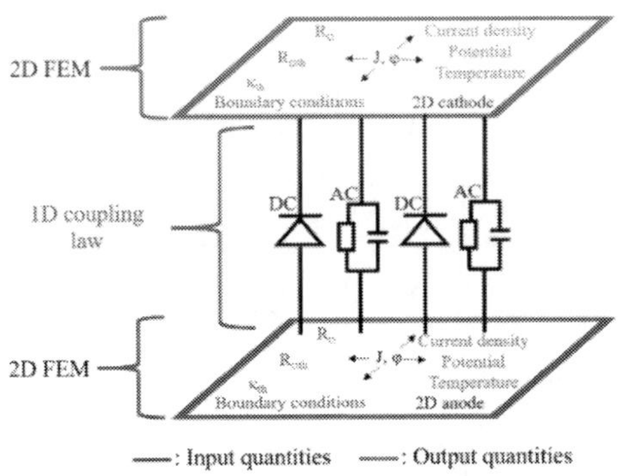

Figure 3: Structure of the electro-thermal model of the Laoss software with the coupled 2D domains and 1D coupling laws (DC and AC).

In this study, we intend to simulate the temperature distribution on a solar cell, caused by a sinusoidal small-signal. We restrict ourselves to the unidirectional electro-thermal coupling thereby assuming that the simulated temperature solely depends on electrical device characteristics and not vice versa.

To solve the thermal AC equations, Laoss requires steady-state temperature and voltage distribution results for the small-signal linearisation. The fundamental heat conservation law for the calculation of the temperature distribution on the top and bottom electrodes in steady-state is described as

$$-\nabla \cdot \left(\frac{1}{R_{th}} \nabla T\right) = P_{joule}^{electrode} + P_{joule}^{stack} - P_{ex}^{envir} \quad (2)$$

and is further explained in [14]. Equation (2) is solved for the temperature distribution T on the electrodes which depends on current-induced Joule heating in the electrodes $P_{joule}^{electrode}$, Joule heating in the stack P_{joule}^{stack} caused by vertically flowing currents, which includes also heat conduction between the electrodes, and convective, conductive and radiative heat exchange with the environment P_{ex}^{envir}. The simulated temperature is also influenced by the thermal sheet resistance R_{th} describing the heat conductivity of the top and bottom electrode materials.

The electro-thermal AC model is based on the three-dimensional diffusion equation for transient heat conduction in solids with the source term q_{source} including Joule heating, the specific heat capacity c_p, the density of the electrode ρ and the thermal conductivity λ and can be written as

$$\rho c_p \frac{\partial \mathrm{T}}{\partial \mathrm{t}} + \nabla \cdot (-\lambda \nabla T) = q_{source}. \quad (3)$$

A Fourier transform has been applied to obtain the model in frequency domain. Subsequently, a linearisation of the non-linear Joule heating terms around the DC potential and a reduction of the equations from 3D to 2D electrodes was carried out. The resulting equations for the 2D top and bottom electrodes can then be solved for the desired local AC temperature.

The Laoss simulations are based on a 2D top and bottom geometry which also considers the silver, FTO and carbon layers (transition layers) as subdomains, connecting the active area (s. Figure 1b). An active area of 1.44 cm^2 was assumed and a 4e-4 cm^2 rectangular subdomain was added to quantify the defect. Electrical steady-state simulations require sheet resistance values assigned to the individual materials in addition to the 1D coupling law. A vertical 1D coupling law represents an ideal cell and is in the best case a small-area device without any edge effects and low series resistance. In our case, we did not have a small-area device available, so we had to acquire the coupling law from the device itself. To obtain the DC coupling law for the active area, we therefore measured its current-voltage characteristic in the dark and subsequently fitted it with single-diode model parameters. However, this fitted curve was afflicted with a series resistance caused by the transition layers and the top and bottom electrodes of the large-area device. This series resistance has been removed for the 1D coupling law by setting the single-diode model parameter R_s to zero. Laoss later compensates for the series resistance with the electrode's sheet resistances.

To model the defect, we initially decided to use an ohmic coupling law, which is a typical approach when modelling defects [24]. In our case, however, the measured dark current-voltage characteristic of the PSC did not show an increased current at low potentials and therefore had a sufficiently high parallel resistance. Consequently, we modelled the defect with a diode-like behaviour, which is an often used modelling approach for pinhole defects in thin film solar cells [25], [26].

For the AC coupling law $Y(\omega, \varphi_{DC})$ we used the admittance of a parallel RC circuit consisting of a frequency independent conductance $G(\varphi_{DC})$ and a bias voltage independent susceptance $B(\omega)$. The conductance per area is calculated from the provided DC coupling law $f(\varphi_{DC})$, which is a voltage dependent current density:

$$G(\varphi_{DC}) = \frac{\partial f(\varphi_{DC})}{\partial \varphi_{DC}}. \quad (4)$$

The susceptance per area depends on the relative stack permittivity ε_r, the frequency and the stack thickness d and is calculated as

$$B(\omega) = \frac{\omega \cdot \varepsilon_r \cdot \varepsilon_0}{d} \quad (5)$$

This results in the following expression describing the stack admittance:

$$Y(\omega, \varphi_{DC}) = G(\varphi_{DC}) + iB(\omega). \quad (6)$$

For thermal steady-state simulations, the thermal sheet resistance R_{th} for the top and bottom electrodes must be calculated from the thermal conductivity of the electrode materials. Additionally, the thermal conductivity of the stack and the heat transfer coefficient of the individual subdomains have to be provided. Thermal AC simulations require an additional heat capacity per unit area K_{th}, which can be calculated from the electrode materials as well. The lock-in frequency of the applied voltage signal is 1 Hz for the SS-DLIT simulations, which is consistent with the applied frequency during the measurements.

3 RESULTS

To reproduce the behaviour of the cell with electrical steady-state and thermal AC simulations, we first assigned a resistive behaviour to the defect. However, this led to substantial deviations to the measurements in both the steady-state and the thermal AC simulations. Using a diode model as the coupling law of the defect, we were able to fit the steady-state and thermal AC measurements well.

It is likely that the solar cell has a defect with a semiconductor junction as the analysed sample has been manufactured in ambient air. Consequently, dust particles could cause a clog in the screen-printing mesh, which could lead to pinholes in the layer stack. Pinholes can result in a connection between HTL and ETL and enhance local recombination currents [25]. In our case, a gap in the $mZrO_2$ layer would lead to a connection between the carbon and the $mTiO_2$ layer. Since the finished mesoporous stack is then filled with perovskite precursor, a semiconductor junction with increased recombination currents can form in the pinhole region.

Non-ohmic shunt currents are known to reduce the efficiency of solar cells by increasing the charge carrier recombination and thus increasing the dark saturation current J_0, leading to a shift of the current-voltage characteristic towards lower voltages and consequently also to a lower open-circuit voltage (V_{OC}) in the illuminated case [24], [27]. To reproduce the SS-DLIT and current-voltage measurements and fit the defect coupling law with Laoss, we took the same single-diode model parameters that we used for the active area coupling law but increased J_0 as a fitting parameter.

Solving the small-signal equations requires steady-state results for the linearisation around the bias potential. Since the current induced Joule heating is crucial for the simulated and measured heating in the cell, we fit the electrical current-voltage simulations to the measurement first. Steady-state simulations also provide the local potential distribution, which is exponentially related [28] to electro-luminescence (EL) images owing to the diode nature of the solar cell. This allows for an additional comparison of the measured EL-image with the simulations and also provides information about the accuracy of the device model. The comparison between the measured EL image and the simulations is further explained in the end of this section.

The measured current-voltage characteristic of the investigated device was manually fitted by adjusting the sheet resistance values of the individual electrode materials. The fitted R_{sh} values, which lead to the current-voltage characteristic shown in Figure 4, are in good agreement with the device manufacturer's reference values. Interestingly, varying J_0 on the coupling law of the defect subdomain had a minimal influence on the resulting current-voltage characteristic regarding the entire voltage range. This is mainly due to its small area and the lack of an ohmic component. Therefore, it would be very difficult to quantify this defect with steady-state data alone. Consequently, the fitting of simulations to additional SS-DLIT measurements becomes advantageous.

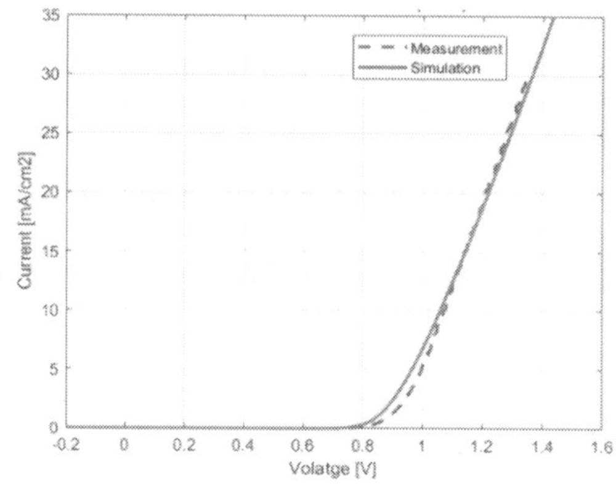

Figure 4: Measured (blue) and simulated (orange) current-voltage characteristic of the perovskite solar cell.

Figure 5 shows the image comparison between the temperature amplitude of the SS-DLIT simulation and the measurement at 0.9 and 1.6V bias. The comparison at 0.9V only shows a hotspot (s. Figures 5a and 5b), while at 1.6 V bias, the active area in the upper end of the cell heats up as well (s. Figures 5c and 5d).

As explained, we increased the dark saturation current on the coupling law of the defect subdomain, until we obtained the simulations with the best fitted AC temperature amplitude (s. Figure 6a). For this purpose, we compared the simulated peak AC temperatures of the active area and the defect with those of the measurements at different bias voltages. The comparison between measured and simulated AC temperature amplitude for the two subdomains is shown in Figure 6b. The measured curves show that the maximum temperature on the active area increases continuously, while the maximum temperature at the defect shows a peak and a subsequent decrease at high voltages.

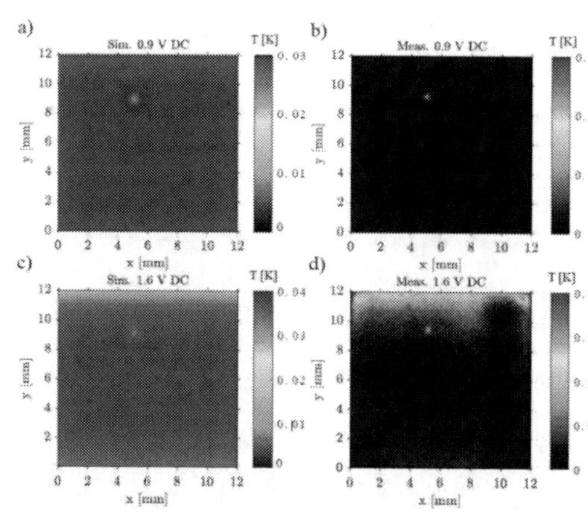

Figure 5: Amplitude image comparison between SS-DLIT simulations at 0.9 V (a) and 1.6 V (c) and measurements at 0.9 V (b) and 1.6 V (d) bias.

The reason for a temperature rise and subsequent drop in the defect subdomain is related to a complex interplay between the two coupling laws and the sheet resistance, which is responsible for a potential drop along the carbon top electrode. At lower voltages, for example at 0.9 V bias, the main current flow and subsequent heating takes place at the defect. At higher voltages, when more current flow is allowed on the active area, more Joule heating occurs in the upper part of the device. The current takes the path of least resistance, which in this case is the better conducting FTO-bottom electrode. The thermal energy is therefore generated to a greater extent on the upper part active area and less on the defect, which is located further in the center of the device. This can be seen very clearly in the SS-DLIT measurement and simulation at 1.6 V bias.

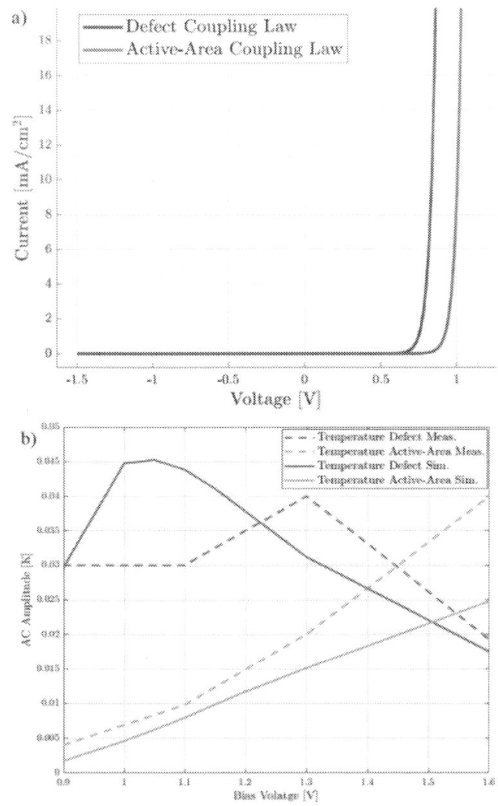

Figure 6: Resulting DC coupling laws of the active-area (orange) and the defect subdomain (blue) in (a) and (b) shows the quantitative comparison between the simulated and measured SS-DLIT temperature amplitude maxima of the defect (orange) and the active area (yellow).

Figure 7 shows the measured and simulated EL image at 1.2 V bias. Defects, such as pinholes, usually show a less radiative dark spot [29], which is clearly visible in the measurement and also in the simulation, where the potential drops in the region of the defect. The EL-measurement shows further interesting artefacts. For example, dark grid lines are visible on the entire right side. After a comparison with the printing mesh sizes used by the manufacturer, it became clear that these originate from applying too much pressure during the screen printing of the carbon layer. Dark spots are also visible on the lower corners. These originate most likely from poor perovskite filling of the porous layers. The qualitative comparison of the EL image with the simulation shows that the luminescence drop caused by the sheet resistance is well reproduced, but the cell has some artefacts such as the poor perovskite filling and the visible screen printing mesh pattern, which have not been modelled for the sake of simplicity.

The grid pattern originating from the screen-printing mesh and dark areas at the lower corners can also be seen in the SS-DLIT measurement at 1.6 V bias, as these regions heat up less due to the lower injection of charge carriers. Therefore, not only hot spots but also cold spots are important to consider when analysing such thermal measurements.

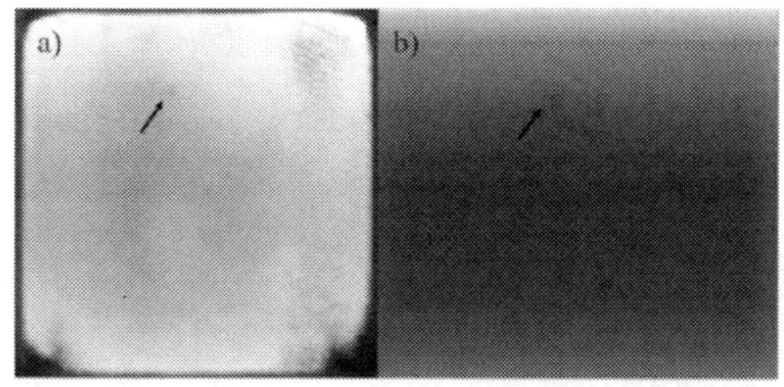

Figure 7: Qualitative comparison between measured EL-image (a) and simulated electro-luminescence based on the potential difference between top and bottom electrode (b), both at 1.2 V bias. The poor perovskite filling and the visible screen-printing mesh pattern have not been modelled for the sake of simplicity.

4 CONCLUSIONS

The prevention of defects during production of large-area perovskite solar cells is a crucial step for their commercialization. For this reason, we modelled a large-area hole-transporter-free carbon-based PSC with a pinhole defect in our FEM simulation tool and fitted the electrical steady-state and thermal AC simulation results to actual measurements (SS-DLIT) in order to characterize the defect and gather insight into its electro-thermal behaviour.

The comparison between SS-DLIT simulations and measurements allowed the characterization of the pinhole defect as we fitted the temperature amplitude of the hotspot and the active area. The SS-DLIT measurement method is worthwhile to compare measurements with simulations in order to better understand the nature of the defects and to quantify them.

This study has shown the potential in detecting nonuniformities in perovskite solar cells from a combination of current-voltage, EL and LIT measurements. These methods, together with semiconductor simulations, complement each other in many ways and can contribute significantly to improve the fabrication process and prevent degradation of solar cells and modules.

REFERENCES

[1] NREL, "Best Research-Cell Efficiencies: Rev. 09-14-2022," *Best Research-Cell Efficiency Chart | Photovoltaic Research | NREL*. National Renewable Energy Laboratory (NREL), p. https://www.nrel.gov/pv/cell-efficiency.html, 2022.

[2] W. H. Nguyen, C. D. Bailie, E. L. Unger, and M. D. McGehee, "Enhancing the Hole-Conductivity of Spiro-OMeTAD without Oxygen or Lithium Salts by Using Spiro(TFSI) 2 in Perovskite and Dye-Sensitized Solar Cells," *J. Am. Chem. Soc.*, vol. 136, no. 31, pp. 10996–11001, Aug. 2014.

[3] A. Mei *et al.*, "A hole-conductor-free, fully printable mesoscopic perovskite solar cell with high stability," *Science (80-.).*, vol. 345, no. 6194, pp. 295–298, Jul. 2014.

[4] R. S. Sanchez and E. Mas-Marza, "Light-induced effects on Spiro-OMeTAD films and hybrid lead halide perovskite solar cells," *Sol. Energy Mater. Sol. Cells*, vol. 158, pp. 189–194, Dec. 2016.

[5] S. Kim *et al.*, "Relationship between ion migration and interfacial degradation of CH3NH3PbI3 perovskite solar cells under thermal conditions," *Sci. Rep.*, vol. 7, no. 1, p. 1200, Dec. 2017.

[6] L. Etgar *et al.*, "Mesoscopic CH 3 NH 3 PbI 3 /TiO 2 Heterojunction Solar Cells," *J. Am. Chem. Soc.*, vol. 134, no. 42, pp. 17396–17399, Oct. 2012.

[7] A. Mei *et al.*, "Stabilizing Perovskite Solar Cells to IEC61215:2016 Standards with over 9,000-h Operational Tracking," *Joule*, pp. 1–15, Oct. 2020.

[8] M. R. Leyden, Y. Jiang, and Y. Qi, "Chemical vapor deposition grown formamidinium perovskite solar modules with high steady state power and thermal stability," *J. Mater. Chem. A*, vol. 4, no. 34, pp. 13125–13132, 2016.

[9] Z. Li *et al.*, "Scalable fabrication of perovskite solar cells," *Nat. Rev. Mater.*, vol. 3, no. 4, p. 18017, Apr. 2018.

[10] L. Rakocevic *et al.*, "Loss Analysis in Perovskite Photovoltaic Modules," *Sol. RRL*, vol. 3, no. 12, p. 1900338, Dec. 2019.

[11] M. Jaysankar *et al.*, "Toward scalable perovskite-based multijunction solar modules," *Prog. Photovoltaics Res. Appl.*, vol. 27, no. 8, p. pip.3153, Jun. 2019.

[12] Fluxim AG, "Large Area Organic Semiconductor Simulation (LAOSS)," 2022. [Online]. Available: https://www.fluxim.com/laoss.

[13] C. Kirsch *et al.*, "Electrothermal Simulation of Large-Area Semiconductor Devices," *Int. J. Multiphys.*, vol. 11, no. 2, pp. 127–136, Jun. 2017.

[14] M. Diethelm *et al.*, "Finite element modeling for analysis of electroluminescence and infrared images of thin-film solar cells," *Sol. Energy*, vol. 209, no. May, pp. 186–193, Oct. 2020.

[15] E. L. Comi *et al.*, "Sinusoidal Small-Signal (AC) and Steady-State (DC) Analysis of Large-Area Solar Cells," *Sol. Energy Adv.*, p. 100003, Oct. 2021.

[16] G. Burwell, N. Burridge, E. Bond, W. Li, P. Meredith, and A. Armin, "Parameterization of Metallic Grids on Transparent Conductive Electrodes for the Scaling of Organic Solar Cells," *Adv. Electron. Mater.*, vol. 7, no. 6, p. 2100192, Jun. 2021.

[17] Z. Chen *et al.*, "An underestimated photoactive area in organic solar cells based on a ZnO interlayer," *J. Mater. Chem. C*, 2021.

[18] M. Siegloch, B. E. Pieters, A. Gerber, and U. Rau, "Small-signal lock-in thermography at the maximum power point of an a-Si solar mini-module," *Phys. status solidi - Rapid Res. Lett.*, vol. 8, no. 11, pp. 894–897, Nov. 2014.

[19] O. Breitenstein, W. Warta, and M. Langenkamp, *Lock-in Thermography*, vol. 10. Berlin, Heidelberg: Springer Berlin Heidelberg, 2010.

[20] O. Breitenstein, J. P. Rakotoniaina, and M. H. Al Rifai, "Quantitative evaluation of shunts in solar cells by lock-in thermography," *Prog. Photovoltaics Res. Appl.*, vol. 11, no. 8, pp. 515–526, Dec. 2003.

[21] Fluxim AG, "Platform for all-in-one characterization (PAIOS)," 2022. [Online]. Available: https://www.fluxim.com/paios.

[22] J. M. Cave *et al.*, "Deducing transport properties of mobile vacancies from perovskite solar cell characteristics," *J. Appl. Phys.*, vol. 128, no. 18, p. 184501, Nov. 2020.

[23] G. T. Koishiyev and J. R. Sites, "Impact of sheet resistance on 2-D modeling of thin-film solar cells," *Sol. Energy Mater. Sol. Cells*, vol. 93, no. 3, pp. 350–354, Mar. 2009.

[24] A. Ren *et al.*, "Spatially Resolved Identification of Shunt Defects in Thin Film Solar Cells via Current Transport Efficiency Imaging Combined with 3D Finite Element Modeling," *Sol. RRL*, vol. 3, no. 5, p. 1800342, May 2019.

[25] P. Kaienburg *et al.*, "How Contact Layers Control Shunting Losses from Pinholes in Thin-Film Solar Cells," *J. Phys. Chem. C*, vol. 122, no. 48, pp. 27263–27272, Dec. 2018.

[26] S. Agarwal and P. R. Nair, "Pinhole induced efficiency variation in perovskite solar cells," *J. Appl. Phys.*, vol. 122, no. 16, p. 163104, Oct. 2017.

[27] J. Tao *et al.*, "Investigation of electronic transport mechanisms in Sb2Se3 thin-film solar cells," *Sol. Energy Mater. Sol. Cells*, vol. 197, no. February, pp. 1–6, Aug. 2019.

[28] A. Helbig, T. Kirchartz, R. Schaeffler, J. H. Werner, and U. Rau, "Quantitative electroluminescence analysis of resistive losses in Cu(In, Ga)Se2 thin-film modules," *Sol. Energy Mater. Sol. Cells*, vol. 94, no. 6, pp. 979–984, Jun. 2010.

[29] J. Raguse, J. T. McGoffin, and J. R. Sites, "Electroluminescence system for analysis of defects in CdTe cells and modules," in *2012 38th IEEE Photovoltaic Specialists Conference*, 2012, pp. 000448–000451.

ACKNOWLEDGEMENT

Financial support for the Innosuisse project AIPV (project no. 37304.1 IP) is gratefully acknowledged as well as the support from Hartmut Nussbaumer, Severin Weiss, Alexander Meier and Franz Baumgartner from the Institute of Energy Systems and Fluid Engineering (IEFE) at ZHAW School of Engineering while carrying out EL imaging measurements. Furthermore, we would also like to thank Frank Nüesch, Jakob Heier, David Martineau and Anand Verma from Empa and Solaronix for providing the perovskite solar cells and the fruitful discussions about the measurement results. Finally, the authors would like to thank Thomas Hocker and Sandro Ehrat from the Institute of Computational Physics (ICP) at ZHAW School of Engineering for providing the IR camera.

UNDERSTANDING PEROVSKITE EXPERIMENTAL DEGRADATION PATHWAYS COMBINING EXPERIMENTAL MEASUREMENTS AND SIMULATIONS

Arthur Julien[1], Jean-Baptiste Puel[1,2], Jean-François Guillemoles[1,3]
1. IPVF, Institut Photovoltaïque d'Ile-de-France, 30 RD 128, 91120 PALAISEAU, France
2. EDF R&D, 30 RD 128, 91120 PALAISEAU, France
3. IPVF, UMR 9006, Air Liquide, Chimie ParisTech, CNRS, EDF, IPVF SAS, Total, Ecole Polytechnique, Institut Polytechnique de Paris, 30 RD 128, 91120 PALAISEAU, France
arthur.julien@ipvf.fr

ABSTRACT: Perovskite solar cells have been reported to be sensitive to various stressors, potentially inducing numerous chemical or physical degradation processes, making it challenging to identify causes of performance losses. Here we take advantage of electrical measurements done along aging experiments to identify the degradation processes causing the loss of performances. Correlations between electrical parameters are used: by avoiding time dependence, they define characteristic pathways of the degradation mechanisms. Experimental and simulated degradation paths can therefore directly be compared. To do so, a procedure has been developed to reproduce the initial performances of a given device, based on a randomized approach. It allows implementing several unitary degradation mechanisms, and their pathways are compared to experimental results. An aging experiment performed in-house is investigated and results show the ETL might be the cause of failure. Furthermore, experimental records available in literature are investigated. Causes attributed by authors and even the kinetics approach employed can be discussed, as results at different temperature are analyzed independently. The method, furthermore, allows considering sequential combinations of mechanisms. It is demonstrated how simulations can support the enhancement of perovskite solar cells stability by distinguishing their degradation processes.
Keywords: Perovskite, Degradation, Simulation

1 INTRODUCTION

Perovskites solar cells now provide highly efficient and easy to produce devices. Yet, improving their stability and understanding underlying degradation processes is a main concern [1]. First, these cells are mostly based on metal-halide perovskite materials, usually containing organic cations. Also, fully organic materials are frequently used as hole transport layers (HTL). As a result, devices show instability towards elevated temperature, water, and oxygen coupled with light [2]. Furthermore, ionic migrations induce modifications of material properties, phase segregations and interface degradations, as well as hysteresis behaviors [3]. Finally, even electron transport layers (ETL) face stability issues and have been reported to enhance perovskite material degradation [4]. As a result, potential mechanisms and environmental stressors are numerous and often difficult to distinguish.

The first widely used approach to investigate perovskite solar cells degradation is to precisely characterize the materials and interfaces composing the device before and after degradation [5]. Characterization techniques include X-ray diffraction (XRD) for phase identification, time-of-flight secondary ion mass spectrometry [6] and X-ray photoelectron spectroscopy to determine the composition of a layer [7] and photoluminescence to probe the semi-conductor behavior of the materials [8]. The combination of such techniques can allow for a deeper understanding of the degradation mechanisms taking place. However, it is often time-consuming and costly, and this approach only demonstrates the coincidence between material evolution and photovoltaic conversion efficiency degradations, not the causality. Furthermore, such approaches do not take advantage of measurements that can be performed during ageing of the devices.

A second approach that is frequently used is to record the evolution of parameters related to electrical performances such as short circuit current (Jsc), open circuit voltage (Voc), fill factor (FF) and power conversion efficiency, during ageing [9–15]. Such data are, however, often used only to determine an end-of-life time, or the duration needed to reach a given lower limit. Several stability tests are defined by simply measuring such time under various conditions [1]. Kinetics studies can allow to determine degradation rates [12,16,17], but the dependences on various potential activation processes makes it difficult to determine the causes of degradation.

Here we take advantage of electrical measurements done during aging experiments (short circuit current, open circuit voltage, etc.) to identify the causes of performance losses. We use these results without considering their time dependence, but their correlations. They define pathways that do not depend on any activation process and can be reproduced with drift-diffusion simulations. Unitary degradation mechanisms are implemented, their associated characteristic pathways determined, and compared to experimental records. Finally, this allows to identify the dominant degradation mechanisms impacting the solar cell performance.

In this paper, the procedure developed to reproduce performances of a given device and to implement degradation mechanisms is first presented. Then, results from degradation experiments performed in-house are analyzed, demonstrating the capabilities of the novel method. Finally, experimental measurements reported in literature, by Lim *et al.* [12] are investigated. Results presented here allow one to question the methodology itself and the possibility to conduct kinetics studies when numerous degradation mechanism can be activated simultaneously within the same device. With this example, it is shown that in some cases, degradation measurements can be separated into several steps as a function of time. Interpretation with distinct degradation mechanisms becoming dominant in succession can therefore be discussed. These analyses of two different degradation experiments show that focusing on the cause of performances degradation, instead of concomitance with variations of material properties, can provide the knowledge necessary to focus on the right material or interface to further improve performance stability. As a result, the methodology presented here provides a valuable tool for the study of perovskite solar cells stability and tackles a major goal of the field.

2 METHODS

In this work, perovskite solar cells are simulated with a drift diffusion simulation tool, SCAPS 1D, as a stack of three materials (ETL, perovskite and HTL), top and bottom contacts, and with defects at interfaces and in the perovskite layer. In order to analyze experimental results, it is necessary to define coherent simulations with the real device used in the experiments. However, all properties such as defect densities, doping levels and carrier mobilities are not always known, and some can even vary a lot with fabrication. Therefore, a procedure has been developed to determine suitable values for these material parameters, aiming at reproducing the current-voltage characteristic of the real device before degradation.

A randomized approach based on a genetic algorithm is implemented in several steps. First, sets of material parameters (i.e., drift diffusion simulation inputs) are randomly generated, within realistic physical ranges for the considered materials. Current voltage characteristics are simulated, and depending on Voc, Jsc and FF, suitable parameters sets are retained. Then, steps expanding the selected group by randomly perturbating the selected input sets are used to generate new potential inputs that are simulated and again, selected if close enough to the investigated experimental data. This procedure provides a large number of material parameter sets reproducing experimental results, despite the size and high dimensionality of the topological space to investigate.

Then, inputs generated in the randomized procedure are used to implement nine unitary degradation mechanisms: reduction of doping in ETL or HTL, reduction of carrier mobilities, increase of defect densities at interfaces or in perovskite layer. Each mechanism is implemented on all material parameters set, allowing to compare the impact on the performances of each of them. Finally, the average evolution of Voc, Jsc and FF, as well as confidence interval are computed along degradation for each mechanism.

Importantly, results are represented in terms of correlations through Jsc vs Voc, FF vs Voc and FF vs Jsc graphs. This allows to define degradation pathways in these two-dimensional planes that are not dependent on time. Therefore, the speed of degradation in the simulated scenarios or in experimental results does not influence the interpretation, the only determining factor of these paths being the nature of the mechanism.

Finally, this independence on environmental conditions makes it directly possible to compare experimental and simulated degradation pathways. The closest pathway can be identified as the most probable mechanism having caused the measured degradation.

3 RESULTS AND DISCUSSION

The first part of the procedure described above concern the determination of input parameters to accurately reproduce the JV characteristic of the studied device before degradation. Figure 1 shows the experimental (top left) and simulated (top right) characteristics in the example of a device fabricated and analyzed at IPVF (details are given in the next section). In this example, the genetic algorithm provided 112 sets of inputs parameters, all approaching experimental Voc, Jsc and FF values within 2% error. In fact, several combinations of material parameters can be used to reproduce the same electrical behavior. It is therefore necessary to consider numerous combinations, all reproducing the same measured JV curve, in order to avoid any unfounded assumption on material parameters values.

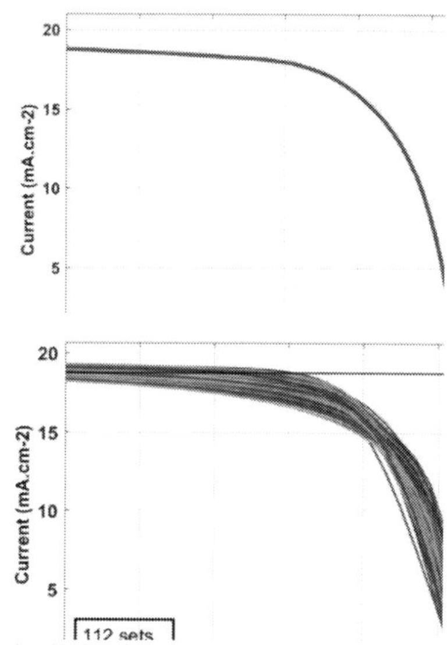

Figure 1: Top: STC current voltage characteristic measured experimentaly on the fresh device, before degradation. Bottom: Simulated current voltage characteristics reproducing the experimental measurements.

Then, each combination (set of material parameters) is used to simulate the response to the nine unitary degradation mechanisms mentioned in the previous section. Average behavior (and confidence interval) per mechanism is computed and results are finally represented as pathways in Figure 2, showing the correlation of Jsc with Voc (top left), FF with Voc (top right) and Jsc with FF (bottom). It appears that most degradation mechanisms follow distinguishable pathways. Importantly these representations provide an intrinsic cartography of the degradation mechanisms, independent on degradation conditions and potential activation processes.

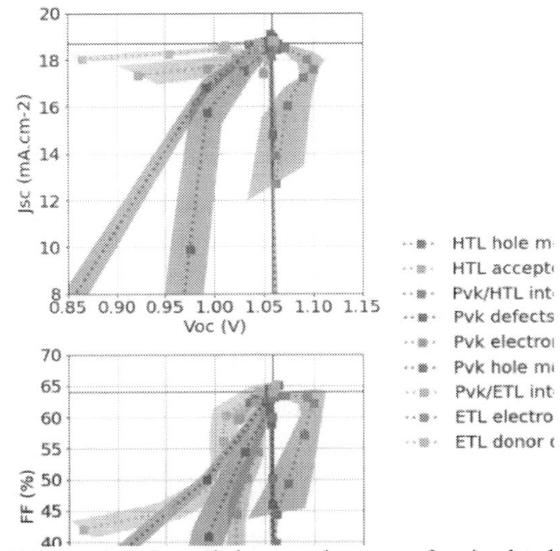

Figure 2: Degradation pathways of simulated degradation mechanisms in Jsc/Voc (top, left), FF/Voc (top, right) and FF/Jsc (bottom) planes. For each mechanism, average behavior (over the 112 finally selected sets) is represented with squares and dashed lines, shaded areas represent the 95 % confidence interval. Pathways allow to distinguish the mechanisms, while being independent from activation processes and environmental conditions.

3.1 Degradation of $Cs_{0.05}(MA_{0.17}FA_{0.83})_{0.95}Pb(Br_{0.17}I_{0.83})_3$ based devices measured at IPVF

The first measurements analyzed here were performed at IPVF on cells based on $Cs_{0.05}(MA_{0.17}FA_{0.83})_{0.95}Pb(Br_{0.17}I_{0.83})_3$ material. The structure is n-i-p (illumination through ETL side) with PTAA HTL and TiO2 ETL. Aging was recorded under constant one sun illumination, air and ambient temperature during more than 1000 hours.

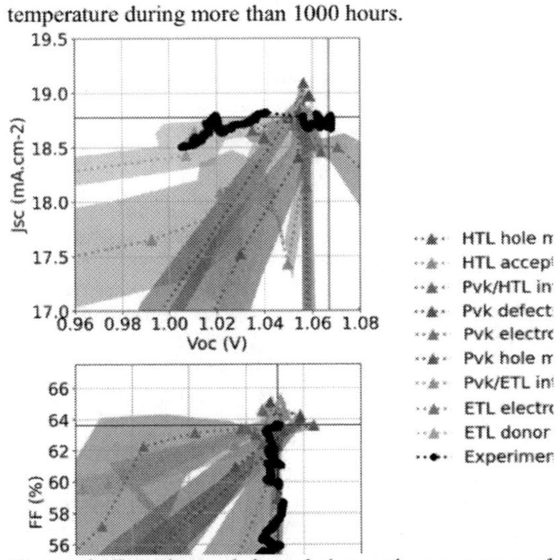

Figure 3: Experimental degradation pathways measured at IPVF: perovskite devices were aged under one sun, air and ambient temperature. Associated simulated pathways (average over 112 material parameters sets used to reproduce initial performances) are superposed. A loss of ETL doping is the closest simulated mechanism.

As explained above, 112 sets are used to reproduce the initial performances and to implement the unitary degradation mechanisms.

Experimental and simulated degradation pathways are superposed in Figure 3. It appears that the experimental pathway resembles to a reduction of doping in the ETL layer (yellow), with a major loss of FF and Voc. Reduction HTL hole mobility (blue) is also a close mechanism on FF vs Jsc graph (right) but would induce almost no reduction of Voc, which is not compatible on the other plane.

This first example shows that the analysis performed here can be employed to support the developpement and fabrication of perovskite solar cells. By identifing the degradation cause, it helps to focus on the appropriate layer or interface to further enhance the stability of the devices.

3.2 Degradation of $FA_{0.95}MA_{0.05}PbI_{2.85}Br_{0.15}$ based devices published by Lim *et al.*

The next measurements analyzed were published by Lim et al. [12]. Devices are semi-transparent encapsulated cells, with n-i-p structure: $FA_{0.95}MA_{0.05}PbI_{2.85}Br_{0.15}$ perovskite material, spiro-OMeTAD HTL and SnO2 ETL. They were exposed to air under one sun at constant temperature, ranging from 25 to 70 °C. Authors investigated the kinetics and thermal activation of measured degradation. Furthermore, authors attributed the loss of performances to a degradation of the HTL by using complementary characterization (X-ray diffraction and absorption measurements).

Details on the results reported in this section are in the following paper, submitted by the same authors: *Distinction of Mechanisms Causing Experimental Degradation of Perovskite Solar Cells by Simulating Associated Pathways.*

The procedure described in the previous section is applied and results show that experimental pathways at different temperatures are not all similar, indicating that potentially different mechanisms occurred. This questions the kinetics study: measuring thermal activation of a chemical process relies on the assumption that the same process is involved and activated at all investigated temperatures.

In fact, degradation recorded at 40 and 55°C can be associated to a decrease of hole mobility in the perovskite layer. On the other hand, experimental data recorded at 70°C follows a distinct path, closer to the reduction of electron mobility in the HTL.

This shows how the procedure developed here can complete analyses based on characterizations: the observed degradation of the HTL could have happened simultaneously with deterioration of the perovskite layer itself, and the latter might be responsible for the performance losses at lower temperature.

Finally, the experimental degradation noticeably shows a two-steps trend, indicating that potentially two distinct mechanisms did impact the performances, one after the other. Under this assumption, it is possible to study the second part as a new distinct device: with a partially degraded starting point, and initial characteristics that can be reproduced through the same method as a non-degraded device.

Results show that the first step of degradation can be explained by a loss of perovskite electron mobility.

Interestingly, the same mechanism might have occurred at all temperatures at this step, being compatible with a kinetics approach.

For the second part of degradation, two distinct experimental pathways are distinguishable, similarly to the previous analysis. Previous results are confirmed, with a loss of hole mobility at lower temperatures, and a loss of doping of HTL at higher temperature.

Finally, the analysis shows the strength and versatility of the method. It is possible to apply it on results available in literature, and it can give complementary insights. Importantly, it is possible to discuss kinetics studies and give complementary insights on the validity of this approach. Here we showed that the assumption that the same process occurs at several temperatures is not necessarily validated.

4 CONCLUSIONS

In this work, a simulation method to help identifying the dominant degradation mechanism has been developed. Correlations of optoelectrical parameters are used to characterize degradation and produce pathways that are intrinsic characteristics of the mechanisms. In fact, avoiding time dependence and related potential activation effects allows to be independent from environmental conditions. Therefore, simulations and experimental results are directly comparable.

A procedure has been established to reproduce the initial performances of any given device with a statistical approach. The developed genetic algorithm provides numerous sets of parameters, and unitary degradation mechanisms are implemented on each of them. Finally, average response can be derived, and directly compared to experimental data in terms of correlation pathways.

This shows that it is possible to analyze degradation of perovskite solar cells through the evolution of studied optoelectrical parameters and provide insights on causes. A first example, based on solar cells fabricated and measured at IPVF, is investigated. Results show the ETL is responsible for the performance degradation. This provides valuable insights to further develop the devices and enhance their stability. The second example analyzed here is extracted from results published in literature by Lim *et al.* Here, the degradation causes are distinguished between temperatures: the perovskite might be in cause at 40 and 55°C, and the ETL at 70 °C. Interestingly, this aspect questions the approach employed by the authors, aiming at measuring the thermal activation of the degradation and relying on the same mechanism being thermally activated. Finally, the versatility of the method is showed, by possibly analyzing sequentially the experimental results.

Finally, the novel method developed here shows that it possible to use simulations to tackle the degradation of perovskite solar cells. Importantly, it directly focuses on the causality of materials deterioration onto performances losses. This is a significant advantage on using heavy characterization techniques, that can merely show the concomitance of material and performance degradation. As a result, the work done here aims at providing a valuable tool to understand the degradation of perovskite solar cells and further develop their stability.

4 ACKNOWLEDGEMENTS

This project has been supported by the French Government in the frame of the program of investment for the future (Programme d'Investissement d'Avenir – ANR-IEED-002-01).

5 REFERENCES

[1] M. V. Khenkin, E.A. Katz, A. Abate, G. Bardizza, J.J. Berry, C. Brabec, F. Brunetti, V. Bulović, Q. Burlingame, A. Di Carlo, R. Cheacharoen, Y.B. Cheng, A. Colsmann, S. Cros, K. Domanski, M. Dusza, C.J. Fell, S.R. Forrest, Y. Galagan, D. Di Girolamo, M. Grätzel, A. Hagfeldt, E. von Hauff, H. Hoppe, J. Kettle, H. Köbler, M.S. Leite, S. (Frank) Liu, Y.L. Loo, J.M. Luther, C.Q. Ma, M. Madsen, M. Manceau, M. Matheron, M. McGehee, R. Meitzner, M.K. Nazeeruddin, A.F. Nogueira, Ç. Odabaşı, A. Osherov, N.G. Park, M.O. Reese, F. De Rossi, M. Saliba, U.S. Schubert, H.J. Snaith, S.D. Stranks, W. Tress, P.A. Troshin, V. Turkovic, S. Veenstra, I. Visoly-Fisher, A. Walsh, T. Watson, H. Xie, R. Yıldırım, S.M. Zakeeruddin, K. Zhu, M. Lira-Cantu, Consensus statement for stability assessment and reporting for perovskite photovoltaics based on ISOS procedures, Nat. Energy. 5 (2020) 35–49.

[2] J. Hidalgo, C.A. Perini, A.-F. Castro-Mendez, D. Jones, H. Köbler, B. Lai, R. Li, S. Sun, A. Abate, J.-P. Correa-Baena, Moisture-Induced Crystallographic Reorientations and Effects on Charge Carrier Extraction in Metal Halide Perovskite Solar Cells, ACS Energy Lett. 5 (2020) 3526–3534.

[3] S.P. Dunfield, L. Bliss, F. Zhang, J.M. Luther, K. Zhu, M.F.A.M. Van Hest, M.O. Reese, J.J. Berry, From Defects to Degradation : A Mechanistic Understanding of Degradation in Perovskite Solar Cell Devices and Modules, Adv. Energy Mater. (2020) 1904054.

[4] C.C. Boyd, R. Cheacharoen, T. Leijtens, M.D. McGehee, Understanding Degradation Mechanisms and Improving Stability of Perovskite Photovoltaics, Chem. Rev. 119 (2019) 3418–3451.

[5] M. cheol Kim, S.Y. Ham, D. Cheng, T.A. Wynn, H.S. Jung, Y.S. Meng, Advanced Characterization Techniques for Overcoming Challenges of Perovskite Solar Cell Materials, Adv. Energy Mater. 11 (2021) 2001753.

[6] J.A. Christians, P. Schulz, J.S. Tinkham, T.H. Schloemer, S.P. Harvey, B.J. Tremolet De Villers, A. Sellinger, J.J. Berry, J.M. Luther, Tailored interfaces of unencapsulated perovskite solar cells for >1,000 hour operational stability, Nat. Energy. 3 (2018) 68–74.

[7] D. Messou, S. Cacocivh, S. Bechu, A. Yaiche, J. Rousset, P. Schulz, M. Bouttemy, Stability of triple cation halide perovskites layers: study of the chemical evolution after light soaking thanks to XPS analysis, in: 2020 47th IEEE Photovolt. Spec. Conf., 2021: pp. 1628–1631.

[8] H.J. Lin, S. Cacovich, A. Rebai, J. Rousset, C. Longeaud, Influence of Environment and Light-

Stress on the Optoelectronic Properties of Triple-Cation Perovskite Thin Films, ACS Appl. Mater. Interfaces. 12 (2020) 19495–19503.

[9] J. Peng, D. Walter, Y. Ren, M. Tebyetekerwa, Y. Wu, T. Duong, Q. Lin, J. Li, T. Lu, M.A. Mahmud, O.L.C. Lem, S. Zhao, W. Liu, Y. Liu, H. Shen, L. Li, F. Kremer, H.T. Nguyen, D.Y. Choi, K.J. Weber, K.R. Catchpole, T.P. White, Nanoscale localized contacts for high fill factors in polymer-passivated perovskite solar cells, Science (80-.). 371 (2021) 390–395.

[10] N. Li, Y. Luo, Z. Chen, X. Niu, X. Zhang, J. Lu, R. Kumar, J. Jiang, H. Liu, X. Guo, B. Lai, G. Brocks, Q. Chen, S. Tao, P. David, H. Zhou, Microscopic Degradation in Formamidinium-Cesium Lead Iodide Perovskite Solar Cells under Operational Stressors, Joule. 4 (2020) 1743–1758.

[11] M. Chen, S.R. Sahamir, G. Kapil, A.K. Baranwal, M.A. Kamarudin, Y. Zhang, K. Nishimura, C. Ding, D. Liu, D. Hirotani, Q. Shen, S. Hayase, Inverted CsPbI2Br perovskite solar cells with enhanced efficiency and stability in ambient atmosphere via formamidinium incorporation, Sol. Energy Mater. Sol. Cells. 218 (2020) 110741.

[12] J. Lim, M. Kim, H.H. Park, H. Jung, S. Lim, X. Hao, E. Choi, S. Park, M. Lee, Z. Liu, M.A. Green, J. Seo, J. Park, J.S. Yun, Kinetics of light-induced degradation in semi-transparent perovskite solar cells, Sol. Energy Mater. Sol. Cells. 219 (2021) 110776.

[13] B.X. Zhao, C. Yao, K. Gu, T. Liu, Y. Xia, Y.L. Loo, A hole-transport material that also passivates perovskite surface defects for solar cells with improved efficiency and stability, Energy Environ. Sci. 13 (2020) 4334–4343.

[14] S.C. Ezike, A.B. Alabi, A.N. Ossai, A.O. Aina, Stability-improved perovskite solar cells through 4-tertbutylpyridine surface-passivated perovskite layer fabricated in ambient air, Opt. Mater. (Amst). 112 (2021) 110753.

[15] A.A. Zaky, E. Christopoulos, K. Gkini, M.K. Arfanis, L. Sygellou, A. Kaltzoglou, A. Stergiou, N. Tagmatarchis, N. Balis, P. Falaras, Enhancing efficiency and decreasing photocatalytic degradation of perovskite solar cells using a hydrophobic copper-modified titania electron transport layer, Appl. Catal. B Environ. 284 (2021) 119714.

[16] J. Kim, N. Park, J.S. Yun, S. Huang, M.A. Green, A.W.Y. Ho-Baillie, An effective method of predicting perovskite solar cell lifetime–Case study on planar CH3NH3PbI3 and HC(NH2)2PbI3 perovskite solar cells and hole transfer materials of spiro-OMeTAD and PTAA, Sol. Energy Mater. Sol. Cells. 162 (2017) 41–46.

[17] J.P. Bastos, G. Uytterhoeven, W. Qiu, U.W. Paetzold, D. Cheyns, S. Surana, J. Rivas, M. Jaysankar, W. Song, T. Aernouts, J. Poortmans, R. Gehlhaar, Model for the Prediction of the Lifetime and Energy Yield of Methyl Ammonium Lead Iodide Perovskite Solar Cells at Elevated Temperatures, ACS Appl. Mater. Interfaces. 11 (2019) 16517–16526.

This is a short version of the paper submitted to the journal Progress in Photovoltaics.

It has been selected by the Committee of the WCPEC-8 for submission to Progress in Photovoltaics.

PROCESS ENGINEERING AND INTERFACIAL DEFECTS PASSIVATION OF LARGE AREA PEROVSKITE SOLAR MODULES

Luigi Vesce[1], Maurizio Stefanelli[1], Stefano Razza[1], Luigi Angelo Castriotta[1], Francesco Di Giacomo[1], Aldo Di Carlo[1,2]
[1]CHOSE – Centre for Hybrid and Organic Solar Energy,
Department of Electronic Engineering, Tor Vergata University of Rome, Rome, Italy
[2]CNR-ISM – Institute for Structure of the Matter, National Research Council, Rome, Italy
vesce@ing.uniroma2.it

ABSTRACT: The industrial and economical transition of the perovskite (PVSK) photovoltaic (PV) technology is feasible by scaling-up effective PVSK recipe and post-treatment from small area cells (< 1 cm^2) to sub-modules (≥ 200 cm^2). In this work, we reported on the ambient air fabrication of a low temperature planar n-i-p PVSK sub-module (total area 320 cm^2, aperture area 201 cm^2, 93% GFF) exploiting the defect passivation strategy, the PVSK composition and a hybrid blade/slot-die coating technique assisted by air and green antisolvent quenching method. Finally, the sub-module record efficiency is 16.13% with 7% and 0% losses with respect to small area cell and mini-module, respectively. Further details, characterizations, life cycle assessment (LCA) and stability test are in the full paper submitted to Progress in Photovoltaics.
Keywords: photovoltaic, perovskite, module manufacturing, scaling-up, deposition, LCA

1 INTRODUCTION

The organometal halide perovskite (PVSK) physical-chemical properties and the solution-based fabrication process led to power conversion efficiencies higher than any thin film photovoltaic (PV) [1–4]. Researchers reached efficiencies of 25.7% for the single junction and 31.25% when used in tandem with silicon thanks to the synergy of material compositions fine tuning, device architectures design and fabrication processes optimization [3,4]. The real exploitation at the market level of PVSK PV technology can be feasible only if the laboratory cells (area < 1 cm^2) results will be transferred to mini (area < 200 cm^2) and sub-module devices (200 cm^2 ≤ area ≤ 800 cm^2) in ambient air environment by scalable printing/coating techniques [5–7]. In the scaling-up route, the high quality large area PVSK films, the low losses induced by front contact and the cells interconnections are the main factors to getting reproducible and reliable high efficiency modules. [8–12] Layers inhomogeneity rises when transferring to module devices processes and material compositions optimized for the small area cells [13,14]. In this context, the polycrystalline nature of solution-processed PVSK layers induces defects, such as at grain boundaries and vacancies during the fabrication process [15,16].

In literature few attempts showed mini- and sub-modules fabricated in ambient air by scalable printing/coating techniques. Regarding the mini-modules, different PVSK compositions ($MAPbI_3$ and multi-cation) were deposited by blade-coating and slot-die coating techniques with 20.2% maximum efficiency [6,17–20]. Besides the efforts on mini-modules, the development of scaled-up manufacturing processes for PVSK solar sub-modules is crucial to provide an expeditious and economical transition into mass production of such devices as PVSK solar panels and arrays [8]. Recently, Bu et al. reported 15.3% efficiency on 205 cm^2 module by blade coated FACs PVSK with methylammonium chloride additives in glovebox, but with high losses with respect to spin-coated small-sized cell (more than 36%) and mini module (more than 25%) [21].

In this work, we scaled up from lab scale cell to sub-module device by performing module designing and related interconnections optimization, by exploiting the interfacial defects passivation and PVSK solvent addiction strategies, and by introducing a hybrid blade/slot-die coating meniscus coating process out of the glovebox with green antisolvent quenching. To the best of our knowledge, this is the first time that a sub-module was fabricated outside the glovebox by scalable deposition technique. The demonstrated homogeneity of the full stack permitted to reach a record efficiency of 16.13% (7% losses with respect to small area cell) The performance is equal to the blade-coated mini-module fabricated in our recent publication [6]. Further details, characterizations, life cycle assessment (LCA) and stability test are in the full paper submitted to Progress in Photovoltaics.

2 MATERIALS AND METHODS

The n-i-p planar module is fabricated in ambient air as follow (Fig. 1). The planar configuration guarantees a low-temperature processed device [22]. Fluorine-doped tin oxide (FTO)-glass substrates (7 ohms/sq, Pilkington) were patterned (P1) to form 16 series connected cells by an UV nanosecond raster scanning laser. A colloidal solution of tin oxide diluted in water was deposited by slot-die coating technique to form the ETL (electron transporting layer). Then the film was annealed in air at 150°C for 30 min. The substrate was treated under an UV-oven ozone lamp to improve the wettability prior to the PVSK deposition. The solvent engineered 3C CsMAFA PVSK was deposited by hybrid blade/slot-die coating (ambient air, 30% RH) technique on a thermal plate at 40 °C. The blade coating deposition process of PVSK precursor was assisted by a heated (40°C) airflow (220 l/min) to pre-dry the PVSK surface. Then, antisolvent deposition was carried out by slot-die coating isopropanol and the substrate was annealed at 100°C on a hot plate for 20 minutes. Phenethyl ammonium iodide (PEAI) solution in IPA was slot-die coated on top of PVSK assisted by dry-air quenching at RT. The HTM (hole transporting material) doped Spiro-OMeTAD solution in chlorobenzene was coated by an air-assisted slot-die method in ambient air.

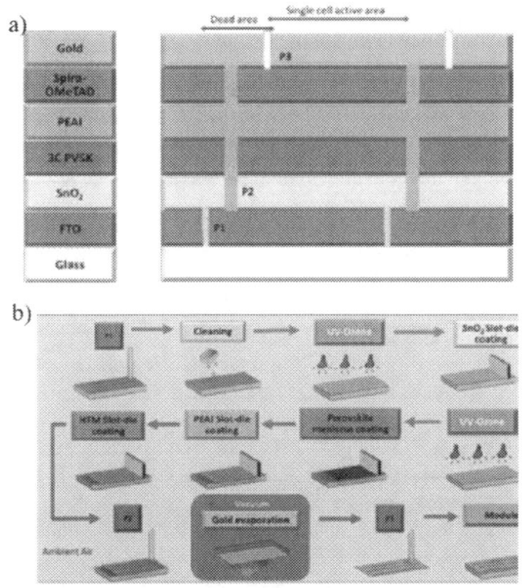

Figure 1: a) Perovskite module stack and b) fabrication full process.

The next step is the removal (P2) by the UV laser the full stack (ETL/PVSK/PEAI/HTL) deposited on FTO from the vertical connection areas to series connect two adjacent cells with the subsequent electrode deposition. Afterwards, 120 nm thick Au counter-electrode is thermally evaporated in a high vacuum chamber. Finally, the counter electrode P3 laser ablation is performed to obtain the electrical insulation between the counter electrodes of adjacent cells.

The photovoltaic characteristics are measured with a class A sun simulator (Sun 2000, Abet) at AM 1.5 $1000 W/m^2$ calibrated with an SKS 1110 sensor (Skye Instruments Ltd., Llandrindod Wells, UK); the system is equipped with a 2612 source meter (Keithley Instruments Inc., Cleveland, OH, USA) and a LabVIEW interface. The EL imaging on module adopts a Silicon CCD camera. The illumination intensity was kept constant for the different measurements.

3 RESULTS AND DISCUSSION

In the developed hybrid meniscus coating technique, the isopropanol antisolvent is disbursed by slot-die coater during the full process by a continuous flow rate forming a steady state meniscus. The synergic effect of the antisolvent and gas quenching causes instant supersaturation of precursor solution and grants fast nucleation and rapid removal of solvents, promoting the homogeneity of PVSK film in the whole substrate. [23]

We pursued two strategies to boost performance and get superior quality of such large area devices. The first one is related to the surface defects passivation of PVSK layer and the second one to the PVSK additive engineering to improve nucleation and quality of the absorber material. The PVSK surface passivation by phenethylammonium iodide (PEAI) is one of the most efficient method to suppress nonradiative recombination losses and to improve charge carrier extraction and photovoltage, by forming a shielding layer in such a way more effective than 2D PEA_2PbI_4 PVSK. [24–26] Besides the surface passivation, we engineered the PVSK precursor solution with a solvent additive to control intermediate phase of PVSK wet film and get superior crystal quality reducing defects and controlling the nuclei formation. Our strategy is focused on adding high vapor pressure non-coordinating ACN solvent in PVSK solution that can uniform the evaporation rate and the supersaturated phase of the absorber film on large substrate area.

The advantages related to the device design, fabrication procedure, PVSK engineering composition and passivation addiction led to a champion efficiency of 16.13% with Voc=18.15 V, Isc=263.42 mA, Jsc=22.65 mA/cm^2, and FF=62.8% (Fig. 2a). The performance are comparable to that obtained with the mini-module reported in our recent work and 7% less with respect to 0.5 cm^2 cells [6]. Preliminary results showed no shunts defects on the module by electroluminescence (EL) analysis with an injection current of 40 mA (Fig. 2b). Further details, characterizations, life cycle assessment (LCA) and stability test are in the full paper submitted to Progress in Photovoltaics.

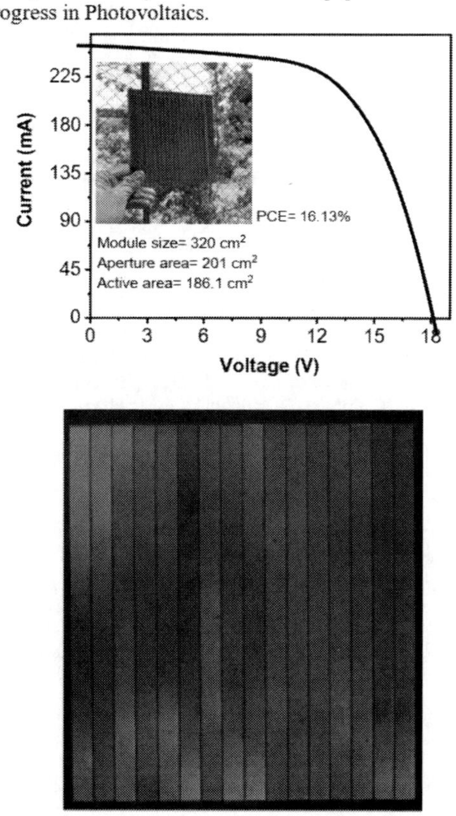

Figure 2: Sub-module a) current-voltage curve and b) EL imaging.

4 CONCLUSION

In conclusion we presented the efforts to improve the sub-module efficiency state of art and reducing the gap with lab scale cells and mini-modules by improving structural design, material chemistry, and process (coating and laser ablation). We demonstrated a planar PVSK sub-module (201 cm^2 aperture area, 93% GFF) with 16.13% efficiency fully fabricated outside the glovebox (first time in literature) by meniscus coating techniques in real manufacturing conditions. We adopted the passivation strategy to suppress non radiative recombination losses and to improve charge carrier extraction and photovoltage. Moreover, we engineered the PVSK precursor solution with acetonitrile to control the intermediate phase of the absorber film. Different characterization techniques (SEM, PL, UV-Vis) testified the homogeneity of the deposited layers on the full device area. EL reveals the homogeneity and the interface quality of the meniscus coated PVSK layer without any shunt defects. The reported simple and reproducible method represent an outstanding demonstration of the PVSK technology exploitation. Further details, characterizations and stability test are in the full paper submitted to Progress in Photovoltaics.

ACKNOWLEDGEMENTS

The authors were supported by the European Union's Horizon 2020 Framework Program for funding Research and Innovation under grant agreements no. 881603 (Graphene Flagship Core 3), no. 826013 (IMPRESSIVE) and no. 764047 (ESPResSo). The authors acknowledge the project UNIQUE, supported under the umbrella of SOLAR-ERA.NET_cofund by ANR, PtJ, MUR (GA 775970), MINECOAEI, SWEA, within the European Union Framework Program for Research and Innovation Horizon 2020 (Cofund ERANET Action, No. 691664).

REFERENCES

1 A. Kojima, K. Teshima, Y. Shirai and T. Miyasaka, *J. Am. Chem. Soc.*, 2009, **131**, 6050–6051.
2 H. S. Kim, C. R. Lee, J. H. Im, K. B. Lee, T. Moehl, A. Marchioro, S. J. Moon, R. Humphry-Baker, J. H. Yum, J. E. Moser, M. Grätzel and N. G. Park, *Sci. Rep.*, 2012, **2**, 1–7.
3 NREL, Best research-cell efficiency chart, https://www.nrel.gov/pv/cell-efficiency.html, (accessed 25 May 2022).
4 M. A. Green, E. D. Dunlop, J. Hohl-Ebinger, M. Yoshita, N. Kopidakis and X. Hao, *Prog. Photovoltaics Res. Appl.*, 2022, **30**, 3–12.
5 M. A. Green, E. D. Dunlop, J. Hohl-Ebinger, M. Yoshita, N. Kopidakis, K. Bothe, D. Hinken, M. Rauer and X. Hao, *Prog. Photovoltaics Res. Appl.*, 2022, **30**, 687–701.
6 L. Vesce, M. Stefanelli, J. P. Herterich, L. A. Castriotta, M. Kohlstädt, U. Würfel and A. Di Carlo, *Sol. RRL*, 2021, **5**, 1–11.
7 NREL, Champion photovoltaic module efficiency chart.
8 D. K. Lee and N. G. Park, *Sol. RRL*, 2021, **2100455**, 1–25.
9 C. Ashworth, *Nat. Rev. Mater.*, 2021, **6**, 293.
10 L. Vesce, M. Stefanelli, L. A. Castriotta, A. Hadipour, S. Lammar, B. Yang, J. Suo, T. Aernouts, A. Hagfeldt and A. Di Carlo, *Sol. RRL*, , DOI:https://doi.org/10.1002/solr.202101095.
11 J. Barichello, L. Vesce, F. Matteocci, E. Lamanna and A. Di Carlo, *Sol. Energy Mater. Sol. Cells*, 2019, **197**, 76–83.
12 L. Vesce, A. Guidobaldi, P. Mariani, A. Di Carlo, M. L. Parisi, S. Maranghi and R. Basosi, in *World Scientific Reference of Hybrid Materials*, 2019, vol. 2, pp. 423–485.
13 Z. Yang, C. C. Chueh, F. Zuo, J. H. Kim, P. W. Liang and A. K. Y. Jen, *Adv. Energy Mater.*, 2015, **5**, 1–6.
14 L. Vesce, M. Stefanelli and A. Di Carlo, *2019 Int. Symp. Adv. Electr. Commun. Technol.*, 2020, **764047**, 1–5.
15 S. G. Motti, D. Meggiolaro, A. J. Barker, E. Mosconi, C. A. R. Perini, J. M. Ball, M. Gandini, M. Kim, F. De Angelis and A. Petrozza, *Nat. Photonics*, 2019, **13**, 532–539.
16 L. Vesce, M. Stefanelli, L. A. Castriotta, A. Hadipour, S. Lammar, B. Yang, J. Suo, T. Aernouts, A. Hagfeldt and A. Di Carlo, *Sol. RRL*, 2022, **2101095**, 1–8.
17 X. Dai, Y. Deng, C. H. Van Brackle, S. Chen, P. N. Rudd, X. Xiao, Y. Lin, B. Chen and J. Huang, *Adv. Energy Mater.*, 2020, **10**, 1–7.
18 Y. Deng, S. Xu, S. Chen, X. Xiao, J. Zhao and J. Huang, *Nat. Energy*, 2021, **6**, 633–641.
19 S. Chen, X. Dai, S. Xu, H. Jiao, L. Zhao and J. Huang, *Science (80-.).*, 2021, **373**, 902–907.
20 L. A. Castriotta, F. Matteocci, L. Vesce, L. Cinà, A. Agresti, S. Pescetelli, A. Ronconi, M. Löffler, M. M. Stylianakis, F. Di Giacomo, P. Mariani, M. Stefanelli, E. M. Speller, A. Alfano, B. Paci, A. Generosi, F. Di Fonzo, A. Petrozza, B. Rellinghaus, E. Kymakis and A. Di Carlo, *ACS Appl. Mater. Interfaces*, 2021, **13**, 11741–11754.
21 T. Bu, L. K. Ono, J. Li, J. Su, G. Tong, W. Zhang, Y. Liu, J. Zhang, J. Chang, S. Kazaoui, F. Huang, Y.-B. Cheng and Y. Qi, *Nat. Energy*, , DOI:10.1038/s41560-022-01039-0.
22 E. Calabrò, F. Matteocci, B. Paci, L. Cinà, L. Vesce, J. Barichello, A. Generosi, A. Reale and A. Di Carlo, *ACS Appl. Mater. Interfaces*, 2020, **12**, 32536–32547.
23 K. Liu, Q. Liang, M. Qin, D. Shen, H. Yin, Z. Ren, Y. Zhang, H. Zhang, P. W. K. Fong, Z. Wu, J. Huang, J. Hao, Z. Zheng, S. K. So, C. S. Lee, X. Lu and G. Li, *Joule*, 2020, **4**, 2404–2425.
24 M. R. et al Yoo, J.J., Seo, G., Chua, *Nature*, 2021, **590**, 587–593.
25 J. Suo, B. Yang and A. Hagfeldt, *Energies*, , DOI:10.3390/en14164836.
26 Q. Jiang, Y. Zhao, X. Zhang, X. Yang, Y. Chen, Z. Chu, Q. Ye, X. Li, Z. Yin and J. You, *Nat. Photonics*, 2019, **13**, 460–466.

INTERFACE BETWEEN POLY-SI ON OXIDE PASSIVATED SI BOTTOM AND PEROVSKITE TOP CELLS – SEARCH FOR THE LEANEST AND BEST WORKING LAYER STACK

R. Peibst[1,2], C. Schwarz[1,2], Y. Larionova[1], S. Wolter[1], M. Rienäcker[1], T. Wietler[1], M. Diederich[1], R. Brendel[1,3]
[1]Institute for Solar Energy Research Hameln (ISFH), Am Ohrberg 1, 31860 Emmerthal, Germany
[2]Institute for Electronic Materials and Devices, Leibniz University Hannover, Schneiderberg 32, 30167, Germany
[3]Institute for Solid State Physics, Leibniz University Hannover, Appelstr. 2, 30167 Hannover, Germany

ABSTRACT: Regarding the interface between Si bottom cells with poly-Si on oxide (POLO) junctions as the uppermost layer and the lowest layer of a perovskite (Pk) top cell (i.e., the hole-selective layer HTL in pin configuration), many questions are still open, including the one which layers are actually required for low-resistive subcell interconnection. In this work, we compare recombination-layer (RL) and tunneling junction (TJ) - based subcell interconnection schemes, namely spiro-TTB/*n*+ poly-Si, spiro-TTB/p^+ poly-Si/n^+ poly-Si, and spiro-TTB/ITO/n^+ poly-Si. The first is – to our knowledge for the first time – investigated experimentally. We prepare dedicated Pk/Si test structures, which are opaque co-evaporated $MAPbI_3$ top cells on Si substrates with the uppermost layers of interest. Dark-*IV* and frequency-dependent impedance measurements are applied. We reveal an upper limit for the contact resistance of 14 mΩcm^2 for the p^+ poly-Si / n^+ poly-Si tunnel junction with Cox&Strack measurements. We observe a reduction of the current densities measured on our Pk/Si test structures upon biasing at more than 1 V, that stabilizes at a factor of ~1/300 for the spiro-TTB/ITO and of ~ 1/10 for both spiro-TTB/poly-Si interfaces as compared to the initial measurement. After stabilization, our *IV* measurements show that the spiro-TTB/n^+ poly-Si recombination junction conducts comparable current densities as the spiro-TTB/p^+ poly-Si interface. Within a framework of a three-element equivalent circuit for impedance measurements, the spiro-TTB/X interface affects the voltage dependent parallel resistance. The comparable current densities for the spiro-TTB/n^+ poly-Si and the spiro-TTB/p^+ poly-Si interface gives hope that n^+-type poly-Si can indeed work as a recombination layer itself, although we did not succeed in a quantification of the specific resistance of the spiro-TTB/X interface. This indication has to be verified on full tandem cells.

Keywords: perovskite, silicon, poly-Si, hole transport layer, tandem solar cells, tunnel junction, recombination junction, interface

1 INTRODUCTION

Perovskite/Si tandem cells, in particular those based on a "PERC-like" Si bottom cell passivated by a poly-Si on oxide (POLO) front junction [1, 2], are very attractive in terms of efficiency potential and levelized cost of electricity (LCOE) [3]. The interface between both subcells is, as in every monolithic tandem device, essential for electrical and optical performance [7]. Recombination layers based on TCOs also work for "PERC-like" Si bottom cells [2], but TCO deposition on the poly-Si needs to be performed after fast firing of the rear-side metallization, and is then causing sputter damage on the POLO junctions that is difficult to cure completely. Tunneling junctions between the n^+-type poly-Si layer from the electron-collecting POLO junction and an additional p^+-type poly-Si layer may prevent the necessity of TCO sputtering on the poly-Si if the HTL is deposited by other methods (e.g. evaporation) and forms a low resistive junction to the p^+ poly-Si. Meanwhile, these p^+ poly-Si / n^+ poly-Si tunneling junctions have proven to exhibit abrupt doping profiles and low inherent contact resistances after firing [1, 8, 9]. However, little is known so far about the interface between the poly-Si as the uppermost layer of the bottom cell and the hole transport layer as the first layer (in pin configuration) of the top cell. Although well-working "all-low-temperature processed" Pk/Si tandem cells with a direct contact between *p*+ nanocrystalline (nc) Si:H and 2,2′,7,7′-tetra(N,N-di-tolyl)amino-9,9-spiro-bifluorene (spiro-TTB) as HTL have been demonstrated [4,5], the best HTL / poly-Si combination still needs to be found. It is not even certain that the additional p^+ -type poly-Si layer is indeed necessary. Perhaps, a recombination junction could be formed directly between the n^+ poly-Si and the HTL (see Fig. 1). This would offer the leanest possible processes flow.

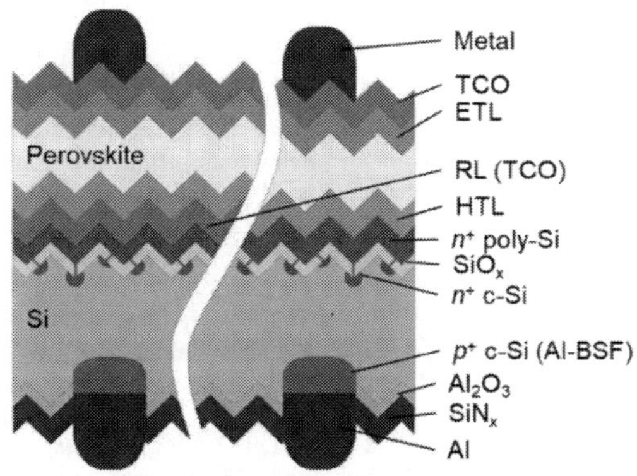

Fig. 1. Schematic drawing of the targeted cell structure. An interconnection of the subcells via a HTL/TCO recombination junction is shown left. It would be attractive to omit the TCO (ITO) layer – as well as further layers like *p*+ -type doped poly-Si - if possible (right).

In this work, we therefore evaluate such a spiro-TTB/n^+-type poly-Si recombination junction and compare it with a spiro-TTB/p^+-type poly-Si junction as used in previous works on p^+ -type nc-Si:H [4,5]. As reference, we also include the spiro-TTB/ITO interface in our split (see Fig. 2).

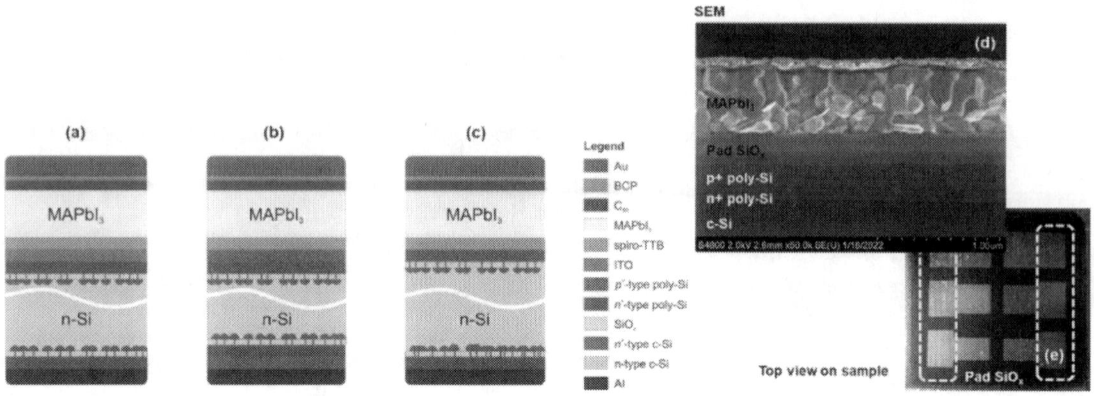

Fig. 2. Schematic drawings of the opaque Pk/Si test structures with the HTL (spiro-TTB) in contact with (a) ITO, (b) p^+ poly-Si, (c) n^+ poly-Si. (d) Scanning electron microscopy cross section image of the structure shown in (b), but in the perimeter regions with SiO_x pads underneath the Pk stack to prevent shunting by probes. (e) Photograph of Si substrate with six diodes of different sizes. The SiO_x pad regions are indicated by the white dashes.

2 EXPECTATIONS BASED ON BAND DIAGRAMM (ANDERSON MODEL)

Fig. 2 shows the band diagrams constructed in the framework of the Anderson model [6] and the photoelectron spectroscopy (UPS) results for spiro-TTB from Ref. [5] for the different spiro-TTB/X interfaces investigated. The work function for ITO is set to 5 eV according to the UPS measurements in Ref. [10], the band gap narrowing in the (poly-) Si is calculated according to [11]. At the interface of the spiro-TTB/ITO recombination junction, an energy difference of only 96 meV needs to be dissipated for electrons and holes to recombine (Fig. 3 (a)). For spiro-TTB/p^+-type poly-Si, an ohmic junction for holes can be expected (neglecting the small valence band offset of 9 meV). The hole current from the top cells finds it's continuation in the electron current from the bottom cell via band to band tunneling at the n^+ poly-Si / p^+ poly-Si interface (Fig. 3 (b)). For the spiro-TTB/n^+-type poly-Si junction, however, the Anderson model predicts the necessity to dissipate at least 596 meV (Fig. 3 (c)). This is very disadvantageous even if the high density of defects within the poly-Si (minority carrier lifetimes in the pico-second rage [1]) might be sufficient to efficiently mediate such a recombination. Other HTL materials with higher work function like NiO_x could mitigate this issue. In face of the limitations of the Anderson model and the uncertainty regarding interface charges and input parameters, we investigate the spiro-TTB/n^+ poly-Si interface experimentally.

3 EXPERIMENTAL APPROACH

Since we anticipated that the investigation of this specific interface within a full tandem cell would be complicated, we base our study on simplified test structures (see Fig. 2). Here, the Si substrate is an ohmic contact to the representative top layers (n-type poly-Si (+ITO), p^+-type /n^+-type poly-Si TJ). We use n-type Cz wafers with a resistivity of 2 Ωcm. On both sides, we form n+-type poly-Si / SiO_x junctions. The wet chemically grown interfacial oxide is broken-up to a large extend during an excessive high temperature anneal in order to blind out the poly-Si/SiO_x/Si junction resistance. The n^+-type poly-Si is the uppermost layer of the Si substrate for the split shown in Fig. 2(c). For the split group shown in Fig. 2(b), a highly p-type doped poly-Si layer is deposited on top of the n^+-type poly-Si by low pressure chemical vapor deposition (LPCVD) at 600 °C. Electrochemical capacitance voltage (ECV) measurements (not shown) reveal an abrupt profile of dopants with concentrations > $2 \cdot 10^{20}$ cm^{-3} on both sides of the p^+-type poly-Si / n^+-type poly-Si tunneling junction. Two-terminal IV measurements according to the Cox&Strack method on sister samples with evaporated, differently-sized Al pads result in an upper limit of 14 mΩcm^2 for the specific resistance of the p^+-type poly-Si / n^+-type poly-Si tunneling junction. For the split group shown in Fig. 3(a), 20 nm ITO are sputtered on top of the n^+-type poly-Si. Al is evaporated as the rear-side metallization for all split groups. Prior to the deposition of the HTL (spiro-TTB) via evaporation, native SiO_x is removed from the surface of the poly-Si by HF. Since a separate contacting of the HTL could imply additional resistive contributions and shunting issues that are not representative for the tandem cell, we "contact" the HTL - as in the final device - via a perovskite ($MAPbI_3$) layer. For contacting of the latter, we apply our standard stack of C_{60}, BCP and Au. Thus, opaque all-evaporated perovskite diodes are formed on top of the Si.

The test structures are characterized by dark IV measurements using a Keithley 4200 analyzer with current preamplifiers. When biasing our test structures to voltages > 1 V, the current density for a fixed voltage reduces. This effect saturates (after several minutes) at a factor of ~1/300 for the spiro-TTB/ITO and of ~1/10 for both spiro-TTB/poly-Si interfaces as compared to the initial current density. We speculate that ion migration is the root cause for change in the IV characteristic as compared to the unbiased sample. We thus decide to characterize our test structures also by frequency-dependent impedance measurements [11] at voltages that do not alter the electrical behavior (e.g. 0.5 V). We use an Agilent 4294A Precision Impedance Analyzer. Although we acknowledge that impedance measurements can reveal many complex aspects in perovskite solar cells [11], we expect that a three-element equivalent circuit (voltage-dependent parallel capacitance and conductance

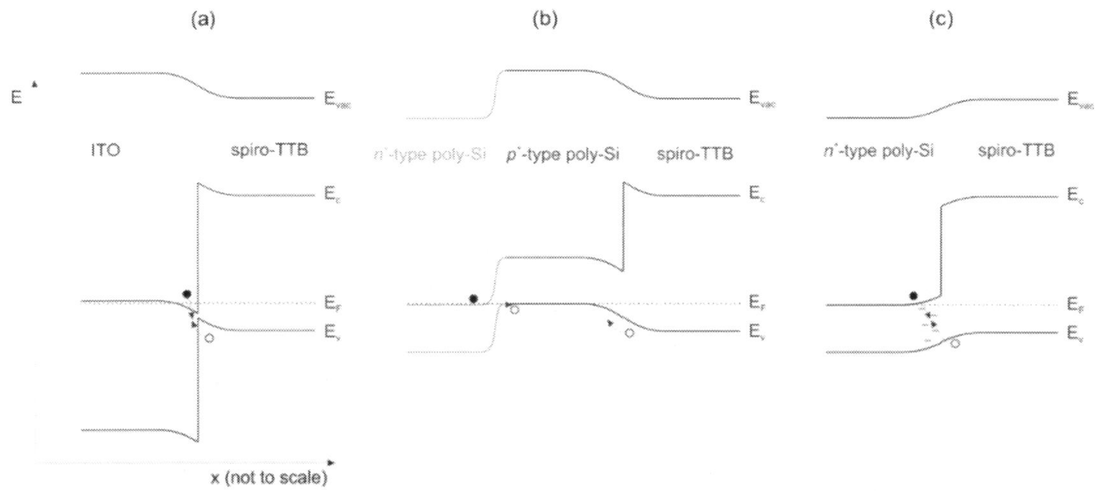

Fig. 3. Band diagrams sketched for (a) the spiro-TTB / ITO, (b) spiro-TTB/ p^+ poly-Si and (c) spiro-TTB / n^+ poly-Si interface according to the Anderson model [6] and Ultraviolet Photoelectron Spectroscopy data for ITO [10] and spiro-TTB [5]. The bandgap of the highly doped Si layers is taken from Ref. [11]. Note that while the energetic alignment far away from interface as well as the band offsets are quantified, while no calculation of the depletion zone widths (which would require the knowledge of the density of states for all materials and a corresponding solution of Poisson's equation) is performed.

$G_P=1/R_P$, voltage-independent resistance R_s in series) can describe the frequency behavior of the impedance in the frequency range investigated here (50 Hz – 500 kHz).

4 RESULTS

4.1 Dark-*IV*

Fig. 4 shows the dark-*IV* curves for representative samples after stabilization. As expected for the Pk diode, all split groups show an exponential voltage dependence. Surprisingly (in the view of the expectations from the Anderson model), the spiro-TTB/n^+-type poly-Si conducts comparable current densities as the spiro-TTB/p^+-type poly-Si interface. The current densities measured on the samples with the spiro-TTB/ITO interface are higher by one order of magnitude. On none of our samples, the current density is – after stabilization - in the range of the short circuit current density (20 mA/cm²) at a voltage that is in the range of the open circuit voltage of the perovskite cell (1057 mV) as measured on a 17 % efficient opaque Pk single junctions cell on ITO covered glass with this stack. However, at least for the spiro-TTB/ITO interface, this was the case for the initial measurement. Since our test structures do not allow a quantification of the voltage drop across the perovskite diode and across other resistive components such as the spiro-TTB/X interface, and since the degradation effects prevent a further increase of the voltage to enable us to quantify the series resistance from the IV curve (as in the Si world [12]), we attempt to extract the latter from impedance measurements at lower voltages.

4.2 Impedance spectroscopy

Fig. 5 shows the Nyquist plots of all three split groups for 0.5 V (open squares) and 2.5 V (filled circles). Within the framework of the three-element equivalent circuit, the locus curves should describe semi-circles in the third square of the coordinate system. The series resistance R_s corresponds to the intersection of the curves with the real axis close to the origin. Although it was not fully reached at our upper frequency limit of 500 kHz, one can conclude that R_s is < 10 Ω (area-related < 2 Ωcm²) for all split groups (see insets in Fig. 5). Within our resolution limits, no significant effect of the spiro-TTB/X interface was found in R_s. The diameters of the semi-circles are given by R_P. They depend on the working point of the Pk diode. For a working point of

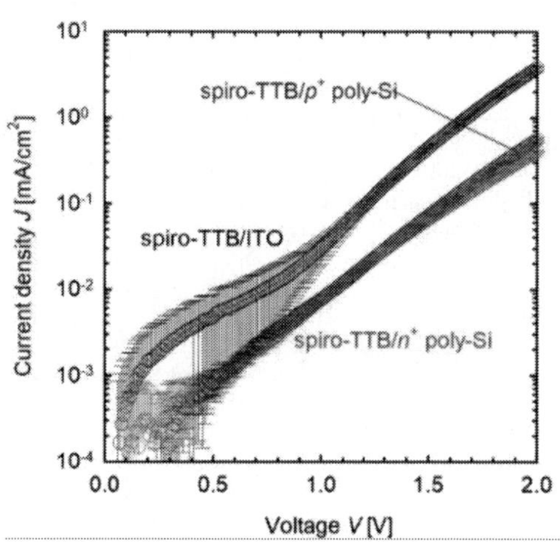

Fig. 4. Dark-*IV* curves for representative samples after preconditioning with spiro-TTB/- ITO (black); p^+ poly-Si (blue); n^+ poly-Si (red) interface. Each data point corresponds to the average value of 10 measurements after stabilization. Error bars refer to the standard deviation. Pad size is 0.2 cm². The reproducibility of the measurement is verified on numerous pads and samples.

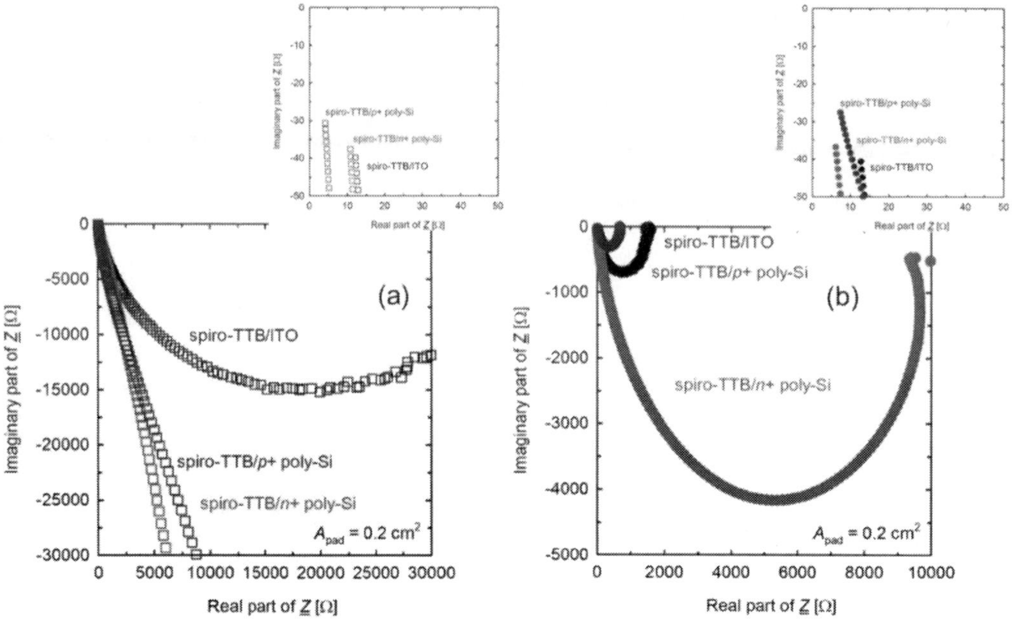

Fig. 5. Nyquist plots for representative samples from all split groups (color coded as in Fig. 4), either (a) at V_{DC} = 0.5 V or (b) V_{DC} = 2.5 V (filled circles). The insets shows the impedance for highest frequencies (≤500 kHz).

V_{DC}=0.5 V, R_p is large, and only a small section of the semi-circle is described for frequencies > 50 Hz. As expected from the differential resistance of the perovskite diode, R_p is reduced at 2.5V, and local curves roughly correspond to the expected semi-circles. Interestingly, R_p strongly differs between the split groups. It has values in the range of ~700 Ω for the spiro-TTB/p^+-type poly-Si interface (area-related ~140 Ωcm²), of 1400 Ω for the spiro-TTB/ITO interface (area-related ~280 Ωcm²), and of ~10 kΩ for the spiro-TTB/n^+-type poly-Si interface (area-related ~2000 Ωcm²). Thus, we interpret our results within the framework of the equivalent circuit shown in Fig. (6) where the spiro-TTB/X interface effects R_p rather than R_s. Whether the influence of the spiro-TTB/X interface on R_p is due to a different conductivity across this interface or indirectly via a change in quasi-Fermi level splitting in the perovskite (due to the different voltage drop across the different spiro-TTB/X interfaces) is the subject of ongoing investigations. For the spiro-TTB / n^+-type poly-Si interface, the parallel resistance is around one order of magnitude higher than for the other split groups. Obviously, the impedance measurements show a small signal behavior that is different from the large signal behavior shown by the *IV* measurements (since the latter indicates similar conductivity for the spiro-TTB/- p^+-type poly-Si and n^+-type poly-Si interface interface, as well as a lower conductivity of the spiro-TTB / p^+-type poly-Si than for the spiro-TTB / ITO interface.

5 CONCLUSIONS

To summarize, *IV* and impedance measurements consistently show that the spiro-TTB/X interface (X=n^+-type poly-Si, p^+-type poly-Si, ITO) is not a simple ohmic contribution but rather affect the *IV* and $\underline{Z}(f)$ in the entire measurement range. Most remarkably, we obtain similar *IV* curves on the samples with spiro-TTB/p^+-type poly-Si and with spiro-TTB/n^+-type poly-Si interface. This is despite the fact that the Anderson model predicts an ohmic contact for the first and a recombination junction with a large energy dissipation for the latter. Reality seems to be more complex. In contrast to similar (large signal) *IV* curves, the (small signal) $\underline{Z}(f)$ curves shows a one order of magnitude higher parallel resistance for the spiro-TTB/n^+-type poly-Si interface than for the with spiro-TTB/p^+-type poly-Si interface. A full tandem cell will have to show whether this recombination junction is suited for an interconnection of the subcells.

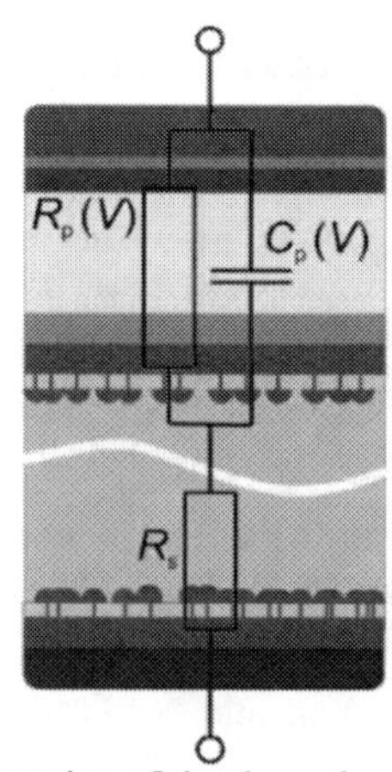

Fig. 6. Interpretation of the three-element equivalent circuit for our test structures. Obviously, the spiro-TTB / X interface affects the voltage dependent parallel resistance R_p rather in the voltage-independent R_s.

6 ACKNOWLEDGEMENTS

We would like to thank Renate. Winter, Sarah Spätlich, Heike Kohlenberg, David Sylla, Mircea Turcu, Jessica Strey and Guido Glowatzki for their contributions to sample processing, and Jan Krügener, Guido Wetzel and Hannah Genath for valuable discussions. This work was founded by the German Ministry for Economic Affairs and Climate Action (FKZ 03EE1056A) and the federal state of Lower Saxony.

7 REFERENCES

[1] R. Peibst, M. Rienäcker, B. Min, C. Klamt, R. Niepelt, T. F. Wietler, T. Dullweber, E. Sauter, J. Hübner, M. Oestreich, and R. Brendel, "From PERC to Tandem: POLO- and p+/n+ Poly-Si Tunneling Junction as Interface Between Bottom and Top Cell," *IEEE Journal of Photovoltaics*, vol. 9, no. 1, pp. 49–54, Jan 2019.

[2] S. Mariotti, K. Jäger, M. Diederich, M. S. Härtel, B. Li, K. Sveinbjörnsson, S. Kajari-Schröder, R. Peibst, S. Albrecht, L. Korte *et al.*, "Monolithic perovskite/silicon tandem solar cells fabricated using industrial p-type polycrystalline silicon on oxide/passivated emitter and rear cell silicon bottom cell technology," *Solar RRL*, vol. 6, no. 4, p. 2101066, 2022.

[3] C. Messmer, B. S. Goraya, S. Nold, P. S. Schulze, V. Sittinger, J. Schön, J. C. Goldschmidt, M. Bivour, S. W. Glunz, and M. Hermle, "The race for the best silicon bottom cell: Efficiency and cost evaluation of perovskite–silicon tandem solar cells," *Progress in Photovoltaics: Research and Applications*, vol. 29, no. 7, pp. 744–759, 2021.

[4] F. Sahli, J. Werner, B. A. Kamino, M. Bräuninger, R. Monnard, B. Paviet-Salomon, L. Barraud, L. Ding, J. J. Diaz Leon, D. Sacchetto *et al.*, "Fully textured monolithic perovskite/silicon tandem solar cells with 25.2% power conversion efficiency," *Nature materials*, vol. 17, no. 9, pp. 820–826, 2018.

[5] Y. Li, B. Shi, Q. Xu, L. Yan, N. Ren, Y. Chen, W. Han, Q. Huang, Y. Zhao, and X. Zhang, "Wide bandgap interface layer induced stabilized perovskite/silicon tandem solar cells with stability over ten thousand hours," *Advanced Energy Materials*, vol. 11, no. 48, p. 2102046, 2021.

[6] R. Anderson, "Germanium-gallium arsenide heterojunctions," *IBM Journal of Research and Development*, vol. 4, no. 3, pp. 283–287, 1960.

[7] M. De Bastiani, A. S. Subbiah, E. Aydin, F. H. Isikgor, T. G. Allen, and S. De Wolf, "Recombination junctions for efficient monolithic perovskite-based tandem solar cells: physical principles, properties, processing and prospects," *Materials Horizons*, vol. 7, no. 11, pp. 2791–2809, 2020.

[8] F.-J. Haug, A. Morisset, P. Wyss, M. Lehmann, A. Hessler-Wyser, Q. Jeangros, A. Ingenito, C. Ballif, C. S. Kumar, S. Eswara *et al.*, "Passivating polysilicon recombination junctions for crystalline silicon solar cells," *physica status solidi (RRL)–Rapid Research Letters*, vol. 15, no. 9, p. 2100272, 2021.

[9] C. Luderer, M. Penn, C. Reichel, F. Feldmann, J. C. Goldschmidt, S. Richter, A. Hähnel, V. Naumann, M. Bivour, and M. Hermle, "Controlling diffusion in poly-si tunneling junctions for monolithic perovskite/silicon tandem solar cells," *IEEE Journal of Photovoltaics*, vol. 11, no. 6, pp. 1395–1402, 2021.

[10] C. Pan and T. Ma, "High-quality transparent conductive indium oxide films prepared by thermal evaporation," *Applied Physics Letters*, vol. 37, no. 2, pp. 163–165, 1980.

[11] "PV lighthouse: Band-gap calculator," Sep. 2022. [Online]. Available: https://-www.pvlighthouse.com.au/bandgap

[12] E. Von Hauff and D. Klotz, "Impedance spectroscopy for perovskite solar cells: characterisation, analysis, and diagnosis," *Journal of Materials Chemistry C*, vol. 10, no. 2, pp. 742–761, 2022.

[13] D. K. Schroder, *Semiconductor material and device characterization*. John Wiley & Sons, 2006, chapter 4, p. 185 ff.

MODELLING AND OPTIMIZATION OF PEROVSKITE-SILICON TANDEM DEVICES FOR ENHANCED FIELD ENERGY PRODUCTION

Špela Tomšič, Benjamin Lipovšek, Marko Jošt, Kristijan Brecl, Marko Topič
University of Ljubljana, Faculty of Electrical Engineering
Tržaška 25, 1000 Ljubljana

ABSTRACT: An important step in the evolution of every novel photovoltaic (PV) technology before it emerges on the market is to move beyond standard test conditions (STC) and strive to maximize the production of electrical energy in realistic operating conditions. In this work we use an in-house developed modelling approach for accurate calculation of the energy yield (*EY*) as the most important figure-of-merit in determining the capabilities of PV devices in realistic operating conditions. We first investigate how the installation of the tandem device at different geographical locations (selected from distinct Köppen-Geiger-Photovoltaic climate zones) affects the annual energy yield as well as the optimal bandgap of the perovskite sub-cell ($E_{G,PK}$). We show that the energy yield values are dominated by the location-conditioned annual insolation, whereas the optimal $E_{G,PK}$ is not significantly influenced by the climate zone of the device operation. However, all optimal $E_{G,PK}$ values obtained under realistic operating conditions are at least 20 meV higher than the one determined under STC conditions. Finally, we perform the sensitivity analysis of *EY* modelling. We show that the inclusion of realistic temperature and especially realistic spectral distribution of incident irradiance is crucial for obtaining accurate results, both in terms of absolute *EY* as well as optimal $E_{G,PK}$ values.
Keywords: tandem, perovskite, modelling, optimization, realistic operating conditions

1 INTRODUCTION

The field of photovoltaics, which offers a low cost and sustainable source of electrical energy, has experienced a nearly exponential growth in the recent years and decades. The majority (~95 %) of the PV market today is taken-up by single-junction mono- and multi-crystalline silicon (c-Si) wafer based solar cells [1], which are slowly approaching the famous Shockley-Queisser (SQ) conversion efficiency limit [2] practically constrained to 29 % [3]. In order to push the SQ limit further up to 45 % [4], double-junction or tandem solar cells have also been developed steadily over the years, with the novel perovskite-based tandems gaining the most attention recently. With the record efficiency of monolithic perovskite-silicon (PK-Si) tandem solar cells just pushing over the 30% barrier [5], and with the stability improving, we are gradually nearing the point of transition from the laboratory testing to the field and rooftop operation. In support of this goal, there has recently been a pronounced tendency in the photovoltaic research field to approach experimental characterization as well as numerical modelling (in form of optical optimization and energy yield calculations) of tandem PV devices by taking the realistic operating conditions fully into account [6]–[10]. This is crucial for proper design and optimization of large-area modules and eventually also large-scale systems to achieve an effective increase in the long-term performance as well as for extraction of fundamental information such as project energy payback times, degradation rates, etc. [11], [12].

However, when considering realistic installation and operation of the device, its long-term behavior may become significantly influenced by the various varying installation and environmental factors (e.g. incident power density, angle of incidence, ambient temperature, etc.). Therefore, with the transition from standard test conditions to realistic operating conditions, further development and optimization of solar cells under STC alone no longer suffice. This compels us to consider all the physical effects associated with the outdoor operation of the PV devices in order to improve their long-term energy yield and ultimately contribute to the increase of the global cumulative installed PV capacity [1].

In this contribution, a specialized, in-house developed modelling algorithm was used for reliably assessing the predicted annually generated (area-normalized) electrical energy – the energy yield (*EY*), given in kWh/m^2 – of two-terminal (2T) encapsulated PK-Si tandem solar cells operating in different geographical locations selected from distinct Köppen-Geiger-Photovoltaic (KGPV) climate zones [13]. The algorithm, based on realistic (measured) optical, electrical, thermal and environmental input parameters, also served us as the figure-of-merit in determining the optimal bandgap of the perovskite sub-cell ($E_{G,PK}$) for each of the selected locations. Finally, we focused on studying how the *EY* and optimal $E_{G,PK}$ are affected by two key aspects: i) spectrum of the incident irradiance and ii) operating cell temperature. From that we were able to propose certain guidelines, that may lead to increased long-term outdoor performance.

2 ENERGY YIELD MODELLING

For the purpose of this work, we used an in-house developed EY calculation algorithm that is based on three modelling approaches – optical, electrical, and thermal, which are coupled to determine the current density-voltage (*J-V*) characteristics of the investigated tandem solar cell at any given time and conditions.

Optical modelling is performed by coupled wave-ray optics methods in the simulator CROWM [14]. The results are then linked with the collected realistic irradiance (*G*) data to assess the short-circuit current density (J_{sc}) of each sub-cell in the tandem device. The operating cell temperature (T_{cell}), required both for electrical modelling as well as for including the effects of temperature-induced bandgap variations, is calculated based on the steady-state Duffie-Beckman thermal model [15] as a function of measured irradiance, ambient temperature (T_a) and wind speed (*WS*). The electrical model embraces measured *J-V* characteristics of individual perovskite and silicon single-junction solar cells obtained under well-controlled laboratory environment and, if required, extrapolated to a broader G and T_{cell} range, as necessary by the geographical location. The J-V curve of the investigated tandem device together with the corresponding maximum power is obtained at any given time by graphical summation of the corresponding J-V curves of individual sub-cells.

3 RESULTS AND DISCUSSION

In the scope of this study we focused on an encapsulated 2-terminal PK-Si tandem device presented schematically in Figure 1. For the top perovskite sub-cell it has been selected the "inverted" design (p-i-n), of which the absorber layer is comprised of a multiple cation, multiple halide composition. The bottom sub-cell is a double-side textured rear-emitter n-type silicon heterojunction (SHJ) cell, which has been optimized for the tandem architecture as reported previously [7], [16]. Additionally, encapsulation stack – glass (3.2 mm) / ethylene-vinyl acetate-EVA (500 µm) – is placed above the described tandem device, also demonstrated in Figure 1. While the bandgap of the perovskite absorber was optimized with respect to the long-term energy yield, its thickness was fixed to 1000 nm.

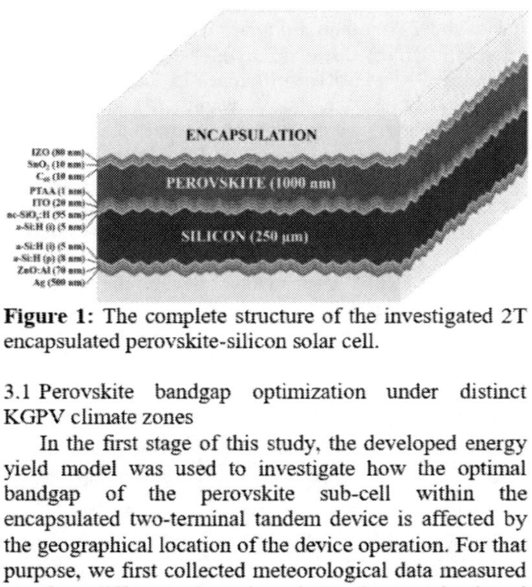

Figure 1: The complete structure of the investigated 2T encapsulated perovskite-silicon solar cell.

3.1 Perovskite bandgap optimization under distinct KGPV climate zones

In the first stage of this study, the developed energy yield model was used to investigate how the optimal bandgap of the perovskite sub-cell within the encapsulated two-terminal tandem device is affected by the geographical location of the device operation. For that purpose, we first collected meteorological data measured at nine different Central and North America-based geographical locations from distinct Köppen-Geiger-Photovoltaic (KGPV [13]) climate zones. In this work, realistic meteorological data (T_a, WS, spectrally-resolved global tilted (GTI) and direct normal irradiance (DNI) measurements) were acquired from the National Solar Radiation Database [17] in hourly intervals over the entire typical meteorological year (TMY [18]).

For each of the nine geographical locations we then varied the value of $E_{G,PK}$ in the range of 1.65 – 1.80 eV and calculated the annual energy yield of the device in the optimal free-rack orientation (location-dependent [19]). The maximal achievable energy yield values (EY_{max}) and the corresponding effective conversion efficiencies (EY/G_{tot}; with G_{tot} representing the total solar energy that reaches the device surface over the entire TMY) were extracted from the calculations for each location and are presented in Figure 2. The power conversion efficiency (PCE) obtained under STC conditions is also indicated in the figure.

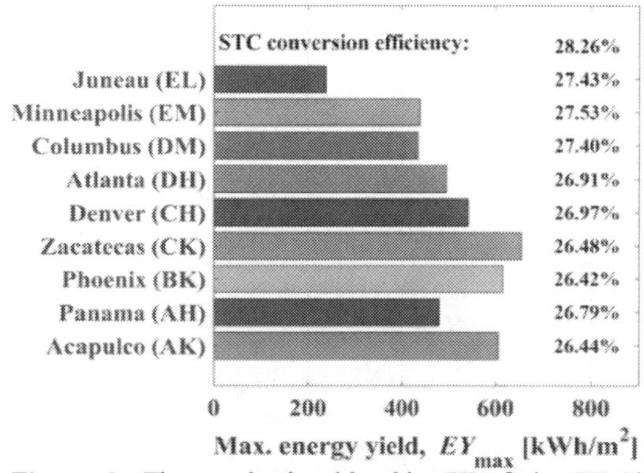

Figure 2: The maximal achievable EY of the PK-Si tandem device located at all nine selected geographical locations. Also indicated are the corresponding climate zone designations (in parentheses) and the effective conversion efficiencies (EY/G_{tot}).

It can be observed that tandem device installed at Zacatecas generates the most electrical energy, since Zacatecas represents a location with the highest yearly insolation, followed closely by Acapulco and Phoenix. However, it can also be noticed that the PV device located at these three locations exhibits the lowest conversion efficiencies due to higher operating temperatures specific for these climates zones (AK, BK and CK) with high G_{tot} and T_a. Since Minneapolis is one of the coolest location with relatively high G_{tot}, the device operating there exhibits the highest effective efficiency of 27.53%.

Finally, the optimal $E_{G,PK}$ values that rendered the maximal EY were calculated for all selected geographical locations as well as for the STC conditions. The results are presented in Figure 3. It can be clearly seen that the optimal $E_{G,PK}$ values obtained under location-specific realistic conditions differ from the optimal value determined under STC, and are in fact higher in all cases. Despite the strong variety in climate conditions, the impact of geographical location on the optimal $E_{G,PK}$ is only minor, since the difference between the highest (Panama, 1.724 eV) and the lowest (Denver, 1.704 eV) $E_{G,PK}$ is only 20 meV.

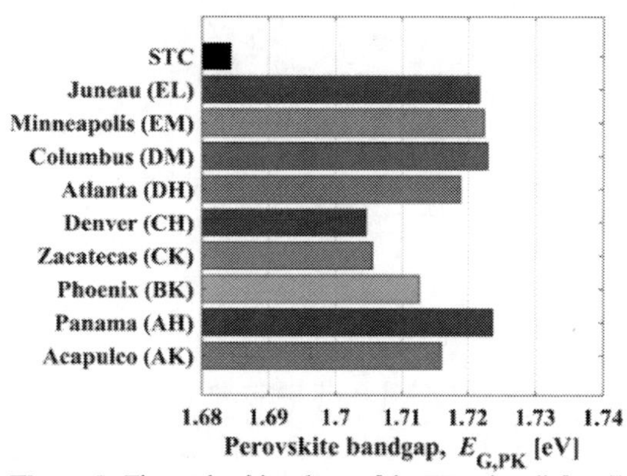

Figure 3: The optimal bandgap of the PK sub-cell for all selected geographical locations. Also indicated are the corresponding climate zone designations (in parentheses).

3.2 Sensitivity analysis of *EY* modelling

In the second stage of our work, our goal was to study the interdependencies of the incident irradiance spectrum and the operating device temperature. The analysis was in this case focused only on one geographical location – Atlanta, Georgia, USA – , since it represents the DH KGPV [13] that covers the Southern Europe and also a large area of the USA.

From the point of view of *EY* modelling and modelling-assisted design of tandem devices, the aspect of the spectral content of incident irradiance is important since in many cases environment monitoring at the target location is performed by measuring only the total irradiance, while its spectral dependency is not recorded and is often assumed to be that of the reference AM1.5 spectrum. Therefore, to disentangle the impact of the spectrum we applied both (i) measured and (ii) AM1.5 spectral distribution in our *EY* calculations. The aspect of the operating temperature, on the other hand, is important to accurately include all the temperature-induced effects, which influence the device performance. In our study, we investigated its impact by including three cases: (i) by assuming a constant cell temperature of 25 °C, and by calculating T_{cell} via the Duffie-Beckman model, where the upper and bottom side of the PV module are subjected to (ii) efficient forced convective cooling due to wind and (iii) very little to no wind cooling.

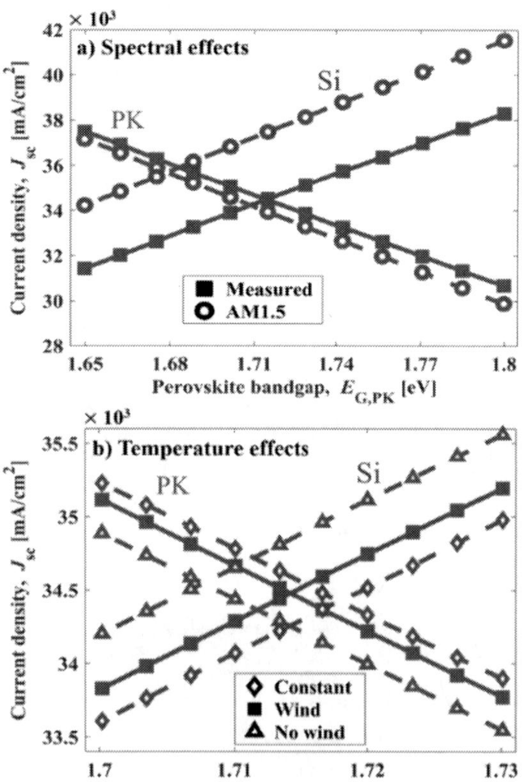

Figure 4: The annually-integrated J_{sc} of the PK and Si sub-cell in the tandem device installed in Atlanta: a) The influence of applied solar spectrum (realistic, AM1.5), and b) the influence of cell temperature modelling: i) constant 25 °C and Duffie-Beckman model ii) with wind cooling, iii) without wind cooling.

To observe how and to which degree the spectrum and the operating temperature affect the optical behavior of the device, we first varied the perovskite bandgap in the range of 1.65 – 1.80 eV and extracted the values of annually-integrated J_{sc} for all combinations of the above-mentioned aspects. The results are presented in Figures 4a) and 4b). (For better visualization, the results in Figure 4b) are shown within a narrower $E_{G,PK}$ range.)

As expected, we can see that both solar spectrum and T_{cell} influence the amount of J_{sc} in each individual sub-cell and thus change the position of the current matching point. The discrepancies are generally more pronounced for the silicon sub-cell, especially when assuming the reference AM1.5 spectrum instead of the measured one, whereas the effect of T_{cell} is somewhat lower (note the different ranges of the vertical axes). The actual spectral distribution of the incident illumination is significantly different from the fixed AM1.5 spectrum, particularly in the NIR region, which substantially affects the optical behavior of the studied tandem device. In Figure 4b) it can be observed that with increasing T_{cell} the point of current matching is moving towards lower PK bandgaps, due to the opprosing trends of bandgap temperature dependencies of perovskite and silicon sub-cell.

In addition, we also show the annual *EY* as a function of the perovskite bandgap in Figure 5, where vertical lines indicate the optimal $E_{G,PK}$ values. As predicted based on the J_{sc} results, we can see that both the spectral data and T_{cell} notably influence the calculated amount of the annually generated energy and the optimal PK bandgap. A significant difference in the average photon energy (ΔAPE = 55 meV; where *APE* designates the average photon energy that characterizes the spectrum of incident irradiance) between the measured and AM1.5 spectra causes a ~35 meV shift of the optimal perovskite bandgap and, moreover, a 2 % discrepancy in calculated EY_{max}. On the other hand, as already seen in Figure 4b), Figure 5 illustrates that different T_{cell} calculation appraches have a much smaller effect on the results, be it J_{sc}, *EY* or optimal $E_{G,PK}$ values.

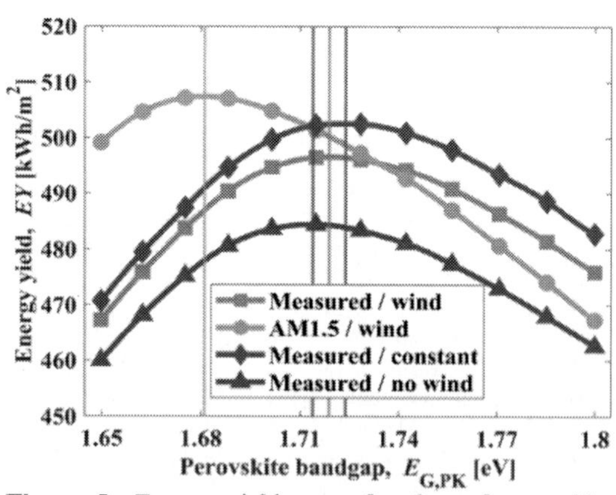

Figure 5: Energy yield as a function of perovskite bandgap for the tandem device installed in Atlanta, calculated by applying different spectral distributions of incident irradiance and cell temperature models. The optimal $E_{G,PK}$ values are indicated by vertical lines.

4 CONCLUSIONS

We used an in-house developed numerical model for accurate calculation of the long-term electrical energy produced by two-terminal encapsulated tandem PV device under realistic operating conditions from different Köppen-Geiger-Photovoltaic climate zones. The calculated energy yield served us to determine the optimal bandgap of the perovskite sub-cell within the tandem device.

We demonstrated that tandem device installed at Zacatecas is subjected to the highest yearly insolation and consequently generates the most electrical energy, however exhibits the lowest effective conversion efficiency due to high operating device temperatures. The results also showed that the optimal PK bandgap is higher under all selected location-specific realistic environmental conditions compared to the one obtained under STC. However, the impact of the geographical location on the $E_{G,PK}$ optimal value is very small.

Finally, in the scope of sensitivity analysis of *EY* modelling, we showed that inclusion of realistic device temperature and, more importantly, the exact spectral distribution of incident irradiance is crucial for accurate design and optimization of tandem devices. If applying the reference AM1.5 spectrum instead of the measured one, we observed a substantial shift (cca. 35 meV) of the optimal $E_{G,PK}$, and around 2 % discrepancy in the maximal *EY* between the two cases. Inclusion of realistic cell temperature by application of different thermal models, on the other side, had a much smaller effect on the results.

ACKNOWLEDGMENTS

The authors acknowledge the financial support from the Slovenian Research Agency (program P2-0415 and project J2-1727). Š. Tomšič thanks the Slovenian Research Agency for her Ph.D. funding. We also thank the Helmholtz Association for funding the project TAPAS (Tandem Perovskite And Silicon solar cells – Advanced optoelectrical characterization, modelling and stability) within the EU partnering program.

REFERENCES

[1] "Photovoltaics Report - Fraunhofer ISE," *Fraunhofer Institute for Solar Energy Systems ISE*. https://www.ise.fraunhofer.de/en/publications/studies/photovoltaics-report.html (accessed Apr. 12, 2021).

[2] W. Shockley and H. J. Queisser, "Detailed Balance Limit of Efficiency of *p-n* Junction Solar Cells," *J. Appl. Phys.*, vol. 32, no. 3, pp. 510–519, Mar. 1961, doi: 10.1063/1.1736034.

[3] R. M. Swanson, "Approaching the 29% limit efficiency of silicon solar cells," in *Conference Record of the Thirty-first IEEE Photovoltaic Specialists Conference, 2005.*, Lake buena Vista, FL, USA, 2005, pp. 889–894. doi: 10.1109/PVSC.2005.1488274.

[4] M. H. Futscher and B. Ehrler, "Efficiency Limit of Perovskite/Si Tandem Solar Cells," *ACS Energy Lett.*, vol. 1, no. 4, pp. 863–868, Oct. 2016, doi: 10.1021/acsenergylett.6b00405.

[5] "Best Research-Cell Efficiency Chart." https://www.nrel.gov/pv/cell-efficiency.html (accessed Mar. 15, 2021).

[6] M. H. Futscher and B. Ehrler, "Modeling the Performance Limitations and Prospects of Perovskite/Si Tandem Solar Cells under Realistic Operating Conditions," *ACS Energy Lett.*, vol. 2, no. 9, pp. 2089–2095, Sep. 2017, doi: 10.1021/acsenergylett.7b00596.

[7] M. Jošt *et al.*, "Textured interfaces in monolithic perovskite/silicon tandem solar cells: advanced light management for improved efficiency and energy yield," *Energy Environ. Sci.*, vol. 11, no. 12, pp. 3511–3523, 2018, doi: 10.1039/C8EE02469C.

[8] M. Langenhorst *et al.*, "Energy yield of all thin-film perovskite/CIGS tandem solar modules," *Prog. Photovolt. Res. Appl.*, vol. 27, no. 4, pp. 290–298, Apr. 2019, doi: 10.1002/pip.3091.

[9] J. Lehr *et al.*, "Energy yield of bifacial textured perovskite/silicon tandem photovoltaic modules," *Sol. Energy Mater. Sol. Cells*, vol. 208, p. 110367, May 2020, doi: 10.1016/j.solmat.2019.110367.

[10] F. Gota, R. Schmager, A. Farag, and U. W. Paetzold, "Energy yield modelling of textured perovskite/silicon tandem photovoltaics with thick perovskite top cells," *Opt. Express*, vol. 30, no. 9, p. 14172, Apr. 2022, doi: 10.1364/OE.447069.

[11] B. Lipovšek, M. Jošt, Š. Tomšič, and M. Topič, "Energy yield of perovskite solar cells: Influence of location, orientation, and external light management," *Sol. Energy Mater. Sol. Cells*, vol. 234, p. 111421, Jan. 2022, doi: 10.1016/j.solmat.2021.111421.

[12] M. Roffeis *et al.*, "New insights into the environmental performance of perovskite-on-silicon tandem solar cells – a life cycle assessment of industrially manufactured modules," *Sustain. Energy Fuels*, p. 10.1039.D2SE00096B, 2022, doi: 10.1039/D2SE00096B.

[13] J. Ascencio-Vásquez, K. Brecl, and M. Topič, "Methodology of Köppen-Geiger-Photovoltaic climate classification and implications to worldwide mapping of PV system performance," *Sol. Energy*, vol. 191, pp. 672–685, Oct. 2019, doi: 10.1016/j.solener.2019.08.072.

[14] B. Lipovšek *et al.*, "Detailed optical modelling and light-management of thin-film organic solar cells with consideration of small-area effects," *Opt. Express*, vol. 25, no. 4, p. A176, Feb. 2017, doi: 10.1364/OE.25.00A176.

[15] A. H. M. Smets, K. Jäger, O. Isabella, R. van Swaaij, and M. Zeman, *Solar energy: the physics and engineering of photovoltaic conversion, technologies and systems*. Cambridge, England: UIT Cambridge, 2016.

[16] A. Al-Ashouri *et al.*, "Monolithic perovskite/silicon tandem solar cell with >29% efficiency by enhanced hole extraction," *Science*, vol. 370, no. 6522, pp. 1300–1309, Dec. 2020, doi: 10.1126/science.abd4016.

[17] M. Sengupta, Y. Xie, A. Lopez, A. Habte, G. Maclaurin, and J. Shelby, "The National Solar Radiation Data Base (NSRDB)," *Renew. Sustain. Energy Rev.*, vol. 89, pp. 51–60, Jun. 2018, doi: 10.1016/j.rser.2018.03.003.

[18] F. E. Vignola, A. C. McMahan, and C. N. Grover, "Bankable Solar-Radiation Datasets," in *Solar Energy Forecasting and Resource Assessment*, Elsevier, 2013, pp. 97–131. doi: 10.1016/B978-0-12-397177-7.00005-X.
[19] "JRC Photovoltaic Geographical Information System (PVGIS) - European Commission." https://re.jrc.ec.europa.eu/pvg_tools/en/#TMY (accessed Aug. 23, 2021).

This is a short version of the paper submitted to the journal Progress in Photovoltaics.

It has been selected by the Committee of the WCPEC-8 for submission to Progress in Photovoltaics.

SUSTAINABLE MATERIAL EVALUATION FOR SILICON BASED TANDEM SOLAR CELLS AT TERAWATT SCALE PRODUCTION

Li Wang[1], Yuchao Zhang[1], Moonyong Kim[1], Matthew Wright[2], Robert Underwood[1], Ruy Sebastian Bonilla[2] and Brett Hallam[1]
[1]UNSW Sydney, NSW 2052, Australia
[2]Department of Materials, University of Oxford, Oxford, OX1 3PH, United Kingdom

ABSTRACT: In accordance with "Net zero emissions" goal, annual photovoltaic (PV) production must be expanded to terawatts (TW) scale around 2030. This goal not only accelerates the requirement of improvement on efficiencies of solar cells, but also intensifies the consumption of scarce materials contained, especially in the high efficiency Si-based tandem solar cells. This abstract provides a sustainability analysis of various potential silicon-based tandem solar cell designs in heading towards terawatt scales of annual PV production by 2030 on the path towards net-zero emissions by 2050. It investigates the current and target consumption values of critical elements for future silicon-based tandem devices in the absorbing materials (eg. III-V, CIGS, CdTe and perovskite), and bottom cell (PERC, TOPCon and SHJ) candidates and requirement for metal electrodes, as well as lateral conductors/buffer layers typically based on indium-tin-oxide (ITO). We show that III-V, CIGS and CdTe top cells will not support TW scales of manufacturing, with perovskites provide the most sustainable top cell, if it is indium-free. For the bottom cell, PERC could allow significantly lower silver consumption than TOPCon and SHJ cells.
Keywords: PV Materials, Sustainable, tandem

1 INTRODUCTION

In order to significantly mitigate climate change, the annual photovoltaic (PV) module production rate must dramatically increase from approximately 135 gigawatts (GW) in 2020 to roughly 3 terawatts (TW) around 2030 [1] on its path to net-zero emissions by 2050. A key concern for such dramatic PV production expansion, is the intensively increased material requirements, both for scarce materials due to potential depletion of material resources at the cell/module level, as well as abundant materials used at the module/systems level, due to the significant environmental and social impacts. One key method to reduce material consumption and the associated impacts is to increase cell efficiency, through substantial savings for module and balance of systems (BOS) components (eg. aluminium, copper, steel and concrete). Tandem devices have the opportunity to reach module efficiencies >30%, greatly assisting with the potential of reducing the material consumption.

As silicon solar cells are currently dominating the PV market (PERC, TOPCon and SHJ technologies), Si-based tandem solar cells are expected to be a next-generation representative with industrial compatibility. However, there are various top cell candidates that can be integrated into a tandem structure, and many of these rely on scarce materials. III-V solar cells have consistently achieved high efficiencies, from 24.2% InP cell [2] to 29.1% efficient GaAs cell [3] then 35.9% efficient GaInP/GaInAsP/Si [4] triple junction solar cell, which is the highest reported efficiency to date for silicon-based multijunction solar cell technologies. As an established technology, CdTe cells have achieved 22.1% efficiency [5]. Due to the toxicity of Cd, Cd-free CIGSe cells have been developed and hit a higher efficiency of 23.35% [6]. The rapid development of perovskite cells has realised world record efficiencies of 25.8% [7] and 29.8% [8] for single junction perovskite and perovskite/Si tandem devices, respectively.

Therefore, it is of great importance to understand the material usages and limitations for future tandem devices. This work investigates the critical material usage for various top cell candidates (III-V, CIGS, CdTe and Pervoskite), Si bottom cells candidates (PERC, TOPCon and SHJ) and associated silver metal requirement, and lateral transport/buffer layers typically based on in indium tin oxide (ITO) for SHJ and perovskite solar cells. From this, for sustainable PV manufacturing capacity (SMC) is estimated to provide valuable consideration for future sustainable tandem designs.

2 METHODOLOGY

Various cell types with world-record efficiencies above 22% were selected as the top cell candidates as listed in Table I. In order to obtain a range of element consumption in mg/W, three representatives from III-V groups were selected and a range of thicknesses for those from CIGS, CdTe and Perovskite groups were chosen from the literatures. A material utilization rate of 80% was assumed for the material growth for CIGS, CdTe and Perovskite candidates, whereas utilization rates of 48-60% was assumed for Si considering material losses during the mineral production and purification. With current usage (in mg/W) and the assumption that using only 20% of 2020 global primary material supply (in tonnes/yr) based on US Geological Survey [9] for each critical material listed above. Sustainable manufacturing capacity (SMC_x), for each cell technology was then calculated (in GW/yr or TW/yr), where x represents the element with the lowest SMC (i.e. the limiting material) of the corresponding cell technologies. What's more, the target usage (in mg/W) for each element was also obtained by dividing 20% global supply by 3 TW assumed annual production to identify the gaps between current and target usages. Table I summarizes assumptions and parameters used for this study.

3 RESULTS AND DISCUSSION

The target usages (in mg/W) for each element were obtained using the method detailed above and are shown in Fig. 1. From the calculated target usage consider the case of GaAs cells requiring gallium and arsenic. If such technology could generate annual production of 3TW and only consume 20% of 2020 global annual supply of gallium and arsenic, respectively, the 'target material consumptions' for gallium and arsenic under this ideal scenario must be below 0.0218 mg/W and 4 mg/W, respectively. Same rules applied to CIGS (copper, indium, gallium, selenium), CdTe (cadmium, tellurium), perovskite (lead, iodine) and silicon cells (silicon). For target usage of metal conductor silver, it is assumed that all the cells contributing to the 3 TW production contain silver metal contacts.

Table I: Cell type, efficiency and material thickness used for estimating the consumption of critical elements

Cell type	*Material*	*Efficiency (%)*	*Thickness (nm)*	*Ref*
III-V, III-V/Si	GaAs	29.1	1000	[3]
	InP	24.2	5200 (InP)	[2]
			100 (AlInAs)	
	GaInP/GaInAsP/Si	35.9	530 (Ga0.51In0.49P)	[4]
			1680 (Ga0.93In0.07As0.87P0.13)	
			700 (Al0.06Ga0.94As)	
CIGS	CuIn.9Ga.1Se2	23.35	1000-2500	[6]
CdTe	CdTe	22.1	2000-8000 (CdTe)	[5, 10]
			60-200 (CdS)	
Perovskite, Pervoskite/Si	FAPbI3	25.8	300-700. $FAPbI_3$ and $Cs_{0.05}(FA_{0.77}MA_{0.23})_{0.95}Pb(I_{0.77}Br_{0.23})_3$	[7, 8]
	In in ITO layer		40-285 (ITO) on Perovskite only	
PERC	Silicon, Ag	23.83	Si: wafer thickness 160-175 μm, cell sizes 16.6×16.6= 275.56 cm^2 and 21×21=441 cm^2; Ag: cell metallisation patterns as our previous work, 9 busbars for 275.56 cm^2 and 12 busbars for 441 cm^2; In: 80-150nm ITO for each side of SHJ cells, 275.56 and 441 cm^2,	[11]
TOPCON	Silicon, Ag	24.58		
SHJ	Silicon, Ag, In	25.11		

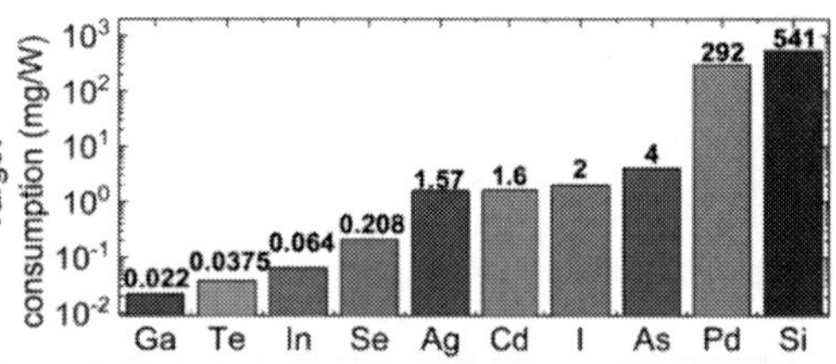

Figure 1: Target consumptions for these materials with assumed annual PV production of 3TW and only using 20% of annual primary material supply from USGS 2020 data

3.1 Sustainable top cell options

The resulting materials consumption and SMC are illustrated in Fig. 2. where (a), (b), (c) illustrate the range of material consumption in mg/W for each material, and (d), (e), (f) show the associated range of sustainable manufacture capacity in GW/year when employing these cell technologies. In order to directly compare the target and actual usage for each element, dash lines in Fig. 2 (a, b, c) indicate the target usage previously shown in Fig. 1.

For the III-V group the target consumption for gallium (Ga), indium (In), and Arsenic (As) are approximately 0.022 mg/W, 0.064 mg/W and 4 mg/W, respectively. In practice GaAs and GaInP/GaInAsP/Si cells consume 12-34 mg/W of Ga and 12-17 mg/W of As, as shown in Fig. 3 (a). These are at least ×545 and ×3 times the target usages, respectively. The In usage from InP and GaInP/GaInAsP/Si cells ranges from 6 to 104 mg/W, which is ×94 to ×1625 times larger than the target values. The corresponding SMC for Ga, In and As are 2-6 GW/year, 2-32 GW/year, and 720-1000 GW/year, respectively, as shown in Fig. 2 (b). This shows that SMC of III-V group is limited by Ga, with SMCGa as low as 2-6 GW/year, which means, even if 20% of Ga global annual supply could be employed to fabricate III-V solar cells with the efficiencies as high as 35.9% as listed in Table I, the maximum production capacity of such solar cells is up to 6 GW/year, which is significantly below the TW/year scale.

CIGS candidates utilise 1.6-4 mg/W of Ga, 13.7-34.3 mg/W of Selenium (Se), and 14.5-15.5 mg/W of In as illustrated in Fig. 2 (b). These are over ×70, ×86 and ×223 times of the target values shown Fig. 1 for the three materials. As a result, the corresponding SMC of these three materials are 16-41 GW/year, 18-45 GW/year and 12-13 GW/year, respectively, as shown in Fig. 2 (e). Therefore, the SMC of CIGS solar cells is constrained by Se with a SMC_{Se} of 16-41 GW/year which is dramatically below the TW/year production goal. Similar limitations arise when comparing the current usage to the target usage of Cadmium (Cd) and Tellurium (Te) as shown in Fig. 2 (b). CdTe candidates consume 20 to 83 times the target value of Cd, and 940-3750 times the target value of Te. The SMC for CdTe technology is hence limited by Te with a SMC_{Te} as low as 0.8-3.2 GW/year as shown in Fig. 2 (e).

For perovskite cells, the consumption of lead (Pb) and iodine (I) were considered as the critical elements of the absorbing layers in the forms of $FAPbI_3$ and $Cs_{0.05}(FA_{0.77}MA_{0.23})_{0.95}Pb(I_{0.77}Br_{0.23})_3$. As shown in Fig. 2 (c), the consumption of Pb is 1.5-3.6 mg/W which is significantly lower than the target value of 292 mg/W, so that the SMC of Pb is 190-470 TW/year as labelled in Fig. 2 (f). The consumption of I is 1.9-7 mg/W which is 1-3.5 times as that of the target value of 2 mg/W as listed in Fig. 2 (c), so that the SMC of I is 714-2490 GW/year, as labelled in Fig. 2 (f). If the usage of I can be controlled below 6 mg/W, the corresponding SMC of I will be higher than 1 TW/year. However, Perovskite cells require TCO layers to provide lateral conductivity. For the perovskite candidates selected in this work the ITO thickness ranges from 40-285 nm. With the highest efficiency of 25.8% for single junction perovskite cells, indium (In) consumption in this ITO layer equals to 0.8-5.9 mg/W, such that SMC of In for perovskite cells are 26-186 GW/year, as shown in Fig. 2 (e). Therefore, TW/year goal is achievable for perovskite cells only if they are In-free and the I usage can be well controlled.

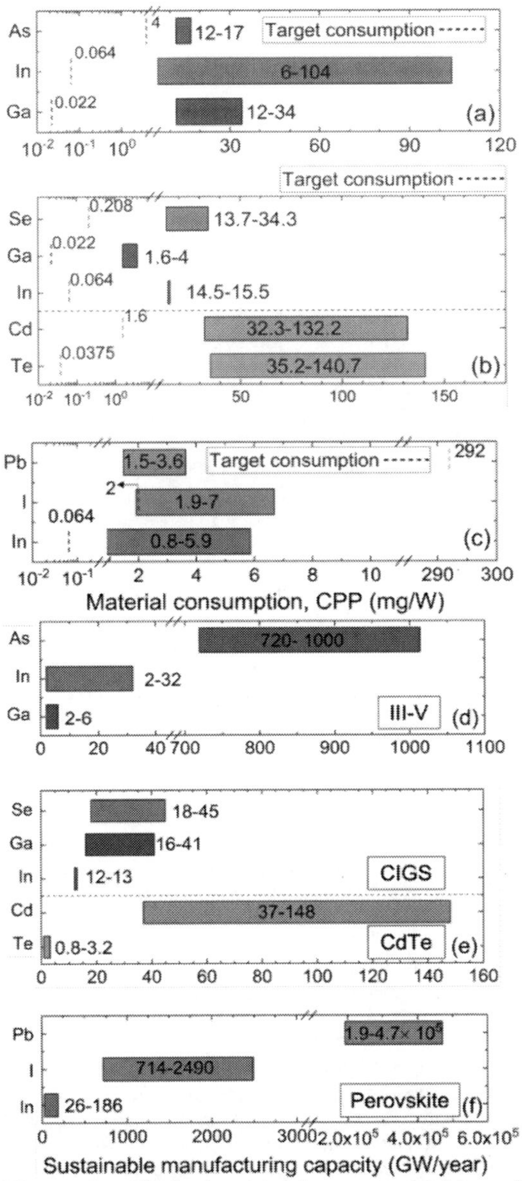

Figure 2: Obtained material usages (a)-(c) and corresponding SMC (d)-(f) for top cell options of III-V, CIGS, CdTe and perovskites, based on assumptions and parameters described in Table I.

3.2 Sustainable bottom cell options

Among three dominant silicon solar cell technologies, PERC, TOPCon and SHJ solar cells involve different structures and fabrication process, and hence they consume different materials quantities. In addition, adaptations of these three technologies have been used as Si bottom cells in a number of perovskite/Si tandem devices. Therefore, it is worth making comparison among material usages not just for PERC, TOPCon, and SHJ cells, but also Si bottom cells used in reported perovskite/Si tandem cells. Similar to the material evaluation on top cell candidates, critical material consumptions on Si and silver (Ag) were obtained using the methodology above with cell thicknesses, efficiencies, Ag metallisation geometries, and cell details listed in Table I. The resulting material consumption and sustainable capacities are illustrated in in Fig. 3

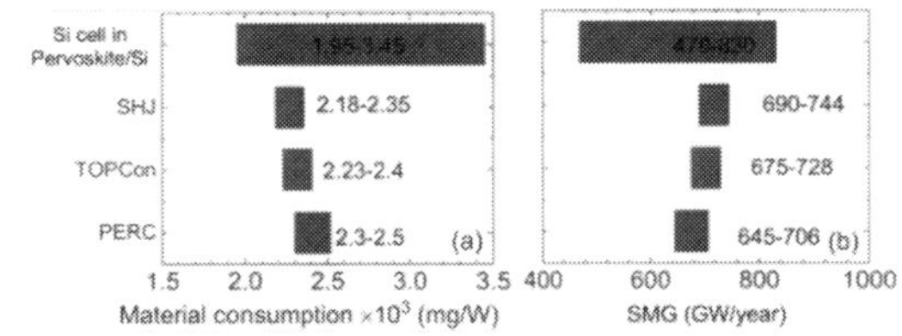

Figure 3: Obtained Si material usages (a) and corresponding SMC (b) for single junction PERC, TOPCon and SHJ cells, and Si subcells in tandem structures. Based on assumptions and parameters described in Table I

From Fig. 3 (a) it is clear that the Si consumptions for PERC, TOPCon and SHJ are similar, ranging from approximately 2.2 g/W for SHJ to 2.5 g/W for cells, with the difference primarily attributed to the differences in efficiency as listed in Table I. On one hand, with the same Si thickness of around 170 μm, the advantage of higher efficiency on the reduction of material consumption can be observed from Fig. 3. Especially, the lowest predicted consumption for Si used in tandem devices is less than 2 g/W, which can save up to ~17% and 23% of Si material compared to SHJ and PERC, respectively. On the other hand, nearly 5% absolute efficiency gain in tandem devices (29.8%) compared to SHJ devices (25.1%) cannot outweigh the disadvantage of the thicker Si wafers that are used for research scale tandems, which give an upper limit to material consumption as high as 3.45 mg/W. The reason is that the majority of Si bottom cell in the lab scale perovskite/Si tandem structures are fabricated on Float Zone (FZ) wafers with a thickness up to 300 μm. This leads to higher Si consumption for tandem devices, which can be more than 40% of Si material than used for single junction SHJ cells (2.4 g/W, 25.1% efficient, 175 μm Si). If the bottom cell is fabricated on Czochralski (CZ) wafers with a thickness of ~170 μm similar to single junction Si cells, the value could be < 2 g/W. Cz wafers are hence expected to be employed at large scale for Si-based tandem production, in addition to their compatibility with industrial PV processing in terms of size, thicknesses, and cost-effectiveness. This highlights that thickness reductions on silicon cells will continue to be sought after in future PV technology. As shown in Fig. 3 (b), the SMC_{Si} of PERC, TOPCon and SHJ cells also shares a similar range from 650 - 740 GW/year, whereas the SMC_{Si} for the Si-based tandems shows a wider range of 470-830 GW/year. Unfortunately, none of these can achieve TW/year scale. Therefore, despite the increasing trend on poly-Si production, these figures indicate that research efforts should be put in reducing silicon wafer thickness and increasing material utilisation efficiency.

3.3 Silver electrodes

Fig. 4 (a) shows Ag consumptions SHJ, TOPCon and PERC single junction cells to be 25.6-39 mg/W, 22.8-34.5 mg/W and 12.8-18.8 mg/W, respectively. SHJ and TOPCon consume ~2 times and ~1.8 times more Ag than that of PERC. As a result, the corresponding SMC of Ag are 250-367 GW/year for PERC, and less than ~200 GW/year for SHJ and TOPCon, as shown in Fig. 4 (b). Even if the tandem cell efficiency is nearly 30%, the SMC of Ag can only be improved by 1.2-1.3 times, which are equivalent to 313-460

GW/year, 120-183 GW/year and 165-250 GW/year for tandems using Ag grid designs in PERC, SHJ and TOPCon, respectively. This manufacturing capacities are dramatically lower than the TW/year goal, and current usage is considerably higher than the target usage of 1.6 mg/W (Fig. 1) for 3 TWp.a. on the basis of using 20% of global Ag supply. Therefore, significant Ag usage reduction is urgently required in order to prevent silver shortage.

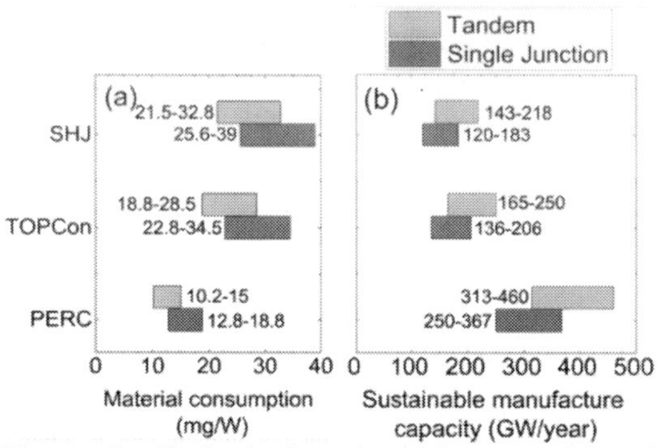

Figure 4: Obtained silver material usages (a) and corresponding SMC (b) for bottom cell options of PERC, TOPCon and SHJ cells in a single junction and tandem structures. Based on assumptions and parameters described in Table I, and (h) target usages for each material using 20% 2020 global production supply.

3.4. Indium in ITO layers

Indium is largely used in ITO and zinc doped indium oxide (IZO) layers as a transparent conductor for perovskite and SHJ solar cells, as well as a recombination layer between the subcells in 2T perovskite/Si tandems. The In consumption for SHJ cells ranges from 3.7-5.3 mg/W as shown in Fig. 5 (b) which is 58-83 times larger than the 0.064 mg/W target value calculated using only 20% of annual global indium supply. The corresponding SMC of In for SHJ is only 36-52 GW/year (Fig. 5 (c)), which is lower than the SMC of Ag as shown in Fig. 5 (b). As a result, the SMC of SHJ is constrained by indium which makes the TW/year scale production goal of SHJ cells even more difficult to hit. As discussed earlier, the consumption of In in 40-285 nm thick ITO layers required for perovskite cells is 0.8-5.9 mg/W, such that SMC of In for perovskite cells are 26-186 GW/year. These are also plotted in Fig. 5 in addition to single junction SHJ cells and perovskite/Si tandem cells. The ITO layer thicknesses for 2T perovskite/Si tandem cells ranges from 40-320 nm. It consists of 40-150 nm on the perovskite top cells, 0-150 nm on the silicon bottom cells depends on the type of bottom cells, and 0-40 nm as the recombination layer between the two subcells. Details on these thickness calculations are included in invited paper for this work [12]. With efficiencies of 24.5% to 29.8%, the obtained In usages are 0.87-6.7 mg/W, so that the corresponding SMC of In for 2T perovskite/Si tandems are 23-176 GW/year. By contrast, although a recombination layer between the two subcells is not required in a 4T tandem structure, 4T perovskite/Si tandems still consume 145-480 nm ITO, including 80-285 nm on perovskite top cells and 0-230 nm for Si bottom cells. With efficiencies of 25.2% to 28.3%, the obtained In usages are 2.7-9 mg/W, which is 1.3-3 times higher than that of 2T counterpart. Consequently, the corresponding SMC of In for 4T perovskite/Si tandems are as low as 17-57 GW/year. These figures show that with present In usage it is not feasible to realise TW/year target even if tandem devices are nearly 30% efficient.

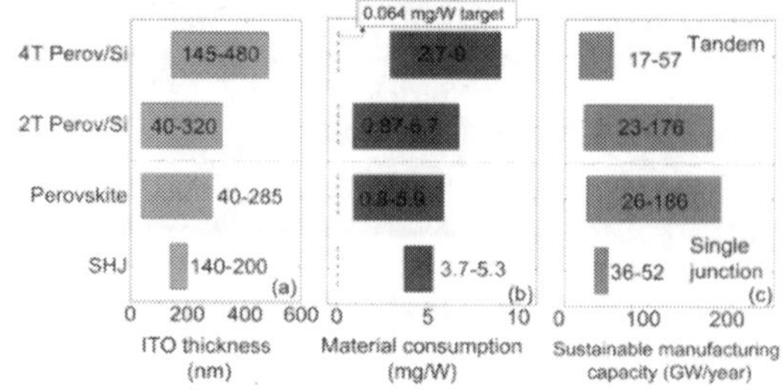

Figure 5: (a) ITO thicknesses currently used, (b) calculated In consumption ranges and (c) SMC_{In} for single junction SHJ and perovskite cells, 2T and 4T perovskite/Si tandem cells.

4 CONCLUSION

Based on our conservative evaluation, SMC of III-V top cell, CIGS and CdTe are below 6 GW/year (limited by Ga), 13 GW/year (limited by In), and 3.2 GW/year (limited by Te), respectively. Therefore, none of these high efficiency top cell candidates are able to achieve TW scale production. On the other hand, perovskite solar cells can have SMC values over 1 TW/year, provided their TCO layers are indium-free and usage of iodine can be well controlled. Si consumption in PERC, TOPCon and SHJ are similar and the corresponding SMCSi are all below 740 GW/year. The majority of the present perovskite/Si tandem cells utilize thick FZ silicon wafers, so that SMCSi in the tandem cells is 470 GW/year. This will increase to 830 GW/year if using CZ wafers. The SMC of Ag for PERC single junction cells is up to 367 GW/year, and below 200 GW/year for TOPCon and SHJ single junction cells. Such SMC can be improved to up to 460 GW/year, 183 GW/year and 250 GW/year for 2T tandems using PERC, SHJ and TOPCon as the bottom cells with present Ag grid designs, respectively. The gain in capacity is largely due to the gain in efficiency and the low J_{mp}/V_{mp} ratio of 2T tandem devices. Indium is the limiting material in the SMC for SHJ single junction cells and perovskite/Si tandem cells, with SMC 32G W/year for SHJ cells, and 57 GW/year and 136 GW/year for 4T and 2T perovskite/Si tandem configurations, respectively. Therefore, in order to realized TW/year target, reductions on silicon wafer thickness and improvement on material utilization efficiency must be realized. This include innovative screen-printing metallization design and Ag-lean pastes, as well as In-free TCO layers such that cell technologies that significantly reduce the demand on Si, Ag and In materials are developed.

Although some materials are potentially available in the crust and annual supply can be expanded, the environmental impact of such expansion including mining and purification, must be evaluated. Additionally, due to the 25-30 years life-span of PV modules, materials used in the PV module have to wait for such a period of time to be recycled if they can be, and there will be material losses during the recycling process. Based on this work, we note that PV is being developed at such a tremendous scale, that more research focus should be shifted from purely improving the efficiency towards improving the sustainable deployment of PV. This must include reducing both material consumptions and the impacts on environment, society, and human wellbeing. A compromise of efficiency gain and resources depletion might be a key to open the gate towards TW era photovoltaics deployment.

5 ACKNOWLEDGEMENTS

This work has been supported by the Australian Centre for Advanced Photovoltaics (ACAP) and the Australian Government through the Australian Renewable Energy Agency (ARENA). Responsibility for the views, information or advice expressed herein is not accepted by the Australian Government.

6 RERENCE

[1] P. J. Verlinden, J. Renewable Sustainable Energy 12 (5), 053505 (2020).
[2] https://patents.google.com/patent/US9590131B2/en
[3] https://optics.org/news/9/12/19
[4] Schygulla, P et al. Prog Photovolt Res Appl. 2021; 1-11. doi:10.1002/pip.3503
[5] MA. Green et al. ,"Solar cell efficiency tables (version 58)", Prog Photovolt Res Appl 29, 657– 667 (2021).
[6] Nakamura et al. IEEE Journal of Photovoltaics, 9, 6, (2019)
[7] H. Min et al.,Nature 598, 444–450 (2021)
[8] P. Tockhorn et al., TandemPV workshop 2022
[9] U.S. Geological Survery, Mineral Commodity Summary. (2022)
[10] https://www.nrel.gov/pv/cadmium-telluride-solar-cells.html
[11] Y. Zhang et al., En. Environ. Sci, 14, 5587 (2021).
[12] L. Wang et al., invited by PiP, under review

FLEXIBLE GAAS//CU(IN_{1-X}, GA_X)(S_Y, SE_{1-Y})$_2$-BASED THREE-JUNCTION SOLAR CELLS USING MODIFIED SMART STACK TECHNOLOGY

Kikuo Makita[1], Hiroshi Tomita[2], Hidenori Mizuno[1], Ryuji Oshima[1], Yasushi Shoji[1], Ralph Müller[3], David Lackner[3], Frank Dimroth[3], and Takeyoshi Sugaya[1]
[1]National Institute of Advanced Industrial Science and Technology (AIST), Tsukuba, Japan
[2]Idemitsu Kosan Co., Ltd., Atsugi, Japan
[3]Fraunhofer Institute for Solar Energy Systems ISE, Freiburg, Germany
AIST Tsukuba, Central 2, Umezono 1-1-1, Tsukuba, Ibaraki, 305-8568, JAPAN
Phone: +81-29-861-3263, Fax: +81-29-861-5615, E-mail: kikuo-makita@aist.go.jp

ABSTRACT: Multijunction (MJ) solar cells achieve very high efficiencies by effectively utilizing the solar spectrum. Previously, we constructed a GaAs//$CuIn_{1-y}Ga_ySe_2$ (CIGSe) MJ solar cell using a unique mechanical stacking technology with Pd nanoparticle and silicone adhesive at the bonding interface (modified smart stack technology). The modified smart stack was developed to increase the bonding strength of the GaAs//CIGSe interface. The efficiency of 28.06% with a InGaP/AlGaAs//CIGSe three-junction solar cell was the best value among GaAs//CIGSe tandem solar cells as two-terminal. In this report, we fabricated a flexible InGaP/AlGaAs//Cu(In_{1-x},Ga_x)(S_y,Se_{1-y})$_2$ (CIGSSe) three-junction solar cell using a flexible CIGSSe bottom cell. Appling the modified smart stack, the total efficiency with 24.3% for aperture area (26.7% for active area) was realized under the global standard AM1.5G spectrum. This is the first demonstration as a flexible GaAs//CIGSe-based MJ solar cell. The superior performance can be achieved by improving the reflection loss at the bonding interface. We predict that the efficiency could be boosted to 35% by minimizing the thickness of transparent conducting oxide layer and the surface roughness of CIGSSe cell. The results demonstrate the potential of GaAs//CIGSe-based MJ solar cell as next-generation solar cells for application such as vehicle-integrated photovoltaics.
Keywords: III-V Semiconductors, CIGS, Multijunction Solar Cell

1 INTRODUCTION

Multijunction (MJ) solar cells have achieved very high efficiencies (> 30%) by effectively utilizing the sun's energy. MJ solar cells based on mechanical stacking technology have been developed as next-generation solar cells. Mechanical stacking technology allows the most appropriate combination of different cells, enabling the construction of efficient and low-cost solar cells. A realistic combination for this purpose is the GaAs/Si MJ solar cell. Recently, efficiencies exceeding 30% were reported in mechanically stacked GaAs//Si MJ solar cells [1, 2]. Against this background, we previously developed a unique mechanical stacking method with Pd nanoparticle alignment, named smart stack technology [3, 4]. Using smart stack, an InGaP/AlGaAs//Si with 30.6% efficiency [5] was realized. Recently, a GaAs//$CuIn_{1-y}Ga_ySe_2$ (CIGSe) multi-junction solar cell attracts attention [6-8]. The CIGSe cell is another possible substitute for the Si bottom cell, as mass production technology and its fabrication cost. In addition, CIGSe allows flexible substrate [9] such as thin metal plate and polymer film. The bendability is advantageous in mobility application. We previously demonstrated an InGaP/AlGaAs//CIGSe three-junction solar cell with 28.06% via a modified smart stack, which involves Pd nanoparticle and adhesive material for the bonding interface [10, 11]. The bonding for CIGSe cell is generally difficult because of large surface roughness (Rq-100nm). The feature of modified smart stack is the reinforcement of bonding strength using adhesive. The value of 28.06% was the highest efficiency ever reported for two-terminal GaAs//CIGSe-based MJ solar cells [12].

In this study, we reported a flexible InGaP/AlGaAs//Cu(In_{1-x},Ga_x)(S_y,Se_{1-y})$_2$ (hereafter referred as CIGSSe) three-junction solar cell using modified smart stack. This device has a bendability using CIGSSe cell formed on flexible substrate. The obtained efficiency for the cell size of 1 cm^2 was 24.3% for aperture area (26.7% for active area) at 1 sun. The results demonstrate the potential of GaAs//CIGSe-based MJ solar cell as next-generation solar cells for applications such as vehicle-integrated photovoltaics.

2 DEVICE APPROACH

Figure 1 is a diagram of the fabricated InGaP/AlGaAs//CIGSSe three-junction solar cell. The upper cell is a two-junction cell comprising p-InGaP (Eg-1.89 eV) and p-$Al_{0.06}Ga_{0.94}As$ (Eg-1.50 eV) absorption layers connected by a tunneling layer. The bottom cell is a CIGSSe solar cell which was developed by Idemitsu Kosan Co.,Ltd.. The CIGSSe bottom cell is formed on a flexible substrate with the device stack consisting of CIGSSe absorption layer and transparent conducting oxide (TCO) layer. The CIGSSe layer was formed by SAS (Sulfurization after Selenization) method [13]. The bandgap energy of CIGSSe absorption layer is adjusted to be approximately 1.0 eV by controlling Ga distribution in CIGSSe absorption layer. These cells were connected via modified smart stack [10, 11]. Figure 2 shows a stacking process. First, the Pd nanoparticle arrays were formed on CIGSSe bottom cell using a thin-film template of a self-assembled block copolymer [3, 4]. Next, silicone adhesive was applied on CIGSSe cell to increase the bonding strength. The thickness of silicone adhesive is controlled to be thin by through dilution with toluene and the speed of spin coating. On the other hand, GaAs-based cell was separated from the growth substrate using the epitaxial lift-off (ELO) technique. In this study, the detachment and the bonding for large area becomes possible using slightly adhesive sheet for reinforcement. Finally, the upper and bottom cells are connected by a light load (5 N/cm^2). In here, The CIGSSe bottom cell was not prepared through a bother and expensive flattening process such as chemical mechanical polishing (CMP). Although the surface roughness of CIGSSe cell was over 100 nm, the bonding stability was reinforced notably using modified smart stack.

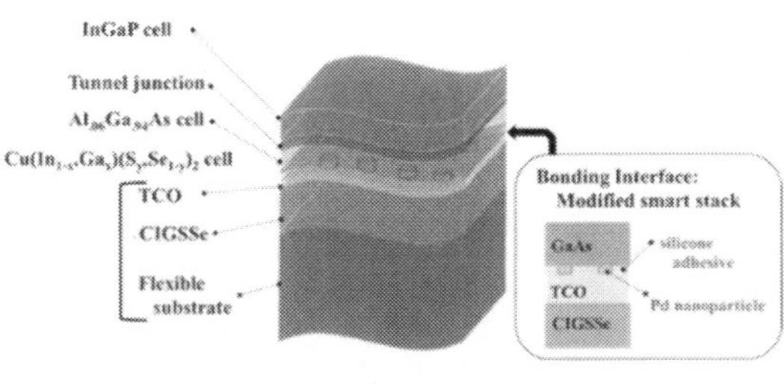

Figure 1: Structure of a flexible InGaP/AlGaAs//CIGSSe three-junction solar cell via modified smart stack technology.

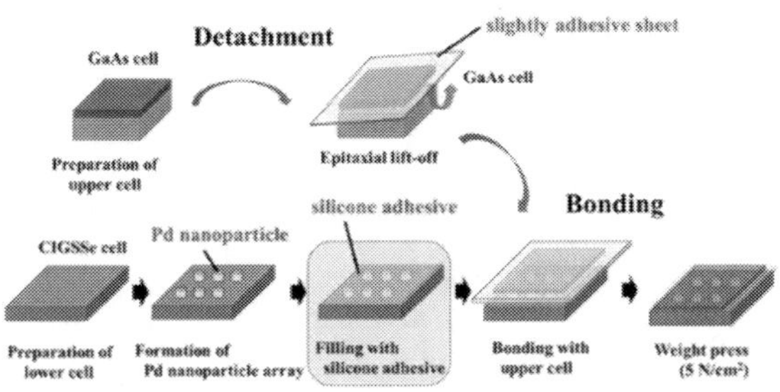

Figure 2: Schematic of the fabrication process of the tandem solar cell using the modified smart stack technology.

3 EXPERIMENT AND DISCUSSION

Figure 3 is a photograph of the fabricated flexible InGaP/AlGaAs//CIGSSe three-junction solar cell. The InGaP/AlGaAs two-junction upper cell with the size of 1 cm^2 was bonded to the flexible CIGSSe bottom cell with size of 2.5 cm^2 using modified smart stack.

Figure 4 shows the I-V curve of the fabricated cell. In the measurement, the solar cell was covered with a shadow mask to estimate the accurate current densities. The performance of the solar cells was determined for the aperture area (S=0.932cm^2). At 1 sun, the total efficiency (η) was 24.3% for aperture area. An open-circuit voltage, short-circuit current density, and fill factor were 2.97 V, 10.72 mA/cm^2, and 76.4%, respectively. For active area (excluding electrode area), the efficiency of 26.7% was estimated. Figure 5 shows external quantum efficiencies (EQE) spectra for each cell. In here, the photo current densities for each cell are 12.0 mA/cm^2 (InGaP). 10.2 mA/cm^2 (AlGaAs), and 11.2 mA/cm^2 (CIGSSe). The performance is limited by the photo current of the middle AlGaAs cell. For high efficiency, the middle cell should be changed from AlGaAs to GaAs (to have a narrow bandgap energy). In this case, the low EQE (nearly 0.7) of CIGSSe bottom cell is concerned. In our speculation, the deterioration of EQE is mainly due to the reflection loss, which is caused by the configuration of the bonding interface. First, because the bonding interface is mainly composed of GaAs/TCO/CIGSSe, an inevitable reflection loss occurs at the bonding interface, which depends on the thickness of the TCO layer with a low refractive index. Secondly, the relatively large roughness of the CIGSSe cell (~100 nm) should be considered. The surface roughness leads to an increase in the effective gap width at the bonding interface. In the future, by minimizing the thickness of the TCO layer (< 20 nm) and the reduction of surface roughness of CIGSSe bottom cell (< 10 nm), the optical loss can be reduced to less than 5%. Such a design leads to an increase of the EQE in the CIGSSe bottom cell. The high EQE allows the constructed GaAs middle cell and contributes to improving the current-matching level, resulting in high potential efficiency. According to our simulation, a predicted efficiency for InGaP/GaAs//CIGSSe three-junction solar cell is 35% [10].

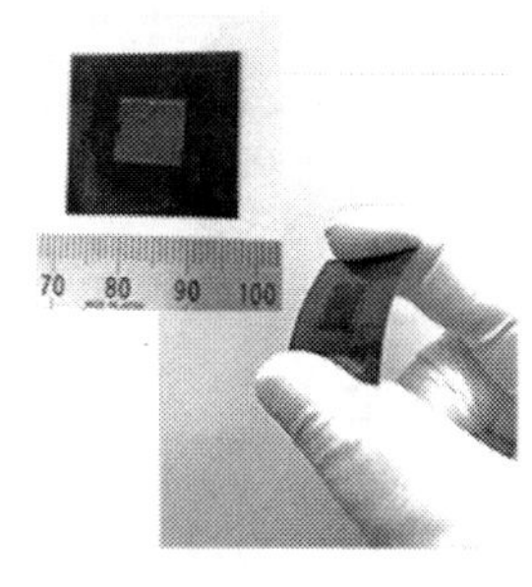

Figure 3: Photograph of a flexible InGaP/AlGaAs //CIGSSe three-junction solar cell.

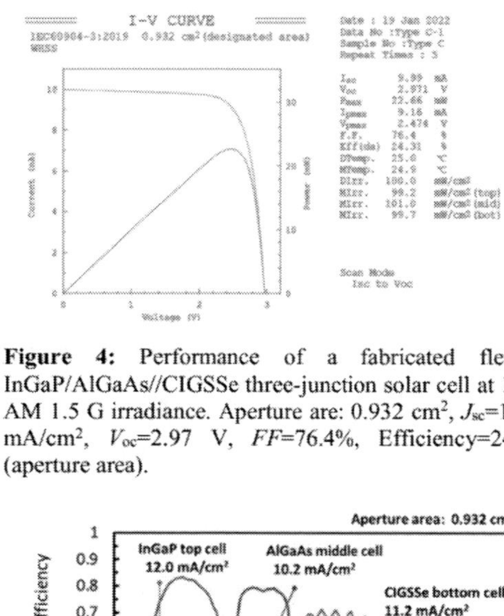

Figure 4: Performance of a fabricated flexible InGaP/AlGaAs//CIGSSe three-junction solar cell at 1 sun AM 1.5 G irradiance. Aperture are: 0.932 cm^2, J_{sc}=10.72 mA/cm^2, V_{oc}=2.97 V, FF=76.4%, Efficiency=24.3% (aperture area).

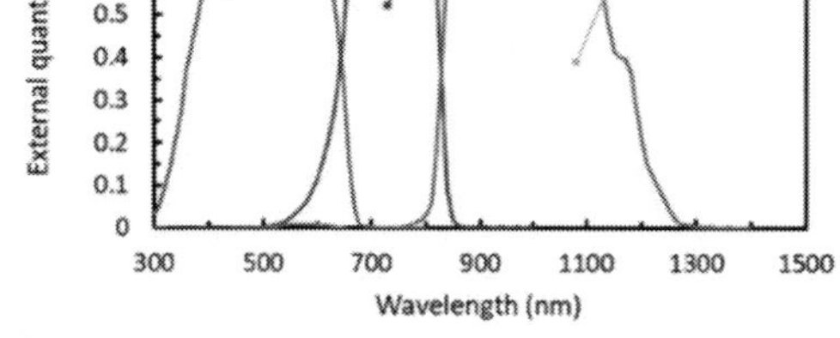

Figure 5: Spectra of the external quantum efficiencies for each cell.

4 CONCLUSIONS

We fabricated a flexible InGaP/AlGaAs//CIGSSe three-junction solar cell using modified smart stack, achieving a maximum efficiency of 24.3%. This is the first demonstration as a flexible GaAs//CIGSe-based MJ solar cell using mechanical stack technology. The results suggest that GaAs//CIGSe tandem solar cells are promising solutions for innovative and economical next-generation solar cells such as vehicle-integrated photovoltaics, and that production engineering can be realized through modified smart stack technology.

The development of the III-V cell structures at Fraunhofer ISE was supported by the German Federal Ministry for Economic Affairs and Energy under the project PoTaSi (FKz. 0324247). This work is based on results obtained from a project, JPNP20015, commissioned by the New Energy and Industrial Technology Development Organization (NEDO) under the Ministry of Economy, Trade, and Industry (METI), Japan.

References

[1] S. Essig, C. Allebé, T. Remo, J.F. Geisz, M.A. Steiner, K. Horowitz, L. Barraud, J.S. Ward, M. Schnabel, A. Descoeudres, DL. Young, M. Woodhouse, M. Despeisse, C. Ballif, A. Tamboil, "Raising the one-sun conversion efficiency of III–V/Si solar cells to 32.8% for two junctions and 35.9% for three junctions," Nat Energy, Vol. 2 (2017) Art. no. 17144.

[2] P. Schygulla, R. Müller, D. Lackner, O. Höhn, H. Hauser, B. Bläsi, F. Predan, J. Benick, M. Hermle, S. Glunz, and F. Dimroth, "Two-terminal III-V//Si triple-junction solar cell with power conversion efficiency of 35.9% at AM1.5g," Prog. Photovolt. Res. Appl., Vol. 30 (2022) 869.

[3] H. Mizuno, K. Makita, K. Matsubara, "Electrical and optical interconnection for mechanically stacked multijunction solar cells mediated by metal nanoparticle arrays," Appl. Phys. Lett., Vol. 101 (2012) Art. no. 191111.

[4] H. Mizuno, K. Makita, T. Sugaya, R. Oshima, Y. Hozumi, H. Takato, K. Matsubara, "Palladium nanoparticle array-mediated semiconductor bonding that enables high-efficiency multi-junction solar cells," Jpn. J. Appl. Phys., Vol. 55 (2016) Art. no. 025001.

[5] K. Makita, H. Mizuno, T. Tayagaki, T. Aihara, R. Oshima, Y. Shoji, H. Sai, H. Takato, R. Müller, P. Beutel, D. Lackner, J. Benick, M. Hermle, F. Dimroth, T. Sugaya, "III - V//Si multijunction solar cells with 30% efficiency using smart stack technology with Pd nanoparticle array," Prog. Photovolt. Res. Appl., Vol. 28 (2020) 16.

[6] S. Kawakita, M. Imaizumi, K. Makita, J. Nishinaga, T. Sugaya, H. Shibata, S. Sato, T. Oshima, "High efficiency and radiation resistant InGaP/GaAs//CIGS stacked solar cells for space applications," in Proc. 43rd IEEE Photovolt. Spec. Conference, Portland, USA, (2016) Paper 2574.

[7] D.W. Kang, Y. Takiguchi, P. Sichanugrist, M. Konagai, "InGaP//GaAs//CIGS 3-junction spectrum-splitting solar cells with low-concentration ratio," Phys. Status Solidi A, Vol. 213 (2016) 1535.

[8] M. Imaizumi, T. Takamoto, H. Sugimoto, T. Oshima, S. Kawakita, "Preliminary study on super radiation-resistant mechanical-stack triple-junction space solar cell: PHOENIX," in Proc. 46th IEEE Photovolt. Spec. Conference, Chicago, USA, (2019) 1495. [Online]. Available: https://doi.org/10.1109/PVSC40753.2019.8980537

[9] J. Ramanujam, D. M. Bishop, T. K. Todorov, O. Gunawan, J. Rath, R. Nekovei, E. Artegiani, A. Romeo, "Flexible CIGS, CdTe and a-Si:H based thin film solar cells: A review", Prog. Mater. Sci., Vol. 110 (2020) Art. no. 100619.

[10] K. Makita, Y. Kamikawa, H. Mizuno, R. Oshima, Y. Shoji, S. Ishizuka, R. Müller, P. Beutel, D. Lackner, J. Benick, M. Hermle, F. Dimroth, T. Sugaya, "III-V/$Cu_xIn_{1-y}Ga_ySe_2$ multijunction solar cells with 27.2% efficiency fabricated using modified smart stack technology with Pd nanoparticle array and adhesive material," Prog. Photovolt. Res. Appl., Vol. 29 (2021) 887.

[11] K. Makita, Y. Kamikawa, H. Mizuno, R. Oshima, S. Ishizuka, R. Müller, D. Lackner, F. Dimroth, T. Sugaya, "GaAs//$CuIn_{1-y}Ga_ySe_2$ three-junction solar cells with 28.06% efficiency fabricated using a bonding technique involving Pd nanoparticles and adhesive," IEEE J. Photovolt., Vol. 12 (2022) 639.

[12] M.A. Green, E.D. Dunlop, J. Hohl-Ebinger, M. Yoshita, N. Kopidakis, X. Hao, "Solar cell efficiency tables (Version 58)," Prog. Photovolt. Res. Appl., Vol. 29 (2021) 657.

[13] M. Nakamura, K. Yamaguchi, Y. Kimoto, Y. Yasaki, T. Kato, H. Sugimoto, "Cd-Free $Cu(In,Ga)(Se,S)_2$ Thin-Film Solar Cell With Record Efficiency of 23.35%", IEEE J. Photovolt., Vol. 9 (2019) 1863.

STUDY OF CDSETE/CDTE DEVICES FABRICATED BY THERMAL EVAPORATION

Elisa Artegiani, Prabeesh Punathil, Solidea Zanetti[1], Narges Torabi and Alessandro Romeo

LAPS-Laboratory for Applied Physics, Department of Computer Science, University of Verona

Ca' Vignal 1, Strada Le Grazie 15, 37134 Verona, Italy, tel.: +39-045-8027974, e-mail address: alessandro.romeo@univr.it

[1]also at Isopan S.p.A., via Giona 5, Trevenzuolo, Italy

ABSTRACT: Our group fabricates and studies CdTe based solar cells deposited by thermal evaporation method at low substrate temperature (below 450 °C).
In this paper we present and study CdSeTe/CdTe cells, where the CdSeTe compound is formed by evaporating in sequence a CdSe and a CdTe layer, subsequently mixed by a $CdCl_2$ activation treatment. These devices, with an absorber thickness of only 2 µm, show current densities above 27 mA/cm^2.
Earlier we also introduced an innovative method to develop CdSeTe by treating a thin CdTe layer at high temperature in selenium atmosphere (selenization), which gets subsequently mixed with a thicker CdTe layer.
We study the structure of the new devices completely manufactured by evaporation and we compare them with the samples fabricated by selenization. This reveals how the manufacturing method of the CdSeTe layer affects the device properties.

Keywords: CdTe, CdSe, Thin Film Solar Cell.

1 INTRODUCTION

In the last years the efficiency of CdTe panels has jumped over 19 %. The efficiency improvement relates to the introduction of Se into the absorber layer, forming a CdSeTe compound at the junction. With a band gap narrower than CdTe, it allows to absorb photons at longer wavelengths and to remove the CdS layer and its parasitic absorption, making an optimal junction with wide band gap buffer layers such as SnO_2.

In our laboratory CdSeTe/CdTe devices totally fabricated by evaporation have reached current densities above 27 mA/cm^2. With this method, to obtain a suitable CdSeTe compound, it is very important to calibrate the thicknesses of the CdSe and CdTe layers and the amount of $CdCl_2$ together with the temperature of the activation treatment. These parameters affect the structure and the defects of the devices.

Moreover, our group already presented a study on CdSeTe/CdTe devices fabricated by selenization of a thin CdTe layer [1][2]. Therefore, we compare the morphology, crystal structure and efficiency parameters of samples fabricated with different methods. This highlights how the manufacturing method of the CdSeTe layer affects the properties of the device.

2 EXPERIMENTAL

2.1 Preparation of the devices

In our laboratory, CdTe solar cells in superstrate configuration are deposited by thermal evaporation method at low substrate temperature (below 450 °C) [3].

We have already presented an innovative method to fabricate CdSeTe/CdTe junction, by treating a thin CdTe layer (between 150 and 300 nm) at high temperature in selenium atmosphere (selenization), which gets subsequently mixed with a thicker CdTe layer by the activation treatment [1][2].
In this paper we compare this fabrication process with a more standardized procedure where CdSe is directly deposited by thermal evaporation in order to address how the CdSeTe manufacturing method affects the properties of the device.

Specifically, on TEC12 (SnO_2:F/SnO_2) commercial coated glasses a CdSe layer followed by a CdTe layer is deposited at a substrate temperature of 340 °C in vacuum (10^{-4} Pa). The total thickness of the absorber is around 2 µm. Subsequently, as activation treatment, a solution of $CdCl_2$-methanol (MeOH) is drop casted on the surface and then the stack is annealed at 400 °C in air. This, in addition to re-crystallizing the absorber, allows the mixing of the CdSe and CdTe layers generating the desired CdSeTe layer at the junction.

The surface is later cleaned from the $CdCl_2$ residues by etching it with a Br-MeOH solution. This step also allows the formation of CuTe compounds at the back contact, improving the ohmic contact and the stability of the device [4]. Finally, a copper/gold bilayer is deposited by thermal evaporation, and the finished device is annealed at 200 °C in air to favor the diffusion of copper in the absorber [5]. Alternatively, Cu can be inserted in the absorber by applying a $CuCl_2$ solution on the top of the CdTe layer and subsequently annealing the stack [6].

2.2 Characterization

The morphology of the absorbers is analyzed by a NT-MDT Solver Pro atomic force microscopy (AFM). The crystal structure is studied by X-ray diffraction (XRD) analysis with a Thermo ARL X'TRA powder diffractometer (in Bragg-Brentano geometry, equipped with a Cu-anode X-ray source (Kα, λ=1.5418 Å) and a Peltier Si (Li) cooled solid state detector). Current density-voltage (JV) characteristics are measured with a Keithley Source Meter 2420, under an AM 1.5 spectrum at 100 mW/cm^2, using a LOT Quantum Design Europe solar simulator LS0306.

3 DEVICES CHARACTERIZATION

To evaluate the influence of CdSe and CdTe thicknesses on the properties of the devices, we have fabricated samples with a CdSe thickness of 75 nm, 150 nm, 225 nm and 300 nm, keeping the total thickness of the absorber fixed at around 2 µm.

The surface morphology of the as-deposited CdTe analyzed by AFM images (Figure 1) is a first striking example of how these thicknesses affect the absorber structure. In fact, by increasing the CdSe thickness from 75 nm to 300 nm, the diameter of the grains of the as-deposited CdTe reduces from around 6-8 µm to 2 µm. Fig. 1A (CdSe 75 nm) and Fig. 1B (CdSe 150 nm) show big grains of CdTe with diameters up to 6-8 µm. From Fig. 1A to Fig. 1B the presence of small grains with a diameter of 1-2 µm seems to increase. By increasing the thickness of the CdSe to 225 nm the proportions are reversed: the surface of the CdTe reveals the presence of a few large grains surrounded by small grains. By further increasing the thickness to 300 nm, the large grains disappear, giving way to a more homogeneous surface with small grains having a diameter of about 2 µm.

The influence of CdSe thickness on the absorber structure is further confirmed by X-ray diffraction analysis (Fig. 2A) on the treated samples. CdTe films deposited on 75 nm and on 300 nm thick CdSe layer present a strong (111) peak. However this orientation is strongly predominant for the thinner CdSe layer, while the polycrystals are more randomly oriented (increase of the (311) and (422) peaks) when the thickness of the CdSe increases to 300 nm.

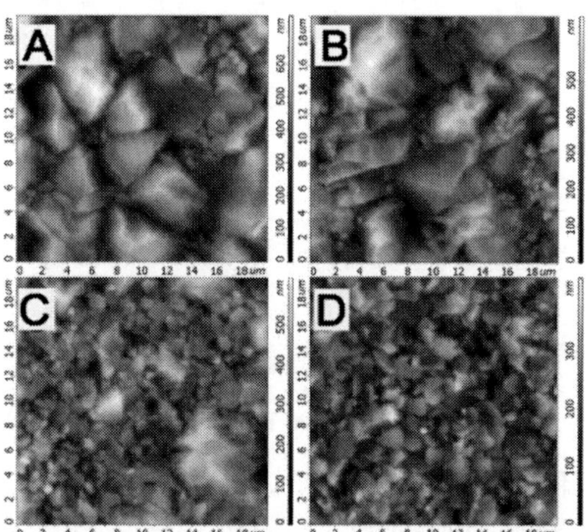

Figure 1: AFM images of the surface of CdTe deposited on a: A) 75 nm, B) 150 nm, C) 225 nm, D) 300 nm thick CdSe layer.

Fig. 2B compares the XRD spectra of treated absorbers fabricated by CdSe evaporation (225 nm thick) and by selenization of a 150 nm thick CdTe layer. Again the analysis highlights a different crystal structure: the selenized device shows the predominance of the CdTe (311) orientation on the (111) and (422). This can only be related to the different method to incorporate selenium, since the manufacturing method, and in particular the total thickness of the absorber and the activation treatment, are the same.

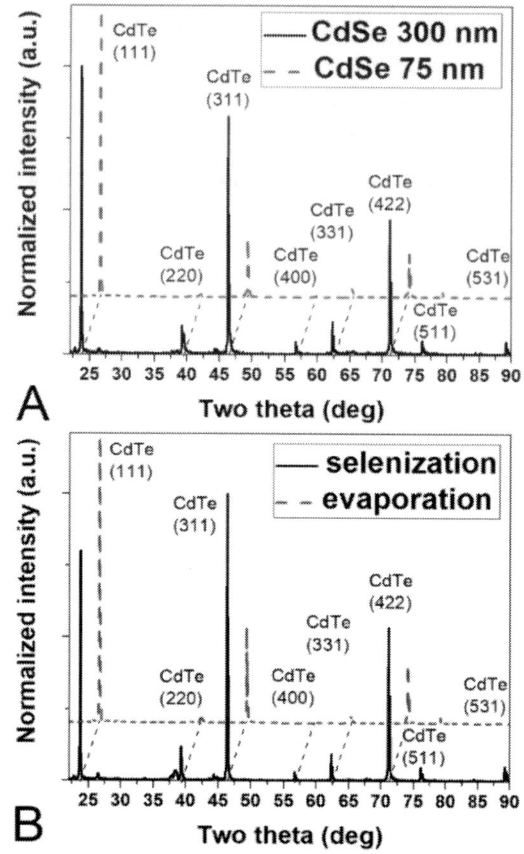

Figure 2. A) XRD spectra of treated absorbers fabricated with a 300 nm (black-solid line) and a 75 nm (red-dash line) thick CdSe layer. B) XRD spectra of treated absorbers fabricated by selenization of a CdTe layer (black line) and by CdSe evaporation (red line). In both cases the red spectra are shifted 3° to the right to improve the comparison.

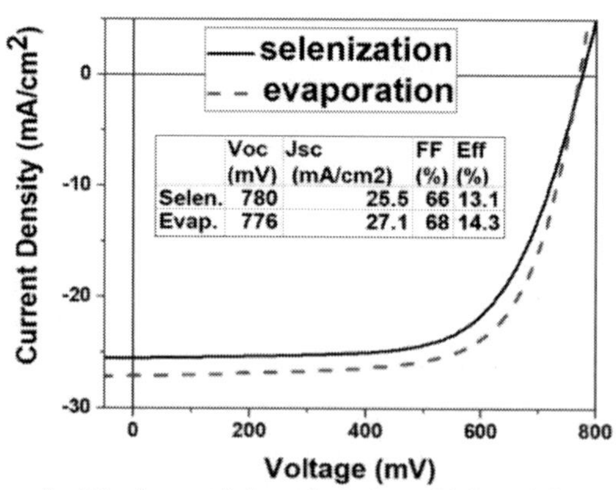

Figure 3. JV characteristics of devices fabricated by selenization and by evaporation

Regarding the actual operation of the finished solar cells, Figure 3 compares the JV characteristics of a typical device with CdSeTe by selenization and of a device with CdSeTe by evaporation. Moreover, the efficiency parameters of the cells are specified in the inset of Fig. 3.

Compared to the samples subjected to selenization, the cells fabricated totally by evaporation show similar open circuit voltage (Voc) and slightly higher fill factor (FF). However, what stands out the most is that this latter manufacturing process leads to much improved short circuit current (Jsc) values. In fact, the samples show Jsc above 27 mA/cm^2, which is a remarkable achievement especially considering the ultra-thin thickness of the absorber (about 2 µm). The increase in current leads in turn to a good improvement in efficiency of more than 1 %.

4 CONCLUSIONS

The introduction of selenium into the absorber of CdTe based solar cells has led to a large increase in current density and consequentially in efficiency. Generally CdSeTe is made by deposition of CdSe followed by a layer of CdTe, subsequently mixed through the activation treatment. However the physical properties of the absorbers deposited on this buffer layer and the mechanism leading to the formation of $CdSe_xTe_{1-x}$ still need to be addressed further.

By studying CdSeTe/CdTe devices totally fabricated by thermal evaporation, we have showed how CdSe thickness affects the morphology and the crystal structure of the whole absorber layer. The crystal structure is also affected by the method used to incorporate selenium, as we highlighted by comparing CdSeTe/CdTe devices fabricated through a selenization step with samples manufactured by evaporating a CdSe layer. The latter demonstrate an efficiency improvement of more than 1 %, mainly thanks to a remarkable current gain of almost 2 mA/cm^2, reaching Jsc values above 27 mA/cm^2. This is a remarkable result considering the ultra-thin absorber thickness of about 2 µm.

5 REFERENCES

[1] E. Artegiani et al., "A new method for CdSexTe1-x band grading for high efficiency thin-absorber CdTe solar cells," Solar Energy Materials and Solar Cells, vol. 226, p. 111081, Jul. 2021, doi: 10.1016/j.solmat.2021.111081.

[2] E. Artegiani et al., "Effects of CdTe selenization on the electrical properties of the absorber for the fabrication of CdSexTe1-x/CdTe based solar cells," Solar Energy, vol. 227, pp. 8–12, Oct. 2021, doi: 10.1016/j.solener.2021.08.070.

[3] A. Romeo, E. Artegiani, and D. Menossi, "Low substrate temperature CdTe solar cells: A review," Solar Energy, 2018. doi: 10.1016/j.solener.2018.02.038.

[4] I. Rimmaudo et al., "Improved stability of CdTe solar cells by absorber surface etching," Solar Energy Materials and Solar Cells, 2017, doi: 10.1016/j.solmat.2016.12.044.

[5] E. Artegiani, J. D. Major, H. Shiel, V. Dhanak, C. Ferrari, and A. Romeo, "How the amount of copper influences the formation and stability of defects in CdTe solar cells," Solar Energy Materials and Solar Cells, vol. 204, p. 110228, Jan. 2020, doi: 10.1016/j.solmat.2019.110228.

[6] E. Artegiani et al., "Analysis of a novel CuCl2 back contact process for improved stability in CdTe solar cells," Progress in Photovoltaics: Research and Applications, vol. 27, no. 8, pp. 706–715, 2019, doi: 10.1002/pip.3148.

THEORETICAL STUDY ON THE IMPROVED BI-LAYER ABSOBER OF SILVER INCORPORATED SB2SE3 DEVICES

Sanghyun Lee[a], Michael McInenery[b]
[a]Department of Engineering Technology, University of Kentucky, Lexington, KY 40506, USA
[b]Department of Physics and Optical Engineering, Rose-Hulman Institute of Technology, Terre Haute, IN 47803, USA

ABSTRACT: We have studied Ag incorporated $Sb_2(S_xSe_{1-x})_3$ bi-layer devices by modeling and simulation to optimize the absorber layer. Four different device structures are extensively investigated to optimize the device configuration. Firstly, as a frame of reference, simple single-layer devices are studied with the absorber film of each $AgSb(S_xSe_{1-x})_3$ and $Sb_2(S_xSe_{1-x})_3$ as the Sulfur composition increases. Following the study of reference devices, the impact of Sulfur composition on bi-layer devices has been modeled and simulated with two different absorber configurations such as $Sb_2(S_xSe_{1-x})_3/AgSb(S_xSe_{1-x})_3$ and $AgSb(S_xSe_{1-x})_3/$ $Sb_2(S_xSe_{1-x})_3$. Compared with any single-layer devices, the absorber configuration of $AgSb(S_xSe_{1-x})_3/$ $Sb_2(S_xSe_{1-x})_3$ bi-layers shows best performance at the Sulfur composition, x = 0.4. This is ascribed to the reduced effective Schottky barrier by 126 mV at the back contact, improving the roll-over effect by 4.3 times at 1 V. The resulting efficiency is 18.6 %, Voc, is 778.9 mV, Jsc is 37.2 mA/cm2, and FF is 64.0 %. Once the bi-devices with the highest efficiency and the optimal Sulfur composition are identified, a set of different device parameters have been studied to optimize these bi-layer devices further. The best efficiency is 21.4 % with Voc (822 mV), Jsc (38.1 mA/cm2), and FF (68.4 %) based on this study.

Keywords: Thin film solar cells, Sb_2Se_3, bi-layer devices, silver

1 INTRODUCTION

Antimony chalcogenide, $Sb_2(S_xSe_{1-x})_3$ thin-film solar cells have been emerging as third-generation photovoltaic devices in the past several years. $Sb_2(S_xSe_{1-x})_3$ utilizes earth-abundant and non-toxic absorber materials with tunable bandgap (1.0 – 1.6 eV), stable upon exposure to sunlight under ambient conditions [1-5]. $Sb_2(S_xSe_{1-x})_3$ shows a high absorption coefficient at visible light (> 10^5 cm^{-1}), good carrier mobility (< 15 cm^2/Vs), long carrier lifetime (< 67 ns), and simple binary compound with high vapor pressure and low melting point (550 C) [1-5].

To further explore the absorber material, Silver (Ag) Antimony, $AgSb(S_xSe_{1-x})_3$ has been studied with similar tunable bandgap (0.7 - 1.9 eV), doping concentration (<10^{16}), and high absorption coefficient (>10^4 cm^{-1}) [6-8]. There have been several studies about photovoltaic structures using $AgSb(S_xSe_{1-x})_3$ with a single absorber layer by depositing $Sb_2(S_xSe_{1-x})_3$, Ag, and Se thin-film precursors, followed by the heat treatment [4]. The reported bandgap ranges from 0.71 - 1.1 eV ($AgSbSe_3$) to 1.7 - 1.9 eV ($AgSbS_3$) [4], making an ideal absorber candidate with tunable bandgap. However, most studies have focused on performance improvement based on different bandgap energies of the single absorber film by varying the concentration of elements. With this favorable optical bandgap of $AgSb(S_xSe_{1-x})_3$ (x=0.53, 0.58, and 0.61), a progress of device performance has been made, showing efficiency approximately >2.77 %, Voc (410 - 520 mV), Jsc (5.2 - 9.7 mA/cm2), and FF (32 - 50 %) [3]. In another approaches, $AgSbSe_3$ devices demonstrate <0.1 %, Voc (370 mV), Jsc (0.5 mA/cm2), and FF (<50 %) [4], whereas $AgSbS_3$ devices show approximately 0.27 %, Voc (500 mV), Jsc (1.1 mA/cm2), and FF (48 %) [5].

To fully utilize the promising properties of both $Sb_2(S_xSe_{1-x})_3$ and $AgSb(S_xSe_{1-x})_3$ films, we have theoretically studied Antimony/Silver Antimony bi-layer devices, consisting of $Sb_2(S_xSe_{1-x})_3$ and $AgSb(S_xSe_{1-x})_3$ absorber layers. With developed device models from our preliminary studies, we have used in-house MATLAB modeling suites and Sentaurus TCAD simulators [9-13]. From electric band structures perspectives, we will focus on identifying the efficiency limiting factors and mechanism. Cumulatively, we will optimize the device parameters and configuration with proposed device parameters.

2 DEVICE STRUCTURE AND METHODOLGY

2.1 Device Structure

The proposed device structures are shown in Fig. 1. The first device structures are two different configurations with a single-film absorber: ($Mo/MoS_xSe_{1-x}/$ $Sb_2(S_xSe_{1-x})_3$/CdS/i-ZnO/ZnO:Al) in Fig. 1a and ($Mo/MoS_xSe_{1-x}/AgSb(S_xSe_{1-x})_3$/CdS/i-ZnO/ZnO:Al) in Fig. 1b. These single-layer devices serve as a reference for the comparison with bi-layer devices. Molybdenum (Mo) is a back contact, and MoS_xSe_{1-x} is and a thermodynamically stable phase, which is always present in contact with chalcogenide films with Sulfur or Selenium [2,9]. Beyond the absorber layer, transparent conductive oxide (TCO), front window, and buffer layers are ZnO:Al (ZnO), intrinsic-ZnO (i-ZnO), and CdS films, respectively. Antimony/Silver Antimony bi-layer devices have two different configurations. Fig. 1c is the absorber film with $AgSb(S_xSe_{1-x})_3/Sb_2(S_xSe_{1-x})_3$ bi-layer devices and Fig. 1d is the absorber film with $Sb_2(S_xSe_{1-x})_3/AgSb(S_xSe_{1-x})_3$ bi-layer devices.

After Mo metal back contact (workfunction - 4.9 eV), we added an inevitable interlayer (MoS_xSe_{1-x} between the

Fig. 1. Device structures for this study, (a) a refernece - $Sb_2(S_xSe_{1-x})_3$ single-layer devices (thickness 2 um), (b) $AgSb(S_xSe_{1-x})_3$ single-layer devices (thickness 2 um), (c) bi-layer devices: $AgSb(S_xSe_{1-x})_3/Sb_2(S_xSe_{1-x})_3$ (thickness 1 um/1 um), and (d) bi-layer devices: $Sb_2(S_xSe_{1-x})_3/$ $AgSb(S_xSe_{1-x})_3$ (thickness 1 um/1 um).

Table 1. Key modeling parameters used in the work

Parameters	ZnO	CdS	Sb_2S_3	Sb_2Se_3	$AgSbSe_3$	$AgSbS_3$
Thickness (um)	0.07	0.05	2	2	1	1
bandgap (eV)	3.3	2.4	1.62	1.06	1.03	1.74
Electron Affinity (eV)	4.45	4.2	3.7	4.08	4.08	3.28
Relative permittivity, ϵ_r	9	10	7.08	9.86	9.86	9.21
Donor concentration, N_D (cm^{-3}), 10^{17}	10	1.0	na	na	na	na
Acceptor concentraton, N_A (cm^{-3}), 10^{14}	na	na	57.2	10	0.1	19

absorber and Mo metal back contact throughout this study. The defect concentrations of each layer are maintained as 2×10^{14} cm^{-3}.

Once the simulation of single-layer devices is conducted, Antimony/Silver Antimony bi-layer devices are modeled and simulated based on the same set of single-layer devices' parameters except for inserting a Silver Antimony layer and updating each bi-layer thickness to 1 um to keep the total thickness 2 um. As the bandgap of $Sb_2(S_xSe_{1-x})_3$ and $AgSb(S_xSe_{1-x})_3$ depends on the composition (Sulfur concentration), we have used the empirical equation of bandgap (Eq. 1) and electron affinity with a bowing parameter as below. The rest of the parameters are modeled with the linear relation of composition variation (Fig 2a).

$$E_g(x)^{Sb_2(S_xSe_{1-x})_3} = xE_g^{Sb_2S_3} + (1-x)E_g^{Sb_2Se_3} + b^{Sb_2(S_xSe_{1-x})_3}x(1-x)$$

$$E_g(x)^{AgSb(S_xSe_{1-x})_3} = xE_g^{AgSbS_3} + (1-x)E_g^{AgSbSe_3} + b^{AgSb(S_xSe_{1-x})_3}x(1-x)$$

$$E_g(x)^{Mo(S_xSe_{1-x})_2} = xE_g^{MoS_2} + (1-x)E_g^{MoSe_2} + b^{Mo(S_xSe_{1-x})_2}x(1-x)$$

Equation 1. Modeling formula for band gap that used in this study.

3 RESULTS AND DISCUSSION

3.1 Modeling of single-layer devices

Fig. 2b shows the efficiency of $AgSb(S_xSe_{1-x})_3$ and $Sb_2(S_xSe_{1-x})_3$ as Sulfur (x) composition increases. The best performed devices are $AgSb(S_{0.4}Se_{0.6})_3$, demonstrating the optimal composition, x = 0.4. Once Sulfur composition is optimized, we have simulated two different absorber configurations of bi-layer devices, $AgSb(S_xSe_{1-x})_3$/ $Sb_2(S_xSe_{1-x})_3$ (fig 1c) and $Sb_2(S_xSe_{1-x})_3$/$AgSb(S_xSe_{1-x})_3$ (fig. 1d) using the same device parameters except the film thickness. To keep the total thickness 2 um, each bi-layer film's thickness is adjusted to 1 um for the consistency.

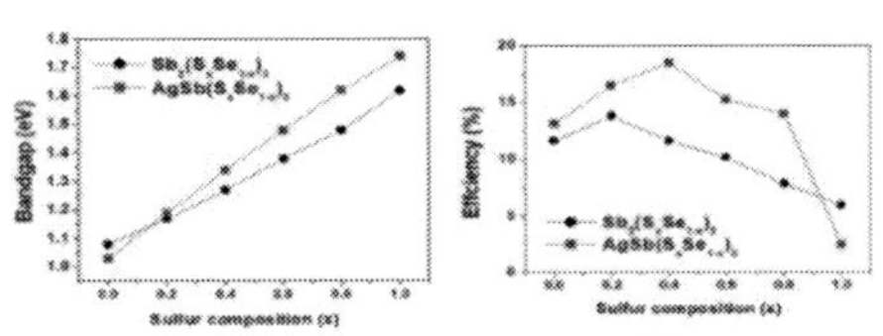

(a) (b)

Fig. 2. (a) Band gap energy increases as Sulfur (x) concentration increases, (b) baseline single-layer devices efficiency

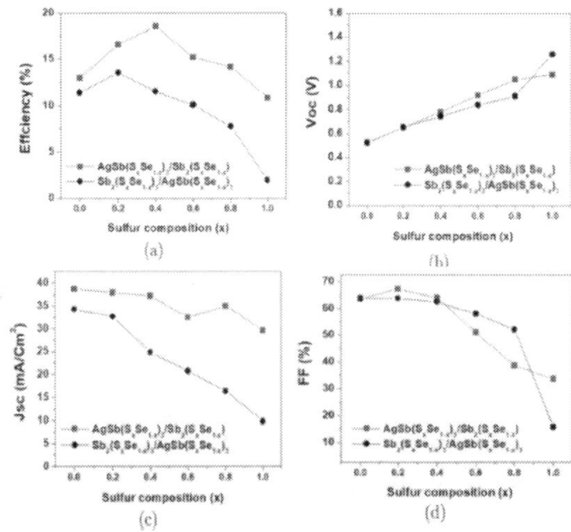

Fig. 3. Bi-layer devices per Sulfur composition, (a) efficiency, (b) Voc, (c) Jsc, and (d) FF

3.2 Modeling of Bi-layer devices

In Fig. 3, similar to single-layer devices, Sulfur composition at x = 0.4 shows the highest efficiency, 18.6 %, Voc (778.9 mV), Jsc (37.2 mA/cm^2), and FF (64.0 %) for $AgSb(S_{0.4}Se_{0.6})_3$/ $Sb_2(S_{0.4}Se_{0.6})_3$ devices while $Sb_2(S_{0.4}Se_{0.6})_3$/$AgSb(S_{0.4}Se_{0.6})_3$ shows 13.6 % efficiency. Moreover, $AgSb(S_{0.4}Se_{0.6})_3$/ $Sb_2(S_{0.4}Se_{0.6})_3$ bi-layer devices show highest efficiency than the best single-layer devices of $AgSb(S_{0.4}Se_{0.6})_3$ by 0.2 % without modifying any parameters. In Fig. 4, current-voltage (J-V) characteristics are demonstrated for each $AgSb(S_{0.4}Se_{0.6})_3$ (single-layer devices) and $AgSb(S_{0.4}Se_{0.6})_3$/ $Sb_2(S_{0.4}Se_{0.6})_3$ (bi-layer devices). The 0.2 % efficiency increase is driven by higher Voc by 1.6 mV and FF by 0.5 %, compared to single-layer devices. What is interesting is that both devices' J-V curves under one sun (100 mW/ cm^2) show the current saturation at the higher forward biases, so-called the roll-over effect in Fig. 4. However, the roll-over effect is far more prominent for single-layer devices. To elucidate the governing factor for the improvement, we have investigated an energy band structure in Fig. 5a and 4b.

In Fig. 5a and 5b, the effective Schottky barrier of single-layer devices is greater than bi-layer devices by 126 meV even with the same MoS_xSe_{1-x} interlayer. Hence, the slight modification of the absorber layer with 1 um of $Sb_2(S_{0.4}Se_{0.6})_3$ in bi-layer devices clearly provides better back contact characteristics. Despite the increase being

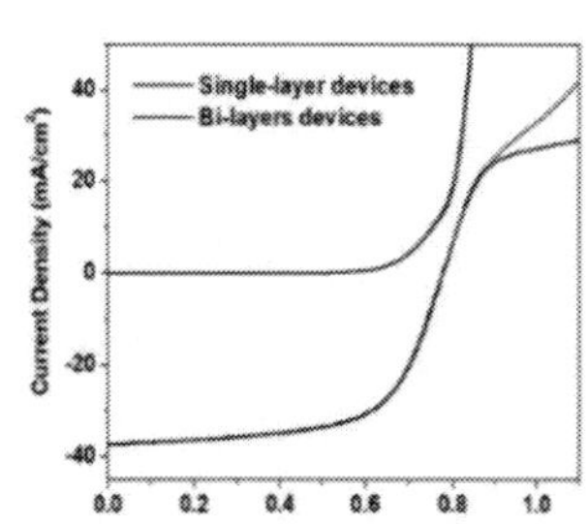

Fig. 4. J-V for $AgSb(S_{0.4}Se_{0.6})_3$ single-layer and $AgSb(S_{0.4}Se_{0.6})_3$/ $Sb_2(S_{0.4}Se_{0.6})_3$ bi-layer devices

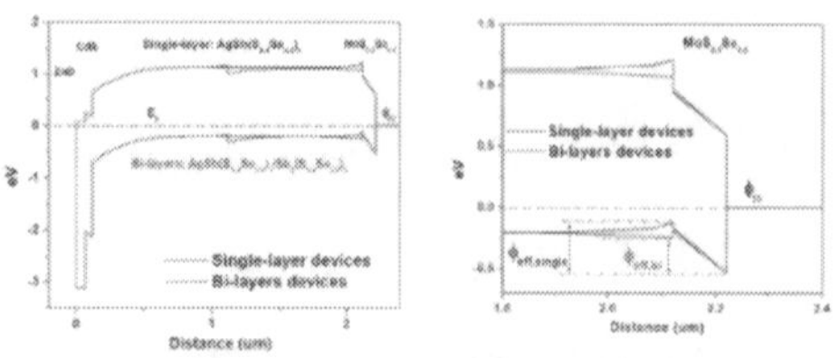

(a) (b)

Fig. 5 (a) Band diagram for two devices of $AgSb(S_{0.4}Se_{0.6})_3$ single-layer and $AgSb(S_{0.4}Se_{0.6})_3$/ $Sb_2(S_{0.4}Se_{0.6})_3$ bi-layer devices, (b) Zoom-up of (b) at the back contact interface, showing the reduced Schottky barrier of bi-layer devices.

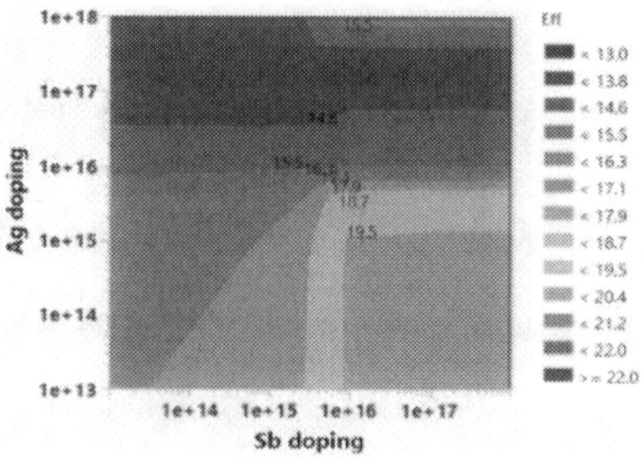

Fig. 6 Efficiency of bi-layer devices with various doping concentrations of $AgSb(S_{0.4}Se_{0.6})_3$ and $Sb_2(S_{0.4}Se_{0.6})_3$ absorber films

relatively small 1.6 mV (Voc) and 0.5% (FF), the so-called back-to-back diode structure due to the Schottky barrier at the back contact tends to deteriorate the roll-over of J-V curves far greater (4.3 times at 1 V) for single-layer devices as the current increases. Moreover, we expect the higher Schottky barrier of single-layer devices at the back contact to worsen at different operating temperatures.

In Fig. 6, the efficiency gain of $AgSb(S_{0.4}Se_{0.6})_3$/ $Sb_2(S_{0.4}Se_{0.6})_3$ bi-layer devices can be maximized (21.4 %) by increasing $Sb_2(S_{0.4}Se_{0.6})_3$ doping more than 1 x $10^{16}/cm^2$ while keeping $AgSb(S_{0.4}Se_{0.6})_3$ doping below 1 x $10^{15}/cm^2$.

5 CONCLUSIONS

In conclusion, we have studied the electronic structure of Antimony and Silver Antimony bi-layer devices. The best devices are $AgSb(S_{0.4}Se_{0.6})_3$/ $Sb_2(S_{0.4}Se_{0.6})_3$ bi-layer absorber devices with 18.6 %, higher than any single-layer devices due to the improved Schottky barrier at the back contact. Moreover, this bi-layers devices can be further enhanced by optimizing each bi-layers' doping concentration independently.

REFERENCES

[1] A. Mavlonov, T. Razykov, F. Raziq, J. Gan, J. Chantana, Y. Kawano, et. al., A review of Sb2Se3 photovoltaic absorber materials and thin-film solar cells, Sol. Energy 201 (2020) 227–246.

[2] S. Lee, M. McInerney, Modeling of Ag incorporated $Sb_2(S_xSe_{1-x})_3$ bi-layer devices: Enhanced bi-layer absorber configuration, Solar Energy, 238 (2022) 363-370

[3] J. O. Gonzalez, S. Shaji, D. Avellaneda, G. A. Castillo, T. K. D. roy, B. Krishnan, Photovoltaic structures using AgSb2(SxSe1-x)3 thin films as absorbers, Appl. Phys. A 116 (2014) 2095 -2105

[4] J. Garza, S. Shaji, E. Arato, A. Rodriguez, T. Roy, B. Krishnan, Chemically Deposited Silver Antimony Selenide Thin Films for Photovoltaic Applications, Mater. Res. Soc. Symp. Proc. 1165 (2009) 1165-M08-25.

[5] F. Xie, H. Tang, X. Zhao, M. Li, L. Li, J. Cheng, Thin-Film Solar Cells using a Selenized Silver Antimony Sulfide Absorber Prepared by Spray Pyrolysis Deposition, Phys. Status Solidi RRL 2000514 (2021) 1-9

[6] K. Bindu, J. Campos, M. Nair, A. Sanchez, P. Nair, Semiconducting AgSbSe2 thin film and its application in a photovoltaic structure, Semicond. Sci. Technol. 20 (2005) 496–504.

[7] T. Naiel, J. Henry, K. Mohanraj, G. Sivakumar, $AgSbS_2$ and Ag_3SbS_3 absorber materials for photovoltaic applications, Mat. Chem. and Phys. 181 (2016) 415–421.

[8] Y. Ho, M. Lee, $AgSbS_2$ semiconductor-sensitized solar cells, Electro. Comm. 26 (2013) 48–51.

[9] S. Lee, K. Price, E. Saucedo, S. Giraldo, Engineering of effective back-contact barrier of CZTSe: Nanoscale Ge solar cells: $MoSe_2$ defects implication, Sol. Energy 194 (2019) 114-120

[10] S. Lee, K. Price, D-H Kim, Two local built-in potentials of H2S processed CZTSSe by complex impedance spectroscopy, Sol. Energy, 225 (2021) 11-18

[11] S. Lee, K. Price, E. Saucedo, Estimation of hetero- and back contact junctions of CZTSe:Ge bi-layers solar cells by impedance and modulus spectroscopy, J. Phys. D: Appl. Phys., 54 (2021) 335501

[12] S. Lee, K. Price, E. Saucedo, Interface models of CZTSe: Ge bilayers solar cells with Impedance Spectroscopy, IEEE Proceedings of PVSC 47th Photovoltaic Specialists Conference, vol 1 (2020) 1 – 3, DOI: 10.1109/PVSC45281.2020.9300607

[13] 15. S. Lee, Kent Price, E. Saucedo, S. Giraldo, Improved Back Contact Barrier of CZTSe solar cells by incorporating nanoscale Ge bi-layers, Proceedings of 7th World Conference on Photovoltaic Energy Conversion, vol. 1 (2018) 1-3, DOI:10.1109/PVSC.2018.8547770

INSIGHTS INTO THE LIGHT TRAPPING MECHANISMS IN 100 NM-THICK NANOPATTERNED GAAS SOLAR CELLS

D. N. Micha*[a,b], M. Giteau[b], A. Cattoni[a,b], J.-F. Guillemoles[a,c], S. Collin[a,b]
[a]Institut Photovoltaïque d'Ile de France (IPVF), Palaiseau, 91120, France
[b]Centre de Nanosciences et de Nanotechnologies (C2N), CNRS, Université Paris-Saclay, Palaiseau, 91120, France
[c]UMR IPVF 9006, CNRS, Ecole Polytechnique Institut Polytechnique de Paris, PSL Chimie ParisTech, IPVF SAS, Palaiseau, 91120, France
*daniel.micha@ipvf.fr

ABSTRACT: By maximizing the interaction of sunlight with the absorber, light trapping allows for more than one order of magnitude reduction in the solar cell thickness with minor impact on performance. This possibility has been experimentally demonstrated for mature technologies, such as Si and GaAs ultrathin solar cells, with very high short-circuit currents (J_{SC}). In this work, we optimize the optical properties of a 100 nm-thick GaAs solar cell using a periodic subwavelength diffraction grating and demonstrate an optically equivalent J_{SC} of 27.2 mA/cm^2. Along with the calculations of the optical properties and a detailed analysis of the losses, we present a step-by-step evolution of the light trapping design from a simple back mirror to an advanced design combining a semiconductor diffraction grating, a spacer layer and a flat mirror. Finally, we go one step further in the optimization using *OptiPV*, an in-house developed software. It combines optical simulations with electrical simulations using SCAPS, through the calculation of the carrier generation rate in stacks that include photonic nanostructuration, allowing to consider practical aspects of the device operation.

Keywords: III-V semiconductors, Gallium Arsenide based cells, Modelling, Optical properties, Thin film.

1 INTRODUCTION

Recent advances in light trapping technologies allow for the reduction of material usage in a photovoltaic (PV) device. The so-called ultrathin solar cells (UTSC) are a new class of devices with thicknesses order(s) of magnitude smaller than their thick counterpart. The reduced thickness brings several benefits, such as lower fabrication costs, flexibility, low-weight and endurance to harsh environments, making these devices appropriate to several applications [1], such as agri-PV, vehicle-integrated-PV and, especially, space-PV. In the latter case, ultrathin devices show a major benefit as they are more resilient to particle radiation in space due to a lower interaction volume [2]. In such ultrathin devices, the thickness reduction must be compensated by mechanisms that enhance the light path inside its volume to avoid a drastic reduction in the short-circuit current (J_{SC}). A very effective and technologically reproducible approach for light trapping is the combination of back side reflectors and subwavelength periodic patterning (either in the front or back side or both) to confine sunlight by diffracting it beyond the total internal reflection angle, exciting guided modes in the waveguide-like structure. With this approach, 0.86 µm-, 2.5 µm- and 10 µm-thick Si- and 0.2 µm- and 0.26 µm-thick GaAs-solar cells have been experimentally demonstrated with J_{SC} surpassing the double pass absorption limit [3-7]. In addition, several theoretical works also indicate ways to exceed that limit [8,9], exhibiting absorption even higher than the maximum values set by the Lambertian scattering model [1,10-12].

In this work, we show a theoretical study on the optimization of a multilayer structure with a subwavelength nanopatterned diffraction grating for a 100 nm GaAs solar cell. We present our results in a didactic approach by increasing the light trapping complexity step-by-step and showing the gains of each innovation to the absorption enhancement. The best back side architecture is made of a semiconductor/dielectric grating, a dielectric spacer layer and a flat back mirror that shows a state-of-the-art J_{SC} of 27.2 mA/cm^2. In addition, we consider practical aspects of the device operation using *OptiPV*, an in-house software that combines both optical and electrical simulations.

2 OPTIMIZATION OF LIGHT TRAPPING IN ULTRATHIN SOLAR CELLS

2.1 Ultrathin Solar Cell: Optical Design

The investigated UTSC designs are presented in Fig. 1. The design is based on fully flat front side layers composed of a double-layer anti-reflection coating (DLARC), the active semiconductor layers (Fig. 1a) and a textured back side mirror with different architectures (Fig. 1b). The materials used in the layers were chosen to maximize the extracted current and to minimize losses by reflection and parasitic absorption. The period p and the height H of the square diffraction grating ($p_x = p_y$) and the side L of its square pattern were optimized using a genetic algorithm, as described in the next section.

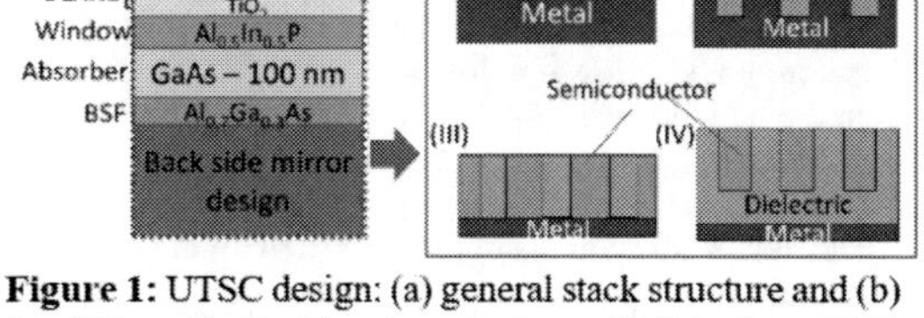

Figure 1: UTSC design: (a) general stack structure and (b) the different back side mirror designs - (I) flat mirror (FM) with a dielectric spacer (green), (II) metallic diffraction grating with a FM, (III) semiconductor diffraction grating with a FM and (IV) semiconductor diffraction grating with a spacer and a FM.

2.2 Optimization Method

We consider as a figure of merit for optimization the optical equivalent short-circuit current density $J_{SC,eq}$, which is the J_{SC} assuming that all photogenerated carriers in the GaAs absorber are extracted (perfect collection efficiency):

$$J_{SC,eq} = q \int_0^{\infty} A_{abs}(\lambda)\Phi(\lambda)d\lambda, \qquad (1)$$

in which q is the electron charge, A_{abs} is the absorptance of the absorber layer and Φ is the incident photon flux (AM1.5G solar spectrum). The calculation of the absorptance for all structures shown in this work was performed using the rigorous coupled wave analysis – RCWA (with 11 x 11 Fourier orders for the optimization runs and 21 x 21 Fourier orders for the final spectra presented here), as implemented in Reticolo [13]. The optimization of the structures with diffraction gratings was done using a genetic algorithm (ga toolbox in *Matlab*) using the $J_{SC,eq}$ as the objective function.

2.2 Evolution of the Light Trapping Architecture

In a GaAs slab of 100 nm, only around 40% of the solar photons in the relevant wavelength range (λ < 870 nm) can be absorbed in a single pass (assuming no reflection in the front surface). For a realistic structure with a DLARC (see Fig. 1a), a $J_{SC,eq}$ of only 11.5 mA/cm^2 is obtained whereas full absorption of useful photons would lead to 33 mA/cm^2. In Fig. 2, we show the best results obtained for the different light trapping designs shown in Fig. 1.

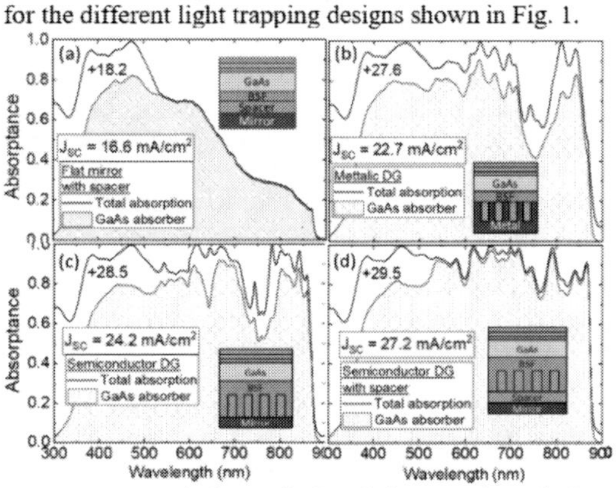

Figure 2: Evolution of the light trapping design: absorptance curves for (a) a flat mirror (FM) with a spacer, (b) a metallic diffraction grating with a FM, (c) a semiconductor diffraction grating with a FM and (d) a semiconductor diffraction grating with a spacer and a FM.

The simplest light trapping design is the use of a flat reflector at the back side of the structure right (Fig. 1b-I). In this work, we use an Ag mirror and a MgF_2 spacer layer. The $J_{SC,eq}$ reaches 16.6 mA/cm^2, as shown in Fig. 2a, with still important reflection losses (14.8 mA/cm^2), especially in the long wavelength range. When a subwavelength diffraction grating is introduced, losses are drastically reduced, and $J_{SC,eq}$ is increased to 22.7 mA/cm^2, and parasitic absorption occurs in particular in the metal (3.3 mA/cm^2). It is well known that when a metal is nanostructured, it gives rise to plasmonic resonances in the visible-near infrared range depending on their size and shape [14]. Therefore, a way of preventing such losses is to avoid metallic nanostructures, using a more transparent diffraction grating, such as the one of Fig. 2b-III. With such architecture, a higher absorption in the absorber is obtained, with lower parasitic absorption in the metal, 2.6 mA/cm^2. Finally, in order to further reduce losses, we implement a low refractive index spacer layer (MgF_2) to have the metal far from the grating. After optimization, we reach a $J_{SC,eq}$ of 27.2 mA/cm^2 with the optimized parameters, the highest short-circuit current predicted numerically for this thickness range.

2.3 *OptiPV* and the study of practical aspects

Few simple simulation tools for fully optimizing an optoelectronic multilayered ultrathin device considering accurately its optical and electrical properties are available. For this purpose, we built *OptiPV*, an in-house set of codes in *Matlab* to integrate existent available solutions and perform in-depth simulation of nanopatterned solar cells. For the calculation of optical properties, we use an in-house written transfer matrix model algorithm and the *Reticolo* software for RCWA. After obtaining the field distribution, the Poynting vector and the absorption in each layer of the structure are calculated. For a chosen irradiation spectrum, the generation rate profile is determined and, with the structure electrical parameters, is the input for the drift-diffusion model implemented in SCAPS [15]. From SCAPS, we obtain the device IV curve and its figures of merit as well as energy diagrams and carrier concentration profile.

In Fig. 3, we show the simulation results for our best structure obtained with *OptiPV*: (a) the generation rate profile, (b) the energy band profile with quasi-Fermi level distribution at the maximum power point (MPP), (c) the carrier concentration profiles at MPP and (d) the IV curve. The best structure was found after an optimization of the doping levels under the radiative limit to further optimize the J_{SC}.

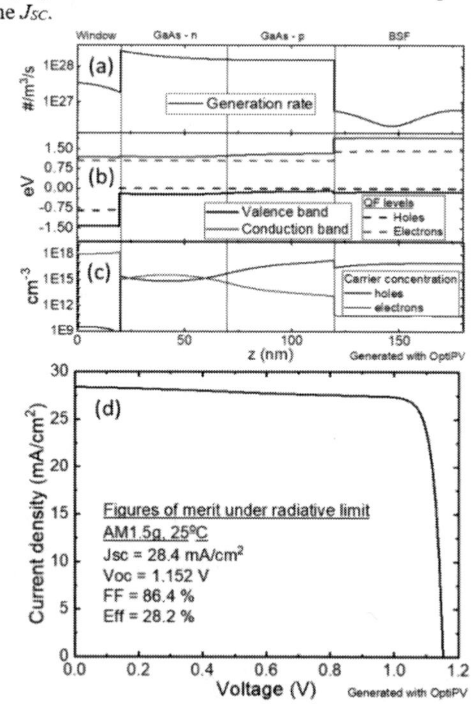

Figure 3: Optoelectronic properties calculated with *OptiPV* for the best light trapping architecture: (a) generation rate profile, (b) energy band profile with quasi-Fermi level distribution at MPP, (c) IV curve and (d) carrier concentration profiles at MPP. The figures of merit are for ideal conditions (radiative limit).

From the generation rate (Fig. 3a), we see an almost flat profile indicating homogeneity of carriers along the absorber. With the carrier concentration profile (Fig. 3c), it is possible to calculate the sheet resistances of the front and back sides to infer the total series resistance of the device and optimize front and back side electrical contact spacing (not shown here). Finally, the IV curve in Fig. 3b shows the figures of merit of the device under AM1.5g under the radiative limit.The result for the J_{SC} shows that a contribution as high as 1.2 mA/cm^2 can be expected from the window and BSF layers.

3 CONCLUSIONS AND PERSPECTIVES

In this work, we present a theoretical study on the optimization of the light trapping design for a 100 nm GaAs solar cell. We used a step-by-step approach evolving from the simplest light trapping strategy, namely a flat back side mirror, to the most complex one involving a semiconductor/dielectric grating and a flat mirror with an optimized spacer layer to thoroughly understand the improvement in the absorption properties and the reduction of losses. The best structure shows an optically equivalent J_{SC} of 27.2 mA/cm^2 under AM1.5g, the highest ever reported in this thickness range. In addition to the optical optimization, we presented results for the electrical properties of the device calculated using an in-house software called *OptiPV*. The generated curves allow for a more in-depth analysis of the device operation condition. They suggest routes for further optimization of the structure, especially regarding the doping levels and the electrical contact spacing at the front and back sides. An additional increase in J_{SC} of up to 1.2 mA/cm^2 can be expected from generation in the cladding layers.

4 ACKNOWLEDGEMENTS

The authors acknowledge IPVF/SUNLIT team members for fruitful discussions, Mark Burgelman for the technical support with SCAPS, the LIA NextPV currently IRP NextPV-II, a CNRS-University of Tokyo collaboration, and financial support from the French Government through the ANR project ICEMAN (ANR-19-CE05-0019), and in the framework of the "Programme d'Investissement d'Avenir" (ANR-IEED-002-01). M.G. is currently a Severo Ochoa Excellence postdoctoral fellow at the Institut de Ciencies Fotoniques (ICFO), Barcelona, Spain.

5 REFERENCES

[1] I. Massiot, A. Cattoni and S. Collin, "Progress and prospects for ultrathin solar cells," Nature Energy 5 (2020) 959-972.

[2] L.C. Hirst, M. K. Yakes, J. H. Warner, M. F. Bennett, K. J. Schmieder, R. J. Walters and P. P. Jenkins, "Intrinsic radiation tolerance of ultra-thin GaAs solar cells," Applied Physics Letters 109 (2016) 033908.

[3] V. Depauw, C. Trompoukis, I. Massiot, W. Chen, A. Dmitriev, P. Roca i Cabarrocas, I. Gordon and J. Poortmans, "Sunlight-thin nanophotonic monocrystalline silicon solar cells," Nano Futures 1 (2017) 021001.

[4] A. Gaucher, A. Cattoni, C. Dupuis, W. Chen, R. Cariou, M. Foldyna, L. Lalouat, E. Drouard, C. Seassal, P. Roca i Cabarrocas and S. Collin, "Ultrathin epitaxial silicon solar cells with inverted nanopyramid arrays for efficient light trapping," Nano Letters 16 (2016) 5358-5364.

[5] M. S. Branham, W.C. Hsu, S. Yerci, J. Loomis, S. V. Boriskina, B. Hoard, S. E. Han and C. Gang, "15.7% Efficient 10-μm-thick crystalline silicon solar cells using periodic nanostructures," Advanced Materials 27 (2015) 2182-2188.

[6] H.-L. Chen, A. Cattoni, R. de Lépinau, A. W. Walker, O. Hoehn, D. Lackner, G. Siefer, M. Faustini, N. Vandamme, J. Goffard, B. Behaghel, C. Dupuis, N. Bardou, F. Dimroth and S. Collin, "A 19.9%-efficient ultrathin GaAs solar cell with a silver nanostructured back mirror," Nature Energy 4 (2019) 761-767.

[7] J. Buencuerpo, T. Saenz, M. Steger, M. Young, E. Warren, J. Geisz, M. Steiner and A. Tamboli. "Efficient light-trapping in ultrathin GaAs solar cells using quasi-random photonic crystals," Nano Energy 96 (2022) 107080.

[8] J. Buencuerpo, J. Llorens, J. Ripalda, M. Steiner and A Tamboli. "Engineering the reciprocal space for ultrathin GaAs solar cells," Optics & Laser Technology 142 (2021) 107224.

[9] E. C. Abad, H. Joyce and L. Hirst, "Transparent Quasi-Random Structures for Multimodal Light Trapping in Ultrathin Solar Cells with Broad Engineering Tolerance," ACS Photonics 9 (2022) 2724–2735.

[10] S. Bhattacharya, I. Baydoun, M. Lin and S. John, "Towards 30% power conversion efficiency in thin-silicon photonic-crystal solar cells," Physics Review Applied 11 (2019) 014005.

[11] S. Eyderman et al. "Light-trapping optimization in wet-etched silicon photonic crystal solar cells," Journal of Applied Physics 118 (2015) 023103.

[12] S. Eyderman, S. John, S and A. Deinega, "Solar light trapping in slanted conical-pore photonic crystals: beyond statistical ray trapping," Journal of Applied Physics 113 (2013) 154315.

[13] J. P. Hugonin and P. Lalanne, RETICOLO software for grating analysis, Institut d'Optique, Orsay, France (2005), arXiv:2101:00901.

[14] H. Nishi and T. Tatsuma, "Full-Color Scattering Based on Plasmon and Mie Resonances of Gold Nanoparticles Modulated by Fabry–Pérot Interference for Coloring and Image Projection," ACS Appl. Nano Mater. 2 (2019) 5071–5078.

[15] M. Burgelman, P. Nollet, S. Degrave, "Modelling polycrystalline semiconductor solar cells" Thin Solid Films 361-362 (2000) 527-532.

EVALUATION OF PERFORMANCE MEASUREMENTS OF DIFFERENT TYPE PEROVSKITE DEVICES AT DIFFERENT MEASUREMENT CONDITIONS

V. Paraskeva[1], M. Hadjipanayi[1], M. Norton[1], A. Aguirre[2,3,4], A. Hadipour[5], A. Krishna[2,3,4], G.E. Georghiou[1]
[1]FOSS Research Centre for Sustainable Energy, Department of Electrical and Computer Engineering, University of Cyprus, 75 Kallipoleos St., Nicosia, 1678, Cyprus
[2] Imec, imo-imomec, Thin Film PV Technology – partner in Solliance, Thor Park 8320, 3600 Genk, Belgium
[3]EnergyVille, imo-imomec, Thor Park 8320, 3600 Genk, Belgium
[4] Hasselt University, imo-imomec, Martelarenlaan 42, 3500 Hasselt, Belgium
[5] Kuwait University, Department of Physics, Condensed Matter Physics Group, University city Shadadiya campus, Kuwait

ABSTRACT: Perovskite solar cell devices are driving a re-evaluation of photovoltaic measurement techniques due to their inherent instabilities that complicate the application of traditional characterization approaches. Being a relatively novel technology, the experiences of measuring these devices contribute to their advancements in production. In this context, the work aims to contribute to the growing body of indoor test experiences by presenting indoor measurement results obtained from different perovskite technology modules. Three different types of perovskite technology mini-modules have been tested with different measurement performance strategies to investigate factors that might improve the reliability of performance measurements of perovskite devices. Impact of sweep rate, scan direction priority, time delay between forward-reverse scans, MPP and light preconditioning are some of the factors studied at the different technology perovskite devices. Measurements revealed that $MAPbI_3$ perovskite-based devices are more sensitive to sweep rate, scan direction priority and time delay between forward and reverse scans. An investigation of (IV) traces at different sweep rates is highly recommended in order to give an insight regarding the impact of ion migration effects in the perovskite devices under study and for the selection of the appropriate sweep rate.
Keywords: Characterization, Perovskite, Performance

1 INTRODUCTION

Due to the hysteretic behavior of most perovskite cells the indoor and outdoor characterization of those devices present several challenges. Their photocurrent density – voltage (J-V) demonstrates anomalous dependence on the voltage scan direction/rate/range, voltage conditioning history and device configuration [1]. For this purpose, the IEC has developed some guidelines on how to measure perovskite devices. This work is described in IEC standard 63228 which presents the measurement protocols for photovoltaic devices based on organic, dye-sensitized or perovskite materials. Although a large body of research work has demonstrated the impact of various measurements factors on perovskite performance [2], [3], more experience of those studies in several types of perovskites is still needed. The collection of IV traces at different sweep rates will demonstrate the interplay between ion migration and capacitance effects and the impact of perovskite composition on the response of the device will be established. This will lead to improvement of the existing measurement strategies for the measurement of perovskites with specific active layer compositions.

To investigate in more detail other factors that may improve the reliability of performance measurements, current-voltage (IV) characteristics of perovskite devices of different technologies have been studied under different conditions. Particularly, three different types of perovskites have been utilized for these studies: $MAPbI_3$ (methylammonium lead iodide) (type A module), and 2 different types of two-cation perovskite CsFaPbIBr: type B module and type C module. Different measurement strategies have been applied to give insight into the impact of ion migration on the performance curves of different perovskite active layers. Impact of sweep rate, scan direction priority, time delay between forward-reverse scans, MPP and light preconditioning are some of the factors studied in this paper during the performance measurements of the different technology perovskites.

Measurements have revealed the effect of sweep rate and scan direction priority in $MAPbI_3$ perovskite-based devices. Using a lower sweep rate minimizes hysteresis and brings those devices towards their steady-state operating conditions. Moreover, the scan direction priority in these devices determines their power output. Different steady-state conditions exist during forward-first and reverse-first studies even at low sweep rates. On the other hand, type B perovskite devices are not affected by sweep rate or scan direction priority. Finally, in type C perovskite devices, a reduction of hysteresis was found at higher sweep rates indicating the dominance of ion migration of slow-moving ions. An investigation of (IV) traces at different sweep rates is highly recommended in order to give an insight regarding the impact of ion migration effects in the perovskite devices under study and for the selection of the appropriate sweep rate.

2 EXPERIMENTAL APPROACH

The module performance was measured by using a Sol3A Class AAA solar simulator from Newport using a light power at 1000W/m^2. The current at the contacts of the module was collected using a Keithley 2430 source meter driven by a python script. A temperature-controlled base was utilized to keep the temperature of the samples at 25°C. Different perovskite devices have been utilized for the investigation of the appropriate testing techniques for performance measurements. This is necessary in order to understand how the different testing conditions affect each perovskite structure and to extract a general conclusion for the parameters that affect the performance measurements of those devices. Perovskite modules were preferred over cells for these performance studies as they exhibited better stability over time. The different types of perovskites studied are the following: $MAPbI_3$ (type A module), and two different types of 2-cation perovskite CsFaPbIBr: type B module and type C module. The type A modules consist by six (6) cells and the overall dimensions of the modules are 8.5 x 3.5 cm^2. The type B and type C modules consist by seven (7) cells and their overall dimensions are 3.17 cm^2.

3 RESULTS AND DISCUSSION

3.1 Impact of sweep rate

The impact of sweep rate and voltage sweep order was initially investigated in type A modules ($MAPBI_3$). Forward-first and reverse-first measurements have been applied during those studies to spot differences in the IV curves using different scan direction priorities. The measurements (Figure 1) indicate that reverse and forward curves differ significantly in devices that exhibit high levels of hysteresis. Reverse IV sweeps provide higher maximum power output in agreement with previous reports[4]. The lowest sweep rate in both forward-first and reverse-first approach presented the lowest level of hysteresis and was found to move both reverse and forward curves towards the expected steady state performance and seems to be the best choice for performance measurements. The MPP value at steady state conditions in the forward-first approach (indicated by a green dot in Figure 1a) was found by collecting a full forward IV curve at very low sweep rate. Particularly, at each voltage point, the waiting time was set to 5 minutes which corresponds to a sweep rate of 0.001 V/sec. This point (green dot in Figure 1a) approaches the MPP of both forward and reverse curves in the forward-first approach at low sweep rates.

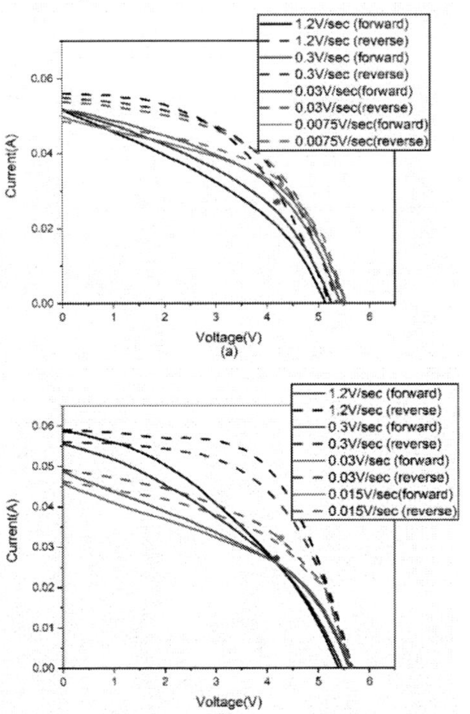

Figure 1: Performance measurements of perovskite type A at different sweep rates and in (a) forward-first and (b) reverse-first approach.

In the same manner, the MPP value at steady state conditions during reverse-first measurements (indicated by a red dot in Figure 1) was found by collecting a full reverse IV curve at the same sweep rate (0.001 V/sec). At this sweep rate, the maximum power in reverse-first curves was found to be smaller than the maximum power found in the forward-first ones. Therefore, it can be deduced that the steady state conditions are determined not only from the sweep rate but also from the voltage sweep direction. Forward-first measurements at low sweep rates seems to provide higher output power compared to reverse-first measurements at the type A samples. Moreover, the voltage sweep rate, which minimizes hysteresis effects and brings the samples towards their steady state conditions in both cases (forward-first and reverse-first), is found to be lower than 15 mV/sec. The impact of scan direction priority was also found during the outdoor testing of those devices as well [5].

In type B devices, no change of the current-voltage characteristics is obtained at different sweep rates and at different scan direction priority (see Figure 2). Moreover, both forward- first and reverse-first measurements have been applied but no difference was obtained between the data.

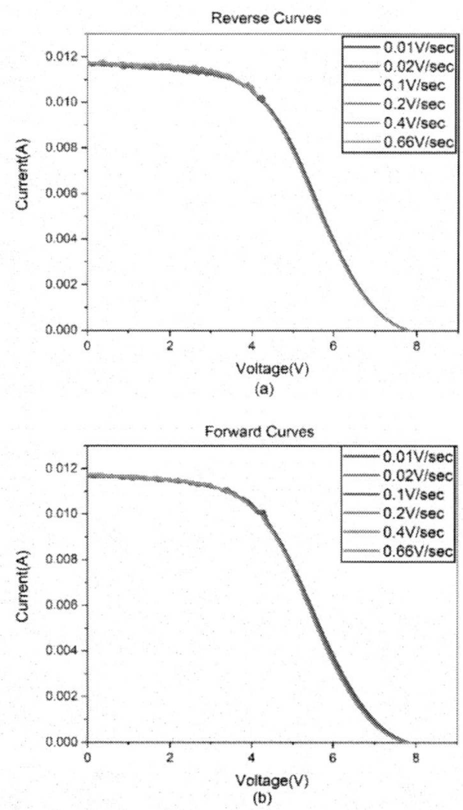

Figure 2: (a) Reverse and (b) forward curves of perovskite type B device at different sweep rates. Reverse-first approach was used.

Finally, a type C perovskite module was studied at different sweep rates in order to explore their effect on another type of perovskite device. Only forward-first measurements have been conducted on that device. As can be obtained from Figure 3, in this type of module hysteretic behavior is obtained at lower sweep rates.

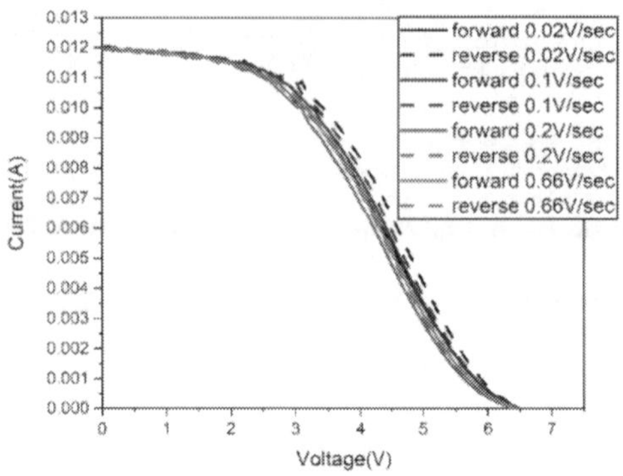

Figure 3: Performance measurements of type C device at different sweep rates. The results indicate the dominance of ion migration of slow-moving ions.

Particularly, hysteresis is obtained at sweep rates lower than 0.1 V/sec. This effect shows opposite behavior compared to the data obtained for type A perovskite as presented above. It is believed that ion migration of slow-moving ions might be the dominant mechanism for this observation. What is interesting in this case is the same power output obtained from the perovskite module at the different sweep rates, which indicates that the application of different sweep rate does not affect the overall output of the device.

3.2 Impact of time delay between forward-reverse curves

Time delay between forward and reverse curves has been introduced to see its impact on the current-voltage curves. Forward-first measurements have been applied in this case. In the forward-first approach the time delay between forward and reverse curves is expected to affect only the reverse curves.

Initially, type A device was studied. The reverse curves at different time delays have been collected and compared with the steady state curve of the cell. The current in this curve was collected at very low sweep rates (0.001V/sec) and it is assumed that those conditions are the steady state conditions for the sample. Time delay between forward-reverse curves in forward-first measurements move the reverse curve away from the steady state condition (see Figure 4).

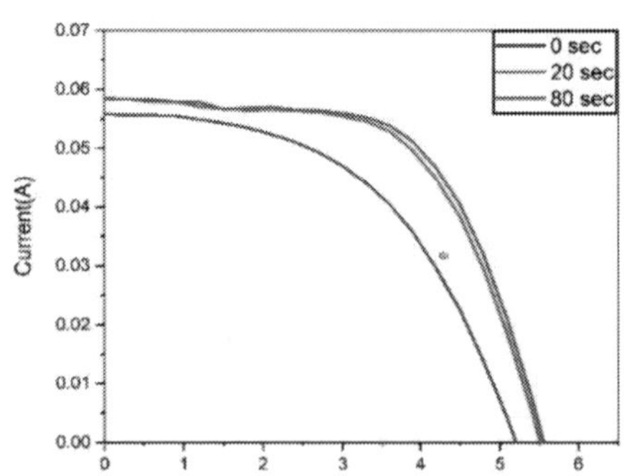

Figure 4: Reverse curves of type A module at different time delays between forward and reverse curves in forward-first measurements.

The same procedure was applied in type B devices. Time delay between forward and reverse curves was introduced and reverse curves were collected at each case. The results are depicted in Figure 5. No significant change is obtained at the different time delays applied between forward and reverse curves indicating that the addition of time delay between forward and reverse curves does not have an impact in this type of perovskite devices.

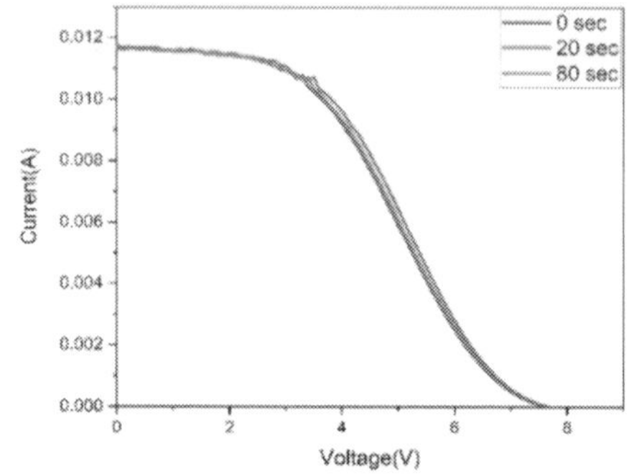

Figure 5: Reverse curves of type B module at different time delays between forward and reverse curves in forward-first measurements.

3.3 Impact of MPP and light Preconditioning

The impact of MPP and light preconditioning on device output was investigated next. Type B perovskite devices were studied in this case. During those studies, an initial IV curve was collected and then the device was exposed to conditions close to MPP for 5 minutes under light and then the performance measurements were repeated again. In this manner the devices were exposed to light soaking for another 5 minutes and in the presence of MPP voltage and then output of the device was collected. The maximum power point voltage (Vmp) of the device under test found from previous performance measurements was around 4.2V and this have been applied in the device during MPP preconditioning tests. In that voltage the corresponding current has been collected.

Table 1: Major electrical parameters extracted from reverse-first curves at different sweep rates in the presence and absence of MPP and light preconditioning.

	Imp (A)	Vmp (V)	Pmax (W)
	Sweep rate: 0.66V/sec		
No MPP	0.01009	4.2	0.04238
MPP	0.01006	4.5	0.04527
	Sweep rate: 0.4V/sec		
No MPP	0.01015	4.2	0.04262
MPP	0.01004	4.5	0.04516
	Sweep rate: 0.2V/sec		
No MPP	0.01014	4.2	0.04257
MPP	0.01007	4.5	0.04532
	Sweep rate: 0.1 V/sec		
No MPP	0.01011	4.2	0.04245
MPP	0.01001	4.5	0.04506
	Sweep rate: 0.02 V/sec		
No MPP	0.00991	4.3	0.04260
MPP	0.01004	4.5	0.04516
	Sweep rate: 0.01 V/sec		
No MPP	0.01007	4.3	0.04329
MPP	0.01018	4.5	0.04580

During that procedure, it was obtained that the maximum power point obtained before MPP and light preconditioning was no longer valid and a new maximum power point value was exhibited from the measurements which was larger than before. Reverse-first measurements at different sweep rates have been collected in the presence and absence of MPP and light preconditioning. Table 1 summarizes the main parameters obtained from reverse-first curves at different sweep rates in the presence and absence of MPP and light preconditioning.

As can be obtained in Table 1 at all sweep rates the output power of the device shifts towards higher values in the presence of MPP and light preconditioning. This result was confirmed through repeatable measurements and indicates that the combination of MPP and light preconditioning in those devices affects their output.

3.4 Maximum power-point (MPP)-first measurements

In an attempt to study if keeping the perovskite device at maximum power-point before current-voltage measurements can result to more reliable performance collection, the samples have been initially set to MPP conditions (for 5 minutes) and then started to collect their IV curves at different sweep rates starting from that point. Type A devices were studied with this approach. In almost all cases studied in the past, the starting point of the IV curves was either at voltage 0 V (which corresponds to short-circuit conditions) either the Voc (open-circuit voltage) [4]. The first case corresponds to forward-first measurements and the last one to reverse-first measurements. In our case, MMP-first measurements have been applied. Since the starting point is at MPP, different scan directions should be applied. Figure 6 represents the possible scan directions in the case where the initial point in the IV curve is the MPP. In Figure 6(a) the currents in forward direction have been collected first and then a complete reverse curve was captured. The second case (Figure 6b) starts from MPP and the currents are collected in reverse direction first and then collect currents in forward curve. In this case the reverse curve is collected first and the forward curve is collected second. In all cases the MPP is selected to be the value found in forward-first measurements at low sweep rates presented above (section 3.1).

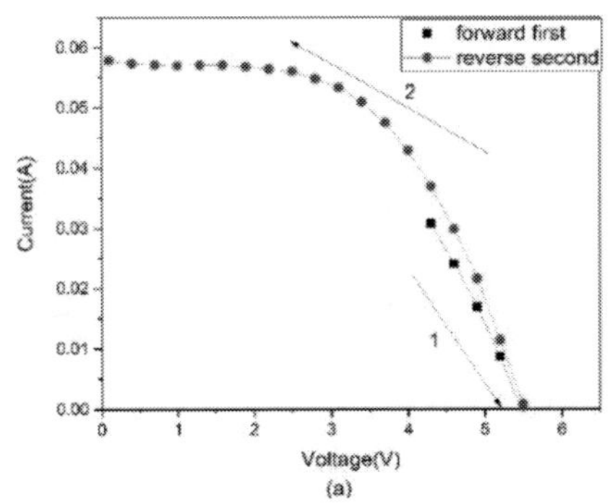

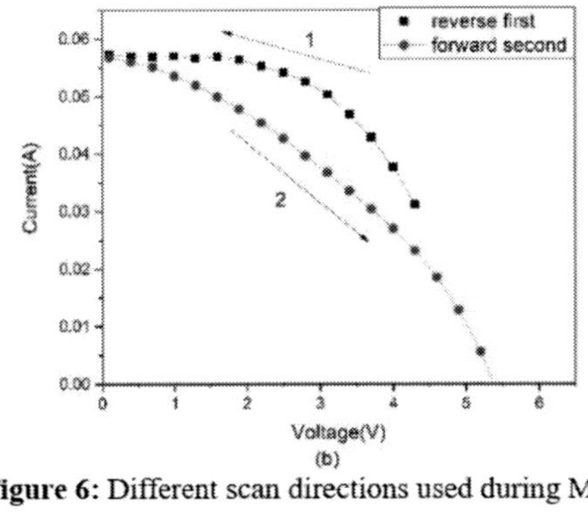

Figure 6: Different scan directions used during MPP-first measurements of type A devices.

The IV curves at the different directions have been studied at different sweep rates in an attempt to find if one of the curves is approaching the steady state curve. The steady state curve of the perovskite device corresponds to the forward first IV curve at very low sweep rates (0.001V/sec).

In the case of high sweep rates (2V/sec), the steady state curve of the device is not approached by any of the collected curves indicating that the performance measurements in both scan directions is far from the real one. The results are depicted in Figure 7.

Then the same procedure was repeated at lower sweep rate (Figure 8). As the sweep rate becomes lower all the curves are approaching each other: the steady state curve is approaching the forward second curve at low voltage biases and reverse second curve at higher voltage biases. However, none of those curves does not represent the steady state condition of the device at all voltage biases indicating that this testing technique is not appropriate for reliable performance testing of perovskites.

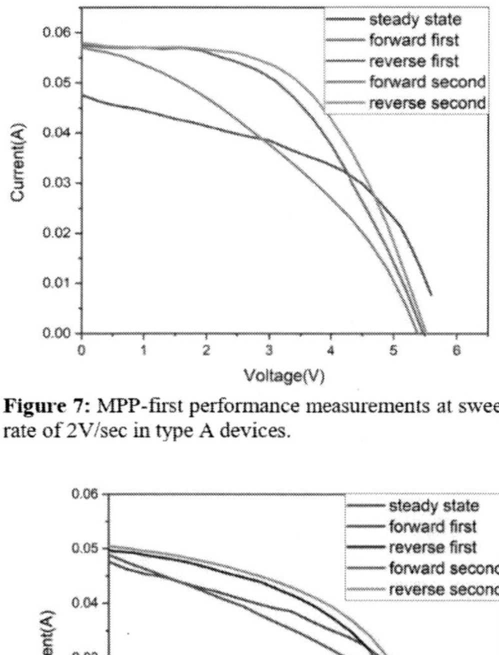

Figure 7: MPP-first performance measurements at sweep rate of 2V/sec in type A devices.

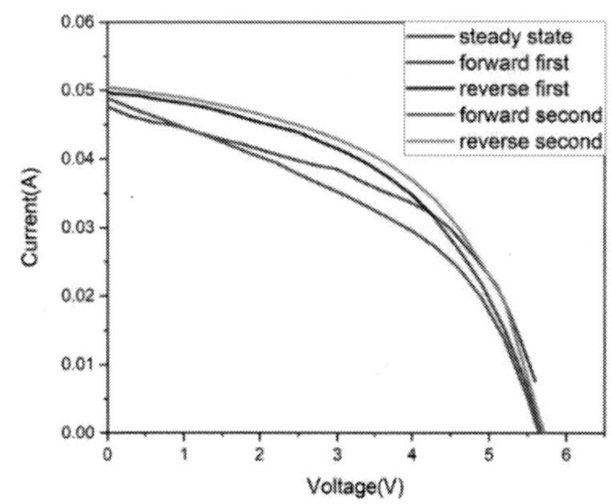

Figure 8: MPP-first performance measurements at sweep rate of 0.03V/sec in type A devices.

4 CONCLUSIONS

After a detailed study of several factors that might affect the performance measurements of different perovskite modules, it can be concluded that the most critical factor that affects the performance measurements is the sweep rate. By studying different types of perovskites, it is demonstrated that sweep rate in $MAPbI_3$ devices should be as low as possible in order to reach the device the steady state conditions. However, this behavior is not obtained in all perovskite devices since it was found that in CsFaPbIBr devices the sweep rate is not critical and the same output is obtained at different sweep rates. A careful study of the performance of the perovskite at different sweep rate is highly recommended for reliable performance studies. Regarding the scanning direction in $MAPbI_3$ devices (type A) the scan direction was found to affect the MPP of the devices. In our investigations it was found that the MPP during reverse first scan was lower in those devices and this is obtained even at very low sweep rates where steady-state conditions exist.

The impact of time delay between forward-reverse scans and vice-versa was found to affect mainly the $MAPbI_3$ devices (type A). Time delay between forward-reverse scans was found to move the efficiency of the type A perovskite away from the steady state conditions and should be avoided. Time delay between forward and reverse scans was not found to affect type B devices. CsFaPbIBr devices (type B) performance was found to be affected by the MPP and light preconditioning. Performance of those devices was shifted to higher values after MPP and light preconditioning for some minutes prior the IV collection.

Finally, MPP first measurements were not found to be an option for reliable performance studies, since none of the IV curves collected by applying this method didn't approach the steady state condition curves at all voltage biases in devices with strong hysteresis (type A).

ACKNOWLEDGMENT

This work was funded through the European Regional Development Fund and the Republic of Cyprus in the framework of the project "DegradationLab" with grant number INFRASTRUCTURES/1216/0043.

5 REFERENCES

[1] B. Chen, M. Yang, S. Priya, and K. Zhu, *J. Phys. Chem. Lett.*,7, (2016), 905–917.

[2] J. Christians, J. Manser, and P. Kamar, *J. Phys. Chem.* 6, (2015), 852–857.

[3] R. B. Dunbar *et al.*, *J. Mater. Chem. A*, 5, (2017) 22542.

[4] F. Wu, R. Pathak, K. Chen, and G. Wang, *ACS Energy Lett.*, 3, (2018), 2457–2460.

[5] M. Norton, M. Kohlstädt, M. Hadjipanayi, V. Paraskeva, U. Würfel, and G. E. Georghiou, Proceedings 37th European Photovoltaic Solar Energy Conference (2020), 606.

PEROVSKITE-BASED SOLAR INK FOR DIFFERENT FABRICATION PROCESSES

Anjusree Shyla[1], Deepak Thrithamarassery Gangadharan[1], Erin Moloney[2], Sahar M. Sam[1], Sonal Prajapati[1] & Makhsud Saidaminov[2]
[1]Solaires Entreprises Inc., Victoria, British Columbia, Canada, & [2] University of Victoria, Victoria, British Columbia V8P 5C2 Canada

ABSTRACT: Perovskite solar cells (PSCs) have made tremendous progress in the past years with an efficiency of over 25% in lab-scale devices. Scalability of the perovskite technology is one of the major challenges that restrict perovskites from entering the market. In this research, we performed a systematic experimentation study to demonstrate the scalability and stability of our perovskite-based Solar Ink™ and demonstrated its compatibility with different coating processes. The goal is to address the industry need of improving the performance of the perovskite solar cells by having a stable perovskite ink compatible with air-processing and ambient atmosphere, and reduce the fabrication cost by eliminating the requirement of a controlled environment. The aim of this work is to demonstrate the stability and scalability of our novel perovskite ink (Solar Ink™) with high efficiency and stability, and its compatibility with a variety of scalable manufacturing processes.

1 INTRODUCTION

Perovskite technology has emerged as the low-temperature solution processing photovoltaic technology with high efficiency because of its outstanding properties, including strong light absorption, tunable bandgap and long diffusion length [1-4]. Perovskite solar cells (PSCs) can be easily fabricated by solution-based processing coating methods [5-6]. Although the number of publications and reports on the efficiency of PSCs is increasing, almost all reports are based on small area device fabrication techniques, like spin coating, which is not feasible for large-scale manufacturing of solar panels [7-8].

Scalability of the perovskite technology is one of the major challenges that restrict perovskites from entering the market [9]. The issue of its scalability is defined in two categories: 1- perovskite material and 2- fabrication process for PSCs. Although several industries and research institutions are focused on scaling up the fabrication process [10-12], the scalability of perovskite material, the so-called perovskite ink, has been overlooked. The scalability of the perovskite ink is largely dependent on its stability i.e., whether the perovskite ink is capable of retaining its structural integrity throughout the fabrication process. Most of the perovskite ink formulations are stable only for a day or 2 which limits the storage of ink for large-scale manufacturing. To investigate the properties of ink and its stability various rheological properties have been studied [13].

The present work demonstrates and validates the capability of our Solar Ink for industrial-scale deposition techniques like inkjet printing and blade coating as well as for conventional lab-scale techniques like spin-coating used for the fabrication of fully air-processed PSCs. The Solar Ink™ was also tested with the different device architectures.

2 EXPERIMENTAL DETAILS

2.1 Materials

ITO-coated glass substrates were purchased from Ossila with a sheet resistance of 10 Ohm/Sq and dimensions of 25 mm x 25mm. FAI, MACl, and MABr were purchased from Great Cell, PbI_2 from TCI chemicals, and $PbBr_2$ was purchased from Alfa Aesar as perovskite precursors. All the solvents for the perovskite precursor solution were purchased from Sigma-Aldrich. SnO_2 15% in H_2O colloidal dispersion solution was purchased from Alfa Aesar.

The SnO_2 solution was made by diluting 15% in the ratio of 1:6 in water. The diluted SnO_2 solution was sonicated for 30 min before deposition.

Perovskite-based Solar Ink™ was made by mixing the precursor powders in DMF and NMP solvents as described in a previous publication [14].

The spiro solution was made using 0.1g Spiro-OMeTAD, 39μl tBP, 1.1 ml chlorobenzene, 25μl Li-TFSI solution dissolved in acetonitrile (540 mg/ml), and 10μl cobalt salt dissolved in acetonitrile (376 mg/ml).

2.2 Device fabrication

ITO-coated glass substrates as received were cleaned by sonication with DI water, acetone, and isopropanol, respectively, each for 15 minutes. The substrates were then dried with an N_2 gas gun and were kept in a UV-Ozone cleaner. The ozone treatment was performed for 30 minutes. After that, the substrates were immediately spin-coated with 150μl SnO_2 solution at 3000 rpm for 30 seconds. After deposition, all the films were thermally annealed at 150°C for 30 minutes. For the next step, the substrates were again kept for 30 minutes of ozone treatment and then 100μl of the perovskite-based Solar Ink™ was spin-coated in a one-step process on top of SnO_2 and thermally annealed at 150°C for 3 minutes for crystallization. Before thermally annealing an antisolvent step was performed by dipping the perovskite coated sample in antisolvent. The next hole transport layer (HTL) with 75μl Spiro-OMeTAD solution was deposited with a dynamic spin coating method at 2000 rpm for 30 seconds. Finally, a thin layer of ~80 nm of gold was evaporated on top of the film for front metal connections.

For blade coating, the cleaning step and ozone treatment for the ITO coated substrates with the dimension of 3inch x 1.5inch was performed as described for spin coating and after that 45 μl of the same SnO_2 solution was blade coated on the heated substrate at 80°C. A gap of 150 micrometers and a speed of 10 mm/s were used for deposition. After deposition, it was annealed for 30 min at 150 C. After this, 30 minutes of ozone treatment was performed, and then perovskite-based Solar Ink™ was blade coated using a gap of 150 micrometers and speed of 7 mm/s s and the films were annealed at 150 C for 30 minutes for uniform crystallization. Lastly, the spiro-OMeTAD solution was spin-coated onto the perovskite film by dynamic spinning and the top electrodes of gold were thermally evaporated.

The inkjet-printed PSCs were HTL-free, meaning their device architecture was different from spin and blade coated. The FTO-coated glass substrates were used for the inkjet printed PSCs. As an electron transport layer (ETL) compact TiO2 was deposited on the FTO-coated glass substrates by using optimized inkjet printed parameters. Perovskite-based Solar Ink™ was deposited using inkjet printing on TiO2-coated substrates and then annealed at 140 C for 10 minutes. Finally, the carbon layer was inkjet printed as the top electrode.

All three fabrication processes were carried out in fully ambient air-processed conditions.

3 RESULTS AND DISCUSSION

3.1 Formulation to improve the stability of the perovskite ink

We carefully formulated a stable perovskite ink that is compatible with major industrial scalable coating techniques. To investigate the stability of Solar Ink™, XRD of perovskite films deposited from aged ink were monitored. Figure 1 below shows the perovskite films were formed with 104 day old Solar Ink™with the characteristic X-ray diffraction (XRD) peaks 14° and 28.2°.

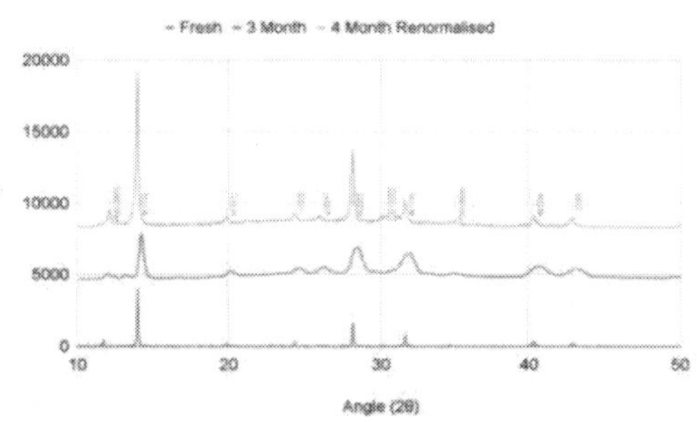

Figure 1: XRD of films made from aged perovskite-based Solar Ink™

We also fabricated solar cells made from aged Solar Ink™, and tracked power conversion efficiency (PCE), as shown in Figure 2. The devices made with Solar Ink™ after aging for 90 days showed similar performance compared to fresh ink This shows the stability of the ink in the device as well as its shelf life.

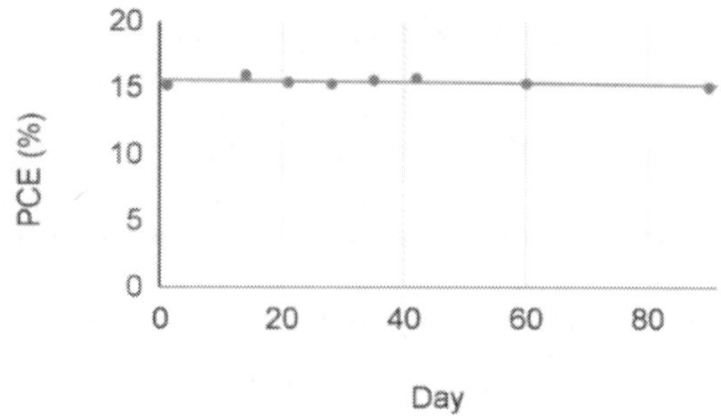

Figure 2: PCE of devices fabricated with aged ink.

3.2 Fabrication of perovskite solar cells via different coating techniques

Spin coating

The initial optimization of the device was done using lab-scale coating technique, spin coating. To test and demonstrate the performance of Solar Ink, PSCs were fabricated with an indium tin oxide (ITO) / SnO_2 / perovskite/ Spiro-OMeTAD / Gold (Au) device structure. The PCE of the tested devices was approximately 20% over an active area of 0.38 cm^2 shown in Table I. The I-V curve and SEM for these devices can also be seen in Figure 3 below.

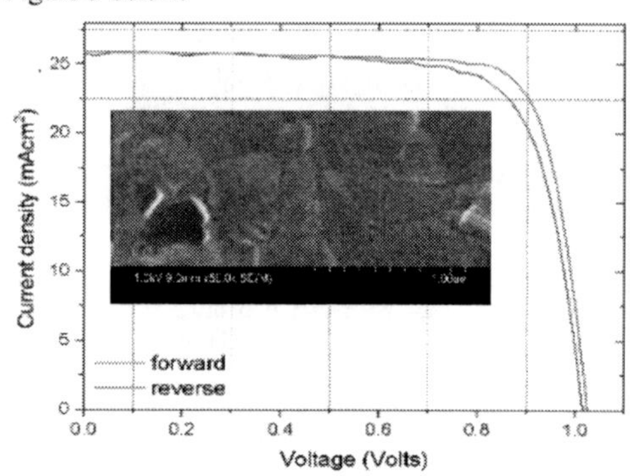

Figure 3: I-V curve and SEM for the spin-coated device with an active area of 0.38 cm^2

Table I: Device characteristics of the spin-coated device with Solar Ink of an active area of 0.38 cm^2

Bais	Efficiency (%)	FF (%)	Jsc (mA/cm2)	Voc (V)
Reverse	20.88	78.76	25.85	1.02
Forward	19.52	74.53	25.79	1.01

Blade-coating

One of the major challenges of upscaling perovskite coating is getting a uniform, pin-hole-free, dense film coating over a large area. Here, we demonstrate that Solar Ink™ can produce a dense, pin-hole-free film using the blade-coating technique (Figure 4b). The large area (1 cm^2) devices fabricated with blade-coated perovskite film showed a PCE of 15.39%, a comparable performance to the devices made from the spin-coating technique and is shown in Figures 4a & b.

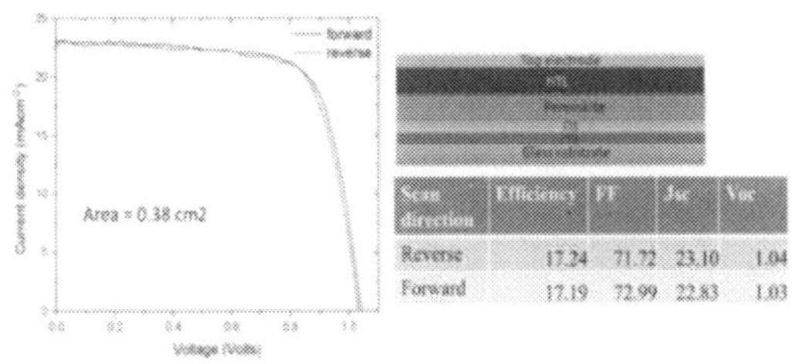

Figure 4a: Device architecture for blade-coated device and I-V curve with an active area of 0.38 cm2.

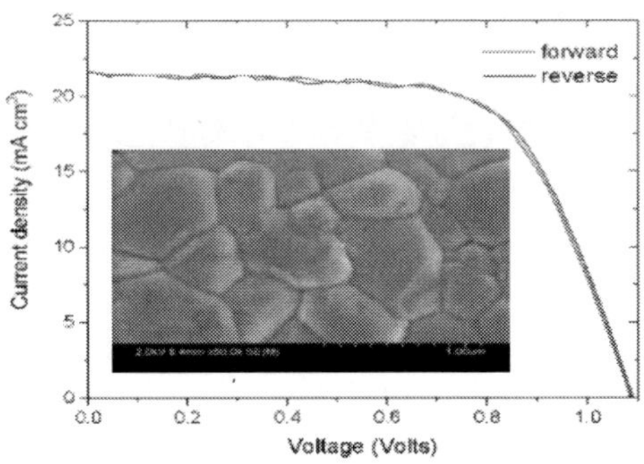

Figure 4b: I-V curve and SEM for blade-coated with an active area of 1.03 cm^2.

Table II: Device characteristics of the blade-coated device with Solar Ink™ with an active area of 1.03 cm^2

Scan direction	Efficiency (%)	FF (%)	Jsc (mA/cm2)	Voc (V)
Reverse	15.37	65.61	-21.51	1.08
Forward	15.39	65.92	-21.51	1.08

Inkjet printing

We also demonstrated the capability of ink coating using inkjet printing in a carbon-based HTL-free device architecture which also showed a performance of approximately 10% over a 1cm^2 active area. It is also difficult to incorporate an antisolvent step for the perovskite film formation for the inkjet printing process. The annealing of the perovskite film was performed at two different temperatures 150 C for 3 minutes, resulting in 6.94% efficiency and 140 C for 10 minutes with 9.97% efficiency. The performance and architecture of the inkjet-printed devices are shown below in Figure 5. It should be noted that this is a very preliminary result and we are optimizing various device fabrication parameters to improve performance further.

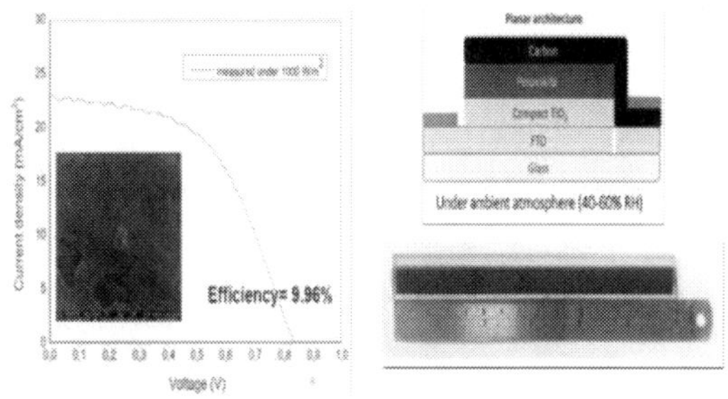

Figure 5: Inkjet-printed perovskite solar cell and device architecture

A comparison of spin-coated, blade-coated, and inkjet-printed devices is presented in Table III, showing that while the efficiency varies depending on deposition method and size, all cells display adequate performance.

Table III: Device characteristics of perovskite solar cells with Solar ink by the different coating methods

Device Characteristic	Spin	Blade	Inkjet
Active area of cell (cm^2)	0.38	1	1
Efficiency (%)	20.88	15.39	9.97

3.3 Stability of the fabricated devices using Solar Ink™

Perovskite solar cells are sensitive to UV light, high temperature, oxygen, and humidity and tend to degrade faster in their presence [15-16]. Hence, it is necessary to hermetically seal the PSCss with encapsulant material so it does not come in contact with the external environment. In the present study, the blade-coated devices were encapsulated with cover glass and epoxy and efficiency was measured at regular time intervals for 90 days. It was found that an unencapsulated blade-coated PSC degraded after 20 days while the encapsulated device was operational up to 90 days, shown in Figure 6. The blade-coated and inkjet-printed devices were tested for stability under ambient conditions.

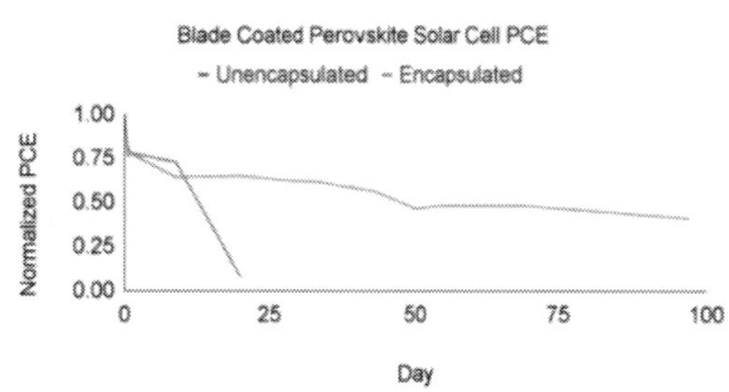

Figure 6: Stability of blade-coated encapsulated and unencapsulated devices.

The inkjet printed device with solar ink was tested for 1000 hours (Figure 6) and was found to be stable under ambient conditions.

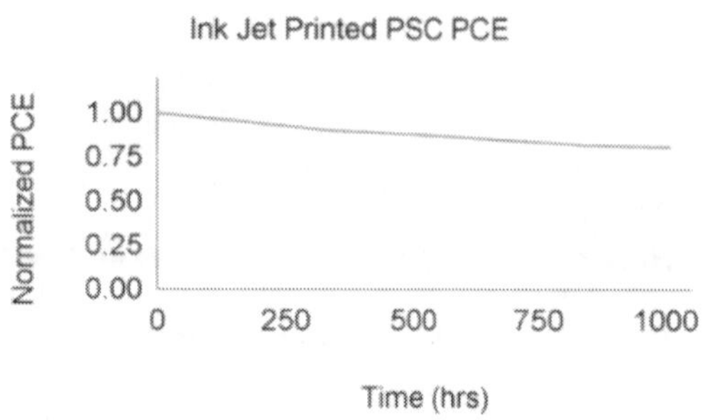

Figure 7: Stability of inkjet printed devices using Solar Ink™.

4 CONCLUSION

In conclusion, we have demonstrated a unique perovskite ink formulation (Solar Ink™) that is stable for more than 3 months in ambient conditions without any effect on the performance of the devices fabricated using this formulation. Additionally, the same Solar Ink™ formulation was successfully tested for different scalable coating processes and device architectures. The stability of Solar Ink™ with longer shelf life and scalable processing provides a gateway for the transition from lab scale to industrial scale.

5 REFERENCES

[1] J You, Z Hong, Y Yang, Q Chen, M Cai, T Song, C Chen, S Lu, Y Liu, H Zhou, Y Yang, ACS Nano 8 (2014) 1674.

[2] J Huang, Y Yuan, Y Shao, Y Yan. Nat Rev Mater. 2 (2017) 1.

[3] C Chen, Y Jiang, Y Wu, J Guo, X Kong, X Wu, Y Li, D Zheng, S Wu, X Gao, Z Hou. Solar RRL 4 (2020) 1900499.

[4] A. R. Uhl, Z. Yang, A. K. Y. Jen and H. W. Hillhouse, J. Mater. Chem. A, 2017, 5, 3214–3220

[5] B. Li, D. Binks, G. Cao, J. Tian, Small 15 (2019) 1903613

[6] L. Chao, Z. Wang, Y. Xia, Y. Chen, W. Huang, J. Energy Chem.27 (2018) 1091

[7] H. Min, M. Kim, S.-U. Lee, H. Kim, G. Kim, K. Choi, J. H. Lee, S. I. Seok, Science (Washington, DC, U.S.) 366 (2019) 749.

[8] Q. Jiang, Y. Zhao, X. Zhang, X. Yang, Y. Chen, Z. Chu, Q. Ye, X. Li, Z. Yin, J. You, Nat. Photon. 13 (2019) 460.

[9] J Yan, T Saveniji, L Mazzarella, O Isabella Sustainable Energy & Fuels 6 (2022) 243.

[10] L Fagiolari, F Bella Energy Environmental science 12 (2019) 3437.

[11] https://sauletech.com/product/

[12] J Benitez-Rodriguez, D Chen, M Gao, R. Caruso RRL Solar 5 (2021) 2100341.

[13] E Berger, M Bagheri, S Asgari, J Zhou, M Kokkonen, P Talebi, J Luo, A Nogueira, T Watson, S Hashmi Sustainable Energy Fuels 6(2022) 2879.

[14] E Moloney, D Gangadharan, V Yeddu, D Zhang, S Moradi, A Askar, M Adachi, D Leitch, M Saidaminov Chem. Mater. 34 (2022) 4394.

[15] C Aranda, L Caliò and M Salado. Crystals 11 (2021) 519.

[16] A Senocrate, T Acartuerk, G Kim, R Merkle, U Starke, M Grätzel, J Maier. J. Mater. Chem. A 6 (2018) 10847.

TESTING EVA, PMMA AND PVDF ENCAPSULATED PEROVSKITE SOLAR CELLS IN A CLIMATIC CHAMBER BY FOLLOWING THE INTERNATIONAL SUMMIT ON ORGANIC PHOTOVOLTAIC STABILITY (ISOS-T) PROTOCOLS.

L. Ocaña[1,2], C. Montes[1,2], B. González-Díaz[2], S. González-Pérez[3], E. Llarena[1].
[1]Instituto Tecnológico y de Energías Renovables, S. A. (ITER)
Pol. Industrial de Granadilla, s/n, E 38600 Granadilla de Abona, Spain.
Ph. +34 922 747 700 / Fax +34 922 747 701 / E-mail lmgonzalez@iter.es
[2]Departamento de Ingeniería Industrial. Universidad de La Laguna.
Camino San Francisco de Paula, s/n, 38206 San Cristóbal de La Laguna. S/C de Tenerife. Spain.
[3]Departamento de Didácticas Específicas. Universidad de La Laguna.
c/ Pedro Zerolo, s/n, Edificio Central Planta 2. Apartado 456, 38200 San Cristóbal de La Laguna. S/C de Tenerife. Spain.

ABSTRACT: The Solar Cell Laboratory from the Instituto Tecnológico y de Energías Renovables (ITER), with the support of the La Laguna University (ULL), has been working on the development of low temperature encapsulation methods, suitable to protect Perovskite Solar Cells (PSCs). This document summarizes the work carried out to evaluate the performance of perovskite solar cells, based on $MAPbI_3$ absorbers, which were encapsulated at low temperature and then the International Summit on Organic photovoltaic Stability (ISOS-T) protocols were applied. Three types of polymer materials were studied, those composed of Ethyl Vinyl Acetate (EVA), those composed of Poly(methyl methacrylate) (PMMA), and those composed of Polyvinylidene fluoride (PVDF) combined with EVA. The degradation of the samples was characterized by imaging microscope, optical ellipsometry and the electrical performance was simulated by the SCAPS-1D software. The results indicate EVA encapsulated samples remained stable after the 200 thermal cycles, keeping the samples a 15.06% PCE compared to the 15.96% PCE of the pristine ones, with a reduction of 5% of total efficiency, meanwhile the PMMA and EVA-PVDF samples lost 59% and 35% PCE respectively.
Keywords: Perovskite, degradation, absorption, encapsulated, Polymer Film.

1 INTRODUCTION

Perovskite solar cells (PSCs) have made significant improvement in recent years, but progress is still required to reach commercial stage. The main efforts of the scientific community have focused on efficiency, stability, reducing lead leakage and developing eco-friendly lead-free perovskites [1]. Currently the PSCs have obtained competitive power conversion efficiency (PCE) values, accomplishing 25.7% for perovskite cells [2], 29.8% for perovskite / Si tandem [2] and 17.9% for perovskite modules [3]. Besides, progress has also been made regarding the stability of the devices. Noteworthy are the contributions made by Anyi Mei et al, who got a printable PSC filled with $(5\text{-AVA})XMA_1\text{-}XPbI_3$ that worked for more than 9,000 h at a maximum power point of 55°C ±5°C without obvious decay [4], the ones made by Yen-Hung Lin et al, who maintained the efficiency of their encapsulated cells for 1200 hours at 85°C [5] and the work of Zonghao Liu et al, who maintained approximately 86% of the initial performance of their encapsulated perovskite solar modules, after continuous operation for 2,000 h under AM1.5G light illumination [6]. Despite of that, the stability values for the encapsulated PSCs are still far from those typically found in silicon solar cells.

This work continues the line of research carried out by ITER and ULL in order to develop a low-temperature encapsulation method to obtain stable perovskite solar cells when exposed to light and environment agents. Thus, the first generation of the encapsulated devices withstood temperatures up to 150°C [7] and 350 hours exposure under full solar radiation [8,9]. The second generation, in which the curing times in the encapsulating process were optimized, resulted on an improvement of the cells' resistance to temperature, remaining stable after 200 thermal cycle test, performed as per the norm IEC-61215 [10]. In this paper we present the results obtained after the application of encapsulation methods to include, besides EVA based sheets (Encapsolar PC-135 A, from Stevens Urethane, Easthampton, MA, USA), other polymers, such as Poly(methyl methacrylate) (PMMA; Alfa Aesar, Thermo Fisher Kandel GmbH) and Polyvinylidene fluoride (PVDF; X 3M 8590M Deutschland GmbH). The encapsulated were tested following the procedures proposed at the International Summit on Organic photovoltaic Stability (ISOS-T) protocols [11,12].

2 EXPERIMENTAL

To know the behavior of the different encapsulating materials, we have developed three sample types, all of them of a size of 25 mm x 25 mm (see table 1): Samples composed by just a layer of the encapsulating polymer material, sandwiched by two glass substrates (G_EVA; G_PMMA: G_PVDF), samples with perovskite thin films, based on $CH_3NH_3PbI_3$ absorbers [7], encapsulated with the polymer material and a glass substrate (Pk_EVA; Pk_PMMA; Pk_PVDF) and, finally, samples fabricated with fully functional PSCs, encapsulated with the polymer material and a glass substrate with the polymer material and a glass substrate (PSC_EVA; PSC_PMMA; PSC_PVDF). These sample sets allowed to evaluate the transmittance and the absorbance of the encapsulating polymer film in the first set, the behavior of the perovskite layer in the second and the effect on the complete devices.

The samples were fabricated inside an ISO 7 cleanroom in ambient conditions [7]. The EVA and PVDF sheets were placed on the perovskite film, at its commercial state, without any previous treatment, while the PMMA solutions were prepared with a concentration of 10:90 in weight (wt%) by adding 0.874 g of PMMA and 7 ml of chlorobenzene, which then were deposited on the perovskite films by dropping 200 µl of solution. Subsequently, the solar cells were encapsulated following the low temperature encapsulation method previously reported [10].

Table I: Research samples

ID	Description	Characterization
G-EVA	Glass-EVA-Glass	Transmittance
G-PMMA	Glass-PMMA-Glass	
G-EVA-PVDF	Glass-PVDF-Glass	
Pk-EVA	Glass-$CH_3NH_3PbI_3$-EVA-Glass	Absorbance
Pk-PMMA	Glass- $CH_3NH_3PbI_3$-PMMA-Glass	
Pk-EVA-PVDF	Glass- $CH_3NH_3PbI_3$-PVDF-Glass	
PSC-EVA	Glass-PSC-EVA-Glass	Simulation current voltage (J-V) curves
PSC-PMMA	Glass-PSC-PMMA-Glass	
PSC-EVA-PVDF	Glass-PSC-PVDF-Glass	

Following the parameters defined in ISOS-T-3 [11], we carried out the aging processes in a climatic chamber (CTC256, Manufactured by MEMMERT), in dark conditions, establishing 6 to 7 thermal cycles daily from -40 to 85 °C (Table 2), being the relative humidity (RH) set to 55% at temperatures above 40 °C (and without RH set point below that point). We selected the maximum temperature rate under 100°C/h. The temperature of the samples remained stable at least 10 min. The total duration of the cycle never exceeded 6 hours, after 200 cycles.

The purpose of the thermal cycling test was to determine the ability of the device to resist thermal imbalances, fatigue or other stresses caused by repeated changes in temperature. This test is especially relevant for any outdoor-destined PV technologies because they simulate realistic conditions of humidity [11]. In the case of the PSCs, this test acquire a special importance to study the degradation processes, attributed to the ion accumulation at the contacts [11,13], when the temperature is varied, instead of the application of constant extreme temperatures.

Table II: Thermal cycle

Parameter	Start	Ramp	Stand	Ramp	Stand	Ramp
Time per step (min)	0	45	15	90	15	45
Temp. (°C)	25	-40	-40	85	85	25

3 CHARACTERIZATION

To analyze the encapsulation materials (EVA, PMMA and PVDF) performance against temperature and humidity, we divided the characterization process into two phases: (i) the characterization of the samples before the aging process, to know their pristine state and (ii) the same measurements after the ageing processes, to analyze the thermal cycling effects on the perovskites encapsulated with the studied polymers and therefore, to feed the gathered information to carry out simulations to evaluate its effect on completed perovskite devices.

The measurements performed to carry out the characterization of the samples were: The search for macroscopic defects on the surface of the absorbing layers, using a digital microscope [14], the examination of the absorbance curves by means of spectroscopic ellipsometer, M-2000X, 75 W Xenon lamp, with a focused beam size of 500 µm, to detect variations related with material degradations. In fact, previous studies have shown that the effect of exposure to high temperatures affect the absorbance curve in the spectral range between 600 and 800 nm, reducing the perovskite band gap [7]. Thus, the absorbance values and the band gap were determined from transmittance measurements on a spectral range from 300 to 1000 nm [15].

Finally, the electrical parameters (Jsc, Voc, FF, PCE and Pmax) during the aging process were obtained for the encapsulated PSCs by using the SCAPS-1D simulation software [16], since a reduction in these parameters allows to evaluate their degree of stability.

4 RESULTS AND DISCUSSION

4.1 Digital microscope

Figure 1 depicts images of the samples before and after the thermal cycling test. The first column corresponds to samples composed of a layer of perovskite encapsulated with EVA, the second column shows samples comprised by a layer of perovskite encapsulated with PMMA and the third column corresponds to samples consisting of a layer of perovskite encapsulated with EVA and PVDF. As it can be seen, after 200 thermal cycles, the sample encapsulated with EVA maintained their original appearance and did not show macroscopic defects, which indicate a high degree of stability throughout the test. However, the samples encapsulated with the other polymers (PMMA and EVA with PVDF) showed discoloration right after the encapsulation process took place, with a further loss in color by the end of the thermal cycles, suggesting that these materials are probably not suitable to be used over perovskite layers, nor offer enough protection against temperature changes in humid environments.

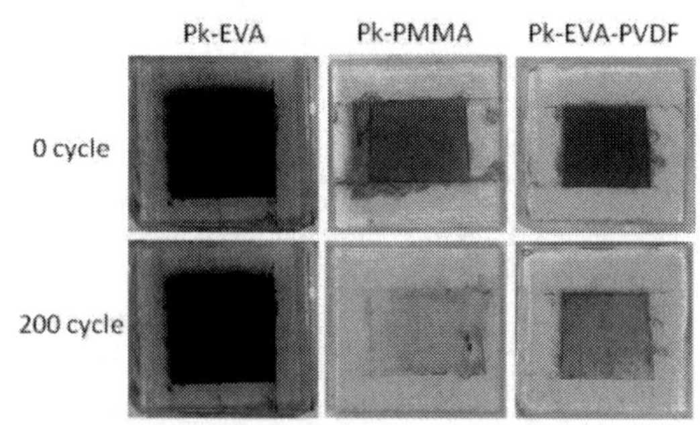

Figure 1. - Images of the samples before and after de thermal cycling test performed inside the climatic chamber.

4.2 Spectroscope ellipsometer

First, the transmittance of the three encapsulating polymeric materials was analyzed by using the G-EVA, G-PMMA and G-EVA-PVDF samples (see Table I). As it can be seen in figure 2, the PMMA allowed-s the light to pass from 300 nm, while the encapsulates that include an EVA layer successfully cut the irradiation above the

400 nm. Therefore, the latter materials should be more suitable for protecting perovskite layers from the UV part of the spectrum. Also, the solo-EVA encapsulation showed more transmittance than the EVA plus PVDF one, which should allow the perovskite to absorb more light and, thus, to be more efficient.

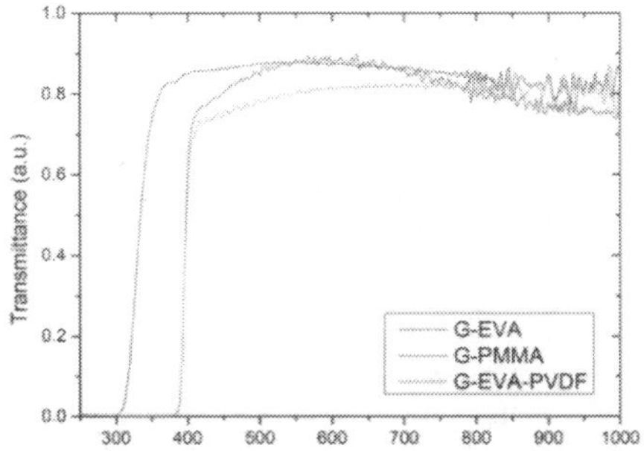

Figure 2. - Transmittance of the EVA, PMMA and EVA-PVDF after having been encapsulated

To evaluate the effect that the EVA, PMMA and EVA-PVDF encapsulation on the perovskite layers during the thermal cycling test, the absorbance curves of the samples were measured before and after the degradation process .

As it can be seen in Figure 3, the samples encapsulated with EVA present lower reduction of the absorbance curve, less than 15%, between 400 and 760 nm (therefore, below perovskite's bandgap, 1.6 eV). However, the samples encapsulated with PMMA and EVA-PVDF presented unfavorable results, reaching 90% and 80% decreases in absorbance, respectively. In fact, for the case of PMMA, we observed that decrease was similar throughout the spectrum, while the samples encapsulated with EVA and PVDF had their greatest degradation around 450-550 nm.

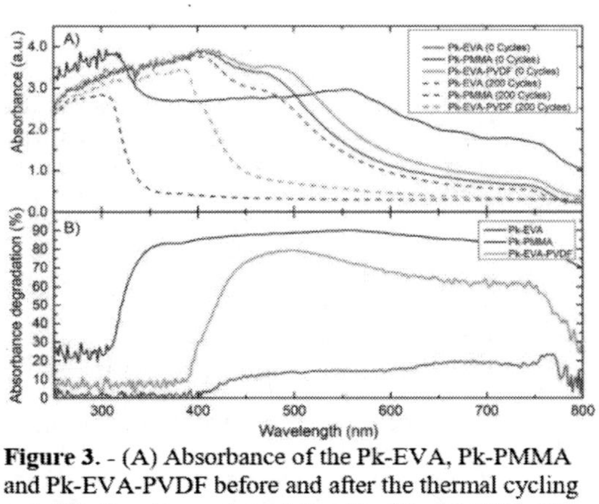

Figure 3. - (A) Absorbance of the Pk-EVA, Pk-PMMA and Pk-EVA-PVDF before and after the thermal cycling test. (B) Absorbance degradation % of the Pk-EVA, Pk-PMMA and Pk-EVA-PVDF after 200 thermal cycles.

4.3 Encapsulated PSC simulations

To evaluate the effects the aforementioned degradation on the the perovskite layers would have on complete cells encapsulated with the materials under study, we have performed simulations with the obtained data using the SCAPS-1D simulation software. This software has been developed by the Department of Electronics and Information Systems (ELIS) of the University of Ghent, Belgium, being widely reported as an accurate tool for the study of PSCs [17–22]. Thus, the thickness, the band gap and the curve of the absorbance of the device layers allowed us to extrapolate the degradation of the samples after the thermal cycling test radiation to the SCAPS 1-D software, determining to estimate their influence on the short-circuit current (Jsc), the open circuit voltage (Voc), the fill factor (FF) and the energy efficiency (PCE) [10].

Figure 4 compares the IV curve of the samples encapsulated with EVA, PMMA and EVA+PVDF (or PSC-EVA, PSC-PMMA and PSC-EVA-PVDF, respectively) before and after undergoing the thermal cycling test. As it can be seen, the samples encapsulated with EVA maintained the power conversion efficiency after the test, with a reduction of only 0.9%. However, the samples encapsulated with PMMA go from a 20.20% to an 8.28% while the samples encapsulated with EVA + PVDF, go from 17.06% to an 11.10% (See table III).

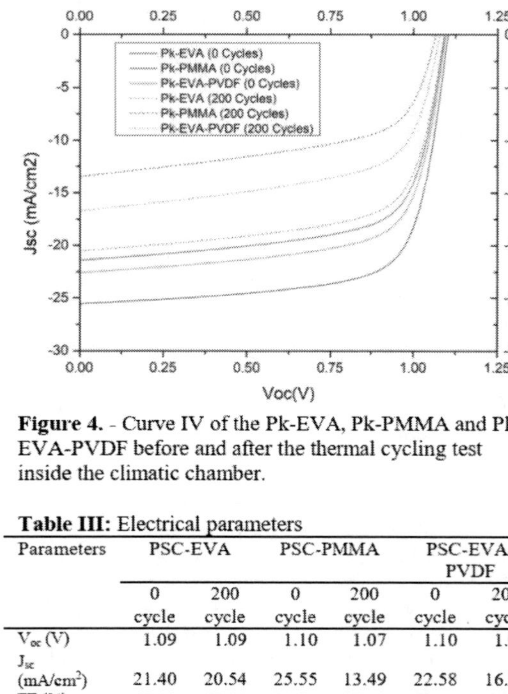

Figure 4. - Curve IV of the Pk-EVA, Pk-PMMA and Pk-EVA-PVDF before and after the thermal cycling test inside the climatic chamber.

Table III: Electrical parameters

Parameters	PSC-EVA		PSC-PMMA		PSC-EVA-PVDF	
	0 cycle	200 cycle	0 cycle	200 cycle	0 cycle	200 cycle
V_{oc} (V)	1.09	1.09	1.10	1.07	1.10	1.08
J_{sc} (mA/cm^2)	21.40	20.54	25.55	13.49	22.58	16.72
FF (%)	68.24	67.25	71.73	57.58	69.34	61.58
PCE (%)	15.96	15.06	20.20	8.28	17.16	11.10
V_{mp} (V)	0.92	0.91	0.93	0.88	0.92	0.90
J_{mp} (mA/cm2)	17.42	16.48	21.82	9.39	18.65	12.37

5 CONCLUSIONS

In this work, the performance of three different polymers, EVA, PMMA and EVA combined with PVDF, as encapsulation materials to protect perovskite films against temperature and humidity variations have been evaluated. To this purpose, the international summit on organic photovoltaic stability (ISOS-T) protocols have been applied.

After the tests were carried out, it was possible to

ascertain that the encapsulation made with solo-EVA is a far better option, since it showed superior results that the ones performed with the PMMA and EVA combined with PVDF, both from the point of view of its protection against UV radiation, since it was capable of cutting below 400 nm (figure 2), as well as from its ability to protect against dramatic changes in temperature, resulting on absorbance curves that decreased by only 15%, while the absorbance observed for the samples encapsulated with the other polymers were reduced between 80% and 90%.

From these results, the absorbance, band gap and thicknesses of the perovskite layers were used to simulate the effect of subjecting the encapsulated perovskite solar cells to the mentioned thermal cycles, obtaining the electrical parameters of EVA encapsulates, as power conversion efficiency, fill factor, short circuit intensity and open circuit voltage, remain stable after 200 thermal cycles, keeping an efficiency of 15.06% compared to the initial 15.96%.

6 ACKNOWLEDGEMENTS

This work was developed within the MACLAB-PV project framework, which has been co-financed by the INTERREG Madeira-Azores-Canarias Territorial Cooperation Programme (MAC) 2014-2020. 2nd Call. Axis 1. Enhancing research, technological development and innovation.

7 REFERENCES

1. Wu, T.; Qin, Z.; Wang, Y.; Wu, Y.; Chen, W.; Zhang, S.; Cai, M.; Dai, S.; Zhang, J.; Liu, J.; et al. The Main Progress of Perovskite Solar Cells in 2020–2021. *Nano-Micro Lett.* 2021, *13*.
2. Laboratory National Renewable Energy, (NREL) Best Research-Cell Efficiency Chart Available online: https://www.nrel.gov/pv/cell-efficiency.html.
3. National Renewable Energy Laboratory, (NREL) Champion Photovoltaic Module Efficiency Chart Available online: https://www.nrel.gov/pv/module-efficiency.html.
4. Mei, A.; Sheng, Y.; Ming, Y.; Hu, Y.; Rong, Y.; Zhang, W.; Luo, S.; Na, G.; Tian, C.; Hou, X.; et al. Stabilizing Perovskite Solar Cells to IEC61215:2016 Standards with over 9,000-h Operational Tracking. *Joule* **2020**, *4*, 2646–2660, doi:10.1016/j.joule.2020.09.010.
5. Lin, Y.-H.; Sakai, N.; Da, P.; Wu, J.; Sansom, H.C.; Ramadan, A.J.; Mahesh, S.; Liu, J.; J Oliver, R.D.; Lim, J.; et al. *A Piperidinium Salt Stabilizes Efficient Metal-Halide Perovskite Solar Cells*;
6. Liu, Z.; Qiu, L.; Ono, L.K.; He, S.; Hu, Z.; Jiang, M.; Tong, G.; Wu, Z.; Jiang, Y.; Son, D.Y.; et al. A Holistic Approach to Interface Stabilization for Efficient Perovskite Solar Modules with over 2,000-Hour Operational Stability. *Nat. Energy* **2020**, *5*, 596–604, doi:10.1038/s41560-020-0653-2.
7. Ocaña, L.; Montes, C.; De Sousa-Vieira, L.; Moreno-Ramírez, J.S.; González-Pérez, S.; González-Díaz, B.; Hernández-Rodríguez, C.; Friend, M.; Cendagorta, & M. On the Stability of Planar CH3NH3PbI3 Perovskite Solar Cells Produced in Ambient Conditions by Using Polymer Encapsulates. In Proceedings of the 36th European Photovoltaic Solar Energy Conference and Exhibition; Marseilles, 2019; pp. 721–727.
8. Ocaña, L.; Montes, C.; González-Pérez, S.; González-Díaz, B.; Friend, M.; Cendagorta, & M. Improvements on the Stability of MAPbI3 Thin Films against UV Radiation by a Low Temperature Encapsulation Method That Uses Commercial Ethylene-Vinyl Acetate Sheets. In Proceedings of the 37th European Photovoltaic Solar Energy Conference and Exhibition; EU PVSEC, Ed.; 2020; pp. 659–662.
9. Ocaña, L.; Montes, C.; González-Pérez, S.; González-Díaz, B.; Llarena, E. Characterization of a New Low Temperature Encapsulation Method with Ethylene-Vinyl Acetate under UV Irradiation for Perovskite Solar Cells. *Appl. Sci.* **2022**, *12*, 5228, doi:10.3390/app12105228.
10. Ocaña, L.; Montes, C.; González-Pérez, S.; González-Díaz, B.; Llarena, E. Testing Encapsulated Perovskite Solar Cells in a Climatic Chamber by Following the IEC 61215 and IEC 61646 Standards for the Thermal Cycling Test. In Proceedings of the 38th European Photovoltaic Solar Energy Conference and Exhibition; 2021; pp. 406–410.
11. Khenkin, M. V.; Katz, E.A.; Abate, A.; Bardizza, G.; Berry, J.J.; Brabec, C.; Brunetti, F.; Bulović, V.; Burlingame, Q.; Di Carlo, A.; et al. Consensus Statement for Stability Assessment and Reporting for Perovskite Photovoltaics Based on ISOS Procedures. *Nat. Energy* **2020**, *5*, 35–49, doi:10.1038/s41560-019-0529-5.
12. Reese, M.O.; Gevorgyan, S.A.; Jørgensen, M.; Bundgaard, E.; Kurtz, S.R.; Ginley, D.S.; Olson, D.C.; Lloyd, M.T.; Morvillo, P.; Katz, E.A.; et al. Consensus Stability Testing Protocols for Organic Photovoltaic Materials and Devices. *Sol. Energy Mater. Sol. Cells* **2011**, *95*, 1253–1267, doi:10.1016/j.solmat.2011.01.036.
13. Schwenzer, J.A.; Rakocevic, L.; Gehlhaar, R.; Abzieher, T.; Gharibzadeh, S.; Moghadamzadeh, S.; Quintilla, A.; Richards, B.S.; Lemmer, U.; Paetzold, U.W. Temperature Variation-Induced Performance Decline of Perovskite Solar Cells. *ACS Appl. Mater. Interfaces* **2018**, *10*, 16390–16399, doi:10.1021/acsami.8b01033.
14. Digital Microscope U500X, Focus Range 15mm - 40mm, Range 50X-500X by Unimake.
15. Montes, C.; Ocaña, L.; De Sousa-Vieira, L.; Moreno-Ramírez, J.; González-Pérez, S.; González-Díaz, B.; Hernández-Rodríguez, C.; Friend, M.; Cendagorta, M. Producing Uniform and Smooth Thin Layers of Perovskite under Ambient Conditions by Adjusting to the Existing Levels of Moisture. In Proceedings of the 35th European Photovoltaic Solar Energy Conference and Exhibition; Brussels, 2018; pp. 933–937.
16. SCAPS-1D Simulation Software, Developed by the Department of Electronics and Information Systems (ELIS) of the University of Ghent Has Been Used, Belgium.
17. Chowdhury, M.S.; Shahahmadi, S.A.; Chelvanathan, P.; Tiong, S.K.; Amin, N.; Techato, K.;

Nuthammachot, N.; Chowdhury, T.; Sukluenng, M. Effect of Deep-Level Defect Density of the Absorber Layer and n/i Interface in Perovskite Solar Cells by SCAPS-1D. *Results Phys.* 2020, *16*.

18. Slami, A.; Belkaid, A.B. *Numerical Study of Based Perovskite Solar Cells by SCAPS-1D*;
19. Mandadapu, U.; Vedanayakam, S.V.; Thyagarajan, K. Simulation and Analysis of Lead Based Perovskite Solar Cell Using SCAPS-1D. *Indian J. Sci. Technol.* **2017**, *10*, 1–8, doi:10.17485/ijst/2017/v10i11/110721.
20. Lin, L.; Jiang, L.; Li, P.; Fan, B.; Qiu, Y.; Yan, F. Simulation of Optimum Band Structure of HTM-Free Perovskite Solar Cells Based on ZnO Electron Transporting Layer. *Mater. Sci. Semicond. Process.* **2019**, *90*, 1–6, doi:10.1016/j.mssp.2018.10.003.
21. Bansal, S.; Aryal, P. Evaluation of New Materials for Electron and Hole Transport Layers in Perovskite-Based Solar Cells through SCAPS-1D Simulations. In Proceedings of the Conference Record of the IEEE Photovoltaic Specialists Conference; Institute of Electrical and Electronics Engineers Inc., November 18 2016; Vol. 2016-November, pp. 747–750.
22. Chakraborty, K.; Choudhury, M.G.; Paul, S. Numerical Study of Cs2TiX6 (X = Br−, I−, F− and Cl−) Based Perovskite Solar Cell Using SCAPS-1D Device Simulation. *Sol. Energy* **2019**, *194*, 886–892, doi:10.1016/j.solener.2019.11.005.

OPTIMIZATION OF COATING PARAMETERS FOR ENABLING HOMOGENEOUS PEROVSKITE THIN FILMS VIA UPSCALABLE HYBRID EVAPORATION – SLOT-DIE COATING DEPOSITION ROUTE

Van Son Nguyen*[1], Elisa Grépin[1], Pia Dally[1,2], Iwan Zimmermann[1], Elise Bruhat[3], Olivier Dupré[3], Muriel Bouttemy[2], Solenn Berson[3], Matthieu Manceau[3], Jean Rousset[1,4]

[1] IPVF, Institut Photovoltaïque d'Ile-de-France, 18 Boulevard Thomas Gobert, 91120 Palaiseau, France.
[2] Institut Lavoisier de Versailles (ILV), Université de Versailles Saint-Quentin-en-Yvelines, Université Paris-Saclay, CNRS, UMR 8180, 45 avenue des Etats-Unis, 78035 Versailles Cedex, France.
[3] University of Grenoble Alpes, CEA, LITEN, INES, 50 Avenue du Lac Léman, 73375 Le Bourget du Lac, France.
[4] EDF, Electricité de France R&D, 30 Route Départementale 128, 91120 Palaiseau, France.

ABSTRACT: We report the use of slot-die coating as an up-scalable technique to convert a co-evaporated $CsBr/PbI_2$ precursor layer into a triple-cation mixed halide perovskite. We optimize this two-step deposition process by varying the coating parameters such as the evaporation rates in the first step and the solution concentration and dispense rates in the second step. We observe that the co-evaporation of CsBr and PbI_2 at 0.1 A s^{-1} and 1.0 A s^{-1}, respectively, results in a porous layer which is desired for the conversion by slot-die coating. In the second step, with an optimization of solution concentration (40.7 mg/mL) and solution dispense rate (3.3 µL s^{-1}), uniform perovskite film with large grain size on the 5 x 5 cm^2 substrates is achieved. The homogeneity, composition and stoichiometry of the converted perovskite film were studied using X-ray Photoelectron Spectroscopy (XPS), from surface to bulk, using in depth profiling by sequential Ar^+ sputtering. The perovskite layer has been implemented into a single-junction device, yielding a champion efficiency of 13.5% power conversion efficiency with hysteresis (average 10 - 11%), which is the first cell efficiency obtained using this hybrid deposition method based on the dual-source $CsBr/PbI_2$ thermal evaporation and slot-die coating.

Keywords: Perovskite solar cells, thermal co-evaporation, slot-die coating, hybrid perovskite deposition

1 INTRODUCTION

Organic-Inorganic metal halide perovskite materials have been evolving rapidly to one of the most promising photovoltaic absorbers leading to the current record Power Conversion Efficiency (PCE) of 25.7% since its first apparition in solar cell application in 2009.[1–3] Such record and most of the highly efficient perovskite solar cells have been realized with small active areas (< 1 cm^2) which is not eligible for moving towards commercial mass production. A common and simple method but very suitable for small area perovskite thin film deposition is spin-coating. However, such method is not suitable for large area deposition due to the inhomogeneous issues and large waste of materials. To overcome these issues, alternative coating methods have been utilized such as blade coating [4], slot-die coating [5,6], or inkjet printing [7]. Amongst these techniques, slot-die coating appears to be a very promising candidate, which can be used to deposit large area perovskite.[6,8,9]

Recently, research has focused on improving the efficiency of the perovskite solar devices together with finding the way to upscale them. By implementing the perovskite cell on top of the crystalline silicon (c-Si) bottom cell, tandem solar cell today reaches 31.3% PCE.[10] Still, most of these tandems have been realized on polished c-Si surface. For industrial fabrication, the planar c-Si cells are not favored due to the high polishing cost. In addition, planar devices exhibit higher light reflection thus lower photogeneration current compared to their non-planar counterparts. However, depositing perovskite on non-planar surface is challenging. One-step coating such as spin-coating or slot-die coating could not ensure the conformal coating on standard c-Si surfaces. Therefore, hybrid two-step deposition method has been employed to deposit a conformal perovskite layer on the non-planar surface. This technique combines a thermal vacuum evaporation of an inorganic layer (usually PbX_2 or/and CsX with X = I, Br, Cl) and followed by a solution process such as spin-coating to convert that layer into perovskite. Thereby, Sahli et al. fabricated a 2-terminal perovskite/tandem solar cell, yielding 22.7% PCE on a 1.43 cm^2 area, by evaporating PbI_2 and CsI coupling with subsequent spin-coating of a FAI/FABr containing solution.[11] Using a similar approach, Schulze et al. obtained 13.8 % single junction cell device efficiency.[12] However, these works still employ spin-coating for the conversion process. Here, we use slot-die coating to convert the dual-source evaporated CsBr and PbI_2 layers into mixed halide perovskite. This hybrid perovskite deposition, coupling thermal co-evaporation and slot-die coating, is believed to enable large area deposition with conformal and high throughput of the perovskite films on non-planar surface.

The hybrid evaporation/slot-die coating technique was recently realized for the first time by our group.[13] This work demonstrated a successful conversion of PbI_2 precursor layer into perovskite using slot-die coating. Compared to the co-evaporation, the single source deposition is much easier to control. However, the converted perovskite does not present Cs^+ cations because it cannot be introduced in the second conversion step since Cs compounds are not soluble in isopropanol solvent. Because Cs^+ cation doping is proven to enlarge the bandgap and to stabilize the perovskite,[14] we therefore incorporated Cs^+ cation by evaporating CsBr together with PbI_2 in the first step. Regarding this dual-source evaporation combination for the hybrid dry-wet method, only few publications can be found in literature. Qiu et al. evaporated $CsBr/PbI_2$ precursor layer then converted it into perovskite using chemical vapor deposition (CVD) of FAI, resulting in a 13.3% efficient cell and 10% efficiency on 91.8 cm^2 large-area module.[15] Sahli and coworkers spin-coated FAI/FABr solution on an evaporated $CsBr/PbI_2$ film, leading to 11.35% and 25.2% PCE of a semitransparent single junction cell and a fully textured 2-terminal perovskite/c-Si tandem cells, respectively. [16] In one other paper, Werner et al. also reported the use of this evaporation combination but followed by FAI spin-coating, yielding 14.0% PCE of a 3- terminal perovskite/c-Si tandem cell device, with conformal perovskite layers on top of c-Si pyramids.[17]

Herein, we coated a mixed MA^+ and FA^+ cation-based solution using slot-die coating on top of the $CsBr/PbI_2$ co-evaporated precursor layer. We study the film composition and verify the film uniformity and stoichiometry using XPS. With the optimization of the evaporation rates in the first step and the solution concentration and dispense rate in the second step, we obtained a single-junction device with 13.5% PCE (with hysteresis).

2 EXPERIMENTALS

2.1 Preparation of perovskite solar cells

Commercial Fluorine doped Tin Oxide (FTO) substrates were purchased from Solems. Formamidinium bromide (FABr), Formamidinium iodide (FAI), Methylammonium iodide (MAI) and FK-209 Co(III) TFSI Salt (Co-FK-209, 98%) were purchased from Greatcell Solar. Lead Iodide (PbI_2, 99.99%) was purchased from Tokyo Chemical Industry (TCI) and 2,2',7,7'-Tetrakis [N,N-di(4-methoxyphenyl) amino]-9,9'-spirobifluorene (Spiro-OMeTAD, 99.8%) from Borun New Material Technology Ltd. Cesium Bromide (CsBr, 99.99%), methylammonium chloride (MACl, > 98%), Isopropanol (IPA, 99.5%, anhydrous), chlorobenzene (99.8%), 4-tert butylpyridine (TBP, 98%) and Lithium bis(trifluoromethanesulfonyl)imide (LiTFSI, 99.95%) were purchased from Sigma Aldrich. Zinc powder was purchased from Alfa Aesar. All materials were used as received without further purification.

Conductive FTO coated glass substrates (3 mm, 10 $\Omega\ sq^{-1}$) were etched with Zn powder and concentrated HCl (4M) to obtain the electrode patterns. Then substrates were sonicated in an ultrasonic bath with detergent RBS 35 solution (2%) for 60 min at 60°C, followed by subsequentially sonicating in acetone and isopropanol (IPA) for 30 min each. Substrates were then treated under UV-ozone for 15 min to remove organic residues.

The devices were stacked in n-i-p configuration, as shown in Fig.1. Details of SnO_2 as Electron Transport Layer (ETL) deposited by Chemical Bath Deposition (CBD) method, Spiro-OmeTAD (doped with TBP, LiTFSI and Co-FK-209) as Hole Transport Layer (HTL) by spin-coating and gold electrode by evaporation can be found in [18]. For hybrid method, the 5x5 cm^2 SnO_2/FTO substrates were first loaded in the thermal evaporator (MBraun-ProVap-5G) to co-grow a thin film of CsBr and PbI_2 at high vacuum of 3.10^{-7} mbar. The deposition rates and thicknesses of CsBr and PbI_2 films can be independently controlled during the evaporation. Then the substrates were quickly transferred to Slot-die glovebox for the conversion (controlled N_2 atmosphere with H_2O and O_2 levels below 10 ppm). Then, we inject a solution containing organic cations (0.08 mol MAI, 0.06 mol FAI, 0.12 mol FABr, and 0.05 mol MACl dissolved in IPA) via slot-die head of a commercial slot-die coater from nTact. The solution infiltrates into and reacts with the $CsBr/PbI_2$ layer to form a wet perovskite film. A post annealing on a hotplate at 150 °C for 15 minutes was carried out to dry out the film and crystallize perovskite. The samples were then cut into smaller pieces of 2.5×2.5 cm^2 for film characterizations and cell finalization. Full cell fabrication process is presented in Fig.1.

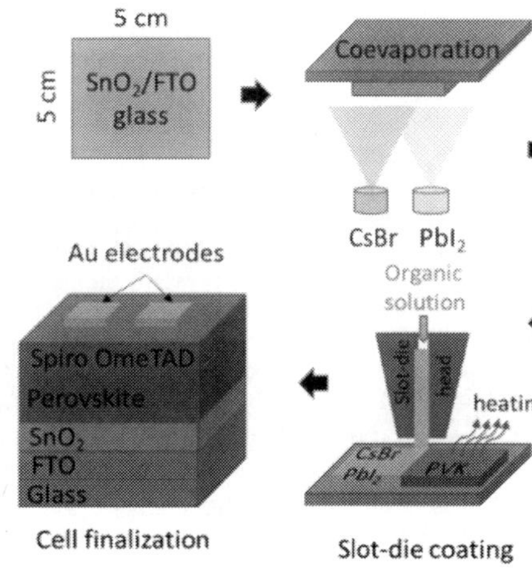

Figure 1: Schematics of device fabrication process by hybrid method and final cell structure.

2.2 Characterization methods

X-ray diffraction (XRD) was performed with an Empyrean system (PANalytical) using Cu Kα radiation. Scans were recorded within the 2θ-range from 5° to 60°. Scanning Electron Microscope (SEM) analysis was carried out with a Zeiss Merlin VP Compact at 5 and 15 kV acceleration voltage for top surface and cross-section, respectively. Absorption and transmission of perovskite layers were measured with a Cary 5000 UV-Vis-NIR spectrophotometer (Agilent Technologies). Spectra were recorded from 300 nm to 900 nm in steps of 1 nm. Current-voltage (J-V) characteristics of solar devices were measured using an Arkeo system from Cicci Research under standard AM1.5G one sun (100 mW cm^{-2}) illumination (Xenon lamp). The light intensity and spectrum were calibrated using a Si reference cell from Newport. A shadow mask was used to define an active area of 0.09 cm^2. The scan speed and voltage step were kept at 20 mV s^{-1} and 20 mV, respectively. All the measurements were carried out in ambient air (with humidity ~ 40-50%).

XPS surface analyzes were carried out using a Thermo Electron K-Alpha+ spectrometer using a monochromatic Al-Kα X-Ray source (1486.6 eV) with an X-ray spot size of 400 µm. Calibration of the spectrometer was done using Cu and Au samples following the ASTM-E-902-94 standard procedure. A Constant Analyzer Energy (CAE) mode was used for the acquisition of High energy resolution spectra with a pass energy of 20 eV and an energy step size of 0.1 eV. For surface analysis, the exposure to X-ray was optimized and limited to 1 hour per spot measurement. The stability of the perovskite layer was verified by comparing two survey spectra on the same sport, at the beginning and the end of the acquisition sequence. For depth profiles, mono-atomic Ar^+ sputtering was used with an energy of 1000 eV, a current density of 10 mA and an ion gun orientation of 30° with respect to the sample surface. Thermo Fisher scientific Avantage© data system was used for data treatment and peak fitting. Sample transfer was realized under controlled atmosphere (transfer tube sealed in the glovebox) to preserve the perovskite layer chemistry.

3 RESULTS AND DISCUSSION

CsBr and PbI_2 were thermally co-evaporated on top of SnO_2 / FTO substrate, with CsBr rates varying from 0.1 to 0.5 Å s^{-1} while PbI_2 rate was kept unchanged at 1.0 Å s^{-1}. We first analyzed the phase composition of the evaporated thin film. Fig. 2a presents the X-rays diffractograms of the films deposited with three different CsBr evaporation rates. At the lowest rate of 0.1 Å s^{-1}, we observed only high intense characteristic peak (001) of PbI_2 located at 12.7° because of a small introduction of CsBr content into the film. The color of the film was yellowish, appeared to be the typical color of the evaporated PbI_2 thin film. The non-perovskite delta phases and the inorganic perovskite phases are formed when the rates increase to 0.3 Å s^{-1} and 0.5 Å s^{-1} respectively, as indicated by the characteristic peak (100) located at 14.1°. This evolution is coherent with the gradual change of the typical color of the film from yellowish to brownish. The formed inorganic perovskite phase does not ease the organic perovskite conversion in the slot-die step thus this must be avoided. On the other hand, regarding the film morphology, Fig.1b shows the SEM images of CsBr/PbI_2 films with the varying compositions. The film with the lowest CsBr rate appeared porous with round-shaped grains while at the highest CsBr rate, the perovskite grains are formed leading to a compact film. Therefore, to facilitate the slot-die conversion step, low CsBr rate of 0.1 Å s^{-1} and high PbI_2 rate of 1 Å s^{-1} are chosen since the porous scaffold layer allows better solution infiltration into the CsBr/PbI_2 film, resulting in a fast conversion.

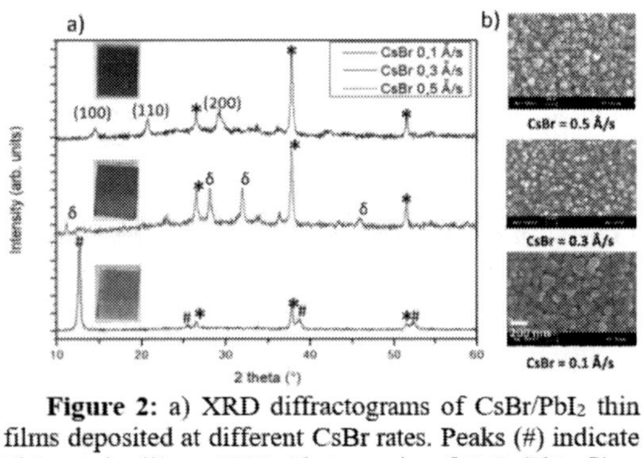

Figure 2: a) XRD diffractograms of CsBr/PbI_2 thin films deposited at different CsBr rates. Peaks (#) indicate PbI_2, peaks (*) are FTO. Photographs of CsBr/PbI_2 films corresponding to each deposition rate composition; b) SEM images showing morphology of three samples.

Once the evaporation rates are determined, in the second step, we transform the CsBr/PbI_2 precursor layer into perovskite by slot-die coating. The organic solution containing MA^+ and FA^+ cations is distributed via the slot-die head with a constant coating gap of 60 µm (the distance between the head tip-end and the sample's surface) and a coating speed of 2.5 mm s^{-1}. To crystalline perovskite, the converted film was then annealed at 150°C for 30 min.

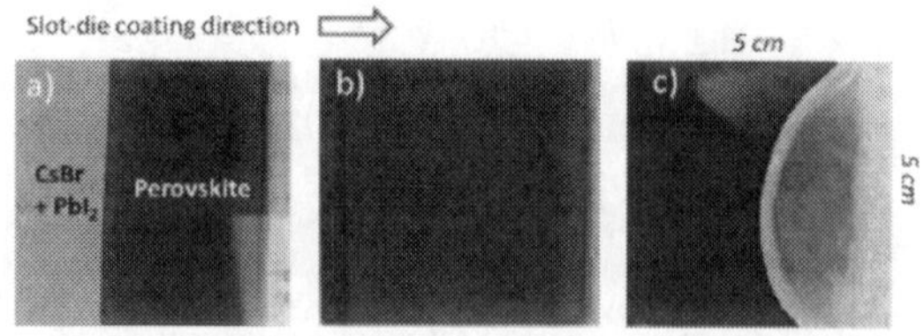

Figure 3: Three scenarios of slot-die coating depending on solution dispense rates: a) 2.9 µL s^{-1}, b) 3.3 µL s^{-1} and c) 3.7 µL s^{-1}. The thickness of CsBr/PbI_2 film is 200 nm. The coating direction from left to right.

The solution dispense rate must be controlled so that the meniscus of the solution is fully formed under the tip-end of the head near sample surface. This parameter is critical to ensure a full conversion perovskite on large-area substrate, in this case, of 5x5 cm^2. Fig.3 exhibits three scenarios that could happen during the slot-die coating on the evaporated film. As the dispense rate is low (of 2.9 µL s^{-1}), there is a lack of solution, the meniscus formed too late, thus the coating starts near the middle of the substrate, left behind the CsBr/PbI_2 film unconverted. If too much solution is injected (3.7 µL s^{-1}), the large amount of the remained solution accumulated at the end of the substrate is flooding back to redeposit on the converted film, cause the white circular shape on the substrate, as observed in Fig. 3a. Only with an appropriate solution amount applied on the film, about 3.3 µL s^{-1}, full meniscus is formed but not too much, the conversion is therefore homogeneous on the full surface (Fig.3b).

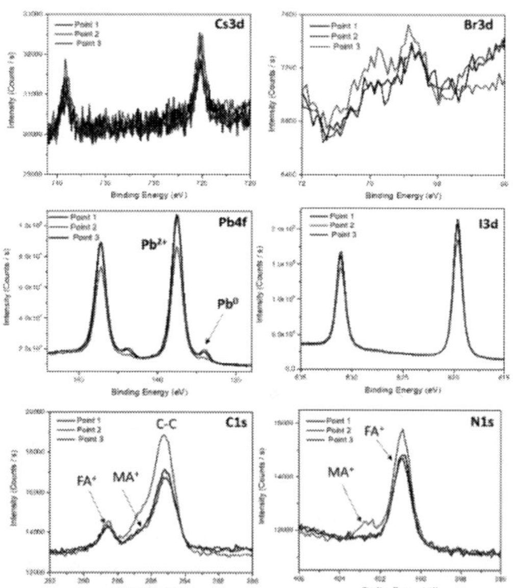

Figure 4: High-energy-resolution XPS spectra of Cs 3d, Br 3d, Pb 4f, I 3d, C 1s and N 1s of the perovskite film (350 nm thick) on FTO glass substrate obtained for three different spots randomly distributed at the film surface (point #1(black), point #2(red) and point #3(blue)).

To further study the chemistry of the converted perovskite film we performed XPS analysis. The surface composition homogeneity was assessed by comparing the XPS response obtained on three different spots distributed at the perovskite surface and spaced from almost 1 cm distance. It is crucial to highlight that the XPS probed depth is around 10 nm bringing inherently information of the extreme surface chemistry. All the constitutive elements (Cs, Br, Pb, I, C and N) are present. Regarding the organic compounds, the N 1s spectra showed the presence of FA^+ compound located at 400.9 ± 0.2 eV and MA^+ at 402.4 ± 0.2 eV. The C 1s spectra also showed the presence of MA^+ and FA^+ compounds at 286.4 eV and 288.3 ± 0.2 eV, respectively. The peak at 284.8 eV is attributed to the C – C bond linked to the presence of adventitious carbon contamination [13]. There is also an evidence that a small Cs^+ content is well introduced from the evaporation, together with a weak signal of Bromine, according to only 10% of CsBr rate (compared to PbI_2). We first observed that the XPS spectra fingerprint of three locations on the film surface are found quite identical for all the inorganic elements (Cs 3d, Br 3d, Pb 4f, I 3d). This laterally homogeneous element distribution might be resulted from the uniform coatings of evaporation and the perovskite transformation in slot-die step. The Pb 4f peak shows, in addition to the Pb^{2+} peak at 138.6 eV, a metallic Lead (Pb^0) contribution around 136.9 eV. A small discrepancy is observed for point 2 presenting a slightly higher MA^+ and FA^+ content (see N 1s spectra) but less Pb^0 (degradation product) compared to point 1 and 3. The inhomogeneity of organic cations can be attributed to the volatility nature of organic compounds together with the high annealing temperature (in this case, 150°C) as reported elsewhere.[19] Anyway, whatever the spots peaks shapes and energy positions are consistent with usual values we already measured on triple cation perovskite [20], confirming the perovskite phase formation.

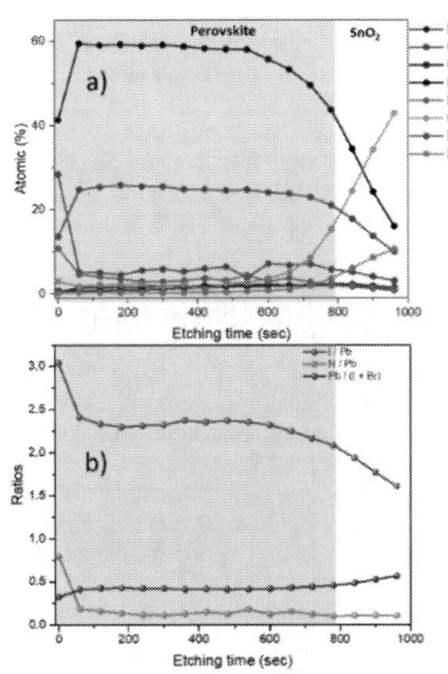

Figure 5: a) XPS in-depth profiling of the {Perovskite / SnO_2 / TiO_2 / FTO / Glass} stack showing the evolution of the atomic percentages of elements in function of the etching time. b) Plots of element ratios evolution across the film.

To evaluate the homogeneity of perovskite layer in bulk, XPS in-depth profiling was performed using monoatomic Ar^+ bombardment. The reliability and limits of this technique were assessed in a previous work on conventional triple cation perovskite.[21] Figure 5 presents the evolution of the atomic percentage in depth measured for of our perovskite layer from the surface to the interface with the underneath SnO_2 layer. The interface area is reached after 800 s abrasion and marked with the appearance of Sn and O elements representative of the SnO_2 layer.

As already observed, after the first 60 s of etching, a partial loss of the organic compound is observed (decrease of N and C %) with a concomitant increase of I and Pb. The evolution of all elements remains then quasi stable in the perovskite bulk until reaching the SnO_2 interface. This result confirms the composition uniformity across the film. Indeed, as we already demonstrated that there is no cumulative effect of bombardment artefacts.

The final stochiometric surface compositions (element ratios) calculated from XPS atomic % of each spot are gathered in Table I. For example, the stoichiometry composition experimentally obtained for the spot #2 is $Cs_{0.04}MA_{0.32}FA_{64}Pb(I_{0.97},Br_{0.03})_3$. As compared to the theoretical film composition expected from the precursors which is $(Cs_{0.02}MA_{0.35}FA_{0.63}Pb(I_{0.87},Br_{0.13})_3$, they are all quite well-coherent except there is an excess of Iodine ($I_{0.97}$ versus $I_{0.87}$) at the interface, possibly due to the PbI_2 excess.

Table I: Details of the film compositions of three investigated surface spots compared to the expected theoretical values for $Cs_{0.02}MA_{0.35}FA_{0.63}Pb(I_{0.87},Br_{0.13})_3$ converted perovskite.

Ratios	Theory	Point #1	Point #2	Point #3
I/(I+Br)	0.87	0.99	0.97	0.98
Cs/Pb	0.02	0.02	0.04	0.03
Cs/(I+Br)	0.01	0.01	0.01	0.01
Pb/(I+Br)	0.33	0.38	0.30	0.33
I/Pb	2.61	2.60	3.23	2.95
C/N	1.60	1.80	1.70	1.90
MA/FA	0.60	0.20	0.50	0.20

To reduce the PbI_2 excess in the perovskite, we optimize the solution concentration. As shown in Fig.6a, we obtained a higher quality perovskite film with large grain sizes (up to 800 nm) with higher solution concentration. The corresponding XRD peaks in Fig.6b show that the PbI_2 (001) peak intensity centered at 12.7° is lowered and smaller than the perovskite (100) peak at 14.1°. This confirms that the excess amount of PbI_2 is mitigated with 40.7 mg/mL concentration. As the results, higher absorption and lower transmission are also observed in the UV-Vis spectrum (Fig.6c). The on-set absorptions of these two films are around 750 nm, thus the bandgap of these films is about 1.65 eV wide.

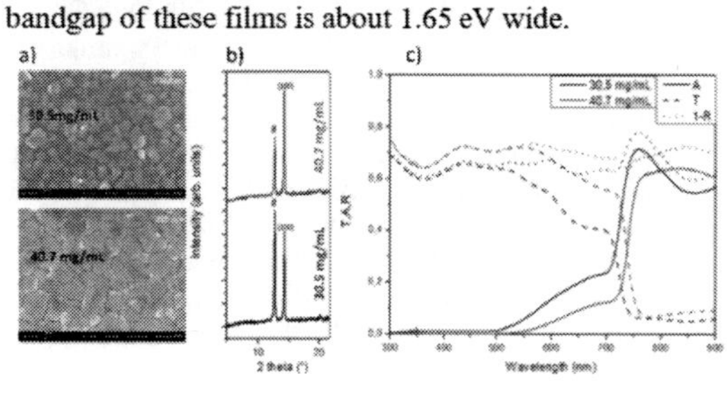

Figure 6: (a) SEM image of the perovskite surface with solution concentration: 30.5 mg/mL and 40.7 mg/mL, (b) the corresponding XRD, Peaks (#) indicate PbI_2, peaks (*) are FTO; and (c) the UV-Vis spectrum of the films (T = Transmission, A = Absorption, R = Reflection).

The optimized layer was then stacked in a n-i-p cell configuration, with SnO_2 and Spiro-OmeTAD as electron and hole transporting layers, respectively, and finalized with gold electrodes. The cell performance is shown in Fig.7. We obtained a champion device with 13.5% PCE under reverse scan (with a large hysteresis). The average PCE obtained of other cells is between 10 to 11 %. It is well-noted that the main factor that limits the device's performance is the open-circuit voltage V_{OC}, which is relatively low (lower than 1.0 Volt). This s possibly due to defects at the interfaces between perovskite and transporting layers which can detriment the Voc. Therefore, some solution treatments (such as KCl or KOH) at the Perovskite/SnO_2 interface is proposed to reduce defects, passivate the contact and stabilize the device [22].

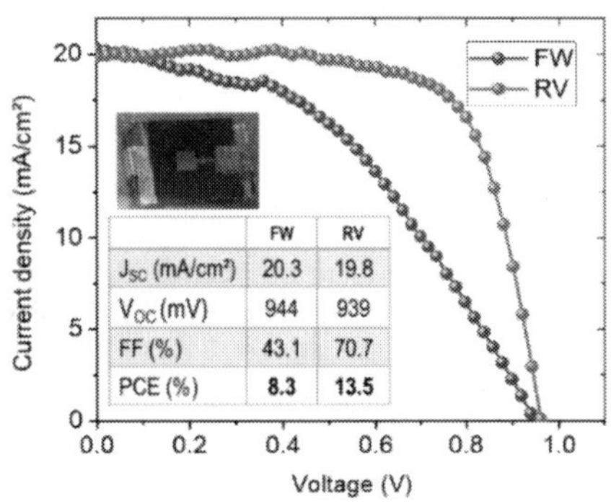

Figure 7: J-V characteristics of the optimized perovskite cell. Insets a photograph of the opaque cell with 0.09 cm^2 active area and a table summarizing cell performance parameters (FW = Forward scan, RV = Reverse scan).

4 CONCLUSIONS

In conclusion, we use the hybrid sequential deposition technique which conjoins CsBr-PbI_2 thermal co-evaporation and slot-die coating for achieving a triple cation wide-bandgap perovskite. This work demonstrates an achievement of perovskite conversion with homogeneous coating thanks to the optimization of deposition conditions such as evaporation rates, solution dispense rate and solution concentration on the large substrates of 25 cm^2. The perovskite is fully converted with 0.1 Å s^{-1} CsBr and 1.0 Å s^{-1} PbI_2 in the first step together with 3.3 μL s^{-1} dispense rate and 40.7mg/mL solution concentration. We also investigated the compositions of the converted perovskite using XPS, showing small discrepancies the organic cations (MA^+ and FA^+) is observed on the surface, however, elemental in-depth distribution is uniform across the layer. The film is found stoichiometrically correlated to the theoretical film but there is an excess of Iodine at the surface. The cell integrated the hybrid perovskite layer delivers 13.5% PCE under reverse scan, which is the first high efficiency obtained using CsBr/PbI_2 thermal co-evaporation coupled with slot-die coating. As perspective, to further increase the V_{OC} which is currently an issue which limits the cell efficiency, a treatment at the interface of Perovskite and charge-carrier transporting layer can be used to mitigate the defects in the perovskite thin film. Finally, optimization of the coating on the larger area is currently carried out in our group with an effort to demonstrate that this hybrid fabrication method paves the way to fabricate up-scalable textured two-terminal perovskite/silicon tandem solar cells.

ACKNOWLEDGEMENT

We acknowledge the support from the French Government in the Programme d'Investissement d'Avenir ANR-IEED-002-01. This work is also supported by CEA internal fundings.

REFERENCE

[1] A. Kojima, K. Teshima, Y. Shirai, T. Miyasaka, Organometal halide perovskites as visible-light sensitizers for photovoltaic cells, J. Am. Chem. Soc. 131 (2009) 6050–6051. https://doi.org/10.1021/ja809598r.

[2] https://www.nrel.gov/pv/cell-efficiency.html, (n.d.). https://www.nrel.gov/pv/cell-efficiency.html.

[3] J. Jeong, M. Kim, J. Seo, H. Lu, P. Ahlawat, A. Mishra, Y. Yang, M.A. Hope, F.T. Eickemeyer, M. Kim, Y.J. Yoon, I.W. Choi, B.P. Darwich, S.J. Choi, Y. Jo, J.H. Lee, B. Walker, S.M. Zakeeruddin, L. Emsley, U. Rothlisberger, A. Hagfeldt, D.S. Kim, M. Grätzel, J.Y. Kim, Pseudo-halide anion engineering for α-FAPbI3 perovskite solar cells, Nature. 592 (2021) 381–385. https://doi.org/10.1038/s41586-021-03406-5.

[4] Z. Ouyang, M. Yang, J.B. Whitaker, D. Li, M.F.A.M. van Hest, Toward Scalable Perovskite Solar Modules Using Blade Coating and Rapid Thermal Processing, ACS Appl. Energy Mater. 3 (2020) 3714–3720. https://doi.org/10.1021/acsaem.0c00180.

[5] M. Fievez, P.J. Singh Rana, T.M. Koh, M. Manceau, J.H. Lew, N.F. Jamaludin, B. Ghosh, A. Bruno, S. Cros, S. Berson, S.G. Mhaisalkar, W.L. Leong, Slot-die coated methylammonium-free perovskite solar cells with 18% efficiency, Sol. Energy Mater. Sol. Cells. 230 (2021). https://doi.org/10.1016/j.solmat.2021.111189.

[6] S. Huang, C. Guan, P. Lee, H. Huang, C. Li, Y. Huang, W. Su, Toward All Slot-Die Fabricated High Efficiency Large Area Perovskite Solar Cell Using Rapid Near Infrared Heating in Ambient Air, Adv. Energy Mater. 10 (2020) 2001567. https://doi.org/10.1002/aenm.202001567.

[7] Z. Li, P. Li, G. Chen, Y. Cheng, X. Pi, X. Yu, D. Yang, L. Han, Y. Zhang, Y. Song, Ink Engineering of Inkjet Printing Perovskite, ACS Appl. Mater. Interfaces. 12 (2020) 39082–39091. https://doi.org/10.1021/acsami.0c09485.

[8] I. Zimmermann, M. Al Atem, O. Fournier, S. Bernard, S. Jutteau, L. Lombez, J. Rousset, Sequentially Slot-Die-Coated Perovskite for Efficient and Scalable Solar Cells, Adv. Mater.

Interfaces. 8 (2021) 1–9. https://doi.org/10.1002/admi.202100743.

[9] Z. Yang, W. Zhang, S. Wu, H. Zhu, Z. Liu, Z. Liu, Z. Jiang, R. Chen, J. Zhou, Q. Lu, Z. Xiao, L. Shi, H. Chen, L.K. Ono, S. Zhang, Y. Zhang, Y. Qi, L. Han, W. Chen, Slot-die coating large-area formamidinium-cesium perovskite film for efficient and stable parallel solar module, Sci. Adv. 7 (2021). https://doi.org/10.1126/sciadv.abg3749.

[10] Chart of Best Research-Cell Efficiencies Provided by NREL, (n.d.). https://www.nrel.gov/pv/assets/pdfs/best-research-cell-efficiencies-rev220630.pdf.

[11] F. Sahli, B.A. Kamino, J. Werner, M. Bräuninger, B. Paviet-Salomon, L. Barraud, R. Monnard, J.P. Seif, A. Tomasi, Q. Jeangros, A. Hessler-Wyser, S. De Wolf, M. Despeisse, S. Nicolay, B. Niesen, C. Ballif, Improved Optics in Monolithic Perovskite/Silicon Tandem Solar Cells with a Nanocrystalline Silicon Recombination Junction, Adv. Energy Mater. 8 (2018) 1701609. https://doi.org/10.1002/aenm.201701609.

[12] P.S.C. Schulze, K. Wienands, A.J. Bett, S. Rafizadeh, L.E. Mundt, L. Cojocaru, M. Hermle, S.W. Glunz, H. Hillebrecht, J.C. Goldschmidt, Perovskite hybrid evaporation/ spin coating method: From band gap tuning to thin film deposition on textures, Thin Solid Films. 704 (2020) 137970. https://doi.org/10.1016/j.tsf.2020.137970.

[13] V.S. Nguyen, I. Zimmermann, E. Grépin, S. Jutteau, F. Donsanti, E. Bruhat, A. Duchatelet, S. Berson, J. Rousset, Up-scalable hybrid perovskite deposition process coupling Thermal Evaporation and Slot-Die Coating. [Unpublished manuscript], n.d.

[14] M. Saliba, T. Matsui, J.-Y. Seo, K. Domanski, J.-P. Correa-Baena, M.K. Nazeeruddin, S.M. Zakeeruddin, W. Tress, A. Abate, A. Hagfeldt, M. Grätzel, Cesium-containing triple cation perovskite solar cells: improved stability, reproducibility and high efficiency, Energy Environ. Sci. 9 (2016) 1989–1997. https://doi.org/10.1039/C5EE03874J.

[15] L. Qiu, S. He, Y. Jiang, D. Son, L.K. Ono, Z. Liu, T. Kim, T. Bouloumis, S. Kazaoui, Y. Qi, Hybrid chemical vapor deposition enables scalable and stable Cs-FA mixed cation perovskite solar, (2019) 6920–6929. https://doi.org/10.1039/c9ta00239a.

[16] F. Sahli, J. Werner, B.A. Kamino, M. Bräuninger, R. Monnard, B. Paviet-Salomon, L. Barraud, L. Ding, J.J. Diaz Leon, D. Sacchetto, G. Cattaneo, M. Despeisse, M. Boccard, S. Nicolay, Q. Jeangros, B. Niesen, C. Ballif, Fully textured monolithic perovskite/silicon tandem solar cells with 25.2% power conversion efficiency, Nat. Mater. 17 (2018) 820–826. https://doi.org/10.1038/s41563-018-0115-4.

[17] J. Werner, F. Sahli, F. Fu, J.J. Diaz Leon, A. Walter, B.A. Kamino, B. Niesen, S. Nicolay, Q. Jeangros, C. Ballif, Perovskite/Perovskite/Silicon Monolithic Triple-Junction Solar Cells with a Fully Textured Design, ACS Energy Lett. 3 (2018) 2052–2058. https://doi.org/10.1021/acsenergylett.8b01165.

[18] I. Zimmermann, M. Provost, S. Mejaouri, M. Al Atem, A. Blaizot, A. Duchatelet, S. Collin, J. Rousset, Industrially Compatible Fabrication Process of Perovskite-Based Mini-Modules Coupling Sequential Slot-Die Coating and Chemical Bath Deposition, ACS Appl. Mater. Interfaces. 14 (2022) 11636–11644. https://doi.org/10.1021/acsami.1c24558.

[19] K. Liao, J. Yang, C. Li, T. Li, F. Hao, Off-Stoichiometric Methylammonium Iodide Passivated Large-Grain Perovskite Film in Ambient Air for Efficient Inverted Solar Cells, ACS Appl. Mater. Interfaces. 11 (2019) 39882–39889. https://doi.org/10.1021/acsami.9b12829.

[20] S. Cacovich, P. Dally, G. Vidon, M. Legrand, S. Gbegnon, J. Rousset, J.-B. Puel, J.-F. Guillemoles, P. Schulz, M. Bouttemy, A. Etcheberry, In-Depth Chemical and Optoelectronic Analysis of Triple-Cation Perovskite Thin Films by Combining XPS Profiling and PL Imaging, ACS Appl. Mater. Interfaces. 14 (2022) 34228–34237. https://doi.org/10.1021/acsami.1c22286.

[21] S. Cacovich, D. Messou, A. Bercegol, S. Béchu, A. Yaiche, H. Shafique, J. Rousset, P. Schulz, M. Bouttemy, L. Lombez, Light-induced passivation in triple cation mixed halide perovskites: Interplay between transport properties and surface chemistry, ACS Appl. Mater. Interfaces. 12 (2020) 34784–34794. https://doi.org/10.1021/acsami.0c06844.

[22] T. Bu, J. Li, F. Zheng, W. Chen, X. Wen, Z. Ku, Y. Peng, J. Zhong, Y.B. Cheng, F. Huang, Universal passivation strategy to slot-die printed SnO 2 for hysteresis-free efficient flexible perovskite solar module, Nat. Commun. 9 (2018) 1–10. https://doi.org/10.1038/s41467-018-07099-9.

CHARACTERIZATION AND DEGRADATION OF PEROVSKITE MINI-MODULES

R. Ebner[1], G. Ujvari[1], A. Mittal[1], M. Hadjipanayi[2], V. Paraskeva[2], G. E. Georghiou[2], A. Hadipour[3], A. Aguirre[4,5,6], T. Aernouts[4,5,6], T. Fontanot[7], S. Pechmann[7], S. Christiansen[7,8]
[1]AIT Austrian Institute of Technology, Center for Energy, Giefinggasse 2, 1210 Vienna, Austria, T +43 50550-6628, F +43 50550-6390, rita.ebner@ait.ac.at, www.ait.ac.at
[2]University of Cyprus, 1 Panepistimiou Avenue, 2109 Aglantzia, Nicosia, Cyprus, www.foss.ucy.ac.cy
[3]Kuwait University, Department of Physics, Condensed Matter Physics Group, University city Shadadiya campus, Kuwait
[4]Imec, imo-imomec, Thin Film PV Technology – partner in Solliance, Thor Park 8320, 3600 Genk, Belgium
[5]EnergyVille, imo-imomec, Thor Park 8320, 3600 Genk, Belgium
[6]Hasselt University, imo-imomec, Martelarenlaan 42, 3500 Hasselt, Belgium
[7]Fraunhofer Institute for Ceramic Technologies and Systems IKTS, Äußere Nürnberger Str. 62, 91301 Forchheim, Germany, www.ikts.fraunhofer.de
[8]Max Planck Institute for the Science of Light, Günther-Scharowsky-Strasse 1, 91058 Erlangen, Germany, mpl.mpg.de

ABSTRACT: With their exceptional optoelectronic properties, compatibility with low-cost and large-scale fabrication methods, organic-inorganic hybrid metal halide perovskites are poised to revolutionize the next generation of photovoltaics. The leap forward in the power conversion efficiency (PCE) enabled by lead halide perovskites is unprecedented, with PCEs emerging from 3.8% in its first study to a current certified value of 25.5% in single-junction and 29.52% in perovskite-silicon tandem devices [1-3]. The main challenge for the successful commercialization of perovskite solar cells is to achieve high stability at the module level. The commercially available solar modules undergo a series of characterization procedures that analyze their properties and ensure their quality. However, these procedures and protocols cannot unambiguously be applied to perovskite solar modules (PSM). To this end, more advanced characterization methods are needed to understand the degradation mechanisms in the PSM. In this context, optical and electrical measurement methods are effectively employed in quality control and development support and are essential characterization tools in industry and research.

KEYWORDS: Perovskite, Optical and electrical characterization, Degradation

1 INTRODUCTION

In the proposed work, we employed optical and electrical characterization methods to understand the degradation of perovskite mini-modules. Optical techniques such us electroluminescence (EL), photoluminescence (PL), and dark lock-in thermography (DLIT) are non-destructive measurement techniques and provide high-resolution images showing a two-dimensional distribution of the characteristic features of PV cells and allowing the investigation of cracks, defects, shunts, and stacking faults in the cells [4]. Furthermore, electrical measurements like current-voltage (IV) characteristics and external quantum efficiency (EQE) can provide information on the power output and other device parameters, which could be used to identify the possible degradation route.

2 EXPERIMENTS

A - double cation-double halide perovskite active layer with the composition **$Cs_{0.18}FA_{0.82}PbI_{2.82}Br_{0.18}$** was used. To make large-area devices, so called mini-modules were produced by laser scribing, to generate 7 sub-cells connected in series. To prevent penetration of metallic particles of the top electrode into the soft perovskite layer, ITO (indium-tin-oxide) was used. ITO was also selected as a top electrode to obtain semi-transparent modules. The module stack was as follows: glass/ITO/Hole transport layer (HTL)/550nm 2C Perovskite with bandgap of 1.6 eV/Electron transport layer (ETL)/ITO/glass. Fig.1 depicts the structure of the perovskite sub-cell and Fig. 2 illustrates the cross-section of the mini-module. Fig.3 shows two images of the front and back side of one perovskite mini-module (substrate size: 3cm x 3cm and module size: 2cm x 2cm). 4 perovskite mini-modules ("S9", "S10", "S11" and "S12") were produced and characterized by applying DLIT-, EL- and PL methods alongside IV and EQE measurements. The results are presented in section 3.

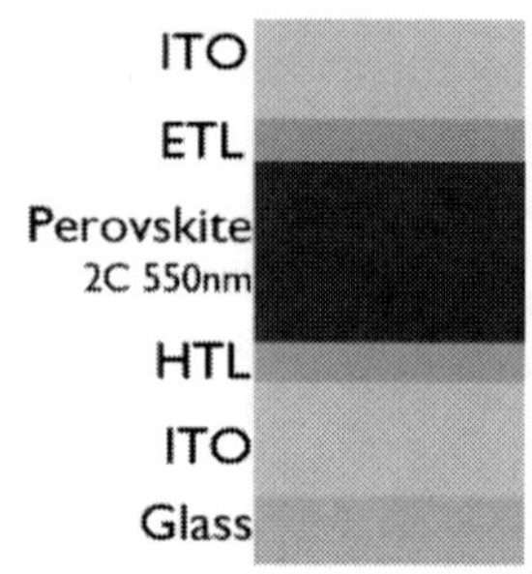

FIG.1 PIN-structure of each sub-cell.

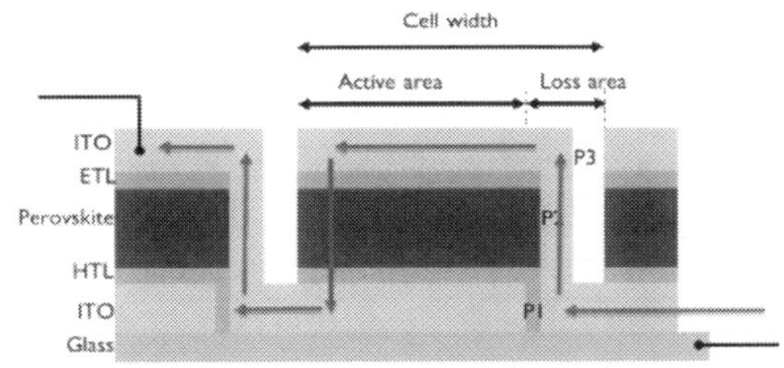

FIG.2 Cross-section of the mini-module.

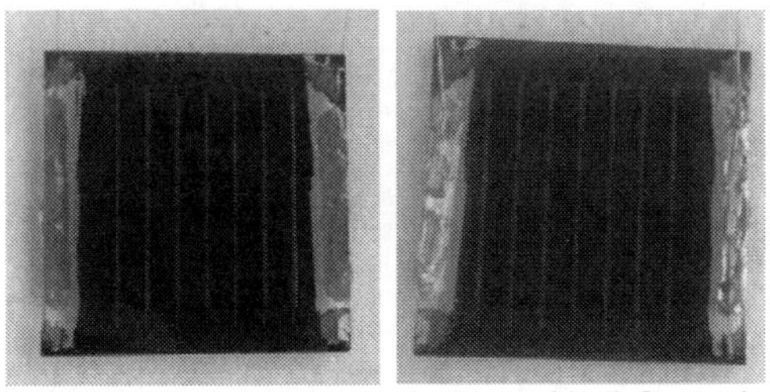

FIG.3 Perovskite mini-module: Front side (left), Back side (right).

3 RESULTS

3.1 Mini-module "S9"

Fig.4 shows the EL images (front side and back side) of mini-module "S9". All cells are active. The IV measurement results of mini-module "S9" are presented in Table 1, while the EQE measurement results are shown in Fig.5.

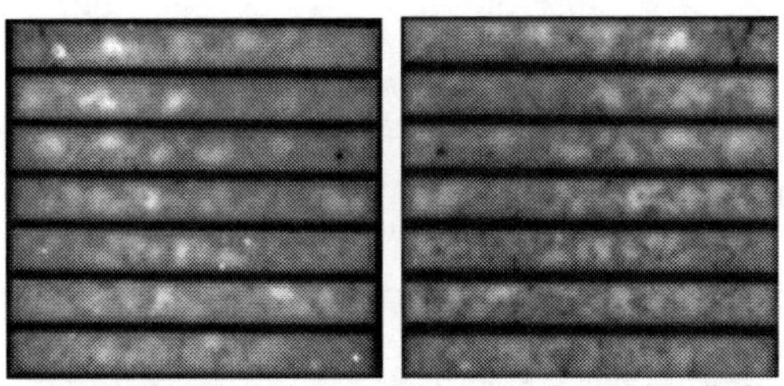

FIG.4 Perovskite mini-module "S9": Front side (left) and back side (right).

Table 1:
IV-measurement results of the mini-module "S9"

I_{SC} [mA]	V_{OC} [V]	FF [%]	P_{MPP} [mW]	Jsc [mA/cm²]
10.26	7.789	42.91	34.29	13.10

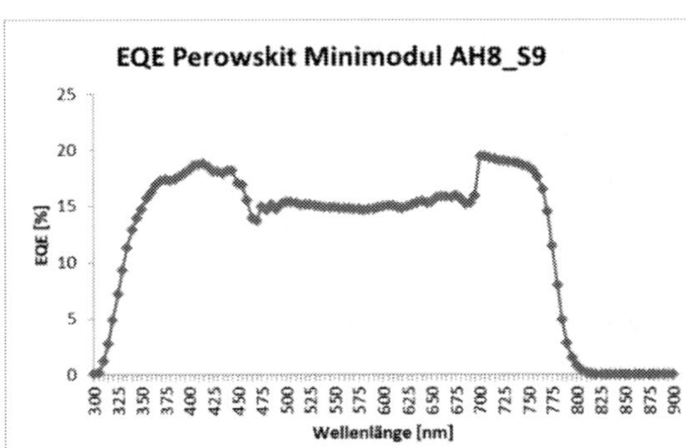

FIG.5 EQE of the perovskite mini-module "S9"

3.2 Mini-module "S10"

Fig.6 shows the EL images of mini-module "S10" with two inactive cells. The IV measurement results of minimodule "S10" are listed in Table 2. The EQE measurement results are shown in Fig.7.

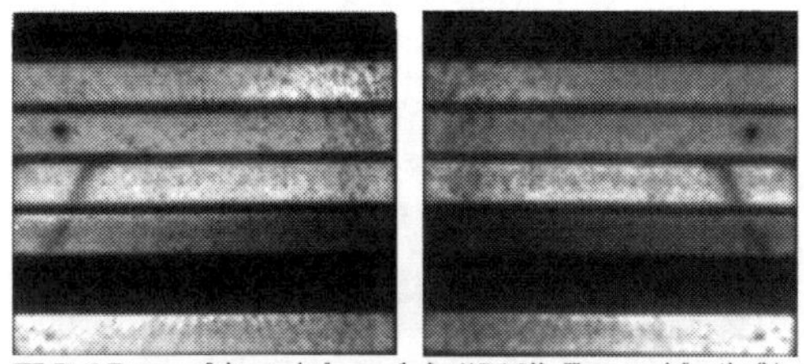

FIG.6 Perovskite mini-module "S10": Front side (left) and back side (right).

Table 2:
IV-measurement results of the mini-module "S10"

I_{SC} [mA]	V_{OC} [V]	FF [%]	P_{MPP} [mW]	Jsc [mA/cm²]
10.56	6.106	39.70	25.59	13.48

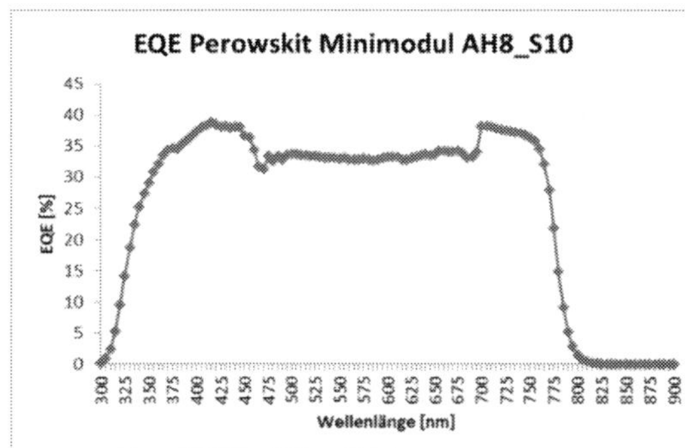

FIG.7 EQE of the perovskite mini-module "S10"

3.2.1 DLIT-measurements

DLIT measurements were carried out on the selected mini modules. However, it soon became evident that these measurements put a great strain on the modules, with the possibility to induce degradation, interrupt electrical contact or even re-establish broken contacts.

In the case of the mini-module "S10", it can be clearly seen in the EL image (see Fig.8), that an interrupted contact was reactivated, after the DLIT measurement performed on 10/2021 was carried out. After DLIT measurement there was only one inactive subcell instead of two.

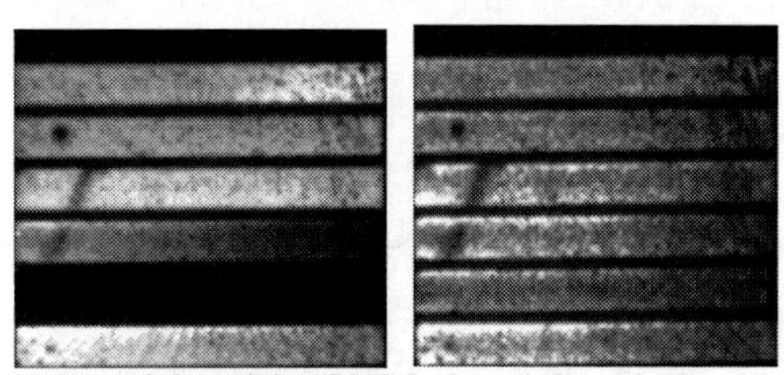

FIG.8 Mini-module "S10" before (left) and after (right) DLIT- measurement

The IV measurement results also showed a significant increase in performance after the DLIT measurement was performed (see Table 3).

Table 3: IV-measurement results of the mini-module "S10", before and after DLIT measurement (10/21).

Mini-Modul	I_{SC} [mA]	V_{OC} [V]	FF [%]	P_{MPP} [mW]	Jsc [mA/cm²]
S10	10.56	6.106	39.70	25.59	13.48
S10 DLIT	10.39	6.756	39.81	27.94	13.27

However, there was already a drop in the measurement performed on 11/2021 and a subsequent return to the initial values.

3.3 Mini-module "S11"
Fig.9 shows the EL recordings of the minimodule "S11" with three inactive cells. The IV measurement results of the minimodule "S11" are presented in Table 4. The EQE measurement results are depicted in Fig.10.

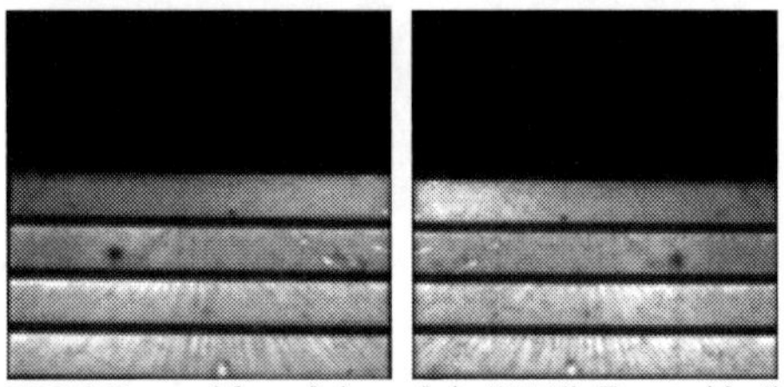

FIG.9 Perovskite mini-module "S11": Front side (left) and back side (right).

Table 4:
IV-measurement results of the mini-module "S11"

I_{SC} [mA]	V_{OC} [V]	FF [%]	P_{MPP} [mW]	Jsc [mA/cm²]
10.37	5.260	37.60	20.50	13.24

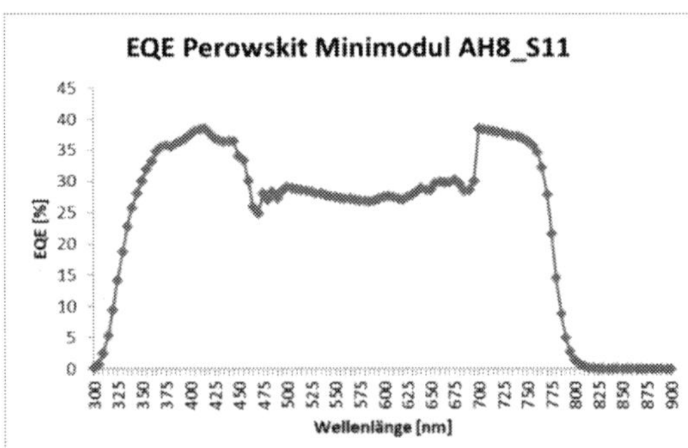

FIG.10 EQE of the perovskite mini-module "S11"

3.4 Mini-module "S12"
Fig.11 shows the EL images of the minimodule "S12" with two inactive cells.

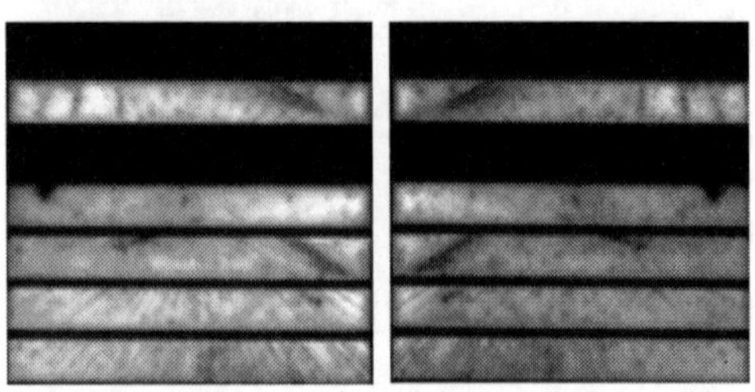

FIG.11 Perovskite mini-module "S12": Front side (left) and back side (right).

The IV measurement results of the minimodule "S12" are listed in Table 5. The EQE measurement results are shown in Fig.12.

Table 5:
IV-measurement results of the mini-module "S12"

I_{SC} [mA]	V_{OC} [V]	FF [%]	P_{MPP} [mW]	Jsc [mA/cm²]
10.44	6.993	40.60	29.64	13.33

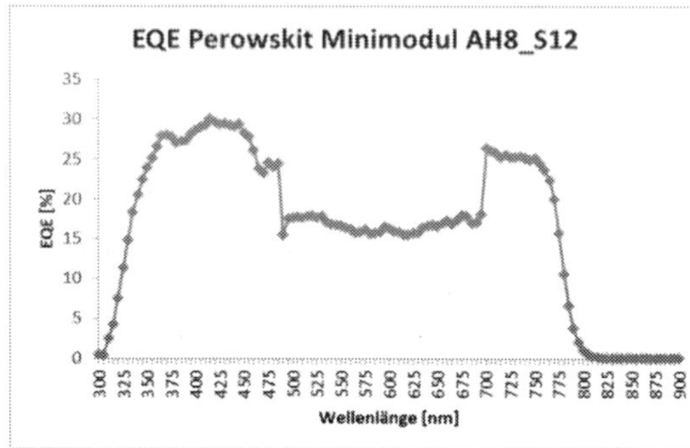

FIG.12 EQE of the perovskite mini-module "S12"

3.5 Aging behavior of the perovskite mini-modules S9, S10, S11 and S12
For the perovskite mini-modules, IV measurements as well as EL measurements were performed regularly from **July 2021 (07/21) to February 2022 (02/22)** to determine the **aging behavior** of the mini-modules. The modules were stored in the dark between the measurements. As can be seen in **Fig.13**, the mini-modules behaved in a very stable manner over the selected period.

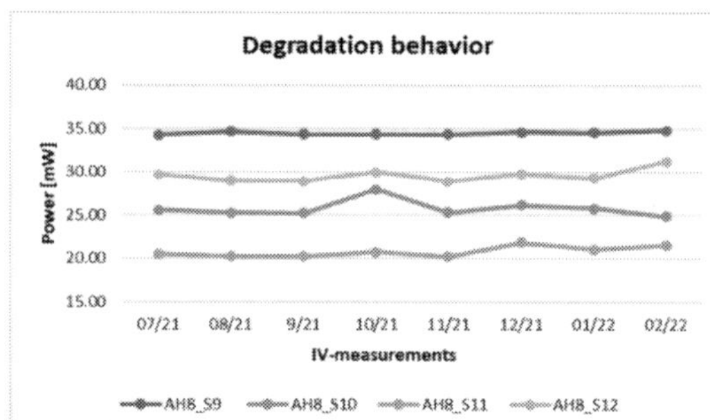

	AH8_S9	AH8_S10	AH8_S11	AH8_S12
7/21	34.29	25.59	20.50	29.64
8/21	34.68	25.24	20.25	28.96
9/21	34.33	25.20	20.23	28.93
10/21	34.31	27.94	20.75	30.00
11/21	34.35	25.30	20.25	28.88
12/21	34.61	26.18	21.84	29.75
1/22	34.53	25.78	21.08	29.26
2/22	34.81	24.89	21.56	31.27

FIG.13 Aging behavior of the perovskite mini-modules

Only in the case of mini-module "S10" a significant increase in power was detectable in 10/2021 (after a DLIT measurement was carried out). However, already in November 2021 there was a decrease in power and a return to the initial values.

4 CONCLUSIONS

First defects and shunts were identified by EL, PL and DLIT measurements. The EQE measurements gave very high values for perovskite cells (e.g. about 90%, cell size:0.13cm^2, see Fig.12), but still too low values for perovskite mini modules (Fig.5).

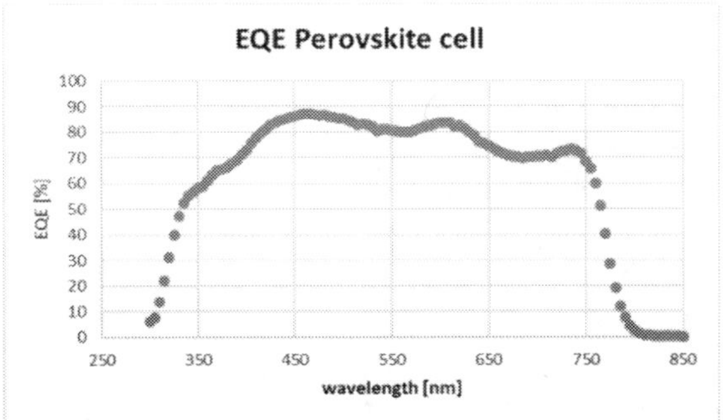

FIG.12 EQE of a perovskite cell

Further indoor tests and outdoor tests of differently structured perovskite samples and perovskite tandem cells have to be performed in order to identify more defects and thus be able to improve the perovskite cell and module structure. The aging behavior of the perovskite mini-modules will also be further analyzed. In addition, measurement protocols for indoor and outdoor tests of perovskites will be established.

5 REFERENCES

[1] Antonio Urbina 2020 J. Phys. Energy 2 022001

[2] "Oxford PV retakes tandem cell efficiency record",https://www.pv-magazine.com/2020/12/21/oxford-pv-retakes-tandem-cell-efficiency-record/

[3] Enzheng Shi et al., "Two-dimensional halide perovskite lateral epitaxial heterostructures", *Nature*, 2020; 580 (7805): 614

[4] R. Ebner et al., "Non-destructive techniques for quality control of PV modules", 39th Annual

Acknowledgement:
This work was funded through the European Regional Development Fund and the Republic of Cyprus in the framework of the project "DegradationLab" with grant number INFRASTRUCTURES/1216/0043.
http://www.foss.ucy.ac.cy/degradationlab/

PEROVSKITE SOLAR CELLS PREPARED IN AMBIENT CONDITIONS BASED ON AN HTM-FREE ARCHITECTURE THAT HAS A BACK CONTACT DEPOSITED FROM A PASTE MADE WITH GRAPHITE, BLACK CARBON, MAI AND PVDF DISPERSED IN CHLOROBENZENE

C. Montes[1,3], L. Ocaña[1,3], B. González-Díaz[3], S. González-Pérez[2], E. Llarena[1].
[1]Instituto Tecnológico y de Energías Renovables, S. A. (ITER)
Pol. Industrial de Granadilla, s/n, E 38600 Granadilla de Abona, Spain. Ph. +34 922 747 700. Fax +34 922 747 701. E-mail: cmontes@iter.es
[2]Departamento de Didácticas Específicas.
C/ Pedro Zerolo, s/n
Edificio Central Planta 2
Apartado 456. Código postal 38200
San Cristóbal de La Laguna. S/C de Tenerife. Spain.
[3]Departamento de Ingeniería Industrial.
Universidad de La Laguna.
Camino San Francisco de Paula, s/n
38206 S/C de Tenerife, Spain

ABSTRACT: In this work an improved carbon-based composite and a simplified architecture has been developed to produce and deposit a back contact to fabricate HTM-free Perovskite Solar Cells in ambient conditions. A batch of cells were produced that had efficiencies stabilized at about 1%, with a record value of 1.646%. Estimations of the devices' serial and shunt resistances were also obtained, resulting on 2kΩ and 200Ω respectively, which indicates that, on the one hand, they are still too inefficient at allowing the flow of the photo-generated electrons and, on the other, that alternative recombination pathways yet exist, which further hamper their efficiencies. Also, the observed phenomena of IV curve hysteresis (between the forward and reverse measurements), together with a yellowing of the perovskite layers after completing a set of tests under continuous illumination, suggest that substantial ionic shift happens in the crystal lattice, that also contributed to its degradation. In order to compensate these effects, different strategies were proposed, which will be tested in future iterations.
Keywords: Perovskites, Contact, Metallization, Graphite

1 INTRODUCTION

The Solar Cell Laboratory from the Instituto Tecnológico y de Energías Renovables (ITER), with the support of the La Laguna University (ULL), has been working for several years on perovskite solar cells (PSCs), based on the use of methylammonium lead triiodide (MAPbI3), while focusing, on the one hand, on fabricating these devices in ambient conditions [1]–[5] and, on the other hand, on developing ink-based solutions, in order to provide the means for producing their back contacts as an alternative to thermal evaporation [6]–[11]. Also, considerable efforts have been dedicated to find solutions for encapsulating perovskite layers by using lamination techniques that apply ethylene-vinyl acetate (EVA) as encapsulate agents [12]–[14]. By using the knowledge acquired with these three initiatives and inspired by the work of Guanglan Liao et Al [15] and of Xingzhong Zhao et Al [16] who developed carbon-based back electrodes for Hole Transport Material Free (HTM-free) planar heterojunction PSCs, an agglomerate of graphite and black carbon in an EVA in toluene solution was developed, in order to be applied as back contact electrode by means of an innovative cell architecture, so as to produce HTM-free perovskite solar cells [17] which, further inspired by the work of Xingzhong Zhao et Al [18], was refined in order to introduce a Methyl Ammonium Iodide (MAI) rich sub layer in a carbon in Polyvinylidene Fluoride (PVDF) based back contact, which led to devices with had Power Conversion Efficiencies up to 1.32% [19]. Furthermore, the IV curves performed to these devices indicated that there is an important range to improve the electrode thickness to perform efficiently and to reduce the series resistance of the devices. Besides, it was found that the use of a MAI dispersed in the carbon layer located above the perovskite one led to a patent improvement on the PCE of the devices, after having subjected them to a mild thermal process of 100 ºC during 30 minutes. As a result, it was concluded that this phenomenon was in accordance with a scenario where adding excess MAI helped to relieve the uniformity problems already affecting the perovskite layers, thus improving the shunt resistance of the devices, as well as providing optimized interface energetics [20] and reduce the PbI_2 impurities [21]. Unfortunately, this effect quickly faded after the devices were further tested.

This paper summarizes the results obtained by continuing this line of research, this time focusing mainly on simplifying the perovskite solar cell architecture.

2 EXPERIMENTAL

2.1 Preparation of the powder of graphite, black carbon and PVDF

Solutions of PVDF (Kureha KF W#1700 powder) in DMSO (Sigma Aldrich 276855) at 15:85% by weight percentage (wt%) were made by mixing 1.008 g of PVDF in 5 ml of DMSO. The mixing method was first using a spatula and then a magnetic stirrer (3D, by IBX Instruments), set up at 300 rpm during 5 minutes. Then the mixture was taken to a magnetic stirrer aided with a heating element (BasicMagMix, by OVAN), where it was further stirred during 60 minutes at 75ºC and 450rpm. As soon as PVDF comes into contact with DMSO, it forms a whitish dispersion (like aspirin in water) which, as it heats up, thickens and gels, turning to a light yellowish color.

4.875 g of PVDF in DMSO solution, 9.098 g of graphite (Graphite powder SC 20 OS, Graphite Kropfmühl GmbH) and 3.534 g of carbon black were mixed, producing 18.184 g of mixture (28:52:20% in wt%). To mix such an amount of conductive powder, an extra 20 ml of DMSO had to be added, bringing the final weight percentages of the mixing components to 2:66:23:9% (PVDF:DMSO:Graphite:Black Carbon, wt%) before the mixture was taken to the three-roll mill (ESSG65, by Shangai Espread), where 5 passes were made.

To evaporate the DMSO, this mixture was spread with a spatula on a glass and then put in a stove (CONTERM 19L, by J.P.Selecta.) for seven hours at 130°C. Finally, the dry product was scraped from the glass using a steel spatula and the mixture was placed in a coffee grinder, in order to pulverize it.

After this process, with all the DMSO evaporated from the powder, we have obtained a mixture made up of PVDF:Graphite:Black Carbon with weight percentages 6:68:26% (which represents 94.4% of conductive powders).

Figure 1: (a) Agglomerate of PVDF with Graphite and Black Carbon in DMSO, spread on a glass and, ready to be put in the stove. Mix of dried PVDF with Graphite and Black Carbon before (b) and (c) after having been pulverized in a coffee grinder.

2.2 Preparation of the Graphite, Black Carbon and MAI with PVDF in chlorobenzene paste.

To a small plastic beaker, 1.043 g of PVDF, Graphite and Black Carbon powders, prepared as indicated in the previous section, were added together with 0.241 g of methylammonium iodide (CH_3NH_3I or MAI) (Sigma Aldrich 793493), followed by 4 ml of chlorobenzene (C_6H_5Cl) (Sigma Aldrich 284513). Then it was mixed by using a spatula for 5 minutes. The resulting mixture was covered and allowed to stand until required for use.

2.3 Preparation of the HTM-free Perovskite Solar Cells (PSCs)

The fabrication process of the device was carried out inside an ISO 7 cleanroom in ambient conditions, at relative humidity (RH) of about 40%, and at a temperature set to 23°C.

First, a Fluorine doped Tin Oxide (FTO) coated glass (Sigma Aldrich 735167) was cut in pieces with sizes 2.5 x 2.5 mm, which were cleaned sequentially by ultrasonication in a 2% aqueous solution of soap, deionized water, acetone and finally in 2-propanol. Then, the pieces were dried with nitrogen gas flow. Coming up next, an UV Ozone Cleaner (UV Ozone Cleaner, model L2002A2-UK by Ossila) was used to remove any remaining contaminant from the surface of the samples.

Secondly, a compact layer of Titanium dioxide (TiO_2) was deposited by spin coating (NT12000 by Navson Technologies) over the FTO layer, at a speed of 2000 rpm for 60 seconds. The titanium oxide precursor solution was prepared by dropwise adding a solution of 175 µl of Titanium isopropoxide ($Ti\{OCH(CH_3)_2\}_4$) (Sigma Aldrich 205273 Titanium(IV) isopropoxide, 97%) in 1.25 ml of Ethanol (EtOH) (Sigma Aldrich 493546) into a solution of 17.5 µl of 2 M hydrochloric acid (HCl) (Sigma Aldrich 258148) in 1.25 ml of EtOH [22], [23]. This layer was cured in a furnace at 500 ° C for 60 min. Prior to their deposition, the produced solutions were passed through a 0.22 µm Polytetrafluoroethylene (PTFE) filter.

Third, thin layers of perovskite, from the methylammonium lead iodide kind ($CH_3NH_3PbI_3$ or $MAPbI_3$), were fabricated in ambient conditions. The perovskite precursor solution was produced by mixing lead iodide (PbI_2) (Sigma Aldrich 900168) and methylammonium iodide (CH_3NH_3I) (Sigma Aldrich 793493) in an equimolar ratio of 1:1, with a solvent mixture of dimethylformamide (DMF) (Sigma Aldrich 227056) and DMSO, prepared with a 9:1 volume ratio. The solution was mixed in a magnetic stirrer with heating capabilities by set it at 300 revolutions per minute (rpm) and 75 °C during 50 minutes. Then the heating was interrupted, leaving the solution to cool down to room temperature, while keeping the stirring at 1050 rpm overnight. Prior to its deposition, the produced solution was also passed through a 0.22 µm PTFE filter. The deposition proper was made by spin-coating 100 µl of the perovskite precursor solution at 5000 rpm for 30 seconds so that, 6 seconds after the start of the spinning, 200 µl of a non-polar solvent, in this case chlorobenzene, was dropped over the substrates. Throughout the spin-coating process the RH inside the chamber was maintained at a range of 6-8 %, depending on the existing ambient humidity, by using a constant flow of clean dry air. Finally, after the deposition process took place, the samples were cured at 100 °C for 10 minutes inside a stove furnace and stored in a vacuum desiccator (DE05-250-001 by LBG), so as to avoid their degradation.

In order to incorporate the carbon based layer as the back contact for the HTM-free perovskite solar cells, a new cell architecture was developed (see Figure 2), in which the EVA sheet was used both as an encapsulating material and to hold in place the copper adhesive tape that is required for interfacing with both, front and back electrodes, as a way to transport the charges generated by the device when illuminated (see Figure 3).

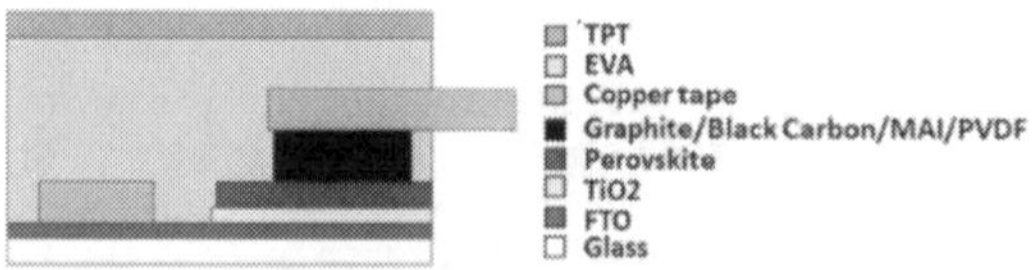

Figure 2: Schematic view of the developed HTM-free PSC architecture.

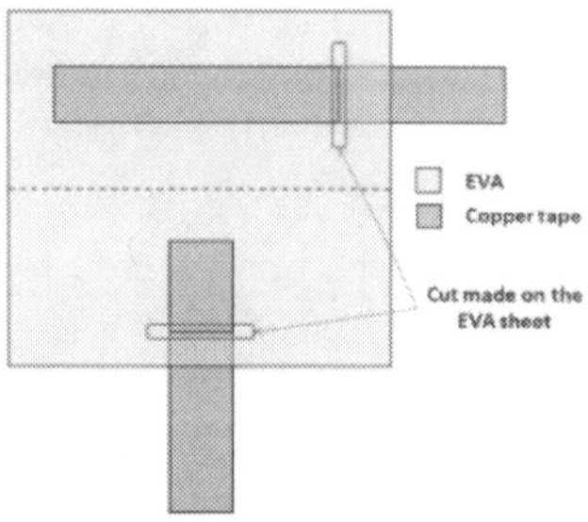

Figure 3: Diagram showing the arrangement of the copper tapes glued on the EVA sheet.

Subsequently, the carbon-based electrode was deposited over the perovskite layer. This process was carried out with the purpose of that, once cured, it became as thin as possible (which is why the mixture was so much diluted), while at the same time, as spread and homogenous as possible. Unfortunately, the developed mixture had a very low cohesive properties and did not allow to do both things at the same time. Therefore the deposition was carried out by dropping a little amount of paste with a spatula, avoiding splashing the FTO area, while trying to get it spread over the perovskite layer, being careful not to exert too much pressure to avoid scratching it. Next, to evaporate the chlorobenzene, the substrates were placed in a stove for 30 minutes at 100 ºC (see Figure 4). As a result, the electrodes had 69.1% by weight percentage of conductive powders.

Figure 4: Image showing the carbon electrodes deposited and ready for the curing stage.

Next, an EVA sheet with the pre-assembled copper tapes arranged as per Figure 3 was placed on each substrate, so that one extreme of the copper tape was placed on the carbon electrode (previously deposited and cured) and the other on the FTO. To close the encapsulation, a TPT sheet was added on top of each sample and the sandwich so produced was secured in place with Kapton tape (see Figure 5).

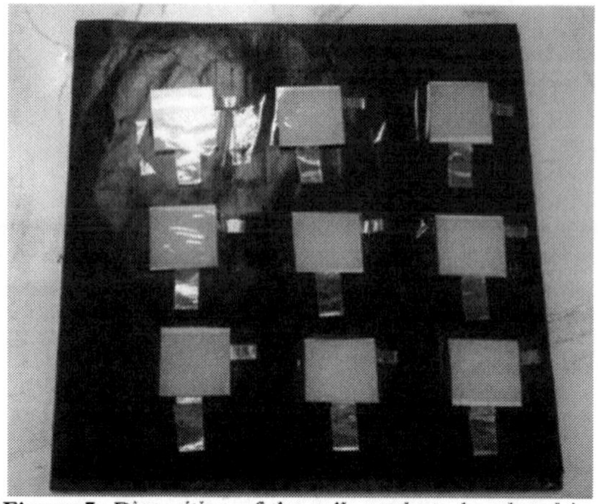

Figure 5: Disposition of the cells ready to be placed in the stove in order to complete the encapsulation process.

Next, a thick steel plate was placed over the devices, in order to exert uniform pressure during the last thermal process, also performed in a stove furnace set at 100 ºC, but this time lasting 165 minutes, to take into account the thermal inertia of the steel plate. Finally, the devices were allowed to cool down for another 125 minutes, before removing the steel plate.

2.4 HTM-free PSCs characterization

First, a visual inspection of the produced samples was carried out by using a digital microscope (U500X, by Unimake), to evaluate their morphology. Special attention was paid to detect encapsulation failures and to look for signs that suggested that damage had been caused to any of the many elements that comprised the devices (see Figure 6).

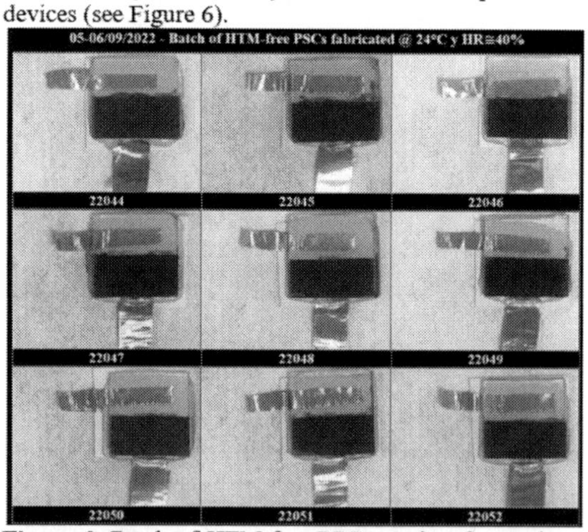

Figure 6: Batch of HTM-free PSCs produced at a clean room at 23ºC and relative humidity (RH%) about 40%.

Then, the devices were characterized by using a semiconductor characterization system (SCS-4200 by Keithley) in combination with a solar simulator (ABET-10500 by ABET Technologies), illuminating through a mask with an open surface of 0.38 cm^2 (see Figure 7), for determining their electrical properties.

The measurements were performed consecutively and in the following order:

1. A fast IV sweep in the forward direction.
2. A fast IV sweep in the reverse direction.

3. An IV-sweep performed at cell stability conditions in the forward direction.
4. An IV-sweep performed at cell stability conditions in the reverse direction.
5. Three fast IV sweeps in the forward direction.
6. Three fast IV sweeps in the reverse direction.
7. An IV-sweep performed at cell stability conditions in the forward direction
8. An IV-sweep performed at cell stability conditions in the reverse direction.

The first fast measurements were made to ensure that the tip of the probes were well positioned, as well as to obtain information on how the initial response of the cell was. Next, the "cell stability" measurements were made to get data from their operation in such conditions. As the measurements continue to progress, under continuous light exposure, the effect of light soaking [24]–[26] was also evaluated.

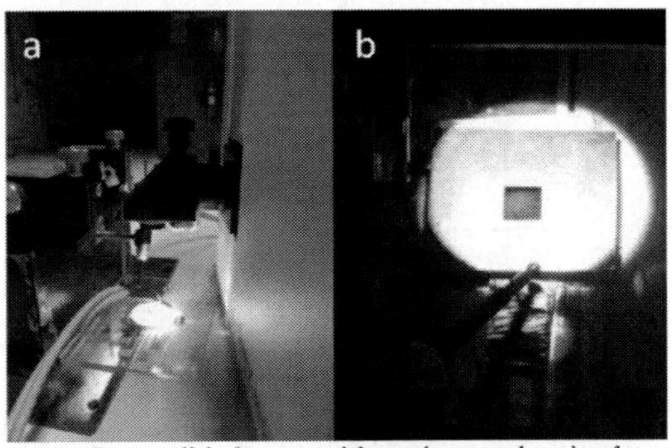

Figure 7: A cell being tested by using a solar simulator set at STC conditions (a) by iluminating trough a 0.38 cm^2 mask (b).

3 RESULTS AND DISCUSSION

Table I summarizes the most relevant data measured from the best produced cell in the manufactured batch. Figures 8 to 11 show the corresponding IV curves.

Table I: Summary with the values of short circuit current (Isc), open circuit voltage (Voc), fill factor (FF) and power conversion efficiency (PCE) obtained from the measurements made at the sample 22052.

Measurement	Isc (mA)	Voc (V)	FF (%)	PCE (%)
Fast forward	-0,099	0,675	n/a	0,130
Fast reverse	-1,505	0,700	49,626	1,383
Cell stability forward	-0,386	0,623	n/a	0,607
Cell stability reverse	-0,742	0,623	54,884	0,672
Fast forward	-0,409	0,690	n/a	0,598
Fast forward	-0,522	0,700	n/a	0,699
Fast forward	-0,583	0,705	n/a	0,723
Fast reverse	-1,728	0,725	46,723	1,548
Fast reverse	-1,788	0,730	47,655	1,646
Fast reverse	-1,467	0,685	46,261	1,230
Cell stability forward	-1,323	0,719	42,768	1,076
Cell stability reverse	-0,940	0,623	52,279	0,810

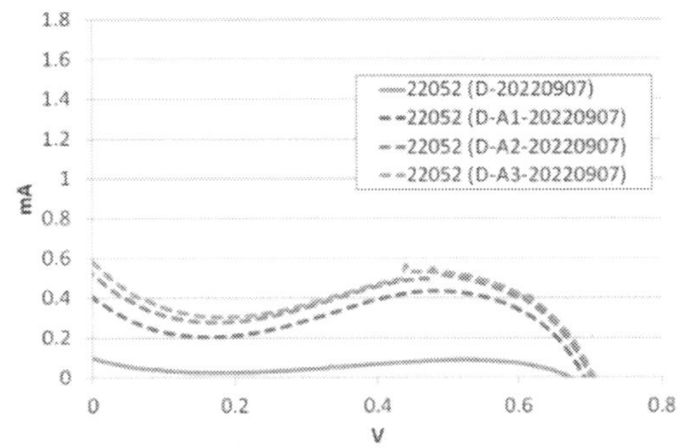

Figure 8: Fast sweep IV curves measured in the forward direction.

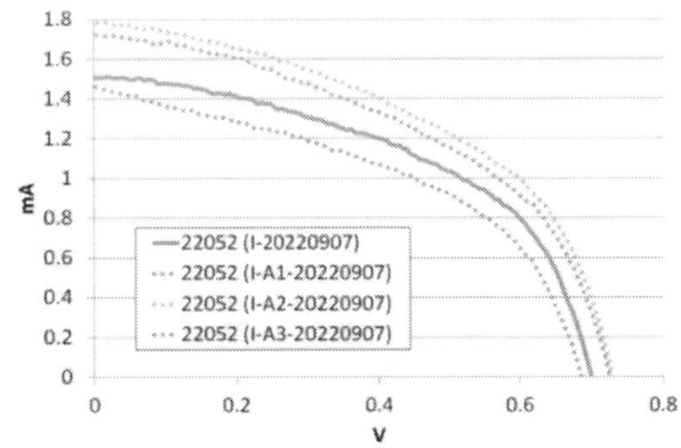

Figure 9: Fast sweep IV curves measured in the reverse direction.

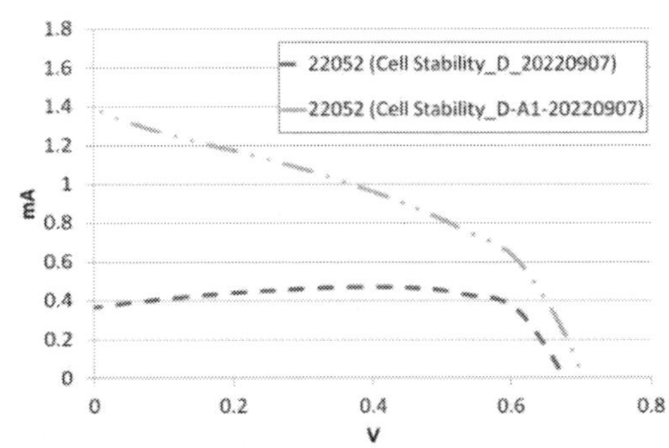

Figure 10: IV curves measured in the forward direction at cell stability conditions.

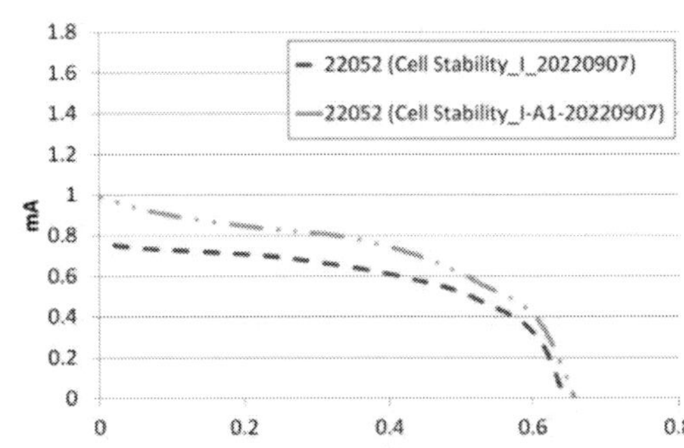

Figure 11: IV curves measured in the reverse direction at cell stability conditions.

As it can be seen, the best results were obtained in fast sweeps in the reverse direction, reaching the highest PCE obtained for this kind of cell architecture by our laboratory so far, with **1.646%** efficiency, and about 1% on average at cell stability conditions, a value that, considering the measurements made for the rest of the cells in the batch, seems to be the one expected for the developed technology (in the absence of manufacturing errors).

Also, Figure 9 curves were used to estimate the device's serial and shunt resistances (Rs and Rsh) [27]. Thus, the values obtained for the serial resistance was about **2kΩ** (or 760 Ωcm^2) and about **200Ω** (or 76 Ωcm^2), for the shunt resistance, that is, Rs that were two orders of magnitude above of ones found in the literature and Rsh that were one order of magnitude below reported ones [28]–[32]. Having such a high series resistance indicate that the devices were very inefficient at allowing the flow of the photo-generated electrons. This could be related to the thickness and quality of the developed carbon electrode, meaning that, despite the efforts made, it was still too thick and little conductive. However, before concluding this line of research, other options could be attempted, such as to modify the cell's architecture in order to include a mesoporous TiO_2 layer above the compact one, so as to increase the contact area of the perovskite [33], and to incorporate a proper HTM layer (possibly an inorganic one [34]). In addition, having such a low shunt resistance indicates that it is necessary to reduce alternative recombination pathways taking place in the devices. In this sense, in a short time it is expected to start manufacturing the perovskite layer in a glove box in our laboratory, in the hope that, in the absence of moisture and oxygen, both during the solution preparation stage and (especially) during the deposition and curing stage, this layer will be free of trapped degrading agents [30], [35].

Finally, as it can be seen in the IV curves, there was significant hysteresis between the forward and reverse measurements, as well as an increase in the open-circuit voltage (Voc) and fill factor (FF) (see Figures 8 to 10), while the short circuit current (Isc) experienced a quick increase and then a decrease upon light exposure (see Figure 9). These phenomena indicated a decrease in the charge accumulation at the electrode interfaces as well as a substantial ionic shift in the crystal lattice of $MAPbI_3$ which, in turn, could explain the degradation observed in this layer after completing a set of measurements under continuous light exposition [36]–[38]. To improve this situation, it might be necessary to incorporate multi-cation mixtures, such as formamide [FA; HC(NH2)2+], cesium (Cs) or rubidium (Rb) to act as passivation agents [39]–[41].

Figure 12: Image depicting a PSC after going through a set of measurements, notice that there is a yellowing in the perovskite sheet right where the opening in the mask was located.

4 ACKNOWLEDGEMENTS

This work has been developed within the MACLAB-PV project framework, which has been co-financed by the INTERREG Madeira-Azores-Canarias Territorial Cooperation Programme (MAC) 2014-2020. 2nd Call. Axis 1 – Enhancing research, technological development and innovation.

5 REFERENCES

[1] L. Ocaña *et al.*, "Adaptation of a Crystalline Silicon Solar Cell Laboratory to Produce Perovskite Solar Devices," *EU PVSEC Proc.*, no. 2015, pp. 1138–1143, doi: 10.4229/EUPVSEC20152015-3BV.5.40.

[2] L. Ocaña *et al.*, "Optimizing the Deposition of Thin Layers of Organic-Inorganic Hybrid Perovskite Methylammonium Lead Iodide (CH3NH3PbI3) on Large Surfaces through Their Optical Properties," *EU PVSEC Proc.*, pp. 1316–1323, 2016, doi: 10.4229/EUPVSEC20162016-3DV.2.26.

[3] M. F. & M. C. C. Montes, L. Ocaña, C. Quinto, S. González-Pérez, B. González-Díaz, C. Hernández-Rodríguez, "A One-Step Deposition Method Assisted with Non Polar Washing Solvent Treatment to Produce Uniform Thin Layers of Perovskite Validated through Ellipsometry," in *EU PVSEC Proceedings*, 2017, pp. 1181–1185. doi: 10.4229/EUPVSEC20172017-3DV.2.71.

[4] C. Montes *et al.*, "Producing uniform and smooth thin layers of perovskite under ambient conditions by adjusting to the existing levels of moisture," in *35th European Photovoltaic Solar Energy Conference and Exhibition*, Nov. 2018, pp. 933–937. doi: 10.4229/35THEUPVSEC20182018-3BV.3.36.

[5] M. C. C. Montes, L. Ocaña, L. De Sousa-Vieira, S. González-Pérez, B. González-Diaz, J.S. Moreno-Ramírez, C. Hernandez-Rodriguez, M. Friend, "Fabrication of Smooth, Mirror-Like and PbI2-Free Thin Film Perovskite Layers in Ambient Conditions," in *EU PVSEC Proceedings*, 2019, pp. 717–720. doi: 10.4229/EUPVSEC20192019-3BV.2.8.

[6] C. Quinto *et al.*, "Screen Printing for Perovskite Solar Cells Metallization," *EU PVSEC Proc.*, pp. 1144–1148, 2015, doi: 10.4229/EUPVSEC20152015-3BV.5.41.
[7] C. Montes *et al.*, "Conductive Inks With Epoxy Resin Based Vehicles for Perovskite Screen Printing," *EU PVSEC Proc.*, pp. 1303–1309, 2016, doi: 10.4229/EUPVSEC20162016-3DV.2.20.
[8] M. C. C. Montes, L. Ocaña, C. Quinto, S. González-Pérez, B. González-Diaz, M. Friend, "Conductive Inks with Epoxy Resin Based Vehicles for Perovskite Screen Printing Metallization as a Viable and Low-Cost Alternative to Thermal Evaporation," in *EU PVSEC Proceedings*, 2017, pp. 1158–1162. doi: 10.4229/EUPVSEC20172017-3DV.2.56.
[9] C. Montes *et al.*, "Exploring the use of collodion as a binder in screen printing vehicles for perovskite solar cells metallization," in *35 th European Photovoltaic Solar Energy Conference and Exhibition*, Nov. 2018, pp. 938–932. doi: 10.4229/35THEUPVSEC20182018-3BV.3.34.
[10] C. Montes *et al.*, "Studying the use of mixed binders made with epoxy resin and collodion for producing conductive inks for the metallization of perovskite solar cells via screen printing techniques," in *36th European Photovoltaic Solar Energy Conference and Exhibition*, 2019, pp. 742–745. doi: 10.4229/EUPVSEC20192019-3BV.2.17.
[11] M. C. C. Montes, L. Ocaña, S. González-Pérez, B. González-Diaz, M. Friend, "Technical Analysis of Mixed Binders Based on Epoxy Resin and Ethylene-Vinyl Acetate to Produce Conductive Inks for the Metallization of Perovskite Solar Cells via Screen Printing Technique," in *EU PVSEC Proceedings*, 2020, pp. 663–666. doi: 10.4229/EUPVSEC20202020-3BV.1.23.
[12] M. Cendagorta *et al.*, "On the Stability of Planar CH3NH3PbI3 Perovskite Solar Cells Produced in Ambient Conditions by Using Polymer Encapsulates," *36th Eur. Photovolt. Sol. Energy Conf. Exhib.*, pp. 721–727, Oct. 2019, doi: 10.4229/EUPVSEC20192019-3BV.2.10.
[13] L. Ocaña, C. Montes, S. González-Pérez, B. González-Díaz, M. Friend, and & M. Cendagorta, "Improvements on the stability of MAPbI3 thin films against UV radiation by a low temperature encapsulation method that uses commercial ethylene-vinyl acetate sheets," in *37th European Photovoltaic Solar Energy Conference and Exhibition*, Oct. 2020, pp. 659–662. doi: 10.4229/EUPVSEC20202020-3BV.1.22.
[14] L. Ocaña, C. Montes, S. González-Pérez, B. González-Díaz, and E. Llarena, "Testing encapsulated perovskite solar cells in a climatic chamber by following the IEC 61215 and IEC 61646 standards for the thermal cycling test," in *38th European Photovoltaic Solar Energy Conference and Exhibition*, Nov. 2021, pp. 406–410. doi: 10.4229/EUPVSEC20212021-3BV.1.4.
[15] Z. Liu, T. Shi, Z. Tang, and G. Liao, "A large-area hole-conductor-free perovskite solar cell based on a low-temperature carbon counter electrode," *Mater. Res. Bull.*, vol. 96, pp. 196–200, Dec. 2017, doi: 10.1016/J.MATERRESBULL.2017.03.069.
[16] F. Li *et al.*, "Fully Air-Processed Carbon-Based Efficient Hole Conductor Free Planar Heterojunction Perovskite Solar Cells With High Reproducibility and Stability," *Sol. RRL*, vol. 3, no. 2, p. 1800297, doi: 10.1002/solr.201800297.
[17] M. F. & M. C. C. Montes, L. Ocaña, S. González-Pérez, B. González-Díaz, "Developing an Agglomerate of Graphite and Black Carbon in an Ethylene-Vinyl Acetate in Toluene Solution for Producing Electrodes for HTM-Free Perovskite Solar Cells," in *EU PVSEC Proceedings*, 2020, pp. 667–672. doi: 10.4229/EUPVSEC20202020-3BV.1.27.
[18] Z. Yu *et al.*, "Stable Organic-Inorganic Perovskite Solar Cells without Hole-Conductor Layer Achieved via Cell Structure Design and Contact Engineering," *Adv. Funct. Mater.*, vol. 26, no. 27, pp. 4866–4873, Jul. 2016, doi: 10.1002/adfm.201504564.
[19] C. Montes, L. Ocaña, S. González-Pérez, B. González-Díaz, and E. Llarena, "Exploring the use of methylammonium iodide to improve the uniformity of the MAPBI3 layer in HTM-free perovskite solar cells equipped with back contacts based on pastes of graphite and black carbon in a polyvinylidene fluoride in toluene solution," in *38th European Photovoltaic Solar Energy Conference and Exhibition*, 2021, pp. 398–402. doi: 10.4229/EUPVSEC20212021-3BV.1.2.
[20] Z. Hawash, S. R. Raga, D. Y. Son, L. K. Ono, N. G. Park, and Y. Qi, "Interfacial Modification of Perovskite Solar Cells Using an Ultrathin MAI Layer Leads to Enhanced Energy Level Alignment, Efficiencies, and Reproducibility," *J. Phys. Chem. Lett.*, vol. 8, no. 17, pp. 3947–3953, Sep. 2017, doi: 10.1021/ACS.JPCLETT.7B01508/ASSET/IMAGES/LARGE/JZ-2017-015085_0003.JPEG.
[21] W. Lee *et al.*, "Controllable deposition of organic metal halide perovskite films with wafer-scale uniformity by single source flash evaporation," *Sci. Reports 2020 101*, vol. 10, no. 1, pp. 1–10, Nov. 2020, doi: 10.1038/s41598-020-75764-5.
[22] S. Pathak *et al.*, "Atmospheric influence upon crystallization and electronic disorder and its impact on the photophysical properties of organic-inorganic perovskite solar cells," *ACS Nano*, vol. 9, no. 3, pp. 2311–2320, 2015, doi: 10.1021/nn506465n.
[23] P. Vivo *et al.*, "Influence of TiO2 compact layer precursor on the performance of perovskite solar cells," *Org. Electron.*, vol. 41, pp. 287–293, Feb. 2017, doi: 10.1016/J.ORGEL.2016.11.017.
[24] C. Zhao, B. Chen, X. Qiao, L. Luan, K. Lu, and B. Hu, "Revealing Underlying Processes Involved in Light Soaking Effects and Hysteresis Phenomena in Perovskite Solar Cells," *Adv. Energy Mater.*, vol. 5, no. 14, Jul. 2015, doi: 10.1002/aenm.201500279.
[25] W. Tress *et al.*, "Interpretation and evolution of open-circuit voltage, recombination, ideality factor and subgap defect states during reversible

light-soaking and irreversible degradation of perovskite solar cells," *Energy Environ. Sci.*, vol. 11, no. 1, pp. 151–165, Jan. 2018, doi: 10.1039/c7ee02415k.

[26] C. L. Watts *et al.*, "Light soaking in metal halide perovskites studied via steady-state microwave conductivity," *Commun. Phys.*, vol. 3, no. 1, pp. 1–10, Apr. 2020, doi: 10.1038/s42005-020-0350-2.

[27] M. Diantoro, T. Suprayogi, A. Hidayat, A. Taufiq, A. Fuad, and R. Suryana, "Shockley's equation fit analyses for solar cell parameters from I-V curves," *Int. J. Photoenergy*, vol. 2018, 2018, doi: 10.1155/2018/9214820.

[28] P. H. Huang, Y. H. Wang, J. C. Ke, and C. J. Huang, "The effect of solvents on the performance of CH3NH3PbI3 perovskite solar cells," *Energies*, vol. 10, no. 5, pp. 1–8, 2017, doi: 10.3390/en10050599.

[29] D. Liu, M. K. Gangishetty, and T. L. Kelly, "Effect of CH3NH3PbI3 thickness on device efficiency in planar heterojunction perovskite solar cells," *J. Mater. Chem. A*, vol. 2, no. 46, pp. 19873–19881, 2014, doi: 10.1039/c4ta02637c.

[30] C. Bracher, B. G. Freestone, D. K. Mohamad, J. A. Smith, and D. G. Lidzey, "Degradation of inverted architecture CH3NH3PbI3-xClx perovskite solar cells due to trapped moisture," *Energy Sci. Eng.*, vol. 6, no. 1, pp. 35–46, 2018, doi: 10.1002/ese3.180.

[31] K. Nishi, T. Oku, T. Kishimoto, N. Ueoka, and A. Suzuki, "Photovoltaic Characteristics of CH3NH3PbI3 Perovskite Solar Cells Added with Ethylammonium Bromide and Formamidinium Iodide," *Coatings*, vol. 10, no. 4, 2020, doi: 10.3390/coatings10040410.

[32] T. Oku, Y. Ohishi, and A. Suzuki, "Effects of SbBr3 addition to CH3NH3PbI3 solar cells," *AIP Conf. Proc.*, vol. 1807, no. January, 2017, doi: 10.1063/1.4974789.

[33] J. J. Choi, X. Yang, Z. M. Norman, S. J. L. Billinge, and J. S. Owen, "Structure of methylammonium lead iodide within mesoporous titanium dioxide: Active material in high-performance perovskite solar cells," *Nano Lett.*, vol. 14, no. 1, pp. 127–133, Jan. 2014, doi: 10.1021/NL403514X/SUPPL_FILE/NL403514X_SI_001.PDF.

[34] R. Singh, P. K. Singh, B. Bhattacharya, and H. W. Rhee, "Review of current progress in inorganic hole-transport materials for perovskite solar cells," *Applied Materials Today*, vol. 14. Elsevier, pp. 175–200, Mar. 01, 2019. doi: 10.1016/j.apmt.2018.12.011.

[35] K. Yamamoto *et al.*, "Degradation Mechanism of Planar Heterojunction Perovskite Solar Cells for High Stability," *{ECS} Meet. Abstr.*, vol. {MA}2016-02, no. 15, p. 1398, Sep. 2016, doi: 10.1149/ma2016-02/15/1398.

[36] H. J. Snaith *et al.*, "Anomalous hysteresis in perovskite solar cells," *J. Phys. Chem. Lett.*, 2014, doi: 10.1021/jz500113x.

[37] E. L. Unger *et al.*, "Hysteresis and transient behavior in current-voltage measurements of hybrid-perovskite absorber solar cells," *Energy Environ. Sci.*, vol. 7, no. 11, pp. 3690–3698, Nov. 2014, doi: 10.1039/C4EE02465F.

[38] Y. Yuan and J. Huang, "Ion Migration in Organometal Trihalide Perovskite and Its Impact on Photovoltaic Efficiency and Stability," *Acc. Chem. Res.*, vol. 49, no. 2, pp. 286–293, Feb. 2016, doi: 10.1021/ACS.ACCOUNTS.5B00420/ASSET/IMAGES/MEDIUM/AR-2015-00420D_0003.GIF.

[39] Z. Yang *et al.*, "Effects of formamidinium and bromide ion substitution in methylammonium lead triiodide toward high-performance perovskite solar cells," *Nano Energy*, vol. 22, pp. 328–337, Apr. 2016, doi: 10.1016/J.NANOEN.2016.02.033.

[40] M. Saliba *et al.*, "Incorporation of rubidium cations into perovskite solar cells improves photovoltaic performance," *Science (80-.).*, vol. 354, no. 6309, pp. 206–209, Oct. 2016, doi: 10.1126/SCIENCE.AAH5557/SUPPL_FILE/SALIBA.SM.PDF.

[41] M. Saliba *et al.*, "Cesium-containing triple cation perovskite solar cells: improved stability, reproducibility and high efficiency," vol. 9, no. 6, pp. 1989–1997, Jun. 2016, doi: 10.1039/C5EE03874J.

INVESTIGATION ON PEROVSKITE SOLAR CELLS STABILITY THROUGH IMPEDANCE SPECTROSCOPY AND DARK CURRENT- VOLTAGE CHARACTERIZATIONS

Ilaria Matacena*[1], Laura Lancellotti[2], Pierluigi Guerriero[1], Brigida Alfano[2], Eugenia Bobeico[2], Antonella De Maria[2], Vera La Ferrara[2], Lucia V. Mercaldo[2], Mara Miglietta[2], Tiziana Polichetti[2], Gabriella Rametta[2], Gennaro V. Sannino[2], Paola Delli Veneri[2], Santolo Daliento[1]
[1]University of Naples Federico II, Naples, Italy,
[2]ENEA Portici Research Center, Portici, Italy

ABSTRACT: The aim of this work is the study of the degradation of two different PSCs. A PSC with undoped SnO_2 ETL was compared to a PSC with ETL made of G-SnO_2. Degradation was monitored by performing both impedance spectroscopy and dark I-V measurements. Experimental data were collected just immediately after fabrication and after one week. It was observed that the degradation of the solar cells affects both the AC behavior of the devices, modifying associated Nyquist plots, and the DC behavior, observable from dark current-voltage curves. From the Nyquist plots a quantitative analysis was performed by comparing the parameters of a proper equivalent circuit model. Results were coherent with those achieved with DC analysis, thus providing that the arising from these two techniques allows to gain a deeper knowledge of the degradation phenomena.

Keywords: perovskite, impedance spectroscopy.

1 INTRODUCTION

Interest in perovskite solar cells (PSCs) has continuously grown in the last decade due to the sharp increase in the efficiency of this new type of solar cells. A record efficiency of 25.5% has been achieved in 2021 [1]. Such impressive result, achieved in few years, has attested to this new emerging class of solar cells as a protagonist of the next generation of photovoltaics.

In spite of rapid progresses, there are some challenges to overcome. Among others, the degradation of the perovskite is of foremost importance since it affects the stability of the solar cell. Some works attribute material modifications to external causes, such as humidity and heat exposure, some others relate them to internal phenomena arising in the material due to its chemical composition. Surely, a relevant effect on material stability is given by the quality of the interface between HTL (hole transport layer)/perovskite and ETL (electron transport layer)/perovskite. In particular, ETL are often made of SnO2 [1], deposited through low-temperature procedures [2], such processes may generate traps in the SnO2 [3], which lower the quality of the perovskite/ETL interface [4]. In order to diminish trap concentration graphene has been proposed [5] as material for SnO2 dopant, demonstrating that ETL modification by graphene has the result of enhancing carrier collection and improving solar cell stability, without any storage in low humid air and vacuum which is needed to prevent degradation in the case of undoped SnO2 ETL [6].

In this work, the effect of graphene nanoplatelets addition to SnO_2 on PSC degradation has been investigated. Experiments, performed by means of both impedance spectroscopy analysis [7]-[10] and current-voltage (I-V) dark characteristics, have shown that PSC [6], where ETL is made by undoped SnO_2, suffers from performance degradation just after 1 week from fabrication. Conversely, PSC embedding graphene nanoplatelets in SnO_2 (G-SnO_2) achieves better stability. Performance of PSC with G-SnO_2 are evaluated as fabricated and after one week and compared to PSC with only SnO_2 by means of both impedance spectroscopy analysis and I-V dark characteristics. More in detail, Section II contains experimental setups employed for impedance spectroscopy analysis and dark I-V curves. Results arising from both analyses are also reported in the same section, while conclusions are drawn in Section III.

2 EXPERIMENTAL SETUP AND RESULTS

2.1 Impedance spectroscopy

The Impedance data have been acquired using a Solartron 1260 Impedance Analyzer. Frequency has been swept from 1 MHz to 100 mHz, with a logarithmic sweep. The amplitude of the superimposed AC signal has been set to 100 mV. The DC bias has been changed from -0.2 V to 1 V. Nyquist plots achieved for a DC bias of 0.4 V are shown in Figure 1 (PSC with G-SnO_2 (#1)), and Figure 2 (PSC with undoped SnO_2 (#2)). Each plot shows two curves: the curve acquired just after sample fabrication, (dashed line), and the curve acquired after one week (full line). From Figure 1 can be clearly inferred that the behavior of #1 substantially doesn't modify with time. The bigger circumference corresponds to the medium-low frequency range (in this case such range is [700 Hz, 100 mHz]), while the smaller circle is associated to the high frequency range, a magnified view of the high frequency arc is given in Figure 3. Nyquist plots can be interpreted through an equivalent circuit configuration. More in detail, each arc can be associated to the parallel of a resistance and a capacitance (RC pair), while the shift on the x-axis from the origin is equal to the series resistance. This means that in the case of Figure 1, the equivalent circuit (inset in Figure 1) consists of two RC pairs and a series resistance. Both these arcs keep the same behavior with time, indicating that the performance of #1 didn't change in the period considered. Similar analysis has been applied to #2. Figure 2 shows the Nyquist plots corresponding to #2 measured just after fabrication and one week after. Also in this case, it is possible to recognize two different arcs in the impedance plot, which means that also in this case the equivalent circuit associated to Nyquist plots consists of two RC pairs and a series resistance, as shown in the inset in Figure 2.

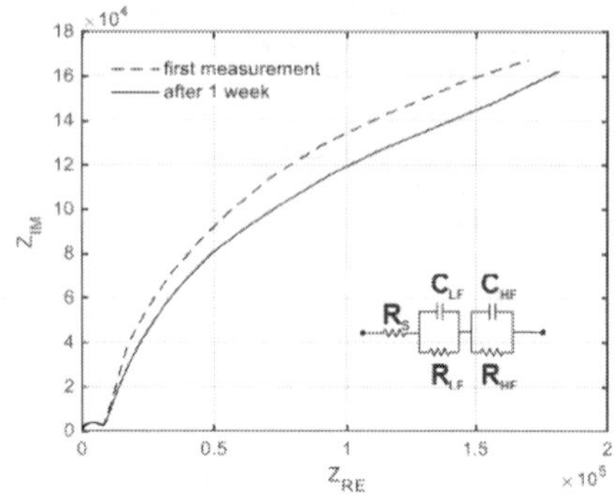

Figure 1: Nyquist plot obtained for #1(PSC with G-SnO_2) with a DC bias of 0.4 V, after fabrication (dashed line) and after 1 week (full line).

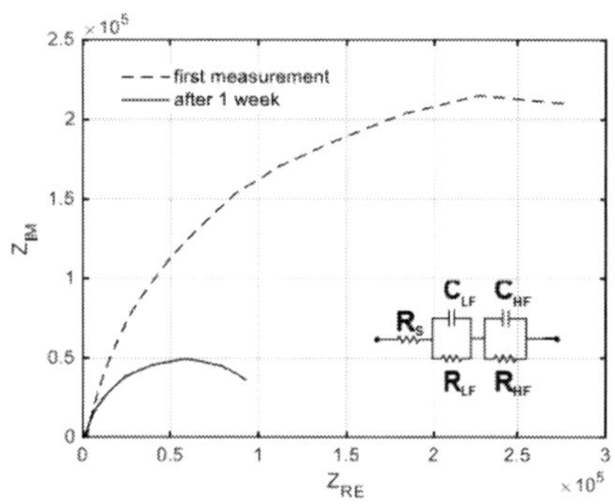

Figure 2: Nyquist plot obtained for #2(PSC with undoped SnO_2), with a DC bias of 0.4 V, after fabrication (dashed line) and after 1 week (full line).

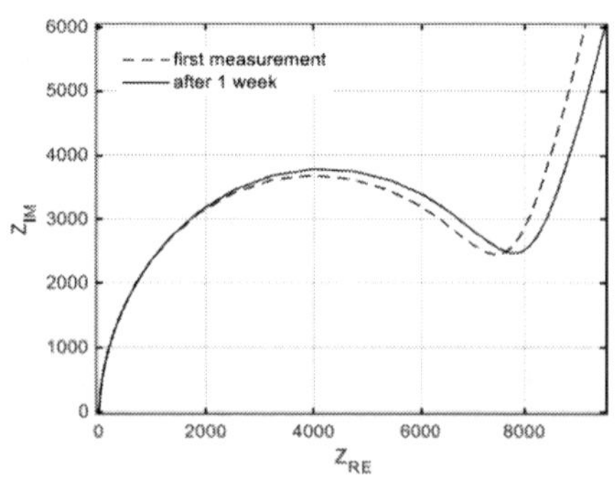

Figure 3: Zoom of Nyquist plot of Figure 1 at high frequencies.

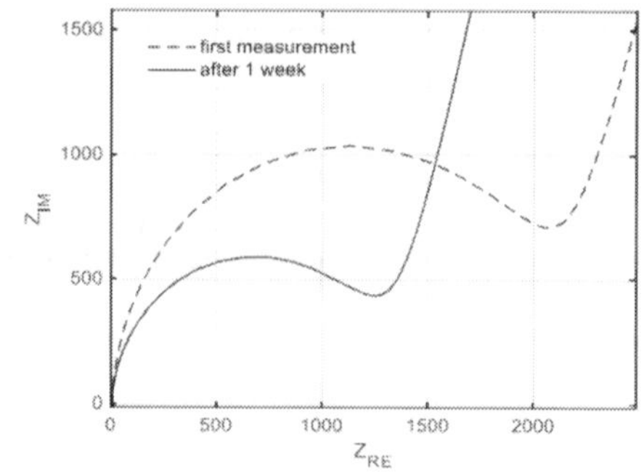

Figure 4: Zoom of Nyquist plot of Figure 2 at high frequencies.

Unlike the previous case, the behavior of Nyquist plots with time appears clearly different. The medium-low frequency arc radius strongly reduces with time. This reduction is also appreciable in the high frequency (Figure 4). Such behavior suggests that performance of #2 considerably modifies with time. Each RC pair is identified by a time constant, τ, related to the physical mechanisms arising in the device. This means that two time constants can be found from the experimental data, one is associated to phenomena taking place in the high frequency region, τ_{HF}, while the other is related to physics in the low frequency region, τ_{LF}. The time constants can be calculated as:

$$\tau_{HF} = R_{HF} \cdot C_{HF} \tag{0.1}$$

and

$$\tau_{LF} = R_{LF} \cdot C_{LF} \tag{0.2}$$

In the point where Z_{IM} is maximum, the following equation can be written:

$$\tau \cdot \omega = 1 \tag{0.3}$$

where $\omega = 2\pi f_{MAX}$

Hence in the two frequency ranges, equation (0.4) can be translated in:

$$\tau_{HF} \cdot \omega_{HF} = 1$$
$$\tau_{LF} \cdot \omega_{LF} = 1 \tag{0.5}$$

Time constants calculated for both PSCs are presented in Table 1. From this analysis it is possible to assert that τ_{LF} and τ_{HF} appear unchanged for #1, while they noticeably reduce for #2. As R_{HF} reduces after one week, τ_{HF} also changes and, more in detail, it becomes lower. Same consideration can be done about τ_{LF}. This seems to suggest that recombination mechanisms existing both in low and high frequency range become more evident with time.

Table 1: Time constants obtained from impedance spectroscopy.

	Sample #1	*Sample #2*
	first measurement	
τ_{LF}_[s]	2.51	8
τ_{HF}_[s]	$1 \cdot 10^{-3}$	$4 \cdot 10^{-4}$
	after 1week	
τ_{LF}_[s]	2.51	1
τ_{HF}_[s]	$1 \cdot 10^{-3}$	$1.6 \cdot 10^{-4}$

2.2 Current-voltage curves

Dark I-V curves have been measured using a Keithley 236 Source Measure Unit. Solar cell under test has been placed in a darkened box which was electrically connected to the multimeter. Voltage has been swept from - 0.5 V to 1.5 V with a step of 5 mV. Each point has been acquired after a time of 300 ms. Resulting I-V curves are shown for #1 and for #2 in the form of semilogarithmic I-V plots in Figure 5 and Figure 6, respectively. For each sample, the I-V curves have been measured as fabricated and after 1 week of ageing. Typically, in low forward bias, current curve has a shape which is mainly due to shunt resistance and recombination mechanisms. On the other side, in the high voltage range, current is affected by series resistance. Comparing I-V curves of both solar cells acquired in different time, it is possible to assert that solar cell embedding G-SnO_2 ETL has more stable I-V characteristics with respect to the PSC with undoped SnO_2. More in detail, the standard solar cell (#2) I-V curve modifies with time, especially in the low voltage range, Figure 6. These two curves tend to overlie when voltage increases, overlapping for a voltage greater than 0.9 V. This behavior suggests that only phenomena visible in the low voltage range changed with time, while all the other mechanisms remained unvaried. As preannounced, the predominant factors occurring in the low voltage range consist in the shunt resistance and recombination processes. Shunt resistance can be easily detected with a symmetric slope around zero voltage. Since I-V curves behavior is not completely symmetric for all the cases considered, then it is possible to assert that recombination mechanisms are dominating in the standard solar cell.

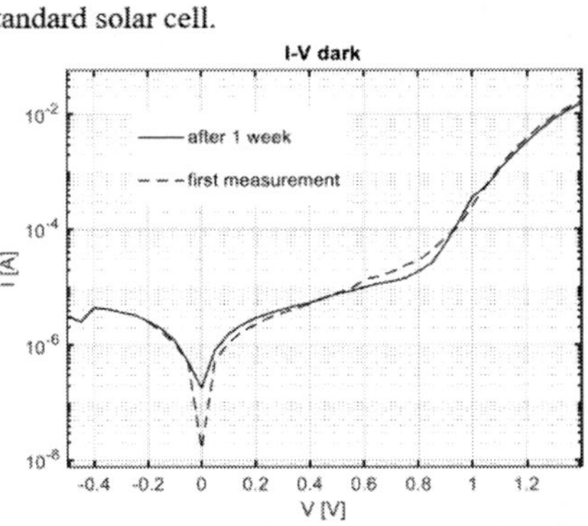

Figure 5: I-V curve of the # 1 (perovskite solar cell with G-SnO_2).

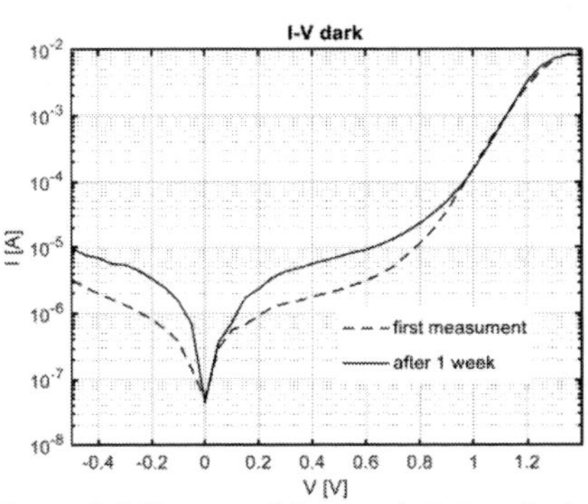

Figure 6: I-V curve of the sample 2 (standard perovskite solar cell).

3 CONCLUSIONS

The innovation of this work is twofold. The first aspect consists in the validation of the beneficial effects of graphene nanoplatelets on SnO_2 ETL in terms of PSCs degradation. In addition, an innovative approach to characterize the behavior of PSCs over time, exploiting both impedance spectroscopy (AC technique) and dark I-V curves (DC analysis) is proposed. Data coming from AC and DC analysis suggest that embedding graphene nanoplatelets in SnO_2 has a beneficial effect on the degradation of PSCs, inhibiting the increase in recombination mechanisms at the interface perovskite/ETL.

4 REFERENCES

[1] Green, Martin A., et al. "Solar cell efficiency tables (Version 58)." *Progress in Photovoltaics: Research and Applications* 29.7 (2021): 657-667.

[2] Bai, Yang, et al. "Low temperature solution-processed Sb: SnO2 nanocrystals for efficient planar perovskite solar cells." *ChemSusChem* 9.18 (2016): 2686-2691.

[3] Godinho, Kate G., Aron Walsh, and Graeme W. Watson. "Energetic and electronic structure analysis of intrinsic defects in SnO2." *The Journal of Physical Chemistry C* 113.1 (2009): 439-448

[4] Zuo, Lijian, et al. "Tailoring the interfacial chemical interaction for high-efficiency perovskite solar cells." *Nano letters* 17.1 (2017): 269-275.

[5] Dai, Xinchen, et al. "Focussed Review of Utilization of Graphene-Based Materials in Electron Transport Layer in Halide Perovskite Solar Cells: Materials-Based Issues." *Energies* 13.23 (2020): 6335.

[6] La Ferrara, Vera, et al. "The effect of storage cycle on improvement in the photovoltaic parameters of planar triple cation perovskite solar cells." *Materials Advances* 2.16 (2021): 5396-5405.

[7] Suresh, M. S. "Measurement of solar cell parameters using impedance spectroscopy." *Solar Energy Materials and Solar Cells* 43.1 (1996): 21-28.

[8] Pascoe, Alexander R., et al. "Insights into planar CH3NH3PbI3 perovskite solar cells using impedance spectroscopy." *The Journal of Physical Chemistry C* 119.9 (2015): 4444-4453.

[9] Matacena, Ilaria, et al. "Impedance spectroscopy for the characterization of the all-carbon graphene-based solar cell." Energies 13.8 (2020): 1908.

[10] Adachi, Motonari, et al. "Determination of parameters of electron transport in dye-sensitized solar cells using electrochemical impedance spectroscopy." The Journal of Physical Chemistry B 110.28 (2006): 13872-13880.

STABILITY ASSESSMENT OF PEROVSKITE CELLS WITH H3PP MOLECULE ADDITIVES AT DIFFERENT LOADING CONDITIONS

V. Paraskeva[1], M. Hadjipanayi[1], M. Norton[1], S. C. Hayes[2], N. Charalambous[2], C. Pereyra[3], M. Lira[3], G.E. Georghiou[1]
[1]FOSS Research Centre for Sustainable Energy, Department of Electrical and Computer Engineering, University of Cyprus, 75 Kallipoleos St., Nicosia, 1678, Cyprus
[2] Molecular Spectroscopy Laboratory, Department of Chemistry, University of Cyprus, University of Cyprus, 1 University Ave, 2109, Nicosia, Cyprus
[3]Catalan Institute of Nanoscience and Nanotechnology (ICN2), CSIC and the Barcelona Institute of Science and Technology (BIST), Building ICN2, Campus UAB E-08193, Bellaterra, Barcelona, Spain

ABSTRACT: Additive engineering in perovskite devices has been explored in the last few years towards enhancing photovoltaic performance and stability. Control of the morphology, grain boundary, grain size, charge recombination and density of defect states in the perovskite layer are important for the optimization of the perovskite-based devices. In this work, we study the stability testing of perovskite cells with and without 3-phosphonopropionic acid (H3pp) additives in the active perovskite layer at different loading conditions in an attempt to detect differences in their photo-current and External Quantum Efficiency (EQE) spectrum due to the different loading conditions. Raman spectroscopy and spatially-resolved Electroluminescence (EL) techniques were applied before and after degradation of the samples to detect differences in their molecular composition and defect evolution respectively. Measurements revealed the higher stability of the perovskite with H3pp additives and the larger impact of loading conditions in those cells. EQE measurements established higher reduction of the EQE values in the control cells (cells without additives). Finally, spatially-resolved EL measurements indicated larger recombination losses in control cells while Raman measurements demonstrated changes in the Formamidinium (FA) cation composition in the control cells.
Keywords: Characterization, Perovskite, Stability, Electroluminescence, Degradation

1 INTRODUCTION

In less than a decade, perovskite technology has emerged with immense promise as potentially the cheapest alternative to the present commercially available photovoltaic technologies. However, despite an expeditious rise in their efficiency perovskite-based devices are still far from commercialization because of inadequate long-term stability. Defect density at the surface and grain boundaries, significantly affect both operational stability and power conversion efficiency (PCE) of perovskites. The introduction of additives into the halide perovskite layer is a method that has been effectively applied to passivate defects and to increase the efficiency of the solar cell [1]. One of the organic molecules used recently to improve the stability of perovskites is the 3-phosphonopropionic acid (H3pp) additive. Stability analysis of perovskites with H3pp showed negligible loss of the initial efficiency compared to their counterparts that lack additives in the perovskite layer [2].

In this work, we study the stability of perovskite cells with and without H3pp additives in the active perovskite layer at different loading conditions. Short-circuit current losses in both types of cells (with H3pp and without) were studied by applying open-circuit voltage (Voc) and maximum power point (MPP) loading conditions between IV scans during stability testing. Loading condition is an important parameter to access the operational stability of the perovskite solar cells. Several studies in the past have shown the impact of loading during stability testing of perovskites [3]. Most of the references address the importance of the MPP load during stability testing as this result to the least degradation of the perovskite devices. Voc load was found to cause severe degradation mainly because of the fill factor and open-circuit voltage losses. Bias loading conditions were found also to affect the performance recovery mechanisms in the perovskite devices [4]. Absence of load was found to reduce the reversible processes in the devices.

Apart from the performance measurements external quantum efficiency (EQE) measurements have been implemented before and after stability testing in order to establish the change of the EQE due to light soaking of the samples at the different loading conditions. Raman spectroscopy measurements were conducted before and after degradation in order to detect changes in the molecular composition of the samples. Finally, this work studies long-term trends in defect evolution by comparing spatially-resolved electroluminescence (EL) images taken before and after stability testing. Scanning Electron Microscope (SEM) method was applied in the samples after degradation for microscopic characterization of the perovskite layer. This method will establish vital information regarding the chemical composition and crystalline structure of the samples under test.

2 EXPERIMENTAL APPROACH

The two different types of perovskite cells were tested under continuous irradiation conditions at 1000 W/m^2, 50% relative humidity air and 25°C for two hours. A scan rate of 25mV/sec was performed in all measurements. Stability testing was performed at different loading conditions between scans (i.e. Voc and MPP loading) while Raman measurements, EL images and EQE spectra of the devices were collected before and after the two hours of stability testing.

Alongside the collection of IV traces, spatially-resolved EL measurements were taken from the cells before and after light soaking to look for changes occurring in EL images after performance degradation thus enabling the study of defect evolution at the two different types of cells. A high-performance CCD matrix camera at -5°C was used during the EL studies. Dedicated software was utilized to collect the EL images at the same testing parameters for comparable results.

Furthermore, EQE measurements have been implemented before and after light soaking at the two different types of cells in order to report the changes in the EQE spectrum and thus short-circuit current due to light application at different loading conditions. The EQE measurement set-up consists of a steady-state Quartz-Tungsten-Halogen light source in series with a monochromator to produce the monochromatic light input, which is then chopped by a chopper, superimposed over continuous bias light and measured by digital lock-in-amplifiers.

Raman spectroscopy technique was applied to detect differences in the vibrational modes of molecules within the two types of perovskites. A custom-made Raman microscope (Raman AlphaSource RLS4032, NanoBioAnalytics) was employed for resonant excitation of the perovskite cells at 532 nm. All measurements were acquired under air in standard conditions, employing low excitation powers (10 mW) to minimise sample degradation.

SEM cross-section of the degraded samples were collected with the system FEI Quanta 650F ESEM. Electrons were accelarated at 10kV at a distance of 10mm and using secondary electron detector. Electrons were accelarated at 10kV at a distance of 10 mm.

Two different types of cells have been used. The one type of cell includes H3pp molecule additives in the halide perovskite layer. Both types have the configuration FTO/c-TiO2/m-TiO2/HP/Spiro-OMeTAD/Au based on the multication perovskite absorber with the nominal formula $Rb_{0.05}Cs_{0.05}MA_{0.15}FA_{0.75}Pb_{1.05}(I_{0.95}Br_{0.05})_3$ (RbCsMAFA). All the devices are un-encapsulated and present similar performance with open-circuit voltage at around 1.1V and efficiency 18-19%. The short-circuit current density is around 22 mA/cm^2. The dimensions of the samples are 1.5 cm x 2.5 cm while the active perovskite region has dimensions of 0.5 cm x 0.5 cm. A SEM cross-sectional image of the devices with H3pp molecule additives under study is shown in Figure 1.

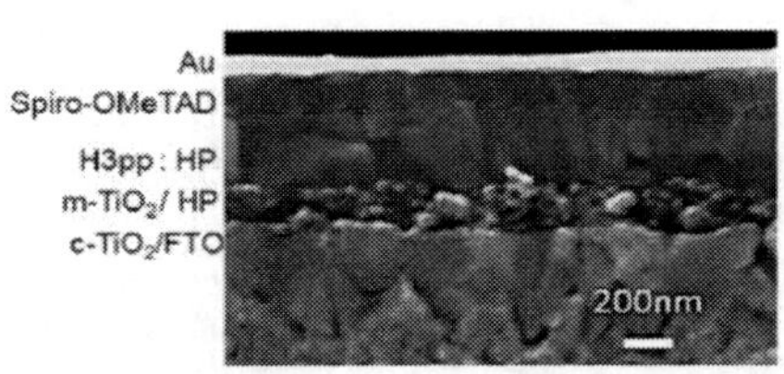

Figure 1: SEM cross-sectional image of the perovskite devices with the H3pp: HP material as the absorber.

3 RESULTS AND DISCUSSION

3.1 Stability testing

Stability testing was applied in 4 devices under test and lasted for two hours. During the light soaking period, the current-voltage curves of the devices were collected every 10 minutes. In two of the samples, MPP load was utilized between IV curves while in the remaining two devices Voc load was utilized. The samples without H3pp molecules in the perovskite layer are labeled as *control* samples. The evolution of the short-circuit current of the four samples under test over time is depicted in Figure 2.

The current losses were calculated for all the 4 samples under test. Short-circuit current measurements at the beginning and at the end of the light soaking were taken into consideration for this calculation. The data are provided in Table 1. Table 1 demonstrates that in both open-circuit and MPP loading conditions the control cells, which lack H3pp molecules in the halide perovskite layer, present larger current losses compared to their counterparts with H3pp additive molecules. While similar current losses are obtained in cells with and without H3pp additives at open-circuit loading conditions, there is a significant difference between current losses of both types of cells at MPP loading conditions. At MPP loading conditions cells with H3pp molecules present very low current losses compared to the case where open-circuit voltage load is applied giving evidence of the larger impact of loading conditions in cells with molecular additives.

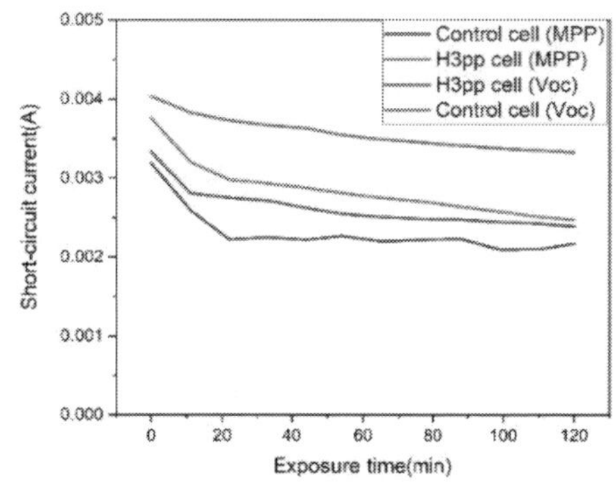

Figure 2: Short-circuit current stability curve for all devices under test.

	Control sample	H3pp sample	Control sample	H3pp sample
Voc load			34.5%	28.4%
MPP load	32.2%	17.6%		

Table 1:Short-circuit current losses of 4 devices under test at different loading conditions after two hours of stability assessment.

3.2 EQE measurements

EQE measurements were implemented before and after stability testing of the devices. The EQE results from samples before and after stability testing at MPP loading conditions are depicted in Figure 3. EQE curves after light soaking demonstrate more reduction of the EQE values at the near-infrared (NIR) region of the spectrum in agreement with previous works [5]. This is obtained in cells tested with both loading techniques (Voc & MPP). In addition, cells with H3pp molecule additives present enhanced EQE values after stability testing at low wavelengths (<550 nm). EQE losses due to stability testing (at MPP conditions) were then calculated for the control sample and for the sample with H3pp additives (Figure 4).

EQE losses up to 31% were found in the control sample in the NIR part of the spectrum. On the other hand, EQE losses in the sample with additives were significantly lower over the entire response region of the cells. EQE losses up to 15% were detected in this type of cells. Both types of cells present higher EQE losses in the NIR.

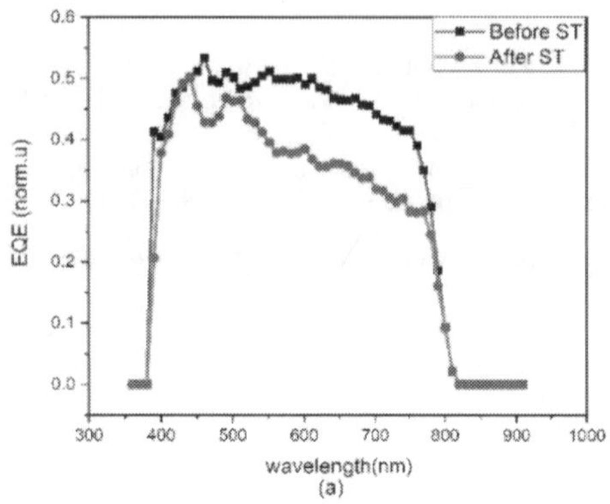

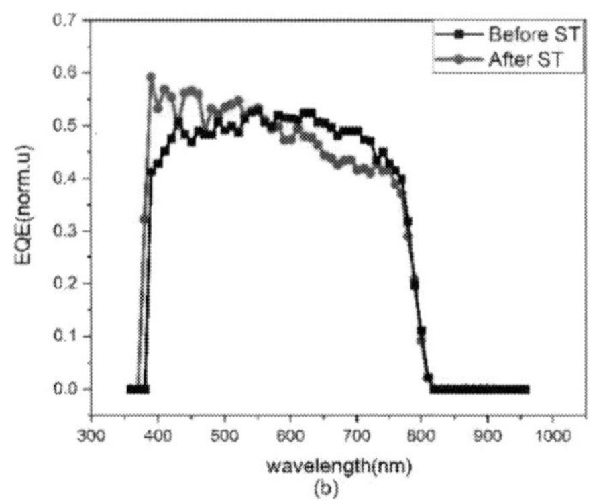

Figure 3: EQE curves before and after stability testing at MPP loading conditions for (a) a control sample and (b) a sample with H3pp molecule additives.

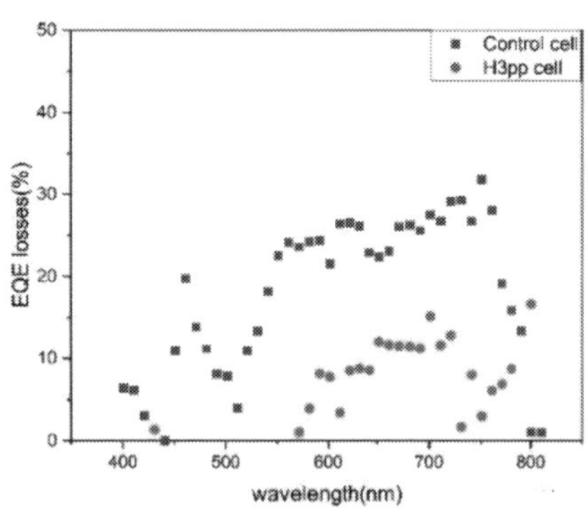

Figure 4: Spectrally-resolved EQE losses for a control sample and for a sample with H3pp molecule additives due to stability testing at MPP loading conditions.

3.3 Resonance Raman spectroscopy

Resonance Raman spectroscopy method was applied on the samples before and after stability testing to detect differences in the vibrational modes of molecules within the two types of perovskites. The Raman spectrum for a control sample and for a sample with H3pp additives before and after stability testing at MPP load is depicted in Figure 5. Figure 5 demonstrates the Raman spectrum in the region of perovskite at around 250-280 cm^{-1}. The Raman peak at 250 cm^{-1} is associated with the Formamidinium (FA) based compounds and particularly is assigned to the symmetric and asymmetric out-of-plane bending of H_2N-C-NH_2 [6]. Absence of the Raman peak at 250 cm^{-1} is obtained in the degraded control sample after stability testing indicating FA cation degradation. In contrast, the respective peak is still apparent in the case of the perovskite sample with additives.

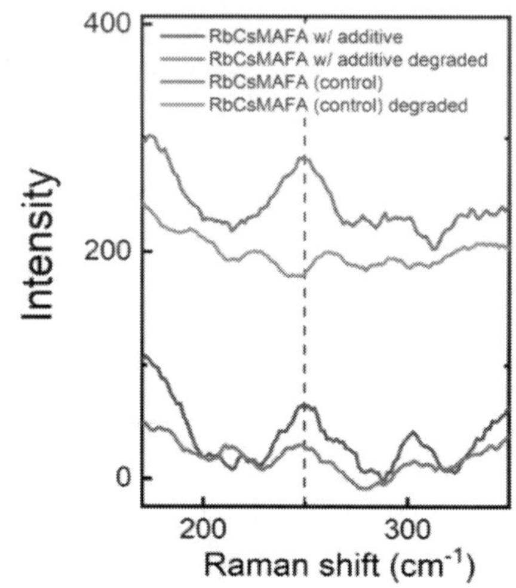

Figure 5: Raman spectrum of a perovskite cell with H3pp molecule additives and of a control module before and after stability testing in the region of perovskite.

3.4 Spatially-Resolved Electroluminescence (EL)

For the investigation of trends in radiative emission and defect evolution, spatially-resolved EL imaging was used prior and after stability testing of the samples of different structure. Same bias current was applied in the devices for comparison purposes. The results are depicted in Figures 6 and 7 respectively. Similar EL image patterns were obtained in the two cells prior stability testing. Larger dark areas were obtained in the control cells after light soaking indicating larger losses in radiative recombination in those devices. The results are in agreement with the current losses, EQE and Raman spectroscopy studies.

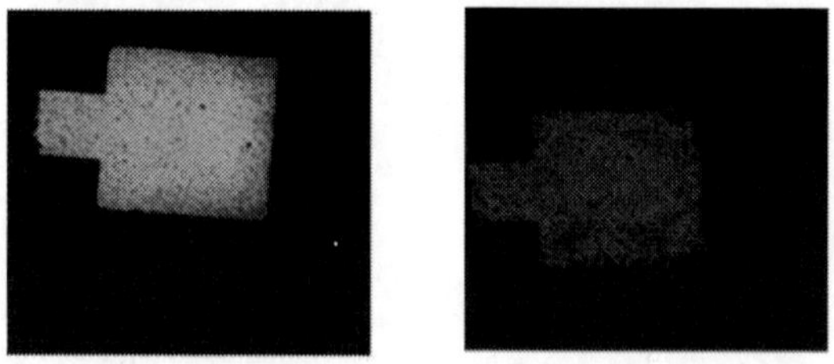

Figure 6: Spatially-resolved EL images before (left) and after (right) two hours of stability testing for a sample with H3pp molecule additives in the perovskite layer.

Figure 7: Spatially-resolved EL images before (left) and after (right) two hours of stability testing for a control sample.

3.4 Scanning Electron Microscopy (SEM)

Scanning Electron Microscopy technique was finally applied to the degraded samples for their microscopic characterization. Different images have been taken alongside the film to show different spots of the cell. This is because one SEM image is not enough to detect degradation from the sample. SEM images of two degraded samples with H3pp molecule additives that have been tested at different loading conditions are obtained in Figure 8. In the images all the layers (FTO: Meso-TiO2: Perovskite: Spiro; gold) in the cell configuration can be identified easily.

SEM images of the two identical perovskite samples indicate different perovskite degradation in their active region. Particularly, the sample tested at MPP load presents no perovskite layer segregation while the sample tested at Voc load presents perovskite segregation. This observation gives evidence of the impact of load on the structural characteristics of the cells.

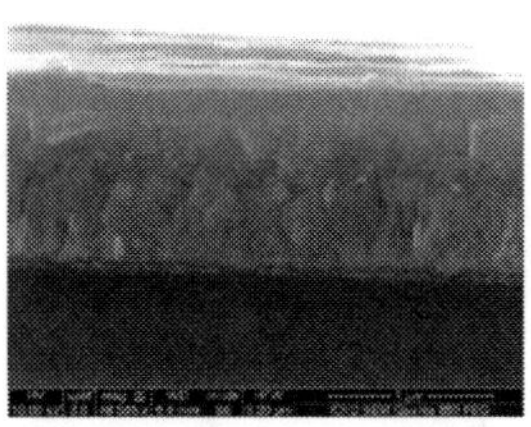

(a)

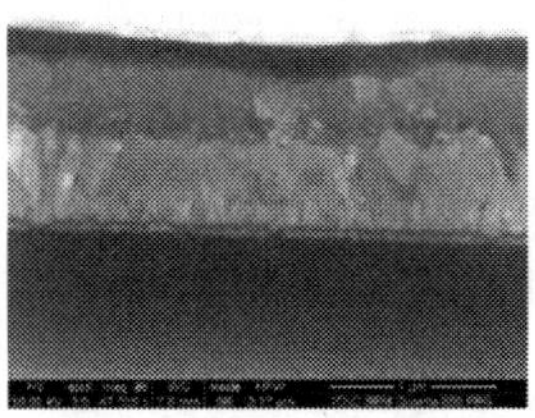

(b)

Figure 8: SEM images of samples with H3pp molecule additives after stability testing at (a) MPP and (b) Voc loading conditions.

4 CONCLUSIONS

In summary, stability testing of samples with and without H3pp molecule additives in the perovskite layer was studied at different loading conditions (Voc and MPP tracking). Results demonstrated the higher stability of the perovskite with H3pp additives and the larger impact of loading conditions in those cells. EQE measurements established higher reduction of the EQE values in the control cells at both loads under study. EQE losses were found to be more severe in the NIR part of the spectrum at all cells under study. Spatially-resolved EL technique was applied on the cells before and after stability testing. EL images indicated larger recombination losses after light soaking in control cells which lack H3pp molecules. Resonance Raman spectroscopy method was applied on the samples before and after light soaking demonstrating changes in the FA cation composition in the control cells. Finally, SEM images after stability testing demonstrated severe perovskite degradation in cells with H3pp molecule additives tested at Voc load.

ACKNOWLEDGMENT

This work was funded by the University of Cyprus through the project 'PERSI-PV' under the ONISILOS funding scheme for post-doctoral researchers.

5 REFERENCES

[1] S. Liu, Y. Guan, Y. Sheng, Y. Hu, Y. Rong, and A. Mei, *Adv. Energy Mater.*, 10, (2019), 1902492.

[2] H. Xie *et al.*, *Joule*, 5, (2021), 1.

[3] K. Domanski, E. A. Alharbi, A. Hagfeldt, M. Grätzel, and W. Tress, *Nat. Energy*, 3, (2018), 61.

[4] M. Prete *et al.*, *Appl. Energy Mater.*, 4, (2021), 6562.

[5] P. Hierrezuelo-Cardet, A. F. Palechor-Ocampo, and J. Caram, *J. Appl. Phys.*,127,(2020), 235501.

[6] J. Ibaceta-jan, H. Mirhosseini, D. Heinrich, M. R. Wagner, T. D. Ku, and B. Szyszka, *Phys. Chem. Chem. Phys.*, 22, (2020), 5604.

ROLE OF ION MIGRATION ON THE S-TYPE J-V CHARACTERISTICS OF PEROVSKITE SOLAR CELLS

Dhyana Sivadas, Pradeep R Nair
Department of Electrical Engineering, Indian Institute of Technology Bombay, India
dhyana@iitb.ac.in, prnair@ee.iitb.ac.in

ABSTRACT: Many reports attribute the origin of S-type characteristics in the thin film solar cells to interface recombination and energy level mismatch between materials. However, along with interface degradation and non-ideal energy levels of materials, ion migration is an unavoidable phenomenon in perovskite solar cells whose impact on the S-type characteristics is not well established. Here, we use a double diode approach to study the formation of S-type characteristics, in which the first diode characteristics is due to the energy level offset, interface recombination and ion distribution. The second one is the conventional diode which represents the drift and diffusion current in the device. Our findings suggest that the deactivation of first diode with applied bias due to the ion redistribution results in the formation of S-type J-V characteristics. Insights from the study also provide theoretical explanation on different shapes of S kink seen in the J-V characteristics using double diode approach.
Keywords: Perovskite, interfaces, ion migration

1 INTRODUCTION

Perovskite solar cells (PSCs) exhibit remarkable power conversion efficiencies of ~25% [1]. However, PSCs face several roadblocks to commercialization. Long-term stability, S-type J-V characteristics, hysteresis are some of the anomalies observed in PSCs [2]–[5]. The formation of the S-type J-V characteristics has been attributed to interface recombination and energy level mismatch between perovskite material and transport layers in PSCs [6]–[11]. However, a critical analysis on the influence of ion migration on the same is lacking. Here, with the help of numerical simulations, we deduce the impact of ion redistribution in the formation of S-type J-V characteristics. Also, a qualitative explanation for these peculiar current characteristics is provided using double diode approach. Insights provided in the study elucidate the trends in J_{SC}, V_{OC} and also correlate the shape of S-type curve with ion density and energy level offset.

2 MODEL SYSTEM

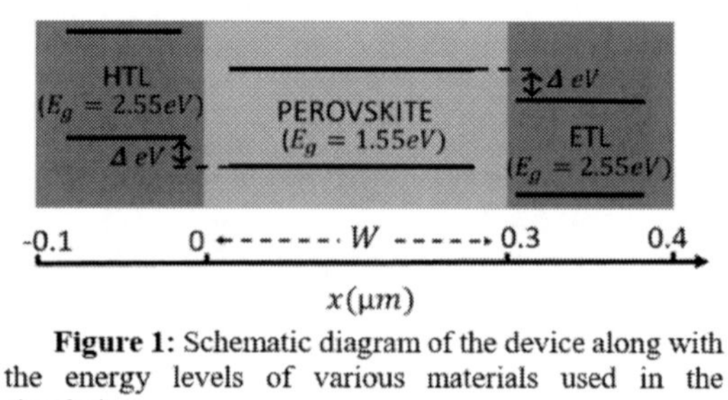

Figure 1: Schematic diagram of the device along with the energy levels of various materials used in the simulation.

In this report, we study the impact of ion migration on the S-type features in the J-V characteristics of perovskite solar cells with the help of numerical simulations. The schematic of the device is shown in Fig. 1. We use a self-consistent drift-diffusion model which solves the Poisson equation (eq. 1), continuity equations for electrons (eq. 2), holes (eq. 3) and ions for analysing the steady state behaviour of perovskite solar cells.

$$\frac{d^2V}{dx^2} = -\frac{q}{\varepsilon}\left(p - n - N_I(x) + N_{dop}\right) \quad (1)$$

$$\frac{\partial n}{\partial x} = \frac{1}{q}\frac{\partial J_n(x)}{\partial x} + G - R \quad (2)$$

$$\frac{\partial p}{\partial x} = -\frac{1}{q}\frac{\partial J_p(x)}{\partial x} + G - R \quad (3)$$

Here, V is the potential, x is the spatial cordinate, q is the charge, ε is the dielectric permittivity, p is the hole density and n is the electron density. N_{dop} denotes doping density of the material (for hole transport layer, HTL, $N_{dop} = -N_A$, for perovskite $N_{dop} = 0$, and for electron transport layer, ETL, $N_{dop} = N_D$) whereas, J_n and J_p indicates the electron and hole current. Generation rate of $G = 5.21 \times 10^{21} cm^{-3}s^{-1}$ is used, which corresponds to a J_{sc} of 25 mA/cm^2 and the recombination of carriers (R) is modelled by SRH, Auger, Radiative recombination[12]. The interface recombination is taken into account via recombination velocity (S_v).

Further, the ionic charge neutrality principle[13]–[15] leads to

$$\int_0^W \left(N_{I,p}(x) - N_{I,n}(x)\right) dx = 0 \quad (4)$$

where, $N_{I,p}$ is positive immobile ions ($N_{I,p}(x) = I_{den}$, with I_{den} is a constant which denotes the average ionic density), $N_{I,n}$ is negative mobile ions[16]–[18] and W is the width of perovskite layer (see Fig. 1). Under steady state conditions, the analytical solution to continuity equation for mobile ions could be written as $N_{I,n}(x) = Ke^{\frac{V(x)}{V_T}}$, (where K is prefactor and V_T is the thermal voltage) using Einstein relation. By substituting these equations of $N_{I,p}$ and $N_{I,n}$ in eq. 4, we would obtain ion distribution equation as [19],

$$I_{den} \times W - \int_0^W Ke^{V(x)/V_T}\, dx = 0 \quad (5)$$

The above-mentioned system of equations (i.e., eqs. 1-3 and eq. 5) is solved using finite difference method with Scharfetter-Gummel discretization scheme [16], [19], [20]. Moreover, the parameters used for the simulations are shown in Table I.

Table I: Parameters used for simulations[19].

Parameters	HTL/ETL	Perovskite
$N_{dop}(cm^{-3})$	10^{18}	0
ε_r	4	24
Nc,Nv(cm^{-3})	10^{20}	10^{19}
μ (cm^2/Vs)	0.012	10
EA (eV)	3/4	4

3 RESULTS AND DISCUSSION

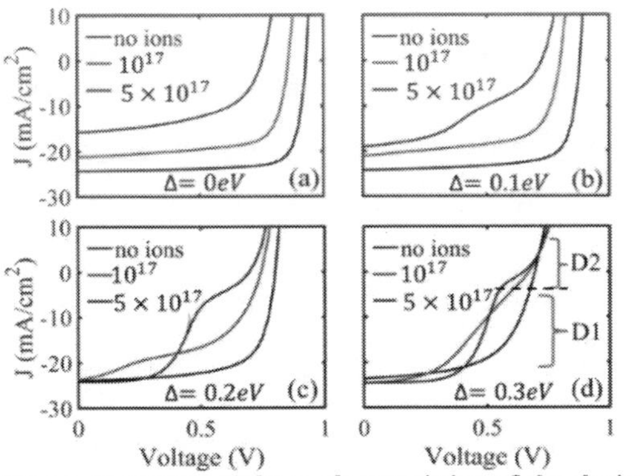

Figure 2: Current-Voltage characteristics of the device with different ion densities (cm^{-3}) with surface recombination velocity (S_v) of 10^5 cm/s for different band offset values (a) $\Delta = 0eV$, (b) $\Delta = 0.1eV$, (c) $\Delta = 0.2eV$, (d) $\Delta = 0.3eV$. (Under uniform generation rate of 5.21×10^{21} cm^{-3}/s).

Figure 2 shows the J-V characteristics of perovskite solar cells under 1 sun illumination with different energetic offsets at ETL ($\Delta_{ETL} = E_{c,per} - E_{c,ETL}$) and HTL ($\Delta_{HTL} = E_{v,HTL} - E_{v,per}$). Here, we assume a symmetric offset condition where $\Delta_{ETL} = \Delta_{HTL} = \Delta$ (see Fig .1). The effect of ion density [13], [21], [22] on these devices is also plotted in the same figure. Note that, significant interface recombination ($S_v = 10^5$ cm/s) is considered for all the devices. Further, the current characteristics for the PIN device generally follow an exponential dependence with bias [1], [23], [24]. However, an S-type characteristics or double diode behaviour is prominently seen for devices with high ion density and offset. For example, the devices devoid of ions do not show evidence of S-type characteristics for any band offset (see black curves in Fig. 2a-d). But, with $\Delta = 0.1eV$, the device with an ion density of $5 \times 10^{17} cm^{-3}$ shows the S-type current characteristics (see blue curve in Fig. 2b), whereas, for $0.2eV$ and $0.3eV$ band offset, double diode behaviour is observed for the device with both low ($1 \times 10^{17} cm^{-3}$) and high ($5 \times 10^{17} cm^{-3}$) ion densities (see blue and red curves in Fig. 2c,d).

Specifically, for the device in the presence of ions and adequate band offset, a rapid decrease of current near lower biases are observed. However, at higher biases, a slow variation of current is seen (for example, see blue curve after $\sim 0.5V$ in Fig. 2c and $\sim 0.55V$ in Fig. 2d). Here, we refer the lower biases characteristics to first diode (D1) and the slow variation of current at higher biases to second diode (D2) characteristics (see, Fig. 2d). Note that, the second diode behaviour appears at a later voltage with an increase in ion density (see red and blue curve in Fig. 2c,d) and offset (see blue curves in Fig. 2b-d).

For more insight, we examine the energy band diagrams and carrier profiles of the device in the presence and absence of ions with an offset of $0.3eV$ (see, Fig. 3). Here, the energy band diagrams Figs. 3 a,c corresponds to no ion case where S-type characteristics are absent and Fig. 3 b,d represents the band diagram for the device with ions which shows the double diode behaviour (see black and blue curves of Fig. 2d). As the devices shown here are having high interface recombination (which indeed is the bottleneck in current extraction), a detailed analysis of interface properties is essential to elucidate the above-mentioned effects.

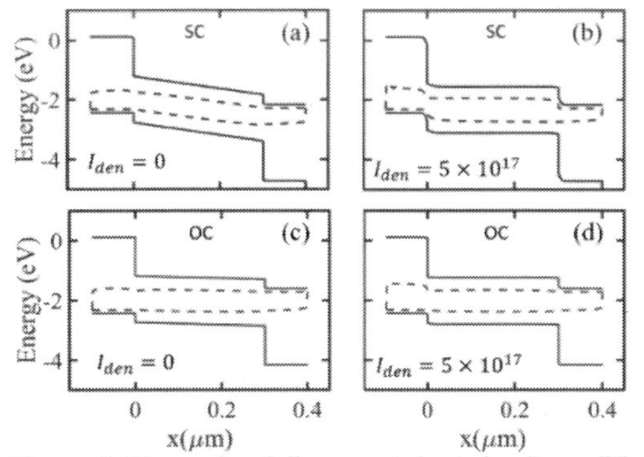

Figure 3: Energy band diagram at short circuit condition for the device with in the (a) absence of ions and (b) in the presence of ions. Energy band diagram at near open circuit condition for the device in the (c) absence of ions and (d) in the presence of ions. All devices have symmetric energy offset, $\Delta = 0.3eV$, generation rate of 5.21×10^{21} cm^{-3}/s and $S_v = 10^5$ cm/s. Ion density of $5 \times 10^{17} cm^{-3}$ is considered for the device in the presence of ions.

The energy alignment mismatches with the transport layer introduce an injection barrier for electrons near ETL and holes near HTL. As a result, the concentration of both carriers near interfaces would be on the lower side as compared to the device with an ideal energy alignment[24] ($\Delta = 0$). For the device with ions, the negative mobile ions accumulate in the perovskite near ETL interface and expose the immobile positive ions in the perovskite near HTL interface. As a result, the electric field inside the active layer is reduced [25], [26] (see, Fig. 3b). The presence of these ions at the interface and the corresponding band bending further reduces the electron and hole concentration near ETL and HTL interface respectively (see Fig.3a,b, ($E_c - E_{fn}$) near ETL interface and ($E_{fp} - E_v$) near HTL interface, which are greater for the device with ions). For significant offset and ionic concentration, the electrons become minority carrier near ETL interface while holes are the minority carriers near the HTL interface. This is in contrast to the device in the absence of ions, where electrons and holes are the majority near ETL and HTL interfaces.

Figure 4a shows the carrier density profile of device in the absence of ions near HTL interface and ETL interface (for the biases where S type transition is seen in blue curve of Fig. 2d). Here, the application of bias for the device without ions results in the increase of both type of carriers. Interestingly, for the device in the presence of ions, the change in bias cause redistribution of mobile ions which significantly influence the band bending near interfaces. As a result, a substantial change in carrier profile is expected (see Fig. 4b). With bias, for the device with appreciable ion density and offset, the electron concentration (minority) near the ETL interface increase and hole concentration decrease due to ion redistribution. Similarly, near HTL interface hole concentration (minority) increases and electron concentration reduces with bias (see Fig. 4b hole variation from dark blue to light blue and electron variation from dark red to light red). This change in the ion concentration with bias prompt drastic change in the minority concentration which in turn results in the first diode (D1) characteristics (see current charcteristics till ~$0.55eV$ in Fig. 2d).

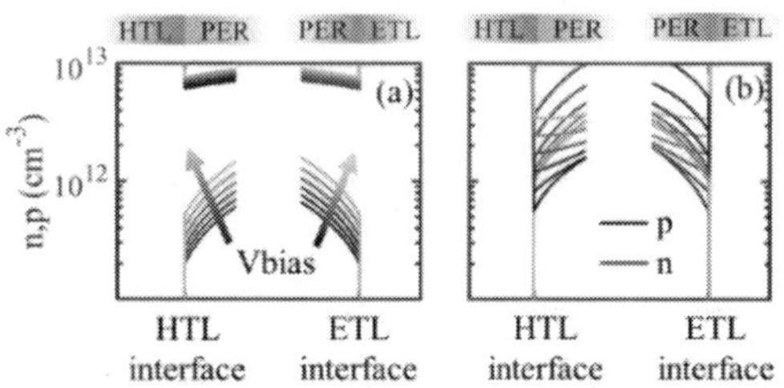

Figure 4: Carrier density profile near HTL interface and near ETL interface for the device in the (a) absence of ions and (b) in the presence of ions (under the application of bias, V_{bias} from $0.55V$ to $0.66V$). All devices have symmetric energy offset, $\Delta = 0.3eV$, generation rate of $5.21 \times 10^{21}\ cm^{-3}/s$ and $S_v = 10^5\ cm/s$. Ion density of $5 \times 10^{17} cm^{-3}$ is considered for the device in the presence of ions.

At medium biases (near V_{bias}~V_{BI}), due to the low ion accumulation (i.e., redistribution) near the interfaces, electrons become majority carriers near ETL interface while the holes are the majority carriers near HTL interface. Under such conditions, the internal electric field is very small and hence the electron and hole densities near the interfaces remain invarient with applied bias. As a result, the current due to diode D1 remains unchanged and the rest of the device J-V characteristics is dictated by diode D2. Hence, the device in the presence of ions shows similar behaviour as the device in the absence of ions at higher bias (see Fig. 3c,d , similar band diagrams at open circuit conditions) and the current characteristics follow diode D2 behavior.

The increase in energetic offset results in an increase in the short circuit current (J_{SC}) for the device in the presence of ions (see blue and red curves in Fig 2a-d). However, the J_{SC} of device in the absence of ions are nearly constant with offset (see black curves in Fig 2a-d) due to the presence of sufficient electric field. As mentioned earlier, with offset the minority carriers (electrons near ETL and holes near HTL) decrease with ion density. This leads to an improvement in the J_{SC} with offset for device with ions ($15 - 24.5\ mA/cm^2$). In addition, the V_{OC} which shows a significant change with offset for device in the absence of ions ($0.92V - 0.69V$) whereas the reduction in V_{OC} for the device with ions are not quiet drastic ($0.75V - 0.65V$ for ion density of $5 \times 10^{17} cm^{-3}$, even though they are on lower side). Here, the device without ions have linear electric field and they reduce directly with a reduction in V_{BI} (reduces with offset [8]). This enables interface recombination and shows a huge variation in V_{OC}. However, for the devices with sufficient ions, bands are flat and a significant recombination is expected irrespective of offset.

4 CONCLUSIONS

In summary, with the help of numerical simulation, we elucidate the influence of ion density and energy alignment mismatch between perovskite and transport layers on the formation of S-type characteristics (or double diode behavior) under significant interface degradation. The findings also qualitatively correlate the variation of V_{OC} and J_{SC} with ion density and energy level offset. Also, we provide a clear explanation for different shapes seen in the current characteristics using double diode model. These findings help us to understand the impact of ion distribution on the charge transport in the perovskite solar cells.

5 ACKNOWLEDGEMENTS

This work was supported by Science and Engineering Research Board(SERB, project code: CRG/2019/003163), Department of Science and Technology (DST), India. The authors acknowledge the support from National Centre for Photovoltaic Research and Education (NCPRE), IITB Nanofabrication Facility(IITBNF), and DST Visvesvaraya Young faculty Research Fellowship.

6 REFERENCES

[1] J. J. Yoo *et al.*, "Efficient perovskite solar cells via improved carrier management," *Nature*, vol. 590, no. 7847, pp. 587–593, 2021, doi: 10.1038/s41586-021-03285-w.

[2] J. A. Christians *et al.*, "Tailored interfaces of unencapsulated perovskite solar cells for >1,000 hour operational stability," *Nature Energy*, vol. 3, no. 1. pp. 68–74, 2018. doi: 10.1038/s41560-017-0067-y.

[3] J. Wu *et al.*, "Quantifying the Interface Defect for the Stability Origin of Perovskite Solar Cells," *Adv Energy Mater*, vol. 9, no. 37, Oct. 2019, doi: 10.1002/aenm.201901352.

[4] Z. Lin *et al.*, "Interface studies of the planar heterojunction perovskite solar cells," *Solar Energy Materials and Solar Cells*, vol. 157, pp. 783–790, Dec. 2016, doi: 10.1016/j.solmat.2016.07.045.

[5] A. Guerrero *et al.*, "Interfacial degradation of planar lead halide perovskite solar cells," *ACS Nano*, vol. 10, no. 1, pp. 218–224, Jan. 2016, doi: 10.1021/acsnano.5b03687.

[6] L. X. Shi, Z. S. Wang, Z. Huang, W. E. I. Sha, H. Wang, and Z. Zhou, "The effects of interfacial recombination and injection barrier on the electrical characteristics of perovskite solar cells," *AIP Adv*, vol. 8, no. 2, 2018, doi: 10.1063/1.5021293.

[7] S. E. Chiang *et al.*, "Origins of the s-shape characteristic in J-V curve of inverted-type perovskite solar cells," *Nanotechnology*, vol. 31, no. 11, 2020, doi: 10.1088/1361-6528/ab5a02.

[8] N. Tessler and Y. Vaynzof, "Insights from Device Modeling of Perovskite Solar Cells," *ACS Energy Lett*, vol. 5, no. 4, pp. 1260–1270, 2020, doi: 10.1021/acsenergylett.0c00172.

[9] L. Xu, R. Molaei Imenabadi, W. G. Vandenberghe, and J. W. P. Hsu, "Minimizing performance degradation induced by interfacial recombination in perovskite solar cells through tailoring of the transport layer electronic properties," *APL Mater*, vol. 6, no. 3, 2018, doi: 10.1063/1.5021138.

[10] Q. Wang, C. Bi, and J. Huang, "Doped hole transport layer for efficiency enhancement in planar heterojunction organolead trihalide perovskite solar cells," *Nano Energy*, vol. 15, pp. 275–280, 2015, doi: 10.1016/j.nanoen.2015.04.029.

[11] S. Shao *et al.*, "Efficient Perovskite Solar Cells over a Broad Temperature Window: The Role of the Charge Carrier Extraction," *Adv Energy Mater*, vol. 7, no. 22, 2017, doi: 10.1002/aenm.201701305.

[12] V. Nandal and P. R. Nair, "Predictive Modeling of Ion Migration Induced Degradation in Perovskite Solar Cells," *ACS Nano*, vol. 11, no. 11. pp. 11505–11512, 2017. doi: 10.1021/acsnano.7b06294.

[13] S. van Reenen, M. Kemerink, and H. J. Snaith, "Modeling Anomalous Hysteresis in Perovskite Solar Cells," 2015, doi: 10.1021/acs.jpclett.5b01645.

[14] P. Calado *et al.*, "Evidence for ion migration in hybrid perovskite solar cells with minimal hysteresis," *Nat Commun*, vol. 7, pp. 1–10, 2016, doi: 10.1038/ncomms13831.

[15] G. Richardson *et al.*, "Can slow-moving ions explain hysteresis in the current-voltage curves of perovskite solar cells?," *Energy Environ Sci*, vol. 9, no. 4, pp. 1476–1485, 2016, doi: 10.1039/c5ee02740c.

[16] T. Saketh Chandra *et al.*, "Ion mobility independent large signal switching of perovskite devices," *Appl Phys Lett*, vol. 119, no. 2, pp. 1–5, 2021, doi: 10.1063/5.0051342.

[17] B. Rivkin, P. Fassl, Q. Sun, A. D. Taylor, Z. Chen, and Y. Vaynzof, "Effect of Ion Migration-Induced Electrode Degradation on the Operational Stability of Perovskite Solar Cells," *ACS Omega*, vol. 3, no. 8, pp. 10042–10047, 2018, doi: 10.1021/acsomega.8b01626.

[18] M. de Bastiani *et al.*, "Ion migration and the role of preconditioning cycles in the stabilization of the J-V characteristics of inverted hybrid perovskite solar cells," *Adv Energy Mater*, vol. 6, no. 2, pp. 1–9, 2016, doi: 10.1002/aenm.201501453.

[19] D. Sivadas, S. Bhatia, and P. R. Nair, "Efficiency limits of perovskite solar cells with n-type hole extraction layers," *Appl Phys Lett*, vol. 119, no. 20, 2021, doi: 10.1063/5.0059221.

[20] H. K. Scharfetter, D. L., & Gummel, "Large-signal analysis of a silicon read diode oscillator," *IEEE Trans Electron Devices*, vol. 16(1), pp. 64–77, 1969.

[21] L. Bertoluzzi *et al.*, "Mobile Ion Concentration Measurement and Open-Access Band Diagram Simulation Platform for Halide Perovskite Solar Cells," *Joule*, vol. 4, no. 1, pp. 109–127, 2020, doi: 10.1016/j.joule.2019.10.003.

[22] L. Bertoluzzi, R. A. Belisle, K. A. Bush, R. Cheacharoen, M. D. Mcgehee, and B. C. O. Regan, "In Situ Measurement of Electric-Field Screening in Hysteresis-Free," 2018, doi: 10.1021/jacs.8b04405.

[23] S. Agarwal *et al.*, "On the uniqueness of ideality factor and voltage exponent of perovskite-based solar cells," *Journal of Physical Chemistry Letters*, vol. 5, no. 23, pp. 4115–4121, 2014, doi: 10.1021/jz5021636.

[24] S. Agarwal and P. R. Nair, "Device engineering of perovskite solar cells to achieve near ideal efficiency," *Appl Phys Lett*, vol. 107, no. 12, 2015, doi: 10.1063/1.4931130.

[25] P. Lopez-Varo *et al.*, "Effects of Ion Distributions on Charge Collection in Perovskite Solar Cells," *ACS Energy Letters*, vol. 2, no. 6. pp. 1450–1453, 2017. doi: 10.1021/acsenergylett.7b00424.

[26] A. Pockett and M. J. Carnie, "Ionic Influences on Recombination in Perovskite Solar Cells," *ACS Energy Lett*, vol. 2, no. 7, pp. 1683–1689, 2017, doi: 10.1021/acsenergylett.7b00490.

GROWTH OF ZNO FILMS BY PLASMA ASSISTED REACTIVE EVAPORATION WITH IMPROVED THICKNESS HOMOGENEITY

G. Gordillo[1], J.C. Pena[2], J. Piña[1],
[1] Departamento de Física, Universidad Nacional, Bogotá, Colombia
[2]Fundación Universitaria Agraria de Colombia

ABSTRACT: This work focuses on the development and evaluation of a system for the deposition of ZnO thin films with improved thickness homogeneity and suitable optoelectrical properties to be used as TCO layer in solar cells, using the Plasma Assisted Reactive Evaporation (PARE) method. This development made it possible to significantly improve the inhomogeneity of the thickness of the ZnO films that were deposited using a previous version of the PARE system developed by our Group. Through modeling and theoretical simulation of the processes that take place during the deposition of ZnO thin films by the PARE method, information was obtained that allowed the implementation of a reactor with a novel design that allowed the growth of large-area ZnO films with good homogeneity in thickness. By optimizing the preparation parameters, it was possible to deposit large area ZnO thin films with similar values of resistivity and transmittance than those obtained with small area ZnO films deposited using the previous version of the reactor. ZnO films with resistivities varying between $9.5*10^{-4}$ and $2*10^{4}$ Ωcm and transmittances greater than 85% (in the visible region) were achieved

1. INTRODUCTION

Zinc oxide (ZnO) is a semiconductor wide-band-gap (3.37 eV) which is the most used for the manufacture of solar cells, as transparent conducting oxide (TCO) layer in solar cells based on Cu(In,Ga)Se2 (CIGS) [1]] and CuZnSnS4 (CZTS) [2]. Several methods have been used to grow thin ZnO films to be used for different applications [3-9], been the magnetron sputtering RF method the most widely used at industrial level in the manufacture of photovoltaic (PV) modules. However, it has some disadvantages such as high production costs and limitation for homogeneous deposition in large areas due to the use of ceramic targets with extrinsic doping.

The method proposed for this research is plasma assisted reactive evaporation, which has advantages over the RF sputtering method, reflected in terms of costs and ease of homogeneous deposition in large areas. Taking into account these advantages of the PARE method, we consider that the implementation of ZnO thin film production systems based on this method could be useful for industrial purposes since large area ZnO thin films with good thickness homogeneity and high degree of reproducibility of its opto-electric properties could be achieved by incorporating an adequate design of de Zn evaporation source and an electronic system for automatic control of the deposition process.

2. EXPERIMENTAL DETAILS

The main objective of this work is the development of a system for the deposition of ZnO thin films with improved thickness homogeneity and suitable optoelectrical properties to be used as TCO layer in solar cells, using the Plasma Assisted Reactive Evaporation method. For that, special emphasis was placed on the implementation of a deposition system that allowed to significantly improve the homogeneity in thickness of the ZnO films through a novel design of the reactor, but also achieving high reproducibility of optoelectric properties by incorporating a system with facilities to automatically control the deposition process of ZnO films.

Fig. 1 schematically shows the system implemented to deposit ZnO thin films by the PARE method, controlling the deposition process automatically. The system includes the following units:
- Glow Discharge (GD) Reactor to ionize the precursor species (Zn, O_2) and get them to react quickly and thus induce rapid growth of the ZnO films.
- Zn evaporation source manufactured with a new design based on the Knudsen cell concept that allows larger surface films to grow with good thickness homogeneity.
- Zn Sensor designed and implemented to generate a signal that alerts the moment in which the evaporation of Zn begins. The knowledge of the moment in which the evaporation of Zn begins is critical to be able to guarantee a good reproducibility of the optoelectrical properties of the ZnO films.
- Unit control with facilities to control electronically, the alert of start of Zn evaporation, Knudsen cell temperature, oxygen flow and the partial pressure change DP within the chamber induced by the reaction of Zn and oxygen that gives rise to the formation of ZnO within the glow discharge (GD) zone. The magnitude of these two variables is controlled, by controlling the Zn^{2+}/O^{2-} ratio, which is done by keeping the oxygen flow constant and controlling the change in partial pressure DP. The DP control allow the electrical resistivity of the ZnO films to be controlled in a reproducible way.

The acquisition, monitoring and control algorithms were developed in the LabVIEW ™ programming environment. This environment is based on the graphical programming language and data flow, which allows the developer to write his code in the form of a block diagram with which it is possible to perform the control and monitoring of equipment and processes. The software is based on the development of virtual instruments (VI), which fulfill the functions of individual physical instruments or set of instruments.

The hardware used for the monitoring and control of the PARE system is the following:

- National Instruments Compact FieldPoint interface module (cFP 1804). The data acquisition and generation cards that were connected to this module are:

- cFP-TC-120, 8 Channel Thermocouple Input Module. (TC, mV)
- cFP-AIO-610, Multi- function I/O module for Compact FieldPoin. I (mA, mV, V), O(V).
- cFP-AI-110, 8 ch Analog Input Module (mA, mV, V)
- NI USB 6008 card used as a module for analog and digital inputs and outputs.

With this system it was possible to prepare ZnO thin films with a high degree of reproducibility of their optoelectric properties.

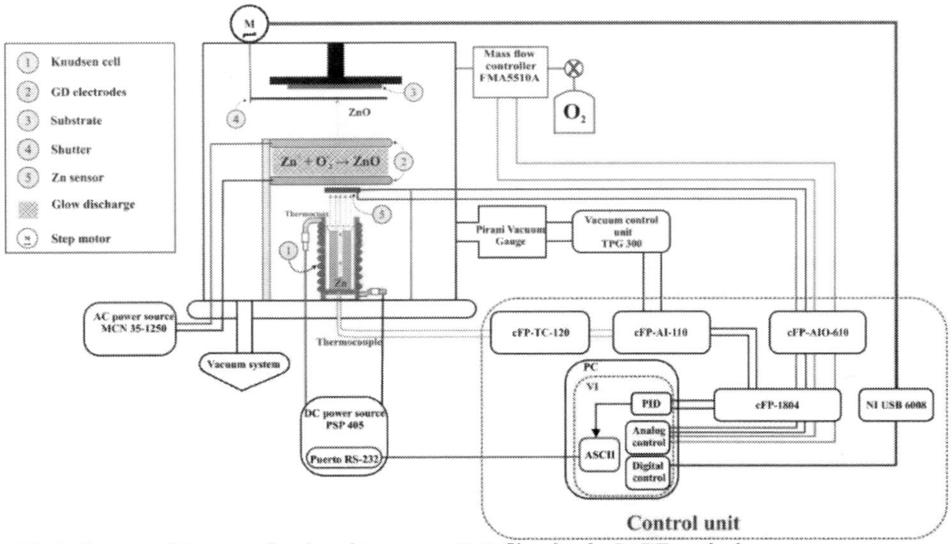

Fig. 1: Block diagram of the setup developed to prepare ZnO films by the PARE method.

It was found experimentally that ZnO films deposited on a 2.5 x 3.0 cm substrates by the PARE method using a conventional Knudsen cell, present a thickness profile as shown in Fig.2.

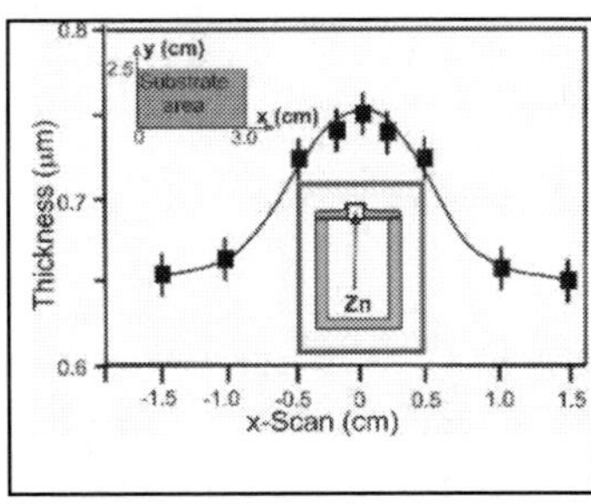

Fig.2. Experimental thickness profile of ZnO films, deposited using a conventional Knudsen cell.

This behavior is attributed to the fact that the Zn vapor flow is greater in the central region of the Knudsen cell because it has a single exit hole in the central part. In order to improve the inhomogeneity in thickness of the ZnO films, the design of the Zn evaporation source was modified based on a study carried out previously to solve the limitation of inhomogeneity in thickness that this type of deposition systems typically present. This study was carried out by modeling and simulating the plasma-assisted reactive evaporation process (see details in Ref.[3]). The theoretical simulation also indicated that the inhomogeneity of the thickness exhibited by the ZnO layers deposited by the PARE method can be considerably improved using a Knudsen cell that includes several exit holes distributed in such a way that the sum of their surfaces is similar to that of the substrate. The thickness profile predicted by theoretical simulation for ZnO films deposited on larger area substrates by the PARE method assuming a Knudsen cell with the proposed new design is shown in Fig. 3.

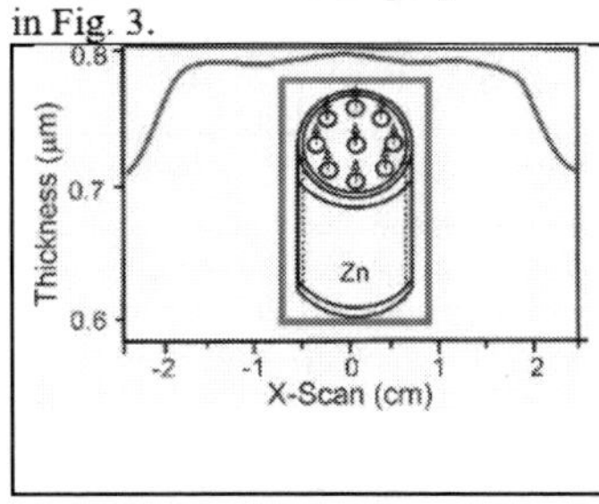

Fig. 3: Thickness profile of ZnO film theoretically simulated for a Knudsen cell with the new design.

Fig. 4 shows a schema of the Zn evaporation source built taking into account the new design, that guarantees the growth of ZnO films with improved thickness homogeneity. This was manufactured by integrating two elements built separately. The first one is a cylindrical stainless steel crucible with an internal diameter of 4 cm, which constitutes the base of the cell and the second is a conical stainless steel piece that is integrated into the base of the crucible through a thread system (see inset of Fig. 4). By varying the angle of inclination of the conical piece, it is possible to increase the exit area of the Zn vapor and thus the area of deposition of the ZnO, guaranteeing improvement of thickness homogeneity. The conical piece has a small hole (3 mm cross section diameter) in the lower part through which the Zn vapor exits and in the upper part a lid with multiple holes (each 2 mm cross section diameter) is placed, which is tightly fitted to the conical piece to achieve a high Zn vapor pressure. The crucible is heated by an oven made up of a stainless steel cylinder that is heated by means of a spiral-wound thermocoax type heating element.

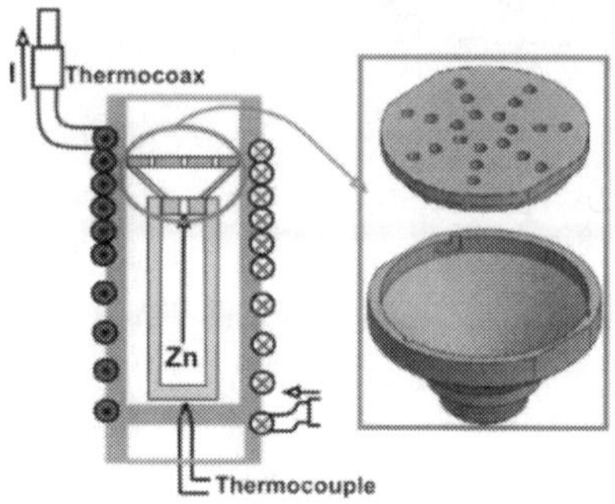

Fig. 4: scheme of the Zn evaporation source designed and implemented to deposit ZnO films with improved thickness homogeneity by the PARE method.

3. RESULTS AND DISCUSSION

3.1 Test of the ZnO growth system and verification of the improvement of thickness homogeneity.

The developed automated ZnO growth system was tested by depositing thin films of ZnO. Through a study of the influence of the main deposition parameters (pressure change DP, glow discharge current and oxygen partial pressure controlled by controlling the O_2 flow) on resistivity and transmittance, conditions were found to deposit thin films of ZnO highly transparent with resistivities varying in a wide range in such a way that they could be used as a TCO layer (n+-ZnO) and as an interdiffusion barrier (i-ZnO) in solar cells. It was found that the transmittance and resistivity of the ZnO films are affected by the three mentioned deposition parameters, however, the decrease in DP pressure that is observed when Zn starts its reaction with oxygen to form ZnO molecules is the parameter that most strongly affects both transmittance and resistivity. Fig. 5 shows the typical routine used for the deposition of highly transparent ZnO films, whose resistivities can be varied between $2x10^4$ Ωcm (i-ZnO) and $9.5x10^{-4}$ Ωcm (n+-ZnO).

Fig. 6 shows the effect that pressure change DP, produces both on transmittance T (measured at a wavelength $\lambda = 800$ nm) and on electrical resistivity ρ of the ZnO films. It is observed that the resistivity of the ZnO films decreases abruptly when DP increases from 0.02 to 0.1 mB. This behavior is mainly caused by an increase in the density of free carriers associated with an increase in oxygen vacancies generated as a consequence of an increase in the Zn^+/O^- ratio. It is also observed in Fig. 6 that transmittance decreases with increasing DP, apparently due to the generation of photon absorption centers attributed to the presence of oxygen vacancies.

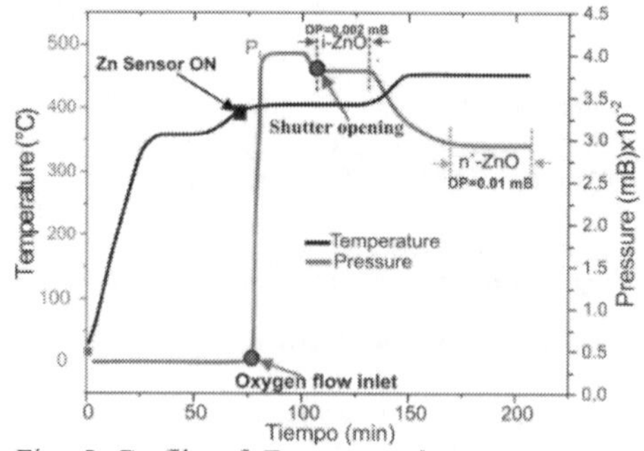

Fig. 5. Profile of Zn evaporation temperature and partial pressure change DP used to deposit highly transparent ZnO thin films with adequate resistivities to be used as TCO layer and interdiffusion barrier of solar cells.

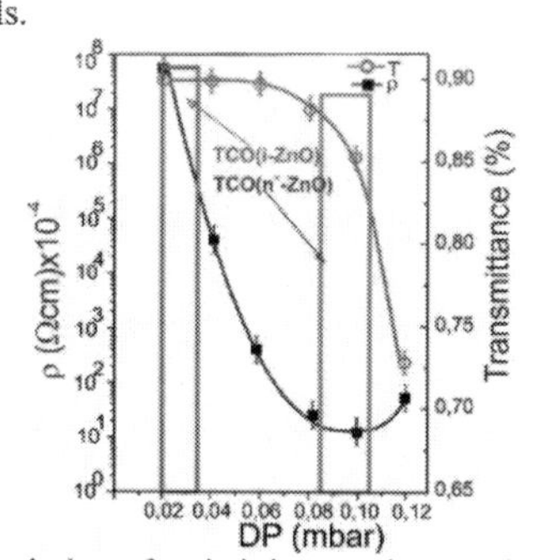

Fig. 6: Variation of resistivity ρ and transmittance T as a function of pressure change DP (ZnO films deposited keeping the current at 6 mA and PO_2 at 0.15 mB

The two regions framed in red and green rectangles (Fig. 6) indicate the DP values that give rise to ZnO films with optoelectric properties suitable to be used as i-ZnO layer (green region) and n+-ZnO (region red). The i-ZnO/n+-ZnO double layer typically used as part of the optical window of CIGS and CZTS-based solar cells, can be deposited In-Situ (without interrupting the process) by properly controlling the Zn evaporation routine.

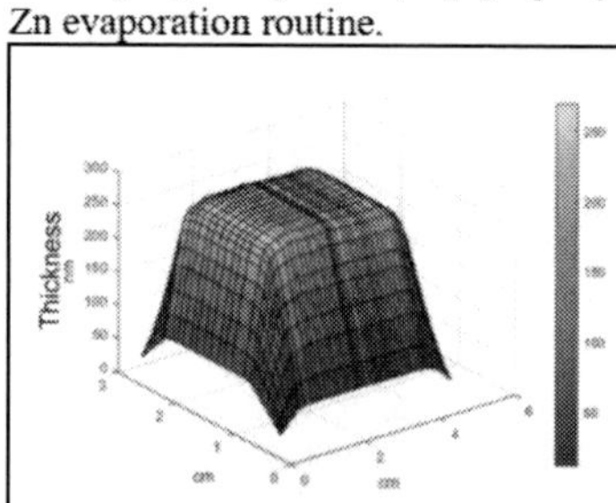

Fig. 7: Thickness profile of ZnO films deposited using the Knudsen cell implemented with a new design.

To verify if the new designed Knudsen cell allows improving the thickness homogeneity, ZnO films were deposited on 5.2 cm x 4.0 cm glass substrates and their thickness was measured with a Veeco Dektak 150 surface profiler making scans across the length and width of the ZnO samples. Fig. 7 shows the results of the thickness homogeneity study carried out with ZnO films deposited using the Knudsen cell designed and built in this work. These results show that the evaporation source built with the new design, effectively lead to a significant improvement of the homogeneity in thickness of the ZnO films.

4.2 Optical characterization of ZnO thin films

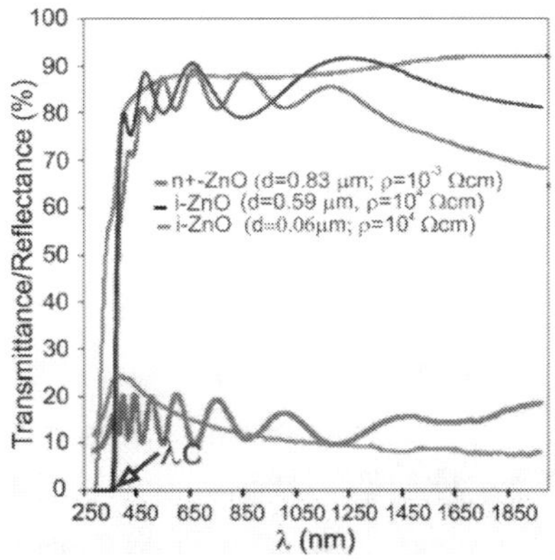

Fig.8: Spectral transmittance and reflectance curves of n+-ZnO (0.83 μm) and i-ZnO (0.59 and.06 μm) films.

Figure 8 shows spectral transmittance and reflectance spectra corresponding to ZnO films typically used as TCO layer (0.83 μm, 10^{-3} Ωcm) and diffusion barrier (60 nm, 10^{4} Ωcm) in solar cells. The transmittance and reflectance spectra were obtained by using a Varian–Cary 5000 spectrophotometer. It is observed that the transmittance of very thin i-ZnO film shifts towards the ultraviolet with respect to the transmittance of thicker n+-ZnO and i-ZnO film, indicating that the film thickness is not responsible for this displacement of λc. This behavior has been attributed to the strain in the film introduced during the deposition [4] as well as to a band gap shrinkage effect, induced by the change of both the nature and strength of interaction potentials between donors and the host lattice [5]. These results are consistent with those reported by other authors [6–10].

4. CONCLUSIONS

Highly transparent and conductive ZnO thin films with improved thickness homogeneity were obtained by plasma-assisted reactive evaporation method using a knudsen cell as crucible for Zn evaporation, built with a novel design. The new design of the Zn evaporation source was carried out using results of a theoretical simulation of the PARE process. The performance of the new implemented reactor was tested depositing a large number of ZnO films. It was found that the ZnO films grow with quite good thickness homogeneity and with suitable transmittances and resistivities to be used as TCO layer and as diffusion barrier in solar cells, without using extrinsic doping. The low resistivity of ZnO films is mainly caused to a high density of free carriers generated by oxygen vacancies.

ACKNOWLEDGEMENTS

Este trabajo fue financiado por la Fundación para la Promoción de la Investigación y Tecnología del Banco de la República y la Universidad Nacional de Colombia-Sede Bogotá-Facultad de Ciencias-Departamento de Física- Grupo GMS&ES- Cra 45 #26-85, Bogotá, Cod. Postal 111321- Colombia

REFERENCES

[1] T. Kato, "Cu(In,Ga)(Se,S)2 solar cell research in Solar Frontier: Progress and current status," *Jpn. J. Appl. Phys.*, vol. 56, 2017.

[2] A. Gueddim, N. Bouarissa, A. Naas, F. Daoudi, and N. Messikine, "Characteristics and optimization of ZnO/CdS/CZTS photovoltaic solar cell," *Appl. Phys. A*, vol. 124, p. 199, Feb. 2018, doi: 10.1007/s00339-018-1626-1.

[3] A. A. Ramirez, I. Gil, G. Gordillo, and A. M. Latifi, "Analysis of a plasma-assisted reactive evaporation process for preparation of ZnO thin films: Modeling and experimentation," *Thin Solid Films*, vol. 698, p. 137846, 2020, doi: https://doi.org/10.1016/j.tsf.2020.137846.

[4] S. Al-Khawaja, B. Abdallah, S. Abou Shaker, and M. Kakhia, "Thickness effect on stress, structural, electrical and sensing properties of (0 0 2) preferentially oriented undoped ZnO thin films," *Compos. Interfaces*, vol. 22, no. 3, pp. 221–231, Mar. 2015, doi: 10.1080/15685543.2015.1002259.

[5] A. P. Roth, J. B. Webb, and D. F. Williams, "Band-gap narrowing in heavily defect-doped ZnO," *Phys. Rev. B*, vol. 25, no. 12, pp. 7836–7839, Jun. 1982, doi: 10.1103/PhysRevB.25.7836.

[6] P. Y. Dave, K. H. Patel, K. V Chauhan, A. K. Chawla, and S. K. Rawal, "Examination of Zinc Oxide Films Prepared by Magnetron Sputtering," *Procedia Technol.*, vol. 23, pp. 328–335, 2016, doi: https://doi.org/10.1016/j.protcy.2016.03.034.

[7] J. J. Lu, Y. M. Lu, S. I. Tasi, T. L. Hsiung, H. P. Wang, and L. Y. Jang, "Conductivity enhancement and semiconductor metal transition in Ti-doped ZnO films," *Opt. Mater. (Amst).*, vol. 29, pp. 1548–1552, Jul. 2007, doi: 10.1016/j.optmat.2006.08.002.

[8] A. Jazmati and B. Abdallah, "Optical and Structural Study of ZnO Thin Films Deposited by RF Magnetron Sputtering at Different Thicknesses: a Comparison with Single Crystal," *Mater. Res.*, vol. 21, Apr. 2018, doi: 10.1590/1980-5373-mr-2017-0821.

[9] S. maki, A. Shaban, and S. hussain, "Thickness Influence on Structural and Optical Properties of ZnO Thin Films Prepared by Thermal Evaporation," *Ibn AL-Haitham J. Pure Appl. Sci.*, vol. 31, p. 79, Sep. 2018, doi: 10.30526/31.2.1946.

[10] M. Suchea, S. Christoulakis, M. Katharakis, N. Vidakis, and E. Koudoumas, "Influence of thickness and growth temperature on the optical and electrical properties of ZnO thin films," *Thin Solid Films*, vol. 517, pp. 4303–4306, Jun. 2009, doi: 10.1016/j.tsf.2008.11.143.

EFFECT OF THE Sb/Se RATIO ON THE STRUCTURAL, MORPHOLOGICAL AND ELECTRICAL PROPERTIES OF Sb_xSe_y FILMS

T.M.Razykov[1], A.Bosio[2], S. Pasini[2], *B.A.Ergashev[1], D.Z.Isakov[1], K.M.Kouchkarov[1], R.R.Khurramov[1], M.A.Makhmudov[1], A.A.Mavlanov[3], A.Romeo[4], N.Romeo[2], Sh.B.Utamuradova[5]

[1]Physical-Technical Institute, Uzbek Academy of Sciences, Chingiz Aytmatov Street 2B, Tashkent 700084, Uzbekistan
[2]University of Parma, G.P Usberti 7/A - 43124 Parma, Italy
[3]Ritsumeikan Global Innovation Research Organization, Ritsumeikan University, 1-1-1 Nojihigashi, Kusatsu, Shiga 525-8577, Japan
[4]Universita' di Verona, Ca' Vignal 2- Strada Le Grazie 15, 37134 Verona, Italy
[5]Institute of Semiconductors Physics and Microelectronics, Yangi Olmazor Street 20, Tashkent 100057, Uzbekistan

*Corresponding author: tel. +998931890304, bobur_7007@yahoo.com

ABSTRACT: At the present, attention of many researchers is focused on Sb_2Se_3, Sb_2S_3 and $Sb_2(S_x, Se_{1-x})_3$ chalcogenide compounds. This is due to the physical properties of these materials, such as *p*-type conductivity, band gap Eg = 1.1÷1.8 eV, high absorption coefficient $\alpha > 10^5 cm^{-1}$ (for visible solar radiation), low melting point (823K and 885K for Sb_2Se_3 and for Sb_2S_3 respectively) and high partial pressure, which make these materials suitable for use as an absorbing layer in high efficiency thin film solar cells. In this work, we presented the effect of the Sb/Se ratio on the structural, morphological, and electrical properties of Sb_xSe_y films. Sb_xSe_y films were fabricated by chemical molecular beam deposition (CMBD) method from Sb and Se precursors at a substrate temperature of 500°C at atmospheric pressure hydrogen flow. The scanning electron microscope and X-ray diffraction analysis show that all films have a crystallite size of 2÷3 μm and an orthorhombic crystal structure. The electrical conductivity of the films varies in the range of ~10^{-5}÷ 10^2 (Ohm • cm)$^{-1}$ depending on Sb/Se ratio which was measured by van der Pauw method.

Keywords: Thin film solar cells, Sb/Se ratio, CMBD, p-type, n-type, electrical properties

1 INTRODUCTION

At the present, attention of many researchers is focused on the use of Sb_2Se_3, Sb_2S_3 and $Sb_2(S_x, Se_{1-x})_3$ chalcogenide compounds solid solution, as an effective absorbing layer for thin-film solar cells [1-3]. This is since the physical properties such as high absorption coefficient $\alpha > 10^5 cm^{-1}$ (visible solar radiation), *p*-type conductivity, band gap Eg = 1.1÷1.8 eV, low melting point (823K and 885K for Sb_2Se_3 and for Sb_2S_3 respectively) and high partial pressure of these materials are very close to the properties of the dominant materials $Cu(In,Ga)(Se,S)_2$ and CdTe. But it is known, that $Cu(In,Ga)(Se,S)_2$ and CdTe have also disadvantages for large-scale production [4-5]. Due to the significant shortcomings of existing solar cells, scientists began to explore new materials to be used in thin film solar cells. New materials must conform to their physical properties, as well as their abundance in the earth's crust. The following material fully meets these requirements. It is known that Sb and Se elements are relatively inexpensive, earth abundant, resistant to external influences and have low toxicity. In the future, this will make it feasible to manufacture environmentally friendly and efficient solar modules, as well as arise the way for their large-scale production.

2 EXPERIMENTAL DETAILS

In this work we presented the physical properties of Sb_2Se_3 thin films, obtained by the chemical molecular beam deposition (CMBD) method from Sb and Se precursors at a substrate temperature of 500 °C. High purity (99.999%) Sb and Se materials were used as the source materials, which were evaporated in a hydrogen flow at atmospheric pressure. The evaporation temperature ranges were (800 ÷ 900) °C and (400 ÷ 500) °C for Sb and Se respectively. The hydrogen carrier gas flow is ~20 cm^3/min. Depending on the required film thickness, the duration of the deposition process is ranged from 30 to 60 min The thickness of the deposited Sb_2Se_3 films of various compositions is 2÷3 μm. Soda-lime glass was used as a substrate. The composition of the resulting Sb_2Se_3 films was varied by controlling the ratio of the partial pressures of Sb and Se in the vapor phase during the growth. Thin films of various compositions obtained by the CMBD method is described in detail in [6]. Structural studies of the samples were performed on a Panalytical X'pert PRO X-ray diffractometer with a Cu tube (Kα = 1.5406 Å). The measurements were carried out at room temperature in the range of 2θ angles from 20° to 80° in the step-by-step scanning mode with a step of 0.02 degrees and a signal accumulation time at each point of 5 s. Morphological studies were performed using a SEM-EVO MA 10 scanning electron microscope, and the film compositions were determined using an EDX (Oxford Instrument, Aztec Energy Advanced X-act SDD) energy-dispersive elemental analyzer. The electrical conductivity of the films was measured by the van der Pauw method.

3 RESULTS AND DISCUSSION

The effect of the Sb/Se ratio on the structural, morphological and electrical properties of Sb_2Se_3 films has been investigated. Fig. 1 shows X-ray diffraction patterns of thin films with different Sb/Se ratios. All thin films have prominent peaks that are in good agreement with the standard JCPDS map (No. 15–0861) of the orthorhombic Sb_2Se_3 phase without any detectable impurities. As it is shown in picture 1, with an increase in the amount of selenium in the vapor phase, the direction of dominance (*hkl*, *l*=1), referred to the (211) (221) intensity is enhanced. At the same time (hkl, l≠1), a decrease in the intensity of the (120) peak is observed.

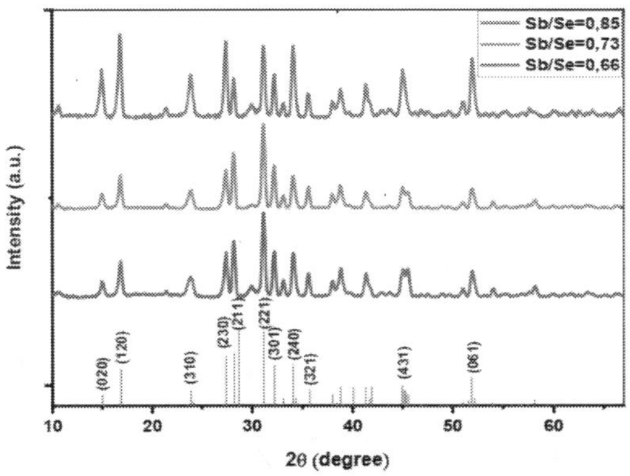

Fig. 1. Typical X-ray diffraction pattern of a Sb_2Se_3 film of various compositions on glass coated with a molybdenum layer.

These results showed that the ratio of Sb and Se in the vapor phase plays a decisive role in the choice of orientation in the solid phase. Moreover, thin films of Sb_2Se_3 characterized with preferential orientations (*hkl, l* = 1) (221) showed good conduction properties of the charge carriers [7].

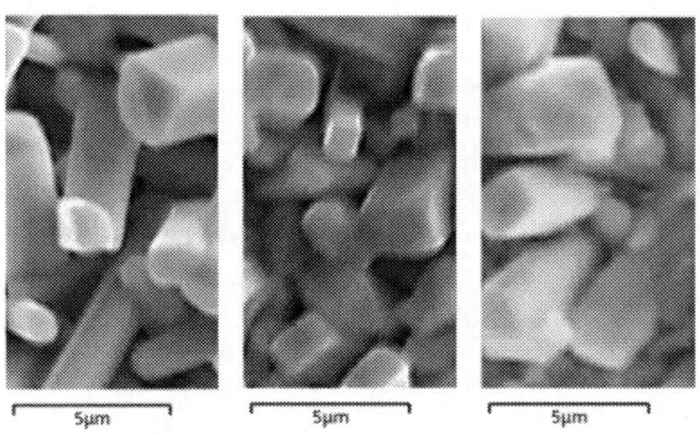

Sb/Se = 0.66 Sb/Se = 0.7 Sb/Se = 0.8

Fig. 2. Scanning electron microscope images of Sb_2Se_3 films with various compositions deposited at a temperature of 500°C.

As can be seen from Fig. 2, the microcrystals of the Sb_2Se_3 films are evenly distributed over the substrate surface. From the figure, the crystal sizes for all samples are 2÷3 μm with a long rod-like shape. When the Sb/Se ratio of the film compositions is in the range of 0.6÷0.8, the crystal growth is perpendicular to the substrate. The electrical conductivity of the films was measured by the van der Pauw method. With an increase in the partial pressure ratio in the range Sb/Se = 0.3÷0.66, the electrical conductivity (~10^{-5} (Ω cm)$^{-1}$) of the Sb_2Se_3 films almost did not change. On the contrary, a sharp increase of the conductivity from 10^{-5} (Ω cm)$^{-1}$ to 10^{2} (Ω cm)$^{-1}$ was observed when Sb/Se ≥ 0.66. The samples showed *p*-type conductivity in the composition range 0.3 ≤ Sb/Se ≤ 0.7 while, at 0.8 ≤ Sb/Se ≤ 1.22 exhibited *n*-type conductivity. An inversion of the conduction type was found at Sb/Se = 0.7, which corresponds to the stoichiometric ratio of Sb_2Se_3.

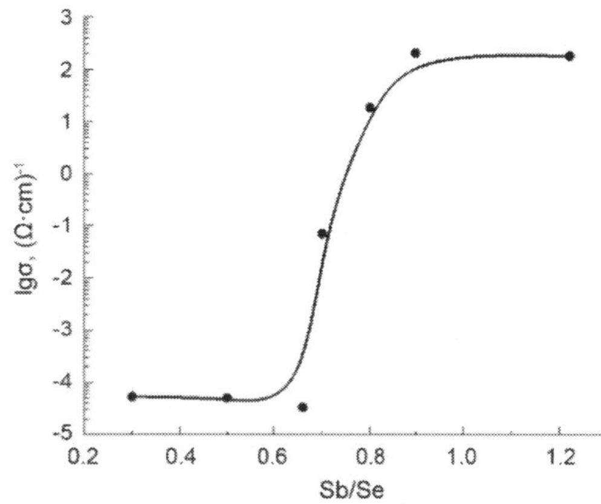

Fig. 3. The dependence of the conductivity of Sb_xSe_y films on the Sb/Se ratio.

4 CONCLUSIONS

The scanning electron microscope data and the X-ray diffraction analysis show that all films have a crystallite size of 2÷3 μm with an orthorhombic crystal structure characterized by predominant (211) and (221) orientations. The electrical conductivity of the films varies in the range of ~10^{-5}÷ 10^{2} (Ω cm)$^{-1}$ depending on Sb/Se ratio.

5 REFERENCES

[1] Z. Li, X. Liang, G. Li, H. Liu, H. Zhang, J. Guo, J. Chen, K. Shen, X. San, W. Yu, R.E.I. Schropp, Y. Mai, 9.2%-efficient core-shell structured antimony selenide nanorod array solar cells, Nat. Commun. 10 (2019) 125. https://doi.org/10.1038/s41467-018-07903-6

[2] Choi, Y.C., Lee, D.U., Noh, J.H., Kim, E.K., and Seok, S.I. (2014). Highly improved Sb_2S_3 sensitized-inorganic-organic heterojunction solar cells and quantification of traps by deeplevel transient spectroscopy. Adv. Funct. Mater. 24, 3587–3592. https://doi.org/10.1002/adfm.201304238

[3] Wang, X., Tang, R., Jiang, C., Lian, W., Ju, H., Jiang, G., Li, Z., Zhu, C., Chen, T. Manipulating the Electrical Properties of $Sb_2(S, Se)_3$ Film for High - Efficiency Solar Cell. Advanced Energy Materials 2020, 10 (40), 2002341. https://doi.org/10.1002/aenm.202002341

[4] Jixiang Zhou and Changyu Li. Research on Copper Indium Gallium Selenide (CIGS) Thin-Film Solar Cells. E3S Web of Conferences 267, 02031 (2021) https://doi.org/10.1051/e3sconf/202126702031

[5] T.M.Razykov, C.S.Ferekides, D.Morel, E.Stefanakos, H.S.Ullal, H.M.Upadhyaya. Solar Photovoltaic Electricity: Current Status and Future Prospects (Review). Solar Energy 85 (2011) 1580–1608. https://doi.org/10.1016/j.solener.2010.12.002

[6] T.M.Razykov Chemical molecular beam deposition of II–VI binary and ternary compound films in a gas flow. Applied Surface Science Volumes 48–49, 1991, Pages 89-92 https://doi.org/10.1016/0169-4332(91)90311-7

[7]__Rong Tang, Xingye Chen, Yandi Luo. Controlled Sputtering Pressure on High-Quality Sb_2Se_3 Thin Film for Substrate Configurated Solar Cells. Nanomaterials 2020, 10, 574; https://doi.org/10.3390/nano10030574

CPEPh-Na/MoO3 BI-STACK HOLE INTERFACIAL LAYERS FOR ENHANCING THE PERFORMANCE OF NONFULLERENE ORGANIC PHOTOVOLTAICS

Enas Moustafa*, Josep Pallarès and Lluis F. Marsal*
Department of Electric, Electronic, Engineering and Automatic, Universitat Rovira i Virgili
43007, Tarragona, Spain.
E-mail: enas.moustafa@urv.cat, lluis.marsal@urv.cat

ABSTRACT: The interface-associated organic photovoltaics (OPVs) devices consider a vital issue that affects their overall performance. Accordingly, in the present work, we demonstrated an efficient interfacial modification through using conjugated poly electrolytes (CPE) to modulate the barrier for charge injection into the semiconducting layers and in turn improve the device performance Herein, we report an aqueous conjugated poly electrolyte of PCPDTPhSO$_3$-Na as a buffer interlayer to modify the interface between the hole transporting layer (HTL) of MoO$_3$ and the active blend of PM6:Y7 in an inverted device architecture of ITO/ZnO/PM6:Y7/HTL/Ag. This bi-stack modulates the interfacial traps through diminishing the accumulated charges and in turn, enhances the transfer and the collection of the generated current. Accordingly, the PCPDTPhSO$_3$-Na/MoO$_3$ based devices exhibited higher short circuit current density than the MoO$_3$ based ones by 8 %, and the fill factor was further enhanced by 5 %, consequently, the power conversion efficiency improved by 14 %.
Keywords: interracial layers, nonfullerene organic photovoltaics, conjugated poly electrolytes.

1 INTRODUCTION

It is well known that the organic photovoltaics (OPVs) are assembled from a complex set of layers of different materials, each layer playing a distinct role within the device[1]. Accordingly, several strategies were investigated to enhance the performance of the OPVs through modyfing each layer. One of the avenus was concern about improving the bulk hterojunction active blend using differnet donor and acceptor materials, additives, and further treatments[2]–[4]. Hence, the power conversion efficiency of OPVs has been boosted by the current development of the nonfullerene acceptors (NFA) approaching the 20 % for a single junction devices[3], [5]–[7]. Recently, Zou and co-workers synthesized a new N3 acceptor materials that modified the Y6 acceptors[8], [9].

Morover, other approches were investingaing different interfacial layers between the active blend and the contacts[10]–[13]. Accordingly, hole transporting layers (HTLs) play a crucial role in OPVs, not only enhancing the performance but also the stability behaviour[14], [15]. HTL materials used in OPVs can be describes into two head categories, which are inorganic HTL and organic HTL materials. Despite OPVs with inorganic HTL materials can get saisfied PCE, but they are not suitable for large scale production due to the inescapable high-temperature vacuum evaporation process[14]. Hence, large numbers of latest research succefully developed organic HTL materials for OPVs[14]. One of the used organice HTL materials was the conjugated poly electrolytes (CPE) such as PCPDTPhSO3-Na/K ones[16]. CPEs used as an interfacial material showing a significant improvement of the PCE through solution processing[16].

Hence, in the present work, the CPEPh-Na was used as an additional buffer interlayer to fabricate inverted nonfullerene organic photovoltaics (iNF-OPVs) with devices configuration of ITO/ZnO/PM6:Y7/CPEPh-Na/MoO$_3$/Ag as presented in **Fig.1a** to benfit from the dual advantages of organic and inorganic HTLs. The control devices with only MoO$_3$ as single hole transporting layer were also conducted for comparison.

2 EXPERIMENTAL METHODS

2.1 Materials

The transparent conducting oxide use d was indium tin oxide (ITO) patterned glass substrate with a resistivity of 10 Ω sq^{-1} purchased from PsiOTec. The donor polymer PM6, the non-fullerene acceptor Y7 and the buffer layer of conjugated poly electrolytes material PCPDTPhSO$_3$-Na (CPEPh-Na) were supplied from One-Material. To synthesize the Zinc oxide (ZnO) as Electron transporting layer (ETL) precursor solution, zinc acetate dihydrate 99.999% (A-Aldrich), 2-methoxyethanol 99.9% (Sigma Aldrich) and ethanolamine 99.5% (A-Aldrich) were purchased. Vanadium oxide (MoO$_3$) used as hole transporting layer (HTL) was delivered from Sigma Aldrich and the 99.999% purity Ag from Testbourne Ltd. The used solvents of methanol, chlorobenzene (CB), diiodo octane (DIO) and chloronapthelene (CN) were purchased from Sigma Aldrich.

2.2 Device Fabrication

The inverted NF-OPVs were fabricated in the traditional sandwich structure of ITO/ZnO/PM6:Y7(NF)/HTL/Ag, the HTLs were MoO$_3$ for the control devices and CPEPh-Na/MoO$_3$ bi-stack for the investigated iNF-OPVs. The device archeticture was illustrated in **Fig.1a** along with the bandgab cascade of the corresponding layers in **Fig.1b** and the chemical structure of the CPE buffer layer demonstrated in **Fig. 1c**.

The ITOs were cleaned by acetone, methanol and

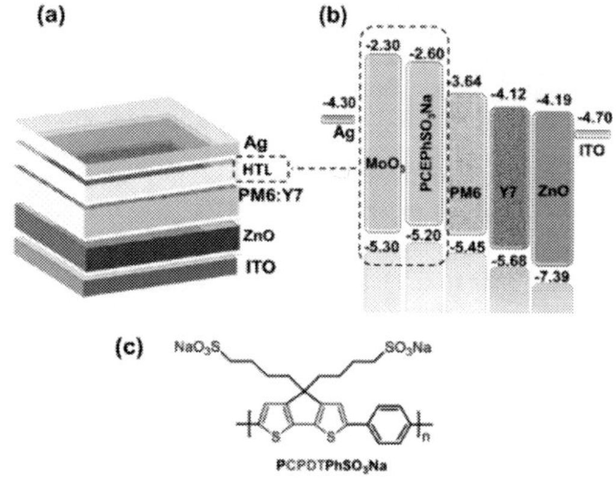

Fig.1. (a). The architecture of the inverted fabricated NF-OPVs, (b) Energy band cascade of the fabricated inverted NF-OPVs, and (c) the chemical structure of the CPE.

isopropanol then treated by UV-Ozone for 20 min. The ZnO film were fabricated by depositing the ZnO precursor solution on the pre-cleaned ITO, then lefft forl h annealing at 200 °C to produce thin film with 30 nm thickness [2], [17]. The ZnO precursor solution was synthesized by dissolving 0.68M of zinc acetate dihydrate and 0.46M of ethanolamine in 2-methoxyethanol, then the mixture was left for 1 h under vigorous stirring at 70 °C. Furthermore, the prepared solution was diluted by methanol 1:1 v/v ratio then filtered and spun at 3000 rpm [2], [17].

20 mg/ml blend solution of PM6 donor and Y7 acceptor was synthesized by dissolving 1:1 ratio in CB with 1 % v/v CN solvent additive. Then continuously stirred at least for 3 h at 80 °C. Subsequently, the prepared solution was filtered by 0.2 µm PVDF filter and directly spun-coated at 2000 rpm for 40 s over the ZnO layer and thermally treated at 100 °C to form a film thickness of 120 nm. Finally, a bilayer of MoO_3 (8 nm) covered by Ag (100 nm) were thermally evaporated under high vacuum conditions ($\leq 1 \times 10^{-6}$ mbar) for the control device.

The $PCPDTPhSO_3$-Na (CPEPh-Na) buffer layer was synthetized by dissolving 0.1 mg in 1 ml of methanol. The prepared solution was deposited over the active blend film at 3000 rpm for 40 s, then evaporating the MoO_3 and Ag as mentiond previously.

In the fabrication processes, the top electrod evaporated using shadow mask with device active area of 0.09 cm². It is worth to mention that all the deposition steps (after the ZnO films) were carried out inside the glove box under nitrogen environment.

2.2 Device Characterization and Measurements

The thickness of the deposited films in the fabricated devices was measured via the surface profilometer (AMBIOS TECHNOLOGY-XP-1). Current density-voltage (J-V) characteristics of the fabricated devices were conducted at room temperature with Keithley 2400 source-measure unit under illumination and dark conditions. The illumination was obtained from solar simulator (Abet Technology model 11000 class type A, Xenon arc). Calibration for the solar simulator intensity was operated with a Fraunhofer certified silicon photovoltaic cell to yield a 100 mW/cm² and AM1.5 spectrum. Impedance spectroscopy measurements were carried out under simulated AM 1.5G illumination using an HP4192A impedance analyzer.

3 RESULTS AND DISCUSSIONS

The CPEPh-Na was used as an additional buffer interlayer to fabricate iNF-OPVs with devices configuration of ITO/ZnO/PM6:Y7/CPEPh-Na/MoO_3/Ag as presented in **Fig. 1a**, named as B-CPEPh-Na/MoO_3 devices. Control devices with only MoO_3 as single hole transporting layer were conducted for comparison, named

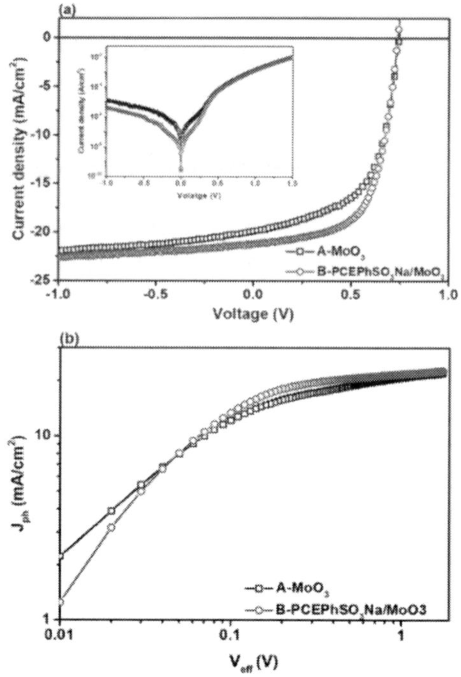

Fig.2. (a) Current density-voltage (J-V) characterisrics under AM 1.5G illnmination, the inset figure for the J-V cureves at dark condition of the iNF-OPVs (b) J_{ph} vs V_{eff} curves of the devices.

as A-MoO_3 devices. The photovoltaics parameters of the fabricated devices were listed in **Table I**.

Fig. 2a displays the representative current density-voltage (J-V) characteristics of the fabricated devices incorporating various buffer interlayer under simulated AM 1.5G illumination condition (100 mW/cm²). The A-MoO_3 based devices exhibited a power conversion efficiency (PCE) of 8.88 % with an open circuit voltage (V_{OC}) of 0.74 V, a short circuit current density (J_{SC}) of 20.64 mA/cm², fill factor (FF) of 0.59, series (R_S) and shunt (R_{Sh}) resistances of 2.23 and 270 Ω, respectively. While using CPEPh-Na/MoO_3 as a buffer interfacial layer enhanced the J_{SC} to 22.21 mA/cm² and the FF to 0.62, due to lower obtained R_S of 1.88 Ω and almost doubled shunting resistance of 484 Ω, resulting in a better PCE of 10.16 %. Moreover, the inset J-V dark curve in **Fig. 2a** showed a lower leakage current observed for the B-CPEPh-Na/MoO_3 based devices, revealing an improved shunt resistance, indicating superior charge selectivity and charge transport ability of the anode interlayer[18]. This observation might explain the enhanced J_{SC} obtained for the B-CPEPh-Na/MoO_3 based devices (**Table I**).

Table I: The Performance Parameters of the Fabricated Inverted Non-Fullerene Organic Photovoltiac Devices.

HTLs	V_{OC} (V)	J_{SC} (mA/cm²)	FF	R_S (Ω.cm²)	R_{SH} (Ω.cm²)	PCE (%)
A- MoO_3	0.74	20.64	0.59	2.23	270	8.88
B-PCEPh-Na/MoO_3	0.74	22.21	0.62	1.88	484	10.16

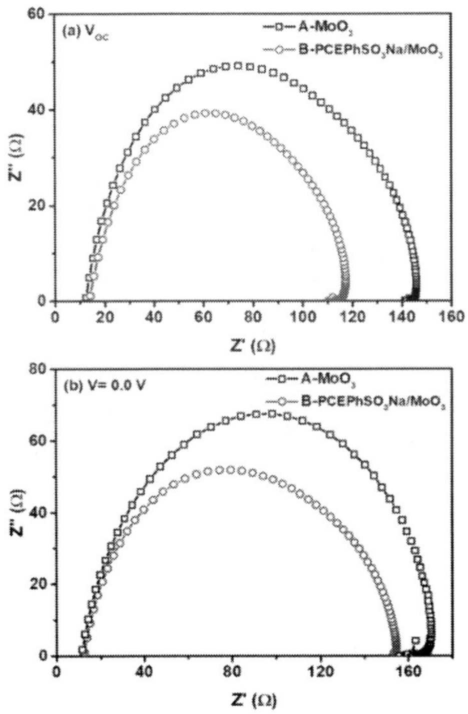

Fig.3 Cole-Cole plots under AM 1.5G illumination of the fabricated iNF-OPVs at (a) open circuit voltage V_{OC}, (b) short circuit condition 0.0 V.

To gain insight into the charge extraction process, we conducted the photocurrent density (J_{ph}) as a function of the effective voltage (V_{eff}) as presented on **Fig. 2b**. We could calculate the exciton dissociation probability (P_{diss}) of the fabricated devices as reported[2], [19], which show values of 97.5 % and 98.3 % for the A-MoO_3 and B-CPEPh-Na/MoO_3 based devices, respectively. This higher P_{diss} value for the B-CPEPh-Na/MoO_3 based devices, disclosing the improved charge extraction, leading to increased J_{SC}, the high FF and as a sequence PCE[19].

Fig. 3 shows the Cole-Cole plot of the impedance spectra for the fabricated iNF-OPVs. Performing impedance spectroscopy on photovoltaics was a great challenge for the new material properties and complex device designs. The impedance spectroscopy measurements were carried out under illumination condition of AM 1.5G, frequency sweep from 1 Hz to 10 MHz at V_{OC} bias voltage (**Fig. 3a**) and short circuit conditions of 0.0 V(**Fig. 3b**), 50 mV amplitude of AC signal. It was interesting to notice that the arc size of the plot was smaller for the B-CPEPh-Na/MoO_3 based devices than the A-MoO_3 ones in both cases of applied voltage (V_{OC} and 0.0V), indicating the lower resistance[20] possessed by the B- CPEPh-Na/MoO_3, confirming the R_S obtained from the J-V characteristics in **Fig. 2a** and **Table I**. In addition, the smaller arc diameter at short circuit condition defined that the B-CPEPh-Na/MoO_3 devices have a rapid charge extraction whereas A-MoO_3 devices have a slower charge extraction, which is contributed to the J_{SC} values of the devices that agrees with the result from **Table I**. Moreover, it is worth to notice that the B-CPEPh-Na/MoO_3 based devices exhibited lower low arc frequency, disclosing the termination of the charge accumulated at the interfaces that cannot be extracted by the contacts[21]. Accordingly, B-CPEPh-Na/MoO_3 based device showed higher output generated current.

4 CONCLUSIONS

In summary, we demonstrated that CPEPh-Na buffer layer can significantly enhance the performance of the iNF-OPVs. This might be attributed to modifying the interfaces between the active blend and the thermally evaporated MoO_3 layer through diminishing the interfacial traps and improving the charge extraction process.

5 Reference

[1] O. A. Abdulrazzaq, V. Saini, S. Bourdo, E. Dervishi, and A. S. Biris, "Organic solar cells: A review of materials, limitations, and possibilities for improvement," *Particulate Science and Technology*, vol. 31, no. 5, pp. 427–442, 2013.

[2] E. Moustafa, A. A. A. Torimtubun, J. Pallarès, and L. F. Marsal, "Effect of Additives and Annealing on the Performance of Nonfullerene-Based Binary and Ternary Organic Photovoltaics," *Solar RRL*, vol. 2100480, pp. 1–16, 2021.

[3] J. Lv *et al.*, "Additive-induced miscibility regulation and hierarchical morphology enable 17.5% binary organic solar cells," *Energy & Environmental Science*, vol. 14, no. 5, pp. 3044–3052, 2021.

[4] M. Günther *et al.*, "Increasing Photostability of Inverted Nonfullerene Organic Solar Cells by Using Fullerene Derivative Additives," *ACS Applied Materials and Interfaces*, vol. 13, no. 16, pp. 19072–19084, 2021.

[5] Q. Liu *et al.*, "18% Efficiency organic solar cells," *Science Bulletin*, vol. 65, no. 4, pp. 272–275, 2020.

[6] Y. Li *et al.*, "Mechanism study on organic ternary photovoltaics with 18.3% certified efficiency: from molecule to device," *Energy & Environmental Science*, vol. 15, pp. 855–865, 2022.

[7] Y. Cui *et al.*, "Single-Junction Organic Photovoltaic Cells with Approaching 18% Efficiency," *Advanced Materials*, vol. 32, no. 19, pp. 1–7, 2020.

[8] J. Yuan *et al.*, "Single-Junction Organic Solar Cell with over 15% Efficiency Using Fused-Ring Acceptor with Electron-Deficient Core," *Joule*, vol. 3, no. 4, pp. 1140–1151, 2019.

[9] K. Jiang *et al.*, "Alkyl Chain Tuning of Small Molecule Acceptors for Efficient Organic Solar Cells," *Joule*, vol. 3, no. 12, pp. 3020–3033, 2019.

[10] E. Moustafa, M. Maria, J. Pallar, and L. F. Marsal, "Solar Energy Materials and Solar Cells Low temperature based PDINO cathode interlayer for high operational photostable inverted non-fullerene organic solar cells," *Solar Energy Materials and Solar Cells*, vol. 248, no. January, p. 111985, 2022.

[11] E. Moustafa, L. F. Marsal, and J. Pallarès, "Significant Stability Improvement of Fullerene

Organic Photovoltaics via ZnO Film Modification through the Intermittent Spray Pyrolysis Technique," *ACS Applied Energy Materials*, 2022.

[12] X. Xu *et al.*, "Interface-enhanced organic solar cells with extrapolated T80 lifetimes of over 20 years," *Science Bulletin*, vol. 65, no. 3, pp. 208–216, 2020.

[13] H. Zhang, Y. Li, X. Zhang, Y. Zhang, and H. Zhou, "Role of interface properties in organic solar cells: From substrate engineering to bulk-heterojunction interfacial morphology," *Materials Chemistry Frontiers*, vol. 4, no. 10, pp. 2863–2880, 2020.

[14] H. Xu *et al.*, "Hole transport layers for organic solar cells: Recent progress and prospects," *Journal of Materials Chemistry A*, vol. 8, no. 23, pp. 11478–11492, 2020.

[15] B. Ecker *et al.*, "Degradation effects related to the hole transport layer in organic solar cells," *Advanced Functional Materials*, vol. 21, no. 14, pp. 2705–2711, 2011.

[16] S. D. C. Huiqiong Zhou , Yuan Zhang , Cheng-Kang Mai and and A. J. H. Thuc-Quyen Nguyen , Guillermo C. Bazan, "Advanced Materials - 2013 - Zhou - Conductive Conjugated Polyelectrolyte as Hole-Transporting Layer for Organic Bulk.pdf," *Advanced Materials*, vol. 26, no. 5, pp. 780–785, 2013.

[17] E. Moustafa, J. G. Sánchez, L. F. Marsal, and J. Pallarès, "Stability Enhancement of High-Performance Inverted Polymer Solar Cells Using ZnO Electron Interfacial Layer Deposited by Intermittent Spray Pyrolysis Approach," *ACS Applied Energy Materials*, vol. 4, no. 4, pp. 4099–4111, 2021.

[18] M. Cui *et al.*, "A Cost-Effective, Aqueous-Solution-Processed Cathode Interlayer Based on Organosilica Nanodots for Highly Efficient and Stable Organic Solar Cells," *Advanced Materials*, vol. 32, no. 38, pp. 1–8, 2020.

[19] V. D. Mihailetchi, L. J. A. Koster, J. C. Hummelen, and P. W. M. Blom, "Photocurrent generation in polymer-fullerene bulk heterojunctions," *Physical Review Letters*, vol. 93, no. 21, pp. 19–22, 2004.

[20] E. Von Hauff, "Impedance Spectroscopy for Emerging Photovoltaics," *Journal of Physical Chemistry C*, vol. 123, no. 18, pp. 11329–11346, 2019.

[21] B. Arredondo, M. B. Martín-López, B. Romero, R. Vergaz, P. Romero-Gomez, and J. Martorell, "Monitoring degradation mechanisms in PTB7:PC71BM photovoltaic cells by means of impedance spectroscopy," *Solar Energy Materials and Solar Cells*, vol. 144, pp. 422–428, 2016.

THE EFFECT OF HEATED DEPOSITIONS ON THE EFFICACY OF MOLYBDENUM NITRIDE BARRIER LAYERS FOR CIGS BACK CONTACT APPLICATIONS

Liam M Welch, Nada Benhaddou, Jake W Bowers
Loughborough University
CREST, Wolfson School of Mechanical, Electrical and Manufacturing Engineering, Loughborough, LE11 3TU, UK
01509635305, L.Welch@Lboro.ac.uk

ABSTRACT: Molybdenum nitride barrier layers have been previously shown to be effective in preventing the excessive ingress of selenium into the back contact of CIGS solar cells during a high temperature selenisation. Heated depositions have also been shown to decrease the sheet resistance of sputtered molybdenum, with the potential to produce superior back contacts. However, the effect of temperature on the molybdenum nitride barrier layers is unknown. In this work, heated depositions and vacuum annealing of molybdenum and molybdenum nitride barrier layers were carried out to investigate the effects on the barrier layer. While heated depositions and vacuum anneals were found to be beneficial for the molybdenum layers, reducing the resistivity, they were found to lower the efficacy of the molybdenum nitride barrier layer. A survivability study found that heated depositions and vacuum annealing dramatically increased the chance of the selenium diffusion barrier failing, followed by the excessive formation of molybdenum selenide.
Keywords: Back contact, CIGS, CZTS, Diffusion

1 INTRODUCTION

$Cu(In,Ga)(S,Se)_2$ (CIGS) is a well-established and commercially deployed thin-film photovoltaic technology reaching high laboratory efficiencies (23.35%) [1]. Non-vacuum processed CIGS, typically grown using solution chemistry. Solution processing typically has a high material utilization, and is capable of a high-throughput, and doesn't require high capital-cost vacuum equipment. This makes solution processing an attractive low-cost and environmentally friendly alternative to vacuum-based technologies. Non-vacuum processed CIGS typically require a high temperature annealing treatment, in the presence of selenium vapor (called a selenisation treatment), which converts the solution-deposited precursor film into the final desired film. This typically improves the morphology of the CIGS absorber, producing larger CIGS crystals, leading to a higher efficiency solar cell [2]. Whilst high temperature selenisations are necessary to convert the precursor film to the desired chalcopyrite phase of the absorber, the Mo back contact typically reacts with Se leading to the formation of $MoSe_2$. A thin layer of $MoSe_2$ is beneficial to the performance of CIGS device, creating an ohmic contact with the absorber [3]. However, the formation of excessive amounts of $MoSe_2$ is detrimental to the solar cell performance, since it is highly resistive, increasing the series resistance of the device. In extreme cases, the formation of $MoSe_2$, and the subsequent expansion of the back contact, can lead to fatal delamination of the cell during the selenisation, which can be a common occurrence in solution-processed precursor films which are typically quite porous.

Therefore, a diffusion barrier can be used to prevent the excessive formation of $MoSe_2$. Our group has previously shown that MoN_x can easily be deposited as a barrier layer using reactive sputtering, by introducing N_2 to the sputtering gas during the deposition of Mo [4]. By depositing the barrier layer within the back contact, a sacrificial layer can then be deposited above the barrier, that will form a thin layer of beneficial $MoSe_2$.

Heated depositions can lead to an increase in crystallinity in sputtered Mo films, resulting in a back contact with a decreased sheet resistance [5]. Therefore, a heated deposition could be used to increase the performance of the back contact, resulting in a device with a reduced series resistance. However, a highly amorphous structure is desirable in the barrier layer, since a columnar grain structure can provide a path for the Se to diffuse through [6]. Therefore, it is possible that a heated back contact deposition could be beneficial to the Mo layers, while being detrimental to the performance of the MoN_x barrier. The Mo and MoN_x barrier layer are likely to be deposited in a single deposition, since it can be expected to be prohibitively expensive to leave the sample to cool in the deposition area.

Since sheet resistance of a back contact is irrelevant if the back contact is destroyed during processing, this work aims to check the viability of heated back contact depositions by examining the effect of different heating regimes on the survivability of the back contact.

2 EXPERIMENTAL

2.1 Molybdenum back contact depositions

Thin film molybdenum back contacts were deposited by DC magnetron sputtering (AJA International ATC 2200-V) on Na-free EAGLE XG (Corning) glass. The full Mo back contact stack used in the survivability section of this work (section 3.2) is composed of 4 layers; an initial layer deposited at a high sputtering pressure designed to produce a strong adhesion to the substrate (5 mTorr sputtering pressure in Ar, 1.79 Wcm^{-2} sputtering power density, 300 nm film thickness). This is followed by a layer deposited at a low sputtering pressure that is designed to have a low sheet resistance (1 mTorr sputtering pressure in Ar, 1.79 Wcm^{-2} sputtering power density, 750 nm film thickness). Then a MoN_x barrier layer is deposited (1 mTorr sputtering pressure, 2/1 Ar/N_2 gas ratio, 1.79 Wcm^{-2}, 100 nm film thickness). Finally, a thin sacrificial layer of Mo is deposited on top of the barrier (1 mTorr sputtering pressure in Ar, 1.79 Wcm^{-2} sputtering power density, 25 nm film thickness).

In section 3.1, individual layers of molybdenum and molybdenum nitride were deposited for the purpose of layer characterisation. Individual layers were deposited onto glass according to their normal deposition parameters, but to a thickness of 750 nm.

2.2 Heated depositions and annealing

In order to test the effect of heating on the MoN_x layer, two heating regimes were trialled alongside a control:

Heated deposition (HD) - The deposition was performed with a substrate temperature of 300 °C. The substrate was soaked at 300 °C for 5 minutes prior to deposition.

Vacuum anneal (VA) - Following a room temperature deposition, the substrate and layer was heated to 300 °C for 10 minutes in the deposition chamber.

Room temperature (RT) - The deposition was performed at room temperature, as a control. No further annealing was performed.

To minimise oxidation of the Mo, the substrates were allowed to cool below 100 °C in the evacuated chamber prior to removal.

The low-pressure layers are given the suffix "L-Mo", while the MoN_x layers are given the suffix "MoN_x"

2.3 Characterisation

Layer thicknesses were measured using scanning electron microscopy (SEM), using a Jeol 7800F FEGSEM with an accelerating voltage of 5 kV. Sheet resistance measurements were performed on a 4-point probe with a probe spacing of 0.635 mm. X-ray diffraction spectroscopy (XRD) was performed using a Bruker D2 Phaser benchtop diffractometer using a Cu-Kα source. Samples were cut to approximately $1cm^2$. X-ray Photoelectron Spectroscopy (XPS) was performed using a Fisher Scientific K-alpha XPS surface analysis tool. An electron flood gun was used to reduce charging to avoid peak shifts. Charge was corrected to the C1s peak at 284.8 eV. To minimise the effect of surface contamination, Ar ion milling was used to perform the analysis at a depth of 21 nm.

3 Results and Discussion

3.1 Characterisation of single layers

In order to observe any differences caused by the different deposition regimes, single layers from the back contact stack are deposited onto the glass substrates individually. Single layers of Mo and MoN_x were deposited in the three deposition regimes for characterisation, as shown below in Table I.

Table I: Experimental layout for the heated deposition and vacuum anneal layers

Sample Name	Deposition Temperature (°C)	Anneal Temperature (°C)	Resistivity (Ω·cm)
RT-L-Mo	-	-	$11x10^{-6}$
VA-L-Mo	-	300	$9x10^{-6}$
HD-L-Mo	300	-	$8x10^{-6}$
RT-MoN_x	-	-	$112\ x10^{-6}$
VA-MoN_x	-	300	$103x10^{-6}$
HD-MoN_x	300	-	$50x10^{-6}$

The resistivity of the L-Mo layers decreases with both the HD and VA deposition regimes relative to the RT control. HD produces the greatest decrease in resistivity. This is likely due to the increased crystallinity expected with heated depositions. This suggests that a heated deposition could yield a superior back contact for CIGS

The MoN_x shows a similar trend, with a dramatic decrease in resistivity associated with the HD deposition regime. The very moderate decrease with the VA deposition suggests that a post deposition anneal at 300 °C may not be sufficient to cause a significant increase in the crystallinity of the MoN_x.

Cross-sectional SEM of the L-Mo samples reveals no significant morphological difference between the samples deposited using the three regimes (Fig. 1a). All three samples show a crystalline structure with columnar grains. This is in keeping with the resistivity measurements, which show only marginal differences between the deposition regimes. However, the MoN_x layer (Fig. 1b) shows a significant morphological change when deposited in the HD regime compared to the VA and RT regimes. The HD-MoN_x layer's structure contains large columnar grains, similar to those observed within pure Mo. VA and RT regimes produced samples with an amorphous MoN_x layer. This suggests that the HD deposition regime could produce an inferior Se diffusion barrier, since a crystalline structure is undesirable.

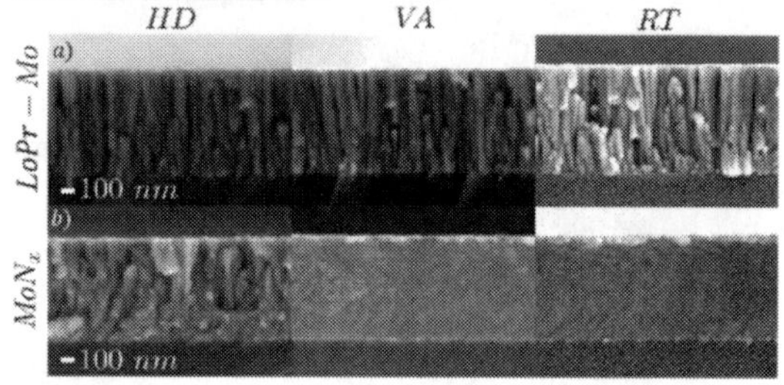

Figure 1: Cross-sectional SEM images of a) the L-Mo and b) MoN_x layers deposited using the three deposition regimes.

XRD spectra of the MoN_x layers reveals that the HD-MoN_x layer has a strong and relatively sharp Mo (110) peak, as well as (200), (211) and (220) peaks (Fig. 2). This suggests that there is a large amount of Mo with a large crystallite size present. The resistivity is still high for a pure Mo sample (approximately 6 times higher than HD-L-Mo), suggesting that there are still MoN_x species present. Therefore, it appears that in the heated deposition, the high temperature is causing a stoichiometric change instead of a simply morphological one.

The samples deposited using the VA and RT regimes show extremely broad Mo_5N_6 (113) and Mo_2N (004) peaks, as well as a broad Mo (110) peak. This suggests that these samples have a larger proportion of MoN_x than the HD-MoN_x sample. There are minimal differences between the VA and RT samples, supporting the theory that the vacuum anneal has a limited effect on the MoN_x layer.

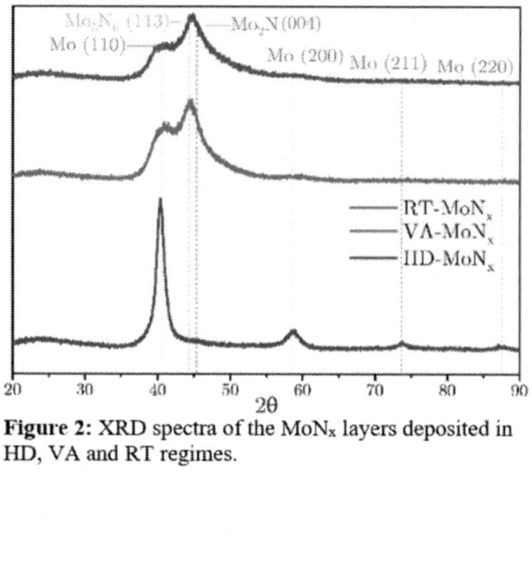

Figure 2: XRD spectra of the MoN_x layers deposited in HD, VA and RT regimes.

XPS analysis was performed on the three samples at a depth of 21 nm. The stoichiometry of the samples was determined by calculating the area of the Mo 3d, N 1s and O 1s peaks, and is listed in table II.

Table II: Stoichiometry as determined by XPS

Sample Name	Ratio (XPS) Mo:N	Mo:O
RT-MoN_x	1:0.22	1:0.13
VA-MoN_x	1:0.20	1:0.12
HD-MoN_x	1:0.11	1:0.12

The HD-MoN_x sample has a Mo:N ratio of 1:0.11. This represents a dramatic increase in the Mo:N ratio compared to the VA-MoN_x and RT-MoN_x samples. This suggests that the heated deposition inhibited the incorporation of N into the film. This could be due to the heated surface preventing nitrogen species from adsorbing onto the substrate, decreasing the amount that is included in the final film. This is expected to decrease the efficacy of the MoN_x barrier layer, since there will be a reduction in the amount of MoN_x species present.

The VA-MoN_x and RT-MoN_x samples have a similar Mo:N ratio, suggesting that only a small amount of N was lost during the vacuum anneal. This is likely because any N that is present in the film after the deposition is strongly bound to Mo atoms, and therefore no longer labile.

Additionally, there seems to be a relationship between the Mo:N ratio and the resistivity of the film, with a high Mo:N ratio associated with a higher resistivity film. This is expected, since N in the structure disrupts the lattice and reduces the crystallinity, while also making the overall bonding less metallic [7].

The Mo:O ratio is approximately equal for all three samples, suggesting that the heating in the vacuum system did not cause any additional oxidation. The amount of oxidation for all samples is still high, considering the low base pressure of the system (10^{-7} Torr). It is likely that the oxidation came from months of storage in a dry atmosphere, and that the measurement depth of 21 nm was not deep enough to avoid the oxidised surface.

3.2 Back contact stack survivability testing

To test the efficacy of MoN_x barrier layers deposited using the different deposition regimes, a survivability study was carried out. Full back contact stacks with MoN_x barrier layers of varying thicknesses (100-40 nm) were deposited using HD and VA regimes and compared against an RT deposition. The back contact stacks were selenised in a rapid thermal processing (RTP) furnace at 550 °C for 40 minutes. Following the selenisation, samples were visually inspected and assigned an outcome:

Pass → Pinholing → Ingress → Delamination.

A result of "Pass" means that the sample appears as it did before the selenisation – without pinholes and exhibiting specular reflection. "Pinholing" means that the barrier failed in a few isolated locations, leading to the formation of pinholes in the film. A result of "Ingress" means that the sample lost specular reflection and became a dull grey colour. "Fail" means that the sample delaminated from the substrate. Examples of these results are shown in figure 4.

Figure 3: Photographs of a) "Pinholing" result, b) "Delamination" result, c) "Ingress" result and d) cross-sectional SEM of an "Ingress" result.

There is a clear trend of worse outcomes associated with reduced barrier layer thicknesses. For the RT deposition regime, reducing the barrier layer thickness from 80 nm to 40 nm changes the outcome from "Pass" to "Pinholing" to "Ingress".

There is additionally a trend of worse outcomes for the same barrier layer thickness between the different deposition regimes. At a MoN_x thickness of 80 nm, RT achieves a "Pass", VA receives "Ingress" and HD "Fails". This trend continues at all MoN_x thicknesses, with HD always receiving a worse outcome than VA, and VA always receiving a worse outcome than RT. The poor performance of the HD regime stack is likely the result of two factors:

The increased crystallinity of the MoN_x barrier layer providing a pathway for Se atoms along the grain boundaries:

And the decrease in the N content, resulting in a lower amount of MoN_x species forming, and more Mo. This Mo can then be itself converted into $MoSe_2$.

Together these dramatically reduce the efficacy of the barrier, so that even thick barriers fail to prevent the ingress of Se.

Table III: Survivability Grid Considering Deposition Regime and MoN_x Layer Thickness

Deposition Regime	MoN_x Thickness 100 nm	80 nm	60 nm	40 nm
RT	Pass	Pass	Pinholing	Ingress
VA	Pinholing	Ingress	Ingress	Ingress
HD	Ingress	Fail	Fail	Fail

Sheet resistance measurements were also taken for the surviving back contact stacks post-selenisation (Table IV). Since $MoSe_2$ is highly resistive, the sheet resistance is expected to increase dramatically if the MoN_x barrier fails and the back contact is converted to $MoSe_2$.

The results of Table IV largely corroborate with the visually inspected performance outcomes displayed in Table III. Once an "Ingress" result is achieved, the sheet resistance of the back contact stack increases dramatically. The exception is VA-80, which visually appears to have suffered from Se ingress, but still has a low sheet resistance. SEM-shows that there is no excessive formation of $MoSe_2$ in VA-80 (Fig. 4a), but in VA-60 the entire back contact has been converted to $MoSe_2$ (Fig. 4b). This is evident by the great increase in thickness, as well as in the morphology of the back contact stack.

Table IV: Back contact stack sheet resistances following selenisation

	Sheet Resistance (Ω/□)			
Deposition	MoN_x Thickness			
Regime	100 nm	80 nm	60 nm	40 nm
HD	**15776**	X	X	X
VA	0.106	0.121	**7297**	**5183**
RT	0.124	0.160	0.187	**8823**

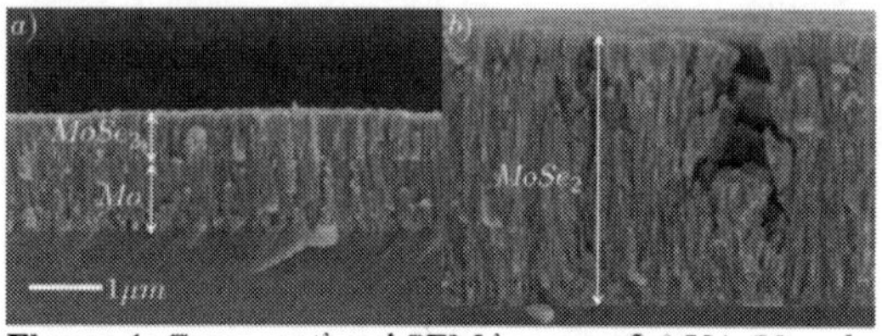

Figure 4: Cross-sectional SEM images of a) VA-80 and b) VA-60 following selenisation. In the VA-80 sample, only the sacrificial Mo layer was converted to $MoSe_2$. The MoN_x barrier layer failed in the VA-60 sample, resulting in the conversion of the entire back contact to $MoSe_2$.

4 CONCLUSIONS

Heating, in the form of a heated deposition or vacuum anneal, was found to be slightly beneficial on the resistivity of the pure molybdenum layer, suggesting that the technique has the potential be industrially applicable.

However, a heated deposition was found to dramatically decrease the efficacy of the MoN_x barrier layer. The heated deposition was shown to decrease the amount of nitrogen in the film while increasing the crystallinity. These effects were demonstrated to increase the likelihood of catastrophic selenium ingress into the back contact.

Vacuum annealing of the MoN_x layer was found to have more modest effects, cut still decreased the efficacy of the barrier layer.

Since the negative effects of the heated deposition and vacuum anneal on the MoN_x barrier layer outweigh the positive effects on the pure molybdenum layer, the process is unlikely to be industrially relevant if the layers are deposited sequentially.

5 REFERENCES

[1] Solar Frontier, "Press Release: Solar frontier achieves world record thin-film solar cell efficiency of 23.35%," 2019. .

[2] W. Wang, S. Y. Han, S. J. Sung, D. H. Kim, and C. H. Chang, "8.01% $CuInGaSe_2$ solar cells fabricated by air-stable low-cost inks," *Phys. Chem. Chem. Phys.*, vol. 14, no. 31, pp. 11154–11159, 2012, doi: 10.1039/c2cp41969f.

[3] K.-J. Hsiao, J.-D. Liu, H.-H. Hsieh, and T.-S. Jiang, "Electrical impact of $MoSe_2$ on CIGS thin-film solar cells," *Phys. Chem. Chem. Phys.*, vol. 15, pp. 18174--18178, 2013, doi: 10.1039/C3CP53310G.

[4] S. Uličná *et al.*, "Deposition and application of a Mo-N back contact diffusion barrier yielding a 12.0% efficiency solution-processed CIGS solar cell using an amine-thiol solvent system," *J. Mater. Chem. A*, vol. 7, no. 12, pp. 7042–7052, 2019, doi: 10.1039/c8ta12089g.

[5] H. Zhao *et al.*, "Effects of heating mode and temperature on the microstructures, electrical and optical properties of molybdenum thin films," *Materials (Basel).*, vol. 11, no. 9, 2018, doi: 10.3390/ma11091634.

[6] D. J. Kim, Y. T. Kim, and J. W. Park, "Nanostructured Ta-Si-N diffusion barriers for Cu metallization," *J. Appl. Phys.*, vol. 82, no. 10, pp. 4847–4851, 1997, doi: 10.1063/1.366346.

[7] I. Jauberteau *et al.*, "Molybdenum Nitride Films: Crystal Structures, Synthesis, Mechanical, Electrical and Some Other Properties," pp. 656–687, 2015, doi: 10.3390/coatings5040656.

ANALYSIS OF SELENIZATION TEMPERATURE FOR THE PERFORMANCE IMPROVEMENT OF SPIN COATED CZTSSe SOLAR CELLS

Prabeesh Punathil, Solidea Zanetti[1], Elisa Artegiani, Narges Torabi and Alessandro Romeo*

LAPS-Laboratory for Applied Physics, Department of Computer Science, University of Verona

Ca' Vignal 1, Strada Le Grazie 15, 37134 Verona, Italy, tel.: +39-045-8027974,

*e-mail address: alessandro.romeo@univr.it

[1]also at Isopan S.p.A., via Giona 5, Trevenzuolo, Italy

ABSTRACT: In our laboratory, $Cu_2ZnSn(S,Se)_4$ (CZTSSe) films are prepared by selenization of spin coated CZTS precursors. In this study, different temperatures are tested for the selenization process and their effects on the device are investigated. The influence of the selenization temperature on the Se incorporation and structural properties of the absorber have been analyzed by X-ray diffraction, Raman spectroscopy and scanning electron microscopy methods. Current-voltage analysis reveals a fill factor reduction with increasing temperature, concurrently with an improvement in short circuit current and efficiency. SLG/Mo/CZTSSe/CdS/iZnO/ITO/Au cells fabricated with selenization process at 550 °C reach an efficiency of 6.3 %.

Keywords: CZTSSe, Selenization, Thin Film Solar Cells.

1 INTRODUCTION

The current requirement for high efficiency, low cost, non-toxic and stable solar cells drives the research efforts on thin film solar cells. Polycrystalline $Cu_2ZnSn(S,Se)_4$ (CZTSSe) thin films have the potentiality to reduce PV production costs due to the simple production process, availability and low toxicity of the elements used. Recently, by regulating crystal growth via organic lithium salt additive in a solution-based cell, Zhou et al. improved the efficiency of CZTSSe devices to a record efficiency of ~13 % [1]. However, based on the Shockley-Queisser limit, the theoretical efficiency of CZTSSe solar cells is ~33 %. The main limiting factors of the efficiency of CZTSSe solar cells are a large deficit in open-circuit voltage (V_{OC}) and a relatively low fill factor (FF). Sn-based defects control in the CZTSSe system is one way to address this large V_{OC} deficit. Several approaches have been adopted for the fabrication of CZTSSe solar-cell devices, which include electrodeposition spray pyrolysis, spin coating, dip coating, successive ionic layer absorption and reaction (SILAR), sputtering, thermal evaporation, etc[2][3]. However, high-efficiency CZTSSe solar cells are also fabricated utilizing non-vacuum methods. Among these techniques, the selenization of metallic precursor layers is a convenient, high efficiency, and simple process. Additionally, this method avoids the usage of expensive metal targets and equipment during production stage.

2 EXPERIMENTAL

2.1 Preparation of the devices

In our laboratory we fabricate CZTSSe solar cells by spin coating technique followed by low temperature (450 °C) selenization. With this method we regularly obtain CZTSSe solar cells with an efficiency of 5.4 % [4], more recently with an additional Ge incorporation an efficiency exceeding 7% has been obtained [5]. The selenization process is an effective recrystallization step for the growth of CZTSSe thin films. It also affects the phase purity and defect states in the system. Most of the reported works employ high temperature annealing (above 500 °C) in order to increase the crystallite quality and to improve the optoelectronic properties of the absorber [6] [7].

In this work we present a study on the influence of the selenization temperature on the structural, morphological and electrical properties of the CZTSSe. In the interest of improving the properties of the final compound, a large number of absorber layers have been grown and subsequently processed into solar cells.

The substrates consist of Mo-coated soda lime glass in the size of 3×3 cm^2, fabricated by RF magnetron sputtering with a typical thickness of 1.6 µm. Subsequently, CZTS precursors are deposited by spin coating technique followed by the above mentioned selenization process. In order to investigate the effects of selenization temperature, a 30 minutes-selenization was carried out at various temperatures (450 °C, 500 °C, 550 °C). After this process step, a 60 nm thick CdS film is applied by chemical bath deposition (CBD) from an aqueous solution containing 30 mM $Cd(CH_3COO)_2{:}2H_2O$, 0.6 M thiourea and 25ml of NH_4OH (20 %) (for 250 ml total volume). Subsequently 80 nm thick intrinsic ZnO and 1.2 µm thick ITO are deposited by RF magnetron sputtering. Finally, a gold metal grid is deposited through an aperture mask by thermal evaporation. Precursor solution preparation and detailed device fabrication stags are reported in our previous works [5][4]. A total cell area of 0.13 cm^2 is defined by mechanical scribing.

2.2 Characterization

The crystal structure is studied by X-ray diffraction (XRD) analysis with a Thermo ARL X'TRA powder diffractometer (in Bragg-Brentano geometry, equipped with a Cu-anode X-ray source (Kα, λ=1.5418 Å) and a Peltier Si (Li) cooled solid state detector) and by Micro-Raman spectroscopy experiments in backscattering geometry with Thermo Scientific DXR2 Raman Microscope, over the Stokes-shifted region between 100 cm^{-1} to 350 cm^{-1} by a diode laser (l_{exc}=633 nm). Current density-voltage (JV) characteristics are measured with a Keithley Source Meter 2420, under an AM 1.5 spectrum at 100 mW/cm^2, using a LOT Quantum Design Europe solar simulator LS0306.

3 DEVICES CHARACTERIZATION

3.1 Structural properties

CZTS thin films selenized at various temperatures (450°C, 500°C and 550°C) show a tetragonal chalcopyrite crystal structure with a dominant peak along (112) reflection plane (see figure 1). The diffractogram also shows the presence of (101), (110), (103), (121), (220), (132) characteristic diffraction peaks (ICDD PDF # 01-082-9159) showing the growth of a CZTSSe material with an ordered cation sublattice. By comparing the position of (112) plane with the standard ICSD pattern file, S/Se ratio can be predicted as 0.25, indicating a tendential Se rich nature of the film. Furthermore, the lattice parameters and crystallite size of the films have been calculated, and are a=b=5.6515 Å, c=11.3143 Å, 30 nm, a=b=5.6439 Å, c=11.3339 Å, 36 nm and a=b=5.6515 Å, c=11.3032 Å, 40 nm for films selenized at 450 °C, 500 °C and 550 °C respectively [8]. By increasing selenization temperature a slight enlargement of the size of the crystallites is denoted.

In addition to CZTSSe phase, all films show peaks attributed to Mo, in particular the (110) orientation. More interesting, for the 500°C-case a peak around 31° arises, this can be attributed to the secondary phase of $SnSe_2$. Instead when the selenization is done at 550°C the secondary phase is reduced. The arise and decrease of this peak by increasing selenization temperature is consistent with the position of (112) peak in the three XRD patterns. As a matter of fact, a shift of the (112) peak can be observed, for this reason in Figure 2 we highlight it by magnifying the spectra in the two theta range between 26° and 29°. Typically pure CZTSe presents a peak positioned at 27.16°; as sulphur is replacing Se the peak shifts towards higher 2θ values up to 28.44 °, which corresponds to pure CZTS [9]. So as the blue peak (500 °C) shifts to the left of the graph at lower wavenumbers, the layers is richer in Se compared to the 450°C- and 550°C-cases (red and black peaks).

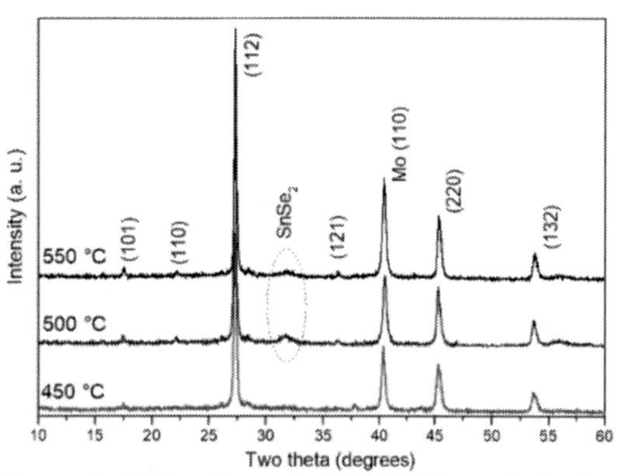

Figure 1. XRD pattern of CZTSSe films selenized at different temperatures.

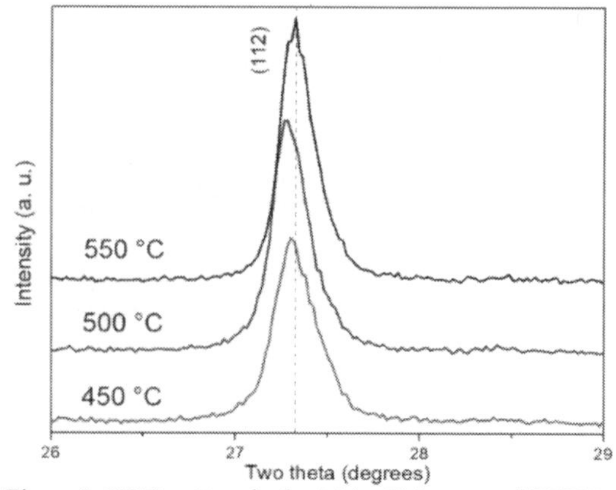

Figure 2. XRD pattern in the 26 ° - 29 ° range of CZTSSe films selenized at different temperatures.

Figure 3 presents the Raman spectra of the films selenized at different temperatures. All the films show the characteristic CZTSSe peaks at 56 cm^{-1}, 80–82 cm^{-1}, 174–177 cm^{-1}, 196–198 cm^{-1}, 235–250 cm^{-1} and 329–332 cm^{-1} corresponding respectively to the E, A, A, B and A vibrational modes of the CZTSSe phase [5]. Also in this case, by comparing the three spectra, a shift of the peaks can be observed. Similarly to the XRD, also for Raman the shift towards higher wavenumbers results in an increase of sulfur vs selenium.

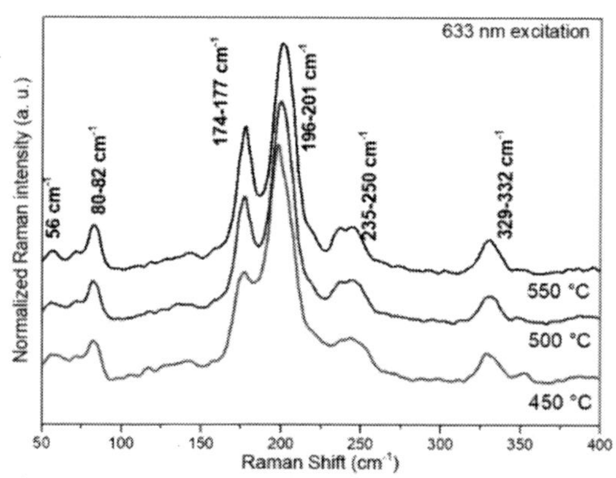

Figure 3. Raman spectra at 633 nm laser excitation of CZTSSe films selenized at different temperatures.

In Figure 4 we highlight the shift of the 196–198 cm^{-1} peak by magnifying the spectra in the 180 cm^{-1} - 225 cm^{-1} range. In this case the right-shift of the peak is accentuated with increasing temperature. The different behaviour, compared to XRD, can be explained by the fact that the Raman is mainly a surface analysis. So the shifts indicate a reduced amount of Se at the CZTSSe surface, demonstrating that the higher the annealing temperature the larger is the Se loss on the top of the absorber, where CdS is deposited.

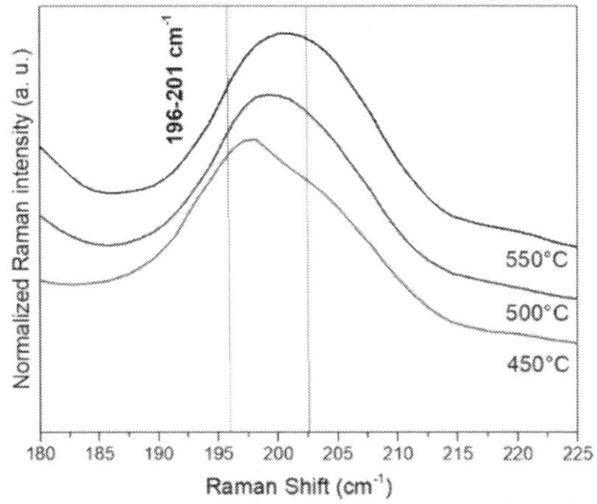

Figure 4. Raman spectra at 633 nm laser excitation, in the 180 cm^{-1} - 225 cm^{-1} range, of CZTSSe films selenized at different temperatures.

3.2 Performance parameters

As already mentioned the absorbers selenized at different temperatures were used to realized finished devices and their performance were measured.

Figure 5 shows the JV curves of three typical solar cells based on CZTSSe absorber layers selenized at 450°C, 500°C, and 550°C, the respective efficiency parameters are presented in Table I.

By increasing the selenization temperature, the efficiency of the devices improves, mainly thanks to a gain in Jsc. On the other hand, the FF gradually decreases, while the Voc demonstrates its maximum value at 500 °C and it is slightly lower at the other temperatures.

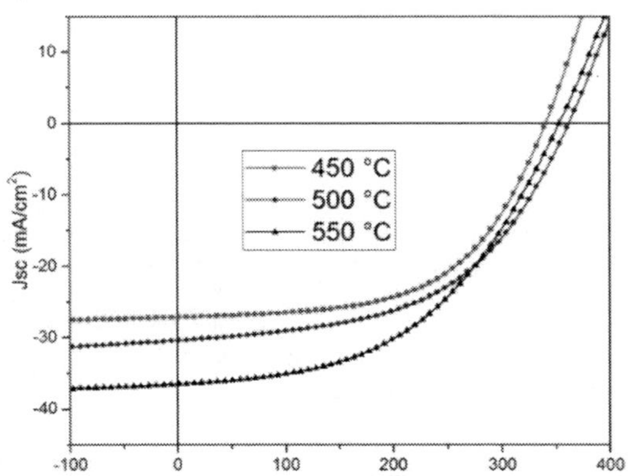

Figure 5. JV characteristics of CZTSSe devices fabricated through different selenization temperatures: 450° C red line, 500 °C blue line, 550 °C black line.

Table I: Efficiency parameters of the devices presented in Figure 5.

Selenization temperature	Voc (mV)	Jsc (mA/cm2)	FF (%)	η (%)
450 °C	340	27	56	5.2
500 °C	362	31	52	5.9
550 °C	353	37	48	6.3

4 CONCLUSIONS

The Jsc gain is observed for the 500°C-case and it is much larger for the 550 °C-case, we could speculate that as the selenization temperature increases the crystal quality improves (attested by the XRD analysis), however the measured excess of Se in the bulk for 500 °C limits the Jsc increase. With the 550 °C selenization, the Jsc reaches the maximum value: the stoichiometry is adjusted and the secondary phases reduced.

The reduction of FF might be attributed to the $MoSe_2$ that is identified by the small peak at 57° (see Figure 1) as well as to the loss of Se at the surface observed by the Raman analysis.

In conclusion, increasing the selenization temperature improves the CZTSSe quality, however it can differently affect Se content in the bulk and at the surface. One possible solution could be to control the Se flux during the annealing process in order to compensate the Se loss at higher temperature and the Se excess at lower temperatures.

5 REFERENCES

[1] J. Zhou *et al.*, "Regulating crystal growth via organic lithium salt additive for efficient Kesterite solar cells," *Nano Energy*, vol. 89, no. PB, p. 106405, 2021.

[2] S. H. Hadke *et al.*, "Synergistic Effects of Double Cation Substitution in Solution-Processed CZTS Solar Cells with over 10% Efficiency," *Adv. Energy Mater.*, vol. 8, no. 32, pp. 1–9, 2018.

[3] P. Prabeesh, P. Saritha, I. P. Selvam, and S. N. Potty, "Fabrication of CZTS thin films by dip coating technique for solar cell applications," *Mater. Res. Bull.*, vol. 86, pp. 295–301, 2017.

[4] P. Punathil, S. Zanetti, E. Artegiani, V. Kumar, and A. Romeo, "Analysis of the drying process for precursors of Cu2ZnSn(S, Se)4 layers by low cost non vacuum fabrication technique," *Sol. Energy*, vol. 224, no. July, pp. 992–999, 2021.

[5] P. Punathil, E. Artegiani, S. Zanetti, L. Lozzi, V. Kumar, and A. Romeo, "A simple method for Ge incorporation to enhance performance of low temperature and non-vacuum based CZTSSe solar cells," *Sol. Energy*, vol. 236, no. March, pp. 599–607, 2022.

[6] K. H. Kim and I. Amal, "Growth of Cu2ZnSnSe4 thin films by selenization of sputtered single-layered Cu-Zn-Sn metallic precursors from a Cu-Zn-Sn alloy target," *Electron. Mater. Lett.*, vol. 7, no. 3, pp. 225–230, 2011.

[7] P. Prabeesh, I. P. Selvam, and S. N. Potty, "Effect of annealing temperature on a single step processed Cu2ZnSnS4 thin film via solution method," *Thin Solid Films*, vol. 606, pp. 94–98, 2016.

[8] P. Prabeesh, I. Packia Selvam, and S. N. Potty, "Structural properties of czts thin films on glass and mo coated glass substrates: A rietveld refinement study," *Appl. Phys. A Mater. Sci. Process.*, vol. 124, no. 3, pp. 1–6, 2018.

[9] P. M. P. Salomé *et al.*, "Growth and characterization of Cu 2ZnSn(S,Se) 4 thin films for solar cells," *Sol. Energy Mater. Sol. Cells*, vol. 101, pp. 147–153, 2012.

DEFECT ANALYSIS OF GRADED $CdSe_xTe_{1-x}$ / CdTe THIN-FILM SOLAR CELLS BY MEANS OF THERMAL ADMITTANCE SPECTROSCOPY

Maximilian Schönau[1], Michel Nicklaus[2], Bernd Hüttl[2], Georgios Papageorgiou[1], Robert Arndt[1], Bettina Späth[1], Bastian Siepchen[1]
[1]CTF Solar GmbH, Manfred-von-Ardenne-Ring 4, 01099 Dresden, Germany
[2]Department of Electrical Engineering and Computer Sciences, Coburg University, Friedrich-Streib-Straße 2, 96450 Coburg, Germany
E-mail adress: Maximilian.Schoenau@ctf-solar.com

ABSTRACT: Thermal admittance spectroscopy (TAS) is often used to analyze trap states in CdTe thin-film photovoltaics. Specifically, the apparent energy level and the concentration of defects can be calculated using this method. In this work, we present TAS measurements on $CdSe_xTe_{1-x}$ / CdTe photovoltaic devices with varying Se concentrations and power conversion efficiencies from 12 % to 20 %. The results show two main trap signatures. Trap signature 1 has an apparent energy of about 0.43 eV above the valence band and is observed in all samples. A second trap signature was measured at about 0.10 eV for samples with low Se concentrations and 1.2 eV for samples with high Se concentrations. The defects concentrations are evaluated using the emission parameter instead of the typically used capture cross-section. When using a fit to an Arrhenius plot to characterize a trap signature, errors in the measurements lead to a compensation of the calculated apparent energy level and the emission parameter. After correcting the measurements by this effect, plots of PV performance versus the emission parameter of trap signature 1 show a good correlation, indicating that the emission parameter is suitable for determining concentration differences between the various samples.

1 INTRODUCTION

CdTe-based thin-film photovoltaics, such as $CdSe_xTe_{1-x}$ / CdTe solar cells, are a matured technology and have better performance stability over temperature and under low irradiance than silicon PV [1]. In addition, low production costs make them economically viable for tandem solar cells [2]; and their degradation was substantively reduced by replacing Cu-doping with As-doping [1].

As a result, CdTe PV may become a significant contributor to the worldwide effort to phase out conventional fuels. For single junction dot cells, efficiencies of up to 22.1 % were reached [3], with the theoretical limit at about 32 % [4] offering potential for improvement. However, the cell performance is highly limited by the Shockley-Read-Hall recombination of impurities in the cell absorber layer [5].

With Thermal Admittance Spectroscopy (TAS), these impurities can be characterized, enabling a better understanding of the defects in the research cells of CTF Solar GmbH. In this paper, we perform TAS measurements on $CdSe_xTe_{1-x}$ / CdTe photovoltaic devices with varying Se concentrations and power conversion efficiencies from 12 % to 20 %.

TAS and the similar Deep Level Transient Spectroscopy (DLTS) are comparatively difficult to apply to thin-film devices. Since the measurement can get affected by many interfering and sometimes unknown influences, the energy level and concentration of a trap are hard to obtain for thin film photovoltaics using TAS or DLTS [6–9]. Especially for calculating a trap's concentration, much semiconductor and defect-specific knowledge is required [7]. To get an indication of a trap's concentration nevertheless, the emission parameter [8] is used in this work instead of the commonly used capture cross-section.

2 METHODOLOGY

2.1 Fabricated Samples

CTF Solar GmbH prepared 10 graded $CdSe_xTe_{1-x}$ / CdTe solar cells. Samples with a wide range of power conversion efficiencies η from 12 % to 20 % with high and low Se concentrations [Se] were selected. The replacement of the "traditional" CdS / CdTe configuration with the $CdSe_xTe_{1-x}$ / CdTe enabled efficiencies exceeding 20 %, with significantly improved short circuit current densities. Samples 1 to 5 have low [Se] and efficiencies of 12 % to 17 %. Samples 6 to 10 have high [Se] and efficiencies of 18.5 % to 20 %.

2.2 Thermal Admittance Spectroscopy

The application of TAS is described extensively in the literature (e. g. [7, 10]). In a typical TAS measurement, the admittance $\underline{Y} = G + j\omega C$ is measured over a range of frequencies f and temperatures T. Then, the capacitance (or conductance) is plotted over temperature, and the capacitance inflection points (or the admittance maxima) are extracted. At these extrema, the thermal emission rates of the deep traps e_p resonate with the applied angular frequency $\omega = 2\pi f$. Therefore, m specific temperatures T_{1-m} can be acquired through the extrema of the admittance, with m being the number of deep traps. In short, the set of the emission rates' temperatures as a function of the angular frequency can be described with:

$$T_{1-m}(\omega) = \left\{T \in \mathbb{R}^+ : \frac{\partial^2 C(T,\omega)}{\partial T^2} = 0\right\} = \left\{T \in \mathbb{R}^+ : \frac{\partial G(T,\omega)}{\partial T} = 0\right\} \quad (1)$$

The temperatures are calculated with Eq. (1) for every applied frequency. Then, the emission rates associated with the calculated temperatures are derived with [7]:

$$e_p = \frac{1}{2} \cdot \omega \quad (2)$$

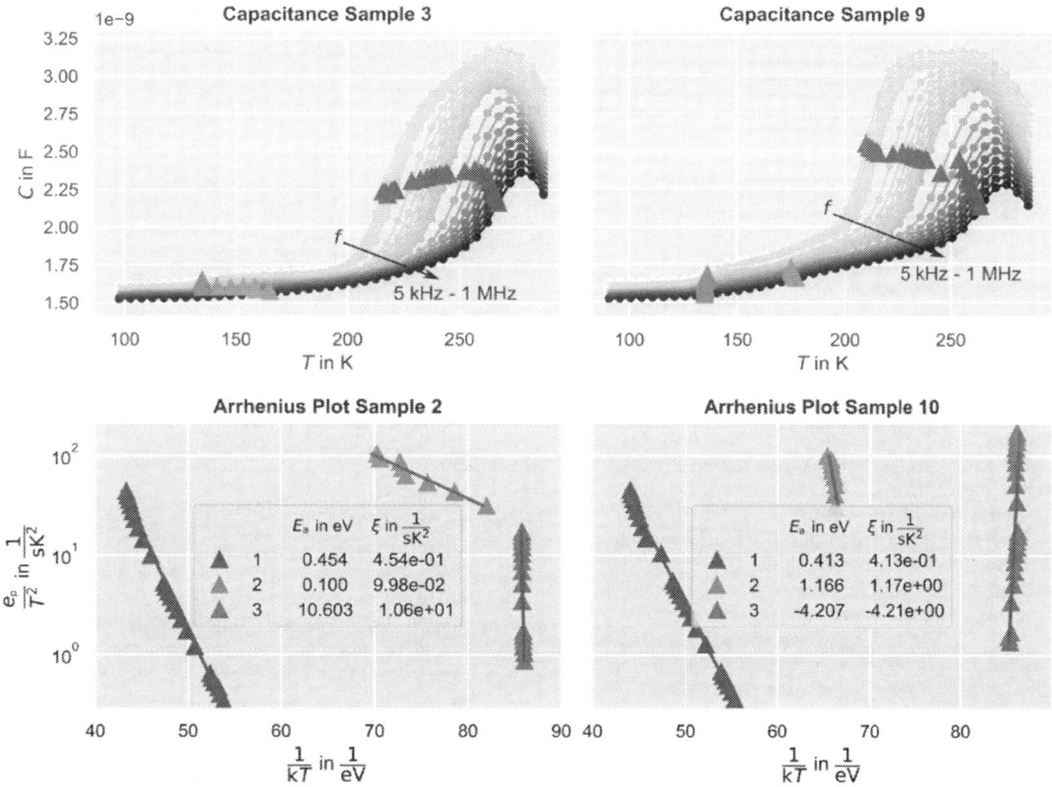

Figure 1: Capacitance spectrum over temperature and Arrhenius plots of Sample 3 (low [Se]) and Sample 9 (high [Se])

Thus a number of m sets of temperature-dependent emission rates called "trap signature" are created, which each should follow [6, 8]:

$$e_{\mathrm{p}}(T) = \xi T^2 \exp\left(-\frac{E_{\mathrm{a}}}{\mathrm{k}T}\right) \quad (3)$$

Where the emission parameter ξ is calculated with:

$$\xi = N_{\mathrm{V}} \langle v_{\mathrm{p}} \rangle T^{-2} \sigma_{\mathrm{a}} \quad (4)$$

With k being the Boltzmann constant, N_{V} the effective density of the valence band states, $\langle v_{\mathrm{p}} \rangle$ the average thermal velocity of holes and σ_{a} the apparent capture cross-section.

When displaying Eq. (3) semilogarithmically in an Arrhenius plot, the emission parameter follows a straight line, with the slope being the apparent energy level of the defect E_a. The line intersects the y-axis at the emission parameter. This parameter can be considered temperature-independent because $N_V\langle v_p \rangle \sim T^2$ which results in $N_V\langle v_p \rangle \cdot T^{-2}$ being constant [6]. Both the apparent energy and the apparent capture cross-section can differ from the actual energy level of a defect E_T and the capture cross-section σ [7], which limits TAS in its accuracy.

To get a reasonable estimation of the concentration of a defect, one must first calculate the capture cross-section. However, this requires much semiconductor and defect-specific knowledge [7]. Thus, it is often challenging to calculate the concentration of a defect when performing TAS or DLTS, especially for the characterization of thin-film PV, where those calculations are particularly complicated [6]. As an alternative, the emission parameter may be carefully used to get qualitative information about the concentration of a defect when comparing the same defect of similar behaving samples.

2.3 Capacity Measurements

Both the capacity and conductance measurements can be used to create Arrhenius plots. Our analysis was done using capacity measurements, which seem to be the better choice for thin film devices, being less affected by series and shunt resistance [11]. We applied a reverse bias of 0.2 V on our samples and measured with frequencies from 5 kHz to 1 MHz.

The capacity was measured with a Keithley 4200A-SCS Parameter Analyzer configured with two Keithley 4210 SMUs. A cryostat was used to temper the samples from 80 K to 290 K. Because of the thermal settling times of our cryostat system, an extended measurement interval of about 10 min was chosen, for which thermal coupling between the cryostat and sample could be demonstrated experimentally. However, with the long measurement duration, the sample rate of the capacity measurements over temperature could not be chosen very high. Therefore, we settled on measurements of about every 5 K after ensuring that higher resolutions do not lead to significantly better TAS results.

The measurement must be interpolated to find the inflection points of the capacitance as accurately as possible with this sample rate. Simple interpolation methods were insufficient for our analysis, particularly because two differentiations of the capacitance must be calculated to get its inflection points. However, good results were achieved using the "MS filter" of M. Schmid et al. [12].

RESULTS AND DISCUSSION

3.1 Trap Signatures of $CdSe_xTe_{1-x}$ / CdTe Cells

Figure 1 shows capacitance versus temperature and extracted Arrhenius plots of an exemplarily low and a high [Se] sample. Three trap signatures per sample can be observed, with energies matching various types of defects reported elsewhere [5, 13]. Fewer trap signatures were found than expected in CdTe-based PVs (e. g. [9, 14–16]). This may result from our observable trap signatures overshadowing additional deep states in our $CdSe_xTe_{1-x}$ / CdTe Cells.

All samples demonstrate a trap signature at an apparent energy of about 0.43 eV (trap signature 1). A second trap signature was measured at about 0.10 eV for samples with low [Se] and 1.2 eV for high [Se] samples. The trap signature 3 of samples 3 and 9 are anomalous because of their very high and even negative apparent energy, respectively. Such trap signatures were discussed elsewhere (e.g. [17]) and are not the focus of this work. Because trap signature 1 could be measured reliably for all samples and correlated with our cells' performance, it is discussed here further.

To be sure about the origin of this trap signature, an additional feature, the back-contact Schottky barrier, which is typical in CdTe devices, should be considered. Since such a Schottky barrier may appear in TAS as a trap signature [17], we performed temperature-dependent dark current density-voltage measurements for all samples. Because the Schottky barrier was hardly detectable, we attribute trap signature 1 to a defect. However, there is not enough data to completely deconvolute the two effects. Trap signature 1 is further analyzed using the emission parameter.

3.2 Correction of the Emission Parameter

An exponential relationship between the calculated apparent energy and emission parameter of trap signature 1 can be observed (Figure 2). This relationship has been reported in the literature and was sometimes attributed to the kinetic compensation effect or Meyer–Neldel rule (e.g. [6]). However, a mathematical origin of this effect seems much more plausible [8, 18]:

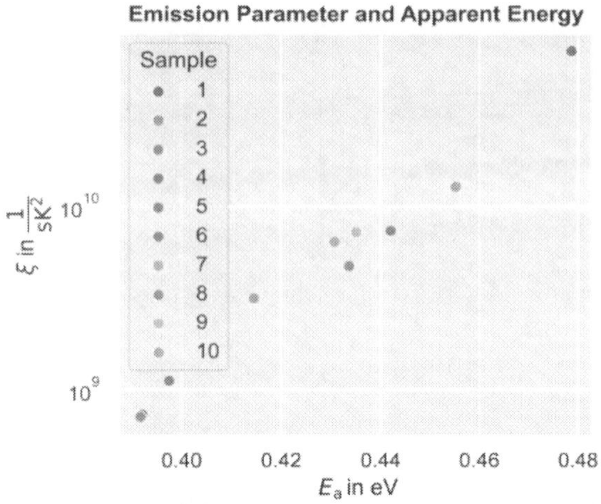

Figure 2: Emission parameter over apparent energy of trap signature 1

Because the apparent energy and the emission parameter are fitted with Eq. (3) simultaneously, their calculation is not independent. Thus, errors in the calculated apparent energy (the slope in the Arrhenius plot) are propagated to the calculated emission parameter (the intersection of the straight line with the y-axis of the Arrhenius plot). Therefore, whenever measurement errors lead to a higher energy level than the "true" value, the resulting emission parameter is too high, and when the energy level is calculated too low, the emission parameter gets calculated too low as well. Random measurement errors are to be expected mainly because of the tempering of the samples in the cryostat, which is limited in accuracy.

In this work, we compensated for this effect by refitting the emission parameter while using a fixed value for the apparent energy level of trap signature 1. The linear fits in the Arrhenius were recalculated using the expected value of the apparent energy levels of $\bar{E}_a = 0.43$ eV. This correction enabled consistent results and a correlation between the emission parameter and PV performance.

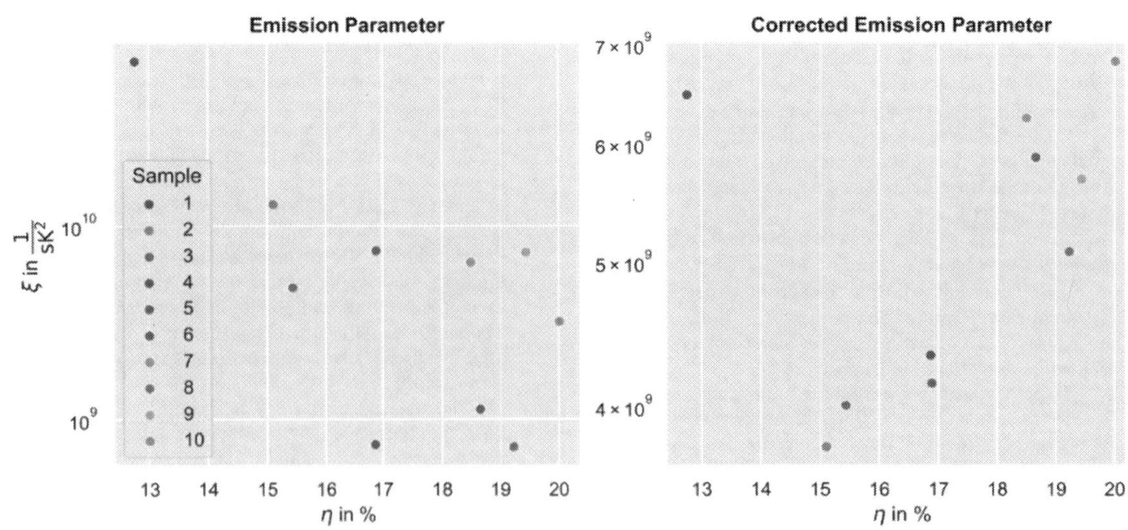

Figure 3: Emission parameter and corrected emission parameter (with $\bar{E}_a = 0.43$ eV) over efficiency of trap signature 1

3.3 Interpretation of the Emission Parameter

Figure 3 shows plots of the emission parameter and corrected emission parameter versus power conversion efficiency of all ten samples. A good correlation between the corrected emission parameter and efficiency can be observed. Sample 1 is an outlier in this correlation; the low performance can be addressed by processing errors. Assuming that the emission parameter is a metric of the defect's concentration, higher concentrations of defect 1 result in higher PV performance. Additionally, the defect's concentration is higher for samples with high [Se] than those with low [Se]. One explanation could be that the measured defect acts as a p-dopant.

This interpretation of the corrected emission parameter is not indisputable; different effects could explain the discussed correlation. As displayed in Eq. (4), the emission parameter does not only depend on the trap's concentration. However, interpreting the emission parameter as an indicator of a defect's concentration seems plausible.

4 CONCLUSION

We performed TAS on $CdSe_xTe_{1-x}$ / CdTe photovoltaic devices with varying Se concentrations and power conversion efficiencies from 12 % to 20 %. The results show two main trap signatures. Trap signature 1 has an apparent energy of about 0.43 eV and is observed in all samples. The concentrations of this defect were evaluated using the emission parameter. After correcting the measurements by the kinetic compensation effect, plots of PV performance versus the emission parameter of trap signature 1 show a good correlation, indicating that the emission parameter is suitable for determining concentration differences between the various samples. Thus, the corrected emission parameter could be a helpful complementary indicator for characterizing the defects of a thin-film device.

LITERATURE

[1] W. K. Metzger *et al.*, "As-Doped CdSeTe Solar Cells Achieving 22% Efficiency With −0.23%/°C Temperature Coefficient," *IEEE J. Photovoltaics*, pp. 1–4, 2022.

[2] A. C. Tamboli, D. C. Bobela, A. Kanevce, T. Remo, K. Alberi, and M. Woodhouse, "Low-Cost CdTe/Silicon Tandem Solar Cells," *IEEE J. Photovoltaics*, vol. 7, no. 6, pp. 1767–1772, 2017, doi: 10.1109/JPHOTOV.2017.2737361.

[3] NREL. "Best Research-Cell Efficiency Chart." https://www.nrel.gov/pv/assets/pdfs/best-research-cell-efficiencies-rev211117.pdf (accessed Feb. 2, 2022).

[4] G. Xiong and W. Metzger, "Thin Film Cadmium Telluride Photovoltaics," in *Earth systems and environmental sciences*, Elsevier, Ed., Elsevier, 2021.

[5] D. N. Krasikov, A. V. Scherbinin, A. A. Knizhnik, A. N. Vasiliev, B. V. Potapkin, and T. J. Sommerer, "Theoretical analysis of non-radiative multiphonon recombination activity of intrinsic defects in CdTe," *Journal of Applied Physics*, vol. 119, no. 8, p. 85706, 2016, doi: 10.1063/1.4942529.

[6] D. Abou-Ras, T. Kirchartz, and U. Rau, *Advanced characterization techniques for thin film solar cells*, 2nd ed. Weinheim: Wiley-VCH Verlag, 2016.

[7] P. Blood and J. W. Orton, *The electrical characterization of semiconductors: majority carriers and electron states* (Techniques of physics 14). London: Academic Press, 1992.

[8] A. Gerhard, "Elektrische Defektspektroskopie an $CuGaSe_2$ und verwandten Halbleiterdünnschichten," Dissertation, Freie Universität Berlin, Berlin, 2000. [Online]. Available: https://refubium.fu-berlin.de/handle/fub188/10950

[9] J. V. Li, S. W. Johnston, X. Li, D. S. Albin, T. A. Gessert, and D. H. Levi, "Discussion of some "trap signatures" observed by admittance spectroscopy in CdTe thin-film solar cells," *Journal of Applied Physics*, vol. 108, no. 6, p. 64501, 2010, doi: 10.1063/1.3475373.

[10] A. R. Peaker, V. P. Markevich, and J. Coutinho, "Tutorial: Junction spectroscopy techniques and deep-level defects in semiconductors," *Journal of Applied Physics*, vol. 123, no. 16, p. 161559, 2018, doi: 10.1063/1.5011327.

[11] C. León, J. M. Martin, J. Santamaría, J. Skarp, G. González-Díaz, and F. Sánchez-Quesada, "Use of Kramers–Kronig transforms for the treatment of admittance spectroscopy data of p - n junctions containing traps," *Journal of Applied Physics*, vol. 79, no. 10, pp. 7830–7836, 1996, doi: 10.1063/1.362391.

[12] M. Schmid, D. Rath, and U. Diebold, "Why and How Savitzky-Golay Filters Should Be Replaced," *ACS measurement science Au*, early access. doi: 10.1021/acsmeasuresciau.1c00054.

[13] F. H. Seymour, V. Kaydanov, T. R. Ohno, and D. Albin, "Cu and CdCl2 influence on defects detected in CdTe solar cells with admittance spectroscopy," *Appl. Phys. Lett.*, vol. 87, no. 15, p. 153507, 2005, doi: 10.1063/1.2099515.

[14] J. Versluys, P. Clauws, P. Nollet, S. Degrave, and M. Burgelman, "DLTS and admittance measurements on CdS/CdTe solar cells," *Thin Solid Films*, vol. 431-432, pp. 148–152, 2003, doi: 10.1016/S0040-6090(03)00202-5.

[15] F. H. Seymour, V. Kaydanov, T. R. Ohno, and D. Albin, "Cu and CdCl2 influence on defects detected in CdTe solar cells with admittance spectroscopy," *Appl. Phys. Lett.*, vol. 87, no. 15, p. 153507, 2005, doi: 10.1063/1.2099515.

[16] Y. Y. Proskuryakov, K. Durose, B. M. Taele, G. P. Welch, and S. Oelting, "Admittance spectroscopy of CdTe/CdS solar cells subjected to varied nitric-phosphoric etching conditions," *Journal of Applied Physics*, vol. 101, no. 1, p. 14505, 2007, doi: 10.1063/1.2402961.

[17] Y. Y. Proskuryakov, K. Durose, B. M. Taele, and S. Oelting, "Impedance spectroscopy of unetched CdTe/CdS solar cells—equivalent circuit analysis," *Journal of Applied Physics*, vol. 102, no. 2, p. 24504, 2007, doi: 10.1063/1.2757011.

[18] P. J. Barrie, "The mathematical origins of the kinetic compensation effect: 1. The effect of random experimental errors," *Physical chemistry chemical physics : PCCP*, early access. doi: 10.1039/c1cp22666e.

A STUDY OF THE ROLE OF THE ANNEALING TEMPERATURES AND TIMES ON C2uZnSnS4 NANOCRYSTAL THIN FILMS PROPERTIES

Asaad Alluhaybi [1,2], Douglas Halliday [2], Marek Szablewski [2]

[1] Jeddah University, Khulais College of Science and Arts, Department of Physics, Khulais, 21921, Saudi Arabia, Email: asaad.m.alluhaybi@durham.ac.uk, aalluhaybi@uj.edu.sa

[2] Durham University, Department of Physics, Centre for Materials Physics, South Road, Durham, DH1 3LE, United Kingdom

ABSTRACT: Cu_2ZnSnS_4 (CZTS) nanoparticle thin films were produced by hot-injection spin-coating technique, and successfully deposited on soda lime glass substrates (SLG). The impact of different annealing temperatures and annealing times were investigated by using X-ray diffraction (XRD), Focus Ion Beam Microscopy (FIB), Energy-dispersive X-ray spectroscopy (EDS), Raman spectroscopy and Ultraviolet-visible spectroscopy (UV-vis). The XRD measurements confirmed the formation of kesterite structure in the CZTS thin films. From the optical properties, the films demonstrate improvement in uniformity and crystallinity of CZTS with annealing. This study shows that annealing conditions of 585 °C in a N_2 and H_2S environment for one hour is suitable step to produce CZTS absorber layers for successful CZTS solar cell device fabrication. Keywords: CZTS, Hot-injection synthesis, Thin-films

1 Introduction

In recent years, Copper-based quaternary semiconductor material Cu_2ZnSnS_4 (CZTS) has been investigated as it might be one of the alternative thin film materials to use in low-cost thin film solar cells devices because of its promising properties. Its high absorption coefficient (>10^4 cm^{-1}) and the direct band gap of around 1.5 eV which are ideal properties for solar cell applications give promising properties [1]. In addition to these properties, all elements of CZTS are earth abundant, environmentally friendly, and cheap to produce in contrast to Copper Indium Gallium Selenide (CIGS) and Cadmium Telluride (CdTe), which have scarce elements in them like indium (In) and tellurium (Te) as well as having high prices in their production [2-4].

CZTS can be fabricated by technologies based on vacuum condition, such as sputtering [4], coevaporation [5, 6], and epitaxial methods [7] or based on non-vacuum condition, such as hot-injection method [8,9]. As the vacuum approaches seem to have disadvantages such as the high-cost, complicated process and difficulty in scaling up production, the non-vacuum methods can be used here in this study for its lower cost and more suitability for mass production.

There are several published papers of the annealing treatment work, however, in this study a different range of annealing temperatures is investigated [10,11]. In addition, UV-vis is used here compared to the other published papers.

2 Experimental details

Our approach is based on the procedures in the published paper by Guo et al. 2009 [12]. The CZTS nanocrystal inks were successfully synthesised with Cu-poor Zn-rich composition (Cu/Zn+Sn) = 0.87, (Zn/Sn) = 1.34 and (S/Cu+Zn+Sn) = 1.29. In this method, the nanoparticles were prepared as follows: 1.44 mmol (0.3769 g) of copper (II) acetylacetonate, 0.94 mmol (0.2478 g) of zinc (II) acetylacetonate, 0.70 mmol (0.2714 g) of tin (IV) bis(acetylacetonate) dichloride and 10 ml of oleylamine (OLA) were added into a three-neck flask with a small stirring. A heating mantle was used to heat the mixture to around 140 °C under vacuum and degassed for 30 minutes and purged with Ar several times. The temperature was then increased to 250°C where a solution of 4 mmol (0.1283 g) of sulphur in 4 ml of OLA was injected into the mixture with holding the same temperature for 30 minutes. Then the mixture was left to cool down to 50°C by removing the heating mantle. Then, 5 ml of toluene and 40 ml of isopropanol were added into the solution, and the nanoparticles were collected using the centrifuge at 10000 rpm for 10 minutes. After that, the supernatant was decanted. Then, 20 ml of toluene and 10 ml of isopropanol were added to the mixture and centrifuged again, but at 8000 rpm this time for 10 minutes and the supernatant was decanted for the second time (this process was repeated twice). The created CZTS powder was then dried under vacuum overnight and prepared to form the ink solution by adding in hexanethiol as a solvent (250 mg of CZTS in 1 ml of hexanethiol).

The CZTS nanoparticle inks were deposited on glass substrate via spin-coating technique. The glass first was cleaned and washed by using acetone, methanol and distilled water for ten minutes for each of them in ultrasonic bath. Then, a nitrogen gun was used to dry the glass substrate. After that, the spin-coater was set up for the suitable parameters: 2500 rpm for the spinning speed for 15 seconds and spin acceleration index 010. The dried glass was put into the spin-coater and drops (2-10 drops) of the ink (40 microlitre per drop) were dropped into the centre of the glass with the rotated speed mentioned above making sure of achieving uniformed thin films. Two drops of thin films are used for FIB, EDS and UV-vis measurements while 10 drops producing thicker films are used for XRD and Raman measurements. The glass was then heated for 300 °C by heating mantle for 30 seconds in air to remove any solvent after each drop and produce stable thin films.

The prepared CZTS thin films on substrates undergo sulphurisation annealing at a suitable atmosphere (N_2 and H_2S) to improve the crystallinity. The two ends of the furnace's tube caps were connected to different valves: a vacuum pump, exhaustion line, nitrogen and H_2S supplies. The annealing process was applied under the conditions of set temperatures at 560 °C, 585 °C, 610 °C, 635 °C and (585 °C with no H_2S atmosphere) for one hour at ramping rate of 10 degrees C per minute. After that the furnace was left to cool naturally overnight. When cooled to room temperature, the furnace tube finally was flushed for several times with nitrogen gas and vented.

3 Results and discussion

3.1 The impact of annealing temperatures

For the structural measurement, X-ray diffraction (XRD) is used to determine the crystal structure of CZTS nanoparticles thin film as well as the size of the scattering domain. The XRD peak positions are determined from the fitting of the spectra by using Lorentzian curves.

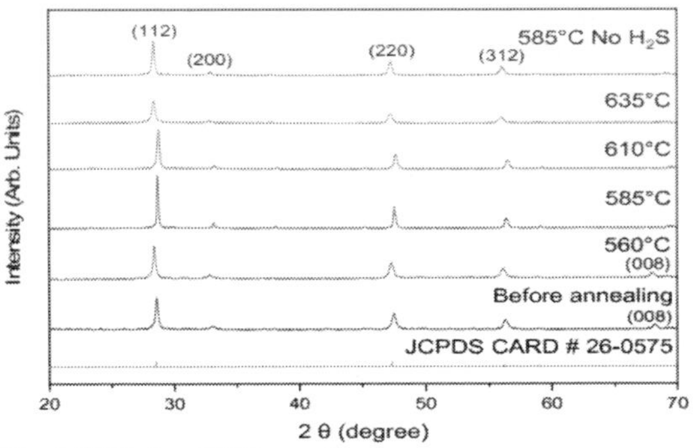

Fig. 1 XRD for CZTS samples.

The XRD measurement in fig. 1 confirmed the formation of kesterite structure in CZTS thin films for all the annealing temperatures used as the four main diffraction peaks (112), (200), (220) and (312) appeared in the location corresponding to JCPDS card number 26-0575. In addition (008) peak observed in the samples of before annealing and 560°C which also correspond to JCPDS card number 26-0575. However, because of the similarity of these peaks' locations with the secondary phases of ZnS (PDF card 65-1691) and Cu_2SnS_3 (PDF card 27-0198), it is difficult to distinguish between them and the presence of CZTS [13]. Therefore, Raman spectroscopy is used to identify the possible presence of any secondary phases figures 3 and 8 below [14]. The peaks for the (112) structure planes were located at 28.47°, 28.34°, 28.66°, 28.74°, 28.35° and 28.34° for the annealing temperatures at before annealing, 560 °C, 585 °C, 610 °C, 635 °C and (585 °C with no H_2S atmosphere) respectively. The values of *d*-spacing were 0.3133 nm, 0.3143 nm, 0.3115 nm, 0.3106 nm, 0.3139 nm and 0.3147 nm respectively. The values of lattice parameter *a* were 0.5417 nm, 0.5435 nm, 0.5409 nm, 0.5399 nm, 0.5440 nm and 0.5441 nm respectively. The values of lattice parameter *c* were 1.0895 nm, 1.0929 nm, 1.0738 nm, 1.0690 nm, 1.0868 nm and 1.0944 nm respectively. The *d*-spacing, *a* and *c* values of the 585°C show similar values with the literature [15]. The possible secondary phases peaks such as SnS_2 at 33.02° (JCPDS card 83-1707), SnS at 45.80° (JCPDS card 83-1758) and Cu_3SnS_4 at 37.27°, 39.81°, 51.80° (JCPDS card 36-0217) were not observed in Fig. 1 [16, 17].

The scattering domain size in table I was calculated from Debye-Scherrer formula as follow

$$L = \frac{k\lambda}{D\cos\theta} \qquad (1)$$

Where k is a constant = 0.9, λ is the x-ray wavelength (for Cu Kα, this is 0.154 nm), D is the full width at half maximum in radians (FWHM) and θ is the Bragg angle of diffraction (peak position) [18].

Table I: Values of the size of scattering domain

Temperatures (°C)	Scattering domain (nm)
Before annealing	15.85
560	28.28
585	32.47
610	42.71
635	53.27
585 no H_2S	11.68

The size of scattering domain showed a clear trend which increases with the rise of annealing temperatures except when the H_2S was absent at 585°C with a low value of 11.68 nm.

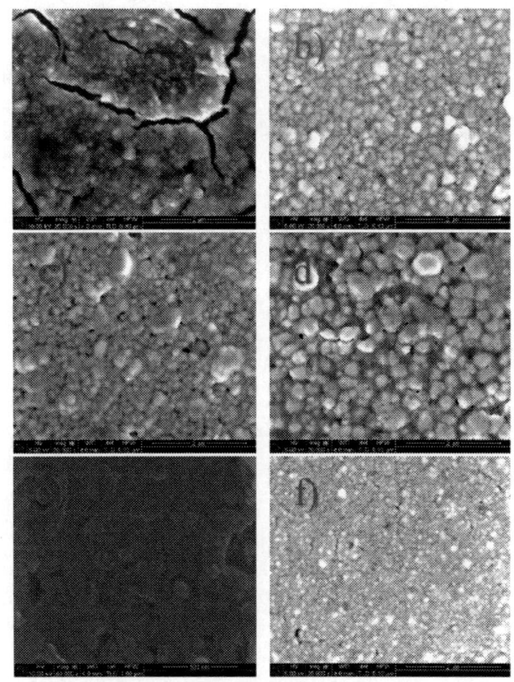

Fig. 2 FIB images for CZTS nanoparticles as a) Before annealing, b) 560°C, c) 585°C, d) 610°C, e) 635°C and f) 585°C No H_2S.

In fig. 2, it is clear that the annealing temperature plays an important role on the uniformity of the CZTS nanoparticles as this improved compared to before annealing. In addition, the gaps and cracks in the film before annealing almost disappeared in all other images which might indicate the development of the crystallinity of the CZTS nanoparticles and the formation of a continuous thin-film. In images b, d and f, there were brighter nanoparticles than the rest owing to the excessive amount of Zn, while there were less of them in image c. The dark areas in image e were because of the absence of Sn and Zn as a result of annealing at higher temperatures.

In table I-I, the ratios of 560 °C and 585 °C were similar to the starting composition of Cu-poor Zn-rich in the experimental details. However, the temperature of 585 °C with no H_2S showed different ratios than the original films due to the absence of the H_2S atmosphere.

Table I-I: Composition ratios of CZTS nanoparticles by EDS

Temperatures (°C)	Cu/(Zn+Sn)	Zn/Sn	S/(Cu+Zn+Sn)
Before annealing	0.65	0.67	1.17
560	0.80	1.15	1.07
585	0.83	1.17	1.32
610	1.05	1.75	0.85
635	1.15	2.25	0.67
585 no H_2S	0.41	2.41	0.80

The Raman peak fitting is done by using Lorentzian curves. Raman main and strong peaks of CZTS in fig. 3 are located at 334 cm^{-1} for before annealing and 560 °C, 337 cm^{-1} for 585 °C and 635 °C, 335 cm^{-1} for 610 °C and 339 cm^{-1} for 585 °C with no H_2S. These all locations are in good agreement with the literature papers [19-21]. Another weak peak appears at 290 cm^{-1} for 585 °C with no H_2S which seems to be a secondary phase of ZnS [22,23].

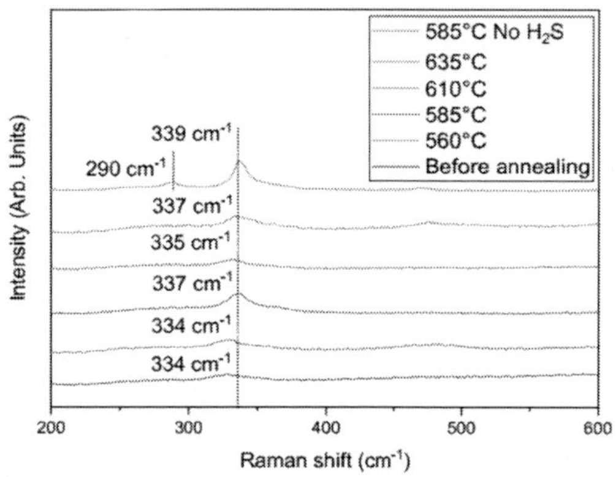

Fig. 3 Raman spectra of CZTS samples with different annealing temperatures for 60 minutes of annealing times.

For the optical measurement, UV-vis spectrophotometer was used to determine the band gap of CZTS nanoparticles thin film from the optical absorption spectrum.

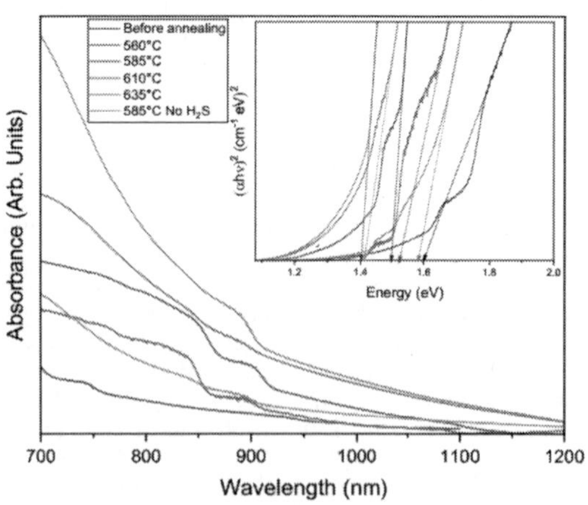

Fig. 4 Absorbance spectrum of CZTS samples and the inset shows Tauc plot to determine the band gap of CZTS.

The energy band gap can be determined by extrapolating straight line of the plot $(Ahv)^2$ versus Eg as shown in fig. 4 by using the following equation:

$$(\alpha h v) = \sqrt{A(hv - Eg)} \quad (2)$$

Where A is a constant determined by the electronic band structure, h is the Planck constant, v is the frequency of the light, and Eg is the band gap of the material [24]. Fig. 4 showed the energy band gaps as 1.60, 1.52 ± 0.005, 1.50 ± 0.021, 1.41 ± 0.028, 1.40 ± 0.010 and 1.58 eV for annealing temperatures at 560 °C, 585 °C, 610 °C, 635 °C and 585 °C with no H_2S respectively. All these energy band gaps correspond well with that reported in literature of between 1.40-1.60 eV [25-30]. It is observed that there is a proportional relationship between the energy band gap and the annealing temperatures except for the sample annealed at 585°C with no H_2S atmosphere as fig. 5 shows below.

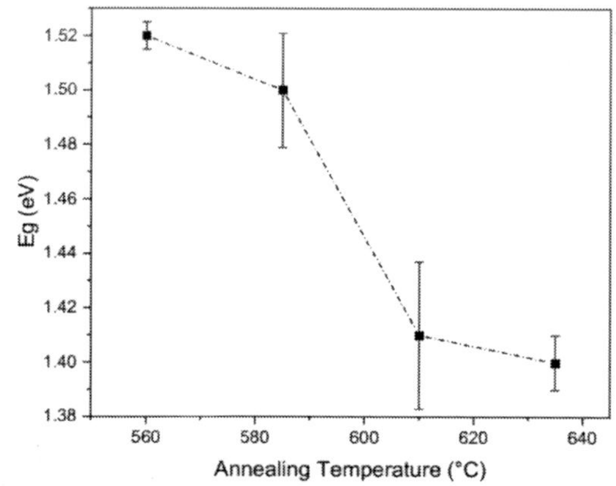

Fig. 5 The trend of the band gap of CZTS samples vs different annealing temperatures.

According to the analysis of the annealing temperatures by using XRD, FIB, EDS and UV-vis as above, the temperature of 585°C was used to study the impact of the annealing times.

3.2 The impact of annealing times

The formation of kesterite structure in CZTS thin films were confirmed by XRD for all the annealing times as it is shown in fig. 6.

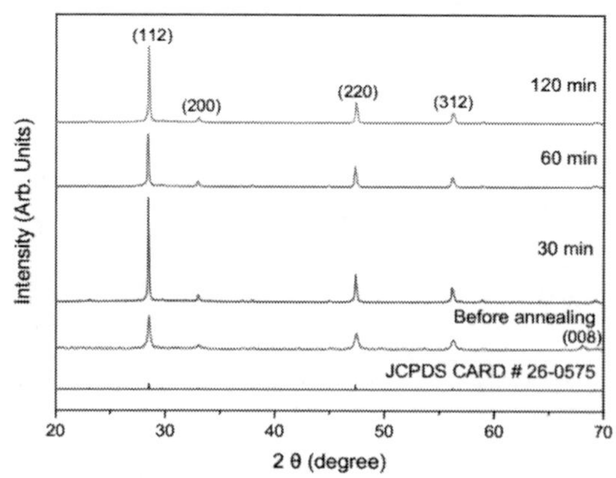

Fig. 6 XRD for CZTS samples with different annealing times.

The four main diffraction peaks (112), (200), (220) and (312) appeared at the location correspond to JCPDS card number 26-0575. In addition (008) peak appears in the samples before annealing which also correspond to

JCPDS card number 26-0575. The peaks for the (112) structure planes were located at 28.47°, 28.44°, 28.66° and 28.45° for the annealing times at before annealing, 30 minutes, 60 minutes and 120 minutes respectively. The values of *d*-spacing were 0.3133 nm, 0.3136 nm, 0.3115 nm and 0.3134 nm respectively. The values of lattice parameter *a* were 0.5417 nm, 0.5426 nm, 0.5409 nm and 0.5421 nm respectively. The values of lattice parameter *c* were 1.0895 nm, 1.0892 nm, 1.0738 nm and 1.0884 nm respectively. The d-spacing, *a* and *c* values of 60 min show similar values with the literature papers [15].

Table I-I-I: Values of the size of scattering domain for different annealing times

Times (minutes)	Scattering domain (nm)
Before annealing	15.85
30	66.62
60	32.47
120	52.43

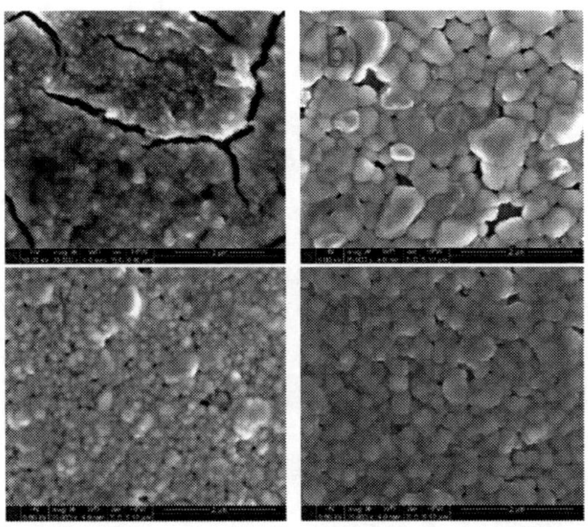

Fig. 7 FIB images for CZTS nanoparticles with different annealing times as a) Before annealing, b) 30 min, c) 60 min, d) 120 min.

The annealing times also played an important role as the temperatures did on the uniformity of the CZTS nanoparticles. In image c, the development of the crystallinity and uniformity of the CZTS nanoparticles was much better than on other samples, displaying a consistently uniform range of particle sizes.

Table I V: Composition ratios of CZTS nanoparticles with different annealing times by EDS

Times (minutes)	Cu/(Zn+Sn)	Zn/Sn	S/(Cu+Zn+Sn)
Before annealing	0.65	0.67	1.17
30	0.76	1.40	0.93
60	0.83	1.17	1.32
120	0.91	1.57	1.47

The ratios of 60 min were similarly matching to the Cu-poor Zn-rich ratios in the experimental details.

Raman main and strong peaks of CZTS in fig. 8 are located at 334 cm^{-1} for before annealing, and 339 cm^{-1} for 30 minutes, 337 cm^{-1} for 60 minutes and 338 cm^{-1} for 120 minutes. These locations are all in good agreement with the literature papers [19-21]. Another peak appears at 288 cm^{-1} for 120 minutes which seems to be a secondary phase of MoS_2 [31,32] or SnS [33].

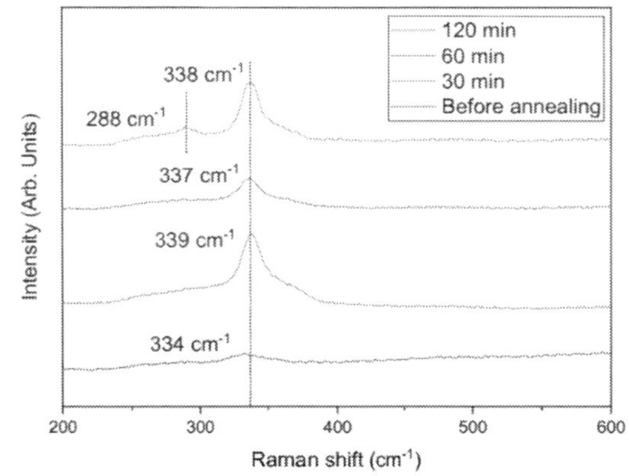

Fig. 8 Raman spectra of CZTS samples with different annealing times for 585°C.

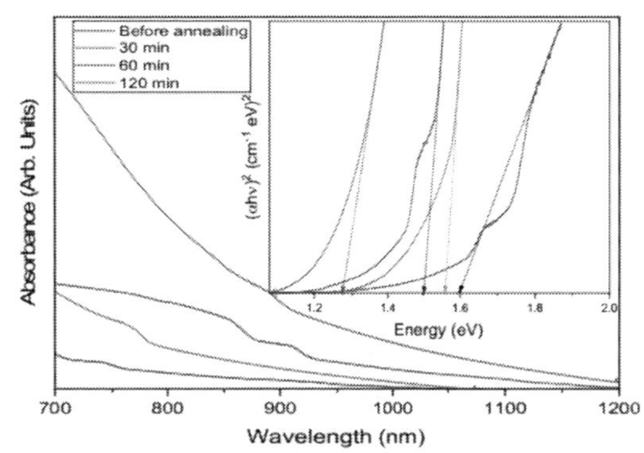

Fig. 9 Absorbance spectrum of CZTS samples with different annealing times and the inset shows the Tauc plot.

Fig. 9 showed the energy band gaps as 1.60, 1.28, 1.50, and 1.56 eV for annealing times at before annealing, 30 minutes, 60 minutes, and 120 minutes respectively. The observed band gap of the 60 and 120 minutes time period (1.50 eV and 1.56 eV) correspond well with several literature reports [25-30]. It seems that too short annealing time was not preferable to have the suitable energy band gap for CZTS as the 30 minutes sample showed a smaller band gap of 1.28 eV.

4 Conclusion

In conclusion, CZTS nanoparticles inks were prepared via hot-injection and deposited into (SLG) glass substrates by spin-coating. A range of different annealing temperatures 560 °C, 585 °C, 610 °C, 635 °C and 585 °C with no H_2S atmosphere were investigated to find out the most suitable one for the fabrication of CZTS solar cell devices by using several characterization instruments such as XRD, FIB, EDS and UV-vis. Based on the analysis of data collected, 585°C was picked up to be suitable for the annealing temperature. In addition, a range of different annealing times (30 min, 60 min, and 120 min) were investigated to study the duration effect on CZTS thin films. The 60 min sample showed a better performance on matter of energy band gap, uniformity and crystnilty among other times. Therefore, annealing temperature 585°C for 60 min might be the most suitable to improve the performance of a CZTS solar cell device.

Acknowledgments

Authors are thankful to Durham University, UK and Jeddah University, Saudi Arabia for support. The authors acknowledge Professor A. Beeby for the use of Raman facilities.

4.1 References

[1] M. Islam, et al, Journal of Advanced Research in Fluid Mechanics and Thermal Sciences 2021. 81(1): p. 73-87.
[2] H. Zhou, et al, Energy & Environmental Science 2013. 6(10).
[3] W. Liu, et al, Journal of Materials Chemistry A 2013. 1(9).
[4] L. Peter, L.M, Philos Trans A Math Phys Eng Sci 2011. 369(1942): p. 1840-56.
[5] T. Dhakal, et al, Solar Energy 2014. 100: p. 23-30.
[6] R. Moreno, et al, Journal of Physics Conference Series, 2016. 6875.
[7] T. Tanaka, et al, Physica Status Solidi c 2006. 3(8): p. 2844-2847.
[8] X. Zhang, et al, Nanomaterials (Basel) 2019. 9(3).
[9] H. Wei, et al, Materials Letters 2010. 64(13): p. 1424-1426.
[9] S. Rondiya, et al, Journal of Alloys and Compounds 2022. 890
[10] Y. Altowairqi, et al, Materials Today: Proceedings, 2019. 18: p. 473-486.
[11] S. Jain, et al, Inorganic Chemistry Communications 2021. 133.
[12] Q. Guo, et al, Am Chem Soc 2009, 131, 33, p. 11672–11673.
[13] M. Wei, et al, Materials Letters 2012. 79: p. 177-179.
[14] V. Dzhagan et al, Electronic Materials 2022. 3.1 p. 136-153.
[15] Q. Tian, et al, Cryst Eng Comm 2012. 14(11).
[16] J. Raiguru, et al, National Conference on Processing and Characterization of Materials 2017, IOP Publishing Ltd: Bristol.
[17] S. Bahramzadeh, et al, Journal of Alloys and Compounds 2015. 642: p. 124-130.
[18] VD. Mote, et al, Journal of Theoretical and Applied Physics 2012. 6(1): p. 1-8.
[19] G. Babu, et al, Semiconductor Science and Technology 2008. 23(8).
[20] P. Fernandes, et al, Thin Solid Films 2009. 517 p. 2519.
[21] M. Altosaar, et al, Phys. Status Solidi (a) 2008. 205 p. 167.
[22] Y.C. Cheng et al, Journal of Applied Physics 2009. 106 (12).
[23] W.G. Nilsen, Physical Review 1969. 182 (3), p.838.
[24] X. Lin, et al, Thin Solid Films 2013. 535: p. 10-13.
[25] G. Babu, et al, Semicond Sci Technol 2008, 23. 085023.
[26] R. Adhi, et al, Physica Status Solidi 2007, 204(10): p. 3373-3379.
[27] G. Zoppi, et al, Progress in Photovoltaics: Research and Applications 2009. 17(5): p. 315-319.
[28] G. Babu, et al, Journal of Physics D: Applied Physics, 2008. 41(20).
[29] M. Hiroaki, et al, Journal of Crystal Growth 2000. 208: p. 416-422.
[30] K. ITO, John Wiley & Sons, Ltd. 2015.
[31] S.J. Sandoval et al, Physical Review 1991. B 44 (8), p.3955.
[32] P. Fernandes, et al, Journal of Alloys and Compounds 2011. 509(28): p. 7600-7606.
[33] L.S. Price et al, Chemistry of Materials 1999. 11 (7), p.1792.

PRESSURE-AIDED GROWTH APPROACH OF HIGHLY EFFICIENT TANDEM PEROVSKITE SOLAR CELL VIA SANDWICH EVAPORATION TECHNIQUE

Jia-Ci Jhou[1], Chung-Han Chang[1], Ashish Gaurav[1], Hsin-Ting Lin[2], Ching-Fuh Lin[1,3,4*]
1. Graduate Institute of Photonics and Optoelectronics, National Taiwan University,
No. 1, Sec. 4, Roosevelt Road, Taipei 106319, Taiwan
2. Graduate Institute of Advance Technology, National Taiwan University,
No. 1, Sec. 4, Roosevelt Road, Taipei 106319, Taiwan
3. Graduate Institute of Electronics Engineering, National Taiwan University,
No. 1, Sec. 4, Roosevelt Road, Taipei 106319, Taiwan
4. Department of Electrical Engineering, National Taiwan University,
No. 1, Sec. 4, Roosevelt Road, Taipei 106319, Taiwan
*lincf@ntu.edu.tw

ABSTRACT: The perovskite layer was fabricated using sandwich structure principle. Initially, 1st layer (MAI) was spin-coated and PbI_2 as 2nd layer was thermally evaporated. MAI was evaporated at low pressure by sandwich evaporation technique (SET) using double inter-diffusion step to make the perovskite react completely. Changing the pressure flow of the growth chamber can result in different physical and electrical properties of the formed film. SET chamber pressure was varied to study the effect of pressure change on the surface morphology and electrical properties of the perovskite film. At pressure 3 torr, the semitransparent perovskite solar cell (ST-PSC) showed the higher efficiency of 15.03%. Finally, four terminal (4T) tandem configuration was established having a power conversion efficiency (PCE) of 22.35%.
Keywords: perovskite, sandwich evaporation technique, double inter-diffusion, pressure-aided growth approach, tandem solar cells

1 INTRODUCTION

In the last few years, inorganic-organic hybrid perovskite ($MAPbI_3$) has shown its potential in the photovoltaic field due to attractive properties like low non-radiative recombination rate, strong absorption coefficient and facile synthesis technology and has good characteristics like low cost, efficiency, stability along with enhanced functionality [1]. As a result of these features and properties, various kinds of solar cells have been designed. Among the essential growth parameters of the perovskite thin film, pressure is one of the crucial parameters [2]. Pressure-aided growth approach describes the dependence of pressure on the ease of thin film perovskite fabrication, as in the high-pressure range, due to collision with air particle, the mean free path for MAI increases, resulting in lesser MAI deposition on the perovskite surface and vice versa for the low-pressure range. Generally, during the sandwich evaporation technique (SET) process, nitrogen is introduced to isolate water and oxygen. We have investigated the effect of pressure change on the electrical property of the different perovskite film through J-V curve and further examined its application in tandem based perovskite solar cell which can be classified as two terminals and four terminals.

2 EXPERIMENTS

Pre-cleaned ITO/glass substrate subjected to UV-Ozone treatment for 30 mins was used here. PEDOT:PSS layer was spin-coated at 4000 rpm for 30 secs as HTL, followed by annealing at 120°C for 10 mins. The $CH_3NH_3PbI_3$ (MAI) layer was formed with sandwich like structure fabricated by first spin-coating MAI at 3000 rpm; annealed at 120°C for 10 mins. Second layer of PbI_2 was deposited via thermal evaporator, which was also annealed at 120°C for 10 mins. Third, final layer of MAI was deposited with handmade chamber with evaporation time for 30 mins at different SET pressure. This was followed by rinsing residual MAI with IPA at 4000 rpm and annealed at 120°C for 5 mins to smoothen the surface. The sample was kept in vacuum for 12 hrs after spin-coating $PC_{60}BM$ powder dissolved in chlorobenzene (CB) (20 mg/mL) at 3000 rpm for 60s. For the electrodes, 5 nm bathocuproine (BCP) and 120 nm Ag were deposited via thermal evaporator, layered over by a semi-transparent Ag thin film electrodes. The tandem configuration was achieved by amalgamation of perovskite with Si cell via four terminal methodology.

The crystallinity and morphology of the thin film was studied through X-ray diffraction (XRD) and field emission scanning electron microscopy (FE-SEM). The absorption spectrum of the fabricated film was analyzed from UV-vis spectroscopy. Current density and voltage curves were measured by solar simulator connected with Keithley 2400 under light source condition AM 1.5G, intensity 100 mW/cm^2.

3 RESULTS AND DISCUSSION

Figure 1 shows the XRD patterns of perovskite thin films for different SET pressure. From the result, the main diffraction peak of perovskite is located at $2\theta = 14.1°$ with (110) plane, the secondary diffraction peak is located at $2\theta = 28.1°$ with (004) plane and $2\theta = 28.4°$ with (220) plane. Investigating the PbI_2 diffraction peak at $2\theta = 12.4°$, the peak intensity first decreased and then increased with increase in SET pressure and the lowest PbI_2 intensity was found to be at 3 torr, indicating that most of the PbI_2 reacted with MAI to form perovskite.

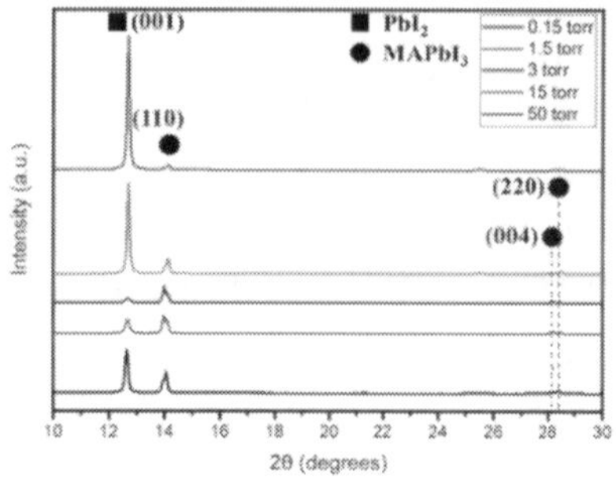

Figure 1: XRD patterns of perovskite thin films for different SET pressure.

Figure 2(a) shows the absorption spectrum of perovskite formed under different SET pressure for the wavelength range of 400-800 nm. The absorption increases with the increasing pressure up to 3 torr; giving the maximum absorption of 61.33%, and beyond this pressure it decreases significantly. Figure 2(b) shows the physical appearance of the perovskite thin film formed at different pressure. Film formed at low pressure was found to be dark brown in color as compared to film fabricated at high pressure, with the darkest at 3 torr, indicating that it is most suitable pressure for perovskite reaction.

(a)

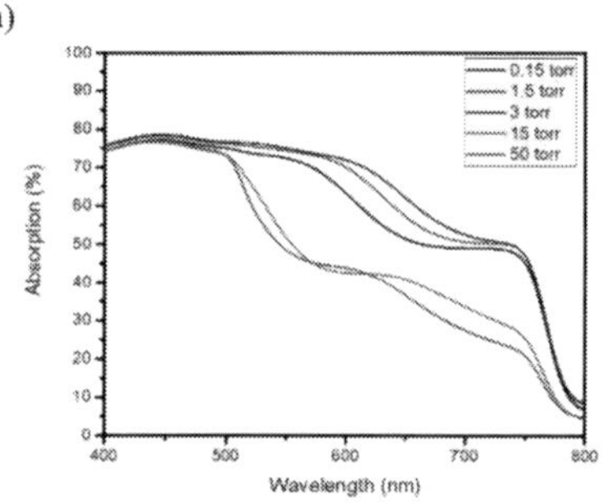

(b)

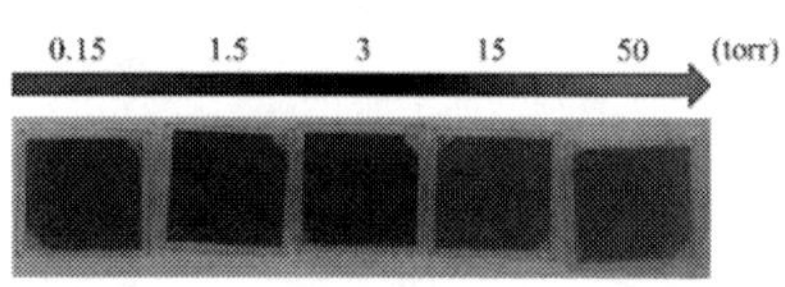

Figure 2: (a) UV-vis absorption spectrum (b) Thin films appearance of perovskite for different SET pressure.

Additionally, we investigated the morphology of perovskite thin films via SEM measurement. Fig. 3 shows cross-section SEM images of perovskite thin films at different SET pressure. Through SET, the sandwich structure can cause double inter-diffusion of ions, which not only control the diffusion distance of MAI ions more effectively, but also increases the actual effective reaction thickness. Fig. 3d shows that about 120 nm of PbI_2 reacted with MAI into perovskite and PbI_2 content is lower in content after formation. Therefore, through the pressure-aided growth approach, under the appropriate pressure at 3 torr, the uniformity and flatness of the resulting film are significantly improved, and the reduction of discontinuous boundaries facilitating the transmission of charge carriers along with the removal of residual PbI_2.

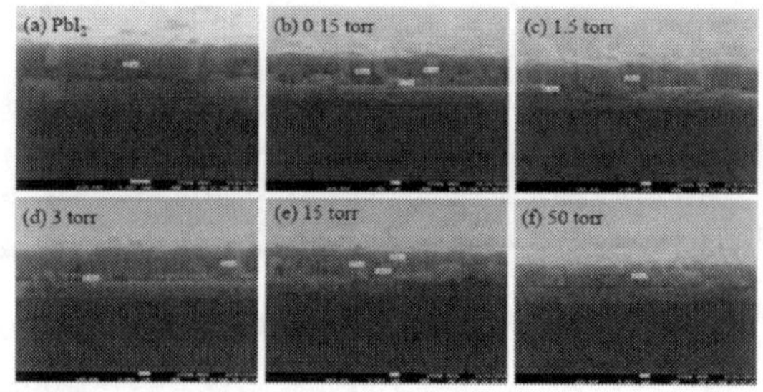

Figure 3: Cross-section SEM images of perovskite thin films for (a) PbI_2 and SET pressure at (b) 0.15 torr (c) 1.5 torr (d) 3 torr (e) 15 torr (f) 50 torr.

As shown in Figure 4(a)-(b), semitransparent perovskite solar cells (PSCs) with Ag transparent electrodes exhibited PCE of 15.03%. The PCE (%) of the 4T tandem solar cell was enhanced to 22.35% through the addition of perovskite top cell PCE (%) with filtered c-Si bottom cell, resulting in boosting the PCE (%) to above 20%, as compared to the single opaque PSCs.

(a)

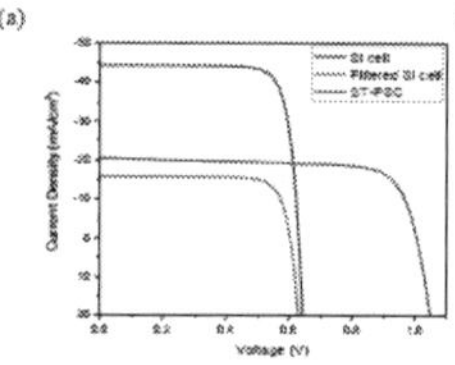

(b)

Parameter	Area (cm²)	Voc (V)	Jsc (mA/cm²)	FF (%)	PCE (%)
Si cell	2.25	0.64	44.3	77.75	22.04
ST-PSC	0.08	1.01	20.52	72.72	15.03
Filtered Si cell	2.25	0.61	15.8	76.3	7.32
Sum	0.08	---	---	---	22.35

Figure 4: (a) J-V curves of semitransparent PSC, and silicon solar cell with and without the semitransparent PSC filter (b) Summary of photovoltaic parameters of semitransparent perovskite cell, silicon cell, filtered silicon cell, and the summed 4T perovskite/silicon tandem cell.

4 CONCLUSION

In this paper we have successfully resolved the residual problem of PbI_2 in the evaporation method by combining the pressure aided growth approach with sandwich evaporation technique. The pressure flow of nitrogen gas was controlled during the fabrication process which enhanced the reaction between the MAI and PbI_2 and improved the performance of the cell. From characterization, the most suitable pressure was found to be 3 torr, which results in efficiency of 15.03% and combining the fabricated perovskite top cell with the filtered c-Si bottom cell resulted in PCE (%) up to 22.35% through the realization of 4T tandem configuration.

5 ACKNOWLEDGMENT

The authors gratefully acknowledge Ministry of Science and Technology in Taiwan for supporting the research with the contract numbers: MOST 108-2221-E-002-032-MBK, MOST 108-2221-E-002-145-MY3, MOST 107-2221-E-002-172-MY3. Besides, we would like to thank Ms. Yuan-Tzu Lee of the Instrumentation Center at National Taiwan University for their assistance with the SEM operations.

6 REFERENCES

[1] Azhar Fakharuddin, Rajan Jose, Thomas M. Brown, Francisco Fabregat-Santiago, and Juan Bisquert, Energy Environ. Sci. 7 (12), 3952 (2014).

[2] G. Xiao, C. Zhu, Y. Ma, B. Liu, G. Zou, and B. Zou, Angew Chem Int Ed Engl 53 (3), 729 (2014).

UPSCALING OF PEROVSKITE-SILICON TANDEM SOLAR CELLS

O. Schultz-Wittmann[1], P. S. C. Schulze[1], O. Er-Raji[1,3], R. Efinger[1], Ö. Ş. Kabaklı[1], M. Heydarian[1,3], K. A. McMullin[1], O. Fischer[1,3], A. J. Bett[1], D. Erath[1], S. Pingel[1], U. Heitmann[1], Z. Kiaee[1], M. Kohlstädt[1], B. P. Kore[1,3], H. Nagel[1], M. C. Schubert[1], M. Bivour[1], J. C. Goldschmidt[1,2], J. Borchert[1,3], M. Hermle[1], S. W. Glunz[1,3]
[1] Fraunhofer Institute for Solar Energy System (ISE), Heidenhofstr. 2, 79110 Freiburg, Germany
Email: oliver.schultz-wittmann@ise.fraunhofer.de,
[2] Philipps-University Marburg, Renthof 7, 35032 Marburg, Germany
[3]INATECH, University of Freiburg, Emmy-Noether Str.2, 79110 Freiburg, Germany

ABSTRACT: We are working on upscaling of perovskite-silicon tandem solar cells from small lab size areas of several square millimeters to full commercial wafer size. This requires the use of processing techniques that can replace the usual spin coating process employed for small lab size areas. As alternatives for large-area depositions spray coating, inkjet printing and slot-die coating are investigated. For highest specific energy yield under field conditions perovskite-silicon tandem solar cells should be textured on the front side. To ensure conformal coating of the random pyramid textured silicon bottom cell we follow a hybrid process route. In a first step, the inorganic components are thermally evaporated to form a scaffold allowing conformal coating of pyramids of arbitrary size and shape. In a second step, the conversion into perovskite is accomplished by infiltration of the scaffold with organic salt solutions. A thermal anneal completes the conversion into a polycrystalline perovskite layer.
In addition to upscaling of the cell area we work on replacing the evaporated silver contact grid by screen printed ultra-low temperature silver paste. First results on planar perovskite-silicon tandem solar cells, still fabricated with spin coating processing but with screen-printed front contacts, resulted in a certified stabilized conversion efficiency of 22.5% on 104 cm^2 total cell area.
Keywords: Perovskites, Tandem, Hybrid

1 INTRODUCTION

Perovskite-silicon tandem solar cells with highest specific energy yield under field conditions should be textured on the front side [1]. Since the standard lab procedures of wet-chemical processing like spin-, spray- or blade coating do not yield conformal coating of µm-sized textured surfaces, we follow a hybrid route to deposit mixed cation mixed halide perovskite absorbers $[HC(NH_2)_2]_{1-x}Cs_xPb(I_{1-y}Br_y)_3$ [2,3]. In a first step, the inorganic components are thermally evaporated to form a scaffold allowing conformal coating of pyramids of arbitrary size and shape. In a second step, the conversion into perovskite is accomplished by infiltration with organic salt solutions. For the second step, several deposition techniques are evaluated which are already used in other industrial fields: a) spray coating b) inkjet printing c) slot-die coating (Figure 1). An annealing process finalizes formation of a polycrystalline perovskite.

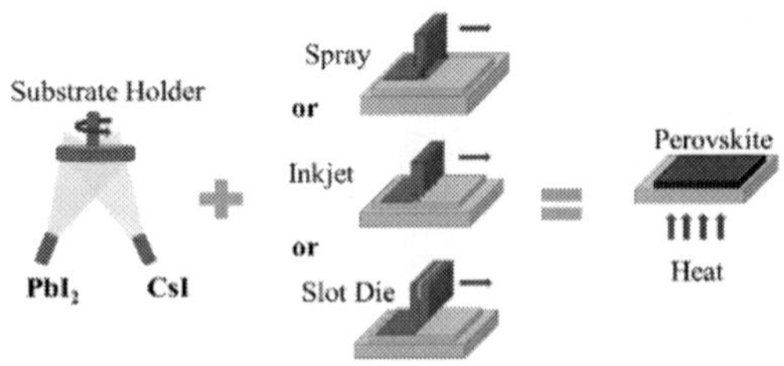

Figure 1: Illustration of hybrid route used for perovskite deposition onto µm-sized textured substrates. We evaporate a porous PbI_2/CsI scaffold (yellow) onto a silicon bottom cell and infiltrate this layer with an organic salt solution (red). After an annealing process a polycrystalline perovskite (black) is formed.

Sintering of screen-printed silver paste in a high-temperature firing process is the standard industrial process for metallization of single-junction silicon solar cells. However, perovskites are temperature-sensitive, hence for this approach ultra-low temperature pastes that cure below 150°C are required [4] and will be used in our experiments.

On medium size solar cells of 4 cm^2 full area, we analyze the impact of shunts that are distributed across the entire cell area. Such shunts are often not present in small area (1 cm^2) champion cells because of the ability to select small shunt-free cells for analysis.

2 EXPERIMENTAL

2.1 Tandem Cell Structure

As silicon bottom cells we used 250 µm thick boron-doped Float Zone wafers of 1 Ω cm resistivity and amorphous silicon heterojunction surfaces [5]. The rear electrode consists of evaporated silver whereas on the front side a ca. 20 nm thick ITO completes the bottom cell preparation. The top cells were prepared by spin coating of hole transport layers (HTL) followed by spin coating of perovskite solution for the case of planar front or by co-evaporation of PbI_2/CsI for textured surfaces. For this hybrid route subsequent infiltration of the inorganic scaffold was accomplished by either spin coating of organic salt solutions as our reference process or by spray coating or inkjet printing as our target process sequence. Subsequently, a thin layer of C_{60} was evaporated followed by atomic layer deposition of SnO_x and sputtering of ITO. The front electrode was either prepared by evaporation of silver through a shadow mask or screen-printing of an ultra-low temperature silver paste. Finally, a MgF_2 anti reflection layer was evaporated on top (Figure 2).

We used two different approaches for the deposition of the perovskite layers: For planar front sides established spin coating processing was used while for textured surfaces we used a hybrid approach. For cell areas of 5x5 mm^2, 10x10 mm^2 and 20x20 mm^2 we used evaporated silver through a shadow mask as the method for front metallization. For even larger areas of 50x50 mm^2 and 102x102 mm^2 we used screen printing of silver paste, that can cure at very low temperatures of about 130 °C.

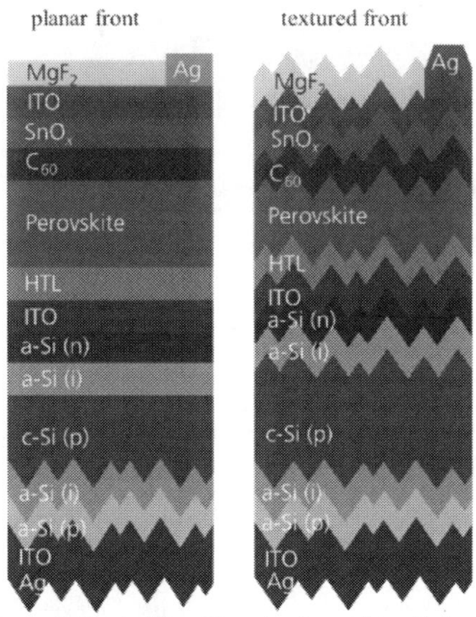

Figure 2: Perovskite-silicon tandem solar cell structure with planar or textured front.

2.2 Characterization

After certain processing steps we used photoluminescence imaging to assess the material quality and homogeneity of the prepared layers [6]. Using blue generation light of 450 nm allowed to selectively excite the perovskite cell. Also, we used lock-in thermography [6,7] to identify shunted areas.

For *IV* measurements the definition of the active cell area is of major importance. The layout for the different solar cell structures is shown in Figure 3 where the small cell design consists of four 5x5 mm^2 area with a U-shape metallization on a 25x25mm^2 substrate. In such a design the *IV* measurements are made on the "designated area" (d.a.) which is not metallized [8]. For 10x10 mm^2 and 20x20 mm^2 cells the metallization fingers are part of the active cell area but the broad circumferent metallization busbar is not considered part of the active area, again resulting in a "designated area" measurement. For large-area cells we have cut out the cell boundaries by laser scribing and cleaving. Hence, the entire remaining wafer, including the metallization grid, is considered being part of the active area and the IV measurement is described as a "total area" (t.a.) measurement.

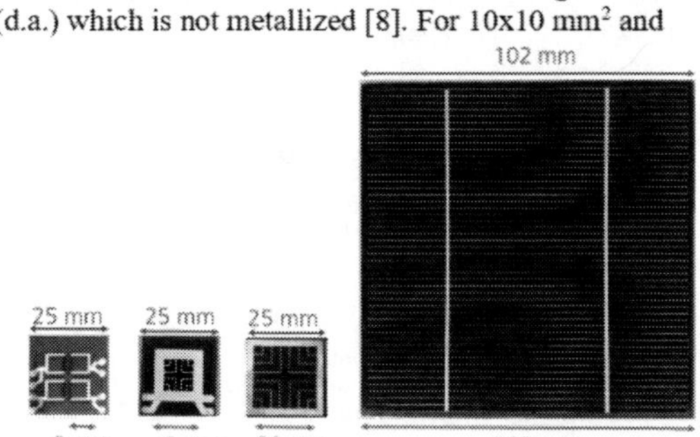

Figure 3: Tandem cell layout and area definitions for *IV* measurements. Red: Substrate dimension. Green: Active cell dimension.

3 RESULTS AND DISCUSSION

3.1 Spin Coating of Perovskite Absorber

When scaling up the area of perovskite-silicon tandem cells homogeneity of layer quality as well as density and severity of shunts are key performance indicators. In the following experiments processing of the perovskite was performed with spin coating, which is the established lab process for small areas. However, we also used it for larger areas to test the limit for this approach. We have observed that with adjustment of the processing parameters (spin speed, volume of liquid etc.) reasonably homogeneous layer properties could be obtained in the center of 100 mm diameter wafers. This can be concluded from Figure 4 where a photoluminescence image irradiated with 450 nm wavelength with an intensity of about 587 Wm^{-2} is shown for a wafer with 100 mm diameter containing seven individual cells of 20x20 mm^2 active area. The corresponding *IV* measurements reveal similarly small spread of power conversion efficiency.

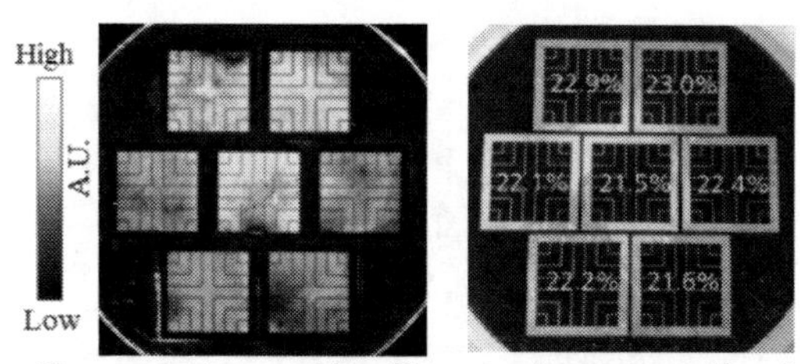

Figure 4: 7 perovskite-silicon tandem solar cells of 20x20 mm^2 active area on a 100 mm diameter wafer. Left: Photoluminescence image excited with 450 nm light. Right: Picture of the same wafer with corresponding power conversion efficiencies averaged from IV measurements in forward and reverse direction.

Figure 5 Left: Illuminated Lock-In Thermography (ILIT) image of the perovskite sub cells using a blue 470 nm LED for the excitation of charge carriers. Right: ILIT image of the silicon sub cells using an infrared 950 nm LED for excitation. Bright spots within the active area of the solar cells mark regions with increased heat dissipation. Consequently, in these regions ohmic shunts or increased non-radiative recombination can be found.

In Figure 5 a measurement of the same wafer with Illuminated Lock-In Thermography (ILIT) is shown. The use of blue light for excitation of charge carriers allows measurement of the top cell. Likewise, infrared light is used to measure the silicon bottom cell. Bright spots within the active area reveal locations of increased heat dissipation, hence one can find ohmic shunts (or increased non-radiative recombination) in those locations. While these spots are frequently found in the perovskite top cell, they are also found in the silicon bottom cell. This stresses the need to have high quality for both sub cells that combined work as a tandem cell.

We made further attempts to increase the area on planar cells with spin coating of the perovskite absorber. The PL imaging results for a large cell of 102x102 mm^2 perovskite-silicon tandem cell can be seen in Figure 6a. Excitation with 450 nm wavelength allows for selectively detecting the signal of the top cell and strong inhomogeneities are revealed. This indicates that such large area is beyond the current process maturity of the spin coating process for the perovskite absorber. Hence, we tested the hybrid route for the case of inkjet printing (Figure 6b) and spray coating (Figure 6c). Though for the inkjet printed sample 3 mm wide stripes are observed, overall homogeneity is greatly improved. Also, these stripes can be avoided by a set-up in which the entire area will be printed in a single swath rather than in multiple passes covering 3 mm on each pass. Similarly, the spray coating sample also exhibits a much more homogeneous PL signal. A small gradient from top to bottom is observed but these results support the confidence into the hybrid route approach for large-area perovskite-silicon tandem solar cells.

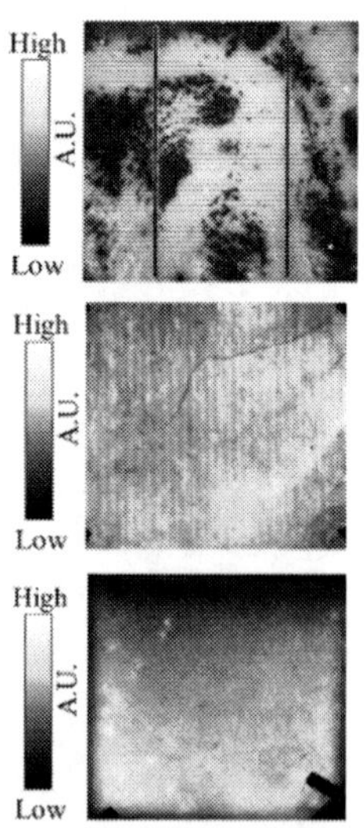

a) 102x102 mm^2 tandem cell prepared by spin coating. The resulting top cell shows a strongly inhomogeneous PL signal.

b) 100x100 mm^2 top cell prepared by hybrid route with inkjet printing. While stripes are visible overall homogeneity is improved.

c) 100x100 mm^2 top cell prepared by hybrid route with spray coating. While a gradient is observed overall homogeneity is improved.

Figure 6: PL images of perovskite layers on planar surfaces excited by a 450 nm laser.

3.2 Hybrid Route

In order to obtain a conformal coating of the surface also for textured bottom cells we follow the hybrid route. In a first step, the inorganic components are thermally evaporated to form a scaffold allowing conformal coating of pyramids of arbitrary size and shape. A cross section SEM image is shown in Figure 7 (top). In a second step, the conversion into perovskite is accomplished by infiltration of the porous scaffold with organic salt solutions.

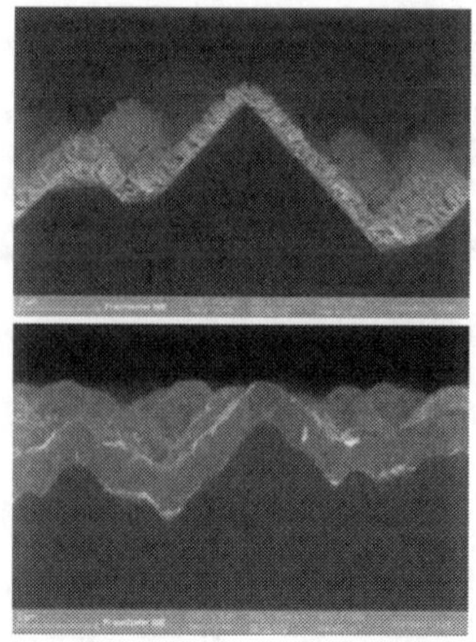

Figure 7: SEM images of evaporated PbI_2/CsI scaffold (top image) and after infiltration with organic salt solution and crystallization (bottom image).

In the SEM image in Figure 7 (bottom) this was accomplished by spin coating but inkjet printing and spray coating yield the same results and a conformal coating of the textured surface is obtained. This texture on the front side is intended to not only increase the power conversion efficiency of the cell but also the performance ratio over the year in a fixed tilt installation [1]. The change in the relative EQE shape for a textured cell produced by the hybrid route and a planar cell is shown in Figure 8. Compared to the planar front the EQE becomes flatter.

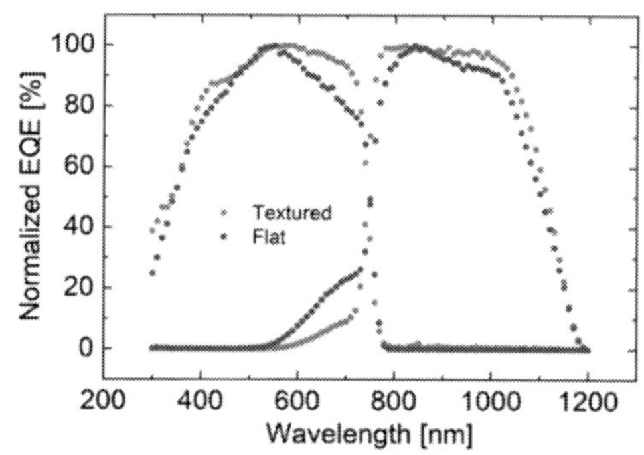

Figure 8: Relative EQE measurements of a textured and a planar perovskite-silicon tandem cell normalized to 100% on the respective highest measurement data points. The change in the shape of the EQE is related to the front texture.

3.3 *IV* Results

In Table 1 *IV* results of perovskite-silicon tandem solar cells made at Fraunhofer ISE are shown. The upper rows show measurements in forward direction, the lower rows in reverse and η_{steady} is the stabilized steady state efficiency value.

Cell A is textured and was produced with the hybrid route where the infiltration of the inorganic scaffold was accomplished by spin coating of organic salts. Cells B and C have a planar front and the perovskite was produced by spin coating. With a total area of 104 cm^2 cell C is the largest cell we have produced so far and has a screen-printed silver metallization of the front.

Table 1: *IV*-parameters of recent perovskite-silicon tandem solar cells of varying size made at Fraunhofer ISE. The upper rows show measurement in forward direction, the lower rows in reverse.
*Calibrated measurement at Fraunhofer ISE CalLab
A: textured front, hybrid route, evaporated metal
B: planar front, evaporated metal
C: planar front, screen-printed metal

#	*Area* [cm²]	η_{steady} [%]	j_{sc} [mA cm^{-2}]	V_{oc} [mV]	*FF* [%]	η [%]
A	$1.0_{d.a.}$	25.2	18.0	1786	77.4	24.9
			18.1	1786	77.6	25.0
B	$4.0_{d.a}$	23.0*	17.0	1771	75.1	22.7
			17.0	1771	75.1	22.6
C	$104_{t.a}$	22.5*	17.8	1772	68.2	21.5
			17.9	1771	70.5	22.4

4 SUMMARY AND CONCLUSIONS

We fabricated textured perovskite-silicon tandem solar cells with a hybrid processing route. This resulted in the formation of a top solar cell that conformally coats the µm-sized random pyramid texture of the bottom solar cell. Initial results on 1 cm² designated area reached 25.2% where the infiltration of the inorganic scaffold was accomplished by spin coating of organic salts.

Increasing the area on planar surfaces using spin coating for the perovskite formation was successful and for very large cells of 104 cm² total area we reached certified 22.5% stabilized steady state efficiency. Furthermore, we managed to integrate screen-printing of a low-temperature silver paste curing at 130 °C for the front metallization. However, inhomogeneities in the top cell have been detected by photoluminescence as the area increased. Hence, we work on replacement of the spin coating process. Inkjet printing as well as spray coating show promising initial results on large areas.

As a next step we will replace the spin coating process in the hybrid route by inkjet printing or spray coating. We expect this to enable cell efficiencies in the range of 28% for large area, textured perovskite silicon tandem cells with printed metallization.

5 ACKNOWLEDGEMENTS

This work was partially funded by the German Federal Ministry for Economics Affairs and Climate Action under contract number 03EE1086A (PrEsto) and 03EE1087A (Katana) as well as a Fraunhofer LIGHTHOUSE PROJECT (MaNiTU). Ö. Ş. Kabaklı gratefully acknowledges scholarship from the Dr. Ruth Heerdt Stiftung.

6 REFERENCES

[1] Tucher, N., Höhn, O., Murthy, J.N., *et al.*: 'Energy yield analysis of textured perovskite silicon tandem solar cells and modules', *Optics Express*, 2019, **27**, (20), A1419

[2] Sahli, F., Werner, J., Kamino, B.A., *et al.*: 'Fully textured monolithic perovskite/silicon tandem solar cells with 25.2% power conversion efficiency', *Nat Mater*, 2018, **17**, (9), pp. 820–826

[3] Schulze, P.S.C., Wienands, K., Bett, A.J., *et al.*: 'Perovskite Hybrid Evaporation/ Spin Coating Method: From Band Gap Tuning to Thin Film Deposition on Textures', *Thin solid films*, 2020, **704**, p. 137970

[4] Kamino, B.A., Paviet-Salomon, B., Moon, S.-J., *et al.*: 'Low-Temperature Screen-Printed Metallization for the Scale-Up of Two-Terminal Perovskite–Silicon Tandems', *ACS Applied Energy Materials*, 2019, **2**, (5), pp. 3815–3821

[5] Schulze, P.S.C., Bett, A.J., Bivour, M., *et al.*: ' 25.1% High - Efficient Monolithic Perovskite Silicon Tandem Solar Cell with a High Band Gap Perovskite Absorber' , *Solar RRL*, 2020, **4**, (7), p. 2000152

[6] Schubert, M.C., Mundt, L.E., Walter, D., Fell, A., Glunz, S.W.: 'Spatially Resolved Performance Analysis for Perovskite Solar Cells', *Adv. Energy Mater.*, 2020, p. 1904001

[7] Mundt, L.E., Heinz, F.D., Albrecht, S., *et al.*: 'Nondestructive Probing of Perovskite Silicon Tandem Solar Cells Using Multiwavelength Photoluminescence Mapping', *IEEE Journal of Photovoltaics*, 2017, **7**, (4), pp. 1081–1086

[8] Green, M., Dunlop, E., Hohl - Ebinger, J., Yoshita, M., Kopidakis, N., Hao, X.: 'Solar cell efficiency tables (version 57)' , *Prog Photovoltaics*, 2021, **29**, (1), pp. 3–15

APPLICATION OF ADVANCED HYDROGENATION PROCESS ON SIGE/SI SOLAR CELLS

Zhenyu Sun[1] Fukun Lei[1] Shude Zhang[2] Brett Hallam[1] Li Wang[1]
[1]School of Photovoltaic and Renewable Energy Engineering, UNSW Sydney, NSW 2052, Australia
[2]Suzhou Talesun Solar Technologies Co., Ltd. Jiangsu 215542. China

ABSTRACT: The lattice mismatch is a significant issue for monolithic III-V/Si tandem solar cells due to the creation of islanding and misfit dislocations at the interface. $Si_{1-x}Ge_x$ buffer layers have demonstrated promising results in mitigating the lattice-mismatch between III-V and Si materials. However, dangling bonds and dislocations can still exist among the $Si_{1-x}Ge_x$ buffer layers, causing significant recombination losses. Herein, we apply an advanced hydrogenation process on $Si_{1-x}Ge_x$ cells grown on Si with $Si_{1-x}Ge_x$ graded buffer layers in between, aiming to reduce the buffer layer and base region defects, hence improving the cell's performance. SiN_x through PECVD has been applied as both a hydrogen source and an anti-reflection coating on SiGe/Si cells with four different Ge concentrations (i.e., 78%, 82%, 85% and 88%). After accumulated laser treatment of 30 seconds, V_{oc} and efficiency of SiGe cells (78%, 82% and 85% Ge) have been improved by 2-4%$_{REL}$ (equivalent to 6-12 mV) and 5-17%$_{REL}$, respectively. These results demonstrate the effectiveness of hydrogenation passivation on $Si_{1-x}Ge_x$ buffer layers and SiGe base defects, also indicating opportunities for broad applications of the advanced hydrogenation technique on other tandem structures.
Keywords: Germanium, Hydrogen, Laser Processing, Passivation, Tandem

1 INTRODUCTION

The world record efficiency for single-junction silicon (SJ-Si) solar cells is 26.7% [1], with a theoretical efficiency limit of 29.1% [2,3]. To further surpass the theoretical efficiency limit of SJ-Si solar cells, we must rely on the so-called Multi-junction techniques. Amongst all types of solar cells, the highest power conversion efficiency is 47.1%, achieved by an III-V multi-junction solar cell (MJSC) [4]. However, this type of cell is currently limited to extraterrestrial applications mainly due to its expensive material and fabrication costs. An obvious solution to this problem is to replace their conventional gallium arsenide (GaAs) or germanium (Ge) substrates with much cheaper Si material via direct epitaxial growth of III-V materials on Si wafers to fabricate III-V/Si MJSCs.

However, epitaxial growth of high-quality III-V materials onto Si substrates is difficult when introducing materials with varying lattice constants [5]. The lattice mismatch is a significant issue, which can create islanding and misfit dislocations at the III-V/Si interface. As a solution, $Si_{1-x}Ge_x$ (silicon germanium) buffer layers with a tunable lattice constant have been demonstrated to be an appropriate material for realising lattice-constant transition in monolithic III-V/Si tandem devices [6]. During the growth of $Si_{1-x}Ge_x$, the Ge concentration gradually increases until reaching the aimed composition. Such graded buffer layers can mitigate the defects originating from the lattice mismatch but cannot eliminate them thoroughly. Dangling bonds and dislocations can still exist among the $Si_{1-x}Ge_x$ buffer layers and the $Si_{1-x}Ge_x$/Si interface, causing significant recombination losses. Therefore, it is necessary to introduce an extra passivation approach to increase the buffer layers' quality further.

Hydrogen passivation, which has been shown as a promising and effective passivation technique on Si solar cells [7–9], is investigated in this study to passivate the $Si_{1-x}Ge_x$ buffer layers. Here, we deposit a hydrogenated silicon nitride (SiN_x) layer on SiGe samples through PECVD (plasma-enhanced chemical vapour deposition). This SiN_x layer functions not only as a hydrogen source but also as an anti-reflection coating (ARC) [10,11]. After the SiN_x deposition, a thermal treatment is performed to release hydrogen ions into the sample, functioning as a passivation source. However, although the hydrogen ions released through thermal treatments can provide effective passivation for $Si_{1-x}Ge_x$ buffer layers, the passivation efficacy reduces when the Ge concentration increases (especially when greater than 85%). This is due to the fact that the majority of hydrogen ions generated during the firing process are negatively charged, and the dangling bonds in $Si_{1-x}Ge_x$ (with a high Ge concentration) are also negatively charged, resulting in an ineffective interaction and inefficient passivation. Therefore, advanced hydrogenation treatment involving laser illumination has also been applied to increase the amount of positively charged hydrogen ions. Laser illumination has been shown to be able to manipulate the hydrogen charge state in Si solar cells [14] (from negatively charged to positively charged), hence enabling more efficient passivation for Ge-rich SiGe/Si solar cells.

Here we apply the SiN_x as both a hydrogen source and ARC and combine it with an advanced hydrogenation process involving annealing and laser illumination (which are widely deployed as an industrial process for silicon solar cells [15]) to achieve significant hydrogen passivation results on SiGe/Si solar cells. SiGe/Si cells with four different Ge concentrations (i.e., 78%, 82%, 85% and 88%) are prepared and fired at 660 °C initially, followed by laser treatment for 30 s accumulatively. Significant hydrogen passivation effects have been observed for all cells with Ge% below 88%, verified by the EQE (external quantum efficiency) and light I-V results.

2 EXPERIMENTAL METHODS

2.1 Sample preparation

CZ Si substrates with a 6° offcut were prepared for this experiment. After a standard cleaning process, $Si_{1-x}Ge_x$ buffer layers and SiGe cells with four different Ge compositions (i.e., 78%, 82%, 85% and 88%) were grown in an RPCVD (reduced pressure chemical vapour deposition) reactor. Then SiN_x was deposited via a plasma-enhanced chemical vapour deposition (Roth & Rau MAiA XS PECVD System), followed by a firing process with a peak temperature of 660 °C in a furnace (Roth & Rau CAMiNI). Photolithography processes combined with a buffer HF etching were conducted to partially etch through the SiN_x, creating finger openings for the cell. A stack of Ti/Pd/Ag (titanium, palladium and silver) and a full-area Al (aluminium) were evaporated as the front and back contacts, respectively, through e-beam evaporation (Lesker PVD75). Finally, samples were edge isolated in an RIE machine (reactive-ion etching). Schematics of the process flow and cell structure are shown in Fig. 1.

2.2 Advanced hydrogenation process

An advanced hydrogenation process was performed using a DR hydrogenation laser with a chuck temperature of 170 °C, a laser power of 1226 W (wavelength of 1064 nm) and an illumination time of 30 s, an optimised recipe from previous work [16]. As the thermal budget plays a vital role in this process, plus the fact that laser illumination can dramatically increase the samples' temperature, we gradually increased the thermal budget by treating samples for accumulated 10 s, 20 s and 30 s. EQE and light I-V measurements were performed before and after each laser treatment, respectively.

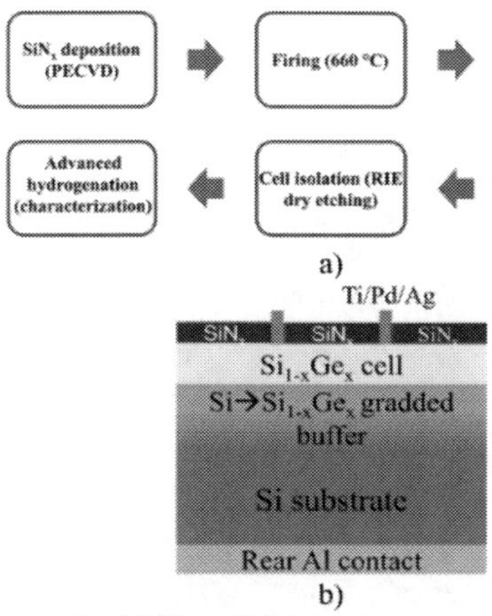

Figure 1: a) SiGe cell fabrication process flow chart and b) a schematic of the SiGe/Si solar cell (cross-sectional view).

3 RESULTS DISCUSSION

EQE results are shown in Fig. 2 with EQE-integrated J_{sc} labelled. Cells with 78% and 82% Ge obtained similar improvement in EQE (1 and 1.3 mA/cm^2 increase in EQE integrated J_{sc}, respectively), whereas 88% Ge cell only reached 0.3 mA/cm^2 increase. The EQE improvement mainly from the wavelength range of ~800-1180nm indicating hydrogen passivation in the SiGe base and buffer layers region. The 85% Ge cell obtains the greatest improvement in EQE response with ~2.4 mA/cm^2 increase in EQE-integrated J_{sc}, attributing to the thick SiGe base (2.75 μm) (cells with other Ge% have a 0.75 μm thick base), which also provides evidence of hydrogenation passivation since thicker base region contains more volume of defects during the material growth.

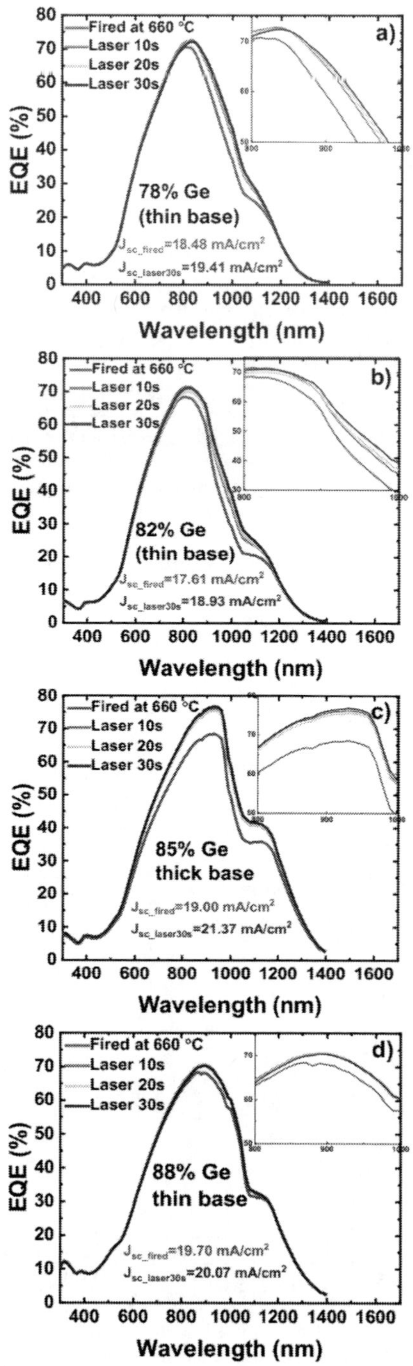

Figure 2: EQE of a) 78%, b) 82%, c) 85%, and d) 88% SiGe/Si cells as fired and after advanced hydrogenation treatment for accumulated 10 s, 20 s and 30 s, respectively. It should be noted that the samples with a Ge concentration of 78%, 82% and 88% have a 0.75 μm Ge base, while the one with 85% Ge has a thicker base of 2.75 μm.

Light I-V measurements are also performed on the samples, and relative changes in V_{oc} and efficiency results are shown in Fig. 3. After accumulated laser illumination of 30 s, the cell with 85% Ge obtains the largest relative improvement in V_{oc} (3-4%$_{REL}$ equivalent to 9-12 mV) and efficiency (12-17% $_{REL}$). Again, for the other three cells with a 0.75 μm SiGe base, the cell with 82% Ge performs the best, achieving a relative increase on V_{oc} of ~3% $_{REL}$ and efficiency of ~15% $_{REL}$, which agrees with the results observed in EQE. The better performance of cells with a thicker base indicates a significant passivation impact on the SiGe bulk of the advanced hydrogenation technique.

The reason why $Si_{1-x}Ge_x$ cells with lower Ge% achieved more obvious improvement after laser treatment could be the transition of $Si_{1-x}Ge_x$ alloy property from Si-like to Ge-like at approximately 85% Ge composition [17]. The charge states of both hydrogen and defects can significantly influence the passivation efficacy [18]. Both the charge-state transition levels for dangling bonds and interstitial hydrogen in Si are within the conduction band minimum and valance band maximum [19]. In contrast, the charge state transition levels of dangling bonds in Ge are lower than the valance band maximum [20], [21]. This suggests the dangling-bonds defects will always be negatively charged in Ge. Hence, a vast majority of the dangling bonds in Ge-rich SiGe/Si cells are also negatively charged. Therefore, hydrogen passivation through only thermal treatment will be inefficient since most of the hydrogen ions released are also negatively charged [21].

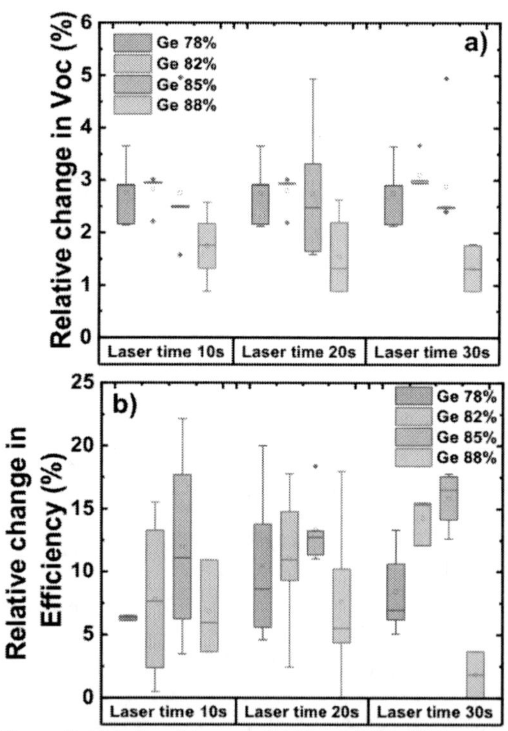

Figure 3: Relative changes in (a) V_{oc} and (b) efficiency of 78%, 82%, 85% and 88% SiGe/Si cells after accumulated 10 s, 20 s and 30 s advanced hydrogenation treatment.

Furthermore, after laser processing, a further cell performance improvement was seen for the cell with 88% Ge concentration. The reason could be the hydrogen charge state manipulation by laser process [18], and thus providing further passivation for Ge-rich samples (> 85%).

4 CONCLUSION

Herein, we have demonstrated an implementation of hydrogen passivation on SiGe/Si solar cells through an advanced hydrogenation technique involving annealing and laser treatment with SiN_x as a hydrogen source. Significant performance improvements have been observed, with relative improvements of 2-4%$_{REL}$ (equivalent to 6-12 mV) in V_{oc} and 5-17%$_{REL}$ in efficiency achieved on 78%, 82% and 85% SiGe/Si cells. The results also indicated that SiGe/Si cells with a thicker base and Ge% ≤ 85% obtain more dramatic improvement in device performance, indicating an effective passivation impact on the SiGe bulk from the advanced hydrogenation technique. Future work will include material characterisations and in-depth analyses of hydrogen passivation effects.

5 ACKNOWLEDGEMENTS

This work has been supported by the Australian Centre for Advanced Photovoltaics (ACAP) and the Australian Government through the Australian Renewable Energy Agency (ARENA). Responsibility for the views, information or advice expressed herein is not accepted by the Australian Government. This work is partially completed at Australian National Fabrication Facility (ANFF) and AmberWave Inc.

6 REFERENCES

[1] K. Yoshikawa, H. Kawasaki, W. Yoshida, T. Irie, K. Konishi, K. Nakano, T. Uto, D. Adachi, M. Kanematsu, H. Uzu, K. Yamamoto, Silicon heterojunction solar cell with interdigitated back contacts for a photoconversion efficiency over 26%, Nat. Energy. 2 (2017) 17032. https://doi.org/10.1038/nenergy.2017.32.

[2] M.J. Kerr, A. Cuevas, P. Campbell, Limiting efficiency of crystalline silicon solar cells due to Coulomb-enhanced Auger recombination, Prog. Photovoltaics Res. Appl. 11 (2003) 97–104. https://doi.org/10.1002/pip.464.

[3] A. Richter, M. Hermle, S.W. Glunz, Reassessment of the limiting efficiency for crystalline silicon solar cells, IEEE J. Photovoltaics. 3 (2013) 1184–1191. https://doi.org/10.1109/JPHOTOV.2013.2270351.

[4] M.A. Green, E.D. Dunlop, D.H. Levi, J. Hohl-Ebinger, M. Yoshita, A.W.Y. Ho-Baillie, Solar cell efficiency tables (version 54), Prog. Photovoltaics Res. Appl. 27 (2019) 565–575. https://doi.org/10.1002/pip.3171.

[5] A.K. Okyay, A.M. Nayfeh, K.C. Saraswat, T. Yonehara, A. Marshall, P.C. McIntyre, High-efficiency metal-semiconductor-metal photodetectors on heteroepitaxially grown Ge on Si, Opt. Lett. 31 (2006) 2565. https://doi.org/10.1364/ol.31.002565.

[6] D.M. Isaacson, C.L. Dohrman, A.J. Pitera, S. Gupta, E.A. Fitzgerald, Mid-105 cm-2 threading dislocation density in optimised high-Ge content relaxed graded SiGe on Si for III-V solar on Si, 836 (2005) 2–7.

[7] P. Hamer, B. Hallam, S. Wenham, M. Abbott, Manipulation of hydrogen charge states for passivation of P-type wafers in photovoltaics, IEEE J. Photovoltaics. 4 (2014) 1252–1260. https://doi.org/10.1109/JPHOTOV.2014.2339494.

[8] M. Sheoran, A. Upadhyaya, A. Rohatgi, Bulk lifetime and efficiency enhancement due to gettering and hydrogenation of defects during cast multicrystalline silicon solar cell fabrication, Solid. State. Electron. 52 (2008) 612–617. https://doi.org/10.1016/j.sse.2007.10.001.

[9] L. Song, A. Wenham, S. Wang, P. Hamer, M.S. Ahmmed, B. Hallam, L. Mai, M. Abbott, E.R.

Hawkes, C. Chong, S. Wenham, Laser enhanced hydrogen passivation of silicon wafers, Int. J. Photoenergy. 2015 (2015). https://doi.org/10.1155/2015/193892.

[10] Aberle , Armin G, Overview on SiN surface passivation of crystalline silicon solar cells, Sol. Energy Mater. Sol. Cells. 65 (2001) 239 - 248.

[11] F. Jiang, M. Stavola, A. Rohatgi, D. Kim, J. Holt, H. Atwater, J. Kalejs, Hydrogenation of Si from SiNx(H) films: Characterisation of H introduced into the Si, Appl. Phys. Lett. 83 (2003) 931–933. https://doi.org/10.1063/1.1598643.

[12] D. Li, Material Characterization and Device Optimization of Silicon Germanium Solar Cells Grown on Silicon Substrates, (2016).

[13] J.R. Weber, A. Janotti, C.G. Van De Walle, Dangling bonds and vacancies in germanium, Phys. Rev. B - Condens. Matter Mater. Phys. 87 (2013) 1–9. https://doi.org/10.1103/PhysRevB.87.035203.

[14] P. Hamer, S. Wang, B. Hallam, S. Wenham, C.M. Chong, A. Wenham, M. Abbott, Laser illumination for manipulation of hydrogen charge states in silicon solar cells, Phys. Status Solidi - Rapid Res. Lett. 9 (2015) 111–114. https://doi.org/10.1002/pssr.201409495.

[15] M. A. Green, "The Passivated Emitter and Rear Cell (PERC): From conception to mass production," Sol. Energy Mater. Sol. Cells, vol. 143, pp. 190–197, 2015.

[16] F. Lei, Z. Sun, B. Hallam, X. Zhang, and L. Wang, "Advanced Hydrogen Passivation on SixGe1-x solar cell grown on Si substrate." Poster presentation at PVSEC-31 Conference 13-15 Dec. 2021.

[17] R. Braunstein, A. R. Moore, and F. Herman, "Intrinsic optical absorption in germanium-silicon alloys," Phys. Rev., vol. 109, no. 3, pp. 695–710, 1958.

[18] P. Hamer, B. Hallam, S. Wenham, and M. Abbott, "Manipulation of hydrogen charge states for passivation of P-type wafers in photovoltaics," IEEE J. Photovoltaics, vol. 4, no. 5, pp. 1252–1260, 2014.

[19] C. Herring, N. M. Johnson, and C. G. Van de Walle, "Energy levels of isolated interstitial hydrogen in silicon," Phys. Rev. B - Condens. Matter Mater. Phys., vol. 64, no. 12, pp. 1252091–12520927, 2001.

[20] J. R. Weber, A. Janotti, P. Rinke, and C. G. Van De Walle, "Dangling-bond defects and hydrogen passivation in germanium," Appl. Phys. Lett., vol. 91, no. 14, pp. 2005–2008, 2007.

[21] J. R. Weber, A. Janotti, and C. G. Van De Walle, "Dangling bonds and vacancies in germanium," Phys. Rev. B - Condens. Matter Mater. Phys., vol. 87, no. 3, pp. 1–9, 2013.

NEAR-INFRARED LIGHT MANAGEMENT IN SILICON HETEROJUNCTION SOLAR CELLS FOR SILICON-BASED III–V MULTIJUNCTION TANDEM CELL APPLICATIONS

Hyunju Lee[1,2*], Tappei Nishihara[2], Yoshio Ohshita[3], and Atsushi Ogura[1,2]
[1]Meiji Renewable Energy Laboratory, Meiji University, 1-1-1 Higashi-mita, Tama-ku, Kawasaki 214-8571, Japan
[2]Toyota Technological Institute, 2-12-1 Hisakata, Tempaku-ku, Nagoya 468-8511, Japan
[3]School of Science and Technology, Meiji University, 1-1-1 Higashi-mita, Tama-ku, Kawasaki 214-8571, Japan
*Author for correspondence: Tel. +81-52-809-1877, Fax +81-052-809-1879, E-mail: sollee@meiji.ac.jp

ABSTRACT: The silicon heterojunction (SHJ) cell technology dominates in Si-based tandem cell research, and also the world record Si-based III–V multijunction tandem cell is based on SHJ bottom cells. Despite this exciting performance, the tandem cell was still limited by the SHJ bottom-cell photocurrent. The limited photocurrent is owing to significant losses in the near-infrared (NIR) part of the spectrum on the rear side of SHJ bottom cells. This contribution demonstrates an improved NIR response of the tungsten-doped indium oxide (IWO)/Ag rear contact of rear-junction SHJ solar cells. The carrier concentration and the thickness of the rear IWO layer are optimized in order to minimize the free-carrier and the plasmonic absorption losses without detrimentally affecting the selectivity and the electrical transport properties of SHJ solar cells. The NIR light reflectance of rear IWO/Ag reflectors are shown to be significantly impacted by the carrier concentration of IWO layers within the investigated ranges. As a result of these optimizations, a significant decrease of the parasitic absorption loss in the NIR part of the spectrum is achieved, leading to a champion device with a short-circuit current density of up to 40.8 mA/cm^2 and an efficiency of 22.4%.
Keywords: TCO Transparent Conducting Oxides, Silicon Solar Cell, Optical Losses, Heterojunction, Tandem

1 INTRODUCTION

Silicon (Si)-based III–V multijunction tandem cells have been attracting significant amounts of attention owing to their cost effectiveness. The recorded maximum efficiency of a Si-based III–V multijunction tandem cell has well exceeded 35% [1] and thus has attracted attention as a next-generation high-efficiency low-cost solar cell. Currently, the silicon heterojunction (SHJ) bottom cell technology, which provides high open-circuit voltage values of up to 750 mV, dominates in Si-based tandem cell research [2,3], and also the world record Si-based III–V multijunction tandem cell is based on SHJ bottom cells [1]. Despite this exciting performance, the tandem cell was still limited by the bottom-cell photocurrent. For instance, the total photocurrent density of the world record GaInP/GaAs//SHJ tandem cell was only 38.7 mA/cm^2 [1]. For comparison, 42.7 mA/cm^2 were reached in the world record SHJ cell [4], which is very close to the theoretical maximum for silicon of about 43.3 mA/cm^2 [5]. An important optical loss mechanism in this tandem cell, in comparison to an optimized single-junction silicon cell, is the increased reflection and reduced absorption in the near-infrared (NIR) part of the spectrum, hence reduced photocurrent in SHJ bottom cells. In particular, on the rear side of SHJ bottom cells, an unoptimized transparent conducting oxide (TCO)/metal contact is the origin of significant losses in the NIR part of the spectrum. These losses come from plasmonic absorption at the TCO/metal interface and from free carrier absorption in the TCO [6,7]. Therefore, the NIR light management on the rear side of SHJ solar cells becomes increasingly important as short-circuit current in the NIR part of the spectrum has to be increased owing to the increasing interest of using SHJ solar cells as bottom cells in Si-based III–V multijunction tandem cells.

In this paper, tungsten-doped indium oxide (IWO) is optimized for enhancing the NIR response of IWO/Ag rear contacts as the thickness of the rear IWO layer and the oxygen content in the deposition gas flow are varied to find the optimal combination leading to the highest rear reflectance of IWO/Ag rear contacts in the NIR regions and the maximum Jsc of SHJ solar cells.

2 EXPERIMENTAL DETAILS

Samples with an IWO layer, were fabricated on a mirror-polished Czochralski p-type (100) oriented crystalline Si (c-Si) substrate (ρ of 1000 Ω cm) by reactive plasma deposition (RPD) technique. 160-nm-thick IWO layers were deposited from a ceramic target which is made of 99 wt% In_2O_3 and 1 wt% WO_3. O_2/Ar mixed gases with varied ratios, R from 23 to 54 % (R = $[O_2]/[Ar+O_2] \times 100$ %), were utilized as process gases. Some of the deposited IWO layers experienced post-deposition annealing at 220 °C for 5 min in air.

Samples with an IWO/Ag rear reflector were fabricated on double-side textured Czochralski n-type (100) oriented c-Si substrates (ρ of 1–5 Ω cm). Front 120-nm-thick IWO layers were fabricated with a 54% mixed gas. 70–400-nm-thick rear IWO layers were fabricated with a 23% mixed gas for studying the effect of thicknesses on the NIR reflectance of the samples and 160-nm-thick rear IWO layers were fabricated with 19–54% mixed gases for studying the effect of oxygen concentrations on the NIR reflectance of the samples. Rear 200-nm-thick silver (Ag) layers were thermally evaporated. Some of the fabricated samples with a rear IWO/Ag reflector experienced post-deposition annealing at 220 °C for 5 min in air.

SHJ solar cells were fabricated with double-side textured Czochralski n-type (100)-oriented c-Si wafers (ρ of 1–5 Ω cm). Front i/n stack and rear i/p stack thin-film amorphous silicon layers were prepared by plasma-enhanced chemical vapor deposition (PECVD). IWO layers were deposited on the front and rear sides of the wafers, respectively. The photolithography process was used to define the contact area for metallization. Front and rear metal contacts were thermally evaporated Ag. The fabricated SHJ solar cells experienced post-deposition annealing at 220 °C for 5 min in air.

The opto-electrical properties of the IWO layers were evaluated with Hall effect measurements, UV-Vis-

NIR spectroscopy, and spectroscopic ellipsometry. Current-voltage (I–V) characteristics of the complete SHJ solar cells were measured using a solar simulator under 1-sun illumination conditions (100 mW/cm^2, AM 1.5G). External quantum efficiency (EQE) measurements were performed to evaluate wavelength-dependent optical response of the SHJ solar cells. The experimental procedure and the structures of the prepared samples are shown in Fig. 1.

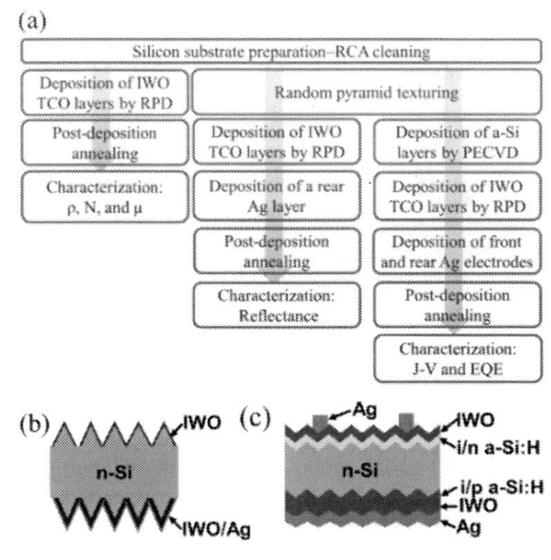

Figure 1: (a) Experimental procedure flowchart of this study. Schematic illustration of (b) the samples with an IWO/Ag reflector and (c) the SHJ solar cells.

3 RESULTS AND DISCUSSION

3.1 Electrical properties of 160-nm-thick IWO layers on Si substrates

Hall effect measurements demonstrate that the IWO layers are n-type semiconductor. Fig. 2 shows their carrier concentration (N_e), carrier mobility (μ_e), and resistivity (ρ) change as a function of O_2/Ar flow ratio, R ($R = [O_2]/[Ar+O_2] \times 100$ %). For increasing R from 23% to 54%, carrier concentration monotonously decreases, while carrier mobility is almost stable. Consequently, the resistivity monotonously increases. After post-deposition annealing carrier concentration and mobility increase, and thus resistivity decreases. The change of carrier concentration is related to the oxygen vacancy (Vo) and metal substitution such as W^{6+} substituted In^{3+} in the IWO layers. However, Vo may decrease by oxidation during post-deposition annealing in air, and therefore the Vo is not the predominant mechanism in our IWO layers. Therefore, the increase of activated W^{6+} dopants is responsible for the increase of the carrier concentration and thus the decrease of resistivity. The change of carrier mobility is related to the scattering mechanisms such as the grain boundary scattering and ionized impurity scattering. The ionized impurity scattering is mainly influenced by the carrier concentration in the layers [8,9]. However, the carrier concentration increases slightly after post-deposition annealing, and therefore the ionized impurity scattering is not the predominant scattering mechanism in our IWO layers. Therefore, the reduction of grain boundary scattering is responsible for the increase of carrier mobility and thus the decrease of resistivity. The highest mobility of 65 cm^2/Vs and the low carrier concentration of 1.1×10^{20}/cm^3 can be obtained at R of 54% after post-deposition annealing.

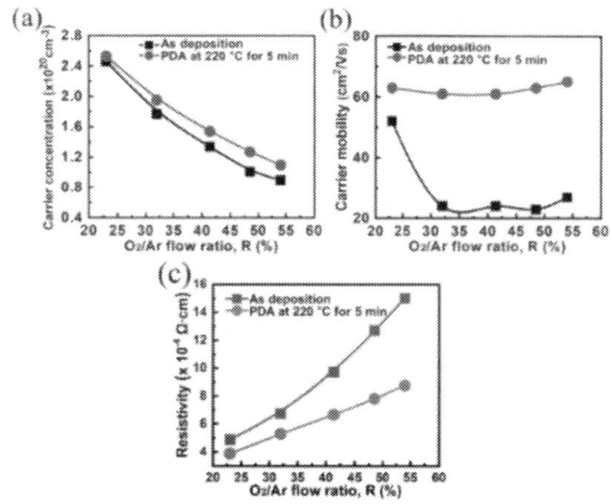

Figure 2: (a) carrier concentration (N_e), (b) carrier mobility (μ_e), and (c) resistivity (ρ) changes of the IWO layers as a function of O_2/Ar flow ratio, R.

3.2 Effect of the IWO thickness on the reflectance of IWO/Ag rear reflectors

The reflectance of IWO/Ag rear reflectors with different IWO thicknesses and a constant oxygen content (R of 23%) in the IWO deposition gas flow is shown in Fig. 3. It is oberved that there is less near-infrared (NIR) parasitic absorption in the samples with thicker rear IWO layers, resulting in higher NIR reflectance. The maximum reflectance of the samples is achieved with the 160 and 200-nm-thick IWO layers before and after post-deposition annealing, respectively. These thick IWO layers may reduce the amount of evanescent waves reaching the Ag layer by increasing the internal reflectance at the rear surface, which yields to the reduction of plasmonic absorption (PA) in the Ag layer, and thus the increase of the NIR reflectance. For the samples with the 400-nm-thick IWO layer, however, the reflectance largely decreases because free carrier absorption (FCA) in the such thick IWO layers negates the optical gains by the reduction of PA in the Ag layer.

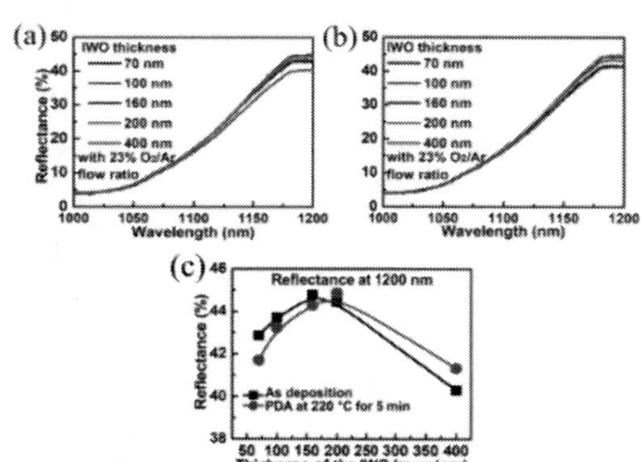

Figure 3: Reflectance change of the samples with an IWO/Ag reflector (a) before and (b) after post-deposition annealing. The rear IWO layers are fabricated with different thicknesses. (c) Reflectance change of the samples at 1200 nm.

3.3 Effect of the oxygen content in the IWO deposition gas flow on the reflectance of IWO/Ag rear reflectors

The reflectance of IWO/Ag rear reflectors with different oxygen content in the IWO deposition gas flow and a fixed thickness of 160 nm is shown in Fig. 4. Apparently, it is observed that the NIR reflectance largely increases with the oxygen content. Increasing the R from 19% to 54%, the reflectance increased from 39% to 57% for as-deposited samples, and from 42% to 53% for annealed samples, respectively. As shown in Fig. 2, the IWO carrier concentration decreases as the R increases. Therefore, decreasing IWO carrier concentration reduces the FCA losses in the IWO layers, resulting in enhaced NIR reflectance of the samples. The reflectance is saturated at the R of 48% and 41% before and after post-depositon annealing, respectively. Meanwhile, the reflectance of some annealed samples is slightly lower than that of as-deposited samples. This may be owing to the oxidation of Ag at the IWO/Ag interface during post-depositon annealing. The degree of oxidation seems to be related to the R, and thus the oxygen content in the IWO layers.

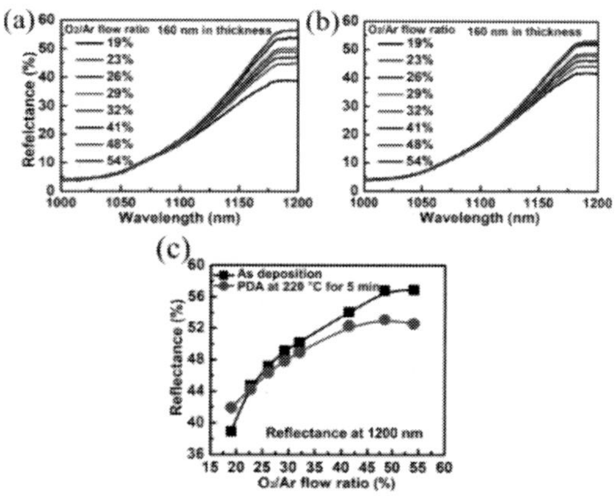

Figure 4: Reflectance change of the samples with an IWO/Ag reflector (a) before and (b) after post-deposition annealing. The rear IWO layers are fabricated with different O_2/Ar flow ratio, R. (c) Reflectance of the samples at 1200 nm.

3.4 Effect of the oxygen content in the IWO deposition gas flow on the performance of SHJ solar cells

The J_{sc} and EQE of SHJ solar cells with different oxygen content in the IWO deposition gas flow and a fixed thickness of 160 nm, are shown in Table. 1 and Fig. 5. Apparently, it is observed that the J_{sc} largely increases with the oxygen content. Increasing the R from 23% to 54% increases the J_{sc} by 0.7 mA/cm^2, leading to the increased power conversion efficiency by 0.7%. Consistently, the corresponding EQE curve is higher in the 1000–1200 nm spectral range for higher oxygen contents. This is well consistent with the observe result from reflectance measurements of the samples with an IWO/Ag rear reflector. As shown in Fig. 2 and 4, the reflectance at 1200 nm increases for higher oxygen contents and lower carrier concentrations. Therefore, we can conclude that the increase in oxygen content reduces the FCA in the IWO layer, resulting in an enhanced NIR rear reflector and higher J_{sc}.

Table I: Open-circuit voltage, short-circuit current density, fill factor, and power conversion efficiency of SHJ solar cells with R of 23% and 54%.

O_2/Ar flow ratio, R	Voc (mV)	Jsc (mA/cm^2)	FF (%)	PCE (%)
23%	701	40.1	77.2	21.7
54%	703	40.8	78.2	22.4

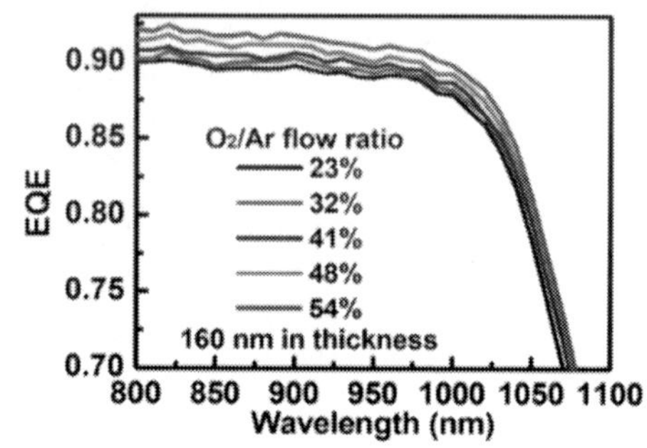

Figure 5: External quantum efficiency (EQE) change of the SHJ solar cells as a function of O_2/Ar flow ratio, R.

4 CONCLUSIONS

We optimized IWO layers by adjusting the thickness and the oxygen content in the IWO deposition gas flow for NIR light management in SHJ solar cells. The opto-electrical properties of the TCO layers were found to be strongly dependent on the oxygen content in the IWO deposition gas flow and the post-deposition annealing process. The highest mobility of 65 cm^2/Vs and the low carrier concentration of 1.1×10^{20}/cm^3 can be obtained at R of 54% after post-deposition annealing. The NIR reflectance of samples with an IWO/Ag reflector largely increases with increasing the oxygen content in the IWO deposition gas flow because decreasing IWO carrier concentration by increasing the oxygen content reduces the FCA losses in the IWO layers, resulting in enhaced NIR reflectance of the samples. Finally, the J_{sc} of SHJ solar cells largely increases by 0.7 mA/cm^2 after optimization of IWO layers. This is consistent with the enhaced EQE in the NIR region of SHJ solar cells after optimization of IWO layers. Our study will be beneficial to ehhance the performance of SHJ bottom cells in Si-based III-V multijunction tandem cells.

ACKNOWLEDGEMENT

This study is supported by NEDO, Japan.

REFERENCES

[1] S. Essig, C. Allebé, T. Remo, J. F. Geisz, M. A. Steiner, K. Horowitz, L. Barraud, J. S. Ward, M. Schnabel, A. Descoeudres, D. L. Young, M. Woodhouse, M. Despeisse, C. Ballif, A. Tamboli, Nature Energy 2 (2017) 17144.

[2] M. Taguchi, A. Yano, S. Tohoda, K. Matsuyama, Y. Nakamura, T. Nishiwaki, K. Fujita, E. Maruyama,

IEEE Journal of Photovoltaics 4 (2014) 96.
[3] Y. Liu, Y. Li, Y. Wu, G. Yang, L. Mazzarella, P. Procel-Moya, A. C. Tamboli, K. Weber, M. Boccard, O. Isabella, X. Yang, B. Sun, Materials Science and Engineering R 142 (2020) 100579.
[4] K. Yoshikawa, H. Kawasaki, W. Yoshida, T. Irie, K. Konishi, K. Nakano, T. Uto, D. Adachi, M. Kanematsu, H. Uzu, K. Yamamoto, Nature Energy 2 (2017) 17032.
[5] A. Richter, M. Hermle, S. W. Glunz, IEEE Journal of Photovoltaics 3 (2013) 1184.
[6] Z. C. Holman, S. de Wolf, C. Ballif, Light Science and Applications, 2 (2013) e106.
[7] Z. C. Holman, M. Filipic, A. Descoeudres, S. De Wolf, F. Smole, M. Topic, C. Ballif, Journal of Applied Physics 113 (2013) 013107.
[8] S. Calnan, A. N. Tiwari, Thin Solid Films 518 (2010) 1839.
[9] A. J. Leenheer, J. D. Perkins, M. F. A. M. van Hest, J. J. Berry, R. P. O'Hayre, D. S. Ginley, Physical Review B 77 (2008) 115215.

Optimization and characterization of indium zinc oxide TCOs for perovskite-silicon tandem solar cell applications

Volker Sittinger1, Hunter King[1], Özde Ş. Kabakli[2], Patricia S. C. Schulze[2], Jan Christoph Goldschmidt[2,3]

* Corresponding author: Volker Sittinger, Fraunhofer IST, Bienroder Weg 54E, 38108 Braunschweig, Tel.: +49-531-2155-512; E-Mail address: volker.sittinger@ist.fraunhofer.de

[1] Fraunhofer Institute for Surface Engineering and Thin Films IST, Bienroder Weg 54E, 38108 Braunschweig, Germany
[2]Fraunhofer Institute for Solar Energy Systems, Heidenhofstr. 2, 79110 Freiburg, Germany
[3]Philipps-University Marburg, Renthof 7, 35032 Marburg, Germany

ABSTRACT: This work details the development and optimization of indium zinc oxide (IZO) as top transparent conductive oxide (TCO) for use in perovskite-silicon tandem solar cells, focusing on the determination of application-specific ideal film stoichiometry. The indium zinc oxide films were produced by means of a serial reactive co-sputtering process with a zinc planar target and an indium rotatable target. The indium was sputtered with argon as sputter gas, and oxygen as reactive gas. The IZO films were optimized for conductivity, carrier mobility and absorption. The measured absorption of the films was weighted with global standard spectral data (AM1.5G) in order to more closely mimic the effect of absorption on the resulting solar cell efficiency. Based on previous work, a process window varying Zn target power (100 – 140 W) and oxygen flow rate (30 – 34 sccm, influencing total process pressure, and In target power at constant O_2 partial pressures) was selected. The optimized IZO films were then deposited onto perovskite-silicon tandem cells where the power conversion efficiencies (PCE), short-circuit current densities (Jsc), open-circuit voltages (Voc), and fill factors (FF) were characterized and compared with tandem cells with a reference ITO as top TCO. The best performing cells showed a PCE of 25.8 %, with process parameters corresponding to 34 sccm O_2 flow, and 100 W Zn target power.

Keywords: TCO Transparent Conducting Oxide, IZO, Sputtering, Perovskite, Silicon, Tandem

1. INTRODUCTION

While silicon solar cells are the current industry standard, they are reaching their limits in efficiency [1]. In order to continue to improve solar cell efficiencies a shift to new materials and technologies is required. Perovskite-based PV cells have shown significant promise as use as a new absorber material with higher achievable efficiencies than traditional silicon cells. Perovskite-silicon tandem cells utilized the well-established silicon bottom absorber with a perovskite-based top absorber material. However, for this technology to make a market impact, the efficiencies > 30 % and long-term stability is required in order to reduce the overall levelized cost of energy delivered.

In perovskite-silicon tandem cells, a significant number of films can play a role in the resulting efficiency of the solar cell. However, outside of the absorber layers, the top TCO can play an especially significant role as tunnel and front contact (Figure 1). Additional details to the cell structure and the individual films can be found in [2].

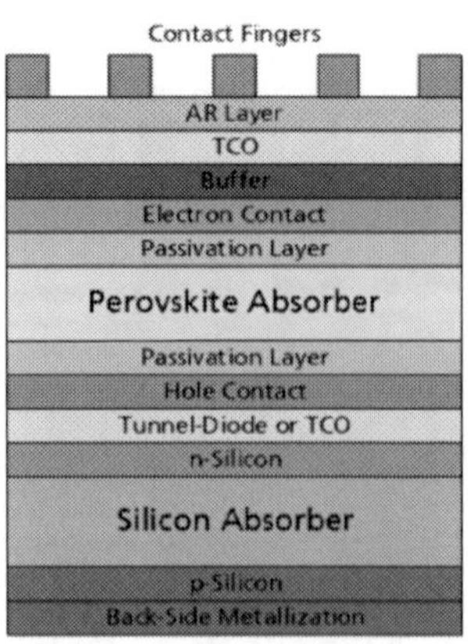

Figure 1: Example structure of a perovskite-silicon tandem solar cell (layers not to scale).

This work continues previous work by King et al. [2], and further investigates the influence of film stoichiometry on the resulting cell performance metrics of perovskite-silicon tandem cells (specifically J_{SC}, V_{OC}, FF, and PCE).

2. EXPERIMENTAL DETAILS

IZO films were produced by means of a reactive co-sputtering process utilizing a planar zinc target, and rotatable indium target.
Indium/zinc concentrations were varied by means of a variation in sputter power of the zinc cathode towards the indium target material. The parameters were selected based on previous work by King et al. [3] with 3 sets of process parameters investigated. Additionally, 3 processes were performed with each parameter set to investigate effects of process drift. The complete summary of process parameters for the system are shown in Table I.

Table I: Range of process parameters

Working Pressure	0.35 Pa
Working Gas	Ar = 100 sccm
Reactive Gas	O_2 = 30, 34 sccm
Base Pressure	< $2*10^{-6}$ Pa
Power Range (In)	0.5 – 1.2 kW
Power Range (Zn)	100 – 140 W
Rotatable Target (In)	In (4N), d = 143 mm x 400 mm
Planar Target (Zn)	Zn (4N), 88 mm x 500 mm
Substrate Temperature	50 °C
Target-Substrate Distance	95 mm
Substrate Material	EagleXG glass 100 x 100 mm² and tandem cells

The optical and electrical properties of the films were characterized by means of optical spectroscopy (transmission, reflection), 4-point resistivity, and Hall-effect measurements to confirm the carrier concentration and mobility of the films. The optical absorption data was

then weighted against solar spectral data, to increase measurement relevance. The films were then integrated as front electrodes in perovskite-silicon tandem solar cells, and the resulting power conversion efficiency, fill factor, short-circuit current density, and open-circuit voltage were determined.
All films were deposited at low target powers and low substrate temperatures (< 80 °C) in order to prevent damaging the underlying sensitive perovskite absorber [2].

3. RESULTS

The IZO films were optimized as a function of reactive gas (oxygen) flow rate, and zinc target power (corresponding to varying indium/zinc stoichiometry). An optimum between weighted film absorption and conductivity was found, while maintaining deposition temperatures at 50 °C. The IZO films were optimized by maintaining operation of the deposition system in the sputtering transition zone between metallic and oxide operation. In this range the films were optimized to minimize weighted film absorption and maximize conductivity. Reflectance and transmittance of IZO films deposited on glass were measured which were deposited in the same run as the tandem cells. Figure 2 shows the results of absorption measurements of the IZO films.

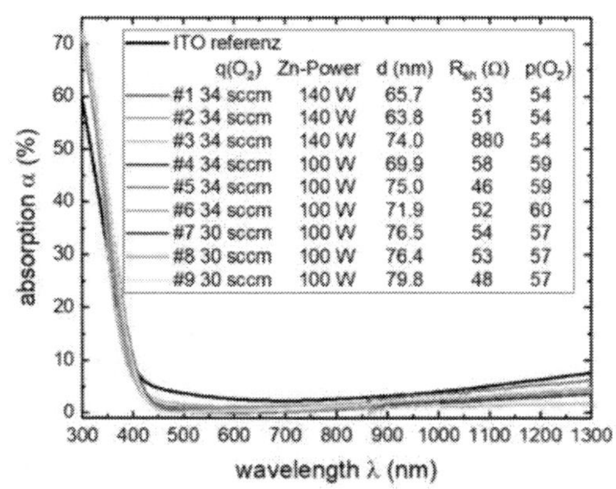

Figure 2: Measured absorption as function of oxygen flow and Zn-Target power. Film thickness was determined by fitting of the optical spectra. Sheet resistance was determined by 4-point-measurement.

The ITO reference is also plotted in the graph. All IZO samples produced in this work showed lower absorption values when compared with the reference ITO films over the entire spectrum (until the UV zone).

Figure 3 shows the results of the IZO optimization of resistivity and weighted absorption as a function of oxygen flow rate, and zinc target power. All samples show improved weighted absorption values when compared with the reference ITO film (2.8 % – 3.4 % vs. 4.6 %, respectively), and all but one sample (#3 which shows significant variation due to process issues) show improvement compared to the resistivity of the reference ITO (345 – 415 μΩcm vs. 513 μΩcm, respectively).

These improvements are further reflected in the Hall effect measurement results as a function of oxygen flow rate and zinc target power shown in Figure 4.

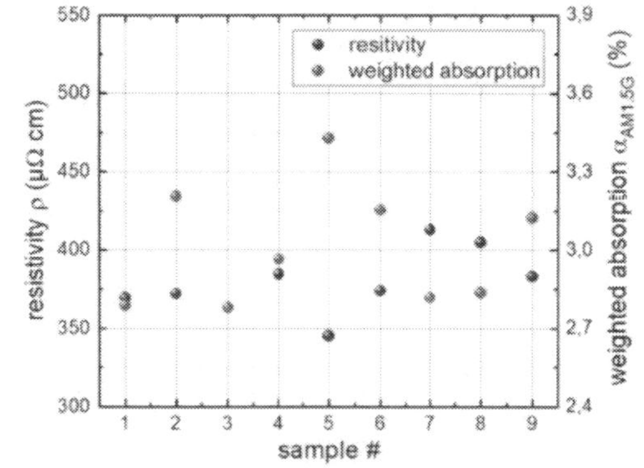

Figure 3: Resistivity and weighted absorption (convoluted with AM 1.5 spectrum for relevant wavelength range) as function of oxygen flow and Zn-Target power. Process parameters for each sample listed in Figure 1.

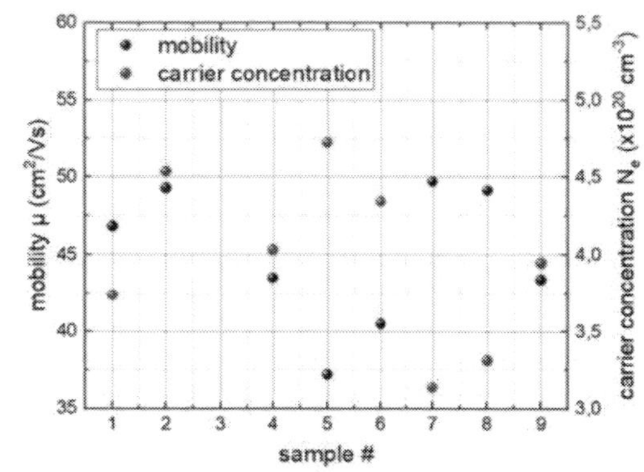

Figure 4: Mobility and carrier concentration as function of oxygen flow and Zn-Target power. Process parameters for each sample listed in Figure 1.

While standard ITO films have mobilities around 30 cm^2/Vs at similar deposition temperatures, IZO films can achieve significantly higher mobilities, in this work up to 50 cm^2/Vs (as can be seen in Figure 4). This increase in mobility allows for equivalent, and higher, film conductivities to be achieved, with lower carrier concentrations.

One important trend to note is the increased carrier concentration and absorption of the samples within each series (shown in Figure 4). Here three samples for each set of process parameters were produced (group 1: samples 1-3, group 2: samples 4-6, and group 3: samples 7-9, as described in the legend in Figure 1). Each group of samples varied oxygen flow rate, and zinc target power, and corresponded to 34 sccm / 140 W for group 1, 34 sccm / 100 W for group 2, and 30 sccm / 100 W for group 3, respectively. This drift is likely due to an increase in zinc atom incorporation into the indium target over time. Although the target was pre-sputtered for at least 1 hour between each process, the same film properties could not be achieved with identical process conditions.

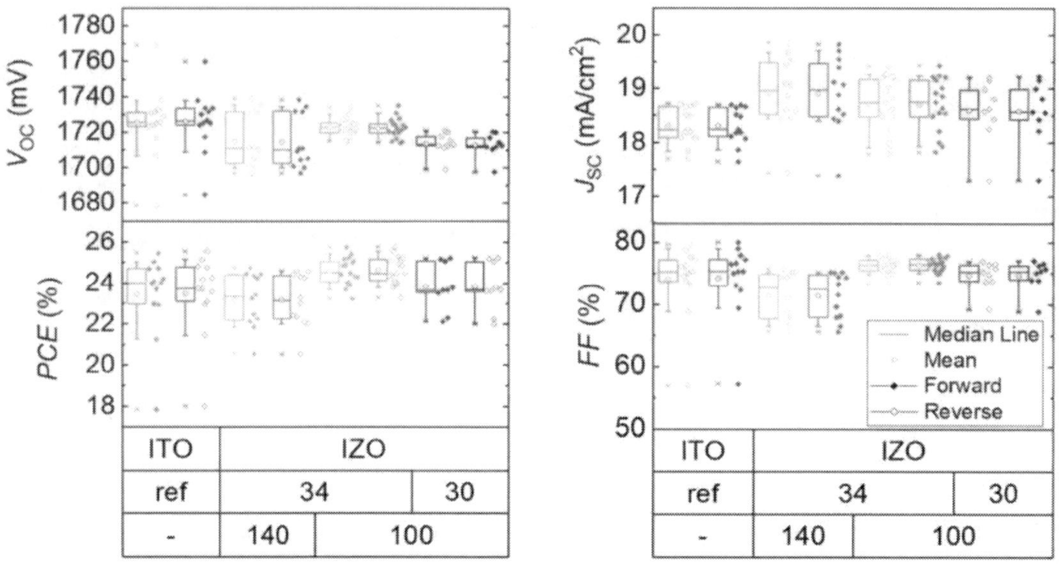

Figure 5: Cell parameters Jsc, Voc, FF and PCE compared with standard ITO reference front contact.

The influence of the deposited IZO which was in parallel deposited on to perovskite-silicon tandem cells is shown in Figure 5. It was observed that higher oxygen flow rate and zinc target power allowed for reaching higher short circuit current density (J_{SC}) of the tandem solar cells. However, these parameters also led to lower fill factor values, but in general, higher overall power conversion efficiencies. A maximum solar cell efficiency of 25.8 % could be reached with the optimized IZO layers.

4. OUTLOOK

In future work, this is planned to be corrected by utilizing an optical emission spectroscopy (OES) based control system to maintain sputtered film characteristics independent of chamber and target state. Additionally, electron probe micro analysis (EPMA) measurements are planned to compare the trend of increasing carrier concentration to the stoichiometric composition of the film.

REFERENCES

[1] A. Richter, M. Hermle, S.W. Glunz, IEEE J. Photovolt. (2013), 4, 1184-1191.

[2] P. S. C. Schulze et al., Sol. RRL (2020), 4, 2000152

[3] H. King, V. Sittinger, T. Harig, D. Stoll, Proceedings of 38th EUPVSEC (2021), 511-514.

[4] V. Sittinger, H. King, Proceedings of 37th EUPVSEC (2020), 769 -771.

ACKNOWLEDGEMENT

This work was supported by the Fraunhofer Lighthouse project (MaNiTU). Özde Ş. Kabakli gratefully acknowledge scholarship support from the Dr. Ruth Heerdt Stiftung. The authors would like to thank Materion Advanced Materials Germany GmbH for providing the rotatable In target.

AlGaAs/InGaP HETEROJUNCTION SOLAR CELL BASED ON TEMPERATURE-GRADED GROWTH

Gan Li[1], Hassanet Sodabanlu[2], Kentaroh Watanabe[2],
Masakazu Sugiyama[1,2], Yoshiaki Nakano[1]

1. The University of Tokyo, Bunkyo-ku, Tokyo, 113-8656, Japan
2. Research Center for Advanced Science and Technology, Meguro-ku, Tokyo, 153-8904, Japan

ABSTRACT: In this study, a temperature-graded growth layer (TGL) is newly introduced to achieve continuous growth between AlGaAs and InGaP grew at 700°C and 560°C, respectively. It is observed that the deterioration of V_{oc} is successfully mitigated after this grading layer is applied. Absolute electroluminescence and atomic force microscopy measurements are further performed to confirm an enhanced ERE and improved interface topography, indicating that the TGL can effectively suppress oxygen incorporation at AlGaAs/InGaP heterojunction, thereby improving the smoothness of the epitaxial layer, which in turn prohibits nonradiative recombination. In all, the TGL is verified to enhance the quality of AlGaAs/InGaP heterojunction solar cells and may serve as a general routine for AlGaAs-based heterogeneous growth with metal-organic vapor phase epitaxy (MOVPE).

1 INTRODUCTION

InGaP is a competent material for the topcell in III-V multi-junction solar cells (MJSCs) due to its adequate bandgap and lattice-matched growth on the GaAs substrate. However, the performance of cells suffers from a low minority carrier mobility in p-type InGaP [1], which leads to insufficient carrier collection and brings down current density. Meanwhile, indium scarcity also limits the massive application of this material.

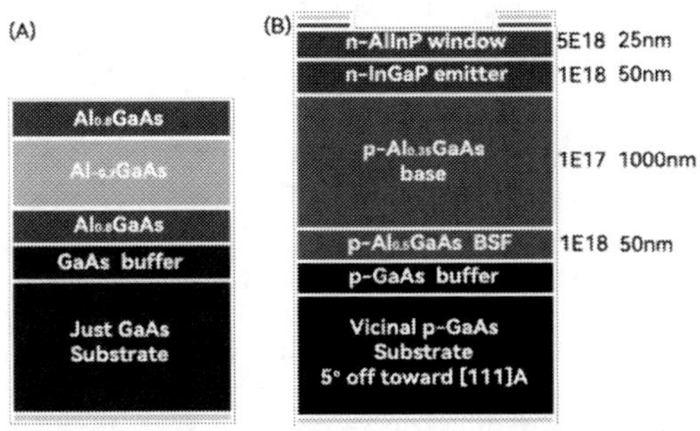

Figure 1: Structures of (A) a double-hetero device and (B) a solar cell device

The above problems can be solved using p-$Al_{0.35}$GaAs to replace the p-InGaP base (Figure 1(A)). Due to these two materials' closed bandgap and lattice constant. This heterojunction combination is supposed to offer higher carrier mobility without changing the bandgap of a solar cell. Besides, indium use can be significantly reduced since only a thin n-InGaP emitter is required.

A similar structure was investigated by Slimane *et al.*[2] using molecule beam epitaxy (MBE). However, the degrading growth quality as Al fracture increases made them compromise to use $Al_{0.25}$GaAs in solar cell devices, of which the bandgap is less than the required one for three or four-junction devices. We expected high-quality $Al_{0.35}$GaAs, with the exact same bandgap as InGaP, to be grown using MOVPE, as it is well known that MOVPE offers better growth quality than MBE accredited to the higher growth temperature.

This study investigates the growth of AlGaAs/InGaP heterojunction cells in our MOVPE system. We found that the offset between two materials' preferred growth temperature is the bottleneck to building high-quality InGaP/AlGaAs heterojunction because of a temperature ramping interval required. After inserting a TGL during temperature adjustment, the device shows significantly improved voltage performance, symbolizing defects at the interface that might be suppressed. Absolute electroluminescence (EL) and atomic force microscopy (AFM) are applied to confirm the improvement of growth quality.

2 EXPERIMENTAL DETAILS

Epitaxial growth of solar cells with a thin InGaP emitter and thick AlGaAs base was performed by planetary MOVPE reactor (Aixtron, AIX2000HT) using tertiarybutylarsine (TBAs), tertiarybutylphosphine (TBP), trimethylgallium (TMGa) and trimethylindium (TMIn) as precursors. Dimethylzinc (DMZn) and diethyltellurium (DETe) were chosen for n-dopant and p-dopant, respectively.

This study chose 700°C as the growth temperature for AlGaAs and 560°C InGaP in MOVPE growth. A novel TGL was employed to achieve continuous growth at the abovementioned temperature offset. This TGL was carried out from 700 °C to 560 °C in 8 minutes, with a low growth rate of 0.6 μm/h as a quarter to which of the bulk to reduce the layer thickness. The grown samples were processed into solar cell devices with standard metal evaporation, photolithography, and wet etching. The highly doped front n-GaAs contact layer was removed by chemical etching for photovoltaic performance evaluation. The current-voltage (J–V) characteristics were measured with a commercial Xenon light source solar simulator (Bunkoukeiki, CEP-2000LS) under 100 mW/cm^2 AM1.5G illumination, and the external quantum efficiency (EQE) spectrum was measured with 2.5 mW/cm^2 monochromatic light. Dark J-V was fitted using an open-source two-diode model [3]. All the samples were measured without antireflection coating (ARC)

Absolute EL measurement and AFM (Agilent, 5420 AFM) were utilized to examine the improvement of the AlGaAs/InGaP interface. In particular, two heterojunction samples with similar structures to the cell samples were reproduced with a skinny InGaP cladding layer (25 nm) over the AlGaAs base for AFM evaluation. This cladding layer is aimed to protect the AlGaAs surface from oxygen while inheriting the interface morphology.

3 RESULTS AND DISCUSSION

In this study, we first prepared a series of AlGaAs double hetero (DH) structure samples for photoluminescence (PL), as shown in Figure 1(B). These devices are grown at 590 °C, 620 °C, 650 °C, 680 °C, and 700 °C to optimize growth temperature for the AlGaAs absorber.

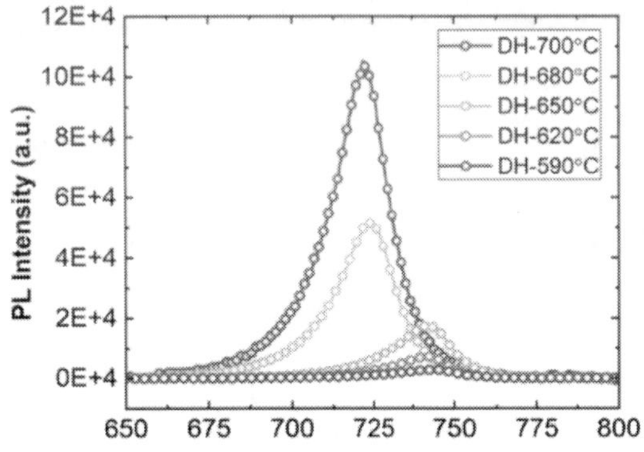

Figure 2: Continuous-wave PL of AlGaAs DH structures with different growth temperatures.

Figure 2 shows an incremental photoluminescence intensity with increasing temperature, which suggests a better crystal quality of AlGaAs grown at 700°C for its less probability of non-radiative recombination. The shift of bandgap is caused by adjustment of Al partial pressure and should not affect the conclusion. On the other hand, InGaP presents a smooth surface under AFM at low growth temperatures, according to previous research [4]. A 140 °C offset between two materials' preferred growth temperatures causes an inevitable temperature ramping process between the growth of two layers and consequently exposes the $Al_{0.35}GaAs$ surface for several minutes in MOVPE growth. Driven by the solid Al-O bonding, incorporating residual oxygen from the reactor chamber or sources will form defects in AlGaAs [5], especially when the Aluminum mole fraction is large. These undesired defects, behaving as non-radiative recombination centers, would degrade cell performance by reducing carriers' diffusion length.

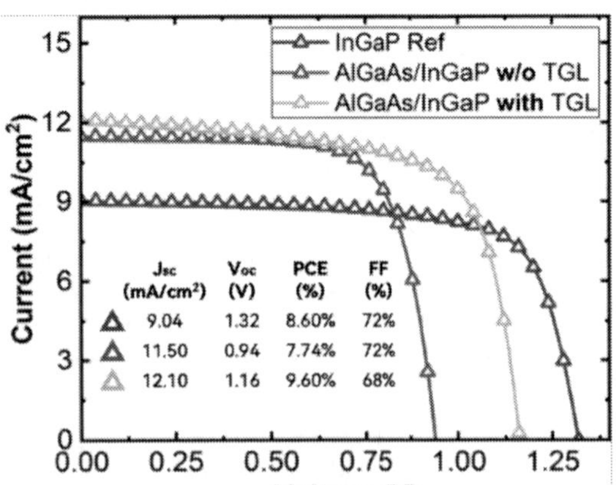

Figure 3: J-V characteristics with AM1.5G (100mW/cm^2) illumination. Inset gives performance parameters.

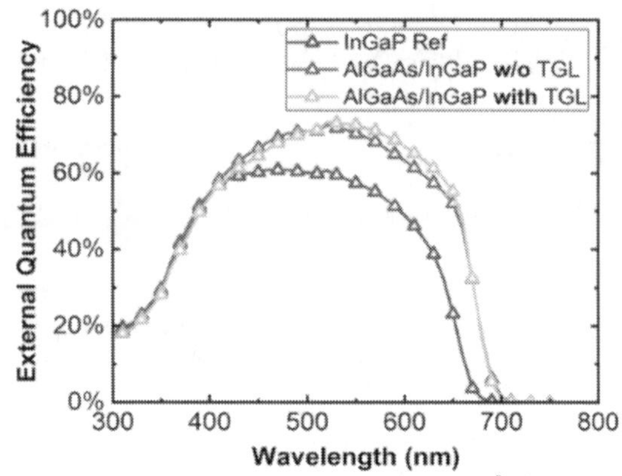

Figure 4: EQE spectrum using 2.5mW/cm^2 chromatic illumination

Two AlGaAs/InGaP heterojunction solar cell samples, without and with TGL and one InGaP homojunction reference sample, were characterized and compared systematically. Both AlGaAs/InGaP heterojunction cells have increased short-circuit current density (J_{sc}) with a diminished open-circuit voltage (V_{oc}) compared with InGaP one, see Figure 3. External quantum efficiency (EQE) reveals that the increment of J_{sc} comes from two parts: better extraction of carriers and a narrower energy bandgap than setting value. The bandgap deviation is probably caused by an inappropriate growth profile and will be modified in the later study. The enhanced carrier extraction aligns with our expectations as p-AlGaAs have better carrier mobility than p-InGaP.

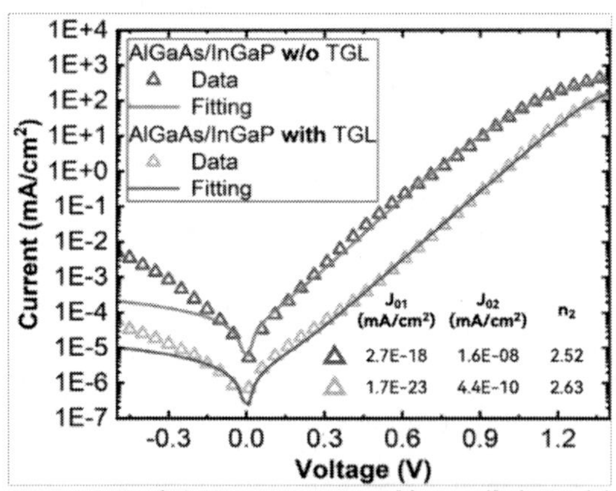

Figure 5: Dark J-V measurement with two-diode model fitting, with parameters of the fitting curves in inset.

For voltage performance, the AlGaAs/InGaP samples first show unsatisfactory V_{oc} of 0.94V, with an enormous $W_{oc} = E_g/q - V_{oc} = 0.87V$, which can be attributed to the temperature ramping interval between AlGaAs and InGaP layers that induces oxygen incorporation. After inserting a TGL, the V_{oc} of the device was significantly improved to 1.16V with no weakened current, indicating that the temperature-graded growth strategy succeeded. Dark J-V measurements are shown in Figure 4. The dark J-V curve was fitted using a two-diode model [3] with the ideal factor $n_1 = 1$ for band-to-band recombination and a non-preset ideal factor n_2 for other types of recombination. Both decreased J_{01} and J_{02} suggest that lower recombination rate in the device with a TGL, which agrees with the increment in V_{oc}. We also note a slightly higher current density in the device with a TGL.

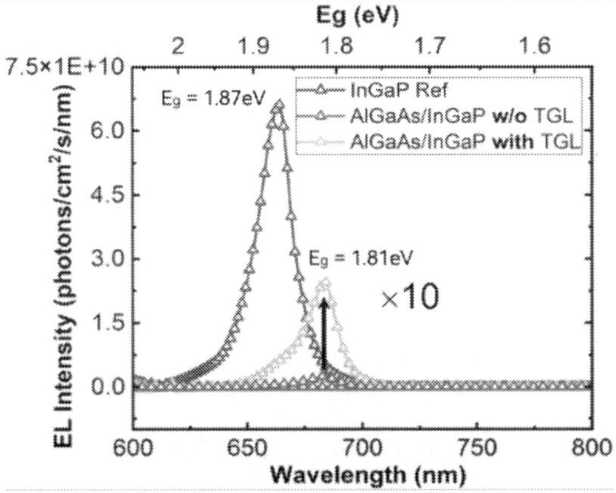

Figure 6: Absolute electroluminescence (EL) intensity with J_{sc} injected. The intensity of the heterojunction cell with TGL is tenfold for clarity, while the signal from the one without TGL is below the sensitivity.

Absolute EL is further performed to obtain external radiative efficiency (ERE) to evaluate the non-radiative recombination of the samples. According to the reciprocity relation given by Rau [6], the difference between V_{oc} and open-circuit voltage at the radiative limit, V_{oc}^{rad}, is a negative exponential function of ERE:

$$\Delta V_{oc} = V_{oc}^{rad} - V_{oc} = \frac{-k_B T}{q} \ln(ERE)$$

which gives a good benchmark for the material quality and its voltage performance, even for materials with different E_g.

	InGaP ref	AlGaAs/InGaP w/o TGL	AlGaAs/InGaP **with** TGL
ERE	3.6E−3 %	~8.6E−11 %	~6.8E−5 %
V_{oc}	1.32V	0.93V	1.16V
V_{oc}^{rad}	1.60V	~1.53V	1.53V
$\boldsymbol{\Delta V_{oc}}$	**0.28V**	**~0.60V**	**0.37V**

Table I: ERE, V_{oc}^{rad} and ΔV_{oc} obtained by absolute PL intensity measurement.

Figure 6 plots the absolute EL intensity of the samples. The hetero-structure sample without a TGL has too weak an EL signal to be detected even after prolonged exposure: only the other two samples were successfully measured as the ERE and V_{oc}^{rad} are calculated in Table 1. The boost from grimmer to vigorous intensity could indicate that non-radiative recombination can be effectively suppressed when TGL is applied. Besides, we confirmed that the EL result agreed with the EQE spectrum that an E_g deviation of 0.06 eV exists between the AlGaAs/InGaP samples and the InGaP sample, which can also contribute to the diminished V_{oc}. Therefore, ERE and ΔV_{oc} turned out to be more useful merit of figures regarding the different E_g. By assuming that both AlGaAs/InGaP cells have the same bandgap (they should have for their similar structure except TGL), we estimated that the dim sample without TGL has an approximate ΔV_{oc} of 0.60V. The TGL was verified to have a beneficially substantial effect on the cells' performance seeing the reduction of ΔV_{oc}, which denotes improvement of AlGaAs/InGaP interface quality. Nevertheless, compared with the InGaP reference cell, the ERE of the heterojunction device with TGL is still two orders of magnitude lower. This passable voltage suggests further modification of AlGaAs crystal quality in future work.

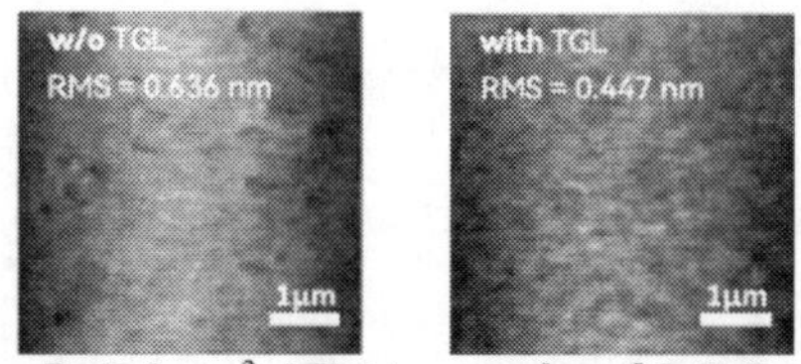

Figure 7: 5×5 μm² AFM topography of InGaP emitter 25nm above the heterojunction of the sample without (left) and with (with) TGL

In the end, AFM was employed to confirm the improvement of growth quality to evaluate the roughness of the AlGaAs/InGaP interface. Figure 7 displays the 5×5 μm² AFM images and the root mean square (RMS) values of the surface roughness of a 25 nm-thick InGaP layer clapping on AlGaAs. The InGaP surface evenness became ameliorated after the TGL was applied, which can be explained by the reduction of external interstitials incorporated into the AlGaAs surface. With improved ERE and interface evenness, it is proved that the graded growth effectively suppresses oxygen incorporation at AlGaAs/InGaP heterojunction and thereby prohibits non-radiative recombination, enhancing V_{oc}.

4 SUMMARY

We proposed a new growth strategy of TGL to improve the performance of heterojunction solar cells, where optimal growth temperatures are different for AlGaAs and InGaP layers, which aimed to prohibit oxygen incorporation by continuous growth. It is verified that solar cell devices with a TGL can achieve higher V_{oc} and J_{sc}. Moreover, AFM and absolute EL measurements proved that, by applying TGL, the interface could be smoothened along with a suppressed non-radiative recombination. On the other hand, AlGaAs/InGaP heterojunction solar cells, even with TGL, still show performance inferior to InGaP homojunction ones regarding their ΔV_{oc}, demanding systematic research on improving its growth quality in the future.

5 ACKNOWLEDGEMENTS

A part of the results of this study is based on a project, JPNP20015, commissioned by the New Energy and Industrial Technology Development Organization (NEDO). G. LI himself is supported by the Quantum Science and Technology Fellowship Program (Q-STEP) of the University of Tokyo.

References:

[1] M. Ikeda, K. Kaneko, Journal of applied physics, 66 (1989) 5285
[2] A. Ben Slimane, A. Michaud, O. Mauguin *et al.*, Progress in Photovoltaics: Research and Applications, 28 (2020) 393
[3] S. Rißland, O. Breitenstein, Energy Procedia, 38 (2013) 167
[4] M. Asami, K. Watanabe, Y. Nakano *et al.*, physica status solidi (b), 259 (2022) 2100305
[5] S. Heckelmann, D. Lackner, C. Karcher *et al.*, IEEE Journal of Photovoltaics, 5 (2014) 446
[6] U. Rau, Physical Review B, 76 (2007) 085303

Applying GaAsP with Ultra-high Phosphorous Composition to InGaAs/GaAsP Strain-balanced Quantum Structure Solar Cells to Achieve High-efficiency Multi-junction Solar Cells

Meita Asami[1], Kentaroh Watanabe[2], Yoshiaki Nakano[1], Masakazu Sugiyama[1, 2]
[1]School of Engineering, The University of Tokyo, [2]Research Center for Advanced Science and Technology, The University of Tokyo
[1]7-3-1 Hongo, Bunkyo-ku, Tokyo 113-8656, [2]4-6-1 Komaba, Meguro-ku, Tokyo 153-8904

ABSTRACT: This study was aimed at determining a suitable strain-balanced quantum structure to further improve the energy conversion efficiency of conventional Ge-based lattice-matched triple-junction solar cells. We investigated the voltage losses in several types of quantum-structure solar cells using the detailed balance theory and optoelectronic reciprocity relation. Based on the voltage loss analysis, we found that the combination of thick InGaAs quantum well layers and thin GaAsP layers was suitable for high-performance quantum-structured solar cells. To maintain the strain-balance condition for a quantum structure with thick InGaAs, we proposed the use of ultrahigh phosphorous composition GaAsP quantum barrier layers. The results of this study are significant for the design of future quantum-structure solar cells.

Keywords: Quantum Well, Multi-junction Solar Cell, Recombination, III-V Semiconductors

1 INTRODUCTION

Bandgap adjustment is indispensable for improving the energy conversion efficiency of conventional dual-junction and triple-junction solar cells. To achieve this, in previous studies, quantum structures have been inserted into the GaAs sub-cells of InGaP/GaAs dual-junction solar cells and those of InGaP/GaAs/InGaAs inverted metamorphic (IMM) triple junction solar cells [1, 2]. Thus, the highest energy conversion efficiencies for dual-junction and triple-junction solar cells have been achieved [3]. Bandgap adjustment using quantum structures is a promising method to increase the efficiency of solar cells.

Both Ge-based lattice-matched triple-junction solar cells and IMM triple-junction solar cells have been used in space applications. In these triple-junction solar cells, the amount of current generated in the middle cell, that is, the GaAs subcell, is the lowest, and the total current flowing through the entire cell is restricted by this current. Therefore, previous studies inserted quantum structures into GaAs sub-cells to adjust the effective bandgap for current matching [2, 4, 5]. The advantage of Ge-based lattice-matched triple-junction solar cells over IMM triple-junction solar cells is that lattice-matched solar cells do not require thick buffer layers, which are necessary for metamorphic growth.

However, although the world's highest energy conversion efficiency has been achieved for IMM triple junction solar cells with quantum structures, high-performance Ge-based lattice-matched triple-junction solar cells with quantum structures have not been reported to date [2]. This is because superlattices cannot easily be grown on 6° off misoriented substrates [6]. To avoid the formation of an anti-phase domain on GaAs grown on Ge, GaAs and superlattices should be grown on 6° off misoriented substrates [7, 8]. Notably, a superlattice can be undulated on misoriented substrates, and an undulated superlattice termed Wire-on-Well (WoW) can be formed as a result of step-bunching [9]. The schematic cross-sectional scanning transmission electron microscopy (STEM) images of a planar superlattice (PSL), which is a normal superlattice, and the WoW are shown in Fig. 1. The actual STEM images were reported in [10]. The open-circuit voltage of GaAs solar cells with WoW tends to be lower than that of other GaAs-based solar cells. To fabricate Ge-based lattice-matched triple-junction solar cells with high energy conversion efficiency, the open-circuit voltage of the GaAs subcell with WoW needs to be enhanced. In this study, to elucidate the cause of the low voltage, we quantitatively evaluated the type of voltage loss that occurs in WoW solar cells. Based on the evaluation and analysis, we proposed a new quantum structure to reduce the voltage loss.

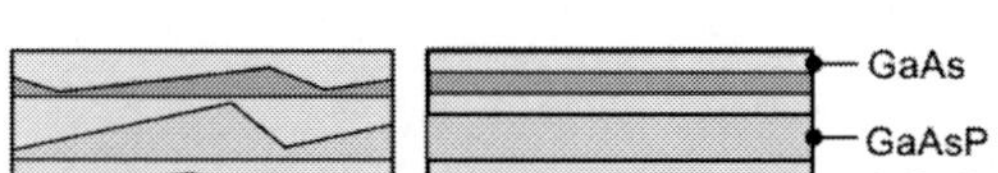

Figure 1: Schematic cross-sectional image of each superlattice.

2 EXPERIMENTAL DETAILS

The structures of the samples fabricated and evaluated in this study are shown in Fig. 2. The samples were labelled GaAs ref., 20 PSL, 50 PSL, 20 WoW, and 50 WoW, respectively; for example, a 20 PSL is a GaAs p-i-n junction solar cell with 20 layers of PSL. Solar cell samples were grown using metal-organic vapour phase epitaxy (MOVPE) in a reactor AIX200/4 (AIXTRON). The precursors used in this study are described in [11]. Two types of GaAs wafers were used in this study. GaAs (0 0 1) 2-in wafers without misorientation were used for solar cells with planar superlattices, and GaAs (0 0 1) 2-in wafers with a 6° off misorientation toward <1 1 1>B were used for solar cells with WoW. The detailed growth conditions have been described in our previous paper [12].

We performed a voltage loss analysis based on the detailed balance theory [13] and optoelectronic reciprocity relation [14]. We can calculate the Shockley–Queisser (SQ) limit open-circuit voltage, V_{SQ} using the detailed balance theory. We have developed a new method for determining the bandgap of quantum structure solar cells. More information on this method is presented in our current study [12]. The open circuit voltage at the radiative limit can be calculated using an optoelectronic reciprocity relation (1) [14].

$$V_{OC}^{rad} = \frac{kT}{q}\ln\left(\frac{q\int_0^\infty Q_e\phi_{sun}dE}{q\int_0^\infty Q_e\phi_{bb}dE}\right), \qquad (1)$$

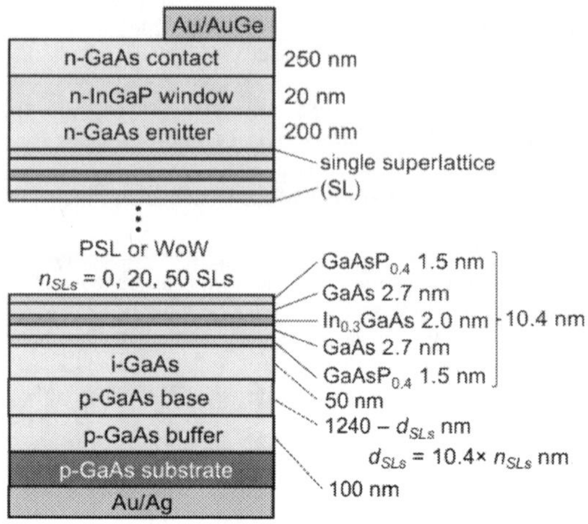

Figure 2: Structure of samples fabricated using metal-organic vapor phase epitaxy. For GaAs reference sample (n_{SLs} = 0), 50-nm-thick p-GaAs is grown instead of 50-nm-thick i-GaAs. The thickness of the p-GaAs base layer is adjusted so that the total thickness of each sample is the same while changing the number of superlattices (SLs) n_{SLs}. The samples with planar superlattice are grown on the GaAs (0 0 1) wafers without misorientation, and the samples with Wire on Well (WoW) are grown on the 6°-off wafers.

where k is the Boltzmann constant, and T is the temperature, E is the energy, Q_e is the external quantum efficiency (EQE), ϕ_{sun} is the solar irradiance flux, and ϕ_{bb} is the ambient blackbody radiation flux. As no non-radiative recombination occurs at the radiative limit, the external radiative efficiency is 100%. A non-radiative recombination voltage loss can be evaluated using Equation (2).

$$\Delta V_{nonrad} = V_{OC}^{rad} - V_{OC}^{IV}, \qquad (2)$$

where V_{OC}^{IV} is the actual open-circuit voltage obtained from the I-V measurements. The radiative recombination voltage loss can be calculated using Equation (3).

$$\Delta V_{rad} = V_{SQ} - V_{OC}^{rad} \qquad (3)$$

The validity of the application of the optoelectronic reciprocity relation to quantum structure solar cells has been discussed in detail in the literature [15].

3 RESULTS AND DISCUSSION

Table I and Fig. 3 show the results of the I-V and EQE measurements. Fig. 4 shows the results of the voltage-loss analysis. The nonradiative recombination voltage loss ΔV_{nonrad} is significantly larger than the radiative recombination voltage loss ΔV_{rad}. However, the difference in ΔV_{nonrad} for each sample is small, and the difference in the total voltage loss depended on ΔV_{rad}. To reduce ΔV_{nonrad}, we must improve the crystal quality. However, this is challenging because it requires improvement of the reactor conditions and thorough adjustment of the crystal growth conditions. In this study, we addressed the reduction in voltage loss due to radiative recombination.

We investigated factors determining ΔV_{rad}. Because

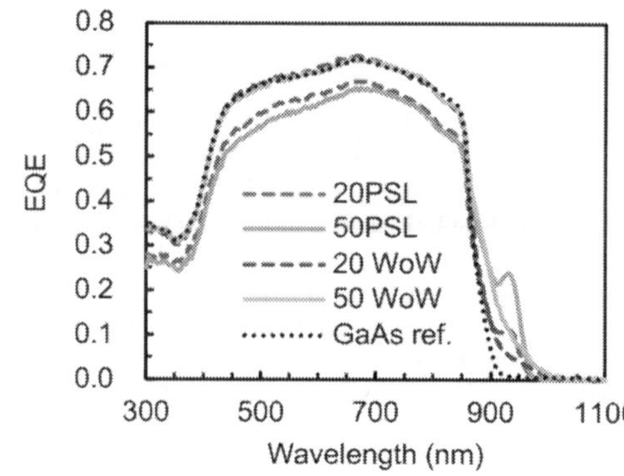

Figure 3: Results of EQE measurements

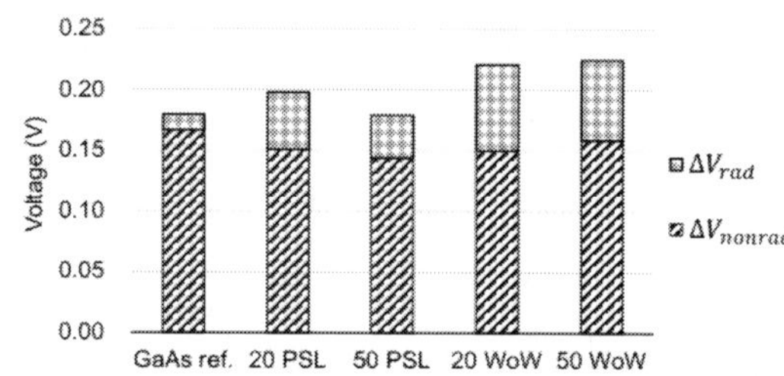

Figure 4: Results of voltage loss analysis based on detailed balance theory and optoelectronic reciprocity relation.

the voltage at the SQ limit is determined by the bandgap, V_{OC}^{rad} must be high to reduce ΔV_{rad}. V_{OC}^{rad} is determined by the flux of blackbody radiation, the flux of sunlight, and EQE, as shown in Equation (1). The blackbody radiation and solar fluxes are universal in each sample. As the only factor that varies with the samples is the EQE value, the shape of the EQE curve determines V_{OC}^{rad}.

To investigate the effect of the shape of the EQE curve on V_{OC}^{rad}, a double-step EQE model was used in this study. In this model, the EQE over the range of 800–900 nm wavelength, $EQE_{800\text{-}900nm}$, varies from 0 to 1, as shown in Fig. 5. Fig. 6 shows the change in ΔV_{rad} as $EQE_{800\text{-}900nm}$ varies from 0 to 1. ΔV_{rad} increases rapidly when the value of $EQE_{800\text{-}900nm}$ is small.

The 50-WoW sample has a gradually declining EQE tail, and the value of the EQE is small within a wide range of wavelengths. The reason for this gradually declining EQE tail is that the quantum well width is not constant in the WoW. The non-uniformity of the quantum well width accounts for the non-uniformity of the energy levels. We calculated the relationship between the quantum well width and the energy levels in the quantum well using the effective mass approximation. The results are shown in Fig. 7. In the simulation, we assumed that the quantum structure contains the 3-nm-thick $GaAsP_{0.8}$ quantum

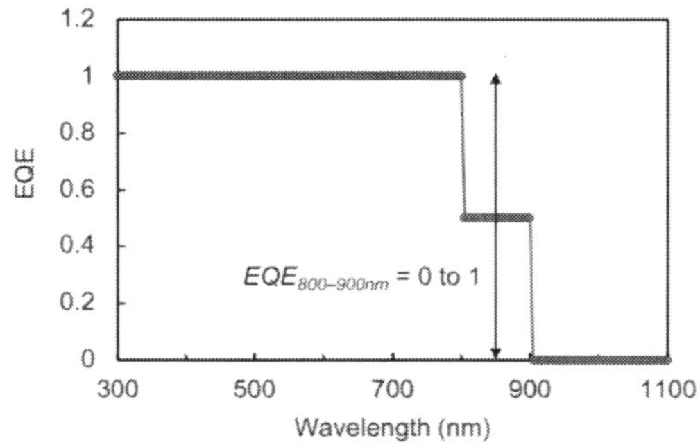

Figure 5: Double step EQE model.

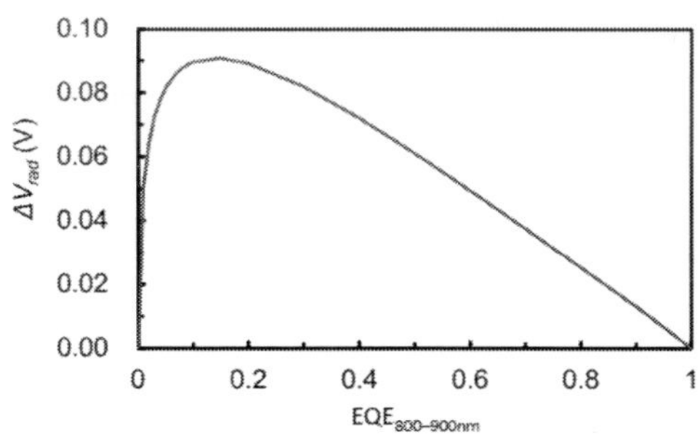

Figure 6: Double step EQE model.

barrier layers and x-nm-thick $InGaAs_{0.09}$ quantum well layers. The graph of the effective bandgap becomes more gradual as the quantum well width becomes thicker. For example, as shown in Fig. 7, the ±2-nm-thick fluctuation at a 4-nm-thick quantum well results in a 51-meV fluctuation of the energy level, whereas at a 14-nm-thick quantum well, the fluctuation of the energy level is only approximately 18 meV. Therefore, as the thickness of the quantum well layer increases, the energy level fluctuation with the quantum well width decreases. Thus, thicker quantum wells are suitable to realize sharp EQE edges, which are necessary to reduce the radiative recombination voltage loss.

Therefore, in this study, we grew samples containing thick InGaAs quantum well layers on a 6° off misoriented substrate to suppress the radiative recombination voltage loss. However, if the InGaAs quantum well layer is thickened, usually the GaAsP barrier layer must also be thickened to achieve a strain-balanced condition. The maximum P composition of the GaAsP barrier layer used in previous studies was 40% [1, 2, 5]. In this study, we used a 13.2-nm-thick $In_{0.09}GaAs$ quantum well layer. If we use 40% P composition GaAsP, the thickness of the $GaAsP_{0.4}$ layer must be approximately 6 nm to compensate for the crystal strain caused by these 13.2-nm-thick $In_{0.09}GaAs$ well layers. However, a previous study has shown that the thickness of a $GaAsP_{0.4}$ barrier layer must be less than 3 nm to maintain a high carrier transport efficiency [16]. Therefore, in this study, we decided to increase the P content of the GaAsP barrier layer to 80%. The fabricated quantum structure is illustrated in Fig. 8. Because many crystal defects are expected to occur at the interface between ultra-high P composition GaAsP and InGaAs owing to the difference in lattice constants, a $GaAsP_{0.09}$ intermediate layer was inserted between $GaAsP_{0.8}$ and $In_{0.09}GaAs$. Notably, carriers can easily

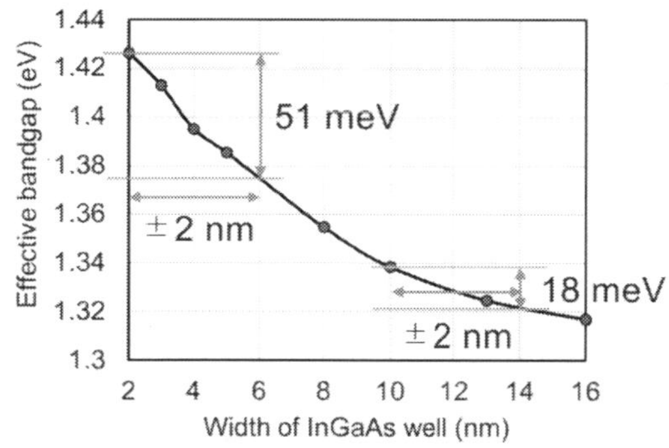

Figure 7: Relation between Effective bandgap formed in InGaAs wells and width of InGaAs well.

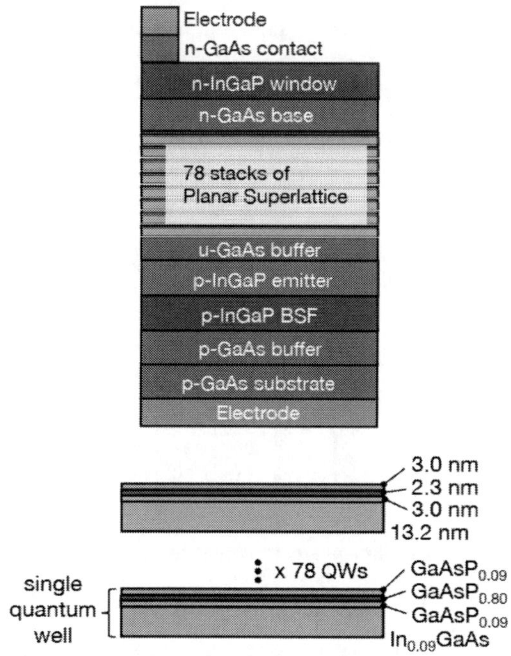

Figure 8: Schematic of the new solar cell and the new quantum structure.

escape from the InGaAs well layer to the $GaAsP_{0.09}$ intermediate layer via thermal escape. The results of the EQE and I-V measurements are shown in Fig. 9. The open-circuit voltage of this new quantum-structure solar cell is higher than that of our conventional quantum-structure solar cells. The current density is lower than that of the WoW solar cells because of the shift of the absorption edge to the short-wavelength side and the low EQE in the short-wavelength region. The low EQE in the short-wavelength region in the new quantum structure solar cell is due to the poor conditioning of the n-GaAs base layer and the n-InGaP window layer rather than the quantum structure. This EQE degradation at short wavelengths is also currently observed in normal GaAs solar cells grown in our MOVPE reactor, and the growth conditions are reviewed.

The voltage loss due to the radiative recombination in the new quantum structure solar cell is shown in Fig. 10. The ΔV_{rad} of the new quantum-structure solar cell was significantly smaller than that of our conventional

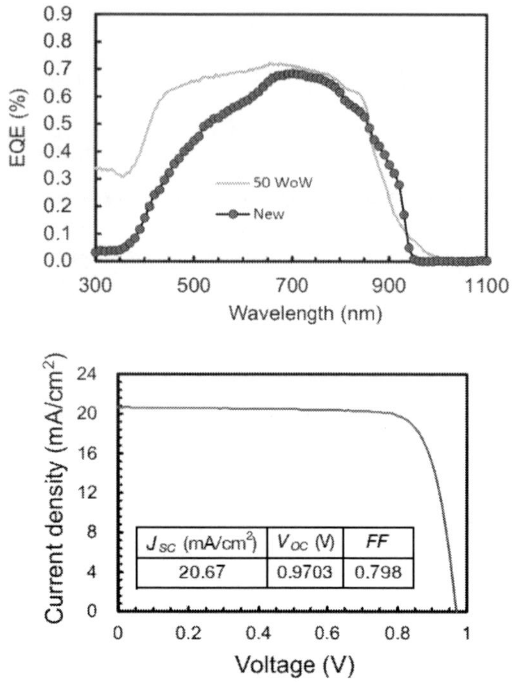

Figure 9: EQE and I-V measurements results of the new quantum structure solar cell.

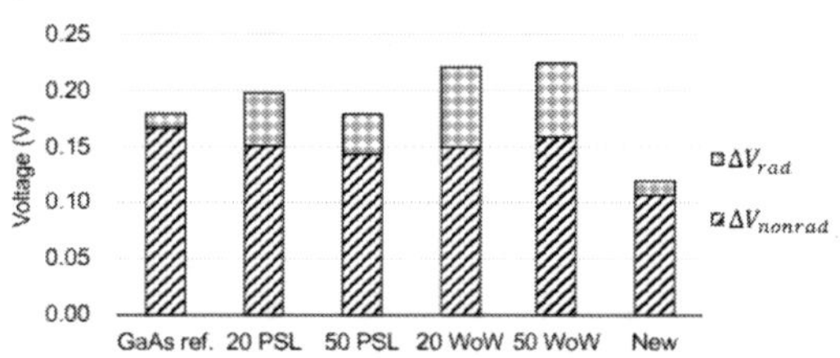

Figure 10: Voltage loss in each solar cell including the new quantum structure solar cell.

quantum-structure solar cells. In addition, ΔV_{nonrad} caused by crystal defects due to the insertion of ultra-high P composition GaAsP barrier layers was a concern. However, it was successfully suppressed to a lower level than that of conventional quantum structure solar cells. Although the degradation of carrier transport caused by the barrier height of ultra-high P composition GaAsP was also a concern, the fill factor value was high; furthermore, the EQE results showed that the carriers excited in the quantum wells could be efficiently extracted.

4 CONCLUSIONS

In this study, we optimised the quantum structure grown on 6° off misoriented substrates based on voltage loss analysis. We found that the in-plane fluctuation of the thickness of the well layers in WoW result in large voltage losses. We also found that a thicker well layer could reduce the thickness fluctuation effect. Based on these results, we fabricated a new quantum structure solar cell with thicker InGaAs well layers. The crystal strain of thicker InGaAs well layers was compensated for by ultra-high P composition GaAsP barrier layers; the P composition was 80%. This ultra-high P composition of GaAsP enables us to realise strain-balanced conditions with thick InGaAs well layers and efficient carrier transport simultaneously. A solar cell with this new quantum structure succeeded in reducing the radiative recombination voltage loss and achieved a high open-circuit voltage. The results of this study are significant for the design of future quantum-structure solar cells.

References

[1] M. A. Steiner, R. M. France, J. Buencuerpo, J. F. Geisz, M. P. Nielsen, A. Pusch, W. J. Olavarria, M. Young, and N. J. Ekins-Daukes, "High efficiency inverted GaAs and GaInP/GaAs solar cells with strain-balanced GaInAs/GaAsP quantum wells," Adv. Energy. Mater. 11, 2002874, 2021.

[2] R. M. France, J. F. Geisz, T. Song, W. Olavarria, M. Young, A. Kibbler, and M. A. Steiner, "Triple-junction solar cells with 39.5% terrestrial and 34.2% space efficiency enabled by thick quantum well superlattices," Joule 6, pp. 1121–1135, 2022.

[3] NREL, "Best Research-Cell Efficiency Chart," https://www.nrel.gov/pv/cell-efficiency.html, (accessed : 23rd Sept. 2022).

[4] N. J. Ekins-Daukes, J. M. Barnes, K. W. J. Barnham, J. P. Connolly, M. Mazzer, J. C. Clark, R. Grey, G. Hill, M. A. Pate, and J. S. Roberts, "Strained and strain-balanced quantum well devices for high-efficiency tandem solar cells," Sol. Energy Mater. and Solar Cells 68, pp. 71–87, 2001.

[5] H. Fujii, K. Toprasertpong, Y. Wang, K. Watanabe, M. Sugiyama, and Y. Nakano, "100-period, 1.23-eV bandgap InGaAs/GaAsP quantum wells for high-efficiency GaAs solar cells: toward current-matched Ge-based tandem cells," Prog. Photovolt.: Res. Appl. 22, pp. 784-795, 2014.

[6] H. Fujii, H. Sodabanlu, M. Sugiyama, and Y. Nakano, "Low-temperature MOVPE using TEGa for suppressed layer undulation in InGaAs/GaAsP superlattice on vicinal substrates", J. Cryst. Growth 414, pp. 3-9, 2015.

[7] L. Lazzarini, L. Nasi, G. Salviati, C. Z. Fregonara, Y. Li, L. J. Giling, C. Hardingham, and D. B. Holt, "Antiphase disorder in GaAs/Ge heterostructures for solar cells," Micron 31, pp. 217–222, 2000.

[8] P. -H. Wu, Y. -K. Su, I.-L. Chen, S. -F. Chen, C. -H. Chiou, S. -H. Guo, J. -T. Hsu, and W. -R. Chen, "Research of surface morphology in Ga(In)As epilayers on Ge grown by MOVPE for multi-junction solar cells," J. Cryst. Growth298, pp. 767–771, 2007.

[9] H. Sodabanlu, Y. Wang, K. Watanabe, M. Sugiyama, and Y. Nakano, "Growth of InGaAs/GaAsP multiple quantum well solar cells on mis-orientated GaAs substrates," J. Appl. Phys. 115, 233104, 2014.

[10] M. Sugiyama, H. Fujii, T. Katoh, K. Toprasertpong, H. Sodabanlu, K. Watanabe, D. Alonso-Álvarez, N. J. Ekins-Daukes, and Y. Nakano, "Quantum wire-on-well (WoW) cell with long carrier lifetime for efficient carrier transport Prog. Photovolt.: Res. Appl.24, pp. 1606–1614, 2016.

[11] M. Asami, K. Toprasertpong, K. Watanabe, Y. Nakano, Y. Okada, and M. Sugiyama, "Comparison of effective carrier mobility between wire on well and planar superlattice using time-of-flight

measurement," IEEE J. Photovolt. 10, pp. 1008–1014, 2020.
[12] M. Asami, K. Watanabe, Y. Nakano, and M. Sugiyama (to be submitted).
[13] W. Shockley, and H. J. Queisser, "Detailed balance limit of efficiency of p–n junction solar cells," J. Appl. Phys. 32, pp. 510–519, 1961.
[14] U. Rau, "Reciprocity relation between photovoltaic quantum efficiency and electroluminescent emission of solar cells," Phys. Rev. B 76, 085303, 2007.
[15] M. Asami, K. Watanabe, Y. Nakano, and M. Sugiyama, "Evaluation of crystal quality of solar cells using voltage metrics based on optoelectronic reciprocity relation and detailed balance theory" (submitted to a journal).
[16] K. Toprasertpong, H. Fujii, Y. Wang, K. Watanabe, M. Sugiyama, and Y. Nakano, "Carrier escape time and temperature-dependent carrier collection efficiency of tunneling-enhanced multiple quantum well solar cells," IEEE J. Photovolt. 4, pp. 607–613, 2014.

This is a short version of the paper submitted to the journal EPJ Photovoltaics.

It has been selected by the Committee of the WCPEC-8 for submission to EPJ Photovoltaics.

EPITAXY AND CHARACTERIZATION OF INP/INGAAS TANDEM SOLAR CELLS GROWN BY MOVPE ON INP AND SI SUBSTRATES

Stefano Soresi,[a] Mattia da Lisca,[b,c,d] Claire Besancon,[a] Nicolas Vaissiere,[a] Alexandre Larrue,[a] Cosimo Calo,[a] José Alvarez,[b,c,d] Christophe Longeaud,[b,c,d] Ludovic Largeau,[e] Pablo Garcia Linares,[f] Eric Tournié,[g] Jean-Paul Kleider,[b,c,d] Jean Decobert[a]

[a] III-V Lab, 1 Avenue Augustin Fresnel, 97167 Palaiseau, France

[b] Université Paris-Saclay, CentraleSupélec, CNRS, Laboratoire de Génie Electrique et Electronique de Paris, 91192, Gif-sur-Yvette, France

[c] IPVF, Ile de France Photovoltaic Institute, 30 route départementale 128, 91120 Palaiseau, France

[d] Sorbonne Université, CNRS, Laboratoire de Génie Electrique et Electronique de Paris, 75252, Paris, France

[e] C2N/CNRS-Universite Paris-Saclay, Route de Nozay, 91460, Marcoussis, France

[f] Instituto de Energía Solar, Universidad Politécnica de Madrid, E.T.S.I. Telecomunicación, Av. de la Complutense 30, 28040 Madrid, Spain

[g] IES, Univ. Montpellier, CNRS, F- 34000, Montpellier, France

ABSTRACT: The integration of III-V multi-junction solar cells on Si substrates is one of the most encouraging prospects to combine high photovoltaic performance with a reduction of the manufacturing costs.

We propose a prospective study for the realization of an InP/InGaAs tandem solar cell lattice-matched to InP on a commercially available Si template by direct MOVPE growth. The InP top cell and the InGaAs bottom cell were firstly separately grown and optimized using InP substrates, which exhibited conversion efficiencies of 13.5% and 11.4%, respectively. The two devices were then combined in a tandem device by introducing an intermediate InP/AlInAs lattice-matched tunnel junction, showing an efficiency of 18.4%. As an intermediate step towards the realization of the tandem device on Si, the InP and InGaAs single junction solar cells were grown on top of a commercial InP/GaP/Si template. Each cell was electrically characterized by external quantum efficiency and illuminated current-voltage under solar simulator: efficiencies of 3.6% and 2.0% for the InP and InGaAs solar cells were obtained, respectively. The material quality was also analyzed by means of X-ray diffraction, Atomic-Force Microscopy, Transmission Electron and Scanning Electron Microscopy.

Keywords: III-V Semiconductors; Epitaxy; Multijunction Solar Cell

1 INTRODUCTION

During the last decade, III-V materials based multi-junction solar cells (MJSC) have demonstrated to be one of the most efficient solutions to reach high conversion efficiencies. In particular, InP present a wide range of alloys that are suitable for photovoltaic (PV) use under solar spectrum, covering the entire energy range from 0.76 eV ($In_{0.53}Ga_{0.47}As$) up to 1.42 eV ($Al_{0.48}In_{0.52}As$). However, in order to render III-V semiconductors competitive at industrial level, a substantial cost reduction is required [1]. In this concern, the use of Si substrates as 1.1 eV bandgap bottom cells [2] has already demonstrated being an excellent solution [3,4].

In this article, we propose a prospective study for the realization of an InP/InGaAs tandem solar cell lattice-matched to InP on a commercially accessible Si template by direct MOVPE growth. In particular, in the first part of this work, the primary purpose is the experimental demonstration of an InP/InGaAs tandem solar cell (TSC) on an InP wafer. InP/InGaAs TSCs provide a bandgap combination very close to the optimum efficiency for a double-junction solar cell with AM1.5D spectrum at high concentration: considering the bandgaps of InP (1.34 eV) and $In_{0.53}Ga_{0.47}As$ (0.76 eV) constituting the two sub-cells, a theoretical limiting efficiency of ~ 32% can be reached for the tandem cell [5].

We have first fabricated and characterized the two single junction cells (SJCs) grown on InP substrates. Then, we have combined them into a tandem device by including an intermediate, transparent InP/AlInAs tunnel junction formerly presented in [6].

In the second part, we aimed to reproduce the full process on commercially available InP/GaP/Si templates from NAsP company [7], with the purpose of evaluating the effect of the metamorphic InP buffer relaxation onto device performance. This latter was structurally characterized, and the PV performance of the resulting solar cells was evaluated and compared with the previous devices developed on InP substrates.

A commercial template has been selected to demonstrate the feasibility of realization of III-V TSCs on low-priced and easily accessible substrates with a view to propose new strategies for the development and reducing the production cost on 300 mm Silicon platform.

2 DESIGN OF THE STRUCTUES

Table I and Table II show the epitaxial layer structures of the InP and $In_{0.53}Ga_{0.47}As$ p^+n SJCs with the relative doping concentrations, respectively. The epitaxial layer structures were grown on (n) InP substrates. Both devices present an (n^+) InP layer of 700 nm and 400 nm for the InP and InGaAs SJCs, respectively, which acts both as back surface field (BSF) and as buffer layer. An (n) InP and (n) $In_{0.53}Ga_{0.47}As$ base with a thickness of 3 μm was used. The emitter consists in (p^+) InP and (p^+) $In_{0.53}Ga_{0.47}As$ layers with a thickness of 150 nm and 200 nm, respectively. A window layer of (p^+) $Al_{0.48}In_{0.52}As$ (1.44 eV) and (p^+) $Al_{0.31}In_{0.53}Ga_{0.16}As$ (1.2 eV) has been included in the InP and InGaAs SJCs structure, respectively. In both devices, a (p^+) InP etch-stop layer was introduced to protect the window layer during the etch of the (p^{++}) $In0_{.53}Ga_{0.47}As$ contact layer. On top of InP solar cells is deposited an anti-reflecting coating (ARC) of Si_3N_4 with a thickness of 115 nm. The optimal thickness was calculated in order to minimize the reflection at the wavelength of maximum absorption of InP (925 nm) [8]. The two devices were then combined in a tandem device by introducing an intermediate (n^{++})InP/(p^{++})$Al_{0.48}In_{0.52}As$ tunnel junction. A sketch of the TSCs on InP is reported in Figure 1.

Table I: Structure of the p⁺n single junction InP cell

Layer	Material	Thickness (nm)	Doping level (cm^{-3})
Contact	(p^{++}) $In_{0.53}Ga_{0.47}As$	200	$+2x10^{19}$
Etch-stop	(p^{+}) InP	10	$+1.5x10^{18}$
Window	(p^{+}) $Al_{0.48}In_{0.52}As$	25	$+1.5x10^{18}$
Emitter	(p^{+}) InP	150	$+1x10^{18}$
Base	(n) InP	3000	$-1x10^{17}$
BSF	(n^{+}) InP	700	$-2x10^{18}$
Substrate	(n) InP	/	$-2x10^{18}$

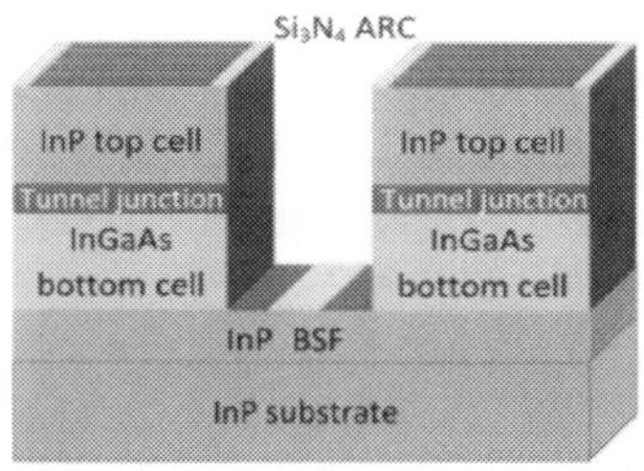

Figure 1: Tandem solar cell fabricated on InP.

3 EXPERIMENT

The samples presented in this paper were all grown in a vertical MOVPE Aixtron Close-Coupled-Showerhead reactor at a 150 mbar pressure. The substrates were two inches (0 0 1) n-type InP wafers or InP/GaP/Si templates diced from a 300 mm commercial wafer [7,9]. The used precursors were trimethylindium (TMIn), trimethylgallium (TMGa), trimethylaluminum (TMAl), phosphine (PH_3) and arsine (AsH_3) under hydrogen (H_2) flow. The n-type dopant was Si (from disilane, Si_2H_6, precursor), whereas the p-type dopant was Zn (from diethylzinc, DEZn, precursor). The growth temperatures were 630 °C for the n-type layers and 600 °C for the p-type layers. The growth conditions for the tunnel junction are somewhat different from the rest of the cell and they have been thoroughly described elsewhere [6]. The InP/AlInAs tunnel junction allows to reach a high peak tunneling current of 1570 A/cm² at voltages between 120 mV and 160 mV, thus indicating very low losses under tandem cell operation.

A proper device design and processing procedure was developed for the solar cell fabrication. The p-type front contact was obtained using Pt/Au deposited by sputtering. The contact resistivity, evaluated by Transmission Line Measurements (TLM), was $3x10^{-6}$ Ωcm^2. Wet etching of the mesas (5 mm side squares) for the cells grown on InP substrates was performed with a $HBr:H_2O_2:H_2O$ (15:1:15) non-selective solution. The InGaAs contact layer and the InP etch-stop layer were etched by selective $H_3PO_4:H_2O_2:H_2O$ (3:1:40) and HCl: H_3PO_4 (1:3) solutions. The Si_3N_4 ARC layer was deposited by plasma-enhanced chemical vapor deposition (PECVD) at 245°C. The Ti/Pt/Au n-type rear metallic contact was directly deposited by sputtering on the back surface of the InP substrate.

The NAsP template, presented in Table III, was grown on a 300 mm (0 0 1) Si substrate. After a chemical etching followed by a high temperature annealing directly in the VPE chamber to deoxidize the substrate, a 200 nm thick Si buffer was grown on top of the substrate using SiH_4 as a precursor. A 60 nm thick GaP layer was then grown by MOVPE using tertiarybutylphosphine (TBP) and triethylgallium (TEGa) as precursors [9]. For our template, a 1 μm thick InP layer was directly grown by the supplier on the GaP layer by MOVPE. The latter was added in order to adapt the lattice parameter of the template to that of InP and aiming to lower the unavoidable threading dislocation density (TDD).

All the samples were characterized by X-ray diffraction (XRD) in order to evaluate the quality of the structures. Furthermore, the samples made on Si templates underwent Atomic-Force Microscopy (AFM) and Scanning Electron Microscopy (SEM) characterizations to evaluate the effects of the lattice-mismatched substrate on the morphology.

Cross-sections of solar cells grown on Si template have been prepared by Focused Ion Beam (FIB) on a FEI Strata DB400 system and observed by Transmission Electron Microscopy (TEM) on a FEI Tecnai Osiris 200kV system equipped with X-EDS and a GIF Quantum detectors to assess their structural quality.

Current density-voltage (J-V) measurements were performed using 4-wire configurations, using an Oriel Instruments 81192 solar simulator (the maximum illuminated area of the simulator being 100 x100 mm², with a nominal power of 1 kW and adjusted to provide a solar spectrum similar to AM1.5G. The measurements were taken with Keithley 2400 and 2450 SourceMeters. The Spectral Response (SR)/External Quantum Efficiency (EQE) measurements were performed by a modified Fourier Transform InfraRed Spectrometer (FTIR) [36].

4. RESULTS AND DISUCSSION

4.1 Solar cells on InP substrate

The three solar cells presented in this section belong to the devices fabricated on InP substrates. The J-V characteristics acquired at room temperature under one-sun AM1.5G illumination for the single junction

Table II: Structure of the p⁺n single junction InGaAs cell

Layer	Material	Thickness (nm)	Doping level (cm^{-3})
Contact	(p^{++}) $In_{0.53}Ga_{0.47}As$	250	$+2x10^{19}$
Etch-stop	(p+) InP	10	$+1.5x10^{18}$
Window	(p^{+}) $Al_{0.31}In_{0.53}Ga_{0.16}As$	30	$+1.5x10^{18}$
Emitter	(p^{+}) $In_{0.53}Ga_{0.47}As$	200	$+1x10^{18}$
Base	(n) $In_{0.53}Ga_{0.47}As$	3000	$-1x10^{17}$
BSF	(n^{+}) InP	400	$-2x10^{18}$
Substrate	(n) InP	/	$-2x10^{18}$

This is a short version of the paper submitted to the journal EPJ Photovoltaics.

It has been selected by the Committee of the WCPEC-8 for submission to EPJ Photovoltaics.

Table III: Layer structure of the NAsP template.

Layer	Material	Thickness	Doping level (cm^{-3})
Buffer	InP	1000 (nm)	$-2x10^{18}$
Nucleation	GaP	60 (nm)	--
Si buffer	Si	200 (nm)	--
Substrate	Si	775 (μm)	$1x10^{15}$

InP (red line) and InGaAs (green line) cells are presented in Figure 2, together with that of the complete TSC (blue line). Table IV summarizes the main PV parameters of the devices.

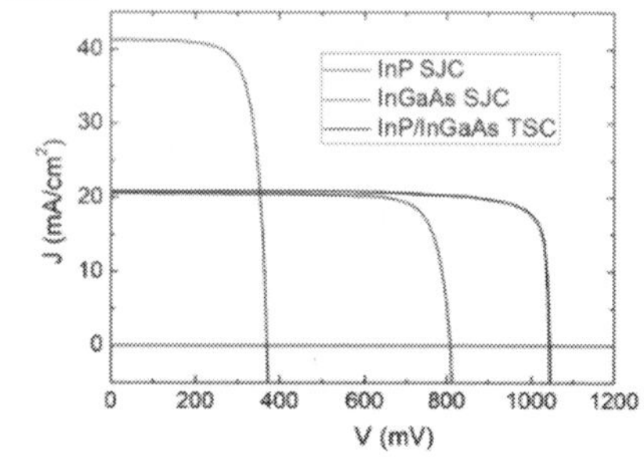

Figure 2: J-V characteristics under 1-sun illumination of the obtained InP/InGaAs tandem cell (blue line) compared to the single J-V characteristics of InP top cell (red line) and InGaAs bottom cell (green line).

4.1.1 Single junction solar cells

The InP cell presented an open-circuit voltage (V_{OC}) of 809.3 mV and a fill factor (FF) of 81.4 %, which are similar with the best InP cell performance reported so far. Conversely, the short-circuit current density (J_{SC}) of 20.6 mA/cm^2 can be further improved up to values of 31.2 mA/cm^2 as calculated by Jain et al. [10] or 29.5 mA/cm^2 as experimentally reached with the current InP world record cell [11]. The efficiency η was then limited to 13.5%. One of the possible mechanisms that reduces J_{SC} is probably related to the base thickness, limited to 3 μm in this study. In addition, problems in carrier collection, and non-optimized ARC and window layer, could also be responsible for such a reduction. The InGaAs bottom cell showed an excellent PV performance: J_{SC} of 41.2 mA/cm^2, V_{OC} of 369.9 mV and FF of 74.5%, providing an η of 11.4% which is near to the best InGaAs cell performance reported so far.

4.1.2 Tandem solar cell

The TSC, characterized under the same conditions, presented values of J_{SC} of 20.8 mA/cm^2, V_{OC} of 1.05 V, FF of 83.9% and η of 18.3%. From the comparison with the PV parameters of the SJCs, we can assume that the overall photo-generated current is limited by the InP top cell. The V_{OC} is around 100 mV lower than expected. The V_{OC} loss is most probably caused by a non-optimal passivation of the tandem cell interfaces. Nonetheless, the excellent quality of the device can be inferred from the noticeably high V_{OC} and FF values. The overall high η is a promising result and comparable to the ones available in the literature [11].

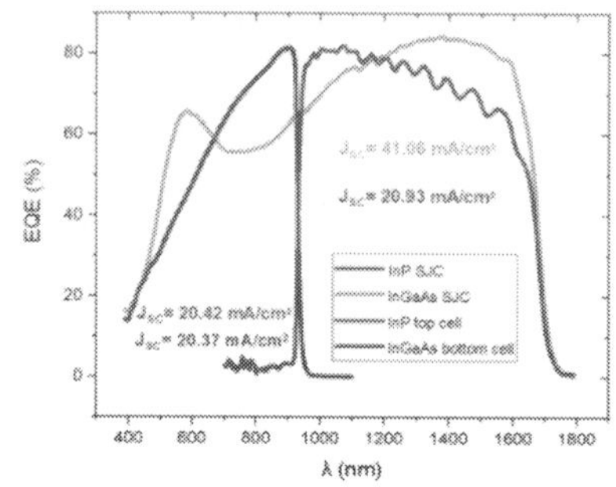

Figure 4: EQE characteristic of the InP/InGaAs tandem solar cell. It was obtained by separately measuring the contributions of the top (violet line) and bottom (orange line) cells. The results are compared with the ones obtained from the single cells (red line for InP and green line for InGaAs).

Figure 4 represents the EQE curve of the TSC, compared to those obtained from the constituting InP (red line) and InGaAs (orange line) SJCs. In order to evaluate the individual contributions of the top (violet line) and the bottom (green line) cells to the overall EQE, they were separately saturated using a proper bias light source, namely a 785 nm laser source and a properly filtered halogen lamp, for the top and bottom cell, respectively. As expected, the InP top single junction device and the one integrated in the tandem presented similar behaviors. Conversely, the integrated bottom cell presented a different trend compared to the single junction cell. This is evident in the 1200-1600 nm range, where the clear discrepancy is partially caused by the higher reflectivity of the device at long wavelengths since the ARC thickness is optimized for the InP top cell.

4.2 Solar cells on Si substrate

4.2.1 Material characterization of the Si template.

The quality of the NAsP template was evaluated by performing a series of characterizations such as optical microscopy, XRD, AFM and TEM. Furthermore, in order to assess the impact of the initial annealing taking place during the growth of the solar cell structure, the same characterizations were carried out after ten minutes annealing at 700 °C/150 mbar and under a flow of phosphine (PH_3) in the MOVPE reactor. The NAsP template presented a mirror-like surface morphology, not altered by the high temperature annealing, as shown in the images reported in Figures 5a and 5b obtained from optical microscopy. The XRD scans in Figure 5c present three sharp peaks. The most intense peak corresponds to the silicon substrate around 34.6°; afterwards, the GaP

Table IV: Main PV parameters of the tandem cell compared with those of the two single-junction individual cells.

	InP	InGaAs	Tandem
J_{SC} (mA/cm^2)	20.6	41.2	20.8
V_{OC} (mV)	809.3	369.9	1045.9
FF (%)	81.4	74.5	83.9
η (%)	13.5	11.4	18.3

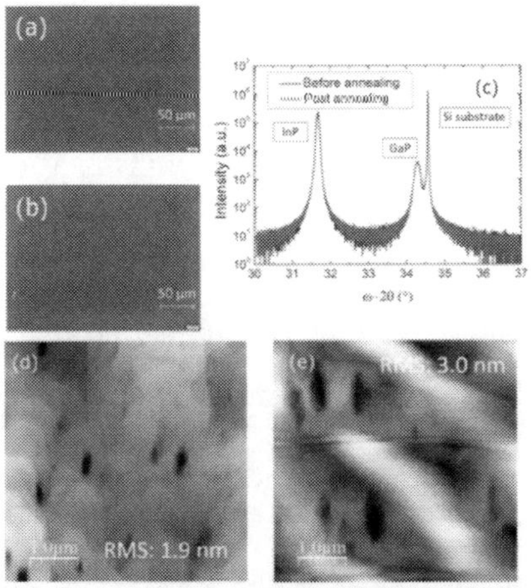

Figure 5: NAsP sample before and after the 700°C annealing: morphology a) before and b) after from optical microscopy; (c) XRD diffractograms; AFM (d) before and (e) after.

layer growth on the top of the substrate is around 34.3° and the last epitaxied layer of InP is around 31.7°. These three peaks are located at the same place before and after the annealing step revealing that the crystal quality of the template is not affected by this procedure. Thanks to the previous values, the peak mismatched compared to the silicon substrate can be extracted: 0.77% corresponds perfectly to fully strained GaP, whereas the one at 8.38% matches with fully relaxed InP. The AFM scans performed on the surface of the template presented a root mean square (RMS) roughness of 1.9 nm before annealing and 3.0 nm after annealing, which is acceptable for our purposes, as reported in Figures 5d and 5e. The AFM scans revealed the presence of surface irregularities which can be visualized as black spots in the topography.

A deeper evaluation of crystalline properties of (0 0 1) NAsP templates was carried out by performing a series of STEM and TEM scans towards the (1 1 0) direction. Figure 6 presents HAADF-STEM scans of the GaP/InP interface. The orange arrow indicates the presence of periodical misfit dislocations at the interface which arises from the different lattice parameters of GaP and InP (a_{GaP} = 5.4505 Å and a_{InP} = 5.8687 Å, respectively).

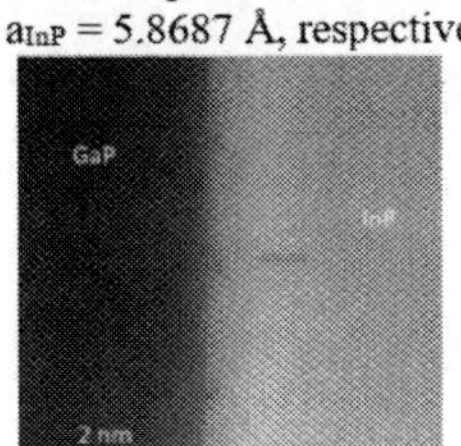

Figure 6: HAADF-STEM image of the (0 0 1) NAsP template towards the (1 1 0) direction showing the GaP/InP interface.

A misfit dislocations periodicity of 7.17 nm was evaluated by FFT of the image treated with a band-pass filter by ImageJ software. This value is within the experimental error limit of the theoretical value of 7.63 nm. The presence of such a misfit periodicity and the absence of irregular rearrangements at the interface tend to confirm the high quality of the template.

Finally, Figure 7 shows two TEM images with different magnification of the Si/GaP/InP template. Figure 7a reveals the excellent crystalline quality of the GaP layer and the presence of the expected periodic misfit dislocations at the GaP/InP interface. An image treatment permitted to find a misfit periodicity of 7.38 nm, which is in good agreement with the value calculated from the SEM image. Figure 7b presents a larger zone, emphasizing the presence of threading dislocations developing from the mismatched structure, the density of which tends to decrease with increasing the thickness.

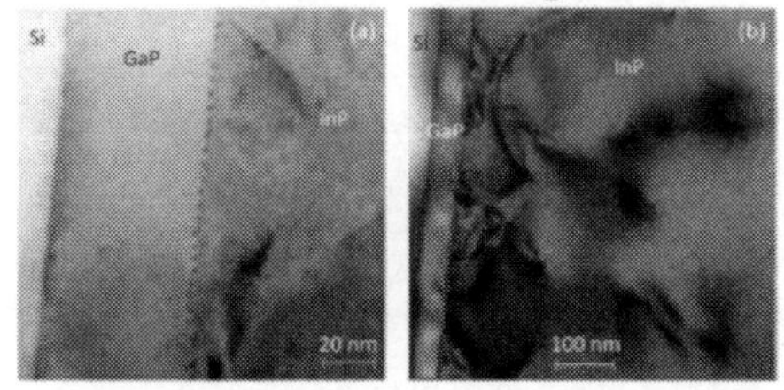

Figure 7: TEM images of the cross section showing the Si/GaP/InP template. a) points out the evaluation of misfits periodicity, b) illustrates some crystalline defects arising from the high mismatch

4.2.2 SJCs on Si Substrate: material characterization and PV performance

Before studying the PV performance in detail, the InP SJC on Si was characterized by the same techniques used for the NAsP template in the previous section. In particular, the surface morphology, shown in Figure 8, despite being completely mirror like to the naked eye, appears slightly rough at both the center (8a) and edges (8b). The AFM image in Figure 8c shows that the RMS decreases from 3.0 nm (measured after annealing at 700 °C) to 1.5 nm, a value lower than that we found in the unannealed template since the increasing thickness of the layer tends to level out the roughness. The thermal mismatch induced a series of cracks in the structure which can be visualized as vertical stripes on the whole figure surface.

The XRD diffractogram in Figure 8d demonstrates that a longer and thicker deposition of InP did not affect the overall crystal quality of the structure since the positions of the peaks are not shifted with respect to the XRD scan on the bare template, also reported in Figure 8d as comparison. Additionally, the InP peaks present a thin and comparable full width at half maximum as further evidence of a good quality crystal of the entire structure. Note that, the intensity of the InP related peak is higher since the XRD is now detecting all the InP stack constituting the device and not simply the 1 µm InP buffer layer of the template.

The InP and InGaAs cells integrated on Si were characterized under the same conditions presented in Section 4.1. The J-V curves obtained at room temperature under illumination are presented in Figure 9, whereas the PV parameters are summarized in Table V.

As expected, the integration on a Si substrate had a limiting effect on the PV performance. By comparing the J-V curves in Figure 9 with that of Figure 3, we can identify a higher influence on both R_{sh} (in particular, for the InGaAs cell) and R_s which has a negative impact on FF. All the electrical parameters decreased: both cells

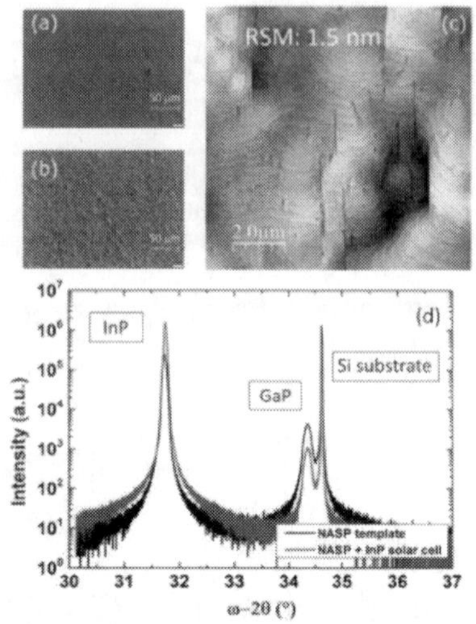

Figure 8: InP solar cell grown onto NAsP template: morphology (a) at the center and (b) at the edge of the sample from optical microscopy; (c) AFM; (d) XRD scans of NAsP template (black line) and NAsP + InP solar cell (red line).

presented a ~ 40% J_{SC} reduction, a V_{OC} reduction of 22% and 40% and a FF reduction of 43 % and 50%, for the InP and the InGaAs devices, respectively. Consequently, the efficiency dropped to 3.6% and 2.0% for the InP and InGaAs cells, respectively. There are examples in literature where J_{SC} and V_{OC} have been maintained after the fabrication of the III-V on Si cells [12-13]. Conserving a good FF has been, in general, more challenging, due to the R_s and R_{sh} variation caused by the cell fabrication. In our case, this implies that there is still an important scope for improvement for all our structures.

To get into more details regarding the decrease of the PV performance, in Figure 10 are reported SEM and TEM images for the InP and InGaAs SJCs, respectively. In particular, Figure 10a presents the surface and the cross-section of the InP devices and it allows to distinguish the InP layer from the Si substrate. The irregularities at the interface on the InP side are probably due to the local disorder induced by the presence of dislocations making the cleaving process challenging. The image presents an inset treated with a bandpass filter by ImageJ to better expose the irregularities. Figure 10b presents a TEM image of the InGaAs solar cell device grown onto the template.

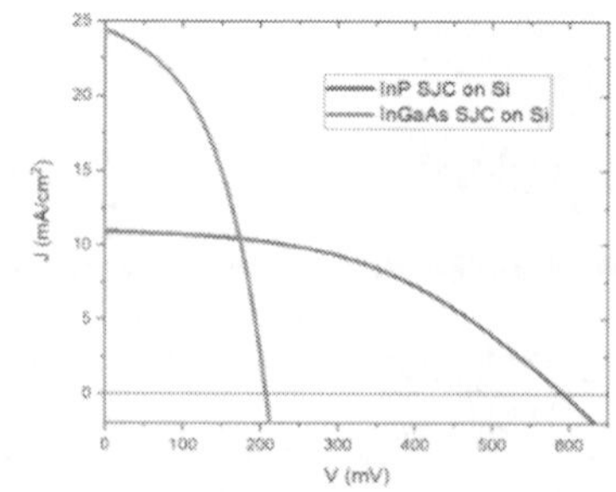

Figure 9: J-V characteristics under 1-sun illumination of the two SJCs integrated on Si. The InP cell is represented in red, whereas the InGaAs cell in green.

Table V: Parameters of the two solar cells on Si.

	InP	**InGaAs**
J_{SC} (mA/cm²)	12.1	24.7
V_{OC} (mV)	634.0	220.2
FF (%)	46.4	36.9
η (%)	3.6	2.0

Some dislocations clearly propagate in the InGaAs layer and the evaluation of the TDD along the growth direction indicates a mean TDD of 8.8x10^8 cm^{-2} for the InP buffer that decreases to 2.0x10^8 cm^{-2} in the InGaAs base. This value is higher with respect to similar devices present in literature [14]; it has a direct impact on the minority carrier (electron) lifetime, which is considerably degraded, worsening the PV performance.

The most probable explanation for the differences between the PV parameters of the cells grown on InP and on Si template is related to the presence of structural defects due to the lattice mismatch. In particular, as shown in Figure 8c, the thermal mismatch caused cracks in the structure which can act as charge carrier traps and recombination centers. In this sense, the worsening of the quality of the material impacts the minority carriers lifetime and the corresponding diffusion length. This is responsible of increasing the losses, which is in line with the results determined by TEM.

Figure 11 presents the EQE characteristic of the best single solar cells grown on the Si template (black full line), compared to the best result obtained for the same device grown on the InP substrate (red dashed line). Figures 11a and 11b illustrate the InP and the InGaAs solar cells, respectively. Both solar cells show a significant reduction of the overall quantum efficiency,

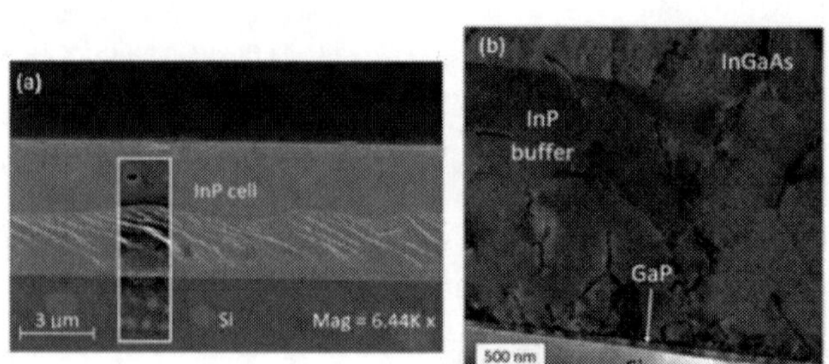

Figure 10: (a) SEM images of the InP cell cross-section grown onto NAsP template; the image presents an inset showing band-pass filter treatments performed with ImageJ software to highlight the irregularities. (b) TEM images of the cross section of the grown InGaAs single solar cell displaying a part of the InGaAs base with the presence of dislocations.

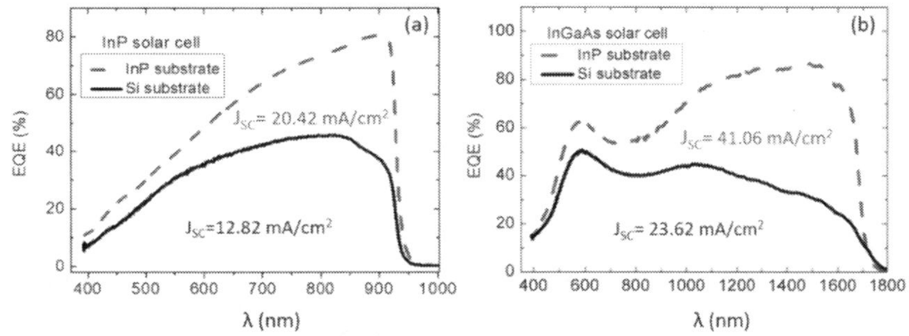

Figure 11: EQE characteristic of the best InP a) and InGaAs b) solar cells grown on Si template (black line), compared to the same result obtained on a typical InP substrate (dashed red line).

notably in the spectral range associated to the absorption in the base: 700-920 nm and 1000-1700 nm, respectively for the InP and InGaAs devices.This observation clearly points out deficiencies of the diffusion length at the base as well as surface recombination issues at the rear. The rest of the spectral range (λ<700 nm for InP and λ<1000 nm for InGaAs) also shows a lowering of the EQE due to similar issues taking place this time at the emitter and at the front surface. The EQE spectra confirm a general reduction of carrier collection compared to the devices fabricated on InP.

Finally, it is worth reminding that TEM characterizations have indeed evidenced the presence of structural defects, due to the lattice mismatch, which are well known to impact the charge carriers transport and recombination processes. Such defects are frequently the origin of parallel current paths for the photocurrent producing shunts that reduce FF and V_{OC}.

5 CONCLUSION

We have first presented the realization of high-performance InP and InGaAs solar cells lattice-matched to InP which demonstrated efficiencies of 13.5% and a 11.4%, respectively. The combination of the two devices through a InP/$Al_{0.48}In_{0.52}As$ tunnel junction, allowed the fabrication of a TSC which showed an efficiency of 18.3%, with little influence from shunt and series resistances.

In a second step, a prospective study in view of integrating the previous III-V SJCs on Si substrates through direct MOVPE growth has been accomplished. The commercial InP/GaP/Si template from NAsP company and the MOVPE-grown cells were characterized by XRD, SEM and TEM pointing out specifically a fully relaxed InP buffer and the presence of structural defects that can propagate to the cell grown atop. J-V characterizations revealed that these metamorphic cells kept a diode-like trend and produced photocurrent, although the efficiencies were reduced to 3.6% and 2.0% for the InP and InGaAs single junctions, respectively.

These preliminary results demonstrate the potentiality of the proposed methodology although the realization of the final tandem device on the Si substrate was postponed until further advancements in the quality and PV performance of the SJCs will be reached.

Several paths to improve the presented results are feasible. For instance, in the epitaxial approach on InP substrates, the thickness of the base for the tandem top cell must be optimized reaching the typical 5 µm reported in literature.

Finally, regarding the integration on Si substrate, the reduction of threading-dislocations at the InP/GaP interface seems a key concern to improve the material quality and the PV performance of the final device.

Future improvements in this regard may open a new generation of multijunction solar cells on Si substrates both more efficient and more cost-effective.

6 REFERENCES

[1] K.A.V Horowitz et al., National Renewable Energy Laboratory, Technical Report NREL/TP-6A20-72103 (2018)
[2] R. Cariou et al., Nat Energy **3,** 326–333 (2018)
[3] R. Cariou et al., Journal of Photovoltaics. **7**, 1 (2017)
[4] S. Fan et al., Cell Reports Physical Science. **1,** 9 (2020)
[5] J.P. Connolly et al., Progress in Photovoltaics:Research and Applications. **22,** 7 (2014)
[6] S. Soresi et al., Physica status solidi(a). **215**, 8 (2018)
[7] https://www.nasp.de/gap-on-si.html
[8] L.G. Meiners et al., Journal of Vacuum Science and Technology **19**, 373 (1981)
[9] K. Volz et al., Journal of Crystal Growth. **315,** (2011)
[10] R.K. Jain et al., IEEE Transactions on Electron Devices. **40**, 1 (1993)
[11] M.A Green et al., Solar cell efficiency tables (version 59) (2021)
[12] T. Jimbo et al., Science and Technology of advanced Materials. **6**, 1 (2005)
[13] V. Raj et al., Applied Materials Today. **18**, 100503 (2014)
[14] A. Freundlich et al., Journal de Physique Colloques. **50,** C5 (1989)

IMPROVEMENT OF INGAP SOLAR CELLS GROWN WITH TBP IN PLANETARY MOVPE REACTOR

Hassanet Sodabanlu,[1] Gan Li,[2] Kentaroh Watanabe,[1] Yoshiaki Nakano,[2] Masakazu Sugiyama[1,2]

1. Research Center for Advance Science and Technology, The University of Tokyo, Tokyo, 153-8904, Japan
2. Department of Electrical Engineering and Information Systems, The University of Tokyo, Tokyo, 113-8656, Japan

ABSTRACT: The carrier lifetime in unintentionally doped InGaP was significantly deteriorated with an increase in growth (surface) temperature from 560 to 590 °C. With a temperature higher than 600 °C, the growth of InGaP was failed with extremely rough surface morphology. Meanwhile, a larger amount of H_2 carrier gas led to a faster TBP flow velocity and an improvement of InGaP surface roughness. These results suggested that, due to fast TBP decomposition processes, the depletion of TBP and insufficient P partial pressure for surface chemical reactions were major problematic issues for the growth of P-related materials in a large planetary MOVPE reactor. The quality of p-type InGaP grown in this reactor with TBP was not as good as that of n-type InGaP indicated by Hall mobility. In this work, InGaP rear homo junction (RJ) and rear hetero junction (RHJ) solar cells utilizing of thick n-InGaP layers exhibited much better photo-carrier collection and performance than the traditional n-on-p solar cell comprising of a thick p-InGaP base layer. Moreover, the quality of InGaP was enhanced by the mis-orientation of GaAs (001) toward (111)A, while InGaP grown on GaAs (001) with a mis-cut toward (111)B showed worse results in term of carrier lifetime and solar cell performance.

1 INTRODUCTION

The past few decades have witnessed the tremendous progress on III-V semiconductor solar cells owing to the development of fabrication technologies, including the crystal growth by metalorganic vapor phase epitaxy (MOVPE) method. Even tough extensive researches regarding MOVPE growth have been carried out up to the present, there is no notable report on the growth of InGaP solar cells in a planetary MOVPE reactor using tertiarybutylphosphine (TBP) which has a relatively lower intrinsic toxicity than PH_3. According to few available references, InGaP solar cells grown with TBP generally fall behind PH_3 samples in terms of quality and performance [1]. In this work, we develop and optimized growth conditions of InGaP solar cells ined a planetary MOVPE using TBP and their structures to address this quality issue.

With regard to our previous report on the growth of InGaAsP solar cells on InP, the growth temperature is the most important factor to realize high quality epitaxial layers and solar cells in our planetary MOVPE [2]. Hence, the first parameter to be investigated would be the growth temperature. After the optimization of growth temperature, the effects of gas flow velocity would be studied by varying of amount of H_2 carrier gas. Then, the electrical properties of both p-InGaP and n-InGaP would be investigated, together with fabrication of InGaP solar cells with three different structures. Last, the impacts of substrate mis-orientation on the quality and properties of InGaP will be examined.

2 EXPERIMENTAL DETAILS

Epitaxial growth in this work was carried out in a planetary MOVPE reactor (Aixtron, AIX2000HT) using trimethylgallium (TMGa), trimethylaluminium (TMAl), trimethylindium (TMIn), tertiarybutylarsine (TBAs) and tertiarybutylphosphine (TBP) as precursors for Ga, Al, In, As and P, respectively, with H_2 carrier gas. Dimethylzinc (DMZn) and diethyltellurium (DETe) were utilized for p-dopant and n-dopant, respectively. The in-situ monitoring was conducted using a Laytec Epicurve DA IR T. This equipment allows us to observe the accumulating strain inside epitaxial layers, surface roughness, and wafer's surface temperature during epitaxial growths by monitoring wafer bowing, surface reflectance, and emissivity-corrected pyrometry, respectively. The reactor pressure during the growth was stabilized at 100 mbar. The growth rate of InGaP can be in-situ monitored by resolving distances between oscillation peaks of reflectance spectra. In this work, the growth rate and the V/III ratio of InGaP were 1.2 μm/h and 20, respectively. Atomic force microscopy (AFM) was utilized to observe the surface morphology of epitaxial layers. The carrier lifetime in InGaP bulk layer was evaluated by time-resolved photoluminescence (TRPL) excited with 405-nm laser.

For the solar cell fabrication, Ag/Au and AuGe/Au metal electrodes were used for ohmic contact with p-GaAs substrates and n-GaAs contact layers, respectively. After cell fabrication with a mesa area of 0.5x0.5 cm^2, current-voltage (I-V) characteristic and external quantum efficiency (EQE) were evaluated using a solar simulator under AM1.5G (100 mW/m^2) and an EQE measurement setup with ZnS/SiO_2 antireflection coating (ARC), respectively.

3 RESULTS AND DISCUSSION

First, the carrier lifetimes in InGaP double hetero (DH) samples, as depicted in Fig. 1(a), grown at different temperatures on GaAs (001) substrates were evaluated by TRPL, and their transient decay spectra were plotted in Fig. 1(b). The decay spectra showed that the carrier lifetime was decreased with a higher growth temperature. The reason was not clear but could relate to increases in (1) background carrier concentration (n-type) measured by

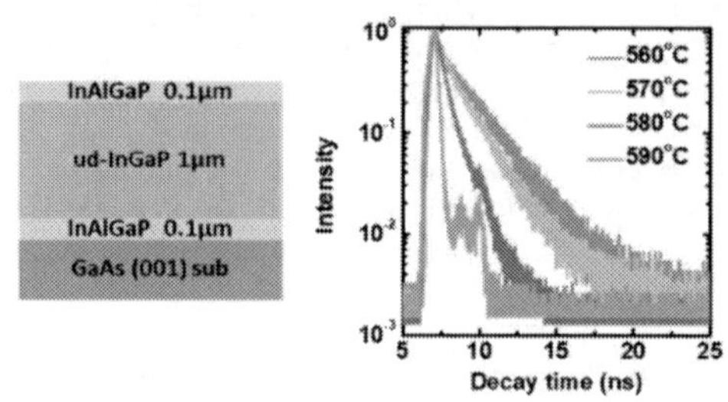

Figure 1: (a) double hetero structure (DH) and (b) transient decay spectra of InGaP DH samples grown at different temperatures.

ECV method, (2) indium segregating, (3) surface/interface roughness, (4) parasitic gas-phase reactions, and (5) degree of spontaneous ordering with higher growth temperatures. The AFM images of InGaP epitaxial layers grown on GaAs (001) at 560 and 590 °C. The amount of H_2 gas carrying TBP precursor was varied from 2.4 to 5.6 and 12.2 slm. The surface roughness of InGaP was remarkably improved with a faster gas flow velocity via a larger H_2 supply, especially for high temperature condition. The TRPL and AFM results can lead to a conclusion that a major problematic issue for the growth of P-related materials is the P partial pressure or actual V/III ratio for chemical surface reactions. Considering the 100%-decomposition temperature of TBP at 550 °C [3], a low growth temperature and a fast TBP flow velocity are indeed preferred for InGaP growth in a large planetary MOVPE reactor. However, with the H_2 flow amount of 12.2 slm, the growth rate of InGaP was reduced from the designated value of 1.2 µm/h to 1.0 µm/h. The reason was most likely the poor gas phase diffusion of group III-MO through the chemical boundary layer. In addition, the composition of InGaP was not altered by the amount of H_2 flow.

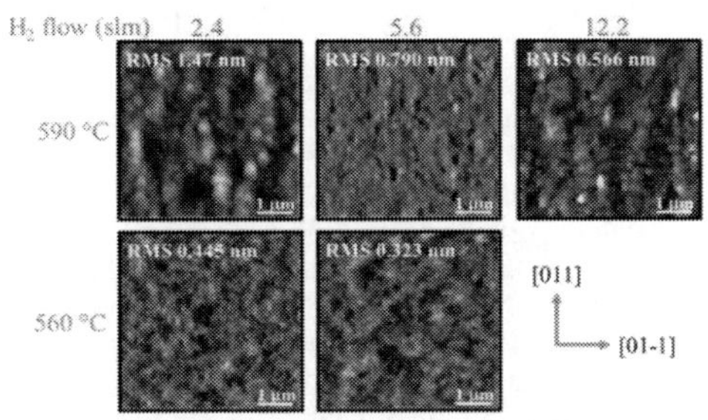

Figure 2: 5x5 µm² AFM images of InGaP grown at 560 and 590 °C with amount of H_2 carrier gas for TBP of 2.4, 5.6 and 12.2 slm.

With the optimized growth temperature of 560 °C and the H_2 flow of 5.6 slm, the electrical properties of both p-doped and n-doped InGaP were examined by Hall Effect measurement. Figs. 3(a) and 3(b) show the dependences of carrier mobility on the doping concentration of p-type and n-type InGaP, respectively. The red fitting lines were calculated from the empirical low-field mobility model [4]. For p-InGaP, the hole mobility seemed to be limited at 25-35 cm²/Vs range, regardless of the doping concentration. A plausible explanation was the n-type background doping in InGaP, which required a certain

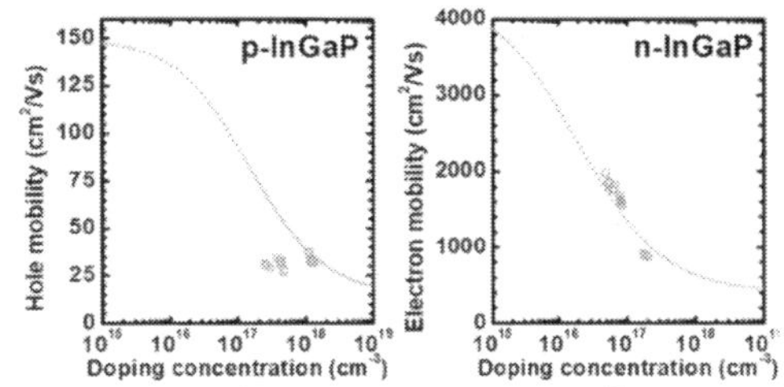

(a) (b)

Figure 3: Dependences of (a) hole mobility in p-type InGaP and (b) electron mobility in n-type InGaP on the doping concentrations. The dot lines represent the theoretical value given by the empirical low-filed mobility model.

amount of Zn dopant for compensation, leading to defects and traps. This behavior was not observed in case of n-InGaP, suggesting that the quality of n-InGaP grown in this reactor was reasonably better than that of p-InGaP.

According to the Hall mobility results in Fig. 3, InGaP solar cells utilizing of thick n-InGaP base layers are proposed in order to realize high conversion efficiency. Fig. 4(a) shows the traditional n-on-p (np) InGaP solar cell with a thick p-InGaP base layer. It was found that by employing highly doped np junction, the open circuit voltage (V_{oc}) of np InGaP solar cells could be enhanced by decreasing of recombination in space charge region. This highly doped junction approach was inherited to the rear homo-junction (RJ) and rear hetero-junction (RHJ) devices, shown in Figs. 4(b) and 4(c), respectively, which thick n-InGaP layers were utilized. The position of pn junction are hi-lighted in each figures.

After cell fabrication including ARC deposition, the I-V characteristics under AM1.5G illumination and external quantum efficiency (EQE) spectra of these cells were displayed in Figs. 5(a) and 5(b), respectively. Suffering from poor electrical properties, the np solar cell had poor carrier extraction leading to low fill-factor (FF), shunt-like I-V curve and low short circuit current density (J_{sc}). It is also clearly evidenced by the degradation of EQE at long wavelength range. The RJ and RHJ, owing to superior electrical properties in n-InGaP, exhibited good carrier extraction resulting in high J_{sc}, FF and EQE. It is noticeable that the RHJ solar cell had a bit lower V_{oc} than the others possibly caused by poor InAlGaP/InGaP hetero-interface. For MOVPE process, there exists an interruption period, typically for few seconds, when the growth is changed from one layer to another layer, in order to adjust and stabilize the bubbling conditions of MO precursors.

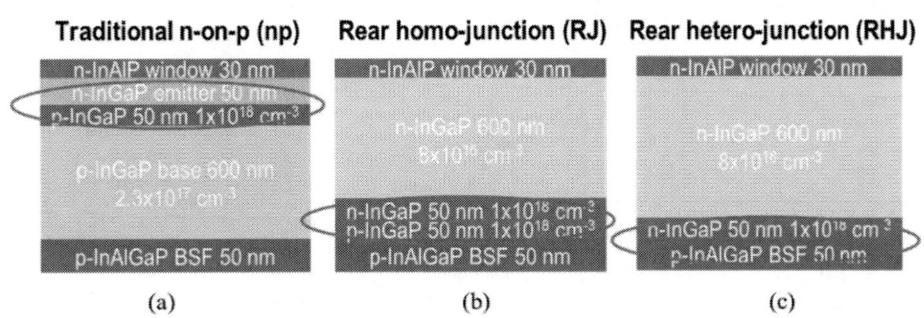

(a) (b) (c)

Figure 4: Structures of InGaP solar cell investigated in this work: (a) traditional n-on-p, (b) rear homo-junction and (c) rear hetero-junction, denoted as np, RJ and RHJ respectively. The positions of p-n junction are hi-lighted in the figures.

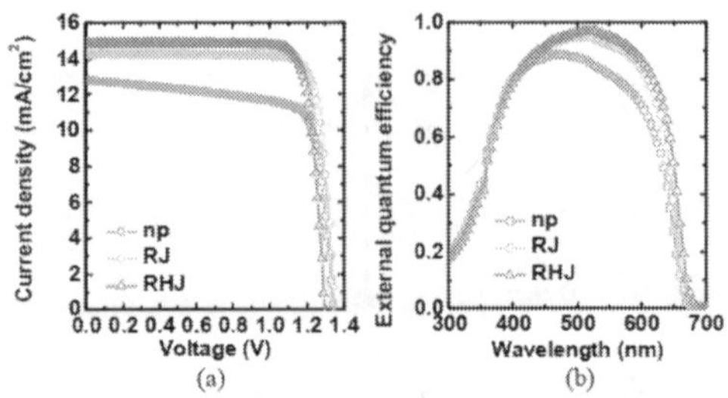

Figure 5: (a) I-V characteristics under AM1.5G (100 mW/cm^2) illumination and (b) EQE spectra of InGaP solar cells with different structures shown in Fig. 4

However, during this period, the surface of InAlGaP is vulnerable to the oxygen contamination that induces defects and traps at the InAlGaP/InGaP hetero-interface. Removing of this interruption period results in an improvement of hetero-interface, revealing in the next experiment section.

Next, the effects of GaAs substrate mis-orientation on the properties of InGaP epitaxial layers were investigated. GaAs (001) wafers with four different mis-orientations were utilized: exact, 5° toward (110), 5° toward (111)A and 6° toward (111)B, denoted as exact, 5(110), 5A and 6B, respectively. The TRPL and PL spectra of InGaP DH samples grown on GaAs with various mis-orientations, employing the above mentioned new gas flow sequences, are displayed in Figs. 5(a) and 5(b), respectively. Obviously, the carrier lifetime in InGaP grown on exact substrate was much improved from the result in Fig. 1(b), owing to the improvement of hetero-interfaces. The carrier lifetime was the longest for 5A and the shortest for 6B, while the other two samples had a very similar lifetime, as shown in Fig. 5(a). The indium composition was not much altered by the substrate mis-orientation as confirmed by in-situ curvature and ex-situ X-ray diffraction measurements. However, the PL peak positions shifted to lower energies for the samples on 5(110) and 6B substrates, ascribing to the accenting of spontaneous ordering in InGaP. The 5A sample had the strongest PL intensity and closely followed by the exact sample, but the 5(110) and 6B had only one fifth and one tenth of 5A intensity, respectively. These results suggested that the quality InGaP grown 5A substrate was possibly the best, and the degree of ordering in InGaP was considerably the smallest.

Figs. 7(a), 7(b) and 7(c) showed the I-V characteristic under AM1.5G (100 mW/cm^2) illumination, dark I-V curves and EQE spectra of InGaP RHJ solar cells (Fig.

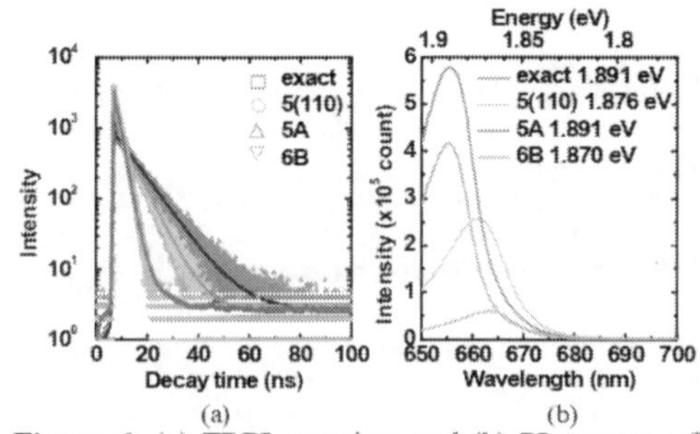

Figure 6: (a) TRPL transient and (b) PL spectra of InGaP DH samples grown on GaAs (001) with various mis-orientations

4(c)) grown on GaAs (001) wafers with various mis-orientations. Clearly, due to the better InAlGaP/InGaP hetero-interface, the V_{oc} of solar cell on exact substrate was much improved from the previous result in Fig. 5. The 6B solar cell had a low V_{oc} of about 1.30 V, compared to 1.38 V of exact and 5A cells. The explanation could be the quality of InGaP on 6B which was deteriorated from the accented ordering effect. The 5(110) solar cell had also a degraded V_{oc} of 1.36 V, but this decrease was not as much as 6B case. Although the carrier lifetime in InGaP on 5A was much longer than that on exact GaAs (Fig. 6(a)), their V_{oc} were almost similar. Hence, there should be other factors, for example, hetero-interfaces, in addition to the quality of InGaP that limit the cell efficiency.

4 SUMMARY

The growth of InGaP in a planetary MOVPE reactor using TBP as P source were thoroughly investigated and optimized. There are four key parameters revealed in this work that enhance the quality of InGaP and the solar cell performance. First, the growth of InGaP need a relatively lower growth temperature than GaAs in this reactor in order to prevent pre-decomposition of TBP and parasitic gas-phase reactions. Second, owing to the excellent quality of n-InGaP, the solar cells with thick n-InGaP absorber layers, i.e. RJ or RHJ, are recommend to realize good performance. Third, the gas flow sequences during the growth of InAlGaP/InGaP hetero-interface is modified to minimize the damage from oxygen contamination. Last, GaAs (001) wafers with a mis-orientation toward (111)A is favorable for InGaP growth in this work, since InGaP epitaxial layers and solar cells exhibit the best properties and performance. The best InGaP RHJ solar cell grown

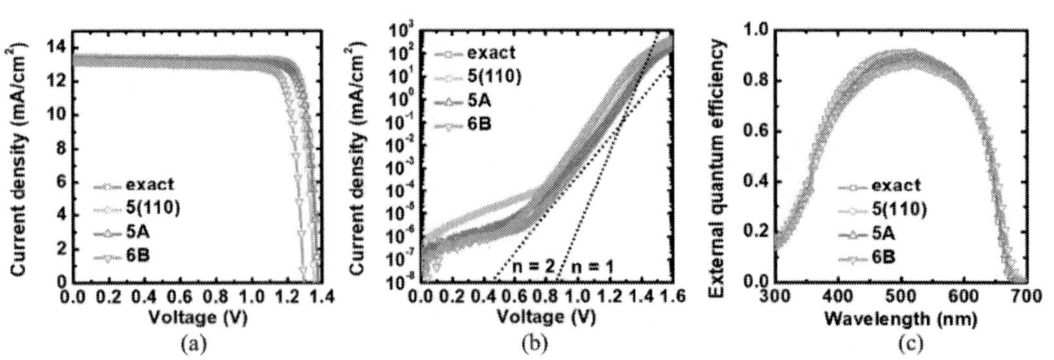

Figure 7: (a) I-V characteristics under AM1.5G (100 mW/cm^2) illumination, (b) dark I-V curves and (c) EQE spectra of InGaP RHJ solar cells grown on GaAs (001) with different mis-orientations.

with TBP in this work is comparable to that grown using PH_3 in the view point of cell efficiency [5]. This achievement encourages the fabrication of InGaP-based devices using TBP in a planetary MOVPE reactor.

5 ACKNOWLEDGEMENTS

A part of the results of this study is based on a project, JPNP20015, commissioned by the New Energy and Industrial Technology Development Organization (NEDO).

References:

[1] I. Garcia et al., J. Cryst. Growth 310, 5209 (2008)
[2] H. Sodabanlu et al, Jap. J. Appl. Phys. 57, 08RD09 (2018)
[3] A. C. Jones, P. O'Brien, CVD of compound semiconductors, Chapter 3, Wiley (2008)
[4] M. Sotoodeh et al., J. Appl. Phys. 87, 2890 (2000)
[5] J. F. Geiszz et al., Appl. Phys. Lett. 103, 041118 (2013)

FLEXIBLE SOLAR CELL ASSEMBLIES- TESTING IN HARSH CONDITIONS

Roberta Campesato[1], Mariacristina Casale[1], Giuseppe Gabetta[1], Emilio Fernandez Lisbona[2]
[1]CESI S.p.A.; [2] ESA/ESTEC
[1]Milan, via Rubattino 54, Italy; [2] 2200 AG Noordwijk, Keplerlaan 1, The Netherlands

ABSTRACT: This paper describes the results achieved during the qualification campaign conducted on Solar Cell Assemblies CTJ30-80-SCA manufactured by CESI using the qualified CTJ30-80 thin (80 µm) bare solar cells.
The results of a set of specific engineering tests grouped under the denomination of "Subgroup M" aimed at verifying the flexibility and the reliability of thin (80 µm) solar cells / solar cell assemblies mounted on curved surfaces or on flexible supports are also summarized. The use of solar cell assemblies based on thin solar cells and flexible coverglasses in the fabrication of non-planar solar arrays or flexible structures requires the assessment of the bending capabilities of the solar cell assemblies. The aim of this activity was the identification of procedures and methods to test the cell behavior in bending configuration in BOL and EOL conditions.

Keywords: Space Cells, III-V Semiconductors, Flexible Substrate, Qualification and Testing

1 INTRODUCTION

New generation of flexible solar arrays need flexible and lightweight solar cells and solar cell assemblies.

To this purpose triple junction solar cells based on CESI product CTJ30 have been thinned down to 80 micrometers obtaining a thin and flexible device with efficiency class 29% AM0, 1 sun. The so called CTJ30-80 solar cell was specifically developed and qualified for this application [1].

This paper discusses the test campaign that was designed to verify the bending properties of CESI solar cells CTJ30-80 at bare and SCA level in BOL and EOL conditions.

This activity may provide the necessary background if a demand for standardization of the flex properties of solar cells/ solar cell assemblies is raised by flex solar array manufactures.

The approach to the test plan definition was iterative: a first test plan was designed and applied to flexible solar cells (bare and SCA). Subsequently, the test methods and conditions were optimized to reach the final test plan version that completely characterizes the flex properties of the devices BOL and EOL.

This campaign was suitable to prove that a bending radius as low as 10 cm can be successfully reached by bare solar cells and SCAs.

2 DESCRIPTIONS OF FLEXIBLE SOLAR CELLS

2.1 Bare solar cells

Triple-junction space solar cells with a thickness of 80 µm (CTJ30-80). were manufactured by CESI with an InGaP/InGaAs/Ge epitaxial structure.

The solar cell structures were grown on 100 mm diameter and 140µm thick Ge substrates. After the epitaxial growth the Ge substrate was thinned down to 80 µm using a suitable wet etch process. Two solar cells, with 26.5 cm^2 size and typical 1.3g mass were manufactured on each wafer. The front and rear contacts were made by evaporated silver 5 µm-thick, capped with 200nm Gold. The broad band AR coating was an optimized dual-layer Ta_2O_5/SiO_2.

About 500 CTJ30_80 solar cells (Fig. 1) were manufactured with an average efficiency of about 29% in AM0, 25°C. Figure 1 reports a picture of a CTJ30-80 solar cell bent by an operator.

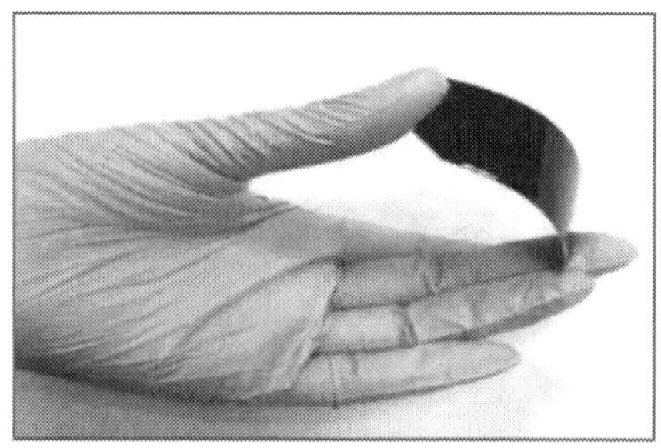

Figure 1: CTJ30-80 solar cell in bent configuration

2.2 Solar Cell Assemblies

Flexible CTJ30-80 solar cells were used to manufacture flexible solar cell assemblies.

The Front and Rear Interconnects were Invar coated silver with a flash of gold and were welded by parallel gap technology on thin solar cells.

A flexible coverglass 100 µm thick from Qioptiq and a thin 100 µm thick silicon bypass diode from Azur Space were also procured to complete the assembly.

The flexible solar cell assembly (SCA) process was developed in the frame of an ESA project supported by the Italian Space Agency. A flexible SCA is shown in figure 2.

Figure 2: Flexible Solar Cell Assembly based on CTJ30-80 solar cell

3 TEST PLAN FOR FLEXIBLE SOLAR CELLS

The path towards very high power solar array (> 35 kW) with limited mass foresees the use of flexible arrays based on flexible and lightweight solar cells and solar cell assemblies. To date, for space applications no methods /procedures have been defined to assess the flexibility and reliability of flexible and thin solar cells/solar cell assemblies. The fundamental questions to be answered are:

- To what extent are thin solar cells, namely CTJ30-80, flexible?
- Is the flexibility maintained in EOL conditions?
- Flexible solar cell assemblies using an innovative flexible coverglass [2] withstand space environment?

To answer these questions a special test campaign was designed and performed to verify the bending properties of solar cells at bare and SCA level in BOL and EOL conditions. Moreover, some standard test like electron irradiation, UV test and thermal cycling were performed to assess the reliability of flexible SCAs.

This activity may provide the necessary background if a demand for standardization of the flex properties of solar cells/ solar cell assemblies is raised by flex solar array manufactures.

The approach to the test plan definition was iterative: a first test plan was designed and applied to flexible solar cells (bare and SCA). Subsequently, the test methods and conditions were optimized to reach the final test plan version.

The set of test methods and procedures was called "Group M". Such group was divided in several subgroups to test different flexible/ bendable properties of solar cells at bare and SCA level

At bare solar cell level three different subgroups were introduced to test the flexibility of CTJ30-80:

- Subgroup M1/BSC – Long-term Bending properties at Bare cell level.
- Subgroup M2/BSC – Shrinking test and ultimate bending.
- Subgroup M3/BSC – Ultimate Bending properties at Bare cell level.

At SCA level the following subgroups have been introduced:

- Subgroup M1/SCA – Long-term Bending properties at SCA level
- Subgroup M2/SCA – shrinking test at SCA level on flexible blanket
- Subgroup M3/SCA – Ultimate Bending properties at SCA level
- Subgroup B – UV test
- Subgroup C – Electron irradiation 1MeV, 1 $*10^{15}$ e^- /cm^2 fluence
- Subgroup E –Thermal cycling

Subgroup M3 was aimed at evaluating the minimum bending radius that solar cell bare and SCA can withstand before breakage of samples BOL. Subgroup M1 was aimed at reproducing the effects of the long storage in bent configuration, typical of "roll-up" solar array with a diameter of 10 cm. Subgroup M2 was dedicated to assess the mechanical robustness of the thin solar cells submitted to severe temperature variation like sun/eclipse condition. At SCA level this test was carried out also on a few samples preventively glued on a flexible blanket.

The ultimate bending test was performed at last,after all subgroups to demonstrate that the flexibility of solar cells remained unchanged in EOL

At SCA level the UV test, electron irradiation and thermal cycling were also performed to verify the suitability of the innovative flexible coverglass.

3.1 Long term bending test

The test was aimed at proving that thin solar cells/solar cell assemblies in bent configuration (i.e. wrapped around a substrate of known curvature radius) did not electrically degrade and did not retain memory of the assembly in bent configuration. The test consisted in keeping 10 solar cells / Solar Cell Assemblies in bent configuration around a 11 cm-radius cylinder for 8 months as shown in figure 3.
The electrical performances and the flatness of the samples were recorded before and after the test.

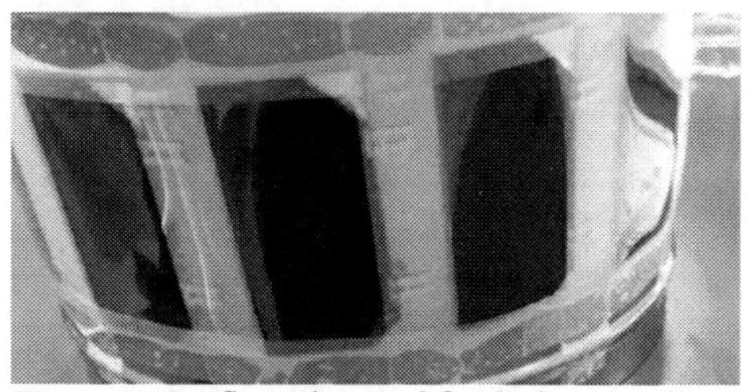

Figure 3: Configuration used for the long bending test of flexible CTJ30_80-SCA.

3.2 Shrinking test

The objective was to evaluate how the bare solar cells and the SCAs adapted to the abrupt temperature changes by themselves or when mounted on a blanket (solar cell assemblies only). The test consisted in performing a thermal shock test on samples free to shrink. A steep temperature gradient was applied making 20 cycles between -197°C/+130°C each cycle lasting 2 minutes. The test was carried out on bare cells, SCA, and SCA on a flexible blanket. The blanket was a carbon reinforced polyamide foil 50μm thick

The requirement of this test was that samples would not break or electrically degrade (ΔI_L <3%).

3.3 Standard bending

Bare solar cells and solar cell assemblies were bent around a cylinder having radius of 11 cm to verify that no breakages were introduced by this condition. The bending test consisted of wrapping the samples using a tape according to the following four modes:

Face-front, long side
Face front, short side
Face- rear, long side
Face-Rear, short side.

3.3 Ultimate bending

The bare solar cells and the solar cell assemblies were bent until the breakage condition was reached to determine the minimum bending radius the bare cells and the SCAs can withstand to without breaking. The bending test consisted in wrapping the samples around cylindrical surfaces of progressively decreasing radiuses of curvature according to the following four modes:

Face-front, long side
Face front, short side
Face- rear, long side
Face-Rear, short side.

4 EXPERIMENTAL RESULTS

4.1 Bare solar cells

Bare solar cells were submitted to the various bending tests as reported in chapter 3. Before and after each bending test the evaluation of the electrical performances, the electroluminescence imaging, the visual inspection and bendability to 11 cm radius cylinder was carried out.

No breakage in any CTJ30-80 solar cell was noticed after the long bending test and the shrinking test.

The electrical performances of the solar cells at bare level were measured under an AM0 dual source solar simulator from Wacom. Measurements were performed in BOL (20 samples) and after long bending test (10 samples) and shrinking test (10 samples). The average measurement results were reported in Table I; no degradation of electrical performances was observed.

Table I: Performances of CTJ30-80 bare solar cells after bending test (area 27 cm^2)

	Isc (mA)	Voc(V)	Pm(W)
BOL	476	2.55	1.04
After long bending test	476	2.55	1.03
After Shrinking test	476	2.55	1.03

The ultimate bending was measured on CTJ30-80 solar cells in BOL, after long bending test and after shrinking test. The majority of the solar cells break at bending radiuses around 4-5 cm no matter which test sequence was applied to the samples before, and the preferred orientation leading to breakage was "front side up" on the long side (figure 4). In figure 5 a typical distribution of the bending radius at breakage is reported.

Figure 4: Typical breakage of bare CTJ30-80 solar cell on 4 cm dia cylinder

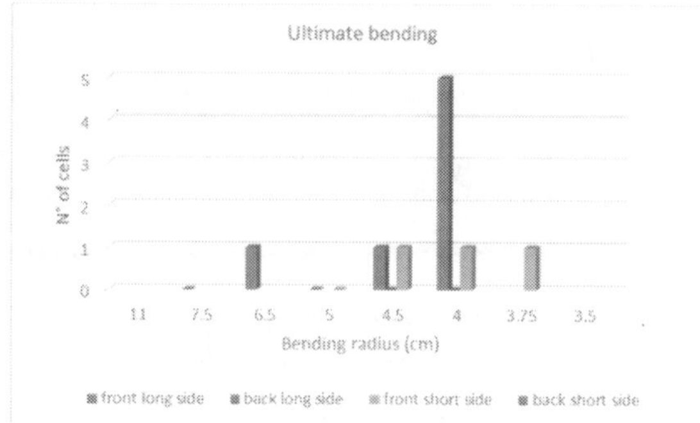

Figure 5: Ultimate bending of bare CTJ30-80 solar cells

4.2 Solar cell assemblies

Flexible solar cell assemblies were also submitted to a test plan foreseeing the execution of all the bending tests of chapter 3, each test preceded and followed by the measurement of the electrical performances, the electroluminescence imaging, the visual inspection and the bending around a 11 cm radius cylinder. This last test was performed to verify that the space environment effects like radiation, temperature variation, storage did not affect the samples bendability. Furthermore, at the end of each test sequence the ultimate bending was performed to verify whether any variation in flexibility was induced by the environmental tests.

Table II summarizes the average electrical performances results obtained on the SCAs after long bending test and shrinking test. Very important it is to mention that the electroluminescence images and the visual inspection did not evidence any breakage of flexible SCAs after long term bending and shrinking tests at SCA level.

Table II: Average performances of CTJ30-80 flexible SCA after bending test (area 27 cm^2)

	Isc (mA)	Voc(V)	Pm(W)
BOL	480	2.54	1.03
After long bending test	478	2.55	1.02
After Shrinking test	478	2.55	1.03

The ultimate bending was measured on CTJ30-80 solar cell assemblies in BOL, after long bending test and after shrinking test. The majority of the SCAs broke when the bending radius was around 6-7 cm no matter which test sequence the SCA were submitted. The breakage mode in this case depends on the sample orientation, in fact, only the solar cell or the coverglass broke: when the SCA was oriented front-up, it was the coverglass to break whereas when the SCA was bent face-down, the breakage affected the solar cell.
In figure 6 a typical distribution of the radius of curvature that led to breakage is reported.

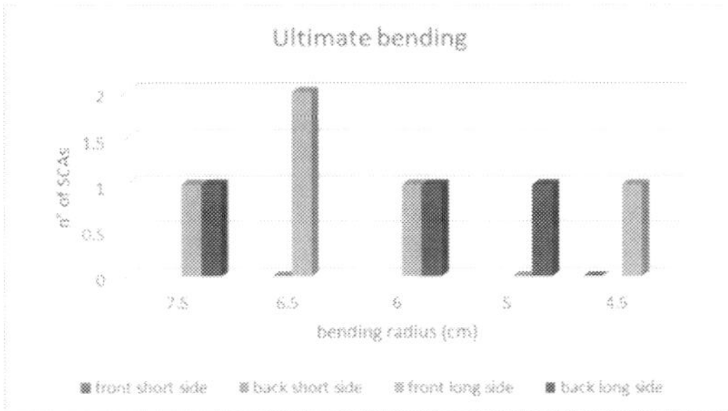

Figure 6: Ultimate bending of CTJ30-80 solar cell assemblies

The flexible SCAs were submitted to electron irradiation (1 MeV electron with a fluence of 1e15 e^-/cm^2) to verify that the assembly did not change the radiation resistance of CTJ30-80 solar cells. Table III compares the remaining factors obtained on CTJ30-80 samples bare and on samples equipped with a flexible CMG and a standard CMG. No differences are noticeable therefore the flexible CMG coverglass is not deteriorating under electron irradiation.

UV test was also performed on 3 samples equipped with the flexible coverglass: no degradation of short circuit current was observed as shown in figure 7. Absorptance and emittance were also comparable to standard SCA and did not degrade after UV test.

Table II: Performances of CTJ30-80 flexible SCA after bending test (area 27 cm^2)

	Isc (mA)	Voc(V)	Pm(W)
BOL	480	2.54	1.03
After long bending test	478	2.55	1.02
After Shrinking test	478	2.55	1.03

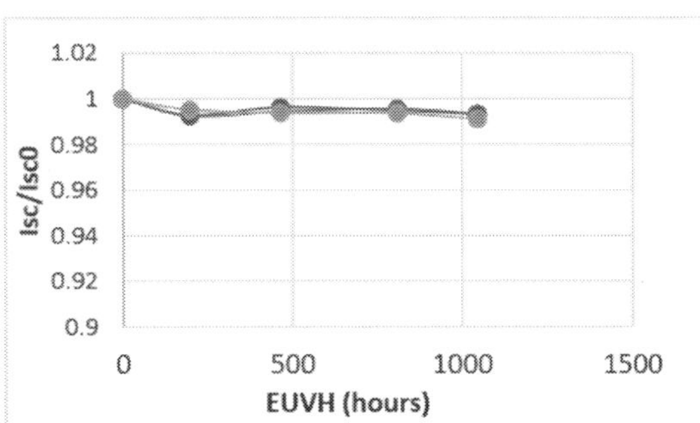

Figure 7: UV test on flexible solar cell assemblies

A Thermal cycling test simulating one year operation in LEO environment was carried out. It consisted in 6000 cycles between -140°C and +140°C in inert atmosphere. In this case too, no degradation and no breakages were detected on thin flexible solar cell assemblies CTJ30-80 SCA.

4.3 Solar cells on coupon

Flexible bare solar cells and solar cell assemblies were mounted on a carbon reinforced polyimide, 50 micron thick and submitted to Shrinking Test (20 cycles -197/+130°C free to shrink). Figure 7 shows a flexible SCAs on a blanket ready for shrinking test.The samples were glued on the foil by a thin (50µm) Elastosil RTV-S691 space approved silicone.

- The behavior of the bare solar cells and of the SCAs was very different: The shrinking test on bare solar cells mounted on blanket led to the breakage of all samples and was not deemed representative of a real situation.
- The shrinking test on SCAs was partially successful, no solar cell breakage on 10 samples have been noticed but 2 coverglasses cracked.

The test was also performed on SCAs mounted on the polyimide blanked and fixed to an external frame to prevent the torsion of the foil under thermal stress. This configuration is deemed to better approach the orbital behavior of the flexible solar arrays. In this case no breakages were detected.

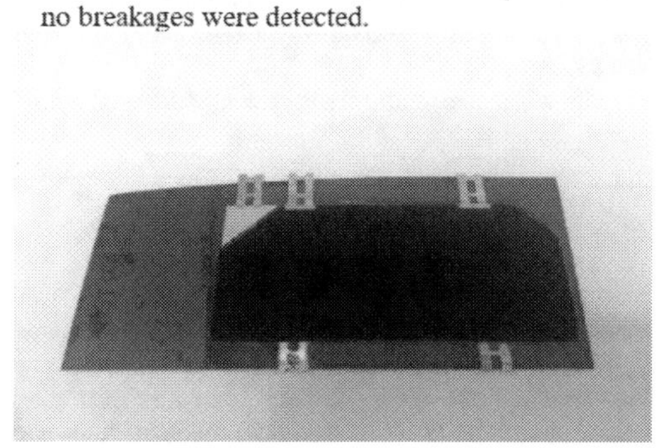

Figure 7: Flexible SCA mounted on a flexible blanket of carbon reinforced polymide

5 CONCLUSIONS

The specific engineering tests described in this report allowed to collect information on the reliability of the thin (80 µm) solar cells CTJ30_80 at bare level and in SCA configuration with flexible coverglass.

- flexible CTJ30_80 SCAs have similar EOL properties of the qualified CTJ30_80 SCAs after electron irradiation, UV test and thermal cycling.
- the electrical properties of CTJ30_80 (both bare cells and SCAs) after long and multiple bending are in line with standard BSC/SCA,;
- The ultimate average bending radius at bare level is in the range 4-5 cm in all tested subgroups;
- The ultimate average bending radius at SCA level is in the range 6-7 cm in all tested subgroups.

REFERENCES

[1] R. Campesato, M. Casale, G. Gabetta, E. F. Lisbona, L. D'Abrigeon "Thin and Flexible Triple Junction cells 30% efficient: qualification results and future space applications", Proceeding IEEE PVSC 2017

[2] Gray et Al. "Flexible Cover Glass for next-generation solar arrays" in Proceedings 10th EU Space Power Conference (2014).

OPTICAL CHARACTERIZATION OF THE INGAASP SOLAR CELLS WITH ABSOLUTE ELECTROLUMINESCENCE OBSERVATION

Kentaroh Wawtanabe[1], Riko Yokota[2], Meita Asami[2], Hassanet Sodabanlu[1], Yoshiaki Nakano[2], Masakazu Sugiyama[1,2]
[1] Research Center for Advance Science and Technology, The University of Tokyo, Japan
[2] Department of Electrical Engineering and Information Systems, The University of Tokyo, Japan
[1]4-6-1 Komaba, Meguro-ku, Tokyo 153-8904, [2]7-3-1 Hongo, Bunkyo-ku, Tokyo 113-8656

ABSTRACT: The rear-hetero structure InGaAsP (1.04 eV) single junction (RHJ) solar cells are fabricated and investigated in this study. The rear-hetero structure in III-V semiconductor based photovoltaic cell is focused their substantial enhancement of open-circuit voltages (Voc). Because there still a few examples of low bandgap material have been developed for an improved mechanical stacked multijunction solar cells, it is important to qualify the material of lattice-match InGaAs and InGaAsP grown on InP substrate. In addition to the standard light I-V and quantum efficiency (QE) measurement, an absolute measurement of electroluminescence (EL) under the current injection equivalent to the photo-carrier generation with AM1.5G illumination is demonstrated. Owing to the reciprocal relationship between the light absorption and emission in the photovoltaic devices, the device performance is qualified as a luminescent efficiency. Adjusting the lattice matching condition, the obtained RHJ InGaAsP solar cell reached to 0.653V of Voc and the corresponding external radiative efficiency (ERE, η_{ext}) is measured as η_{ext} ~3.0%.
Keywords: Electroluminescence, Multijunction Solar Cell, Heterojunction, III-V Semiconductors

1 BACKGROUND

The improvement in multijunction solar cells rapidly increased in recent years. There is two ways of producing remarkable efficiency improvement in the III-V based multijunction solar cells, one is the metamorphic epitaxial growth and the other is mechanical stacking. The current maximum conversion efficiency 47.1% under the concentrated illumination is recorded by inverted metamorphic (IMM) 6-junction structure [1]. Another strategy for multi-layer integration is mechanical stacking. As an example of the direct wafer bonded 4-junction with InGaP/GaAs//InGaAsP/InGaAs cell has been recorded 46% of efficiency under the concentration [2]. One of the advantages of such direct wafer bonding is a high material quality owing to the lattice matched epitaxial growth on each different lattice constant substrate with GaAs and InP, respectively. In contrast to the metamorphic growth, improved crystal quality in each subcell with lattice matched epitaxial growth can be expected by reduced defect and then resulted in higher open circuit voltage (Voc). The dual junction InGaP/GaAs grown on the GaAs substrate is well studied and realized the high-performance devices. However, there still a small number of reports about the narrow bandgap material grown on the InP substrate have been investigated as a photovoltaic (PV) application. In previous research, we reported the $In_{0.53}Ga_{0.47}As$ lattice matched single junction PV with enhanced Voc ~0.42 V by applying rear hetero structure. In that case, the absolute electroluminescence (EL) was also measured and then resulted in enhanced external radiative efficiency (ERE, η_{rad}) ~8.2% at 1SUN equivalent injection current [3]. These results indicated the bottom subcell candidate is improved by a rear hetero structure of n-InGaAs/p-InP junction. In this study, we also attempted to fabricate and investigate the rear hetero solar cell with n-InGaAsP absorber layer which has bandgap energy around 1.05 eV as a candidate of the 3rd subcell in the mechanically stacked 4-junction device.

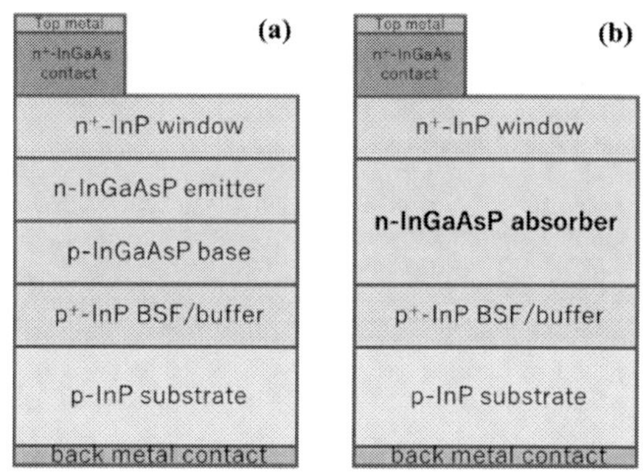

Figure 1: The schematic structure of front homo junction type of device (a) and rear hetero junction type device (b) with InGaAsP main absorption layer.

2 EXPERIMENTAL

2.1 Device fabrication

All of the samples tested in this research are grown by the planetary-type metal organic vapor phase epitaxy (MOVPE) reactor (AIX2000HT, AIXTRON) on the 2'' size InP (001) substrate at 550°C. As the source materials, the TMGa, TMIn, TBAs and TBP are used. The DMZn MO source and gas phase DETe are also used as the acceptor and donor impurity source, respectively. After metallization of planer rear side and grid-type front metal contacts, individual solar cell devices in 5.5 × 5.5 mm^2 sizes are passivated their side edge by wet chemical treatment. The cross-sectional sample structure is shown in Fig. 1. The standard front homo junction (FJ) type samples are fabricated as Fig.(a). And the rear hetero structure type samples (b) are composed by n-InGaAsP absorber and rear p-InP back surface field (BSF) layer. The doping concentration in the minority carriers blocking n-InP window and p-InP BSF layer is $>5\times10^{18}$ cm^{-3}, respectively. The doping concentration in the main absorber of p-InGaAsP for FJ and n-InGaAsP for RHJ is set around 3×10^{17} cm^{-3}. In the present work, the FJ and RHJ samples with same thickness (1500 nm in InGaAsP layer) and FJ samples grown at 550°C and 560°C with the strain balanced InGaAsP layer with same thickness are fabricated for comparison.

2.2 Device characterization

The fabricated InGaAsP solar cell was evaluated their performance by current-voltage (I-V) and external quantum efficiency (EQE) measurements under the AM1.5G standard sunlight illumination with 100 mW/cm^2 power density. The optical efficiency, including the spectrometer, InGaAs array detector, and optical system was calibrated by standard light source to make a numerical characterization from the observed EL spectra. An absolute optical measurement is the effective evaluation method for PV devices. There is the reciprocity relationship between the open circuit voltage under light illumination and the internal radiative efficiency under current injection [4]. Using this relationship, the observed radiative efficiency can indicate the equivalent Voc.

Table I: Summarized cell performance of the InGaAsP cells with RHJ and FJ structure grown at 550°C and the FJ cell grown at 560°C.

Structure, Growth Temp.	Jsc (mA/cm²)	Voc (V)	FF (%)	η (%)
RHJ, 550°C	32.8	0.637	64.9	13.7
FJ, 550°C	33.5	0.640	61.3	13.1
FJ, 560°C	27.1	0.646	61.3	11.6

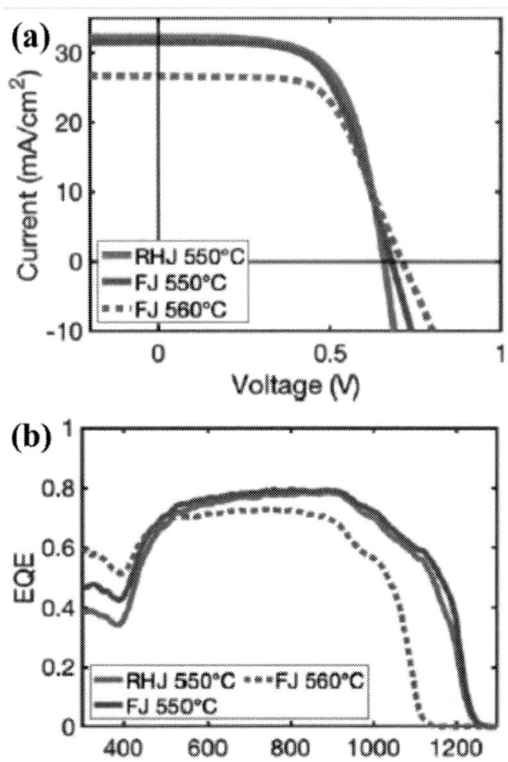

Figure 2: Observed I-V characteristics (a) EQE (b) under the AM1.5G illumination for RHJ and FJ structure InGaAsP cell grown at 550°C, respectively. The dashed line shows the measurement result for the FJ sample grown at 560°C.

3 RESULTS AND DISCUSSION

3.1 Comparison of the RHJ and FJ structure.

The observed I-V characteristics and EQE for the InGaAsP cell with RHJ and FJ structure grown at 550°C and FJ cell grown at 560°C are shown in Figure 2. Cell performance estimated from measurement is also summarized in Table I. Comparing the growth temperature for samples of the same FJ structure and thickness (1500 nm in total emitter and base layer), the growth temperature at 560°C showed decreased short circuit current density (Jsc) and increased Voc. And there is no significant difference between the samples of RHJ and FJ structures. These results are also supported by absolute EL measurement results shown in Figure 3 and Table II. For both structures, the similar bandgap energy ~1.04 V for InGaAsP grown at 550°C is observed. The estimated external radiative efficiency (η_{EL}) ~1.3 and ~1.8% are obtained from RHJ and FJ samples, respectively. The FJ sample grown at 560°C showed much less η_{EL} than the samples grown at 550°C and increased bandgap energy ~1.14 eV and shorten the cutoff wavelength ~1100 nm in EQE. The 10°C differences in growth temperature changed the composition ratio in quaternary system material and increased non-radiative recombination despite of the lattice matching condition. The composition ratio of $In_{0.79}Ga_{0.21}As_{0.46}P_{0.54}$ is estimated by external X-ray diffraction measurement and EL peak wavelength for the sample grown at 550°C, revealed to be appropriate bandgap energy for 4 junction stacking.

Table II: Summary of estimated bandgap energy and radiative efficiency from absolute EL measurement.

Structure, Growth Temp.	E_g (eV)	EL Photon Flux (Photons/s/cm²)	η_{EL} (%)
RHJ, 550°C	1.04	2.7×10^{15}	1.3
FJ, 550°C	1.03	3.5×10^{15}	1.8
FJ, 560°C	1.14	2.0×10^{14}	0.12

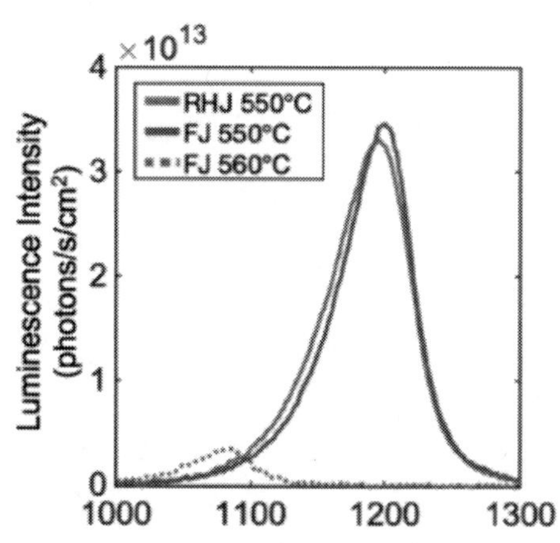

Figure 3: Absolute EL luminescence spectra for RHJ and FJ InGaAsP cells under 1SUN equivalent

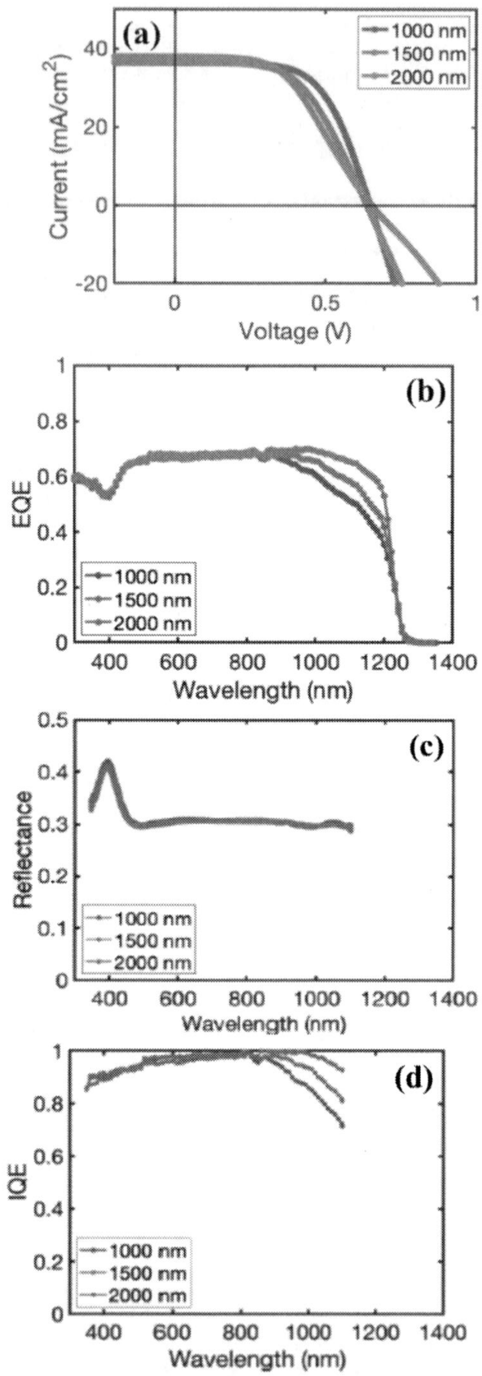

Figure 4: Characterization results of the thickness dependency for RHJ InGaAsP solar cells with 1000-2000 nm in absorber. The I-V characteristics (a), EQE (b), reflectance spectra (c) and IQE (d) is evaluated, respectively.

3.2 RHJ cell with different absorber thickness

We also fabricated the RHJ structure InGaAsP cells with a different absorber thickness of 1000, 1500 and 2000 nm. The evaluation results are shown in Figure 4 and Table III. The highest external radiative efficiency η_{EL} ~ 3.0% and Voc=0.653 V are recorded by RHJ sample with 2000 nm in absorber thickness, an substantial advantages in Voc are suggested in this result compared to previous reported InGaAsP cells [5-7]. And there was no significant thickness dependency on the Voc η_{EL} values. In this thickness range, simply improved Jsc value with increased optical absorption is obtained. This result indicated that there is no significant shortage for minority carrier for dominant absorber in the cell structure. Therefore, a diffusion length of the n-type absorber layer should be larger than 2000 nm. This result is also indicated by internal quantum efficiency (IQE) estimated from the measurement results of the EQE and reflectance (R) spectra. Here, the simple assumption of the IQE=EQE/(1-R) is applied. The IQE indicated the carrier extraction efficiency is sufficiently high with larger than 90% for optically thick wavelength range. There is improvement in IQE in the wavelength range in λ>800 nm accompanied with absorber thickness, directly influenced to the Jsc improvement in the I-V curve. From this result, it is possible to estimate the available Jsc as a 3rd subcell in multijunction device. Figure 5 shows the estimation of IQE under the 3-μm-thick GaAs optical filter simulating the existence of InGaP/GaAs dual junction overlay on the InGaAsP cell. According to this calculation, Jsc=14.7, 15.8 and 17.0 mA/cm^2 will be expected for each absorber thickness with 1000, 1500 and 2000 nm under the ideal anti-reflection coating (ARC), respectively. Because the Jsc value around 14 mA/cm^2 is typically expected to the lattice matched InGaP/GaAs/Ge triple junction solar cell, The RHJ InGaAsP 3rd subcell should be adjusted its thickness regarding to the realistic optical efficiency of the ARC and absorption in 4th $In_{0.53}Ga_{0.47}As$ subcell for 4 junction devices.

Table III: Summary of RHJ InGaAsP solar cell with different thickness.

Absorber Thickness	Jsc (mA/cm^2)	Voc (V)	FF (%)	η (%)	η_{EL} (%)
1000 nm	36.3	0.649	61.1	14.4	2.7
1500 nm	36.4	0.637	54.8	12.7	1.3
2000 nm	37.8	0.653	48.8	12.1	3.0

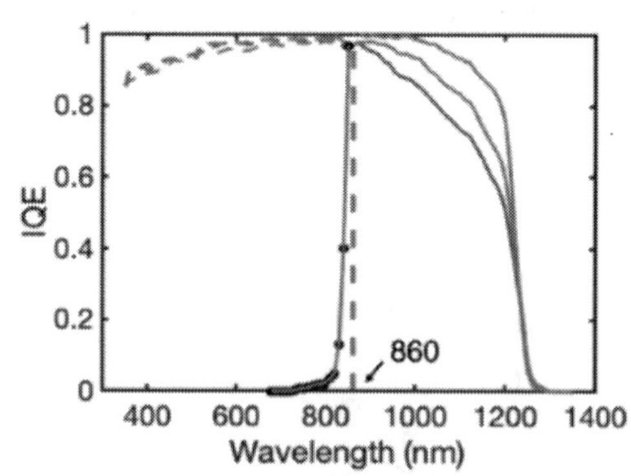

Figure 5: Estimated IQE for 3rd subcell of InGaAsP RHJ device under the GaAs optical filter.

4 CONCLUSIONS

In this research, the rear hetero structure of InGaAsP solar cell lattice matched to the InP substrate is demonstrated for application of the mechanically stacked multijunction device. Unlike in the case of InGaAs device, there is no significant difference from the conventional front junction structure is observed in InGaAsP cell. This result indicates that the no obvious reduction in n-type InGaAsP absorber layer appeared compared to the p-type base layer. The external radiative efficiency evaluation of absolute EL measurement is also performed to numerical qualify. Typically, over 1% of radiative efficiency is observed in the InGaAsP cell with Eg~1.04 eV epitaxially grown by MOVPE at 550°C. Finally, the sufficient value of Jsc as the candidate of subcell in multijunction is obtained from estimation by IQE under the GaAs optical filter with various thickness of the n-InGaAsP absorber in RHJ structure.

Acknowledgement

A part of this article is based on results obtained from a project subsidized by the New Energy and Industrial Technology Development Organization (NEDO).

References

[1] John F. Geisz, Ryan M. France, Kevin L. Schulte, Myles A. Steiner, Andrew G. Norman, Harvey L. Guthrey, Matthew R. Young, Tao Song, and Thomas Moriartyusch, W. J. Olavarria, M. Young, and N. J. Ekins-Daukes, "Six-junction III-V solar cells with 47.1% conversion efficiency under 143 suns concentration," Nature Energy, **5**, pp. 326–335, 2020.
[2] F. Dimroth et al., "Four-Junction Wafer-Bonded Concentrator Solar Cells," IEEE Journal of Photovoltaics, **6**, pp. 343-349, 2016.
[3] K. Watanabe, H. Xu, R. Yokota, H. Sodabanlu, Y. Nakano, M. Sugiyama, "Absolute Electroluminescence Characterization for InGaAs PV Cell Lattice-Matched to InP Substrate," 29th International Photovoltaic Science and Engineering Conference (PVSEC-29), Xi'an, China, 2019.
[4] U. Rau, "Reciprocity relation between photovoltaic quantum efficiency and electroluminescent emission of solar cells," Phys. Rev. B 76, 085303, 2007.
[5] N. Dharmarasu, et al., "High-radiation-resistant InGaP, InGaAsP, and InGaAs solar cells for multijunction solar cells," Appl. Phys. Lett., **79**, pp. 2399-2401, 2001.
[6] L. Ji, et al., "Investigation of InGaAsP solar cells grown by solid-state molecular beam epitaxy," Sol. Eng. Mater. Solar Cell, **137**, pp. 68-72, 2015.
[7] H. Sodabanlu, et al., "Growth of InGaAs(P) in planetary metalorganic vapor phase epitaxy reactor using tertiarybutylarsine and tertiarybutylphosphine for photovoltaic applications," Jpn. J. Appl. Phys., **57**, p. 08RD09, 2018.

THIN-FILM SOLAR CELLS WITH MGF_2/AG BACK MIRROR PATTERNING FOR IMPROVED NEAR-IR REFLECTANCE

Lara Barros Rebouças [1*], Gerard J. Bauhuis [1], Jens Olhmann [2], Jeroen Maasen [1],
Elias Vlieg [1] and John J. Schermer [1]
1 Institute for Molecules and Materials, Radboud University, Nijmegen, the Netherlands
2 Institute for Solar Energy Systems ISE, Freiburg, 79110, Germany
*l.reboucas@science.ru.nl

ABSTRACT: Combining epitaxial lift-off with a dielectric-metal back mirror boosts III-V solar cells efficiencies by enhanced photon recycling without sacrificing the costly growth wafer. In this work, GaAs and GaInP/GaInAs solar cells are produced with a patterned MgF_2/Ag mirror. The sub-bandgap reflectance increases by 2.1% and 2.6%, respectively, compared to the devices with a full Ag back mirror. We investigate the effect of increased sub-bandgap reflection on the temperature-induced open-circuit voltage degradation during operation in low-pressure environments.
Keywords: epitaxial lift-off, back mirror, MgF_2, infrared reflectance

1 INTRODUCTION

Producing thin-film solar cells by allows for the application of a back mirror that reflects emitted photons back into the absorber layers, increasing the probability of internal photon recycling and avoiding parasitic absorption losses to the substrate. As a result, the device's open-circuit voltage (V_{OC}) increases [1]. More recent studies show that the back mirror design can be further optimized with the addition of a low refractive index dielectric layer between the semiconductor and back metal contact. To allow electric contacting of the rear side, the dielectric layer is typically applied in a pattern to allow metal deposition directly onto the contact layer. ZnS [2], SiO_2 [3], SU-8 photoresist [4], MgF_2 [5], and a MgF_2/AlO_x stack [6] are shown to improve the standard performance of III-V solar cells.

In a quasi-planar configuration, the back mirror could reduce the operating temperature of single and multi-junction solar cells by reflecting the unused near-infrared photons. That is specially interesting for space applications, where thermal radiation is the main heat dissipation mechanism.

Further, if the thin-films are produced by epitaxial lift-off (ELO) the possibility of wafer reuse could lower the production costs of these high-efficiency III-V devices [7]. However, due to the interaction of HF with dielectric materials, combining ELO processing with a dielectric-metal reflector is challenging. Oxides such as silica are readily dissolved, whereas photoresists such as SU-8 are found to peel off easily when immersed in HF 48% solution [8].

In this study, we investigate the influence of the patterned dielectric/metal back-side mirror on the operating temperature of solar cells by measuring the V_{OC} of devices with a patterned MgF_2/Ag back mirror and a standard full Ag mirror during a 10 min simulated operation. For comparison, a device with a low-reflective back support is also demonstrated. Additionally, study the interaction of MgF_2 and HF and demonstrate the production of a dual-junction GaInAs/GaInP solar cell with a patterned MgF_2/Ag mirror in which the underetching of the MgF_2 is prevented by a simple silver/gold protective edge rim.

2 MATERIALS AND METHODS

2.1 Growth and processing of solar cells

The single junction GaAs solar cell structures, shown in figure 1, were grown in an inverse order to facilitate wafer removal. The solar cells were grown using an Aixtron 200 reactor on 2-inch (100) GaAs substrates with 2° off to (110) orientation.

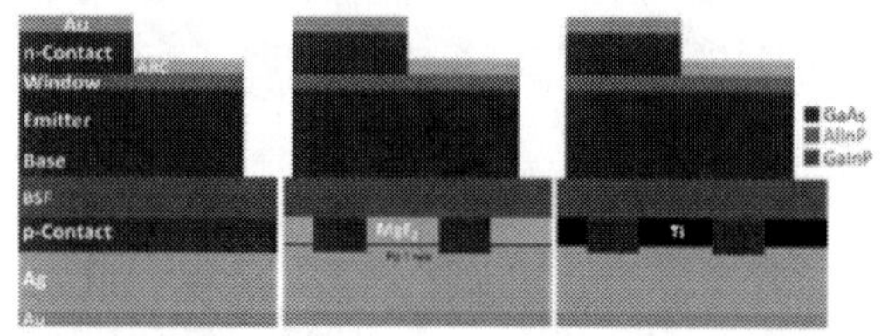

Figure 1: Schematic representation of the single-junction GaAs with (a) standard Ag mirror, (b) a patterned MgF_2/Ag back mirror, and (c) a low reflective Ti back as a substrate-based equivalent for comparison.

The GaInP/GaInAs solar cell structure, shown schematically in figure 2, was grown in an AIX2800 G4 TM MOVPE reactor on 4-inch (100) p-Ge wafers with a 6° offcut towards <111> direction.

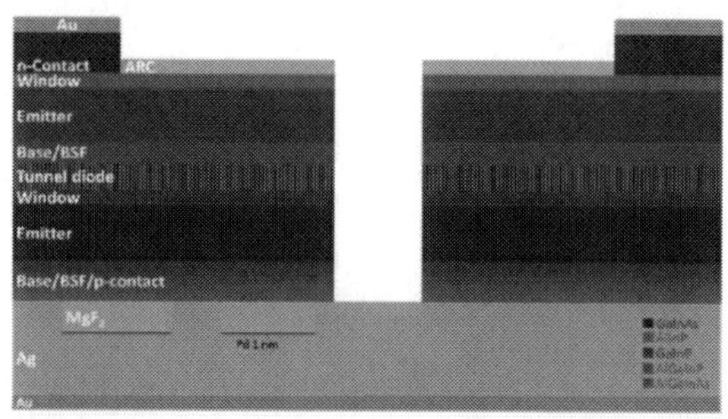

Figure 2: Schematic representation of dual junction GaInP/GaInAs solar cells with a patterned MgF_2/Ag back mirror and a reference Ag mirror processed on the same wafer.

The back mirror pattern was defined by using standard photolithography consisting of regularly spaced contact points with a 127 μm diameter covering 10% of the solar cell area. A 220 nm MgF_2 layer was deposited, followed by 1 nm Pd to improve adhesion dielectric/metal adhesion. Note that this was only performed on half of the wafer to allow comparison with 100% Ag mirror cells produced from the same wafer. After resist removal of the masked contact points, a 3 μm Ag layer was evaporated in the whole wafer. The back mirror was capped by 50 nm Au to prevent oxidation. For the low reflective sample, 220 nm of Ti was deposited instead of MgF_2.

For the single junction GaAs, substrate etching was

performed by etching in [1:3] $NH_4OH:H_2O_2$. ELO was performed by etching a 10 nm AlAs sacrificial layer in [1:1] 25% HF:Acetone. For the dual-junction GaInP/GaInAs.

The released thin-films were processed in a similar procedure; after the removal of the buffer and etch stop layers, a 200 nm Au layer was evaporated in a photo-lithographically defined front grid (16.6% metal coverage). Metal lift-off was performed in acetone, followed by selective etching of the cap layer between the grid fingers. The cell areas are defined by MESA etching with HCl (37%) for the phosphorous-containing layers and [1:1:8] $H_3PO_4:H_2O_2:H_2O$ for the GaAs-based layers. 44 nm ZnS and 94 nm MgF_2 are applied as a top anti-reflection coating (ARC).

2.1 Suitability for ELO processing

To test the stability of MgF_2 during ELO, a 100 nm thick layer was evaporated onto the AlGa(In)As p-contact and exposed to concentrated HF 48%. Half of the sample was masked during evaporation to allow surface profile step height measurements. Optical microscopy images were obtained using a Reichert-Jung Polyvar MET microscope.

2.3 Solar Cells Characterization

Wavelength-dependent reflectance spectra from the top of the processed devices were measured using a ReRa SpeQuest system with ReRa Photor 3.1 software using a step size of 5 nm coupled with an integrating sphere (Bentham DTR6).

Standard AM1.5G JV characteristics of the solar cells were measured at 25 °C with an ABET Technologies Sun 2000 Class AAA solar simulator, which provides illumination with a power density of 1000 W m^2 over a 10 x 10 cm^2 area, a Keithley 2401 source meter, and ReRa Tracer 3.0 software. The same setup was used to allow continuous measurement of V_{OC} over a 10 min operating cycle. Data acquisition was performed by an in-house developed software. During the measurement, the solar cells are kept in a enclosed box at 0.006 mBar. The ambient temperature under illumination was monitored with a type K thermocouple and sufficient time was given in between measurements to allow for complete cooling of the system, ensuring similar starting conditions.

3 RESULTS AND DISCUSSION

3.1 MgF_2 in HF

The 100 nm MgF_2 layer exposed to HF for 30, 60, and 120 seconds showed no significant step height variation in the contact profilometer measurements. However, as shown in figure 3, the dielectric layer is under etched and detaches from the semiconductor surface in smaller pieces.

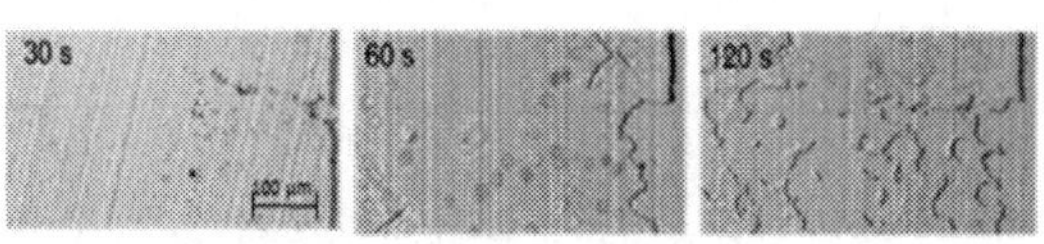

Figure 3: Polarized optical microscopy of a 100 nm MgF_2 layer on AlGa(In)As after exposure to 48% HF for 30, 60, and 120 seconds.

The observed under etching seems to occur initially in the edge of the deposited layer and readily propagates through the structure. Therefore, a 5 mm mask was applied to cover the edge rim during MgF_2 and Pd deposition to process an epitaxial lift-off solar cell with a patterned mirror. This protective edge is then fully covered with silver during the back contact evaporation. Subsequently, the ELO of a GaInP/GaInAs demonstrated that this scheme prevented structural damage to the released thin-film.

3.2 Thin-film characteristics

The JV-characteristics of thin-film solar cells produced by ELO and substrate etching are given in table I. The parameters are given by the average of the three best-performing devices.

Table I: AM1.5G JV-characteristics in room temperature.

Solar Cell - Back mirror	J_{SC} mA.cm^{-2}	V_{OC} V	FF %	η %
SJ - Ti	21.0	1.03	85.2	18.5
SJ - Ag	21.2	1.03	84.9	18.5
SJ - MgF_2	22.2	1.04	83.3	19.3
ELO DJ - Ag	10.2	2.31	83.6	19.8
ELO DJ - MgF_2	10.2	2.31	84.2	19.8

For the single junction GaAs, a gain in open-circuit voltage from the patterned MgF_2/Ag mirror device compared to the Ag and Ti solar cells indicates enhanced photon recycling. The more complex dual junction solar cells, however, show no significant improvement is for the dielectric-metal mirror compared with a standard Ag mirror. Indicating that further optimization is required. Nevertheless, no deterioration occurs from the added MgF_2 layer processed by epitaxial lift-off. Demonstrating the feasibility of combining ELO with a MgF_2/Ag mirror.

2.3 Sub-bandgap reflectance

Figures 4 and 5 show the total reflectance measured from the top of the fully processed solar cell structures.

As expected, a strong absorption is observed for wavelengths below the GaAs and the GaInAs bandgap. In the longer wavelengths (lower energy photons). For the sub-bandgap region, the reflectance increases when a MgF_2/Ag mirror for both solar cell structures. An increase of 2.1% in the range of 900-1100 nm is observed for the single junction GaAs and 2.6% in the range of 945-1100 nm for the GaInP/GaInAs dual junction. The sample with a Ti back shows a reflectivity of 30.7% between 900-1100 nm, and was included to simulate a substrate-based solar cell, with no mirror.

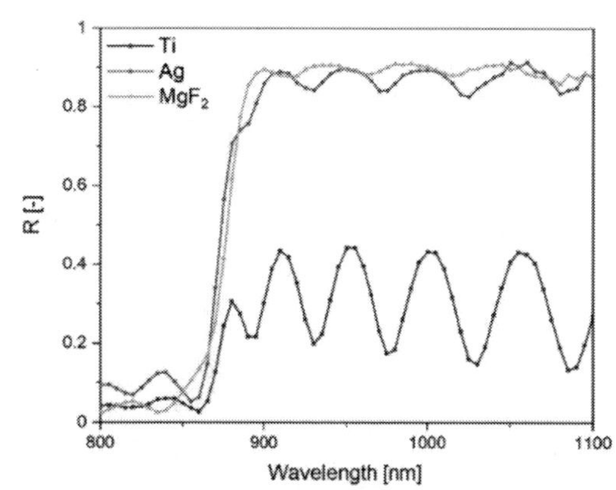

Figure 4 Wavelength-dependent reflectance of single-junction GaAs with different back mirrors.

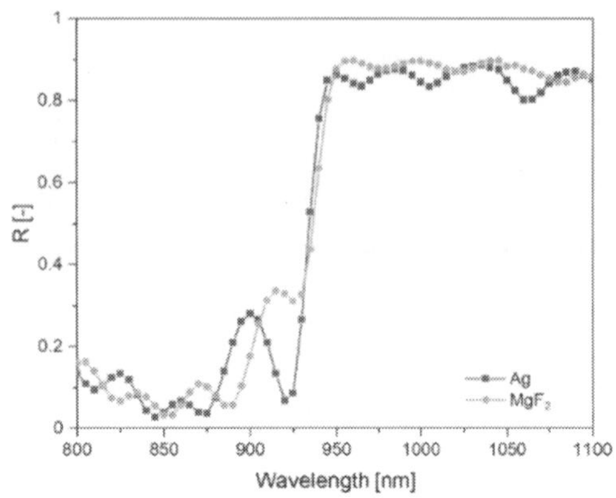

Figure 5 Wavelength-dependent reflectance of dual junction GaInP/GaInAs solar cells produced via ELO with a patterned MgF_2/Ag back mirror, a standard Ag back mirror.

3.4 Operating V_{OC}

III-V solar cells, however, rarely operate in a controlled room temperature environment. Elevated temperatures have a negative impact on the efficiency of solar cells. This degradation is mainly apparent in the V_{OC}. Figure 6 shows the variation in V_{OC} during a 10 min cycle under simulated AM1.5G at 0.006 mBar

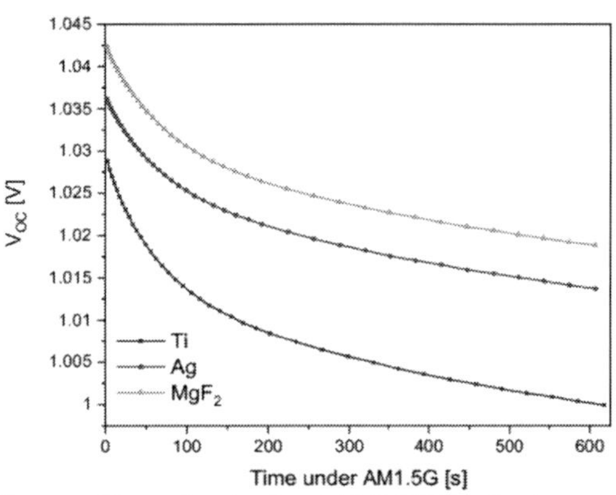

Figure 6 V_{OC} under simulated AM1.5G illumination over 10 min at 0.006 mBar for single junction GaAs solar cells with patterned MgF_2/Ag back mirror, a standard full Ag mirror, and a low reflective Ti back coverage (substrate equivalent).

A significant decrease in V_{OC} is observed for the low-reflective Ti solar cell, which represents a 14 °C increase in temperature during the 10 min operation. Although starting from a higher V_{OC}, the difference between the Ag mirror and the patterned MgF_2 mirror is not significant since variation ±1°C is typically observed in cells from the same batch. The calculated temperature variation for the Ag mirror device is 11°C and 11.7°C for the patterned MgF_2/Ag mirror. This result shows that the enhanced sub-bandgap reflection plays a role in reducing the heat load of the solar cell when the difference in reflection is large, comparing thin-films with substrate-based solar cells, for example. This effect is expected to have a higher impact in space applications, where waste heat is only removed through thermal radiation.

4 CONCLUSIONS

The solar cells produced with a MgF_2/Ag mirror showed enhanced sub-bandgap reflection. For single junction GaAs produced an increased V_{OC} was also achieved, indicating an enhanced photon recycling for devices with a patterned dielectric-metal mirror. The presence of a highly reflective mirror was shown to reduce the operating temperature of the device, as observed in the V_{OC} measurement during a 10 min operating cycle at 0.006 mBar. However, no significant difference was observed in the operating temperature of thin-film solar cells with different mirror designs. Further, thin film solar cells with a patterned MgF_2/Ag back mirror can be produced by epitaxial lift-off with the use of a protective edge rim to prevent the delamination of the thin-film from its carrier during ELO. A dual junction GaInAs/GaInP realized by this method showed no significant damage to the released film and device operation when compared to a standard full metal mirror processing.

REFERENCES

[1] O.D. Miller, E.Yablonovitc, and S.R. Kurtz, Strong internal and external luminescence as solar cellsapproach the Shokley-Queisser limit. IEEE J. Photovolt. 2012;2(3):303-311.

[2] N. Gruginskie, S.C.W. van Laar, G.J. Bauhuis, et al., Increased performance of thin-film GaAs solar cells by rear contact/mirror patterning. Thin Solid Films. 2018;660:10-18.

[3] D.Alonso-Álvareza, C.Weiss, J.Fernandez et al., Assessing the operating temperature of multi-junction solar cells with novel rear side layer stack and local electrical contacts. Solar Energy Materials and Solar Cells. 2019;200:110025.

[4] M.K. Arulanandam, M.A. Steiner, E.J. Tervo et al., GaAs thermophotovoltaic patterned dielectric back contact deviced wit improved sub-bandgap reflectance. Solar Energy Materials and Solar Cells. 2022;238:111545.

[5] C.L. Schilling, O Höhn, D.N, Micha, et al., Combining Photon Recycling and Concentrated Illumination in a GaAs Heterojunction Solar Cell, IEEE J. Photovolt. 2018;8(1):348–354.

[6] H. Helmers, E. Lopez. O. Höhn, et al. 68.9% efficient GaAs-based photonic power conversion by photon recycling and optical resonance. Phys. Status Solidi. 2021;15:2100113.

[7] G. J. Bauhuis, P. Mulder, and J. J. Schermer, Thin-film III-V solar cells using epitaxial lift-off, in: High-efficiency solar cells-physics, Springer, Chap 203 pp 623-643.

[8] D.C.S. Bien, P.V. Rainey, S.J.N. Mitchell, et al., Characterization of masking materials for deep glass micromachining. Journal of Micromechanics and Microengineering. 2003;13(4):34-40.

ACKNOWLEDGEMENTS

This work was supported by the Regeneration project (grant agreement 17043), funded by EIT Raw Materials from the European Institute of Innovation and Technology (EIT), an independent body supported by the European Union's Horizon 2020 program. The authors would also like to thank Daan van der Woude for his support with the epitaxial lift-off processing, Wiebe Leenders for carrying out the latest measurements, and Wil Corbeek.

ULTRATHIN GAAS SOLAR CELLS WITH ROUGH GAP LIGHT-TRAPPING STRUCTURE AND EPITAXIAL LIFT-OFF

[1]Daan van der Woude, [1]Gerard Bauhuis, [1]Luc van der Krabben, [1]Tom Daemen, [2]Jae Jin Kim, [1]Peter Mulder, [3]Joost Smits, [1]Elias Vlieg and [1]John Schermer

[1]Radboud University, Nijmegen, 6525 AJ, The Netherlands
[2]Shell International Exploration and Production Inc., Houston, USA
[3]Shell Global Solutions International B.V., Amsterdam, Netherlands

ABSTRACT: In this work, ultrathin GaAs solar cells are presented with a rough backscattering structure based on lattice mismatched gallium phosphide, grown with MOCVD. This way, a simple and reproducible light-trapping scheme is introduced with low parasitic absorption and full compatibility with hydrofluoric acid-based epitaxial lift-off. Solar cells with two different GaP surface morphologies are demonstrated and shown to be compatible with epitaxial lift-off without any complications or performance loss. An increase in short-circuit current density of up to 9% is observed, a direct result of the increased photon absorption induced by the scattering of the rough GaP texture.
Keywords—ultrathin, GaAs, epitaxial lift-off, light-trapping

1 INTRODUCTION

By increasing the photonic pathlength using a suitable light-trapping scheme, ultrathin GaAs solar cells can be made optically thick and achieve current-densities matching those of conventional thin film (2-3 μm) GaAs cells [1]. The thinner absorber in this type of cell give a higher resistance for proton and electron radiation induced defects [2], and might offer better performance in cells that are grown in ways that do not necessarily ensure high crystalline quality, such as hydride vapor phase epitaxy (HVPE) [3] or III-V on silicon (III-V/Si) [4].

In this work, we present ultrathin GaAs solar cells with a 300 nm thick absorber and a rough backscattering structure based on lattice mismatched gallium phosphide (GaP), grown directly on the device layers with MOCVD. Growing GaP with MOCVD has the benefit of simply growing the light-trapping structure continuously after the solar cell growth, with a high level of reproducibility. The wide, indirect bandgap (2.24 eV) of GaP means it is expected to have even less parasitic absorption than aluminum-based light-trapping layers, while offering high resistance towards hydrofluoric acid (HF) for the sacrificial layer etching during epitaxial lift-off (ELO). We show two different GaP morphologies that are incorporated in the same ultrathin solar cell design, as well as a reference with similar cell structure but with a planar silver mirror. Finally, the compatibility with ELO is demonstrated by producing one of the GaP-textured cell structures using this method.

2 METHODS

2.1 Device growth and characterization

Both the planar and GaP textured inverted ultrathin GaAs homojunction solar cell structures, as shown in table I, were grown using low-pressure MOCVD on 2'' (100) GaAs substrates, 2° off to the (110) orientation. Three different solar cell structures are examined in this study; a reference cell with planar Ag back mirror and two GaP-textured solar cells, each GaP layer grown 10 min. in total but at different temperatures. One is kept at a constant 700 °C as previously reported [5], while the other is grown by a linear ramp to 700 °C after 30 s at 630 °C. These GaP surfaces will from here on be referred to as GaP Constant and GaP Ramp, respectively. The inverted cell structure means that the GaP is the last layer to be grown, ensuring that the dislocations that roughen this layer have no effect on the epitaxial layers and the device performance.

Table 1: Layer structure of the GaP-textured ultrathin GaAs solar cells used in this study. Equivalent planar cells have also been produced that do not incorprate the final etch stop and GaP texture.

Layer	Material	Thickness [nm]	Dop. Conc [cm^{-3}]
n-contact 1	GaAs	40	Te: 1.0E19
n-contact 2	GaAs	200	Si: 5.0E18
window	AlInP	25	Si: 1.8E17
emitter	GaAs	200	Si: 1.0E17
base	GaAs	100	Zn: 1.0E18
BSF	InGaP	100	Zn: 5.0E17
p-contact 1	$Al_{0.3}GaAs$	100	Zn: 2.0E18
p-contact 2	$Al_{0.1}GaAs$	20	C: 1.3E20
etch stop	InGaP	50	-
texture	GaP	~500	-

2.2 Device processing

Local Ohmic contact points were defined using standard photolithography and etched through the GaP layer using a solution of 0.15 M I_2 and 0.3 M KI in a pH 13 0.132 M NaOH and 0.05 M KCl solution at 25 °C. After the InGaP etch stop removal in HCl, a 3 μm Ag back contact/mirror was evaporated on the sample and capped with 50 nm of Au. The growth substrate of the solar cell structure with GaP Constant was destructively removed by chemical etching in 1:3 NH_4OH (32%) and H_2O_2 (30%), while the substrate of the cells with GaP Ramp was recovered using a lateral sacrificial etch of an AlAs release layer in a 1:1 solution of HF (24%) and acetone during ELO.

Both substrate etch and ELO expose the front side of the film, onto which a 200 nm Au contact grid (16.6% coverage) was defined on each cell by lithography and evaporated. After the cap-etch in a 1:2:10 solution of NH_4OH (32%), H_2O_2 (30%) and H_2O, an array of 5×5 mm^2 cells was defined and mesa etched using HCl and NH_4OH for the window

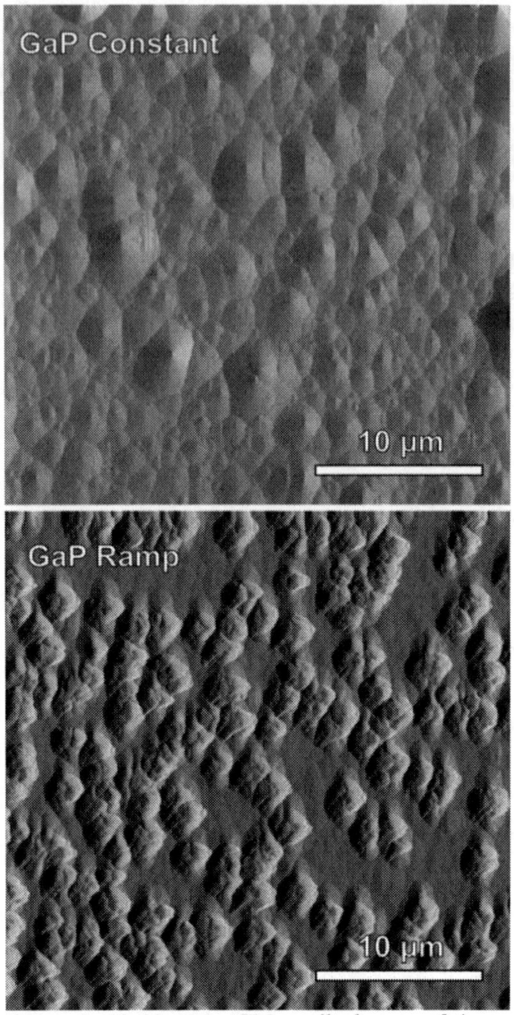

Figure 1: 30×30 µm2 AFM amplitude map of the as-grown GaP Constant (up) and GaP Ramp (down) surfaces.

and emitter respectively. Finally, a 44/94 nm thick ZnS/MgF_2 bilayer anti-reflection (ARC) coating was applied to all structures.

2.3 Device characterization

AFM images were made using a Dimension 3100 system, operated using a Nanoscope (v6.12r1) Digital Instruments control unit. Current Density-Voltage (JV) characteristics of the solar cells were measured with an ABET Technologies Sun 2000 Class AAA solar simulator, which simulates the AM1.5G spectrum with a power density of 1000 $W \cdot m^{-2}$, a Keithley 2601B source meter and ReRa Tracer 3.0 software. A ReRa SpeQuest system with ReRa Photor 3.1 software using a step size of 5 nm was used to measure the wavelength-dependent reflectance spectra in combination with a Bentham DTR6 integrating sphere.

3 RESULTS & DISCUSSION

3.1 GaP surface morphology

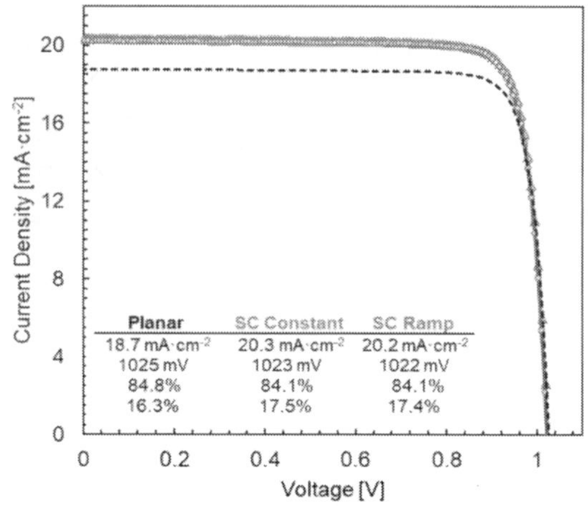

Figure 2: Current Density-Voltage characteristics of the planar, GaP Constant and GaP Ramp textured ultrathin GaAs solar cells used in this work. A front grid coverage of 16.6% and a ZnS/MgF2 ARC.

Figure 1 shows the 30×30 μm^2 AFM amplitude maps of the GaP Constant and GaP Ramp surfaces. The hillocks of GaP Constant form as a result of lattice strain relaxation as the GaP layer surpasses the critical layer thickness of around 1-2 nm. During the duration of the layer growth (10 min.), the hillocks grow in height and volume until a surface is created with a mean inclination around 6° to 9° and root-mean-square roughness (S_q) of 80 nm. The surface of GaP Ramp demonstrates areas with similar morphology, interspersed between semi-random features that originate from accelerated growth at the top of the largest hillocks. These feature result in a much greater variation in surface inclination and higher roughness, with a mean inclination around 25° and S_q of 230 nm.

3.2 Solar cell performance

Figure 2 shows the illuminated JV characteristic of the best performing solar cells (SC) with GaP Constant and GaP Ramp, compared to the best performing equivalent cell with a planar back mirror for reference. Both GaP-textured cells show a similar increase of 8% (SC Ramp) and 9% (SC Constant) in short-circuit current density (J_{SC}) compared to the planar reference. This is a direct result of the increased photon absorption induced by the scattering off the rough GaP texture and silver mirror. The open-circuit voltage (V_{OC}) and fill-factor (FF) above 84% demonstrate that the structures do not suffer from material degradation as a result of the lattice-mismatched GaP growth. Moreover, the equal performance of the substrate etched device and the one obtained with ELO demonstrates that the intended goal of an HF-resistant light-trapping scheme has been achieved.

Figure 3 shows the total and diffuse reflection of GaP-textured thin films with a similar layer structure as the solar cells, but without the GaAs emitter/base, to simulate the scattering at the rough GaP/Ag interface. Photons up to 650 nm can be absorbed in the InGaP (BSF and etch stop) and AlGaAs (p-contact) layers, resulting in a low reflection signal. Beyond 650 nm, the total reflection of both structures has a steep increase but the GaP Constant reaches a significantly higher level before it levels off, around 2x the maximum of GaP Ramp at the GaAs bandgap. The diffuse part of the reflected light of

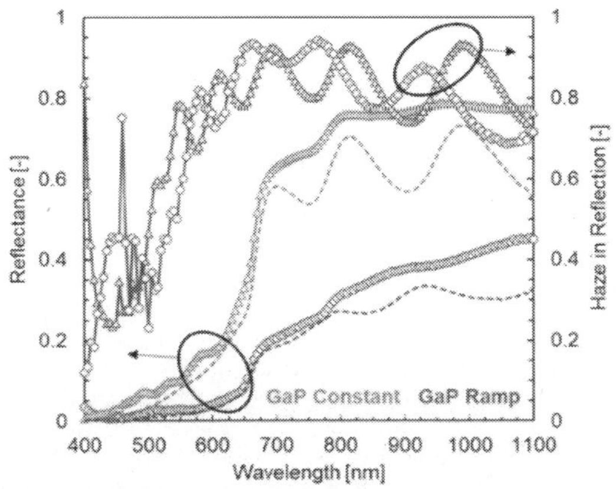

Figure 3: Total and diffuse reflectance, as well as the haze in Reflection of the GaP Constant and GaP Ramp surfaces in an equivalent cell structure without the GaAs absorber.

both structures scales with their respective total reflectances, resulting in a haze (i.e., the ratio of diffuse and total reflectance) that is very similar between both structures, as shown in Figure 3.

The lower signal obtained for the GaP Ramp sample can be attributed to plasmonic losses at the GaP/Ag interface [6]. Plasmonic losses are a direct result of the rough texture of the light-trapping scheme. Because GaP Ramp has a much higher average surface roughness, it can be expected that the plasmonic losses here are more severe. It is expected that this loss can be reduced by adding a dielectric layer between the rough GaP and the Ag mirror.

4 CONCLUSION & FUTURE WORK

In this work, ultrathin GaAs solar cells are demonstrated with a rough back scattering layer based on lattice mismatched GaP. This way, a simple and reproducible light-trapping scheme is introduced with low parasitic absorption and compatibility with HF-based epitaxial lift-off. This compatibility was demonstrated by comparing a substrate-etched GaP-textured cells with one that has been produced with epitaxial lift-off, with both cells performing equally well. Two different GaP morphologies have been studied and found to increase the short-circuit current density by 8 and 9% compared to the equivalent cell with a planar mirror. Future experiments will incorporate a dielectric layer in between the rough GaP and silver mirror, which is expected to reduce plasmonic loss at the silver mirror and increase performance of the GaP textured solar cells.

ACKNOWLEDGMENT

Acknowledged financial support by Shell Global Solutions International B.V. under contract number CWS-149392.

REFERENCES

[1] J. Buencuerpo, M. A. Steiner, and A. C. Tamboli, "Optically-thick 300 nm GaAs solar cells using adjacent photonic crystals," Optics Express, vol. 28, no. 9, p. 13845, Apr. 2020, doi: 10.1364/oe.391737.

[2] S. J. Polly, G. T. Nelson, J. R. D'Rozario, R. Tatavarti, and S. M. Hubbard, "Radiation Effects in Thinned GaAs Photovoltaics Incorporating DBRs for Improved Radiation Tolerance of Multijunctions," in Conference Record of the IEEE Photovoltaic Specialists Conference, 2019, pp. 2818–2821.

[3] J. Simon et al., "Upright and inverted single-junction GaAs solar cells grown by hydride vapor phase epitaxy," IEEE J. Photovoltaics, vol. 7, no. 1, pp. 157–161, Jan. 2017.

[4] M. Feifel, D. Lackner, J. Ohlmann, J. Benick, M. Hermle, and F. Dimroth, "Direct Growth of a GaInP/GaAs/Si Triple-Junction Solar Cell with 22.3% AM1.5g Efficiency," Sol. RRL, vol. 3, no. 12, p. 1900313, Dec. 2019.

[5] D. van der Woude, L. van der Krabben, G. Bauhuis, et al. "Ultrathin GaAs solar cells with a high surface roughness GaP layer for light-trapping application," Prog Photovolt Res Appl. 2022; 1- 10. doi:10.1002/pip.3534

[6] F.J. Haug, T. Söderström, O. Cubero, V. Terrazzoni-Daudrix, and C. Ballif, "Plasmonic absorption in textured silver back reflectors of thin film solar cells", Journal of Applied Physics 104, 064509 (2008) https://doi.org/10.1063/1.2981194

IMPROVEMENT OF PASSIVATION PERFORMANCE OF SILICON NANOCRYSTAL/SILICON OXIDE COMPOUND LAYER BY HYDROGEN PLASMA TREATMENT

* Masashi Matsumi[1], **Kazuhiro Gotoh[1], Markus Wilde[2], Yasuyoshi Kurokawa[1], Katsuyuki Fukutani[2] and Noritaka Usami[1]
[1] *Graduate School of Engineering, Nagoya University, Nagoya, Japan*
[2] *Institute of Industrial Science, The University of Tokyo, Tokyo, Japan*
* *Masashi Matsumi (matsumi.masashi@e.mbox.nagoya-u.ac.jp)*
** *Kazuhiro Gotoh (gotoh.kazuhiro@material.nagoya-u.ac.jp)*

ABSTRACT: Solar cells using stacks of a polycrystalline silicon layer and an ultra-thin silicon oxide layer have commanded considerable attention due to their high conversion efficiency. However, thinning the silicon oxide film to transport carriers leads to a decrease in passivation performance. Therefore, we have developed the NAnocrystalline Transport Path in Ultra-thin dielectrics for Reinforced passivation (NATURE) contact that combines both thickness and conductivity. Hydrogen plasma treatment (HPT) is an effective method to improve passivation performance. We previously reported that the implied open circuit voltage (i-V_{OC}) was significantly improved after HPT at 500 °C. For further improvement, we here investigate the impact of the HPT duration and the H_2 pressure on the i-V_{OC} values and on the hydrogen depth distribution. The i-V_{OC} was measured by the quasi-steady state photo-conductance method and the hydrogen depth distribution was characterized by resonant nuclear reaction analysis (RNRA). This revealed that hydrogen atoms are sufficiently incorporated into the NATURE contact for HPT duration of 15 min at a H_2 pressure of 0.5 Torr, yielding the highest i-V_{OC} of 724 mV.
Keywords: Hydrogen, Hydrogen Plasma Treatment, Polycrystalline Silicon on Oxide Junction, SiO_2, Passivation

1 INTRODUCTION

Photovoltaics is regarded as a power source in order to continue sustainable development while meeting the world's growing demand for electricity. This has led to the development of low-cost, high-performance, stable, and environmentally friendly photovoltaic (PV) devices [1-5]. Among them, crystalline silicon (c-Si)-based PV devices are being intensively studied because of their high-power generation efficiency and commercial-scale production [6-8].

Recently, c-Si solar cells incorporating poly-Si (poly-Si)/silicon oxide (SiO_X) passivation contacts have attracted considerable interest as one of the most promising ways to overcome the efficiency limitations of passivated emitter rear cell (PERC) technology for c-Si solar cells [9-11]. On a laboratory scale, efficiencies of 26.1% have been demonstrated for poly-Si/SiO_X interdigitated back-contact (IBC) cell structures [12]. Among them, tunnel oxide film passivation contact (TOPCon) solar cells have attracted more interest because of their various advantages, such as easy availability of raw materials, easy process steps, and expected high efficiency [13,14].

In TOPCon solar cells, the SiO_X film thickness is limited to 1-2 nm to extract carriers from the c-Si substrate due to the very high resistivity of the SiO_X layer [15]. Thick SiO_X films have excellent passivation performance and are preferred for long-term stabilization [16-18]. In addition, the time-dependent dielectric breakdown has been reported to be suppressed in metal-oxide-semiconductor field-effect transistor devices with silicon oxide thicker than 4 nm [19]. Therefore, if thick SiO_X layers can be applied to TOPCon solar cells, the performance of TOPCon solar cells can be further improved.

Then, the NAnocrystalline Transport Path in Ultra-thin dielectrics for Reinforced passivation (NATURE) contact, in which conductive silicon nanocrystals (Si NCs) are embedded in the SiO_X layer and carriers are transported through them, has been studied to achieve both depositions of thick SiO_X layer and carrier transport [20]. Figure 1 shows the structure of NATURE contact. Currently, NATURE contact records a silicon oxide film of 7.4 nm, which is more than three times thicker than the conventional film and is expected to have high passivation performance and long-term reliability. However, a large number of dangling bonds on the surface of Si NCs in the NATURE contact may limit the passivation performance of the NATURE contact. Therefore, we performed hydrogenation treatment on the NATURE contact to improve the passivation performance. Hydrogenation is a process that introduces hydrogen into the sample and bonds it to the silicon dangling bonds and is an effective means of improving passivation performance [21-23].

In this manuscript, we searched for the most efficient conditions for hydrogenation of NATURE contacts by varying HPT duration and H_2 pressure and using passivation performance as a measure.

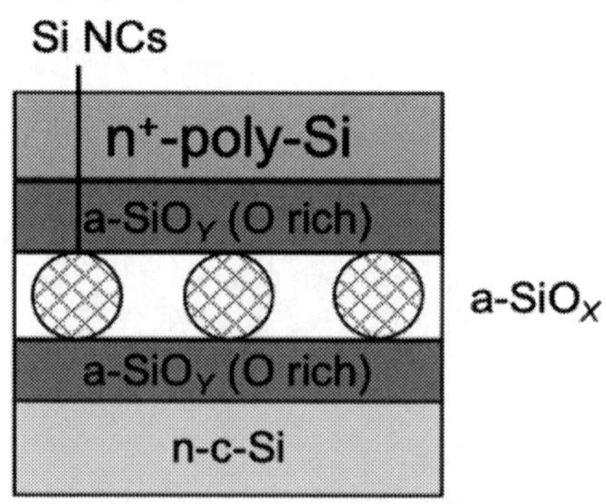

Figure 1: Schematic structure of the NATURE contact

2 EXPERIMENTAL DETAILS

Czochralski (CZ) n-type monocrystalline wafers (orientation of (100), resistivity of 3.1 Ω cm, and thickness of 200 μm) were cleaned in an ultrasonic bath using Semicoclean-23 (Furuuchi Chemicals Co.) for 6 min, followed by immersion in 5% HF for 1 min to remove native oxides on the substrate. Then, NATURE contact and n-type hydrogenated amorphous silicon (n^+-a-Si:H) layers were deposited on the c-Si substrate by plasma-enhanced chemical vapor deposition (PECVD; ULVAC Inc., CME-200J). Silane (SiH_4), hydrogen (H_2), and phosphine (PH_3) precursor gases were employed during the PECVD, along with deposition parameters such as a frequency of 27.12 MHz, working pressure of 25 Pa, a plasma source power density of 32.5 mW/cm^2, and a substrate temperature of 180 °C. The NATURE contact was deposited by sandwiching a Si-rich a-SiO_X:H layer with an O-rich a-SiO_Y:H layer ($X < Y$). The thicknesses of the bottom a-SiO_Y:H, middle a- SiO_X:H, and top a-SiO_Y:H layers were fixed at 1 nm, 5 nm, and 1.4 nm, respectively. The layer thickness was controlled by changing the deposition time of a- SiO_X:H and a-SiO_Y:H. The SiH_4 and CO_2 gas flow rates for the a-SiO_X:H layer were 10 and 10 sccm, respectively, and those for the a-SiO_Y:H layer were 6 and 50 sccm, respectively (sccm = standard cubic centimeters per minute). Next, a 30 nm thick n^+-a-Si:H layer was deposited using SiH_4, H_2, and PH_3 gases at flow rates of 40, 400, and 20 sccm, respectively. Forming gas annealing was then performed at 750°C for 30 minutes using a lamp furnace (ADVANCE RIKO, MILA-5050). At this time, the n^+-a-Si:H layer crystallized into an n^+-poly-Si layer, and silicon nanocrystals were formed in the a-SiO_X layer. The samples were then hydrogenated by a hydrogen plasma treatment (Katagiri Engineering Co., Ltd.). For electrical measurements, aluminum electrodes were deposited by vacuum evaporation (ULVAC Inc., VPC-260F).

Quasi-steady-state photoconductance (QSSPC; Sinton Instruments, WCT-120TS) was used to measure i-V_{OC} and evaluate sample passivation performance [24]. The root mean square (RMS) of the sample surface roughness was evaluated using an atomic force microscope (AFM; Hitachi High-Tech Corp., AFM100 Plus) [25], and the contact resistivity (ρ_c) was evaluated by current-voltage (I-V) measurements using the Cox-Strack configuration [26,27]. The layer thickness of each sample was measured by variable angle spectroscopic ellipsometry (SE; J. A. Woollam, M-2000DI). The depth profile of hydrogen was measured at RNRA by the resonant ^1H (^{15}N, $\alpha\gamma$)^{12}C reaction [28,29]. $^{15}N^{2+}$ ion beams were produced at 20-35 nA at the MALT bandegraph tandem accelerator at the University of Tokyo.

3 RESULTS AND DISCUSSION

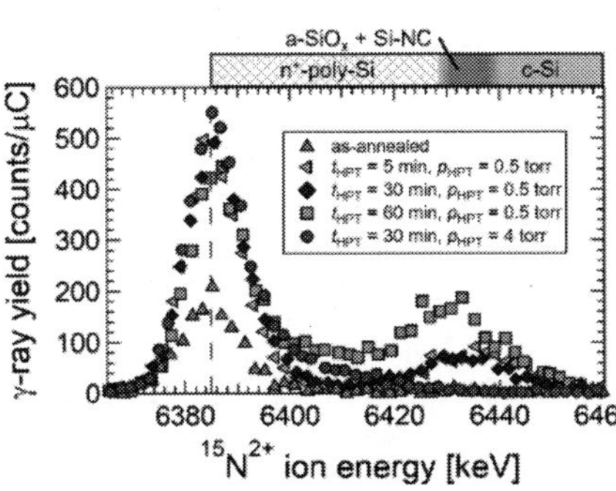

Figure 2: RNRA γ-ray yield curves of the samples after annealing at 750 °C, and after HPT in different durations and H_2 pressures. Schematic structure is shown at the top of the figure.

Figure 2 shows the RNRA γ-ray yield curves for different t_{HPT}. The schematic structure at each depth is shown at the top of the figure. The hydrogen plasma treatment time (t_{HPT}) was varied from 5 to 60 minutes The HPT temperature and H_2 pressure were fixed at 500 °C and 0.5 Torr, respectively. The hydrogen depth distributions of the samples for t_{HPT} between 5 min and 30 min were almost the same, showing H incorporation into the NATURE contact region. On the other hand, after HPT for 60 min, a larger amount of hydrogen is observed, also in the n^+-poly-Si layer.

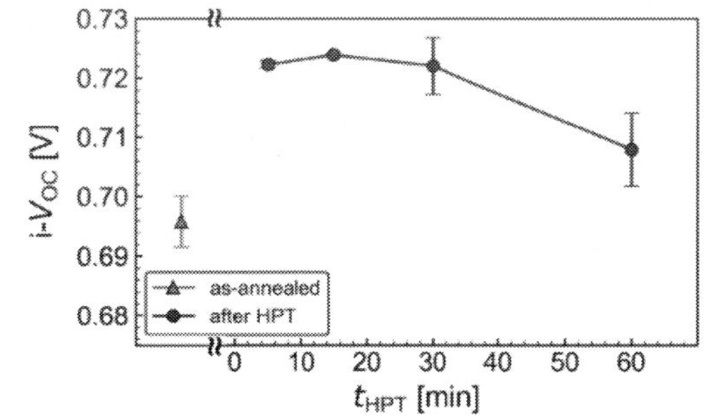

Figure 3: Implied V_{OC} of the samples after annealing at 750 °C and HPT in 0.5 Torr H_2 as a function of HPT duration.

Figure 3 shows the t_{HPT} dependence of i-V_{OC}. Almost identical i-V_{OC} values were observed for the samples HP-treated from 5 min to 30 min, whereas the i-V_{OC} decreased after a t_{HPT} of 60 min.

These results suggest that the hydrogen inside the structure was already saturated by 5 min HPT, and that the equilibrium between hydrogen introduction and desorption was maintained up to t_{HPT} = 30 min, so that the hydrogen distribution and i-V_{OC} may not have changed. The higher i-V_{OC} measured for this t_{HPT} range suggests that a sufficient amount of hydrogen was introduced to effectively terminate the dangling bonds on the Si NC surfaces already after HPT for 5 min. However, at t_{HPT} = 60 min, hydrogen was introduced in excess. According to Slaoui *et al*, the excess hydrogen introduced cleaves weak Si-Si bonds and forms Si-H complexes [30]. Excess hydrogenation also creates sub-surface defects due to the accumulation of diatomic hydrogen (H_2^*). These may have resulted in the formation of dangling bonds, which may have lowered the i-V_{OC}. Then, hydrogen bonds with the new dangling bonds, resulting in the presence of a large amount of hydrogen inside the structure.

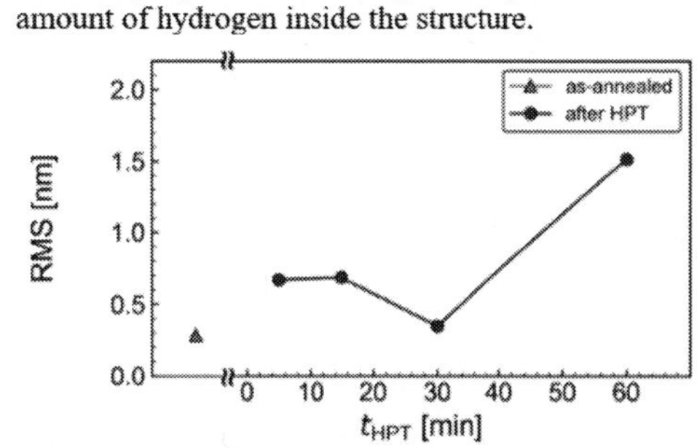

Figure 4: Surface roughness of the samples after annealing at 750 °C and HPT in 0.5 Torr H_2 as a function of HPT duration.

Figure 4 shows the t_{HPT} dependence of root mean square (RMS) of surface roughness. At $t_{HPT} \leq 30$ min, almost no change in RMS before and after HPT was observed, while RMS increased at $t_{HPT} = 60$ min. This result suggests that at $t_{HPT} = 60$ min, the sample surface is etched due to the excessive introduction of hydrogen, suggesting the formation of dangling bonds in the sample.

From this experiment, the HPT time was found to be closely related to the amount of hydrogen introduced into the sample. Sufficient hydrogenation of the NATURE contact is realized already after HPT for 5 min and should not exceed 30 min.

Next, we investigated how pressure changes in hydrogen plasma treatment affect the hydrogenation of NATURE contact. Figure 2 shows the RNRA γ-ray yield curves for different p_{HPT}. The hydrogen plasma treatment pressure (p_{HPT}) was varied from 0.5 to 4 Torr, and the electrical properties and hydrogen depth distribution were measured. In this case, the HPT temperature and HPT time were fixed at 500 °C and 30 min, respectively. Hydrogen was effectively incorporated into the NATURE contact region at a p_{HPT} of 0.5 Torr. At $p_{HPT} = 4$ Torr, hydrogen atoms are not contained in the NATURE contact region but only in the n^+-poly-Si layer.

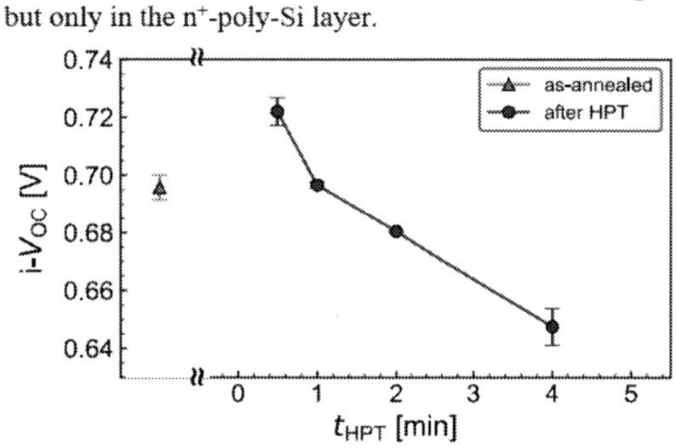

Figure 5: Implied V_{OC} of the samples after annealing at 750 °C and HPT for 30 min as a function of H_2 pressure during HPT.

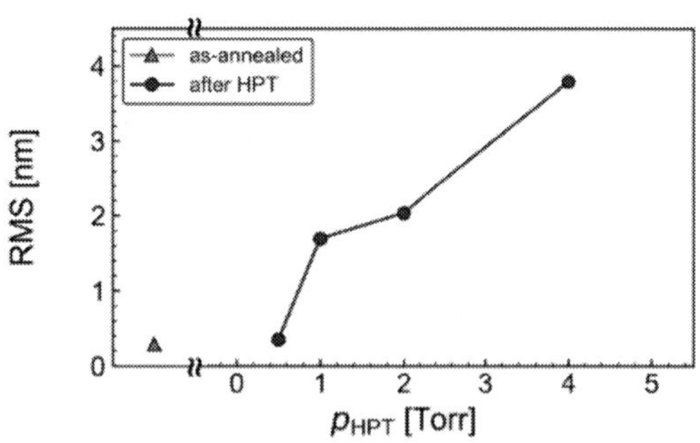

Figure 6: Roughness of the samples after annealing at 750 °C and HPT for 30 min as a function of H_2 pressure during HPT.

Figures 5 and 6 show the i-V_{OC} and RMS of the NATURE contact as a function of the p_{HPT}, respectively. As the p_{HPT} increases, i-V_{OC} decreases while RMS increase.

Nunomura *et al* reported that increasing hydrogen pressure in HPT also increases the hydrogen radical density [31]. Based on this fact and the results of this experiment, collisions between the excess radicals and the sample surface may have occurred more frequently in the high-pressure plasma. The sample surface would then have been etched, resulting in increased surface roughness and the formation of large amounts of dangling bonds, which would have degraded passivation and conductive performance. This consideration is consistent with previous studies that have shown that high-density hydrogen plasma treatment introduced plasma damage to the sample surface [32]. In addition, collisions between radicals may have occurred more frequently in high-pressure plasma. As a result, the energy of the hydrogen radicals may have been reduced, resulting in hydrogen incorporation only at shallow locations on the sample.

From this experiment, we know that HPT pressure is closely related to the damage that occurs when hydrogen is introduced into the sample. The lower the HPT pressure, the less damage to the sample, and at $p_{HPT} = 0.5$ Torr, only the passivation performance was greatly improved with almost no damage to the sample.

4 CONCLUSIONS

To further improve the passivation performance of NATURE contact, which is expected to have high passivation performance and long-term reliability, hydrogen plasma treatment was performed at various durations and H_2 pressures. As a result, a short time (5 to 30 min) and low pressure (0.5 Torr) HPT resulted in a sufficient amount of hydrogen incorporation into the NATURE contact, and the passivation performance was greatly improved. On the other hand, HPT at a long time (60 minutes) introduced excessive hydrogen into the sample, resulting in a decrease in passivation performance. This may be due to the fact that the excess hydrogen may have created dangling bonds inside the sample, which reduced the passivation performance. The sample employing HPT at high pressure (4 Torr) showed a significant decrease in passivation performance, and the sample surface was very rough. This is thought to be because the excess hydrogen radicals produced under high pressure frequently collided with the sample surface and generated many dangling bonds on the sample surface.

In conclusion, NATURE contact showed the most efficient hydrogen passivation when HPT was performed for 15 minutes at 0.5 Torr, and the highest i-V_{OC} value of 724 mV was recorded. The HPT under appropriate conditions is very important to improve the passivation performance because some HPT conditions can deteriorate the passivation performance.

ACKNOWLEDGEMENT

The work was supported by Japan Society of the Promotion Science (JSPS) KAKENHI Grant number 18H05951, 20K15127, and Grant-in-Aid for Scientific Research on Innovative Areas, MEXT, Japan (Hydrogenomics), JP18H05514 and JP18H05518.

REFERENCES

[1] N. Asim, K. Sopian, S. Ahmadi, K. Saeedfar, M.A. Alghoul, O. Saadatian, S. H. Zaidi. A review on the role of materials science in solar cells. *Renew. Sustain. Energy Rev.*, **2012**, 16, 5834–5847.

[2] Y. Lee, C. Park, N. Balaji, Y.J. Lee, V.A. Dao. High-efficiency Silicon Solar Cells: A Review. *Isr. J. Chem.*, **2015**, 55, 1050-1063.
[3] P. Mandal. Application of Plasmonics in Solar Cell Efficiency Improvement: a Brief Review on Recent Progress. *Plasmonics*, **2022**, 17, 1247-1267.
[4] H. Zhang, X. Ji, H. Yao, Q. Fan, B. Yu, J. Li. Review on efficiency improvement effort of perovskite solar cell. *Sol. Energy*, **2022**, 233, 421-434.
[5] M. Yang, J. Wu, Z. Lan, J. Lin, M. Huang, L. Fan. Hotspots, frontiers, and emerging trends of tandem solar cell research: A comprehensive review. *Int. J. Energy Res.*, 2022, 46(1), 104-123.
[6] M.A. Green, E.D. Dunlop, J. Hohl-Ebinger, M. Yoshita, N. Kopidakis, X. Hao. Solar cell efficiency tables (version 59). *Prog. Photovolt. Res. Appl.*, **2022**, 30(1), 3–12.
[7] M.Z. Rahman. Advances in surface passivation and emitter optimization techniques of c-Si solar cells. *Renew. Sustain. Energy Rev.*, **2014**, 30, 734–742.
[8] A. ur Rehman, M. Z. Iqbal, M. F. Bhopal, M. F. Khan, F. Hussain, J. Iqbal, M. Khan, S. H. Lee. Development and prospects of surface passivation schemes for high-efficiency c-Si solar cells. *Sol. Energy*, **2018**, 166, 90–97.
[9] M.A. Green. Photovoltaic technology and visions for the future. *Prog. Energy*, **2019**, 1, 013001.
[10] F. Feldmann, M. Simon, M. Bivour, C. Reichel, M. Hermle, S.W. Glunz. Carrier- selective contacts for Si solar cells. *Appl. Phys. Lett.*, **2014**, 104, 181105.
[11] F. Feldmann, M. Bivour, C. Reichel, H. Steinkemper, M. Hermle, S.W. Glunz. Tunnel oxide passivated contacts as an alternative to partial rear contacts. *Sol. Energy Mater. Sol. Cell.*, **2014**, 131, 46–50.
[12] F. Haase, C. Hollemann, S. Schäfer, A. Merkle, M. Rienäcker, J. Krügener, R. Brendel, R. Peibst. Laser contact openings for local poly-Si-metal contacts enabling 26.1% efficient POLO-IBC solar cells. *Sol. Energy Mater. Sol. Cells*, **2018**, 184–186.
[13] F. Feldmann, M. Bivour, C. Reichel, M. Hermle, S.W. Glunz. Passivated rear contacts for high-efficiency n-type Si solar cells providing high interface passivation quality and excellent transport characteristics. *Sol. Energy Mater. Sol. Cells*, **2014**, 270–274.
[14] D.K. Ghosh, S. Bose, G. Das, S. Acharyya, A. Nandi, S. Mukhopadhyay, A. Sengupta. Fundamentals, present status and future perspective of TOPCon solar cells: A comprehensive review. *Surfaces and Interfaces*, **2022**, 30, 101917.
[15] A.S. Kale, W. Nemeth, S. P. Harvey, M. Page, D. L. Young, S. Agarwal, P. Stradins. Effect of silicon oxide thickness on polysilicon based passivated contacts for high-efficiency crystalline silicon solar cells. *Sol. Energy Mater. Sol. Cells*, **2018**, 185, 270–276.
[16] M. J. Kerr, A. Cuevas. Very low bulk and surface recombination in oxidized silicon wafers. *Semicond. Sci. Technol.*, **2002**, 17, 35–38.
[17] H. Nagayoshi, Y. Onozawa, M. Ikeda, M. Yamaguchi, Y. Yamamoto, T. Uematsu, T. Saitoh, K. Kamisako. Effect of hydrogen-radical annealing for SiO2 passivation. *Jpn. J. Appl. Phys.*, **1996**, 35, L1047–L1049.
[18] D. L. Young, W. Nemeth, S. Grover, A. Norman, H. Yuan, B.G. Lee, V. LaSalvia, P. Stradins. Carrier selective, passivated contacts for high efficiency silicon solar cells based on transparent conducting oxides. *Energy Procedia*, **2014**, 55, 733–740.
[19] J.W. McPherson, R.B. Khamankar. Molecular model for intrinsic time-dependent dielectric breakdown in SiO2 dielectrics and the reliability implications for hyper-thin gate oxide. *Semicond. Sci. Technol.*, **2000**, 15, 462–470.
[20] R. Tsubata, K Gotoh, M Matsumi, M Wilde, T Inoue, Y Kurokawa, K Fukutani, N Usami. Silicon Nanocrystals Embedded in Nanolayered Silicon Oxide for Crystalline Solar Cells. *ACS Appl. Nano Mater.*, **2022**, 5, 2, 1820-1827.
[21] A. Descoeudres, L. Barraud, S. De Wolf, B. Strahm, D. Lachenal, C. Guérin, Z.C. Holman, F. Zicarelli, B. Demaurex, J. Seif, J. Holovsky, C. Ballif. Improved Amorphous/Crystalline Silicon Interface Passivation by Hydrogen Plasma Treatment. *Appl. Phys. Lett.*, **2011**, 99, 123506.
[22] G. Dingemans, W. Beyer, M.C.M. van de Sanden, W.M.M. Kessels. Hydrogen Induced Passivation of Si Interfaces by Al2O3 Films and SiO2/Al2O3 Stacks. *Appl. Phys. Lett.*, **2010**, 97, 152106.
[23] E. Cartier, J.H. Stathis, D.A. Buchanan. Passivation and Depassivation of Silicon Dangling Bonds at the Si/SiO2 Interface by Atomic Hydrogen. *Appl. Phys. Lett.* **1993**, 63, 1510–1512.
[24] R. Sinton, S. Cuevas. Contactless determination of current-voltage characteristics and minority-carrier lifetimes in semiconductors from quasi-steady-state photoconductance data. *Appl. Phys. Lett.*, **1996**, 69, 2510–2512.
[25] G. Binnig, C.F. Quate, Ch. Gerber. Atomic Force Microscope. *Phys. Rev. Lett.*, **1986**, 56, 930-933.
[26] R. H. Cox, H. Strack. Ohmic contacts for GaAs devices. *Solid- State Electron*, **1967**, 10, 1213–1218.
[27] R. P. Gupta, J. B. White, O. D. Iyore, U. Chakrabarti, H. N. Alshareef, B. E. Gnade. Determination of contact resistivity by the Cox and Strack method for metal contacts to bulk bismuth antimony telluride. *Electrochem. Solid-State Lett.*, **2009**, 12, H302–H304.
[28] M. Wilde, K. Fukutani. Hydrogen detection near surfaces and shallow interfaces with resonant nuclear reaction analysis. *Surface Science Reports*, **2014**, 69, 196-295.
[29] K. Fukutani, M. Wilde, M. Matsumoto. Nuclear-reaction analysis of H at the Pb/Si(111) interface: Monolayer depth distinction and interface structure. *Physical Review B*, **2001**, 64, 245411-1-6.
[30] A. Slaoui, E. Pihan, I. Ka, N.A. Mbow, S. Roques, J.M. Koebel. Passivation and etching of fine-grained polycrystalline silicon films by hydrogen treatment. *Sol. Energy Mater. Sol. Cells*, **2006**, 90, 2087–2098.
[31] S. Nunomura, H. Katayama, I. Yoshida. Hydrogen atom kinetics in capacitively coupled plasmas. *Plasma Sources Sci. Technol.*, **2017**, 26, 055018.
[32] S. Rattanapan, H. Yamamoto, S. Miyajima, T. Sato, M. Konagai. Hydrogen plasma treatment for improving bulk passivation quality of c-Si solar cells. *Curr. Appl. Phys.*, **2010**, 10, S215-S217.

COMPARISON OF CDS AND CDSE BUFFER LAYERS FOR ANTIMONY SELENIDE (SB_2SE_3) SOLAR CELLS BY VACUUM EVAPORATION

Torabi N., Zanetti S., Artegiani E., Piccinelli F.[+], Punathil P., and Romeo A*.
LAPS-Laboratory for Applied Physics, Department of Computer Science, University of Verona,
Ca' Vignal 1, Strada Le Grazie 15, 37134 Verona, Italy, tel.:+39-045-8027974, *e-mail address:
alessandro.romeo@univr.it
[+]Department of Biotechnology, University of Verona, Ca' Vignal 1, Strada Le Grazie 15, 37134 Verona, Italy

ABSTRACT: In this work, we study the growth of Sb_2Se_3 deposited by thermal evaporation on different buffer layers for superstrate configuration thin film solar cells. A comparison between CdS which has been widely used as a buffer layer and CdSe is represented. Selenium vacancies in Sb_2Se_3 thin film act as recombination centers, and according to the literatures, applying an excessive amount of Se improve the application of solar cell. By using CdSe as a buffer layer, selenium could diffuse into the film and it is supposed to increase the efficiency. In our case both devices show the same efficiency of around 3.5% and higher J_{sc} for CdSe device. Also, morphology results indicate slightly higher crystal quality for CdSe/Sb_2Se_3 thin film and the solar cells perform almost the same degradation results.

Keywords: Sb_2Se_3, buffer layer, thin film solar cells

1 Introduction

Outstanding optoelectronic properties of antimony selenide like favorable band gap, crystal structure, high absorption coefficient, good carrier mobility, low melting point, earth-abundant elements, and low toxicity have served it as a promising photovoltaic material [1]. Its one-dimensional crystal structure limits the charge carrier loss in grain boundaries and its band gap (1.2 eV) allows the theoretical efficiency of 30%. However, the highest efficiency that has been reported up to now is under 10% [2,3]. Therefore, achieving the 1D structure to improve the charge transport is a critical issue and the influence of the buffer layer in this aspect is fundamental for this kind of device [4]. Yan et. al [5] studied and compared CdS and CdSe buffers for Sb_2Se_3 thin film by close-spaced sublimation (CSS) deposition. They have reported that applying CdSe improves the stability of Sb_2Se_3 solar cells by Se diffusion. Also, we found from our previous study that co-selenization improves the crystal quality and enhances the efficiency by diffusion of excessive amount of selenium [6]. In this research, Sb_2Se_3 is grown on CdS and for the first time on CdSe buffer layer at low temperature by vacuum evaporation, in order to study the morphology and photovoltaic properties of these devices.

2 Experimental

Sb_2Se_3 solar cells are prepared in superstrate configuration -glass/SnO_2:F(FTO)/SnO_2(TO)/buffer layer/Sb_2Se_3/Au- as shown in figure 1. Two different buffer layers of CdS and CdSe are deposited on FTO/TO coated glasses (TEC 12D). CdS thin film is grown by chemical bath deposition containing 0.03 M aqueous solution of cadmium acetate ($Cd(O_2CCH_3)_2(H_2O)_2$), 0.6 M aqueous solution of thiourea ($SC(NH_2)_2$) and 32 ml of ammonia (NH_3). The substrate is immersed in the chemical bath at a temperature of 60 °C for 16 min. After deposition thin film is cleaned by isopropyl alcohol for 1 min in an ultrasonic bath. CdSe layer is deposited by thermal evaporation at a substrate temperature of 340 °C in a pressure of $1*10^{-5}$ mbar. Different thicknesses of CdSe have been studied and the best results are available for a thickness of 100 nm. Sb_2Se_3 is evaporated from a graphite crucible at a temperature of 500-600 °C at a pressure of

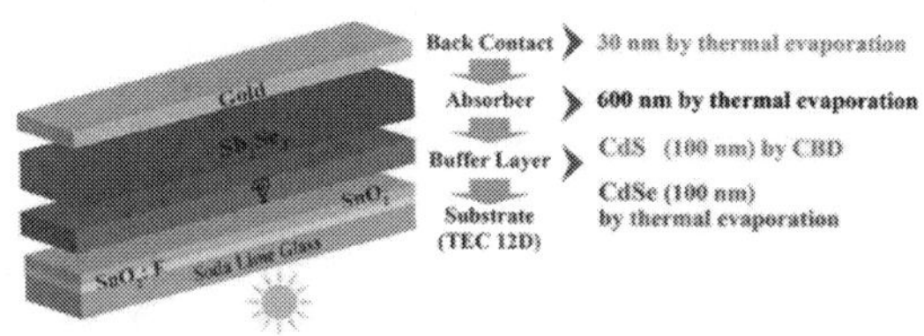

Figure 1: Schematic of Sb_2Se_3 solar cells in superstrate configuration.

$3*10^{-6}$ mbar with an evaporation rate of 0.15 nm/sec. After some optimization, the thickness of the absorber was fixed at 600 nm. After deposition the stacks are annealed in vacuum at 350 °C for 30 min in the same chamber. For back contact 30 nm of gold is thermally evaporated on top of Sb_2Se_3.

The morphology, crystal structure, stoichiometry of both buffers as well as of the absorber layer deposited on the different buffers are characterized. Analysis were carried out by NT-MDT Solver Pro atomic force microscopy (AFM) and by X-ray diffraction (XRD) with a Thermo ARL X'TRA powder diffractometer (in Bragg-Brentano geometry, equipped with a Cu-anode X-ray source (Kα, λ=1.5418 Å) and a Peltier Si (Li) cooled solid state

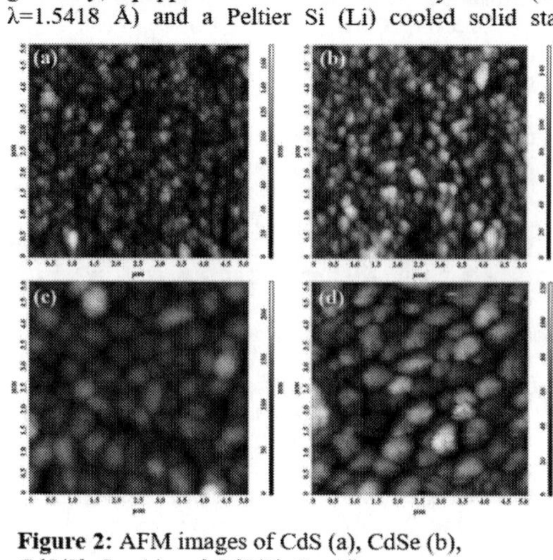

Figure 2: AFM images of CdS (a), CdSe (b), CdS/Sb_2Se_3 (c) and CdS/Sb_2Se_3 (d).

detector). Low-frequency micro-Raman experiments were conducted on Sb_2Se_3 samples by means of He–Ne laser (λ_{exc} = 633 nm) coupled to a single-stage Raman spectrometer (Thermo Scientific DXR2) and a charge-coupled device, thermoelectrically cooled.

3 Results

Figure 2 illustrates AFM images of the buffer layers and of Sb_2Se_3 on these two layers. The analysis of CdS and CdSe shows the same root mean square (RMS) roughness of 22 nm and a slight bigger particle size for CdSe. On the other hand the RMS roughness of Sb_2Se_3 thin film is 25 nm and 18 nm on CdS and CdSe layers, respectively. Moreover, the absorber grows columnar on both buffer layers with a slightly bigger grain size on CdSe.

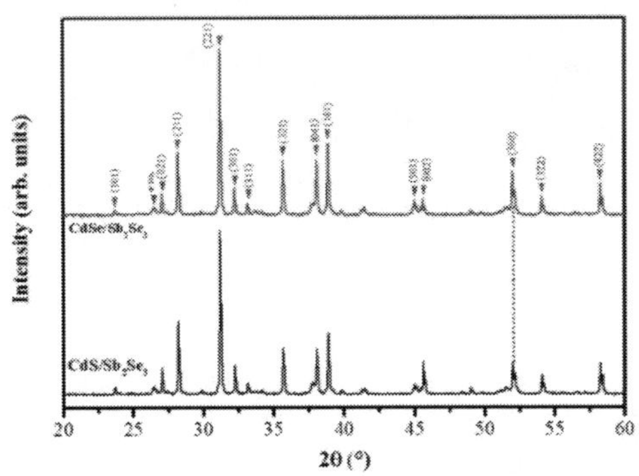

Figure 3: X-ray pattern of CdS/Sb_2Se_3 and $CdSe/Sb_2Se_3$ thin films.

As shortly mentioned in the introduction, Sb_2Se_3 is an anisotropic material with a ribbon-like structure where the electronic transport is strongly affected by the orientation of the ribbons. The predominance of the (hk1) orientation improves the carrier transport mechanism since it implies ribbons perpendicularly oriented on the substrate, while (hk0) peaks indicate that the ribbons are parallel to the substrate. The XRD pattern for Sb_2Se_3 thin films on CdS and on CdSe buffer layers is shown in figure 3. In both cases, all detected peaks belong to Sb_2Se_3 and are well matched with the JCPDS card n. PDF 15 0861. The ribbons of two Sb_2Se_3 thin films are mainly oriented along (hk1) directions, in particular (211), (221), (301), (321), (041), and (141). Rietveld refinement has been performed

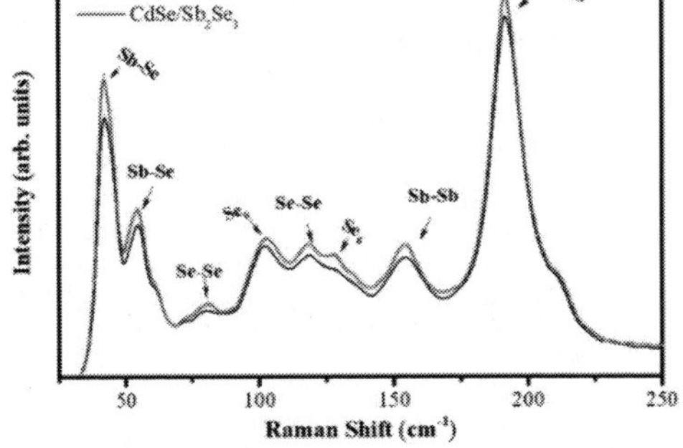

Figure 4: Raman spectra of CdS/Sb_2Se_3 and $CdSe/Sb_2Se_3$ thin films.

to calculate lattice parameters of these thin film. It indicates a modest increase in the C lattice parameter and size of the cell volume for $CdSe/Sb_2Se_3$ film. Considering the radius of S and Se, this could suggest that sulphur diffuses into the absorber reducing the volume of the crystal cell. On the other hand, the intensity of the

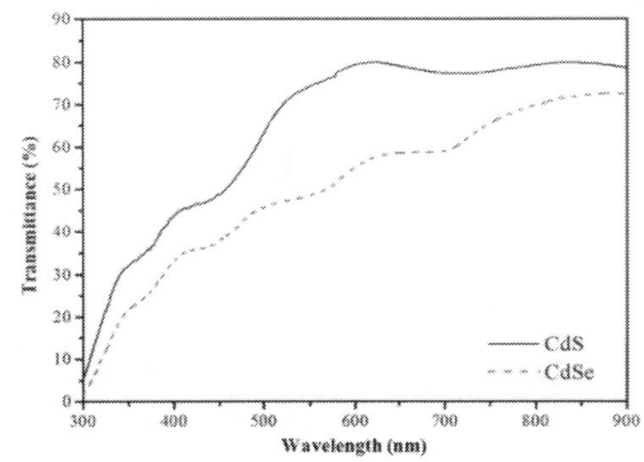

Figure 5: Transmittance spectra of CdS and CdSe layers.

undesired peak (360) is higher for Sb_2Se_3 on CdSe layer.

Raman spectra of Sb_2Se_3 thin films are illustrated in figure 4. The Raman peaks around 153 and 191 cm^{-1} have been attributed to the Sb_2Se_3 phase. These peaks show higher intensity for CdSe which is related to crystal quality. The sharper peaks of $CdSe/Sb_2Se_3$ at 103 and 129 cm^{-1} that are assigned to the Se_6 ring of rhombohedral Se, as well as the peaks at 82 and 118 cm^{-1} which are related to Se-Se bond demonstrate the presence of more selenium in the absorber. Also, broad peaks of CdS/Sb_2Se_3 imply a lower crystal quality that could suggest the presence of more defects in the crystal structure [7,8].

Figure 5 illustrates the transmittance spectra of CdS and CdSe layers: the CdSe layer shows lower transparency in the whole spectrum. This is in agreement with the band gap of CdSe (1.7 eV) which is narrower than the one of CdS (2.4 eV). So we would expect that solar cell with CdSe perform lower current than CdS case.

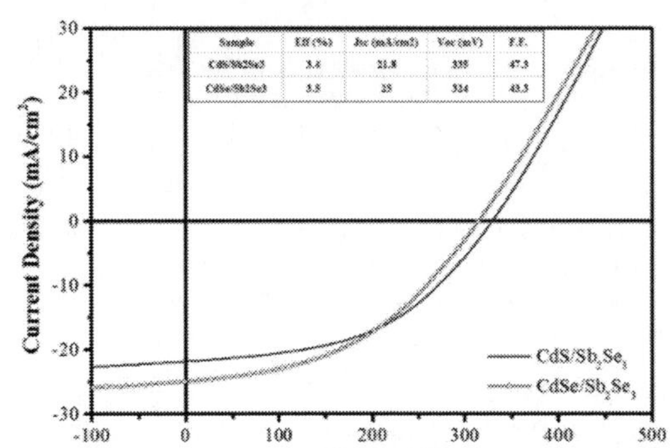

Figure 6: J-V curves of CdS/Sb_2Se_3 and $CdSe/Sb_2Se_3$ solar cells.

The J-V results of CdS/Sb_2Se_3 and $CdSe/Sb_2Se_3$ solar cells is illustrated in figure 6 and both devices perform almost the same efficiency. Despite of CdSe transmittance spectrum and band gap, $CdSe/Sb_2Se_3$ solar cell shows higher short circuit current (J_{sc}) of 25 mA/cm^2. Lower open circuit voltage (V_{oc}) of $CdSe/Sb_2Se_3$ solar cell could refer to the band alignment of CdSe and Sb_2Se_3 at the junction [6].

The EQE results are presented in figure 7, for CdSe device the response is higher in short wavelength region that is in consonance with the J_{sc} values. The response of EQE in this region is related to the defects density in thin film. Accordingly, this implies a higher defect density in CdS/Sb_2Se_3 which it might explain the reason for lower J_{sc}.

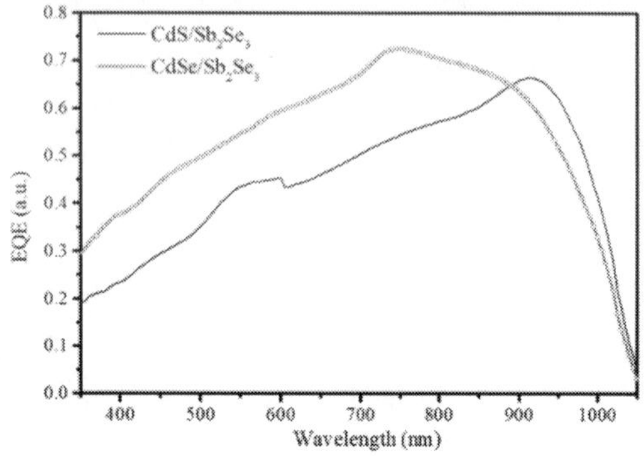

Figure 7: EQE results of CdS/Sb_2Se_3 and $CdSe/Sb_2Se_3$ solar cells.

Figure 8 illustrates the degradation study of the photovoltaic parameters of these devices, for reasons of space we present here only the graph referring to the efficiency values. The efficiency is slightly more stable for CdS/Sb_2Se_3 solar cell, though it is enhanced in the first hours under stress condition for CdSe case. The degradation of the fill factor is the reason for the efficiency behaviour in the $CdSe/Sb_2Se_3$ device; while J_{sc} and V_{oc} almost keep the initial values even after a long period of stress test.

4 Conclusion

As already mentioned, both devices have the same efficiency. However, $CdSe/Sb_2Se_3$ solar cells possess higher current density, despite the lower CdSe band gap and the larger lattice mismatch in this case. This can be attributable to a lower defect density of the absorber that is not affected by sulfur diffusion.

According to the degradation, devices with CdSe buffer layer shows a little lower stability than with CdS, in contrast to what has been reported by Guo et al [5]. In particular, Guo's devices fabricated with CdS showed a strong instability under accelerated lifetime test just after 120 hours (a reduction to 70% of the initial conversion efficiency), which is not observed in our case where the efficiency did not change. On the other hand for the CdSe case we observe, after 1000 hours, a reduction to 80% ± 10% while no change is reported by Guo et al (after 120 hours).

We believe that the different behaviour can be explained by the different absorber deposition technique. For Guo et al. a high substrate temperature CSS deposition is used and this enhances the sulfur diffusion from buffer to absorber. Instead our substrate temperature, during deposition, does not exceed 300 °C resulting in a limited intermixing the CdS and Sb_2Se_3.

In any case CdSe shows a higher current density and suggests an improved stoichiometry with possibly reduced defect density, resulting as a good alternative for Sb2Se3-based devices.

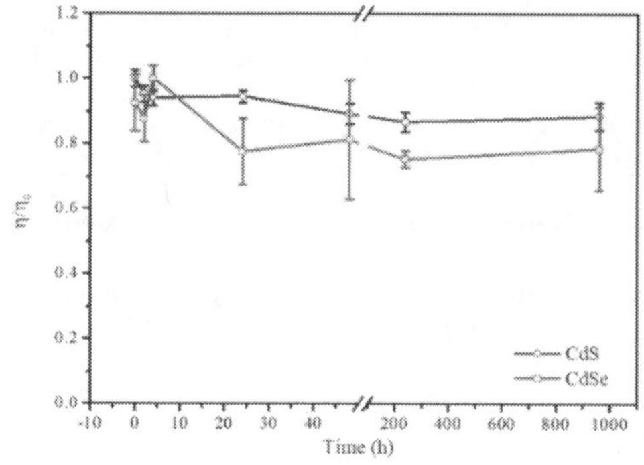

Figure 8: Degradation results of CdS/Sb_2Se_3 and $CdSe/Sb_2Se_3$ solar cells.

5 References

[1] A. Mavlonov, T. Razykov, F. Raziq, J. Gan, J. Chantana, Y. Kawano, T. Nishimura, H. Wei, A. Zakutayev, T. Minemoto, X. Zu, S. Li, L. Qiao, A review of Sb2Se3 photovoltaic absorber materials and thin-film solar cells, Sol. Energy. 201 (2020) 227–246. doi:10.1016/J.SOLENER.2020.03.009.

[2] Z. Li, X. Liang, G. Li, H. Liu, H. Zhang, J. Guo, J. Chen, K. Shen, X. San, W. Yu, R.E.I. Schropp, Y. Mai, (No Title), (n.d.). doi:10.1038/s41467-018-07903-6.

[3] W. Wang, Z. Cao, L. Wu, G. Chen, J. Ao, J. Luo, Y. Zhang, Interface Etching Leads to the Inversion of the Conduction Band Offset between the CdS/Sb 2 Se 3 Heterojunction and High-Efficient Sb 2 Se 3 Solar Cells, Cite This ACS Appl. Energy Mater. (2022). doi:10.1021/acsaem.1c04078.

[4] K. Zeng, D.-J. Xue, J. Tang, Antimony selenide thin-film solar cells, Semicond. Sci. Technol. 31 (2016) 063001. doi:10.1088/0268-1242/31/6/063001.

[5] L. Guo, C. Grice, B. Zhang, S. Xing, L. Li, X. Qian, F. Yan, Improved stability and efficiency of CdSe/Sb2Se3 thin-film solar cells, Sol. Energy. 188 (2019) 586–592. doi:10.1016/J.SOLENER.2019.06.042.

[6] V. Kumar, E. Artegiani, P. Punathil, M. Bertoncello, M. Meneghini, F. Piccinelli, A. Romeo, Analysis of Se Co-evaporation and Post-selenization for Sb 2 Se 3-Based Solar Cells, (2021). doi:10.1021/acsaem.1c02301.

[7] I. Efthimiopoulos, C. Buchan, Y. Wang, Structural properties of Sb2S3 under pressure: Evidence of an electronic topological transition, Sci. Rep. 6 (2016). doi:10.1038/srep24246.

[8] A. Kumar, V. Kumar, A. Romeo, C. Wiemer, G. Mariotto, Raman Spectroscopy and In Situ XRD Probing of the Thermal Decomposition of Sb 2 Se 3 Thin Films, J. Phys. Chem. C. 125 (2021) 2022. doi:10.1021/acs.jpcc.1c05047.

EVALUATION OF HNO3 DOPING IN GRAPHENE/SILICON SOLAR CELLS BY MEANS OF CAPACITANCE - VOLTAGE CURVES AND IMPEDANCE SPECTROSCOPY

Ilaria Matacena[1], Laura Lancellotti[2], Pierluigi Guerriero[1], Eugenia Bobeico[2], Nicola Lisi[3], Paola Delli Veneri[2], Santolo Daliento[1],
[1]University of Naples Federico II, Naples, Italy,
[2]ENEA Portici Research Center, Portici, Italy,
[3]ENEA Casaccia Research Center, Rome, Italy

ABSTRACT: The aim of this work is the study of beneficial effects of HNO_3 doping for G/Si SJSCs by means of capacitance versus voltage (C-V) curves and impedance spectroscopy characterizations. Capacitance curves in forward bias allowed to characterize defects distribution at graphene/silicon interface, proving induced defect passivation of molecular doping. Impedance spectra both at reverse and forward bias enforced achieved results. Degradation after storing the solar cell in air was also monitored using the same techniques. AC analyses performed were confirmed by dark current-voltage (I-V) curves collected before doping, after doping and 2 weeks after doping storing cell in air.

Keywords: graphene, capacitance, Schottky.

1 INTRODUCTION

Schottky junction solar cells (SJSC) have been extensively studied in literature in last decades. Among these devices, SJSC based on graphene/silicon Schottky junction have gained deep interest due to the twofold nature of graphene [1]-[2] which can be used as top transparent layer and as active layer for electron-hole separation. The main shortcoming is that the graphene/silicon interface is highly affected by recombination of charge carriers which strongly influences electrical performance. The strategies to reduce recombination dwell on the addition of an interfacial layer [3] or on treatments [4] aiming at interface defects saturation. Amongst the latter, treatments of graphene film with volatile oxidants, such as HNO_3, HCl, H_2O_2, $SOCl_2$, have been demonstrated to ameliorate the efficiency of graphene/silicon solar cells [5]. These treatments not only dope graphene but also passivate interface defects. Doping effects have been demonstrated through Raman spectra [6] while interface defects passivation has never been proved.

In this work, the quality of graphene/silicon interface is analyzed in order to demonstrate the effectiveness of HNO_3 doping in G/Si SJSC. The approach based on C-V curves analysis in forward bias operating condition before and after HNO_3 exposure resulted particularly suitable. Actually, the capacitance measured at the device terminals can be directly associated to junction properties and possible unexpected capacitance behavior, such as double peaks occurrence, can be related to interface defects and, hence, recombination. Further knowledge has been acquired by studying Nyquist plots collected before and after doping. Results coming from these techniques are supported by dark I-V characteristics. Even though, the effectiveness of molecular doping in G/Si SJSC was proved by experimental analysis, this treatment showed time variation. Solar cell degradation after 2 weeks storage in air has also been monitored, through C-V curves, IS analysis and dark I-V characterizations. In Section 2, details about the fabrication of solar cell under test are reported, together with the experimental setup and results. Conclusions are then drawn in Section 3.

2 EXPERIMENTAL CASE OF STUDY

The solar cell under test consists of few graphene layers (FLG) grown through CVD and transferred on a n-type commercial silicon substrate (1 Ω cm), Figure 1. Front contacts are made by graphitic glue, while back contact is formed by evaporated Al.

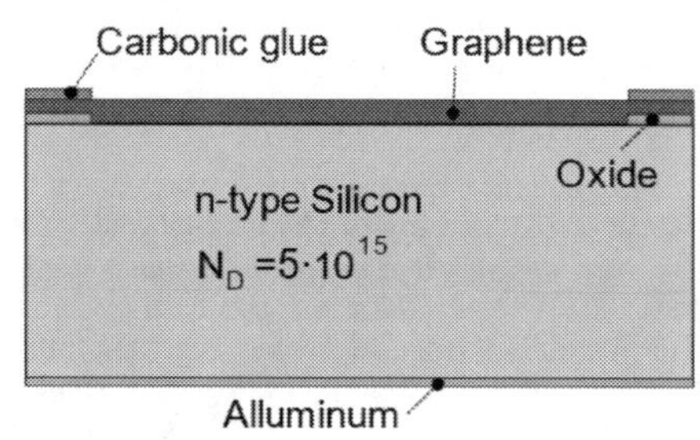

Figure 1: Structure of the G/Si solar cell under test.

Capacitance versus voltage (C-V) curves are obtained using Solartron 1260 Impedance Analyzer. Experimental results for f_1 = 5 kHz, f_2 = 6 kHz and f_3 = 7 kHz are presented in Figure 2. In this figure, capacitance (full line) exhibits a lower peak for a voltage close to 0.2 V and a dominant peak for a voltage near to 0.7 V. This behavior is unexpected since the capacitance curve of a such structure, with a single active junction, should display single peak. In fact, multiple peaks in capacitance behavior can be ascribable *i)* to the existence of multiple junctions or *ii)* to non-uniform properties along the interface [7]. Since only one junction is embedded in the solar cell under test, multiple peaks in capacitance are explained with an uneven defect distribution at the active interface. It should be noted that defectivity of G/Si interface notably affects the overall solar cell performance. This drawback of this technology is solved through molecular doping. The device under test was doped by exposing the top side, the graphene side, of the cell to HNO_3 (65%) acid vapors for 5 minutes after heating the cell at 180 °C for 6 minutes. Observing capacitance after doping (dashed line in Figure 2), it is clear that the dominant peak at higher voltage completely disappears while the first peak is mainly unaffected. This capacitance behavior is evidence of the capability of HNO_3 doping in passivating interface.

Subsequently the sample was stored in air for 2 weeks and C-V curves were repeated. Capacitance curves measured after 2 weeks (markers in Figure 2) showed that the second peak appears again, confirming the degradation of HNO_3 doped G/Si SJSC stored in air (nota bene in absence of encapsulation).

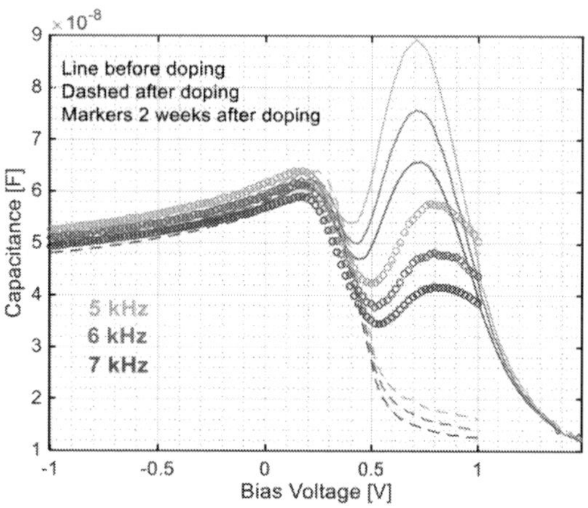

Figure 2: C-V curves of the cell of Fig.1 before doping (full lines), after doping (dashed lines) and 2 weeks after doping and storage in air (markers). Curves are collected at 5 kHz (yellow traces), 6 kHz (orange traces) and 7 kHz (blue traces).

Impedance spectroscopy is carried out at the same steps of C-V curves. Results for a DC bias of -0.5 V and 0.7 V are shown, respectively, in Figure 3 and Figure 4, where the effect of HNO_3 doping is clearly visible as a significant reduction of the radius the of Nyquist plot after doping strongly reduces. Taking into account the data obtained 2 weeks after doping, the beneficial effect of doping partially fades. Thus, the Nyquist plots tends to recover the pristine shape. This effect is more evident at V_{DC} = 0.7 V, while it seems that reverse bias behavior of the solar cell after doping (either just after or 2 weeks after) experiences a poorer variation. This confirms results coming from C-V analysis where the capacitance curves substantially change only at high forward bias.

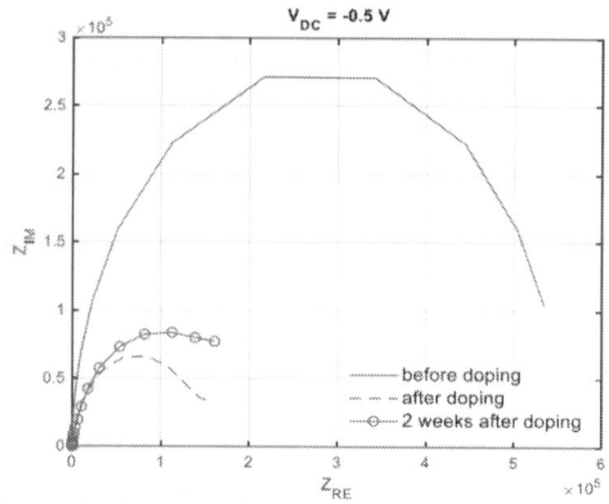

Figure 3: Nyquist plot for a DC bias V_{DC} = -0.5 V, before doping (full line), after doping (dashed line) and 2 weeks after doping and storage in air (markers).

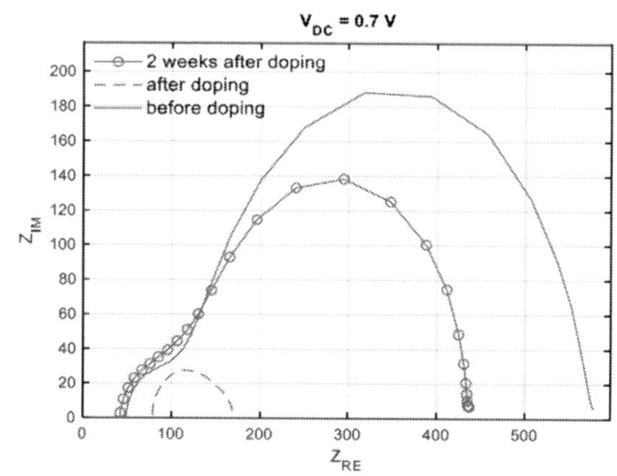

Figure 4: Nyquist plot obtained for a DC bias V_{DC} = 0.7 V, before doping (full line), after doping (dashed line) and 2 weeks after doping and storage in air (markers).

In order to enforce these results, current-voltage (I-V) characterization in dark condition are carried out. In Figure 5, it is shown that doping has effect on two key parameters of the solar cells. On one side it reduces reverse current; on the other side it increases the equivalent shunt resistance which is visible as a lower slope near origin.

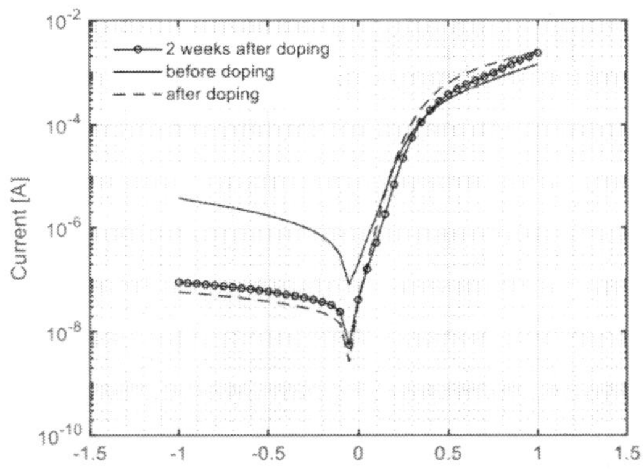

Figure 5: Dark I-V characteristics obtained before doping (full line), after doping (dashed line) and 2 weeks after doping and storage in air (markers).

3 CONCLUSIONS

In conclusion, this work validates the effectiveness of nitric acid molecular doping to passivate defects at G/Si interface in SJSCs. Nevertheless, there is a critical aspect of the implementation of HNO_3 treatment related to partial fading of beneficial effects when the samples are stored in air.

The innovation of this work consists in the validation of nitric acid molecular doping to passivate defects at G/Si interface in SJSCs, thus enabling further progress in such promising technique. Furthermore, the adopted technique based on the analysis of the forward bias capacitance and impedance spectroscopy is proposed as an effective approach in interface characterization to be applied in solar cells and, more in general, in semiconductor devices. Moreover, the work highlights a critical aspect of the implementation of HNO_3 treatment which is related to undesired solar cells performance degradation when the samples are stored in air.

4 REFERENCES

[1]. Zhu, Y., Murali, S., Cai, W., Li, X., Suk, J.W., Potts, J.R., Ruoff, R.S., 2010. Graphene and graphene oxide: synthesis, properties, and applications. Adv. Mater. 22 (35), 3906–3924.

[2]. Neto, A.H Castro, et al., 2009. The electronic properties of grapheme. Rev. Modern Phys. 81 (1), 109. Ocak, Y.S., Kulakci,

[3]. Gnisci, A., Faggio, G., Lancellotti, L., Messina, G., Carotenuto, R., Bobeico, E., Delli Veneri, P., Capasso, A., Dikonimos, T., Lisi, N., 2019. The Role of Graphene-Based Derivative as Interfacial Layer in Graphene/n-Si Schottky Barrier Solar Cells. Physica Status Solidi (A) 216 (3), 1800555.

[4]. Cui, Tongxiang, Lv, Ruitao, Huang, Zheng-Hong, Chen, Shuxiao, Zhang, Zexia, Gan, Xin, Jia, Yi, Li, Xinming, Wang, Kunlin, Wu, Dehai, Kang, Feiyu, 2013. Enhanced efficiency of graphene/silicon heterojunction solar cells by molecular doping. J. Mater. Chem. A 1 (18), 5736.

[5]. Chandramohan, S., Durairaj, S., Seo, T.H., Ryu, B.D., Hong, C.-H., 2021. Observation of dopant-dependent efficiency in chemically doped graphene/silicon solar cells and prospects for MoOx to overcome the stability and efficiency limits. J. Appl. Phys. 129 (1), 013101.

[6]. Lancellotti, L., Bobeico, E., Castaldo, A., Veneri, P.D., Lago, E., Lisi, N., 2017. Doping of multilayer graphene for silicon based solar cells. In: 2017 6th International Conference on Clean Electrical Power (ICCEP). IEEE, pp. 509–512

[7] Matacena, Ilaria, et al. "Forward bias capacitance investigation as a powerful tool to monitor graphene/silicon interfaces." Solar Energy 226 (2021): 1-8.

This is a short version of the paper submitted to the journal EPJ Photovoltaics.

It has been selected by the Committee of the WCPEC-8 for submission to EPJ Photovoltaics.

LIGHT SEPARATION FOR THE LATERAL SYSTEM OF LUMINESCENT SOLAR CONCENTRATORS

Mykola R. Kulish[1], Anatoly V. Sachenko[1], Vitaliy P. Kostylyov[1], Igor O. Sokolovskyi[1], and Anatoli I. Shkrebtii*[2]

[1] V. Lashkaryov Institute of Semiconductor Physics (ISP), NAS of Ukraine, 45 Pr. Nauky, Kyiv 03028, Ukraine

[2] Ontario Tech University, 2000 Simcoe St. N., Oshawa, ON, L1G 0C5, Canada

*Corresponding author: anatoli.chkrebtii@ontariotechu.ca

ABSTRACT: Luminescent solar concentrator (LSC) is a device with waveguide (or set of the waveguides) to transport sunlight to a photovoltaic cell attachment, where electricity is generated. We offer a new LSC design with the spectrum splitting on lateral plates. Such design substantially decreases the reflection losses, even though the escape cone losses increase due to absence of sequential utilization of descending photons. Ultimately, the lateral LSC can be substantially more efficient compared to the stacked LSC system. The efficiency of the lateral LSC design is determined by the dispersion system. In this research, we analyse four cases, namely: (i) Ideal solar spectrum separation without losses, considered as the limiting case. (ii) Dispersive prism system with the modeled efficiency above 80%. (iii) The set of dispersive mirrors. The experimental efficiency for every mirror can exceed 90%, but the losses are accumulated as for the stack of the LSC plates. (iv) Diffraction gratings. They are highly efficient for polarized light, but not for the unpolarized solar spectrum in the wide wavelength range. As we established, the dispersive prism and mirrors systems demonstrate the photon transport efficiency exceeding 57% for high-refractive index matrices. Photoconversion efficiency was assessed based on silicon solar cells (SCs).

Keywords: Photoluminescence, Concentrators, Lateral

1 INTRODUCTION

A luminescent solar concentrator (LSC) main component is a luminophore doped, rectangular in this case plate made of transparent material. Luminophore absorbs the solar photons and radiates luminescent photons in a narrow band. Luminescent photons are next transported by the total internal reflection to the edges of the plate, where solar cells (SCs) convert the power into electricity.

Luminescent solar concentrators are recently attracting a lot of interest [1,2]. To maximize their photoconversion efficiency, photon management approach has to be implemented [2] to utilize the photoactive solar photons with minimal thermalization losses. The results for the photon transport in the stack of concentrator plates in a tandem LSCs were recently reported [3]. Further improvement of multicomponent LSCs requires optimization of the solar spectral energy redistribution between the plates.

LSC with the plates in the stack [3] is advantageous due to partial reuse of the luminescent photons, lost to the escape cone, but with detrimental reflection increase on the plates' boundaries. While the plates boundaries are optimized for the total internal reflection, the light transmission is not optimized, causing an increase in the reflection losses.

To achieve maximal photoconversion efficiency, a photovoltaic system should maximize the utilization of solar photons by minimizing the thermalization losses. The LSCs both concentrate the solar energy to increase the voltage and decrease the solar cells semiconductor material use. Fig. 1 illustrates the concept of the single plate LSC. In the tandem luminescent solar concentrator [1,3], the plates (waveguides) are stacked on top of each other, similar to the tandem solar cells [2]. On the other hand, an alternative and natural way of the solar photon management in LSC is to split the solar spectrum using several laterally placed elements, and each element utilizes its part of the spectrum.

The results on the photon transport efficiency for the stack of concentrator plates in the tandem LSC were recently reported [3]. Fig. 2 demonstrate the light separation concept for the vertical tandem LSC design. In this research, we offer a new concept of parallelly (laterally) arranged concentrator plates and evaluated the theoretical maximal photons transport efficiency for such luminescent concentrator systems. We focused here mainly on the photon transport, even though the resultant photoconversion efficiency also depends on the luminescence efficiency, reabsorption and the efficiency of the photovoltaic solar cells (SCs), attached to the edges [3]. When the photon transport efficiency is known, the formalism allows straightforward extension to calculate the resultant photoconversion efficiency of the whole LSC device.

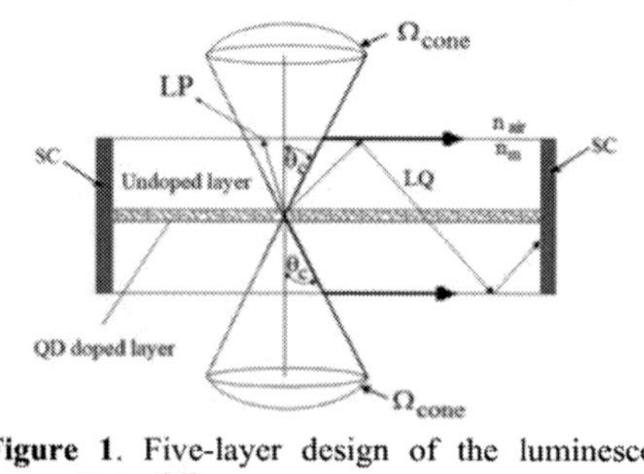

Figure 1. Five-layer design of the luminescent solar concentrator [4].

2 FORMULATION OF THE PROBLEM

Further improvement of multicomponent systems with luminescent solar concentrators requires optimization of the solar spectrum separation between the plates. It should be noted that an obvious tandem design with the plates in the stack (Fig. 2) provides both an advantage of partial reuse of luminescent photons lost to the escape cone and a disadvantage of sequential light reflection on the plates' boundary. As the boundaries are optimized for the total internal reflection, the light transmission in the stack is not optimized, leading to the cascade increase in the reflection losses.

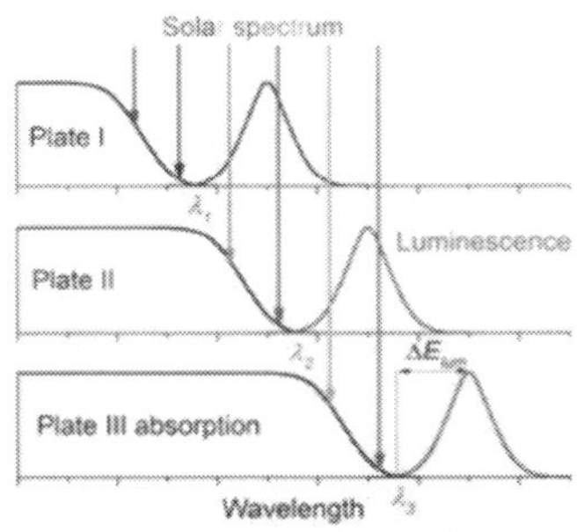

Figure 2. Spectrum separation for the stack of three concentrator's plates in tandem LSC.

On the other hand, the spectrum separation for the laterally exposed to the sunlight plates (see Fig. 3) increases escape cone losses since there is no serial utilization of downward moving photons. The efficiency of this design is determined by the properties of the dispersion system.

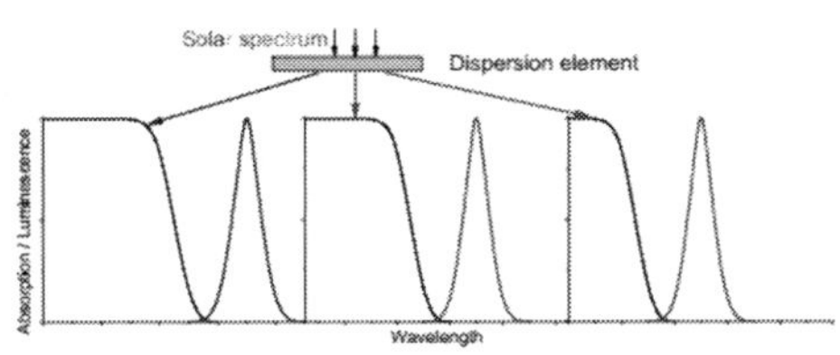

Figure 3. Solar spectrum separation in the concentrator's plates of the three-plate lateral LSC. Blue, green and red Gaussian-like curves are the emission spectra of the luminophores and the black line is the luminophores' absorption.

Power density of the AM1.5 solar spectrum in three-plate concentrator is demonstrated in Fig. 4.

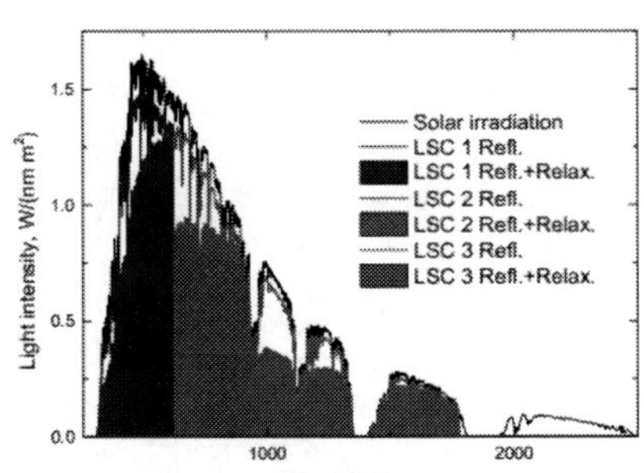

Figure 4. Power density of the AM1.5 solar spectrum in a wide wavelength range (black line), the density entering the concentrator plates after reflection ("Refl.", blue, green and red lines) and with thermalization ("Refl. + Relax.", blue, green and red filled area) for the ideal lateral three-plate luminescent solar concentrator.

Two promising designs with the sunlight separation by dispersive mirrors and dispersive prisms are shown in Fig. 5, while Fig. 6 shows diffraction gratings systems. In additions to the designs in Figs. 5 and 6, in this research, we study four cases in total.

1. Ideal spectrum separation without losses. This is referred as the limiting case.
2. Dispersive prism system, Fig. 5. A photon transport efficiency exceeding 80% was reported for such a system ([2], 83.5%, modelling).
3. The set of dispersive mirrors, Fig. 4. The efficiency for every mirror can exceed 90% ([3], experimental), but the losses for the system are accumulated in the same way as for the stack of the tandem LSC plates. Here we used 95% efficiency for the light angle of incidence, close to 45°.
4. Diffraction gratings. They are highly efficient for polarized light, but their efficiency is low for unpolarized solar light in the wide wavelength range [5]. Therefore 50% efficiency is considered.

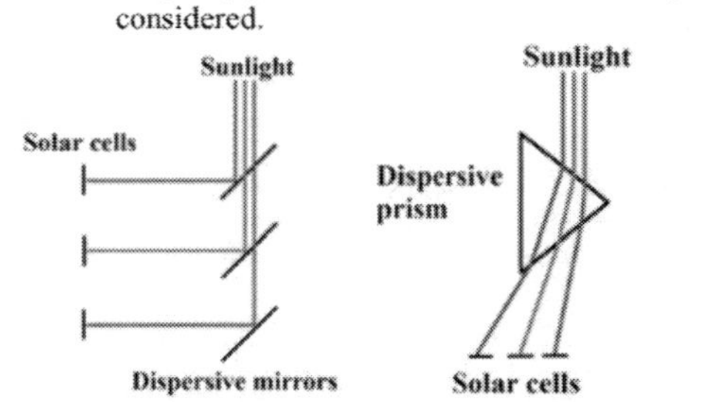

Figure 5. Sunlight separation by dispersive mirrors (at the left) and by dispersive prism (at the right).

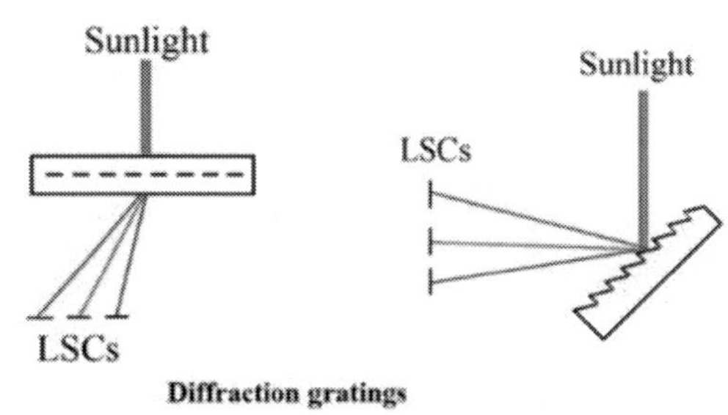

Figure 6. Sunlight separation by diffraction grating.

Next, we discuss the efficiencies of the three designs (shown in Figs. 5 and 6) and the ideal reference case.

3 RESULTS

3.1 Optical efficiency

We modified the formalism for the tandem LSC, developed in [3], to model the lateral LSC design. In all the considered cases, except the system of dispersive mirrors, a single-reflection model was considered.

We have proven that the LSC designs with dispersive prism and mirrors demonstrate the photon transport efficiency exceeding 57% for high-refractive index matrices (such as, *e.g.*, N-LASF9 glass). The ideal limiting efficiency for such design is close to 70%, higher than the maximum efficiency of about 60% for stacked fused silica plates [3]. On the other hands, the photon transport efficiency for diffraction grating is below 35% (because of 50% losses in the diffraction grating).

Fig. 7 shows the transport efficiency for all the cases considered. The system of dispersive mirrors demonstrates the best efficiency for low numbers of the plates, aligning with dispersive prism efficiency for the higher plate numbers due to the serial increase in the reflection losses. The system of diffraction gratings has low efficiency due to the limitations described above.

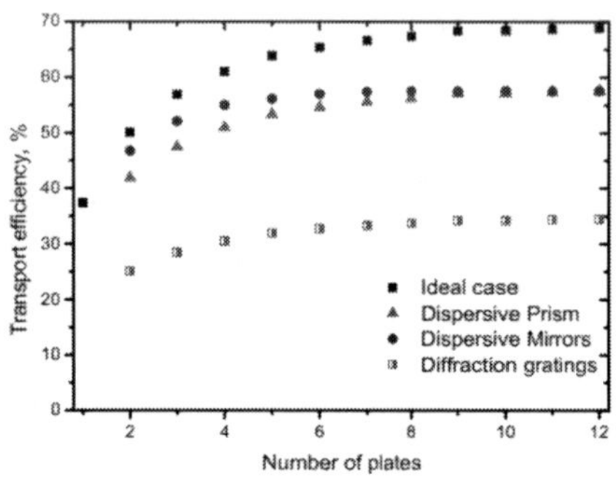

Figure 7. Photon transport efficiency vs. number of the plates: the ideal limiting case (black squares); LSCs with dispersive mirrors (blue circles); dispersive prisms (green triangles) and gratings (red squares).

It should be noted that for the LSC plates made from materials with lower refractive index (such as most of the polymers) all the efficiencies considerably decrease when compared to the plates with the optical materials with high refraction (*e.g.,* common fused silica or N-LASF9 glass). Vice versa, for the materials with higher refractive index the efficiencies increase since the detrimental plate reflection losses accumulation is not present.

3.2 Electrical efficiency

To estimate the overall efficiency of the LSC system, one needs to calibrate optical efficiency to the efficiency of solar cells on the edges of the plates. Using known 23% efficient silicon SC [6] as a sample and applying our [7] modified approach, we've got 46.9 % efficiency for monochromatic 100 mW/cm^2 illumination. The optimal wavelength for this case is 993 nm. As we mentioned, the luminescence wavelength of QD can be varied by the choice of the core material and its size. SC efficiency can be modified by the optimization of illumination, thickness, doping etc. One can wait for an analogous situation for other SC materials.

Therefore, electrical efficiency for the systems with different dispersion elements and 50% photoconversion efficiency of SCs is shown in Fig. 8.

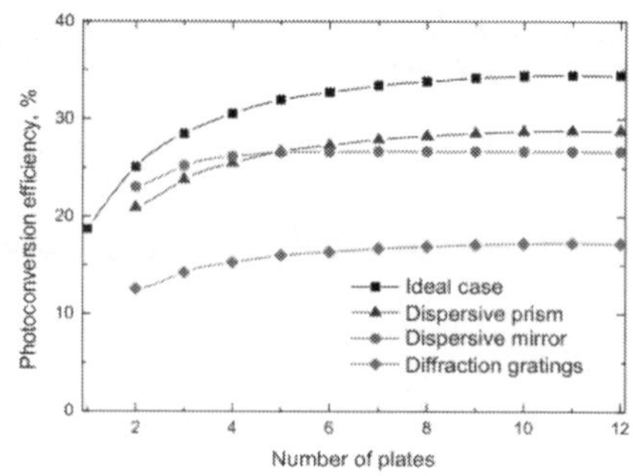

Figure 8. Total photoconversion efficiency vs. number of the plates.

4 CONCLUSIONS

We offer the lateral design of the tandem luminescent solar concentrator (LSC) and developed a formalism to calculate the photon transport efficiency. We proved that the lateral design of the luminescent solar concentrator potentially provides higher transport efficiency, close to 70%, as compared to about 60% for the stacked LSC plates [3]. The efficient lateral LSC structure requires the material with the highest refractive index, which decreases the escape cone losses. Practical use of this structure, however, might be challenging due to complexity of the design.

Overall photoconversion efficiency of the system was also estimated using silicon SC efficiency under monochromatic illumination as a reference.

Detailed discussion and numerical formalism used has been presented in [8]

5 REFERENCES

[1]. K. Wu, H. Li and V. I. Klimov, Nature Photonics **12**, pp.105–110 (2018).
[2]. C. Chen, S. Zheng, and H. Song, Chem. Soc. Rev., **50**, pp. 7250-7329 (2021).
[3]. I.O. Sokolovskyi, et al., 38th EU PVSEC, 1BV.3.2, pp. 45-48 (2021).
[4]. M. Kulish, A. Shkrebtii, V. Kostylyov, I. Sokolovskyi and A. Sachenko, 2021 IEEE 48th Photovoltaic Specialists Conference (PVSC), 2021, pp. 0687-0692, doi: 10.1109/PVSC43889.2021.9518629.
[5]. M. Stefancich, A. Zayan, M. Chiesa, et al., Opt. Express **20**, 9004-9018 (2012).
[6]. H. Sai, T. Oku, Y. Sato, M. Tanabe, T. Matsui, K. Matsubara, Prog Photovolt Res Appl. 2019; 27: 1061–1070. https://doi.org/10.1002/pip.3181
[7]. A. V. Sachenko et al., 2021 IEEE 48th Photovoltaic Specialists Conference (PVSC), 2021, pp. 0544-0550, doi: 10.1109/PVSC43889.2021.9518764.
[8]. M. R. Kulish, A. V. Sachenko, V. P. Kostylyov, I.O. Sokolovskyi, and A. I. Shkrebtii. EPJ, submitted.

CHARACTERIZATION OF HIGHLY EFFICIENT SILICON-BASED SOLAR CELLS WITH FLAT AND TEXTURED SURFACES

A.V. Sachenko[1], V.P. Kostylyov[1], V.M. Vlasiuk[1], G. Perinparajah[3], I.O. Sokolovskyi[1], M. Evstigneev[2], and A.I. Shkrebtii[3*]
[1] V. Lashkaryov Institute of Semiconductor Physics, NAS of Ukraine, 03028 Kyiv, Ukraine
[2] Department of Physics and Physical Oceanography, Memorial University of Newfoundland, St. John's, NL, A1B 3X7, Canada
[3] Ontario Tech University, Oshawa, ON, L1G 0C5, Canada

*Corresponding author: anatoli.chkrebtii@ontariotechu.ca

ABSTRACT: Since the photoconversion efficiency η of the silicon-based solar cells (SCs) is approaching their theoretical fundamental limit close to 29%, further improvement of their performance requires theoretical modeling and/or numerical simulation to optimize the SCs parameters and their design. For several types of high-performance silicon-based SCs with both flat and textured top surfaces, we (i) analyzed experimentally key characteristics, namely external quantum efficiency (*EQE*), short-circuit current density $J_L(d)$, open-circuit voltage V_{OC}, *I-V* curves and the efficiency η; (ii) modeled these characteristics by a semi-analytical approach, developed by the authors, and (iii) numerically simulated using Quokka, SunSolve and other dedicated software. Analyzing and comparing the combined experimental, modelled and simulated results provide valuable information to further improve the design and enhance the SCs performance. The experimental $EQE(\lambda)$ curves for high-efficiency silicon planar and textured SCs have been analyzed in terms of the base thickness dependent short-circuit current density J_L, calculated using the empirical formula for $EQE(\lambda)$, and the same parameters were simulated using Quokka 3. The theoretical and modeling results, dependent on the SCs type and structure, recombination parameters, radiation intensity, *etc*., agree well with the experiment for the considered IBC type flat and textured SCs.
Keywords: silicon, solar cell, texture, efficiency, recombinations, optimization.

1 INTRODUCTION

Crystalline silicon (c-Si) remains a dominant photovoltaics material in solar cell industry. Currently, scientific and technological advances make possible producing the c-Si solar cells (SCs) with the efficiency of 26.7% (see, e.g., [1-3]), below the fundamental limit only by two percent. This record efficiency of the SCs in the lab has been achieved in 2018 using the heterojunction intrinsic thin layer on a thin *a*-Si passivating layers and interdigitated back contact (IBC) design on *n*-type silicon [1]. Also, the efficiency of the wafer-based silicon modules has reached 24.5% [1]. Finally, the commercial SC module efficiency with *n*-type Si interdigitated back contact technology is projected over 21.8%, while the average module efficiencies of 17.4%, 18.4%, and 19.4% are projected for the conventional *p*-type multi-, mono-, and mono-passivated emitter and rear cell (PERC), respectively [4]. Different design of the Si based SCs [5] have also been offered to produce highly efficient devices. In such case, to progress in the SCs field use of the theoretical and numerical simulation tools become ubiquitous [2,3,6,7]. Such tools allow to optimize the SCs under consideration by combining the experiment and the theory. To demonstrate the significance and practicality of such approach, in this work, we first analyzed the experimentally measured SCs key characteristics, namely external quantum efficiency (*EQE*), short-circuit current density $J_L(d)$, open-circuit voltage V_{OC}, and the photoconversion efficiency η for high-performance silicon-based SCs with both flat and textured top surfaces. Next, we (i) modeled these characteristics by a semi-analytical approach, developed by the authors, [6,8] and (ii) numerically simulated the SCs, using Quokka, SunSolve and other dedicated solar cell software, see, *e.g.*, [7]. We demonstrated that analyzing and comparing the combined experimental, modelled and simulated results provide valuable information to further improve the design and enhance the SCs performance.

We focused in the paper on a comprehensive characterization of three IBC type [5,9] solar cells, both flat and textured. Compared to the planar SC design [10], the texturized surfaces provide the light-trapping interfaces to reduce optical loss and thus increase the efficiency of solar cells [11]. This enhancement can be attributed to the decrease of the front surface reflectance for short-wavelength range and the increase of the light path for long-wavelength range. We considered namely (i) planar 19% efficient SC [10], and textured both (ii) a commercial SunPower 21% efficient SC [9,12], as well as (iii) the efficiency record breaking 26.7% efficient SC [1]. While we measured in the lab the SunPower SC key parameters for (ii), including *I-V* characteristics and $EQE(\lambda)$, we also used experimental characteristics given in the corresponding research papers, namely, [10] for the flat SC, case (i), and [1] for the record efficient solar cell, case (iii).

The theoretical modeling and numerical simulations were aimed to find out the SCs optical characteristics, namely the internal and external quantum yields (efficiencies) as well as J_L(d), V_{OC}, and η. The *EQE* for SCs with flat and textured surfaces was calculated using the formalism from [13], while for the key photoelectric parameters we followed [8]. The experimental $EQE(\lambda)$ curves for high-efficiency silicon planar SCs [10] have been analyzed in [6].

Using proposed in [6] empirical formula for $EQE(\lambda)$, the dependence of the short-circuit current density J_L on the base thickness d for the textured SCs was calculated. Next, the same parameters were simulated using Quokka 3 software with the input, dependent on the SCs type and structure, recombination parameters, radiation intensity, *etc*. Good agreement, achieved for the flat and textured IBC type SCs considered, indicates that combining theory, simulations and experiment brings more comprehensive description of the solar cells under consideration.

2 TECHNICAL DETAILS

We characterize in this paper the highly efficient crystalline Si-based solar cells (SCs). The experimental measurements of the key SC parameters have been complemented by the results from the theoretical semi-analytical formalism and computer simulations using a commercial software. This combined approach allowed us to both determine the SCs key characteristics and to further optimize the SCs performance.

The paper uses, firstly, a physics-based semi-analytical approach [8], to model the high-efficiency silicon SCs with both flat (planar) and textured surfaces. Secondly, the above results were compared to the numerical simulations, using several existing programs, including in particular, commercial Quokka 3 software [14] and SunSolve solar cell software, see, *e.g.*, [7,15].

We refer the readers to [8] for the details of the formalisms and experiment, outlining here only the main points. To calculate the photoconversion efficiency η of the silicon SCs considered, a self-consistent approach was used [8]. It is valid when (i) the diffusion length of the minority charge carriers L_d exceeds significantly the base thickness d; and (ii) the surface recombination rate S is small enough to guarantee the inequality $S << D_A/d$, where D_A is the ambipolar diffusion coefficient. All the above criteria are satisfied for the high efficiency SCs considered here. In this approach, the short-circuit current density $J_L(d)$ is calculated first using the empirical expressions for $EQE(\lambda)$. Then the open circuit voltage V_{OC}, light IV curves and $J_L(V)$ are calculated (here V is the applied voltage). At the maximum power point (MPP) the condition $d[J_L(V)\cdot V]/dV=0$ is used to determine the efficiency η and the IV curve filling factor (FF). Comparing the calculated V_{OC} and η with the experiment, the surface recombination rate S and the series resistance R_s are found.

3 THEORY AND SIMULATION FORMALISMS

In the semi-analytical formalism [8] (we refer this as theory), we included into consideration the Shockley-Read-Hall (SRH) recombination, surface recombination, radiative recombination, recombination in the space charge layer, nonradiative exciton Auger recombination via deep recombination center and interband Auger recombination. The surface recombination rate S dependence on the doping level is also considered and analyzed. Therefore, our semi-analytical formalism accounts for all the existing recombination mechanisms in silicon, as included in the general expression for the effective recombination lifetime, e.g., in Si with the n-type SCs base:

$$\tau_{eff}(n)=\left[\frac{1}{\tau_{SRH}(n)}+\frac{1}{\tau_{exc-Auger}(n)}+\frac{S_{S0}}{d}\left(1+\frac{\Delta n}{n_0}\right)^r+\frac{1}{\tau_r(n)}+\frac{1}{\tau_{Auger}(n)}+\frac{S_{SC}(n)}{d}\right]^{-1}, \qquad (1)$$

where $n=n_0+\Delta n$ is the total majority carrier concentration, consisting of the equilibrium, n_0, and excess Δn contributions; τ_{SRH} is Shockley-Read-Hall (SRH) lifetime; $S_{S0}\sim n_0$ is the net recombination velocity on the front and rear surfaces at low injection level, usually the dimensionless coefficient $r=1$; $\tau_{exc-Auger}(n)=\tau_{SRH}n_x/n_0$ is the exciton Auger recombination time with n_x=$8.2\cdot 10^{15}$ cm^{-3} [16]; τ_r is radiative recombination time, and τ_{Auger} is band-to-band Auger recombination time [17]. The SRH lifetime in an n-type base is given by:

$$\tau_{SRH}(n)\cong\frac{\tau_{p0}(n_0+\Delta n+n_1)+\tau_{n0}(p_1+\Delta n)}{n_0+\Delta n}, \qquad (2)$$

where $\tau_{p0}=\left(C_pN_t\right)^{-1}$ and $\tau_{n0}=(C_nN_t)^{-1}$ are electron and hole life-times, n_1 and p_1 are the equilibrium electron and hole concentrations when Fermi level coincides with the deep-impurity energy.

It is shown that for the complete characterization of the SCs under investigation the following key experimental data are required: $EQE(\lambda)$ dependencies, light current-voltage characteristics, dependences of the short-circuit current on the open-circuit voltage $J_L(V_{OC})$ in sufficiently wide range of V_{OC}.

For the numerical simulations we used a free Quokka2 software together with OPAL2 [18,19] to consider both flat and structured Si surfaces with and without antireflecting coating (ARC); and Quokka3 with SunSolve commercial solar cell software [7,15,19], (we refer this as simulation). Regarding the parameter choice for the semi-analytical formalism [8] compared to the simulations [14], one has to mention, that in the simulation the SC parameters set, described below, provided the best fit of the experimental measurements. However, unlike the formalism of [8], Quokka software (as the most today's simulations) does not consider non-radiative exciton recombination. As it was demonstrated in [8], a contribution of the non-radiative exciton recombination can be substantial. In our opinion, this contribution is masked by the SRH and surface recombination. Indeed, re-definition of the recombinations parameters can be done as: the modified SRH time τ^*_{SRH}=$\tau_{SRH}/(1+n_0/n_x)$ and the surface recombination velocity S^*=$S_0+(S_0+dn_0/(\tau_{SRH}\cdot n_x))\Delta n/n_0$. This, however, can lead to a wrong impression that introduction of the non-radiative exciton Auger recombination is unnecessary. Note that for $n_0=4.9\cdot 10^{15}$ cm^{-3}, the value of τ^*_{SRH} is smaller than τ_{SRH} by about 60%. It should also be noted that an expression for the effective lifetime in Si similar to the formula for τ^*_{SRH} was used in some previous works [20] with $n_x=7.1\cdot 10^{15}$ cm^{-3}. Therefore, the Quokka recombination parameters from the fit should not necessarily be the same as in the analytical modelling.

4 RESULTS

We compared the experimental results, measured in our lab for commercial SunPower SCs [9], as well as available from the literature, with those calculated by the semi-analytical formalism in [6,8] and Quokka simulations. Three types of IBC Si-based solar cells, namely (i) planar [10] and textured both (ii) commercial SunPower and (iii) the efficiency record breaking [1] SCs. For the numerical simulations as well as for the theoretical modeling, we analyzed the quantum efficiency and IV characteristics of both planar and textured SCs. The light reflection from the SCs surface was calculated using the Opal 2 Software, provided by PV Lighthouse [21]. The output was then used with Quokka 2. Using SunSolve, we determined the key optical characteristics of the solar cells, using the output file in Quokka 3 to find the I-V characteristics.

Here we first demonstrate application of the combined

theory/simulation approach using the textured commercial SunPower SCs. In this simulation, a 65 nm thick SiN_x antireflecting film and 20 nm thick SiO_2 film with a random upright pyramid structure with the base of 180 μm was considered on top of a flat silicon substrate. *IV* and External Quantum Efficiency simulations were carried out using the Quokka software [15].

Experimentally, we characterized the SunPower SCs with rear metallization, with a photoactive area of 154.7 cm^2, the base thickness $d = 165$ μm.

In the Quokka simulation we used the following bulk recombination parameters: the fixed bulk recombination time is 1250 μs, Auger recombination is "Richter 2012", as described in [17], radiative recombination time is $4.73\cdot10^{-15}\,\frac{cm^3}{s}$. The Shockley-Read-Hall defect has the density of $7\cdot10^{9}\,cm^{-3}$, with two defect level energies, one is at 0.21 eV, electron capture cross section equals $1.7\cdot10^{-16}$ cm^{-2}, hole capture cross section is $1.1\cdot10^{-14}$ cm^{-2}. The second SRH defect level at -0.01 eV has the electron capture cross section equals to $6.3\cdot10^{-15}$ cm^{-2}, hole capture cross section is $2.4\cdot10^{-15}$ cm^{-2}. This corresponds to the gold impurity. The simulation uses a 3 cm/s surface recombination velocity for both the front and rear part of the cell. Fig. 1 shows *EQE*(*λ*), calculated for a 165 μm thick 180 mm × 180 mm cell with a doping level of $9\cdot10^{14}$ cm^{-3}. To simulate the *IV* characteristics (Fig. 2), a shunt resistance of $1\cdot10^{7}$ $\Omega\cdot cm^2$ and series resistance of 0.25 $\Omega\cdot cm^2$ were used.

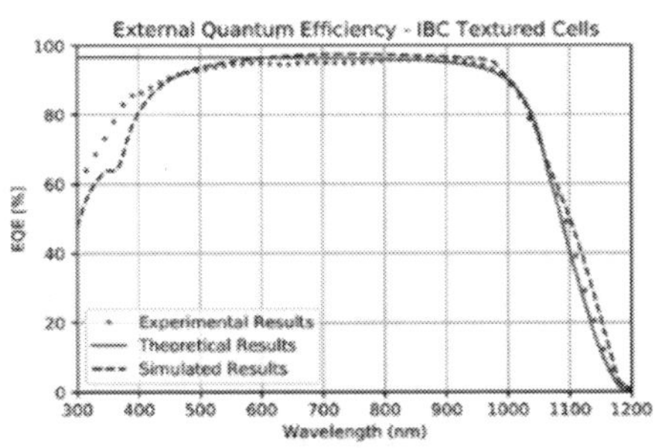

Figure 1. External Quantum efficiency *EQE* for the commercial 21.3% efficient textured Si-based SC from SunPower [9]. Experimental *EQE*(λ): red circles. Green solid line (theoretical results) shows *EQE*(λ) calculated using the empirical formula [6,8]. Blue-dashed line is the simulated $EQE(\lambda)$ using Quokka.

Very good agreement for the $EQE(\lambda)$ dependence has been achieved in the photoactive part of the solar spectrum for both the theoretical (using the empirical formula) and the simulated results. Even though the empirical formula does not describe accurately $EQE(\lambda)$ in the short-wavelength region of the solar spectrum, its application in our theoretical formalism leads to practically identical *IV* characteristics, see the next Figure 2. Therefore, we can provide consistently accurate SC efficiencies, the final goal for the combined experiment, theory and simulations.

Furthermore, even though our theoretical EQE modelling, being less accurate than numerical simulation, it gives an important thickness dependence of the absorption edge. When the short-wave part can be measured, thickness-dependent current can be modelled using our approach by combining experimental short-wavelength and theoretical long-wavelength parts.

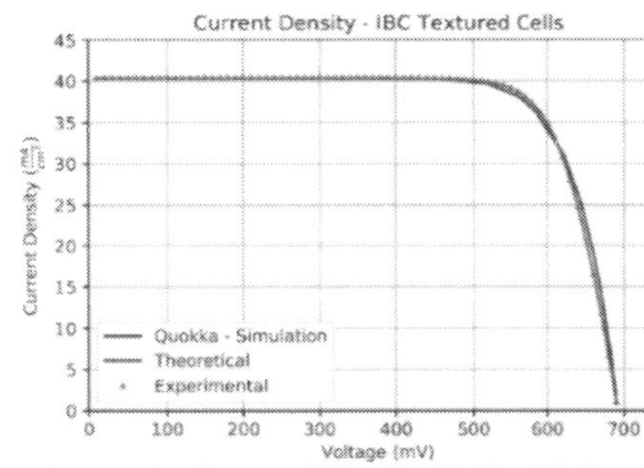

Figure 2. *IV* characteristics for 21.3% efficient SunPower SC [9]. Experiment: red circles. Green line is the *IV* from the theoretical modeling, blue-dashed line is the simulated *IV* curve.

Practically ideal fit of the experimental curve by both the theory and simulations has been achieved, indicating the accuracy of the theoretical formalism proposed.

Next, we characterized the record-breaking SC [1], using the experimental *EQE*(λ) curve, presented in the article. This experimental *EQE*(λ) curve is superimposed with those both theory and simulation in Fig. 3.

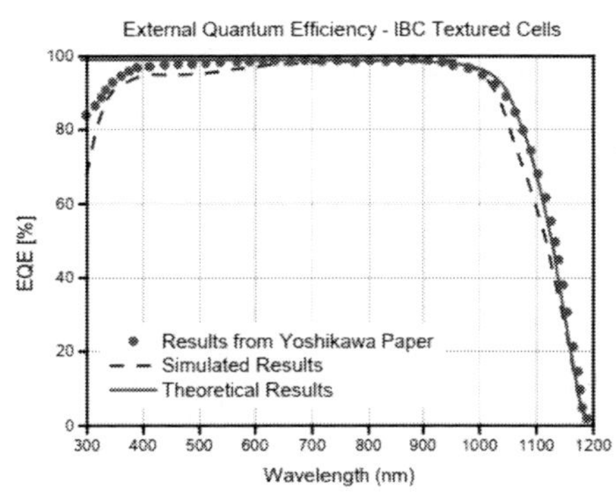

Figure 3. External Quantum efficiency *EQE* for the record-breaking 26.7% efficient textured Si-based SC [1]. Experimental *EQE*(λ): red circles. Green solid line (theoretical results) shows *EQE*(λ) calculated using the empirical formula [6,8]. Blue-dashed line is the simulated $EQE(\lambda)$ using Quokka.

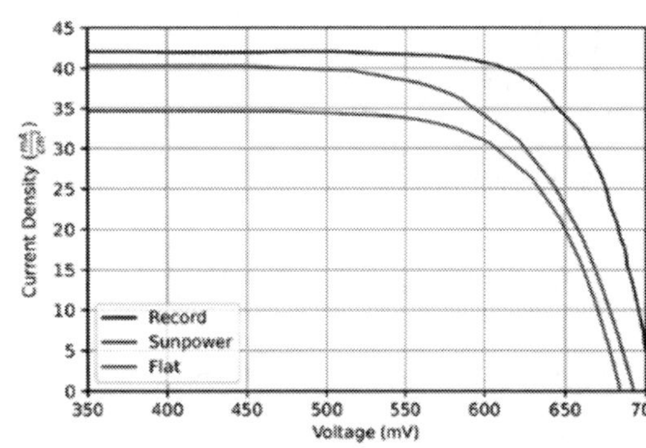

Figure 4. Calculated *IV* characteristics of the three SCs under consideration. Flat SC: red line. Green line is the *IV* from the textured SunPower SC, blue-dashed line is the *IV* curve for the record-breaking SC.

5 CONCLUSIONS

In conclusion, we analyzed and compared combined experimental, modelled and simulated results for the high-efficiency silicon-based SCs. Three IBC types of the SCs considered, including both planar and textured systems. Good agreement between the modeling, experiment and analytical formalism for the commercial SCs and from the research lab SCs brings a valuable information, required to further improve the SC design and enhance the SCs performance.

It is shown that in the cases considered, the experimental results agree well with the modeling and simulations. Such comprehensive combined characterization makes possible not only to further optimize the solar cells under consideration but to also determine the SCs parameters when fitting the experimental curves to the semi-analytical and simulated results.

6 ACKNOWLEDGEMENTS

This work was partially supported (V. Kostylyov and V. Vlasiuk) by National Research Foundation of Ukraine by the state budget finance (project 2020.02/0036 "Development of physical base of both acoustically controlled modification and machine learning–oriented characterization for silicon solar cells") and (G. Perinparajah and A. Shkrebtii) by Ontario Centres of Excellence (OCE) and IntriEnergy Canada Labs Inc.

7 REFERENCES

[1] K. Yamamoto, K. Yoshikawa, H. Uzu, and D. Adachi, Japanese Journal of Applied Physics **57**, 08RB20 (2018).

[2] A. Richter, R. Müller, J. Benick, F. Feldmann, B. Steinhauser, C. Reichel, A. Fell, M. Bivour, M. Hermle, and S. W. Glunz, Nature Energy **6**, 429 (2021).

[3] L. C. Andreani, A. Bozzola, P. Kowalczewski, M. Liscidini, and L. Redorici, Advances in Physics: X **4**, 1548305 (2019).

[4] Y. Chen, P. P. Altermatt, D. Chen, X. Zhang, G. Xu, Y. Yang, Y. Wang, Z. Feng, H. Shen, and P. J. Verlinden, IEEE Journal of Photovoltaics **8**, 1531 (2018).

[5] J. Liu, Y. Yao, S. Xiao, and X. Gu, J. Phys. D **51**, 123001 (2018).

[6] A. Sachenko, V. Kostylyov, I. Sokolovskyi, B. Arzhang, and M. Evstigneev, 2020 47th IEEE Photovoltaic Specialists Conference (PVSC), 0715 (2020).

[7] A. Richter, J. Benick, F. Feldmann, A. Fell, M. Hermle, and S. W. Glunz, Solar Energy Mater. Solar Cells **173**, 96 (2017).

[8] A. Sachenko, V. Kostylyov, V. Vlasiuk, I. Sokolovskyi, and M. Evstigneev, in *IEEE 48th Photovoltaic Specialists Conference (PVSC)*, 2021, p. 0590.

[9] Q. Pengcheng and Q. Pengxiang, IOP Conference Series: Earth and Environmental Science **621**, 012067 (2021).

[10] G. Zin, et al. Proceedings 49th AuSES, Sydney, Australia, p. 1-7 (2011).

[11] E. Y. Plotnikova, A. V. Arsentiev, and M. E. Harchenko, IOP Conference Series: Materials Science and Engineering **1035**, 012002 (2021).

[13] A. Sachenko, V. Kostylyov, V. Vlasiuk, I. Sokolovskyi, and M. Evstigneev, in *2021 IEEE 48th Photovoltaic Specialists Conference (PVSC)*, 2021), p. 0532.

[14] A. Fell, IEEE Trans. Electron Devices **60**, 733 (2013).

[15] A. Fell, *Quokka3, a software to simulate (silicon) solar cells in 3D*, 2022), 5.

[16] A. V. Sachenko, Diamond and Related Materials, 141 (2016).

[17] A. Richter, S. W. Glunz, F. Werner, J. Schmidt, and A. Cuevas, Phys.Rev.B **86**, 165202 (2012).

[18] K. R. McIntosh and S. C. Baker-Finch, in *2012 38th IEEE Photovoltaic Specialists Conference*, 2012), p. 000265.

[19] ARC simulator, *OPAL 2 calculator OPAL 2 (pvlighthouse.com.au), https://www2.pvlighthouse.com.au/calculators/OPAL%202/OPAL%202.aspx*, 2020).

[20] J. G. Fossum, Solid-State Electronics **19**, 269 (1976).

[21] *https://www2.pvlighthouse.com.au/resources/courses/altermatt/The%20Solar%20Spectrum/The%20global%20standard%20spectrum%20(AM1-5g).aspx*, 2016).

OPTIMIZING THE POSITIONS OF QUANTUM DOT LAYERS FOR LIGHT ABSORPTION ENHANCEMENT IN QUANTUM DOT SOLAR CELLS WITH LIGHT TRAPPING

Yusuke Oteki[1,2], Naoya Miyashita[1], Maxime Giteau[1], Kento Kitahara[1,2], Yoshitaka Okada[1,2]
1. Research Center for Advanced Science and Technology (RCAST), The University of Tokyo, 4-6-1 Meguro-ku, Tokyo, 153-8904, Japan
2. Graduate School of Engineering, The University of Tokyo, 7-3-1 Hongo, Bunkyo-ku, Tokyo 113-8656, Japan

We have investigated the effect of light trapping on the photoabsorption in quantum dot (QD) solar cells. Numerical simulations show that optimizing the positions of quantum dot layers can lead to an absorption enhancement using Fabry-Perot (FP) resonance and guided diffracted modes. Further, we experimentally demonstrate that positioning the InAs/GaAs quantum dot layers at the intensity peaks of FP resonance results in an increase of EQE and hence photocurrent. This effect can be used to improve the performance of QD solar cells and also photodetectors.
Keywords: Quantum Dots, Thin Film Solar Cell, Light Trapping

1 INTRODUCTION

The conversion efficiency of single-junction solar cells is fundamentally limited to ~33% under 1-sun illumination, known as the Shockley-Queisser limit [1]. This limit can theoretically be overcome using several concepts, for example by embedding the quantum dot (QD) layers within a host absorber material [2,3]. QDs can increase the output current of an intermediate band solar cell (IBSC) with a maximum theoretical efficiency of 63.2% at full sunlight concentration *via* two-step photon absorption (TSPA) process [4]. One main limitation is that a single QD layer absorbs only a small fraction of incident light. One solution is to stack multiple QD layers [5], but this could lead to generation of increased defect density [6]. To reduce the number of layers, several works have considered nanostructures to increase the absorption through optimized light trapping [7]. For quantum well (QW) solar cells, we previously reported that the absorption can be further increased by adjusting the positions of QW layers to match with Fabry-Perot (FP) resonances [8]. In this work, we show that a similar approach can be applied to QD solar cells [9]. We first optimized the positions of QD layers to increase the absorption of specific wavelengths using FP resonance and then experimentally investigated this effect on the external quantum efficiency (EQE) of QD solar cells with a FP resonance structure. We show that this extra absorption enhancement can help improve the efficiency of QD solar cells.

2 SIMULATIONS

The spectral power density absorbed in the QD layers at a depth z (in W.cm^{-3}.nm^{-1}) can be directly related to the local intensity of the electric field ($\boldsymbol{E}$) as [8, 9].

$$p(\lambda, z) = \frac{1}{2} c\varepsilon_0 n(\lambda, z)\alpha(\lambda, z)|\boldsymbol{E}(\lambda, z)|^2 \quad (1)$$

, where c is the speed of light in vacuum, ε_0 is the dielectric permittivity of vacuum, n is the refractive index, and $\alpha(\lambda) = 4\pi\kappa(\lambda)/\lambda$ is the absorption coefficient, respectively. For simplification, we consider very thin QD layers (≈1 nm), such that the total absorption of QD layers is obtained by integrating the absorbed power over the thickness of each QD layer, relative to the intensity of the incident field ($\boldsymbol{E}_{\mathbf{inc}}$). Thus, the normalized field intensity (F) can be considered homogeneous within a QD layer, leading to

$$F(\lambda, z) = n(\lambda, z)\frac{|\boldsymbol{E}(\lambda, z)|^2}{|\boldsymbol{E}_{\mathbf{inc}}(\lambda, z)|^2} \quad (2)$$

We then calculated the $F(\lambda, z)$ in an 1800 nm-thick GaAs layer on an Au mirror using the rigorous coupled wave analysis (RCWA) method (Fig. 1(a)). The effect of a small number of QD layers on the simulation is negligible since the host and QDs have the almost same real part of the refractive index. The bulk refractive index of GaAs and InAs were used in this work. Five resonant modes with many lobes were observed in the region beyond the bandgap of GaAs, with a maximum $F(\lambda, z)$ of as high as 13 times. Fig. 1(a) shows that the $F(\lambda, z)$ varies depending

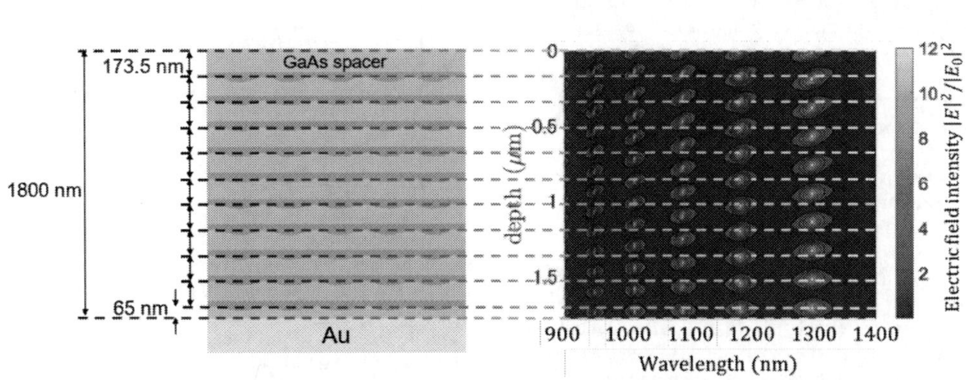

Fig. 1. (a) Normalized field intensity (F) as a function of the wavelength and position in the 1800 nm-thick GaAs on an Au mirror, (b) Schematic structure assumed to simulate for several resonant modes: an 1800 nm-thick GaAs layer with 10 QD layers on an Au back mirror.

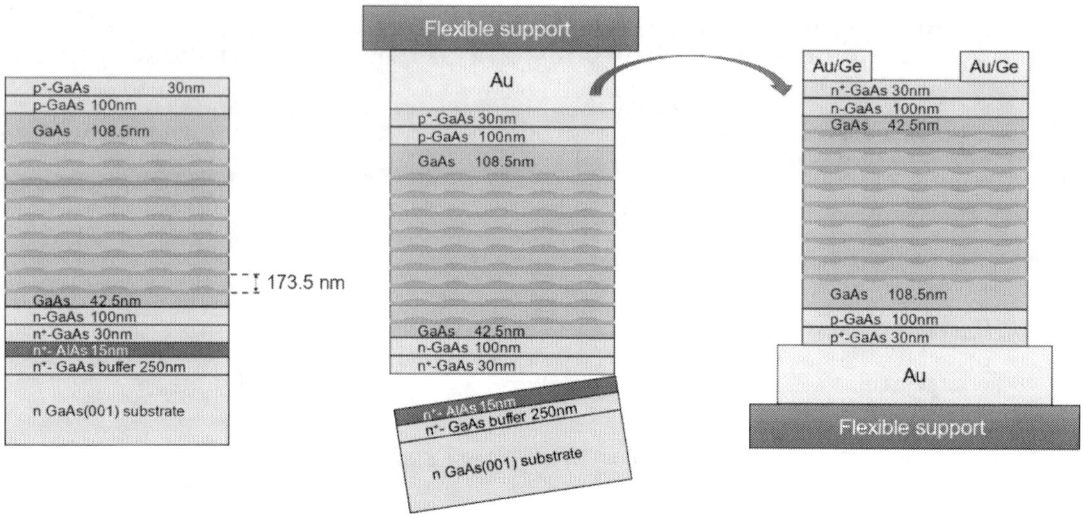

Fig. 2. Schematic device structure of thin-film Fabry-Perot QDSC fabricated in this work.

on the depth even at the same wavelength. Therefore, it is an important issue to adjust the positions of the QD layer to the peak position of the electric field intensity. We focus on the resonant mode centered around λ=1.192 μm, which shows 10 intensity resonance peaks (not counting the one at the surface). The peaks appear every 173.5 nm from the surface. By introducing 10 QD layers into the slab, one at each peak as shown in Fig. 1(b), we can fully benefit from the resonant absorption of this resonant mode.

3 EXPERIMENTAL

Figure 2 shows the schematic sample structure and flow of fabrication adopted in this work. First, an n^+-GaAs buffer layer of about 300 nm was grown, followed by a 15 nm thick n^+-AlAs release layer on a GaAs (001) substrate by molecular beam epitaxy (MBE). Following the growth of an n-GaAs region, an 1800 nm-thick InAs/GaAs QDSC containing 9 layers of QDs was grown. The position of the QD layers was optimized for resonant absorption around 1.192 μm. InAs QD layers were grown after every 173.5 nm of GaAs from the top of AlAs layer, upside down from Fig. 1(b). There were 10 QD layers in Fig. 1(b). On the other hand, 9 QD layers were grown in the device structure of Fig. 2 since both the upper and lower regions were doped. GaAs layers were grown at a growth rate of 1.0 ML/s under As_2 pressure (2.0×10^{-4} Pa), and InAs QD layers were grown to a thickness of 1.8 monolayer (ML) at a growth rate of 0.04 ML/s under As_2 pressure (1.0×10^{-4} Pa), respectively. Finally, p-GaAs layers were grown. The sample was then processed to have a FP light trapping structure. An Au mirror (800 nm) was deposited on the front of the epitaxial stack, using electron beam evaporation. The sample was then transferred to a flexible support film. The GaAs substrate was removed by the epitaxial lift-off (ELO) method [10,11]. Finally, the electrodes were patterned and formed after depositing AuGe alloy using thermal evaporation. The cell size was ~ 4 mm×4 mm, and no anti-reflection coating was used.

4 RESULTS AND DISCUSSION

The photoluminescence (PL) spectra were obtained for the fabricated sample before and after ELO, for the same laser excitation (λ =782 nm, P=100 W.cm^{-2} with a spot diameter of 100 μm). The excitation energy is slightly above the bandgap of GaAs, so the carriers are generated in the host GaAs. Some of these carriers then get trapped and recombine in the QDs, leading to the measured PL signal. The photon flux spectrum emitted by a semiconductor per unit solid angle can be described by the generalized Planck law [10].

$$\Phi(h\nu) = A(h\nu)\frac{2(h\nu)^2}{h^3c^2}\left[\exp\left(\frac{h\nu-\Delta\mu}{k_BT}\right)-1\right]^{-1} \quad (3)$$

, where $\Delta\mu$ is the quasi-Fermi level splitting between electrons and holes, $h\nu = hc/\lambda$ is energy, and T is the lattice temperature. This expression can be extended to describe the emission of multiple QD layers provided these layers are identical and all share the same $\Delta\mu$. The ratio $R(\lambda)$ between the PL spectra from the sample after

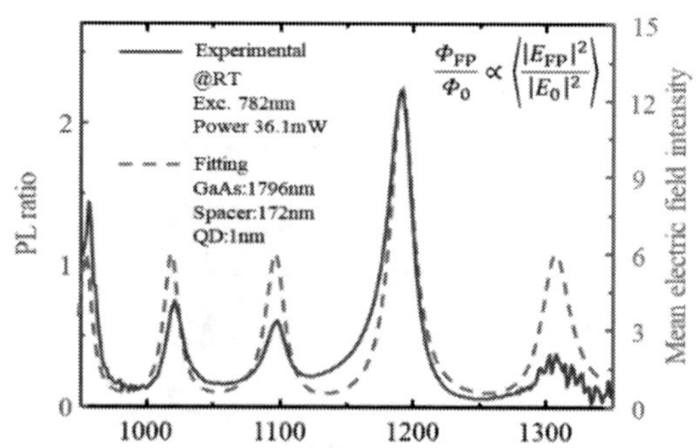

Fig. 3. The ratio of the room-temperature PL spectra (red line, left axis) and light path enhancement simulated for a 1796-nm-thick QDSC with 9 QD layers positioned to maximize the absorption for the FP resonance at 1.192 μm (dashed blue line, right axis).

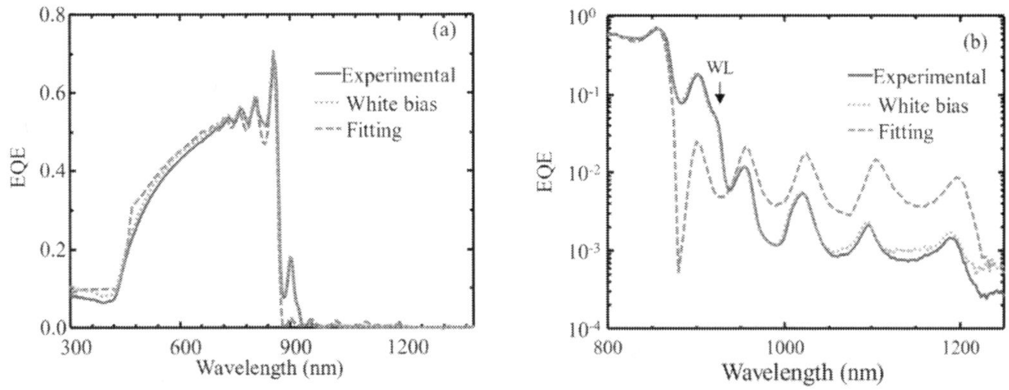

Fig. 4. EQE spectra measured under dark (red line) and 1 sun light bias (green dotted line); (a) over whole spectrum, and (b) enlarged in the near infrared region, respectively. The light path enhancement simulated for a 1818-nm-thick QDSC with 9 QD layers positioned to maximize the absorption for the FP resonance at 1.192 μm is also plotted in dashed blue line).

processing Φ_{FP} and before processing Φ_0 thus takes the expression

$$R(\lambda) = \frac{\Phi_{FP}(\lambda)}{\Phi_0(\lambda)} \propto \frac{E_{FP}(\lambda)}{E_0(\lambda)} \quad (4)$$

The PL ratio $R(\lambda)$ was shown in Fig. 3, in red line. We observed 4 peaks at 1.020 μm, 1.097 μm, 1.190 μm and 1.300 μm, with a significantly higher intensity for the third peak. We adjusted the total thickness to 1796 nm in the model to compute the $F(\lambda)$ shown in Fig. 3, in blue dashed line. We observe a good match in the peak position between both curves, confirming that the oscillations in the $R(\lambda)$ originate from the FP resonances $F(\lambda)$. Also, $R(\lambda)$ is relatively independent of the laser intensity, confirming that this difference in the relative intensity of the FP peaks cannot be attributed to nonlinear effects in the QDs.

Furthermore, the peak for which the QD layer position is optimized shows an enhancement of more than twice that of the other peaks. For the three non-optimized resonant modes, the $F(\lambda)$ is only about a half the optimal value. Therefore, by optimizing the position of QD layers, we believe we can further increase the absorption at least by a factor 2 is possible.

The EQE spectrum measured at room temperature is shown in red solid line in Fig. 4(b) shows EQE in the range of near infrared wavelengths from 0.9 μm to 1.2 μm. We adjusted the total thickness to 1818 nm in the model to fit the result of EQE measurement shown in Fig. 4 (blue dashed line). We performed a theoretical analysis of IBSC by the self-consistent drift diffusion method [12] considering the optical electric field intensity distribution. The generation rate at each wavelength (λ) at a depth (z) is given by

$$g(\lambda) = \frac{p(\lambda)}{h\nu}, \quad (5)$$

, where $h\nu = hc/\lambda$ is the photon energy. Using Equations (1) and (5), we find:

$$g(\lambda) = \frac{2\pi\varepsilon_0}{h} n(\lambda)\kappa(\lambda)|E(\lambda)|^2. \quad (6)$$

We approximate the density of states for the intermediate band (IB) as a Dirac delta function in energy. Assuming optical transitions with non-overlapping spectral bands, the combination of bands excited by the photocarriers is determined by the energy of the photons. The photoluminescence (PL) wavelength peak, which originates from the VB-to-IB transition of InAs/GaAs QDs, occurs around 1.2 μm in our samples in Fig. 3. Therefore, we assume that photons with wavelengths beyond 1.2 μm are absorbed by the IB-to-CB transition, while that between 0.87 μm (GaAs bandgap) and 1.2 μm is absorbed by the VB-to-IB transition (we assume that the absorption of the different transitions does not overlap spectrally). The photocurrent extracted from the device was calculated from the generation rate of Eq. (6). EQE was calculated using the solar spectrum under AM1.5G. For the drift-diffusion analysis of the IBSC, the Poisson equation, current continuity equation, and balance equation are solved self-consistently for the electrons in the IB using COMSOL Multiphysics®:

$$G_{CI}(z) - R_{CI}(z) = R_{IV}(z) - R_{IV}(z) \quad (7)$$

, where $G_{CI}(z)$ is the generation rate of CB–IB, $G_{IV}(z)$ that of IB–VB, $R_{CI}(z)$ is the radiative recombination rate of CB–IB, and $R_{IV}(z)$ that of IB–VB. $G_{CI} = 0$ in the wavelength range 0.87-1.2um while $G_{IV} = 0$ in the wavelength region longer than 1.2 μm due to the assumption of non-overlapping absorption bands. The radiative recombination rates are defined as those found in Ref. 13. The radiative lifetime of InAs QDs is assumed to be 1 ns.

The small difference in the total thickness between PL and EQE is due to nonuniformity in the plane and the different measurement positions. We again observe a good fit in the peak wavelengths between both curves. The FP resonance hence results in an increase of photocarrier generation between the valence band and the QD states/intermediate band (IB). On the other hand, the EQE signal becomes smaller at wavelengths longer than 1.2 μm where the light absorption mainly corresponds to exciting the carriers from IB to conduction band. It can be clearly observed that EQE increases in this wavelength region when measured under a white light bias condition (green dotted line). This is because the carrier occupation of IB under a bias illumination will increase the rate of second photo-excitation step and hence current. The additional short-circuit current due to QD contribution is estimated to be ~ 0.39 mA/cm^2 under AM1.5 spectrum. There is still a difference between the EQE measurement results and the fitting in the near infrared region. This gap results from taking the refractive index of InAs bulk. We further observe a shoulder in the measured EQE at a wavelength of ~0.924 μm which is due to the wetting layer (WL) contribution.

The calculated EQE is lower than the experimental value in this region since the simulation does not take into consideration for the light absorption coefficient of WL. It is expected that the difference can be reduced by accurately measuring the refractive index of QDs. We observed a good fit in the EQE peak wavelengths between the measurement and simulation.

5 CONCLUSION

In this study, a Fabry-Perot light-trapping structure for a QDSC of was designed and fabricated. The five resonant modes with many lobes were observed in the simulation of the electric field intensity distribution in an 1800 nm-thick GaAs layer on an Au mirror in the wavelength region from 0.9 to 1.3μm. We focused on the resonant mode centered around λ=1.192 μm. The resonant absorption of this resonant mode could be benefited by introducing 10 QD layers into the slab, each layer at the position of resonance peak from the PL measurement. We observed a good fit in the peak wavelengths between the measurement and simulation.

6 ACKNOWLEDGMENTS

This work is supported by the National Research and Development Agency, New Energy, and Industrial Technology Development Organization (NEDO), and Ministry of Economy, Trade and Industry (METI), Japan.

REFERENCES

[1] W. Shockley and H. J. Queisser, J. Appl. Phys. **32**, 510 (1961).
[2] A. Luque and A. Martí, Phys. Rev. Lett. **78**, 5014 (1997).
[3] D. Das, D. Panda, B. Tongbram, J. Saha, V. Deviprasad, H. Rawool, S. Singh, V. Chavan, and S. Chakrabarti, Sol. Energy Mater. Sol. Cells **185**, 549 (2018).
[4] A. Luque, A. Martí, and C. Stanley, "Understanding intermediate-band solar cells", Nature Photonics, **6**, 146-152 (2012).
[5] T. Sugaya, O. Numakami, R. Oshima, S. Furue, H. Komaki, T. Amano, K. Matsubara, Y. Okano, and S. Niki, Energy Environ. Sci. **5**, 6233 (2012).
[6] A. Takata, R. Oshima, Y. Shoji, K. Akahane, and Y. Okada, 35th IEEE Photovoltaic Specialists Conference 001877 (2010).
[7] T. Aho, F. Elsehrawy, A. Tukiainen, S. Ranta, M. Raappana, R. Isoaho, A. Aho, A. Hietalahti, F. Cappelluti, and M. Guina, Appl. Opt. **59**, 6304 (2020).
[8] B. Behaghel, R. Tamaki, N. Vandamme, K. Watanabe, C. Dupuis, N. Bardou, H. Sodabanlu, A. Cattoni, Y. Okada, M. Sugiyama, S. Collin, and J.-F. Guillemoles, Appl. Phys. Lett. **106**, 081107 (2015).
[9] M. Giteau, Y. Oteki, K. Kitahara, N. Miyashita, R. Tamaki, and Y. Okada, J. Photonics for Energy, **12**, 022203 (2022).
[10] M. F. Bennett, Z. S. Bittner, D. V. Forbes, S. R. Tatavarti, S. P. Ahrenkiel, A. Wibowo, N. Pan, K. Chern, and S. M. Hubbard, Appl. Phys. Lett. **103**, 213902 (2013).
[11] T. Sogabe, Y. Shoji, P. Mulder, J. Schermer, E. Tamayo, and Y. Okada, Appl. Phys. Lett. **105**, 113904 (2014).
[12] K. Yoshida, Y. Okada, and N. Sano, J. Appl. Phys. **112**, 084510 (2012).
[13] A. S. Lin, and J. D. Phillips, IEEE Trans. Electron Devices **56**, 3168–3174 (2009).

Scanning Tunneling Spectroscopy Study of Quantum Dot sized SiGe Alloy Nanocrystals prepared with Nonthermal Capacitively-Coupled Plasma

Md. Seraj Uddin[a], **Salma Khatun**[c], **C. Vijayan**[a] **and J. K. Rath**[a,b,*]
[a]*Department of Physics, Indian Institute of Technology Madras, Chennai 600036, India*
[b]*Utrecht University, Debye Institute, P.O. Box 80000, 3508 TA Utrecht, The Netherlands*
[c]*School of Physical Sciences, Indian Association for the Cultivation of Science, Kolkata 700032, India*
*Corresponding author: jkr@iitm.ac.in

ABSTRACT: We report on the Scanning Tunneling Spectroscopy (STS) study of quantum dot sized SiGe alloy nanocrystals (NCs) to explore its bandgap energies. These alloy NCs have been prepared in a multichamber very high frequency plasma enhanced chemical vapor deposition (VHF PECVD) reactor using a capacitively-coupled plasma technique. The STS study yielded a density of states (DOS) spectrum of NCs in which the conduction band (CB) and valance band (VB) edges were located, in addition, even midgap defect states in the forbidden region in some of the NC samples. The measured bandgap of SiGe alloy NCs in different samples reveals Ge content dependence. Though there was no dopant introduced intentionally during the growth, a shallow p-type nature of alloy NCs has been observed in all the samples even which may be attributed to charged grain boundary defects in high Ge content alloy NCs. The high bandgap value of NCs compared to bulk alloy indicates quantum confinement in the SiGe alloy NCs.

Keywords: VHF PECVD, SiGe alloy NCs, Scanning Tunneling Spectroscopy (STS)

1 INTRODUCTION

Bandgap engineering in semiconducting materials is the demand of device industry to widen its application further to existing [1–5] as well as new technologies [6–10]. To that end, one of the measures is to synthesze nanoparticles (NPs) of size below or of the order of exciton Bohr radius, also called quantum dots (QDs) [11] which can have size tunable bandgaps below the exciton Bohr radius. The synthesis of QDs with accurate control over the size is difficult or non-reproducible due to very small Bohr exciton radius, e.g., 5 nm for Si [12, 13]. Therefore, another more pronounced method utilized for bandgap tuning is the compositional modification of elements in alloy nanocrystals (NCs) [14]. The control over size of NCs becomes easier if the Bohr radius is comparatively high, e.g. alloying the Si with Ge. This offers a control over optical and electronic properties and making inroads for application in the optoelectronic devices. Alloys of group IV elements in periodic table especially of Si and Ge (i.e, $Si_{1-x}Ge_x$; $0<x<1$) are widely recognized due to its potential applications in transistors [15], photovoltaic [16] and infrared detection [17].

SiGe alloy NCs combine the high Bohr exciton radius of Ge (24.3 nm) with good electrical properties of Si, therefore, results in a more pronounced quantum confinement [18]. A high extinction coefficient and controlled bandgap tuning makes SiGe a suitable candidate as a gain medium in bottom cell of a QDs based tandem solar cell [16, 19]. However, an accurate knowledge of energy band gap of SiGe QDs is required to effectively incorporate for device application. Though many techniques have been used for the band gap determination such as Optical spectroscopy [20, 21], Cyclic Voltammetry (CV) [22, 23] and Conducting AFM [24], they all have some limitations. Optical spectroscopy probes electronic transition between two levels of the incorporate molecule or QD which satisfy selection rules. Therefore, not all energy levels of the system are reflected in the optical resonances. In addition, the spatial resolution of optical spectroscopy is limited to one fourth of the wavelength, i.e., up to few hundred nanometer [25] and so, it is usually not possible to probe a single nano-object. CV study has limitation of doing measurement in solution phase and AFM has its limit due to tip resolution. However, the scanning tunneling spectroscopy (STS) [26], considering its highly localized nature, can efficiently study the individual nanostructures [27, 28] and interfaces [29], and, can directly measures the position of density of state (DOS) of CB and VB edge. The STS has been successfully employed previously to obtain the band gap of SnS quantum dots by *C. Prastani et al.* [30].

In this paper, we report the STS study on SiGe alloy NCs, synthesized in a non-thermal VHF PECVD reactor. The synthesis and other characterisation of these SiGe alloy NCs have already been reported [31]. The growth process of these alloy NCs takes place in gas phase through complex process steps called nucleation, agglomeration and coalescence. The analytic model of Fridman and Bouchoule [32] best describes these phenomena. The SiGe NCs reported here are nano-crystallites embedded in an amorphous matrix similar to polymorphous film [33, 34]. The crystallinity of these non-thermal plasma deposited NCs results from the heating and cooling events on the nanoparticle surface through electron-ion recombination, dangling bond passivation, hydrogen abstraction and hydrogen recombination processes [35]. From STS study, we report on the bandgap measurement and the semiconducting nature (p-type or n-type) of SiGe alloy NCs in a dopant free deposition environment, and, also the existence of midgap defect states in the NCs. In addition, a composition dependence of bandgap in alloy NCs has also been reported. Previously, Motaki *et al.* [36] has reported a similar study and showed the p-type nature of SiGe alloy films, prepared in dopant free condition using Hall measurements, and also established its dependence on the point defects and Ge content.

2 EXPERIMENTAL

The SiGe alloy NCs have been prepared in one of the chambers of a custom made multi-chamber very high frequency PECVD reactor (ASTER). This chamber is especially designed to grow well-defined quantum dot (QD) size nanoparticles in single step. A detailed study of this NP chamber has been reported [37]. Unlike the other chambers of ASTER, NP chamber accommodates one extra grounded electrode in between the bottom showerhead RF electrode and the substrate holder attached floating electrode, this results in a triode configuration. The grounded electrode of NP chamber has a grid like structure which confines the plasma in the vicinity of the grid-holes with a high electron density [38]. This design of NP chamber makes it unique for the nanoparticle synthesis in gas phase. The triode configuration of the chamber constrains the plasma in between the grounded and showerhead electrode and therefore protects the deposited samples from direct exposure to the plasma. The significance of this synthesis technique is that the properties of NPs can be tuned just by varying the plasma parameters which ultimately depends on the inter-electrode distance, pulsing of applied power and the precursor gas flow rate in the chamber.

The synthesis of SiGe NCs has been processed by passing the mixture of precursor gases such as silane, germane, argon and hydrogen (SiH_4 + GeH_4 + Ar + H_2) into the chamber through the bottom showerhead electrode. These precursor gases were dissociated in the chamber by applying an oscillating very high frequency (VHF) 60MHz electric field. An input power of 150W was applied to the bottom showerhead electrode to create and sustain the plasma in the chamber and a pressure of 0.8 mbar was maintained in the chamber during the nanoparticle synthesis. The holes in grid (grounded) electrode allowed the formed NPs to escape the plasma region, and the phenomena called thermophoresis effect [34, 35] helped the NPs to deposit on the substrate maintained at ground potential. A series of depositions were carried out at different silane-to-germane gas flow (F) ratios (r_g) (i.e, r_g=F(silane)/F(germane)) in both H_2 and (Ar+H_2) gas dilution conditions, and in continuous wave (CW) mode as well as pulsed mode [41] of plasma with ON time (T_{ON}) 250 msec and OFF time (T_{OFF}) 1 sec. To obtain the pulsed mode, a modulation in the VHF power was done using a square wave modulator. The NCs were synthesized in dust and collected on a Corning glass (Eagle XG) mounted on the substrate holder, by utilizing the thermophoresis force originating from the temperature gradient between the plasma gas and the substrate, and, by the applied drag force on the gases due to pump, connected back side of the substrate.

Here, out of the several deposited samples, the four samples prepared in different growth regimes are reported for the STS study. They are named as NP_S1, NP_S2, NP_S3 and NP_S4;

1. Ar : H_2 : SiH_4 : GeH_4 = 25 sccm : 25 sccm : 4 sccm : 4 sccm (r_g =1; CW plasma mode)
2. Ar : H_2 : SiH_4 : GeH_4 = 25 sccm : 25 sccm : 4 sccm : 4 sccm (r_g =1; Pulsed plasma mode, 250 msec.)
3. Ar : H_2 : SiH_4 : GeH_4 = 25 sccm : 25 sccm : 4 sccm : 2 sccm (r_g =2; CW plasma mode)
4. H_2 : SiH_4 : GeH_4 = 50 sccm : 4 sccm : 4 sccm (r_g =1; CW plasma mode)

The STS study of the samples has been carried out in Nanosurf easyScan2 STM system in ambient condition. For this, the alloy NC samples were prepared on a highly oriented pyrolytic graphite (HOPG) substrate, serves as a base electrode for STS. For STS, the SiGe NCs were first dissolved in ethanol solvent and sonicated for 20 minutes to disperse them. A small drop of diluted SiGe NCs were casted on the HOPG substrates and kept in vacuum to dry. To get the tunneling current through the samples, a bias voltage was applied to the tip of the microscope which is made with a mechanical cut of a Pt/Ir (80:20) wire of diameter 0.25 mm. A tip-approach voltage 1.0 V was initially applied to get the corresponding set-point current 0.1 nA. The set-point current optimizes tip-sample separation to match the tunneling condition.

In STS, the instrument records tunneling current through the NC as a function of tip-bias voltage. A derivative of the tunneling current, i.e, dI/dV, has correspondence to the density of states (DOS). From dI/dV spectra, the CB (conduction band) and VB band (valance band) edge of NC are identified in the form of first peak at both the negative and positive bias voltages to the tip respectively, considering the Fermi energy to be aligned at 0 V. This infers the fact that the first peak at a positive tip-bias voltage in dI/dV spectrum implies the withdrawal of electrons from the NC and hence denote the VB edge. Similarly, the first peak at negative tip-bias voltage in the dI/dV spectrum implies the injection of electrons into the NP and hence denote the CB edge.

3 RESULTS AND DISCUSSION

To investigate the energy band structure of SiGe alloy NCs samples, the individual NCs were first identified on the substrate using STM topography. Then, the STM tip was stabilized on top of the desired NC, and a voltage sweep was applied to the tip w.r.t the NC to get the tunneling current. The tunneling current enables us to obtain the DOS spectra of NCs. Figure 1(a) shows the topography image of the alloy NC for sample NP_S1. The white spot in the image is seen as individual NC of size around 20 nm and morphology slightly distorted from a sphere. Figure 1(b) shows the DOS spectrum of the NC of sample NP_S1 obtained from derivative plot of the tunneling current. In the DOS spectrum, a peak at positive bias voltage results from the tunneling of electrons from NC to the tip and therefore, this positive bias voltage peak close to 0 V represents the VB-edge. Similarly, a peak at negative bias voltage results from the electron tunneling from tip to the NC, therefore, this negative bias voltage peak close to 0 V represents the CB-edge. The subsequent extra peaks before CB edge and after VB edge result from quantum confinement. The tunneling current measurement in STS is an extremely localized phenomenon, therefore, the DOS spectra of the SiGe NC were recorded at many different points on the NC. The CB and VB energies obtained from around 50-60 different points on the NC are summed up in the histogram as shown in Figure 1(c). The energy distribution of CB and VB edge in histogram may result from the presence of defects and disordered state in the sample. The CB and VB edge position of NC in dI/dV spectrum in Figure 1(b) shows a shallow p-type character which we consider to have resulted from charged grain boundary defects of Ge atom rich NC [42]. The bandgap of NC in NP_S1 sample estimated from CB and VB edge is E_g = 1.0 eV.

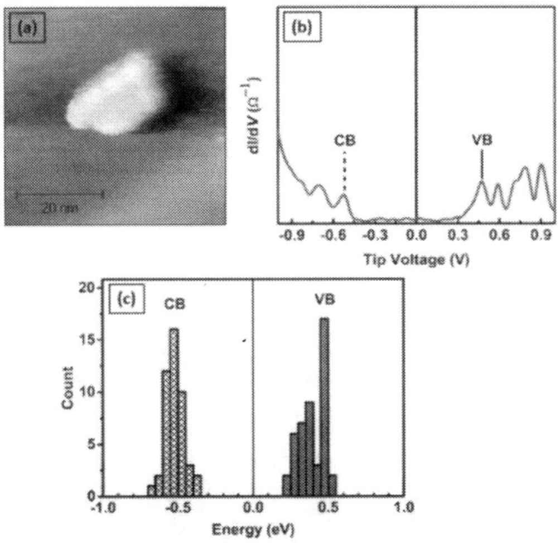

Figure 1: (a) Topography of the NC from sample NP_S1, (b) Typical DOS spectrum for NC in sample NP_S1, (d) associated histogram of CB and VB energies of SiGe alloy NC in sample NP_S1.

A similar study was also performed on the NCs of NP_S2 sample. Figure 2(a) shows the topography of the NCs in sample NP_S2. An ensemble of individual NCs of different sizes ranging from 5 nm-10 nm dispersed on the substrate can be seen in the image as white spots with well-defined morphology. The image also shows some extra unwanted white patches which might have resulted from the solvent and artifacts. Figure 2(b) shows the DOS spectrum of SiGe alloy NC, obtained from tunneling current from a set of different NCs circled in Figure 2(a). The DOS spectrum gives energy band edge positions of CB and VB in the negative and positive side of the Fermi level (0 V) respectively. There are some extra small peaks also observed in DOS spectrum near the Fermi level (0 V) and therefore, it is difficult to distinguish the VB edge position. However, the actual VB edge position can be found from VB histogram shown in Figure 2(c). The bandgap of NC obtained from CB and VB edges is E_g = 0.90 eV. The extra peaks in the forbidden region presumably result from the deep defect states in SiGe alloy NC. However, this is a very unexpected result from a QD sized NC, as the QDs are self-purifying media that diffuse out the intrinsic defect due to high surface area to volume ratio [43]. However, a recent computational study [44] has reported that the surface defects such as oxide and dangling bonds may lead to deep level defect states. This needs to be further investigated experimentally for the SiGe QDs. The histogram of CB and VB energies in Figure 2(c) has been plotted for about 50–60 different points on the NCs due to localized nature of STS measurement. The CB and VB edge position in dI/dV spectrum of sample NP_S2 is again showing a shallow p-type character.

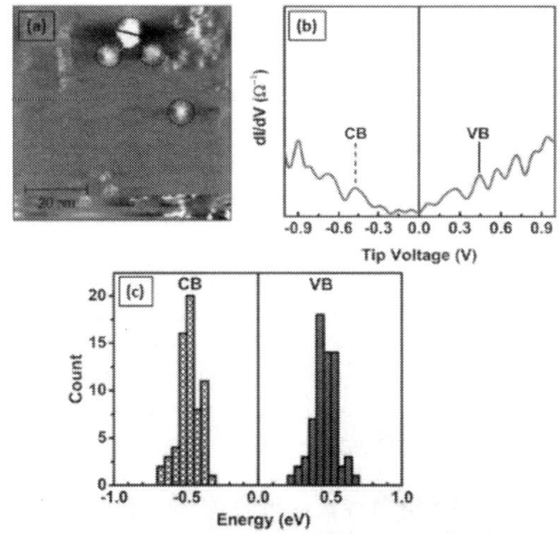

Figure 2: (a) Topography of NPs for sample NP_S2, (b) Typical DOS spectrum for NC in sample NP_S2, (c) associated histogram of CB and VB energies of SiGe alloy NC in sample NP_S2.

Figure 3(a) shows the topography of SiGe alloy NCs in sample NP_S3. The bright spots featured in the figure is identified as individual NCs of size around 10 nm-12 nm. The topography image also shows the different layers of HOPG substrate. Figure 3(b) shows the DOS spectrum of the SiGe alloy NCs (circled in Figure 3(a)). The DOS spectrum gives the energy band edge position of CB and VB for the SiGe NC in sample NP_S3. The bandgap value obtained from the CB and VB edge position is $E_g = 0.95$ eV. A histogram of the CB and VB energies has been plotted in Figure 3(c) by taking around 50–60 different points on the NPs. The CB and VB edge position in DOS spectrum of sample NP_S3 is again showing a p-type character. However, compared to the sample NP_S1 and NP_S2, it has a stronger p-type nature and also carry some extra peaks in the forbidden gap.

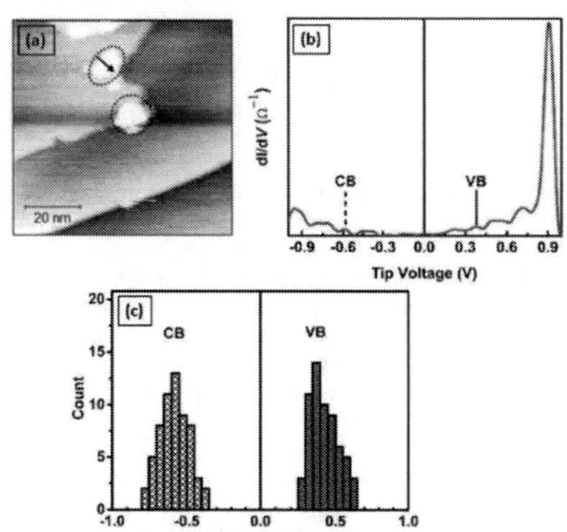

Figure 3: (a) Topography image of NPs for the sample NP_S3, (b) Typical DOS spectrum, (d) associated histogram of CB and VB energies of SiGe alloy NC in sample NP_S3.

Figure 4(a) shows the topography of SiGe NCs in sample NP_S4. The NCs of size-range from 4 nm-7 nm are observed as white spots in the image. Figure 4(b) shows the DOS spectrum, obtained from the derivative plot of tunneling current w.r.t tip voltage from a set of different NCs circled in Figure 4(a). The DOS spectrum gives the edge value of CB and VB. The bandgap value evaluated from the DOS spectrum is E_g = 0.77 eV. A histogram of CB and VB energy, plotted for around 50–60 different points on the circled NPs, is shown in Figure 4(c). The CB and VB edge positions in DOS spectrum of sample NP_S4 demonstrate a p-type character and there are a few week peaks also observed in the forbidden gap. However, the p-type character is much weaker compared to the other samples.

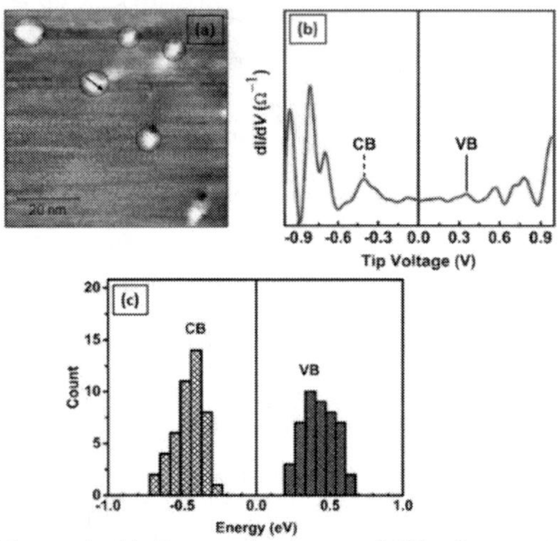

Figure 4: (a) Topography image of NPs for sample NP_S4, (b) Typical DOS spectrum and (d) associated histogram of CB and VB energies of SiGe alloy NC in sample NP_S4.

The bandgap of SiGe alloy NCs obtained from the STS measurement is shown in **Table 1.** The bandgap value is well within the expected range [18, 45]. The bandgap of alloy NCs can be compared with the bandgap of bulk $Si_{1-x}Ge_x$ alloy, where 0<x<1, calculated by formula [46];

$$E_g(x) = (1.17 - 0.896x + 0.396x^2)\ eV \qquad (1)$$

Here, 'x' is the Ge content in bulk $Si_{1-x}Ge_x$ alloy.

The Ge content in SiGe alloy NCs of each samples mentioned in this paper have already been reported [31]. For a given Ge content, the bandgap of alloy SiGe NC measured from STS is compared with its bulk part as shown in **Table 2**. The bandgap of SiGe alloy NCs is greater than that of respective bulk SiGe alloys which indicates the quantum confinement. The Ge contents in NCs of sample NP_S1 and NP_S2 have same value but showing different bandgaps. This is possible as the bandgap value of alloy NCs depends on other deposition parameters, such as growth under pulsed plasma condition. However, the bandgap in NC samples is found to vary with Ge content, as in the case of the bulk alloy.

Table 1. Bandgap of SiGe alloy NCs in different samples obtained from STS.

Sample ID	Ge (at.%) in NCs	CB edge (eV)	VB edge (eV)	SiGe NC bandgap (eV)
NP_S1	0.66	-0.525	0.475	1.00
NP_S2	0.66	-0.475	0.425	0.90
NP_S3	0.57	-0.575	0.375	0.95
NP_S4	0.73	-0.405	0.365	0.77

One interesting observation is that the samples made with CW plasma have comparatively cleaner mid gap states, whereas the samples made with pulsed plasma (sample NP_S2) seem to have a number of peaks which may be correlated with agglomeration of small particles (as may be observed from topography) with many defects. A further analysis of the midgap states in the reported NCs needs to be done.

Table 2. Bandgap comparison of SiGe alloy NCs with its bulk.

Sample ID	Ge (at.%) in NCs	SiGe NC Bandgap (eV)	Bulk SiGe Bandgap(eV)
NP_S1	0.66	1.00	0.75
NP_S2	0.66	0.90	0.75
NP_S3	0.57	0.95	0.79
NP_S4	0.73	0.77	0.73

4 CONCLUSION

In conclusion, a systematic analysis of the band gap is performed in SiGe alloy NC samples using STS. The bandgap values in all the samples indicate quantum confinement when compared with bulk alloy. The SiGe alloy NCs have composition dependence bandgap and they all show a p-type character even though there is no intentional impurity doping during growth. The p-type nature of SiGe alloy NCs may attribute to the Ge atom rich nature of NCs. Extra peaks in the forbidden region were also observed in a few samples, notably for QDs made with pulsed plasma. Further work using atom probe analysis is currently underway in order to identify the defects.

5 ACKNOWLEDGMENT

The author would like to thank Department of Physics, IIT Madras for providing the research grants and also thanks to Prof. Amlan J. Pal, IACS Kolkata for giving access to the STM facility. Authors also acknowledge DSEHC which has been funded by Department of science and technology (DST), Government of India, through grant No.DST/TMD/SERI/HUB/1(C).

REFERENCES

[1] Y. Li, C. Cui, J. Song, Q. Liu, S. Yuan, C. Zeng, J. Xia, Nanotechnology. 31 (2020) 385603.

[2] D. V. Talapin, J.S. Lee, M. V. Kovalenko, E. V. Shevchenko, Chem. Rev. 110 (2010) 389–458.

[3] A. Torres, M. Moreno, A. Kosarev, A. Heredia,

J. Non. Cryst. Solids. 354 (2008) 2556–2560.
[4] Q. Guo, S.J. Kim, M. Kar, W.N. Shafarman, R.W. Birkmire, E.A. Stach, R. Agrawal, H.W. Hillhouse, Nano Lett. 8 (2008) 2982–2987.
[5] M. Baira, B. Salem, N.A. Madhar, B. Ilahi, Superlattices Microstruct. 117 (2018) 31–35.
[6] H. Liu, W. Winkenwerder, Y. Liu, D. Ferrer, D. Shahrjerdi, S.K. Stanley, J.G. Ekerdt, S.K. Banerjee, IEEE Trans. Electron Devices. 55 (2008) 3610–3614.
[7] I.B. Akca, A. Dâna, A. Aydinli, R. Turan, Appl. Phys. Lett. 92 (2008) 052103.
[8] S.Wiss, X. Michalet, F. F. Pinaud, L. A. Bentolila, J. M. Tsay, S. Doose, J. J. Li, G. Sundaresan, A. M. Wu, S. S. Gambhir, Science, 538 (2013) 538–545.
[9] S. Buckley, K. Rivoire, J. Vučković, Reports Prog. Phys. 75 (2012) 126503.
[10] P. Bardella, L.L. Columbo, M. Gioannini, Opt. Express. 25 (2017) 26234.
[11] E.G. Barbagiovanni, D.J. Lockwood, P.J. Simpson, L. V. Goncharova, Appl. Phys. Rev. 1 (2014) 011302.
[12] D. Carolan, Prog. Mater. Sci. 90 (2017) 128–158.
[13] I. Dolan, M.C.M. Van De Sanden, Plasma Process. Polym. 13 (2016) 19–53.
[14] M.D. Regulacio, M.Y. Han, Acc. Chem. Res. 43 (2010) 621–630.
[15] S. Koelling, R.C. Plantenga, H.I.T. Hauge, Y. Ren, A. Li, M.A. Verheijen, S. Conesa Boj, S. Assali, P.M. Koenraad, E.P.A.M. Bakkers, ECS Trans. 75 (2016) 751–760.
[16] A. Dey, D. Das, J. Phys. Chem. Solids. 154 (2021) 110055.
[17] I. Stavarache, C. Logofatu, M.T. Sultan, A. Manolescu, H.G. Svavarsson, V.S. Teodorescu, M.L. Ciurea, Sci. Rep. 10 (2020) 1–9.
[18] J.E. Yang, C.B. Jin, C.J. Kim, M.H. Jo, Nano Lett. 6 (2006) 2679–2684.
[19] A. Mohan, F.D. Tichelaar, M. Kaiser, M.A. Verheijen, R.E.I. Schropp, J.K. Rath, Chem. Phys. Lett. 661 (2016) 185–190.
[20] P. Makuła, M. Pacia, W. Macyk, J. Phys. Chem. Lett. 9 (2018) 6814–6817.
[21] S. Sohila, M. Rajalakshmi, C. Ghosh, A.K. Arora, C. Muthamizhchelvan, J. Alloys Compd. 509 (2011) 5843–5847.
[22] E. Kucur, J. Riegler, G.A. Urban, T. Nann, J. Chem. Phys. 119 (2003) 2333–2337.
[23] P.P. Ingole, G.B. Markad, D. Saraf, L. Tatikondewar, O. Nene, A. Kshirsagar, S.K. Haram, J. Phys. Chem. C. 117 (2013) 7376–7383.
[24] D. Moerman, N. Sebaihi, S.E. Kaviyil, P. Leclère, R. Lazzaroni, O. Douhéret, Nanoscale. 6 (2014) 10596–10603.
[25] T. Ichimura, T. Jin, H. Fujita, H. Higuchi, T.M. Watanabe, Front. Physiol. 5 (2014) 273.
[26] S.U. Nanayakkara, J. Van De Lagemaat, J.M. Luther, Chem. Rev. 115 (2015) 8157–8181.
[27] B. Kundu, A.J. Pal, J. Phys. Chem. C. 122 (2018) 11570–11576.
[28] A. Arramel, T.C. Pijper, T. Kudernac, N. Katsonis, M. Van Der Maas, B.L. Feringa, B.J. Van Wees, J. Appl. Phys. 111 (2012) 083716.
[29] T.Y. Chien, J. Chakhalian, J.W. Freeland, N.P. Guisinger, Adv. Funct. Mater. 23 (2013) 2565–2575.
[30] C. Prastani, SnS nanoparticles to boost CuInS2 solar cells, PhD thesis, Utrecht University, 2015.
[31] M.S. Uddin, C. Vijayan, J.K. Rath, EPJ Appl. Phys. 91 (2020) 20801.
[32] A.A. Fridman, L. Boufendi, T. Hbid, B. V. Potapkin, A. Bouchoule, J. Appl. Phys. 79 (1996) 1303–1314.
[33] A. Mohan, M. Kaiser, M.A. Verheijen, R.E.I. Schropp, J.K. Rath, J. Cryst. Growth. 467 (2017) 137–144.
[34] N. Chaâbane, P. Roca i Cabarrocas, H. Vach, J. Non. Cryst. Solids. 338–340 (2004) 51–55.
[35] N.J. Kramer, R.J. Anthony, M. Mamunuru, E.S. Aydil, U.R. Kortshagen, J. Phys. D. Appl. Phys. 47 (2014) 075202.
[36] M. Satoh, K. Arimoto, K. Nakagawa, S. Koh, K. Sawano, Y. Shiraki, N. Usami, K. Nakajima, Jpn. J. Appl. Phys. 47 (2008) 4630–4633.
[37] A. Verkerk, Plasma deposition of thin film silicon at low substrate temperature and at high growth rate, PhD thesis, Utrecht University, 2009.
[38] A. Mohan, M.M. De Jong, I. Poulios, R.E.I. Schropp, J.K. Rath, J. Phys. D. Appl. Phys. 48 (2015) 375201.
[39] G.M. Jellum, J.E. Daugherty, D.B. Graves, J. Appl. Phys. 69 (1991) 6923–6934.
[40] U. Bhandarkar, U. Kortshagen, S.L. Girshick, J. Phys. D. Appl. Phys. 36 (2003) 1399–1408.
[41] A. Bouchoule, A. Plain, L. Boufendi, J.P. Blondeau, C. Laure, J. Appl. Phys. 70 (1991) 1991–2000.
[42] T. Matsui, C.W. Chang, K. Mizuno, Y. Takeuchi, M. Kondo, Jpn. J. Appl. Phys. 51 (2012) 091302.
[43] J. Li, S.H. Wei, S.S. Li, J.B. Xia, Phys. Rev. B - Condens. Matter Mater. Phys. 77 (2008) 1–4.
[44] N.P. Brawand, M.B. Goldey, M. Vörös, G.Galli, Chem. Mater. 29 (2017) 1255–1262.
[45] G.G. Pethuraja, R.E. Welser, A.K. Sood, C. Lee, N.J. Alexander, H. Efstathiadis, P. Haldar, J.L. Harvey, Mater. Sci. Appl. 03 (2012) 67–71.
[46] T. Ando, Y. Arakawa, K. Furuya, S. Komiyama, H. Nakashima, Mesoscopic Physics and electronics, Springer, 2006.

INVESTIGATION OF PHOTOLUMINESCENT PIGMENTS FOR ULTRAVIOLET LIGHT WAVELENGTH SHIFTER FOR SOLAR CELLS

Oliveira, O. J.[1], Crestani, T.[2], Coutinho, N. F.[1], Freitas, G. T. D.[1], Alvarez, H. S.[3], Villalva, M. G.[2], Marques, F. C.[1].

[1] 'Gleb Wataghin' Institute of Physics, UNICAMP, Campinas, Brazil
[2] School of Electrical and Computer Engineering, UNICAMP, Campinas, Brazil
[3] BYD, Campinas, Brazil

ABSTRACT: Although silicon and perovskite solar cells have achieved efficiency of up to 26 %, their efficiency is still limited by several factors, such as the reduced efficiency in the ultraviolet region of the solar spectrum. One of the proposed mechanisms to overcome this problem is to use wavelength shifters. The objective of this work was to investigate optical and structural properties of thin films of common photoluminescent pigments used in several areas, such as coatings, graphic arts, cosmetics and adhesives. Thin films of the pigments and tin dioxide (SnO_2) films with embedded pigments were prepared by spin coating. The films excited by ultraviolet laser present emission in the visible region, thus functioning as wavelength shifters, which can potentially be used to improve the absorption of ultraviolet light in solar cell.
Keywords: wavelength shifter, photoluminescence, solar cells.

1 INTRODUCTION

The world demand for electricity has grown in recent decades, which has led to a more expressive search for renewable and sustainable energy sources. Consequently, solar cells have received attention for their ability to convert sunlight into electrical energy through the photovoltaic effect. Although silicon and perovskite solar cells have achieved efficiencies as high as 26% [1], it is still possible to increase these efficiencies by combining materials that convert high-energy photons, above the bandgap of the absorbing material, into lower-energy photons that are better absorbed by solar cells. Thus, in addition to improving the use of high-energy photons, additional benefits are obtained by reducing the thermalization process, increasing the efficiency of solar cells [2]. In addition, the solar glass used in solar modules is not sufficiently transparent in the ultraviolet region, so some of the high-energy UV does not reach the solar cells inside the module, as it is absorbed into the glass. The use of an appropriate wavelength shifter on top of the solar module, which converts UV into visible light, can be used to supply extra photons to the solar cells within the module, thus increasing the module's efficiency.

Photoluminescent additives have already been used in transparent polymers used in CIGS solar cells, resulting in increased efficiency [3]. Here we investigate the properties of low-cost photoluminescent pigment thin films currently used in coatings, graphic arts, cosmetics and adhesives. The objective was to evaluate the ability to shift ultraviolet light to the wavelength where silicon solar cells are most efficient (around 600 nm). By shifting high energy wavelength photons to lower energy, the probability of losing the electron-hole pair created due to surface recombination, Auger and Shockley-Read-Hall recombination in the subsurface region is reduced. With the use of photoluminescence, a higher short-circuit current, J_{SC}, a slight increase in the open-circuit voltage, V_{OC}, and therefore a higher efficiency of the solar cell is expected [4].

2 MATERIALS AND METHODS

2.1 Deposition of pigment thin films

Precursor solutions were prepared by dissolving different pigments (green, orange, pink, red) on ethanol at a concentration of 28 mg/mL. The deposition of thin films was performed by the spin-coating technique with rotation of 3000 rpm for 30 seconds. After each deposition, the substrates were placed on a heating plate (50°C) for 10 minutes in order to evaporate the solvent. Ten cycles of deposition were performed for every film. In order to incorporate the pigment on SnO_2 thin films, 10 mg of pigment were added on a solution of $SnCl_4{:}5H_2O$ in 10 mL ethanol (0.05M) and deposited by spin coating, following the same procedure adopted for the films prepared with pigment only.

2.2 Characterization

Pigments (powder and thin films) were characterized by UV-Vis, AFM, photoluminescence and X-ray diffraction. Photoluminescence emission was measured using UV excitation laser (400 nm) and the PL signal detected by an Ocean optical spectrometer. Transmittance measurements were performed with a Perkin Elmer UV-Vis Lambda 1050 spectrometer. The AFM measurements were performed on an AFM - Bruker Innova, using a TESP, Bruker tip, in tapping and scanning mode with a size of 5 µm. X-ray diffractometry (XRD) was performed using the PANanalytical X'Pert PRO MPD using CuK radiation, with 40 kV, 30 mA and integration time of 1s from 10° to 80°.

3 RESULTS AND DISCUSSIONS

3.1 Pigment powder

Figure 1 shows XRD diffractograms of fluorescent pigment powders (orange, red, green and pink). The spectra are compatible with the expected structure of organic pigments. They present an amorphous characteristic and small differences between the different colors of the pigments.

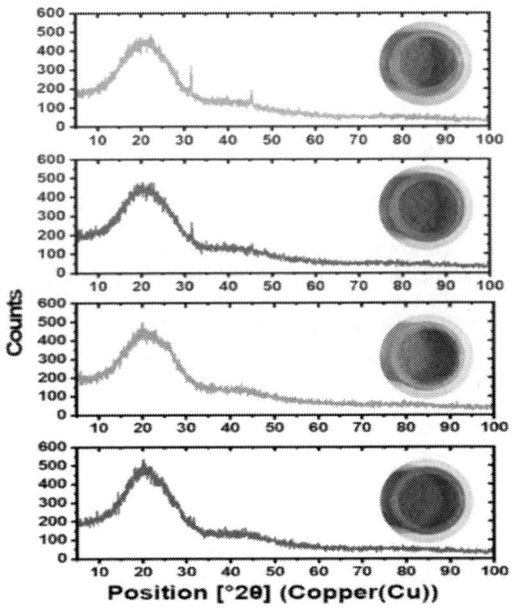

Figure 1: XRD patterns of the orange, red, green and pink fluorescent pigments.

Photoluminescence measurements of pigment powders are shown in Fig. 2. The PL emission covers a range of 500 – 650 nm, which is in the region of highest efficiency of silicon solar cells. The green and orange pigments have relatively high emission compared to the other two pigments.

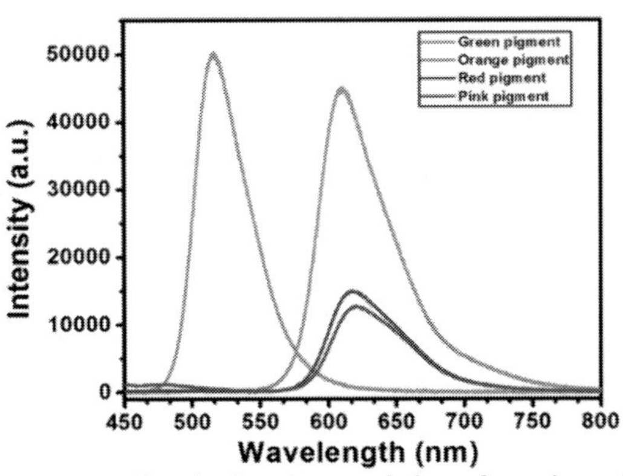

Figure 2: Photoluminescence emission of powders of orange, red, green and pink fluorescent pigments.

3.2 Thin films of fluorescent pigments

Thin films deposited by spin coating using solutions with green, orange, red and pink pigments are shown in Figure 3a, showing good uniformity. The UV-Vis spectra show some absorption bands related to specific structures of pigments and/or impurities. The intensity of the PL signal can be correlated with the absorption (reduced transmission) at the wavelength of the PL excitation laser (400 nm) - the greater the absorption, the greater the PL intensity, as inferred by verifying the transmittance at 400 nm (dotted blue line).

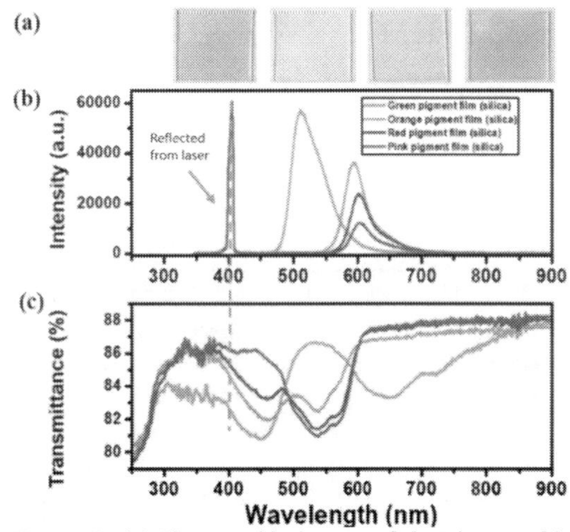

Figure 3: (a) Photographic images of the pigment thin films deposited by spin coating on silica substrates, (b) photoluminescence spectra of the films, excited by a laser at 400nm. The signal at 400nm is due to reflection of the laser beam; (c) UV-Vis transmittance of the films.

Figure 4 shows the absorption and emission spectra of the orange pigment film. The absorption (blue curve), occurs in the wavelength range from 300 to 700 nm, being more intense in the UV range. This absorption corresponds to that associated with the conversion of UV light into visible light. PL emission, on the other hand, is shifted to the range of 550 to 700 nm, as indicated by the red curve. This result shows that the film made of pigment, deposited by spin coating, works as a wavelength shifter.

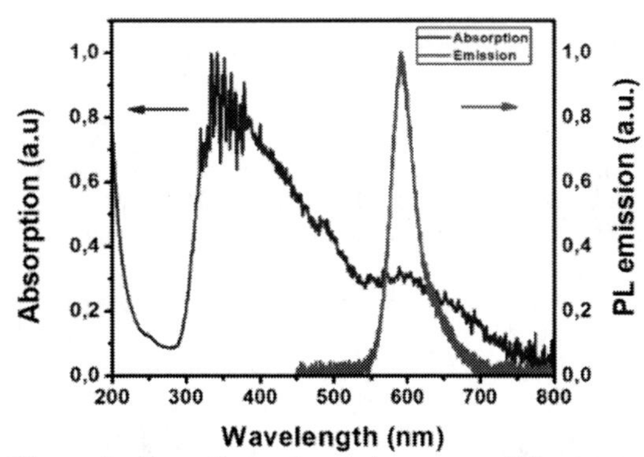

Figure 4: Absorption and emission curves of the orange pigment thin film.

To analyze the thermal stability of the pigment, the degradation process was carried out on the film as a function of temperature. The red pigment film deposited on the glass substrate was annealed in the range of 100 to 300 °C, Fig. 5. The annealing steps were performed in air on a hot plate for 10 minutes at each temperature and photoluminescence measurements were performed after each annealing temperature. The film is reasonably stable up to 180°C, starting to degrade faster at higher temperatures, probably related to the degradation of the pigment structure.

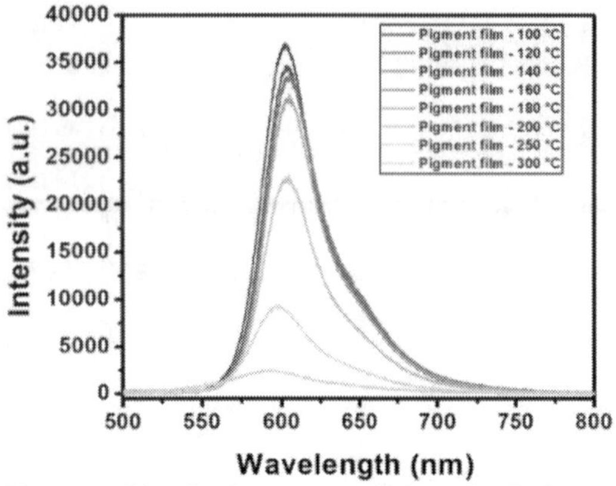

Figure 5: Photoluminescence emission of red pigment thin film as a function of annealing temperature.

3.3 Fluorescent pigment in SnO_2 thin films

Figure 6 shows the PL of three different films: 1) SnO_2; 2) orange pigment and 3) SnO_2 with orange pigment incorporated. The insect shows an AFM image of the orange pigments embedded in the thin film of SnO_2. The SnO_2 film with the pigment was prepared to have the same amount of pigment as the film deposited with the pigment alone. The UV-Vis of all films show that they have very good transmittance in the visible region of the solar radiation spectrum. The PL emission of SnO_2 is not observed as it was excited with photon energy (~3eV) below its band gap of ~3.5 eV, as determined by UV-Vis. On the other hand, the film made of pure pigment has an intense PL emission. The SnO_2/Pigment film also shows PL emission in the same range as that observed in pure pigment film. Thus, the SnO_2/Pigment structure can also potentially be used as a wavelength shifter matrix. Its PL intensity is lower than expected, considering that the film contains the same amount of pigments as the film made with pigment alone. One of the reasons may be associated with the chemical interaction of the SnO_2 precursor with the pigment molecules. However, this type of procedure may be an option to incorporate a wavelength shifter into an anti-reflective coating, which can be used directly on top of the solar cell. More investigations are needed to evaluate the effect of this film structure on a solar cell.

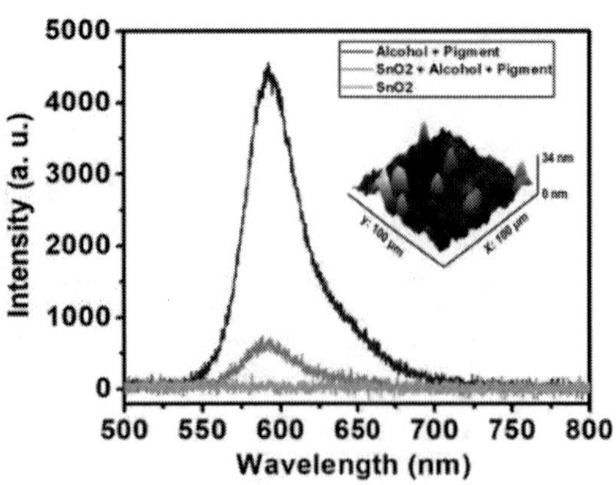

Figure 6: Photoluminescence emission of pristine pigment, SnO_2 and SnO_2/Pigment films. The inset is an AFM image of the SnO_2/Pigment film.

4 CONCLUSION

Optical and structural properties of low-cost fluorescent pigment thin films have demonstrated their function as wavelength shifters. The incorporation of the pigment in an SnO_2 matrix produces a film with similar characteristics, also functioning as a wavelength shifter. These results suggest the potential for using wavelength-shifting compounds incorporated into an anti-reflective coating for application in solar cells.

5 ACKNOWLEDGMENT

The authors gratefully acknowledge support from BYD Energy Brazil through the PADIS/MCTI program, Project No. 5779/FUNCAMP, FAPESP (São Paulo Research Foundation, Project number 2017/11986-5), INCT/INES/CNPq (grant 465423/2014-0), CNPq (grants 302349/2021-9, 313703/2021-3, 435260/2018-9), the multi-use laboratory of the Gleb Wataghin Physics Institute (LAMULT) and LARX-UEL.

REFERENCES

[1] Best Research-Cell Efficiency Chart, available at www.nrel.gov/pv/cell-efficiency.html. Accessed on 09/12/2022.

[2] T. Trupke, M. A. Green and P. Würfel, Journal of Applied Physics 94 (3) (2002) 1668-1674.

[3] Borshchev, O. V., Surin, N. M., Skorotetcky, M. S. and S. A. Ponomarenko, Ineos Open 2 (4) (2019) 112–123.

[4] C. Strümpel, M. McCann, G. Beaucarne, V. Arkhipov, A. Slaoui, V. Švrček, C. Del Canizo, and I. Tobias, Sol. Energy Mater. Sol. Cells 91(4) (2007) 238–249.

ASSESSING THE EFFECTIVENESS OF ENCAPSULATION SCHEMES FOR PEROVSKITE DEVICES USING PHOTOLUMINESCENCE IMAGING

George Koutsourakis[1*], Simone M. P. Meroni[2], James C. Blakesley[1], Ross Dix-Peek[3], Richard Dixon[4], Ernest Van Dyk[3], Pufinji Obene[5], Ian Arnold[5], Tateos Tvapanyan[5], Jacqueline Crozier McCleland[3], Roelof Roodt[3], Frederik Vorster[3], Alexander G. R. Howe[4], Trystan M. Watson[2], Fernando A. Castro[1]

[1]National Physical Laboratory (NPL), Hampton Road, Teddington, Middlesex, TW11 0LW, United Kingdom
[2]SPECIFIC, Swansea University, Bay Campus, Swansea SA1 8EN, United Kingdom
[3]Department of Physics, Nelson Mandela University, Port Elizabeth, 6011, South Africa
[4]Dycotec Materials, Unit 6 Stainer Road, Porte Marsh Industrial Estate, Calne, Wiltshire, SN11 9PX, United Kingdom
[5]Precision Varionic Intl. Ltd. Langley Road, Hillmead, Swindon, SN5 5WB, United Kingdom

ABSTRACT: Photovoltaic (PV) devices based on metal halide perovskite technology are typically more sensitive to environmental conditions, such as moisture and oxygen, than conventional inorganic PV materials like silicon. This means scalability and commercialization of perovskite PV technology are reliant upon highly effective and reliable encapsulation schemes. Evaluation of edge seal materials and encapsulation effectiveness can be challenging, especially if contactless schemes for characterisation are required. In this work, a contactless technique to evaluate encapsulation methods for perovskite devices is presented, based on photoluminescence (PL) imaging. Perovskite devices exhibit a strong PL emission signal when illuminated, while there are dramatic changes in the local PL emission when the devices degrade due to interactions with moisture or oxygen. Such changes in PL signal present an effective way to monitor the effectiveness of encapsulation schemes for perovskite devices. Triple-mesoscopic carbon perovskite mini modules were encapsulated with different schemes and subjected to damp heat aging tests at 85 °C/85 % relative humidity in the dark. PL imaging was used at different times during ageing to monitor the spatial PL emission of samples. In addition, encapsulated devices were placed outdoors and were monitored by PL imaging at frequent intervals in order to compare outdoor exposure with accelerated tests. It has been demonstrated that encapsulation failure can be identified at different times during ageing, with an observed decrease in measured short circuit current of the aged device. Novel encapsulation schemes can be tested using this methodology and compared with more common encapsulation methods.
Keywords: Perovskites, Photoluminescence, Characterisation, Reliability

1 INTRODUCTION

Photovoltaic (PV) devices based on perovskite solar cells (PSCs) are typically much more sensitive to environmental conditions such as moisture and oxygen compared to conventional inorganic PV materials like silicon [1]. As a result, scalability and commercialisation of this technology are dependent on highly effective and reliable encapsulation schemes [2]. Low quality encapsulation configurations usually used in research laboratories are insufficient to provide protection to PSC devices for outdoor use or standard reliability testing [3]. Evaluation of edge seal materials and encapsulation effectiveness can be challenging [4], especially if contactless schemes for characterisation are required for fast, inline inspection.

In this work, a contactless technique to evaluate encapsulation methods for perovskite devices is presented, based on photoluminescence (PL) imaging. Perovskite devices exhibit a strong PL emission signal when illuminated, and PL signal has already been demonstrated for monitoring perovskite crystal formation [5]. It has also been shown that dramatic changes occur in the local PL emission when the devices degrade due to interactions with moisture or oxygen [6]. Such changes in PL signal present an effective way to monitor the effectiveness of encapsulation schemes for perovskite devices.

Mini modules based on triple-mesoscopic carbon perovskite solar cells (mCPSC) [7] were encapsulated with different schemes and damp heat aging tests were performed at 85 °C/85 % R.H. in the dark. PL imaging was used at different time-points during ageing to monitor the spatial PL emission of samples. In addition, encapsulated devices were placed outdoors at Nelson Mandela University (NMU) facilities, South Africa, in February (summer month for South Africa), and were monitored by PL imaging at frequent intervals, in order to compare outdoor exposure with indoor accelerated tests. It is shown that encapsulation failure can be observed at different time-points during ageing. This demonstrates that large area PL imaging is an appropriate tool to evaluate encapsulation of perovskite devices.

2 EXPERIMENTAL SETUP

A multispectral, large area PL imaging system has been developed at the National Physical Laboratory (NPL) along with a nearly-identical system developed in parallel at NMU. The PL imaging system consists of 16 multispectral LED arrays that include 4 different excitation wavelengths to allow excitation of a range of samples, from single cell to tandem samples, up to 200 mm × 200 mm. The available nominal excitation wavelengths of the system are 380 nm, 450 nm, 625 nm and 850 nm. CMOS cameras are used for both the NPL and NMU systems, with a configuration of filtering and lens settings depending on the samples under study. For the experiments of this work, 450 nm (blue) or 625 nm (red) LED excitation were used at different intensity levels. A GaAs reference cell was used to ensure stable and repeatable irradiance intensity at the sample plane during all measurements. The PL images captured always included both the PSC device to be measured and the GaAs reference, so that the PL signal from the GaAs cell is used to normalise the PSC device PL images. In this way the PL images can be compared with each other regardless of changes in exposure time settings. This normalisation process has been applied to all images presented in this work. A 700 nm longpass filter was used for filtering, as this rejects LED illumination but transmits perovskite PL emission, the peak of which for these samples is expected between 750 nm and 800 nm. The schematic of the PL imaging system and a picture of the sample and reference platform under illumination at the PL imaging system at NPL are presented in **Figure 1**.

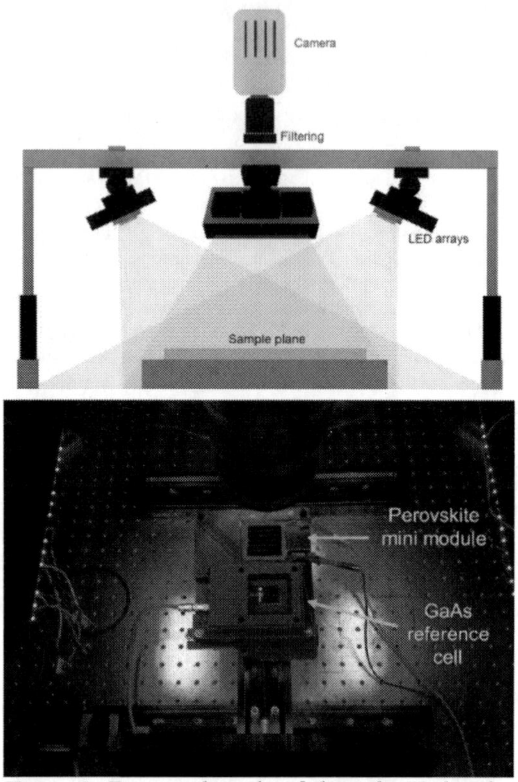

Figure 1. Top, a schematic of the twin PL imaging systems built at NPL and NMU. Bottom, a PSC mini module next to the GaAs reference, placed on a 3D printed sample platform under the LED PL imaging system at NPL

The mini modules tested in this work consist of 6 cells in series and are based on screen printed layers of mesoporous TiO_2, ZrO_2, and carbon, which are then infiltrated with the perovskite solution, as described in [8]. The modules have a size of 50 mm × 50 mm. Samples of the same batch were sent both to NPL and NMU so that indoor and outdoor degradation could be compared. During each degradation period- either indoor damp heat testing in the dark or outdoor exposure - a reference mini module of the same batch of samples was also kept in a desiccator in the dark and tested with PL imaging at the same intervals as the exposed samples. Samples presented in this work were encapsulated with a simple method at SPECIFIC (glass/glass), with a foil of polyurethane, edge sealed with an ethylene-vinyl acetate (EVA) foil and a glass sheet from the top with a second edge sealant of epoxy glue. No temperature control was applied during measurements.

3 RESULTS

Results of PL imaging with different excitation wavelengths, at intervals during damp heat exposure at NPL are presented in Figure 2 using red light, and Figure 3 using blue light for excitation. The measured irradiance level at the sample plane was 0.5 suns equivalent in terms of induced current for the GaAs reference cell. The PSC mini module sample was taken out of the damp heat chamber for PL measurements at each interval. In the measurements presented here, after the 1 h and 4 h duration exposures, the sample was kept in the dark in the desiccator overnight, and a small regeneration of performance was observed each time.

Different excitation wavelengths have a different penetration depth within the material, which can show slightly different spatial features. In the results below it can be observed that the encapsulation has failed already in the first degradation period. In the initial PL images of Figure 2 and Figure 3 the PL emission is rather uniform for all cells, with some local defects being visible. After 1 hour 85 °C/ 85 % R.H the PL image intensity distribution of the sample for both red and blue excitation has changed dramatically, with an increase in PL signal of different magnitude for various areas of the sample. After another 4 hours of damp heat exposure a PL emission decline can be observed with some areas providing no emission at all. It has been already reported in the literature that moisture exposure significantly increases the intensity of PL emission in metal halide perovskite materials [9]. Further exposure to both moisture and O_2 simultaneously enhances the overall degradation of the perovskite lattice, causing a decrease in PL signal, which also agrees with previous work [10].

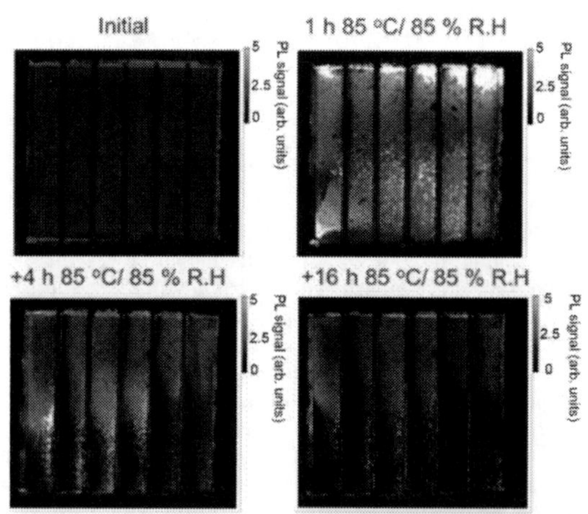

Figure 2. PL imaging of a mPSC mini module, at specific intervals during damp heat exposure using the 625 nm (red) LED arrays. Irradiance levels were at 0.5 suns equivalent in terms of induced current for the GaAs reference cell.

The measured short-circuit current (I_{SC}) of the device under the 625 nm (red) excitation, at 0.5 suns equivalent current for the GaAs reference, is presented in Figure 4. It can be observed that there is a significant decrease in I_{SC} even after the first hour of damp heat exposure, which indicates loss of active material, which is consistent with reduced PL intensity. There is evidence of weak regeneration of the device when the sample is kept in the desiccator in the dark overnight, but it is much smaller than the permanent degradation of the sample. After 21 h total damp heat exposure at 85 °C/ 85 % R.H there is approximately a 75 % loss of photocurrent for the device. The reference mini module kept in the dark in the desiccator was also measured during these runs and exhibits no decrease in I_{SC} or noticeable change in PL (not presented here). Optical images of the degraded and reference mini modules at the end of the ageing tests are presented in Figure 5.

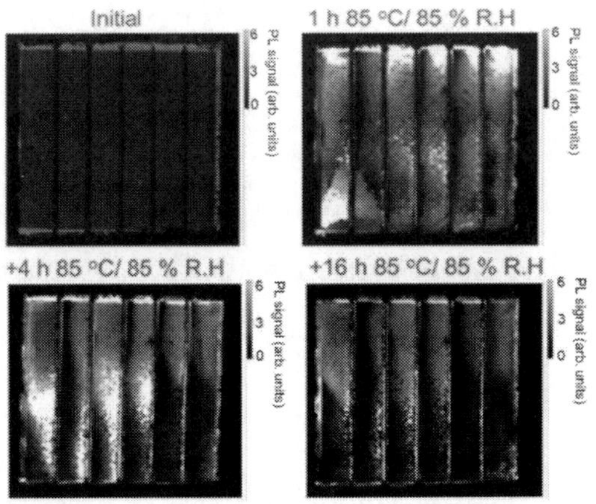

Figure 3 PL imaging of a mPSC mini module, at specific intervals during damp heat exposure using the 450 nm (blue) LED arrays. Irradiance levels were at 0.2 suns in terms of equivalent induced current for the GaAs reference cell.

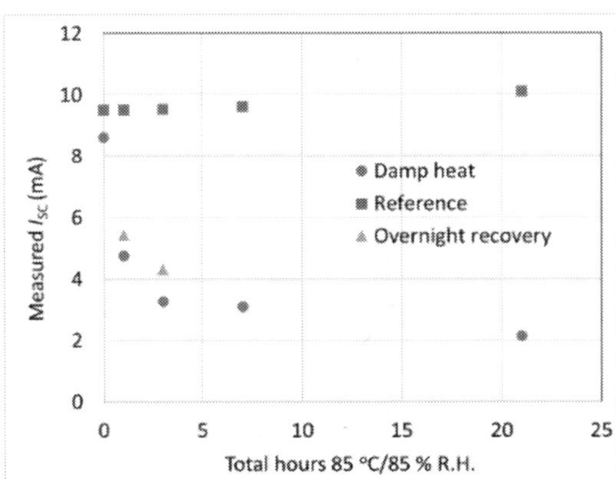

Figure 4. The measured I_{SC} current of the degraded and reference devices under the 625 nm (red) excitation, at 0.5 suns equivalent current for the GaAs reference cell.

Similar behaviour for the PL imaging results is observed with outdoor exposure of a PSC sample of the same batch at NMU. The outdoor setup that the mini-module was mounted upon is presented in **Figure 6**, a zoomed in image of the PSC mini-module sample outdoors is included in the same figure. The device temperature, ambient relative humidity and irradiance during outdoor exposure were measured and logged with 1 s resolution during outdoor exposure, the data are presented in **Figure 7**. The sample was taken indoors between the exposure intervals for PL imaging measurements.

Figure 5. An optical image of the degraded (left) and reference (right) mini modules.

Figure 6. Outdoor setup at NMU, South Africa, zoomed-in picture shows the perovskite mini-module.

Outdoor exposure was performed on different days. As can be observed in Figure 7, there is some variability of the outdoor conditions during the outdoor exposure. The irradiance in the plane of the array varies during the day, and on the last 4 hour interval there is a significant increase in relative humidity with a small drop of temperature compared to previous intervals.

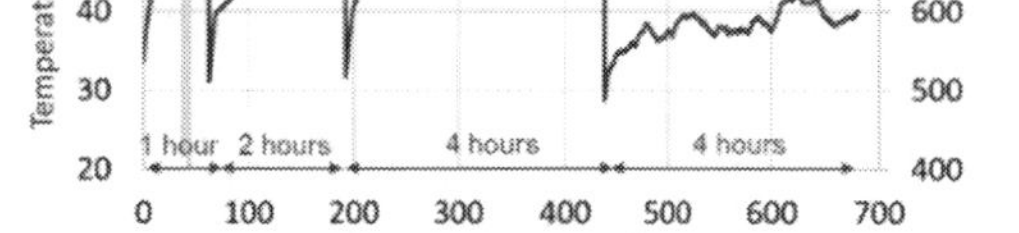

Figure 7. Device temperature, relative humidity and irradiance measurements during the outdoor exposure of the perovskite sample.

PL imaging results from the system at NMU at specific time-intervals during outdoor exposure are presented in Figure 8. Results using 625 nm (red) LED excitation are presented. The irradiance levels at the sample plane for excitation are set in the same way as the NPL system, using a GaAs reference cell. Degradation is obvious in the PL images, with a large portion of the active area in the centre of the sample degraded at the end of the exposure, and the timescales are comparable with the damp heat tests at NPL, regarding visible spatial degradation (loss of active area). The NMU PL imaging setup uses a different camera model, resulting in a different dynamic range for the images. The progression of spatial degradation is different than in the case of indoor damp heat testing, as it expands from the centre rather than from the edges as in the damp heat exposure case. Nevertheless, a statistical analysis with more samples is necessary for meaningful conclusions and for determining the causes of degradation.

From the results of both indoor accelerated testing (damp heat) and outdoor exposure of PSC samples, it is shown that PL imaging can be used to efficiently evaluate the effectiveness of encapsulation for PSC devices. Although just one sample was used in each case of indoor and outdoor degradation, it is apparent that any spatial changes due to degradation of PSC samples will be captured by PL imaging.

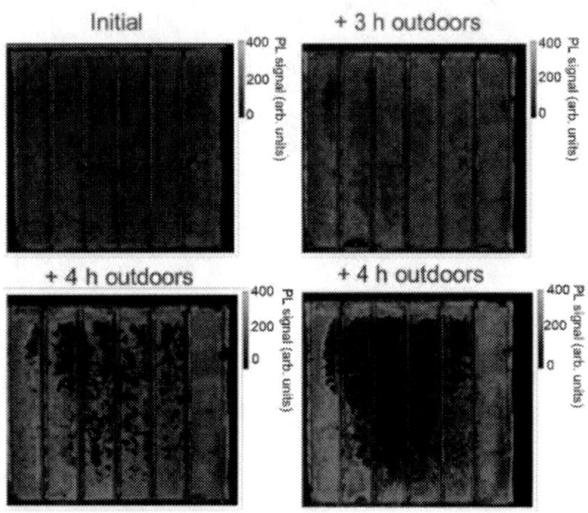

Figure 8. PL imaging of a mPSC mini module, at specific intervals during outdoor exposure, using the 625 nm LED arrays. Irradiance levels were at 0.5 suns in terms of equivalent induced current for the GaAs reference cell.

4 CONCLUSIONS

A multispectral, large area PL imaging system has been developed at NPL, with a twin system set up in parallel at NMU. The systems were used to perform PL imaging measurements on PSC samples undergoing damp heat exposure at NPL or outdoor exposure at NMU.

The results demonstrate that PL imaging can be used as a contactless method to monitor encapsulation effectiveness of perovskite devices. PL imaging results have been presented for both indoor damp heat ageing and outdoor exposure, demonstrating similar timescales for changes in PL images to occur. Although evidence of degradation is clearly observed, experiments with a larger number of samples and variable conditions are necessary to draw conclusions regarding the dominant causes of degradation and in order to decouple the effects of temperature, moisture, UV exposure and irradiance.

ACKNOWLEDGMENTS

This work was co-funded by UK Research and Innovation (UKRI) through the Innovate UK Energy Catalyst project "USPV Ultrasealing for Photovoltaics" (grant number 105907) and by the UK Government's Department for Business, Energy and Industrial Strategy (BEIS) through the UK's National Measurement System programmes.

REFERENCES

[1] S. Zhang and G. Han, "Intrinsic and environmental stability issues of perovskite photovoltaics," *Prog. Energy*, vol. 2, no. 2, p. 022002, May 2020, doi: 10.1088/2516-1083/ab70d9.

[2] C. A. Aranda, L. Caliò, and M. Salado, "Toward commercialization of stable devices: An overview on encapsulation of hybrid organic-inorganic perovskite solar cells," *Crystals*, vol. 11, no. 5, pp. 1–16, 2021, doi: 10.3390/cryst11050519.

[3] Q. Emery *et al.*, "Encapsulation and Outdoor Testing of Perovskite Solar Cells: Comparing Industrially Relevant Process with a Simplified Lab Procedure," *ACS Appl. Mater. Interfaces*, vol. 14, no. 4, pp. 5159–5167, Feb. 2022, doi: 10.1021/acsami.1c14720.

[4] M. Owen-Bellini, D. B. Sulas-Kern, G. Perrin, H. North, S. Spataru, and P. Hacke, "Methods for In Situ Electroluminescence Imaging of Photovoltaic Modules Under Varying Environmental Conditions," *IEEE J. Photovoltaics*, vol. 10, no. 5, pp. 1254–1261, Sep. 2020, doi: 10.1109/JPHOTOV.2020.3001723.

[5] L. Wagner *et al.*, "Distinguishing crystallization stages and their influence on quantum efficiency during perovskite solar cell formation in real-time," *Sci. Rep.*, vol. 7, no. 1, p. 14899, Dec. 2017, doi: 10.1038/s41598-017-13855-6.

[6] T. Kirchartz, J. A. Márquez, M. Stolterfoht, and T. Unold, "Photoluminescence-Based Characterization of Halide Perovskites for Photovoltaics," *Adv. Energy Mater.*, vol. 10, no. 26, p. 1904134, Jul. 2020, doi: 10.1002/aenm.201904134.

[7] S. M. P. Meroni, C. Worsley, D. Raptis, and T. M. Watson, "Triple-Mesoscopic Carbon Perovskite Solar Cells: Materials, Processing and Applications," *Energies*, vol. 14, no. 2, p. 386, Jan. 2021, doi: 10.3390/en14020386.

[8] S. M. P. Meroni *et al.*, "Scribing method for carbon perovskite solar modules," *Energies*, vol. 13, no. 7, 2020, doi: 10.3390/en13071589.

[9] K. P. Goetz, A. D. Taylor, F. Paulus, and Y. Vaynzof, "Shining Light on the Photoluminescence Properties of Metal Halide Perovskites," *Adv. Funct. Mater.*, vol. 30, no. 23, p. 1910004, Jun. 2020, doi: 10.1002/adfm.201910004.

[10] N. Aristidou, C. Eames, M. S. Islam, and S. A. Haque, "Insights into the increased degradation rate of $CH_3NH_3PbI_3$ solar cells in combined water and O_2 environments," *J. Mater. Chem. A*, vol. 5, no. 48, pp. 25469–25475, 2017, doi: 10.1039/C7TA06841G.

EFFICIENT CHARGE CARRIER EXTRACTION IN PEROVSKITE-SILICON TANDEM SOLAR CELLS INVESTIGATED BY NUMERICAL SIMULATION

Christoph Messmer[1,2], Jonas Schön[1,2], Uli Würfel[2,3], Patricia S. C. Schulze[2], Martin C. Schubert[2], Martin Bivour[2], Stefan W. Glunz[1,2], Martin Hermle[2]
[1] INATECH, University of Freiburg, Emmy-Noether-Str. 2, 79110 Freiburg, Germany
[2] Fraunhofer Institute for Solar Energy Systems ISE, Heidenhofstr. 2, 79110 Freiburg, Germany
[3] Materials Research Center FMF, University of Freiburg, Stefan-Meier-Str. 21, 79104 Freiburg, Germany

ABSTRACT: This work presents a comprehensive opto-electrical simulation model in Sentaurus TCAD for perovskite-silicon tandem devices which incorporates sophisticated physical models that allow for an in-depth study of efficient charge carrier extraction in tandem devices.

Keywords: photovoltaics, perovskite-silicon tandem solar cells, simulation

1 INTRODUCTION

Perovskite-silicon tandem devices exceeded the 30% efficiency threshold in laboratory [1] and are supposed to be suitable for low-cost mass-production in the near future [2, 3]. Besides the experimentally remarkable success within the last years, the main focus of simulation work was put into optical modelling, whereas electrical modelling somewhat lagged behind. However, electrical modeling could play an important role in understanding and systematically reducing the limitations in current tandem solar cells.

The aim of this and future work is to understand the present limitations of recently published tandem devices and to help overcoming the limitations by discrete guidelines which boost the tandem efficiency.

2 SIMULATION MODEL

We elaborated a full opto-electrical simulation model in Sentaurus TCAD [5] that is capable to describe state-of-the-art perovskite-silicon tandem solar cells [4, 6]. Figure 1 shows the full layer stack of the perovskite-silicon tandem device. The bottom cell is a silicon heterojunction (SHJ) with textured rear side. The planar front side features the a-Si electron contact followed by the ITO recombination junction towards the top cell's hole transport layer (HTL) which is made up of a 1-nm-thick 2PACz layer acting as self-assembled monolayer. The top cell is featuring a triple cation perovskite absorber with a C_{60}/SnO_x stack as electron transport layer (ETL) that is followed by an ITO for lateral conductance towards the silver fingers. The MgF_2 layer is used to minimize the reflection of the device.

Optical modelling is performed using raytracing in crystalline silicon and transfer-matrix-method (TMM) for the top and bottom cell thin-film layer stacks (including the perovskite absorber), which was experimentally validated in a previous publication [2] for the 25.1%-efficient perovskite-silicon tandem cell of Schulze *et al.* [6]. The optical model was used to improve the tandem devices by a proposed optical roadmap which was capable to describe the most recently published perovskite-silicon tandem device by Heydarian *et al.* [4].

The electrical model for the silicon bottom cell and tunneling at the TCO recombination junction is used as published in several previous publications [2, 7]. The innovation of this paper is the extension of electrical model of the perovskite top cell including all relevant layers shown in Figure 1. It includes a drift-diffusion model for the mobile anions and cations within the perovskite absorber and allows for an in-depth study of the charge-carrier extraction at the electron and hole transport layer (in this study: C_{60} and 2PACz, respectively) of the perovskite top cell.

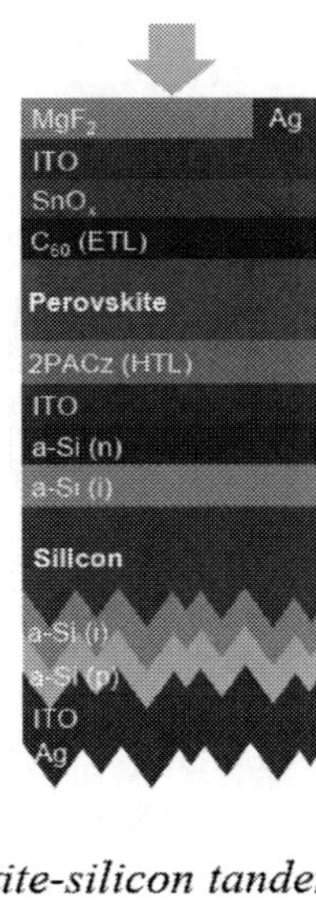

Figure 1: *Layer stack of the perovskite-silicon tandem solar cell as published in [4].*

3 RESULTS

3.1 Investigation of efficient charge carrier extraction

Requirements for efficient charge carrier extraction at the top cell's electron and hole contact layers are investigated by assessing the band diagrams of the perovskite-silicon tandem device as shown in Figure 2. Figure 2(a) shows the band diagram at open-circuit conditions V_{oc}. One can see that SnO_x/C_{60}/perovskite interface at which efficient electron extraction is required. The parameters of the C_{60} and perovskite were chosen according to measurements following the publication of [8]. At the C_{60}/perovskite interface, the high non-radiative recombination reported in literature [4, 8] is accounted for by Shockley-Read-Hall recombination varying the hole recombination velocity $S_{0,h}^{ETL}$ (shown in gray) at this interface. One can see that for very high $S_{0,h}^{ETL}$, that the internal voltage of the perovskite top cell iV_{oc}^{Pero} drops significantly revealing a passivation loss. Furthermore, the Quasi-Fermi level of the electrons shows a significant drop towards the ETL/Pero interface which is highlighted in Figure 2b as ΔV_{oc}, revealing further extraction losses due to the highly recombination active ETL/Pero interface.

The hole extraction of the top cell 2PACz layer was modelled by a 1-nm-thick layer with dipole moment of $5 \cdot 10^{13}$ cm^{-2} which effectively changes the ITO work function WF_{ITO} (from around 4.8 eV to 5.4 eV according to [9] facilitating the hole extraction. The 2PACz/ITO junction allows for band-to-band tunneling of holes from the 2PACz side with electrons from the ITO side which electrically interconnects the top and bottom cell of the tandem device. Figure 2a also shows the electron contact of the silicon bottom cell featuring amorphous silicon (i.e. an a-Si(i)/a-Si(n) stack). Electron tunneling at the a-Si(n)/ITO barrier is accounted for in the model.

The anion and cation density within the perovskite top cell is shown in Figure 2 in green and light blue, respectively (see right axis). For the negatively charged anions, a diffusion coefficient of $1 \cdot 10^{-12}$ cm²/s was assumed whereas the diffusion coefficient of the positively charged cations is assumed to be one order of magnitude lower according to measurement reported in literature concerning triple cation perovskites [10]. In quasi-steady state (i.e. holding the transient simulations sufficiently long) the negatively charged anions accumulate at the ETL (shown in green), whereas the positively charged cations accumulate at the HTL (shown in blue). This increases the requirements on efficient charge carrier extraction since anions and electrons (cations and holes) compete at the ETL (HTL) interface.

Figure 2b shows the band diagram at short-circuit conditions j_{sc} (i.e., at zero voltage) after a 90 second long transient sweep from V_{oc} (Figure 2a). One can see that the distribution of the less mobile cations barely changed with respect to Figure 2a, whereas the mobile anions moved further towards the ETL due to the electric field within the perovskite absorber. The higher anion concentration at the ETL/Pero interface and lower anion concentration at the Pero/HTL interface raises the requirements for efficient electron and hole extraction at the respective interfaces which is typically visible in the forward scan from j_{sc} to V_{oc} (shown in the next subsection).

3.2 Influence on the JV parameters

Figure 3 shows the simulated transient JV curves of the bottom (red), top (blue) and tandem device (green) for the reverse (i.e., V_{oc} to j_{sc}, solid line) (i.e., j_{sc} to V_{oc}, dashed line), respectively. One can see a high current density of about 19.6 mA/cm² of the current-matched device according to the experimental devices presented by Heydarian *et al.* [4]. The hysteresis of the perovskite top cell (i.e., difference between reverse and forward scan) is almost negligible which is in line with [4]. However, the potential of more efficient charge carrier extraction can be visualized by plotting the internal voltage curves (see dotted lines for perovskite top cell and tandem device). At V_{oc}, the difference ΔV_{oc} between the internal voltage iV_{oc} and external voltage is shown (orange). This can be attributed to the high SRH recombination at the ETL/Pero interface which is currently being investigated to further improve the electron extraction at this interface. At the MPP the ΔV_{mpp}. Here, a reduction of the effective series resistance of the device and more selective ETL and HTL layers could further improve the charge carrier extraction which is subject to future research.

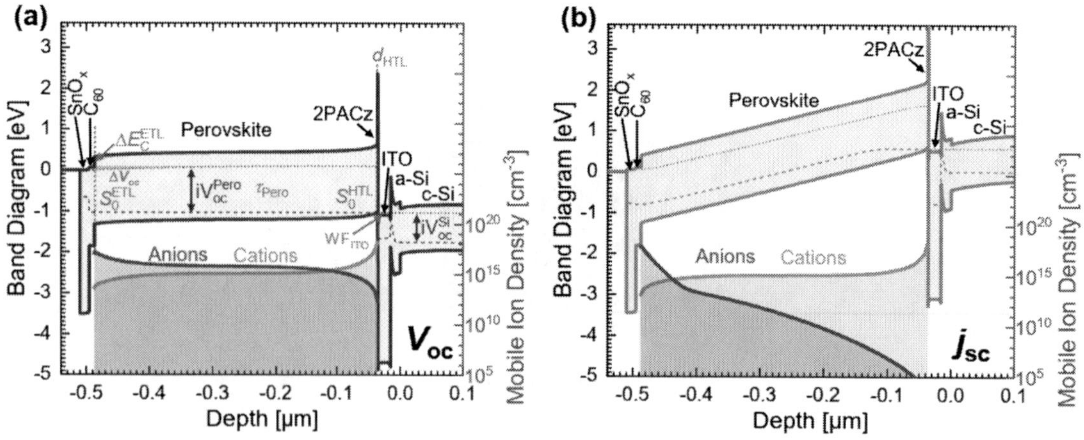

Figure 2: ***(a)*** *Band diagram at* V_{oc} ***(b)*** *at* j_{sc}. *Right axis shows the density of mobile anions and cations within the perovskite absorber.*

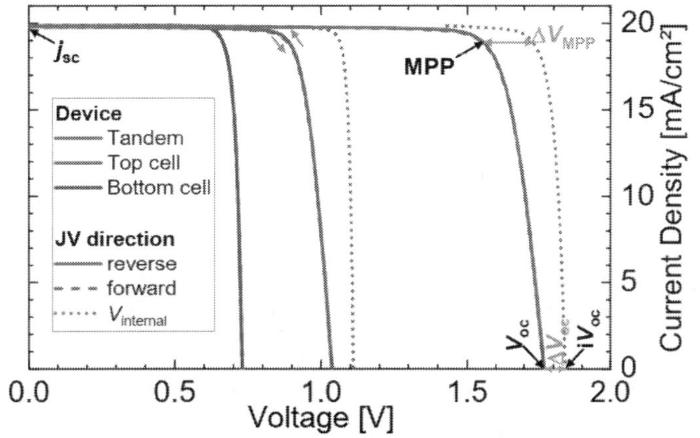

Figure 3: *Simulated JV curves of top (blue), bottom (red) and tandem (green) solar cell for reverse and forward scan as solid and dashed curves, respectively. The dotted curves show the internal voltage curves (i.e., the current density versus the averaged Quasi-Fermi level splitting within the respective absorber(s))*

4 CONCLUSION

In this work, we presented a full opto-electrical simulation model in Sentaurus TCAD which incorporates sophisticated physical models to investigate the charge carrier extraction in perovskite-silicon tandem cells. This comprehensive model allows investigate the charge carrier transport in perovskite-silicon tandem devices and to assess the opto-electrical properties of the device including band diagrams and the transient JV sweeps including migration of anions and cations within the perovskite absorber. It will be applied in future works to interpret experimentally observed research topics and will be further used to optimize perovskite-silicon tandem devices for higher efficiency and stability.

5 ACKNOWLEGEMENTS

This work was supported by the German Federal Ministry of Economic Affairs and Climate Action within the research projects "KATANA" (contract number 03EE1087A and 03EE1087B) and "PrEsto" (contract number 03EE1086A) and the Fraunhofer LIGHTHOUSE PROJECT (MaNiTU).

6 REFERENCES

[1] EPFL, *New world records: perovskite-on-silicon-tandem solar cells: EPFL and CSEM smash through the 30% efficiency barrier for perovskite-on-silicon-tandem solar cells—setting two certified world records.*, 2022.

[2] C. Messmer, B. S. Goraya, S. Nold, P. S. Schulze, V. Sittinger, J. Schön, J. C. Goldschmidt, M. Bivour, S. W. Glunz, and M. Hermle, "The race for the best silicon bottom cell: Efficiency and cost evaluation of perovskite–silicon tandem solar cells," *Prog. Photovolt: Res. Appl.*, vol. 29, no. 7, pp. 744–759, 2021.

[3] L. A. Zafoschnig, S. Nold, and J. C. Goldschmidt, "The race for lowest costs of electricity production: techno-economic analysis of silicon, perovskite and tandem solar cells," *IEEE Journal of Photovoltaics*, vol. 10, no. 6, pp. 1632–1641, 2020.

[4] M. Heydarian, C. Messmer, A. J. Bett, M. Heydarian, D. Chojniak, Ö. S. Kabakli, L. Tutsch, M. Bivour, F. Schindler, G. Siefer, M. C. Schubert, J. C. Goldschmidt, M. Hermle, S. W. Glunz, and P. S. Schulze, "Maximizing Current Density in Monolithic Perovskite Silicon Tandem Solar Cells," *Solar RRL*.

[5] Synopsis, "Sentaurus TCAD: Release Q-2019.12," http://www.synopsys.com, 2019.

[6] P. S. C. Schulze, A. J. Bett, M. Bivour, P. Caprioglio, F. M. Gerspacher, Ö. Ş. Kabaklı, A. Richter, M. Stolterfoht, Q. Zhang, D. Neher, M. Hermle, H. Hillebrecht, S. W. Glunz, and J. C. Goldschmidt, "25.1% High-Efficiency Monolithic Perovskite Silicon Tandem Solar Cell with a High Bandgap Perovskite Absorber," *Sol. RRL*, vol. 4, no. 7, p. 2000152, 2020.

[7] C. Messmer, J. Schön, S. Lohmüller, J. Greulich, C. Luderer, J. C. Goldschmidt, M. Bivour, S. W. Glunz, and M. Hermle, "How to make PERC suitable for perovskite–silicon tandem solar cells: A simulation study," *Prog. Photovolt: Res. Appl.*, https://onlinelibrary.wiley.com/doi/full/10.1002/pip.3524, 2022.

[8] D. Menzel, A. Al-Ashouri, A. Tejada, I. Levine, J. A. Guerra, B. Rech, S. Albrecht, and L. Korte, "Field Effect Passivation in Perovskite Solar Cells by a LiF Interlayer," *Advanced Energy Materials*, p. 2201109, 2022.

[9] J. Hu, W. Fu, X. Yang, and H. Chen, "Self-assembled monolayers for interface engineering in polymer solar cells," *Journal of Polymer Science*, 2022.

[10] B. Li, C. Kan, P. Hang, Y. Fang, L. Zuo, L. Song, Y. Zhang, D. Yang, and X. Yu, "Understanding the Influence of Cation and Anion Migration on Mixed-Composition Perovskite Solar Cells via Transient Ion Drift," *Physica Rapid Research Ltrs*, vol. 15, no. 9, p. 2100225, 2021.

This is a short version of the paper submitted to the journal EPJ Photovoltaics.

It has been selected by the Committee of the WCPEC-8 for submission to EPJ Photovoltaics.

IDENTIFYING DEFECTS ON SOLAR CELLS USING MAGNETIC FIELD MEASUREMENTS AND ARTIFICIAL INTELLIGENCE TRAINED BY A FINITE ELEMENT MODEL

Kjell Buehler[1], Kai Kaufmann[2,3], Markus Patzold[2], Mawe Sprenger[2], Stephan Schoenfelder[1]
[1] Leipzig University of Applied Sciences, Faculty of Engineering, Leipzig, Germany
[2] DENKweit GmbH, Halle, Germany
[3] Hochschule Anhalt – University of Applied Sciences, Köthen, Germany

Renewable energies have an increasing share in the energy supply. In order to ensure the security of this supply, the reliability of the systems is therefore increasingly important. In photovoltaic modules or in manufacturing defective solar cells due to broken busbars, cross-connectors or faulty solder joints must be detected quickly and reliably. This paper shows how the magnetic field imaging method can be used to detect defects in solar cells and modules without contact during operation. For the evaluation of the measurement data several neural networks were used, which were trained with the help of results from finite element simulations. Different training data sets were set up in the simulation model by varying the electrical conductivities of the different parts of the solar cell. There is an advantage of a combination of simulated and experimental training data, which is presented and discussed.
Keywords: Defects, Magnetic Field, Neural Networks, Machine Learning, Finite Element Analysis

1 INTRODUCTION

In Germany, photovoltaics (PV) account for the second-largest share of electricity generation systems among renewable energies after wind power. Renewable energies accounted for 42.3% of the total amount of electricity fed into the grid in Germany in 2019 [1]. In order to ensure a safe power supply, the reliability of PV systems is becoming increasingly important. Typical defects of PV modules are defect solder joints, busbars or cross connectors as well as broken solar cells, which are all leading to a decreasing yield. The mentioned defects are typically detected and analyzed by electroluminescence (EL) or thermography. Both methods have disadvantages. With EL, the measurements of solar parks can only take place at night to provide shading of the modules and the modules must be supplied with external power. Thermography measurements can also be carried out on a cloudless day, but since the voltage drop in the p-n junction is primarily considered here, minor defects such as broken solder points cannot be detected. As another analyzing technique, DENKweit, a startup from Halle (Saale), used the Magnetic Field Imaging (MFI) method and developed the 'B-Tech' technology which is a magnetic field measurement system for PV cells and modules which allows to measure contactless the amount and direction of electric currents changed due to defects [6], [7] [8]. The magnetic field can either be measured manually or automatically.

If solar modules are measured in production or even in field, a big amount of data will be generated and needs to be analyzed reliably and automatically. Thus, DENKweit provides a platform for vision AI. The AI can detect different defects, based on the training data set. In order to optimize the data set, the results of a finite element (FE) simulation were used for the AI training, which has become an increasingly common practice in recent years as in [2], [3], [4], [5].

In this work, an electromagnetic FE model of a four busbar solar cell is presented, which is used to generate training data for artificial neural networks which are then used to analyze real measurement data. The advantages of using a finite element model for the data generation are:

- increasing number of training data leading to an improved and more accurate prediction by the AI
- saving time for measuring, but having a sufficient amount of training data
- a wide variation of possible defects regarding size or position can be considered
- getting data before a technical system is in operation and measurable (e.g. new solder or busbar concept).

Thus, the FE simulation model is applied in this work and results are presented and discussed.

2 MATERIALS AND METHODS

2.1 Magnetic Field Imaging and Experimental Setup

Under working conditions or when power is supplied, solar modules show a characteristic distribution of electric current. The current density is greatest in the busbars. Electric current generates a magnetic field, which can be described by a fundamental relationship in magnetostatics, the Biot-Savart law,

$$d\vec{B}(\vec{r}) = \frac{\mu_0}{4\pi} I d\vec{l} \times \frac{\vec{r} - \vec{r}'}{|\vec{r} - \vec{r}'|^3}$$

where B is the magnetic flux density, I is the current $d\ell$, an element of the conductor, the constant μ_0 is the permeability of free space, and $\vec{r}'$and $\vec{r}$ the locations of the conductor and sensor respectively. Especially in the case of busbars, which can be approximated as line conductors, this model can be used. Defects due to broken solder joints then lead to changes in the magnetic field due to the locally changed current flow, which in turn can be detected.

In [6], [7] and [8] the magnetic field analysis of a solar cell or module using a line sensor was presented. The publications give an overview of the detectable defects and their effects on the resulting magnetic field. Therefore, the sensor technology must cover the measurement range from a few µT to a few mT, the usually generated magnetic flux densities due to current strengths in solar modules.

For the investigation of the magnetic field of solar modules, two cases, the measurement in the laboratory under forward voltage and the measurement in the field under illumination, needs to be distinguished. In the laboratory, a forward voltage is applied to the module and an electric current of 8 A is applied which roughly equals 1000 W/m². In the field, the electric current is generated by solar irradiation. The magnetic field generated by this is directly proportional to the electric current, i.e. directly proportional to the irradiance. The validation measurement shown in this paper was made in the laboratory using a DENKweit B-Lab laboratory instrument. Here a 16 cm wide line sensor with 64 data points along the line is integrated. The measuring frequency is 30 Hz. Field measurements for testing the performance of the neural networks are carried out with a hand-held device. The sensor technology corresponds to that in the B-Lab.

Figure 1 shows the example of a solar cell with a broken connector to the cross-connector. The connection failure means that no electrical current can flow from the cross connector into the busbar. In the magnetic field image, this is shown by a reduced flux density in the busbar concerned. Since the cell has cross-conductivity due to the grid fingers, the current density in the busbars is equalized over the length of the cell.

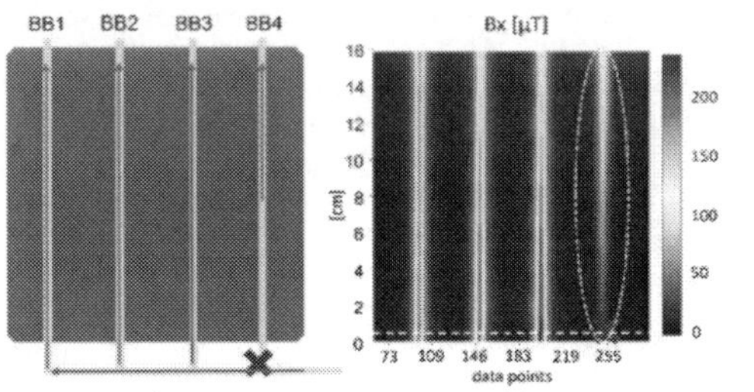

a) b)

Figure 1: As an example a four busbar cell with broken cross connector (left) does not allow current to flow through the faulty region, this leads to a reduced magnetic field in the corresponding busbar (right).

2.2 SIMULATION

In order to generate data for training the neural network, a model for a magnetostatic analysis was set up using the finite-element software ANSYS Mechanical. A four-busbar solar cell (156x156mm²) was modelled in consideration of the real layer stack consisting of the busbars, solder connections, grid fingers, the silicon wafer as well as a metallization of the entire back side (see Figure 2b). For simplification, the grid fingers were not explicitly modeled, but substituted by one volume with the same thickness as the original grid fingers. In this volume the electrical resistivity was reduced and the conductivity was restricted to the x and z direction to represent the electrical properties correctly. This procedure has already been shown in [9]. For busbar defects, the busbars are constructed from different volumes with corresponding electrical conductivities. As shown in Figure 2c exemplarily for busbars 1 and 2, the resistance of all busbars can be adjusted individually at the solar cell inlet and before and after each solder pad on the cell's backside. The cross-connectors were modeled with a better conductivity than the busbars to ensure that the electric current in all four busbars is the same and not distorted by ohmic losses. An additional ambient surrounding volume was modeled with the material properties of air to calculate the magnetic field around the solar cell. For more information about the dimensions, material and properties of the layers see Table I. The model is based on ohmic resistors and neglects solar cell physics like p-n junction or diodes. The relative permeability of all bodies was set to be 1.

The electric boundary conditions were applied on low-resistance cross connectors (CC) connecting the upper and lower busbars respectively. On the top CC, a current of 8A (approximating 1 sun) was set, on the bottom CC the electrical potential was set to 0V (see Figure 2a). A parallel magnetic flux on the outer surfaces of the model was defined as the magnetic boundary condition. The global mesh size was set to be 3mm, however, each line was divided into at least two elements.

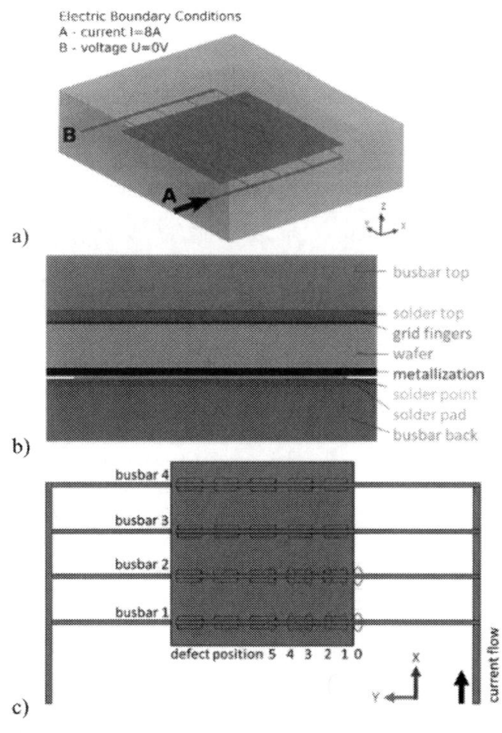

Figure 2: FE model of a four busbar solar cell showing the electric boundary conditions (a), a cross section of the layer stack (b) as well as an overview of the numbering of busbars and defect positions (c).

Table I: Information about the material properties and dimensions of the layers used in the FE-Model.

Body	**Specific El. Resistance [Ω·m]**	**Thickness (z) [µm]**	**Width (x) [mm]**
Busbars	$1.7\cdot10^{-8}$	200	3
Cross Connector	$1.7\cdot10^{-11}$	200	5
Solder	$1.45\cdot10^{-9}$	top=50 pad=30 point=10	top=3 pad=5 point=3
Grid Finger	$4.5\cdot10^{-8}$	10	
Metallization	$1.0\cdot10^{-6}$	40	

The model described above is used for both, validation and AI training data generation. For this purpose, variations of the model are calculated and combined into simulation sets according to their usage. The electrical conductivity of some solar cell components was adjusted in the simulation set in accordance with real solar cells, which can exhibit electrical and therefore magnetic behavior that deviates from that of a perfect solar cell. This may happen due to variation in manufacturing, subsequent assembly or slight damage during operation. In this way, for example, faulty solder points or defects can be modeled. The following simulation sets were set up and calculated:

For Validation of the Finite Element Model

1. **Solar cell with one defect busbar**
   For further FE model validation, the solar cell was calculated with a defective busbar (busbar 1 at the cell inlet "defect position 0", see Figure 2c).

For AI Training

2. **Solar cell without defects**
   To train the AI for anomaly detection simulations without busbar defects were performed. In this set, the electrical conductivities of the individual busbars, the grid fingers and the various solder connections were changed. The values from Table I were varied randomly in a range of 90...110%.
3. **Solar cell with busbar defects**
   In order to train the AI to recognize busbars and defective busbars (feature detection) simulations were carried out in which the electrical conductivity of the busbars was increased at various points (see Figure 2c). In addition, the input resistors of the busbars into the cell, the electrical conductivities of the individual busbars, the grid fingers and the various solder connections were varied randomly from 33...300%.

The used specific electrical resistances in Ω•m for the different cell components are shown in Table II.

Table II: Variation of the used specific electric resistances of the model parts in the simulation sets.

Specific El. Resistance [Ω•m]	Set 1	Set 2	Set 3
Busbars	$1.7 \cdot 10^{-8}$	$1.5...1.9 \cdot 10^{-8}$	$6.5...51.0 \cdot 10^{-9}$
Grid Finger (x & z)	$4.5 \cdot 10^{-8}$	$4.1...5.0 \cdot 10^{-8}$	$1.5...13.5 \cdot 10^{-8}$
Solder	$1.45 \cdot 10^{-9}$	$1.3...1.6 \cdot 10^{-9}$	$5.0...43.5 \cdot 10^{-10}$
Busbar Input	$1.7 \cdot 10^{-8}$	$1.5...1.9 \cdot 10^{-8}$	$1.0...2.4 \cdot 10^{-8}$
Busbar Defects	1.0	-	1.0

2.3 AI: Data Selection and Training

2.3.1 Datasets for Training and Testing

In order to be able to apply networks trained with artificial data to real data, the training data must have a similar variation with respect to typical characteristics as the real measured data. In the case of solar modules, this is the current strength in each busbar, which then determines the strength of the magnetic flux density in the measurement. In addition, measurements with a hand-held device are affected by variations due to operator influence. This includes a change in the measurement distance, which can result from tilting or a slight lifting of the measurement system. The training data were varied as described in chapter 2.2. The variation of the material parameters in the simulation model resulted in different magnetic flux densities in the busbars. In the data sets used for the training, the component B_x varied in a range of 38%. With these variations, effects that occur in the field due to local shading are covered as well.

In order to evaluate the influence of this variation, a data set of 75 images was created from set 2. The dataset was duplicated and supplemented with a few real measurement data to test the training with mixed data (simulation and MFI measurement data).

For the training of the neural networks for busbar defect detection, a data set with one broken busbar at the cell edge was used. In total, this data set consists of 82 images. Here two classes, "busbars okay" and "busbars broken" were introduced to distinguish both cases. Table III summarizes the datasets.

Table III: Training data sets

Dataset *S = Simulation data* *M = MFI data*	Training *S*	Training *M*	Validation *S*	Validation *M*
BB detection Simulation data only	50	0	25	0
BB detection Simulation + MFI data	50	7	25	4
BB defect detection Simulation data only	55	0	27	0

In order to test the busbar detection networks, 58 MFI measurements of single cells from the same solar module were used. The measurement was done with a handheld device. Some measurements contained 4, some three busbars, the total amount of visible busbars on the test images was 212. For testing the defect detection networks, a dataset consisting of 6 MFI images was used, each showing a solar cell with a single broken busbar, resulting in 6 defect BBs. Again, all measurements were performed with a handheld device (see Table IV).

Table IV: Test data sets, corresponding numbers of busbars and defect busbars.

	MFI Images	Sum of BBs	Defect BBs
BB detection	58	212	0
BB defect detection	6	16	6

2.3.2 Data pre-processing

In the field, especially with hand-held measurements, measurements might be slightly rotated with respect to each other, i.e. that the busbars are not parallel to the sensor bars. This results in the B_x component losing intensity with increasing rotation, but B_y gaining intensity. B_z shows invariance to rotations and is therefore used for neural network training. A Gaussian gradient filter is also applied to the image, to allow easy labeling (see Figure 3a). The result is then rotation invariant and represents qualitatively the position of electric conductors on the image. This filter also ensures that measurements of adjacent strings on a module are comparable. Otherwise, the reversed current flow leads to reversed measurement results. The current paths appear as line-like structures and are well suited to generate the necessary markers for training. The training data set is additionally artificially augmented by reflections, random rotations and contrast variations to account for effects such as a rotated sensor or differences in illumination.

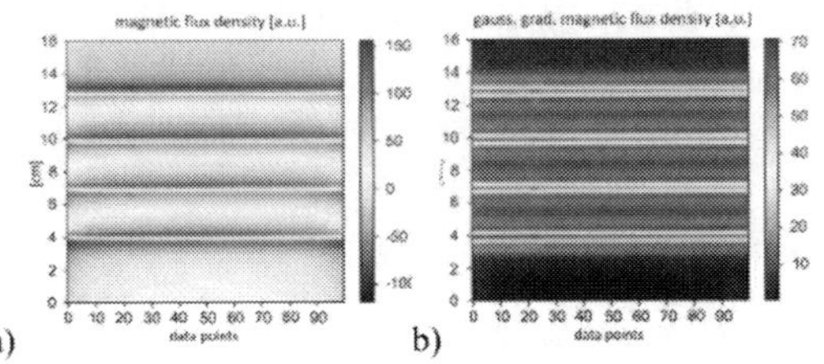

a) b)

Figure 3a): Magnetic flux density B_z component of a simulated solar cell, b): the same simulation with an applied Gaussian gradient filter.

2.3.3 Training

An instance segmentation network was trained using the DENKweit vision AI platform. This mesh type evaluates the image at pixel level, i.e. each pixel is assigned whether it belongs to an object class or not. In the second step, the pixels are separated to objects, something like a busbar. Training took place on a Nvidia RTX 2080Ti and took approximately 30 minutes.

3 RESULTS

3.1 Model Validation

In the first step, a validation of the FE model is carried out to check whether the results of the simulation agree with the experimental measured values. As described in chapter 2.2, a model with a defective busbar in the area of the cell entrance was calculated.

Figure 4 shows the magnetic flux density in the x direction 1.5mm above the upper busbars as a heatmap (Figure 4a) as well as the results along the shown path across the solar cell (Figure 4b). Simulation and experiment fits fairly well. One can clearly see the defective busbar (1), as well as the three intact ones (2-4). The first busbar shows no magnetic field at the inlet to the solar cell (y=0mm) which is caused by no electric current at this position. Busbars 2-4 have increased currents at this position. Along the busbars, the electrical currents equalize due to the cross-conductivity of the grid fingers. The wave-like structures along the busbars of the experimental data (see Figure 4c) being caused by the solder points are less visible in simulation.

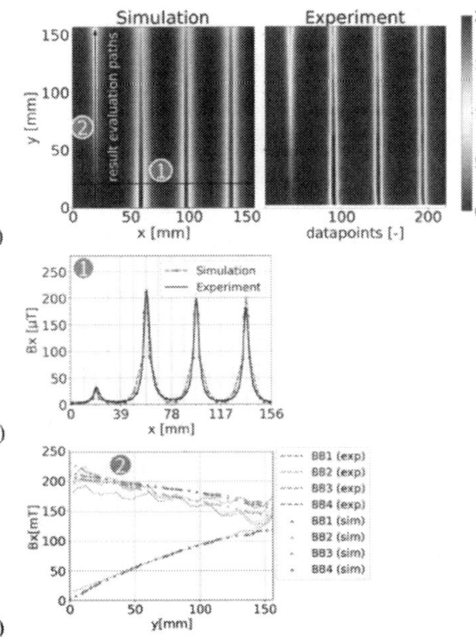

a) b) c)

Figure 4: X-component of the magnetic field of a solar cell with a defect first busbar as heatmaps (a), as a lineplot perpendicular to (b) and along (c) the busbars comparing the simulation and the experiment.

3.2 Automatic Data Analysis by AI

3.2.1 Detection of Busbars

The effectiveness of the neural networks is tested by application to the test data set. The AI marks objects on the images and assigns them a probability. Results with a probability below 30% were ignored. The number of busbars found that had a probability greater than 30% was then determined. The results can be found in Table V. Using simulation data only for training the neural network 98.11% of the busbars were labeled correctly. The additional real measurement data increases the correctly labeled busbars up to 100%. Thus, simulation data shows a good training effect reaching very good results. The networks can be made significantly more robust by adding real data.

Table V: Percentage of correctly labeled busbars

	Correctly labeled [%]	**Training type**
BB detection	98.11	Simulation data only
	100.00	Simulation + MFI data
BB defect detection	100.00	Simulation data only

For certainty of detection, the associated probabilities were statistically evaluated for each instance found. The result is summarized in the box plot in Figure 5. Adding real measurement data on a small scale of 14% (see Table 3), leads to an increase in detection probability (see Figure 5). It is noticeable that the network trained with the data mix show more outliers in the boxplots. The reason for this is that the probability distributions become narrower, but still contain some low values. Since the wisker limits are determined by the interquartile intervals, narrow distributions lead to more outliers.

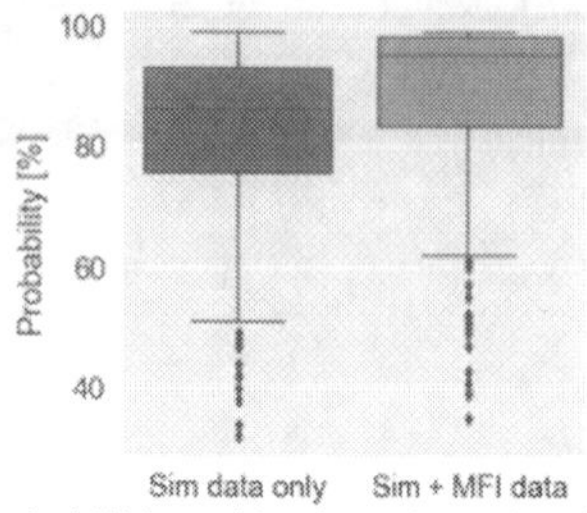

Figure 5: Probabilities of busbars detection for the neural network depending on the training data sets. Real measurement data increases the probabilities of detection.

3.2.2 Detection of Defect Busbars

Defects were reliably detected with the "busbars okay" and "busbars broken" class. 100% of the broken busbars were marked on the images. All busbars without defects were also foundFigure 6 shows an example result of a measurement evaluated by the neural network. The defect busbar was correctly marked, as well as the functioning busbars. The classification probabilities are in the range above 90% for both the intact busbars and the broken busbars.

a) 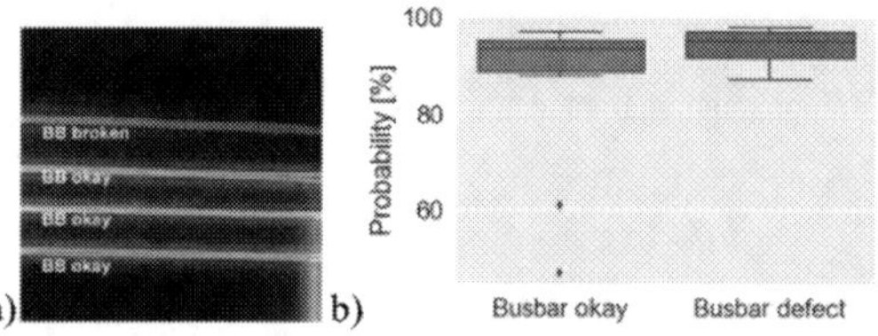 b)

Figure 6 a): example result with both intact and broken busbars labeled by the neural network, b) detection probabilities for intact and broken busbars.

4 DISCUSSION

The simulation model being used could be validated by the MFI measurements. The deviations between simulation and experiment were small. The effect of a defective busbar could be mapped realistically although the model was meshed relatively coarsely to keep the simulation times as short as possible. This is essential for an efficient creation of the AI training data sets.

The results show, it is possible to train neural networks by simulation data to analyze MFI magnetic field data. Though, it is crucial that the simulation data show a similar variation in typical properties as the real measured data. The data sets from the simulation can be extended with a few measurement data sets, if they are available. This leads to an increased number of features found as well as an increased probability to detect these features by the neural networks. Thus, simulation can help to train an AI if no or not enough measurement data are available. With simulation models the variance of the results can be adjusted quickly and purposefully, allowing the training region to be extended in a physically correct way, even if no measurement data are available for these regions. The simulation also helps to improve the understanding of the relationship between defects and the resulting magnetic field being measured. Furthermore, the demonstrated method can be used to model and train other solar cell defects such as defective solder points or short circuits in the silicon wafer in order to extend the detection capabilities.

5 SUMMARY

This paper presented the Magnetic Field Imaging method and showed how it can be used to find features and defects on solar cells and modules. The AI being used for this was trained by the results of FE simulations. For this purpose, an electromagnetic FE model of a four busbar solar cell was presented in this work. This model has been successfully validated with MFI measurements. Subsequently, the model was used for various defect scenarios providing defined and large number of training data.

By varying the electrical conductivities of individual cell components a training data set was generated. This set was partially supplemented by a few real measurements. With these training data sets a neural network was trained. By using a test data set, the recognition probability was analyzed. The results were very good on correctly labeled features. Extending the simulation training data with a few measurement data, the recognition rate could be even increased. Thus, FE models for generating artificial and physically correct data for AI training are beneficial for reliable AI usage.

ACKNOWLEDGMENT

This work was funded by the German Federal Ministry for Economic Affairs and Climate Action as part of the 'Central innovation programme for small and medium-sized enterprises (ZIM)' within the mobiInspec - MagPV project (funding number: 16KN083023).

REFERENCES

[1] Statistisches Bundesamt (Destatis), 2021, "13 Umwelt, Energie und Mobilität", Auszug aus dem Datenreport 2021

[2] T. Gantala, K. Balasubramaniam, "DPAI: A Data-driven simulation-assisted-Physics learned AI model for transient ultrasonic wave propagation", 2022, Ultrasonics Volume 121, 106671, https://doi.org/10.1016/j.ultras.2021.106671

[3] N. Ødegaard, A. O. Knapskog, C. Cochin and J. Louvigne, "Classification of ships using real and simulated data in a convolutional neural network," *2016 IEEE Radar Conference (RadarConf)*, 2016, pp. 1-6, doi: 10.1109/RADAR.2016.7485270.

[4] H Y Hsiao, K N Chiang, "AI-assisted reliability life prediction model for wafer-level packaging using the random forest method", 2020, Journal of Mechanics Volume 37, https://doi.org/10.1093/jom/ufaa007

[5] Yun Gao, Xiaoyang Liu, Jiawei Xiang, "FEM Simulation-Based Generative Adversarial Networks to Detect Bearing Faults", 2020, IEEE Transactions on Industrial Informatics Volume 16, Issue: 7, doi: 10.1109/TII.2020.2968370

[6] Paduthol A., Kunz, O., Kaufmann, K., Patzold, M., Lausch, D. and Trupke, T., "Magnetic Field Imaging: Strengths and limitations in characterising solar cells", 2019 IEEE 46th Photovoltaic Specialists Conference (PVSC), 2019, pp. 0822-0824

[7] Kaufmann, K., Lausch, D., Lin, C., Rudolph, M., Hahn, D. and Patzold, M., „Evaluation of the Quality of Solder Joints within Silicon Solar Modules Using Magnetic Field Imaging", (2021), Phys. Status Solidi A, 218: 2000292.

[8] D. Lausch, M. Patzold, M. Rudolph, C.-M. Lin, J. Fröbel, K. Kaufmann, in Proc. of the EU-PVSEC, Brussels 2018, pp. 1060–1064.

[9] U. Zeller, D. Lausch, M. Pander, K. Kaufmann, S. Slaby, S. Schoenfelder, "Comparison of magnetic field imaging (MFI) and magnetic field simulation of silicon solar cells", AIP Conference Proceedings 2147, 020021 (2019), https://doi.org/10.1063/1.5123826

EVALUATION INDOOR LIGHTING SIMULATOR FOR EMERGING PV

Yean-San Long, Min-An Tsai, and Teng-Chun Wu
Center for Measurement Standards, Industrial Technology Research Institute
Hsinchu, Taiwan

ABSTRACT: The measured I-V hysteresis is complicated by the vast array of different materials, lighting sources, and device architectures. In our study, we used the RTOS method, is shown call asymmetrical method, for I-V then the results showed better accuracy by eliminating in real-time the acceptance effect. We also used this method to compare emerging PV cells with the latter cells that exhibited promising power conversion efficiency and no hysteresis behavior under an indoor lighting simulator. In this study, we defined a performance ratio to compare indoor lighting and solar and the results will be shown different performance trends for the different types of PV in indoor/outdoor, which depends on the device. That will be shown what's the type of PV that has energy harvesting for the indoor lighting measurement system.
Keywords: Larger solar cell, Indoor simulator, Solar simulator, Performance ratio.

1 INTRODUCTION

In general indoor lighting, a lot of different light sources exist. The International Commission on Illumination (CIE) is responsible for publishing binding standards for different types of light sources, called illuminants. A standard illuminant represents a mathematical table of relative energy versus wavelength, used for colorimetric calculations. It is a (theoretical) source of visible light with a set spectrum, determined by convention and therefore provides a worldwide basis for comparing images under different lighting. Which illuminants are useful standards for indoor illumination ? We consider some typical, widely used, light sources for indoor environments (residential and commercial) and relate them to an appropriate illuminant (standard). First, we will propose a standard for the common incandescent light bulb. The CIE has agreed upon a standard for the incandescent light bulb, called "illuminant A". This illuminant is intended to represent typical, domestic, tungsten-filament lighting. Its relative spectral power distribution is that of a Planckian radiator at a temperature of approximately 2856 K. In other words, a typical light bulb can be characterized by a black body at temperature T = 2856 K. Technically, illuminant A is only defined over the spectral region from 300 nm to 830 nm. However, solar cells often absorb longer wavelengths. Therefore, we will not use illuminant A itself to represent the spectrum of the incandescent light bulb, but we will use Planck's law for a black body at a temperature of 2856 K. In the 300–830 nm range, the difference between Planck's law at this temperature and illuminant A never exceeds 1%. However, In indoor lighting, the International Commission on Illumination is an organization devoted to international cooperation and exchange of information between its member countries on all matters relating to the science and art of lighting. In accordance with an agreement between ISO and CIE, standards are published as double logo standards by ISO. Standards produced by the CIE are a concise documentation of data defining aspects of light and lighting, for which international harmony requires a unique definition. Therefore, the standard follow that the CIE has defined some lighting spectrums for different indoor colorful applications and situations. However, the requirements for color testing do not apply to related indoor energy harvesting under different lighting uniformity and stability, for performance measurement of emerging PV and device. Furthermore, countless types of indoor lighting patterns are required to do performance characteristics measurement for different application contexts. The most common performance indicator is the PV efficiency under standard test conditions (STC). It is the maximum electrical power divided by the total irradiance. For indoor lighting illuminance values are measured used a lux meter or spectradiometer. Therefore, we find (a) Lux is a typical unit for measuring indoor lighting. (b) For outdoors, irradiance is measured in W/m2 and there is an approximate conversion of 0.0079 W/m2/ Lux for the sun. Irradiance includes the power from all wavelengths weighted equally, whereas illuminance weights the power from each wavelength in proportion to the sensitivity of the human (Fig. 1). Although these conditions are seldom achieved in practice (except in the lab), this characterization provides a reasonable guideline for comparing different solar cell types under outdoor conditions. However, the STC are not relevant for indoor applications. Accurate characterization of emerging PV requires level lighting consideration on each very slow temporal response in the I-V curves of the emerging PV are clearly dependent on the voltage sweep direction, even when the sweep time is the order of seconds [1]. Furthermore, the temporal response is dependent on different level lighting consideration. In our study, RTOS (real-time one-sweep) method can real-time remove the capacity effect under solar simulator for I-V successfully and proposes the forward/backward schematic reaction mechanism for DSC [2]. In this study, RTOS method [3, 4] practices for measuring perovskite solar cell performance showed having better accuracy results by eliminating acceptance effect instantly. Additionally, RTOS method will be useful to measuring cell performance more accurately and rapidly when evaluating emerging PV performance in indoor lighting condition. Therefore, we also defined a performance ratio to compare between indoor lighting and solar simulator.

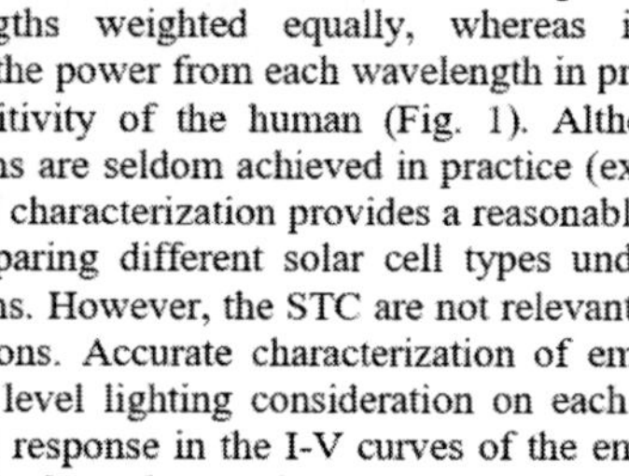

Figure. 1. Illumination spectra of IEC 60904/CIE Standard Lighting [11, 12]

3 EXPERIMENTS

The samples Sample size ranges from 0.1 cm2 to 1 cm × 5 cm. These I-V curves were done on a Keithley Source Meter (Model 2460), AM1.5G solar simulator, and Indoor Multi-Light Source System (see Fig. 4). Its meet follow as:

1) Spectral match, available methods are the use of spectroradiometer. Standard lighting spectrum is defined by the deviation from CIE[12]. For three wavelength intervals of interest, the percentage of total irradiance is specified in Table I. Calculate the spectral match for each wavelength interval, which is the ratio of calculated percentage for the CIE spectrum and the lighting simulator spectrum (Table I). The data comparison with the solar spectrum shall indicate the spectral match classification as per the following as Table II.

2) Non-uniformity of irradiance in the test plane, the lux meter is recommended to be used as uniformity detector for determining the non-uniformity of irradiance in the test area of the lighting simulator. The uniformity detector shall have a spectral response appropriate for the simulator. The linearity and time response of the uniformity detector shall conform to the characteristics of the simulator being measured. And the designated test area divided by 25. The area covered by the detector measurements should be 100% of the designated test area. The measurement positions should be distributed uniformly over the designated test area. Therefore, the maximum and minimum irradiance are those measured with the detector(s) over the designated test area. The calculation of non-uniformity is defined as Eq. (1).

3) Temporal instability of irradiance, the lux meter is recommended to be used as temporal instability detector for determining the temporal instability of irradiance in the test area of the lighting simulator. For lighting simulator the temporal instability is related to the irradiance change of measured data sets during the time of data acquisition (Fig. 2).

The Indoor Multi-Light Source included D65 (6500 K, Average North USA sky daylight) / TL84 (4100 K, European shop fluorescent) / CWF (4150 K, cool white fluorescent, shop lighting) / U30 (3000 K, shop lighting) / A (2856 K, typical home lighting), meeting CIE Standard illuminants. The adjustable light intensity (ranging from 0 to 2500 lux) is supplied with an ultraviolet light source for aging testing, with non-uniformity of less than 2% (at 20 cm × 20 cm) and temporal instability of less than 2%. During the I-V measurement in the dark and under Indoor Multi-Light Source System, we followed the testing flowchart in Fig. 5; the scan direction was forward and backward, the sample temperature should be stable at 25 °C with a fluctuation of less than 1 °C, and the irradiance and luminance were determined using a NIST-traceable spectroradiometer (StellarNet) and an NML (National Measurement Laboratory, Taiwan)-traceable lux meter (HIOKI, model FT3424).

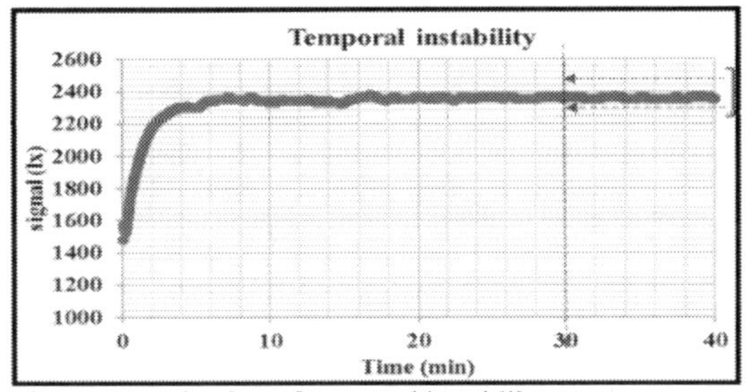

Figure 2. Evaluation of temporal instability

TABLE I THE RATIO OF CALCULATED PERCENTAGE FOR THE CIE SPECTRUM

Wavelength	Bandwidth	Percentage of total irradiance (300 nm – 900 nm)				
Band		A	D65	U30	CWF	TL84
1	300 - 450	3.78	20.31	11.40	17.99	15.18
2	450 - 650	40.15	45.35	81.72	76.73	79.13
3	650 - 900	56.06	34.34	6.88	5.29	5.69

TABLE II SPECTRAL DEFINITION OF LIGHTING SIMULATOR CLASSIFICATIONS

Classification	Percentage
A	1 %
B	3 %
C	5 %

$$Non-uniformity(\%) \equiv \frac{(max-min)}{(max+min)} \times 100\%$$

(1)

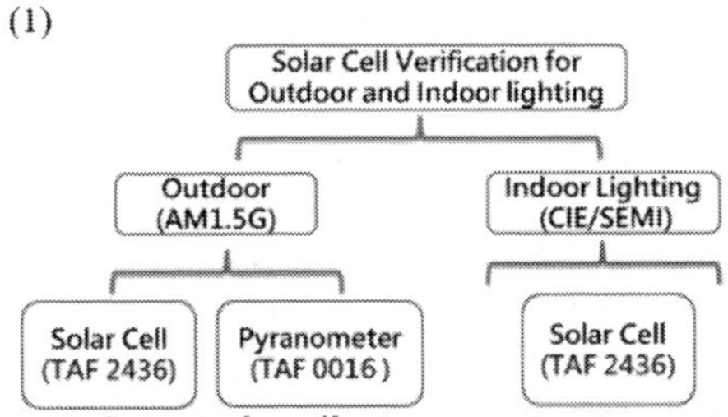

Figure 3. Larger solar cell measurement system custom-made by CMS/ITRI for solar cell measurement.

4. TEST AND ANALYSIS

In this study, we follow the testing flow [3,4] in the I-V measurement. We used real-time removal of capacity effect; the RTOS method shows the best repeatability. A proposed list of steps [2,5] for the measurement of a sample is given in Fig. 8. For most traditional semiconductor PV devices, the current will respond to small changes in bias voltage in less than a millisecond. Sweeping the voltage from open-circuit voltage (V_{oc}) to short-circuit current (I_{sc}) in a few seconds at a constant illumination for such a device will yield accurate currents. Many emerging PV cells respond far more slowly, and consequently, the sweep time must be increased.

For indoor lighting measurement system light source evaluation is performed according to SEMI PV80-0218, the check cycle is one year, and the light source evaluation results are recorded to confirm that the uncertainty of the system light source conditions falls within the scope of this document. Sources of uncertainty for maximum power measurements of the emerging PV include type A uncertainty assessed by statistical methods and type B uncertainty assessed by non-statistical methods:

TABLE III UNCERTAINTY SOURCE FOR INDOOR LIGHTING MEASUREMENT SYSTEM

Type A		Type B	
u_{A1}	Repeatability	u_{B1}	Effective illumination
u_{A2}	Stability	u_{B2}	Temperature effect
		u_{B3}	Accuracy for electrical measurement equipment
		u_{B4}	Measurement condition correction

Therefore, we used the RTOS Method [2] following the testing flowchart (see Fig. 4) in the I-V measurement. applied bias voltage is changed from I_{sc} to V_{oc} (forward scan) or in the reverse direction (backward scan, V_{oc} to I_{sc}). When a capacitance effect occurs, this measurement process produces the phenomenon. In the forward scan, an overshot current appears immediately after an abrupt increase of applied voltage and then the current gradually decreases to an equilibrium state. Sweeping to determine maximum power (P_{max}) may not be valid if the device is metastable because of the long times that it sits near I_{sc} or V_{oc} before getting to P_{max} confirm the "near-true" P_{max} of emerging PV, especially for samples that depend strongly on sweep direction or rate. This also shows that the asymptotic I-V method may be the only reliable way to obtain efficiency.

In the light intensity under artificial lighting conditions found in offices and factories is less than 5 W/m^2 as compared to 100–1000 W/m^2 under outdoor conditions, depending on the type of light source and its distance. Moreover, the spectrum can be totally different from the outdoor solar reference spectrum (e.g. AM1.5G). The spectrum depends not only on the type of light source, but also on the presence of reflected and diffused light. Therefore, we shown a definition of performance (η) between indoor lighting and solar simulator below as

$$\eta \equiv (P_{max}/Area)/P_{in}. \qquad (2)$$

Where (a) P_{max} means maximum power of device, (b) Area means area of device, (c) P_{in} means Illuminant power of lighting source. Therefore, we also defined a normalized power density (performance ratio, η_n) to compare between indoor lighting and solar simulator below as

$$\eta_n \equiv \eta_{Indoor} / \eta_{AM1.5G} \qquad (3)$$

Where (a) η_{Indoor} means performance of device at each indoor lighting (A/TL84/D65/U30/CWF), (b) $\eta_{AM1.5G}$ means performance of device at solar simulator.

7 DISCUSSION

The characterization after testing showed that one must measure performance under all typical artificial light sources. The 1000 lx of the system uncertainty is shown as

TABLE IV UNCERTAINTY BUDGET FOR INDOOR LIGHTING MEASUREMENT SYSTEM

	Uncertainty Sources	I_{sc}	I_{mp}	V_{oc}	V_{mp}	P_{max}	FF
u_{B1}	Effective Irradiance	1.54	1.54	0.0921	0.0921	1.54	0.185
u_{B2}	Temperature	0.0268	0.0268	0.131	0.131	0.134	0.190
u_{B3}	Electrical Accuracy	0.00405	0.00405	0.00247	0.00247	0.00474	0.00670
u_{B4}	Correction to 1000 lx/25 °C	0.0572	0.0572	0.140	0.140	0.152	0.214
u_{A1}	Repeatability	0.0957	0.215	0.0293	0.256	0.0698	0.164
u_{A2}	Stability	0.235	0.653	0.143	0.438	0.829	0.532
Effective degree of freedom		8982	213	52	9	101	11
Confidence Level		95 %					
Coverage factor k		1.96	1.96	2.01	2.26	1.98	2.2
Relative Combination Standard Uncertainty (%)		1.56	1.69	0.258	0.551	1.76	0.653
Expanded Uncertainty (%)		**3.1**	**3.4**	**0.52**	**1.3**	**3.5**	**1.5**

Therefore, we propose that each solar cell should be measured varied indoor and outdoor lighting condition for its characterization by RTOS method. In Fig. 5a, we will be compare the P_{max} of the different solar cells in light levels in indoor environments to show linearity under near range lighting level, which depends on device. In Fig. 5b shown nearly a constant from 400 lx to 1000 lx, it will be get an average of normalized power density and follow the defined of equation (3). Therefore, the results will be show the different solar cells in light levels in indoor environments to show linearity under near range lighting level, which depends on device. From Fig. 10b, it is also shown the different solar cells in indoor lighting and solar environments to show different performance, ex. DSSC have more 0.26 times for TL84 than D65. We are also known STC under outdoor application, therefore, we also define a STC for indoor application, that is " LS/ 25 °C/ 1000 lx " , LS mean: lighting source, ex. A/D65/TL84/U30/CWF.

V. CONCLUSION

In this work, we defined performance and normalized power density, is called performance ratio, shown a comparison for different emerging PV in indoor lighting and solar environments. Therefore, the results also find OPV have more 0.26 times for TL84 than D65, but OPV/a-Si have less 0.5 times for TL84 than D65. That's shown OPV have more indoor application than others in this case study. When we will be follow through CIE of the defined some lighting spectrums for different indoor colorful applications and situations. However, the requirements for color testing do not apply to related PV indoor performance measurement under different lighting uniformity and stability, for performance measurement of emerging PV and device, and Its characteristics will be having a classification. Furthermore, countless types of indoor lighting patterns are required to do performance characteristics measurement for different application contexts. We should be define a different STC than STC in AM1.5G, which we will be define a STC in indoor lighting as below as "I-V measurement for Indoor STC condition: 1000 lx @ 25 °C and Illuminant A". Furthermore, our goal is to determine which light source is relevant for a new international standard test method.

Upon the indoor characterization of solar cells, it is the spectrum range of the light source that decides the lighting performance, not the light intensity, which is present and affects the cell equally through different light sources.

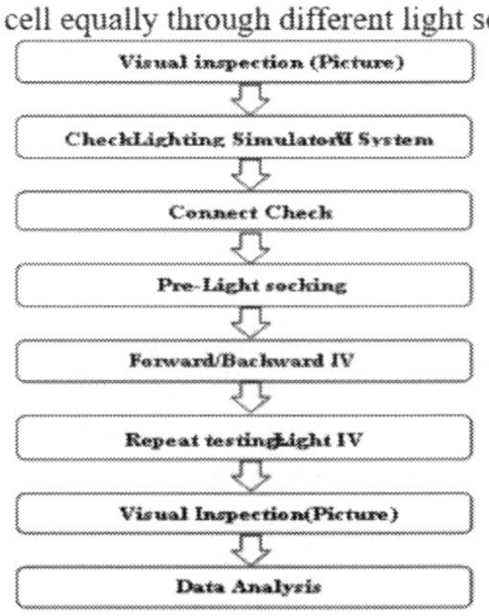

Fig. 1. Testing flowchart [2,5].

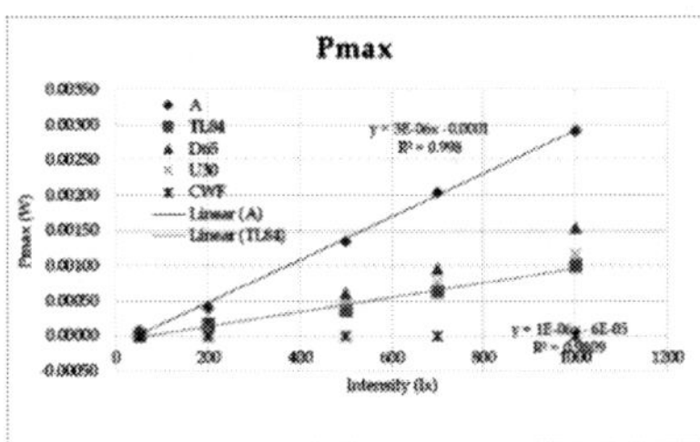

(a) P_{max}

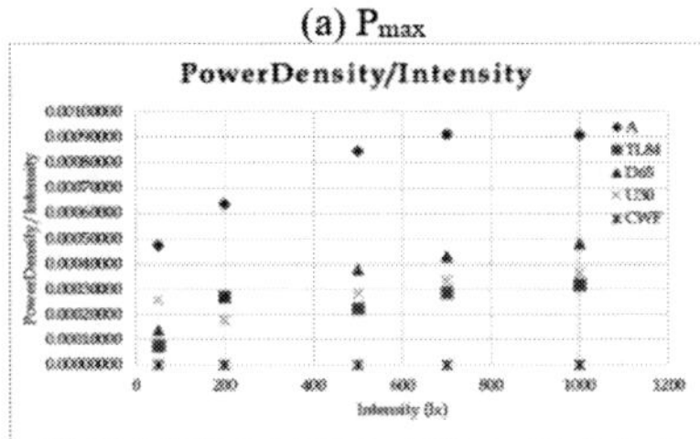

(b) Normalized Power density

Fig. 2. Light linearity of OPV measured by indoor multi-light source system under varied lighting condition (I-V using RTOS method).

REFERENCES

[1] Y. S. Long, S. T. Hsu, and T. C. Wu, "Induction of Internal Capacitance Effect in Performance Measurement of OPV (Organic Photovoltaic) Device by RTOSM (Real-Time One-Sweep Method)," *Journal of Energy and Power Engineering*, Vol 8, pp1059-1066, 2014.

[2] Y. S. Long, S. T. Hsu, and T. C. Wu, "Capacitance Effect from High to Low Level Lighting by Real-Time One-Sweep Method for OPV Device," *IEEE PVSC42*, 2015.

[3] Y. S. Long, S. T. Hsu, and T. C. Wu, "A Study of Capacitance Effect in DSC from High to Low Level Lighting by Real-Time One-Sweep Method," *OPTIC2014*.

[4] SEMI PV57-1214, "Test Method for Current-Voltage (I-V) Performance Measurement of Organic Photovoltaic (OPV) and Dye-Sensitized Solar Cell (DSSC)."

[5] SEMI PV80, "Specification of indoor lighting simulator requirements for emerging PV "

[6] IEC 60904-1, "Photovoltaic Devices – Part 1: Measurement of Photovoltaic Current-Voltage Characteristics," 2006.

[7] CIE/ISO 10526, "CIE Standard Illuminants for Colorimetry," 2007.

[8] K. Emery, S. Winter, S. Pinegar, and D. Nalley, "Linearity Testing of Photovoltaic Cells," *4th World PV Conf.*, 2006.

[9] ASTM D 1729, "Standard Practice for Visual Appraisal of Colors and Color Differences of Diffusely-Illuminated Opaque Materials";

[10] Handbook of Photovoltaic Science and Engineering, Chapter 16

ELECTROLUMINESCENCE RADIANCE MAPS BASED ON MULTIPLE EXPOSURE IMAGES FROM INGAAS CAMERAS

Claire Mantel[1], Sergiu V. Spataru*[2], Rodrigo Del Prado Santamaría[2], Peter Behrensdorff Poulsen[2], Gisele Alves dos Reis Benatto[2], Søren Forchhammer[1]
[1]DTU Fotonik, Kgs. Lyngby, Denmark, [2]DTU Fotonik, Roskilde, Denmark

ABSTRACT: This paper presents a method to increase the dynamic range of the EL signal for diagnostics of defective areas in PV modules. It applies a computational method for reconstructing a high dynamic range radiance map from a stack of standard dynamic range images at multiple exposure times designed by Debevec and Malik for visible light cameras (RGB). Increasing the precision of the EL signal has two interests: more quantitative information is then available for automatic analysis, and better visual quality is available for visual inspection in case of a small amplitude EL signal, such as acquired in an outdoor daylight setup using lock-in EL. Both are demonstrated via two experiments during which a PV module is electrically biased and imaged at 8 different exposure times with an InGaAs camera: one indoor and one outdoor in daylight. The exposure times are combined to reconstruct an EL image of higher precision. To measure the increased precision, the number of unique pixel values is used: it increases by a factor larger than 2.5 for the indoor case and a factor higher than 100 in the outdoor case. In the outdoor case the improvement towards visual inspection is also clearly apparent.

Keywords: Electroluminescence, Defects, Experimental methods, Evaluation, Reliability

1 INTRODUCTION

This paper focuses on the usage of electroluminescence (EL) images from InGaAs cameras for diagnostic of photovoltaic (PV) cells in a PV module. Since the first presentation of EL imaging for inspection of PV modules [1], advancements have been made allowing its usage outdoor in daylight [2] or towards faster inspection using for example drones [3].

The purpose is here to achieve a better precision of the EL signal in terms of amplitude at each pixel. Achieving a finer granularity on the EL signal provides more information on the variation of the EL signal and is interesting in two regards: added information will help automatic analysis and diagnostic algorithms and, in the case of a low EL signal amplitude, increasing the signal granularity will also help for visual inspection by making more details visible. That second aspect is for example relevant for outdoor daylight EL-based inspection of solar installations [2], for example using drones to acquire images [3]. Indeed, in such cases, the range of the EL image obtained by computing the difference between electrically biased and background images can be rather small. For instance, it has been shown to be lower than 100 values for images taken outdoor under global horizontal irradiance (GHI) of approximately 670 W/m^2 [4].

As a finer granularity in a signal can be seen as a larger dynamic range, we use an approach designed for visible light cameras that aims at retrieving the high dynamic range radiance map of a scene from a stack of standard dynamic range images acquired at different exposure times [5]. That approach operates in two steps. First, the response function of the camera is determined, modeling the relation between the product of the radiant flux incident on the sensor and exposure time and the pixel value. In a second step, a high dynamic range image of the radiant flux at each pixel is reconstructed, up to a scaling factor. An advantage of using such data-based methods is to limit the a priori knowledge on the camera system and environment.

The article is organized in the following way: first the original method is presented in Sec. 2. It is then applied on datasets acquired during two experiments: one performed indoor (without noise due to sunlight) and one performed outdoor using electrical modulation and image processing to retrieve the EL signal [2]. The experiments are presented along with the obtained results in Sec. 3 for the indoor setup and Sec. 4 for the outdoor setup, respectively.

2 PREVIOUS WORK

One usage of EL images of PV modules is to estimate optoelectronic characteristics of the modules such as the diffusion length of minority carriers [1] or series resistance [6]. Some models exist describing the relationship between the characteristics of a solar cell and an EL image of it [6]. In [6], the authors include in their model the camera system using a priori knowledge such as the quantum efficiency or gain of the camera. The method presented here is another way to estimate the emitted flux incident on the sensor using several images with multiple exposure times but no other prior knowledge about the camera. The result achieved is determined up to a scaling factor and therefore needs a reference point for absolute calibration (e.g. a specific light source like in [7]). As the sensitivity of a camera depends on environmental factors such as its temperature, this data-based approach can adapt more easily to the current condition of the camera and is therefore quite flexible.

In practice, we have adapted image processing techniques that reconstruct a large dynamic range radiance map using multiple images taken with standard dynamic range at multiple exposure times. Those techniques were developed for visible light cameras (RGB), as opposed to InGaAs cameras in our case, and are assuming that the content of the scene imaged is static. The principle of this

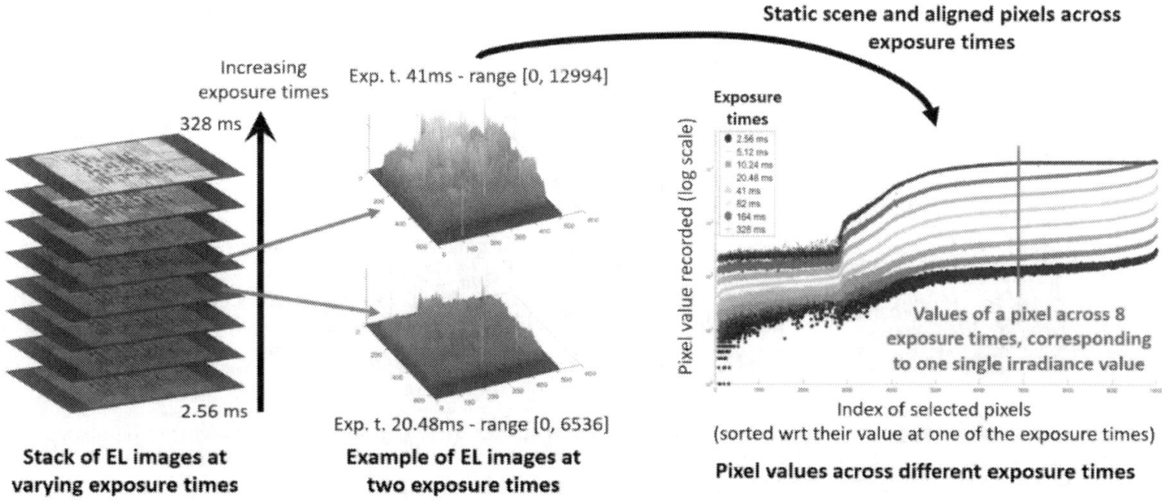

Figure 1 Data structure of multiple exposure time EL images and its utilisation in the reconstruction method from [5]

process was first presented by Debevec and Malik in [5]. The way the stack of images at multiple exposure times are used is illustrated in Figure 1.

The main idea is that for a specific pixel, the digital value issued by a camera depends on the camera response, the radiant flux received on the sensor and the exposure time of the sensor. Using the assumption that the radiant flux on each pixel is constant, it is possible to compute the response of a camera to the product of a specific radiant flux by the exposure time. Once this camera response function is computed, it is possible to retrieve, up to a scaling factor, the radiant flux incident on each pixel with a higher precision. To deduce fully the emitted flux, one also needs to account for other factors modifying the flux before it is incident on the sensor, such as vignetting of the lens of the camera or the transmittance of the atmosphere [6].

The modeling of the pixel response Z_{ij} in [5] is: $Z_{ij} = f(\varphi_i \times t_j)$ where f is a non linear function (the camera response), φ_i is the irradiance (radiant flux received by the sensor area) and t_j the exposure time.

By posing $g = ln(f^{-1})$, the problem can be reformulated as solving for g that minimizes the following cost, C:

$$C = \sum_{i=1}^{N}\sum_{j=1}^{P}\left[g(Z_{ij}) - \ln(\varphi_i) - \ln(\Delta t_j)\right]^2 + \lambda \sum_{z=Z_{min}+1}^{Z_{max}-1} g''(z)^2 \tag{1}$$

where N is the number of pixels, P the number of different exposure times and Z_{min} and Z_{max} the smallest and largest possible pixel values, respectively. The second term is added for regularization, with λ controlling the smoothness of the response. The logarithm of the inverse response function, g, needs to be resolved for the values between Z_{min} and Z_{max}.

The assumption that the radiant flux is constant on each pixel implies both that the scene is static across the acquisitions at all exposure times and that the pixels are registered across all exposure times.

3 INDOOR EXPERIMENT

3.1 Dataset

For the indoor experiment, a dataset of EL images at 16 exposure times was acquired. The module imaged is a 36 cells mechanically stressed multi-crystalline silicon module that was current biased at 100% I_{SC}. The InGaAs camera used for the experiment is a Raptor OWL 640 which has the following characteristics: resolution 640x512, spectral sensitivity 0.4µm to 1.7µm, pixel pitch 15mm x 15mm and readout noise (RMS) at low gain: <195 electrons. It uses active TEC cooling to keep the sensor at stabilized temperature. The lens mounted on it is a SWIRECON 25 mm SWIR lens.

The exposure time can be set to a minimum of 1 µs. To scan the whole dynamic range of the scene, 16 different exposure times were used for acquisition, with a x2 factor between each: [0.8, 0.16, 0.32, 0.64, 1.28, 2.56, 5.12, 10.24, 20.48, 41, 82, 164, 328, 656, 1312, 2624] ms. The acquisition was performed indoor without any daylight and 5 repeats of each exposure time were acquired. The lowest exposure times present a high amount of noise and the highest exposure times present saturated pixels. Only the 8 best exposure times 6 to 13 [2.56, 5.12, 10.24, 20.48, 41, 82, 164, 328] ms are included in the analysis below as their combination achieves the best result.

The input images were averaged over the 5 repetitions to remove part of the acquisition noise and then rounded to the nearest integer. The results are depicted in Figure 2. The images are scaled to the displayable range [0, 255] for better visualization whereas the actual range of values in each image is indicated above. It should be noted that there can be large differences between displayable and actual range of values.

3.2 Results

In practice, we solved Equation (1) for a subset of 10000 pixels randomly selected within the module and a smoothing parameter λ=3. The implementation was that available in [5], that is based on Singular Value Decomposition to solve the equation system, modified to handle input images with a bitdepth higher than 8 bits.

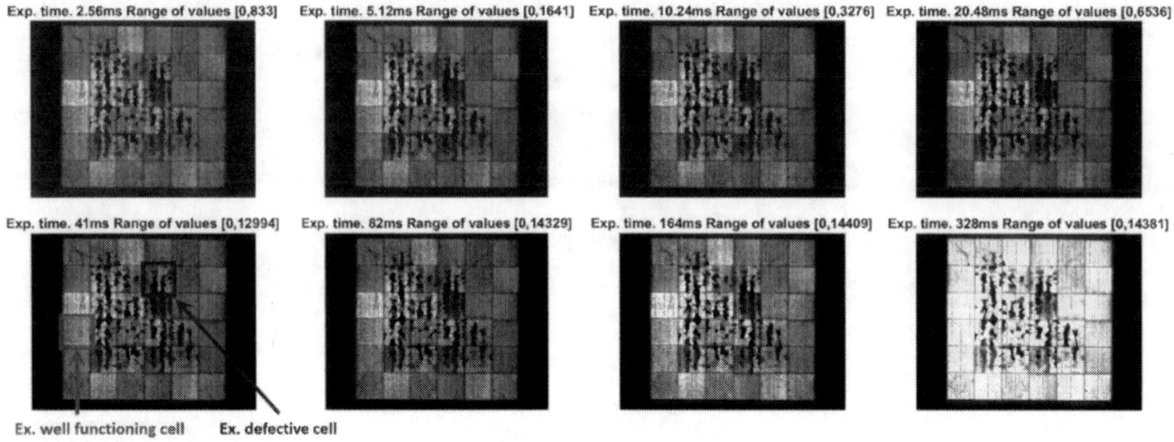

Figure 2: EL indoor images at the exposure times selected. The acquired images have been averaged over the 5 repeats and rounded. The range of values is indicated above each image and they are scaled between their 0.001 and 0.999 quantiles for visualization. The two cells selected for illustrating results are highlighted: in red for the well-functioning one and blue for the defective example.

Once the response function of our InGaAs camera is computed, the radiance map is reconstructed (using functions from [8] modified for higher bitdepth).

Results are shown in Figure 3 for a well-functioning cell, the one highlighted in red in Figure 2, and in Figure 4 for a cell presenting important defects, the one highlighted in blue in Figure 2 .

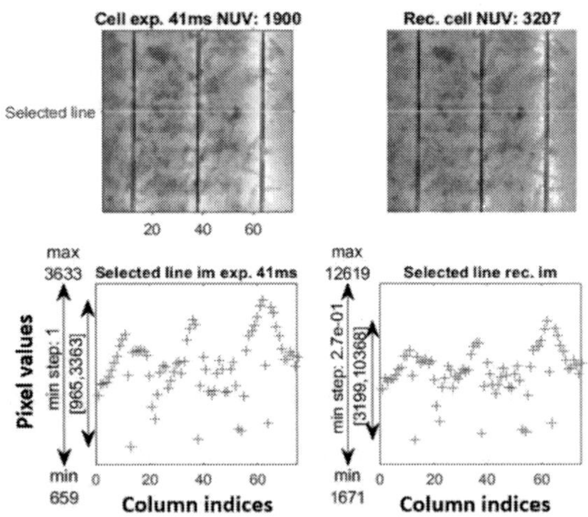

Figure 3: Example of best acquired (i.e. 41 ms exposure time, left) and reconstructed (right) images for a well-functioning cell (highlighted in red on Figure 1) with the corresponding number of unique value (NUV). In the second row, the profile of a selected line illustrates the interest of the method with regard to the minimum step size estimated and total cell variation.

In both cases, both a camera acquired image and the reconstruction are shown. The camera acquired image used here is the "best case scenario": i.e. the one with the highest possible exposure time that does not saturate (exposure time of 41 ms in our experiment which reaches a maximum value of 12994). It should be noted that the number of pixel values represented in Figure 3 and Figure 4 is limited to 256 (as [0, 255] is the standard range of displays), which is much lower than the real range of pixel values present in the images (up to 5500).

3.3 Analysis of results

The results are evaluated by comparing the EL image at 41 ms, i.e. the "best case" directly acquired EL, and the reconstructed EL image with a larger dynamic range. It is done both qualitatively through visual inspection of the EL images and quantitatively through the number of unique pixel values (NUV) and minimal step size present in the images.

As noted, the range of the figures presented here is much smaller than the full range of the images ([0, 255] as opposed to [0, >5000]). As it is also more than what the human eye is able to perceive without adaptation, the images at the top row of both Figure 3 and 4 appear similar. In this context of indoor imaging, when it is possible to acquire EL images with a large dynamic range, the method presented is not useful for visual inspection when looking at the whole cell. In case of zoom over a smaller portion of the image as done in the second row of Figure 4, the higher dynamic range can allow seeing ore details in both bright and dark areas. It here visible hereon the close up on a defective area (images 1 and 3 of the second row): the EL image at 41ms presents less details than the reconstructed EL (their dynamic range are [88, 211] and [114, 615], respectively).

It is however useful for automatic detection and analysis of defects. Indeed, the graphs on the bottom row of both Figures 3 and 4 show a higher number of NUV (increased by more than 2.5 times) and smaller step sizes obtained for the case of the reconstructed images (to the right). Automatic detection algorithms will leverage the finer step size to separate more accurately areas with optimal performance from those not functional in defective solar cells. In the middle row of Figure 4, an example of how automatic selection of inactive cell area based of EL signal could use the added information, both to identify defective areas and evaluate the level of defect within such a defective area.

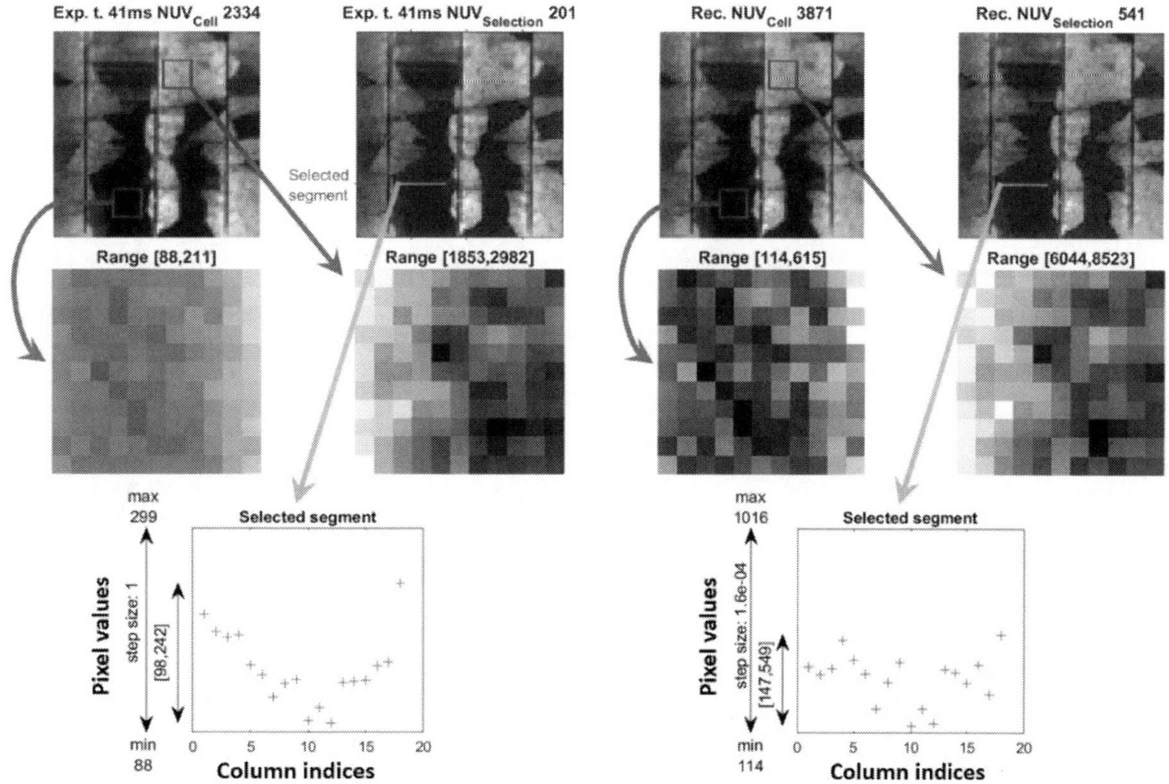

Figure 4: Example of best acquired (i.e. 41 ms exposure time, two leftmost columns) and reconstructed (two rightmost columns) images for a cell presenting significant defects (highlighted in blue on Figure 1). The first row depicts the images and a selected area representing the largest loss in terms of EL signal (highlighted in blue) and indicates their respective NUV. The second row focuses on zooms within the cell to illustrate the interest of higher dynamic range: finer granularity of values to make details visible in both dark and bright areas. The third row presents the profile of the selected segment to illustrate the interest of the method with regard to the minimum step size estimated and variation within the selected area.

4 OUTDOOR EXPERIMENT

4.1 Dataset

For the outdoor experiment, the same module as for the indoor experiment was imaged with the same InGaAs camera and lens. The camera was operating at 120Hz.

As the acquisition was performed in daylight (GHI was measured at 639 W/m^2 and the irradiance in the plane of array 71 W/m^2), the panel was electrically modulated for reconstruction [2] and a bandpass filter from Edmund Optics was used (25 nm band width, 1150 center wavelength, optical density 4). The module was biased at 100% I_{SC} with a square wave signal at 30Hz. The acquisition consisted in 2 electrically biased images followed by 2 unbiased images, also called background. 200 images were acquired at each exposure time, yielding 100 biased images and 100 background images.

Some processing was necessary to retrieve the electroluminescence signal. First, the corners of the module were detected, and the image cropped to the area of interest [9]. Then the images sequence was separated into biased and background images so the averaging over each category could be performed, before computing the difference [2].

Series of 200 images were acquired at the following 16 exposure time: [0.05, 0.08, 0.12, 0.18, 0.27, 0.398, 0.589, 0.87, 1.28, 1.9, 2.82, 4.17, 6.17, 9.13, 13.5, 20] ms. The exposure times were calculated as 14 points on a logarithmic scale between the lower bound representing a signal visible but noisy and the higher bound presenting saturation in the module area. An example of an image acquired and the retrieved EL image are depicted in Figure 5.

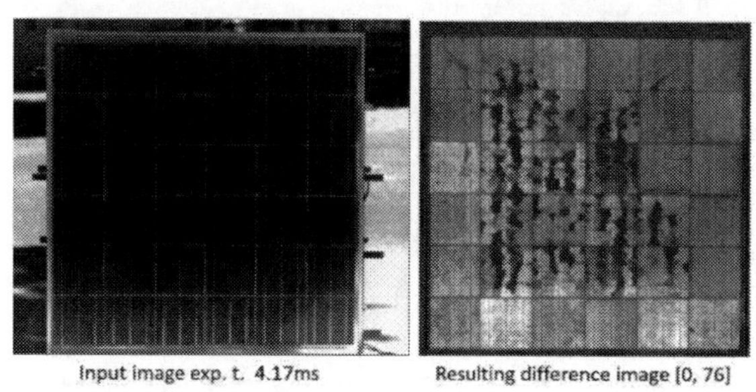

Figure 5: Example of outdoor SWIR image at exposure time 4.17 ms (left) and resulting EL image after module detection, BEL/BG identification, averaging and difference (right).

Due to the relative strength of the EL signal with regards to the noise and sun variation, a subset of 8 exposure times were selected to achieve the best results. Table I below presents the exposure times selected and the corresponding range of values present in the acquired SWIR image and the resulting EL images.

Table I: Selected exposure times and corresponding range of values in difference image for outdoor DS

Exp. Time (ms)	SWIR Image range	Difference image range
0.398	[396,4150]	[0, 20]
0.589	[546,6109]	[0, 21]
0.87	[791,9025]	[0, 29]
1.28	[1156,12970]	[0, 41]
1.9	[1587,14479]	[0, 53]
2.82	[1586,14479]	[0, 56]
4.17	[1634,14622]	[0, 76]
6.17	[1714,14655]	[0, 104]

4.2 Results

The same process was applied to reconstruct a larger dynamic range image for the 8 EL images at various exposure times: random selection of 10 000 pixels, solving of Equation (1) with smoothing parameter λ=3 using code adapted from [5] and reconstruction of EL image with adapted code from [8]. The obtained EL image is depicted in Figure 6.

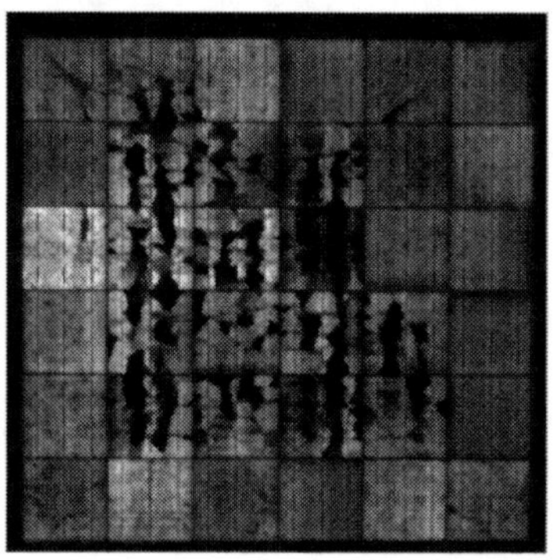

Figure 6: Larger dynamic range EL image obtained from 8 EL images at exposure times ranging from 0.398 ms to 6.17 ms.

Detailed results for the same cells as for the indoor experiment (Sec. 3) are presented hereafter: in Figure 7 for the well-functioning cell and in Figure 8 for the defective cell.

4.3 Analysis of results

The usefulness of the method is in that case evident both in terms of NUV but also visually. Indeed, the best achieved EL signal obtained through averaging and differencing (left column of Figure 7 and 8) presents clearly less information for visual inspection than the reconstructed larger dynamic range EL image using 8 exposure times (right column of Figure 7and 8). The second row of Figure 8 illustrates the interest of higher dynamic range for visual inspection when the input EL signal obtained by difference has a low amplitude by showing that more details are visible either in dark or bright areas.

When comparing with results from Sec. 3, it is apparent that the difference lies mainly in the quality of the EL images acquired outdoor. Indeed, the maximal amplitude of the EL image obtained through averaging and differencing between the electrically biased images and the background images is only around 100, as opposed to more than 14000 for the indoor experiment. The main reason for that limitation is that the amplitude of the EL signal in the SWIR images is low in comparison of the rest

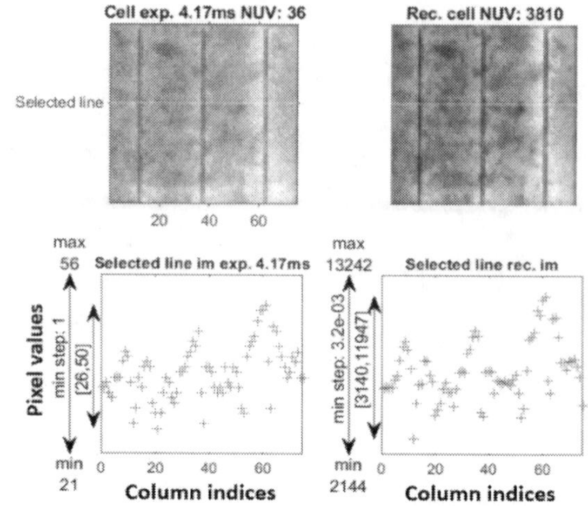

Figure 7: Example of best acquired (i.e. 4.17 ms exposure time, left) and reconstructed (right) images for a well-functioning cell (highlighted in red on Figure 1) with the corresponding NUV. In the second row, the profile of a selected line illustrates the interest of the method with regard to the minimum step size estimated and total cell variation.

of the image.

In the same way as shown for the indoor case, the increase in NUV and decrease in step size indicate that these results will allow more accurate automatic cell defect and fault detection in daylight EL images. In this case of low amplitude EL input signal, the NUV is increased by a factor higher than 100.

5 CONCLUSION AND FUTURE WORK

This paper applies a method designed for visible light cameras that combines images of a static scene acquired at different exposure times to retrieve an irradiance map of the scene with a larger dynamic range.

Reconstructing a larger dynamic range version of an EL image can be used to increase the available information for analysis and diagnostic of EL images of PV modules in the case of both automatic and visual inspections.

Both cases are illustrated through two experiments and evaluated through both the quantitative characteristics of the EL images and their visual aspect.

An indoor experiment using images acquired at 8 different exposure times is first presented. In that case the full range of the camera is used for the EL signal, which then presents a large range of values (more than 14000 in our case). The presented method is interesting for automatic inspection as the precision of the EL signal is increased, as measured by an increase factor of 2.5 to 5 in NUV and decrease 10^{-4} for the minimal variation present in the image.

An outdoor experiment allows illustrating the interests for both automatic and visual inspection. In this case the range of the camera is used for the complete SWIR image, i.e. both the EL signal and the unwanted noise due to the sunlight. Therefore, the EL signal obtained through averaging and differencing from EL lock-in input presents a rather small range of values (at most 104 for our experiment performed at 639 W/m^2 global irradiance and 71 W/m^2 POA). The image reconstructed using the presented method produces then a clear interest both for visual inspection and in terms of number of unique pixel values (multiplied by more than 100 in that case).

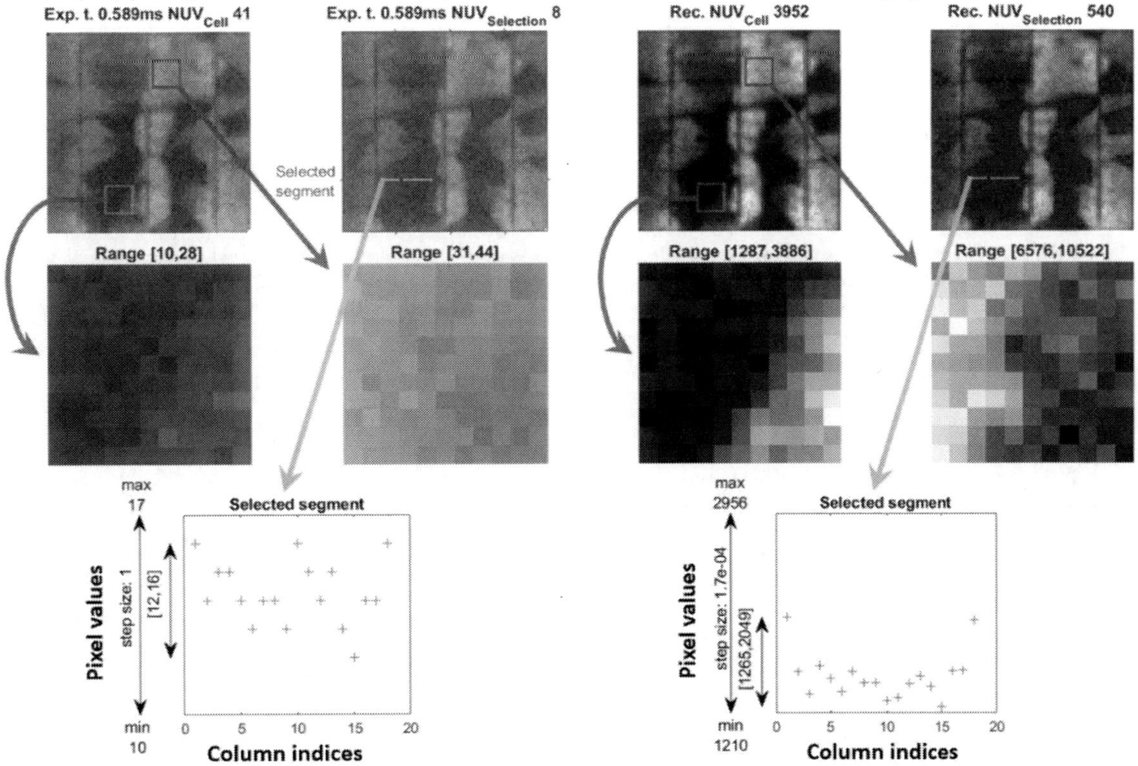

Figure 8: Example of best acquired (i.e. 4.17 ms exposure time, left) and reconstructed (right) images for a cell presenting significant defects (highlighted in blue on Figure 1). The first row depicts the images and a selected area representing the largest loss in terms of EL signal (highlighted in blue) and indicates their respective NUV. The second row focuses on zooms within the cell to illustrate the interest of higher dynamic range: finer granularity of values to make details visible in both dark and bright areas. The third row presents the profile of the selected segment to illustrate the interest of the method with regard to the minimum step size estimated and variation within the selected area.

6 ACKNOWLEDGMENTS

This research has been carried in the project DronEl2 project under the EUDP - Energy Technology Development and Demonstration Program with grant number 64019-0603.

7 REFERENCES

[1] T. Fuyuki et al., "Photographic surveying of minority carrier diffusion length in polycrystalline silicon solar cells by electroluminescence," Applied Physics Letters, 2005

[2] L. Stoicescu, et al., "Daysy: luminescence imaging of pv modules in daylight," in 29th European photovoltaic solar energy conference and exhibition proceedings (EU PVSEC), 2014

[3] G. Benatto et al., "Drone-Based Daylight Electroluminescence Imaging of PV Modules". IEEE Journal of Photovoltaics, 2020

[4] Mantel, C et al., "SNR Study of Outdoor Electroluminescence Images under High Sun Irradiation", In Proceedings of 7th World Conference on Photovoltaic Energy Conversion, 2018

[5] P. E. Debevec, J. Malik, "Recovering high dynamic range radiance maps from photographs", in: Proceedings of ACM SIGGRAPH, 1997

[6] D. Ory et al., "Extended quantitative characterization of solar cell from calibrated voltage-dependent electroluminescence imaging", Journal of Applied Physics, 2021

[7] Mochizuki et al., "Solar-cell radiance standard for absolute electroluminescence measurements and open-circuit voltage mapping of silicon solar modules", Journal of Applied Physics, 2016

[8] Banterle, F. et al., "Advanced High Dynamic Range Imaging (2nd Edition)", AK Peters (CRC Press), July 2017

[9] Mantel, C., et al., "Method for Estimation and Correction of Perspective Distortion of Electroluminescence Images of Photovoltaic Panels", IEEE Journal of Photovoltaics, 2020

A NEW EXPLICIT FIVE-PARAMETER SOLAR CELL MODEL

Andreea Sabadus[1], Marius Paulescu[2]
[1]Institute for Advanced Environmental Research, West University of Timisoara, Timisoara, Romania
[2]Faculty of Physics, West University of Timisoara, Timisoara, Romania

ABSTRACT: The most common equivalent electrical circuit of a real solar cell is probably the one-diode model. Throughout the years, numerous research have focused on the extraction of model parameters from measurements done in standard test conditions (STC) with the goal of replicating the current-voltage characteristics (I-V). The equation that describes the I-V characteristics is highly nonlinear, and also an implicit equation. There are many attempts to write an explicit equation, achieving only limited success. This study proposes an explicit Shockley equation for the five-parameter solar cell model. The form of the equation is new, consisting of the ideal Shockley equation and an additive term. The results confirm the feasibility of the proposed approach, indicating a promising way to a simple-explicit five-parameter solar cell model.
Keywords: solar cell, one-diode model, explicit model

1 INTRODUCTION

The current-voltage (I-V) characteristic of a photovoltaic (PV) module result from the superposition of the I-V characteristics of its constituent solar cells. A key element in evaluating the performance of a solar cell is the I-V characteristics measured in standard test conditions (STC) (normal incoming solar radiation at $1000\ W/m^2$ under AM1.5G spectrum and a cell temperature of $T = 25°C$), this being a proxy for computing the fill factor and cell efficiency. Even at STC, due to its highly nonlinear nature, the mathematical modeling of the I-V characteristic of a solar cell is far from a trivial task.

To describe the operation of PV module, almost all mathematical models are rooted in Shockley theory of the illuminated p-n junction [1]. The one-diode model (Fig. 1) is the most common and mainly used due to its apparent simplicity.

Based on the one-diode equivalent circuit from Fig. 1, probably the most popular equation for the I-V characteristic of a solar cell is written as:

$$I = I_L - I_S\left[exp\left(\frac{q\left(V+IR_S\right)}{mk_BT}\right)-1\right]-\frac{V+IR_S}{R_P} \qquad (1)$$

where q is the electron charge, k_B is the Boltzmann constant, and T is the cell temperature. The photocurrent I_L, diode saturation current I_S, diode ideality factor m, serial resistance R_S, and shunt resistance R_P form a set of five parameters (specific for a cell), which give the name of the model, respectively the five-parameter model. Besides being a highly nonlinear equation, Eq. (1) is also an implicit equation. There are many attempts over time to write an explicit equation, achieving only limited success.

Numerous studies for extracting the model parameters from measured data at STC, have been developed over the last decades, ranging from classic (e.g. Newton-Raphson [2]) to exotic artificial intelligence-based approaches (e.g. flower pollination algorithm [3]). Remarkably, several studies report significantly distinct sets of parameters for the same solar cell, while all reproducing the measured I-V characteristics with high accuracy. We addressed this subject in a recent work [4] and emphasized that there is an infinity of possible sets of parameters that can accurately replicate the I-V characteristic of a solar cell.

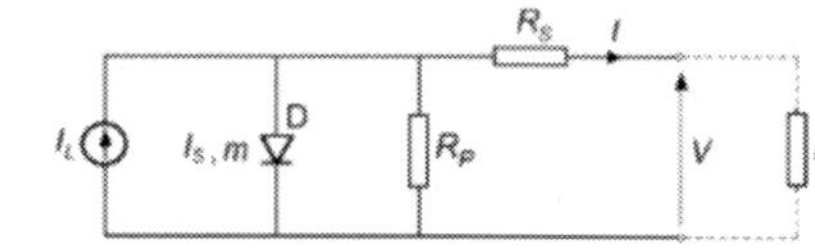

Figure 1: One-diode equivalent circuit for a solar cell.

The purpose of this study is to revise the implicit Shockley equation associated to the one-diode solar cell model, aiming to write an explicit equation. A blend among theory and empirical procedures is proposed for finding an additive term to the ideal Shockley equation of a solar cell.

The paper is organized as follows. Section 2 presents the new explicit equation for the five-parameter model. The feasability of the new equation is discussed in Section 3. Section 4 presents I-V characteristics of the new explicit equation. Section 5 gathers the main conclusions of this study.

2 EXPLICIT MODEL

The one-diode model parameters are extracted using an algorithm, regardless of its nature, that deals with highly non-linear equations derived from Eq. (1). However, there is no universally accepted procedure for extracting the one-diode model parameters.

The novelty of this study compared to other studies on this subject is the proposal for an additive function $g(V)$ to Shockley's ideal diode equation.

$$I = I_L - I_S\left[exp\left(\frac{qV}{mk_BT}\right)-1\right]-\frac{V}{R_P}+g(V) \qquad (2)$$

Eq. (2) represents an explicit equation, which simplifies the problem of parameter extraction from measured data, contributing to avoiding an algorithm to be trapped in a local minimum.

The form of $g(V)$ is obtained from Eq. (2) and Eq. (1). By writing the Taylor series for the exponential term and retaining only the first two terms we obtain the additive term $g(V)$. After further simplifying the calculus, the final form for $g(V)$ is obtained Eq. (3).

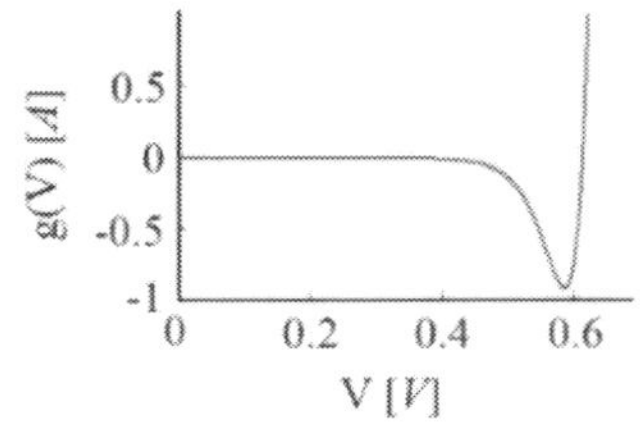

Figure 2: Graphical representation of the additive function $g(V)$.

Table I. Chosen reference sets of parameters.

Parameters	Reference Ref. [5]	Ref. [6]
I_L [A]	8.25	8.21
I_S [A]	$6.39 \cdot 10^{-8}$	$9.83 \cdot 10^{-8}$
m	1.28	1.30
R_S [Ω]	0.0024	0.0042
R_P [Ω]	2.49	10.98

$$g(V) = \left[-I_S \frac{qR_S}{mk_BT} exp\left(\frac{qV}{mk_BT}\right) - \frac{R_S}{R_P} \right] \cdot \frac{\left[I_L - I_S \left[exp\left(\frac{qV}{mk_BT}\right) - 1 \right] - \frac{V}{R_P} \right]}{1 + \frac{R_S}{R_P} + I_S \frac{qR_S}{mk_BT} exp\left(\frac{qV}{mk_BT}\right)} \quad (3)$$

Replacing Eq. (3) in Eq. (2) we remain with an explicit equation for the current equation. A simple, elegant, expression is obtained Eq. (4).

$$I = \frac{\left[I_L - I_S \left[exp\left(\frac{qV}{mk_BT}\right) - 1 \right] - \frac{V}{R_P} \right]}{1 + \frac{R_S}{R_P} + I_S \frac{qR_S}{mk_BT} exp\left(\frac{qV}{mk_BT}\right)} \quad (4)$$

Eq. (4) represents an explicit equation, which simplifies the problem of parameter extraction from measured data.

3 MODEL ASSESSMENT

The additive function $g(V)$ can be seen in Fig. 2. $g(V)$ is different from zero only for voltages higher than the threshold value from where the `knee` of the I-V characteristic begins. This is an expected behavior, considering that the additive term is a correction to the ideal Shockley equation.

Two reference sets of parameters [5-6] were chosen in order to verify the feasibility of the proposed model. The parameters were obtained for the Kyocera KC200GT PV module.

Using the two sets of parameters we replicated the measured I-V characteristic of the Kyocera PV module using the implicit Eq. (1) and explicit Eq. (2) equations. The coefficient of determination Eq. (5) is a measure of how well a model fits a set of data.

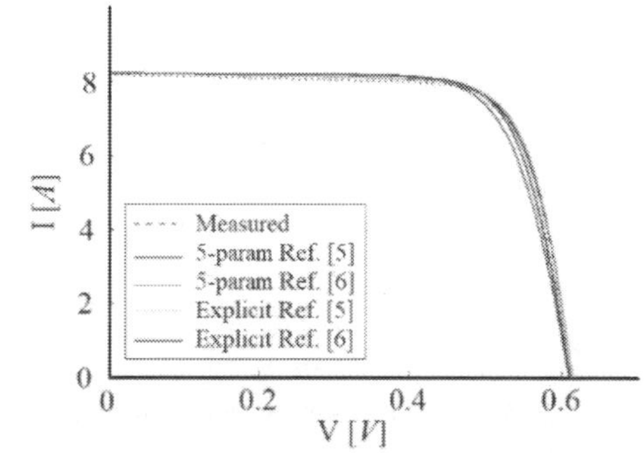

Figure 3: The I-V characteristics obtained with both implicit Eq. (1) an explicit Eq. (4) equations for the Kyocera KC200GT PV module.

Table II. The coefficient of determination R^2.

Parameters from	Implicit equation Eq. (1)	Explicit equation Eq. (2)
Ref. [5]	0.991	0.986
Ref. [6]	0.980	0.996

$$R^2 = 1 - \frac{\sum_{i=1}^{N} \left[I_{measured_i} - I_{computed_i} \right]^2}{\sum_{i=1}^{N} \left[I_{measured_i} - \overline{I_{measured}} \right]^2} \quad (5)$$

where $I_{measured}$ and $I_{computed}$ refer to the measured and computed values for the current, while N is the sample size. For each curve R^2 was calculated in order to assess the accuracy of the new explicit equation.

In both cases, the coefficient of determination shows a very good approximation of the measured I-V characteristic. Considering the parameters from Ref. [6], the explicit equation is an almost perfect approximation.

The I-V characteristics confirm the feasibility of the proposed approach, indicating a promising way to a simple-explicit five parameter solar cell model.

4 I-V CHARACTERISTICS

Fig. 4 – Fig. 7 show the relative effect of the one-diode model parameters (diode saturation current, diode ideality factor, serial resistance and shunt resistance) on the I-V characteristic and the additive term, respectively. In all of the figures each of the curves given by Eq. (4) and Eq. (3) were plotted varying only one of the parameters and keeping all the others constant.

Fig. 4 and Fig. 6 display a common picture of the parameters influence on the I-V characteristics of a solar cell. Varying the saturation current in the range of $[10^{-6}\,A, 10^{-9}\,A]$ changes the characteristics visibly (Fig. 4a). The increase in I_S determine a decrease in the estimated open-circuit voltage. Fig. 5a highlights that the behavior of the additive term is the opposite, which confirms again the corrective nature of it. In Fig. 4b, the ideality factor acts the opposite way as the saturation current, while in Fig. 5b, the behavior of both parameters is similar.

Overall, Fig. 4 shows that an increase in the diode saturation current that implies a decrease in the open-circuit voltage is compensated by an increase in the diode

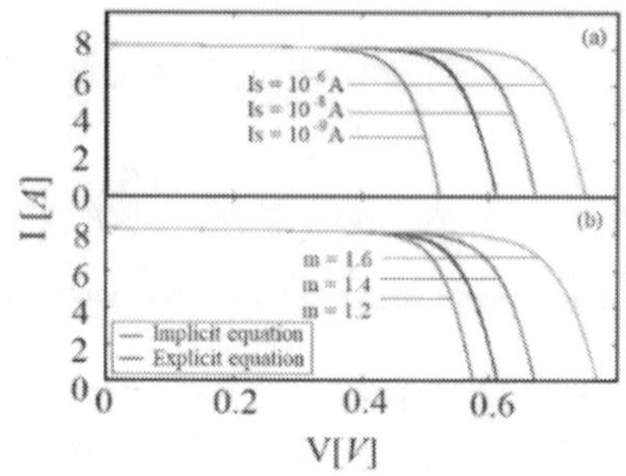

Figure 4: Solar cell I-V characteristic dependence on the parameters in Eq. (4). (a) Diode saturation current I_S, (b) Diode ideality factor m.

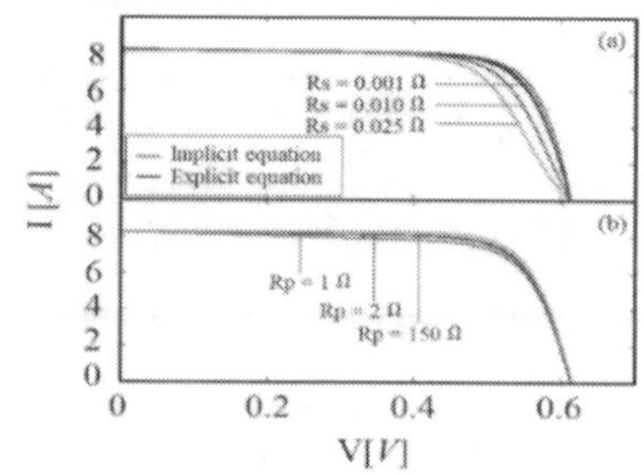

Figure 6: Solar cell I-V characteristic dependence on the parameters in Eq. (4). (a) Serial resistance, (b) Shunt resistance.

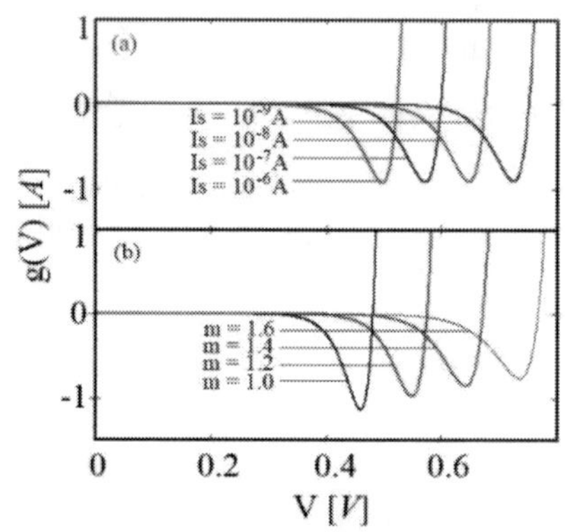

Figure 5: Additive term $g(V)$ dependence on the parameters in Eq. (3). (a) Diode saturation current I_S, (b) Diode ideality factor m.

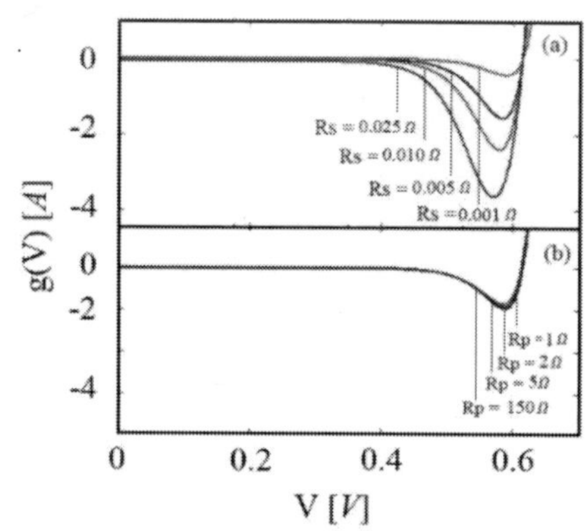

Figure 7: Additive term $g(V)$ dependence on the parameters in Eq. (3). (a) Serial resistance, (b) Shunt resistance.

ideality factor. The open-circuit voltage is a physical parameter, so compensation is a mandatory internal action of any numerical algorithm. A discussion on the relationship between I_S and m is presented in [3]. Using the new equation Eq. (4), this compensation is implicitly encapsulated in the additive term.

The variation of the serial resistance and shunt resistance from Fig. 6 are not changing the open circuit voltage. The same behavior for the resistances can be seen in Fig. 7.

5 CONCLUSIONS

There are several methods dealing with the extraction of parameters of the one-diode model. For solar engineering applications, accuracy and computational accessibility are the two key factors taken into account when choosing a procedure for extracting the parameters of the one-diode model. However, the resulted set of parameters may be far from their values resulting from measurements [4]. The new explicit equation Eq. (4) proposed in this study may present a solution for this problem, avoiding an algorithm to be trapped in a local minimum, thus finding a global solution.

The results confirm the feasibility of the proposed approach, indicating a promising way to a simple-explicit five-parameter solar cell model. Eq. (4) represents an explicit equation, which simplifies the problem of parameter extraction from measured data. The equation form is new, consisting of the ideal Shockley equation and an additive term.

Further studies are needed to see the influence of retaining more terms of the Taylor series for the exponential term on the accuracy in replicating the measured I-V characteristics. Also, a study for the feasibility of the explicit equation for the two-diode model has to be discussed.

6 ACKNOWLEDGMENTS

This work was supported by a grant from PN-III-P2-2.1-PED-2019-3942 All-sky imager-based solar power forecasting system for smart-grid operation.

7 REFERENCES

[1] Nelson J. Imperial College Press: London, UK, 2013.
[2] Appelbaum J, Peled, A. Sol. Energy Mater. Sol. Cells 2014, 122, 164–173.
[3] Ram JP, Babu TS, Dragicevic T, Rajasekar N. Energy Convers. Manage. 2017, 135, 463-476.
[4] Sabadus A, Paulescu M. Energies 2021, 14, 3974.
[5] Alam DF, Yousri DA, Eteiba MB. Energy Convers. Manage. 2015, 101, 410–422.
[6] Ali EE, El-Hameed MA, El-Fergany AA, El-Arini MM, Sustain. Energy Technol. Assess. 2016 17: 68-76.

VALIDATION OF PLASMONIC LUMINESCENT SOLAR CONCENTRATOR MODELLING SOFTWARE

Glenn, A.*, Chandra, S., McCormack, S.
Trinity College Dublin
College Green, Dublin 2, Ireland
aglenn@tcd.ie

ABSTRACT: This study aims to validate theoretical modelling of Plasmonic Luminescent Solar Concentrators (PLSC) software so that it can be used to optimize the PLSC fabrication process. The model will also be used to optimize the overall PLSC output in terms of solar power conversion, optical efficiency, dye concentration and device configuration. The two-stage process involves initially validating the LSC simulations by matching the modelling results with experimentally observed outputs, before moving onto subsequent PLSC model validation which requires further additional inputs, the metal nanoparticle extinction spectra, and the photon density enhancement factor (PDEF). The metal nanoparticle extinction spectra were measured experimentally and as the PDEF was the final remaining variable, it was taken from the literature and altered until it fit the particles being used in this study. The results of these simulations were then compared with the indoor and outdoor experimental characterizations and after noting that the trends and key data points correlated between the two methods of testing, the model was deemed to be validated.

Keywords: Modelling, Building Integrated PV (BIPV), Concentrators, Photovoltaic, Characterisation.

1 INTRODUCTION

Global warming and climate change induced droughts, flooding, heatwaves, and other extreme weather events are becoming more and more frequent [1]. This is, and will continue to be, a huge issue for civilization. A survey with a sample size of over 10,000 people performed in 2019 spanning 22 different countries found that climate change is considered the most commonly top-ranked problem that needs immediate reaction in the world at this moment [2].

The most prevalent cause of this change in climate is the incessant reliance on fossil fuel materials for energy production. When fossil fuels are processed, they emit greenhouse gases (GHGs) such as carbon dioxide into the atmosphere, which then traps heat and contributes to this warming effect in the earth's atmosphere [3]. In effect, the atmosphere with these GHG additives is acting like a greenhouse in that it is trapping the incoming solar radiation that would usually escape and is therefore causing a long-term heating effect, with the issues mentioned above being an unfortunate outcome. 9 out of the 10 hottest years on record since 1880 have occurred since 2005 with each of the top 5 happening since 2015 [4]. Aside from behavioural changes, one proven and critically important solution to this problem is the research and development of renewable energy technologies to reduce this reliance on the finite fossil fuel resources. The main objective of this study in particular is to demonstrate the ability of a modelling tool which simulates a novel type of plasmonic luminescent solar concentrator device which is a clean solar energy technological alternative.

Luminescent Solar Concentrators (LSCs) are devices that capture, redirect, and concentrate the incoming light towards photovoltaic (PV) cells that are situated along the edges of the layer. This effect is demonstrated in figure 1. They absorb the incoming solar energy using luminescent materials (i.e Quantum Dots, dye) of choice and re-emit it at longer wavelengths which confines the photons between the surfaces of the layer. The photons are transported along the device until they reach the edge PV cells through total internal reflection (TIR) [5].

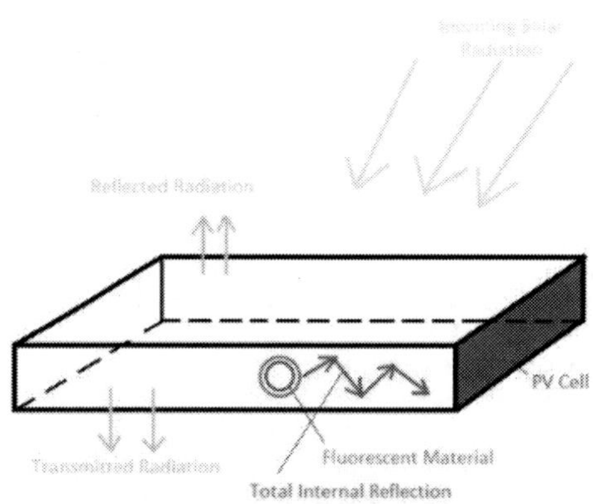

Figure 1 – Luminescent Solar Concentrator

LSCs were first proposed in 1976 and were referred to as 'greenhouse collectors' because of their working mechanisms [6]. The technology has had a resurgence in interest in recent years due to increased availability in the range of luminescent materials and their high quantum efficiency [7]. The design of an LSC minimises the amount of PV material required to produce electrical energy by concentrating the incoming solar radiation onto a smaller PV surface area. This is an important concept for building integration as it means that the technology does not need to track the sun and is therefore not just limited to a specific type of roof and façade installation. The technology could replace a window-type building element or be included in the façade of a construction as the technology is tuneable in terms of its transparency and colouration.

Figure 2 is an example of different LSCs in one panel displaying the variety of different shapes and colours available depending on the brief or outputs required (11). Some advantages to LSCs are that they perform well in both direct and diffuse lighting environments, they are static and do not need to track the sun, and their properties can be tuned to suit the task at hand as demonstrated below [8][9][10].

Figure 2 – Various LSC panels [11]

Plasmonic luminescent solar concentrators (PLSC) take this technology a step further with the inclusion of an additional plasmonic component such as metal nanoparticles which enhances the optical efficiency of the device [12]. These additional components compliment the organic dyes to further increase the efficiencies of the device by concentrating the photons in the subwavelength range that the dye particles can capture and re-direct [13]. MNPs have a very useful property in that their optical absorption properties can be tuned and therefore set to compliment the dye of choice in a PLSC [14]. The addition of MNPs to a LSC introduces a phenomenon known as surface plasmon resonance (SPR) in which the striking of an MNP by an incoming photon causes an oscillation to begin which will then induce an electric dipole. This oscillation between the coulombic and electromagnetic forces creates a localized SPR where the optical absorption of each dye particle is enhances as there are more electrons nearby for the particles to absorb [15].

This study focuses on the validation of LSC simulations initially before subsequently confirming the PLSC modelling capabilities.

2 METHODOLOGY

2.1 The PLASMON Model

The model used in this study was developed by the Solar Energy Applications Group (SEAG) within Trinity College Dublin [16]. It is a hybrid model that utilizes a ray tracing algorithm and finite-difference time-domain methods (FDTD) to simulate the optical processes that take place within LSC/PLSC devices [17]. These processes include transmission, reflection, refraction, absorption, emission, scattering and wave-guiding, as well as the plasmonic interactions which will be discussed later [18]. The graphical user interface (GUI) of the model is shown in figure 3.

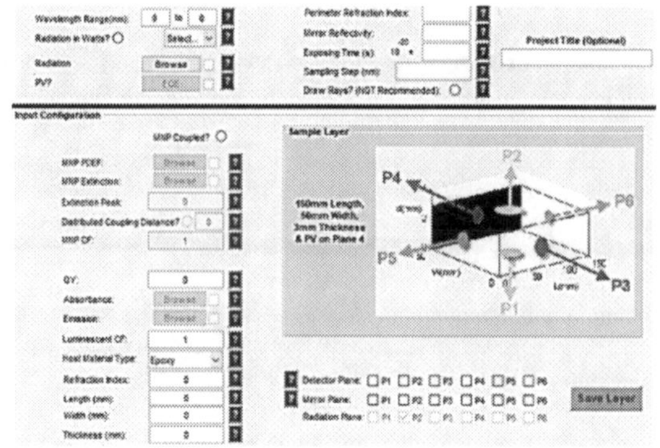

Figure 3 – GUI of the PLSC/LSC Modelling Software

The model implements a Monte Carlo ray tracing approach in which light rays are represented by photons that are traced until their final fate. The photons go through various optical processes, and their outcome is decided via probability theory. A constant 203,000 ray sample size test was performed throughout all simulations for comparative purposes as this was the upper limit of the PCs' hardware capabilities.

2.2 Luminescent Solar Concentrator Validation

LSCs were produced of varying thicknesses, dimensions, and dye concentrations so that they could be tested for their optical characteristics in the lab and compared with the model for validation.

Simulations were performed with different layer thicknesses, dimensions and dye concentrations also. To perform LSC simulations, the following inputs are required;

- Solar Radiation Spectrum in the same wavelength range as specified
- Absorption Spectrum of luminescent material
- Emission Spectrum of luminescent material
- Fluorescence Quantum Yield (FQY)
- Background absorption of the polymer
- Refractive index of the polymer
- Exposure duration

The model also requires the dimensions of the layer along with the configuration of PV cells and reflective surfaces. For example, if a configuration is to have 4-edge PV cells, then P3, P4, P5, and P6 must be selected in the detector plane section (see figure 3). Many of the inputs should remain constant for comparison purposes; It is important to note that the spectra input should be set in wavelength intervals of 1nm.

2.3 Plasmonic LSC Validation

The validation process for the PLSC modelling follows the same approach as for the LSC discussed earlier, with the exception that two additional input spectra are required for the model;

- MNP Photon density enhancement factor (PDEF)
- MNP extinction spectra

For the purposes of this study, an LSC sample of known concentration was produced in the lab and tested for its optical characteristics. The MNPs optical properties, the absorption and emission spectra were measured and input within the model. This value was subdivided to find the optimal concentration, with the previous red dye concentration remaining constant. This meant that the PDEF input was the final variable and therefore could be manually altered from another set of known MNPs taken from literature until the results matched the lab observations [19]. After it was determined that the model successfully simulated the layer already produced, it meant that it was suitable to simulate other concentrations and find an optimal overall. This optimal PLSC layer was fabricated in the lab and tested then compared.

3 RESULTS

3.1 Luminescent Solar Concentrator Validation

Included in figure 4 is the resulting output of the simulations of a 1-edge PV LSC layer of varying Lumogen F Red 305 dye concentration. The concentration was varied between 10 and 300 ppm.

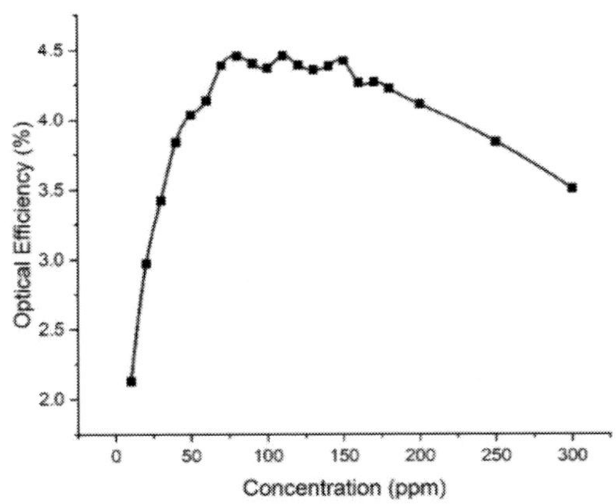

Figure 4 – Dye Concentration vs Optical Efficiency Simulation 1-Edge PV

Figure 5 displays the same test process as above, but in this simulation the configuration has 4 PV cells along the edges instead of 1.

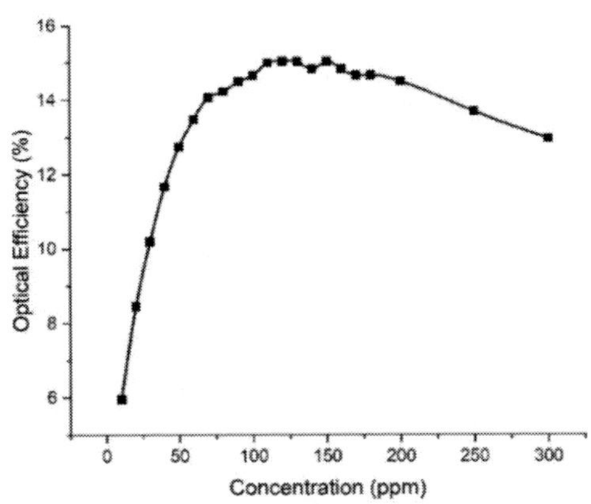

Figure 5 – Dye Concentration vs Optical Efficiency Simulation 4-Edge PV

Shown below in figure 6 is the integrated edge emission for a 1-edge LSC device measured using a BLK-C-SR-50 Concave Grating Spectrometer (StellarNet) connected to a 6-inch integrating sphere (RT-060-Sf, Labsphere) illuminated by a DH-2000-BAL UV-VIS-NIR light source (Ocean Optics).

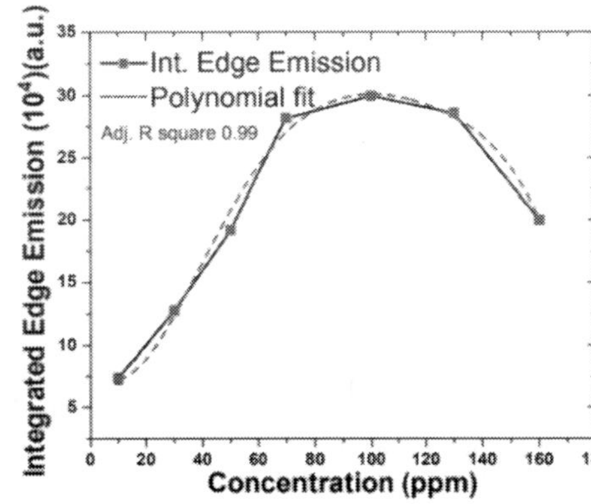

Figure 6 – Emission at the Edges of LSC Layer with Regards to Dye Concentration [20]

3.2 Plasmonic Luminescent Solar Concentrator Validation

Figure 7 shows the change in optical efficiency with varying MNP concentration when the lumogen red dye concentration had been set to a constant 70ppm. Figure 8 shows the experimental results of the MNP concentration with optical enhancement factor.

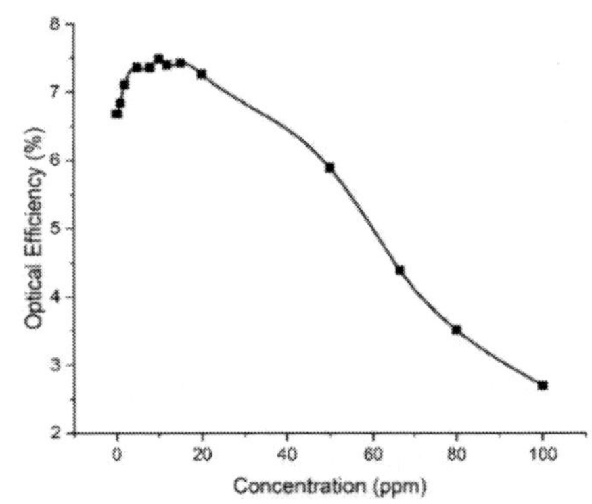

Figure 7 - MNP Concentration vs Optical Efficiency Simulation 1-Edge PV

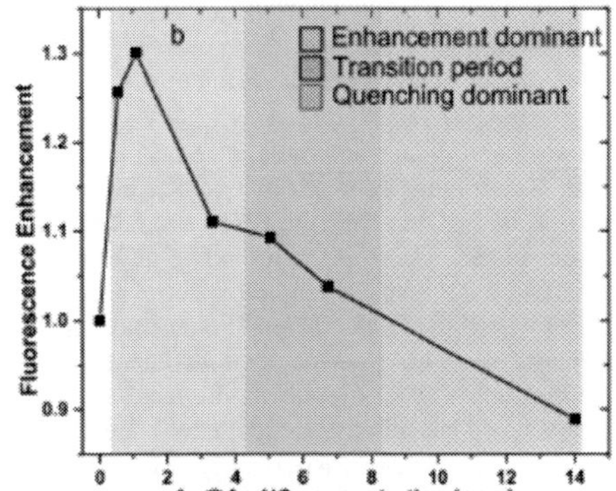

Figure 8 – MNP Concentration vs Enhancement Factor [20]

Figure 9 displays the final variable that had to be determined, the PDEF.

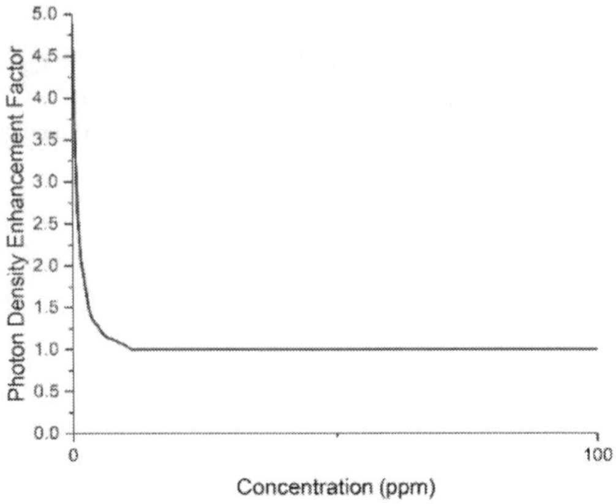

Figure 9 – Photon Density Enhancement Factor for Increasing MNP Concentration.

4 DISCUSSION

4.1 Luminescent Solar Concentrator Validation

Figures 4 - 6 demonstrate the process of the LSC simulation validation. Figures 4 and 5 show how the change in dye concentration affects the optical efficiency of the LSC device. The parameters of the simulation were set as to be identical in form to the fabricated layers in the lab with the only variable being the lumogen red dye concentration. The dye concentration was varied between 10-300 ppm throughout each analysis. The optimal optical efficiency output occurs at a dye concentration of 70ppm, and at concentrations any higher than that saturation begins to cause a plateau in efficiency before it begins to decline. This trend remained the same within both the single edge PV test and the 4 edge PV. The method in which the early fabricated devices were tested for their efficiencies in the lab was by measuring the emission at the edges of the layers using an integrated sphere and spectrometer. It can be seen that the output trend matches with the simulations and that an optimal yield occurs at a concentration of 70ppm.

After the LSC simulations had been validated the model could be used to predict optimal layers with other variables in mind such as the surface area of the device or the layer thickness. This greatly reduced the amount of time and materials needed to find and produce the optimal layer of each type.

4.1 Plasmonic Luminescent Solar Concentrator Validation

PLSC modelling requires two additional inputs when compared to LSC modelling, MNPs extinction spectrum and the PDEF. As this study uses gold core silver nanocuboid MNPs, the extinction spectra were measured and then varied to test the different concentrations. Figure 7 displays the change in optical efficiency when the concentration of MNPs is varied. Figure 8 demonstrates this same trend wherein the 3 stages of MNP enhancement, diminishing returns and quenching become clear.

The final variable to obtain for future PLSC simulations was the PDEF of the setups, included in figure 9, and as this was the last variable, it meant that a general PDEF from any gold/silver MNP could be taken from the literature (17) and then altered until the simulation results matched what had been observed in the lab. The exponential decay trend is common throughout all PDEFs as the distance from the particle is the critical factor. It was observed that a PDEF of 5 matched our MNPS best.

5 CONCLUSIONS

The objective of this study was to validate the developed combined ray trace and FDTD model (15) for PLSC device optimisation. Luminescent Solar Concentrator modelling has been successfully validated as the results of the simulations match with what has been measured experimentally. As well as this, an optimal output in terms of optical efficiency has been observed at a 70ppm lumogen red dye concentration in both cases.

With regards to PLSC modelling, two additional input variables had to be acquired, MNP extinction spectra and the PDEF. The MNP extinction spectra were obtained through experimental measurement and the PDEF was taken from the literature and adjusted to fit the MNPs of choice. A PDEF of 5 was observed as the best fit for the gold core silver coated nano-cuboid MNPs used in this study. Lumogen red dye concentration was determined to be optimal at 70ppm, whilst the MNP concentration has an optimal concentration of 10ppm.

The validation of this modelling software was successful as the simulated results match the trends and key measurements that have been observed experimentally. The model can now be used to simulate a vast array of different LSC & PLSC device configurations which will significantly reduce the amount of time, costs, and materials required in the optimization process.

6 REFERENCES

[1] - Meehl, G. A., Zwiers, F., Evans, J., Knutson, T., Mearns, L., and Whetton, P. (2000). Trends in Extreme Weather and Climate Events: Issues Related to Modeling Extremes in Projections of Future Climate Change. *Bulletin of the American Meteorological Society* 81, 3, 427-436, available from: < https://doi.org/10.1175/1520-0477(2000)081<0427:TIEWAC>2.3.CO;2> [Accessed 17 August 2022]

[2] – Climate change ranks highest as vital issue of our time. Amnesty International. December 10th, 2019.

[3] - Hoel, M., Kverndokk, S. Depletion of fossil fuels and the impacts of global warming, Resource and Energy Economics, Volume 18, Issue 2, 1996, Pages 115-136, ISSN 0928-7655, https://doi.org/10.1016/0928-7655(96)00005-X.

[4] – MacMillan, A., Turrentine, J. Global Warming 101. Natural Resources Defence Council. April 7th, 2021.

[5] - Li, Y., Zhang, X., Zhang, Y., Dong, R., and Luscombe, C. K. 'Review on the role of polymers in

luminescent solar concentrators'. 2019. J. Polym. Sci Part A 57, 201–215. doi: 10.1002/pola.29192
[6] - Weber, W. H., & Lambe, J. (1976). 'Luminescent greenhouse collector for solar radiation'. In APPLIED OPTICS (Vol. 15, Issue 10).
[7] - Green, M. A. 'How Did Solar Cells Get So Cheap?'. Joule, Volume 3, Issue 3,2019, Pages 631-633, ISSN 2542-4351, https://doi.org/10.1016/j.joule.2019.02.010.
[8] – Banal, J. L., Zhang, B., Jones, D. J., Ghiggino, K. P., & Wong, W. W. H. (2017). 'Emissive molecular aggregates and energy migration in luminescent solar concentrators'. Accounts of Chemical Research, 50(1), 49–57. https://doi.org/10.1021/acs.accounts.6b00432
[9] – Müller, M. F., Indoor Photovoltaics, Materials, Modeling and Applications. Chapter 6.1, Page 134. 2020.
[10] – Meinardi, F., Bruni, F., & Brovelli, S. (2017). 'Luminescent solar concentrators for building-integrated photovoltaics'. In Nature Reviews Materials (Vol. 2). Nature Publishing Group. https://doi.org/10.1038/natrevmats.2017.72
[11] - Van Sark, W., Moraitis, P., Aalberts, C., Drent, M., Grasso, T., L'Ortije, Y., Visschers, M., Westra, M., Plas, R., & Planje, W. (2017). 'The "Electric Mondrian" as a Luminescent Solar Concentrator Demonstrator Case Study'. Solar RRL, 1(3–4). https://doi.org/10.1002/solr.201600015
[12] – Chandra, S., Doran, J., McCormack, S. J., Kennedy, M., Chatten, A. J. 'Enhanced Quantum Dot Emission for Luminescent Solar Concentrators using Plasmonic Interaction'. Journal of Solar Energy Materials & Solar Cells, Volume 98, Page 385-390, 2012.
[13] – Chandra, S. Ahmed, H., Doran, J. McCormack, S. J. 'Plasmonic quantum dot solar concentrator', Proc. SPIE: Physics, Simulation, and Photonic Engineering of Photovoltaic Devices VI, 10099, Page 77-84, 2017.
[14] – Nolan, J.P., Sebba, D.S. 'Surface-Enhanced Raman Scattering (SERS) Cytometry'. Methods in Cell Biology, Academic Press, Volume 102, 2011, Pages 515-532, https://doi.org/10.1016/B978-0-12-374912-3.00020-1.
[15] – Kelly, K. L. The Optical Properties of Metal Nanoparticles: The Influence of Size, Shape, and Dielectric Environment. 2002. 107: 668–677. doi:10.1021/jp026731y.
[16] – Rafiee, M. '3D Modelling Tool to Optimise Advanced Plasmonic Luminescent Solar Devices for Building Integrated Photovoltaic Applications'. Trinity College Dublin. School of Engineering, 2019.
[17] – Rafiee, M., Chandra, S., Ahmed, H., Barnham, K., McCormack, S. J. 'Small and large scale plasmonically enhanced luminescent solar concentrator for photovoltaic applications: modelling, optimisation and sensitivity analysis'. Optics Express (Energy and Environmental Optics Express) 29, page 15031-15052, 2021.
[18] -
[19] - Thomas, R., Kumar, J., Swathi, R. S., & Thomas, K. G. Optical effects near metal nanostructures: towards surface-enhanced spectroscopy. 2012. Vol. 102, Issue 1.
[20] - Sethi, A., Chandra, S., Ahmed, H., & McCormack, S. (2019). Broadband plasmonic coupling and enhanced power conversion efficiency in luminescent solar concentrator. Solar Energy Materials and Solar Cells, 203. https://doi.org/10.1016/j.solmat.2019.110150

CHARACTERIZATION OF THE ELECTRICAL BEHAVIOR OF A BIFACIAL PHOTOVOLTAIC SOLAR MODULE OBTAINED THROUGH STATIC AND TRANSIENT ANALYSES

Dr. Valdivia-Lefort Patricio, patricio.valdivial@usm.cl
Ing. Gonzalez Valentina, valentina.gonzalezbe@sansano.usm.cl
Dr. Rodrigo Barraza, Rodrigo Barraza <Rodrigo.Barraza@usm.cl>
Universidad Técnica Federico Santa Maria
Av Vicuña Mackenna 3939, San Joaquin, Santiago - Chile

ABSTRACT: In the last 18 years PV solar power has increased from 0 to 500 GWp approximately as indicated by International Energy Agency Photovoltaic Power Systems Program IEA PVPS [13]. However, an issue that has been left aside is the maintenance of the constituent elements of a solar PV plant, this mainly because the elements are still under warranty. In particular, one of the fundamental players in a solar PV plant is the Photo Voltaic Solar Module PVSM, which usually has a guarantee of around 20 years. However, it has been proven that in practice a large number of modules have presented unexpected failures, e.g., power degradation faster than what the manufacturer considered when defining the warranty. Considering an average annual degradation of 1%, it would be expected that within the first 10 years, the PV modules of a PV plant present a 10% degradation in their performance.
In this work it is carried out a characterization of the electrical behavior of a bifacial photovoltaic solar module obtained through static and transient analyses operating under no-fault and fault conditions at controlled conditions using the Flasher IV A+A+A+ Solar Simulator of Eternalsun. Later, the response of photovoltaic solar modules, for their healthy and faulty operation, were studied.
As a result, it is obtained a characterization of the electrical behavior of bifacial photovoltaic solar module under no-fault and fault conditions at controlled level.

1 *INTRODUCTION*

The Chilean PV industry started in 2014 with a progressive insertion of Photovoltaic solar Plants, starting with 1.2% [9], and by 2022 there is already 20% of penetration [9]. Solar energy is one of the most promising in our country since, according to data from the Ministry of Energy, the Great North of Chile would receive the largest amount of average daily solar radiation, estimating that an area of 20 [km2] of photovoltaic panels could meet the energy demand of the entire country [2] Given the conditions of desert in the great north of Chile, a technology that is especially promising is the bifacial, since it takes advantage of the radiation reflected from the ground to illuminate the rear face of the modules and thus increase the generation, making a more effervescent.

Together with the development of technology, it is necessary to create techniques that allow evaluating the performance of this, it is here where the need to generate methodologies of characterization. In the case of monofacial modules, there are a considerable number of methodologies recorded in the literature given its settlement and stability over time, since that this type of device has a long history, however, for new technologies, such as bifacials, there are more challenges associated with new behaviors that must be studied, such as their performance in different scenarios, degradation mechanisms, failures etc.

The characterization of photovoltaic modules can be carried out using different methods, among them is the characterization with electrical variables from IV curves, optical characterization, characterization by electroluminescence, among others. The present study will specifically address the characterization by means of electrical variables, electroluminescence and thermography. The characterization, in addition to providing information on the current performance of the measured device, allows detection of possible degradations or failures.

The degradation of photovoltaic modules is a problem that has been present in the industry for a long time, and understanding this process allows manufacturers to improve the materials and processes used in the production of these devices [1]. Some degradation mechanisms are mentioned in [10]:

1.1 Packaging material degradation

This mechanism occurs when the materials that protect the cell are damaged. Some examples of this degradation are the breakage of the glass that covers the modules (Figure 1), blocking diode failure, potting discoloration (Figure 2), delamination (Figure 3) and cracking.

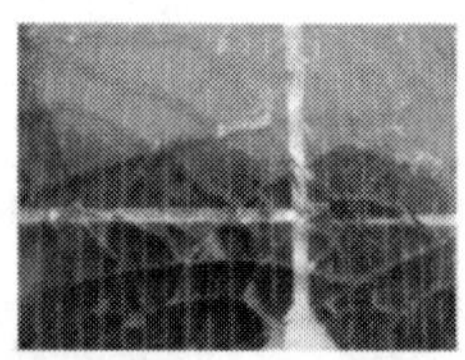

Figure 1: Glass breakage in a module [1].

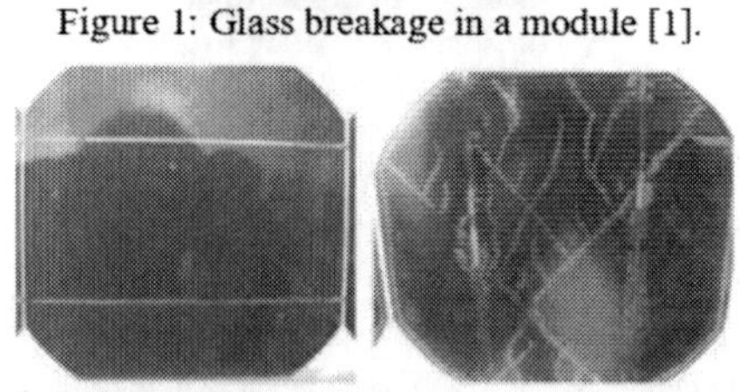

Figure 2: Blocking diode failure, potting discoloration [1]

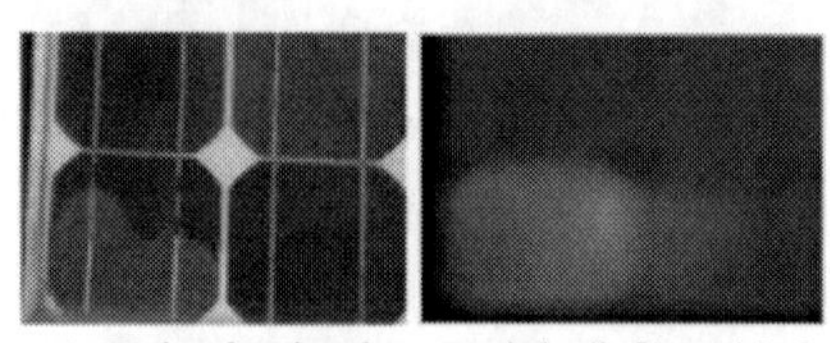

Figure 3: Delamination in a module (left) and infrared image of the area (right) [1]

1.2 Moisture intrusion:
The humidity that reaches the module from the rear face or through the edges of the sheets causes corrosion (Figure 4) and increases leakage currents. it attacks metallic components in the case of c-Si modules, while in modules thin film attacks the semiconductor layers, causing losses in their electrical performance.

Figure 4: Corrosion in interconnections and metallic contacts of two different modules [1].

In this work it is carried out a characterization of the electrical behavior of a bifacial photovoltaic solar module obtained through static and transient analyses operating under no-fault and fault conditions at controlled conditions using the Flasher IV +AAA Solar Simulator of Ethernalsun. Later, the response of photovoltaic solar modules, for their healthy and faulty operation, were studied.

1.3 IV Curve abnormalities
The shape of the IV curves allows to identify anomalies in a photovoltaic cell. The standard IEC 62446-1 [5] contains some keys for the interpretation of the different forms that these curves can present.

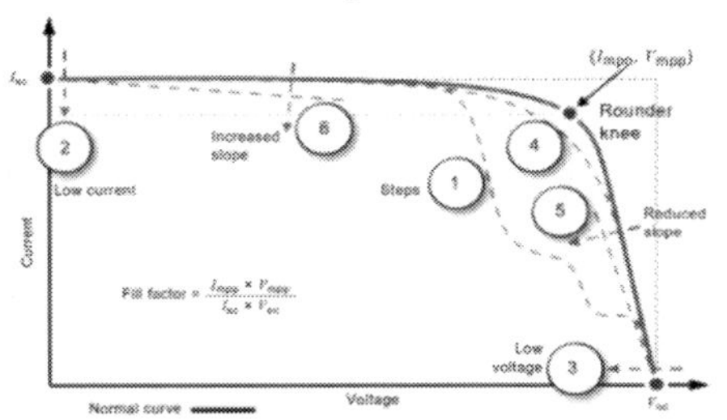

Figure 5: Some of the variations that IV curves can have [5].

1.1. Steps or crevices: They can indicate faults between the different areas of the measured device. This deviation indicates that the blocking diode turned on, so some current is being redirected, bypassing some internal cells protected by the diode. This can occur due to partial shading, dirt or soiling of the module, damaged cells, or shorted blocking diode.

1.2. Curve with low current: To identify that the current is lower than expected, the reference is rated short-circuit current. Some causes of this decrease may be uniform dirt, shadow fringing, degradation, poor sensor calibration of irradiance, measurement problems, or irradiances that are too low.

1.3. Curve with low voltage: To identify that the voltage is lower than expected, the reference is rated open circuit voltage. Some causes can be the diodes in conductor shunt or shorted, potential induced degradation (PID), significant and uniform shading throughout the cell, no correspondence with the actual temperature of the device and the value at which the curve is plotted.

1.4. Curve with soft knee: Rounding of the knee of the IV curve may be a manifestation of the process module aging. This applies to curves that have changed their shape, since there are some technologies that have this form intrinsically and not necessarily corresponds to deterioration.

1.3.5. Reduced slope in vertical leg: Corresponds to the slope that includes the section from the point of maximum power (where Vmpp is located) and the open circuit voltage (Voc). This slope is influenced by the series resistance of the module, where the higher the resistance, the lower slope incline. Some Causes of Increased Series Resistance can be damage to the wiring or faults in the interconnections of modules or substrings, which can be caused by corrosion, degradation, or factory errors. In this case, thermographic inspection can be a useful tool when it comes to identify high-resistance faults.

1.5. Greater slope in horizontal leg: This variation may be related to low parallel resistances or parallel current fluxes, differences in the short-circuit current of the cells of a module or string (mismatch), shading or punctual soiling. Shunt current corresponds to deviations due to localized defects in the cell or in the interconnections, which can cause localized hot spots that can be identified through thermographic inspection.

2 SET UP

To carry out the characterization of the bifacial photovoltaic modules, three different techniques were utilized: IV curves, electroluminescence and thermography; the equipment used for each case is detailed as follow:

2.1. PV Modules
The main element for the measurement is the test object, in this case two types of bifacial modules.

Risen RSM72 Bifacial Modules
For the measurement, a total of 10 Risen RSM72 bifacial modules PERC+ were used, their data Nominal values are presented in Table 1

Brand	Risen Energy Co., Ltd.
Model	RSM72-6-370BMDG
Dimensions [mm]	1984x992x25
Pmax [W]	370
Vmpp [V]	39,60
Impp [A]	9,35
Voc [V]	48,15
Isc [A]	9,90
Tolerance Isc	±3%
Tolerance Voc	±4%
Bifaciality	75%

Table 1: Specifications of the Risen RSM72 bifacial modules [3].

Also, three mini bifacial modules of (hetero junction) HJT technology, coming from the CEA INES laboratory were used. Its nominal data is specified in Table 2.

	Módulo $N°11$	Módulo $N°12$	Módulo $N°13$
ID	BOM3-09	BOM3-09B	BOM1-16
I_{sc} $[A]$	8,828	8,829	8,776
V_{oc} $[V]$	2,925	2,911	2,918
P_{max} $[W]$	19,216	19,047	18,990
I_{mpp} $[A]$	8,172	8,179	8,117
V_{mpp} $[V]$	2,352	2,329	2,340
FF	74,42	74,12	74,15

Table 2: Specifications of BOM CEA INES bifacial mini modules.

2.2 Curves IV: TCLF

The Temperature Controlled Lab Flasher (TCLF) solar simulator is a device developed by Eternalsun Spire, its purpose is to perform indoor characterization measurements of photovoltaic modules from IV curves, allowing measurements to be made under standard conditions STC, but also under controlled conditions of both temperature and irradiance.

Figure 6: Equipment TCLF

The Flasher Spire measures photovoltaic modules under AM 1.5G conditions following the ASTM G173 standard [5]. Their technical specifications are found in Tables 3 and 4.

Dimensions max. module	2140 x 13540 mm
Effective test area	2100 x 1300 mm
Classification	A+A+A+ (IEC 60904-9)

Table 3: TCLF General specifications

Lamp type	SLP filtered Xenon tube
Pulse duration	110 ms
Pulse type	single pulse
Spectral range	300 – 1200 nm
Spectral Match	< 12,5% (A+)
Irradiance non-uniformity	< 1% (A+)
STI-LTI	0.10% (A+)
Irradiance range	100 – 1200 Wm−2
Time between flashes	< 20 s
Guaranteed lamp life	200,000 flashes
Typical Lamp Life	> 400,000 flashes

Table 4: Light Specifications

2.3 Electroluminescence

The electroluminescence measurement is performed on all modules (N°1 – N°13). The maximum current supplied for RISEN RSM72 PV modules was 6 [A]. In the case of the mini bifacial modules, the maximum current used was lower than Isc since having a transparent structure, the light emitted by the reverse reflected off the supporting wall, contaminating the measurement results; consequently, the level of current chosen to be able to appreciate in a good way the emitted spectrum, and at the same time decrease the effect of reflection was 1 [A].

In both cases images captured of both, the front face and the rear face of the modules, were always preventing the connections to interfering with the measurement.
From this test, only the capture of images was performed at the current level indicated previously, since in this way it was possible to obtain as much information as possible to qualitative analysis. The connection used for this test is shown in Figure 7.

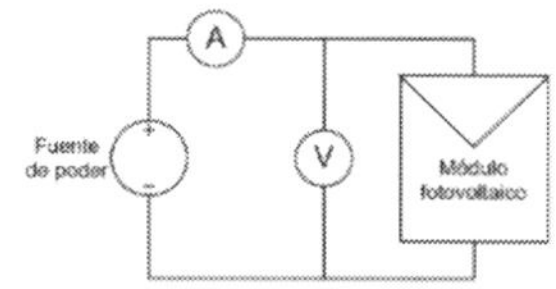

Figure 7: Connection circuit for the electroluminescence test

Brand	Ele-tech
Model	HY5020E
Dimensions [mm]	1984x992x25
Pmax [W]	1000
Voltage [V]	0-50
Current [A]	0-20

Table 5: Technical specifications of the power source used for the test of electroluminescence

Brand	Nikon
Model	D750
Sensor	CMOS de 26 MP
Filtered wavelengths	< 850[nm]

Table 6 includes the details of the camera, lens, and filter used for the electroluminescence assay.

2.3 Thermography

The last test carried out is the capture of IR images of the modules. Are used two different connections depending on the type of module being worked on, in the case of Risen RSM72 bifacial modules, these are connected according to Figure 8.

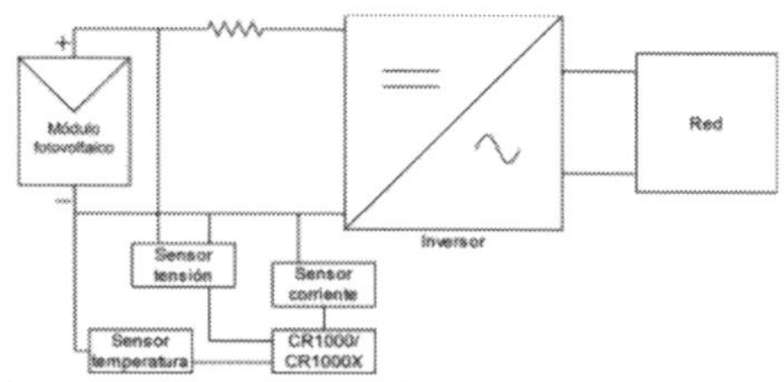

Figure 8: Risen RSM72 bifacial photovoltaic module connection diagram.

On the other hand, in the case of the bifacial mini modules, the PV tests were connected with a rheostats as indicated in Figure 9. When connected only a mini module, the generated current was around 1 [A], for this reason a connection of modules in series, to increase the voltage at the terminals. Connecting the three mini bifacial modules in series, and using the minimum resistance value of the rheostats, a current value of 97 [A] was obtained.

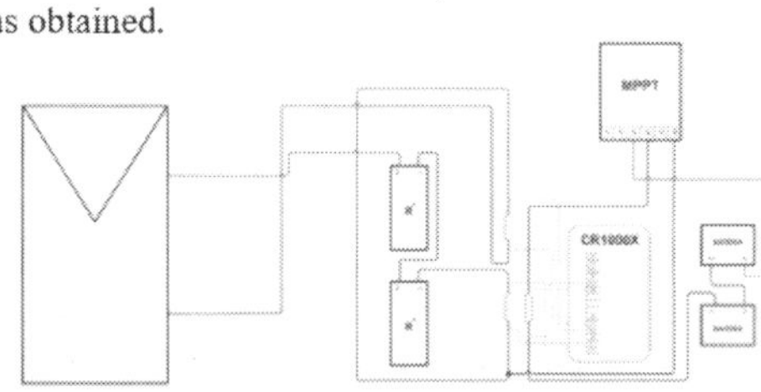

Figure 9: Connection diagram of the BOM CEA INES mini bifacial photovoltaic modules [1].

Brand	Fluke
Model	Ti400
Lens	FLK-LENS/TELE2
CDVI	1.31 mrad
Minimal focus distance	15 cm
focus system	Automatic

Table 7: Specifications of the thermal imaging camera used for the Thermography test

3 METHODOLOGIES

To characterize bifacial photovoltaic modules, the following tests are proposed to make: I-V curves, electroluminescence, thermography, and other such visual inspection [1]

1 IV CURVES:

1.1 Measurements at constant temperature of 25°C:

After the visual inspection, the measurement of IV curves is carried out with the solar simulator TCLF. For each module, both front and back lighting are tested, using temperature correction at 25∘C and varying the irradiance level from100 [Wm-2] to 1000 [Wm-2], with a step of 100 [Wm−2] between each measurement.With the data obtained, the IV and PV curves are plotted both with illumination front and rear for irradiance levels of 100 − 500 − 1000 [Wm−2] of all the modules that correspond to the same type, that is, the first group contains modules from 1 to 10, corresponding to Risen RSM72 bifacial modules and the second group from 11 to 13, corresponding to bifacial mini modules BOM CEA INES. From the IV curves, the data is processed to obtain:

- Graph of electrical variables (Isc, Voc and Pmax) as a function of irradiance, for the front face and back face.
- Bifaciality coefficients
- BiFi, PmaxSTC, PmaxBiF i100 and PmaxBiF i200

1.2 Temperature controlled measurements

Once the IV curves corrected at 25∘C have been obtained, the tests are carried out with control temperature at 25 − 50 − 75∘C, and a fixed irradiance of 1000[Wm−2]. From the Risen RSM72 bifacial modules, modules No. 1, No. 8 and No. 10 were chosen. From the BOM CEA INES mini bifacial modules, N∘11 was chosen.

With the data obtained, a graph of electrical variables (Isc, Voc and Pmax) is made depending on the temperature.

For module N∘10, the test of both rise and fall in temperature is recorded, therefore, temperatures of 50−25∘C are also added to the original points.

4 RESULTS

4.1 IV Curves

IV and PV curves are plotted at irradiance levels of 100 [Wm−2], 500 [Wm−2] and 1000 [Wm−2], which represent a low, intermediate, and high level, and thus observe if there are changes in the shape of the curves depending on this factor. In this section are only included the graphs for the irradiance level of 1000 [Wm−2].

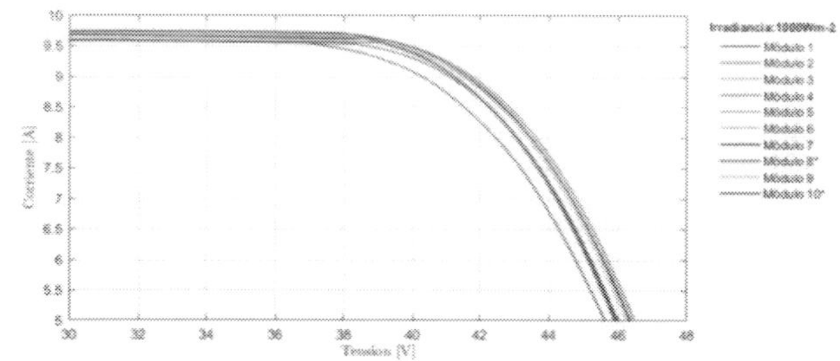

Figure 10: IV curves for the 10 Risen RSM72 bifacial modules at an irradiance of 1000[Wm−2] on the front face.

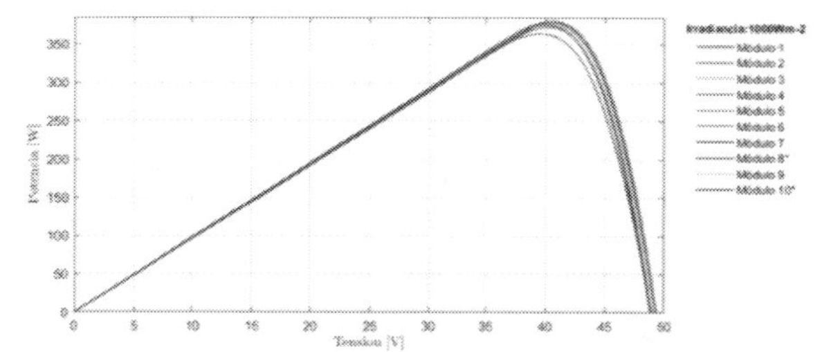

Figure 11: IV curves for the 10 Risen RSM72 bifacial modules at an irradiance of 1000[Wm−2] on the front face.

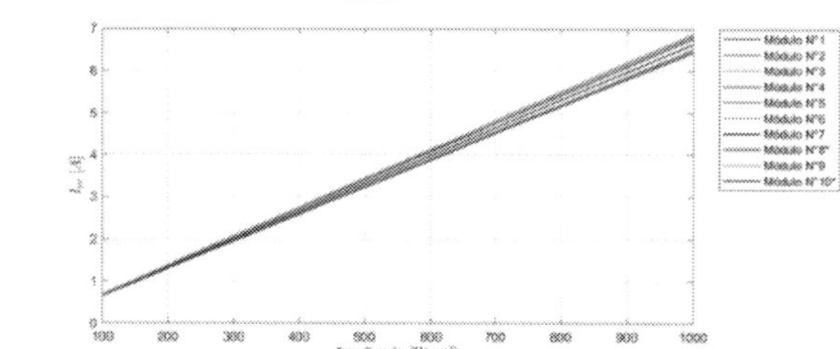

Figure 12: Short circuit current of the 10 Risen RSM72 bifacial modules for irradiances from 100 to 1000[Wm−2] on the rear face.

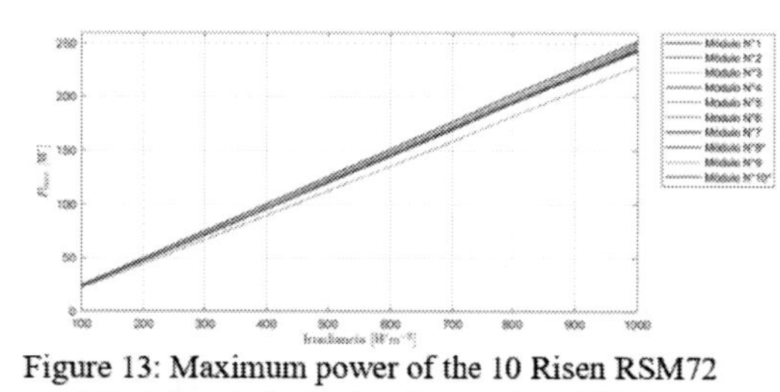

Figure 13: Maximum power of the 10 Risen RSM72 bifacial modules for irradiances from 100 to 1000[Wm−2] on the rear face.

When the irradiances are low, the dispersion is not appreciable, however, as the irradiance level increases, the change in slopes is noticeable. Minors slopes are associated with modules N∘3, N∘5, and N∘10, at an intermediate point finds module N∘7, and the largest slopes are associated with modules N∘1.

In the same way as for the Risen RSM72 bifacial modules, for BOM CEAN INES PV modules, the graphics are included of IV and PV curves for the irradiance level of 1000 [Wm−2].

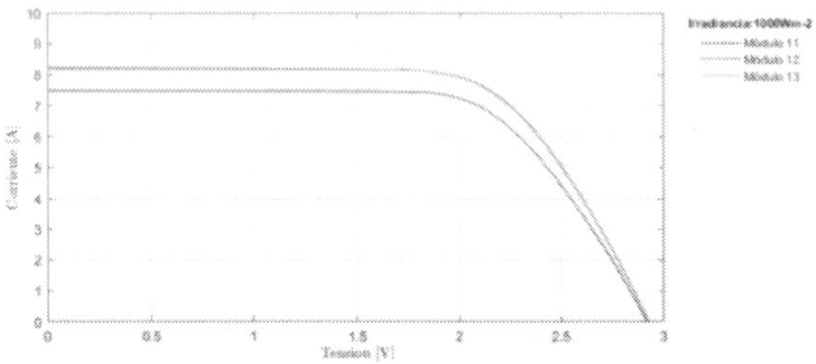

Figure 14: IV curves for the 3 bifacial mini-modules at an irradiance of 1000[Wm−2] on the back face

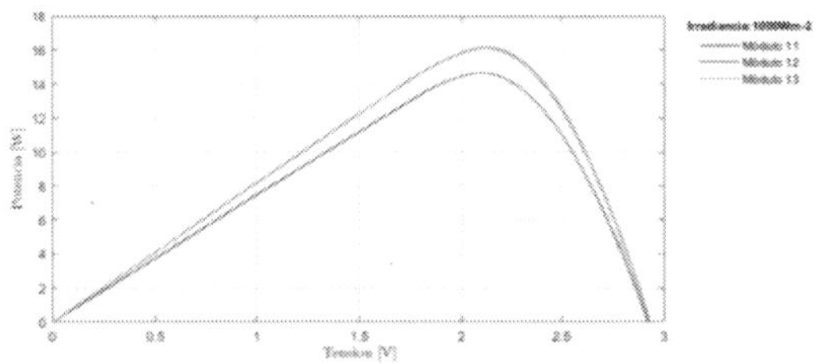

Figure 15: PV curves for the 3 bifacial mini-modules at an irradiance of 1000[Wm−2] on the rear face.

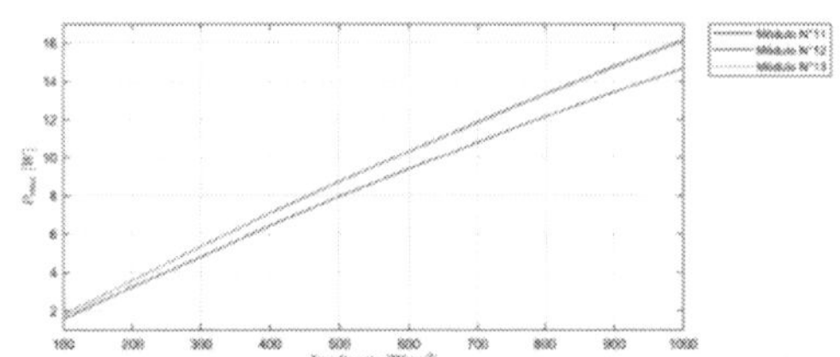

Figure 16: Maximum power of the 3 mini bifacial modules for irradiances from 100up to 1000 [Wm2] on the rear face.

4.2 BIFI

The BiFi corresponds to the slope of the linear regression of the maximum power in function of the irradiance of the rear face of the module, this corresponds to the slope of the first points obtained in Figure 13. In this case, the levels of irradiance of 100, 200 and 300[Wm−2], given the recommendations of the IEC standard TS 60904-1-2:2019 [6]. The values of BiFi, PmaxSTC, PmaxBiF i100 and PmaxBiF i200 are presented in Table 8.

$N°$ módulo	$BiFi$	P_{maxSTC}	$P_{maxBiFi100}$	$P_{maxBiFi200}$
1	0.2478	-1.5619	23.2201	48.0022
2	0.2449	-1.5229	22.9644	47.4516
3	0.2256	-0.9794	21.5785	44.1364
4	0.2447	-1.2942	23.1798	47.6537
5	0.2433	-1.1149	23.2113	47.5374
6	0.2387	-0.7622	23.1059	46.9740
7	0.2434	-1.4954	22.8416	47.1787
8	0.2544	-1.4001	24.0361	49.4722
9	0.2496	-1.9960	22.9633	47.9225
10	0.2432	-1.6505	22.6696	46.9898

Table 8: BiFi obtained for each Risen RSM72 bifacial module and their respective powers PmaxSTC, PmaxBiF i100 and PmaxBiF i200.

6 Electroluminescence

The images obtained in the electroluminescence measurement for RISEN RSM72 PV Modules are shown in Figure 17.

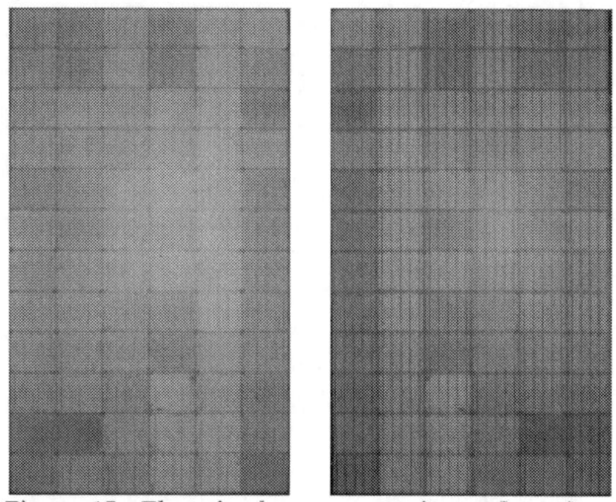

Figure 17: Electroluminescence results performed on bifacial module No. 1. a) Front face. (b) Back face.

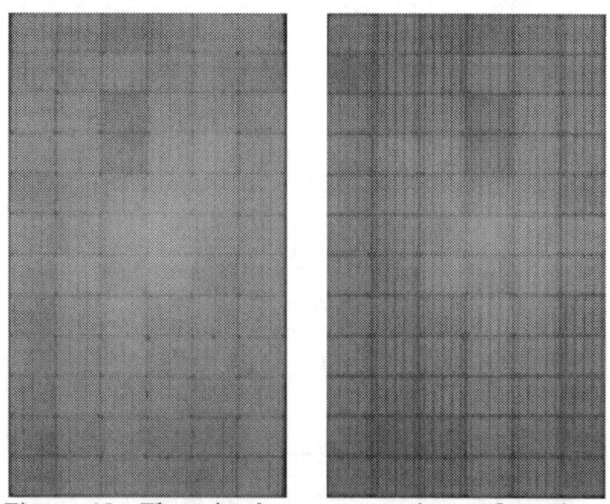

Figure 18: Electroluminescence results performed on bifacial module No. 10. a) Front face. (b) Back face.

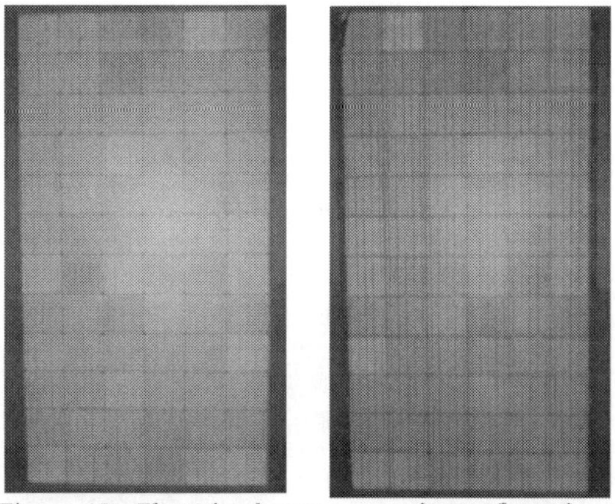

Figure 19: Electroluminescence results performed on bifacial module No. 8. a) Front face. (b) Back face.

Figure 20: Electroluminescence results performed on bifacial module No. 11.

The front face of the bifacial mini module BOM CEA INES N◦11 shows a series of dark spots similar to stripes, which are observed both from the front (Figure 20 left) and back (Figure 20 right). These spots can be associated with the decrease in time of life of the minority carriers, since its structure resembles the impurities seen in Figure 15, as well as cell contamination, as seen in Figure 15. Figure 14, where line-shaped defects appear. The light emitted by the cells looks uniform, without appreciable differences. Electroluminescence shows no indication of the delamination mentioned on visual inspection.

4.4 Thermography

The images obtained in the thermography are analyzed based on the numbering shown in Table 8. Modules N◦1, N◦2, N◦3, N◦4, N◦6 and N◦7 was found positioned vertically, while the modules N◦5, N◦8. No.9 and No.10 were positioned horizontally.

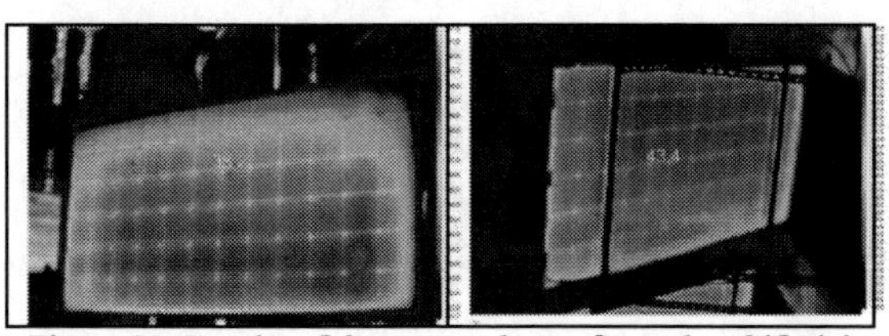

Figure 21: Results of thermography performed on bifacial module No. 1.

Figure 22: Results of thermography performed on bifacial module No. 8.

Figure 23: Results of thermography performed on bifacial module No. 10.

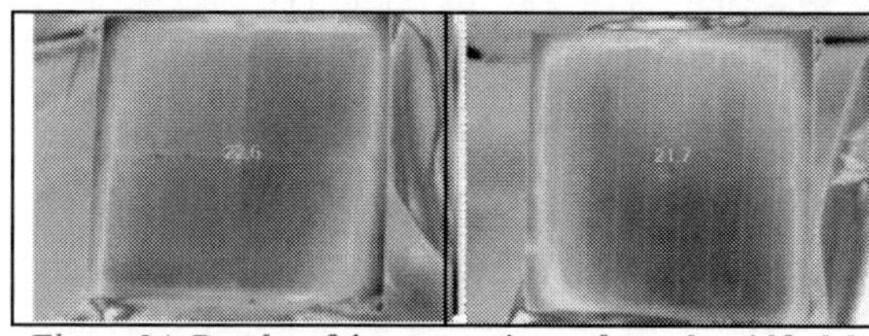

Figure 24: Results of thermography performed on bifacial BOM CEA INED module No. 11.

4 DISCUSSIONS

At level of irradiance of 1000[Wm−2], the difference in performance can be seen clearly, in particular, the IV curve of module N◦8 is less pronounced when it approaches the knee, which according to [5] can occur due to processes aging of the module and is consistent considering that the module has been in operation. As for the. others, module No. 1 appears within the minor maximum power levels, even reaching lower powers than in the case of module N◦10, which has been previously exposed. The other curves have tightened to the previously mentioned behaviors for the irradiance levels of 100 [Wm−2] and 500 [Wm−2].

As the irradiance level varies, in Figure.12, the short-circuit current grows linearly. Qualitatively, it is observed that for the front face the value of Isc does not vary much. The module that showed higher short-circuit currents when illuminated from its front face was the module N◦10, while the one that reached the minors was the module N◦8. In this case, no particular behavior is observed that differentiates the exposed modules of the unexposed. On the other hand, when the lighting is done on the back a clear separation is observed in the short-circuit current levels reached.

In the case of BOM CEA INES MODULES, it can be identified an abnormality in module 11, probably caused by a delamination, which is confirmed with the IV Curves and thermography test.

Table 8 shows that the minimum BiFi value corresponds to module No. 3, result that is evident when observing Figure 13, since this is the one that has the lower peak power levels when lit from the back. On the other hand, the module that had the best maximum power performance was the N∘8, and it coincides with the highest BiFi value. This quantitative result also allows corroborating some performances that had been previously observed on the charts, for example, the highest BiFi value corresponds to module #6.

For EL test, Figure 17, shows the electroluminescence results of module No. 1. In general, the first thing that is most evident is the difference in illumination of the cells, where the most illuminated is column E, contrasting with column F since it is darker. Some cells that are also particularly dark are (1,B), (2,B), (1,D), (2,D) and (11,B), which according to [7] could be due to a decrease of the lifetime of the minority carriers of the semiconductor.

In the case of module N∘8 (Figure 17) dark spots are observed on the edges of the last cells of columns D and E. Due to the color of the stains could be associated with type C cracking, which is associated with the disconnection of regions from the rest of the circuit, and therefore, decreasing the power performance, which was seen clearly in Figure 10, where this module had the lowest performance when increasing the irradiance level of the front face. In addition, darker cells are observed, such as (3,C) and (4,C), which show faded black spots. All Finds Viewed by the front face are also replicated when observing the spectrum emitted by the back face.

An important detail is that, although the modules that present cracks show a low in power performance, this is only apparent when lit from the face front, since on the back the behavior is totally different, reaching even to observe that the module N∘8 reaches the highest levels of maximum power. In module N∘10 (Figure 17) some more noticeable darkening are observed in cells (2,C), (7,B) and (8,C), which can be caused by the decrease in lifetime of minority carriers.

Module N∘10, during Thermography test, shows uniform temperature distributions, without anomalies on its front face (Figure 22a), in the case of its rear face a fringe pattern with a lighter color (Figure 22b), which is attributable to reflections metal structure product.

Figure 24 a shows that the front face of the mini bifacial module No. 11 has some hot spots in the area of the junction between cells (1,A) and (2,A). hot spots evidenced fit with the areas that have presence of dirt on the surface. For another side, Figure 24b shows the temperature distribution on the rear face of the module, which has an increment in the cells of the second row, which is consistent with identified delamination. Also, the back side shows some hot spots between cells (1,A) and (1,B), as well as at the junction between cells (1,B) and (2,B), which fits with the presence of dirt on the surface.

As a synthesis, a block diagram is made with the routines carried out and diagnoses based on the findings that can be evidenced. This considers the test suite was i) IV curves ii) electroluminescence iii) thermography.

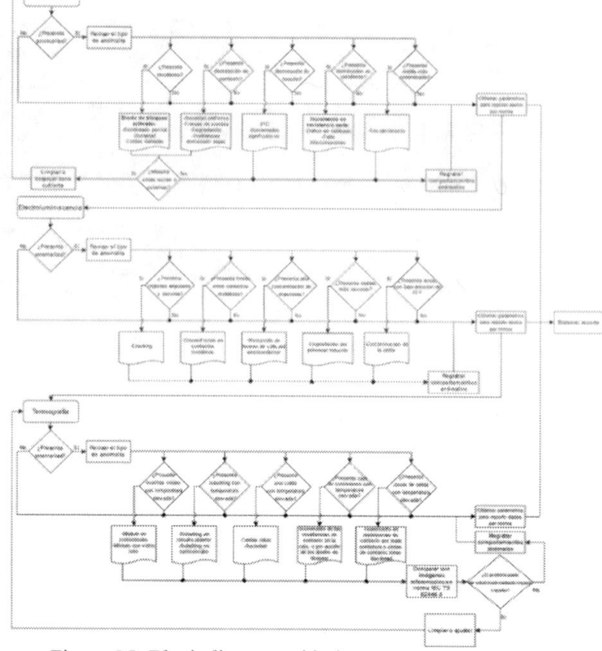

Figure 25: Block diagram with the measurement methodology and its possible diagnoses.

5 CONCLUSIONS

The main contribution of this work is the creation of a methodology for the characterization for bifacial photovoltaic modules from three tests: plotting curves IV, electroluminescence and thermography, in addition to the detection of anomalies through complementary visual inspection. Within this, one of the points of greatest emphasis was the use of the TCLF equipment, a solar simulator specialized in the characterization of photovoltaic devices from IV curves in the laboratory. Later, a guide was generated with procedures for measurement with and without temperature control. One of the main advantages that this device showed is the fast data acquisition, the possibility of working under standardized conditions, and finally the possibility of controlling the variables of temperature and irradiance.

Regarding the characterization based on electrical variables, it was observed that the IV curves were clear in showing abnormalities in three main cases i) the presence of delamination ii) cracking iii) shading. In the case of delamination, there was a noticeable decrease in the short-circuit current of the affected device, while the cracking was shown to from the smoothing of the knee of the curve, which in the PV characteristic is evidenced as a decrease in the maximum power point. Finally, the shading generated current steps, decreasing the maximum power performance of the devices. This phenomenon was observed on the rear face of all Risen bifacial modules RSM72.

With the parameters extracted from the IV curves, graphs were made based on the irradiance for both the front and back face. These charts help complement the information provided by the IV curves, showing that the decreases more noticeable in the slope for the case of Isc and Pmax (BIFI) are related to degradation.

Regarding the electroluminescence tests, these provided key information to identify type A and B cracking modes, but they did not serve to detect delamination or scratches on the surface of the modules. This assay also showed cells more darkened, which could be associated with shorter lifespans of carriers minority. Contrasting these latter patterns, no marked relationship was observed between the existence of dimmer cells and the shape of the IV curves.

IR image capture enabled identification of hot spots for the case of Risen RSM72 modules. The latter did not show significant variations in their IV curves, however, considering the recommendations of the standard, a follow-up of these points to prevent this type of anomaly from becoming security risk. In the case of the mini modules, this essay provided information very important for the identification of overheating in some specific cells and the junction box. In addition, with thermography it was possible to identify by means of a high temperature zone the presence of delamination in module No. 11, which reaffirms that the different tests can be complemented to obtain a characterization full of modules.

Finally, it can be affirmed that carrying out the three proposed tests provides a characterization complete, since each brings a different perspective that can complement the results obtained. On the one hand, electroluminescence tests and thermography help to identify the degradation mechanisms present in the devices measured, which complement the diagnosis through IV curves, where it is observed the performance of the modules when exposed to light.

6 ACKNOWLEDGEMENTS

We acknowledge our funding institution ANID (FONDECYT PROYECT N°11220697)

7 REFERENCES

[1] Shashwata Chattopadhyay et al. "Visual Degradation in Field-Aged Crystalline Silicon PV Modules in India and Correlation With Electrical Degradation". En: IEEE Journal of Photovoltaics 4.6 (2014), págs. 1470-1476. doi: 10.1109/JPHOTOV.2014. 2356717.

[2] Ministerio de Energía. Explorador Solar. Ver en http://solar.minenergia.cl/inicio.

[3] High Performance Bifacial PERC Monocrystalline Module. RSM72-6-365BMDG- 385BMDG. risen solar technology. Mar. de 2019.

[4] Classification of solar simulator characteristics. Standard. International Electrotechnical Commission, sep. de 2020.

[5] Grid connected systems - Documentation, commissioning tests and inspection. Standard. International Electrotechnical Commission, ene. de 2017.

[6] Measurement of current-voltage characteristics of bifacial photovoltaic (PV) devices. Standard. International Electrotechnical Commission, ene. de 2019.

[7] Electroluminescence of photovoltaic modules. Standard. International Electrotechnical Commission, ago. de 2018.

[8] Photovoltaic modules and plants – Outdoor infrared thermography. Standard. International Electrotechnical Commission, jun. de 2017.

[9] Coordinador Eléctrico Nacional. Reporte Anual de Desempeño del SEN (Art. 72°-15). Ver en https : / / www . coordinador . cl / reportes / documentos / reporte - anual-de-desempeno-del-sen/. 2021.

[10] M.A. Quintana et al. "Commonly observed degradation in field-aged photovoltaic modules". En: Conference Record of the Twenty-Ninth IEEE Photovoltaic Specialists Conference, 2002. 2002, págs. 1436-1439. doi: 10.1109/PVSC.2002.1190879.

[11] M. Smith, A. Miller, Proceedings 17th European Photovoltaic Solar Energy Conference, Vol. I (2002) 903.

[12] K.W. Boer, Solar Cells 16 (1996) 591.

[13] ISBN 978-3-907281-35-2: Trends in Photovoltaic Applications 2022. REPORT IEA PVPS T1-43:2022

COST-EFFECTIVE FLUORENE AND THIOPHENE CONTAINING HOLE CONDUCTORS TOWARDS SEMI-TRANSPARENT Sb_2S_3 ABSORBER-BASED SOLAR CELLS

Sreekanth Mandati[1,*], Nimish Juneja[1], Atanas Katerski[1], Aistė Jegorovė[2], Sarune Daskeviciute-Geguziene[2], Raitis Grzibovskis[3], Aivars Vembris[3], Nicolae Spalatu[1], Artiom Magomedov[2], Smagul Karazhanov[4], Vytautas Getautis[2], Malle Krunks[1], and Ilona Oja Acik[1]

[1]Department of Materials and Environmental Technology, Tallinn University of Technology, Ehitajate tee 5, 19086 Tallinn, Estonia.

[2]Department of Organic Chemistry, Kaunas University of Technology, Kaunas LT-50254, Lithuania.

[3]Institute of Solid State Physics, University of Latvia, Kengaraga Str. 8, Riga, Latvia.

[4]Institute for Energy Technology (IFE), P.O Box 40, NO 2027, Kjeller, Norway.

*corresponding author: sreekanth.mandati@taltech.ee

ABSTRACT: Building integrated photovoltaics (PV) is among the major PV technologies being pursued to align with the energy efficiency buildings and net-zero emission policies. Semi-transparent solar cells could be key components to achieve these targets and efforts are underway to develop novel PV absorbers and devices with the priority of materials' abundance, cost and stability. Sb_2S_3 thin films are promising PV absorbers owing to their superior optoelectronic properties, stability and availability of raw materials. However, the conventional expensive hole conductors like P3HT and Spiro-OMeTAD coupled with their post-processing needs and stability issues could be obstacles for Sb_2S_3 solar cells. In this study, cost-effective and transparent fluorene and thiophene-based hole transport materials (HTM) are explored in Sb_2S_3 solar cells. HTMs are spin coated to fabricate solar cells in the superstrate configuration (Glass/FTO/TiO_2/Sb_2S_3/HTM/Au) wherein TiO_2 and Sb_2S_3 are ultrasonic spray deposited. Solar cells with new HTMs have yielded similar power conversion efficiencies (3.8-3.9 %) compared to P3HT-based cells (3.7 %) while exhibiting 20-25 % enhanced average visible transmittance (400-800 nm) in the device stack without the metal contact. Optical properties and band energetics of individual layers demonstrate that the new cost-effective HTMs exhibit favourable band off-set for application in semi-transparent Sb_2S_3 solar cells.

Keywords: antimony sulfide, hole transport material, semi-transparent, solar cells

1 INTRODUCTION

Photovoltaics (PV) is a continuously evolving technology for energy generation. Considering the world is progressing towards artificial intelligence, smart cities and to align with the UNO's clean energy policies such as net zero emissions and energy efficiency buildings, building integrated PV has become one of the major PV technologies being pursued. Semi-transparent solar cells that could be integrated into smart windows are the need of the hour and efforts are underway towards developing novel PV absorbers and device architectures while with the priority of materials abundance, cost and their stability [1]. Thin films solar cells based on Sb_2S_3 are promising and is an emerging materials class with continuously dedicated efforts towards improving the efficiency with evolving device architectures as the absorber promises with earth abundant, non-toxic constituents and long-term stability [2]. Sb_2S_3 possesses desired optoelectronic properties like direct bandgap ($\approx$ 1.7 eV) and large optical absorption coefficient ($\approx 10^5$ cm^{-1}) for solar cells applications [3]. In addition, owing to its relatively wider bandgap, Sb_2S_3 is a suitable candidate for semi-transparent solar cells [4] and also, for tandem junction cells. Despite possessing the advantages of environmentally friendly materials and long-term stability, the practical realization of Sb_2S_3 solar cells is yet to be seen due to the low power conversion efficiencies. In a recent effort, a multi-sulfur source approach is explored in chemical bath deposition with CdS as buffer layer to obtain 8 % efficiency in planar Sb_2S_3 solar cells wherein 2,2′,7,7′-tetrakis[*N*,*N*-di(4-methoxyphenyl)amino]-9,9′-spirobifluorene (Spiro-OMeTAD) is used as hole transport material (HTM) [5]. In addition to the reasonably low power conversion efficiencies, the use of expensive HTMs like Spiro-OMeTAD, poly(3-hexylthiophene) (P3HT), poly{4,4-dialkyl-4*H*-cyclopenta[2,1-*b*;3,4-*b*′]dithiophene-2,6-diyl-*alt*-2,1,3-benzothiadiazole-4,7-diyl} (PCPDTBT), etc. could limit the practical realization of Sb_2S_3 solar cells [6,7]. Further, the conventional HTMs possess stability issues to moisture and HTMs like P3HT also require a post-processing treatment at elevated temperatures (> 150 °C) and also exhibits parasitic absorption losses, which hinder the overall transparency of the devices, a parameter that is essential in semi-transparent PV [8]. HTM is an important layer in Sb_2S_3 solar cells that efficiently collects and transports photogenerated holes while blocking the electrons to suppress the recombination losses at the absorber-HTM surface and also prevents the direct contact of metallic layer with the absorber. Therefore, it is unavoidable in the solar cells to particularly attain high efficiencies. Hence, the need for cost-effective and transparent HTMs has been increasing and so are the efforts to synthesize novel materials through cost-effective approaches and improved yields [9].

The present study focuses on exploring fluorene and thiophene-based HTMs V1236 and V808, which are economic owing to their simplified synthesis processes coupled with higher yields. Furthermore, the thiophene terminated HTMs are tailor made for Sb_2S_3 solar cells as they are expected offer better interaction plausibility with the Sb_2S_3 surface owing to the S-S anion interaction from the absorber and thiophene unit of HTMs. These low-cost and transparent HTMs are spin coated onto glass/FTO/TiO_2/Sb_2S_3 surface in the fabrication of Sb_2S_3 solar cells. For effective validation of new HTMs, devices are also fabricated using conventional P3HT as HTM. Characterization of individual layers and solar cells demonstrate that the new HTM-based cells perform on par with P3HT containing devices with favourable band off-set with Sb_2S_3 absorber and exhibit enhanced transparency in the visible region over P3HT.

2 EXPERIMENTAL

The procedure adopted for the fabrication of Sb_2S_3 solar cells started with the cleaning of glass/FTO substrates thoroughly with deionized water, ethanol and methanol followed by boiling in deionized water for 15 min. The substrates were dried with nitrogen prior to the deposition. Indigenously developed ultrasonic spray pyrolysis (USP) was used for the deposition of TiO_2 and the absorber (Sb_2S_3). The precursor of TiO_2 was prepared by mixing 0.2 M TTIP with 0.2 M acetylacetone in ethanol. TiO_2 electron transport layers were deposited with a spray rate of 2.5 ml/min for 30 min at the deposition temperature of 340 °C. The as-deposited TiO_2 samples were annealed at 450 °C for 30 min in air. For the deposition of Sb_2S_3, 60 mM of antimony chloride ($SbCl_3$) was mixed with 180 mM of thiourea in methanol. The same USP setup was used for Sb_2S_3 where the substrate temperature is 200 °C with a spray rate of 2 ml/min for 30 min. The as-deposited Sb_2S_3 thin films on glass/FTO/TiO_2 substrates were annealed at 250 °C for 5 min in N_2 atmosphere. The HTMs – P3HT, fluorene and thiophene-based materials were coated using spin coating from the respective precursors dissolved in chlorobenzene. P3HT was activated by heat treatment in vacuum at 170 °C for 5 min whereas V1236 and V808 did not require this activation step. The detailed synthesis process of V1236 (N^2,N^2,N^7,N^7-tetrakis[2,2-bis(4-methoxyphenyl)vinyl]-9,9-dihexyl-9*H*-fluorene-2,7-diamine) is explained in our previous report [8], while V808 (N^2,N^2,N^7, N^7-tetrakis(4-methoxyphenyl)-9-(thiophen-2-ylmethylene)-9*H*-fluorene-2,7-diamine) also involves a similar process. Finally, Au metal contacts were deposited by thermal evaporation. The active area of the fabricated solar cells was measured to be 7.06 mm^2.

The optical studies were performed using a Jasco V-670 ultraviolet-visible spectrophotometer (UV-VIS) in the range of 300-1000 nm. For the band diagram construction, ionization potential and electron affinities of the functional layers were measured using photoelectron emission spectroscopy as detailed previously. The work function of Au contacts was measured using a Kelvin probe. The current–voltage (I–V) characteristics of the fabricated solar cells were measured using Wavelabs LS-2 LED solar simulator with AM1.5G (100 mW cm^{-2}) light source. A Newport 69911 system with a 300 W Xenon lamp was used to measure the external quantum efficiency (EQE) of the fabricated solar cells.

3 RESULTS AND DISCUSSION

The quality, crystallinity and optical properties of electron transport layer (TiO_2) and Sb_2S_3 absorber deposited by ultrasonic spray pyrolysis deposition are similar to the absorber used in our previous report [8]. In brief, the Sb_2S_3 absorber exhibits stibnite crystal structure devoid of any unwanted phases such as antimony oxide and possesses a bandgap of ≈ 1.8 eV. The conventional P3HT, fluorene (V1236) - and thiophene-terminated (V808) - HTMs are spin coated onto glass/FTO/TiO_2/Sb_2S_3 surface followed by the thermal evaporation of Au metal contacts to complete the solar cell fabrication. The current density-voltage (J-V) characteristics of solar cells with P3HT, V1236 and V808 are shown in Figure 1a. As the curves are amply suggestive of the fact that solar cells with different hole conductors exhibit similar characteristics and the extracted power conversion efficiencies are 3.7, 3.9 and 3.8 % for devices with P3HT, V1236 and V808, respectively. Further, it may be noted the devices exhibit almost identical open circuit voltage (V_{OC}) while the short circuit current density (J_{SC}) is slightly higher for the solar cells with new hole conductors. To further unravel the reasons for enhanced J_{SC}, the external quantum efficiency (EQE) curves are recorded for the solar cells and the normalized EQE spectra are shown in Figure 1b. As it may be observed, the EQE spectrum of solar cell with P3HT as HTM exhibits a dip in the range of 550-700 nm, which is absent for cells with new hole conductors (V1236 and V808). This could be due to the fact that P3HT, owing to its similar bandgap value (≈ 1.8 eV) as that of Sb_2S_3 (≈ 1.7 eV) exhibits parasitic absorption and thus, responsible for the slight loss of J_{SC}. Parasitic absorption losses occur due to the absorption of reflected light from the metal back contact by the HTM, which otherwise is reabsorbed by the absorber to contribute to the photocurrent [10]. Such absorption by an HTM is possible when the thickness of HTM is comparable to that of absorber and/or the bandgaps of HTM and absorber are similar, which is the case in the present study. The new fluorene (V1236) and thiophene (V808) HTMs possess bandgap values of 2.6 and 2.9 eV, respectively, and are much thinner to obtain high efficiency of solar cells, as previously reported [8], and thereby do not exhibit the parasitic absorption losses thus producing slightly enhanced J_{SC} of the solar cells.

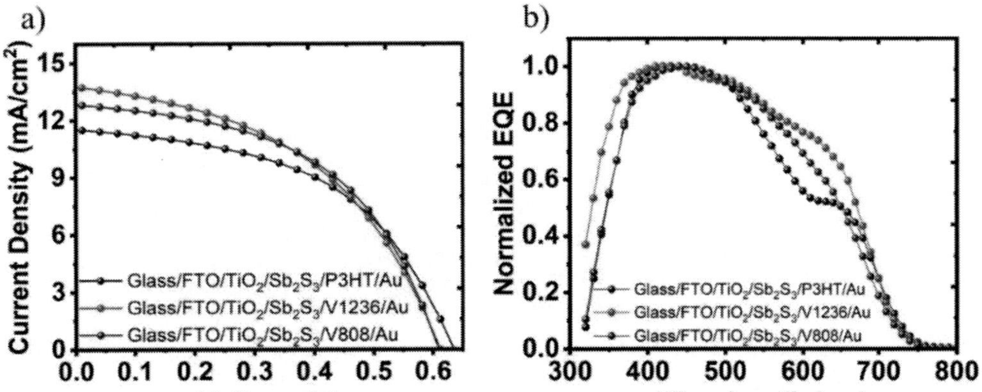

Figure 1: a) J-V characteristics and b) Normalized EQE curves of Sb_2S_3 solar cells containing P3HT, V1236 and V808 as hole transport materials.

The optical properties of individual HTM layers are studied where the transmittance of glass/FTO/P3HT, glass/FTO/V1236 and glass/FTO/V808 are recorded as shown in Figure 2a and the curve of glass/FTO is also included for the reference. The transmittance curves of new HTMs is almost similar to that of glass/FTO indicating complete transparency. The transmittance of P3HT, on the other hand, shows a significant drop in the 450-700 nm region, which corresponds to the absorption of P3HT and responsible for the loss of transparency but also affirms the fact pertaining to parasitic absorption. Furthermore, the overall transparency of Sb_2S_3 solar cells with different hole conductors is studied as shown in Figure 2b, devoid of the top metal contact. The transmittance curves clearly demonstrate enhanced transparency of solar cell stack for devices with new HTMs compared to the P3HT and the average visible transmittance (AVT), calculated in the range of 400-800 nm, is computed to be 21, 27 and 33 %, respectively, for solar cell stack with P3HT, V1236 and V808 HTMs. The AVT values further confirm the improved transparency of solar cell stack with new hole conductors.

In addition, the ionization potential (I_p) and electron affinity of individual solar cell layers are determined using photoelectron emission spectroscopy to obtain the band edge positions, which are then used to construct the band diagram as shown in Figure 3. As the diagram clearly shows, the valence band edge (VBE) positions of new HTMs are closer to the Sb_2S_3's VBE compared to P3HT, which ascertains that the holes are collected more efficiently. On a similar note, the conduction band edge (CBE) values of new HTMs are much higher than that of P3HT in comparison to that of CBE, which in other words poses a larger potential barrier for the electrons thereby blocking the electrons effectively to reduce the recombination probability at the absorber-HTM interface. The band energetics suggest the favourable band alignment of new HTMs over P3HT with the Sb_2S_3 absorber that further validates the applicability of fluorene and thiophene-containing hole conductors as the more suitable HTMs for application in semi-transparent Sb_2S_3 solar cells.

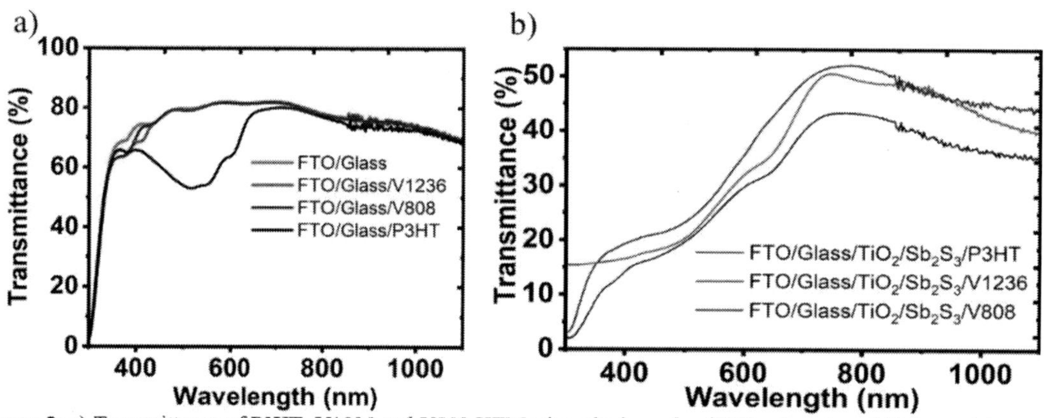

Figure 2: a) Transmittance of P3HT, V1236 and V808 HTMs deposited on glass/FTO substrate and b) transmittance of the respective Sb_2S_3 solar cell stack without Au

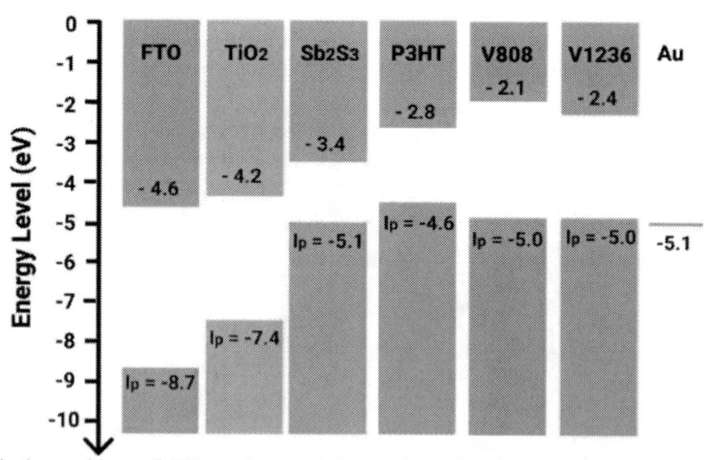

Figure 3: Band edge positions of different layers of Sb_2S_3 solar cells with new hole conductors (V1236 and V808)

4 CONCLUSIONS

In an effort to fabricate Sb_2S_3 solar cells for semi-transparent applications, the study herein explores the use of cost-effective and transparent hole transport materials containing fluorene and thiophene units. The new hole conductors are significantly cheaper owing to their simplified synthesis methods coupled with higher yields. The new hole conductors are thus demonstrated in Sb_2S_3 solar cells which exhibit similar power conversion efficiencies in comparison with conventional materials like P3HT. The solar cell stack with fluorene and thiophene-based hole transport materials displayed enhanced visible transparency over P3HT devices and the band energetics present the favourable band alignment of new hole conductors with Sb_2S_3 over P3HT, thereby demonstrating their efficacy for application in semi-transparent Sb_2S_3 solar cells.

5 ACKNOWLEDGEMENTS

The "Development of Semi-Transparent Bifacial Thin Film Solar Cells for Innovative Applications" benefits from a 999372 € grant from Iceland, Liechtenstein and Norway through the EEA Grants. The aim of the project is to develop new approach based on novel materials and structures and production technologies, which are the key to further increase the share, and range of application of PV in areas with sub-average sunlight, including Baltic and Nordic countries. Therefore, development of resource saving, cost-effective and efficient PV devices is a primary challenge of this project. Project contract with the Research Council of Lithuania (LMTLT) No is S-BMT-21-1(LT08-2-LMT-K-01-003). Department of Materials and Environmental Technology, Tallinn University of technology has received funding Estonian Research Council project PRG627 "Antimony chalcogenide thin films for next-generation semi-transparent solar cells applicable in electricity producing windows", the Estonian Centre of Excellence project TK141 (TAR16016EK) "Advanced materials and high-technology devices for energy recuperation systems", the European Union's Horizon2020 programme under the ERA Chair project 5GSOLAR grant agreement No 952509 and PSG689 "Bismuth Chalcogenide Thin-Film Disruptive Green Solar Technology for Next Generation Photovoltaics". Institute of Solid-State Physics, University of Latvia has received funding from the European Union's Horizon 2020 Framework Programme H2020-WIDESPREAD-01-2016-2017-TeamingPhase2 under grant agreement No. 739508, project CAMART².

6 REFERENCES

[1] S.-J. Lee, S.-J. Sung, K.-J. Yang, J.-K. Kang, J.Y. Kim, Y.S. Do, D.-H. Kim, Approach to Transparent Photovoltaics Based on Wide Band Gap Sb2S3 Absorber Layers and Optics-Based Device Optimization, ACS Appl. Energy Mater. 3 (2020) 12644–12651. https://doi.org/10.1021/acsaem.0c02552.

[2] U.A. Shah, S. Chen, G.M.G. Khalaf, Z. Jin, H. Song, Wide Bandgap Sb2S3 Solar Cells, Adv. Funct. Mater. 31 (2021) 2100265. https://doi.org/https://doi.org/10.1002/adfm.202100265.

[3] R. Kondrotas, C. Chen, J. Tang, Sb2S3 Solar Cells, Joule. 2 (2018) 857–878. https://doi.org/https://doi.org/10.1016/j.joule.2018.04.003.

[4] J.S. Eensalu, A. Katerski, E. Kärber, L. Weinhardt, M. Blum, C. Heske, W. Yang, I.O. Acik, M. Krunks, Semitransparent Sb2S3 thin film solar cells by ultrasonic spray pyrolysis for use in solar windows, Beilstein J. Nanotechnol. 10 (2019) 2396–2409.

[5] S. Wang, Y. Zhao, B. Che, C. Li, X. Chen, R. Tang, J. Gong, X. Wang, G. Chen, T. Chen, J. Li, X. Xiao, A Novel Multi-Sulfur Source Collaborative Chemical Bath Deposition Technology Enables 8%-Efficiency Sb2S3 Planar Solar Cells, Adv. Mater. n/a (2022) 2206242. https://doi.org/https://doi.org/10.1002/adma.202206242.

[6] R. Nie, S. Il Seok, Efficient Antimony-Based Solar Cells by Enhanced Charge Transfer, Small Methods. 4 (2020) 1900698. https://doi.org/https://doi.org/10.1002/smtd.201900698.

[7] R. Yaoguang, H. Yue, M. Anyi, T. Hairen, S.M. I., S.S. Il, M.M. D., S.E. H., H. Hongwei, Challenges for commercializing perovskite solar cells, Science (80-.). 361 (2018) eaat8235. https://doi.org/10.1126/science.aat8235.

[8] N. Juneja, S. Mandati, A. Katerski, N. Spalatu, S. Daskeviciute-Geguziene, A. Vembris, S. Karazhanov, V. Getautis, M. Krunks, I. Oja Acik, Sb2S3 solar cells with a cost-effective and dopant-free fluorene-based enamine as a hole transport material, Sustain. Energy Fuels. (2022). https://doi.org/10.1039/D2SE00356B.

[9] S. Daskeviciute, C. Momblona, K. Rakstys, A.A. Sutanto, M. Daskeviciene, V. Jankauskas, A. Gruodis, G. Bubniene, V. Getautis, M.K. Nazeeruddin, Fluorene-based enamines as low-cost and dopant-free hole transporting materials for high performance and stable perovskite solar cells, J. Mater. Chem. A. 9 (2021) 301–309. https://doi.org/10.1039/D0TA08452B.

[10] E. Zimmermann, T. Pfadler, J. Kalb, J.A. Dorman, D. Sommer, G. Hahn, J. Weickert, L. Schmidt-Mende, Toward High-Efficiency Solution-Processed Planar Heterojunction Sb2S3 Solar Cells, Adv. Sci. 2 (2015) 1500059. https://doi.org/https://doi.org/10.1002/advs.201500059.

PRIMARY CALIBRATION OF SOLAR MODULES WITH DIRECT SUNLIGHT

Stefan Riechelmann*[1], Dirk Friedrich[1,2], Michael Müller[1], Frank Schmaljohann[1], Hendrik Sträter[1], Stefan Winter[1]
*Corresponding author: stefan.riechelmann@ptb.de
[1]Physikalisch-Technische Bundesanstalt (PTB), Bundesallee 100, 38116 Braunschweig, Germany
[2]now at Bundesamt für Sicherheit in der Informationstechnik (BSI) Hüttenstraße 14, 01705 Freital, Germany

ABSTRACT: One of the most accurate primary calibration methods for solar cells is the direct sunlight method (DSM), using the direct light of the sun as light source. We present a measurement device that scales the DSM method to the size of PV modules. The setup provides a light field of the size of 2.5 x 2.5 m² and is therefore capable of measuring large-sized PV modules based solar cells with an area of 210 mm x 210 mm. Technically, the setup consists of a rectangular tube with a size of 4 m x 4 m x 7.2 m, capable of tracking the sun. In addition, the PV module holder inside the tube can be tilted relative to the sun, enabling us to perform an angle of incidence measurement of the device under test. We present our first angular measurements of a half-cut solar module and a power measurement as first step towards a primary calibration of the same module to showcase the capabilities of our new setup.

Keywords: Characterisation, Energy Rating, PV Module, AOI modifier, Primary calibration

1 INTRODUCTION

In the field of solar cell calibration, the DSR (differential spectral responsivity) method that uses monochromatic radiation delivers the lowest measurement uncertainties [1], but this can hardly be scaled to a large light field necessary for primary PV module calibration. Using direct natural solar radiation for a primary calibration of solar cells is the second most accurate calibration method. In this so-called Direct Sunlight Method (DSM) [2], the solar cell is placed in an approximately 70 cm long tube with a black interior that blocks out all diffuse radiation from the sky. The measurements are traced back to a cavity radiometer, which is an electrical substitution radiometer, providing a radiometric standard for direct solar irradiance with an uncertainty of typically 0.2 %. Overall, if all components are accurately characterized, a measurement uncertainty of 0.6 % can be achieved using this method. However, this method has never been used for PV modules so far because of the much larger tube needed to narrow the viewing angle of a PV module to prevent diffuse light from reaching its surface. We therefore present a measurement setup, that can perform a DSM calibration with PV modules. In addition, a rotating stage is implemented in the measurement device, that enables incident angle measurements of solar modules, which are usually performed in the lab. Compared to artificial light sources, the direct sunlight supplies a perfectly collimated and uniform light field, enabling us to precisely determine the incidence angle modifier of any tested module. Reducing both, the uncertainty of PV module calibrations and incidence angle modifier calibrations, will help to bring down overall system uncertainties and thus lead to a reduced financial uncertainty of PV systems.

2 SETUP

The new DSM measurement setup for PV modules shares many features of its smaller version for solar cells: A tube with an aperture limits the field of view (FOV) of the device under test. Since the illuminated area shall have a size of 2.5 x 2.5 m², an at least slightly larger aperture of the same size is needed to illuminate the measurement plane. The FOV is now determined by the distance between aperture and measurement plane, as shown in Figure 1. A FOV of 5°, comparable to typical FOVs of cavity radiometers, would result in a tube length of 28.63 m, which is technically not feasible. We make a compromise between FOV and instrument size by choosing a distance between aperture and measurement plane of 4.5 m. This results in a FOV of 31°. Nevertheless, only 7.2 % of the sky is visible in comparison to an unrestricted FOV, efficiently reducing the incident diffuse irradiance.

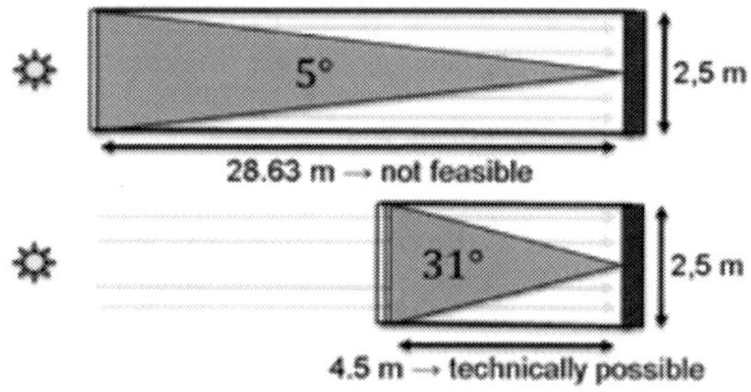

Figure 1: A tube for PV modules with a FOV of 5° would result in a 28.63 m long apparatus. Limiting the distance between aperture and PV module to 4.5 m, a FOV of 31° can be achieved.

Figure 2: Solar module tube setup in operation at PTB.

The tube is mounted on a rotary stage and equipped with two synchronized linear stages at each side to align the aperture of the setup towards the current sun position (Figure 2). In addition, the PV module is mounted on a 2-axis holder that can tilt the module in relation to the incident sunlight, this way the angular responsivity of a device can also be measured with this setup (Figure 3).

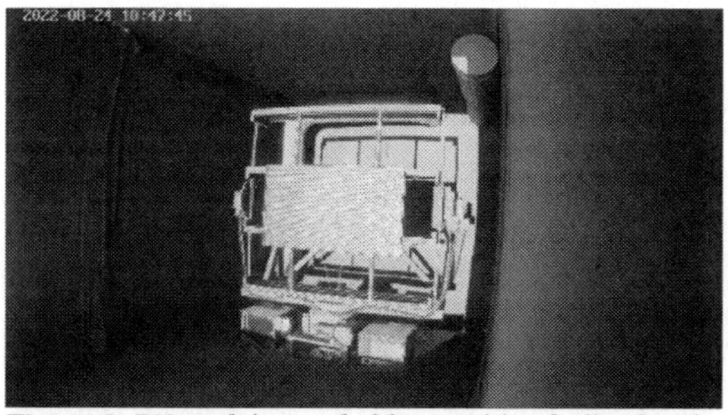

Figure 3: PV module on a holder capable of adjusting the tilt angle in relation to the incident sunlight.

3 FIRST RESULTS

We conducted our first measurements on a half-cut module on a sunny and cloudless day of August 2022 (Figure 4).

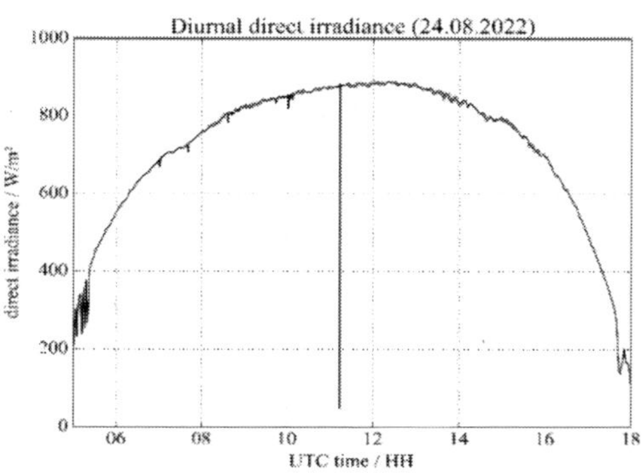

Figure 4: Diurnal direct irradiance on August 24, 2022. The time periods when measurements were conducted are marked red.

In addition to measurements conducted with our solar module tube, diurnal measurements of direct irradiance have been derived with three SHP1 Kipp & Zonen pyranometers. Spectral irradiance from 280 nm to 4000 nm is provided by a Spectrafy SolarSIM-D2 filter radiometer. Note that this is a preliminary setup with simple auxiliary measurements, in the future we will use three Davos Instruments PMO6/PMO8 electrical substitution radiometers for direct irradiance measurements and an Instrument Systems CAS spectroradiometer setup for spectral direct irradiance measurements.

3.1 Angular responsivity

The spectral responsivity of a half-cut PV module is derived by angular scans from -95° to 95° in steps of 5°. At each step an IV curve is measured and I_{SC} is derived from these curves. I_{SC} values are then corrected for temperature drifts by temperature measurements conducted on the rear side of the module as stated in IEC 61853-2 [3]. The set of angular transmission measurements shows a standard deviation of less than 1 % (σ = 2) for all angles less than 80 %. By fitting the analytical function for relative light transmission given by IEC 61853-2 [3], we derive an a_r factor of 0.1495. A low a_r factor relates to less irradiance reflected back to the sky by the device under test, the value is in accordance with anti-reflective-coated solar modules according to literature values [4].

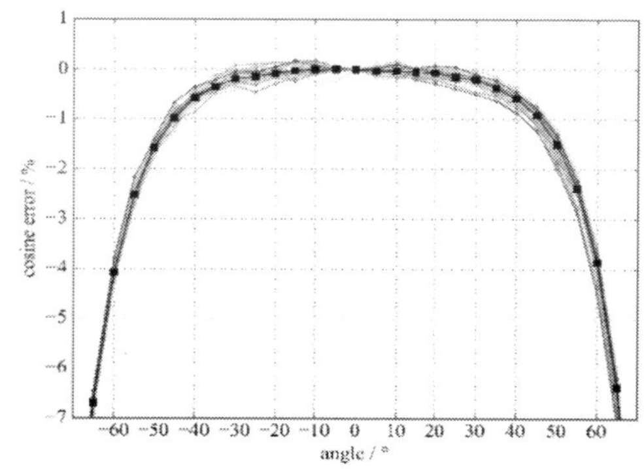

Figure 5: The colored lines represent 25 angular measurements with the solar module tube setup. The square points connected by a black line is the mean value the individual measurements.

3.2 Primary module calibration

To showcase the capability of our setup to conduct primary module calibrations we derive from each angular responsivity measurement the IV curves at 0°. Each IV curve is corrected for spectral mismatch according to IEC 60904-7 [5]. In addition, we calculate the equivalent cell temperature (ECT) of the module under test according to IEC 60904-5 [6] and perform a temperature and irradiance correction of all IV curves towards STC according to IEC 60891 [7]. The corrected values of a half-cut module with a nameplate maximum power point value (P_{MPP}) of 365 W are shown in Figure 6.

3.3 Comparison to lab measurements

The half-cut module measured with the DSM outdoor measuring site has been measured with an LED-type solar simulator [8] in the lab. Figure 7 shows the zoomed IV curves around the P_{MPP} of both setups. The solar simulator has a non-uniformity of less than 2 % [9] which translates to class A after IEC 60904-9 [10], nevertheless a step in module current of 0.2 A is apparent at around 10 V. The outdoor measurement shows due to the excellent uniformity only an insignificant current step of less than 0.05 A at around 20 V.

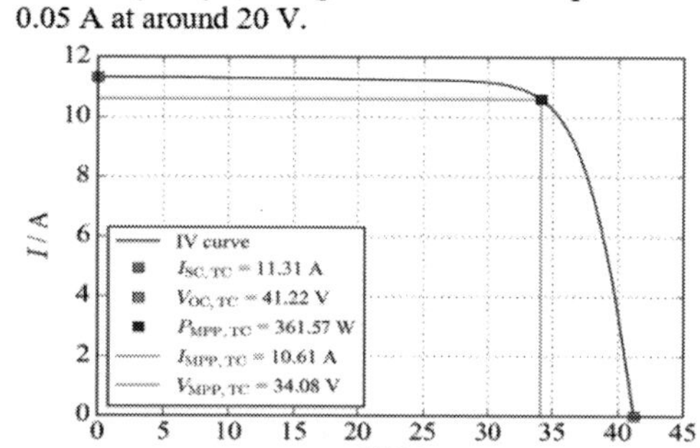

Figure 6: IV curve of a half-cut module with a nameplate P_{MPP} value of 365 W measured within the new DSM setup for solar modules.

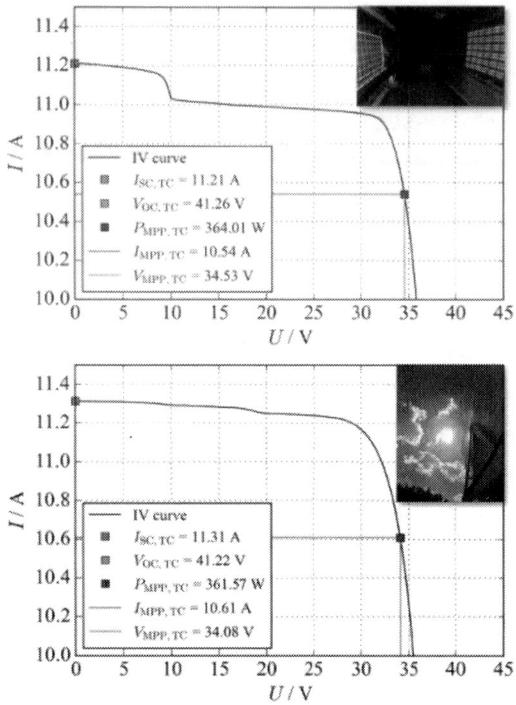

Figure 7: Comparison of non-uniformity features of IV curves measured with a solar simulator indoors (upper plot) and the DSM setup outdoors (lower plot), respectively. Note that the y-axis starts at 10 A.

4 CONCLUSIONS

We measured the angular responsivity and the electoral properties of a half-cut PV module to show the capabilities of our new "Solar Module Tube" measuring site. The angular measurement yielded plausible results in line with literature values. The electrical properties derived from the IV curves of the measurements performed at an angle of 0° are also plausible and show a superior non-uniformity behavior compared to lab measurements with a class A solar simulator. There are still technical features that need to be added to the measuring site and to the evaluation:

- Module temperature control will be added to the setup provided by a cooling unit and a fast-moving stage curtain, that keeps the module in the dark until the desired temperature is reached.
- A correction factor for the field of view difference between the Solar Module Tube and the instruments used for direct sunlight measurement will be determined. Various solutions will be tested, a pyranometer mounted inside the tube with a shading ball for deriving the diffuse part of the 31° FOV may be a practical solution.

5 ACKNOWLEDGEMENT

The solar module tube setup is developed within the project „**MetroKomPV**", which is funded by the *Federal Ministry for Economic Affairs (BMWi)*, Germany.

6 REFERENCES

[1] Winter S, Wittchen T, Metzdorf J et al. (2000) Primary Reference Cell Calibration at the PTB Based on an Improved DSR Facility. 16th European Photovoltaic Solar Energy Conference and Exhibition 2198–2201
[2] International Electrotechnical Commission Photovoltaic devices - Part 4: Photovoltaic reference devices - Procedures for establishing calibration traceability (IEC 60904-4:2019)
[3] International Electrotechnical Commission Photovoltaic (PV) module performance testing and energy rating - Part 2: Spectral responsivity, incidence angle and module operating temperature measurements 61853-2(IEC 61853-2:2016)
[4] Herrmann W, Schaaf S, Rimmelspacher L et al. (2019) Angular Response Measurement of Thin-Film PV Modules with Solar Simulators. 5 pages / 36th European Photovoltaic Solar Energy Conference and Exhibition; 1192-1196
[5] International Electrotechnical Commission Photovoltaic devices - Part 7: Computation of the spectral mismatch correction for measurements of photovoltaic devices (IEC 60904-7:2019)
[6] International Electrotechnical Commission Photovoltaic devices - Part 5: Determination of the equivalent cell temperature (ECT) of photovoltaic (PV) devices by the open-circuit voltage method (IEC 60904-5:2011)
[7] International Electrotechnical Commission Photovoltaic devices - Procedures for temperature and irradiance corrections to measured I-V characteristics (IEC CD 60891:2019)
[8] Riechelmann S, Sträter H, Winter S (eds) (2020) Determination of a PV module power matrix with an LED solar simulator
[9] Sträter H, Riechelmann S, Neuberger F et al. (2019) LED-Based Differential Spectral Responsivity Measurements of PV Modules. 36th European Photovoltaic Solar Energy Conference and Exhibition; 1121-1126
[10] International Electrotechnical Commission Photovoltaic devices - Part 9: Classification of solar simulator characteristics (IEC 60904-9:2020)

This is a short version of the paper submitted to the journal EPJ Photovoltaics.

It has been selected by the Committee of the WCPEC-8 for submission to EPJ Photovoltaics.

ADVANCED UV-FLUORESCENCE IMAGE ANALYSIS FOR EARLY DETECTION OF PV-POWER DEGRADATION

Lukas Neumaier*[1], Gabriele C. Eder[2], Yuliya Voronko[2], Karl A. Berger[3], Gusztáv Újvári[3], Karl Knöbl[4]

[1] SAL Silicon Austria Labs, Europastraße 12, 9524 Villach, Austria
[2] OFI Austrian Research Institute for Chemistry and Technology, Franz-Grill-Straße 5/Objekt 213, 1030 Vienna, Austria
[3] AIT Austrian Institute of Technology GmbH, Giefinggasse 4, 1210 Vienna, Austria
[4] University of Applied Sciences Technikum Wien, Höchstädtplatz 6, 1200 Vienna, Austria

ABSTRACT: A major component in the growth of developing PV markets around the world is the reliability and longevity of photovoltaic (PV) modules. The polymeric encapsulation materials' chemical and physical stability directly affects reliability. Ultraviolet Fluorescence (UVF) imaging is a technique that can detect the impacts of the polymeric encapsulant's aging directly on site at a PV plant. This research focuses on advanced imaging analysis of UVF pictures and the subsequent correlation to electrical characteristics of PV modules that have undergone long-term, accelerated aging processes particular to a certain climate. For establishing a correlation, a so called UVF area ratio was established, resulting from the typical fluorescence patterns of the encapsulant material, which arise due to stress impact (e.g.: water vapor ingress, elevated temperature, irradiation) and aging/degradation processes. The results of the data analysis demonstrate a distinct correlation between the electrical parameters and the UVF area ratio with increasing aging time. Long-term test series with different climate-specific aging processes confirm UVF area ratio correlations with power and series resistance. Hence, calculating the UVF area ratio can be used to predict the service life and electrical power dissipation of individual PV modules within a PV array, under the assumptions of the utilization of the same kind of polymeric encapsulation and backsheet materials and a similar climate.
Keywords: Reliability, Characterization, System Performance, Monitoring, Performance

1 INTRODUCTION

The growth of emerging PV markets around the world is significantly influenced by the dependability and durability of photovoltaic (PV) modules. The polymeric encapsulation materials' chemical and physical stability directly affects reliability. Ultraviolet Fluorescence (UVF) imaging is one technique that can identify ageing effects of the polymeric encapsulant directly in the field [1-6]. The performance and several electrical parameters of PV mini-modules, that underwent long-term, climate-specific accelerated aging procedures, are discussed in this work along with an enhanced imaging analysis of UVF images. The goal is to establish a correlation between these parameters, which would be suitable for operation and maintenance approaches in terms of predicting power dissipation of PV modules.

Based on data measured within the Austrian (flagship) project INFINTY and current data processing within the follow-up project ADVANCE! a so called UVF area ratio was established. It is resulting from typical fluorescence patterns of the encapsulant material, which arise due to stress impacts, such as water vapor ingress, elevated temperature, and irradiation, as well as material aging/degradation processes. In general, the UVF pattern gives information on the extent of stress-induced changes of the encapsulant. The encapsulant material changes because of the impact of climatic stresses with increasing lifetime (natural weathering) or storage duration in a climate chamber (accelerated aging). The architecture and bill of materials of the PV-module, such as the permeability of the backsheet material, and the type of encapsulant, are secondary influences on the UVF pattern [2,6,7,8]. Within this work, only PV modules with an identical bill of material and module design were used to exclude material related effects. Hence, aging induced changes of the encapsulant material, visualized by its UVF pattern, were converted into a scalar quantity (UVF area ratio).

2 APPROACH

The paper shows the correlation of UVF images and electrical parameters of 6-cell test modules (EVA encapsulation, PET-based backsheet) exposed to climate-specific long-term accelerated aging processes up to 6000 h [2]. In the context of this work, the analyses of two specific accelerated aging procedures (Table I) are shown.

Table I: Test sequences for climate specific aging

	Duration	Temperature	Humidity	Interval
Tropical 1	6000 h	85 °C	85 %	500 h
Tropical 2	3000 h	90 °C	90 %	500 h

2.1 UV-Fluorescence Imaging

UVF imaging is an established inspection tool for PV modules. Especially when a fast, non-destructive, on-site characterization method is required for detecting compound aging effects [4,6,9-13] and/or cell crack detection [14-18]. PV modules do not show distinctive fluorescence effects in the original state, but with increasing lifetime, the encapsulant material ages due to the interaction with climatic stress factors (mainly irradiation by sunlight, elevated temperature, and humidity) by forming fluorophores (Figure 1).

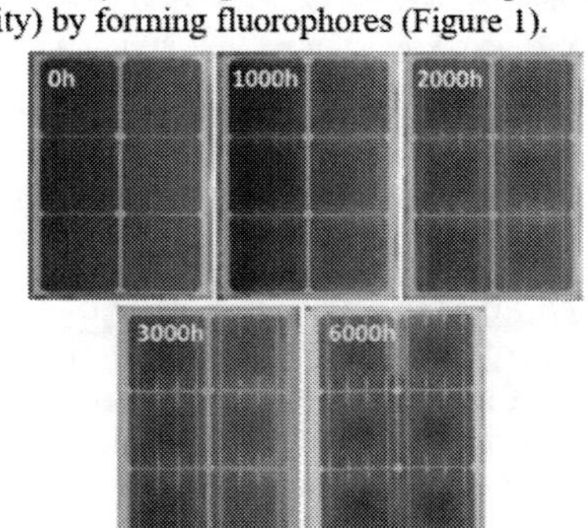

Figure 1: UVF-images of the Tropical 1 time series

UVF images were recorded in dark environment by UV light illumination of the PV modules (self-made UV-lamp with three power- tunable LED-arrays as light source; emission maximum at 365 nm (50 W, tunable); low pass filter to cut off visible light) and subsequent detection of the fluorescing light in the visible region by utilizing a standard camera system (with long pass UV filter). From the obtained UVF-patterns, the so called UVF area ratio was derived. It is defined as the ratio of the area with low UVF in the cell center divided by the cell area (Fig.1). The ratio is set to 1 when no pattern was detected. The UVF area ratio was manually extracted from the UVF images using an image manipulation software (GIMP 2.10.30). To calculate the ratio, we divided the masked UVF pattern area by the total cell area of the six solar cells. Middle and edge regions were excluded from the calculations. Finally, we calculated the average UVF area ratio from six cells to extract correlation parameters for each minimodule (Figure 2). This procedure was repeated for all aging steps to obtain time resolved UVF area ratios.

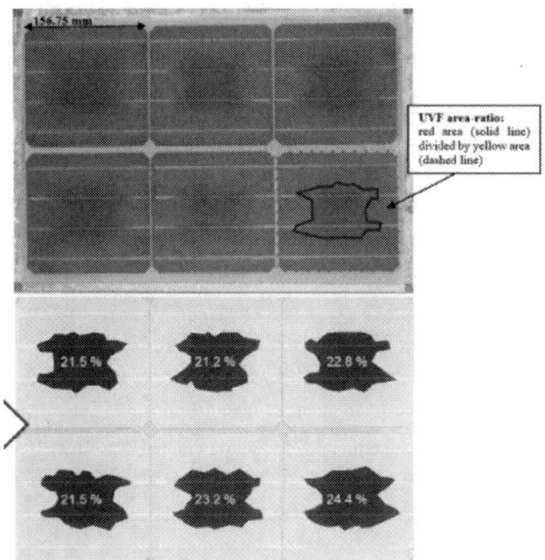

Figure 2: Typical UVF pattern after 3000 h of (accelerated) aging (Tropical 1); Manual extraction of UVF patterns on cell/module level

2.2 Electrical Characterization

Electrical characterization data was measured at standard laboratory conditions (STC; 25°C; 1000 W/m2 with AM 1.5 spectral distribution) utilizing a PASAN HighLIGHT VLMT A+A+A+ flash tester (IEC60904-1 & IEC60904-9). These time-resolved characterization parameters represent relevant electrical properties of PV modules, which were recorded at every intermediate step of the specific accelerated aging procedures.

3 RESULTS

Figure 3 shows the UVF ratio as a function of the aging duration for two accelerated aging procedures Tropical 1 (Damp-Heat, 85°C, 85% r.H.) and Tropical 2. Here, the UVF area ratio significantly declines with aging duration. Figure 4 shows the correlation between the UVF area ratios and electrical module parameters. Analyzing the results in detail revealed an increase in the series resistance (R_S) and a concomitant decrease in the output power (P_{MPP}) while the short-circuit current (I_{SC}) stays unchanged. Increasing moisture penetration through the backsheet with exposure time was found to be responsible for (i) grid finger/busbar corrosion effects, reflected in changes in R_S and P_{MPP} and (ii) changes in UVF area ratio. The analysis of the yellowness index and transparency of the encapsulant showed no significant changes with age, consistent with the negligible reduction in I_{SC}.

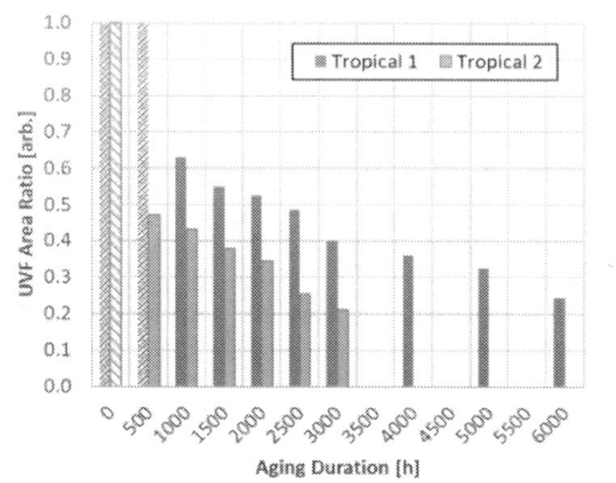

Figure 3: UVF area ratio analysis for accelerated aging procedures Tropical 1 & Tropical 2

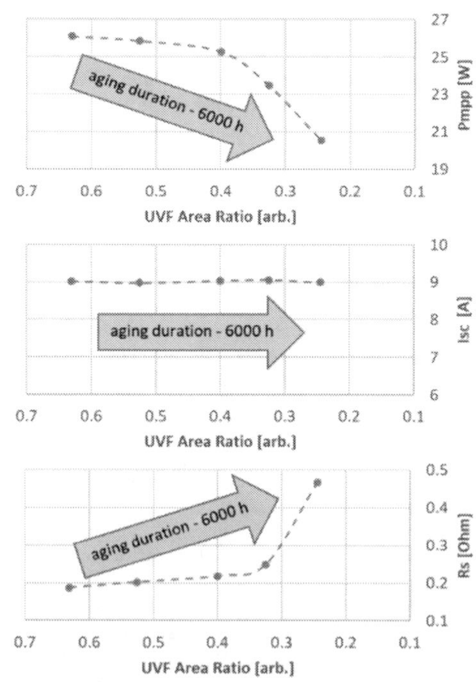

Figure 4: Correlation of electrical parameters and UVF area ratio for Tropical 1 (85°C, 85 % r.H.)

4 CONCLUSION

Concluding we can state that there is a clear correlation between the UVF area-ratio and the R_S (and consequently P_{MPP}) which could be utilized for a fast, on-site and easy applicable performance evaluation/

approximation based in UVF images, without directly interacting with the PV plant. Especially for early detection of power degradation on site related to operation and maintenance approaches, this could be of great interest. The findings are confirmed by extensive long-term test series with different climate-specific accelerated aging procedures.

5 ACKKNOWLEDGEMENT

This work was conducted as part of the Austrian "Energy Research Program" project ADVANCE! (Nr. 881133), funded by the Austrian Climate and Energy Fund and the Austrian Research Promotion Agency (FFG), both are gratefully acknowledged.

6 REFERENCES

[1] F. J. Pern and S. H. Glick, "Fluorescence analysis as a diagnostic tool for polymer encapsulation processing and degradation", AIP Conference Proceedings, vol. 306, p. 573, 1994.

[2] M. Köntges, A. Morlier, G. Eder, E. Fleiß, B. Kubicek and J. Lin, "Review: Ultraviolet Fluorescence as Assessment Tool for Photovoltaic Modules", in IEEE Journal of Photovoltaics, vol. 10, no. 2, pp. 616-633, 2020.

[3] G.C. Eder, Y. Voronko, Ch. Hirschl, R. Ebner, G. Újvári and W. Mühleisen, "Non-destructive failure detection and visualization of artificially and naturally aged PV modules", Energies, vol. 11, no. 5, p. 1053, 2018.

[4] J. Schlothauer, S. Jungwirth, B. Röder, M. Köhl, "Fluorescence imaging-a powerful tool for the investigation of polymer degradation in PV modules", Photovoltaics International, vol. 10, pp. 149 154, 2010.

[5] W. Mühleisen, C. Hirschl, G. Brantegger, L. Neumaier, M. Spielberger, H. Sonnleitner, B. Kubicek, G. Ujvari, R. Ebner, M. Schwark, G.C. Eder, Y. Voronko, K. Knöbl, L. Stoicescu," Scientific and economic comparison of outdoor characterisation methods for photovoltaic power plant", Renewable Energy, vol. 134, pp. 321-329, 2019.

[6] G.C. Eder, Y. Voronko, P. Grillberger, B. Kubicek, K. Knöbl, "UV-Fluorescence measurements as tool for the detection of degradation effects in PV-modules", Proceedings of the European Weathering Symposium EWS, Vienna, Austria 09.2017; Herausgeber: Gesellschaft für Umweltsimulation e.V. GUS.

[7] Y. Voronko, G.C. Eder, M. Knausz, G. Oreski, T. Koch, K.A. Berger, "Correlation of the loss in photovoltaic module performance with the ageing behaviour of the backsheets used", Prog. Photovolt: Res. Appl. 23, pp. 1501–1515, 2015.

[8] A. Omazic, G. Oreski, M. Halwachs, et al. "Relation between degradation of polymeric components in crystalline silicon PV module and climatic conditions: a literature review", Sol Energy Mater Sol Cells. 192, pp.123-133, 2019.

[9] G.C. Eder, Y. Voronko, S. Dimitriadis, et al. "Climate specific accelerated ageing tests and evaluation of ageing induced electrical, physical, and chemical changes", Prog Photovolt Res Appl.; 27, pp. 934– 949, 2019, https://doi.org/10.1002/pip.3090.

[10] F. J. Pern, "Factors that affect the EVA encapsulant discoloration rate upon accelerated exposure", Sol. Energy Mater. Sol. Cells, vol. 41–42, pp. 587–615, 1996.

[11] F. J. Pern, "Polymer encapsulants characterized by fluorescence analysis before and after degradation", Photovoltaic Specialists Conference, 1993., Conference Record of the 23rd IEEE; DOI:10.1109/PVSC.1993.346968

[12] J. Schlothauer, S. Jungwirth, M. Köhl, B. Röder, „Degradation of the encapsulant polymer in outdoor weathered photovoltaic modules: Spatially resolved inspection of EVA ageing by fluorescence and correlation to electroluminescence", Solar Energy Materials and Solar Cells, vol 102, pp. 75-85, 2012.

[13] A. Morlier, M. Siebert, I. Kunze, S. Blankemeyer, and M. Köntges, "Ultraviolet fluorescence of ethylene-vinyl acetate in photovoltaic modules as estimation tool for yellowing and power loss", in Proc. IEEE 7th World Conf. Photovolt. Energy Convers., pp. 1597–1602, 2018.

[14] M. Köntges, S. Kajari-Schröder, I. Kunze, "Cell Cracks Measured by UV Fluorescence in the Fields", in 27th European Photovoltaic Solar Energy Conference, pp. 3033–3040, 2012.

[15] M. Köntges, S. Kajari-Schröder, and I. Kunze, "Crack Statistic for Wafer-Based Silicon Solar Cell Modules in the Field Measured by UV Fluorescence", IEEE J. Photovoltaics, vol. 3, no. 1, pp. 95–101, 2013.

[16] M. Köntges et al., Report IEA-PVPS T13-01:2014: Review of Failures of PV Modules, 2014; ISBN 978-3-906042-16-9.

[17] W. Mühleisen, G.C. Eder, Y. Voronko, K. Knöbl, M. Spielberger, H. Sonnleitner, R. Ebner, G. Uijvari, Ch.Hirschl, "Outdoor Detection and Visualization of Hailstorm Damages of Photovoltaic Plants", Renewable Energy 118, pp. 138-145, 2018.

[18] A. Morlier, M. Siebert, I. Kunze, G. Mathiak, M. Köntges, "Detecting Photovoltaic Module Failures in the Field during Daytime with Ultraviolet Fluorescence Module Inspection", IEEE Journal of Photovoltaics, vol. 7,no. 6 , 1710, 2017.

MACHINE AIDED ESTIMATION OF SOLAR CELL CRACK CAUSED POWER LOSS FROM ELECTROLUMINESCENCE IMAGES

Rodrigo Del Prado Santamaría, Thøger Kari Hass, Gisele A. Dos Reis Benatto, Jacob Wilder Ng, Peter Behrensdorff Poulsen, Sergiu V. Spataru
Department of Electrical and Photonics Engineering, Technical University of Denmark,
Frederiksborgvej 399, 4000 Roskilde, Denmark

ABSTRACT: In this work, we present and validate a machine learning approach for estimating power loss in mono-crystalline silicon PV cells due to cell cracks. The primary focus is selection and evaluation of robust and accurate prediction-features derived from electroluminescence (EL) images of solar cells. To achieve this, we have experimentally created training and validation EL and I-V measurements datasets of cells with two to up to five busbars, with different extents of cell crack degradation. We have evaluated simple features derived from the cells' luminescence histogram, which are easy to compute; these are the cell inactive area, the mean, standard deviation (STD), skewness, and kurtosis. Learning models are then trained against cell I-V measurements and validated. This cell-level approach has the advantage of being scalable so that it could be used to estimate the power loss of a full PV module based on a module EL image that is segmented in cell images. Similarly, this bottom-up approach could facilitate the development of more generalizable machine algorithms, where just one model can be used to simulate power no matter the size of the module, number, layout, or type of the cracked cells. As a baseline, a linear regression approach was attempted, which yielded a R^2 score of 0.85 and a maximum error of 0.26. These results were subsequently improved by applying a 2nd degree polynomial linear regression and a custom non-linear regression to 0.869 of R^2 and 0.17 of maximum error. The model selection has been chosen from a statistical exploratory analysis of the dataset, where it has been demonstrated that mono-crystalline cells of with different number of busbars behave differently in terms of histogram features correlations with the output power, and that adding a classifier variable as the busbar number as a predictor it is possible to account for the non-linearity of each cell type and model it accordingly.
Keywords: Monocrystalline Silicon, Cell Cracks, Machine Learning, Power Loss

1 INTRODUCTION

Year after year, the photovoltaic (PV) installed capacity has increased, adding 133 GW of new installed capacity at the end of 2021 [1]. Following that trend, inspection of the PV fields is becoming an important matter for maintaining the PV assets deployed in the field. The importance of PV reliability lies on the development of tools that facilitate the inspection and assessment of PV plants.

When deployed in the field, PV modules are subject to stress, that can be during installation, transport, weather conditions, etc. These stresses can cause failures in the modules, among which, cell cracks are frequent [2].

Cracks in the PV cells cause power loss and risk of hotspots, furthermore, they can evolve over time, turning into a more severe power loss and other potential safety risks for the PV installation [3], [4].

To detect and monitor failures in the modules such as cracks, electroluminescence (EL) inspection is often used, which is an imaging technique that allows identifying material faults in solar cells and PV modules [5].

Most commonly, EL images of PV modules are evaluated qualitatively to detect defects. However, in recent years, machine learning methods have been presented that can automatically identify PV module defects and degradation such as microcracks and other major disconnections, most of them, using machine and deep learning approaches [6]–[11]. Furthermore, these approaches have been expanded to be able to classify automatically failures appearing in EL images of both PV cells and modules [12]–[15].

These models, while useful in field inspection and for fault detection, still lack a quantification in terms of power output based on the classified faults.

Current state-of-the-art methods for estimating PV module power (or power loss) from EL images are either based on analytical luminescence models of the solar cells or empirical machine learning models. Analytical models rely on calculating the distributed series resistance map of the cells in a module and can yield very accurate simulations with a solid theoretical background [16]. However, they are sensitive to correct initial parametrization of the model.

Statistical analysis of correlation between I-V parameters and EL images has been investigated successfully [17]–[19], setting the base for machine learning models that are able to predict the efficiency of cells and I-V characteristics of PV modules [20]–[23].

These machine learning approaches present a significant advantage to analytical models in field applications because of their robustness. However, obtaining a good quality model training dataset of EL images is not trivial, as it requires a long time for electrical characterization of modules and EL image recording that needs to cover a wide range of PV module types and degradation extents. To increase the ability of the machine learning model to generalize to different module and cell types, the prediction features must be designed to be cell- and module-type agnostic.

The purpose of this work is to investigate the process of simulating the power output of mono-crystalline PV cells affected by cracks by using a machine learning approach with cell level EL images as input.

An important feature of the model is that it should be able to deal with different cell types and image quality, in order to build a functional and complete module power loss estimation model, that is able to generalize to a better degree than current models, and with less training data. An exploratory analysis has been carried to find appropriate image features and study their relationship towards each type of cell included in the dataset, aiming to model the different types of cells from one single dataset.

One of the main advantages of a cell-based approach is that, in comparison to other models, we can validate the simulation on the cell level instead of in the whole module, facilitating granularity that can be taken advantage of when combined with knowledge of a PV-module's electronics such as bypass-diodes. This work aims to set the basis for a future bottom-up machine learning approach that can predict power output on module level with EL images taken in field inspections.

2 METHODOLOGY

2.1 Cell selection and characterization

A collection of monocrystalline silicon PV cell samples of different types have been tested and characterized to build a dataset that contains cells with different busbars and amount of cracks.

Table I: Type, number of cell samples and number of measurements of each cell type included in the dataset.

Cell Type	Nº of samples	Nº of measurements
IBC Zebra	4	13
IBC SunPower	4	16
2 Busbars	42	104
3 Busbars	4	15
4 Busbars	4	15
5 Busbars	16	59
Total	**74**	**222**

Table I presents the different cell types included in the dataset, the number of samples manufactured for each cell and the number of measurements that, after the characterization process, includes the dataset. A total number of 74 cracked solar cell samples, were used to create the dataset. Each cell sample consist of a single PV cell encapsulated using EVA with a back sheet and a flexible transparent front sheet that allows mechanically stressing the cell more easily.

The dataset consists of multiple monocrystalline silicon cells with different number of busbars, such as 2, 3, 4 and 5 busbars, plus some Zebra and SunPower back contacted IBC cells. Figure 1 shows examples of each type of the analyzed cells EL images.

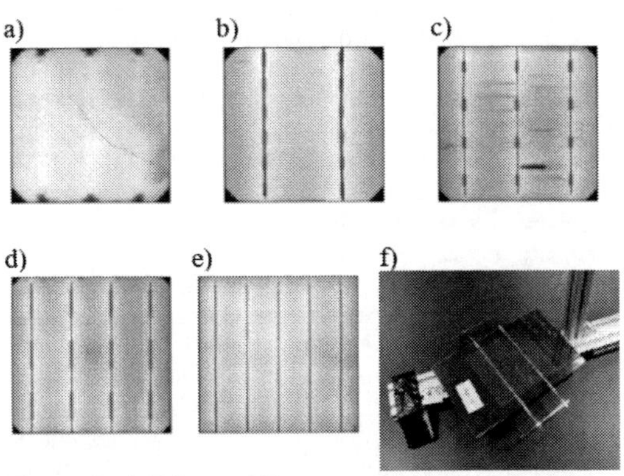

Figure 1: a) Zebra and Sunpower IBC cells EL. b) 2-busbar cells EL. c) 3-busbar cells EL. d) 4-busbar cells EL. e) 5-busbar cells EL. f) Example of a manufactured cell sample

All samples have been characterized and mechanically stressed over several degradation stages to cause cell cracks, as previously described in [17], where after an initial characterization, the cell mini-modules were mechanically stressed and re-characterized after checking for cracks. To have a generalizable and spread-out dataset, multiple levels of cracked areas were induced, with data that includes relatively undamaged cells to inactive area up to 50% of the total cell area. The characterization consisted of light I-V measurements at STC conditions (1000 W/m^2, 25 ºC and 1.5A M) using an EcoSun 10L REV E class AAA Sun simulator, whereas EL images were acquired using a Nikon DSLR 24.2MP camera in indoor lab conditions at 100% I_{SC} bias.

After the characterization process, the main cell electrical parameters (P_{MPP}, I_{SC}, V_{OC}, I_{MPP}, V_{MPP}) were extracted from the I-V measurements.

The final dataset consists of 222 distinct EL images and I-V curves of the cells at different degradation stages. Figure 2 displays the distribution of the dataset. At the moment, the majority of the analyzed cells are 2 busbar cells followed by 5 busbar cells. The significant lower amount of 3 and 4 busbar cells must be taken into consideration when evaluating the models and the results.

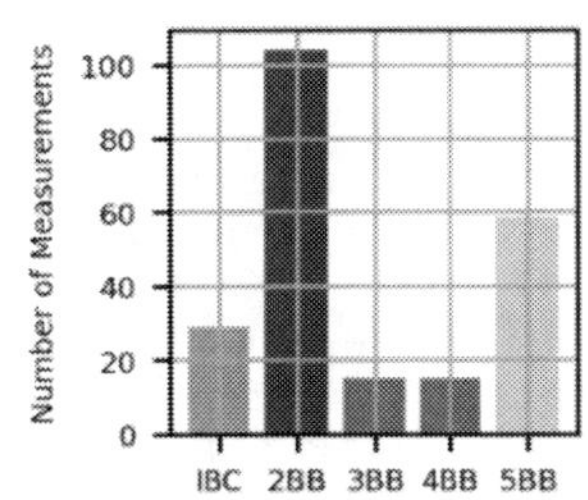

Figure 2: Dataset measurements distribution.

Next, the EL images have been processed to create comparable datasets. The raw recorded images have been cropped and orientation corrected. Furthermore, a mask has been applied to the edges and busbars of the cells to remove the regions without luminescence. Moreover, the raw EL images have been down sampled to 200x200. Since the future aim is to develop a tool that can be validated in a module level, extracting the EL of the cells of a module will not be as high resolution as the resolution of a DSLR 24.2MP camera. This consideration also applies to images taken with lower resolution InGaAs cameras, and potentially outdoor EL images.

Similarly, the electrical parameters obtained from the I-V curves have been normalized to a reference cell of each of the different groups of cells analyzed, to facilitate generalization of the models. These references correspond to the average of the electrical parameters of 2 or 3 undamaged cells of each type.

2.2 Cell image histogram-derived predictor features

After the characterization and data normalization process, an investigation on what image features could be obtained from the EL images was carried out.

The feature selection for this work is focused on modelling the cell crack impact on the cell's power loss from the point of view of the cell's EL image histogram. The goal was to extract simple image features that are easy to compute and robust to image noise. Furthermore, these features should be sensitive to different extents of cells cracks and their severity in terms of power loss.

Previous research proposes a way of quantifying cell cracks percentage by using 100% and 10% I_{SC} bias EL images [24]. By applying an automatic threshold on the high bias image to separate the dark areas of the cell, one can calculate their percentage over the total cell area. By following this approach, one image feature was extracted for each cell, total inactive area. This corresponds to the total percentage of B and C cracks combined in the cell.

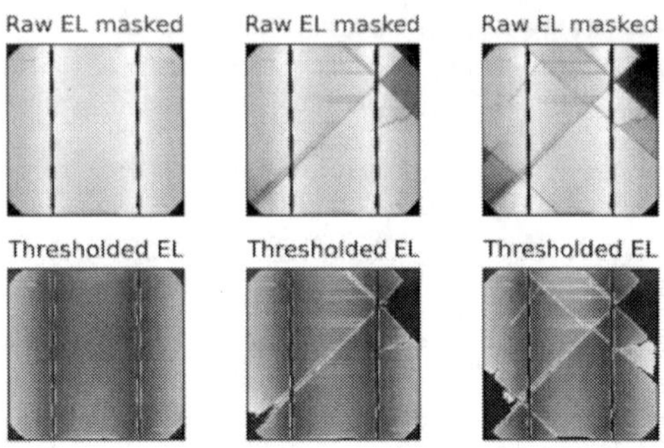

Figure 3: From left to right, initial to second degradation stage of a 2-busbar cell EL image. The top image shows the raw EL image with the busbar and edge mask. The bottom image shows in purple the inactive areas (B and C cracks) of the cell by applying the algorithm introduced in [24].

Furthermore, it has been shown that the histogram parameters of EL images are correlated to the appearance of cracks [25]. Therefore, mean, standard deviation (STD), kurtosis and skewness of the EL image histogram have also been selected as image features for the model. These parameters have been calculated on all the cell's Electroluminescence Intensity Distribution (ELID) or $p_{EL}(i)$ as defined in [25]. Where i is the occurrence of each grey pixel or intensity level. Figure 4 shows an example of an ELID histogram of a 2-busbar cell in its initial state (blue) and after two rounds of mechanical degradation (orange), where the above-mentioned histogram parameters have been calculated. The induced cracks change the shape of the image histogram, leading to a change in magnitude of the proposed image features. In this example, and as a general trend, the mean, skewness and kurtosis of the image will decrease, and the STD of the image histogram will increase due to the appearance of new dark areas in the cell.

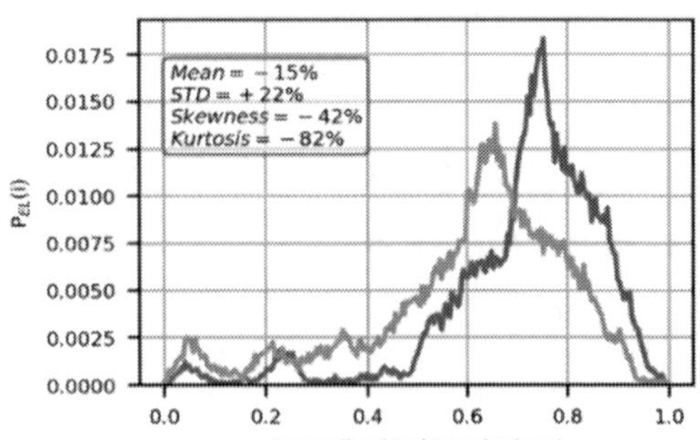

Figure 4: Example of histogram parameters (mean STD, skewness and kurtosis) evolution calculated on a ELID histogram of a cell EL image before (blue) and after (orange) mechanical stress.

Like the electrical parameters, the mean and the STD of each cell image has been normalized to a relative value corresponding to the same reference cells mentioned in the previous section. This measure accounts for the different type of EL emission the different cells have, for example, IBC cells usually have a higher EL mean. The rest of the image features are, by definition, comparable to each other between the different cells.

2.3 Model validation

The models analyzed in this section will be evaluated using the Root Mean Square Error (RMSE), Mean and Maximum Absolute Error (MAE and MaxAE) and the R^2 score defined below, where R_i is the real measured power and S_i is the simulated or predicted power for n measurements. Since these scores will be calculated on the normalized power, they will range between 0 and 1.

$$RMSE = \sqrt{\frac{1}{n}\sum_{i=1}^{n}(R_i - S_i)^2} \quad (1)$$

$$MAE = \frac{1}{n}\sum_{i=1}^{n}|R_i - S_i| \quad (2)$$

$$MaxAE = \max(|R_i - S_i|) \quad (3)$$

$$R^2 = 1 - \frac{\sum_{i=1}^{n}(R_i - S_i)^2}{\sum_{i=1}^{n}(R_i - \bar{R}_i)^2} \quad (4)$$

Furthermore, to validate the result of the models, a K-fold cross validation has been performed in the dataset using K = 8. This validation procedure consists of randomly shuffling the dataset and then splitting it into K equal sets. Then for each iteration, K-1 sets are used to train a model, while the remaining one is used to test it. Once that is done, the model scores are calculated, and the process is repeated until all sets have been used for validation.

Examples of real vs. simulated electrical parameters will be shown as an extra validation of the models, for that, the dataset was split into a training and a test set consisting of 75% and 25% of the data respectively.

3 RESULTS AND ANALYSIS

3.1 Linear regression approach

For a better understanding of the dataset and its distribution, we first looked at the correlation of the different variables to the maximum power of the cell.

Figure 5 shows as scatter plots the relationship between the P_{MPP} and the EL image histogram parameters and inactive area of the cells. Here, one can notice the apparent linearity of some of the variables with the power. For example, the inactive cell area shows a very characteristic downward trend, similar to the STD of the dataset. The appearance of cracks in the cells have an influence on the power output, but also on the image STD, since the dark regions in the image spread out the histogram. The trend of the skewness and kurtosis resembles more a logarithmic trend, whereas the mean increases with the cell power since the luminosity of the cells is generally brighter in non-damaged states.

Figure 5f shows the correlation coefficients that represent the linearity of the relationship between the different variables in the dataset. As previously mentioned, we observe a strong inverse correlation of the inactive area and the STD, and positive correlation of the mean with the P_{MPP} define the dataset.

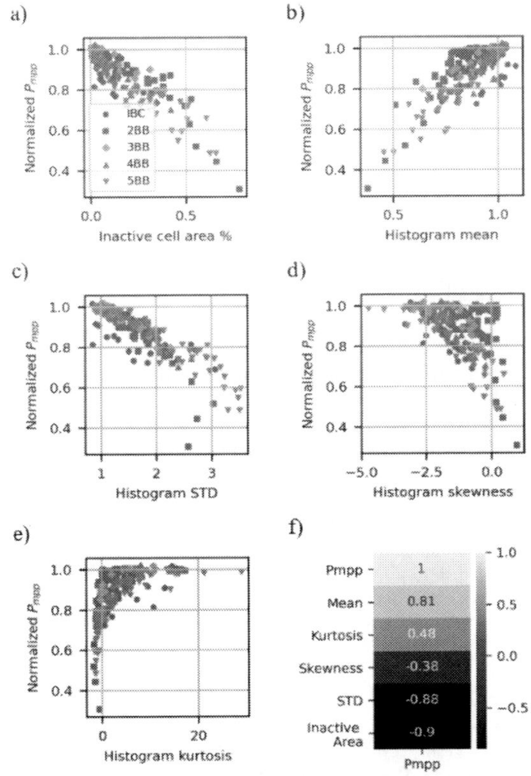

Figure 5: a-e) Impact of the image histogram features on P_{MPP}. Color legend is the different busbar cell category. f) correlation coefficients of the EL image features to P_{MPP}.

Due to the linearity of most of the features, a linear regression model was selected for a first test and to set a benchmark for the rest of the future approaches.

Table II: Linear regression validation scores with K-folds cross validation (K=8).

	MAE	MaxAE	RMSE	R^2
Max.	0.036	0.264	0.068	0.964
Min.	0.017	0.041	0.020	0.709
Avg.	0.027	0.127	0.041	0.850

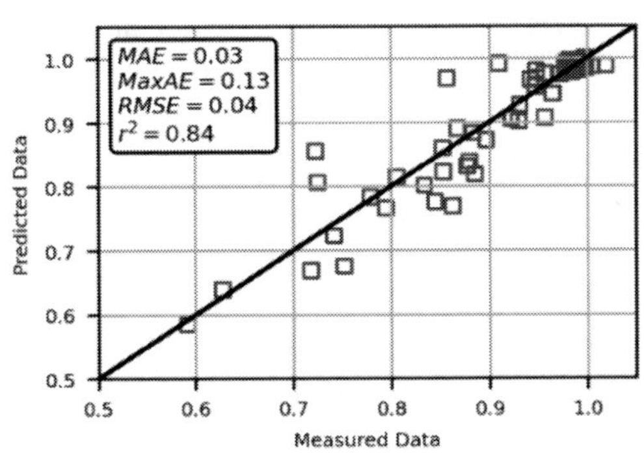

Figure 6: P_{MPP} predicted vs measured using a linear regression model.

The validation of this model (Table II) presents a considerable maximum absolute error of 0.26, which considering that the simulated power is normalized between 0 and 1, means that this model inconsistency in simulation can yield big deviations if a certain part of the dataset is used to train this model. This behavior suggests that the general trends of the dataset are not robust enough to simulate certain P_{MPP} outliers, which could be that certain group of cells used for training does not represent well cells with a different number of busbars.

Figure 6 shows as an extra validation example, a power simulation using a random test/train set. The scores resemble the K-fold cross validation results, but present a much lower maximum error, which proves the inconsistency of this model.

The results from this first simulation present a promising baseline that can be used to evaluate more complex approaches, but the occurrence of random errors given a different test/train distribution cannot be neglected.

3.2 2nd degree polynomial regression approach

Looking at Figure 5 again, one can see that a simple linear regression model will be limited by the skewness and kurtosis image features. Furthermore, looking at the mean, STD and Inactive Area, certain types of cells follow slightly different trends, with different slopes, not a completely linear response in general. Therefore, a second model has been tested by applying a 2nd degree polynomial linear regression.

Table III: 2nd degree Polynomial regression validation scores with K-folds cross validation (K=8).

	MAE	MaxAE	RMSE	R^2
Max.	0.038	0.187	0.059	0.953
Min.	0.020	0.052	0.024	0.630
Avg.	0.028	0.103	0.038	0.859

Table III presents the validation of this model over the K = 8 folds cross validation of the randomized dataset. This model shows a better result than the previous linear regression model, with an increase of 0.09 of R^2 average score and mainly a big decrease of the maximum error over the validation, that went from 0.26 to 0.187. However, this model is still not reliable, as the minimum R^2 value has decreased, meaning that it is still inconsistent and relies too much on the training data to be able to model certain outliers.

This model seems to fit better the different trends of the dataset in terms of average R^2 and maximum error, but improvements are still possible regarding the variability of the maximum and minimum scores.

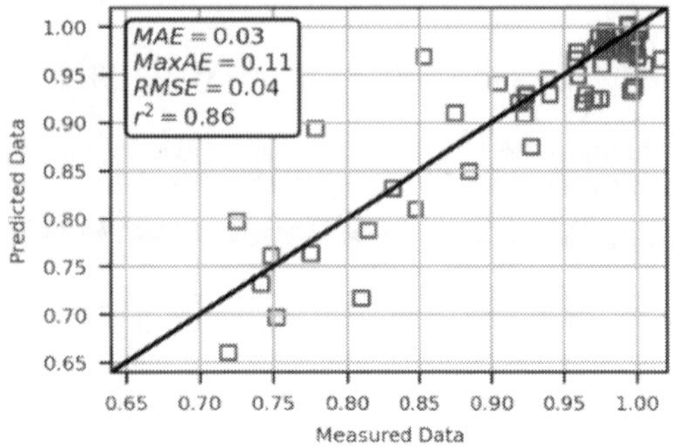

Figure 7: P_{MPP} predicted vs measured using a polynomial 2nd degree linear regression model.

Figure 7 shows the measured vs. simulated power from a random state of the dataset test/train split simulation, where the average results are similar to the validation results. The maximum error in this case is lower, however, variations are expected when randomly testing the model.

3.3 PCA and statistical analysis

To better gauge the impacts and relationships between the features and apply them into potentially more advanced models, the variables in the dataset were subjected to a Principal Component Analysis (PCA).

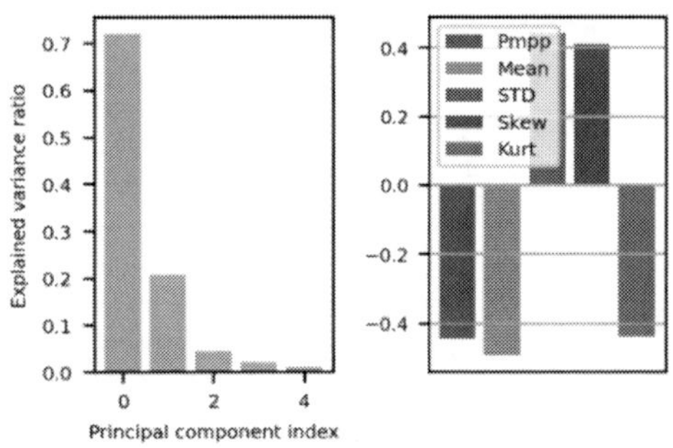

Figure 8: Left: PCA result, explained variance of each dataset component. Right: First principal component features contribution.

Figure 8 shows the principal component proportions of the dimensions covering of the main statistical metrics of the image histograms, plus the P_{MPP}. The eigenvalues of the two first principal components, PC_0 and PC_1 explain together more than 90% of the variance in the dataset. In both, the P_{MPP} has a strong presence. Naturally, this is a good sign in terms of the explanatory power of the variables, but on the other hand, it is clear from the eigenvalues, that there is a high correlation, positive and negative between all the input variables.

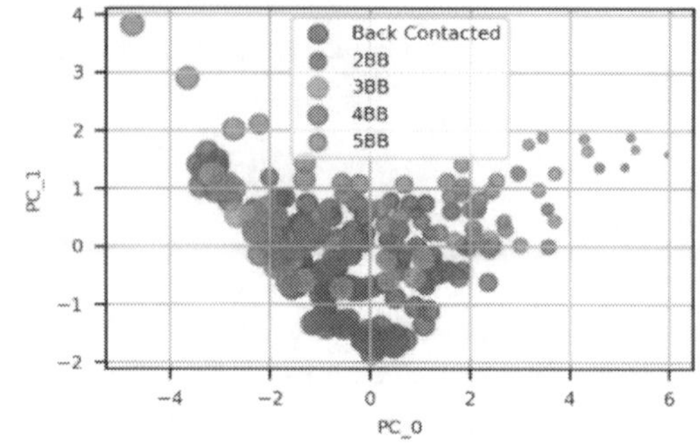

Figure 9: Projection of first and second component of the PCA containing P_{MPP}, mean, STD, skewness and kurtosis. Dot size is proportional to P_{MPP}.

By transforming the dataset and projecting it onto the two main principal components (Figure 9) a few things become apparent. Observing the size of the plot points, which is set to be proportional to the P_{MPP}, it's clear that the first principal component is the main predictor of P_{MPP}, with the P_{MPP} increasing along it. For the second axis, however, there is no longer any clean trend, and it should be remembered that P_{MPP} is also part of this axis.

In addition, it's clear that, when the dataset is grouped by the number of busbars, some amount of clustering of the data is occurring for each busbar category.

Taking this information into account, the obvious next step is to add the busbar numbers and run a PCA again. This time, only the histogram mean was added, both in order to reduce interference to any conclusions, and because both the 1st and 2nd eigenvectors showed high correlation between the mean and the other metrics, meaning that it should be sufficient to draw further conclusions.

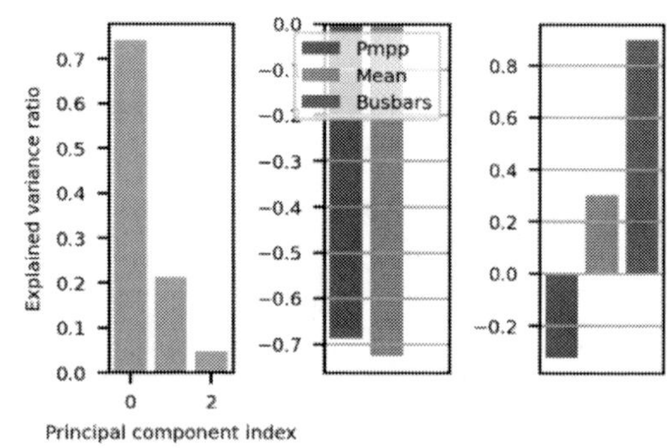

Figure 10: Left PCA result using P_{MPP}, mean and Busbars. Middle: First principal component features contribution. Right: Second principal component features contribution.

Running a PCA of the dataset against the histogram mean, P_{MPP} and number of busbars of each cell, the result is a first principal component that explains almost 75% of the dataset variance (Figure 10), while the first two components together explain >95%. In this case, the busbar effect appears greatly in the second component, which is not surprising since it reflects the category of the cells. When projecting onto the first two principal axes (Figure 11), the clustering observed before can be much better appreciated by adding this variable.

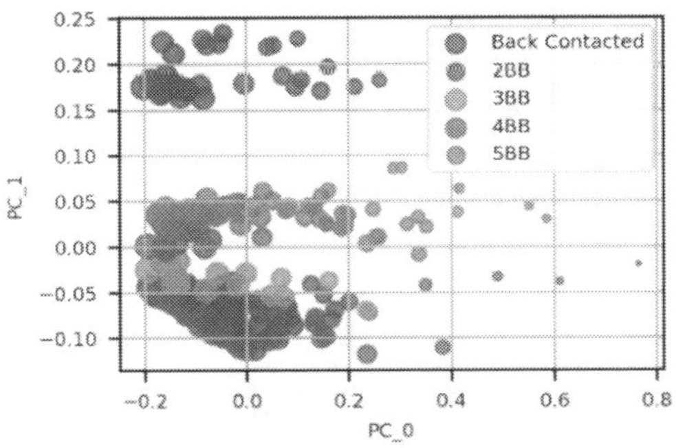

Figure 11: Projection of first and second component of the PCA containing P_{MPP}, mean and Busbar number. Dot size is proportional to P_{MPP}.

The busbar-groupings of the plot clearly have different relationships to the histogram mean. If the busbar numbers were purely a classification, this should not occur.

Recollecting the eigenvectors from Figure 10, the busbar number has a minor correlation to the P_{MPP}, but it correlates even less to the mean. Given this, the different slopes of the groupings are quite distinctive. Furthermore, the first axis is again the main P_{MPP} predictor, which in this case contains the mean. The different set of variables that can influence the power have therefore a different behavior in terms of the trend that each cell type follows, meaning that one should model the dataset accordingly.

The interpretation to take away from this is, that the number of busbars on the cells have an impact on the relationship between the P_{MPP} and the histogram parameters, thereby introducing non-linearity. The extent of this is difficult to gauge from the PCA analysis, since the busbar number is intentionally dominant for visualization purposes, and the third principal component is only indirectly visible via the size of the plot points, as it correlates mostly with the P_{MPP}.

One solution to this non-linearity could be to create five separate models. However, some of the cell types have a low number of data points, making the quality of any sub-model questionable. By instead introducing the busbar number as a non-linear factor, the overall trends from all the data can be shared, while the busbar-dependent behavior can be interpolated between cell types.

3.4 Non-Linear regression approach

Following the reasoning explained in the PCA, a non-linear approach was examined, considering the nonlinearity that a different number of busbars presents. Furthermore, observing the relative improvement of the linear second-degree polynomial model, a second-degree non-linear equation was formulated:

$$f(x_0, x_1, BB) = c_0 + c_1 BB + c_2(1 + c_3 BB)x_0 + c_4(1 + c_5 BB)x_0^2 + c_6(1 + c_7 BB)x_1 + c_8(1 + c_9 BB)x_1^2 \quad (5)$$

As observed in the PCA, and for the sake of simplicity, a model using three independent variables was formulated, where the non-linear factor comes from the addition of the busbars category.

Different combinations of features were also tested to find the ones that fit this model the best. By combining the inactive area metric and the STD of the cells as x_0 and x_1, an improvement to the previous models was found.

Table IV shows the error metrics of the K-folds cross validation, with an average R^2 of 0.869, the best one out of the tested models, and a maximum absolute error of 0.17, a significant reduction from the previous two tested models. An important fact that makes this a better model is that, on top of improving all average scores, the distance between maximum and minimum has been reduced. Furthermore, this model presents the highest minimum R^2 scores in validation, which makes it the most robust out of the tested models so far.

Table IV: Non-linear regression validation scores with K-folds cross validation (K=8).

	MAE	MaxAE	RMSE	R^2
Max.	0.037	0.170	0.056	0.963
Min.	0.020	0.051	0.025	0.720
Avg.	0.027	0.100	0.037	0.869

This result is proof that the linear models, although sufficient for a proof-of-concept, are limited. The non-linear models can fit better the particularities of the dataset when having multiple types of cells.

Figure 12 shows the validation of predicted vs measured power using the non-linear model, but in this case using the whole dataset to train the model. This is the expected result of using the whole information in the dataset to train the model, and as expected, it shows a better R^2 score than the average of the validation, but a higher maximum error. This result shall be the baseline of future approaches, not forgetting about the real worst-case scenario of the minimum scores of the model validation.

The findings of this custom model are especially interesting because they prove that different types of cells can be modelled in the same dataset and with just one model if the non-linearity of the dataset is investigated and a model is built around it. Therefore, it opens the possibility of including more cells with different characteristics in the future, cells with more busbars or even half or third cut cells.

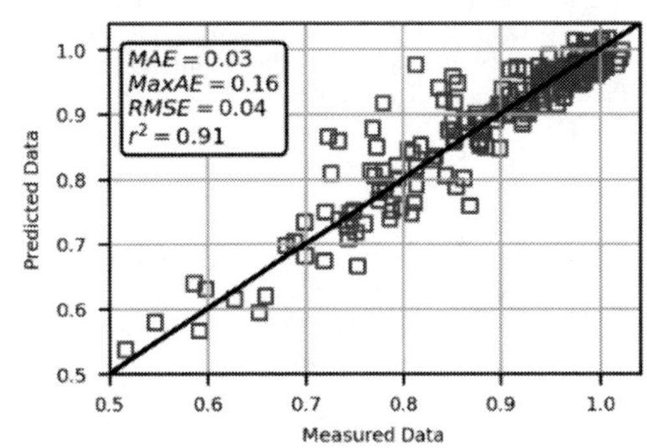

Figure 12: P_{MPP} predicted vs measured using a custom non-linear regression model for the whole dataset.

4 SUMMARY AND CONCLUSIONS

In this work a machine learning approach to simulate P_{MPP} of monocrystalline silicon cells from EL images has been proposed.

The proposed set of image features consist of cell EL image histogram parameters such as mean, STD, skewness, kurtosis and the inactive cell area.

The statistical analysis carried out for the dataset showed certain linear trends in the correlation between the power and the inactive area and the mean and STD of the EL images, thus, in order to set a baseline model, a linear regression was used. This yielded a result of average 0.85 R^2 scores and maximum error of 0.26 over a K-fold cross validation of K = 8. Next, to try to improve this baseline and given that the features correlation to the power were not completely linear, a polynomial linear regression of second-degree was applied to the dataset. Using this model, already an improvement can be noticed, with an average R^2 score of 0.859 and a reduced maximum error 0.187.

In order to check in more detail the correlation between the features, and given the fact that the linear models were not enough to fit the data due to the different set of cell types, a PCA was applied. Results from the PCA demonstrated that the number of busbars influences the image features correlations, creating clusters that behave differently depending on the type of cell. The result of this exploratory analysis is proof that modelling the power output of different types of cells is possible from one single dataset without the need of different modelling for each type of cell individually and that data from different cells is useful input to the model. Finally, to validate these findings a second-degree custom non-linear model was tested. This model aims to account for the non-linearity of the dataset that was observed when using the busbars as a classification feature. The result of this model yielded improved results compared to the linear models, with 0.869 R^2 score, reduced minimum R^2 and absolute error, validating the PCA analysis and fitting better the different cell outliers in each different validation step.

The results presented in this work set a benchmark more advanced models in the future, that can work with a dataset that includes cells of different types and number of busbars. These results can be limited by the modelling or by the image features, therefore, further investigation is needed in the feature side, where features that account for the location and crack connectivity are included in the dataset. On top of that, the procedure presented in this work can also be applied to other electrical parameters of the I-V characteristic in order to model them accordingly.

5 REFERENCES

[1] A. Lebedys *et al.*, "Renewable Energy Statistics 2022," Jul. 2022.

[2] D. C. Jordan, T. J. Silverman, J. H. Wohlgemuth, S. R. Kurtz, and K. T. VanSant, "Photovoltaic failure and degradation modes," *Progress in Photovoltaics: Research and Applications*, vol. 25, no. 4, pp. 318–326, Apr. 2017, doi: 10.1002/pip.2866.

[3] S. Kajari-Schršder, I. Kunze, and M. Kšntges, "Criticality of Cracks in PV Modules," *Energy Procedia*, vol. 27, pp. 658–663, 2012, doi: 10.1016/j.egypro.2012.07.125.

[4] M. Köntges, I. Kunze, S. Kajari-Schröder, X. Breitenmoser, and B. Bjørneklett, "The risk of power loss in crystalline silicon based photovoltaic modules due to micro-cracks," *Solar Energy Materials and Solar Cells*, vol. 95, no. 4, pp. 1131–1137, Apr. 2011, doi: 10.1016/j.solmat.2010.10.034.

[5] T. Fuyuki, H. Kondo, T. Yamazaki, Y. Takahashi, and Y. Uraoka, "Photographic surveying of minority carrier diffusion length in polycrystalline silicon solar cells by electroluminescence," *Appl Phys Lett*, vol. 86, no. 26, p. 262108, Jun. 2005, doi: 10.1063/1.1978979.

[6] M. Sun, S. Lv, X. Zhao, R. Li, W. Zhang, and X. Zhang, "Defect Detection of Photovoltaic Modules Based on Convolutional Neural Network," 2018, pp. 122–132. doi: 10.1007/978-3-319-73564-1_13.

[7] D.-M. Tsai, S.-C. Wu, and W.-C. Li, "Defect detection of solar cells in electroluminescence images using Fourier image reconstruction," *Solar Energy Materials and Solar Cells*, vol. 99, pp. 250–262, Apr. 2012, doi: 10.1016/j.solmat.2011.12.007.

[8] J. Fioresi *et al.*, "Automated Defect Detection and Localization in Photovoltaic Cells Using Semantic Segmentation of Electroluminescence Images," *IEEE J Photovolt*, vol. 12, no. 1, pp. 53–61, Jan. 2022, doi: 10.1109/JPHOTOV.2021.3131059.

[9] H. R. Parikh *et al.*, "Solar Cell Cracks and Finger Failure Detection Using Statistical Parameters of Electroluminescence Images and Machine Learning," *Applied Sciences*, vol. 10, no. 24, p. 8834, Dec. 2020, doi: 10.3390/app10248834.

[10] S. Spataru, P. Hacke, and D. Sera, "Automatic detection and evaluation of solar cell micro-cracks in electroluminescence images using matched filters," in *2016 IEEE 43rd Photovoltaic Specialists Conference (PVSC)*, Jun. 2016, pp. 1602–1607. doi: 10.1109/PVSC.2016.7749891.

[11] C. Mantel *et al.*, "Machine learning prediction of defect types for electroluminescence images of photovoltaic panels," in *Applications of Machine Learning*, Sep. 2019, p. 1. doi: 10.1117/12.2528440.

[12] M. Y. Demirci, N. Beşli, and A. Gümüşçü, "Efficient deep feature extraction and classification for identifying defective photovoltaic module cells in Electroluminescence images," *Expert Syst Appl*, vol. 175, p. 114810, Aug. 2021, doi: 10.1016/j.eswa.2021.114810.

[13] A. M. Karimi *et al.*, "Automated Pipeline for Photovoltaic Module Electroluminescence Image Processing and Degradation Feature Classification," *IEEE J Photovolt*, vol. 9, no. 5, pp. 1324–1335, Sep. 2019, doi: 10.1109/JPHOTOV.2019.2920732.

[14] S. Deitsch *et al.*, "Automatic classification of defective photovoltaic module cells in electroluminescence images," *Solar Energy*, vol. 185, pp. 455–468, Jun. 2019, doi: 10.1016/j.solener.2019.02.067.

[15] Y. Wang *et al.*, "Adaptive automatic solar cell defect detection and classification based on absolute electroluminescence imaging," *Energy*, vol. 229, p. 120606, Aug. 2021, doi: 10.1016/j.energy.2021.120606.

[16] T. Kropp, M. Schubert, and J. Werner, "Quantitative Prediction of Power Loss for

Damaged Photovoltaic Modules Using Electroluminescence," *Energies (Basel)*, vol. 11, no. 5, p. 1172, May 2018, doi: 10.3390/en11051172.

[17] R. del Prado Santamaría *et al.*, "Characterization of Electrical Parameters of Cracked Crystalline Silicon Solar Cells in Photovoltaic Modules," in *2021 IEEE 48th Photovoltaic Specialists Conference (PVSC)*, 2021, pp. 846–853. doi: 10.1109/PVSC43889.2021.9519081.

[18] C. M. Whitaker, B. G. Pierce, R. H. French, and J. L. Braid, "Properties of PV Cell Fractures and Effects on Performance of Al-BSF and PERC Modules," in *2021 IEEE 48th Photovoltaic Specialists Conference (PVSC)*, Jun. 2021, pp. 1779–1785. doi: 10.1109/PVSC43889.2021.9519030.

[19] J. S. Fada *et al.*, "Correlation of I-V Curve Parameters with Module-Level Electroluminescent Image Data Over 3000 Hours Damp-Heat Exposure," in *2017 IEEE 44th Photovoltaic Specialist Conference (PVSC)*, Jun. 2017, pp. 2697–2701. doi: 10.1109/PVSC.2017.8366659.

[20] A. M. Karimi *et al.*, "Generalized and Mechanistic PV Module Performance Prediction From Computer Vision and Machine Learning on Electroluminescence Images," *IEEE J Photovolt*, vol. 10, no. 3, pp. 878–887, May 2020, doi: 10.1109/JPHOTOV.2020.2973448.

[21] M. Hoffmann *et al.*, "Deep-learning-based pipeline for module power prediction from electroluminescense measurements," *Progress in Photovoltaics: Research and Applications*, vol. 29, no. 8, pp. 920–935, Aug. 2021, doi: 10.1002/pip.3416.

[22] Y. Buratti, A. Sowmya, R. Evans, T. Trupke, and Z. Hameiri, "Half and full solar cell efficiency binning by deep learning on electroluminescence images," *Progress in Photovoltaics: Research and Applications*, vol. 30, no. 3, pp. 276–287, Mar. 2022, doi: 10.1002/pip.3484.

[23] S. Rodrigues, C. Buerhop, B. Doll, J. Hauch, C. Brabec, and I. M. Peters, "PREDICTING MODULE I-V CURVES FROM ELECTROLUMINESCENCE IMAGES WITH DEEP LEARNING," Sep. 2020.

[24] S. Spataru, P. Hacke, D. Sera, S. Glick, T. Kerekes, and R. Teodorescu, "Quantifying solar cell cracks in photovoltaic modules by electroluminescence imaging," in *2015 IEEE 42nd Photovoltaic Specialist Conference (PVSC)*, Jun. 2015, pp. 1–6. doi: 10.1109/PVSC.2015.7356124.

[25] S. Spataru, H. Parikh, P. Hacke, G. Alves dos Reis Benatto, D. Sera, and P. B. Poulsen, "Quantification of Solar Cell Failure Signatures Based on Statistical Analysis of Electroluminescence Images," in *Proceedings of the 33rd European Photovoltaic Solar Energy Conference and Exhibition*, 2017, pp. 1466–1472. doi: 10.4229/EUPVSEC20172017-5CO.8.5.

LONG-TERM PERFORMANCE ANALYSIS OF PV MODULES INSTALLED IN OPEN-RACK AND BIPV MOUNTING CONFIGURATIONS

Ebrar Özkalay[1,2], Gabi Friesen[1], Alessandro Virtuani[2,3], Andrew Fairbrother[2], Mauro Caccivio[1], Christophe Ballif[2,3]
1 – University of Applied Sciences and Arts of Southern Switzerland (SUPSI-PVLab), Via Flora Ruchat-Roncati 15, 6850 Mendrisio, Switzerland
2 – École Polytechnique Fédérale de Lausanne (EPFL), Institute of Electrical and Micro Engineering (IEM), Photovoltaics and Thin-Film Electronics Laboratory, Rue de la Maladière 71b, 2002 Neuchâtel, Switzerland
3 – CSEM PV-Center, Rue de Jaquet-Droz 1, 2002 Neuchâtel, Switzerland
Email: ebrar.ozkalay@supsi.ch

ABSTRACT: BIPV (Building Integrated Photovoltaics) systems play an important role in generating electricity for net-zero energy buildings. Since BIPV modules and systems are fully integrated into the building envelope, their rear side is exposed to no or low ventilation, leading to increased operating temperature. Therefore, the performance of the modules – which decreases with increasing temperatures – is reduced, and the durability of the module – and its components (e.g. encapsulant, backsheet etc.) – may be impacted. Over the years, three different BIPV mock-ups have been installed at SUPSI (in Southern Switzerland) in different configurations: open-rack, BIPV-ventilated (partial ventilation) and BIPV-insulated (no ventilation) to investigate the impact of the operating temperature on the long-term performance of different BIPV modules. In this work, first, we show that since larger thermal stress accelerates UV-induced degradation of EVA (discoloration), there could be loss in I_{sc} in BIPV configurations. Secondly, due to larger thermo-mechanical stress, there could be damaged cells and metallization, which causes loss in FF. Additionally, operating in a BIPV configuration does not mean that there will be faster degradation. There could be faster degradation only if there are larger thermal and-or thermo-mechanical stresses.
Keywords: Long-term outdoor performance, Performance loss rate (PLR), Building Integrated Photovoltaics (BIPV)

1 INTRODUCTION

A deeper understanding of the module performance and performance degradation during real-world operation is critical to predict the long-term reliability of building integrated photovoltaics (BIPV) systems. It is difficult to estimate a lifetime for the BIPV products since they are mostly custom-made (produced in small volumes), operate at harsher conditions, and there is less experience in BIPV. Therefore, analysis of the long-term performance of monitored BIPV systems is crucial.

Due to reduced or restricted rear-side ventilation, BIPV modules are typically operating at elevated temperatures (inducing larger thermal stresses) and are exposed to larger diurnal temperature variations (which may induce larger thermo-mechanical stresses) [1]. Despite the fact that the literature on the subject is still limited, some studies have started showing that [2, 3] the exposure of the modules to higher operating temperatures and larger diurnal temperature changes may lead to higher degradation rates (e.g. higher rate of encapsulant discoloration, damaged interconnections, solder joints, cells etc.) and a faster occurrence of wear-out-failures that shorten the lifetime of a PV module.

In this work, we investigate the long-term performance of BIPV modules by analyzing their real-world data and indoor measurements. In this way, more information regarding the behavior of modules and its materials under more challenging BIPV operating conditions can be better understood. Furthermore, these findings provide valuable information in order to design BIPV modules with increased reliability and indoor accelerated reliability and safety tests suitable to the operating conditions of BIPV modules.

2 EXPERIMENTAL SETUP

2.1 Outdoor BIPV Test Stands

Three BIPV outdoor test stands were monitored at SUPSI in Canobbio, Switzerland. The modules on the test stands were installed in open-rack and BIPV configuration, as summarized in Table 1. In the BIPV-ventilated configuration, there is a ventilation chamber between the module and the insulation layer, which ensures partial ventilation on the rear side of the module. The BIPV-insulated configuration has no air gap between the module and the insulation layer. Hence, the rear-side ventilation is restricted. The modules in the BIPV configuration operated at higher temperatures than the same module type in open-rack [1]. Each module's maximum power (P_m) and IV curve parameters (I_{sc}, V_{oc} and FF) were individually monitored using a maximum power point (MPP) tracker at 1-minute intervals and IV curve tracer at 5-minute intervals, respectively. The plane of array irradiance (G_{POA}) (measured by a pyranometer) and module temperature (measured on the back of modules) data were collected every minute.

2.2 Long-term Data Analysis and Indoor Measurements

In this work, performance loss rate (PLR) analyses of the outdoor performance of the modules are performed using the averaging PLRs calculated from different PLR methods in [4] as suggested by [5] since there is no consensus on a PLR method. Indoor measurements, namely visual inspection (VI), electroluminescence (EL) imaging and performance measurement (P_m, I_{sc}, V_{oc}, FF) at STC were performed before and after outdoor exposure of the modules. The changes on the modules' performance and EL images were observed.

Table I: Information about the three Test Stands, the modules used and the typology of installation.

Test Stand	Cell and Module Technologies	Installation Configuration	Azimuth (North = 0°) / Tilt Angles	Duration	Monitored Parameters
#1	Al-BSF - G/EVA/BS Commercial module	• Open-rack • BIPV-Insulated	176° / 6° (Roof)	51 months	• G_{POA} (every 1 minute) • Module Temperature (Pt100 on rear-side of modules) (every 1 minute) • Module performance (MPPT/IV-tracer) (every 1/5 minutes)
	Al-BSF - G/PVB/G Commercial module				
#2	c-Si - G/G Prototype module	• Open-rack • BIPV-Ventilated	176° / 20° (Roof)	53 months	
	c-Si - G/EVA/BS Commercial module	• BIPV-Ventilated			
#3	PERC - G/PVB/G Custom module	• BIPV-Ventilated	176° / 90° (Façade)	27 months	
Al-BSF: Aluminum back surface field, PERC: Passivated Emitter and Rear Contact, G: Glass, BS: Backsheet, EVA: Ethylene-vinyl acetate, PVB: Polyvinyl-butyral, MPPT: Maximum power point tracker and G_{POA}: Plane of array irradiance.					

3 RESULTS AND DISCUSSION

3.1 Test Stand # 1

Table II shows that the G/BS BIPV-insulated module degraded faster than the G/BS open-rack module (Open-rack: -0.03%/yr and BIPV-insulated: -0.33%/yr). As shown in Table III, which represents the difference between indoor measurements at STC before and after outdoor monitoring, the performance loss of the BIPV-insulated module is mainly driven by losses in I_{sc} (Open-rack: 0.1% and BIPV-ventilated: -0.9%) and losses in FF (Open-rack: -0.5% and BIPV- insulated: -1.6%). Figure 1 shows the EL images of the G/BS BIPV module before and after outdoor monitoring. Because of the larger diurnal temperature changes, this module was exposed to larger thermo-mechanical stress, and therefore, there are damaged metallization and cell cracks that cause loss in FF. In addition, due to larger thermal stress, UV-induced degradation of EVA was accelerated, and discoloration occurred. Hence, the G/BS BIPV module has loss in I_{sc}. On the contrary, the open-rack module do not have any discoloration. The G/BS module in the BIPV-insulated configuration degraded faster than the same module technology in the open-rack configuration due to larger thermal and thermo-mechanical stresses.

Table II: Performance loss rates (PLR) of PR and IV parameters of the modules on Test Stand #1 by applying averaging PLRs method in [4].

Module Technology (duration of monitoring)	**Al-BSF – G/EVA/BS** (4+ years of operation)		**Al-BSF – G/PVB/G** (4+ years of operation)	
Mounting Configuration	**Open-rack**	**BIPV-Ins.**	**Open-rack**	**BIPV-Ins.**
PR [%/yr]	-0.03	-0.33	-0.05	0.25
I_{sc} [%/yr]	-0.09	-0.03	-0.04	0.48
V_{oc} [%/yr]	0.03	-0.07	-0.07	0.03
FF [%/yr]	0.03	-0.29	0.01	-0.15

The outdoor data analysis (Table II) shows that the G/G BIPV-insulated module's performance improved due to a gain in I_{sc}. However, the indoor measurements show that the performance of both G/G modules (Open-rack and BIPV-insulated) degraded due to loss in I_{sc}. This may be related to the behavior of PVB encapsulant at high temperatures. Thus, PVB encapsulant's effect at high temperatures has been under further investigation. The G/G BIPV-insulated module degraded slightly more than the open-rack module (Open-rack: -1.7% and BIPV-insulated: -1.9%), but the difference is not significant.

Table III: Change between the initial and final indoor IV measurements of the modules on Test Stand #1.

Module Technology (duration of monitoring)	**Al-BSF – G/EVA/BS** (4+ years of operation)		**Al-BSF – G/PVB/G** (4+ years of operation)	
Mounting Configuration	**Open-rack**	**BIPV-Ins.**	**Open-rack**	**BIPV-Ins.**
ΔP_m [%]	-0.3	-3	-1.7	-1.9
ΔI_{sc} [%]	0.1	-0.9	-1.7	-1.7
ΔV_{oc} [%]	0.2	-0.3	-0.05	-0.15
ΔFF [%]	-0.5	-1.6	0.1	-0.1

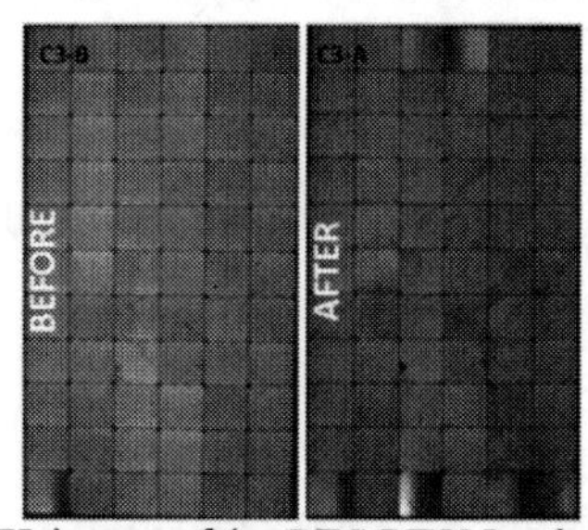

Figure 1: EL images of the G/BS BIPV-Insulated module on Test Stand #1 before and after outdoor monitoring.

3.2 Test Stand # 2

Table IV shows a noticeable loss in performance ratio for the prototype modules (c-Si, G/G) in both configurations (Open-rack: -0.83%/yr and BIPV-ventilated: -1.83%/yr). The performance losses are mainly driven by the losses in I_{sc} (Open-rack: -0.28%/yr and BIPV-ventilated: -0.97%/yr) and FF (Open-rack: -0.51%/yr and BIPV-ventilated module: -0.81%/yr). During the visual inspection, we observed discoloration of the encapsulant of both prototype modules (c-Si, G/G) in open-rack and BIPV-ventilated configurations. The discoloration of the BIPV-ventilated module is visually more intense than the discoloration of the open-rack module. Higher temperature seems caused more intense discoloration. Discoloration of the encapsulant explains the observed loss in I_{sc} over time. Discoloration on the interconnect ribbon is noticeable for both open-rack and BIPV-ventilated (more intense) modules. Since the open-rack prototype module operated at lower thermal stresses, performance loss over time is less than the BIPV-ventilated module. Nevertheless, no change is observed on EL images of the modules (not shown).

Table V shows the difference between the modules' initial and final indoor measured IV parameters on Test Stand #2 at STC. The performance losses after the outdoor deployment of the open-rack and the BIPV-ventilated prototype modules are significant (Open-rack: -4.4% and BIPV-ventilated: -11.6%). The PLR results of the open-rack and BIPV-ventilated modules are mostly in line with the indoor STC results considering the uncertainty of the statistical analyses of outdoor measurement.

Table IV: Performance loss rates (PLR) of PR and IV parameters of the modules on Test Stand #2 by applying averaging PLRs method in [4].

Module Technology (duration of monitoring)	**c-Si – G/G Prototype module** (4+ years of operation)		**c-Si – G/EVA/BS Commercial module** (4+ years of operation)
Mounting Configuration	**Open-rack**	**BIPV-Vent.**	**BIPV-Vent.**
PR [%/yr]	-0.83	-1.83	-0.01
I_{sc} [%/yr]	-0.28	-0.97	-0.10
V_{oc} [%/yr]	-0.11	-0.01	-0.06
FF [%/yr]	-0.51	-0.81	0.07

The commercial module (c-Si, G/EVA/BS) in the BIPV-ventilated configuration on Test Stand #2 barely showed any change in its performance (Table IV). Table V shows that the module's performance only dropped -0.8%, mainly driven by the loss of FF by -0.7% after over four years of operation. No visible change was observed on the module. According to [1], this module did not operate at elevated temperatures (thermal stress) and large diurnal temperature changes (thermo-mechanical stress). This shows that if a ventilation chamber of a BIPV-ventilated module is well designed, there will not be high thermal and thermo-mechanical stresses and accelerated degradation mechanisms due to those additional stresses.

Table V: Change between the initial and final indoor IV measurements of the modules on Test Stand #2.

Module Technology (duration of monitoring)	**c-Si – G/G Prototype module** (4+ years of operation)		**c-Si – G/EVA/BS Commercial module** (4+ years of operation)
Mounting Configuration	**Open-rack**	**BIPV-Vent.**	**BIPV-Vent.**
ΔP_m [%]	-4.4	-11.6	-0.8
ΔI_{sc} [%]	-2.7	-6.2	-0.4
ΔV_{oc} [%]	-0.1	-0.2	0.2
ΔFF [%]	-1.5	-0.81	-0.7

3.3 Test Stand # 3

The BIPV-ventilated façade module only had larger thermo-mechanical stress [1]. Tables VI and VII show that the module did not degrade, except for a slight decrease in FF from indoor measurements. This module needs to be monitored for longer durations (3 to 5 years) to be able to say something about the possible effects of larger thermo-mechanical stress.

Table VI: Performance loss rates (PLR) of PR and IV parameters of the modules on Test Stand #3 by applying averaging PLRs method in [4].

Module Technology (duration of monitoring)	**PERC – G/EVA/BS** (2+ years of operation)
Mounting Configuration	**BIPV-Ven. - Façade**
PR [%/yr]	-0.02
I_{sc} [%/yr]	-0.04
V_{oc} [%/yr]	0.02
FF [%/yr]	-0.01

Table VII: Change between the initial and final indoor IV measurements of the modules on Test Stand #3.

Module Technology (duration of monitoring)	**PERC – G/EVA/BS** (2+ years of operation)
Mounting Configuration	**BIPV-Ven. - Façade**
ΔP_m [%]	0.01
ΔI_{sc} [%]	0.07
ΔV_{oc} [%]	0.2
ΔFF [%]	-0.26

4 CONCLUSION

This work shows the performance changes of the modules in open-rack and typical BIPV installation configurations (BIPV-ventilated for reduced rear-side ventilation and BIPV-insulated for restricted rear-side ventilation). We analyzed the change in the modules' performance by performing PLR analysis using the monitored data during the outdoor exposure and STC power measurements at indoor before and after the outdoor exposure. First, we showed that since larger thermal stress accelerates UV-induced degradation of EVA, there could be loss in I_{sc} in BIPV configurations. In addition, due to larger thermo-mechanical stress, there could be damaged cells and metallization that cause FF loss.

PV modules may have issues originated from their production, e.g. encapsulant material susceptible to UV and thermal stresses. In this case, the results from Test Stand #2 (c-Si, G/G, BIPV-ventilated prototype module) show that those issues could not be important for the safety and performance of conventional PV; however, for BIPV applications, imperfect production could be more significant since the operating conditions are more challenging than conventional PV.

Modules operating in a BIPV configuration do not have to degrade faster. There could be faster degradation only if there are larger thermal and-or thermo-mechanical stresses. As shown from Test Stand #2 (c-Si, G/BS), if a ventilation chamber of a BIPV-ventilated module is well-designed, there will not be higher thermal and thermo-mechanical stresses and accelerated degradation mechanisms related to those stresses. Last of all, the BIPV-ventilated façade module did not degrade after 2 years. However, it needs to be monitored for longer durations to be able to see possible effects of larger thermo-mechanical stress.

5 ACKNOWLEDGEMENT

The project called "Reliability of PV Systems integrated into the built environment" (REBIPV) is supported by the Swiss National Science Foundation under COST IZCOZ0_182967 and the Swiss Federal Office of Energy. In addition, the COST Action PEARL PV (CA16235) provided additional (collaborative) support. Finally, the authors would like to thank SUPSI-PVLab (University of Applied Sciences and Arts of Southern Switzerland), the BIPV-Innovative Building Skin Team researchers, and all the staff members who contributed to supplying the relevant data for this work.

6 REFERENCES

[1] E. Özkalay, G. Friesen, M. Caccivio, P. Bonomo, A. Fairbrother, C. Ballif and A. Virtuani, "Operating temperatures and diurnal temperature variations of modules installed in open-rack and typical BIPV configurations," *IEEE Journal of Photovoltaic,* 2022.

[2] D. Jordan, S. Kurts, K. VanSant and J. Newmiller, "Compendium of photovoltaic degradation rates," *Progress in Photovoltaics,* pp. 978-989, 2016.

[3] H. Hu, W. Gambogi, K. R. Choudhury, L. Garreau-Iles, T. Felder, S. MacMaster, O. Fu and T.-J. Trout, "Field analysis and degradation of modules and components in distributed PV applications," in *35th European Photovoltaic Solar Energy Conference and Exhibition*.

[4] E. Özkalay, A. Virtuani, A. Fairbrother, A. Skoczek, G. Friesen and C. Ballif, "Quantifying performance loss rates of photovoltaic modules using ground-based vs satellite-based meteorological data," in *EUPVSEC*, 2021.

[5] S. Lindig, D. Moser, A. J. Curran, K. Rath, A. Khalilnejad, R. H. French, M. Herz, B. Müller, G. Makrides, G. Georghiou, A. Livera, M. Richter, J. Ascencio-Vásquez, M. van Iseghem, M. Meftah, D. Jordan, C. Deline, W. van Sark, J. S. Stein, M. Theristis, F. Baumgartner and W. Luo, "International collaboration framework for the calculation of performance loss rates: Data quality, benchmarks, and trends (towards a uniform methodology)," *Progress in Photovoltaics: Research and Applications,* vol. 29, no. 6, p. 573– 602, 2021.

AN APPROACH FOR A SHADING RESISTANCE CLASSIFICATION OF PV MODULES

Hendrik Sträter[1], Stefan Riechelmann[1], Stefan Winter[1]
[1]Physikalisch-Technische Bundesanstalt (PTB), Bundesallee 100, 38116 Braunschweig, Germany

ABSTRACT: Shading of PV modules is known as a common reason for underperforming of PV installations, especially in roof top installations. With increasing total installation goals necessary for the energy transition, more and more less favorable (roof) installation sites come into focus. There is an awareness about the impact of shading situations on the overall output power and an increased number of PV modules claimed to be shading resistant are found on the market. Nevertheless, shading resistance is not well defined and lacks a fair, understandable, and reproducible comparability measure. In our contribution, we present a robust and pragmatic classification for the shading resistance of PV modules. We therefore developed a rating scheme and use it in a way that can easily be conducted with a standard solar simulator or an outdoor test stand, and commercially available shading materials, respectively. In addition, we propose classification intervals in a way that typical and already available technologies fall within reasonable classes.
Keywords: Performance, PV Module, Shading

1 INTRODUCTION

With increasing awareness of problems caused by shading of PV modules, different mitigation strategies have been developed to increase the shading resistance of PV systems. The shading resistance of a PV system can be improved already in the planning phase and on the inverter level or by using individual power optimizers directly connected to individual PV modules, respectively. On the PV module level, certain technologies as installation of bypass diodes [1], shingled solar cells [2] or dividing the module in parallel connected [3] submodules have also been developed. At this point it is important to keep in mind that shading resistance is always a compromise between shading resistance, maximum power and cost.

It is understandable that innovative module designs are advertised for their improved shading resistance. But comparison of different module technologies and designs gets complicated if shading resistance as a property of a PV device is not well defined, and no standardized means for testing a PV module for its shading resistance compared to common technologies are available.

With this work we propose a classification scheme for a better comparison of shading resistance of typical PV modules and demonstrate by indoor and outdoor measurements how commercially available PV modules of different design would be classified under this scheme. We designed the scheme to be quick and easy in execution, and to cover only the most relevant shading scenarios. As a starting point we limited our measurements to Silicon-based PV modules of typical dimensions.

2 THE CLASSIFICATION SCHEME

We evaluated a classification scheme which rates the power loss of a PV module under different shading patterns where the power loss is solely determined by the shaded area of the PV module. Ideally, the power loss would be equivalent to the size of the shaded area, e.g. if 20 % of the module area is shaded, a perfect shading resistant device would only lose 20 % of module power. Any additional loss (AL) larger than this expected loss is mainly due to the circuit design of the PV module. The additional loss is defined as:

$$AL = \left(1 - \frac{P_{\mathrm{mpp,shaded}}}{P_{\mathrm{mpp}}} - \frac{A_{\mathrm{shaded}}}{A_{\mathrm{total}}}\right) \cdot 100\ \%$$

By definition, $AL \geq 0$ % if we neglect potential non-linear effects and has the best case of $AL = 0$ %, meaning that the shading does not reduce power generation of the unshaded part of the module. The worst case is $AL = \left(1 - \frac{A_{\mathrm{shaded}}}{A_{\mathrm{total}}}\right) \cdot 100$ %, meaning that the shaded area of the module causes a total loss of power generation of the unshaded part of the module.

Although different methods for evaluation of different shading scenarios already exist [4], e. g. complete blocking a full cell or moving a shading pattern along the PV module [5], we consider three different, stationary shading patterns:

- shading along the long side of the PV module
- shading along the short side of the PV module
- spot shading within one solar cell,

as shown in Figure 1. Many more shading patterns have been discussed and evaluated for their impact but could be reduced to the proposed patterns.

The shaded areas applied on the module shall have the following dimensions:
$A_{\mathrm{long\ side}} = L_{\mathrm{Module}} \times 25$ cm for the long side shading,
$A_{\mathrm{short\ side}} = W_{\mathrm{Module}} \cdot 25$ cm for the short side shading,
and $A_{\mathrm{spot}} = 9\ \mathrm{cm} \cdot 3\ \mathrm{cm}$ for the spot shading.

This ensures that for the first two shading patterns even for the new generation of M12 solar cells at least one cell row is completely within the shaded area. For the spot shading rating, it is important that the shading material is placed only on the surface of one solar cell and as less metallization as possible is covered to maximize the shading effect. For simplicity the shading should be done with a material with a transmittance of 0 %. The shading material for the long side and short side test shall be placed flush with the module edge and not with the cell area. Otherwise, inconsistencies between framed and frameless modules may occur in rating.

Note that we propose to use identical shading patterns for all tested PV modules no matter what their total size, solar cell size, number of bypass diodes and interconnection pattern actually is. The reason for this is the idea, that for a fair rating the shade does not depend on the module properties, and it is an intrinsic property of the PV module how to perform for a given shade situation.

The first two shading patterns can be understood as worst case scenarios, e. g. snow or vegetation covering one of the outermost rows or columns of the PV module over a longer period of time and thus decreasing the module power significantly. A typical mitigation technique for a shading along the long side is the existence of usually three bypass diodes. For standard PV modules, a shading along the short side of the PV module will result in a total loss of power and a potential damage of the PV module. Dividing the PV module in two parallel connected submodules (butterfly design), the shade resistance along the short side is improved, although it is only possible to achieve the same current and voltage of the PV module if the solar cells are cut in half.

Since we may not know the exact circuit diagram of the PV module, each shading measurement should be performed at both sides of the module and the worst result of both shading scenarios should be taken.

The spot shading emulates shading by soiling caused by leaves, bird droppings [6] or similar types of total irradiance blocking. Here, the power loss of the module depends on the solar cell properties as size, shunt resistance and reverse behavior. Without further knowledge about the solar cell characteristics, it is difficult to estimate how spot shading reduces the module power.

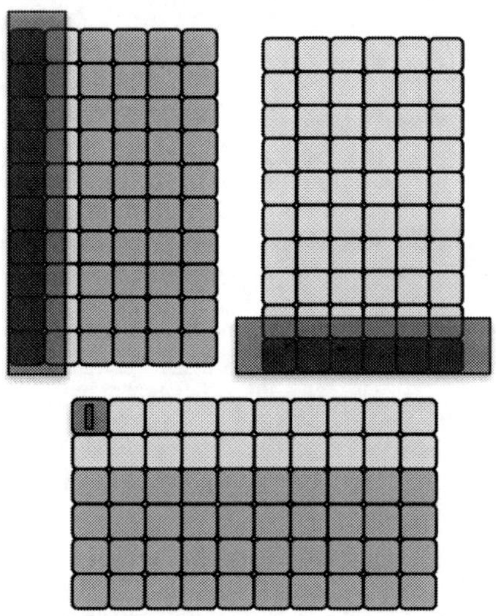

Figure 1: Typical shading patterns on a standard 60 cell PV module denoted as long-side (top left), short-side (top right) and spot (bottom) shading. Dark red cells are completely shaded, light red areas are partially or indirectly affected by the shading and blue areas are not affected by the shade. For larger modules, the shading material would only be adapted by its length to shade the module over the whole length and width. The width of the shading would be kept constant.

The classification of PV modules according to their additional loss *AL* for the different shading patterns is listed in Table I. We define 4 classes from A (very good) D (very bad). Since the long and short side shadings are very similar, they also have the same class limits. The spot shading causes way smaller *AL* values compared to the other patterns, therefore different class limits are chosen.

Table I: Classification scheme for the shading resistance depending on the additional loss *AL*.

Class	long side shading / %	short side shading / %	spot shading / %
A	$AL < 5$	$AL < 5$	$AL < 2$
B	$5 \leq AL < 25$	$5 \leq AL < 25$	$2 \leq AL < 5$
C	$25 \leq AL < 50$	$25 \leq AL < 50$	$5 \leq AL < 10$
D	$AL \geq 50$	$AL \geq 50$	$AL \geq 10$

3 MEASUREMENTS

3.1 Samples

We examined 5 different PV modules and classified their shading resistance according to the proposed scheme. The length and width of the modules ranges from around 169 cm to 177 cm and 99 cm to 114 cm, respectively.

Table II: Samples used to evaluate the classification.

Module	solar cells	bypass diodes
Standard	60	3
Half-cut cells	120	3
IBC cells	104	3
Shingle	340	2
Integrated diodes	60	60

All tested modules offer different mitigations to increase their shading resistance for certain shading patterns. The first module tested is a 60 cell standard module with 3 bypass diodes. The second module is a half-cell module with 120 half-cut cells in total where the module is divided into two parallel connected parts.

The third module is a PV module with 104 interdigitated back contact (IBC) cells arranged in 8 strings á 13 cells. The n-type solar cells are smaller and show a different reverse characteristic compared to standard solar cells [7] enabling an improved shade resistance for small shadings [8]. According to the data sheet it is equipped with 3 bypass diodes with no further explanation if and how the solar cells are divided in individual substrings.

The fourth module is a shingle module with 10 stripes consisting of 34 small solar cells. There are many possible interconnection patterns regarding shingle modules, in this case we assume that there are only two bypass diodes, and that the module is divided in in two parts connected in series while each of the two parts consists of 5 stripes connected in parallel. Due to its size the module does not fit into the light field provided by our solar simulator, therefore the module is only examined by outdoor measurements.

In the fifth module, each solar cell is equipped with a small bypass diode in parallel. These bypass diodes are of unknown type and smaller than typical bypass diodes installed in a junction box. In theory, this module should show the best shading resistance. The data sheet gives no information about additional string bypass diodes.

3.2 Shading material

We used opaque material made from black PVC with the dimensions 200 cm × 25 cm, 120 cm × 25 cm for the shading along the long and short side, and a 9 cm × 3 cm polished aluminum stripe for the spot shading, respectively. Using PVC for the spot shading is not recommended as the solar cell can heat up significantly and deform the shading material. The shading material has been fixed exactly at the edges of the PV module.

3.3 Indoor experiments

For the indoor experiments, we used an LED-type solar simulator of class AAA+. The measurement time of the IV curve was set to 300 ms. Measurements were performed close to standard test conditions with monitor correction using a WPVS reference cell. We did not perform a spectral mismatch correction since we estimated the impact on this relative measurement to be negligible.

The spot shading has been examined for each solar cell of all modules except the shingle module due to its larger size. Before each measurement, an unshaded reference measurement has been performed. The reference measurement has been repeated at the end of the measurement series to check for significant drifts.

3.4 Outdoor experiments

The outdoor measurements have been conducted using two electronic loads with a measurement time of 300 ms. One PV module served as reference while the other module has been shaded using the three shading patterns. The measurements of the long and short side shading have been repeated multiple times without any significant difference in the measurement.

For the spot shading, multiple measurements over several minutes and at multiple positions on the module have been recorded while we measured only at close to clear sky conditions. We did not notice a significant temperature dependence on the calculated *AL*.

Since the test and the reference module can have a different power in the unshaded case, we measured the power of both modules without shade simultaneously and calculated the power ratio to use it as an additional correction factor.

4 RESULTS

4.1 Standard PV module

Figure 2 shows the I-V and P-V curve of the standard PV module under the three different shading situations, but with two different positions for the spot shading. As expected, the power of the standard module decreases by around 33 % if the long side is shaded and 100 % if the short side is shaded.

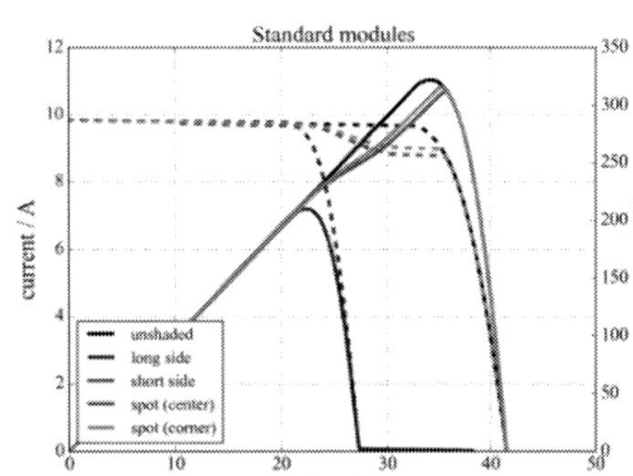

Figure 2: I-V and P-V characteristics of the standard PV module for different shading patterns. As expected, the module does not generate any power if the short side is shaded.

A further examination of the additional loss due to spot shading as function of shaded cell has shown that the *AL* can differ more than 3 % depending on the position in the PV module, see Figure 3, and which can lead to different classifications of the spot shading.

The observed *AL* pattern can be explained with the known irradiance non-uniformity of our solar simulator which shows almost the exact same pattern for a 60 cell module, as shown in Figure 4. The non-uniformity of the irradiance for a 60 cell module determined according to IEC 60904-9:2020 [9] accounts to 1.7 %.

1.0	2.0	2.5	3.0	3.3	3.3	3.1	2.8	2.4	1.3
1.6	2.5	2.8	3.1	3.3	3.4	3.1	2.9	2.5	1.8
1.6	2.4	2.8	3.1	3.3	3.2	3.0	2.7	2.7	1.8
1.9	2.7	3.0	3.2	3.5	3.3	3.3	3.0	2.6	2.0
1.9	2.9	3.2	3.4	3.6	3.7	3.7	3.2	2.7	2.0
2.2	2.8	3.3	3.9	4.2	4.1	4.0	3.3	2.7	1.6

Figure 3: Additional loss as function of position in the standard PV module. The minimum and the maximum AL are shown in white.

We explain residual differences with an additional unknown non-uniformity of the solar cell properties in the PV module. We find this pattern for all examined modules except the IBC cell module, which can be explained by their improved shade resistance for small shadings [8].

1013	1004	995	988	987	987	989	995	1002	1010
1013	1008	1000	997	994	994	997	1001	1009	1014
1015	1010	1005	1001	997	997	1002	1007	1012	1016
1015	1010	1005	1002	997	998	1002	1005	1013	1017
1013	1006	1000	997	994	995	998	1004	1012	1017
1008	1000	995	988	984	984	989	999	1007	1016

Figure 4: Non-uniformity of the light field of our solar simulator for a light field of about 160 cm × 100 cm. The left and right edges and decreased towards the center as well as to the top and bottom edge.

Table III: Experimental results for the standard PV module of the indoor measurement.

	long side	short side	spot
Power loss	34.8 %	100.0 %	1.2 %
Shaded area	24.5 %	13.4 %	0.2 %
AL	10.3 %	86.6 %	1.0 %
Class	B	D	A

4.2 Half-cell module

Figure 5 shows the I-V and P-V curve of the half- cell module under the same shading situations. The effect of the two parallel connected halves results in a power loss of only 50 % compared to 100 % of the standard module if the PV module is shaded along the short side, as shown in Table IV. For the long side shading, the half-cell module is comparable to the standard module.

The *AL* ranges from 3.6 % to 5.5 %. The mean *AL* accounts to 4.4 %. Again, the *AL* non-uniformity has the same pattern as the irradiance non-uniformity of the solar simulator.

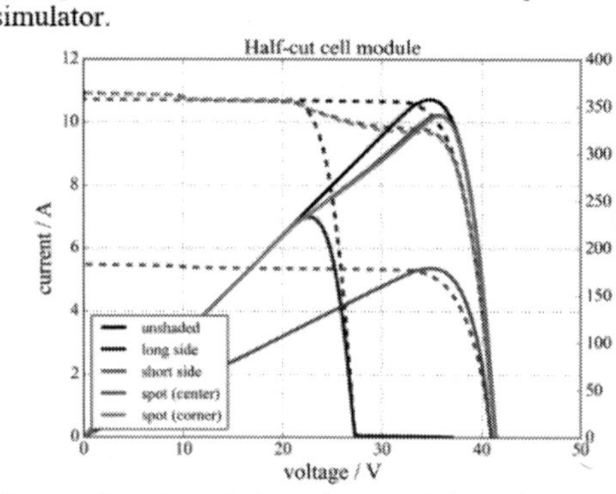

Figure 5: I-V and P-V characteristics of the half-cut module with two and parallel connected halves.

Table IV: Experimental results for the half-cut cell module of the indoor measurement.

	long side	short side	spot
Power loss	34.7 %	49.7 %	3.8 %
Shaded area	23.8 %	14.1 %	0.2 %
AL	10.9 %	35.6 %	3.6 %
Class	B	C	B

4.3 IBC cell module

The I-V and P-V curves of the IBC cell module are shown in Figure 6. This module has a much better and almost ideal shading resistance for the long side shading while the short side shading behaves only slightly better compared the standard module, although an output power of about 15 % of the unshaded case can be observed due to the different reverse characteristics of the solar cells. The spot shading is not as good as one would expect but can be explained with the significantly smaller solar cells compared to a standard module.

The corresponding power losses, calculated additional losses and resulting shading resistance classes are shown in Table V.

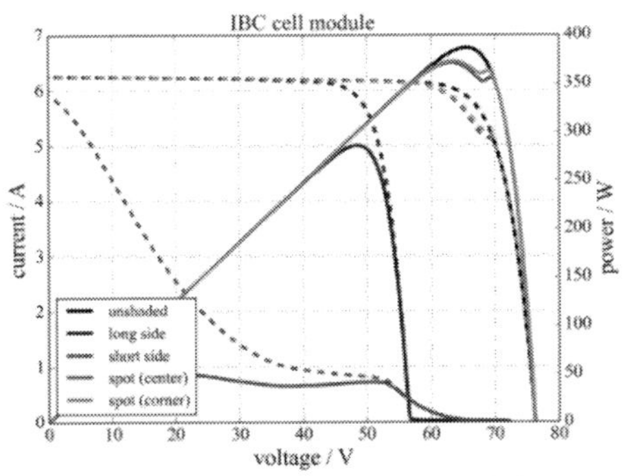

Figure 6: I-V and P-V characteristics of the IBC cell module.

Table V: Experimental results for the IBC cell module of the indoor measurement.

	long side	short side	spot
Power loss	26.1 %	86.7 %	3.8 %
Shaded area	23.9 %	14.8 %	0.2 %
AL	2.2 %	71.9 %	3.6 %
Class	A	D	B

4.4 Shingle module

The I-V and P-V curves of the shingle module are shown in Figure 7. This module shows a shading resistance behavior along the long and short side which is inverse to that of a 60 cell standard module. For the long side shading the power loss accounts to 100 % while the short side shading results in a loss of only 23 %, see Table VI. This demonstrates how this rating scheme can be used to identify the optimum module type for a given mounting situation.

Due to the significantly smaller cell size compared to a full cell, the spot shading leads to a larger power loss resulting in a higher *AL* value of more than 5 %. This is only one possible interconnection of a shingle module and a better shade resistance is e.g. possible by adding more bypass diodes.

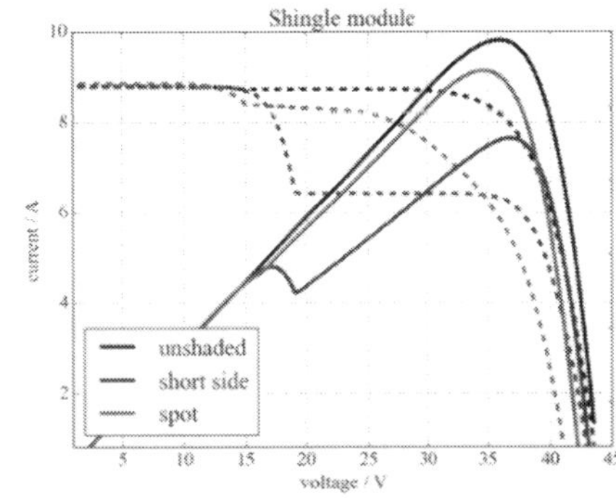

Figure 7: I-V and P-V characteristics of the shingle module. For a better visualization of the characteristics under different shading scenarios, all curves have been scaled towards the Isc of the reference measurement.

Table VI: Experimental results for the shingle module of the outdoor measurement.

	long side	short side	spot
Power loss	100.0 %	22.7 %	5.6 %
Shaded area	21.9 %	14.5 %	0.2 %
AL	78.1 %	8.2 %	5.4 %
Class	D	B	C

4.5 Integrated diodes module

The integrated diodes module shows the best shading resistance of all examined modules as can be seen from the I-V and P-V curves in Figure 8. Although it remains unclear what the properties of this small bypass diodes are and if there are any side effects under unshaded operation, they increase the shading resistance significantly.

The laterally resolved spot shading resistance also resembles the irradiance non-uniformity of the used solar simulator.

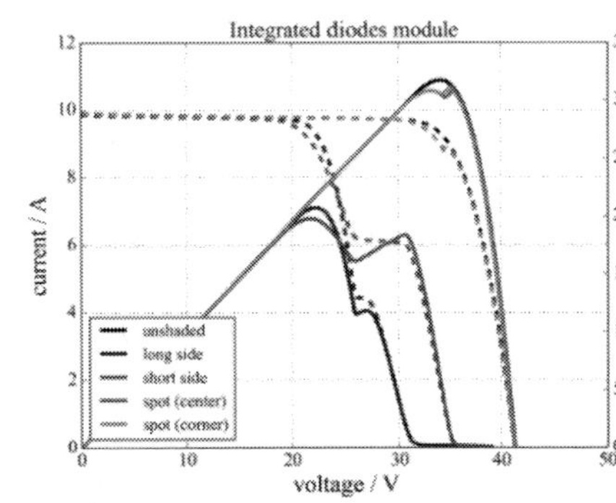

Figure 8: I-V and P-V characteristics of the integrated diodes module.

Table VII: Experimental results for the module with integrated diodes. The first value indicates the indoor, the second value the outdoor measurement.

	long side	short side	spot
Power loss	34.8 %	38.1 %	1.2 %
Shaded area	25.1 %	14.8 %	0.2 %
AL	9.7 %	23.3 %	1.0 %
Class	B	B	A

4.5 Time series of spot shading

We performed time series of the spot shading using natural sun light with assumed perfect uniformity to calculate the *AL*, see Figure 9, to get a better statistic. The resulting *AL* values are within the measured *AL* values of the indoor measurement and suspiciously close to the lowest *AL* of the laterally resolved measurement of the spot shading.

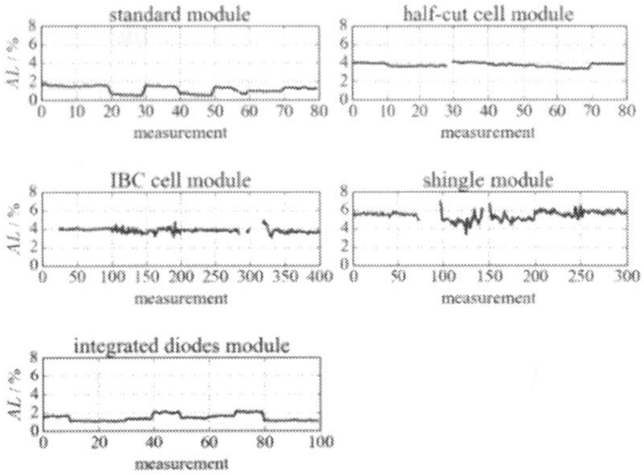

Figure 9: Additional loss *AL* for the spot shading determined from outdoor measurements over a longer period of time and.

The spot shading, as proposed in this metric, aims towards a shading resistance due to solar cell characteristics and its behavior under reverse bias is rated.

Comparing the *AL* for spot shading determined from indoor and outdoor measurements we find a suspiciously good agreement between the lowest *AL* from the indoor measurement and the mean value from the outdoor measurement, for all examined modules. Right now, it remains unclear if the irradiance non-uniformity leads to an increase of *AL* and the lowest *AL* is the correct one, or if more elaborated measures are necessary, e. g. analytically correct the experimental data using the actual irradiance non-uniformity.

We nevertheless think that this type of spot shading is important to rate a PV module regarding its resistance against small area soiling. Our experiments show that different module and cell technologies behave differently for spot shading. Even a spot shading less than 0.2 % of the area leads to a power loss of more than 5 %. Since we keep the spot shading constant (as the dirt does not get smaller when the cells under it are tinier), a smaller solar cell size does not lead to an increased shade resistance.

5 SUMMARY

We presented a classification based on a rating scheme for the additional power loss *AL* of a PV module under shading. We measured the different shading resistance of PV modules by performing indoor and/or outdoor measurements. We showed that with our proposed metric different technologies and designs aimed for an improved shading resistance can be classified accordingly. The corresponding class limits were defined in such a robust way that, on the one hand, slightly different measurement results, e.g. due to measurement uncertainties, do not lead to different classes, but, on the other hand, technological improvements leading to a significant increase in shadow resistance are reflected in a better classification.

Since the power loss for the first two shading patterns is caused by the circuit diagram of the PV module and the implemented bypass diodes, we find a very good reproducibility and agreement between the indoor and outdoor measurement, as can be seen in Figure 10, where all results for indoor and outdoor measurements are shown. The resulting shading classes for all examined modules are listed in Table VIII.

The small observable differences between indoor and outdoor measurements can be explained with the use of different modules for indoor and outdoor measurements, varying temperature and irradiance, small discrepancies while placing and mounting the shading material, and the use of different measurement equipment, respectively. But these differences did not lead to different classifications.

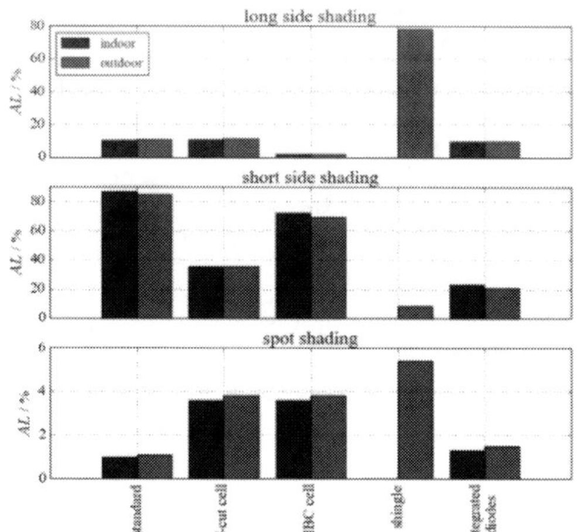

Figure 10: I-V and P-V characteristics of the shingle module. For a better visualization of the characteristics under different shading scenarios, all curves have been scaled towards the Isc of the reference measurement.

Table VIII: Overview of the classification results for all examined PV modules.

	long side	short side	spot
Standard	B	D	A
Half-cell	B	C	B
IBC cell	A	D	B
Shingle	D	B	C
Integrated diodes	B	B	A

Due to the irradiance non-uniformity of the solar simulator and a potential interaction with solar cell properties, determination of the spot shading of one solar cell is challenging. Determined additional losses can vary several percentage points depending on the position within the PV module and leading to an inconclusive classification for spot shading.

The reason for this is the non-uniformity of the solar simulator. But this drawback can be circumvented by making multiple measurements or by taking advantage of the perfect uniformity of natural solar radiation.

Up to now, only Si-based, monofacial PV modules have been examined, our next steps include measurements of thin film, bifacial PV modules and matrix module. In addition, impact on small strings of several PV modules connected in series will be examined.

6 REFERENCES

[1] R. G. Vieira, F. M. U. de Araújo, M. Dhimish, M. I. S. Guerra: *A Comprehensive Review on Bypass Diode Application on Photovoltaic Modules*,

[2] H. Hanifi, J. Schneider, J. Bagdahn: *Reduced Shading Effects on Half-Cell Modules – Measurement and Simulation*, 31st EU PVSEC, 2015

[3] N. Klasen, D. Weisser, T. Rößler, D. H. Neuhaus, A. Kraft: *Performance of shingled solar modules under partial shading*, Prog. Photovolt. Res. Appl. **30**, 2022

[4] A. G. Galeano, M. Bressan, F. J. Vargas, C. Alonso: *Shading Ratio Impact on Photovoltaic Modules and Correlation with Shading Patterns*, Energies **11**, 2018

[5] A. Dolara, G. C. Lazarroiu, S. Leva, G. Manzolini: *Experimental investigation of partial shading scenarios on PV (photovoltaic) modules*, Energy **55**, 2013

[6] A. K. Sisodia, R. K. Mathur: *Impact of bird dropping deposition on solar photovoltaic module performance: a systematic study in Western Rajasthan*, Environmental Science and Pollution Research **26**, 2019

[7] H. Chu , L. J. Koduvelikulathu, V D. Mihailetchi, G. Galbiati, A. Halm, R. Kopecek: *Soft breakdown 1behavior of interdigitated-back-contact silicon solar cells*, Energy Procedia 77, 2015

[8] H. Chu , A. Halm, V D. Mihailetchi, G. Galbiati, R. Kopecek: *Shade performance of a back-contact module assembled with cells featuring soft breakdown characteristic*, Energy Procedia **92**, 2016

[9] International Electrotechnical Commission Photovoltaic devices - Part 9: Classification of solar simulator characteristics (IEC 60904-9:2020)

7 ACKNOWLEDGMENTS

The project 19ENG01 (Metro-PV) leading to this publication has received funding from the EMPIR programme co-financed by the Participating States and from the European Union's Horizon 2020 research and innovation programme.

This is a short version of the paper submitted to the journal Progress in Photovoltaics.

It has been selected by the Committee of the WCPEC-8 for submission to Progress in Photovoltaics.

PV MODULE PERFORMANCE MEASUREMENTS – STATISTICAL ANALYSIS OF TECHNOLOGICAL TRENDS

Ulli Kräling[1], Paul Gebhardt[1], Martin Kaiser[1], Daniel Philipp[1]
[1]Fraunhofer Institute for Solar Energy Systems
Heidenhofstr. 2, 79110 Freiburg, Germany

ABSTRACT: Determining the performance of photovoltaic (PV) modules accurately and in different operational states is key for evaluating their value and basic requirements for yield assessments and bankability considerations. In this publication, we share our experience from the last 10 years of performance measurements and module calibrations at the CalLab PV Modules of Fraunhofer ISE. Our dataset of STC results comprises 61060 modules from 739 different manufacturers (3548 different module types) from North America, Europe, and Asia. We thereby present a valuable reference for various stakeholders in the PV community summarizing changes in performance and efficiency over time. Furthermore, we present and discuss technology trends, including the effect of module design aspects such as number of ribbons, wafer size, number of cells and cell technology on the performance at standard conditions, module parameters such as temperature coefficient and low light performance.
Keywords: PV module, efficiency, low light behavior, temperature coefficient, technology trends

1 INTRODUCTION

Determining the performance of photovoltaic (PV) modules accurately and in different operational states is a key quality indicator and a basic requirement for yield assessments and bankability considerations. These measurements are carried out according to standards IEC 60904-1 [1], IEC 60891 [2], and IEC 61853-1 [3]. At CalLab PV Modules, located at the Fraunhofer Institute for Solar Energy Systems ISE in Freiburg, Germany, we have carried out these measurements on a large scale for more than a decade for different purposes like calibration and quality control.

Here, we present performance data of a large set of PV modules from the last 10 years, as measured according to international standards IEC 60904-1 [1], IEC 60891 [2], and IEC 61853-1 [3]. This data could potentially serve as reference when evaluating or modelling the effect of different module technologies, or help estimating the effect on the energy rating or energy yield, since the discussed data is typically the input data for energy rating calculations.

The given data and analyses can serve as reference for different stakeholders of the PV community: Researchers can use it as indication of performance and weak light behavior for current and past modules, e.g., for theoretical considerations. Likewise, the presented data can be used to validate modelled/projected technology improvements [4].

2 METHODOLOGY

Fraunhofer ISE's accredited laboratory CalLab PV Modules produces measurement results for its customers with highest accuracy since more than a decade [5]. Beside the measurement results, basic characteristics of each module, e.g., manufacturer, module type, module dimensions, and cell material are recorded. More details were added in the recent years (e.g., number of cells, cell dimensions, and ribbons). So, for recent measurements, more module characteristics are available than for measurements 5 years ago.

All these results are used in an anonymized way to analyze the development of the discussed module characteristics in the past ten years.

2.1 Initial data set

The evaluation of the data is based on the period 2011/12 – 2022/08. Within this period, all measurement results were evaluated using the same inhouse developed software.

Different types of measurements are considered in this analysis: Measurement at Standard Test Conditions (25 °C, 1000 W/m²) according to IEC 60904-1 [1]; measurement at 25 °C and varying irradiance levels from 100 W/m² to 1100 W/m² according to IEC 60904-1 [1] and IEC 61853-1 [3]; measurement of temperature coefficients at 1000 W/m² according to IEC 60891 [2] and IEC 61853-1 [3]. Measurement types and conditions and the used abbreviations are shown in Table I.

Table I: Measurement types, abbreviations, and measurement conditions

Measurement type	abbr.	Temperture [°C]	Irradiance [W/m²]
Standard test condition	STC	25	1000
Irradiance dependency	G	25	100 - 1100
Temperature coefficient	TC	15 - 75	1000

The measured modules considered in this analysis are property of Fraunhofer ISE's customers and belong to different types of projects with different objectives. This includes large sets of identical modules of the same type, e.g., from a PV power plant or single modules, e.g., golden modules in a production line. Beside the number of modules per type (i.e. product name), also the number of measurements for an individual module varies because the evaluation of light-induced degradation (LID), light- and elevated temperature-induced degradation (LETID) [6], potential-induced degradation (PID), and others, requires module measurements at different states of degradation.

To make measurement results comparable, several filters were applied. Thus, only measurements of the same state are extracted and the imbalance between large and small groups of identical measurement objects is eliminated. So, the final data set still represents the variety of all modules measured at Fraunhofer ISE's module calibration lab CalLab PV Modules. The filtering steps are described in detail in the following section.

2.2 Inconsistency filter

The initial filter sorts out measurement results with inconsistent data. E.g., target temperature and target irradiance must align with the designated measurement type. The serial number of the device under test (DUT) in the results information is mandatory to identify repeating measurements. Measurements of the rear side of bifacial modules are also excluded. Finally, data sets with physically impossible results are excluded, like negative short circuit current I_{SC}, negative open circuit voltage V_{OC} or fill factor beyond 100 %. Those results might come from measurements of defective modules (e.g., due to transportation damage).

2.3 Initial measurement filter

The 2nd filter only preserves the first measurement – of a specific measurement type like STC – of a DUT within a project. The intention is to get only results at initial state of the module. This is the only comparable state over a large set of data because the intermediate state of a module depends on the individual schedule and focus of a project and therefore differs for each project. This also means that the herein presented data was acquired before any stabilization, e. g. by light soaking.

2.4 First appearance filter

The 3rd filter removes data sets of recalibrations and other re-measurements. Typically, golden modules are sent to our lab to be recalibrated in regular time intervals. Then, the identical module refers to different projects respectively orders. To avoid double counting of these modules, only the first appearance of a serial number of a specific module is valid and all subsequent remeasurements of a module are sorted out.

2.5 Imbalance filter

In the next step, from each module type in a project, only the measurement results closest to the average power (respectively the average TC_{Pmpp} for TC measurements) for this module type is considered. Hence, the imbalance due to different numbers of modules per project is eliminated. In addition, this paper is focused on general and representative trends in PV module development and not in the presentation of highest possible results.

3 RESULTS AND DISCUSSION

3.1 Filter characteristics

61060 STC measurements, 1382 irradiance dependence measurements (G) and 1381 temperature coefficient measurements (TC) in the period 2011/12 – 2022/08 were analyzed. By different filtering algorithms described in the previous section, data sets were significantly reduced to 9.7 % for STC, 40.9 % for G and 58.1 % for TC measurements compared to the initial available data sets. The available data sets after the individual filtering steps are shown in Table II.

Table II: Number of available data sets after several filtering steps for STC, irradiance dependence (G) and temperature coefficients (TC) measurements and ratio of filtered data.

Filter steps	STC	G	TC
0 Initial data set	61060	1382	1381
1 Inconsistency filter	58282	1243	1320
2 Initial measurements	37066	979	1217
3 First appearance filter	36357	964	1204
4 Imbalance filter	5940	565	803
Ratio (filtered/initial)	9.7 %	40.9 %	58.1 %

The filtering was also analyzed for mono-Si and poly-Si modules separately (Table III). The filter ratio for mono-Si and poly-Si modules was nearly identical for STC. For G and TC, the ratio for mono-Si was about 5 % points lower than for poly-Si.

Table III: Filtering results for mono-Si and poly-Si modules for different measurement types.

Standard test conditions (STC)	poly-Si	mono-Si
Available	26848	28155
Filtered	2618	2740
Ratio	9.8 %	9.7 %

Irrandiance dependency (G)	poly-Si	mono-Si
Available	544	682
Filtered	243	271
Ratio	44.7 %	39.7 %

Temperature coefficient (TC)	poly-Si	mono-Si
Available	612	652
Filtered	377	371
Ratio	61.6 %	56.9 %

In Figure 1, the filtering of the data is shown for STC results for mono-Si modules. The different color grades indicate the different intermediate filtering steps, from initial data set (light gray) to filtered data (dark blue). After filtering, a clear trend over the years towards higher efficiencies is already visible as will be discussed in detail later.

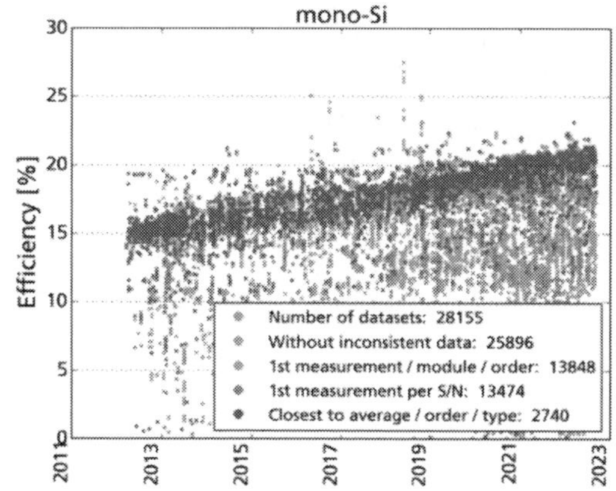

Figure 1: Visualization of data filtering for mono-Si PV modules. Measured efficiencies at STC are shown vs.

measurement date. The different colors indicate the different filtering levels.

As indicated in Table II, filtering reduced the number of data sets, e.g., for STC to 9.7 % of the initial data sets. Nevertheless, more than 92 % of module types and producers are still available in the filtered data sets as shown in Table IV. Hence, the diversity of producers and modules could be preserved after filtering and the imbalance of different sample sizes and re-measurements was eliminated. 686 different PV module producers and 3508 different PV module types remain after filtering for evaluation. The inconsistency filter causes the reduction of producers and module types in the data set.

Table IV: Available number of producers and module types in the data sets before and after filtering for different measurement types

Producers	STC	G	TC
Available	739	156	194
Filtered	686	145	187
Ratio	92.8 %	92.9 %	96.4 %

Module types	STC	G	TK
Available	3548	511	693
Filtered	3508	479	687
Ratio	98.9 %	93.7 %	99.1 %

3.2 Technology trends

Beside the measurement results, the documented module characteristics give an overview of the technological development over the past 10 years. As an independent accredited calibration and testing lab, CalLab PV Modules is not involved in the module development and has therefore only limited access to detailed information about the module components and related technologies. Therefore, the analysis is mainly focused on visible characteristics and information stated on the module label or datasheet values provided by the customer. The presented results in this section are based on the filtered STC results.

The number of ribbons per cell was monitored from 2016 on. Figure 2 shows the share of different ribbon numbers for each year. Until 2020, modules with 5 or less ribbons were dominating for mono-Si modules with a share of more than 50 %. Since then, the share drastically dropped and modules with 9 or 10 thinner ribbons or round wires took over the lead [7].

For poly-Si modules, the 5-ribbon technology is still dominating. The reason for this difference to mono-Si modules could be the lack of process development for poly-Si modules: The transition to new interconnection technologies is most likely hindered by the generally smaller focus and efforts on multi-Si modules.

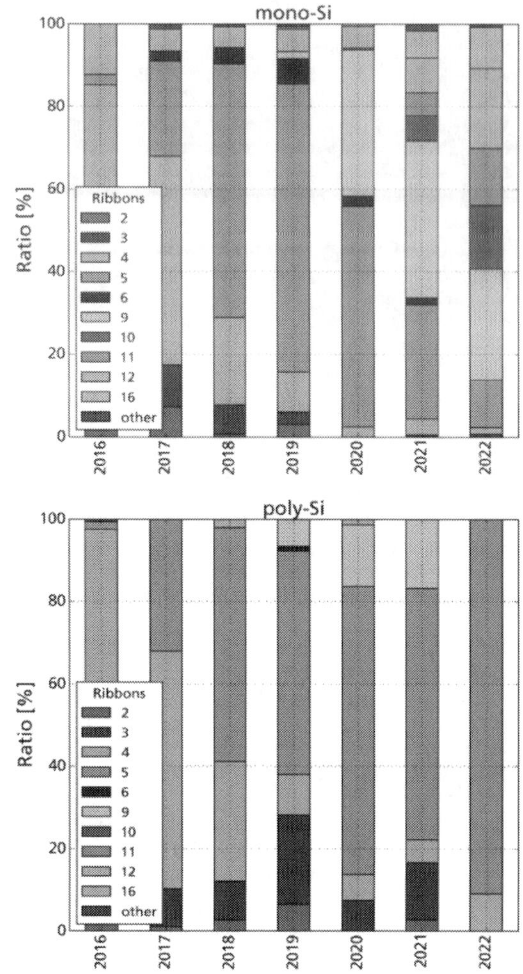

Figure 2: Share for different ribbon numbers per cell for mono-Si and poly-Si PV modules

The development of the module formats and cell count did not change for a longer period. The share of 60-cell and 72-cell modules was almost constant from 2012 until 2017 for mono-Si modules at 80 % (Figure 3). Other module layouts with typically less than 60 cells played a minor role. But since 2017, the dominating formats were steadily replaced by half-cell modules with 120 and 144 cells. The overall share of these full-cell and half-cell modules remains above 80 % until 2021. In the last months, many diversified module concepts with various cell counts and sizes arose.

For poly-Si modules, the classical layout with 60 and 72 full-cells is still dominating with more than 70 % to date. The share of the corresponding half-cell modules with 120 and 144 cells increased over the years. Other formats for poly-Si modules are barely relevant.

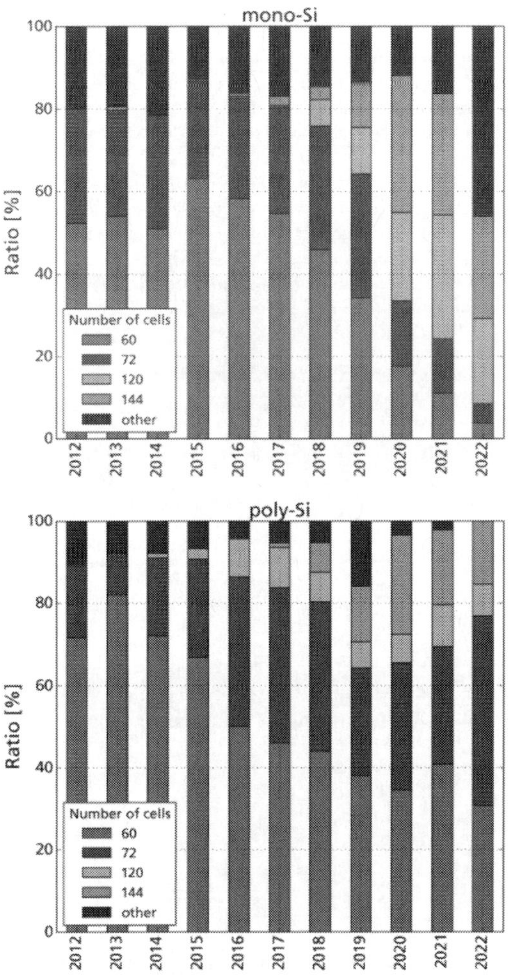

Figure 3: Share for modules with different number of cells for mono-Si and poly-Si PV modules

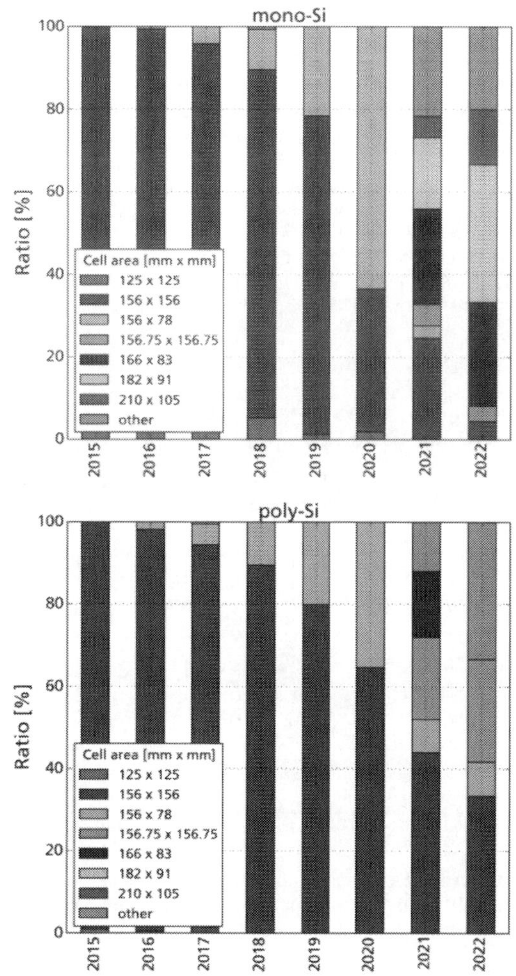

Figure 4: Share for cell dimensions for mono-Si and poly-Si PV modules

The cell sizes used in PV modules changed drastically in the last years (Figure 4). While for both, mono-Si and poly-Si PV modules, the standard cell size of 156 mm and 156.75 mm (M0 and M2 wafer) dominated with a share of about 80 % until 2019, the trend towards half-cut cells and larger wafer geometries changed the share. In 2020, the share of half-cut cells of 156 mm x 78 mm (M0/M2 wafer) raised and was in the following years replaced by half-cut cells of 166 mm x 83 mm (M6 wafer), 182 mm x 91 mm (M10 wafer) and 210 mm x 105 mm (M12 wafer) size.

For poly-Si modules, the full-cell formats of M0 wafers and M2 wafers (156.75 mm x 156.75 mm) still have a share greater 50 % to date.

The share of other cell formats is raising up to about 20 % share, but despite the larger diversity, none of the module types with e.g., M3 or M4 wafers or modules with third-cut cells have a significant share. So, all these formats are summarized in category "other".

3.3 STC results

5358 STC measurements of mono-Si and poly-Si PV modules were filtered and analyzed. Mono-Si modules were divided in 2 subcategories: Silicon-Heterojunction (SHJ) and others, which is indicated by "not specified". There is no differentiation between passivated emitter and rear cell (PERC), aluminum back surface field (Al BSF), TOPCon, MWT, IBC or other cell-technologies because the information was not available for most of the modules.

Figure 5 and Figure 6 show the development of module efficiency and fill factor (FF) over the last 10 years. The lines indicate the measured median for each category per year. The colored band indicates the efficiency interquartile range (IQR) from 25 % percentile to 75 % percentile, which represents 50 % of the results. The mono-Si modules improved almost constantly from 15 % in 2012 to 20 % in 2022, which is an efficiency improvement of 0.5 %-points per year. The efficiency of mono-Si SHJ Modules is even about 0.6 % higher over the last years.

The efficiency gain over the years for poly-Si modules is less than for mono-Si modules and converged at around 17 % since 2020. From 2012 to 2020, the efficiency gain was 0.3 %-points per year. A reason for the stagnation of the efficiency could be the switch of manufacturers towards mono-Si modules and their higher potential towards higher efficiencies. Hence, the investments in the development of poly-Si modules were reduced. This is supported by the share of measured poly-Si PV module types which decreased drastically since 2018 whereas the number of mono-Si module types in the lab is constantly high.

FF increased from about 75 % in 2012 to almost 80 % in 2022 for mono-Si modules. The values are slightly higher for mono-Si SHJ modules. For poly-Si modules, FF remains below 78 % in the last years. The use of half-cut cells [8] is one example how the FF was improved in the last years.

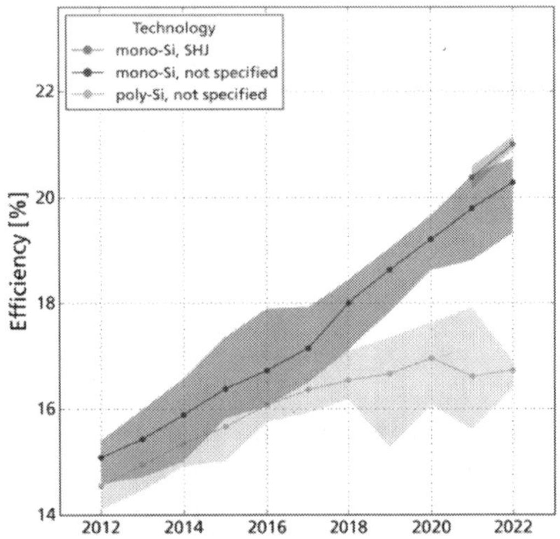

Figure 5: Development of the module efficiency over the last 10 years; the median and the interquartile range (IQR) are shown for mono-Si SHJ, mono-Si, and poly-Si modules

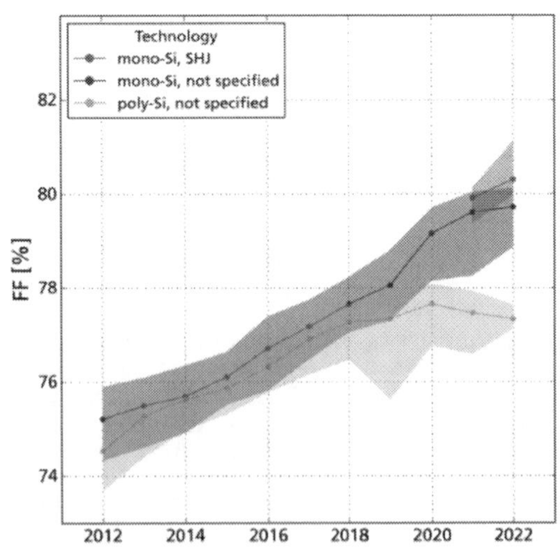

Figure 6: Development of the module FF over the last 10 years; the median and the interquartile range (IQR) are shown for mono-Si SHJ, mono-Si, and poly-Si modules

Figure 7 shows the V_{OC} for mono-Si modules. A trend towards higher V_{OC} is visible for all types. The cell interconnection of half-cell modules yields to the same V_{OC} as for full-cell modules. Therefore, 60-cell and 120-cell modules respectively 72-cell and 144-cell modules have nearly the same V_{OC}. The gap between those two V_{OC} levels can be explained by the difference of in-series connected cells, which define the module V_{OC}.

The larger fluctuations mainly of the 120-cell modules in the last years can be explained by the mixture of SHJ modules and other technologies in this graph. The SHJ technology yields higher V_{OC} values about 44.5 V (742 mV per cell) compared to 41.3 V for other technologies (688 mV per cell) [9].

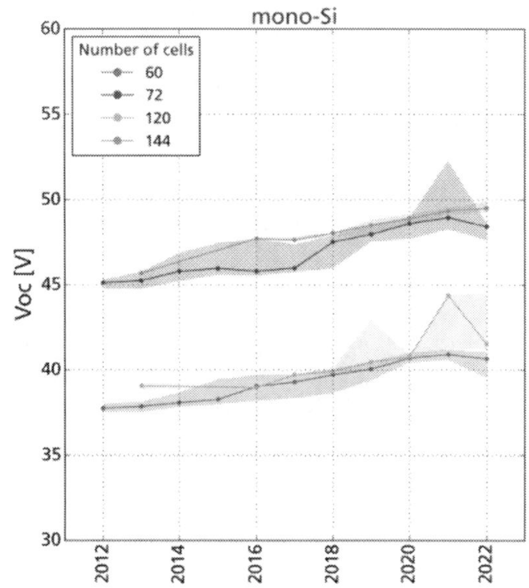

Figure 7: Development of the module open circuit voltage (V_{OC}) over the last 10 years; the median and the interquartile range (IQR) are shown for mono-Si modules with 60, 72, 120, and 144 cells

Beside the trends over time, it is interesting to filter the measured data by some module characteristics. In Figure 8, a boxplot shows the module efficiencies for different number of ribbons for mono-Si and poly-Si modules. The boxplot shows the median, the interquartile range (IQR), and the whiskers. The lower whisker limit Q1 - 1.5 * IQR and the upper whisker limit Q3 + 1.5 * IQR define a range of 99.3 % of the measured data.

With increasing number of ribbons, the module efficiency increases. This trend correlates with the efficiency over time as presented before because the cell design was developed over the years starting from 2-ribbon cells towards 9- or 10- ribbon cells respectively multiwire technology with more than 10 wires nowadays. Of course, the number of ribbons is not affecting the increase in efficiency directly. Larger cells as well as optimized cells yield to higher currents but resulting in increased electrical losses in the cell interconnections. Therefore, the optimization of the current distribution in the cell lead to more ribbons [4]. In addition, the underlying cell technology changed as well over time from BSF cells towards e.g., PERC cells and SHJ cells. Hence, the number of ribbons is not the cause for a gain in efficiency but comes along with the cell improvement and the optimization of the module design e.g., by the cell-to-module analysis [10]. But the number of ribbons clearly indicates the development stage of the module respectively the cell technology.

The graph also shows that the diversity of used ribbons for poly-Si modules is less compared to mono-Si modules. The multiwire approach with 10 or more round wires did not occur for poly-Si modules.

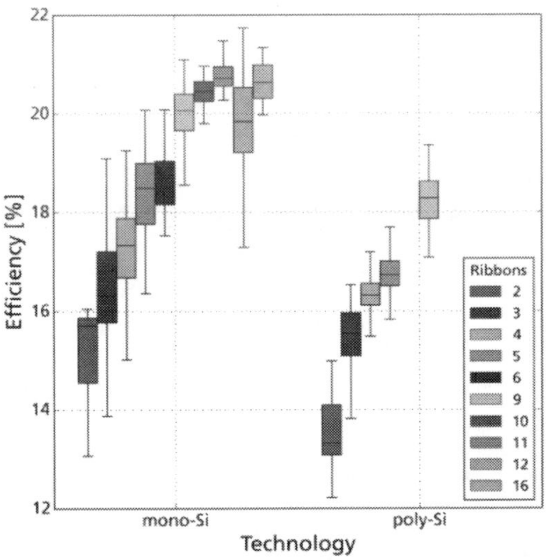

Figure 8: Boxplot of the module efficiency for mono-Si and poly-Si for different number of cell ribbons

In Figure 9, it is shown that modules with half-cut cells (120 and 144 cells) yield higher module efficiencies compared to full cell modules with 60 and 72 cells for both, mono-Si and poly-Si modules. The reason are the lower resistive losses in the module due to reduced internal currents per cell [8]. The large difference in the whisker size for full-cell mono-Si modules can be explained by the fact that full-cell modules were measured in 2012 and are still available in 2022. So, the efficiency gain over the years is summarized in the boxplot. Whereas half-cell modules entered the market about 5 years ago and where already higher efficiencies were state-of-the-art. Nevertheless, the potential for higher efficiencies for half-cell modules is also indicated by the upper whisker limit.

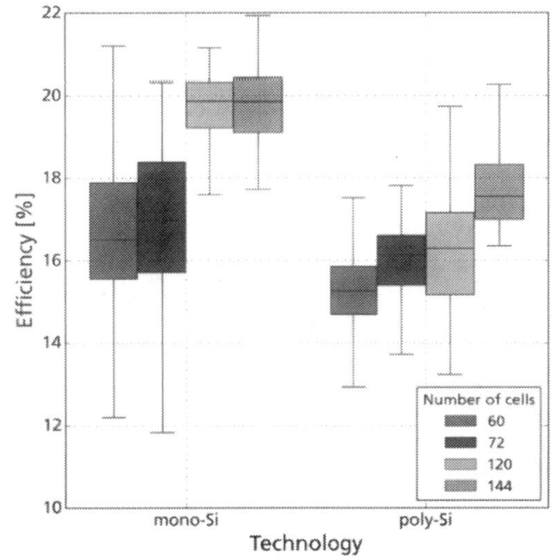

Figure 9: Boxplot of the module efficiency for mono-Si and poly-Si for different number of cells

3.4 Low light behavior

The trend of low light performance relative to STC over the years shows a negative development for mono-Si modules at first glance (Figure 11). However, as the comparison with the absolute efficiencies shows (Figure 10), the effect of generally higher efficiencies of modules from recent years clearly outweighs the low light behavior – meaning that newer modules, on average, outperform older modules despite seemingly worse low light behavior due to their absolute efficiency increase. This is important to consider when comparing PV module datasheets.

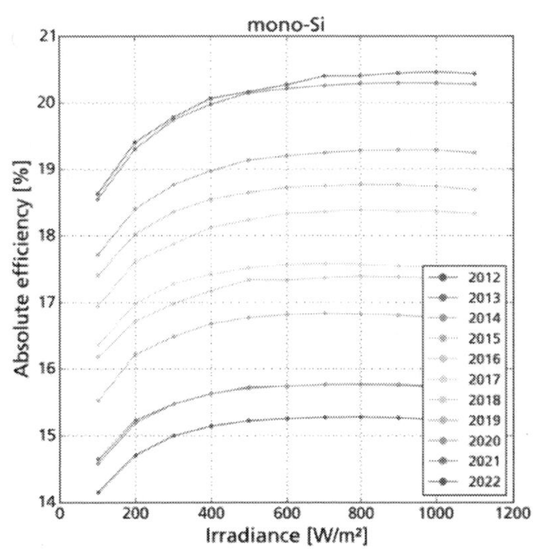

Figure 10: Yearly median of absolute efficiencies vs. irradiance for mono-Si modules

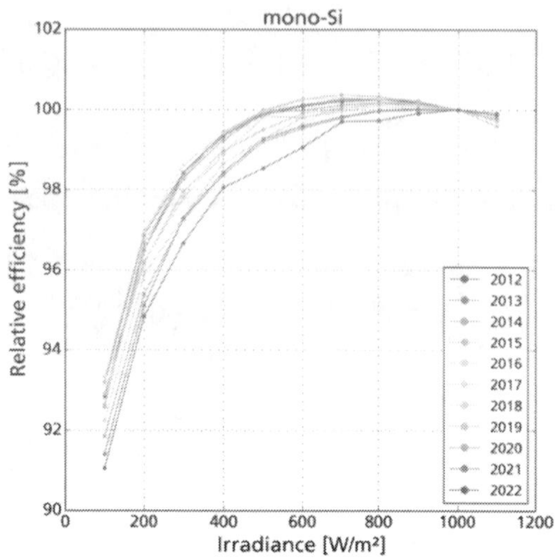

Figure 11: Yearly median of relative efficiencies vs. irradiance for mono-Si modules, normalized to the efficiency at STC

The absolute efficiencies for poly-Si modules are lower compared to mono-Si modules (Figure 12). However, the same negative trend for relative low light performance is visible (Figure 13).

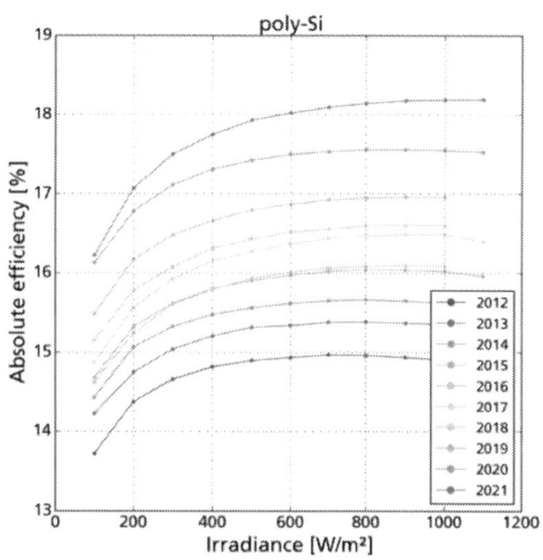

Figure 12: Yearly median of absolute efficiencies vs. irradiance for poly-Si modules

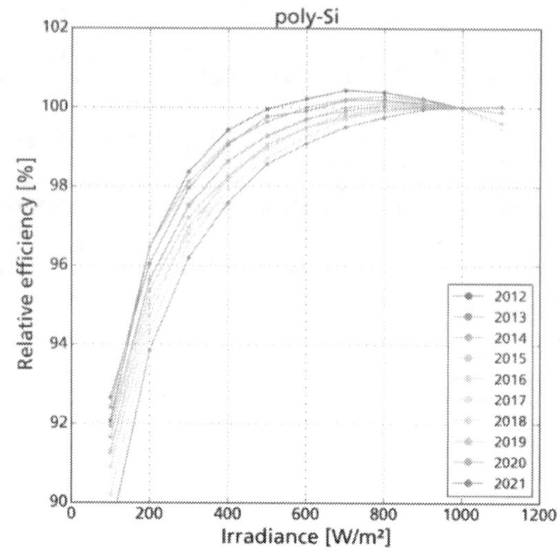

Figure 13: Yearly median of relative efficiencies vs. irradiance for poly-Si modules, normalized to the efficiency at STC

3.5 Temperature coefficient results

The temperature coefficient of power at maximum power point TC_{Pmpp} (Figure 14) shows an absolute improvement over the years. This trend is most likely connected to the usage of newer cell technologies. Modules with SHJ cells, which appeared in the last years, have an even better performance than other mono-Si modules. 10 years ago, the poly-Si modules outperformed the TC_{Pmpp} of mono-Si modules. This trend changed in 2017/18.

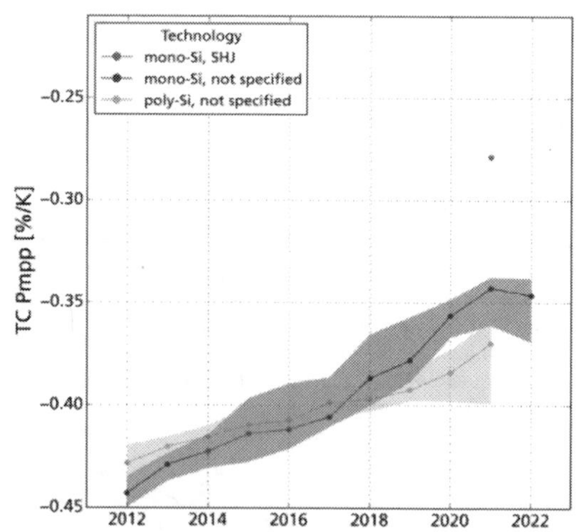

Figure 14: Development of the relative TC_{Pmpp} over the last 10 years; the median and the interquartile range (IQR) are shown for mono-Si SHJ, mono-Si, and poly-Si modules

The temperature coefficient of the open-circuit voltage TC_{Voc}, which is also strongly related to the cell development [9], did also improve over the years for mono-Si modules. For poly-Si modules, an improvement is also visible but on a lower level. The SHJ modules outperform the other technologies with the highest TC_{Voc}.

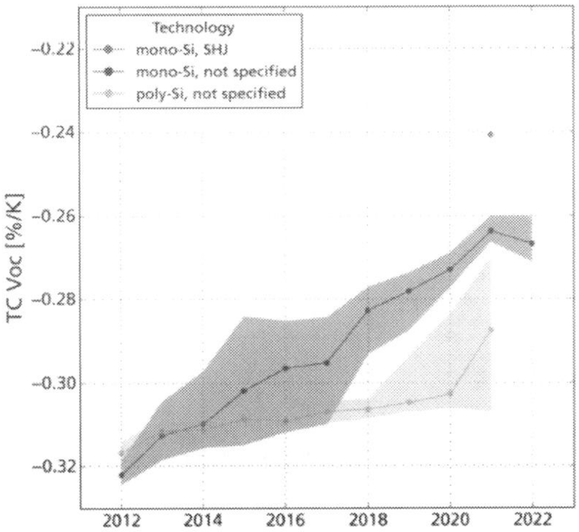

Figure 15: Development of the relative TC_{Voc} over the last 10 years; the median and the interquartile range (IQR) are shown for mono-Si SHJ, mono-Si, and poly-Si modules

Temperature coefficients for short-circuit current TC_{Isc} are very small and therefore the degree of uncertainty is normally higher than for the other temperature coefficients. Nevertheless, a clear distinction between mono-Si and poly-Si modules with higher coefficients is visible. It is noticeable, that from 2018/19, the trend towards higher coefficients was reversed.

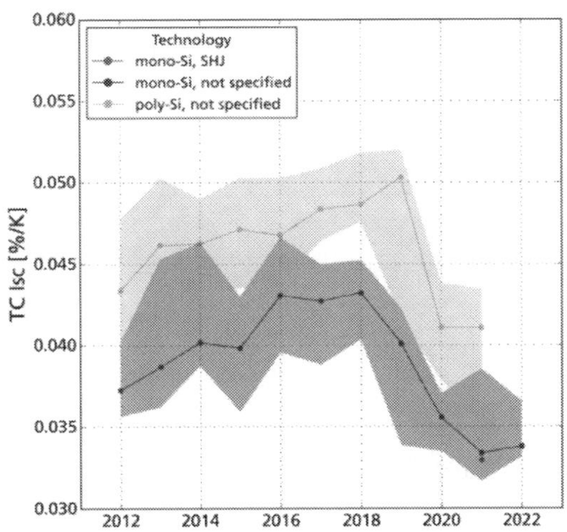

Figure 16: Development of the relative TC_{Isc} over the last 10 years; the median and the interquartile range (IQR) are shown for mono-Si SHJ, mono-Si, and poly-Si modules.

4 CONCLUSION

In this work, we presented a methodology to make a large data set of measurements with high variability comparable. Multiple filters were applied to reduce the imbalance introduced by the variability of the data that was accumulated of our day-to-day business to convert it to a more expressive dataset. For STC, only 9.7 % of the data were available after filtering (40.9 % for G, and 58.1 % for TC). Nevertheless, the diversity of module types and producers in the filtered data set was maintained with 92.8 % (92.9 % for G, and 96.4 % for TC). Due to the large number of module types and producers in this evaluation, the shown graphs can therefore be interpreted as quite representative although all measurements were performed only in one laboratory which does certainly not cover the full market.

The data was analyzed to identify technological trends as far as client confidentiality allowed. Various technological trends were shown like the development from 2-ribbon cell interconnection towards 9-ribbon and multiwire interconnection replacing older technologies. The module layout with 60 or 72 full cells which was the standard for years, was successively replaced by half-cut cells with 120 and 144 cells per module. These modules represent the major share now. But increasing wafer formats and new module concepts are going to yield modules with wider variations in the number of cells in the upcoming years especially or mono-Si modules.

A constant increase of the absolute efficiencies even at low light conditions was shown for both, mono-Si and poly-Si modules. In contrast, the relative low light efficiencies decreased over time. This fact plays an important role e.g., in judgement and comparison of datasheets of different modules.

Along with the continuous development over cell level (from BSF to the introduction of passivation layers and SHJ), the TC for the modules improved. Modules with SHJ cells outperform other crystalline modules regarding the TC values.

We hope that the shown data and discussed trends can serve as reference, e.g., to evaluate if a module is state-of-the-art or fits to a certain year, to classify or rank module. Surely care must be taken with drawing general conclusions, especially regarding later trends of technical innovations. The fact that not all technological innovations could be resolved in this study and not all new trends reached the same grade of technical optimization must be considered. The study does not replace the individual characterization and evaluation of the module type of interest.

Furthermore, the data can be used as input for simulations to compare modules, e.g., CSER, which we plan to focus on in a follow-up publication.

In contrast to other publications on technological trends, such as the ITRPV roadmap [11], our work is not limited to mere estimations and opinions of the questionnaire participants but based on actual measurements and lessons learned from our daily lab work. Although inducing its own limitations (e. g. due to the large but still limited dataset) and remaining uncertainty about how representative our data is to the whole market, it still supplies a valuable resource from a different viewpoint to forecast future technology trends. We thus allow an alternative approach to support these considerations, because analyzing the speed and effects of technological changes that occurred in this industry's past can also educate assumptions on future developments.

5 ACKNOWLEGEMENTS

The authors of this work would like to thank all coworkers from Fraunhofer ISE's CalLab PV Modules, present and past, for performing the numerous measurements.

6 REFERENCES

References

[1] International Electrotechnical Commission (IEC). Photovoltaic devices – Part 1: Measurement of photovoltaic current-voltage characteristics. 3rd ed. (IEC 60904-1: 2020-09). Geneva, Switzerland: International Electrotechnical Commission (IEC); 2020.

[2] International Electrotechnical Commission (IEC). Photovoltaic devices – Procedures for temperature and irradiance corrections to measured I-V characteristics. 3rd ed. (IEC 60891: 2021-10). Geneva, Switzerland: International Electrotechnical Commission (IEC); 2021.

[3] International Electrotechnical Commission (IEC). Photovoltaic (PV) module performance testing and energy rating - Part 1: Irradiance and temperature performance measurements and power rating. 1st ed. (IEC 61853-1: 2011-01). Geneva, Switzerland: International Electrotechnical Commission (IEC); 2011.

[4] Tummalieh A, Pfreundt A, Mittag M. Trend Tracking of Efficiency and CTM Ratio of PV Modules. In: 37th EU PVSEC 2020; 2020.

[5] Fraunhofer Institute for Solar Energy Systems ISE. Fraunhofer ISE's CalLab PV Modules Improves Measurement Uncertainty to Record Value of 1.1 %. Freiburg, Germany; 2020.

[6] Fokuhl E, Philipp D, Mülhöfer G, Gebhardt P. LID and LETID evolution of PV modules during outdoor operation and indoor tests. EPJ Photovolt. 2021; 12: 9, DOI: 10.1051/epjpv/2021009.

[7] Walter J, Rendler LC, Ebert C, Kraft A, Eitner U. Solder joint stability study of wire-based interconnection compared to ribbon interconnection. Energy Procedia 2017; 124: 515–25, DOI: 10.1016/j.egypro.2017.09.288.

[8] Guo S, Singh JP, Peters IM, Aberle AG, Walsh TM. A Quantitative Analysis of Photovoltaic Modules Using Halved Cells. International Journal of Photoenergy 2013; 2013: 1–8, DOI: 10.1155/2013/739374.

[9] Steinkemper H, Geisemeyer I, Schubert MC, Warta W, Glunz SW. Temperature-dependent modeling of silicon solar cells—Eg, ni, recombination, and VOC. IEEE J. Photovolt. 2017; 7(2): 450–7, DOI: 10.1109/JPHOTOV.2017.2651822.

[10] Hädrich I, Eitner U, Wiese M, Wirth H. Unified methodology for determining CTM ratios: Systematic prediction of module power. Sol Energ Mat Sol C 2014; 131: 14–23, DOI: 10.1016/j.solmat.2014.06.025.

[11] ITRPV. International Technology Roadmap for Photovoltaic (ITRPV): 13th edition, 2021 Results; 2022.

MODELLING AND MEASURING THE POWER OUTPUT OF VERTICAL BIFACIAL SOLAR PANELS IN NORDIC CONDITIONS

Sami Jouttijärvi[1], Mehmet Tok[1], Lauri Karttunen[1], Hugo Huerta[2], Samuli Ranta[2], Kati Miettunen[1]
[1]Department of Mechanical and Materials Engineering, University of Turku, Vesilinnantie 5, 20500 Turku
[2]New Energy group, Turku University of Applied Sciences, Joukahaisenkatu 7, 20520 Turku

ABSTRACT: Predicting the power production of solar photovoltaic (PV) panels is necessary for successful commissioning and operation of PV power systems. Lifetime estimations are required to analyze commercial viability of the planned PV system, whereas hourly and daily production estimations are important considering operation in the electricity market. Here, a specific setup, consisting of vertically mounted bifacial solar panels in a high-latitude location, is monitored. Measured power production is compared to modelled productions achieved by several simple modelling approaches. The aim is to analyze how specific and high-quality data is actually needed to model accurately the total energy production in several temporal scales, ranging from one-minute to multi-year. The results help to understand the required complicity for pre-installation production estimations and thus contribute to increasing bankability of vertical bifacial PV systems.
Keywords: Energy performance, Modelling, PV module

1 INTRODUCTION

Predicting accurately power production of solar photovoltaic (PV) systems in both long- and short-term is crucial to advance commissioning of PV. This work focuses on a specific setup of vertically mounted bifacial PV (VBPV) system located in Turku, Finland (60°N, 22°E). The main benefit of VBPV is the change in power production profile: while monofacial, south-facing PV panels (MPV) produce one high peak around noon, VBPV produces two smaller peaks, at morning and evening, as shown in Figure 1. This provides better match with electricity consumption and maximizes the PV power production during high-price hours [1].

The power production of the studied panel is measured directly and modelled with several different approaches. The most important variable is the incident irradiation. Two different approaches are used: 1) measured global tilted irradiance (GTI) and 2) modelled GTI based on measured global horizontal irradiance (GHI). The irradiance-to-power conversion is done a) with 6k regression model [2] and b) with temperature-corrected standard test condition (STC) front- and rear-side efficiencies reported by the manufacturer ($\eta_{STC,F}$ = 17.7% and $\eta_{STC,R}$ = 15.9% respectively). The bifaciality factor (BF), defined as a fraction of rear- and front-side efficiencies, was 90%.

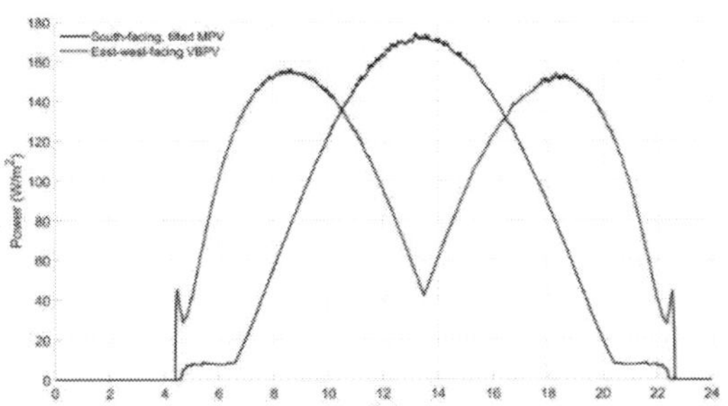

Figure 1: Power production profiles of south-facing, tilted (30°) MPV and east-west-facing VBPV panels on a sunny summer day in Turku, Finland.

The one-minute modelled power production datasets gained with different approaches are compared to measured production. Aggregated energy production in time scales varying from one-hour to multi-year are compared as well. The aim is to determine how large additional errors will be caused by insufficient data availability and how much these errors are reduced with data aggregation.

Figure 2: The vertical bifacial PV test setup located in Turku, Finland. The monitored panel is shown with the blue rectangle. Green circles show the locations of the monitor cells (top) and thermocouple (at front cover of the panel). Original photograph: Nyberg/UTU.

2 METHODS

2.1 Data collection

Data was collected in Turku with two different setups located at each other's close proximity. Weather station measured ambient temperature (T_A), wind speed (WS) and global horizontal irradiance (GHI). The VBPV test setup (Figure 2) measured power production (P_{MEAS}) of a vertically mounted bifacial solar panel (PrismSolar Bi60-375BSTC). Moreover, the panel temperature ($T_{M,MEAS}$) was measured with a thermocouple and global tilted irradiance (GTI) incident on east- and west-oriented vertical surfaces ($I_{E,MEAS}$ and $I_{W,MEAS}$) with monitor cells. The resolution of the raw data was one minute. The setup is shown in Figure 2 and described in more detail elsewhere [3].

2.2 Data processing

The datasets were combined so that only the period when all variables mentioned in Section 2.1 were available (Jan 2018 – Sep 2020) was included and the timestamps were full minutes (e.g. 2018-07-25 17:35:00). This was done by linear interpolation of two adjacent datapoints. If at least one of the adjacent datapoints in the raw data was missing, the converted datapoint was marked as missing. Only those timestamps which had valid datapoints for all variables were accepted to the set containing valid datapoints (named as "Valid points").

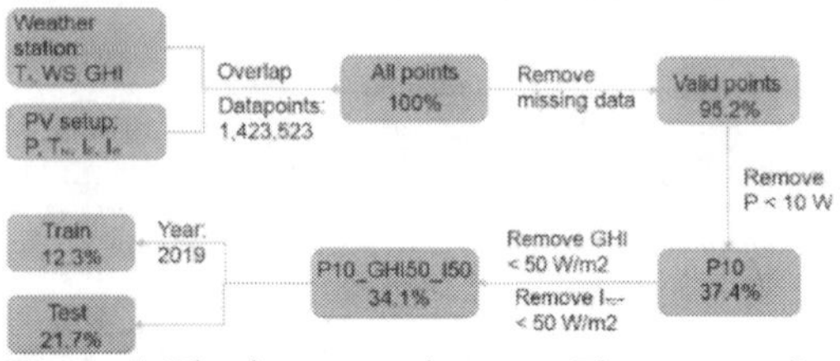

Figure 3: The data processing steps. The percents show how large fraction of the initial data ("All points", total 1,423,523 datapoints) is left.

The set with valid datapoints was filtered as follows. First, all datapoints with $P < 10$ W were removed, to exclude nighttime data and very low irradiance conditions ("P10"). Then, the filtering was done also to datapoints with GHI < 50 W/m^2 and $I_{TOT} = I_{E,MEAS} + I_{W,MEAS} < 50$ W/m^2 (P10_GHI50_I50). Finally, the set was split according to the year: 2019 data was used to train the model ("Train") and non-2019 data was used to test the trained model ("Test"). Table I shows the size of the datasets and how large fraction of the total produced electricity is cut due to filtering. E_{MEAS} is the sum of produced energy covering the whole dataset. Figure 3 visualizes the data processing and filtering procedure.

Table I: The data filtering process. E_{MEAS} = 100% corresponds to the total produced electricity during the "Valid points" dataset.

Set name	# of datapoints	E_{MEAS} (%)
Valid points	1355603	100
P10	532222	98.7
P10_GHI50_I50	484904	97.1
Test	309308	63.6
Train	175596	33.5

2.3 Power production modelling

Solar irradiation was obtained via different methods. Measured GTI was used as input as it is. GTI modelled based on GHI was obtained by using Starke decomposition model [4] with "dfc" climate zone coefficients and Perez transposition model with parameters defined in [5] (1994 version). Modelling was done with PVLIB [6].

The temperature of the front cover of the module was approximated with the Sandia model [7]:

$$T_{M,MOD} = T_A + I_{TOT} \cdot e^{a+b \cdot WS} \tag{1}$$

where a and b are regression coefficients. The cell temperature was approximated as:

$$T_C = T_{M,MOD} - \frac{I_E}{I_{STC}} \cdot \Delta T, \quad if\ I_E > 2 \cdot I_W \tag{2a}$$

$$T_C = T_{M,MOD} + \frac{I_W}{I_{STC}} \cdot \Delta T, \quad if\ I_W > 2 \cdot I_E \tag{2b}$$

$$T_C = T_{M,MOD}, \quad otherwise \tag{2c}$$

where I_{STC} = 1000 W/m^2 and ΔT = 3°C. Three different regions corresponds to sunshine to the front side of the panel where the thermocouple is located (Eq.(2a)), sunshine to the rear side (Eq.(2b)) and periods which are cloudy or the sunshine is coming from south (Eq.(2c)).

The 6k regression model is defined as [2]:

$$P_{MOD} = I_{REL} \cdot (I_{STC} \cdot n_{STC} + k_1 \cdot \ln(I_{REL}) + k_2 \cdot \ln(I_{REL})^2 + k_3 \cdot T_{REL} + k_4 \cdot T_{REL} \cdot \ln(I_{REL}) + k_5 \cdot T_{REL} \cdot \ln(I_{REL})^2 + k_6 \cdot T_{REL}^2) \tag{3}$$

where k_1, …, k_6 are regression coefficients and

$$I_{REL} = \frac{I_E + BF \cdot I_W}{I_{STC}} \tag{4}$$

$$T_{REL} = T_C - T_{STC} \tag{5}$$

where T_{STC} = 25 °C.

Alternatively, a simple power production model, which based on only the incident irradiation, cell temperature and STC efficiency, was applied, i.e.:

$$P_{MOD} = n_{STC} \cdot (1 + \beta \cdot T_{REL}) \cdot (I_E + BF \cdot I_W) \tag{6}$$

Two GTI inputs and two irradiance-to-power conversion methods resulted as four different power production datasets. The sets were numbered (1-4) as presented in Table II.

Table II: Different power production datasets. GTI is obtained either by direct measurement (monitor cell) or by modelling based on measured GHI.

Set	GTI	Irradiance-to-power
Set1	Measured (mon. cell)	6k
Set2	Measured (mon. cell)	Simple
Set3	Modelled (GHI)	6k
Set4	Modelled (GHI)	Simple

The degradation of the solar panels during the operation was considered by approximating the annual performance loss rate (PLR) to be 1.4%. This value is high compared to expected performance loss during the lifetime of the panel, but it is realistic due to rapid performance loss during the infancy phase of the installation.

2.4 Error analysis

To compare different modelling approaches, several error indicators were used. Absolute errors during the whole span of the dataset showed the total difference between the modelled and measured energy production. Normalized root mean square error (nRMSE)

$$nRMSE = \frac{\sqrt{\sum[(P_{MOD,i} - P_{MEAS,i})^2]/n}}{\sum P_{MEAS,i}/n} \tag{7}$$

where n is the number of datapoints was chosen as one error indicator since it is a common indicator and enables direct comparison with other studies. Considering the power production analysis, a more representative indicator is relative power-weighted average bias error (PWABE), which is defined as:

$$PWABE = \frac{\sum(P_{MEAS,i} \cdot |P_{MOD,i} - P_{MEAS,i}|/P_{MEAS,i})}{\sum P_{MEAS,i}} = \frac{\sum(|P_{MOD,i} - P_{MEAS,i}|)}{\sum P_{MEAS,i}} \tag{8}$$

and reveals the expected error of the modelled power, weighting the high-power moments.

3 RESULTS

3.1 Parametrization of models

The thermal model was parametrized by fitting Eq. (1) to the "Train" dataset. Parameters a and b were optimized so that the sum of squared errors (SSE) of the modelled temperature was minimized. Values of $a = -3.4422$ and $b = -0.1181$ s/m were obtained. These values were used for all simulations, although it should be noticed that they are based on the measured irradiance and may differ from optimal values with modelled irradiance.

The 6k model was parametrized by minimizing the SSE between P_{MOD} (Eq.(3)) and P_{MEAS}. Parametrization was done separately for each different solar irradiation input method. The parameters are shown in Table III. There is significant difference between the parameters obtained with measured and modelled GTI. The values obtained with measured GTI are close to what was expected based on the original 6k publication [2], indicating that the 6k fit with modelled GTI may not be very accurate.

Table III: Parameters of the 6k model.

GTI method	k_1	k_2	k_3
Measured	-0.08754	-6.81361	-0.46184
Modelled	12.7976	3.87095	-0.61999
	k_4	k_5	k_6
Measured	-0.21365	-0.34551	-0.00819
Modelled	1.69512	0.32382	0.03343

3.2 Testing parametrized models

The "Test" dataset was used to study the accuracy of the models (Eqs. (3)&(6)). The parameters shown in Section 3.1 were used for all simulations. The quality of the models was compared by using error indicators described in Section 2.4.

The analysis was done with one-minute resolution (single datapoints) and aggregated data. Three different aggregation levels were used: one hour, one day and one month. Analyzing both aggregated and initial data allowed to compare production estimates in temporal scales that are relevant for electricity market operation (hour and day), seasonal variation (month) and the lifetime production estimation of the solar panels (multi-year). The key error indicators are presented in Figure 4. The total produced electricity and its relative error are presented in Table IV.

Table IV: The total electricity production and relative error (compared to the measured value) for the "Test" dataset.

Set	Total energy (kWh)	Relative error (%)
Measured	549.22	0.00
Set1	548.95	-0.05
Set2	564.45	2.77
Set3	544.01	-0.95
Set4	547.00	-0.40

Two obvious major trends are observed. First, more advanced methods (6k vs. simple method for power calculation) and more accurate initial data (measured vs. modelled GTI) decrease the error. Second, aggregating data to larger quantities reduces the error. This was expected: when data is aggregated, the positive and negative errors of individual datapoints partly cancel each other. All data shown in Figure 4, follow these major trends.

Although the total errors in energy production were relatively small in all cases, excluding Set2, which showed significant overproduction, the errors in minute- and hour-scales were significant. This can be problematic especially when considering energy systems with high share of PV: minute-scale fluctuations can affect to the grid stability and hour-scale fluctuations makes effective economic operation, such as selling PV electricity to the day-ahead and intra-day market, difficult.

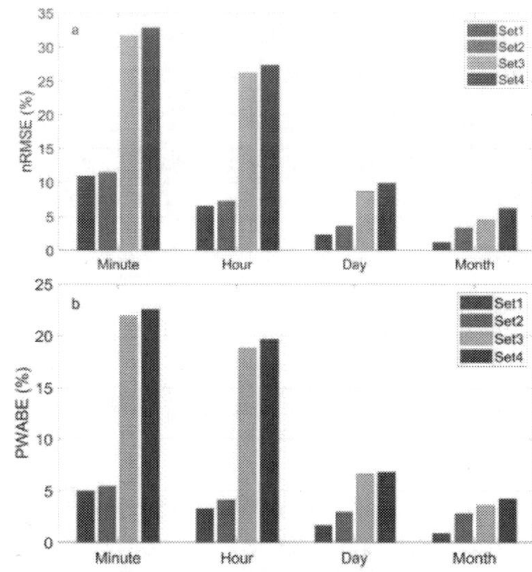

Figure 4: nRMSE (a) and PWABE (b) of the different power production modelling methods with "Test" dataset.

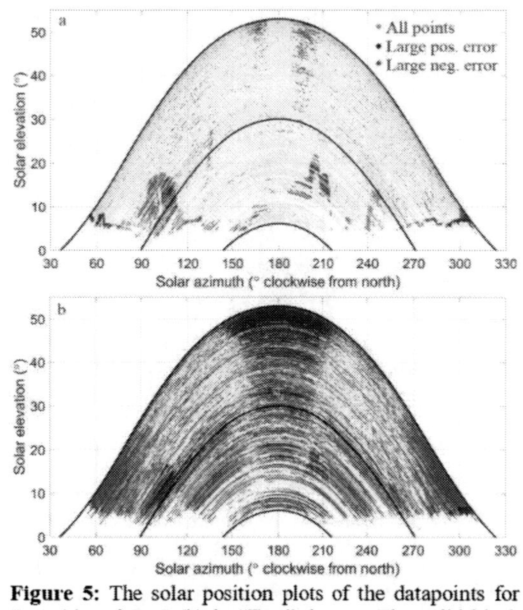

Figure 5: The solar position plots of the datapoints for Set1 (a) and Set3 (b) in "Test" dataset. The solid black lines show the solar path during solctices and equinoxes.

To further analyze the possible error sources, the datapoints with major errors (>20% and >10 W) were plotted according to the solar position. The plots for

Sets1&3 are shown in Figure 5. The plots for Sets2&4 were similar to shown plots and thus excluded.

In Figure 5a, the most significant errors originate near the horizon, close to solar noon and from the object located at east (a tree). The high positive errors due to the tree and when the solar elevation angle is low results from the location of the monitor cells at the top of the panel: the panel can be partly shaded even if the monitor cells are not. The high errors close to solar noon can originate from high angle of incidence, resulting as a high reflection from the cover glasses.

In Figure 5b, the amount of large positive errors is increased significantly and large negative errors are a major factor as well. This highlights that the simple modelling approach used here provides unreliable results, especially in minute- and hour-levels.

Finally, similar error analysis was done for the "P10" dataset, i.e also the datapoints with low irradiance but reasonable power production were included. The results are shown in Figure 6 and Figure 7. These are almost identical to results achieved with the "Test" set. This was expected, since majority of the data in "P10" set was also included in the "Test" set. However, this demonstrates that the models parametrized with typical data filtering procedures, especially GHI < 50 W/m^2, keep their performance level when they are applied in low-GHI conditions. This is especially important for VBPV, since its production can be notable even in low-GHI conditions due to east-west orientation of the panel.

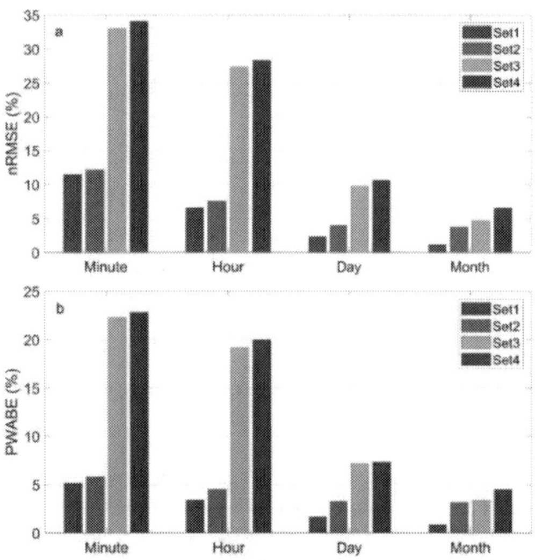

Figure 6: nRMSE (a) and PWABE (b) of the different power production modelling methods with "P10" dataset.

Overall, the results showed that estimating the energy production of a VBPV panel in a high-latitude location, which is considered as a challenging setup to model, can be estimated with a reasonable accuracy, even with very simple data input, when the aggregation level of the data is high enough. This allows to accurately analyze the economic viability of the planned system in advance. However, for more sophisticated analysis, such as producing accurate estimations in hour-level, which is crucial for the operation in the electricity market, very accurate, high-resolution GTI data is needed.

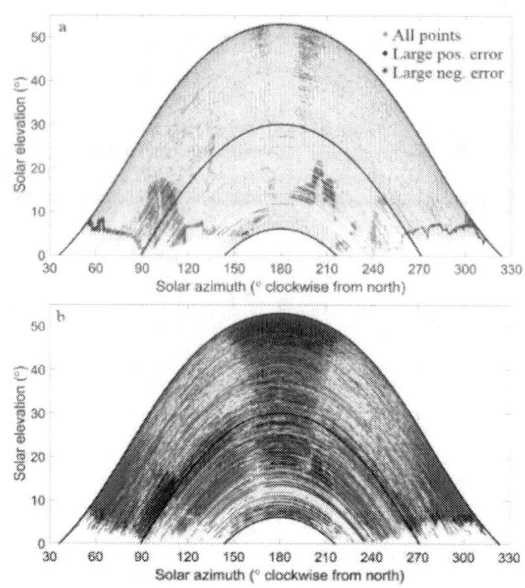

Figure 7: The solar position plots of the datapoints for Set1 (a) and Set3 (b) in "P10" dataset. The solid black lines show the solar path during solctices and equinoxes.

Finally, it should be noted that the modelling procedures presented here assume that the input irradiation value (GTI or GHI) is measured and accurate. Thus, especially considering hour- and minute-level modelling, the error resulting from the inaccuracy of the irradiation forecast is neglected. Considering monthly or lifetime production, this effect is not significant, since predicting weather for time periods up to years is impossible and thus historical data can be used as an accurate input.

4 CONCLUSIONS

The power production estimate based on the measured irradiance showed good accuracy. This demonstrates that, besides multi-year and lifetime estimations of PV production, also high-resolution estimations can be made based on only one year of data. This is important considering e.g. operation in day-ahead electricity markets, where the electricity is sold on hourly basis. However, especially for lifetime estimations, detailed information on the PLR of the panel is required. Here, the PLR is based on a linear estimation for the first years after commissioning of the setup and thus it is expected to decrease during operation.

The accuracy of the modelled irradiation -based power production estimation was low, especially with high temporal resolution (minute, hour). The most significant reasons for this are 1) the challenge of modelling solar irradiation with low solar elevation angles and 2) the location of the GHI sensor compared to the PV panel: the shading conditions are different. Thus, the simple irradiation modelling used here should be improved to consider shading. However, estimating the production of the VBPV panel in timescales from month to lifetime, even this method can be suitable.

Overall, this work contributes to expand the existing literature on PV power modelling to cover also the specific setup of VBPV at a high-latitude location, which is challenging to model. This contribution highlights that on-site measured irradiance values are needed to make accurate predictions for the panel energy production, especially when high-resolution data is needed. For lower resolution, such as multi-year and lifetime predictions, also simpler datasets can be sufficient. This indicates that a procedure where the decision of commissioning of a PV system is done based on simple data input and estimations, and the system is monitored during use to increase the accuracy and resolution of the power production estimations can be recommended.

ACKNOWLEDGEMENTS

The work was funded by Emil Aaltonen Foundation, Erasmus+ organization, Academy of Finland (project BioEST, no. 336577) and the Finnish Cultural Foundation.

REFERENCES

[1] S. Jouttijärvi, G. Lobaccaro, A. Kamppinen, and K. Miettunen, *Renewable and Sustainable Energy Reviews*, vol. 161, p. 112354, Jun. 2022, doi: 10.1016/J.RSER.2022.112354.

[2] T. Huld *et al.*, *Solar Energy Materials and Solar Cells*, vol. 95, no. 12, pp. 3359–3369, Dec. 2011, doi: 10.1016/j.solmat.2011.07.026.

[3] S. Ranta, J. S. Stein, E. Whitney, H. Huerta, and A. Heinonen, *Conf. Proc. 36th PVSEC,* 2020. doi: 10.4229/EUPVSEC20192019-6CO.15.5.

[4] A. R. Starke, L. F. L. Lemos, J. Boland, J. M. Cardemil, and S. Colle, *Renew Energy*, vol. 125, pp. 472–484, Sep. 2018, doi: 10.1016/J.RENENE.2018.02.107.

[5] M. P. Utrillas and J. A. Martinez-Lozano, *Solar Energy*, vol. 53, no. 2, pp. 155–162, Aug. 1994, doi: 10.1016/0038-092X(94)90476-6.

[6] J. S. Stein, W. F. Holmgren, J. Forbess, and C. W. Hansen, *Conference Record of the IEEE Photovoltaic Specialists Conference*, Nov. 2016, vol. 2016-November, pp. 3425–3430. doi: 10.1109/PVSC.2016.7750303.

[7] D. L. King, W. E. Boyson, and J. A. Kratochvil, "Photovoltaic array performance model," 2004. doi: 10.2172/919131.

MODEL TO ESTIMATE THE PERFORMANCE OF FULL-COLORED SOLAR PANELS

Lenneke H. Slooff-Hoek [1], Antonius R. (Teun) Burgers [1], Anna J. Carr [1] Tom Minderhoud [2], Ger Gijzen [2], Thijs Sepers [3], Wouter van Strien[4]
[1]TNO, P.O.Box 15, 1755 ZG Petten, The Netherlands, tel: +31 650009631, email: lenneke.slooff@tno.nl
[2] UNStudio,, PO Box 75381, 1070 AJ Amsterdam, The Netherlands
[3]TS Visuals, Schuit 14, 1724 VD Oudkarspel, The Netherlands
[4]Solar Visuals, Schuit 14, 1724 VD Oudkarspel, The Netherlands

ABSTRACT: In this paper we will present a model that is used to calculate the performance of colored modules. Especially in building integrated photovoltaic systems, the use of colored modules continues to increase for aesthetic reasons. The manufacturers often offer a standard product portfolio as well as client specific colors and prints. The effect of the color or print on the performance is in most cases determined by making a sample and measuring the performance or by estimating the performance based on experience. Here we present a model that calculates the effect of the print on the current density of the cells. With this, the overall module performance can be determined. The model parameters are determined by fitting the model to measurements on single cell laminates with a monochromatic color. Next the model is validated using the print files and measured data of full color prints and interconnected cells. The results show that a very good match is obtained between the model and the measurements.
Keywords: Modelling, Building Integration, Façade, Aesthetic, Performance

1 INTRODUCTION

In the last decade, the development of colored photovoltaic (PV) modules has resulted in various products for building integrated applications. modules [1-10] The products range from homogeneously colored modules to modules with a full-color print and can differ in the way the color is applied to the module. Some manufacturers add an extra layer inside the module on which the color or image is applied. Others print on the front glass, mostly at the inside, to protect the color from degradation by sunlight. These techniques make use of more or less standard digital printing techniques. However, there are also products where the color is generated by multilayer coatings. The coloring technique might differ, but in all cases, the application of color in the module will reduce the amount of light that reaches the solar cells in the module. The magnitude of the reduction depends strongly on the colors and the technology and affects the power output of the module. Most suppliers have performed tests to determine the power output of the products in their portfolio, but for new full-color PV modules or custom-made colors, often real life samples must be made to establish the performance. For a manufacturer it is therefore interesting to have a model that is able to determine the module output without the need to make test samples. There are a few reports in literature that describe models to calculate the output of colored modules. The first one by Halme and Mäkinen describes a model to calculate the efficiency limit of a colored solar cell for a color described by its CIE 1931 X-, Y- and Z-coordinates.[11] It is based on ideal abortion spectra to represent the colors. As such, the model is not suitable for realistic predictions of the module output. Peharz and Ulm used a similar approach based on narrow band constant reflection spectra and an ideal c-Si solar cell. But they also calculated the power loss for measured spectra (300-1200 nm) to indicate the expected losses in real modules. McKintosh *et al.* used a ray tracing approach to simulate the performance of modules in which the color is derived from changes in thickness of the anti-reflection coating. Their model can also take into account different front side texture and different observer angles.[12] The last report again uses measured reflection spectra and model spectra of solar modules to estimate the losses due to the color. The spectra are also converted to XYZ and CIELab color coordinates to show trends in the loss.[13] All the models mentioned above only work for homogeneously colored modules. Here we present a model to determine the loss in current for full color solar panels with any kind of print. It is based on the image file of the intended print. In addition, the model can also be used to optimize the output of the module by varying certain parameters and calculating the differences in performance for these variations. Besides that it can help establish if there are too large differences in light intensities on the cells that could cause problems in current matching between the strings.

The model calculates the relative amount of light that reaches the individual cells in a full size colored module. This individual illumination is used in subsequent model to calculate the overall module performance. The overall model result is validated for small laminates with homogenous colors as well as for full size full color PV modules.

2 MODEL FOR CELL ILLUMINATION

The model described below is only applicable for PV modules in which the image or color is applied by a digital printing technique. The basic idea behind the model is that it determines the transmission of each pixel in the final print that is applied to the actual module.

In digital printing the original image file is usually a pdf file with the image in RGB color space. This file is converted into CYMK color space upon printing. To understand this conversion better we start with the CIE xyY color space that covers all the colors that can be seen by the human eye. See Fig. 1. It is good to realize that each color space can contain different color models. RGB for example contains a.o. sRGB, plotted by the black lined triangle in Fig 1., AdobeRGB plotted by the blue lined triangle and ProPhotoRGB plotted by the red lined triangle. They do overlap to some extent, but not in all areas. If we then look at the CYMK color space, see white pentagram in Fig. 1., it overlaps only partly with the RGB color space. As long as the color in RGB space overlaps with the CYMK color space, the conversion is easy. But when the color is outside the CYMK space, or out of gamut as it is called, it cannot be easily converted. In that case a conversion algorithm needs to be chosen for the conversion. This conversion algorithm is a print parameter setting in the software called intent. There are 4 options: Perceptual, Relative colorimetric, Absolute colorimetric and saturation. [14]. Which one to choose depends on the application of the print. Upon printing, the print software converts this information into the color space of the printer, which is CYMK based. So for each new ink set and/or printer, the final CYMK values will be different.

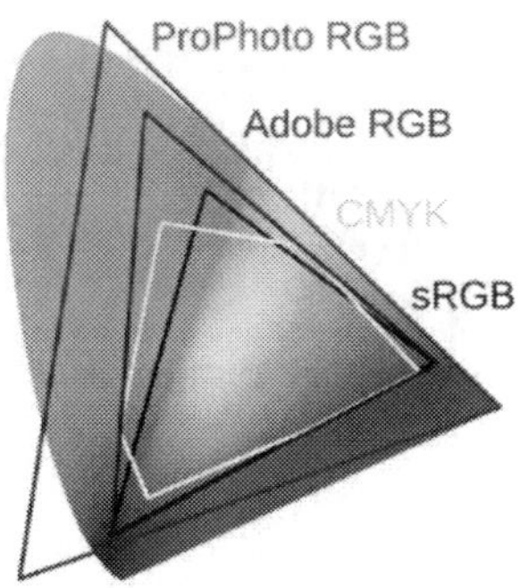

Figure 1: presentation of the CIE xyY color space together with the different RGB and CYMK color spaces [15].

In the model this conversion needs to be done as well, because in the end, the amount of ink of each color will determine the absorption in the ink. The model starts with the pdf file containing the color image for the PV module. Each pixel in the image is described by a certain combination of RGB values that represents the color of the pixel. For the images below it is in sRGB color space. In Acrobat Pro, the pdf color space is converted into a CYMK color space for printing which in this case is Coated Fogra 39. The file is then saved as tiff file. The rest of the conversion is done in a home-built Matlab program. It reads the tiff file of the image and uses standard Matlab image processing functions to convert the Coated Fogra 39 CYMK values into the printer and ink specific CYMK values, making use of icc files. For this it is necessary to have the printer and ink specific icc file from the manufacturer of the printer. See Fig. 2. [16]. The Coated Fogra 39 or other standard color space icc files can easily be found online. In the Matlab conversion function also the intent parameter can be set.

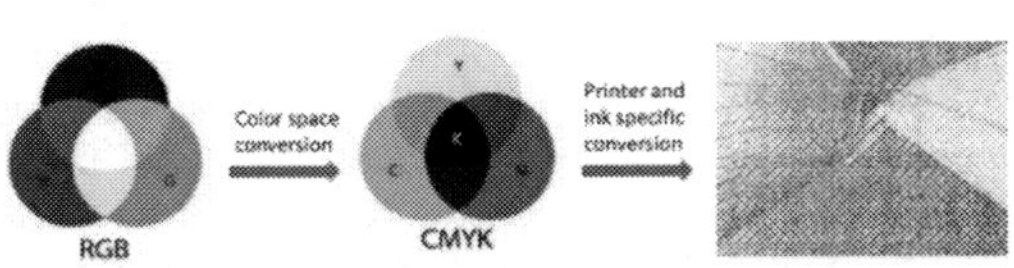

Figure 2: schematic of the RGB to ink and printer specific CYMK conversion.

It is important that the conversion profile and intent settings of the algorithm that are chosen are consistent with the conversion profile and settings that are used when sending the file to the actual printer. Table I shows the results of the RGB-CYMK color conversion for the pure RGB colors (red, green and blue). An sRGB color conversion profile was used and a few variations of the parameter settings for source intent and destination intent were analyzed. As can be seen, the resulting CMYK values can vary substantially with these parameter settings, showing the importance of using the correct conversion settings. These settings must be documented at the time of printing of the file.

Table I: results of the RGB-CYMK color conversion for the pure RGB colors (red, green and blue) for different combinations of source intent (src_intent) and destination intent (dest_intent). Here Abs. Col. is absolute colorimetric, Percept. is perceptual and Rel. Col. is relative colorimetric.

src_intent	dest_intent	rgb	c	m	y	k
Abs.Col.	Abs.Col.	Blue	243	222	0	0
Percept.	Abs.Col.	Blue	243	222	0	0
Percept.	Percept.	Blue	222	215	0	0
Percept.	Rel. Col.	Blue	239	229	0	0
Percept.	Saturation	Blue	249	231	1	0
Abs. Col.	Abs. Col.	Green	106	0	209	0
Percept.	Abs. Col.	Green	106	0	209	0
Percept.	Percept.	Green	115	0	209	0
Percept.	Rel. Col.	Green	112	0	227	0
Percept.	Saturation	Green	155	0	255	0
Abs. Col.	Abs. Col.	Red	0	236	254	0
Percept.	Abs. Col.	Red	0	236	254	0
Percept.	Percept.	Red	0	238	240	0
Percept.	Rel. Col.	Red	0	240	255	0
Percept.	Saturation	Red	0	245	254	0

In the program the cell distribution (See Fig. 3) for an example) is specified and the program determines which pixels of the image file coincide with the cells.. Then for each of those pixels the converted CYMK values are determined. In the program, for all these colors a certain absorption is assigned which in the end determines how much light reaches the cells. The program can in principle be used for various front side materials. We have applied it to modules with a flat glass front side and a lightweight polymer front side by changing the parameter for the front side reflection. Other materials can be used as well. For the front side reflection of glass we used a fixed reflection of 4%.

For calculating the transmission, two different optical models have been implemented and validated for performance. The first model assumes a linear dependence of the transmission of the pixel on the absorption of the individual colors. This means that a photon of a specific color can be absorbed by all the different color pixels.

$$T = 1 - R_{fixed} - w * \frac{A_w}{255} - c * \frac{A_c}{255} - m * \frac{A_m}{255} - k * \frac{A_k}{255}$$

Here R_{fixed} is the reflection of the front side, w,c,m,y and k are the W, C,Y,M, and K values of the pixel as determined from the tiff file, the A values are the absorption parameters for the different colors at maximum W, C,Y, M and K values of 255.

The second model is a non-linear model, where it is assumed that cyan, yellow and magenta have non-overlapping absorption spectra and the absorption parameters need to be summed to determine the overall transmission:

$$T = (1 - R_{fixed}) * T_w * T_b * (T_c + T_m + T_y - 2)$$

Here R_{fixed} is the reflection of the front side, the T values are the transmission for the different colors and the other layers in the module, e.g $T_c=(1- c*A_c/255)$.

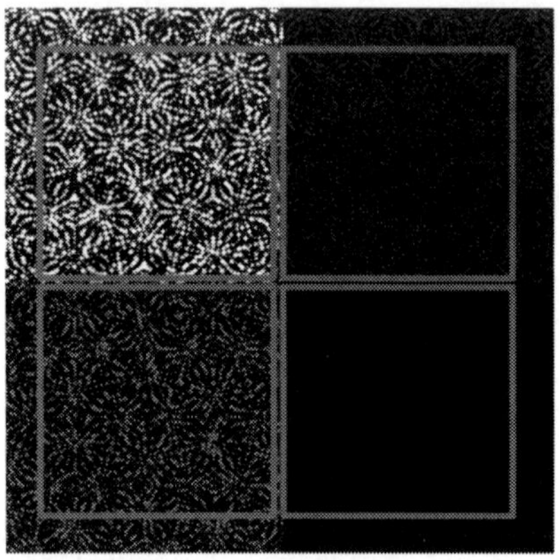

Figure 3: example of a print pattern of the laminates that were used to fit the parameters in the model. In red the cell distribution in the laminate Each cell could be individually connected to measure the I-V characteristics.

3 FITTING OF ILLUMINATION MODEL

As mentioned above, the model contains parameters for the absorption of the CYMK colors. These parameters were determined by fitting the model to measurement results on single cell laminates. For these laminates a dot-like color pattern containing a homogeneous color was used. An example is given in Fig. 4 for the magenta laminate.

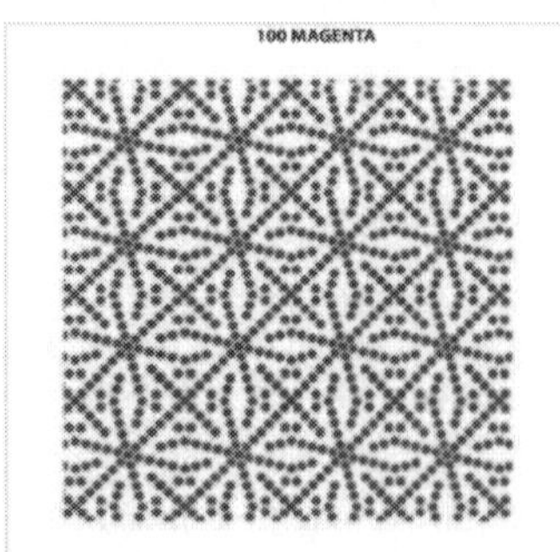

Figure 4: example of the print pattern of the C, Y, M, K laminates that were used for fitting the absorption parameters.

The laminate technology used for validating the model is based on our previous work on colored PV modules [16-19] of which the layer sequence is shown in Fig. 5. In between the different layers an EVA encapsulant was used. Of each color 2 laminates were made.

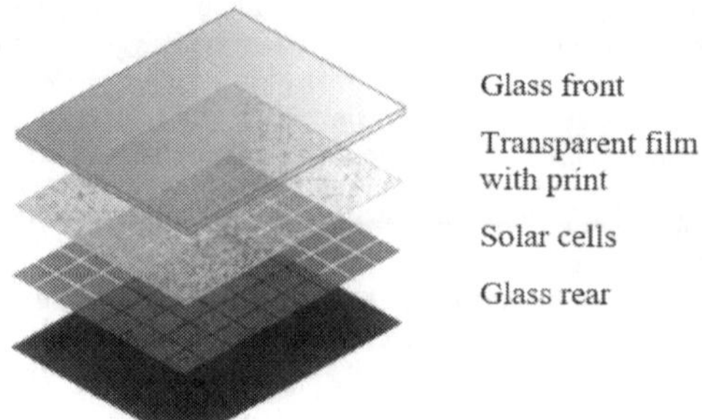

Figure 5: schematic of the different layers in the module

Current-voltage measurement of the laminates were performed using a Pasan HighLIGHT Lab Flash tester 4.4.1. Before lamination the cells were first measured in our WACOM solar simulator under standard test conditions to be able to determine the reduction in cell current. The resulting relative currents for the different colors are shown in Table II.

Table II: Measured and calculated relative current of colored PV laminates wrt the cell current of the cell in the laminate for the linear optical model.

Color	Measurement	Calculation	error
	0.689	0.690	0.2%
	0.788	0.783	0.8%
	0.825	0.831	0.7%
	0.833	0.834	0.1%
	0.844	0.842	0.2%
	0.887	0.893	0.6%
white	0.905	0.899	0.6%

As can be seen, the white laminates show the lowest reduction in cell current, whereas the black shows the highest reduction in cell current.

Table II also shows the relative currents generated by the model assuming the linear model. They all lie within 1% of the measured values, showing good agreement between the model and the measurements. If the gray laminates as shown in Fig. 2 are also included in the fit, the error increases slightly to an average of 1.3%, see Table III.

Table III: resulting absorption parameters for the individual colors (A_c, A_m, A_y and A_k for cyan, magenta, yellow and black respectively) for the two optical models. Not shown in the table is the absorption by the other layers, but it is taken into account in the number for the Total Absorption.

Optical model	Resid. error	Total Abs.	A_c	A_m	A_y	A_k
linear	1.3%	1.2	0.23	0.14	0.08	0.47
non-linear	1.8%	0.93	0.22	0.04	0.01	0.83

However, this fit result shows non-physical absorption parameters as the total absorption is above 1. The non-linear model on the other hand has a somewhat higher, but still low, residual error of 1.8% and does show physical absorption parameters. Based on these results it was decided to continue with the non-linear model.

4 VALIDATION OF ILLUMINATION MODEL

To validate the model a number of full size modules with different patterns have been characterized in the Pasan Lab Flash tester (described earlier), the resulting performance (IV curves) are then compared to predicted results using the illumination model.

6.a. Pattern #1

6.b. Pattern #2

6.c. Pattern #3

6.d Pattern # 4

Figure 6: Print designs used in the validation.

To begin, the module layout, number of cells, dimensions and cell position behind the pattern are defined. See Fig. 7. The color tif file of the print is then analyzed and the transmission determined on a cell by cell basis thereby determining the light incident on each cell. Figure 8 shows the transmission values as heatmaps for each of the patterns. Pattern number 4 has the highest transmission and pattern 2 the lowest.

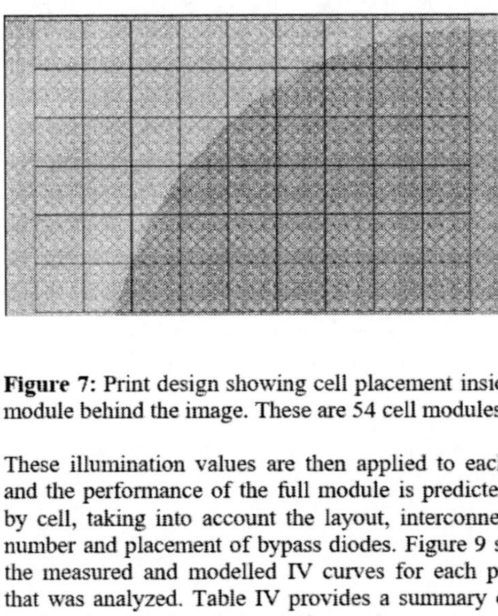

Figure 7: Print design showing cell placement inside the module behind the image. These are 54 cell modules

These illumination values are then applied to each cell and the performance of the full module is predicted cell by cell, taking into account the layout, interconnection, number and placement of bypass diodes. Figure 9 shows the measured and modelled IV curves for each pattern that was analyzed. Table IV provides a summary of the results, showing the relative difference between measured and modelled module performance parameters, where there is at most a 2% difference, indicating very good agreement with the model.

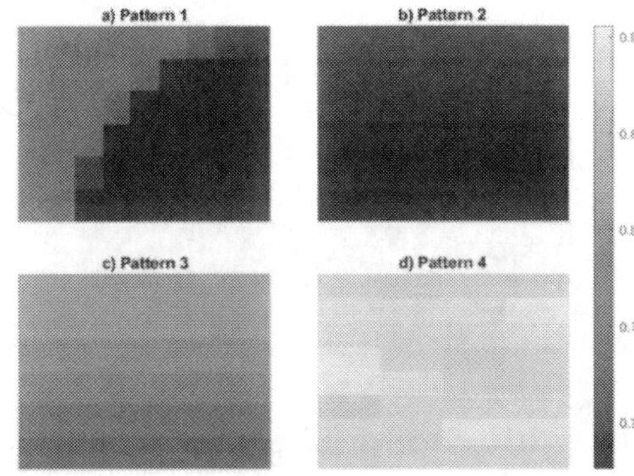

Figure 8: Heatmaps for all designs showing the modelled transmission across the active area of the modules.

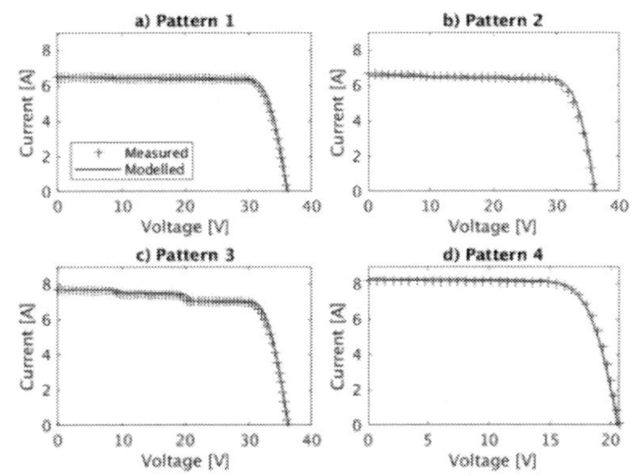

Figure 9: Comparison of measured and modelled IV curves for all designs. The curves show a very goog match between measured and modeled results. Pattern 3 has pronounced steps in the IV curve.

Table IV: Current-voltage comparison between modelled and measured results. The values are the relative [%] differences between modelled and measured values.

Pattern	Pm [%]	Isc [%]	Imp [%]	Voc [%]	Vmp [%]	FF [%]
1	0.9	0.8	-1.0	0.1	1.9	0.0
2	1.0	-1.7	-1.3	0.1	2.3	2.6
3	2.0	0.0	0.0	-0.1	2.0	2.0
4	-0.5	1.5	0.6	-0.9	-1.1	-1.1

The steps in the IV curve of pattern 3 indicate that the mismatch between the 3 strings is substantial and is causing bypass diodes to be activated. By using the Illumination model to adjust the transmission we have made a first attempt at reducing these steps and thereby designing a safer module. See Fig. 10. Figure 11 shows the result of this first adjustment, the steps have been reduced but also the overall performance, due to the large drop in short circuit current (Isc).

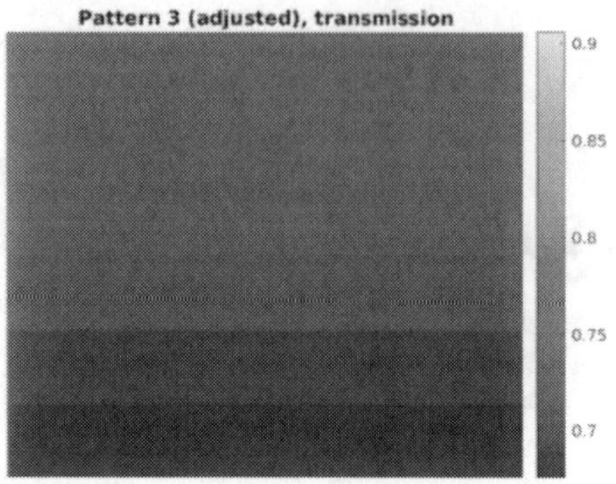

Figure 10: Adjusted transmission for pattern #3. Compared to Figure 8c we can see an overall reduction in transmission.

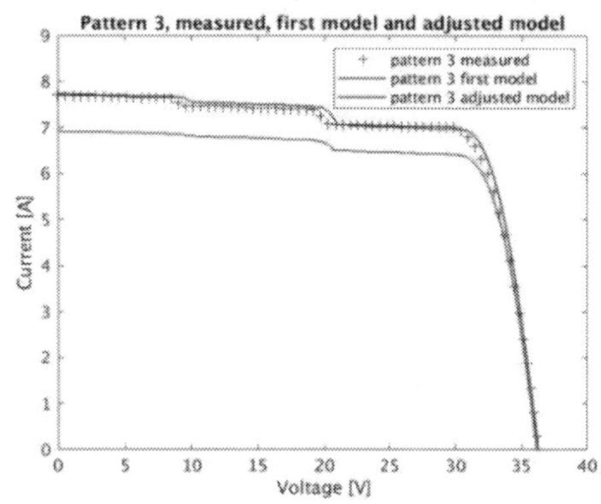

Figure 11: Comparison of measured and modelled IV curves for pattern 3. The adjusted model shows that the first 'step' in the IV cure has almost been removed, and the second has been reduced.

Adjusting patterns to optimize module performance will be an iterative process between creating a print with specific parameters, and then modelling the predicted IV curve. The results of the modelling help in advising what changes may be required in the print design to ultimately have a safe and optimized performance.

5 CONCLUSIONS

In this paper we present a model that can be used to estimate the performance of colored PV modules. The model first determines the loss in individual cell performance due to the print and subsequently calculates the overall module performance. The model parameters are determined by fitting the model to measurements on single cell laminates with a monochromatic color. The model is validated using the print files and measured data of full color prints and interconnected cells. The results show a very good agreement between the model and the measurements. Besides the performance estimates, the model can also be used to determine and solve intrinsic current mismatches due to the print as well as optimization prints for aesthetics and performance.

ACKNOWLEDGEMENTS

This work has been supported by the Dutch Government through the Topsector Energy Subsidies of the Ministry of Economic Affairs of the Netherlands.

REFERENCES

[1] https://www.solarvisuals.nl/products/
[2] https://kameleonsolar.com/sample-shop/
[3] https://www.zigzagsolar.com/
[4] https://hermanstechnisolar.com/products/
[5] https://www.soluxa.solar/
[6] https://www.swissinso.com/
[7] https://metsolar.eu/metsolar-bipv/colored-solar-panels/
[8] https://www.futurasun.com/en/products/monocrystalline-pv-panels/coloured-silk-pro-colour/
[9] http://www.issol.eu/colourful-pv/
[10] https://www.pixasolar.com/pixacolor
[11] J. Halme, P. Mäkinen, Energy Environ. Sci. **12** (2019) 1274
[12] K. R. McIntosh, M. Amara, F. Mandorlo, et al, AIP Conference Proceedings 1999 (2018) 020017
[13] A. Røyset, T. Kolås, B. P. Jelle, Energy and Buildings, Volume 208 (2020) 109623
[14] Color Management: Color Space Conversion (cambridgeincolour.com)
[15] https://learn.zoner.com/srgb-prophoto-rgb-and-more-do-you-know-your-color-spaces/
[16] J.A.M. van Roosmalen, L.H. Slooff, L.A.G. Okel, M.N. van den Donker, T. de Vries, T.Minderhoud, N. Wang, G. Gijzen, T. Sepers, M. Rietbergen, L. Polinder, E.M.B. Heller, A. Versluis, F.Frumau, 12th Conference on Advanced Building Skins, Bern, Switzerland, 2017.
[17] L.H. Slooff, J.A.M. van Roosmalen, L.A.G. Okel, T. de Vries, T. Minderhoud, G. Gijzen, T. Sepers, A. Versluis, F. Frumau, M. Rietbergen, L. Polinder, E.M.B. Heller, F. de Vries, 34th EU-PVSEC conference, Amsterdam, The Netherlands, 2017.
[18] L.H. Slooff, J.A.M. van Roosmalen, L.A.G. Okel, T. Minderhoud, G. Gijzen, L.C. Polinder, F. Goethals, T. Sepers, 35th EU-PVSEC conference, Brussels, Belgium, 2018.
[19] L.H. Slooff, A.R. Burgers, K.M. De Groot, N.J.J. Dekker, T. Minderhoud, G. Gijzen, T. Sepers, Y. de Groot, W. van Strien, J.A.M. van Roosmalen, 38th EU-PVSEC conference, online, 2021.

This is a short version of the paper submitted to the journal EPJ Photovoltaics.

It has been selected by the Committee of the WCPEC-8 for submission to EPJ Photovoltaics.

DEEP LEARNING TEMPERATURE ESTIMATION MODEL FOR PV MODULES

Anastasios Kladas, Bert Herteleer, Jan Cappelle
KU Leuven Research Group ELECTA Ghent,
Gebroeders De Smetstraat 1, 9000 Ghent, Belgium
Keywords: PV temperature, Deep learning, PV modelling

ABSTRACT: Simulations and modelling are very important in order to estimate the produced energy of photovoltaic systems, which helps its management and optimization. This work presents the development of a reliable deep learning PV temperature prediction model based on experimental data. The model performance is evaluated from the data of two different sites with different climates and compared with state-of-the-art models. The results show that the developed model provides very accurate and stable estimations for 1-minute data (MAE 1.03°C, RMSE 1.34°C) on the first site, but it shows weaker performance when tested in a different climate and different mounting conditions, especially during weather patterns that have not been included in its training. The comparative analysis shows that the developed model outperforms the other examined models on PV temperature estimations, yet the coefficients translate poorly for other climates: the developed model must be calibrated for each site for optimal results. Data-driven models developed for universal usage must contain a wide range of possible meteorological cases.

1 AIM AND APPROACH

After irradiance, temperature is the largest uncertainty component in estimating system performance. Furthermore, PV module temperature used for Performance Ratio corrections [1], with contractual consequences for owner and system operator. This work develops a deep learning temperature estimation model, and validates and compares with other state-of-the-art temperature prediction models. The data used for the model development and validation is described in Table 1.

Table 1: Datasets and locations used for model development and validation. T_{amb} = ambient temperature, G_{PoA} = plane of array irradiance, WS = wind speed, while T_{CELL} and T_{BS} are the PV temperature measured against solar cell and backsheet respectively.

	KU Leuven [1]	NIST [2]
Location	Ghent, Belgium	Maryland, USA
PV array mounting	18° tilt, 180° (due S), rooftop free-back c-Si	10° tilt, 180° (due S), rooftop c-Si with wind deflectors
Data period & resolution	2021, 1-minute resampled from 1-s measured	2017, 1-minute resampled from 10-s measured
Data & sensors	T_{amb} (4-wire RTD), G_{POA} (CMP11), T_{CELL} (4-wire RTD), WS (cup anemometer)	T_{amb} (4-wire RTD), G_{POA} (CMP11), T_{BS} (4-wire RTD), WS (Vaisala WMT52 ultrasonic anemometer)

2 METHODS

The PV module temperature can be calculated via a thermodynamic approach on the PV module based on the incident irradiance, ambient temperature and wind speed [3]. PV temperature models can be divided into four categories: Physics-based models; steady-state models [4], [5]; dynamic models [6], [7]; and machine learning models.

Machine learning models are models trained according to the data given in order to estimate the target values based on the predictors given. Using artificial neural networks the algorithm can be trained in order to identify the patterns between the predictors (input) and the target value (output). Machine learning techniques can be "easily" implemented to monitor existing systems consisting tailor-made temperature models that can capture issues, either by measurement equipment (broken sensors), or used to sense-check values from measured data. By considering the measurements from prior timesteps and benefiting of the supervised learning techniques that neural networks are using, a very accurate model can be developed, capable of providing reliable estimations for testing the validity of set up's assessed measurements.

The deep learning model of this work is implemented in a Python environment, by using Keras API from the TensorFlow library. As the target value is a continuous label, a regression algorithm is used. It consists of the input layer with 15 neurons (predictors) and two hidden layers before the output layer with 75 and 25 neurons respectively. On each layer, the activation function used is a rectified linear unit, as it can be nonlinear and has the ability to not use all the neurons at the same time, making the network more efficient. The model time resolution is one minute, with the inputs of the model being one-minute averaged values for the plane of array irradiance, ambient temperature and wind speed. The previous four minutes are also used as recent historical data.

3 RESULTS AND DISCUSSION

To evaluate the stability of the model during the examined year, the Pearson r coefficient is calculated between the daily mean absolute error (MAE) and the examined weather factors. Those factors have been chosen to be the total solar insulation of the examined day, the irradiance fluctuation calculated by the average value of every five-minute interval's standard deviation, the mean wind speed, the standard deviation of wind speed and the mean ambient temperature. MAE fluctuates around 1 °C throughout the year, without showing any high correlation with the weather factors (person $r < 0.4$). The above indicate a reliable model for the temperature estimations of the KU Leuven site. As can be seen in Table 2, the accuracy between the trained and the test data similar, indicating that the model is not overfitted.

Table 2: Errors of the deep learning model on training and testing dataset as well as the overall errors of both sites. The training (70%) and testing (30%) data are subsets of the KU Leuven dataset. Before splitting, the dataset was shuffled to ensure that both sites. The training (70%) and testing (30%) data are subsets of the KU Leuven dataset. Before splitting, the dataset was shuffled to ensure that both datasets contain data from all seasons.

	MAE (°C)	RMSE (°C)	r^2
Training dataset (KUL)	1.03	1.34	0.98
Test dataset (KUL)	1.02	1.34	0.98
KUL all	1.03	1.35	0.98
NIST	3.21	4.3	0.93

When evaluating the universality of the deep learning model by testing it on another dataset, it seems that it performs 3 times worse when used on the NIST data. Figure 1 shows how the developed model tends to follow the patterns of the measured PV temperature, yet with relatively large deviations during the day, especially when the PV module temperature drops below zero. This is likely due to the difference in mounting methods, with the NIST rooftop array having wind deflectors which hamper convective cooling. Moreover, the NIST site sees a wider range of ambient and PV temperatures than the KUL data on which the deep learning model was trained.

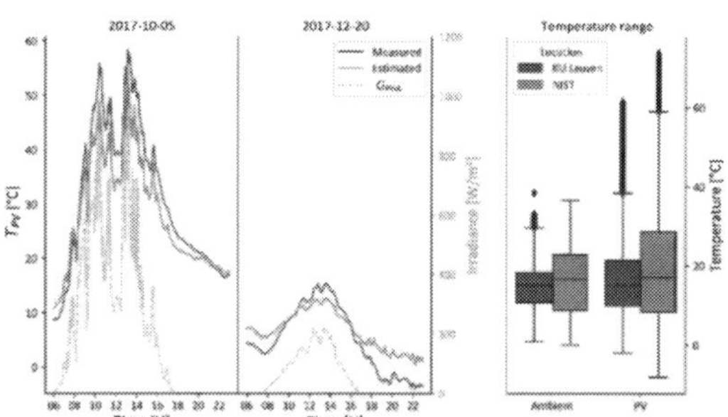

Figure 1: Time-series view of the KUL-data trained model performance during two selected days from NIST datasets, (left), as well as the distributions of ambient and PV temperature on the two datasets (right).

To rank the developed deep learning model, its performance is compared against two steady-state [4], [5] and one dynamic [6] temperature model, which are described in Table 3.

Table 3: List of the examined models for comparison. ΔTs,a is the deviation between the sky and the ambient temperature, C' is a normalization constant. Coefficient k is equal to 10-6 and τ is the time constant. Coefficients u1, u2 and u3 have been set to 25, 7, and 0.23 respectively.

State-of-the-art comparison models	Ref
$T_{PV} = T_{amb} + \frac{0.32 * G_{PoA}}{(8.91 + 2 * (\frac{WS}{0.67})}$	Skoplaki et al[4]
$T_{PV} = \frac{(20.9 + 3.32 * WS) * T_{amb} + G_{PoA} * 0.59}{(20.9 + 3.32 * WS) - 0.0005 * G_{PoA}}$	Mattei et al[5]
$T_{PV}(t_{n-k}) = \frac{1}{C'} * \sum_{K=0}^{N} \exp\left(-k * \frac{\Delta T}{\tau}\right) * \left[T_{amb}(t_{n-k}) + \frac{G_{PoA}(t_{n-k})}{u_1 + u_2 * WS(t_{n-k})} + u_3 \Delta T_{s,a}(t_{n-k})\right]$	Barry et al[6]

After the implementation of the models on the examined datasets, the developed deep learning model dominates in accuracy and stability in comparison with the other examined models for the KU Leuven set-up, as can be seen from Figure 2. The two steady-state models have similar accuracy and uncertainty, while the dynamic model slightly improves the accuracy and uncertainty versus the steady-state models. Using the NIST data, the dynamic model from Barry et al has the best accuracy and stability, performing slightly better than the developed deep learning model, while the steady-state models show poorer performance, as expected for the 1-minute time resolution.

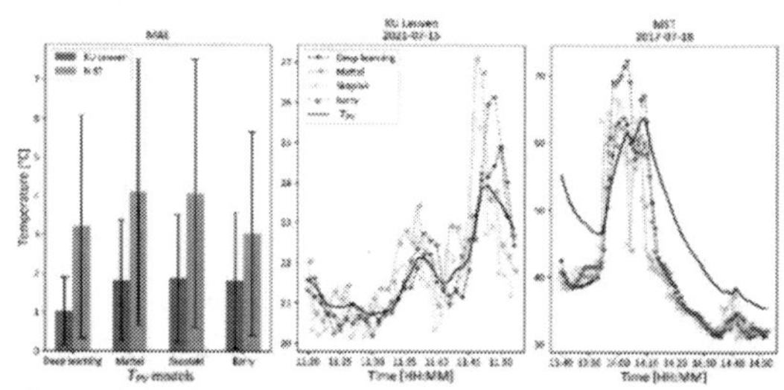

Figure 2: Left panel:Accuracy and uncertainty comparison of the models on the two sites. Middle & right panels: Demonstrations of the time mismatch issue between the models and the measured temperature on two sites. The NIST dataset is more challenging for all more challenging for all models than the KUL dataset.

4 CONCLUSIONS AND NEXT STEPS

The deep learning model achieves a high accuracy for the dataset on which it was trained with an MAE of 1.03°C and 1.34°C RMSE, and significantly reduces the time mismatch compared to often-used steady-state temperature models, yet the direct applicability to other datasets as shown on the NIST data still needs more work. For the full paper, the translation of the reverse direction, i.e. using the NIST-trained deep learning model on the KU Leuven dataset will be examined, and evaluated against an additional site's data. The influence of wind direction will also be investigated.

5 CONCLUSIONS AND NEXT STEPS

The deep learning model achieves a high accuracy for the dataset on which it was trained with an MAE of 1.03°C and 1.34°C RMSE, and significantly reduces the time mismatch compared to often-used steady-state temperature models, yet the direct applicability to other datasets as shown on the NIST data still needs more work. For the full paper, the translation of the reverse direction, i.e. using the NIST-trained deep learning model on the KU Leuven dataset will be examined, and evaluated against an additional site's data. The influence of wind direction will also be investigated.

6 REFERENCES

[1] B. Herteleer, B. Huyck, F. Catthoor, J. Driesen, and J. Cappelle, "Normalised efficiency of photovoltaic systems: Going beyond the performance ratio," Solar Energy, vol. 157, pp. 408–418, Nov. 2017, doi: 10.1016/j.solener.2017.08.037.

[2] M. Boyd, "Performance data from the NIST photovoltaic arrays and weather station," Journal of Research of the National Institute of Standards and Technology, vol. 122, 2017, doi: 10.6028/JRES.122.040.

[3] D. Faiman, "Assessing the outdoor operating temperature of photovoltaic modules," Progress in Photovoltaics: Research and Applications, vol. 16, no. 4, pp. 307–315, Jun. 2008, doi: 10.1002/pip.813.

[4] E. Skoplaki, A. G. Boudouvis, and J. A. Palyvos, "A simple correlation for the operating temperature of photovoltaic modules of arbitrary mounting," Solar Energy Materials and Solar Cells, vol. 92, no. 11, pp. 1393–1402, Nov. 2008, doi: 10.1016/j.solmat.2008.05.016.

[5] M. Mattei, G. Notton, C. Cristofari, M. Muselli, and P. Poggi, "Calculation of the polycrystalline PV module temperature using a simple method of energy balance," Renewable Energy, vol. 31, no. 4, pp. 553–567, Apr. 2006, doi: 10.1016/j.renene.2005.03.010.

[6] J. Barry et al., "Dynamic model of photovoltaic module temperature as a function of atmospheric conditions," Advances in Science and Research, vol. 17, pp. 165–173, Jul. 2020, doi: 10.5194/asr-17-165-2020.

[7] M. Prilliman, J. S. Stein, D. Riley, and G. Tamizhmani, "Transient Weighted Moving-Average Model of Photovoltaic Module Back-Surface Temperature," IEEE Journal of Photovoltaics, vol. 10, no. 4, pp. 1053–1060, Jul. 2020, doi: 10.1109/JPHOTOV.2020.2992351.

[8] M. G. Chowdhury, A. Kladas, B. Herteleer, J. Cappelle, and F. Catthoor, "Sensitivity analysis of the state of the art silicon photovoltaic temperature estimation methods over different time resolution," in Conference Record of the IEEE Photovoltaic Specialists Conference, Jun. 2021, pp. 1340–1343. doi: 10.1109/PVSC43889.2021.9519068.

VIRTUAL ENERGY RATING: A METHOD FOR OPTIMIZING MODULE PERFORMANCE THROUGH CELL-TO-MODULE ANALYSIS

A. Protti, J. Shahid, M. Mittag, D. H. Neuhaus, U. Kräling, M. Kaiser
Fraunhofer Institute for Solar Energy Systems ISE
Heidenhofstr. 2, 79110 Freiburg, Germany
alexander.aguilar.protti@ise.fraunhofer.de

ABSTRACT: The implementation of energy rating of photovoltaic modules based on IEC-61853 imposes challenges to developers of PV module technology. In response, we present how a simulation-based analysis methodology, adapted for non-STC conditions, can be used to obtain all information needed for determining a virtual Climate Specific Energy Rating (vCSER), namely: a power matrix, the spectral response, the angular loss, and heat transfer coefficients. A conventional full-cell module and a "butterfly" half-cell topology are used to exemplify how the proposed method can be used to identify gain and loss factors of a module design affecting the performance of each module in terms of rated energy yield per area and Climate Specific Energy Rating (CSER). Recommendations for improving the module design are ranked and discussed, based on a sensitivity analysis on eleven parameters.
Keywords: Energy Rating, CTM, Simulation, PV Module, Design, Performance, Loss Analysis

1 INTRODUCTION

Photovoltaic (PV) modules are to some extend priced based on their rated power, measured at Standard Test Conditions (STC: 1000 W/m^2 irradiance, 25°C module temperature and AM1.5G spectrum [1]) which are rarely found at installation sites. However, environmental conditions and how a module responds to them defines the energy yield of a module and the economic revenue that can be obtained from it. As solar modules constitute one of the main cost of PV systems [2], improving the way in which modules are rated can foster innovation and the bankability of PV projects.

In response, the International Electrotechnical Commission (IEC) introduced the IEC 61853 series: Photovoltaic (PV) module performance testing and energy rating [3–6], which provides an international standard for the PV industry to characterize and compare modules based on an estimation of their performance for typical climatic regions, without the effect of the PV system (shading, inverter, AC cables, transformer, etc.) or module degradation.

As the IEC 61853 gains relevance in the PV market [7–11], module producers might adapt the design and the bill of materials (BOM) of their modules with focus on the energy rating. Nonetheless, various issues have been pointed out regarding the costs of the additional measurements needed for characterizing the modules [12] and the difficulties in achieving an unanimous interpretation and application of the standard [13,14]. In other words, the IEC 61853 implies additional challenges that might increase the time-to-market of new technologies and slow down the overall development of the PV industry.

To overcome this barrier, the current work presents how a cell-to-module (CTM) analysis model [15–17] extended for non-STC simulations [18] can be used to calculate a virtual Climate Specific Energy Rating (vCSER). This model, which requires as input the module layout and the physical properties of the module components, allows to reduce expenses from prototyping, field testing and the additional measurements required by the standard [3,4], during the development phase of a module. Furthermore, it enables a rapid assessment of the impact of module design parameters on its performance, facilitating the usage of the energy rating as a tool in technology development and benchmarking. It also allows to give a forecast or a quick update of the energy rating of a module for higher power cell classes.

A similar methodology has been implemented by Blakesley [12] for optimizing the costs of the CSER, however his analysis aims at minimizing the costs of the measurements needed for the IEC 61853 and the electrical losses in the module design are not explicitly contemplated in his model. Haedrich [19], Thomson [20] and Mittag [21] have also applied CTM analysis for yield simulations. However, their work does not follow the IEC 61853 procedure. Furthermore, Haedrich [19] and Thomson [20] make use of various independent models. In contrast, the current work has been implemented in a consistent bottom-up multi-physics loss channel analysis ("SmartCalc.Module") [22,23], whose graphical user interface facilitates the usage of the model for rapid assessment of module technology.

In the next section, the details of IEC 61853 relevant for explaining our approach are presented. Then, the vCSER approach and its application are shown. This paper finalizes with a summary of the main aspects that were discussed.

2 OVERVIEW OF THE ENERGY RATING BASED ON IEC 61853

The IEC 61853 series (from now on, "the standard", for simplicity) consists of four parts which establish the characterization methods [3,4], a reference weather data set [6] and the algorithm [5] for calculating the energy output of the module [Wh] for each time step and for a whole year (rated power or yield, $E_{mod,year}$ [$Wh/year$]):

$$E_{mod,year} = \sum_{j=1}^{j=8760} P_{mod,j} \cdot 1\ hour \tag{1}$$

where $P_{mod,j}$ [Wp] is the module maximum power point at the given conditions for each timestep j. The standard also defines the Climate Specific Energy Rating (CSER, dimensionless), which corresponds to the annual performance ratio (PR) of the PV module for each climatic region:

$$CSER = \frac{E_{mod,year}/H_p}{P_{mpp,STC}/G_{ref}} \quad (2)$$

Here H_p [Wh/m^2] is the total of the hourly global in-plane irradiation for the reference period, G_{ref} is the irradiance at STC (1000 W/m^2), and $P_{mpp,STC}$ [Wp] the corresponding maximum power at STC.

2.1 Inputs for the IEC 61853 algorithm

The algorithm of the standard [5] requires four inputs: a power matrix, the spectral response, the angular dependence, and the thermal coefficients for calculating the module operating temperature. These inputs are sequentially used for obtaining discount coefficients associated to each environmental factor, which are then used to calculate the energy, generated by the module at each time step, given the weather data [6].

2.1.1 Power matrix (PM)

The PM accounts for the response of the module to light intensities and temperatures different from STC. It consists of a set of P_{MPP} values obtained from IV curve measurements at the temperatures and irradiances shown in Table I. The full matrix consists of 22 data points. The measurements can be done either outdoor or indoor. Details for the interpolation and extrapolation of the data are part of the algorithm [5].

Table I: Temperatures and irradiances needed for the power matrix according to IEC 61853

Irradiance	Module temperature [°C]			
[W/m^2]	15	25	50	75
100			NA	NA
200			NA	NA
400				NA
600				
800				
1000				
1000	NA			

2.1.2 Spectral response (SR)

The SR is used for obtaining the spectral mismatch of the module and for correcting to the reference spectral conditions during module characterization. It must be measured following the procedure from IEC 60904-8 [24].

2.1.3 Angular dependence

This input accounts for the changes in the module I_{SC} at 25 °C with respect to the angle of incidence (θ). Hence, it is used to calculate the effective light transmission (τ) in the module at different incidence angles, which affects how the module responds to both direct and diffuse light. For simplicity of computation, a fitting of the measurements to the Martin-Ruiz model [25] is suggested by the standard, so that the measurements can be summarized by the angular loss coefficient (a_r, also known as the angle of incidence response [4]). The following equation applies:

$$\tau(\theta) = \frac{Isc(\theta)}{Isc(0)\cos(\theta)} = \frac{1 - e^{\frac{-\cos(\theta)}{a_r}}}{1 - e^{\frac{-1}{a_r}}} \quad (3)$$

2.1.4 Coefficients for calculating the module operating temperature

These coefficients are obtained from fitting the Faiman model [26] to outdoor measurements. Here, the sample module and additional reference modules are required. The data used in the fitting must include measurements from at least ten different days which contain no less than ten valid points before and after noon. The valid data points must fulfill steady state requirements within irradiance and wind speed limits further detailed in the standard [4]. The model for relating the irradiance on the module (G [W/m²]), the wind speed (v [m/s]), the module operating temperature (T_{mod} [°C]) and the ambient temperature (T_{amb} [°C]) has the form:

$$\frac{G}{T_{mod} - T_{amb}} = u_0 + u_1 v \quad (4)$$

Here u_0 [W/m²/°C] describes the influence of the irradiance at no-wind conditions and u_1 [W-s/m³/°C] describes the wind impact.

2.1 Standard reference climatic profiles

The reference weather data set provided in the standard [6] contains tabulated data of hourly values over a full year for six different climatic profiles:

Table II: Locations and climatic profiles used in the IEC 61853

Latitude	Type
1°S	Tropical Humid
33°30'N	Subtropical arid (desert)
33°22'N	Subtropical coastal
56°N	Temperate coastal
34°N	High elevation (above 3000 m)
57°N	Temperate continental

The given weather variables are a time stamp, local solar time, ambient temperature, wind speed at the module height, solar angles, direct and global horizontal irradiance, direct and global in-plane irradiance, and spectrally resolved global in plane-irradiance for a set of discrete band intervals.

3 VIRTUAL ENERGY RATING MODEL

The calculation procedure consists of two stages. In the first stage, the model relies on a physical characterization of the individual module components to calculate all the inputs needed for the IEC 61853 algorithm by using a CTM model [15–17], which has been extended and validated for non-STC conditions as explained in [18]. At a second stage, the algorithm from the standard [5] is used to estimate the yield and a virtual CSER of the module. For ease of use, the model has been implemented in the software SmartCalc.Module [22,23], which can be downloaded from www.cell-to-module.com.

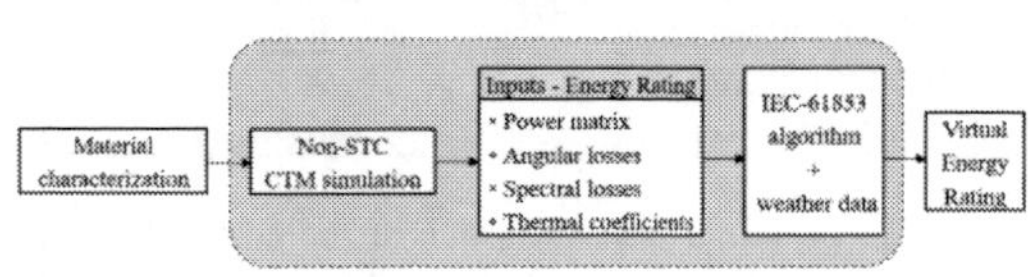

Figure 1: Flow diagram of the virtual Energy Rating

For calculating the power matrix, a linear behavior of the photo-current from the one-diode-model is assumed. Temperature coefficients for P_{MPP}, I_{SC} and V_{OC} are considered independent from the irradiance.

Fresnel equations as presented in [18] are used to make angularly and spectrally resolved calculations of the optical losses due to the module cover and encapsulants. With respect to the optical coupling, the gains from reflective back sheets can also calculated in detail as described by [17]. The optical coupling gains from cell metallization, interconnection ribbons and the cell-encapsulant coupling are considered as a constant value for all angles and wavelengths. This way, a synthetic value of the SR and a_r that responds to changes in the optical properties of the module is generated.

A one-dimensional thermodynamical model described by Mittag [27] is used to obtain the temperature under different operating conditions. A sample of the weather data from the standard [6] that fulfills the requirements from [4] is used as weather input for the calculations. Finally, u_0 and u_1 are obtained with the least-squares method. It is worth mentioning that the standard [4] warns about potentially large variation in the values of these coefficients, depending on the location and the season. In contrast, the proposed approach is less sensitive to these conditions because the data points selected are chosen using a pseudo-random sequence, which allows direct repeatability of the weather data.

4 APPLICATION OF THE VIRTUAL ENERGY RATING

To illustrate the application of the virtual energy rating procedure, two exemplary modules are considered. Module A is constituted by 60 M4 full square cells connected in series. Module B contains 120 half-cut cells connected in a "butterfly" topology (2 parallel blocks of 60 cells connected in series). The module layout follows the specifications presented in Table III. Here, the top margin is 10 *mm* less in module B because the string interconnector is placed between the two parallel blocks of cells. Hence, module B has 10 *mm* in its center, between both cell blocks. However, module B is 22.5 *mm* longer, because it has nine additional cell spaces along its length. The rest of the layout dimensions are the same for both module designs.

Table III: Module layout

Characteristic	Module A	Module B
String connection type	Series	Butterfly
Cells per string	10	10
Strings per module	6	12
Middle distance [*mm*]	0	10
Top margin [*mm*]	20	10
Bottom margin [*mm*]	10	10
Side margins [*mm*]	10	
Cell distance [*mm*]	2.5	
String distance [*mm*]	2.5	
Frame width [*mm*]	13	
Module length [*mm*]	1696	1718.5
Module width [*mm*]	1029	1029
Module area [m^2]	1.745	1.768

The IV curve characteristics of the cells, taken from [28], are shown in Table IV. Here, the cell cutting generates a reduction in cell efficiency of 0.32%$_{abs}$. However, since both cell types have practically the same efficiency, a clean comparison between module designs can be done.

Table IV: Cell specifications

Characteristic	Module A	Module B
Cell type	Full cells	Half cells with edge losses
Length [mm]	161.75	80.875
Width [mm]	161.75	161.75
Pseudo-square diameter [mm]	0	0
Area [cm^2]	261.631	130.815
Number of busbars	6	6
I_{SC} [*A*]	10.46	5.23
V_{OC} [*V*]	0.683	0.681
I_{MPP} [*A*]	10.00	4.97
V_{MPP} [*V*]	0.585	0.580
P_{MPP} [*Wp*]	5.85	2.88
Fill factor [%]	81.88 %	80.93 %
Efficiency	22.36%	22.04 %
Metallized area (front) [mm^2]	888	444

4.1 Comparison of module designs

4.1.1 CTM analysis

The cell to module analysis is shown in Figures 2 and 3. With respect to the efficiency, the module margin (k1) and cell spacing (k2) are the main cause of efficiency loss in both modules. The additional cell spaces of module B contribute to increasing the total active area loss (k1+k2) by 0.24%$_{abs}$ with respect to module A, which also leads module B to having 0.06%$_{abs}$ greater cover coupling gains (k11). As a result, the geometrical and optical losses in module B are in total 0.18%$_{abs}$ greater than in module A.

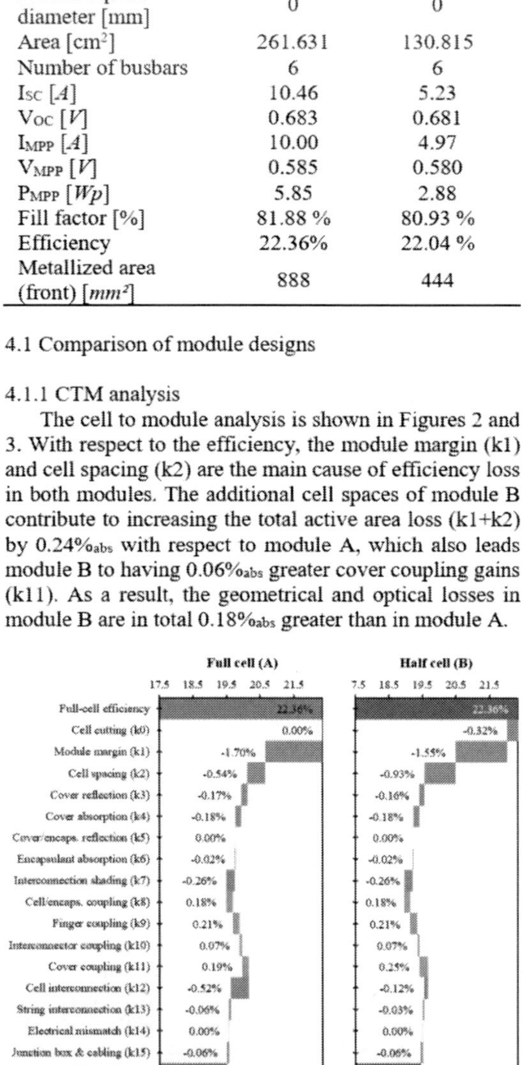

Figure 2: CTM analysis in terms of efficiency of the module design with full cells (A) and half cells (B)

With respect to the electrical factors, module B has about four times less efficiency losses in the cell interconnectors (k12) because half as much current flows through them ($\Delta P = I^2 * R$). Also, half of the current flows through about two thirds of the string interconnector length, leading module B to having half the losses in these elements (k13). The resulting advantage of 0.43%$_{abs}$ less electrical losses (k12-k15) is 0.25%$_{abs}$ greater than the difference in the optical factors. However, it is not enough to overcome the 0.32%$_{abs}$ losses due to cell cutting, leading the butterfly topology (module B) to have and efficiency of 19.45%, which is 0.06 %$_{abs}$ or 3%$_{rel}$ smaller than module A (19.51% efficiency).

Regarding the CTM in terms of module power (Figure 3), the inactive area only plays a role in the optical gains (k11), but not in the losses. Therefore, module B has a rated power of 343.88 *Wp*, which is greater than the 340.57 *Wp* of module A by a difference of 3.31 *Wp* (less than 1%).

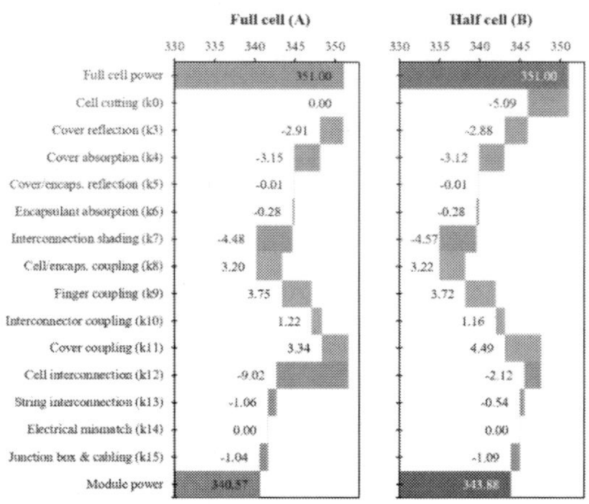

Figure 3: CTM analysis in terms of power of the module design with full cells (A) and half cells (B)

4.1.2 Results from the virtual Energy Rating

When looking at the results from the virtual Energy Rating (Figures 4 and 5) module A has an advantage over module B for all climatic regions. In terms of yield per area and virtual CSER, module A is on average 1.19% and 1.44% better than module B.

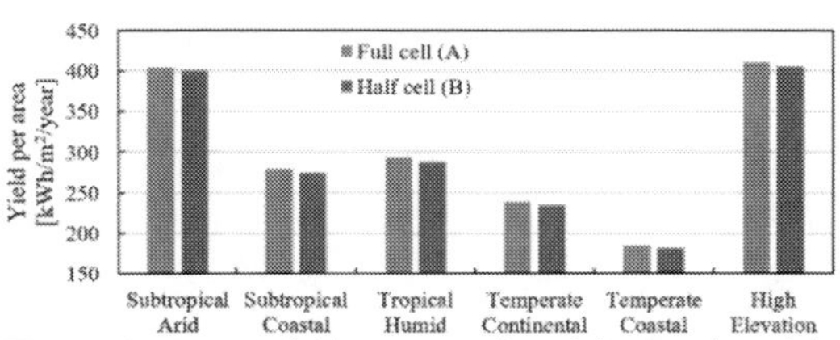

Figure 4: Energy yield per area of module A and module B for six climatic regions

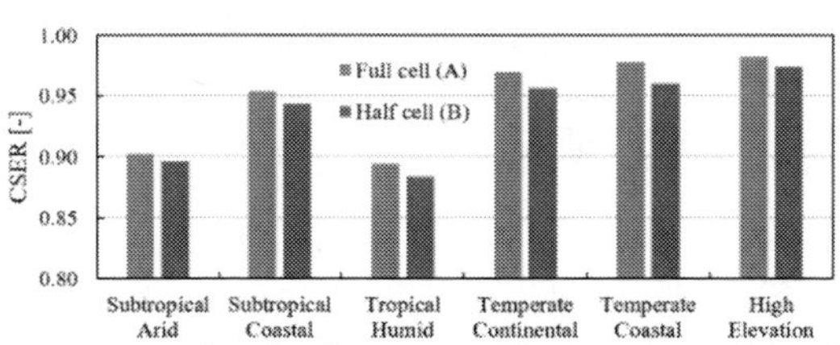

Figure 5: Virtual Climate Specific Energy Rating of module A and module B for six climatic regions

With respect to the causes for these results, Figures 6 and 7 show that half-cut cells module has a worse low light response [28], which it a competitive disadvantage with respect to the simulated full cell module when partial shading at the installation site is not expected [29].

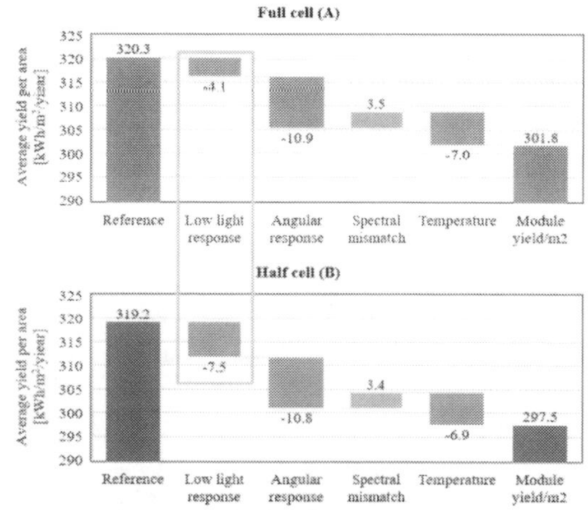

Figure 6: Comparison of environmental factors affecting the CSER of module A and B

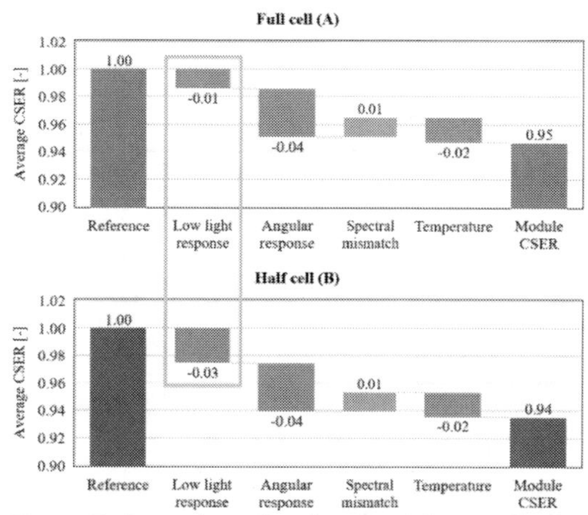

Figure 7: Comparison of environmental factors affecting the average yield per area of module A and B

4.2 Sensitivity analysis on module design parameters

We now illustrate how the energy rating of module B could be improved by changing its design. For narrowing the analysis, a selection of eleven parameters is done based on the CTM analysis. The results are presented in Figure 8 and Figure 9 in the form of spider plots. For comparison, the changes are expressed as percentual change with respect to the initial reference value.

Regarding the yield per area, increasing the number of strings per module has the most important effect, followed by the number of cells per string. When changing these parameters, it is assumed that spacings and margins are kept the same, but more cells or strings are added (or removed), meaning that the number of cells and module dimensions are increased (decreased). In principle, adding either cells or strings improve in a similar way the active area share. However, having more cells in a string would also increment the cell interconnection losses (k12), whereas the strings per module couple with the string connector losses, which are lower due to the larger cross section of the string connector ribbon.

Next in line, comes increasing the thickness of the cell and string interconnector. These actions would not have a direct impact on any other variable, than themselves and the downstream electrical losses.

Increasing the width of the string interconnector has an equivalent effect to increasing its thickness, since both improve its cross-section area and reduce its resistance. The sensitivity for these two parameters is clearly non-linear because the dependency between solar cells and the series resistance follows the diode model and because the irradiance distribution is skewed towards low light conditions, giving a lower marginal benefit to having lower resistance at higher irradiance conditions.

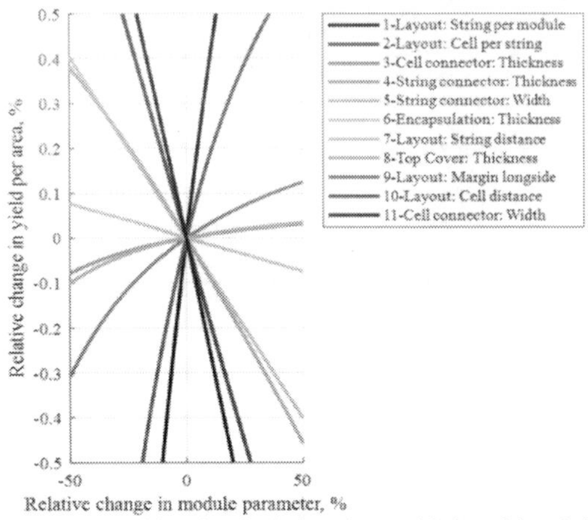

Figure 8: Spider plot depicting the sensitivity of the yield per area of module B with respect to module design parameters

The small negative effect of increasing the thickness of the encapsulant means that this parameter could be adjusted if having thicker interconnectors caused problems in the lamination process.

Then, reducing the string distance would increase the module packing density and decrease the total module area. This would also shorten the string connectors. A similar effect on the yield per area is observed when the glass thickness is reduced.

Following, a reduction in the module margin and the cell distance would reduce the module area, improving the yield per area. Here, a decrease in the losses due to the cell connectors when reducing the cell distance, gives this aspect a greater impact.

Finally, the shading caused by the cell connector could be reduced by making it narrower. Any increments on the resistance, are less important.

Before looking at the results from the sensitivity analysis on the CSER, it is worth noticing that a module designed by CSER optimization is different from a module optimized for yield per area, because the area units in the definition of the CSER (Equation 2) are cancelled out. In other words, the CSER can be understood as a normalized ratio between rated yield and rated power (G_{ref} and H_p are constant).

Consequently, in Figure 9, we observe that increasing the module margin, string distance and cell distance has a positive impact on the CSER, because that would increment the gains from the internal light recycling, the back cover coupling (k11).

Also, we see that the cell interconnector width result has a U-shape, being the current width close to the minimum CSER. In other words, a different width will improve the yield-to-power ratio.

In the case of the string interconnector, increasing either its width or thickness will also have a negative effect on the CSER, since the benefits on the rated power are lower compared to the yield gains. Nonetheless, a reduction in the encapsulation or in the front cover thickness would have a beneficial impact on the yield-to-power ratio.

In the case of the cells per string and the strings per module, the associated losses from the interconnectors have a negative impact on the CSER.

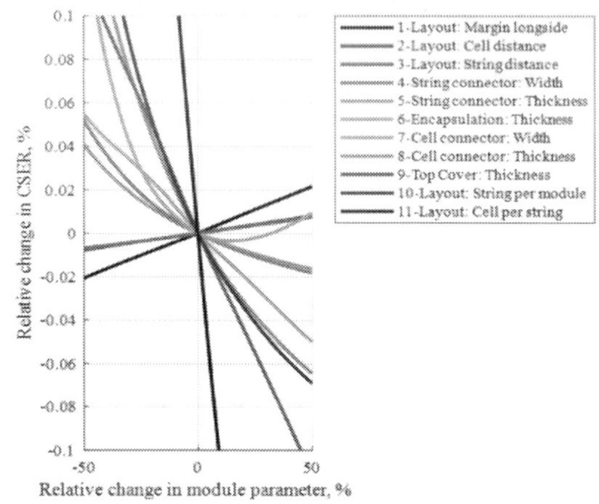

Figure 9: Spider plot depicting the sensitivity of the CSER of module B with respect to module design parameters

Finally, the effect on the yield per area of improving five selected parameters by 20% in the design of module B can be observed in Figure 10. The yield per area of the changed design is 301.8 W/m²/year, which is 1.45% better than the initial design and is now as good as the reference full cell module. Regarding the rated efficiency and rated power at STC, they are now 19,74% and 345,0 Wp, respectively, which are both better than the reference values of both modules.

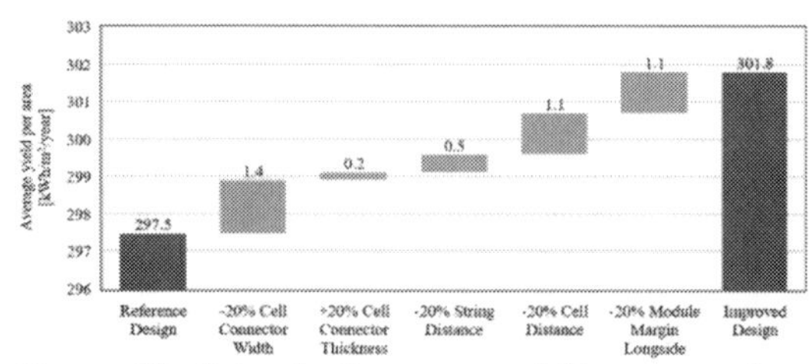

Figure 10: Gains in the average yield per area from sequential changes in the design of module B

4 CONCLUSIONS

A methodology for studying and quantifying the effect of module design parameters on the rated yield of a PV module, using Cell-to-Module simulations, together with the weather data and algorithm from the IEC 61853, is presented. It is also shown how the vCSER can be used for relating the rated power with the rated yield of a module.

It is expected that the vCSER approach assists module manufacturers in their decision making and speeds up the development of photovoltaic modules with an improved performance, by reducing the need for prototyping and testing specially in the early stages of product development.

6 AKOWLEDGEMENT

We would like to thank the German Federal Ministry for Economic Affairs and Climate Action (FKZ 0324289 and 03EE1028) for its funding.

7 REFERENCES

[1] International Electrotechnical Commission, Photovoltaic devices – Part 1: Measurement of photovoltaic current-voltage characteristics, 3rd ed., International Electrotechnical Commission (IEC), Geneva, Switzerland, 2020.

[2] IRENA, Renewable Power Generation Costs in 2020 (2021).

[3] International Electrotechnical Commission, Photovoltaic (PV) module performance testing and energy rating – Part 1: Irradiance and temperature performance measurements and power rating, 1st ed., IEC Central Office, Geneva, Switzerland, 2011.

[4] International Electrotechnical Comission, Photovoltaic (PV) module performance testing and energy rating – Part 2: Spectral responsivity, incidence angle and module operating temperature measurements, 1st ed., IEC Central Office, Geneva, Switzerland, 2016.

[5] International Electrotechnical Commission, Photovoltaic (PV) module performance testing and energy rating – Part 3: Energy rating of PV modules, 1st ed., IEC Central Office, Geneva, Switzerland, 2018.

[6] International Electrotechnical Commission, Photovoltaic (PV) module performance testing and energy rating – Part 4: Standard reference climatic profiles, 1st ed., IEC Central Office, Geneva, Switzerland, 2018.

[7] N. Dodd, N. Espinosa, Preparatory study for solar photovoltaic modules, inverters and systems, 2019.

[8] A. Wade, et al., The Product Environmental Footprint (PEF) of photovoltaic modules-Lessons learned from the environmental footprint pilot phase on the way to a single market for green products in the European Union, Prog Photovolt Res Appl 26 (2018) 553–564. https://doi.org/10.1002/pip.2956.

[9] D. Polverini, et al., Building Criteria for Energy Labeling of Photovoltaic Modules and Small Systems, Solar RRL 6 (2022) 2100518. https://doi.org/10.1002/solr.202100518.

[10] European Commission. Joint Research Centre., Preparatory study for solar photovoltaic modules, inverters and systems: final report, Publications Office, 2020.

[11] A. Wade, et al., The Product Environmental Footprint (PEF) of photovoltaic modules-Lessons learned from the environmental footprint pilot phase on the way to a single market for green products in the European Union, 2021.

[12] J.C. Blakesley, et al., Accuracy, cost and sensitivity analysis of PV energy rating, Progress in Solar Energy 1 203 (2020) 91–100. https://doi.org/10.1016/j.solener.2020.03.088.

[13] N. Riedel-Lyngskær, et al., Interlaboratory comparison of angular-dependent photovoltaic device measurements: Results and impact on energy rating, Prog. Photovolt: Res. Appl. 29 (2021) 315–333. https://doi.org/10.1002/pip.3365.

[14] M.R. Vogt, et al., Interlaboratory Comparison of the PV Module Energy Rating Standard IEC 61853-3. 5 pages / 37th European Photovoltaic Solar Energy Conference and Exhibition; 811-815 (2020). https://doi.org/10.4229/EUPVSEC20202020-4BO.13.2.

[15] I. Hädrich, et al., Unified methodology for determining CTM ratios: Systematic prediction of module power, Sol Energ Mat Sol C 131 (2014) 14–23. https://doi.org/10.1016/j.solmat.2014.06.025.

[16] M. Mittag, et al., Electrical and thermal modeling of junction boxes, in: Proceedings of the 33rd European Photovoltaic Solar Energy Conference and Exhibition (EU PVSEC), Amsterdam, Netherlands, 2017, pp. 1501–1506.

[17] A. Pfreundt, et al., Rapid Calculation of the Backsheet Coupling Gain Using Ray Groups, in: 35th European Photovoltaic Solar Energy.

[18] A. Pfreundt, M. Mittag, J. Shahid, Cell-To-Module Analysis Beyond Standard Test Conditions, in: 47th IEEE Photovoltaic Specialists Conference 2020.

[19] I. Haedrich, D.C. Jordan, M. Ernst, Methodology to predict annual yield losses and gains caused by solar module design and materials under field exposure, Solar Energy Materials and Solar Cells 202 (2019) 110069. https://doi.org/10.1016/j.solmat.2019.110069.

[20] A. Thomson, et al., Impact of PV module configuration on energy yield under realistic conditions, Opt Quant Electron 49 (2017) 329. https://doi.org/10.1007/s11082-017-0903-0.

[21] M. Mittag, et al., Approach for a Holistic Optimization from Wafer to PV System, in: Proceedings of the 7th World Conference on Photovoltaic Energy Conversion, Hawaii, 2018.

[22] Fraunhofer Institute for Solar Energy Systems ISE, SmartCalc.Module. http://www.cell-to-module.com (accessed 1 September 2022).

[23] M. Mittag, M. Ebert, Systematic PV module optimization with the cell-to-module (CTM) analysis software, Photovoltaics International (2017) 97–104.

[24] International Electrotechnical Commission, Photovoltaic Devices - Part 8: Measurement of Spectral Response of a Photovoltaic (PV) Device, International Electrotechnical Commission (IEC), Geneva, Switzerland, 1998.

[25] N. Martín, J.M. Ruiz, A new model for PV modules angular losses under field conditions, International Journal of Solar Energy 22 (2002) 19–31. https://doi.org/10.1080/01425910212852.

[26] D. Faiman, Assessing the outdoor operating temperature of photovoltaic modules, Prog. Photovolt: Res. Appl. 16 (2008) 307–315. https://doi.org/10.1002/pip.813.

[27] M. Mittag, et al., Thermal Modelling of Photovoltaic Modules in Operation and Production, in: 36th European Photovoltaic Solar Energy Conference and Exhibition (EUPVSEC) 2019.

[28] M. Mittag, et al., Techno-Economic Analysis of Half Cell Modules - The Impact of Half Cells on Module Power and Costs, 2019.

[29] N. Klasen, et al., A Comprehensive Study of Module Layouts for Silicon Solar Cells under Partial Shading, IEEE Journal of Photovoltaics (2022). https://doi.org/10.1109/JPHOTOV.2022.3144635.

DIFFERENTIATION OF PV DEVICES ACCORDING TO ENERGY RATING IEC 61853

Ana María Gracia-Amillo*[1], Robert P. Kenny[1], Diego Pavanello*[1]
European Commission, Joint Research Centre (JRC), Ispra (VA), Italy
e-mail: *[1]agracia@cener.com (now with Cener); [1]robert.kenny@ec.europa.eu;
*[1]Tel: +39 0332786449 diego.pavanello@ec.europa.eu

ABSTRACT: The energy rating IEC 61853 Standard series "*Photovoltaic (PV) module performance testing and energy rating*" provides a standardized measure of the performance of PV devices considering real working conditions of ambient temperature and irradiation, representative of worldwide reference climates relevant for PV installations. The *Climatic Specific Energy Rating* (*CSER*) parameter defined in the standard could serve as a tool for the comparison of PV technologies and devices. The scope of this paper is the validation of the ability of the *CSER* to differentiate between PV technologies and devices of the same technology, evaluating how much of the observed differences are due to uncertainties in the estimation chain or input data other than the result of the intrinsic behavior of the considered devices. Ten devices representative of commercially available PV devices have been analyzed. The observed differences, higher than the associated uncertainty in the estimation of the *CSER* parameter, indicates the ability of the standard to differentiate between different technologies and devices of the same technology. Furthermore, the variability in the estimated performance depends on the climatic conditions.
Keywords: Energy Rating, Evaluation, PV Module, Qualification and Testing

1 INTRODUCTION

PV modules are currently compared, rated and priced based on their power output or their efficiency measured under Standard Test Conditions (STC) [1, 2], even though these specific conditions of module temperature and irradiance (broadband and spectral content) hardly ever occur when PV modules are installed outdoors. Despite this accepted and extended rating parameter, the most significant parameter for PV users would be the expected energy yield over the lifetime of the device. However, at present, there are no standards defining a methodology for the estimation of such lifetime performance, which depends on many factors. On the one hand, factors related to the device being evaluated, like spectral response, thermal or light behavior or the degradation rate. And on the other hand, factors relative to the specific location like the long-term climatic characterization, presence of shadows or soiling; and factors relative to the installation and configuration, like the inclination and orientation angles, or the operation and maintenance activities.

In opposition to the maximum power output measured under the specific STC ($P_{max,\,STC}$), the IEC 61853 Standard series "*Photovoltaic (PV) module performance testing and energy rating*" [3-6] describes a methodology to evaluate the performance of PV devices under real working conditions. These are defined by six distinctive climatic profiles, which represent the climatic conditions that most PV installations worldwide will be subjected to. In addition to the diverse weather conditions used in the simulations in terms of irradiance, ambient temperature and wind speed, the methodology to estimate the power output of the PV device takes into consideration as well, the device's characteristics such as the angle of incidence or spectral response and its behavior under the varied temperature and irradiation conditions of the climatic profiles.

The output of the standard, the *Climatic Specific Energy Rating* (*CSER*), is a dimensionless parameter which relates the estimated energy yield of a PV device under the aforementioned predefined climatic conditions over one year to the energy yield that would have been obtained if the module efficiency would have constantly been the one measured under STC. Therefore, the *CSER* could serve as a tool to compare and rate PV devices in a more meaningful way than by just considering the power at STC ($P_{max,\,STC}$) thanks to all the factors considered in the estimation of the *CSER*.

To confirm this hypothesis, the aim of the present paper is the validation of the ability of the energy rating standard and the *CSER* to discriminate between different currently commercially available PV technologies, as well as between devices of the same PV technology. In particular, this paper will analyze how diverse the *CSER* values, calculated based on detailed module measurements, could be. The present paper will also evaluate how much of these differences are due to the devices' characteristics or to uncertainties in the *CSER* estimation, by comparing the magnitude of the observed differences in relation to the uncertainties.

Even though the IEC 61853 Standard series was completed in 2018 with the publication of Part 3 and 4 [5, 6], to our knowledge, it hasn´t been implemented by PV manufacturers yet, as the *CSER* has not been included in the PV module information datasheets. In this regard, in the framework of the Ecodesign Directive [7] and the Energy Labelling Regulation [8], the European Commission is working on the implementation of these tools to PV modules and systems [9]. The energy label, for example, could be defined based on the estimated lifetime energy yield from the PV device under consideration. According to the proposed methodology, this lifetime performance would be estimated from the *CSER* values in the various climates relevant to Europe. Therefore, if this Regulation was implemented, manufacturers could be requested to provide the *CSER* parameter or the necessary input data for its estimation in the PV module datasheet. Just like the A to G energy labeling scheme provides stakeholders an information tool to properly evaluate the "energy efficiency" of PV modules and enable an informed decision when selecting among various products, the *CSER* parameter could also serve to that purpose allowing the selection of the most suitable PV device for the location of interest.

2 IEC 61853 STANDARD SERIES

The International Electrotechnical Commission (IEC) Standard series 61853 "*Photovoltaic (PV) module performance testing and energy rating*" describes a methodology to estimate the performance of PV devices in terms of the yearly energy output ($E_{mod,year}$, kWh) and the *CSER*. These are obtained assuming no degradation, one single year of climatic data and using a fixed configuration of ground mounted fixed rack facing the equator with an inclination angle of 20°. Simulation conditions that do not make this procedure suitable for the estimation of the energy yield of a specific module and PV installation. For this task specific information about the PV installation is required.

The IEC 61853 Standard series consists of four parts as described in the following:

2.1 IEC 61853-1. Irradiance and temperature performance measurements and power rating

The first part of the Standard series [3] describes the methodology to obtain the so-called power matrix, which is a collection of power output measurements over a range of PV installation relevant in-plane irradiance (from 100W/m^2 to 1100W/m^2) and module temperature (from 15°C to 75°C) levels.

2.2 IEC 61853-2. Spectral responsivity, incidence angle and module operating temperature measurements

Part 2 of the Standard series [4] describes the experimental procedures to determine the set of parameters required to quantify the various effects needed for the estimation of the PV power output. These include the angle of incidence (AOI) effect by means of parameter *ar*, the spectral effects by means of the spectral response of the device (*SR*) and the thermal behavior of the device by means of two parameters (U_0 and U_1).

2.3 IEC 61853-3. Energy rating of PV modules

The third part of the Standard series [5], describes the complete model, concatenation of various steps, to estimate the yearly energy yield of the PV device under test and the *CSER*. It requires as inputs the parameters obtained in Part 1 and 2 and the climatic datasets of Part 4.

2.4 IEC 61853-4. Standard reference climatic profiles

The last part of the Standard series [6] contains the six standard reference climatic profiles named: high elevation above 3000m, subtropical coastal, temperate continental, subtropical arid, temperate coastal and tropical humid.

Every climatic profile is described by a time series of hourly values over one year of ambient temperature, wind speed and irradiance at 2m above the ground. Global and direct horizontal and in-plane broadband irradiance are defined, along with the spectrally resolved global in-plane irradiance integrated into 32 spectral bands. The angle of incidence between the sun and the normal to the module's surface is also reported, as it is required for the AOI effect estimation.

Figure 1 shows, for each of the six reference climates, the monthly average global horizontal (*G*) and beam (*Gb*) irradiance (W/m^2) and the monthly maximum, minimum and average ambient temperature (°C). There are significant differences between the reference climates. Not only the amount of monthly irradiation is significantly different between climates, like between the subtropical arid and the temperature coastal, but also the proportion of this monthly irradiation being direct irradiance can differ substantially. While the irradiation received in the subtropical arid climate is mainly direct, this is mainly diffuse in the tropical humid climate or in the temperate coastal. There are also significant differences in the ambient temperatures. While temperatures are very stable and elevated all through the year in the tropical humid region, the temperature range reaches 40°C in the high elevation climate, when modules will be submitted to considerable thermal gradients. This is also the case of the temperate continental climate where maximum and minimum monthly ambient temperatures can vary by as much as 30°C.

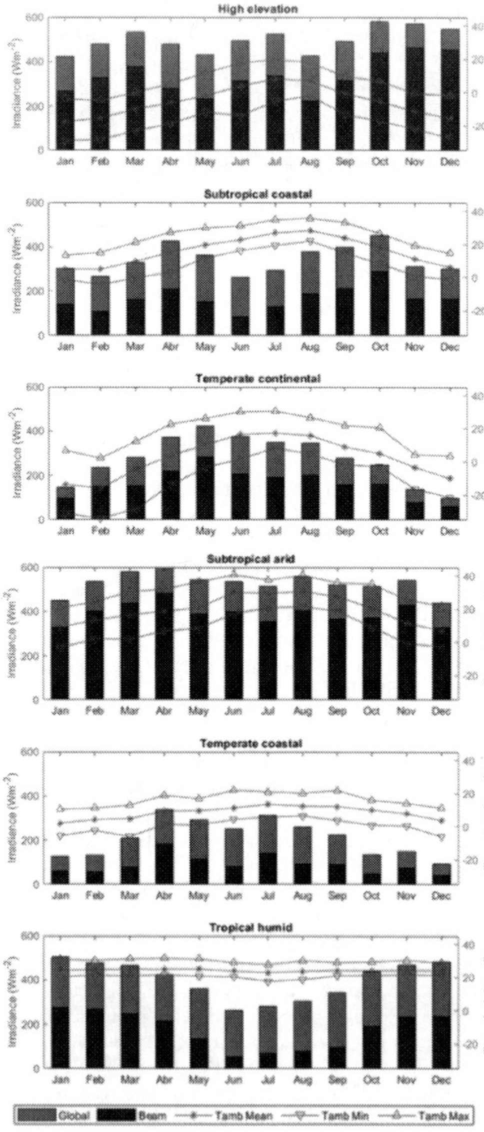

Figure 1: Average monthly *G* and *Gb* irradiance and temperature range for each reference climatic dataset.

These working conditions provide information about the performance of the PV device. For example, technologies whose performance is affected by temperature will tend to perform worse in the subtropical arid and tropical humid climates, characterized by high temperatures, compared to cooler climates. And devices with low performance under low light conditions, would tend to perform poorly in locations of the temperate coastal region where there are low levels of irradiation which is also mainly diffuse.

2.5 Uncertainties on the energy rating estimation

The *CSER*, like any other estimated value has an associated uncertainty that should be reported. This can derive from possible errors in the determination of the input data and from the estimation models used. The determination of the *CSER* uncertainty is not straightforward due the amount of input data used and the various calculation steps applied. In fact, how the uncertainty in the various input data and applied models is propagates throughout the different steps of the energy rating methodology into the final *CSER* value is not defined nor quantified in the IEC 61853 Standard [10].

One of the main factors that explain the different performance of PV devices is their irradiance and temperature dependence which is modelled, according to the energy rating standard, based on power matrix measurements defined in Part 1 and the linear interpolation and extrapolation procedure defined in Part 3. Each power matrix measurement is subjected to uncertainties and these often depend on the target conditions, which are commonly optimized for STC. Therefore, measurements at other points in the power matrix will most likely present higher uncertainty values. Other input data, such as the spectral response, the coefficients to model the thermal behavior of the PV device or the angle of incidence effects are also subjected to errors, which are then propagated in the estimation chain to calculate the final *CSER* value. On the contrary, the input data relative to the climatological conditions provided in Part 4 are fixed reference datasets and are not considered to contribute to uncertainty [10, 11]. In addition to the uncertainty in the input data, the models integrated in the IEC 61853-3 methodology may not be equally accurate for all PV technologies and devices due their intrinsic characteristics, which may derive in additional errors for some technologies. However, the standard is assumed to be technology agnostic, so it does not penalize some technologies over others. Therefore, the only source of uncertainty in the energy yield and *CSER* estimation, according to the Standard series, is from uncertainties in the input parameters described in Part 1 and 2 [11].

An exhaustive uncertainty study was carried out in [11] analyzing the sensitivity of the uncertainty of the energy rating output to different sources of uncertainty in the input measurements. Authors used a generic measurement uncertainty model, adding the uncertainties in measurements of Part 1 and 2 parameters declared by accredited laboratories employing a Monte Carlo approach. This analysis considered the same uncertainty in the determination of each parameter for all module types, i.e., assuming there is no difference in measurement uncertainty between modules types and that all technologies can be measured as accurately as c-Si.

The study concluded that the main uncertainty in the *CSER* arises from the measurement of the in-plane irradiance during the power matrix determination (due to the instrument calibration chain), followed by the definition of the two parameters used to estimate the module temperature (U_0 and U_1), especially in the warmest climates. Due to the low sensitivity of the *CSER* to the a_r parameter or the SR values, errors on these parameters are not significant on the final *CSER* value.

According to [11] based on information provided by accredited laboratories, the average uncertainty in the measurement of $P_{max,\ STC}$ is 1.88%, while the average uncertainty in the estimation of the energy yield and *CSER* in the different climates are respectively 2.2% and 1.9%. The *CSER*, calculated normalizing by $P_{max,\ STC}$ has a reduced impact of the irradiance measurement errors. According to this, for devices whose *CSER* values differ by less that this threshold, it may not be possible to define whether differences are due to their intrinsic characteristics and performance or due to uncertainties in the estimation of the *CSER* values.

The uncertainties of the energy yield and *CSER* obtained in the analysis [11] for the different six reference climates are shown in Table I.

Table I: Uncertainties (%) in the energy yield (EY) and *CSER* estimation at the six reference climates as defined in [11]

Reference climate	Uncertainty EY (%)	Uncertainty *CSER* (%)
High elevation	2.21	2.04
Subtropical coastal	1.99	1.62
Temperate continental	2.17	1.88
Subtropical arid	2.18	1.94
Temperate coastal	2.19	1.82
Tropical humid	2.35	2.08
Average	2.18	1.90

3 *CSER* ANALYSIS

3.1 Considered PV devices

Eight crystalline silicon and two CdTe PV modules, representative of commercially available PV devices, have been characterized in accordance to Part 1 and 2 of the IEC 61853 Standard series. Not all devices could be fully characterized for AOI and thermal temperature, and the a_r, U_0 and U_1 parameters used in the *CSER* estimation are obtained from a subset of these devices and applied the crystalline silicon modules and CdTe modules appropriately.

Table II shows the power output at STC ($P_{max,\ STC}$), the module's area and efficiency at STC (Eff_{STC}) of the analyzed PV devices, which varies between 11.32% of one of the CdTe devices and 22.07% of one c-Si based device.

Table II: Main characteristics of the eight c-Si and two CdTe devices analyzed in the present study.

Device - Tech	$P_{max,\ STC}$ (W)	Area (m^2)	Eff_{STC} (%)
D1 c-Si	330	1.844	17.89
D2 c-Si	330.3	1.844	17.91
D3 c-Si	343.9	1.709	20.13
D4 c-Si	342.5	1.709	20.05
D5 c-Si	389.8	1.766	22.07
D6 c-Si	303	1.937	15.64
D7 c-Si	329.9	1.674	19.70
D8 c-Si	271.4	1.633	16.61
D1 CdTe	81.4	0.719	11.32
D2 CdTe	391.5	2.398	16.33

3.2 Input data for the *CSER* estimation

The *CSER* values have been calculated for each of the six reference climatic profiles, considering the power matrix and *SR* of each device. However, it has not been possible for this analysis to obtain the AOI and thermal coefficients for each device. Only two c-Si devices and one CdTe have been fully characterized and their set of parameters have been applied to the other devices of the same technology.

As a result of the available a_r and U_0 and U_1 parameters from two devices, each c-Si device could be simulated four times by combination of the two a_r values and two sets of U_0 and U_1 parameters. The a_r values differ by 12.5%, while the U_0 and U_1 parameters differ by 5.6% and 8.6% respectively. Due to these differences, in our simulations the main impact on the EY and *CSER* estimation derives from the a_r value used, even though the sensitivity to this parameter is lower [11]. In fact, IEC 61853-2 allows omitting the tests to obtain the a_r parameter for flat uncoated front glass plate modules and use standard values instead without significantly affecting the accuracy of the energy rating. According to our simulations, changes by as much as 20% on the a_r value affect the *CSER* estimated value by 1.37% in the most affected climate, temperate coastal. In the subtropical arid reference climate this impact is reduced to 0.89% when the a_r value is reduced by 20%. The same 20% variation is applied for the U_0 value, and the *CSER* changes by as much as 1.6% in the tropical humid climate and only 0.39% in the temperate coastal. Similarly for the U_1 value, representative of the wind effect on the module temperature, the *CSER* changes by 0.57% in the subtropical arid and 0.09% in the high elevation climate. Uncertainties in the U_0 determination would have a higher impact on the *CSER* values than the U_1 parameter.

As an example of the sensitivity analysis performed, Figure 2 shows the variation, in percentage, of the *CSER* and AOI corrected in-plane diffuse (*GTdAOI*), direct (*GTbAOI*) and global (*GTAOI*) irradiance as the initial value of a_r is modified by as much as 20%, for one of the c-Si devices in the temperate coastal region. These values give an approximation of the impact of possible uncertainties in the definition of the input data on the final *CSER* values.

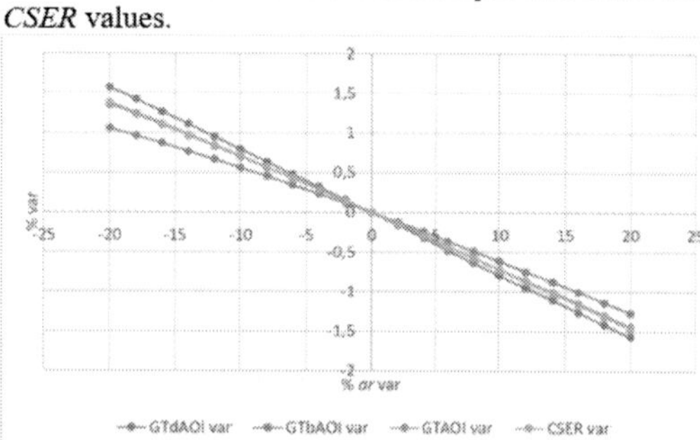

Figure 2: Variation on the in-plane AOI corrected irradiance and *CSER* values as a_r parameter is modified in the temperate coastal region.

Figure 3 shows the spectral response values of the 10 devices. There are differences between technologies and between devices of the same technology mainly at the UV wavelengths and between 900 and 1100nm for the c-Si devices.

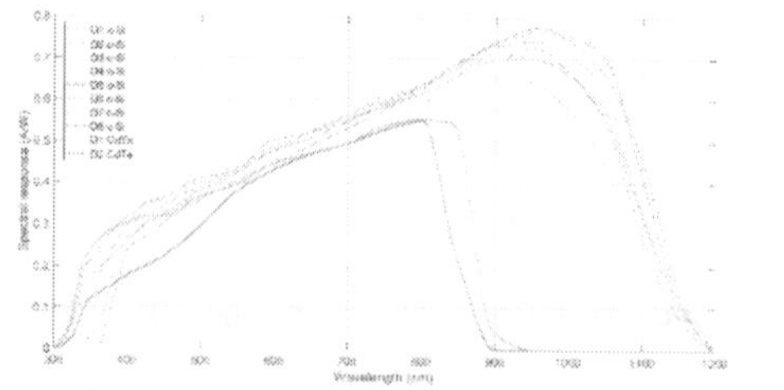

Figure 3: Spectral response of the considered PV devices.

As an example of the power matrix measurements, Figure 4 shows the measurements at 1000W/m^2, normalized to the value at 15ºC of the different devices. The different thermal behavior is noticeable especially as the module temperature increases.

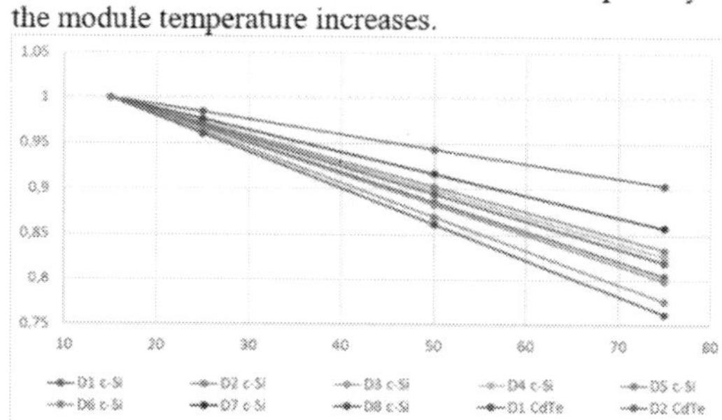

Figure 4: Variation of the normalized power output registered at 1000W/m^2 at the different measured module temperatures for the different devices.

3.2 Estimated *CSER*

For simplicity, only the *CSER* values obtained using one set of a_r and U_0 and U_1 parameters will be shown and discussed.

Table III shows the obtained *CSER* values for the two CdTe devices, along with the relative difference between the two for the different reference climates. The same comparison with the efficiency at STC is also shown. Even though the second device has an STC efficiency 44.25% higher than the first device, the *CSER* values are lower in all considered climates. However, differences in the energy rating parameter are lower, ranging between 3.78% and 9.15%, depending on the reference climate.

Table III: *CSER* estimated values for the two considered CdTe devices, efficiency at STC and relative difference between both devices.

Reference climate	D1 CdTe	D2 CdTe	Diff (%)
High elevation	0.9814	0.9443	-3.78
Subtropical coastal	1.0238	0.9448	-7.72
Temperate continental	1.0178	0.9410	-7.55
Subtropical arid	0.9658	0.9004	-6.76
Temperate coastal	1.0446	0.9490	-9.15
Tropical humid	1.0207	0.9297	-8.91
Eff STC (%)	11.3	16.3	44.25

Due to the lack of specific input data about the AOI and thermal effects, the AOI corrected irradiance and the estimated module temperature, which depends on this AOI corrected irradiance, would be the same for both devices. As a result, the observed differences in performance derive from the different spectral responses and power matrix measurements of each CdTe device. The performance of D2 is more homogeneous in the various climates than D1.

Besides, the device D2 is more affected by temperature than D1, as observed from the power matrix measurements (example case shown in Figure 4) and the *CSER* values of the warmest climatic regions.

Similarly to the results of the CdTe devices, even though there were more input data available regarding the AOI and thermal effects, the results shown here for the c-Si devices have been obtained assuming the same *ar*, U_0 and U_1 parameters for all eight c-Si devices. Therefore, the differences on the *CSER* derive from the power matrix measurements and the spectral response data.

Figure 5 shows the obtained *CSER* values for the eight c-Si devices in the six reference climates. Compared to the CdTe devices, the device D1 (CdTe) has *CSER* values in the range of those obtained by the c-Si devices or even higher in certain climates. While D2 (CdTe) always results in lower *CSER* values than the c-Si devices.

The c-Si device D6, which has the lowest STC efficiency (15.64%) obtains in all reference climates the lowest *CSER*. Conversely, the c-Si device with the highest STC efficiency, D5 (22.07%), does not outperform the rest of devices, in terms of *CSER*, in any reference climate, in contrast with what might be expected from the STC data alone. In fact, device D8, with 16.61% STC efficiency has higher *CSER* values than the rest of devices in the High elevation reference climate, whose conditions of high irradiation and low temperatures benefit the performance of c-Si technology generally, but this one the most. Depending on the reference climate, one c-Si device or another, could be considered the most suitable one as the device with the highest *CSER* value. For example, D4 has the highest performance under the tropical humid conditions, D7 has the highest *CSER* under the subtropical regimes, and D3 has the best performance under the temperate climates, as shown in Figure 5.

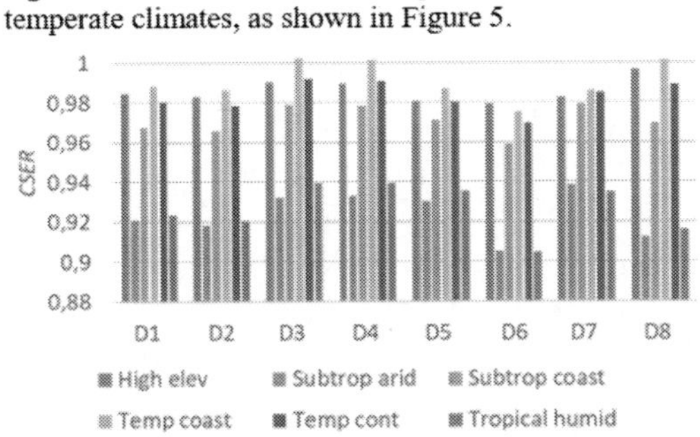

Figure 5: Estimated *CSER* values for the analyzed c-Si devices on the six reference climates.

As observed in Figure 5, the performance of the c-Si devices changes significantly between the different climatic conditions. The differences between the *CSER* values in the most favorable and least favorable climate conditions ranges between almost 10% in the case of the D8 device and 5.5% for the D7.

However, within each reference climate the *CSER* values of the different c-Si devices are similar, as shown in Figure 6. The differences between the maximum and minimum *CSER* value in every climate goes between 1.74% in the high elevation climate to 3.82% in the tropical humid climatic conditions.

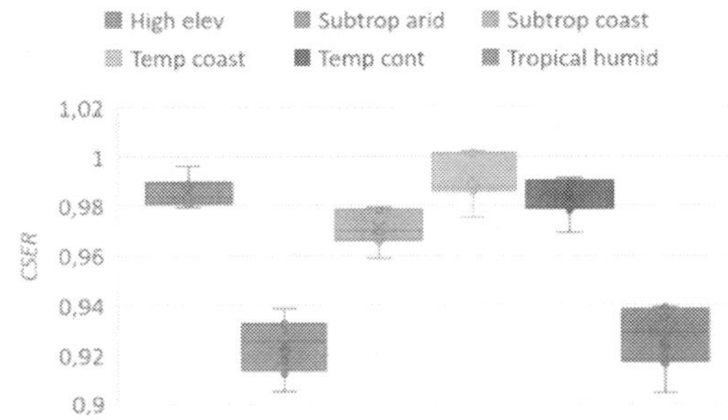

Figure 6: Estimated *CSER* values for the analyzed c-Si devices on the six reference climates.

4 CONCLUSIONS

The *Climatic Specific Energy Rating* (*CSER*) parameter of the IEC 61853 energy rating Standard series could serve end-users when selecting the most suitable PV device for the location of interest, in addition to the commonly used power output or efficiency measured under STC. Despite this application, the standard hasn´t to our knowledge been widely implemented by PV manufacturers. However, if the proposal of the European Commission to include PV modules in the Energy Labelling Regulation is approved, since this A to G classification may be based on the *CSER* parameter, it could make it necessary for manufacturers to include this parameter or the input data needed for its estimation into the PV module datasheet.

Considering this possible future implementation, the aim of this paper is to validate the capacity of the *CSER* parameter to distinguish between devices of the same and/or different PV technology. Additionally, these differences are compared to the *CSER* uncertainty to define whether the observed differences are robust. Eight c-Si devices and two CdTe devices representative of commercially available devices have been analysed according to the IEC 61853 Standard series. The complete characterization of all devices was not possible so the effect of the AOI and the thermal behaviour of every device could not be analysed separately for each of them. The estimated *CSER* values based on resulting from the different spectral response data and the power matrix for each particular device, show significant differences. These differences could help end-users select the most appropriate device for each location. The differences in performance vary with climatic region, the high elevation reference climate where c-Si device CSER values are more similar. However, under warmer climates, the c-Si devices show higher variability. As has been observed, a PV device classification based on the STC efficiency does not always correlate with a classification based on the *CSER* parameter, which provides a better estimate of the performance of the device under real working conditions.

A detailed analysis on the uncertainty associated to the estimation of the *CSER* value [11], concluded that the main sources derive from the irradiance measurements on the power matrix and from the definition of the two thermal coefficients. This uncertainty was estimated as 1.9% as the average for the six reference climates. However, the uncertainty depends on the reference climate as indicated on Table I. Except for the high elevation climate, the uncertainty is lower than the variability in the observed *CSER* values for the eight c-Si devices.

Considering as well the two CdTe devices, uncertainties are significantly lower than the observed *CSER* variability between the ten PV devices analysed. This indicates the ability of the *CSER* parameter to distinguish between devices, as the observed differences tend to be higher than the uncertainty in the estimation of the *CSER* parameter.

5 FUTURE LINES OF RESEARCH

It has not been possible for the present paper to fully characterize all considered PV devices according to the complete IEC 61853 Standard series. In particular, it was not possible to define the AOI and thermal parameters (a_r, U_0 and U_1) for all devices. Once data relevant to these two effects become available, the study presented in this paper could be repeated and extended using the complete characterization of every device. Differences on the *CSER* values are expected to increase when all effects are considered, which would contribute to the differentiation of devices.

Additionally, extending this study considering a higher number of devices would be beneficial for the PV community. Besides, as new PV devices and technologies are developed, they should be analysed according to the IEC 61853 Standard series.

Further research on the uncertainty of the *CSER* parameter should be performed, especially considering the fact that, according to [11], it would be possible to perform the energy rating analysis with a lower number of measurements in the power matrix, without losing accuracy on the final *CSER* estimation. Since these power matrix irradiance measurements were identified as one of the main sources of the *CSER* uncertainty, this new approach could reduce the uncertainty on the *CSER* parameter and therefore enabling a better differentiation of PV devices.

The other main source of uncertainty in the *CSER* estimation derives from the determination of the two thermal parameters (U_0 and U_1) [11] defined in Part 2 of the IEC 61853 Standard series. Future revisions of Part 1 and 2 could simplify the methodology for obtaining these two parameters and the power matrix, which could decrease their uncertainty and consequently the *CSER* uncertainty.

6 BIBLIOGRAPHY

[1] IEC TS 61836:2016 ED3. Solar photovoltaic energy systems: Terms, definitions and symbols. Edition 3.0 International Electrotechnical Commission (2016).

[2] IEC 60904-3:2019 ED4. Photovoltaic devices – Part 3: Measurements principles for terrestrial photovoltaic (PV) solar devices with reference spectral irradiance data. Edition 4.0 International Electrotechnical Commission (2019).

[3] IEC 61853-1. Photovoltaic (PV) module performance testing and energy rating – Part 1: Irradiance and temperature performance measurements and power rating. Edition 1.0 International Electrotechnical Commission (2011).

[4] IEC 61853-2. Photovoltaic (PV) module performance testing and energy rating – Part 2: Spectral responsivity, incidence angle and module operating temperature measurements. Edition 1.0 International Electrotechnical Commission (2016).

[5] IEC 61853-3. Photovoltaic (PV) module performance testing and energy rating – Part 3: Energy rating of PV modules. Edition 1.0 International Electrotechnical Commission (2018).

[6] IEC 61853-4. Photovoltaic (PV) module performance testing and energy rating – Part 4: Standard reference climatic profiles. Edition 1.0 International Electrotechnical Commission (2018).

[7] Directive 2009/125/EC of the European Parliament and of the Council of 21 October 2009 Establishing a Framework for the Setting of Ecodesign Requirements for Energy-Related Products, OJ L285, 31.10.2009, p. 10–35, https://webstore.iec.ch/home.

[8] Regulation (EU)2017/1369 of the European Parliament and of the Council of 4 July 2017 Setting a Framework for Energy Labelling and Repealing Directive 2010/30/EU, OJ L198, 28.7.2017, p. 1–23, https://webstore.iec.ch/home.

[9] Polverini, D., Gracia Amillo, A., Taylor, N., Sample, T., Salis, E., Dunlop, E.D. Building criteria for energy labeling of photovoltaic modules and small systems. *Solar RRL* (2021) 2100518.

[10] IEA PVPS Task 13. Performance, Operation and Reliability of Photovoltaic Systems. Climatic rating of photovoltaic modules: different technologies for various operating conditions. Report IEA-PVPS T13-20:2020. ISBN 978-3-907281-08-6.

[11] Blakesley, J.C., Huld, T., Müllejans, H., Gracia-Amillo, A., Friesen, G., Betts, T.R., Hermann, W. Accuracy, cost and sensitivity analysis of PV energy rating. *Solar Energy* (2020), 203, 91-100.

COMPARISON OF ENERGY PERFORMANCE OF VERTICAL AND TILTED BIFACIAL PV MODULES IN ARID ENVIRONMENTS

Maulid Kivambe*, Giovanni Scabbia, Amir Abdallah, Benjamin Figgis and Juan Lopez-Garcia
Qatar Environment and Energy Research Institute (QEERI), Hamad Bin Khalifa University (HBKU)
P.O. Box: 34110 Doha, Qatar

ABSTRACT: This work reports on field data results of energy performance for bifacial crystalline silicon (c-Si) photovoltaic (PV) modules in a desert climate. Energy yields of south-facing tilted (22°) and vertical (90°) E-W oriented modules were compared. The main objective is to study energy yield in real operating conditions for soiled and cleaned module and investigate bifacial module especially the vertical configuration as a soiling mitigation approach. For tilted modules, soiling losses increased with time up to 60% and 45% for monofacial and bifacial modules respectively. Generally, in average clean tilted bifacial module slightly (<5.2%) outperformed clean vertical module. However, vertical module outperformed the tilted module by up to 8% during the summer. No significant soiling losses were observed for the vertical module. The results show that bifacial PV is a natural soiling mitigator and vertical module installation is feasible and has a strong potential as a soiling mitigation strategy in desert climates.

Keywords: Bifacial, c-Si, Performance, PV Module.

1 INTRODUCTION

Bifacial crystalline silicon (Si) photovoltaic (PV) devices have become a competitive technology because they can produce additional output energy in comparison to conventional monofacial modules. A market share of 40% is expected by 2025 [1]. Desert climates present some advantages for PV modules such as high irradiation and high natural albedo but also challenges such as high soiling rates [2]. The rear-side of bifacial module experiences little or no soiling compared to the front-side. Furthermore, bifacial PV modules allow for vertical East-West orientation [3-6] that brings interesting advantages compared to conventional technologies installed in a tilted plane and might be considered as a passive soiling mitigation approach that achieves reduced soiling losses and cleaning cost [7]. In addition, the vertical mounting allows for shared land use (agroPV [8], sound barriers and integration in building facades [9]), harvesting the solar radiation available when the sun is low at sunrise (East) and sunset (West), higher electricity value derived from the production at "off-peak" times (early morning and late afternoon) [10, 11]. Nonetheless, due to the multiple variables that influence the performance of such devices, their characterization lacks consistency. In particular, for bifacial systems operating under real operating conditions improvements can be made in achieving meaningful characterization for the long-term energy yield assessment in specific harsh desert climates. This work reports on energy yield data for soiled and cleaned south-facing tilted and vertical E-W oriented bifacial module and provide valuable insights on the significant potential of bifacial modules, particularly vertical installation, for soiling mitigation and reduction of cleaning costs in desert climates.

2 EXPERIMENTAL

Commercial bifacial c-Si PV modules with Pmax bifaciality (φPmax) 90% were installed in the Outdoor Test Facility (OTF) at QEERI, in Doha (Qatar, 25°19'37.36" N, 51°25'58.44" E, at 15 m above sea level) in a harsh desert climate. The test setup is shown in **Figure 1**. Two n-type 60 cells frameless glass-glass modules without any partial self-shading in the rear were installed vertically E-W oriented, about 0.8m ground clearance, and two were installed in conventional tilt (22° facing) about 1.2 m above ground level. In addition, two 22° tilted 60 cells frameless monofacial c-Si modules were used as references. One module of each configuration was weekly cleaned and the other was left intentionally naturally soiled. The variability of operating conditions - ground albedo, the in-plane front and rear side irradiances and module temperature were monitored with a series of sensors. The in-plane front and rear side irradiances were measured with calibrated c-Si reference cells and pyranometers in both bifacial modules, i.e., vertical and tilted. The DC power of each module (operated in MPPT) was recorded every 1 minute and filtered to remove outliers.

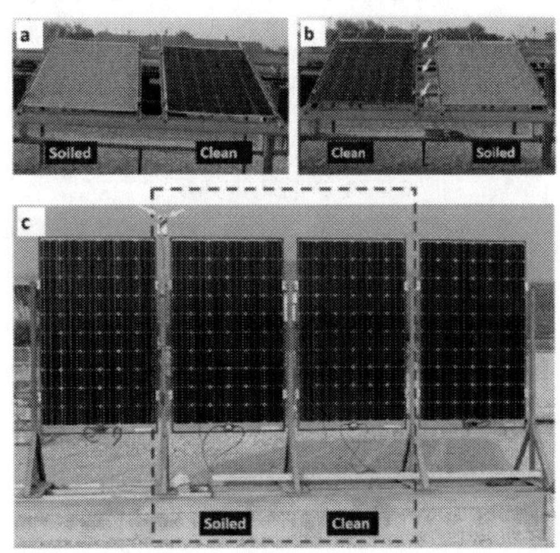

Figure 1: Image of the tilted (**a**) monofacial soiled and clean modules, (**b**) bifacial clean and soiled and (**c**) vertically mounted East-West oriented modules. Soiling is evident in the tilted modules, but not in vertical modules.

The location of radiation sensors are indicated by arrows.

3 RESULTS

The hourly and daily average albedo for the site during the study period (October 2019 – August 2022) was 0.45±0.03. **Figure 2** shows Irradiance for a clear sky day in January 2022. It is observed that the rear of tilted modules receive a small fraction of the plane of array irradiance, (G_POA) while both East and West sides of vertical modules receive significantly higher and similar irradiation.

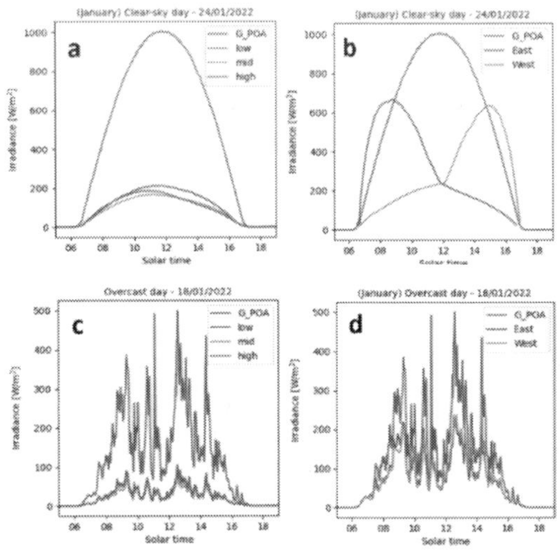

Figure 2: Irradiance for a clear sky day **(a, b)** and overcast day **(c, d)**, including the plane of array (G_POA), the low, mid and high rear irradiance for the tilted modules and the East and West irradiance for the vertical modules.

Comparing yield loss, i.e. energy difference in % of the soiled module from clean reference for the tilted modules (**Figure 3**), it is observed that the losses are cumulative. Between March and December 2021 Monofacial losses reached >60%. Bifacial losses were significantly lower, reaching ~45%. This shows that bifacial PV is naturally a soiling mitigator. Bifacial gain ranged from 20% (January) to 35% (July).

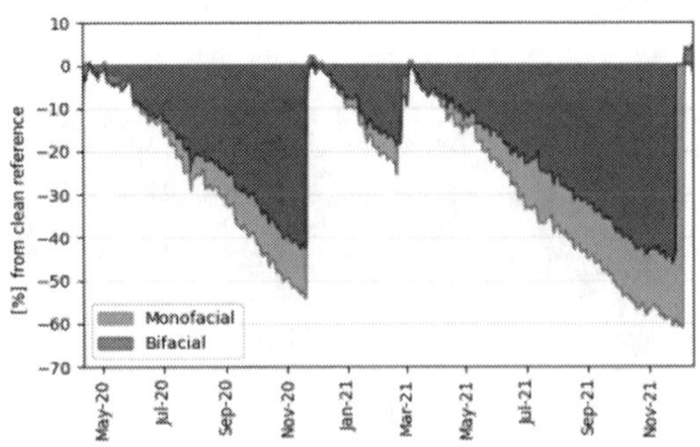

Figure 3: Difference between the monofacial soiled and clean and the bifacial soiled and clean. Values at 0 correspond to clean modules (when the soiled modules were completely cleaned, for example by rain events.

The energy yield comparison of clean and soiled bifacial tilted modules is shown in **Figure 4.**

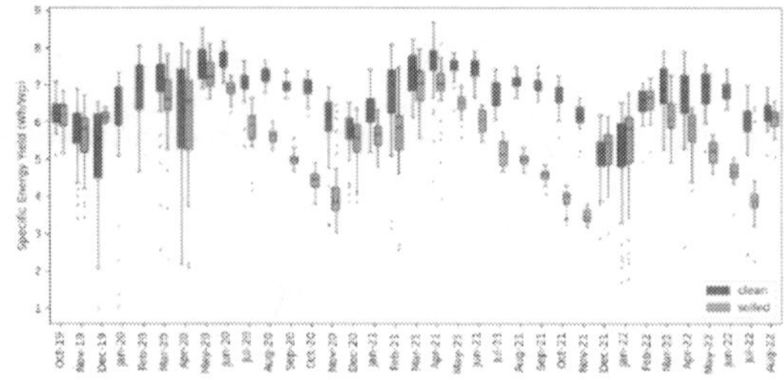

Figure 4: Daily specific energy yield for the clean and soiled tilted bifacial modules. Boxplots represent the median and the quartiles. The whiskers are 1.5 times the interquartile length.

Unlike the tilted modules, vertical module showed little soiling loss compared to the vertical clean reference. In 2022 for example, the soiled module produced a maximum monthly mean loss of less than 2%.

Figure 5a shows the specific energy yield for clean tilted and vertical modules. In average over the test period, the tilted module slightly (by <5.2%) outperformed the vertical module. However, during the summer, the vertical module outperformed the tilted module by up to 8%, as shown in **Figure 5b**. This is significant because both consumption and soiling are highest during summer months.

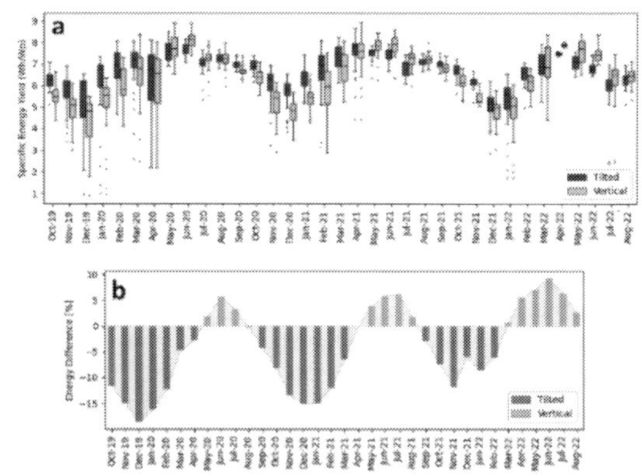

Figure 5: Daily specific energy yield (**a**), and difference for the clean tilted and vertical bifacial modules (**b**).

Comparing the vertical and tilted soiled modules, **Figure 6,** the vertical module yield is higher in average. However, during the winter when natural cleaning events such as rain occur, the tilted module slightly outperforms the vertical module. During the summer, the vertical module strongly outperforms the tilted module.

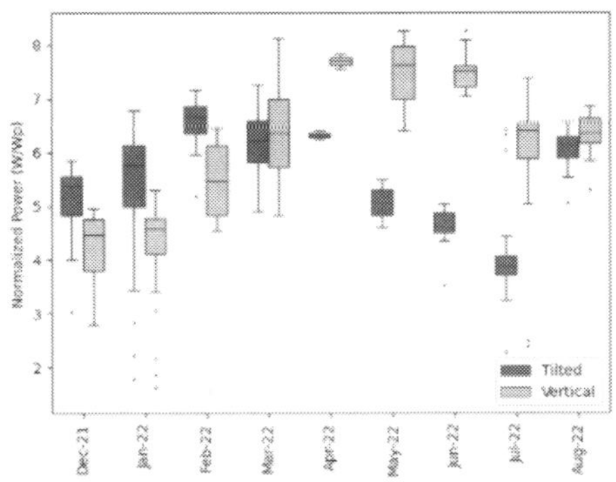

Figure 6: Normalized power for the soiled tilted and vertical bifacial modules.

4 CONCLUSIONS

Soiling can reduce energy yield of monofacial modules by up to 60% during the summer. The losses are significantly lower (by ~15%) for bifacial modules, considering a short circuit current bifaciality factor (φIsc) of 88.6%. The specific energy yield of clean vertical bifacial modules outperformed the tilted bifacial module during the summer (May-August) with maximum of 8%. However, in average the tilted module outperformed the vertical module by up to 5.2%. The bifacial gain ranged from 20% (January) to 35% (July), slightly higher than other reported values. The results show the significant potential of bifacial modules, particularly vertical installation, for soiling mitigation and reduction of cleaning costs in desert climates.

5 REFERENCES

[1] ITRPV-2021, International Technology Roadmap for Photovoltaic (ITRPV) 12th edition, 2021.

[2] K. Ilse, et al., Techno-Economic Assessment of Soiling Losses and Mitigation Strategies for Solar Power Generation, Joule, 3 (2019) 2303.

[3] A. Asgharzadeh, C. Deline, J. Stein, F. Toor, A Comparison Study of the Performance of South/North-facing vs East/West-facing Bifacial Modules under Shading Conditi, 7th IEEE WCPEC 2018, 2018, pp. 1730.

[4] L. Bodlak, H. Hernandez, J. Bierbaum, J. Kang, S. Nold, R. Preu, Price-bifaciality relationship of bifacial modules in vertical east-west oriented PV systems, 35th EUPVSEC, Brussels, Belgium, 2018, pp. 2134.

[5] M.R. Khan, A. Hanna, X. Sun, M.A. Alam, Vertical bifacial solar farms: Physics, design, and global optimization, Appl. Energy, 206 (2017) 240.

[6] J. Lopez Garcia, R. Urraca, R.P. Kenny, T. Sample, Comparison of the energy yield of vertical (e-w orientation) and tilted (equator facing) bifacial pv module arrays, 37th EUPVSEC, Online 2020, pp. 883.

[7] S. Bhaduri, A. Kottantharayil, Mitigation of Soiling by Vertical Mounting of Bifacial Modules, IEEE J. Photovoltaics, 9 (2019) 240.

[8] H. Hildebrandt, 3 MWp vertical E-W oriented system in Germany, 5th PV bifacial Workshop, Konstanz, Germany, 2017.

[9] B. Soria, E. Gerritsen, P. Lefillastre, J.E. Broquin, A study of the annual performance of bifacial photovoltaic modules in the case of vertical facade integration, Energy Science and Engineering, 4 (2016) 52.

[10] J. Bonilla-Castro, M. Herz, C. Monokroussos, M. Schweiger, Energy yield comparison between biacial and monofacial PV modules: real world measurements and validation with bifacial simulations, 35th EUPVSEC, Brussels, Belgium, 2018, pp. 1236.

[11] C. Deline, S. Macalpine, B. Marion, F. Toor, A. Asgharzadeh, J.S. Stein, Assessment of Bifacial Photovoltaic Module Power Rating Methodologies-Inside and Out, IEEE J. Photovoltaics, 7 (2017) 575.

INFLUENCE OF IR SOLDERING PROFILE ON INDUSTRIAL SILICON HETEROJUNCTION SOLAR CELLS

Angela De Rose, Cristina Rosado Alberdi, Achim Kraft

Fraunhofer Institute for Solar Energy Systems ISE
Heidenhofstraße 2, 79110 Freiburg i.Br., Germany

*Corresponding author, e-mail to: angela.de.rose@ise.fraunhofer.de

ABSTRACT: Within this work, the influence of infrared (IR) light on bifacial *n*-type silicon heterojunction (SHJ) solar cells is investigated, with the focus on the industrial IR soldering process. The EL images and *I-V* parameters of industrial SHJ cells are compared before and after treatment with IR light. The temperature-time-profile of this thermal treatment is varied according to established soldering profiles used in silicon photovoltaics, reaching peak temperatures above 200 °C. As a result, a reliable process window for the interconnection process of SHJ solar cells is identified (max. 250 °C, max. 6 s), where thermal degradation is avoided and even slight improvements of the cell performance are observed (ΔV_{oc} = +1.9 mV, ΔFF = +0.3 %, $\Delta\eta$ = +0.1 %). Overall, these investigations demonstrate the feasibility of the established IR soldering process for SHJ solar cells during module integration and therefore extend the range of usable process options for a broad deployment of SHJ modules.

Keywords: silicon heterojunction (SHJ), temperature, soldering, interconnection, infrared light

1 INTRODUCTION

For the interconnection of silicon heterojunction (SHJ) solar cells, several challenges have to be faced, as reported by different groups [1–5]. Previous work has shown that reliable and damage-free infrared (IR) soldering can be used to interconnect SHJ solar cells and build full size solar modules [4]. Today, commercially available SHJ modules on the marked use both options, multi-wire interconnection and soldering flat ribbons [6]. All of them use process temperatures between 200 °C and 250 °C for interconnection. Beside a precise alignment of ribbon/wire and busbars/pads, the success of the soldering result strongly depends on the used temperatures within the whole production chain.

Due to their temperature sensitivity, the excellent passivation of the hydrogenated amorphous silicon (*a*-Si:H) layer of the SHJ solar cell could be damaged at high temperatures [7]. In addition, a too high thermal budget leads to an increase of the contact resistance at the heterojunction [8]. Hence, during cell processing, low-temperature processes are used for *e.g.* deposition of the *a*-Si layers (~200 °C) [5, 9], sputtering of the transparent conductive oxide (< 210 °C) [10, 11], and curing of the metallization (< 240 °C) [12, 13]. Also, additional annealing steps and light soaking during solar cell manufacturing have shown to be beneficial for the performance [14, 15]. Different SHJ cell manufacturer use slightly different temperatures, materials and cell thicknesses to optimize their cell technology. Therefore, the soldering process has to be adapted carefully, for a given fabrication sequence or cell supplier.

Within this work, the power of the IR radiator is varied to consequently change the peak temperature of the soldering profile between 200 °C and 320 °C, and the effect on the *I-V* parameters of industrial SHJ solar cells is evaluated. To optimize the whole process chain for production of SHJ solar modules, the effects of IR light and cell temperature have to be evaluated. The aim of this work is to find a reliable process window for IR soldering while minimizing the process time to ensure a high throughput for industrial manufacturing.

2 EXPERIMENTAL

To define a suitable temperature range for soldering SHJ solar cells, the influence of IR light on the SHJ cell has been studied by a systematic temperature variation during IR light treatment and selected solar cell characterization.

2.1 Infrared light treatment

A semi-automatic IR soldering station is used for thermal treatment of the industrial SHJ solar cells (*n*-type, monocrystalline, 156.75 mm × 156.75 mm, 5 busbars, bifacial, rear emitter). The experimental setup consists of an IR lamp (λ_{max} = 1.2 µm) and a hotplate with down-holding unit. One of the SHJ solar cells is used to develop and control the temperature profiles, given in Figure 1. For this, a thermocouple is placed in the center on the cell's front side beside the middle busbar. A hotplate is used for pre-heating to 120 ° C. After the alignment of SHJ cell and down holder on the hotplate, the IR treatment is performed, and the cell surface temperature is monitored against time.

Seven different temperature profiles are used. The peak temperature is varied between 200 °C and 320 °C with ΔT = 20 K, while the "soldering" time is kept constant at $t_s = t_{FWHM} \approx 6$ s. To reliably vary the peak temperature up to 320 °C and ensure comparable results, a constant soldering time t_s is used. Each variation is tested on N = 5 SHJ cells.

In addition to IR light, the influence of alternative heat sources on the SHJ cells are tested. To imitate contact soldering with a solder iron, a solder tip is used to locally induce heat into the cell metallization. Heat transfer by convection is tested with an oven at ambient air for different temperatures.

In addition to IR light, the influence of alternative heat sources on the SHJ cells are tested. To imitate contact soldering with a solder iron, a solder tip is used to locally induce heat into the cell metallization. Heat transfer by convection is tested with an oven at ambient air for different temperatures.

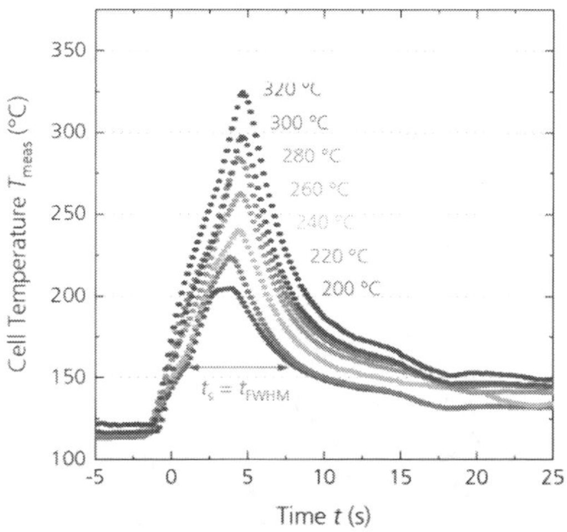

Figure 1: Measured solar cell temperature T_{meas} on the surface in the middle of an SHJ solar cell during IR treatment with different peak temperatures. The "soldering" time t_s is given by the full width at half maximum t_{FWHM} and kept constant.

During all heat treatments of the SHJ cells, no solder materials (*i.e.* solder alloy, flux) are used to focus on the effect of the IR treatment and eliminate other factors as *e.g.* possible material interactions.

2.2 Solar cell characterization

Solar cells are electrically characterized by electroluminescence (EL) imaging and *I-V* measurements, both before and after IR treatment of the cells. EL is done at 10 A. An inline *I-V* tester is used to measure the *I-V* curve under STC conditions. A reference group, consisting of five SHJ solar cells of the same batch, are used as measurement reference for EL and *I-V*. The systematic uncertainties (2σ) of the *I-V* measurements are: $\sigma_{Voc} = \pm 0.09$ %, $\sigma_{Isc} = \pm 0.15$ %, $\sigma_{FF} = \pm 0.17$ %, $\sigma_{pFF} = \pm 0.12$ %, $\sigma_{Pmpp} = \pm 1.62$ %, $\sigma_{\eta} = \pm 0.28$ %.

Selected samples are further characterized by Suns-V_{oc} measurements, using $J_{sc} = 37$ mA/cm², $\rho_B = 4$ Ω·cm and 165 µm cell thickness as input parameters, performing eight measurements per busbar.

2.3 Microscopic imaging

For high resolution imaging, scanning electron microscopy (SEM) is conducted in top view mode with 5 kV acceleration voltage, using an in-lens detector to measure the secondary electrons.

3 RESULTS AND DISCUSSION

In the following, the results of the IR light treatment of SHJ solar cells are presented. They are referred to a certain peak temperature, introduced during IR treatment. All qualitative and quantitative results are compared to the initial state of the cells before treatment.

3.1 Electrical characterization by EL and *I-V*

EL imaging is a fast and reliable method to detect defects (*e.g.* cracks, cell damage), introduced during cell interconnection. Therefore, we use EL images to evaluate possible local damage of the cells after the IR treatments.

Figure 2 exemplarily shows the EL images of one of the five solar cells for each peak temperature. The images are scaled to the same limits of the EL signal for better comparison. For a peak temperature of 260 °C and higher, local damage of the cell can be observed, indicated by a reduced EL signal (dark areas). These defect areas are more pronounced for higher temperatures. The position and exact shape of these spots strongly correlates to the arrangement of the IR radiators and the used downholding unit, both influencing the light propagation.

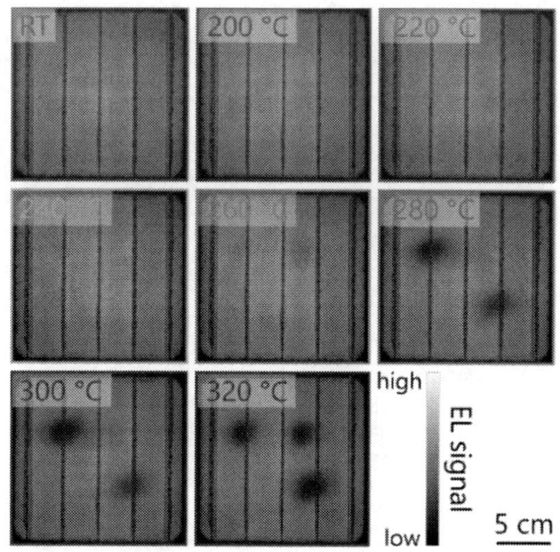

Figure 2: EL images of one example for each peak temperature, visualizing a local degradation for temperatures above ~260 °C.

To interconnect solar cells with a Sn60Pb40 solder alloy, peak temperatures of 230 °C – 240 °C have to be reached for a time of at least 1 s – 2 s. The EL images in Figure 2 let assume that no damage of the SHJ layers up to 240 °C for at least 6 s can be expected.

To quantify the damage of the SHJ cells for higher temperatures and define an upper limit for soldering, *I-V* measurements for all groups are performed. Figure 3 shows the open circuit voltage V_{oc} (top) and the fill factor *FF* and pseudo fill factor p*FF* (bottom) for the reference group, stored at room temperature (RT) and each peak temperature up to 320 °C.

By increasing the peak temperature up to 260 °C, a gain in V_{oc} of +(0.2 ± 0.1) % (absolute: +(1.9 ± 0.8) mV) is found. As reported by other researcher, investigating the impact of light soaking on SHJ solar cells, this is due to improved hydrogen passivation within the SHJ layers [15] and further healing of sputter damage. This shows that the soldering process could be used to further optimize the overall cell performance, when considering the whole production chain from cell to module.

After a certain threshold temperature is reached, the passivation and therefore V_{oc} degrades. Hydrogen effusion leads to depassivation and a local increase of the dark saturation current j_0 can be found [4]. Temperatures above 260 °C should be avoided during soldering to not damage the SHJ solar cell.

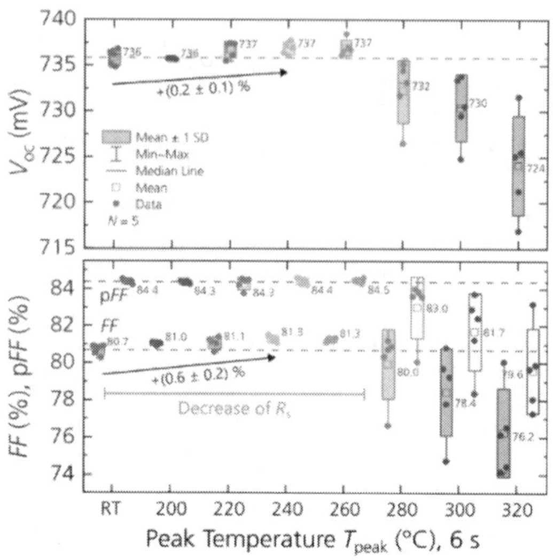

Figure 3: Open circuit voltage V_{oc}, fill factor FF and pseudo fill factor pFF of SHJ solar cells after IR treatment with different peak temperatures T_{peak} for a constant time of about 6 s. The horizontal dashed lines indicate the mean value of the reference group without treatment (RT = room temperature). Up to 260 °C, a gain in V_{oc} and FF is measured, whereas higher temperatures induce degradation of the SHJ solar cell.

The same trend is found for the fill factor FF, improving by +(0.6 ± 0.2) % (absolute: +(0.3 ± 0.1) %) for increased temperature up to 260 °C. In this temperature range, the pseudo fill factor pFF remains constant, whereas a decrease of the series resistance R_s is measured (not shown here), yielding to a boost in FF. The improvement of R_s can be attributed to post-curing of the metallization paste, which will be discussed in section 3.3.

Above 260 °C, pFF decreases as well, resulting in a decrease of FF. Here, several effects are overlapping: in addition to the decrease of R_s due to "post-curing", reduced charge carrier transport at the *a*-Si:H/TCO interface leads to a local increase of R_s [4, 8]. Since the overall change in Rs remains nearly constant (*cf.* Table 1), the decrease in pFF is dominated by local depassivation.

A summary of the change of the relevant *I-V* parameters is given in Table 1.

Table 1: Change of the *I-V* parameters, showing the mean change of $N = 5$ SHJ cells after the IR treatment with 240 °C for 6 s. The values are corrected by the *I-V* data of a control group of additional five SHJ cells (without IR treatment).

Parameter	Abs. change	Rel. change
V_{oc}	(1.9 ± 0.8) mV	(0.2 ± 0.1) %
J_{sc}	(0.0 ± 0.0) mA/cm²	(0.0 ± 0.0) %
pFF	(0.1 ± 0.1) %	(0.1 ± 0.1) %
FF	(0.3 ± 0.1) %	(0.6 ± 0.2) %
R_s	(-0.03 ± 0.02) Ω·cm²	(-10.2 ± 3.9) %
η	(0.10 ± 0.02) %	(0.7 ± 0.1) %

Similar overlapping effects, as described above, are found for the temperature around 260 °C. Here, the EL image already shows local depassivation (*cf.* Figure 2), whereas an overall gain of the *I-V* parameters (esp. V_{oc}, pFF, FF) is found (*cf.* Figure 3). These results show that global and spatial electrical analysis should be used complementary to provide a better understanding of the influence of IR soldering on SHJ solar cells.

3.3 SEM analysis of metallization

The IR light treatment does not only affect the layers of the SHJ cell structure, but also the metallization. The electrodes of the investigated cells consist of screen-printed and cured fingers and busbars, using an Ag low-temperature (LT) metallization paste. Figure 4 shows top view SEM images of the finger, before (left) and after IR light treatment with 340 °C peak temperature for 6 s (right, finger of degraded region).

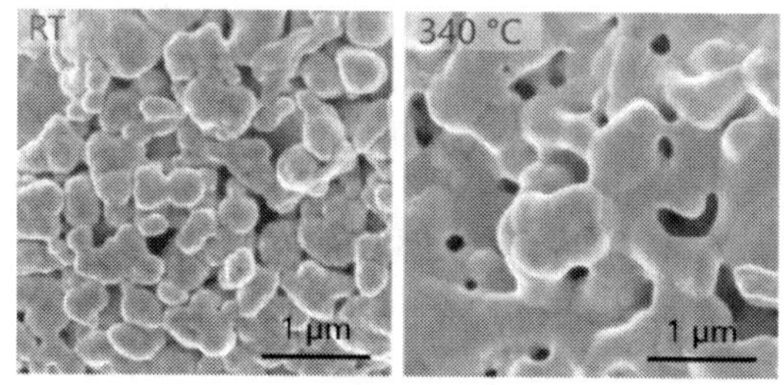

Figure 4: Top view SEM images of finger metallization (LT paste) of front side of industrial SHJ solar cell before (left) and after (right, degraded region) IR treatment with 340 °C for 6 s.

The IR light induces "post-curing" of the finger paste, leading to further sintering of the Ag particles. This results in a lower series resistance R_s, which causes an increase of the fill factor FF (*i.e.* Figure 3).

3.4 Local analysis of degradation by Suns-V_{oc}

By *I-V* measurements, mean electrical parameters of the whole SHJ solar cell are extracted. Local information over the wafer can be obtained by *e.g.* luminescence imaging. Another method to obtain local quantitative results for the quality of the cell is Suns-V_{oc}. One SHJ solar cell, treated with IR light at 340 °C for 6 s, is characterized by Suns-V_{oc}. The V_{oc}, determined by *I-V*, is 736 mV in the initial state and 706 mV in the degraded state after IR treatment, giving a loss of ΔV_{oc} = -30 mV.

Figure 5 shows the result of the Suns-V_{oc} measurement of this SHJ cell. The underlying EL image correlates well to the Suns-V_{oc} measurement. A maximum degradation of -75 mV is found at the position where the EL image shows the lowest signal. Also in the outer cell range, which seemed to be unaffected in the EL image, losses around 0.5 mV to 3 mV are observed.

The Suns-V_{oc} measurement confirms that the IR radiation leads to damage of the passivation layers, resulting in local degradation of up to ΔV_{oc} = -75 mV, correlating with a decrease in pFF (not shown here). In contrast, the V_{oc}, extracted from *I-V* measurements, only represents a mean value of the cell and is therefore less useful to qualify the damage. For the temperature range between 220 °C and 280 °C, these local changes play a major role; in this range, soldering is carried out. As electrical gains and losses are overlapping in this regime, as seen from our experiments, careful optimization is important.

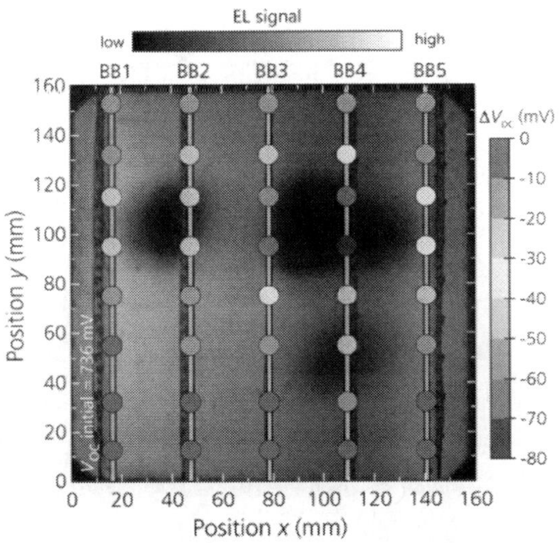

Figure 5: Suns-V_{oc} measurement of an SHJ solar cell after IR treatment for 6 s at 340 °C, correlated to its EL image. The measurement is performed at the marked positions along the five busbars.

3.4 Alternative heat sources for interconnection

Concerning SHJ solar cells, several annealing experiments have already been carried out, most of them using hotplates to control temperature and time precisely [7, 8, 10, 16]. However, for SHJ string and module production, other heat sources are more relevant. Therefore, we investigate the influence of different heat sources on the SHJ solar cells. Contact heat, as it is applied during contact soldering or lamination, convection heat induced by an oven or *e.g.* hot air soldering and, as reference, IR radiation. For each condition, an EL image after heat treatment is exemplarily shown in Figure 6.

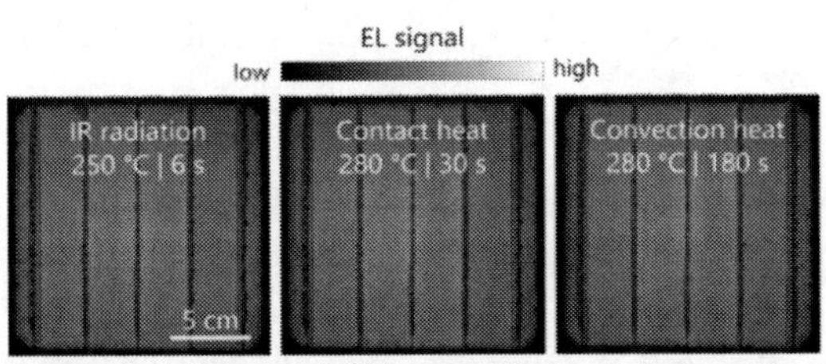

Figure 6: Influence of different soldering techniques (heat sources) on SHJ solar cells. For each technique, five SHJ cells are tested, whereas the temperature maximum T_{max}, tested for a time of t_{max} without inducing cell damage, is given. Both are set-parameters.

The aim of this variation is to identify a maximum thermal budget (temperature T_{max} for a certain time t_{max}), without inducing damage to the solar cell. The boundaries of these processes are identified using the same analysis by EL and *I-V* measurements as presented above, before and after thermal treatment. For all tested heat sources, the maximum suitable temperature is between 250 °C and 280 °C. The values for T_{max} and t_{max}, given in the EL images, refer to a maximum heat treatment, without degrading the electrical parameters of the cells (not shown here). Especially for IR radiation, one has to keep in mind the more inhomogeneous temperature distribution over the cell, compared to *e.g.* convection heating. Therefore, the given temperature of 250 °C can underestimate the local cell temperature in some regions.

For conventional soldering with Sn60Pb40, peak temperatures between 220 °C and 250 °C are used for soldering times < 2 s. During low-temperature soldering with Pb-free solder alloys, even lower temperatures < 220 °C are required [17]. Therefore, the identified values are far above the required temperatures and times for soldering solar cells.

Nevertheless, the maximum thermal budget of SHJ cells during module production has to be evaluated for each SHJ cell structure, since it is influenced by the thermal budget used during SHJ cell processing. However, a broad temperature range can be used for interconnection, allowing also peak temperatures up to 260 °C for 6 s (*e.g.* for IR soldering).

4 CONCLUSION AND OUTLOOK

In the past, studies have shown the high sensitivity of SHJ solar cells to elongated thermal treatment [7, 10]. Therefore, most work has taken a conservative approach and limiting the temperature range below 210 °C. Today, most processes have reached a high maturity and mass manufacturing is built up. Now, the limits of the thermal treatments have to be tested, to enable high throughput manufacturing.

In this work, we investigate the influence of different temperatures, induced by IR light, on industrial SHJ solar cells to identify a guideline for soldering of SHJ solar cells for manufacturing PV modules. Our evaluation shows no degradation of the SHJ solar cells after IR treatment for T_{max} < 260 °C and t_{max} = 6 s, as well tested for other heat sources (contact, convection). Indeed, the IR light causes a light soaking effect of the cell [15]. A slight gain of +0.2 % in V_{oc} and +0.6 % in *FF* is observed due to enhanced passivation quality and a post-curing of the metallization after IR treatment. SEM images of the finger metallization confirm further sintering of the Ag particles within the paste. These gains in cell performance could be further quantified after climate chamber testing of modules to proof the stability after further heat or light exposure.

For temperatures above ~260 °C for 6 s, local degradation can be observed by EL imaging and quantified by *I-V* measurements. Suns-V_{oc} measurements have shown to be suitable to identify local gains and losses in V_{oc}, also helpful to achieve homogenous solder joint formation over the cell. Since SHJ cells are sensitive to temperature and light, any small adjustment of the soldering process (*e.g.* temperature, belt speed, downholder *etc.*) can have a large influence on the soldering result and the SHJ cell. Furthermore, the temperatures given in this study, most probably underestimate the actual cell temperatures in the hottest regions during IR soldering. Therefore, the given thresholds (temperature, time) could also be understood as a lower limit for the onset of cell degradation.

During string production, EL imaging can be used as fast and reliable method to optimize the soldering result. However, the images must be evaluated quantitatively or directly correlated to *I-V* or Suns-V_{oc} measurements, since slight degradation may not be visible. These techniques are not jet implemented inline into module manufacturing lines before lamination, since a fast and damage-free transport of the strings is challenging.

Given the presence of further thermal treatments during module manufacturing, the whole process chain for SHJ should be optimized to enable the leanest fabrication path. This is focus of ongoing work at our institute.

5 REFERENCES

[1] A. Faes *et al.*, "Smartwire solar cell interconnection technology," in *Proceedings of the 29th European Photovoltaic Solar Energy Conference and Exhibition (EU PVSEC)*, Amsterdam, Netherlands, 2014, pp. 2555–2561.

[2] A. De Rose, T. Geipel, D. Erath, A. Kraft, and U. Eitner, "Challenges for the interconnection of crystalline silicon heterojunction solar cells," *Photovoltaics International*, vol. 40, pp. 78–86, 2018.

[3] C. Kaiser, V. Nikitina, T. Geipel, and A. Kraft, "Reduction of ECA amount for the ribbon interconnection of heterojunction solar cells," in *Proceedings of the 37th European Photovoltaic Solar Energy Conference and Exhibition (EU PVSEC)*, online, 2020, pp. 1086–1090.

[4] A. De Rose, T. Geipel, D. Eberlein, A. Kraft, and M. Nowottnick, "Interconnection of Silicon Heterojunction Solar Cells by Infrared Soldering - Solder Joint Analysis and Temperature Study," in *Proceedings of the 36th European Photovoltaic Solar Energy Conference and Exhibition (EU PVSEC)*, Marseille, France, 2019, pp. 229–234.

[5] S. De Wolf, A. Descoeudres, Z. C. Holman, and C. Ballif, "High-efficiency Silicon Heterojunction Solar Cells: A Review," *Green*, vol. 2, no. 1, pp. 7–24, 2012, doi: 10.1515/green-2011-0018.

[6] S. K. Chunduri and M. Schmela, "Heterojunction Solar Technology: 2022 Edition," 2022.

[7] B. A. Korevaar, J. A. Fronheiser, X. Zhang, L. M. Fedor, and T. R. Tolliver, "Influence of annealing on performance for hetero-junction a-Si/c-Si devices," in *Proceedings of the 23rd European Photovoltaic Solar Energy Conference and Exhibition*, Valencia, Spain, 2008, pp. 1859–1862.

[8] C. Luderer, C. Messmer, M. Hermle, and M. Bivour, "Transport Losses at the TCO/a-Si:H/c-Si Heterojunction: Influence of Different Layers and Annealing," *IEEE Journal of Photovoltaics*, vol. 10, no. 4, pp. 952–958, 2020, doi: 10.1109/JPHOTOV.2020.2983989.

[9] U. K. Das, M. Z. Burrows, M. Lu, S. Bowden, and R. W. Birkmire, "Surface passivation and heterojunction cells on Si (100) and (111) wafers using dc and rf plasma deposited Si:H thin films," *Appl Phys Lett*, vol. 92, no. 6, p. 63504, 2008, doi: 10.1063/1.2857465.

[10] J. Haschke *et al.*, "Annealing of Silicon Heterojunction Solar Cells: Interplay of Solar Cell and Indium Tin Oxide Properties," *IEEE J. Photovoltaics*, pp. 1–6, 2019.

[11] L. Tutsch *et al.*, "Implementing transparent conducting oxides by DC sputtering on ultrathin SiOx / poly-Si passivating contacts," *Sol Energ Mat Sol C*, vol. 200, pp. 1–5, 2019, doi: 10.1016/j.solmat.2019.109960.

[12] S. Pingel *et al.*, "Low-temperature Ag-Paste Screening for Silicon Heterojunction Solar Cells and Modules," in *Proceedings of the 37th European Photovoltaic Solar Energy Conference and Exhibition (EU PVSEC)*, online, 2020, pp. 508–511.

[13] D. Erath *et al.*, "Fast screen printing and curing process for silicon heterojunction solar cells," in *Proceedings of the 9th Workshop on Metallization and Interconnection for Crystalline Silicon Solar Cells*, Genk, Belgium, 2021, p. 20006.

[14] E. Kobayashi *et al.*, "Increasing the efficiency of silicon heterojunction solar cells and modules by light soaking," *Sol Energ Mat Sol C*, vol. 173, pp. 43–49, 2017, doi: 10.1016/j.solmat.2017.06.023.

[15] A. Moldovan *et al.*, "Improved layer properties combined with light soaking enabling for 23% efficient silicon heterojunction solar cells," in *Proceedings of the 37th European Photovoltaic Solar Energy Conference and Exhibition (EU PVSEC)*, online, 2020, pp. 218–222.

[16] S. De Wolf and M. Kondo, "Nature of doped a-Si:H/c-Si interface recombination," *Journal of Applied Physics*, vol. 105, no. 10, 103707, 2009, doi: 10.1063/1.3129578.

[17] D. Güldali and A. De Rose, "Material Joint Analysis of Lead-Free Interconnection Technologies for Silicon Photovoltaics," in *45th International Spring Seminar on Electronics Technology (ISSE)*, 2022.

DEVELOPMENT OF SHINGLE MATRIX TECHNOLOGY FOR INTEGRATED PV APPLICATIONS

[1]Daniel von Kutzleben, [1]Torsten Rößler, [1]Max Mittag [1]Julian Weber, [1]Sanjeev Sigdel, [1]Nils Klasen, [2]Philipp Zahn, [1]Achim Kraft, [1]Holger Neuhaus
[1]Fraunhofer Institute for Solar Energy Systems ISE, Freiburg, Germany
[2]M10 Solar Equipment GmbH, Freiburg, Germany

ABSTRACT: Linear shingling of solar cells has already been known for many years. This work evaluates a new approach of arranging the shingle cells, called matrix shingling and discusses advantages and challenges of this technology. An overview of the most crucial process steps to build matrix modules is presented. Cell-to-module (CTM) simulations are used for a profound loss/gain analysis from host cell to module. A comparison between half-cell modules and matrix modules shows the potential of the matrix technology. Glass facade elements for building integrated PV with red MorphoColor® coating are demonstrated to show the potential of this growing field of application.
Keywords: photovoltaic module, shingling, ECA, BIPV, CTM

1 INTRODUCTION

The most recent part of the IPCC report [1] stated that the most effective way to avoid greenhouse gas emissions is the deployment of wind and solar energy, since it results in the lowest cost per ton of CO_2 saved. Large-scale PV installations require immense areas of land that compete with other uses such as agriculture, forestry, or built-up areas. Therefore, multiple uses of space are necessary. Building integrated photovoltaics (BIPV) is one of those solutions. Architects and building owners, however, are keen not to make their building look like a power plant but to use the solar modules as a design and creative element.

For the growing market of integrated photovoltaics, modules with high requirements to the aesthetic appearance are demanded. Yet, an appealing cell interconnection technology is missing. As shingle interconnection of solar cells enables highly uniform PV modules it particularly fulfills these aesthetical demands. Shingling is known for a long time as interconnection method for solar cells [2]. Companies like Sunpower/Maxeon, Solaria and Tongwei have offered shingled solar modules as commercial products for many years [3–5]. Several manufacturers have placed patents, protecting their intellectual property on some aspects along the production chain. The general idea of shingling solar cells is used since the 1960s and cannot be part of any patent leaving the possibility open for new "players" to enter the BIPV market.

In this scientific work an advanced concept of shingling solar cells is discussed that is called shingle matrix technology [6]. Shingling in the matrix arrangement was already introduced in 1990 [7]. This approach not only contains the advantages of linear shingling but further increases active area in the module and brings a high shading resilience for an increased energy harvest in partial shading situations, which occur more likely in integrated PV [8, 9]. Figure 1 shows a matrix module with red MorphoColor® [10, 11] used to demonstrate colored facade elements.

Section 2 gives an overview about the principle of shingle matrix interconnection, the laser separation process and the shingle matrix stringer followed by a simulated comparison between a half-cell module design and matrix modules. In section 3, produced shingle matrix modules are presented and analyzed in terms of their cell-to-module losses and gains.

Figure 1: Full-size glass-glass matrix module with red MorphoColor® coating as a glass facade element suitable for building integration. The upper left corner is without MorphoColor® to show the matrix cell arrangement.

2 SHINGLE MATRIX TECHNOLOGY

2.1. Shingle matrix interconnection

The matrix technology is an innovative and challenging approach to further improve shingling technology. Although small-scale, lab-type "matrix shingle" modules were shown in the past [7], only Fraunhofer ISE and M10 Solar Equipment have upscaled the shingle matrix technology to improve aesthetics, efficiency [12] and circumvent patent blockades of the conventional shingle interconnection. As a result of this project an industrial stringer machine is ready to bring this innovative approach to the PV market [13]. Fraunhofer

ISE sees a high demand for PV modules that visually satisfy the highest expectations for the market of integrated PV.

The concept of matrix shingling is based on the principle of linear shingling, depicted in Figure 2, but includes an additional lateral offset in every second row by inserting half shingles in the outermost columns.

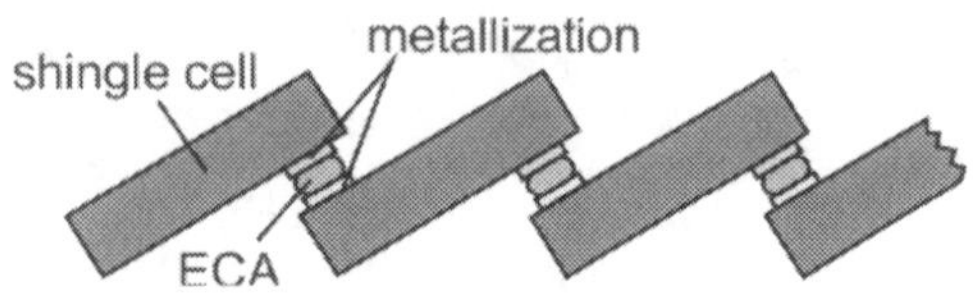

Figure 2: Principle of shingling solar cells. The cells overlap with their busbar metallization and are mechanically and electrically bonded by an electrically conductive adhesive (ECA).

It is the same principle as of a masonry wall. The result is a matrix of flexible length and width, where the shingle cells of one row are connected in parallel by using the busbars of the cells overlapping this row. The rows are in series connection with each other. A drawing of a matrix can be seen in Figure 3. High flexibility in length and width of the shingle matrix is one of its core advantages especially interesting for building integrated applications with predefined glass sizes.

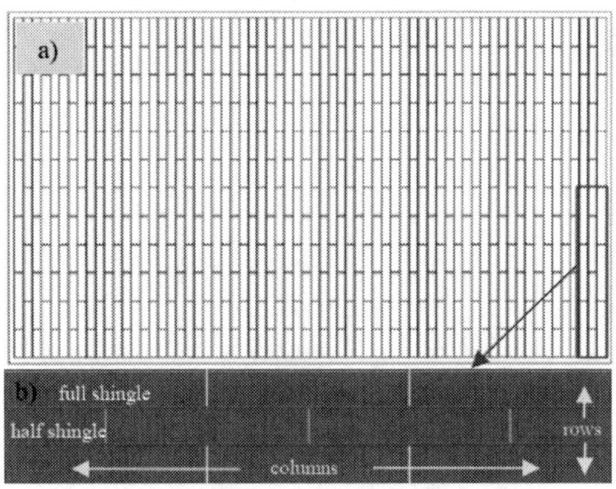

Figure 3: Drawing of a) a full-size matrix module and b) part of the matrix with three full shingle cells in parallel and three rows in series interconnection.

Matrix shingling offers the same advantages as linear shingling, namely lead-free cell interconnection, high active module area and a homogeneous appearance resulting from the absence of ribbons and not existing or minimized cell gaps, respectively. The biggest advantage over linear shingling, however, is the high shading resilience [8, 9]. Partial shading is an issue that must be considered, especially in the field of integrated PV. Matrix modules with a significantly higher power output in situations of shading caused by traffic signs, bird droppings or leaves are predestined for this application. In the case of a matrix module, shaded cells can be circumvented by the current using the lateral current paths through the busbars of adjacent solar cells [14]. In case of diagonal shading, a matrix module can reach up to 73.8% and in the case of random shading 96.5% more power output compared to a module with linear shingle strings [8, 9]. The implementation of the shingle matrix technology comes along with significant technology challenges such as parallel processing of shingles, terminal and intermediate connections and handling of the matrix.

2.2. Laser separation processes

To obtain shingle cells, host wafers must be cut into 1/5th to 1/7th stripes of the original cell. A common way to separate shingle host cells into shingle cells is by means of a laser. There are two industrially relevant laser cutting processes: Laser scribing and mechanical cleaving (LSMC) [12, 15] and thermal laser separation (TLS) [12, 16]. Both separation processes result in unwanted recombination of charge carriers at the cut edges. However, LSMC is known to lead to larger recombination related efficiency losses compared to TLS when separating the intact host wafer. In Figure 4 is an overview of efficiency of the initial host wafer and cut shingles. The shingles are 1/5th of a G1 PERC host cell with the size of 158.75×31.75 mm^2. The datapoints of the host wafer are divided by five to make the comparison more comprehensible. The shingles are subdivided into pseudo-square (PSQ) and full-square (FSQ) for LSMC and TLS. PSQ shingles are the two outer shingles of a host cell with only one laser separated edge while the FSQ have two.

The host wafers have a power of 5.5 W_p measured with GridTouch (1.1 Wp if, naively, divided by 5). FSQ/LSMC shingles show a power of 1.04 W_p, while FSQ/TLS shingles feature 1.06 W_p. The corresponding PSQ cells show slightly higher efficiency. The shingles are measured with pin bars between front- and rear side busbar [12]. The difference between host cell and shingle can be separated into two effects. One is a higher series resistance due to increased finger length resulting from the different measurement methods between host cell and shingle, explained in [17]. The second effect is edge recombination on the unpassivated lasered edges. In case of a full-size matrix module with a number of 324 cells, the difference in power output between LSMC FSQ and TLS FSQ cells would be 6.5 W_p. Passivated edge technology (PET) developed by Fraunhofer ISE can recover up to $0.1\%_{abs}$–$0.4\%_{abs}$ of the efficiency losses due to recombination for PERC solar cells [18].

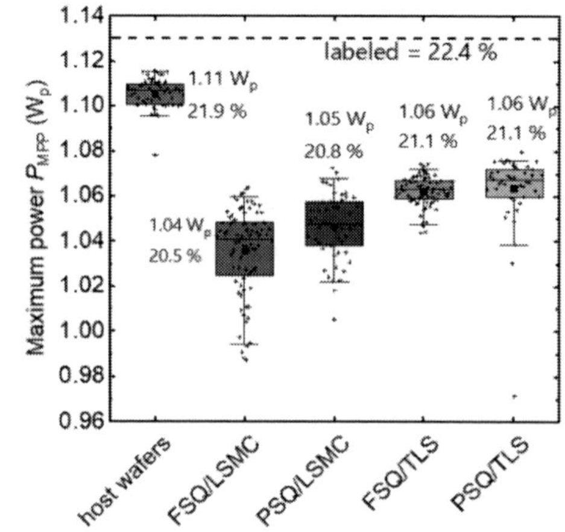

Figure 4: Effect of laser separation processes on the maximum power P_{MPP} of a shingle. LSMC cut shingle

cells show slightly higher cell division losses than TLS cut shingles. For better comparison the host wafer power is divided by 5.

2.3. Shingle matrix stringer

For the matrix modules produced in this work cell arrangement and interconnection are performed on a shingle matrix stringer built by M10 Solar Equipment (Figure 5). Cut shingle cells are used as input material. Controlled by a vision system, the cells are placed row wise in the matrix configuration. Dispensing units are used to apply electrically conductive adhesive (ECA) on the rear side busbar of each shingle after which the rows are placed with the specified offset on an assembling tray. A curing unit using infrared lamps follows. The matrix is transferred by a full-size vacuum gripper and placed on a glass-ethylene-vinyl acetate (EVA) stack. In the last step before the lamination process, terminal and intermediate connectors are attached and prepared for connection with the junction boxes, followed by an EVA and backsheet layer. The stringer is flexible in width and length of the matrices to be produced. This opens up design possibilities for BIPV applications.

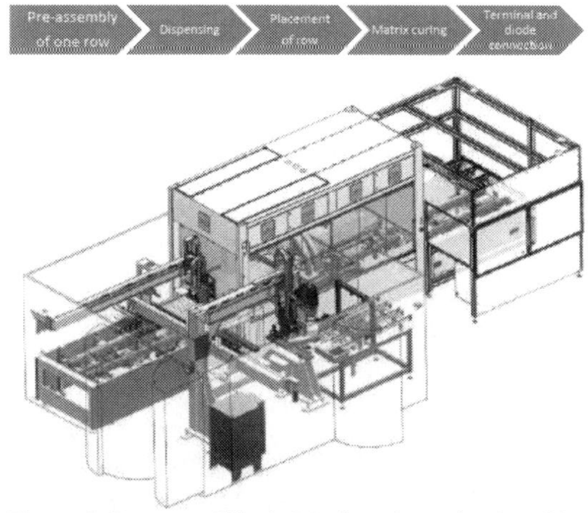

Figure 5: Prototype Shingle Matrix Stringer developed in the project SHIRKAN by M10 Solar Equipment.

Table 1: Comparison of two scenarios for an optimized matrix module with a half-cell module (drawings not to scale).

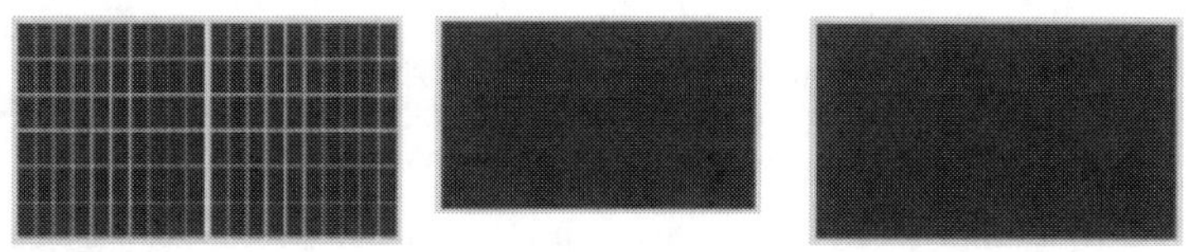

	Unit	Half-cell	Scenario 1 Same host wafer	Scenario 2 Same module area
Dimensions	mm^2	1680.5 × 1002.5	1601.5 × 995.0	1693.0 × 995.0
Cell spacing/overlap	mm	1	−1	−1
String spacing/lateral offset	mm	2.0	0.5	0.5
Edge spacing	mm	17	17	17
Top/bottom spacing	mm	34.5	28	28
Center spacing	mm	12.5	-	-
Interconnection	-	5 busbars	ECA	ECA
Module area	m^2	1.685	1.592	1.685
Number of host wafers	-	60	60	63.6
Separated cell efficiency	%	21.6	21.1	21.1
Input power cell level	W_p	326	319	338
Module efficiency	%	18.6	19.2	19.3
Module power	W_p	313	306	324

2.4. Host-cell-to-module prediction

If the shingle matrix technology wants to take a foothold on the market, it must be able to compete with the established half-cell module designs. To find out if and under which conditions matrix technology has advantages in terms of electrical performance, a comparison between a standard half-cell module and two different scenarios for the shingled matrix technology is made by means of cell-to-module (CTM) SmartCalc.Module simulations. Table 1 gives a comprehensive overview about the input and output data for the simulation. The first scenario is using the same number of G1 host wafers as in the half-cell module, namely 60. If those host wafers are cut into shingle cells and interconnected with an overlap the module area is smaller than the one of the half-cell module. The second scenario shows the case of the same module area as the half-cell module by using more host wafers in the shingle module. Starting point for the simulation is the shingle host wafer efficiency of 21.9% corresponding to a maximum power of 1.11 W_p (5.55 W_p for the full wafer) in Figure 4. In case of the shingle module simulations, shingle cells with an efficiency of 21.1% corresponding to the FSQ/TLS cut shingles with a power of 1.06 W_p taken from Figure 4 are used as input for the SmartCalc.Module calculations [19, 20]. To estimate the half-cell efficiency starting with the same 21.9% host wafer efficiency a reduction of $-0.2\%_{abs}$ to consider cutting losses was found to be reasonable [21]. The increased finger resistance of a pin-based measurement in comparison to the GridTouch measured cell was considered to lead to a $-0.1\%_{abs}$ loss [17]. The half-cell input efficiency for the SmartCalc.Module simulation was therefore 21.6%, corresponding to 326 W_p of cell power. The module output is 313 W_p with an efficiency of 18.6% on an area of 1.685 m^2.

Scenario 1: If the same host wafer equivalent number of cells is interconnected in matrix configuration the module area can be reduced to 1.592 m² to avoid unreasonably high unused space. The output power is 7 W_p lower than the power of the half-cell module due to the cell overlap of 1 mm as well as the missing cell and string distances. However, efficiency is enhanced by $0.6\%_{abs}$ to 19.2% because of higher packing density.

Scenario 2: The module area is kept constant in comparison to the half-cell module and thus the number of host wafers is increased to 63.6 (additional 18 shingle cells). The input power on cell level is therefore 12 W_p higher, 338 W_p in total. The module ends up with an efficiency of 19.3% and a power of 324 W_p. This leads to an increase in efficiency of $0.7\%_{abs}$ compared to the half-cell module and to an 11 W_p higher power output on the same module area.

This study shows mainly two things. First, either module area or number of host cells have to be adapted going from half-cell to shingle module. Second, shingling does not necessarily achieve high CTM_{power} ratios (the shading in the overlap is typically a significant power loss factor). However, shingling reaches large power densities: Power output considering a certain module area, which means it delivers higher module efficiencies.

In case of half-cell module technology, the drop in efficiency from the GridTouch measured host wafer to five busbar interconnected half-cell is relatively low (finger resistance and edge recombination) while it is high from half-cell to module (cell and string spacing, loss in the ribbons). In shingling it is rather opposite with a high efficiency loss from host wafer to shingle cell (finger resistance and edge recombination) and a moderate drop from shingle cell to module (no or almost no cell and string spacing, no ribbons). To improve the efficiency of matrix shingle technology we see the introduction of edge passivation as an important next step.

3 EXPERIMENTS AND RESULTS

3.1. Full size modules

Seven full format shingle matrix modules made from industrial G1 PERC shingle cells are manufactured, one of those modules is depicted in Figure 6. Two different ECAs are used. The modules were laminated as glass-backsheet stack in an industrial laminator from Bürkle. The encapsulation materials are EVA and a black polyethylene terephthalate (PET) backsheet. An antireflection coated solar glass with the dimensions of 1700 × 1000 × 3.2 mm³ is used. The reached power and efficiency for all modules is depicted in Figure 7. The power ranges between 316 W_p and 322 W_p. The best module shows a shingle-to-module power loss of −4.5% with a cell power of 337 W_p and a module power of 322 W_p. Efficiency changes from 20.5% at cell level to 18.8% at module level. A gain and loss analysis for the manufactured modules follows in the next chapter.

3.2. Shingle module CTM analysis

For the full-size matrix modules with the size of 1708 × 1006 mm² (glass and frame) having 6 columns and 54 full-shingle-equivalent rows, power and efficiency gains and losses are analyzed. Due to practical reasons, changes in the module design are made in comparison to the presented optimized matrix modules from section 2.4. The main differences are bigger glass margins due to available glass sizes and the usage of one inactive shingle cell row which is used to implement the end contact. Although TLS cut shingle cells potentially lead to higher module efficiency, LSMC cut shingles are used since an automated TLS process was not accessible by the time of module manufacturing. The shingle overlap is increased to 1.3 mm to meet the metallization design of the used shingle solar cells. Theoretical module efficiency (19.3%, from Table 1) and experimental results (18.8%, from Figure 7) show therefore a deviation of $0.5\%_{abs}$. In the next paragraph the power and efficiency loss from shingle cell to module using real data from the built modules are explained.

Figure 6: Full size matrix module with the dimensions of 1708 × 1006 mm² manufactured on the Shingle Matrix Stringer. The module features a power of 322 W_p and an efficiency of 18.8%.

Table 2 gives an overview on the *I-V* performance beginning with the host wafer values measured with GridTouch at Fraunhofer ISE. Industrial shingle host cells are labeled with a certain power by the manufacturer, but it stays unclear which measurement setup was used. Therefore, the datasheet values can drastically differ from measurements conducted at Fraunhofer ISE. A GridTouch measurement of host wafers can be a practical workaround, but highly overestimates the performance of separated shingles. The separated shingle cells, measured by pin arrays on the front side and rear side busbar lead to a more realistic value since the current paths are the same as in the interconnected state as well as the effect of edge recombination losses are included in the measurement. A thorough study about this topic can be found in [17]. According to Table 2 the pin array measured LSMC/FSQ shingles have an efficiency of 20.5%, which is 1.4%$_{abs}$ lower than the host cell efficiency. The values measured for separated shingles are used to analyze the shingle-to-module losses of the built modules with a SmartCalc.Module calculation. The detailed gains and losses in power and efficiency are shown in Figure 8.

The biggest drop in efficiency results from very high glass margins (k1, −1.8%). Interconnection shading (k7) of blue cell area inside the overlap leads to a power loss of 9.5 W_p and an efficiency loss of 0.55%$_{abs}$, respectively. The loss due to the resistance induced by interconnection (k12) is calculated based on assuming a specific contact resistance between ECA and busbars of 0.2 mΩcm². This value is within the range given in [22] and based on current investigations at Fraunhofer ISE. As a result, the ohmic power loss proposed by the Smartcalc.Module simulation is 2.5 W_p and −0.1%$_{abs}$ efficiency loss. Note that the factors k1 to k15 that yield a large power loss at the same time correspond to the greatest potential to improve module efficiency.

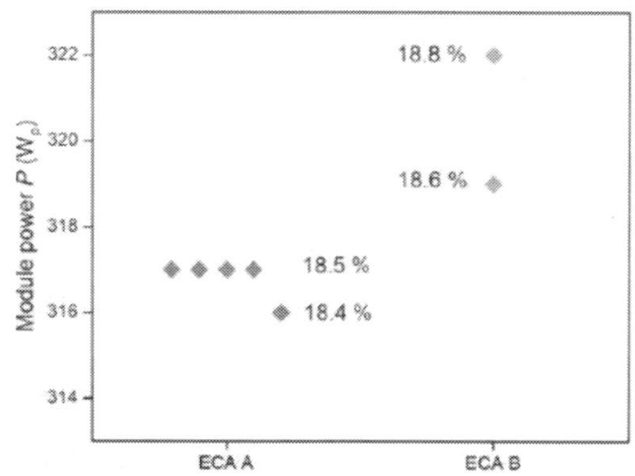

Figure 7: Module power and efficiency of full format matrix modules with two different ECAs.

Table 2: Overview of measured values from host cell to module for fill factor (*FF*), maximum power (P_{MPP}) and efficiency (η).

Description	***FF* (%)**	**P_{MPP} (W_P)**	**η (%)**
Host cell based on Fraunhofer ISE GridTouch measurement	80.8	5.55	21.9
Shingle measured with pin arrays	76.9	1.04	20.5
Produced module (non-optimized) (1.718 m²)	76.3	322	18.8
Optimized module (SmartCalc.Module simulation) (1.685 m²)	See scenario 2 in Table 1		

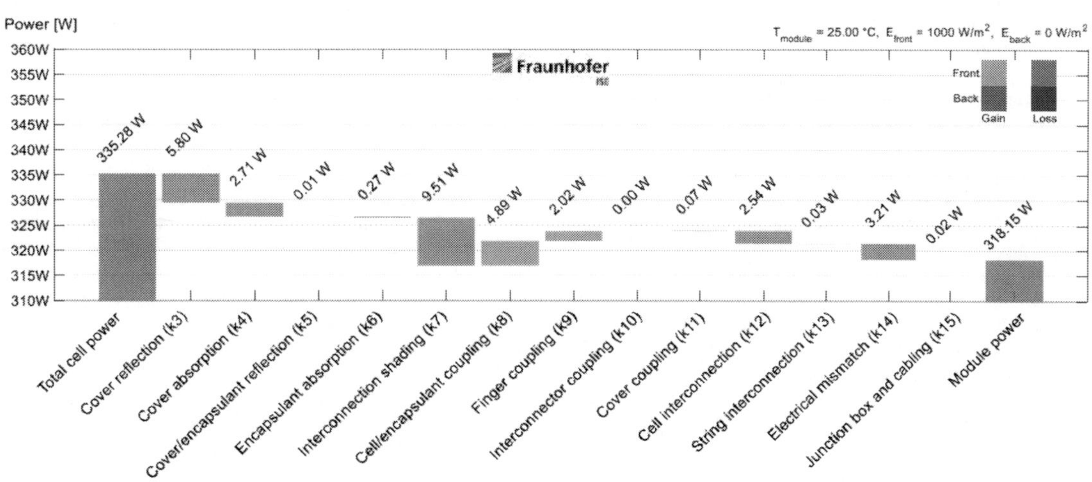

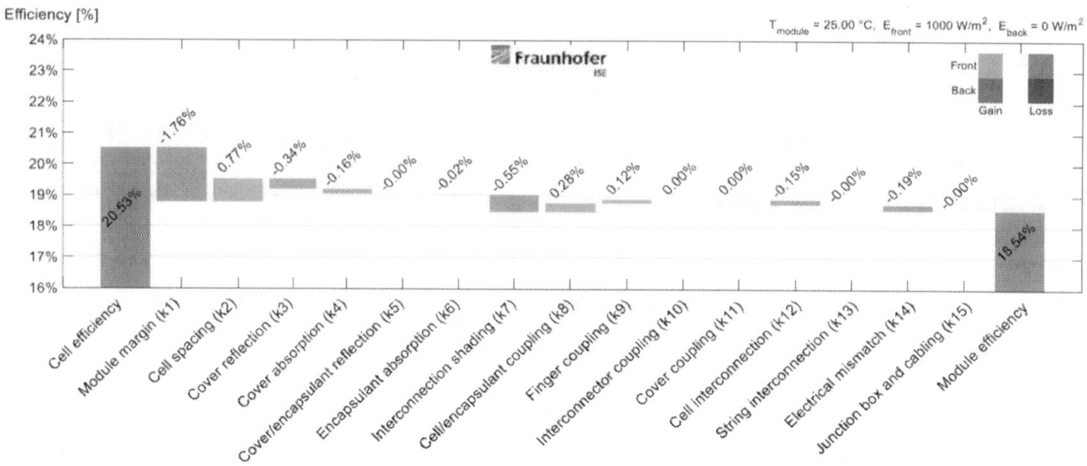

Figure 8: Shingle to module gains and losses diagram for a full-size shingle matrix module as depicted in Figure 6 with G1 1/5th LSMC cut FSQ shingles as input. The factors relevant to the matrix technology are discussed in section 3.2. A detailed description of all factors can be found in [20].

4 CONCLUSION AND OUTLOOK

In this work matrix shingling was analyzed. This technology brings many advantages for BIPV, but for standard modules as well. The absence of ribbons and the bricked wall like arrangement ensure a very homogeneous appearance. High power outputs during partial shading compared to conventional modules ensure high energy yields. Scalability makes integration into predefined glass sizes easy and maintains excellent area efficiency. Better coverage of the module area with cells leads to higher module efficiency. Fraunhofer ISE has developed and tested this technology while an industrial stringer was brought to the market by M10 Solar Equipment. The key points for designing of high efficiency matrix modules are a low damage laser cutting process, optimized glass margins and reduced shingle overlap. A comparison between a common half-cell module and matrix modules shows that either module area or number of host cells must be adjusted for an optimized matrix module. The matrix technology reaches higher module efficiency (19.3% compared to 18.6% with a half-cell module). However, more silicon is used per module. This shows that – simply speaking – shingle matrix technology does not get the best out of the host wafer but out of the module area. Therefore, shingle matrix module technology is an attractive alternative to wire soldering with half-cut cells for BIBV as well as for standard modules.

Seven matrix modules were built on a glass area of 1700 × 1000 mm². They feature a power up to 322 W_p and an efficiency of up to 18.8%. In experiment as well as in a detailed shingle-cell-to-module simulation, a relative power loss of −4.5% is obtained. The most important points in order to build highly efficient matrix modules, namely low glass margins, low damage cut shingles and small shingle overlap are presently under development at Fraunhofer ISE.

Long-term stability for modules with similar material was already shown for linear shingled modules by Fraunhofer ISE [17]. A batch of 10 modules is currently under test to show the same for matrix interconnected shingles. The test sequences according to IEC 61215 [23] includes TC200, humidity freeze, DH1000, mechanical load and hot-spot endurance.

Matrix integrated into a MorphoColor® façade element was presented which shows the aesthetic attractiveness of such modules. Technology wise matrix shingling offers a solution for upscaling flexible BIPV productions with a very high design flexibility.

ACKNOWLEDGEMENT

This work was funded from the Federal Ministry of Economic Affairs and Climate Action (BMWK) under the contract number 03EE1026A with the acronym Shirkan.

5 REFERENCES

[1] P.R. Shukla, J. Skea, R. Slade, A. Al Khourdajie, R. van Diemen, D. McCollum, M. Pathak, S. Some, P. Vyas, R. Fradera, M. Belkacemi, A. Hasija, G. Lisboa, S. Luz, J. Malley, (eds.), "IPCC, 2022: Summary for Policymakers. In: Climate Change 2022: Mitigation of Climate Change. Contribution of Working Group III to the Sixth Assessment Report of the Intergovernmental Panel on Climate Change," 2022.

[2] Jr Donald C Dickson, "Photo-voltaic semiconductor apparatus or the like," May 31, 1960.

[3] *Solaria: Solar Panels for Home & Business.* company homepage. [Online]. Available: https://www.solaria.com/solar-panels (accessed: Aug. 18 2022).

[4] *SunPower Performance Solar Panels.* [Online]. Available: https://sunpower.maxeon.com/int/solar-panel-products/performance-solar-panels (accessed: Aug. 18 2022).

[5] Tongwei Solar, "Shingled monofacial module: TH635~660PMB6, 68 SC datasheet," 2021. [Online]. Available: http://www.tw-solar.com/en/Products/PVModules/

[6] A. Mondon, N. Klasen, M. Mittag, M. Heinrich, and H. Wirth, "Comparison of Layouts for Shingled Bifacial PV Modules in Terms of Power Output, Cell-to-Module Ratio and Bifaciality," in

Proceedings of the 35th European Photovoltaic Solar Energy Conference and Exhibition (EU PVSEC), Brussels, Belgium, 2018, pp. 1006–1010.
[7] W. Schmidt and K.-D. Rasch, "New interconnection technology for enhanced module efficiency," *IEEE Trans. Electron Devices*, vol. 37, no. 2, pp. 355–357, 1990, doi: 10.1109/16.46366.
[8] N. Klasen, D. Weißer, T. Geipel, D. H. Neuhaus, and A. Kraft, "Performance of shingled solar modules under partial shading," *Prog. Photovolt: Res. Appl.*, pp. 1–14, 2021, doi: 10.1002/pip.3486.
[9] N. Klasen, F. Lux, J. Weber, T. Roessler, and A. Kraft, "A Comprehensive Study of Module Layouts for Silicon Solar Cells Under Partial Shading," *IEEE J. Photovoltaics*, vol. 12, no. 2, pp. 546–556, 2022, doi: 10.1109/JPHOTOV.2022.3144635.
[10] B. Blasi, T. Kroyer, T. Kuhn, and O. Hohn, "The MorphoColor Concept for Colored Photovoltaic Modules," *IEEE J. Photovoltaics*, vol. 11, no. 5, pp. 1305–1311, 2021, doi: 10.1109/JPHOTOV.2021.3090158.
[11] C. Kutter *et al.*, "Decorated Building-Integrated Photovoltaic Modules: Power Loss, Color Appearance and Cost Analysis," 2018, doi: 10.4229/35THEUPVSEC20182018-6AO.8.6.
[12] P. Baliozian *et al.*, "PERC-based shingled solar cells and modules at Fraunhofer ISE," *Photovoltaics International*, vol. 43, pp. 129–145, 2019.
[13] Solarthemen Media GmbH, *Schindel-Solarmodul von ISE und M10 ist serienreif.* [Online]. Available: https://www.solarserver.de/2022/05/10/photovoltaik-stringer-zur-fertigung-von-matrix-schindelmodulen-geht-in-serie/ (accessed: Sep. 13 2022).
[14] Nils Klasen, Julian Weber, Achim Kraft, "Lateral Currents in Shingle Solar Modules Detected by Magnetic Field Imaging," *submitted to Journal of Photovoltaics*, 2022.
[15] M. Oswald, M. Turek, J. Schneider, and S. Schoenfelder, *Evaluation of Silicon Solar Cell Separation Techniques for Advanced Module Concepts*, 2013.
[16] P. Baliozian *et al.*, "Thermal Laser Separation of PERC and SHJ Solar Cells," *IEEE J. Photovoltaics*, vol. 11, no. 2, pp. 259–267, 2021, doi: 10.1109/JPHOTOV.2020.3041251.
[17] T. Rößler *et al.*, "Progress in shingle interconnection based on electrically conductive adhesives at Fraunhofer ISE," *Aip Conf Proc*, 2022.
[18] P. Baliozian *et al.*, "Postmetallization "Passivated Edge Technology" for Separated Silicon Solar Cells," *IEEE J. Photovoltaics*, vol. 10, no. 2, pp. 390–397, 2020, doi: 10.1109/JPHOTOV.2019.2959946.
[19] M. Mittag and M. Ebert, "Systematic PV module optimization with the cell-to-module (CTM) analysis software," *Photovoltaics International*, no. 36, pp. 97–104, 2017.
[20] M. Mittag, T. Zech, M. Wiese, D. Blaesi, M. Ebert, and H. Wirth, "Cell-to-Module (CTM) analysis for photovoltaic modules with shingled solar cells," in *44th IEEE PV Specialist Conference PVSC.*
[21] J. Schneider, H. Hanifi, D. Daßler, M. Pander, F. Kaule, and M. Turek, "Half-cell solar modules: The new standard in PV production?," *Photovoltaics International*, vol. 2019.
[22] M. Devoto Acevedo, T. Timofte, A. Halm, and D. Tune, *Improved Measurement of the Contact Resistivity of ECA-Based Joints*, 2021.
[23] *IEC 61215-2 Terrestrial photovoltaic (PV) modules – Design qualification and type approval – Part 2: Test procedures*, 2016.

6 ANNEX

SmartCalc.Module half-cell simulation

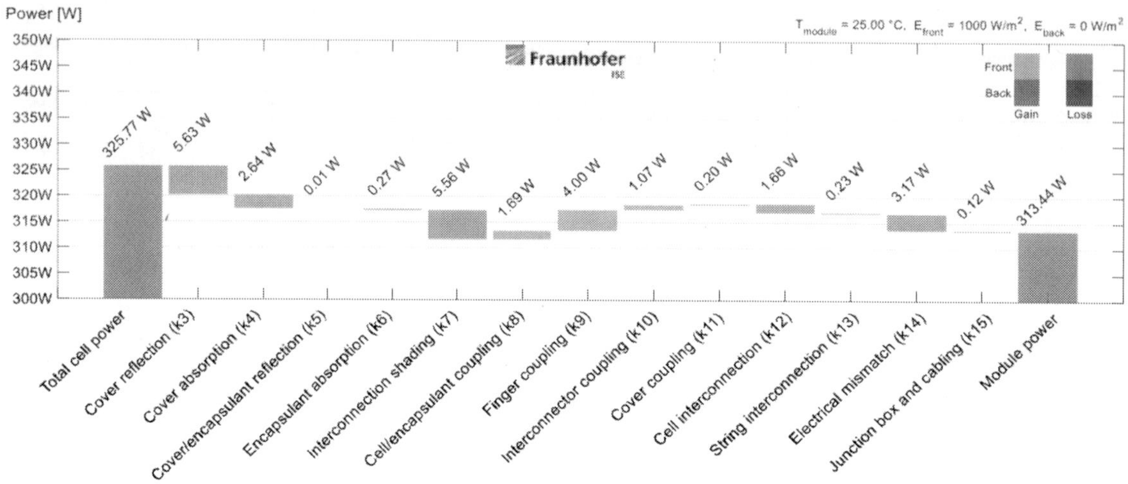

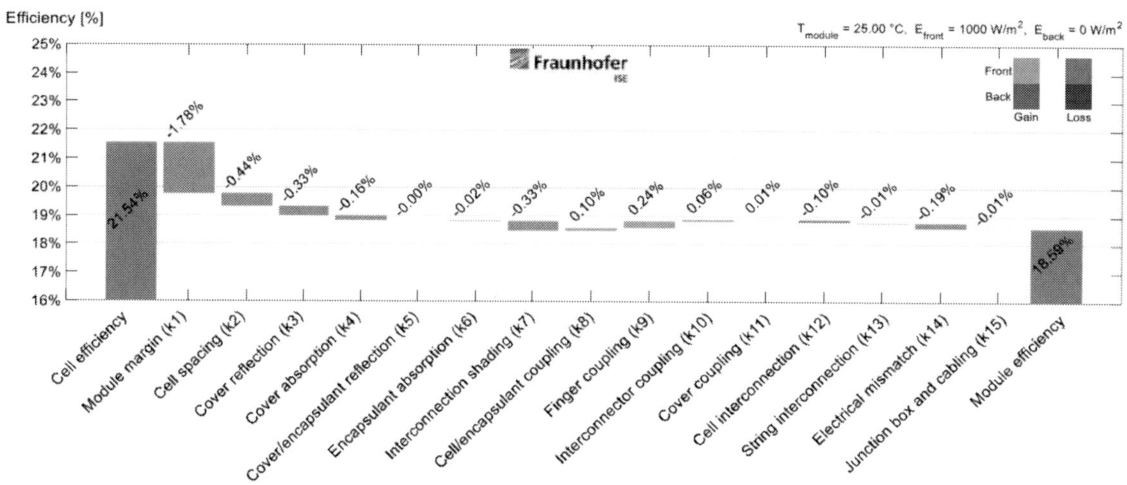

SmartCalc.Module matrix simulation: Scenario 1

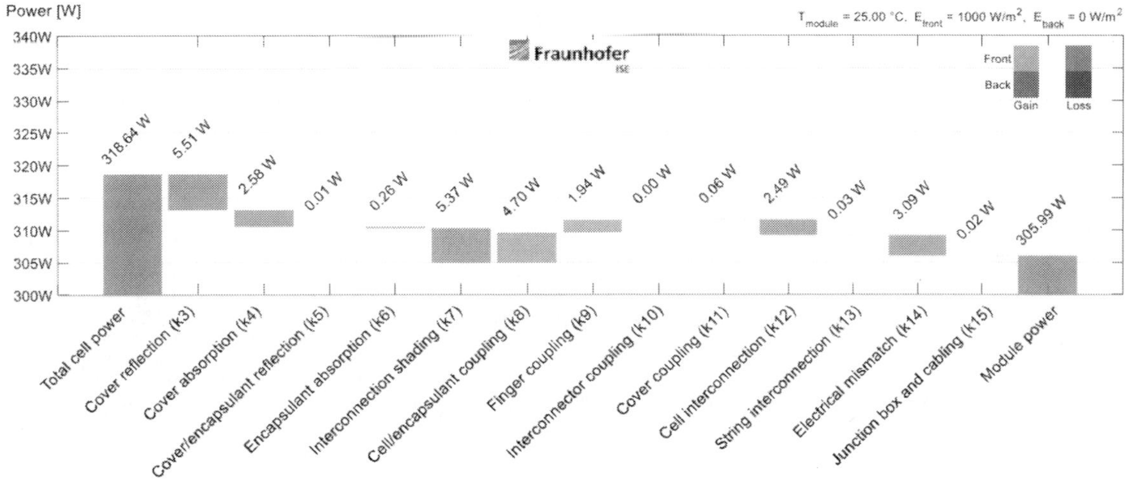

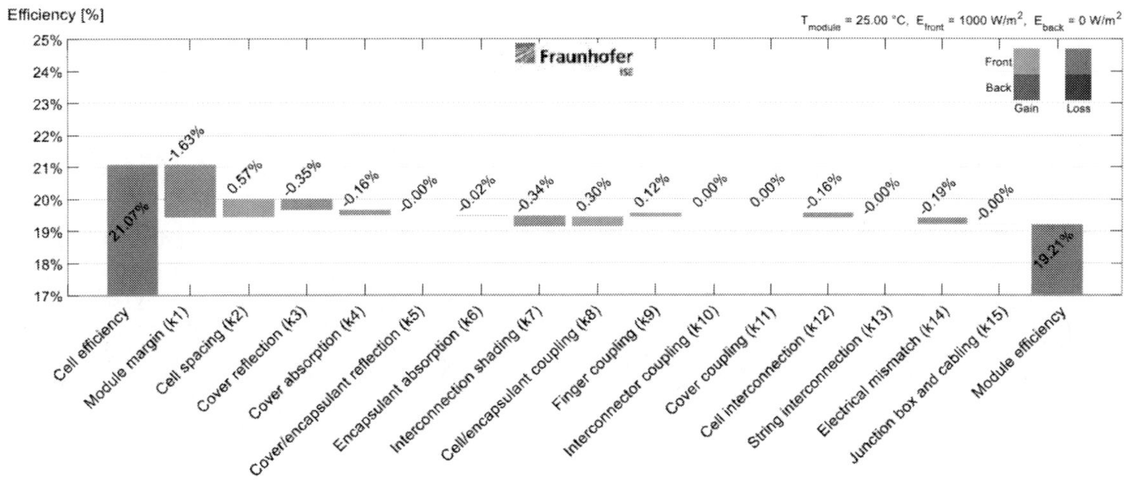

SmartCalc.Module matrix simulation: Scenario 2

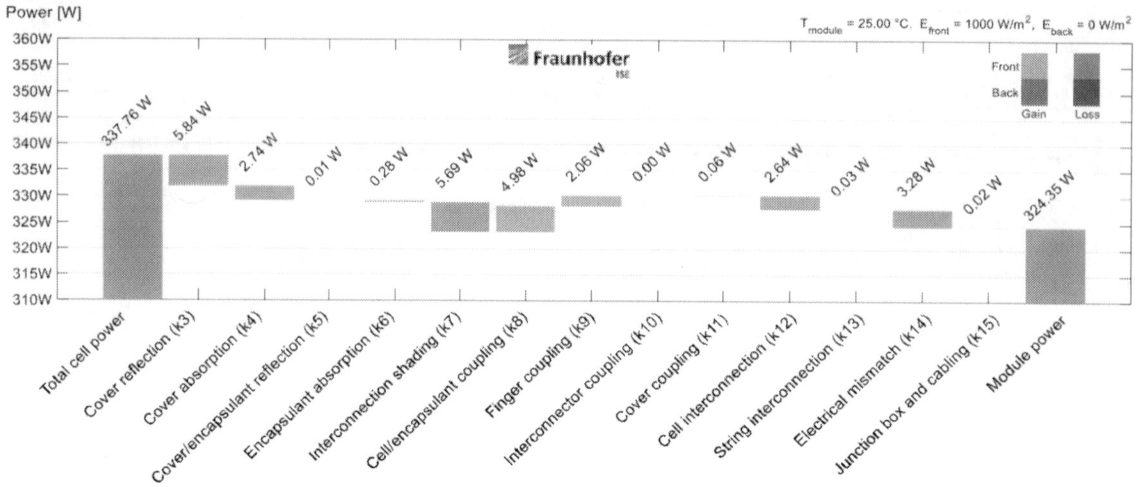

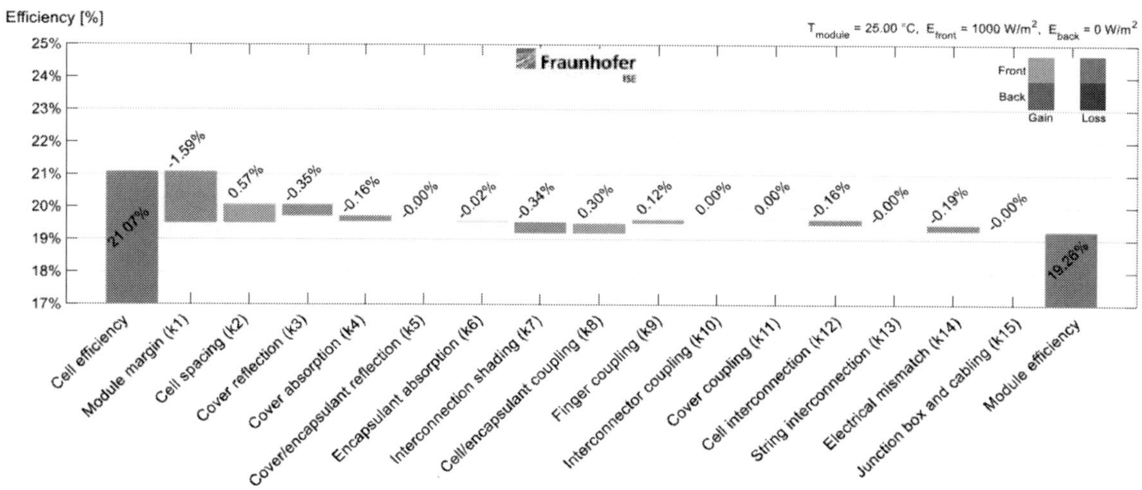

FEASIBILITY STUDY OF SYNERGISTIC BACK-CONTACT SILICON SOLAR CELLS METALLIZATION AND ELECTRICALLY CONDUCTIVE ADHESIVE INTERCONNECTION APPROACH

Tudor Timofte[1], Stephan Großer[2], Matthias Pander[2], Thomas Fischer[3], Robert Wade[3,5], Christof Erban[4], Maria Ignacia Devoto[1,6], Joris Libal[1], Karl Wienands[1], Tobias Meßmer[1], Daniel Tune[1], Andreas Halm[1]
[1] ISC Konstanz, Rudolf-Diesel-Str. 15, 78467 Konstanz, Germany
[2] Fraunhofer-Center für Silizium-Photovoltaik CSP, Otto-Eißfeldt-Str. 12, 06120 Halle (Saale), Germany
[3] teamtechnik Maschinen und Anlagen GmbH, Planck Str. 40, 71691 Freiberg, Germany
[4]Sunovation Produktion GmbH, Glanzstoffstr. 21, 63820 Elsenfeld, Germany
[5]Solarnative GmbH, Oeserstr. 3C, 65934 Frankfurt am Main, Germany
[6]Hochschule Anhalt / Anhalt University of Applied Sciences,
Bernburger Straße 55, 06366, Köthen, Germany

ABSTRACT: Besides technical and reliability aspects, the successful industrialization of a solar cell technology must also consider the optimization of material and process costs. The aim of this study is to push the boundaries of silver reduction in the solar cell metallization as well as in the interconnection process with electrically conductive adhesive (ECA). The feasibility of replacing the silver busbars of the ZEBRA IBC cell [1] by an ECA interconnection line is investigated, as well as realizing the interconnection through ECA pads [2][3]. These variations are tested in an established industrial process environment [4] (TeamTechnik stringers: TT 1400ECA for ECA interconnection and TT2100 for soldered strings) for relevant mass production results. The produced strings were investigated on process, material, string and mini-module level regarding the extent of ECA curing, polymer compatibility, mechanical stability of the interconnection, electrical performance and reliability, by using methods like differential scanning calorimetry, 180° ribbon peel test, electroluminescence and I–V measurement, cross-section analysis of interconnection, lock-in thermography, magnetic field imaging [5], mechanical load test [6] and accelerated ageing tests (TCT and DHT). Through this comprehensive investigation, a clear identification is obtained of reliable configurations that are worth following at further ZEBRA IBC interconnection process developments, as well as the experimentally determined material saving for realistic cost of ownership assessment.
Keywords: Back Contact, Cost Reduction, Module Manufacturing

1 INTRODUCTION

Electrically conductive adhesive (ECA) interconnections confers several opportunities as: switching to lead-free interconnection, implementation of ECA as busbar replacement, and the possibility of tuning the ECA consumption and interconnection geometry by screen print design or dispensing pattern. A cost of ownership (CoO) model calculation was conducted for the ZEBRA IBC cell and this experiment setup, considering several assumptions for industrialization. By opting for an ECA interconnection, additional silver consumption increases the production costs by approximately 2.9 USDct / Wp (**Fig. 1**).

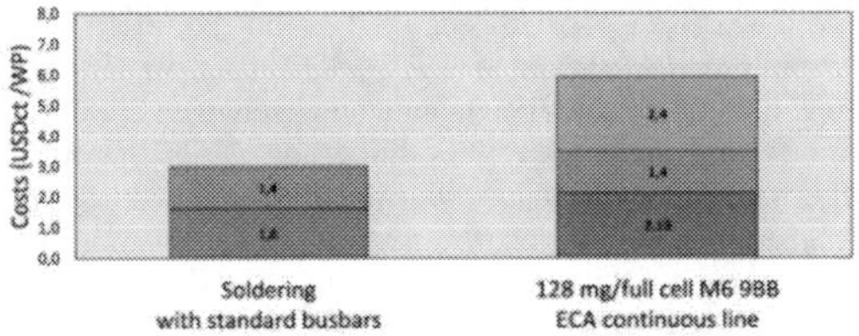

Figure 1: CoO for ZEBRA IBC cell interconnection with standard soldering or ECA (BB = busbar).

To reduce the ECA consumption, two screen print variations for pad printing are considered: a 0.4 x 0.5 mm ECA pad size with potential silver reduction of approximately 77% and a 0.6 x 0.5 mm ECA pad geometry with potential Ag reduction of approximately 42% (both versions calculated for M6 6BB cell layout). The ECA interconnection proved to be a mostly reliable interconnection for the ZEBRA IBC cell, as it was tested with two commercially available, highly silver filled ECAs [4]. These two ECAs (ECA01 1K acrylic-based and ECA02 2K epoxy-based), since already qualified and proven reliable, were used also for the tests presented in this work. The investigation consists of a systematic reduction of ECA quantity per cell by printing ECA pads on cell busbars instead of a continuous line. Additionally, also replacing the busbar metallization by ECA during the interconnection process is considered (printing ECA on busbar-less cells).

For this experiment, ZEBRA IBC cells of same size and with same metallization layout were used: a test group with cells with full standard ZEBRA IBC metallization and a test group with same metallization (paste, firing or drying process and layout geometry) as for the previous group, but without the busbar metallization.

For all tests, half-cell strings with 1–2 mm cell–cell distance were used. The test strings with ECA variations were produced on a TeamTechnik TT1400 ECA stringer with integrated screen printer, with a fully automated process, whereas the soldered strings used for certain tests as comparison group were produced on a TeamtTechnik TT2100 stringer with a semiautomatic process.

An intensive qualification schedule was planned for identification of a robust solution: verification of curing process and material compatibility (since for the groups with busbar-less cells feature a new contact between ECA and electrical insulating paste), characterization regarding connection geometry, mechanical stability, performance and reliability.

2 EXPERIMENTAL SETUP

A test plan targeting maximal, yet still functional ECA and silver reduction for ZEBRA IBC cell interconnection was created. Considering previously reported results and process experience [4], five ECA contacting variations (CV) were investigated, as shown in **Figure 2**.

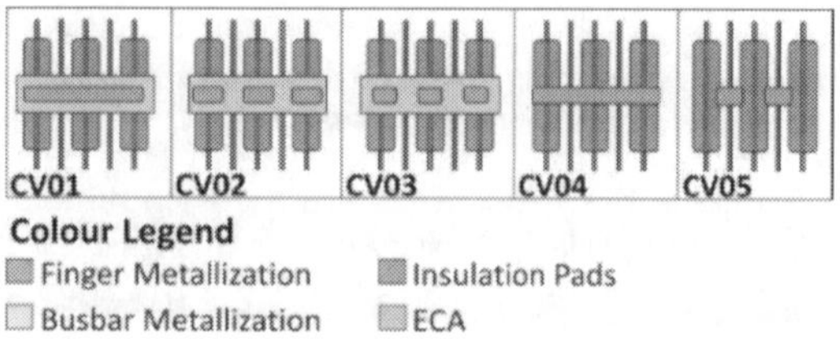

Figure 2: The ZEBRA IBC contacting variations studied.

This variation scheme contains three ECA contacting variations for the cell with busbars and two contacting variations for the busbar-less cell, whereas for CV01 / CV04 and for CV02 / CV05 the same screen prints were used. Therefore, for the above-mentioned testing variations only three screen prints were necessary. The geometric features of the screen print openings are depicted in **Figure 3** and listed in **Table I**. The screen print type, as well as screen print process was reported previously [4]

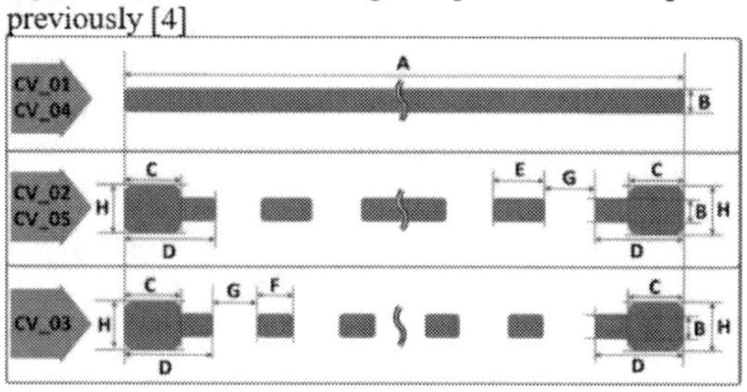

Figure 3: Print screen opening design for the three screen print variations

Table I: Opening dimensions for the tested screen prints

Dimension (mm)	CV01 CV04	CV02 CV05	CV03
A	77.50	77.37	77.37
B	0.5	0.5	0.5
C	-	0.6	0.6
D	-	1.1	1.1
E	-	0.6	-
F	-	-	0.4
G	-	0.5	0.5
H	-	1.1	1.1

The semiautomatic process flow was as follows: cell fluxing (fluxing pen with industrial, low residue flux), cell positioning on transport belt and ribbon positioning on the busbars. The ribbons for semiautomatic process were cut to required dimension and stretched by 0.3%. The ribbons have following features for all groups: 1.5 x 0.20 mm straight copper core, silver coated for ECA gluing process and $Sn_{60}Pb_{40}$ coated for the soldering process. The nominal silver coating thickness was 0.5 µm (homogeneous over the entire surface of ribbon), whereas the $Sn_{60}Pb_{40}$ coating nominal thickness was 16–24 µm, measured in the middle of ribbon width (since the coating is elliptic). Both ribbon were of soft type, with offset yield strength ($Rp_{0.2}$) in range of 70–80 MPa. The soldering and gluing processes were presented in more detail previously [4], with following features in the current test setups:

- Gluing process
  - Setup curing temperature: 180°C
  - Setup heating ramp: 15s
  - Dwell duration at setup temperature: 35s
  - No IR heating during curing
- Soldering process
  - Setup soldering temperature: 210°C for heating plate
  - Setup soldering temperature: 230°C for IR heating
  - Setup heating ramp: 10.8s
  - Soldering time (with IR heating): 1.6s / cell
  - Set cooling ramp to 80°C: 13.44s

The acrylic-based ECA01, delivered as 1K frozen cartridges, was thawed and degassed at room temperature using a laboratory glass desiccator under vacuum (150 mbar) for approximately 30 min prior to use. The epoxy-based ECA02 was delivered as 2K, mixed according to producer specification and also degassed for 30 min. before use. For all tests, G1 (158.75 x 158.57 mm) ZEBRA IBC solar cells with same metallization layout were used, with contacting finger pitch 1.1 mm and four continuous busbars per polarity, 1.5 mm wide each. For two testing variations the busbar metallization was omitted. Furthermore, all processes on stringers were set for half-cell process, therefore the samples created on the stringers with ECA or solder consist of two or four half-cell strings with a cell–cell distance of 2–3 mm. To avoid undesired measurement effects, only cells from one batch were used for the peel tests. The cells used for cell-to-module (CTM) loss calculation, I–V performance and climate chamber testing were metallized at ISC Konstanz from one available G1 solar cell precursor batch (all ZEBRA IBC processing steps completed up to metallization) with two variations: half of the precursors from the batch were fully metallized according to standard ZEBRA IBC processes and half of the precursors were metallized with the same standard ZEBRA IBC process, but with the last metallization step (busbar printing) omitted. All the cells were segmented into half-cells by an established laser scribing and mechanical cleavage process. For peel tests, two half-cell strings were segmented to quarter cell segments and the ribbons were peeled at 180° on a digital force testing machine (ZwickRoell BT1-FR0.5TN.D14). For investigation of ECA curing degree and for verification of compatibility between polymer components, a differential scanning calorimeter (Netzsch 214 Polyma) was used with aluminium crucibles (Netzsch Concavus® type). Nitrogen served as protective and purge gas (protective N_2 flow rate: 40 ml/min; purge N_2 flow rate: 60 ml/min) and a 10 K/min heating and cooling rate was implemented.

For further testing, two types of samples were produced: glass–backsheet (GB) mini-modules, using a commercially available PET-based transparent 3-layer (PET–PET–Primer) backsheet, and glass-glass (GG) mini-modules. For the glass–backsheet mini-modules a commercially available thermoset (cross linking) poly-olefin-based encapsulation material with low UV cut-off was used (one layer of 450 µm thickness film on each cell side). Lightly textured 3.2 mm solar glass without anti-reflective coating (ARC) was used for GB modules, and 2.0 mm lightly textured solar glass without ARC was used for GG modules. The ID code of each cell was tracked back at the mini-modules, furthermore using the same cell halves of a full cell at each two half-cell mini-module. The bussing of mini-modules was made with commercially available $Sn_{60}Pb_{40}$ coated copper core (5–6 mm width / 0.3 mm thickness) ribbon. Glass–backsheet mini modules were laminated at a laboratory laminator, with lamination program set to exceed 75% cross linking degree of encapsulant. The glass–glass mini-modules were produced by Sunovation using a proprietary liquid encapsulation (silicone-based) technique. The cells were measured at an AAA h.a.l.m. GmbH flasher, using 4-point contacting at STC (flasher irradiance 1000 W/m², 25°C ± 3°C). The mini-modules were measured at two AAA h.a.l.m. GmbH flashers, using 4-point contacting at STC conditions. Control modules (from each testing group) were considered for the measurements to extract eventual not corrected measurement drifts and offsets. For CTM losses, performance and accelerated ageing tests (DHT and TCT), only two half-cell mini-modules were used for all testing configurations. For lock-in thermography (LIT), an InfraTec ImageIR 9400 tool was utilized. For magnetic field imaging (MFI), a home-build system at FhG CSP equipped with an MFI-sensor from DENKweit was used. LIT and MFI were measured at samples without any accelerated ageing. Cross section analysis was made using contrast mode microscopy (bright field / dark field) at metallographic polished two half-cell strings. One string of each group was assigned for cross section analysis, and for each string the cross section was planned at four defined locations (**Fig.4**): one at center of one half-cell and three close to the cell edge, at approximately 6 mm distance to cell edge, attempting to cover most of the process variation for thickness and width of interconnection.

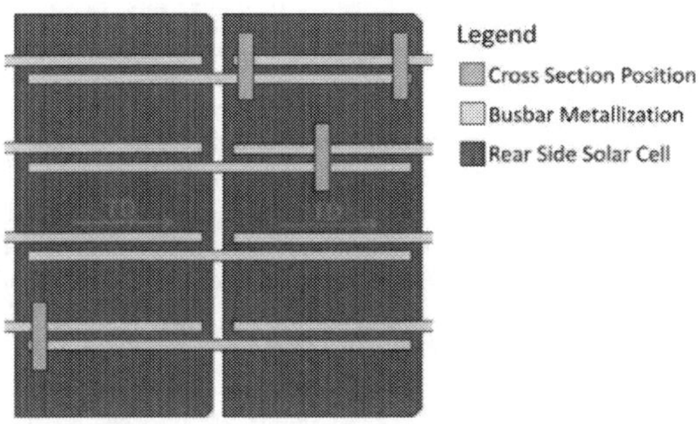

Figure 4: String drawing with two half-cells / inter-connection side towards viewer (TD = transport direction at stringer).

For mechanical load test (MLT) bigger size test module samples were required, therefore a 900 x 360 x 4 mm glass size was used for all samples. The testing modules were produced in glass–backsheet configuration with polyolefin-based encapsulation material. At each mini-module for mechanical load test, two strings, each containing four half-cells, were joined in series (2 mm cell–cell distance in both directions of string matrix). The two half-cell strings (each containing four half-cells) were positioned parallel to the 360 mm edges of glass and the cell matrix was centered on glass surface. From each test configuration a mini module was set under cyclic mechanical loading test through a 4-point-bending setup at a Zwick Z400 universal testing machine with electro-luminescence monitoring during and after mechanical load. The setup and load range (100 N – 600 N) was chosen to simulate similar string strain amplitude at full module size in accordance with testing standards [6]. The accelerated ageing conditions for damp heat test (DHT: + 85°C; 85 % relative humidity) and temperature cycling test (TCT: - 40°C / + 85°C) comply with the IEC 61215-2 norm procedure, with the exception, that no current injection at samples was used during TCT testing.

3 RESULTS

For verification of curing degree at ECA, an aluminium crucible with 10–11 mg uncured ECA, was set at center of a half-cell, localized at standard position on transport belt and sent through the TeamTechnik TT1400ECA stringer curing path. For both ECAs no residual curing peaks could be detected, meaning full curing for both ECAs was achieved (**Fig.5** for ECA01; **Fig.6** for ECA02).

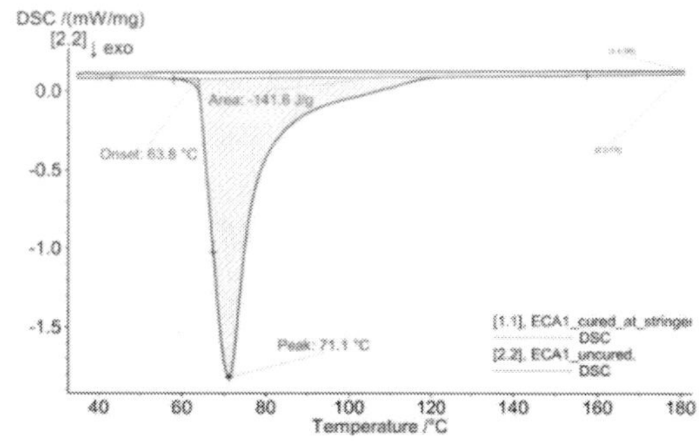

Figure 5: 1st DSC heating curve for ECA01: blue line after curing process; the red line = DSC of uncured ECA01.

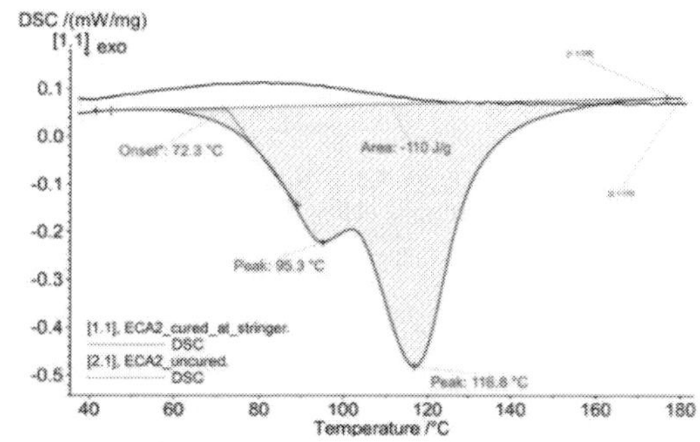

Figure 6: 1st DSC heating curve for ECA02: blue line after curing process; the red line = DSC of uncured ECA02.

At **Figure 6** a broad peak can be noticed at the cured sample, which is related to absorbed moisture, being slowly released during heating step [7].
For interconnection variations CV04 and CV05, a new material interface situation occurs, since the absence of busbars leads to direct contact between insulation paste and ECA (**Fig.7**). DSC can be used also to verify the chemical compatibility and stability at this interface.

Figure 7: Interconnection section at edge of a busbar-less IBC ZEBRA cell with ECA pads applied on fingers between insulation pads (light brown colored stripes)/ CV05.

The DSC data revealed no additional peaks than the curing peaks of the ECAs [4], therefore, the compatibility between insulation pad polymer paste and both ECAs could be confirmed in range + 25°C to + 200°C (**Fig.8**). A light endothermic peak could be observed at the DSC measurement of the previously cured insulation paste in the range + 50° to + 70°C, which might correspond to release of moisture.

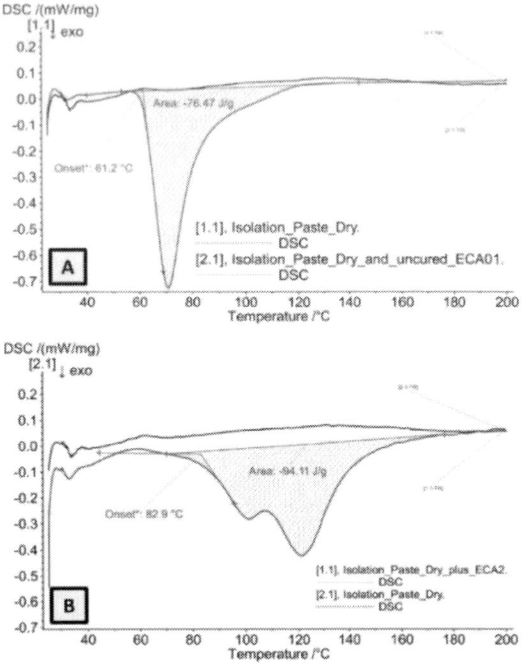

Figure 8: DSC diagrams with 1st heating of cured insulation paste without (blue line) and with (red line) uncured ECA01 **(A)** and ECA02 **(B)**.

After production of strings, for each variation of interconnection (**Fig.3**) and ECA type, one string with two half-cells was used for cross section analysis (a soldered string was also considered). The cross section was investigated by bright / dark field light microscopy regarding condition of the bond, geometry and eventual defects, at the marked positions with green rectangle at **Figure 4**. The exemplary mostly observed types of good and failed interconnection are presented at **Figure 9**.

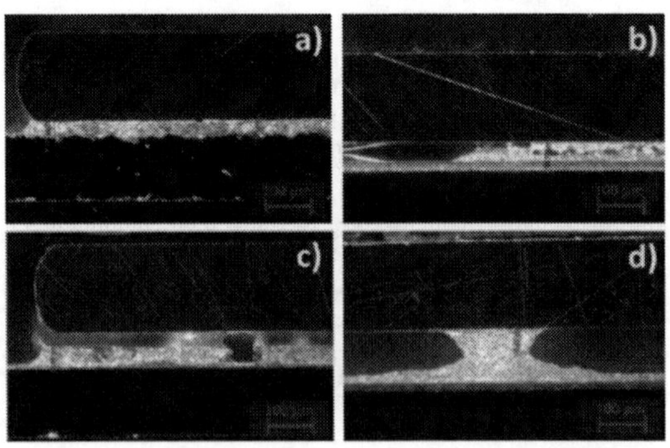

Figure 9: Cross sections at interconnections with ECA02, revealing: ***a)*** *good interconnection,* ***b)*** *cohesive fracture in the ECA interconnection,* ***c)*** *detachment of ribbon from ECA and bubble within ECA and* ***d)*** *necking of ECA.*

Ribbon detachments and ECA cracks were observed for both tested ECAs at regions starting from both cell edges and propagating toward cell middle over a distance of more than 10 mm at each cell side, caused by thermo-mechanical stress upon cooling after inter-connection [8]. ECA squeeze out is neglected, since observed rarely at this test. Most of ribbon detachments in cell-edge range were noticed at groups with ECA pads and at groups with busbar-less cells. The distribution of thickness is depicted in **Figure 10**.

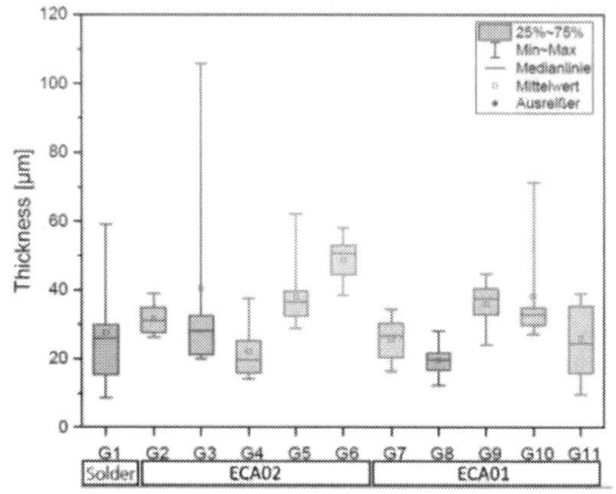

Figure 10: Distribution of interconnection thickness for solder and ECA joints (at least 10 measurements made at each testing group).

Within each group and between the groups no correlation or systematic tendency could be noticed between individual inspected cross sections. The median thickness is at 27 µm, considering all groups with ECA01 (minimum/ maximum thickness: 10 µm / 71 µm) and for the groups with ECA02 the median thickness is at 31 µm (minimum/ maximum thickness: 14 µm / 106 µm). The broad distribution of thickness observed at several groups, without correlation to interconnection type, is caused by a combination of factors, such as: thermo-mechanical stress due to different coefficients of thermal expansion between copper ribbon and solar cell [8], as well as process features, such as down holder type, pressing on ribbons at six equidistant positions. The measured width of interconnection (distribution of width at groups depicted in **Figure 11**) is used to calculate the force per mm width.

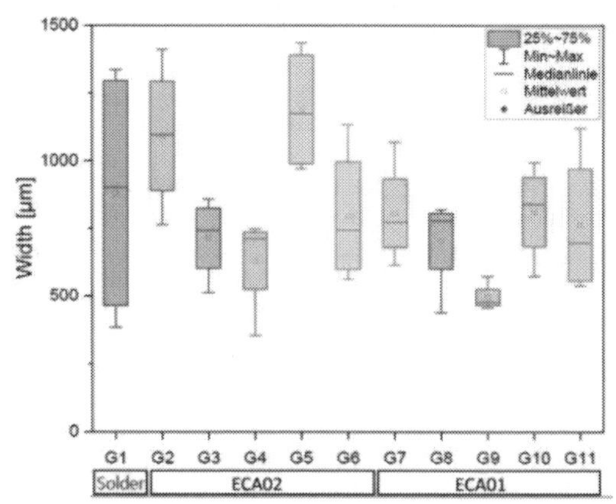

Figure 11: Distribution of interconnection width for solder and ECA joints (at least 4 measurements made per testing group).

The groups with busbar-less cells tend to have slightly wider effective width, but also broader spread of values. The evaluation of adhesion proved difficult, since at many samples with ECA pads a full detachment of ribbon during first peeled centimeter occurred. Therefore two questions occur:

1. Which is the length of retained interconnection?
2. Which is the suitable segment from peel diagram at all groups, which allows a comparison, excluding measurement artefacts or the effect of premature detachment of ribbons?

To answer first question, an observation of peel force through the entire peel path is recommended (**Fig.12**).

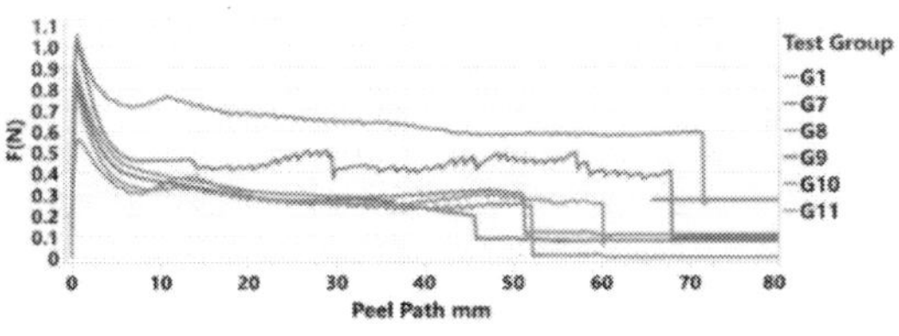

Figure 12: Force - peel path diagram for one exemplary ribbon of each testing group of soldered (G1) and glued connection variations with ECA01 (G7 – G11)

Whereas at groups 1 and 7 almost the complete cell length is interconnected, providing peel force data for analysis, at groups 8, 9, 10 and 11 a clear impairment of peel force data results from premature detachment of ribbons (**Fig.12**). Worth mentioning is the constant adhesion and higher peel force of ECA01 at CV01 interconnection (G7) compared to soldering group (G1), showing that certain ECAs might clearly outperform the soldering interconnection.

To approach question number two, a data cut is necessary: the initial part of diagram with artefacts and last considerable segment of diagram (because of above mentioned issue) are excluded from all data sets, leading to the peel path segment: 15 mm–30 mm used for peel force comparison. For a better visualization and analysis, the data sets of testing groups in the above-mentioned peel path segment are represented as bars with superposed thin box plots; for each testing group all selected peel data were merged in one box plot per polarity (**Fig.13**). For simplifying the representation, the peel force bar of soldered group was omitted, mentioning that both polarities of this group have peel force (median) in range 0.70 – 0.78 N, which is comparable with the values of ECA reference groups (contacting versions CV01).

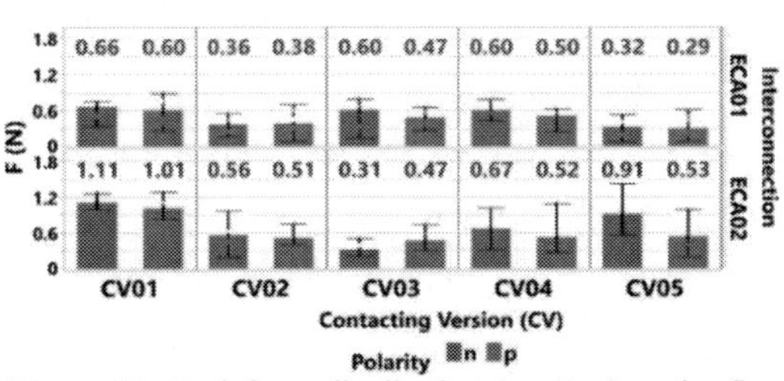

Figure 13: Peel force distribution (**median**) at the five contacting versions for ECA01 and ECA02.

At contacting variations made by ECA pads (CV02, CV03, CV05), it is expected that the peel force as average or median would be lower than for groups with ECA line interconnection, since lower adhesion at ECA pads at front and end region of each pad occur.

Contacting version variation CV01 would be considered reference, since cells with busbars are interconnected by uninterrupted soldering or ECA adhesive lines. Contacting versions variations with pads (CV02, CV03 and CV05) reveal, as expected, lower peel force values, with 40% less median peel force per group for both ECAs at CV02 in comparison with CV01.

Considering the line interconnections CV01 and CV04, the peel force at busbar-less cells is lower than at cells with busbars: for ECA01 the difference is 15% less peel force for busbar-less cells at positive polarity, whereas at ECA02 the peel force drop is 48.5% for positive polarity. Furthermore, ECA02 tends to reveal higher differences than ECA01 between polarities at each interconnection variation. This is probably a surface wetting related issue or might be also caused by rheological differences between the two ECAs. The above-mentioned observations do not consider the differences at the effective width of interconnection. By implementing the determined median effective width for each group at the measured median peel force, the data sets can be better compared, in N/mm (**Fig.14**).

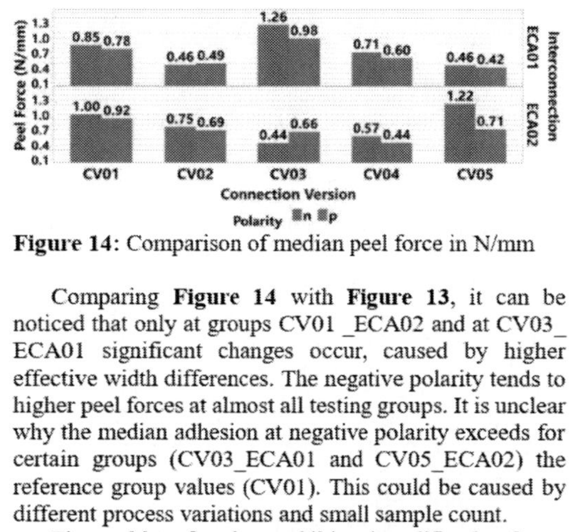

Figure 14: Comparison of median peel force in N/mm

Comparing **Figure 14** with **Figure 13**, it can be noticed that only at groups CV01 _ECA02 and at CV03_ ECA01 significant changes occur, caused by higher effective width differences. The negative polarity tends to higher peel forces at almost all testing groups. It is unclear why the median adhesion at negative polarity exceeds for certain groups (CV03_ECA01 and CV05_ECA02) the reference group values (CV01). This could be caused by different process variations and small sample count.

The peel interface is an additional qualification factor for understanding the interconnection issues (**Fig.15**).

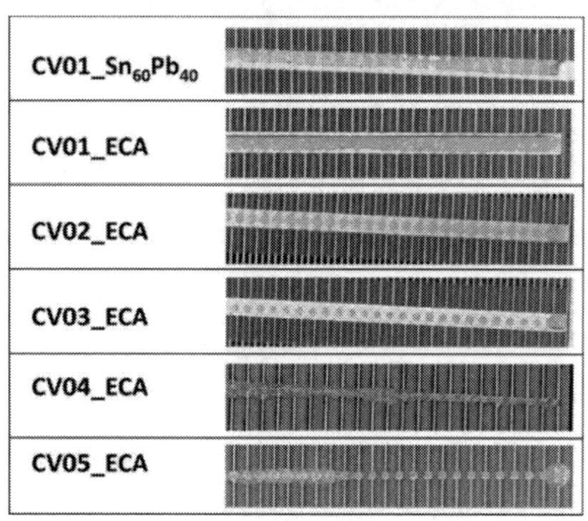

Figure 15: Typical peel interface pictures of tested interconnection variations, representative for both tested ECAs and peeled polarities.

The peel interface is similar at groups with cells with busbars and of same contacting geometry, but with different ECAs. An inhomogeneous pressing before or during curing at ECA is particularly visible at groups with busbar-less cells in form of considerable variation at the effective width of ECA pattern over complete busbar length (**Fig. 15** / CV04 and CV05). A clear difference between ECAs at peeled samples with busbar-less cells could be noticed: ECA01 presents clearly more **adhesive breakage** between ECA and insulation pad than ECA02. Furthermore, ECA01 reveals a better homogeneity regarding the shape of peeled ECA pad polarities, whereas at ECA02 the positive polarity had a poor mechanical contacting compared to the negative polarity (ECA pads at positive polarity were not or less pressed during interconnection process, than the ECA pads of negative polarity). This is in accordance with the differentiation of peel force for polarities at CV05 and ECA02, with considerably less peel force (approximately 41%) for positive than for negative polarity (**Fig.14**). Separation surface of peeled samples for all groups is usually characterized by cohesive breakage within ECA, combined with adhesive breakage to ribbon, respectively to cell (at groups with busbar-less cells) for both ECAs.

Continuing the qualification at laminate level, three mini modules per testing group with two half-cells each were characterized after production by EL, LIT and MFI imaging, to inspect the homogeneity of interconnection. For each group, the three mini modules reveal almost identical characteristic EL, LIT and MFI images, concluding that retrieved image patterns are representative for tested variations and no arbitrary effects. At groups with lower ECA or without busbars considerable inhomogeneities at interconnection and not connected regions were detected. At **Figure 16** the extreme cases of our study are presented: a good, homogeneous interconnection by ECA01 with CV01 (cell with busbars and ECA line) and a bad interconnection by ECA01 with CV05 (busbar-less cell with ECA pads). Similar images reveal ECA02 at same contacting variations. All other contacting variations (CV02, CV03 and CV04) present EL, LIT and MFI images (not shown here) with intermediate states between the good and bad example shown below (**Fig.16**).

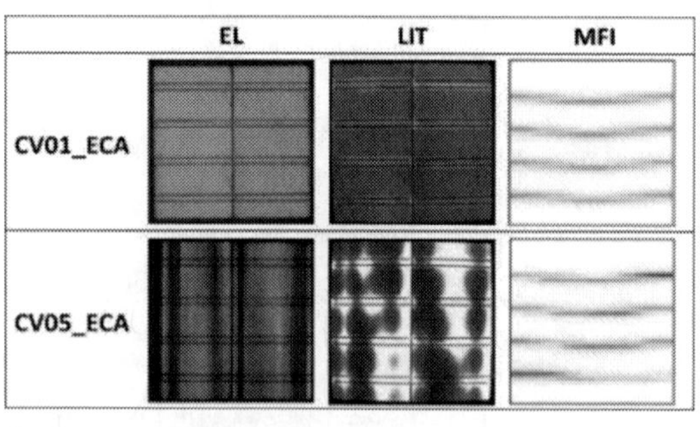

Figure 16: Good (CV01_ECA) and bad (CV05_ECA) interconnections characterized by EL, LIT and MFI imaging for ECA01.

The EL picture reveals at groups with ECA pads or at groups with busbar-less cells interruption over the cell surface; the most electrically separated areas are located at cell edge regions with interconnectors (dark regions at EL image for CV05 in **Figure 16**). Due to inhomogeneous interconnection observed by EL imaging, the current injection during LIT test is distributed inhomogeneously, therefore the lock-in thermography image reveals an inhomogeneous thermal power loss in the solar cell, as well. MFI images show the 1D magnetic flux density in perpendicular direction caused and dominated by currents in the interconnection ribbons. MFI image provides therefore spatially resolved mapping for cell interconnection, a versatile method to determine inhomogeneous current densities at cell interconnections, since the measured magnetic flux density is directly proportional to passing current at the inspected sites [5]. Current flow interruptions or inhomogeneous current flow (due to interruptions, cracks in ECA joints or inhomogeneous interconnection) can be visualized by absence of signal (no line) or by intensity variation of the signal, presented as colored line (**Fig.16**, MFI image for CV05_ECA). The 4-busbar solar cell of CV01 exhibited comparable currents within the parallel busbars observable by equal intensities (**Fig.16**, MFI image for CV01_ECA). Through the combination of these three characterization techniques, a reliable and detailed qualification of the interconnection can be achieved for complete string surface. Differentiation of interrupted regions from eventual local shunts, as well as inhomogeneous distribution of current in parallel ribbons at electrically active regions can be mapped at a resolution in mm^2 scale. Table II depicts the detected defects at testing matrix, considering the EL, LIT and MFI inspection findings.

Table II: Overview of main contacting defects related to interconnection variations, detected by EL, LIT and MFI (x means that defect is present)

Imaging Method			EL	LIT	MFI
Recorded Interconnection Issue			not connected cell areas	inhomogeneous thermogram / (cell hot spots excluded)	uneven current path distribution along interconnection direction
Contacting Version	CV01	$Sn_{60}Pb_{40}$			
		ECA01		x	x
		ECA02		x	x
	CV02	ECA01	x	x	x
		ECA02	x	x	x
	CV03	ECA01	x	x	x
		ECA02	x	x	x
	CV04	ECA01	x	x	x
		ECA02	x	x	x
	CV05	ECA01	x	x	x
		ECA02	x	x	x

Most of contacting variations, even the ones made at cells with busbars and uninterrupted ECA line reveal inhomogeneities at interconnection. The affected surface at strings and the contrast of inhomogeneities with respect to properly contacted cell areas is what differentiate the groups, with the extent of inhomogeneously interconnected area at the string raising from CV01 to CV05 (same trend for both ECAs).

To investigate the impact of ECA pads and removal of cell busbar on module performance, a CTM analysis was made. Since cells and mini-modules were contacted differently at IV flasher, an absolute determination of CTM losses was not subject of this study. However, measuring the cells and the mini-modules with same setup, a relative comparison between groups is possible. The mini-modules from groups with busbar-less cells and ECA pads performed poor, being either electrically inactive after production, or delivering such low values, that an IV measurement could not be properly obtained. It was also noticed at IV measurements, that groups with busbar-less cells and continuous ECA interconnection generate considerably less power than groups with cells with busbars, therefore, the relative Pmpp (power at maximum power point) CTM comparison was made only at the groups with cells with busbars (**Fig.17**).

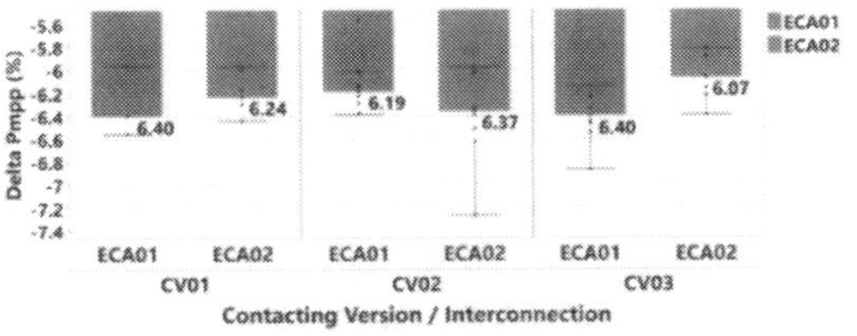

Figure 17: Relative Pmpp CTM loss (%) at contacting variations with busbar cells.

The CTM power losses are in range of 6.1– 6.4% for all three tested variations and two ECAs, with slightly broader distribution at the groups with contacting ECA pads. For ECA groups with contacting pads, no higher CTM was recorded, even remarkably for CV03, which uses the shorter 0.4 mm ECA pads. At (functional) group with busbar-less cells and ECA interconnection line, a broad distribution of mini-module power was observed (**Fig.18** / CV04). Both ECAs with the CV04 interconnection reveal CTM losses in range of 24 % for ECA01 and 41 % for ECA02. These broad power losses are caused by lowered fill factor (FF), thus increased resistive losses and by lowered short circuit current (Isc), due to interrupted cell areas.

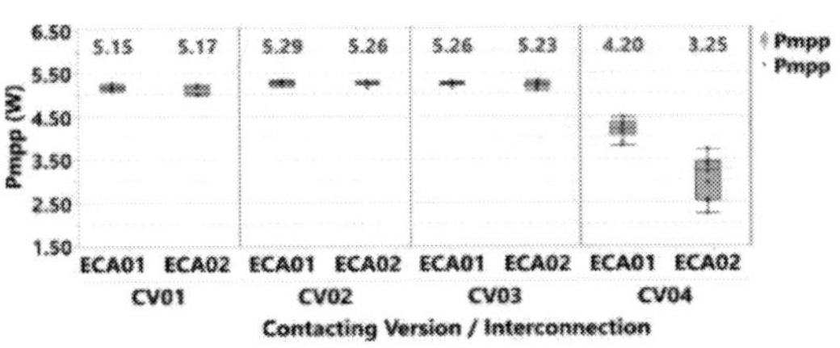

Figure 18: Pmpp (W) of mini-modules manufactured with contacting variations CV01 to 04 and both ECAs (6 minimodules for each variation and ECA type)

A dynamic mechanical loading test (MLT) was performed to detect mechanical susceptibilities of the tested interconnections. After 1000 mechanical load cycles with a load between 100 N and 600 N, considerable power losses were registered at groups with busbar-less cells (**Fig.19**).

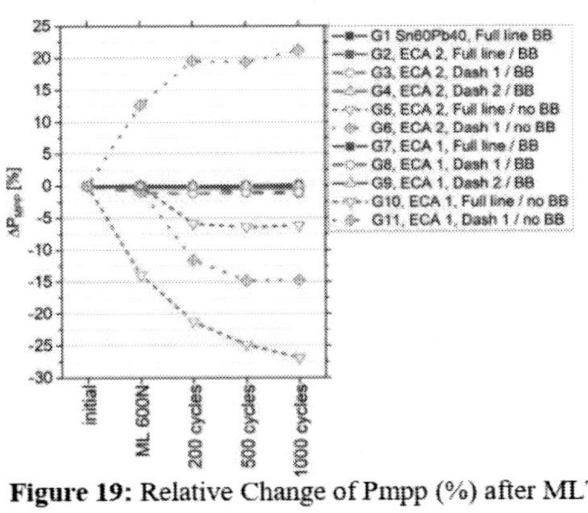

Figure 19: Relative Change of Pmpp (%) after MLT.

At sample from group 11 (ECA01 pads on busbar-less cell) a power increase to initial value was detected after the MLT, which is most probably caused by temporary improvement of contacting between ribbons, ECA pads and fingers through the mechanical load displacements. Most of the measured losses are caused by occurred cell cracks during the test. The modules from the groups with cells with busbars reveal Pmpp changes in range of only ± 1%. An EL inspection (not presented here) at tested modules after MLT reveals clearly the cause of the high power losses at groups with busbar-less cells. In case of cell cracks at mini-modules with busbar-less cells after MLT, broad inactive cell areas occur in the cracked regions. This is not the case for mini-modules made with cells with busbars, which also reveal certain cell cracks after MLT, with the difference that the broken cell segments at cell crack region are mostly electrically active.

In addition, accelerated ageing testing was performed to investigate the reliability of tested interconnection variations. Three mini modules were assigned for each test variation and climate chamber testing. The preliminary ageing results reported here consist of 100 cycles TCT and 650h DHT. For a solid material–process qualification at least 400 cycles TCT and 2000 h DHT are considered necessary. The first 650 h DHT (+ 85°C / 85% rel. humidity) reveal already susceptibilities at several test setups (**Fig.20**).

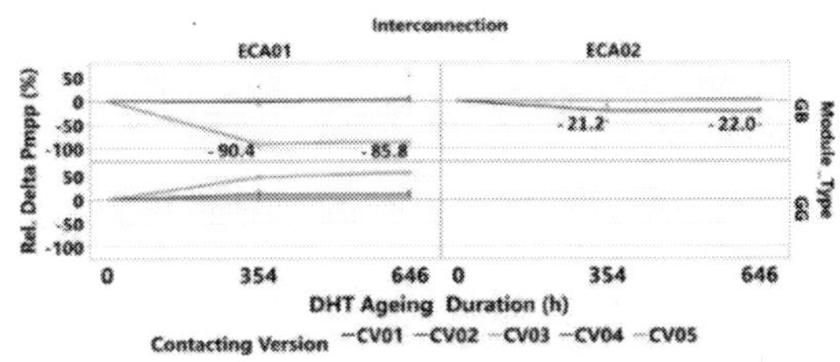

Figure 20: Relative Pmpp changes (%) after DHT ageing.

The glass–glass modules (GG), tested only with ECA01, present a stable behavior for all groups. The increase in power compared to initial value at CV05 (green line in **Figure 20** for GG mini-modules) is associated with a similar mechanism noticed at MLT, only that for this case mechanical work is of thermo-mechanical nature. For glass–backsheet modules (GB) considerable power losses, up to 22% for CV04 with ECA2 and up to 90.4 % for CV05 with ECA1 were recorded. The CV05 data for ECA02 is missing at **Figure 20,** because the produced mini-modules were initially electrically inactive or with barely measurable IV parameters. For this test setup, ECA02 is not suitable for CV04 and CV05, at least in glass–backsheet module configuration. For both ECAs at glass–backsheet configuration the setups CV01, CV02 and CV03 are stable up to 650 h DHT.

The mini-modules were aged up to 100 cycles TCT, without current injection. For almost all groups power drops beyond the pass-fail limit of 5% were recorded (**Fig.21**). CV04 and CV05 variations (both groups with busbar-less cells) in glass–backsheet configuration have power losses in range of 35-100 % for both ECAs. ECA01 seems to be more susceptible to degradation at all silver saving configurations (CV02-CV05) in glass–backsheet format, since even for contacting variation CV02 (0.6 mm long ECA pads on busbars) the power drop after only 100 cycles TCT is close to 5%. ECA02 at same glass-backsheet configuration is at least up to 100 cycles TCT stable for CV02 configuration. All other testing variations with ECA02 (except reference CV01) reveal power losses close to or far beyond 5%.

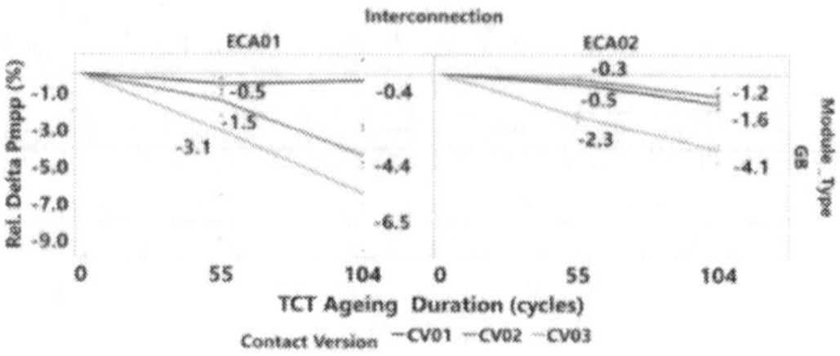

Figure 21: Relative Pmpp change (%) after TCT ageing for glass–backsheet (GB) module type.

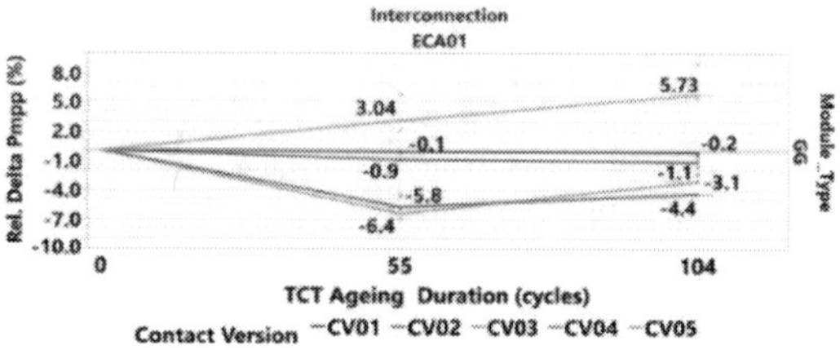

Figure 22: Relative Pmpp change (%) after TCT ageing for glass–glass (GG) module type.

Glass-glass module setup protects better the cell interconnection from thermo-mechanical stress during TCT ageing (only ECA01 tested), however, even at this test a stable behavior seems to provide only CV02 contacting variation, from all tested silver saving setups **(Fig.22)**.

4 SUMMARY

Four silver saving interconnection variations were tested for the ZEBRA IBC solar cell with two previously qualified ECAs, one acrylic and the other epoxy-based. The experimental setup showed that it is possible to reduce ECA consumption by 42 to 77 wt. %, mentioning that only stable configuration of this test with ECA pads was the one with 42 wt. % less ECA02 material, passing at least 650h DHT and 100 cycles TCT. Another possibility to reduce silver metallization of ZEBRA IBC cell by approximately half weight, is replacing the busbars with ECA (synergistic effect). This option proved unstable for both ECAs with the utilized screen prints and tested process. Through the observations and experimental results of this test it can be concluded that the tested variations have potential to pass the qualification tests, if following aspects are considered for further process optimizations:

a) accuracy improvement for ECA pattern alignment to cell metallization;

b) down holder design for more homogeneous pressing on interconnecting ribbons;

c) wider and longer ECA pads (than tested) in range of cell edges to withstand thermo-mechanical stress after stringing;

d) use of ribbons with textured surface toward ECA for enhanced adhesion.

Adhesive breakage between ECA01 and insulation pads should be further investigated. A better wetting through ECA01 of the insulation pad surface should be reached by chemical functionalization or by surface processing.

If the busbars cannot be dismissed from the metallization scheme of ZEBRA IBC cell, because of reliability issues, the next approach could be to test busbar pads in combination with ECA line or ECA pads.

Current tested sequence during this reported work is a reliable tool to qualify interconnection variations regarding processing possibility, as well as performance and reliability aspects, being able to detect inapplicable interconnection variations.

It must be mentioned, that reported results claim validity only for tested materials and processes, without being generally transferable to any epoxy or acrylic-based ECAs available or to any other tested component.

5 ACKNOWLEDGEMENTS

The authors like to thank German Federal Ministry of Economic Affairs and Energy in the frame of Zquadrat Project (contract nr.: 03EE1005) for support, as well as to TeamTechnik, Fraunhofer CSP, Sunovation and all partners and colleagues, which contributed, for great cooperation.

6 REFERENCES

[1] A.Halm, et.all – doi: 10.4229/27thEUPVSEC2012-2AO.2.1

[2] T.Geipel, et.all – doi: 10.1109/JPHOTOV.2018.2828-829

[3] C.Kaiser, et.all – doi: 10.4229/EUPVSEC20202020-4AV.2.6

[4] T.Timofte, et.all – doi: 10.4229/EUPVSEC20212021-4BO.5.2

[5] D.Lausch, et.all – doi: 10.4229/35thEUPVSEC2018-2018-5BO.11.5

[6] M.Pander, et.all – doi: 10.4229/28thEUPVSEC2013-4CO.10.3

[7] R. Artiga, et.all – doi: 10.1016/j.tca.2004.11.016

[8] L.C. Rendler, et.all–doi: 10.4229/EUPVSEC20162016-1CO.11.2

This is a short version of the paper submitted to the journal Progress in Photovoltaics.

It has been selected by the Committee of the WCPEC-8 for submission to Progress in Photovoltaics.

THERMOMECHANICAL DESIGN RULES FOR PV MODULES

Andreas J. Beinert[1,2,*], Pascal Romer[1], Martin Heinrich[1], Jarir Aktaa[2] and Holger Neuhaus[1]
[1] Fraunhofer Institute for Solar Energy Systems ISE, Heidenhofstrasse 2, 79110 Freiburg, Germany
[2] Karlsruhe Institute of Technology (KIT), Institute for Applied Materials, Hermann-von-Helmholtz-Platz 1, 76344 Eggenstein-Leopoldshafen, Germany
*Corresponding author: e-mail to: andreas.beinert@ise.fraunhofer.de

ABSTRACT: We present a set of thermomechanical design rules to support and accelerate future PV module developments. The design rules are derived from a comprehensive parameter sensitivity study of different PV module layers and material properties by finite element method simulations. We develop a three-dimensional FEM model, which models the PV module geometry in detail from busbar and ribbons up to the frame including the adhesive. The FEM simulation covers soldering, lamination and mechanical load at various temperatures. The FEM model is validated by mechanical load tests on three 60-cell PV modules. Here, for the first time, stress within a solar cell is measured directly using stress sensors integrated in solar cells (SenSoCells®). The results show good accordance with the simulations.

The parameter sensitivity study reveals that there are two critical interactions within a PV module: (1) Between ribbon and solar cell. (2) Between front/back cover and interconnected solar cells. Here, the encapsulant plays a crucial role in how the single layers interact with each other. Therefore, its mechanical properties are essential and four design rules are derived regarding the encapsulant. Also, four design rules concern front and back side and three address the solar cells. Finally, two design rules each deal with module size and frame, respectively. Altogether we derive a set of 15 thermomechanical design rules.

As a rule of thumb of how well a bill of material will work from a thermomechanical point of view, we introduce the concept of specific thermal expansion stiffness $\hat{E}_\alpha = E \cdot \alpha \cdot A_\mathrm{j} \cdot h$ as the product of Young's modulus E, coefficient of thermal expansion α, joint area A_j and materials height h. The difference between two materials is a measure of how much thermal strain one material can induce in another. A strong difference means that the material with the larger value will induce thermal strain in the other.

Keywords: PV Module, Design, Module Manufacturing, Modelling, FEM Simulation, Durability, Reliability

1 INTRODUCTION

Stress in solar cells plays a crucial role in the reliability of photovoltaic (PV) modules. Influences on stress are as diverse as the number of different materials in a PV module and become more and more complex with growing variety of PV modules for different applications. Nevertheless, the bill of materials development for new PV modules is often a try and error, which is very time and cost consuming. The finite element method (FEM) in combination with material characterization offers an approach for digital prototyping, which has the potential to reduce the amount of needed physical prototypes significantly. Accordingly, FEM simulations are frequently used in research as the review from Nivelle *et al.* [1] shows. The aim of this work, which is conducted within a dissertation [2], is to derive general thermomechanical design rules for PV modules on the base of a parameter sensitivity study performed with a detailed 3D FEM model of a 60-cell PV module, including ribbons and frame. Additionally, a factor for the quick and straight forward thermomechanical assessment of materials, called specific thermal expansion stiffness $\hat{E}_\alpha$ is introduced. The full version of this paper is published in Progess on Photovoltaics [3].

2 METHOD

The basis of the design rules is a comprehensive parameter sensitivity study conducted with a detailed 3D FEM model of a 60-cell PV module using COMSOL Multiphysics. Intermediate versions of this FEM model have been published previously [4–6]. It covers a simplified soldering process (cooling from 179 ° to 25 °C), lamination (cooling from 150 °C to 25 °C) and homogeneous surface load of 2400 Pa at different temperatures. The stress tensor from each study is transferred to the subsequent study at the temperature at which the subsequent study starts. For more details, we refer to [2].

The parameter sensitivity study consists of 72 parameter combinations covering the material properties shown in Table 1, different frame materials (aluminum, steel, wood), solar cell dimensions and formats as well as module sizes for glass-foil and glass-glass modules. The results of the size variation have been published previously [6]. The results are evaluated according to stress, cell fracture probability P_f and deflection d. For a detailed description of used parameters as well as simulation results, we refer to the authors open access dissertation [2]. In this article, we focus on the design rules themselves.

Table 1: Variation of PV module layer material properties.

	Front/back cover	Encapsulant	Solar cells
Young's modulus E	x	x	
CTE α	x		
Height h	x	x	x

3 THERMOMECHANICAL DESIGN RULES

We find two major stress causes:

1. The mismatch of thermal expansion when exposed to temperature differences.
2. The curvature of PV module and solar cells when exposed to mechanical loads.

Table 2: Derived Thermomechanical design rules clustered by component. The magnitude of the influence on the stress within the solar cells is given in the last column, with increasing magnitude from + to +++.

Component	Design Rule	Influ-ence
Encapsulant	Sharp melting point with a sudden increase of the Young's modulus E_e.	+
	Sharp glass transition with a sudden increase of the Young's modulus E_e.	+
	The encapsulants specific thermal expansion stiffness $\hat{E}_{\alpha,e}$ should be between the ribbon $\hat{E}_{\alpha,r}$ and solar cell $\hat{E}_{\alpha,c}$ value: $\hat{E}_{\alpha,r} < \hat{E}_{\alpha,e} < \hat{E}_{\alpha,c}$.	++
	Low thickness, given that there is sufficient material between ribbon and front- and backsheet and that the critical stress originates from the solar cell-ribbon interaction.	+
Front/Back Cover	At least one stiff layer is needed with a minimum thickness. For soda-lime glass around 3 mm.	++
	The CTE of stiff layers should have a value between the one of the solar cell and ribbon: $\alpha_c < \alpha < \alpha_r$. In symmetric module designs more critical.	+++
	A larger CTE of the backsheet is advantageous for push loads: $\alpha_{bs} > \alpha_{fs}$.	+
	For minimal bending stress: Place the solar cells in the neutral axis, e.g. by a symmetrical module design.	++
Solar Cells	High solar cell thickness.	++
	Small solar cell edge length.	+
	Split cells: Alignment of the shorter side along the higher curvature.	+
PV Module Size	Smaller module area decreases stress.	+
	Module aspect ratio: Longer modules instead of wider modules (for mounting on long side and non-extreme ratios).	+
Frame	The higher the frames stiffness, the better.	+
	Frame design has to be adapted to specific module design.	+

The first one is responsible for compressive stress of solar cells after lamination as well as high tensile stress at the end of the ribbon, which dominates the cell fracture probability P_f.

The second one is responsible that corner solar cells show tensile stress although the deflection from mechanical load is minimal (see Figure 1). For different mounting solutions, like frameless modulus with laminate clamps it could be shown that the highest tensile stress even occurs at the position of the largest curvature instead of the largest deflection [5].

Both causes interact with each other and should not be analyzed separately. Regarding the layer interfaces, we have identified two critical interactions:

1. Between ribbon and solar cell.
2. Between front/back cover and interconnected solar cells.

The thermomechanical design rules reflect these influences. They are summarized in Table 2. For more details we refer to the full paper version [3] as well as the authors open access dissertation [2].

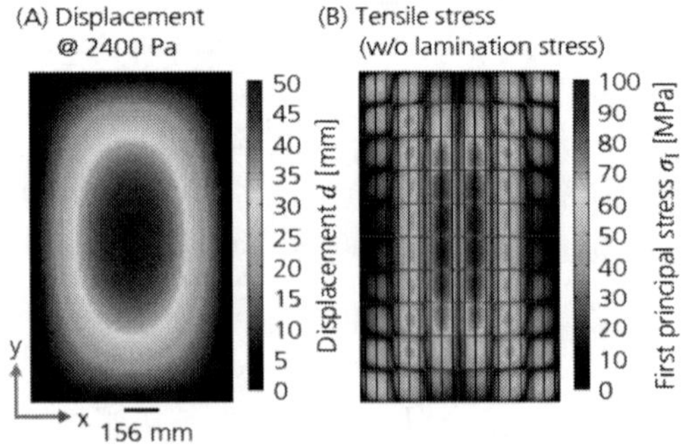

Figure 1: Displacement d at 2400 Pa mechanical load (A) with corresponding first principal stress σ_I in solar cells, neglecting stress after lamination (B). Modified after [2].

4 CONCLUSION

There are many and often entangled influences on stress within solar cells, which we summarize in the derived 15 Thermomechanical Design Rules. Additionally, we introduce the specific thermal expansion stiffness $\hat{E}_\alpha = E \cdot \alpha \cdot A_j \cdot h$ to characterize PV module materials easily and straightforwardly according to their influence on the thermomechanics.

We derive the design rules from a comprehensive parameter sensitivity study using a manifold approach: FEM simulations complemented by solar cell integrated sensors as validation. The developed FEM model covers the PV modules geometry in detail, from the busbar metallization until the frame. The analysis of the parameter sensitivity study results show that some phenomena are only revealed by this multi-scale approach.

ACKNOWLEDGEMENTS

This work was supported by a PhD scholarship from the Cusanuswerk in Bonn, Germany.

The authors would like to thank the TestLab PV Modules, especially Heinrich Berg and David Hottenrott, for conducting the mechanical load tests, Christoph Herzog for support in the PV module production and Jan Benick for production of the SenSoCells.

REFERENCES

[1] P. Nivelle, J.A. Tsanakas, J. Poortmans, M. Daenen, Renewable and Sustainable Energy Reviews 145 (2021), 1–18, 10.1016/j.rser.2021.111022.

[2] A. Beinert Thermomechanical Design Rules for the Development of Photovoltaic Modules. Dissertation, Karlsruhe (2021), 10.5445/IR/1000142510.

[3] A.J. Beinert, P. Romer, M. Heinrich, J. Aktaa, D.H. Neuhaus, Prog. Photovoltaics Res. Appl. (Progress in Photovoltaics: Research and Applications) (2022), 1–13, 10.1002/pip.3624.

[4] A.J. Beinert, A. Büchler, P. Romer, V. Haueisen, L.C. Rendler, M.C. Schubert, M. Heinrich, J. Aktaa, U. Eitner, Sol Energ Mat Sol C 193 (2019), 351–360, 10.1016/j.solmat.2019.01.028.

[5] A.J. Beinert, M. Ebert, U. Eitner, J. Aktaa, Proceedings of the 32nd European Photovoltaic Solar Energy Conference and Exhibition (2016), 1833–1836, 10.4229/EUPVSEC20162016-5BV.1.14.

[6] A.J. Beinert, P. Romer, M. Heinrich, M. Mittag, J. Aktaa, H. Neuhaus, IEEE J. Photovoltaics 10 (2020), 70–77, 10.1109/JPHOTOV.2019.2949875.

This is a short version of the paper submitted to the journal Progress in Photovoltaics.

It has been selected by the Committee of the WCPEC-8 for submission to Progress in Photovoltaics.

THE BEAUTY AS WELL AS THE BEAST: TOWARDS APPEALING AESTHETICS AND RELIABLE PHOTOVOLTAICS FOR VEHICLE INTEGRATION AND BEYOND

Jonathan Govaerts[1,2,3], Tom Borgers[1,2,3], Bin Luo[1,2,3,4], Rik Van Dyck[1,2,3,4], Arvid Van der Heide[1,2,3], Bart Reekmans[1,2,3], Luc Vastmans[1,2,3], Reinoud Moors[1,2,3], Geert Doumen[1,2,3], Loïc Tous[1,2,3], Jozef (Jef) Poortmans[1,2,3,4]

[1]imec, imo-imomec, Genk, Belgium
[2]EnergyVille, imo-imomec, Genk, Belgium
[3]Hasselt University,imo-imomec, Hasselt, Belgium
[4]KULeuven, Leuven, Belgium

ABSTRACT: First we give some historical background and trends in PV module technologies with a focus on the growing motivation for increased aesthetics, and the related ongoing developments. Then we elaborate on two areas where we are active. In terms of multi-wire interconnection, the experiments show promising results on extended reliability testing, including thermal cycling, damp heat and humidity freeze as well as UV exposure and high-temperature storage. In terms of encapsulation, we introduce our approach for curved surfaces using a double-membrane laminator and present results on fabricating curved modules, targeting on the one hand glass-glass sunroofs, on the other hand lighter-weight bonnets.
Keywords: c-Si, Reliability, Module Integration

1 BACKGROUND

Historically, PV modules have typically been implemented in a way to maximize energy generation and/or in applications to (ostentatively or not) indicate and promote its owner as sustainable and environmentally-conscious. In both cases, this implies the PV modules are very visible, with the typical pattern of blue (semi-)squares of the solar cells on a white module background. Gradually however, PV has grown to be more mainstream and slowly becomes ubiquitous in urban environments, in particular in the developed world and the built environment where PV is either required or profitable (or both) in new projects, while also the necessary prefinancing is typically available.

All this means that PV is ever more present in our daily lives and its typical appearance is experienced as uncomfortably unattractive by some to even downright unacceptable by others. Because of this, in recent years, there is a growing trend towards a more unobtrusive appearance. As low-hanging fruit that was picked up very early on, a black backsheet was implemented to replace the white background, immediately subduing the "blue squares" pattern that is visible from afar. In the same makeover, also the bare Al frame was replaced with a blackened version. This could already resolve most of the concerns of building owners related to the aesthetics of their roofs and facades that are far enough away. After this, and coming closer to the PV module, the interconnect wiring is the next eyecatcher coming into view. To address this effect, a recent trend that is gaining traction in industry is to distribute the cell interconnect wiring more evenly across the cell surface (more ribbons but with smaller cross-section) to create a more uniform appearance on the cell area (with additional benefits in terms of resistive and/or optical gains in a trade-off with the cell metallization cost). The next step, or alternative step, is to hide the visible metallic wiring, in and outside of the active area. This can be done by implementing a black absorber in front, either by painting the ribbons (before or after stringing) or by adding a black material in front within the lamination stack. Both approaches imply an increased process complexity and cost, for only an aesthetical gain, compared to the previous adaptations, that either were simply a replacement (backsheet or frame), or combined with an improved performance/cost (wiring redistribution).

Outside the scope here, but for completeness: in parallel, back-contact modules offer the inherent benefits of improved aesthetics and performance, due to the minimization of metallization on the frontside of the cell, albeit at the cost of process complexity and associated cost.

In a later evolution currently picking up speed, the challenge is to hide the modules in plain sight, even at the cost of a reduced performance and increased material cost. The underlying idea is that additional surfaces may thus be unlocked for PV electricity generation. Options in various stages of development include colouring of glass [1] or foils [2,3] in front of the cells, up to assembling even stone or wood veneer in front [4].

Beyond the hiding of the PV cells behind or blending them uniformly in the module's colours, the next level of PV integration entails the form factor (curved surfaces). For this, the interconnection should allow of course the imposed curvatures and stresses, but the most challenging is probably the encapsulation process. The current standard technology for PV encapsulation is based on single membrane flat-plate lamination [5]. However, as such, it does not allow for curved surfaces. To this end, a range of approaches are being implemented. Staying close to standard lamination, one option is to keep the same (flat) process and bend the module after lamination, either while it is still hot, or after cool-down. Depending on the bill-of-materials (type and thickness), the module will need additional fixation (gluing, frame clamping, ...) to keep its (curved) form [6]. A second option is to adapt the lamination equipment. A typical adaptation in this approach is to include a mould support during the lamination [7]. The drawback here is the complexity related to its influence on the thermal behaviour (speed and uniformity of heating and cooling). From another perspective, with more affinity to the glass industry, people are looking into vacuum bagging and autoclaving approaches [7].

Depending on the application, also weight limitations

have to be considered for the further integration of PV. Dominating the weight of standard PV modules, glass is the first component to reduce. Thinner glass can be used, though this will also impact the mechanical stability and fragility. To avoid the use of glass sheets, possible alternatives include the use of polymer sheets with or without fibre reinforcement (composites) [8], honeycomb structures [8] and metal, all of which are endemic in vehicle applications.

Of course, for final integration additional application-specific requirements will be required, ranging from reliability and lifetime to mechanical and electrical safety. This is beyond the scope of the current abstract.

In this background of a growing demand for seamless (and invisible) integration of PV everywhere close to the consumer, we discuss in the following sections some progress that has been made at imec in the fields of interconnection and encapsulation technologies towards this goal.

2 MULTI-WIRE INTERCONNECTION

In previous publications [9] we have reported on a multi-wire approach based on hybrid stitch-weaving of wires into so-called interconnect foils, as illustrated in Figure 1. This allows for more uniform aesthetics, as illustrated in Figure 2.

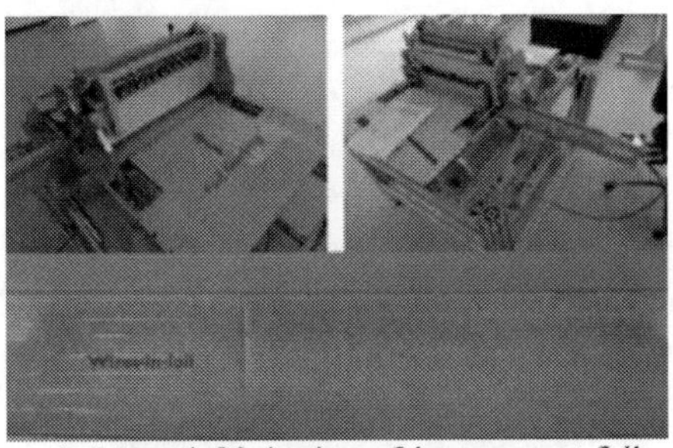

Figure 1: Manual fabrication of interconnect foils: tool operation (top) and resulting foil (bottom)

Figure 2: Improvement with multi-wire aesthetics

In an updated version, we have now improved and extended reliability results with regards to thermal cycling (TC), damp heat (DH), high-temperature storage (HT; 500h@105°C and 10h@120°C), UV (through Xenon irradiation) exposure (UV) and humidity freeze cycling (HF).

We fabricated glass-glass samples with three versions of updated interconnect foils, either 2x2-cell (TC) or 1-cell (DH, HT, UV and HF) laminates. In addition, four types of 2x2-cell reference laminates were included each time, two different encapsulants for both a reference multi-wire approach and a standard 5BB buildup. Table 1 shows typical starting performance values of the different types and their respective interconnect cross-sections.

	MW182 (test)	MW183 (test)	MW184 (test)	SWCT (ref)	5BB (ref)	SWCT-POE (ref)	5BB-POE (ref)
I_{sc} [A]	8.56	8.49	8.55	8.38	8.59	8.61	8.78
V_{oc} [V]	2.926	2.889	2.926	2.921	2.626	2.925	2.658
FF [%]	78.1	76.6	78.8	80.1	78.2	78.7	77.7
P_{mpp} [W]	19.571	18.784	19.701	19.613	17.641	19.808	18.136
Section [mm²]	0.63	0.63	0.63	0.88	1.00	0.88	1.00

Table 1: Representative starting performance of test and reference 2x2-cell samples

Figures 3 to 7 indicate the degradation of the different sample types for each test, both in terms of Pmpp throughout the test and electroluminescence (EL) comparison between start and end of the test. While most samples hold up relatively well, it seems the MW182 is performing the worst of the 3 variations. MW183 has a consistently lower starting performance due to shunting that is both visible in the (pseudo-)fill factor values and in the EL signal. Interestingly, this shunt can be recovered with a single exposure to -40°C as in the TC and HF test.

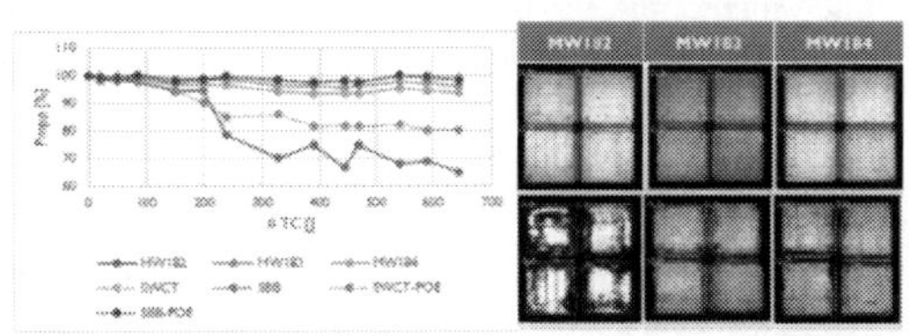

Figure 3: Pmpp (left) and EL (right) degradation throughout TC

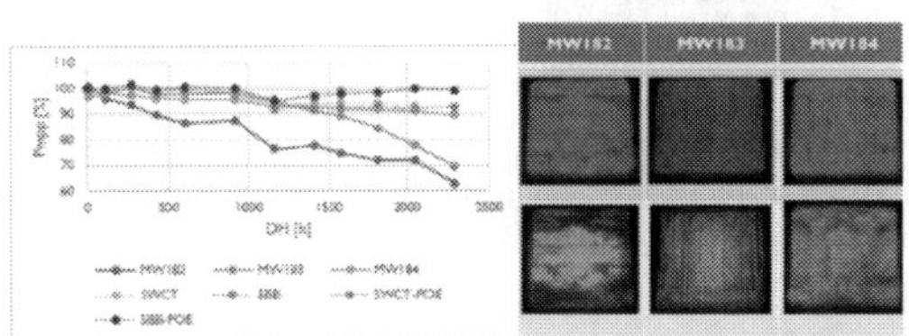

Figure 4: Pmpp (left) and EL (right) degradation throughout DH

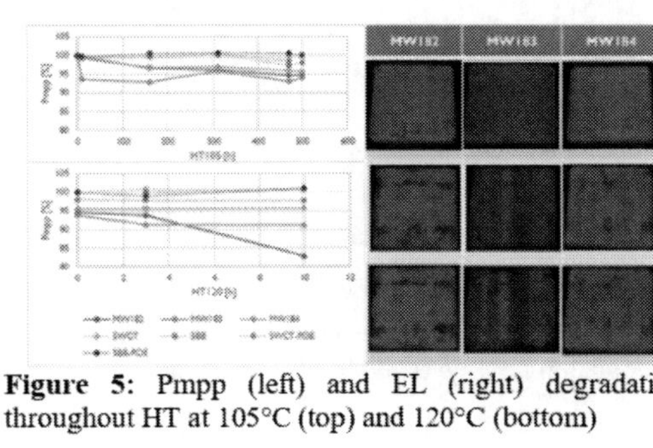

Figure 5: Pmpp (left) and EL (right) degradation throughout HT at 105°C (top) and 120°C (bottom)

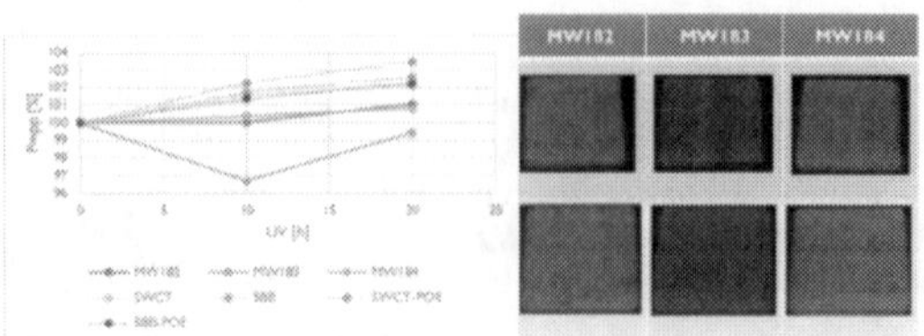

Figure 6: Pmpp (left) and EL (right) degradation throughout UV

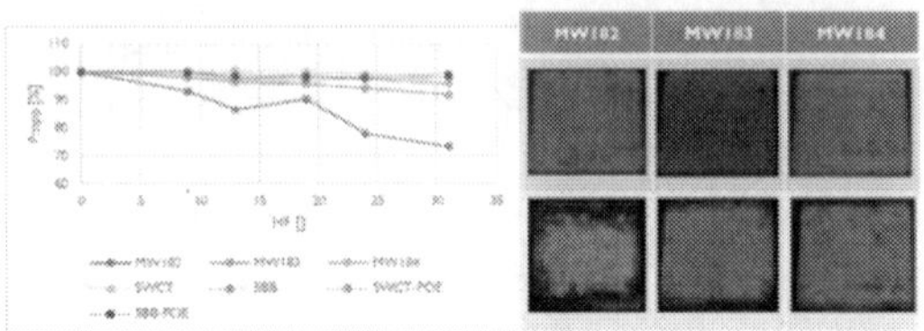

Figure 7: Pmpp (left) and EL (right) degradation throughout HF

3 DOUBLE-MEMBRANE LAMINATION

To address curvature requirements, we are looking into a hybrid approach, based on a standard laminator that is equipped with an additional bottom membrane. Figures 8 and 9 illustrate how simultaneously building up the pressure from both the top and bottom membrane allows to laminate complex curvatures without the need for a supporting mould. In fact, such a process is a step in the direction of autoclaving. While the pressure does not rise to the same levels though, the heating rate is also reduced because of the air all around (as opposed to the direct contact between glass and hotplate in a standard single-membrane flat-plate lamination. The absence of a mould allows for more flexibility and tolerance in form factor.

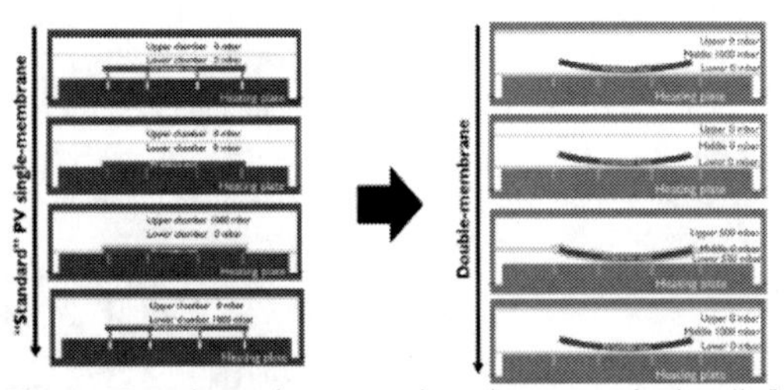

Figure 8: schematic comparison between single- (left) and double-membrane (right) lamination

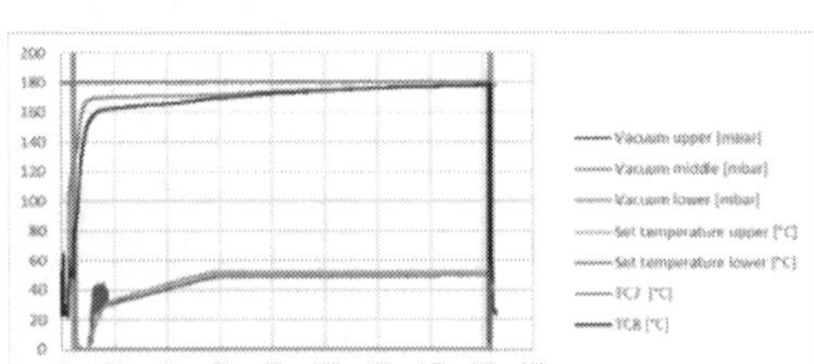

Figure 9: example of lamination recipe with symmetrical buildup of pressure and heating

In the following subsections, we apply this type of lamination for 2 types of VIPV applications.

3.1 Glass-glass sunroof

In a first application, the goal is to integrate PV in the existing sunroof of a car. This involves a glass-glass buildup with 2D curvature. In first instance, we implement 6 standard strings of 11 half cells (PERC-type) on an area of roughly 1.15 x 1.15 m2. Some issues with frame height and temperature and pressure (distribution) were addressed by modifying the tool, the process and the module buildup. Figure 10 illustrates how at a certain point glass breakage was observed, but without inducing cracks on the incorporated solar cells. After solving this, Table 2 reports on the performance parameters of such PV sunroofs. The improvement in current for the 2nd trial is due to the front glass having a higher transparency.

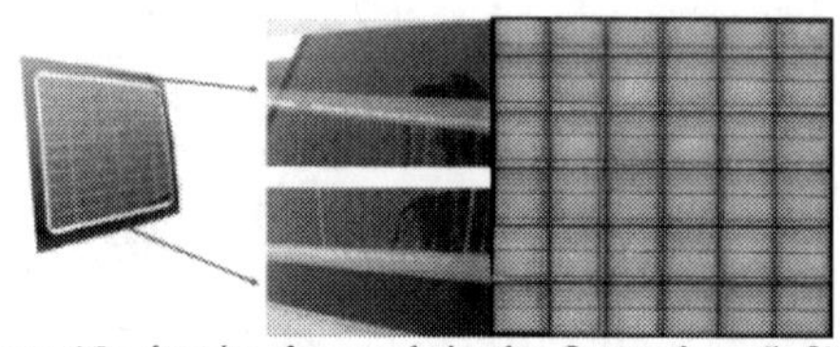

Figure 10: despite the crack in the front glass (left), the solar cells inside are not affected (right) (curved glass courtesy of AGP eGlass)

	Trial 1	Trial 2	increase
I_{sc} [A]	4.344	4.717	+8.6%
V_{oc} [V]	45.048 (66x 0.683)	45.070 (66x 0.683)	+0.0%
FF [%]	81.66	81.48	-0.2%
P_{mpp} [W]	159.8 (66x 2.421)	173.2 (66x 2.625)	+8.4%

Table 2: Performance of 2 PV sunroofs indicating the increase (in current) due to the use of a more transparent front glass

3.2 Light-weight bonnet

Apart from the sunroof's PV integration, where anyway glass-glass was required as a boundary condition, many VIPV applications require minimal weight. In other publications [10,11] we have elaborated our approach towards fibre reinforced encapsulants (GFRE) and back covers (GFRB), including results on reliability testing as TC, DH and HF as well as UV exposure and hail impact testing. Figure 11 indicates also the minimal additional weight of integrating PV on top of the existing metal (Al) pieces. In this work we implement our GFRB approach next to a direct integration on the Al surface of a car bonnet as in Figure 12. In first instance we tested cutout pieces for the lamination.

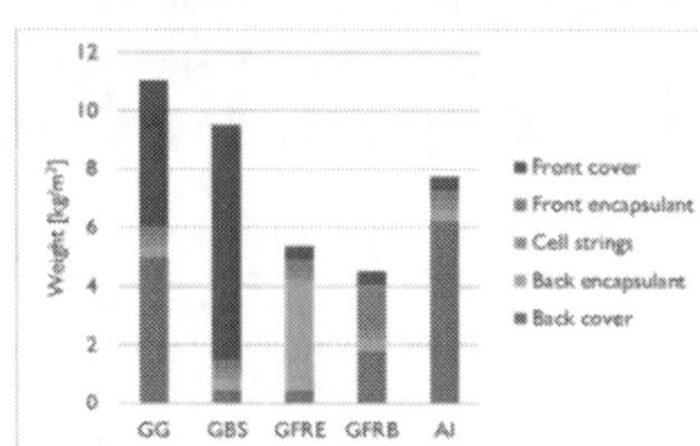

Figure 11: typical weight distribution of various module buildups

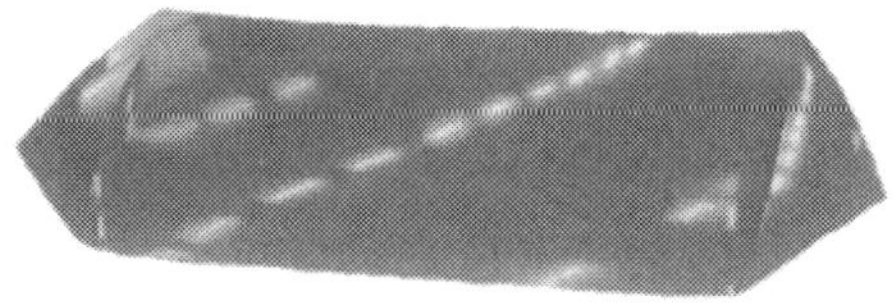

Figure 12: car bonnet for PV integration

Figure 13 shows the fabrication and result of integrating a 9BB string of 6 half PERC cells on top of the Al. For aesthetical reasons, we included a black backsheet and a transparent frontsheet.

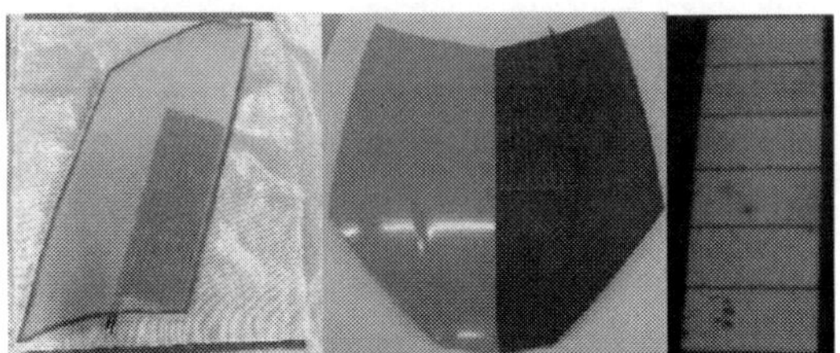

I_{sc} [A]	V_{oc} [V]	FF [%]	P_{mpp} [W]	Active area [cm²]
4.69	4.17	80.9	15.84	~768

Figure 13: PV on an Al bonnet piece: from (top) left to right layup (top left), comparison before/after (top middle), resulting EL (top right) and electircal performance parameters (bottom)

Similarly, we also fabricated a version where we replaced the Al with a GFRB back cover. First, we "duplicated" the bonnet piece using the Al part as a mould in a separate double-membrane lamination. Afterwards we used a shingled string to fabricate the PV laminate. This is illustrated in Figure 14.

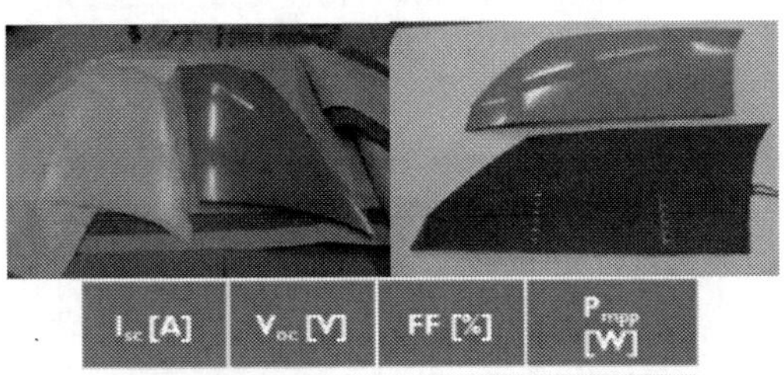

I_{sc} [A]	V_{oc} [V]	FF [%]	P_{mpp} [W]
1.25	8.84	79.3	8.78

Figure 14: PV on a GFRB bonnet piece: GFRB duplication of the bonnet piece (top left), comparison before/after (top right) and electrical performance parameters (bottom)

Finally, Figure 15 illustrates some upscaling that is ongoing. Further details, upscaling efforts and testing results are being considered for publication [12].

ACKNOWLEDGMENT

The authors gratefully acknowledge the Flemish government for its financial support through the funded ICON projects SNRoof, DAPPER and CSP+ and related project partners, as well as the European Union's Horizon 2020 Programme for research, technological development and demonstration for funding part of this work under Grant Agreement no. 857793 (HighLite).The authors also would like to acknowledge the partial funding by the Kuwait Foundation for the advancement of Sciences under project number P115-15EE-01.

REFERENCES

[1] E.g. Soltech, Issol, Sunovation, Onyx,...
[2] Escarre et al., IEEE PVSC 2015
[3] van Roosmalen et al., Advanced Building Skins 2017
[4] Morlier et al., EUPVSEC2021
[5] Cattaneo et al. (2015). Lamination process and encapsulation materials for glass–glass PV module design. Photovoltaics International. 82.
[6] E.g. Galuppi et al., International Journal of Solids and Structures 67–68 (2015) 231–243
[7] Sato et al., 4th PVTEC webinar 2021
[8] A.C. Martins, PhD thesis
[9] Govaerts et al. EUPVSEC2020
[10] Govaerts et al., EPJ-PV Volume 13, 2022
EU PVSEC 2021: State of the Art and Developments in Photovoltaics
[11] Bin EUPVSEC2022 (accepted)
[12] Govaerts et al. PiP (submitted)

IMPACT OF MODERN CELL CONCEPTS AND MODULE LAYOUTS ON HOT SPOT TEMPERATURES IN SOLAR MODULES

K. Ramspeck, C. Müller and M. Meixner
halm elektronik gmbh
Friesstrasse 20, 60388 Frankfurt/M.
e-mail: k.ramspeck@halm.de

ABSTRACT: In our contribution we investigate the effect of modern cell concepts and module layouts on the formation of hot spots in partly shaded cells in solar modules under operational conditions. We set up a thermo-electrical model to calculate reverse biasing conditions and the resulting temperature distribution for cells with and without localized reverse current flow for different cell types and module layouts. As an example for a localized hot spot we examine a local shunt with 50 Ω resistance. For the thermal model we assume a glass-glass module with 180 µm thick silicon solar cells. We investigate the effect of module size and layout for CZ-BSF, PERC, TopCon and HJT cells to evaluate the impact, the evolution of solar cell architecture has on the achievable module temperatures as well as the effect of increasing module size. We find, that local hot spot temperatures even of moderate hot spots can easily exceed 200 °C. Such high temperatures will be able to cause melting of encapsulants and may damage interconnection of solar cells in the module in case of low temperature cell concepts. Furthermore, the presence of such high temperature in a module underlines the importance of temperature dependency of reverse current flow for the hot spot behavior in solar modules. We examine the temperature coefficients of reverse resistance of cells with localized hot spots and deduce its impact on final module temperature in the field.

Keywords: Characterization, Reliability, Shading, Shunts, PV-Module

1 INTRODUCTION

Hot spots in PV modules are a topic in photovoltaics since decades [1-4]. Yet, still they are one of the major concerns when it comes to system failures in the field [5]. Several authors addressed module temperatures caused by localized reverse currents (I_{rev}) and predicted from inline thermography measurements of solar cells in production [6-9]. Other publications examine module temperatures under partial shading conditions [7,10-12]. We examine the impact of modern cell concepts, increased module size and varying module layouts on the formation of hot spots due to I_{rev} and partial shading. We present a thermo-electric model and investigate the temperature a 50 Ω point-like shunt causes on a partially shaded cell for different module types, cell technologies and geometries. We take up the topic of reverse current temperature coefficients (T_{CIrev}) previously discussed in [13,14], present measurements of reverse resistance temperature coefficients (T_{CR}) and examine their impact on module temperatures using our thermal model.

2 THERMO-ELECTRIC MODEL

In order to model the behavior of hot spots in solar modules, we calculate voltages and heat source power from measured solar cell *IV*-curves and use the resulting heating power as input for a simplified thermal module simulation. Changing the *IV*-curves used in the electric calculation will thus allow to model the effect different cell concepts have on the resulting hot spot temperatures.

The reverse voltage, a shadowed cell in an externally short-circuited string with bypass-diode is biased with, can be calculated as follows:

$$V_{rev} = (n-1) \cdot V(T, I, \Phi) + V_{diode} \quad (1)$$

Here, n ist the number of cells connected in series in the string and V_{diode} is the voltage the bypass diode requires to effectively bypass the current amount of $I_{MPP} - I$. $V(T, I, \Phi)$ is the voltage one cell in the string provides at the remaining current I that is generated by the limiting cell, the operating temperature T and irradiance Φ. Throughout this paper, we assume an irradiance Φ of 1000 W/m². Having calculated the reverse bias using equation (1), an assumption on a hot spot resistance yields the hot spot power to be considered in the thermal model. Figure 1 (a) shows the resulting hot spot power for a 50 Ω shunt for different module configurations and cell types. Depending on the number of cells connected in series and the cell type, large differences can be seen ranging from 2.5 W hot spot power for a 60 cell PERC module up to > 6.5 W for a 84 cell HJT module.

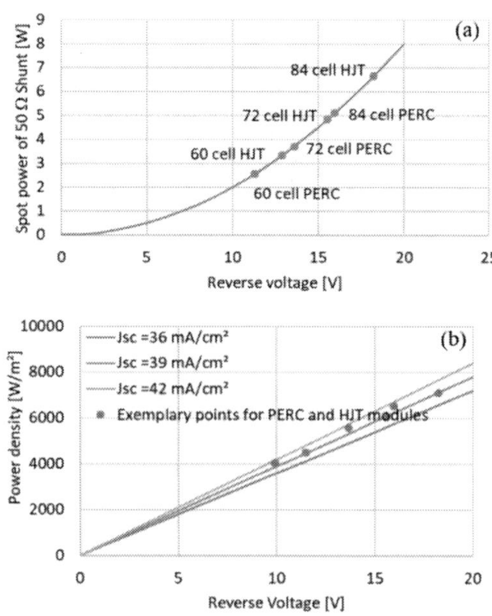

Figure 1 (a): Hot spot power for a 50 Ω shunt as a function of reverse voltage with points marking different module types, (b) heating power density for different short circuit current densities with points for HJT and PERC modules.

Using the short circuit current density from the IV-curve of the cell, the heating power density (P_{ill}) in an illuminated region of a partially shadowed cell can be calculated

$$P_{ill} = J_{SC} \cdot V_{rev} \quad (2)$$

This heating power density is considered in the thermal model for an illuminated region with radius r_{ill} around a hot spot in a partially shadowed cell. Figure 1(b) shows its value as a function of reverse bias for 36 mA/cm², 39 mA/cm² and 42 mA/cm² short circuit current density. Dots in the graph represent values for different module configurations for PERC and HJT modules. Typical heating power densities observable for modern cells in modules will accordingly range in between 4000 W/m² and 7000 W/m².

The thermal modeling is performed in a radial symmetric way to achieve a low dimensionality of the model. This simplification as well as further simplifications will be discussed in the appendix.

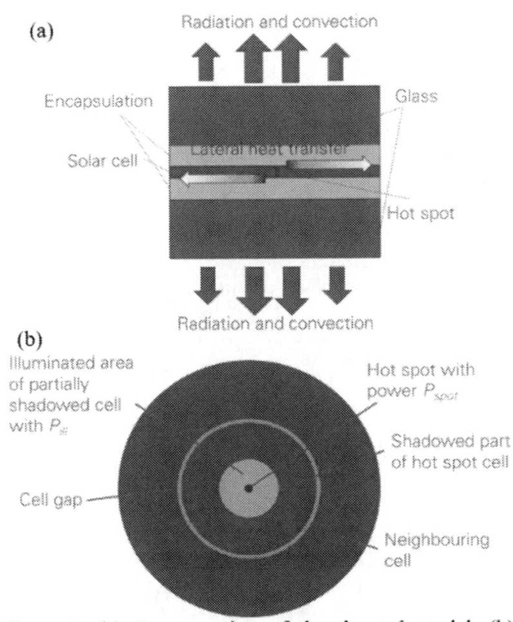

Figure 2: (a) Cross section of the thermal model, (b) topview of the thermal model.

Figure 2 (a) shows a cross section, Figure 2 (b) a topview on the model. A localized heat source with radius r_{spot} is placed in the center of the model, surrounded by an area representing the illuminated part of the partially shadowed cell with radius r_{ill}. These two regions contain the heat sources of the model as one boundary condition. Heat is transported laterally with one heat conductivity composed by a parallel connection of the materials the module is composed of, mainly dominated by silicon and glass. Lateral heat conductivity of encapsulation materials is neglected. At the cell gap, the contribution of silicon to the heat conductivity is set to zero. In vertical direction, heat losses to the surrounding are calculated considering radiation and convection with a fixed heat transfer coefficient for the convective part and a fixed emissivity for the radiation part of the heat losses. No vertical temperature gradients within the module materials are accounted for as a third simplification. Both processes, lateral heat transfer and vertical losses are calculated using the following equations:

$$\frac{dT}{dr} = \frac{P(r)}{\lambda(r)} \quad with$$

$$\lambda(r) = 2\pi r \left(\lambda_{Si} \cdot w_{Si}(r) + \lambda_{glass} \cdot w_{glass}(r)\right) \quad (3)$$

and

$$\frac{dP}{dr} = P_{gen}(r) - 2\varepsilon\sigma(T(r)^4 - T_{BG}^4) - 2h_{glass/air}(T(r) - T_{BG}) \quad (4)$$

Here $P_{gen}(r)$ is the generated power at distance r from the center, which is composed of the spot power and the power through illumination. Far away from the heat sources, the remaining lateral power transport shall reduce to zero and the temperature shall converge to the operating temperature of the module which gives the boundary conditions for the calculation of the center temperature. To solve the model, the center temperature is varied and it is checked whether the boundary conditions are fulfilled. Numeric values used for the parameters in the calculation are given in Table I. These values are used throughout the paper, unless otherwise stated.

Table I: Standard parameters for thermal calculation.

Parameter	Value	Unit
I	4	A
Φ	1000	W/m²
n	24	Cells per string
λ_{Si}	150	W/mK
λ_{glass}	1	W/mK
w_{Si}	180	µm
w_{glass}	2.3	mm
$h_{glass/air}$	20	W/m²K
Emissivity ε	0.85	Unitless
R_{spot}	50	Ω
P_{ill}	5470	W/m²
Cell type	PERC (2018)	
r_{spot}	0.5	mm
r_{ill}	45	mm
r_{gap}	60	mm
Gap width	2	mm
V_{rev}	-13.64	V
T_{module}	50	°C

3 TEMPERATURE CALCULATIONS

Using the model described above, we calculate the temperature field using the conditions shown in Table I as a standard case to start with. As illustrated in Figure 3, a hot spot temperature of 191 °C in spot position results, with a temperature field falling below 100 °C at $r = 25$ mm. This result shows that a 50 Ω shunt can already yield a critically high module temperature under partial shading conditions for a 72 cell PERC module.

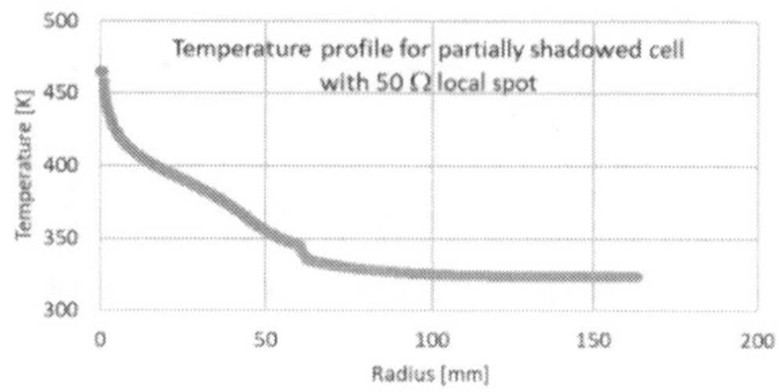

Figure 3: Temperature profile around a 50 Ω spot at the conditions shown in Table I.

In order to examine effects modern cell concepts and different module layouts have on the resulting spot temperature, we vary a number of parameters of the model, keeping the rest of them fixed, namely the cell type, string length, size of illuminated region, size of the hot spot, effect of cell gaps (r_{gap} and gap-width) and the cell thickness. The results of these variations are given in Figure 4. We observe a large impact of all of these parameters on the resulting spot temperature. Namely, we find that using most recent cell technology like TopCon or HJT, a further increase of the spot temperature from 191 °C up to 221 °C results.

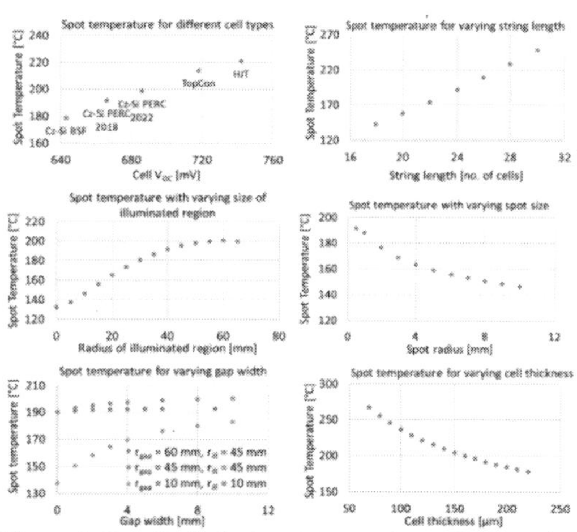

Figure 4: Results for hot spot temperature when varying important parameters in our thermo-electrical model.

Similar, a temperature increase by about 30 K is observed for every increase of 4 additional cells in string length. Varying the size of the illuminated region of the partially shadowed cell shows, that a maximum temperature of 200 °C is observed for r_{ill} = 60 mm. For larger illuminated regions, the increasing residual current of the cell reduces the reverse voltage and the hot spot temperature starts to decrease. Please note, that 6 inch cells are the basis for the calculation of the reverse current as a function of residual current, for different sizes of the cells (or part cells), the radius for the hottest temperature will be different. For a variation of the spot size, r_{spot}, larger spots lead to a significant decrease of the center temperature of the spot. For a spot radius of 10 mm, the center temperature reduces by 40 K, showing how sensitive the spot temperature depends on the intensity of local heating. A cell gap, far away from the spot has only little impact on the spot temperature, larger impact is only observed if the gap is in close proximity to the spot, limiting at the same time the possible size of the illuminated region in our model. A variation of the cell thickness leads to a change in lateral heat conductivity and has large impact on the spot temperature. Assuming a cell thickness reduction from 180 µm to 120 µm, the spot temperature increases from 191 °C to 221 °C.

4 REVERSE-IV TEMPERATURE COEFFICIENTS

As we observe that hot spot temperatures exceed even 200 °C under many conditions, their critical operation takes place at conditions, differing strongly from the measurement conditions of reverse-*IV* measurements in solar cell production. It is well known from literature [13,14] that reverse current flow exhibits significant temperature dependency. The authors of these studies measured temperature coefficient values in the range of ±2 %/K locally on solar cells using thermographic methods. But even in the simplest case, an ohmic shunt caused by a metallic conduction path, the temperature dependency of metallic current conduction would be expected to yield a resistance change of 0.4 %/K at conditions of sufficiently low temperature. To gather experimental data on the temperature coefficient of reverse-*IV*, we examine three groups of cells with reverse current issues from different manufacturers (all PERC) and measure their reverse *IV*-curves at 18 °C, 35 °C and 50 °C. From the data we derive the respective temperature coefficients of reverse resistance (T_{CR}). The measurements are performed in a commercially available cetisPV-Celltest4-BF system with temperature control option, allowing for bifacial *IV*-characterization of all Si-solar cells at temperatures from 15 °C to 60 °C. Figure 5 displays the temperature coefficients measured on the cells totaling a number of 42. Temperature coefficients were evaluated at -12 V and -15 V reverse bias.

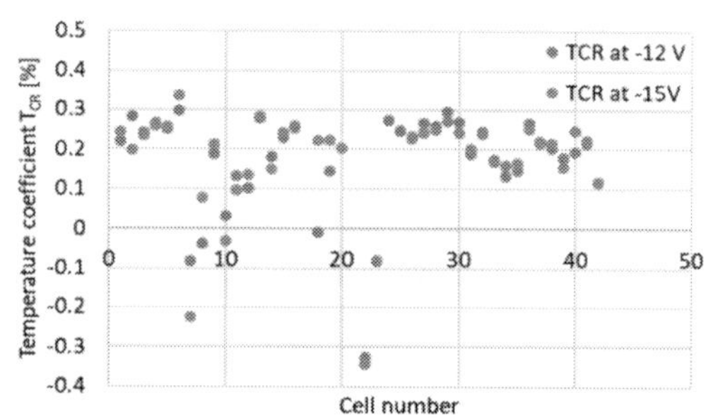

Figure 5: Reverse resistance temperature coefficients for 42 PERC solar cells from three different manufacturers.

We observe a resistance temperature coefficient (T_{CR}) in the range between 0.1 %/K and 0.3 %/K for the majority of the cells. A few cells showing negative coefficients all have very small I_{rev} values which do not cause a significant hot spot. Some cells for which no value is visible in figure 5 for the measurement at -15 V exhibited very early breakdown and thus very high reverse current at -15 V. The breakdown voltage shifted to higher reverse bias for all breakdown effects observed on these cells, leading to a much higher temperature coefficient for the reverse resistance, reaching values of up to 2 %/K. This latter value reached in case of reverse breakdown is in good agreement with the local measurements described in literature.

As a result of the reverse temperature coefficient measurements, we can state that for the cell batches we examined, all relevant hot spots exhibit positive reverse resistance temperature coefficients, thus the hot spot current reduces at higher temperature. However, this result cannot be generalized to different cell concepts or all PERC cells in general. Furthermore, no examination whether this temperature coefficient changes at higher temperatures could be performed so far.

The increase of resistance, associated with a decrease in current, with increasing temperature leads to a saturating behavior of the hot spots and has been implemented into our thermal model by including a temperature dependent resistance of the spot (*Rspot(T)*). Varying the spot power @ 25 °C and the T_{CR} value, we investigate the effect of the temperature coefficient on the hot spot temperature leaving all other parameters at their standard values shown in Table I. The results for hot spot power P_{spot} @ 25 °C ranging from 0 W to 10 W is shown in Figure 6 (a). Figure 6 (b) displays the effective hot spot power at operational conditions for the same set of calculations. We observe, that positive T_{CR} values lead to a saturation of the hot spot temperature and allow to tolerate much higher spot power at 25 °C. While a 10 W spot without temperature coefficient achieves a spot temperature of 315 °C in the calculation under the standard conditions used here, with a T_{CR} of 0.4 %/K it's temperature is reduced to 210 °C and the spot power acting is reduced from 10 W to below 5 W.

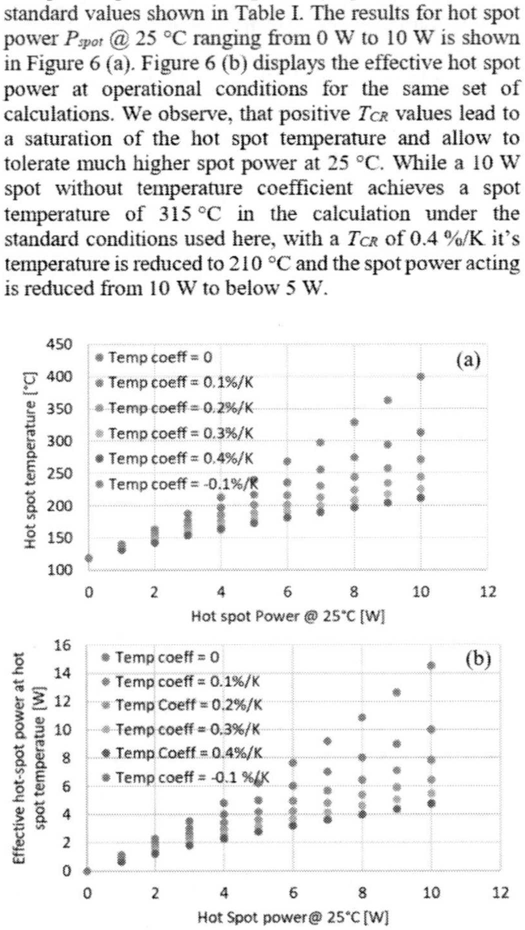

Figure 6 (a) Temperature of hot spots including reverse resistance temperature coefficient and effective acting spot power (b) for the same conditions.

Assuming a negative temperature coefficient for the reverse resistance quickly leads to a thermal catastrophe in the calculation. While for -0.1 %/K a spot power > 14 W and a spot temperature above 400 °C result, for larger values the calculation quickly doesn't converge anymore demonstrating, that a linear constant decrease of the resistance with temperature is not possible up to such high temperature.

5 TANDEM AND THINFILM CELLS

While we concentrate mostly on current high-efficiency cell concepts, an outlook to next generation technologies is worthwhile. Further development activities in the PV sector are more and more dedicated to Tandem-cell architectures to allow for even higher efficiency. While future concepts for such cells are not yet fully clear, two different, important candidates are interesting to discuss from the point of view of hot spot effects. A Si-wafer based tandem cell (with its thermal behavior being very much comparable to Si-single junction cells) and thin film approaches were both subcells are processed directly on glass. For the latter case, we examine the effect of a vanishing thermal conductivity of Si on the results of the calculation. While our assumptions in the model will cause some larger errors in this case, the general trend will be well described and becomes obvious in Figure 7 – a much more pronounced hot spot temperature peak is caused by the low internal heat conductivity, leading to 470 °C hot spot temperature in our standard case calculation. Changes in source power are not yet accounted for in this calculation. Furthermore, it should be noted that pure thin-film based approaches will tend to have a very different geometry and an adapted model will be required for this.

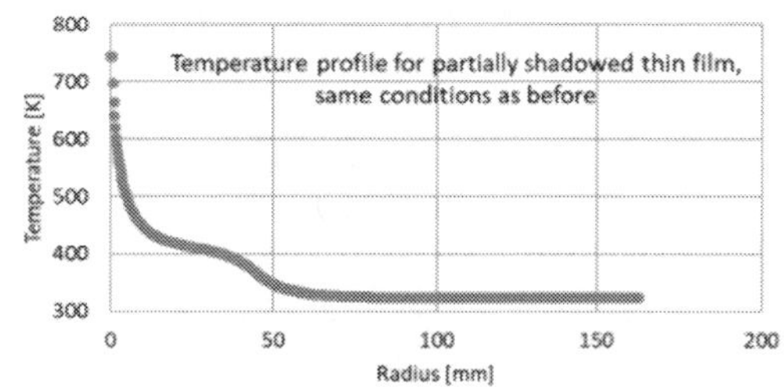

Figure 7: Temperature profile resulting for the conditions given in Table I if silicon thickness is reduced to 3 μm and thermal conductivity in the model is thus strongly reduced.

Considering a Si-based tandem cell, the main change is not within the thermal model, but applies to the calculation of reverse bias and spot power. Assuming two-terminal approaches, the voltage of both subcells add to each other, while the current is reduced by about 50 %. With the voltage nearly being tripled in an ideal case, an additional point added into Figure 1(a) if tandem cells are considered might be what is shown in Figure 8.

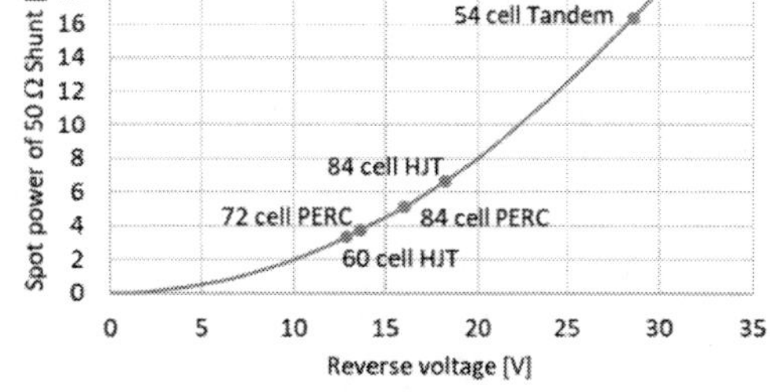

Figure 8: Hot spot power for a 50 Ohm shunt as a function of reverse voltage and points marked for different module types now including an approximate value for a Tandem module.

Even for a number of only 18 cells in a string, 28 V reverse voltage could be applied and lead to excessive heat in a 50 Ω spot. In order to prevent such critical reverse bias conditions, Tandem-Cell modules will require a new module layout or need to be equipped with electronics which allow to switch a cell string to V_{OC} conditions when it becomes limiting in current due to a shadow or a crack. In other cases, either very high reverse voltage resilience is required for tandem cells (both subcells) or complicated module interconnection with a large amount of bypass diodes is required to reduce the voltage to a value each subcell can stand. Please note that reverse *IV*-measurements need to be done for both subcells individually to determine critical hot spots in 2-terminal tandem approaches.

6 CONCLUSIONS AND DESIGN RULES

In this paper we have demonstrated how a simplified thermal model of a PV-module allows to examine the effect of different parameters and conditions on the hot spot behavior and temperature. We demonstrated that high temperatures are to be expected in many different configurations especially for string lengths of 24 cells and above. With hot spot operating temperatures being close to 200 °C we propose to lay focus on the temperature dependency of reverse-current flow in Si-solar cells and demonstrate the effect different temperature coefficients have on module temperature using our thermal model. Comparing with measured values for the reverse resistance temperature coefficient we could show, that in many cases, hot spots tend to limit themselves partly at elevated temperatures, but further investigation of their behavior up to 200 °C is still necessary. In order to limit the effect of partial shadowing several design rules for cells, modules and/or systems can be set up.

1.: Perform thermography measurements on cell level at an appropriate voltage to investigate the lateral distribution of reverse current flow and sort out cells with localized spots.

2.: Avoid long cell strings in modules, a maximum of 24 cells should be bypassed by one diode

3.: Higher efficiency leads to higher module temperatures under reverse operation – optimize module materials and cell technology for the appropriate temperature

4.: Use electronics in the module which allow to switch limiting cell strings to V_{OC} condition if no save operation can be guaranteed by cell and module design

5.: In further development, especially for tandem cells – incorporate the question of partial shadowing early in the development process of cells and/or modules to allow for safe operation later on.

REFERENCES

[1] L. C. Larue and E. du Trieu, 3rd EUPVSEC, Cannes (1980), p. 490ff

[2] O. Breitenstein, J. P. Rakatoniaina, M.H. Al Rifai, Prog. In Phot. 11 (2003) 515 - 526

[3] O. Breitenstein, J.P. Rakatoniaina, M. Kaes, S. Seren, T. Pernau, G. Hahn, W. Warta and J. Isenberg, 20th EUPVSEC Barcelona (2005), p. 590 - 593

[4] K. Ramspeck, P. Altermatt, N.P. Harder and R. Brendel., 21st EUPVSEC, Dresden (2006), p. 1471 – 1474

[5] D.C. Jordan, T.J. Silverman, J. H. Wohlgemuth, S.R. Kurtz and K.T. VanSant, Prog. In Phot. 25 (2017) 318 - 326

[6] I. Geisemeyer, F. Fertig, W. Warta, S. Rein and M. C. Schubert, in 29th EUPVSEC Amsterdam (2014), p. 2429 - 2434.

[7] K. Ramspeck, S. Schenk, D. Duphorn, A. Metz and M. Meixner, Energy Procedia 55 (2014) 133

[8] S. Wasmer, N. Rajsrima, I. Geisemeyer, F. Fertig, J.M. Greulich and S. Rein, J. Appl. Phys. 123 (2018) 093105

[9] T. Roth, R. Siebert and K. Meyer, Energy Procedia 55 (2014) 504 - 508

[10] S. Deng, Z. Zhang, C. Ju, J. Dong, Z. Xia, X. Yan, T. Xu and G. Xing, Energy Procedia 130 (2017) 77 - 86

[11] H.J. Solheim, H.G. Fjaer, E. A. Sorheim and S.E. Foss, Energy Procedia 38 (2013) 183 – 189

[12] R. Witteck, M. Siebert, S. Blankemeyer, H. Schulte-Huxel and M. Köntges, IEEE Journal of Photovoltaics, vol. 10, no. 6 (2020) 1828-1838

[13] O. Breitenstein, J. Bauer, K. Bothe, W. Kwapil, D. Lausch, U. Rau, J. Schmidt, M. Schneemann, M.C. Schubert, J.M. Wagner and W. Warta, J. Appl. Phys. 109 (2011) 071101

[14] F. Fertig, C. Willibald, I. Geisemeyer, M.C. Schubert and S. Rein, Energy Procedia 38 (2013) 32 – 42

APPENDIX

Simplifications of the thermal model and their impact on modeling results will be discussed in this Appendix. The Relevant simplifications are radial symmetry, vertically constant module temperature and neglecting of lateral heat transfer by encapsulation materials:

1. Radial symmetry

The assumption of radial symmetry in our calculation leads to an error whenever the temperature field reaches the end of cell. Within a cell, around a point-like hot spot a circular symmetric temperature field will evolve. Errors due to cell edges will be small, if spots are far enough from the cell edge. The most critical case where the error will become larger is a spot in a cell corner with that cell being positioned at the corner of the module. Such spots will show higher center temperatures than calculated in our model. However, the focus of our investigation lies in the comparison of different cell types and module layouts and errors caused by the assumption of radial symmetry are going to be comparable for all different cases. If an estimate of the worst case condition with very low heat conductivity in the corner of a module is required, this will be similar to a drastically reduced cell thickness in our model. Generally, the assumption of radial symmetry will be best for spots in the center of cells and thus rather underestimate critical hot spot temperatures but not overestimate them.

2. Vertical constant module temperature

The assumption of vertical constant module temperature neglects the temperature gradient between cell and the module/air boundary. This temperature gradient is governed by the thermal transfer through glass and encapsulant and the power to be transferred. The maximum power transfer in our standard calculation is slightly below 5000 W/m^2 in hot spot position at 191 °C. The thermal conductivity of glass at 1 W/mK and of the encapsulant of about 0.35 W/mK (for the case of EVA) together with the thickness of glass (2.3 mm) and encapsulant (maximum 0.3 mm) yield a heat transfer capability of 316 W/m^{2}K. The hot spot temperature will thus tend to be up about 16 K hotter than estimated in our simple model. Please note that our result will however be close to the glass surface temperature of a module under hot spot operation.

3. Neglecting of lateral thermal conductivity of encapsulation materials

The encapsulation material has a low heat conductivity and is much thinner than glass. Its contribution to the lateral heat conductivity is about 22 times smaller than the contribution of the glass and about 257 times smaller than the contribution from the silicon solar cell with its high heat conductivity. Neglecting its effect in lateral heat transfer is thus well justified.

This is a short version of the paper submitted to the journal Progress in Photovoltaics.

It has been selected by the Committee of the WCPEC-8 for submission to Progress in Photovoltaics.

A LOW CONCENTRATION APPROACH TO SPECTRAL SPLITTING OPTICS FOR BIFACIAL 4-T MODULES

Andrea Farina[1], Alice Carlotto[1,2], Stefano Varas[2], Alessandro Chiasera[2], Silvia Maria Pietralunga[*1]
[1]CNR, IFN - Institute for Photonics and Nanotechnologies, Piazza Leonardo da Vinci, 32 20133 MILANO, Italy
[2]CNR, IFN-CNR CSMFO Lab. and FBK Photonics Unit, Povo-Trento, Italy

ABSTRACT: We numerically evaluate the optical performance of an asymmetric low-gain concentrator for 4T bifacial solar modules operating in spectral-splitting (SS) configuration. The optical element is a right-angle-wedged (RAW) dielectric prism that operates as a lightguide, by trapping light through dichroic mirrors and total-internal-reflection. Concentration is active on the visible portion of the solar spectrum. A wide-gap solar cell of suitable technology is supposed to convert visible light, while a bifacial silicon cell converts the near infrared. By a ray-tracing simulation, values for spectrally integrate optical efficiency up to 88.7% and up to 62% are respectively estimated for direct irradiation and sky-diffused light. Solar modules based on the RAW-SS geometry can work trackless. Proper orientation of the modules minimizes ground shading and is therefore appealing for bifacial operation. Possible applications of the single module are in sub-watt indoor PV and Internet-of-Things power suppliers. Multiple replicas of the optical element can be integrated in larger panels for utility-scale plants and building-integrated solutions.

Keywords: bifacial modules, optical concentrators, PV module, four-terminal, spectral-splitting.

1 INTRODUCTION

Renewable energy sources, including solar, are forcefully acquiring importance in the general quest for decarbonization and sustainable energy production to support life and progress on planet earth [1]. Recent international events have also stressed their role as strategic assets for the economic and political independence of nations. In this perspective, there is a renewed and empowered sense of confidence in the key role that photovoltaic (PV) energy conversion production can play in the global energetic challenge [2], [3]. The main objective of PV innovation is to improve the Power Energy Conversion (PCE) of modules, to get the most out of the irradiation conditions at installation sites. A second main objective is of course to find technical solutions that are also economically sustainable, meaning that the Levelized Cost of Energy (LCoE) of the specific kind of solar plant must be minimized. The LCoE is a complex parameter that provides the specific annual cost of the energy produced by a specific technology in one selected environment. It can be generally expressed as LCoE [€/kWh] = α C_P/E, where α coefficient accounts for the sum of the fixed charge rate of the investment and the O&M (operation and maintenance) yearly cost coefficient, Cp (€) is the total plant cost and E (kWh) is the average energy produced by the plant in one year. Therefore, to minimize the LCoE, PV solutions must not only be highly efficient but also cost-effective and with low O&M charges [4].

To boost the PCE of PV modules beyond the Shockley-Queisser limit of a single junction solar cell [5], in principle a multijunction approach can be used, where every PV cell is best matched in energy to one selected band of the solar spectrum [6], [7]. In fact, this is a classical way to get around the problem and much valuable efforts are being devoted to this. The most classical solution is represented by the tandem configuration where the cells are stacked and electrically connected in series so that a two terminal (2T) device results. The topmost cell converts the most energetic portion of sunlight (in the blue and visible range) and transmits downwards the less energetic bands, which are converted by the cell at the bottom of the stack. Most efficient tandem solutions involve the heteroepitaxial growth of multiple junctions in III-V semiconductors [8], but they cannot be intended for the commodity market, due to their high price. In view of utility-scale use, cheaper solutions must be developed. The most direct way is coupling a bottom cell in silicon and a wide bandgap (WG) top cell in tandem configurations [9], [10]. A good number of different technologies for the top cell are being studied in these times. In any case, efficient tandem structures need a reliable and durable current matching and this still represents a crucial issue. A flexible alternative to standard 2T tandem cells is the four-terminal (4T) electrical connection. In it two cells are stacked and optically "in series", but not electrically and there is no need for monolithic integration [11]–[13]. In this way, technological constraints are somewhat relaxed and the flexibility in the choice of the materials for the WG cell is improved. In any case, still performant cells of different technology must match in area. In the case of wide silicon bottom cells, this implies that the technology of fabrication of the top cell must assure high performance, reliability and durability of large area devices. And also this is still an open challenge. As an alternative to cell stacking, Spectral Splitting (SS) configurations provide separate optical paths for light in different bands by means of some kind of spectral selective optical components [14], [15]. In this way, matching in area of the cells is no more necessary and many technological constraints are relaxed. However, one major drawback of SS is represented by its typical poor land occupation, because the modules are typically wider than standard ones, to spatially separate the spectral bands. The impact on land occupation can be reduced by introducing optical concentration. If Low Concentration PV (LCPV) solutions are chosen, the use of expensive double-axes sun trackers is not required [16], [17].

Another strategy to improve the LCoE in utility-scale plants is by bifacial modules, which collect additional energy from albedo radiation at their rear surfaces[18], [19]. A market share for bifacial PV exceeding 30% by 2030 is forecasted by The International Technology Roadmap for PhotoVoltaic (ITRPV) [19]. Merging bifaciality and SS in 4T connection is therefore appealing for maximizing PV energy conversion.

In this work we present the conceptual design of an optical building block for 4T SS solar modules operating in hybrid LCPV. The optical core of the module is a Right Angle Wedged (RAW) prism coupled to dichroic mirrors and providing Total Internal Reflection (TIR). The VIS fraction of light is concentrated and guided to a WG cell at right angle with respect to the top surface, so that no increase in land use exists with respect to a stacked 4T configuration. The LCPV optics can operate trackless without much loss in PEC. At best orientation RAW-SS promotes the collection of albedo irradiation from the ground by reducing self-shading effect, which is an asset for bifacial operation and a peculiar feature of the solution. Therefore, the geometry here presented can be considered as a building block for large panels in utility-scale plants as well as in building-integrated solutions.

2 METHOD

2.1 Design and modelling

The optical core of the SS-4T module is a Right-Angled-Wedge (RAW) prism of apex angle α that performs a low gain optical concentration in the visible (VIS) portion of the spectrum by lightguiding. Light is confined by Total Internal Reflection (TIR) at the top surface of the wedge, while the SS function is performed by a pair of dichroic filters. The RAW-SS optics has been designed using a commercial 3D non-sequential ray-tracing software (ZEMAX OpticStudio®). Dielectric data of commercially available bulk materials, coatings and filters are available from embedded libraries. For the test case here reported, the prism is made of PMMA (Polymethyl-methacrylate), that is a typical material for lightweight plastic optics, with a refractive index around n=1.5. Zeonex® glass has been considered for the 3 mm-thick protective top layer. Near-perfect optical matching at interfaces is also supposed in this preliminary evaluation. Ideal dichroic mirrors are modelled as surface coatings with cut-off wavelength λ_D = 735 nm, a 20-80% transition region over 2 nm and 0-100% spectral selection. The specific value of λ_D compatible with a 4T solution that uses a silicon solar cell and a hybrid perovskite WG cells with a bandgap Eg = 1.63 eV [20]. In order to estimate the amount of light reaching the solar cells, the model considers detectors optically coupled to the bottom and rear face of the optical prism. The detectors are pixelated to evaluate the uniformity of power distribution.

The prism is sketched in Figure 1. To simulate direct solar illumination at incidence angle θ_i a rectangular source of parallel rays has been created such that all rays hit the top surface. The condition of diffused irradiation is instead simulated by using an isotropic rectangular source placed close to the top surface of the wedge and of equal area. Spectral distribution for the solar irradiation, both direct and diffused, has been set based on the typical distribution AM1.5G.

2.2 Principle of operation of RAW-SS

As it is shown in Figure 1, the RAW-SS optics guides VIS light to the WG cell coupled to its rear side. This effect is provided by the combination of TIR at the top interface and the dichroic reflections at the bottom and rear sides. The requirement for TIR to occur sets a constraint on the angle of acceptance for input light, in that only light incident at angles $\theta_i > \theta_{iL}$ experiences TIR and is confined into the RAW prism, where $\sin\theta_{iL} = n\sin(\theta_L - 2\alpha)$ being n the refractive index of the prism and of the top solar glass, n=1 the refractive index of air and θ_L the limit angle for TIR at the internal prism-air interface. Therefore, if we measure incidence angles from the normal to the top surface of the prism, we can define a "dead angle" θ_{iL} such that incident light at $\theta_i < \theta_{iL}$ is not guided and is re-emitted from the top surface, after being reflected at the bottom face.

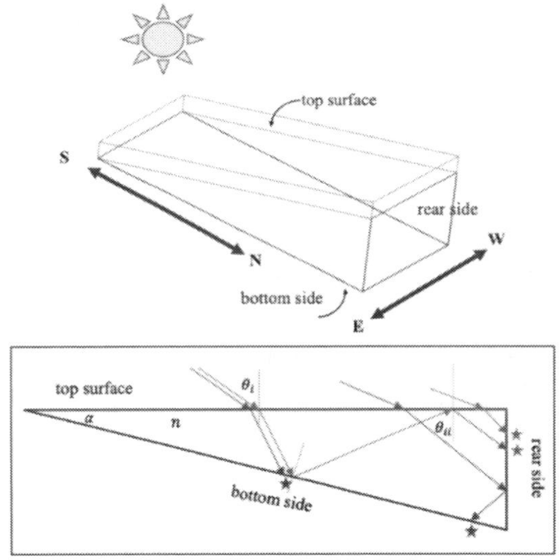

Figure 1: A sketch of the RAW-SS optical element, as modelled by ray-tracing software. Top: the element is oriented in the NS direction and its top surface makes right angles with the rear side and the lateral sides. Bottom: blue and red rays indicate possible trajectories for VIS and NIR light. VIS light is reflected at bottom side and absorbed at the rear side and vice versa for the NIR light.

However, this limitation can be overcome if one considers that the angular elevation of the sun over the horizon has a defined excursion and reaches its maximum at summer solstice. Let ϕ_s be the angle of sun elevation at summer solstice and β be the tilt angle of the module between the plane of the top surface and the horizontal plane (ground). The module can be aligned so that $\beta = (90° - (\phi_s + \theta_{iL}))$. In turn, the value for θ_{iL} is related to the prism parameters n and α that contribute to define the limit for possible optical concentration of the prism.

The RAW-SS is an asymmetric and hybrid concentrator, working as LCPV for the VIS portion of light and it does not concentrate the NIR portion. Therefore an analytical description of its working principle can be separately provided for the two spectral ranges. In any case, optical concentration brings limitations in the angular acceptance for incoming light, according to the thermodynamical principle of conservation of the étendue [21]. In this work we have carried out the calculations by considering an asymmetric angular admittance θ_{MAX} measured from the horizon and up to the solar elevation angle. The limit in optical concentration becomes:

$$tan(\alpha) \geq \frac{\cos^2\theta_{MAX}}{2n^2} \quad (1)$$

In the following, a trade-off geometrical concentration gain for VIS light $C_{G-VIS} = 1/\tan\alpha \approx 4$ ($\alpha = 14°$) is considered, which implies that $\theta_{MAX} \approx 48°$. Such an acceptance is comparable with the angular excursion of the position of the sun in the sky at noon along the year. Besides, in wintertime it allows to couple sunlight right from $\phi = 0°$. In summertime, the limited acceptance in angle introduces a penalty that depends on the latitude of installation. Only elevation angles higher than $\phi = (\phi_s - \theta_{MAX})$ above the horizon contribute to light collection. At $\phi_N = 45°$ $\phi \approx 20°$ and the limitation still copes with most effective installation conditions without introducing much additional penalty (it does not exceed by much the elevation of 10° from the horizon that is generally considered as a common operational limit to light collection). The penalty becomes eventually negligible at higher latitudes of installation. On the contrary, at $\phi_N < 45°$ the penalty increases up to $\phi \approx 42°$ for sites at the equator at the equinoxes or at tropical locations at solstices. However, we will discuss in the next sub-section the reason why, in case $\phi_N < 45°$ a proper alignment of the RAW-SS design strongly favors bifacial operation and this fact can at least partially compensate the loss in angular acceptance in the visible range. Moreover, no angular limitation is set for NIR irradiation.

It also has to be stated that so far we have discussed of irradiation in the sagittal plane of the prism. However, this does not introduce any loss of validity to the description of the operation of the RAW-SS, practically in all possible cases when glass and polymeric optical materials are used. Actually, in the case of $n \geq \sqrt{2}$ optical concentration is also provided for skew light rays entering from the top surface, by means of reflections at the lateral faces of the prism. A module aligned north-south (NS) has its rear face supporting the WG cell than looks south. Optical performance ratio P_{ro} is defined as the ratio between the output power from the optics that enters the solar cells and the incident power at the top surface. In our case of two different optical paths for the VIS and NIR light, we can define it as $P_{ro} = (P_{out,VIS} + P_{out,NIR})/P_{in} = P_{ro-VIS} + P_{ro-NIR}$. It depends on the incidence angle for light and can be evaluated for NS-oriented module without losing generality. When considering the angular admittance in the NS direction and the full admittance in the EW direction, it is evident that a trackless operation of the module is foreseeable with acceptable levels of PCE. Clearly, nothing prevents tilt repositioning of the module twice a year at different angles β_w and β_s in winter and in summer, or the use of an elevation sun tracking NS and of an azimuth tracker EW, to achieve even higher efficiencies despite at higher plant costs.

2.3 Possible installation and alignment

Figure 2 reports three possible orientations of the module for outdoor installation, in dependence on the latitude of the site, supposed to be in the northern hemisphere.

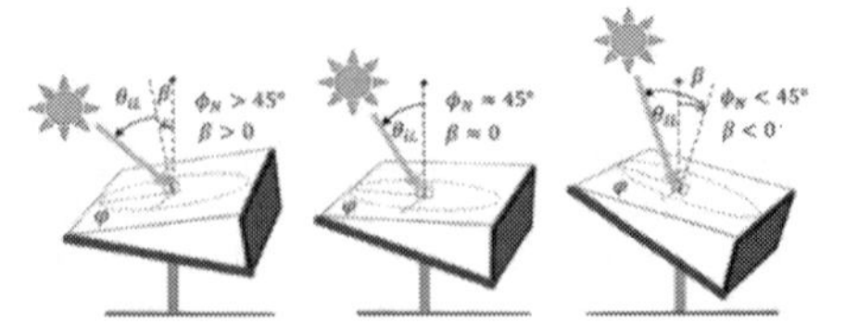

Figure 2: Possible alignment of the RAW-SS module, at three different latitudes. The TIR condition combines with maximum sun elevation to define the optimal tilt angle.

In the reference case with the parameters described above, $\theta_{iL} \approx 21°$. As the latitude of installation site decreases, ϕ_s increases, so that the tilt of the module is adjusted for best efficient operation and decreases until it can eventually turn negative. For instance, in case of a module installed at latitude $\phi_N = 45°$ (Lombardy-Italy) ϕ_s =68°26' and $\beta \approx 0°$ and the module would be aligned almost horizontally. If $\phi_N < 45°$, $\beta < 0°$ and the top surface of the module is slightly pointing north-facing. In case $\phi_N > 45°$, $\beta > 0°$ as for standard flat panels, however the tilt in this case is lower than for standard installations. Therefore, in all cases the alignment of RAW-SS decreases the effect of self shading of the panel with positive impact on the amount of albedo irradiation from the ground. A second positive asset of RAW-SS design is shown in Figure 3. Thanks to its right-angle geometry, the RAW-SS solution does not increase the land coverage. Since the VIS solar cell is located in the shadow cone of the module and illuminated by guided light, it never contributes additional ground shading. Therefore, the widening of land coverage which is usually caused by SS PV solutions is fully avoided in the present case. With reference to Figure 3, the length of the shaded area at ground is L" such that:

$$L" = L'/\sin(\phi) = L \cdot \sin(\phi + \beta)/\sin(\phi) \qquad (2)$$

In the case of standard panels, $\beta \geq 0°$ and $L" \geq L$. In this case, the tilt of RAW-SS module is tipically lower than for standard modules, so that the increase of the shaded area is less severe. Moreover, the RAW-SS design also admits negative tilt $\beta < 0°$, bringing to L"<L, as shown in Figure 3.

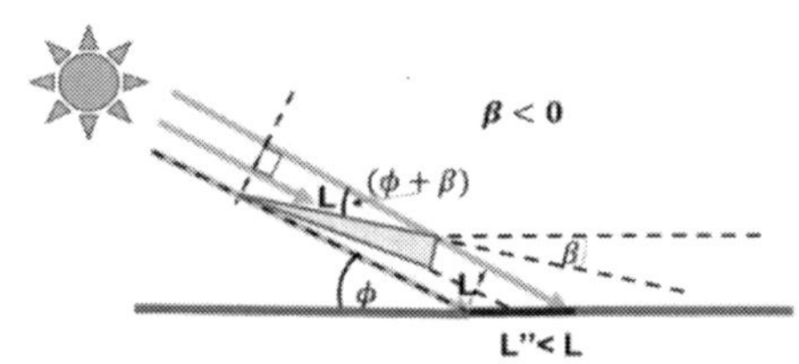

Figure 3: Example of installation and alignment of the RAW-SS design under direct irradiation, showing the reduction in ground shading with respect to a standard panel of equal dimension.

From what described above, it is clear that the RAW-SS design is particularly appealing if operated bifacially, by coupling a bifacial silicon solar cell to its bottom side, to benefit from the retroreflected irradiation from the ground. Bifaciality may increase PCE of the module by the 20%-30%, in strong dependence on installation and operating conditions[22]–[24]. Detailed evaluations of bifacial gain are made for specific locations, ground types and average weather conditions and are outside the scope of this work.

3 RESULTS AND DISCUSSION

An example of the performance of a RAW-SS geometry is reported in Figure 4, respectively for direct irradiation, in dependence on its angle of incidence angle in the sagittal plane, and for isotropically diffused radiation. Plotted ray trajectories are issued by ray-tracing software. By diffused irradiation one can also statistically represent the general case of direct skew irradiation and non sagittal rays, as it typical happens in trackless operation. It is evident that the EW component of rays is confined, while back-reflection is responsible for the loss in optical power in dependence on the angle of incidence of the NS component.

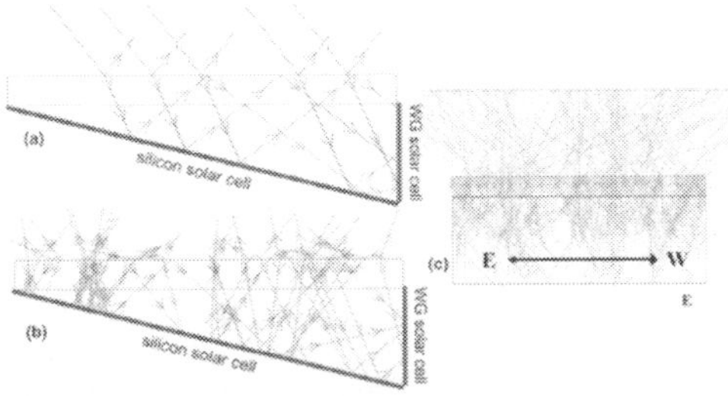

Figure 4: Simulated trajectories of rays in the RAW-SS optics, for visible light (blue lines) and near-infrared light (red lines). (a): direct irradiance; (b): isotropically diffused light, also statistically representing skew incidence; (c): an EW view of the wedge that highlights the light confinement in the EW direction.

A plot of P_{ro} as a function of the angle of incidence θ_i in the sagittal plane is reported in Figure 5. Simulations have been performed with 1 million rays striking the top surface. The blue solid line is the angle-dependent efficiency of collection P_{ro-VIS} of the VIS light that is concentrated at the rear side of the wedge. Its drop for $\theta_i < \theta_{iL} \approx 20°$ is due to the TIR cutoff. However, a certain degree of optical efficiency in the VIS can still be appreciated also beyond the cutoff. It is contributed by the small fraction of rays that impinge at top surface very close to the rear surface and reach it either directly or upon just being reflected once by the cold mirror, without the need for TIR. The red line represents P_{ro-NIR} and is much less angle-dependent since no TIR occurs and only the common contribution by Fresnel reflection at the top interface is effective. Most of NIR rays entering the module either directly reach the bottom surface or are directed to it after just one reflection by the hot mirror. The total P_{ro} is contributed by both VIS and NIR bands and it gradually decreases beyond the TIR cutoff to around 48%. The upper limitation in angular acceptance for the concentrating optics, due to the étendue theorem, is also smoothed and at $\theta_i = 80°$ $P_{ro-VIS} \approx 57\%$. This can be explained by the fact that a portion of VIS light at grazing angles of incidence directly reaches the rear surface according to the laws of refraction, without undergoing reflection at the bottom surface. The specific conditions for these extra contributions are strictly dependent on the values for the refractive index of the optical wedge and on the choice of the concentration gain. Once again, the NIR spectral component provides its contribution to global optical efficiency for incidence angles beyond the angular acceptance for LCPV in the VIS. In sum, a spectrally integrated P_{ro}>80% is obtained over a whole angular acceptance $\theta_A \approx 48°$, with $P_{ro,MAX}$=88.7% at an optimal angle $\theta_i = 37°$. In case of isotropically diffused light (same spectrum) the optical efficiency is independent on panel orientation and can be evaluated as $P_{ro,diff} = P_{ro,diff-VIS} + P_{ro,diff-NIR} = 17\% + 45\% = 62\%$.

4 DISCUSSION

The RAW-SS optics is conceived to be coupled to different solar cells, optimized in PEC respectively for the visible and near-infrared. In the present reference case, the pair of cells is thought to be contributed by a bifacial silicon solar cell and a thin-film hybrid perovskite solar cell, of a specific composition and bandgap. For a balanced 4T electrical connection, the number of solar cells to be used is that which balances the value for the output voltage V_{oc} of the module.

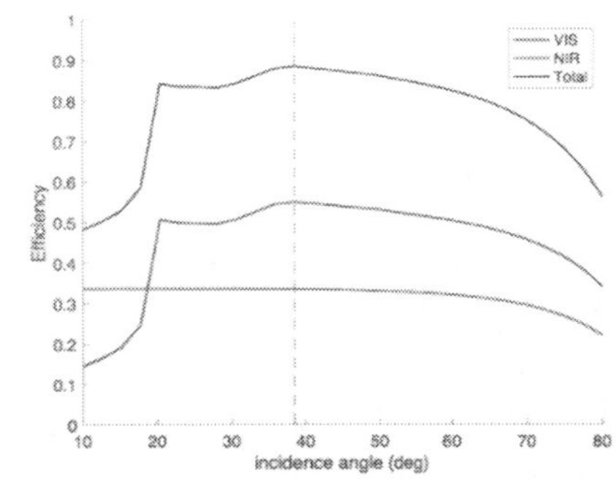

Figure 5: Spectrally integrated ptical efficiency of the RAW-SS concentrator, computed for the VIS and NIR bands in case of direct irradiation at different angles of incidence in the NS direction.

For instance, in the present case, the silicon and perovskite cells feature $V_{oc,NIR} \approx 0.7V$ and $V_{oc,VIS} \approx 1{,}05V$ respectively. A common value of $V_{oc} = 2.1V$ can be obtained by connecting in series three silicon cells and two perovskite cells. The absolute dimensions of the prism must therefore accommodate the cells and must also cope with any technological constraint relative to the fabrication of the cell themselves, besides considering also the need for economical sustainability of the solution. The geometrical parameters of the solution designed and simulated in this work are a valuable trade-off. The height of the rear side can measure about 2 cm and the length of the top surface in the sagittal plane can be around 8 cm. In this way, the whole geometry would also be compatible in thickness with standard panels. Trackless operation is also compatible with standard infrastructures, so that no specific increase in the Balance of System is expected for a utility-scale installation of the RAW-SS solution. Instead, what is evident is that a geometry like the one here proposed, which copes with small dimensions of the WG cell in LCPV, necessarily implies the use of small silicon cells and is constituted by relatively small optical elements. As it is, a single element might provide sub-watt supply and cope with demands in Internet-of-Things applications. Multiple replicas of the core element might be arranged together to configure optics for wider panels of required topology and power. The fabrication of large patterns of dielectric wedges, by printing and casting techniques and by optically coupling dichroic mirrors grown by sputtering industrial facilities is possible and the whole manufacturing process would be at the reach of highly automated production lines.

5 CONCLUSIONS

In this work we have reported about the design of a PV optical element, suitable for connecting solar cells in double-junction 4T configuration. The optical element is a right-angled-wedge (RAW) prism, coupled to dichroic mirrors for spectral splitting (SS) operation in the VIS and NIR bands. Lightguiding effect concentrates a great portion of VIS light. The NIR light is not concentrated. Therefore, the 4T connection can be established between silicon solar cells and cells operating in the visible range, highly performant but of small area. It is envisaged that a single optical wedge element measures in the centimeter range and multiple replicas can be arranged to configure a wider panel. Optical performance analysis for the single element has been ruled using commercial ray-tracing software under direct irradiance as well as isotropically diffused. A trade-off geometry is obtained by setting the geometrical concentration gain in the VIS at $C_{G-VIS} \approx 4$ ($\alpha = 14°$), which implies that $\theta_{MAX} \approx 48°$. By considering a wedge of refractive index 1.5<n<1.56, the angular acceptance required by concentration is compatible with trackless operation of the module. By considering PMMA as the dielectric medium for the optical wedge and by assuming ideal dichroic mirrors as the spectral splitting elements, in case of direct sunlight irradiation at AM1.5G standard spectrum, the overall optical efficiency of the module is P_{ro}>80% over the whole angular NS acceptance, reaching $P_{ro,MAX}$=88.7% at an angle of incidence of 37°. The overall optical efficiency for isotropically diffused light is $P_{ro,diff}$=62%, ($P_{ro,diff-NIR}$=45%, $P_{ro,DIFF-VIS}$=17%). This condition brings to align the panels at tilt angles smaller than those of standard flat panels installed at same latitude. Negative values for the tilt may even be required. In this way, the self-shading of the panels is reduced, with positive effects in terms of promoting albedo irradiance and its collection by the bottom cell. The right-angle shape also ensures that spectral splitting is performed without impairing land occupation. Therefore, the present solution can be considered as specifically appealing for 4T operation in bifacial configuration.

References

[1] R. Madurai Elavarasan, R. Pugazhendhi, M. Irfan, L. Mihet-Popa, P. E. Campana, and I. A. Khan, Appl. Energy 307 (2022) 118173.
[2] N. S. M. N. Izam, Z. Itam, W. L. Sing, and A. Syamsir, Energies 15 (2022) 1.
[3] D. R. Myers, Compr. Renew. Energy 1 (2012) 213.
[4] G. Grasso, A. Righetti, M. C. Ubaldi, F. Morichetti, and S. M. Pietralunga, Sol. Energy 86 (2012) 1725.
[5] W. Shockley and H. J. Queisser, J. Appl. Phys. 32 (1961) 510.
[6] M. E. Nell and A. M. Barnett, IEEE Trans. Electron Devices 34 (1987) 257.
[7] A. Luque and S. Hegedus, Handbook of Photovoltaic Science and Engineering, 2nd Ed. John Wiley & Sons Ltd., 2011.
[8] M. Yamaguchi, F. Dimroth, J. F. Geisz, and N. J. Ekins-Daukes, J. Appl. Phys. 129 (2021)
[9] H. Liu, Z. Ren, Z. Liu, A. G. Aberle, T. Buonassisi, and I. M. Peters, Opt. Express 23 (2015) A382.
[10] S. Akhil et al., Mater. Des. 211 (2021) 110138.
[11] E. Raza and Z. Ahmad, Energy Reports 8 (2022) 5820.
[12] S. Rühle et al., J. Renew. Sustain. Energy 1 (2009) 013106.
[13] Z. Liu et al., 2017 IEEE 44th Photovolt. Spec. Conf. PVSC 2017 (2017) 1.
[14] A. Mojiri, R. Taylor, E. Thomsen, and G. Rosengarten, Renew. Sustain. Energy Rev. 28 (2013) 654.
[15] G. Martinelli, M. Stefancich, and A. Antonini, US 2008/0149162 A1, 2008
[16] F. Morichetti, A. Righetti, G. Grasso, M. C. Ubaldi, and S. M. Pietralunga, in Conference Record of the 2012 38th IEEE Photovoltaic Specialists Conference, 2012, 000957.
[17] G. Grasso, F. Morichetti, A. Righetti, S. M. Pietralunga, and M. C. Ubaldi, in Conference Record of the 2012 38th IEEE Photovoltaic Specialists Conference, 2012, 002048.
[18] R. Guerrero-Lemus, R. Vega, T. Kim, A. Kimm, and L. E. Shephard, Renew. Sustain. Energy Rev. 60 (2016) 1533.
[19] X. Sun, M. R. Khan, C. Deline, and M. A. Alam, Appl. Energy 212 (2018) 1601.
[20] S. M. Pietralunga *et al.*, Adv. Mater. Interfaces 7 (2020) 1.
[21] G. Smestad, H. Ries, R. Winston, and E. Yablonovitch, Sol. Energy Mater. 21 (1990) 99.
[22] S. Kim *et al.*, Sci. Rep. 11 (2021) 1.
[23] S. Pal and R. Saive, Conf. Rec. IEEE Photovolt. Spec. Conf. (2021) 1159.
[24] P. Jain, G. Raina, S. Mathur, and S. Sinha, Renew. Energy Sustain. Growth Assess. (2022) 181.

ACKNOWLEDGMENTS

• Authors acknowledge partial funding by European Union under grant - FESR, PON Ricerca e Innovazione 2014-2020, Project n. ARS01_00519.
• Italian Patent application N. 102022000013840, filed on June 30th, 2022

EXTENSIVE MODULE BACKSHEET QUALIFICATION PROTOCOL CORRELATED TO SHJ MODULE RELIABILITY

Timea Bejat*[1], Marion Sérasset[1], Eszter Voroshazi[1], Romain Soulas[1], Lorenzo Carbone[2], Lionel Sicot[1], Vincent Barth[1], Marion Vite[1], Gaetano Izzo[2], Antonino Ragonesi[2], Francesco Rametta[2], Alessandro Fucile[2], Cosimo Gerardi[2]
[1]Univ. Grenoble Alpes, CEA, Liten, INES, Le Bourget du Lac, France,
[2]ENEL Green Power, Contrada Blocco Torrazze, Catania, Italy

ABSTRACT: In the race of ultra-high performance modules and the integration of novel cell architectures for bifacial configuration, many efforts are required in order to ensure module reliability and maintain low BOM cost, leading to research of new material families and configuration among PV module backsheet. Although, PVF and PET based backsheets are in use since several decades with well-established testing methods [1], new materials, surface treatments and manufacturing methods, such as co-extrusion, are rapidly emerging in the PV backsheets field. Therefore, the new backsheet samples require new testing methods and characterization strategy since their structure and physical properties differ from the commonly used backsheets in PV industry. In this paper, we present a three-stage investigation strategy on nine commercial backsheets intended to be integrated in heterojunction (SHJ) PV modules manufactured by 3SUN [2]. We characterize three different groups of backsheets: first group containing a coextruded PP backsheet, the second group covers PET based backsheets containing aluminum laminated layer and the third group about laminated PET based backsheets without aluminum layer. First stage is material level testing (optical, mechanical, adhesion, water absorption and tensile strength) enabling rapid ranking based on backsheet intrinsic properties. Second stage is accelerated aging testing in mini-module configuration while third stage is full-sized module testing. The material characterization allowed the conclusion that adhesion properties seem to have the most influence on final module behavior while we need to highlight that only the WVTR of backsheet material cannot predict the PV module behavior predictions in DH testing and adhesion strength plays a critical role.
Keywords: Durability, Reliability, PV Materials, PV Module

1 AIM AND APPROACH

The purpose of the study herein proposed is to establish a reliable methodology for material qualification of durable PV modules backsheets and to investigate the correlation between backsheet material properties and the module reliability in accelerated damp heat testing conditions as one of the most critical test of the IEC61215 PV reliability standard [3]. Indeed, the aim of this investigation is to assist PV manufacturing companies in accelerating the integration of novel materials in innovative PV module and saving time/money within research and development activities. Therefore, we present an extensive material and mini module testing sequence and its validation. Our study presents the backsheets selection process for a final module made by 144 SHJ half-cell with glass-backsheet configuration.

Upon exhaustive market survey, we chose to study nine backsheets (Table 1) intended to serve as rear face material in SHJ modules suitable for the next generation of photovoltaic product of 3SUN EGP [4]. First, we conducted an initial material testing sequence to characterize their following properties (on 3 samples/material). Shrinkage ratio tests were carried out, in order to evaluate the sample size variation under temperature. We measure the dimensional variations in both machine and cross directions, with expected values below 1%. Optical tests correspond to reflectivity measurements on small backsheet samples before and after aging under UV light (Xenon lamps) for 15 kWh/m² [3]. Water adsorption measurements were carried out submerging laminated backsheet samples into deionized water for at least 8 weeks [5]. Regular weight monitoring of the water uptake by the backsheet were evaluated and only the 0.5% of mass increase was consider acceptable. The methodology for tensile stress – strain measurements of laminated backsheets was performed on dumbbell shaped samples and repeated along both material structure orientation i.e. the machine and cross directions,[6] before and after 200h of damp heat (DH) treatment. We also determined Young modulus of each samples. These tests were done on a standardized experimental device (Instron) [6]. For adhesion testing special samples were manufactured to assess adhesion between the backsheet and the encapsulant. The peel force was measured in 180° configuration [3]. We took into account three possible peel interfaces: at backsheet/encapsulant interface, within the encapsulant layer, at glass encapsulant interface.

Following this extensive material testing sequence, we made a first backsheet selection. We identified three groups of backsheets according to their physical behavior. We retained backsheets from each group with the lowest WVTR values and those with lowest price and availability for mass manufacturing. Thereafter, we fabricated mini-modules with eight half-cells connected in series and a module stack configuration based on glass/POE/ SHJ cells/POE/BS with unique difference in the backsheet material. For each configuration, three samples were fabricated and underwent accelerated DH testing for 3000 h following IEC61215 with intermediate characterization of current-voltage (IV) measurement and electroluminescence (EL) imaging. We used glass/glass modules as reference. Lamination process qualification to reach the required level of material cross-linking and adhesion has been conducted prior to the module fabrication. The selected POE has been qualified too in an earlier study to pass 3000h DH testing with limited power loss in GG configuration. The backsheets with power loss < 5% upon 1000h of DH are then integrated in industrial size modules with 144 half-cells configuration with 3 central bypass diodes for final qualification in damp heat testing. This closed loop testing enables correlation between final module results, material property and mini-module level testing.

2 PREPARING THE MANUSCRIPT

Several authors proposed water vapor transmission rate (WVTR) as a determining factor to extrapolate module performances towards damp heat testing [1]. However, as following reported in the presented study, the aluminum containing backsheets (D and I) and backsheet F report the lowest WVTR values, but after 2000 h of DH testing, only backsheets D and I showed small decrease in electrical performance (Pmax variation less than 2%) while backsheet F, lost more than 5% of power output. This result highlights that only WVTR evaluation is not enough for PV module behavior predictions. The material characterization identified that adhesion properties seem to have the most influence on final module behavior. Indeed, the backsheets from the third group, showing poor adhesion performances reflect the poorest performance in terms of Pmax among studied materials. We have notable outlier about backsheet A which showed poor performance during initial testing stage, and presents less than 5% power loss upon 2000 h of DH in mini module configurations. We recorded a surprising recovery between 1500 h and 2000 h of DH tests for backsheets of the third group (F, H). This behavior will be explained by further material analysis.

3 RESULTS AND CONCLUSIONS

The first test at material level is the shrinkage ratio measurement. All tested backsheet pass this test except backsheet A, which showed strong distortions and not allowed to determine its dimension variation. All upcoming test used laminated samples. During optical tests no notable changes have been reported, as all studied backsheets have a special UV protective layer supposed to prevent UV light induced degradation. Water absorption tests gave results below 0.5% of water uptake for group two and three. Backsheet A showed a very different behavior compared to all others. It had more than 2% of mass increase. Backsheet F has an intermediate behavior, slightly above the targeted value. We decided to continue the testing of both these backsheets. Tensile strength results show 3 groups upon mechanical behavior (Figure 1), backsheet A [1] shown very high strain at low tensile stress. We measured up to 525% of strain while other backsheet do not reach 110%. We noticed delamination within both aluminized multilayer structures samples, D and I, at each test. Their results correspond to the tensile stress of last layer. Despite of systematic delamination their tensile performance remained very good. The last group includes backsheets with elastic/plastic behavior, with rather high values for tensile strength, reached at the end of each test without delamination.

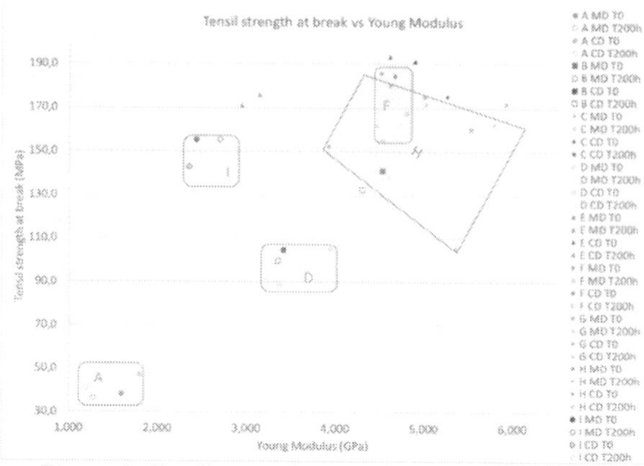

Figure 1: Tensil strength versus Young Modulus of studied backsheets in machine (MD) and cross directions (CD)

Table I: Studied BSs' composition and properties

BS	Composition (P=primer layer)	Remarks N° of group (W=white; TR=transparent)
A	PP	1, W, Co-extruded
B	Kynar(PVDF)/PET/P	3, W, laminated
C	Tedlar(PVF)/PET/P	3,TR, laminated
D	PET/Al/PET/P	2, W, laminated
E	coating/PET/SiOx-PET/P	3, TR, laminated
F	ETFE/PET+/PET/P	3, TR, laminated
G	ETFE/SiOx-PET/PET/P	3, TR, laminated
H	coating/PET/PET+/PET/P	3, TR, laminated
I	PET/Al/PET/POE	2, W, laminated
REF	Glass	-

During adhesion testing, we mainly focused on the backsheet/encapsulant interface, which is considered the crucial surface for the peeling force test. Backsheet rupture occurred systematically for multilayered laminated structures with aluminium layers, meaning that adhesion forces between backsheet and encapsulant are higher than internal material adherences. We measured the peel strength at T0 and after 200h of damp heat exposure of all backsheets (5 bands of 10mm width each time). The backsheets which shown the poorest performance (E and G) were discarded for further characterization. Table 1 report the summary of tested BS with material composition. Material level testing results showed that, in some cases, backsheets were close to the given thresholds described above, thus we decided to keep them for mini-module testing. Afterwards we have fabricated mini-modules with the selected backsheet. The average initial performance of mini-modules with white backsheet has been the following: Pmax=20.45 Wp, Isc=4.57A, FF of 75.7 % and Voc of 5.91V while for samples with transparent backsheet of Pmax=19.67 Wp, Isc=4.37A, FF of 76.4 % and Voc of 5.89V. Figure 2 presents the normalized Pmax variation of mini-modules with various backsheet and reference module. In concordance with earlier work, the degradation in DH conditions is mostly attributed to the FF related to the degradation of TCO, metallization layer and/or their interface [7].

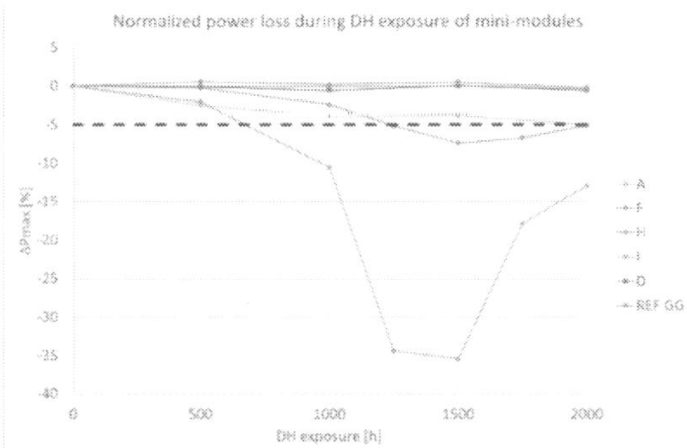

Figure 2: Normalized mini-module power output variation in function of DH exposure duration

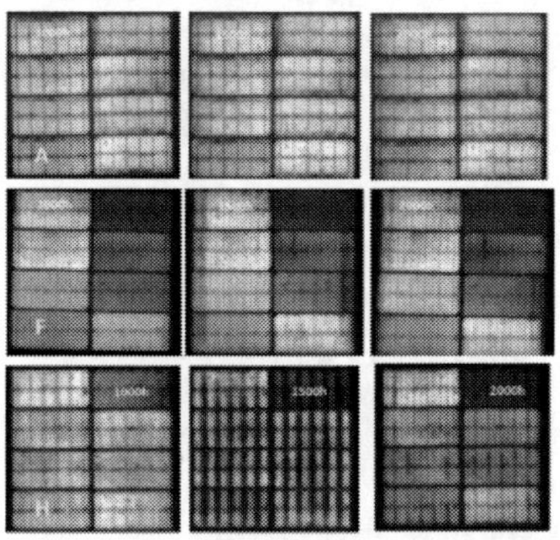

Figure 3: EL images of mini-modules during DH tests

Figure 3 presents electroluminescence images of modules with backsheets A, F and H at 1000h, 1500h and 2000h which corroborate the FF loss with the gradual appearance of dark areas around the busbars. Unlike earlier studies exhibiting darkening at edge areas linked to side moisture ingress, we observe a homogenous degradation of the cells clearly confirming humidity ingress from the backsheet side. Our initial hypothesis is that moisture penetration triggers a decrease of TCO conductivity within the interface degradation (further investigation are ongoing). Interestingly we notice recovery effect at 2000h DH suggesting a partially reversible degradation process, potentially drying mechanisms for higher permeability backsheets of the third group. Based on the durability of mini-modules in DH study, we selected the backsheets D, F, I for qualification of full-sized modules. Initial results on full size modules up to 1000h ageing corroborate the results of the mini-modules and degradation mechanisms with -1.04%, -16.75% and -1.02% power loss upon 1000h for backsheets D, E, and I, respectively (Figure 4).

4 REFERENCES

[1] G. Oreski et al. Designing new materials for photovoltaics: Opportunities for lowering cost and increasing performance through advanced material innovations IEA PVPS Task 13, Report IEA-PVPS T13-13:2021, April 2021 ISBN 978-3-907281-02-4.

[2] G. Condorelli et al., "High Efficiency Hetero-Junction: From Pilot Line To Industrial Production," 2018 IEEE 7th World Conference on Photovoltaic Energy Conversion (WCPEC) (A Joint Conference of 45th IEEE PVSC, 28th PVSEC & 34th EU PVSEC), Waikoloa Village, HI, 2018, pp. 1970-1973, doi: 10.1109/PVSC.2018.8548197.

[3] IEC 61215: Crystalline silicon terrestrial photovoltaic (PV) modules–Design qualification and type approval (2005, 2nd edition)

[4] G. Condorelli et al., "Initial Results of Enel Green Power Silicon Heterojunction Factory and Strategies for Improvements," 2020 47th IEEE Photovoltaic Specialists Conference (PVSC), Calgary, OR, 2020, pp. 1702-1705, doi: 10.1109/PVSC45281.2020.9300806

[5] ISO 62:2008 Plastics – Determination of water absorption.

[6] ISO 527-3:2018 Plastics – Determination of tensile properties – Part3: Test conditions for films and sheets

[7] W. Liu, L. Zhang, X. Yang, J. Shi, L. Yan, L. Xu, Z. Wu, R. Chen, J. Peng, J. Kang, K. Wang, F. Meng, S. De Wolf, Z. Liu, Joule, Volume 4, Issue 4, 2020, pp 913-927

ARE PV-MODULE BACKSHEETS WITH AN OUTER FLUOROPOLYMER LAYER THE BETTER SOLUTION?

Jonathan Kriening, Maximilian Engel, Dieter Geyer, Dirk Stellbogen, Peter Lechner
Zentrum für Sonnenenergie- und Wasserstoff-Forschung Baden-Württemberg (ZSW)
Meitnerstr. 1, 70563 Stuttgart, Germany

ABSTRACT: The aim of this work is to determine if polyethylene terephthalate (PET)-core backsheet laminates with an outer weather protection layer based on fluoropolymers are advantageous or necessary to guaranty adequate long-term stability and electrical safety of PV-modules. After observing a manifold of severe issues with coextruded polyamide backsheets (trade name AAA) [1, 2, 3] fluoropolymer protected backsheets gain back significant popularity. However, it is increasingly under discussion, whether fluoropolymers are really of low concern for human and environmental health [4]. More specifically, concerns are expressed regarding the safe disposal of fluorochemical polymers at the end of their life cycle and the rather limited options for recycling. Our approach was comparative testing of field aged modules with and without fluorochemical outer layer backsheets. In the test laboratory Solab of ZSW Stuttgart the modules were visually inspected, the STC power and electrical insulation was measured and the embrittlement of the backsheet was evaluated. While we observed some differences concerning the visual anomalies, the STC power and the electrical insulation of the modules as well as the embrittlement of the related backsheets indicate that fluorine-free outer layer polymers protect the PET core of the backsheet just as well as fluorochemical polymers.

1 MOTIVATION

The starting question is whether fluorochemical outer layer polymers are needed in order to protect the PET core of the backsheet to provide a reliable backsheet for a long service life of the PV modules. With respect to a sustainable production and environmental concerns it would be preferable to use materials that are free of fluorine polymers, thus not hazardous with a simpler end-of-life treatment.

For a sustainable PV business with a production of all the needed components in Europe it is important to not only look at competitive production costs but to also consider further factors, in particular a life cycle assessment of the produced components. The CO_2 equivalents, the toxicity and possible recycling solutions should also be assessed. Fluorine polymers are usually of higher cost in comparison to other polymers. The manufacturing is hazardous and the disposal of fluorochemical polymers is a difficult process. The usual approach is thermochemical pyrolysis or incineration where many hazardous agents can arise [7, 8]. The recycling is an even more challenging process. Therefore, it would be beneficial from a financial and environmental perspective to avoid fluorochemical polymers for a more sustainable PV business.

Figure 1 shows a schematic structure of a typical 3-layerd backsheet. The inner, cell facing layer is responsible for a good bond between the backsheet and the encapsulant .The core layer, which is the thickest layer in the backsheet, provides electrical insulation and mechanical stability. The outer, air-facing layer protects the core layer from environmental influences and is the focus of this work. A comparison between fluorochemical and fluorine-free outer layer polymers will be presented.

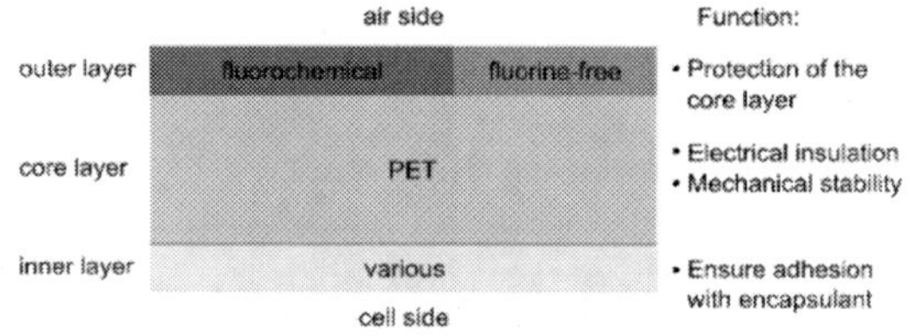

Figure 1: Typical structure of a 3-layered backsheet

2 APPROACH

Our approach was a comparative test of field aged PV modules in our PV laboratory Solab of ZSW Stuttgart. The state of health was analyzed for two groups of PV modules. One group with fluorochemical outer layers was compared to a second group with fluorine free outer layers in their backsheet layer composition.

The following test were performed in the laboratory:

- visual inspection with regard to backsheet related anomalies
- power measurement at STC conditions
- insulation testing by wet leakage current measurements
- evaluation of the embrittlement by elongation at break testing of backsheet samples

2.1 Investigated Modules

In this work, 143 modules which were in operation between four and eleven years in 31 different European photovoltaic parks were analyzed. The mean operation time for the modules with fluorochemical outer layer polymers was 8.4 years and 8.3 years for the modules with a fluorine-free outer layer. Figure 2 shows the service life distribution of the investigated PV modules.

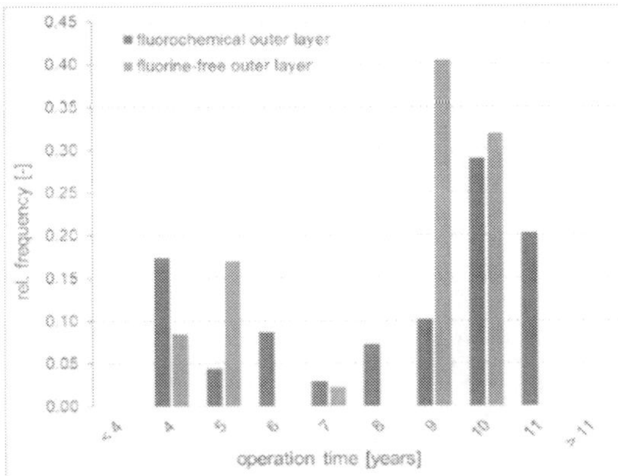

Figure 2: Operation time of tested PV modules with a fluorochemical (green) and a fluorine-free (red) outer layer polymer

2.2 Identification of polymers

The different types of polymers were identified by ATR-FTIR (Attenuated Total Reflection Fourier Transform Infrared spectroscopy). As the ATR measurements are surface sensitive, only the outer layer polymer could be identified directly. The entire structure of the backsheet was either identified by surface grinding and consecutive ATR-FTIR measurements or by microscopic ATF-FTIR spectra measurements of the cross-section samples. Figure 3 shows a sample of an exemplary cross-section of a backsheet layer composition with a fluorochemical PVDF outer layer, a PET core layer and a modified PE inner layer.

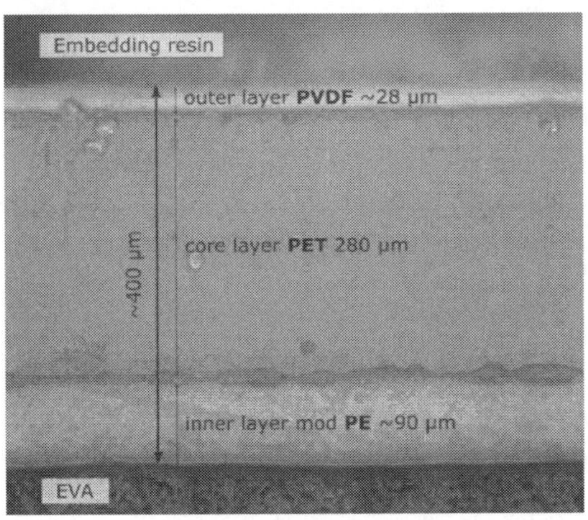

Figure 3: Cross-section of a PET based backsheet with an outer fluorochemical PVDF layer

Table 1 shows the distribution of the identified outer layer polymers. Of all 143 analyzed modules, 60 % had backsheets with fluorochemical and 40 % fluorine-free outer layer polymers.

Table 1: Identified different outer layer polymers of the field aged modules

	outer layer polymers		Freqency [%]
fluorochemical	PVF	Polyvinyl fluoride, Tedlar	21.5%
	PVDF	Polyvinylidene fluoride	25.9%
	FC	Fluorine coating	4.4%
	THV	Tetrafluoroethylene Hexafluoropropylene Vinylidene Fluoride	6.7%
	ETFE	Ethylene tetrafluoroethylene	1.5%
fluorine-free	PET	Polyethylene terephthalate	15.6%
	PA	Polyamide	24.4%

3 RESULTS

The following sections show the findings of the conducted lab tests for analyzed PV modules.

3.1 Visual inspection

As criteria for the visual inspection of the PV modules, four categories for backsheet related anomalies were chosen.

Corrosion: Chemical reactions between metallic components of the PV module and mainly related to acidic acid and resulting in reduced conductivity of the corroded elements. Consecutive failures can include the loss of electric contacts, the formation of hotspots, and loss of insulation.

Delamination: Failure due to reduced adhesion between backsheet and encapsulant that can result in moisture ingress and subsequent insulation problems.

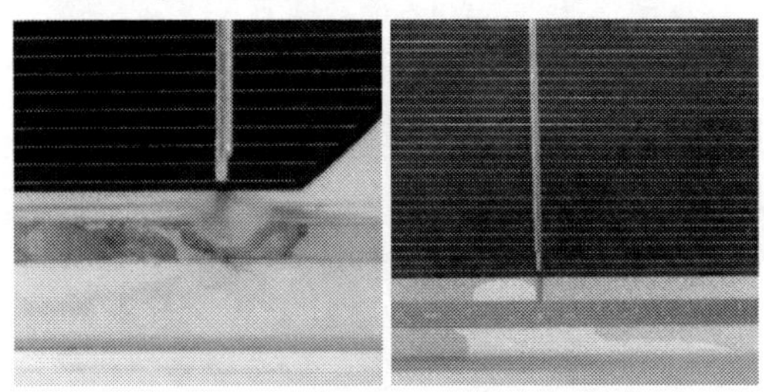

Figure 4: Exemplary visual anomalies of analyzed modules with corrosion (left) and delamination (right).

Backsheet cracks: Mechanical tear of the backsheet or parts of the backsheet. In the two shown examples (see Figure 5) only the outer, air facing polymer layer is torn off while the underlying PET core layer shows no cracks.

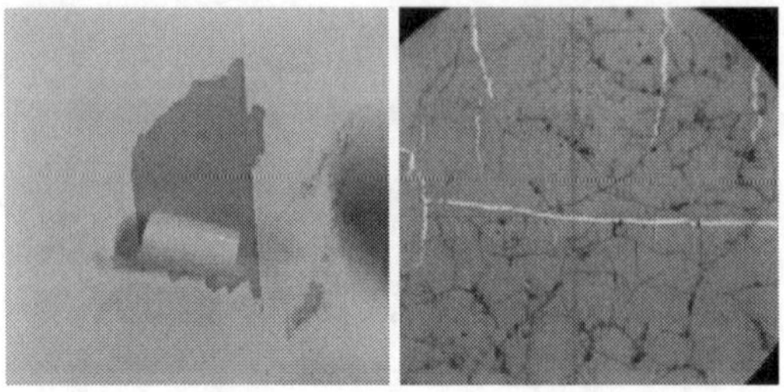

Figure 5: Superficial cracks of a backsheet with a fluorochemical PVDF (left) and a fluorine-free polyamide (right) outer layer polymer.

Chalking: White TiO_2 powder that is used as a UV protection are partially released due to a photocatalytic process.

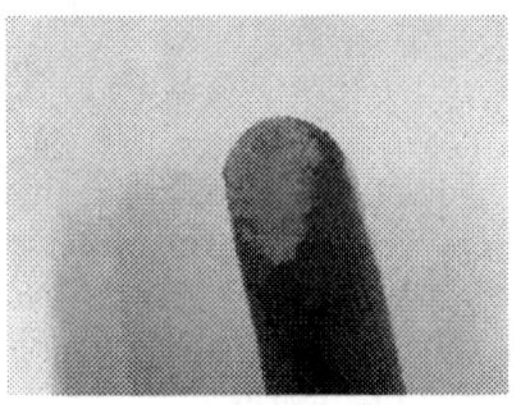

Figure 6: White TiO_2 particles abrasion of a fluorine-free polyamide outer layer on a black glove

Figure 7 shows the frequency of the visual anomalies. While nearly 60 % of the modules with a fluorine-free outer layer show chalking and superficial cracks, they are almost free of corrosion and delamination. Of these, the modules with chalking and superficial cracks all have a polyamide outer layer.

The modules with fluorochemical outer layer polymers in comparison are affected by corrosion, delamination, superficial backsheet cracks and chalking in about 10 % of the cases.

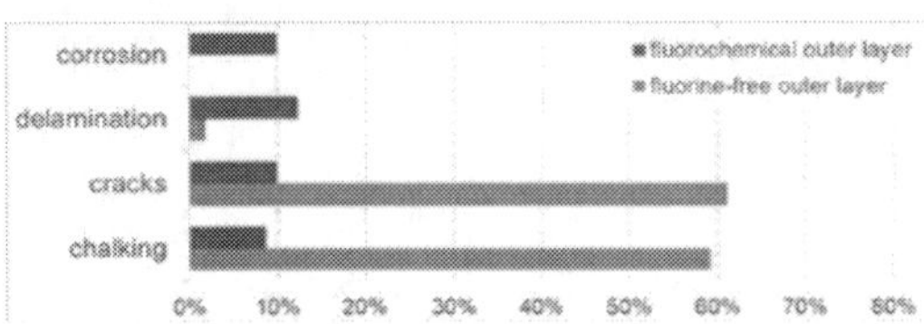

Figure 7: Backsheet related visual anomalies for modules with fluorochemical (red) and fluorine-free (green) outer layer polymers

3.2 Power measurements

After visual inspection and cleaning of the modules, the highest module power under STC conditions was measured. Figure 8 shows the distribution of the relative STC power of all modules. The relative STC power is the measured power under STC conditions divided by the corresponding nameplate power. Whilst the relative STC power of some modules is low, there is no correlation with backsheet related issues. In further step, data from modules with PID, bypass diode failures and glass breakage where filtered out. Figure 9 shows the resulting STC power distribution of modules with fluorochemical and fluorine-free outer layer polymers with suspected backsheet issues. The two distributions show no significant difference. The mean relative. STC Power is 2.5 % below nameplate power for the modules with fluorochemical outer layer polymers and 3.2 % below nameplate power for the modules with fluorine-free outer layer polymers.

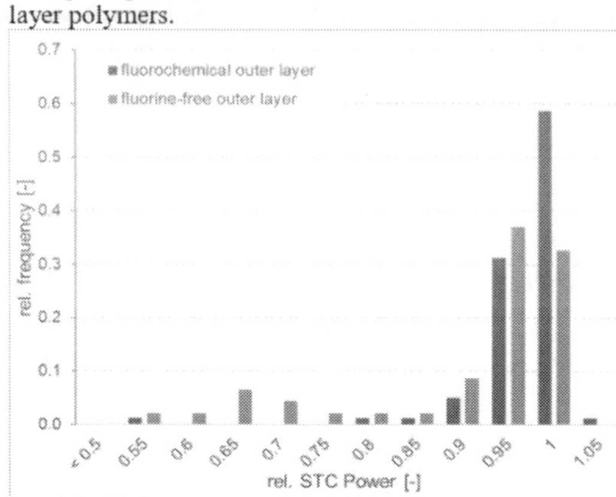

Figure 8: Distribution of the rel. STC Power of all modules

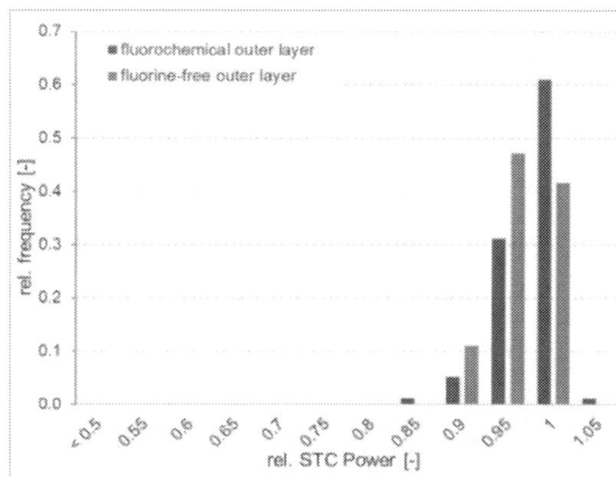

Figure 9: Distribution of the rel. STC Power of modules with backsheet related issues

3.3 Insulation

The insulation of the modules was measured by a wet leakage current test according to IEC 61215-2 MQT 15. The pass/ fail criterion is at 40 $M\Omega \cdot m^2$.

Of all the tested modules only five failed the insulation tests. The distribution between the fluorochemical and fluorine-free outer layer polymers is quite similar. No general insulation problem can be observed.

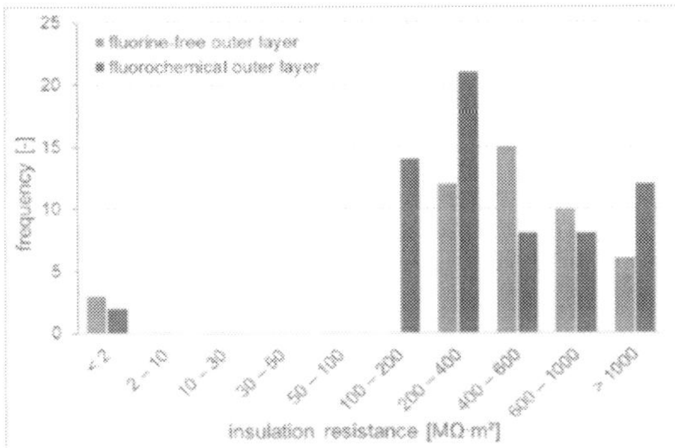

Figure 10: Distribution of the measured insulation value of modules with fluorochemical (red) and fluorine-free (green) outer layer polymers

3.3 Embrittlement of the backsheets

As a direct mechanical property, the embrittlement of the backsheet was measured by elongation at break (EaB) testing. For the EaB measurement test stripes were cut out of the backsheet and subjected to the tensile testing machine. In a previous work with coextruded polyamide backsheets (AAA) [3] we found a good correlation between the occurrence of BS cracks and the corresponding EaB value. For EaB values above 30 % the backsheets showed no cracks. Below EaB of 30 % the risk of crack formation increases rapidly with declining EaB values. For EaB values below 10 % over 60 % of the modules showed cracks in the backsheet [4]. If we apply the findings from AAA backsheets to the PET bases backsheets, there is an area, where we would expect cracked backsheets. Figure 11 shows the distribution of the EaB values for backsheet samples with and without fluorochemical outer layer polymers. There are only few measurements in the area of risk with EaB values below 30 %. None in the very critical area with EaB values below 10 %. This is consistent with our visual observation where we have not found any cracks in the PET core layer. The EaB distribution of the two groups is practically identical which indicates that fluorine-free outer layer polymers protect the PET core just as well as fluorochemical outer layer polymers.

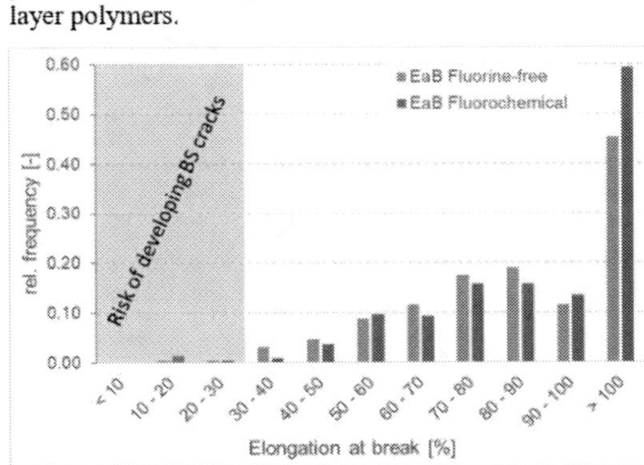

Figure 11: Distribution of elongation at break (EaB) values for PET based backsheets with fluorochemical (red) and fluorine-free (green) outer layer polymers.

4 SUMMARY

In this study, we compared the state of health of two groups of field aged modules from different European PV power plants after an average service live of 8.3 years. One group with fluorochemical - and the other group with fluorine-free outer layer polymers.

Laboratory tests show no significant difference in STC power, electrical insulation or embrittlement of the backsheets which indicates that fluorine-free outer layer polymers protect the PET core just as well as fluorochemical outer layer polymers.

The visual inspection of the modules showed some differences. While the modules with a fluorochemical outer layer showed corrosion, delamination, superficial cracks or chalking in 10 % of the cases, modules with a fluorine-free outer layer were almost not affected by corrosion or delamination. On the other hand, 60 % of those modules showed chalking and superficial cracks in the backsheet outer layer. Further investigations are necessary in order to establish whether chalking and especially superficial cracks of outer layers have any effect on the PET core in the remaining operation time. As most of the modules originate from a more temperate climate, data for other climate zones would be important to validate these findings for other climatic regions.

4 REFERENCES

[1] Eder, Gabriele & Voronko, Yuliya & Oreski, Gernot & Muehleisen, Wolfgang & Knausz, Marlene & Omazic, Antonia & Rainer, Alois & Hirschl, Christina & Sonnleitner, Horst. (2019). "*Error analysis of aged modules with cracked polyamide backsheets*". Solar Energy Materials and Solar Cells.

[2] G.M. Wallner, K.J. Geretschläger, M. Dankl, I. Hintersteiner (2013). "*Aging Behavior of Polyamide Based Backsheet Films*", 28th European Photovoltaic Solar Energy Conference and Exhibition

[3] P. Lechner, H. Wirth, J. Schnepf, S. Hummel, D. Geyer, B. Weinreich, R. Haas (2019). "*Analysis of Fielded PV-Modules with Backsheet Issues*", 36th European Photovoltaic

[4] R. Lohmann, I. T. Cousins, J. C. DeWitt, J. Gluege, G. Goldenmann, D. Herzke, A. B. Lindstrom, M. F. Miller, C. A. Ng, S. Patton, M. Scheringer, X. Trier, Z. Wang; "*Are Fluoropolymers Really of Low Concern for Human and Environmental Health and separate from Other PFAS?*", Environ. Sci. Technol. 2020, 54, 12820 – 12828

[5] J. Schnepf, P. Lechner, H. Wirth, S. Hummel, D.Geyer (2021) "*Analysis of Field Aged PV Modules with PET Based Backsheets Regarding their Mechanical Stability*", 37th European Photovoltaic Converence online

[6] Rong Deng, Nathan L. Chang, Zi Ouyang, Chee Mun Chong, "*A techno-economic review of silicon photovoltaic module recycling*", Renewable and Sustainable Energy Reviews, Volume 109, 2019, Pages 532-550, ISSN 1364-0321

[7] Aryan, Venkat and Font-Brucart, Mercè and Maga, Daniel, "*A comparative life cycle assessment of end-of-life treatment pathways for photovoltaic backsheets*", Progress in Photovoltaics: Research and Applications, Volume 26, Number 7, 2018, Pages 443-459

[8] Danz P, Aryan V, Möhle E, Nowara N. "*Experimental Study on Fluorine Release from Photovoltaic Backsheet Materials Containing PVF and PVDF during Pyrolysis and Incineration in a Technical Lab-Scale Reactor at Various Temperatures.*" Toxics. 2019 Sep 18;7(3):47.

DETAILED PERFORMANCE & DEGRADATION ASSESSMENT OF DIFFERENT PV TECHNOLOGIES WITH MORE THAN 10 YEARS LIFETIME

Sascha Lindig[1,*], Atse Louwen[1], Lukas Koester[1,2], Alexander Astigarraga[1], David Moser[1]
[1]Institute for Renewable Energy - EURAC Research, Viale Druso 1, 39100 Bolzano, Italy
[2]Dipartimento di Energia, Politecnico di Milano, Via Lambruschini 4, 20156 Milano, Italy
[*]Phone: +39 0471 055 715; E-mail: sascha.lindig@eurac.edu

ABSTRACT: In this paper, performance losses and correlated degradation of five different PV systems representing five different PV technologies are investigated. Here, we describe the overall performance loss of a PV system in greater detail to better understand which underlying reasons trigger the performance decline of the investigated PV systems. The plants were installed in 2010, and almost 12 years of system operation are considered. Based on best practices developed by the IEA PVPS Task 13, the linear performance loss rates of all PV systems have been calculated. The results are correlated with corresponding nonlinear performance losses, voltage and current evolution, I-V curves and infrared and luminescence images captured in the field. Overall, it was shown that crystalline-based PV systems experience a lower overall performance loss compared to thin-film systems. The losses in crystalline systems stem from cell mismatches within modules and gradually developing degradation modes such as snail trails and cell cracking. The thin-film systems under investigation are subject to higher performance loss rates, (most) likely triggered by a degradation of the active PV material. Furthermore, glass-glass thin-film modules are additionally exposed to physical stresses by the mounting clamps leading to cracked glass, which may jeopardize the mechanical integrity of the modules.
Keywords: PV system, performance degradation, system performance

1 INTRODUCTION

Photovoltaic (PV) modules, and the materials they are composed of, are subject to continuous cycles of temperature, humidity, irradiation, and mechanical stress, as well as other external influences. These stresses affect the performance of PV systems, depending on the underlying mechanisms, reversibly or irreversibly. The combined effects on the performance can be quantified by the performance loss rate (PLR). The PLR describes the overall annual performance reduction of a PV system. It is the combination of degradation, which describes the performance loss due to emerging degradation modes (DM), and other, usually reversible, performance impairing effects [1]. DM are effects that reduce the output power of a PV system irreversibly and might create safety problems [2].

Advanced PV system models predict the lifetime energy yield of PV systems attempting to consider all performance impairing effects. In order to better understand performance loss and degradation in PV modules, environmental stress tests are performed in indoor laboratories, and the results are integrated into physical PV system and degradation models as a bottom-up approach. The problem is that stresses are usually applied in isolation. In a recent study, a combined stress test setup has been developed to create more realistic ageing conditions to better understand the interplay of varying stressors and thereby varying aging mechanisms [3].

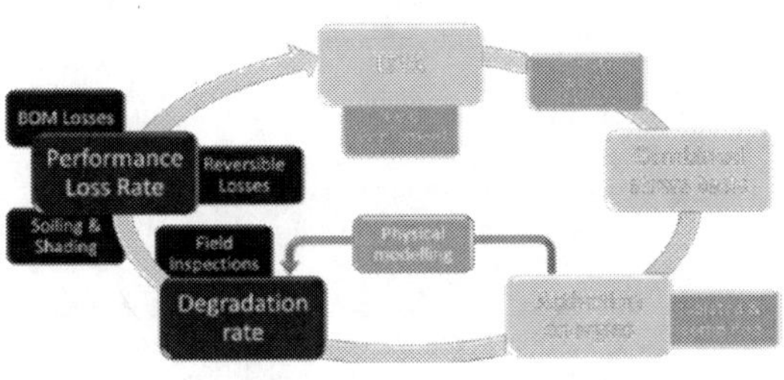

Figure 1: Degradation modelling workflow, performed analysis is highlighted

Figure 1 shows the connection between theoretical modelling of performance losses and yield estimates, indoor stress tests and measured performance of operating PV systems. In this work, we follow a top-down approach where we investigate the performance evolution of five PV systems operating for 12 years. Thereby, the calculated PLR is the starting point. In theory, the PLR should contain various contributing factors, such as actual degradation, soiling, shading and others. By applying state-of-the-art investigation techniques, we attempt to better understand physical processes within PV modules during outdoor operation.

In Section 2, we introduce the PV systems investigated in this study. The methodologies used are elaborated in Section 3, including linear and nonlinear performance loss calculations, maximum power point current (I_{MPP}) and voltage (V_{MPP}) as well as I-V curve evaluation and the application of different imaging techniques. The results are summarized in Section 4 and the paper closes with a discussion and a summary in Section 5.

2 PV SYSTEM

The PV systems in question are part of a larger experimental PV installation located at the Airport Bolzano Dolomiti (ABD) in South Tyrol/Italy. The latitude of the location is 46.4625°N and the longitude 11.3299°E. The plant is installed at a fixed tilt of 30°, an orientation of 8.5° west of south and is in operation since August 2010. The prevailing climate according to Köppen-Geiger [4] is classified as Cfb (temperate, without dry season, warm summer) and according to a PV sensitive climate classification temperate with medium irradiation [5]. The technologies considered in this study are mono-crystalline silicon (c-Si), poly-crystalline silicon (pc-Si), single junction amorphous silicon (1j-a-Si), cadmium telluride (CdTe) and copper indium selenide (CIS). The main characteristics of the systems are summarized in TABLE I. DC- and AC-side electrical parameters are measured every 15 minutes by commercial SMA inverters that ensure a high level of accuracy in current and voltage with an average difference of less than 5% and 2%, respectively.

At the test side, a dedicated weather station was installed in February 2011. The ambient temperature (T_{amb}), wind speed (ws), plane-of-array irradiance (G_{POA}) and other irradiance components are recorded with a measurement frequency of 1 min. The weather data are averaged on a 15 min interval to coincide the electrical signals. On the rear side of each system the module temperature is measured. The sensors are systematically cleaned and periodically calibrated in order to comply with the standard IEC61724-1:2017 [6]. Due to the delayed start of recording weather data, the investigated period runs from February 2011 to August 2022.

Because of an electrical fault and the inability to fix the issue on-site due to the pandemic, weather data of 22 months from January 2020 to October 2021 have been lost. Fortunately, back up sensors are in place. These sensors are not calibrated periodically but are a good source of training data to impute the missing measurements. Ambient temperature and wind speed values were directly used, as the values agreed well with the ones of the weather station. Irradiance data were imputed using ExtraTree regression [7], implemented with the Python library scikit-learn [8], followed by a polynomial irradiance-based correction. Missing module temperature values were modelled based on the Sandia module temperature model [9] according to:

$$T_{mod}[°C] = G_{POA}(e^{a+b*WS}) \quad (1)$$

a and b are empirical parameters depending on mounting configuration, module backside materials as well as solar cell material.

TABLE I: Main parameter of PV systems. Number of modules N_{mod}, nominal Power per module $P_{nom,mod}$, total array power $P_{nom,tot}$ & power temperature coefficient γ. c-Si: mono-crystalline silicon; pc-Si: poly-crystalline silicon; 1j-a-Si: single junction amorphous silicon; CdTe: Cadmium Telluride; CIS: Copper Indium Selenide.

Tech	Model	N_{mod}	$P_{nom,mod}$ [W]	$P_{nom,tot}$ [kW]	γ [%/K]
c-Si	Solarwatt M230-96	8	220	1.76	-0.5
pc-Si	Trina TSM 230-PC05	9	230	2.070	-0.45
1j-a-Si	Schott Solar Asi TM 100+	12	100	1.2	-0.2
CdTe	First Solar FS 277	90	77.5	6.975	-0.25
CIS	Würth WSG0036E80	14	80	1.12	-0.36

3 METHODOLOGY

3.1 Linear Performance Loss Rate Calculations

The linear PLR of the systems is calculated based on best practices defined by the International Energy Agency (IEA) Photovoltaic Power System Program (PVPS) Task 13 [10]. The calculation steps include:

0) Data analysis including data quality assessment
1) Data cleaning and filtering
2) Aggregation of several performance metrics
3) Optional: time series feature correction
4) Statistical model application to calculate ensemble of PLR
5) Filtering calculated ensemble
6) Averaging results to final PLR

The quality of the system data is defined based on the number of outliers, the amount of missing data and the longest data gap. The datasets are of high quality with average values of outliers of 2.8%, missing data of 2.2% and the average longest gap is 8 days. The soiling ratio is around 1% across all systems [11], keeping in mind that the systems are not cleaned as we want to see natural aging effects without human intervention.

According to the IEA PVPS Task 13, the most reliable way to calculate PLR is to use several different approaches to calculate individual PLR values, filter the results based on a 95% confidence interval and average the ones remaining after filtering. Thereby, a combination of five different filters, the daily and monthly aggregation of three performance metrics and the application of three time series decomposition and statistical model calculation combinations resulted in 90 individual PLR values, which were filtered and consequently averaged to calculate the final PLR of each individual PV system, given in units of %/a, or %/year.

Additionally, a simplified linear PLR calculation methodology is applied on I_{MPP} and V_{MPP} values to detect if the overall PLR is driven by losses in one or the other parameter [12].

3.2 Nonlinear Performance Loss Rate Calculations

The performance evolution of a PV system and consequently the PLR is commonly assumed to be linear, and a single PLR value represents the entire lifetime. This representation is useful as a first assessment of the health status of a PV plant. Nevertheless, detailed performance patterns remain hidden because the performance decline of PV systems is usually highly nonlinear [13, 14, 15]. Instead, a nonlinear PLR, described by a nonlinear function, would describe the performance evolution in greater detail but would be poorly interpretable. The approach applied here divides a nonlinear performance trend into multiple subsets, which are evaluated independently in a linear fashion. The algorithm provides thereby several PLR values for one PV system, which are divided by breakpoints. The so-called multistep PLR (MS-PLR) calculation algorithm used in this study was developed by Lindig et al. [16].

The calculation pipeline follows a similar pattern compared to the one in Section 3.1 of input data assessment, data filtering, performance metric selection, data aggregation, and the application of a statistical model to calculate the multistep PLR. Thereby, the multistep performance loss model contains three steps: performance metric time series (monthly temperature corrected Performance Ratio ($PR_{T_{corr}}$)) decomposition [17], determination of breakpoints [18], and construction of final multistep performance loss model. The results are the timestamp of the breakpoints, and the performance loss values between two individual breakpoints in %/a.

3.3 Field measurements

Additionally, apart from the performance loss estimates, on-site measurements have been performed to link the calculation results with degradation modes detected in the field.

I-V curves of the PV systems have been recorded using the HT Solar I-V500w solar tracer. The curves have been compared to the initial curves at measurement and Standard Test Conditions (STC - T_{STC} = 25°C, G_{STC} =

1000 W/m², air mass of 1.5 [19]). The initial curves were modelled according to De Soto et al. [20] and implemented in pvlib [21]. To easier compare the modelled curves with the measured ones, all values have been normalized to 1. This has been done according to:

$$I_{norm} = \frac{I_{meas}(I_{mod})}{I_{SC} * N_{parallel}} \quad (2)$$

$$V_{norm} = \frac{V_{meas}(V_{mod})}{V_{OC} * N_{series}} \quad (3)$$

Here, I_{norm} and V_{norm} are the normalized current and voltage values. The numerator in the equations is the measured or modelled current (equation 2) and voltage (equation 3), I_{SC} the short circuit current, $N_{parallel}$ the number of modules connected in parallel, V_{OC} the open circuit voltage and N_{series} the number of modules in series. The I_{SC} is the point in the diagram where the curve crosses the y axis (zero voltage) and in turn the V_{OC} is the point where the curve crosses the x axis (zero current).

Infrared (IR) images are taken with a Fluke 435 Series II infrared camera for a qualitative interpretation. The emissivity of the PV module front glass was set to 0.9. The minimum irradiance threshold set by the IEC 62446-3 [22] was respected.

Additionally, daylight photoluminescence (dPL) inspection was performed on both crystalline silicon PV systems. dPL uses the sunlight as excitation source resulting in emission of near-infrared radiation from the (silicon) solar cells, while the module is in open circuit. The luminescence signal is detected by subtracting high current images from the open circuit images. Therefore, dPL images were recorded as described by Koester et al. [23], using combiner box switches to alter the PV modules operation point, combined with batched image acquisition.

4 RESULTS

TABLE II summarizes the calculated linear PLR, losses in I_{MPP} and V_{MPP} as well as the resulting MS-PLR for all systems. The intercept shown in the table refers to the MS-PLR intercept with the y-axis, which expresses the temperature corrected performance ratio (PR_{Tcorr}) value at the beginning of the lifetime of the plants. Additionally, Figure 2 shows the MS-PLR evolution of the system performance. The images of the on-site inspection and corresponding data are stored in the Appendix. When looking at the results in TABLE II, it is visible that the overall PLR is lowest for crystalline silicon systems and higher with increasing order for 1j-a-Si, CdTe and CIS.

The PLR of the crystalline systems roots back primarily to a loss in MPP current while both, MPP current and voltage, are affected considering the thin-film systems under investigation. When looking at the multistep performance loss evolution, it is visible that the performance of the thin-film systems decreases throughout the time of observation, whereas the performance of the two crystalline systems shows an apparent "improvement" at the beginning and end of the time series. It is suspected that the initial gain is based on a combination of PV technology behaviour and favourable weather conditions. In the summer of 2013 and the subsequent winter high temperature corrected PR values were recorded before a measurable reduction in performance was detected. As efficiency of crystalline systems decreases under low irradiance conditions due to higher related resistance losses, high average irradiance conditions under constant temperature yields high efficiency values and, therefore, an increase in performance [24, 25]. These high performance values can be connected to a very sunny July, in which the highest monthly yields were recorded, followed by a mild and sunny winter where the PR reaches very high values of almost 100%. A similar reason is expected for the positive curve gradient at the end of the time series. The observation leads to the assumption that the usage of a temperature corrected PR does not entirely exclude weather dependent effects.

According to Jordan et al. [26], degradation in older crystalline PV systems in moderate climates is dominated by encapsulant discoloration, delamination, cell cracks, hot spots and circuity discoloration, the latter caused by corrosion. Thin-film systems are more likely to be subject to glass breakage and absorber or transparent conductor oxidation. In the following, the systems are discussed individually regarding the detected DM.

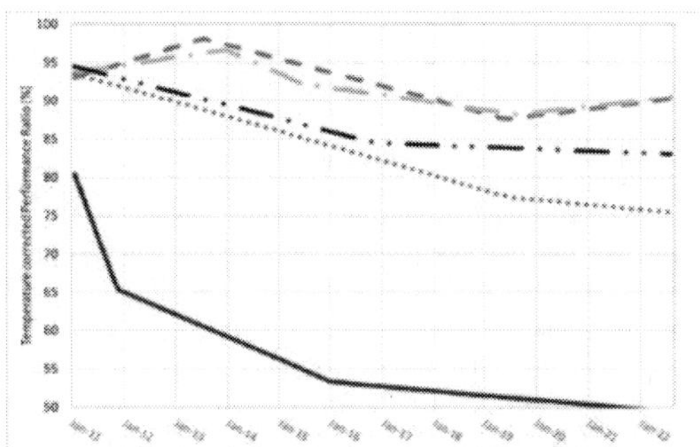

Figure 2: Multistep performance loss rate time series

TABLE II: Intercept of PR at beginning of operation, linear system performance loss rate, performance loss rate in MPP current & MPP voltage, multistep performance loss rate values with corresponding breakpoints.

	Intercept	PLR_{lin}	PLR I_{MPP}	PLR V_{MPP}	PLR_1		PLR_2		PLR_3		PLR_4
	[%]	[%/a]	[%/a]	[%/a]	[%/a]	BP1	[%/a]	BP2	[%/a]	BP3	[%/a]
c-Si	93.1	-0.77	-0.88	-0.28	2.05	07/13	-1.79	06/19	0.85		
pc-Si	93.7	-0.65	-0.93	0.06	1.02	12/13	-2.94	06/15	-0.97	08/19	0.77
1j-a-Si	94.5	-1.06	-0.76	-0.51	-1.73	11/16	-0.25				
CIS	80.4	-2.43	-0.82	-2.44	-18.59	11/11	-2.94	01/16	-0.59		
CdTe	93.5	-1.67	-1.31	-0.66	-1.92	07/19	-0.63				

PLR: Performance Loss Rate; BP: breakpoint

c-Si

The monocrystalline silicon/c-Si system has a moderate to high linear PLR with -0.77%/a. The recorded I-V curve of the entire c-Si system (see green line in Figure 5) shows a similar, albeit decreased, trend compared to the initial curve at measurement conditions. Therefore, we cannot identify any extraordinary performance loss effects, instead a rather balanced overall decrease in the I-V curve parameter. Possibly, a slight decrease in the shunt resistance can be assumed due to a slightly lifted gradient of the curve close to the I_{SC}. Furthermore, the IR image (Figure 3) as well as the PL image (Figure 4) shows that one cell string in one module in the upper part of the system is bypassed. The effect is also clearly visible in the recorded I-V curve of that module (see red line in Figure 3). It is visible that this particular module has a lower overall performance, especially at higher voltage. The upturn of the curve close to the V_{OC} might be caused by the bypass diode allowing the current to flow at low current conditions.

pc-Si

Similar to the c-Si system, the pc-Si system under investigation shows elevated but not extremely high linear PLR values and no degradation of the V_{MPP}. Visual inspections (as shown in Figure 6) revealed the presence of snail trails. It has been observed that this effect spread across many cells of the system and they in turn create visible cell cracks, burn marks across the contact fingers and corrosion of the contact fingers close to the affected areas. An example of the cell cracks leading to disconnected cell parts and hot cells can be seen in the IR and PL images (Figure 7 and Figure 8). Through regular site visits, it can be confirmed that this effect has worsened over time. In general, developing snail trails have been observed in several poly- and mono-crystalline systems installed in the late 2000's and early 2010's. While this developing degradation mode has a major visual impact, and triggers the appearance of other DM, it does not seem to have a strong impact on performance over the period of operation and until now. Figure 9 shows the recorded I-V curve of the plant compared to the initial curve at measurement conditions and the I-V curve of the module, which is affected most by the DM listed above. While we can detect a balanced decrease of all I-V curve parameters, the main difference between the curves for the system and the individual module is a higher shunt resistance, expressed through a steeper gradient close to I_{SC}. This type of progressing degradation should be checked and observed regularly to ensure that no safety related issues or an increasing performance loss appear.

1j-a-Si

The PLR of system 1j-a-Si is with -1.06%/a the lowest among the three investigated thin-film systems, but still above the manufacturer's performance guarantee of a three percent loss in the first year followed by an annual reduction of 0.5%/a for the remaining part of the 25 year performance guarantee. The system experiences balanced losses between I_{MPP} and V_{MPP}. When looking at the MS-PLR it is visible that the system experienced higher performance losses of -1.73%/a for the first six years of operation which smoothed out in the last six years with a value of -0.25%/a. The shape of the I-V curve (see Figure 12) variates from the initial curve only close to the V_{OC}, presenting a stronger decrease at low current conditions.

CIS

The CIS system experiences the highest performance losses among all systems and technologies investigated in this study, with a linear PLR of -2.43%/a. Additionally, the initial performance is already the lowest with a starting value of 80.4%, whereas the average of the other systems is 93.7%. Looking at the MS-PLR, it is visible that a strong, continuous, performance loss appeared very early under operation with a PLR_1 value of -18.59%/a for the first year. After, the performance loss settled at -2.94%/a for four years before it further decreased to -0.59%/a. Looking at the losses in current and voltage in TABLE II as well as the recorded I-V curve in Figure 15, it is visible that the overall performance loss is primarily driven due to a loss in voltage (both V_{OC} and V_{MPP}). Additionally, the curve has a pronounced gradient close to the I_{SC}, suggesting considerable shunting losses. It is expected that the high voltage losses are connected to a degradation of the active PV material (absorber layer). Figure 14 shows an EL image of one CIS module under initial conditions, and one taken after 10 years of operation. It is visible that the EL signal decreased considerably over time, resulting in a very dark image. The EL signal intensity is directly proportional to the module voltage and thereby directly affected [27].

CdTe

The CdTe system is part of a larger commercial PV plant installed next to the experimental PV plant. With 90 modules, it is the largest system under consideration. The overall linear PLR is with -1.67%/a quite high. The modules installed are the second generation of First Solar CdTe modules, which have been known to have stability issues. In contrast, the manufacturer offers warranties with the lowest annual degradation rate on the market with a value of -0.2%/a for its current 6th generation of CdTe modules [28]. Based on that, it seems that the material dependent stability issues have been resolved. According to the MS-PLR results, the temporal performance loss is quite homogenous, with a decreasing value from 2019 onwards. Looking at MPP current and voltage, it is visible that larger losses are observed for the I_{MPP} by a 2:1 ratio. Instead, when looking at the recorded I-V curves (see Figure 17), it is visible that the V_{OC} decreased significantly while the I_{SC} is still at initial conditions. In the graph, three recorded I-V curves are shown, one for each string in the system. Each string consists of five parallel circuits of each six modules connected in series. While two curves are very similar, the third one deviates notably with a lower V_{OC} and a higher gradient close to the I_{SC}, indicating large shunting losses (similar to the CIS system). These shunting losses are triggered by two broken modules within that string, visible in Figure 16. The modules are glass-glass modules, and it seems that the mounting clamps apply continuous physical stress to the modules which ultimately resulted in glass cracks and breakage in several modules. Similar problems have been observed for other modules of this plant and for other PV systems with similar setups elsewhere. Additionally, as stated above, glass breakage was found to be one of the dominating degradation modes for thin film PV modules in a study by Jordan et al. [26] underlining the importance of the rigorous application of best practices in PV module installation.

5 SUMMARY & CONCLUSION

In this study, the performance of five PV systems, representing five different PV technologies, two crystalline silicon and three thin-film systems, has been investigated. The systems are part of a larger experimental PV plant operating in Bolzano, Italy for 12 years running. To get a broader perspective on the performance evolution of the systems, linear performance loss rates have been calculated together with nonlinear performance losses as well as losses in maximum power point current and voltage time series. These key performance indicators are very useful to track the overall performance of a PV system. To better understand what triggered certain performance trends, the results were supported by field inspections, including visual inspection, I-V curve acquisition, as well as the application of infrared and luminescence imagery.

In general, the two crystalline silicon-based systems experience a lower overall PLR of around -0.7%/a, compared to the investigated thin-film systems with an average of -1.7%/a (amorphous silicon, copper indium selenide and cadmium telluride). The losses for the crystalline systems stem primarily from a loss in I_{MPP} while a loss, both, in current and voltage, was detected for the thin-film systems.

Crystalline systems show increasing shunting losses due the development of snail trails leading to cell cracks. One module of the crystalline system has an bypassed cell string resulting in an overall lower performance with a great impact on the I-V curve at higher voltage values.

The nonlinear PLR evolution shows that the three thin-film technologies experience an initial steeper performance loss followed by a lower PLR. It is expected that the higher recorded losses of the thin-film systems are based on a degradation of the active PV material, resulting in a decreasing open circuit voltage. While the copper indium selenide system experienced by far the highest losses with a very high degradation in the first year of operation, the cadmium telluride system with glass-glass modules has two modules which have broken glasses due to the applied physical stress of the clamps.

Studies such as this help to get a better idea of common degradation pathways depending on technology and climate. Thereby, they can provide complementary information to indoor testing where, specific climate stressors and their effects are investigated. The connection between observed performance losses and degradation will enrich the common knowledge about the occurrence and severity of degradation modes and other technical risks in specific climates, which is important for both the development of new products and a better estimation of lifetime yield assessments.

ACKNOLEDGEMENTS

The authors would like to thank Julián Ascencio-Vásquez for his support in imputing irradiance data.

REFERENCES

[1] S. Lindig et al., "International collaboration framework for the calculation of performance loss rates: data quality, benchmarks and trends (towards a uniform methodology)," *Under Review*, 2020.

[2] M. Köntges, G. Oreski, M. Herz, U. Jahn, P. Hacke and K.-A. Weiß, "Assessment of Photovoltaic Module Failures in the Field," IEA PVPS Task 13, Subtask 3 Report IEA-PVPS T13-09:2017, 2017.

[3] M. Owen-Bellini, P. S. S. Hacke, D. Sulas, N. Hannah, D. C. Miller and M. D. Kempe, "Combined-Accelerated Stress Testing for Advanced Reliability Assessment of Photovoltaic Modules," National Renewable Energy Lab. (NREL), Golden, CO (United States), 2020.

[4] M. Kottek, J, Grieser, C. Beck, B., Rudolf and F. Rubel, "World Map of the Köppen-Geiger climate classification updated," *Meteorol. Z.*, pp. 259-263, 2006.

[5] J. Ascencio-Vásquez, K. Brecl and M. Topič, "Methodology of Köppen-Geiger-Photovoltaic climate classification and implications to worldwide mapping of PV system performance,," *Solar Energy*, vol. 191, p. 672–685, 10 2019.

[6] International Electrotechnical Commission, "IEC 61724-1:2021 Photovoltaic system performance - Part 1: Monitoring," International Electrotechnical Commission, Geneva, Switzerland, 2021.

[7] P. Geurts, D. Ernst and L. Wehenkel, "Extremely Randomized Trees," *Machine Learning*, vol. 63, pp. 3-42, 2006.

[8] F. Pedregosa, G. Varoquaux, A. Gramfort, V. Michel, B. Thirion, O. Grisel, M. Blondel, A. Müller, J. Nothman, G. Louppe, P. Prettenhofer, R. Weiss, V. Dubourg, J. Vanderplas, A. Passos, B. M. Cournapeau, M. Perrot and E. Duchesnay, "Scikit-learn: Machine Learning in Python," *J. Mach. Learn. Res.*, vol. 12, p. 2825.2830, 2011.

[9] J. A. Kratochvil, W. E. Boyson and D. L. King, "Photovoltaic Array Performance," Sandia Report, United States, 2004.

[10] R. H. French, L. S. Bruckman, D. Moser, S. Lindig, M. van Iseghem, B. Müller, J. S. Stein, M. Richter, M. Herz, W. van Sark, F. Baumgartner, J. Ascencio-Vásquez, D. Bertani, G. Maugeri, A. J. Curran, K. Rath, J. Liu and A. Khalilnejad, "Assessment of Performance Loss Rate of PV Power Systems - IEA PVPS T13-22:2021," IEA PVPS Task 13, 2021.

[11] M. G. Deceglie, L. Micheli and M. Muller, "Quantifying Soiling Loss Directly From PV Yield,," *IEEE Journal of Photovoltaics*, vol. 8, no. 2, pp. 547-551, 2018.

[12] S. Lindig, P. Ingenhoven, G. Belluardo, D. Moser and M. Topic, "Evaluation of Technology-Dependent Maximum Power Point Current and Voltage Degradation in a Temperate Climate," in *35th EU PVSEC*, Brussels, 2018.

[13] M. Köntges et al., "Review of failures of photovoltaic modules - IEA PVPS T13-03:2014," IEA PVPS Task 13, 2014.

[14] A. Virtuani et al., "35 years of photovoltaics: Analysis of the TISO-10-," *Prog Photovolt,* vol. 27, no. 4, pp. 328-339, 2019.

[15] D. C. Jordan, T. J. Silverman, B. Sekulic and S. R. Kurtz, "PV degradation curves: non-linearities and failure modes," *Progress in photovoltaics, Research and application,* vol. 25, no. 7, pp. 583-591, 2017.

[16] S. Lindig, A. Louwen, D. Moser and M. Topic, "New PV Performance Loss Methodology applying a self-regulated Multistep Algorithm," *IEEE Journal of Photovoltaics,* vol. 11, no. 4, pp. 1087-1096, 2021.

[17] R. B. Cleveland, W. S. Cleveland, J. E. McRae and I. Terpenning, "STL: A Seasonal-Trend Decomposition Procedure Based on LOESS," *Journal of Official Statistics,* vol. 6, no. 1, pp. 3-33, 1990.

[18] V. Muggeo, "segmented: an R Package to Fit Regression Models with Broken-Line Relationships," 2008. package version 3.6.1. [Online]. Available: https://cran.r-project.org/doc/Rnews/.

[19] International Electrotechnical Commission, "IEC60904-3:2016 Photovoltaic Devices—Part 3: Measurement Principles for Terrestrial Photovoltaic (PV) Solar Devices," International Electrotechnical Commission, Geneva, Switzerland, 2016.

[20] W. De Soto, S. A. Klein and W. A. Beckman, "Improvement and validation of a model for photovoltaic array performance," *Solar Energy,* vol. 80, pp. 78-88, 2006.

[21] W. F. Holmgren, R. W. Andrews, A. T. Lorenzo and J. S. Stein, "PVLIB Python 2015," in *42nd Photovoltaic Specialists Conference*, New Orleans, 2015.

[22] "IEC62446-3:2017 Photovoltaic (PV) systems - Requirements for testing, documentation and maintenance - Part 3: Photovoltaic modules and plants - Outdoor infrared thermography," International Electrotechnical Comission, Geneva, CH, 2017.

[23] L. Koester, A. Astigarraga, S. Lindig and D. Moser, "Development of Daylight Photoluminescence Technique for Photovoltaic Modules and Investigation of Temperature Dependency," in *37th EU-PVSEC*, Online, 2020.

[24] F. Mavromatakis, F. Vignola and B. Marion, "Low irradiance losses of photovoltaic modules," *Solar Energy,* vol. 157, pp. 496-506, 2027.

[25] I. de la Parra, M. Muñoz, E. Pigueiras, M. García, J. Marcos and F. Martínez-Moreno, "PV performance modelling: A review in the light of quality assurance for large PV plants," *Renewable Sustain. Energy Rev.,* vol. 78, pp. 780-797, 2017.

[26] D. C. Jordan, T. J. Silverman, J. Wohlgemuth, S. R. Kurtz and K. VanSant, "Photovoltaic failure and degradation modes," *Progress in photovoltaics, Research and application,* vol. 25, pp. 318-326, 2017.

[27] C. Han, "Analysis of moisture-induced degradation of thin-film photovoltaic module," *Solar Energy Materials and Solar Cells,* vol. 210, 2020.

[28] T. Lombardo, "engineering.com - First Solar Achieves Industry's Best PV Degradation Rate," 03 05 2021. [Online]. Available: https://www.engineering.com/story/first-solar-achieves-industrys-best-pv-degradation-rate. [Accessed 09 09 2022].

APPENDIX

c-Si

Figure 3: Infrared image c-Si system

Figure 4: Photoluminescence image c-Si system

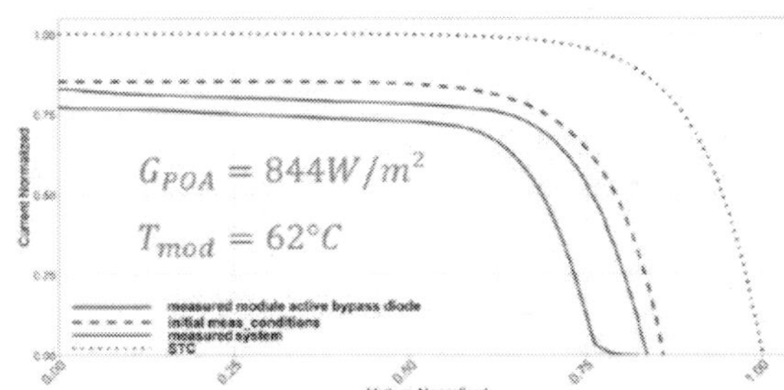

Figure 5: I-V curve c-Si system, Curves - black dotted: modelled STC, dark-red dotted: modelled at measurement conditions, green: measured system, red: measured module with active bypass diode

pc-Si

Figure 6: Image pc-Si module

Figure 7: Infrared image pc-Si module

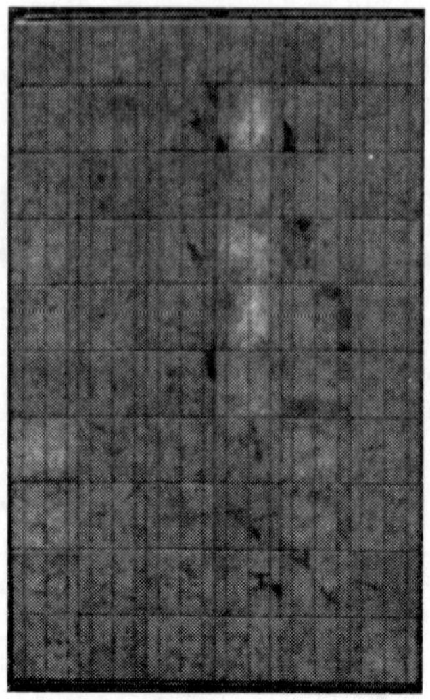

Figure 8: Photoluminescence image pc-Si module

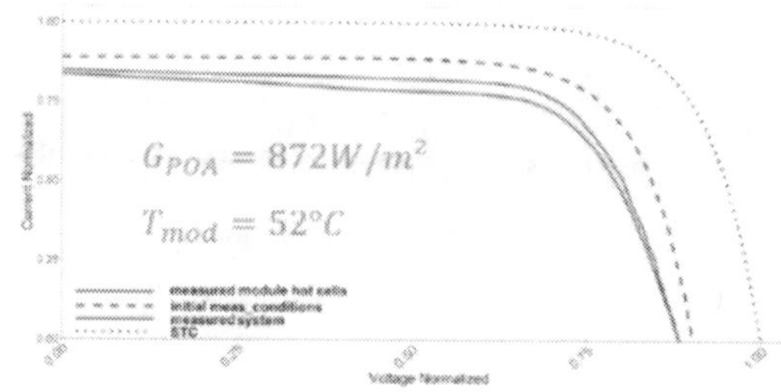

Figure 9: I-V curve pc-Si system, Curves - black dotted: modelled STC, dark-red dotted: modelled at measurement conditions, green: measured system, red: measured module with strong development of snail trails

1j-a-Si

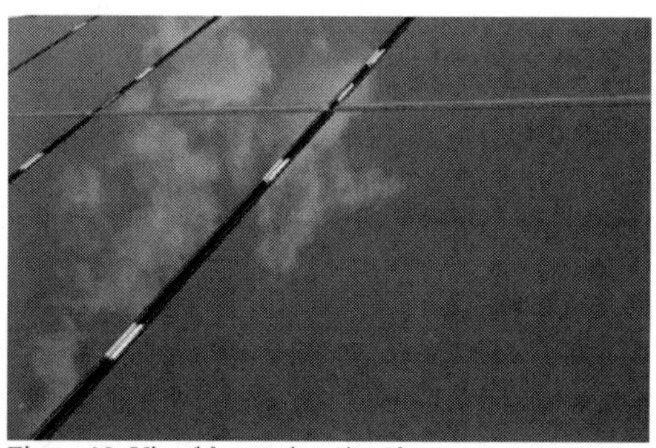

Figure 10: Visual inspection 1j-a-Si system

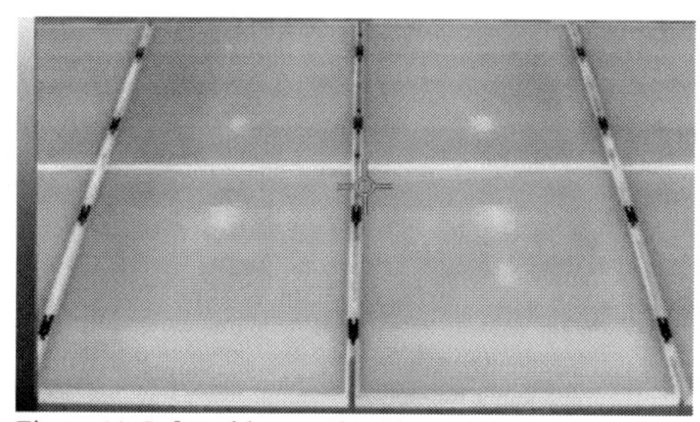

Figure 11: Infrared image 1j-a-Si system

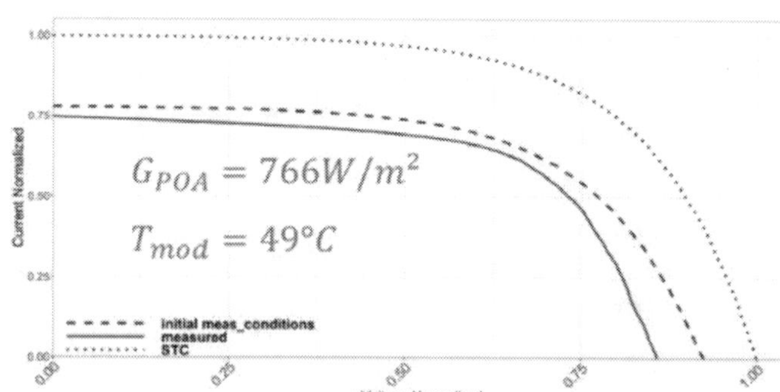

Figure 12: I-V curve 1j-a-Si system, Curves - black dotted: modelled STC, dark-red dotted: modelled at measurement conditions, green: measured system

CIS

Figure 13: Infrared image CIS system

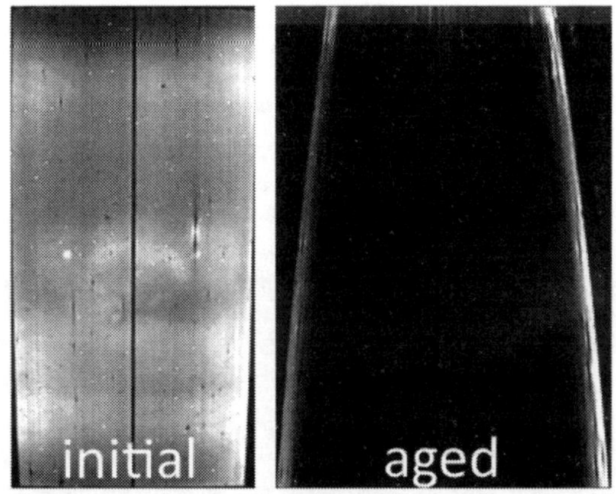

Figure 14: EL image CIS module; left - module under initial conditions; right - aged module

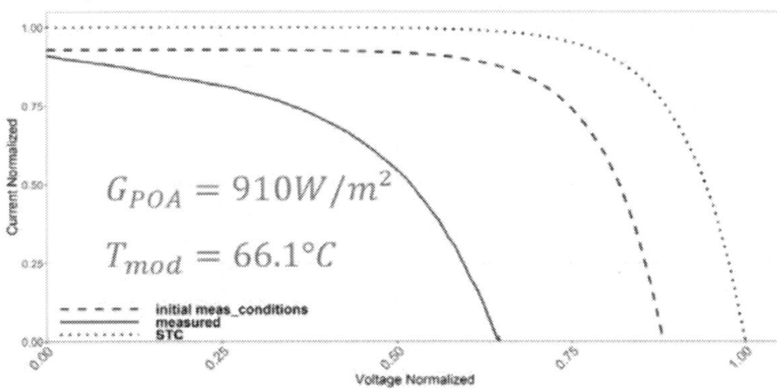

Figure 15: I-V curve CIS system, Curves - black dotted: modelled STC, dark-red dotted: modelled at measurement conditions, green: measured system

CdTe

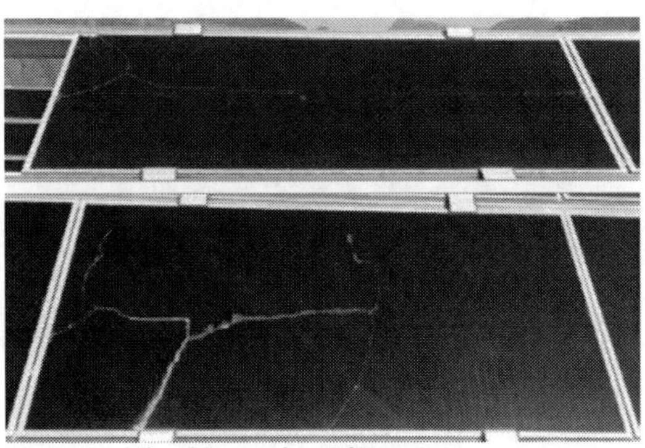

Figure 16: Visual inspection CdTe system

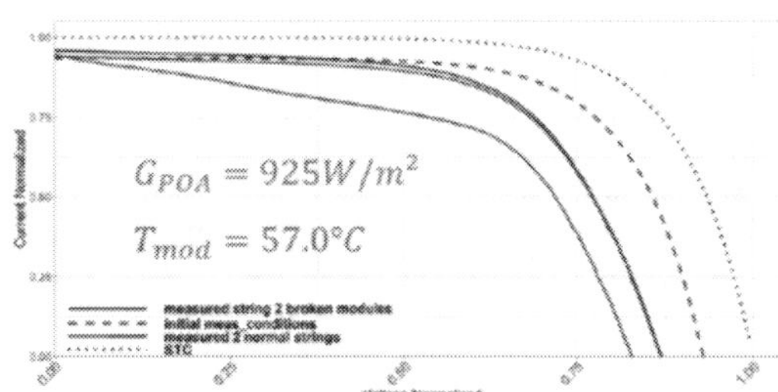

Figure 17: I-V curve CIS system, Curves - black dotted: modelled STC, dark-red dotted: modelled at measurement conditions, blue, green and red curves are individual module strings of 30 modules each

A STUDY OF MODULE DEGRADATION FROM A 10-YEAR INSTALLATION IN THE URBAN TROPICAL ENVIRONMENT

Mauro Pravettoni[1], Mike Mahesh S/O Sundar Das[1], Amit Singh Rajput[1], Wan Li[2], and Selvam Valliappan[3]
[1]National University of Singapore, Solar Energy Research Institute of Singapore, 7 Engineering Dr 1, Singapore 117574
[2]Vidyasirimedhi Institute of Science and Technology, Rayong, Thailand
[3]Green Building Technology, Environmental Sustainability Group, Building Construction Authority (BCA), Singapore

ABSTRACT: The solar photovoltaic installation capacity in Southeast Asia is growing widely in the region at the fastest pace among other renewables. Today's market requires performance warranties that are decades long, and although much effort is put nowadays into module qualification and study of new and ever more aggressive degradation mechanisms, fewer studies are available from the field experience in the tropical climate compared to the continental one. This work aims to present one of the first comprehensive reliability studies of a statistically relevant sampling of PV modules after 10 years of installation in the Tropical region. Degradation modes are listed and analysed, and their occurrence is correlated to power degradation. A total of 72 modules were sampled from the original plant of 750 modules in one of Singapore's central districts. Soiling was observed to be one of the sources of power degradation: even self-cleaning from the frequent tropical rains is ineffective and cementation takes place. All modules have been analysed and various degradation modes are observed. Their correlation with the observed deviations from the nameplate of the electrical parameters is presented.
Keywords: Degradation, Lifetime, Optical Losses, Performance, Reliability.

1 INTRODUCTION

Southeast Asia accounts for a 680 million population, almost one-tenth of the world, nearly half of which live in urban areas. Hosting most of the fastest-growing economies, it has become a major manufacturing and trading hub. The energy demand is subsequently growing and expected to grow further, as still around 10% of the region's population has no access to electricity and larger numbers lack access to modern cooking fuels [1]. Ensuring an affordable and environmentally sustainable supply of energy for businesses and for all citizens is a key objective for the region's governments.

The solar photovoltaic (PV) installation capacity is growing widely in the region at the fastest pace among renewables. Solar accounts for about 60% of Southeast Asia's installed capacity of non-hydro renewable electricity, the areas that receive the strongest solar intensity being concentrated in Thailand, Myanmar, Cambodia, and Vietnam. But the whole region offers a variety of installation potentials, from the conventional rooftop and utility-scale plants to building-integrated PV (BIPV), floating PV (both on reservoirs and offshore), agrivoltaics, and in the future also likely vehicle integrated PV (VIPV). However, reliability concerns arise from the hot and humid tropical climate of a large part of the region and from soiling, of either natural or anthropic origin [2].

Today's market requires performance warranties that are decades-long [3]. Recently, Maxeon Solar (the Singaporean head-quartered spin-off of SunPower) announced that it will offer a world-record 40-year performance warranty for its Interdigitated Back Contact (IBC) product line. However, recent works have shown how midlife failures may be worryingly frequent in tropical regions [4]. Most of the reliability studies in the literature focus on continental climate, while data and statistics from the tropical hot and humid climates are still poorly available. The combination of tropical climate and the urban environment of Southeast Asia megacities is the best testbed for challenging a performance warranty of just more than 20 years. Studying the performance and history of degradation mechanisms in the most aggressive environmental conditions for PV performance is therefore becoming essential to allow achieving outstanding reliability targets.

This work aims at representing one of the first comprehensive reliability studies of a statistically relevant sampling of PV modules after 10 years of installation in the tropical region. Degradation modes are listed and analysed, and their occurrence is correlated to the power degradation measured at SERIS' ISO/IEC 17025 accredited testing laboratory.

2 THE CASE-STUDY: ZERO ENERGY BUILDING OF THE BCA ACADEMY IN SINGAPORE

2.1 Description of the case-study

The modules analysed in this work belong to a system of 750 modules decommissioned from the roof of the Zero Energy Building (ZEB, [5]) of the Building Construction Authority (BCA) Academy in Singapore (Fig. 1, total capacity: 142.5 kWp; Module type: multi-crystalline Si; year of installation: 2009).

Figure 1: The case-study: the roof of the Zero Energy Building (ZEB) of the Building Construction Authority (BCA) Academy in Singapore [5].

The first 18 months of operation were analysed and published [6], showing an initial good performance ratio of 0.81 and a final yield of 3.12 h/d averaged over all 22 arrays. These values were fluctuating very little over the year, due to the limited weather seasonality of the region. The system was decommissioned in 2019, after 10 years of operation.

2.2 Sampling and module's specifications

A total of 72 PV modules were sampled based on the string performance from the decommissioned original plant of 750 modules. Fig. 2 indicates the location and number of the 72 sampled modules; five different locations were selected to map the best, the worst and the average performing strings, with different tilts and orientations.

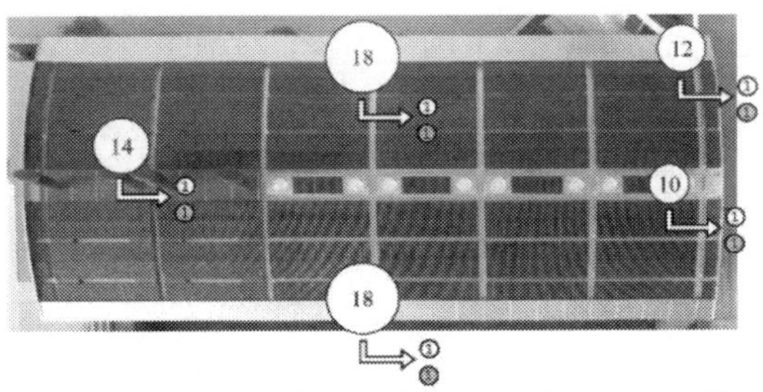

Figure 2: Sampling of 72 modules from 5 locations. The locations were selected to map the best, worst and average performing strings and different tilts and orientations of the installed modules. The small circles indicate the subsampling of 2 modules, the visually cleanest and dirtiest, per location.

Two modules per location where subsampled (the cleanest and the dirtiest at visual inspection, respectively): these were analysed in detail for performance degradation due to soiling.

The modules were manufactured by one of the top ten module manufacturers of polycrystalline Silicon (poly-Si) technology of the time [7]; they have 1658 × 834 mm size, 50 cells (M0): PET backsheet; nominal performance: P_{max} = 190 W, I_{SC} = 8.23 A, V_{OC} = 30.8 V. The technical information available at the time of commercialization reported "unknown potential induced degradation (PID)" risk.

3 DEGRADATION ANALYSIS

3.1 Visual inspection

Fig. 3 lists the frequency of occurrence of the most important degradation mechanisms observed on the tested modules.

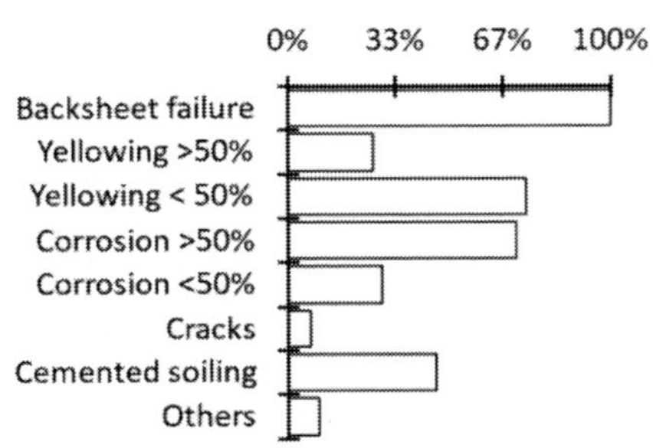

Figure 3: Results of visual inspection: frequency of occurrence of the various degradation mechanisms observed.

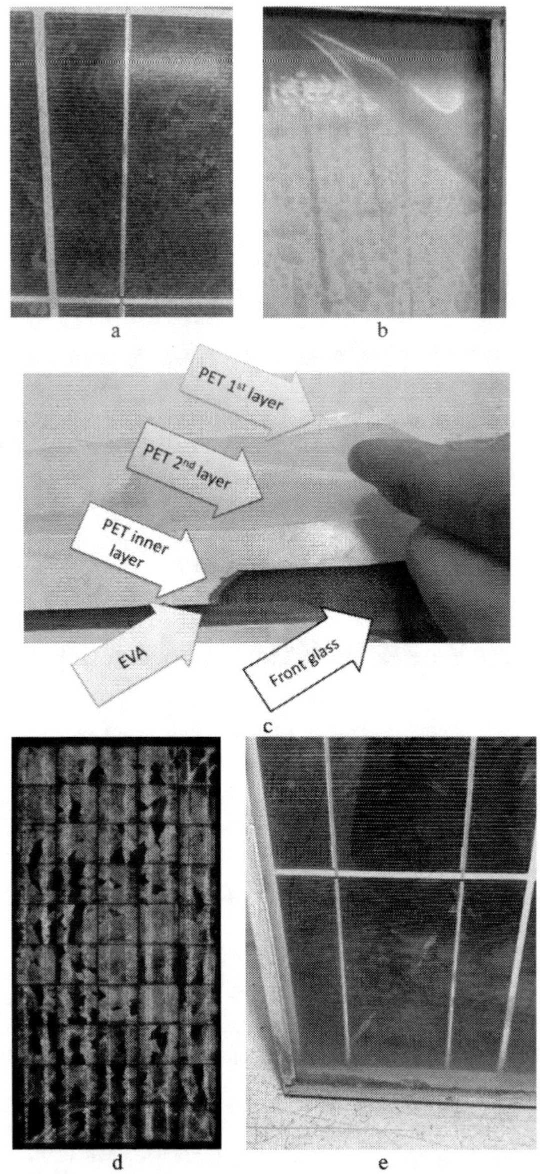

Figure 4: Examples of the most frequent degradation mechanisms observed: a) Yellowing of the backsheet; b) Backsheet failure, possibly causing moisture ingress and corrosion; c) The subsequent delamination of the three PET layers; d) Microcracks causing the electrical isolation of parts of the cells; e) Cemented soiling.

Yellowing (Fig. 4a) is clearly visible in all modules: some modules (26%) the backsheet and encapsulant are yellowed almost over the entire module, while the remaining modules have only portions that are visually yellowed. The effect of yellowing is analysed also in the transmittance results of section 3.4.

Backsheet failures (Fig. 4b) resulted in all 72 modules in the delamination of the backsheet, with formation of bubbles and in most cases with severe detachment of the 1st layer of PET (Fig. 4c). These failures are the possible cause of moisture ingress [8] which determine extensive corrosions in the metallization in all modules (in 71% of the cases, distributed over the entire module).

Cracks and microcracks causing the electrical isolation of parts of the cells (Fig. 4d) were observed on few modules (7%), while cemented soiling (Fig. 4e) was overall observed in nearly half of the cases (46%).

Other failures observed include delamination of the front encapsulant with the formation of bubbles, junction box and connector failures.

3.2 Electroluminescence

Fig. 5 shows the electroluminescence (EL) images of the best performing module (Fig. 5a), of the worst performing module (Fig. 5b) and of the average case (Fig. 5c).

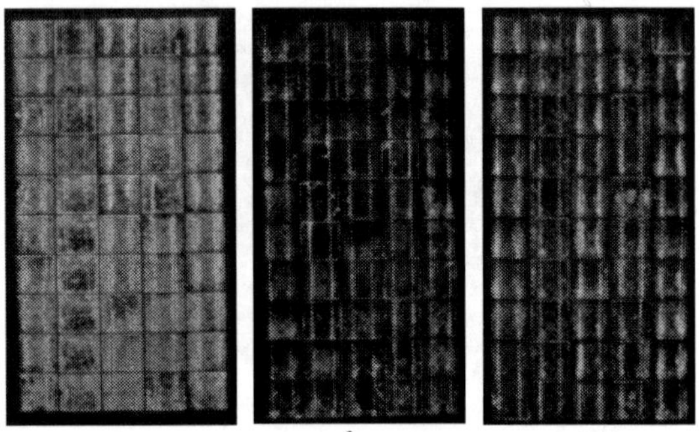

a b c

Figure 5: Electroluminescence imaging: a) The best performing module; b) The worst performing module; c) The average case.

All EL images highlighted an advanced status of corrosion of the metalization, and effects of acetic acid corrosion. On the contrary, the typical pattern of PID [9] was not observed on any module.

3.3 Soiling analysis

Soiling was analysed in detail in a subsampling of 10 modules and reported in [2]. There, it was observed that tropical storms were generally only able to partially remove soiling in the dirtiest modules (up to 5%), while manual cleaning with detergents can achieve up to 10% short-circuit current increase.

The presence of metals (Fe, Al, Zr, etc.), alongside carbon, silicon and oxygen was observed: these are elements that are commonly present in soil [10-11]. It is remarkable, however, the absence of nitrogen and poor presence of sulphur, suggesting that the transport pollution is well confined in the urban environment of Singapore.

The Thermo-Gravimetric Analysis (TGA) highlighted the presence of smoke points between 200 and 300°C, which may signify the presence of oil or grease. This may be consistent with the frequent cases of haze events occurring in the area, mainly caused by the burning of crops in the Borneo Forest and in the surrounding region.

3.4 Transmittance losses

The transmittance of the front glass and of the front EVA was measured on a module with average degradation (M61, relative deviations from the nameplate values: $\Delta I_{SC} = -16\%$, $\Delta V_{OC} = -2\%$, $\Delta FF = -54\%$, $\Delta P_{max} = -62\%$). The result is shown in Fig. 6, which highlights the yellowing of the EVA (caused by the prolonged UV exposure) and of the front glass (caused mainly by soiling). The chart shows also the transmittance of a non-significantly yellowed portion of the EVA: yellowing is shown to cause up to 20% absolute loss in transmittance in the visible, thus expecting to affect significantly the measured I_{SC}.

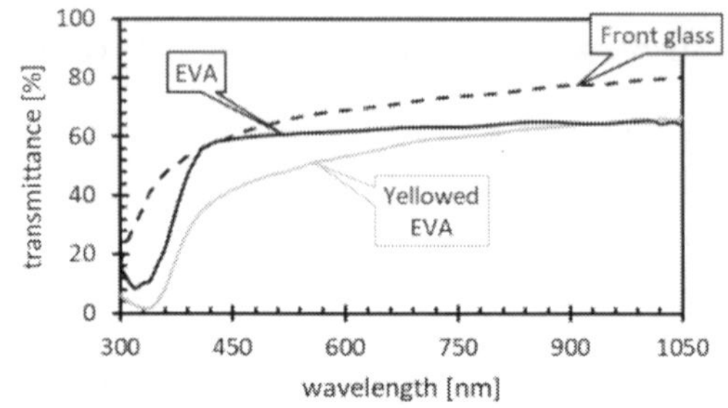

Figure 6: Spectral transmittance measured on an averagely degraded module (module M61: I_{SC} deviation from the nameplate: −16%). Yellowness of the EVA (likely caused by UV exposure) and of the front glass (likely caused by soiling) are shown.

3.4 Performance losses

The electrical performance of the 72 sampled modules was measured via current-voltage (I-V) characterization at Standard Test Conditions (STC). Measurements were performed at SERIS' ISO/IEC 17025 accredited laboratory for PV module testing, on an A+A+A+ solar simulator. Each module was gently removed of the dust before measurements, but no additional detergent was used to remove more persistent soiling, apart from the ten subsampling of section 3.3.

Fig. 7a compares the I-V curve of the best performing and of the worst performing module with the theoretical I-V curve of the module at the time of initial installation ("nameplate"), as a benchmark for the degradation analysis. In the absence of a control module, this I-V curve (the dashed line in Fig. 7a) has been reconstructed based on the nameplate values of I_{SC}, V_{OC} and P_{max}. The best performing module shows a remarkable –4% loss in I_{SC}, no change in V_{OC} and −18% loss in FF, for a total –22% power loss, corresponding to a degradation rate of –2.2%/yr.

. The worst performing module shows the most severe degradation in FF (−58%) and I_{SC} (−52%), a marginal degradation in V_{OC} (−3%) and −81% power degradation (−8.1%/yr).

In general (Fig. 7b), performance degradation of all parameters is in good approximation normally distributed, with an average degradation rate of approximately –2%/yr in I_{SC}, no significant degradation in V_{OC}, ~–4.5%/yr degradation in FF and –5.5%/yr maximum power loss. The power degradation seems therefore to be attributed to yellowing and soiling losses (in minimum part) and to corrosion (in larger part). That corrosion is the larger contribution to maximum power degradation is also confirmed by Fig. 7c, which shows the trend of the power degradation rate as a function of the series resistance (obtained directly from I-V curve measurements).

In summary, not a single module of the 72 sampled satisfy the threshold of 10% maximum degradation after 10 year of operation (corresponding to −1%/yr, in linear approximation).

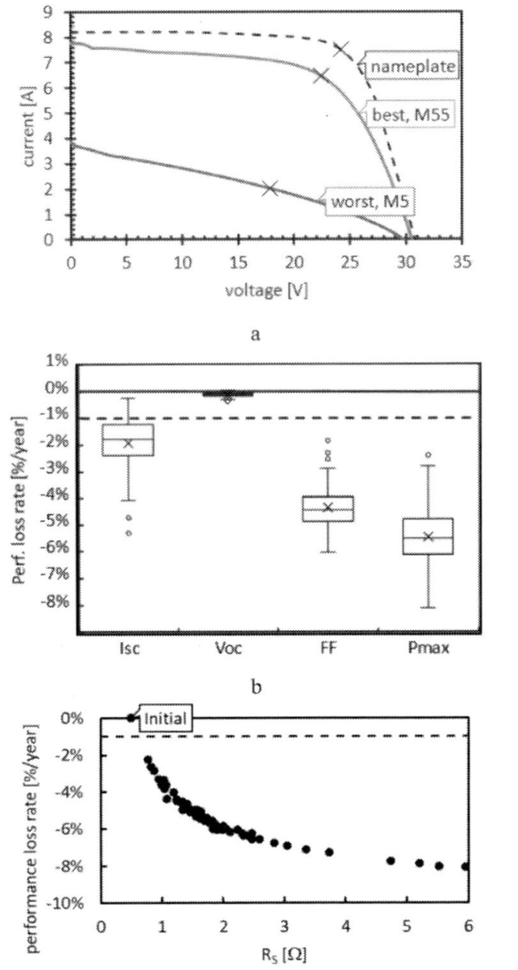

Figure 7: Performance losses: a) Current-voltage (I-V) characteristics of the best and worst performing modules, compared with the initial I-V; b) Loss rate (in %/year) for I_{SC}, V_{OC}, FF and P_{max}; c) Loss rate (P_{max}) as a function of the series resistance.

4 ADDITIONAL CONSIDERATIONS ON SIMILAR SYSTEMS IN THE ISLAND

Soiling represents a degradation mechanisms that can be particularly severe in Singapore: Fig. 8a shows the status of soiling from a selection of a 230 kWp roof at CleanTech One, Jurong West district. The Suntech modules (210 W, glass-glass, poly-Si) have been installed in 2010 and present cemented soiling which is visually evident even to a distant observer.

To the same building and, installed shortly afterwards, belong the system of REC235PE (BLK) poly-Si, glass-backsheet modules shown in Fig. 8b. In this case, the integrity of the backsheet seems to be preserved, and backsheet failures was not visually detected.

Figure 8: Examples from roof systems in Singapore after ~10 year operations: a) Cemented soiling; b) Integrity of the backsheet.

5 CONCLUSIONS

A sampling of modules decommissioned after 10 years of installation in Singapore (central district) showed severe soiling and corrosion as the main causes of premature failure.

While the frequent tropical storms have shown to only partially remove soiling (up to 5%), manual cleaning with detergents can achieve up to 10% short-circuit current increase in the dirtiest case. The presence of metals (Fe, Al, Zr, etc.), alongside carbon, silicon and oxygen, was reported: the absence of N and only traces of S indicates that transport and industrial pollution cannot be considered among the main causes of soiling. Thermo-gravimetric Analysis (TGA) may also confirm the presence of oil/grease, but natural and anthropogenic origin.

Severe transmittance losses due to soiling and yellowing (both glass and EVA) were reported. Backsheet failure likely causes moisture ingress that accelerated power degradation

No V_{OC} degradation suggests that modules did not suffer PID, which finds confirmation in the EL images that do not show the typical pattern of PID.

Finally, while soiling cementation is generally observed, examples of strong resistance to climate severity are available.

6 ACKNOWLEDGMENTS

SERIS is a research institute at the National University of Singapore (NUS). SERIS is supported by NUS, the National Research Foundation Singapore (NRF), the Energy Market Authority of Singapore (EMA) and the Singapore Economic Development Board (EDB).

7 REFERENCES

[1] "Renewable Energy Market Analysis: South East Asia", IRENA, 2018.

[2] W. Li, A. Singh Rajput, M. Pravettoni, *SiliconPV 2022, The 12th International Conference on Crystalline Silicon Photovoltaics*, in press.

[3] N. Taylor, and A. Jäger-Waldau, "Photovoltaics Technology Development Report 2020", EUR 30504 EN, Publications Office of the European Union, Luxembourg 2020

[4] W. Luo, et al., *Renewable Energy* **177** (2021) 327-335.

[5] S. K. Wittkopf, A. Pandey, A. K. Seng, P. Poh, *Proceedings of the 25th conference on passive and low energy architecture (PLEA)*, Dublin (2008).

[6] S. K. Wittkopf, S. Valliappan, L. Liu, A. K. Seng, S. C. J. Cheng, *Renewable Energy* **47** (2012) 9-20.

[7] www. renewableenergyworld.com.

[8] O. K. Segbefia, A. G. Imenes, T. O. Sætre, *Solar Energy* **224** (2021) 889–906.

[9] W. Luo, Y. S. Khoo, P. Hacke, V. Naumann, D. Lausch, S. P. Harvey, J. P. Singh, J. Chai, Y. Wang, A. G. Aberle, S. Ramakrishna, *Energy Environ. Sci.* **10** (2017) 43-68.

[10] K. K. Ilse, B. W. Figgis, V. Naumann, C. Hagendorf, J. Bagdahn, *Renewable and Sustainable Energy Reviews* **98** (2018) 239–254.

[11] J. Tanesab, D. Parlevliet, J. Whale, T. Urmee, *Renewable Energy* **111** (2017) 105-115.

INVESTIGATING LONG-TERM UV DURABILITY OF GLASS/TRANSPARENT BACKSHEET LAMINATES FOR BIFACIAL PHOTOVOLTAICS – CROSS-SECTIONAL ANALYSIS OF BACKSHEET DEGRADATION

Xiaohong Gu, Soshana Smith, Stefan Mitterhofer, Stephanie Moffitt, Song-Syun Jhang, Stephanie Watson, and Li-Piin Sung
Engineering Laboratory, National Institute of Standards and Technology, Gaithersburg, MD 20899, USA

ABSTRACT: Bifacial modules with glass/transparent backsheet (G/CB) structures offer lighter weight and higher hail resistance over traditional glass/glass (G/G) modules. However, research on long-term durability of transparent backsheets and their usage in bifacial modules is lacking. In this study, the G/CB laminated coupons were constructed with polyolefin elastomer encapsulant (POE) and three types of fluoropolymer-based transparent backsheets. The coupons were aged using an integrating sphere-based weathering device at 65 °C and 50 % relative humidity (RH) for 3600 h, with an approximately 1800 MJ/m^2 UV dose (295 nm to 400 nm), followed by 200 cycles of thermal cycling from 85 °C to -40 °C. Cross-sectional characterizations of optical, chemical, and mechanical degradation were conducted on the transparent backsheets of the exposed coupons using confocal fluorescence microscopy, microscale infrared spectroscopy, and nanoindentation, respectively. All three types of backsheets show one or more degradation modes including yellowing, surface/bulk cracking, or delamination. The greatest yellowing and cracking are seen on a fluoroethylene vinyl ether (FEVE)/polyethylene terephthalate (PET)/ethylene-vinyl acetate (EVA)-based transparent backsheet. The results also show changes in fluorescence intensity are consistent with chemical and mechanical changes across the thickness of the PET core layer in the FEVE/PET/EVA backsheet.
Keywords: glass/transparent backsheet; bifacial modules; accelerated laboratory testing; cross-section; chemical, optical and mechanical properties; fluorescence; cracking; nanoindentation

1 INTRODUCTION

Bifacial modules have increasingly gained market share in recent years due to the growing affordability. According to a report by Wood Mackenzie Power & Renewables [1], the size of the bifacial market is anticipated to increase tenfold from 2019 to 2024. Compared to traditional monofacial modules, one major advantage of bifacial modules is their capability of absorbing light and generating electrical power from both front and rear sides, leading to reduced levelized cost of energy (LCOE) of photovoltaic (PV) generated electricity [2, 3]. The additional power yield of bifacial modules can be as high as 30 % to 40 % by optimization of various configurations [4, 5], such as installation conditions (e.g., height, angle, and direction), ground albedo, tracking, and the bifaciality of the solar cells.

Traditional bifacial modules consisting of a double glass (G/G) structure are resistant to moisture, fire, scratching and sand abrasion. However, with the trend of larger silicon cells, the G/G bifacial modules have become too heavy, making the installation and transportation costly [3, 6]. The elevated operating temperature of G/G bifacial modules can also reduce power generation and increase the degradation rate of modules. By comparison, the glass/transparent backsheet (G/CB) modules have advantages of lighter weight, high light transmittance, smaller heat capacity, higher stain resistance and corrosion resistance, and lower risk for hail [7-9]. Despite these advantages, only a few reports are available on the reliability and long-term performance of transparent backsheets and their bifacial modules [9-11].

Our previous study [11] on the durability of free-standing films of three fluoropolymer-based transparent backsheets after exposure to UV/75 °C/50 % RH showed cracking could occur once transparent backsheets films become brittle due to degradation. Note that local microenvironment (e.g., moisture, oxygen concentration, residual stress/strain, etc.) could be different between a free-standing backsheet and the same backsheet laminated to the glass substrate through an encapsulant. A follow-up study has been conducted based on the glass/polyolefin elastomer (POE)/transparent backsheet coupons. In this study, the specimens were exposed to NIST SPHERE (Simulated Photodegradation via High Energy Radiant Exposure) at 65 °C and 50 % RH for up to 3600 h (1800 MJ/m^2), followed by 200 cycles of thermal cycling. Three types of fluoropolymer-based backsheets, i.e., polyvinyl fluoride (PVF)/polyethylene terephthalate (PET) /fluoroethylene vinyl ether (FEVE), polyvinylidene fluoride (PVDF) /PET/FEVE, and FEVE/ PET/ ethylene vinyl acetate (EVA), were selected for the lamination. The air side (backsheet side) of the G/CB coupons faced the UV light source. The cross-sections of the aged G/CB transparent backsheets were examined by confocal fluorescence microscopy, microscale infrared spectroscopy, and nanoindentation for their optical, chemical, and mechanical degradation, respectively. For a comparison, the degradation of glass/POE/glass and glass/EVA/glass coupons aged at the same conditions was also studied.

2 EXPERIMENTAL*

2.1 Materials

Glass/POE/transparent backsheet coupons (i.e., G/CB1, G/CB2, G/CB3) were prepared using a 3 mm thick polished fused silica wafer with a transmittance of around 95 % between 250 nm to 800 nm. Materials and structures of these backsheets can be found in our previous work [11]. CB1, CB2 and CB3 are primarily PVF/PET/FEVE, PVDF/PET/FEVE and FEVE/PET/FEVE, respectively. The top and bottom encapsulants used for lamination were a polyolefin elastomer UV-transparent (POE-T) and a POE with UV cut-off at ≈ 380 nm (POE-UVA), respectively. Glass/POE/glass (G/POE/G) coupons were laminated with two fused silica wafers and the same types of encapsulants were used in all G/CB coupons. Considering the effect of topological and thermo-mechanical stress on backsheet cracking [12], 0.2 mm thick and 0.9 mm wide tabbing ribbons were laminated into the samples, creating a small ridge in the backsheet. The orientation of the ribbons was either parallel or perpendicular to the machine direction of the backsheet. For comparison, glass/EVA/glass coupons were prepared with a UV-transparent EVA (EVA-T) and a UV-cut off EVA (EVA-UVA) for the top and bottom encapsulants, respectively. An illustration of the coupon structures can be seen in Figure 1. The ribbons can be seen in Figure 2.

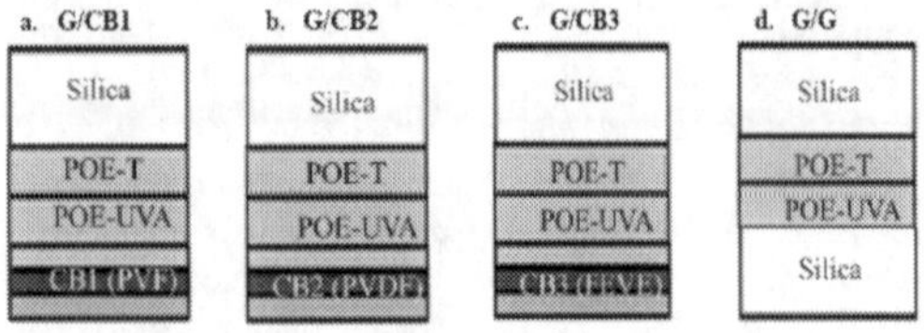

Figure 1. Illustration of coupons used for this study

2.2 Accelerated Laboratory Weathering

Laboratory accelerated weathering was performed using the NIST SPHERE *(Simulated Photodegradation via High Energy Radiant Exposure)* at 65 °C/50 % RH for up to ≈ 3600 h. The average UV irradiance at the sample locations was approximately 140 W/m^2 (295 nm to 400 nm). G/CB coupons of 75 mm × 100 mm were mounted on top of a ≈ 2 mm thick Teflon sheet in a black anodized sample holder with the airside of the backsheet facing the UV light. Samples were periodically removed from the SPHERE for non-destructive optical, chemical and microscopic characterization. Thermal cycling tests (TC) were performed using an ESPEC BTX-475 environmental chamber between – 40 °C and 85 °C with a dwell of 10 min at each temperature maxima and a ramp rate of 100 °C/h after UV exposure. Samples were subjected to two rounds of 100 cycles of TC tests after UV exposure, one was at approximately 2000 h (≈1000 MJ/m^2 UV) and the other at approximately 3600 h (≈1800 MJ/m^2 UV).

Two samples of each backsheet were removed with a surgical knife and potted in epoxy with their cross-sections exposed and then cured at ambient conditions. They were then cut with an EM UC7 microtome (Leica Biosystems Nussloch GmbH, Nussloch, Germany) to achieve a smooth and flat surface. The specimens were then used for cross-sectional characterizations.

2.3 Characterization

2.3.1 UV-Vis

Optical degradation of the aged and unaged G/CB and G/G coupons were measured in transmission mode using a Lambda 900 spectrophotometer (PerkinElmer, Inc., Waltham, MA) equipped with an integrating sphere in the range of 190 nm to 820 nm with a resolution of 1 nm. The yellowness index (YI) was calculated based on ASTM E313. [13] All measurements were done on one sample in at least three different locations to calculate the mean and standard deviation.

2.3.2 Laser Scanning Confocal Microscope (LSCM)

A Zeiss model LSM800 (Carl Zeiss Microscopy, White Plains, NY) laser scanning confocal microscope (LSCM) was used to image the cross-sections of the aged and unaged backsheets of G/CB coupons. The sample surface was first brought into focus using reflection mode and then switched to fluorescence mode for mapping. All LSCM images shown in this study are 2D projection images. The reflection mode was captured with a laser wavelength of 405 nm using objectives of 5×, 20×, and 50×. Each frame consisted of 1024 pixels × 1024 pixels. The fluorescence mode was performed using a 5× objective with a diode laser (405 nm) to excite the sample. Laser power was set to 8.5 % while the pinhole size was 90 µm.

2.3.3 Attenuated Total Reflectance Fourier Transform Infrared Spectroscopy with Microscope (Micro-ATR-FTIR)

Cross-sectional chemical characterization of the aged and unaged backsheets of G/CB coupons was performed using a Thermo Scientific Nicolet iN10 MX Infrared imaging microscope (Thermo Fisher Scientific, Madison, WI) equipped with a liquid nitrogen-cooled mercury-cadmium-telluride (MCT) detector in an ATR mode. A germanium conical tip with an aperture size of 150 µm × 150 µm was used for all measurements. Spectra were collected in the range of 4000 cm^{-1} to 650 cm^{-1} with 16 scans. For all samples, at least 3 locations were measured. All acquired spectra were baseline corrected.

2.3.4 Cross-sectional Nano-Indentation

Young's moduli of the individual layers of the aged and unaged backsheets of G/CB coupons were investigated with a G200 nano-indenter system (KLA, Milpitas, CA) using a 1 µm radius, 60° diamond cone indenter. Fives lines of nanoindentations at an angle of 45°, with a spacing of 15 µm between each indent, were performed to evaluate the depth-dependency of the mechanical properties of the backsheet due to exposure. Additional measurements were taken in the thin layers of the backsheets. The indent marks were then imaged in the LSCM in reflection mode. Reported moduli of the samples are the averaged values obtained between the depths of 1 µm to 2 µm for a minimum of 5 indents, with more measurements on the thicker layers. As this was a destructive test method, it was only carried out with unexposed and the fully exposed specimen after the full ≈ 3600 h UV exposure and 200 thermal cycles.

3 RESULTS AND DISCUSSION

3.1 Optical Changes of G/CB and G/G Coupons after UV/TC Exposure

After ≈ 3600 h of UV exposure at 65 °C/50 % RH and 200 thermal cycles, notable optical changes are observed on all three G/CB coupons (Figure 2). Compared to those without exposure, all coupons have yellowed to varying degrees. G/CB3 appears to be dark yellow. Very little yellowing is seen on G/CB1 and G/CB2. The quantitative change in yellowing is plotted over UV doses shown in Figure 2. Among all three coupons, G/CB3 has shown the highest rate of yellowing. This result is consistent with those observed in the free-standing transparent backsheets [11], indicating the optical degradation of CB3 is faster than those of CB1 and CB2, whether in a free film or being laminated in a G/CB coupon. In addition to yellowing, the delamination of the backsheet has been observed near the edges of G/CB2 and G/CB3 coupons, implying the weak adhesion of the multilayers in these two backsheets as a result of degradation.

A visible crack is also observed from the backsheet side of G/CB3 when the UV dose reached ≈ 1800 MJ/m^2, followed by an additional 200 TC. It is found that thermal cycling has played a great impact on the formation of the cracks for this coupon because no cracks were observed before 200 TC. In addition to the large crack (Figure 3a), relatively shorter and thinner cracks are also revealed by laser scanning images (Figures 3b and 3c). These numerous thin cracks cover almost the entire sample, with an orientation leaning toward the machine direction. Interestingly, by a careful examination of LSCM images, it is found that these thin cracks are not on the FEVE surface; they are located beneath the FEVE outer layer, ≈ 10 μm down into the PET core layer. This finding is consistent with those observed in the free-standing CB3 backsheet [11], suggesting the cracking of this backsheet can be largely resulting from the brittleness of the PET core layer after UV/TC exposure.

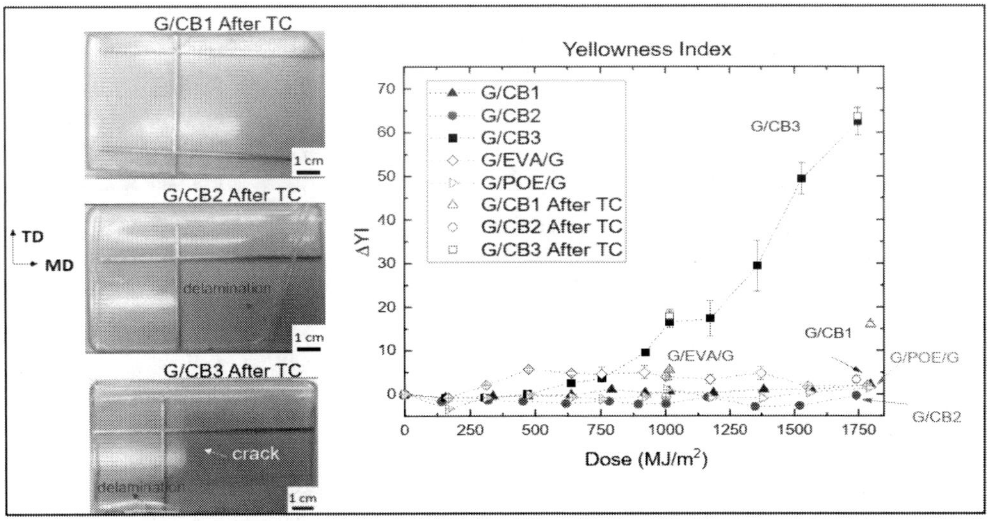

Figure 2: (Left) Photos of three coupons of G/CB1, G/CB2 and G/CB3 subjected to UV dose of 1800 MJ/m² and 200 thermal cycles; (Right) The increase in yellowness index (YI) of the G/CB coupons as well as reference G/POE/G and G/EVA/G coupons a function of UV dose. Each data point represents the average value from the three replicate measurements, while error bars show one standard deviation.

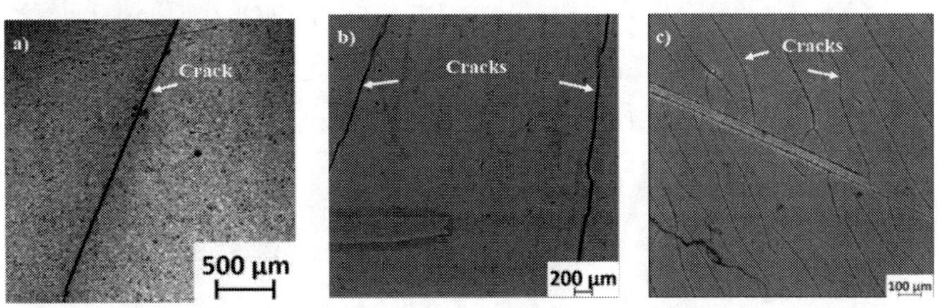

Figure 3. LSCM images of airside (FEVE) of G/CB3 backsheet after 1800 MJ/m² UV exposure and 200 TC, showing various types of cracks. Note that the thin cracks shown in Figure 3c are located beneath the FEVE outer layer, approximately 10 μm down into the PET core layer.

3.2 Cross-Sectional Optical, Chemical, and Mechanical Degradation Characterization of Transparent Backsheets of Aged G/CB Coupons

Cross-sectional LSCM images of the epoxy-embedded transparent backsheets obtained from three aged G/CB coupons after exposure to 1800 MJ/m² UV and 200 TC are displayed in Figure 4. Two magnifications are exhibited for each coupon in order to show both overall backsheet structures and the fine cracking features near the aged outer layers. A number of cracks are seen in the outer layer of CB1 (Figures 4a and 4b). The higher magnification image of this region shows that these cracks are less than ≈ 10 μm long from the UV-exposed air surface, limited to the outermost portion of the PVF outer layer. No cracks are seen in the adhesive layer or the PET core layer of CB1. For comparison, no visible cracking is seen in the whole cross-section of CB2 after exposure (Figures 4c and 4d).

LSCM images of a cross-section of the aged G/CB3 backsheet show a few interesting features (Figures 4e and 4f): 1) delamination of the FEVE outer layer from the PET core layer, 2) cracking in the PET core layer, and 3) no visual cracks in the EVA inner layer. Although the delamination of the FEVE outer layer appears to occur during the preparation of cross-sectional samples, it suggests that the adhesion between the FEVE outer layer and the PET core layer has been weakened after exposure (no delamination is observed for the unexposed counterpart). In addition, two large cracks are seen across the entire PET layer, as shown in Figures 4e and 4f. Furthermore, numerous microcracks are observed near the interface between the FEVE outer layer and the PET core layer, extending approximately 8 μm deep to the PET layer. Unlike the outer and core layers of CB3, the EVA inner layer doesn't show any cracks, which is different from the results observed in the aged free-standing CB3 films [11], signifying the degree of degradation of this layer might be reduced in the laminated G/CB3 coupon.

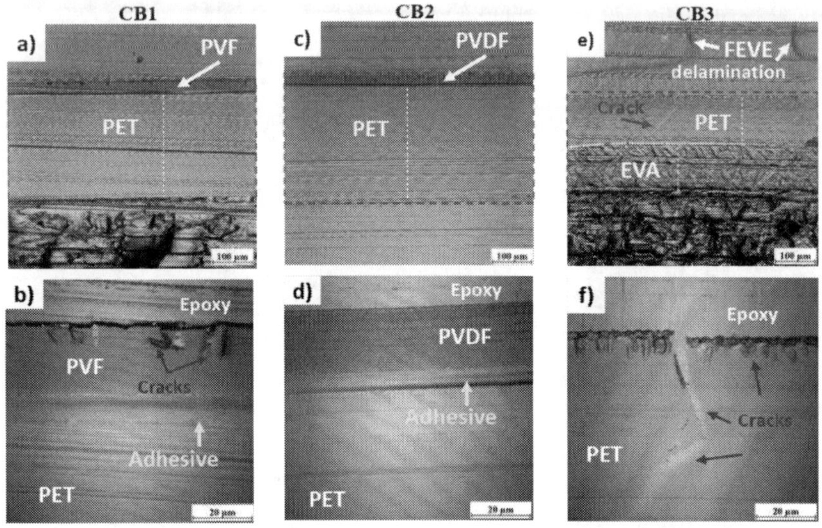

Figure 4. LSCM images of the cross-sections of CB1 (a, b), CB2 (c, d), and CB3 (e, f) backsheets obtained from G/CB coupons after exposure to 1800 MJ/m^2 UV and 200 TC. The higher magnification images (b, c, f) are focused on the UV-exposed outer layer, adhesive layer and partial PET core layer of the three transparent backsheets.

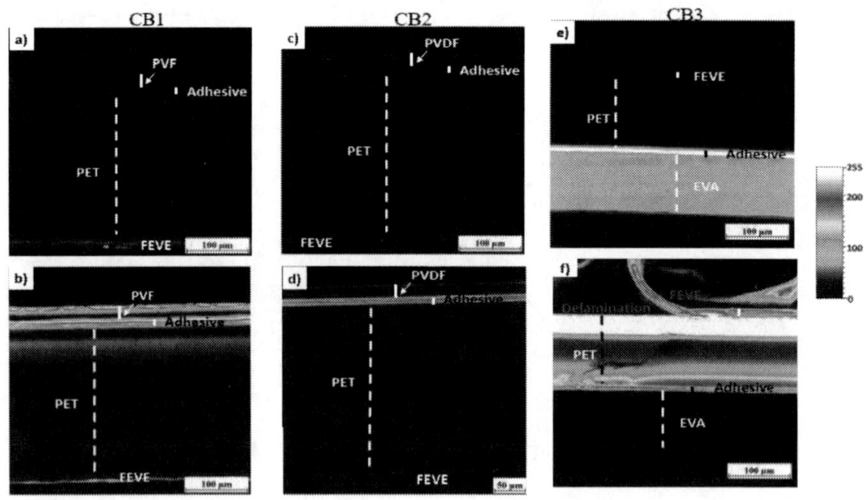

Figure 5. Confocal fluorescence mapping of the cross sections of unexposed (a, c, e) and exposed (b, d, f) backsheets of the G/CB1, G/CB2 and G/CB3 coupons after 1800 MJ/m^2 UV and 200 TC.

Figure 5 shows the corresponding confocal fluorescence images of three G/CB backsheets before and after exposure. Similar changes in fluorescence were observed on CB1 and CB2 backsheets after exposure. First, both adhesive layers show a significant increase in the intensity of fluorescence. Second, both PET core layers show a similar level of fluorescence. Compared to the very slightly fluorescent PVDF outer layer in the aged CB2, the top portion of the PVF outer layer in CB1 shows an obviously increased fluorescence, suggesting that degradation has taken place near the skin of the PVF outer layer due to direct UV exposure (Figure 5b), which is consistent with the microcracks shown in Figures 4a and 4b. On the other hand, substantial changes are shown in the confocal fluorescence images of CB3 after the same exposure. The fluorescence seen in the unexposed EVA inner layer has disappeared, while apparent fluorescence can be seen in the FEVE outer layer, the upper and bottom portions of the PET core layer, and the adhesive layer. The loss of the fluorescence in EVA might be related to the loss of the additives in the EVA inner layer as a result of UV exposure. High fluorescence of the aged FEVE has made it possible to locate the FEVE layer in the cross-sectional samples, even though the layer has been partially delaminated from the PET core layer. One can clearly see the outer FEVE layer detaching from the PET core layer and curving upwards (Figure 5f). The increase in the fluorescence of the FEVE layer indicates that this outer layer has been degraded, and this degradation seems to weaken the bonding between the outer layer and the underlying PET layer, leading to delamination.

One can find a very strong fluorescent area (the white region in Figure 5f), approximately 30 μm thick, lying just below the FEVE outer layer in the PET layer. This region corresponds to the area with numerous cracks shown in Figure 4f, indicating severe degradation has occurred. The degradation of this region could contribute to the strong yellowing of G/CB3 seen in Figure 2. To better understand these features, the chemical changes in this region were examined by the micro-ATR-FTIR spectroscopic mapping on the G/CB3 backsheet before and after aging. Results from the unexposed sample show there is little variation in the IR spectra across the width of the PET layer (Figures 6a and 6b). However, IR results of the degraded sample show a difference depending on the location within the sample (Figures 6c and 6d). Position 1 in Figure 6d corresponds to approximately 30 μm from the FEVE layer; this is a similar region to the white areas seen in Figure 5f. This area shows markedly different behavior compared to the other positions. At this position, the signal of PET is coupled with the signal of the mounting epoxy. This indicates there were small microcracks in the top layer of the PET layer prior to embedding, as shown in Figure 4f. Positions 2-6 show little to no evidence of epoxy co-mingling, however, comparing these positions to the unexposed PET shows a broadening of carbonyl peaks in the lower wavenumbers seen in Figure 6d, indicating the formation of degradation products such as carboxylic acid dimers, TPA, etc. [14, 15]. Figure 6e shows the change in FTIR intensity of the peak at 1700 cm^{-1}, which is attributed to the carbonyl stretching of carboxylic acids [14]. It appears the intensity of carboxylic acids is sensitive to the position within the sample. The top PET region (Position 1) near the FEVE outer layer shows the highest level of carboxylic acid formation. As the distance from the light source increases, Position 8, which is furthest away from the light source, shows levels similar to that of the unexposed coupon (circled).

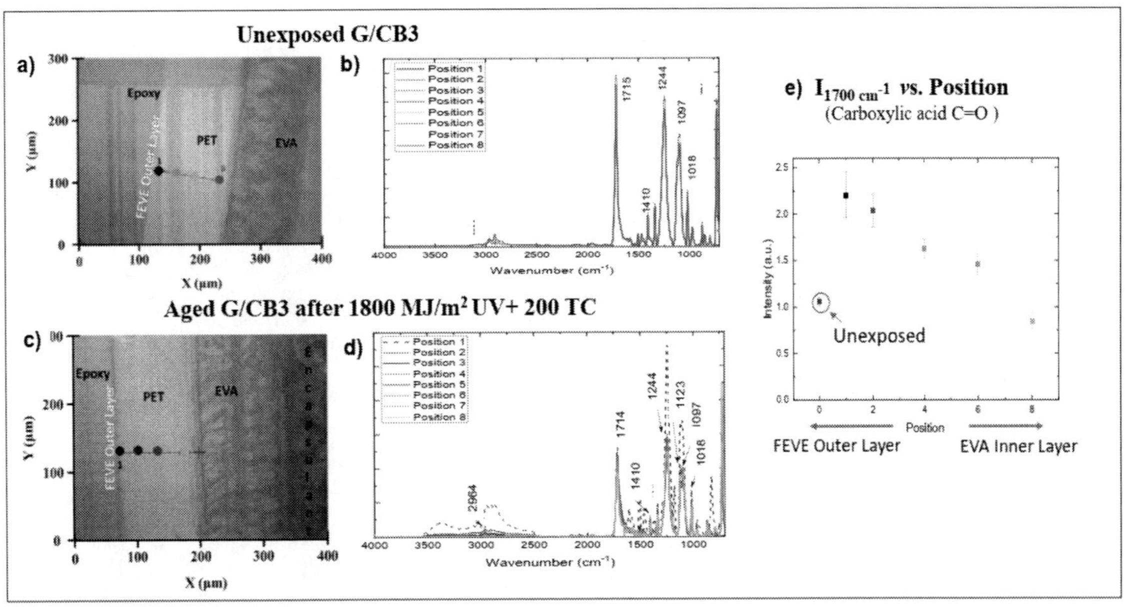

Figure 6. Micro-ATR-FTIR optical microscope images of (a) unexposed and (c) exposed G/CB3 with indications of areas measured within the PET core layer. Position 1 is approximately 30 μm from the FEVE outer layer. The spacing between two adjacent positions is 15 μm. (b) and (d) are corresponding IR spectra of locations indicated in (a) and (c), respectively. (e) Evolution of peak intensity at ≈ 1700 cm^{-1} (carboxylic acid C=O) *vs.* measurement position in the PET core layer. Each data point in e) represents the average value from the three replicate measurements, while error bars show one standard deviation.

Cross-sectional nanoindentation of the G/CB3 backsheet shows an obvious increase in modulus of the PET region near the FEVE outer layer (Figure 11). This region corresponds to the very high fluorescent region (white) in Figure 5f, as well as Position 1 in Figure 6c which has the highest formation of carboxylic acids across the PET core layer. This correlation indicates that severe degradation leads to the brittleness of the top PET portion underneath the FEVE layer, and the subsequential cracking. Combining the above results, one can find that, despite being protected from UV light by the outer layer, the PET core layer in CB3 is still prone to high levels of photodegradation. Therefore, designing a more UV-protective transparent outer layer or a more UV- stabilized PET core layer will greatly promote the long-term reliability of transparent backsheet-based bifacial modules.

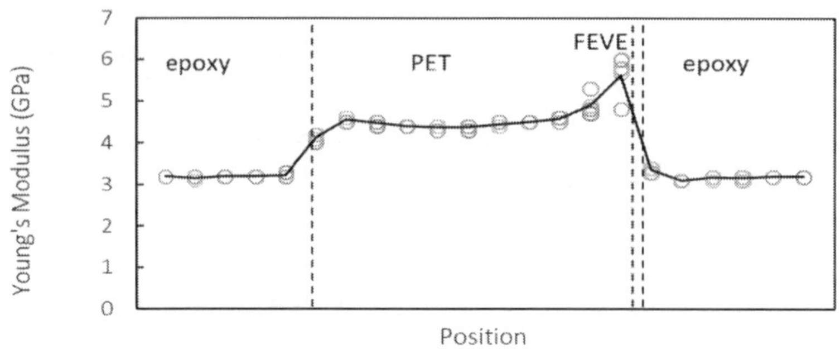

Figure 7. Young's modulus of a partial cross-section of G/CB3 after aging, showing the increased modulus in the PET layer close to the FEVE outer layer (the EVA inner layer was missing due to delamination from the PET core layer). The black solid line shows the average of all data from that position. The vertical dashed lines denote the interfaces between the layers.

4 SUMMARY

The cross-sections of transparent backsheets of three G/CB coupons, G/CB1, G/CB2, and G/CB3, before and after exposure were characterized by nanoindentation, micro-IR, and LSCM in both reflection and fluorescence modes. Results show that a superficial outer layer cracking occurred on PVF-based CB1 backsheet, and local delamination appeared near a corner of PVDF-based CB2 backsheet. Among the three types of transparent backsheets, FEVE-based CB3 exhibited the most optical, chemical, and morphological changes after exposure. In addition to large cracks across the entire PET layer, numerous microcracks were observed underneath the FEVE outer layer, down to the PET core layer. The closer to the light source, the higher level of the chemical degradation of the PET was observed. This study has indicated that, despite being protected from UV light by the outer layer, the PET core layer is still prone to photodegradation. Thus, the UV durability of the transparent backsheets is of importance to the long-term reliability of G/CB bifacial modules.

*DISCLAIMER:

Certain commercial products or equipment are described in this paper to specify adequately the experimental procedure. In no case does such identification imply recommendation or endorsement by the National Institute of Standards and Technology, nor does it imply that it is necessarily the best available for the purpose.

5 ACKNOWLEDGMENTS

We greatly acknowledge the support from a NIST/Industry consortium on "Characterization and Modeling of Polymers for Photovoltaic Systems".

6 REFERENCES

[1] https://www.woodmac.com/news/editorial/bifacial-2019/

[2] A. Cuevas, A. Luque, J. Eguren, and J. Del Alamo, "50 Percent more output power from an albedo-collecting flat panel using bifacial solar cells," Solar Energy, 29 (1982) 419.

[3] E. Urrejola, et al., "bifiPV2020 Bifacial Workshop: A Technology Overview," NREL/TP-5K00-77817. Oct 2020; National Renewable Energy Laboratory: Golden, CO, USA, 2020.

[4] X. Sun, M. R. Khan, C. Deline, and M. A. Alam, "Optimization and performance of bifacial solar modules: A Global Perspective," Appl. Energy, 212 (2018) 1601.

[5] G. W. Wilson, et al., "The 2020 Photovoltaic technologies roadmap," J. Phys. D: Appl. Phys, 53 (2020) 493001.

[6] A. Sinha, et al., "Glass/glass photovoltaic module reliability and degradation: a review," J. Phys. D: Appl. Phys. 54 (2021) 413002.

[7] M. H. Saw, Y. S. Khoo, J. P. Singh, and Y.Wang, "Enhancing optical performance of bifacial PV modules," Energy Procedia, 124 (2017) 484.

[8] https://www.pv-tech.org/industry-updates/jinkosolar-transparent-backsheet-vs-dual-glass-advantages-and-disadvantages

[9] K. Arihara, et al., "Reliability and long term durability of bifacial photovoltaic modules using transparent backsheet," Japanese Journal of Applied Physics, 57 (2018), 08RG15.

[10] K. R. Choudhury, "Performance and Long-term Durability of Transparent Backsheets for Bifacial PV Applications," bifiPV2020, Golden, CO, USA, 2020.

[11] S. Smith, et al., "Transparent backsheets for bifacial photovoltaic modules: material characterization and accelerated laboratory testing," Progress in Photovoltaics: Research and Applications, 10/2021, doi.org/10.1002/pip.3494

[12] M. D. Kempe, T. Lockman, and J. Morse, "Development of testing methods to predict cracking in photovoltaic backsheets," 2019 IEEE 46th Photovoltaic Specialists Conference (PVSC), 2019, pp. 2411-2416.

[13] ASTM E313-20. "Standard Practice for Calculating Yellowness and Whiteness Indices from Instrumentally Measured Color Coordinates" ASTM International: West Conshohocken, PA, 2020.

[14] F. J. Horne, J. J. Liggat, W. A. MacDonald, and S. W. Sankey, "Photo-oxidation of poly(ethylene terephthalate) films intended for photovoltaic backsheet," Journal of Applied Polymer Science, 137 (2020) 48623.

[15] T. Grossetête, et al., "photochemical degradation of poly(ethylene terephthalate)-modified copolymer," Polymer 41 (2000) 3541.

INFLUENCE OF PRODUCTION PROCESSES ON PID-S SENSITIVITY OF C-SI MODULES AND NOVEL MITIGATION STRATEGIES

Alexander Stauffenberg[1], Bengt Jäckel[1], Matthias Pander[1], Jens Fröbel[1], Christof Erban[2]
[1]Fraunhofer Center for Silicon Photovoltaics CSP, Otto-Eißfeldt-Str. 12, 06120 Halle (Saale), Germany
[2]SUNOVATION Produktion GmbH, Glanzstoffstr. 21, 63820 Elsenfeld, Germany
Contact: +49-345-5589-5215, alexander.stauffenberg@imws.fraunhofer.de

ABSTRACT: Potential-induced degradation (PID) is one of the major degradation mechanisms of PV modules, reducing their power and useful time of operation. Novel mitigation strategies for the shunting type, PID-s, on module level are investigated by varying production parameters and materials in modules. This includes variations of encapsulant material (silicone and EVA), silicone mixing ratio and degree of EVA cross-linking as well as different front and back cover constructions using glass and transparent polymers. Modules are stressed under IEC standard conditions for PID-s and characterized using I-V-characteristics and electroluminescence imaging.
Modules with silicone encapsulant are resistant to degradation after 300+ hours of PID-s stress exposure, but are not with damp heat preconditioning. A higher mixing ratio between silicone and hardener might benefit PID-s resistance. Degradation of modules with EVA encapsulant varies strongly with parameters: A PID insensitive EVA with high degree of cross-linking reduces PID-s sensitivity, but ultimately can't stop degradation. Recovery from PID-s by inverting the potential is possible but limited, depending on the degree of degradation. Modules with a polymer sheet in front instead of glass show significantly reduced PID-s susceptibility, leading to a new mitigation strategy.
Keywords: PV Module, Degradation, Reliability, c-Si, Encapsulation

1 INTRODUCTION

Potential induced degradation (PID) was found in the field more than 10 years ago [1][2] and intensive research has been conducted since to understand the underlying mechanisms [3][4]. For PID-s (shunting) and PID-d (delamination), standard procedures were developed to screen PV modules for their resistance against potential induced degradation phenomena [5][6].

A conventional p-type c-Si module with grounded frame and negatively biased cells is prone to the shunting type PID-s. Induced leakage currents from module frame to cells take different possible pathways through front or back surface, along interfaces or directly through encapsulant and lead to module power loss. The pathway through the front surface has been determined as most relevant factor of degradation since the surface conductivity of front glass increases with contact to water due to rain and humidity, with soiling and in a salty environment. Since the leakage currents pathway may vary, the occurrence and severance of PID-s depend on applied potential and module construction. Although PID-s can be partially recovered by thermal or reverse-biased voltage recovery, there are no fully PID-free technologies available on the market. With an increased maximum system voltage to 1500 V, the PID problem will be more severe in the future [7].

The current transition to larger wafer sizes, new cell technologies with higher efficiencies can support more environmentally friendly production throughout the value chain [8]. It is known that even small variations in cell production processes can lead to an increase in PID sensitivity. Additionally, the module production processes are also subject to change due to material variations [9]. This study focuses on the production process influences of encapsulant materials, their mixing ratios and degree of cross-linking as well as cover materials used for module construction, to further develop the understanding on new and innovative materials and to uncover alternative mitigation strategies.

2 MATERIAL AND METHODS

2.1 Samples and variations

In this series of experiments, specially manufactured highly PID-s sensitive p-type multicrystalline silicon solar cells (156.75x156.75 mm, 4.5 W) were used to build 80 mini-modules with three cells each. Variations include encapsulant material, production parameters for encapsulants and module construction (see Table I). Materials used as encapsulant are a two-component silicone, PID sensitive and insensitive ethylene-vinyl acetate (EVA). The mixing ratio between silicone and hardener was varied, resulting in four variations using glass/glass modules built by Sunovation. For each EVA material, the degree of cross-linking was varied, resulting in the two variations partly and fully cured using glass/backsheet modules. For module construction, the front and back cover were varied using glass and commercially available transparent polymer sheets, resulting in the four variations glass/backsheet (BS), glass/glass, frontsheet (FS)/glass and an additional setup of glass/backsheet modules with a polymer frontsheet on top of the glass (FS + glass). A PID sensitive, fully cured EVA was used for these module variations.

Table I: Specimen matrix.

	Front	Back	Encapsulant	Production parameter
1a	glass	glass	silicone	1:2
1b	glass	glass	silicone	1:1
1c	glass	glass	silicone	2:1
1d	glass	glass	silicone	4:1
2a	glass	BS	ins. EVA	partly cured
2b	glass	BS	ins. EVA	fully cured
2c	glass	BS	sens. EVA	partly cured
2d	glass	BS	sens. EVA	fully cured
3	FS	glass	sens. EVA	fully cured
4	glass	glass	sens. EVA	fully cured
5	FS + glass	BS	sens. EVA	fully cured

2.2 Test procedure

Modules were characterized initially using a Berger AAA class Xenon solar simulator for I-V-characteristics at 200 and 1000 W/m² and a greateyes Lumisolar system for electroluminescence imaging at I_{sc} and 0.1 I_{sc}. Before the PID stress, some samples from each module variation were preconditioned with damp heat exposure for 200 hours at 85 °C, 85% rh in a CTS climate chamber to induce humidity and trigger or accelerate PID-s. All modules were stressed to test for PID-s according to IEC 62804-1:2020 [6] standard conditions at 60 °C and 85% rh with the module front face additionally covered in aluminum foil (see Figure 1).

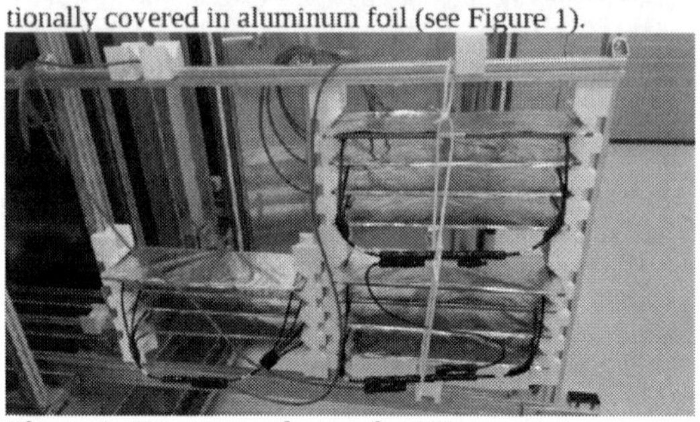

Figure 1: Experimental setup for PID stress.

Because of the known high PID-sensitivity of cells a temperature of 60 °C instead of 85 °C was used. A stress potential of –1000 V was applied between cell circuitry and aluminum foil for at least 12 hours. The values of V_{oc}, I_{sc}, P_{mpp} and FF were determined from I-V-characteristics and compared to the initial values by calculating the percentage deviation due to degradation. Depending on degree of degradation, modules were further stressed in the climate chamber up to a total test duration of > 300 hours. An inverted potential (+1000 V) was applied to some quickly degrading modules with sensitive EVA encapsulant to investigate a potential recovery.

3 RESULTS AND DISCUSSION

As reference, the initial module parameters were I_{sc} = 9.17±0.15 A, V_{oc} = 1.95±0.01 V, P_{mpp} = 12.94±0.53 W and FF = 72.41±2.48%, showing a narrow distribution.

3.1 Silicone variations

Shown in the following figures is the percentage change in power compared to the initial value at 0 hours measured at STC (1000 W/m², 25 °C) over the PID exposure time. The glass/glass modules with silicone as encapsulant material provided by Sunovation have shown no significant power loss without damp heat preconditioning (see Figure 2). Variations of ±1% result from measurement inaccuracies, contact resistance and temperature deviations from STC (25 °C). Mixing ratio between silicone and hardener does not have a significant effect on the maximum power during PID testing. It must be noted, that mixing ratio will impact long term reliability of modules due to other effects.

However, under damp heat preconditioning, PID-s sensitivity is increased. The preconditioning on its own resulted in a power loss of -1.2 to -4.3%. Under PID stress, these modules degraded from -3.0% up to -27.9% after 300 h. The glass/glass module with a 1:1 mixing ratio stands out as it degraded more than the other modules due to darkening effects observed in electroluminescence imaging (see Figure 2A). A trend can be observed that the PID-sensitivity decreases with higher mixing ratio, meaning that an increased amount of silicone in the two-component silicone can reduce degradation.

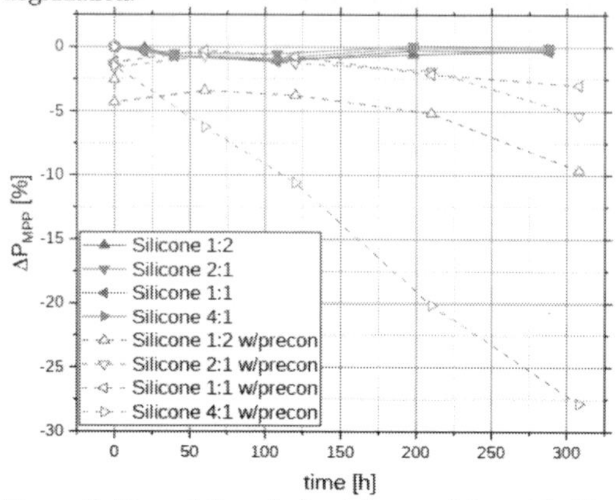

Figure 2: Degradation of glass/glass modules with silicone encapsulant mixing ratio variations with and without preconditioning at STC.

3.2 EVA variations

The glass/backsheet modules with variations of EVA as encapsulant material show greater differences than the modules with silicone (see Figure 4). As expected, modules with PID insensitive EVA show a significantly lower degradation than those with PID sensitive EVA. After 346 h of PID stress, maximum power decreased -8.5% for fully and -24.8% for partly cured EVA (see Figure 3B after 256 h). Compared to the modules using a silicone encapsulant, the power loss is clearly higher. Preconditioning seems to accelerate the degradation effect for PID insensitive modules [10]. However, this effect does not apply for modules with PID sensitive EVA. After only 12 h, maximum power in the modules without preconditioning decreased over -85%. The preconditioned modules with PID sensitive EVA show a significantly lower degradation. Especially the fully cured EVA module stands out, losing only -9.6% after 12 h and -35.2% after 190 h. The partly cured EVA module, with a degradation of -58% after 24 h, was recovered to -8.1% with an inverted voltage for 168 h. Although it was PID-stressed for 90 h, it degraded only slightly to -11.9%. Recovery on already degraded glass/backsheet modules with EVA encapsulant is possible but limited.

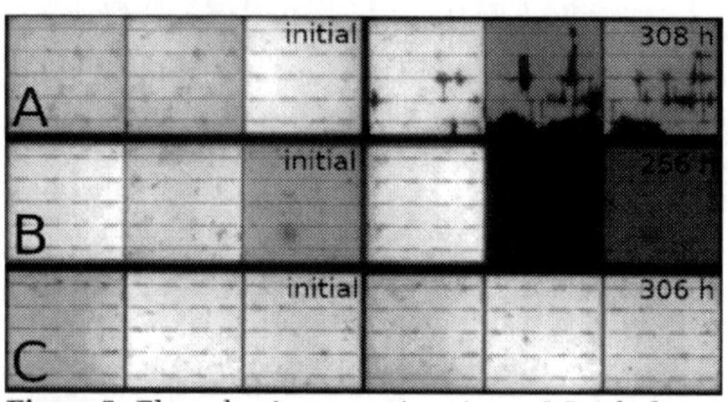

Figure 3: Electroluminescence imaging at 8.5 A before and after PID-stress. (A) glass/glass with silicone 1:1

w/ precon (see Figure 2), (B) glass/backsheet with partly cured, insensitive EVA (see Figure 4) and (C) frontsheet/ glass module (see Figure 5).

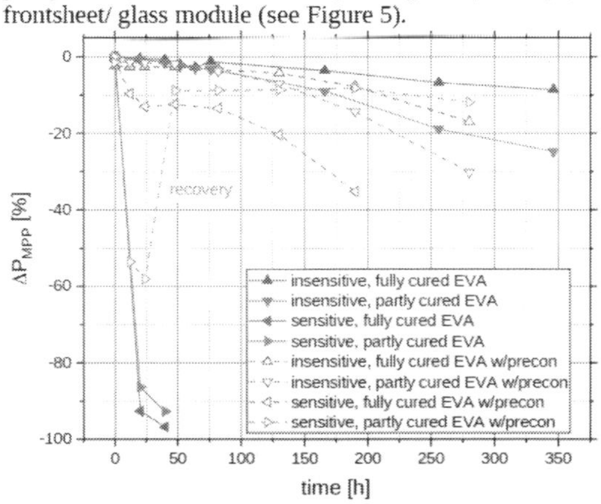

Figure 4: Degradation and recovery of glass/backsheet modules with EVA encapsulant variations with and without preconditioning at STC.

3.3 Module construction and cover materials

Cover material variations were tested on modules with a PID sensitive, fully cured EVA encapsulant (Figure 5). Modules with a front glass degrade within 24 h of PID stress exposure, regardless of the back cover material being glass or a polymer sheet. In contrast, modules with a polymer sheet as front cover did not degrade more than -2.9% after 306 h of exposure, also visible in electroluminescence (see Figure 3C). The preconditioning of frontsheet/glass modules resulted in a power loss of -1.4% initially and -4.9% after over 300 h of PID exposure. Since the polymer sheet instead of a glass as front cover provided a reduction in PID sensitivity, another set of glass/backsheet modules was produced with an additional polymer sheet on the front glass to profit from the glass's mechanical resistance and the polymers PID resistance.

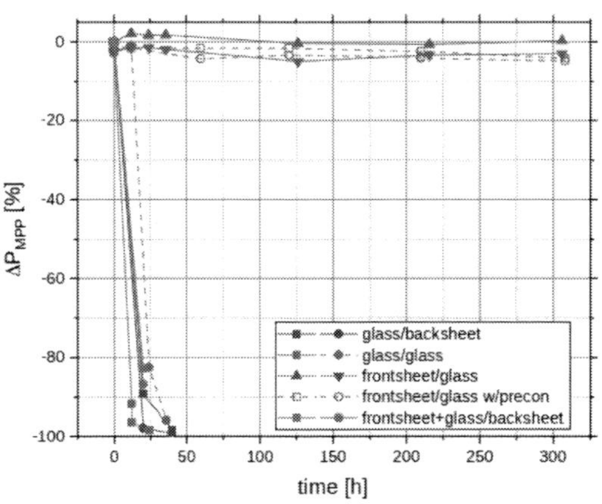

Figure 5: Degradation of PID sensitive EVA modules with construction variations at 200 W/m², 25 °C.

However, this frontsheet did not make a difference in terms of degradation as it lost over -85% of its power after 12 h of exposure. This degradation was probably caused by the contact between glass and wrapping aluminum foil on the modules sides which is a possible leakage current pathway according to Luo et al. [7]. This problem might be solved by either covering not only the surface but the entire front glass including the edges with a polymer sheet or by having a polymer sheet between glass and EVA. This way, the source of diffusing ions from the glass still exists, but the setup will show significant increased resistance, resulting in a lower applied potential close to the solar cell.

4 SUMMARY AND CONCLUSION

This work investigated the influence of different production parameters and materials on PID-s sensitivity and showed that the commonly used materials used to mitigate PID-s are not sufficient. Although silicone is a novel potent encapsulant material, damp heat preconditioning increased PID-s sensitivity. The physical interaction between silicone and the damp heat conditions needs to be further investigated. A higher mixing ratio between silicone and hardener in a two-component silicone is a possible mitigation strategy against PID-s. Every standard module with EVA as encapsulant degraded eventually, regardless of insensitive material or a high degree of cross-linking. Therefore, it is currently safe to say that EVA cannot provide a sufficient solution to completely suppress PID-s. Damp heat preconditioning accelerated the PID-s sensitivity of all constructions. For still PID-s sensitive solar cells even highly "PID-resistant EVAs" will not fully solve the problem. New mitigation strategies and materials against PID-s still need to be developed, especially to cover the impact of moisture. However, test protocols must also be adopted to this more in depth understanding as moisture ingress is currently very limited within the typical 96 hours.

By inverting the potential, recovery from PID-s is possible, depending on the degree of degradation. Regarding construction setups, the back cover material did not impact PID-s sensitivity, although this statement might only be applicable for monofacial and not for modules utilizing bifacial solar cells. A polymer sheet in front instead of glass can be used to reduce PID-s sensitivity significantly. A possible mitigation strategy would be to have a polymer sheet between the front glass and EVA instead of only covering the surface. A stack of glass/ polymer sheet/ EVA could reach a high PID resistance while still keeping the mechanical resistance of front glass. This novel way to have a PID-s resistant PV module while still using a front glass will be investigated further at Fraunhofer CSP to ultimately ensure long lifetime and durability of PV modules.

5 ACKNOWLEDGEMENTS

The authors gratefully acknowledge the financial support by the German Federal Ministry for Economic Affairs and Climate Action for financial support of the project "ProModul" under grant no. 0324354.

6 REFERENCES

[1] S. Pingel et al., "Potential induced degradation of solar cells and panels", 2010 35th IEEE Photovoltaic Specialists Conference, pp. 002817–002822, 2010. DOI: 10.1109/

PVSC.2010.5616823.

[2] M. Schütze et al, "Laboratory study of potential induced degradation of silicon photovoltaic modules", 2011 37th IEEE Photovoltaic Specialists Conference, pp. 000821-000826, 2011. DOI: 10.1109/PVSC.2011.6186080.

[3] P. Hacke et al., "Acceleration factor determination for potential-induced degradation in crystalline silicon PV modules", 2013 IEEE International Reliability Physics Symposium (IRPS), pp. 4B.1.1-4B.1.5, 2013. DOI: 10.1109/IRPS.2013.6532009.

[4] V. Naumann, "Sodium decoration of PID-s crystal defects after corona induced degradation of bare silicon solar cells", Energy Procedia, vol. 77, pp. 397–401, 2015. DOI: 10.1016/j.egypro.2015.07.055.

[5] B. Jäckel et al., "PV module degradation under high potentials – a comparative study between test setups", Proceedings of the 28th European Photovoltaic Solar Energy Conference and Exhibition, 2013.

[6] IEC TS 62804-1-1. "Photovoltaic (PV) Modules – Test Methods for the Detection of Potential-Induced Degradation - Part 1-1: Crystalline Silicon - Delamination", 2020.

[7] W. Luo et al., "Potential-induced degradation in photovoltaic modules: A critical review," Energy & Environmental Science, 2017. DOI: 10.1039/C6EE02271E.

[8] A. O. Gnocchi, "Encapsulant selection for PID resistant modules made with heterojunction solar cells", Proceedings of the 37th European Photovoltaic Solar Energy Conference and Exhibition, 2020.

[9] S. Schulze et al., "Development of a novel technique for EVA-curing by means of UV-radiation", Proceedings of the 27th European Photovoltaic Solar Energy Conference and Exhibition, 2012.

[10] V. Naumann et al., "Influence of soiling and moisture ingress on long term PID susceptibility of photovoltaic modules", AIP Conference Proceedings 2147, 090005, 2019. DOI: 10.1063/1.5123873.

WORLDWIDE MOISTURE INGRESS EVALUATION IN PHOTOVOLTAIC MODULES USING FINITE ELEMENT SIMULATIONS AND MACHINE LEARNING

Stefan Mitterhofer[1], Julián Ascencio Vásquez[2,3], Xiaohong Gu[1]
[1]National Institute of Standards and Technology, 100 Bureau Drive, Gaithersburg, MD 20899, United States
[2]Envision Digital, 203 Redwood Shores Pkwy, Redwood City, CA 94065, United States
[3]ASVA Consulting, c/ Amalia Alayón 11, Santa Cruz de Tenerife, 38650, Spain

ABSTRACT: Ambient water can be absorbed by photovoltaic (PV) modules, affecting their degradation. This paper presents a two-step approach to evaluate the moisture profile inside fielded PV modules around the world. The first step consists of moisture ingress simulations at more than 17000 global locations using a cluster, running up to 160 simulations in parallel. Results of the simulations show a low R^2-value of 0.65 when correlating the ambient relative humidity and the saturation moisture concentration in front of the cell, emphasizing the need for such simulations to provide more accurate inputs for degradation modelling. In the second step, we train a machine learning algorithm to obtain a model finding a better correlation between the important climatic parameters and the simulation results. The ML algorithm can accurately find such a correlation, shown by the R^2-value of 0.98. We use the algorithm to further predict water ingress and interpolate the results to other global locations in a higher spatial resolution of the climatic input parameters.
Keywords: moisture ingress, finite element method, machine learning

1 INTRODUCTION

As the photovoltaic (PV) market grew rapidly over the past few decades, durability and reliability of the technology are becoming increasingly important topics [1]. Water vapor is one of the most crucial climatic factors affecting the degradation of PV modules [2]–[4]. Most current degradation models take the ambient relative humidity *RH* as input to predict either general loss of output of the module [5], [6] or module behavior caused by individual prominent degradation modes [7], [8]. However, most module degradation modes are affected by the moisture inside the PV modules, not by the ambient *RH*. It can ingress through the polymeric backsheet or edge seals and further diffuse through the encapsulant. This is commonly simulated using the finite element method (FEM) and the Fickian diffusion model [4], [9]. Such simulations require knowledge of both the climatic and material parameters. The former is not only limited to *RH* in the air and the ambient temperature T_{amb}. Depending on the model used to calculate the module temperature T_{mod}, further requirements can include the irradiance of ultraviolet and wind speed [10], [11]. The latter includes material parameters directly affecting diffusion, such as the diffusion coefficient D or the solubility of water in the material S, but also others affecting T_{mod}, commonly condensed down to fitting parameters for the module in a specific tilt and installation configuration.

Due to the large number of required parameters, such simulations have only been carried out either for modules in controlled conditions in climatic chambers or in well monitored outdoor testing sites [4], [12]–[15]. Furthermore, experimental verification of these calculations inside PV modules has been lacking. Only recently, such methods have been developed and applied to PV modules to confirm the simulation results, for example through miniature temperature and *RH* sensors laminated in the module [16]. While not all polymers used for PV encapsulation exhibit strictly Fickian behavior [17], [18], the model is still the most commonly used and accepted in such materials, with its applicability being proven repeatedly under controlled conditions [4], [14], [15], [19]–[21]. Furthermore, recent results showed a good fit between Fickian simulation results and measurements in some of the most common materials in fielded PV mini-modules, such as ethylene vinyl acetate (EVA), and polyethylene terephthalate (PET) based backsheets, such as a Tedlar©-PET-Tedlar© (TPT) or PET-PET-EVA (PPE) [3].

This work presents an approach to extend the above simulations [3] in crystalline silicon modules to a worldwide scale, using re-analysis data and finite element simulations. We use the obtained dataset to train a machine learning (ML) model, connecting climate input data and important simulation outputs. The ML model enables fast evaluation of even more locations, using a finer spatial resolution of the climate input data. The methodology and the simulation results will be presented in section 2. Section 3 will show and discuss the results.

2 METHODS

2.1 Climate data

We use the ERA5 dataset, the fifth generation reanalysis dataset of the European Center for Medium-Range Weather Forecasts (ECMWF), to obtain the required climatic parameters in a $1°\times1°$ grid for the entire world, excluding Antarctica [22], which equals 17,674 locations. Figure 1 shows an example of points in North America.

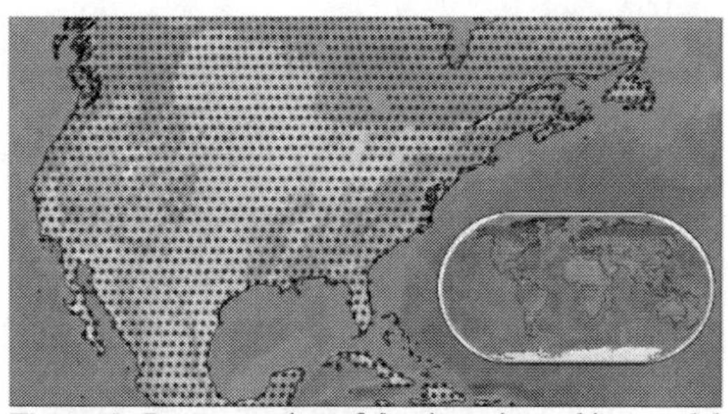

Figure 1: Representation of density points with a spatial grid resolution equal to $1°\times1°$ for North America.

The extracted parameters are T_{amb}, *RH*, wind speed *WS* and global horizontal irradiance (GHI). The typical meteorological year (TMY) for each location is calculated using the Sandia method [23]. For this specific study, a novel set of TMY weights have been implemented, given in Table I. For PV modules the standard TMY is calculated using 80 % or more for GHI, because the typical interest in TMY's is to run long-term yield assessments of PV systems. However, in this study, we have selected an arbitrary lower weight for GHI and added a 20 % in RH due to the importance of this variable for our research interest.

The operating conditions of the PV modules were calculated under a scenario of fixed tilted racking type, assuming the inclination equals to the latitude of the site. The plane-of-array irradiance *PoA* is calculated using Erbs method for decomposition and Perez method for transposition [24], [25]. The module temperature (T_{mod}) is then calculated for the TMY with the Faiman model (equation 1) [11].

Table I: Weights for determination of TMY

Parameter	Weight
Global horizontal irradiance	50 %
Ambient temperature	20 %
Relative humidity	20 %
Wind Speed	10 %

$$T_{mod} = T_{amb} + \frac{PoA}{u_0 + u_1 \cdot WS} \quad (1)$$

The values of the model fitting parameters u_0 and u_1 are 32.6 W/m²/°C and 3.8 Ws/m³/°C, respectively, taken after fitting the measurements of a poly-crystalline PV module.

2.2 Finite element simulations

We use COMSOL Multiphysics 5.6 for 2-dimensional Fickian diffusion simulations (equation 2).

$$\frac{\partial C(x,y,t)}{\partial t} = D \cdot \Delta C(x,y,t) \quad (2)$$

Here, C is the moisture concentration at position (x,y) and time t. A Dirichlet boundary condition (BC) sets $C(t)$ on the backsheet-air interface. The value of this BC is calculated from the T_{amb}, RH and T_{mod} assuming the same partial water vapor pressure in the air in the environment and close to the backsheet using the Arden-Buck-Equation [26]. For every material we presume Henry's Law of sorption, i.e., a linear sorption isotherm. Both D and S are taken from the literature for an EVA encapsulant and a PPE backsheet [3]. The PV cell and front glass cover are moisture barriers. The initial value of C in the entire module is presumed to be 0 kg/m³.

Due to the large number of required simulations, several initial tests at different climates are carried out to evaluate the impact of possible simplifications on the results. The climates include two moderate (Scotland, Germany), one Mediterranean (Italy), two desert (Niger, Chile), two tropical (Indonesia, Brazil) and two polar (Russia, Greenland). The interest of these locations was highlighted in [27]. The investigated parameters are:

- The upper limit of the timestep taken in the simulations, including 2 h, 4 h, 8 h, 12 h and 1 day.
- Total simulation time. Because diffusion is temperature-dependent, modules in warmer climates require less time to reach equilibrium with the environment and can be simulated during an accordingly short timeframe.
- Model geometry.
- Various simulation specific parameters, for example mesh density (the lower limit being the pre-defined setting "Very Coarse" in COMSOL), additional finite elements on material interfaces, or time-stepping tolerance (a parameter defining a criterium for the BC, possibly adding additional simulations at shorter intervals if changes are too fast)

After evaluation of these simplifications, which are described in detail in section 3.1, a model is created to keep the simulation time to a minimum while not adding a significant systematic error. This error is defined as the root mean square error between the simulation results using a possible simplification and the results using the most extensive settings. The latter are, for example, the finest evaluated mesh in addition to boundary elements, or the lowest time-step. While the model simulates the moisture profile around an entire cell in the module, C is saved only at four points of interest, shown in Figure 2. One point is between cell and backsheet, another is between two cells, and the other two are between cell and glass, in which one is in front of the center of the cell, and the other is in front of the point halfway between cell center and cell edge.

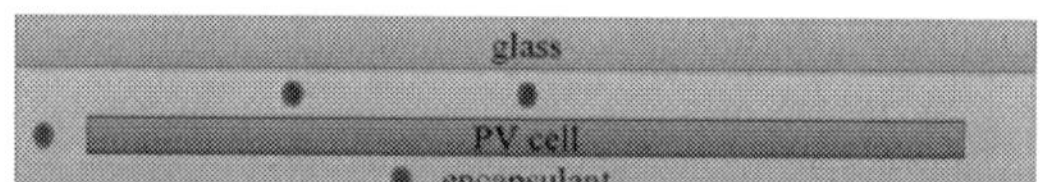

Figure 2: Points in which C is extracted, marked by red dots, before evaluating possible simplifications of the geometry. The figure is not true to scale due to the very low thickness of the layers in comparison to their other dimensions. Only half the cell is simulated due to symmetry.

2.3 Data evaluation

We extract two important parameters from the results. The first is C in front of the cell in equilibrium, C_{eq}. As C still shows yearly variations depending on seasonal changes, we define this value as the yearly average in the last simulated year. The second result is the required time to reach a certain percentage of thisvalue. While a value as high as possible, i.e., as close to the equilibrium value as possible, gives more useful information, going too high can reduce the accuracy of the algorithm due to small variations every leap year. Thus, we chose 95 % in this evaluation.

2.4 Machine learning for higher spatial resolution

We use an ML approach to find a correlation between the climatic input parameters and the simulation results, and replicate the COMSOL model results for new locations and input datasets with a faster computational process. The chosen method is the Light Gradient Boosting Machine (LGBM) [28] due to the high accuracy and very low processing time compared to similar approaches [29]. The LGBM is an ensemble model based on the gradient boosting decision tree (GBDT) approach, in which predictors are trained sequentially, making a stronger prediction model at each iteration [30]. This model is first trained and tested on the dataset obtained from the simulations. The split between training and testing is thereby 80/20. It is then applied to a different climate dataset, available in a higher spatial resolution: ERA5-Land [27]. We retrieve the climatic parameters of interest for the United States with a spatial resolution of 9 km (~0.08°).

3 RESULTS AND DISCUSSION

3.1 Simulation simplifications

Both mesh density as well as inclusion of boundary elements do not result in a notable change of the results. While a higher mesh density on the backsheet-air interface can influence the initial conditions, which are extrapolated from C in the module and the Dirichlet BC at the interface, the differences disappear after the first few simulated days. This result is not surprising, as the differential equation leads to no singularities or other problematic areas in the module geometry, which would require a fine mesh. Furthermore, changes of the moisture profile within the module only happen over a long time compared to the timesteps taken in the simulations.

Furthermore, C between cell and backsheet shows no dependency on the lateral position. Lateral diffusion only begins to influence the results close to the cell edge, contrary to the region between cell and glass, where it dominates. Thus, a large part of this area can be removed to achieve faster simulations. A 10 mm wide area from the cell edge is sufficient to keep the error negligible. Figure 3 shows the simulated area and extracted points according to these simplifications.

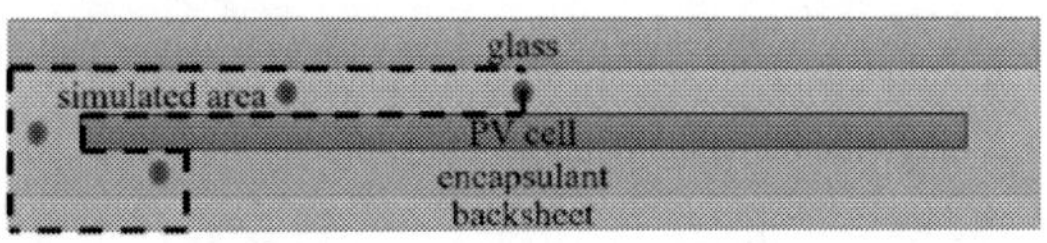

Figure 3: Simulated area and evaluated points after simplifications of the geometry.

Other parameters, however, show an obvious impact on the moisture ingress. Figure 4 displays an example of simulations using 2-h and larger time-steps in Brazil. These differences are caused by the daily changes of the climatic conditions. In general, T_{mod} is higher and RH is lower during the day than during the night. Diffusion in the module is therefore slower when RH is high. Taking an average of the climatic conditions over a time-span of 4 h or more and simulating accordingly long time-steps misses these changes. It is thus not surprising that the longer time-steps cause, in general, an overestimation of the moisture content in the module. We thus stick to a 2-h time-step in further simulations.

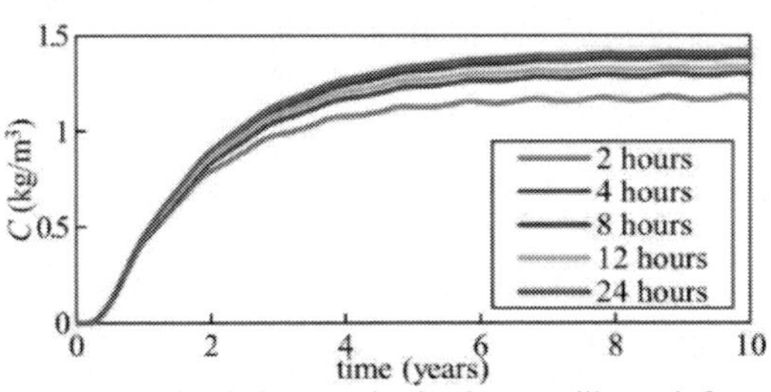

Figure 4: Simulation results in the Brazilian rainforest using different time-steps.

The strong temperature-dependence of D also influences the required total simulation time. The lower end is 15 years in the hottest climates on the Earth, for example in the tropics and some hot deserts such as the Sahara. The upper end is ~250 years, only required in central Greenland.

3.2 Simulation results

Figure 5 shows a map of the C_{eq}. Not surprisingly, PV modules in coastal and tropical regions are expected to take more moisture than in drier regions. The occurrence for different values of C_{eq} is shown in Figure 6.

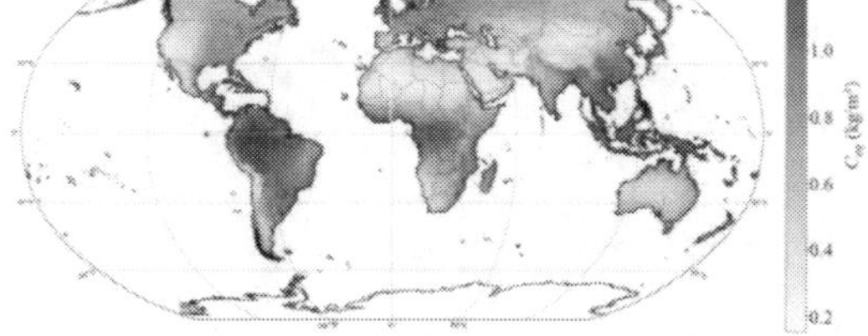

Figure 5: Worldwide mapping of C_{eq} in kg/m³.

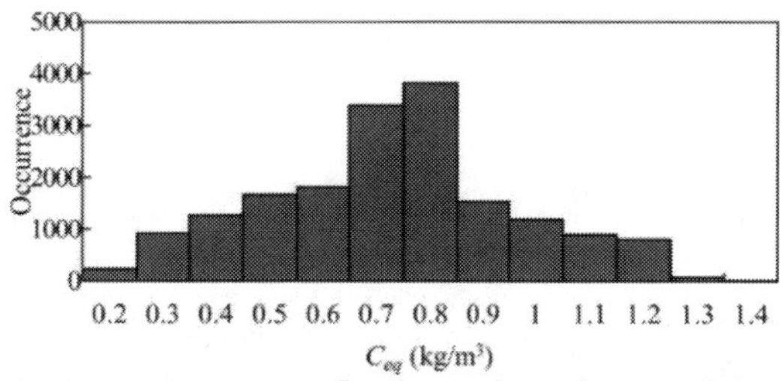

Figure 6: Occurrence of various values of C_{eq}. Each bar corresponds to an interval of 0.1 kg/m³.

While there is a correlation between the average ambient RH and C_{eq}, the spread of the data is very wide, shown in Figure 7. This result suggests that only taking RH in degradation modelling may induce a possible error for degradation modes being influenced by C. Reasons for this wide spread can include temperature differences between the climates. The solubility of encapsulant and backsheet depend on T: in hotter climates, the polymers can take up more water, thus directly impacting the results for C_{eq}. Furthermore, a strong seasonality or daily variations may impact the results. Pronounced cycles can increase drying out the module during the day or during hot and dry seasons, adding another possible discrepancy between the C_{eq} and the average RH.

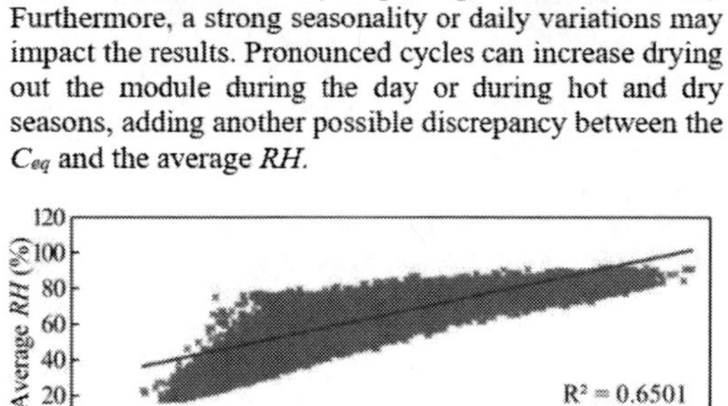

Figure 7: Correlation between C_{eq} and the average ambient RH. The R^2-value is calculated from the linear fit, which is presumed due to Henry type of sorption.

The time till equilibrium is reached shows a better correlation to the average module temperature, shown in Figure 8. The corresponding map is shown in Figure 9.

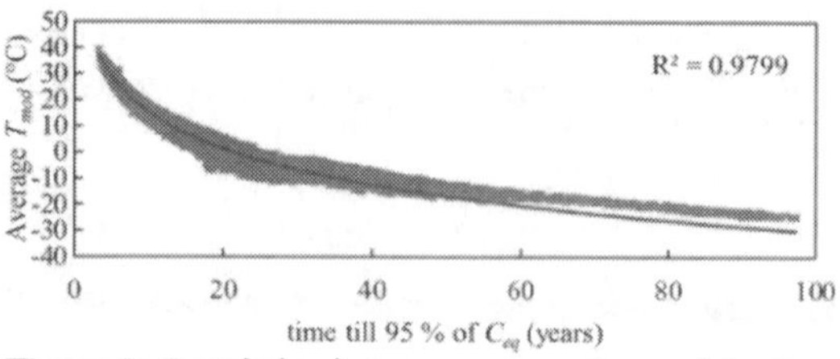

Figure 8: Correlation between average T_{mod} and the time till 95 % of C_{eq} in front of the cell is reached. The R^2-value is calculated from the logarithmic fit, which is presumed due to the Arrhenius dependence of D from T.

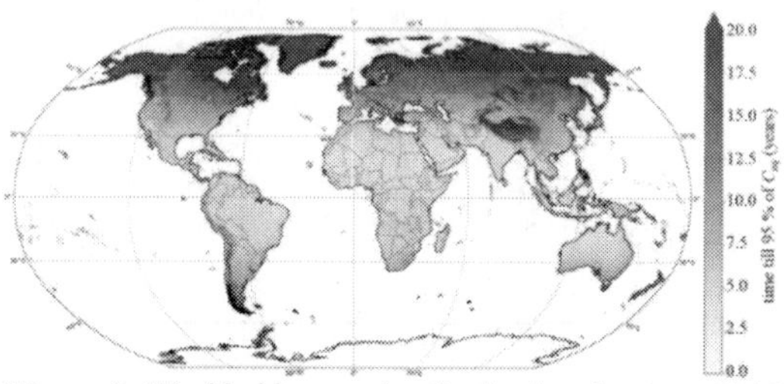

Figure 9: Worldwide mapping for the time in years till 95 % of C_{eq} is reached.

3.3 Machine learning results

Due to the large number of simulation results given for training, the coefficient of determination of the pair ML-COMSOL is 0.98. The ML approach using the LGBM method can accurately predict the parameters obtained from the simulations using only the climatic data. Figure 10 shows the results and comparison of the time till 95 % of C_{eq} for the United States in the higher resolution. This improvement can enable a more local assessment considering microclimates and including coastlines.

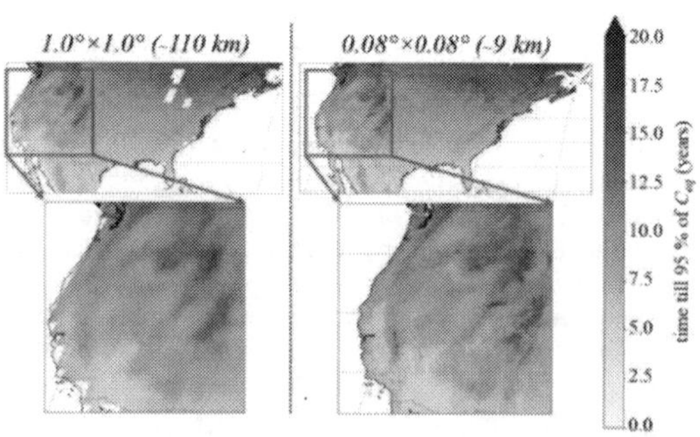

Figure 10: Comparison between lower (left) and higher (right) resolution maps for the United States replicating the COMSOL simulations on ERA5-Land data using LGBM.

4 CONCLUSION

This paper presented worldwide Fickian simulations of moisture ingress in PV modules. Comparison of the results with the ambient conditions, such as *RH*, emphasize the need to evaluate the moisture profile within PV modules for more accurate degradation predictions, even after equilibrium is reached. The time till 95 % of C_{eq} is reached can thereby be several years in hot and several decades in cold climates, further adding to the discrepancy between C and RH. The ML approach has shown to be a viable alternative to simulations to obtain some parameters, if the dataset to train the ML model is large enough. Future work in this topic will include further evaluation of the requirements of the ML model, expansion to other materials and further simulations to improve accuracy. The latter will, for example, include additional two-dimensional simulations in front of the cell, which have shown to give more accurate results in that region [3]. Expansion to other materials may have to include a more in-depth evaluation of non-Fickian behavior in said materials. As of now, EVA, TPT and PPE are the only PV encapsulant and backsheets, in which a good accuracy of the Fickian model has been experimentally verified in the field [3].

ACKNOWLEDGEMENT

The authors acknowledge the help of Dr. Nicos Martys and Xiaoxu Guan for their help in setting up the simulations on Hercules.

DISCLAIMER

Certain commercial products or equipment are described in this paper to specify adequately the experimental procedure. In no case does such identification imply recommendation or endorsement by the National Institute of Standards and Technology, nor does it imply that it is necessarily the best available for the purpose.

This paper contains modified Copernicus Atmosphere Monitoring Service information (2022). Neither the European Commission nor ECMWF is responsible for any use that may be made of the Copernicus information or data it contains [32].

REFERENCES

[1] K.-A. Weiss, L. S. Bruckman, R. H. French, G. Oreski, and T. Tanahashi, "Service Life Estimation for Photovoltaic Modules," IEA PVPS Task 13 Report, Jun. 2021.

[2] O. K. Segbefia, A. G. Imenes, and T. O. Sætre, "Moisture ingress in photovoltaic modules: A review," *Sol. Energy*, vol. 224, pp. 889–906, Aug. 2021, doi: 10.1016/j.solener.2021.06.055.

[3] S. Mitterhofer, "Modelling and monitoring of moisture transport phenomena in polymeric materials and photovoltaic modules," PhD Thesis, University of Ljubljana, Ljubljana, Slovenia, 2021. [Online]. Available: https://repozitorij.uni-lj.si/IzpisGradiva.php?id=132446&lang=eng

[4] M. D. Kempe, "Modeling of rates of moisture ingress into photovoltaic modules," *Sol. Energy Mater. Sol. Cells*, vol. 90, no. 16, pp. 2720–2738, 2006.

[5] A. B. Subramaniyan, R. Pan, J. Kuitche, and G. TamizhMani, "Quantification of environmental effects on PV module degradation: A physics-based data-driven modeling method," *IEEE J. Photovolt.*, vol. 8, no. 5, pp. 1289–1296, 2018.

[6] I. Kaaya, M. Köhl, A. P. Mehilli, S. de Cardona Mariano, and K. A. Weiß, "Modeling Outdoor Service Lifetime Prediction of PV Modules: Effects of Combined Climatic Stressors on PV Module Power Degradation," *IEEE J. Photovolt.*, vol. 9, no. 4, pp. 1105–1112, 2019.

[7] P. Hacke *et al.*, "Accelerated testing and modeling of potential-induced degradation as a function of

temperature and relative humidity," *IEEE J. Photovolt.*, vol. 5, no. 6, pp. 1549–1553, 2015.

[8] D. Carlson *et al.*, "Corrosion effects in thin-film photovoltaic modules," *Prog. Photovolt. Res. Appl.*, vol. 11, no. 6, pp. 377–386, 2003.

[9] A. Fick, "Ueber Diffusion," *Ann. Phys.*, vol. 170, no. 1, pp. 59–86, 1855.

[10] R. G. Ross Jr., "Interface design considerations for terrestrial solar cell modules," in *12th Photovoltaics Specialists Conference (PVSC)*, Baton Rogue, LA, 1976, pp. 801–806.

[11] D. Faiman, "Assessing the outdoor operating temperature of photovoltaic modules," *Prog. Photovolt. Res. Appl.*, vol. 16, no. 4, pp. 307–315, Jun. 2008, doi: 10.1002/pip.813.

[12] P. Hülsmann and K.-A. Weiß, "Simulation of water ingress into PV-modules: IEC-testing versus outdoor exposure," *Sol. Energy*, vol. 115, pp. 347–353, 2015.

[13] P. Hülsmann, K.-A. Weiß, and M. Köhl, "Temperature-dependent water vapour and oxygen permeation through different polymeric materials used in photovoltaic-modules," *Prog. Photovolt. Res. Appl.*, vol. 22, no. 4, pp. 415–421, 2014.

[14] D. Wisniewski, R. Lv, S. V. Nair, J.-N. Jaubert, T. Xu, and H. E. Ruda, "Measurement and modelling of water ingress into double-glass photovoltaic modules," *Prog. Photovolt. Res. Appl.*, vol. 27, no. 2, pp. 144–151, 2019.

[15] A. Dadaniya and N. V. Datla, "Water diffusion simulation in photovoltaic module based on the characterization of encapsulant material using in-situ gravimetric technique," *Sol. Energy Mater. Sol. Cells*, vol. 201, p. 110063, Oct. 2019, doi: 10.1016/j.solmat.2019.110063.

[16] M. Jankovec *et al.*, "In-Situ Monitoring of Moisture Ingress in PV Modules Using Digital Humidity Sensors," *IEEE J. Photovolt.*, vol. 6, no. 5, pp. 1152–1159, Sep. 2016, doi: 10.1109/JPHOTOV.2016.2583779.

[17] S. Mitterhofer, C. Barretta, L. F. Castillon, G. Oreski, M. Topič, and M. Jankovec, "A Dual-Transport Model of Moisture Diffusion in PV Encapsulants for Finite-Element Simulations," *IEEE J. Photovolt.*, vol. 10, no. 1, pp. 94–102, 2019.

[18] M. D. Kempe, A. A. Dameron, and M. O. Reese, "Evaluation of moisture ingress from the perimeter of photovoltaic modules," *Prog. Photovolt. Res. Appl.*, vol. 22, no. 11, pp. 1159–1171, 2014.

[19] M. D. Kempe, D. Panchagade, M. O. Reese, and A. A. Dameron, "Modeling moisture ingress through polyisobutylene-based edge-seals," *Prog. Photovolt. Res. Appl.*, vol. 23, no. 5, pp. 570–581, 2015.

[20] A. Dadaniya and N. V. Datla, "Multilayer Backsheet Characterization Using Diffusion Experiments and Optimization Method for Water Diffusion Simulation Inside the Photovoltaic Module," *IEEE J. Photovolt.*, vol. 10, no. 1, pp. 306–314, 2019.

[21] P. Hülsmann, "Characterization and simulation of water vapor ingress into PV-modules," PhD Thesis, Universitaet Linz, Linz, Austria, 2016.

[22] Copernicus Climate Change Service (C3S), "ERA5: Fifth generation of ECMWF atmospheric reanalysis of the global climate." Copernicus Climate Change Service (C3S), 2017. [Online]. Available: https://cds.climate.copernicus.eu/cdsapp#!/home

[23] Sandia National Laboratories, "Sandia TMY method." [Online]. Available: https://github.com/TerriaJS/aremi-tmy/tree/master/sandia

[24] D. G. Erbs, S. A. Klein, and J. A. Duffie, "Estimation of the diffuse radiation fraction for hourly, daily and monthly-average global radiation," *Sol. Energy*, vol. 28, no. 4, pp. 293–302, 1982, doi: 10.1016/0038-092X(82)90302-4.

[25] R. Perez *et al.*, "A new operational model for satellite-derived irradiances: description and validation," *Sol. Energy*, vol. 73, no. 5, pp. 307–317, Nov. 2002, doi: 10.1016/S0038-092X(02)00122-6.

[26] A. L. Buck, "New Equations for Computing Vapor Pressure and Enhancement Factor," *J. Appl. Meteorol.*, vol. 20, no. 12, pp. 1527–1532, Dec. 1981, doi: 10.1175/1520-0450(1981)020<1527:NEFCVP>2.0.CO;2.

[27] J. Ascencio-Vásquez, K. Brecl, and M. Topič, "Methodology of Köppen-Geiger-Photovoltaic climate classification and implications to worldwide mapping of PV system performance," *Sol. Energy*, vol. 191, pp. 672–685, Oct. 2019, doi: 10.1016/j.solener.2019.08.072.

[28] Microsoft and LightGBM Contributors, "LightGBM." [Online]. Available: https://lightgbm.readthedocs.io/en/v3.3.2/#

[29] J. Ascencio-Vásquez, J. Bevc, K. Reba, K. Brecl, M. Jankovec, and M. Topič, "Advanced PV Performance Modelling Based on Different Levels of Irradiance Data Accuracy," *Energies*, vol. 13, no. 9, p. 2166, May 2020, doi: 10.3390/en13092166.

[30] L. Mason, J. Baxter, P. Bartlett, and M. Frean, "Boosting Algorithms as Gradient Descent," in *Advances in Neural Information Processing Systems*, 1999, vol. 12. [Online]. Available: https://proceedings.neurips.cc/paper/1999/file/96a93ba89a5b5c6c226e49b88973f46e-Paper.pdf

[31] Copernicus Climate Change Service, "ERA5-Land hourly data from 2001 to present." ECMWF, 2019. doi: 10.24381/CDS.E2161BAC.

[32] "Copernicus Products - License agreement." Accessed: Sep. 21, 2022. [Online]. Available: https://cds.climate.copernicus.eu/api/v2/terms/static/licence-to-use-copernicus-products.pdf

This is a short version of the paper submitted to the journal Solar RRL.

It has been selected by the Committee of the WCPEC-8 for submission to Solar RRL.

QUANTIFYING THE ENERGY IMPACT OF SOILING THANKS TO THE TOOL SOILRATIO

Sébastien Arbaretaz*[1], Eric Pilat[2], Hervé Colin[2], Elias Urrejola[3]
[1]Univ. Savoie Mont Blanc, CEA, Liten, INES
[2]Univ. Grenoble Alpes, CEA, Liten, Campus INES
[3]ATAMOSTEC

[1,2]150 avenue du lac Léman, 73375 Le Bourget du Lac, France
[3]Avenida Angamos 601, Antofagasta, 1270300, Chile

ABSTRACT: Herein, a soiling measurement tool, termed "SoilRatio," and its software is presented. The originality of this fully integrated device lies on its standalone feature and its sensors on a module scale. It is able to quantify precisely, quickly, and regardless the type of soiling, the production loss of a photovoltaic plant. Most soiling measuring devices consist of two similar modules mounted side by side, one is cleaned frequently and the other is left soiled. Other systems use a photodiode optical sensor that has to be calibrated according to the type of soiling targeted or a cell protected and another exposed as sensors. The sensor surface is too small to generate results representative of commercial modules and should therefore, be processed and interpreted. The concept chosen by the CEA is to use a pair of modules and cover one of them to keep it clean. In this way, it does not require any cleaning, which ensures better accuracy and lowers operating costs. Various experiments are performed to assess its performance and improve the SoilRatio device. The specifications document of the next version is already written to integrate modules with multiple measurement areas so as to identify the soiling patterns.

1 INTRODUCTION

Soiling is the deposition of dust or pollutants on the modules surface, reducing the amount of light reaching solar cells. Therefore, it leads to an output power reduction. The increase in installed PV capacity in regions with high soiling conditions like arid or tropical climates leads to greater soiling losses on a global scale. Even when considering optimal cleaning, the soiling losses are expected to reach 4-7% of worldwide solar power production in 2023, which amount to 4-7 billion € [1]. The challenge is to find a balance between the benefit of energy savings and cleaning costs. To achieve this goal, soiling monitoring is an important stage to select the most appropriate mitigation strategy and assess the optimal cleaning frequency.

Estimation shows that almost 60,000 sensors would be needed to monitor the global installed PV capacity considering an intermediate sensors density. For the year 2024 itself, more than 15,000 sensors would be required to monitor the newly installed capacity [2]. These numbers give a representation of the size of the soiling sensor market.

Three main types of device are currently available to monitor soiling. First, the soiling stations consist of two similar modules mounted side by side. One is cleaned frequently and the other one is left exposed to soiling. The cleaning can be done manually or automatically with a rotating brush or water spray [3-4]. The soiling ratio (SR) is calculated as the ratio between the soiled module current and the clean module current. Measuring power in addition to current allows to detect non-uniform soiling events [5]. However, the lack of reproducibility of the cleaning can affect the accuracy and they cannot be considered as a stand-alone device. The Arizona State University (ASU) developed a soiling station with a pair of half-cells exposed and another pair protected by a glass which opens at solar noon during short-circuit current measurement [4]. The device is fully autonomous. However, they refer only to a surface the size of a cell and cannot measure the impact of non-uniform dust distribution that can affects commercial modules.

Second, systems use a photodiode optical sensor that has to be calibrated with the targeted type of soiling. In the Kipp&Zonen's DustIQ, a photodiode sensor measures the reflection of a monochromatic light on the backside of a soiled glass [2]. The Mars Soiling Sensor of Atonometrics takes images of the soiling deposit on a glass, illuminated by sunlight or LEDs, with an optical microscope and measures the brightness of the images [6]. The authors noted that the accuracy of this method is reduced for high soiling rates, for which the image background is difficult to distinguish from the particles [6]. These systems are less expensive and require less maintenance. However, the sensor surface is too small like the soiling station of ASU and measurements should therefore be processed and interpreted.

Third, it consists in using soiling extraction algorithms. The principle is to measure the decline in efficiency from the data of electrical output power and irradiance [7-8]. No soiling sensors are required for these methods, but soiling must be accurately distinguished from others failures and cleaning events must be detected correctly. Besides, soiling monitoring sensors are still necessary to validate soiling extraction algorithms.

Two promising methods are still under development. First, the soiling image analysis consists in taking images of panels and identify the surface cover by soiling and the brightness using statistical or machine learning algorithms [9-11]. The results can be converted into transmission or energy loss. Second, estimation models are algorithms that deduce the soiling ratio from environmental parameters, site and system characteristics. They are based on linear regression, artificial neural network, numerical simulation or physical model to estimate and predict annual, monthly or daily soiling loss.

The CEA-INES has been working since 2015 on a specific soiling station concept. Instead of cleaning the reference module, this one is prevent from soiling by covering it between each measurement in order to keep it as clean as possible. This ensures a better accuracy and lower the operating costs. All the electronics are integrated for an easy installation. The measurement processing is manage by an algorithm. to evaluate the energy losses due to soiling. A first demonstrator was manufactured in 2017 at a module scale and conclusive tests were carried out. A significant step was reached in 2021 with the manufacturing and testing of an industrial prototype, termed "SoilRatio", which is the topic of this paper.

2 DESCRIPTION OF THE "SOILRATIO" SYSTEM

The SoilRatio (Figure 1) device is a soiling station system. It assesses the soiling ratio by comparing the short-circuit current of a dirty module with that of a similar clean module. Modules are composed of 36 half-size Si cells. The tilt angle can be set to the value of the plant tilt between 0 ° and 50 ° (Figure 1.C).

Figure 1: SoilRatio (A) with the drawer closed to protect the clean module, (B) with the drawer open to measure the soiling ratio, (C) back of the SoilRatio with the adjustment of the tilt angle

The soiled module is exposed all the time and the clean module is protected in a drawer. It opens only for the measurement. Using a module instead of only a cell allows evaluating more precisely the energy loss due to soiling on commercial modules. Indeed the sensors must be big enough to take into account the distribution of soiling and spots such as bird droppings. The measurement of the short-circuit current is fully automatized. The drawer opens during 10 minutes to allow the measurement once a day around noon, the measurement can be done up to 3 times a day. It opens only if the current of the soiled module is above 50 % of its value at Standard Test Conditions. This condition avoids exposing the clean module to dirty conditions such as sandstorms or rainy days. A 5 minutes delay before measurement is set to allow the clean module to heat up and have a temperature close to the exposed module. Measurements are made every 5 seconds during the last 5 minutes of opening. This is a compromise between opening the drawer long enough to stabilize the temperature of the protected module, a long enough measurement period to avoid artifacts such as temporary shading, and not exposing the protected module for too long to avoid soiling.

The short-circuit current is the measured electrical parameters because it is proportional to the incident photon radiation captured by the cells of the photovoltaic panels. It has the advantage of being less sensitive to temperature variations than voltage and therefore power, so we avoid a correction that could be a source of uncertainty. An argument for using power instead of current is the mismatch loss and bypass diode activation. However, our modules have no by-pass diode so they are not affected by mismatch loss. SoilRatio stand-alone version includes the possibility to add a battery and a 4G/ Wi-Fi router to be operational even in remote locations. In case it is located in a black spot, data can be saved and downloaded later via an USB port. It operates with a software that makes the calibration and calculates the soiling ratio.

3 SOFTWARE DESCRIPTION

An algorithm that calculate the soiling ratio automatizes the measurement. It integrates some corrections on the short-circuit current to remove differences between the protected and the exposed modules due to temperature and angle of incidence.

The angle of incidence of the solar radiation has an impact on the generated current. In order to correct the slight parallelism error between the two modules, a self-calibration process calculates the correction coefficients. This self-calibration takes also into account the different optical characteristics between the two modules. It is made at the beginning of the measuring period and can be made when necessary, for example after a cleaning considered as perfect. It has been observed that the soiling ratio depends on the angle of incidence, which varies during the day [2]. The soiling loss tends to be lower for small angle of incidence, it means in the middle of the day. Therefore, the IEC61724-1 standard recommends using only measurements within two hours around solar noon for calculating the soiling ratio on fixed tilted system [12]. The optimal time of each day for the minimum angle of incidence is calculated according to the location of the system. This information is transmitted to the system in order to make the measurement at this specific moment.

First, the short-circuit current of the exposed module is corrected according to the temperature gap between the two modules. Even if a delay is set before the measurement in order to allow the protected module to rise in temperature, a slight temperature difference could be observed. Then the short-circuit current is corrected with the incident angle using the previously calculated coefficients. Finally, the daily soiling ratio is calculated as the ratio of the short-circuit current of the exposed module to that of the protected module. Three different ways are used to average all the measurements of each day, the mean, the median and the affine regression. The current experimental results have not made it possible to determine which formula is the most appropriate yet.

4 FIRST RESULTS

4.1 Experimental settings

The SoilRatio system was operated at a 30 ° tilt angle between July and mid-October 2021 at the CEA-INES outdoor platform named "INCAS", located at Le Bourget-du-Lac in France. The natural soiling rate during this period was very low, the soiling ratio was above 0.99. Because of this low soiling rate, various experiments were performed using a translucent film and artificially deposited dust.

4.2 Tests with a translucent film

First, a translucent film with a measured spectral hemispherical transmittance loss of 62.1 % was applied on the entire surface of the module. The average short-circuit current loss on two consecutive days was 39.9 %. Then, a measurement was made with only a strip at the bottom of the module masking one and a half cells.

This distribution is often observed as the consequence of self-cleaning phenomenon due to humidity or rain leading to solid deposit at the bottom of modules. The measured loss was slightly lower, 30.5 %, than for the whole covered module. The module is a single string module. Its current is limited by the less efficient cell, which means that if a cell is covered the current is governed by this cell. In our case, we should have expected the same level of loss between the completely covered module and just a strip. The difference between both experiments can be justified by the assumption that the cells covered in the second experiment are not among the least efficient cells. Thus, in the first experiment, the film affected the least efficient cell but not in the second, which explains the lower loss. This soiling pattern will be further studied with the next version of the soiling ratio, which has been designed especially for quantifying non-uniform soiling loss. These first experiments have validated the proper operation of the system and the software.

4.3 Tests with Arizona test dust

A large mass of 0-10 µm Arizona Test Dust was thrown on the exposed module. This dust is mainly composed of silica and silicate particles of sizes lower than 10 µm in diameter. The short-circuit current was reduced by 47.1 % just after deposition and 43.6 % after 2 days of exposure. The module appeared completely cleaned after the first rain with a soiling ratio nearing 1. Nevertheless, a small amount of fine dust was still visible on the surface of the module, which did not affect the short-circuit current. This observation would tend to show that this type of dust does not stick to the surface.

It confirms our previous observations in indoor test bench that no cementation process occurs with this type of dust, which can be explained by the absence of soluble compound. Thereby, cleaning without any mechanical action such as a waterjet, blowing or vacuuming would be sufficient.

A second artificial soiling was made with a smaller amount of Arizona Test Dust (Figure 2.A). It resulted in a measured loss of 4.3 % just after deposition and 3.9 % after 2 days of exposure. The recovery in the days following these depositions may be due to wind or dew formation at the end of the night. Indeed, we can observe that the dust had spread over the surface of the module and drops formed by dew (Figure 2.B). As the dewdrops rolled down the module, they carried away some of the dust performing a partial cleaning. It resulted in a dust accumulation at the bottom of the module, which shows the importance of the frame design not to retain the dust. The module was completely washed by the rain that occurred a few days later.

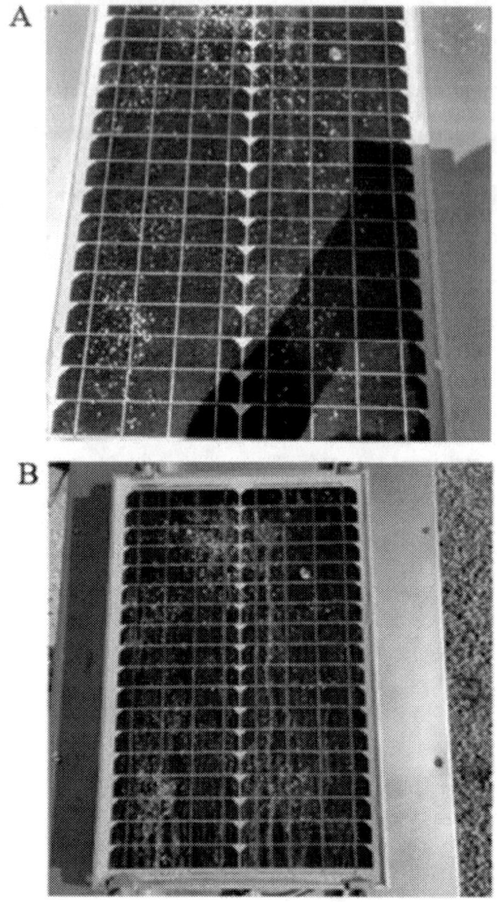

Figure 2: (A) Just after the projection of a light amount of Arizona Test Dust and (B) after one night of exposure

4.4 Tests with Atacama Desert's dust

Finally, an experiment was conducted with dust collected from the Atacama Desert, more precisely on the ATAMOS-TEC test platform (Atacama Desert Solar Platform, PSDA). The process of cementation has been studied for this location, it appears that the gypsum naturally present in this dust is responsible for the cementation [13].

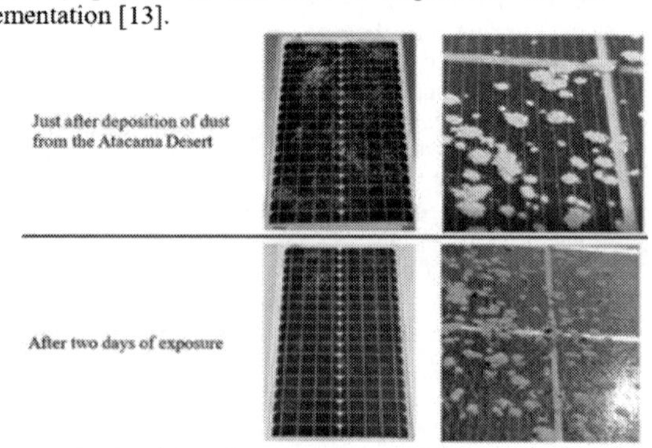

Figure 3: Module after projection of dust from the Atacama Desert and after two days of exposure

A light deposit of this dust generated a loss of 5.5% (Figure 3), a slight recovery was observed after two days of exposure, the remaining loss was 5.3 %. Unlike the Arizona Test Dust, dewdrops did not roll down and dust did not spread. The dust aggregated and adhered strongly to the surface (Figure 3); this could explain the lower recovery.

5 PERSPECTIVES

5.1 Implementation in the Atacama Desert

Soilratio prototype is currently implemented in Chile in the Atacama Desert, more precisely on the ATAMOS test platform (PSDA), so as to characterize and quantify the soiling phenomena in this so sunny and thus energy attractive location (Figure 4). The monthly soiling rate was calculated using the performance ratio in a previous study to be around 4 % per month and the soiling loss reached 35 % after 10 months of exposure [14]. SoilRatio's results are expected in the next months to provide information that is more representative.

Figure 4: SoilRatio installed at the PSDA, Atacama Desert, Chile

5.2 New configuration of the "SoilRatio"

The new configuration will integrate specific "multi areas" modules, each area being an independent sensor, in order to be able to identify the soiling pattern (i.e. the distribution over the module surface). The current version of SoilRatio is fine for measuring uniform soiling but lacks precision for measuring non-uniform soiling.

Soiling can be distributed in different patterns that affect the production differently. In the case of a homogeneous distribution of particles on the surface, the current loss is directly proportional to the optical transmission loss, the relation depends mainly on the spectral response of the cells combined with the spectral filter generated by the soiling. Soiling can also be divided into several spots, this is the case of bird droppings. The current loss depends on several parameters such as the size and opacity of the spots and the bypass diodes configuration. A frequently reported pattern that was observed during our experiments at the CEA-INES platform is a solid deposit at the bottom of modules. This distribution is often observed as the consequence of self-cleaning phenomena due to humidity or rain, in this situation also, the current decrease depends strongly of the bypass diodes configuration. In case of non-uniform distribution of soiling, short-circuit current loss can underestimate real power loss.

The next improved version of SoilRatio will identify the soiling pattern and calculate the energy loss according to the electrical architecture. The new modules will be made with four by four PERC full M2 cells. It will be divided into five areas and each area will have a junction box and a thermocouple. These sensors will provide the current and temperature measurements by areas (Figure 5). As the bottom of the module is a critical area, it will be divided into two measurement areas. An algorithm will take into account all those data, firstly to determine the soiling pattern and secondly calculate the current loss on each area. Therefore, we will be able to calculate the corresponding energy loss for a whole array by using the open access software PVMismatch and taking into account the electrical architecture of the whole system [15].

Figure 5: Module configuration of the next version of the SoilRatio

6 CONCLUSION

The SoilRatio is a stand-alone soiling station. The specificity of this fully integrated and fully automatized device lies on the use of a pair of modules and the covering of one of them to keep it clean. Instead of cleaning frequently the reference module, it is protected in a drawer and exposed only during the measuring time. In this way, it does not require any cleaning, which ensures better accuracy, lowers operating costs and works even in remote locations. The short-circuit current is measured every day for both modules when the angle of incidence is minimum. A self-calibration is made at the beginning of the measuring period to correct the slight difference between both modules. The SoilRatio system was operated between July and mid-October 2021 at the CEA-INES outdoor platform located in Le Bourget-du-Lac, France. Various experiments were performed in order to assess its performance. The experiments were conducted using a translucent film, and artificially deposited Arizona Test Dust and dust from the Atacama Desert. Several phenomena were observed such as accumulation of dust at the bottom of the module due to dew and aggregation of dust. The rain cleaned the module, allowing the soiling ratio to rise to one although a thin layer of dust was still visible. Currently, this prototype is implemented in Chile in the Atacama Desert so as to characterize and quantify the soiling phenomena in this region. The next improved version is being developed to identify the soiling pattern and calculate the energy loss according to the electrical architecture. It will integrate specific multi areas modules to characterize non-uniform soiling.

7 ACKNOWLEDGEMENTS

This work was partially supported by the French National Program "Programme d'Investissements d'Avenir - INES.2S" under Grant ANR-10-IEED-0014-01.

8 REFERENCES

[1] K. Ilse, L. Micheli, B.W. Figgis, K. Lange, D. Daßler, H. Hanifi, F. Wolfertstetter, V. Naumann, C. Hagendorf, R. Gottschalg, J. Bagdahn, "Techno-Economic Assessment of Soiling Losses and Mitigation Strategies for Solar Power Generation", Joule 2019, 3 (10), 2303, https//doi.org/10.1016/j.joule.2019.08.019.

[2] J.G. Bessa, L. Micheli, F. Almonacid, E.F. Fernández, "Monitoring photovoltaic soiling: assessment, challenges, and perspectives of current and potential strategies", iScience 2021, 24 (3), 102165, https//doi.org/10.1016/j.isci.2021.102165.

[3] S. Toth, M. Hannigan, M. Vance, M. Deceglie, "Predicting Photovoltaic Soiling From Air Quality Measurements", IEEE Journal of Photovoltaics 2020, 10 (4), 1142, https//doi.org/10.1109/JPHOTOV.2020.2983990.

[4] T. Curtis, S. Tatapudi, G.S. Tamizh Mani, "Design and Operation of a Waterless PV Soiling Monitoring Station", in Proc. 2018 IEEE 7th World Conf. Photovolt. Energy Conversion, Waikoloa, June 2018, 3407, https//doi.org/10.1109/PVSC.2018.8547820.

[5] M. Gostein, T. Düster, C. Thuman, "Accurately Measuring PV Soiling Losses With Soiling Station Employing Module Power Measurements", in Proc. 2015 IEEE 42nd Photovolt. Spec. Conf., New Orleans, June 2015, https//doi.org/10.1109/PVSC.2015.7355993.

[6] M. Gostein, S. Faullin, K. Miller, J. Schneider, B. Stueve, "Mars Soiling SensorTM", in Proc. 2018 IEEE 7th World Conf. Photovolt. Energy Conversion, Waikoloa, June 2018, 3417, https//doi.org/10.1109/PVSC.2018.8547767.

[7] A. Kimber, L. Mitchell, S. Nogradi, H. Wenger, "The effect of soiling on large grid-connected photovoltaic systems in California and the Southwest Region of the United States", in Proc. 2006 IEEE 4th World Conf. Photovolt. Energy Conversion, Waikoloa, May 2006, 2391, https//doi.org/10.1109/WCPEC.2006.279690.

[8] M.G. Deceglie, L. Micheli, M. Muller, "Quantifying Soiling Loss Directly from PV Yield", IEEE Journal of Photovoltaics 2018, 8 (2), 547, https//doi.org/10.1109/JPHOTOV.2017.2784682.

[9] S. Mehta, A.P. Azad, S.A. Chemmengath, V. Raykar, S. Kalyanaraman, "DeepSolarEye: Power Loss Prediction and Weakly Supervised Soiling Localization via Fully Convolutional Networks for Solar Panels", in Proc. 2018 IEEE Winter Conf. Appl. Comput. Vision, Munich, March, 2018, 333, https//doi.org/10.1109/WACV.2018.00043.

[10] W.K. Yap, R. Galet, K.C. Yeo, "Quantitative Analysis of Dust and Soiling on Solar PV Panels in the Tropics Utilizing Image-Processing Methods", presented at 2015 Asia-Pacific Sol. Res. Conf., Queensland, December, 2015.

[11] M. Yang, J. Ji, B. Guo, "Soiling Quantification Using an Image-Based Method: Effects of Imaging Conditions" IEEE Journal of Photovoltaics 2020, 10 (6), 1780, https//doi.org/10.1109/JPHOTOV.2020.3018257.

[12] International Electrotechnical Commission, 2017. Photovoltaic System Performance – Part 1: Monitoring (IEC 61724-1, Edition 1.0, 2017-03). CENELEC.

[13] D. Olivares, P. Ferrada, A. Marzo, J. Llanos, C. Miranda-Ostojic, V. del Campo, S. Bravo, E. Fuentealba, "Microstructural analysis of the PV module cementation process at the Solar Platform of the Atacama Desert", Sol. Energy Mater. Sol. Cells 2021, 227 (February), 111109, https//doi.org/10.1016/j.solmat.2021.111109.

[14] F. Araya, P. Ferrada, J. Rabanal, A. Marzo, E. Fuentealba, "Performance Analysis of Photovoltaics Systems Installed At Different Sites in The Atacama Desert", in Proc. 2007 32nd European Photovolt. Sol. Energy Conf., Munich, June, 2007, 1635, https//doi.org/10.4229/EUPVSEC20162016-5CO.14.6.

[15] Mark Mikofski, Bennet Meyers, Chetan Chaudhari, SunPower Corporation, Richmond, CA, "PVMismatch Project: https://github.com/SunPower/PVMismatch". 2018.

ANALYSIS OF ENCAPSULANT DISCOLORATION EFFECT ON THE TEMPERATURE OF PHOTOVOLTAIC MODULE

Arti Pareek, Roopmati Meena*, and Rajesh Gupta
Department of Energy Science and Engineering, Indian Institute of Technology Bombay, India
*Corresponding email id: meenaroopmati@iitb.ac.in

ABSTRACT: Encapsulant discoloration is a major reliability issue in a photovoltaic (PV) module that significantly affects its power generation. In addition, the discoloration can also affect the temperature of a PV module. High temperature negatively affects the reliability and power output of PV module. In this work, the effect of encapsulant discoloration on the temperature of PV modules has been quantitatively analyzed using simulation performed through electro-thermal modelling. The loss in power output obtained by the electrical model has been used to calculate the heat source for the thermal model to estimate the temperature. Temperature has been estimated for two case of uniform and non-uniform discoloration. The results show that uniform discoloration can increases the temperature by a few degrees compared with a pristine module. While non-uniform discoloration can cause a rise in temperature of upto 37 °C.

Keywords: PV module, Encapsulant, Discoloration, Electro-thermal modelling, Module temperature

1 INTRODUCTION

The reliability of a multilayer structured crystalline photovoltaic (PV) module, consisting of backsheet, glass, and solar cells gluing with the help of polymer encapsulant, is critical for the smooth functioning of the PV module. The most widely used polymer encapsulant is based on ethylene-vinyl acetate (EVA) co-polymer [1]. It provides electrical isolation, supports the mechanical structure, and maintains optical coupling while ensuring the electrical integrity of PV module. However, stresses in field-operating conditions such as moisture, ultraviolet radiation, and temperature affect the properties of EVA and prevent its reliable functioning [2]. The discoloration of EVA encapsulant from transparent to yellow or brown as shown in Fig. 1, is a common reliability issue in commercial PV modules that can cause a significant loss in power generation [3]. Discoloration of EVA is reported under all climatic conditions caused by the degradation of its polymer structure [4].

Figure 1: Discoloration in cells of the module.

Discoloration prevents solar radiation from reaching solar cells and reduces short circuit current (I_{sc}) while absorbtion of solar irradiance by EVA increases [5]. Non-uniform discoloration further causes power mismatch within the cells of the module. Ultimately, discoloration causes a decrease in the power generation of the PV module. Solar radiation incident on the module and not converted into output power due to discoloration converts into heat and remains within the PV module. The accumulated heat affects the PV module temperature. Thus, in addition to electrical attributes, the discoloration can also affect the thermal attributes of a PV module.

PV module performance is sensitive to high temperature as it can harm module performance as well as long-term reliability. High temperature is the main factor for various degradation mechanisms such as broken interconnect, broken cell, failure of the junction box, and encapsulant degradation (discoloration and delamination) [6]. Thus, it becomes important to study the effects of discoloration from temperature aspect. In this work, the effect of discoloration on the temperature of the PV module has been analysed through modelling and simulation.

Modelling and simulation provide ease of analyzing discoloration in a noise-free environment. Wherein, electro-thermal modelling approach has been used to model the effects of discoloration on the temperature of the PV module by using the electrical and thermal model of solar cells. The loss in electrical output due to discoloration has been calculated from the electrical model and used as input in the thermal model while considering the optical losses. Various cases of uniform and non-uniform discoloration have been modelled to understand their impact on the module temperature.

The layout of the paper is as follows. The methodology is given in the Section 2, results and discussion are discussed in detail in the Section 3. The conclusion is given in the Section 4.

2 METHODOLGY

In this paper, the effect of EVA-based encapsulant discoloration on the module temperature has been studied through modelling and simulation. For this purpose, the PV module with 36 cells has been modelled by an electro-thermal model of solar cell [7]. Wherein, the combination of electrical modelling and thermal modelling has been used in tandem considering the energy balance of the system. The schematic of the process used in this study is shown in Fig. 2 along with the electro-thermal model.

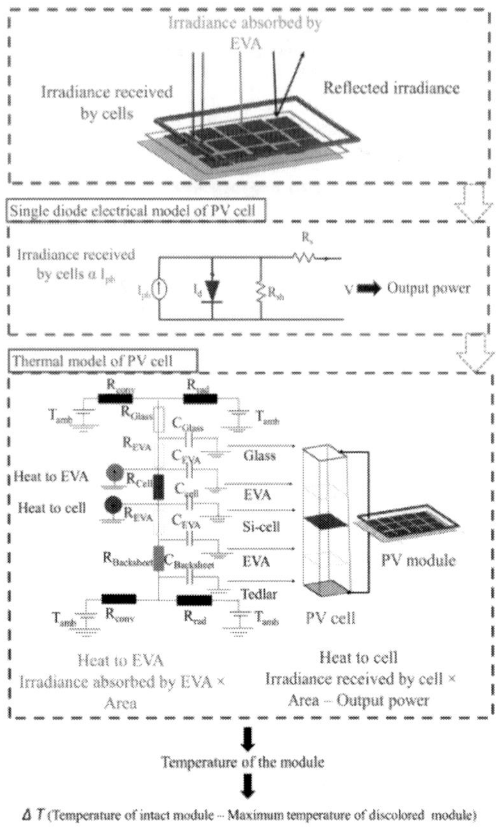

Figure 2: Schematic of the methodology used in this work.

The electrical modelling has been performed using a single diode model (SDM) of a solar cell. It has been defined using five parameters namely, reverse saturation current (I_d) of the diode, I_{sc}, ideality factor (n) of diode, R_s, and shunt resistance (R_{sh}). In this work, electrical parameters have been experimentally obtained from the multi-crystalline PV module through the current-voltage (I-V) measurements Experimently extracted electrical parameters are given in table 1.

Table 1. Parameters of SDM of the solar cell used in the simulation

Parameters	Numerical values
Light generated current (I_{ph})	10.97 A
Shunt resistance (R_{sh})	0.729 Ω
Series resistance (R_s)	0.0032 Ω
Ideality factor (n)	1.1295
Reverse saturation current (I_d)	1.04 E-08 A
Cell dimensions	156×156 mm^2

Here, loss in I_{sc} has been used to simulate the extent of discoloration in the simulation. The maximum loss in I_{sc} due to discoloration has been taken from available literature as 40 % [4]. Electrical modelling has been carried out to estimate the power output of good and discolored cells which have further been used to calculate the heat source for the thermal model to estimate the effect of EVA discoloration on the temperature of the PV module. The thermal modelling has been done by using the analogy between Ohm's law for flow of electric current and Fourier's law for the flow of heat.

Herein, each layer of the PV module (glass, front EVA, cell, back EVA, and backsheet) has been expressed by their respective thermal resistance (R_t) and thermal capacitance (C_t). The R_t and C_t have been obtained from equations 1 and 2 using dimensions and material properties of respective layers of PV module as given in [7]. Numerical values of the thermal conductivity (k), thickness (t), density (ρ), and specific heat capacity (c) of different layers are given in Table 2.

$$R_{th} = \frac{t}{kA} \tag{1}$$

$$C_{th} = c\rho At \tag{2}$$

A in equations 1 and 2 is the cross-section area of respective layers in the direction of heat flow.

Table 2: Numerical values of material properties [7]

	Glass	EVA	Cell	Backsheet
Thermal Conductivity (W/m-K)	1.8	0.35	148	0.2
Heat Capacity (J/kg-K)	500	2090	677	1250
Thickness (mm)	3	0.5	0.225	0.1
Density (kg/m^3)	3000	960	2330	1200

In the thermal model, resistance for radiation and convection heat transfer has been calculated using equations 3 and 4 respectively, where h_{rad} and h_{conv} are coefficients of radiation and convection heat transfer respectively.

$$R_{rad} = \frac{1}{h_{rad}A} \tag{3}$$

$$R_{conv} = \frac{1}{h_{conv}A} \tag{4}$$

Optical losses of solar radiations due to reflection and absorption by EVA have been inculcated in the thermal model to increase the accuracy of the results. [4]. Both electrical and thermal models have been simulated in the PSpice circuit simulator.

3 RESULTS AND DISCUSSION

In this section, the results showing the effect of uniform and non-uniform discoloration on the module temperature have been discussed in detail. The change in temperature (ΔT) has been studied through simulations using the electro-thermal model. Herein, ΔT has been defined as the difference between the maximum temperature of the discolored and the non-discolored PV module.

To examine the temperature under uniform discoloration, the extent of discoloration has been simultaneously varied in all the cells of the PV module. The change in the extent of discoloration has been modelled as loss of I_{sc}, in the step of five up to the maximum loss of 40 %, where 40 % loss in I_{sc} represents the 100 % extent of discoloration over a solar cell. In Fig. 3 ΔT with different extent of discoloration is shown. It can be observed from Fig. 3 that the uniform discoloration can increase the temperature of the module around to 2 °C. This shows that the uniform discoloration may not severely affect the PV module in aspects of temperature rise. It can be understood from the fact that the discolored EVA absorbs the incoming radiations and increase the temperature of EVA while the solar cells under it receive less light and are heated up less. Thus, the overall module temperature remains almost unchanged in the case of uniform discoloration.

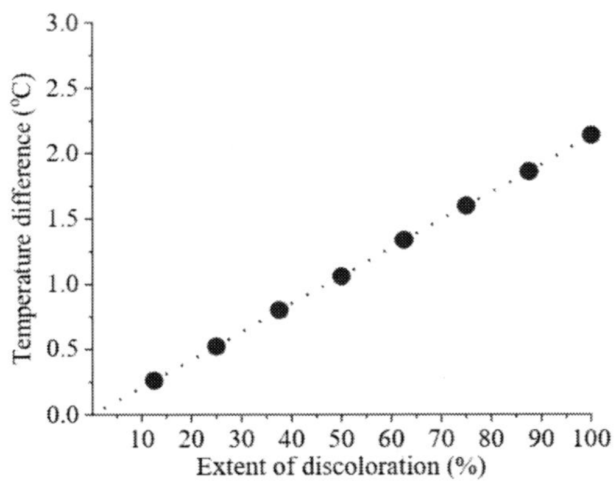

Figure 3: Temperature of the PV module with different extents of uniform discoloration.

In addition to uniform discoloration, PV modules also suffer from non-uniform EVA discoloration under field-operating conditions [8]. To study the effect of non-uniform EVA discoloration, simulation has been performed by varying discoloration extent only for one cell of the module by reducing I_{sc} in the steps of five up to 40 %. Discoloration extent has been varied for one solar cell in the module to enable the maximum thermal mismatch due to discoloration. Fig. 4 shows the ΔT with the non-uniform discoloration in the PV module. It can be observed that ΔT increases with an increasing the extent of discoloration of one of the cells. However, ΔT increases more rapidly after a discoloration extent of 25 %. It can be attributed to the current mismatch between the discolored cell and healthy cells, causing the discolored solar cell to operate under reverse bias conditions. Due to reverse bias, the excess power generated by healthy cells gets dissipated in the discolored solar cell leading to an increase in its temperature. For the scenario, where discoloration extent was less than 25 %, the current mismatch has not been enough to cause reverse bias operation of the discolored solar cell. However, the cells with less than 25 % discoloration generate electricity at a lower efficiency, which caused a slight increase in their temperature. From Fig. 4, it can be observed that non-uniform discoloration can cause a temperature rise of up to 37 °C in a discolored cell compared to healthy cells of the PV module.

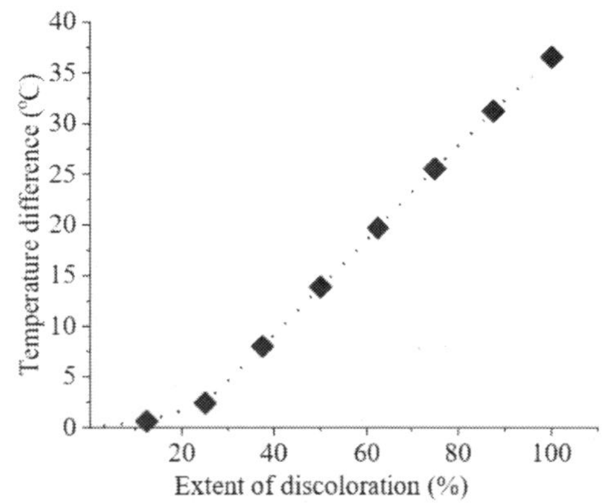

Figure 4: Maximum temperature difference (ΔT) between discolored and non-discolored PV module with the extent of discoloration in one cell of the module.

Furthermore, the non-uniform discoloration can also be simulated with variation in the number of discolored solar cells (N_d). To study the effect of non-uniform discoloration with changing N_d, it is assumed that discolored cells have a 100% discoloration extent resulting in a 40 % reduction in I_{sc}. Fig. 5 shows the ΔT with a change in N_d in the module. It can be observed that with the increase in the N_d, ΔT is decreasing. However, for more than 4 N_d, ΔT is almost negligible and unchanged. It can be understood from the fact that for up to 3 N_d, the discolored cells are in reverse bias due to the current mismatch in the PV module. As such, the unused electrical power gets dissipated in the discolored cell. With the increase in N_d, the current mismatch reduces and the discolored cells start to contribute to power generation rather than dissipation. Thus, the ΔT remains almost constant.

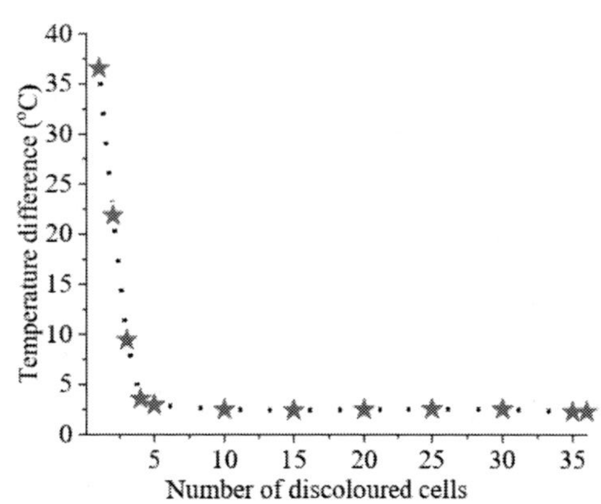

Figure 5: Maximum temperature difference (ΔT) between discolored and non-discolored PV module along the change in the number of discolored cells (N_d).

Based on the simulation results, it can be observed that uniform discoloration in the PV module does not have a significant impact on module temperature. Whereas the non-uniform discoloration can cause a temperature difference as high as 37 °C and create a thermal mismatch within the module. Non-uniform discoloration may lead to localized heating in the solar cells which can lead to the formation of hotspots causing permanent damage to cells in the long run. These effects of EVA discoloration on module temperature can become detrimental to long-term reliability and the performance under field-operating conditions.

4 CONCLUSIONS

In this work, the effect of uniform and non-uniform encapsulant discoloration on PV module temperature has been analyzed by electro-thermal modelling using PSpice circuit simulator. The results have shown that the uniform discoloration causes less effect on module temperature (about 2 °C temperature rise). In comparison, the non-uniform discoloration has affected the module temperature severely. The non-uniform discoloration caused a mismatch in the power output of the module's solar cells. As a result, cells with a high degree of discoloration dissipated the excess power generated by the good cells and caused a significant temperature rise. In the considered cases of non-uniform discoloration, the temperature of the discolored cells was increased up to 37 °C more as compared to the good cells of the module.

REFERENCES

[1] M. C. C. de Oliveira, A. S. A. Diniz Cardoso, M. M. Viana, and V. de F. C. Lins, "The causes and effects of degradation of encapsulant ethylene vinyl acetate copolymer (EVA) in crystalline silicon photovoltaic modules: A review," *Renew. Sustain. Energy Rev.*, vol. 81, no. June 2017, pp. 2299–2317, 2018, doi: 10.1016/j.rser.2017.06.039.

[2] J. H. Wohlgemuth, M. D. Kempe, and D. C. Miller, "Discoloration of PV encapsulants," *Conf. Rec. IEEE Photovolt. Spec. Conf.*, pp. 3260–3265, 2013, doi: 10.1109/PVSC.2013.6745147.

[3] A. Bouaichi *et al.*, "Experimental evaluation of the discoloration effect on PV-modules performance drop," *Energy Procedia*, vol. 119, pp. 818–827, 2017, doi: 10.1016/j.egypro.2017.07.107.

[4] R. Meena, S. Kumar, and R. Gupta, "Comparative investigation and analysis of delaminated and discolored encapsulant degradation in crystalline silicon photovoltaic modules," *Sol. Energy*, vol. 203, no. November 2019, pp. 114–122, 2020, doi: 10.1016/j.solener.2020.04.041.

[5] A. Kraft *et al.*, "Investigation of Acetic Acid Corrosion Impact on Printed Solar Cell Contacts," *IEEE J. Photovoltaics*, vol. 5, no. 3, pp. 736–743, 2015, doi: 10.1109/JPHOTOV.2015.2395146.

[6] E. Kaplani, "Detection of degradation effects in field-aged c-Si solar cells through IR thermography and digital image processing," *Int. J. Photoenergy*, vol. 2012, 2012, doi: 10.1155/2012/396792.

[7] H. Mohammed, M. Kumar, and R. Gupta, "Bypass diode effect on temperature distribution in crystalline silicon photovoltaic module under partial shading," *Sol. Energy*, vol. 208, no. August, pp. 182–194, 2020, doi: 10.1016/j.solener.2020.07.087.

[8] A. Sinha, O. S. Sastry, and R. Gupta, "Nondestructive characterization of encapsulant discoloration effects in crystalline-silicon PV modules," *Sol. Energy Mater. Sol. Cells*, vol. 155, pp. 234–242, 2016, doi: 10.1016/j.solmat.2016.06.019.

OPTIMAL METHODOLOGY FOR FORECASTING PERFORMANCE LOSS RATE OF FIELDED PHOTOVOLTAIC SYSTEMS USING ROBUST PRINCIPAL COMPONENT ANALYSIS AND AUTO REGRESSIVE INTEGRATED MOVING AVERAGE

Anna Michail*, Andreas Livera, Andreas Kyprianou and George E. Georghiou
PV Technology Laboratory, FOSS Research Centre for Sustainable Energy, University of Cyprus
Panepistimiou 1 Avenue P.C. 2109, Aglantzia, P.O. Box 20537, 1678, Nicosia, Cyprus
*Corresponding author email: michail.m.anna@ucy.ac.cy

ABSTRACT: In this paper, an optimal methodology for forecasting the performance loss rate (PLR) of photovoltaic (PV) systems is presented using robust principal component analysis (RPCA) and seasonal auto regressive integrated moving average (SARIMA) methods. The SARIMA models were identified based on four different criterions (i.e., information and empirical criterions) using three different cases for the values of the seasonal and non-seasonal differencing orders of the models. The SARIMA models were used to forecast the performance ratio (PR) time series of fielded PV systems. Then, the RPCA method was used to estimate the PLR based on the forecasted PR values. The results demonstrated that the forecasted PLR values using the different cases and criteria were in good agreement with the PLR values extracted from actual field data, exhibiting absolute error (AE) below 1.03%/year. Therefore, the RPCA method can be used to estimate the PLR, even when using forecasted PR values. Moreover, for almost all PV systems, the models with the lower p-values yielded lower values for the RMSE between the actual and forecasted PR time series. The optimal methodology was finally chosen based on a comparative analysis focusing on the forecasting accuracy of both PR and PLR, statistical significance, and method's simplicity.
Keywords: Degradation, Performance Loss Rate, Performance Ratio, Seasonal Auto Regressive Integrated Moving Average, Robust Principal Component Analysis

1 INTRODUCTION

Photovoltaic (PV) systems exhibit different degradation modes resulting to different performance loss rates (PLR). The degradation modes can be attributed to environmental and internal factors (e.g., temperature, humidity, soiling) as well as on PV module/system performance loss causes (shading, module/cell cracks and mismatches, etc.) [1]. The aforementioned factors result in reduction of PV module electricity production, which must be quantified through the estimation of the PLR.

The estimation of PLR provides the capability of determining the decay of output power of a PV system over its lifetime [2], [3]. Different statistical and comparative approaches have been reported in literature for PLR estimation including linear regression (LR), classical seasonal decomposition (CSD), year-on-year (YoY), robust principal component analysis (RPCA) and seasonal auto regressive integrated moving average (SARIMA) methods [1], [4]. Each method results to different PLR values, that in turn impart uncertainty to the financial value of a PV project [5].

Even though, different approaches have been reported in the literature, the selection of the most robust methodology for PLR estimation is an area that is not fully explored. Up to date, a standardised PLR estimation method does not exist [1]-[5], mainly due to the lack of a statistically significant model that can accurate capture the behaviour of the performance ration time series [2], [6].

In this work, an optimal methodology for forecasting the PLR of grid connected PV systems is presented using RPCA and SARIMA methods. Specifically, SARIMA models, were identified using four different criterions in order to model and forecast future values of the PR time series of PV systems installed in Cyprus. Three of them were based on information criterions, namely the Akaike Information Criterion (AIC), the modified AIC for small sample size (AICc) and the Bayesian Information Criterion (BIC). The fourth one was based on an empirical methodology , referred as empirical criterion (EC) in this paper, by utilising the autocorrelation function (ACF), partial ACF (PACF) and residuals analysis [2]. Then, the RPCA method was used to estimate the PLR based on the forecasted PR values. Finally, the optimal methodology was derived using a comparative analysis based on the forecasting accuracy for both PR and PLR, statistical significance, and method's simplicity.

2 METHODOLOGY

2.1 Experimental setup

At the outdoor test site of the University of Cyprus (UCY), 11 grid-connected PV systems of different technologies (mono-c-Si, multi-c-Si and thin films) and approximately 1 kW_p capacity, were installed in June 2006. The PV systems were installed in an open-field arrangement due to South, at a fixed tilt angle of 27.5°. The PV performance and the prevailing meteorological conditions were recorded according to the IEC 61724-1 [7], at a resolution of 1 second and stored as 15-min averages with the use of a monitoring platform. The acquired meteorological data include the in-plane irradiance (G_I), ambient temperature (T_{amb}), wind speed (W_S) and direction (W_a). The PV operational data include the back surface module temperature (T_{mod}), the array DC current (I_A), voltage (V_A) and power (P_A), and AC power to the utility grid (P_{TU}).

2.2 PV datasets of observed measurements

The measurement of P_A and G_I were initially checked for invalid data and outliers [8] and then used to construct the monthly PR time series (see Figure 1) for 11 test PV arrays over an 8-year period (June 2006 to May 2014). The temperature corrected PR [9] was not selected, since the statistical method used in the present study is suitable for time series with a seasonal behaviour [6]. The PR time series exhibit seasonality and whose values on average decrease over the years, indicating that the systems experience performance losses. This on average decrease of the values over time can be thought, in statistical analysis terms, as non-stationary behaviour. The optimal statistical model for times series that exhibit both seasonal and non-stationary behaviour is the SARIMA model [6].

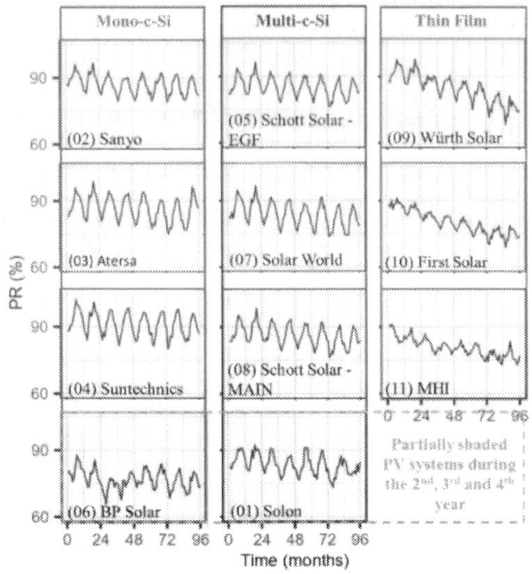

Figure 1: PR time series for the 11 PV systems installed in Cyprus categorized by module technology (i.e., 1st column: mono-c-Si, 2nd column: multi-c-Si and 3rd column: thin film). The BP Solar and the Solon PV systems suffered from partial shading during the 2nd, 3rd, and 4th years of operation (i.e., 4th row). The numbering corresponds to the numbering used in Table I.

2.3 Performance loss rate estimation approach

The 8-year monthly PR time series (i.e., 96 values) were separated into train and test subsets. The first 5 years were used as the train set, aligning with the minimum period requirements for accurate PLR estimation [10]. The following 3 years were used as the test set. The train set was used to identify and estimate the SARIMA models for each PV system using the "forecast::auto.arima" function in R [11]. Monthly PR forecasts were then generated for the next 3 years using the estimated SARIMA models ("astsa::sarima.for" function [12] in R) based on four different criterions. Three of them were based on information criterions (the AIC, AICc and the BIC). The fourth criterion was based on an empirical evaluation of the ACF, partial PACF and residuals analysis as depicted in [2]. The 3-year forecast time series were then combined with the train set to create an 8-year time series of monthly PR for each PV system. These models are linear; thus, it can be said that the estimation of the PLR is based on a linear assumption.

The RPCA method [13], was then used to obtain the PLR at the end of the 6th, 7th and 8th year for the actual and forecasted data of each PV system. For RPCA implementation, the forecasted and actual data were divided into 8×12-month time series and collected in a data matrix ($\boldsymbol{D}$), for each PV system. Finally, the SARIMA models based on the four criterions were compared in order to identify the optimal one for estimating the PR and the PLR.

The comparative analysis was based on the forecasting accuracy for both PR and PLR, statistical significance, and method's simplicity. Specifically, the forecasting accuracy of PR time series was based on mean value of the root mean square error (RMSE) [14] between the actual and forecasted PR time series. The forecasting accuracy for the PLR was based on the mean value of the absolute error (AE) [14] between the actual and forecasted PLR for the last 3 years. Moreover, four different values were used for both the mean RMSE of PR and AE of PLR for each PV technology, the product of which constitutes the forecasting accuracy score. For evaluating the statistical significance, the adequacy of the models was based on the residuals' behaviour. In particular, the statistical significance score was based on the p-values of the portmanteau lack-of-fit test proposed by Ljung and Box [6]. Furthermore, the EC methodology was assumed to be a more empirical approach and a good understanding of the ARIMA theory was needed to identify the SARIMA models. In contrast, with the criteria based on the three different information criterions, simple and automatic functions were utilised for the identification of the SARIMA models, therefore they were assumed simpler than EC. However, it is worth mentioned that even if automatic functions can identify the SARIMA models, a "manual" check of the results is needed to verify the statistical adequacy of the models. Lastly, the total score of each methodology is the product of the score for the forecasting accuracy, statistical significance, and method's simplicity, with the preferred methodology exhibiting the lowest total score.

3 RESULTS

3.1 Identification of SARIMA models orders

Since monthly PR aggregation was chosen, the seasonality s was set equal to 12 and three different cases were used for the values of the differencing, d, and seasonal differencing, D. In the present study, $d \neq 0$ and $D \neq 0$ can be translated as that the monthly and yearly (since $s = 12$) PR values on average decay over time, respectively. In case 1 (C1), the d value was chosen automatic from the algorithm (either 0 or 1) and the D was set equal 1. In case 2 (C2), both d and D were chosen automatic from the algorithm. Lastly, in case 3 (C3) both d and D were set equal 1, which values were also used for the application of the EC.

Table I summarizes the derived SARIMA models based on different criterions and cases. The SARIMA models identified in case 1 and 2 derived identical results, for this reason from now on C1 and C2 will be both referred as C1. In C1 the identified SARIMA models using the different information criterions yielded identical results for 7 out of 11 PV systems (expect for the BP solar, SolarWorld, Schott Solar (MAIN) and Würth Solar PV systems), however, none of the derived models was in agreement with the ones derived from the EC.

On the other hand, in C3 the resulted models yielded identical results for 6 out of 11 PV systems (expect for the Atersa, Schott Solar (EGF), BP solar, SolarWorld and Schott Solar (MAIN) PV systems). However, the model identified in C3 using the AIC yielded to the same model EC for the Atersa PV system. Furthermore, the models identified by the three cases using the information criterions resulted to identical models for the Sanyo, First Solar and Mitsubishi Heavy Industries (MHI) PV systems.

Focusing on the estimated parameters (summation of p, q, P and Q) of each model, it can be observed that the models of C3 and EC in most cases have higher summed value than that of C1. Lastly, the BIC criterion yields

Table I: Identified SARIMA models $(\boldsymbol{p}, \boldsymbol{d}, \boldsymbol{q}), (\boldsymbol{P}, \boldsymbol{D}, \boldsymbol{Q})_s$ for the test PV systems based on the different cases and the AIC, AICc, BIC and EC criteria.

a/a	Manufacturer	**C1** ($d = arbitrary, D = 1, s = 12$) **or** **C2** ($d = arbitrary, D = arbitrary, s = 12$)			**C3** ($d = 1, D = 1, s = 12$)			EC
		AIC	AICc	BIC	AIC	AICc	BIC	
01	Solon *	$(1,0,0)(1,1,0)_{12}$	$(1,0,0)(1,1,0)_{12}$	$(1,0,0)(1,1,0)_{12}$	$(0,1,1)(0,1,1)_{12}$	$(0,1,1)(0,1,1)_{12}$	$(0,1,1)(0,1,1)_{12}$	$(1,1,1),(1,1,1)_{12}$
02	Sanyo	$(0,1,1)(1,1,1)_{12}$	$(0,1,1)(1,1,1)_{12}$	$(0,1,1)(1,1,1)_{12}$	$(0,1,1)(1,1,1)_{12}$	$(0,1,1)(1,1,1)_{12}$	$(0,1,1)(1,1,1)_{12}$	$(2,1,1),(1,1,1)_{12}$
03	Atersa	$(0,0,0)(1,1,0)_{12}$	$(0,0,0)(1,1,0)_{12}$	$(0,0,0)(1,1,0)_{12}$	$(2,1,1)(1,1,1)_{12}$	$(2,1,0)(1,1,0)_{12}$	$(0,1,1)(1,1,0)_{12}$	$(2,1,1),(1,1,1)_{12}$
04	Suntechnics	$(0,0,0)(1,1,0)_{12}$	$(0,0,0)(1,1,0)_{12}$	$(0,0,0)(1,1,0)_{12}$	$(2,1,0)(1,1,1)_{12}$	$(2,1,0)(1,1,1)_{12}$	$(2,1,0)(1,1,1)_{12}$	$(3,1,1),(1,1,1)_{12}$
05	Schott Solar (EGF)	$(0,0,0)(1,1,1)_{12}$	$(0,0,0)(1,1,1)_{12}$	$(0,0,0)(1,1,1)_{12}$	$(2,1,0)(1,1,0)_{12}$	$(2,1,0)(1,1,0)_{12}$	$(0,1,1)(1,1,0)_{12}$	$(3,1,1),(1,1,1)_{12}$
06	BP Solar *	$(0,1,2)(1,1,0)_{12}$	$(2,1,0)(1,1,0)_{12}$	$(0,1,0)(1,1,0)_{12}$	$(0,1,2)(1,1,0)_{12}$	$(2,1,0)(1,1,0)_{12}$	$(0,1,0)(1,1,0)_{12}$	$(3,1,2),(1,1,0)_{12}$
07	SolarWorld	$(0,0,0)(1,1,1)_{12}$	$(0,0,0)(1,1,1)_{12}$	$(0,0,0)(1,1,0)_{12}$	$(2,1,1)(1,1,0)_{12}$	$(2,1,0)(1,1,0)_{12}$	$(2,1,0)(1,1,0)_{12}$	$(3,1,1),(1,1,1)_{12}$
08	Schott Solar (MAIN)	$(0,0,0)(1,1,1)_{12}$	$(0,0,0)(1,1,1)_{12}$	$(0,0,0)(1,1,0)_{12}$	$(2,1,0)(1,1,0)_{12}$	$(2,1,0)(1,1,0)_{12}$	$(0,1,1)(1,1,0)_{12}$	$(3,1,1),(1,1,1)_{12}$
09	Würth Solar	$(1,0,0)(1,1,0)_{12}$	$(0,0,0)(1,1,0)_{12}$	$(0,0,0)(1,1,0)_{12}$	$(3,1,0)(1,1,0)_{12}$	$(3,1,0)(1,1,0)_{12}$	$(3,1,0)(1,1,0)_{12}$	$(3,1,1),(1,1,1)_{12}$
10	First Solar	$(1,1,0)(1,1,0)_{12}$	$(1,1,0)(1,1,0)_{12}$	$(1,1,0)(1,1,0)_{12}$	$(1,1,0)(1,1,0)_{12}$	$(1,1,0)(1,1,0)_{12}$	$(1,1,0)(1,1,0)_{12}$	$(3,1,1),(1,1,1)_{12}$
11	MHI	$(0,1,1)(1,1,0)_{12}$	$(0,1,1)(1,1,0)_{12}$	$(0,1,1)(1,1,0)_{12}$	$(0,1,1)(1,1,0)_{12}$	$(0,1,1)(1,1,0)_{12}$	$(0,1,1)(1,1,0)_{12}$	$(3,1,1),(1,1,1)_{12}$

models with estimated parameters equal or less than the other information criterions, which was expected due to its formula [6].

3.2 Model adequacy assessment

In the present study the forecasted PR values were used to forecast the yearly PLR. Thus, it is vital to capture as best as possible the actual behavior of the PR time series of each PV system. The evaluation of the goodness of fit of the SARIMA models was divided into two pillars, (a) diagnostic checks of the residuals of the identified models, and (b) evaluation of the estimation of PLR and the PR time series based on performance metrics (i.e., RMSE and AE).

After the SARIMA models were identified, diagnostic checks (pillar (a)) were needed to be applied to the fitted models to assure the quality of fitting. The most used and appropriate checks are applied to the residuals of the fitted models, which can indicate necessary modifications to the selected model. Box et al. in [6] proposed two diagnostic checks for the model's residuals, the autocorrelation check and the cumulative periodogram of the residuals. In the present study both diagnostic checks were used, however the results presented herein focus on the autocorrelation of the residuals.

Autocorrelation check refers to the visual check of the ACF graphs of the residuals, as showed in Figure 2.a. The approximate upper and lower bounds used for the standard error of a single autocorrelation is $(-1/n) \pm (2/\sqrt{n})$ (where n is the number of observed residuals). If the residuals' autocorrelation coefficients fall within these bounds, then the model is assumed to be adequate.

Table II summarises the analysis of the ACF graphs for each case. C1 and specifically the models based on BIC yielded the highest mean value of 1.5 for out-falling lags per model; indicating that the model does not fully capture the PR behaviour [6]. On the contrary, the models based on EC resulted in a significant lower value for out-falling lags per model with a mean value of 0.4.

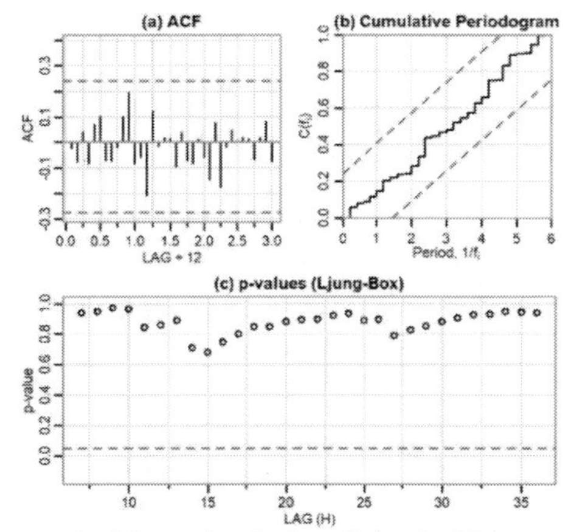

Figure 2: Diagnostic checks of the SARIMA model obtained using the EC for the BP Solar PV system including the (a) ACF graph, (d) cumulative periodogram, and (c) p-values of the Ljung-Box portmanteau lack-of-fit test.

Table II: Position (Pos.) and number (No.) of out-following lags of the residual ACF boundaries for the SARIMA models of each PV system based on the different cases and the AIC, AICc, BIC and EC criteria.

a/a	Manufacturer	**C1** ($d = arbitrary, D = 1, s = 12$) **or** **C2** ($d = arbitrary, D = arbitrary, s = 12$)						**C3** ($d = 1, D = 1, s = 12$)						EC	
		AIC		AICc		BIC		AIC		AICc		BIC			
		Pos.	No.	Pos.	No.	Pos.	No.	Pos.	No.	Pos.	No.	Pos.	No.	Pos.	No.
01	Solon *	-	0	-	0	-	0	-	0	-	0	-	0	-	0
02	Sanyo	5	1	5	1	5	1	5	1	5	1	5	1	$r_5^2(\hat{a})$	1
03	Atersa	3 and 24	2	3 and 24	2	3 and 24	2	-	0	-	0	3 and 24	2	-	0
04	Suntechnics	3, 21 and 24	3	3, 21 and 24	3	3, 21 and 24	3	-	0	-	0	-	0	-	0
05	Schott Solar (EGF)	3	1	3	1	3	1	-	0	-	0	3, 21 and 24	3	-	0
06	BP Solar *	-	0	14	1	2	1	-	0	14	1	2	1	-	0
07	SolarWorld	3	1	3	1	3	1	-	0	-	0	-	0	-	0
08	Schott Solar (MAIN)	3	1	3	1	3, 21 and 24	3	-	0	-	0	3, 21 and 24	3	-	0
09	Würth Solar	24	1	4 and 24	2	4 and 24	2	24	1	24	1	24	1	-	0
10	First Solar	-	0	-	0	-	0	-	0	-	0	-	0	-	0
11	MHI	7, 13 and 21	3	7, 13 and 21	3	7, 13 and 21	3	7, 13 and 21	3	7, 13 and 21	3	7, 13 and 21	3	7, 13 and 21	3
Mean			**1.2**		**1.4**		**1.5**		**0.5**		**0.5**		**1.3**		**0.4**

Table III: Mean values of p-values of Ljung-Box portmanteau lack-of-fit test using the (using the "stats::Box.test" function in R, at lags between 1 to 36) for the SARIMA models of each PV system using the different cases and the AIC, AICc, BIC and EC criteria.

a/a	Manufacturer	**C1** ($d = arbitrary, D = 1, s = 12$) **or** **C2** ($d = arbitrary, D = arbitrary, s = 12$)			**C3** ($d = 1, D = 1, s = 12$)			EC
		AIC	AICc	BIC	AIC	AICc	BIC	
01	Solon *	0.6794	0.6794	0.6794	0.4804	0.4804	0.4804	0.8254
02	Sanyo	0.4234	0.4234	0.4234	0.4234	0.4234	0.4234	0.4028
03	Atersa	0.1142	0.1142	0.1142	0.7352	0.6182	0.2560	0.7352
04	Suntechnics	0.1020	0.1020	0.1020	0.7554	0.7554	0.7554	0.9469
05	Schott Solar (EGF)	0.1317	0.1317	0.1317	0.6015	0.6015	0.0202	0.7779
06	BP Solar *	0.7702	0.3550	0.1449	0.7702	0.3550	0.1449	0.8707
07	SolarWorld	0.2942	0.2942	0.1650	0.4744	0.5101	0.5101	0.6603
08	Schott Solar (MAIN)	0.3256	0.3256	0.0676	0.6082	0.6082	0.0761	0.9079
09	Würth Solar	0.5515	0.1045	0.1045	0.8877	0.8877	0.8877	0.9241
10	First Solar	0.8524	0.8524	0.8524	0.8524	0.8524	0.8524	0.8417
11	MHI	0.1612	0.1612	0.1612	0.1612	0.1612	0.1612	0.0757
Mean for mono-c-Si PV		0.2132	0.2132	0.2132	0.6380	0.5990	0.4783	0.6950
Mean for multi-c-Si PV		0.2505	0.2505	0.1215	0.5614	0.5733	0.2021	0.7820
Mean for thin film PV		0.5217	0.3727	0.3727	0.6338	0.6338	0.6338	0.6138
Mean for partial shaded PV		0.7248	0.5172	0.4121	0.6253	0.4177	0.3127	0.8480
Mean for all PV		**0.4005**	**0.3221**	**0.2679**	**0.6136**	**0.5685**	**0.4153**	**0.7244**

To quantify the autocorrelation of the residuals a second check was performed as well, based on the portmanteau lack-of-fit test proposed by Ljung and Box [6]. Figure 2.c illustrate the p-values of the Ljung-Box portmanteau test of the SARIMA model obtained using the EC for the BP Solar PV system. Table III summarises the mean of the p-values for each SARIMA model per PV system using the different cases and criteria. It can be observed that the results of Ljung-Box portmanteau test were consistent with the residuals' autocorrelation coefficients.

In particular, the p-values of the EC models were higher (in most cases) compared with other criterions, indicating that behaviour of their residuals had higher possibility to follow a white noise behaviour. On the other hand, the p-values of the C1-BIC models exhibited the lowest values. However, the p-values for all the models (except the C3-BIC model for the Schott Solar (EGF) PV system) were higher than the significance level of $a = 0.05$, indicating that the fitted SARIMA models are adequate to model the PR time series (no lack of fit).

For pillar (b), during the model adequacy assessment, all criterions showed good fitting for the forecasted PR values, with RMSE values below 6% (see Figure 3). In general, all the cases resulted to similar RMSE values for each PV system. However, for the BP Solar (no. 6) PV system the RMSE values vary from 3.4% to 5.8%, with the BIC criterion yielding the lower value in contrast to the results of the Ljung-Box portmanteau lack-of-fit test where BIC had the lowest p-values.

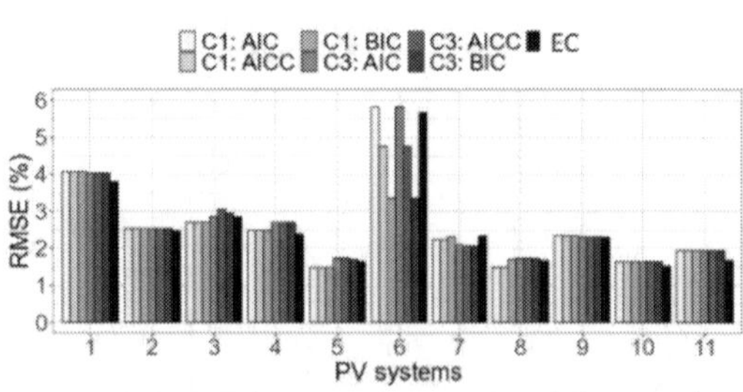

Figure 3: RMSE between the actual and forecasted PR data for the identified SARIMA model of each PV system based on the different cases and criteria.

This contradicts the theory that a model is adequate when its residuals behave as a white noise. Comparing the results of Table III and Figure 3 it is observed that in almost all cases the models with the lower p-values, hence lower probability for their residuals to follow a white noise, yield better RMSE values. However, there are some cases, such as for Solon and Würth Solar PV system (no. 1 and 9, respectively) where the model based on EC resulted to the highest mean p-value and the lowest RMSE value.

3.3 Estimation of the PLR

The RPCA data was used to define the PLR of the PV systems. Specifically, the PLR was calculated using the actual and forecasted PR timeseries at the end of the 6th, 7th, and 8th year for all the 11 PV systems. Figure 4 shows the PLR calculated from the RPCA actual dataset for these three years of the 11 PV systems and the maximum, mean, median and minimum value categorized by PV technology.

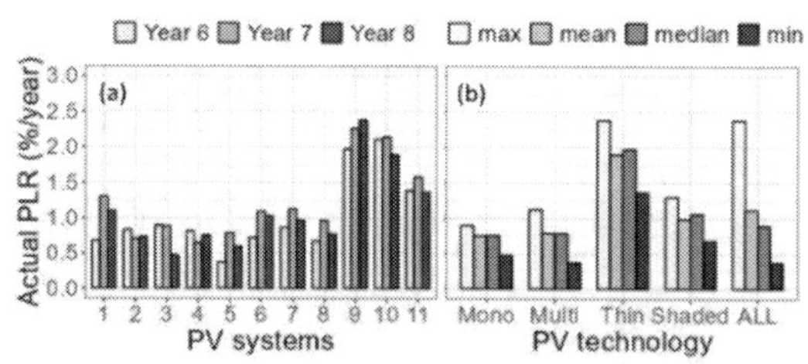

Figure 4: PLR for the RPCA actual datasets (a) for the 6th 7th and 8th year for each PV system and (b) maximum, mean, median and minimum values over the last 3 years categorized by PV technology.

The median and mean PLR values fall within the reported values in literature, i.e., median, and mean value between 0.5 – 0.6%/year and 0.8 – 0.9%/year, respectively for c-Si and with median value for thin-film greater than 1%/year [3]. These results confirm that the RPCA analysis can be used for accurate linear PLR estimations.

By comparing the actual and forecasted PLR values, it was observed that all the derived models manage to estimate the PLR sufficiently, with AE below 1.03%/year (Figure 5.a). Generally, all the criterions exhibited higher values of AE of PLR for the partially shaded PV systems. The EC showed better results for thin-film technology with mean AE equal to 0.16%/year, while the C1: BIC achieved AE < 0.51%/year and mean AE < 0.34%/year for over all the PV technologies (Figure 5.b).

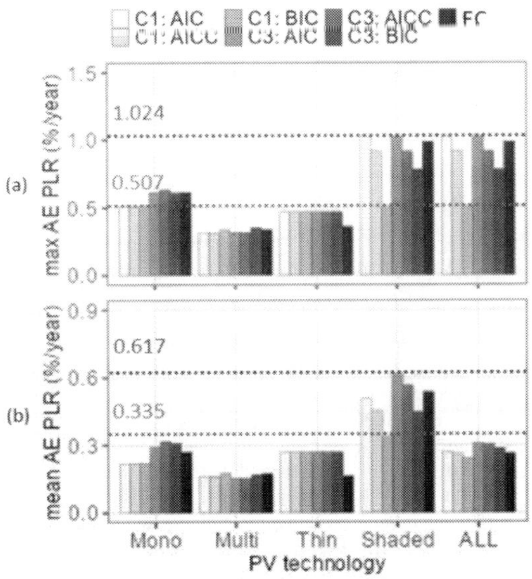

Figure 5: (a) Maximum and (b) mean value of the absolute error (AE) of the PLR values between the RPCA actual and forecasted data based on the different cases and criteria categorised to each PV technology.

3.4 Comparative analysis

To indicate the optimal model for PLR estimation, a comparative analysis was performed, and the results are summarised in Table IV. All parameters are presented in manner that the lowest values indicate the most preferable method. The optimal method for forecasting the PLR of PV systems is the one achieving the lowest score.

Based on the forecasting accuracy, it can be observed that all the criteria of C1 resulted to better forecast results than the ones of C3. This indicates that the assumption that all the PV systems have a non-seasonal, hence monthly, differencing ($d = 1$) used in the automated function was not the most appropriate. However, this assumption functioned better using the EC, where the orders and parameters of each SARIMA model were estimated based on the empirical methodology focused strictly on ACF, PACF and residuals analysis.

Furthermore, the BIC criterion yielded better results in comparison to the AIC and AICc in both C1 and C3 cases. The most preferable forecasting accuracy score was achieved by the models of C1-BIC with score of 0.117, followed by C1-AICc and EC with 0.164 and 0.168, respectively. On the contrary, the models of C3-AIC achieved the least preferable score of 0.350.

Even if the C1-BIC achieved the most preferable score in forecasting accuracy, its statistical significance scores were quite high. Based on the statistical significance score, the models of EC yielded the most preferable score of 1.380. However, the simplicity of the EC was set equal to 4 as this method was assumed to be a more empirical approach, where a more deeply understanding of the ARIMA theory was needed. Finally, the models of C1-BIC achieved the most preferable total score of 0.872, followed by EC with 0.929.

4 CONCLUSIONS

An optimal methodology for estimating accurately the PLR of fielded PV systems using SARIMA and RPCA methods was presented in this paper. Four different criterions and three different cases were used (and compared) for deriving the SARIMA models.

During the model adequacy assessment, all the methods showed good fitting of the forecasted PR values, with the RMSE values being below 6%. The analysis of the residuals, and the inspection of the residuals' autocorrelation coefficients showed that the models of C1-BIC had the highest mean value, 1.5 of out-falling lags per model. On the contrary, the models of EC resulted in a significant lower value of out-falling lags per model with a mean value of 0.4.

Additionally, the Ljung-Box portmanteau lack-of-fit test showed results that were consistent with the residuals' autocorrelation coefficients. In particular, the p-values of the models of EC were higher in most cases compared with the other methods, indicating that behaviour of their residuals has higher possibility to follow a white noise behaviour. On the other hand, the p-values of the models of C1-BIC exhibit the lowest values. However, the p-values for almost all the models were higher than the significance level of $a = 0.05$, indicating that the fitted SARIMA models are adequate and do not lack of fit.

The results also showed that for almost all PV systems the models with the lower p-values yielded better values for the RMSE between the actual and forecasted PR time series. Nevertheless, this contradicts the theory that a model is adequate and considered good when its residuals

Table IV: Comparative analysis of the different cases based on the AIC, AICc, BIC and EC criteria to identify the SARIMA models and the total score of each method. All parameters are presented in manner that the lowest values indicate the most preferable method.

Criterion	Forecasting accuracy									Statistical significance	Simplicity	**Total score**
	Mean value of RMSE of the PR (%) time series				Mean value of AE of the PLR (%/year) value for the 6th, 7th, and 8th year				Score	1/Mean (p–values)		
	Mono-c-Si	Multi-c-Si	Thin Film	Partial Shaded	Mono-c-Si	Multi-c-Si	Thin Film	Partial Shaded				
C1: AIC	2.559	1.723	1.966	4.945	0.220	0.162	0.269	0.505	0.206	2.497	2.000	**1.031**
C1: AICc	2.559	1.723	1.963	4.413	0.220	0.162	0.268	0.451	0.164	3.104	2.000	**1.020**
C1: BIC	2.559	1.826	1.963	3.719	0.220	0.174	0.268	0.335	0.117	3.733	2.000	**0.872**
C3: AIC	2.676	1.852	1.956	4.927	0.292	0.152	0.268	0.617	0.350	1.630	2.000	**1.142**
C3: AICc	2.750	1.840	1.956	4.395	0.316	0.152	0.268	0.563	0.316	1.759	2.000	**1.112**
C3: BIC	2.716	1.825	1.956	3.701	0.307	0.169	0.268	0.447	0.223	2.408	2.000	**1.074**
EC	2.557	1.875	1.830	4.735	0.269	0.172	0.163	0.535	0.168	1.380	4.000	**0.929**

behave as a white noise. Moving to the results of the PLR, the actual and forecasted PLR values were in agreement with the linear PLR estimation reported in literature for the c-Si and thin-film technologies, confirm the reliability of the RPCA analysis even by using forecasting PR values. Comparing the actual and forecasted PLR values it was observed that all methods manage to estimate the PLR sufficiently with AE below 1.03%/year.

Finally, the comparative analysis showed that the values of the total score of all methods were very closed (up to 0.27 difference). The models of C1-BIC had the lowest scores for the forecasting accuracy, but the statistical significance scores of the method was quite high. However, this method achieved the most preferable score with a total score of 0.872 followed by EC with 0.929. Future work will focus on provide insights on how and if the diagnostic checks can improve the models' forecasting accuracy and the statistical significance between their results. Lastly, the methodology proposed in this work should be tested on different PV systems topologies and locations, as well as compared with other linear and non-linear methodologies reported in literature.

5 ACKNOWLEDGEMENTS

This work was funded by the ROM-PV project, which is supported under the umbrella of SOLAR-ERA.NET Cofund 2 by the Research and Innovation Foundation of Cyprus (RIF, P2P/SOLAR/0818/0009), the Spanish Ministry of Economy, Industry and Competitiveness – State Research Agency (MINECO-AEI, PCI2019-111852-2) and the Greek General Secretariat for Research and Technology (GSRT, T11EPA4-00023). SOLAR-ERA.NET is supported by the European Commission within the EU Framework Programme for Research and Innovation HORIZON 2020 (Cofund ERA-NET Action, N° 786483).

6 REFERENCES

[1] A. Phinikarides, N. Kindyni, G. Makrides, and G. E. Georghiou, "Review of photovoltaic degradation rate methodologies," *Renew. Sustain. Energy Rev.*, vol. 40, pp. 143–152, Dec. 2014, doi: 10.1016/j.rser.2014.07.155.

[2] E. Pieri, A. Kyprianou, A. Phinikarides, G. Makrides, and G. E. Georghiou, "Forecasting degradation rates of different photovoltaic systems using robust principal component analysis and ARIMA," *IET Renew. Power Gener.*, vol. 11, no. 10, pp. 1245–1252, 2017, doi: 10.1049/iet-rpg.2017.0090.

[3] D. C. Jordan and S. R. Kurtz, "Photovoltaic Degradation Rates - An Analytical Review," *Prog. Photovoltaics Res. Appl.*, vol. 21, no. 1, pp. 12–29, Jan. 2013, doi: 10.1002/pip.1182.

[4] I. Romero-Fiances *et al.*, "Impact of duration and missing data on the long-term photovoltaic degradation rate estimation," *Renew. Energy*, vol. 181, pp. 738–748, Jan. 2022, doi: 10.1016/j.renene.2021.09.078.

[5] R. H. French *et al.*, "IEA PVPS T13-22:2021: Assessment of Performance Loss Rate of PV Power Systems," 2021. [Online]. Available: www.iea-pvps.org.

[6] G. E. P. Box, G. M. Jenkins, G. C. Reinsel, and G. M. Ljung, *Time Series Analysis Forecasting and Control*, 5th ed. John Wiley & Sons Inc., Hoboken, New Jersey, 2015.

[7] IEC 61724-1:2017, "Photovoltaic system performance - Part 1: Monitoring," vol. 32, pp. 2846–2855, 2017.

[8] A. Phinikarides, G. Makrides, and G. E. Georghiou, "Estimation of the Degradation Rate of Fielded Photovoltaic Arrays in the Presence of Measurement Outages," in *32nd European PV Solar Energy Conference and Exhibition*, pp. 1754–1757.

[9] T. Dierauf, A. Growitz, S. Kurtz, J. L. B. Cruz, E. Riley, and C. Hansen, "Weather-Corrected Performance Ratio," Golden, CO (United States), Apr. 2013. doi: 10.2172/1078057.

[10] G. Makrides, B. Zinsser, M. Schubert, and G. E. Georghiou, "Performance loss rate of twelve photovoltaic technologies under field conditions using statistical techniques," *Sol. Energy*, vol. 103, no. May, pp. 28–42, 2014, doi: 10.1016/j.solener.2014.02.011.

[11] Rob J. Hyndman and Yeasmin Khandakar, "Automatic Time Series Forecasting: The forecast Package for R," *J. Stat. Softw.*, vol. 27, no. 3, p. 22, 2008, [Online]. Available: http://www.jstatsoft.org/%0Ahttp://www.jstatsoft.org/v27/i03/paper.

[12] D. Stoffer, "Applied Statistical Time Series Analysis: Package 'astsa.'" 2021, [Online]. Available: https://cran.r-project.org/web/packages/astsa/astsa.pdf.

[13] A. Kyprianou, A. Phinikarides, G. Makrides, and G. E. Georghiou, "Definition and Computation of the Degradation Rates of Photovoltaic Systems of Different Technologies with Robust Principal Component Analysis," *IEEE J. Photovoltaics*, vol. 5, no. 6, pp. 1698–1705, 2015, doi: 10.1109/JPHOTOV.2015.2478065.

[14] S. Theocharides, G. Makrides, G. E. Georghiou, and A. Kyprianou, "Machine learning algorithms for photovoltaic system power output prediction," in *2018 IEEE International Energy Conference (ENERGYCON)*, Jun. 2018, pp. 1–6, doi: 10.1109/ENERGYCON.2018.8398737.

ESTIMATING SOLAR CELL OPERATING TEMPERATURE VIA DEEP NEURAL NETWORKS

Th. Katsaounis*[,1,2], G. Papadomichelakis[2], K. Kotsovos[3], I. Gereige[3], M. Abdullah[3], A. Jamal[3] and A. Tzavaras[4]

[1]Institute of Applied and Computational Mathematics (IACM), Foundation for Research and Technology - Hellas (FORTH), Heraklion, Greece

[2]Dept. of Math. & Applied Mathematics, Univ. of Crete, Heraklion, Greece

[3]Saudi Aramco R&D Center, Carbon Management Division, Dhahran, Saudi Arabia

[4]Computer Electrical and Mathematical Science & Engineering (CEMSE), King Abdullah University of Science and Technology (KAUST), Thuwal, Saudi Arabia

*Corresponding author: IACM-FORTH & Univ. of Crete, Heraklion, Greece, email: thodoros.katsaounis@uoc.gr

ABSTRACT: Saudi Arabia has great potential in clean energy production from photovoltaic installations. There are two main obstacles towards this goal which are related to the local climate: sandstorms and high ambient temperatures. Sandstorms are quite frequent in the region; scattered particles reduce solar irradiance and soil deposition on solar panels surface is significant, requiring regular cleaning. Furthermore, for most part of the year daily ambient temperatures can reach as high as 45 – 55 °C which are well above the range of STC for solar panels. These two factors can reduce the energy yield output of PV installations significantly. A priori knowledge and quantification of this efficiency loss can help design and assist in strategic planning to compensate and reduce the effects of local climate characteristics in energy yield output. In this work we aim on the estimation of the operating temperature of solar cells which is one of the key parameters for the assessment of the actual performance of photovoltaic panels. The proposed methodology is based on machine learning techniques and on Deep Neural Networks. The models provide excellent results and approximate solar cell operating temperature with less that 1 °C of accuracy.

1 INTRODUCTION

Over the past decades the world-wide market for solar photovoltaic (PV) technology has grown impressively. Every year the cumulative photovoltaic increases and it's estimated to be sufficient to supply about 1.5% of global electricity demand. Saudi Arabia belongs to a group of countries with the highest insolation worldwide, therefore, is an attractive market for PV implementation. The average energy from sunlight falling on Saudi Arabia is 2200 kWh/m^2, which allows a very competitive energy production cost. Even though Saudi Arabia receives a high level of solar insolation around the year, successful deployment of PV technologies can be challenging due to local weather conditions. Dust storms are frequent in the region, causing soling on the panel surface and reducing sunlight radiation intensity. In addition, the combination of increased ambient temperatures and high levels of solar radiation pose additional challenges in PV module energy yield performance. These two factors can reduce the energy yield output of PV installations significantly. A priori knowledge and quantification of this efficiency loss can help design and assist in strategic planning to compensate and reduce the effects of local climate characteristics in energy yield output.

In the literature there are several models for estimating the PV module operating temperature. These models are mathematical formulas usually derived from a data fitting process and use various parameters to provide a prediction. A set of models use electrical characteristics of the module, e.g., V_{oc} and/or I_{sc}. Another group of models, use parameters which, in general, are available a priori, such as air temperature (T_{air}), solar irradiance (G_{irr}), wind velocity (W_{vlc}) and efficiency of the module (η_{ref}). For a comprehensive review of these models, we refer the reader to [2,3] and the references therein. In [4] the authors used a some of these models to forecast module temperature by a nonlinear least square fitting process.

In this study, we propose two different methods for estimating solar cell operating temperature. The main ingredient in our approach is to use machine learning techniques. Our model is based on Deep Neural Networks (DNN). The input vector in the DNN consists of solar cell parameters and/or environmental factors which are known either apriori or aposteriori. The sole output of the DNN is the forecasting solar cell temperature.

2 EXPERIMENTAL SETUP – DATA ACQUISITION

The testing site is the New Energy Oasis (NEO) test field near the Red Sea coast (22.30 N, 39.10 E), located in KAUST, Thuwal, Saudi Arabia. This is a challenging location for PV installation since ambient temperature can reach values as high as 45-50 °C, which are far beyond the standard testing conditions (STC) range. The system consists of a monofacial PV panel of AlBSF technology an IV measuring system with radiation sensors and individual thermocouples which measure the temperature at the back of the module. The module consists of 60 solar cells, nominal efficiency is 14.5% at 45.7 °C NOCT, and maximum power of 240W. The measurements cover a period of ten days 28/03 – 06/04/2018, with samples every 10min.

3 METHODOLOGY

We propose a machine learning approach to estimate the solar cell operating temperature. The base model that we use is a Deep Neural Network (DNN) with an input vector consisting of five components, several hidden layers, and neurons and with a single output. We propose two methods that they differ in the kind of input parameters used in DNN. The first method (Method 1) uses as input the ambient temperature (T_{air}, °C), the solar irradiance (G_{irr}, W/m^2), wind velocity (W_{vlc}, m/s), open circuit voltage (V_{oc}, V) and short circuit current (J_{sc}, A). This method uses actual/online measurements of solar cell's performance. In the second method (Method 2) we replace V_{oc} and J_{sc} by two manufacturing parameters of the panel namely, the T_{NOCT} value and module efficiency η_{ref}. The main advantage of the second method is that all input values can be known apriori. The sole output of both DNNs is the forecasting solar cell temperature.

We have tested the two DNNs depending on the input parameters used: (T_{air}, G_{irr}, W_{vlc}, V_{oc}, J_{sc}) for Method 1 and (T_{air}, G_{irr}, W_{vlc}, T_{NOCT}, η_{ref}) for Method 2. In both methods the first three parameters are the same and refer to atmospheric factors which can be obtained from well-known databases e.g., [1] in case local meteorological data are not available. The last two parameters in Method 1 are V_{oc}, J_{sc} which under operating conditions they might not be possible to acquire, while Method 2 uses T_{NOCT} and η_{ref} which are provided by the solar cell manufacturer. In [5] the authors consider a single hidden layer DNN with a six-component input vector which consists of with the same parameters as Method 1 plus the value of the produced power. The authors in [6] use also a single hidden layer DNN with a two-component input vector consisting only with (T_{air}, G_{irr}).

4 RESULTS

Before feeding the data to the DNN they were scaled-normalized so they all mapped to [0,1]. The DNN was built using Python 3.8 with an in-house code using stochastic gradient descent as the optimizing algorithm and a fixed learning rate. The number of hidden layers and neurons in the DNN they were kept as parameters to test their effect in the accuracy of the model. The training set was comprised by 70% of the data and the rest was used for validation-testing. It's worth mentioning that even with 30% of the data as a training set the Mean Absolute Error (MAE) in the testing set was below 1 °C, showing the robustness of the method. Tables I, II show the MAEs on the testing set for both methods for a range of values of hidden layers (HL) and neurons (NR).

Table I : MAE on the testing set for Method 1

	Method 1: MAE's (°C)			
NR/HL	2	3	4	5
5	0.711	0.623	0.602	0.609
10	0.589	0.578	0.532	0.447
20	0.619	0.629	0.392	0.396
40	0.567	0.530	0.322	0.229

Table II : MAE on the testing set for Method 2

	Method 2: MAE's (°C)			
NR/HL	2	3	4	5
5	1.233	1.182	1.074	1.109
10	1.106	0.863	0.945	0.806
20	0.928	0.609	0.564	0.471
40	0.963	0.604	0.475	0.424

The first observation from Tables I, II is that for both methods the MAE decreases as the number of HL and neurons increase reaching a limiting value. The second observation is that MEA in Method 2 is larger that of Method 1. This expected since Method 1 uses actual values of V_{oc}, J_{sc} accounting for the actual cell operating conditions, in contrast to Method 2 which all its parameters are known apriori. Figure 1 shows the comparison between the prediction (blue dotted line) of Method 1 with 2 HL and 5 neurons on each node and the experimental values (solid red line) for the whole dataset. In this case we have MAE=0.889 °C. The training time for the DNN with 5HL and 40 neurons was less than 1min for an MAE of about 1 °C. Smaller training times can be achieved using one of the standard accelerating techniques for the stochastic gradient descent algorithm, e.g., Adam, [8]. The same DNN was build using *Tensorflow* computational framework, [7] for purely comparative purposes. For the same set of parameters, the in-house code and *Tensorflow* provided very similar results differing only on the execution time.

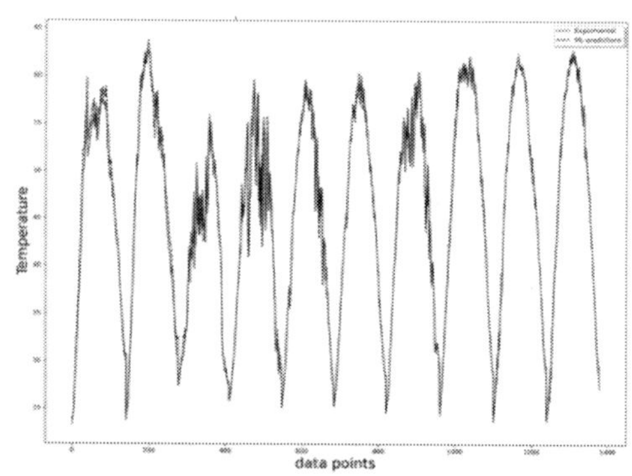

Figure 1: Method 1: 2HL, 5NR, Experimental Data (Red), DNN Model (Blue).

5 CONCLUSIONS

In this study we have developed machine learning techniques for estimating the operating temperature of solar cells. We have used DNNs with various number of layers and neurons. Two different approaches were used in terms of the kind of input data for these DNNs. Both methods produced excellent results in predicting solar cell's operating temperature. The first method that uses ambient factors and electrical characteristics produces the best results. The second method uses ambient data and parameters known apriori produced comparable results.

Both methods performed well with the differences between model predictions and actual measurements were small, less than one degree Celsius, for both methods.

ACKNOWLEDGEMENTS

The authors acknowledge the support of KAUST and Saudi Aramco R&D Center - Carbon Management Division for their financial support in devel- oping this work. This work was partially supported by grant OSR–2020 – 4433.02 from KAUST and Saudi Aramco.

REFERENCES

[1] https://re.jrc.ec.europa.eu/pvg_tools/en/
[2] A.Q. Jakhrani, A.K. Othman, A.K., A.R. Rigit, S.R. Samo, World Applied Sciences Journal (2011), 14, 1–8.
[3] C. Coskun, U. Toygar, O. Sarpdag, Z. Oktay, Journal of Cleaner Production (2017), 164, 1474–1485.
[4] Th. Katsaounis, K. Kotsovos, I. Gereige, A. Basaheeh, M. Abdullah, A. Khayat, E. Al-Habshi, A. Al-Saggaf, A.E. Tzavaras, A. E. Renewable Energy (2019), 143, 1285–1298.
[5] G. Ciulla, L.V. Brano, E. Moreci, International Journal of Photoenergy 2013, (2013), 1–10.
[6] I. Ceylan, O. Erkaymaz, E. Gedik, A.E. Gürel,. Case Studies in Thermal Engineering (2014), 3, 11–20.
[7] https://www.tensorflow.org/
[8] M.J. Kochenderfer, T.A. Wheeler, Algorithms for Optimization; (2019); pp. 1–521, MIT Press.

DEVELOPMENT OF A NEW CHARACTERIZATION METHOD OF SOILING ON PV MODULE BASED ON OPTICAL MICROSCOPY

Sébastien Arbaretaz*[1], Christophe Ménézo[2], Emilie Planès[3], Eric Pilat[4]
[1]Univ. Savoie Mont Blanc, CEA, Liten, INES
[2]Univ. Savoie Mont Blanc, CNRS, LOCIE
[3]Univ. Grenoble Alpes, Univ. Savoie Mont Blanc, CNRS, Grenoble INP, LEPMI
[4]Univ. Grenoble Alpes, CEA, Liten, Campus INES

[1,4]150 avenue du lac Léman, 73375 Le Bourget du Lac, France
[2]73376 Le Bourget du Lac, France
[3]38000 Grenoble, France

ABSTRACT:
Herein, a new characterization method is developed based on optical microscopy. Images are acquired with transmitted light microscope at two magnifications and using image combination on vertical axis. The image analyzing is done with ImageJ software, using gray scale threshold function. Size of particles is analyzed with the equivalent diameter taken as size parameter. The particle morphology is also analyzed. Similar methods in literature use the shape factor to describe morphology of particles. However, the shape factor shows deviation due to the perimeter estimation of pixelated forms. Therefore, new morphological factors are defined without using the perimeter. Cementation process is reproduced in an indoor soiling test-bed with dust from the Atacama Desert. This process is study with this new characterization method. Particle and agglomerate size are smaller after cementation than just after dust deposition. Also, particle shape seems rounder after cementation. However, it requires further study about shape analysis before conclusions can be drawn.

1 INTRODUCTION

Soiling reduces the amount of light reaching solar cells and leads to a decrease in module output. Different type of material could be deposited depending on location: dust, pollution, pollen, bird dropping or agriculture emission. Many environmental parameters influence soiling like relative humidity, wind, temperature and particulate matter. Among these parameters, humidity has a strong effect on dust adhesion. Several phenomena such as cementation, caking and capillary aging enhance adhesion of particles on the surface. The method described here aims to analyze the influence of these parameters on soiling at a microscopic scale. It consists in acquiring images with an optical microscope coupled to an image analysis software to study dust deposition.

Similar methods have already been developed. Ilse et al. [1] used an optical microscope with dark field at 20× magnification to measure proportion of surface covered using the auto threshold function of ImageJ software. Smestad et al. [2] measured particle size and proportion of surface covered on images acquired at 100× and 500× magnification excluding particles smaller than 1 pixel. They suggested that analysis of this type should employ multiple magnifications and image combination. Olivares et al. [3] and Ferrada et al. [4] developed a method acquiring images at 20× magnification to analyze dust size and morphology with the shape factor (Eq. 1).

$$Shape\ Factor = Perimeter^2/(4.\pi.Area) \qquad \text{Equation 1}$$

The shape factor quantifies the compactness of a figure. It is equal to one for a circle and increase with the ratio of perimeter to area for other shape. Although the shape factor describes well the compactness of a figure, it is based on perimeter. However, there are a deviation in the perimeter estimation due to pixel in image. It is illustrated with pixelated circles of different size (see Fig. 1). Below 15 px, the perimeter is underestimated and overestimated above. It results that the shape factor is not equal to one even though the analyzed shapes are circles.

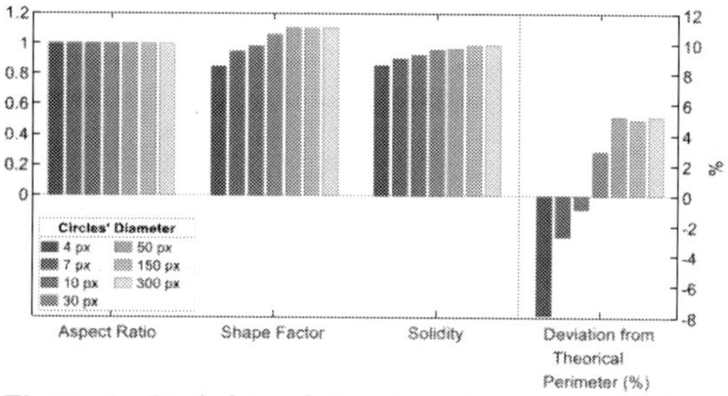

Figure 1: Deviation of the shape factor compared to other factors for circles of different sizes

The objective of this study is the development of an innovative characterization method of dust deposition based on optical transmission microscopy. Images were acquired at two magnifications and were the results of image combination on vertical axis. In addition, new factors are defined to describe shape without using perimeter. This method enables understanding at the microscopic scale the behavior of soiling which leads to a decrease in module output. The specific phenomena study in this paper are cementation and caking.

2 METHOD

2.1 Images acquisition

Images are taken with an optical transmission microscope, (ZEISS Axio Imager.Z2 Vario). The transmitted light microscopy give better result than reflected light microscopy, the particles stand out better against the illuminated background. In this paper, the term particle includes isolated particles and agglomerates of particles because the particles within agglomerates cannot be distinguished from each other.

Laser diffraction analysis were conducted on the dust used for the experiments, non-agglomerates particle size was in the range of 0.3 to 140 µm. Observations at several magnifications are therefore necessary to cover this wide particle size range. The 200x and 500x magnifications are selected. Images are acquired at three locations on sample to ensure a good representativeness of all the surface of sample. For each location, one image is acquired at 200x magnification which covered a surface around 0.39 mm^2. As images at 500x magnification cover only 0.06 mm^2, three are acquired for each location. The number of images taken is a compromise between having enough images to improve the statistics of the study and keep a reasonable acquisition time.

Because the focal depth is smaller than maximum particle height, all images are the resulted of image combination on vertical axis to have a clear image over the entire height of particle. This process is execute with the microscope software, images are taken at an equidistant depth and combined.

2.2 Images processing

The images are then processed with ImageJ software to detect particles and measure their properties. Particles are separated from the background with a gray scale threshold function, the threshold value is the default value provide by ImageJ, which gives good results.

A minimum particle size is set for 500x magnification, only particle bigger than 0.7 µm diameter (9 px^2) are kept for analysis. Smaller particles are too small to be analyzed accurately. These smaller particles can be in fact artifacts in the image that can introduce errors. Nevertheless, particles under 0.7 µm represent only 4% of total particle volume according to the laser diffraction analysis. Similarly, only particles bigger than 15 µm diameter (662 px^2) are kept for analysis at 200x magnification.

2.3 Particle size analysis

First, a particle size analysis is made to study the distribution of particle size. The equivalent diameter is taken as particle size parameter, it is the diameter of a perfect circle with the same area than the particle (see Equation 2).

$$\textit{Equivalent diameter} = 2.\sqrt{(Area/\pi)} \qquad \text{Equation 2}$$

For each magnification, only the particles whose size is within the analysis range are retained. At 500x magnification this range is between 0.7 and 20 µm diameter. At 200x magnification, only particles bigger than 15 µm diameter are retained.

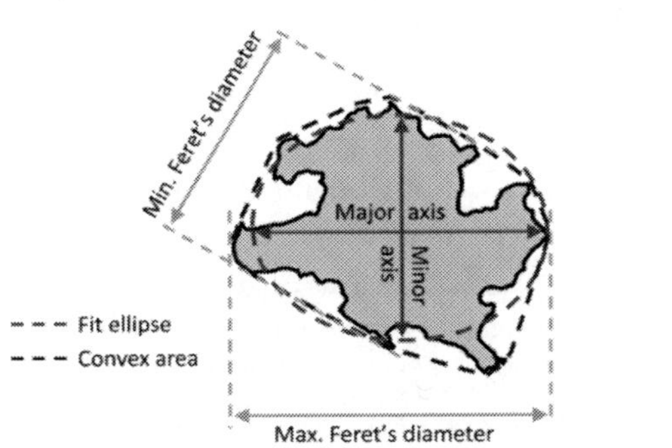

Figure 2: Parameters directly measured with ImageJ software

The particles are arranged in size classes to describe the distribution of particle size. For each image, the ratio of surface covered for each size class is calculated. Thus, this provides a distribution that can be compared to those of other images regardless of the magnification. Moreover, the average distribution for each location, and then for each sample, can be calculated. The median, first and third quartile are calculated from particle size distribution to compare results.

2.4 Particle morphology analysis

As previously mentioned, the shape factor gives a biased estimation of the particle shape. Therefore, other factors are defined to describe shape without using the perimeter (Fig. 3).

$$Solidity = \frac{Area}{Convex\ Area} \qquad \text{Equation 3}$$

$$Inverse\ Aspect\ Ratio = \frac{1}{Aspect\ Ratio} = \frac{Minor\ Axis}{Major\ Axis} \qquad \text{Equation 4}$$

$$Feret\ Ratio = \frac{Minimum\ Feret's\ Diameter}{Maximum\ Feret's\ Diamete} \qquad \text{Equation 5}$$

$$Major\ Feret\ Ratio = \frac{Major\ Axis}{Maximum\ Feret's\ Diameter} \qquad \text{Equation 6}$$

$$Minor\ Feret\ Ratio = \frac{Minor\ Axis}{Minimum\ Feret's\ Diameter} \qquad \text{Equation 7}$$

Figure 3: Equations of newly defined morphological factors

Among these new defined factors, ImageJ software directly calculates the Solidity and the Aspect Ratio. The others are based on parameters measured with ImageJ software: the major and minor axis of the fit ellipse and the maximum and minimum Feret's diameter (illustrated in Fig. 2). The Inverse Aspect Ratio is chosen instead of the Aspect Ratio to have a value between zero and one like the other factors. The Aspect Ratio or the Inverse aspect Ratio quantifies the radial symmetry of particles. The others factors quantify the particle compactness.

3 EXPERIMENTS

3.1 Case study: the Atacama Desert

The soiled sample were obtained with dust from the Atacama Desert, more precisely on the ATAMOS-TEC test platform (Atacama Desert Solar Platform, PSDA). It appears that this dust naturally contains gypsum that cause cementation [3]. A typical soiling day in the Atacama Desert begins with temperature rising in the morning and relative humidity decreasing. Dust is bring by wind during the afternoon. Then temperature decreases during the night when relative humidity increases. At the end of the night, ambient temperature increases while panel temperature is low [3].

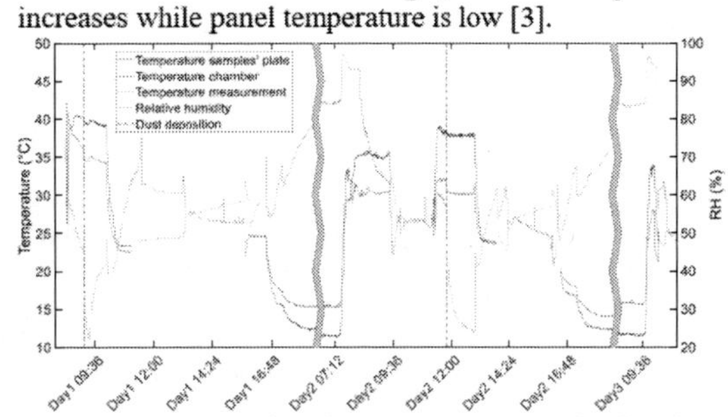

Figure 4: Record of measurements of chamber temperature, sample temperature and relative humidity

This temperature difference with high relative humidity induces the formation of dew in which the soluble compounds dissolve. The surface is drying in the morning and these compounds recrystallize leading to compaction and cementation. As known that cementation enhances adhesion of particle, this phenomenon was reproduced in an indoor dirt test-bed to use it as case study with this new characterization method.

3.2 Indoor dirt test-bed

The indoor dirt test-bed (Fig. 5) is composed of a chamber where ambient temperature, sample temperature and relative humidity can be control and monitor. A powder dispersion generator projected dust on a deflector to generate a homogeneous cloud inside the chamber.

Figure 5: Indoor dirt test-bed used for artificial soiling

3.3 Cycles of dew formation and evaporation

Three samples of glass of 8 by 4 cm were used for this study. The experiment was conducted during 3 days (Fig. 4). The typical soiling day describe earlier can be simplified as a cycle: deposition of dust at high temperature and low humidity, followed by the formation of dew and finally its evaporation. Two successive cycles were performed. First, dust deposition is made at high temperature; the sample temperature is higher than chamber temperature like when the PV modules are in production. Second, the sample and chamber temperature is reduced to room temperature to output samples for microscopic image acquisition. The samples are put back in the test-bed after image acquisition. During the night, chamber temperature is decrease and relative humidity is increase. The sample temperature is lower than the chamber temperature to enable dew to form. In the early morning, the sample and chamber temperature are increased to dry the samples. Finally, the sample and chamber temperature is reduced to room temperature and samples are taken out for microscopic image acquisition.

4 RESULTS

4.1 Size

The size distribution of particles and agglomerates was measure for all samples as describe in the Method section. Results are shown in Figure 6. The dust deposition is not completely uniform between the different locations on samples, differences in the particle size distribution are visible. There is a greater difference between points after dew formation than just after deposition. In addition, the particle size is smaller after dew formation, indicating a compaction of the agglomerates. These observations are valid for both cycles.

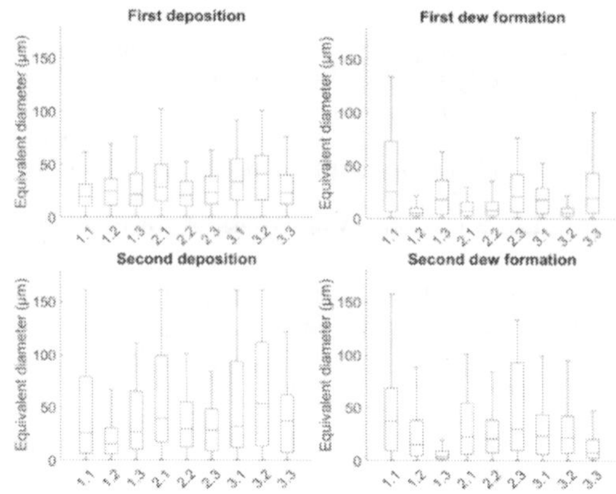

Figure 6: Results of the size analysis for each location on samples

It can also be seen from the sample averages that the particle size is smaller after dew formation than just after dust deposition (Fig. 7). This indicates a compaction of agglomerate as it could have been expected. In addition, the second deposition has increased the size of the particles, which shows that newly deposited particles accumulated to the already formed agglomerates.

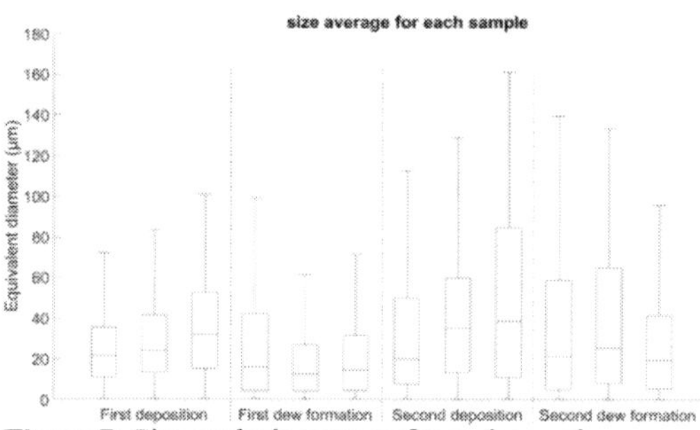

Figure 7: Size analysis average for each sample

These results demonstrate that the method allows the analysis of the size distribution of dust. The purpose of this analysis is to show the behavior of the particles: the particles tend to spread out on the surface in an isolated way or they gather in agglomerates.

4.2 Shape

For the five morphological factors, a value of one corresponds to a circle and the value decrease towards zero with the compactness. The evolution of particle shape was compared between the different stages of observation (Fig 8).

There is a slight trend that emerges; there are more rounded particles after the dew formation than just after dust deposition. Moreover, the range of value of the factors is greater after dew formation. These results corroborate with what is observed on the images, the dew compacts the agglomerates into more rounded shapes. Crystal shapes are also observed on sample that could be formed by recrystallization of soluble compounds.

However, the difference are not significant and none of the morphological factors seems to perform better than the others. More experiments need to be performed in different conditions and with different type of dust in order to determine if it comes from the methods or the experimental conditions did not bring enough differences in particle and agglomerate shapes to be observed by this method.

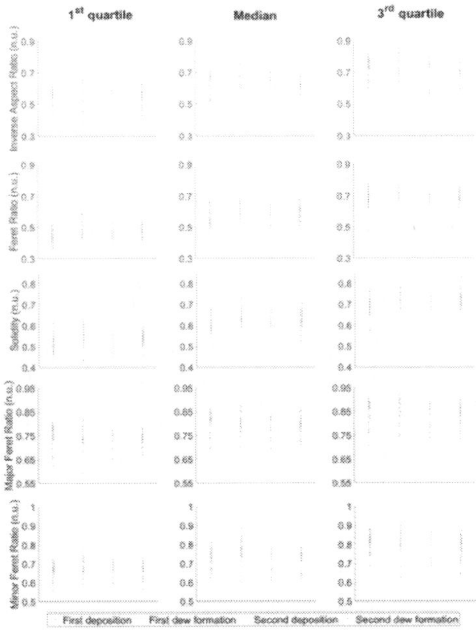

Figure 8: Results for the shape analysis

5 CONCLUSION

A new characterization method is developed using transmitted light microscope and ImageJ software for image analysis. Observations at two magnifications, 200x and 500x, are used to cover a wider range of particle sizes. Moreover, images are the result of image combination on vertical axis to have a clear image over the entire height of particle.

The method is applied on artificial soiling samples. They are produced in an indoor dirt test-bed by reproducing simplified typical soiling day in Atacama Desert with dust from on site. It consists in high temperature and low humidity during the day and inversely during the night. Dew forms at the end of the night before evaporating in the morning. This region is known for cementation that enhances particle adhesion on module. The necessary conditions for cementation are reproduced in order to study this phenomena at the microscopic level.

The results of size analysis for both magnifications are combine to obtain the particle size distribution overall range of size. As the shape factor shows deviation, new morphological factors are defined to describe particle and agglomerate morphology.

Particle and agglomerate size are smaller after dew formation than just after dust deposition. In addition, particle and agglomerate shape are rounder after dew formation. It seems that dew compacts the particles into agglomerates with a rounded shape. However, measured differences between each stage of the experiment are small. It requires further study before conclusions can be drawn.

6 ACKNOWLEDGEMENTS

This work was partially supported by the French National Program "Programme d'Investissements d'Avenir - INES.2S" under Grant ANR-10-IEED-0014-01.
Some authors are supported by the French National Program "Programme d'Investissements d'Avenir - EUR" ANR-18-EURE-0016 - Solar Academy.

7 REFERENCES

[1] K. K. Ilse et al., "Comparing indoor and outdoor soiling experiments for different glass coatings and microstructural analysis of particle caking processes," IEEE J. Photovoltaics, vol. 8, no. 1, pp. 203–209, 2018, doi: 10.1109/JPHOTOV.2017.2775439.

[2] G. P. Smestad et al., "Modelling photovoltaic soiling losses through optical characterization," Sci. Rep., vol. 10, no. 1, pp. 1–13, 2020, doi: 10.1038/s41598-019-56868-z.

[3] D. Olivares et al., "Microstructural analysis of the PV module cementation process at the Solar Platform of the Atacama Desert," Sol. Energy Mater. Sol. Cells, vol. 227, no. February, p. 111109, 2021, doi: 10.1016/j.solmat.2021.111109.

[4] P. Ferrada et al., "Physicochemical characterization of soiling from photovoltaic facilities in arid locations in the Atacama Desert," Sol. Energy, vol. 187, no. April, pp. 47–56, 2019, doi: 10.1016/j.solener.2019.05.034.

DECOUPLING PERFORMANCE GAINS OF SILICON HETERO-JUNCTION BIFACIAL MODULES

Hugo Quest*[1, 2], Alessandro Virtuani[1], Luca Gnocchi[1], Alejandro Borja Block[1], Nicolas Wyrsch[1], Christophe Ballif[1, 3]
[1]EPFL, Institute of Electrical and Micro Engineering (IEM), Photovoltaics and Thin-Film Electronics Laboratory (PV-Lab), Rue de la Maladière 71b, CH-2002 Neuchâtel, Switzerland
[2]3S Swiss Solar Solutions AG, Schorenstrasse 39, CH-3645 Gwatt (Thun), Switzerland
[3]CSEM, Sustainable Energy Centre, Rue Jaquet-Droz 1, CH-2002 Neuchâtel, Switzerland

ABSTRACT: Silicon heterojunction (SHJ) solar cells and bifacial modules are expected to rapidly grow in the PV market, with their high power conversion efficiency, good performance at low irradiation and better temperature coefficients (TCs) allowing for improved energy yields. It is therefore increasingly relevant to have comparative analyses with conventional monofacial modules and market leading PERC cells. The aim of this contribution is to assess the performance of five PV modules on a rooftop monitoring station in Neuchâtel: two bifacial SHJ, two monofacial SHJ and one benchmark monofacial PERC. These three module types allow for detailed comparisons of performance enhancing factors both on the module structure and solar cell technology levels, allowing to decouple individual performance effects. After almost two years of monitoring, results show that bifacial modules yield 7.2% energy gains over both monofacial technologies, and 5% improved performance ratio (PR). For the corrected PR, current bifacial corrections may not fully describe the angular, optical and seasonal effects, as a 2.2% bifacial gain is still observed after temperature and rear irradiance corrections, leaving room for improvement in the current bifacial PR definition. In terms of temperature dependence, field-based TCs are found to be better for SHJ cells than PERC (-0.16%/°C vs. -0.31%/°C at STC, respectively), although performance gains also vary with irradiance conditions.
Keywords: Silicon Heterojunction, Bifacial, Monitoring, PERC, Performance Ratio

1 INTRODUCTION

With their high energy conversion efficiencies, low temperature coefficient (TC) and intrinsic bifacial nature, silicon heterojunction (SHJ) solar cells are soon expected to become competitive with Passivated Emitter and Rear Contact (PERC) cells, the current go-to technology [1, 2]. As SHJ cells become more prominent in PV markets, it is increasingly relevant to have robust analyses of their performance and characteristics when compared to industry standards, evaluating their potential gains and advantages. This contribution aims to provide an up-to-date and in-depth comparison of three commercial technologies, focusing on bifacial vs. monofacial applications and SHJ vs. PERC solar cells, which will allow to decouple the contributions to increased energy yields and performance. The bifaciality of the SHJ modules raises important questions for PV system designers, installers and users, given that rear irradiation level also needs to be taken into account. The promise of dual-sided PV modules is that they can make use of more incoming irradiance than typical modules by collecting additional light at the rear side, however quantifying the gains is not straightforward and depends on many factors: the substrate behind the system (e.g. white panels, grass, gravel,...), the system geometry, and solar cell types. Having the two cell technologies and three module structures in the same monitoring conditions is therefore ideal to establish the performance differences. For example, it is expected that SHJ cells will perform better in clear-sky and elevated temperature conditions due to their lower temperature coefficients, and that the bifacial modules will have improved yields and performance in diffuse irradiance conditions.

The main outcomes of this work will therefore be: (1) detailed data analysis pipelines for reproducible and standardised measurements, (2) a practical comparative analysis of PV module technology performances, focusing on energy yield, performance metrics and temperature coefficients, leading to (3) proposals for improved bifacial system modelling and monitoring guidelines.

2 EXPERIMENTAL

The monitoring station is situated in Neuchâtel, Switzerland (temperate oceanic climate, Cfb in Köppen climate classification), and modules are installed with a 15° tilt and 175° orientation (South-facing) on aluminum mounting racks at 0.1 m from the gravel-covered rooftop. Five modules are monitored: two bifacial SHJ, two monofacial SHJ, and one PERC monofacial. **Table I** summarises the characteristics and nominal electrical parameters of the installed modules (nameplate manufacturer values) measured at standard test conditions (STC), where P_0 [Wp] is the rated power, I_{sc} [A] the short-circuit current, V_{oc} [V] the open-circuit voltage and γ [%/°C] the maximum-power temperature coefficient.

Figure 1 shows photographs of the setup with the three module types. The monitoring station setup measures the module-level I-V curves at 180 sec. timesteps to retrieve electrical outputs (keeping modules at maximum power point (MPP) between acquisitions), and back-surface sensors collect module temperatures. For the bifacial modules, a dummy glass-glass module is setup in the module array to retrieve module temperature without having to shade the rear of the measured modules. Additionally, global horizontal, diffuse irradiance and a back-surface irradiance sensor (albedometer) measure irradiation at 10 min. intervals. Data is available starting January 2021, however the rear-side irradiance measurements only start in February 2022.

3 METHODOLOGY

The available data allows for a two-fold comparative analysis: (i) comparing module topologies with the bifacial and monofacial layouts (based on the two types of SHJ panels), and (ii) investigating cell technologies by comparing the performance of SHJ to the benchmark PERC cells. The following subsections will describe the metrics, methods and filtering approaches used in this work.

Table I. Characteristics of the studied modules (Mo=Monofacial, Bi=Bifacial) at standard test conditions (STC).

ID	Bifacial	P_0 [W_p]	I_{sc} [A]	V_{oc} [V]	γ [%/°C]
SHJ-Bi-1	✓	380	53.4	9.17	-0.25
SHJ-Bi-2	✓	380	40.6	10.4	-0.25
SHJ-Mo-1	×	380	53.4	9.17	-0.25
SHJ-Mo-2	×	380	53.4	9.17	-0.25
PERC-Mo	×	330	53.4	9.17	-0.37

Figure 1. Photographs of the outdoor test facility and monitored devices.

3.1 Performance metrics

The main metric used for the analysis is the performance ratio (PR), defined as the quotient between the system's final and reference yield. Following the IEC 61724-1 guidelines, the PR can be further adjusted for temperature coefficients (denoted PR'), and bifaciality (denoted PR'_{bi}). This notation will be followed to differentiate the various correction levels of PR. Equations (1), (2) and (3) summarise the metric definition:

$$PR'_{bi} = \frac{\left(\sum_k P_{mp,k} \cdot \tau_k\right)}{\left(\sum_k \frac{\left(C_{k,25°C} \cdot P_0\right) \cdot G_{i,k} \cdot BIF_k \cdot \tau_k}{G_{i,ref}}\right)} \quad (1)$$

$$C_{k,25°C} = 1 + \gamma \cdot (T_{mod,k} - 25°C) \quad (2)$$

$$BIF_k = (1 + \varphi_{max} \cdot \rho_i) \quad (3)$$

Where for the k timesteps with recording interval τ, P_{mp} [W] is the measured power at MPP, P_0 [Wp] the rated power, G_i [Wm^{-2}] the measured in-plane irradiance, $G_{i,ref} = 1000$ Wm^{-2} the irradiance at which P_0 is determined (STC), $C_{k,25°C}$ the power rating temperature adjustment factor, T_{mod} [°C] the module temperature, γ [%/°C] the relative maximum-power TC, BIF the bifacial irradiance factor, φ_{max} the maximum-power bifaciality coefficient and ρ_i the in-plane rear-side irradiance ratio. For PR', the bifacial irradiance factor is set to 1, and for PR both bifacial and temperature correcting terms are set to 1. The energy yield Y is also used to compare the module performances (sometimes referred to as specific yield), and is defined as the module energy output (DC) E [Wh] per rated installed W_P.

For a more detailed analysis, the change rates of available electrical parameters are also evaluated. The maximum power P_{mp} and short-circuit current I_{sc} are corrected for irradiance and normalised by the reference values of **Table I**, while V_{oc} is simply normalised [3].

3.2 Temperature coefficients

It is well established that solar cell performance declines with increasing temperature, primarily due to the reduction of open-circuit voltage and FF [4]. Silicon heterojunction solar cells are considered ideal for bifacial modules due to their good low-light performance and lowest temperature coefficients amongst silicon solar cells [5]. Moreover, it has been shown that TCs are in reality nonuniform, and vary depending on the environmental and operating conditions [3], [6]. With the available datasets, it is possible to study the temperature dependency of cell technology performance, as well as the variability of the TCs, usually defined at STC. This will allow both the comparison of module performances and the data-based characterisation of TCs. To do so, the methodology outlined in [6] is followed: the DC Power is plotted as a function of module temperature for different irradiance conditions and filters, and a linear regression is used to extract the irradiance-dependent TCs. This analysis method also allows to extrapolate the STC power from field data using the 1000 Wm^{-2} irradiance corresponding to STC, which can be compared to nominal values.

3.3 Data filtering

In terms of filtering, three procedures were applied for the performance metrics: (i) an irradiance filter applied to the G_i measurements removing very low (< 200 Wm^{-2}) irradiance values to filter night-time and variable cloudy conditions and very high (> 1200 Wm^{-2}) irradiance (i.e. due to extreme cloud reflections), (ii) for daily or higher aggregations, a day type classification algorithm was applied to filter out overcast days, where high weather variability would generate noise in the data, (iii) for PR measurements, an additional threshold filter excludes outlier values due to shading or nonuniform irradiance conditions ($0.75 \leq PR \leq 1.2$).

For the temperature coefficient vs. irradiance level analysis, data is filtered for ten irradiance steps $G_i = 50$, 100, 200, 300, …, 800, 1000 Wm^{-2} with $\pm 5\%$ tolerance, where 900 is excluded due to outlier values. To avoid further outliers in the power measurements, the PR filter used for the performance metrics is applied to remove shading faults.

4 RESULTS AND DISCUSSION

4.1 Performance gains analysis

As a first step, the performance metrics described in section 3.1 are used to quantify the differences between the studied module technologies and structures. **Figure 2** shows the production profiles of the five modules during a clear-sky day, where nonuniform behaviours are minimised. The specific power (normalisation by the rated power), standard PR and corrected PR'_{bi} are represented, allowing to compare the different capacity modules through various lenses. Starting with the specific power, a zoom in the midday hours shows the performance gains observed for the monofacial SHJ modules over the PERC module, as well as the bifacial gains. Over the entire day, the SHJ monofacial modules outperform the PERC module by 1.7%, and the bifacial SHJ modules show a 6.1% gain over the monofacial SHJ. Moreover, looking at individual timesteps, peak gains for specific power (up to 3%) for the SHJ vs. PERC were observed around midday, directly correlated to peak irradiance and temperature. From this simple clear-sky day comparison, we therefore observe and quantify both the temperature and bifaciality performance effects. Looking at the entire monitoring period, the energy yield gain due to bifaciality (comparing with SHJ and PERC monofacial modules) is 7.2%, with the bifacial modules producing ~2200 Wh/Wp from January 2021 – July 2022. Monofacial SHJ modules also produced ~1.5% more than PERC, likely due to the TC. A potential caveat is the measured value of the nominal power, which in this case is the manufacturer rated value, which is notoriously unreliable. Ideally, the rated power should be measured in an accredited lab for optimal accuracy, in which case the difference between SHJ and PERC would likely increase.

Looking at the standard PR, one can further quantify the gains, although without decoupling the individual effects leading to increased performance, as there is no temperature or rear-irradiance correction. Considering most commercial systems will only have access to the standard PR, it nevertheless remains a valuable metric for standardised comparisons of modules. Here, during the studied clear-sky day, we observe 4.8% and 8.7% gains for bifacial SHJ over monofacial SHJ and PERC, respectively. The advantage of SHJ is notably only at high temperature and irradiance conditions, as the PR equalises for the monofacial modules in the morning and evening hours.

Finally, in order to further detail and decouple the bifacial and cell technology gains, the corrected performance ratio PR'_{bi} is considered, and the measured rear-irradiance G_{rear} [Wm^{-2}] is represented for the studied clear-sky day. The rear irradiance profile shows two peaks in the morning and evening, likely due to the sensor angular response or reflections at low sun angles. The PR'_{bi} profiles also show an angular trend: the performance ratios match well around noon, with the PERC module even outperforming the SHJ modules, which can be attributed to the TC difference. Interestingly, bifacial gains are still present in the morning and evening, even after correction, indicating that the applied rear-irradiance correction is not sufficient to account for all bifacial enhancements. This effect is likely due to environmental factors decreasing the measured rear-irradiance (e.g. sensor shaded by aluminum frame or surroundings). Another possible explanation is the sun angle dependency of both the solar cells and albedometer. Firstly, given that the glass-glass bifacial modules are frameless, internal light reflections at low sun angles coming from the module sides could contribute to increased current. Secondly, during morning and evening hours with lower sun angles, the rear-irradiance sensor likely does not capture the total irradiance reaching the module backside, which includes not only diffuse and reflected components, but also direct irradiance. The combination of these two factors, along with other optical effects possibly linked to the installation environment, could lead to the bifacial enhancements observed, even after correction.

In order to confirm the angular dependency, the relative difference between SHJ bifacial and monofacial PR'_{bi} is aggregated and averaged per hour of the day during the period of February-July 2022, as shown in **Figure 3**. A clear pattern of low angular gains is isolated, confirming the seasonal and sun angle-dependent bifacial behaviour. This result shows that a simple rear-irradiance correction model, as described in IEC 61724-1, might not be sufficient for capturing all bifacial enhancement factors, or that special attention should be placed on the rear-irradiance measurements. A future improvement for the standard could therefore be an angular-corrected bifacial factor.

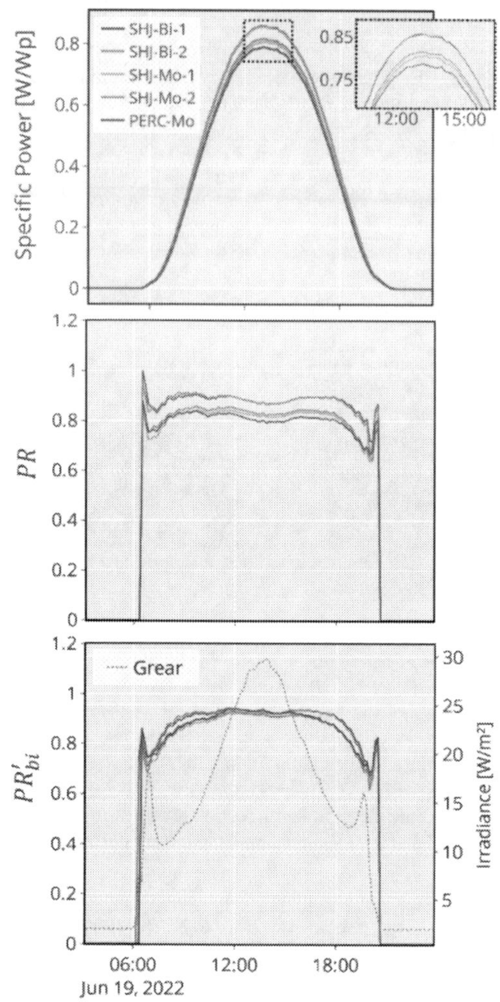

Figure 2. Daily production profile during a clear-sky day, for the specific power (normalised with rated power), standard PR, and temperature and rear irradiance corrected PR'_{bi} (shown with rear-irradiance measurement).

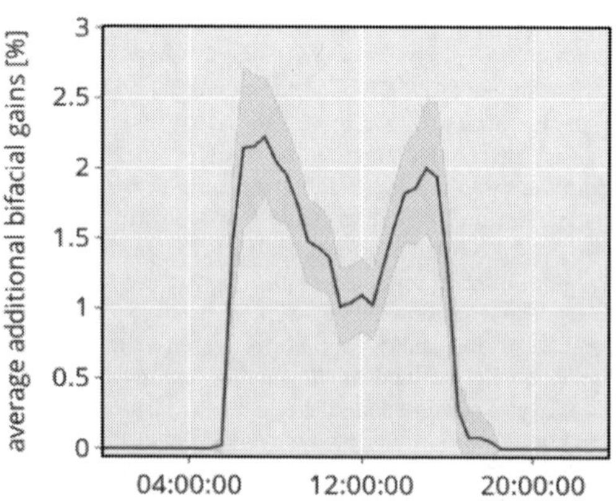

Figure 3. Hourly aggregated and averaged additional SHJ bifacial gains after temperature and rear-irradiance corrections when compared to SHJ monofacial module performance, isolating the remaining angular-dependent

gains. Standard deviation is used for uncertainty (shaded area). A pattern of morning and evening gains is observed.

Long-term temporal trends of the normalised power (equivalent to standard PR) and I-V curve parameters are shown in **Figure 4**. Normalisation, filtering and analysis procedures are described in sections 3.1 and 3.3. Results show the impact of bifaciality on the PR over the monitoring duration, with a ~5% performance gain over the SHJ monofacial modules, while the SHJ cells show a ~1% gain over PERC. The observed temporal trends of the electrical parameters show a better short-circuit current performance for the bifacial vs. monofacial SHJ modules. Given that current is linearly correlated to light intensity reaching the cells [7], the difference is simply explained by the increased irradiance due to bifaciality.

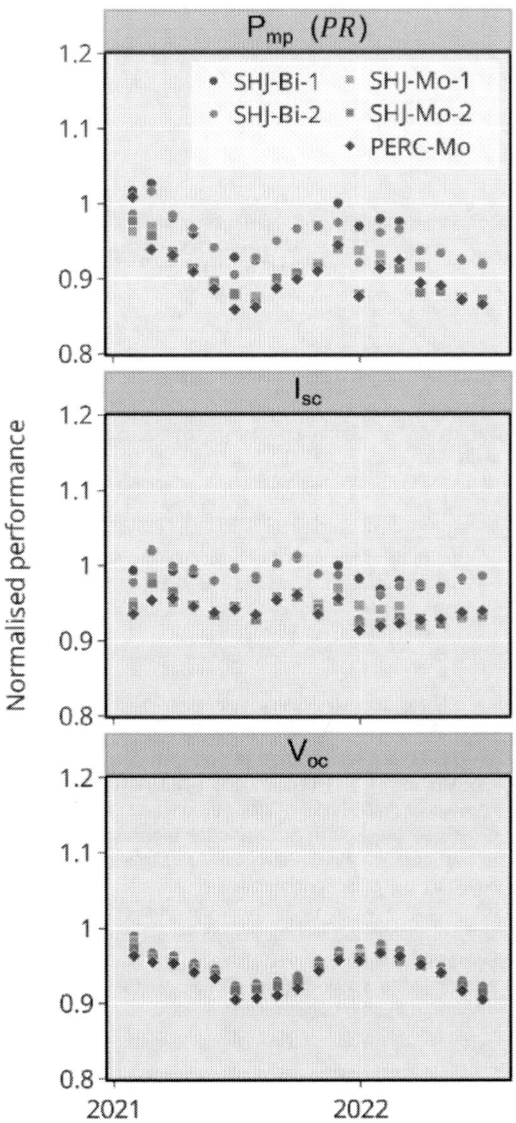

Figure 4. Normalised monthly performance rates of the DC power (equivalent to standard $\boldsymbol{PR}$) and I-V parameters for the studied modules.

Table II. Average performance ratios (Feb. – Jul. 2022).

ID	Mean PR	Mean PR'	Mean PR'_{bi}
SHJ-Bi-1	0.932	0.949	0.927
SHJ-Bi-2	0.930	0.948	0.925
SHJ-Mo-1	0.883	0.908	0.908
SHJ-Mo-2	0.880	0.904	0.904
PERC-Mo	0.882	0.916	0.916

As discussed with the corrected performance ratio, the rear irradiance is highly dependent on the sun angle and environmental conditions, as it can be composed of a combination of sky diffuse, reflected or direct back-surface irradiance. At low sun angles, the glass-glass structure of the bifacial modules could also contribute to additional internal reflective gains, leading to the improved current. The open-circuit voltage trend is stable for all modules, however the SHJ initial degradation rate is steeper than the PERC cells. A long-term analysis (>2 years) would be necessary to conclude, however it has been observed in literature that SHJ cells could degrade more in V_{oc} than typical c-Si cells in long-term field exposure [8].

Having seen how the performance metrics can be used to analyse and decouple the bifacial gains, the next step is to focus on solar cell technolgy differences, which mostly come down to temperature coefficients.

4.2 Temperature effects on performance and TC variability

The methods described in 3.2 are applied to the available datasets to study the temperature dependence of performance, as well as the variability of the TCs for different irradiance conditions. **Figure 5** shows heatmaps of the standard PR as a function of the irradiance and module temperature. From this visualisation, it is clear that bifaciality offers the most gains in performance, especially in mid to low light conditions. Moreover, the differences in TCs between SHJ and PERC cells are also visible, with lower performance ratios over a wider array of high irradiance and temperature values for the PERC cells: for module temperatures above 55°C, the SHJ monofacial outperform PERC by ~2.5%.

As a next step, the maximum-power TC variability is quantified for different irradiance conditions. **Figure 6** shows an example of the irradiance level splitting and linear fitting for the SHJ-Bi-1 module data. The linear regressions coefficients, once normalised by the module rated power at each given condition, give the TCs at the various irradiance levels. Results for all modules are shown in the lower part of **Figure 6**, where the preliminary results from the heatmaps and performance metrics analyses are confirmed: SHJ solar cells have better TCs, with all four SHJ modules converging to -0.16 %/°C at STC conditions, while the PERC module TC is estimated at -0.31 %/°C. Interestingly, the TCs only diverge for the different cell technologies for mid to high irradiance values, confirming the nonuniform temperature behaviours. The TCs tend to higher values for low irradiance conditions, which could be a data artefact due to angular dependencies or measurement errors, although such effects have been previously observed and discussed in literature [6, 9, 10]. Nevertheless, TCs are often considered uniform in PV modelling tools, which could lead to errors and uncertainties, which is why field-based measurements could bring valuable additional insights.

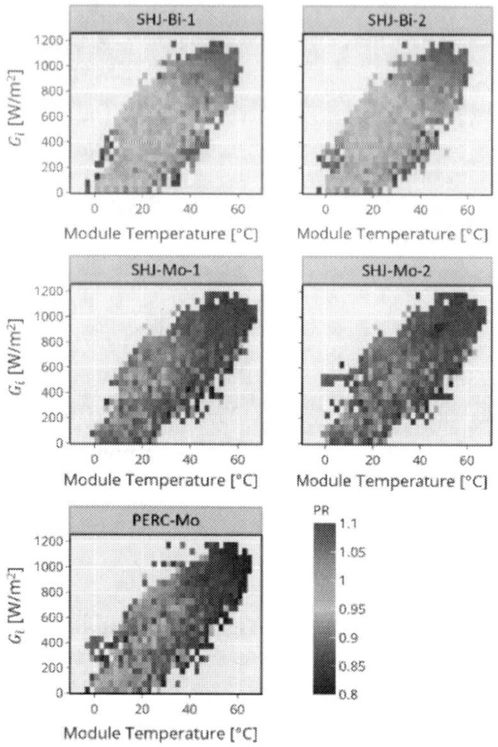

Figure 5. Performance ratio heatmaps of the irradiance vs. module temperature relationship. ***PR*** is averaged over intervals of 50 W/m² and 2 °C.

5 CONCLUSIONS

Bifacial modules and SHJ solar cells are rapidly entering the PV market, creating the need for more advanced understanding of their advantages over conventional, widely studied PERC cells. This contribution aimed at decoupling the performance gains due to bifaciality (module structure) and differing solar cells (module technology), using data from five monitored modules in Neuchâtel, Switzerland. Based on the analysis of performance metrics such as the performance ratio, as well as temperature and irradiance dependencies, the following main outcomes are deduced:

- The bifacial SHJ modules showed an average 7.2% increased energy yield compared to both monofacial technologies over the monitoring period.
- The standard, non-corrected performance ratio is a valuable metric to quantify bifacial and cell technology gains. Overall, bifaciality improves the *PR* by ~5% when comparing the bifacial and monofacial SHJ modules.
- The IEC 61724-1 guidelines for *PR* corrections may not account for all bifacial enhancements, as a 2.2% average gain is still found after correction. This is likely due to angular dependency, which is confirmed with morning and evening peak gains, along with seasonal variability. A future proposal may include angular corrections to account for all rear irradiance components.
- SHJ solar cells are found to have significant advantages over PERC in terms of maximum-power temperature coefficients, although the gains are not uniform depending on the irradiance conditions, with the highest impact of cell technology found at mid to high irradiation. The field-based TC values at STC are found to be -0.16%/°C and -0.31%/°C for SHJ and PERC respectively.

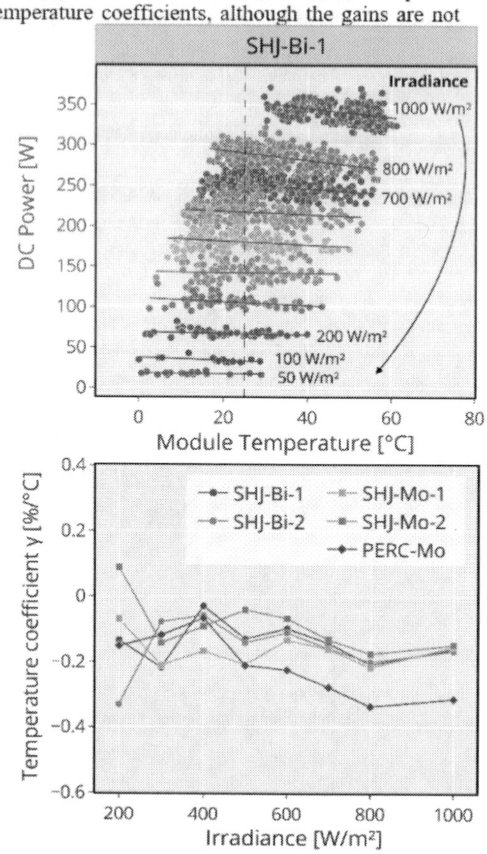

Figure 6. Top: module power vs. temperature relationship, with irradiance level filtering. Bottom: Irradiance dependence of the temperature coefficients.

6 ACKNOWLEDGEMENTS

This work was funded by the European Commission as part of the Horizon 2020-GOPV Project (grant agreement # 792059). The authors are grateful to all GOPV partners participating in the monitoring campaign, in particular to Enel Green Power (Lorenzo Todaro) for sourcing the solar modules and to CEA (Romain Soulas) for taking care of solar module post-processing. The authors also acknowledge the financial support from 3S Swiss Solar Solutions AG. Finally, the authors gratefully acknowledge support from EPFL PV-Lab staff for the installation and outdoor testing facility.

7 REFERENCES

[1] X. Li *et al.*, 'Silicon heterojunction-based tandem solar cells: past, status, and future prospects',

Nanophotonics, vol. 10, no. 8, pp. 2001–2022, Jun. 2020, doi: 10.1515/nanoph-2021-0034.
[2] J. Haschke, O. Dupré, M. Boccard, and C. Ballif, 'Silicon heterojunction solar cells: Recent technological development and practical aspects - from lab to industry', *Solar Energy Materials and Solar Cells*, vol. 187, pp. 140–153, Dec. 2018, doi: 10.1016/j.solmat.2018.07.018.
[3] A. Gok, E. Ozkalay, G. Friesen, and F. Frontini, 'The Influence of Operating Temperature on the Performance of BIPV Modules', *IEEE Journal of Photovoltaics*, vol. 10, no. 5, pp. 1371–1378, Sep. 2020, doi: 10.1109/JPHOTOV.2020.3001181.
[4] M. A. Green, 'General temperature dependence of solar cell performance and implications for device modelling', *Progress in Photovoltaics: Research and Applications*, vol. 11, no. 5, pp. 333–340, 2003, doi: 10.1002/pip.496.
[5] X. Li *et al.*, 'Reassessment of silicon heterojunction cell performance under operating conditions', *Solar Energy Materials and Solar Cells*, vol. 247, p. 111951, Oct. 2022, doi: 10.1016/j.solmat.2022.111951.
[6] B. Zinsser, G. Makrides, M. Schubert, G. Georghiou, and J. Werner, *Temperature and Irradiance Effects on Outdoor Field Performance*. 2009, p. 4086. doi: 10.4229/24thEUPVSEC2009-5BV.2.19.
[7] K. Emery and C. Osterwald, 'Measurement of photovoltaic device current as a function of voltage, temperature, intensity and spectrum', *Solar Cells*, vol. 21, no. 1, pp. 313–327, Jun. 1987, doi: 10.1016/0379-6787(87)90130-X.
[8] D. C. Jordan *et al.*, 'Silicon Heterojunction System Field Performance', *IEEE Journal of Photovoltaics*, vol. 8, no. 1, pp. 177–182, Jan. 2018, doi: 10.1109/JPHOTOV.2017.2765680.
[9] B. R. Paudyal and A. G. Imenes, 'Investigation of temperature coefficients of PV modules through field measured data', *Solar Energy*, vol. 224, pp. 425–439, Aug. 2021, doi: 10.1016/j.solener.2021.06.013.
[10] C. Berthod, R. Strandberg, G. Yordanov, H. G. Beyer, and J. Odden, 'On the Variability of the Temperature Coefficients of mc-Si Solar Cells with Irradiance', *Energy Procedia*, vol. 92, pp. 2–9, Aug. 2016, doi: 10.1016/j.egypro.2016.07.002.

STUDY OF OPTICAL TRANSMISSION LOSSES OF SATINATED PV GLASS

Markus Babin, Anna Bertomeu i Baldé, Sergiu V. Spataru, Michael L. Jakobsen, Sune Thorsteinsson
Department of Electrical and Photonics Engineering, Technical University of Denmark
Frederiksborgvej 399, 4000 Roskilde, Denmark

ABSTRACT: Satinated glass is an increasingly important component of PV modules, especially in the building integrated (BIPV) sector, where it can contribute to more uniform appearance as well as significantly reduce the risk of glare. Different satination methods, such as wet chemical etching or mechanical abrasion, are available, which can impact the PV-relevant optical parameters in different ways. In order to characterize the associated performance losses, in this study eight different satinated glass types are compared in terms of spectral transmittance, surface morphology, relative transmission loss and angular dependent reflection losses.
Measurements show only insignificant transmission losses of < 1 % for acid-etched glass compared to reference PV glass at normal incidence, with significant gains at high incidence angles > 60°. No significant differences between different acid-etched samples can be observed, all featuring homogeneous appearance. Other satination technologies, such as sandblasting or laser-etching show significantly higher transmission and reflection losses at low incidence angles, making them less viable for use in PV applications.
Keywords: Optical Losses, Performance, PV Materials, Texturisation

1 INTRODUCTION

Satinated glass is becoming an increasingly used component in PV modules, especially for building-integrated photovoltaics (BIPV). [1] Compared to standard PV cover glass with either a smooth or lightly textured surface, AR-coated glass or even deeply structured glass, satinated glass can significantly reduce specular reflections responsible for glare. [2] Additionally, satination of the glass surface can lead to a more uniform appearance. In colored PV products based on thin-film interference coatings, satinated glass is even an essential component to reduce iridescence and achieve a homogeneous appearance. [3]

Apart from its effect on reflections and appearance, satinated glass also has an impact on the angular dependent performance measured through the incidence angle modifier (IAM). Especially at high incidence angles, satinated glasses are known to show higher IAM values than standard PV cover glasses. On the other hand, at lower incidence angles, satinated glasses are usually slightly outperformed by standard glass solutions. At normal incidence, textured glass has been shown to have increased light transmission compared to smooth glass, as reflected light can have a second chance to enter the glass and be transmitted to the cell. [4]

While satinated glass can be produced through either mechanical abrasion (i.e. sandblasting) or wet chemical etching, the latter method is the more commonly used. Nevertheless, even this method can employ a variety of HF-based acids, etch rates and environmental conditions, leading to significant variation in surface morphology. [5][6]

Nevertheless, the specific effects of different satination methods on the PV-relevant optical properties of glass have not been investigated until now. This study presents a comparison of transmission loss in a number of acid-etched samples as well as satinated glasses produced from alternative methods, both at normal incidence and higher incidence angles.

2 METHODOLOGY

A total of fourteen different glass types are sourced from the market and compared using total transmittance measurements, short-circuit current (Isc) measurements before and after encapsulation in PV mini-modules to determine relative transmission losses, as well as incidence angle modifier (IAM) measurements. An overview of the samples with short descriptions can be seen in Table I. Sample #1 is considered a good example of standard PV cover glass and is subsequently used as a reference when comparing satinated glasses. Glass types #2-6 are standard acid-etched samples sourced from different manufacturers, while glass types #7-10 are provided by a single manufacturer.

Table I: Glass types and number of individual samples tested

#	Description	Samples
1	Lightly structured low-iron float glass	4
2	#1 sandblasted	3
3	#1 laser-satinated	3
4	Acid-etched low-iron float glass	3
5	Acid-etched low-iron float glass	3
6	Acid-etched float glass	3
7a	Flat low-iron float glass	1
7b	Flat float glass	1
8a	#7a lightly acid-etched	1
8b	#7b lightly acid-etched	1
9a	#7a acid-etched variant 1	1
9b	#7b acid-etched variant 1	1
10a	#7a acid-etched variant 2	1
10b	#7b acid-etched variant 2	1

2.1 Glass characterization

Total spectral transmittance measurements on glass samples are carried out by illuminating samples with collimated light from a laser-driven light source (LDLS) in front of a 6" integrating sphere coupled to an OceanOptics QE65000 spectrometer using an optical fiber. An OceanOptics DH2000 halogen-deuterium light source is used as a correction light source (CLS) to compensate for back-reflections from the glass sample to the inside of the integrating sphere. This setup and the required correction procedure have been described in greater detail in earlier publications. [7][8]

In order to characterize the surface structure of different samples, through-light optical microscopy is carried out perpendicular to the glass surface.

2.2 Relative transmission losses

PV mini-modules are laminated from the glass samples using 2-busbar mono-Si cells (Al-BSF), EVA as encapsulant as well as a polymeric black backsheet.

To determine the PV-relevant relative transmission losses, IV curves are measured on all cells before and after lamination using a Pasan cell tester. Subsequently, short-circuit currents (Isc) are extracted and compared to determine the relative transmission losses, and EL images are taking throughout the process to ensure no processing damage.

2.3 Angular dependent reflection losses

To determine angular dependent reflection losses, measurements of the incidence angle modifier (IAM) are performed on the mini-modules. For these measurements, the collimated light from the LDLS is contained within the cell area and Isc is measured for incidence angles in 5° steps and subsequently averaged over positive and negative angles. This methodology and its validation have also been described in previous publications. [7]

Finally the relative Isc losses are multiplied with the measured IAM to determine the angular dependent effective transmittance of different glass types, in order to enable a direct comparison.

3 RESULTS

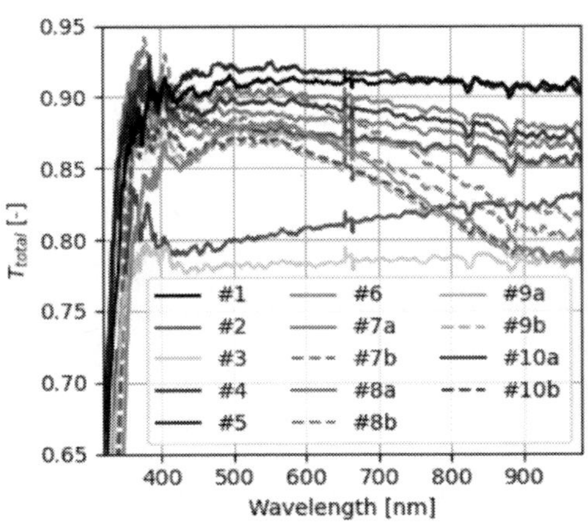

Figure 1: Total transmittance spectra of all glass samples

3.1 Glass characterization

Figure 1 shows the total transmittance spectra of all glass samples, with average transmittance between 78-92 %. While the highest values conform to the expected Fresnel reflection losses of glass in air (approximately 4 % for each air-glass interface), the lower values indicate significant additional reflection or absorption losses.

However, due to the strong scattering of light in case of highly satinated glass, some amount of transmitted light may not be captured by the integrating sphere in the total transmittance measurements, leading to higher uncertainty regarding the magnitude of the measured optical transmittance. Subsequent analysis therefore focuses on the spectral shape rather than magnitude.

Samples #6 and #7b, #8b, #9b, #10b show near-infrared transmission losses characteristic for iron-containing float glass, normally not used in PV encapsulation. Conversely, sandblasted glass (#2) shows a transmittance spectrum weighted more towards the near-infrared, however with overall significantly reduced transmittance compared to the reference. This relative increase in near-infrared transmittance is in line with literature. [B]

The surface structure of the different glass types is shown in Figure 2, featuring through-light microscopy images of four selected samples at varying magnifications. The images show significant differences in the surface structure, with acid-etched glass featuring quasi-circular dips surrounded by higher ridges, while sandblasted glass has a highly irregular surface structure with sharp peaks and ridges as well as small patches of almost smooth glass.

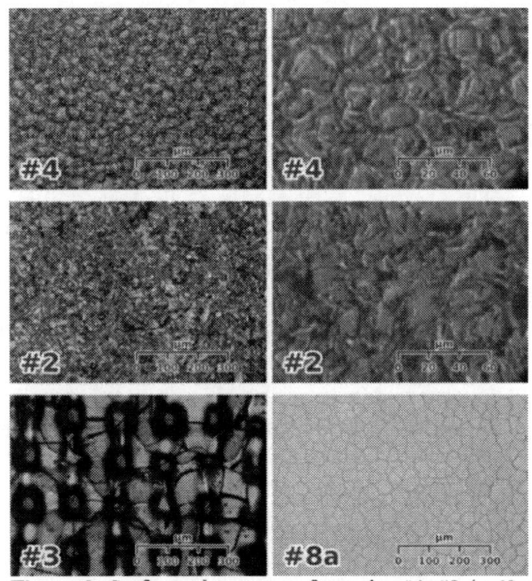

Figure 2: Surface microscopy of samples #4, #2 (~ 43x and 216x optical magnification) and #3, #8a (~ 43x optical magnification)

No significant differences can be observed between the individual acid-etched glass samples (#4-6, #9-10), neither upon visual inspection nor under through-light microscopy. While the size of microstructures varies slightly, no clear change in morphology is apparent.

Sandblasted glass (#2), however, has a significantly less uniform appearance, looking generally more white and opaque. Furthermore, continuous abrasion of the sharp surface structures leads to a perpetual dirty-white appearance.

The laser-satinated samples (#3), while uniform, exhibit a slightly glittery appearance, most likely due to the deep wells in the glass leading to a limited number of characteristic reflection angles. Glass #8 only features very light satination leading to a slightly hazy appearance with a surface structure characterized by wide, shallow dips.

3.2 Relative transmission losses

The relative loss in short circuit current (Isc) of the mini-modules compared to the unencapsulated PV cells is shown in Figure 3. For glass types #1-6, the $\mu \pm 3\sigma$ interval over all samples is shown, whereas for glass types #7-10 data points represent individual samples.

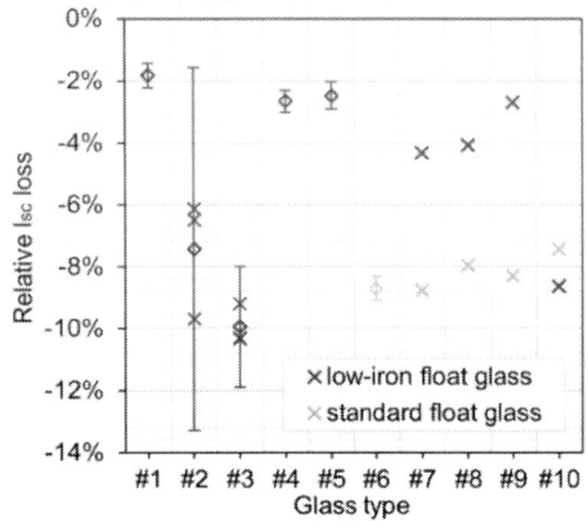

Figure 3: Relative loss in short-circuit current compared to unencapsulated PV cells

While the acid-etched glass types show relative Isc losses of < 1 p.p. beyond that of the lightly textured reference glass, losses are significantly higher and more varying for the sandblasted glass. Similarly, the laser-etched glass shows additional losses of > 5 p.p. beyond the reference. Due to the limited number of samples for glass types #7-10, accuracy of these measurements cannot be ascertained, however the individual data points indicate that satinated glass performs similar or even better than completely flat glass.

3.3 Angular dependent losses

The relative incidence angle modifier (IAM) compared to reference glass #1 is shown in Figure 4 for the different glass types.

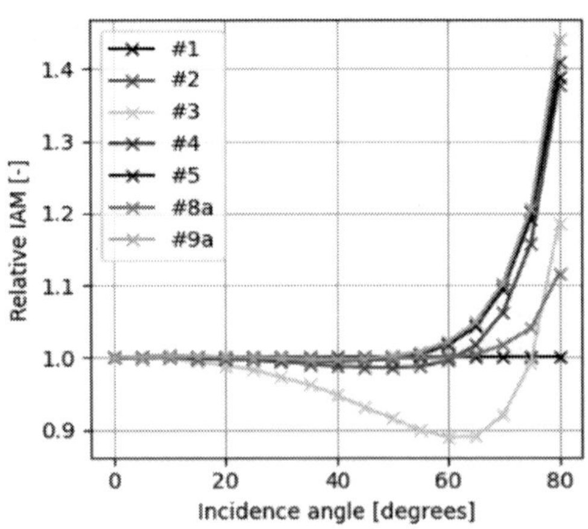

Figure 4: Relative IAM

The measurements show a significant increase in angular dependent transmittance at incidence angles > 55° for acid-etched samples compared to the lightly structured reference glass. Differences in IAM between these samples can only be observed at incidence angles > 70° and are within the measurement uncertainty of the setup.

Both the sandblasted and laser-etched glasses only show a higher IAM compared to the reference at significantly higher incidence angles, performing especially badly in the case of the laser-satinated glass. The lightly satinated glass (#8a) also slightly outperforms the reference at high incidence angles without reductions in IAM at lower angles.

Combining the measured IAM with the relative Isc loss results in the graph shown in Figure 5, giving the effective relative transmittance compared to unencapsulated cells at normal incidence. It shows beneficial irradiance conditions for acid-etched glass for incidence angles > 60°, with only minor reductions in effective transmittance at lower angles. Due to their low transmittance at normal incidence, both sandblasted and laser-etched glasses surpass the reference only at > 70° and 80°, respectively.

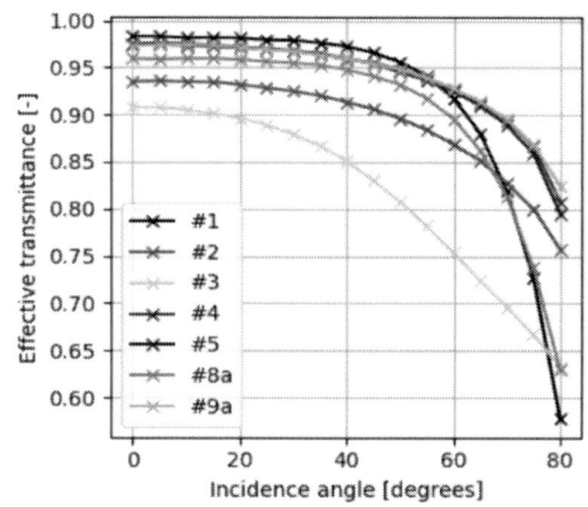

Figure 5: Angular dependent effective transmittance relative to unencapsulated cells at normal incidence

4 DISCUSSION

The results in Figure 5 indicate that while having slightly lower transmittance at low incidence angles, acid-etched satinated glass does not perform substantially worse than standard PV glass with slightly structured surfaces. For fixed tilt installations with frequent occurrence of high incidence angles and/or large fractions of diffuse light, satinated glass can actually be beneficial for the overall annual yield. This is for example the case in building-integrated (BIPV) installations, where non-optimal orientations are quite common.

Regarding the uncertainty of the presented results, due to the limited number of samples per glass type, a full statistical analysis could not be performed. The measurements indicate, however, that there is a significantly higher variance between samples for the sandblasted and laser-satinated glass compared to the references and acid-etched glass.

5 CONCLUSIONS

In general it can be concluded that all acid-etched glass samples investigated in this study very closely resemble each other both in appearance and performance, indicating that manufacturer or specific product choice is of secondary concern when selecting satinated glass for PV applications.

From relative transmittance and IAM measurements the following conclusions can be made regarding different glass satination techniques for PV applications:

Acid-etched glass performs best, with transmittance losses < 1 p.p. compared to standard PV glass at normal incidence, surpassing the reference at incidence angles > 55°. Sandblasted glass has lower visual uniformity and higher variance in transmission losses, possibly leading to increased cell and module mismatch. The relative IAM gain is also lower than that of acid-etched glass. Due to the jagged surface structure, continuous abrasion leading to dirty-white appearance can also be observed.

Laser-satinated glass is unsuitable for PV applications due to very high transmission and IAM losses, most likely due to the deep wells in the glass surface.

Further studies will be required to verify the measurement accuracy, however this work indicates that acid-etched satinated glass leads to insignificant performance reductions at low incidence angles and significant gains at high incidence angles. Its use can therefore be generally recommended for PV applications, especially in non-optimal fixed-tilt orientations and glare sensitive areas, such as in building integration. Satinated glass produced by other technologies, such as sandblasting or acid-etching, should be advised against, however.

6 ACKNOWLEDGEMENTS

The authors would like to acknowledge the support from EUDP as a funding agency for project "64021-1079 UnitSun".

7 REFERENCES

[1] G. Eder, G. Peharz, R. Trattnig, P. Bonomo, E. Saretta, F. Frontini, C. S. P. López, H. R. Wilson, J. Eisenlohr, N. M. Chivelet, S. Karlsson, N. Jakica, A. Zanelli, Coloured BIPV: Market, Research and Development, Report, IEA PVPS Task 15 (2019)

[2] M. Babin, S. Thorsteinsson, M.L. Jakobsen, S.V. Spataru, Glare Potential Evaluation of Structured PV Glass Based on Gonioreflectometry, IEEE J. Photovolt. (2022)

[3] SwissINSO, Kromatix™ Technology, online, https://www.swissinso.com/technology (accessed: 13-Sep-2022)

[4] M. Duell, M. Ebert, M. Muller, B. Li, M. Koch, T. Christian, R.F. Perdichizzi, B. Marion, S. Kurtz, D.M.J. Doble, Impact of Structured Glass on Light Transmission, Temperature and Power of PV Modules, Proc. 25th EUPVSEC (2010) pp. 3867-3872

[5] G.A.C.M. Spierings, Wet chemical etching of silicate glasses in hydrofluoric acid based solutions, J. Mat. Sci., vol. 28 (1993) pp. 6261–6273

[6] M. Starý, F. Novotny, M. Horák, M. Stará, V. Hotař, O. Matúšek, Summary of the properties and benefits of glass mechanically frosted with an abrasive brush, Constr. Build. Mater., vol. 206 (2019) pp. 364-374

[7] M. Babin, S. Thorsteinsson, A.A. Santamaria Lancia, P.B. Poulsen, A. Thorseth, C. Dam-Hansen, M.L. Jakobsen, Dependency of IAM Losses in Colored BIPV Products on the Refractive Index of Colorants, Proc. 38th EUPVSEC (2021) pp. 583-588

[8] A. Tirpak, R. Young, Accurate transmission measurements of translucent materials, Report, Optronic Laboratories (2008)

APPLICATION OF ELECTROLUMINESCENCE CURRENT-INTENSITY RELATION FOR ESTIMATION OF PID IMPACT IN CRYSTALLINE SILICON PHOTOVOLTAIC MODULE

Ravi Kumar*, Vishal Puranik, and Rajesh Gupta
Department of Energy Science and Engineering, Indian Institute of Technology Bombay
Powai, Mumbai, Maharashtra-400076, India
(E-mail: ravikr@iitb.ac.in*, vishal.puranik@iitb.ac.in, rajeshgupta@iitb.ac.in)

ABSTRACT: Potential induced degradation (PID) is a severe degradation mechanism in crystalline silicon photovoltaic (PV) modules which results in significant power loss in a very short duration. Regular monitoring of PID is required for taking preventive actions. Currently, current-voltage (*I-V*) characteristics and shunt resistance measurement are used for PID impact quantification, while the electroluminescence (EL) imaging technique is used for qualitative investigation only. *I-V* characteristics and shunt resistance measurement is a hectic task in a PV field as it requires an auxiliary instrument for each module. A new approach to PID impact quantification is investigated in this study which is based on a quantitative analysis of EL images. EL is a spatial technique, so the cell-level investigation is also possible. EL current and intensity are related with a current calibration constant whose value changes with PID progression in a PV cell. This current calibration constant is used as a PID index to estimate the PID impact on a crystalline silicon PV cell.

Keywords: Photovoltaic, Electroluminescence, Potential-induced degradation

1 INTRODUCTION

Potential induced degradation (PID) is a severe degradation mechanism that can lead up to 40% loss in the performance of a solar module within a few months of installation, especially in a humid and hot environment [1], [2]. In p-type crystalline silicon PV module, sodium ions present on the surface of a solar cell drift towards the pn junction under high voltage stress and accumulate inside the crystallographic defects crossing the pn junction, which results in shunting [3]. This shunted path provides an additional sinking path for photogenerated current; hence the performance of the PV module drops. Currently, current-voltage (*I-V*) characteristics, electroluminescence (EL) imaging, and shunt resistance (R_{sh}) measurements are commonly used techniques for PID detection [4]–[6]. However, *I-V* and R_{sh} measurements need auxiliary instrumentation per module, which is a tedious job in field conditions. Also, it is not useful in the spatial investigation of PID and assessing its severity level since cell level *I-V* measurement is not possible for an intact module. Whereas EL is an image-based technique being used in PID detection, and in recent days, drone-based applications of EL imaging can inspect large PV fields in a short duration. So far, EL imaging has not been used beyond the qualitative detection of PID.

In this work, an EL imaging-based method is proposed to estimate the impact of PID in the crystalline silicon PV module. EL image is captured by exciting PV cell or module electrically. Current flowing through the bulk region of the cell results in EL emission due to band-to-band recombination. The presence of PID affects the intensity of EL image as fractional current flows through the leakage paths induced due to PID, known as shunting. The current flowing through the shunt paths does not contribute to the EL signal. Hence EL intensity of a cell is inversely proportional to the extent of PID within a PV cell. This work studies the relationship between the current and EL intensity. It was observed that EL current follows a linear relationship with EL intensity on a log-log scale. For healthy cells, the EL intensity-current curve of a cell appears linear, but as the PID progresses, it becomes non-linear in nature, as shown in Fig. 3. From this current-EL intensity curve, the current calibration constant is extracted and used as a PID index. Using simulations, an empirical relationship is found between this PID index and relative power loss. Later the experimental work supported the simulations satisfactorily and showed the possibility of using this PID index as a determinant of PID for estimating the impact on the performance of a PV module.

2 THEORY

2.1 Analysis of Double Diode Model (DDM)

A PV cell with PID can be represented by a DDM, as shown in Fig. 1. It is a model with seven parameters where J_{ph} represents the photogenerated current, diode 1 parameters (J_{01}, n_1) represent the bulk properties, diode 2 parameters (J_{02}, n_2) represent the recombinative shunting, R_{sh} represents the shunt resistance, and R_s represent the series resistance. Among these parameters, three parameters (J_{02}, n_2, R_{sh}) are affected by PID progression. Ohmic and non-ohmic shunting in a PV cell depend on the domination of shunt resistance or recombinative diode [7].

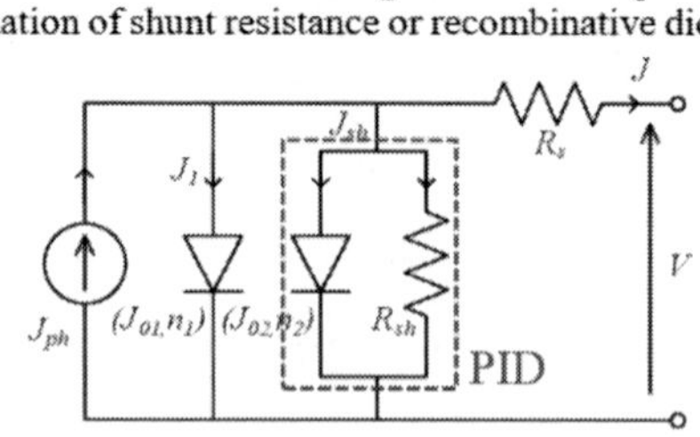

Figure 1: Double diode model of a PV cell

The output current (in terms of current density) from a PV cell can be given as:

$$J = J_{ph} - J_1 - J_2 - J_{R_{sh}} \quad (1)$$

$$J = J_{ph} - J_{01} \exp\left[\frac{V+IR_s}{n_1 V_t}\right] - J_{02} \exp\left[\frac{V+IR_s}{n_2 V_t}\right] - \frac{V+JR_s}{R_{sh}} \quad (2)$$

$$V_t = \frac{kT}{q} \quad (3)$$

Here, J_{ph} is photocurrent generated due to photovoltaic action, J_{01} and J_{02} are saturation currents, while n_1 and n_2 are ideality factors of diode 1 and diode 2, respectively, R_{sh} is shunt resistance, and R_s is series resistance.

The power output is a function of all seven parameters shown by (4), in which three parameters (J_{02}, n_2, R_{sh}) are affected by PID progression, influencing the shunt current, J_{sh}.

$$P = [J_{ph}, J_{01}, n_1, J_{02}, n_2, R_{sh}, R_s] \quad (4)$$

2.1 Current Calibration Constant as a PID Index

In a PV cell, EL emission results due to recombination of charge carriers when current is supplied externally. The emitted intensity lies in the infra-red region; therefore, a dark environment is required with a sensitive camera to detect emission. An image is formed based on spatial intensity variation on a PV module. The circuit diagram of a PV cell under dark EL conditions is shown in Fig. 2.

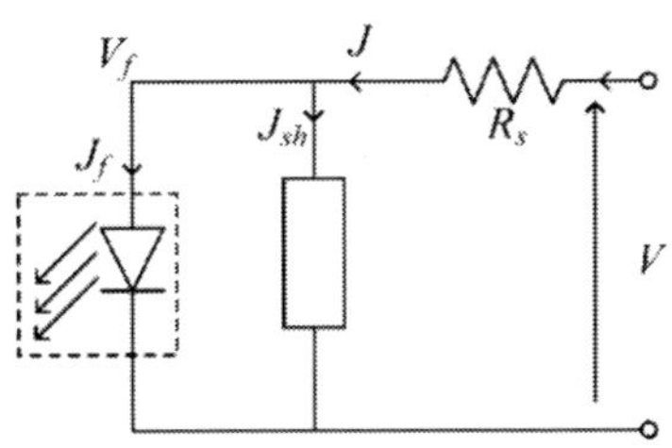

Figure 2: Circuit diagram for EL imaging under dark conditions

The relation between the intensity emitted from a PV cell and input current density can be given as [8]:

$$\log(\phi_{EL}) = A'' + n_1 \log(J_{EL}) \quad (5)$$

The relation (refer eq. 5) resembles the straight-line equation (y = mx + c). From Fig. 2 and eq. 5, it can be observed that if the current through the shunted path, J_{sh} will increase with PID progression, and the EL intensity will decrease. This increase in J_{sh} will affect the A" which is known as the current calibration constant. This constant is being used as a PID index in this study. The non-linearity of the EL-current curve serves the motivation behind this study. This non-linearity changes the constant A" named as the current calibration constant. The variation in the EL intensity (ϕ_{EL}) curve with PID progression is shown in Fig. 3.

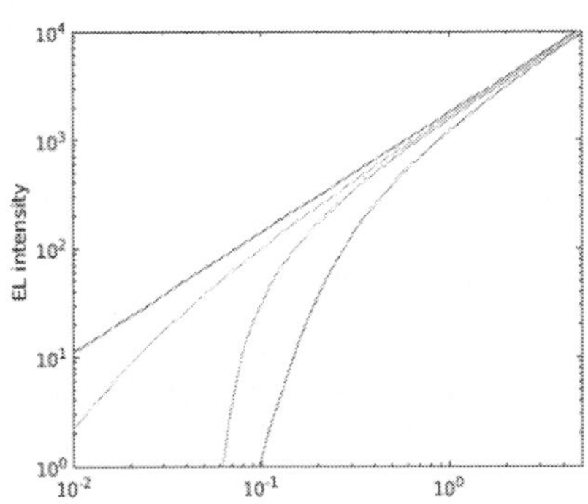

Figure 3: Variation in the EL intensity with current at different stages of PID progression

For the simulation purpose, the EL intensity (ϕ_{EL}) can be calculated theoretically by measuring the voltage across the PV cell by using eq. 6 [9].

$$\phi_{EL} = A\exp\left(\frac{V_f}{V_t}\right) \quad (6)$$

Here A is voltage calibration constant, and V_f is forward bias voltage. The EL intensity variation with the applied voltage across the PV cell is shown in Fig. 4. As the voltage increases, the EL intensity also increases, which is exponentially in nature.

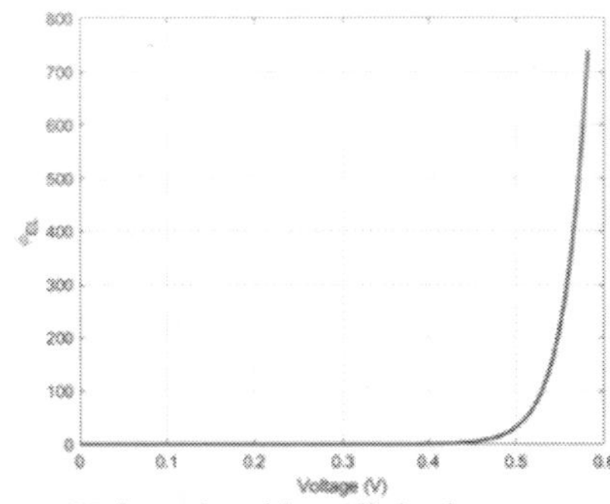

Figure 4: EL intensity with applied voltage

A (voltage calibration constant) and A" (current calibration constant) depend on the material properties. A is related to the terminal voltage, which remains constant even in the presence of a shunt. It remains the same for a module despite PID, as it does not affect the bulk region (J_{01},n_1). As PID get induced (refer to Fig. 2.), the leakage path sinks current ($J = J_f + J_{sh}$), and J_{EL}-ϕ_{EL} curve becomes non-linear with PID progression. The relative difference in *A"* values between a cell and a reference cell at low EL current (20% of I_{SC}) is used as a PID index. Here, the reference cell is considered the one with the highest power output among all the cells. This PID index is calculated as:

$$\Delta A'' (\%) = \frac{A''_{ref} - A''_{cell}}{A''_{ref}} \times 100 \quad (7)$$

Relative change in current calibration constant (%A") reflects the degree of shunting and can be used as a sole indicator of PID. A" variations purely reflect shunting; hence it enables PID detection even in the presence of other degradation. It purely shows the existence of a leakage path; therefore, in the presence of any defects related to R_s as well as bulk parameters (J_{01}, n_1), PID can be detected efficiently and reliably.

4 METHODOLOGY

The approach used in the simulation study is shown in Fig. 5. DDM of a healthy cell is considered for this study. The current-voltage *(I-V)* characteristics of the cell is shown in Fig. 6, and the parameters used in simulations are given in Table I.

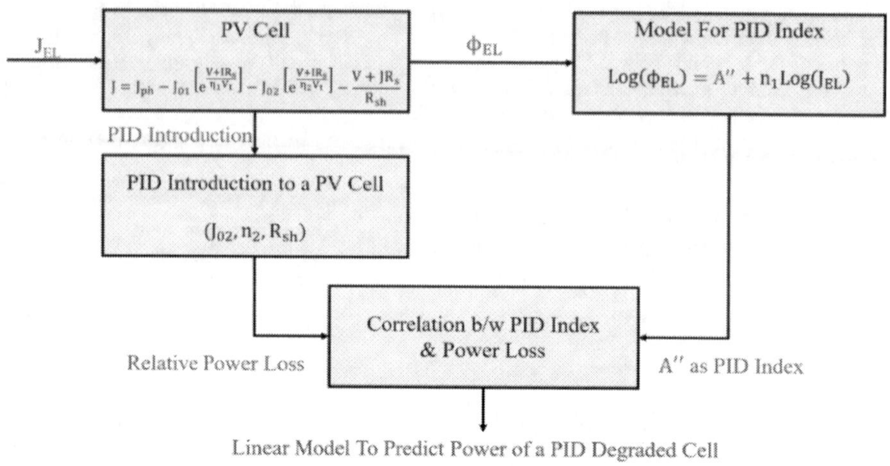

Figure 5: Block diagram showing work approach

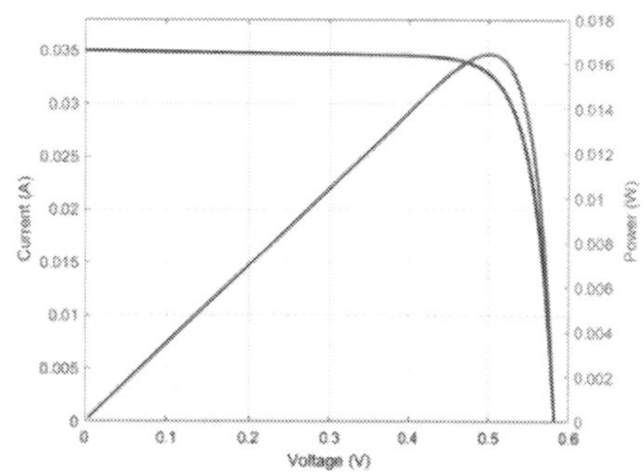

Figure 6: I-V characteristics of a healthy cell used in the study

Table I: Parameters of healthy cell used in simulations

Parameter	Value
J_{sc}	0.035 A/cm^2
R_s	0.0001 Ω/cm^2
J_{01}	10^{-11} A/cm^2
n_1	1.03
J_{02}	10^{-09} A/cm^2
n_2	2
R_{sh}	1000 Ω/cm^2

In simulations, PID was introduced by randomly changing the three PID-influenced parameters (J_{02}, n_2, R_{sh}) and power output, P_{MPP} for each case was calculated. This random variation includes both types of PID-shunting, ohmic and non-ohmic. On the other hand, EL intensity, Q_{EL} was estimated for each PID case using (6) and current calibration constant, A" was estimated using (5). Using P_{MPP} and A", a correlation is found, which governs the possibility of using A" as a potential PID index to estimate the PID impact using EL imaging technique as shown in Fig. 7(a).

To support the simulation work, some experiments were carried out. Using EL setup, 20% of I_{sc} current is passed as required, and the EL image is taken. The intensity value is extracted from this EL image and using (5), the current calibration constant is calculated at the cell level. Cell level measurement is done by assessing cells after cutting the backsheet. The power is measured and plotted with the current calibration constant for every cell, as shown in Fig. 7(b).

5 RESULTS AND DISCUSSION

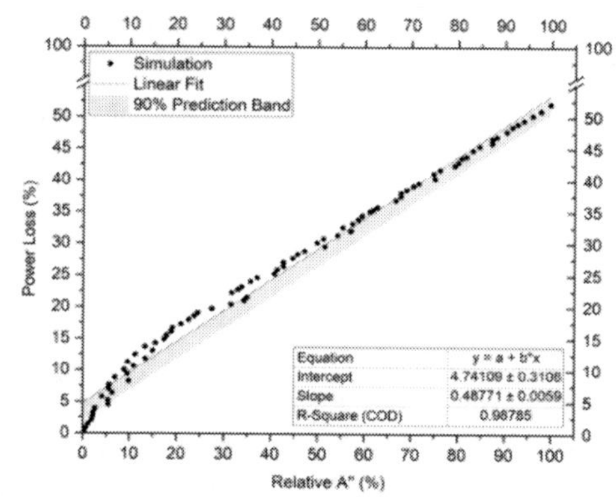

(a) Simulation

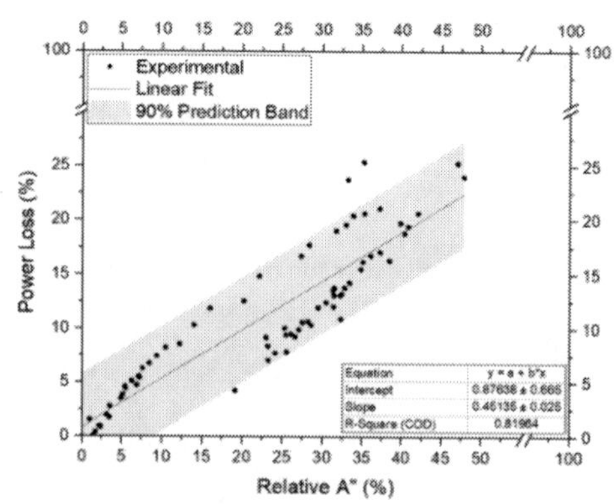

(b) Experimental

Figure 7: (a) Simulation and (b) Experimental linear fit between PID index and power loss at low EL current (20% of I_{SC})

Based on the simulations, Fig. 7(a) shows a good correlation between ΔA"% and ΔP_{MPP}% at low current (20% of I_{sc}). Here all the data points lie in a straight line, and a linear model is fitted with more than 98% R-square value.

Experimental data are fitted in linear relation with more than 80% R-square value, as shown in Fig. 7(b). The trends obtained in simulation work are supported by experimental work satisfactorily.

Using the obtained relation between the power loss and relative change in the current calibration contant, power loss can be predicted satisfactorily in the initial to moderate phase of PID. This method requires only one EL image and is much more suitable for field application as drones equipped with EL systems are used in PV fields.

Limitations. There are a few limitations associated with this method which are given as follows:

a) Any severe pre-existing shunted cell (other than PID) also qualifies for PID, which can not be segregated from PID using this method.
b) It might fail in severely affected PID cells since no useful EL emission occurs from these cells.

4 CONCLUSIONS

In this paper, EL intensity is used to quantify the impact of PID on the performance of a crystalline silicon PV module. The relation between input current and EL intensity under dark conditions is used to extract a parameter known as the current calibration constant, which is used as a PID index. This PID index is correlated with power loss. Both simulations and experimental work show a direct relationship between the PID index and relative power loss. Power loss can be estimated satisfactorily in the initial to moderate phase of PID using this empirical relationship. PID within a cell can be easily judged without estimating any other parameter and knowing datasheet parameters.

REFERENCES

[1] S. Pingel *et al.*, "Potential induced degradation of solar cells and panels," in *Conference Record of the IEEE Photovoltaic Specialists Conference*, 2010, pp. 2817–2822. doi: 10.1109/PVSC.2010.5616823.

[2] W. Luo *et al.*, "Potential-induced degradation in photovoltaic modules: A critical review," *Energy and Environmental Science*, vol. 10, no. 1. Royal Society of Chemistry, pp. 43–68, Jan. 01, 2017. doi: 10.1039/c6ee02271e.

[3] V. Naumann, C. Hagendorf, S. Grosser, M. Werner, and J. Bagdahn, "Micro structural root cause analysis of potential induced degradation in c-Si solar cells," in *Energy Procedia*, 2012, vol. 27, pp. 1–6. doi: 10.1016/j.egypro.2012.07.020.

[4] I. M. Kwembur, J. L. Crozier McCleland, E. E. van Dyk, and F. J. Vorster, "Detection of Potential Induced Degradation in mono and multi-crystalline silicon photovoltaic modules," *Phys. B Condens. Matter*, vol. 581, Mar. 2020, doi: 10.1016/j.physb.2019.411938.

[5] V. E. Puranik and R. Gupta, "Analysis and insight of electroluminescence imaging in the assessment of potential-induced degradation in crystalline silicon photovoltaic module," *Eng. Fail. Anal.*, vol. 134, Apr. 2022, doi: 10.1016/j.engfailanal.2022.106027.

[6] V. E. Puranik and R. Gupta, "Novel Quantitative Electroluminescence Method for Detailed Performance Analysis of PID-s Affected Crystalline Silicon PV Module," *IEEE J. Photovoltaics*, vol. 11, no. 6, pp. 1470–1478, Nov. 2021, doi: 10.1109/JPHOTOV.2021.3108764.

[7] R. Gupta, P. Somasundaran, and D. K. Nandi, "Electrical simulation and characterization of shunts in solar cells," *Appl. Mech. Mater.*, vol. 110–116, no. 1, pp. 2453–2457, 2012, doi: 10.4028/www.scientific.net/AMM.110-116.2453.

[8] K. Takashi, Fuyuki; Hayato, Kondo; Yasue, "Analytic findings in the electroluminescence characterization of crystalline silicon solar cells," *J. Appl. Phys.*, vol. 023711, no. January, 2007, doi: 10.1063/1.2431075.

[9] T. Fuyuki and A. Kitiyanan, "Photographic diagnosis of crystalline silicon solar cells utilizing electroluminescence," *Appl. Phys. A Mater. Sci. Process.*, pp. 189–196, 2009, doi: 10.1007/s00339-008-4986-0.

ADDING THE MLFM TO PVPMC/PVLIB

Steve Ransome[1] and Juergen Sutterlueti[2]
[1]Steve Ransome Consulting Ltd, #99 KT2 6AF, U.K. mailto:steve@steveransome.com
[2]Gantner Instruments GmbH, 6780 Schruns, Austria

ABSTRACT: PVPMC/PVLIB is a global collaborative project headed by Sandia to develop a verified open-source PV modelling library in python. The MLFM (Mechanistic Loss Factors Model) combines the Loss Factors Model LFM (orthogonal, normalised loss calculations from MPPT measurements, iv curves or IEC 61853-like matrices) with the Mechanistic Performance Model MPM (robust, orthogonal, normalised fitting coefficients to LFM losses). The MLFM is being added to PVLIB, this paper describes the new functionality as shown by these models and how to interpret the graphs and datafits. This paper also describes some of the procedures used to add code to PVLIB and aims to encourage others to submit their own PV modelling code following PVLIB's styles, nomenclature, naming conventions, formatting, guidelines, designs, tests etc.
Keywords: Energy Rating; Energy Performance; Modelling

1 INTRODUCTION

The MLFM equations and procedures have proven accurate and useful in PV module measurement characterization, identifying performance limits and loss mechanisms, deriving performance and temperature coefficients, analyzing degradation rates and causes and/or seasonal annealing, and comparing fit accuracies with other common models [1][2][3].

Their functionality is now being added to PVLIB which is the industry leading python open-source modelling library.

MLFM modelling algorithms can be used with existing PVLIB algorithms such as atmospheric data, TMY, solar position and angle of incidence, tracking, inverter modelling and the performance compared with other models in PVLIB (such as the 1-diode model, PVUSA and SAPM. [2])

The sequence, methods and procedures used to add code has been documented to help others contribute.

MLFM algorithms have been added as a Jupiter Notebook tutorial 'mlfm.ipynb' with a library file 'mlfm.py' to PVLIB The code is available as a downloadable python tutorial mlfm.ipynb in PVLIB[]. Three sample data files contain either indoor measurements (CFV IEC 61853) or outdoor (NREL and Gantner Instruments) with 4 or 6 datapoints.

A flow chart of the mlfm procedures added to PVLIB including graphs as from the tutorial is shown in figure 1.

Each box is identified by a letter [A] to [K] for later discussion in the paper.

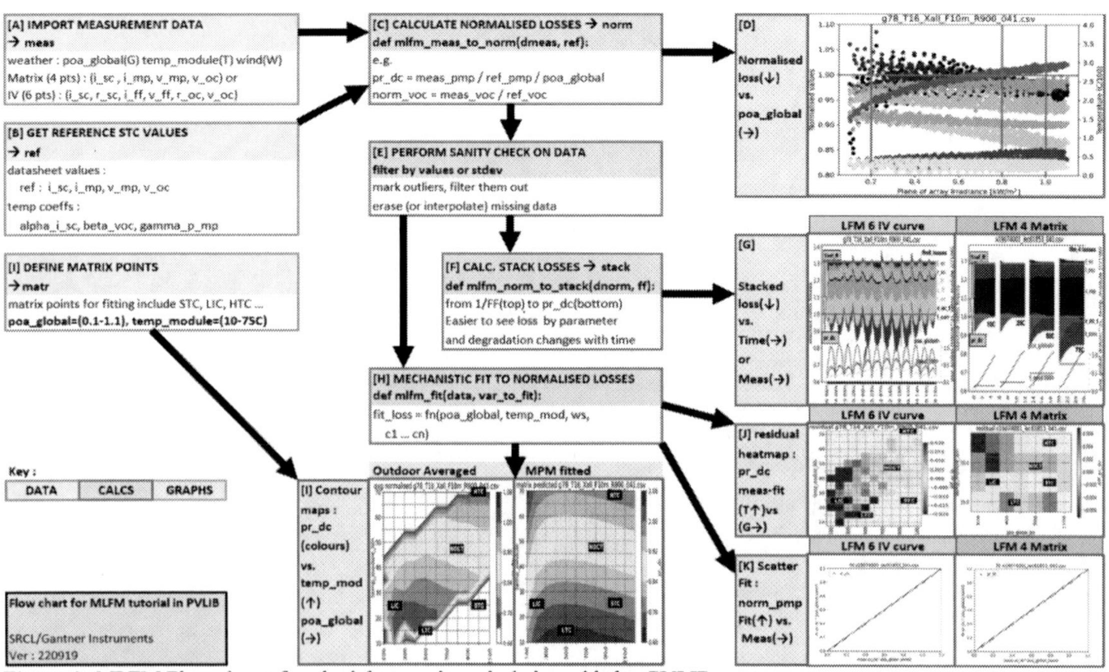

Figure 1: MLFM Flow chart of methodology and graphs being added to PVLIB

2 FURTHER DETAILS

The Loss Factors Model derives meaningful, orthogonal, and normalised loss factors from the shape of IV curves compared with reference STC values as illustrated in figure 2.

These coefficients show the magnitude of losses attributable to each of the factors i_sc to v_oc.

The number of normalised loss factors depends on whether measurements of iv parameters include r_sc and r_oc slopes (1/dI/dV at i_sc or v_oc respectively).

The product of the loss coefficients equation (1) multiplies the datasheet module ref_p_mp point ① to measured meas_pmp point ⑥ using the equation for pr_dc6 (with r_sc and r_oc as in a full iv curve) or pr_dc4 (without r_sc and r_oc e.g. as with an IEC 61853 matrix).

(Spectral and reflectivity vs. aoi are not included here).

Normalised LFM loss equation:

$$pr\_dc_n = \frac{1}{ref\_ff} \times \prod_{i=1}^{n} norm\_lfm_{(i)} \qquad (1)$$

Where:

norm_lfm$_{(1..6)}$=
['i_sc','r_sc','i_ff','v_ff','r_oc','v_oc']

or

norm_lfm$_{(1..4)}$=
['i_sc', 'i_mp', 'v_mp', 'v_oc']

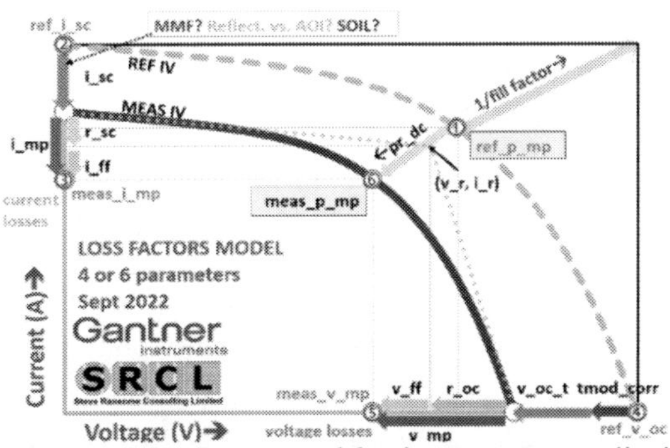

Figure 2: Loss Factors model using 4 or 6 normalised losses (plus optional temperature correction on v_oc) to calculate to ⑥meas_p_mp from ①ref_p_mp.

The names and standard LFM colours of the losses are listed in table I.

Table I: LFM_6 and LFM_4 parameters and colours (with optional temperature correction on v_oc).

LFM_6 e.g. IV curves (with r_sc and r_oc) # Param	 Colour	LFM_4 e.g. IEC 61853 matrix (no r_sc, no r_oc) # Param	 Colour
Ref Lossless =① * 1/ref_ff		Ref Lossless =① * 1/ref_ff	
i_sc	Purple	i_sc	Purple
r_oc	Orange		
i_ff	Lime	i_mp	Green
v_ff	Cyan	v_mp	Blue
r_oc	Pink		
v_oc(_t)	Brown	v_oc(_t)	Brown
(t_corr)	Red	(t_corr)	Red
⑥ Meas = Pr_dc		⑥ Meas = Pr_dc	

3 MLFM PROCEDURES AND GRAPHS

Each box and graph from figure 1 is identified by letter [A] to [K] and is described with some python code below.

Sample python code from pvlib is included but its whitespace (tabbing) may have been modified for clarity.

Note: The code within pvlib is correctly formatted.

2.1 [A] 'Import measurement data → meas'

The MLFM analysis process imports measured weather and iv point or slope data "-1/(di/dv)" from csv files into a dataframe 'dmeas' with the following essential variable names which follow the pvlib naming convention where possible.

Optional parameters such as 'temp_air' can be included but aren't analysed as standard.

```
# Python code [A]

dmeas :
# essential measured values
#                  [unit]  # comment
# weather
    poa_global     [W/m^2]
    temp_module    [C]
    wind_speed     [ms^-1] # 0 if indoor
# iv points
    i_sc           [A]
    v_oc           [V]
    i_mp           [A]
    v_mp           [V]
    p_mp           [W]     # p_mp = i_mp * v_mp
# optional
    r_sc           [Ohm] #-1/IVslope@Isc
    r_oc           [Ohm] #-1/IVslope@Voc
```

Three sample data sources (Gantner 6 parameter and NREL 4 parameter outdoor plus an indoor CFV Matrix 4 parameter) are included but more can easily be added in the required meas csv format above.

2.2 [B] 'Get reference stc values → ref'

STC Reference data 'ref' (from datasheets) is imported for the module chosen in (A) again following pvlib naming conventions.

```
# Python code [B]

ref :
# Essential Reference values at STC.
#                  [unit]  # comment
# electrical
    i_sc           [A]
    i_mp           [A]
    v_mp           [V]
    v_oc           [V]
    p_mp           [W]     # = i_mp * v_mp
# temperature coefficients
    gamma_p_mp     [1/C]   # pmp temp. coeff.
    beta_v_oc      [1/C]   # voc temp. coeff.
    alpha_i_sc     [1/C]   # isc temp. coeff.
```

2.3 [C] 'Calculate multiplicative normalised losses → norm'

Normalised multiplicative losses are calculated from meas and ref using equation (1) for either LFM_6 or LFM_4 parameter measurements.

Voltage values are normalised as 'norm=meas/ref', Current values are normalised as 'norm= meas/ref/irradiance_suns'.

Further equations are used to derive normalised losses from the r_sc and r_oc using i_r and v_r which is the intercept of the r_sc and r_oc lines in figure 2.

```
# Python code [C]

def mlfm_meas_to_norm(dmeas, ref):
```

```
# Convert measured power, current and voltage to
# normalized values e.g.

dnorm['i_sc'] = dmeas['i_sc'] \
    / (dmeas['poa_global'] / G_STC) / ref['i_sc']

dnorm['v_oc'] = dmeas['v_oc'] / ref['v_oc']

# where
# i_r = ((i_sc*r_sc - v_oc)/(r_sc - r_oc))
# v_r = ((r_sc*(v_oc - i_sc*r_oc)/(r_sc - r_oc)))

if lfm_6: # full iv curve with r_sc and r_oc
    # calculate normalised resistances r_sc, r_oc

    dnorm['r_sc'] = i_r / dmeas['i_sc']
    dnorm['r_oc'] = v_r / dmeas['v_oc']

    # calculate remaining fill factor losses
    # partitioned to i_ff, v_ff

    dnorm['i_ff'] = dmeas['i_mp'] / i_r
    dnorm['v_ff'] = dmeas['v_mp'] / v_r

if lfm_4: # matrix no r_sc or r_oc
    dnorm['i_mp'] = dmeas['i_mp'] / dmeas['i_sc']
    dnorm['v_mp'] = dmeas['v_mp'] / dmeas['v_oc']
...
return dnorm
```

2.4 **[D] 'Graph of normalised lfm : loss(↓) vs. poa_global(→)'**

Figure 3 illustrates the 6 different normalised loss parameters for a cSi module at Gantner Instruments' OTF in Tempe, AZ vs. poa_global.

Loss parameters i_ff, v_ff, r_sc and r_oc are all quite narrow and smooth indicating good module behaviour and well measured. The LFM analysis can often identify temperature coefficients of terms such as r_sc and r_oc [JAPAN Tues]

i_sc is a little more widely scattered (as it has uncorrected soiling and spectral losses, particularly at lower light levels where aoi/reflectivity differ from clear skies to diffuse).

Because v_oc is quite temperature sensitive it is usually better plotted temperature corrected by beta_v_oc.

```
# Python code [D]

dnorm['v_oc_temp_corr'] = dnorm['v_oc'] \
    * (1 - ref['beta_v_oc'] \
    * (dmeas['temp_module'] - T_STC))
```

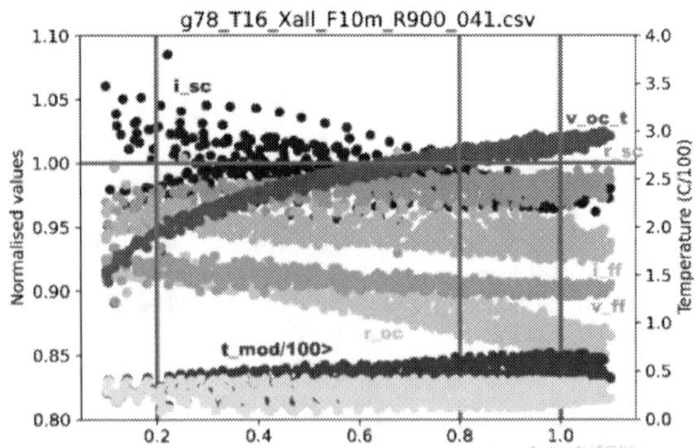

Figure 3: Normalised LFM_6 losses(↓) vs. Irradiance(→) typical c-Si module at Tempe, AZ.

Pr_dc is the product of all these loss parameters as in equation (1). Therefore if any of the coefficients change with irradiance or temperature it will affect the overall pr_dc performance and can be quantified as in figure 3 .

- Low light pr_dc drop(↙) is caused by r_sc (r_shunt) and v_oc (which depends on ln(poa_global)).
- High light pr_dc drop(↘) is caused by r_oc (r_series).
- Coefficients that are almost 'flat' with irradiance = i_ff, v_ff.

2.5 **[E] 'Perform sanity checks on data'**

Normalised data should be checked for missing values, errors, and outliers.

Both normalised limits (e.g. '0.7 < pr_dc < 1.3') and standard deviations (e.g. < 3σ) can be used for outdoor data, whereas normalised indoor matrix measurements would likely be remeasured if there were any outliers.

The limits needed will depend on the accuracy and variability of the measurements.

```
# Python code [E]

# remove values outside limits e.g. <0.5 or >1.5

norm = norm[((norm['pr_dc'] > 0.5) &
             (norm['pr_dc'] < 1.5))]

# remove all mlfm values outside x~3 stdevs
if qty_mlfm_vars == 6:
    stdevs = 3
    # remove data > x stdevs
    for lfm in ('i_sc','r_sc','i_ff',
                'v_ff','r_oc','v_oc'):
        norm = norm[
            ((norm[lfm] - norm[lfm].mean()) /
                norm[lfm].std()).abs() < stdevs
        ]
```

2.6 **[F] 'Calculate stacked losses → stack (subtractive)'**

Plots of multiplicative LFM losses (1) as shown in figure 3 overlap, sometimes it's difficult to discern relative losses and changes.

It can be useful to mathematically transform them to subtractive losses which can be represented more easily on a stack plot as in figs 4 and 5 using equation (2).

```
# Python code [F]

def mlfm_norm_to_stack(dnorm, fill_factor):
'''Converts normalised values to stacked
    subtractive normalized losses.
    Normalized values can reveal losses via
    scatter plots vs. irradiance or temperature.
    Stacked subtractive losses can show relative
    loss proportions.
    Stacked losses partition the difference
    between the normalized power and the power
    that corresponds to the reference fill
    factor. '''
...
return dstack
```

Stack loss equation

$$pr\_dc_n = \frac{1}{ff} - \sum_{i=1}^{n} stack\_lfm_{(i)} \qquad (2)$$

Where:

```
stack_lfm(1..6) =
    ['i_sc','r_sc','i_ff','v_ff','r_oc','v_oc']
```

or

```
stack_lfm(1..4) =
    ['i_sc',    'i_mp',       'v_mp',    'v_oc']
```

2.7 **[G] 'Plot stacked subtractive lfm : loss(↓)**

vs. time(outdoor) or measurement(indoor)(→)'

Stacked losses for 6 and 4 parameters are indicated by their colours from top to bottom as listed in Table I. The height of each loss is proportional to its magnitude. Loss colours are from Table I.

Losses fall from from 1/ref_ff down to pr_dc. Irradiance (green) and temp_mod/100 (red) are shown both on the right y axis.

Figure 4 (left) plots one clear day a month (Jan – Dec for a year) for a Gantner c-Si module in Tempe AZ. There are six independent losses plus temperature corrections in a stacked format from a lossless limit 1/ref_ff (top), subtracting each loss value in turn until it reaches pr_dc (bottom).

```
# Python code [G]
```

```
# Stacked loss values(y) vs. date and time
#(or matrix measurement)(x)

fig_stack = plot_mlfm_stack(
    dmeas=meas, dnorm=norm, dstack=stack,
        # dataframes measurements
    xaxis_labels=12,
        # show #x_labels or 0=show all
    is_i_sc_self_ref=False,
        # is i_sc self referenced?
    Is_v_oc_temp_module_corr=True
        # is v_oc temp corrected?
)
```

If IV curves are not available to find r_sc and r_oc (e.g. just indoor matrices IEC 61853) then similar stacked loss plots with just 4 parameters can be generated.

Figure 4(right) plots losses for LFM_4(i_sc, i_mp, v_mp, v_oc_t, temperature correction) for a cSi module IEC 61853 matrix measured by CFV.

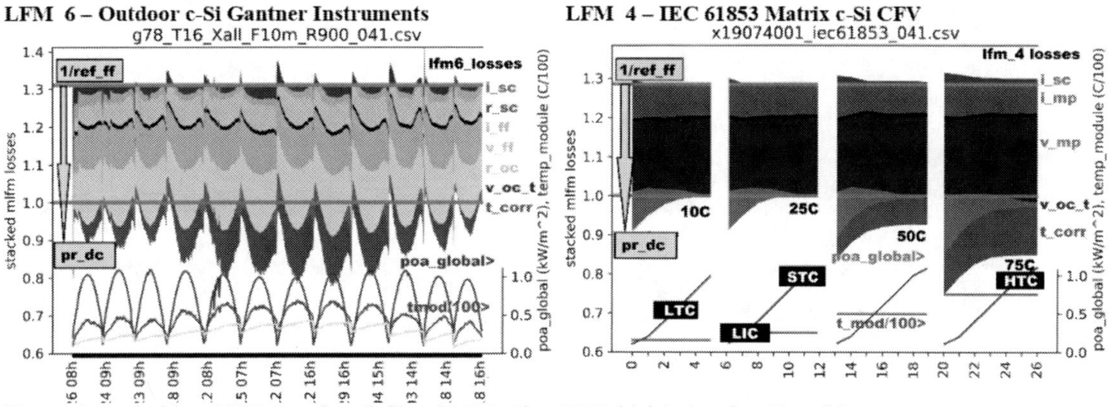

Figure 4: Stacked loss LFM_6 outdoor (left) and LFM_4 iec 61853 (right) plots for cSi modules

Explanations for figure 4 left:

- 1 clear day per month Jan to Dec Tempe, AZ
- Lowest pr_dc at middle of days due to highest temperature and roc losses (~r_series).
- Lowest pr_dc summer months due to temperature.
- This module has much higher r_oc (~r_series) loss than r_sc (~ r_shunt).
- Isc loss is non-zero, cause is soiling and spectral?

Explanations for figure 4 right:

- IEC 61853 matrix measurements indoor
- i_sc loss (purple) ~0 (measured at AM1.5, aoi=0 and soil = 0 so no corrections are needed).
- i_mp loss is almost constant, less than v_mp.
- v_mp increases with temp. and irradiance (→ ~r_series).
- v_oc loss worst at low light (↙left of each stack).
- temp_module losses worst at high temp. (↘right=75C).

2.8 [H] 'Mechanistic fit to normalised losses'

A mechanistic performance model MPM [ref] is used to fit any of the LFM parameter behaviours with irradiance and module temperature (3).

This algorithm minimizes rmse using python's optimize.curve_fit.

$$mpm = c_1 + c_2 * (temp\_module - 25) + c_3 * log_{10}(poa\_global) + c_4 * poa\_global + c_5 * ws + c_6 \,/\, poa\_global \qquad (3)$$

```
# Python code [H]

# setup initial values and initial boundary
# conditions

# initial    c1    c2    c3    c4    c5    c6<0
p_0      = (1.0, 0.01, 0.01, 0.01, 0.01, -0.01)
# boundaries
bounds = ([ -2,   -2,   -2,   -2,   -2,   -2],
          [  2,    2,    2,    2,    2,    0])
```

The mpm gives meaningful, normalised and orthogonal coefficients unlike other models.

For indoor measurements c_5 (wind_speed) is 0, most modules can be fitted with a c_6 =0 but if not, it must be <0 as the boundary conditions constrict it.

Some typical example coefficients are given in table II

Table II : Typical MPM coefficients

Coeff	Value	Unit	Comment
C_1	+106.8%		Overall quality
C_2	-0.45%	/K	Temp. coeff. gamma
C_3	+0.48%		Low light Voc. Rsh
C_4	-7.03%	/(kW/m^2)	High light Rs
C_5	-0.063%	/(ms^{-1})	Windspeed
C_6	-1.54%	*(kW/m^2)	Extra Low light

2.9 [I] 'Plot contour maps: pr_dc(colours) vs. temp_mod(↑) and poa_global(→)'

Figure 5 left shows the measured (left) vs. mpm fitted (right) pr_dc per irradiance bin (100W/m^2→) and temp_module bin(5C↑) from a Gantner Instruments cSi module in Tempe, AZ.Figure 5 right gives the IEC 61853 Matrix by CFV.

Both show close MPM fits as there are close similarities against temp_module(C↑) and irradiance(W/m^2→).

Some of the test condition points are shown as the Matrix measurement does not have linear irradiance and temperature values.

```
# Python code [I]
# Measured vs. MPM Fitted vs. poa_global(→)
# and temp_module(↑)
contour_plot = plot_contourf(
    df=matr2,
    y_axis='temp_module',
    x_axis='poa_global',
    z_axis=mlfm_sel,
    title='matrix predicted ' + mlfm_meas_file,
    vmin=0.7,
    vmax=1.05,
    levels=9
```

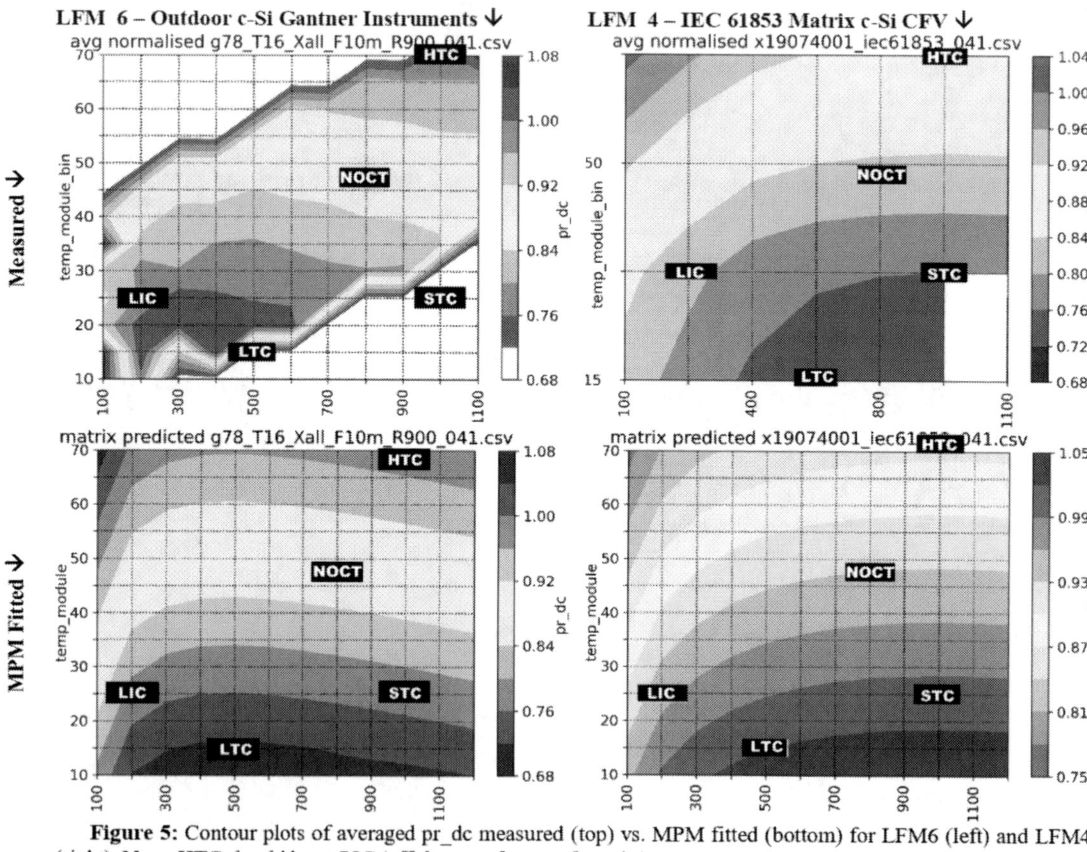

Figure 5: Contour plots of averaged pr_dc measured (top) vs. MPM fitted (bottom) for LFM6 (left) and LFM4 (right). Note: HTC should be at 75C (off the top of most of graphs)

2.10 [J] 'Plot residual fit heatmap: pr_dc 'meas – fit' vs. temp_mod(↑), irradiance (→)'

Figure 6 plots residuals '(fit – meas pr_dc)' against irradiance (W/m^2→) and module temperature (C↑) bins for LFM_6 (left) and LFM_4 (right).

Apart from some extreme outdoor weather conditions i.e. the top and bottom of the weather distribution most weather bins have the 'meas – fit' within ±-1% (light cyan to light orange).

Note the fit to the indoor measurements is even better at mostly ± 0.4%. The 50C lines is all overestimated a little, it's possible that this chart has just identified a slight measurement inaccuracy at 50C as they are all ~+0.3% rather than -0.1%. This will be investigated further.

```
# Python code [J]

# Residual MLFM fit heatmap(colours)
# vs. poa_global(x), temp_module(y)
heatmap_plot = plot_heatmap(
    dnorm=norm,
    dmeas=meas,
    fit=mlfm_sel,
    y_axis='temp_module_bin',
    x_axis='poa_global_bin',
    z_axis='diff_' + mlfm_sel,
    title='residual ' + mlfm_meas_file
)
```

LFM_6 – Outdoor c-Si Gantner Instruments **LFM_4 – IEC 61853 Matrix c-Si CFV**

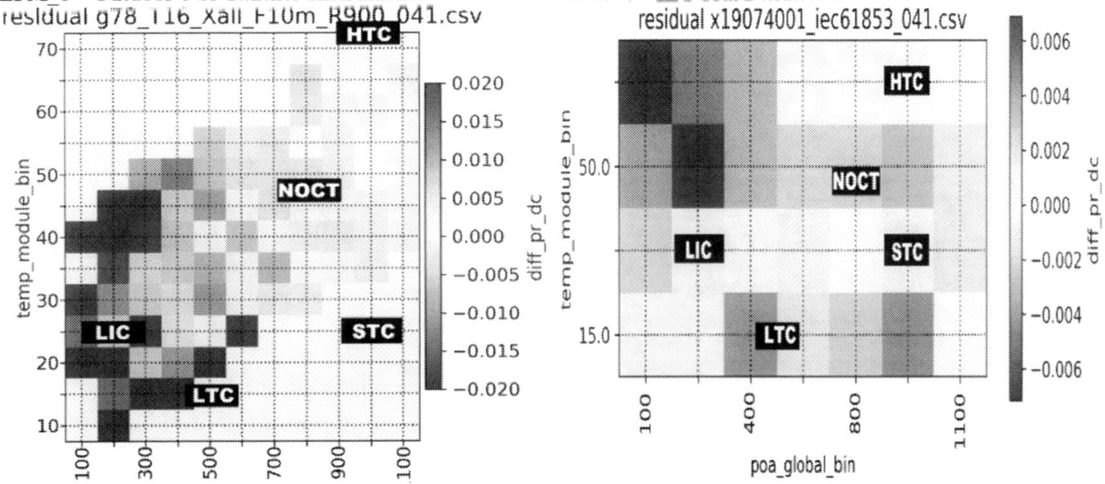

Figure 6: Residual error heatmaps pr_dc 'meas – fit' for LFM6 (left) and LFM4 (right)

2.11 **[K] 'Scatter plot: norm pmp fit(↑) vs. meas(→)'**

Figure 7 gives the mpm fitted vs. measured normalised p_mp (= meas_p_mp / ref_p_mp) with a 1:1 line) indicating a very good fit for both LFM_6 and LFM_4.

```
# Python code [K]

# plot fit vs. measured,
# include a 1:1 line
fit_plot = plot_fit(
    dmeas=meas,
    dnorm=norm,
    fit=mlfm_sel,
    title='fit ' + mlfm_meas_file
)
```

LFM_6 – Outdoor c-Si Gantner Instruments **LFM_4 – IEC 61853 Matrix c-Si CFV**

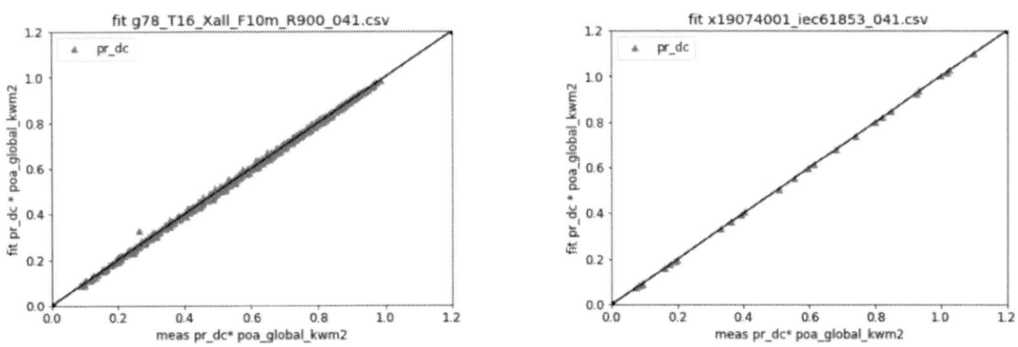

Figure 7: Scatter plot showing normalised p_mp fitted (↑) vs. measured (→)

It is hoped that the mlfm can be put into pvlib soon. Please see the pull [4] request and help if you can.

4. CONTRIBUTING

Table III summarises links and information on helping adding code to PVLIB.

Table III: How to contribute to PVLIB

INFORMATION :

• Stack Overflow :	http://stackoverflow.com/questions/tagged/pvlib
• Google groups :	https://groups.google.com/forum/#!forum/pvlib-python
• GitHub issues :	https://github.com/pvlib/pvlib-python/issues

• Pull requests : https://github.com/pvlib/pvlib-python/pulls
• Jupyter Notebook tutorials : https://github.com/pvlib/pvlib-python/tree/master/docs/tutorials
• Follow PEP8 : https://www.python.org/dev/peps/pep-0008/. (Max line length 79 chars)
• Add your project : https://github.com/pvlib/pvlib-python/wiki/Projects-and-publications-that-use-pvlib-python.
• Variables and Symbols : https://pvlib-python.readthedocs.io/en/stable/user_guide/variables_style_rules.html#variables-style-rules
• Documentation style : https://pvlib-python.readthedocs.io/en/stable/contributing.html#documentation

Coding :
• PVSystem and Location classes provide convenience wrappers around the core PVLIB functions.
• Remove logging calls and print statements.
• Include documentation and Comprehensive unit tests
• Functions must return the desired output for all inputs see https://github.com/pvlib/pvlib-python/issues/394

PVPMC : https://pvpmc.sandia.gov/
Document Library : https://pvpmc.sandia.gov/resources-and-events/documents/
Code of conduct : https://github.com/pvlib/pvlib-python/blob/master/CODE_OF_CONDUCT.md
Cite this paper: https://doi.org/10.21105/joss.00884

5. Acknowledgements

https://doi.org/10.21105/joss.00884

Gantner Instruments https://www.gantner-instruments.com/ and NREL for outdoor data,

CFV for indoor data
https://pvpmc.sandia.gov/download/7701/

6. CONCLUSIONS

The MLFM is being added to PVLIB python

The robust modelling of the LFM and MPM have been described

These are proving useful in industrial projects from single module OTFs to large power plants

Information has been given on how other coders can contribute their projects

7. REFERENCES

[1] "Modelling of PV modules and systems" VIRTUAL PVCOST Training School 2021 Romania http://www.steveransome.com/pubs/2021_07_PVCOST_Romania_Ransome_210706t11tobepresented.pdf
[2] "Benchmarking PV performance models with high quality IEC 61853 Matrix measurements (Bilinear interpolation, SAPM, PVGIS, MLFM and 1-diode)" PVSC 49 Jun 2022 Philadelphia, USA http://www.steveransome.com/pubs/2206_PVSC49_philadelphia_4_presented.pdf
[3] "Improving IEC 61853 Energy Yield Modelling with LFM and MPM models" 2022 PVPMC Salt Lake City USA https://pvpmc.sandia.gov/download/8494/
[4] MLFM pull request https://github.com/pvlib/pvlib-python/pull/1354
[5] "Quantifying and analysing the variability of PV module resistances Rsc and Roc to understand and optimise kWh/kWp modelling" Asian PVSEC-27 2017 Shiga Japan.
http://www.steveransome.com/PUBS/1711_7TuPo.225_27PVEC_Ransome_3.pdf
[6] William F. Holmgren, Clifford W. Hansen, and Mark A. Mikofski. "pvlib python: a python package for modeling solar energy systems." Journal of Open Source Software, 3(29), 884, (2018).

OUTDOOR ENERGY PERFORMANCES FOR STANDARD AND BI-FACIAL MODULES AS WELL ON THE FAILURE MODES OBSERVED IN OUTDOOR CONDITIONS

Ottanà A.(1, 3); Rametta F. (1); Gangemi W. (2); Colletti C. (1); Di Stefano A. (2); Canino A. (2); Foti M. (1); Gerardi C.(1); Bizzarri F. (3);
1) EGP - Enel Green Power, Contrada Blocco Torrazze Zona Industriale 95121, Catania, Italy, 2) EGP - Enel Green Power, Contrada Passo Martino Zona Industriale 95121, Catania, Italy, 3) EGP - Enel Green Power Viale Regina Margherita 125, 00198 Roma, Italy

ABSTRACT: In this article is reported the outdoor performance evaluation of 3SUN bifacial HJT PV modules carried out in the Enel Green Power Innovation Lab of Catania – Passo Martino. In the analysis in terms of Performace Ratio, a monofacial PV technology is used as a reference, and in order to ensure a fair comparison, different commercial module technologies have been selected and monitored. The present paper aims at demonstrating the superior production of the heterojunction modules due to their higher passivation, better thermal coefficients and highlight the bifacial gain.
Keywords: Heterojunction, Bifacial, Outdoor Performance, Performance Ratio

1 INTRODUCTION

In this article is presented the outdoor performance analysis of 3SUN "BEST-4U Initial Generation" bifacial HJT PV modules conducted at the Enel Green Power Innovation Lab of Catania – Passo Martino.
The present work aims to demonstrate an increased energy yield up to 5% in sunny compared to Al-BSF solar cells, and to demonstrate that bifaciality can bring energy yield in line with expectation of albedo, and module spacing (i.e. typically more 15-20% more energy than an equivalent monofacial module).
For this purpose, the performance evaluation is carried out in comparison with a monofacial PV module technology, considered in this document as a "world widespread" reference. Moreover, in order to ensure a fair comparison with the 3SUN modules, different commercial bifacial module technologies have been selected and tested. In the following paragraphs 3SUN "BEST-4U Initial Generation" PV modules will be named "Bifi HJT 3SUN" in order to allow an easier comparison with the other modules under test by highlighting their main characteristics.

2 TESTING FACILITY AND PV MODULES UNDER TEST

The test facility dedicated to double-sided modules was designed and built in the Innovation Lab of Catania – Passo Martino in 2015.
As it can be seen in Fig. 1, the module support structures do not have metal bars on the back.

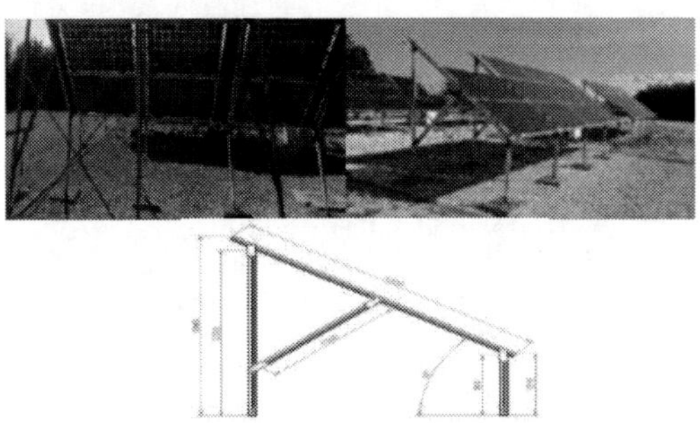

Figure 1: Supporting structures for bifacial modules

For the test, monofacial Poly p-Si modules are used as a "world widespread" reference and, in addition to 3SUN dual-glass bifacial modules based on ECA-busbar interconnections, the outdoor performance has been evaluated also for different commercial module technologies in order to ensure a fair comparison. Other bifacial dual-glass modules were tested: two based on HJT technology (one of which with Smart Wire Connection Technology) and one based on PERT solar cells technology. Here below the list of names used to identify the different technologies:

- Bifi HJT 3SUN
- Bifi HJT SWCT
- Bifi HJT A
- Bifi PERT
- Monofacial Poly p-Si

All the exposed modules have carried out the same process of characterization imposed by the 61215 IEC standard: visual inspection (MQT 01) and electroluminescence, Performance at STC (MQT 6.1), initial stabilization (MQT 19.1), electrical insulation test in a dry and wet environment (MQT 03, MQT15), which will be repeated at the end of the bench test period to evaluate the structural degradation of the modules of each technology.
According to the standard it is necessary to expose outdoor the modules in two steps of 5 kWh/m2 each. At the value of about 5 kWh/m2 of absorbed solar radiation, the modules have been dismounted from the bench and brought to the sun simulator laboratory for the execution of a new STC power rating measurement.
The STC maximum power Pmax is compared with the Pmin and if (Pmax – Pmin)/ Paverage ≥ 1% (where Paverage is (Pmax + Pmin) /2), the modules must continue the outdoor exposure in order to absorb additional 5 kWh/m2 until the following stabilization condition is verified:

$$(P_{max} - P_{min}) / P_{average} < 1\% \qquad (1)$$

Here, Pmax, Pmin and Paverage are defined as two consecutive output power measurements, P1 and P2, taken from a sequence of alternating stabilization and measurement steps using the results of the STC power rating measurement of the solar simulator. The previous condition (1) proves the stabilization of the modules. The integration of solar radiation should be calculated over the time interval when the irradiance is higher than 500 W/m2. Table 1 and Figure 2 show the measured values for all the tested technologies.

Table I: Pmpp gain after modules stabilization (MQT 19.1 IEC 61215-2)

Module Type	Avg Pmpp @ t0 [W]	Avg Pmpp @ tf [W]	Gain after stab. [%]
Bifi HJT 3SUN	358.5	360.4	0.55
Bifi HJT SWCT	355.8	358.1	0.63
Bifi HJT A	306.8	307.9	0.36
Bifi PERT	301.1	301.4	0.09
Monofacial Poly p-Si	275.9	273.7	-0.78

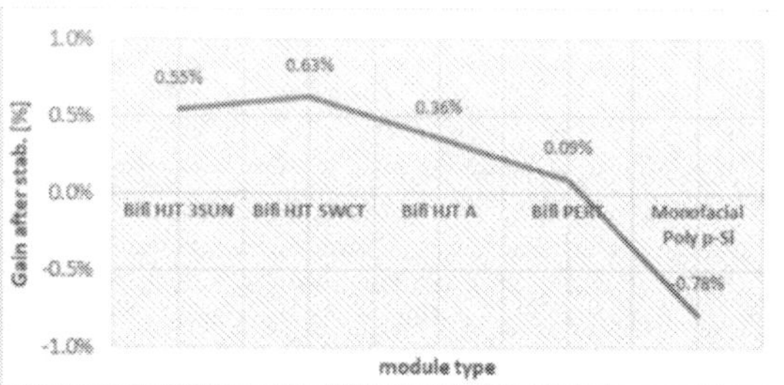

Figure 2: Pmpp gain after modules stabilization (MQT 19.1 IEC 61215-2)

After the initial performance test, the modules have been exposed outdoor and connected to a fixed resistive load or to micro-inverters in order to impose operation conditions close to the maximum power point at STC. For each module technology, independent systems of approximately 3 kW have been created, using a single-phase inverter with 1 MPP tracker for each string and dedicated transducers for detecting the current and voltage in DC Systems. The number of modules present in each string depends on the available place on the structures. As depicted in Fig. 3, the following technologies have been installed adjacent to each other for direct comparison:

- Bifi HJT A / Bifi PERT
- Bifi HJT 3SUN / Bifi HJT SWCT

The Monofacial Poly p-Si string is positioned in the top row of the installation and, for reasons of space, it is divided into 2 rows belonging to the same inverter.

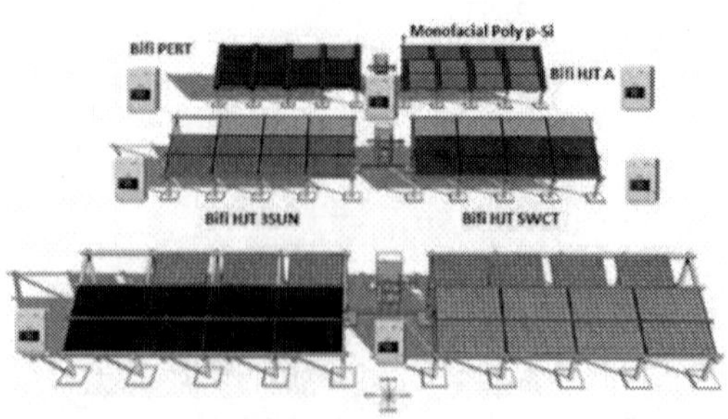

Figure 3: Installation final configuration

In Table 2 is reported the number of PV modules installed and monitored at Passo Martino for each technology.

Table II: PV modules tested in Passo Martino

Module type	Date of installation	Number of modules	Number of cells
Monofacial Poly p-Si	May-19	11	60
Bifi HJT SWCT	May-19	8	72
Bifi HJT 3SUN	May-19	8	72
Bifi PERT	May-19	10	60
Bifi HJT A	Oct-19	10	60

Note: HJT – heterojunction; SWCT – smart wire

The systems began to acquire data from the middle of 2019.

2.1 Monitoring system

The monitoring equipment is composed of:

- 8 module temperature sensors
- 1 front side pyranometer
- 1 back side pyranometer
- DC voltage and current sensors for each system

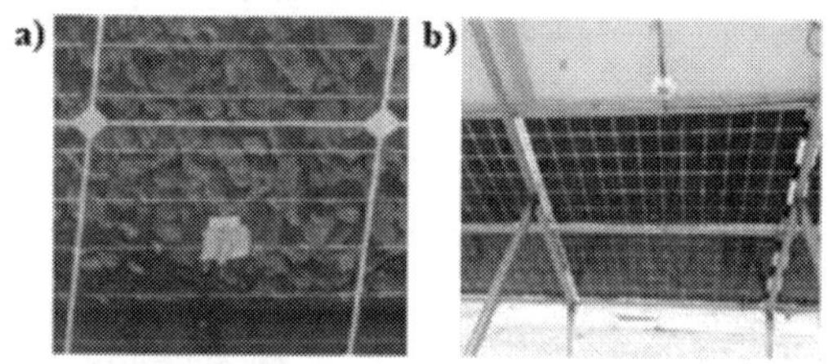

Figure 4: a) Temperature sensor at the back side of a module; b) Back side irradiance sensor.

Monitoring is entrusted to precision instrumentation that sends data 24 hours a day to a server, directly accessible via web portal. Here you can find a great variety of measurements both in DC and AC, radiometric data, temperature etc..

3 ANALYSIS OF OUTDOOR PERFORMANCE AT ENEL INNOVATION LAB PASSO MARTINO

In this section is reported the outdoor performance evaluation of the different PV module technologies, carried out at the Enel Green Power Innovation Lab of Catania – Passo Martino.

The monitoring period goes from May 2019 to April 2020, with the exception of Bifi HJT A modules, which were installed in October 2019. The elaboration of the acquired data has been carried out during 2020, once the collection of the complete dataset was concluded.

3.1 Performance indicators

The acquisition of data for all the technologies involved in the comparison is standardized through i-EM systems, which can be retrieved online at any time and processing is still ongoing. The performance evaluation of each single technology can be carried out by following two apparently different paths, which in reality are closely connected to each other. The first involves calculating the Performance Ratio (PR), given by the ratio between the energy produced and the expected energy. Specifically, the PR is calculated with the total energy produced by both the front and rear sides of the module, divided only by the radiation captured by the front of the module in order to compare the performance of the single and double-sided modules, according to the formula:

$$PR = \frac{\frac{E_{tot}}{P_{nom}}}{\frac{E_{Irr,front}}{Irr_{STC}}} \quad (2)$$

The second performance calculation system evaluates the equivalent hours, i.e. the actual hours of production of the strings. A unique daily value (and a monthly sum of equivalent hours) is obtained for each system tested according to the formula:

$$EOH = \frac{E_{tot}}{P_{nom}} \quad (3)$$

3.2 Analysis from May-19 to April-20

Figure 5 and Figure 6 show the average PR and EOH that go from the installation period of the various technologies involved in the test bench until the beginning of April.
It is necessary to consider the gap of a module in the upper first row in the installation of the Bifi HJT 3SUN system (see Figure 3) due to a pre-existing project unrelated to the one discussed in this document, which could represent an advantage over the Bifi HJT SWCT string.

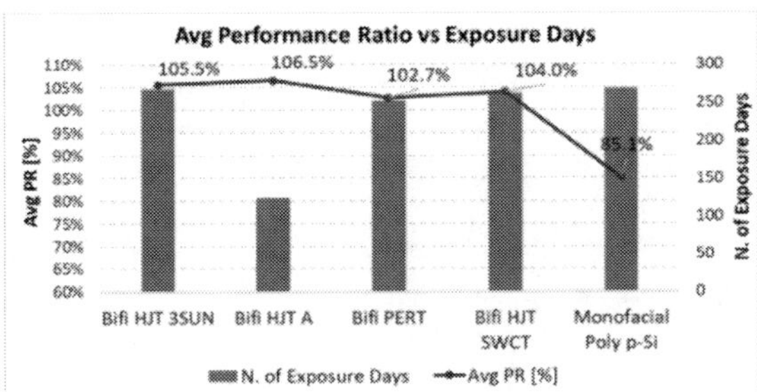

Figure 5: Average PR and exposures days from May 2019 to April 2020

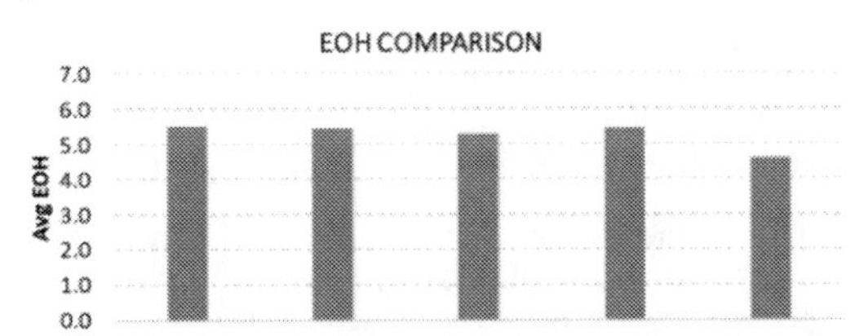

Figure 6: Average EOH from May 2019 to April 2020

The performance results of the different PV technologies are now presented considering Monofacial Poly p-Si as the reference technology. A direct comparison is also shown between technologies installed adjacent to each other.
Figure 7 and Figure 8 show the PR trend of the 3SUN technology and Bifi HJT SWCT technology compared to the Monofacial Poly p-Si reference technology. The outlined scenario is consistent with expectations, considering the similar installation conditions and the same installation period that characterizes the two strings of bifacial modules.

Figure 9 shows a comparative graph of the average gain in PR over the Monofacial Poly p-Si reference, both of Bifi HJT 3SUN and Bifi HJT SWCT technologies.

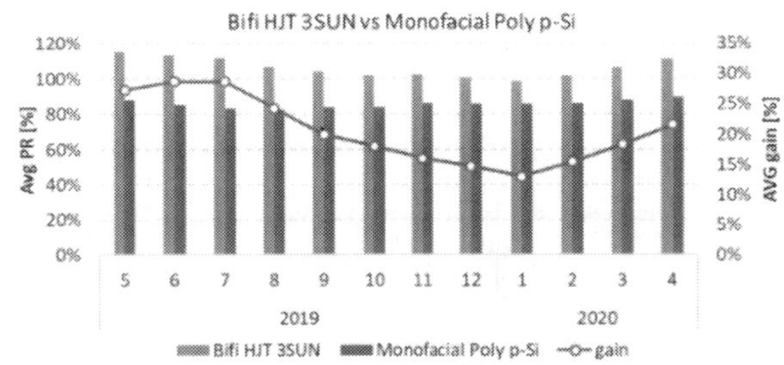

Figure 7: PR and gain comparison between Bifi HJT 3SUN and Monofacial Poly p-Si from May 2019 to April 2020

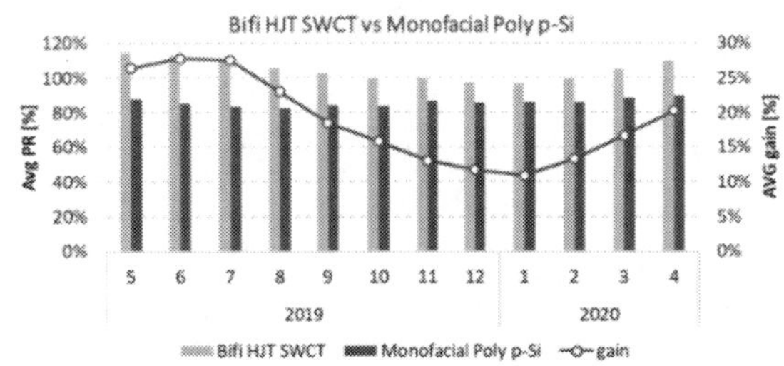

Figure 8: PR comparison and gain between Bifi HJT SWCT and Monofacial Poly p-Si from May 2019 to April 2020

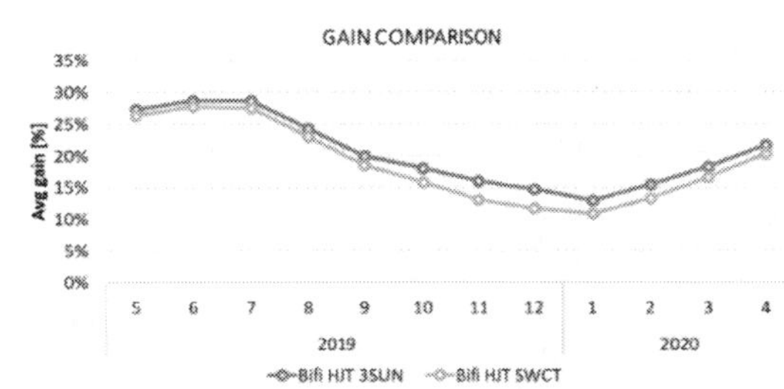

Figure 9: Comparative graph on the average gain of Bifi HJT 3SUN and Bifi HJT SWCT vs Monofacial Poly p-Si

The PR comparison and gain between Monofacial Poly p-Si modules and the Bifi PERT and Bifi HJT A technologies are shown in Figure 10 and Figure 11, respectively. Figure 12 displays the direct comparison of the average gain of the two technologies over the reference one.

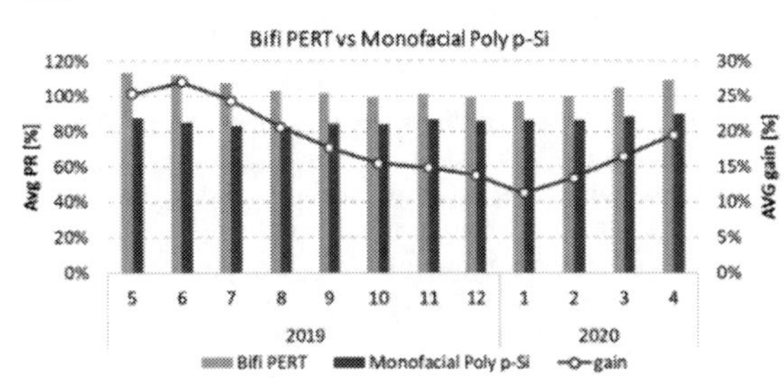

Figure 10: PR comparison and Gain between Bifi PERT and Monofacial Poly p-Si from May 2019 to April 2020

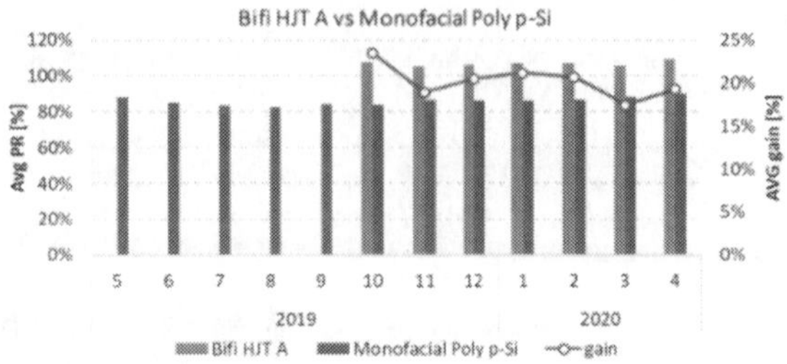

Figure 11: PR comparison and Gain between Bifi HJT A and Monofacial Poly p-Si from May 2019 to April 2020

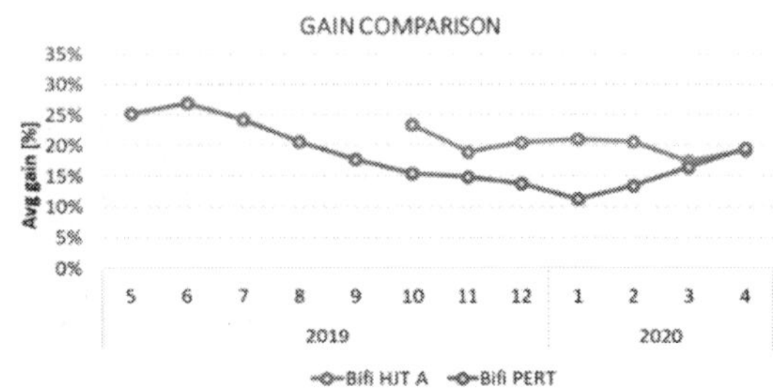

Figure 12: Comparative graph on the average gain of Bifi HJT A and Bifi PERT vs Monofacial Poly p-Si

The average gain of all the four bifacial technologies tested in comparison with the Monofacial Poly p-Si reference is reported in Figure 13.

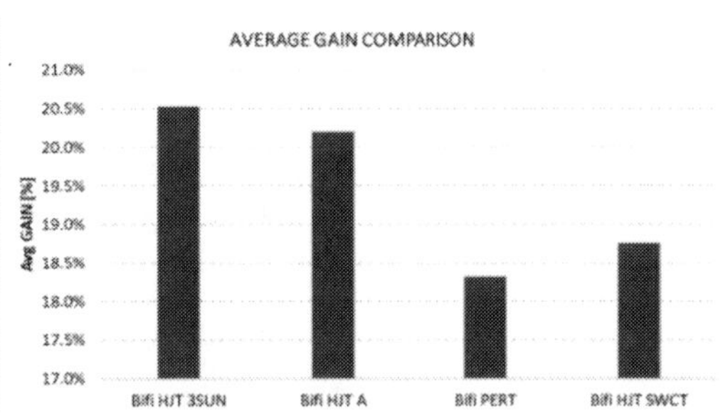

Figure 13: Average gain of each module set vs Monofacial Poly p-Si

3.3 Module failure modes

For the arrangement of the modules and the period of their installation, at the moment there is a fairly linear performance trend, but the most reliable result can be obtained at the end of the annual cycle, where the weather conditions that influence the data so much, with good approximation will align to the initial ones and real degradation will be seen. Installation differences affect how much the modules, in the back and in the front, perceive sunlight. In fact, when the solar angle is zero with respect to the normal of the module surface, the acquired low irradiance data become relevant.

The installation of the following modules occurred quite in the same period, May 2019:

- monofacial polycristalline p-Si
- bifacial HJT with Smart Wire
- bifacial HJT 3SUN
- bifacial PERT

while the following modules were added in October:

- bifacial HJT A

In the following graphs it is shown the performance trend closely related to temperature changes.

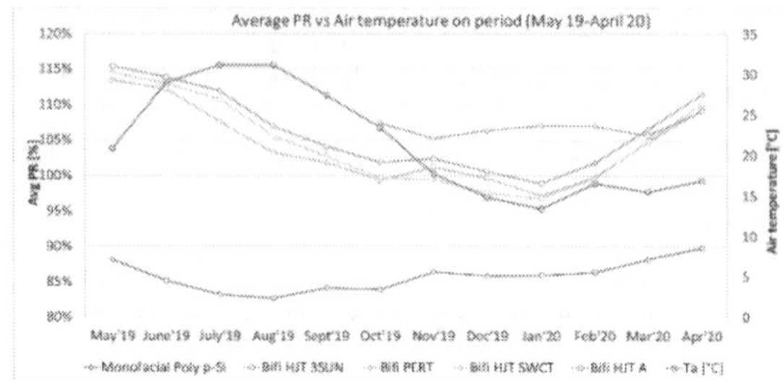

Figure 14: Average PR in relation to the ambient temperature in the period May'19-April'20

Figure 15 reports the average PR of all the technologies in relation to the ambient temperature. From this graph, Bifi HJT A technology appears to be the one with the highest performance. However, the lowest degradation related to its installation must be considered. Therefore, taking into account the installation timeline, we can say that the most performing technology is HJT 3SUN.

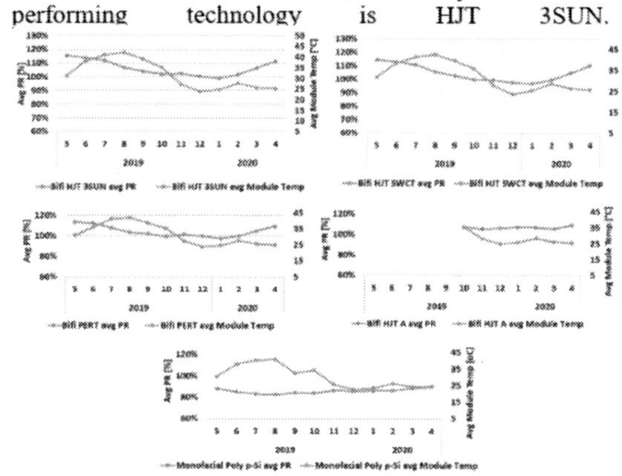

Figure 15: Average PR related to temperatures measured on the back of the modules.

4 CONCLUSION

All modules technologies have been installed and monitored over almost a year, and the outdoor performance evaluation is conducted in terms of comparison the Performace Ratio parameter.

The initial goal of demonstrating an increased energy yield up to 5% in sunny regions compared to Al-BSF solar cells, and to demonstrate that bifaciality can bring energy yield in line with expectation of albedo, and module spacing (i.e. typically more 15-20% more energy than an equivalent mono-facial module) has been achieved. Indeed, the average gain of bifacial 3SUN "BEST-4U Initial Generation" modules with respect to the "world widespread" monofacial reference technology is 20.5%.

In conclusion, it has been successfully demonstrated the superior energy production of the 3SUN "BEST-4 Initial Generation" bifacial heterojunction modules.

ACKNOWLEDGMENTS

This work has been supported by the PON Ricerca e Innovazione 2014-2020 project BEST4U, funded under Decreto Direttoriale di concessione dell'agevolazione del 21-05-2019 prot. n. 991, contract code ARS01_00519.

REFERENCES

[1] W. Favre et al, "25% efficient large area silicon solar cell: paving the way for premium PV manufacturing in Europe" in EU PVSEC, 11 September 2020

[2] G. Condorelli et al., "High Efficiency Hetero-Junction: From Pilot Line To Industrial Production," 2018 IEEE 7th World Conference on Photovoltaic Energy Conversion (WCPEC) (A Joint Conference of 45th IEEE PVSC, 28th PVSEC & 34th EU PVSEC), Waikoloa Village, HI, 2018, pp. 1970-1973, doi: 10.1109/PVSC.2018.8548197.

[3] G. Condorelli et al., "Initial Results of Enel Green Power Silicon Heterojunction Factory and Strategies for Improvements," 2020 47th IEEE Photovoltaic Specialists Conference (PVSC), Calgary, OR, 2020, pp. 1702-1705, doi: 10.1109/PVSC45281.2020.9300806

[4] G. Razongles, L. Sicot, M. Joanny, E. Gerritsen, P. Lefillastre, S. Schroder, andP. Lay, "Bifacial photovoltaic modules: Measurement challenges",Energy Procedia, vol. 92, pp. 188–198, 2016.

[5] J. S. Stein, D. Riley, M. Lave, C. Hansen, C. Deline, F. Toor, "Outdoor Field Performance from Bifacial Photovoltaic Modules and Systems," 2017 IEEE 44th Photovoltaic Specialist Conference (PVSC), 2017, pp. 3184-3189, doi: 10.1109/PVSC.2017.8366042.

[6] H. Park, S. Chang, S. Park, W. K. Kim, "Outdoor performance test of bifacial n-type silicon photovoltaic modules," Sustainability,vol. 11, no. 22, 2019, Art. no. 6234, doi: 10.3390/su11226234.

[7] Janssen, G.J.; Van Aken, B.B.; Carr, A.J.; Mewe, A.A. "Outdoor Performance of Bifacial Modules by Measurements and Modelling," Energy Procedia2015, 77, 364–373, doi:10.1016/j.egypro.2015.07.051.

DYNAMIC SIMULATION MODEL OF AN AIR-BASED BIPVT ROOF TILE SYSTEM

L. Clasing, P. H. Yam, C. Ziller, U. Blieske
Cologne University of Applied Sciences, Faculty for Process Engineering, Energy and Mechanical Systems, Cologne Institute for Renewable Energy
Betzdorfer Str. 2, 50679 Cologne, Germany

ABSTRACT: The application of building-integrated PV roof tiles on residential buildings is gaining increasing attention in recent years. Besides many advantages of these systems, they have one major disadvantage: compared to standard PV systems, there is a lack of rear ventilation and thus the temperature of the PV cells increases. This temperature increase leads to a reduction in the electrical performance and also the lifetime of the modules decreases. In order to reduce the PV cell temperature in such systems, a novel air-based building-integrated PV solar thermal roof tile with active rear ventilation is presented. The aim of this work is to develop and validate a dynamic simulation model for a novel air-based PV and solar thermal roof tile system. The model is developed in order to optimize and assess the overall annual energy yield under real operating conditions. Furthermore, the simulation model is validated by measured experimental data conducted on a test roof with the solar roof tiles under investigation in Cologne, Germany. The Root Mean Square Error of 15 days of validation in June shows a good agreement for outlet air and PV module temperature with 1.78 °C and 2.95 °C, respectively. Finally, time-series simulations are performed with the developed model and an optimized control strategy for a case study. In this case study the solar roof tile systems air-duct outlet is coupled to an air-source heat pump. The simulation results reveal seasonal performance factor of 3.4 compared to a reference case with the ambient air inlet of 3.3. The presented model allows us to predict a realistic thermal energy yield of the solar roof tile under real operating conditions and thus assessment of the economic viability.
Keywords: BIPVT, Modelling, Thermal Performance

1 INTRODUCTION

Building integrated PV and thermal systems (BIPVT) combine many advantages such as PV cell cooling, waste heat utilization for residential heating and custom design options. In this work, a simulation model of a novel air-based BIPVT roof tile with active rear ventilation is presented [1]. In order to provide realistic predictions of the overall energy yield and economic viability to professional architects and potential investors, an accurate energy simulation model of this BIPVT roof tile is lacking. Recent works presented simplified quasi steady-state thermal models of different BIPVT collector designs [2]. In this work a dynamic thermal model for the BIPVT system with active rear ventilation coupled to an air source heat pump (ASHP) is presented.

Transient computational fluid dynamics (CFD) simulation with dynamic boundary conditions reveals detailed insights, but requires high calculation times. On the contrary, the calculation time of a dynamic simulation in a block diagram environment of MATLAB Simulink is significantly lower, but the input parameters regarding the airflow need to be identified for the modeling process. Therefore, this work uses an approach, in which the input parameters are identified by steady-state CFD simulations and used in a dynamic simulation model in MATLAB Simulink. This approach can represent the outlet air and PV cell temperature due active to rear ventilation more accurately, allowing realistic estimations of the overall power output of the designed system.

2 BIPVT SOLAR ROOF TILE

The novel air-based BIPVT solar roof tile (Figure 1) consists of two main parts. A custom-designed glass-glass module with mc-Si PV cells and size with length x width x height of 339 mm x 293 mm x 7.1 mm [1].

Figure 1: Cross-sectional profile of the BIPVT roof tile with a rear ventilation system

An aluminum frame allows the glass-glass module to be moveable and provides the air-duct (Figure 2) for active rear ventilation.

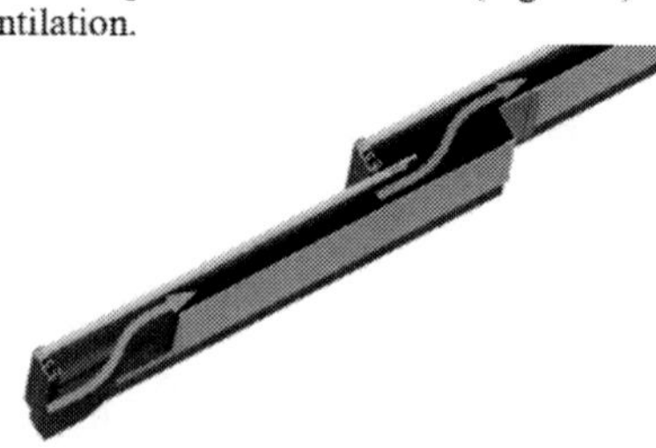

Figure 2: Cross-sectional profile of the BIPVT roof tile with a rear ventilation system

3 SIMULATION MODEL

3.1 Solar irradiance and electrical PV model

Effective irradiance is calculated considering angular losses for direct and diffuse irradiance components by angular loss parameters according to Martín and Ruiz model [3,4]. Furthermore, diffuse irradiance on the tilted plane is calculated by Klucher diffuse sky irradiance transposition model [5].

The electrical PV model is based on De Soto et al. one-diode equivalent circuit five parameter model [6]. Therefore, IV-curve measurements are conducted under standard test conditions (STC, 1000 W/m², AM1.5g, 25 °C). These measurement data are used to calculate the parameters required for the model and the electrical efficiency under real operating conditions. The used irradiance and electrical models are described more in detail by Bartkowski et al. [7].

3.2 Thermal model

The thermal simulation model is developed in MATLAB Simulink by means predefined blocks for physical systems of the Simscape library. The model of a single BIPVT roof tile collector consists of a fluid and a thermal physical network.

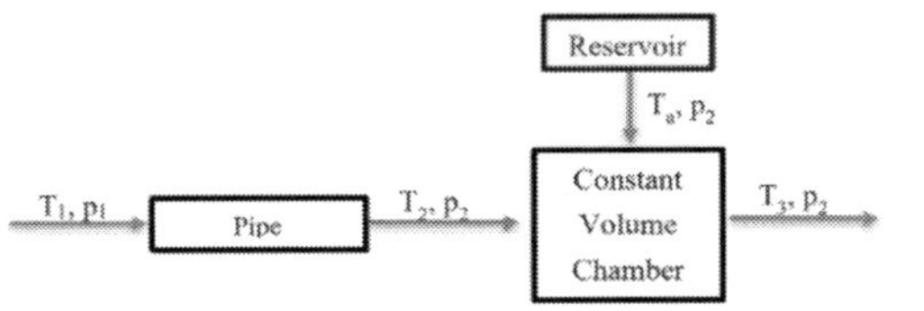

Figure 3: Fluid network of the BIPVT roof tile subsystem

Figure 3 shows the structure of the fluid network of a BIPVT roof tile subsystem. The rear ventilation system is modelled as a pipe, which calculates the convective heat transfer between the airflow and the PV glass-glass module as well as the pressure loss. A constant volume chamber represents an air gap, where cold air flows in through the leakages and mixes with the warm air from the ventilation system. In the system simulation model 12 BIPVT roof tile subsystems are connected in series.

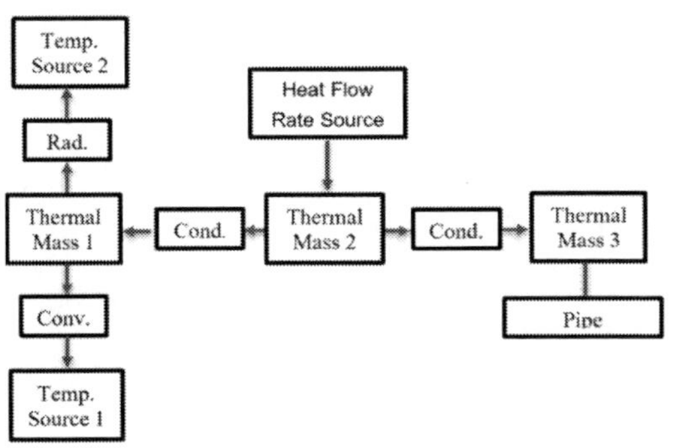

Figure 4: Thermal network of the BIPVT roof tile subsystem

An energy balance for a multi-node solar collector model according to Helmers and Cramer is applied to the thermal model for the BIPVT roof tile [8]. Figure 4 shows the structure of the thermal network. The PV glass-glass module is modelled in three parts: the solar glass on top of the solar cells (thermal mass 1), the solar cells (thermal mass 2) and the glass under the cells (thermal mass 3). The solar heat energy, which was calculated from the input weather data, flows from a heat flow rate source to the PV cells and to the glasses through conduction.

Thermal convective and radiative front losses are approximated with empirical relationships based on wind speed and ambient temperature Tamb [9,10]. These heat losses from the solar glass is modelled in relation to the ambient temperature (Temp. Source 1) and the sky temperature (Temp. Source 2). The bottom glass is directly connected to the thermal port of the pipe, so that it is treated as the pipe wall and its heat transfer with the airflow in the pipe can be calculated. The required input parameters: the heat transfer coefficients, the air flow rate through the leakages as well as the pressure loss are identified by computational fluid dynamics (CFD) simulations (Figure 5).

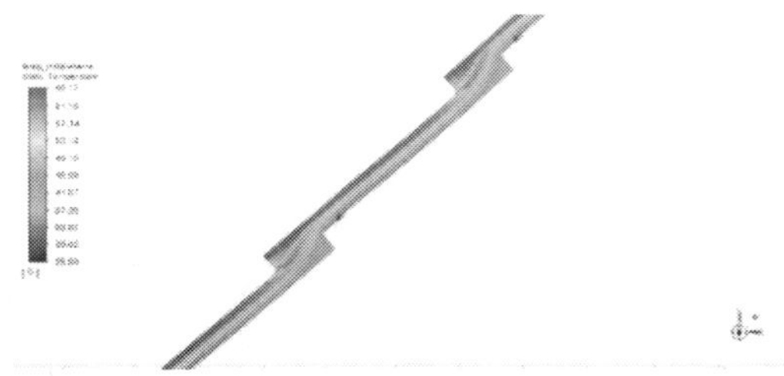

Figure 5: Temperature contour plot of the cross-section of the BIPVT solar roof tile air-duct system

A parametric study was conducted with the CFD model and the results show that these parameters vary with the air flow rate in the ventilation channel. Thus, the output data from CFD simulations with varying flow rate are interpolated and the resulted functions are built into the model.

4 EXPERIMENTAL DETAILS

To validate the developed simulation model, field measurements are conducted on an experimental test roof (Figure 6) covered with the novel BIPVT roof tiles in Cologne, Germany. The test roof is configurated with 12 parallel air-duct systems with each 12 solar roof tiles in series. This results in PV installation capacity of 2.78 kWp and 19.8 m² irradiated area. An external fan for active rear ventilation is installed in the center of the test roof. A pipe with different inlets provides approximately equal mass flow distribution for each air-duct system.

Figure 6: Experimental BIPVT solar roof tile system located in Cologne, Germany

Measurement data of outlet mass flow, module temperature, outlet air temperature, plane-of-array (POA) solar irradiance, in-plane wind speed and ambient temperature is recorded in 20-second intervals.

The weather data recorded during the experiment are used as input parameters in the simulation and the calculated outlet air temperature in the manifold, as well as the PV module temperature, are compared.

5 CASE STUDY

A case study is conducted to assess the usable overall annual energy of the novel BIPVT roof tile based on the simulation model presented in this work. Therefore, a BIPVT roof tile system based on the test roof with 2.78 kWp PV installation capacity and 19.6 m² irradiated area (12 parallel air-duct systems) is simulated. Typical meteorological year (TMY) data provided by PVGIS from 2007 – 2016 at the location of Cologne, Germany [11]. Domestic hot water (DHW) profile generated by means the tool DHWcalc [12] are used. A hot water storage in the thermal system is not considered in this work. Seasonal performance factor (SPF) of ASHP with P=3.5 kW is calculated in dependency of air outlet temperature by means of a model based on averaged heat pump efficiencies according to Ruhnau et al. [13].

6 RESULTS AND DISCUSSION

The dynamic simulation model of the novel BIPVT roof tile is validated with measurement data of an experimental test roof (Figure 7). To assess the prediction accuracy, Mean Bias Error (MBE) and Root Mean Square Error (RMSE) is calculated for residual error of measurement and simulation. MBE shows up with 0.84 °C and -1.07 °C for PV module temperature T_M and outlet air temperature T_F for five days at clear-sky irradiance conditions in June. In comparison, RMSE is observed with 2.84 °C and 1.94 °C for the identical dataset.

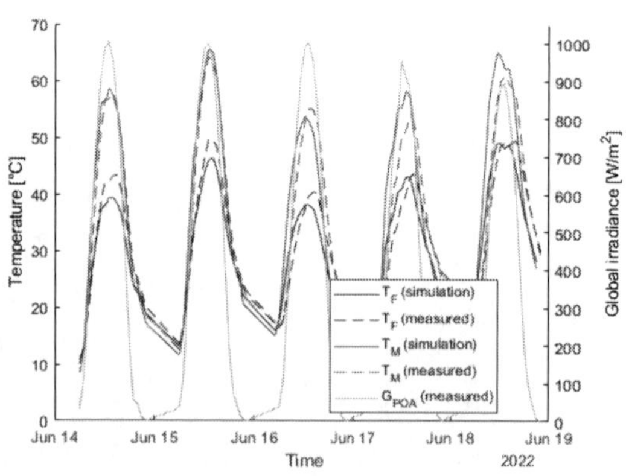

Figure 7: Simulated and measured PV module T_M and outlet air temperature T_F for five days at clear-sky irradiance conditions in June 2022 (V=25 m³/h)

Table I shows the calculated residual error statistics for 15 days of model validation in summer season at 25 m³/h volume flow rate in each of the 12 air-duct systems of the BIPVT roof tile. The dynamic thermal model shows a good agreement in terms of module and outlet air temperature. However, module temperature is slightly overestimated and outlet air temperature slightly underestimated. Hence, the following possible sources of error are analyzed:

1. Overestimate of heat loss to surroundings due to a simplified convective and radiative heat transfer model
2. Inaccurate modeling of leakages in the CFD model geometry
3. Inaccurate turbulence model parameter in the CFD simulations

Table I: Residual error of 15 days in June 2022 (V=25 m³/h x 12 air-duct systems)

Residual Error	T_M	T_F
MBE (°C)	1.08	-0.69
RMSE (°C)	2.95	1.78

The SPF of the simulated ASHP coupled to the BIPVT systems air outlet, shows 3.4 and 3.3 for the reference with ambient temperature T_{amb}. In comparison, the SPF of the BIPVT system including the energy consumption of the external fan can be overserved with 3.37. The temporal discrepancy between the availability of solar irradiance at the location and the DHW profile lead to comparatively low gains in the SPF for the energy concept used.

7 CONCLUSIONS

The presented approach enables realistic thermal energy yield predictions of the novel BIPVT solar roof tile and thus assessment of the economic viability. In future work, simulations with more suitable demand profiles and energy concepts are needed to optimize thermal energy utilization. Furthermore, refinement of the model to improve accuracy and expansion of the data base for model validation is necessary.

8 ACKNOWLEDGEMENTS

This project has been financed by the European Fonds for Regional Development North Rhine-Westphalia (EFRE.NRW, Code: EFRE-0801561, Reference no. EU-2-2-048B)

9 REFERENCES

[1] L. Clasing, C. Dick, U. Blieske, P. Hakenberg, S. Leyer, and J.-R. Hadji-Minaglou, "Analysis of a Novel Improved BIPV/T Roof Tile Design", Proceedings of the *15th Conference on sustainable development of energy, water and environment systems (SDEWES)*, Cologne, Germany, September 2020.

[2] C. Brosig, L. Clasing, U. Blieske, and E. Waffenschmidt, "Modeling a novel building-integrated PVT-air-collector system coupled to a heat pump using open source libraries", Proceedings of the *16th Conference on sustainable development of energy, water and environment systems (SDEWES)*,

Dubrovnik, Croatia, October 2021.
[3] N. Martin and J. M. Ruiz, "Calculation of the PV modules angular losses under field conditions by means of an analytical model", Solar Energy Materials & Solar Cells, vol. 70, pp. 25-38, 2001.
[4] N. Martin and J. M. Ruiz, "Corrigendum to 'Calculation of the PV modules angular losses under field conditions by means of an analytical model'", Solar Energy Materials & Solar Cells, vol. 110, pp. 154, 2013.
[5] Klucher, T. M. (1979): Evaluation of models to predict insolation on tilted surfaces. In: Solar Energy 23 (2), S. 111–114. DOI: 10.1016/0038-092X(79)90110-5.
[6] W. D. Soto, S. A. Klein, and W. A. Beckman, "Improvement and validation of a model for photovoltaic array performance," Solar Energy, vol. 80, no. 1, Art. no. 1, Jan. 2006, doi: 10.1016/j.solener.2005.06.010.
[7] M. Bartkowski, H. R. Saleemi, L. Clasing, C. Brosig, E. Waffenschmidt, U. Blieske, „Development and validation of a performance model for simulation of an air-based BIPVT roof tile system using open-source libraries", Proceedings of the *8th World Conference on Photovoltaic Energy Conversion (WCPEC)*, Milan, Italy, September 2022.
[8] H. Helmers and K. Kramer, "Multi-linear performance model for hybrid (C)PVT solar collectors," Solar Energy, vol. 92, pp. 313–322, Jun. 2013.
[9] W. H. McAdams, Heat Transmission, vol. 120. Kogakusha: McGraw-Hill New York-London, 1954.
[10] W. C. Swinbank, „Long-wave radiation from clear skies", Q.J.R. Meteorol. Soc., Jg. 89, Nr. 381, S. 339–348, 1963.
[11] Photovoltaic Geographical Information System (PVGIS), "TMY Data: PVGIS-SARAH: 2007 2016", European Commission Joint Research Centre.
[12] U. Jordan and K. Vajen, „Influence of the DHW load profile on the fractional energy savings: A case study of a solar combi-system with TRNSYS simulations".
[13] O. Ruhnau, L. Hirth, and A. Praktiknjo, "Time series of heat demand and heat pump efficiency for energy system modeling," Scientific Data, vol. 6, no. 1, Art. no. 1, Oct. 2019.

DATA PROCESSING FOR PHOTOVOLTAIC PERFORMANCE LOSS ANALYSIS IN NORDIC CLIMATE CONDITIONS

Aapo Poskela[1,2], Lauri Karttunen[1], Heikki Palonen[1], Hugo Huerta Medina[3], Samuli Ranta[3], Kati Miettunen[1]
[1]Solar Energy Materials and Systems, University of Turku | [2]Solar Finland Ltd. | [3]Turku University of Applied Sciences
[1]Vesilinnantie 5, 20500 Turku, Finland | [2]Salorankatu 5-7, 24240 Salo, Finland | [3]Joukahaisenkatu 3, 20520 Turku, Finland

ABSTRACT: The aim of this work is to demonstrate a photovoltaic performance loss analysis workflow for a rooftop system installed in Nordic climate. There is a variety of methods for evaluating the performance loss of a PV system, but they give different results, making comparisons between studies challenging. This study uses several combinations of filters and statistical methods to calculate the performance loss rate, giving values between 1.5 and 3.6 %/year. The range of values is large and steps should be taken to minimise the variance and reach a value that can be compared with other studies. One method to reduce variance in the calculated performance loss rate is to improve data quality, and thus reduce the impact the data filters will have on the analysis. Data quality can be improved by taking great care in the design of the measurement setup and by monitoring the data quality as the measurement is ongoing. Comparability can also be improved during the data analysis by performing it with multiple methods and data filters, as in this work, thus generating a distribution of performance loss rates that can be handled with statistical methods or compared individually with other studies using similar methods.
Keywords: PV module, lifetime, degradation, modelling

1 INTRODUCTION

Photovoltaic (PV) technologies are responsible for ever larger share of global energy production.[1] As PV is becoming more vital in the energy market, it is important to be able to accurately forecast the performance losses and lifetime of PV panels so that energy demand can be met and interruptions in supply avoided. Common method for studying degradation of PV panels is through the use of performance loss rates (PLR), which are calculated from data collected over a long period of time from an active photovoltaic installation. The data collection period is usually several years.[2]

PV performance data automatically recorded during a long period of time is rarely accurate enough to be used for the calculation of performance loss rate as is, and requires processing that removes outliers and matches the produced power to the irradiance conditions. This work also covers generating data that was not directly measured, such as plane-of-array irradiance (POA) and cell temperature, by using physical models to calculate them from other results. We discuss the use of different filters to remove outliers and how the filters impact the calculated performance loss rate. Finally, we demonstrate that even the statistical method used to calculate the performance loss has an effect on the final result. We give our recommendations on how to perform a performance loss calculation based on these findings.

This study analyses data from a rooftop installation in Turku, southern Finland. Finland and the Nordic countries in general are often not considered to be prime locations for PV deployment, due to their association with long and dark winters. However, the winters are balanced by long daylight hours during summer and lower temperatures which improve silicon PV efficiency. This contribution provides an overview of performance loss analysis in Nordic conditions and discusses what to consider when performing such analysis.

2 METHODS

2.1 PV setup

The PV performance data was collected from 18 south-facing (15° tilt) silicon solar panels (KD-P250, Zhejiang Kingdom Solar Energy Technic Co.) on a rooftop in Turku, southern Finland (60°N, 22°E). The panels were installed and data collection began in the beginning of 2018. The data was logged with 5-minute intervals. Saved data included current, voltage and energy produced by the panels, while nearby weather station logged wind direction and speed, rainfall, relative humidity, ambient temperature and global horizontal irradiance (GHI). The panels and weather station were not disturbed nor cleaned during the test, which means full effects of soiling and snow cover are visible in the data. This is the most realistic scenario for a rooftop installation. Previous study by Pettersen et al.[3] suggests that soiling is negligible in the Nordics.

2.2 Initial data handling

The weather and power generation data spans from the system installation in January 2018 to September of 2020. Initial clean-up of the data included merging the weather and power datasets and adjusting the timestamps so that all time was UTC+2 i.e. daylight-saving time was converted to the UTC+2.

Since the system lacked a reference cell for measuring plane of array (POA) irradiance, it was calculated from the GHI data measured by the weather station. Functions from the pvlib library for Python [4] were used in the calculation. Initially, the solar zenith angles were obtained for the timestamps in the measured data. Using the zenith angles and GHI values from the weather station, the direct normal irradiation (DNI) and diffuse horizontal irradiation (DHI) were calculated with Erbs model [5]. The estimate for the POA irradiance was then calculated from the GHI, DNI, DHI and panel surface angles.

Modelling the theoretical power generation by the solar panels requires cell temperature. Since the cell temperature was not measured, it had to be estimated from the ambient temperature, irradiance conditions and wind speed. The cell temperature was calculated both with empirical heat loss factor model PVsyst [6] and Sandia

Array Performance Model [7]. After comparison of the cell temperature model results (data not shown here), it was determined that the difference between the models had an insignificant impact on the results of the overall analysis. In the end, data produced by Sandia Array Performance Model was used for the degradation analysis.

The measured power of the PV installation was normalized with a modelled power. The modelled power did not account for performance degradation over time to allow it to be visible in the normalized data. The model used for calculating theoretical power was PVWatts [8], 4500 Wp was used as the nameplate value for the whole installation (18×250 Wp) and -0.408 %/C° was the temperature coefficient according to the panel datasheet provided by the manufacturer. The normalized power calculated with PVWatts is also called power performance index, which is one of the most commonly used performance indices for solar power, and can be expressed as

$$PI = \frac{P_{measured}}{P_0 \frac{G}{G_{STC}}[1+\gamma(T_{cell}-T_{STC})]}. \quad (1)$$

$P_{measured}$ is the measured power, P_0 is the nameplate power, $\frac{G}{G_{STC}}$ is the ratio of measured POA to the POA in STC conditions, T_{cell} is the temperature of the cell and γ is the temperature coefficient.

2.3 Data filtering

Normalized power time series is further filtered to remove outliers caused by low light intensity (night or heavy cloud cover) or erroneous data recording. The impact of applying different filters with varying settings was compared. The simplest filters are the power filter and POA filter which removes all data outside of certain range of values for normalized power and POA, respectively. A clipping filter was also utilized, but its impact was minimal since there was no significant clipping in the data.

Effects of a power/POA filter are also studied. Power/POA calculates the average ratio of power and POA irradiation for the panels, and removes all datapoints of measured power that are a certain distance from the average. [9]

Final filter showcased in this study is the clear-sky filter. It compares the measured GHI to a calculated value that assumes the day is cloudless. If the measured GHI differs 20% from the theoretical sunny value, it is removed. Therefore, clear-sky filter should remove values measured while the weather was overcast. [4]

2.4 Calculating performance loss

Finally, the rate of performance degradation must be calculated from the filtered data. This study applies two different methods: year-on-year (YOY) and least squares regression.

Linear least squares regression is a common method in statistics that fit a line to given data by minimising the squared error between the measured data and the linear fit.

YOY method [10] calculates several performance loss rates by calculating the loss from datapoints a whole year apart, repeating for each time that has a corresponding measurement year later. This method reduces the effect of seasonal variation, since comparisons are made for datapoints collected in the same time of a year.

3 RESULTS AND DISCUSSION

The PV performance data after initial processing is displayed in Figure 1. There are clear gaps in the data during the winter months, revealing that either the irradiance conditions or snow cover on the panels prevented the collection of accurate data during winter periods. The lack of data from winter months makes it more challenging to quantify the exact effects winter months have to the performance of PV panels, but data from the rest of the year can be used to determine the overall loss of PV performance. The effects of snowy winters on solar panels might require more specialised and controlled tests to account for the effects of partial shading, snow weight-load and moisture ingress from melting snow.

Based on Figure 1, the modelled POA and panel power are accurate, since the majority of the normalised datapoint gather close to the 1.0 value. Minor inaccuracies in normalisation are also not an issue, since they are likely to manifest as a systematic error and we are studying the rate of change in performance which is not affected by a constant error.

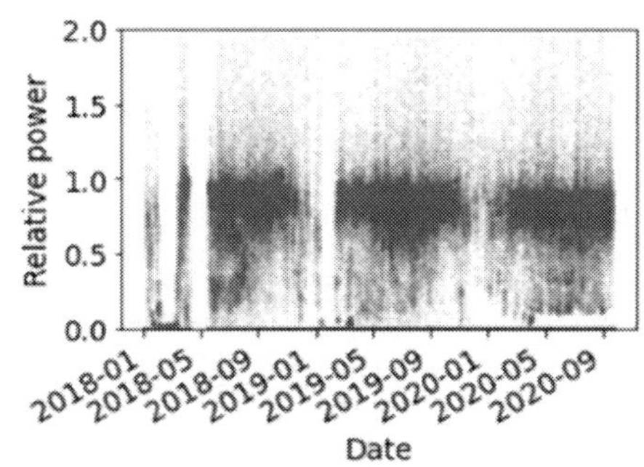

Figure 1 Measured power relative to the power simulated with PVWatts. Before filtering and aggregation there is over 360 000 data points.

Application of filters greatly reduced the amount of datapoints. In the example in Figure 2a the clipping filter, POA filter of 300...1200 W/m^2 and power filter for 0.5...1.25 of the relative power, over 90 % of the datapoints were filtered out. The high amount of filtering is to be expected since it includes all points measured during night-time and heavy cloud cover. The first filtering step also removes all times when the panels were under a snow cover. Increasing the pass-range in the POA and power filters would decrease the amount of data filtered out, but it has its downsides: the calculation of performance relies on models to calculate the theoretical power generation which likely becomes less accurate further away from the STC conditions.

Next, the clear-sky filter (Figure 2b) once again removes most of the remaining datapoints. The clear-sky filter is effective in removing data recorded during overcast periods, which can also result in bias to the outcome. If all datapoints are recorded only during full sunshine, some effects on the PV performance during lower light intensity may be overlooked. On the other hand, clear-sky filter removes datapoints where the irradiance measurements are erroneous [11]. It is therefore recommended to perform the performance analysis both with and without clear-sky filter.

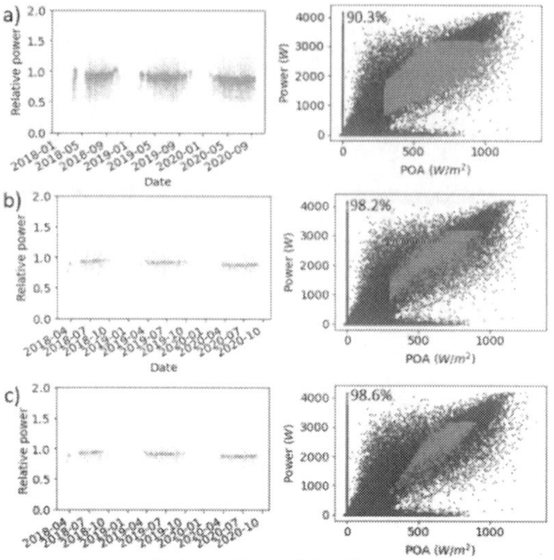

Figure 2 Cumulative effects of the filters on the raw data. Left-hand figures show the normalised power datapoints and right-hand figures display what data remains (orange) of the whole dataset (blue). The percentage in the corner of the right figures shows the share of filtered datapoints. a) Clipping filter, POA 300-1200, Power 0.5-1.25. b) Clear-sky filter added. c) Power/POA filter added.

Final applied filter is the Power/POA filter (Figure 2c). In the example it cuts datapoints with low and high POA, but the slope of the cut does not follow the high datapoint density, and can be considered undesirable way of filtering data. Since the Power/POA filter is calculated from the data it filters, its usefulness depends largely on the quality of the data. The data in this study has a large number of outliers, which leads to the Power/POA filter to perform poorly.

The performance loss rates calculated with YOY method with five different filter combinations are shown in Figure 3. Depending on the used filters, the calculated performance loss for the PV panels is between 2.59 % and 2.96 % per year. Generally, the performance loss gets smaller when the number and strictness of the filters is increased.

As discussed earlier in this chapter, Power/POA filter appears to have the smallest impact on the calculated performance loss rate. To improve the accuracy of Power/POA filter, the limits of the filter should be adjusted to be suitable for each separate dataset, a process which is not practical for continuous use.

The clear-sky filter performs slightly better, removing some of the outliers in normalised energy, but as discussed earlier, clear-sky filter may introduce unwanted bias into the result and should be always compared with a case where it is not used. Applying Power/POA filter after the clear-sky filter does not affect the performance loss rate calculation.

The last example in Figure 3 uses a high cut-off for the POA filter lower limit (800 W/m^2). It was manually chosen based on visual examination of the data, removing the datapoints that are not clustered around 1.0 relative power. In this case the regression line fits better into the clustered datapoints, since deviation in the dataset is lower.

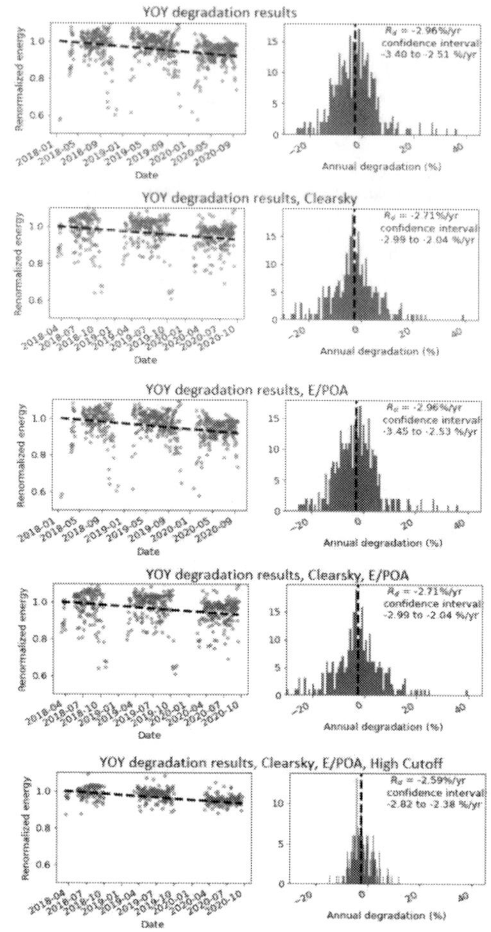

Figure 3 Showcase of degradation rates calculated using YOY.

Figure 4 collects the degradation rates calculated with different methods and filters. The general trend suggests that using stricter filters improves the calculated performance loss rate, which is to be expected since most of the outliers in the data are below the largest cluster of datapoints around 1.0 normalised power. All of the degradation rates are higher than expected for silicon solar cells, but during an inspection of the panels in late 2021 it was discovered that some of the modules had suffered cracking, possibly due to a heavy snow load. Some indication of more severe than expected soiling was also detected. In addition, the data covers only the first three years since the panel installation and it has been shown that the degradation rate slows after few years. [12]

At first glance, it would appear that choosing a filter that ensures most data outside of clusters is filtered would give the most accurate result, but it can be also considered to introduce bias. With high filtering, the data that remains is measured when the conditions are close to STC, which would be akin to measuring degradation of efficiency of the panels in laboratory conditions. However, changes in panel efficiency are not always the goal of the study, many studies and especially monitoring of real PV installations is more interested in the overall performance of the whole system. Too generous application of filtering may obfuscate effects such as soiling, tracking equipment malfunctions or degradation phenomena with nonlinear response to irradiance.

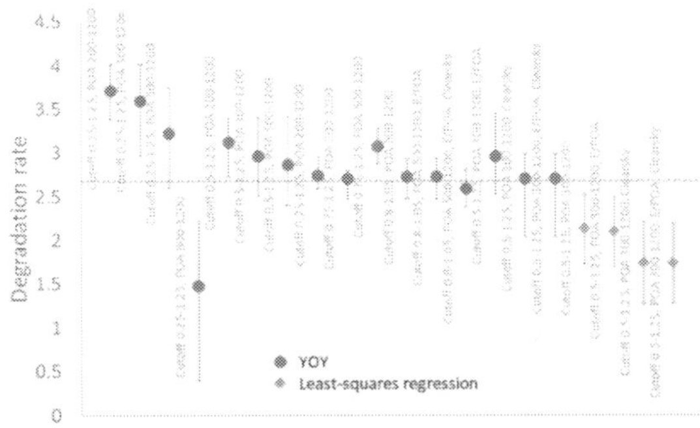

Figure 4 Collection of degradation rates calculated with different combinations of filters and statistical methods. The dashed vertical line is the average degradation rate.

This study demonstrates that the calculated performance loss rate in Nordic conditions varies greatly depending on the methods used. There are several measures that can be taken to mitigate or handle the variance before or during the measurements and data analysis. Best method to ensure accurate analysis is to improve the data quality, which can be achieved by making sure the weather data, especially temperature and irradiance is measured as accurately as possible, and as close to the PV panels themselves as possible. It is also important to monitor the data quality, especially early in the measurement so that any issues can be dealt with as soon as possible. In Nordic conditions, if data is desired from the winter months, accumulated snow should be cleared from the panels, although it should be remembered that panels are likely to be cleaned only in the largest installations in real conditions.

The reliability of the calculated performance loss rate can be improved by performing the analysis with multiple methods, as in this study. Care should be taken to also consider if the used methods are suitable for the current study, or introduce unwanted bias by the choice of too strict or lenient filters.

4 CONCLUSIONS

To accurately determine the performance loss rate of a PV system in Nordic conditions several factors should be considered. The obvious difference with more southern locations is the low irradiance and possibility of snow cover during winter months, which can lead to complete lack of datapoints for several months. The number of viable winter datapoints can be increased by cleaning accumulated snow from the panels, but it should be noted that snow removal does not represent a realistic rooftop installation.

More general factors to consider in every PV performance study include the methods used for data processing. To successfully remove outliers from the dataset without biasing the result, a correct set of filters with suitable strictness should be used. Similarly, there is a large number of different methods for calculating the performance losses themselves. A suggested method for finding an estimate for the performance loss rate is to perform the analysis with multiple different filters and performance metrics and use the distribution of results to quantify the performance loss rate.

6 REFERENCES

[1] IEA, "Solar PV," 2021. https://www.iea.org/reports/solar-pv.

[2] D. C. Jordan *et al.*, "Reducing Interanalyst Variability in Photovoltaic Degradation Rate Assessments," *IEEE J. Photovoltaics*, vol. 10, no. 1, pp. 206–212, 2020, doi: 10.1109/JPHOTOV.2019.2945191.

[3] A. Pettersen, H. Pedersen, J. K. Selj, and E. S. Marstein, "PV modules in Nordic climate : Effects of soiling and snow," p. 2015, 2015.

[4] W. F. Holmgren, C. W. Hansen, and M. A. Mikofski, "Pvlib Python: a Python Package for Modeling Solar Energy Systems," *J. Open Source Softw.*, vol. 3, no. 29, p. 884, 2018, doi: 10.21105/joss.00884.

[5] D. G. Erbs, S. A. Klein, and J. A. Duffie, "Estimation of the diffuse radiation fraction for hourly, daily and monthly-average global radiation," *Sol. Energy*, vol. 28, no. 4, pp. 293–302, 1982, doi: 10.1016/0038-092X(82)90302-4.

[6] D. A. R. Barkhouse, O. Gunawan, T. Gokmen, T. K. Todorov, and D. B. Mitzi, "Yield predictions for photovoltaic power plants:empirical validation,recent advances and remaining uncertainties," *Prog. Photovoltaics Res. Appl.*, vol. 20, no. 1, pp. 6–11, 2015, doi: 10.1002/pip.

[7] D. L. King, W. E. Boyson, and J. A. Kratochvil, "Photovoltaic array performance model, SANDIA Report SAND2004-3535," *Sandia Rep. No. 2004-3535*, vol. 8, no. December, pp. 1–19, 2004.

[8] A. P. Dobos, "PVWatts Version 5 Manual (NREL/TP-6A20-62641)," *Natl. Renew. Energy Lab.*, no. September, p. 20, 2014, [Online]. Available: http://www.nrel.gov/docs/fy14osti/62641.pdf.

[9] S. Lindig *et al.*, "International collaboration framework for the calculation of performance loss rates: Data quality, benchmarks, and trends (towards a uniform methodology)," *Prog. Photovoltaics Res. Appl.*, vol. 29, no. 6, pp. 573–602, 2021, doi: 10.1002/pip.3397.

[10] A. J. Curran, C. Birk Jones, S. Lindig, J. Stein, D. Moser, and R. H. French, "Performance Loss Rate Consistency and Uncertainty Across Multiple Methods and Filtering Criteria," *Conf. Rec. IEEE Photovolt. Spec. Conf.*, pp. 1328–1334, 2019, doi: 10.1109/PVSC40753.2019.8980928.

[11] D. C. Jordan, C. Deline, S. R. Kurtz, G. M. Kimball, and M. Anderson, "Robust PV Degradation Methodology and Application," *IEEE J. Photovoltaics*, vol. 8, no. 2, pp. 525–531, 2018, doi: 10.1109/JPHOTOV.2017.2779779.

[12] D. C. Jordan and S. R. Kurtz, "Photovoltaic Degradation Rates-an Analytical Review," *Prog. Photovoltaics Res. Appl.*, vol. 21, no. 1, pp. 12–29, Jan. 2013, doi: 10.1002/pip.1182.

UNIFORM AND NON-UNIFORM SOILING OF PV MODULES: A COMPREHENSIVE STUDY OF PERFORMANCE AND TEMPERATURE EFFECTS

Daniel Sena Braga*, Antonia Sonia A.C. Diniz*, Suellen C.S. Costa, Vinícius Camatta Santana*, and L.L. Kazmerski*†
*Pontifícia Universidade Católica de Minas Gerais (PUC Minas), Belo Horizonte, Brazil
†Renewable and Sustainable Energy Institute (RASEI), University of Colorado Boulder, USA

ABSTRACT: The science and engineering of soiling of solar devices is now well into its eighth decade, with more than 400 research papers this past year. The interests are technologically diverse, ranging from performance monitoring to mitigation methodologies. The derating of the energy output from a PV installation is a concern for both technology reliability and operating costs. This paper examines aspects of the relationships between the soiling of PV modules and the effects of this environmental process on the resulting performance and temperature characteristics. Thin-film CdTe and crystalline-Si modules are evaluated and compared for both uniform and non-uniform soiling patterns. The comprehensive studies document and compare laboratory simulations with actual outdoor cases for framed and frameless module structures. There is some emphasis on edge-buildup patterns, which are among the most common non-uniform deposition configurations, for both portrait and landscape module orientations. These non-uniform accumulations not only cause decrease in the produced power- output but result in *shading* that can cause increased confine areal heating ("hot spots") that can lead to module degradation. This paper further documents effects of both uniform and non-uniform soiling patterns on the J-V characteristics, documenting changes in the shape of these characteristics with the geometry and thickness distributions on the module surfaces. The special focus is to evaluate the specific effects of the non-uniform distribution on module temperature distributions using IR thermography and complementary techniques. These soiling patterns, performance parameters, and temperature distributions/levels are determined for modules in several Brazil climate zones. Temperature differences can exceed 40°C in excess normal operating temperatures for cells due to non-uniform soiling for the crystalline-Si cases; and corner soiling can lead to hot-spot formation no matter the module orientation..
Keywords PV modules, soiling, uniform, non-uniform, degradation

1 INTRODUCTION

The integrity of the first surface of interaction for the incoming solar photons remains a priority for PV module reliability and performance though this issue has been examined for >8 decades [1]. Soiling of PV-module surfaces is a growing concern because of the derating of energy output from solar installation. With the opening of the solar markets in sunbelt and desert areas of the world, areas that have not only the highest solar resource availability but also the most critical dust issues, the *understanding and mitigation* of this problem become crucial for avoiding any showstoppers for rapid and widespread deployment. The soiling issue is not, of course, confined to these harsh-climate areas. Significant losses can also be encountered in areas that have tropical and temperate environments. The importance and diversity of this problem is highlighted by the growing volume of research reports, exceeding over 400 in 2021.

This aims to contribute within the worldwide efforts toward examining, understanding, and adding to the knowledge base of the effects and fundamental interaction of these soiling particles with the PV module surface. We build on our previous reports and observations of the soiling process, adhesion mechanisms, and performance effects. We focus on our research in Brazil, working in the 5-climate areas in which we have deployed Si and thin-film soiling monitoring stations.

This paper further reports effects of both uniform and non-uniform soiling patterns on the J-V characteristics, documenting changes in the shape of these characteristics with the geometry and thickness distributions on the module surfaces. The special focus is to evaluate the specific effects of the non-uniform distribution on module temperature distributions using IR thermography and complementary techniques. These soiling patterns, performance parameters, and temperature distributions/ levels are determined for modules in several Brazil climate zones. Some correlations of the soiling distributions to the chemistry and composition of the soiling particles and the adhesion mechanism to the module glass surfaces. Major results of these investigations are: (1) uniform soiling distributions affect mainly the short-circuit current in both Si and CdTe technologies; (2) non-uniform soiling can significantly change the shape of the IV characteristics of Si modules, giving evidence that the P_{max} is a more accurate monitoring parameter for soiling losses as previously reported; (3) non-uniform soiling can have significant effects on the temperature distributions in Si modules. The non-uniform distributions of the soiling build-ups can cause temperature differences in single cells and cell combinations exceeding 30°-40°C in excess of the normal module operating temperatures. The severity is compared for modules oriented in portrait (preferred) and landscape positioning; (4) thin-film modules do not exhibit significant 'hot spots' due to soiling; and (5) laboratory simulations correlate well with actual operating (field) module measurements. The purpose is to validate some critical relationships between the module performance parameters and the soiling process, important for modeling, simulations, predictions, and potential minimizations or eliminations of these externally-produced reliability issues.

2 DETAILS AND EXPERIMENT METHODS

Current-voltage characteristics were recorded using our soiling stations (Atonometrics), monitoring the short-circuit current, maximum power, and when required, the total IV curve [2]. The stations were verified for accuracy by comparing results with for the same modules in a class-AAA solar simulator. The solar irradiance was monitored using

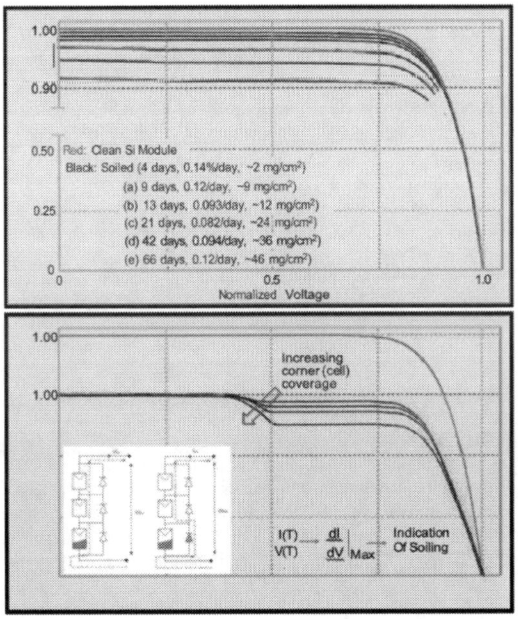

Figure 1: Normalized IV characteristics for c-Si modules – UPPER: Uniform soiling; LOWER: Bottom edge with lower corner soiling. Red curves indicate clean module characteristic. Inset shows condition of bypass diodes causing the progressive distortion in the curves.

calibrated reference cells (and checked periodically against a separate reference cell and a calibrated pyranometer for consistency). The stations incorporated weather stations to record all meteorological conditions for the given location (climate zone). Module temperature was recorded with an RTD on the back surface of the module (and with calibrated thermocouples). Module temperature mapping was accomplished using IR cameras (NEC/Sanei and Seek). Temperatures were also independently (regularly) checked using IR detectors and thermocouples. All data are corrected for temperature and irradiance levels. Silicon (Canadian Solar 2650W and Kyocera 185-W) modules and First Solar CdTe Series-4 modules were used for the *Crystalline-Si* and *Thin-Film Monitoring Stations*, respectively.

Figure 2: Soiling buildup along frame and at corners.

Table I: Correlation and estimation of soiling area covered (corners) using the slope of the IV-characteristic and the difference in the two levels of soiling buildup.

Cell area covered (%)	10	22	36	62	78
ΔArea (covered by soiling-ratio)	-	2.2	1.64	1.72	1.19
$(dI/dV)_T$	7.6	3.6	2.2	1.3	1.1
Δ_L (difference in levels)	0.04	0.11	0.19	0.48	0.62

3 SUMMARY RESULTS AND DISCUSSION

Previously, we have provided initial evaluations of the importance of soiling patterning on the measured IV characteristics and indicated potential issues relating to temperature distributions and hot-spot generation in some module types [2,3]. The issue of hot-spots is an important one for PV reliability— and is a major focus of this paper.

3.1 Soiling and IV characteristics

It is common to evaluate the soiling effects on module performance by monitoring the short-circuit current. It has been shown that it is can be critical to monitor the maximum power output as well because non-uniform distributions can possibly lead to misleading results [3-5]. Investigations of the non-uniform dust accumulation on PV modules surfaces has been shown to affect the shape of the IV-characteristics and possibly generate hot spots due to shading [5-7]. We have observed these effects and have categorized the shapes of the IV characteristics for Si (framed) and thin-film CdTe (unframed) modules on several identified soiling patterns.

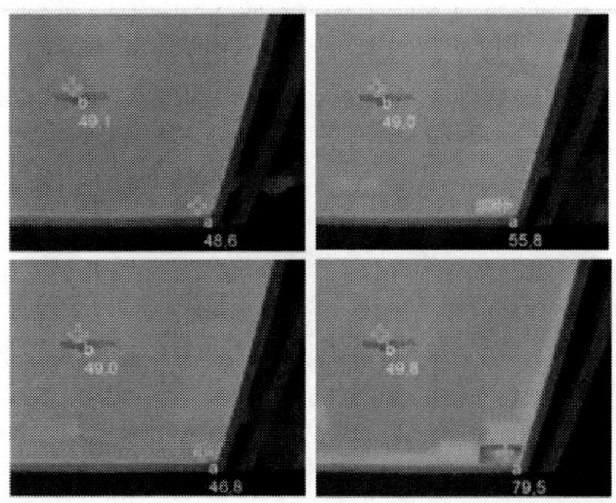

Figure 3: IR-mapping of c-Si module with soiling buildup at corners. As exposed to sun, there is initial cooling of the cell (bottom left) followed by heating of that region as shaded cell is exposed to incoming sunlight.

We associate soiling pattern with interpretion of the shapes and levels of the resulting module IV characteristics. Initially, this generates a database of distinctive characteristics that can then be used in the inverse fashion—to identify the major soiling patterns or combinations that are occurring. These can then be used in various modeling and simulation activities. Our approach compares *field-observed* soiled modules with various soiling patterns and "laboratory" *simulations*

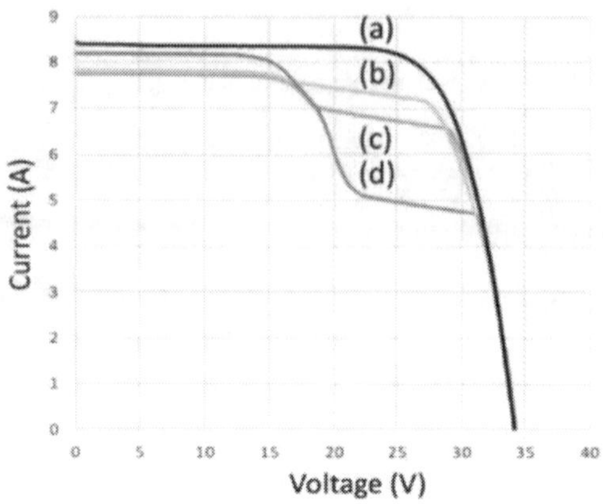

Figure 4: IV characteristics measured for 4-module soiling conditions as presented in Fig. 2 and 3, with corner cell change in temperature.

using both calibrated (for transmittance and spectral characteristics) papers/organics and artificially produced dust patterns using dust samples collected from modules.

Figure 1 provides examples of several generated IV-characteristics, illustrating the effects of the soiling patterns found at these locations. The first set is for the case of uniform (i.e., nearly evenly distributed thickness or g/cm^2 layers) deposits over the entire module surface. Loss of the short-circuit current dominates as a function of the soiling deposit thickness and soiling particle characteristics. The more complex non-uniform distribution can occur far more commonly in the field.

Figure 1 (lower) shows a series of IV characteristics for edge-soiling build-up along the frame bottom edge and corner (Fig. 2) of a Si module—comparing these IV shapes to the edge. The characteristic shape is affected by both the incursion of the soiling from the edge into the module and the thickness of the build-up. The excessive corner buildup causes the same effect for either portrait or landscape orientations. We have documented these effects for both field-deployed modules and laboratory-produced test cases. The correlation of these provides the interpretations and the predictions. The dust generally builds on the sides and bottom module edges along the frame, and most important, at the corners. The shape of these patterns depends on the bypass diode configuration and the extent of soiling. This paper provides some initial IV spectroscopy combined with image analysis to give indication of the soiling distribution (with one summary shown in Table I). (These edge and corner build-ups were also simulated in laboratory conditions with nearly identical results.)

3.1 Soiling patterns and temperature mapping

A priority for this study is to evaluate the relationships between soiling patterns and hot spot generation. For selected module soiling patterns, we have carefully monitored the temperature distribution along with the IV characteristics. IR mapping of the temperature clearly shows the development of hot regions (as a function of exposure to the sun's radiation). An example of one such temperature map is shown in Fig. 3 for a crystalline Si module with bottom edge and one-corner soiling. The module was held near normal operating temperature (in this case ~49°C) (Fig. 3a). Initially, the regions beneath the soiled regions are slightly cooler because they are shaded from the sun's radiation (Fig. 3b) and benefit from some heat removal by the close metal frame. As time-exposure to the sun increases, the cells increase in temperature directly with soil build-up regions on the module surface. Adjacent regions are heated by conduction from this cell. We monitor the module glass surface temperature as a function of time—as well as the cell temperature. As time increases, the temperature of those cells shaded by the soiling increases with the increasing temperature. We have recorded realtime videos of the process showing the changes in tempertaure distributions. The effect on the module performance is shown in Fig. 4, corresponding to the temperature captures of Fig. 3. The corner dust accumulation (Fig. 2)_that results in the hot-spot formation likewise degrades the performance significantly.

3 SUMMARY AND HIGHLIGHTS

This paper extends and supports the results of several previous investigations of soiling effects on module performance, in additions to some first-time correlations among the performance parameters, dust patterns and properties, and the resulting modules temperature distributions. Specifically, this paper has been concerned with (major contributions):

- In-depth evaluations of the effects of several specifically identified and defined soiling patterns on the module IV characteristics. The studies are reported for both field experience and laboratory simulations. Examples have been provided for all three pattern types, with specific identification of the effects on the shapes of the IV patterns;
- Correlation of these soiling patterns with the time-dependent module temperature distributions are confirmed, and possible "hot-spot" formations are observed (with temperature increases of more than 30°C-40°C over 'normal' regions). The module temperatures are mapped using high resolution IR camera techniques. The direct effect of the rising temperature on the IV characteristics is also reported and animated in real time.
- The orientation of the module (i.e., portrait or landscape positioning) is important [8]. For crystalline-Si, portrait is preferred for series-connected modules. For thin-films (CdTe), landscape is better for soiling considerations. This ensures minuimum loss due to the electrical circuit configurations.
- For all module (no matter the module orientation), corner soiling is a primary concern because it leads to hot-spot formation which can not only result in performance loss, but also lead to permanent module degradation.

The studies have been carried out in Brazil's 5-climate zones in which our monitoring stations are deployed.

4 ACKNOWLEDGEMENTS

The authors gratefully acknowledge the support of Brazil's CAPES for part of this work through "Pesquisa Visitante Estrangeiro," and Ciência sem Fronteiras that supported the students and postdocs involved with our Brasil soiling reliability project. We also acknowledge also the help and advice of the National Renewable Energy Laboratory (NREL), which maintains strong leading research and development programs in PV soiling & reliability research.

5 REFERENCES

[1] Suellen C.S. Costa, et al., Solar energy dust and soiling R&D progress: Literature review update for 2016. Renewable & Sustainable Energy Rev. (2018). [http://dx.doi.org/10.1016/j.rser.2017.09.015]

[2] L.L. Kazmerski, et al. Interrelationships between non-uniform soiling disbributions and PV module performance parameters, climate conditions, and soiling particle and module surface properties. Proc. 44th IEEE Photovoltaic Spec. Conf.-Washington, DC (IEEE, New York; 2017).

[3] Ghosh, S., Roy, J.N., & Chakraborty, C. A model to determine soiling, shading and thermal losses from PV yield data. Clean Energy 6, 372-391 (2022). https://doi.org/10.1093/ce/zkac014

[4] Li, C., Yang, Y., Fan, F., Xia, L., Peng, P., Wang, Y., Zhang, K., & Wei, H. A novel methodology for partial shading diagnosis using the electrical parameters of photovoltaic strings. IEEE Journal of Photovoltaics 12, 1027-1035 (2022). DOI: 10.1109/JPHOTOV.2022.3173723

[5] M. Gostein, B. Littmann, J.R. Caron, & L. Dunn, Comparing PV power plant soiling measurements extracted from PV module irradiance and power measurements. Proc. 39th IEEE Photovoltaic Spec. Conf. (IEEE, NY: 2013). DOI:10.1109/PVSC.2013.6745094

[6] L. Dunn, B. Littmann, J.R. Caron, & M. Gostein, PV module soiling measurement uncertainty analysis. Proc. 39th IEEE Photovotltaic Spec. Conf., (IEEE, NY; 2013). DOI: 10.1109/PVSC2013.6744236

[7] E. Lorenzo, R. Moretón, & I. Luque, "Dust effects on PV array performance: In-field observations with non-uniform patterns. Progr. in Photovoltaics 22, 666-670 (2013).

[8] K.A. Bapurao, et al., A review of static reconfiguration techniques of solar PV to mitigate mismatch loss and minimize partial shading effect. IETE J. Res. DOI: 10.1080/03772963/2021.1987992

Very Accurate Outdoor Estimation of V_{OC}'s Temperature Coefficient from Equivalent Cell Temperature Sensitivity Analysis?

Georgi Hristov Yordanov[1,2*] and Stan Herthogs[1,†]
[1]University of Leuven (KU Leuven), Dept. of Electrical Engineering (ESAT), Div. Electrical Energy Systems and Applications (ELECTA), Kasteelpark Arenberg 10, 3001 Leuven-Heverlee, Belgium;
[2]EnergyVille, Thor Park 8310, 3600 Genk, Belgium;
*Corresponding author; Email: georgi.yordanov@kuleuven.be; †Now with EcoTechnics, Belgium.

ABSTRACT: We explore a novel method to potentially very accurately estimate the temperature coefficient (TC) $\boldsymbol{\beta}$ of a PV module's Open-circuit voltage $\mathbf{V_{OC}}$ in outdoor characterization. The goal is to improve the accuracy of the classical methods for outdoor $\boldsymbol{\beta}$ evaluation which can have significant uncertainties, causing a less accurate Equivalent Cell Temperature (**ECT**) evaluation. The 'RMSE method' is based on sensitivity analysis of **ECT** with respect to the error in the used $\boldsymbol{\beta}$ value. The difference between the calculated **ECT** and measured $\mathbf{T_{back}}$ is plotted over the effective in-plane irradiance, resulting in a $\mathbf{\Delta T}$ scatter plot. Our decomposition of $\mathbf{\Delta T(\boldsymbol{\beta}')}$ very accurately describes how the $\mathbf{\Delta T}$ plot changes with an erroneous value $\boldsymbol{\beta}'$. We refute the initial hypothesis, namely, the $\boldsymbol{\beta}'=\boldsymbol{\beta_R}$ corresponding to minimal scatter is not the true $\boldsymbol{\beta}$ (but for most c-Si PV, $\boldsymbol{\beta_R}$ and the classic value $\boldsymbol{\beta_o}$ differ within a few %). Our accurate **RMSE** fitting for c-Si and thin-film PV suggests correct mathematical modeling but the corresponding scatter term in the decomposition is independent on the true $\boldsymbol{\beta}$. The 'linear' and 'vertical-axis intercept' terms can be investigated in future works. The **ECT** evaluation procedure has also been adapted to bifacial (BiFi) c-Si PV modules in earlier works, and therefore the present line of research is also applicable to that technology.
Keywords: PV Module; Outdoor Testing; Temperature Coefficient; Equivalent Cell Temperature; Modelling.

1 ECT FROM V_{OC} VS. BACK-OF-MODULE T

Significant (in some cases, >10°C) and steady spatial non-uniformities of a PV module's operating temperature can persist under full solar forcing, in particular in floating PV (FPV). For outdoor PV performance testing, the effective operating temperature is needed in order to calculate the temperature coefficients and to depict the temperature dependence of e.g., the relative efficiency with respect to STC. The IEC Standard 60904-5 (1st withdrawn Edition from year 1993 [1]; 2nd current Edition from 2011 [2]) defines the Equivalent Cell Temperature (**ECT**) as the uniform PV device T that would lead to the same measured Open-circuit voltage, $\mathbf{V_{OC}}$ at the same (uniform) in-plane irradiance. The Standard (originally developed in the monofacial-PV era) prescribes a procedure for calculating the **ECT** from the measured $\mathbf{V_{OC}}$ and in-plane irradiance. A priori knowledge of the $\mathbf{V_{OC}}$'s TC, $\boldsymbol{\beta}$, and the Open-circuit ideality factor, $\mathbf{n_{OC}}$ (or simply **n**), is required. Next, there are other requirements and limitations such as: the constancy of the ideality factor; negligibly small current leakage via p-n junction shunts; largely uniform illumination; as well as irradiance not less than 200 W/m^2 (0.2 suns) below which the procedure is compromised by, according to the Standard, variation of $\boldsymbol{\beta}$. In fact, the **ECT** calculation procedure only uses the $\boldsymbol{\beta}$ after the measured $\mathbf{V_{OC}}$ is first translated from the measured irradiance to the STC one (or another reference level; we always use STC, while a reference irradiance of 0.7 or 0.8 sun may be more suitable for façade BIPV). We have developed a modified procedure for the special case of linear dependence of the $\mathbf{n_{OC}}$ on both the irradiance and the PV device's back temperature for some rare cases [3]. In earlier works, we also showed that it was in fact the variation of the $\mathbf{n_{OC}}$ at low irradiance that led to increased errors in the calculated **ECT** [4, 5]. We studied scatter plots of the difference between the calculated **ECT** and the measured back **T** over irradiance ($\mathbf{\Delta T}$ plots) from up to 16 back T sensors per PV module and found that some of those plots were 'benign', i.e., mostly linear at high and intermediate irradiance, sometimes all the way down to very low irradiance; see Fig. 1. From sensitivity analysis of the $\mathbf{\Delta T}$ plots, we obtained the low-irradiance behavior of the $\mathbf{n_{OC}}$ as a by-product [5, 6]. There was excellent agreement of both the $\mathbf{n_{OC}}$ behavior and the corrected **ECT** estimates after linearizing four benign $\mathbf{\Delta T}$ plots (based on four slim-Pt100 back T sensors) at low irradiance for the same PV module. That approach was also successfully applied to several *bifacial* c-Si PV modules when using a simple effective irradiance, namely the sum of the measured front and rear in-plane irradiance.

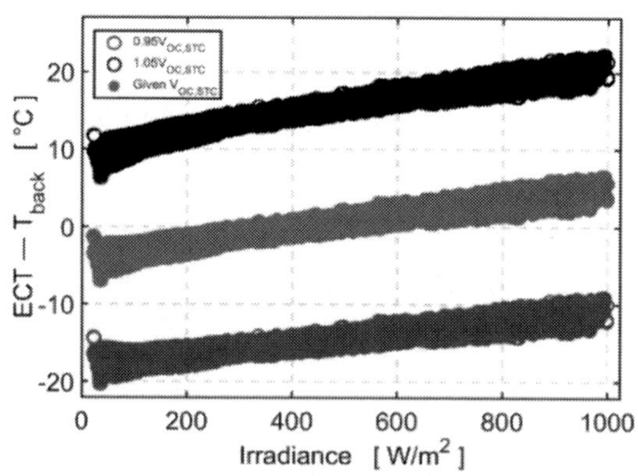

Figure 1: Three $\mathbf{\Delta T}$ scatter plots of a mc-Si PV module calculated with the given STC value and ±5% of the $\mathbf{V_{OC}}$ [4]. The measured $\mathbf{T_{back}}$ data were used in each. Since we expect that $\mathbf{ECT \rightarrow T_{back}}$ without any solar forcing (at near-zero irradiance), a $\mathbf{\Delta T}$ plot with zero vertical-axis intercept should give the *true* $\mathbf{V_{OC,STC}}$.

2 THE RMSE METHOD

In a most recent work, we focused on the **ECT** sensitivity to errors in the $\boldsymbol{\beta}$ value derived from either indoor or outdoor PV testing [5]. Such errors change the slope, the overall spread, and the vertical-axis intercept of the best linear fit to the **ΔT** scatter plot; see Fig. 2. All the other involved parameters such as the $\mathbf{n_{OC}}$ and the rated $\mathbf{V_{OC,STC}}$ are assumed to be absolutely correct. A method to correct the latter was explained in the caption of Fig. 1 while in past works, we have presented methods to accurately determine the local ideality factor's value along most of the I-V curve and at Open circuit in particular [4, 7, 8].

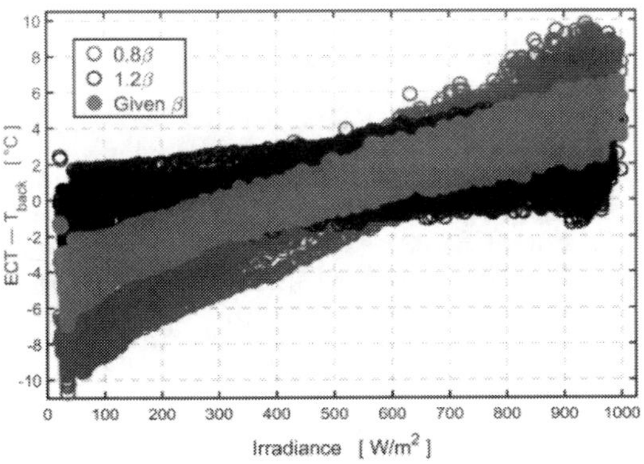

Figure 2: Three **ΔT** scatter plots created with the indoor-measured value and ±20% of the ***β*** [4]. Note that the rated $\mathbf{V_{OC,STC}}$ value has not been corrected, yet.

In [5], we used a simplified expression for the **ECT** error (from its partial derivative with respect to ***β***) and arrived at an approximate yet quite accurate model of the **RMSE(*β'*)** function of the trial TC value ***β'***. The empirical plots had a quasi-parabolic shape with a well-defined minimum; see Fig. 3.

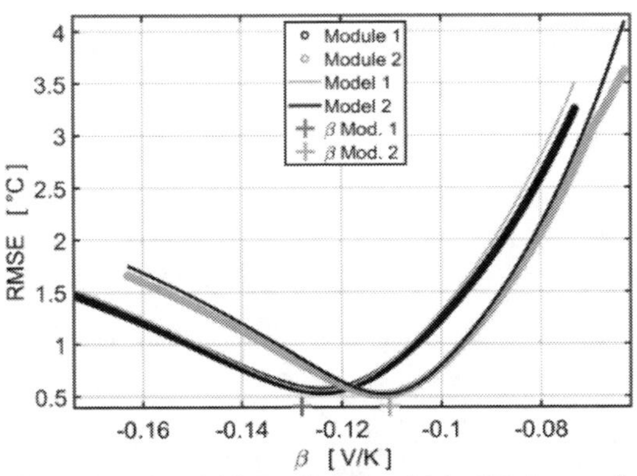

Figure 3: **RMSE** of linear fits to multiple **ΔT** plots made with trial values of ***β*** from [4] for two monofacial mc-Si PV modules. The solid lines are models based on Eq. (4) in [5] and the plus symbols denote the ***β*** values obtained from fitting outdoor $\mathbf{V_{OC}}$ data over $\mathbf{T_{back}}$ at 1.00±0.05 sun.

We derived an approximate proportional relationship between the true ***β*** and the **RMSE** function's minimum location, $\boldsymbol{\beta_{RMSE}}$ (from now on, $\boldsymbol{\beta_R}$) [5]:

$$\beta \approx \frac{\beta_{RMSE}}{\left[1+\frac{\sum r(r+R)}{\sum R(r+R)}\right]}, \qquad (1)$$

where the summation goes over the total number of data points, **N**; **r** are the residuals from the linear fit to the **ΔT** plot corresponding to the true ***β***; and **R** are the residuals from the linear fit of $\mathbf{T_{back}}$ over the irradiance, which are independent on the trial value ***β'*** and are typically of a much higher magnitude than the **r**. (Note that the indices *i*=1..**N** of both **r** and **R** are omitted for brevity.) The 2nd term in the square brackets in Eq. (1) ranged within ±3% for most of the tested c-Si PV modules and equaled –2.0% for the CIGS module. The new estimates of the true ***β*** differed by only ±1-2% (median –0.62%, sigma 1.63%) from those obtained in the classic way (by linear fitting of the $\mathbf{V_{OC}}$ over module's $\mathbf{T_{back}}$ at irradiance of about 1.0 sun). In fact, those values and the corresponding deviations were computed with some bugs present in our MatLab code. After debugging, we obtained values for ***β*** that were even closer to the outdoor estimates, within ±0.4% (median –0.043%, sigma 0.23%) [9].

The present work aims at deriving exact expressions of the error **ΔECT** due to an error **Δ*β***, of the resulting model of **RMSE(*β'*)**, and of the relationship between the true ***β*** and $\boldsymbol{\beta_{RMSE}}$. The ultimate goal is to give the outdoor PV testing community a method for very accurate evaluation of the true ***β***, with an uncertainty of less than 1%. One challenge to that is that we start with an imperfect classic estimate, and thus the residuals **r** in an equation similar to Eq. (1) would differ (albeit slightly) from those corresponding to the true ***β***. Both residuals must be compared, and the resulting (supposedly small) error must be evaluated. Here, we summarize and discuss the latest results found in [9].

3 ΔECT AND RMSE MODEL DEBUGGING

We re-analyzed our previously derived expressions for the sensitivity **ΔECT** with respect to the value used for ***β*** [4, 5] and arrived at a correct, exact formula:

$$\Delta ECT_\beta = -\frac{\Delta\beta}{\beta' + \frac{kN_S n}{q}\ln\left(\frac{G}{G_{STC}}\right)}\left(ECT(\beta) - 25°C\right) \qquad (2)$$

where the denominator term is an update to [4, 5] adding irradiance dependence to the error **ΔECT**:

$$\beta' + \frac{kN_S n}{q}\ln\left(\frac{G}{G_{STC}}\right) = \beta' + B(G) \qquad (3)$$

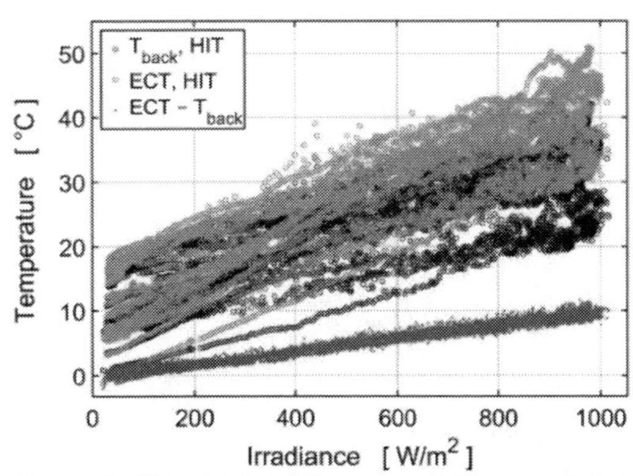

Figure 4: Calculated **ECT**, measured $\mathbf{T_{back}}$, and their difference **ΔT** plotted over the in-plane irradiance for a HIT module. The data are from [4].

Eq. (2) is then used to re-formulate the decomposition

of the corresponding **ΔT(*β'*)** scatter plot into a 'linear', 'vertical-axis intercept', and 'scatter' terms (where the last one is used to calculate the **RMSE** of the best linear fit). This is illustrated in Figs. 4 and 5, and the decomposition of **ΔT(*β'*)** is given in Eq. (4).

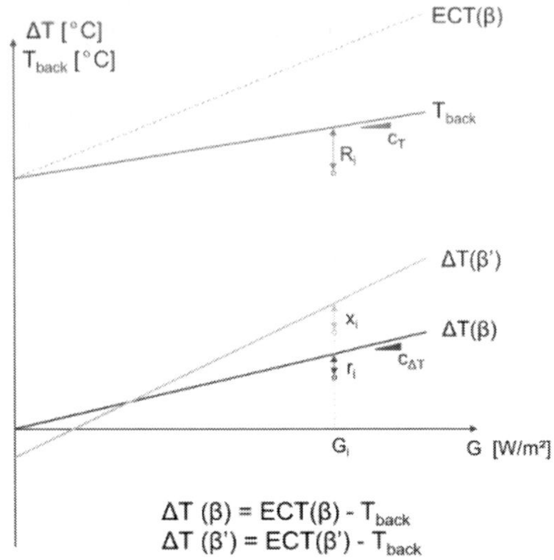

Figure 5: An illustration of the **ΔT(*β'*)** scatter plot decomposition from [9] based on Figs. 2 and 4. The lines are best linear fits to the corresponding scatter plots, and the residuals of interest are $\mathbf{r_i}$, $\mathbf{R_i}$, and $\mathbf{x_i}$. The slopes of the fitted lines to the **ΔT(*β*)** and $\mathbf{T_{back}}$ scatter plots are $\mathbf{c_{\Delta T}}$ and $\mathbf{c_T}$, respectively. So, for example, $\Delta T(\beta)_i = c_{\Delta T} G_i + r_i$. The residuals $\mathbf{x_i}$ are used to calculate the empirical **RMSE** curves for various trial values ***β'***.

$$\Delta T(\beta')_i = \left[\left(1-\frac{\Delta\beta}{\beta'+B_i}\right)c_{\Delta T} - \left(\frac{\Delta\beta}{\beta'+B_i}\right)c_T\right]G_i + \frac{\Delta\beta}{\beta'+B_i}\left(25°C - T_{back,G=0}\right) + \left[r_i - \frac{\Delta\beta}{\beta'+B_i}(r_i+R_i)\right] \qquad (4)$$

$$x_i = r_i - \frac{\Delta\beta}{\beta'+B_i}(r_i+R_i) \qquad (5)$$

$$RMSE = \sqrt{\sum_{i=1}^{N}\frac{x_i^2}{N}} \qquad (6)$$

The 1st term in Eq. (4) resembles the linear term in the much simpler expression for $\mathbf{\Delta T(\beta)_i}$ (see the caption of Fig. 5) but its 'slope' factor is irradiance-dependent due to the variable $\mathbf{B_i}$; see Eq. (3). The overall slope of the best linear fit to the **ΔT(*β'*)** scatter plot will therefore be defined by the combined contributions of all the data points. The 2nd term can be regarded as defining the vertical-axis intercept of the best linear fit. Finally, the 3rd term is the only one containing residuals, and therefore it represents the individual residuals $\mathbf{x_i}$ of the best linear fit to the scatter plot **ΔT(*β'*)** (see Eq. (5)) that are used to calculate the **RMSE** corresponding to each ***β'***; see Eq. (6).

Fig. 6 shows the excellent agreement of the new model with the empirical **RMSE** curves (obtained from fitting many **ΔT(*β'*)** scatter plots where **ECT** was calculated with a mock value ***β'***), much better than what was achieved in Fig. 3 with the old, approximate model [5]. Both models are very accurate in the vicinity of the curve's minimum at $\beta' = \beta_R$, but the newer model's superior performance far away from it supports the correctness of the mathematical treatment of the **ΔT(*β'*)** decomposition done in [9].

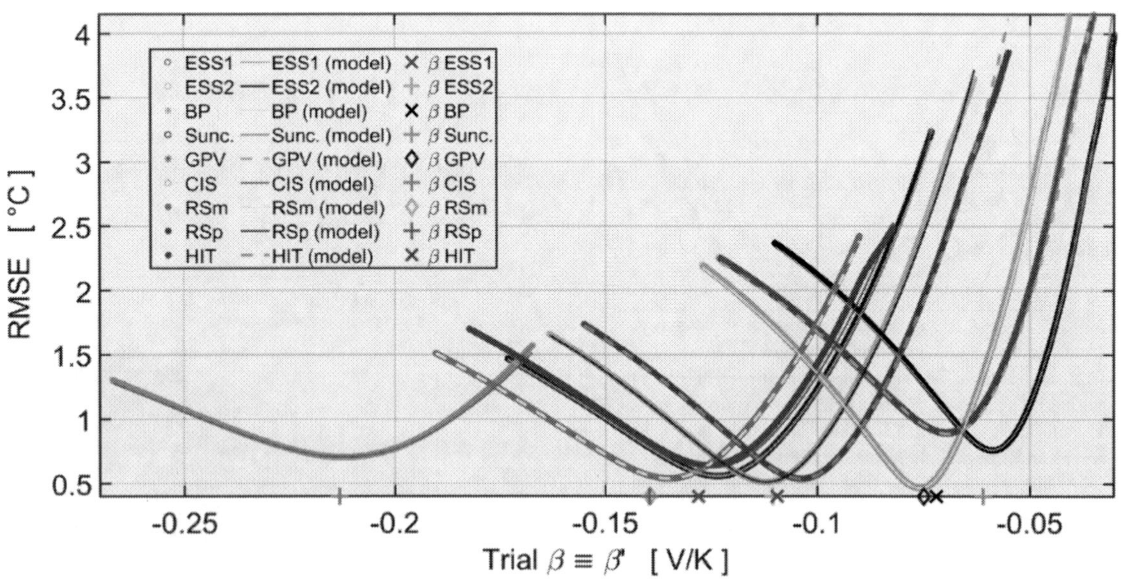

Figure 6: Empirical (discrete symbols) and modeled with Eqs. (5) and (6) (lines) **RMSE** curves for 8 new and aged mono- and mc-Si PV modules, and one new CIGS module. Raw data are from [4]. The discrete symbols along the horizontal axis mark the classic outdoor estimates $\boldsymbol{\beta_o}$.

Fig. 6 also shows that the classic outdoor estimates β_o lie close to the **RMSE** curve's minimum location at $\beta'=\beta_R$. Our next task is, based on the exact expressions in Eqs. (5) and (6), to derive an exact expression linking the true β to the β_R similar to the approximate one in Eq. (1). This is done by differentiating the **RMSE** with respect to β' and imposing the requirement for an extremum by setting the derivative equal to zero (which is satisfied for $\beta'=\beta_R$). The obtained algebraic equation must be solved for β. Since the model fits the empirical curves very accurately, and especially so in the vicinity of the minimum, the β_R can be regarded as an empirical constant.

4 THE ULTIMATE EQUATION – IMPLICATIONS

Important intermediate step in solving the algebraic equation for the **RMSE**(β') minimum is to eliminate the residuals r_i that by definition correspond to the true (and unknown) β. We expressed every r_i and (r_i+R_i) term by means of the variables β, β_R, R_i, B_i, and the x_i corresponding to $\beta'=\beta_R$ (that is, the residuals from the best linear fit to the $\Delta T(\beta_R)$ scatter plot). The detailed analysis can be found in [9]. In the end, the algebraic equation for the function's minimum location boils down to:

$$\sum_{i=1}^{N}\left[\frac{x_i(R_i+x_i)}{\beta_R+B_i}\right]=0 \qquad (7)$$

The unexpected (and unpleasant!) result means that the true β is eliminated completely, and that there cannot be an exact relationship of it to the β_R similar to the approximate Eq. (1).

Eq. (7) suggests that a minimum scatter of the $\Delta T(\beta')$ plot (**RMSE** minumum) is reached for the (set of) residuals x_i that, in combination with the (fixed) residuals R_i (from fitting the T_{back} scatter plot) and with the irradiance-dependent parameters B_i result in a zero sum. The corresponding $\beta'=\beta_R$ is unrelated to, and independent from, the true β. The scatter in the $\Delta T(\beta)$ plot, the set of residuals r_i, is likely caused by other physical phenomena such as thermal transients and instrument noise.

In view of the above key finding, the fact that the approximation in Eq. (1) results in values for β that are suspiciously very close to the classic outdoor estimates β_o suggests that the approach was wrong from the beginning. Indeed, that relationship involves the residuals r_i corresponging to β which are unknown apriori, leaving us with their best estimate, those corresponding to β_o. Thus, both hypotheses are refuted in the present work, namely hypothesis 0 from [4] ($\beta=\beta_R$) and hypothesis 1 from [5] expressed in the approximate Eq. (1).

5 CONCLUSION AND OUTLOOK

Despite the promising accuracy of the mathematical model and the excellent fitting of the empirical **RMSE** curves, the latter turned out to lack any info about the true β. The two other terms in the $\Delta T(\beta')$ decomposition can be investigated next. Nevertheless, it is an important observation that, for most of the tested PV modules, the classically estimated values β_o are within ± a few % from the curve's minimum location, β_R. This makes the latter a good rough estimate, but the accuracy level (up to several per cent as for one of the tested modules) is considered dissatisfactory, likely exceeding that of the classic β_o.

REFERENCES

[1] IEC Standard (60)904-5:1993 "Photovoltaic devices - Part 5: Determination of the equivalent cell temperature (ECT) of photovoltaic (PV) devices by the open-circuit voltage method," 1st Edition, withdrawn.
[2] IEC Standard 60904-5:2011 "Photovoltaic devices - Part 5: Determination of the equivalent cell temperature (ECT) of photovoltaic (PV) devices by the open-circuit voltage method," current 2nd Edition (under revision).
[3] G.H. Yordanov *et all.*, "Equivalent cell temperature calculation for PV modules with variable ideality factors," Proc. 38th IEEE PVSC, 2012, pp. 000505–000508.
[4] G.H. Yordanov, "Characterization and analysis of photovoltaic modules and the solar resource based on in-situ measurements in southern Norway," Doctoral thesis, NTNU, Trondheim, 2012.
[5] G.H. Yordanov *et all.*, "Beyond-standard Equivalent Cell Temperature evaluation," Proc. 38th EU PVSEC, 2021, pp. 820–824.
[6] R. D'Hoore, "Validation of the PVGIS model for the performance modeling of modern monofacial and bifacial c-Si PV modules," Master's thesis, KU Leuven, Leuven, 2021.
[7] G.H. Yordanov *et all.*, "Series resistance determination and further characterization of c-Si PV modules," *Renewable Energy* **46** (2012), pp. 72–80.
[8] G.H. Yordanov *et all.*, "Differential analysis of PV module I-V curves," Proc. 28th EU PVSEC, 2013, pp. 3207–3211.
[9] Stan Herthogs, "Alternative evaluation of the temperature coefficient of a PV module's V_{OC} from outdoor testing data," Master's thesis, KU Leuven, Leuven, 2022.

A STUDY ON USING MICRO-COMPUTED TOMOGRAPHY FOR PV INTERCONNECTION TECHNOLOGY DEVELOPMENT

Rik Van Dyck[1,2,3,4], Alexandros Prapavesis[1], Tom Borgers[2,3,4], Jonathan Govaerts[2,3,4], Loic Tous[2,3,4], Jef Poortmans[2,3,4,5], Aart Willem van Vuure[1]
[1]KU Leuven Departement of Materials Engineering - Oude Markt 13, 3000 Leuven. Belgium
[2]Universiteit Hasselt, imo-imomec - Martelarenlaan 42, 3500 Hasselt, Belgium
[3]imec, Departement of Energy - Thorpark 8320, 3600 Genk, Belgium
[4]EnergyVille, imo-imomec - Thorpark 8320, 3600 Genk, Belgium 5KU Leuven Departement of Electrical Engineering - Oude Markt 13, 3000 Leuven. BelgiumAddress(es)

ABSTRACT: Current more novel interconnection methods for Si solar cells such as shingling, multi-wire, conductive backsheet, and multi-ribbon approaches, are structurally more complex than conventional 2-5 busbar stringing. The most common techniques used for the development of these concepts are electrical characterization (current-voltage measurements and electroluminescence imaging) in combination with a cross-section analysis using SEM and EDX. Other characterization methods, such as scanning acoustic microscopy (SAM) are implemented, especially to search for voids and delamination within the module. Using micro-computed tomography (μCT) is reported as a viable method to inspect new module concepts in the past, however, very few effective results are published. Current state-of-the-art μCT devices are capable to achieve a lower voxel size and higher resolution on larger samples than ever before. In this work, the multi-ribbon interconnection for back-contact cells under development at imec is being used as a case study to investigate the potential of these new μCT devices for PV research.
A 20 x 20 cm2 one-cell module is electrically characterized before, during and after reliability tests in the form of thermal cycling and humidity-freeze cycling. The results show an unstable shunt being present in the module. Before and after the reliability tests, a TESCAN UniTOM XL scanner is used to scan the one-cell module using μCT.
Due to a large amount of metal in the module, artifacts such as photon scattering and shadowing in the region around the ribbons were observed. The use of binning and different metal filters was tested as a way to decrease these artifacts. Using a 1 mm tin filter in combination with binning seems to achieve the best results when scanning (part of) a PV module.
Keywords: Interconnection, characterization

1 INTRODUCTION

In the current development of novel interconnection and module concepts, evaluation is mainly based on two methods: post lamination inspection and reliability testing combined with electrical characterization. The former one usually consists of a cross-section analysis with optical or scanning electron microscopy (SEM), optionally with additional EDX or Raman spectroscopy mapping. For the latter one, the used reliability tests are based on the intended application of the module, such as terrestrial usage, space PV, and vehicle- or building-integrated PV.

The widely used cross-section analysis requires time-consuming sample preparation and is a destructive test. Less commonly used non-destructive inspection techniques exist. These are scanning acoustic microscopy (SAM) [1,2,3] and micro-computed tomography (μCT) [2,4]. SAM has proven it is suited to detect failure modes in such as delamination and undesired voids in a broad range of research fields including PV [1,3,5,6]. However, it lacks the ability to generate a 3D image of solder joints, metallization print, and other interconnection structures. Overcoming this limitation, μCT has the great potential to be used to evaluate a solder joint formation. This is especially important in the development of module concepts using a combined lamination and soldering approach.

μCT was already introduced as a new diagnostic tool for the development of PV module technologies [7], also, it is further used to inspect modules using the conductive backsheet approach for back-contact solar cells at TNO/ECN [8]. Yet further elaboration on implementing μCT in PV research is still required, especially for newer scanning devices with increased performance. This work attempts to implement and further investigate how industrial μCT can benefit PV research with the three-dimensional multi-ribbon interconnection as a case study.

2 MATERIALS AND METHODS

2.1 Sample preparation

Previous publications reported on a multi-ribbon interconnection for back-contact solar cells based on a 3D woven fabric of interconnection ribbons in an encapsulant matrix [9]. Reliability testing on sample-sized modules has shown the potential of this technology [10]. To enhance the industrial relevance of the concept, further tests on the fabrication processes are required. Especially the solder joint formation is an important aspect in evaluating new production methods and materials. For this reason, a sample-sized module using this interconnection technology will be used as a case study in this work.

The one-cell module is fabricated similar to the samples made in previous publications related to the interconnection concept. The explanation is beyond the scope of this abstract (but can be included later). An image of the front- and backside of the sample module is given in Figure 1. The 20 x 20 cm2 module with an interdigitated back-contact cell consists of two 2-mm-thick glass plates and a glass-fiber reinforced polyolefin encapsulant. The interconnection ribbons have a 0.8 x 0.07 mm2 copper core and are coated with a 5- to 25-μm-thick layer of a SnBiAg solder. The ribbons are directly soldered to the fingers using a combined soldering and encapsulation step.

Figure 1: Image of the front- and backside of one-cell module using the 3D interconnection method.

2.2 Micro-computed tomography

The module was mounted in a TESCAN UniTOM XL scanner, equipped with a 230 kV/300 W microfocus x-ray tube, with a tungsten reflection target and a 2856 × 2856 pixels detector with a 150 μm pixel pitch. The scans were performed at 230 kV and 50 W with a voxel size of 25 μm while 2×2 pixel binning was enabled on the detector, thus resulting in a 50 μm voxel size. Binning is a technique in which the charge for several pixels on the detector is added together. 2×2-pixel binning results in a doubling of the voxel size, but with a trade-off in a better signal-to-noise ratio, shorter exposure times, smaller projections, and reconstruction of datasets, implying higher quality, faster reconstruction, and analysis. Furthermore, a 1 mm thick metal beam filter was installed on the source to remove the longer-wavelength (or 'soft') part of the polychromatic x-ray spectrum. This reduces artifacts like photon scattering and beam hardening. Both a copper and a tin filter were tested. In total, 2600 radiographic projections were acquired over a 360° angle sample rotation, each with an exposure time of 420 ms and frame averaging set of 5 reducing the random noise. The total acquisition time was 86 minutes. Cross-sectional imaging reconstruction was thereafter performed by applying a filter back-projection algorithm in the TESCAN reconstruction software Panthera.

2.3 Reliability testing

First, the sample is electrically characterized using electroluminescence (EL) imaging and current-voltage (IV) measurements. Then, a μCT scan was taken using the procedure as described. Reliability experiments were performed in the form of thermal cycling (TC) and humidity-freeze (HF) according to the IEC61215 standard [11]. At intermediate intervals, the module was electrically characterized. After 550 thermal cycles and 16 humidity-freeze cycles, also a new μCT scan was performed.

3 RESULTS AND DISCUSSION

3.1 Electrical characterization

Figure 2 shows the IV measurement results of the module at various stages during the reliability testing. A low initial Fill Factor (FF) would indicate that there is either a high series resistance or a short circuit (shunt). The pseudo-Fill Factor (pFF) is also low, implying that most likely, a shunt is present in the module causing the low FF.

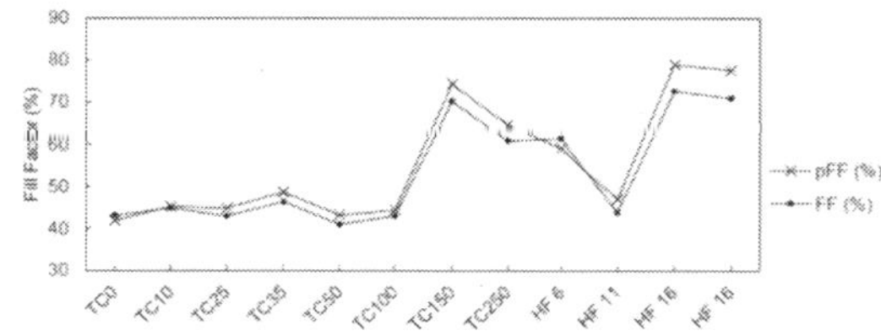

Figure 2: Graph showing the pseudo-Fill Factor (pFF) and Fill Factor (FF) of the one-cell module during thermal cycling (TC) and humidity-freeze (HF) reliability testing. TCX and HFY refers to the sample after X thermal cycles and Y humidity-freeze cycles.

The EL-images of the module at six stages during the reliability testing are given in Figure 3. As the FF and pFF from the IV measurements suggest, there is barely any change in the module performance during the first 100 thermal cycles. This is also demonstrated by the first two EL-images. Another 50 cycles seem to improve the performance, as both FF and pFF increase. This is most likely due to the shunt being removed due to internal thermomechanical strain, generated during TC. Another 100 thermal cycles and 11 HF cycles reduce again the pFF and FF close to the original values. Also, the EL-image shows this performance drop. This dropping and increasing of the performance are repeated in the last thermal and HF cycles, indicated in both the IV measurements and EL-images. Although one could argue that the sample is not appropriate for further testing since it is unpredictable and yields unreliable results, this unstable shunt is a great opportunity to test the capabilities of μCT scanning.

Figure 3: Electroluminescence (EL) images of the one-cell module at different stages during reliability testing.

3.1 Micro-computed tomography

Figure 4 (top) shows a partial volume rendering of a scan before the module underwent reliability tests. Only the voxels of the metal (metallization, interconnection ribbons and solder) are shown. Removing the glass, encapsulant and Si cell from the image allows a better interconnection evaluation. Since the ribbons are long and continuous metal structures, they form barriers for the x-ray photons to penetrate, yielding artifacts in the form of photon scattering and shadowing.

Figure 4 (bottom) shows similar volume renderings for a scan taken after the reliability tests. For this scan, the source filter material was changed from copper to tin. This removes more of the lower energy photons, which are anyway unable to penetrate the metals in the module. Also, binning is implemented, further improving image quality. This enables us to further identify if a solder joint has been established between fingers and ribbons.

Figure 4: Three-dimensional volume renderings of a part of the module (top) before the reliability testing, using a copper filter and no binning and (bottom) after reliability testing using a tin filter and binning.

By applying a voltage in reverse bias to the module, the current will start to flow through the unwanted shunts. Because of the high current through these shunts (10 A), they will dissipate heat to their surroundings. By taking a picture with an infrared-sensitive camera, we could pinpoint the location of potential shunts. Figure 5 shows such an image, with increased temperatures at the top right corner and left edge of the module.

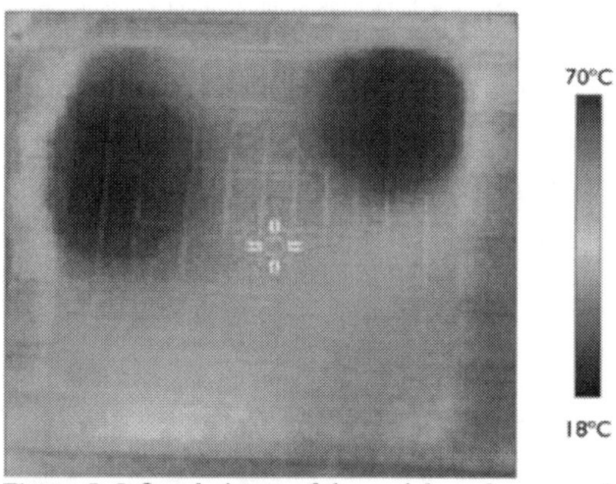

Figure 5: Infrared picture of the module under reversed bias showing two shunt locations.

Using the above-mentioned locations, we could pinpoint the short-circuit location. Figure 6 shows a connection between a positive polarized ribbon and negative finger. This is most likely the result of uncontrolled solder flow during the lamination process.

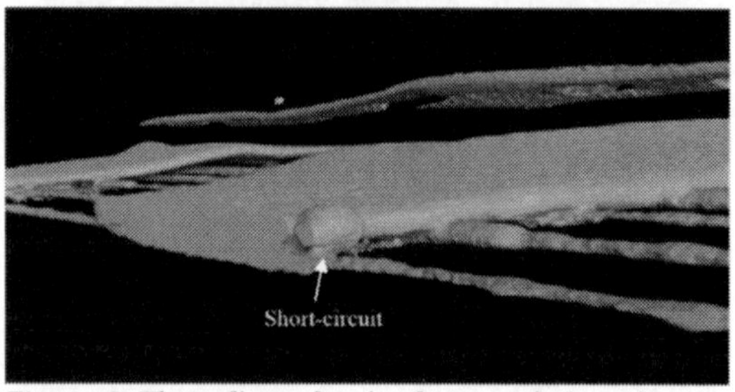

Figure 6: Three-dimensional volume rendering of a scan on a 2x2 cm2 part of a module.

4 CONCLUSION

Reliability testing on a one-sample module was combined with micro-computed tomography (μCT) as an additional visual inspection technique. The result of this work illustrates the potential of μCT for PV research. They specify that using binning and a tin filter to remove lower energy photons from the source yields the best results for optical analysis of the interconnection structures. Shunts (short-circuit) and disconnections within the module could be pinpointed and visualized. In futher work, the compatibility with cross-section analysis using SEM.

5 ACKNOWLEDGEMENTS

The authors gratefully acknowledge imec's SiPV industrial affiliation programme and its partners, Fonds Wetenschappelijk Onderzoek, through the SB Ph.D. Fellowship under grant number 1S14120N, and their large infrastructure I013518N project for the financial support of the X-ray infrastructure. The KU Leuven XCT Core facility is acknowledged for the 3D image acquisition and quantitative post-processing tools (https://xct.kuleuven.be/), and ISC Konstanz for providing the ZEBRA IBC solar cells.

6 REFERENCES

[1] K. Dressler *et al.*, "Characterisation of Rear Local Contacts Including BSF Formation Using Raman and Scanning Acoustic Microscopy," *27th EU PVSEC.* 2012.

[2] K. Dressler, M. Rauer, M. Kaloudis, S. Dauwe, A. Herguth, and G. Hahn, "Nondestructive characterization of voids in rear local contacts of PERC-Type solar cells," *IEEE J. Photovoltaics*, 2015.

[3] D. E. Mansour, S. Kotterer, D. Philipp, and P. Gebhardt, "Pv Backsheet Failure Analysis by Scanning Acoustic Microscopy," *37th EU PVSEC*, 2020.

[4] A. De Rose, T. Geipel, D. Eberlein, A. Kraft, and M. Nowottnick, "Solder Interconnection of Silicon Heterojunction Solar Cells by Infrared Soldering - Solder Joint Analysis and Temperature Study," *36th EUPVSEC*, 2019

[5] F. Bertocci, A. Grandoni, and T. Djuric-Rissner, "Scanning acoustic microscopy: A robust method for defect detection during the manufacturing process of ultrasound probes for medical imaging," *Sensors*, 2019.

[6] S. L. Gao and J. K. Kim, "Scanning acoustic microscopy as a tool for quantitative characterisation of damage in CFRPs," *Compos. Sci. Technol.* pp. 345–354, 1999.

[7] E. C. Quintana, M. A. Quintana, K. D. Rolfe, K. R. Thompson, and P. Hacke, "Exploring diagnostic

capabilities for application to new photovoltaic technologies," *IEEE Photovolt. Spec. Conf.*, 2009.
[8] D. Veldman, I. J. Bennett, B. Brockholz, and P. C. De Jong, "Non-destructive testing of crystalline silicon photovoltaic back-contact modules," *IEEE Photovolt. Spec. Conf.*, 2011.
[9] R. Van Dyck, T. Borgers, J. Govaerts, J. Poortmans, and A. W. van Vuure, "Three-dimensional multi-ribbon interconnection for back-contact solar cells," *Prog. Photovoltaics Res. Appl.*, 2021.
[10] R. Van Dyck *et al.*, "Three-Dimensional Multi-Ribbon Back-Contact Interconnection : Latest Results on Reliability Testing," *38th EU PVSEC*, 2021
[11] "IEC 61215-1 Terrestrial photovoltaic (PV) modules – Design qualification and type approval," 2021.

THERMAL MODELLING OF PV SYSTEMS BASED ON INDOOR AND OUTDOOR PV MODULES TESTING

Oume Lgheit Rhazi*[1], Julien Dupuis[1], Emmanuel Boyère[1], Geoffrey Rachinel[2], Floriane Gay[2], Daline Nap[2]

[1]EDF R&D, EDF Lab Les Renardières, Avenue des Renardières, 77250 Moret Loing et Orvanne, France
[2]EDF Renouvelables, 100 Esplanade du Général de Gaulle, 92932 Paris La Défense Cedex, France
*Corresponding Author: oume-lgheit.rhazi@edf.fr

ABSTRACT: EDF R&D has developed its in-house 3D PV modelling tool PVNOV®. It is using reverse raytracing for a more efficient calculation of irradiance on the cells/modules of PV plant. The energy yield is calculated in the aftermath of the irradiance estimation through an improved single-diode model and an advanced thermal model. The thermal model considers the thermal inertia of the module, as well as the convective and radiative heat exchange with environment. This model is challenged and recalibrated based on experimental data for several module technologies and various setups that are presented in this work: monofacial and bifacial modules, full-cell and half-cell modules, glass-glass or glass-backsheet modules, in addition to less-common white ceramics glass-glass modules. The indoor and outdoor thermal measurements are performed using several temperature probes on the back of the module and the open-circuit voltage according to the method described in the IEC 60904-5:2011 standard adapted to bifacial modules. A comparison of the thermal model performance with 2-years measurements from an instrumented testbed in an EDF's outdoor test zone is finally introduced.

Keywords: modelling, modules, thermal behavior, cell temperature, back temperature, monofacial, bifacial, glass-glass, glass-backsheet, full-cell, Half-cell,

1 INTRODUCTION

Although several GW of bifacial systems were installed over the last years, the need for accurate modelling tools able to correctly assess the bifacial yield in various conditions is decidedly still relevant. Therefore, we aim at improving yield modelling by considering all the phenomena that come into play in PV systems' production for any design choices, and within reasonable time. Among other physical parameters, getting an accurate model to predict the cell temperature during operation is of foremost importance due to the thermal dependency of the electrical yield.

2 PVNOV® MODELLING

2.1 A raytracing software

PVNOV® is a raytracing-based 3D tool developed at EDF R&D [1-3]. This software is dedicated to accurate prediction of PV systems yield. Its main application is for building numerical models of PV plants. It allows predicting the theoretical yield of the future project before it is built depending on the equipment chosen. It can consider the performance of different kind of modules, inverters and study the irradiance, thermal and electrical phenomena that affect the yield estimate. Its development was initiated to assess the performance of innovative technologies, which were not accurately estimated by commercial tools, such as bifacial technology, then half-cell modules, floating PV or Agro PV.

PVNOV®'s accuracy was also validated through several benchmarks such as the 2021 IEA Task13 modelling tools benchmark [4] and compared to several experimental measurements at 10 min time step. Results have shown a consistency with other PV simulation tools based on raytracing such as the one developed by Fraunhofer ISE [5].

2.2 PVNOV® Thermal Model

The thermal model is one of the three main components allowing the determination of pv power output. Once optical model and ray-tracing methods have predicted the irradiance reaching PV panels, thermal and electrical models are then called out to estimate the part of the solar radiation converted into electricity. The remaining part heats the panel and reduces the efficiency of photovoltaic-to-electricity mechanisms.

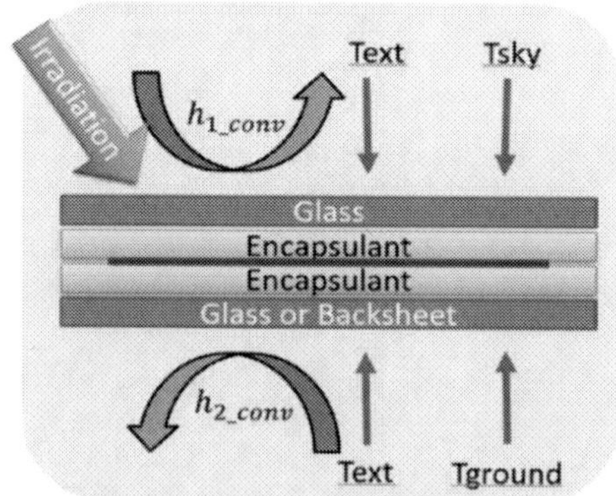

Figure 1: Heat transfer in PV modules

The thermal model used in PVNOV® is a physics compliant realistic model which represents the solar panel as an equivalent thermal capacity. This model is constructed with Dymola/Modelica [6] and considers thermal heat exchanges through:

- Conduction: between the PV module layers.
- Convection: The photovoltaic panels are subject to the convective effects of the wind [7-9] and natural convection [10].
- Long-wave radiation: considered between the PV modules, the environment and the sky.
- Short-wave radiation: considered between the PV modules and the sun.

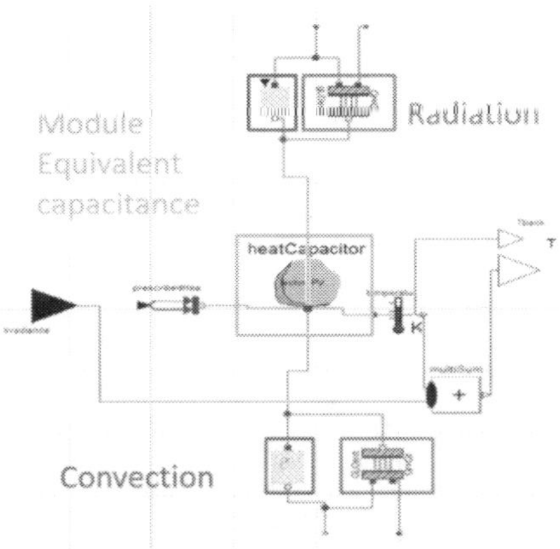

Figure 2: Thermal model assembly under Dymola

The thermal balance between the incoming fluxes, the heat exchanges and the produced electricity is built. This allows to determine the temperature of the equivalent thermal capacity. The equivalent cell temperature is therefore derived from a correction of the thermal capacity using the effective irradiation [11]. This model was initially calibrated and validated with monofacial modules and the first bifacial modules in the market. As manufacturers are now proposing many kinds of bifacial module architectures, an update of the few model empirical parameters is required through experimental data on several module technologies.

3 EXPERIMENTAL

3.1 Configuration of the test facility

The literature agrees on the fact that it is complicated to perform a thermal model validation due to the difficulty in accurately measuring the cell temperature. Actually, cell temperature can be either estimated using a probe laminated in the module during the manufacturing phase, or using an extrapolation of the module back temperature probe or deduced from the open-circuit voltage using the IEC 60904-5:2011 standard.

Thus, indoor and outdoor measurements performed in this study alternate between Maximum Power Point Tracking (MPPT) and open-circuit voltage measurement sequences.

3.2 Tested photovoltaic modules

Nine modules with different characteristics were tested in indoor and outdoor conditions (see Table 1).

Module	Bifacial	Cell type	Back side of the module	White ceramics
Module 1	No	Full-Cell	White Backsheet	No
Module 2	Yes	Full-Cell	Transparent Backsheet	No
Module 3	No	Half-Cell	Glass	No
Module 4	Yes	Half-Cell	Glass	Yes
Module 5	Yes	Full-Cell	Glass	No
Module 6	No	Full-Cell	White Backsheet	No
Module 7	No	Half-Cell	White Backsheet	No
Module 8	Yes	Half-Cell	Transparent Backsheet	No
Module 9	Yes	Full-Cell	Glass	Yes

Table I: Tested modules characteristics

3.3 Indoor measurements

The tests took place in the indoor laboratory of EDF R&D (PVLab) and monitored the thermal transient history of the different panels under controlled conditions at several levels of irradiation (600W/m², 800W/m², 1000W/m², 1200W/m²), ambient temperatures (25°C, 40°C) and for 3 fan speeds. In addition to the chamber temperature measurement, 9 temperature probes were used on the rear side of each module (Figure 3).

Several thermal probes were glued to the back of the modules with several layers of aluminum adhesive tape to limit the sensitivity of the sensor to other external fluxes [12] as shown in Figure3.

The estimation of the cell-temperature from the short-circuit voltage measurement required an additional measurement of the thermal diode voltage of each module, deduced from flashes at the same temperature (25°C) but different irradiation levels.

Figure 3: Indoor setup for the thermal experiment

3.4 Outdoor measurements

The outdoor tests were carried out in the EDF PV-Zen outdoor test zone located in EDF lab les Renardières, in the south of Paris-France to take into account the real climatic conditions and constraints exerted on several modules as well as to keep track of the transient thermal behavior of each module.

Two pyranometers were used and allowed the measurement of the irradiation on the front and back faces of the tested modules. The speed and the direction of the wind were recorded with an anemometer installed behind the modules and at the same height. Several temperature sensors were glued to the rear side of the modules. The mounting structure used is bifacial-compatible, and wooden panels were installed to simulate the proximity of other modules.

The tests were performed at open-circuit voltage and MPP per pair of modules with a reference module each time to highlight a given characteristic. For instance, the test including module 1 and module 2 emphasizes the thermal behaviour of a bifacial module compared to a monofacial one while the test with module 2 and module 8 reveals the difference between a full-cell and a half-cell module from a thermal point of view.

Figure 4: Outdoor setup for the thermal experiment

4 RESULTS

4.1 Modules Setups comparison

The results presented in this section are based on cell temperature measurements using IEC 60904-5:2011 standard for monofacial as well as for bifacial modules, by taking into account each module's thermal voltage, the front and rear irradiation measured in addition to the bifaciality coefficient for bifacial modules. They were cross-checked with the measurements from thermal sensors placed on the back of modules to detect any inconsistency.

4.1.1 Monofacial Modules

4.1.1.1 Glass-Glass vs Glass White Backsheet

The glass-white backsheet module is cooler than the glass-glass one showing a mean bias difference (MBE) of 1.6°C. It indicates that over the whole test period, the glass-glass module's cell temperature was in average 1.6°C above the glass-backsheet one. This is in accordance with the higher amount of irradiation absorbed in the rear side of the module due to the glass transparency but without any additional power production because of the monofacial cells.

4.1.1.2 Half-Cell module vs Full-Cell module

As expected, the Half-Cell (HC) monofacial module is slightly cooler than the Full-Cell (FC) module (MBE= 1.2°C) thanks, among other parameters, to the electrical losses reduction due to current cut in half-cells.

4.1.2 Bifacial vs monofacial module

When it comes to the comparison between monofacial and bifacial module, it is key to distinguish between glass-glass and glass-backsheet modules.

4.1.2.1 Glass-Backsheet modules

The module 1 using a white backsheet is cooler than the module 2 using a transparent backsheet showing a Mean Bias difference of 1.5°C module. The trend is highlighted in the Figure 5. The higher temperature observed on the module 2 can be attributed to the high reflection of the white backsheet that obstructs a major amount of light reaching, on the contrary, the bifacial module and increasing its temperature as a consequence of the efficiency limitation.

Despite this higher temperature, the bifacial module delivers more energy as its bifaciality allows an additional production of 5% to 10% in the tested conditions.

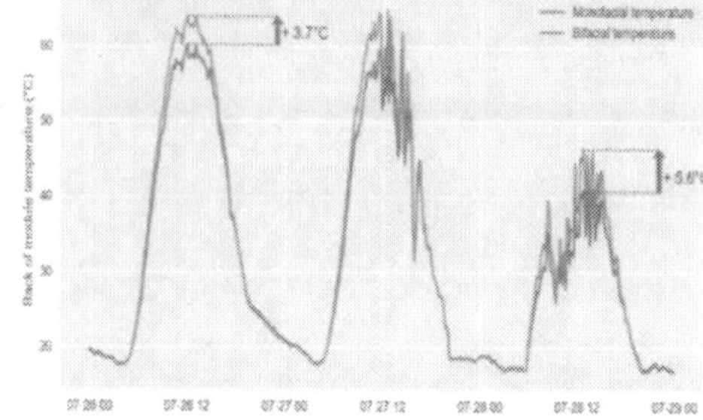

Figure 5: Bifacial and Monofacial Backsheet modules

4.1.2.2 Glass-Glass modules

Contrary to the previous case, here we can consider the same amount of rear irradiation reaching the module cells for both monofacial and bifacial modules. However, a part of this irradiation is converted into electricity for the bifacial module and hence works as a heat sink. This was proved through measurements that reveal a lower cell temperature for the bifacial glass-glass module compared to the monofacial module exhibiting an MBE of 1.8°C over the test period.

4.1.3 Bifacial modules

4.1.3.1 Bifacial glass- backsheet module vs glass-glass module

The module 2 using transparent backsheet is showing a lower temperature compared to the module 5 using a glass/glass structure (MBE = 2.5°C), due to the higher thermal resistance of the glass that prevents the dissipation of heat into the environment leading therefore to lower cell operating temperature. M.Mittag et al have highlighted this glass characteristic through a detailed layer analysis of the PV module in [13].

Transparent backsheets have already demonstrated a better yield than glass-glass module with 1% to 2% more producible in some benchmarks [14]. Nevertheless, this gain may not be related only to the temperature difference and degradation aspects are to be considered in the balance.

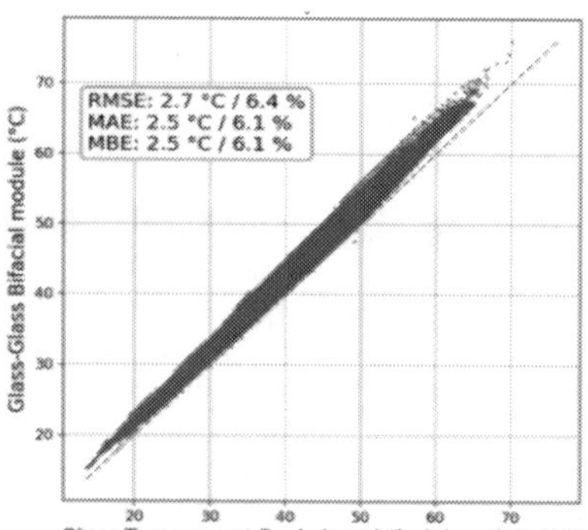

Figure 6: Bifacial and Monofacial Backsheet modules comparison

4.1.3.2 Glass-glass with white ceramics

The glass-glass with white ceramics increases the internal reflections in the module, causing additional irradiation to be absorbed by the cell which is also an additional heat source due to cell efficiency limitation. Whereas for the standard glass-glass module, more light passes through module layers. The measurements exhibit a 0.7°C average difference.

4.1.3.3 Half-Cell (HC) vs Full-Cell (FC) module

Surprisingly during outdoor tests, HC module was warmer compared to the FC bifacial module (MBE=1.3°) although the electrical losses diminution. The outdoor tests were indeed performed during winter period in low irradiation conditions that do not showcase the advantages of HC technology. Nevertheless, indoor tests have confirmed the tendency. New flashes were conducted to re-evaluate the efficiency of the tested HC bifacial module compared to the FC one. They revealed, as expected, a lower efficiency of the HC module of 0.88% at STC conditions and up to 1.0% at 400 W/m^2 irradiation. Further investigations are ongoing to understand this efficiency noticeable gap.

4.2 Evaluation of the temperature difference between the back of the module (Tback) and the real cell temperature (Tcell)

Figure 7 highlights the difference between the estimated cell temperature (Tcell) and the measured mean back module (Tback) does not evolve in a linear way with the received irradiance and depends on the module technology. The average bias between Tback and Tcell for the tested modules is in the [0.6°C - 3.5°C] range.

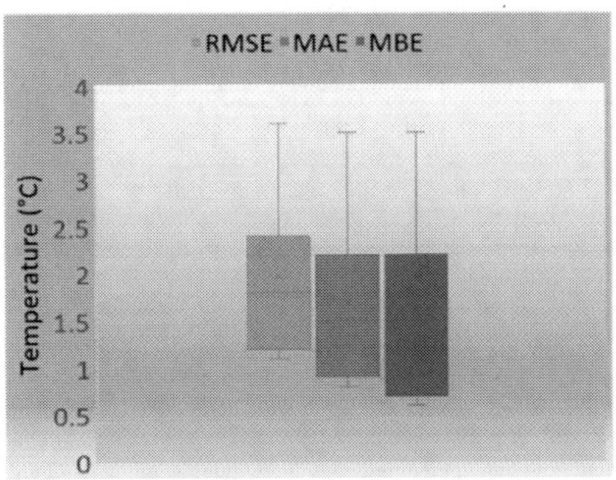

Figure 7: Root Mean Square Error, Mean Absolute Error and Mean Bias Error on temperature Tback vs Tcell

Besides, depending on their position underneath the module, the technology, and the wind/fan speed, a significant mismatch is noticeable between the temperature probes, up to 3°C.

Moreover, in a real power plant, the string configuration of the modules may deeply impact the effective modules temperature. Measurements from a real PV plant are meant to be used to validate the model.

5. VALIDATION

As explained previously in this paper, the main objective of the performed thermal testing is to recalibrate the PVNOV® thermal model and enhance its accuracy for several module architectures.

The used testbed for the following results is a 15° tilted fixed system installed on a concrete ground. It is composed of 2 S-cabled bifacial strings of 15 modules and a monofacial string of 6 modules. Each of the strings is on an independent MPPT. Bifacial and monofacial modules are glass-backsheet.

Figure 8: A PVzen testbed and its PVNOV® model

The presented results in Figure 9 and Figure 10 are a comparison of the simulated back module temperature with PVNOV® and the measured temperature on this PV testbed for years 2018 and 2019.

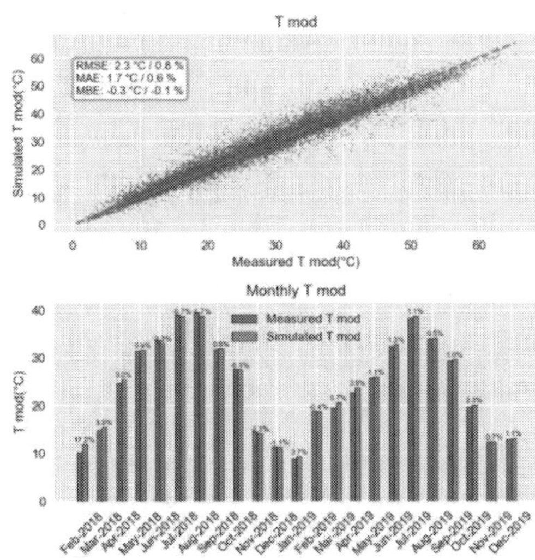

Figure 9: 10 min time steps correlation of measured/modelled module temperature for a monofacial white backsheet module

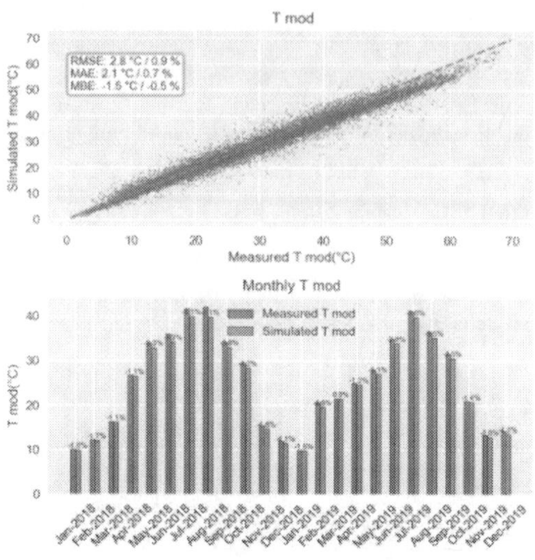

Figure 10: 10 min time steps correlation of measured/modelled module temperature for a transparent backsheet bifacial module

The thermal model seems slightly underestimating the module temperature. The annual average error is in fact of 0.3°C for the monofacial and -1.5°C for the bifacial system, a deeper analysis of this difference is in progress and a sensitivity analysis allowed determining the most impacting parameters.

6 CONCLUSIONS

PVNOV® thermal model shows a global good agreement with measurements for several types of modules. Nevertheless, it still needs to be improved based on the data collected from the presented tests revealing the impact of the module design on the thermal behavior of the system.

The cell temperature estimation from back module temperature should undoubtedly consider wind speed, module technology, in addition to irradiation to ensure more robust results. Moreover, the measurements have underlined a temperature mismatch on the rear side of the module exhibiting the necessity of using several thermal probes for module temperature measurement for accurate assessment.

This noticed mismatch phenomenon is undoubtedly more pronounced on large power plants and need to be deeply studied. Needless to say, a part of this mismatch is already estimated in the electrical mismatch caused by irradiation discrepancy. Nonetheless, it may also occur owing to the difference in thermal fluxes caused by the wind speed or direction and further tests will be performed in order to estimate its impact on the module behavior, the string and of the PV plant performance.

7 ACKNOWLEDGMENTS

The authors are very grateful to the EDF R&D laboratory team for the help provided in performing module testing.

8 BIBLIOGRAPHIE

[1] A. Lindsay, M. Chiodetti, P. Dupeyrat, D. Binesti, E. Lutun, K. Radouane, Proceedings of the 31st European Photovoltaic Solar Energy Conference, Hamburg, Germany (2015) 1764-1769

[2] A. Lindsay, M. Chiodetti, D. Binesti, S. Mousel, E. Lutun, K. Radouane, J. Christopherson, Proceedings of the 32nd European Photovoltaic Solar Energy Conference, Munich, Germany (2016) 1610-1617

[3] M. Chiodetti, A. Lindsay, P. Dupeyrat, D. Binesti, E. Lutun, K. Radouane, S. Mousel, Proceedings of the 32nd European Photovoltaic Solar Energy Conference, Munich, Germany (2016) 1449-1455

[4] Bifacial Photovoltaic mdules and systems: Experience and results from international research and pilot applications 2021, Report IEA-PVPS T13-4:2021 pp.78-109

[5] M. Chiodetti, J. Kang, C. Reise, A. Lindsay, Proceedings of the 35th EUPVSEC (2018), Brussels, Belgium, 1623-1629

[6] M. Dempsey, "Dymola for Multi-Engineering Modelling and Simulation", Proceedings 2006 IEEE Vehicle Power and Propulsion Conference, pp. 1-6, 2006.

[7] S. Armstrong and W.G. Hurley, A thermal model for photovoltaic panels under varying atmospheric conditions, Applied Thermal Engineering.

[8] J.A. Palyvos, A survey of wind convection coefficient correlations for building envelope energy systems' modeling, Applied Thermal Engineering.

[9] R.C. Lessmann, A. Johary, Heat transfer during wind flow over rectangular bodies in the natural environment, F.L. Test. Journal of Heat Transfer.

[10] A.D. Jones, C.P. Underwood, A thermal model for photovoltaic systems, Solar Energy

[11] King et al. « Photovoltaic Array Performance Model

[12] K. Nishioka, K. Miyamura, Y. Ota, M. Akitomi, Y. Chiba, A. Masuda, (2018). Japanese Journal of Applied Physics, 57(8S3), 08RG08

[13]M.Mittag,L.Vogt,C.Herzog,A.Pfreundt,J.Shahhid,D. Holger,H.Wirth, Thermal modelling of photovoltaic modules in operation and production.

[14] W.Gambogi, M.Demko, S.MacMaster, B-L.Yu, A Borne, H.Hu, Z.Pan, Assessing the reliability of transparent polymeric backsheets for durable bifacial modules,

MODULE QUALITY CONTROL USING VERY LOW-COST MOBILE NEAR-INFRARED SPECTROSCOPY DEVICE: APPLICATION TO BACKSHEET TYPE IDENTIFICATION

Julien Dupuis[1], Christine Abdel-Nour[1], Romain Bodeux[2], Paul Lefillastre[3], Axel Becker[3]
[1]EDF R&D, EDF Lab Les Renardières, Avenue des Renardières, 77250 Moret Loing et Orvanne, France
[2]EDF R&D – IPVF, 18 boulevard Thomas Gobert, 91190 PALAISEAU, France
[3]EDF Renouvelables, 100 esplanade du Général de Gaulle, Cœur Défense - Tour B, 92932 Paris La Défense Cedex

ABSTRACT: This article addresses the problematic of having a cost-effective material quality control for photovoltaic modules with a focus on backsheet. A low cost near infrared (NIR) spectroscopic equipment is used to determine the backsheet type present on modules with unknown material. An initial calibration is carried out using RAMAN spectroscopy to associate NIR spectra signature with five different backsheet types. RAMAN equipment allows to identify the global backsheet composition, but some uncertainties can remain linked to modified backsheet composition with similar spectrum form. The depth of analysis is also limited: the external and core layers are accessible, but the inner layer remains almost out of range. Therefore, we can't always be completely sure of the exact backsheet composition. Destructive analysis using physical or chemical identification may be required to detect all elements unambiguously if needed. After this calibration, the NIR equipment was able to identify the main backsheet types from eight out of ten modules with unknown backsheet in a few seconds. The two remaining have a high gloss which seems to have disrupted the signal giving non reproducible spectra. Also, it has to be noted that some infrared signatures seem to be sufficiently unique to detect unambiguously the core and outer layer composition (polyamide in AAA backsheet in particular) but some other ones have very similar spectrum and the device may not be always able to identify them correctly (PVDF/PET/PVDF ("KPK") backsheets in particular with EVA/PET/PET ("PPE")). More module types are needed to validate the method robustness but the already obtained results are quite impressive regarding the simplicity of such a tool. In addition, the NIR spectrometer software allows to share the reference infrared spectra to other users. This can be used on existing and new power plants to test module's backsheet type with the same references. Further works on NIR measurements using other backsheet types or dirty and aged modules will also be carried out to validate the possibility of using these devices on modules from the field or aged artificially. Testing is also in progress on the module junction box composition and the silicon glue used for fixing it to the rear side.

Keywords: Near-Infrared Spectroscopy, PV module backsheet, calibration, infrared signatures, non-destructive tools.

1 INTRODUCTION

Polymers are one of the main components impacting the lifetime and performance of photovoltaic modules. Controlling the exact polymers types within incoming modules for PV plants under construction is relevant as some manufacturers might use polymers different from the ones agreed upon during the module's order.

Moreover, it is known that polymers in PV modules are one of the main components impacting the lifetime and performance of photovoltaic modules when degrading. Some reliability issues have also been reported on polyamide backsheet (AAA) [1], controlling existing modules where the backsheet type was not mentioned can help in anticipating possible future issues.

Figure 1: RAMAN Spectroscopy characterization [2]

Spectroscopic measurements are very powerful non-destructive tools to evaluate the type and state of polymers within photovoltaic modules but are often too expensive to be deployed such as RAMAN spectroscspy (Figure 1) on all PV power plants under construction.

2 AIM AND APPROACH

The objective of this work is to have a robust and non-destructive material quality control solution for PV power plants at reasonable time and cost.

For that purpose, a low cost near infrared (NIR) spectroscopic equipment (Figure 2) is used to determine the backsheet type present on modules with unknown material

This paper is focusing on backsheet properties, but the same approach is possible for other parts of the module containing polymers (junction-box material, silicon glue, encapsulant…).

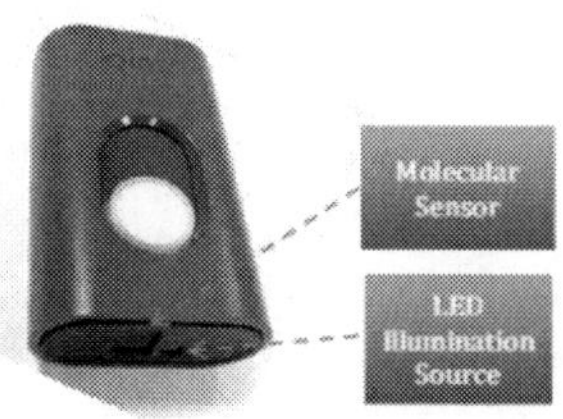

Figure 2: Low cost near infrared spectroscopic equipment

3 METHODOLOGY

3.1 Experimental setup

Fifteen silicon photovoltaic modules using white backsheet (excluding glass) from various manufacturers and manufacturing dates were tested using stored modules at EDF R&D (noted from A to N). The objective was to identify the main types of backsheets on modules where the BOM was not available.

RAMAN measurements were done using two types of equipment:

- A Horiba LabRAM HR RAMAN spectrometer using 532nm laser, 100µm spot with a x50 enlargement (wavelength from 200 to 3500cm-1).
- A Bruker BRAVO RAMAN spectrometer using two lasers (785 an 852nm) (wavelength from 300 to 3200cm-1).

The low-cost mobile NIR device tested here is the SCiO from Consumer physics (Figure 2), already evaluated among other NIR spectrometers mainly in food related studies [3], [4], [5]. The choice of evaluating such low-cost device is linked to the important number of existing PV power plants and PV power plants under construction. As budget for standard mobile RAMAN or NIR equipment is usually several thousand euros, low-cost sensors are of high interest.

The NIR device is small and simple to use: a user-friendly graphical interface is used to perform the measurements and process the collected data using a mobile phone with Android or Apple application. The analyzed spectrum covers the 700 to 1100nm wavelength range but the device, aiming mainly to introduce spectroscopy application, can only plot the absorption in function of wavelength without indicating axis units and ranges ("spectral fingerprint") as it can be seen on the Figure 5).

NIR measurements were first performed on five modules with different backsheets types using previously measured RAMAN modules as references to record the known backsheets signatures. Then, RAMAN and NIR measurements were both done on ten other modules to compare the results.

In order to use the device, the first steps have consisted in recording the signature of known backsheet with initial RAMAN measurements in the NIR mobile application. To do that, the NIR spectrometer is measuring several times the reference backsheet (10 times) to get accurate reference spectra. The Figure 3 is showing 5 types of recorded backsheet signatures associated to different types of backsheet. Ten measurements were recorded for the reference spectrum to get an accurate backsheet signature using different parts of the backsheet. We can notice some "spread" of the reference spectrum due to the multiple measurements performed.

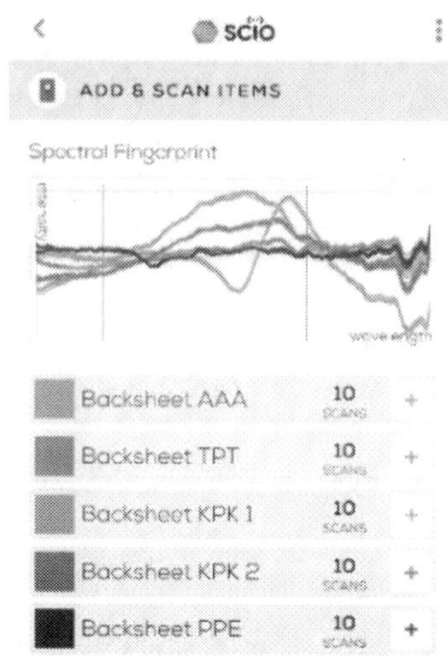

Figure 3: Mobile software interface with 5 types of recorded backsheet signatures

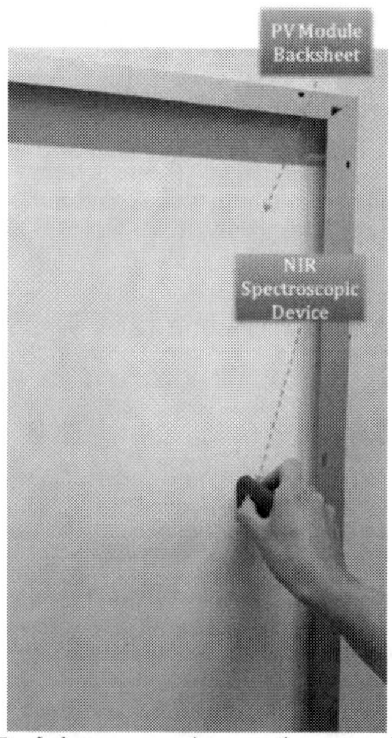

Figure 4: Backsheet scanning and spectral data processing

After the calibration and characterization procedures, the backsheet scanning is performed on 15 PV modules (Figure 4). The scanning results are processed and analysed through the NIR mobile application.

4 RESULTS AND DISCUSSIONS

4.1 Results

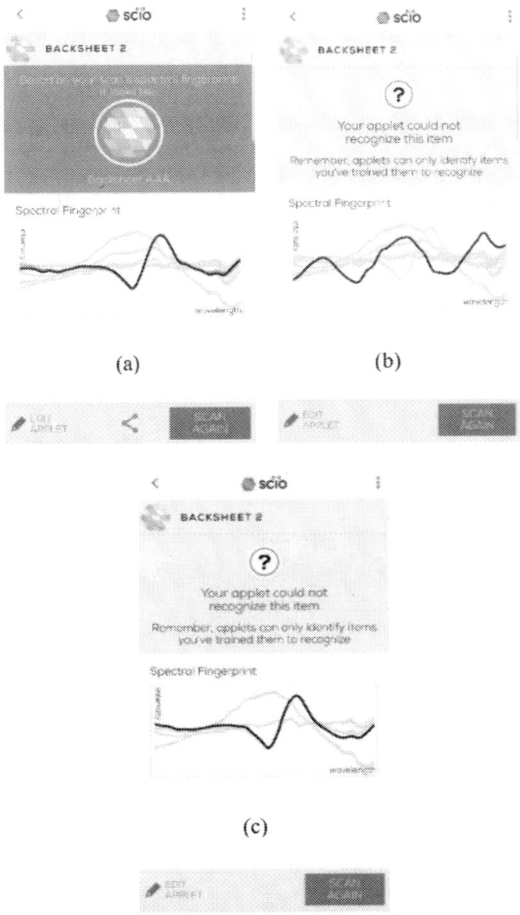

(a) (b) (c)

Figure 5: Infrared spectral signature identification:
a) Backsheet AAA type, b) and c) Ununified spectral signature

Figure 5 shows the infrared spectral signature identification for 3 scanned samples on the NIR mobile application. Figure 5a represents infrared signature identification of the "N" module with AAA backsheet type. In Figure 5b, the infrared signature of the "H" module is shown: the high reflectivity of the backsheet is preventing the NIR device to have a reproductible spectrum signature.

Even with a RAMAN spectrometer, it is not always possible to be fully accurate on the material detection as RAMAN spectra may present very similar forms for some kinds of materials, especially if the module is not destroyed (example: TPT and TPE which differ only by their inner layer, too deep to be characterized by the RAMAN spectroscopy). Moreover, generally mentioned backsheet types may differ in term of composition depending on the backsheet manufacturer and some elements detected may not clearly be associated with a regular backsheet type, leaving some identification uncertainties (like KPK backsheet in this study). Similar RAMAN results were found using lab or mobile RAMAN device with sufficiently accurate spectra for our application. The main drawback is still the depth of analysis limited to the outer layer and a part of the core layer as the device is used in a non-destructive way.

The Table I summarizes the results and identified backsheet types.

Table I: Backsheet type according to RAMAN and NIR devices measurements. The * indicate the use of RAMAN measurements to calibrate the NIR spectral signature

Modules	Detected elements (RAMAN)	Potential backsheet type (RAMAN)	Potential backsheet type (NIR)
A	EVA, PE, PA, TiO2	AAA	AAA*
B	PVF, PET and/or PE, PVF, TiO2	Tedlar© based (TPT or TPE)	TPT or TPE*
C	PVDF, PET, TiO2	Kynar© based (KPK or KPE)	KPK or KPE*
D	PVDF, PET, TiO2	KPK or KPE	KPK or KPE*
E	PET, TiO2	PPE	PPE*
F, N	PA, TiO2	AAA	AAA (based on "A")
J, L, M, O	PVDF, TiO2, PET	KPK	KPK (based on "C")
H, K	PVDF, PET, TiO2	KPK	Not identified (high backsheet reflectivity)
G, I	PVF, TiO2, PET or PE	TPT or TPE	TPT and/or TPE (based on "B")

4.2 Discussion

As shown in Table I, nearly all the modules were able to give as accurate results with the NIR spectrometer measurements compared to RAMAN characterisation except two modules (H and K) where the spectral signature was not reproducible (Figure 5b). We suppose that the high gloss of these backsheets led to signal disruption and prevented its identification.

All the other characterizations performed were successful but not always with the automatic detection system. Even if the signature obtained is similar to one of the references recorded, the system may not recognise it. In particular AAA backsheet and KPK backsheet were not always recognized (Figure 5c) but the spectral signature is sufficiently typical to be directly identified by the user.

This identification can be more difficult for PPE and TPT backsheets as their signatures are very alike but surprisingly, the NIR device works well on the measured modules and the results obtained were in line with RAMAN ones.

5 CONCLUSION AND PERSPECTIVES

A nondestructive method for a robust quality control of PV power plants was presented in this work. The main advantage of this method is a reasonable time and cost solution without having to transport the PV panels to the laboratory in order to test them.

Similar RAMAN results were found using lab or mobile RAMAN device with sufficiently accurate spectra for our application.

Backsheet types may differ in term of composition depending on the manufacturer and some elements detected may not clearly be associated with a regular backsheet type, leaving some identification uncertainties.

The backsheet database created in the NIR spectrometer software was successfully shared through the application allowing future users in worldwide PV plants to access to the same calibrated spectral signature. As no graphical scales or software solution to export the measured spectra in the basic software exists, further analysis is hardly possible and requires advanced systems (like RAMAN, FTIR, or less basic NIR spectrometer) to analyse and identify detected peaks in some cases.

Nevertheless, the presented NIR spectrometer device can potentially be used as a basic device to identify known materials.

Further work is in progress to verify if this low-cost solution can be applied to other photovoltaic module backsheets or materials types. Moreover, this device represents a potential solution to detect PV modules degradation mechanisms.

5 ACKNOWLEDGMENTS

The authors are very grateful to the EDF R&D and IPVF laboratory teams for the help provided in performing all the experimental work.

6 REFERENCES

[1] Y. Lyu et al., Prog Photovolt Res Appl. 28 (2020) 704–716

[2] https://blue-scientific.com/photovoltaics-raman-spectroscopy

[3] A. Kartakoullis et al., Food Chemistry 278 (2019) 314-321

[4] C. McVey et al., Foods 10 (2021) 956-971

[5] S. Mayr et al., Talanta 223 (2021) 121809-121816

Performance evaluation and comparison of solar cell technologies based on historical data

Yannis Pantazis[1], Evangelia Kalligiannaki[1], Yiannis Kamarianakis[1], Konstantinos Kotsovos[2], Issam Gereige[2], Marwan Abdullah[2], Athanasios Tzavaras[3], Theodoros Katsaounis[1,4]

[1]FORTH, IACM, Heraklion, Greece, [2]ARAMCO, Renewable Energy, Carbon Management Division, Thuwal, Saudi Arabia, [3]KAUST, CEMSE, Thuwal, Saudi Arabia,
[4]University of Crete, Dept. of Mathematics & Applied Mathematics, Heraklion, Greece

ABSTRACT: In the current work we study the performance of three photovoltaic (PV) cell technologies under real operating conditions and five years of observations. The measurements are in Saudi Arabia, a challenging place for solar panels' performance as they often operate at temperatures far beyond the Standard Test Conditions (STC). We perform statistical comparisons between the three solar cell technologies (Back-Contact, Hetero-Junction, and Aluminum Back Surface Field) and estimate each PV cell's efficiency. Towards this purpose, we employ statistical and machine learning tools. We use two approaches for cell technology comparisons. The first is non-parametric and depends only on the energy yield output. The second approach is parametric, and a linear model between the energy yield and the irradiance is learned, with stationary and time-varying coefficients. The energy output differences between the cell technologies were small but statistically significant. We conclude that the highest average power output is given by the BC technology, followed by the HJT and the AlBSF technologies. Additionally, we quantify the deterioration of the cell performance over time and the variations due to seasonal factors.

1 INTRODUCTION

The great majority of Saudi Arabia's landscape is located inside the sun belt resulting in a significant opportunity for renewable energy from solar panels [1,2]. However, the development of solar plants is quite limited, with the contribution to the national energy mix being about 0.5% for 2020. Given the high potential and the future need for renewable energy, this study aims to highlight the performance differences between various PV cell technologies located in Saudi Arabia. We furthermore elucidate their behavior over a long period of time under the challenging weather conditions of Saudi Arabia, such as the high operating temperatures and dust storms.

Specifically, the technologies we compare are the Aluminum Back Surface Field (AlBSF), the Hetero-Junction (HJT), and the Back-Contact (BC). AlBSF is one of the first solar cell architectures used in the PV industry. The BSF technology is used primarily for reducing the surface recombination velocity, thus increasing solar cells' performance. HJT solar cells combine two different technologies into one cell: a crystalline silicon (monocrystalline or polycrystalline) cell sandwiched between two layers of amorphous silicon. The main idea behind the BC solar cells is to move all or almost all of the front contact grids to the rear of the device. This results in larger surfaces exposed to light and reduced shading effects, thus increasing the efficiency of the solar cell [2,3].

In work [4], authors compare the cell architectures under realistic conditions based on a combination of numerical simulations and statistical correction. In the present work, we employ machine learning and statistical tools [5, 6] to explore further cell technologies' performance based solely on observations. results and conclusions rely solely on the observations. We first, in sec. 2, present the data and the pre-processing procedure, then in sec. 3, we discuss the methodology, in sec. 4, we present and discuss the results, and we conclude in sec. 5.

2 HISTORICAL DATA LOCATION AND PRE-PROCESSING

The location of the PV solar cells for which historical observations are collected is the New Energy Oasis (NEO) test field near the Red Sea coast (22.30 N, 39.10 E), KAUST, Thuwal, Saudi Arabia. The available historical data consists of irradiance (IR) and energy yield (EY) observations from the three cell technologies (AlBSF, HJT & BC) for the 2015-2019 calendar years. For each day, an observation is recorded every 10 minutes, starting at 6:00 and ending at 18:50. Thus, the maximum number of daily observations, at times $t_n > 0$, $n = 1,...,T$, is $T = 78$; see, e.g., Fig. 1 for one day of observations.

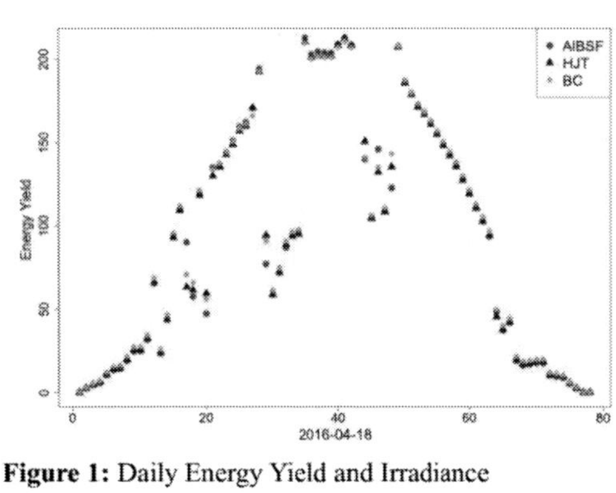

Figure 1: Daily Energy Yield and Irradiance

We initially performed an exploratory data analysis and visualization. We identified a small percentage of the records which did not follow the overall sample distributions. Irradiance and EY measurements had few problematic values, which we consider outliers, and we applied all the necessary data cleaning and preprocessing steps. Moreover, we further reduce the noise in the data by excluding measurements with an EY value below *10 W/m²*.

3 METHODOLOGY

We employ two approaches for comparing the three solar cell technologies. The first approach is a straightforward calculation of the EY difference between two technologies measured at the same time points. The second one is parametric in the sense that it incorporates the estimation of the efficiency coefficient for each architecture.

3.1 Direct, non-parametric comparisons

The first approach is a straightforward calculation of the EY difference between two technologies measured at the same time points. This "synchronous" comparison is required because the variability in the EY values stemming from weather or maintenance factors or simply the existence of missing data could affect the statistics of the quantity of interest.

The quantity of interest is the *relative performance difference* between architectures *A* and *B*, defined by

$$d_{AB}(t_n) = \frac{EY_A(t_n) - EY_B(t_n)}{1/2[EY_A(t_n) + EY_B(t_n)]} \quad (1)$$

for instant times $t_n > 0$, n=1,2,... .

Here A (and B) denotes the type of the architecture being AlBSF, HJT, or BC. We study the relative performance d_{AB} statistics on a monthly and annually basis and for all the available observations.

3.2 Indirect, parametric comparisons

The parametric approach is defined via a linear model

$$EY_A(t_n) = c_A IR(t_n) + z(t_n), \; t > 0 \quad (2)$$

where c_A is the slope and $z(t)$ the residual error.

The linear model is motivated by the estimated correlation coefficient between irradiance and EY is above 92%, even when we consider all records.

The linear irradiance model (2) has only one parameter, the slope c_A, and it is directly interpretable as the efficiency of the solar cell. Given that the relationship between irradiance and energy yield is not stationary for the whole five-year period, we slice the data monthly. Then, we compute the slope using two estimation methods: Ordinary Least Squares (OLS) and Robust Least Squares (RLS) with Huber weights [5].

We then compare the estimated slopes and utilize statistical testing to assess the significance of the differences.

4 RESULTS AND DISCUSSION

The EY differences between cell technologies were small but distinctive. The BC technology achieved the highest average power output, followed by the HJT and the AlBSF technologies. We next report the results supporting this conclusion while we quantify the differences. Also, we present the relative performance results for non-parametric (NP) and parametric (P) approaches.

4.1 Results on the non-parametric comparisons

Tables I and II report the average relative performance gain (or loss), $E[d_{AB}]$, between the three solar cell technology pairs per year and in total, respectively. We observe that the BC outperforms the other two technologies for each year from 2015 to 2019 and, on average for 2015-2019. Moreover, the relative performance difference between the BC and AlBSF has increased over the years. Similar is the behavior for the BC and HJT, except for 2018. Table II also verifies that HJT has better performance than AlBSF.

Table I: Average relative performance difference, $E[d_{AB}]$, between the three cell technologies. Positive values imply that the first cell technology is better.

Year	2015	2016	2017	2018	2019
BC vs AlBSF	0.5	4.4	15.6	13.2	18.8
BC vs HJT	1.5	2.4	5.4	3.0	4.8
HJT vs AlBSF	-1.0	2.1	12.4	-	-

Table II: Relative performance difference for 2015 - 2019

d_{AB} 100% / *A* vs *B*	$E[d_{AB}]$	95% C.I.
BC vs AlBSF	5.84	(5.75, 5.94)
BC vs HJT	3.11	(3.06, 3.16)
HJT vs AlBSF	2.53	(2.42, 2.64)

Figs. 1 and 2 compare the EY average value for the BC vs. the HJT and AlBSF technologies yearly and monthly. The BC technology achieved the highest average power output for all years and months. Similar is the behavior for the HJT and AlBSF pair, which we do not report here for ease of presentation.

The mean average difference of deviations for the EY output is 0.9 W/m² for the BC vs. HJT and 5.1 W/m² for the BC vs. AlBSF. These measurements ensure that the HJT performance is closer to the BC than the AlBSF.

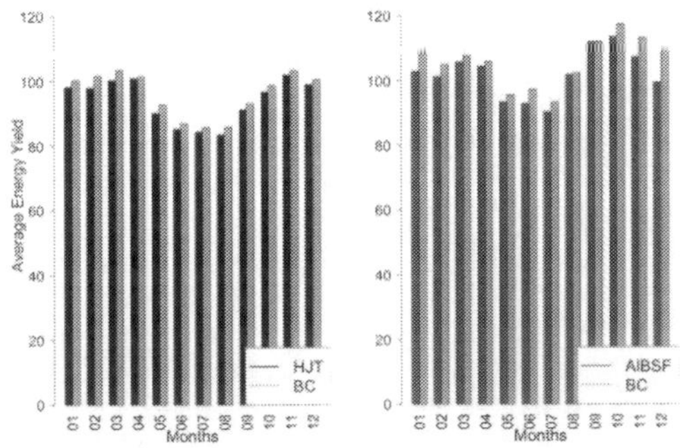

Figure 2: BC vs. AlBSF, monthly average EY.

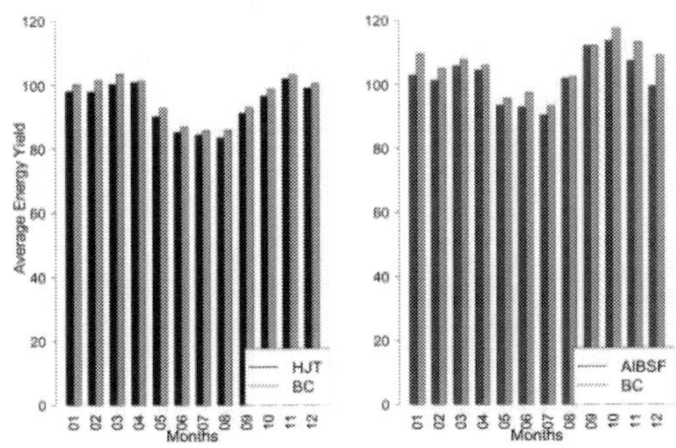

Figure 3: BC vs. HJT and AlBSF, yearly average EY.

4.2 Results of the parametric approach

In this section, we present the results of the parametric approach; see section 3.2. We note that both OLS and RLS estimators provided similar results; thus, we only present the results with OLS. Table III depicts the relative performance difference corresponding to the parametric approach. Comparing the results reported in Tables I and III, we observe that both approaches report qualitatively similar results. However, the parametric indirect approach is more conservative than the direct non-parametric approach. One possible explanation for this difference is that the parametric approach puts more weight on measurements with high irradiance values, while the non-parametric approach treats all measurements equally. Given that more power output is produced when irradiance is higher, the parametric approach is expected to provide more consistent comparisons. Despite being indecisive for 2015, it is evident that BC technology outperforms the other two for the subsequent calendar years.

Table III: Parametric approach. The relative performance between the three cell technologies.

Year / $E[d_{AB}]$	2015	2016	2017	2018	2019	All
BC - AlBSF	-1.8	1.9	12.8	9.0	18.3	5.5
BC - HJT	-0.4	0.6	4.8	1.0	3.5	2.1
HJT - AlBSF	-1.4	1.3	4.6	-	-	3.2

The upper plot of Fig. 3 shows the estimated slopes for the three cell technologies on a monthly basis. The time-varying behavior of the slope, hence, the solar cell efficiency, is evident. The lower panel of Fig. 3 shows the efficiency difference between pairs of solar cell technologies. This plot is a higher resolution presentation of the efficiency differences relative to Table III. These differences in performance are statistically significant as quantified by the p-values of the t-test reported in Table IV.

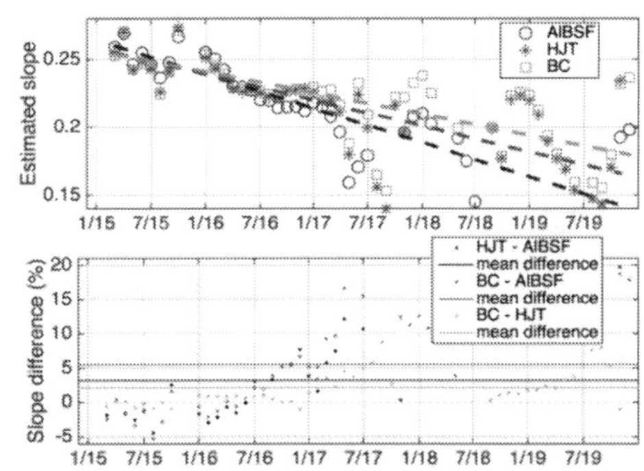

Figure 4: Upper panel: Monthly estimated slope for the three cell technologies. The slope is a direct estimator of cell efficiency. Thick dashed lines correspond to the linear trend of the slopes for each cell type and reveal the drop in efficiency over time. Lower panel: the efficiency difference for each pair (dots) and the respective average difference (solid lines).

Table IV: p-values for the t-test.

	BC–AlBSF	BC–HJT	HJT–AlBSF
p-value	0.000362	0.000008	0.027369

Moreover, it is evident from the upper panel of Fig. 3 that there is an annual performance deterioration in all cell technologies, which is different for each technology. We model this decay with a linear regression model, and the inferred model corresponds to the dashed lines in the upper panel of Fig. 3. For BC, the *average rate of deterioration* was 5.6%, for HJT, the rate was 6.9%, and for AlBSF, was 8.4%. To put these values into perspective, given that AlBSF and BC cells started with the same efficiency at the beginning, the efficiency of BC after 3 years will be the same as the efficiency of AlBSF after only 2 years!

Finally, we model the temporal evolution of the efficiency with a linear regression model that not only incorporates the decay trend but also the seasonal variation of the efficiency. It is obvious from Fig. 5 (dashed lines) that there are strong seasonal phenomena. Moreover, the efficiency is, on average, about 25% lower during the summer months relative to winter ones. This large difference shows the adversity of the weather conditions in desert environments which directly impact the production of solar energy.

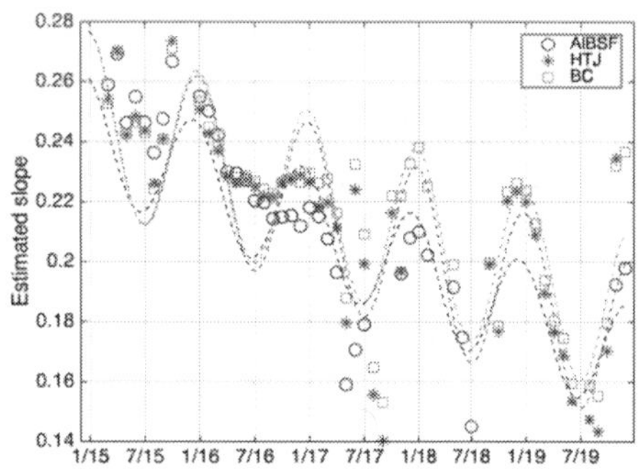

Figure 5: Same as the upper plot of Fig. 4 but with a regression model that considers the efficiency's seasonal trend.

5 CONCLUSIONS

BC was, on average, 5.5-5.8% more efficient than AlBSF and 2.1-3.1% more efficient than HJT. Interestingly, for 2015 the ranking is different (see Tables I and III). The deterioration over time is also evident: we estimated a 5.6-9% annual drop in power output efficiency depending on the cell technology. Data also revealed that the most dramatic performance difference was observed during seasonal alteration. The relative efficiency drop in summer relative to winter is up to 40%, eliminating the gains from increased irradiance. Therefore, informed decision-making for PV installation projects should take into special account the temperature coefficient of the PV cells. At the same time, mitigation measures such as cooling may be financially viable and worth considering.

We observe a temporal variation in the cell performance (the slopes). We attribute it to the heat; since the temperature in the afternoon (when the sun falls from its azimuth) is higher than in the morning (when the sun rises towards its azimuth).

6 ACKNOWLEDGMENTS

The authors acknowledge the support of KAUST and Saudi Aramco R&D Center - Carbon Management Division for their financial support in developing this work. This work was partially supported by grant #OSR–2020-4433.02 from KAUST and Saudi Aramco.

REFERENCES

[1] A.H. Almasoud and H.M. Gandayh, J of King Saud University - Engineering Sciences (2015) **27**, 153-157

[2] Al Khuffash, K., Lamont, L.A. & El Chaar, L. *Appl. Sol. Energy* 50, 215–220 (2014)

[3] S. S.Rashwan, A. M. Shaaban, F. Al-Suliman, Renewable and Sustainable Energy Reviews, 80, 313-318 (2017).

[4] Katsaounis, T., et al. Renewable Energy, 143, 1285–1298, (2019).

[5] P.J. Huber and E. Ronchetti, *Robust Statistics; 2nd Ed.*, John Wiley & Sons (2009) New Jersey, USA

[6] Casella, G. and Berger, R. L., Statistical Inference, Thomson Learning, (2002).

LINEARITY MEASUREMENTS OF PHOTOVOLTAIC MODULES WITH ATTENUATING IRRADIANCE FILTER TECHNIQUE

Bazkir Ozcan, Meric Seval
National Metrology Institute (TUBITAK-UME), P.O.Box.54, 41470 Kocaeli/Türkiye
ozcan.bazkir@tubitak.gov.tr, seval.meric@tubitak.gov.tr

ABSTRACT: For the accurate evaluations of energy rating, the electrical performance parameters obtained at STC need to be transferred to real conditions by the use of well determined linear equations defined in IEC 61829 standard. One of the most important requirements for this is the determination of the linearity of electrical performance parameters of PV modules at various irradiance levels.
In this work, the linearity measurements of a mono-crystalline, a polycrystalline, a CIGS thin film and a bifacial PV module were investigated using a xenon light based solar simulator system with irradiance attenuating filters. To measure the linearity of PV modules accurately a comprehensive analysis of all factors that may affect the linearity were done. In this context, initially the solar simulator was optically characterized according to the IEC 60904-9 standard requirements in terms of spectral match, spatial non-uniformity and temporal instability. Next, an effect of irradiance attenuating filters on non-uniformities of simulator was determined. Then, PV modules were electrically stabilized to remove any fluctuations in their electrical performance parameters. Finally, the electrical performance parameters of PV modules were measured at a range of irradiance levels from 100 W·m^{-2} to 1000 W·m^{-2} and their linearity were determined using the least-squares fit method.
Keywords: Attenuating Filters Linearity, Photovoltaic Modules, Responsivity, Solar Simulator

1 INTRODUCTION

In general, the electrical performance parameters of PV devices (short-circuit current, I_{sc}; open-circuit voltage, V_{oc}; maximum power, P_{max}) are determined at Standard Test Conditions (STC) so as to ensure their standardization. In IEC TS 61836 standard, the STC is defined as 1000 W·m^{-2} irradiance, 25 °C device temperature and AM1.5 spectral irradiance distribution [1]. However, in real applications the PV devices can be subjected to various irradiance levels, typically from few mW·m^{-2} to more than 1100 W·m^{-2} which are different from STC conditions irradiance level (1000 W·m^{-2}). Therefore, determination of wide range of irradiance conditions on the electrical performance parameters of PV devices is one of the most important requirements for the accurate performance analyses. In such conditions, in order to get information about the performances of PV devices at different radiation levels, the knowledge of the linearity of PV devices is used. The linearity of the electrical performance parameters of PV devices can be derived from the measurements carried out at a range of irradiance levels and at a fixed temperature.

There have been many indoor and outdoor works on linearity measurements of PV devices [2, 3]. For the measurements in both indoor and outdoor, there are important conditions that must be met so as to do measurements.

In this work, to perform the linearity measurements indoor measurement technique was used. For the indoor measurements, the solar simulator is used to irradiate the PV devices and measure their electrical performance parameters. According to the IEC 60904-9 standard the solar simulator to be used for this purpose has to fulfill some requirements like; spectral match, spatial non-uniformity of irradiance and temporal instability [4].

In solar simulators the range of irradiance levels needed for the linearity measurements can be provided by using attenuation masks or mesh irradiance adjusting equipments. However, these irradiance adjusting equipments in accordance with IEC 60904-10 standard must not disturb the spectral match, spatial non-uniformity of irradiance and temporal instability of the simulator system.

For this reason, in this work before the linearity measurements, optical characterizations of both the simulator and the irradiance adjusting equipments were carried out so as to check the requirements specified in the IEC 60904-9 IEC 60904-10 standards.

2 CHARACTERIZATION OF MEASUREMENT SYSTEM

2.1 Characterization and Calibration of Solar Simulator

The solar simulator system shown in Fig.1, used as the irradiance source for the linearity measurements of PV modules. Before the measurements, the solar simulator system was characterized in terms of spectral match, spatial non-uniformity of irradiance and temporal instability according to the IEC 60904-9 standard's requirements [4]. The measurement results as shown in Table I indicate that the solar simulator's spectral match, spatial non-uniformity and temporal instability according to the IEC 60904-9 standard's classification requirements corresponds to A+ A+ A+ classes. After the characterization, the solar simulator was calibrated using the reference PV cell (2 cm x 2 cm) which is traceable to PTB DSR measurement system. Since the reference PV cell was calibrated at STC conditions (1000 W·m^{-2}, 25 °C and AM1.5) the calibration of the solar simulator system was also performed at the same conditions [5]. To set the reference PV cell to 25 °C, the thermal chamber integrated with the simulator system was used. The temperature of thermal chamber was controlled such that the temperature non-uniformity of the less than ± 2 °C over the 2 m x 2 m area was obtained, which is almost two times better that the IEC 60891 standard requirements (± 2 °C).

Table I. Solar simulator class parameters

Parameters	TUBITAK UME Simulator		IEC 60904-9 Standard	
	Class	Value	Class	Value
Spectral match	A+	0,991 - 1,028	A+	0,875 - 1,125
Non-uniformity	A+	≤ % 1	A+	≤ % 1
Temporal instability	A+	≤ % 1	A+	≤ % 1

Figure 1. Thermal chamber integrated solar simulator system

2.2 Transmittance of Filters

In this work, solar simulator's irradiance levels from 800 W·m^{-2} to 1100 W·m^{-2} were adjusted using the solar simulator's control cell and from 100 W·m^{-2} to 800 W·m^{-2} were adjusted using totally eight filters with 30 cm x 30 cm dimensions. For the measurements they were placed to the filter holder present just in front of the light source of the simulator system. The important parameter to be considered when using these filters are the temperature-dependent transmittance of the filters and the homogeneity of the transmitted light from the filters.

The fan placed in front of the filter holder prevents the heat generated on the filters due to the chamber of the light source.

The filters used for the used for attenuation of irradiance for the linearity measurements shouldn't change the simulator system's spatial non-uniformity of irradiance. To control this before the linearity measurements the spatial non-uniformities of the solar simulator system's light source with filter were measured. For this, using specially designed 16 cm x 16 cm PV cell an area of 2 m x 2 m was scanned and current at 64 points were measured (Fig. 2). From the recorded values and using the Eqn1 the spatial non uniformities were evaluated for each filter. The measurement results as shown in Table II indicates filters used for the attenuation of irradiance does not change the solar simulator's spatial non-uniformity.

$$Nonunif. = 100 * \left[\frac{(Max\ Irrad - Min\ Irrad)}{(Max\ Irrad + Min\ Irrad)}\right] \quad (1)$$

Table II. Spatial non-uniformity of solar simulator with filters

Simulator with attenuation filters	Irradiance non-uniformity, (%)
Simulator + F_{100} at 100 W·m^{-2}	0,57
Simulator +F_{200} at 200 W·m^{-2}	0,59
Simulator +F_{300} at 300 W·m^{-2}	0,45
Simulator +F_{400} at 400 W·m^{-2}	0,56
Simulator +F_{500} at 500 W·m^{-2}	0,46
Simulator +F_{600} at 600 W·m^{-2}	0,47
Simulator +F_{700} at 700 W·m^{-2}	0,35
Simulator +F_{800} at 800 W·m^{-2}	0,39
Simulator +No filter at 900 W·m^{-2}	0,35
Simulator +No filter at 1000 W·m^{-2}	0,34

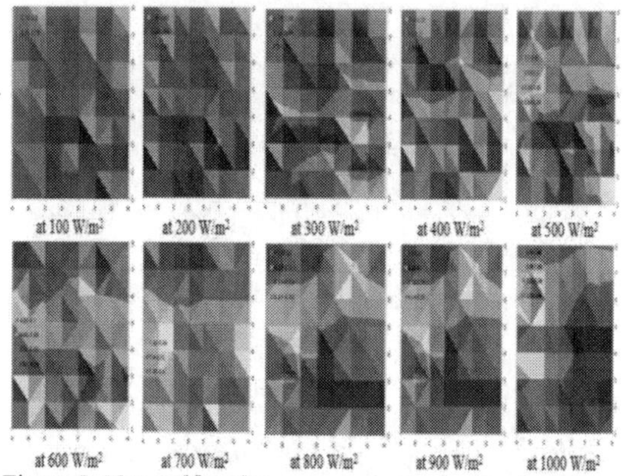

Figure 2. Non uniformity measurement of solar simulator as a function of irradiance

After the measurements of spatial non-uniformity of irradiance the total transmittance values of these filters were measured in the simulator system. In the measurements, firstly when the filters were not attached to the filter holder of the simulator, the solar simulator system was adjusted to give 1000 W·m^{-2} irradiance via measuring the short circuit current value produced by the PV device in the sample holder of the simulator. Then, the filters respectively were placed in the filter holder of the simulator and the short circuit current of the PV device were measured again. Taking the ratio of the measurement results obtained with and without filters the transmittance values were obtained (Table III).

Table III. Transmittance values of filters

Filter	Transmittance, (%)	Filter	Transmittance, (%)
F_{100}	10,25	F_{500}	51,35
F_{200}	20,50	F_{600}	62,00
F_{300}	30,84	F_{700}	72,56
F_{400}	41,09	F_{800}	82,76

3 LINEARITY MEASUREMENTS OF PV MODULES

In this work, the linearity measurements of a monocrystalline, a polycrystalline, a CIGS thin film and a bifacial PV modules were measured. For accurate linearity measurements all the factors that could affect the linearity has to be eliminated. Especially at low irradiance ranges any fluctuations in the electrical performance parameters of PV modules can hide their nonlinear behaviors. In order to reduce or to eliminate these fluctuations the PV modules under test were electrically stabilized according to the IEC 61215-2 standard requirements [6]. After each PV device reached the stability requirements of related standards the linearity measurements were performed. The stability and linearity measurements were performed as follows.

3.1 Stabilizations of PV Modules

Since the PV modules are made from semiconductor materials they can be electrically unstable due to possibility of defects in their internal structures. These devices especially at low irradiance ranges can have fluctuations in their electrical performance parameters which can hide the nonlinear behaviors. Therefore, in order to reduce or to eliminate these fluctuations, before the linearity measurements the PV modules under test were electrically stabilized according to the IEC 61215-2 standard requirements using the stabilization system shown in the Fig. 3. In this system, the irradiance level was set to 1000 $W \cdot m^{-2}$ and the PV modules were respectively exposed 10$kWh \cdot m^{-2}$, 5$kWh \cdot m^{-2}$ and 5$kWh \cdot m^{-2}$ irradiation doses and their maximum powers were measured. hen, the stabilities of PV modules were evaluated using $(P_{max}-P_{min})/P_{av}<x$ relation [7]. These irradiance exposure and maximum power measurement processes were repeated until the x values of PV modules defined in related technology specific parts of IEC 61215 is reached [8-11].

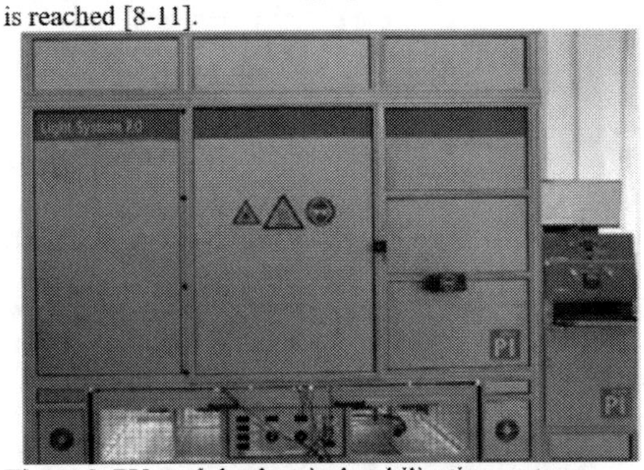
Figure 3. PV module electrical stabilization system

3.2 Temperature Non-uniformity of PV modules

In general PV modules constitute series combinations of PV cells. So, in order to get nearly same contribution from each cell to the module's performances, besides the homogeneous irradiation of all the cells the homogeneous distribution of the temperature is also critical parameter. Since the solar simulator used in this work for the irradiation of PV modules is a pulse type simulator with short pulse duration (10 ms) it has a quite low heating effect. However, the variations in the laboratory conditions may cause the temperature to exceed the limits (25 °C ± 2 °C) specified in STC. To keep modules at these limits during the measurements thermal chamber integrated to simulator was used. The thermal chamber both can heat /cool the modules from 15 °C to 75 °C and also at each temperature level can create a homogeneous temperature distribution over the entire surfaces modules. To record the temperature distribution of PV modules, totally eight (8) temperature sensors were mounted to their back sides, directly behind the cells. Each temperature sensor have ± 1 °C accuracy and ± 0.5 °C repeatability. Then, the temperature set to 25 °C and the temperature from each sensor was recorded to check the temperature nun-uniformity. As shown in Table IV the temperature non uniformity of less than ± 1 °C, over the module area was obtained, which is better that the IEC 60891 standard requirements (± 2 °C) [14].

Table IV. Temperature distribution over the whole back surface of PV module

	T1	T2	T3	T4	T5	T6	T7	T8	$\frac{(T_{max}-T_{min})}{(T_{max}+T_{min})}$
Mono. c-Si	25,00	25,05	25,00	24,90	24,95	25,06	24,04	25,01	0.004
Poly c-Si	24,90	25,05	25,02	24,95	25,00	24,95	24,98	25,02	0.003
CIGS	25,00	25,06	25,01	25,07	25,10	24,98	25,05	24,98	0.002
Bifacial	24,99	25,03	25,04	25,00	25,10	25,05	25,10	24,97	0.002

3.3 Linearity Measurements of PV Modules

After the complete characterizations of both the measurement system and the stabilization of PV modules under test (monocrystalline, a polycrystalline, a CIGS thin film and a bifacial PV) the linearity measurements were performed using solar simulator with irradiance attenuation filters. This method is based on measurements of short-circuit current with respect to incident irradiance. The incident irradiance values were calculated from the measured short circuit current (I_{sc}) of the reference cell, and its calibration value at STC (I_{rc}). The temperature correction also was applied to this irradiance value using the temperature coefficient of the reference cell α_{rc} [12].

$$G = \frac{1000 * I_{sc}}{I_{rc}} * [1 - \alpha_{rc}(T_m - 25)] \qquad (2)$$

The irradiance levels from 100 $W \cdot m^{-2}$ to 800 $W \cdot m^{-2}$ were adjusted using totally eight filters and from 800 $W \cdot m^{-2}$ to 1100 $W \cdot m^{-2}$ were adjusted using the solar simulator's control cell. The filters as described in sec B do not affect the irradiance distribution on the PV modules under the test. In order to eliminate the effect of temperature on the linearity measurements arise from the heating variation due to different incident irradiance values, the thermal chamber integrated to solar simulator system was used. In this system, the PV modules under the tests were respectively mounted to the Solar Simulator's module holder and PT-100 temperature sensors were attached to their back sides (Fig. 1). Then the temperature inside the thermal chamber was set to 25 °C. After the desired thermal stability was reached, the PV modules front surfaces at each irradiance levels were irradiated by flash type simulator and their electrical performance parameters were measured. From the short-circuit currents versus open circuit voltages (I_{sc}-V_{op}) and the maximum powers versus open circuit voltages (P_{max}-V_{oc}) graphs the irradiance dependent values of short circuit currents and the maximum power values for all the PV modules were obtained as given in Fig. 4. – Fig. 13.

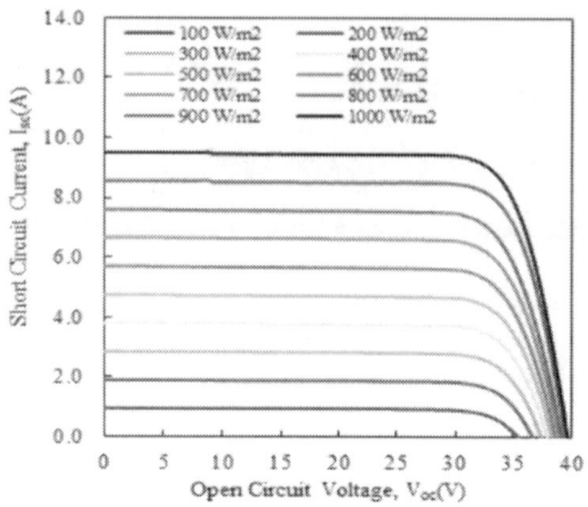

Figure 4. I_{sc}-V_{oc} curves of monocrystalline PV module

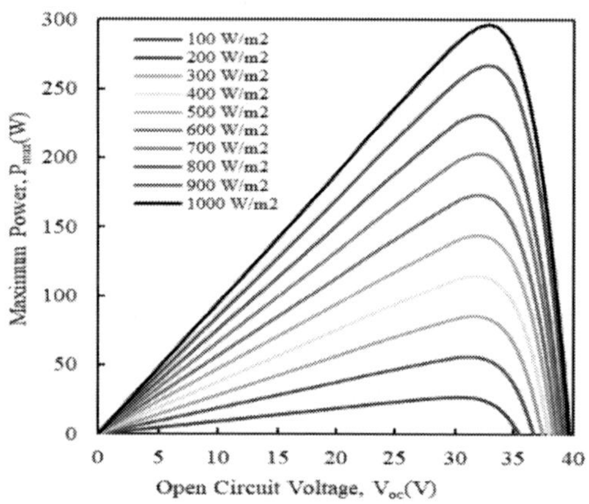

Figure 5. P_{max}-V_{oc} curves of monocrystalline PV module

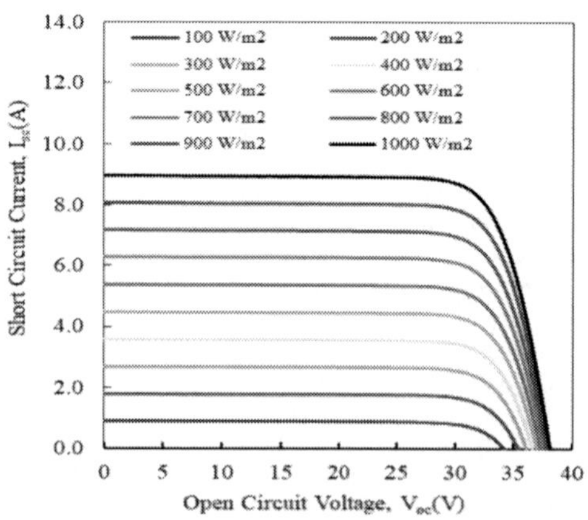

Figure 6. I_{sc}-V_{oc} curves of polycrystalline PV module

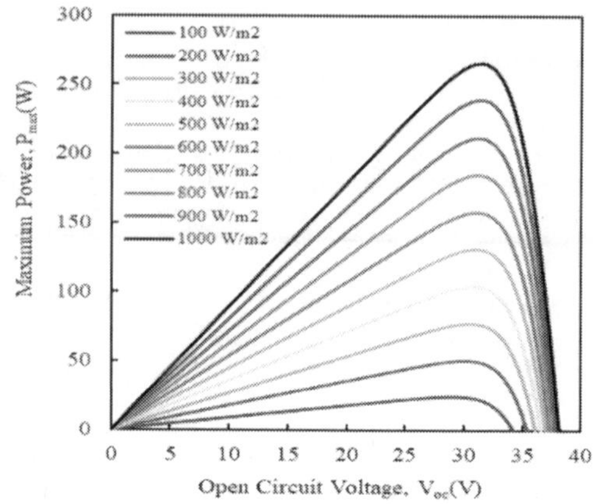

Figure 7. P_{max}-V_{oc} curves of polycrystalline PV module

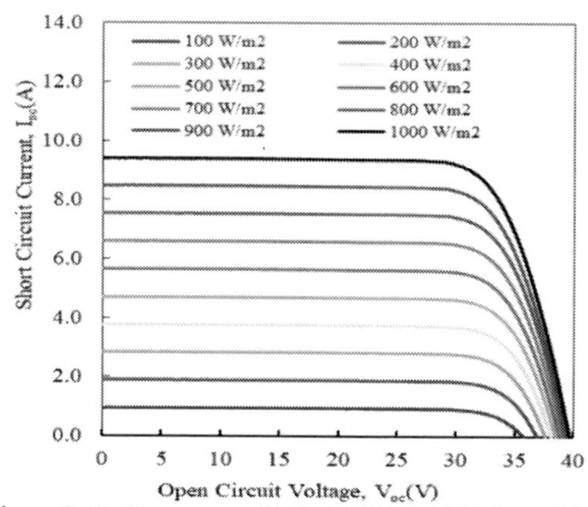

Figure 8. I_{sc}-V_{oc} curves of bifacial PV module front side

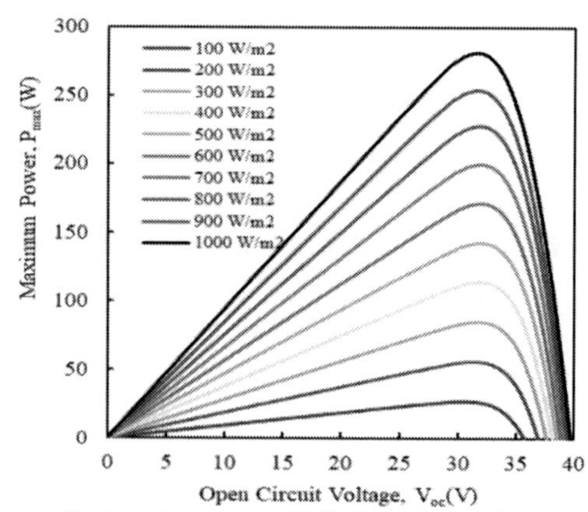

Figure 9. P_{max}-V_{oc} curves of bifacial PV module front side

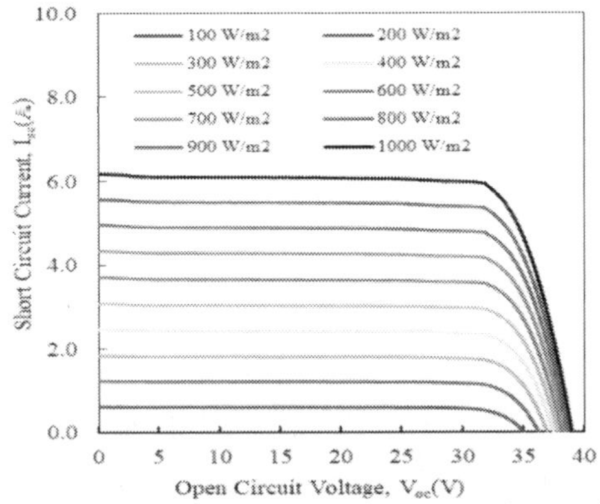

Figure 10. I_{sc}-V_{oc} curves of bifacial PV module rear side

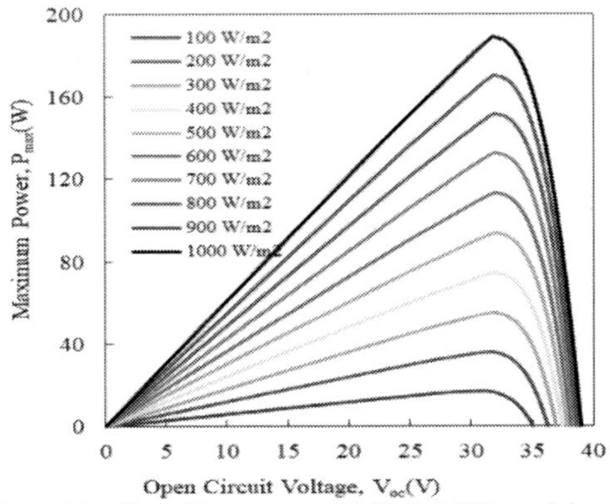

Figure 11. P_{max}-V_{oc} curves of bifacial PV module rear side

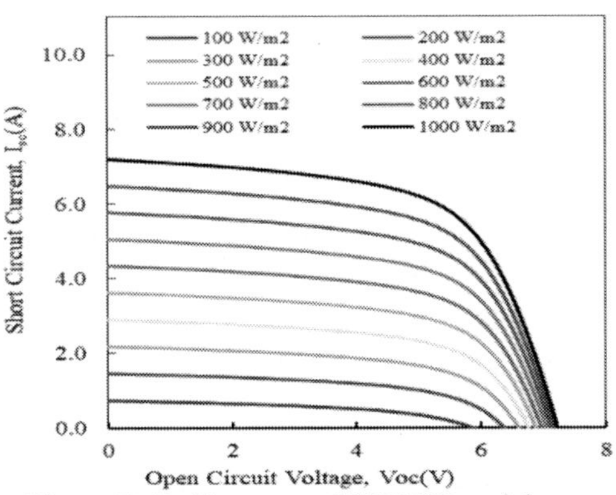

Figure 12. I_{sc}-V_{oc} curves of CIGS PV module

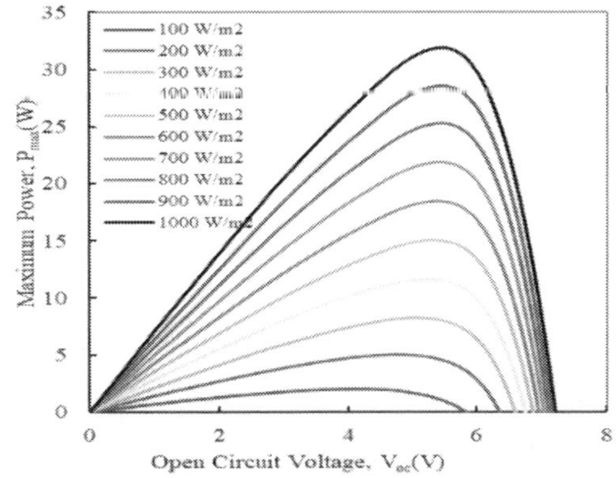

Figure 13. P_{max}-V_{oc} curves of CIGS PV module

From the short-circuit current versus open circuit voltage (I_{sc}-V_{oc}) and the maximum power versus open circuit voltage (P_{max}-V_{oc}) graphs the irradiance dependent values of short circuit current and the maximum power values for the PV modules were obtained as given in the Fig. 14. – Fig. 15.

The linearity of electrical performance parameters of PV modules were evaluated using the least-squares fit method as described in IEC 60904-10 standard was used [13].

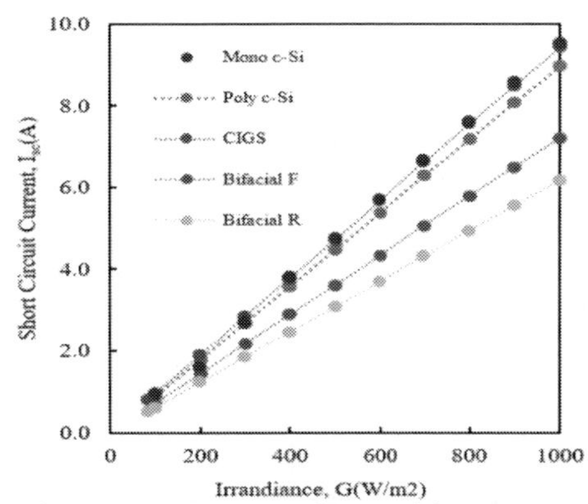

Figure 14. I_{sc}-G curves of PV modules

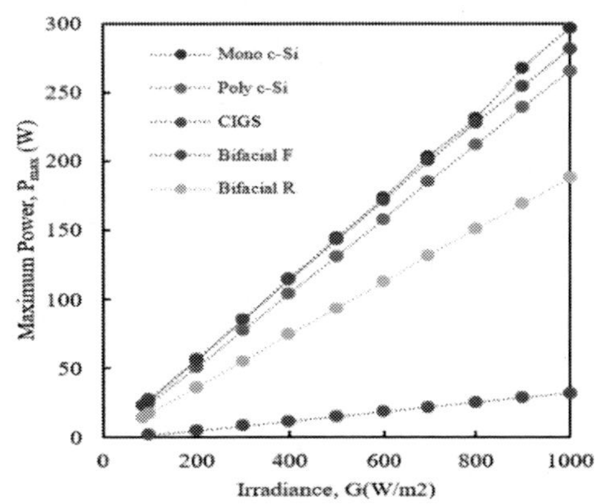

Figure 15. P_{max} -G curves of PV modules

4 RESULTS AND DISCUSSIONS

For the performance evaluations of PV modules at the range of irradiance levels (100 $W \cdot m^{-2}$ to 1000 $W \cdot m^{-2}$) in the field, the electrical performance parameters obtained at STC irradiance level (1000 $W \cdot m^{-2}$) need to be transferred to fields irradiance levels by the use of well determined linear equations defined in IEC 61829 standard. This requires the knowledge of the behaviors (linearity) of the electrical performance parameters of PV modules at various irradiance levels. In this work, the linearity of PV modules (mono-crystalline, a polycrystalline, a CIGS thin film and a bifacial PV module) were studied by analyzing the factors that may affect the linearity. That is, effects of simulator's parameters (spectral match, spatial non-uniformity and temporal instability), irradiance attenuating filters parameters (temperature-dependent transmittance and the homogeneity of the transmitted light), PV modules parameters (electrical instabilities and temperature nun uniformities) on the linearity of PV modules were studied. The behavior of the electrical performance parameters of PV modules were measured at a range of irradiance levels from 100 $W \cdot m^{-2}$ to 1000 $W \cdot m^{-2}$ and their linearity were determined using the least-squares fit method.

ACKNOWLEDGEMENT:

This work is supported by EMPIR Metro PV project

REFERENCES

[1] PD IEC/TS 61836, Solar photovoltaic energy systems – Terms, definitions and symbols, p.41, 2016
[2] K. George, E.Timothy, K. Ingo, C.Blakesly James, "High-resolution linearity measurements of photovoltaic devices using digital light processing projection", Meas. Sci. Technol. 32 055901 (8p), 2021
[3] S. Markus, S. Michalski, J. Ulrike, H. Werner, R. Uwe, "Non-linearity of temperature coefficients, equivalent cell temperature and Temperature behavior of different pv-module technologies", 28th European Photovoltaic Solar Energy Conference and Exhibition,pp.3265-3268, 2014
[4] BS EN IEC 60904-9:2020, Photovoltaic devices-Part 9: Solar Simulator Performance Requirements, pp.9-21, 2020
[5] BS EN 60891:2010, Photovoltaic devices - Procedures for temperature and irradiance corrections measured I-V characteristics, P.13, 2010
[6] IEC 61215-2, Terrestrial photovoltaic (PV) modules - Design qualification and type approval - Part 2: Test procedures, pp.44-46,2021
[7] IEC 61215-1, Terrestrial Photovoltaic (PV) Modules - Design Qualification And Type Approval - Part 1: Test Requirements, pp.20-23,2021
[8] IEC 61215-1-1, Terrestrial photovoltaic (PV) modules - Design qualification and type approval - Part 1-1: Special requirements for testing of crystalline silicon photovoltaic (PV) Modules, p.7,2021
[9] IEC 61215-1-2,Terrestrial photovoltaic (PV) modules - Design qualification and type approval - Part 1-2: Special requirements for testing of thinfilm Cadmium Telluride (CdTe) based photovoltaic (PV) modules, p.7, 2021
[10] IEC 61215-1-3,Terrestrial photovoltaic (PV) modules - Design qualification and type approval - Part 1-3: Special requirements for testing of thinfilm amorphous silicon based photovoltaic (PV) modules, p.7, 2021
[11] IEC 61215-1-4, Terrestrial photovoltaic (PV) modules - Design qualification and type approval - Part 1-4: Special requirements for testing of thinfilm Cu(In,GA)(S,Se)2 based photovoltaic (PV) modules, p.7, 2021
[12] IEC 61853-1, Photovoltaic (PV) module performance testing and energy rating – Part 1: Irradiance and temperature performance measurements and power rating, pp.9-16, 2011
[13] BS EN IEC 60904-10:2020, Photovoltaic devices – Part 10: Methods of linearity measurement, pp.11-16, 2020

COMPUTATION OF FAULTY IV CURVES BASED ON A DISTRIBUTED SOLAR CELL ALGORITHM

Cárdenas-Bravo C.[1,2], Dutykh D.[2,3], and Ha D.[1]
[1]Univ. Grenoble Alpes, INES, F-73375, Le Bourget du Lac, France
[2]Univ. Grenoble Alpes, Univ. Savoie Mont Blanc, CNRS, LAMA, 73000 Chambéry, France.
[3]Mathematics Department, Khalifa University of Science and Technology, PO Box 127788, Abu Dhabi, United Arab Emirates

This work proposes a novel calculus method that goes deeper into the nature of photovoltaic technology. For this purpose, a distributed solar cell model is used to calculate the IV curves of a solar module. The main advance of this approach is that the user can set the topology of the solar module, as well as each parameter of the solar cells that compose it. Also, the proposed calculus method offers the possibility of modifying the protection system model, adding another degree of freedom to the original problem. Finally, similarly to PVmismatch, the proposed method is able to calculate IV curves given the irradiance and cell temperature of each solar cell.
Keywords: Modelling, Photovoltaic, Software.

1 INTRODUCTION

Several explanations have been proposed in the literature to represent defects in solar modules. However, despite the nature of the failure, the available data indicates that defects in a solar module can be modeled by modifying the internal parameters of the solar cells that compose the module. Consequently, having software capable of generating IV curves based on a distributed solar cell model represents an advantage for acquiring in-deep knowledge about the failure phenomena in solar modules. In addition, having access to such algorithms opens the possibility of creating new approaches, such as the generation of a thermal image based on the internal parameters of a solar module cell and environmental data (plane of array irradiance matrix, ambient temperature, and wind speed).

There are different ways to calculate the electrical behavior of a solar module at different levels. One example of this is the well-known PVlib open-source library [1], which is capable of calculating atmospheric variables, as well as DC and AC electrical variables of large PV systems. This library offers the possibility of modifying the irradiance on the plane and the cell temperature assuming a uniform distribution. However, lacks the option of introducing mismatches in a solar array, which is a limiting factor when analyzing faulty scenarios. Therefore, specialized software is required to solve these cases.

Another well-known and widely software used is PVmismatch [2]. This software, as well as PVlib, is able to calculate IV curves at different levels of irradiation and cell temperatures for the distinct sections of a solar plant, i.e., solar cell, solar module, solar string, and solar system level. However, the advantage of PVmismatch is that the user can calculate IV curves when having different irradiances or cell temperatures at the cell level (for a solar module) or module level (for a solar string or solar system). This feature provides a degree of flexibility when studying the faults present in the solar system. Despite these advantages, PVmismatch is not capable of calculating atmospheric variables, or the AC generation of a photovoltaic system. Therefore, a complete analysis should be performed mixing both PVlib and PVmismatch libraries.

The aim of this work is to determine the IV curve of solar modules by using a distributed solar cell model. The proposed algorithm needs to be capable of estimating a wide set of operation conditions, such as uniform variations of the incident irradiance and cell temperature, as well as mismatch relative to irradiance, cell temperature, and internal parameters of the electrical model used.

To solve this problem, a distributed mathematical system is implemented in Python. By solving the equation system, the user has access to all the electrical variables under optimization, i.e., photovoltaic current and photovoltaic voltage of each solar cell, and voltage and current of each protection element.

Some libraries such as PVlib and PVmismatch can calculate the IV curve given a specific set of input data (solar cell model, irradiance and cell temperature). However, the solution proposed in this work offers the possibility to determine the solar cell model given an IV curve. Based on it, internal faults present in the solar module can be estimated using as input the IV curve, the irradiance and the cell temperature. Finally, the internal model of the faulty solar cell helps us to have a better understanding of degradation mechanisms.

2 ABOUT PVLIB AND PVMISMATCH

Understanding the failure mechanisms in solar modules has become a critical issue over the last few years. To solve this problem, open-source programs such as PVlib and PVmismatch have been developed. These programs have become popular because of their capability for estimating a wide operation condition. Table 1 summarizes the main characteristics of these algorithms, where relevant features of the calculus software are indicated. Additionally, features of the proposed software are added.

Despite the advantages of PVlib and PVmismatch, the differences in the internal model of each solar cell that composes the solar module are not considered. Therefore, given this gap in the state-of-the-art, is important to develop a calculus method to determine the IV curves considering these differences present in the solar cells. This work proposes a solution of the electrical equation system using the solar module topology and the environmental variables as input, i.e., number of rows, number of columns, protection diodes, plane of irradiance matrix, and cell temperature matrix.

Note that each IV curve can be described by calculating the cardinal points: Short-circuit current I_{sc}, open-circuit voltage V_{oc}, maximum power point current I_{mpp}, maximum power point voltage V_{mpp}, maximum power P_{mpp}.

Table I: Features available in PVlib, PVmismatch, and the proposed algorithm.

	Pvlib	**Pvmismatch**	**Proposed**
Platform	Python	Python	Python
Electrical model	Single diode model SDM	Double diode model DDM	Single diode model SDM
Solar module type	Concentrated	Distributed	Distributed
Input parameters	- Solar module irradiance - Solar module temperature - Solar module SDM	- Single solar cell irradiance - Single solar cell temperature - Solar cell DDM (all solar cells are equal) - Protection diode model	- Single solar cell irradiance - Single solar cell temperature - Solar cell SDM (**individual for each solar cell**) - Protection diode model
Output	- Solar module IV curve - Solar module generation - Plane of array irradiance	- Solar system/String/module/cell IV curve	- Solar module/sub-string/cell IV curve

2 RESULTS

The results presented come from three independent tests. In the first place, a comparison between the proposed algorithm and PVmismatch under different irradiances and cell temperatures is performed. For this test, both the irradiance and the cell temperature distributions are uniform. Next, the performance of the proposed algorithm is studied when a single solar cell is shaded, comparing its results with using PVmismatch. Finally, real faulty IV curves are replicated using the proposed distributed model.

3.1 Comparison with PVlib and PV mismatch

The solar module under analysis is composed of a 12x6 solar cells array, divided into three equal sub-strings. This solar module uses as a basis the default solar cell DDM included by PVmismatch. However, since PVmismatch does not count with the SDM, it is calculated using an explicit algorithm.

To determinate the error of the proposed method under uniform distributions of irradiance and cell temperature, the cardinal points of 10.000 IV curves are compared with PVmismatch, calculating the error as shown in Equation (1). Here, the error of each cardinal point X is considered independently. For this test, the irradiance varies in the range [100 〚W m〛 ^(-2), 1000 〚W m〛 ^(-2)], while the cell temperature is in the range [0 °C, 75 °C]. Fig. 1(a) illustrates the density plot of the proposed method for different uniform plane of array irradiance and cell temperatures, reaching a mean value equal to 0.59 % and a standard deviation equal to 2.48 %. It is possible to appreciate that both, the mean value and the standard deviation present a low value. This fact indicates that the performance of the proposed method is consistent with PVmismatch simulations when a uniform irradiance and temperature distribution is considered.

Fig. 2(b) shows the error density of the proposed method when estimating 1000 IV curves when the irradiance of a single solar cell is in the range [100 〚W m〛 ^(-2), 1000 〚W m〛 ^(-2)]. The error is calculated by applying the Equation (1). From this figure, it can be appreciated that the mean value is near to zero, which indicates that almost all the estimated cardinal points are similar to the PVmismatch predictions. In addition, the standard deviation equals 1.51 % which is lower than the ones presented in sub-section "uniform distribution comparison". Based on this result, it is confirmed that the proposed method is capable of replicating with high accuracy the results indicated by PVmismatch.

$$Error\ (\%) = 100 \times \frac{x_{proposed} - x_{PVmismatch}}{x_{PVmismatch}} \quad (1)$$

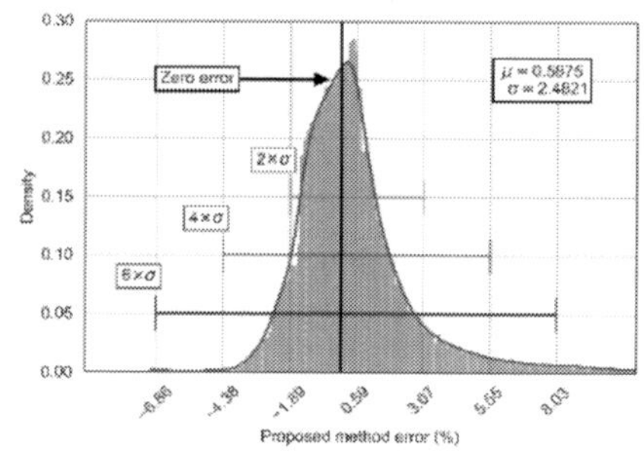

(a) Uniform distribution.

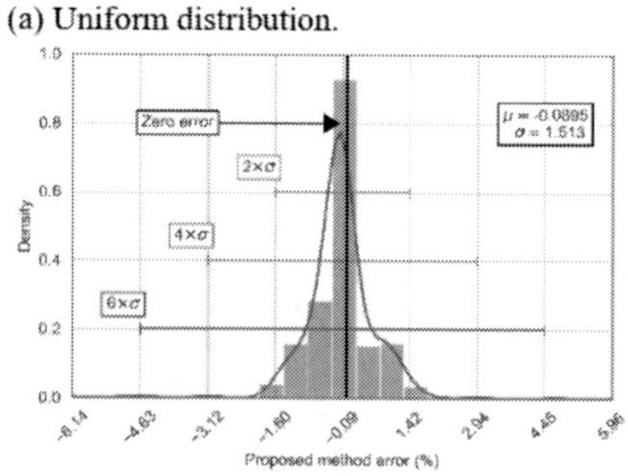

(b) Shading.

Figure 1: Error density when comparing estimaions of the proposed model with PVmismatch.

3.2 Faulty IV Curve fitting

For this test, real measurements are performed on the CEA-INES installation for two faulty solar modules. It should be noted that the solar modules analyzed have multiple faults, modifying the measured IV curves. In order to demonstrate the applicability of the proposed algorithm, the model of each faulty solar cell has been modified to simulate an IV curve similar to the measurements. Fig. 2 shows the IV curves for the two samples under analysis. Here it is possible to appreciate three IV curves: (i) reference curve determined by using the healthy model, (ii) faulty IV curve estimated by modifying the damaged solar cells, and (iii) the measured data. This result suggest that is it possible to estimate a faulty IV curve by modifying the internal parameters of some solar cells. Moreover, by using this approach, internal power estimations can be performed to understand degradation mechanism present on the solar module.

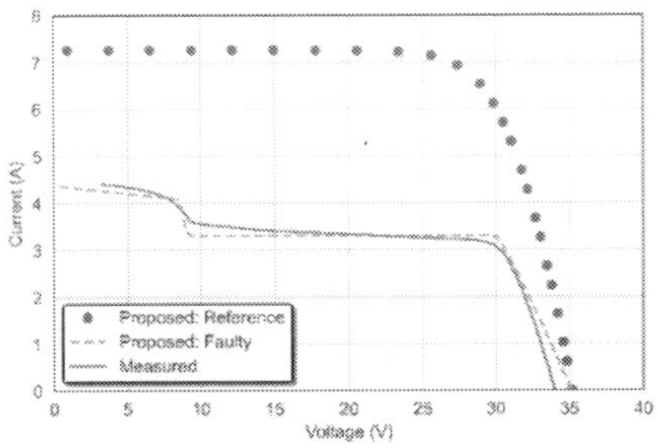

(a) Sample 1.

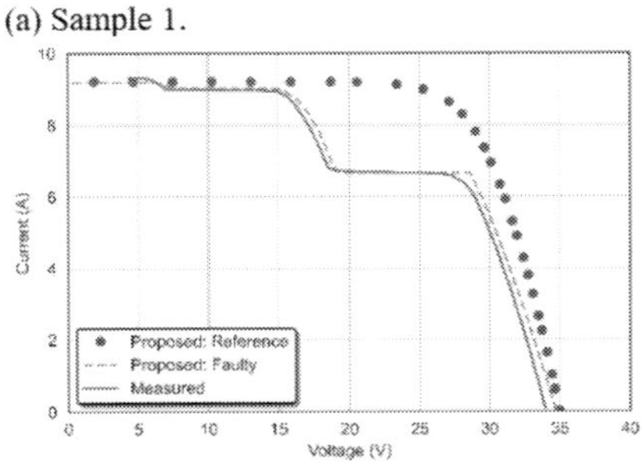

(b) Sample 2.

Figure 2: IV curves measured and simulated using two solar module samples.

4 CONCLUSIONS

A distributed solar cell algorithm is proposed in this work. This algorithm supposes an advantage when analyzing faulty solar modules, since it is possible to manipulate independently each parameter that compose the solar module, as well as the environmental variables. To demonstrate the performance of the algorithm three test are conducted: (i) uniform irradiance & cell temperature, (ii) irradiance mismatch over a single solar cell, and (iii) simulation of the IV curve for two faulty solar modules. Based on this, the results of tests 1 and 2 suggest that the performance of the proposed approach is similar when compared to PVmismatch, achieving a median error of 0.59 % for the first case and -0.09 % for the second test. On the other hand, from test 3 it is appreciated that the both the simulated and measured IV curve presents the same behavior. Therefore, it is possible that adding independent faults on some solar cells replicate a faulty IV curve.

5 ACKNOWLEDGMENTS

The work of D. Dutykh has been supported by the French National Research Agency, through the Investments for Future Program (ref. ANR-18-EURE-0016 - Solar Academy). In addition, C. Cárdenas-Bravo would like to acknowledge the INES.2S French Institute for the Energy Transition, the PhD's contract USMB/2021-454, and the CSMB - Conseil Savoie Mont Blanc for its support.

6 REFERENCES

[1] PVLib, https://github.com/pvlib/pvlib-python.
[2] PVMistmatch, https://github.com/SunPower/PVMismatch.

STEP BY STEP CURRENT-VOLTAGE CURVE DIVISION METHOD FOR THE ESTIMATION OF DEGRADED PHOTOVOLTAIC CELL PARAMETERS

[1]Vishal E. Puranik, [1]Ravi Kumar* and [1]Rajesh Gupta
[1] Department of Energy Science and Engineering, Indian Institute of Technology Bombay, Mumbai, India-400076.
Email ID: vishal.puranik@iitb.ac.in, *ravikr@iitb.ac.in, rajeshgupta@iitb.ac.in

ABSTRACT: Modeling and simulation of a PV cell or module are crucial for diagnosing the cell defects and degradations, estimating resulted losses, and predicting cell or module performance at desired environmental conditions. PV cell is generally modeled using a single diode model (SDM) or double diode model (DDM). For the simulation of these models, accurate cell parameters extraction from an I-V characteristic is essential. Various methods have been reported in the literature to estimate cell parameters; however, most of them discuss parametric extraction of defect free PV cell. Also, most of the existing parametric estimation algorithms offer high accuracy on the cost of complexity and heavy computational burden. In this work, the simplified parametric extraction method is proposed based on the step-by-step division of I-V curve. Analysis of seven PV cell parameters and their region of dominance in I-V curve is studied. Based on that, step by step cell parameters extracted dividing I-V curve in the range of [0, open circuit voltage (V_{oc})]. This method is verified on healthy and defective cells such as potential-induced degradation (PID) affected and cracked solar cells. Results show that I-V curve reproduced by cell simulation based on the extracted cell parameters shows a good agreement with the experimental IV curve with root mean square error < 0.2. The proposed method can effectively estimate seven parameters of a PV cell at any stage of degradation with reduced complexities and computational burden.

Keywords: PV cell, current-voltage characteristic, defect, degradation

1 INTRODUCTION

Extraction of photovoltaic (PV) cell or module parameters is essential for modeling and simulating PV sources. Normally, a PV model generates the current-voltage (I-V) and power-voltage (P-V) curves of a cell, and array. Model parameters define its accuracy. Also, the knowledge of cell parameters is essential in the optimization of the cell fabrication process or assessing defects and degradation. However, these parameters information not available in the module datasheet; also, they change over time. Therefore, it is essential to estimate these parameters from the available information. Generally, a single diode model of a PV cell is commonly used for the performance analysis of a PV cell. However, the field-aged module or cell containing shunt defects and degradation needs an estimation of seven parameters for accurate analysis; in that perspective, limited literature has been reported so far [1]. The single-diode and double-diode models require five and seven parameters, respectively. It is difficult to determine these parameters since the conventional cell equation is implicit and non-linear in nature; hence its solution for seven parameters is not straightforward. Curve fitting and multiple optimization techniques have been reported in the literature; however, most of them are complex to implement, with a lot of computational burdens [2-5].

In this paper, a simple step by step I-V curve division method is proposed for the estimation of seven parameters of healthy to degraded PV cells. I-V curve is analyzed in detail, in which the effect of each parameter on I-V curve and the corresponding region of dominance are studied. Based on this analysis, step by step division of I-V curve is performed for the extraction of seven parameters of a PV cell.

2 METHOD

The double diode model (DDM) of a PV cell is generally used for either higher accuracy or modeling shunt-related defects such as potential-induced degradation (PID) [1], [6]. Fig. 1 shows an equivalent circuit of a PV cell modeled using DDM, where I_{ph} – light generated current, I_{01}, n_1 – bulk diode reverse saturation current and ideality factor, I_{02}, n_2 – recombination diode reverse saturation current and ideality factor, R_{sh}, R_s are the shunt and series of a PV cell. The performance of DDM depends on the parametric values of these seven parameters.

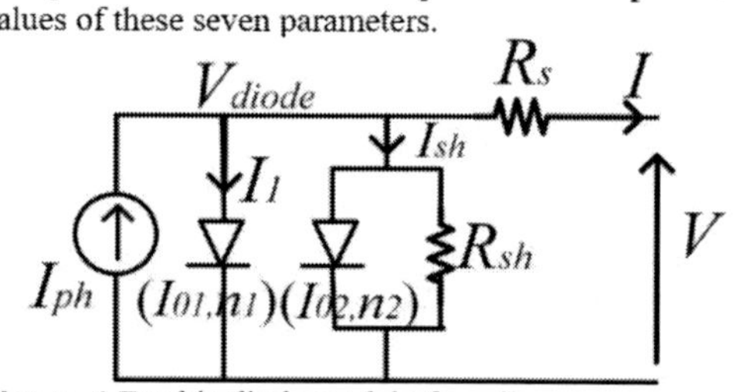

Figure. 1 Double diode model of a cell

$$I = I_{ph} - I_{01}\exp\left(\frac{V+IRs}{n_1 V_t}\right) - I_{02}\exp\left(\frac{V+IRs}{n_2 V_t}\right) - \frac{V+IRs}{R_{sh}} \quad (1)$$

Eq. (1) show the current-voltage relation of a PV cell, which contains seven unknowns (I_{ph},I_{01},n_1,I_{02},n_2,R_s,R_{sh}) whereas V_t is diode thermal voltage. It is very complex to estimate seven cell parameters by solving Eq. (1) since it is implicit and non-linear in nature. Hence, this work aims to propose a simplified parametric extraction method capable of estimating PV cell and module parameters under any stage of degradation. Each of the seven parameters has its dominance region in the IV curve within voltage range [0, open-circuit voltage (V_{oc})]; based on that, The following step-by-step I-V curve division approach is proposed as shown in Fig. 2.

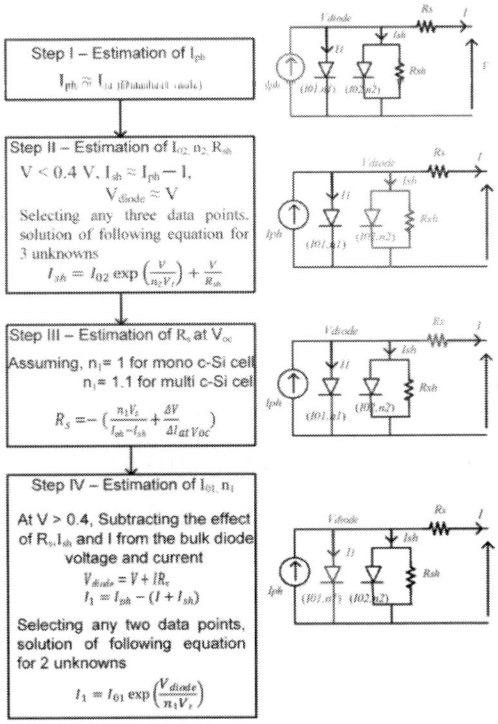

Figure. 2 Flowchart of the proposed step by step parametric estimation method

Step I: Estimation of I_{ph}

Generally, in a healthy cell $I_{ph} \approx I_{sc}$. However, in the case of defective cell $I_{sc} < I_{ph}$. Cell I_{ph} can be obtained from a module datasheet, i.e., $I_{ph} \approx I_{sc}$ (datasheet value). In case of defects or degradation, I_{sc} falls due to other parameters, whereas I_{ph} remains the same. Hence, the initial datasheet value of I_{sc} can be taken as I_{ph}.

Step II: Estimation of I_{02}, n_2, R_{sh}

Parameters (I_{02},n_2) represent the recombination happening in the charge region of a cell. R_{sh} represents the cumulative effect of ohmic leakage paths within a cell. In a healthy cell, the effect of I_{02},n_2,R_{sh} is negligible; however, in the case of cell suffering from PID-s, recombination and ohmic shunting, these parameter dominates and affects cell performance. Fig. 3 and Fig. 4 show the effect of R_{sh} and (I_{02}, n_2) parameters on a cell I-V curve.

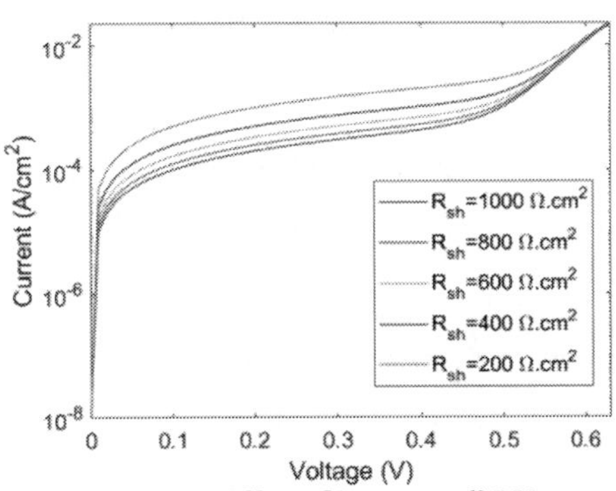

Figure 3: Effect of R_{sh} on a cell I-V curve

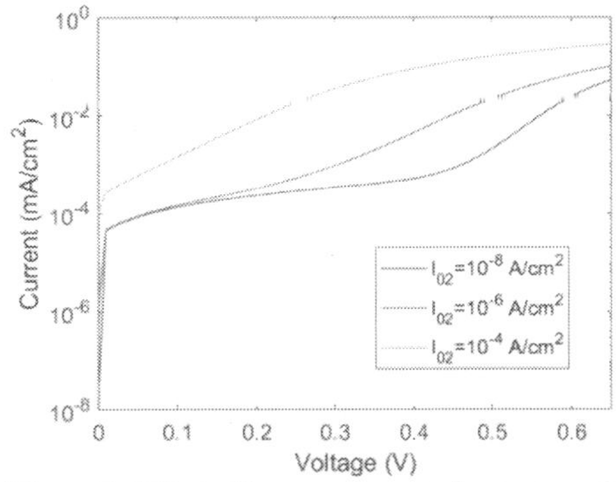

Figure 4: Effect of increased space charge recombination (I_{02}) on a cell I-V curve

Fig. 3 and Fig. 4 show that recombinative diode and shunt parameters (I_{02}, n_2, R_{sh}) dominate in low voltage region V < 0.4 of a I-V curve. Hence, from Fig. 1, the solution of Eq. (2) for any three points of a curve between [0, 0.4 V] gives the estimation of I_{02}, n_2, R_{sh} In low voltage region V < 0.4 V, $I_{sh} \approx I_{ph} - I$ and $V_{diode} \approx V$ since the effect of R_s are negligible in low voltage range. Selecting any three points between 0 < V< 0.4 V and solving the set of non-linear Eq. (2), the value of I_{02}, n_2, and R_{sh} can be obtained.

$$I_{sh} = I_{02}\exp\left(\frac{V}{n_2V_t}\right) + \frac{V}{R_{sh}} \quad (2)$$

Step III: Estimation of R_s

Cell R_s represents the series resistance faced by current that includes the cumulative effect of a finger, busbar, contact, and bulk resistances. Defects like busbar interconnect failure, cracks, finger breakages, and corrosion increases cell R_s. Fig. 5 shows the effect of increasing R_s on a cell I-V curve. It may be observed that R_s dominate between maximum power point (MPP) and V_{oc}.

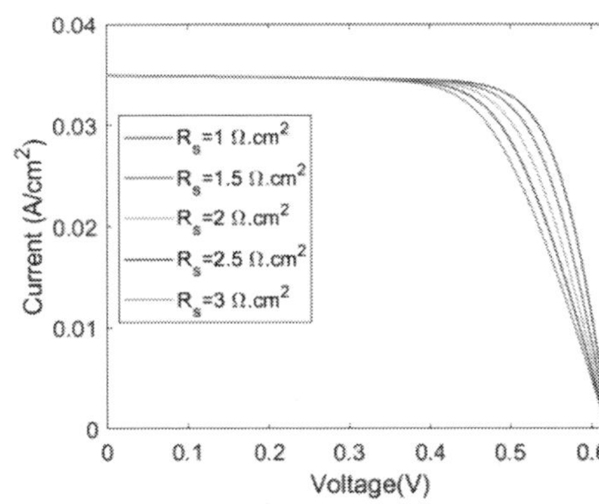

Figure. 5 Effect of R_s on a cell I-V curve

From Fig. 1,

$$V_{diode} = V + IR_s \quad (3)$$

$$n_1V_t\log\left(\frac{I_{ph}-(I+I_{sh})}{I_{01}}\right) = V + IR_s \quad (4)$$

Differentiating Eq. (4) with respect to I, an expression for R_s can be derived as

$$R_s = -\left(\frac{n_1V_t}{I_{ph}-I_{sh}} + \frac{\Delta V}{\Delta I}_{at\,Voc}\right) \quad (5)$$

For a healthy cell, effect of I_{sh} can be ignored; hence Eq. (5) reduces to Eq. (6).

$$R_s = -\left(\frac{n_1V_t}{I_{ph}} + \frac{\Delta V}{\Delta I}_{at\,Voc}\right) \quad (6)$$

Although the value of n_1 is unknown at this stage, for simplicity, it may be assumed as 1 and 1.1 for mono and multi-crystalline cells, respectively.

Step IV: Estimation of I_{01} and n_1

From Fig.1, subtracting the effect of R_s as well as other currents (i.e, I_{sh} and I), only voltage and current flowing through a bulk diode can be estimated using Eq. (7) and (8), respectively.

$$V_{diode} = V + IR_s \quad (7)$$

$$I_1 = I_{ph} - (I + I_{sh}) \quad (8)$$

Bulk diode turns on at a voltage > 0.4 V hence parameters (I_{01}, n_1) dominate in the high voltage region $0.4 < V \leq V_{oc}$. Selecting any two points between (0.4 V, V_{oc}) and solving Eq. (9) gives the value of I_{01} and n_1.

$$I_1 = I_{01} \exp\left(\frac{V_{diode}}{n_1 V_t}\right) \quad (9)$$

3 RESULTS

Fig. 6 shows the results of the proposed method for four test cells. The proposed parametric extraction method is verified on the healthy (Fig. 6(a)), PID-affected cells (Fig. 6 (b) and (c)), as well as a cell containing multiple defects (Fig. 6(d)). Based on the experimental I-V curve, seven parameters of each cell were estimated by implementing the step by step I-V curve division method. DDM of each cell was simulated using the estimated parameters. The simulated I-V curve is compared with experimental I-V, as shown in Fig. 6. Root Mean Square Error (RMSE) between experimental and simulated I-V curve is taken as a figure of merit, and it is calculated using Eq. (10).

$$RMSE = \sqrt{\frac{\sum(I_{experimental} - I_{simulated})^2}{n}} \quad (10)$$

Extracted parameters of the test cell and corresponding RMSE is listed in Table I. Fig. 6 highlights that the simulated IV curve shows a good match with the experimental IV curve with RMSE < 0.2. Results highlight that the proposed method effectively estimates seven cell parameters even with defective conditions.

4. CONCLUSION

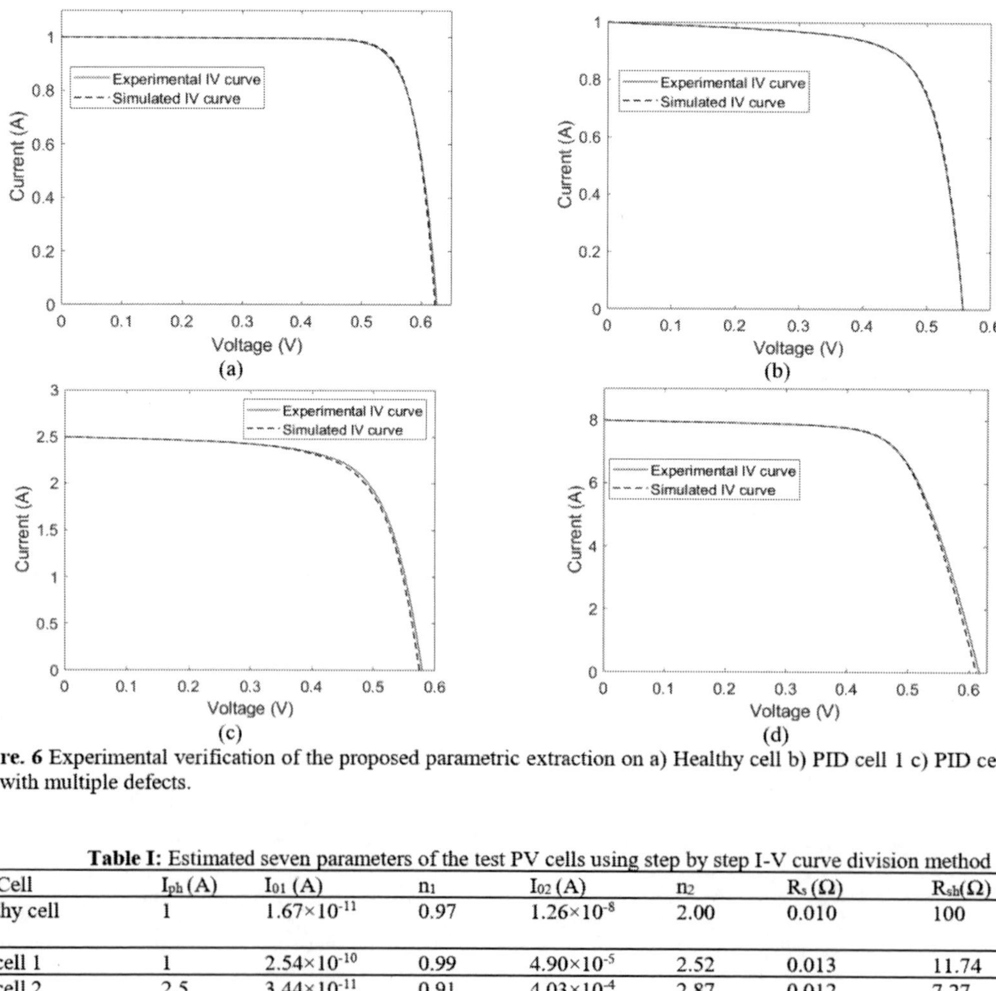

Figure. 6 Experimental verification of the proposed parametric extraction on a) Healthy cell b) PID cell 1 c) PID cell 2 d) Cell with multiple defects.

Table I: Estimated seven parameters of the test PV cells using step by step I-V curve division method

Test Cell	I_{ph} (A)	I_{01} (A)	n_1	I_{02} (A)	n_2	R_s (Ω)	R_{sh}(Ω)	RM
Healthy cell	1	1.67×10^{-11}	0.97	1.26×10^{-8}	2.00	0.010	100	0.03
PID cell 1	1	2.54×10^{-10}	0.99	4.90×10^{-5}	2.52	0.013	11.74	0.05
PID cell 2	2.5	3.44×10^{-11}	0.91	4.03×10^{-4}	2.87	0.012	7.27	0.14
Cell with multiple defects	7.99	6.84×10^{-10}	1.03	16×10^{-4}	4.63	0.016	4.21	0.15

This paper presented the simplified step by step I-V curve division method for estimating healthy and defective cell parameters. The effect of seven cell parameters on I-V curve was studied, and based on their respective region of dominance, this method was developed. Results highlight that the proposed method gives satisfactory performance in estimating cell parameters with the observed RMSE <0.02 between the experimental and simulated I-V curves. Compared with the existing parametric extraction methods based on curve fitting, optimization, and analytical approaches, the proposed method is simple to implement. Also, it is applicable for estimating cell parameters affected by different types of defects.

REFERENCES

[1] G. Petrone, C. A. Ramos-Paja, and G. Spagnuolo, *Photovoltaic Sources Modeling*. 2017.

[2] M. Kumar and A. Kumar, "An efficient parameters extraction technique of photovoltaic models for performance assessment," *Sol. Energy*, vol. 158, no. June, pp. 192–206, 2017.

[3] N. Boutana, A. Mellit, V. Lughi, and A. Massi Pavan, "Assessment of implicit and explicit models for different photovoltaic modules technologies", *Energy*, vol.122, pp. 128-143,2017.

[4] Bana, S., Saini, R., "Identification of unknown parameters of a single diode photovoltaic model using particle swarm optimization with binary constraints", *Renewable Energy*, vol. 101, pp.1299–1310. 2017

[5] A. Gholami, M. Ameri, M. Zandi, R. G. Ghoachani, S. Pierfederici, and H. A. Kazem, "Step-By-Step Guide to Model Photovoltaic Panels: An Up-To-Date Comparative Review Study," *IEEE J. Photovoltaics*, vol. 12, no. 4, pp. 915–928, 2022.

[6] V. E. Puranik and R. Gupta, "Novel Quantitative Electroluminescence Method for Detailed Performance Analysis of PID-s Affected Crystalline Silicon PV Module," *IEEE J. Photovoltaics*, vol. 11, no. 6, pp. 1470–1478, 2021.

ACCURATE MEASUREMENT OF THE ANGULAR DEPENDENCE OF PV MODULE PERFORMANCE USING A DUAL AXIS SOLAR TRACKER

Arvid van der Heide[1,2,3], Arttu Tuomiranta[1,2,3], Michiel Daniels[1,2,3], Niels Capiot[1,2,3], Michaël Daenen[1,2,3], Stefan Wendlandt[4], Loïc Tous[1,2,3]

[1] imec, imo-imomec, Thor Park 8320, 3600 Genk, Belgium
[2] Hasselt University, imo-imomec, Martelarenlaan 42, 3500 Hasselt, Belgium
[3] EnergyVille, imo-imomec, Thor Park 8320, 3600 Genk, Belgium
[4] PI Photovoltaik-Institut Berlin AG, Wrangelstrasse 100, 10997 Berlin, Germany
arvid.vanderheide@imec.be

ABSTRACT: A new approach is presented to determine the angular response of PV modules outdoor, using a dual axis solar tracker and a reference module with known angular response. The method is based on measuring Isc data for both sample and reference module, rotating both modules away from the sun to vary the angle of incidence from 0-90 °. From these data, the Incident Angle Modifier (IAM) curve is calculated. Thanks to the quick rotation (< 1 min), no temperature sensors are needed. For outdoor calibration of reference (mini-)modules, a 1.2 m long tube is used, in which the reference module is rotated while the tube is tracking the sun. In this way, the IAM curve can be determined without diffuse light disturbance. Using such a reference module, IAM curves were determined for several modules. In particular, a mini-module previously measured indoor at PI-Berlin was measured, the maximum difference between the two curves was 1.1%. Compared to the outdoor IAM measurement of PV modules in IEC 61853-2, the method above requires less equipment/sensors and is faster. The method is also accurate, the use of a reference module and the calculation method ensure correction for the varying diffuse irradiation across the sky.
Keywords: Characterisation, Optical properties, Energy rating, Incident angle modifier

1 INTRODUCTION

PV module power is measured with the light beam at normal incidence. However, most PV modules are mounted in a fixed orientation, so the solar irradiation is usually not perpendicular to the surface of the PV module. This results in a lower short-circuit current due to a geometrical effect described by a cosine relation, and an optical effect characterised by the Incident Angle Modifier (IAM). In fact, this is the cosine-corrected value of the normalised I_{sc} (with normalisation on I_{sc} (0)):

$$IAM(\theta) = \frac{I_{SC}(\theta)}{I_{SC}(0) * cos(\theta)}$$

where I_{sc} is the short circuit current, and θ is the angle of incidence.

The procedure to determine the IAM curve for a PV module has been described in the IEC 61853-2 standard [1], both for indoor and outdoor setups.

Indoor setups are using a solar simulator, and the rotation of the module during the measurement leads to a non-uniform irradiance across full-size modules due to the relatively short distance to the light source.

Outdoor setups are using a solar tracker that should be able to rotate the module away from the direct solar beam over a large angular range (90 °).

A clear advantage of measuring the IAM outdoor is that the distance to the sun is so large that rotating of the module does not influence the uniformity of the direct irradiance, even for large ones. On the other hand, the disturbing influence of the diffuse irradiation from the sky (that is also varying in intensity across the sky) needs to be corrected for when measuring outdoor.

To correct for the presence of the diffuse irradiation, the IEC standard requires the use of a pyrheliometer. For the diffuse component, the difference is taken between the global irradiance in the tracker plane and the direct irradiance (measured by the pyrheliometer).

Also, temperature sensors are required to correct for temperature changes of the module during the rotation (due to the reduction of the irradiance).

The approach in this paper, based on the comparison of the module under test to a reference module with known IAM that is placed next to it in the tracker plane, is a simplification of the procedure described in IEC 61853-2. However, it has the promise of being equal in accuracy, it also compensates for the varying diffuse component from the clear sky.

For this approach it is also necessary to have a way to calibrate the reference module, this method that excludes the diffuse irradiation will also be presented in this paper.

IAM curves obtained indoor (at PI-Berlin) and outdoor (at imec) for the same module will be compared.

Finally, measurements obtained for different mini-module from an IAM round-robin will be presented (though the comparison with the other institutes is still pending).

2 OUTDOOR IAM MEASUREMENT METHOD

2.1 Concept of the measurement method

The concept of the measurement method is based on measuring Isc data for both a sample and a reference module (with known angular dependence), while rotating both modules away from the sun, to vary the angle of incidence θ from 0-90 °. From these data, the IAM curve can be calculated. Because of the quick rotation (< 1 min), no temperature sensors are needed. Of course, it is important to perform an accurate calibration of the reference module since all IAM measurement results will be derived from this IAM calibration. For this purpose, a 1.2 m long tube has been built, in which a (mini-)reference module is rotated while the tube is tracking the sun. In this way, the IAM curve for the reference module can be accurately determined without diffuse light disturbance. More details about the construction of the measurement setup will be given later.

When the tracker is rotated away from the sun to vary θ for both sample and reference, the following parameters are measured and saved every 0.1 s: $I_{sc,sample}$, the short-circuit current of a reference module $I_{sc,ref}$, and the module tilt angle α. At this rotation speed, the 0.1 s time step corresponds to an angular step size of 0.17 °. The value of θ is calculated at each timestep by taking the difference between the actual module tilt angle α (that is being varied during the rotation) and the theoretical tilt angle α_{th} of a plane perpendicular to the direct solar irradiation at that time (it only changes slightly during the 54 s rotation time).

The basic method to obtain $IAM_{sample}(\theta)$ is to take the ratio of the normalised I_{sc} for sample and reference module, $[(I_{sc,sample}(\theta)/I_{sc,sample}(0°))/(I_{sc,ref}(\theta)/I_{sc,ref}(0°))]$, and multiply this ratio for each θ with the known $IAM_{ref}(\theta)$ of the reference module. Of course, to improve the accuracy of the calculation, a correction for the diffuse irradiation component from the sky must be included for both sample and reference module. This correction is made by subtracting $I_{sc,sample}(90°)$ or $I_{sc,ref}(90°)$ from the measured I_{sc} at each θ so that only the direct solar irradiation is taken into account, leading to the following final formula:

$$IAM_{sample}(\theta) = \frac{I_{sc,sample}(\theta) - I_{sc,sample}(90°)}{I_{sc,sample}(0°) - I_{sc,sample}(90°)} \frac{I_{sc,ref}(0°) - I_{sc,ref}(90°)}{I_{sc,ref}(\theta) - I_{sc,ref}(90°)} * IAM_{ref}(\theta)$$

It should be mentioned that the intensity of the diffuse irradiation varies across the sky, so that the mentioned subtraction of the I_{sc} at 90 ° is in principle not sufficient as correction. However, since both the sample and the reference module are sensitive to this variation of diffuse irradiation in (almost) the same way, the "ratio calculation method" presented above essentially eliminates this effect so that only the angular response to the direct beam remains.

2.2 Description of the dual axis solar tracker

For the measurements in this paper, a dual-axis solar tracker has been used that has been built in-house at imec to allow for full flexibility of control and angular range. In this paper only the IAM measurements are described, but it is also in use for outdoor I-V characterisation. It is shown below in Figure 1.

Figure 1: In-house built dual axis solar tracker, here with 2 full-size modules for I-V characterisation

The azimuth axis turns the tracker from left to right and the zenith axis controls the tilt angle of the module. Compared to a commercially available dual-axis tracker, the total range of the zenith angle is much higher, which is necessary to perform the IAM measurements.

The two axes are driven by two Siemens SIMOTICS S-1FL6 servo motors, via a slewing drive from KMI group (type KDE3D-62M-RC) with a drive ratio of 62:1. The servo motors are controlled by drivers that are connected to a Siemens S7-1200 PLC. The PLC and the drivers are placed in a switch cabinet together with the required safety equipment. Initial testing showed that the servo motors did not reach the required torque to turn the tracker when two PV modules are mounted. A gearbox (WITTENSTEIN NP025SMF2-16-1E1-1S) with a gear ration of 16:1 was added to each servo motor to increase the torque.

Siemens TIA portal V15 is used to program the PLC that controls the solar tracker. The PLC uses the highly accurate NREL SPA algorithm [2] to calculate the position of the sun, which is used to control the servo motors to follow the sun. To enable the rotation of the modules away from the sun, it is possible to specify an angular offset for the horizontal axis, usually set to 95 ° when doing an IAM measurement. During the rotation over this offset angle, the necessary parameters (electrical currents and tilt angle) are logged.

The PLC can measure the short-circuit currents from two PV modules simultaneously. This is done with two Rawet PX24 current converters that can handle input currents up to 15 A and a Siemens SM 131 ADC that is plugged into the PLC and reads the small current signals (4-20 mA range) from the current converters. Measuring is done using this current signal instead of a voltage signal since the long cables (~ 25 m) to the ADC could pick up quite some noise when (low) voltage would be used for the signal.

2.3 Equipment for reference module calibration

For the IAM measurement method in this paper, an IAM reference module is required, that has to be calibrated in an accurate way. First the reference module will be described, followed by a description of the calibration hardware.

The reference module used for this paper was a 20 cm x 20 cm module made at imec, using glass with an anti-refelction coating (ARC) at the front and an opaque black back sheet at the rear to avoid any disturbing influence of light entering from the rear. It is shown below in Figure 2.

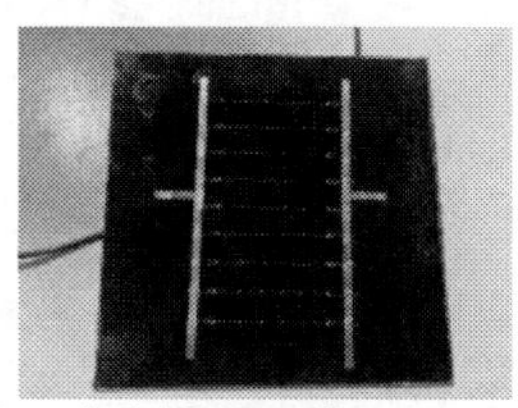

Figure 2: Monofacial IAM reference module with one half cell of crystalline silicon, and glass with ARC

A half cell was used instead of an entire cell to limit the current to minimise the voltage drop across the module. A large voltage drop across the module (due to resistance of cables and measurement hardware) at irradiances still below 1 sun would lead to an unwanted sublinear increase of the module current as a function of irradiance.

The IAM reference module was calibrated inside a 1.2 m long, 22 cm x 22 cm square aluminium tube which is needed to block the diffuse component of the sunlight. It consists of a 1 m long upper part and a 20 cm long lower part. The module is placed at the junction of the two parts so that the module can be mounted easily. The tube is mounted perpendicular to the solar tracker plane while it is tracking the sun, to ensure that the tube is always pointing towards the sun. The tube, mounted on the tracker is shown in Figure 3.

Figure 3: Tube used for calibration of IAM reference modules mounted perpendicularly to the tracker plane

The inside of the tube is covered with a KoPro IR (infrared) Flock Sheet, to absorb possible light reflections inside the tube as much as possible. Compared to many other ultra-black materials, it has the advantage that it also absorbs infrared light. The total reflectance of the IR flock sheet over the range 300 nm - 1200 nm was measured to be in the range of 0.6-0.7%, although the supplier claimed an even lower value, around 0.3%.

In the lower part of the tube, an axis is mounted that enables the rotation of the module to be calibrated. This axis is driven by an RS PRO 2.8V 0.9° hybrid stepper motor. The motor is powered by a TB6600 stepper motor driver. The stepper motor system is powered with a 60 W (12V-5A) DC power supply and is controlled by an Arduino Nano.

To correct for variations in solar irradiance during the measurement, a solar module is placed on the solar tracker next to the tube.

For the current measurement, both the mini-module and the "sun monitor module" next to the tube have a 5.000 mΩ Burster 1240-0.005 precision resistance across them that allows to determine the short-circuit current from the voltage across the resistance. At each angular step of 0.9 °, the voltages are measured simultaneously using two Tektronix DMM 4050 digital multimeters.

The calculation of $IAM_{ref}(\theta)$ is simply done using the formula in the introduction. It should be noted that the measured $I_{sc,ref}(\theta)$ is corrected for the change in the solar irradiance compared to its value at the start of the rotation before inserting it in the equation.

3 RESULTS

3.1 Calibration of the IAM reference module

On June 15th 2022, the calibration of the mini-module took place at clear-sky conditions. The IAM curve was measured 3 times after each other. The resulting IAM curves are shown below in Figure 4.

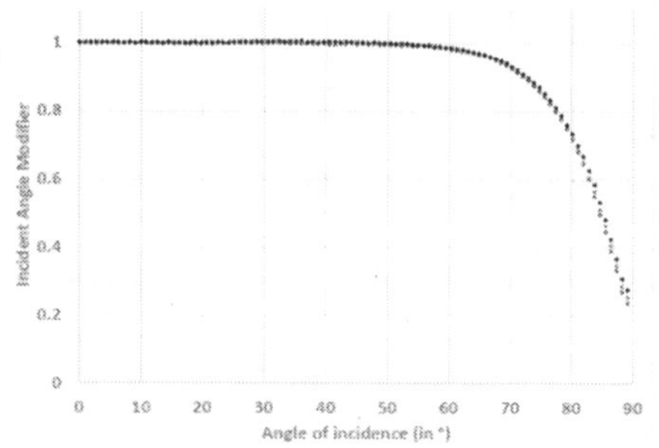

Figure 4: Three IAM curves obtained for reference module using module rotation inside the tube

From this graph, it is clear that the differences between the curves are rather limited. To obtain the best accuracy for the calibration curve, the 3 curves were averaged, the result is shown in Figure 5. The data points for θ up to 40 ° have been set exactly to 1, since in this part, the noise in the measured IAM curves is larger than the real deviation from 1.

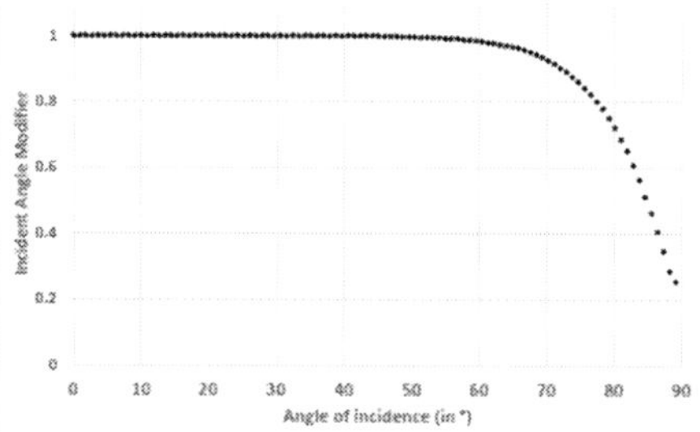

Figure 5: Calibration IAM curve for reference module obtained by averaging 3 curves in Figure 4

3.2 Comparison of indoor and outdoor IAM results

To evaluate the accuracy of the presented outdoor approach, it was decided to make a comparison between IAM results that had already been obtained using an indoor setup at PI-Berlin and outdoor results from the dual axis tracker. The module for this comparison had been made at PI-Berlin, using glass with ARC and containing a few half cells, of which only the one at the centre has been electrically connected.

The IAM measurements also took place on June 15th 2022 at clear-sky conditions. Figure 6 shows both the PI-Berlin module and the IAM reference module mounted next to each other on the tracker.

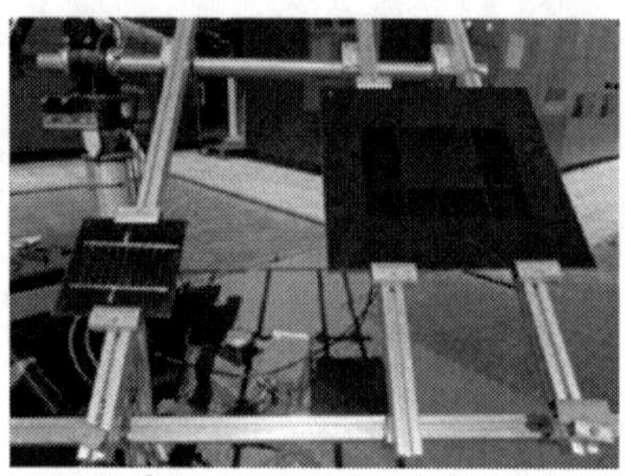

Figure 6: Reference module (left) and PI-Berlin module (right) on the tracker

Since the indoor results were obtained for rather large step sizes of 10 ° (at low θ) or 5 ° (at higher θ), the comparison could only be done at a rather limited number of points, even though the outdoor data are available with higher angular resolution (0.9 °). In Figure 7, the comparison between the indoor and outdoor data is shown.

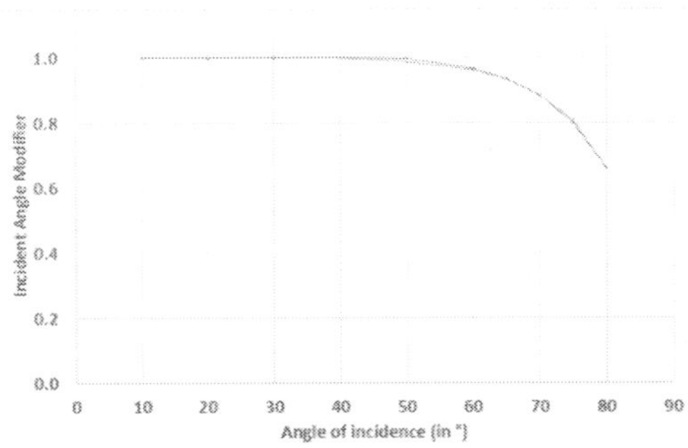

Figure 7: Comparison between AIM data obtained for the PI-Berlin module. Indoor curve (PI-Berlin) is blue and outdoor curve (imec) is orange.

The maximum difference between the IAM values is only 1.1%, which is a very good result. The indoor value of the IAM at 50 ° is probably too high, when looking at other available data for modules that have an ARC on the front glass. For the data points at 75 °, it is more difficult to say anything about the correctness of either measurement.

3.3 IAM curves for different modules from a round-robin

For an even better comparison, it was decided to take part in a round-robin with other institutes concerning IAM measurements, organised by SERIS and TÜV. Within this round-robin, several institutes take part, measuring the IAM curve for several mini-modules that have different angular responses, some of them having coloured glass. The results that have been obtained with the tracker at imec are shown in Figure 8.

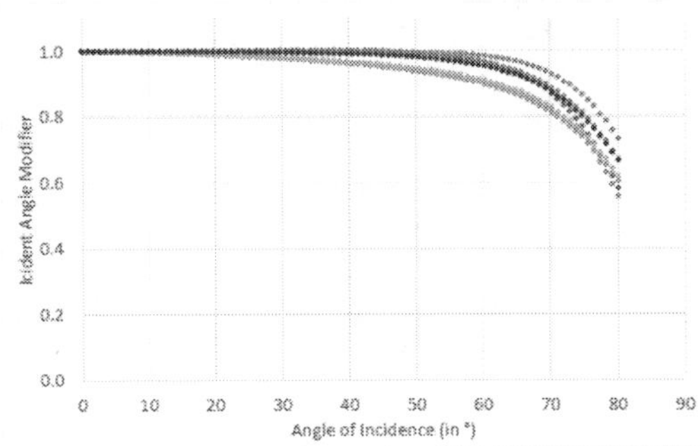

Figure 8: IAM measurements performed for IAM round-robin organised by SERIS and TÜV. The lowest curves were obtained for modules with coloured glass

For the moment, these results have not yet been compared to that of the other institutes. However, it serves as example of the differences in AIM curves that can occur, and also gives and indication of the noise level in the curves. Note that the angular resolution of 0.9 ° is around 5 times better than for the PI-Berlin data in Figure 7.

4 CONCLUSIONS

The outdoor IAM measurement method based on a dual axis solar tracker and an IAM reference module generates the IAM curve in a simplified way compared to IEC 61853-2, while still being accurate. The method includes corrections for the current contribution of the non-uniform diffuse irradiation from the sky.

To enable accurate results, an accurate calibration of the reference module is crucial. The proposed method to do this using rotation of the small reference module inside a tube that blocks diffuse irradiation proved to be accurate and reproducible. The angular resolution of this calibration is higher than usual for IAM measurements, with a step size of 0.9 °.

Very good agreement between outdoor results and indoor results (from PI-Berlin) was obtained, the maximum difference between the curves was found to be 1.1%. This also shows that the calibration method inside the tube is working well.

Finally, IAM measurements have been performed on different types of mini-modules for an IAM round-robin organised by SERIS and TÜV. The comparison of the imec results with other institutes (that is still pending) will still give more insight about the accuracy of the presented IAM measurement approach.

5 ACKNOWLEDGEMENTS

This work has been funded in part by the HIGHLITE and TRUSTPV projects, that have received funding from the European Union's Horizon 2020 research and innovation program under grant agreements 857793 (HIGHLITE) and 952957 (TRUSTPV).

6 REFERENCES

[1] IEC 61853-2, "Photovoltaic (PV) module performance testing and energy rating - Part 2: Spectral response, incidence angle and module operating temperature measurements" 2016

[2] NREL, "Solar Position Algorithm (SPA)." https://midcdmz.nrel.gov/spa/.

IMPACT OF TECHNOLOGICAL TRENDS IN CRYSTALLINE SILICON PV MODULES ON THE ENERGY YIELD PERFORMANCE: PRELIMINARY RESULT OF THE 14TH SUPSI OUTDOOR MEASUREMENT CAMPAIGN

Gabi Friesen, Mauro Caccivio, Ebrar Özkalay, Flavio Valoti
University of Applied Sciences and Arts of Southern Switzerland (SUPSI-PVLab), Via Flora Ruchat-Roncati 15,
Email: gabi.friesen@supsi.ch

ABSTRACT: In order to investigate the advantages and disadvantages of the different technological innovations introduced over the last years, in 2021 the SUPSI PVLab has installed a new outdoor test facility with a set of last generation commercially available PV modules. The focus is here set on the energy performance of monocrystalline silicon glass/backsheet PV modules under optimal and sub-optimal test conditions, where technological differences can have a stronger impact. The results of this study aim to fill the gap of limited field data of these new module technologies, demonstrate the claimed advantages and identify possible degradation problems in the early stage. The performance analysis is performed by means of scientific monitoring data in combination with the characterization of the modules in the laboratory.

1 INTRODUCTION

SUPSI PVLab started in the early 90s with the monitoring and analysis of energy production of commercial PV module technologies. The measurement campaigns (test cycles) of a duration of 2-4 years each, followed the evolution of crystalline silicon PV cell and module production and the increase of efficiencies, especially with the introduction of first generation HJT and IBC technologies. Thin film modules were added to the test cycles at a later stage. Differences in the specific yield [kWh/Wp] of c-Si modules stayed for long time in the range of ±2% shifting the attention of the test laboratories to new generation technologies with higher variability. The last 10 years therefore focused on the numerous thin film modules entering the market (12th test cycle started in 2011) followed by coloured and bifacial technology (13th test cycle started in 2018).

The recent and fast transition to new contact passivation technologies (e.g. PERC/PERT/PERL, TOPCon), in combination with different contacting methods (e.g. multi-bus bar, multi-wire, back contact, shingled solar cells without busbars, tiling ribbon technology), increasing wafer dimensions and half- or multiple-cut cells with new series-parallel connection schemes motivated the initiation of a new measurement campaign, the 14th test cycle. Higher energy yields due to either lower resistive losses, higher power density, lower temperature coefficients or reduced shading losses are claimed by module manufacturers. However, each technological change can bring with it some potential risks which can affect the performance and reliability and the combination of more than one at the same time can lead to unexpected or counter-acting effects.

The new measurement campaign launched in January 2022 aims to demonstrate and analyse the impact of various technological changes on energy yield and to identify possible degradation problems in the early stages. Due to the greater variability of module designs, greater differences in energy yield are expected, especially when installed under partial shading.

2 APPROACH

2.1 Outdoor test facility

Figure 1 shows the three outdoor test facilities installed in December 2021 and operational since January 2022. The outdoor measurements are performed in two configurations, (1) in standard open-rack mounting conditions with optimal inclination (30°) and south orientation without significant shading and (2) in low inclination (10°) to reproduce the typical conditions of flat roofs in Switzerland, with limited air circulation, higher temperatures and increased soiling. The aim is to study the sensitivity of the different technologies to optimal and sub-optimal test conditions.

Figure 1: Pictures of the test stands mounted in Mendrisio (Switzerland): (top and middle) optimal inclination and ventilation conditions (left) low inclination representing typical flat roof conditions.

Each module is connected to a maximum power point tracker (MPPT3000) with integrated I-V tracer. Additionally, to the maximum power point parameters (I_m, V_m), in-plane irradiance (G) and module temperature (T_{mod}) are monitored in 1 minute intervals. The irradiance is measured with some calibrated pyranometers, whereas the temperature with two PT100 sensors attached on the back of the module. The I-V curves are recorded every 5 minutes. The measurements are combined with data from the close by meteo station. The test facility meets the requirements of the recommendations provided in the best practice guideline IEA-PVPS T13-11:2018.

2.1 Test devices

Table 1 shows the seven module types procured for the purpose of this project. A minimum of 3 modules for each type are monitored on the new outdoor test stands (two at 30° and one at 10° inclination). One module is stored in the dark as a reference module and others are dedicated to accelerated indoor and outdoor tests performed within another R&D project.

The module types differ in cell technology, wafer size, cell cuts, cell interconnection and contact technology. The STC power of the modules ranges from 350 to 475 Wp. The test samples was limited to mono-facial glass/backsheet modules and with mono-crystalline silicon cells.

Table I: Overview of module technologies.

Cell technology	Wafer size	Cell cuts	Contact technology	Nom. Power
mono p-type PERC	M6	half-cut	9BB	365 W
mono p-type PERC	M12	third-cut	9BB	385 W
mono p-type PERC	M2	5-cut	shingled	365 W
mono p-type PERC	M2	full size	integrated diodes	320 W
mono p-type TOPCON	M6	half-cut	tiling ribbons	470 W
mono n-type HJT	M2	half-cut	smart wire	375 W
mono n-type IBC	12"	full size	back contact	400 W

2.3 Module characterisation

Prior to outdoor exposure the quality and initial performance of the modules, have been analyzed indoor by acquiring electroluminescence imagines, recording the IV characteristics at standard test conditions, determining the thermal coefficients and the linearity through five different illumination intensities. To quantify any degradation, the modules will be re-measured after 1 year of operation (January 2023). All the here described measurements are performed with our multi-flash approach, which allows to avoid any capacitive measurement artefacts.

Figure 2 shows the measured and nominal P_m temperature coefficient. The datasheet values are all within the limits of the PVLab measurement uncertainty. The TOPCon module couldn't be measured because exceeding the size of our thermal chamber.

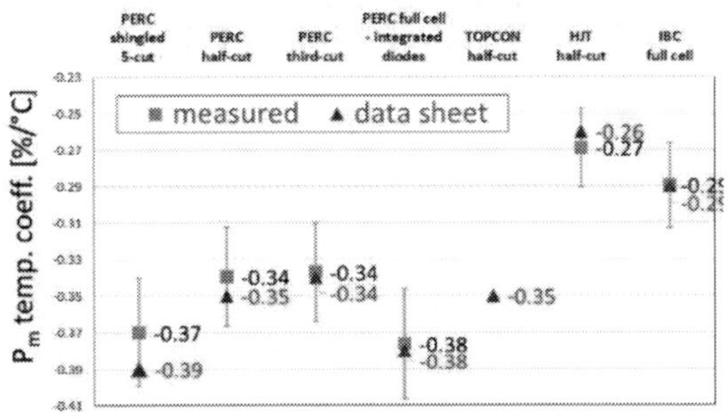

Figure 2: Relative Pm temperature coefficients.

Figure 3 shows the electrical performances at low light levels, which give important feedbacks on the quality of solar cells and of their stringing, highlighting the presence of local or border shunts.

The good performances of the "hotspot free" PERC module with integrated diodes (1 diode per cell) is due to an increase of the series resistance caused by the bypass diodes. At the same time, this results in lower STC and FF power, which is reflected in lower efficiency compared to the other PERC module types.

The shingled PERC modules present the worst low light performance, confirming the quality issues detected also through electroluminescence images.

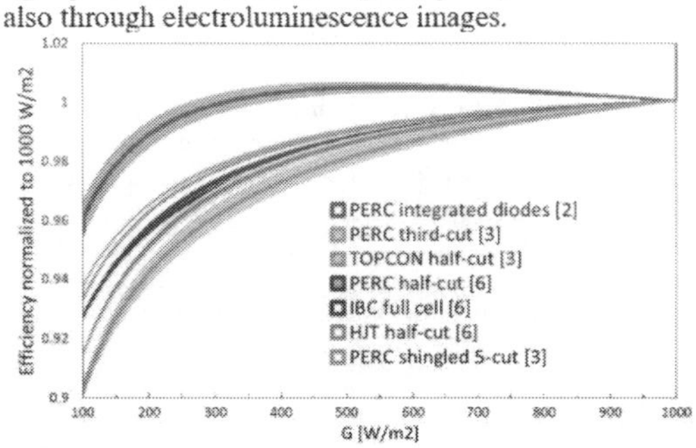

Figure 3: Irradiance dependency of relative module efficiency, measured at 25°C, for all 7 module technologies, together with the variability within the module batch (sample numbers are in brackets).

The STC power of the modules is measured before and after stabilization. The modules, stabilized under a steady state simulator, were therefore connected to a MPP-tracker and the temperature was maintained in the range of 50±10°C. Figure 4 shows the results of the power measured at the begin and after the first (LS1) and second light soaking step (LS2) consisting each in at least 5 kWh/m^2. The first number in the figure gives the degradation rate of the reference module, whereas in brackets are reported the maximum and minimum degradation rates of the other tested modules. The number of tested samples is represented by the last value. A significant variability within the same module types is observed, with HJT and TOPCon showing a positive trend whereas the others a generally characterized by slight negative trends of maximum -0.6%. The only exception a there shingled modules showing slightly higher degradation rates up to -1.36%, with no clear stabilization even after three light soaking periods. In order to speed up the stabilization activities some modules were exposed also outdoors. Differences between indoor and outdoor stabilization were found, that will be further investigated and presented at a later stage.

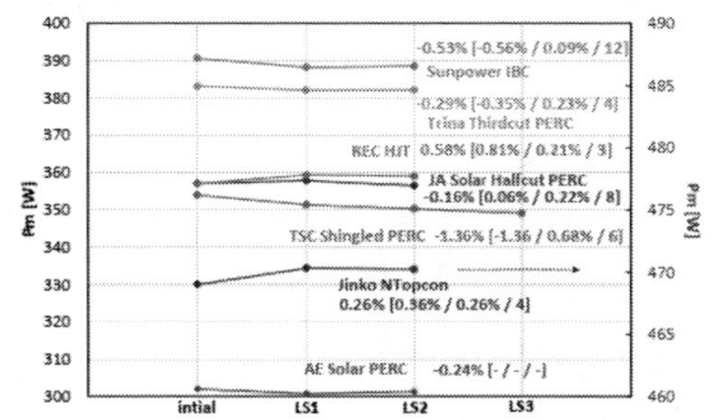

Figure 4: Stabilisation tests on different technologies.

For the purpose of any energy benchmarking a reference power $P_{m,stc}$ has to be chosen for the calculation of the specific energy yield in kWh/Wp or the

Performance Ratio (PR). Due to the observed dependency on the stabilization procedure and an increase of variability within the same module type after light soaking, the initial power was here chosen for referencing. The consideration of LID and LeTID on the benchmarking will be presented with the 1 year results, where more data will be available.

3 RESULTS

3.1 Module benchmarking

The module ranking is here performed by comparing the performance ratio of the modules for the period April to September 2022 (6 month). The first months of exposure had to be excluded due to some missing data. The equation is $PR=\sum P_m \cdot G_{stc}/\sum G \cdot P_{m,stc}$, with G_{stc} being equal to the reference irradiance of 1000 W/m², P_m and G the measured power and irradiance values measured in 1 minute intervals and $P_{m,stc}$ the STC reference power.

Figure 5 shows the 6 months PR of the 6 module technologies mounted under optimal open-rack conditions, together with the relative deviations respect to the best module technology. The average of two modules is here taken.

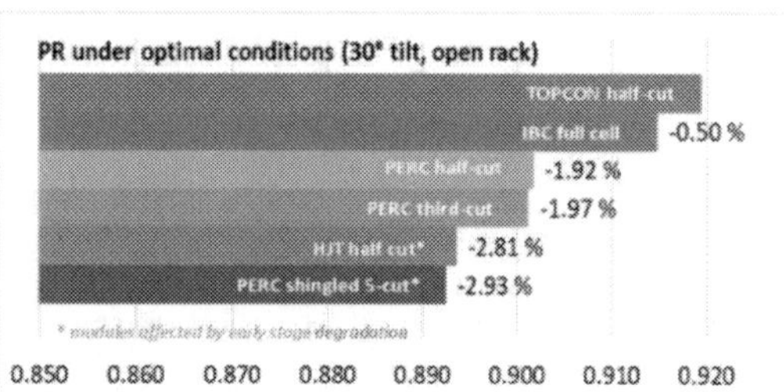

Figure 5: Comparison of the performance ratio of 6 different technologies mounted on the 30° tilted test stand.

Under optimal 30° tilted open-rack and shading free mounting conditions the spread of PR is in the range of ±1.5%, with TOPCON and IBC having the best PR close to 0.92. PERC half-cut and PERC third-cut modules show deviations close to the PVLab STC power meas. uncertainty of ±1.6% [k=2] and can therefore be considered very close to the top modules. The slightly higher deviations of the HJT and PERC shingled modules observed in both test benches can be attributed to an early stage degradation, discussed in more detail in chapter 3.

Figure 6 shows the PR and the relative deviations respect to the best module technology for the modules mounted under sub-optimal conditions. One more technology is here tested, which is the PERC module with integrated diodes.

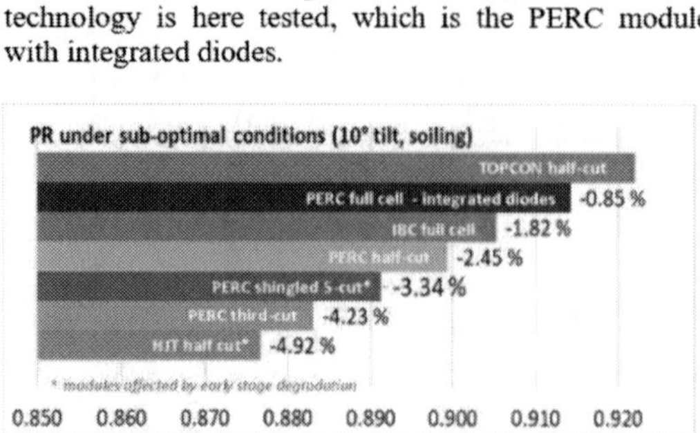

Figure 6: Comparison of the performance ratio of 7 different technologies mounted on the 10° tilted test stand.

For the modules mounted at 10° inclination, representing typical flat roof PV systems in Switzerland, the PR spread is increasing up to ±2.5% (±2.1% excluding modules with early-stage degradation). The TOPCON leads to the highest output followed by the PERC full cell with integrated diodes. For IBC and PERC third-cut the divergence is increasing, whereas it is remaining unvaried for the PERC half-cut module.

This increase can be explained by the different impact that increased soiling at a tilt angle of 10° has on modules with different cell arrangements, interconnections and type of module front design (glass type and frames). In addition to this, the low irradiance performance is having a larger impact. The combination of these two aspects leads to a better PR for the PERC module with integrated diodes than for the other PERC modules tested.

3.2 Degradation figures

The early-stage degradation observed on the HJT modules and PERC shingled modules is still under investigation. Only preliminary results could be reported here.

Figure 7 shows the drop in performance drop detected through the outdoor data. Data measured in the irradiance range of 600-700 W/m2 and 900-1000 W/m2 were therefore corrected to 1000 W/m2 and 25°C. The lower irradiance range allows to get also values for the winter period.

A negative trend could be observed for the HJT modules, which was not visible with the other technologies. The modules were then taken to the laboratory to test them under standard test conditions. A degradation in the range of 4-5% after 10 months of exposure was measured, resulting in PR values that were not representative for the technology. PR values close to those of other high-efficiency technologies, with similar low temperature coefficients would have been expected.

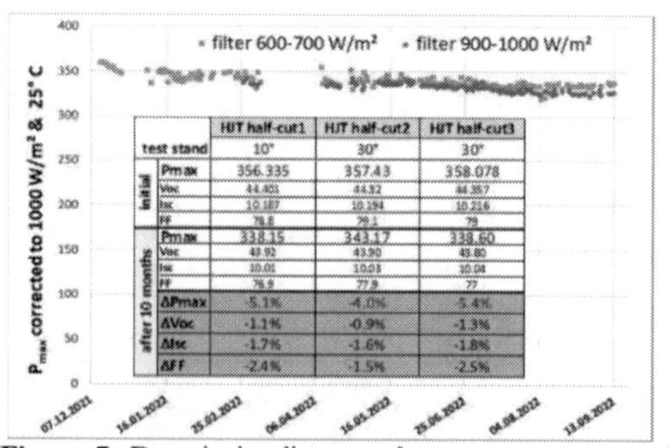

Figure 7: Drop in irradiance and temperature corrected power measured in the field together with indoor measurements under standard test conditions.

To better understand the origin of this early failure, EL and visual inspection images were analysed before and after exposure. The EL image of the new module, shown in Figure 8 a), shows no defects. Figure 8 b), performed after 10 months of outdoor exposure, shows moisture penetration along the cell edges and delamination along the smart wires. Figure 8 c) shows the EL picture after thermal cycling TC200 a new module according to IEC 61215. Both failures could be reproduced. The failure is probably due to the manufacturing process.

Figure 8: EL and visual pictures of new and aged HJT modules. (a) Initial EL. (b) EL and VI picture after 10 months of outdoor exposure and (c) EL picture after TC 200.

The 5-cut shingled PERC modules, which showed lower performance ratios in both test stands, also showed some degradation in the field, which is still being studied. The EL pictures of new modules showed already the evidence of cell mismatch and grid finger problems. Indoor measurements have still to be performed. Higher light-induced degradation rates due to LID/LeTID must be also considered to be responsible for the lower performance.

4 CONCLUSIONS AND OUTLOOK

Under optimal mounting conditions, the new generation sc-Si glass/backsheet mono-facial module technologies, with no obvious degradation problems, showed a comparable PR in the order of ±1%. A performance loss analysis will be carried out after one full year of operation to quantify the impact of temperature, irradiance, spectrum and angle of incidence, including the missing winter months. The impact of LID/LeTID will be further discussed.

Under sub-optimal conditions, with an increase in natural soiling, the impact of different technological solutions (cell layout, contact and module design) is getting increasingly evident, leading to a slight increase in the PR spread up to ±2.1%. Measurements under different shading conditions will be added to better understand the impact that the different layouts have on the PR. The work is carried out within the framework of the European METRO-PV project.

For the half-cut HJT modules, an early-stage degradation in the order of 4-5% over 10 months was measured, which could be attributed to moisture penetration and delamination. The degradation rates of all modules will be reported after 1 year of exposure and then compared with some system data and modules exposed to more severe temperature and shading conditions, within the framework of two other running R&D projects.

5 ACKNOWLEDGEMENT

The here presented results are part of the project ATTRACT funded by the Swiss Federal Office of Energy under the contract SI/502036-01.

DEVELOPMENT AND VALIDATION OF A MODEL FOR PERFORMANCE SIMULATION OF AN AIR-BASED BIPVT ROOF TILE SYSTEM USING OPEN-SOURCE LIBRARIES

M. Bartkowski[1], H. R. Saleemi[2], L. Clasing[1], C. Brosig[3], E. Waffenschmidt[3], U. Blieske[1]
[1] Cologne University of Applied Sciences, Faculty of Process Engineering, Energy and Mechanical Systems,
[2]Cologne University of Applied Sciences, Faculty of Spatial Development and Infrastructure Systems
[3]Cologne University of Applied Sciences, Faculty of Information, Media and Electrical Engineering
Betzdorfer Str. 2, 50679 Cologne, Germany

ABSTRACT: In this work a simulation model for a novel air-based BIPVT roof tile system with rear ventilation is presented. This work contributes to closing the research gap concerning the performance estimation of the electrical and thermal system of such BIPVT systems, also under real operating conditions. Therefore, a simulation model which enables a holistic evaluation of a BIPVT system energy yield and performance is set up and refined. To achieve this, the initial simulation model is adapted to state-of-the-art PV modelling and expanded by still required simulation approaches for system components not investigated yet. Additionally, besides the coupling of insights from previous research, influencing parameters are derived specifically for installed components on site. The calculation results for the installed test site indicate electrical system yields within the range of conventional PV systems and above average performance ratios. Due to the consideration of the solar cooling effect, an efficiency gain for the electric energy yield of 2.0 % is achieved. An increase of the annual performance of the heat pump of 6.0 % at best case is identified by using the solar roof tiles waste heat. In this context, a low mass flow rate of the base-fluid appears to be the operating condition leading to the highest benefits for the overall system yield. A model validation performed with the use of infield measurement data shows an overestimation of 1.3 % concerning the actual electrical yield. Overall, the final model predicts the electric power output before inverter losses with an accuracy of 4.7 % in relative RMSE. It can be concluded that the developed simulation model enables a realistic overall energy yield prediction for the novel BIPVT solar roof tile.
Keywords: BIPVT, Modelling, System Performance

1 INTRODUCTION

Photovoltaic thermal systems, hereinafter referred to as PV/T systems, consist of photovoltaic modules combining the use of solar radiation to produce electric and thermal energy. The hybrid system technology converts the sunlight based on the basic principles of photovoltaic solar cells, while transferring thermal energy via a base-fluid from the module which simultaneously functions as an absorber. Overall, the innovative technology offers the enhanced utilization of the solar spectrum due to the usage of both heat and light [1, 2]. Within the last years, the implementation of PV/T systems into buildings (BIPVT) received enormous attention due to promising advantages as f.e. decreasing overall system costs and savings in primary energy, resulting a lower environment impact. Despite of the rapid growth in interest, the current status of research regarding BIPVT not only demonstrates a lack of evaluations concerning economic and environmental aspects, but also investigations on the electrical side of the systems, as well as extensive studies representing real operating conditions [1–3].

Within the framework of a research project, the investigated system is a grid connected BIPVT system consisting of 192 hybrid solar roof tiles (SRT), located on the roof of the technical university in Cologne, Germany [4]. For simulating the electric and thermal energy yield of the pilot site, a simulation model based on open-source libraries pvlib-python and TESPy, was set up in previous works [5, 6]. The main objective of this work is the refinement and validation of this simulation model in terms of each optical, electrical, and thermal aspects influencing the model behavior and generated results. Besides general code optimization, the integration of computational fluid dynamics (CFD) simulation results concerning the dynamic of the base-fluid within the thermal subsystem is another objective within this research. Furthermore, the accuracy of the final model is to be validated by onsite measurement data.

2 METHODOLOGY

2.1 Electrical and optical model refinements

The optimization of the simulation model with regards to the electric system of the BIPVT plant includes the modification of the solar modules cell temperature estimation, determination of the incoming effective irradiance and the applied model for calculating the electric energy yield. Furthermore, the cooling effect on the solar modules caused by the heat absorption of the base-fluid is considered. In the following, the initial simulation model, which forms the basis for the optimized version, is referred to as the initial model or original model.

2.1.1 Module temperature estimation

The previous method applied in the initial model in terms of the cell temperature estimation is per the Sandia Array Performance Model (SAPM). Due to the adoption to the IEC 61853 standards for performance testing and energy rating, the module temperature determination is now carried out by the usage of the Faiman model. The module temperature (T_{mod}) is derived in expression (1), taking into account variations in wind speed (v) and the plane of array irradiance (H) [6].

$$T_{mod} = T_{mod} + H/(U_0' + U_1' \times v) \qquad (1)$$

Both constant coefficients U'_0 and U'_1 describe the heat loss of the modules in dependency of their optical efficiency as well as wind conditions and can deviate for different module types [6]. In this context, the heat loss coefficients for the solar modules on site are determined by using a least-squares fit approach as described in [6], based on in situ measurement data of the actual module temperature and the corresponding weather and irradiation conditions.

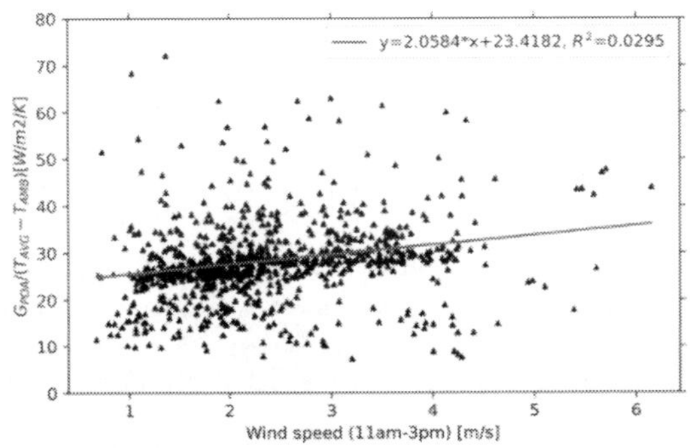

Figure 1: Plot of $G_{POA}/(T_{AVG}-T_{AMB})$ against the daily wind speed for measurement data recorded between the 26th of May to the 30th of June between 11:00 am and 3:00 pm in 5 min intervals, including the least squares linear fit of the data set.

The approach shown in Fig. 1 results in a heat loss coefficient of 23.4182 $Wm^{-2}C^{-1}$ for U'_0 and 2.0584 $Wm^{-2}C^{-1}(m/s)$ for U'_1, where the heat loss coefficients correlate to the slope and intercept of the regression line. It has to be mentioned that these temperature coefficients describe the module temperature considering the cooling effect due to an operating ventilation system. For comparison, a different set of measurement data with a non-operating ventilation system is used to determine the temperature coefficients for representing the module temperature without cooling.

2.1.3 Effective irradiance

The initial simulation model estimates the effective irradiance on the module plane by using the SAPM model as explained in [7]. Optical losses are aspired to be partly considered by the usage of the corresponding modifying factors stated in the Martin and Ruiz model. The basic formula for the effective irradiance (E_e) by SAPM [7] is shown below:

$$E_e = f_1(AM_a) \cdot \left\{ \frac{(E_b \cdot f_2(AOI) + f_d \cdot E_{diff})}{E_o} \right\} \cdot SF \quad (2)$$

For an estimation of the effective irradiance according to SAPM, the reference solar irradiance (E_0) as well as the irradiance components of the diffuse irradiation (E_{diff}) and direct irradiation (E_b) are required. Therefore, E_b is the product of the direct normal irradiance (E_{dni}) and the cosine of the angle of incidence (AOI).

While the solar spectral variation $f_1(AM_a)$ is still derived by the SAPM approach, the solar-angle-of incidence coefficient ($f_2(AOI)$), as well as the equivalent coefficient (f_d) for modifying E_{diff}, are to be brought in via the incident angle modifier by Martin and Ruiz. Again, this model refers to another IEC 61853 standard. Due to the assumption of clean modules, a soiling factor (SF) as an indicator for the module pollution of 1 is assumed. [7]

The estimation of the incident modifier (IAM) as a designation for the angular losses in dependency of the AOI is shown in (5). The model contains the angular losses coefficient (a_r), an empirically estimated parameter which describes the reflectance of different module types with regards to their configuration [8, 9].

$$IAM = \frac{1 - \exp(-\cos(\frac{AOI}{a_r}))}{1 - \exp(-\frac{1}{a_r})} \quad (3)$$

Besides the angular loss factors for the diffuse- and direct irradiation $F_B(AOI)$ and $F_D(\beta)$, the Martin and Ruiz model also considers angular losses affecting the ground reflected (albedo) radiation component. The corresponding factor $F_A(\beta)$ is determined in dependency of the module tilt (β) [8]. The diffuse irradiation components are further distinguished into the sky- and ground diffuse irradiation $E_{diff,sky}$ and $E_{diff,gr}$. Taking into account the parameters explained, the adapted formula applied in the refined model for estimating the effective irradiance is stated as follows:

$$E_e = f_1(AM_a) \cdot \{(E_b \cdot F_B(AOI) + F_D(\beta) \cdot E_{diff,sky} + F_A(\beta) \cdot E_{diff,gr})/E_o\} \cdot SF \quad (4)$$

2.1.4 Electric energy yield

As well as for the module temperature and effective irradiance estimation, the model for determining the electric energy yield is adapted from SAPM to the De Soto Single-Diode Model (CEC).

Due to the requirement for the optimized model to be easily accessible for simulating BIPVT plants with differing system components, the Single-Diode Model appears to be a suitable choice. This is justified with the Single-Diode Model exclusively relying on module data delivered by the manufacturer as in input. In this context it must be mentioned, that the empirical SAPM model requires parameters that need to be distinguished in additional module examinations [10]. While the models accuracy for forecasting the maximum power point (MPP) is proven to be in a similar range as those from the SAPM, the Single-Diode Model is capable of estimating the complete I-V curve of the module instead of only key points [11]. This allows the model to estimate the systems power output under part load operation and relatively large DC/AC ratios [11], which is assumed to be of importance for the investigated system being simulated under varying operating conditions.

The five parameters: diode reverse saturation current ($I_{o,ref}$), light current ($I_{L,ref}$), series resistance ($R_{s,ref}$), shunt resistance ($R_{sh,ref}$) and ideality factor parameter (a_{ref}) - each at reference conditions, are needed to estimate the I-V curve in a first calculation. Based on the I-V curve at reference conditions, the current-voltage relationship can subsequently be determined according to any other operating point as stated in (7) [10].

$$I = I_L - I_o \left[e^{\frac{V+IR_s}{a}} - 1 \right] - \frac{V + R_s}{R_{sh}} \quad (5)$$

where:

$$a \equiv \frac{N_s n_I k T_c}{q} \tag{6}$$

The ideality factor parameter (a) shown in (8), therefore consists of the number of cells in series (N_s), the ideality factor (n_I), the module temperature (T_C), as well as the Boltzman's constant (k) and the electron charge (q) [10]. Within the framework of this research project, the corresponding parameters from a calibration for the modules on site are fitted by the *ivtools.sdm.fit_cec_sam* function of the *PVLIB* library to calculate the mentioned quantities needed for solving the Single-Diode model equation.

Due to the initial BIPVT simulation model's lack of an implementation to calculate the AC power output of the system, the *PVWatts* inverter model from the National Renewable Energy Laboratories (NREL) is added. Briefly explained, the model estimates the inverters efficiency up to the nominal rated inverter efficiency (η_{nom}) as a function of the systems DC power output [12]. For the investigated BIPVT system, two inverters *SUNNY BOY 1.5* of the manufacturer *SMA* with a combined maximum DC input power (P_{dc0}) of 6 kWp and a nominal inverter efficiency of 0.961 are installed [13].

2.2 Thermal model refinements

The investigated system consists of 192 SRTs, each connected in series of 12 SRTs per string with an arrangement of 16 strings in parallel. It is important to mention that this configuration applies for the back ventilation system due to the design of the SRTs, whose air ducts are connected within one string. The individual air volume flows are bundled in one tube bevor entering the ventilator. Furthermore, two strings at each side of the BIPVT plant are ventilated passively, resulting a consideration of the thermal system as 12 strings in parallel each 12 SRTs in series. Diverging to the thermal system, the electrical interconnection of the modules is of a different structure. It corresponds to 96 modules connected in series; each string linked to one inverter.

Adjustments concerning the thermal simulation model relate to a more detailed approach for modelling the pressure loss within the subsystem of the BIPVT plant, as well as the adaption of the SRTs heat loss coefficients. Due to the non-existence within the initial model, a heat pump model of manageable complexity is implemented to enable the performance assessment of a further system component.

2.2.1 Optimized modelling of the systems pressure loss

The initial thermal simulation model, based on the open-source library *TESPy*, foresees a consideration of the systems pressure loss by the usage of a unitary geometry independent friction coefficient (ζ) for expressing the flow resistance within the air vent of the SRT [4].

Due to insights of CFD simulations concerning the behavior of the volume flow of the air as the base-fluid, the necessity of a differentiated view on the leakage mass flow for each SRT became obvious [14]. The leakage mass flow, caused by the air gap between each module, is simulated as a mass flow within an expansion valve, leaving the systems boundaries. One expansion valve exists for each junction between two solar collectors within one string [4]. With the assignment of a leakage mass flow, congruent to the CFD simulation results for each valve, a differentiated integration of the leakage mass flow is ensured. Combined with a corresponding pipe roughness factor (*ks*) as a roughness parameter for the solar collector, the overall pressure drop is then simulated. The pressure drop in dependency of the operating condition again corresponds to the CFD simulation results. The ks values for each solar collector are found with regards to the applied leakage mass flow and by interpolation based on the CFD simulation results.

2.2.2 Adaption of the heat loss coefficients

To consider thermal losses of the SRTs in the simulation model, use of the possibility of assigning heat loss coefficients for the component *solar collector*, representing one SRT, is made of [4]. Due to the unavailability of reference values for solar roof tiles as unconventional absorbers [15], the heat loss coefficents are estimated module specific by a numerical approach. Therefore, a non-linear least squares fit is applied on the equation for the specific collector performance (P_g), specified for the collector surface area (A_c) as shown in (9). The adapted formula to estimate P_g/A_c referring to [16], is than used to estimate the first order heat loss coefficient (a_1) and the second order heat loss coefficients (a_2) numerically.

$$P_g/A_c = [\eta_0 \cdot G - a_1 \cdot (T_m - T_a) - a_2 \cdot (T_m - T_a)^2] \cdot f_P \cdot f_U \cdot f_0 \tag{7}$$

The module temperature (T_m), the total solar irradiation on the collector plane (G) and the ambient temperature (T_a) is based on two days of onsite measurement data in June 2022. While correction factors concerning measurement uncertainty (f_U), unforeseen heat losses (f_0) or heat losses due to pipes and transmission lines (f_P) are neglected, a peak collector efficiency (η_0) of 0.837 is assumed for the SRT, based on material properties.

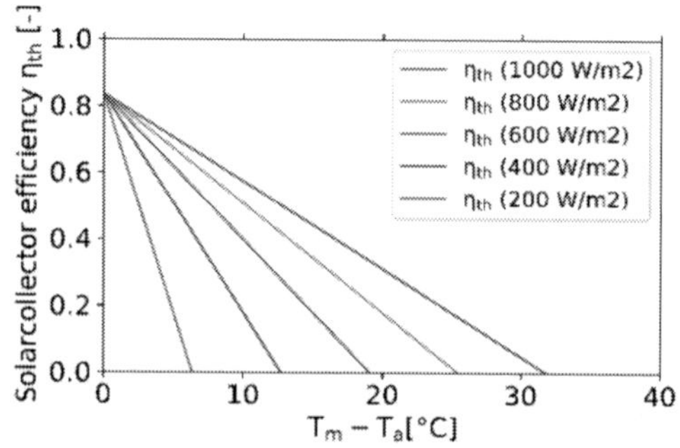

Figure 3: Collector efficiencies of the investigated SRT for different irradiances in dependency of the temperature difference between module (T_m) - and ambient temperature (T_a).

When equation (9) is set equal to zero, the solar collector reaches its stillstand temperature without submitting any thermal power [17]. This sets the prerequisite to determine both heat loss coefficients by applying the nonlinear least squares fit, here with the usage of the *curve_fit* function derived from the open-source library *scipy*. The calculations result in a first order heat loss coefficient a_1 of 26.3 W·K^{-1}·m^{-2} and a second order heat loss coefficient a_2 of 0 $W \cdot K^{-2} \cdot m^{-2}$.

By applying the coefficients in the collector efficiency equation stated in [16], various curves for the efficiency for different irradiances in dependency of the temperature range between the module- and ambient temperature can be derived: Those are shown in figure 3. In comparison, to investigated PV/T modules in a study for hybrid PV/T systems development in 2014 (compare [18]), the SRT shows relatively high collector efficiencies at low temperature differences for 1000 Wm^{-2}. Therefore, the efficiency decreases more rapidly with increasing temperature differences compared to the PV/T modules taken for reference.

2.2.3 Heat Pump Model

Initially, the simulation model calculated the thermal energy output as the merged heat flux from 16 parallel arranged strings of SRTs transferred by a fan located behind the intake pipe. To enhance the expressiveness of the results of the thermal model and to comply with the system design given on site, a model for determining the coefficient of performance (COP) and furthermore the annual performance factor of a heat pump is implemented into the thermal simulation model. The approach refers to the general context of the heat pump's COP linked to the heat source- and sink temperatures, which are again connected to technical properties and atmospheric conditions. The given formula for the COP shown in (10) relies on the quadratic regression of manufacturer data for several heat pumps with regards to the COP for various temperature differences between any combinations of heat- sources and sinks (ΔT). [19]

$$COP = 6.08 - 0.09 \cdot \Delta T + 0.0005 \cdot \Delta T^2 \qquad (8)$$

While the heat sink temperature is set to 50 °C for domestic hot water preparation as the intended use of thermal energy, the merged heat flux of the SRTs behind the fan is adopted as the heat source.

2.3 Performance Evaluation

To evaluate the overall systems performance for the thermal and electrical part, performance parameters are added to the simulation model. Besides the annual performance factor and the COP as performance indicators for the thermal model, use of the DIN EN IEC 61724-1 (VDE 0126-25-1):2021-03 standard is made to evaluate the performance of the electrical system.

According to the referred standard, the performance ratio *PR* as the ratio between the final system yield (net energy yield) (Y_F) and the ideal system yield (Y_R) is determined for the investigated system. Furthermore, the performance ratio under standard test conditions (PR'_{STC}) and for the annual temperature equivalent ($PR'_{annual\text{-}eq}$) are determined under consideration of the reference temperature of 25 °C for PR'_{STC} and the annual average of the module temperature for $PR'_{annual\text{-}eq}$.

While the performance parameters according to the previously stated standard reviews the final electric energy yield considering the inverter losses, the cell efficiency (η_{cell}) based on the module power operating at the maximum power point (P_{MPP}) is calculated as shown below [17]. Here, A_{cell} cell to the solar cell area.

$$\eta_{cell} = \frac{P_{MPP}}{E_e \cdot A_{cell}} \qquad (9)$$

The same applies for the thermal efficiency (η_{cell}) of the system, an efficiency parameter derived from the context of η_{cell}. In this research project, η_{th} is therefore defined as the ratio of the base-fluids heat output to the product of the effective irradiance and the solar cell area. The parameter referred to for evaluating the performance of the heat pump in this case, is the annual performance factor $\mathcal{E}_a$, which is determined as follows.

$$\mathcal{E}_a = \frac{\sum_a \dot{Q}_{ab}}{\sum_a P_{ASHP}} \qquad (10)$$

The annual performance factor is known as the averaged COP, defined as the ratio of the annual sum of the usable thermal energy ($\sum_a \dot{Q}_{ab}$) against the annual electric energy consumption of the heat pump ($\sum_a P_{ASHP}$), in this case of the compressor [17].

3 RESULTS

3.1 Effects of the model optimizations on simulation results

The period for assessing the effects of the model refinements and furthermore the system performance is set to one year. The utilized weather data is based on a data set derived from the PVGIS database of the *European Commissions Photovoltaic Geographical Information system* for 2012, which is stated to be a typical meteorological year [20]. Due to the adaption of the module temperature coefficients U'_0 and U'_1 corresponding to the Faiman model, a relative drop of the annual DC electrical energy yield of 4.3 % can be seen. By the application of the Single-Diode Model, the yield increases again by approximately 1.2 % in contrast to the SAPM model.

With regards to the adaptions for the optical model, the consideration of the angular losses coefficients results a decrease of the absolute effective irradiance per square meter from approximately 1166.2 $kWh \cdot m^{-2}a^{-1}$ by 2.4 % to 1138.0 $kWh \cdot m^{-2}a^{-1}$. The angular losses coefficient a_r is assumed to be 0.14 for the modified glass surface of the solar roof tiles in this case, which lies below the standard a_r value of 0.169 [8] for standard glass in monocrystalline technologies. Due to lower effective irradiances, the annual electric DC energy yield is decreased by 2.5 % for the Single-Diode Model.

Finally, the consideration of the solar cooling effect forms the connection between the thermal and electrical system, leading to lower annual average module temperatures and higher electric energy yields. The annual average module temperature within hours of effective irradiation therefore decreases from 34.2 °C to 31.9 °C due to the cooling effect, causing an additional gain in the electric DC energy yield of 1.95 %.

Concerning the simulation approach for the thermal model, the effects of the model adaptions have a higher influence on the base-fluids energy content and temperature than on the electric energy output in comparison. By integrating valve specific leakage mass flows, the pressure drop within each string of SRTs can be fitted to the existing values from the CFD simulations, as shown in figure 4.

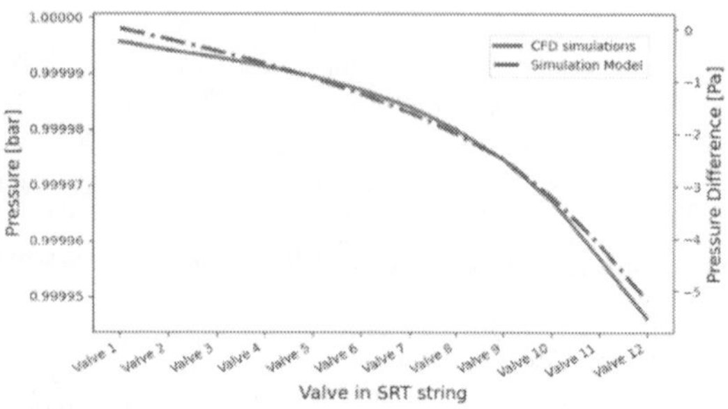

Figure 4: Pressure (bar) within each valve as well as the corresponding pressure difference (Pa) due to the leakage mass flow within one string of SRTs for a base-fluid mass flow of 30 $m^3 \cdot h^{-1}$.

The pressure drop for one string of SRTs at a mass flow rate of 30 $m^3 \cdot h^{-1}$ is shown in figure 4 and can be predicted for each air gap with a root mean squared error (RMSE) of 0.16 Pa with regards to the CFD simulations. The predicted absolute pressure drop of - 5.16 Pa (Valve 12) deviates by 0.36 Pa in comparison to the actual value derived from the CFD simulations for similar operating conditions. Nevertheless, the optimized representation of the pressure drop results a slight reduction of the annual average for the base-fluid temperature and specific annual energy content.

In contrast to this, the adaptation of the first- and second order heat loss coefficients have a higher impact on the thermal energy yield as compared to the refined consideration of the pressure drop. Table I shows the key results of the calculation performed with each initial and adapted coefficients.

Table I: Comparison of the thermal energy output concerning the base-fluid for the initial and adapted first- and second order heat loss coefficients for a mass flow rate of 30 $m^3 h^{-1}$ within a single SRT string.

SRT Heat loss coefficients	Base-Fluid Mean Temperature [°C]	Base-Fluid specific energy [$kWh \cdot m^{-2}$]	Thermal Efficiency [%]
Initial	26.6	110.4	9.7
Adapted	27.7	138.3	12.15

Table I shows that the adapted heat loss coefficients lead to an increase of the base-fluids specific energy content from 110.4 $kWh \cdot m^{-2}$ to 138.3 $kWh \cdot m^{-2}$ per year, related to its temperature increase of 1.1 °C. While the thermal efficiency rises from 9.7 % to 12.15 %, it must be mentioned that the electrical energy yield therefore slightly decreases by approximately 28 $kWh \cdot a^{-1}$.

Ultimately, with the additional integration of the inverter and heat pump model, the final simulation model predicts an electrical energy output of 2901 kWh covering inverter losses for a base-fluid mass flow of 25 $m^3 \cdot h^{-1}$ and the period under consideration. In this context it is important to mention that the results for the heat pumps energy yield and performance evaluation assume a continuously operating heat pump, without considering a thermal storage. The final simulation model's prediction for the specific electric energy yield is 1049 kWh/kWp, which lies around 60 kWh/kWp above the average specific yield for typical monocrystalline photovoltaic power plants of similar alignment and location between 2012 and 2019 [21]. Concerning the solar cell efficiencies, the annual average for η_{cell} within hours of effective irradiation of approximately 14.4 % shows a good agreement with the measured efficiencies from the module calibration report under STC conditions [22].

3.2 Performance estimation for different operating conditions

Figure 5 shows an electric energy gain around 2 % for the final electric energy yield when operating the fan even at a mass flow rate of 25 $m^3 \cdot h^{-1}$ considering the solar cooling effect.

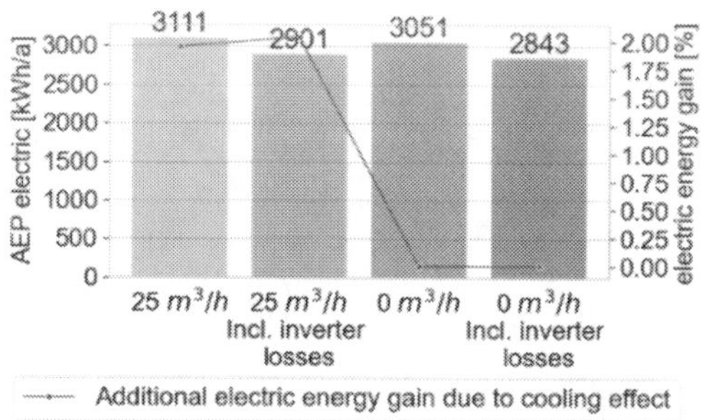

Figure 5: Influence of the solar cooling effect on the annual electric energy production (kWh) before and after inverter losses shown as the electric energy gain (%) for a base-fluid mass flow of 25 $m^3 h^{-1}$.

Concerning the performance of the system, the annual average for the PR is 0.8340 for a mass flow of 25 $m^3 \cdot h^{-1}$, and therefore indicates an above-average performance in comparison to common PV systems (2019) [17, p. 276]. Respectively, the temperature-adjusted $PR'_{annual\text{-}eq}$ reaches 0.8301. In this context, it must be noted that only values for plane of array irradiances above 50 $W \cdot m^{-2}$ are taken into account. This is justified with unreasonable values for the performance ratios at irradiances lower 50 $W \cdot m^{-2}$.

An illustration of the PR for both cases of the fan in and out of operation is shown for a monthly resolution in figure 6. Besides lower system performances in the summer months in comparison to the winter months, it becomes clear that the cooling of the SRTs by a base-fluid holds the potential for increasing the systems electrical performance independent of the time of the year.

While the increase of the PR is with approximately 0.5 % in December at its lowest magnitude, an increase of up to 1 % can be reached in August. In general, the PR ranges between 82.0 % in the summer months and approximately 87.8 in the winter months.

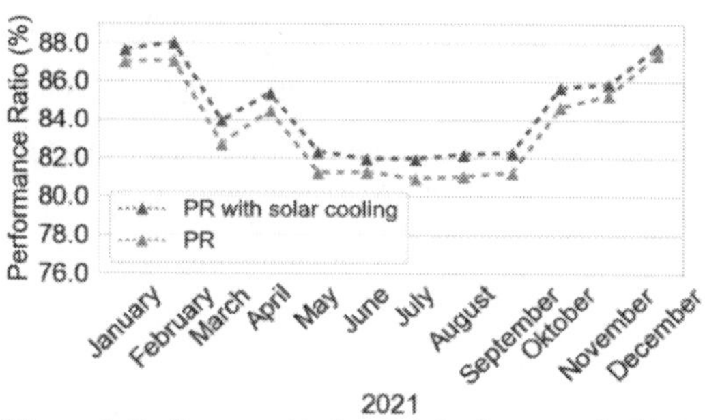

Figure 6: Performance Ratios (%) for the case of a fan in- (with solar cooling) and out of operation for the period

under consideration in a monthly resolution. The base-fluid mass flow is of 25 $m^3 \cdot h^{-1}$.

Analogous to the approximately linear increase of the electrical energy yield with higher mass flows, the base-fluid heat flux, merged after the fan, lies between 53.7 – 138.3 $kWh \cdot m^{-2}$ per year for the same mass flow range as stated in figure 5. In this context, the heat flux can be regarded to as an exergy flow, since it relates to the temperature difference between the heat flux temperature itself and the ambient temperature. The correlation between different mass flow rates, the heat flux temperature and the heat pumps thermal energy output is depicted in Table II.

Table II: Correlation between different mass flow rates and the thermal systems performance, as well as the annual mean base-fluid temperature within hours of effective irradiation. 'HP' corresponds to heat pump.

Mass Flow rate [$m^3 \cdot s^{-1}$]	Base-Fluid Mean Temperature [°C]	HP specific thermal energy [$MWh \cdot m^{-2}$]	Annual Performance Factor HP [-]
10	28.8	5.66	3.54
20	28.2	5.62	3.52
30	27.7	5.60	3.51

In contrast to the correlation of higher mass flow rates and increased efficiencies for the electrical system, the specific thermal energy produced by the heat pump increases with decreasing mass flow rates. This context can be derived also for the base-fluids mean temperature and the heat pumps annual performance factor (ε_a). While ε_a is at 3.51 for a mass flow of 30 $m^3 \cdot h^{-1}$, it reaches a magnitude of 3.54 when reducing the mass flow to 10 m^3. For ensuring a comparability of the results, ε_a for the heat pump operating only with ambient air as a heat source is considered in addition. Ultimately, the heat pump operates with an annual performance factor of 3.34 with ambient air as its heat source and can be increased by approximately 6.0 % when utilizing the solar modules waste heat. With an ε_a of 3.54 at best case, the heat pump is located in the above average performance field, compared to investigated systems for a similar field of application within in a study of the *Fraunhofer ISE* in 2020 [23].

3.3 Performance estimation for different operating conditions

A measurement dataset, recorded for a clear-sky irradiance in June 2022 and partly shown in figure 7, is used to validate the simulation model and evaluate the accuracy of the predicted quantities. To calculate the plane-of-array (POA) irradiance the Klucher model is used [25], which shows the lowest RMSE for the conducted measurements.

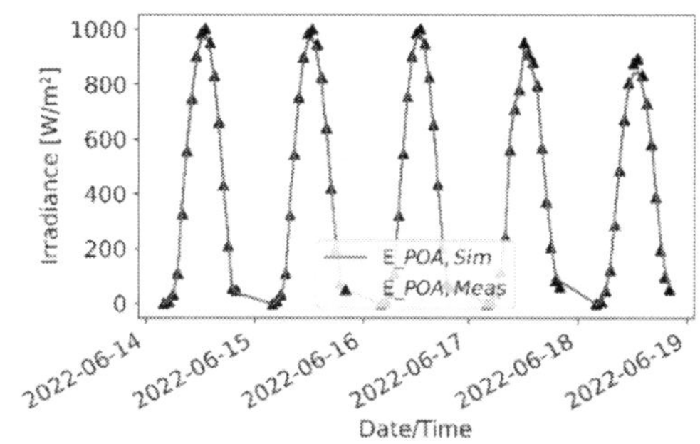

Figure 7: Correlation of measured and simulated irradiances for five days of measurement data in June 2022.

By comparing the measured DC power against the simulated DC power output, a slight overestimation of the electric energy yield with 1.3 % is shown in figure 8.

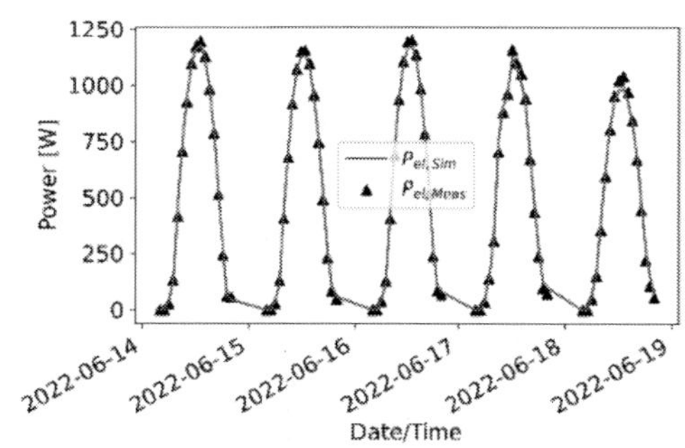

Figure 8: Correlation of measured and simulated power output for five days of measurement data in June 2022.

Ultimately, the final simulation model predicts the DC power output with a RMSE of 25.1 W and a mean absolute error (MAE) of 7.3 W for the period under consideration. Furthermore, the DC electric energy yield is predicted to be 45.9 kWh for the period under investigation, which deviates by 0.6 kWh to the measured energy yield of 45.3 kWh. For the electric energy yield covering inverter losses, the actual electric energy yield of 43.6 kWh is overestimated by 1.2 kWh in a similar order of magnitude as the measured and simulated DC yield.

The predicted and measured module and base-fluid temperature of the SRTs are shown in figure 9. With regards to the module temperatures of the SRTs as well as the base-fluid temperature, figure 9 shows a good agreement in comparison to the measured quantities. For the prediction of the module- and base-fluid temperature, a proportional increase can be seen during the period under consideration.

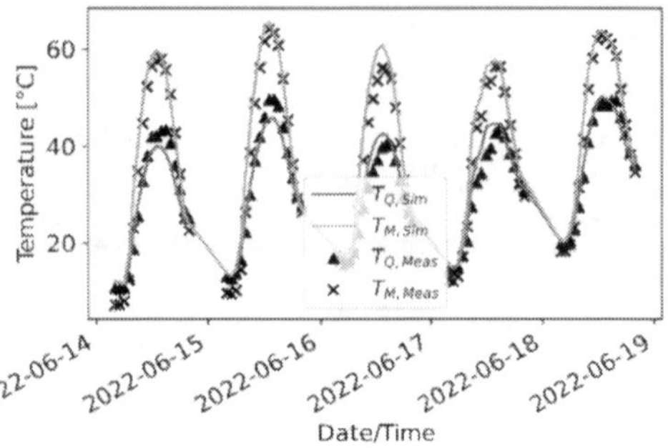

Figure 9: Correlation of measured and simulated temperatures for five days of measurement data in June 2022.

Finally, the highest predicted module temperature ranges up to approximately 65.6 °C, compared to a maximum measured module temperature of around 64.2 °C. The prediction of the module temperatures occurs with an RMSE of 3.4 °C and a MAE of 2.7 °C with regards to the mean module temperature for four measured SRT module temperatures.

For evaluating the prediction accuracy of the base-fluid temperature, the measured air outlet temperature within the backside of an SRT, located in the middle of the first upper row is used for a validation. The corresponding course for the temperature of the base-fluid is shown in figure 9. First, a similar course with a less steep increase of the temperature can be seen for the base-fluid in comparison to the measured module temperatures in the morning. Compared to the actual measurement values, the simulation model slightly overestimates the base-fluid temperature. The model predicts maximum temperatures of 50.3 °C, compared to highest measured base-fluid temperatures of 49.7 °C. Overall, the base-fluids temperature is predicted with a RMSE of around 2.4 °C and a MAE of 0.4 °C, so with a higher accuracy compared to the prediction of the module temperatures.

4 DISCUSSION AND CONCLUSION

With the proposed model, a holistic assessment of a BIPVT plants electrical and thermal energy output, as well as the estimation of the systems overall performance is possible. Due to the optimization of the initial simulation model, essential system components are integrated into the simulation process, whose parameters can be adjusted easily according to their specific properties.

With regards to the investigated system in this research project, the adaptation of the heat loss coefficients according to the Faiman model as well as the consideration of angular losses for the effective irradiance result the highest impact on the electrical yield when it comes to the development of the initial model. Concerning the model adaptions for the thermal model, the fitting of the first- and second order heat loss coefficients shows the biggest influence, in this case for the base-fluids thermal energy. This also shows the importance of an accurate individual specification for the corresponding parameters of the simulated hybrid SRTs to guarantee a reliable statement about the overall systems energy yield.

Furthermore, the solar cooling effect shows a seasonal independent positive influence on the electrical energy yield. The highest additional electric energy gains can be achieved in the summer months, which is expected to be connected to the already lower module temperatures due to low ambient air temperatures in wintertime. Currently, the estimation of the cooled module temperature is only valid for a mass flow of 25 $m^3 h^{-1}$. To investigate a possible variety of the impact of the solar cooling effect on the electrical yield for different operating conditions, further measurements of the module temperature at different mass flow rates are a necessity.

With regards to the thermal energy yield, a small mass flow rate appears to be the most sufficient operating condition to benefit the overall system yield at most. It is important to say that these observations are made for a constant operation of the fan and heat pump. An integration of a part load- and season dependent operation is expected to deliver deeper insights on the systems performance and could be investigations in further research projects. Besides, the used HP model relies only on the temperature difference between heat sink and source. Differing results when considering the base-fluid mass flow are expected.

Concerning the deviation of the predicted against the actual measurands for the five days of measurement data in June, the absolute electric energy yields before and after inverter losses are overestimated by the final model in the same order of magnitude. Out of this, a higher potential for improvements emerges concerning the prediction accuracy of the DC energy yield. Overall, the predicted final electric energy yield deviates for five days by around 3 %.

Similar to these observations, an overestimation of the predicted against the measured temperatures for both SRTs and the systems base-fluid becomes apparent within the course of the model validation process. This fact could also relate to the above average performance of the heat pump, due to the temperature dependence of the used heat pump model according to Ruhnau et al.. A reason for the overestimated module temperature could be the infield measurement data used to find the temperature coefficients according to the Faiman model. Here, a measurement data set is used, that contains windspeeds and their directions which cannot be directly transferred to the wind conditions corresponding to the module plane, even if they were measured at the same location.

Furthermore, a more detailed consideration for the distribution of the module temperatures is conceivable, since the presented model adopts a unitary temperature for each module. Also, the need for accessing the time mismatch with regards to the offset of the module temperatures, assumed to relate to heating and cooling effects of the modules, becomes apparent for reducing the temperature prediction mismatch.

At this point it must be annotated, that the statements concerning the prediction quality of the simulation model is based on five days of measurement data with clear sky irradiance conditions within a specific season of the year. Besides this assessment, the validation of the model for a broader spectrum of measured irradiance conditions over longer time periods could deliver further insights concerning the models prediction accuracy.

5 ACKNOWLEGDEMENTS

This project has been financed by the European Fonds for Regional Development North Rhine-Westphalia (EFRE.NRW, Code: EFRE-0801561, Reference no. EU-2-2-048B)

6 REFERENCES

[1] M. A. Garba and B. YESILATA, "Use of Hybrid Photovoltaic-Thermal (PV/T) Solar Modules for Enhancing Overall System Efficiency", in 3rd International Symposium on Multidisciplinary Studies and Innovative Technologies: Ankara, Turkey, October 11-13, 2019, Ankara, Turkey, 2019, pp. 1–10.

[2] A. H. Al-Waeli, K. Sopian, H. A. Kazem, and M. T. Chaichan, "Photovoltaic Thermal PV/T systems: A review", IJOCAAS, vol. 2, no. 2, pp. 62–67, 2017, doi: 10.24842/1611/0027.

[3] A. Ibrahim, A. Fudholi, K. Sopian, M. Y. Othman, and M. H. Ruslan, "Efficiencies and improvement potential of building integrated photovoltaic thermal (BIPVT) system", Energy Conversion and Management, vol. 77, pp. 527–534, 2014, doi: 10.1016/j.enconman.2013.10.033.

[4] L. Clasing, C. Dick, U. Blieske, P. Hakenberg, S. Leyer, and J.-R. Hadji-Minaglou, "Analysis of a Novel Improved BIPV/T Roof Tile Design", Proceedings of the 15th Conference on sustainable development of energy, water and environment systems (SDEWES), Cologne, Germany, September 2020.

[5] V. Phadke, B. Mukherjee, and L. Clasing, "Performance estimation of hybrid solar roof tiles with effect of incident Angle Modifier using Python", Cologne Institute for Renewable Energy (CIRE), 2021.

[6] C. Brosig, L. Clasing, U. Blieske, and E. Waffenschmidt, "Modeling a novel building-integrated PVT-air-collector system coupled to a heat pump using open source libraries", Proceedings of the 16th Conference on sustainable development of energy, water and environment systems (SDEWES), Dubrovnik, Croatia, October 2021.

[7] D. Faiman, "Assessing the outdoor operating temperature of photovoltaic modules", Prog. Photovolt: Res. Appl., vol. 16, no. 4, pp. 307–315, 2008, doi: 10.1002/pip.813.

[8] J. Kratochvil, W. Boyson, and D. King, "Photovoltaic array performance model", 2004.

[9] N. Martin and J. M. Ruiz, "Calculation of the PV modules angular losses under field conditions by means of an analytical model", Solar Energy Materials and Solar Cells, vol. 70, no. 1, pp. 25–38, 2001, doi: 10.1016/S0927-0248(00)00408-6.

[10] N. Martin and J. M. Ruiz, "Corrigendum to "Calculation of the PV modules angular losses under field conditions by means of an analytical model" [Sol. Energy Mater. Sol. Cells 70 (1) (2001) 25–38]", Solar Energy Materials and Solar Cells, vol. 110, p. 154, 2013, doi: 10.1016/j.solmat.2012.11.002.

[11] W. de Soto, S. A. Klein, and W. A. Beckman, "Improvement and validation of a model for photovoltaic array performance", Solar Energy, vol. 80, no. 1, pp. 78–88, 2006, doi: 10.1016/j.solener.2005.06.010.

[12] J. S. Stein, J. Sutterlueti, S. Ransome, C. W. Hansen, and B. H. King, "Outdoor PV Performance Evaluation of Three Different Models: Single-Diode, SAPM and Loss Factor Model", 2013, doi: 10.4229/28thEUPVSEC2013-4CO.11.1.

[13] A. Dobos, "PVWatts Version 5 Manual", 2014.

[14] SMA Solar Technology AG, "SUNNY BOY 1.5 / 2.5 - The best standard for small PV systems," [Online]. Available: https://files.sma.de/downloads/SBxx-1VL-40-DS-AU-51.pdf

[15] L. Clasing, P. Hin Yam, C. Ziller, and U. Blieske, "Dynamic simulation model of an air-based BIPVT roof tile system", Proceedings of the 8th World Conference on Photovoltaic Energy Conversion (WCPEC), Milan, Italy, September 2022.

[16] C. Brosig, "Energiekonzepte für die Solardachpfanne in Ein- und Mehrfamilienhäusern, sowie für Quartiere", Cologne Institute for Renewable Energy, 2021

[17] F. Bava, J. E. Nielsen, S. Knabl, A. Brunger, S. Furbo, and C. Fink, "Results of IEA SHC Task 45: Large Scale Solar Heating and Cooling Systems. Subtask A: "Collectors and Collector Loop", Energy Procedia, vol. 91, pp. 546–556, 2016, doi: 10.1016/j.egypro.2016.06.197.

[18] V. Quaschning, Regenerative Energiesysteme: Technologie - Berechnung - Klimaschutz, 10th ed. München: Hanser, 2019. [Online]. Available: http://files.hanser.de/Files/Article/ARTK_LPR_9783446461130_0001.pdf

[19] S. Fortuin, M. Hermann, G. Stryi-Hipp, P. Nitz, and W. Platzer, "Hybrid PV-thermal Collector Development: Concepts, Experiences, Results and Research Needs", Energy Procedia, vol. 48, pp. 37–47, 2014, doi: 10.1016/j.egypro.2014.02.006.

[20] O. Ruhnau, L. Hirth, and A. Praktiknjo, "Time series of heat demand and heat pump efficiency for energy system modeling", Scientific data, vol. 6, no. 1, p. 189, 2019, doi: 10.1038/s41597-019-0199-y.

[21] T. Huld, R. Müller, and A. Gambardella, "A new solar radiation database for estimating PV performance in Europe and Africa", Solar Energy, vol. 86, no. 6, pp. 1803–1815, 2012, doi: 10.1016/j.solener.2012.03.006.

[22] J. Schardt and H. te Heesen, "Performance of roof-top PV systems in selected European countries from 2012 to 2019", Solar Energy, vol. 217, pp. 235–244, 2021, doi: 10.1016/j.solener.2021.02.001.

[23] H. Maaroufi, "Kalibrierung von Photovoltaik (PV) Referenzmodulen. Prüfbericht Nr.: 21251142.001", Köln, Nov. 2020.

[24] D. e. Günther, "Wärmepumpen in Bestandgebäuden - ergebnisse aus dem Forschungsprojekt "WPSmart Im Bestand"", Freiburg, Jul. 2020.

[25] Klucher, T. M. (1979): Evaluation of models to predict insolation on tilted surfaces. In: Solar Energy 23 (2), S. 111–114. DOI: 10.1016/0038-092X(79)90110-5.

PROPOSAL FOR THE EXTENSION OF THE ENERGY RATING STANDARD SERIES IEC 61853 TO BIFACIAL MODULES

Robert P. Kenny*[1], Ana María Gracia-Amillo+[1], Teodora Lyubenova[1], Diego Pavanello[1]
[1]European Commission, Joint Research Centre (JRC), Ispra (VA), Italy
*[1]Tel: +39 0332789287, e-mail: robert.kenny@ec.europa.eu; agracia@cener.com (+[1] at present in Cener);
teodora.lyubenova@ec.europa.eu; diego.pavanello@ec.europa.eu

ABSTRACT: Bifacial crystalline (c-Si) photovoltaic (PV) devices are already competing with monofacials in the global PV market given the rear side PV generation capacity and the possibilities of new mounting configurations that would allow them to be integrated with other applications and land uses. The IEC 61853 Standard series "*Photovoltaic (PV) module performance testing and energy rating*" defines a methodology to analyze the performance of PV devices under real working conditions and it could be used to rate and compare devices, providing end-users a tool to select the best suited device for their PV installation. However, the scope of this standard does not include bifacial devices. To that aim, the present paper identifies the required modifications to the standard and proposes a new methodology to extend the energy rating standard to bifacial devices. All four parts of the standard would need to be modified. The proposed methodology has been applied to two c-Si devices under various new mounting configurations relevant to bifacial devices. Results show significant differences, which suggests that the proposed methodology could serve as a tool to compare and select the most appropriate bifacial device for each location and mounting configuration, extending the energy rating Standard series to bifacials.
Keywords: Bifacial, Energy Rating, Qualification and Testing, Simulation, Standardization

1 INTRODUCTION

Bifacial photovoltaic (PV) cells, devices and systems are rapidly overtaking monofacial technologies in the global PV market. Moreover, according to the latest edition of the International Technology Roadmap for Photovoltaic (ITRPV), bifacial modules could reach 60% of the global market share by 2032 [1]. This market penetration can be explained by the fact that, not only bifacial devices are capable of absorbing solar radiation from both sides, but also, by the fact that by applying minor adjustments in the processing line, bifacial cells have reached rear efficiencies close to those of the front side [2]. These characteristics make bifacial devices better suited than monofacials for locations of high albedo such as desert environments where bifacials reach higher production levels under the same installation conditions. Snow covered areas, where monofacial devices suffer considerable losses, are also beneficial for bifacial devices due to the enhanced albedo. In these regions bifacial devices have a significant higher production and experience considerably less losses than monofacial devices [3].

Furthermore, bifacial devices offer the possibility of employing new configurations like vertical East/West oriented PV modules. The power output from this type of installation is better aligned with the normal household consumption, which takes place mainly in the mornings and evenings, than common equator facing PV systems. However, combined they could obtain a production curve similar to that of a solar tracked PV plant, without peaks of production at specific times characteristic of single oriented PV plants [4]. Moreover, the vertical configuration allows PV to be integrated with other land uses, like in agricultural areas as they barely interfere with agricultural activities and growth. Another possible application of bifacial devices in a vertical configuration is their integration in noise barriers. This extensive road infrastructure offers a huge potential for PV deployment which is especially interesting in areas with land scarcity or overpopulation. For example, a detailed simulation for the State of California (USA) then extrapolated to the whole country, estimated the PV potential of the US' existing noise barriers to be between 7 and 9 GW [5]. These features and characteristics of the bifacial devices certainly contribute to the significant increase that they have experienced on the global market in the past decades.

Notwithstanding, the energy output gain of bifacial devices compared to monofacials can vary a lot depending on the bifacial device itself (bifaciality) and the configuration (climatological characteristics, installation and ground albedo) ranging from 10% to up to 50%. This variability could limit the consumers' confidence and therefore hamper bifacial market growth.

PV devices are normally compared, rated and priced according to the power output measured under Standard Test Conditions (STC), even though these specific conditions of irradiance and temperature are hardly ever met outdoors. To overcome this simple, yet extended, approach, in 2018 the International Electrotechnical Commission completed the IEC 61853 Standard Series "*Photovoltaic (PV) module performance testing and energy rating*" which provides an estimate of the performance of PV devices under real working conditions. In opposition to the maximum power output measured at STC, the *Climate Specific Energy Rating* (*CSER*) parameter, output of the aforementioned IEC Standard, enables a more meaningful comparison between devices as it considers the performance under varied climatic conditions. The *CSER* represents the ratio of the energy yield expected from the device under test over one year of real working conditions compared to the expected production if the efficiency would always be the one measured under STC. Therefore, the *CSER* could serve as a decision tool to compare devices and select the best suited one for the climatic conditions at the site of interest. Nevertheless, the scope of the energy rating standard only covers monofacial devices (without restrictions in the PV technology), even though bifacial devices and future investors could benefit from a *CSER* applicable to bifacial devices.

The aim of this paper is to analyze the requirements to extend the scope of the IEC 61853 Standard series to bifacials and to propose a methodology to estimate the

CSER for bifacial devices. The actual Standard series is composed of four parts which would have to be modified in order to be applicable to bifacial devices.

In the present paper, a brief description of the energy rating standard will be presented followed by the proposed modifications and considered assumptions. Besides, an example of the calculated *CSER* for bifacial devices will be presented. The last part of the paper will present future lines of research in order to test and validate the assumptions considered in the presented proposal.

2 IEC 61853 STANDARD SERIES

The International Electrotechnical Commission (IEC) Standard series 61853 "*Photovoltaic (PV) module performance testing and energy rating*" defines an energy rating method (not to be confused with an energy yield estimation model), which takes into consideration various effects like the solar spectrum and the thermal behavior of the PV device. The standard is based on three components: a set of meteorological datasets representative of worldwide climatic conditions, a set of measurements and parameters specific of the PV device under test and a series of models to estimate the performance of PV devices. These components are described in the four parts that form the IEC 61853 Standard series, which are briefly described in the following subsections.

2.1 IEC 61853-1. Irradiance and temperature performance measurements and power rating

The first part of the Standard [6] describes various methodologies, both indoors and outdoors, to obtain the so-called power matrix. This is a set of 22 measurements of the PV module's power output when submitted to various levels of in-plane irradiance and module temperature representative of the working conditions when installed outdoors. The irradiance's range goes from 100 W/m^2 to 1100 W/m^2, while the module temperature varies between 15°C and 75°C. Not all possible combinations are applied. For example, the power output under 100 W/m^2 and 75°C, or 1100 W/m^2 and 15°C are not included as PV modules are only expected to operate under these circumstances in extreme working conditions.

2.2 IEC 61853-2. Spectral responsivity, incidence angle and module operating temperature measurements

In the standard, the performance of the PV device is calculated taking into account various effects linked to the simulated working conditions. In the first place, the irradiance reflected at PV device's cover is accounted for by means of the Angle of Incidence effect (AOI) and the *ar* parameter. Then, the spectral content of the received irradiance and the device's spectral response (SR) are considered in order to estimate the effectively absorbed irradiance. The last considered effect regards the thermal performance of the device for which the temperature of the module needs to be estimated This requires two new parameters, U_0 and U_1.

The second part of the Standard [7] describes the tests to be performed on the PV device in order to define the parameters that describe the considered AOI (a_r), spectral (SR) and thermal effects (U_0 and U_1).

2.3 IEC 61853-3. Energy rating of PV modules

Part 3 of the Standard series [8], contains the complete estimation chain of the PV module energy rating. The concatenation of various models results in the estimation of the expected power output of the device under test and the *CSER*. This requires two sets of input data. The first one describes the device being analyzed by means of the power matrix (Part 1) and the parameters relevant to the various considered effects as described in Part 2. The second set of input data, regards the climatic conditions that represent the real working conditions simulated in the IEC 61853 and are presented in Part 4.

The power output is estimated by interpolation and extrapolation of the power matrix values considering the estimated working conditions defined by the calculated values of effectively absorbed in-plane irradiance and module temperature. This temperature is, in turn estimated considering the ambient temperature, in-plane irradiance corrected for AOI only and the wind speed which can reduce the temperature of the PV device.

2.4 IEC 61853-4. Standard reference climatic profiles

As mentioned in the Introduction, the energy rating standard considers the PV performance under real working conditions, in opposition to STC. These conditions are defined by the six standard reference climatic profiles provided in Part 4 [9], named: tropical humid, subtropical arid, subtropical coastal, temperate coastal, high elevation above 3000m and temperate continental.

Every climatic profile is described by a timeseries which contains hourly values over one year of the following climatic variables: ambient temperature, wind speed and irradiance at 2m above the ground. Global and direct horizontal and in-plane broadband irradiance are included, along with the spectrally resolved global in-plane irradiance integrated into 32 spectral bands. The angle of incidence between the sun and the normal to the module's surface is also reported as it is required for the AOI effect estimation.

A more detailed description of the different parts of the IEC 61853 standard is presented in [10].

3 PROPOSED EXTENSION OF THE IEC 61853 TO BIFACIALS

Due the characteristics of bifacial devices and the new possible mounting configurations that could be applied, all four parts of the Standard series would require modifications. However, our proposal tries to align as much as possible with the actual Standard series premises, as described in [11].

3.1 Bifacial mounting conditions

One single mounting configuration is considered for the energy rating of monofacial devices: ground-mounted free-standing fixed rack facing the equator with an inclination angle of 20°. However, this may not be enough to properly characterize the performance of bifacial devices. Bifacials are likely to be integrated with other land uses by means of a vertical configuration which, in the absence of obstacles, would offer the same insulation conditions to both sides, in opposition to the tilted configuration. Since a standard cannot cover all possible mounting options, the most suitable or representative is to be proposed. Therefore, for the vertical configuration a possible new mounting condition could be the vertical East/West installation.

While the actual standard does not define any ground albedo as the in-plane irradiance is already provided in

Part 4, a value must be considered for bifacial devices. In this regard, in order to maximize the rear side contribution to the device's power output, most bifacial installations use enhanced ground albedo. In order to consider this in the proposed extension, two albedo values could be used: 0.2 representative of grass or concrete pavement and 0.6 typical of white sand or snow.

As a result, three fixed mounting conditions are proposed for bifacial devices, whose assumed location, orientation (Orient), tilt angle and ground albedo are shown in Table I. The two tilted configurations share orientation and tilt with the actual standard. The higher albedo option could represent a bifacial module installed on a flat roof, while for the ground mounted one, an albedo of 0.2 is considered, as for the vertical configuration.

Table I: Bifacial mounting conditions

Configuration	Orient	Tilt	Albedo
Fixed tilt - roof mounted	Equator	20°	0.6
Fixed tilt - ground mounted	Equator	20°	0.2
Fixed vert. E/W - meadow	E/W	90°	0.2

The module clearance above the ground is also a factor to be considered, which is not defined in the actual standard. A module height (from the ground to the bottom edge) of approximately 1m is believed to be appropriate in order to optimize the bifacial gain in a ground mounted open rack configuration, while lower values around 0.1-0.3m are more typical for rooftop systems in order to limit the wind loads.

3.2. Modifications to Part 4.

In order to define the in-plane irradiance on both sides of the bifacial device, the standard reference climatic profiles must be extended to include the irradiance on the rear side of the 20° tilted configuration considering both 0.2 and 0.6 ground albedo, and the in-plane irradiance on the east and west facing vertical planes of the proposed third configuration. The irradiance on the front side of the tilted configuration could be considered the one already available in Part 4.

In the present study, these new global in-plane irradiance values have been obtained applying Perez's anisotropic model, which in addition to the sky dome isotropic diffuse irradiance, considers as well the contribution to the diffuse in-plane irradiance from the horizon band and the circumsolar region [12, 13]. According to this model, the in-plane global irradiance is the sum of this three-component anisotropic diffuse irradiance (sky dome, horizon band and circumsolar), the contribution from the beam irradiance and the ground reflected irradiance. Other models like angular distribution models or ray-tracing could be also considered although, due to the assumed lack of surrounding obstacles and the single module simulation of the energy rating standard, the complexity of these methods may not report much benefit in terms of accuracy compared to the anisotropic model used in the present study.

The new in-plane spectrally resolved global irradiance values on the rear side of the tilted plane and both vertical planes of the E/W configuration have been calculated from the estimated broadband values assuming the same spectral irradiance distribution as in the front side of the tilted configuration (Part 4). However, if deemed necessary and useful compared to the simplified assumption used in the proposed method, it would be possible to calculate the rear side spectrally resolved irradiance taking into account the spectral reflectance of the ground. However, this would require additional input data in this regard.

The Perez model's estimates of broadband in-plane global irradiance have been validated with measurements from two fixed ground mounted bifacial PV arrays: a 20° inclined equator facing array and one E/W vertical array, as shown in Figure 1. The arrays are installed in the outdoor testing field of the European Solar Test Installation (ESTI) of the European Commission in Northern Italy (45°48'43.4''N, 8°37'37.4''E, at 220m above sea level). Due to the lack of spectrally resolved irradiance measurements, the assumption about the spectral irradiance distribution hasn´t been analysed.

Figure 1: Equator facing tilted (top) and E/W vertical (bottom) bifacial arrays

As an example, Figure 2 shows for a clear day the values of the estimated in-plane irradiance and measured ones using calibrated c-Si reference cells. There are 12 on rear side of the tilted rack and 10 on each side of the vertical array. The minimum irradiance measurement and the average are shown. The top graph represents the values of the East plane of the vertical configuration, while the lower graph shows the values on the rear side of the tilted rack. The modelled values (grey curves) were calculated using an albedo value of 0.2 close to that measured in the surroundings of the racks.

There is a good match between modelled and measured values in the vertical system, while for the tilted system, the modelled values on the rear side overestimate the measured in-plane irradiance, especially during the middle of the day. This can be explained by the shade of the module on the ground which reduces the effective albedo. The rear side irradiance was modelled again using a lower value of albedo (20% reduction) and the resulting values (G_{mod} albedo, orange curve) are very similar to the average of all the measured irradiance values, especially during the middle of the day when the shade is more pronounced. At the extremes of the day, when the self-shading does not fall directly below the rear side of the tilted modules, the model considering the measured albedo on the site are closer to the measurements.

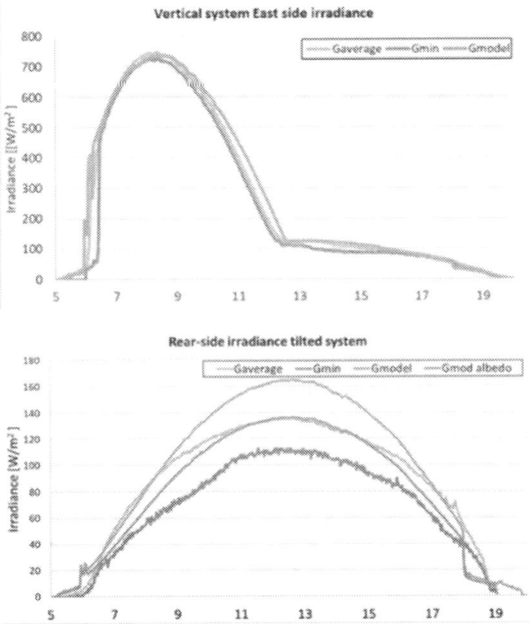

Figure 2: Measured and modelled in-plane irradiance values on the East face of the vertical configuration (top) and on the rear side of the fixed tilted equator facing rack (bottom).

Another important assumption that may not hold true for all new proposed configurations is that the rear side irradiance on the tilted configuration and on the vertical planes is uniform over the module's surface like it is on the front side of the tilted equator facing configuration. The uniformity of the rear-side irradiance is difficult to model and its impact on the module's performance is furthermore difficult to quantify. However, the IEC TS 60904-1-2 [14] indicates, that as long as the non-uniformity is low, i.e. of the order of 10%, an approximation to uniformity may be acceptable. For albedo values of 0.2 this is expected to be a reasonable assumption [15-18], while for albedo of 0.6 this simplification may need further study.

The climatological variables of the original datasets, both the ambient temperature and the wind speed data are assumed at 2m above ground level. While the former does not need to be modified, the original wind speed data (Ws_2) should be modified and adjusted to the height of the new proposed configurations. The wind speed at a different height, h, (Ws_h) could be calculated from the value at 2m, Ws_2 according to Equation 1.

$$Ws\_h = Ws\_2 \cdot (\text{hm}/2\text{m})^{0.2} \qquad \text{(Eq.1)}$$

The angle of incidence between the solar disc and the normal to the new considered surfaces should be included in the extended datasets. This would require three new columns regarding the rear side of the tilted plane and both vertical planes of the E/W configuration.

3.3 Modifications to Part 3

The final output of Part 3 is the *CSER* calculated as shown in Equation 2.

$$CSER = \frac{E_{mod,year} \cdot G_{ref}}{P_{max,STC} \cdot H_p} \qquad \text{(Eq. 2)}$$

Where G_{ref} is equal to 1000W/m^2 as the irradiance applied at STC to measure the maximum power at STC ($P_{max,\ STC}$). H_p is the yearly global in-plane irradiation provided in Part 4 for every climatic region. And $E_{mod,year}$ is the yearly sum of the hourly values of energy output estimated from the effectively absorbed irradiance, the module temperature and the power matrix values.

When working with bifacial devices, the irradiance considered for the energy output estimation should be the equivalent irradiance, calculated summing the effectively absorbed irradiances on both sides, taking into account the bifaciality factor.

The equations defined in Part 3 to estimate the AOI effects, the spectral correction factor and the temperature of the module are assumed valid when applied to bifacial modules. However, the module temperature should be estimated considering the irradiance received by both sides without considering the bifaciality factor. If needed, a wind speed corrected for the corresponding height, for example, for the rooftop configuration, should be used to estimate the module temperature.

When estimating the *CSER* for bifacial devices, the total yearly in-plane irradiation (H_p) considered is the sum of the irradiation received by both sides of the bifacial device, without considering the bifaciality factor. Conversely, the yearly energy yield ($E_{mod,year}$) should be estimated considering the bifaciality factor.

3.4 Modifications to Part 2

The parameters and coefficients relative to the AOI and spectral effects, a_r and SR respectively, should be characterized for both sides. The two thermal coefficients, U_0 and U_1, should be defined using measurements of the device under test in the relevant mounting configuration, for which the NMOT methodology may need to be adjusted, in order to consider heating from both sides.

3.5 Modifications to Part 1

The power matrix should be extended to include the higher equivalent irradiances that are expected at the proposed bifacial mounting configurations and assumed ground albedos. Equivalent irradiance measurements should be performed taking into account guidance from [14]. The bifaciality factor must also be reported, measured following the same Technical Specification, as it is needed in the calculations in Part 3 to obtain the equivalent irradiance.

4 EXAMPLE OF *CSER* ESTIMATION FOR BIFACIALS

4.1 Input data

For the present study, three of the six reference climatic profiles have been selected: temperate coastal, tropical humid and subtropical arid. For each climate, the extended climatic datasets have been calculated using Perez's anisotropic model for the new mounting configurations: fixed tilted with 0.2 and 0.6 ground albedo and the vertical E/W configuration.

Two fully characterized monofacial crystalline silicon modules have been selected. For each module, a corresponding bifacial module has been created assuming a bifaciality factor of 90%, which is representative of commercially available bifacial models. The parameters of the rear side (a_r and SR) are set identical to those of the front side.

4.2 Estimated *CSER* values

The *CSER* values calculated for the different bifacial configurations are reported in Table II. The results for both simulated bifacial devices are shown along the results for the original monofacial devices.

Table II: Estimated *CSER* for two c-Si modules (D1 and D2) for the proposed bifacial mounting conditions: vertical (BV), tilted with an albedo of 0.6 (BT0.6) and tilted with an albedo of 0.2 (BT0.2). For comparison purposes, results for the monofacial devices are also shown (Mono) for the three considered climatic profiles.

		Mono	BV	BT0.6	BT0.2
Temperate coastal	D1	0.982	0.934	0.865	0.930
	D2	0.940	0.893	0.830	0.890
Tropical humid	D1	0.908	0.866	0.767	0.845
	D2	0.880	0.835	0.744	0.819
Subtropical arid	D1	0.904	0.854	0.767	0.847
	D2	0.887	0.831	0.749	0.830

Significant differences are observed between the calculated *CSER* values of the two c-Si modules (D1 and D2), the different bifacial configurations (BV, BT0.6 and BT0.2), and between the monofacial and bifacial modules. The *CSER* values from D1 are always higher than those from D2. The advantage of D1 regarding D2 decreases in warmer climates, from the 4.4% higher *CSER* values of D1, on average for the different configurations, in the temperate coastal climate to the 2.3% difference in the subtropical arid climate. Regarding the various configurations, differences increase in the vertical configuration compared to the monofacial one, especially in the warmer climates. For example, for the subtropical arid climate, when compared to the monofacial configuration, D1's *CSER* values outperforms D2 by 1.89%, while in the bifacial vertical configuration, under our assumptions, this difference increases to 2.8%. However, differences between devices for the bifacial tilted configurations are closer to the differences observed for the monofacial configuration. This supports the consideration of the vertical configuration for the energy rating analysis of bifacial devices in addition to the fixed tilted mounting used in the actual Standard Series.

The *CSER* values for the bifacial modules are lower than for the monofacial devices, especially in the tropical and subtropical climates, characterized by high irradiance and ambient temperatures which lead to high module temperatures. This is especially significant for the tilted configuration with 0.6 albedo, whose *CSER* in these reference climates is almost 15% lower than the monofacial. However, the reduction of the *CSER* values of the tilted configuration with 0.2 albedo compared to the monofacial is around 4%, similar to the one observed for the vertical configuration (7% reduction of the *CSER* values). These reduced *CSER* values are influenced by the higher irradiance levels received by the bifacial devices, which leads to higher module temperatures negatively affecting the module's performance. More significantly, the yearly in-plane irradiation (H_p) used in the estimation of the *CSER* is the sum of front and rear side irradiation and is therefore significantly higher for the bifacial modules. This makes a dominant contribution to decreasing the *CSER* value, even though the yearly energy yield ($E_{mod,year}$) of all bifacial configurations is higher than the energy delivered by the monofacial devices, as shown in Table III. The impact of this latter contribution will depend greatly on the bifaciality of the module and albedo. This behaviour means that CSER values for monofacial and bifacial modules cannot be directly compared, but rather should be used to compare modules of each type independently.

Table III: Estimated yearly energy yield (kWh/kWp) for two c-Si modules (D1 and D2) for the proposed bifacial mounting conditions: vertical (BV), tilted with an albedo of 0.6 (BT0.6) and tilted with an albedo of 0.2 (BT0.2). For comparison purposes, results for the monofacial devices are also shown (Mono) for the three considered climatic profiles (numbers rounded to nearest kWh).

		Mono	BV	BT0.6	BT0.2
Temperate coastal	D1	955	1085	1320	1093
	D2	915	1039	1266	1047
Tropical humid	D1	1523	1523	2084	1716
	D2	1476	1468	2022	1663
Subtropical arid	D1	2076	1998	2709	2296
	D2	2037	1944	2644	2251

Figure 3 shows the distribution of the energy production in the working conditions in terms of module temperature and in-plane irradiance, for the monofacial free standing tilted device and the bifacial roof mounted tilted device with 0.6 ground albedo, in the subtropical arid climate. While the monofacial device produces most of its energy yield at modules temperatures around 57.5-60°C, the bifacial device temperature ranges between 67.5°C and 72.5°C. The vertical bifacial module produces most of its production around a module temperature of 52.5°C, while the temperature of tilted bifacial device with ground albedo of 0.2 is similar to that of the monofacial device.

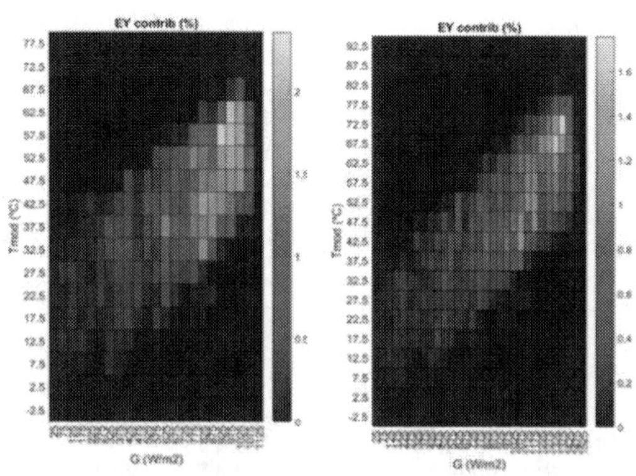

Figure 3: Distribution of the energy production at the subtropical arid climate in the working conditions of module temperature and in-plane irradiance for the monofacial device (left) and the bifacial tilted device with ground albedo of 0.6.

The in-plane irradiance levels (front and rear side combined) are also higher for the bifacial devices compared to the monofacial devices, which leads to the higher energy yield.

5 CONCLUSIONS

The IEC 61853 Standard Series provides a tool to rate and compare monofacial PV devices taking into consideration their performance under real working conditions simulated taking into account various effects (AOI, spectral, irradiance and temperature behaviour), by means of the *CSER* parameter. This parameter could guide users and PV installers in the selection of the most suitable device for the location of interest. Notwithstanding, the scope of this Standard series does not cover bifacial modules, which have the potential to overcome monofacial devices' presence in the global PV market in the forthcoming years.

The scope of the present paper is to propose an extension to the aforementioned Standard series in order to be applicable to bifacial devices. This would imply modifications to the four parts that constitute the standard, as described in the present paper. These changes include considering new mounting options aligned to the possible uses of bifacial devices, such as a vertical configuration or installation in locations with enhanced albedo; extension of the climatic reference datasets in order to provide the irradiance received by both sides of the bifacial device and new measurements to characterize the rear side of the device; in addition to changes in the *CSER* estimation model. In this regard, the in-plane irradiance value to be used in the estimation of the hourly power output values, by interpolation and extrapolation on the power matrix values, would be the equivalent irradiance obtained from the irradiance received on both sides and the bifaciality factor of the device under test. This factor which should be a new input data for the extended standard. However, the temperature of the module would be estimated considering the received irradiance on both sides, without taking into account the bifaciality factor. This would not be considered either in the estimation of the yearly in-plane irradiation needed for the final estimation of the *CSER* value. Due to the expected higher in-plane irradiance levels, the power matrix as described in Part 1 should be extended adding measurements at higher irradiance levels, especially if the configuration with 0.6 albedo is used.

The proposed methodology has been applied to two c-Si bifacial devices, created from monofacial devices considering a bifaciality of 90% and rear side with the same characteristics as the front side, in terms of AOI and spectral response. Results show significant differences between the calculated *CSER* values of the two devices, between the three bifacial configurations and between these and the monofacial configuration which has been simulated according to the actual standard. The *CSER* values from the bifacial devices and configurations are lower than the monofacial values, even though the yearly energy yield of the bifacial devices would be higher. This is partly due to the fact that the bifacial device in all configurations, especially the tilted one with high ground albedo, would be working at higher module temperature compared to the monofacial device, which would result in decreased performance of the c-Si devices. In addition to this, and more significant, the yearly in-plane irradiation is higher for the bifacial configuration which reduces the final *CSER* value, as this is the ratio of the energy yield and this in-plane irradiation.

The observed differences in the estimated *CSER* values of the different devices and configurations suggest that the proposed modifications to the Standard series and the derived *CSER* for bifacial devices could help end-users evaluate and select the most appropriate bifacial device with the relevant mounting configuration and climatic profile according to the location of interest.

5.1 Future lines of research

The proposed methodology to extend the standard to bifacial devices is based on several assumptions, which require further research in order to validate or to modify based on experiments on bifacial devices.

The anisotropic model used to estimate the new in-plane irradiance data of the bifacial devices should be compared with more accurate models like angular distribution models, ray-tracing or smilar as they would allow an estimation the irradiance distribution over the surface of the modules, in opposition to the single value provided by Perez's model in the implementation used in the present study. These other models could be used to describe the non-uniformity on the rear side of the tilted configuration and on the vertical planes, described in [18]. Another aspect that needs further study and validation is the modified ground albedo due to the self-shading of the equator facing tilted modules to the ground below, which reduces the effective albedo and the reflected irradiance. In our current proposal, we used modelled values using the actual ground albedo. Modifications to these values could be considered, perhaps employing two different albedo values, one for the extremes of the day, and a reduced one for the central hours of the day.

The standard is for a single module, but in order to be useful it should provide CSER values reasonable for typical installation conditions. Results from [18] suggest that the method may provide most reliable results for a module located on the top row and at the middle of a very long rack, due to more consistent non-uniformity and shading conditions and it could be defined as such in the standard.

Our proposed methodology also assumes low levels of non-uniformity on the rear-side irradiance, which needs to be quantified, especially at high albedo configurations. Additionally, the non-uniformity impact on the PV performance should be further analysed and quantified. The strong assumption about the uniformity of the rear side irradiance may lead to overestimations of the bifacial gain compared to that experienced in a real system. However, this effect may also be compensated reducing the values of the new in-plane irradiance on the rear side of the extended Part 4.

Measurements of the spectrally resolved irradiance reaching the rear side of tilted bifacial devices and the spectral reflectance of the ground are necessary for the validation of the assumption which considers this distribution equal to that of the front side irradiance.

5 ACKNOWLEDGEMENTS

The research carried out within the project (PV-ENERATE) leading to this application has received funding from the EMPIR programme co-financed by the Participating States and from the European Union's Horizon 2020 research and innovation programme.

6 BIBLIOGRAPHY

[1] ITRPV (International Technology Roadmap for Photovoltaic). 2017–2022. Results. 2016–2021 (8th to 13th editions). VDMA.

https://www.vdma.org/international-technology-roadmap-photovoltaic

[2] IEA-PVPS Task 13. Performance, Operation and Reliability of Photovoltaic Systems. Report T13-12:2021. Bifacial Photovoltaic Modules and Systems: Experience and Results from International Research and Pilot Applications (2021). ISBN 978-3-907281-03-1.

[3] Hayibo, K.S., Petsiuk, A., Mayville, P., Brown, L., Pearce, J.M. Monofacial vs bifacial solar photovoltaic systems in snowy environments. *Renewable Energy* (2022), 193, 657-668. https://doi.org/10.1016/j.renene.2022.05.050.

[4] Gracia-Amillo, A., Lopez-Garcia, J., Kenny, R.P., Sample, T. Energy yield analysis of bifacial PV modules: different technologies and configurations. Proceedings of the 36th European Photovoltaic Solar Energy Conference and Exhibition (EU PVSEC). Marseille 9th – 13th September 2019. doi: 10.4229/EUPVSEC20192019-4AV.2.19.

[5] Wadhawan, S.D., Pearce, J.M. Power and energy potential of mass-scale photovoltaic noise barrier deployment: A case study for the U.S. *Renewable and Sustainable Energy Reviews* (2017), 80, 125-132. http://dx.doi.org/10.1016/j.rser.2017.05.223.

[6] IEC 61853-1. Photovoltaic (PV) module performance testing and energy rating – Part 1: Irradiance and temperature performance measurements and power rating. Edition 1.0 International Electrotechnical Commission (2011).

[7] IEC 61853-2. Photovoltaic (PV) module performance testing and energy rating – Part 2: Spectral responsivity, incidence angle and module operating temperature measurements. Edition 1.0 International Electrotechnical Commission (2016).

[8] IEC 61853-3. Photovoltaic (PV) module performance testing and energy rating – Part 3: Energy rating of PV modules. Edition 1.0 International Electrotechnical Commission (2018).

[9] IEC 61853-4. Photovoltaic (PV) module performance testing and energy rating – Part 4: Standard reference climatic profiles. Edition 1.0 International Electrotechnical Commission (2018).

[10] Huld, T., Gracia Amillo, A., Sample, T., Dunlop E.D., Salis, E., Kenny, R.P. The completed IEC 61853 Standard Series on PV module energy rating, overview, applications and outlook. Proceedings of the 35th European Photovoltaic Solar Energy Conference and Exhibition (EU PVSEC). Brussels 24th – 28th September 2018. doi: 10.4229/35thEUPVSEC20182018-5DO.9.2.

[11] Gracia-Amillo, A.M., Kenny, R. P., Friesen, G., Reise, C., Lopez-Garcia, J., Extension of energy rating to bifacial modules – proposals from the PV-enerate project, JRC122448, 2020.

[12] Perez, R., Ineichen, P., Seals, R., Michalsky, J., Stewart, R. Modelling daylight availability and irradiance components from direct and global irradiance. *Solar Energy* (1990), 44(5), 271-289.

[13] Gracia, A.M., Huld, T. Performance comparison of different models for the estimation of global irradiance on inclined surfaces. JRC81902. EUR 26075 EN. Publications Office of the European Union (2013). doi: 10.2790/91554.

[14] IEC TS 60904-1-2 Ed. 1, Photovoltaic devices - Part 1-2: Measurement of current-voltage characteristics of bifacial photovoltaic (PV) devices (2019).

[15] Lopez-Garcia, J., Gali, S., Grau-Luque, E., Sample, T. Assessment of the Rear Irradiance on Bifacial Silicon PV Modules. Proceedings of the 36th European Photovoltaic Solar Energy Conference and Exhibition (EU PVSEC). Marseille 9th – 13th September 2019. doi: 10.4229/EUPVSEC20192019-4CO.2.6.

[16] Ozkalay, E., Lopez-Garcia, J., Pinero-Prieto, L., Gracia-Amillo, A., Kenny, R.P. Evaluation of the non-uniformity of rear-side irradiance in outdoor mounted bifacial silicon PV module. Proceedings of the 9th International Conference on Crystalline Silicon Photovoltaics, SiliconPV 2019, American Institute of Physics Inc., 2019.

[17] Rauer, M., Schmid, A., Guo, F., Neuberger, F., Hohl-Ebinger, J. Comprehensive Evaluation of IEC Measurement Procedures for Bifacial Solar Cells and Modules. Proceedings of the 37th European Photovoltaic Solar Energy Conference and Exhibition (EU PVSEC). Online 7th – 11th September 2020. doi: 10.4229/EUPVSEC20202020-4CO.2.1.

[18] Calcagnotto Mascarello, M., Kenny, R.P., Lyubenova, T.S., Gracia-Amillo, A., Lopez-Garcia, J. Analyis of the Irradiance non-Uniformity on the Performance of Vertical and Tilted Bifacial PV Module Arrays, Proceedings of the 38th European Photovoltaic Solar Energy Conference and Exhibition (EU PVSEC). Online 6th – 10th September 2021. doi: 10.4229/EUPVSEC20212021-5DO.3.4.

IMPROVEMENT OF A LED BIAS LIGHT SYSTEM FOR THE DOUBLE SIDED CHARACTERIZATION OF BIFACIAL DEVICES

Teodora S. Lyubenova*[1], Robert P. Kenny[1], David Shaw[1], Diego Pavanello[1]
[1]European Commission, Joint Research Centre (JRC), European Solar Test Installation (ESTI), Ispra (VA), Italy
*Tel: +39 0332786574, e-mail: teodora.lyubenova@ec.europa.eu, robert.kenny@ec.europa.eu, david.shaw@ec.europa.eu, diego.pavanello@ec.europa.eu

ABSTRACT: A double-sided solar simulator based on an existing large-area long pulse flash solar simulator and a low-cost LED-based rear bias light has been developed at European Solar Test Installation (ESTI) (Joint Research Centre, European Commission) for studying and validating the double-side indoor testing approach for current-voltage (I-V) characterization of bifacial modules. "Self-reference" methodology is applied using an accredited reference values (ISO/IEC 17025) obtained by single-side illumination. The proposed approach and prototype LED simulator for full size modules has demonstrated adequate performance in comparison with the single-side measurements based on a commercial Class AAA simulator using validated standard methodology. The electrical performance expressed in similar current-voltage (I-V) curves and power output reveals the suitability of the method for characterization of bifacial devices. The LED bias light source could be classified as a "Class C" solar simulator. The analysis of different representative bifacial PV devices available on the market shows different electrical behaviour according to their bifaciality factor.

Keywords: Bifacial, Photovoltaics, Characterization, Double-side, Solar simulator, LED, Standard

1 INTRODUCTION

Industrial production of bifacial crystalline silicon (Si) photovoltaic (PV) devices is increasing due to the attractive enhanced energy output gain and reduction cost associate with extra processing steps. It is expected that the bifacial PV modules will achieve 60% of the total PV market in 2032 [1]. The publication, in 2019, of the Technical Specification (TS) IEC TS 60904-1-2 for the measurement of current-voltage characteristics of bifacial PV devices [2] standardized various methods for measuring bifacial PV devices: natural sunlight, a single-side illumination using the equivalent irradiance method and a double-sided illumination approach. There is a lack of experience reported regarding testing of bifacial PV devices under double-side illumination using two light sources.

The present work is a continuation of previous studies performed for a small size mini module [3] and for a large size commercial PV bifacial device [4]. The results from both studies revealed several drawbacks features related with limitations of the system and the need of additional data for different bifacial technologies that are available on the market. Efforts towards overcoming these shortcomings are addressed in the current research work.

2 MEASURMENT APPROACH

The applied approach is based on the simultaneous illumination of both sides of the bifacial device with two separate different light sources at 1000 W/m^2 on the front and at least two different rear side irradiance levels (100 W/m^2 and 200 W/m^2) (Table I). As front-side illumination (downward facing side), a large-area long pulsed flash solar simulator (Spire SLP4600 with two Xenon linear lamps and pulse duration up to 100 ms, Class A) was used. The rear side of the device is illuminated with low-cost LED-based bias light on the top consisting of a matrix of 55 Chip on Board (CoB) LEDs placed at 51 cm height above the module plane. In Figure 1, a schematic representation measurement set up is shown. The LEDs are generally illuminated for 4 seconds, which facilitates synchronization with the pulsed simulator it is paired with. Black baffles are used to avoid unwanted reflections covering the system and the free space around the device under test (DUT). The single-side illumination method is performed using a pulsed solar simulator PIIIB (PASAN IIIB, 3x3 m^2, flash 10 ms) at the European Solar Test Installation (ESTI), an ISO/IEC 17025 accredited calibration laboratory for PV device testing and calibration. The single - side approach follows the definition of equivalence irradiance [2]. For all measurements involving the PIIIB simulator, the so - called multi-flash (MF) method [5] was employed as the 10 ms pulse is too short for accurate measurement of the high capacitance module under test.

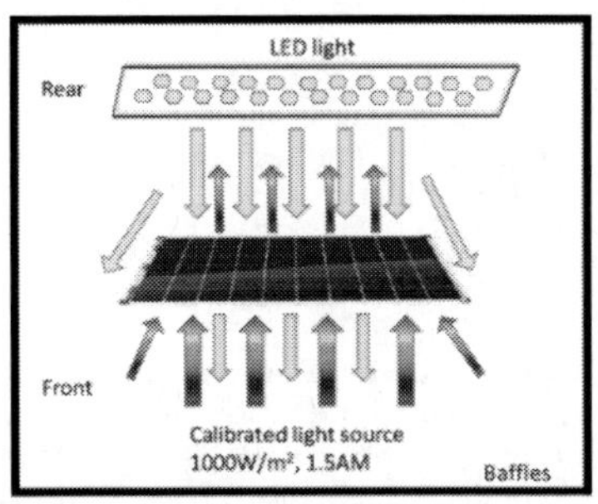

Figure 1: Schematic representation of the double - side measurement set up composed of two different light sources with adjustable irradiance levels.

In this study, several DUTs are characterized by both testing approaches and each one is used as a "self-reference" [4]. The DUTs have been previously tested by the accredited single - side illumination approach IEC TS 60904-1-2) [2] on a Class AAA solar simulator (PASAN IIIB). The values of Isc from the single - side method (Isc_{STC}, Isc_{G1}, Isc_{G2}) are used as a reference to adjust the irradiance at the long - pulse simulator to achieve the same values. The set up parameters are shown in Table I. At G_{STC} only single light source is applied in both configurations using black non- reflective background, while for higher irradiance (G_1 and G_2) additional LED bias light is used in the simultaneous double illumination.

In the present work different bifacial PV modules types representative on the market are considered for characterization in order to obtain more relevant information on their electrical behavior under double-side simultaneous illumination conditions. For this purpose, modules with different bifaciality factor (60% < φ > 90%) are analyzed on the prototype system (Table II).

Table I: Experimental set up parameters of single and double - side approaches.

Method	Irradiance	Front (W/m²)	Rear (W/m²)
Double	G_{STC}	1000	0
	G_1	1000	100
	G_2	1000	200
Single	G_{STC}	1000	0
	G_1	1000 +φ100	0
	G_2	1000 +φ200	0

Table II: c-Si Bifacial PV modules from different manufacturers characterized by both indoor approaches.

DUT	*φ (%)*	*Type*	*Cells*
M58	58	PERC	60
M67	67.6	PERC	60
M86	86	HIT	72
M97	97.6	PERC	60

The spatial rear irradiance non-uniformity of the LED bias light has been obtained by repositioning a calibrated large reference cell across the test plane, which has been divided in sub-areas of 20 cm × 20 cm area each. The resulting measurements are represented as a spatial map. The non-uniformity (NU) of the irradiance on the test plane is calculated according to the standard IEC 60904-9 [6] with the following equation:

$$\text{NU (\%)} = \left[\frac{\max irradiance - \min irradiance}{\max irradiance + \min irradiance}\right] \text{x } 100\%$$

where max *irradiance* and min *irradiance* are, respectively, the maximum and the minimum irradiance values measured on the test plane. NU is given as percentage (%).

3 RESULTS AND DISCUSSION

3.1 I-V measurements

I-V curves of the DUTs characterized by double and single - side illumination approaches are given in Figure 2. Inset graphs show zoomed areas of the I-V curves closed to P_{max} in order to appreciate better the existence of differences in shape and position of the plotted results.

Overall, it can be observed that, the curves are similar and this shows an adequate agreement of the proposed methodology. The I-V curves are smoother and almost overlap in samples M58 and M67 that possess lower bifaciality factors, while the differences become more pronounced in the devices with higher bifaciality. This observation could be related with the increased sensibilities to irradiance non-uniformity fluctuations during measurements.

Results of the current-voltage (I-V) measurements in terms of power output for different bifacial devices are shown in Table III. The power output parameters (P_{max} at G_{STC}, G_1 and G_2) are very similar between the single and double - side approach for the devices M58 and M67. The observed trend is an increasing of P_{max} values when double-side approach is applied, although the Isc at STC conditions has been maintained constant for both methods. However, the devices with lower bifaciality exhibit inferior differences in P_{max} values (form ± 0.02% to ± 0.2%) than the modules with higher bifaciality. The latest, present increased differences having the highest values of almost 2% at G_{STC} for M97 device. The M86 exhibits results contradictory to those observed in the rest of the measured devices (lower P_{max} values) and additional studies to understand the origin of this behaviour have to be performed. This different behaviour may be related to the fact this this module is a HIT type (Heterojunction with Intrinsic Thin - layer, Panasonic Group), whereas the other three are all PERC (Passivated Emitter and Rear Contact), since the HIT technology can be more problematic with single pulse measurements due to its higher capacitance.

Table III: Power output (P_{max}) measured with the equivalent irradiance single-side method and double-side illumination method using the LED simulator.

DUT	*Method*	*P_{STC} (W)*	*P_{G1} (W)*	*P_{G2} (W)*
M58	Single	264.8	279.6	294.4
	Double	265.4	280.3	294.0
	Δ (%)	0.2	0.2	-0.1
M67	Single	297.2	316.3	336.1
	Double	297.4	316.9	336.2
	Δ (%)	0.05	0.2	0.02
M86	Single	209.9	227.1	244.1
	Double	206.6	226.4	240.3
	Δ (%)	-1.6	-0.3	-1.5
M97	Single	250.5	273.4	273.1
	Double	255.4	277.2	276.9
	Δ (%)	1.96	1.4	1.4

The analysed bifacial PV devices with the LED bias light exhibit similar electrical parameters to those obtained with the accredited single - side illumination method and mainly for the lower bifaciality devices. The deviations of the P_{max} could be associated with inherent characteristics of both light sources applied in the double - side approach. However, it is clear from the results in Table III that the predominant influence could be attributed to the inherent characteristics of the long pulse solar simulator that is used to illuminate the front side of the devices, since at G_{STC} no additional bias light is applied.

3.2 Rear-side LED irradiance spatial distribution

Spatial distribution maps of the rear irradiance produced by the LEDs on the test plane at 51 cm distance and for different LED power supply voltages (67 V, 68 V, 70 V and 75 V) that corresponds to irradiance levels at 100 W/m², 135 W/m², 200 W/m² and 300 W/m² are shown on Figure 3. The square represents the module position on the test plane and the axis indicates the dimensions of the solar-simulator illuminated surface. It has to be mentioned that during measurement, although non - reflective baffles are used to avoid parasitic reflections from the non-covered by the module surface (the DUT is smaller than the simulator area), additional light reflections are observed due to the design of the long - pulse solar simulator as it has a glass transmitting surface with a highly reflective interior. In fact, the reflected light (front and rear) from inactive module areas such as gaps between cells and module margins is under ongoing investigation.

From the maps on Figure 3 can be evaluated that the irradiance distribution is not homogeneous along the test area, displaying zones with lower and higher illumination, mostly closed to the borders that varying with irradiance levels with no defined trend.

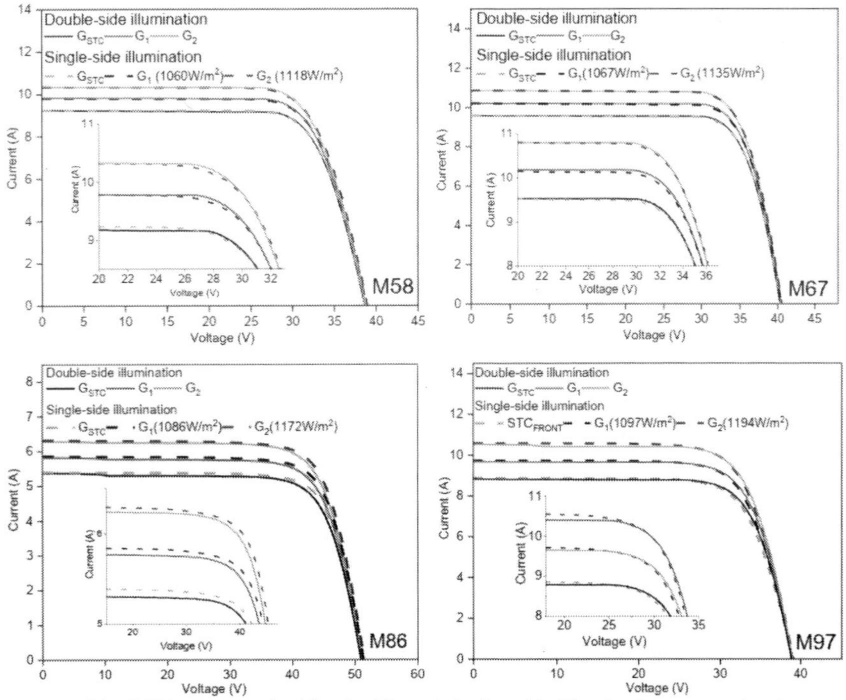

Figure 2: I-V curves of the DUTs characterized by double and single - side illumination approaches. Inset graphs show the zoomed area of the I-V curve closed to P_{max}.

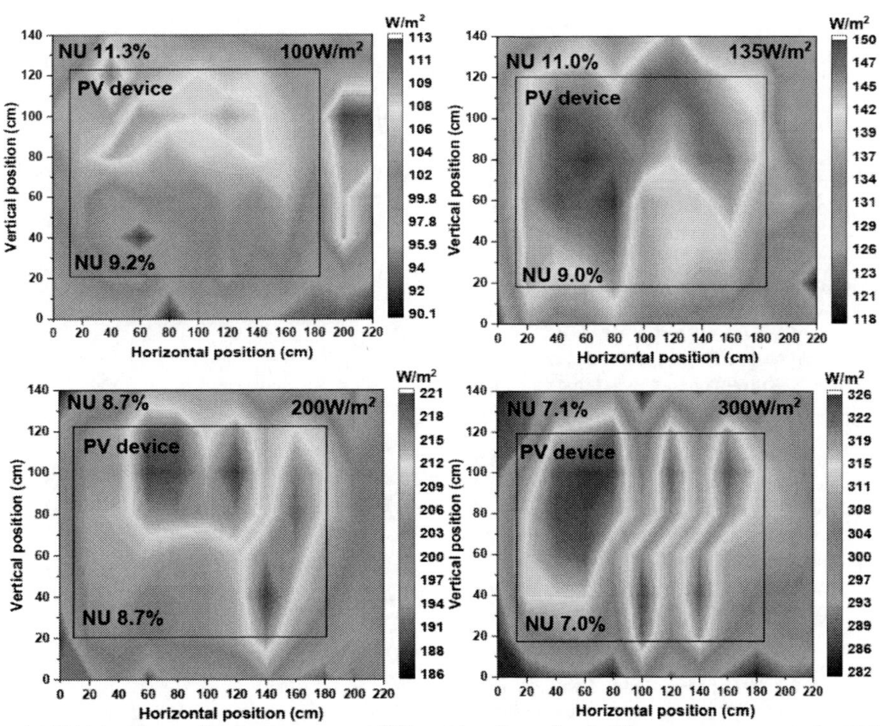

Figure 3: Spatial LED irradiance distribution maps at different irradiance levels. The square represents the DUT position on the test plane.

The rear side NU analysis at different irradiance levels in the double - side illumination system reviled NU values below 10% for higher irradiance levels (200 W/m^2 and 300 W/m^2) classifying the light source in "Class C" according to the IEC 60904-9 [6], while at lower irradiances (100 W/m^2 and 135 W/m^2), the LED bias light exceed this limit considering the full area of the simulator. Although, at higher irradiances the NU is better, a visual inspection of the map at 100 W/m² displays more homogeneous light distribution throughout the plane and a single measurement seems to contribute to the higher calculated NU. Nevertheless, the rear side LED irradiance distribution is found to be under 10% for the working area of the tested modules at all irradiance levels.

3.3 Temporal short-term LED instability of irradiance

Testing the LED flash instability during short period of time (10-15 min) i.e., sufficient for the performing of ordinary I-V measurements has been analyzed (Fig. 4).

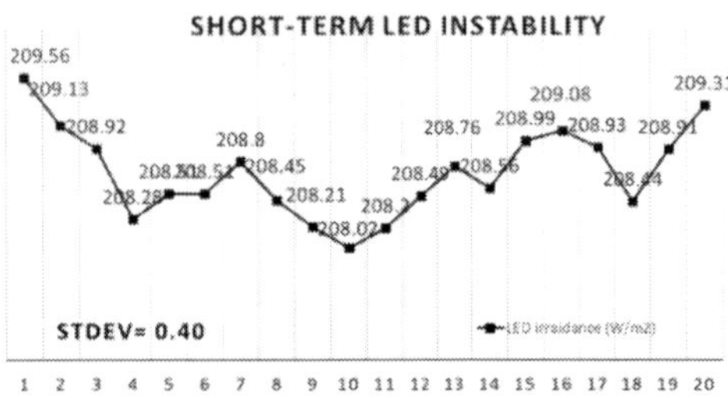

Figure 4: Irradiance changes during consecutive flash to flash testing each 30s.

Multiple consecutive LED flash to flash testing with interval of 30 s have been performed for an arbitrary sample (e.g M58) at 70 V that correspond to 200 W/m² irradiance. It can be observed that the irradiance is ranging from 208.02 W/m² to 209.56 W/m² that stand for 0.4 W/m² of standard deviation between flashes. The LED flashes demonstrate stable temporal behavior for short period and are suitable for I-V measurements.

4 CONCLUSIONS

The present investigation is focused on a double-sided indoor simulator based on LED rear bias light and long-pulsed simultaneous illumination on the front side to provide realistic measurement conditions for different bifacial modules. The proposed approach and prototype LED simulator for full size modules has demonstrated adequate performance in comparison with the single-side measurements based on a commercial Class AAA simulator using validated standard methodology. The electrical performance expressed in similar current - voltage (I-V) curves and power output reveals the suitability of the method for characterization of bifacial devices. A significant proportion of the deviations of P_{max} are associated mainly with inherent characteristics of the long - pulsed solar simulator used to illuminate the front side. The analysis of several different bifacial PV modules types representative of the market shows different behaviour between devices with lower ($70\% < \varphi$) and higher bifaciality ($\varphi > 80\%$), displaying increased sensitivity to irradiance non-uniformity fluctuations for the higher bifaciality samples.

The rear-side non-uniformity (NU) in the double-side illumination system varies with applied LED voltage, classifying the prototype solar simulator in Class C (> 200 W/m²). The rear side LED irradiance distribution is different at varying irradiance levels and is found to be under 10% for the working area of the tested modules. LED flashes demonstrate stable temporal short - term behaviour during measurements.

Overall, the suitability of the proposed method in the IEC TS for the characterization of bifacial PV devices using double sided illumination has been experimentally tested using the developed simulator and light source combination.

Upgrades of the LED system are planned in terms of design and LED electrical control to improve homogeneity of the distribution of the light on the rear side of the device during measurement that should reflect in better measurements of key electrical parameters.

5 ACKNOWLEDGEMENTS

The research carried out within the project (PV-ENERATE) leading to this application has received funding from the EMPIR programme co-financed by the Participating States and from the European Union's Horizon 2020 research and innovation programme.

References

[1] ITRPV-2021 International Technology Roadmap for Photovoltaic (ITRPV) 13th edition, 2022

[2] IEC TS 60904-1-2 Ed. 1 Measurement of current-voltage characteristics of bifacial photovoltaic (PV) devices, 2019

[3] D. Shaw, J. Lopez-Garcia, R. Kenny, L. Pinero-Prieto, E. Ozkalay, Design study of a double-side illumination solar simulator for bifacial silicon PV modules characterisation based on low-cost LED bias light, 35th European Photovoltaic Solar Energy Conference and Exhibition (EUPVSEC), Brussels, Belgium, 2018, pp. 1001 - 1005.

[4] T. S. Lyubenova, R. P. Kenny, D. Shaw, D. Pavanello, J. Lopez-Garcia, Double-sided characterization of full-size bifacial PV modules based on low-cost led bias light, 38th European Photovoltaic Solar Energy Conference and Exhibition (EUPVSEC), 2021, pp. 839 – 844.

[5] J. Lopez-Garcia, B. Haile, D. Pavanello, A. Pozza and T. Sample, "Characterisation of n-type bifacial silicon PV modules" Proc. 32rd European Photovoltaic Solar Energy Conference and Exhibition, pp. 1724-1729, 2017.

[6] IEC 60904-9 Photovoltaic devices - Part 9: Solar simulator performance requirements 2007

DAYLIGHT ELECTROLUMINESCENCE OF PV MODULES IN FIELD INSTALLATIONS: WHEN ELECTRICAL SIGNAL MODULATION IS REQUIRED?

Gisele A. dos Reis Benatto*, Rodrigo Del Prado Santamaria, Thøger Kari Hass, Martin Bartholomäus, Luca Morino, Peter B. Poulsen, Sergiu V. Spataru
Department of Electrical Photonics Engineering, DTU Electro, Technical University of Denmark
Frederiksborgvej 399 4000 Roskilde, Denmark.

ABSTRACT: With the growth of photovoltaic plant installations in quantity and size, the operations & maintenance industry calls for more efficient and accurate ways to inspect the PV modules health. Different country regulations on work hours and installations sites conditions may require that inspections must be done during nighttime, daytime or short time periods, what may limit the use of imaging inspection techniques, such as Infrared Thermography and Electroluminescence (EL). To expand inspection viability, Daylight Electroluminescence (dEL) plays a big role, as EL allows accurate diagnosis of PV panels deep into the cell level. Obtaining EL images in the field during direct sunlight without covers is challenging but effective and robust using a modulated electrical signal and image processing. The thresholds required by the modulation techniques and a detailed description of how the sunlight influences the capture of the EL images are not clearly described in the research or technical literature. In this case study, our goal is to report how sun irradiance in the plane of the array influences dEL image acquisition of PV panels under direct sunlight and to better specify the irradiance limit for when modulation of the forward bias is necessary instead of constant bias. All this considering reaching the minimum acceptable EL image quality for PV diagnosis using previously available and new signal-to-noise ratio (SNR) markers.
Keywords: Characterization, Electroluminescence, PV module, Field inspection

1 INTRODUCTION

The luminescence spectrum of crystalline silicon (c-Si) is a broad band that ranges from 900-1300nm approximately. This range is mostly in the short-wave infrared (SWIR) region of the electromagnetic spectrum. To perform electroluminescence (EL) images of c-Si PV panels, a camera containing a sensor based on silicon (CCD or CMOS) that can detect photons in the entire visible range and partially in the SWIR range, such that long exposure times (in the range of seconds) are required to reach enough quality for the evaluation of the PV module health, including relatively small details such as finger failures and mode-A cracks [1], [2]. This fact unfortunately limits the use of this kind of camera for field EL imaging anytime out of the dusk-to-dawn period. Koch et al. [3] mentions that a boundary irradiance condition to perform EL in the field using cameras with silicon based detectors is 100 W/m^2 due to moon light and sources other than sunlight. During daytime, however, our field experience indicates that this limit doesn't apply for global plane of array irradiance (G_{POA}) for this type of detector, because it saturates the sensor before the EL signal can be captured, even with the use of optical filters. On the other hand, InGaAs sensors can acquire the entire EL spectra range, allowing acquisition of EL images in milliseconds rather than seconds, and making possible the performance of daylight EL (dEL) imaging of PV modules installed in the field, without the need of any cover or enclosure over the panel.

In the literature, it has been explained that the sunlight at 1 sun can be 4 to 5 orders of magnitude more intense than the EL signal coming from a PV panel forward biased at short circuit current (I_{SC}) [4], [5]. In this context, a single background (BG) subtraction is not enough to get rid of the sunlight noise embedded in the images, as short exposure times (below 1 ms) are also a requirement to avoid sensor saturation. With the use of a 1150 nm bandpass filter to block light from most of the sunlight spectrum, the exposure can be increased to 2-4 ms depending on sunlight intensity and camera settings, what can still be short to reach sufficient EL image quality with single BG image subtraction at high irradiance conditions. To enable dEL in such scenario, the acquisition of a sequence of images is required, so the signal-to-noise ratio (SNR) can be enhanced. The image sequence in principle has to be acquired at the same exact position and surrounding conditions with the PV panel biased (EL images) and not-biased (BG images) for later image averaging for noise cancelling and signal enhancement [4], [6].

It has been reported that for measurements done in clear sky conditions [7], [8], the sequence of images can be acquired in batches, i.e. simply taking several images with forward bias ON and then several with bias OFF. Nevertheless, same surrounding conditions can be very challenging to achieve in locations such as United Kingdom and Denmark, where a cloudless sky is rare. Consequentially, the robustness of dEL shall be met using a modulated electrical bias current to the PV sample and fast camera frame rates [9], [10]. PV module bias modulation does require a higher complexity to the EL related equipment, such as the capability of controlling the power supply unit output in a waveform and signal-camera synchronization, but the later (which is the most complex) is only optional if certain image processing methods are used, such as fast Fourier transform (FFT).

In this work, we wish to answer the question: at which sun irradiances it starts to be a requirement to have image sequencing and modulated electrical signal for adequate dEL image quality? Using an InGaAs camera and PERC PV modules installed in a power plant, the image quality was sufficient with single BG subtraction at G_{POA} below 432 W/m^2. Additionally, we introduce a new SNR marker able to indicate the signal quality in a sequence of images intended for performing dEL, along with images processed with an efficient FFT signal detection, to show the correlation with the EL SNR marker developed for single EL-BG subtracted images.

2 METHODOLOGIES

2.1 Experiment details

EL imaging was performed in this work using an InGaAs camera (640×512 pixels sensor) from First Light Imaging (model C-RED 3) installed in a high tripod. The camera, lens, and filter details have been presented previously [5], [11]. During the sequence acquisition, the exposure time was 2 ms, avoiding saturation due to the sunlight. The images were acquired at 480 frames per second. The panels imaged were PERC bifacial modules with 17.7% efficiency, from Trina Solar installed in the tracker facility at Technical University of Denmark (DTU) Risø campus. The modulated electrical signal was provided by a bidirectional DC power supply unit from Delta Elektronika (model SM 1500-CP-30) and a 30 Hz squared waveform generated by an oscilloscope. The power supply was connected to a PV string consisting of 22 panels, while for better image resolution, only two PV panels were imaged per imaged sequence. The same oscilloscope was used to monitor the current running though the PV string. The images acquired from the PV string were taken throughout a bright and cloudless day and before every image sequence acquisition, the G_{POA} was measured using a calibrated reference cell. An image sequence consisted of 300 images acquired at 480 frames per second.

2.2 Image processing and quality metrics

Single BG subtraction consisted of using one image with the EL signal (forward bias signal ON) and subtract it with an identical image without EL signal (forward bias OFF). The image subtraction was performed using the software ImageJ [12]. For EL image quality metrics, SNR_{50} (for individual EL-BG image pairs) and SNR_{AVG} (for image sequences) are markers used in the literature [13], [14]; however, the former proves to carry inconsistencies for dEL imaging [5], [11] and for this reason we did not use it in this work.

SNR_{50} should decrease linearly with sun irradiance, assuming that all the other surrounding conditions during image acquisition are constant. In the IEC 60904-13 technical specification [13], it is suggested that SNR_{50} for outdoor EL images should not be lower than 5 to achieve minimum acceptable quality and we followed this guideline to determinate when modulation starts to be a requirement.

For image sequences processing, an FFT approach was used [15], which consisted in investigating the forward bias signal modulation frequency of the image sequence in phase space. This approach was first mentioned by Guada et al. [7], attesting that it was too computationally demanding to be implemented in large scale. As this approach is suitable to a scenario such as high sun irradiance in addition to camera motion, we focus on it to future implementation to drone based EL imaging. We were able to improve the processing time from 30 min to below one second, making it available for large scale PV inspections and data processing. Together with the image processing FFT approach, a new way to estimate the quality of the image was necessary.

As the FFT approach sees the EL signal in the phase space, we approached the SNR quantification from the signal quality instead of image quality (in contrast with SNR_{50}). First, slow background noise is removed through a filter suitable for the modulation frequency, such as an appropriately sized moving average. Next, the noise level for the time series of a single pixel is calculated as shown in equation (1), and the signal level in equation (2), where the parameters are shown in Figure 1. N_1 and N_2 are the number of samples in each subsample set μ_1 and μ_2. The noise calculation is made for every pixel time-series, p, in an image stack and combined as shown in equation (3), where a weighing scheme is used to mainly capture the SNR level primarily from pixels with a detectable signal. Including both positive and negative values of $\overline{S}$ decreases contributions from areas without signal, as the random offsets cancel out. Thus, the method is able to detect if a signal is present in the image series, as well as its quality, independent of image cropping.

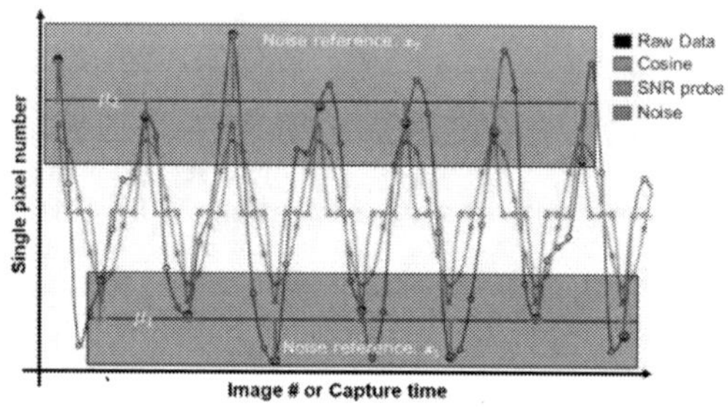

Figure 1: Noise calculation from the modulated electrical signal perspective

$$\overline{N} = \sum|x_{i,1} - \mu_1|/N_1 + \sum|x_{i,2} - \mu_2|/N_2 \quad (1)$$

$$\overline{S} = \mu_2 - \mu_1 \quad (2)$$

$$SNR_{Kari} = \begin{cases} \frac{\sum_p |\overline{S}_p| \overline{S}_p}{\sum_p \overline{S}_p \overline{N}_p}, if \sum_p \overline{S}_p > 0 \\ 0 \end{cases} \quad (3)$$

This approach was named SNR_{Kari} and is used for quality assessment of FFT processed dEL images. The SNR_{Kari} metric for dEL indicates a satisfactory signal for image extraction from $SNR_{Kari} = 1$ and upward. $SNR_{Kari} > 4$ ensures a high quality image, while values below one will result in poor quality.

To capture the most appropriate datapoints for the calculation, the phase and frequency of the modulation signal must be known. It can be found as the dominant frequency and phase in the image series, where the noise cancels itself out.

Near the transition between peak and valley, both interpolation, as well as electrical spikes or ramp-up effects can occur. To mitigate this, the number of samples collected from each peak (or valley) is selected via a simple heuristic:

$$N_{peak} = Max(1, Floor(T_m/2 - Max(4, T_m/4))) \quad (4)$$

, where T_m is the number of samples per modulation period.

Until $T_m = 12$, where $N_{peak} = 2$, only one sample per peak/valley is collected per period, but this is made up for by the large number of pixels in the image. At $T_m = 16$, $N_{peak} = 4$ while $T_m = 32$, provides $N_{peak} = 8$. The relatively narrow selection also allows e.g. sinusoidal signals to be investigated with sufficient precision.

Equation (3) is tailored to square-wave or nearly square-wave modulation signals for performance purposes, but can be generalized to other waveforms.

3 RESULTS AND DISCUSSION

3.1 Batch vs Modulated image sequence acquisition

Figure 2 shows partially a PV module where dEL is performed with batch and modulated image acquisition performed one just after the other at high sun irradiance (G_{POA} 906.8 W/m^2) measured just before the acquisitions. In both cases, the EL signal enhancement was made using image averaging. In the graphs on the left, the variation of the mean pixel number in each image reveals the ambient light variation that is often unavoidable in daylight scenarios. EL and BG indicate the group of images where the EL bias signal was ON and OFF respectively. The images on the right show that dEL is not robust in providing good quality images when the batch acquisition is performed.

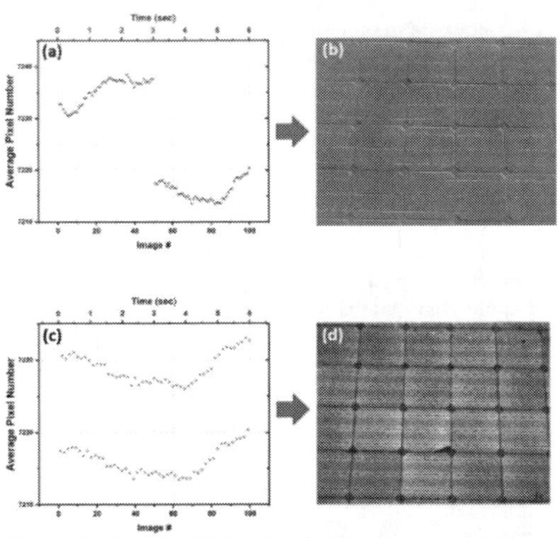

Figure 2: Batch and Modulated dEL image acquisition at G_{POA} 906.8 W/m^2. The graphs on the left show the mean pixel numbers of each image in the sequence and on the right it sshow the final EL image using averaging to improve image quality.

A similar situation was previously described by Bhoopathy et al. [16] for daylight photoluminescence of PV modules, however, the scenario was composed of high efficiency, high luminescence signal PV modules and stationary cameras; a scenario that involves a high luminescence signal suitable to the use of averaging for image improvement. In short, signal enhancement via averaging can be used for both batch image acquisition and modulated acquisition, can only increase the image quality by adding images and is inefficient in enhancing signal when the noise is too high, such as when the camera is in motion.

3.2 Single background subtraction

BG subtraction is commonly used of EL field inspections during nighttime. Being able to perform the simpler image acquisition and image processing with only two images is more straightforward to inspectors to perform it. Additionally, it allows us to use the SNR_{50} marker as a reference for image quality.

In Figure 3, three BG subtracted dEL images of PERC PV modules installed in the field are shown. Two full modules are in the field of view, presenting with several cells displaying issues with the busbar soldering.

The acquisiton was made at G_{POA} 607, 426 and 126 W/m^2. The SNR_{50} was calculated for the images showing lower noise when the SNR_{50} value is higher. At G_{POA} 426 and 126 W/m^2, the EL image is above the minimum acceptable quality for outdoor measurements, what is an indication that for PERC or higher efficency modules, modulation is not required dEL at these irradiances.

It is valid to point out that, even for G_{POA} 607 W/m^2 where the SNR_{50} was below satisfatory (3.50), the main cell defects are still well distinguable. This fact illustrates that an image quality guideline should be developed displaying limits for especific defects, probably by their typical size in a PV cell, rather than only quantifing it in a way that is hard to be interpreted the field during a PV inspection.

3.3 Image sequence processing

In Figure 4, three dEL images of the same modules in Figure 3 are shown. At this time, the image acquisition was made with electrical signal modulation and processed via FFT method. The acquisiton was made at the same G_{POA} as in Figure 3. The SNR_{Kari} was calculated for the images, showing consistensy in relation to the image quality, i. e. the SNR_{Kari} value is higher when the noise in the image is lower, as observed for SNR_{50}. This consistency is not observed with SNR_{AVG} in previous works [5], [14]. The three dEL FFT processed images analyzed here show high visual quality, comparable with EL images acquired indoors.

In contrast with signal enhancement via averaging, the enhancement via FFT can retrieve as much signal from the sequence of images as it contains, has room for numerous filters for signal improvement, making noise addition due to camera motion manageable.

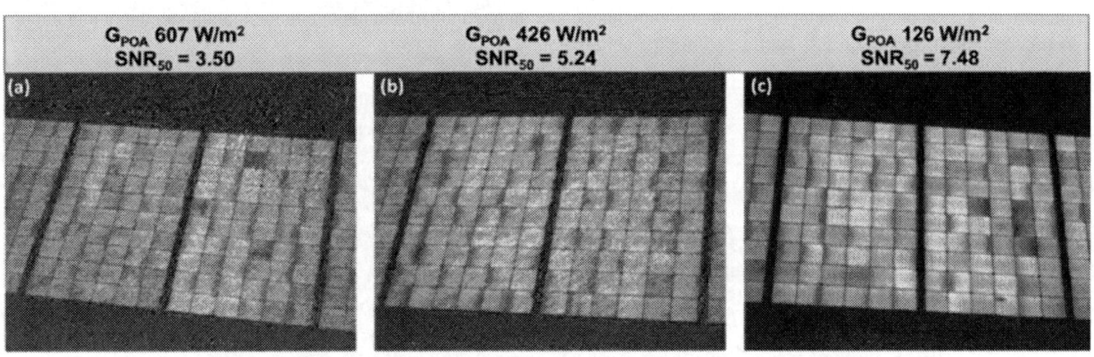

Figure 3: dEL image with single subtracted BG (1 EL image – 1 BG image)

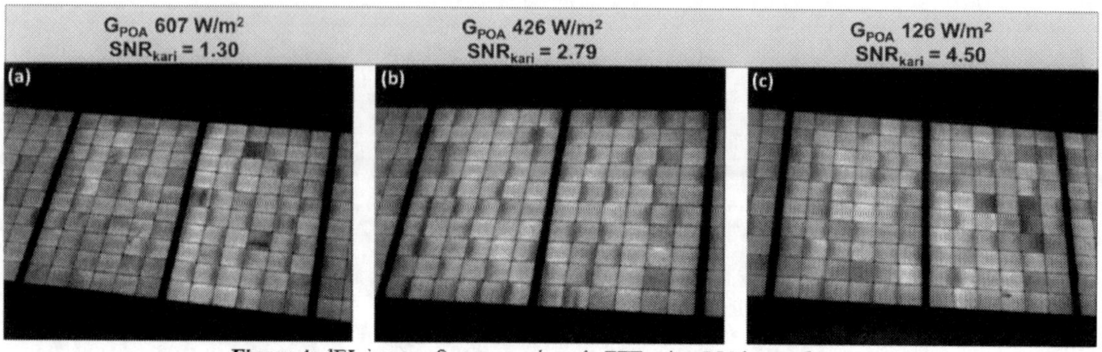

Figure 4: dEL image after processing via FFT using 256 image frames

3.4 SNR markers towards sunlight intensity

Figure 5 shows the results for the SNR_{50} (blue dots) calculated for single BG subtraction dEL images acquired at different G_{POA}. The sun irradiances went from 126 to 607 W/m^2 and a linear fit is also presented (blue dashed line). The SNR decreases with increasing irradiance and the fit crosses the $SNR_{50} = 5$ at 432 W/m^2. That indicates that the quality of the single EL image with subtracted background has satisfactory signal quality for basic diagnosis via EL images at any solar irradiances at G_{POA} below 432 W/m^2. This is valid for this PV panel type imaged at the conditions mentioned above in the methodology section. Figure 3 also shows the results for the SNR_{kari} (red dots) acquired using 256 images of a sequence of 300 images obtained. The linear fit (red dashed line) shows similar trend as SNR_{50}, attesting a good correspondence of the metric. All the dEL FFT processed images analyzed here showed a satisfactory image quality marker with $SNR_{kari} \geq 1$.

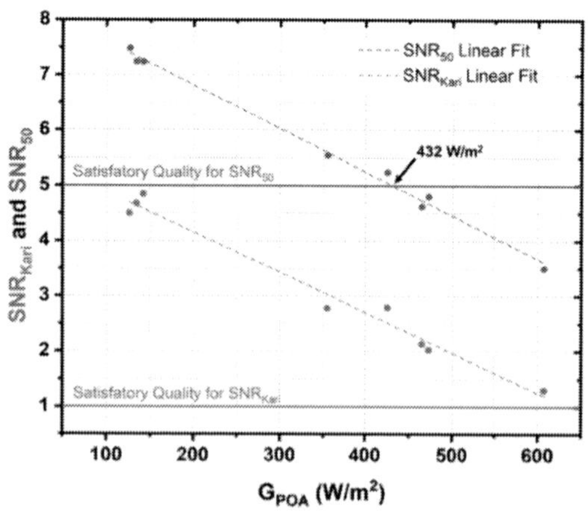

Figure 5: SNR_{50} of single BG subtracted images and SNR_{Kari} calculated for sequences of 256 images with modulated current bias acquired at different solar irradiances at the G_{POA}

4 CONCLUSIONS AND OUTLOOK

This work presents the analysis of Daylight EL SNR and image quality in G_{POA} irradiances ranging from 126 to 607 W/m^2. For the conditions, approaches and samples described, a minimal acceptable dEL image can be achieved for an acquisition made below G_{POA} 432 W/m^2 with simple BG subtraction. This means that signal enhanced using a sequence of images is not required and therefore, no modulation is needed. In this case, the image quality is sufficient to distinguish the defects that occur as large areas in a PV cell, such as partially connected ribbons, large disconnected crack areas, potential induced degradation (PID), among others. These are important features for PV diagnosis that are detectable only or mainly with EL imaging.

A new metric for the SNR of image sequences (SNR_{Kari}) and FFT image processing for dEL are introduced. SNR_{Kari} focuses on the quality of the luminescence signal itself in addition to presenting a good correspondence with visual image quality of dEL images. Both SNR_{Kari} and FFT image processing are applicable for signal detection and dEL acquisition in high noise acquisition conditions, such as high sun irradiances and capture in motion (UAV applications).

For the outlook of this work, the question of the title should be expanded further, considering the important sources of noise to dEL:

a) At which level of sun irradiance variation does modulation start to become a requirement?

b) At which level of noise from camera motion does FFT signal detection start to become a requirement?

These questions were out of the scope of the present manuscript but should be addressed in a near future.

5 ACKNOLEDGEMENTS

This research has been carried in the Eurostars project Automated Daylight Electroluminescence Inspection of Large Photovoltaic Systems (ADELI).

6 REFERENCES

[1] M. Köntges *et al.*, "Review of Failures of Photovoltaic Modules," *IEA-Photovoltaic Power Syst. Program.*, pp. 1–140, 2014, doi: 978-3-906042-16-9.

[2] J. A. Tsanakas *et al.*, *Infrared and Electroluminescence Imaging for PV Field Applications: An Overview of the Latest Report by IEA PVPS Task 13*. 2018.

[3] S. Koch, T. Weber, T. Sobottka, A. Fladung, P. Clemens, and J. Berghold, "Outdoor Electroluminescence Imaging of Crystalline Photovoltaic Modules: Comparative Study between

Manual Ground-Level Inspections and Drone-Based Aerial Surveys," in *32nd European Photovoltaic Solar Energy Conference and Exhibition*, 2016, pp. 1736–1740, doi: 10.4229/EUPVSEC20162016-5DO.12.2.

[4] L. Stoicescu, M. Reuter, and J. . Werner, "Daysy: Luminescence Imaging of PV Modules in Daylight," *29th Eur. Photovoltaics Sol. Energy Conf. Exhib. Amsterdam, Netherlands*, pp. 2553–2554, 2014, doi: 10.4229/EUPVSEC20142014-5DO.16.2.

[5] G. Alves Dos Reis Benatto *et al.*, "Drone-Based Daylight Electroluminescence Imaging of PV Modules," *IEEE J. Photovoltaics*, vol. 10, no. 3, pp. 1–6, 2020, doi: 10.1109/jphotov.2020.2978068.

[6] J. Adams *et al.*, "Non-Stationary Outdoor EL-Measurements with a Fast and Highly Sensitive InGaAs Camera," *32nd Eur. Photovolt. Sol. Energy Conf. Exhib.*, 2016.

[7] M. Guada *et al.*, "Daylight luminescence system for silicon solar panels based on a bias switching method," *Energy Sci. Eng.*, no. June, pp. 1–15, 2020, doi: 10.1002/ese3.781.

[8] R. Bhoopathy, O. Kunz, M. Juhl, T. Trupke, and Z. Hameiri, "Outdoor photoluminescence imaging of photovoltaic modules with sunlight excitation," *Prog. Photovoltaics Res. Appl.*, vol. 26, no. 1, pp. 69–73, 2018, doi: 10.1002/pip.2946.

[9] O. Martínez *et al.*, "Implementation of a Friendly Daylight Electroluminescence System for the Inspection of Solar PV Panels," in *33rd European Photovoltaic Solar Energy Conference and Exhibition*, 2017, pp. 2021–2025, doi: 10.4229/EUPVSEC20172017-6CO.15.4.

[10] G. A. dos Reis Benatto *et al.*, "Development of outdoor luminescence imaging for drone-based PV array inspection," in *2017 IEEE 44th Photovoltaic Specialist Conference (PVSC)*, Jun. 2017, vol. 2009-Janua, pp. 2682–2687, doi: 10.1109/PVSC.2017.8366602.

[11] G. A. dos Reis Benatto *et al.*, "Luminescence Imaging Strategies for Drone-based PV Array Inspection," in *33rd European Photovoltaic Solar Energy Conference and Exhibition*, Nov. 2017, pp. 2016–2020, doi: 10.4229/EUPVSEC20172017-6CO.15.3.

[12] C. A. Schneider, W. S. Rasband, and K. W. Eliceiri, "NIH Image to ImageJ: 25 years of image analysis," *Nat. Methods*, vol. 9, no. 7, pp. 671–675, 2012, doi: 10.1038/nmeth.2089.

[13] IEC, "IEC TS 60904-13 Photovoltaic Devices - Part 13: Electroluminescence of Photovoltaic Modules," http://www.iec.ch/. [Online]. Available: http://www.iec.ch/.

[14] C. Mantel *et al.*, "SNR Study of Outdoor Electroluminescence Images under High Sun Irradiation," in *2018 IEEE 7th World Conference on Photovoltaic Energy Conversion (WCPEC) (A Joint Conference of 45th IEEE PVSC, 28th PVSEC & 34th EU PVSEC)*, Jun. 2018, pp. 3285–3289, doi: 10.1109/PVSC.2018.8548264.

[15] Thøger Kari Hass, S. Spataru, A. A. Santamaria Lancia, H. Parikh, P. B. Poulsen, and G. A. dos Reis Benatto, "Computer Vision Method for Extracting an Induced Electroluminescence Signal from Photovoltaic Modules in Daylight Conditions Using Drone-Captured Images," in *37th European Photovoltaic Solar Energy Conference and Exhibition*, 2020, pp. 1573–1579, doi: 10.4229/EUPVSEC20202020-5CV.3.44.

[16] R. Bhoopathy, O. Kunz, M. Juhl, T. Trupke, and Z. Hameiri, "Outdoor photoluminescence imaging of solar panels by contactless switching: Technical considerations and applications," *Prog. Photovoltaics Res. Appl.*, vol. 28, no. 3, pp. 217–228, 2020, doi: 10.1002/pip.3216.

IMPROVEMENT OF THE AVEARGE PHOTON ENERGY SPECTRAL CORRECTION FUNCTION FOR PHOTOVOLTAIC PERFORMANCE FORECASTING

Rajiv Daxini, Yupeng Wu, Robin Wilson
University of Nottingham Faculty of Engineering
University Park, Nottingham, NG7 2RD, United Kingdom

ABSTRACT: The solar spectrum has a significant impact on the performance of photovoltaic devices, in particular those with a wide bandgap. Accounting for the spectral influence on PV performance requires a mathematical representation of the solar spectrum for analysis. Such representations are traditionally in the form of proxy variables, such as air mass, or parameters calculated directly from measured spectra, such as the Average Photon Energy (APE). The latter is more accurate, but suffers from reduced reliability due to it not being a unique characteristic of the solar spectral distribution. This study proposes a new empirical spectral correction function based on the APE and the depth of a water absorption band, which successfully addresses the APE uniqueness issue. The new model improves the accuracy of the single-variable APE model by up to 10% for a Cadmium Telluride PV device.
Keywords: CdTe, average photon energy, spectral correction, photovoltaic, forecasting

1 INTRODUCTION

Solar irradiance is one of the main factors affecting photovoltaic (PV) performance [1]. However, due to the non-linear spectral response of PV modules, measurements of broadband irradiance alone are insufficient for accurately modelling PV system output.

The impact of spectral irradiance on PV performance is especially significant for wide band gap devices such as amorphous silicon (aSi) and cadmium telluride (CdTe). Its omission from performance models can lead to annual forecasting errors of up to 20% for such devices [2].

The performance of any particular PV device is usually known at a set of reference test conditions (RTC), of which a reference spectrum forms a part. To correct for the difference in performance due to the spectrum under realistic operating conditions (ROC), a spectral correction function (SCF) is used. SCFs are either based on proxy variables used to represent the solar spectral distribution, such as air mass [3], or parameters derived from the spectrum, such as the average photon energy [4].

The parameter selection in proxy-variable SCFs can limit the range of information on the spectrum included, which means the full effect of the spectrum on PV performance is not modelled. On the other hand, the average photon energy offers a more comprehensive characterisation of the spectrum [5] [6]. Daxini et al. (2022) developed and validated an SCF using the APE, which was shown to exceed the accuracy of both the air mass function and a combined air mass and clearness index function [5].

A potential limitation of the APE parameter centres on bijectivity. Research has shown that dips in irradiance in one particular waveband can be countered by increases in irradiance in another waveband, leading to two differently shaped spectra but the same average photon energy. This issue results in an uncertainty of over 3% between 450–900 nm, and 5–11% outside of this range [7]. This would suggest that the APE is not a bijective index of the solar spectrum [8] [9].

Ishii et al. (2012) find that the primary driver for changes in the shape of spectra that maintain the same APE are the negatively correlated depths of water absorption bands and atmospheric windows [10], hereinafter "spectral band" or I_a. They suggest that a solar spectral distribution may therefore be characterised uniquely for all wavelengths with both the APE and I_a.

Since the single-variable APE SCF proposed by [5] was only validated for an aSi device, which has a spectral response range no greater than 800 nm, the bijectivity issue may not have been a problem. This study tests the original APE spectral correction function for a CdTe device, which has a wider spectral response than aSi. The prevalence of the bijectivity issue in the original APE model examined. In addition, a new empirical model is developed using both the APE and I_a. The purpose of the new model is to address this bijectivity issue whilst still retaining benefits of using the APE parameter, in particular its accuracy over spectral proxies.

2 METHODOLOGY

In this section, the methods and parameters used for data collection, processing, and analysis are explained. The methods for characterising PV performance and the spectrum follow those outlined in Ref. [5], but are summarised here for clarity.

2.1 PV performance characterisation

As spectral changes affect the measured short-circuit current of a PV device, I_{sc}, a normalised form of I_{sc} is used to characterise PV performance under the prevailing spectrum. I_{sc} is first translated to a reference temperature (T_r = 25 °C) and reference irradiance (G_0 = 1000 Wm^{-2}) as follows:

$$I'_{sc} = \frac{I_{sc}}{1 + \widehat{\alpha_{I_{sc}}}(T_c - T_r)}\left[\frac{T_r}{G_{poa}}\right]. \qquad [1]$$

$\widehat{\alpha_{I_{sc}}}$ [C^{-1}] is the short-circuit current temperature coefficient, T_c [°C] is the cell temperature, and G_{poa} is the measured plane of array irradiance. T_c is estimated using a one-dimensional thermal conduction model [11] based on the module temperature, T_m, which is measured:

$$T_c = T_m + \frac{G_{poa}}{G_0}\Delta T. \qquad [2]$$

ΔT is the temperature difference between the module and cell and is determined based on the structure of the device under test (DUT), which for the CdTe module in this study is 3 °C.

The purely spectral influence on I_{sc} is expressed by the ratio of I'_{sc} to the reference current, I'_{sc0}, which is the current measured at T_r, G_0, and AM1.5 (reference spectral irradiance):

$$I_{scn} = \frac{I'_{sc}}{I_{sc0}}. \quad [3]$$

2.2 Characterisation of the solar spectrum

The two parameters — APE and I_a — used to characterise the shape of the measured solar spectral distribution are introduced in this section.

2.2.1 Average Photon Energy

The average photon energy, APE (φ), is an indicator of the overall shape of the solar spectrum. It is calculated by dividing the total energy in the spectrum by the number of photons it contains, as follows:

$$\varphi[\text{eV}] = \frac{1}{q}\left(\frac{\int_a^b E_\lambda \, d\lambda}{\int_a^b \Phi_\lambda \, d_\lambda}\right). \quad [4]$$

$\varphi[eV]$ is the average photon energy, $E(\lambda)[\text{Wm}^{-2}\text{nm}^{-1}]$ is the spectral irradiance, $\Phi(\lambda)\,[\text{m}^{-2}\text{nm}^{-1}]$ is the spectral photon flux density, $q\,[\text{C}]$ is the electron charge, and $b\,[\text{nm}]$ and $a\,[\text{nm}]$ are the upper and lower wavelength limits, respectively, of the considered waveband.

2.2.2 Depth of a window or absorption band

The depth of any spectral band is calculated by integrating the spectral irradiance between the lower (c) and upper (d) wavelength limits of the specified band:

$$I_a = \int_c^d E_\lambda \, d\lambda. \quad [5]$$

Figure 1 shows the intercorrelation between the atmospheric window (Band A, 1000–1050 nm) and water absorption band (Band B, 1110–1160) in the solar spectrum, as first observed by Ishii et al. (2012) [10]. The red and blue arrows indicate a higher and lower spectral irradiance in Band A and B, respectively. Therefore, as the SI in Band A increases, that of Band B decreases, and vice versa.

In this study, four different wavebands are tested in conjunction with the APE parameter to determine which offers the best modelling accuracy for a CdTe module.

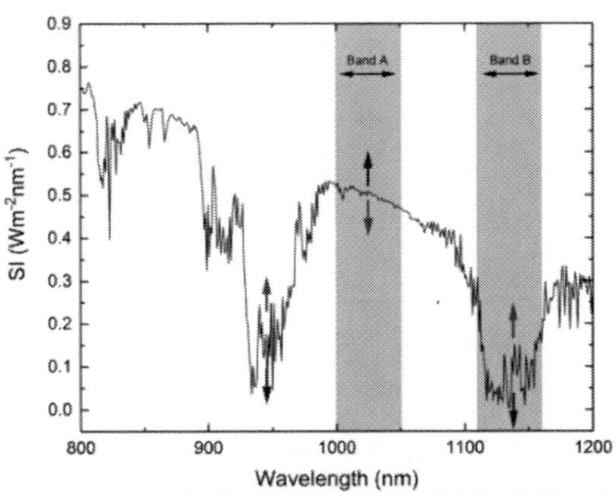

Figure 1 Atmospheric window, Band A (1000–1050 nm), and water absorption band, Band B (1110–1160 nm), as highlighted in Ref. [10]. The red and blue arrows indicate a higher and lower SI in bands A and B, respectively.

These data were measured at the University of Nottingham.

2.3 Device tested

The device tested in this study is a cadmium telluride (CdTe) PV module, the technical specifications of which are summarised in Table I. CdTe devices are especially important for building-integrated photovoltaics applications [12] and, due to their relatively wide band gap, are particularly susceptible to changes in the spectral irradiance.

Table I: Technical specification of the DUT.

Parameter	*Value*
Database identifier	CdTe75669
I_{sc}	1.177 A
V_{oc}	87.66 V
P_{mpp}	65.43 W
$\alpha_{I_{sc}}$	0.04 %/ºC

2.4 Data processing and analysis

The data used in this study are sourced from the Outdoor Test Facility at the National Renewable Energy Laboratory (NREL) [13] [14].

For each of the four wavebands used to plot $I_{scn} = f(\varphi, I_a)$, 22 different surface functions were fitted. The Levenberg-Marquadt algorithm was used with 500 iterations to fit each model to the data. The models were subsequently ranked according to their Akaike (AIC) and Bayesian Information Criterion (BIC) scores. Statistical fit parameters, such as R^2, were also calculated. On average, eight fits were successful whereas the remainder failed to converge. The highest ranked model for each waveband is presented in this work.

The uncertainty on each of the measurements is described in the appendix of Ref. [13]. Errors were propagated to variables such as I_{scn} in the manner described in [5].

3 RESULTS AND DISCUSSION

A new spectral correction function, based on the average photon energy and the depth of a spectral absorption/transmission band, is presented in this section.

3.1 Original APE method

Figure 2 shows the SCF resulting from the original APE method where I_{scn} is assumed to be a function solely of φ. Although the overall fit is reasonably strong with an R^2 value 0.525, which is higher than that calculated for the traditional AM function ($R^2 = 0.319$), the fit is still weaker than the original APE model ($R^2 = 0.870$) derived for an aSi PV module in Ref. [5].

From inspection, it appears that the majority of uncertainty arises from the data points between $1.85 < \varphi[\text{eV}] < 1.90$. In this range, there is a higher density of data points where the same or similar values of φ yield different values of I_{scn}. The implication of this is that spectral distributions that result in different levels of PV performance possess APE values that are insufficiently different to represent the change in I_{scn} to which they lead. This larger spread about I_{scn} for the same or similar values of APE, and thus higher uncertainty of the model $I_{scn} = f(\varphi)$, significantly reduces the reliability of predictions of I_{scn} relative to the performance reported in Ref. [5]. This observation corroborates the findings in Ref. [8] [7] [9] where the bijectivity of the APE index as a means to represent the solar spectrum is questioned.

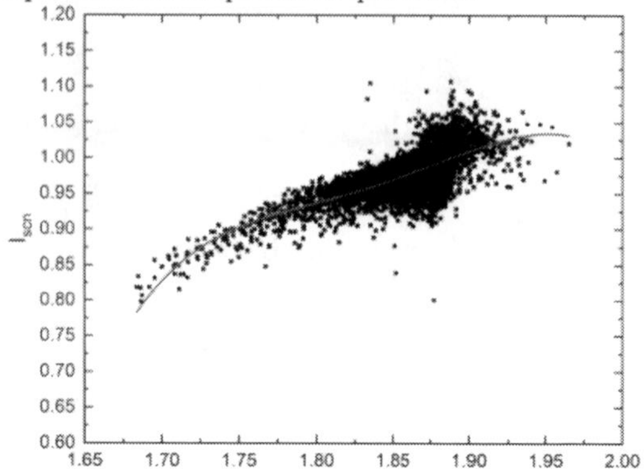

Figure 2 Single-variable SCF, $I_{scn} = f(\varphi)$, based on data measured at the Golden, Colorado, facility between August 2012 and August 2013.

3.2 Modified APE method

Figure 3 shows four new SCFs derived by correlating I_{scn} with the APE and values of I_a derived using four different spectral bands. The highest ranking $f(\varphi, I_a)$ fit for each of the four wavebands investigated is shown. The R^2 values for the fit for each waveband are presented in Table II, which also includes the R^2 value of $f(\varphi)$ for reference.

The points that were previously concentrated around the $1.85 < \varphi[\text{eV}] < 1.90$ range in Figure 2. have been spread into the third dimension created by the additional dependent variable, I_a in Figure 3. It is visually evident that the inclusion of the additional index directly helps address the issue of reduced reliability of I_{scn} predictions within this value range. The spectral distributions with similar APE have now been distinguished sufficiently to be attributed to different values of I_{scn}.

Table II: Coefficient of determination for the highest ranking SCF derived for each of the four wavebands, $f(\varphi, I_a)$, and the single-variable APE SCF, $f(\varphi)$.

I_a	R^2
(a) 750–780	0.567
(b) 920–970	0.547
(c) 950–1000	0.615
(d) 1000–1050	0.568
APE only	*0.525*

Table II shows that regardless of the waveband chosen, an improvement is always achieved relative $f(\varphi)$ by the inclusion of the additional index, I_a. This improvement may be attributed to the benefit achieved through bijective representation of the solar spectrum. However, there is some variation in the fitting accuracy of the different models. The waveband 950–1000 nm generates the most accurate fit, improving on the next highest by almost 5%. This may be explained by the CdTe panel having a particularly high spectral response close to this waveband, making its performance particularly susceptible to changes in the depth of this band [15].

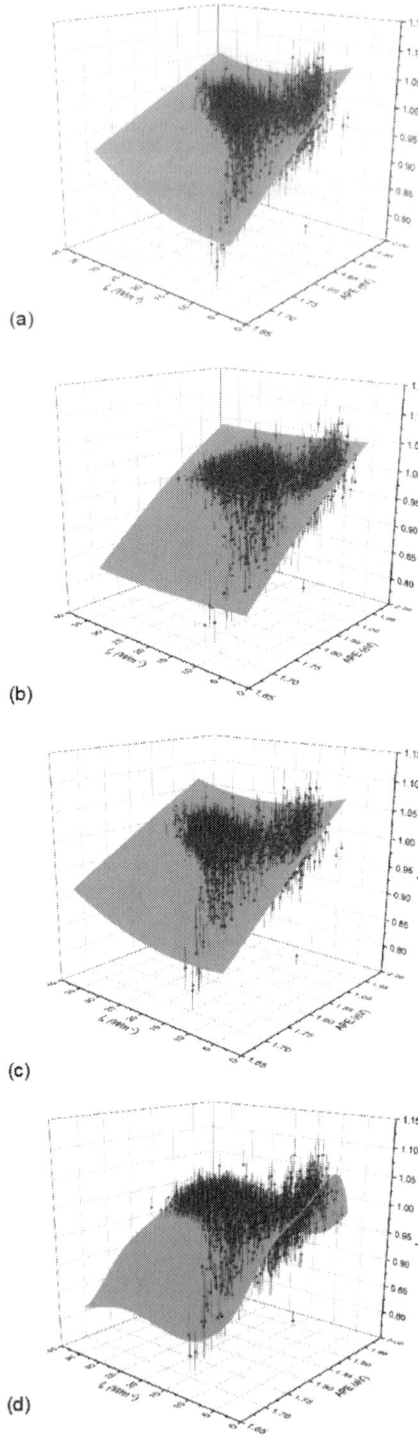

Figure 3 Panels *(a)*, *(b)*, *(c)*, and *(d)* show the highest-ranking fits for the plots of $I_{scn} = f(\varphi, I_a)$, where I_a is the waveband between 750–780 nm, 920–970 nm, 950–1000 nm, and 1000–1050 nm, respectively. R^2 values for each of the fits are presented in Table II.

4 CONCLUSION

Spectral correction functions are a key component of photovoltaic performance models.

Traditional methods have been based on either a variety of proxy variables used to represent the spectrum, or the average photon energy calculated directly from the measured spectrum. The latter is a more accurate method but suffers from significantly reduced accuracy for PV devices with a spectral response greater than 800 nm.

This study has shown that the inclusion of an additional index, namely the depth of a water absorption band or an atmospheric window, improves the accuracy of the original APE SCF for a CdTe module. Four wavebands were investigated, all of which improved on the original function by at least 3% and up to 10%.

5 ACKNOWLEDGEMENTS

This work was supported by the Faculty of Engineering, University of Nottingham, through a PhD studentship awarded to Rajiv Daxini. This work was also supported by the Engineering and Physical Sciences Research Council, UK [grant number EP/W028581/1], and the University of Nottingham Researcher Academy Conference, Travel, and Training Fund.

6 REFERENCES

[1] S. A. Kalogirou, Solar energy engineering: processes and systems, Academic Press, 2013.

[2] Y. Hirata and T. Tani, "Output variation of photovoltaic modules with environmental factors—I. The effect of spectral solar radiation on photovoltaic module output," *Solar Energy*, vol. 55, p. 463–468, 1995.

[3] D. L. King, J. A. Kratochvil and W. E. Boyson, "Measuring solar spectral and angle-of-incidence effects on photovoltaic modules and solar irradiance sensors," in *Conference Record of the Twenty Sixth IEEE Photovoltaic Specialists Conference-1997*, 1997.

[4] C. N. Jardine, T. R. Betts, R. Gottschalg, D. G. Infield and K. Lane, "Influence of spectral effects on the performance of multijunction amorphous silicon cells," in *Proc. Photovoltaic in Europe Conference*, 2002.

[5] R. Daxini, Y. Sun, R. Wilson and Y. Wu, "Direct spectral distribution characterisation using the Average Photon Energy for improved Photovoltaic Performance Modelling," *Renewable Energy*, 2022.

[6] G. Nofuentes, B. García-Domingo, J. V. Muñoz and F. Chenlo, "Analysis of the dependence of the spectral factor of some PV technologies on the solar spectrum distribution," *Applied energy*, vol. 113, p. 302–309, 2014.

[7] G. Nofuentes, C. A. Gueymard, J. Aguilera, M. D. Pérez-Godoy and F. Charte, "Is the average photon energy a unique characteristic of the spectral distribution of global irradiance?," *Solar Energy*, vol. 149, p. 32–43, 2017.

[8] D. Dirnberger, G. Blackburn, B. Müller and C. Reise, "On the impact of solar spectral irradiance on the yield of different PV technologies," *Solar Energy Materials and Solar Cells*, vol. 132, p. 431–442, 2015.

[9] J. Polo, M. Alonso-Abella, J. A. Ruiz-Arias and J. L. Balenzategui, "Worldwide analysis of spectral factors for seven photovoltaic technologies," *Solar Energy*, vol. 142, p. 194–203, 2017.

[10] T. Ishii, K. Otani, A. Itagaki and K. Utsunomiya, "A methodology for estimating the effect of solar spectrum on photovoltaic module performance by using average photon energy and a water absorption band," *Japanese Journal of Applied Physics*, vol. 51, p. 10NF05, 2012.

[11] D. L. King, J. A. Kratochvil and W. E. Boyson, Photovoltaic array performance model, United States. Department of Energy, 2004.

[12] C. Qiu, Y. K. Yi, M. Wang and H. Yang, "Coupling an artificial neuron network daylighting model and building energy simulation for vacuum photovoltaic glazing," *Applied Energy*, vol. 263, p. 114624, 2020.

[13] W. Marion, A. Anderberg, C. Deline, S. Glick, M. Muller, G. Perrin, J. Rodriguez, S. Rummel, K. Terwilliger and T. J. Silverman, "User's manual for data for validating models for PV module performance," 2014.

[14] T. Stoffel and A. Andreas, "NREL solar radiation research laboratory (SRRL): Baseline Measurement System (BMS); Golden, Colorado (data)," 1981.

[15] K. Passow and M. Lee, "Effect of spectral shift on solar PV performance," in *2016 IEEE Conference on Technologies for Sustainability (SusTech)*, 2016.

SEVERITY ANALYSIS OF NON-UNIFORM ENCAPSULANT DISCOLOURATION IN CRYSTALLINE SILICON PHOTOVOLTAIC MODULES USING DIFFERENTIAL CURRENT-VOLTAGE ANALYSIS APPROACH

Roopmati Meena[a,*], Manish Kumar[b], and Rajesh Gupta[a]
[a]Department of Energy Science and Engineering, Indian Institute of Technology Bombay, India, [b]Department of Solar Power Systems, Institute for Energy Technology, Kjeller, Norway
*corresponding email id: meenaroopmati@iitb.ac.in

ABSTRACT: Encapsulant discolouration is a prominent degradation observed in the photovoltaic (PV) module and can have uneven distribution over the cells in the modules. It is often accompanied by other defects and degradations (D&Ds) in module under external operating conditions. Identification of non-uniform encapsulant discolouration in the presence of multiple D&Ds can be a challenge. In this regard, a new, non-destructive, and easy-to-implement analysis method called differential current-voltage (DIV) analysis, has been developed to investigate non-uniform encapsulant discolouration in the presence of other D&Ds. For this purpose, current-voltage (I-V) curve of 20-year-old PV modules have been analyzed while partially shading the consecutive solar cells of the PV module. The normalized percentage change in short circuit current (I_{sc}) and voltage at maximum power point (V_{mpp}) have been studied in detail to analyze the severity/extent of discolouration and determine the type of dominant degradation mode, respectively, in the individual cells of the PV module. The presented approach can be used for identification of encapsulant degradations in the PV modules under outdoor operating conditions.

Keywords: PV module, Differential I-V analysis, Encapsulant discolouration, Defects and degradations.

1 INTRODUCTION

The performance and long-term reliability of crystalline silicon (c-Si) photovoltaic (PV) modules get affected due to various defects and degradations (D&Ds) occurring under external environmental conditions [1,2]. The most commonly observed degradation mode in PV modules has been discolouration in the ethylene vinyl acetate (EVA) based encapsulant under the influence of ultraviolet (UV) radiation, moisture, and high temperature conditions [3–5]. Encapsulant discolouration can have non-uniform distribution with variation in severity and magnitude over the cell areas, within PV modules operating under similar environmental conditions [6]. It can be attributed to the photo bleaching of discoloured encapsulant around cell edges and near busbar region due to the ingression of moisture through backsheet. The presence of moisture can also cause degradation of metallization components in the module namely, fingers, busbar, and interconnect ribbon.

Further, the degradation of EVA leads to release of acetic acid, which again corrodes the metallic components of the PV module [7]. Thus, the discolouration impacts the module output power directly by reducing the incoming irradiation over the cell and indirectly due to metallization corrosion by providing high resistance path for current flow. Early detection of discoloration initiation can prevent its cascading effect on metallization components as well as reduce mismatch losses amongst modules. Thus, there is a need to analyse the encapsulant discolouration through easy-to-implement characterization method under field-operating conditions. However, majority of the characterization techniques available require sophisticated instruments which have limitations for implementation in the outdoor environmental conditions.

In this regard, the commonly available characterization technique, current-voltage (I-V) curve measurement, has been utilised in this study for severity analysis of the uneven encapsulant discoloration over the cells of the PV modules using differential I-V analysis approach. For this purpose, field-aged c-Si PV modules subjected to subtropical climatic conditions for 20-years have been investigated. Wherein, each cell of the module has been partially shaded and the I-V curve of the module has been measured. The normalised change in the electrical parameters of the module, namely, short circuit current (I_{sc}) and voltage at maximum power point (V_{mpp}), have been analysed for the varying extent of encapsulant discoloration related degradation in the cells of the module. Herein, change in I_{sc} has been found to vary based on the extent of encapsulant discolouration in the cells, whereas the V_{mpp} has mainly been found to be affected by the combined effect of encapsulant discolouration and metallization corrosion. The investigated modules have also been pre-characterized using visual inspection, microscopic imaging, and electroluminescence (EL) imaging technique to identify various D&Ds present in the individual cells of the field-aged PV modules. The presented approach of differential I-V analysis can be used for on-field investigation of PV modules for early-stage detection of encapsulant discolouration and related degradations.

2 METHODOLOGY

In this study, the electrical parameters of the PV modules have been systematically analysed using the differential I-V analysis approach for severity analysis of EVA based encapsulant discolouration over the cells of c-Si PV modules. For this purpose, I-V curves of the modules have been analysed while sequentially applying partial shading over its solar cells. The methodology used in this study is given in Fig. 1. Four field-aged c-Si PV modules subjected to subtropical climatic conditions of India for 20 years have been investigated. These modules have been pre-characterized for the presence of various D&Ds through visual inspection, microscopic imaging, and EL imaging technique. The modules predominantly showed presence of discolouration and metallization corrosion over its cells.

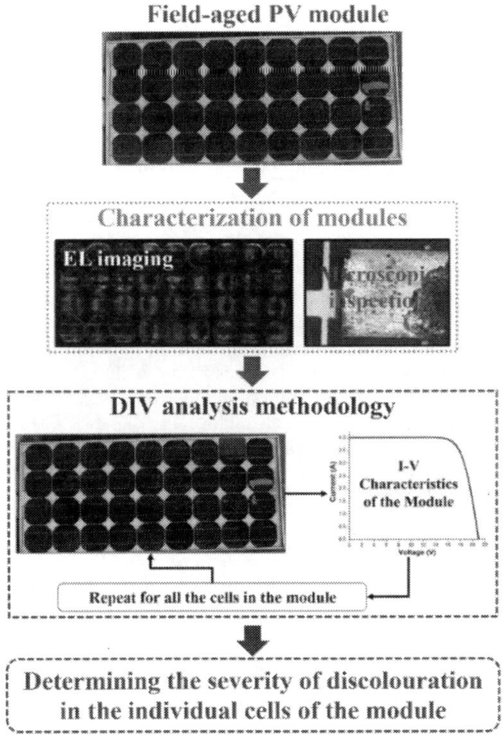

Figure 1: Graphic schematic of the methodology used in this work.

The percentage change in electrical parameters of the module namely, I_{sc} and V_{mpp}, have been calculated with respect to unshaded module parameters with the consecutive partial shading of the cells in a PV module. Further, normalised values of percentage change in I_{sc} and V_{mpp} has been used to examine and differentiate the extent and severity of EVA discolouration from the metallization degradation in the cells of the modules. The modules investigated in this study consisted of 36 series connected cells along with two bypass diodes, each connected across 18 cells. The presence of bypass diodes enabled the independent operation of the two string in the module by bypassing the effect of current mismatch due to D&Ds in another string. Thus, enabling the string wise analysis of non-uniform encapsulant discoloration and corrosion in the cells.

3 RESULTS AND DISCUSSION

The present section discusses the results and analysis from visual inspection, microscopic imaging, and EL imaging performed on field-aged PV modules in detail. Herein, the primary results and analysis of a module, which has been representative of the general observations, have been given.

The images showing results of EL imaging, microscopic imaging, and visual inspection are given in Fig. 2 (a). Several dark regions can be observed over the cells in the EL image of the module. It can be attributed to multiple types of D&Ds, as can be seen in Fig. 2 (b) to Fig. 2 (f). The most commonly observed D&Ds in the investigated modules are encapsulant discolouration, encapsulant delamination, metallization (busbar, finger, and interconnect ribbon) corrosion, cracks, and metallization interruption. These D&Ds effects the module performance and electrical parameter, which primarily depend on the type and extent of the degradation modes.

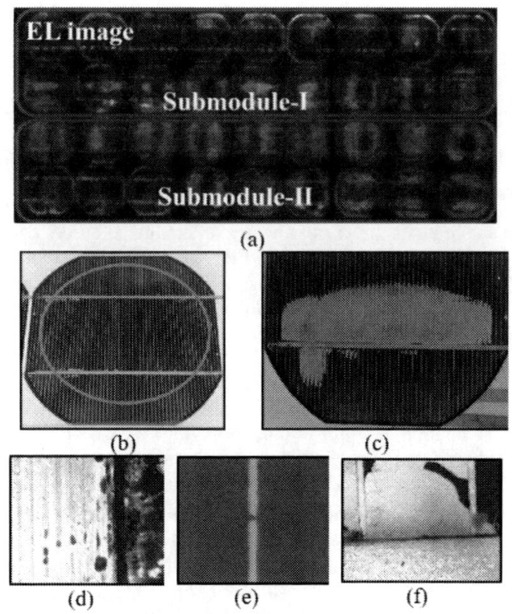

Figure 2: (a) EL image, visual, and microscopic image of D&Ds in the module showing (b) discolouration, (c) delamination, (d) ribbon corrosion, (e) finger interruption, and (f) finger corrosion.

To understand the effect of varying extent of encapsulant discolouration over a solar cell on the module performance, the differential I-V analysis approach has been used. Wherein, the normalised percentage change in the parameters, i.e., I_{sc} and V_{mpp} has been analysed with the partial shading of cells. The graph showing normalised percentage change in the I_{sc} of the module with the shading of each cell is given in Fig. 3.

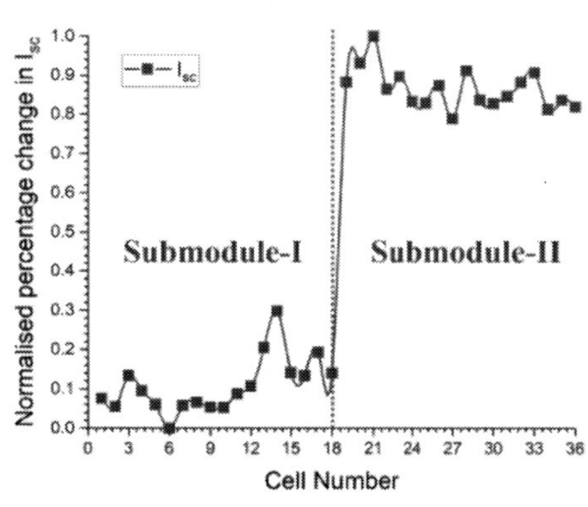

Figure 3: Normalised percentage change with cell number in short circuit current (I_{sc})

It can be observed that there is a distinction in the characteristic pattern of normalised values for a group of cells. These cells have been divided into sub-groups henceforth called submodule-I and submodule-II for cells 1-18 and cells 19-36, respectively. The distinction can be attributed to the presence of bypass diodes across the string of 18 cells, which enable isolated operation of submodule strings. It must be noted that the shift in the magnitude amongst the cells within a submodule has also been different for analysed electrical parameters. The overall change in I_{sc} for submodule-I has been less than in submodule-II, which can be understood using the EL image of the module. Herein, the dark regions over cells in submodule-I has been more than in submodule-II, which can be attributed to the dominance of encapsulant discolouration in submodule-I. When a cell with high extent of encapsulant discoloration is partially shaded, the percentage change in its string current is less compared to the change in current when a less discoloured cell is partially shaded. Thus, the change in current for submodule-II is high due to less extent of discoloration over the cells. Moreover, there has also been variation in I_{sc} change amongst the cells within a submodule. It can be attributed to the difference in the extent of discolouration over the cells within a submodule. The cell with less dark regions in the EL image has shown a greater change in the I_{sc} within a submodule, implying a low extent of encapsulant discolouration over them.

On the other hand, the trend of the change in normalised V_{mpp} in both the submodules has been opposite to that in I_{sc} as shown in Fig. 4. These patterns have been analysed based on the operating efficiency of the cells when subjected to partial shading conditions. The reduction in string current due to partial shading and encapsulant discolouration shifts the operating point of the series-connected cells towards V_{oc}, leading to an increase in the overall voltage of the string and hence, of the PV module. In the case of submodule-I, though the change in I_{sc} has been less during differential I-V analysis (Fig. 3), the cells have permanently been operating at lower efficiencies due to discolouration.

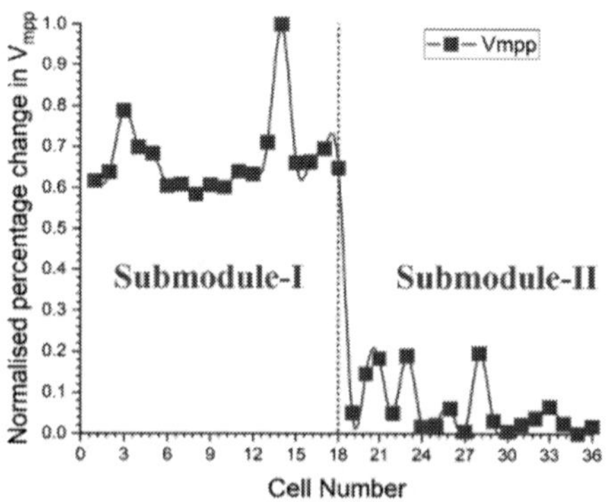

Figure 4: Normalised percentage change with cell number in voltage at maximum power point (V_{mpp}).

Further reduction in current due to partial shading has led to more increase in V_{mpp}. As a result, the overall change in V_{mpp} due to partial shading is higher for cells in submodule-I. Similarly, for the submodule-II, the change in V_{mpp} should have been large for the cells showing a large dark region in EL image following the shift in operating point of cells due to efficiency reduction. However, it must be noted that in submodule-II, some of the cells showing the presence of large dark regions in EL image have not shown a corresponding high change in V_{mpp}, e.g., cells 34-36. Based on the visual inspection and microscopic images, it can be attributed to the metallization corrosion (Fig. 2), resulting from chemical reaction with acetic acid released during encapsulant discolouration and possible moisture ingression due to degradation of EVA. The flow of current through corroded metallization leads to drop in the operating voltage. As a result, the decrease in V_{mpp} due to corrosion acts in opposition to the increase in V_{mpp} due to partial shading and encapsulant discolouration. Thus, the overall change in the module voltage has been less for highly corroded cells with encapsulant discolouration. Moreover, for submodule-II, the overall change in V_{mpp} has been less compared to submodule-I.

Based on the normalised change in I_{sc} and V_{mpp} of a module when its cells are subjected to partial shading conditions, the severity of encapsulant discolouration can be compared within a submodule. The cells with high extent of discoloration show less change in I_{sc} and vice-versa. Further, the dominance of corrosion in metallization components due to acetic acid or moisture ingression in a cell can also be determined through the normalised change in V_{mpp}. Wherein, the cells with high extent of corrosion show comparatively less change in V_{mpp} for a submodule.

4 CONCLUSION

In this work, extent of uneven discolouration in the 20-year-old modules has been investigated using differential current-voltage (DIV) analysis. For this purpose, each cell of the module has been partially shaded while measuring the current-voltage (I-V) curve of the module. The module has been divided into two submodules for analysis. The normalized percentage change in short circuit current (I_{sc}) was found to be dependent on the extent of encapsulant discolouration in the cells. The cells with higher extent of discolouration showed less change in the I_{sc} and vice-versa. To distinguish discolouration from other degradation modes in the cells, voltage at maximum power point (V_{mpp}) have been analyzed. Amongst the two submodules, the submodule having higher extent of discolouration showed higher changes in the V_{mpp}. Further, within a submodule, the cells with higher resistive losses showed less change in the V_{mpp}. The approach used in this work can be utilized to identify the extent and distinguish uneven discolouration in the PV modules under outdoor operating conditions using readily available characterization technique, namely, I-V curve measurement.

REFERENCE

[1] T. Ishii, A. Masuda, Annual degradation rates of recent crystalline silicon photovoltaic modules, Prog. Photovoltaics Res. Appl. 25 (2017) 953–967. https://doi.org/10.1002/pip.

[2] R. Meena, S. Kumar, R. Gupta, Investigation of front-side grey delamination in crystalline silicon photovoltaic modules, in: SiliconPV 2021, 11th Int.

Conf. Cryst. Silicon Photovoltaics, 2022: p. 100002. https://doi.org/10.1063/5.0089273.

[3] R. Meena, S. Kumar, R. Gupta, Comparative investigation and analysis of delaminated and discolored encapsulant degradation in crystalline silicon photovoltaic modules, Sol. Energy. 203 (2020) 114–122. https://doi.org/10.1016/j.solener.2020.04.041.

[4] M.C.C. de Oliveira, A.S.A. Diniz Cardoso, M.M. Viana, V. de F.C. Lins, The causes and effects of degradation of encapsulant ethylene vinyl acetate copolymer (EVA) in crystalline silicon photovoltaic modules: A review, Renew. Sustain. Energy Rev. 81 (2018) 2299–2317. https://doi.org/10.1016/j.rser.2017.06.039.

[5] S. Kumar, R. Meena, R. Gupta, Finger and interconnect degradations in crystalline silicon photovoltaic modules: A review, Sol. Energy Mater. Sol. Cells. 230 (2021) 111296. https://doi.org/10.1016/j.solmat.2021.111296.

[6] A. Sinha, O.S. Sastry, R. Gupta, Nondestructive characterization of encapsulant discoloration effects in crystalline-silicon PV modules, Sol. Energy Mater. Sol. Cells. 155 (2016) 234–242. https://doi.org/10.1016/j.solmat.2016.06.019.

[7] A. Kraft, L. Labusch, T. Ensslen, I. Durr, J. Bartsch, M. Glatthaar, S. Glunz, H. Reinecke, Investigation of Acetic Acid Corrosion Impact on Printed Solar Cell Contacts, IEEE J. Photovoltaics. 5 (2015) 736–743. https://doi.org/10.1109/JPHOTOV.2015.2395146.

MONTE CARLO APPROACH TO CONSIDER POWER PRODUCTION UNCERTAINTY DUE TO RANDOMNESS OF MODULE MANUFACTURING PROCESS

Ali Sohani [a,*], Cristina Cornaro [a], David Moser [b]
[a] Department of Enterprise Engineering, University of Rome Tor Vergata, Rome, Italy
[b] EURAC Research, Bolzano, Italy
[*] Corresponding author, email address: ali.sohani@uniroma2.it

ABSTRACT: This study aims at finding the impact of randomness in manufacturing on the uncertainty of power production by a photovoltaic (PV) module. For this purpose, the module efficiency at the reference condition (η_{ref}) is chosen as the parameter that describes the manufacturing randomness, and using the experimental data, a statistical distribution for that is found. Then, Monte Carlo approach, as a systematic way, is utilized, and the statistical distribution for produced power of the PV module is found. A module from poly crystalline type, with the rated capacity of 120 W, is selected, and the analysis is carried out for that as the case study. The fitting based on the experimental data for 114 PV modules is demonstrated that a normal distribution with the average of 15.8% and standard deviation of 0.498% is the best statistical distribution for η_{ref}. Moreover, according to the obtained results, due to η_{ref} randomness, power production varies in the range of 97 W to 113 W. It means 16 W uncertainty in the power production, which is 13.3% of the nominal capacity.

Keywords: Energy Performance; Manufacturing and Processing; Monte Carlo Approach; PV Module; Uncertainty.

1 AIM AND APPROACH

Power is taken into account as the most important characteristic of a photovoltaic (PV) module. The more accurate PV power is predicted, the better design and control of the system could be achieved. Because of non-ideality in some phenomena, including the structure of the silicon wafers which make a PV module and the cells connections, there is the randomness in manufacturing of a PV module. This randomness leads to uncertainty in the specification of a PV module, and consequently, power production of that. Nonetheless, the issue is not considered in the simulation approaches, which assume a constant value for the specifications of a PV module. It could be identified as a serious gap of the research.

Having found the mentioned gap, in this investigation, the randomness of PV manufacturing process is considered and a method for taking that in the simulation is proposed. In the proposed approach, the efficiency at the reference condition (η_{ref}) is considered as the parameter to describe the manufacturing randomness. Then, using the experimental data, a statistical distribution is found for that. Next, the Monte Carlo approach is employed to find the uncertainty of the power production based on the obtained statistical distribution for η_{ref} manufacturing randomness.

As the case study, a 120W module poly crystalline produced by Yingly company is chosen. After that, using the experimental data for 114 PV modules from that type, the analysis is carried out to find a statistical distribution for η_{ref}, and then, the uncertainty of produced power for the investigated PV module. It should be noted that the experimental data is for the reference condition. The solar radiation and ambient temperature values for the reference condition are 800 W.m^{-2} and 20 °C, respectively.

In the Monte Carlo approach, a specific number of random values is generated (the value is considered 1000 here). Then, the random number is put equal to the cumulative probability of η_{ref} that is predicted by the found distribution. After that, η_{ref} for that random number is found, and power (P) is calculated using Eq. (1):

$$P = GA_{module}\eta_{ref}\left\{1-\beta_{ref}(T_{module}-T_{ref})+\gamma\log_{10}(G)\right\} \quad (1)$$

In which, G, A, and T represent solar radiation, area, and temperature, respectively. The subscripts 'module' and 'ref' also indicate PV module, and reference condition, for which G and T values are 800 W.m^{-2} and 20 °C, respectively. This is also the reference condition for evaluation of module temperature based on the nominal module operating temperature (NOCT) approach. β and γ are also the temperature coefficients for voltage and maximum power, respectively. They are module specifications which are available in the catalogue of the module. In order to calculate module temperature (T_{module}) in Eq. (1), the nominal operating cell temperature (NOCT) method is used here. NOCT suggests Eq. (2) to determine T_{module}:

$$T_{module} = T_{amb} + \frac{G}{G_{ref}}(T_{NOCT} - T_{ref}) \quad (2)$$

All the parameters have been introduced previously except for T_{amb} and T_{NOCT}. The former is the ambient temperature, while the latter is NOCT for the module. Like β and γ, T_{NOCT} is one of the specifications of the module, and the value of that could be found in the catalogue.

This question might come into the mind that plotting power probability curve could be done directly using the available 114 experimental data, and applying Monte Carlo approach is not necessary. As the answer to this question, it should be indicated that, as it has been mentioned in several references like [1], despite direct usage of data, application of Monte Carlo method makes the power probability curve general, and not limited to the limited number of 114 PV modules. This means that it could have a prediction about the uncertainty of power production for 115th, 116th, and other next modules, which have not been tested, as well. For example, if the data is directly used, the approach is not able to predict the probability of power production for a range of η_{ref} outside the training range, i.e., 14.9% to 16.7% (Although the probability of observing η_{ref} outside the range is very low, but it is not zero). This is something similar to prediction by machine learning approaches like artificial neural network (ANN), in which by training based on a limited number of data, a general prediction could be provided.

2 SCIENTIFIC INNOVATION AND RELEVANCE

As indicated, to the best of authors' knowledge, in the employed simulation approaches in the literature, a constant value for each specification of a PV module has been considered. However, due to randomness in the manufacturing of a PV module, there is uncertainty in the values of a parameter like η_{ref}. Not considering the indicated uncertainty is identified as the gap of the research. Therefore, as the novelty, this work finds a statistical distribution for η_{ref} according to experimental data. Then, Monte Carlo approach, as a systematic way, is employed to find an uncertainty for produced power of the PV module because of the indicated uncertainty for η_{ref}. Monte Carlo approach introduces a statistical distribution for produced power as the output.

3 RESULTS AND CONCLUSIONS

By employing the obtained experimental data from the performance of 114 PV modules, all from the poly crystalline type and produced by Yingly company and have the same nominal capacity of 120 W, the probability distribution of the reference efficiency is obtained, which is shown in Figure (1). In this figure, the bin for the histogram is considered 0.2%. Moreover, for better clarity, the figure is provided in a way that values of 14.9% and 15.1% in the horizontal axis represent the ranges of 14.8% to 15.0%, and 15.0% to 15.2%, respectively, and so on.

Figure (1) implies that there is a considerable variation for η_{ref}, which could not be neglected. η_{ref} changes from 14.8% to 16.8%. Among the available items, the normal distribution is found to be the best fit to describe the statistical distribution of η_{ref}. The prediction error of the found distribution for η_{ref} is relatively low, as it is graphically observed in Figure (1). The average and standard deviation, which are the tunning parameters for this distribution, are also 15.8% and 0.498%, respectively.

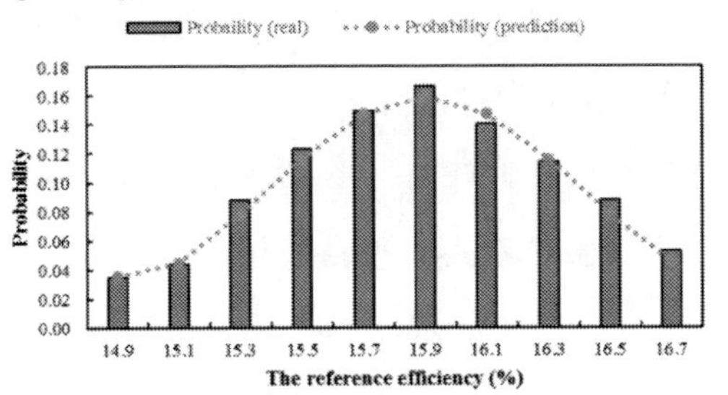

Figure (1): The experimental data for distribution of η_{ref} in addition to the fitted normal distribution

Using the found distribution for η_{ref}, and Monte Carlo approach, the statistical distribution for produced power of the investigated PV module is obtained. It is presented in Figure (2). In order to plot this figure, the bin is considered 2 W. It means that numbers 98 W and 100 W in this figure indicates the range of 97 W to 99 W, and 99 W to 101 W, respectively, and so on.

According to Figure (2), the produced power of the PV module varies from around 97 W to 113 W due to the uncertainty of η_{ref}. This is 16 W variation. The variation is 13.3% of the nominal power of the PV module, i.e., 120 W, which indicates that like η_{ref}, the uncertainty of power is also considerable. The most probability is found for the range of 105 W to 107 W, where the probability is 0.1727. With the value of 0.0633, the lowest probability also belongs to the range of 97 W to 99 W. The average and standard deviation of the found distribution for power are around 105 W and 4.6 W, respectively. It is also a normal distribution, as well.

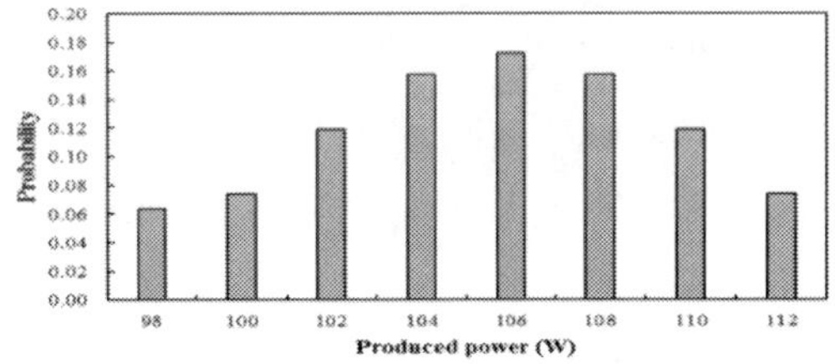

Figure (2): The obtained distribution for produced power of the module

As seen, the obtained results for the conducted analysis have demonstrated that the randomness of PV module manufacturing leads to a considerable uncertainty for power production. Therefore, it has to be considered for analyzing a PV module to achieve a more accurate and realistic condition for the design and control of PV systems.

4 REFERENCES

[1] Mahyar Momen, Mehrdad Shirinbakhsh, Amir Baniassadi, and Ali Behbahani-nia. "Application of Monte Carlo method in economic optimization of cogeneration systems–Case study of the CGAM system." Applied Thermal Engineering 104 (2016): 34-41.

ENERGY-BASED SOILING LOSS MONITORING FOR SOLAR PV SYSTEMS

Pavan Fuke[1,2] (204076012@iitb.ac.in), Shoubhik De[1,2] (shoubhikde@ee.iitb.ac.in), Narendra Shiradkar[1,2] (naren@ee.iitb.ac.in) and Anil Kottantharayil[1,2] (anilkg@ee.iitb.ac.in)
[1]Department of Electrical Engineering, Indian Institute of Technology Bombay, Mumbai, Maharashtra 400076, India
[2]National Centre for Photovoltaic Research and Education, Indian Institute of Technology Bombay, Mumbai, Maharashtra 400076, India

ABSTRACT: The energy loss caused by dust and other particles accumulating on photovoltaic (PV) modules has become one of the most important issues for PV system owners. A site-specific and dedicated soiling monitoring system may be necessary to optimize cleaning schedules and thus improve yearly energy yield. Commercially available electrical soiling measurement kits compare the short circuit current (I_{SC}) or maximum power (P_{max}) of clean and naturally soiled PV modules for a specific time interval and irradiation level. In this paper, we present an energy-based soiling loss monitoring approach. The I_{SC} and P_{max} based methods are benchmarked to the energy based method for soiling data collected from a test station in Mumbai. It is found that the I_{SC} and P_{max} based methods underestimate the soiling losses by about 2.5-3%.
Keywords - photovoltaic modules, soiling, energy loss, soiling ratio

1 INTRODUCTION

The electric yield of solar Photovoltaic (PV) modules are dependent on factors like solar irradiance, temperature, wind speed, precipitation, humidity and dust deposition. Soiling is a commonly overlooked or underestimated external component that might be a deal-breaker for the viability of a solar project. The soiling loss rises linearly in the absence of significant rainfall and regular cleaning. It is projected that the global solar power production loss due to soiling is 3%–4%, which in turn caused 3-4 billion € of revenue losses in 2018 [1]. Various research has been published recently that presents models and methods for monitoring soiling losses [2-5].

Typically, the soiling loss of a PV system is assessed using a pair of PV modules, one of which is kept clean always time, and the other is allowed to soil naturally. The module that is allowed to soil is usually of identical make and model to the ones used for power generation, whereas the clean module can be a reference cell of the same technology or a module similar as the soiled. The output current of the PV module directly depends on light received by the solar cells, which gets reduced by soiling. So, the simplest way to calculate the soiling loss is to compare the soiled and clean modules' I_{SC}. Commercially used soiling monitoring kits such as Atonometrics RDE 300 [3], Campbell Scientific CR-PVS1 Soiling Loss Index RTU [4] and the autonomous soiling monitoring kit developed by Arizona State University [5] compute the soiling losses in terms of I_{SC}. However, dust particles tend to accumulate around the bottom edges of the module due to gravity, precipitation, and wind, resulting in non-uniform soiling. Soiling of PV panels may show significant non-uniformity within a panel, from panel to panel in a string and from string to string within a large powerplant. In the case of non-uniform soling, I_{SC} may not be a good predictor of soiling loss. As a result, some soiling monitoring equipment measures soiling loss in terms of power loss. The Atonometrics RDE 300 can also measure soiling losses in terms of power.

This study proposes an energy-based approach to analyze the soiling losses. The proposed method is applied to analyze the 2-year electrical data obtained from a rooftop PV installation in Mumbai, India. It is shown that the I_{SC} and P_{max} based soil monitoring methods underestimate the energy yield loss due to soiling by 2.5 to 3%.

2 METHODOLOGY

To quantify the soiling loss for Mumbai, India (19.1334° N, 72.9133° E), several field experiments were undertaken. Based on four years of field data, the soiling rate in Mumbai was estimated as 0.43 percent per day during dry seasons [6]. This study analyzed the electrical data of two 327 W_P latitudemounted PV modules installed at Photovoltaic Module Monitoring Station (PVMMS), IIT Bombay, Mumbai, as shown in Fig. 1. The bottom module was cleaned every day, while the top module was permitted to be naturally soiled, as shown in Fig. 2. The modules were connected to a multichannel I–V curve tracer (Daystar MT5 3200). I_{SC}, open circuit voltage, and P_{max} are measured with 1 minute internal and the complete I–V curves are measured once in every 10 minutes.

Fig. 1: PVMMS installed at IIT Bombay, Mumbai

Gostein et. al. [3] measured the soiling ratio in terms of both I_{SC} and P_{max} using Atonometrics RDE 300 setup [4]. The soiling ratio measurement using this method is represented by M1. Campbell Scientific CR-PVS1 Soiling Loss Index RTU and ASU autonomous soiling monitoring kit equipment are configured to measure soiling ratio using I_{SC} [5]-[6]. These soiling monitoring methods are represented by M2 and M3 respectively. Each of these

Fig. 2: Clean (bottom) and soiled (top) modules.

methods analyzed the data only for specific time intervals and for specific solar irradiation range. Table I shows the time and irradiance filters for M1, M2 and M3.

The soiling ratios in terms of current, power, and energy are calculated using equations (1), (2), and (3). It may noted that measured for full day is used for the proposed energy based method, while M1 to M3 use data collected over shorter time intervals. The soiling losses can then be calculated using the soiling ratio as given in equation (4). The soiling ratio difference between the proposed energy-based method and I_{SC} and P_{max} based methods are computed using SR_E as a reference for the comparative study.

$$\text{Soiling ratio in terms of current} = SR_{I_{SC}} = \frac{I_{SC\_soiled}}{I_{SC\_clean}} \quad (1)$$

$$\text{Soiling ratio in terms of power} = SR_{P_m} = \frac{P_{m\_soiled}}{P_{m\_clean}} \quad (2)$$

$$\text{Soiling raio in terms of Energy} = SR_E = \frac{\textit{Integral of power generated by soiled module for the day}}{\textit{Integral of power generated by clean module for the day}} \quad (3)$$

$$\text{Soiling loss} = 1 - \text{Soiling ratio} \quad (4)$$

Table I: Specifications of Different Soiling Monitoring Methods

Soiling ratio monitoring methods	**Specifications**		
	Soiling ratio calculation approach	**Time filter**	**Solar irradiation filter**
M1	I_{SC} and P_m	10 AM to 2 PM	Greater than 500 W/m^2
M2	I_{SC}	11 AM to 1 PM	Greater than 500 W/m^2
M3	I_{SC}	2 minutes at solar noon	-

3 RESULTS AND DISCUSSIONS

Fig. 3 shows the soiling ratios obtained using the 4 methods described. It is clear that if the soiled module is not cleaned, the soiling ratio drops to 0.5 or smaller during the dry seasons in Mumbai, implying more than 50% soiling losses. After heavy precipitation, the soiling ratio recovers to 1 as the module surfaces are naturally cleaned by rain. In the rainy season, the soiling ratio hovers around 1 and the data is generally noisy, more so for the I_{SC} and P_{max}. The energy-based method gives the least noisy data as it is based on the data for the entire day. After analyzing the soiling ratio by considering different approaches, it is found that the soiling ratio obtained using M3 is most noisy. This is because, it calculates the soiling ratio by considering only 2-minute intervals at solar noon, which may often lead to inaccurate measurements. The power-based and current-based soiling ratios obtained as per M1 gives similar results. The energy-based approach showed lower soiling ratio (higher losses due to soiling) values in the summer and winter seasons.

Fig. 4 depicts the differences in different soiling ratio monitoring approaches, using the energy-based soiling ratio as a reference. The current and power-based soiling ratio using M1 shows similar trends, therefore only the power-based soiling ratio is considered for soiling ratio difference calculation. During the summer and winter seasons, it is found that the current and power-based approaches overestimate the soiling ratio, or in other words, they underestimate the soiling losses.

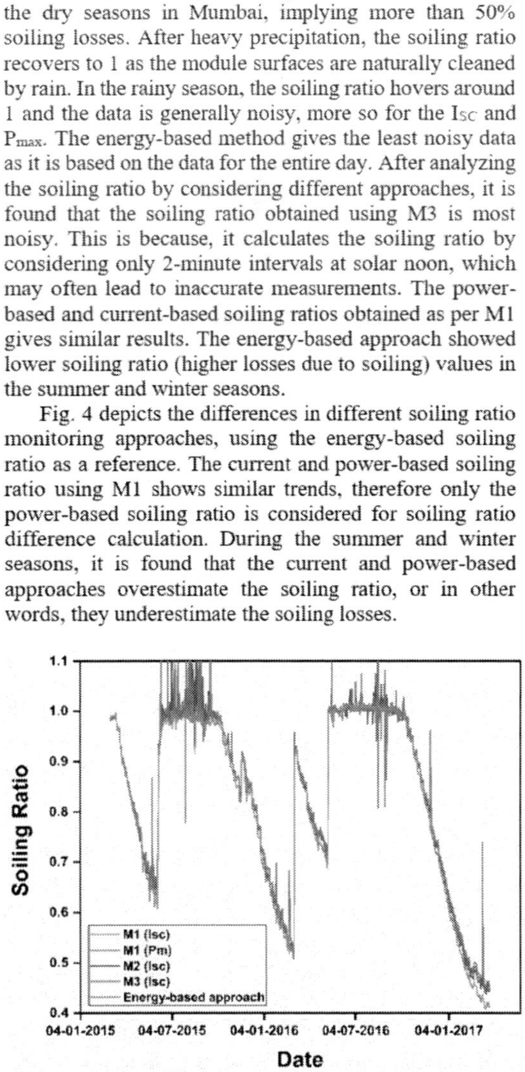

Fig. 3: Soiling ratios obtained using different soiling monitoring methods.

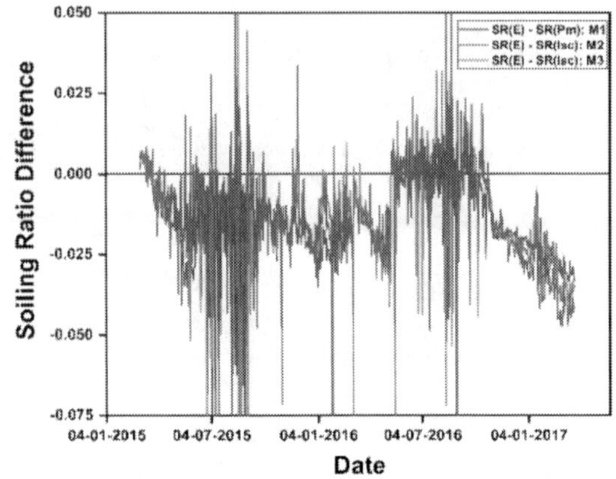

Fig. 4: Soiling ratio differences by considering energy-based soiling ratio as reference.

4 CONCLUSION

We propose an energy-based soiling loss monitoring approach and applied it for analyzing 2 years' of data collected in Mumbai, India. It is found that soiling losses obtained in terms of I_{SC} and P_{max} are lower than the soiling loss in terms of energy by an average of 2.5-3%. So, the current and powerbased soiling loss monitoring approach underestimates the soiling losses.

ACKNOWLEDGEMENT

This work was supported by the National Centre for Photovoltaic Research and Education (NCPRE) at IIT Bombay, funded by the Ministry of New and Renewable Energy of the Government of India through Project No. 31/09/2015-16/ PVSE-R&D dated 15th June 2016.

REFERECES

[1] K. Ilse, L. Micheli, B. W. Figgis, K. Lange, D. Daßler, H. Hanifi, F. Wolfertstetter, V. Naumann, C. Hagendorf, R. Gottschalg, and J. Bagdahn, "Techno-economic assessment of soiling losses and mitigation strategies for solar power generation", Joule, vol. 3, no. 10, pp. 2303-2321, 2019.

[2] M. Gostein, T. Düster, and C. Thuman, "Accurately measuring PV soiling losses with soiling station employing module power measurements", In 2015 IEEE 42nd Photovoltaic Specialist Conference (PVSC), pp. 1-4, 2015.

[3] Atonometrics, "Soiling Measurement System for PV Power Plants." [Online]. Available: http://www.atonometrics.com/products/soiling-measurement-system-for-pv-modules/. [Accessed: 02- Jan-2018].

[4] Campbell Scientific, "SMP100 Solar Module Performance Monitoring System." [Online]. Available: https://www.campbellsci.com/smp100. [Accessed: 02- Jan-2018].

[5] T. Curtis, S. Tatapudi, and G. TamizhMani, "Design and operation of a waterless PV soiling monitoring station", In 2018 IEEE 7th World Conference on Photovoltaic Energy Conversion (WCPEC)(A Joint Conference of 45th IEEE PVSC, 28th PVSEC & 34th EU PVSEC), pp. 3407-3412, 2018.

[6] S. Warade, and A. Kottantharayil, "Analysis of soiling losses for different cleaning cycles", In 2018 IEEE 7th World Conference on Photovoltaic Energy Conversion (WCPEC)(A Joint Conference of 45th IEEE PVSC, 28th PVSEC & 34th EU PVSEC), pp. 3644-3647, 2018.

INFLUENCE OF ENVIRONMENTAL PARAMETERS ON SOILING DEPOSITION FOR PV MODULES

Suellen C.S. Costa*, Dênio Alves Cassini*, Vinicíus Camatta Santana*, L.L. Kazmersk[†]i, & Antonia Sônia A.C. Diniz*
*Pontifícia Universidade Católica de Minas Gerais (PUC Minas), Belo Horizonte, Brazil
[†]Renewable and Sustainable Energy Institute (RASEI), University of Colorado Boulder, USA

ABSTRACT: The soiling of PV modules can have a significant impact on the performance of these solar devices. Soiling losses can be analyzed calculating the soiling ratio (SRatio), the relationship between electrical parameters measured from clean and soiled PV-modules under the same standard operating conditions. But the losses can also be influenced by considering the soiling-layer properties as affected by the specific meteorological conditions. The objective of this paper is to report the effects of soiling deposition for the case study of crystalline-Si PV modules installed in Equitorial climate of Belo Horizonte, Brazil (latitude -19.92°, longitude - 43.99°), and to investigate the correlation of the soiling deposition processes with meteorological variables. The results demonstrate that soiling losses are directly proportional to the state of the deposited particulate matter (density, uniformity), But local climate conditions of higher wind speed combined with low-relative humidity can have a significant, inverse effect, increasing the calculated SRatio and reducing the measured soiling losses.
Keywords: Soiling losses, PV modules, meteorological conditions, climate parameters

1 INTRODUCTION

The solar irradiance incident on a PV module can be significantly affected by the local climate and meteorological conditions. Soiling, in particular, has been shown to be a troublesome concern [1] that can affect not only the electrical power output and associated financial viability, but also the reliability of the system and its components [2,3]. A typical approach is to determine the PV soling losses by comparing the performance between cleaned modules and those left to natural soiling deposition, for example using soiling monitoring stations. This can also be accomplished using performance data from PV plant measured over extended dry periods [4-8]. Soiling losses are specific to each site and system type, which can vary in relation to weather conditions and the structure of the PV plants (tilt, orientation in relation to wind direction, PV technology, device surface condition, among others). Annual energy production can vary between 3% to more than 50% depending on the location and severity of the climate conditions [2,3,9-11]. Accurate knowledge of the factors contolling the soiling q5 the PV power plant's site is important for maintenance scheduling, ensuring the reliability and systems lifetime, and meeting the client's expectations. Brazil is a significant emerging PV market, with installations in 2021 ranked 5th highest in the world (5.7 GW) [12]. Thus, to ensure consumer confidence in these clean tecchnologies, factors that affect the expected electricity production and regulation, including soiling, remain important research and product investments [13].

Direct and on-site investigations of the soiling losses in various climate zones of the world have been reported, with methodolgies to document these power-loss effects [6,8,14]. Many regions, however, exist in which there is insufficient specific soiling information needed for the evaluations. In these cases, the determination of soiling losses through mathematical models can a viable alternative. For example, different models are reported in the literature using environmental parameters as input data [15-19]. But, the environmental parameters. But these may have a different levels of impact on the soiling predications [20-22].

This study evaluates the influence of the environmental parameters on the soiling processes for the case of the tropical region of Belo Horizonte, Brazil. The aim is to help refine the modeling and simulation process to provide a more dependable estimation of the specific location and meteorologial conditions controlling the soiling behavior and losses.

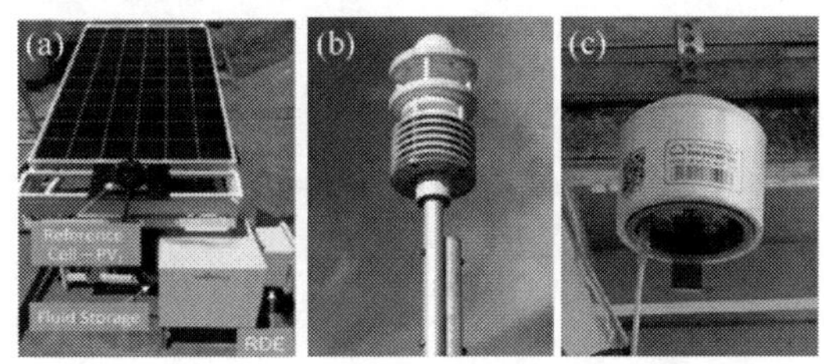

Figure 1: Stations used in this study: (2) Soiling monitoring; (b) Meterological; and (c) Particle Matter (PM)

2 METHODOLOGY

Data collected from meteorological and soiling monitoring stations, both installed at the Pontifical Catholic University of Minas Gerais in Belo Horizonte, Brazil (latitude–19.92°, longitude–43.99°; Equitorial Climate Zone), are used to evaluate the relationship of the meteorological variables relating to the behavior of performance losses of PV modules due to soiling deposition. Data measured during extended dry periods (no precipitation) of the winter (the dry season in this locality) were used..

The PV performance losses due to soiling were evaluated by calculating the soiling ratio (SRatio) [2,3]. SRatio is the relation between output power or short-circuit current from the soiled to the clean state of the module, with the analytical expressions published previously [2,3,5]. This paramater varies from 1 (clean or no soiling) to 0 (complete light attenuation). To calculate the SRatio, data collected by the soiling station, Fig. 1(a), composed of a polycrystalline (or multi-crystalline) Si module left to the natural soiling deposition and by a reference cell that is automatically cleaned periodically, were used. The measurements were performed by a control system providing thermal and electrical parameters from the PV devices. Importantly, all measured data were

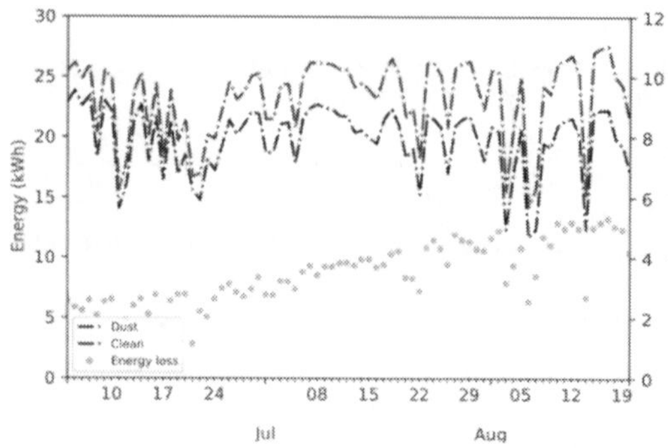

Figure 2: Produced energy from clean (blue line) and soiled (black line) PV module, and energy losses (yellow line) during prolonged dry period in Belo Horizonte, Brazil

corrected to Standard Test Condition (1000 W/m² and 25°C), and the plane-of-array irradiance (POA) was monitored by the soiling station, providing data for the specific tilt and orientation.

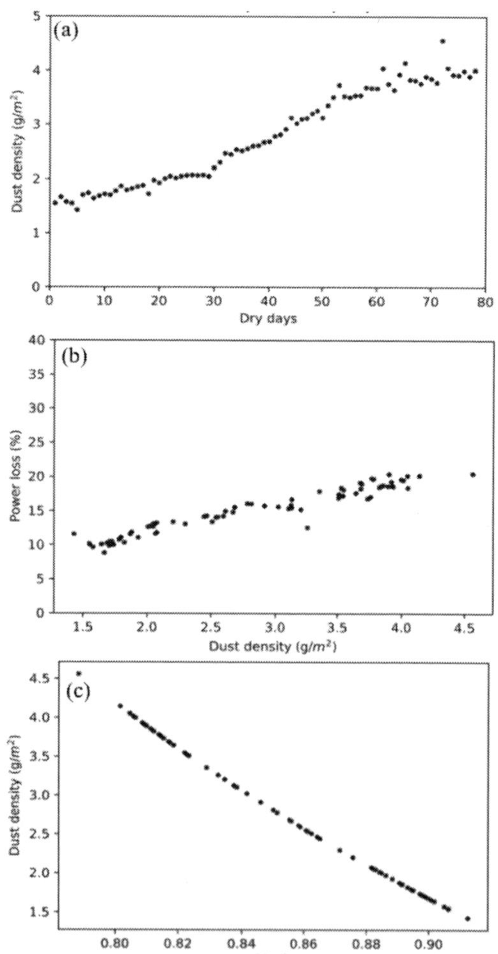

Figure 3: Dust density from silicon PV module installed in Belo Horizonte, Brazil: (a) dust density versus length dry period, (b) power losses versus dust density and (c) dust density versus SRatio

Subsequent to the calculation of the SRatio, two steps are required: (1) analysis of the behavior of the module's electrical parameters through the increase in soiling deposition, and (2) determining the relationship between SRatio and environmental variables. The environmental parameters were monitoried by a compact meteorological station (Fig. 1b), which provides data on solar irradiance incident on a horizontal plane (GHI), ambient temperature, relative humidity, precipitation, wind velocity (speed and direction). The particulate matter (PM) data were monitored using an air-quality sensor, model PA-II (Fig. 1c). Data on PM2.5 (particles with diameters up to 2.5 µm) and PM10 (particles with diameters between 2.5-µm and 10-µm) were collected for the periods of evaluation.

Measured merorological, PM, electrical, and thermal data were collected at one-minute intervals. Data anomalies due to the variation in irradiance were filtered from the analysis.

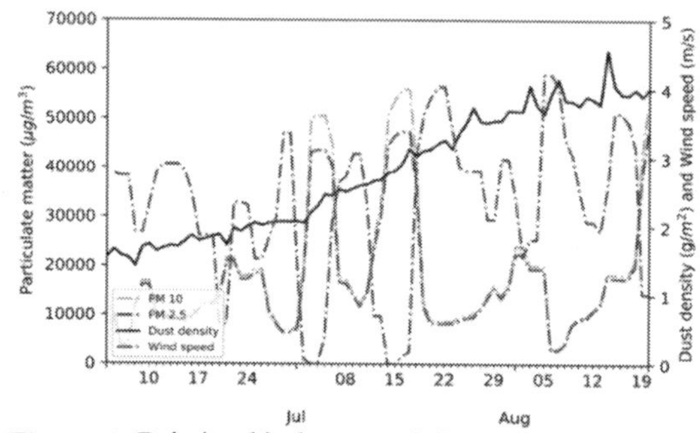

Figure 4: Relationship between daily $PM_{2.5}$ and PM_{10}, daily average wind speed, and daily average dust density to Belo Horizonte, Brazil.

3 RESULTS

Figure 2 presents the daily energy generated from clean and soiled PV module for the a specific dry period evaluated of some 78 days. The difference in energy production for the two devices increases with the length of the dry period, indicating increase performance losses due to soiling.

Figure 3a represents the correlation between the dust density on the Si PV module and the length dry period recorded during this winter, 78-day timeframe without rain. The increase of the particles accumulated on the module increased the proportion of irradiance reflected and not absorbed by the PV cells. Thus, the PV output power tends to decrease with increasing soiling layer density, as shown in Fig. 3b. In other words, the SRatio decreases with increase of the particulates accumulated on the PV module (Fig 3c). Additionally, it is possible to observe variations in the behavior of the dust density throughout the dry period, which can be due to the deposition-resuspension process of soiling. The site is located in the Equatorial climatic zone characterized by having a long dry winter periods and relatively high solar irradiance, which can be due, in turn, to the higher clearness index, causing a decrease in relative humidity, contributing to the reduction of the adhesion forces between the dust particles and the module surface.

As noted, the SRatio reduces with increase of the dust density, resulting in an increase of the power losses. Consequently, it is expected that the dust density increases in periods of high particulate matter index, as shown in Fig. 4. In this evaluated time period, some days with peaks of PM can be observed, especially in July, contributing to a more accentuated increase in dust density (greater slope of the curve). This demonstrates the correlation between PM and dust density (and SRatio). The analysis of relative humidity for the period shows higher values for the first month, decreasing from the following month of July. It is noted that the fluctuations in the dust density curve occur mainly due to the variation of wind speed, and in the first month of analysis, the wind speed had less effect on the soiling deposition on the PV module, which is attributed to higher relative humidity, and consequently, greater adhesion force of the particles.

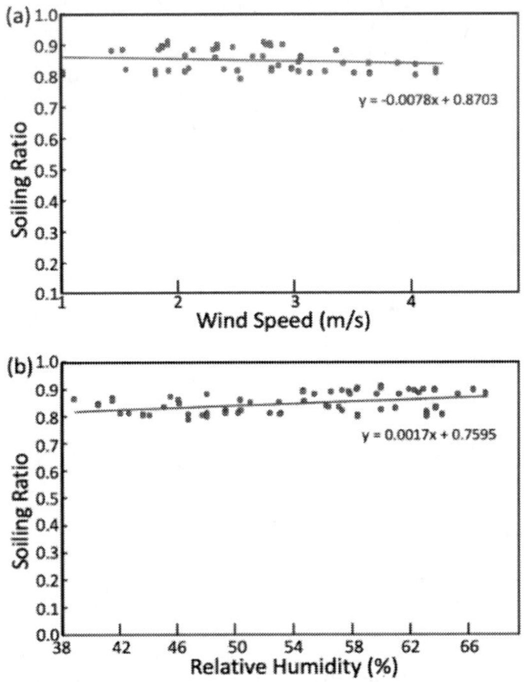

Figure 5: Correlations between SRatio versus: (a) wind speed and (b) relative humidity for Belo Horizonte, Brazil.

From "July data," it is possible to identify that the increase of wind speed associated low-relative humidity can cause a small increase of the dust density (reduction of the SRatio). This is enhanced in the following month, which presented higher wind speeds (Fig. 5). This obervation is different than a previous report this [21,22] that noted that the SRatio increases underr high wind speeds, in this explanation, working as natural cleaning agents. This differenct is dues to the other component of the wind—its direction. This parameter also has to be iincorporated into any interpretation.

The wind speed mainly affects the dust density depending on its direction. The wind rose of Fig 6 shows the wind direction and frequency recorded for this site. This "rose" shows the wind frequency and speed related to different directions. As we move out on the radial scale, the frequency associated with the wind from that direction increases. Each ray is divided by color into wind speed bands. The length of the ray indicates the different wind directions and the ray points to the direction the wind is coming from. In this case, the highest frequency of winds

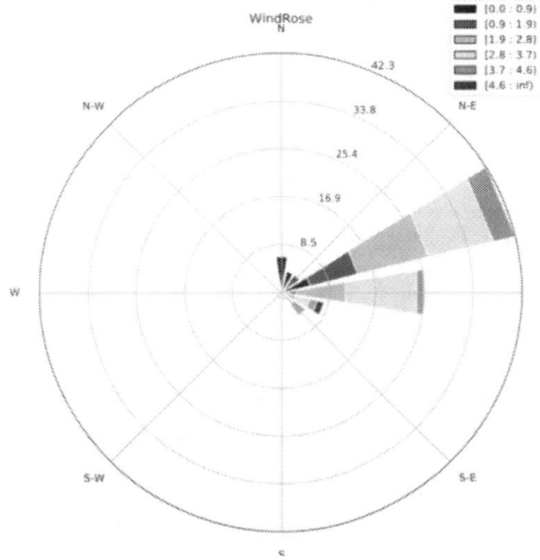

Figure 6: Wind rose for site in Belo Horizonte, Brazil (speed distribution)

comes from the northeast to the east direction, reaching speeds of up to 4.6 m/s. The wind can contribute to the natural cleaning of the module, but for this case, due the direction and wind speed level this variable promoted the increase of the dust density, working as a soiling carrier for periods of low relative humidity.

Thus, the correlation of wind speed with the soiling deposition process cannot be characterized independently, without associating the influence of relative humidity and wind direction and speed. In constrast, particulate matter has a better direct correlation with dust density. These results confirm the report of Micheli and Muller [23] that the particulate matter values are the best soiling predictors.

3 SUMMARY AND CONCLUSIONS

The effects of soiling on electrical losses for a crystalline-Si PV modules operating in the Equitorial Climate Zone of Horizonte, Brazil have been used as a case study to invetigate the various climate and environmental effects. The correlation of these soiling losses with meteorological data were evaluated. For this, data collected by meteorological, soiling, and air quality stations were used and evaluated for a long dry period recorded during winter. The results showed that the performance losses increase with the length of the dry period due to the increase in the particle accumulation (increased dust density). Furthermore, it was found that the SRatio is reduced with the increase in soiling particulate density and with increasing $PM_{2.5}$ and PM_{10} particulate matter.

Dust density and SRatio has some correlation with wind speed, however, this cannot be evaluated separately without considering the effects associated with wind direction and relative humidity.

The results obtained in this study demonstrate the complexity of determining the effects of different environmental parameters on the behavior of the SRatio. This can be justified, due to the combination of effects of different meteorological variables for the soiling deposition process. In addition, it is important to emphasize that the SRatio can be a seasonal factor, and it presents different behavior for different locations and installation types.

4 ACKNOWLEGEMENTS

The authors gratefully acknlwledge Brazil CAPES (Coordenação de Aperfeiçoamento de Pessoal de Nível Superior) and the international institutions NREL (National Renewable Energy Laboratory) and RASEI (Renewable and Sustainable Energy Institute at the University of Colorado Boulder) for supporting the and providing guidance in this research.

5 REFERENCES

[1] J. Cano, J. J. John, S. Tatapudi, and G. TamizhMani, "Effect of tilt angle on soiling of photovoltaic modules," in *2014 IEEE 40th Photovoltaic Specialist Conference (PVSC)*, Jun. 2014, pp. 3174–3176, doi: 10.1109/PVSC.2014.6925610.

[2] S.C. S. Costa, A.S.A.C. Diniz, and L.L. Kazmerski, "Dust and soiling issues and impacts relating to solar energy systems: Literature review update for 2012-2015," *Renew. Sustain. Energy Rev.*, vol. 63, 2016, doi: 10.1016/j.rser.2016.04.059.

[3] S. C. S. Costa, A. S. A. C. Diniz, and L. L. Kazmerski, "Solar energy dust and soiling R&D progress: Literature review update for 2016," *Renew. Sustain. Energy Rev.*, vol. 82, pp. 2504–2536, Feb. 2018, doi: 10.1016/j.rser.2017.09.015.

[4] L. Micheli, M.G. Deceglie, and M. Muller, "Mapping Photovoltaic Soiling Using Spatial Interpolation Techniques," *IEEE J. Photovoltaics*, vol. 9, no. 1, pp. 272–277, Jan. 2019, doi: 10.1109/JPHOTOV.2018.2872548.

[5] S. C. Silva Costa *et al.*, "Comparative Investigations of the Effects of Soiling of PV Modules and Systems in Tropical, Subtropical, and Semi-Arid Climate Zones in Brazil," in *2019 IEEE 46th Photovoltaic Specialists Conference (PVSC)*, Jun. 2019, pp. 2864–2868, doi: 10.1109/PVSC40753.2019.8980505.

[6] M.G. Deceglie, M. Muller, Z. Defreitas, and S. Kurtz, "A scalable method for extracting soiling rates from PV production data," in *2016 IEEE 43rd Photovoltaic Specialists Conference (PVSC)*, Jun. 2016, pp. 2061–2065, doi: 10.1109/PVSC.2016.7749992.

[7] S.C.S. Costa, L.L. Kazmerski, and A.S.A.C. Diniz, "Estimate of Soiling Rates Based on Soiling Monitoring Station and PV System Data: Case Study for Equatorial-Climate Brazil," *IEEE J. Photovoltaics*, vol. 11, no. 2, pp. 461–468, Mar. 2021, doi: 10.1109/JPHOTOV.2020.3047187.

[8] S.C.S. Costa, L.L. Kazmerski, and A.S.A.C. Diniz, "Impact of soiling on Si and CdTe PV modules: Case study in different Brazil climate zones," *Energy Convers. Manag. X*, vol. 10, p. 100084, Jun. 2021, doi: 10.1016/j.ecmx.2021.100084.

[9] M. Piliougine *et al.*, "Comparative analysis of energy produced by photovoltaic modules with anti-soiling coated surface in arid climates," *Appl. Energy*, vol. 112, pp. 626–634, Dec. 2013, doi: 10.1016/j.apenergy.2013.01.048.

[10] M. Mani and R. Pillai, "Impact of dust on solar photovoltaic (PV) performance: Research status, challenges and recommendations," *Renew. Sustain. Energy Rev.*, vol. 14, no. 9, pp. 3124–3131, Dec. 2010, doi: 10.1016/j.rser.2010.07.065.

[11] T. Sarver, A. Al-Qaraghuli, and L.L. Kazmerski, "A comprehensive review of the impact of dust on the use of solar energy: History, investigations, results, literature, and mitigation approaches," *Renew. Sustain. Energy Rev.* 22, 698–733, Jun. 2013. doi: 10.1016/j.rser.2012.06

[12] REN21, 2022 Renewables Global Status Report (REN21 Renewables 2022 Global Status Report; REN21 Secretariat: Paris, France, ISBN 978-3-948393-04-5. https://www.ren21.net/wp-content/uploads/2019/05/GSR2022_Full_Report.pdf

[13] E.deP. Epe and M. de M. e E. Mme, *Expansão da Geração: Empreendimentos Fotovaltaicos*. 2018.

[14] W.J. Jamil, H.A. Rahman, S. Shaari, and M.K.M. Desa, Modeling of Soiling Derating Factor in Determining Photovoltaic Outputs," *IEEE J. Photovoltaics* 10, 1417–1423, Sep. 2020. doi: 10.1109/JPHOTOV.2020.3003815

[15] M. Coello and L. Boyle, "Simple Model for Predicting Time Series Soiling of Photovoltaic Panels," *IEEE J. Photovoltaics*, vol. 9, no. 5,pp. 1382–1387, Sep. 2019, doi: 10.1109/JPHOTOV.2019.2919628

[16] F. Liu *et al.*, "A Method of Calculating the Daily Output Power Reduction of PV Modules Due to Dust Deposition on Its Surface," *IEEE J. Photovoltaics*, vol. 9, no. 3, pp. 881–887, May 2019, doi: 10.1109/JPHOTOV.2019.2903086

[17] L. Micheli, M. G. Deceglie, and M. Muller, "Predicting photovoltaic soiling losses using environmental parameters: An update," *Prog. Photovoltaics Res. Appl.*, vol. 27, no. 3, pp. 210–219, Mar. 2019, doi: 10.1002/pip.3079

[18] S. Toth, M. Hannigan, M. Vance, and M. Deceglie, "Predicting Photovoltaic Soiling From Air Quality Measurements," *IEEE J. Photovoltaics*, vol. 10, no. 4, pp. 1142–1147, Jul. 2020. doi: 10.1109/JPHOTOV.2020.2983990

[19] S. Busquet and J. Kobayashi, "Modelling daily PV performance as a function of irradiation, ambient temperature, soiling, wind speed, and aging - Applied to PV modules operating in Maui," in *2018 IEEE 7th World Conference on Photovoltaic Energy Conversion (WCPEC) (A Joint Conference of 45th IEEE PVSC, 28th PVSEC & 34th EU PVSEC)*, Jun. 2018, pp. 3401–3406, doi: 10.1109/PVSC.2018.8548220

[20] K. Ilse, et al. "Techno-Economic Assessment of Soiling Losses and Mitigation Strategies for Solar Power Generation," *Joule*, vol. 3, no. 10, pp. 2303–2321, Oct. 2019, doi: 10.1016/j.joule.2019.08.019

[21] M. Naeem and G. Tamizhmani, "Climatological relevance to the soiling loss of photovoltaic modules," in *2015 Saudi Arabia Smart Grid (SASG)*, Dec. 2015, pp. 1–5, doi: 10.1109/SASG.2015.7449280

[22] M. Gostein, K. Passow, M. G. Deceglie, L. Micheli, and B. Stueve, "Local Variability in PV Soiling Rate," in *2018 IEEE 7th World Conference on Photovoltaic Energy Conversion (WCPEC) (A Joint Conference of 45th IEEE PVSC, 28th PVSEC & 34th EU PVSEC)*, Jun. 2018, pp. 3421–3425, doi: 10.1109/PVSC.2018.8548049

[23] L. Micheli and M. Muller, "An investigation of the key parameters for predicting PV soiling losses," *Prog. Photovoltaics Res. Appl.*, vol. 25, no. 4, pp. 291–307, Apr. 2017, doi: 10.1002/pip.2860

ANALYSIS OF MEASUREMENT REPRODUCIBILITY FOR PHOTOVOLTAIC MODULES

Xin Huang[1], Chenhui Nan[1], Muyang Lu[1], He Wang[1], Hong Yang[1,*]
[1] MOE Key laboratory for Nonequilibrium Synthese and Modulation of Condensed Matter, School of Physics, Xi'an Jiaotong University, Xi'an 710049, People's Republic of China

ABSTRACT: In this paper, a detailed study on the measurement reproducibility for photovoltaic modules was carried out. It is found that the measurement reproducibility of maximum power current, fill factor and maximum power are larger than that of short circuit current, open circuit voltage and maximum power voltage. The calculated reproducibility of maximum power is 0.147%. This means that if the 0.147% is applied for type approval of photovoltaic modules, degradation rates of modules must be less than 5.14% before and after each test sequence to make sure of passing criteria as per the IEC61215 standard. This work contributes to measurement reproducibility determination and type approval for photovoltaic modules in laboratories.
Keywords: photovoltaic module, measurement reproducibility, IEC61215

1 INTRODUCTION

Performance deterioration of photovoltaic (PV) module is always a big concern for stakeholders. The deterioration mechanisms have been investigated by many papers [1-6]. The degree of performance deterioration can be directly reflected in the decay of output power. For PV product owners, they pay more attention to the real-time output power of solar modules and related power warranty terms. At present, PV module manufacturers provide a linear warranty of 25 years with no more than 20% output power reduction during the warranty period [7]. Due to the existence of uncertainty and reproducibility for measurement, there are always some disputes about output power of PV modules between product owners and manufacturers when determining the remaining maximum power after long-term outdoor operation. In recent years, extensive investigations have been conducted to analyze measurement uncertainty of PV modules. Whitfield et al. [8] proposed an uncertainty estimation procedure for PV electrical performance measurement using natural sunlight and calibrated secondary reference cells. Jordan et al. [9] investigated linear degradation of PV modules caused by different degradation modes within the measurement uncertainty. Kaaya et al. [10] studied different sources of uncertainties in PV degradation rates and lifetime predictions based on physical models. However, almost no one systematically investigated measurement reproducibility of electrical parameters, especially for reproducibility determination in IEC61215:2021 to evaluate the maximum power degradation during approval testing [11]. In this paper, measurement reproducibility for PV modules was studied both theoretically and experimentally for the first time. The electrical performance parameters of the PV module before and after environment tests were measured by a large area pulsed solar simulator model PASAN SunSim 3C. It is found that, the measurement reproducibility of maximum power current (I_{sc}) is the largest. The calculated reproducibility of maximum power (P_{max}) is 0.147% under standard test conditions (STC). If the 0.147% is applied for type approval, degradation rates of PV module must be less than 5.14% before and after each test sequence to make sure of passing criteria based on the IEC61215 standard.

2 EXPERIMENTAL DETAILS

In this experiment, a monocrystalline silicon PV module was used as the testing device. The PV module consists of 72 pieces passivated emitter and rear cells (PERC) with dimensions of 156 mm × 156 mm. The electrical parameters of the module were measured by the solar simulator (PASAN Sunsim 3C) under standard test conditions (STC: 1 kW/m^2 irradiance, 25 °C cell temperature and AM 1.5 global spectrum). The current-voltage tests of the module were carried out for 10 times in the same steps and there was 1 h interval between two consecutive measurements. The measurement results of electrical parameters of the PERC module under STC are showed in Table I.

Table I: The measurement results of electrical parameters of the PERC module under STC

	I_{sc} (A)	V_{oc} (V)	I_{mp} (A)	V_{mp} (V)	FF	P_{max} (W)
1	10.168	41.061	9.6910	33.229	0.7712	322.02
2	10.169	41.063	9.7078	33.211	0.7721	322.41
3	10.174	41.058	9.6899	33.231	0.7709	322.01
4	10.173	41.065	9.7042	33.216	0.7716	322.33
5	10.166	41.064	9.7105	33.208	0.7725	322.47
6	10.177	41.067	9.7005	33.221	0.7711	322.26
7	10.173	41.059	9.6957	33.225	0.7713	322.14
8	10.173	41.064	9.7033	33.217	0.7715	322.31
9	10.174	41.064	9.6976	33.212	0.7710	322.08
10	10.169	41.072	9.7003	33.220	0.7715	322.04

3 RESULTS AND DISCUSSION

3.1 Calculation model of measurement reproducibility for photovoltaic modules

Measurement reproducibility based on the IEC 61215:2021 is defined as Eq. (1)[11].

$$P_{max}(Lab\_GateNo.2) \geq 0.95 \times P_{max}(Lab\_GateNo.1)(1-\frac{r[\%]}{100}) \quad (1)$$

where $P_{max\ (Lab\_GateNo.2)}$ is the maximum power output of each module at the end of each test sequence or for sequence B after bypass diode test, $P_{max\ (Lab\_GateNo.1)}$ is the module's initial measured output power, r denotes measurement reproducibility.

Measurement reproducibility is defined as the consistency between the measurement results for the same value of a quantity, when the individual measurements are made under different conditions of measurement in IEC61215. The definition of measurement reproducibility is of little significance when determining the specific value. In order to deal with this issue, reproducibility limit is considered as the second definition of reproducibility.

We define Y_{pm} as the single measurement result variable of one kind of electrical parameter for PV modules and assume that it follows a normal distribution. Thus, the associated probability density function is expressed as

$$f(y_{pm}) = \frac{1}{\sqrt{2\pi}\sigma}\exp[-\frac{(y_{pm}-\mu)^2}{2\sigma^2}] \quad (2)$$

where μ is the average value and σ is the standard deviation.

We define y^a_{pm} and y^b_{pm} as the two measurement results of the same electrical parameter, respectively. We define y^m_{pm} as the absolute difference between the two results. The normalized absolute difference y'_{pm} can be expressed as

$$y'_{pm} = \frac{y^m_{pm}}{\mu} = \frac{|y^a_{pm} - y^b_{pm}|}{\mu} \quad (3)$$

where y^a_{pm} and y^b_{pm} are the independent variables, but both follow the same normal distribution (Eq. (2)). The corresponding normalized probability density function of y^m_{pm} can be expressed as Eq. (4).

$$f(\frac{y^m_{pm}}{\mu}) = \frac{2}{\sqrt{2\pi}\sqrt{2}\frac{\sigma}{\mu}}\exp[-\frac{(\frac{y^m_{pm}}{\mu})^2}{2(\sqrt{2}\frac{\sigma}{\mu})^2}] \quad (4)$$

From Eq. (4), it can be found that the average value of the normalized distribution is zero and the standard deviation is $\sqrt{2}\sigma/\mu$. The three sigma's law of normal distribution is applicable for the special normal distribution, and the corresponding boundary value can be expressed as Eq. (5).

$$P\left\{0 < \frac{y^m_{pm}}{\mu} < 1.96\frac{\sqrt{2}\sigma}{\mu}\right\} = 95\% \quad (5)$$

Therefore, the expression of measurement reproducibility for PV modules is

$$r = 1.96\frac{\sqrt{2}\sigma}{\mu} \quad (6)$$

3.2 Experimental analysis of measurement reproducibility for photovoltaic modules

Figure 1 shows the histogram of I_{sc}, V_{oc}, I_{mp}, V_{mp}, FF and P_{max} of the sample module under reproducibility measurement condition. From Figure 1, it can be found that the distribution of I_{sc}, V_{oc}, I_{mp}, V_{mp}, FF and P_{max} meet the requirements of normal distribution.

Table II shows the calculated results of measurement reproducibility for electrical parameters. From Table II, it is found the standard deviation of FF is the smallest and the standard deviation of P_{max} is the largest. However, the measurement reproducibility of V_{oc} is the smallest, and the measurement reproducibility of I_{mp} is the largest. This indicates that there is no obvious consistency between standard deviation and measurement reproducibility.

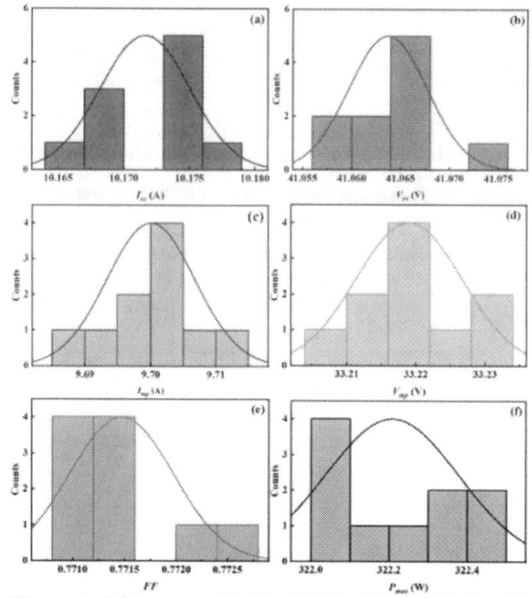

Figure 1: Histogram of (a) I_{sc}, (b) V_{oc}, (c) I_{mp}, (d) V_{mp}, (e) FF and (f) P_{max} of the sample module under reproducibility measurement condition.

Table II: The calculated results of the electric parameters

Parameters	Expectation value	Standard deviation	Measurement reproducibility (%)
I_{sc} (A)	10.172	0.0034	0.093
V_{oc} (V)	41.064	0.0040	0.027
I_{mp} (A)	9.7001	0.0067	0.192
V_{mp} (V)	33.219	0.0077	0.064
FF	0.7715	0.0005	0.180
P_{max} (W)	322.207	0.1704	0.147

Figure 2 shows the measurement reproducibility of I_{sc}, V_{oc}, I_{mp}, V_{mp}, FF and P_{max} of the sample module. The blue dotted line denotes the average value of measurement reproducibility of these six electrical parameters. It can be observed that the measurement reproducibility of I_{mp}, FF and P_{max} are larger than that of I_{sc}, V_{oc} and V_{mp}. This indicates that the consistency of measurement results of I_{mp}, FF and P_{max} are relatively poor. The measurement reproducibility of P_{max} is 0.147%, which means if the 0.147% is used for type approval for PV modules in the laboratory, degradation rates must be less than 5.14% before and after each test sequence. If degradation rates are more than 5.14%, tested modules cannot pass the design qualification and type approval.

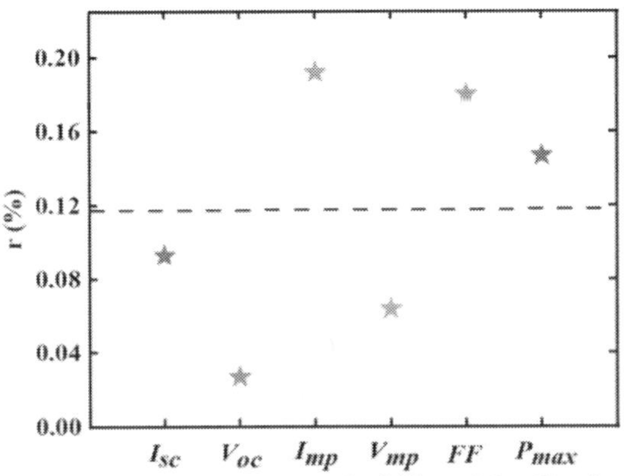

Figure 2: Measurement reproducibility of I_{sc}, V_{oc}, I_{mp}, V_{mp}, FF and P_{max} of the sample module.

4 CONCLUSIONS

In this paper, measurement reproducibility for PV modules was studied both theoretically and experimentally for the first time. The electrical performance parameters of the PV module before and after environment tests were measured by a large area pulsed solar simulator model PASAN SunSim 3C. It is found that, the reproducibility of I_{mp} is the largest. The calculated reproducibility of P_{max} is 0.147% under STC. If the 0.147% is applied for type approval, degradation rates of PV modules must be less than 5.14% before and after each test sequence to make sure of passing criteria based on the IEC61215 standard. This work can provide insight into understanding measurement reproducibility and type approval for photovoltaic modules in laboratories.

5 ACKNOWLEDGEMENTS

The authors would like to thank the support of the Natural Science Foundation of China (Grant No. 61376067 and 61274050). This study was also supported by the National Key R&D Program of China (Grant No. 2018YFB1500700). (*Corresponding author: Hong Yang, Email address: hongy126@126.com)

6 REFERENCES

[1] A. Ndiaye, C. M. Kébé, A. Charki, P. A. Ndiaye, V. Sambou, A. Kobi, "Degradation evaluation of crystalline-silicon photovoltaic modules after a few operation years in a tropical environment." Solar energy 103 (2014): 70-77.

[2] H. Wang, X. L. Cheng, H. Yang, W. S. He, Z. L. Chen, L. H. Xu, D. Y. Song, "Potential-induced degradation: Recombination behavior, temperature coefficients and mismatch losses in crystalline silicon photovoltaic power plant." Solar Energy 188 (2019): 258-264.

[3] F. Bandou, A. H. Arab, M. S. Belkaid, P. Logerais, O. Riou, A. Charki, "Evaluation performance of photovoltaic modules after a long time operation in Saharan environment." International journal of hydrogen energy 40.39 (2015): 13839-13848.

[4] H. Yang, H. Wang, C. K. Chen, D. Y. Cao, H. C. Yu, "Effect of binding force between silver paste and silicon on power degradation of crystalline silicon solar module." Microelectronics Reliability 54.1 (2014): 188-191.

[5] M. Saidan, A. G. Albaali, E. Alasis, J. K. Kaldellis, "Experimental study on the effect of dust deposition on solar photovoltaic panels in desert environment." Renewable Energy 92 (2016): 499-505.

[6] Q. Q. Bai, H. Yang, C. H. Nan, H. Wang, Z. L. Chen, "Analysis of the electrochemical reactions and ions migration for crystalline silicon solar module under high system voltage." Solar Energy 225 (2021): 718-725.

[7] R. Laronde, A. Charki, A. D. Bigaud, "Reliability of photovoltaic modules based on climatic measurement data." International Journal of Metrology and Quality Engineering 1.1 (2010): 45-49.

[8] K. Whitfield, C. R. Osterwald, "Procedure for determining the uncertainty of photovoltaic module outdoor electrical performance." Progress in Photovoltaics: Research and Applications 9.2 (2001): 87-102.

[9] D. C. Jordan, T. J. Silverman, B. Sekulic, S. R. Kurtz, "PV degradation curves: non - linearities and failure modes." Progress in Photovoltaics: Research and Applications 25.7 (2017): 583-591.

[10] I. Kaaya, J. Ascencio-Vásquez, K.-A. Weiss, M. Topič, "Assessment of uncertainties and variations in PV modules degradation rates and lifetime predictions using physical models." Solar Energy 218 (2021): 354-367.

[11] IEC, "IEC 61215:2021, Terrestrial photovoltaic (PV) modules –Design qualification and type approval" International Electrotechnical Commission, International Standard, 2021.

COMPARATIVE STUDY ON TEMPERATURE COEFFICIENT OF PERC AND HIT SOLAR MODULES UNDER DIFFERENT INJECTION CONDITIONS

Chenhui Nan[1], Yuanpang Hao[1], Xiaoqian Li[1], Yunfeng Feng[3], He Wang[1], Hong Yang[1,*]
[1] School of Physics, Xi'an Jiaotong University, Xi'an 710049, People's Republic of China
[2]Gsolar Power Co., Ltd, Xi'an 710000, People's Republic of China

ABSTRACT: In this paper, the differences of temperature coefficient between the bifacial passivated emitter and rear cell (PERC) and the heterojunction with intrinsic thin-layer (HIT) under different module temperatures and irradiance were given. The results show that the power temperature coefficient of HIT is smaller than that of PERC, which is because of the difference of fill factor (FF) temperature coefficient between the two. A new finding is that this phenomenon is more obvious at high irradiance.
Keywords: Temperature coefficients, Heterojunction, PERC

1 INTRODUCTION

Passivated emitter and rear cell (PERC) solar modules have dominated the photovoltaic market in recent years. With the development of heterojunction solar cell technology, the heterojunction with intrinsic thin-layer (HIT) solar modules have gradually entered the market [1]. The temperature coefficient is the main factor affecting the outdoor performance of solar modules, so it is important to study the difference of temperature coefficient for the selection of power station. Many scientists have studied the temperature coefficient of modules. The theoretical formula for temperature coefficients has been derived by some scholars [2-5]. The temperature coefficient of Al-BSF with PERC has been compared in earlier studies [6-7]. Kamkird et al. [8] studied the temperature coefficients of current and voltage of three kinds of solar cells under outdoor conditions. Tian et al. [9] compared the fill factor of Al-BSF and PREC under non-standard test conditions. Bai et al. [10-11] studied effect of temperature on ideality factor for PERC solar module. Tudor et al. [12] studied the temperature coefficient for some types of solar cells. Kasu, M., et al. [13] studied the temperature dependence of Al-BSF, PERC and HIT solar cells. But for different types of solar modules, there is no clear conclusion about which temperature coefficient has the greatest influence on power. At the same time, no one has studied the temperature coefficient under non-standard conditions. So, in this paper, the temperature coefficients of open circuit voltage (V_{oc}), short circuit current (I_{sc}) and fill factor (FF) for two modules were obtained through the solar simulator. The results showed that the power temperature coefficient of PERC is bigger than HIT and the difference becomes greater with increasing irradiance. Through the comparison of various electrical parameters, the conclusion is that the main factor affecting the temperature coefficient of power is FF.

2 EXPERIMENTAL DETAILS

In this paper, two types of solar cells were prepared for testing. The two modules are double-sided photovoltaic modules, and their parameters are shown in Table 1. For bifacial PERC solar cells, both the front and rear sides were passivated with an atomic-layer AlO_X capped with a PECVD SiN_X:H thin film. Standard screen-printed and cofiring were formed on both the front and rear sides. The bifacial PERC solar module consists of a glass superstrate, ethylene-vinyl acetate (EVA), bifacial PERC solar cells, EVA, and a glass superstrate. The module was composed of 5×15+5×15 pieces of 210×70 mm bifacial PERC solar cells. To form a pn heterojunction, the HIT solar cells were made of intrinsic (i-type) amorphous silicon (a-Si) and p-type a-Si deposited on an n-type Czochralski (Cz) crystalline silicon (c-Si) wafer. Transparent conductive oxide (TCO) and metal grid electrodes were prepared on both sides of the doped a-Si layers. All the above operations are performed at a low temperature to avoid thermal damage to the components of solar cells. The module consisted of 6×10+6 × 10 pieces of 210×105 mm bifacial HIT solar cells. The current-voltage (I-V) characteristics of the bifacial PERC solar module and the bifacial HIT solar module were measured by solar simulator with adjustable irradiance levels for single-side illumination under different module temperatures and irradiance in the laboratory.

TABLE 1

Typical Parameters Of Crystalline Silicon Solar Module Under STC

Parameter	V_{oc} (V)	I_{sc} (A)	V_m (V)	I_m (A)	P_m (W)	**Bifaciality (%)**
PERC	51.01	12.46	42.88	11.68	500	70
HIT	44.20	9.82	37.20	9.15	340	>80

STC: 1kW/m^2 irradiance, 25 °C module temperature and AM 1.5 global spectrum.

3 RESULTS AND DISCUSSION

Figure. 1 shows the change of relative temperature coefficient of I_{sc} with temperature under different irradiance. With the increasing temperature, the change of $(dI_{sc}/dT)/I_{sc}$ is not obvious. It can be considered that the temperature coefficient of short-circuit current does not change with temperature and irradiance. The temperature coefficient of I_{sc} is also approximate for the two modules. Therefore, we believe that the I_{sc} temperature coefficients of the two modules are exactly the same.

Figure 2 shows the variation of $(dV_{oc}/dT)/V_{oc}$ with temperatures and irradiance. With the increasing temperature, the absolute values of $(dV_{oc}/dT)/V_{oc}$ for HIT solar module and PERC solar module show an obvious trend of increase. When we compare two types of solar modules, we can see that the average difference in $(dV_{oc}/dT)/V_{oc}$ between a PERC solar module and a HIT solar module decreases from 0.039 %/°C to 0.023%/°C as the irradiance increases from low to high.

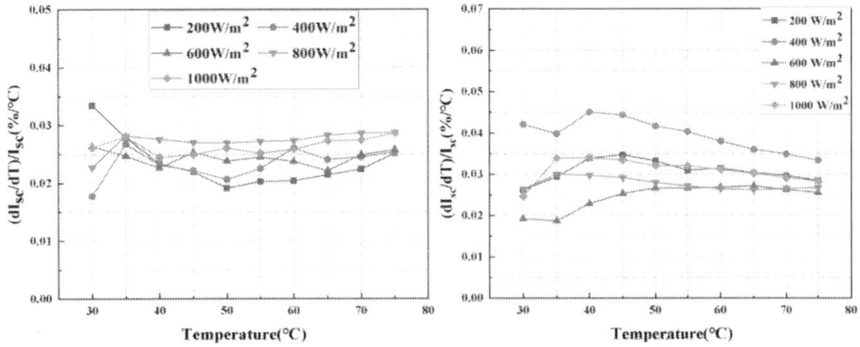

(a) (b)

Figure 1: Measured temperature coefficients of short-circuit current. (a) $(dI_{sc}/dT)/I_{sc}$ of PERC solar module versus temperature under different irradiance. (b) $(dI_{sc}/dT)/I_{sc}$ of HIT solar module versus temperature under different irradiance.

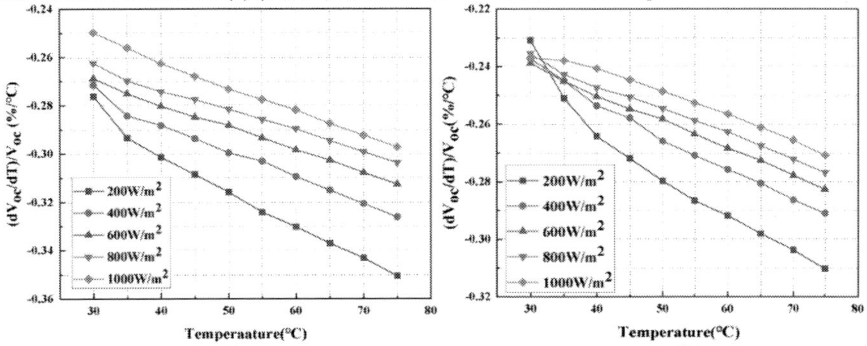

(a) (b)

Figure 2: Measured temperature coefficients of open-circuit voltage. (a) $(dV_{oc}/dT)/V_{oc}$ of PERC solar module versus temperature under different irradiance. (b) $(dV_{oc}/dT)/V_{oc}$ of HIT solar module versus temperature under different irradiance.

Figure 3 shows how (dFF/dT)/FF changes. With the increase of temperature, (dFF/dT)/FF also shows an obvious increase trend for two modules, among which the (dFF/dT)/FF value of HIT solar module is larger. Combining with the correlation formula about (dFF/dT)/FF and the change of $(dV_{oc}/dT)/V_{oc}$ [2][5], we can infer that the variation of R_s for HIT module is very different from that of PERC module and even abnormal. Particularly, the average difference value in (dFF/dT)/FF between PERC solar module and HIT solar module increase from -0.037%/°C to 0.045%/°C, which is a huge difference in the temperature coefficient.

The change of (dPm/dT)/Pm can be seen in Figure 4. The temperature coefficients of both two modules decrease with the increase of temperature. The value of the measured power temperature coefficients is the same as the sum of $(dI_{sc}/dT)/I_{sc}$, $(dV_{oc}/dT)/V_{oc}$ and (dFF/dT)/FF, which proves the accuracy of experimental data from the side. The average difference value increases from 0.008 %/°C to 0.077 %/°C as the irradiance increases from low to high. $(dP_m/dT)/P_m$ of HIT is lower, which means the output performance of HIT solar module is more stable in the face of constant changes in outdoor temperature and irradiance.

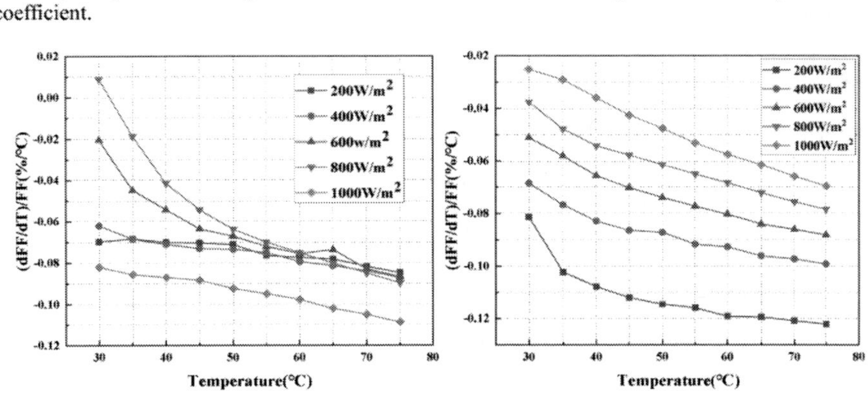

(a) (b)

Figure 3: Measured temperature coefficients of fill factor. (a) (dFF/dT)/FF of PERC solar module versus temperature under different irradiance. (b) (dFF/dT)/FF of HIT solar module versus temperature under different irradiance.

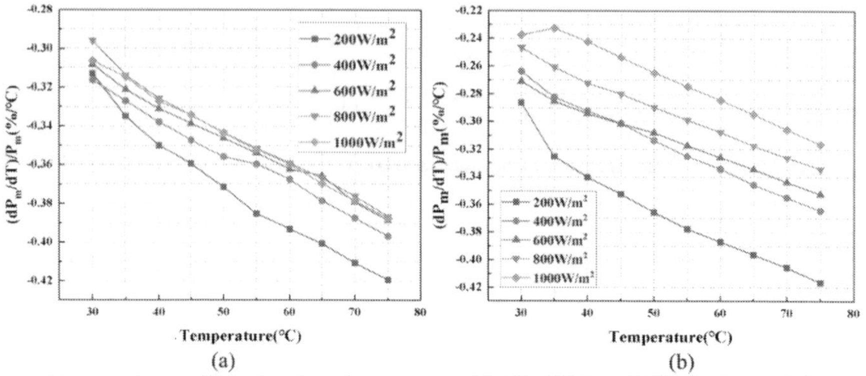

Figure 4: Measured temperature coefficients of maximum power. (a) $(dP_m/dT)/P_m$ of PERC solar module versus temperature under different irradiance. (b) $(dP_m/dT)/P_m$ of HIT solar module versus temperature under different irradiance.

4 CONCLUSION

In this paper, a comparative study on the temperature coefficient of I_{sc}, V_{oc}, FF and P_m was carried out for PERC solar module and HIT solar module. From calculated results, it is found that the temperature coefficient of P_m for HIT solar module is lower than PERC solar module. This phenomenon is more obvious under high irradiance. The main reason for this difference is the difference of temperature coefficient of FF for the two types of solar modules.

ACKNOWEDGEMENT

The authors would like to thank the support of the Natural Science Foundation of China (Grant No. 61274050).

REFERENCES

[1] Shravan K. Chunduri, Michael Schmela. "Heterojunction Solar Technology 2020 Edition", 2020, DOI: 10.13140/RG.2.2.10846.15687.

[2] Dupre, O., R. Vaillon, and M. A. Green. "Experimental assessment of temperature coefficient theories for silicon solar cells." IEEE Journal of Photovoltaics. vol. 6, no. 1, pp. 56-60, 2015.

[3] Green, M. A. "General temperature dependence of solar cell performance and implications for device modelling." Progress in Photovoltaics Research & Applications, vol. 11, no. 5, pp. 333-340, Aug 2003.

[4] O Dupré, R. Vaillon, and M. A. Green. "Physics of the temperature coefficients of solar cells." Solar Energy Materials & Solar Cells, vol. 140, pp. 92-100, Sep 2015, DOI: 10.1016/j.solmat.2015.03.025.

[5] Zhao, J., et al. "Reduced temperature coefficients for recent high-performance silicon solar cells." Progress in Photovoltaics Research & Applications, vol. 2, no. 3, pp. 221-225, 2010, DOI: 10.1002/pip.4670020305.

[6] Wang, H., X Cheng, and H. Yang. "Temperature Coefficients and Operating Temperature Verification for Passivated Emitter and Rear Cell Bifacial Silicon Solar Module." IEEE Journal of Photovoltaics. vol. 10, no. 3, pp. 729-739, May 2020.

[7] Cheng, X., et al. "Comparative Study on Temperature Coefficients of Al-BSF Solar Cells and PERC Solar Cells." 2019 IEEE 46th Photovoltaic Specialists Conference (PVSC), pp. 304-307, 2019.

[8] Kamkird, P., et al. "Investigation on Temperature Coefficients of Three Types Photovoltaic Module Technologies under Thailand Operating Condition." Procedia Engineering. vol. 32, pp. 376-383, 2012.

[9] Tian, H., et al. "Comparative study on fill factor of PERC Silicon Solar cells and Al-BSF Silicon solar cells under non-standard test conditions." 2021 IEEE 48th Photovoltaic Specialists Conference (PVSC), pp. 256-259, 2021, DOI: 10.1109/PVSC43889.2021.9518606.

[10] Bai, QQ., et al. " Effect of temperature on ideality factor for PERC monofacial crystalline silicon solar module." 2021 IEEE 48th Photovoltaic Specialists Conference (PVSC), pp. 244-247, 2021.

[11] Bai, QQ., et al. " Investigation on bifaciality factor and ideality factor of PERC bifacial solar module under different irradiances." 2021 IEEE 48th Photovoltaic Specialists Conference (PVSC), pp. 248-251, 2021.

[12] Tudor, C. D, C. P. Adrian, and M. O. Mihai. "Study of Temperature Coefficients for Parameters of Photovoltaic Cells." International Journal of Photoenergy, 2018, DOI: 10.1155/2018/5945602.

[13] Kasu, M., et al. "Temperature dependence measurements and performance analyses of high-efficiency interdigitated back-contact, passivated emitter and rear cell, and silicon heterojunction photovoltaic modules." Japanese journal of applied physics, 2018, DOI: 10.7567/JJAP.57.08RG18.

INVESTIGATION OF ACCURACY AND CERTAINTY OF ANGLE OF INCIDENCE MEASUREMENTS OF PV MODULES

DOES THE PV OWNER INDUSTRY UNDERSTAND THE IMPACT ON THE YIELD OF THE INCIDENCE ANGLE MODIFIER DURING ENERGY YIELD ASSESSMENTS?

Patricio Sanchez*, Thomas Weber, Mitali Murkar, Benjamin Lippke, Paul Grunow, Steven Xuereb
PI Photovoltaik-Institut Berlin AG, Wrangelstraße 100, 10997 Berlin, Germany
*mail to: sanchez@pi-berlin.com; phone +49 30 814 52 64 -409

ABSTRACT: The aim of this work is to focus on the angle of incidence (AOI) which is described by its angular response curve or incident angle modifier (IAM). AOI measurements are one of the most disregarded problems both during procurement and Energy Yield Assessment (EYA). Hence, this investigation tries to shed some light regarding the ultimate impact of unrealistic angle-modifier curves on the economics of a PV plant by performing the energy yield assessment at different climatic profiles while considering AOI measurements reported by different laboratories. The results showed that unrealistic AOI values could lead to a 4 to 5 % deviation in annual energy yield and most prominently in climatic profiles of higher diffuse light for high angles of incidence, such as Dublin, Hamburg, and Groningen, with 58 %, 55 %, and 56 % Horizontal Diffuse Irradiation (HDI) to Global Horizontal Irradiance (GHI) ratio, respectively. We shed light on the related uncertainties during the IAM determination. Furthermore, we suggest that transmission values higher than 100 % and higher than 70 % shall be secured by the manufacturers' warranty.
Keywords: Energy Performance, Evaluation, Simulation, AOI, PAN-file, Energy Yield Assessment

1 INTRODUCTION

1.1 General

AOI measurements are one of the most disregarded problems both during procurement and Energy Yield Assessment (EYA). EPC companies, banks and any stakeholder involved in the development of a PV project will make use of an energy evaluation as basis for analyzing the feasibility of the project. IAM measurements are used in PAN files to assess the energy yield of the system. However, in literature several interlaboratory comparison have been performed to understand the differences in the AOI measurements setups by comparing between eight scientific institutions in Europe, since the highest uncertainty results from measurements at the highest incidence angles i.e., ±80°. The research work also recommended to investigate the influence of the measurement deviations observed on modelled energy production [1].

The incidence factor (Incidence Angle Modifier "IAM" or Response) corresponds to the decrease of the irradiance that reaches the PV cells' surface, with respect to irradiance under normal incidence. This reduction is largely due to reflections on the glass and increases as the incidence angle increases. Figure 1 shows the IAM loss contributor.

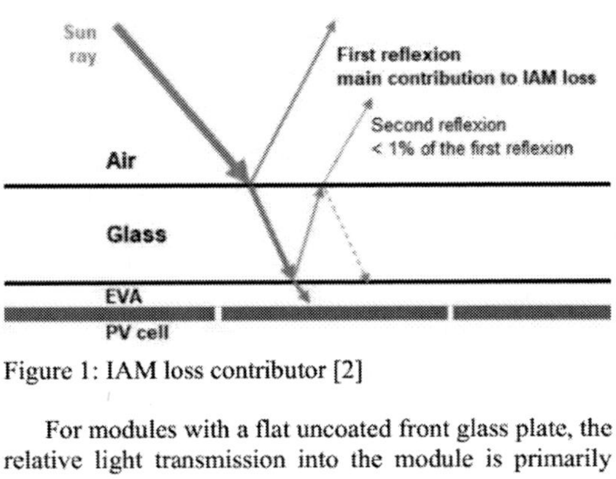

Figure 1: IAM loss contributor [2]

For modules with a flat uncoated front glass plate, the relative light transmission into the module is primarily influenced by the first glass-air interface. For modules with additional optical interfaces, i.e., anti-reflection coatings (ARC), the behavior could have been significantly changed and thus, a validation is recommended for both energy yield and uncertainty values.

Plant owners and operators should use the IAM loss info (together with peak power measurements, temperature coefficients and low-light behavior) as input data for the validation of the so-called *.PAN-files (PAN-files), and use them in simulation software (e.g., PVsyst, SAM, Plant predict) for energy yield assessments. Beyond this, however, the standard proposes an independent calculation of the energy yield and all related procedures (IEC 61853 document group) [3].

Regarding module energy ratings, a well-known problem is the accessibility to the real cell temperature via the module surface during the tests, e.g., for glass-glass-modules. Another problem lies in the full linearity of the irradiance through the reduction of the irradiance level in the flasher. However, one of the most ignored problems in the industry is the accuracy of the angle of incidence (AOI) measurements that determine the IAM values. Even though some laboratories report IAM values of 1.0 at specific angles, according to Fresnel formulas this is not realistic. Transmission values higher than 1 (IAM) at any angle and/or values greater than 0.7 at an angle of incidence of 80° have to be questioned carefully (assuming an uncertainty of ±1 % for this angle). Figure 2 shows the default Fresnel (black) and ARC (orange) IAM-curves for comparison.

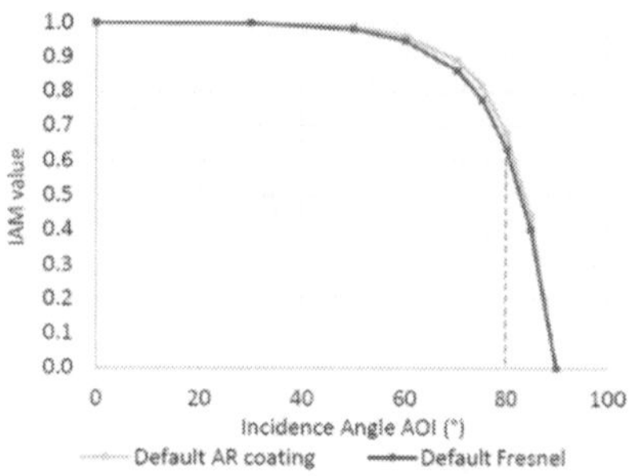

Figure 2: Default Fresnel and ARC curves

This investigation tries to shed some light regarding the ultimate impact of unrealistic angle-modifier curves in the economics of a PV plant. The approach covers the evaluation and comparison of several laboratories whose IAM evaluations are rather questionable, according to physics and the law of reflection of light. There are differences in IAM values at high angles where values above 0.9 were also reported for 80°, though they conflict with optical equations (Fresnel coefficients) while using flat glass surfaces and only one anti-reflective layer (ARC) as module front cover. Hence, this work investigated the ultimate impact of unrealistic angle-modifier curves on the economics of a PV plant by performing the energy yield assessment at different climatic profiles while considering AOI measurements reported by different laboratories.

1.2 Review of measurements

In a previous literature study [4], the Author has reviewed the procedure of AOI measurement provided in IEC 61853-2 and addresses the two main problems i.e. discrepancies in IAM for middle module string vs. one cell and asymmetry of IAM. However, Author suggested one cell method because the string method for measurements of IAM is not suitable due to uneven illumination of front vs. back substrings and following the under-estimated short circuit current of the complete string. To deal with asymmetry of IAM, Author also recommends sensitivity analysis considering offset angle, rotational axis position and carefully positioning of device under test relative to the rotational axes.

Similar practices were also suggested in [5], where it recommended to perform AOI measurements on cell level by performing it on a) Mini-Modules with only one active cell b) separate contacting of one cell through the back sheet, c) partial shading.

As per IEC 61853-2, accurately positioning the module at the specified angles of incidences to an accuracy of ±1°. However, the Author in [5], highlights that the accuracy of AOI measurement is crucial because change in AOI from 81° to 80° leads to an error of 10 %. Therefore, recommends considering accuracy of AOI as crucial.

1.3 Case Study: PAN-File

The IAM is a key component in the creation of the PAN file. It describes the behavior of the PV modules under various irradiance conditions. Thus, we investigated into the PAN file provided by manufacturer.

Manufacturer A sent its standard PAN File (Version 1) during Module Supply negotiations. When the agreement is reached, the manufacturer sent a new set of deliverables, including a new PAN file (Version 2). During the Yield assessment, PI Berlin highlighted that the latest file provides higher IAM losses then the previous. Then, manufacturer provided a third PAN file (Version 3) which then again results in different yield estimates. Comparison of IAM values provided by manufacturer A are depicted in Figure 3.

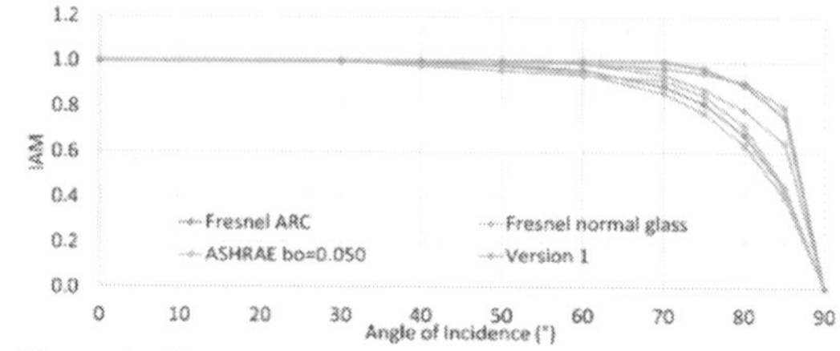

Figure 3: Comparison of IAM values from Manufacturer A

2 METHODOLOGIES

This section describes the methodology used to assess the impact of AOI measurement on energy yield. Firstly, AOI measurement data reported by different laboratories are used for analysis in this research work. Also, the Energy Yield Assessment is performed using PVsyst software for two different locations including Dublin and Murcia. For simplicity of simulation all PV system parameters were fixed, only the AOI data was exchanged, and its economic impact studied. Further, this research work also tries to investigate uncertainty distribution in AOI measurements.

2.1 AOI Measurements

Comparison of AOI measurements reported by different laboratories for various modules were carried out which is illustrated in Figure 4. Data has been taken from one of our customer project at PI Berlin.

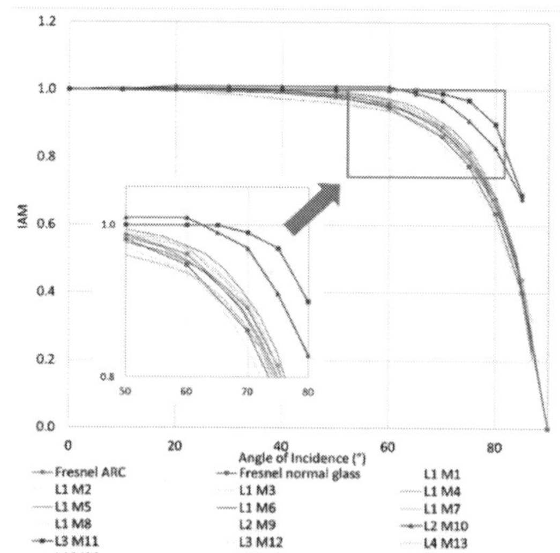

Figure 4: Comparison of AOI measurements reported by different laboratories

3 RESULTS

3.1 Impact on Energy Yield Assessments

The unrealistic values for AOI (out of the range of

Fresnel and ARC values) could lead to large deviations in the performance of a PV module. The results in Figure 5 showed that the yearly variation in energy yield was 4.5 % for Dublin. In a market with rather low margins, differences as high as 5 % in the produced energy could potentially make a PV project unviable. The effect is particularly evident in locations with higher diffuse light for high angles of incidence, such as Dublin, Hamburg and Groningen, with 58 %, 55 % and 56 % Horizontal Diffuse Irradiation (HDI) to Global Horizontal Irradiance (GHI) ratio, respectively.

From all key performance indicators, what matters most is the energy yield of the PV modules and consequently, the PV system. Ultimately, IAM measurements will be used in PAN-files to assess the energy yield of the system. The abovementioned values could lead to differences between 4 to 5 % in the produced energy.

These findings certainly match similar investigations, where it has been reported that the spread of annual angular losses for different glass types range from -1.5 to -2.7 % [6]. For the energy yield evaluation of a specific project, this will have the same effect as using a module with 5 % higher nominal power (Pmax).

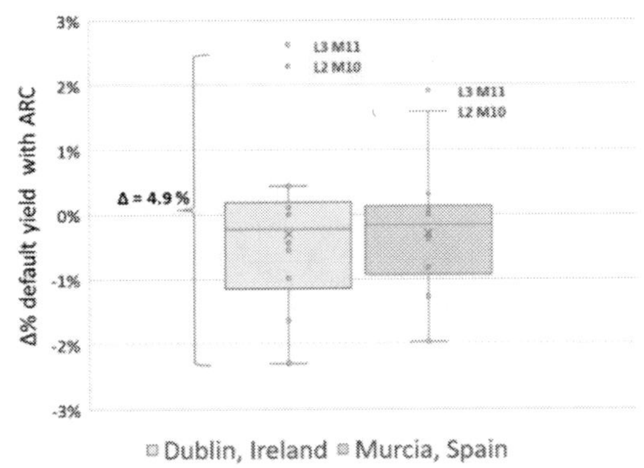

Figure 5: Variation of yearly EYA based on AIMs reported by different laboratories, data taken from Figure 4**Fehler! Verweisquelle konnte nicht gefunden werden.** and simulated with PVsyst. (Reference data is a standard module with ARC).

4 DISCUSSION

4.1 Warranties

Manufacturers likely pick ideal input data from different laboratories for their in-house PAN-files, which does not provide a real benefit for the end customer. Moreover, the existing IEC standard [3] is not robust enough to homogenize the test of the incidence angle modifier.

Due to the expected symmetry, the measurements are normally performed for angles from 0° to 80°. Assuming that 6 % of solar energy is incident in the angular range AOI larger than 70°, the underperformance could have a significant impact on the yield and thus the revenue. Understanding how the effect of the incidence losses (IAM) are verified in the laboratory and specified on datasheets is important for the project's energy yield assessment. Furthermore, transmission values higher that 100 % (at angles 0-40°), and/or higher than 70 % (at angles > 80°) shall be secured by the manufacturers' warranty [7].

4.2 Error Investigation

It is observed that various factors contribute to inaccurate IAM such as measurement error, uncertainty contribution, malfunctioning of rotational device, lack of awareness about best practices.

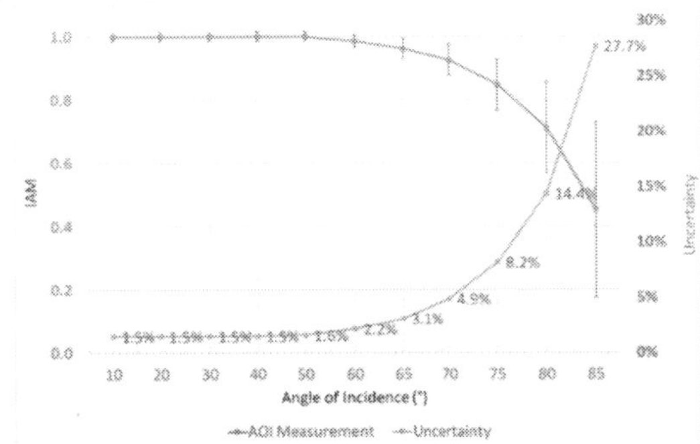

Figure 6: Example of uncertainty distribution of a Lab AOI measurement.

In Figure 6, uncertainty distribution in AOI measurement is plotted. The results showed that, uncertainty distribution is more for higher values of AOI. For an instance, uncertainty increased from 14.4 to 27.7 % for change in AOI from 80° to 85°. Therefore, we recommend verifying the IAM data by third party testing and further investigate in uncertainty which is originated from different sources.

5 CONCLUSION

This IAM overestimation could lead to differences between 4 to 5 % in the produced energy (MWh) and on revenue. These discrepancies are particularly evident in locations with higher diffuse light for high angles of incidence (i.e., North of Europe).

Finally, it can be concluded that the accuracy of AOI measurement is important in energy yield assessment. So, our advice to IEC is to homogenize the test of the incidence angle modifier.

The overestimated AOI measurements could have been originated by a lack of awareness of industry standards (IEC 61853-2) and the best practices of European Metrology Research Programme. The industry requires immediate action regarding a) best practices for IEC 61853-2, b) independent PAN file verification, c) error distribution for IAM factor-ranges.

6 ACKNOWLEDGEMENT

This work is part of the ReliaREN-Pro project and supported by a grant of the BMWi under project number: 03EI4052A. Many people at PI Berlin have contributed to the present work. The authors particularly acknowledge the input by Alexander Drobisch and Romain Penidon.

7 REFERENCES

[1] N. Riedel, et al, "Interlaboratory comparison of methodologies for measuring the angle of incidence

dependence of solar cells," in EUPVSEC, Brussels, 2014.

[2] PVsyst, "Array Incidence Loss," [Online]. Available: https://www.pvsyst.com/help/iam_loss.htm. [Accessed 2022].

[3] IEC 61853-2 Photovoltaic (PV) module performance testing and energy rating – Part 2: Spectral responsivity, incidence angle and module operating temperature measurements, 2016.

[4] PI Berlin, IEC 61853-2: Measurement of Incidence Angle Effects (IAE) A critical review of the measurement procedure, 2018.

[5] European Metrology Research Programme, Good practive guide: Angle of incidence measurements.

[6] L. R. U. J. W. Herrmann, "Optical characterization of PV modules," in WCPEC-6, Kyoto, 2014.

[7] H. e. a. G. Friesen, "Photovoltaic Module Energy Yield Measurements: Existing Approaches and Best Practice," IEA PVPS Task 13, Subtask 3, 2018.

IMPROVED SHADOW FILTERING AND CHANGE-POINT DETECTION METHODS TO EXTRACT SOILING LOSS FROM PV-SCADA DATA

Shoubhik De[1,2] (shoubhikde@ee.iitb.ac.in), Pavan Fuke[1,2] (204076012@iitb.ac.in), Narendra Shiradkar[1,2] (naren@ee.iitb.ac.in) and Anil Kottantharayil[1,2] (anilkg@ee.iitb.ac.in)
[1]Department of Electrical Engineering, Indian Institute of Technology Bombay, Mumbai, Maharashtra 400076, India
[2]National Centre for Photovoltaic Research and Education, Indian Institute of Technology Bombay, Mumbai, Maharashtra 400076, India

ABSTRACT: Soiling of solar panels is becoming a serious issue worldwide. To address this, an important step is to analyze PV-SCADA data so as to extract soiling losses. Previously two models, Fixed Rate Precipitation (FRP) and Stochastic Rate and Recovery (SRR), have been used to extract soiling losses in terms of soiling rates. These models use an irradiance filter and/or a time filter to eliminate data points influenced by shading and angle of incidence effects to calculate a daily Performance Index (PI)/Ratio (PR). However, shadow may be cast during the mentioned time-period on the PV panels or on the irradiance sensor. This work proposes two new filtering techniques to eliminate shadows cast on PV panels, which would result in estimation of true daily PR. Moreover, FRP and SRR assume the soiling deposition rate to be constant between two cleaning events which may not be the case always. The site under study may have changes in climatic conditions due to which the soiling rate may change. Detection of change in the soiling rate would improve soiling rate estimates. Hence, the work also proposes a new algorithm to identify the Change Point(s) (CP) accurately to estimate the true soiling rate.
Keywords: PV, Soiling Loss, Soiling Rate, Temperature-corrected Performance Ratio, Change Point

1 INTRODUCTION

Photovoltaic soiling refers to the accumulation of dust, dirt and other airborne particles on the surface of Photovoltaic (PV) modules. This blocks the sun rays from entering inside the PV modules and thereby reducing its conversion efficiency. PV soiling has been gaining a lot of attention recently, because of the substantial loss of revenue due to reduced electricity generation [1].

This calls for O&M activities to reduce the dust deposition on PV panel surfaces which includes periodic cleaning using water. Research work all around the globe currently is focused on developing soiling mitigation techniques on solar panels. These techniques include robotic cleaning [2], reducing soiling by applying anti-soiling coatings on top of solar panels [3], [4] and using electrostatic-based dust repulsion approaches [4], [5]. Though these mitigation methods are promising, they are costly and their durability in harsh environmental conditions are still to be proven.

The most vital thing in this case is to monitor soiling loss due to dust deposition in large-scale PV power plants. With continuous soiling on PV modules, the power/energy production of the system would go down. Decrease of power beyond a certain limit calls for cleaning so as to restore the system production. To monitor the soiling loss of PV systems, a term called the Soiling Ratio (SR) is used. As per IEC 61724-1, it is the ratio of the outputs of soiled to cleaned PV panels. The output could be the PV panel's power or short-circuit (I_{sc}) current. SR varies from 0 to 1, with 1 indicating a fully clean state. This has led to the development of soiling sensor kits, both optical [6] and electrical [7] to measure the SR. From the daily SR data, the soiling rate can be calculated, giving an idea of the amount of soiling loss for a specific PV installation.

However, there are some problems with these kits. First of all, some of these kits need further field validation. Moreover, it would be challenging to represent the SR of the entire power plant using just one soiling kit. Again, these would be present at a fixed location in the PV plant, and they won't be able to capture the non-uniform SR, which is typical of PV power plants.

An alternative could be to analyze the SCADA data of a PV plant and estimate the soiling rate and hence the soiling loss directly. The PV power plants have weather stations that record various meteorological parameters like solar radiation, daily temperature, wind speed, relative humidity and precipitation. One would get an idea of the extent of soiling losses taking place by understanding the historical trend of SR. Previously, two models have been reported for soiling rate extraction, namely Fixed Rate Precipitation (FRP) model [8] and Stochastic Rate and Recovery (SRR) [9] model. FRP model calculates the soiling rate assuming a constant soil deposition rate between two cleaning events and requires precipitation data. SRR model calculates the soiling rate by identifying positive shifts in the PV time-series data, which is marked as a cleaning event and applying the Thiel-Sen method to the soiling interval to determine the soiling rate. SRR model doesn't require any precipitation data. Both FRP and SRR methods assume the soiling rate between any two cleaning events to be constant.

As indicated by Deceglie et al. [9], a time-series of daily Performance Index (PI) is required to estimate soiling losses. PI is the ratio of actual PV power/energy produced to the modeled PV power/energy. To calculate the daily PI, the SRR model requires filters to eliminate shadow of any form cast on the PV panels. In the literature, an irradiance filter and/or a time filter have been applied to concentrate on solar-noon periods to avoid shadowing and low light effects [9]–[12]. However, any moving object can cast a shadow on the PV panels during solar noon, which may not be eliminated by the filters mentioned above.

In this work, two shadow filtering methods have been described and they have been compared with the existing filtering technique. The daily PI value have been used for comparing the filtering methods. Instead of PI, a different term, Temperature-corrected Performance Ratio ($T_{correc}PR$) has been used in this work.

Moreover, the soiling rate between any two cleaning events may not be constant as the site under study may have changes in climatic conditions. Hence, detecting points where the soiling rate changes may help to determine accurate soiling rates. Recently, a few algorithms have been used to identify the Change Points (CP) in the soiling profile time-series data [12], [13]. This work presents a new Machine Learning (ML)-based algorithm to detect CPs in the soiling profile time-series data to extract accurate soiling rates.

2 METHDOLOGY

2.1 Soiling Data

To study the soiling loss, approximately three months of time-series data of a $100W_p$ PV module installed at the Photovoltaic Module Monitoring Station (PVMMS), IIT Bombay, have been used. And to study the effect of filters on shadows, time-series data of $4.92kW_p$ Evergreen solar installed at the DKA solar centre, Alice Springs, Australia [14], have been used.

2.2 Filtering of Shadows

Shadow cast on PV system refer to the event when a part of the system (panel or group of panels connected in a string) doesn't receive the same solar irradiation in a particular time interval. Shadow cast on PV panels could be due to the following reasons:

- Presence of nearby objects like fences, crops, trees and PV modules.
- Human intervention and flying birds.
- Moving clouds.

A PV plant also has an irradiance sensor to measure the Plane of Array Irradiance (POA), and it may happen that the sensor is shadowed as well. This gives rise to the following conditions:

- Shadowing of the PV system only.
- Shadowing of the irradiance sensor only.
- Shadowing of both.

The scenarios mentioned above are shown in Fig. 1.

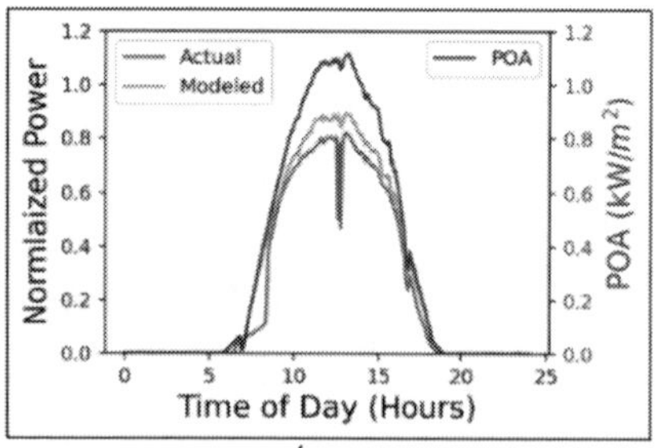

Fig. 1(a): 3rd November 2014

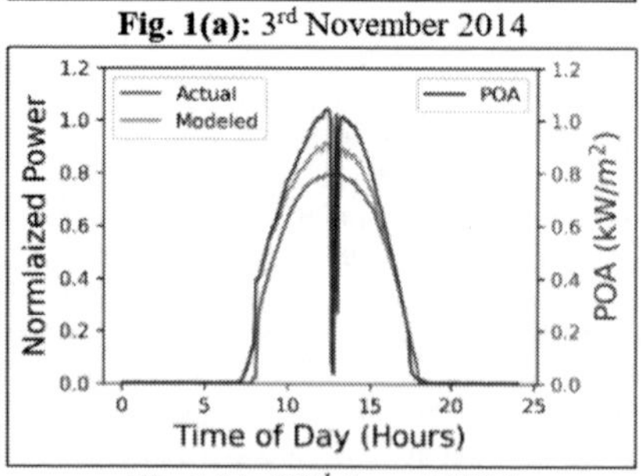

Fig. 1(b): 22nd August 2014

Fig. 1: **(a)** Shadowing of solar PV array during solar noon **(b)** Shadowing of both PV array and irradiance sensor during early morning hours and shadowing of irradiance sensor during solar noon (PV system: $4.92kW_p$ Evergreen solar, Alice Springs, Australia).

Fig. 1 shows the plot of variation of Actual/Measured and Modeled power normalized using the PV system rating and also the POA (as a reference) for 2 specific days. Modeled power has been calculated using NREL's PVWatts DC power model [15].

In Fig. 1(a), it can be seen that the rate of rise of Actual power (blue curve) is different to that of the Modeled power (orange curve) at the early morning hours, which signifies that there is a shadowing on the PV array by the surrounding objects. Again, during the period of solar noon, there is a dip in the Actual power which resembles shadowing on the PV array possibly due to human intervention/flying birds/moving clouds.

In Fig. 1(b), there is shadowing on the irradiance sensor during the solar noon and hence there is also a dip in the Modeled power (since Modeled power is directly proportional to POA). However, no dip is observed in the Actual power. Moreover, during the early morning hours on the same day, it can be seen that the rise of both Actual power and Modeled power suddenly changes indicating shadows cast on both of them.

Therefore, shadowing of these kind need to be removed for accurately evaluating the daily $T_{correc}PR$. Keeping this in mind, two filters have been developed and used for testing along with the filter used in the literature (highlighted above). Description of filters are given below:

- Filter 1: (POA >= $100W/m^2$) and (0.7 <= Ratio <= 1.05) and (ΔRatio <= 0.009)
- Filter 2: (Diff. b/w MP & AP <= 2*x) and (Ratio <= 1.05) and (11AM <= Time Of Day <= 2PM)
- Filter 3: (POA >= $700W/m^2$) and (11AM <= Time Of Day <= 2PM)

Where:

- $\text{Ratio} = \frac{\textit{Measured/Actual PV power}}{\textit{Modeled PV power}}$
- ΔRatio = successive difference of Ratio
- Diff. = Difference between Modeled PV power (MP) and Actual PV power (AP)
- x = median of the difference between MP and AP between 11AM and 2PM

$T_{correc}PR$ for a particular day (where shadows have been cast on PV panels) has been calculated to indicate the performance of the three filters.

2.3 Change Point Detection

The soiling rate between any two cleaning events may not be constant as the site under study may have changes in climatic conditions. For example, a sudden dust storm may alter the soiling rate significantly or an extended period of rain followed by a dry period may impact the soiling rate as well.

As highlighted by Micheli et al. [12], [13], using the SRR model (explained previously), a single line would fit the whole soiling profile (between 2 cleaning events) and the accurate rate of soiling won't be obtained. Therefore, to extract the accurate soiling rate, a number of Change Point (CP) detection algorithms have been described previously such as Piecewise Regression (PCR), Facebook Prophet Algorithm (FPA) [12], [13] and Bayesian Estimation of Abrupt Change, Seasonality, and Trend (BEAST) [12]. These are open-source libraries in Python and R. FPA and BEAST decompose the time series based on seasonal changes to extract change points.

In this work, CP detection has been performed using Python's Linear Tree Regressor (LTR) library. Here the data points are split as per decision rules, and a linear model fits each split. The extracted soiling profiles with

CPs detected using PCR, FPA and LTR have been compared using Coefficient of Determination (R^2) to test the fit.

3 RESULTS AND DISCUSSIONS

3.1 Filtering of Shadows

As mentioned earlier, T_{correc}PR have been used to highlight the performance of different shadow filters (described above). For this, two days were chosen when panels have been shaded.

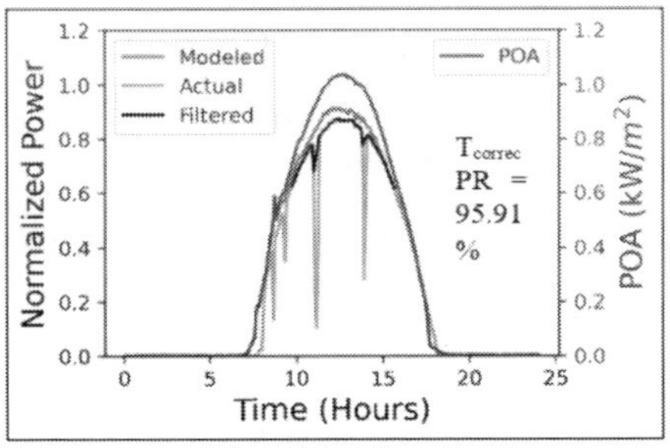

Fig. 2(a): Filter 1

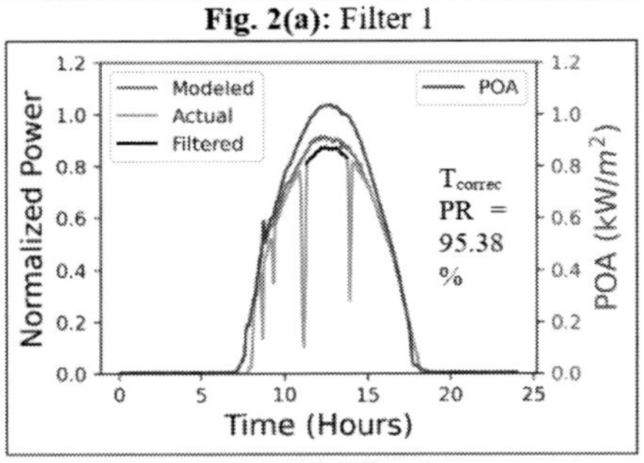

Fig. 2(b): Filter 2

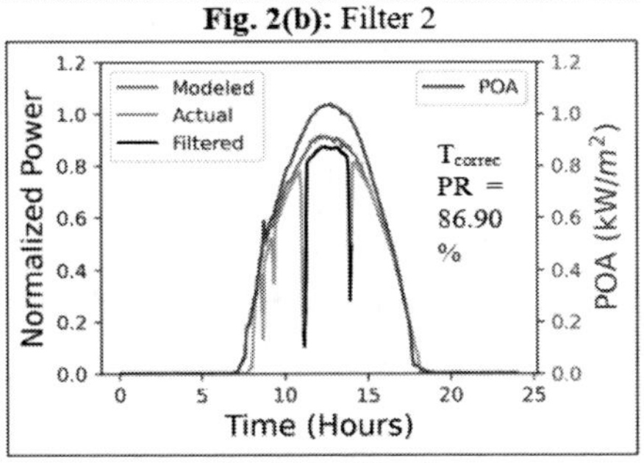

Fig. 2(c): Filter 3

Fig. 2: Performance of (a) Filter 1 (b) Filter 2 (c) Filter 3 on 16th April 2014 (PV system: 4.92kW_p Evergreen solar, Alice Springs, Australia). The T_{correc}PR for each of the three filters have also been shown.

From Fig. 2, it can be seen that the T_{correc}PR for Filter 3 is less than that for Filter 1 and 2. This signifies that Filter 1 and 2 were able to eliminate shadows which were cast on the PV panels on 16th April. The black line on the Actual Normalised PV power curve represents the filtered data points used for evaluating T_{correc}PR.

Filter 2 performs better than Filter 1 especially when the amount of shadow is not that significant (small dip in Actual power). Moreover, the lower limit of Filter 1 (Ratio>=0.75) may create a problem especially when the system is degraded.

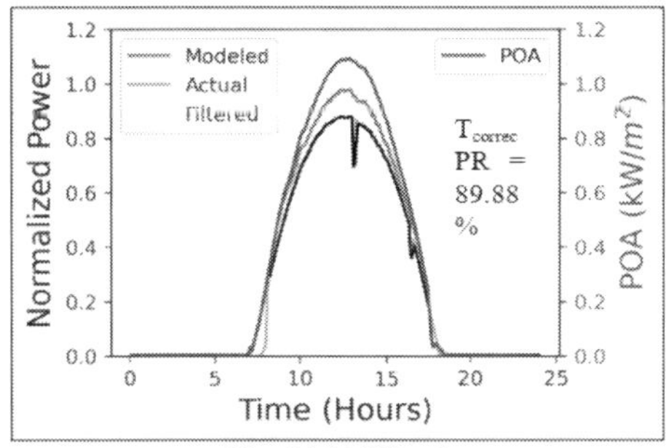

Fig. 3(a): Filter 1

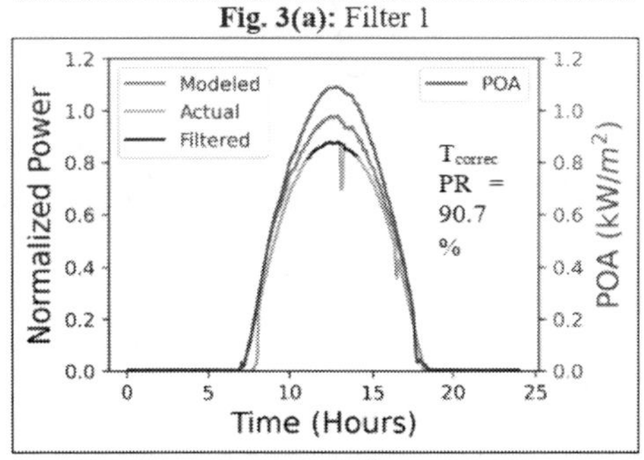

Fig. 3(b): Filter 2

Fig. 3: Performance of (a) Filter 1 (b) Filter 2 on 28th August 2014 (PV system: 4.92kW_p Evergreen solar, Alice Springs, Australia).

From Fig. 3, it is clear that Filter 2 is able to remove the shadow (indicated by the small dip in Normalized Actual Power) cast on the PV panels on 28th August, whereas Filter 1 couldn't. This is because, there is a slight improvement of T_{correc}PR for Filter 2 over Filter 1.

3.2 Detection of Change Point (CP)

Approximately three months of PV time-series data of a 100Wp solar module installed at the PVMMS, IITB have been analyzed to extract the T_{correc}PR. The module has been cleaned once in a month. The plot of T_{correc}PR vs time is shown in Fig. 4. Filter 2 was used for calculating the daily T_{correc}PR.

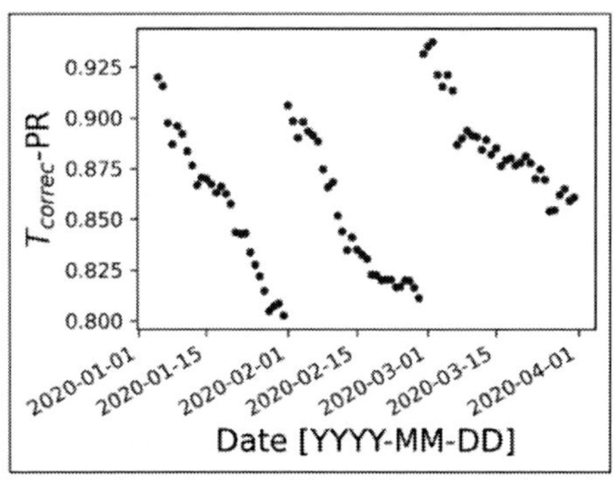

Fig. 4: Daily variation of T_{correc}PR of approximately 3 months of PV data (PV system: 100W_p PV module at PVMMS, IITB).

To estimate the soiling ratio (SR), a clean state of the system has been defined and denoted by $T_{correc}PR_0$. $T_{correc}PR_0$ is the 95th percentile of the T_{correc}PR time-series data [11]. Fig. 5 shows the SR time-series data.

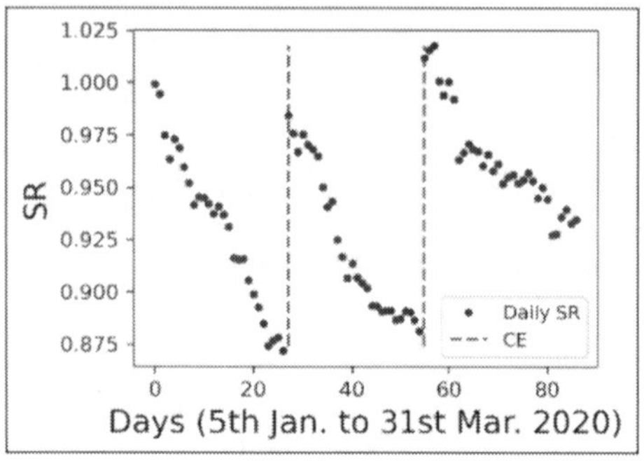

Fig. 5: Daily SR of approximately 3 months of PV data with detected cleaning events (PV system: $100W_p$ PV module at PVMMS, IITB). CE symbolizes Cleaning Event.

In Fig. 5, the cleaning events (day when the PV panel was cleaned) is also shown. To detect the cleaning events, the procedure highlighted in ref. [9] has been followed. Deceglie et al. [9] recommended a running median having a window length of 14 days for detecting the cleaning events. However, a 4-day running median has been found suitable for the dataset used in this paper. Skomedal et al. [11] highlighted that the length of window may vary with dataset. Therefore there are three soiling intervals.

To visualize the CPs, the soiling profile of each of the three soiling intervals have been separately shown for each of the three CP detection algorithms: LTR, PCR and FPA. Figures 6, 7 and 8 show the detected CPs using the above three algorithms for the three soiling intervals respectively. From figures 6(a), 7(a) and 8(a), it can be clearly seen that there are higher number of CPs for each of the three soiling intervals which signifies that soiling profiles with higher accuracy have been extracted.

Table 1 shows the R^2 for each of the three algorithms. Due to higher R^2 value obtained for LTR algorithm, it can detect better CPs than PCR and FPA algorithms and therefore provide a better estimation of the soiling rate.

4 CONCLUSIONS AND FUTURE WORK

This work proposes two new filtering algorithms which help to extract better $T_{correc}PR$ by eliminating shadows cast on the PV panels or the irradiance sensor. This will likely help to find better soiling losses for the concerned PV system.

A new Change Point (CP) detection algorithm (LTR) has also been used to detect CPs in soiling profiles. LTR algorithm has been compared with algorithms (PWR and FPA) which have already been used previously. The higher R^2 obtained via using LTR provides an opportunity in improving the detection of change of soiling rates.

However, the PV time-series data used in this work is a closely monitored and well maintained small PV system (having very low noise). The actual PV power plant data (in MW/GW scale) would have much higher noise and testing LTR algorithm to detect CP would have much higher significance. Hence, the analysis of CP detection would have to be carried for large PV power plants.

5 ACKNOWLEDGEMENT

This work was supported by the National Centre for Photovoltaic Research and Education (NCPRE) at IIT Bombay, funded by the Ministry of New and Renewable Energy of the Government of India through Project No. 31/09/2015-16/ PVSE-R&D dated 15th June 2016.

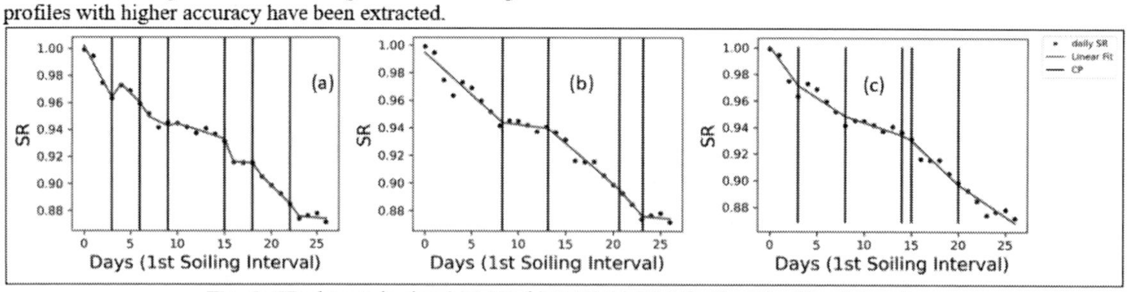

Fig. 6: CPs detected using (a) LTR (b) PWR (c) FPA for the 1st soiling interval.

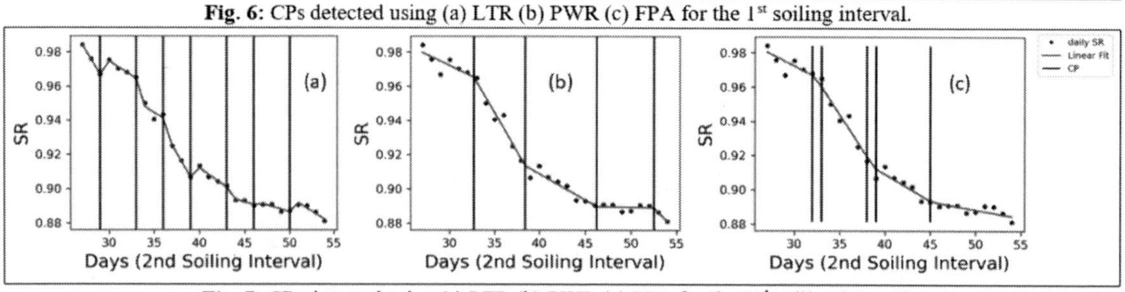

Fig. 7: CPs detected using (a) LTR (b) PWR (c) FPA for the 2nd soiling interval.

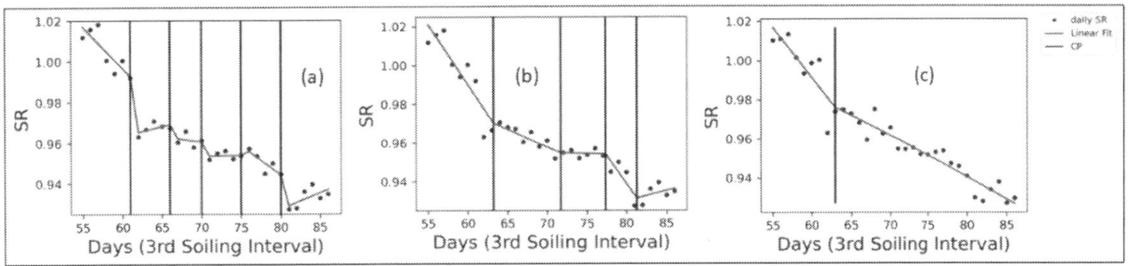

Fig. 8: CPs detected using (a) LTR (b) PWR (c) FPA for the 3rd soiling interval.

Table I: Comparison of LTR, PWR and FPA algorithms in terms of R^2 value

Soiling Interval	LTR	PWR	FPA
First	0.9965	0.9871	0.9860
Second	0.9987	0.9916	0.9902
Third	0.9813	0.9565	0.9406

5 REFERENCES

[1] K. Ilse *et al.*, "Techno-Economic Assessment of Soiling Losses and Mitigation Strategies for Solar Power Generation," *Joule*, vol. 3, no. 10, pp. 2303–2321, 2019, doi: 10.1016/j.joule.2019.08.019.

[2] M. Anderson *et al.*, "Robotic device for cleaning photovoltaic panel arrays," in *Mobile Robotics: Solutions and Challenges - Proceedings of the 12th International Conference on Climbing and Walking Robots and the Support Technologies for Mobile Machines*, 2009, pp. 367–377, doi: 10.1142/9789814291279.

[3] D. Adak *et al.*, "Non lithographic block copolymer directed self-assembled and plasma treated self-cleaning transparent coating for photovoltaic modules and other solar energy devices," *Sol. Energy Mater. Sol. Cells*, vol. 188, no. 15, pp. 127–139, 2018, doi: 10.1016/j.solmat.2018.08.011.

[4] D. Adak *et al.*, "Mitigation of Soiling of Solar Panels by Applying Superydrophobic Aluminum Oxide Thin Film and Dry Cleaning by Electrodynamic Screen," 2022, [Online]. Available: https://doi.org/10.1007/978-981-16-9280-2_9.

[5] S. De *et al.*, "Surface engineering of solar glass covers for soiling related issues by applying Electrodynamic Screens (EDS)," *Surfaces and Interfaces*, vol. 25, 2021, doi: 10.1016/j.surfin.2021.101222.

[6] Á. Fernández-Solas, L. Micheli, M. Muller, F. Almonacid, and E. F. Fernández, "Design, characterization and indoor validation of the optical soiling detector 'DUSST,'" *Sol. Energy*, vol. 211, no. October, pp. 1459–1468, 2020, doi: 10.1016/j.solener.2020.10.028.

[7] "Atonometrics. PV device soiling measurement systems for PV power plants." https://www.atonometrics.com/products/soiling-measurement-system-for-pv-modules/ (accessed Jul. 15, 2021).

[8] A. Kimber, L. Mitchell, S. Nogradi, and H. Wenger, "The effect of soiling on large grid-connected photovoltaic systems in California and the Southwest Region of the United States," in *2006 IEEE 4th World Conference on Photovoltaic Energy Conference*, 2006, pp. 2391–2395, doi: 10.1109/WCPEC.2006.279690.

[9] M. G. Deceglie, L. Micheli, and M. Muller, "Quantifying Soiling Loss Directly from PV Yield," *IEEE Journal of Photovoltaics*, vol. 8, no. 2. pp. 547–551, 2018, doi: 10.1109/JPHOTOV.2017.2784682.

[10] M. G. Deceglie, L. Micheli, and M. Muller, "Quantifying Year-to-Year Variations in Solar Panel Soiling from PV Energy-Production Data," 2018, pp. 1–4, doi: 10.1109/pvsc.2017.8366290.

[11] Å. Skomedal, H. Haug, and E. S. Marstein, "Endogenous Soiling Rate Determination and Detection of Cleaning Events in Utility-Scale PV Plants," *IEEE J. Photovoltaics*, vol. 9, no. 3, pp. 858–863, 2019, doi: 10.1109/JPHOTOV.2019.2899741.

[12] L. Micheli *et al.*, "Improved PV Soiling Extraction through the Detection of Cleanings and Change Points," *IEEE J. Photovoltaics*, vol. 11, no. 2, pp. 519–526, 2021, doi: 10.1109/JPHOTOV.2020.3043104.

[13] L. Micheli, M. Muller, E. F. Fernandez, and F. Almonacid, "Segmentation of Deposition Periods: An Opportunity to Improve PV Soiling Extraction," in *2020 47th IEEE Photovoltaic Specialists Conference (PVSC)*, 2020, pp. 595–598, doi: 10.1109/PVSC45281.2020.9300370.

[14] "Desert Knowledge Australia Centre, Alice Springs. http://dkasolarcentre.com.au/historical-data/download, date accessed: 16/04/2014, 22/08/2014, 28/08/2014, 03/11/2014".

[15] A. P. Dobos, "PVWatts Version 5 Manual (NREL/TP-6A20-62641)," *Natl. Renew. Energy Lab.*, no. September, p. 20, 2014, [Online]. Available: http://www.nrel.gov/docs/fy14osti/62641.pdf.

COMPARISON OF THE YIELD OF MINIATURIZED BIFACIAL PHOTOVOLTAIC MODULES WITH THE YIELD SIMULATION OF BIFACIAL MODULES

Akhildev Pillai*[1], Kevin Depner[1], Conor Brisco Ray[1], Sascha Mebus[1], Eva-Maria Grommes[1], Ulf Blieske[1]
[1]University of Applied Sciences Cologne, Germany
Cologne, Germany
Akhildev.Pillai@smail.th-koeln.de , Kevin.Depner@smail.th-koeln.de , Conor.Ray@smail.th-koeln.de ,
Sascha.Mebus@smail.th-koeln.de , Eva-Maria.Grommes@th-koeln.de , Ulf.Blieske@th-koeln.de

ABSTRACT: Bifacial photovoltaic (PV) is a field that has seen an increase in research and development in the world market in recent years. Increased module efficiency and reduced Levelized Cost of Energy (LCOE) have led to more focus on bifacial panels. This project aims to develop and manufacture miniaturized bifacial solar modules to compare the energy yield of the test module with the yield simulation of bifacial modules from the python simulation software BifacialSimu, developed at the University of Applied Sciences Cologne. The validation of the software for its accuracy has also been conducted. The miniaturized model can easily be transported to different sites for field studies. This makes it possible to record (preliminary) real-time data on the yield of a bifacial PV ground-mounted system and provide a forecast of the solar energy yield under field conditions. For this purpose, miniaturized bifacial glass-glass PV modules were designed and built with a scaling factor of 1:10 and with module dimensions of 100 x 200 mm. Each module consists of a string of six bifacial PERC cells connected in series, each cell with an area of 2098.63 mm^2. The miniaturized modules theoretically deliver the same performance characteristics and behavior as their real-size counterparts under the same test conditions. These modules are assembled into an array, with three rows of three modules and realistic module spacings with sufficient clearance to avoid shading. The I-V characteristics are also recorded under varying ambient conditions. The bifacial module generated a power value between 1.53 W and 2.29 W. The simulated power values calculated showed a deviation from the actual measured values by +2.1 percent and +2.7 percent.
Keywords: Bifacial, Design, Software, Qualification & Testing, Validation

1 INTRODUCTION

In the present scenario, solar photovoltaics has become one of the fastest-growing renewable energy resources in the world. The cost of photovoltaics has also dropped with recent innovations and state-of-the-art manufacturing technologies. In addition to the solar radiation hitting the front side of a solar module, bifacial solar modules can also convert reflected solar radiation reaching the backside of the solar module into electrical energy. This can increase the energy yield per area by 5 to 30 percent, depending on the cell's substrate and installation. This research focuses on optimizing and increasing the bifacial module's energy yield efficiency. Considering this scenario, this project aims to develop a miniaturized bifacial photovoltaic module to compare the yield of a test module with the yield simulation of bifacial modules through python-based BifacialSimu software.

After production, the miniaturized modules were tested in a laboratory and under outdoor conditions. The outdoor measurements were performed on the library rooftop, and energy yield at different tilt angles was recorded. The miniaturized modules are built based on the research made regarding the various manufacturing processes of miniaturized PV modules and the different soldering techniques used. The yield energy of the bifacial modules for different tilt angles is then recorded. A comparison of the results obtained from real-time measurements is made with the one obtained from simulations of bifacial modules using a python program.

2 LITERATURE RESEARCH

The miniaturized bifacial modules enable test measurements with different module setups in various outdoor conditions. This has the advantage of easier handling and lower budgeting. This allows stand-alone and flexible measurements in areas where it is challenging and expensive to test with real-sized modules. Some published scientific studies show miniaturized bifacial module systems and their analysis through measurements, simulations, and evaluations [2]. To produce miniaturized bifacial modules, it was necessary to research the production process with every single step and the function of the unique attributes of bifacial cells in general [3]. Accordingly, both sides should be shadowed as little as possible, and the floor should have a high albedo to get the highest electrical yield of the cells. A low proportion of the shaded area is an essential condition for the production process and the definition of the distances between the rows and the modules [4]. The miniaturized modules are on a defined scale smaller than the original one. This makes the transportation of the modules and the application in terrain with difficult access easier and allows the comparison with real-sized modules of the same cell type.

The miniaturized cells are created through precise laser cutting to get identical surfaces out of the cut cell. Laser cutting is a separation process with a laser application as a fundamental technology that causes the cell to vaporize selectively [5]. Busbars are used to connect the cells physically and electrically with each other. Thereby, during the soldering of the busbars, the cell should not be shadowed from overlapping, e.g., too wide busbars.

The busbars are soldered with flux to get better results in the stability of the connection surfaces. The choice of additional components for the module must be clarified by choosing between glass-glass-modules and glass-foil-modules. Bifacial modules necessitate glass-glass modules. For this module, two layers of encapsulation material and two solar glass sheets are needed in addition to the cells. The encapsulation material is an Ethyl-Vinyl-

Acetate (EVA) foil that contacts the cells directly [6]. It functions as an adhesive with cross-linking, stabilizing and, UV-absorbing properties. It also protects the solar cell against weather, water damage, and humidity.

The extra white glass on the top and the bottom is in touch with the encapsulation. This kind of glass has a lower iron percentage for a higher transmission value up to 91.5 percent and protects the cells against physical effects and weather damage [7, 8].

The lamination machine warms up the components. Once a set temperature has been reached, the EVA will melt. After pressurization, the EVA will bond the glass with the solar cell. The cooling works with a flushing system after a set time [8]. The frame needs suitable row and module distances to achieve as little shadowing as possible for the neighboring modules. If the distances between the modules are bigger, the floor shadowing would be reduced. Otherwise, the whole system needs a larger surface. A higher distance from the floor increases the light's and module's yield. An increase in the height of one meter for bifacial modules can increase the bifacial gain by 30 percent [3, 4, 9].

Setting a steeper angle than a standard monofacial module targets a higher rear side yield and a slightly reduced front side yield. Because of this, it is necessary to find the overarching optimum for the final mounting and measurements. The illumination intensity and homogeneity conditions on the rear side are much more complex than on the front side. This is also the albedo value of the floor below the bifacial modules. The albedo measurements work with a pyranometer or reflectometer mounted with enough space without shadow or other influences around it. The consistency of the measured values during the readings is essential for an accurate comparison [10, 11]. The albedo on the rooftop of the University of Applied Sciences Cologne has been previously measured with a value between 0.238 and 0.263. The deviation is typically about five percent [12]. This influences the setup for the tilt angle. According to outdoor measurements in Switzerland, also with mini bifacial modules, the optimized angle is between 35 and 45 degrees [13].

Furthermore, bifacial modules tilted at a shallow angle are influenced more by diffuse light than monofacial because of the impact on both sides of the modules [14]. The shadowing of the cell surface is also affected by the supporting construction, overlapping frame, crossbars, or cables of the modules. Shadowing because of the installation of these components needs to be kept at a minimum for the installation planning [15].

3 METHODOLOGY

3.1 Design & Construction

To develop the miniaturized solar array, the design was based upon a scaling factor of 1:10, i.e., a 2 x 1 m real-life solar module would be represented by a 0.2 x 0.1 m miniaturized version. The array was designed to have three rows with three modules per row and realistic scaled distancing between modules per the previous literature review. The miniaturized modules are designed and developed at the university laboratory. The cells used have an average efficiency of 22.2 percent, per the datasheet [1]. The module design is based upon a downscaled version of a typical solar array with a scaling factor of 1:10.

Once the dimensions of the modules had been determined, a full-size array could be designed, made of three rows of modules each. This combination allows for consideration of the module shading and effect on the albedo during simulations.

The miniaturized cells were created from single monocrystalline PERC solar cells. The original 166 x 166 mm bifacial solar cell was laser-cut into 12 smaller cells, each with an equal area of 2098.63 mm². The varying shapes and busbar configurations resulted in four cells of 55.3 x 37.95 mm, four cells of 55.3 x 37.2 mm, and two cells of 55.4 x 37.95 mm, and 55.4 x 37.2 mm, respectively.

The laser cutting process effectively scores the cells; therefore, it was still necessary to separate them mechanically afterward. This was achieved by "snapping" the cells apart along the lasered lines. It was essential to pay attention when "snapping" the cells that the pressure was evenly distributed, thereby avoiding unwanted breakages.

The cells could then be soldered together into six cells per module strings with a flat strip of solder-coated wire, with a width of 2 mm. There were difficulties encountered during this phase related to the solder adhesion of the strips and breakages of solar cells from the inherent tension of the solder strip. This is covered later in observations.

The next step involved laminating the glass-glass modules with the following symmetrical construction: glass, EVA, solar cell string, EVA, and glass. These elements were layered upon each other as accurately as possible before being put in the laminator for production. The lamination process is programmable and will run automatically once the parameters have been set. The initial parameters were based on those for another module and were determined experimentally. The same has been listed in Table I. The initial test results were promising, with good adhesion and no obvious cavities or bubbles. As such, these settings were kept for further modules.

Table I: Lamination Device Parameters used for Bifacial Modules

Process	Parameters		
	Temperature (°C)	Time (s)	Time (min.)
Prewarming	-	-	-
Vacuum	110	360	6
Press	110	1	0
Curing	130	1200	20
Cooling	-	-	-
Venting	55	1	0

One issue that occurred during lamination was the cracking of the glass both during and after lamination. This resulted from uneven pressure on the modules within the laminator and the modules themselves. The solution was setting a frame around the modules during lamination to distribute the pressure.

The frame for the array was constructed using an aluminium profile, having been designed in CATIA CAD software beforehand for ease of assembly and confirmation of dimensioning. In addition to the frame, custom bracketry had to be designed and constructed for the modules, as shown in Figure 1. These were produced from CAD models, which were subsequently 3D-printed on-site at the University of Applied Sciences Cologne. The module holders were designed to be fixed to the frame in a horizontal orientation, allowing for tilt adjustment to position the modules at multiple angles.

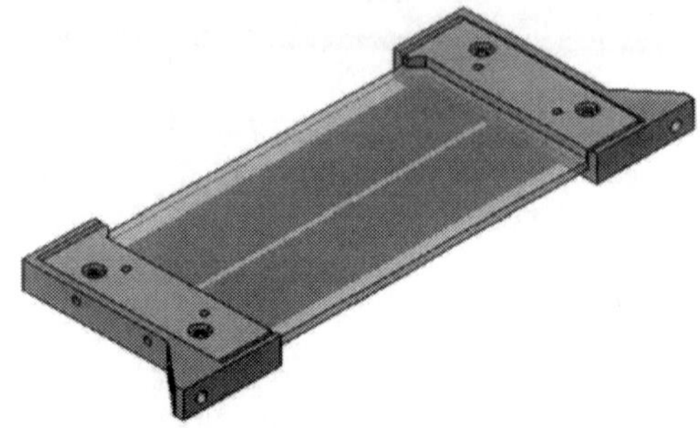

Figure 1: CAD Model of the Miniaturized Bifacial Module and Module Holder

3.2 Measurements

Initial measurements were carried out in laboratory conditions to determine the quality of the final modules and how consistent they performed. These were carried out with a consistent light source directed at each module from a fixed distance. Each module was measured individually, with the short circuit current (I_{SC}) and open circuit voltage (V_{OC}) recorded with which the fill factor was calculated. This was repeated with all modules as stable as possible, from which an assessment could be made as to the efficiency and potential losses.

Once all modules had been evaluated within the laboratory, field measurements could be carried out. For this purpose, short circuit current (I_{SC}), open circuit voltage (V_{OC}), and maximum power point (P_{MPP}) measurements were taken from a single module. This module was in the center of an array of nine modules per the previously described design. These measurements were taken at various angles to provide parameter data for later simulation applications.

Following this, measurements of I_{SC} and V_{OC} were taken every thirty seconds for two hours on two separate days, for a total measurement period of four hours. The I_{SC} and V_{OC} were recorded instantaneously for each moment, i.e., the value that occurred during that second rather than an average. These values were further used for calculating power generated. These period values could then be compared with the simulation results in BifacialSimu.

Instantaneously, the albedo of the surrounding area was also measured and recorded for later application and comparison with the simulation. This was carried out with the use of two pyranometers, one facing upward and another down, which was located at a set height from the ground. The irradiance of each of these is compared and from which an albedo value for the area can be calculated. This measurement took place on the same surface as the Isc measurements, with enough distance between the array and pyranometers to not interfere with each other's readings. The albedo measurements were recorded every six seconds and logged automatically through the LogBoxSE.

3.3 Simulation

The simulation program used to compare the field measurements is an open-source tool from the University of Applied Sciences Cologne called BifacialSimu. Eva-Maria Grommes developed this as an accurate way of proving the advantages of bifacial PV.

The simulation requires the input of a weather file which was possible using weather data from the same period, recorded by the weather station located on the library rooftop. This data was converted into the TMY format to be compatible with the program. Other variables entered included the solar cell characteristics obtained experimentally, as previously described in the measurements section, and from the datasheet of the solar cell manufacturer.

Collating all of this data, it was possible to carry out successful simulations; the irradiance was simulated using View Factors for the front and Ray-Tracing for the rear surface. This could provide remarkably accurate results with the previously described variables. The albedo was simulated according to the averaged values of the measured data as described in the measurements section. Finally, the electrical simulation was based upon the One-Diode Model, which uses separate calculations for the front and rear sides of the module, which are then added to calculate the total energy yield of the system.

4 OBSERVATION AND READINGS

During the production and measurement phases, a few findings were made that will be explained. This starts with the observations and experiences of the production phase. First, the individual cells are broken out of a complete cell. For this purpose, the whole cells are first provided with predetermined breaking points utilizing laser cutting. Despite the predetermined breaking points, clean breaking with a straight edge at the half-cell proves problematic. The whole cell is placed on the table edge to flush with the predetermined breaking point and fixed to the table by applying pressure to the cell with a sufficiently large polystyrene block to obtain a clean breaking edge. The part of the cell that sticks out over the edge of the table and is to be broken into a half-cell is placed between two polystyrene blocks to distribute the force evenly. The protruding part can be bent downwards until the cell breaks cleanly at the predetermined breaking point.

Six half-cells are soldered together in the next step to form a module. When soldering the busbars to the cells for cell connection, it has been shown that the soldering points detach from the intended soldering points when the solder cools down or the cells start to bulge. This can be attributed to the different thermal expansion coefficients and the uneven heating of the cell and the busbars. The busbars contract more than the cell during cooling. If the solder joints hold in some cases, this will cause the cell to bulge when the solder cools to room temperature. This can cause the cell to break during the lamination process. To avoid these stresses after the solder has cooled, the busbars were pre-bent before soldering and only then soldered onto the cell with the bulge facing upwards. As a result, the length of the bus bars between the solder joints is slightly longer than the length of the cell between the solder joints, and thus there is no mechanical tension on the cell after cooling.

The top cover glass repeatedly broke during the subsequent lamination process, with cracks running mainly in the longitudinal direction. The reason for this is assumed to be the rubber mat in the laminator chamber pressing on the modules due to the vacuum. It is assumed that the force is not evenly distributed over the surface of the cover glass. Especially in the edge area, the cover glass is bent downwards, which causes stresses in the glass that can eventually lead to a crack. To distribute the forces evenly over the glass surface, wooden strips were placed around the individual modules before starting the lamination process. These have the same height as the modules in the laminated state. The wooden strips prevent the cover glass from bending at the edges, which results in significantly less glass breakage during the lamination process.

5 DISCUSSION AND RESULTS

The simulation was carried out with the weather data recorded during the actual measurement containing the direct radiation, diffuse radiation, and the albedo, to see how close the measured and simulated power values are. The measured power was recorded in a 30-second interval, whereas only an hourly power can be calculated for the simulation.

For a bifacial module with an area of 12,592 mm^2 per module and Global Horizontal Irradiation (GHI) of 1,510 W/m^2, power values between 1.53 W and 2.29 W could be measured.

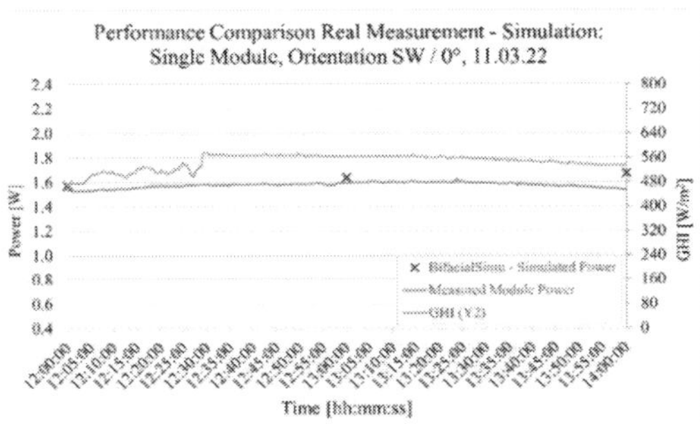

Figure 2: Power measurement and weather data recorded on 11.03.22

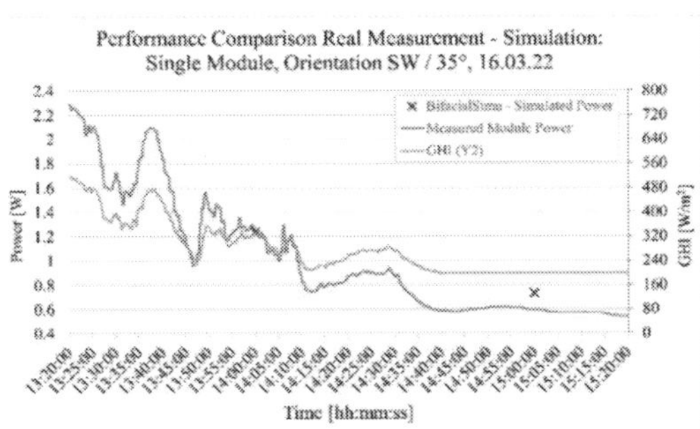

Figure 3: Power measurement and weather data recorded on 16.03.22

As shown in Figure 2, the simulated power deviates from the actual measured values by +2.1 %, +2.7 % and +8.7 %, respectively, when comparing the simulation and actual measurement of the same solar radiation. The discrepancy between measured and simulated performance may be because the cell temperature is assumed constant in the simulation. The highest module output is achieved at a pitch angle of 35°. This is where the most direct radiation hits the more efficient front side of the cells, as the measurement on 16.03.22 in Figure 3 shows. On this day, the sunlight gradually decreased.

Constant measurement uncertainty can be assumed due to the module clamp fastening. Although this was constructed so that it causes as little shadowing as possible, the clamps nevertheless form an opaque frame, which negatively influences the radiation reflected from the ground. This effect cannot be considered in the simulation. The cell connectors used have a width of 3 mm too wide for the 1.6 mm and 1.9 mm wide busbars on the cell, which results in a slight partial shadowing and reduces the short-circuit current of the module. Since the modules were not manufactured by machine but by hand, minor deviations in performance may occur due to manufacturing process steps that are not always carried out consistently. Despite quality control and appropriate sorting of the manufactured modules, deviations between target and actual performance can be assumed. These inaccuracies could also not be determined within the project's time frame and thus could not be considered in the simulation.

6 CONCLUSION

The whole experimental setup was designed and developed in a way to replicate a full-sized PV array. The construction and assembly processes saw many challenges resolved with appropriate solutions. Customized module holders were 3D printed, and the miniaturized bifacial modules were set up on an aluminium frame. The whole setup was prone to sunlight in outdoor conditions on the library rooftop of the University of Applied Sciences Cologne. The albedo effect on the test setup was traced by two pyranometers which showed an average albedo of 0.225. The Isc, and Voc generated by the central module were noted. These values were further used for calculating the power generated for over two hours. These values were compared with those obtained from the BifacialSimu software developed at the University of Applied Sciences Cologne. The comparison showed a very close match to the real-time experimental values, thereby validating the software.

The nine miniaturized bifacial modules fabricated using DMEGC's Mono PERC Bifacial cells showed the same performance characteristics as full-sized modules. The power generated by the real-time module against the results obtained by simulation showed a variation from 2 percent to 7 percent. As a result, the yield of miniaturized bifacial photovoltaic modules matched in close comparison with the yield simulation of bifacial modules.

7 ACKNOWLEDGMENT

First of all, we would like to thank Dipl.-Ing. Rudolf Gecke and Lionel Clasing for their dedicated support with their knowledge and cooperation during the technical preparations and for providing specific equipment like electrical measurement devices and weather data.

We would also like to thank the team of the Maker Space of the University of Applied Sciences Cologne, especially Andreas Käckel, for the opportunity to print our 3D-holders for our module. Also, like to thank Martin Witt and the project group for the aluminium components we used for the array frame. Also, like to thank Sarah Glaubitz for the help and prework for the albedo measurements.

A great thanks go to the company Glasmanufaktur

Brandenburg GmbH with our direct contact Thorsten Schroeter for providing and preparing the PV glasses for the modules.

8 REFERENCES

[1] DMEGC, "DMBD9B166 MONO PERC CELL Datasheet," DM202008, 2021.

[2] Hartmut Nussbaumer, Markus Klenk, Marco Morf, Nicolas Keller, "Energy yield prediction of a bifacial PV system with a miniaturized test array, Solar Energy," vol. 179, pp. 316–325, 2019. [Online]. Available: https://doi.org/10.1016/j.solener.2018.12.042

[3] Energy Experts, "Bifacial solar systems: technology, yield and providers at a glance," energie-experten, 01 Jan., 2022. https://www.energie-experten.org/erneuerbare-energien/solarenergie/solarzelle/bifacial

[4] Hartmut Nussbaumer, Markus Klenk, Marco Morf, Nicolas Keller, "Energy Yield Prediction of a Bifacial PV System with a Miniaturized Test Array," 2019. [Online]. Available: https://digitalcollection.zhaw.ch/bitstream/11475/14510/2/2019_Klenk_Energy_yield_prediction_SolarEnergy_ACCEPTED.pdf

[5] SunWare, Cutting solar cells, 2014. [Online]. Available: https://de.sunware.solar/oem/lasercut

[6] Modulor, "Ethylene Vinyl Acetate (EVA)," 2022. [Online]. Available: https://www.modulor.de/werkstoffbibliothek/kunststoffe/ethylenvinylacetat-eva/#:~:text=Dichte%20und%20Eigenschaften%20richten%20sich,reduziert%20Festigkeit%2C%20Steifigkeit%20und%20Chemikalienbest%C3%A4ndigkeit

[9] Xingshu Sun, Mohammad Ryyan Khan, Chris Deline, Muhammad Ashraful Alam, "Optimization and performance of bifacial solar modules: A global perspective, Applied Energy," vol. 212, pp. 601–1610, 2018. [Online]. Available: https://doi.org/10.1016/j.apenergy.2017.12.041

[10] T. Baumann, M. Klenk, N. Keller, H. Nussbaumer, and F. P. Baumgartner, "Illumination Homogeneity of Bifacial Systems – Outdoor Measurements with Systematically Varied Installation Conditions," 2017, doi: 10.4229/EUPVSEC20172017-5BV.4.28.

[11] S. Modules, Ed., CALCULATING THE ADDITIONAL ENERGY YIELD OF BIFACIAL SOLAR MODULES, 2016.

[12] Grommes, Eva-Maria & Müller-Ost, Julia & Blieske, Ulf, "The impact of albedo measurements on power density calculations of bifacial modules.," 2020. [Online]. Available: https://www.researchgate.net/publication/347521731_The_impact_of_albedo_measurements_on_power_density_calculations_of_bifacial_modules/citation/download

[13] W. Muehleisen et al., "Energy yield measurement of an elevated PV system on a white flat roof and a performance comparison of monofacial and bifacial modules," Renewable Energy, vol. 170, pp. 613–619, 2021, doi: 10.1016/j.renene.2021.02.015.

[14] Hartmut Nussbaumer, Gaby Janssen, Djaber Berrian, Bruno Wittmer, Markus Klenk, Thomas Baumann, Franz Baumgartner, Marco Morf, Antonius Burgers, Joris Libal, André Mermoud, "Accuracy of simulated data for bifacial systems with varying tilt angles and share of diffuse radiation, Solar Energy," vol. 197, pp. 6–21, 2020. [Online]. Available: https://doi.org/10.1016/j.solener.2019.12.071

[15] W. Muehleisen, J. Loeschnig, M. Feichtner, A.R. Burgers, E.E. Bende, S. Zamini, Y. Yerasimou, J. Kosel, C. Hirschl, G.E. Georghiou, "Energy yield measurement of an elevated PV system on a white flat roof and a performance comparison of monofacial and bifacial modules, Renewable Energy," vol. 170, pp. 613–619, 2021. [Online]. Available: https://doi.org/10.1016/j.renene.2021.02.015

PERFORMANCE RATIO AND SOILING RATE CALCULATION OF PV MODULES INSTALLED AT DIFFERENT TILT ANGLES IN UAE DESERT

Jim Joseph John, Kaushal Chapaneri, Shahzada Pamir Aly, Gerhard Mathiak, Vivian Alberts
Research and Development Center, Dubai Electricity and Water Authority,
MBR solar park, Dubai, United Arab Emirates

ABSTRACT: PV modules in desert climate are exposed to environmental stressors cycles such as elevated temperature, UV radiation, sand abrasion and large temperature fluctuations, which can degrade different components of the module and eventually lead to defects. These defects and degradations reduce the power output of the system, as well as the lifetime of PV modules, affecting system reliability. The performance and soiling degradation of 5 different types of PV modules installed in 5 different tilt angles for 5 years (01 July 2017 to 31 July 2022) in desert condition of Dubai have been reported. NREL RdTools has been used to extract the degradation and soiling rate of all the PV modules. The results from the 5×5 heat map matrix of modules show that the highest degradation and soiling is for the modules installed at 0° angle. The average degradation rates in the desert environment are found to be much higher than the industry average reported values of 0.6–0.7%/ year for most commercial PV modules.
Keywords: c-Si, CdTe, Degradation, Durability, Performance, PV Module, Reliability

1 INTRODUCTION

PV systems are expected to be operational for 20-30 years or more. The change in performance of photovoltaic (PV) systems over time has important implications for the economics and value generated by these systems. The system's economic value is influenced by the reduction in energy yield of the PV system, because of its components' aging, degradation, or failures. PV modules installed in the Middle Eastern regions are exposed to relatively harsher weather conditions associated with hot desert conditions. Therefore, improving the risk assessment associated with weather-induced degradation of PV modules will reduce uncertainty in lifetime energy production, which in turn reduces financial risk and improves financing opportunities for projects located in these regions.

The total installed solar capacity in the MENA region is increasing exponentially [1]. By 2024, the expected solar installed capacities could reach up to ~35 GW. With the constant decrease in solar prices and the ambitions of all MENA countries, more projects are expected to be integrated into the grid. However, PV modules in a desert climate are exposed to environmental stressors cycles [2] [3] such as elevated temperature and ultraviolet (UV) radiation, sand abrasion, and large temperature, which can degrade different components of the module and eventually lead to defects such as encapsulant discoloration and delamination, cell cracks, hot spots, finger breakages [4], discoloration of metallization, solder-bond failure, and glass abrasion and breakages. These defects and degradations reduce the power output of the system, as well as the lifetime of PV modules, affecting system reliability.

Many research papers have been published on quantifying degradation rate papers [5] [6] [7] and soiling rates [8]. Notable work has been presented by researchers from NREL using an open-source library RdTools [9]. The degradation rate is defined as the rate at which the PV module/system performance decrease per year, %/year. Most PV module manufacturers give a warranty of 25 years for a maximum of 20% degradation during this time. If a linear degradation is considered, a degradation rate higher than 0.8%/year becomes a problem. Whereas the soiling rate is the daily derate of energy when no cleaning occurs on the PV modules.

In this paper, we use RdTools to estimate the degradation rate and soiling rate of different types of PV module installed at 5 different tilt angles in the desert conditions of United Arab Emirates.

2 DATA ANALYSIS USING RDTOOLS

2.1 Test Setup

A total of 25 PV modules were installed on a multi-tilt structure at the outdoor solar testing facility of the Dubai Electricity and Water Authority (DEWA) R&D center, which is located right next to the planned 5GW Mohammed bin Rashid (MBR) Solar Park in Dubai.

The multi-tilt structure has 5 columns and 5 rows. Where each column has the same module installed at 5 different tilt angles, which are 0°, 15°, 25°, 35°, and 90°. The multi-tilt structure, the different tilt angles, and the module types along with their tags are shown in Figure 1, Figure 2 and Table I. All the PV modules installed in the multi-tilt structure were tested and certified from reputed testing laboratory, before installation in 2017.

Figure 1: PV Module mounting matrix of the test setup that consist of 5 different types of PV modules at 5 different tilt angles

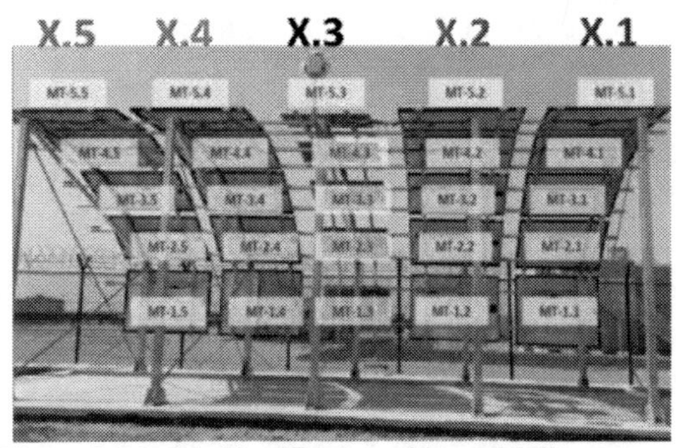

Figure 2: Module identification of the multi-tilt setu

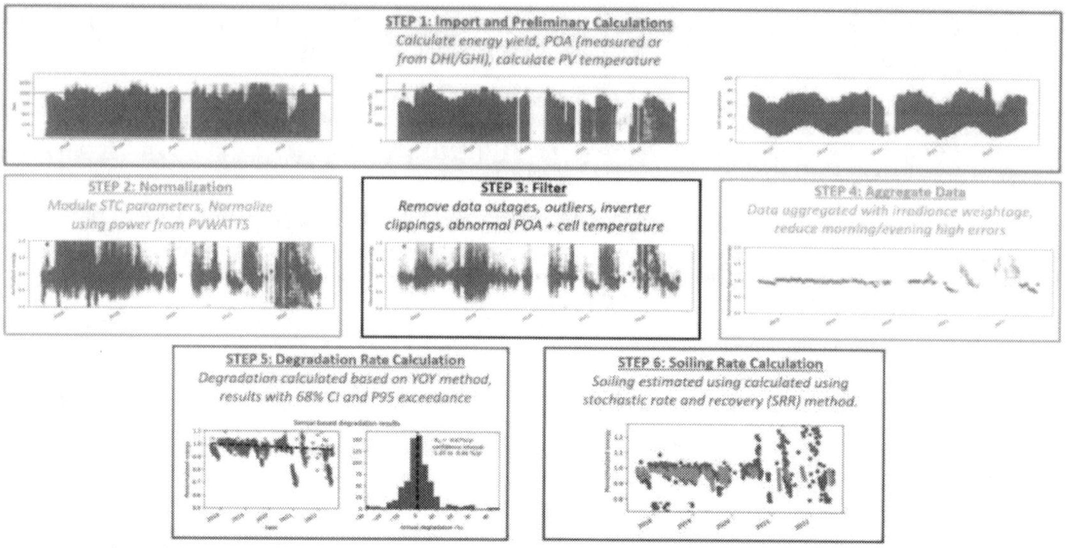

Figure 3. Overview of the NREL/RdTools workflow [8] showing the sample data set used to calculate the degradation and soiling rate

Table I: 5×5 matrix of modules on the multi-tilt structure.

M#	Module Type	90°	35°	25°	15°	0°
X.1	HJT	1.1	2.1	3.1	4.1	5.1
X.2	N-type	1.2	2.2	3.2	4.2	5.2
X.3	CdTe	1.3	2.3	3.3	4.3	5.3
X.4	N-type IBC	1.4	2.4	3.4	4.4	5.4
X.5	P-type PERC	1.5	2.5	3.5	4.5	5.5

2.2 Field Data

The data collection has been ensured to follow the IEC 61724-1:2021 standard instructions. The collected field data, with a frequency of 30 secs, spans for 5 years, starting from 01 July 2017 to 31 July 2022. The measured data consists of the plane of array (POA) irradiances (measured independently for each tilt angle), ambient temperature, and wind speed. The ambient temperature and windspeed is used for calculating the cell temperature. The measured module temperature where not used since the sensors where not properly adhered to the back of the PV modules. Few weeks of missing data in 2020 and 2021 can be seen in Fig 3.

2.3 Data Preparation

Collected field data from the module and weather data was stored in a python data-frame structure as per the RdTools format. Statistical range-based filters were applied on data to remove any outlying data. Pmax limits were set from 0 to 1.3. times rated power of the PV modules. Ambient temperature, windspeed and irradiance filter ranges were set based on historical data. Filtered data was given as input to RdTools for analysis.

2.4 RdTools

RdTools estimates the distribution of the degradation rates using the Year-on-Year (YoY) method which is less sensitive to seasonal effects. Similarly, the soiling rate can be estimated with the stochastic rate and recovery (SRR) method [8].

RdTools can use both the field measured or generated clear-sky data as an input to estimate the performance losses. In this work, we have used the field collected data only. From data collection to performance loss estimates, RdTools can be divided into 6 main steps, simplified in the flow chart shown in **Error! Reference source not found.**. It should be noted that all these steps are sequential.

Step 1 is the import of field data in form of source file data (.csv) and the preliminary calculations, such as cell temperature estimates and the energy yield. Step 2 is the normalization of the PV power using STC parameters and the PVWatts power estimate from Step 1. Step 3 is the data filtering to identify and remove the outliers, such as abnormal POA, and cell temperatures. Step 4 is the data aggregation using irradiance weightages and reducing morning/evening calculation errors. Step 5 is the degradation rate calculations based on the YOY method. The results are reported with a mean value of degradation, along with a 68.2% confidence interval and a P95 exceedance value of this confidence interval. Step 6 is a new addition to the RdTools, which estimates the yearly soiling rates, by renormalizing the energy and the aggregated data from Step 4.

3 RESULTS AND DISCUSSION

The performance losses estimated for all the 25 modules due to degradation and soiling are discussed in the following sections.

3.1 Degradation Rate Calculations

The degradation rate heat map generated for all the 25 modules is shown in Figure 4. The heat map matrix for degradation shows that the highest degradation is seen in the modules installed at 0° tilt, irrespective of the module type. The thin-film CdTe modules seem to be showing the highest degradation, while c-Si technologies are showing a relatively lower average degradation rate (-0.61%/year).

It can be seen from the degrdation heat map in Figure 4 that most of the modules have degradation higher than the usual reported value of approximately 0.6–0.7%/year by most PV module manufacturers. The results indicate that the current module design and technology needs a lot of improvement to be able to withstand the harsh desert weather conditions.

These high degradation rates are financially not viable for the O&M of large PV installations and adversely affect the levelized cost of electricity (LCOE). The X.4 modules (n-type IBC) seem to have the least degradation, though still high compared to the rated 0.5%/year average value for this technology modules, amongst all the tested PV technologies. Further detailed studies need to be carried out to validate this claim and find the exact root causes for higher degradation rate. PV modules at locations 2.1, 2.3 and 1.4 have not been included in the present study due to erroneous data.

3.2 Soiling Rate Calculations

The soiling rate can either be estimated by evaluation of the historical Performance ratio (PR) or by dedicated sensors. In either case, the purpose is evaluating the potential power losses. The daily average soiling ratio can be thus be used to optimize the cleaning frequency of the modules.

The daily average soiling ratio heat map generated for all the 25 modules is shown in Figure . The modules set consisting of 1.X show the expected lower soiling (SR = -0.04%/day) at high tilt angles and high soiling rate (SR = -0.45%/day) at lower tilt angles.

It can be seen from the average soiling rate from heat map that the average soiling rate in Dubai, which is a typical representation for most PV installations in arid areas, is much higher than PV installations in non-arid locations. These daily averages could further be aggregated for different time periods to assess the seasonal variations of the soiling at the given location.

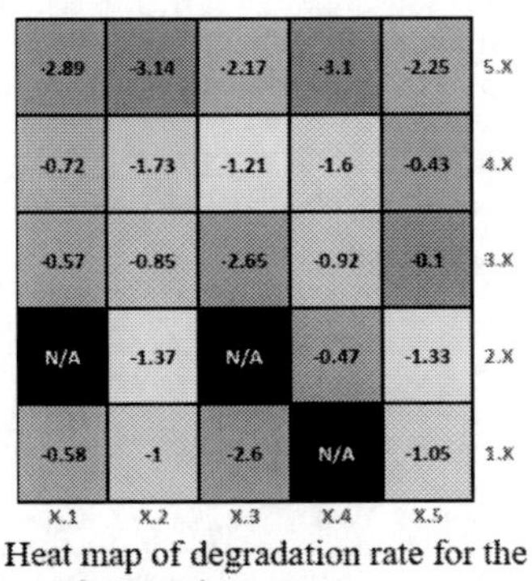

Figure 4: Heat map of degradation rate for the PV modules mounting matrix

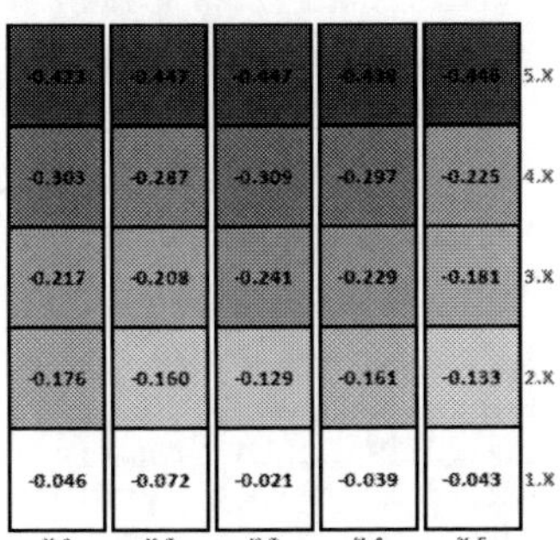

Figure 5: Heat map of average daily soiling rate for the PV matrix for 1 year.

5 CONCLUSIONS

From the statistical data analysis of all the 25 PV modules installed

- Heat maps were generated for the degradation and soiling rates for the entire module mounting matrix.
- The heat map for degradation shows that the highest degradation is seen at the modules installed at 0° tilt, irrespective of the module type.
- The thin-film CdTe modules seem to be showing the highest degradation, while c-Si PV technologies are showing relatively lower degradation rate (-0.61% / year).
- It can be seen from the degrdation heat map in Figure 4 that most of the modules have degradation higher than the usual reported value of approximately 0.5%/year by most PV module manufacturers. The results clearly indicate that the current module design and technology needs improvement to be able to withstand the harsh desert weather conditions.
- The X.4 modules (n-type IBC) and X.1 modules (HJT) seem to have the least degradation, though still high compared to the rated 0.5%/year average value for this technology modules, amongst all the tested PV technologies.
- The modules set consisting of 1.X show the expected lower soiling (SR = -0.04%/day) at high tilt angles and high soiling rate (SR = -0.45%/day) at lower tilt angles.

REFERENCES

[1] Middle East Solar Industry Association, "Solar Outlook Report 2021," MESIA, Dubai, 2021.

[2] S. Kumar, A. Alheloo, H. Hanifi, J. Joseph John and G. Mathiak, "Investigation and analysis of defects and degradations in desert fielded photovoltaic modules,," in IEEE 48th Photovoltaic Specialists Conference (PVSC), 2021.

[3] B. C. A. a. T. H. M. Youness, Z. Kherici, H. Cheghib, N. Kahoul, M. Youness, B. Affari and T. Hazem, "Failure modes of standard photovoltaic modules in Sahara Desert," in IEEE 11th International Renewable Energy Congress (IREC), 2020.

[4] S. Kumar, A. Al Moosawi, B. Adhothu, A. Puthiyapurayil, J. Joseph John, G. Mathiak and V. Alberts, "Identification and analysis of metallization defects in desert operated photovoltaic modules," in IEEE 49th Photovoltaic Specialists Conference (PVSC), 2022.

[5] D. Jordan and S. Kurtz, "Photovoltaic degradation rates—an analytical review," Progress in photovoltaics: Research and Applications, vol. 21, no. 1, pp. 12-29, 2013.

[6] E. Hasselbrink, M. Anderson and Z. Defreitas, "Validation of the PVLife model using 3 million moduleyears of live site data,," in IEEE PVSC, 2013.

[7] J. Singh, J. Belmont and G. TamizhMani, "Degradation analysis of 1900 pv modules in a hot-dry climate: Results after 12 to 18 years of feld exposure," in 39th IEEE PVSC, 2013.

[8] M. Deceglie, L. Micheli and M. Muller, "Quantifying Soiling Loss Directly from PV Yield," IEEE Journal of Photovoltaics, 2018.

[9] M. Deceglie, A. Nag, A. Shinn, G. Kimball, D. Ruth, D. Jordan, J. Yan, K. Anderson and K. Perry, "Github Rdtools version 2.2.0," [Online]. Available: https://github.com/NREL/rdtools.

PV DC YIELD DETERMINED BY DEEP NEURAL NETWORKS: THE CASE OF BUILDING INTEGRATED PV

Eli Shirazi*[1], Ebrar Özkalay[2], Angèle Reinders[1, 3]
1 University of Twente, Enschede, Netherlands,
2 University of Applied Sciences and Arts of Southern Switzerland (SUPSI-PVLab), Mendrisio, Switzerland,
3 Eindhoven University of Technology, Eindhoven, Netherlands

ABSTRACT: PV systems can significantly contribute to achieving the climate neutral goal of the Paris agreement. However, they will present additional challenges to the power grids, due to their intermittent nature. This study aims to model the power output of Building Integrated PV systems (BIPV) on the basis of the application of a Deep Neural Network (DNN) to the following input variables: solar irradiance, module temperature as well as time and location. The DNN has been applied to a dataset containing over four years of data of electrical parameters on module level as well as meteorological data, all at a 5-minutes resolution. The results show that the proposed DNN is able to calculate the PV power output accurately with an R2 score of 0.96 and RMSE of 0.04. Though applied to a BIPV system in this case, the method will be applicable to a myriad of other types of monitored PV systems as well.
Keywords: Building Integrated PV, BIPV, NZEB, Modelling.

1 INTRODUCTION

To achieve a climate neutral world as a goal of the Paris agreement [1] , countries aim to limit greenhouse gas emissions. PV systems can significantly contribute to this aim and realize near zero energy buildings (NZEB). However, they will present additional challenges to the power grids, due to their intermittent nature [2]. This study aims to model PV systems of the category Building Integrated PV systems (BIPV) using machine learning techniques.

Physics-based modelling which covers irradiance, optical, electrical, thermal and systems aspects, is the most applied modelling approach in PV system research. Given all sub-models involved to determine diverse physics or engineering related causes, physics-based modelling of PV systems can become rather complex. For instance, 3D object-shading, system self-shading, complex electric wiring, PV module mismatch and thermal behaviour of PV modules have to be modelled in order to determine the yield of a PV system. With Machine Learning (ML) techniques such as Neural Networks complex 3D physics-based modelling can be avoided saving time and effort when determining power production of PV systems [3]. These approaches are able to model more complex systems in comparison to physics-based approaches [4]without programming exclusively for one specific systems. In this study, we will therefore explore yield assessment of BIPV using a ML approach. In section 2 the proposed method and architecture will be presented, and the results will be discussed in section 3.

2 PROPOSED METHOD

With the development of ML techniques, data driven methods have been extensively used in PV research. One common and robust ML-based approach is Artificial Neural Network (ANN) algorithm, which is capable of mapping many parameters. ANN can be customized through various hyperparameters such as the number of hidden layers, neurons per layer, and the number of training epochs[5]. Each epoch is one cycle where all the training data is used to train the neural network. In shallow architecture, there is only one hidden layer, while in deep architecture, there are more than one layer. Because of the added hidden layer, Deep Neural Networks (DNN) are more robust, and their generalization ability can help to extract nonlinear and more complex patterns in dataset [6]. In this study, a DNN has been applied on a dataset to extract the relation between different inputs and BIPV power.

The hyperparameter of DNN such as number of neurons in each layer, can be optimized though applying various method including empirical one. In this study following equation has been used to optimize the number of neurons in each hidden layer:

$$N_h = \sqrt[N_{hiddenlayers}]{\frac{N_s}{N_i+N_o}} \qquad (1)$$

Where N_h is the number of neurons in each hidden layer, N_s is size of training dataset, N_i is number of inputs and N_o is the number of outputs. The proposed DNN architecture is shown in Figure 1. It consists of an input layer of four nodes, an output layer, two hidden layers of 32 and 64 neurons and epoch size of 200.

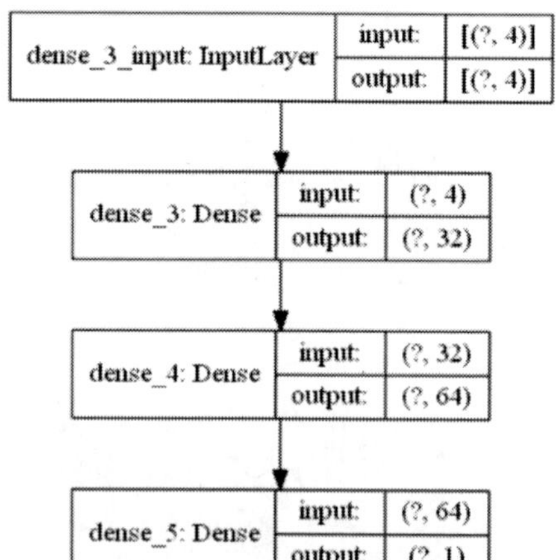

Figure 1. The proposed DNN network architecture

There are various options for the activation function, optimizer, and loss function of DNN as well. In this study Rectified Linear Unit (ReLU) is chosen as the activation function, since it can tackle the problem of vanishing gradient which can prevent DNN from learning effectively. Unlike logistic activation functions, ReLU does not have the saturation problem. Adaptive Moment Estimation (ADAM) is used as the optimizer for DNN, due to its efficiency. ADAM is fast, requires less memory, only needs first-order gradients and fewer parameters for tuning [7]. It is robust to noise and sparsity in data and suitable for large dataset. The pseudocode of the optimizer used in this study is shown in Algorithm I.

Algorithm I. Pseudo code for the optimizer of the proposed DNN

Inputs:
Step size (α)
Exponential decay rates for the moment estimates (β_1, β_2)
Stochastic objective function f_θ (minimize DC yield forecasting error)
Initialization:
Initial parameters vector θ_0 including first m_0 and v_0 second moments vector
Initial timestep t: 0,
While loss function not converged, or the stopping criteria not reached:
Increase timestep t: t+1
Get gradients with respect to stochastic objective at timestep t, $\nabla_\theta f_\theta(\theta_{t-1})$
Update biased first moment estimate, $m_t: \beta_1 . m_{t-1} + (1-\beta_1) . \nabla_\theta f_\theta(\theta_{t-1})$
Update biased second raw moment estimate, $v_t: \beta_2 . v_{t-1} + (1-\beta_2) . (\nabla_\theta f_\theta(\theta_{t-1}))^2$
Compute bias-corrected first moment estimate, $\hat{m}_t: \frac{m_t}{1-\beta_1^t}$
Compute bias-corrected second raw moment estimate, $\hat{v}_t: \frac{v_t}{1-\beta_2^t}$
Update parameters, $\theta_t = \theta_{t-1} - \frac{\alpha . \hat{m}_t}{\sqrt{\hat{v}_t + \epsilon}}$
End while
Return Results θ_t

Finally, Root Mean Squared Error (RMSE) is used as the loss function in this study. RMSE is a criteria to measure the difference between the measured and predicted value.

3 RESULT AND DISCUSSION

The dataset used in this study contains over four years of data (2014-01-29 to 2018-07-10) with five minutes resolution of electrical parameters (V_m, I_m, V_{OC}, I_{SC}) on module level (DC), together with meteorological data (GHI, GPOA and module temperature) have been used. AC power is not included in our study. Both modules were crystalline Silicon (c-Si) with two types of encapsulations: (1) Glass/EVA (ethylene vinyl acetate)/Back sheet and (2) Glass/PVB (polyvinyl butyral)/Glass modules with insulated configuration for no rear side ventilation. There was no shading on PV modules. The modules were installed on the roof of the DACD building at SUPSI's campus in Canobbio, Switzerland, a Cfb Köppen–Geiger climate zone (Figure 2). The latitude and longitude of the location are 46.02°N and 8.91°E, respectively, and the modules were installed at a tilt angle of 6° and a sun azimuth of -4° from the South [8].

Figure 2. Studied modules in SUPSI campus, Switzerland

The dataset has been downloaded from the CKAN data server of the COST Action Pearl PV [9]. The DC parameters under standard test condition (STC) for test modules can be found in TABLE I.

TABLE. I. DC parameters of BIPV modules under STC

Module / Variable	G/PVB/G	G/EVA/BS
Nominal Power [W]	167.09	267.31
Open circuit voltage [V]	40.07	40.99
Short circuit current [A]	5.481	8.701
Voltage of MPP [V]	32.32	32.37
Current of MPP [A]	5.170	8.258
Fill Factor (FF)	76.1 %	74.9 %

3.1 Pre-processing data

The raw dataset consists of more than one million datapoints including nights. It cannot be fed directly into the proposed algorithm and should be changed to an understandable format. To ensure the quality of dataset before applying proposed DNN, the following steps are done. First any timesteps with missing values are removed. Then the outliers are removed using the interquartile range (IQR) which is the difference between first (Q_1) and third (Q_3) quartile of the dataset. Any datapoint outside this range ($Q_1 - 1.5 \times IQR$, $Q_3 + 1.5 \times IQR$) is considered as an outlier. In addition to that if the measured PV power is less than 10% of rated power, it is removed from data set. The last step would be to normalize the datasets, where inputs will be transformed into the range [0,1]. After all the pre-processing the dataset consists of more than 125k data points divided randomly into two sub datasets, training 75% and test 25%. The optimum train and test split could be found through grid search, however, 80%-20%, 75%-25% and 70%-30% are common split percentages used in ML approaches.

3.2 Results

The results of applying proposed the DNN on the SUPSI BIPV dataset are presented in Figure 3 and Figure 4. To evaluate the accuracy of the proposed DNN, two indexes have been considered. R^2 and RMSE. R^2 is the

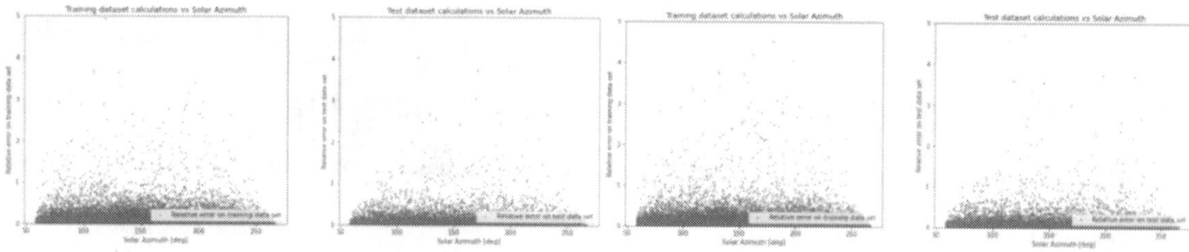

(a) (b) (c) (d)

Figure 3. The relative error of predicted PV power vs solar azimuth on (a) training dataset G/PVB/G, (b) test dataset G/PVB/G, (c) training dataset G/EVA/BS, (d) test dataset G/EVA/BS

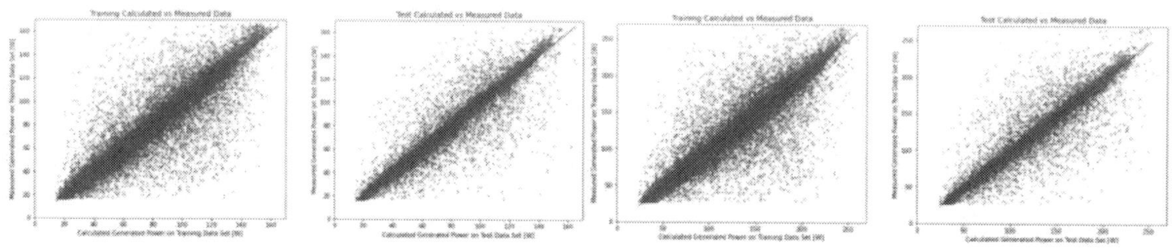

(a) (b) (c) (d)

Figure 4. Calculated vs Measured data on (a) training dataset G/PVB/G, (b) test dataset G/PVB/G, (c) training dataset G/EVA/BS, (d) test dataset G/EVA/BS

proportion of the variance in BIPV DC yield that is predictable from the input variables namely irradiance, temperature, time, and location using the proposed DNN model. The maximum value for R^2 score can be 1 which means that all variations in BIPV yield can be predicted. RMSE is an index to measure the difference between estimated and measured values. The R^2 and RMSE values for assessment of DC yield of two types of BIPV panels are shown in TABLE II.

Having R^2 scores of 0.95 and 0.96 together with the RMSE of 0,04, shows that the proposed DNN can accurately predict the BIPV yield.

TABLE. II. Accuracy indexes for yield assessment of proposed DNN

Index	*G/PVB/G*	*G/EVA/BS*
R^2 score	0.9638	0.959
RMSE	0.0452	0.0485

The RMSEs of each epoch are calculated for both technologies and shown in Figure 5.

As it is already mentioned, BIPV is considered a promising solution to reach NZEB, however there haven't extensive research been done, unlike conventional PV. As the results confirm, the proposed DNN can accurately map different inputs (GHI, GPOA, module temperature and azimuth) to BIPV output. Though applied to a BIPV system in this case, the method will be applicable to a myriad of other types of monitored PV systems as well.

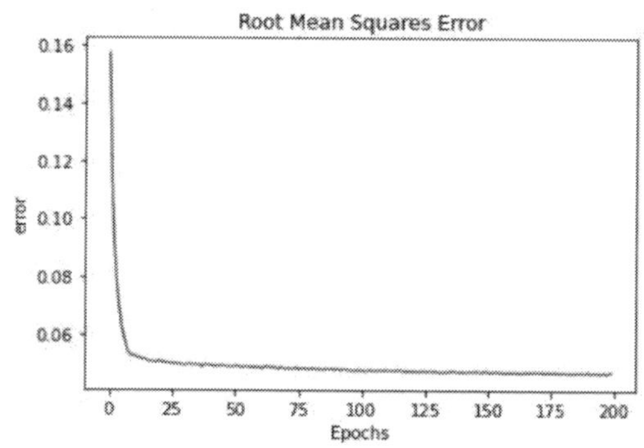

(a)

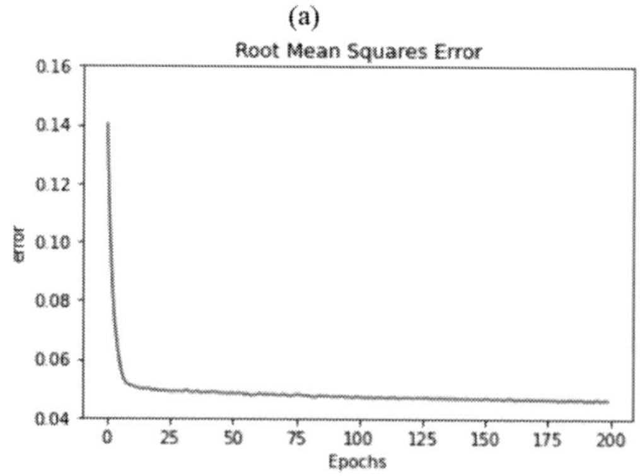

(b)

Figure 5. RMSE of ANN on (a) G/PVB/G, (b) G/EVA/BS

ACKNOWLEDGMENT

This study is based upon work from COST Action

CA16235-Performance and Reliability of Photovoltaic Systems: Evaluations of Large-Scale Monitoring Data – also called PEARL PV, supported by COST (EU Cooperation in Science and Technology): www.cost.eu

The authors would like to thank SUPSI-PVLab and all the staff members who contributed to supplying the relevant data for this work

REFERENCES

[1] U. Nations, "The Paris Agreement | United Nations." [Online]. Available: https://www.un.org/en/climatechange/paris-agreement. [Accessed: 16-Jan-2022].

[2] E. Shirazi, J. Lemmens, M. Gofran Chowdhury, A. Tuomiranta, F. Catthoor, E. Voroshazi, and I. Gordon, "Cloud Detection for PV Power Forecast based on Colour Components of Sky Images," *Conference Record of the IEEE Photovoltaic Specialists Conference*, pp. 2389–2391, Jun. 2021.

[3] G. Hackeling, "Mastering Machine Learning with scikit-learn Apply effective learning algorithms to real-world problems using scikit-learn Mastering Machine Learning with scikit-learn," 2014.

[4] A. Rahman, V. Srikumar, and A. D. Smith, "Predicting electricity consumption for commercial and residential buildings using deep recurrent neural networks," *Appl Energy*, vol. 212, pp. 372–385, Feb. 2018.

[5] N. Halpern-Wight, M. Konstantinou, A. G. Charalambides, and A. Reinders, "Training and Testing of a Single-Layer LSTM Network for Near-Future Solar Forecasting," *Applied Sciences 2020, Vol. 10, Page 5873*, vol. 10, no. 17, p. 5873, Aug. 2020.

[6] J. Gu, Z. Wang, J. Kuen, L. Ma, A. Shahroudy, B. Shuai, T. Liu, X. Wang, L. Wang, G. Wang, J. Cai, and T. Chen, "Recent Advances in Convolutional Neural Networks."

[7] D. P. Kingma and J. Lei Ba, "ADAM: A Method For Stochastic Optimization," in *International Conference for Learning Representations*, 2015, pp. 1–15.

[8] A. Gok, E. Ozkalay, G. Friesen, and F. Frontini, "The Influence of Operating Temperature on the Performance of BIPV Modules," *IEEE J Photovolt*, vol. 10, no. 5, pp. 1371–1378, Sep. 2020.

[9] "SUPSI PVLab Solarbrick Data(2) - Datasets - Pearl PV CKAN Repository." [Online]. Available: https://ckan.pearlpv-cost.eu/dataset/supsi-pvlab-solarbrick-data2. [Accessed: 17-Jan-2022].

PERFORMANCE DEGRADATION BY PID EFFECT FOR DIFFERENT SOLAR MODULES TOPOLOGIES AND PHOTOVOLTAIC CELL TECHNOLOGIES

H. F. Santos[1], H. S. Alvarez[1], R. M. Garcia[1], ***Neves, M. R. M.[2]***

[1]*BYD Energy R&D from Brazil, Campinas, Brazil*
[2] *School of Electrical and Computer Engineering, UNICAMP, Campinas, Brazil*

ABSTRACT: The main objective of this work is to observe the potential induced degradation (PID) effect on four types of single-glass photovoltaic (PV) modules. 1) a M6K monocrystalline PERC full-cell module, 72 cells; 2) a MIK monocrystalline BSF half-cell module, 144 cells; 3) a P6K polycrystalline BSF full-cell module; 4) a PHK polycrystalline BSF half-cell module. All PV modules it was mounted using glass, EVA, solar cells, EVA and PYE SPV L backsheet. Their PID effect was induced by applying a - 1500 V DC power, in a damp heat test chamber, 85 °C, with humidity of 85% for two cycles of 96 h. The modules' electrical parameters were obtained in a PASAN Highlight flash tester AAA+++. At the end of the 192 h of process, the polycrystalline modules P6K and PHK presented a pronounceable PID effect, with the decrease of their relative power (RP) output by 1.3% and 2.1%, respectively, and with the darkening of the cells near their edges. For the monocrystalline modules M6K and MIK, RP only decreased by 1.3% and 0.6 %, with the darkening of their cells happening only for the MIK module.

Keywords: PID degradation, PV modules, degradation PV modules.

1 INTRODUCTION

Photovoltaic solar energy has emerged as an excellent alternative for generating clean energy. This is because the photovoltaic effect does not emit polluting gases - that contribute to global warming - by converting sunlight - a widely available energy source - into electricity. Taking advantage of this environmental appeal, caused by the search for sustainable growth, the participation of the solar matrix has grown 28% between the years 2020 and 2021 [1].

The reduction in the Wp price of photovoltaic devices that occurred in recent years is one of the main factors that leveraged the growth of photovoltaic energy. This fact significantly affected the increase of the solar matrix attractiveness, compared to other energy sources, by having the *Levelized Cost of Energy* (LCOE) as the main benchmark. The LCOE encompasses a series of variables - that have a significant impact on the modules' price - such as the initial system implementation and maintenance costs, and the energy generated by it, the availability and degradation rate of the solar modules used [2].

In the context of photovoltaic plants, potential-induced degradation (PID degradation) is one of the main phenomena that reduce the energy generated. The effects of PID degradation are associated with the electrical potential difference (by the series association of large numbers of panels), that causes leakage currents to flow from the module housing to the solar cells (or vice versa, depending on the module position in a module string). Fig. 1 presents the simplified scheme that shows the potential variation between the modules connected in series [3].

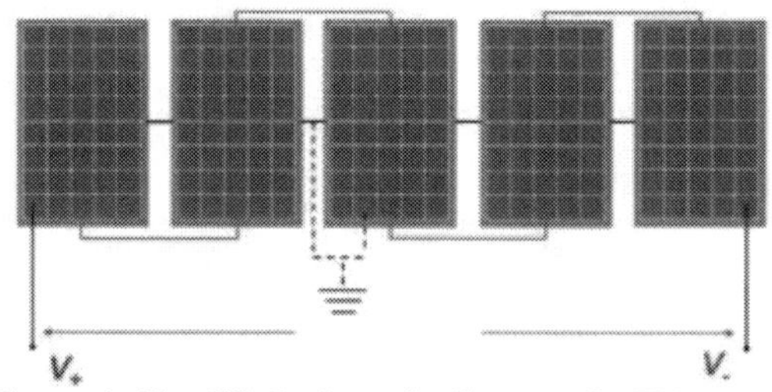

Figure 1: Simplified schematic diagram of a PV system with a floating potential. The modules are serially connected.. (adpated from al. Luo, 2017)

In this context, this work evaluates the performance degradation by PID effect for different topologies of solar modules and photovoltaic cell technologies.

2 METHODOLOGY

2.1 Samples:

The samples used to carry out the tests were provided by BYD. All the cells used in the modules were p-type. All the modules were encapsulated using EVA/ EVA, however, it is not possible to say that both devices use the same weight. The list of topology, technologies, and other characteristics are summarized in Tab. 1.

Table 1: evaluated samples characteristics.

Module	*Cell technology*	*Topology*	*Cell n°*
M6K	MonoPERC	Full	72
MIK	Mono BSF	Half	144
P6K	Poli BSF	Full	72
PHK	Poli BSF	Half	144

2.2 Characterization and testing:

Initially, all samples were exposed to environmental conditions for 120 hours to stabilize the output power with the short-circuited terminals. The electrical parameters were obtained from a solar simulator PASAN HIGHLIGHT SMT AAA+++ of the brand Meyer Burger.

Visual characterization was done using an MPS-EL-AS electroluminescence chamber from Shanghai ASICPY with a current of 3.0 A for the M6K and MIK modules and 7 A for the P6K and PHK modules. Potential-induced degradation tests were performed according to the IEC 61215-2:2016 standard from the MQT 21 test, using a Hototech HTPV-16. The temperature and humidity in the conditioning chamber were 85 °C and 85%. The samples were also negative biased with 1500 V DC.

3 RESULTS:

Looking at the tests performed according to the methodology presented in item 2, Fig.2 shows the results of the electrical characterization performed for the samples over time.

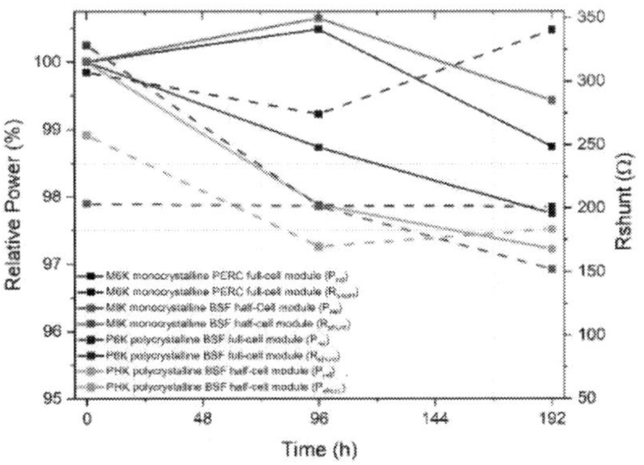

Figure 2: Evolution of electrical parameters throughout the tests.

The initial characterization results (at 0 h) showed a higher shunt resistance value for the full-cell topology samples. This happens due to the non-passivation of the lateral regions of the laser-cut cells - used in the manufacture of the half-cell topology modules - as the cutting process ends up inducing imperfections in the silicon crystals, reducing the resistance between the contacts [4]. From the evolution samples' relative power, in Fig.2, it is possible to observe a higher reduction for the polycrystalline silicon cells samples after 192 h. This phenomenon indicates the susceptibility of the samples with such technology of cells to PID effect, which must also be associated with the higher rate of reduction of shunt resistance after 192 h of testing (evidenced by Table 2). As the sodium ions (Na+) derive through the SiNx anti-reflective coating towards penetrating the crystalline defects and crossing the n+ -p junction [2].

Table 2: Reduction of electrical parameters after 192 h of PID degradation test.

Module	*Relative power loss (%)*	*Relative shunt loss (%)*
M6K	1,3	-10,7
MIK	0,6	25,3
P6K	2,2	38,8
PHK	2,8	28,8

The lowest relative power loss was obtained for the MIK module. However, it was expected that the smaller reduction would be associated with a smaller decrease in shunt resistance, which is the most common type of PID degradation in conventional c-Si modules. There was an increase in the shunt resistance for the M6K module at the end of the 192 h, besides the considerable reduction in its relative power. Therefore, it is possible to infer that the PID degradation route, in this case, was not associated with a reduction in shunt resistance. A very relevant point is that in the case of the M6K, the module is built from MonoPERC silicon cells, and in this case, the resistance fluctuation may be associated with the intrinsic aspects of the cell passivation technology or with its metallization.

Regarding the visual aspects, Figure 3 shows the electroluminescence images of the samples during the degradation test. As observed, one can notice a greater darkening of sample IV (PHK) – concerning the other samples, in the regions close to the frame, expected for the degradation method in a climatic chamber with high temperature and high humidity. The degradation effect at the edges, in this case, may also be associated with the encapsulant resistivity distribution, since the potential variation along the module surface increases as the encapsulation material resistivity decreases.

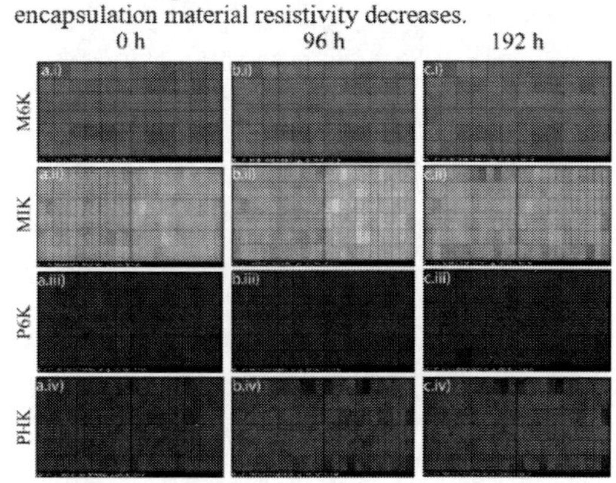

Figure 3: Electroluminescence of samples, before, during and after degradation tests.

Still regarding the images, no evidence of cell darkening was found for the M6K module. As for the MIK module, there was a slight darkening of some cells also located in the edge regions of the device. In the case of P6K, the cells began to darken after 192 h of degradation, showing evidence of the effect of moisture ingress.

4 CONCLUSION

Regarding the tests carried out in this work through the comparison of the constructive topology of photovoltaic modules, as well as different technologies of crystalline silicon photovoltaic cells against the performance qualification as a function of the degradation induced by PID potential, the results showed a greater tendency to degradation of polycrystalline silicon cells, as well as a further reduction in shunt resistance. However, due to the influence of parameters such as encapsulation, moisture vapor transmission rates, BOM, among others, it was concluded that the influence of the modules topology (full-cell x half-cell) must be determined by comparing modules built under the same formulation, which includes the grammage and type of encapsulating material, type of material of the backsheet, in addition to the cell technology, varying just the topology of these devices. In this context, the evaluation of the samples did not allow concluding the influence of the topology of the modules in the face of potential-induced degradation.

5 REFERENCES

[1] GOPI, Ajith et al. Performance modeling of the weather impact on a utility-scale PV power plant in a tropical region. International Journal of Photoenergy, v. 2021, 2021.
[2] VERÍSSIMO, Pedro Henrique Alves et al. Area and LCOE considerations in utility-scale, single-axis tracking PV power plant topology optimization. Solar Energy, v. 211, p. 433-445, 2020.
[3] LUO, Wei et al. Potential-induced degradation in photovoltaic modules: a critical review. Energy & Environmental Science, v. 10, n. 1, p. 43-68, 2017.
[4] SPORLEDER, Kai et al. Root cause analysis on corrosive potential-induced degradation effects at the rear side of bifacial silicon PERC solar cells. Solar Energy Materials and Solar Cells, v. 201, p. 110062, 2019.

6 REGARDINGS

Thanks to the company BYD Energy do Brasil for the continuous investment in research and development. In addition, we reiterate our gratitude to the Ministry of Science, Technology and Innovation of Brazil.

Early and reliable detection of potential-induced degradation in crystalline silicon PV module using electroluminescence imaging

[1]Vishal E. Puranik*, [1]Ravi Kumar and [1]Rajesh Gupta
[1] Department of Energy Science and Engineering, Indian Institute of Technology Bombay, Mumbai, India-400076.
Email ID: vishal.puranik@iitb.ac.in*, ravikr@iitb.ac.in, rajeshgupta@iitb.ac.in

ABSTRACT: Early and reliable detection of PID is essential in a field operation as it poses severe threat to module reliability. Electroluminescence (EL) imaging enables spatial detection of PID based on its characteristic corner pattern, IEC TS 62804-1 and IEC TS 60904-13 have recommended using low and high current EL imaging in PID detection. However, PID detection solely on visual inspection often makes this approach inefficient. In this paper, the method is proposed for early and reliable PID detection using two EL images. The relative difference in cell open-circuit voltage estimated from EL images is used as the PID indicator. The threshold for PID detection is determined equivalent to the pass/fail (5% power loss) criterion used in IEC standards. The proposed method was tested on new and field-aged modules containing multiple defects. Experimental results show that this method effectively detects early phase of PID, which is not possible to detect with the conventional visual inspection approach. Also, reliable detection of PID becomes possible even in the presence of different types of defects and degradation since the used PID index solely indicates shunting defects. Hence, the proposed method would be useful for PID detection in the field as well as in laboratory studies.

Keywords: PV cell, current-voltage characteristic, PID, defect

1 INTRODUCTION

The expected lifetime of a PV module is about 25 to 30 years in a field. Potential-induced degradation (PID) is a severe PV failure mode that rapidly progresses; consequently, increasing a module annual degradation rate causing premature module failure. Shunting gets induced in p-type crystalline silicon modules due to the negative voltage stress experienced by a cell in a field [1]. Shunting defects within a cell decrease cell and module power. PID is detrimental to module lifetime; hence its early detection is essential. Also, precise testing of PID susceptibility of a module before field installations is also essential.

Electroluminescence (EL) imaging is capable of detecting spatial PV defects and it has been widely used for PID detection. Conventionally, high EL current (short circuit current (I_{sc})) is used to detect series resistance related defects and low EL current (10% or 20% of I_{sc}) for shunt related defects detection [2], [3]. However, detection is purely based on visual inspection, which is not sufficient since the early phase of PID defects may not show complete dark appearance [4]. Generally, PID detection is done from low current EL images based on the visual appearance of PID characteristic corner pattern [1]-[3]. Single current-based visual inspection makes PID detection sluggish. Also, low current EL image can not bifurcate pre-existing cell to cell mismatches due to bulk defects or short-circuited cells since they also show pattern similar to PID. Also, characteristic pattern based PID detection is not useful if PID is evenly distributed [4]. Hence, the primary objective of this work is to propose EL method that enables early as well as reliable detection of PID.

In the proposed method, two current (high and low) EL images are used for PID detection. For representing the sole effect of shunting, open circuit voltage (V_{oc}) is used as PID index. The threshold for PID index is defined for effective PID detection.

2 METHOD

PID increases the junction recombination current (reverse saturation current (I_{02}), ideality factor (n_2)) and reduces shunt resistance (R_{sh}) of a PV cell [1] as shown in Fig. 1(a). Fig. 1(b) shows the current-voltage characteristics of healthy and PID-affected cells. PID mainly impacts cell V_{oc} and fill factor as compared with I_{sc}. It may be observed that the difference between V_{oc} of a healthy and PID cell increases at low current since shunting effects are mainly in the low current region.

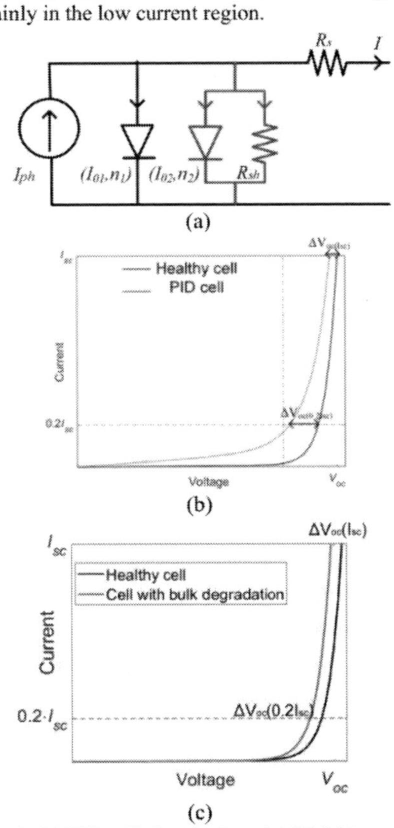

Figure 1: (a) PID cell electrical model (b) I-V curves of healthy and PID cell (c) I-V curve of a healthy and a cell with bulk degradation

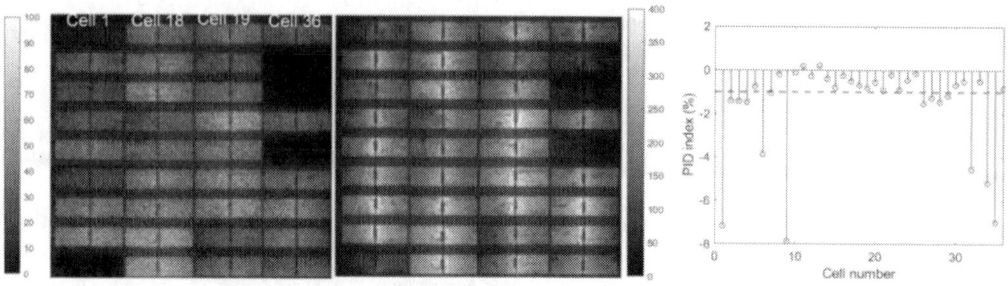

Figure 2: PID detection in 15 W multi c-Si module (a) $0.2{\cdot}I_{sc}$ image (b) I_{sc} image (c) Results

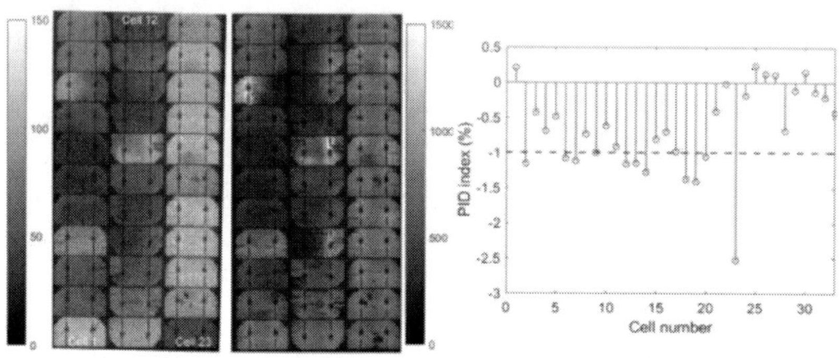

Figure 3: PID detection in field-aged module (a) $0.2{\cdot}I_{sc}$ image (b) I_{sc} image (c) Results

Fig. 1(c) shows the I-V characteristics of the healthy and cell with higher bulk recombination. It may be observed that relative difference between V_{oc} remains almost same irrespective of current. It implies V_{oc} difference at two EL currents can be used for bifurcating PID shunting from other degradation.

In this method, two EL images captured at low ($0.2{\cdot}I_{sc}$) and high EL current (I_{sc}), are used for detecting PID. In EL images cell EL intensity (ϕ) is exponentially related to cell voltage (V) at any EL current, as written in (1) [2].

$$\phi = A \cdot exp(^{V}/_{V_t}) \qquad (1)$$

A- Calibration constant, V_t – thermal voltage of diode
From Fig. 1(b), it is evident that PID progression impacts cell V_{oc}. Hence, fall in cell V_{oc} is taken as the PID indicator as it can be easily estimated from EL images using (1). Open circuit illuminated condition and dark EL condition is equivalent; hence, at I_{sc}, V= V_{oc}, using this condition, module A is determined using Eq. (1) [5]. Healthy or brightest cell within a PID test module is taken as a reference. From I-V curves (Fig. 1(b)), it may be observed that the relative difference between healthy and PID cell voltages (ΔV_{oc}) increases at low EL current, which can be estimated using Eq. (2).

$$\Delta V_{oc} = \frac{log(\frac{\phi_{ref}}{\phi})}{log(\frac{\phi}{A})} \qquad (2)$$

For a detailed understanding of ΔV_{oc}, the simulation study of the PID model (Fig. 1(a)) was conducted, and varying cell parameters corresponding change in cell ΔV_{oc} was observed. It has shown that in the case of (I_{01}, n_1), variation ΔV_{oc} remains almost same at I_{sc} as well as $0.2I_{sc}$. Whereas in case of (I_{02}, n_2, R_{sh}), $\Delta V_{oc(0.2Isc)} > \Delta V_{oc\,(Isc)}$. It implies that ΔV_{oc} estimated at any one current contains the effect of either bulk parameters (I_{01}, n_1), shunt parameters (I_{02}, n_2, R_{sh}), or both. Hence, for reliable detection of PID, the difference between $\Delta V_{oc(Isc)}$ and $\Delta V_{oc(0.2Isc)}$ is used as PID index as it purely indicates the shunting effect.

$$PID\ index = \Delta V_{oc(I_{sc})} - \Delta V_{oc(0.2I_{sc})} \qquad (3)$$

This PID index would also provide the additional advantage of canceling vignetting effect due to subtraction. Hence, EL images without vignetting corrections can also be used directly. For early PID detection, the threshold needs to be defined. Normally, in PID qualification testing 5% power loss is used as the pass/fail criterion [3]. Hence, the magnitude of PID index at which approximately 5% cell power loss results is taken as the threshold for PID detection, which turned out about −1%. Hence, a cell with PID index ≥ −1% qualifies for PID.

3 RESULTS

1. Early detection of PID defects

PID test was conducted on the small size 15 W test module for 96 h. During the test, module was periodically characterized to look for induced PID. Fig. 2 (a)-(b) shows the low and high current EL images of module after 24 h of the damp-heat stress test. Based on the two EL images, cell PID indices were calculated using eq. (2) and (3), and shown in Fig. 2(c). It may be observed that 15 cells are PID affected with PID index > −1%. If the conventional visual PID detection approach is applied, then only 5 cells can be detected based on the PID signature since cells with initial phase PID do not show complete dark appearance, which highlights the effectiveness of the proposed method for detecting initial phase of PID shunting.

2. PID detection in field-aged module

Field-aged module served in the hot and humid climatic conditions of Mumbai for > 15 was investigated using this method. Fig. 3(a)-(b) shows the EL images of the test module, which shows the consistent inhomogeneous pattern in both EL images. The reason behind the inhomogeneous pattern could be the non-uniform degradation and/or presence of shunting. To detect for existence of PID or shunting, the proposed method was applied, and calculated PID indices of the cells are shown in Fig. 3(c). It shows that around 11 cells qualify for PID with PID index $\geq -1\%$. Using the conventional pattern-based method, the presence of shunting in this module cannot be investigated.

The use of threshold enables PID detection in an efficient and reliable way. The result highlights the application of the proposed method for reliable PID detection in field operational modules, which mostly contain different types of simultaneously existing defects and degradations.

4. CONCLUSION

In this paper, the PID detection method was proposed using two EL images. The use of PID index and qualification criterion in the proposed method makes PID detection efficient as compared with the conventional visual inspection approach. Results show that PID in a cell can be detected in its early phase before it gets severe. Also, using subtraction term in the PID index eliminates the effect of other types of mismatches, vignetting, or any other form of offset in EL images. Thus the proposed method ensures the reliable detection of PID in the presence of other defects.

REFERENCES

[1] S. Pingel et al., “Potential Induced Degradation of solar cells and panels,” in 35th IEEE Photovolt. Spec. Conf., Honolulu, HI, 2010, pp. 002817-002822

[2] T. Fuyuki and A. Kitiyanan, “Photographic diagnosis of crystalline silicon solar cells utilizing electroluminescence,” Appl. Phys. A Mater. Sci. Process., vol. 96, no. 1, pp. 189–196, 2009.

[3] Photovoltaic (PV) Modules - Test Methods for the Detection of Potential-Induced Degradation - Part 1: Crystalline Silicon, IEC Standard 62804, 2015.

[4] V. E. Puranik and R. Gupta, “Analysis and insight of electroluminescence imaging in the assessment of potential-induced degradation in crystalline silicon photovoltaic module,” Eng. Fail. Anal., vol. 134, p. 106027, 2022.

[5] V. E. Puranik and R. Gupta, "Novel Quantitative Electroluminescence Method for Detailed Performance Analysis of PID-s Affected Crystalline Silicon PV Module," IEEE Journal of Photovoltaics, vol. 11, no. 6, pp. 1470-1478, 2021.

FULLY BLACK AND RELIABLE PV MODULES
WITH A COST-EFFECTIVE INKJET COATING OF CELL STRINGS

[1]Christian H. Schiller, [1]Stephan Hoffmann, [1]Mike Jahn, [1]Angela De Rose, [1,*]Martin Heinrich

[1] Fraunhofer Institute for Solar Energy Systems ISE,
Freiburg, Germany, Heidenhofstraße 2, 79110 Freiburg i.Br., Germany

*Corresponding author, e-mail to: martin.heinrich@ise.fraunhofer.de

ABSTRACT: The energy transition requires a massive expansion of photovoltaics (PV) on all available areas in the next few years. For integrative photovoltaic applications (XIPV), an aesthetically appealing PV module design for acceptance and a high photoconversion efficiency are desirable. To achieve this goal using common PERC solar cells with five-busbar interconnection, we present an inkjet process that complies with the speed of industrial stringers to blacken the cell connectors after soldering, resulting in an unaffected homogeneous black impression. The optical impression of our manufactured PV modules remains intact after extended (400 temperature cycles (TC), 120 kWh/m^2 ultraviolet radiation (UV), 2000 h damp heat (DH) exposure) and combined (UV-DH, UV-DH+TC) climate chamber tests following IEC standards. The power loss due to printing on cell connectors is 1.4 % compared to PV modules without blackened cell connectors.

Keywords: Inkjet printing, cell strings, integrated PV, black PV modules, ribbons

1 INTRODUCTION

Since 1974 the limits to growth are evident and the discussion on a sustainable future of humanity has emerged [1]. To ensure peaceful human coexistence, we simply must limit the global temperature increase to 1.5 °C. The global economy must be decarbonized. This requires redirecting billions of dollars of global financial flows into sustainable investments (economically efficient, socially just, ecologically sustainable) [2]. As stated specifically in Agenda 2030 [3]: the 7th goal of sustainable development is to ensure access to affordable, reliable, sustainable and modern energy for all [3].

This energy revolution requires a strong expansion of photovoltaics in the next few years [4]. All available areas must be used for this purpose. Photovoltaics shall be integrated in vehicles and on buildings. It may solve land use conflicts in combination with agriculture, water areas and traffic routes [5].

To integrate photovoltaic (PV) modules in façades of buildings or other applications (cars, etc.), it is usually inevitable to achieve an aesthetically pleasing and appealing design for acceptance [6, 7]. In addition to the individual colouring of PV modules as façade elements, the homogeneous appearance for all angles of view and a high photoconversion efficiency are desirable [8]. To achieve this, one approach for coloured PV modules was shown by Bläsi *et al.* whose MorphoColor concept allows the same optical impression from different viewing angles on a PV module [8]. Even for stone façades, a visually appealing proposition was shown to integrate PV modules unobtrusively, however, with some limitations of photoconversion efficiency [9]. Another approach is to use PV modules with the matrix shingle technology that offers a cell interconnection without ribbons, wires or space between cell strings and hence without metallic reflections [10]. However, PV modules with ribbon or wire interconnection are often used as their production technology still prevails [11].

We therefore present an approach that makes the appearance of PV modules for integrative applications containing common PERC solar cells with five-busbar interconnection more homogeneous, while the long-term stability of the module remains unaffected. For this purpose, we develop an inkjet process that complies with the throughput of industrial stringers. We give details about the inkjet process, the optical impression of blackened cell strings in PV modules and how the electrical performance of the PV modules is influenced by our inkjet process.

2 EXPERIMENTAL

2.1 Inkjet Printing for Module Integration

The ink is applied using an industrial inkjet print head based on the continuous inkjet printing (CIJ) process [12]. Therefore, the print head is integrated into a printer platform, which allows movements in all spatial directions (x, y, z, rotation) via computer control. Three-cell strings from industrial monofacial M3 PERC half cells with five cell connectors (0.22 mm × 0.90 mm, Sn60Pb40 solder, industrial infrared soldering on stringer) are used to print on. An industrial inkjet ink based on butanone as solvent is used. The print pattern thus consists of five parallel stripes with a length of 245 mm (three half cells) and a pitch of 31.2 mm (busbar distance). The blackened width is approximately 1.4 mm to cover the vertical edges of the cell connectors and to compensate for the alignment tolerance of the stringer. In order to reduce the complexity of the multidimensional parameter field that affects droplet pitch (modulation frequency of the deflection voltage, jet frequency, printhead-substrate distance, etc.), the printing speed is varied at a fixed modulation frequency of the print head. A sufficient pitch of the droplets on the substrate is adjusted to fully cover the area without pinholes.

2.2 Module Manufacturing

The printed strings are cross-connected. The cross connectors themselves and the cell connectors protruding above the solar cell edges are not blackened in this experiment. Ethylene-vinyl acetate (EVA, 450 µm thickness) without UV cut-off for the front encapsulation and with UV cut-off at 350 nm for the rear encapsulation,

3 mm thick glass and a common black backsheet foil without diffusion barrier layer is used for encapsulation in an industrial laminator.

2.3 Climate Chamber Tests

We perform temperature cycling (TC) and damp-heat (DH) climate chamber tests twice following the IEC standard [13], and ultraviolet (UV) tests following IEC 61730 [14]. That means 400 cycles of TC without current applied, 2000 h exposure to DH, and 120 kWh/m^2 UV radiation. Additionally, we expose similar-built modules to combined tests: UV-DH (120 kWh/m^2) and UV-DH (120 kWh/m^2) followed by 200 thermal cycles, respectively. The IEC tests and the extended tests are performed with two modules containing three-cell-strings each. Two additional modules with inkjet printing and two modules without coating serve as reference. These reference modules are not subjected to ageing tests.

2.4 Module Characterisation

For PV module characterisation, we use three different methods. Scanning the modules is utilised to evaluate the optical impression of the modules initially as well as after passing the accelerated aging tests. We determine the electrical performance of our modules by means of *IV* measurements. Prior to each *IV* measurement, we illuminate the PV modules for 48 h to ensure comparability of the power measurements before and after the ageing processes [15]. Subsequently, we conduct electroluminescence (EL) based upon the determined short circuit currents (I_{SC}) of the modules by *IV* measurements of the modules. Our *IV* measurements show reproducibility within about 0.2 % in I_{SC}, 0.5 % in open circuit voltage (V_{OC}) and 0.7 % in maximum power point (P_{MPP}). The reproducibility for the fill factor (*FF*) measurements is about 1.4 %. Via error propagation, this results in a reproducibility of ± 0.4 % for I_{SC}, ± 1.4 % for P_{MPP}, and ± 2.8 % for *FF* in the relative plot of the *IV* parameters (*cf.* Figs. 4(a) and 6(a)).

3 RESULTS AND DISCUSSION

3.1 Inkjet Printing on Cell Connectors

In this experiment, we determine a sufficiently high droplet density on the surface of cell connectors and busbars using different droplet pitches so that the strings are homogeneously covered in black. The selected pitch of the droplets and the selected printing speed of 80 mm/s with a fixed modulation frequency of the deflection voltage result in a single droplet size of approximately 300 µm and a fully covered area without pinholes (*cf.* Fig. 1(a)). The variation of the printing speed of the print head shows that one cell connector of an M3 half-cell can be printed completely black within about one second in the CIJ process (*cf.* Fig. 1(b)). The inkjet process directly follows the stringing process, which itself thus remains unaffected: No influence on the soldering process and the resulting joints, no further damage of the coating on the cell connectors by additional handling steps within the stringer. In addition, the presented inkjet process blackens the areas of the busbars that are not covered by the cell connectors, because these usually end a few millimetres before the solar cell edge (*cf.* Fig. 1(b)).

We determined the maximum printing speed of our experimental setup to promote the integration of our CIJ process into industrial stringers (soldering process < 2 s per solar cell [16]). As a result, fully covered cell connectors without the appearance of pin holes or other printing defects on M3 cell level have been verified for a printing velocity up to 500 mm/s, limited by the axis system of the used printer platform. The specification of the used CIJ print head indicates a maximum print speed of 10 m/s. This is significantly higher than the typical process speeds of industrial stringers (up to about 6000 solar cells per hour [16]) and therefore does not represent a fundamental limitation.

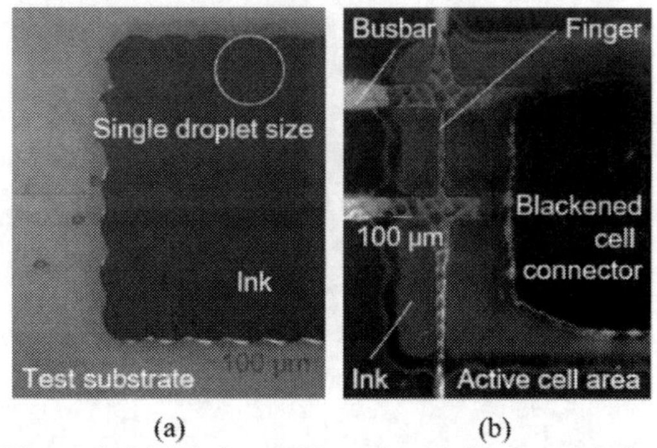

Figure 1: (a) Printing at 80 mm/s with fixed modulation frequency of the deflection voltage on a test substrate and (b) on an interconnected solar cell. The single droplet size is approximately 300 µm. The selected pitch of the droplets results in a fully covered cell connector without pinholes.

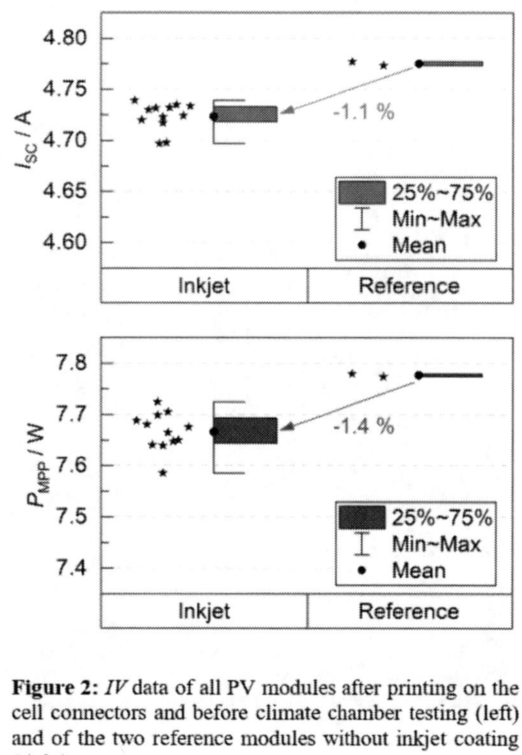

Figure 2: *IV* data of all PV modules after printing on the cell connectors and before climate chamber testing (left) and of the two reference modules without inkjet coating (right).

The initial *IV* data of all PV modules show the influence of the inkjet coating on I_{SC} and P_{MPP}. After printing and module integration of the solar cells, the I_{SC} decreases by 1.1 % ± 0.4 % and the maximum power drops by

1.4 % ± 1.4 %, respectively (*cf.* Fig. 2). The ink-covered surfaces as well as the edges of the cell connectors reduce or even prevent reflection of incident light onto the solar cell surface (*cf.* Fig. 3). We conclude that the selected ink has high opacity; attention must be paid to the print width when printing on the cell connectors to keep I_{SC} losses low but still cover the full connector because of alignment.

3.2 Optical Impression After Aging Tests

Figure 3 shows scans of PV modules in their initial state: (a) before TC test, (b) before UV test, (c) before DH test. The alignment tolerance of the stringer in combination with a narrow print width leads to partially visible busbars (*cf.* Fig. 3(a), (b)) and cell connectors (*cf.* Fig. 3(c)). Nevertheless, these images reveal the major

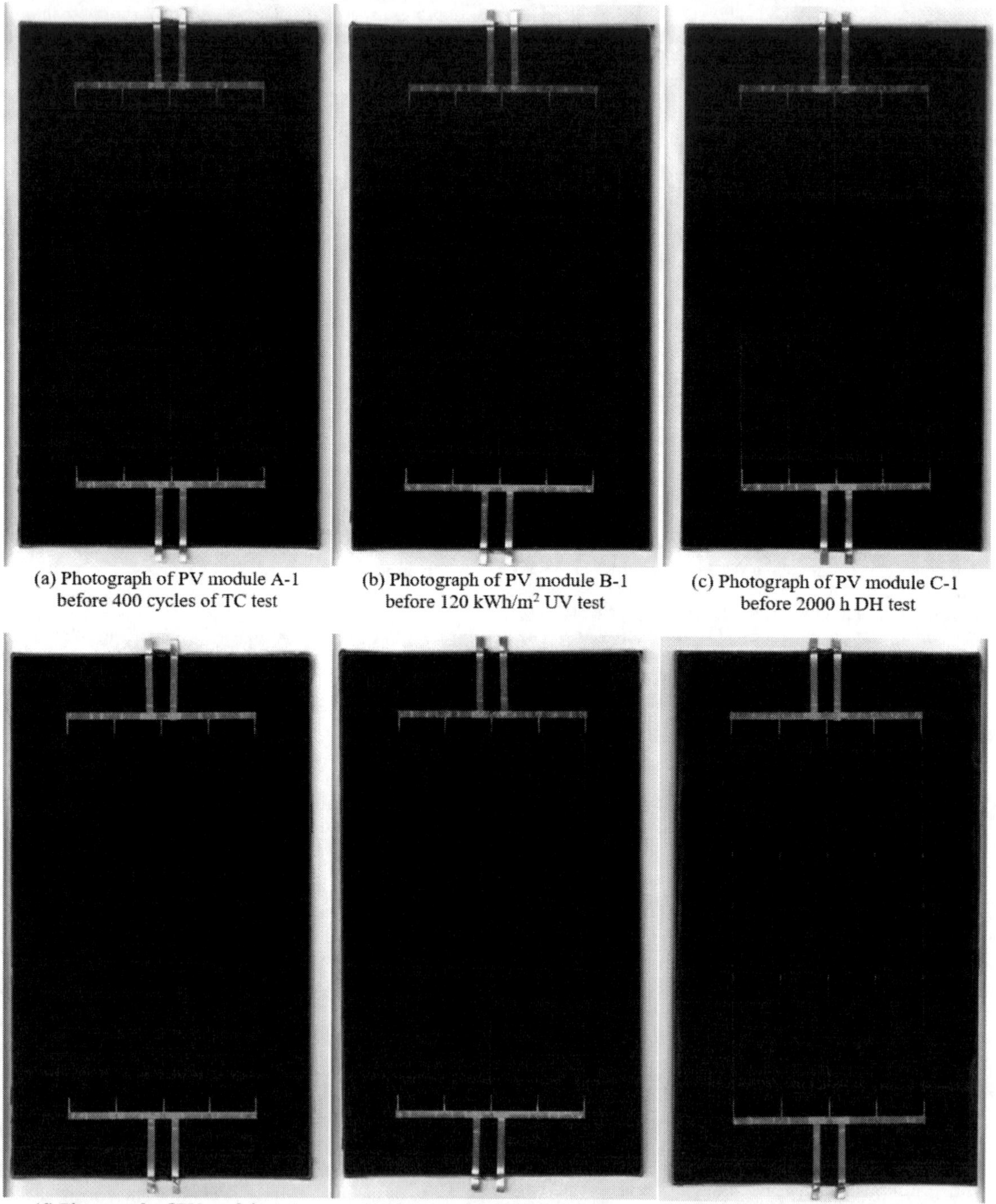

(a) Photograph of PV module A-1 before 400 cycles of TC test

(b) Photograph of PV module B-1 before 120 kWh/m² UV test

(c) Photograph of PV module C-1 before 2000 h DH test

(d) Photograph of PV module A-1 after 400 cycles of TC test

(e) Photograph of PV module B-1 after 120 kWh/m² UV test

(f) Photograph of PV module C-1 after 2000 h DH test

Figure 3: Three exemplary PV modules with inkjet-coated cell connectors in their initial state (a) to (c) and after the double climate chamber test runs according to the IEC standard (d) to (f).

advantage of the CIJ printing process: the pleasing homogeneous black appearance of the cell matrix in PV modules. Fig. 3(d) to (f) shows the same modules after ageing: (d) after 400 cycles of TC, (e) after 120 kWh/m^2 UV, (f) after 2000 h DH. Judging from the visual impression, which is of major importance for XIPV PV modules, a double TC test with 400 thermal cycles seems to have no effect on the stability of the inkjet-printed black coating. The results indicate that the ink coating has good thermomechanical properties; no cracking is visible. After the double UV test, there is also no visible deterioration of the black homogeneous impression. After 2000 h DH test, there is a minimal deterioration of the initial imperfect printing (*cf.* Fig. 3(f)). Especially in the cell gaps, the homogeneous coverage of the cell connectors seems to be reduced to some extent. This deterioration could be related to an interaction with acetic acid formed by the EVA during DH test [17, 18]. A different encapsulation material, which does not form acetic acid under humidity exposure, such as *e. g.* polyolefin (POE) could overcome this issue.

3.3. Module Characteristics After Two Times IEC Tests

Figure 4(a) shows the *IV* data of the PV modules that underwent the TC, UV and DH test procedures twice; each dot represents a tested module. The data are presented in relative form and refer to the initial values, which correspond to unity in the diagrams. Within the measurement uncertainty of the sun simulator, the I_{SC} almost remains constant. TC and DH test procedures seem to have an impact on the P_{MPP} of the PV modules and a minor impact on their *FF*.

However, the inkjet coating does not lower the PV module power after climate chamber tests. But rather certain defects that also occur in PV modules without printed cell connectors, as the EL images show (*cf.* Figs. 4(b) to (d)):

After 400 thermal cycles, finger failures occur because the cell connectors were not optimally aligned over the busbars in the stringing process, but were slightly soldered onto the fingers. As a result, the fine finger metallization tears off (*cf.* Fig. 4(b)). This defect can also be seen in PV modules without printed cell connectors [16]. We conclude that this is an issue of the busbar alignment to the cell but the coating has no influence on this. After UV radiation with 120 kWh/m^2 a dark area in the upper left corner of the top cell is visible. This is a typical error pattern for a faulty cross-connection [19]. After 2000 h

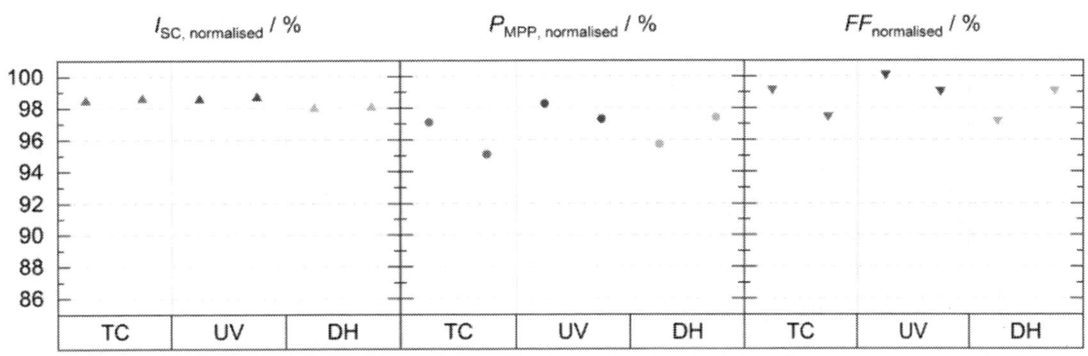

(a) *IV* data of PV modules after climate chamber tests twice following IEC standards. The measurement reproducibility is ± 0.4 % for I_{SC}, ± 1.4 % for P_{MPP}, and ± 2.8 % for *FF*.

(b) EL image of PV module A-2 after 400 cycles of TC test

(c) EL image of PV module B-1 after 120 kWh/m^2 UV test

(d) EL image of PV module C-1 after 2000 h DH test

Figure 4: *IV* data of PV modules after climate chamber tests (a) and EL images of three exemplary PV modules with inkjet-coated cell connectors in their state after the double climate chamber test runs according to the IEC standards (b) to (d).

DH, the visual impression suffers due to coating degradation (*cf.* section 3.2). But cell defects are responsible for the power loss seen in the *IV* data: the inactive outer cell area of the middle cell, and less in the upper cell, is a typical defect of DH (*cf.* Fig. 4(d)) [18].

3.4 PV Module Performance After Combined Tests

The PV modules that were subjected to the combination of test procedures (UV-DH and UV-DH followed by TC) are shown in Figure 5. Judging from the visual impression, the combined test procedures have notable but little impact on the black appearance of the PV modules. As discussed in section 3.3, we ascribe the minimal deterioration of the homogeneous coverage of the cell connectors in the cell gaps to the influence of damp heat (*cf.* section 3.2). The degradation caused by the combined tests is more moderate. The visible effects are less severe than in the individual tests (*cf.* Fig: 6(b), (c)).

The *IV* data along with the EL images of these modules are displayed in Figure 6. The combined climate chamber tests neither significantly degrade the I_{SC}, P_{MPP} nor the *FF* (*cf.* Fig. 6(a)) and the typical failure modes as described in section 3.4 are visible as a combination in the EL images (*cf.* Fig. 6(b) and (c)). This is particularly clear in Fig. 6(c), where DH and TC induced defects appear together: tear off of finger metallization, faulty cross-connection, and inactive outer cell area.

(a) Photograph of PV module D-1 after the combined UV-DH test (120 kWh/m²).

(b) Photograph of PV module E-1 after the combined test UV-DH (120 kWh/m²) followed by TC 200.

Figure 5: Two exemplary PV modules with inkjet-coated cell connectors after the combination of test procedures UV-DH (a) and UV-DH followed by TC (b).

4 CONCLUSIONS

To reach the 7th goal of the sustainable development goals, a strong expansion of photovoltaics is inevitable. To use all available areas, integrated PV is a key technology. An appealing design of PV modules promotes acceptance.

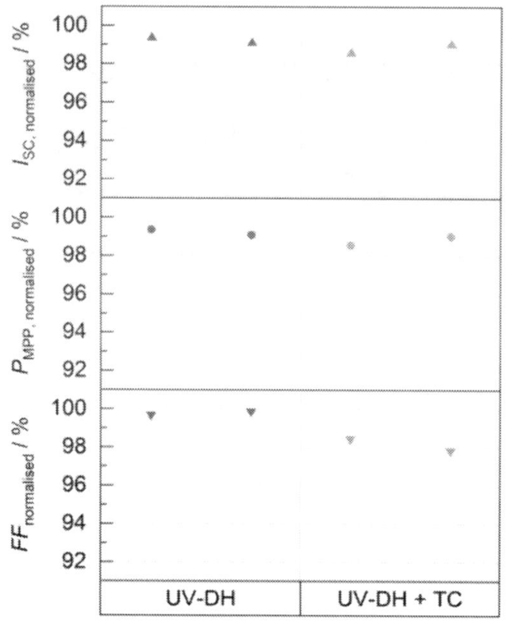

(a) *IV* data of PV modules with inkjet-coated cell connectors after the combination of test procedures UV-DH and UV-DH followed by TC and of the four reference modules. The reproducibility is ± 0.4 % for I_{SC}, ± 1.4 % for P_{MPP}, and ± 2.8 % for *FF*.

(b) EL image of PV module D-1 after the combined test UV-DH (120 kWh/m²).

(c) EL image of PV module E-1 after the combined test UV-DH (120 kWh/m²) followed by TC 200.

Figure 6: *IV* data of PV modules after the combination of test procedures (a) and EL images of the two exemplary PV modules with inkjet-coated cell connectors (b) and (c).

Our CIJ process to blacken cell connectors for homogeneous black appearing PV modules is a promising approach for XIPV. The presented inkjet process results in homogeneous black PV modules while maintaining a high photoconversion efficiency and thus promotes the acceptance of integrated PV by a fast and easy coating process. This uniform optical impression remains constant after 400 thermal cycles, after 120 kWh/m^2 UV radiation exposure and is only slightly affected by 2000 h of DH and the combined climate chamber tests (*cf.* Fig. 7).

Figure 7: Summing up flow chart of the investigated PV modules with inkjet coated cell connectors.

The relative losses in I_{SC} and P_{MPP} of the PV modules due to blackening the cell connectors are about 1.1 % and 1.4 %, respectively. The used industrial inkjet ink is stable after 400 cycles of TC and after 120 kWh/m^2 of UV radiation exposure. The homogeneous coverage of the cell connectors only shows a minor deterioration after 2000 h of DH and the combined tests where DH was applied. The visual deterioration was observed in the cell gaps and may be related to an interaction with acetic acid formed by the EVA during DH test. Another encapsulation material which does not form acetic acid may overcome this issue.

We observed degradation during ageing tests, however the observed degradation mechanisms could all be explained by expected defect patterns that also occur in modules without inkjet coating of the cell connectors. Therefore, we conclude that these are responsible for the lower performance of the PV modules after climate chamber tests.

The continuous inkjet process, presented for blackening the cell connectors on solar cells directly follows the string production, which itself remains unaffected: No influence on the soldering process and the resulting joints, no further damage of the coating on the cell connectors by additional handling steps within the stringer. The specification of the used CIJ print head indicates a maximum print speed of 10 m/s. This is significantly higher than the typical process speeds of industrial stringers. With this and the parallelization of multiple print heads, the CIJ process can be easily adapted to the throughput of PV module production lines. The print width variability also allows an adaption for cells with multiple busbars or wires.

However, a detailed analysis of costs for integrating the CIJ process into existing PV module production lines is still pending. For full-size black PV modules, the cross connectors have to be blackened as well. Depending on the module topology, i.e., the number and positions of the cross-connectors, this presents yet another challenge.

5 ACKNOWLEDGEMENTS

This work was funded by the German Federal Ministry for Economic Affairs and Climate Action under the contract numbers 03EE1049A (PV-Hide).

6 REFERENCES

[1] D. H. Meadows, D. L. Meadows, J. Randers, and W. W. Behrens, *The limits to growth: A report for the Club of Rome's Project on the Predicament of Mankind,* 2nd ed. New York: Universe Books, 1974.

[2] *Paris Agreement*, 2016. [Online]. Available: https://treaties.un.org/Pages/ViewDetails.aspx?src=IND&mtdsg_no=XXVII-7-d&chapter=27&clang=_en

[3] Untied Nations, "Transforming our world: the 2030 Agenda for Sustainable Development: Resolution adopted by the General Assembly on 25 September 2015. Seventieth session, Agenda items 15 and 116," A/RES/70/1, Oct. 2015. [Online]. Available: https://www.refworld.org/docid/57b6e3e44.html

[4] J. Brandes, M. Haun, D. Wrede, P. Jürgens, C. Kost, and H.-M. Henning, "Wege zu einem klimaneutralen Energiesystem: Die deutsche Energiewende im Kontext gesellschaftlicher Verhaltensweisen," Update November 2021: Klimaneutralität 2045, Fraunhofer-Institut für Solare Energiesysteme ISE, Freiburg im Breisgau, Nov. 2021. Accessed: Jun. 24 2022. [Online]. Available: https://www.ise.fraunhofer.de/content/dam/ise/de/documents/publications/studies/Fraunhofer-ISE-Studie-Wege-zu-einem-klimaneutralen-Energiesystem-Update-Klimaneutralitaet-2045.pdf

[5] H. Wirth *et al.,* "Integrierte Photovoltaik: Aktive Flächen für die Energiewende," FVEE Themen 2019 Aktuelle Forschungsfragen - Integrierte Photovoltaik, 2020. [Online]. Available: https://www.google.com/url?sa=t&rct=j&q=&esrc=s&source=web&cd=&ved=2ahUKEwj5nJSWruT5AhUF-qQKHckOB7sQFnoECBAQAQ&url=https%3A%2F%2Fwww.fvee.de%2Fwp-content%2Fuploads%2F2022%2F01%2Fth2019_03_01.pdf&usg=AOvVaw0tJA04bC7GUCGu-6dUINhb

[6] C. Kutter *et al.,* "Decorated Building-Integrated Photovoltaic Modules: Power Loss, Color Appearance and Cost Analysis," 2018, doi: 10.4229/35THEUPVSEC20182018-6AO.8.6.

[7] T. E. Kuhn, C. Erban, M. Heinrich, J. Eisenlohr, F. Ensslen, and D. H. Neuhaus, "Review of Technological Design Options for Building Integrated Photovoltaics (BIPV)," *accepted for publication in the Energy and Buildings special issue review articles from the editors*, 2020.

[8] B. Bläsi, T. Kroyer, T. Kuhn, and O. Höhn, "The MorphoColor Concept for Colored Photovoltaic Modules," *IEEE Journal of Photovoltaics*, vol. 11, no. 5, pp. 1305–1311, 2021, doi: 10.1109/JPHOTOV.2021.3090158.

[9] A. Morlier *et al.,* "Photovoltaic Modules with the Look and Feel of a Stone Façade for Building Integration," *Sol. RRL*, vol. 6, no. 5, p. 2100356, 2022, doi: 10.1002/solr.202100356.

[10] N. Klasen, D. Weißer, T. Geipel, D. H. Neuhaus, and A. Kraft, "Performance of Shingled Solar Modules Under Partial Shading," *Progress in Photovoltaics: Research and Applications*, 2021, doi: 10.1002/PIP.3486.

[11] ITRPV, "International Technology Roadmap for Photovoltaic (ITRPV): 13th edition, 2021 Results," 2022. [Online]. Available: https://www.vdma.org/international-technology-roadmap-photovoltaic

[12] G. Cummins and M. P. Desmulliez, "Inkjet printing of conductive materials: a review," *Circuit World*, vol. 38, no. 4, pp. 193–213, 2012, doi: 10.1108/03056121211280413.

[13] *DIN EN IEC 61215-2 Terrestrial photovoltaic (PV) modules – Design qualification and type approval – Part 2: Test procedures*.

[14] *Photovoltaic (PV) module safety qualification – Part 2: Requirements for testing*, IEC 61730-2:2016, International Electrotechnical Commission (IEC), Geneva, Switzerland, 2016.

[15] A. A. Brand *et al.,* "Ultrafast in-line capable regeneratrion process for preventing light induced degradation of boron-doped p-type Cz-silicon PERC solar cells," in *33rd EU PVSEC*, Amsterdam, The Netherlands, 2017. Accessed: Jul. 4 2019. [Online]. Available: https://www.ise.fraunhofer.de/content/dam/ise/de/documents/publications/conference-paper/33-eupvsec-2017/Brand_2CO95.pdf

[16] A. De Rose, "Evaluation of solder joints on aliminum surfaces for the interconnection of silicon solar cells," Dissertation, Universität Rostock, Rostock, 2021.

[17] I. Duerr, C. Peike, S. Hoffmann, M. Koehl, and K.-A. Weiß, "X-ray study on the damp-heat induced cell degradation (DHID)," in *Proceedings of the 28th European Photovoltaic Solar Energy Conference and Exhibition*, Paris, France, 2013, pp. 3300–3302.

[18] C. Peike *et al.,* "Origin of damp-heat induced cell degradation," *Sol Energ Mat Sol C*, vol. 2013, pp. 49–54, 2013.

[19] M. Köntges, S. Kurtz, C. Packard, U. Jahn, K. A. Berger, and K. Kato, *Performance and reliability of photovoltaic systems: Subtask 3.2: Review of failures of photovoltaic modules : IEA PVPS task 13 : external final report IEA-PVPS.* [Sankt Ursen]: International Energy Agency, Photovoltaic Power Systems Programme, 2014.

APPLICATION OF SHJ AND TOPCON SHINGLE CELLS IN FULL FORMAT AND INTEGRATED MODULES

V. Nikitina[1], D. Reinwand[1], T. Fellmeth[1], S. Neven du Mont[1], D. v. Kutzleben[1], T. Roessler[1], D. H. Neuhaus[1]
[1]Fraunhofer Institute for Solar Energy Systems (ISE)
Heidenhofstr. 2, 79110 Freiburg, Germany

ABSTRACT: Shingle technology is attractive due to the absence of ribbons or wires, large active module area, module design flexibility and aesthetically appealing appearance [1]. This work is looking into the application of silicon heterojunction (SHJ) and tunnel oxide passivated contact (TOPCon) shingle cells in small-scale and full-format modules as well as into the effects of overlap minimization and electrically conductive adhesive (ECA) reduction. Stringer tests with subsequent microscopic analysis showed that precision of the laser cutting process is crucial for the correct alignment of metallization and ECA during the interconnection process. Shunting as well as ECA smearing can be caused by the inconsistent distance of the metallization to the cell edge. The variation of the ECA, printing pattern and encapsulation material in small-scale modules demonstrated no difference in performance after production. However, after accelerated thermal cycling (aTC) 50 the modules with SHJ cells showed $\Delta P_{MPP} = -8.2\%_{rel}$ to $+0.4\%_{rel}$, and TOPCon between $-4.5\%_{rel}$ and $+0.6\%_{rel}$ depending on the used encapsulation material. No difference in performance for the modules with reduced ECA amount in comparison with the reference was detected after aTC50. Full-size solar modules with bifacial TOPCon shingles (1/6 G1 format, 20.7% – 21.3% efficiency after the scribe and break laser process) were manufactured with 50% ECA reduction and the optimal materials based on the previous investigation. Peak power values of up to 392.8 Wp and Eta of 19.84% were demonstrated.

Keywords: PV Modules, Shingling, Heterojunction, TOPCon, Electrically Conductive Adhesive (ECA)

1 INTRODUCTION

Using shingle technology enables creating solar modules of flexible shapes and sizes, which is crucial for the field of integrated photovoltaics. Residential and solar park applications also profit from shingling technology due to the homogeneous appearance of the modules and their large active area [2]. Improved shading resilience of the shingle technology compared to ribbon interconnection is advantageous in all usage scenarios [3]. By reducing the amount of the silver containing electrically conductive adhesive (ECA) in a shingle module, the total cost can be significantly reduced. Carriere et al. demonstrated neither influence on cell-to-module (CTM) loss nor reliability when reducing the ECA amount with silicon heterojunction (SHJ) shingle cells [4]. In this work we investigate the interconnection of SHJ and tunnel oxide passivated contact (TOPCon) cells by shingle technology, targeting full-size and vehicle integrated photovoltaic (VIPV) module production. We aim to cut down the ECA amount as well as maximize the silicon usage by reducing the shingle overlap.

2 METHODS

2.1 Approach

In order to address the issues of shingle overlap minimization, ECA reduction and full-format module integration, a range of experiments was conducted. We investigated the effect of the overlap on the interconnection quality and factors affecting shingle placement precision on a teamtechnik TT1600ECA stringer. The potential to reduce ECA amount as well as different ECAs and encapsulation materials were examined with small-scale modules. The samples were subjected to accelerated thermocycling (aTC) tests [5]. After the most promising material combination had been determined, full-format TOPCon shingle modules were produced. Additionally, a car hood module concept was developed and realized. Evaluation of the string and module performance was done by electroluminescence (EL) and I-V measurements. Metallographic analysis with digital microscopy was used for the shingle overlap investigations. Determination of the series resistance was done according to the procedure described in IEC60891 [6]. Furthermore, Magnetic Field Imaging (MFI) was utilized to look into the terminal connector defects [7].

2.2 Microscopy

A TT1600ECA stringer upgraded for automatic shingle interconnection allows varying overlap in the range of 0.5 mm to 5 mm. Samples for the investigation of overlap reduction were produced on the stringer with the set value of 0.8 mm. Experiments were performed with SHJ shingles, for which the screen design matching with the metallization pads was used. Both front and rear side metallization included rectangular pads every second finger (fig. 1). ECA was applied by screen printing in such a way that every finger-busbar intersection is covered with an ECA rectangle equal to the metallization pad size (0.4 mm × 0.7 mm with 1.4 mm distance, 3.8 mg (ECA-II)). In addition to planar microscopy of the ECA print over the metallization, cross-sections of the joints were prepared according to the procedure described by Eberlein et al [8].

2.3 Fabrication of small-scale modules

Analysis of the small-scale modules was done to obtain the optimal material combination for further full-size module production. Small-scale modules were produced according to the variation in table I with SHJ and TOPCon shingles, respectively. Both shingle types were in 1/6 G1 format (158.75 mm × 26.46 mm). For the TOPCon cells, the metallization of the industrial bifacial precursors was realized at Fraunhofer ISE, using Ag paste for busbars, AgAl paste for front side fingers and Ag paste for the back side fingers. The performance of the shingles after the laser scribe and break process was 20.7% – 21.3% Eta and 0.86 – 0.88 W, measured with pins on the busbars on each side, respectively. For the bifacial

SHJ shingles a low-temperature Ag-based metallization paste with a special print design with reduced paste consumption was used (fig. 1). The cells were cut with laser scribe and break process as well and had efficiency range of 18.4% – 19.0% and P_{MPP} range of 0.78 W – 0.80 W, measured with universal contact probes (UCP) after cutting, respectively.

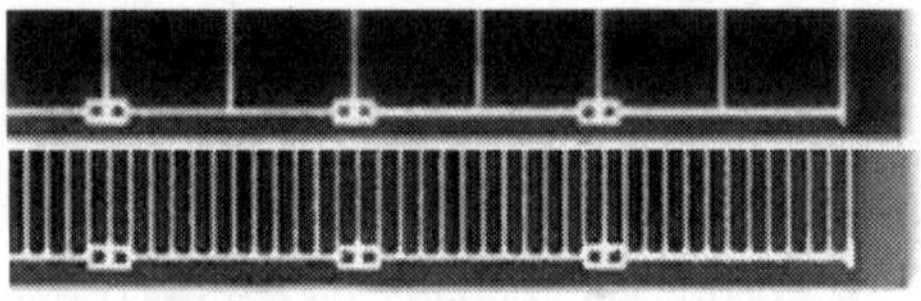

Figure 1: Metallization design on SHJ shingle cells used in this work. Upper – front side, lower – rear side.

Three different ECAs were tested with both cell types: ECA-I (Ag filled acrylate, density 1.9 g/cm³), ECA-II (Ag and Cu filled epoxy, density 4.0 g/cm³) and ECA-III (Ag filled epoxy, density 2.0 g/cm³). For the small-scale modules the adhesives were applied manually as a continuous line of 0.2 mm and as a pad pattern (0.3 mm × 0.5 mm with 0.6 mm distance) through a stencil with 100 µm thickness. The pad design leads down to half of the ECA weight in comparison to the continuous one: 2.5 ± 0.3 (SE) mg vs. 4.8 ± 0.6 mg (measured with ECA-I). The produced small-scale strings (six shingles in a sequence with 1.1 mm overlap) were laminated with polyolefin encapsulation materials from two different manufacturers: PO-I and PO-II. The small-scale modules were laminated in a glass-glass design in a 200 mm × 200 mm size. The glass thickness was 3.2 mm.

Table I: Small-scale module variation. Such sample matrix was used for the SHJ and TOPCon group, respectively; "continuous" and "pads" describe ECA application pattern; "PO-I" and "PO-II" are two types of encapsulation material.

ECA-I				ECA-II		ECA-III	
continuous		pads		pads		pads	
PO-I	PO-II	PO-I	PO-II	PO-I	PO-II	PO-I	PO-II

2.4 Full-size and VIPV module production

With the material choice based on the results of the small-scale module testing, full-format modules (1960 mm × 1010 mm × 2 mm) in glass-glass design with 12 strings 38 shingles each (6 strings in parallel) were produced. Full-size module fabrication was realized with TOPCon shingle cells, ECA-II and PO-I. The TT1600ECA stringer was utilized for the interconnection. The machine used screen printing to apply ECA-II in a pad pattern (4.8 mg per shingle, note the higher density of ECA-II compared to ECA-I). The robot arm system ensured an automatic shingle placement with 1.1 mm overlap, based on the metallization design of the TOPCon shingles (busbar width and position relative to the cell edge). Lamination took place in a Bürkle Ypsator laminator with a 3-step process: plate-membrane (150 °C, 7.5 min), followed by plate-plate (150 °C, 7.5 min) and finishing in cooling press (20 °C).

The vehicle integrated modules in this work (i.e. car bonnets) required an individual solution when it came to material choice and lamination due to their curved shape and metal substrate. The lamination of the bonnet took place in a Bürkle 3D laminator with a custom aluminum negative form. In order to electrically isolate the bonnet and the module, black insulation layer was used. For the VIPV modules, TOPCon shingle cells were interconnected analogically to the strings for the conventional full-size modules. One bonnet module comprised five strings, 30 shingles each, connected in parallel.

2.5 Module testing and characterization

After production, the small-scale modules were measured and compared in terms of *I-V* parameters and uniformity of the electroluminescence (EL) signal. Series resistance was calculated based on the *I-V* curve. Then, the small-scale modules were tested in an accelerated thermocycling (aTC) chamber with 50 cycles followed by 150 cycles. The sample degradation was analyzed. The aTC procedure differs from the thermocycling (TC) testing as described in IEC61215 [9] by the absence of current surge and the use of six times quicker temperature ramps (8 K/min, T_{max} = 85°C, T_{min} = −40°C). This allows identifying differences in the degradation behavior between groups in a short period of time. In order to gain understanding in possible terminal connector defects, Magnetic Field Imaging (MFI) was used.

3 RESULTS

3.1 Microscopic analysis of shingle interconnection

Three principal types of joints were identified during the investigation of the strings with 0.8 mm overlap. Figure 2a shows an ECA printing offset, which is caused by poor screen alignment and can be avoided with an adjustment during production. The second case (fig. 2b) depicts a lower distance from metallization to the shingle edge, causing shunting of the string and negatively affecting the string appearance due to ECA smearing. This can be solved by enhancing the precision of the laser cutting process. Fluctuating distance from the edge to the busbar as a result of unprecise cutting makes it difficult to set up a stable alignment of ECA print and metallization while keeping a constant physical overlap less than 1 mm. Figure 2c demonstrates correct ECA print directly over the metallization pad and precise placement of the shingles above one another creating a shortest possible current path between the shingles.

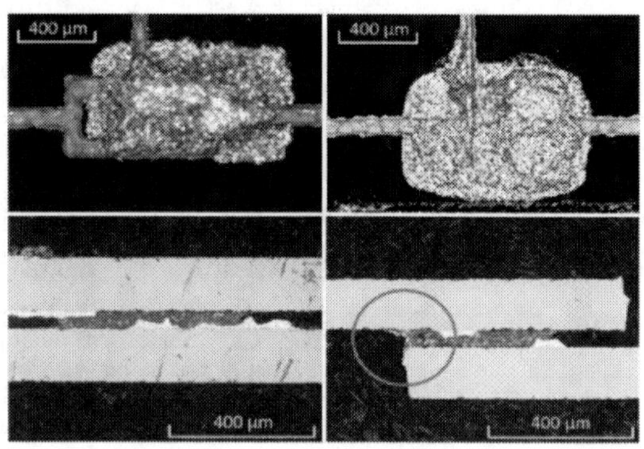

(a) (b)

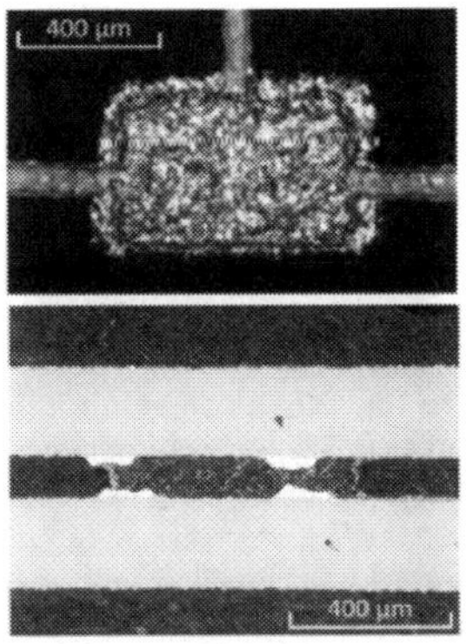

(c)

Figure 2: Top – planar microscopic images of the metallization sections of the SHJ busbar area with printed ECA pads. ECA pad dimensions: 0.7 mm × 0.4 mm. Bottom – cross-sections of the corresponding situations in the shingle joint. (a) ECA is printed with an offset to the cell metallization. (b) Due to the unprecise cutting, the ECA is too close to the edge causing shunting (orange circle). (c) ECA is printed correctly as well as shingles placed precisely, so that metallization of the upper cell lays directly above the metallization of the lower cell with ECA in between.

3.2 ECA reduction and module material optimization

Based on the *I-V* measurement after production (tab. II), no material- or design-dependent differences in the performance of the small-scale modules (fig. 3) were observed.

Figure 3: Small-scale module. Six shingles interconnected in a sequence with a copper terminal connector on each side. Terminal connectors are extended with SnPb-coated bussing ribbons.

Table II: Mean values of the *I-V* parameters of the small-scale modules after lamination.

	V_{OC}	I_{SC}	FF	P_{MPP}
SHJ with PO-I	4.4 V ±0.1%*	1.4 A ±1.4%	77.1 % ±1.5%	4.9 W ±1.8%
SHJ with PO-II	4.4 V ±0.1%	1.4 A ±1.2%	77.1 % ±1.5%	4.9 W ±2.0%
TOPCon with PO-I	4.2 V ±0.1%	1.6 A ±0.8%	78.4 % ±0.7%	5.2 W ±0.5%
TOPCon with PO-II	4.2 V ±0.1%	1.6 A ±3.6%	78.5 % ±1.7%	5.2 W ±2.3%

* standard error (SE)

After the aTC50 test, the small-scale modules were measured again and the loss in *I-V* parameters relative to the initial values was calculated (fig. 4). V_{OC} demonstrated no difference in TOPCon samples, whereas SHJ samples showed average $-3.4\%_{rel}$ across all groups. We do not assume this to be an effect of the shingle interconnection or module fabrication but an instability of the cell itself. I_{SC} is slightly increased for TOPCon ($+0.2\%_{rel}$) and fluctuates around zero change for SHJ. In both cell groups FF and P_{MPP} show strong correlation after aTC50, which implies same governing factors, which in this case is increasing series resistance. TOPCon samples demonstrate no to slight ($-0.5\%_{rel}$) FF or P_{MPP} loss with PO-II whereas PO-I causes up to $-4.5\%_{rel}$ ΔP_{MPP} and $-4.8\%_{rel}$ ΔFF due to string corner breakage (fig. 5) with a strong scattering of the values (fig. 4, green hollow triangles). Usage of a 3.2 mm thick glass for small-scale modules in combination with 450 µm encapsulation material likely causes such an effect. Based on experience with thinner glass (2 mm, as used in full-format modules) no breakage for lower glass-POE thickness ratio is expected.

Within the SHJ group, an overall increase in FF of up to $+4.1\%$ is to be seen. ΔP_{MPP} is dispersed between $-8.2\%_{rel}$ and $+0.4\%_{rel}$, with the majority of the small-scale modules showing no to moderate ($-2.0\%_{rel}$) degradation (Fig. 4, orange hollow circles). Extreme values in this case were caused by the partial detachment of the terminal connectors (not depicted in fig. 4) and degradation around the crack already present after lamination.

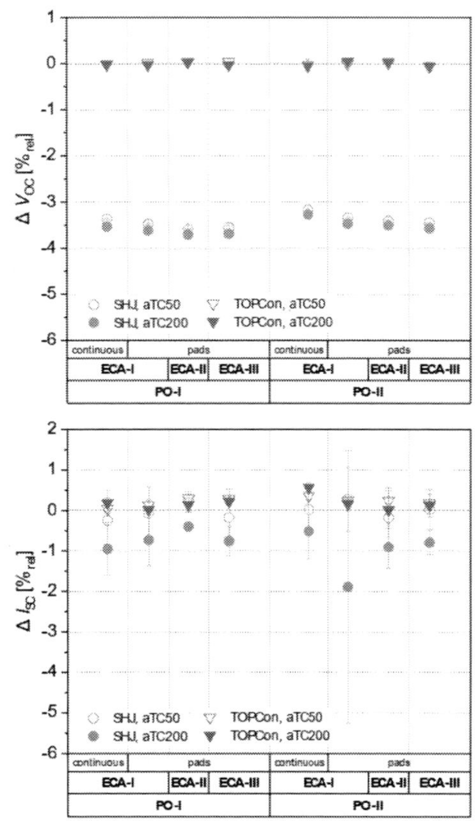

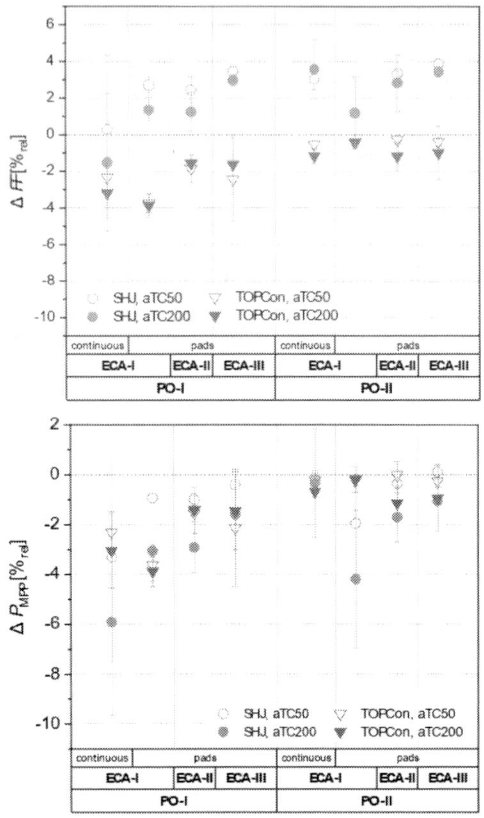

Figure 4: Relative loss in (from top to bottom) V_{OC}, I_{SC}, FF and P_{MPP} of small-scale modules as a result of aTC50 and aTC200 tests. Mean of three samples per variation is plotted, error bar – standard error (SE). Zero level - initial *I-V* characteristics (after lamination). "PO" – encapsulation material, "ECA" – conductive adhesive, "continuous" and "pads" – adhesive printing patterns.

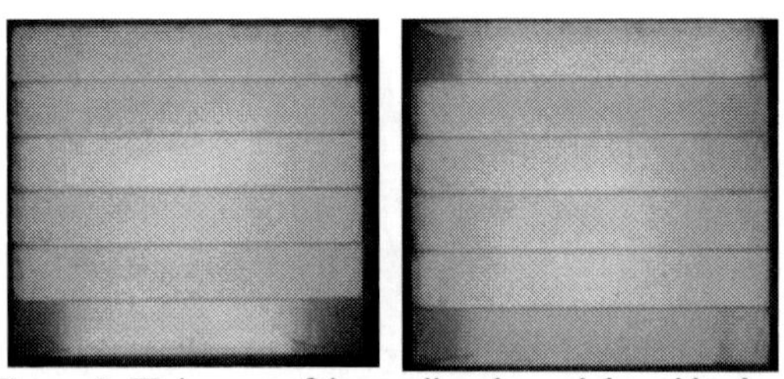

Figure 5: EL images of the small-scale modules with edge string breakage. Measurement setup: current = 3 A, exposure time = 600 ms.

As a result of further 150 cycles in the aTC test, TOPCon modules with PO-II demonstrate stronger degradation than after 50 cycles, whereas samples fabricated with PO-I either keep the same degradation or slightly gain P_{MPP} (fig. 4). Increase in P_{MPP} occurs with the ECA-III most likely due to further curing within the module when exposed to high temperatures (during aTC testing up to 85°C). This is confirmed by the reduced R_S after aTC200 for this adhesive (fig. 6, blue triangles). Similarly to aTC50, ΔFF strongly correlates with ΔP_{MPP}. V_{OC} shows no change after aTC200 in comparison to aTC50. I_{SC} for the samples with continuous print gained up to 0.3%$_{rel}$, whereas ΔI_{SC} for samples with pad application with the same ECA counts +0.1%$_{rel}$ to −0.2%$_{rel}$.

Small-scale modules with SHJ shingles show further V_{OC} decrease across all groups by 0.15%$_{rel}$, which can be attributed to further cell degradation after aTC200. I_{SC} demonstrates 0.3%$_{rel}$ to 2.0%$_{rel}$ lower values in all variations with no trends. ΔI_{SC} correlates with ΔP_{MPP} in samples with PO-II, which is an indication of encapsulation material degradation and its negative effect on the peak power. While ΔP_{MPP} in samples with PO-II is mostly governed by the encapsulation material degradation, P_{MPP} loss in small-scale modules with PO-I are affected by FF loss. ΔFF with SHJ samples after aTC200 produced with PO-I shows the most dependency on the R_S. Samples with PO-II show less change in ΔFF values, due to allegedly less thermomechanical stress, affecting the interconnection.

The series resistance R_S, mostly ranging from 0.1 Ω to 0.3 Ω is slightly different for SHJ (fig. 6, above) and for TOPCon (fig. 6, below) cells. Initial mean R_S values with TOPCon cells are overall lower (0.11 ± 0.03 (SE)) Ω, than with SHJ cells (0.14 ± 0.05) Ω. A comparison of the initial values between the full (continuous) and reduced (pad) ECA application with ECA-I shows that TOPCon cells are robust against the ECA reduction; there is hardly any difference visible (fig. 6, orange triangles). SHJ cells in this case demonstrate a large scattering of the values for the pad ECA application design (orange circles). This indicates that the special metallization design of the SHJ shingles (fig. 1) is sensitive to ECA reduction at least with this particular adhesive. Further reduction of the ECA-I amount on such metallization design may cause worse module performance and reliability. After aTC50 the overall increase in R_S can be observed. Again, TOPCon shingles showed generally better results with average increase by (0.03 ± 0.03) Ω, whereas with SHJ it measured (0.04 ± 0.05) Ω. After aTC200 change in R_s is substantially lower for both cell types: (0.008 ± 0.020) Ω for TOPCon and (0.004 ± 0.026) Ω for SHJ. ECA-II together with TOPCon cells (fig. 6, bottom, middle section) is the best performing combination in terms of R_S values and interconnection stability (data scattering).

Overall, series resistance increase caused by joint degradation and short circuit current loss due to encapsulation degradation are the main causes for the performance loss in small-scale modules. Both of these effects can be avoided if a right material combination (ECA type, ECA application amount, encapsulation type) is used.

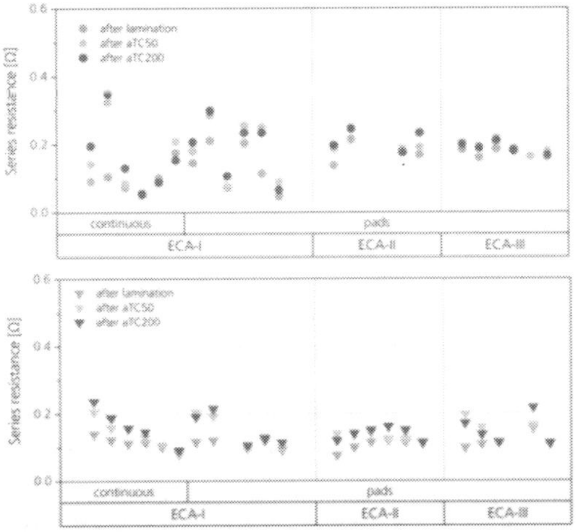

Figure 6: Series resistance of small-scale SHJ (top) and TOPCon (bottom) modules after lamination (orange), aTC50 (grey) and aTC200 (blue). "ECA" – conductive adhesive, "continuous" and "pads" – adhesive printing patterns. Samples with PO-I and PO-II of the same interconnection group are considered together.

Moreover, dendric-shaped features appear in a number of modules with PO-II after aTC200, indicating failure of the encapsulant possibly because of its partial melting and recrystallization (fig. 7).

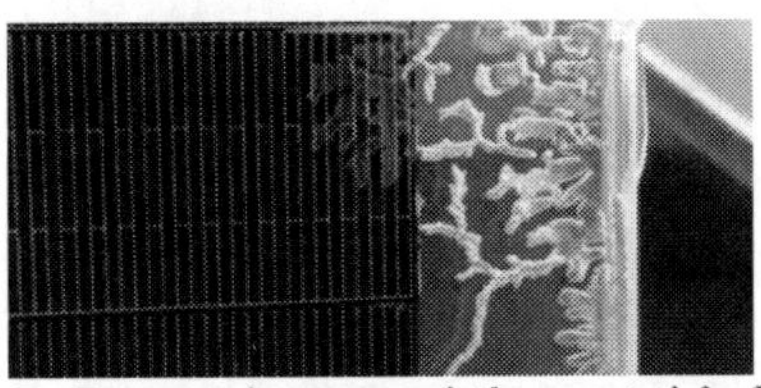

Figure 7: Features in encapsulation material abundant with PO-II samples after aTC200.

3.3 Terminal connection

Two small-scale modules with SHJ cells demonstrated unusually high losses after aTC50 ($-14.2\%_{rel}$ ΔP_{MPP} (fig. 8, first row) and $-4.8\%_{rel}$ ΔP_{MPP} (fig. 8, second row)) while EL images (Fig. 8, left) indicated no breakage but darkening of the whole string with local overexposure in the first cell. MFI images (Fig. 8, right) demonstrated uneven current flow distribution on one pole of each sample, matching with the EL signal. The bottom edge pattern of the MFI images correlates with the positioning of the terminal connector, whereas the current flow on the top is localized in one spot. The busbar of the SHJ cells is more sensitive to terminal connector misplacement or movement during module production than a full busbar metallization design. Visual inspection of the modules confirms the faulty positioning (fig 8, below). Since the data indicates defects in terminal connection, whereas the cell connection within the string appears homogeneous, these samples were not considered in the section 3.2.

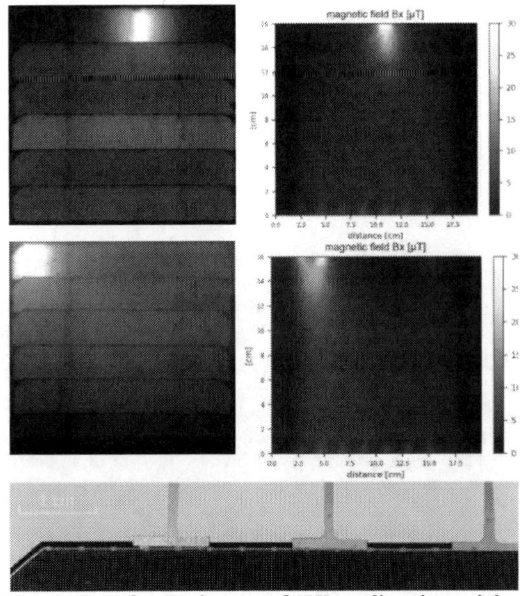

Figure 8: Left – EL images of SHJ small-scale modules with terminal connector defects. EL setup: 0.14 A, 2000 ms. Right – dark MFI of the corresponding samples. MFI setup: 1.7 A. All measurements taken after aTC50. Bottom – photo of the terminal connector defect area.

3.4 Full-format and integrated modules

Based on the results of the small-scale module testing and gained understanding of the degradation behavior of the material combinations an optimal set of TOPCon shingles, ECA-II with pad application and PO-I was chosen for the full-format module production. Three full-format modules were fabricated in a glass-glass configuration. They demonstrated P_{MPP} values of 389.7 W (fig. 9a), 392.8 W (fig. 9b) and 376.8 W (fig. 9c), and full area efficiencies of 19.68%, 19.84% and 19.04%, respectively. This attributes to a CTM of $-2.2\%_{rel}$ to $-6.3\%_{rel}$ in ΔP_{MPP} and $-8.4\%_{rel}$ to $-12.3\%_{rel}$ $\Delta\eta$ (fig. 10). The main portion is related to a loss in short circuit current due to cell shading as a result of overlap. EL images indicate no major defects and homogeneous interconnection.

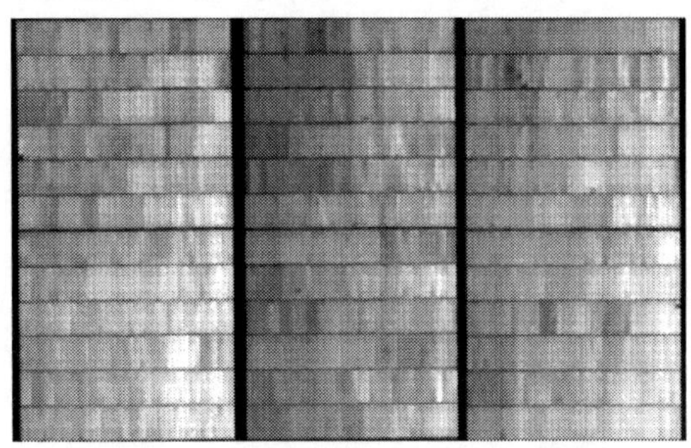

(a) (b) (c)

Figure 9: EL images of the full-format TOPCon modules. Measurement setup: current = 10 A, exposure time = 2000 ms.

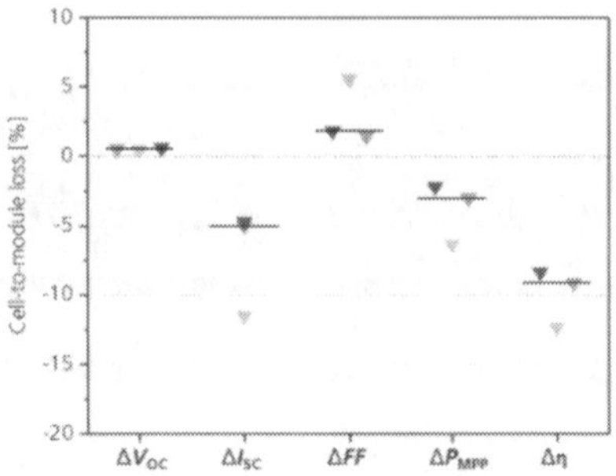

Figure 10: Change of the key *I-V* parameters after the full-format module lamination. Each color depicts one module from fig. 12: cyan – (a), navy – (b), grey – (c).

A prototype of a car bonnet module for Volkswagen Polo was successfully manufactured (fig. 11). Further details to the prototype will be described in in a separate publication and are briefly mentioned in Tous et al. [10]. Such module can serve as a power source in addition to a solar integrated car roof. Another advantage of a solar bonnet is the possibility to upgrade already assembled cars due to the relatively easy bonnet part extraction.

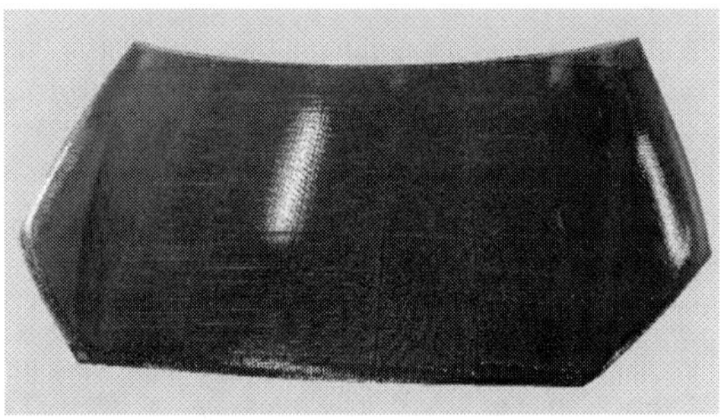

Figure 11: Car bonnet with an integrated solar module developed and produced at Fraunhofer ISE.

4 CONCLUSIONS AND OUTLOOK

4.1 Conclusions

By investigating shingle joints with 0.8 mm overlap, current study demonstrates that reliable interconnection with overlap under 1.0 mm cannot be realized with unprecise cutting, because the metallization print and ECA are too close to the cell edge and may cause shunting.

Based on the small-scale module production and testing with an aTC procedure, following can be concluded. Samples with PO-I after aTC50 show abundant cell breakage. Such breakage is most likely caused by the too low glass-PO thickness ratio and overlay the influence of the ECA. PO-II systematically demonstrates visible defects after aTC200 possibly caused by melting and recrystallization of the polymer. Series resistance of the small-scale modules showed that TOPCon cells with their full busbar have initially lower R_S ($0.11 \pm 0.03\ \Omega$) and are more robust against ECA amount reduction and aTC testing (increase by $0.03 \pm 0.03\ \Omega$) than SHJ shingles ($0.14 \pm 0.05\ \Omega$ and $0.04 \pm 0.05\ \Omega$, respectively). ECA amount reduction by down to 50% does not result in lower *I-V* values neither after the module production nor after aTC. The terminal connection is a critical process that can lead to significant CTM and reliability losses and must be optimized. TOPCon cells interconnected with ECA-II (reduced amount) and laminated with PO-I is the best performing combination after aTC200 (fig. 4. 6).

Full-format glass-glass TOPCon shingle modules were produced and demonstrated P_{MPP} up to 392.8 W with the highest efficiency of 19.84%.

4.2 Outlook

The reduced ECA application design used in this work shows great potential for silver saving, making shingle technology cheaper. Tests according to IEC protocols will be conducted at Fraunhofer ISE in the framework of other projects to demonstrate reliability of shingle interconnection with this application design.

Further optimization work on the shingle string terminal connector is yet to be done. While metallization layouts with wide busbars seem to withstand stress of the reliability tests, metallization paste reduction in the busbar area of the shingle cells requires more robust solutions.

ACKNOWLEDGEMENTS

Current study was prepared in the framework of the project HighLite, which was funded by the European Union's Horizon2020 Programme under Grant Agreement No 857793.

REFERENCES

[1] M. Mittag et al, *Proceedings of the 44th IEEE*, p. 1531–1536 (2017).
[2] T. Rößler, et al, *AIP Conference Proceedings*, accepted (2022).
[3] N. Klasen et al, *IEEE Journal of Photovoltaics*, Vol. 12–2, p. 546–556 (2022).
[4] C. Carriere et al, *Proceedings of the 37th EUPVSEC*, p. 840–844 (2020).
[5] C. H. Schiller et al, *Proceedings of the 36th EUPVSEC*, p. 995–999 (2019).
[6] IEC 60891:2009–12.
[7] R. Lausch et al, *Proceedings of the 35th EUPVSEC*, p. 1060–4 (2018).
[8] D. Eberlein et al, *Practical Metallography*, Vol. 48-5, p. 239–260 (2011).
[9] IEC 61215: 2005.
[10] L. Tous et al, *Progress in Photovoltaics: Research and applications*, submitted (2022).

INTERCONNECTION TECHNOLOGY FOR THE NEXT GENERATION OF (TEMPERATURE-SENSITIVE) SOLAR CELLS SUCH AS HETEROJUNCTION AND C-SI/PEROVSKITE TANDEM

Yonas Zemen[1], Stefan Wendlandt[2], Bernd Litzenburger (†)[1], Lars Podlowski[1]
[1]Solyco Technology GmbH, Baseler Strasse 60, 12205 Berlin, Germany,
[2]PI Photovoltaik-Institut Berlin AG, Wrangelstr. 100, 10997 Berlin, Germany

ABSTRACT: The interconnection technology for solar cells is an important component in the PV module production - not only for module reliability but with respect to the sustainability of the PV industry as well. We are proposing the interconnection technology "TECC Wire" (TECC=Thermoplastically and Electrically Conductive Coating) which is a new low-temperature method – and which avoids the usage of any rare or hazardous materials. As rare or critical materials in the PV industry are rated such as silver, bismuth, lead and indium because global resources are considered to be too limited or too expensive for the terawatt production level. And the usage of lead does not comply with the RoHS (Restriction of Hazardous Substances) directive. Although PV modules are still exempt from this directive, efforts to avoid using lead as a solder alloy are consensus in the industry.
TECC-Wire is conceptually an improved stage of the multi-busbar or multi-wire concept. However, it allows to further reduce the consumption of silver for the solar cells metallization, and it even enables the vision of totally eliminating silver from solar cell production. The TECC wires have a similar copper core that is typically used for multi-busbar soldering of solar cell – but it is coated with an electrically conductive adhesive instead of a solder alloy. The filler material to achieve the electrical conductivity is based on non-silver particles.
The recently built semi-automated stringing machine attaches the TECC wires precisely and in a defined manner. It allows us to build and test standard module configurations. The long-term reliability tests have already passed 3x IEC requirement. After 3000h damp heat test the degradation of the small modules is below 5%. In thermal cycling test the results after TC600 are around 5% for different solar cells. The peel strength is about 1.5 N/mm contact area over the entire solar cell – which is comparable to a solder joints. The homogeneity of the peel strength along the entire surface of the solar cells is remarkable, as the wires adhere to any surface and not only to the printed fingers.
Keywords: Module, Metallization, Low-temperature, Heterojunction

1 INTRODUCTION

Passivated c-Si solar cell technologies such as PERC (Passivated Emitter and Rear Cell) and TOPCon (Tunnel Oxide Passivated Contact) will continue to dominate the market. However, new technologies for higher conversion efficiencies will gain market share. As early as 2024, HJT's (heterojunction technology) share will already be 10%, and by 2032 it will reach 20% market share [1]. Si-based tandem technologies will play a role in mass production beginning from 2024 [1]. The main challenges for HJT in mass production are the enormous silver consumption for the solar cell metallization, the use of rare and expensive indium as TCO (transparent conductive oxide) layer and the necessary low-temperature contacting using tin-bismuth alloy– because HJT solar cells have to be processed at low temperatures <200°C [2], whereas common solar cells such as PERC and TOPCon can be processed and soldered at significantly higher temperatures. In addition to cost and processing concerns, the global deposits of silver, indium and bismuth will conceptually limit the growth of the PV industry in general if no replacements of those critical materials can be established [2].

2 WHAT ARE THE PROPOSED APPROACHES IN THE PV INDUSTRY?

2.1 Reduction silver consumption

Compared to the PERC, the HJT solar cell consumes about twice as much silver [1]. Increasing the number of BBs (busbars) has proven to be an effective means to reduce solar cell paste consumption. While the 12BB layout is already implemented, the industry has begun to increase the BBs up to 16, with some even planning to go more than 18 [3]. As an example, the so-called "SmartWire" technology by Meyer Burger, which contacts busbar-less HJT solar cells, has the lowest silver consumption for screen printed HJT [4]. This is exclusively due to the number of wires and the busbar-less solar cells – up to 22 solder wires are used. However, the technology is no longer available for the market since Meyer Burger claims the technology to be used only for its own module production.

A promising saving potential is to replace silver with copper. Copper plating technologies (with seed layers) have been discussed and tested in the industry for some time – but except for the short period of First Solar´s "Tetrasun" product in 2015/2016 [5] it has not found approval in the PV industry for series production. What is also currently being intensively tested for HJT solar cells is to use metallization pastes with silver-coated copper particle instead of pure silver. Recently a record efficiency of 25.62% on M6 wafers using silver-coated copper pastes has been reported [6]. According to the report it can reduce the silver consumption of the HJT solar cells by 55%. A second approach is to have pure copper pastes and replace silver completely. However, the potential oxidation of the copper surface is a risk factor for reliability, but such HJT solar cells are subject of current R&D. We have recently started the testing of applying the "TECC-Wire" technology to such silver-free solar cells.

2.2 Reduction of indium consumption

Indium is not only very expensive but also belongs to the rare materials. However, indium tin oxide (ITO) is currently the best material for transparent conductive oxide (TCO) layers for HJT solar cells. There are some approaches to replace ITO e.g. by AZO (aluminum zinc oxide) but module reliability is still not sufficient with pure AZO as TCO. A promising approach is the multi-layer TCO with AZO as the bulk material and ITO just as a cap layer. This approach is already delivering promising reliability results and would at least significantly reduce the consumption of indium.

2.3 Bismuth in solder alloys

Bismuth is used as a low-temperature solder alloy for the interconnection of HJT solar cells. As mentioned above bismuth is considered as a critical and limited raw material. In addition, bismuth as a lead substitute is very controversial and was not implemented in the electronics industry when converting to lead-free solder within the framework of the ROHS directive. Currently there is no substitute material for bismuth available for low-temperature soldering, and only a gluing technology can avoid the usage of bismuth.

3. PROPOSED SOLUTION: TECC-Wire

This technology was introduced in 2021 [7]. It bases on copper wires with an electrically conductive thermoplastic coating. It is a low temperature interconnection technology because the melting point of the coating can be engineered in a range between 130°C - 180°C. The technology does not use any rare or critical material such as Ag, Bi or Pb and thus it is fully compliant with the ROHS directive. Since it is a gluing technology, the wires stick to any surface and thus enable the use of busbar-less solar cells with respective savings in silver consumption.

Additional advantages of the "TECC-Wire" technology are:

- there is no limitation to the number of wires used per solar cell. It can use any number of wires – enable reduction of silver consumption
- It does not require any solderable electrical contact surface. It opens the vision to eliminate silver from the solar cells by using very alternative metallization strategies
- No flux is required. Flux is not only "dirty" in module production – but also potential risk factor for field degradation
- Simplification of wire positioning accuracy, since no busbars are required

4. SAMPLE PREPARATION

The semi-automated stringing lab equipment enables to place the wires in a defined manner. Although the stringing machine is still in a lab configuration, it is possible to build strings for full-size modules. Figure 1 shows the semi-automated stinging machine and a full string of half-cut M6 HJT solar cells coming out of the machine.

Figure 1: Semi-automated TEEC-Wire solar cells stringing machine (left), and a full string of half-cut M6 HJT busbar-less solar cells

A full-size module is demonstrated in figure 2. It shows a photo of the module front side (left) and the respective EL image (right). The solar cells are half-cut M6 HJT. Apart from a few minor cracks the EL image is very homogeneous. The cracks appear along the solder joints of the cross-connectors. It shows that more attention when connecting the strings has to been taken by the lab team. The EL image is taken at a current of 9A applied for 1sec.

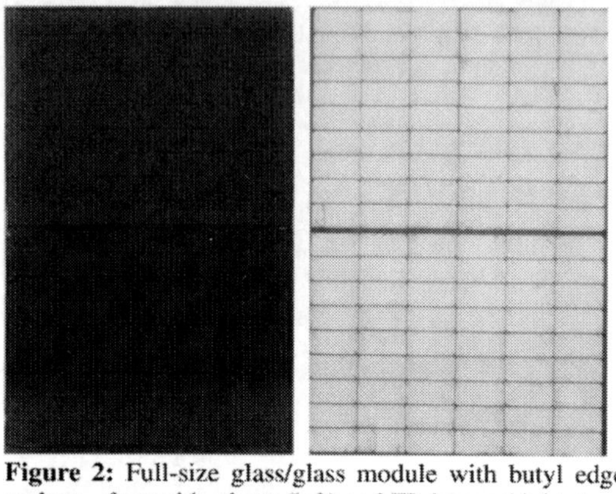

Figure 2: Full-size glass/glass module with butyl edge sealant – front side photo (left) and EL image (right; 1sec @9A). The module was built using the semi-automated TECC wire stringing machine with half-cut M6 HJT solar cells.

5. RESULT AND DISCUSSION

5.1 Peel strength

Figure 3 is showing the peel strength data of busbar-less M6 half-cut HJT solar cell connected using TECC wires. The peel strength test is done using Zwick/Roell machine with a pull angle of 180° and a testing speed of 200mm/min. The measurement results are exemplary for all measured solar cells. The M6 half-cut solar cell was contacted with 16 TEEC wires. The figure shows four wires peeled off from the front side and from the back side, respectively.

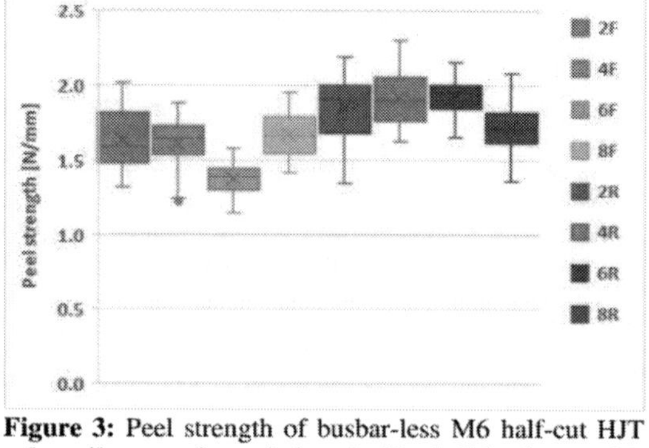

Figure 3: Peel strength of busbar-less M6 half-cut HJT solar cells – e.g., 2 stands for wire No. 2, F for front and R for rear side of the solar cell

What is clearly evident from the boxplot – maximum, minimum, and average value are shown - that no wire adhesion force anywhere gets even close to 0N over the entire solar cell contact area. This demonstrates that the wires stick very well both onto the finger grid as well as onto the solar cell (TCO). The minimum peel strength is always well above 1N/mm contact area, which is comparable to soldering of HJT solar cells with busars [8]. Although the "TECC-Wire" connection technology is more comparable to gluing rather than soldering, we are referencing to the more established data for solar cell soldering technology.

5.2 Damp heat (DH) results

The recent DH test results were made with 1-cell modules (coupons) of M2 HJT solar cells. Some solar cells have been contacted by ECA (Electrically Conductive Adhesives) as reference, and others from the same batch with the "TECC-Wire" interconnection technology. The aim was to compare the TECC wires with the state-of-the-art interconnection technology for HJT solar cells. The solar cells where then encapsulated in glass/backsheet technology, with a POE as encapsulant and a backsheet with moisture barrier.

Figure 4 is showing the results of the ECA and with TECC wires connected coupons after DH3000 – which is equivalent to 3x IEC standards. The data points are the average value for 3 test samples. The power degradation after DH3000 for the ECA coupons is below 2% which is very good results for HJT modules. The "TECC-Wire" coupons reliability is slightly worse to those ECA contacted samples. However, after DH3000 the degradation is around 5% which is still a very good value.

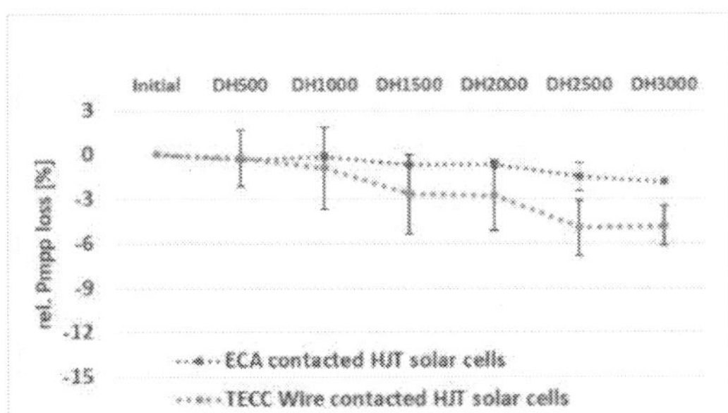

Figure 4: M2 HJT solar cell coupons connected with electrically conductive adhesives (ECA) as reference and the TECC-Wire technology after DH3000. The coupons were encapsulated using a backsheet with moisture barrier

5.3 Thermal cycling (TC) test

With respect to the long-term reliability of a PV module, thermal cycling (TC) is the hardest test to assess the interconnection technology. Using the semi-automated stringing machine and after process optimization, we have reached very good results and even passed 3x the typical IEC requirement. The degradation is below 5% after TC600. Figure 5 is illustrating the results. The M2 HJT solar cells were contacted by using the semi-automated stringing machine, and were then encapsulated in glass/glass technology to 1-cell coupons. The data in figure 5 is showing relative degradation to initial STC (Standard Test Conditions) power measurement.

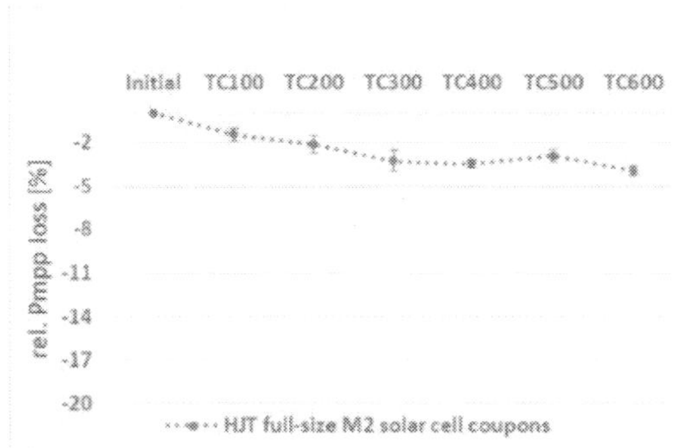

Figure 5: Thermal cycling (TC) test of solar cell coupons (1-cell modules) contacted using TECC wires. Busbar-less M2 HJT solar cells were used, and 22 TECC wires were applied. The data points show an average of min. 10 samples.

6 SUMMARY AND OUTLOOK

New high-efficiency solar cell technologies such as HJT and also c-Si/Perovskite tandems are pushing into the market and will gain market share over the next decade. Currently the PV industry is facing the dilemma that these new technologies i) make solar cells temperature-sensitive and ii) require an increased consumption of critical materials. Critical elements are expensive elements with limited resources such as silver, indium and bismuth. At the same time the PV industry is striving and is forced to minimize consumption of critical elements – which otherwise could slow down the industry's growth to terawatt production level.

The solar cell interconnection technology in PV module production has always been a key process for product quality and reliability, but recently it has become even more important as the interconnection technology determines the solar cell design and thus the usage of critical materials. It decides if lead and/or bismuth is used, and how much silver is consumed for the solar cell grid printing and eventually for ECA.

Having a suitable solar cell interconnection technology is a key factor to introduce new high-efficiency solar cell technologies - and at the same time to reduce the consumption or even eliminate critical materials for the PV industry.

The "TECC-Wire" technology is a very promising approach to meet all future technical and commercial requirements. It is not using lead or bismuth or indium or silver for the wire coating, and it allows low process temperatures which even can be tuned in a range between 130 – 180°C. Additionally it allows a busbarless design for the solar cell front grid, leading to a reduction of silver consumption for the solar cell production. Since it does not require a solderable surface but it achieves a very good mechanical and electrical contact to any surface, it may allow totally new metallization strategies which bases on copper or aluminum or nickel or other cheap metals.

In this paper we have shown reliability test results, and for small modules we have already passed 3x IEC standard requirements with degradation rates below 5%. The results in DH and TC are comparable to small modules made with ECA as a reference technology. The peel strength value of the wires to busbar-less HJT solar cells is above 1 N/mm contact area over the entire solar cell which is comparable to soldered HJT solar cells.

Full-size modules have also been built and will be subject to further extended laboratory testing to verify the long-term reliability of the TECC-Wire technology. One full-size module is already in outdoor testing where the energy yield and possible degradations are monitored in comparison to several commercial products.

7 ACKNOWLEDGMENT

This work was partially supported by a grant from the Federal Ministry for Economic Affairs and Energy (Germany) under contract number 0324293H (Project titel: "Dynasto"), and by a grant from the "ZIM" "Zentrales Innovationsprogramm Mittelstand", program (contract number KK5260601BS1)

8 REFERENCES

[1] International Technology Roadmap for Photovoltaic, 2022, 13th edition
[2] Y. Zhang et. al., "Design Considerations for Multi-terawatt Scale Manufacturing of Existing and Future Photovoltaic Technologies: Challenges and Opportunities Related to Silver, Indium and Bismuth Consumption", UNSW Sydney, 2052 Australia, Energy Environ. Sci., 2021, DOI: 10.1039/D1EE01814K
[3] Taiyang News, Heterojunction Solar Technology, 2022 edition, https://taiyangnews.info/reports/heterojunction-solar-technology-2022-report/, 07. September 2022
[4] J. Rentsch et al., "Technology Challenges for the Industrial Production of Silicon Heterojunction Solar Cells", Fraunhofer ISE, Freiburg, PV HeterojunctionTech virtual Conference 17-18, March 2021
[5] L. Podlowski, "Bifacial High-Efficiency Panels based on First Solar´s TetraCell Technology", First Solar, Workshop on bifacial PV, 2016, Miyazaki (Japan)
[6] Suzhou Maxwell Technologies Co., Ltd., https://www.maxwell-gp.com/en/news-center/maxwell/, 07, September 2022
[7] Y. Zemen et al., "Thermoplastically and Electrically Conductive Coated Wire for the Interconnection of Temperature- Sensitive Solar Cells", Solyco Technology, Berlin, Germany, 2021 IEEE 48th Photovoltaic Specialists Conference (PVSC), DOI: 10.1109/PVSC43889.2021.9518596
[8] A. De Rose et al., "Low-temperature soldering for the interconnection of silicon heterojunction solar cells", Fraunhofer Institute for Solar Energy ISE, 33rd EU PV Solar Energy Conference and Exhibition, September 2017, Amsterdam, Netherlands

PAVING THE WAY FOR LOW BREAKAGE RATES IN INDUSTRIAL PRODUCTION OF N.I.C.E.-WIRE MODULES

Pascal Leibiger[1], Christoph Pönisch[1,2], Thomas Seifert[1], Daniel Kray[1,2]
[1]: Offenburg University of Applied Sciences, Offenburg, Germany, [2]: Institute for Sustainable Energy Systems (INES), Offenburg University of Applied Sciences, Badstraße 24, 77652 Offenburg, Germany, daniel.kray@hs-offenburg.de

ABSTRACT: Significant improvements in module performance are possible via implementation of multi-wire electrodes [1]. This is economically sound as long as the mechanical yield of the production is maintained. While flat ribbons have a relatively large contact area to exert forces onto the solar cell, wires with round cross section reduce this contact area considerably – in theory to an infinitively thin line. Therefore, the local stresses induced by the electrodes might increase to a point that mechanical production yields suffer unacceptably.
In this paper, we assess this issue by an analytical mechanical model as well as experiments with an encapsulant-free N.I.C.E. test setup. From these, we can derive estimations for the relationship between lay-up accuracy and expected breakage losses. This paves the way for cost-optimized choices of handling equipment in industrial N.I.C.E.-wire production lines.
Keywords: Encapsulation, Module Manufacturing, PV Module, Reliability, Sustainable

1 INTRODUCTION

As sustainability aspects get increasingly important in TW_p scale PV production scenarios, module concepts with significant ecological advancements (like Apollon Solar's N.I.C.E.™ [2]) attract more attention. In [3], we introduced the next generation modules using round wires as interconnectors: N.I.C.E.-wire. In case of a positioning offset between wires on the front and on the rear of the solar cell, breakage can occur due to the resulting bending stresses. The maximum allowed offset determines the precision requirement for the automation and has a significant impact on its cost. Thus, we set up an analytical mechanical model and validate it via experiments and comparison to literature. Like this, we calculate the optimum choice in the trade-off between lowest absolute internal module pressure for lowest series resistance and lowest pressing force to maximize mechanical yield.

2 MOTIVATION

Wire electrodes have the potential to increase solar module efficiency compared to flat ribbons [4]. Mechanical yield losses could overcompensate the higher efficiency. Thus, it is highly relevant to assess possible losses by mechanical modeling and experiments. In this paper, we focus on the eco-friendly N.I.C.E.™ technology to pave the way for the industrialization of its next evolutionary step: N.I.C.E.-wire. The design of a production line needs well-defined parameters for pick-and-place accuracies during module lay-up. We elaborate a first guess for the maximum allowed lay-up offset by three elements:

- an analytical mechanical model,
- experiments with a N.I.C.E. lab setup for mechanical testing of N.I.C.E.-wire modules with defined lay-up offset and
- the comparison with literature data.

3 ANALYTICAL MECHANICAL MODEL

Switching standard modules from flat ribbons to round wires can raise mechanical yield questions. In [5], FEM simulations predict local stresses at the solder joints that might exceed the facture strength. However, multi-busbar interconnection technologies are well established and in large-scale production [6]. So, the challenges seem vincible.

In N.I.C.E. technology, the front and rear glass are permanently pressed together due to the lower pressure inside the module. This mechanical force is transmitted via the interconnectors to the solar cells. Obviously, flat ribbons distribute the force over a larger contact area compared to round wires. This might lead to larger local stresses and higher yield losses for N.I.C.E.-wire in case of round wires. To test this hypothesis, we set up an analytical mechanical model, cf. fig. 1. We consciously opted against FEM simulations as analytical models provide much faster calculations and describe correlations as direct functional relationships.

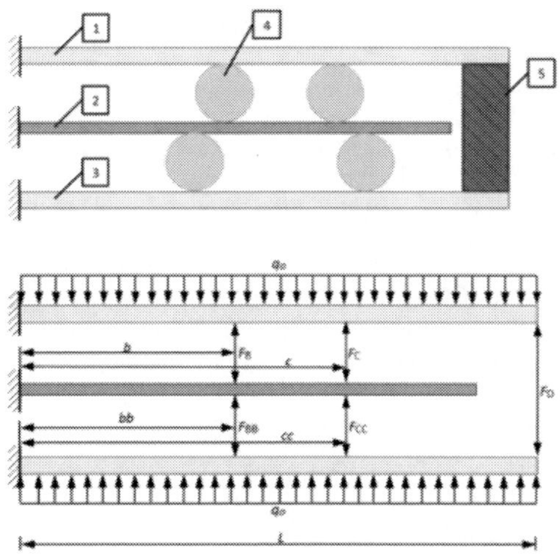

Figure 1: Analytical mechanical model for N.I.C.E.-wire modules. Cross-section of the symmetry element. Top: Three beams (1-3: glass / solar cell / glass) that are coupled via four wires (4), sealant (5). Bottom: Geometry and force parameters.

Using the Euler-Bernoulli bending theory, the evolution of shear force Q, momentum M, bending stress σ, bending angle w' and displacement w along the beam axis can be calculated [7]. For this purpose, the theory of the Föppl bracket is used to describe discontinuities such as jumps or kinks [8]. The upper and lower glass (1 and 3 in fig. 1) as

$$E \cdot I \cdot w'''' = q_o,$$

$$E \cdot I \cdot w''' = q_o \cdot z + C_1 - F_B \cdot \langle z-b \rangle^0 - F_C \cdot \langle z-c \rangle^0 = -Q,$$

$$E \cdot I \cdot w'' = \frac{1}{2} q_o \cdot z^2 + C_1 \cdot z - F_B \cdot \langle z-b \rangle^1 - F_C \cdot \langle z-c \rangle^1 + C_2 = -M, \quad (1)$$

$$E \cdot I \cdot w' = \frac{1}{6} q_o \cdot z^3 + \frac{1}{2} C_1 \cdot z^2 - \frac{1}{2} F_B \cdot \langle z-b \rangle^2 - \frac{1}{2} F_C \cdot \langle z-c \rangle^2 + C_2 \cdot z + C_3,$$

$$E \cdot I \cdot w = \frac{1}{24} q_o \cdot z^4 + \frac{1}{6} C_1 \cdot z^3 - \frac{1}{6} F_B \cdot \langle z-b \rangle^3 - \frac{1}{6} F_C \cdot \langle z-c \rangle^3 + \frac{1}{2} C_2 \cdot z^2 + C_3 \cdot z + C_4.$$

well as the cell (2 in fig. 1) are modeled as beams. The interaction of the beams via the wires (4 in fig. 1) and the sealing (5 in fig. 1) is described by appropriate compatibility conditions. Symmetry is assumed at the left boundary of the model. The differential equation for the Euler-Bernoulli beam assuming a constant line load from pressure is integrated four times to obtain the momentum and, hence, the bending stresses, as well as the displacements. Exemplarily, the integration for beam 1 yields equations (1) above. The z-axis corresponds to the beam axis with zero at the left boundary.

E is Young's modulus and I describes the area moment of inertia obtained from the dimensions of the cross section. For determination of the integration constants C_1 to C_4 of each beam, the corresponding boundary conditions are considered. For beam 1, e.g., we use

$$w(z=0) = 0 \text{ and } w'(z=0) = 0, \qquad M(z=L) = 0 \text{ and } Q(z=L) = -F_D. \quad (2)$$

The loads F_B, F_C, F_{BB} and F_{CC} (fig. 1) are obtained from the compatibility conditions. With these conditions we assume the wires to be rigid, so that the displacement of beam 1 at the upper left wire must be the same as the displacement of beam 2 at the same location. Furthermore, the displacement of beam 2 must be equal to the displacement of beam 3 at the corresponding location and so on ($GL1$, $GL2$: upper and lower glass, W: cell):

$$w_{GL1(z=b)} = w_{W(z=b)}, w_{GL1(z=c)} = w_{W(z=c)}, \qquad w_{W(z=bb)} = w_{GL2(z=bb)}, w_{W(z=cc)} = w_{GL2(z=cc)} \quad (3)$$

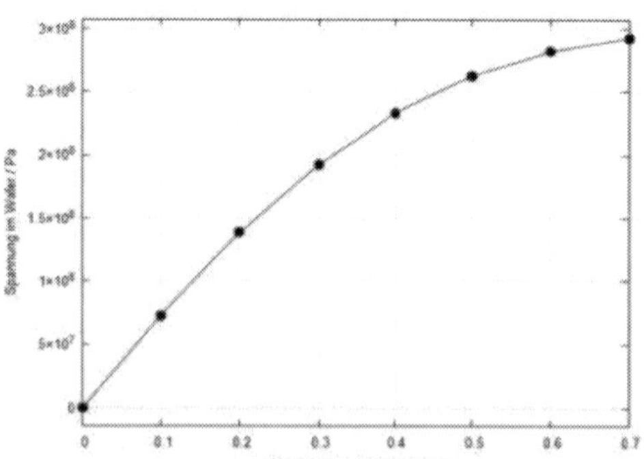

Figure 2: Simulated maximum stress in Pa (y-axis) in the solar cell with varying wire offset in mm (x-axis) for a specific module pressure.

The load F_D (fig. 1) is determined from the stiffness of the sealing C (depending on the height, the area and Young's modulus of the sealing) from the condition

$$F_D = C \cdot \left(W_{GL1(z=L)} - W_{GL2(z=L)} \right). \quad (4)$$

The resulting coupled system of equations is solved using *wxMaxima* which is a GUI for the computer algebra system Maxima. With this program we calculate the shear forces, momentums, bending stresses and displacements for all beams. From the bending stresses in the cell, the probability of failure R_R is calculated using the Weibull distribution [9]:

$$R_R = \int \left(\frac{\sigma}{\sigma_0} \right)^m dV \quad (5)$$

σ_0 is the characteristic fracture stress and m is the Weibull modulus. These parameters are determined later based on experimental results. Moreover, the maximum stresses in the solar cell are calculated depending on the wire offset and the differential pressure between the atmosphere and the module inner volume, cf. fig. 2.

4 EXPERIMENTAL STUDY AND VALIDATION

In order to collect experimental data for model validation, we adapted our re-usable N.I.C.E. test setup [10] as breakage test setup with live electroluminescence (EL) imaging in our flasher system, cf. fig. 3. We prepared solar cells with 5 wires on front and rear attached by precisely dispensed adhesive and dedicated wire offset. These were introduced in the test setup and measured LIV and EL while varying the internal pressure using a controllable vacuum pump. Via EL imaging, we could observe crack initiation and growth, cf. fig. 4.

Clearly, the contact resistance between interconnectors and cell metallization depends on the pressure difference between module interior and atmosphere. The larger this difference, the stronger the forces that the glasses exert on the wires and the latter on the cell, the lower the series resistance, thus the higher the *FF*. As the internal pressure can be chosen freely, one might be tempted to choose the lowest pressure possible. However, this increases mechanical loads on the cells and therefore increases yield losses. The sensible choice of optimum pressure is one of the main foci of this paper.

Figure 3: Test setup with vacuum pump (left) and prepared cell samples with attached wires and tabs (right).

Clearly, the contact resistance between interconnectors and cell metallization depends on the pressure difference between module interior and atmosphere. The larger this difference, the stronger the forces that the glasses exert on the wires and the latter on the cell, the lower the series resistance, thus the higher the *FF*. As the internal pressure can be chosen freely, one might be tempted to choose the lowest pressure possible. However, this increases mechanical loads on the cells and therefore increases yield losses. The sensible choice of optimum pressure is one of the main foci of this paper.

At first, we arbitrarily chose our target module pressure p_t in such a way that *FF* is just 0.2%$_{rel}$ lower than for the lowest achievable pressure. In order to guarantee this pressure at 80°C, we recognized that the module at -40°C would exhibit a lower pressure $p_{t\text{-}40}$. This lowest pressure leads to the highest mechanical loads and is therefore used in the model and the experimental breakage tests. The maximum wire offset w_{max} is thus the offset that allows for acceptable mechanical yield at pressure $p_{t\text{-}40}$. So, the

Figure 4: Crack growth in N.I.C.E.-wire module setup with decreasing internal pressure from top left to bottom right (EL images).

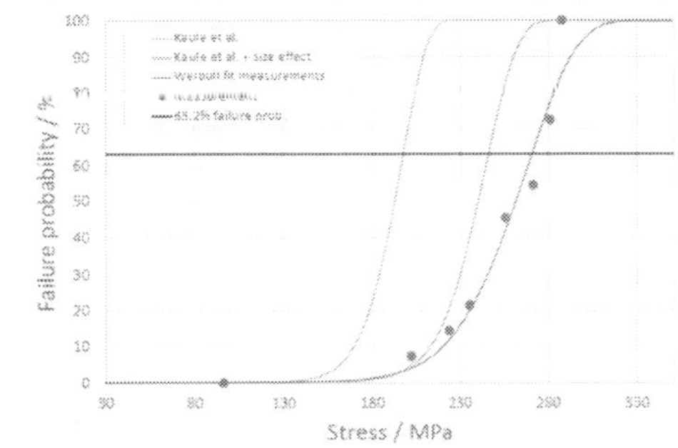

Figure 5: Weibull statistics for cell breakage experiments. Grey: 4-point-bending from [11], orange: Grey curve adapted to our tested volume, blue: this work.

choice of module internal pressure p_{mod} is a trade-off between higher *FF* or higher mechanical yield. This is a N.I.C.E.-specific optimization strategy.

The cell samples with wires were manufactured and put into the N.I.C.E. setup as described above, transferred into the flasher, the vacuum pump was switched on and LIV as well as EL were measured. We decreased p_{mod} step-by-step until visible cell breakage occurred. From the breakage statistics and the modeled stress in the cell, a data set was created. Fitting the data set with the Weibull distribution gave the characteristic fracture strength σ_0 and the Weibull modulus m.

The results for a total of 20 cells with 4 different wire offsets are shown in fig. 5 together with literature results. We see that the fracture strength derived from our model (270 MPa) is close to the extrapolation (from different effective stressed volumes) of the literature data derived from FEM simulations (246 MPa). The experimental data also show a reduced Weibull modulus, most probably due to some cracks introduced during the manual sample preparation. The pretty good match with literature data gives us confidence that the analytical mechanical model's accuracy is sufficient for estimating the necessary pick-and-place accuracy for N.I.C.E.-wire industrial fabrication.

To do so, we need to devise a method to determine the maximum allowable wire offset: Multiplying the survival probability with the relative *FF* gives a suitable optimization parameter in our experiment. In fig. 6, this parameter is plotted against module pressure and wire offset. Red regions indicate suitable production areas where high *FF* are combined with low breakage rates. As a first guess from this graph, we consider 200 µm wire offset as a good requirement for robot placement accuracy. This value depends of course on the diameter and the number of the wires. However, the order of magnitude is such that we are optimistic that industrial production of N.I.C.E.-wire is feasible with good mechanical yield without extreme lay-up precision.

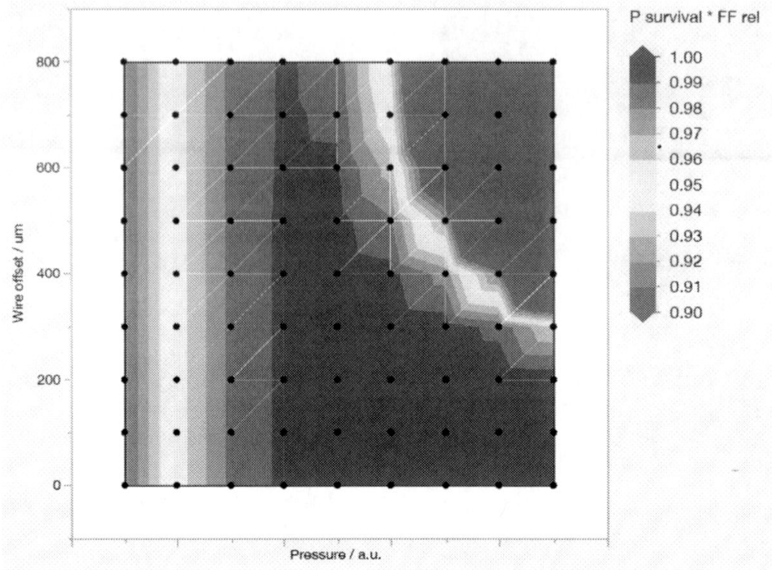

Figure 6: Contour plot of product of survival probability and relative fill factor versus module pressure and wire offset.

5 SUMMARY

The starting TWp age of PV needs sustainable and fully recyclable module technologies. N.I.C.E.-wire is one interesting candidate. The mechanical stresses in the module depend on the front and back wire offset that is determined by the robot pick-and-place accuracy. We have introduced an analytical mechanical model to investigate the maximum allowable offset. The calculated maximum stresses have been validated by experimental and literature data within reasonable differences.
This allowed us to optimize the internal pressure of N.I.C.E.-wire modules for lowest series resistance and highest mechanical yield. We derived a first estimate of 200 µm maximum allowed wire offset for a given module setup. This order of magnitude shows that an industrial fabrication of such modules is conceivable without extreme positioning accuracy.

BIBLIOGRAPHY

[1] S. Braun, G. Micard, and G. Hahn, "Solar Cell Improvement by using a Multi Busbar Design as Front Electrode," *Energy Procedia,* vol. 27, pp. 227-233, 2012, doi: 10.1016/j.egypro.2012.07.056.
[2] J. Dupuis, E. Saint-Sernin, O. Nichiporuk, P. Lefillastre, D. Bussery, and R. Einhaus, "NICE module technology - From the concept to mass production: A 10 years review," in *38th IEEE Photovoltaic Specialists Conference,* Austin, TX, USA, 2012, doi: 10.1109/PVSC.2012.6318254.
[3] D. Kray *et al.*, "N.I.C.E.-Wire: Next Generation Robust Eco-Friendly Bifacial PV Modules With High Efficiency," *IEEE Journal of Photovoltaics,* vol. 12, no. 1, pp. 38-44, January 2022, doi: 10.1109/jphotov.2021.3124168.
[4] A. Mette, "New Concepts for Front Side Metallization of Industrial Silicon Solar Cells," PhD thesis, University of Freiburg, Germany, 2007.
[5] L. C. Rendler, A. Kraft, C. Ebert, S. Wiese, and U. Eitner, "Investigation of Thermomechanical Stress in Solar Cells with Multi Busbar Interconnection by Finite Element Modeling," Proceedings of the 32nd European Photovoltaic Solar Energy Conference and Exhibition, EU PVSEC 2016 : 20 - 24 June 2016, Munich, Germany 2016, doi: 10.4229/EUPVSEC20162016-1CO.11.2.
[6] TaiyangNews, "Advanced Solar Module Technology Report 2018 Edition," *Taiyang News,* 2018.
[7] D. Gross, W. Hauger, J. Schröder, and W. A. Wall, Technische Mechanik 2, 14 ed. Springer, 2021.
[8] D. Gross, W. Hauger, J. Schröder, and W. A. Wall, Technische Mechanik 1, 14 ed. Springer, 2019.
[9] S. Dietrich: Numerische Zuverlässigkeitsuntersuchungen an Siliziumstrukturen. Masterarbeit am Fraunhofer Institut für Werkstoffmechanik, Hochschule für Technik, Wirtschaft und Kultur Leipzig (FH), Leipzig 2007.
[10] D. Reinwand, P. Wiechers, and D. Kray, "New Device for Accurate Measurement of Busbarless Bifacial Solar Cells Using N.I.C.E.TM Technology," Proceedings of the 37th EU-PVSEC (online) 202, doi: 10.4229/EUPVSEC20202020-2CV.1.10.
[12] F. Kaule, W. Wang, and S. Schoenfelder, "Modeling and testing the mechanical strength of solar cells," *Solar Energy Materials and Solar Cells,* vol. 120, pp. 441-447, 2014, doi: 10.1016/j.solmat.2013.06.048.

NOVEL PHOTOVOLTAIC MODULE DESIGN FOR DESERT CLIMATE

(1)Amir A. Abdallah, (1)Kamal Mroue, (1)Ayman Samara, (1)Brahim Aissa, (1)Maulid M. Kivambe, (1)Benjamin W. Figgis, (1)Juan Lopez Garcia
(2)Stephan Großer, (2)Matthias Pander, (2, 3)Bengt Jäckel, (2)Christian Hagendorf, (2)Matthias Ebert, (2, 3)Klemens Ilse
(1)Qatar Environment and Energy Research Institute (QEERI), Hamad bin Khalifa University (HBKU), Qatar Foundation, P.O. Box 34110, Doha, Qatar
(2)Fraunhofer-Center for Silicon-Photovoltaics (CSP), Halle (Saale), Germany
(3)Fraunhofer Institute for Microstructure of Materials and Systems (IMWS), Halle (Saale), Germany

ABSTRACT:
A wide deployment of Photovoltaics (PV) in desert climate conditions requires overcoming the impact of the environmental factors on PV performance and reliability. High Ultra-violet (UV) irradiance, in combination with high module operating temperature that may reach over 70 °C during summer months even for open rack installations, has shown to cause degradation of the PV module material. In this paper, we investigate the impact of UV irradiance on the properties of the PV module back sheet material. The data obtained from real outdoor testing conditions serves as a design guideline to develop a reliable PV module. For instance, FTIR (Fourier Transform Infrared) measurements on the polyamide-based back sheets confirmed the cracking and powdering of the back sheet material after 4 years of exposure in a desert climate.
Keywords: PV Module, Degradation, PV Materials, Durability, Encapsulation

1 INTRODUCTION

With the current PV market evolution with different cell size (M10, M12) [1], full-cell, half-cut or shingled cells, different cell technologies (PERC, SHJ, TopCon and IBC) with multi-busbars, different interconnection technologies and larger module size with a power density of >200 Wp/m^2, the question of which technology suits the desert climate is fundamental.

The PV module consists of solar cell string connected to each other with metallic interconnection and laminated between the front and rear encapsulation materials, as in the case of bifacial technology; the solar cell is then sandwiched between two solar glass sheets. Recently, transparent back sheet has been used for bifacial module.

PV modules are exposed to various stresses, especially in desert climate (e.g., high UV radiation, high temperature, the diurnal and the annual thermal cycles), which may affect module material durability and PV system performance. Many manufacturers of solar modules currently provide performance guarantees for 20 to 30 years. This is only possible when the service lifetime of all relevant components can meet this requirement.

The mostly used type of back sheet in silicon modules consisted of a three-layered "TPT" product: a polyethylene terephthalate (PET) film core layer with pigmented polyvinyl fluoride (Tedlar PVF) [2] on both sides (inner and outer layers). This is still commonly used today, along with a number of alternative materials introduced in the last years. The core layer is the primary providing electrical insulation, mechanical, and moisture barrier properties, and is most commonly a polyester film. The primary function of the outer layers is to protect the core material from environmental stresses and improve the barrier properties of the finished back sheet film. The outer/air side of the back sheet is most commonly a fluoropolymer, materials which are known for their remarkable weatherability due to the high strength of the carbon–fluorine bond which inhibits photolysis and thermolysis along the polymer backbone [3-4]. Common materials include PVF, polyvinylidene fluoride (PVDF), or fluoroethylene vinyl ether (FEVE). Nonfluoropolymer alternatives have also come into use, including PET, polyamide (PA), and polyolefin (PO). The chief advantage of these materials is their lower cost, which often comes at the price of reduced weatherability [1-2], [5]. The inner layer of the back sheet must adhere to the module encapsulation during the lamination process, and must also protect the interior, sun-facing side of the core material.

Therefore, a better understanding of polymeric back sheet degradation by the major environmental stress factors and an indoor testing that simulate the real outdoor testing conditions are required to improve material screening. In this paper, the degradation of back sheet material was studied. The focus was on the changes in polymer morphology and chemical degradation.

2 EXPERIMENTAL PART

Samples were exposed to real operating conditions at the Outdoor Test Facility (OTF) in Doha, Qatar. A 35,000 m^2 outdoor test site where the performance of different PV technologies is monitored. The OTF is equipped with a central weathering station where the meteorological parameters are measured.

Fourier Transform Infrared (FTIR) measurements were carried out in absorbance mode on a Thermo Scientific Nicolet iS50 FTIR spectrometer equipped with an Attenuated Total Reflectance (ATR) sampling accessory with a diamond crystal plate. Spectra were recorded with 32 scans per sample/background in the spectral range of 4000–400 cm^{-1} at 4 cm^{-1} spectral resolution.

Scanning Electron Microscope (SEM) and Energy Dispersive X-ray Spectroscopy (EDS) of the powder samples were dispersed onto adhesive carbon tape and some charging samples are gold coated with 5 nm thickness for electrical conductivity. FEI Quanta650FEG SEM was used for imaging in high vacuum mode at 5KV, while Bruker Quantax EDS was used at 15 kV.

To investigate the crystallographic phases in the samples, powder X-ray diffraction (PXRD) was investigated using the Rigaku SmartLab system (Cu k-α 1.5404 Å). The PXRD scans were performed using Bragg-

Brentano geometry. The powder sample was placed in a standard glass holder (20 x 20 x 0.2 mm³). The 2q scan range was set to be from 3-90̊, with step size of 0.02̊. All indexed phases were found from the Powder Diffraction Files (PDF) by the International Centre for Diffraction Data (ICDD).

The Differential scanning calorimetry (DSC) measurements were carried using TA Instruments (DSC Q2000) which has temperature resolution of 0.1 °C and sensitivity of 0.2 W. Two cycles of DSC test were conducted, Heat > Cool > Heat > Cool, which is needed to clear any thermal history of the material. The heating applied was -50 to 250 °C, and ramping rate of 10 °C/min. Plots of heat flow vs temperature was provided.

The Dynamic Mechanical Analysis (DMA) measurements were carried using TA Instruments (DMA Q800) with temperature resolution of 0.1 °C and force resolution of 10 - 5 N. A tensile stress test was conducted by applying a sinusoidal stretch with amplitude of 10 um and frequency of 1 Hz, while ramping temperature from 30 - 200 °C with 3 °C/min rate. Three DMA plots generated for storage modulus, loss modulus and damping factor (*tan δ*).

3 RESULTS

3.1 Desert climate

One of the main characteristics of desert climate is the high solar irradiance including UV and high operating temperature. From the UV radiometer, the annual UV irradiance of 111 kWh/m² was measured at the Outdoor Test Facility (OTF). On average, a UV-A and UV-B of 557 and 17 kW/m² were measured, respectively. During the summer months of June and July, the UV irradiance reaches its maximum. At the same period, while average ambient air temperature may reach 48 °C, module temperature over 70 °C for open rack installations was recorded during summer months.

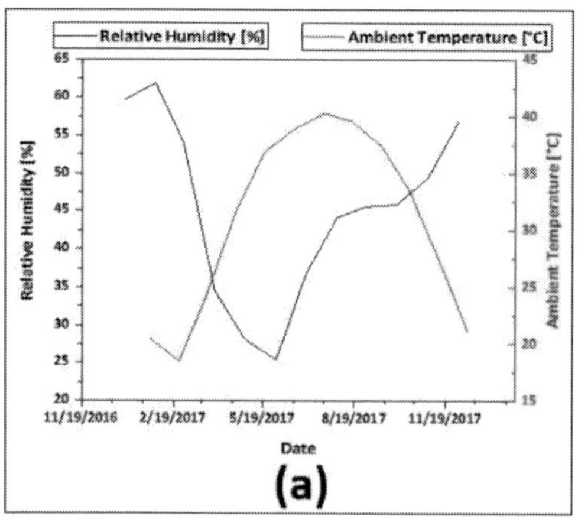

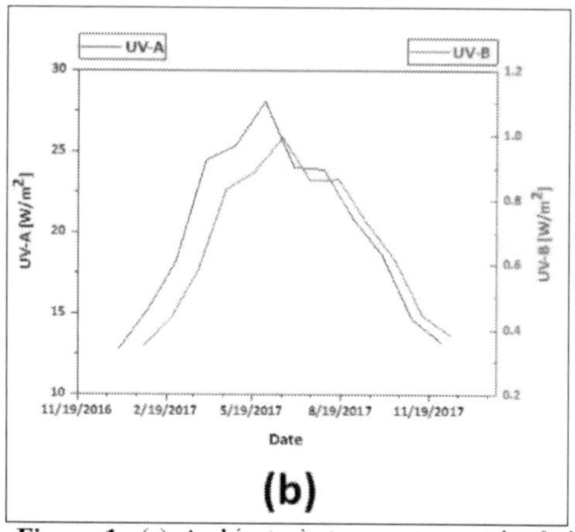

Figure 1: (a) Ambient air temperature and relative humidity, (b) Ground-based measurements of irradiances UVA (320–400 nm) and UVB (280–315 nm) measured at the OTF (Doha, Qatar) during 2017

3.2 Back sheet failure

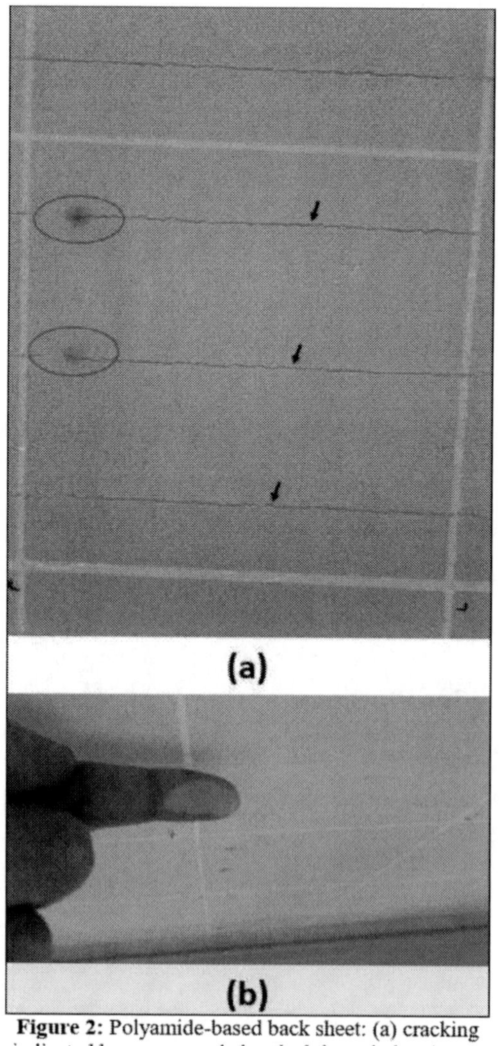

Figure 2: Polyamide-based back sheet: (a) cracking indicated by arrows and chemical degradation (green spots) indicated by red circles, (b) chalking

Various PV module back sheet failure modes were observed after 4 years of operation in a desert climate. Fig. 2a shows a typical back sheet cracking on a polyamide-based back sheet material observed along the busbars, which result in a degradation of the back sheet electrical insulation and barrier properties. In addition to the back sheet cracking and chalking, green points, indicated by red circle in Fig. 3b, were observed. Powder from the white and green points were collected from the degraded back sheet for further indoor analysis.

Firstly, the FTIR spectra of the reference polyamide (Nylon) and the white and green powder (Fig. 3) shows no significant change of the green and the white powder of the polyamide-degraded back sheet.

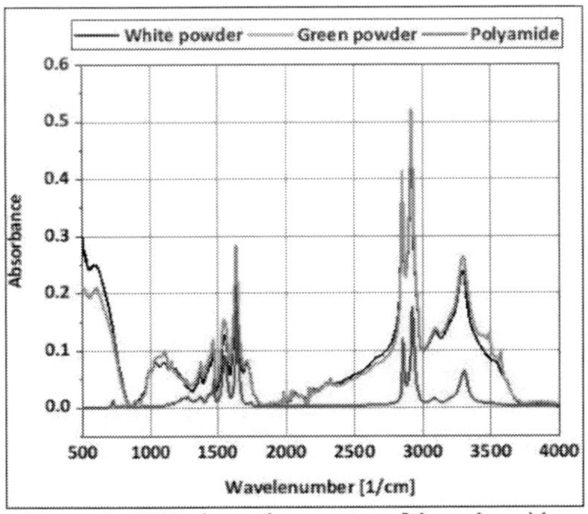

Figure 3: FTIR absorption spectra of the polyamide-based back sheet compared with the white and green powder collected from degraded-back sheet (see Fig. 2b)

Secondly, SEM images of the white and green powder collected from the degraded back sheet were compared as presented in Fig. 4. The white powder shows organic chunks covered with titania nanoparticles (TiOx). The EDS analysis (not shown here) for the white powder confirmed the presence of titania. In addition, a considerable amount of dust with fiber-like shape was also observed in the white powder.

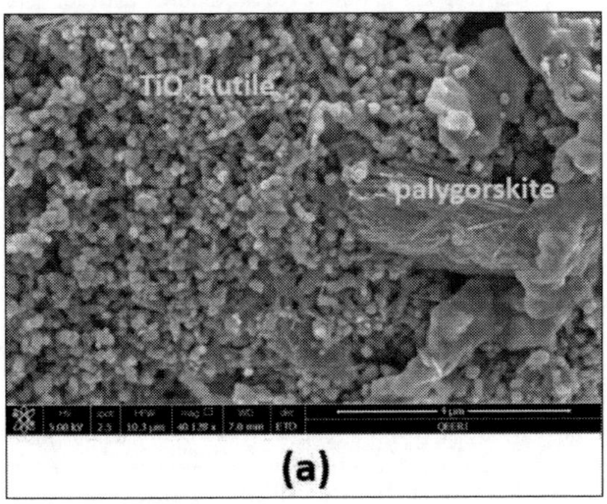

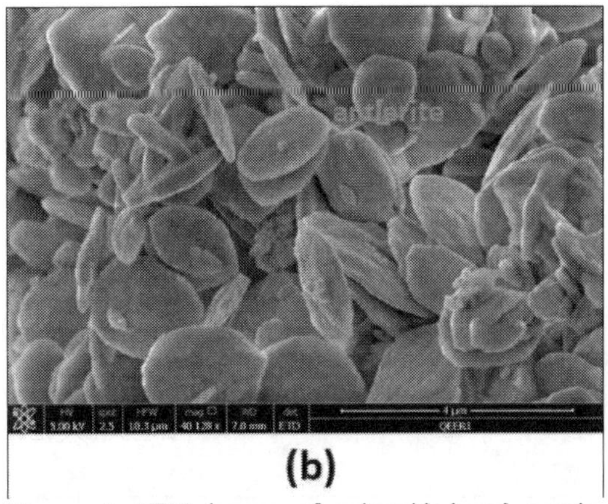

Figure 4: SEM images of polyamide-based powder collected from degraded back sheet: (a) white powder, (b) green powder

On the other hand, the SEM image of the green powder (Fig. 4b) shows antlerite (greenish hydrous copper sulfate mineral $Cu_3(SO_4)(OH)_4$). This has been confirmed by the EDS measurement (not shown here).

Lastly, from the XRD spectra in Fig. 5, the major phase found in both white and green powders is oxide mineral composed of titanium dioxide (TiO_2 Rutile). The broad peak at 2 Theta of 22 [deg] could be attributed to the polyamides. Other minor phases such as palygorskite, calcite, and dolomite are expected to orgiginate from dust accumulated on the back sheet.

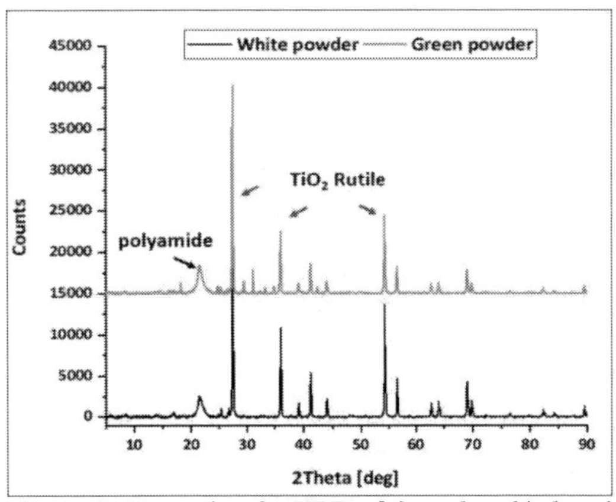

Figure 5: Comparing the XRD of the polyamide-based back sheet white powder and green powder collected from the degraded-back sheet. Polyamide and titanium dioxide (TiO_2 Rutile) are indicated by arrows.

3.3 Back sheet material screening

Different back sheet materials were screened to find the one with most durable properties that suit desert climate. PVL2 (PVDF/Polyester/Primer) and PVL405 (PVDF/Polyester/Primer) were investigated before and after 425 days of outdoor exposure in desert conditions at the OTF.

Fig. 6 shows the DMA plots of the back sheet materials before and after exposure in outdoor. Storage modulus (*E'*), defined as the elastic portion of the viscoelastic complex modulus and the loss modulus (*E"*) defined as the viscous part of the viscoelastic complex modulus, were plotted in Fig. 6 for the reference and aged (after 425 days of exposure at outdoor) back sheets. Initially, PVL405 back sheet shows a reduction in both the E' and the E'' when compared with the reference values, while PVL2 back sheet did not show a significant change in the E' value. However, PVL2 back sheet showed a shift in the E'' curves towards high temperature (Fig. 6b).

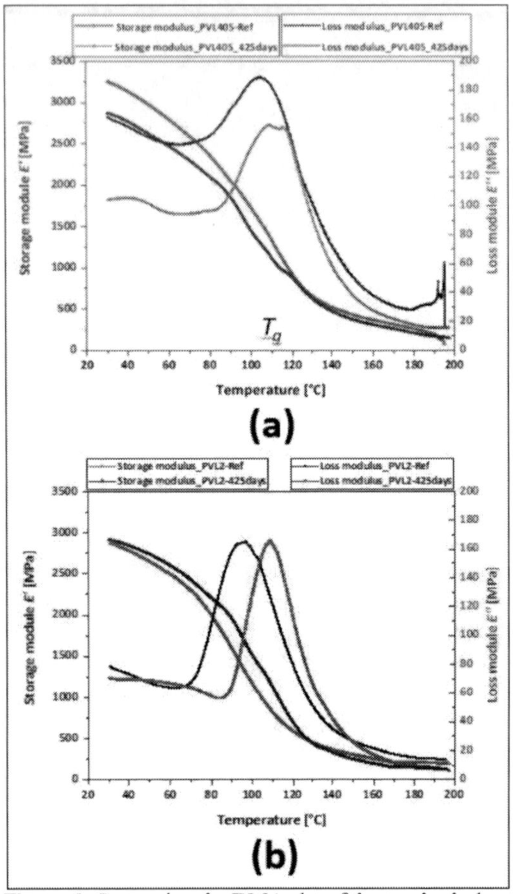

Figure 6: Comparing the DMA plot of the two back sheet materials (a) PVL405 and (b) PVL2.

Fig. 7 shows the DSC measurements for the back sheet materials reference and after outdoor exposure for 425 days. Glass transition (T_g) region starts when the loss modulus starts to have an upward increase, and the T_g temperature is around the peak of the loss modulus curve where the material starts to soften. A good agreement of the T_g region was obtained from both DSC and DMA measurements. For both back sheet materials, the melting point (T_m) is at 165 °C. No changes in the T_g and T_m were observed for the aged back sheets.

4 SUMMARY OF THESE NOTES

- High UV irradiance in combination of high operating temperature and humidity typical to desert climate impact the PV module material durability and PV system reliability.
- Yellowing, cracking, chalking and chemical degradation of a polyamide-based back sheet material were observed after 4 years of operation in desert climate. Further work is required to understand the origin of the Cu in the degraded-back sheet material.

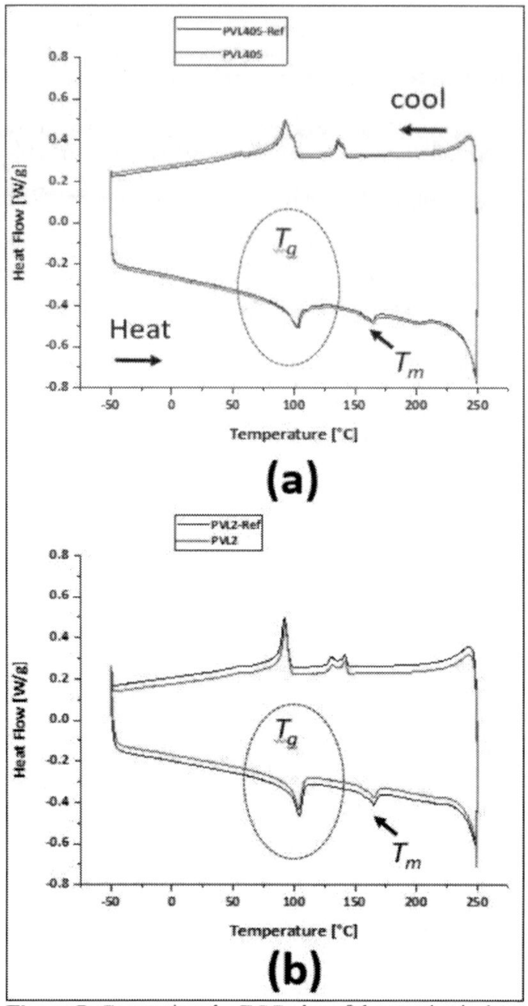

Figure 7: Comparing the DSC plot of the two back sheet materials (a) PVL405 and (b) PVL2. Heat and cool cycles of DSC test were conducted, T_g region is indicated by a circle and melting point T_m is indicated by an arow.

- Back sheet cracking and/or yellowing may affect the electrical insulation of PV module and therefore should be monitored regularly and replaced if necessary.
- Screening of two PVDF based back sheet materials was carried out to investigate material durability to UV. The aging at outdoor conditions does not impact the material properties significantly. However, long-term exposure is required.

5 ACKNOWLEDGEMENT

This publication was made possible by NPRP grant # [NPRP11S-1220-170110] from the Qatar National Research Fund (a member of Qatar Foundation). The findings herein reflect the work, and are solely the responsibility, of the authors. The authors would like to thank the Core Labs at Qatar Environment & Energy Research Institute for material characterization.

6 EFERENCES

[1] International Technology Roadmap for Photovoltaic (ITRPV), 2021 edition.
[2] A. Andrady. Plastics and the Environment. John Wiley & Sons, Inc (2005).
[3] S.H. Hamid, M.B. Amin, A.G. Maadhah S. Ebnesajjad, P.R. Khaladkar, Fluoropolymers Applications in Chemical Processing Industries - the Definitive User's Guide and Data book (First ed.), William Andrew, Inc., Norwich, NY (2005).
[4] Handbook of Polymer Degradation. (Second ed.), CRC Press (2000).
[5] C. Harper Handbook of Plastics, Elastomers, and Composites. (Fourth ed.), McGraw Hill (2002).

2.5 MINUTES LAMINATION PROCESS AND THE INFLUENCES OF THE DEGREE OF CROSS-LINKING AND THE MOISTURE INGRESS ON THE DEGRADATION OF PV MODULES

Aksel Kaan Öz[1], Christine Wellens[1], Max Mittag[1], Martin Wiese[1], Sraisth Sraisth[2], Daniel Klaus[2], Dirk Holger Neuhaus[1], Martin Heinrich[1]
[1]Fraunhofer Institute for Solar Energy Systems ISE, Heidenhofstraße 2, 79110 Freiburg, Germany
[2]Robert Bürkle GmbH, Stuttgarter Straße 123, 72250 Freudenstadt

Corresponding Author: Aksel Kaan Öz | +49 (0) 761 4588 5025 | e-mail: aksel.kaan.oez@ise.fraunhofer.de

ABSTRACT: Faster lamination processes and more sustainable modules will play a critical role to meet the industries demand for green energy. This study presents the influences of short lamination processes on the moisture balance, achieved by increasing the lamination temperature up to 180 °C, and compares these with modules laminated at a common lamination temperature of 150 °C. The investigations are conducted mainly on mini modules, but additionally to those also a standard 60 cell module is built and exposed to the damp heat (DH) test to have a better understanding of the influences of the fast lamination process. The mini modules were exposed to DH for 1500 and 3000 hours and were measured before and after the accelerated ageing in terms of IV measurements and peel tests. We observed no severe degradation caused by the higher lamination temperature in terms of moisture ingress. During the study we also investigated the cost of the 2.5 minutes curing process and discussed the potential benefit for module production facilities. The results of the simulation showed that module manufacturers could reduce their module costs up to 1.3% per module with the faster process.
Keywords: PV Modules, damp heat, degradation, economic analysis, manufacturing

1 INTRODUCTION

The lamination process is one of the most critical steps that influences the quality of a photovoltaic (PV) module in terms of long term stability [1]. In order to provide operational product lifetimes of at least 20 years, the lamination process should ensure a proper adhesion between the laminate layers and a certain degree of crosslinking [2]. Common methods for the quality control of the lamination process is performing peel tests and measuring the degree of crosslinking of the PV modules [3]. However, the lifetime of a module in terms of encapsulation material processing is mainly influenced by the moisture ingress and its resulting consequences that occur during the operational lifetime of a module. The amount of the moisture ingress can be reduced by using edge sealing or encapsulation materials with a high-water vapor transmission rate (WVTR) [4]. Todays most widely used encapsulant is ethylene vinyl acetate (EVA) which has a market share of around 70% [5]. The moisture that enters the module can react with the EVA in the form of hydrolysis and generates acetic acid. The generated acetic acid causes corrosion and leads to a decrease in performance of the modules [6]. Polyolefins are an alternative to EVA and are preferred for laminating SHJ cells or for the glass-glass layup [7]. For the next years, the PV market is expected to keep continue its increasing trend [5]. To meet the demand of PV modules, manufacturers are trying to shorten the lamination process [1]. This can be achieved by using higher lamination temperatures [8]. This study focuses on laminating at 180 °C and the long-term influences of this lamination processes in terms of moisture ingress. To observe the long-term effects, the modules underwent the damp heat (DH) test. The aim of the DH test is to insert moisture into the module to accelerate the ageing process [9,10]. The modules are then evaluated in comparison to modules from the standard temperature lamination processes in regards of performance, and adhesion forces before and after the DH tests. Additionally, three different gel contents and their influence on the degradation behaviour and the productions costs of the 180°C lamination process are analysed.

2 MATERIAL AND METHODS

2.1 Sample Preparation

For this study, a total of 63 mini modules in the sizes of 280 mm x 250 mm were fabricated by using a Bürkle-Ypsator laminator. To protect them from humidity ingress through edges, an edge sealant was used. To analyse the influences of the moisture ingress and its relation to the gel content two different backsheets, one with a high WVTR and one with a low WVTR were chosen. These modules were built with moisture-sensitive heterojunction (SHJ) solar cells at a lamination temperature of 150 °C. Three different lamination duration times were used to get a degree of crosslinking of around 60, 70, and 80%. For the investigations of the short lamination process modules were cured at a peak lamination temperature of 180 °C for again three different lamination duration times. These modules were built with PERC cells and the low WVTR backsheet. In addition to these mini modules a standard-sized module with 60 PERC cells and the same low WVTR backsheet was built and tested in DH1000 to analyse the influence of the 2.5 minutes curing process on a standard module. All modules were laminated with a polyolefin elastomer (POE) encapsulant to protect the moisture-sensitive heterojunction (SHJ) solar cells. The specifications of the materials used in this study are shown in Table 1.

Table 1: List of used encapsulation materials and their specifications

Layer	Material	Thickness [mm]	WVTR [g/m² x day]
glass 1	tempered glass	3	0
glass 2	solar glass	3.2	0
backsheet 1	PET/PET/Primer	0.493	<0.5
backsheet 2	PET/Primer	0.218	<2.2
encapsulant	POE	0.68	n.a

In order to have reliable results, seven prototypes were built for each combination. Three of them were used for the initial measurements, two for DH1500, and the rest for DH3000 testing. The overview of the fabrication and characterization process is shown in Figure 1 and Figure 2.

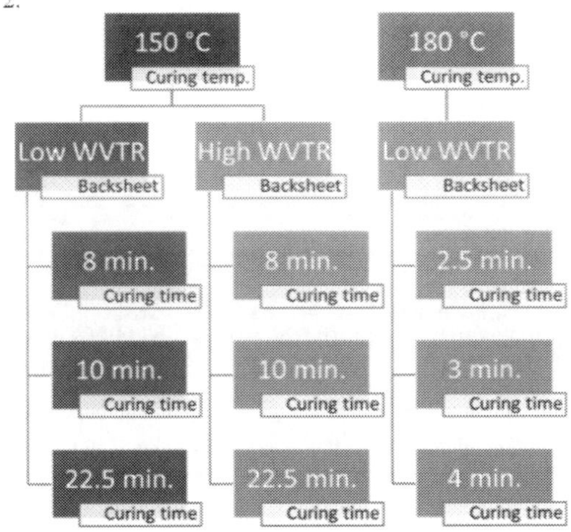

Figure 1: Lamination process specifications of the prepared samples

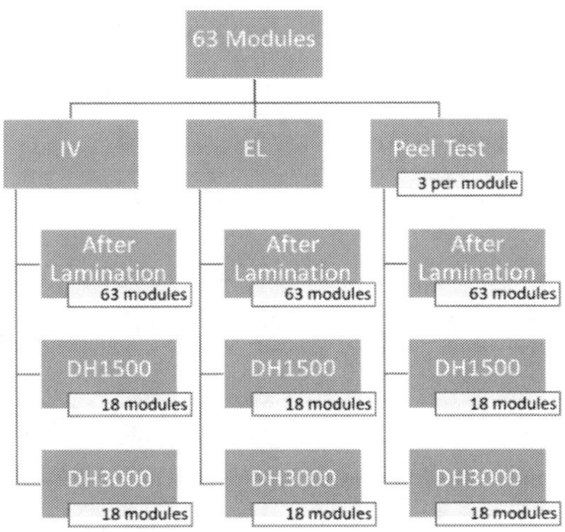

Figure 2: Module characterization methods and number of samples for each degradation condition

2.2 Soxhlet Extraction

To identify initial process durations and determine the degree of crosslinking of the modules, the Soxhlet extraction method was applied based on EN ISO 10147 and EN ISO 6427. It is an easy method that determines the degree of crosslinking by measuring the gel content before and after the extraction. In the PV industry, it is used for the development of new processes and quality control [11]. The serial extraction apparatus from Behrotest shown in Figure 3 was used for the Soxhlet measurements in this study.

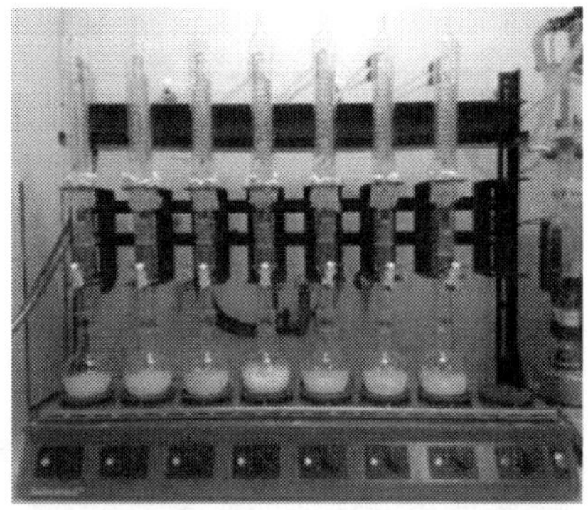

Figure 3: Serial extraction apparatus from Behrotest

2.3 Flasher Measurements

The initial characterization as well as the performance of the modules after the ageing tests were observed by current-voltage (IV) curve measurements with a WaveLabs flasher. The standard-sized module was measured at Fraunhofer ISE TestLab PV Modules according to IEC 60904-1 and IEC 60904-3.

2.4 Electroluminescence (EL) Measurements

To characterize the modules and analyse the influences of the moisture ingress in the cells, EL images of all modules are taken after certain periods. This allows us to make a correlation between the decrease in module performance and the moisture ingress into the module.

2.5 Peel Tests

The peel tests in the study were conducted at the glass-encapsulant interface by using the peel-off instrument manufactured by PSE. All peel test measurements of the mini modules were done according to the BS EN ISO 8510 standard [12]. Three strips with a width of 10 mm each were cut along the module surface and were measured with a peel angle of 90° with a speed of 50 mm/min. The relative maintained peel force (RMPF) which is used for the evaluation is determined by the following formula [8]:

$$RMPF = \frac{Peel\,force\,after\,DH\,3000}{Initial\,peel\,force}\,x\,100\%$$

2.6 Accelerated Ageing

For long-term investigations of the module performance, damp heat (DH) tests were conducted. The DH test is a standard method for ageing polymer materials used in PV. The modules were exposed to DH (85 °C, 85%r.h.) for 1500 hours and 3000 hours. This method was chosen, because it degrades the modules by moisture ingress into the module from the environment and allows us to compare the difference of the low WVTR and the high WVTR backsheets [3].

2.7 Energy Consumption and Cost Calculation

Analyses of the energy consumption of both, the short lamination process and a typical industrial process, (at 155 °C) were conducted by Robert Bürkle GmbH. Ten runs were done with a Bürkle Ypsator laminator for each process. The energy consumption per module was calculated with the following formula:

$$Energy\ per\ module = \frac{energy_{heating} + energy_{\text{switching cabinet}}}{\text{number of modules}}$$

The impact of changes in energy consumption and increased cycle time are analysed using the "SCost" cost model developed by Fraunhofer ISE [13]. The model calculates the Cost of Ownership (COO) for a module or manufacturing site based on module design information (e.g., cells per module, layers, size, etc.), equipment, and process parameters (e.g. cycle times, breakage rate, equipment availability, utility consumption, etc.) and additional information (e.g. taxes, costs for infrastructure, capital costs, etc.). The model is based on SEMI E10 and E35 standards.

Slower and faster lamination process together with their specific energy consumption and other equipment parameters have been modelled and the impact of the changes in the process have been analysed with regard to the impact on the costs of lamination (see Table 2).

Table 2: List of inputs for process cost modelling

	Fast process	Slow process	
Equipment price	610.000	310.000	€
Cycle time	7.5	14	min/module
Electricity consumption	34.2	33.7	kWh/h
Personnel	0.05	0.05	Persons/shift
Floor space	246	246	m²
Equipment uptime	94%	94%	
Breakage rate	0.2%	0.2%	

3 RESULTS

3.1 Soxhlet Extraction

To decide the lamination durationtime and reached gel contents of around 60, 70, and 80% for the 150 °C lamination process Soxhlet measurements were done for certain curing durations. The goal for the 180 °C lamination process was to reach reliable gel contents with a short curing time. The behaviour of the gel content of POE versus the duration time is shown in Figure 4 for the two lamination temperatures of 150 °C and 180 °C.

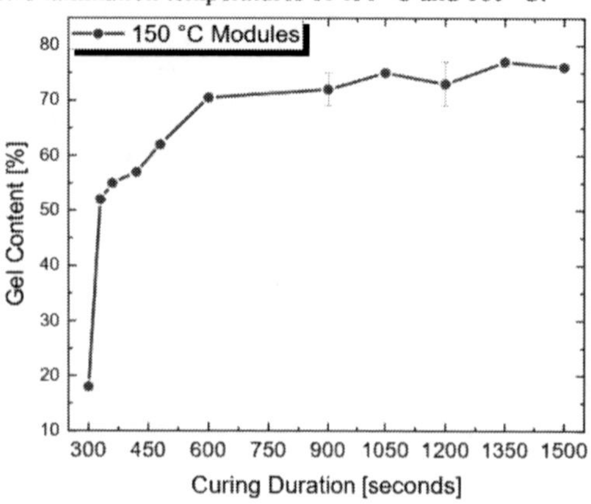

(a)

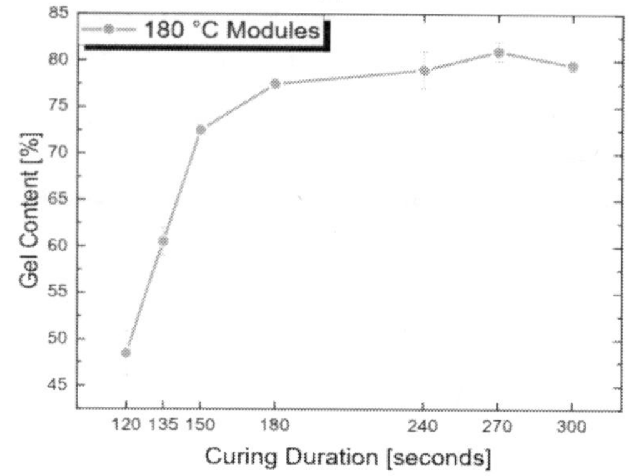

(b)

Figure 4: Gel content measurement results in samples manufactured using different process conditions, (a) 150 °C lamination temperature, (b) for 180 °C lamination temperature

The gel content of the modules laminated at 150 °C in Figure 4 (a) showed a logarithmic behaviour between 300 and 600 seconds. After this period, it became more consistent and reached a gel content of around 75%. To reach approximately the required three gel contents (60, 70, 80%), the curing duration time was set to 480 (62%), 600 (70.5%) and 1350 (77%) seconds, respectively.

For the modules laminated at 180 °C it was observed that the period from 120 to 150 seconds plays a critical role for the degree of cross-linking, whereas after only 180 seconds it roughly reaches its saturation point. With respect to the results shown in the Figure 4 (b), 150, 180 and 240 seconds were chosen as lamination duration times with high potential to pass the further tests.

3.2 Performances before/after Accelerated Ageing

This section analysis the influences of the moisture ingress that took place during the DH test on the performance of the modules, which were laminated under various lamination conditions. The degradation of module performances over time is shown in Figure 5.

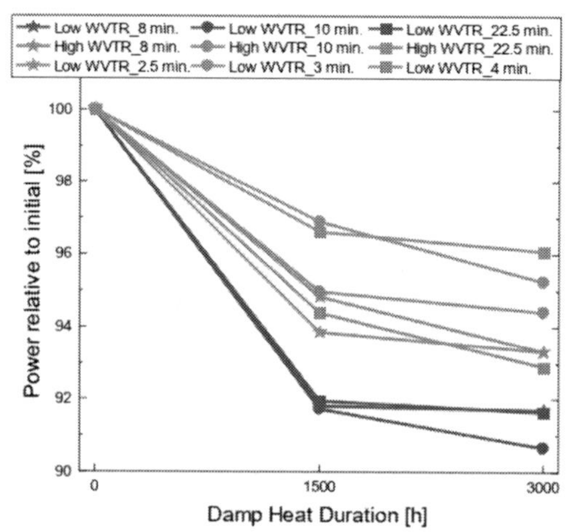

Figure 5: Module performances ratio before to after the DH test. Please note, that the orange curves are obtained

with PERC cells and the blue curves are obtained with SHJ cells.

Figure 5 shows that all mini modules are mainly influenced during the first 1500 hours of DH. It is seen that the degradation rate from DH1500 to DH3000 is much lower compared to the degradation that takes place in the first 1500 hours. It is also observed that there is a trend for the loss in performance with respect to the backsheet. It is found that all modules laminated at 150 °C with the high WVTR backsheet had a higher relative power after both the DH1500 and the DH3000 tests compared to the modules with the low WVTR backsheet. When comparing the results of the low WVTR backsheet for the 150 °C and 180 °C processes, it can be observed, that the 180 °C processes performed better with less degradation values, as expected of the more moisture robust PERC cells used for the 180 °C process. Also, as it is seen in Figure 5, the only modules which show a degradation lower than 5% after DH3000 test were the ones laminated at 180 °C with curing times of 3 and 4 minutes. The only outlying result for the 180 °C processes is the 2.5 min curing process, which has a lower performance to the others.

After evaluating the IV measurements of the mini modules that were exposed to the DH test, visual analysis was conducted on them. The results are shown in Figure 6 below.

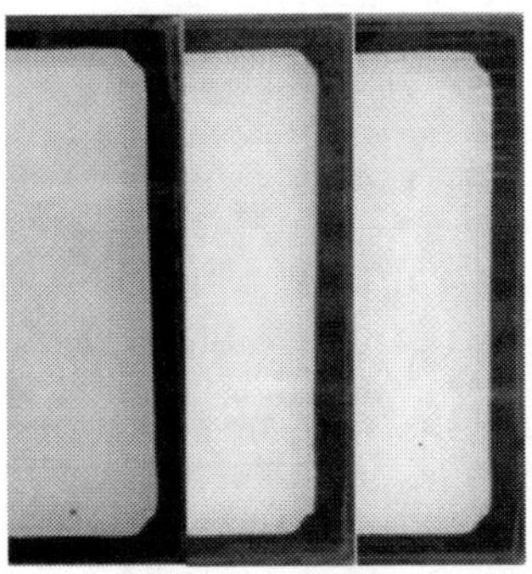

(a)

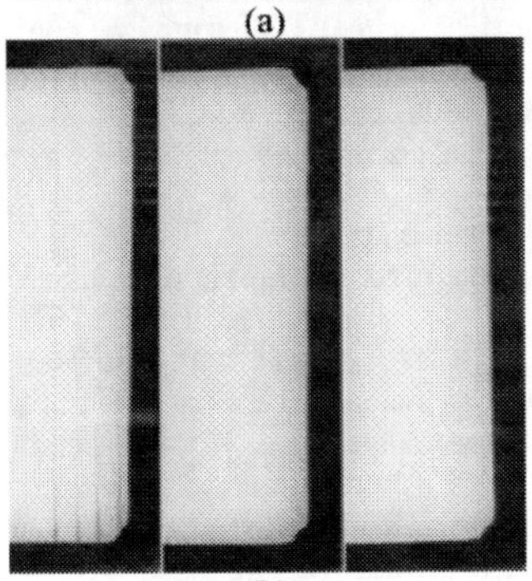

(b)

Figure 6: Discolouration of the 150 °C modules after DH1500, laminated at 8 min, 10 min and 22.5 min, respectively, (a) low WVTR backsheet, (b) high WVTR backsheet

The visual inspections showed that there are differences in terms of discolouration mainly between the two backsheet types, but also between the lamination time, especially for the low WVTR backsheet. Figure 6 (a) which presents the discolouration of the modules laminated with the low WVTR at 8, 10, and 22.5 min curing durations shows that there is a trend in discolouration from low gel content to high gel content. The modules with lower gel content discoloured much more than the modules with higher gel content. The same trend is also obtained in the modules laminated at 180 °C and the ones which used the high WVTR backsheet (not shown here). Whereas at the high WVTR backsheet it is not as clear to see as it is for the low WVTR one. But instead, Figure 6 (b) shows that the moisture ingress probably mainly took place from the edges of the mini modules, because the discolouration at the edges and especially at the corners is stronger compared to the centre of the modules. This phenomenon might also explain, why the module with the 180 °C / 2.5 min process in Figure 5 has significantly lower performance, compared to the ones with higher gel content. Its performance could be lower, because of the higher discolouration and reduction in transparency.

To have a better understanding of the fast lamination process and its long-term influences, a standard 60-cell glass-backsheet module was laminated at 180 °C with the 2.5 min curing process exposed to DH1000. The only difference between this module compared to the mini modules was that it had a frame to protect it from the moisture ingress through the edges instead of the edge sealant. The DH100 results of the Fraunhofer ISE TestLab, which are summarized in Table 3, showed that the degradation in power is less than 1% (below the testing accuracy), which indicates that the standard sized module laminated with the 2.5 min curing process passed the DH1000 test.

Table 3: Performance of the 60-cell module before and after DH1000

		Isc [A]	Voc [V]	Impp [A]	Umpp [V]	Pmpp [W]	FF [%]	Degradation of Pmpp [%]
60 Cell Module	Before DH 1000	10.076	40.609	9.473	32.842	**311.140**	76.038	0.73
	After DH 1000	9.979	40.694	9.370	32.962	**308.867**	76.056	

3.3 Electroluminescence (EL) Measurements

The EL images, of the PERC cells laminated from top to bottom with the 2.5, 3 and 4 minutes curing processes are shown in Figure 7, respectively. The left side shows the EL images before ageing, the right ones after DH3000.

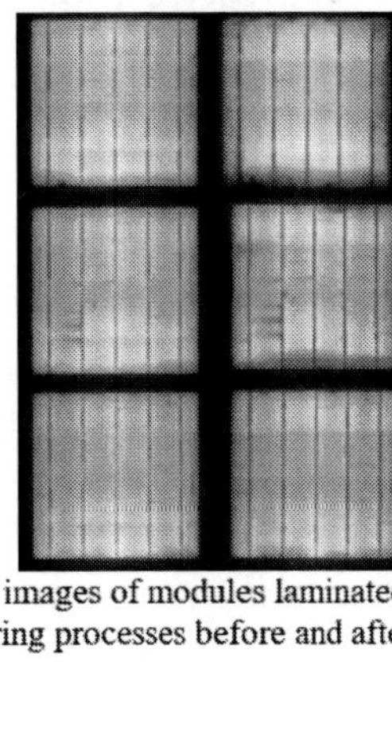

Figure 7: EL images of modules laminated with the 2.5, 3 and 4 min curing processes before and after DH3000

Slight differences in grey scale can be noticed for all modules. The most certain one, however, occurred at the 2.5 min process. Here darker zones at the edges of the module are observed after DH3000. These indicates the ingress of moisture into the module which causes degradation and loss in performance. This result also matches with the one in Figure 5.

The EL images of the standard-sized 60 cell module are shown below. The initial image is represented on the left side, the result after DH1000 on the right.

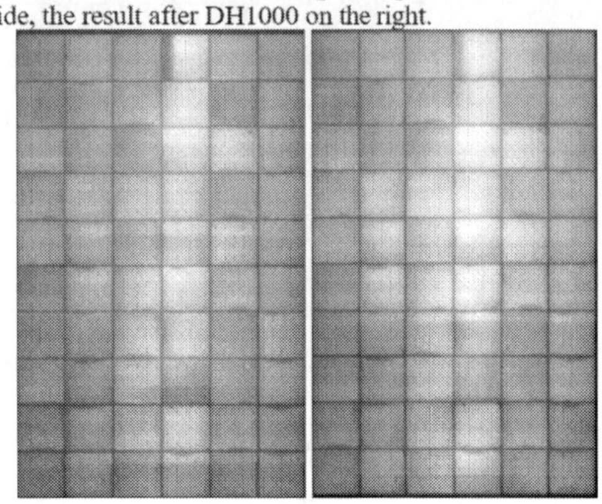

Figure 8: 60 cell module before andafter DH1000

As seen in Figure 8 there are some small defects, but most of them were already there before DH1000. The cells at the edge seem not to be affected that much by moisture compared to the mini modules laminated using the same process.

3.4 Peel Tests

For the peel tests, 3 measurements were done at the glass-encapsulant interface of each module, next to the cells. The overview of the results for the 150 °C processes and the 180 °C processes are shown in Figure 9 (a) and Figure 9 (b), respectively. The time when the measurements were conducted on top of the graph, the duration time of the curing process on the bottom, and the colours of the boxes are done according to the same systems applied in Figure 1.

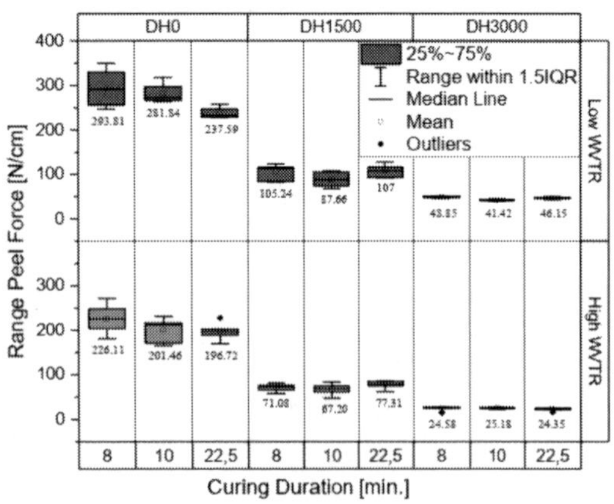

(a)

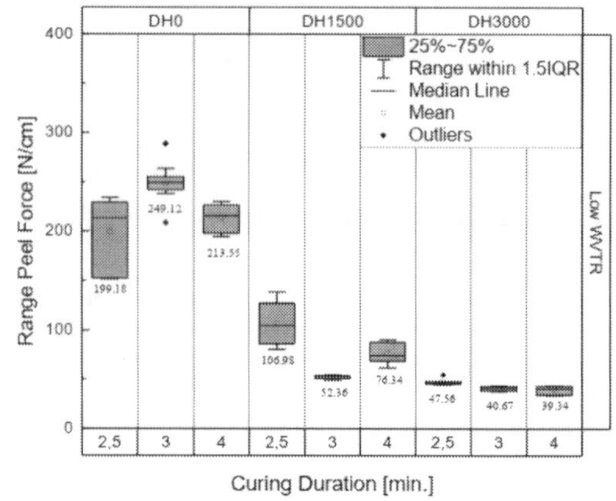

(b)

Figure 9: Peel force at the glass-encapsulant interface before and after DH1500 and DH3000 for (a) the 150 °C lamination process and the (b) 180 °C lamination process

Figure 9 (a) shows that the modules laminated with the low WVTR backsheet have always higher peel forces with respect to the ones laminated with the high WVTR backsheet. The difference between them, however, gets lower after accelerated ageing. Another point is that the highest peel forces, that are observed during the initial measurements for both backsheets are the ones with the lowest gel content (8 min curing).

The RMPF values after DH3000 for the 150 °C, low WVTR modules varied between 14.7 - 19.4% while the RMPF values for the 150 °C, high WVTR modules are with 11.7 - 12.5% lower.The modules laminated with the high WVTR backsheet reach only peel forces of around 25 N/cm.

The peel forces of the modules laminated at 180 °C shown in Figure 9 (b) look more distributed than the 150 °C ones. Initially, lower peel forces at 180 °C lamination as compared to lamination at 150 °C are observed. However, after DH3000, they perform similar to the 150 °C, low WVTR (45.5 N/cm) modules and get peel forces of around 42.5 N/cm. Therefore, we conclude that the lamination temperature does not negatively influence the peel forces after ageing with DH3000. A correlation between the gel content and the RMPF was not observed.

3.5 Energy Consumption

The comparison of the short lamination process at 180 °C with a commonly used lamination process in the industry at 155 °C, showed that, using the new process would cause an increase in energy consumption, due to the higher process temperature. However, through the short process time, the production rate of PV modules would be much higher than with the slow processes. This results in an increased number of modules per day effectively reducing the energy consumption per module. The analysis showed that there is a potential to reduce the energy consumption per module by approximately 46% by using the short lamination process.

The reduction of energy consumption reduces the costs for electricity. While energy costs are responsible for 23% of the total process costs in the slow process, they only make up 17% in the fast process.

Additionally, much higher benefits than energy saving can be found through additional effects on other cost factors, especially infrastructure/facility and labour costs. Those can be considered fix, meaning that they are not significantly impacted by the choice of process. Since the faster process allows for a higher manufacturing output, the lamination costs per module are therefore reduced (see Figure 10).

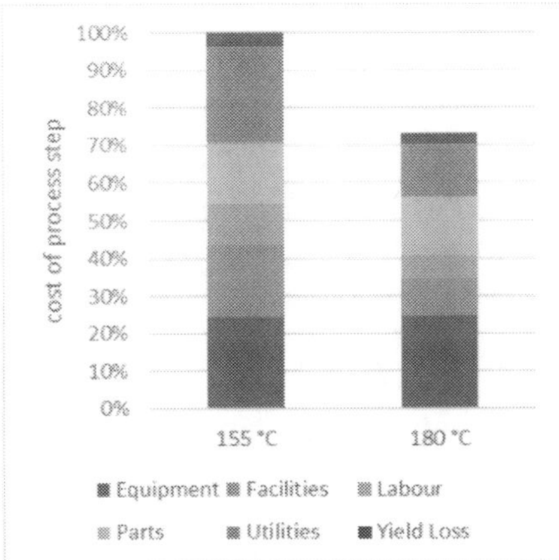

Figure 10: Cost structure of both lamination processes

We find that the increased cycle time effectively reduces the costs of lamination by 27%. Considering that the lamination process step accounts for 5.8% of the total module costs in our calculation, the benefits of the 180 °C, 2.5 min curing process led to a cost advantage of 1.3% in module costs (€/module and €/W_P).

4 CONCLUSION

In this work, we studied the long-term influences of the fast-curing process in terms of moisture ingress, and we simulated the energy consumption and changes in module costs with the "SCost" model of Fraunhofer ISE. Additionally, the influences of the degree of crosslinking and moisture ingress on the degradation of the PV modules were analysed regarding the module's performances and peel forces. The study was conducted on mini modules, which were laminated at different curing duration times to get different levels of gel contents and by using two different backsheets, which had a low and a higher WVTR.

The peel test results indicate that the modules with the low WVTR backsheet have also higher RMPF values after DH3000. When comparing the RMPF values of the 150 °C and 180 °C lamination processes it is seen that they vary between 14.7 - 19,4% and 16.3 - 23.9%, respectively. This means that the peel forces of the 180 °C processes degraded slightly less than for the 150 °C process.

The IV results of the modules laminated at 150 °C showed that the modules with the high WVTR backsheet degraded less than the modules with a low WVTR backsheet. The visual inspection revealed that all modules with the low WVTR backsheet discoloured more than the ones with the high WVTR backsheet. The decrease in reflectivity might be the main cause for the higher loss in performance. The discoloration can be in correlation with the oxygen transmission rate and the WVTR characteristics of the backsheets. It might be that the moisture, that seems to have entered the modules from the edges, could leave the modules with the high WVTR backsheet easier through the backsheet, causing less discoloration. More investigations need to be done to have a better understanding of this phenomenon.

The standard-sized 60-cell module cured for 2.5 minutes showed a loss in performance of less than 1% after DH1000.

The energy consumption and economic analysis showed that the energy consumption of the 180 °C process is a little higher than for the 150 °C process, but due to its fast production rate, the energy consumption per module would reduce by approximately 46%. The simulation also showed that the module costs (€/module and €/W_P) could be reduced up to 1.3% by using the fast lamination process.

It can be concluded that the fast lamination process at 180 °C has many benefits in terms of energy consumption and cost reduction per module. It was also shown that a full-size module manufactured with the process does not show any significant degradation after DH1000.

5 ACKNOWLEDGEMENT

This work was supported by the German Federal Ministry for Economic Affairs and Climate Action (BMWK) under the contract number 0324287C, acronym GEPARD.

References

[1] S. Sraisth, Achieving Faster Lamination Process for Crystalline Photovoltaic Modules by Using Latest Lamination Technologies, 6 pages / 33rd European Photovoltaic Solar Energy Conference and Exhibition; 992-997 (2017).

[2] C. Hirschl, L. Neumaier, S. Puchberger, W. Mühleisen, G. Oreski, G.C. Eder et al., Determination of the degree of ethylene vinyl acetate crosslinking via Soxhlet extraction: Gold standard or pitfall?, Solar Energy Materials and Solar Cells 143 (2015) 494–502.

[3] P. Schenk, M. Pander, U. Zeller, B. Jäckel, M. Ebert, Correlation of Peel Forces to EVA Degree of Cross Linking and Accelerated Weathering, 6 pages / 37th European Photovoltaic Solar Energy Conference and Exhibition; 834-839 (2020).

[4] S. Krauter, R. Pénidon, M. Hanusch, B. Lippke, P. Grunow, PV Module Lamination Durability, in: Proceedings of the ISES Solar World Congress 2011, Kassel, Germany, 2011, pp. 1–9.

[5] M. Fischer, M. Woodhouse, Ss. Herritsch, J. Trube (Eds.), International Technology Roadmap for Photovoltaic (ITRPV), 13th ed., VDMA.e.V., 2022.

[6] A. Masuda, N. Uchiyama, Y. Hara, Degradation by acetic acid for crystalline Si photovoltaic modules, Jpn. J. Appl. Phys. 54 (2015) 04DR04.

[7] M. Fischer, M. Woodhouse, Ss. Herritsch, J. Trube (Ed.), International Technology Roadmap for Photovoltaic (ITRPV), 12th ed., VDMA.e.V., 2021.

[8] A.K. Öz, C. Herzog, C. Wellens, D.E. Mansour, M. Heinrich, A. Kraft, The Impact of the Lamination Process on the Adhesion Properties at the Glass-Encapsulant Interface and Damp Heat Stability of PV Modules, 7 pages / 38th European Photovoltaic Solar Energy Conference and Exhibition; 708-714 (2021).

[9] O.K. Segbefia, A.G. Imenes, T.O. Sætre, Moisture ingress in photovoltaic modules: A review, Solar Energy 224 (2021) 889–906.

[10] D.E. Mansour, C. Barretta, L. Pitta Bauermann, G. Oreski, A. Schueler, D. Philipp et al., Effect of Backsheet Properties on PV Encapsulant Degradation during Combined Accelerated Aging Tests, Sustainability 12 (2020) 5208.
[11] C. Hirschl, M. Biebl–Rydlo, M. DeBiasio, W. Mühleisen, L. Neumaier, W. Scherf et al., Determining the degree of crosslinking of ethylene vinyl acetate photovoltaic module encapsulants—A comparative study, Solar Energy Materials and Solar Cells 116 (2013) 203–218.
[12] BS EN 28510-1:1993, ISO8510-1:1990. Adhesives--peel test for a flexible bonded-to-rigid test specimen assembly-Part 1: 90° peel (1993).
[13] S. Nold, Techno-ökonomische Bewertung neuer Produktionstechnologien entlang der Photovoltaik-Wertschöpfungskette: Modell zur Analyse der Total Cost of Ownership von Photovoltaik-Technologien, Fraunhofer Verlag, Stuttgart, 2019.

INFLUENCE OF THE LAMINATION PRESSURE ON THE ADHESION, DEGREE OF CROSS-LINKING, AND BUBBLE FORMATION OF PV MODULES

Sraisth[1], Aksel Kaan Öz[2], Daniel Klaus[1], Christine Wellens[2], Martin Heinrich[2]
[1]Robert Buerkle GmbH, Stuttgarter Straße 123, 72250 Freudenstadt, Germany
[2]Fraunhofer Institute for Solar Energy Systems ISE, Heidenhofstraße 2, 79110 Freiburg, Germany

ABSTRACT: The lamination parameters can significantly influence the process to make it defect-free and faster or slower, depending on the production need. In this paper, one of these parameters will be discussed, i.e., lamination pressure. This study is focused only on the pressure influence on the glass-backsheet (GB) modules and further compares the differences in two different encapsulant types, Ethylene-Vinyl Acetate (EVA) and Polyolefin Elastomer (POE). The pressure values tested vary from 500 mbar to 1450 mbar. The paper has presented a detailed study of lamination pressure influence on some of the critical characteristics of the PV module, which can impact the module's reliability, namely gel content, peel strength, and bubble formation. The degree of crosslinking for encapsulants was also compared using DSC and Soxhlet measurement methods. The results show that the lamination pressure above atmospheric value significantly influences the visual quality of PV modules.
Keywords: Lamination, Pressure, Peel Test, Gel Content, Bubbles, PV, Glass-Backsheet

1 INTRODUCTION

The lamination process is one of the main steps in module production to produce photovoltaic (PV) modules [1]. The main parameters of the lamination process are temperature, vacuum and vacuum time, pressure, and pressure hold time. Using these parameters and understanding the module lay-up, the process can be designed to laminate the PV modules. A standard lamination process is divided into three steps, excludes the cooling step [2]. First is the evacuation step, where all the unwanted air is evacuated using a vacuum. Then the module is pressed onto the heating plate, having pressure ramp-up time. Lastly comes the curing step, and here, the module is held under the applied pressure. Simply, the lamination pressure is defined as the applied pressure through the membrane or heating plate, depending on the process, on the module to compress all the material to each other and bind them together using heat as a source to melt the encapsulation material. During this pressure phase, the curing of encapsulants happens. The degree of curing depends on all parameters mentioned above.

Since all parameters are equally important for the module's reliability, pressure alone can affect the quality of the product. Poor value can lead to poor reliability of the PV module. Focusing on bubble creation, the lamination pressure greatly influences its formation in the modules [3]. During the lamination process, these bubbles can change in size depending on the pressure used in the process. If the inside pressure of the bubble is higher than the surrounding, the bubbles start growing, but if the inside pressure is lower than the outside pressure, the bubble will begin shrinking [4]. The standard mechanical pressure applied in the lamination of PV modules is around 900-1000 bar, depending on the location. And that's the maximum value a vacuum pump can achieve in the lamination process.

Nowadays, PV module manufacturers are increasing their temperature parameters; this allows them to shorten the lamination process by having a higher throughput [5,6]. However, operating with this higher temperature will also have challenges, like forming air bubbles. This could be overcome by applying a higher pressure than 1000 mbar during the lamination process. Here, Robert Buerkle GmbH emerged as a front-runner in implementing an over-pressure system, patented to use pressure values more than atmospheric up to 1450 mbar. [6].

The initial test results showed that the adhesion and gel content behaved similarly to the changes in pressure. Both were not that much influenced. However, when comparing the peel force of 1000 mbar and 1450 mbar slight changes, dependent on the encapsulant, could be obtained. The peel force for modules with EVA got higher with the increasing pressure, whereas the modules laminated with POE decreased. The formation of bubbles, on the other hand, behaved differently than the others. It became less or vanished by increasing the lamination pressure.

Until now, there has been no study done to understand the exact influence of pressure more than 1000 mbar. And in any other research done before, the pressure value is often not clearly stated in the lamination parameters, assuming all laminations were done around atmospheric pressure values. Only from 750 to 950 mbar is studied, where only the degree of cross-linking is examined [7]. This paper will present a detailed study of the influence of pressure on some of the significant module characteristics, viz. Peel Strength, Gel Content, Bubble Formation, Electroluminescence Imaging. Moreover, using a certain encapsulant or type impacts some of the above features [5,8]. Therefore, for a deeper understanding of the pressure parameter, the current mainstream encapsulants in the industry, Ethylene-Vinyl Acetate (EVA) and Polyolefin Elastomer (POE), are selected for running the trials.

2 EXPERIMENT DESIGN

In this paper, the influence of the lamination process is studied by varying the lamination pressure from 500 mbar up to 1450 mbar using the Buerkle Ypsator Laminator. The process used for the lamination of the PV modules is called the short lamination (SL) process [6]. It is a one-step lamination process plus a cooling step that is used in Buerkle laminators. High temperature and high pressure with shorter process times are used in the SL process.

The press uses a non-silicon membrane to apply the pressure from the top, and the heating is applied from both sides, top and bottom heating plates. Further, the chamber has a spring-loaded pins system to hold the modules 3 mm above the bottom heating surface to avoid direct contact with the heating plate. It prevents any shock to the module and brings homogeneous heating into the module.

For this study, a temperature of 165 °C and a process time of 7 minutes, which includes fixed evacuation and pressing time, are set parameters in the hot press. Later, the cooling step is done by a double-side cooling press with 7 minutes of process time.

Two encapsulant types, EVA and POE (see Table I), for module lay-up were selected to investigate their behavior difference against the pressure further.

For this study, 28 full-size GB modules (1750 mm x 950 mm) were built using EVA and POE and a backsheet, see Fig. 1. As per availability, M10 half-cell strings were used. Because of the glass size constraint, only five strings with 19 half-cut cells were placed in a module.

Figure 1: GB PV Module laminated using five strings

Seven modules for each encapsulant were laminated using seven different pressure settings (with and without cells). Starting the pressure value from 500 mbar, the following values are selected with a difference of 200 mbar. The values in mbar are 500, 700, 900, 1000, 1100, 1300, 1450. As an exception, 1000 mbar is used because it is the most used pressure value in the PV module manufacturing industry. The materials used for laminating PV modules are mentioned in Table I.

Table I: Materials used for PV Module build-up

Layer	Material	Thickness (mm)
Front Glass	Safety Glass	2.8
Encapsulant (EVA)	Supplier A	0.45
Encapsulant (POE)	Supplier B	0.45
Backsheet	PET/PET/Primer 1500V	0.37
Cells	Mono-C, PERC, M6 Half-cut, 9 Busbars	0.18
Interconnected Ribbon	6 mm wide	0.17

2.1 Peel test

The peel strength is measured with a tensile testing machine from PSE AG. The test is performed according to DIN EN 28510-1:2014-07.

In the test, a stripe of 10 mm width is cut into the rear side of the PV module. The cut goes down to the glass's inner surface so that the strip's thickness corresponds to the thickness of two layers of EVA, cells, and the backsheet. Each strip has a length of approximately 25 cm. After the cut, one end of the strip is clamped in the testing machine and peeled off at a speed of 50 mm/min under an angle of 90°.

2.2 Gel content measurement

Gel content or degree of cross-linking is performed to calculate the percentage of encapsulant cured during the lamination process. This study used two methods. The first was a chemical method, measured using the Soxhlet Extraction method, according to IEC 62788-1-6:2017 standard. And the second is a faster thermal method called Differential Scanning Calorimetry (DSC).

For the gel content measurement, the gel sample is cut out from the laminated module without solar cells, and its original weight is measured, W_0. Later, the specimen was cut into several 10 mm x 10 mm pieces and placed in the extraction sleeve in a Soxhlet apparatus for the extraction process. The sample is refluxed in approximately 80 ml of xylene. After about 3 minutes, the solvent is drained off. This procedure is repeated several times for around 16 hours. During this treatment, the non-cross-linked fraction is dissolved in the isomer, and the left is the cross-linked insoluble elastomer or gel. The remaining cured gel residue is dried at 80° C for 8 hours, and net weight is measured, W_1. The resultant gel content is the ratio of the W_1 to W_0 [9].

DSC measures the thermal transition of polymers, including melting, evaporation, glass transition, and exothermic or endothermic reactions [10]. The measurement was carried out on DSC 200 F3 machine from Netzsch GmbH & Co. Holding KG. Circular sample of 5 mm diameter was cut out from the gel sample, and weight was measured and placed in the crucible in the machine.

In measurement, the sample is heated from 20 °C to 200 °C at a constant heating rate of 10 °C/min. The degree of cross-linking is calculated by comparing the reaction enthalpy from an exothermic peak on the heat flow plot in the software of the cured sample to the uncured sample [11].

2.3 Electroluminescence imaging

In electroluminescence (EL) measurement, the module is operated in reverse current. This causes it to emit radiation. The number of emitted charge carriers is proportional to the cell current. Damaged or poorly contacted areas of a cell appear as dark spots on an electroluminescence image.

An electroluminescence measurement can be used to visualize cell breaks, defective backside contacts, and defective current collection fingers (grids). Low-power cells in a network can also be detected.

3 RESULTS AND DISCUSSIONS

3.1 Peel strength

The peel test for the laminated modules was performed at Fraunhofer ISE. In the laminated modules, two locations were chosen, the center and the edge, and two measurements were performed for each position. Each measure gave the maximum, minimum, and continuous values. Fig. 2 shows the peel strength values for POE and EVA encapsulant-based modules at different lamination pressure values.

Peel strength at 700 mbar for POE is probably wrong because the first module was broken while handling, and the new module was laminated separately. The reason could be that the lamination environmental conditions differ from the original one and can affect the adhesion in the interlayers [8], as there was a two-month gap before repeating the module lamination.

As seen from Fig. 2, the peel strength for POE initially decreases with an increase in the pressure until 1000 mbar. However, at 1100 mbar, the adhesion slightly increases, with approximately the same peel strength, ~160 N/cm, as at 500 mbar. Further, from 1100 mbar to 1300 mbar, it reduces marginally by 40 N/cm, but the adhesion increased again for 1450 mbar.

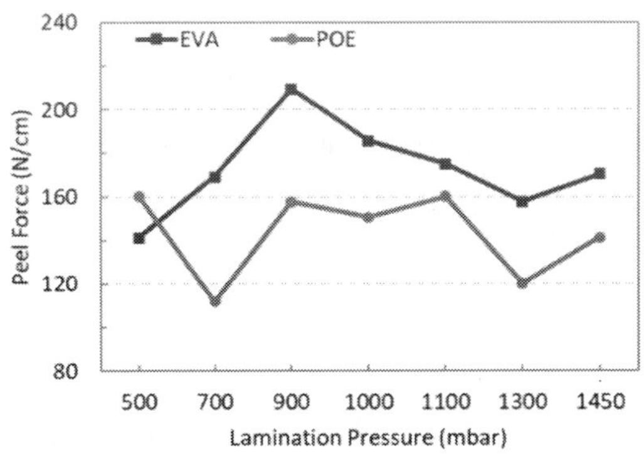

Figure 2: Example of peel force test measurement of POE and EVA-based GB Module laminated at different pressures

Similarly, the peel strength experiment was done on the EVA-based modules. For EVA, the behavior is different from the POE. As seen in Fig. 2, the peel strength increases initially with the increase in the pressure until 900 mbar and reaches the highest peel strength value of ~210 N/cm. This is a significant increase of about 70 N/cm.

However, it shows a downward trend with a further increase in the pressure until 1300 mbar. And like POE, a slight jump in the peel strength is observed at 1450 mbar. Since a similar trend is observed in both POE and EVA, it is likely that peel strength further increases for lamination pressure values beyond 1450 mbar. Comparing the peel strength values of both encapsulants, EVA has higher values compared to POE, see Fig. 3.

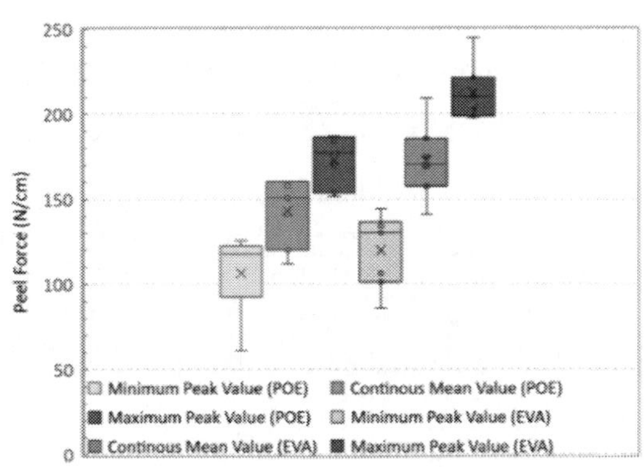

Figure 3: Adhesion between glass and the encapsulant for EVA and POE.

The maximum continuous mean value reaches 160 N and 210 N for POE and EVA, respectively. As seen in Fig. 3, the peel strength values for EVA oscillate in a much wider range, from as low as roughly 85 N to 245 N, whereas POE varies only between 92 N and 187 N, excluding the values for 700 mbar.

However, from the above results, the lamination pressure above 1000 mbar has brought no advantages to the module regarding peel strength.

3.2 Gel content and the degree of cross-linking

The Soxhlet extraction to measure the gel content was done by Fraunhofer ISE. The gel content for EVA has increased by only 2.5% with the increase in pressure from 500 mbar to 1450 mbar (see Fig. 4).

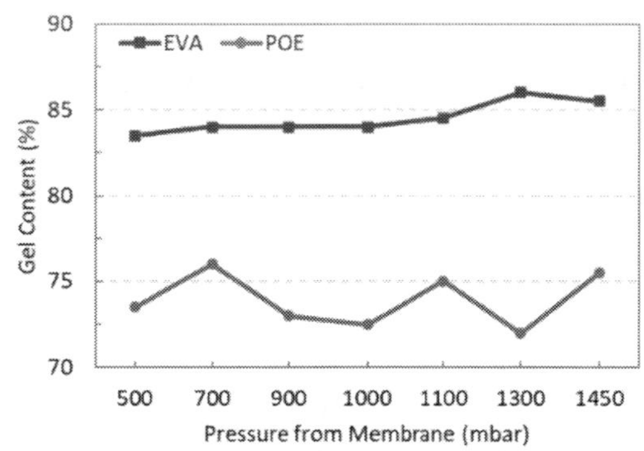

Figure 4: Soxhlet Gel Content value comparison between Eva and POE at different lamination pressures

Whereas POE, unlike EVA, has no trend. The values oscillate around a mean value of 73.9%. The gel value is highest, 76%, at 700 mbar, almost the same as at 1450 mbar. However, the lowest came at 1300 mbar 72%, which is similar to the value at 1000 mbar. The gel value range is good. However, this graph shows that the gel content value for POE does not influence the increase in the lamination pressure. Probably, the gel values are already at the highest level for these lamination parameters, irrespective of the pressure value.

To further study the samples, the faster thermal method for determining the cross-linking of the encapsulant and DSC measurement was done at Buerkle to compare it with the gel content values, see Fig. 5.

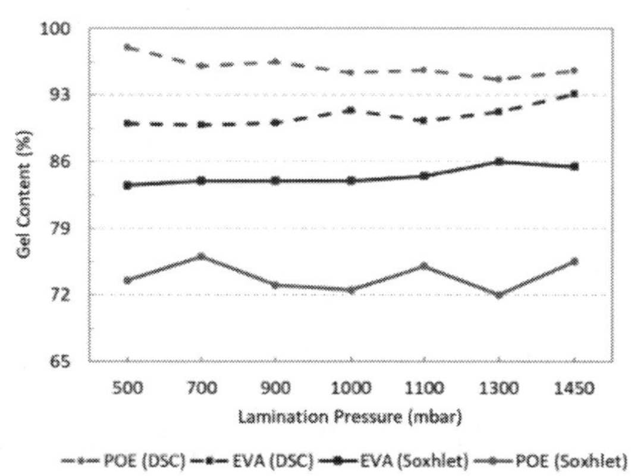

Figure 5: DSC vs. Soxhlet Gel Content value comparison between EVA and POE at different lamination Pressures

Firstly, the DSC values for POE are unimaginably high. It oscillates between 94.6% to 98% and shows no gradual increment or decrement trend with the increase in pressure, see Fig. 5. With every subsequent increment in the pressure value, the DSC value goes up and down, and so on. And the difference between the average values for Soxhlet and DSC for POE encapsulant is 22% (see Fig. 5). It clearly shows that the DSC values are invalid measurements for POE encapsulant.

Whereas the difference in Soxhlet and DSC absolute values for EVA encapsulant is only 6.3% (see Fig. 5). Interestingly, like the gel values, DSC values also increase with an increase in the lamination pressure. The value increases by 3.3% from 500 mbar to 1450 mbar. That is only a difference of 0.8% between the increasing gel and cross-linking values. Here, it can be interpreted that DSC values cannot be neglected and for initial process evaluation can be considered for further improvements, depending upon the cross-linking values or range. In summary, the pressure has minimal influence on gel content. The reason could be that other lamination parameters like high temperature, press time, and double-side heating overcome the pressure value's impact.

3.3 Correlation of peel strength to the gel content

Even though there is no substantial outcome of pressure on the peel strength and gel content. Interestingly, some correlation has been witnessed between peel strength and gel content values. For POE encapsulant, the gel value exactly duplicates the trend of peel strength, excluding the 700 mbar value, see Fig. 6.

It can be interpreted that for POE, these values are interdependent, and by increasing a gel value, one can also increase the adhesion in the POE-based GB PV modules.

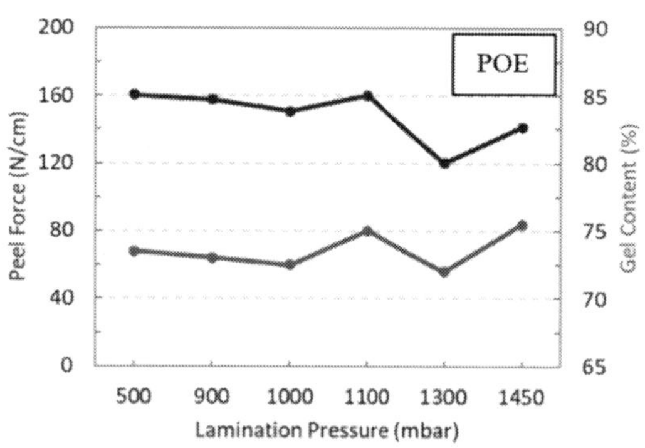

Figure 6: Correlation between peel force and gel content values for POE encapsulant

The trend for EVA is not as perfect as for POE; however, still, the properties correlate. As seen in Fig. 7, until 900 mbar, the peel force increases with the increase in the gel value. However, the trend changes after that; here, properties act oppositely. After 900 mbar, as the gel value increases, the adhesion decreases until 1300 mbar, and when the gel value again reduces at 1450 mbar, the peel force increases.

However, considering other studies [5,12,13], the adhesion is higher for high cross-linked unaged EVA modules, which is different from these results. The modules must be tested through long-term degradation tests to understand pressure influence further.

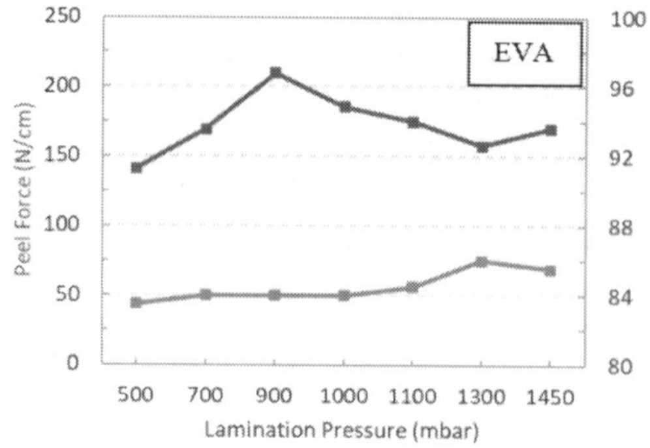

Figure 7: Correlation between peel force and gel content values for EVA encapsulant

3.4 Bubble formation

This section has a direct influence on the quality of the PV module. One extra tiny one mm bubble in the module can displace the module into the B category, depending on the company's quality standards. Bubbles are visible to the naked eye, and it is an easy method to define whether the process is good or bad.

During the experiments, the bubbles were witnessed in almost all modules up to 900 mbar except for EVA-used modules. For EVA and POE at 500 mbar pressure, the bubbles were seen on top of the cells, between the strings, between the cell and the connected ribbon, and between the cells, including the pseudo square area, as seen in pictures, Fig. 8, and Fig. 9. However, latter has bigger bubbles compared to the former.

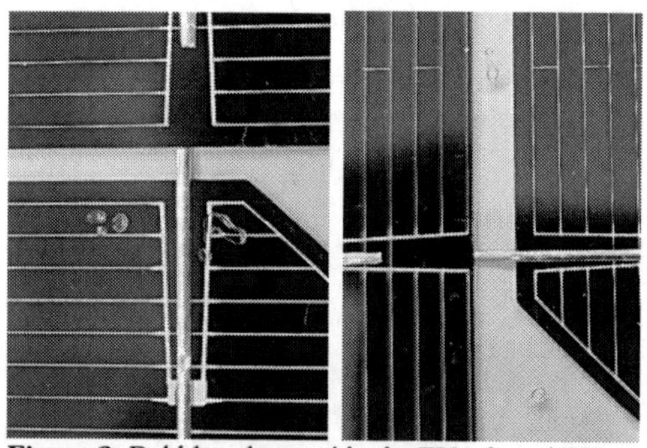

Figure 8: Bubbles observed in the EVA-based PV module laminated at 500 mbar

Figure 9: Bubbles observed in the POE-based PV module laminated at 500 mbar

Some tiny bubbles between the strings and cells were observed at 700 mbar but not on top of the cells. No bubbles were witnessed in EVA-based modules from the pressure of 900 mbar onwards.

Like EVA, modules based on POE have bubbles between the strings and cells but not on the top of the cells at 700 and 900 mbar lamination pressure. At 900 mbar, only two-three bubbles were observed.

However, POE-based modules have bubbles even at 1000 and 1100 mbar pressure values. Even after laminating two times at these pressures, bubbles were observed. These bubbles were tiny as 1 mm or less. At 1300 and 1450 mbar process, no bubbles were witnessed. POE has different chemical characteristics and behavior for the same lamination process as the EVA.

Table II: Bubbles formation in the PV module at different lamination pressure

S. No.	Pressure (mbar)	EVA	POE
1	500	Many bubbles	Many bubbles
2	700	Fewer bubbles than 500 mbar	Fewer bubbles than 500 mbar
3	900	No bubbles (NB)	3-4 bubbles
4	1000	NB	6-7 bubbles*
5	1100	NB	1-7 bubbles*
6	1300	NB	NB
7	1450	NB	NB

*Two modules laminated

Most manufacturers use a lamination process of around 900 mbar or atmospheric pressure because of the limitation of the laminators. And bubbles are formed due to gases formed during lamination and improper timing of application of lamination pressure [3]. During lamination, volatile organic compounds (VOCs), like CO2, 2-ethyl hexanol, and tert-butanol, in the case of EVA, are formed. If the correct pressure is not applied at the right time, bubbles will be present in the module. And in comparison to EVA, POE releases more VOCs during crosslinking [14]. At higher pressure above atm., the chances of having bubbles are diminished because then the pressure applied will be higher than the gas pressure inside the bubbles formed, which would be around atmospheric pressure and, thus, can produce optically suitable modules [3].

Though, alone pressure cannot ensure a bubble-free module. The holding time and timing play a part in that and are interlinked with pressure value. The process time can be reduced with higher pressure and a shorter holding time. Otherwise, with lower pressure, a longer holding time will be required to break down the bubbles and evacuate them during the cycle. This leads to longer process times, which is common in the industry.

3.5 Bubbles formation in backsheet inner layer with POE encapsulant

During the gel sample preparation, an interesting phenomenon was observed in the inner layer of the backsheet, especially during lamination with POE encapsulants. There are bubbles witnessed in the inner layer for processes 500 mbar to 1000 mbar, see Fig. 10. As seen in the picture, the bubbles are following a trend; with the increase in pressure, the number of bubbles is reduced. But, for processes 1100, 1300, and 1450 mbar, there was no bubble observed, see Fig. 11. From these findings, it is pretty clear that pressure influences bubble formation. Higher pressure above atmospheric pressure prevents bubbles from forming during the curing phase.

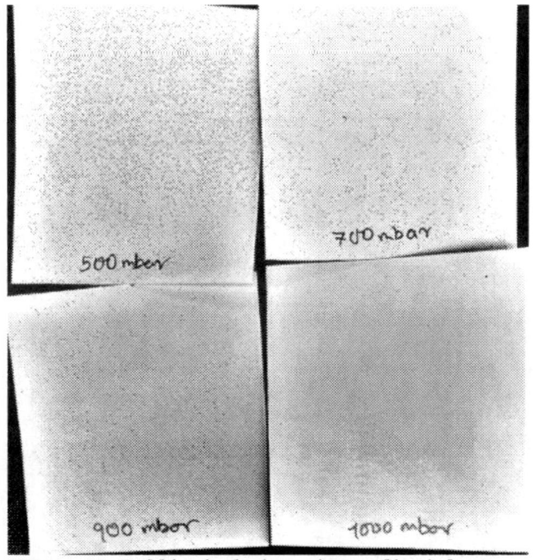

Figure 10: Bubbles in the inner layer of backsheet

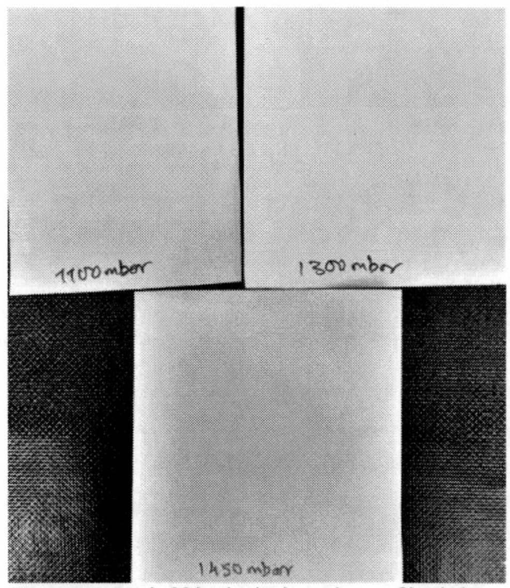

Figure 11: No bubbles in the inner layer of backsheet

3.6 Electroluminescence imaging

Electroluminescence (EL) imaging can help trace micro-cracks or cell-connection defects in the module. In production, there are usually two EL stages, one before lamination and one after lamination. The latter can detect any defects arising due to the lamination process. If defects are seen after the lamination process while comparing the EL images before and after, then one can analyze the process parameters. However, can the pressure of 1450 mbar lead to such cell micro-cracks?

To evaluate this, modules were prepared at 500 mbar and 1450 mbar using both encapsulants, and the EL imaging was done. All the monochromatic pictures shown below in Fig. 12 (EVA) and 13 (POE) are done at a 1 I_{SC} measurement setting, and the colorful ones are imaged at 0.1 I_{SC}. The EL images showed no dissimilarities in the 500 mbar and 1450 mbar lamination processes.

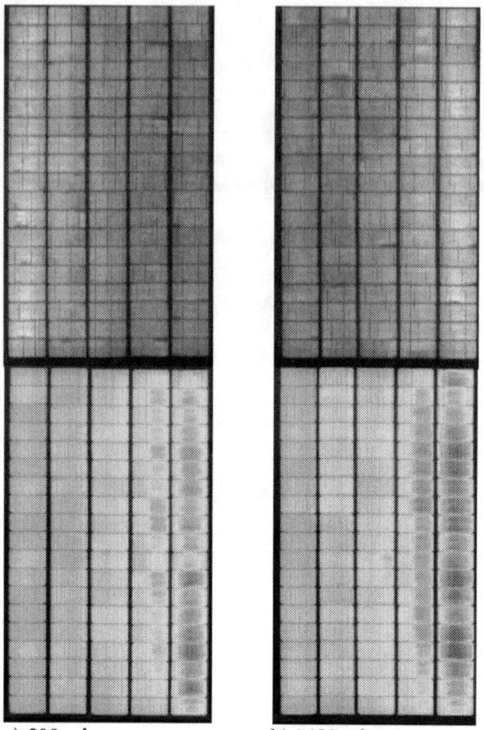

a) 500 mbar b) 1450 mbar

Figure 12: EL imaging for PV modules using EVA

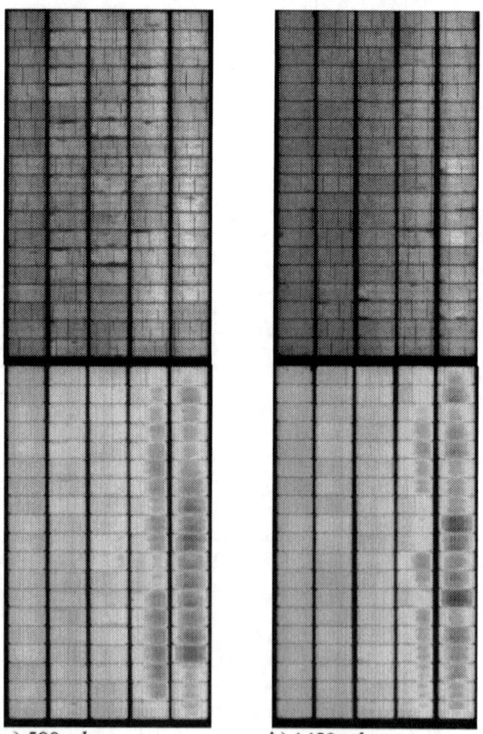

a) 500 mbar b) 1450 mbar

Figure 13: EL imaging for PV modules using POE

From these images, it can be said that the lamination pressure above atmospheric up to 1450 mbar does not create any cell breakage or micro-cracks.

4 CONCLUSIONS

The lamination pressure has minimal influence on the gel content value for PV modules using EVA encapsulants, increasing by 2 % between 1000 mbar and 1450 mbar. And peel strength reduces with the increase in values above atmospheric pressure. In contrast, lamination pressure does not significantly influence adhesion and gel content properties for POE encapsulant-based PV modules.

The lamination pressure dramatically influences the formation of bubbles. No bubbles were observed in modules beyond 900 mbar for EVA and beyond 1100 mbar for POE encapsulant because POE produces more VOCs than EVA. And EL images of the modules show that the higher lamination pressure does not directly impact the cell cracks.

This study observed some correlation between adhesion and gel content while changing only the pressure value. However, EVA and POE both follow different trends. This data can be helpful for manufacturers by analyzing their bill of materials (BOM) to improve their lamination processes.

For higher output, faster lamination processes are quite a need in the current market. Using high temperature combined with high pressure, the process time of the lamination can be reduced.

5 ACKNOWLEDGEMENTS

The authors like to thank Coveme Spa and Voltec Solar for providing the materials for this study. This work was funded by the German Federal Ministry for Economic Affairs and Energy (BMWi) under contract number 0324287C.

6 REFERENCES

[1] A. E. Amrani, A. Mahrane, F. Y. Moussa & Y. Boukennous, Solar Module Fabrication. International Journal of Photoenergy, Article ID 27610 (2007).

[2] R. F. M. Lange, Y. Luo, R. Polo, & J. Zahnd, The lamination of (multi)crystalline and thin film based photovoltaic modules. Progress in Photovoltaics: Research and Applications, 19(2), (2011) 127-133.

[3] L.-E. Perret-Aebi, H.-Y. Li, R. Théron, G. Roeder, Y. Luo, T. Turlings, R. F. M. Lange, C. Ballif, Insights on EVA Lamination Process: Where Do The Bubbles Come From? (2010).

[4] H.-Y. Li, Y. Luo, C. Ballif, L.-E.Perret-Aebi, Modeling of Voids Evolution in the Encapsulation Process of Photovoltaic Modules (2014).

[5] A.K. Öz, C. Herzog, C. Wellens, D.E. Mansour, M. Heinrich, A. Kraft, The Impact of the Lamination Process on the Adhesion Properties at the Glass-Encapsulant Interface and Damp Heat Stability of PV Modules, 7 pages / 38th European Photovoltaic Solar Energy Conference and Exhibition; (2021) 708-714.

[6] S. Sraisth, Achieving Faster Lamination Process for Crystalline Photovoltaic Modules by Using Latest Lamination Technologies, 6 pages / 33rd European Photovoltaic Solar Energy Conference and Exhibition; (2017) 992-997.

[7] K. Harms, L. Neumaier, C. Hirschl, Determination of Depth-Dependent Variations in the Degree of Crosslinking of EVA Due to Changing Lamination

Parameters Using Raman Spectroscopy, 3 pages / 36th European Photovoltaic Solar Energy Conference and Exhibition; (2019) 1042-1044.
[8] F. J. Pern, S. H. Glick, Adhesion Strength Study of EVA Encapsulants on Glass Substrates, NREL/CP-520-33558, 4 pages / National Center for Photovoltaics and Solar Program Review Meeting; (2003).
[9] C. Hirschl et al., Determining the degree of crosslinking of ethylene vinyl acetate photovoltaic module encapsulants - A comparative study. Solar Energy Materials & Solar Cells, (116), (2013) 203-218.
[10] G.W. Ehrenstein, G. Riedel, & P. Trawiel, Thermal Analysis of Plastics: Theory and Practice. Hanser Gardner Publications, (2004) 163-217.
[11] O. Bianchi et al., Assessment of Avrami, Ozawa and Awrami-Ozawa equations for determination of EVA crosslinking kinetics from DSC measurements. Polymer Testing, 27(6), (2008) 722-729.
[12] P. Schenk, M. Pander, U. Zeller, B. Jäckel, M. Ebert, Correlation of Peel forces to EVA Degree of Cross Linking and Accelerated Weathering, 6 Pages / 37th European Photovoltaic Solar Energy Conference and Exhibition; (2020) 834-839.
[13] D. Wu, J. Zhu, D. M-Chicharro, T. R. Betts, R. Gottschlag, Influences of Different lamination Conditions on the Reliability of Encapsulation of PV Modules, 6 Pages / 29th European Photovoltaic Solar Energy Conference and Exhibition; (2014) 3415-3420.
[14] I. Dmitriev et al., Compatibility of crosslinking encapsulants with smart wire connection technology: ways to prevent bubble formation, 3 Pages / 38th European Photovoltaic Solar Energy Conference and Exhibition; (2021) 754-756.

INTEGRATION OF ANTENNA FOR IOT COMMUNICATION IN SMALL SILICON PHOTOVOLTAIC MODULE

Lasse R. Wahlgreen, Sune Thorsteinsson, Peter B. Poulsen*, Martin N. Petersen.
Department of Electrical and Photonics Engineering, Technical University of Denmark
Frederiksborgvej 399, 4000 Roskilde, Denmark
*Corresponding Author: ppou@dtu.dk

ABSTRACT: This study presents proof of concept of a silicon PV module integrated monopole and dipole microstrip antenna made for IoT frequencies in Europe, 868 MHz. Proof of concept for two types of PV integrated antennas is made. A monopole antenna with a length of 50 mm with ground attached to a bike stand and a dipole with each pole the length of 52 mm. Both antennas have been tested and compared to an off-the-shelf stub antenna. The monopole showed close to similar or better readings on SF, RSSI and number of packets. The dipole had better readings on SF and RSSI but had lower number of packets sent. The dipole offers the possibility of easy connection whereas the monopole antenna needs to be connected to both a coax cable and the bike stand.

Keywords: Photovoltaic, antennas, Internet of Things, IoT, ISM band, LoRaWAN

1 INTRODUCTION

IoT solutions are getting more widespread and in the outside environment, an integration with photovoltaics (PV) seems natural. Since a solar module is on the outside of an appliance, it is natural to investigate the possibility of a PV integrated antenna, in order also to maximize the antennas field of view. Several previous studies have proved validity of concept [1] but are yet to be tested or described when implemented in real life applications.

2 AIM AND APPROACH

In this study, a proof of concept is established for a PV integrated antenna made for IoT-frequencies in Europe, 868 MHz. Two antennas are designed, encapsulated, and tested while installed in a bike stand, called Bike Sun by out-sider A/S. This bike stand is made by metal and thus the frame for the antenna is also metal, resulting in significant differences of the antenna function when placed within or outside the bike rack.

The solar module measures 69 x 110 mm and contains two strings, comprised of solar cell elements measuring 18 x 28.5 mm. The solar cells are placed 3 mm from the long edges and in the centre between the two strings, the antenna is placed. The complete solar module is a sandwich build-up consisting of glass (4mm) -EVA-Cell-EVA-PCB (1.6 mm with integrated antenna) construction, and fabricated using standard lamination procedures.

Figure 1: Photo of solar cell module with integrated monopole antenna.

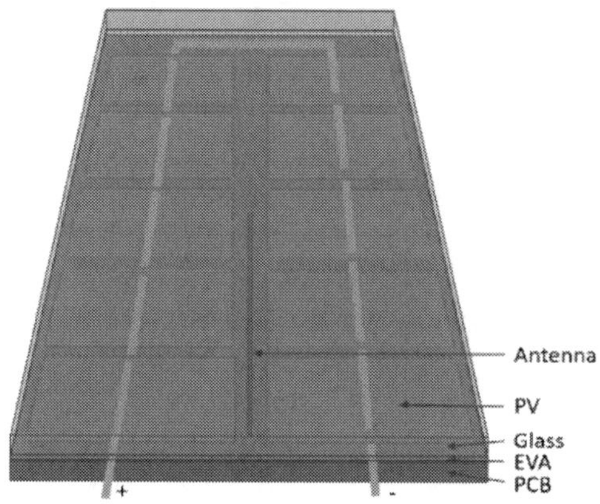

Figure 2: Concept drawing of solar cell module with integrated monopole antenna.

The previous installed solar modules are made by using a PCB as substrate on which cut solar cells are placed using surface mount technology (SMT). Above EVA, glass or another superstrate, is placed on top. As a PCB is already used for the solar modules, this study is also based on a PCB substrate.

In this study two of the antenna designs gave promising results; a monopole microstrip antenna and a dipole microstrip antenna. Both antennas are placed in the centre of the solar module with solar cell strings on each side. The monopole antenna is using the bike stand as ground and the dipole antenna functions without ground making it easier to install.

3 ¼-WAVE MONOPOLE ANTENNA

The length of the monopole antenna is analysed within the bike stand using a N1201SA RF Vector Network Analyzer (VNA). The VNA gives a graphical output on the frequencies as well as the impedance character and the Voltage Standing Wave Ratio (VSWR) and the Reflected Power (S11). The length is found using observations on the VNA as well as an iterative approach to the equation:

$$\lambda_1 \cdot f_1 = \lambda_2 \cdot f_2 \qquad (1)$$

based on the wavelength equation $\lambda = \frac{c}{f \cdot \sqrt{\varepsilon_{eff}}}$ where λ_1 and f_1 is the wavelength and frequency before, and λ_2 and f_2 is the wavelength and frequency after reduction of antenna length. c is the speed of light and ε_{eff} is the efficient (resulting) dielectric constant.

When achieving an antenna length giving a resonance frequency of 868 MHz, solar modules are made to observe the change in resonant frequency because of the changes in dielectric surroundings owing to encapsulation. After encapsulation, the VNA is used to observe the new average resonant frequency of which a new antenna length is found. The encapsulation results in a reduction to the resonance frequency of 270 MHz in average. From this, a shorter antenna length is necessary to have a higher resonant frequency before encapsulation.

A new antenna is made with a resonant frequency of 868 MHz + 270 MHz ≈ 1140 MHz. After encapsulation, measurements are made with the VNA to observe the resulting resonance frequency. Using this resulting frequency, equation 1 is applied for the last design. If the resulting resonance frequency is still unsatisfactory, more iterations can be made.

4 ½-WAVE DIPOLE ANTENNA

A dipole antenna was tested as the monopole antenna reached a length seemingly allowing for a ½-wave dipole. However, due to the surrounding metal frame the VNA could not present any clear resonant frequency. To observe if the encapsulation could change this the dipole antenna was encapsulated. The resulting resonant frequency indicated that the antenna was too short. To change this, another design than a straight line microstrip antenna must be considered, as there is no more room for enlargement in length on the PCB, in this particular product. The dipole antenna did however present a wide bandwidth that covers 868 MHz. With a balun attached, the resulting function of the antenna around 868 MHz might be improved, in particular the impedance to a target value of 50 Ohm which was above 80 Ohm in the measurements.

5 ½-WAVE MEANDER TYPE DIPOLE ANTENNA

To increase antenna length, a meander type dipole was designed. Based on [1] the meander part could be placed below the solar cells without damaging the antenna function. The preliminary tests of antenna length did not give any clear results and thus the PCB with antenna was laminated with solar cells. To be able to reduce antenna length, a lamination with the antenna placed opposite the solar cells was made and compared to the traditional method, see figure 3. The antennas had similar performance around 868 MHz, but with a significant difference of 16 dB around 1 GHz. This however, was assumed to be good enough for a length reduction test. Reducing the length only worsened the performance and further work on this antenna were set aside for later studies.

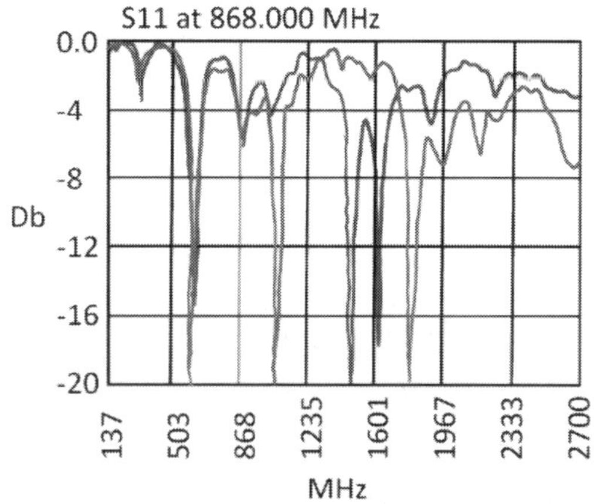

Figure 3: Comparison between two laminations with the antenna. Blue is the traditional method with antenna on the same side as PV. Red is with the antenna on the opposite side.

6 ANTENNA SIGNAL TEST

The two encapsulated antennas are tested towards an off-the-shelf stub antenna. The stub antenna is placed within the bike stand beneath an encapsulated PCB to simulate the possible placement. The test of the antennas is made in the same position and in the same test stand and to minimize the influence of weather the tests are made within one hour per antenna. When installed into the bike stand, the stand is oriented toward each of the four cardinal directions (N, E, S and W) for 15 minutes during tests. This minimised the influence of weather changes. The antennas are connected to an 868 MHz LoraWAN transceiver, and the data itself, along with the number of packets sent, the spreading factor (SF) and the received signal strength indicator (RSSI) is collected through a LoraWAN backend server provided by the Danish LoraWAN provider, Cibicom A/S (www.cibicom.dk)

The collected data is assessed by calculating the average SF and RSSI, and by counting the number of received vs. lost packages in each cardinal direction. By comparing the results between the antennas, a final conclusion on functionality is made.

Figure 4: Photo of the bike stand – Bike Sun, produced by the company out-sider A/S. 3 small silicon solar panels are embedded in the product.

Figure 4 shows the bike-stands that were used for testing. Three solar panels are mounted in it: 1 horizontally in the top, and 2 mounted vertically on the sides in opposite directions. The product provides solar powered light at night to the user and integration of IoT is attractive for online monitoring of the product performance and proving more features to the product and its customers. The antenna integrated photovoltaic module replaces the horizontal module in the top.

6.1 Reducing cell size

To investigate the influence on the size of the solar cell elements and the resulting space between elements where the antenna is placed, a custom-made monopole antenna, fitted in length to an 868 MHz resonance frequency, was encapsulated with cells 2 mm narrower, thus leaving 4 mm of extra space around the antenna. This test was made with two modules and gives indications to what can be expected. The VNA measurements did not deviate notably from the previous case but the signal test resulted in a slightly improved SF and RSSI as well as a slight reduction in lost packages.

6.2 Influence of super- and substrates

An investigation into the effects on the superstrates were made. The influence of the substrate (FR-4) can be found by comparing the wavelength in free air to the wavelength found experimentally.

This can also be calculated by equation 2 from [2] for a patch antenna with ground underneath the substrate:

$$\varepsilon_{eff} = \frac{\varepsilon_{PCB}+1}{2} + \frac{\varepsilon_{PCB}-1}{2} \cdot \left(1 + \frac{12 \cdot h}{W}\right)^{-0,5} \quad (2)$$

where W is the antenna width, which in this project was 1.5 mm. h is the thickness of the substrate measured from antenna to ground. ε_{PCB} represents the dielectric constant of the PCB, and ε_{eff} is the resulting dielectric constant. $W/h > 1$ is a design demand for this formula, but is not upheld, as W is higher than h.

The superstrate influence is here investigated experiment-tally. It is found, that the EVA's influence on the resonant frequency is minimal. There are indications that the PV changes the bandwidth, but further investigations are needed to reach a conclusion. Aside from the substrate FR-4, the glass cover is the major influencing element and reduces the resonant frequency, in this case 270 MHz on average, around 868 MHz. This corresponds to earlier studies [3]–[5].

6.3 Applying an extra layer of PCB and glass

To make a base investigation on the influence of thickness of the sub- and superstrates, tests were made where an extra piece of glass was placed immediately upon the module and a PCB underneath. The influence of both was investigated on several modules, both mono- and dipole.

In figure 5, the measurements on a dipole antenna are seen. The graphs shows S11 which represents the reflected power. -10 dB or less it is generally considered acceptable. The influence of both extra glass and PCB is low, giving an indication that the thickness of the sub- and superstrates are of limited influence.

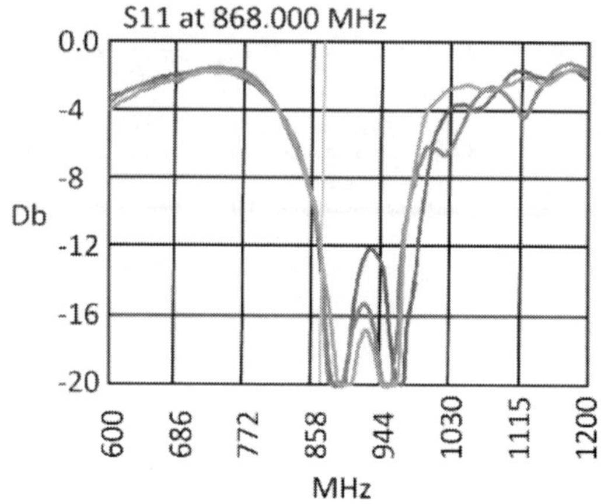

Figure 5: Measurement on a dipole module. Red is the original module. Blue is with extra glass and green with an extra layer of FR-4 PCB material.

6.4 Influence from the bike stand

During the initial project, the bike-stand frame for the module was cut to improve measurements, as the original framing functioned as a Faraday-cage. This was a quick and imprecise cut that left room for improvement. As further measurements were conducted, the frame was adjusted to increase symmetry. In figure 6, an example on the difference in measurements after frame adjustment is seen. Here the reflected power is reduced, even though a slight shift in frequency can be noticed. As the reflected power is dependent on the impedance, which in turn is dependent on the surroundings, the frame adjustment leads to more consistent results.

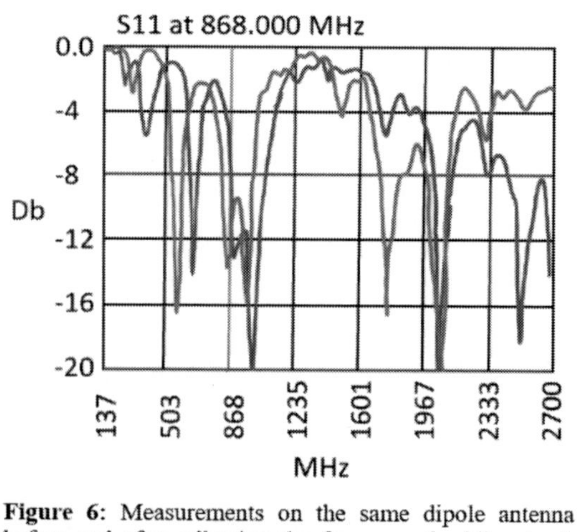

Figure 6: Measurements on the same dipole antenna before and after adjusting the frame on the bike stand. Green is original and red is new.

7 SCIENTIFIC INNOVATION AND RELEVANCE

As this study moves from theoretical analysis to practical PV integrated antennas, it is very applicable in other IoT outdoor solutions. Already several outdoor projects use IoT, for example surveillance of content in trash bins [6], [7]. By presenting proof of concept, inspiration for future use of PV integrated antennas is created.

The solar module in this study is framed by the metal bike stand, but during the experiments the integrated antennas also showed promising results for other use cases. A calibration of antenna length is necessary, as the metal framing had a large influence on the resonance frequency.

The antennas made in this project shows promising results at 868 MHz, and the bandwidth also covers 915 MHZ as seen in figure 7, which is used for LoRaWan in, for example, North America.

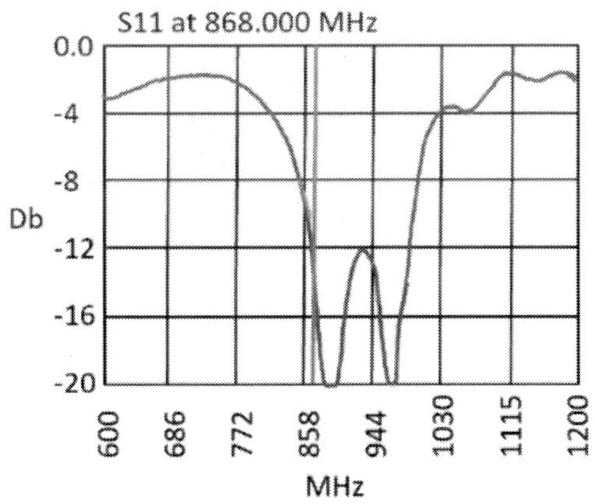

Figure 7: Example of the best dipole antenna. The reflected power is acceptable at 868 MHz with an S11 of more than -12 dB [8].

8 CONCLUSION

Proof of concept for two types of PV integrated antennas is presented; a monopole antenna with a length of 50 mm with ground attached to the bike stand and a dipole with each pole the length of 52 mm. Both antennas have been tested and compared to an off-the-shelf stub antenna. The monopole showed close to similar or better readings on SF, RSSI and number of packets. The dipole had better readings on SF and RSSI but had lower number of packets sent. The dipole offers the possibility of easy connection whereas the monopole antenna needs to be connected to both a coax cable and the ground.

As the antennas in this study are surrounded by the metal frame, they are not directly applicable to other solutions. Most likely, all new integrations require testing prior to production to achieve the correct resonant frequency.

9 ACKNOWLEDGMENTS

This research has been carried in the project ELFORSK 352-045, Solar-powered IoT solutions - Green IoT.

10 REFERENCES

[1] S. B. Thandullu Naganathan and S. Dhandapani, "Patch antenna integrated on solar cells for green wireless communication: A feature oriented survey and design issues," International Journal of RF and Microwave Computer-Aided Engineering, vol. 32, no. 1. John Wiley and Sons Inc, Jan. 01, 2022. doi: 10.1002/mmce.22926

[2] C. A. Balanis, Antenna Theory - Analysis and Design (4th Edition), 4th edition. John Wiley & Sons., 2016.

[3] K. Shafiqul Alam, "Influence of Dielectric Embedding on Antenna Performance," 2016, doi: 10.13140/RG.2.2.31021.08164.

[4] F. Benmeddour, C. Dumond, and E. Kenane, "Effects of Superstrate Layer on the Resonant Characteristics of Annular-Ring printed Antenna," Technology, Engineering & Mathematics (EPSTEM), vol. 7, pp. 147–154, 2019, [Online]. Available: https://hal.archives-ouvertes.fr/hal-02963703

[5] V. Saidulu and K. S. Rao, "Study of the Dielectric Superstrate Thickness Effects on Microstrip Patch Antennas," vol. 11, no. 1, pp. 55–65, doi: 10.9790/2834-11125565.

[6] SWECO, "SWECO IoT trash bin," https://www.sweco.dk/showroom/digital-affaldshandtering-i-rudersdal-kommune/, Feb. 03, 2022. https://www.sweco.dk/showroom/digital-affaldshandtering-i-rudersdal-kommune/ (accessed Feb. 03, 2022).

[7] Nordsense, "Nordsense trash bin," https://nordsense.com/smart-bin-sensors/, Feb. 03, 2022. https://nordsense.com/smart-bin-sensors/ (accessed Feb. 03, 2022).

[8] https://www.thethingsnetwork.org/forum/t/the-big-antenna-s11-catalog/15782 (accessed September 6th 2022).

MITIGATING CELL BREAKAGE LOSSES IN SHINGLED STRINGS AND MODULES

Daniel Tune, Karl Wienands, Ingo Ullmann, Andreas Halm
International Solar Energy Research Centre (ISC) Konstanz
Rudolf-Diesel-Str. 15, 78467 Konstanz, Germany

ABSTRACT: As the shingled module architecture is becoming more widespread and production volumes are increasing, the issue of string breakage and rework is becoming more valuable. In this work, we demonstrate and compare two simple strategies for the rapid rework of broken shingle strings that could be applied in the case when the breakage occurs before lamination. Both strategies are based on the use of conductive patches that are applied to the rear of strings in the area of the damaged cell. One of the strategies is shown to cause losses that are significantly higher than the fractional area loss of the damaged shingle stripe whereas the other is shown to yield only very minor losses.
Keywords: Shingling, String Repair, Module Integration

1 INTRODUCTION

String rework is widely used in small-scale module manufacturing to minimize yield losses due to process and material failures such as incomplete or incorrect soldering and cell breakage. In traditional soldered strings this can be relatively easily achieved by de-soldering and removal of the problem cell(s) in the string, with subsequent re-soldering of a new cell or cells into the string. However, in the shingled module architecture the individual pieces in the shingled string are tightly bonded by conductive adhesive and these bonds are difficult to rapidly and easily break without causing further and more widespread damage to the string. Considering that shingled strings generally consist of at least 32 pieces and may contain many more than this depending on the layout of the module, the losses due to damage of a single piece in the string are multiplied many times when the entire string must be discarded. Rather than trying to remove the damaged cell, this work focusses instead on mitigating the losses due to the damage by using a simple concept of 'patching'. This technique is relevant for production environments in which intermittent or occasional process or material failures, such as a damaged suction cup on a cell or string handling tool, can lead to significant financial losses.

The aim of this work is thus to provide simple, cost-effective and industrially viable strategies for rework of shingled strings in the case of breakages that occur before lamination. The approach for the string rework is to use highly conductive patches placed on the rear of the strings over the region of the broken cell. Two variants of this strategy are investigated as shown in **Figure 1**. The first variant places the patch over the rear of only the broken cell and will be referred to as 'cell patching' whereas the second variant places the patch such that it bridges from the broken cell to the next cell in the string and will be referred to as 'bridge patching'. Small strings of six shingle stripes (one cell equivalent) are used and for simplicity the cell breakage is simulated by a laser cut. The effects of the two variants are compared in terms of current–voltage characteristics as well as through electroluminescence and photoluminescence imaging.

2 RESULTS

Figure 2 shows the results of using the two variants of before-lamination strategy to repair strings. The two variants, referred to as 'cell patch' and 'bridge patch' are illustrated in the schematics (a) and (b). The PL image of the cell-patched string clearly shows the position of the laser cut but is otherwise not dissimilar to before the cut (image not shown).

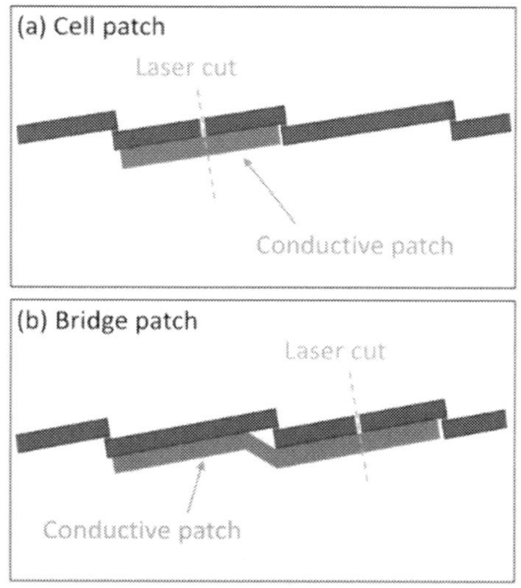

Figure 1 Diagrams illustrating (a) the 'cell patch' string repair strategy and (b) the 'bridge patch' string repair strategy.

However, in the PL image of the bridge-patched string, half of the cut cell exhibits no photoluminescence due to the fact that this cell region is shorted to itself through the bridge-patch and the bond to the adjacent cell in the string. In the EL image of the cell-patched string, one half of the cut cell is dark due to the fact that there no current flowing through it but the whole cut cell is dark in the bridge-patched string, for the same reason.

Figure 3 compares the IV characteristics before and after string repair using the two variants. In the cell-patched string, V_{OC} is unchanged at a little over 4 V since the number of cells in the string is the same. However, I_{SC} is drastically reduced from 1.6 A to less than 0.75 A due to the constriction in the string circuit caused by one of the cells having only half of the electrically active area. In this case, the apparent FF of the repaired string is 0.86 (vs. 0.79 before the string repair), but it is clear that this is artificially inflated due to the severe current restriction. When these factors are combined, the change in string power is significant – going from 5.1 W before breakage to only 2.6 W after the repair attempt, a relative loss of 49 % vs. the fractional loss of the broken cell of 1/6 of the string, or 16.7 %.

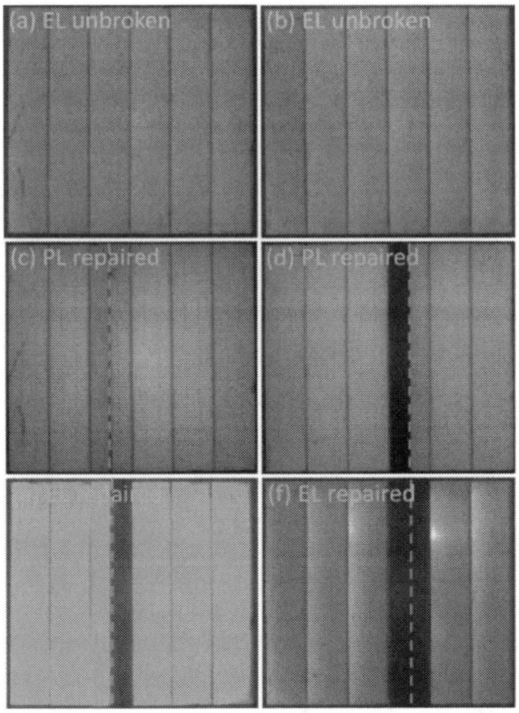

Figure 2 Electro- (EL) and photoluminescence (PL) images taken of strings repaired with either the cell patch (a, c, e) or bridge patch (b, d, f) string repair strategies. EL images of the unbroken strings are shown, as well as PL and EL images of the broken and then repaired strings. The orange dashed lines in the repaired string images indicate the position of the laser cut.

The situation is markedly different in the bridge-patched case, where I_{SC} remains relatively unchanged at a little under 1.6 A and the FF is only marginally reduced from 0.775 to 0.761, but the string voltage is reduced from 4 V to 3.3 V, resulting in a reduction of string power from 4.9 W to 4 W. The relative loss in power is thus only 18.4 % vs. the 16.7 % fractional area loss of the damaged cell – an additional loss of only 1.7 %. The relative losses due to the two variants of the patching strategy can be easily seen by comparing the cell-normalized power, where the cell-patched string exhibits a P_{MPP} per undamaged cell of 0.435 W vs. the bridge-patched string with 0.8 W per undamaged cell. Clearly, the cell patching strategy is not viable for string repair.

However, as shown in **Figure 4**, if a 32-piece shingled string was repaired using the bridge patch strategy, the relative loss in power of the string could be expected to be on the order of around 5 %, resulting in an expected reduction in module power of around 0.5 % (for one repaired string in a 12-string module, considering that the marginally lower voltage of the repaired string will also slightly reduce the power of the other strings that it is in parallel with), or simply 5 % for a module consisting only of repaired strings. This repair strategy could thus allow for losses due to cell breakage to be minimized by creating additional bins of modules with a marginally lower power class. Compared to a 400 W shingled module with 12 undamaged strings of 32 pieces each, projections based on the results of this work show that 398 W could be expected for modules with only one repaired string or 380 W for modules consisting of only repaired strings.

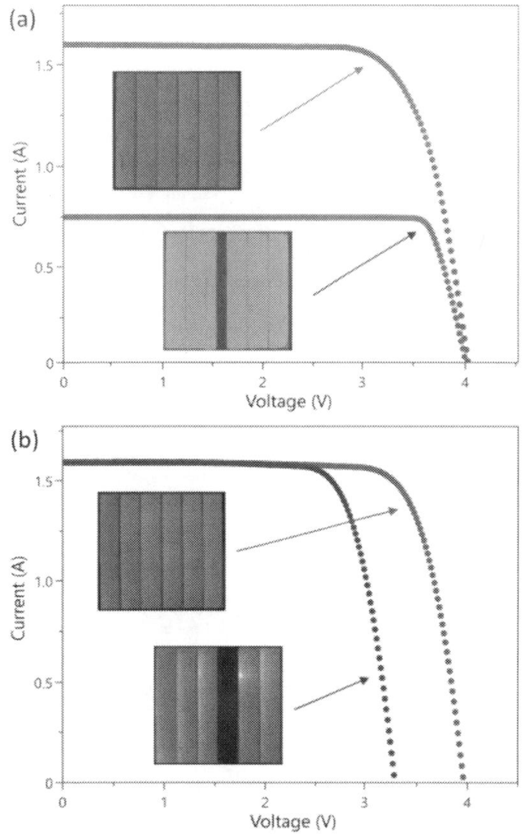

Figure 3 Current–voltage characteristics of 6-piece shingled strings before and after repair with either (a) the cell patch strategy or (b) the bridge patch strategy.

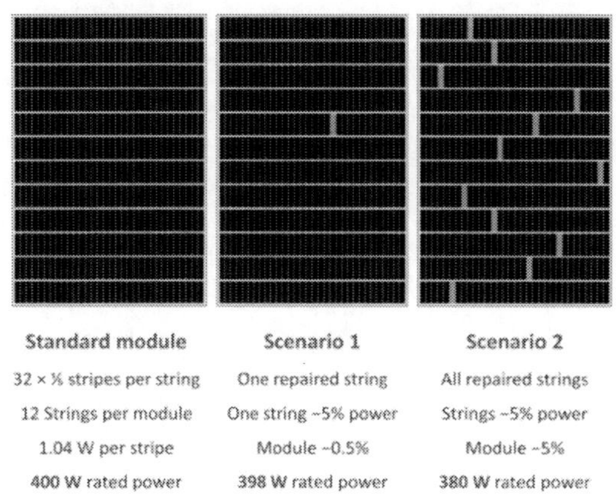

Figure 4 The scenarios considered in the projections of module power obtainable using the bridge patch string repair strategy, compared to the power of a standard module with no repaired strings.

3 SUMMARY

Two variants of a string rework strategy for shingled modules, based on the application of conductive patches to the rear side of the strings in the damaged area, were compared. One variant based on applying the patch to the rear of only the damaged cell was shown not to be viable due to high power losses associated with the constriction of string current through the damaged cell. However, the other variant, based on applying the patch such that it covers the rear of the damaged cell and bridges to the next cell in the string, yielded only very minor losses and could be used in a production environment in situations where large financial losses due to occasional or intermittent process issues would otherwise be incurred.

4 ACKNOWLEDGEMENT

This work was funded by the German Federal Ministry of Economic Affairs and Climate Action (BMWK) as part of the HoSSa project with reference number 03EE1014D.

INTERCONNECTION APPROACH FOR BUSBAR-LESS IBC CELLS
BASED ON PRINTED SOLDER PASTE

Tobias Meßmer[1], Haifeng Chu[1], Valentin Dan Mihailetchi[1], Razvan Roescu[1], Joris Libal[1], Andreas Halm[1]
[1]ISC Konstanz e.V
Rudolf-Diesel-Str. 15, 78467 Konstanz, Germany

ABSTRACT: To decrease the cost for the interdigitated back contact (IBC) cell ZEBRA, a busbar-less interconnection approach is developed in order to reduce the silver consumption. Since the metallization for both polarities is printed on the backside of the cell, a 3D designed metallization layout is used for the ZEBRA cell. Thus, for a busbar-less interconnection approach, the wires must be individually contacted to the fingers between the insulation paste pads, which are printed to insulate the wire from the opposite polarity. Due to the individual contact of the fingers to the wire, the connection area is very small, resulting in a low adhesion between the fingers and the wire, which indicates low reliability of the interconnection approach.
A reliable busbar-less interconnection approach is demonstrated by implementing finger pads below the solder paste pads. Due to the finger pads, the connection area between the wire and the fingers is increased, resulting in a higher adhesion. Moreover, it is shown that the reliability of the interconnection approach without finger pads can be significantly increased by adding an adhesion in order to mechanically attach the wires to the cell.
Keywords: Back Contact, Busbar-less, Module Integration, PV Module

1 MOTIVATION AND AIM

One way to increase the attractiveness of the ZEBRA IBC cell developed at ISC Konstanz [1] is to reduce the cost of the cell concept. Regarding the cell transformation cost, the cost share of the silver pastes is 46 % (Figure 1). Therefore, the costs can be significantly decreased by reducing silver paste.

In a busbar-less interconnection approach the silver paste for busbars is eliminated or at least almost no paste is needed for the busbars. This results in a significant reduction of silver paste and costs.

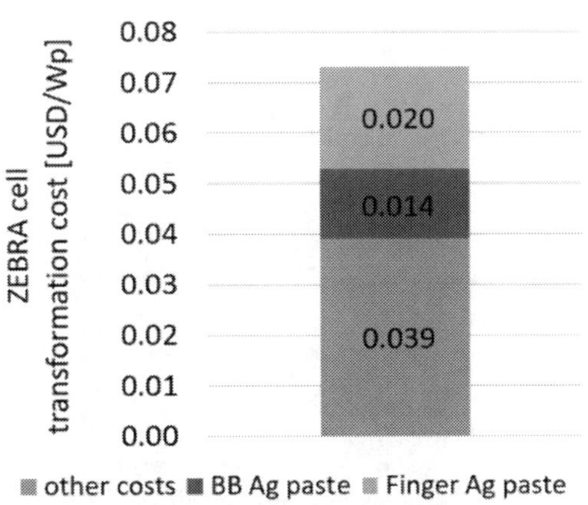

Figure 1: Cell transformation cost of the standard ZEBRA cell concept.

To integrate the newly designed busbar-less IBC cells in a module, a new interconnection approach must be developed as well. Concepts to interconnect busbar-less cells have been proposed [2-6], indicating there is room for improvement by reducing costs and silver consumption. However, since the interconnection approach of the mentioned research work is mostly used for both-side contacted cells, it cannot be implemented for the interconnection of back contact cells, due to the presence of both polarities on the same side of the cell. Van Dyck et al. [7] presented an interconnection approach for busbar-less back contact cells. They use solder-coated wires that are woven into the encapsulant. This gives the encapsulant material a dual function, protection against environmental influences and electrical insulation between opposing polarities on the device.

The present paper shows a new interconnection approach of busbar-less back contact cells, which can be easily implemented in an existing stringer machine and commercially available materials are used.

For back contacted cells, it is expected that higher bifaciality can be achieved by reducing shading due to metallization and thinner wires on the backside. In previous research work [2-6, 8], it has already been shown that the light management in the module is improved by the busbar-less approach when using multiple thin wires for the interconnection. Thus, the interconnection approach presented in this paper includes a multi-wire concept.

2 METHOD

A new bifacial multi-wire module approach to implement quasi busbar-less IBC cells in a module is presented in this paper. Since the wires cannot be connected to the busbars, they must be connected individually to each finger to collect the generated charge carriers from the cell.

The fingers on the ZEBRA cell alternate in the polarity as shown in the schematic in Figure 2. Each second finger belongs to the same polarity. At defined positions, fingers with the same polarity are over-printed with solder paste while the adjacent fingers of opposite polarity are over-printed with an insulation pad. The wires, which are used to connect all the fingers of one polarity are set perpendicular on top of the pads. While the solder paste pads are used to mechanically attach and electrically contact the wire to the fingers of one polarity, the insulation paste pads are used to insulate the fingers of the other polarity from the wire to prevent a short cut. The solder paste pads and the insulation paste pads are located in one row at the location where the wire is connected, so that the wire can be placed straight. Although, it is not a busbar printed, the row of solder paste pads and insulation paste pads looks similar to a busbar. That is why these rows are called quasi busbars.

On the cell, there are several quasi busbars. In order to contact the other polarity as well, on the following quasi busbar the solder paste pads and insulation paste pad are printed the other way around. On the fingers, which are insulated on the previous quasi busbar, a solder paste pad is printed and on the fingers, which were connected to the wire on the previous quasi busbar, an insulation paste pad is printed. The rows of solder paste and insulation pads next to each other, which are connected to the fingers with different polarity, form a pair of quasi busbars as shown in Figure 2. Any number of such pairs can be uniformly distributed across the cell surface.

Since the insulation paste on the cell has a certain height, the solder paste pad must be higher to ensure a physical contact between the solder paste and the wire during the soldering process. For this reason, this metallization concept is referred as 3D design. The schematic of the metallization concept and interconnection approach is shown in Figure 2.

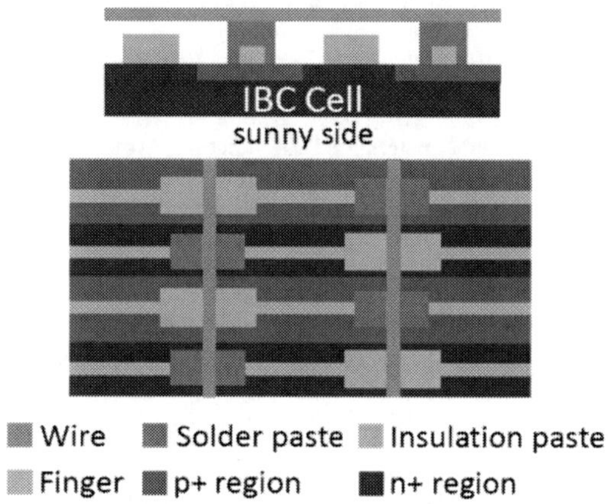

Figure 2: Schematic of the interconnection approach. Top: Cross-section of one row of solder paste and insulation paste pads of the cell. Buttom: Top-view of the back side of the cell. One pair of quasi busbars and the interconnected wires are shown. Each wire contacts the fingers of one polarity.

The wires can cause the cell to break during lamination if they are placed too close to the edge of the cell. Therefore, the wires do not directly contact the fingers at the edges of the cell. To connect the first four fingers counting from the wafer edge, a short busbar is printed at the wafer edge. The wires contact the short busbars on the fourth finger. As a result, a small amount of busbar silver paste needs to be printed for the metallization layout. Compared to the standard ZEBRA cell concept, 85 % of the busbar silver paste consumption is saved due to the busbar-less interconnection approach.

The instead printed solder paste consist from bismuth, tin and silver. The amount of silver that is included in the solder paste is low. 5 – 6 % of the overall silver consumption is used for the solder paste, depending on the metallization layout of the different development steps. Since the solder paste is printed on the fingers, bare copper wires can be used, instead of solder coated ones.

For the interconnection approach described in this paper, a solder paste consisting of tin, bismuth and silver is used. After printing the solder paste is dried in a belt-type furnace. The soldering of the solder paste pads between the fingers and the wires is done manually in the laboratory on two hot plates. To meet the temperature profile suggested by the manufacturer of the solder paste, a preheating step (270 s at 135°C) and a high temperature step (65 s at 180°C) are used for soldering. Additionally a flux is used to pretreat the uncoated copper wires.

Several development steps are taken to develop a reliable busbar-less interconnection approach. For each step mini modules of one half cell or two half cells are assembled to test the reliability of the interconnection approach. These steps are described in the following sub-sections.

2.1 Development step 0 – Replacing busbars by solder paste pads

In Development step 0 the busbars were simply replaced by the solder paste pads. Therefore, the finger silver paste consumption remains the same as for a standard ZEBRA cell. This leads to an overall reduction in the silver consumption of 39 %.

Cells in the wafer format M2 with four pairs of quasi busbars are used in this development step. For the stringing, wires with a cross-section of 1.5 x 0.2 mm² are used. The mini modules for reliability testing include one half cell.

The cells are specially constructed to better observe failures in the mini modules. For this purpose, the fingers are divided into four segments. Each segment is contacted to one pair of quasi busbars. This design makes it easy to recognize the location of a failure when a finger disconnects from the wire during reliability testing.

2.2 Development step 1 – Additional adhesive

In this experiment, additional to the approach in development step 0, an adhesive is used to mechanically attach the wires onto the cell. Therefore, a thermosetting epoxy based adhesive is added at both ends as well as in the middle of each wire. The adhesive is applied in the manner shown in Figure 3. The small lines of the adhesive start on top of the finger and reach the surface of the cell.

The adhesive is fully cured during the soldering process. Thus no further thermal stress is added due to the curing of the adhesive.

Figure 3: Microscopic picture of a sample of the fully cured adhesive (highlighted in blue) integrated into a mini module.

2.3 Development step 2 – Finger pads below solder paste

In development step 2 a new metallization layout was designed to increase the adhesion of the wire to the fingers. Thus, a finger pad is added below the solder paste pad. The size of the finger pad on the p+ region is 0.15 x 1.6 mm² and on the n+ region it is 0.06 x 1.6 mm². Due to the finger pads, the consumption of finger silver paste has increased by 22 % compared to a standard ZEBRA cell as well as compared to the cells used in development step 0 and 1.

However, the saving of busbar silver paste consumption is still more than 85 % compared to the standard ZEBRA cell concept, resulting in an overall reduction in silver consumption of 28 % compared to a standard ZEBRA cell.

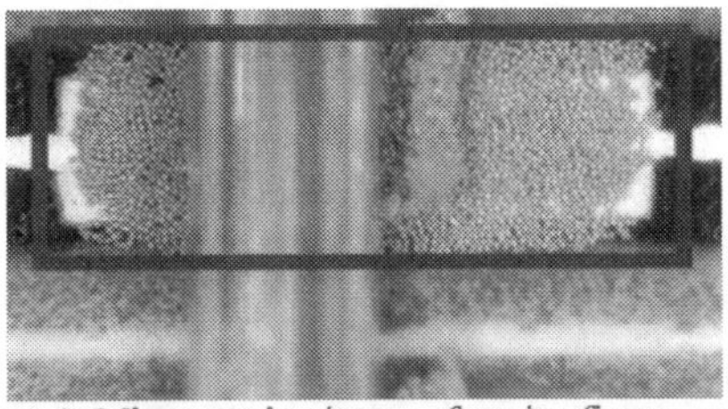

Figure 4: Microscopic picture of a wire-finger-contact with a finger pad (highlited in blue) below the solderer paste pad.

Moreover, a multi-wire approach is implemented in development step 2. After soldering, the different thermal expansions of the used material lead to thermomechanical stresses on the connection. Due to the one-sided contacting of the wires on the cell, a bowing of the cell occurs. To reduce the stress, a multi-wire concept is implemented in this interconnection approach. Since the cross-section per wire can be reduced, a reduction of the thermomechanical stresses in this interconnection approach is expected. For this development step, wires with a cross-section of 0.5 x 0.2 mm^2 are used.

For the development step 2, cells in the wafer format of M6 are used and twelve pairs of quasi busbars are printed on the cell for the multi-wire approach. For the reliability test two half cells are connected in series when assembling the mini modules.

3 RESULTS

As shown in Figure 5 many fingers do not connect to the wire when assembling the modules as described in development step 0. After soldering, all the fingers are connected to the wire but the adhesion is low. For the fingers connected to the p+ region, the adhesion is below 0.1 N/mm, measured in a 180° peel force test. On the n+ region the adhesion is a little higher, where a peel force of 0.25 N/mm is recorded. Due to the low adhesion, some finger disconnect form the wire or disconnect from the cell during lamination already.

During thermal cycling (TC) test most fingers disconnect from the wire or from the cell and are electrically inactive. As shown in Figure 5b as well as in in Figure 8 after TC100 the module significantly degrades. Especially the fingers at the edges of the cell disconnect during TC.

As we believe that due to the low adhesion of the connection, the fingers disconnect from the wire because of the different thermal expansion of the components of the interconnection and the cell. In order to increase the adhesion, an adhesive is used to mechanically attach the wire on to the cell. This approach is implemented in development step 1. To attach the wire mechanically to the cell an adhesion is added as described in chapter 2.3. The additional adhesive shows an adhesion after curing of more than 2 N.

a) b)

Figure 5: Electroluminescence picture of a sample of development step 0. a) Initial picture after module assembly. b) After 100 thermal cycles according to IEC61215.

As shown in Figure 6a after lamination most fingers are connected to the cell. There are some finger interruptions related to bad printing quality and some electrically inactive fingers, which leads to the conclusion that these fingers are not connected to the wire properly at this location.

After 50 thermal cycles it is shown in Figure 6b that there are some finger disconnection related to the TC test. However, the improvement just by adding the adhesive is significant. Although, in the electroluminescence (EL) picture it seems that only a few extra fingers disconnect after TC 50, the power loss (shown in Figure 8) after TC 50 is already 4 %.

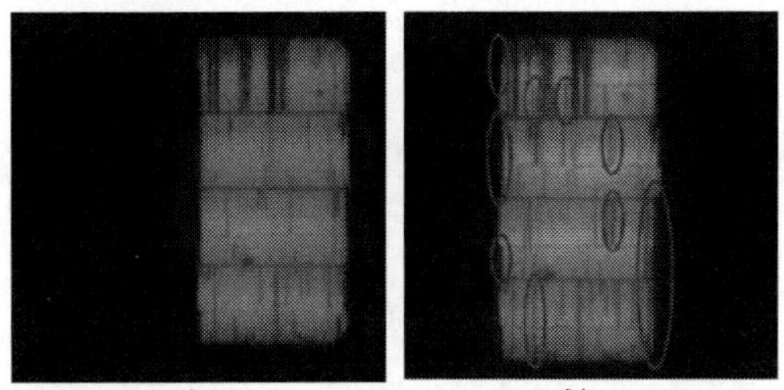

a) b)

Figure 6: Electroluminescence picture of a sample of development step 1. a) Initial picture after module assembly. b) After 50 thermal cycles according to IEC61215.

Walter et al [4] have already shown in their publication, large finger pads on the fingers can lead to higher adhesion. Therefore, in development step 2 this approach was used to implement it to a back contacted IBC cell with a 3 D designed metallization layout.

An adhesion over 0.8 N/mm is achieved when soldering the wire on the finger pad. Since similar adhesion resulted in reliable modules in other interconnect designs, this adhesion is expected to be more reliable in TC tests. For this reason, no additional adhesive is used in this development step.

Soldering the wire to the finger pads results in a mostly homogenously contacted mini module after lamination as shown in Figure 7a. In the top right corner of the EL picture, there is a darker region. This indicates a higher series resistance in this area of the mini module. The electrical connection might not be soldered correctly. However, the fingers are connected to the wire and the area is electrically active.

The comparison of Figure 7a and Figure 7b shows that there is no visible change between the EL pictures. All the fingers remain connected to the finger pads. The TC test does not seem to have a significant effect on the soldered joint at least for 100 thermal cycles. In addition, the power loss after TC100 is low. Figure 8 shows a power loss lower than 0.6 %.

However, it can be observed from the EL picture, that the finger pads (consist of Ag paste) result in higher recombination losses. The pads are visible in the EL pictures and they also result in a lower cells' open-circuit voltage (Voc) in comparison with the standard ZEBRA cell concept as well as for the cell concepts from development step 0 and 1.

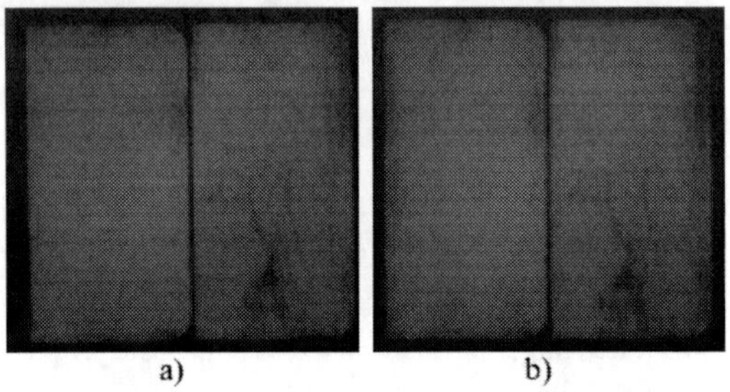

Figure 7: Electroluminescence picture of a sample of development step 2. a) Initial picture after module assembly. b) After 100 thermal cycles according to IEC61215.

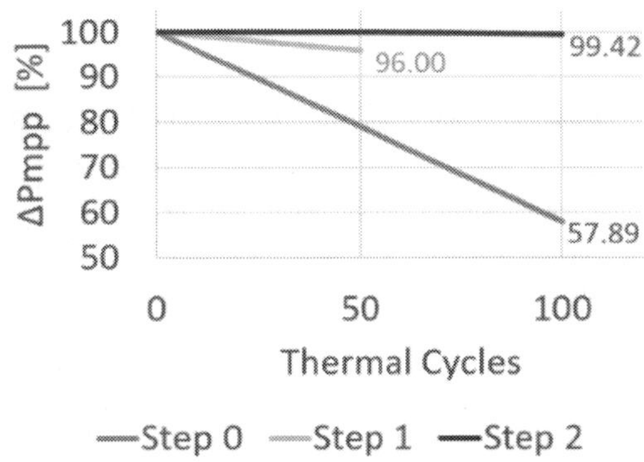

Figure 8: Power loss due to thermal cycles recorded for the best samples of the different development steps.

4 DISCUSSION AND CONCLUSION

In development step 2, a reliable interconnection approach is demonstrated that leads to overall silver saving of 28 % compared to standard ZEBRA cells. Even if the TC test is ongoing and it is done only for up to 100 cycles until now, it is expected to pass TC200 and more with this approach. However, the finger pads implemented in development step 2 lead to two disadvantages that need to be optimized in further development steps.

First, although silver is saved in comparison with the standard ZEBRA cell concept, a higher amount of silver is needed in comparison to development step 0 and 1. This leads to higher costs for development step 2 compared to the other approach introduced in this paper.

Second, because of the finger pads there are cell efficiency losses because of recombination. Since the busbar-less cells cannot be measured precisely at the moment; it is not possible to measure this effect correctly. However, it is expected to lose approximately 2 – 3 % in Voc due to the large finger pads.

Since the connection area between the finger pad and the wire should be rather small to ensure a high Voc, the finger pad must be as small as possible. For the electrical connection the connection area is big enough any way, because the cross-section of the soldered contact is much bigger than the cross section of the finger. However, for the mechanically adhesion, the connection area should be as big as possible to reliably connect the wire to the finger. As shown in the results of development step 0 and 1 the connection area is too small when printing only the finger. This leads to the conclusion that the electrical connection and the mechanical adhesion should be separately developed in a busbar-less approach on a 3D designed back contact cell.

It is already shown in development step 1 that an additional adhesive that mechanically attaches the wire to the cell leads to significant improvements in terms of the reliability of the interconnection approach. For further development of the busbar-less interconnection approach of an IBC cell, the developments from step 1 and 2 should be combined. Thus, a reliable interconnection should be developed that has sufficient mechanical adhesion, a maximal Voc as well as a minimum in silver consumption. Therefore, the implementation of adding an adhesive for improved reliability should be further developed and an adapted finger pad should be developed to minimize the losses due to recombination.

5 SUMMARY

Since the cell transformation cost for the IBC cell ZEBRA is highly dependent on the silver paste consumption, an interconnection approach is developed to reduce the silver paste required for the busbars. This paper describes the development of a busbar-less interconnection approach for a 3D designed back contacted IBC cell based on printed solder pads.

Due to the missing of busbars, the wire has to be connected individually to each finger on the cell. When simply replacing the busbars by solder pads as described in development step 0, the adhesion between the wire and fingers is low. After TC100 the module significantly degrades.

It is shown in development step 1 that adding an adhesive to attach the wire mechanically to the cell leads to significant improvements in reliability. However, the reliability is still not sufficient. The power loss after TC50 is already 4 %.

By implementing finger pads, a reliable connection between fingers and wires is demonstrated. After TC100, the measured power loss is 0.6 %. However, the finger pads cause higher losses in the Voc due to recombination and the silver consumption is increased.

Further research can be focusing on separating the electrical connection and the mechanical adhesion for the busbar-less interconnection approach in order to optimize the reliability as well as the Voc and the silver consumption of the approach.

6 ACKNOWLEDGMENT

This work was partially funded by Solar Era Net and the Federal Ministry for Economic Affairs and Climate Action of Germany as part of the project AMBiPV (grant agreement No. 03EE1036).

7 REFERENCES

[1] A. Halm et al. The Zebra Cell Concept - Large Area n-Type Interdigitated Back Contact Solar Cells and One-Cell Modules Fabricated Using Standard Industrial Processing Equipment, EUPVSEC, 2012.

[2] T. Söderström et al., Smart Wire Connection Technology, EUPVSEC, 2013

[3] A. Faes et al., SmartWire solar cell interconnection technology, EUPVSEC, 2014

[4] J. Walter et al., Multi-wire interconnection of busbar-free solar cells, Energy Procedia 55, 2014, p. 380-388

[5] T. Borgers et al., Combined interconnection and lamination of bifacial busbarless cells trough woven wiring, EUPVSEC, 2017

[6] J. Govaerts et al., Progress in Encapsulant-Integrated Multi-Wire Interconnection, EUPVSEC, 2018

[7] Van Dyck et al., Woven Multi-Ribbon Interconnection for Back-Contact Cells: Extending the Functionality of the Encapsulant, AIP Conference Proceeding, 2019

[8] S. Braun et al., Solar Cell improvement by using a Multi Busbar Design as Front Electrode, Energy Procedia 27, 2012, p. 227-233

INTERCONNECTION TECHNOLOGY IN PV MODULES: IMPACT OF RIBBONS, TAB CONNECTORS AND ELECTRICALLY CONDUCTIVE BACKSHEET ON MODULE PERFORMANCE

A. Tummalieh[1], B. Paritala[2], M. Mittag[1], D. H. Neuhaus[1]
[1]Fraunhofer Institute for Solar Energy Systems (ISE)
Heidenhofstr. 2, 79110 Freiburg, Germany
[2]University of Applied Sciences Jena
Carl-Zeiss-Promenade 2, 07745 Jena, Germany

ABSTRACT: The interconnection technology is one of the aspects that is being continuously researched and developed in photovoltaic (PV) modules [1–4]. The aim of this study is to analyze the impact of the used interconnection technology in the PV module such as ribbons, tab connectors and electrically conductive backsheet (ECB) on cell-to-module (CTM) power losses [5, 6]. To that end, we adapt and develop analytical models that consider the geometry of the contact metallization of the solar cell as well as the geometry and properties of the interconnector being it rectangular ribbon, round ribbon, tab connector or ECB. The study considers different loss and gain channels of the CTM analysis such as power loss due to solar cell's active area shading and ohmic resistance of the interconnection. Simulation results show a maximum PV module power improvement of about 4% by using back contact solar cells interconnected using ECB compared to PV modules with solar cells interconnected using conventional rectangular ribbons. PV module with tab connectors exhibit a power improvement of about 3.4% compared to conventional cell interconnectors. That is a result of clearly less power loss due to ohmic resistance of the tab connectors, because of shorter current path as well as because of the larger cross section area of the current path using ECB. Furthermore, PV modules with back contact solar cells exhibit no loss due to shading of the interconnectors, because of the position of all contact pads on the backside of the solar cell compared to about 6 W power loss in case of rectangular ribbons. CTM analysis shows that the PV module with round ribbons exhibit about 3.6 W power gain due to light coupling by interconnectors compared to 1.5 W for the PV module with rectangular ribbons.

Keywords: PV Modules, Interconnection Technology, Ribbons, Electrically Conductive Backsheet (ECB), CTM

1 INTRODUCTION

The PV market expands and becomes more complex with all components and technologies of the module being permanently optimized to achieve higher power and efficiency [7]. Considering different loss channels in the PV module, the electrical losses due to cell interconnection can be reduced using different approaches and technologies. Typically, solar cells in the PV module are interconnected using rectangular SnPb-coated copper ribbons as shown in Figure 1 (left).

Since the ribbons are commonly placed on the front side of the solar cell, the width of the ribbon should be minimized to reduce the shading effect on the solar cells active area that leads to less power and efficiency. Minimizing the width of the interconnection ribbon decreases its cross-section area and therefore increases its ohmic resistance resulting in higher module electrical losses. Round ribbons as shown in Figure 1 (right) can increase the optical gain due to higher light coupling [1, 2].

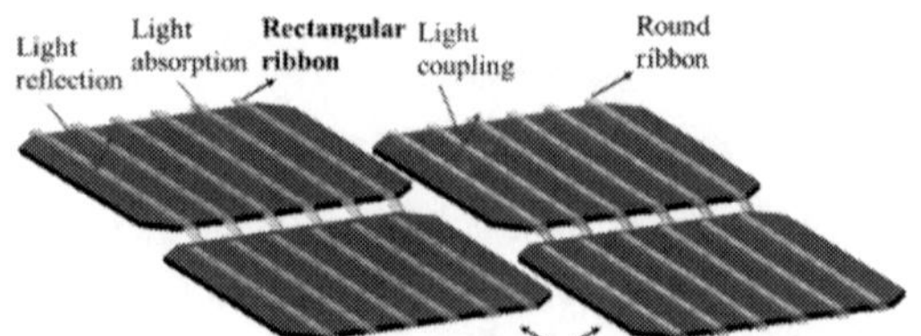

Figure 1: Solar cells interconnected using (left) rectangular and (right) round ribbons.

One way to optimize the PV modules electrical losses without compromising optical losses due to shading is the use of back contact solar cells such as interdigitated back contact solar cells (IBC) or metal wrap through solar cells (MWT) [8, 9], which offer larger active area on the solar cells front side due to missing metallization. Depending on the cell's metallization, such cells can be interconnected using tab connector or an electrically conductive backsheet as shown in Figure 1 (left, right), respectively.

The ECB is normally made of copper, that has a certain pattern to separate negative and positive contacts. Integrating back contact solar cells into PV modules can be realized using specially designed conductive backsheet that exactly matches the metallization pattern of the solar cell [10]. Using electrically conductive adhesive (ECA), back contact solar cells can be simply fixed on the conductive backsheet to be interconnected to each other.

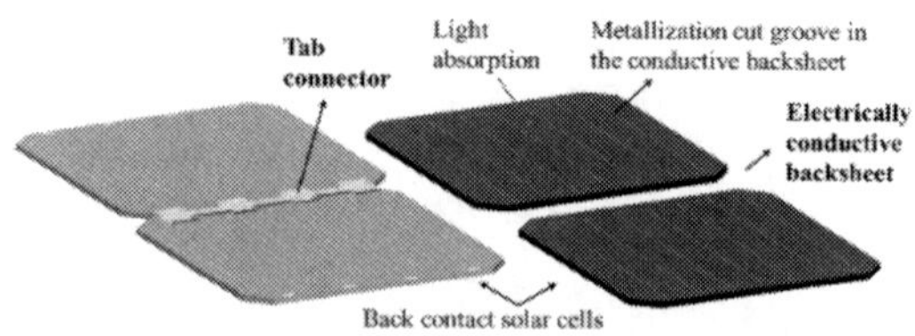

Figure 2: Back contact solar cells interconnected using (left) copper tab connector and (right) electrically conductive backsheet (ECB).

Based on the solar cell's metallization, the ECB pattern, and its thickness, the ohmic resistance of the current path and therefore the electrical loss can be calculated. The pattern of the ECB can be designed in a way that maximizes the width of the busbar current path and therefore minimizes its ohmic resistance. Increasing the thickness of the ECB increases the cross-section area of the current path and therefore decreases the electrical losses.

On the other hand, increasing the thickness of the

conductive backsheet negatively affects the overall price of the PV module resulting in higher specific PV module cost [€/W_P].

2 MODELLING

In this section, the approach used in developing and adapting different models to simulate PV modules using different interconnection technologies is presented. First, the geometry of the solar cell and its metallization is parameterized for further calculations.

Table I: Geometrical parameters and their description of a solar cell with contact pads.

Parameter	Description
l_{cell}	Length of solar cell
w_{cell}	Width of solar cell
d_{psq}	Pseudo-square diameter
l_{pad}	Length of contact pad
w_{pad}	Width of contact pad
d_{pads}	Vertical distance between contact pads
$d_{busbars}$	Horizontal distance between busbars
$d_{left/right}$	Left/right distance between contact/busbar pad and cell edge
$d_{top/bottom}$	Top/bottom distance between contact pad/busbar and cell edge

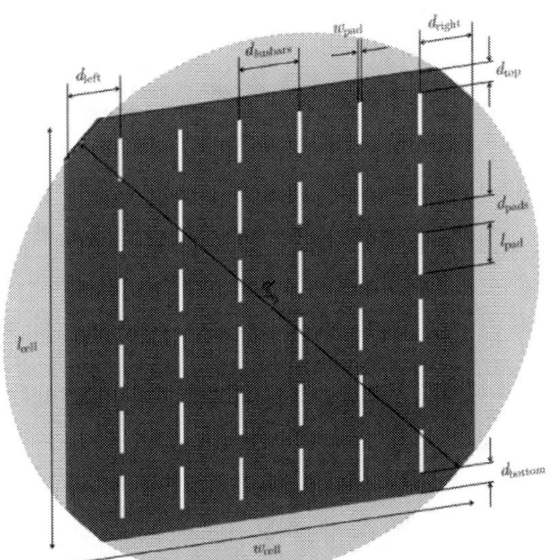

Figure 3: Illustration of the geometry of a solar cell with contact pads.

Basically, the calculation of the power loss P_{loss} due to the ohmic resistance is done using the following equation, where R is the ohmic resistance and I is the flowing electrical current.

$$P_{loss} = RI^2 \quad (1)$$

First, the magnitude of the electrical current collected by each metallization element, (in this case contact pad), is calculated. Since the magnitude of the current can be assumed to be linearly related to the solar cell active area, the calculation is based on the position of the contact pad and therefore on the area of interest ($AOI_{i,j}$) shown in Figure 4. i and j are the indices of the pad and busbar, and N_{pad} and N_{busbar} represent the number of pads and busbars, respectively. In case of continuous busbar i and N_{pad} are equal to 1. In the following equation, the calculation of the current's magnitude of an exemplary contact pad (highlighted in red in Figure 4) is shown.

$$AOI_{2,3} = (l_{pad} + d_{pads})\, d_{busbars} \quad (2)$$

The same approach is used for the front and back side of the solar cell. In case of using back contact solar cells, the geometry of the metallization for both n and p contacts is considered on the back side.

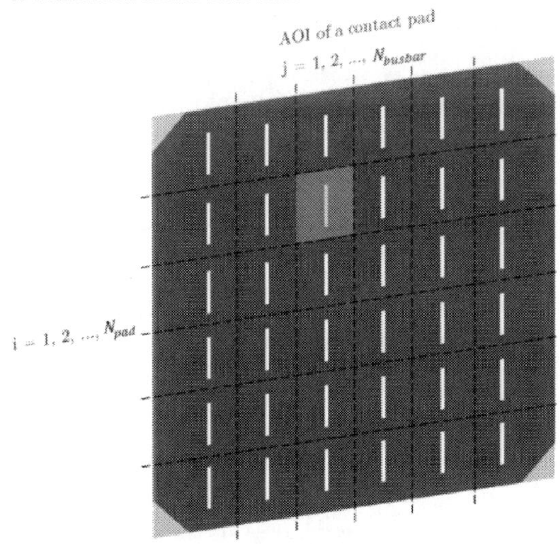

Figure 4: Area of Interest (**$AOI_{i,j}$**) used to calculate the current's magnitude.

To calculate the collected current by a specific metallization element, the following equation is used, where I_{mpp} is the solar cells current at the maximum power point.

$$I_{i,j} = AOI_{i,j}\, I_{mpp} \quad (3)$$

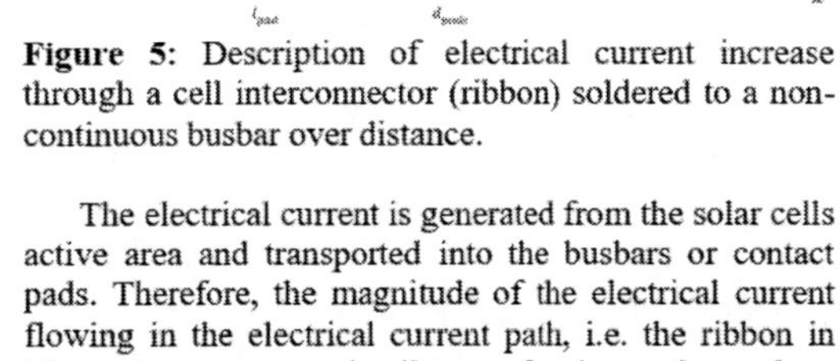

Figure 5: Description of electrical current increase through a cell interconnector (ribbon) soldered to a non-continuous busbar over distance.

The electrical current is generated from the solar cells active area and transported into the busbars or contact pads. Therefore, the magnitude of the electrical current flowing in the electrical current path, i.e. the ribbon in Figure 5, grows over the distance for the sections where the ribbon is soldered onto the metallization. By neglecting the effect of the fingers conducting the current, a constant current for the unsoldered ribbon sections is assumed.

The electrical current increment in a soldered ribbon section can be described by the following relation, where x describes the distance and I_{pad} is the collected current in the contact pad.

$$I(x) = \frac{I_{pad}}{l_{pad}} x \tag{4}$$

Considering solar cells with rectangular ribbons, round ribbons or tab connectors as shown in Figure 6, besides the electrical resistivity ρ, the width w and the thickness t of the cell interconnector are constant and therefore its cross-section area A. That means only the magnitude of the electrical current changes over the distance as explained in Figure 5.

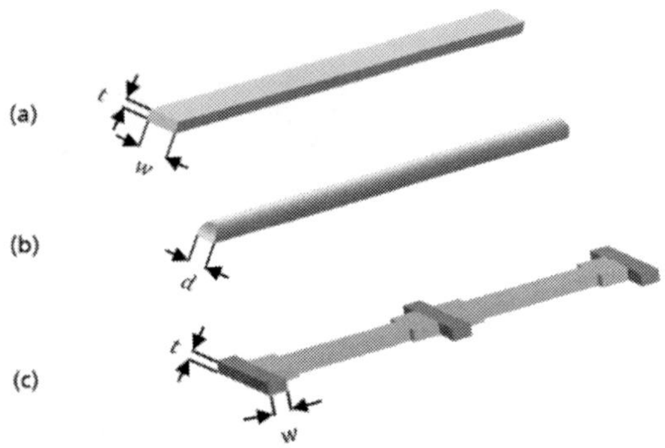

Figure 6: Illustration of different cell interconnectors; (a) rectangular ribbon, (b) round ribbon and (c) tab connector (elements of current path are dark gray).

To assess the ohmic loss in a soldered connector section, it is calculated by integrating the dissipated power in the connector over the length of the metallization element, i.e. the contact pad. The power loss calculation in soldered sections is presented using the following equation [5].

$$P_{\mathrm{loss}} = \frac{\rho}{A} \int_0^{l_{\mathrm{pad}}} I(x)^2 \, dx = \frac{1}{3} \frac{\rho \, l}{A} I_{pad}^2 \tag{5}$$

Since the magnitude of the collected current in unsoldered connector sections is constant (see Figure 5), the power loss in these sections is calculated as shown in the following equation.

$$P_{\mathrm{loss}} = \frac{\rho \, l}{A} I_{pad}^2 \tag{6}$$

Beside the tab connector, back contact solar cells can be interconnected using ECB. Such a backsheet is made of copper and has a certain groove pattern to match the metallization contacts on the solar cell's backside [8]. The cut groove separates the n- and p-contacts from each other to realize the series or parallel connection between solar cells or strings in the PV module as shown in Figure 7.

In addition to the electrical current function, Figure 8 shows the change in the current path width in an electrically conductive backsheet. The width starts from a minimum value of w_0 and increases linearly to reach a maximum value of w. The maximum width w depends on the horizontal distance between two busbars d_{busbars} and the minimum cut groove width w_0 and can be expressed as follows.

$$w = d_{\mathrm{busbars}} - w_0 \tag{7}$$

The width change function of the cut groove $w(x)$ can be described using the following equation, where $d_{\mathrm{ECB,top}}$ and $d_{\mathrm{ECB,bottom}}$ are the top and bottom distance between the cut groove and the edge of the solar cell as illustrated in Figure 9.

$$w(x) = w_0 + \frac{w - w_0}{l_{\mathrm{cell}} - d_{\mathrm{ECB,top}} - d_{\mathrm{ECB,bottom}}} x \tag{8}$$

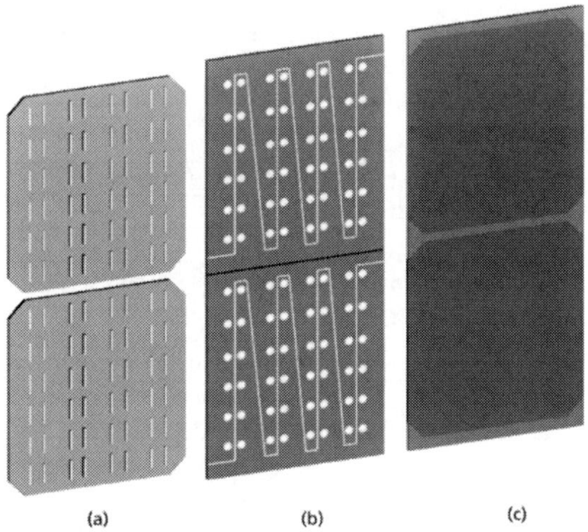

Figure 7: Illustration of (a) two back contact solar cells, (b) an ECB with exemplary cut groove and (c) two back contact solar cells interconnected in series using ECB.

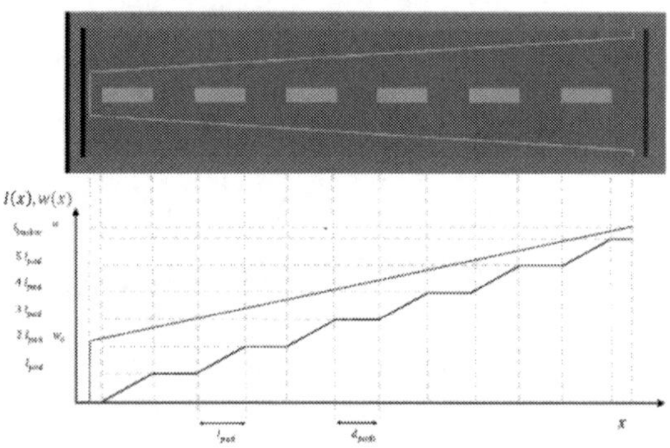

Figure 8: Description of electrical current increase through an ECB section (black) and width of current path increase (red) over distance.

To exemplarily calculate the ohmic power loss in the ECB of the top contact pad section shown in Figure 9, the following equation is used, where ρ and t are the electrical resistivity and the thickness of the ECB, respectively.

$$P_{\mathrm{loss}} = \frac{\rho}{t} \int_{d_{\mathrm{ECB,top}}}^{d_{\mathrm{ECB,top}} + l_{\mathrm{pad}}} \frac{I(x)^2}{w(x)} \, dx \tag{9}$$

Figure 9: Illustration of an ECB section with an exemplary trapezoidal cut groove underneath a back contact solar cell section.

In addition to the power loss due to the ohmic resistance of the cell interconnector, the power loss due contact resistance because of soldering or ECA gluing is considered. To calculate the total ohmic power loss due to interconnection in a solar cell, the power loss is calculated for each connector section and finally summed up. The calculation is done for both p- and n-polarities. To scale up the calculation into a PV module level simulated under standard test conditions (STC), the solar cell power loss is multiplicated by the number of solar cells in the PV module.

3 SIMULATIONS AND RESULTS

To make a comparison between the different interconnection technologies studied in this work, firstly, the common simulation parameters as shown in Table II are set.

Table II: Common simulation parameters of the PV module.

Parameter	
Number of solar cells	60 cells
Solar cell format	Full cells (166×166 mm^2)
Solar cell power	5.5 W
Connection type	Series (10×6)
Type of busbar	Contacts pads
Cell and string distance	2 mm
Interconnector/ECB thickness	0.15 mm
Electrical resistivity	1.8 μΩ cm
Ribbon/tab connector width	1.5 mm
Round ribbon diameter	0.3 mm
ECB cut groove pattern	Trapezoidal (w_0 = 5 mm)

For each interconnection technology, the type of the solar cell and its typical parameters are chosen. For a PV module with rectangular ribbon interconnection, 5 busbars H-pattern solar cells are simulated. In case of round ribbons, multi busbar solar cells with 12 busbars are used. For PV modules with tab connector, back contact solar cells with three terminals are used. PV modules with ECB are simulated with back contact solar cells with four non-continuous busbars.

The simulation is done under standard test conditions (STC) using SmartCalc.Module, a software developed by Fraunhofer ISE by applying a bottom-up multi-physics model [11, 12]. The model used considers loss and gain channels that take place in the PV module when interconnecting the solar cells. That includes geometrical losses, optical losses and gains, and electrical losses [5, 13]. The comparison results are shown in Figure 10.

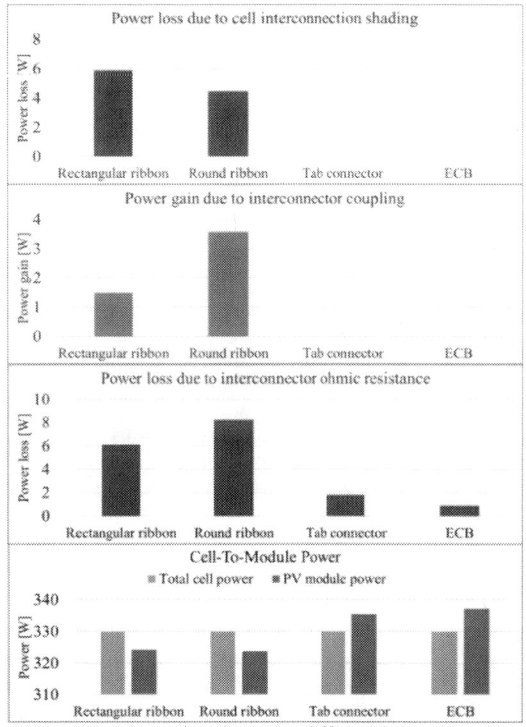

Figure 10: Comparison between different interconnection technologies regarding power loss due to interconnection shading, power gain due to interconnector coupling, power loss due to ohmic resistance and CTM power.

Considering the different loss and gain channels of the CTM analysis, the power loss due to shading of solar cells active area by the interconnection, power gain due to light coupling by the interconnection and the power loss due to ohmic loss are the main channels affected by the interconnection technology in the CTM analysis. Based on the used simulation parameters, the rectangular ribbon has the highest power loss due to shading with about 6 W. Using round ribbons reduces the power loss due to shading to about 4.5 W. Since interconnecting the back contact solar cells using tab interconnectors or ECB does not cover any active area of the solar cell, the power loss due to shading in both cases is 0 W.

Compared to rectangular ribbons, round ribbons reflect more light into the solar cells active area. Therefore, the PV module with round ribbons exhibit more than twice power gain due to light coupling compared to the PV module with conventional rectangular ribbons. Since back contact solar cells are interconnected from the backside, they do not exhibit any power gain due to light coupling on the front side of the solar cell.

Regarding the power loss due to ohmic resistance of the interconnector, the PV modules with IBC solar cells using tab connectors and ECB exhibit a clear lower loss compared to the rectangular and round ribbon. This is due to lower ohmic resistance, because of a shorter current path of the tab connector and the larger current path cross section area of the ECB. The exemplary PV module with ECB shows about 83% less power loss compared to the PV module with the conventional rectangular ribbons.

As a result, and by considering other CTM gain and loss channels [1, 2], the PV module with the ECB shows the highest CTM power with about 337.2 W_p compared to about 335.5 W_p for the PV module with tab connectors, 323.7 W_p for the PV module with round ribbons and 324.3 W_p for the PV module with the conventional rectangular ribbons.

4 CONCLUSIONS AND OUTLOOK

Based on the results shown in the previous section, the PV modules with back contact solar cells exhibit a higher module power compared to the PV modules with conventional H-pattern solar cells. The higher PV module power is mainly due to the lower ohmic resistance of the current path using tab connectors and ECB compared to the rectangular and round ribbons. Furthermore, interconnecting back contact solar cells results in zero loss because of cells active area shading due to cell interconnectors.

On the other side, other aspects should be considered and studied such as the durability and mechanical stability of the solar cell interconnections. Furthermore, production challenges and cost analysis should be studied and evaluated for each interconnection technology to have a holistic comparison between the different interconnection technologies.

ACKNOWLEDGEMENTS

Current study was prepared in the framework of the project HighLite, which was funded by the European Union's Horizon2020 Program under Grant Agreement No 857793.

REFERENCES

[1] I. Hädrich, "Unified methodology for determining CTM ratios: Systematic prediction of module power," 4th Silicon PV, 's-Hertogenbosch, Netherlands, 2014.

[2] M. Mittag *et al.*, "Cell-to-Module (CTM) Analysis for Photovoltaic Modules with Shingled Solar Cells," 2017 IEEE 44th Photovoltaic Specialists Conference (PVSC), Washington, DC, Jun. 2017 - Jun. 2017, pp. 1531–1536.

[3] A. Tummalieh *et al.*, "Trend Tracking of Efficiency and CTM Ratio of PV Modules," in *37th EU PVSEC 2020*, 2020.

[4] Hasnain Yousuf *et al.*, "Cell-to-Module Simulation Analysis for Optimizing the Efficiency and Power of the Photovoltaic Module," *energies*, 2022.

[5] I. Haedrich *et al.*, "Unified methodology for determining CTM ratios: Systematic prediction of module power," *Solar Energy Materials and Solar Cells*, vol. 131, pp. 14–23, 2014.

[6] M. Mittag *et al.*, "Cell-to-Module (CTM) Analysis for Photovoltaic Modules with Shingled Solar Cells: 44th IEEE Photovoltaic Specialists Conference," 2017.

[7] A. Tummalieh *et al.*, "Trend Tracking of Efficiency and CTM Ratio of PV Modules: 37th European Specialist Conference and Exhibition," 2020.

[8] M. Mittag *et al.*, "MOSAIC MODULE CONCEPT FOR COST-EFFICIENT AND AESTHETIC BIPV MODULES," *35th European PV Solar Conference and Exhibition*, 2018.

[9] Max Mittag, Matthieu Ebert, Helen Rose Wilson, Tobias Fellmeth, "Mosaic Module Concept for Cost-efficient and Aesthetic BIPV Modules," in *Proceedings of the 35th European Photovoltaic Solar Energy Conference and Exhibition (EU PVSEC); Brussels, Belgium*, 2018.

[10] I. Bennett *et al.*, "Development of conductive back-sheet for manufacture of PV modules with back-contact cells," *2018 IEEE 7th World Conference on Photovoltaic Energy Conversion (WCPEC) (A Joint Conference of 45th IEEE PVSC, 28th PVSEC & 34th EU PVSEC)*, pp. 3623–3625, 2018.

[11] M. Mittag *et al.*, "Systematic PV module optimization with the cell-to-module (CTM) analysis software," *Photovoltaics International*, no. 36, pp. 97–104, 2017.

[12] Fraunhofer Institute for Solar Energy Systems ISE, *SmartCalc.Module*.

[13] *SmartCalc.CTM*: Fraunhofer Institut for Solar Energy Systems ISE.

AUTHOR INDEX

WIP – Renewable Energies
Sylvensteinstr. 2
81369 Munchen
Germany

ISBN 978-1-7138-7180-4

CURRAN ASSOCIATES INC.
proceedings.com

8th World Conference on Photovoltaic Energy Conversion (WCPEC-8)

Milan, Italy
26 – 30 September 2022

Volume 2 of 2

WIP RENEWABLE ENERGIES

8th World Conference on Photovoltaic Energy Conversion (WCPEC-8)

Milan, Italy
26 – 30 September 2022

Volume 2 of 2

Editors:

A. Scognamiglio
R. Kenny
J. Bergmiller
Y. Ueda
S. Hayase
J. De Gregorio
S. Hubbard
A. Smets

ISBN: 978-1-7138-7180-4

Printed from e-media with permission by:

Curran Associates, Inc.
57 Morehouse Lane
Red Hook, NY 12571

Some format issues inherent in the e-media version may also appear in this print version.

Printed with permission by Curran Associates, Inc. (2024)

For permission requests, please contact WIP – Renewable Energies
at the address below.

WIP – Renewable Energies
Sylvensteinstr. 2
81369 Munchen
Germany

Phone: +49 89 72012735
Fax: +49 89 72012791

wip@wip-munich.de

Additional copies of this publication are available from:

Curran Associates, Inc.
57 Morehouse Lane
Red Hook, NY 12571 USA
Phone: 845-758-0400
Fax: 845-758-2633
Email: curran@proceedings.com
Web: www.proceedings.com

8th World Conference on Photovoltaic Energy Conversion

Proceedings of the International Conference

26 September – 30 September 2022

Edited by:

A. SCOGNAMIGLIO
ENEA - Italian National Agency for New Technologies, Energy and Sustainable Economy
Italy

Y. UEDA
Tokyo University of Science, Japan

S. HUBBARD
RIT- Rochester Institute of Technology
USA

R. KENNY
European Commission Joint Research Centre
Italy

S. HAYASE
Kyushu Institute of Technology
Japan

A. SMETS
TU Delft- Delft University of Technology
The Netherlands

J. BERGMILLER
WIP Renewable Energies
Germany

J. DE GREGORIO
WIP Renewable Energies
Germany

Coordination of the Technical Programme:
European Commission Joint Research Centre
Via E. Fermi 1
21020 Ispra (VA)
Italy

Institutional Support:
European Commission
WCRE – World Council for Renewable Energy

Supporting Organisations:
APVI – Australian PV Institute
ENEA – Italian National Agency for New Technologies, Energy and Sustainable Economic Development
EREF – European Renewable Energies Federation
ESA – European Space Agency
ESMC – European Solar Manufacturing Council
EUFORES –The European Forum for Renewable Energy Sources
EUREC – The Association of European Renewable Energy Research Centres
ETIP PV – European Technology & Innovation Platform PV
Guidehouse Insights
GÜNDER – The Turkish Solar Energy Society
IEA PVPS – International Energy Agency, Photovoltaic Power Systems Programme
INES – National Solar Energy Institute
NUS /SERIS – National University of Singapore / Solar Energy Research Institute of Singapore
ODTÜ-GÜNAM – Center for Solar Energy Research and Applications
REN21 – Renewable Energy Policy Network for the 21st Century

Institutional PV Industry Cooperation:
SolarPower Europe
SOLARUNITED

Supporting Associations:
ABENS – Associação Brasileira de Energia Solar
AREA – African Renewable Energy Alliance
CABA – Continental Automated Buildings Association

WCPEC-8 realised by:
WIP Renewable Energies
Sylvensteinstr. 2, 81369 München, Germany
Tel: +49 89 720 12 735, Fax: +49 89 720 12 791
Email: pv.conference@wip-munich.de
www.wcpec-8.com
www.wip-munich.de

Proceedings produced and published by:
WIP
Sylvensteinstr. 2, 81369 München, Germany
Tel: +49 89 720 12 735, Fax: +49 89 720 12 791
Email: pv.conference@wip-munich.de
www.wcpec-8.com
www.wip-munich.de

ISSN 2196-0992

8th WORLD CONFERENCE ON PHOTOVOLTAIC ENERGY CONVERSION
26 SEPTEMBER – 30 SEPTEMBER 2022

WCPEC-8 COMMITTEES

CONFERENCE ORGANISING COMMITTEE

Conference General Chairs
A. Scognamiglio, ENEA - Italian National Agency for New Technologies, Energy and Sustainable Economy, Italy
Y. Ueda, Tokyo University of Science, Japan
S. Hubbard, RIT, USA

Committee Members
M. Topič, University of Ljubljana, Slovenia
P. Helm, WIP Renewable Energies, Germany
J. Bergmiller, WIP Renewable Energies, Germany
J. de Gregorio, WIP Renewable Energies, Germany
F. Lambert, Genvia, France
N. Pearsall, Northumbria University, United Kingdom
J. Serra, Lisboa University, Portugal
R. Kenny, European Commission, Joint Research Centre, Italy
J. Carabe, CIEMAT, Spain
P. Delli Veneri, ENEA - Italian National Agency for New Technologies, Energy and Sustainable Economy, Italy
M. Kern, WIP Renewable Energies, Germany
A. Schöning, WIP Renewable Energies, Germany
P. Gerlinger, WIP Renewable Energies, Germany
T. Morrison, WIP Renewable Energies, Germany
P. Malbranche, International Solar Aliance, India
H. Ossenbrink, Band Gap, Germany
S. Nowak, NET Nowak Energie & Technologie, Switzerland
A. Grassi, ETA Florence Renewable Energies, Italy
J. Trube, VDMA, Germany
A. Michaelsen, WIP Renewable Energies, Germany
G. Masson, Becquerel Institute, Belgium
R. Drozdowski, IPVF, France
R. Kopecek, ISC Konstanz, Germany
D. Schultz, WIP Renewable Energies, Germany
D. Moser, EURAC, Italy
P. Fath, RCT Solutions, Germany
P Mints, Consultant, USA
M. Woodhouse, NREL, USA
Z. Holman, ASU, USA
P. Verlinden, Amrock, Australia
T. Jester, NEXT Energy Technologies, USA
M. Beck, Department of Energy, USA
K. Hinzer, University of Ottawa
S. Haouka, IPVF, France
P. Frankl, iea, France
I. Weiss, WIP Renewable Energies, Germany
D. Mugnier, Planair, France
L. Großhans, WIP Renewable Energies, Germany
D. Lecoque, ARE, Belgium
C. Pedretti, ARE, Belgium
S. De Iuliis, ENEA - Italian National Agency for New Technologies, Energy and Sustainable Economy, Italy
R. Janssen, WIP Renewable Energies, Germany
R. Blanchard, Loughborough University, United Kingdom
H. M. Albuflasa, University of Bahrain, Bahrain
S. Caneva, WIP Renewable Energies, Germany
G. Ruggieri, Insubria University, Italy
S. Leanza, WIP Renewable Energies, Germany
C. Lins, GWNET, Austria
D. Ngei, GWNET, Austria
W. van Sark, Utrecht University, The Netherlands
R. Egan, UNSW, Australia
W. Sinke, TNO, The Netherlands
C. Ballif, EPFL, Switzerland
J. Poortmans, imec, Belgium
F. Dimroth, Fraunhofer-Institut für Solare Energiesysteme ISE, Germany
P. Szymanski, European Commission, Joint Research Centre, The Netherlands
M. Konagai, Tokyo City University, Japan
Y. Zhao, Nankai University, China
C. Park, Yeungnam University, Korea
B. Hoex, UNSW, Australia
S. Hayase, Kyushu Insititute of Technology, Japan
J.-J. Lee, Dongguk University, Korea
Y. Kurokawa, Nagoya University, Japan
J. Vasi, IIT Bombay, India
A. Kabu, International Solar Aliance, India
J. H. Yun, Korea Institute of Energy Research, Korea
M. K. Tan, SERIS, Singapore
T. Minemoto, Ritsumeikan University, Japan
M. Yamaguchi, Toyota Technological Institute (TTI), Japan
S. Bremner, UNSW, Australia
A. Freundlich, University of Houston, USA
S. Kurtz, Merced University, USA
S. Marsillac, Old Dominion University, USA
A. Smets, TU Delft, The Netherlands
I. Sellers, University of Oklahoma, USA
T. Grassman, Ohio State University, USA
C. Zimmermann, Airbus Defence and Space, Germany
D. Wilt, US Department of Energy, USA
L. Kazmerski, retired from NREL, USA
B. Smith, NREL, USA
K. Hinzer, University of Ottawa, USA
A. Nayfeh, Khalifa University, United Arab Emirates
E. Warren, NREL, USA
P. Bermel, Purdue University, USA
J. McNatt, NASA, USA
S. Polly, RIT, USA
S. Tomasulo, Naval Research Lab., USA
S. Ringel, Ohio State University, USA

INTERNATIONAL ADVISORY COMMITTEE TO THE WCPEC (IAC)

Chairs
H. Ossenbrink, Band Gap, Germany
M. Yamaguchi, Toyota Technological Institute (TTI), Japan
S. Ringel, Ohio State University, USA

Committee Members
M. Topic, University of Ljubljana, Slovenia
W. Sinke, TNO, The Netherlands
J. Serra, University of Lisbon, Portugal
A. Jaeger-Waldau, European Commission, Joint Research Centre, Italy
P. Malbranche, International Solar Alliance, India
S. Nowak, NET Nowak Energy & Technology, Switzerland

J. Poortmans, imec, Belgium
P. Verlinden, Amrock, Australia
F. Ferrazza, ENI, Italy
W. Hemetsberger, SolarPowerEurope, Belgium
P. Helm, WIP Renewable Energies, Germany
A. Scognamiglio, ENEA - Italian National Agency for New Technologies, Energy and Sustainable Economy, Italy
R. Kenny, European Commission, Joint Research Centre, Italy
A. Smets, TU Delft, The Netherlands
M. Konagai, Tokyo City University, Japan
N. Usami, Nagoya University, Japan
B. Hoex, University of New South Wales, Australia
A. Aberle, SERI, Singapore
M. Green, University of New South Wales, Australia
D. Kim, Korea University, Korea
C. Lan, National Taiwan University, Taiwan
Y. Ueda, Tokyo University of Science, Japan
J. Vasi, India India Institute of Technology, India
J. Yi, Korea, SungKyonKwan University, Korea
D. Yang, China, Zheejiang University, China
Y. Zhao, Nankai University, China
S. Hayase, Kyushu Institute of Technology, Japan
A. Freundlich, University of Houston, USA
S. Marsillac, Old Dominion University, USA
A. Reinders, University of Twente, The Netherlands
T. Anderson, University of Massachusetts, USA
S. Bailey, retired from NASA Glenn Research Center, USA
A. Barnett, retired from University of Delaware, USA
R. King, Arizona State University, USA
R. Raffaelle, Rochester Institute of Technology, USA
B. Stanbery, HelioVolt, USA
D. Wilt, Air Force Research Laboratory, USA
L. Kazmerski, retired from NREL, USA
R. Todi, IEEE, USA
S. Kurtz, University of Merced, USA
S. Hubbard, RIT, USA
A. Smets, TU Delft, The Netherlands
Y. Matsumoto, SEES, Mexico
A. Nayfeh, Khalifa University, UAE
K. Hinzer, University of Ottawa, Canada

WCPEC-8 SCIENTIFIC COMMITTEE

Programme Technical Chairs

R. Kenny, European Commission, Joint Research Centre, Italy
S. Hayase, Kyushu Institute of Technology, Japan
A. Smets, TU Delft, The Netherlands

Topic Chairs

Topic 1: Silicon Materials and Cells

S. Glunz, Fraunhofer-Institut für Solare Energiesysteme ISE, Germany
Q. Wang, Jinko solar, China
K. Ding, Forschungszentrum Jülich, Germany

Topic 2: Evolving and Emerging Technologies: Tandems; Thin Film absorbers; III-V; New Materials and Concepts; Advanced Modelling

M. Edoff, Uppsala University, Sweden
U. Bach, Monash University, Australia
C. Hages, University of Florida, USA

Topic 3: Photovoltaic Modules and BoS Components

U. Jahn, VDE Renewables, Germany
A. Lennon, UNSW, Australia
T. Barnes, NREL, USA

Topic 4: PV Systems Engineering, Integrated/Applied PV

David Moser, EURAC, Italy
Thomas Reindl, National University of Singapore (NUS), Singapore
Clifford Hansen, Sandia National Laboratories, USA

Topic 5: Energy Transition – Integration, Storage, Sustainability, Policy, Economics, Energy Poverty, Society

Stefan Nowak, NET Nowak Energy & Technology, Switzerland
Keiichi Komoto, Mizuho Research and Technologies, Japan
Annick Anctil, Michigan State University, USA

Topic Organisers and Paper Review Experts

Topic 1: Silicon Materials and Cells

D. Bredemeier, ISFH Institut für Solarenergieforschung, Germany
S. Riepe, Fraunhofer-Institut für Solare Energiesysteme ISE, Germany
S. W. Glunz, Fraunhofer-Institut für Solare Energiesysteme ISE, Germany
G. Hahn, Universität Konstanz / Department of Photovoltaics, Germany
K. Ding, Forschungszentrum Jülich, Germany
D. Muñoz, CEA / INES, France
P. Roca i Cabarrocas, CNRS-LPICM LPICM Laboratoire de Physique des Interfaces et Couches Minces, France
A. W. Weeber, TNO Energy Transition, The Netherlands
J. Cárabe, CIEMAT, Spain
M. Topic, University of Ljubljana UL-FE, Slovenia
K. Bothe, ISFH Institut für Solarenergieforschung, Germany
F. Ferrazza, ENI S.p.A., Italy
S. Peters, Hanwha Q CELLS, Germany
Q. Wang, Zhejiang Jinko Solar Co., Ltd., China
M. P. Bellmann, STIFTELSEN SINTEF, Norway
M. Benmansour, CEA / INES, France
T. Bruton, Progress in Photovoltaics, United Kingdom
A. Ciesla, UNSW Australia, Australia
C. Hagendorf, Fraunhofer-Center für Silizium Photovoltaik CSP, Germany
K. Hesse, Wacker Chemie AG, Germany
X. Yu, Zhejiang University, China
R. Brendel, ISFH Institut für Solarenergieforschung, Germany
S. Dubois, CEA / INES, France
T. Dullweber, ISFH Institut für Solarenergieforschung, Germany
M. Ernst, ANU - The Australian National University, Australia
M. Hermle, Fraunhofer-Institut für Solare Energiesysteme ISE, Germany
R. G. Naber, Hanwha Q CELLS, The Netherlands
W. Nemeth, NREL - National Renewable Energy Laboratory, USA
R. Preu, Fraunhofer-Institut für Solare Energiesysteme ISE, Germany
B. Terheiden, Universität Konstanz / Department of Photovoltaics, Germany
R. Turan, METU - Middle East Technical University, Turkey
G. Yang, Delft University of Technology, The Netherlands
W. Dauksher, Arizona State University, USA

A. Descoeudres, CSEM - Centre Suisse d'Electronique et de Microtechnique SA, Switzerland
M. Despeisse, CSEM - Centre Suisse d'Electronique et de Microtechnique SA, Switzerland
W. Favre, CEA / INES, France
J. Horzel, Fraunhofer-Institut für Solare Energiesysteme ISE, Germany
Y. Ohshita, Toyota Technological Institute, Japan
J. Rentsch, Fraunhofer-Institut für Solare Energiesysteme ISE, Germany
J. Schmidt, ISFH Institut für Solarenergieforschung, Germany
J. Zhao, CSEM - Centre Suisse d'Electronique et de Microtechnique SA, Switzerland
P. Delli Veneri, ENEA - Italian National Agency for New Technologies, Energy and Sustainable Economy, Italy
S. Gall, Helmholtz-Zentrum Berlin für Materialien und Energie, Germany
T. Matsui, AIST - National Institute of Advanced Industrial Science and Technology, Japan
A. Meeder, Geo-En Energy Technologies, Germany
J. Meier, Meier Technologies, Switzerland
M. Vetter, IPHT Institut für Photonische Technologien Jena e.V., Germany
A. Augusto, Arizona State University, USA
N. Azkona, UPV/EHU - Instituto de Tecnologia Microelectronica (TiM), Spain
D. MacDonald, ANU - The Australian National University, Australia
A. Otaegi, UPV/EHU - Instituto de Tecnologia Microelectronica (TiM), Spain
M. C. Schubert, Fraunhofer-Institut für Solare Energiesysteme ISE, Germany
L. Serenelli, ENEA Casaccia Research Center (SP 064), Italy
K. Sporleder, Fraunhofer-Center für Silizium Photovoltaik CSP, Germany
A. Yamada, Tokyo Institute of Technology, Japan
D. L. Bätzner, Meyer Burger Research AG, Switzerland
A. Danel, CEA / INES, France
P. Fath, RCT-Solutions, Germany
A. Metz, , Germany
H. J. Nonnenmacher, Meyer Burger, Germany
P. J. Verlinden, AMROCK Pty Ltd., Australia
W. Wu, Tongwei Solar (Meishan), China
W. Zhang, Gonda Electronic Technology, China

Topic 2: Evolving and Emerging Technologies: Tandems; Thin Film absorbers; III-V; New Materials and Concepts; Advanced Modelling

J. C. Goldschmidt, Philipps-Universität Marburg, Germany
I. Gordon, imec vzw, Belgium
U. Bach, Monash University, Australia
W. Tress, Zurich University of Applied Sciences, Switzerland
S. Veenstra, TNO Energy Transition, The Netherlands
M. Edoff, Uppsala University, Sweden
C. Hages, University of Florida, USA
A. N. Tiwari, EMPA - Swiss Federal Laboratories For Materials Science and Technology, Switzerland
G. Flamand, imec vzw, Belgium
G. Siefer, Fraunhofer-Institut für Solare Energiesysteme ISE, Germany
A. Martí Vega, UPM - Universidad Politécnica de Madrid, Spain
J. J. Poortmans, imec vzw, Belgium
P. Pearce, UNSW Australia, Australia
I. Ramiro, Universidade Nova de Lisboa, Portugal
S. Albrecht, Helmholtz-Zentrum Berlin für Materialien und Energie G, Germany
G. Coletti, TNO Energy Transition, The Netherlands
F. Dimroth, Fraunhofer-Institut für Solare Energiesysteme ISE, Germany
U. W. Paetzold, KIT Karlsruher Institut für Technologie, Germany
A. Virtuani, EPFL-STI-IMT-PVLAB, Switzerland
A. Walter, CSEM - Centre Suisse d'Electronique et de Microtechnique SA, Switzerland
M. J. Brites, LNEG Laboratório Nacional de Energia e Geologia, Portugal
S. Cros, CEA / INES, France
C. Fell, CSIRO - The Commonwealth Scientific and Industrial Research Organisation, Australia
S. Huang, Macquarie University, Australia
N. Kyranaki, CEA / INES, France
C. F. Lin, National Taiwan University, Taiwan
B. Mihaylov, European Commission JRC, Italy
Y. Saygili, ODTÜ-GÜNAM , Turkey
N. Barreau, University of Nantes, France
V. Bermudez Benito, QEERI- Qatar Environment & Energy Research Institute, Qatar
T. Dalibor, Avancis, Germany
D. Dimova-Malinovska, Bulgarian Academy of Sciences, Bulgaria
S. C. Irvine, Swansea University, United Kingdom
S. Ishizuka, AIST - National Institute of Advanced Industrial Science and Technology, Japan
P. Kratzert, SOLARWATT, Germany
M. Lira-Cantú, CIN2, Spain
M. C. Lux-Steiner, Helmholtz-Zentrum Berlin für Materialien und Energie, Germany
M. Meuris, UHasselt - IMO IMOMEC, Belgium
S. Paetel, Zentrum für Sonnenenergie- und Wasserstoff-Forschung Baden-Württemberg (ZSW), Germany
R. Po, ENI S.p.A., Italy
A. Redinger, University of Luxembourg, Luxembourg
A. Romeo, University of Verona, Italy
V. Sittinger, Fraunhofer-Institut für Schicht- und Oberflächentechnik IST, Germany
M. Theelen, TNO/Solliance, The Netherlands
R. King, Arizona State University, USA
T. Minemoto, Ritsumeikan University, Japan
G. Timò, RSE S.p.A. (Ricerca sul Sistema Energetico), Italy
J. P. Connolly, GeePs-CentraleSupelec, France
J. Govaerts, imec vzw, Belgium
M. A. Hernandez-Fenollosa, UPV - Technical University of Valencia, Spain
K. Hummelen, University of Groningen, The Netherlands
J. P. Kleider, CNRS/GeePs, France
D. König, UNSW - Integrated Materials Design Centre (IMDC), Australia
I. Konovalov, University of Applied Sciences Jena, Germany
Y. Okada, University of Tokyo, Japan
F. Roca, ENEA - Italian National Agency for New Technologies, Energy and Sustainable Economy, Italy
M. Rusu, Helmholtz-Zentrum Berlin für Materialien und Energie, Germany
N. Ekins-Daukes, UNSW Australia, Australia
C. Lévy-Clément, CNRS - Institut de Chimie Matériaux Paris Est, France
B. Rech, Helmholtz-Zentrum Berlin für Materialien und Energie, Germany
R. I. Schropp, Utrecht University, The Netherlands
T. Tayagaki, AIST - National Institute of Advanced Industrial Science and Technology, Japan
W. Warta, Fraunhofer-Institut für Solare Energiesysteme ISE, Germany

Topic 3: Photovoltaic Modules and BoS Components
W. J. Gambogi, DuPont Company, USA
A. Lennon, UNSW Australia, Australia
T. Barnes, NREL, USA
R. Gottschalg, Fraunhofer-Center für Silizium Photovoltaik CSP, Germany
U. Jahn, VDE Renewables, Germany
G. Bardizza, European Commission JRC, Italy
W. Herrmann, TÜV Rheinland Energy, Germany
C. Buerhop-Lutz, Helmholtz-Institut Erlangen-Nürnberg für Erneuerbare Energien HI ERN, Germany
N. Pearsall, Northumbria University, United Kingdom
A. Faes, CSEM - Centre Suisse d'Electronique et de Microtechnique SA, Switzerland
M. Mittag, Fraunhofer-Institut für Solare Energiesysteme ISE, Germany
M. Á. Muñoz-García, Universidad Politécnica de Madrid. UPM-ETSIAAB., Spain
H. Nagel, Fraunhofer-Institut für Solare Energiesysteme ISE, Germany
I. Tsanakas, CEA / INES, France
C. Zhu, LONGi, China
G. Beaucarne, Dow Silicones, Belgium
N. Bosco, NREL - National Renewable Energy Laboratory, USA
C. Camus, LayTec AG, Germany
R. Einhaus, Apollon Solar, France
S. Oktik, Kadir Has University Central Campus, Turkey
G. Oreski, PCCL Polymer Competence Center Leoben, Austria
M. Pander, Fraunhofer-Center für Silizium Photovoltaik CSP, Germany
T. Sample, European Commission JRC, Italy
G. Friesen, SUPSI University of Applied Sciences and Arts of Southern Switzerland, Switzerland
Y. Hishikawa, AIST - National Institute of Advanced Industrial Science and Technology, Japan
J. Lopez-Garcia, QEERI, Qatar
C. Monokroussos, TÜV Rheinland (Shanghai) Co., Ltd., China
H. Müllejans, European Commission JRC, Italy
H. Nussbaumer, Zurich University of Applied Sciences, Switzerland
M. Paverttoni, SERIS, Singapore
J. Roy, IIT Indian Institute of Technology Kharagpur, India
R. Sinton, Sinton Instruments, USA
P. Bauwens, UGhent Ghent University, Belgium
F. Dross, STS, USA
G. Graditi, ENEA - Italian National Agency for New Technologies, Energy and Sustainable Economy, Italy
J. Moschner, KU Leuven Katholieke Universiteit Leuven, Belgium

Topic 4: PV Systems Engineering, Integrated/Applied PV
A. M. Gracia Amillo, Fraunhofer-Institut für Solare Energiesysteme ISE, Germany
W. M. van Sark, Utrecht University, The Netherlands
F. P. Baumgartner, Zurich University of Applied Sciences, Switzerland
C. Hansen, Sandia National Laboratories, USA
J. M. Almeida Serra, University of Lisbon - FCiencias.ID, Portugal
D. Moser, Eurac Research, Italy
G. Mütter, Alternative Energy Solutions, Austria
F. Frontini, SUPSI University of Applied Sciences and Arts of Southern Switzerland, Switzerland
A. Scognamiglio, ENEA - Italian National Agency for New Technologies, Energy and Sustainable Economy, Italy
T. Reindl, NUS National University of Singapore, Singapore
R. E. Valckenborg, TNO Energy Transition, The Netherlands
E. E. Voroshazi, CEA / INES, France
G. Adinolfi, ENEA - Italian National Agency for New Technologies, Energy and Sustainable Economy, Italy
U. Muntwyler, Dr. Schüpbach&Muntwyler, Switzerland
I. Antón Hernández, Universidad Politécnica de Madrid - IES, Spain
K. Araki, University of Miyazaki, Japan
R. Campesato, CESI S.p.A, Italy
S. Taylor, European Space Agency, The Netherlands
P. Blanc, MINES ParisTech, France
A. Louwen, Eurac Research, Italy
T. Oozeki, AIST - National Institute of Advanced Industrial Science and Technology, Japan
C. Protogeropoulos, EEPS, Greece
J. Remund, Meteotest, Switzerland
M. Sengupta, NREL - National Renewable Energy Laboratory, USA
C. Alonso-Tristán, UBU - University of Burgos - Escuela Politécnica Superior, Spain
M. Carbone, ENEL Green Power, Italy
P. Darez, 350 Renewables, Chile
K. Lappalainen, Tampere University, Finland
C. Nyman, Soleco Ltd, Finland
M. Grottke, WIP Renewable Energies, Germany
P. Hacke, NREL - National Renewable Energy Laboratory, USA
K. Kiefer, Fraunhofer-Institut für Solare Energiesysteme ISE, Germany
P. Lechner, Zentrum für Sonnenenergie- und Wasserstoff-Forschung Baden-Württemberg (ZSW), Germany
J. Lin, PV Guider, Taiwan
O. Mayer, Bayern Innovativ - Bayerische Gesellschaft für Innovation und Wissenstransfer, Germany
A. Migan-Dubois, GeePs, France
S. Rodrigues, Forschungszentrum Jülich, (HI-ERN) Helmholtz Institut Erlangen Nürnberg, Germany
J. S. Stein, Sandia National Laboratories, USA
D. Stellbogen, Zentrum für Sonnenenergie- und Wasserstoff-Forschung Baden-Württemberg (ZSW), Germany
P. Suratkar, PV Diagnostics, India
M. Theristis, Sandia National Laboratories, USA
P. Alonso, BWP Baumanagement, Germany
P. Bonomo, SUPSI University of Applied Sciences and Arts of Southern Switzerland, Switzerland
G. C. Eder, OFI, Österreichisches Forschungsinstitut für Chemie und Technik, Austria
I. Hagemann, Architekturbüro Hagemann, Germany
B. Kaempfen, Kämpfen Zinke + Partner, Switzerland
T. Reijenga, BEAR-iD, The Netherlands
E. Román Medina, Fundación Tecnalia Research & Innovation, Spain
K. Sinapis, Trespa International BV, The Netherlands
L. H. Slooff-Hoek, TNO Energy Transition, The Netherlands
L. Tautenhahn, Avancis, Germany
M. Dörenkämper, TNO Energy Transition, The Netherlands
M. Heinrich, Fraunhofer-Institut für Solare Energiesysteme ISE, Germany
R. Peibst, ISFH Institut für Solarenergieforschung, Germany
B. van Aken, TNO Energy Transition, The Netherlands
G. Barchi, Eurac Research, Italy
V. Efthymiou, University of Cyprus, Cyprus
K. Nishioka, University of Miyazaki, Japan
M. Victoria, Aarhus University, Denmark
R. J. Ellingson, Toledo University, USA
V. Khorenko, Azur Space Solar Power, Germany
M. Á. Vázquez Rodríguez, DHV Technology, Spain

Topic 5: Energy Transition – Integration, Storage, Sustainability, Policy, Economics, Energy Poverty, Society

K. P. Kairies, ACCURE Battery Intelligence, Germany
R. Pestana, R&D NESTER, Centro de Investigação em Energia REN - State Grid, Portugal
I. Weiss, WIP Renewable Energies, Germany
A. Anctil, Michigan State University, USA
K. Komoto, Mizuho Research & Technologies, Ltd., Japan
K. Wambach, Wambach-Consulting, Germany
M. Getsiou, European Commission DG RTD, Belgium
P. Malbranche, International Solar Aliance, India
S. Nowak, NET Nowak Energy & Technology, Switzerland
C. Breyer, Lappeenranta University of Technology, Finland
G. Masson, Becquerel Institute, Belgium
H. Ossenbrink, Band Gap, Germany
S. De Iuliis, ENEA Casaccia Research Center (SP 064), Italy
J. Vasi, IIT Bombay, India
N. della Valle, European Commission JRC, Italy
K. Yenneti, University of Wolverhampton, United Kingdom
P. J. Alet, CSEM - Centre Suisse d'Electronique et de Microtechnique SA, Switzerland
M. C. Brito, University of Lisbon - Faculty of Science, Portugal
R. Bründlinger, AIT Austrian Institute of Technology, Austria
F. Carigiet, Zurich University of Applied Sciences, Switzerland
J. S. da Costa Fernandes, Hochschule Offenburg, Germany
B. Gaiddon, HESPUL, France
J. M. Klein, CEA / INES, France
P. Lalanne, HydroClapeyron, France
A. C. Neves Foles, University of Évora, Portugal
E. Schüpbach, Bern University of Applied Sciences BFH, Switzerland
S. Arancón, WIP Renewable Energies, Germany
S. Capaccioli, ETA - Florence Renewable Energies, Italy
V. Fthenakis, Columbia University, USA
G. Heath, NREL - National Renewable Energy Laboratory, USA
W. Palitzsch, LuxChemtech, Germany
E. Parnham, DuPont, United Kingdom
G. G. Arrowsmith, EUREC, Belgium
D. Polverini, European Commission DG GROWTH, Belgium
N. Taylor, European Commission JRC, Italy
T. Haarberg, BNW-Energy AS, Germany
I. Kaizuka, RTS Corporation, Japan
E. Vartiainen, Fortum Growth Oy, Finland
R. Blanchard, Loughborough University, United Kingdom
J. McDonald, ITP Renewables, Australia
A. Nayfeh, Khalifa University, United Arab Emirates
C. Pedretti, ARE, Belgium
S. Caneva, WIP Renewable Energies, Germany
G. Ruggieri, Insubria University, Italy
S. Tay En Rong, National University of Singapore, Singapore
E. Warren, NREL - National Renewable Energy Laboratory, USA

Awards Coordinators

Student Awards Coordinator

A.H.M. Smets, Delft University of Technology, The Netherlands

Poster Awards Coordinator

J. Cárabe, CIEMAT, Madrid, Spain

SUBJECT INDEX

Silicon Materials and Cells

Sessions 1CP.2, 1AO.4, 1AO.5, 1AO.6, 1BO.1, 1BO.2, 1BO.3, 1BO4, 1BO.5, 1DO.11, 1DO.12, 1CV.2, 1CV.3, 1DV.4

Evolving and Emerging Technologies: Tandems; Thin Film absorbers; III-V; New Materials and Concepts; Advanced Modelling

Sessions 2CP.3, 2AO.1, 2AO.2, 2AO.3, 2BO.6, 2BO.7, 2BO.8, 2BO.9, 2BO.10, 2CO.1, 2CO.2, 2CO.3, 2DO.6, 2DO.7, 2DO.8, 2DO.9, 2DO.13, 2AV.1, 2AV.2, 2AV.3, 2BV.2

Photovoltaic Modules and BoS Components

Sessions 3AP.1, 3BO.11, 3BO.12, 3BO.14, 3BO.15, 3CO.4, 3CO.5, 3CO.6, 3DO.17, 3DO.18, 3DO.19, 3EO.1, 3BV.3, 3DV.1, 3DV.3

PV Systems Engineering, Integrated/Applied PV

Sessions 4CP.1, 4AO.7, 4AO.8, 4AO.9, 4BO.16, 4BO.17, 4CO.8, 4CO.9, 4DO.1, 4DO.2, 4DO.3, 4DO.4, 4DO.5, 4DO.16, 4EO.2, 4BV.1, 4BV.4, 4BV.5, 4CV.1

Energy Transition – Integration, Storage, Sustainability, Policy, Economics, Energy Poverty, Society

Sessions 5EP.1, 5CO.10, 5CO.11, 5DO.10, 5DO.14, 5DO.15, 5EO.3, 5DV.2

Topic Code

1 Silicon Materials and Cells
2 Evolving and Emerging Technologies: Tandems; Thin Film absorbers; III-V; New Materials and Concepts; Advanced Modelling
3 Photovoltaic Modules and BoS Components
4 PV Systems Engineering, Integrated/Applied PV
5 Energy Transition – Integration, Storage, Sustainability, Policy, Economics, Energy Poverty, Society

Session Type

P = Plenary Session
O = Oral Session
V = Visual Session

Day Codes

A = Monday, 26 September 2022
B = Tuesday, 27 September 2022
C = Wednesday, 28 September 2022
D = Thursday, 29 September 2022
E = Friday, 30 September 2022

e.g. 1AO.1 ⇒ 1= Silicon Materials and Cells, A=Monday, O=Oral session, 1=Session 1

TABLE OF CONTENTS OF WCPEC-8 PROCEEDINGS PAPERS

Oral PRESENTATIONS 3BO.11 PV Characterization and Imaging Techniques

Oral PRESENTATIONS 3BO.12 Outdoor Performances

Oral PRESENTATIONS 3BO.14 Modelling the Performance

Visual PRESENTATIONS 4BV.4 Solar Resource and Forecasting

FOREWORD

The World Conference on Photovoltaic Energy Conversion (WCPEC) is the world's biggest platform for photovoltaic (PV) research and development, offering a unique possibility of international exchange by combining the three largest and most prominent international PV conferences: the European Photovoltaic Solar Energy Conference (EU PVSEC), the Photovoltaic Specialists Conference (IEEE PVSC) and the International PV Science and Engineering Conference (PVSEC). After being in place for more than a quarter of a century, the WCPEC, being held every four years, is firmly established among the global PV community as the event for the greatest minds and experts in various PV sectors to meet and discuss, to exchange and network.

The WCPEC-8 (combining the 39th EU PVSEC, the 49th IEEE PVSC and the 32nd PVSEC) took place from 26 – 30 September 2022 in Milan, Italy, was great success and highlighted the reality that a global and full-scale PV deployment needs collaboration. In the year of the achievement of the first Terawatt of installed PV capacity, the conference concluded with the vision that with its exponential growth, the multiTW scale PV deployment will be a reality in few years and that despite the many challenges that it presents (the PV research is more than ready to solve them), there are massive opportunities for the PV sector. Research communities around the world and their innovations across the sector, together with the fast deployment and numerous application options are making PV a key pillar of energy production. This year, more than 1500 participants from 60 countries attended this highly inspiring platform delivering over 1100 presentations. It provided a vital forum for information exchange in the field of PV solar energy research, demonstration and applications. In the exhibition 74 companies from all parts of the world welcomed visitors and presented their products and services.

The WCPEC-8 Proceedings give a comprehensive overview of the photovoltaic solar sector, its current status and future prospects in science, research, innovation, development and deployment on almost 3,700 pages. Selection for inclusion in the conference was made by paper review experts and topic organisers (see the listing on pages III-VII), to whom we express our sincere gratitude for their comprehensive review work and overall contribution to the success of the conference. In addition to the 354 submitted papers, the proceedings include 90 presentations (slides) shown during the plenary and oral presentations as well as 140 poster files of the visual presentations. In total this amounts to 584 publications.

The WCPEC-8 maintains its commitment to providing quick and open access to high quality scientific results. They constitute a powerful tool for targeted and quick information search and retrieval, enabling you to search by keyword, paper title and authors. We are sure that these features will help to simplify the use and exploit the full potential of this extensive source of information.

The Conference Proceedings are published as downloadable DVD file and are also fully accessible online. A DOI code (Digital Object Identifier) has been assigned to each paper. This ensures unequivocal and permanent identification and full citability. The WCPEC-8 papers can be viewed and downloaded in a full free open access from the Proceedings website www.eupvsec-proceedings.com.

We are confident that these Proceedings will play an important role in providing a comprehensive overview of the current actors and activities in the global PV sector and that they will disseminate information on the state-of-the-art of technologies and applications. This can generate further research, add momentum to innovation and promote interest in PV worldwide.

We would like to cordially thank all authors and participants of the WCPEC-8 for their contributions and look forward to welcoming you in Lisbon, Portugal from 18 – 22 September 2023 at the EU PVSEC 2023, the 40th European Photovoltaic Solar Energy Conference and Exhibition

The Editors

LIGHTWEIGHT PV MODULES COMPATIBLE WITH ONE STEP LAMINATION PROCESS

Alicia Buceta[a], Miguel Aguirre[a], Ana B. Cueli[a], Mikel Ezquer[a], Elena Llarena[b], Silvia Cal[b], Ana Linares[b], Rocío Castelo[b] and Jaione Bengoechea[a]
[a] National Renewable Energy Center (CENER)
Avenida Ciudad de la Innovación, 7, 31621 Sarriguren (Navarra) SPAIN
Tel.: +34 948 25 28 00 Fax: +34 948 27 07 74
[b] Instituto Tecnológico y de Energías Renovables, S. A. (ITER)
Pol. Industrial de Granadilla, s/n, E 38600 Granadilla de Abona (Canarias), SPAIN
Tel.: +34 922 74 77 00 Fax: +34 922 74 77 01

ABSTRACT: Lab scale PV module designs, aiming to obtain **lightweight solutions**, produced in a conventional lamination process and **capable of withstanding the impact of hail** were studied. To reach this goal, a **polypropylene** (PP) **HoneyComb** (HC) with non-woven polyester skin as **backsheet** with **ethylene tetrafluoroethylene** (ETFE) **or a thin glass** (1 mm thick) **as frontsheet** were used. For this study two different samples of polypropylene HC were tested. The first challenge was adapting the lamination recipe for the HC, since the conventional one caused crushing of the material, especially at the edges. The solution, in order to be able to use one step lamination process, was to use an auxiliary frame to protect the HC during the lamination. Depending on the type of honeycomb the reduction of the pressure was also needed. The samples with ETFE as frontsheet did not resist the hail impact without breaking the solar cell (SC), even if the HC was reinforced. On the other hand, using 1 mm thick glass as frontsheet proved to meet the pass criteria. The preliminary results of temperature and humidity resistance showed that there is some humidity permeability that disappeared after a few days out of the climatic chamber. Even though, the performance was not jeopardized by it.
Keywords: c-Si, lightweight module, honeycomb, lamination

1 INTRODUCTION

Given the global energy situation the need of new solutions that ease the use of renewable energies is of utmost importance. Some approach for the solution include the use of BAPV (Building Applied Photovoltaics), BIPV (Building Integrated Photovoltaics), VIPV (Vehicle Integrated Photovoltaic) and IIPV (Infrastructure-Integrated Photovoltaic) were photovoltaic applications have great potential.[1] For example, the rooftop solar Photovoltaic (PV) installation market is expected to increase in the coming years.[2, 3] Although, one of the bottlenecks for this application can be the modules weight, especially considering light roofs or historical buildings with weight restrictions.[4] Therefore, finding a lightweight solution would increase crystalline silicone photovoltaic module versatility and market applications.[5]

Conventional mono-facial PV module layers are Tedlar®/EVA/SC/EVA/3.2 mm Glass and typically weigh around 11 kg/m^2, where the glass represents 60-75% of the weight.[6, 7] It is considered a c-Si lightweight solution when the module weight is below 6 kg/m^2,[8] and the tendency is to use thinner glasses,[9] or replacing the glass by a lighter material in order to minimize the module weight. One alternative is using polymers like for example ETFE (ethylene tetrafluoroethylene),[10] which can be manufactured as a very thin and light sheet. The limitations will be then the mechanical properties, hence the need of a strong backsheet that can withstand mechanical stress and prevent cells from breaking. Honeycomb (HC) structures stand out for their lightness and their great energy absorption properties,[11] for that reason selecting a HC material as backsheet when using a soft frontsheet could be the key to obtain a sturdy module.[12–14]

The processes and equipment used in the present study for the fabrication of different configurations are the ones currently used in the industry to manufacture conventional c-Si PV modules. Therefore, these new developments can be easily adopted by producers in order to achieve high degree of automation.

2 METHODOLOGY

The selection of the materials was based on previous results obtained in AiSoVol2 project in which this study was developed.[15]

The mechanical and climatic tests performed on the lab scale modules presented in this paper are based on IEC PV module qualification standards.[16, 17]

2.1 Materials

For this study two polypropylene HCs with non-woven polyester skin capable of withstanding punctually temperatures of 140 °C were tested as backsheets. They are opaque white materials, rigid and very light. One was 10 mm thick (HC10) with 8 mm hexagonal cell size and the other sample was 14 mm thick (HC14) with 5 mm hexagonal cell size.

As frontsheet an ETFE matte sheet 101.6 µm thick or a thin glass 1 mm thick were used. The ETFE resists up to 232 °C and has high light transmission (95%) while the 1 mm thick glass used has around 90% transmitance.

An ultra-fast cure EVA (Ethylene-vinyl acetate) was chosen as the encapsulant.

All the tested materials are commercially available.

2.2 Lamination

The laminator used is a PEnergy Lab-Laminator L036LAB, which is an automated module laminator for samples 60 x 60 cm and 3 cm thickness maximum. It has a membrane on the upper chamber and the hot plate on the lower chamber.

2.3 Hail impact test

The hail impact resistance test was performed with an ice ball launcher LBH-25 from ARIES. The experiment was prepared following the IEC 61215-2 standard,[17] therefore 25 mm diameter ice balls were launched at a velocity of 23 m/s ± 2% (MQT 17).

2.4 Climatic chamber

Their resistance to temperature and humidity was tested in a C-70/200 climate chamber from CTS following the IEC 61215-2 standard. The humidity-freeze test (MQT 12) was performed.

3 RESULTS AND DISCUSSION

The objective was to obtain a lightweight solution compatible with current industrial processes. Therefore, all the considered sandwich configurations were chosen for being compatible with lamination. Initially a visual inspection was conducted to detect any major defect such as bent external surfaces or inhomogeneous aspect.[16] Once the samples fulfilled the visual requirements their hail, temperature and humidity resistance were tested.

3.1 Lamination

Once the materials were selected their compatibility with the lamination process was tested. The most challenging one was the HC as the first trials ended up smashing down the samples, especially on the edges (Fig. 1). Different recipes were tried but the results were similar, the edges were the weak point of the HC. Therefore, the need to reinforce the edge led to the use of an auxiliary frame during the lamination process.

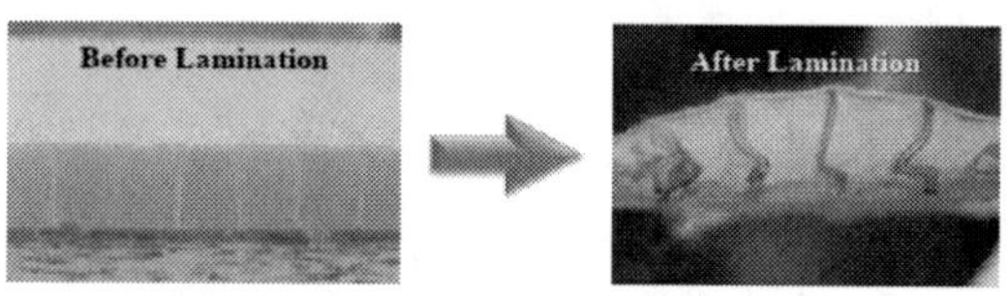

Figure 1: HC10 before (left) and after lamination (right). In the photo to the right, the sample was cut in half after lamination to better show the deformation.

A modular frame was designed so it could be adapted to the sample size, using hard cardboard wrapped in a polymer sheet that could be easily released from the encapsulant once laminated (Fig. 2). Then compatibility between the HC and the EVA was confirmed using a modified recipe to avoid bubble formation. Also, to ensure a proper result, the height of the frame should be as similar as possible to the height of the sandwich that is going to be laminated.

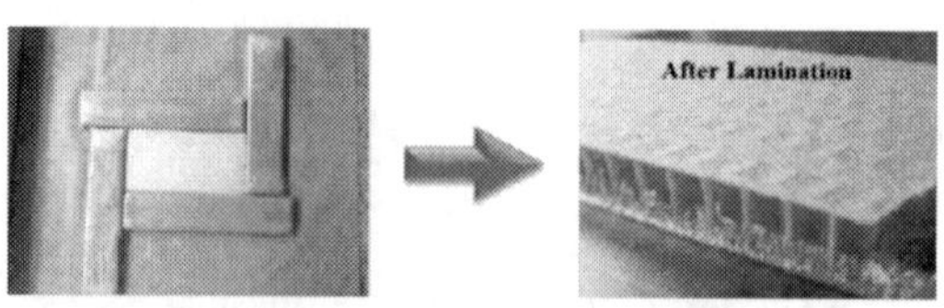

Figure 2: HC10 after a lamination process using an auxiliary frame.

The scalability of the process was tested at lab scale, making samples from one solar cell (23 x 18 cm sample) up to 4 solar cells (40 x 37 cm samples). As it can be observed in Fig. 3, the 4-cell sample laminated with HC14 starts to show some curvature not observed in smaller samples. The non-planarity of the 4-cell samples is quantified measuring the curvature maximum height respect to flat condition; if HC14 is used the sample shows 2 to 3 mm gap while with HC10 it shows 3 to 4 mm gap. Therefore, it will be expected when laminating larger modules to observe higher curvatures.

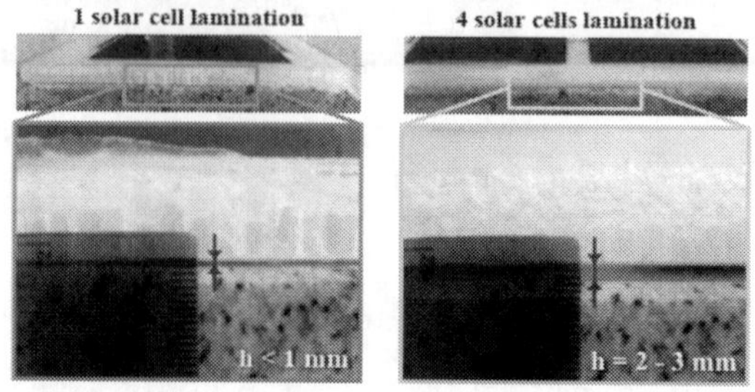

Figure 3: HC14 curvature when larger samples are laminated.

Independently of the frontsheet, the lamination temperature was fixed at 140 °C and the auxiliary frame was always used. However, HC10 exhibited some deformation at its centre when laminating larger pieces. This was probably due to the bigger hexagonal cell size compared to HC14. Consequently, the maximum pressure during the process was lowered from 1000 to 400 mbar.

In Fig. 4 it is possible to observe the consequences of not using an auxiliary frame large enough to protect the whole module perimeter. In this case, the corners of the HC14 crushed during the process.

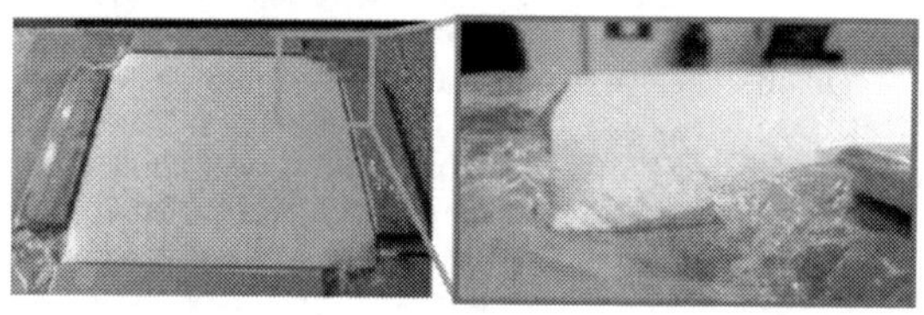

Figure 4: HC14 crushed edges due to poor coverage of the edges with the auxiliary frame.

3.2 Hail Impact Test

All the samples were first tested mounted onto a methacrylate plate. An electroluminescence camera was used to record the ice ball impact in order to detect any potential crack on the cells not visible to the naked eye. In case no cracks were observed, the hail test was conducted again but without any mounting support, just free-standing.

To determine the viability of the modules, the one solar cell samples were tested before upscaling. The first tests were with ETFE as frontsheet. The sandwich that was initially considered was HC/EVA/SC/EVA/ETFE; nevertheless, the solar cell broke after the first hail impact. The next approach was to reinforce the structure adding more EVA at the front of the cell to mitigate the hail impact but no improvement was observed. The HC was then reinforced with a fiberglass based material, but the solar cell did not withstand the impact either (Fig. 5).

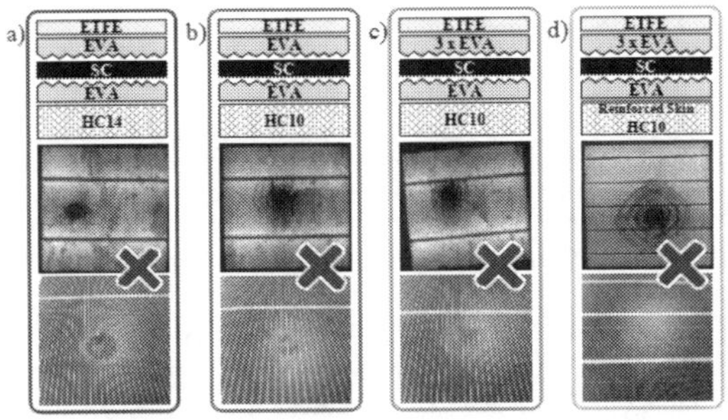

Figure 5: Electroluminescence and surface aspect after hail impact. In all four cases a), b), c) and d) the samples were supported on the methacrylate plate and the SC broke with the first impact.

Hence, further reinforcement is required for samples using honeycomb as backsheet and ETFE as frontsheet to withstand the hail impact without breaking the solar cell. The other candidate material proposed as frontsheet to obtain a lightweight solution is 1 mm thick glass. With this configuration HC/EVA/SC/EVA/1 mm Glass the samples did pass the hail impact test without any damage on the cell when the methacrylate was used and a very small crack appeared when the sample was tested without the support. Therefore, these samples were scaled up to a 4-cell module and tested again for hail impact. To prevent small cracks when tested without the support, another EVA layer was added on the front side of the module: HC/EVA/SC/2xEVA/1 mm Glass. In this case, the modules received 3 hail impacts following a diagonal line, with and without support (6 impacts in total) and no damage was observed in any of the cells (Fig. 6).

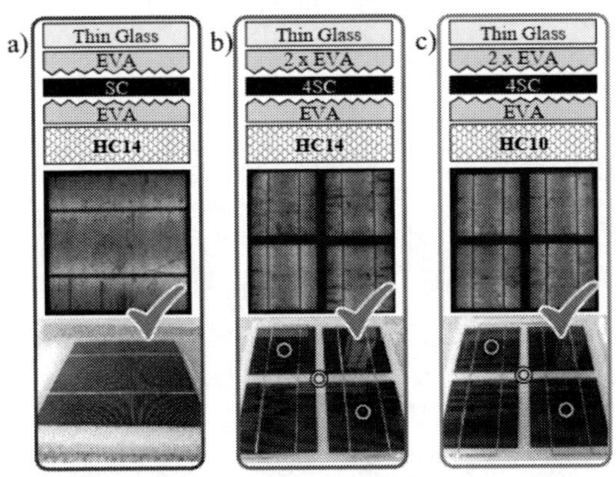

Figure 6 Electroluminescence and surface aspect after hail impact. a) Results after two hail impacts. The SC showed a minor crack when tested without the methacrylate support. In b) and c) No signs of damage were observed after 6 hail impacts: 3 using the methacrylate support and a second round of 3 without it. The yellow rings represent the impact zone.

As a result, it was confirmed the possibility to laminate lightweight modules using HC10 or HC14 that can withstand the hail impact test without any crack on the solar cells.

3.3 Temperature and humidity resistance

The lightweight material configuration was subjected to the humidity-freeze test to determine the ability of the samples to withstand the effects of high temperature and humidity followed by sub-zero temperatures. The samples were tested during 10 cycles of 24 hours each. Each cycle consists of periods of 20 hours at 85 °C and 85% RH followed by 4 hours of descending temperature down to -40 °C. A few days after the test, with the modules outside the climatic chamber, humidity was observed on the tabs (Fig. 7). However, the humidity disappeared over time and the solar cell performance was not affected.

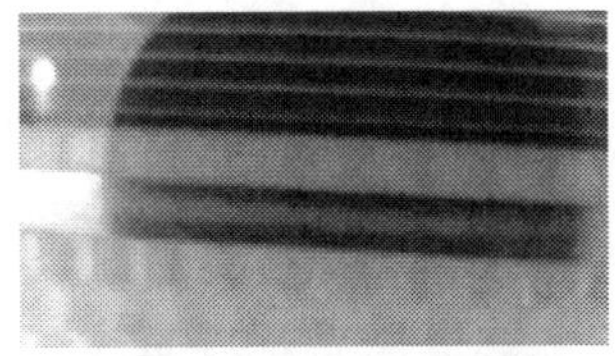

Figure 7: Humidity traces on the tabbing after a few days out of the climatic chamber.

A polymer layer was added to the structure (HC/EVA/Polymer/EVA/SC/EVA/1 mm Glass) as a humidity barrier but no significant improvement was observed. Therefore, further studies on the humidity barrier are still needed.

3.4 Comparison with conventional module

In order to determine the improvements made on the modules, both the weight and the theoretical price are compared with a conventional module configuration. As it is observed in Table I, the proposed solutions are around 40% lighter than the conventional configuration. Nonetheless, the price is over 12 €/m^2 more expensive than the conventional PV module.

Table I: Comparison of weight and theoretical price.

Sample Configuration	Superficial Density (kg/m^2)	*Price (€/m^2)
HC14/EVA/SC/2xEVA/1mm Glass	**6.09**	127.01
HC10/EVA/SC/2xEVA/1mm Glass	**5.63**	124.70
Tedlar®/EVA/SC/EVA/3.2mm Glass (Comercial Conf.)	10.09	**112.43**

*Calculated with the materials bought at lab scale. The SC is not included in the final price.

4 CONCLUSIONS

A frameless self-supporting lightweight solution that withstands the hail impact test was found: HC/EVA/SC/2xEVA/1 mm Glass.

HC10 or HC14 can be used indistinctively, keeping in mind the limitation of HC10 with the pressure during the lamination process. Also, an auxiliary frame was needed during the lamination to avoid deformations on the HC.

It was possible to scale the sample up to 4-cell mini module (40 x 37 cm), although, the larger the module, the higher appreciated curvature.

The price of the lightweight solution would be slightly higher than conventional modules but the weight was reduced by around 40% which makes it suitable for low load-bearing applications.

As future research more resistant HCs materials to temperature and humidity will be tested and scalability to larger devices will be studied.

5 ACKNOWLEDGEMENTS

These results are part of project AiSoVol2, extension of AiSoVol (photovoltaic generation solution as an alternative construction material). It is a project funded by the Ministry of Economy and Competitiveness, within the State Program for R&D&I oriented to meet societal challenge 3: Secure, Clean and Efficient Energy Challenge. The project, which began on 1st June 2020 and will end on 31st December 2023, is composed by the partnership of the Technological Institute of Renewable Energies, SA (ITER) as coordinator and the National Renewable Energy Centre, CENER Foundation.

Special thanks to the entire technical team, and especially to Ibai Abentin, for their invaluable help in carrying out the tests.

6 REFERENCES

[1] D. Singh, R. Chaudhary, A. Karthick, Environmental Science and Pollution Research, Vol. 28 (2021). https://doi.org/10.1007/s11356-021-15349-5.

[2] T. Huld, K. Bodis, I. Pinedo-Pascua, E. Dunlop, N. Taylor, A. Jäger-Waldau, european energy innovation, (2018), 12–15.

[3] S. Philipps, W. Warmuth, *Photovoltaics Report (Updated 24 February 2022)*; Freiburg, (2022).

[4] T. E. Kuhn, C. Erban, M. Heinrich, J. Eisenlohr, F. Ensslen, D. H. Neuhaus, Energy Build., 231 (2021), 110381. https://doi.org/10.1016/j.enbuild.2020.110381.

[5] C. Schmid, Lightweight module durability: Working to address a market barrier https://www.pv-tech.org/lightweight_module_durability_working_to_address_a_market_barrier/ (accessed Aug 22, 2022).

[6] How Much Do Solar Panels Weigh? Actual Facts, kg/sq.m., PSF, Watt/kg https://solarpoweredblog.com/solar-panel-weight/ (accessed Aug 22, 2022).

[7] M. Lunardi, J. Alvarez-Gaitan, J. Bilbao, R. Corkish, A Review of Recycling Processes for Photovoltaic Modules. In *Solar Panels and Photovoltaic Materials*; Alvarez-Gaitan, J. P., Ed.; IntechOpen: Rijeka, 2018. https://doi.org/10.5772/intechopen.74390.

[8] T. Kajisa, H. Miyauchi, K. Mizuhara, K. Hayashi, T. Tokimitsu, M. Inoue, K. Hara, A. Masuda, Jpn. J. Appl. Phys., 53 (2014).

[9] ITPRV. *International Technology Roadmap for Photovoltaic*; Vol. 13th Editi. (2021).

[10] A. Menéndez, A. Martínez, A. Santos, B. Ruiz, K. Moritz, I. Klein, J. Díaz, A. R. Lagunas, T. Sauermann, D. Gómez, Energy Build., 161 (2018), 10–21. https://doi.org/10.1016/j.enbuild.2017.12.023.

[11] Z. Wang, Compos. Part B Eng., 166 (2019) 731–741. https://doi.org/10.1016/j.compositesb.2019.02.011.

[12] A. C. Martins, V. Chapuis, A. Virtuani, C. Ballif, IEEE J. Photovoltaics, 9, (2019) 245–251. https://doi.org/10.1109/JPHOTOV.2018.2876934.

[13] A. C. Martins, V. Chapuis, A. Virtuani, L.-E. Perret-Aebi, C. Ballif, Proceedings of the 33rd European Photovoltaic Solar Energy Conference and Exhibition; (2017) 2604–2608.

[14] M. Heinrich, A. J. Beinert, P. Romer, L. C. Rendler, F. Basler, S. Wendlandt, D. H. Neuhaus, 38th European Photovoltaic Solar Energy Conference and Exhibition; (2021) 761–764. https://doi.org/10.4229/EUPVSEC20212021-4AV.1.26.

[15] CENER. AiSoVol2 project https://www.cener.com/en/areas/photovoltaic-solar-energy-department/outstanding-projects/aisovol-2-adaptive-photovoltaic-generation-solution-for-use-in-building-and-distributed-generation/. (accessed Aug 22, 2022).

[16] IEC. *IEC 61215-1: Terrestrial Photovoltaic (PV) Modules – Design Qualification and Type Approval – Part 1: Test Requirements*, 2nd ed.; International Electrotechnical Commission: Geneva, (2021).

[17] IEC. *IEC 61215-2 Terrestrial Photovoltaic (PV) Modules – Design Qualification and Type Approval – Part 2: Test Procedures*, 2nd ed.; International Electrotechnical Commission: Geneva, (2021).

This is a short version of the paper submitted to the journal Solar RRL.

It has been selected by the Committee of the WCPEC-8 for submission to Solar RRL.

PERFORMANCE ANALYSIS OF POWER OPTIMIZERS BY INDOOR LAB TESTING AND SHADING SIMULATION

Cyril Allenspach*, Fabian Carigiet, Arturo Bänziger, Andrin Schneider, Franz Baumgartner

ZHAW, Zurich University of Applied Sciences, School of Engineering, IEFE
www.zhaw.ch/~bauf, Technikumstr. 9, CH-8401 Winterthur, Switzerland
*phone: +41 58 934 45 76; email: alls@zhaw.ch

ABSTRACT: Photovoltaic systems, which are partially shaded, operate with non-uniform conditions within their module array. Power losses are the consequence in conventional systems as they cannot track each photovoltaic modules in their highest individual maximum power point. In contrast, module-level power electronics can potentially improve the performance of shaded systems, while the increase compared to standard string inverter systems is site-specific. To investigate this, power optimizer and string inverter efficiency measurements were conducted in the ZHAW indoor laboratory. The results were integrated in simulations, with which a module-level power electronics system with power optimizers at every module, and a standard string inverter system was analysed. In detail, the P370 power optimizer and 3500H inverter was used for the power optimizer system and for the standard string inverter setup, the string inverter SUN2000- 3.68KTL-L1, was chosen. The results of the annual simulations with partial shading by a chimney showed a gain of the MLPE system between -0.2% and 0.8% (13-modules), depending on the position of the shading object. Furthermore, the shading adaption efficiency is described, which is a method of quantifying the annual performance for shading situations with weightings to few indoor measured power electronic component efficiencies.
Keywords: Module-Level Power Electronics, MLPE, Power Optimizer, Efficiency, Performance, Testing, Shading

1 INTRODUCTION

The rise of module-level power electronics (MLPE) installations over the past decade has led to the fact that their manufacturers now dominate the residential photovoltaic (PV) market in the USA. But it is not a local phenomenon as they now hold significant market shares in many countries around the world [1]. In detail, nearly an amount of three digit in the millions of MLPE components are now in operation worldwide, whereas their sales have not shown signs of saturation [2]. One of the unique abilities of MLPE is the operation of a PV module in the individual absolute maximum power point (MPP), unrelated on the setup of the respective PV string and especially so, in different shading conditions of each module [3]. Accordingly, it was one of the key selling points and assisted the wide-spread growth of the technology. However, there has not been shown clear evidence in literature, which implicates that an annual performance increase of such PV rooftop systems equipped with MLPE in the double-digit range relative to standard string inverter (SINV) system is feasible. In the few surveys that have been published, nearly all describe the topic of outdoor testing of PV installations [4, 5, 6, 7]. Meanwhile, the performance comparisons of these technologies with outdoor testing show impediments such as their reproducibility and measurement precision, especially when the yield differences of the two compared systems are of the magnitude of a low one-digit number and thus, close to the measurement uncertainties. Further difficulties arise by the need to have an equal diffuse light condition on both installations, MLPE and SINV. And finally, an exact measurement of the related nominal power of all the involved modules is needed, which is a challenging and elaborate task.

The Institute of Energy Systems and Fluid-Engineering (IEFE) at the Zurich University of Applied Sciences (ZHAW),[8] has presented a new method of annual performance assessment of partially shaded PV systems by performing indoor component measurements and optical electrical shading simulation in 2021 [9]. Thus, indoor performance tests on a SolarEdge P370, P405 and Huawei SUN2000-450W-P2 power optimizer under shaded and in unshaded conditions were carried out. The performance tests with the P370 power optimizer revealed that a three-phase power optimizer system had 1.5 % lower annual yield than a string inverter system in unshaded conditions, and 0.9% less energy when partially shaded by a chimney (indoor measurement uncertainty ±0.3 %, k=1) [10, 11]. Similar findings are provided by Bründlinger, stating that the losses of unshaded systems with MLPE are typically 2-3 % higher compared to the conventional string inverter topologies [12]. These findings contrast heavily with manufacturers' datasheet statements claiming that these products ensure greater energy yields with "up to 25 % more energy", or "increase in energy production of more than 36 %" [13, 14]. A further conclusion of the ZHAW IEFE study, was that a quantifiable benefit of a MLPE system must be determined on a case-by-case basis [15]. Although the results strongly depend on the individual efficiency of the power electronic components used, the general rule applies, that in scenarios with extreme shading, the MLPE system provides an advantage, whereas for light shading, lower yields than SINV systems are expected [10].

Furthermore, the case-by-case distinction notably increases the complexity of PV system planning for four reasons. Firstly, PV simulation tools for system planners, which analyse partial shading on the cell level of each PV module with appropriate loss models for MLPEs do not exist yet [9]. Secondly, shading objects would need to be accurately measured in the field and remodelled in 3D, which would increase the planning time. Third, the comparison of MLPE components is challenging, due to the absence of standards for reporting weighted average conversion efficiencies [4]. Power optimizer producers report either peak or weighted conversion efficiencies, but not details on the performance test conditions [12, 16]. And finally, the total costs of MLPE systems are generally higher than string inverter solutions, due to the higher number of components and the additional time required for the installation [17, 18]. Accordingly, it is difficult for planners to calculate when (or if) a potentially higher energy yield justifies higher system cost.

This is a short version of the paper submitted to the journal Solar RRL.

It has been selected by the Committee of the WCPEC-8 for submission to Solar RRL.

2 INDOOR LABORATORY TESTING

In the IEFE laboratory at the ZHAW, a measurement and testing setup for MLPE was developed, which can be operated as a state-of-the-art residential photovoltaic plant consisting of 10 PV-Modules (shown in Figure 1).

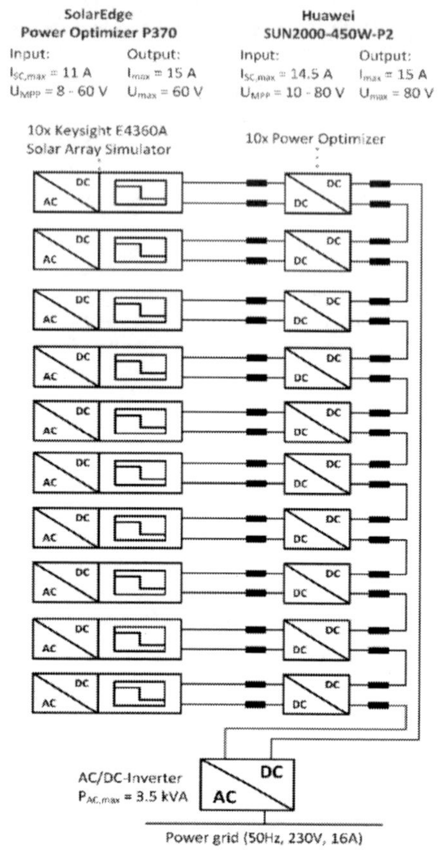

Figure 1: Electrical setup of the analysed MLPE system with 10 power optimizers.

The measurement system can mimic unshaded and shaded PV modules with the use of 10 solar array simulators (SAS) by providing different current-voltage (IV) curve settings, according to the irradiance and temperature conditions on the PV module level. Every workstation has several types of MLPE by different manufacturers installed, which are powered by a Keysight E4360A SAS and measured with a Power Analyzer PPA1530 at the electrical input and output terminals [19, 20]. During testing, the desired operating points of the PV modules are entered in the SAS devices, which calculate respective IV-curves for the given inputs to each MLPE. Accordingly, the SAS generate DC-currents and voltages for the device-under-test based on I-V curves of PV-modules. Further descriptions of the measurement setup are given in "Performance analysis of shaded PV module power electronic systems" and "Performance of New Photovoltaic System Designs" [9, 21].

The losses in the power electronic converters occur in the individual components such as ohmic losses in the inductors and capacitors, ohmic and switching losses in the semiconductor switches and the socket losses due to the power supplies of all active components, e.g., drives, control and communication circuits. Equation (1), (2), (3) and (4) describe the power loss of the buck-boost type power optimizer, which is used in the simulation as underlying model.

$$D_{cycle} = \frac{I_{IN}}{I_{OUT}} \tag{1}$$

$$P_{loss,buck} = a + b \cdot U_{IN} \cdot I_{IN} \cdot D_{cycle} + c \cdot I_{OUT}^2 + d \cdot U_{IN}^2 \tag{2}$$

$$P_{loss,boost} = a + b \cdot U_{IN} \cdot I_{OUT} \cdot |1 - D_{cycle}| + c \cdot I_{IN}^2 + d \cdot U_{IN}^2 \tag{3}$$

$$P_{loss,passthrough} = a + c \cdot I_{IN}^2 \tag{4}$$

With the use of high-precision power analyzers, the DC/DC-efficiency of the MLPE are determined with an accuracy in the range of 0.2 - 0.5 % (k=1) in the ZHAW IEFE indoor laboratory. Accordingly, the whole operating range of buck-boost converter-type Power Optimizers were measured. In Figure 2, the DC/DC-efficiencies of the type P370 for current ratios: $k_I = I_{OUT} = I_{IN}$ ranging from 0.6 to 1.65, and input power: P_{IN} ranging from 0 W to 360 W, are visualized. In detail, the resulting efficiencies $\eta_{DC/DC}$ are between 95 %, at low input power $P_{IN} \approx 18$ W, and 99.5 %, at $P_{IN} = 98$ W and $k_I = 1$. However, distinct differences in efficiency can be identified between, the region where $k_I = 1$, and all other operating points. In detail, in the narrow range around $k_I = 1$ (see red dots in Figure 2), the efficiencies are more than 0.8 % higher than everywhere else, due to the reason that the converter is not operating either in buck or boost mode, in the very few operation points. Therefore, the MOSFETs are inactive, and no switching losses occur. In other words, the input is directly connected to the output and only ohmic losses of the lines, inductor and the MOSFET $R_{DS(on)}$ will occur.

The efficiency mapping in Figure 2 shows that values above 98 % only occur in a limited field of the operating range, while the values mentioned by the manufacturers are either, the maximum efficiency with 99.5 % or the weighted efficiency of 98.9 % [16]. Furthermore, the latter mentioned value could not be reached in the buck- or boost-mode, for which the component was designed.

The measurement results of the 450W-P2 buck-type power optimizer, illustrated in Figure 3 show a similar picture at $k_I = 1$ compared to the P370 results in Figure 2, especially with efficiency values below 97 % at about one tenth of full load. Finally, it is important to realize that for determining the total MLPE system performance, the DC/AC efficiencies of the inverter would be needed to be multiplied with the values presented in this section, which would result in lower overall efficiencies.

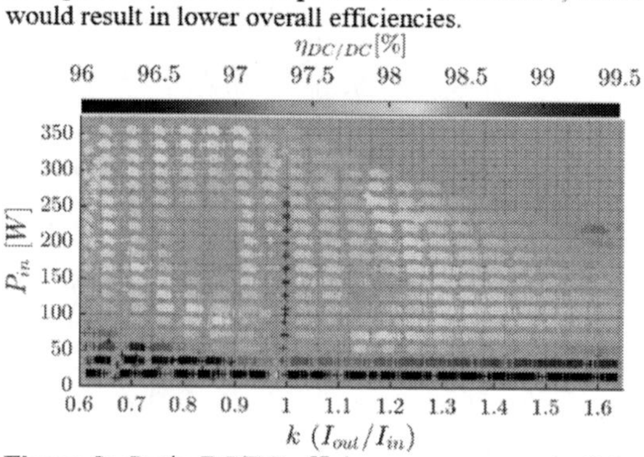

Figure 2: Static DC/DC-efficiency measurement of the buck/boost-type power optimizer P370 at static input voltage U_{IN} = 32 V relative to input/output current ratio.

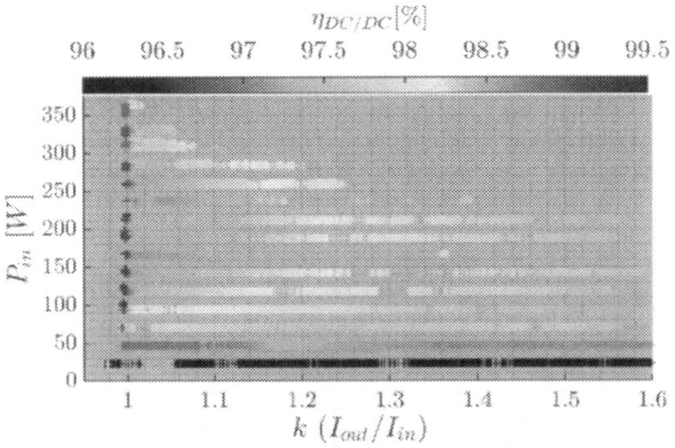

Figure 3: Static DC/DC-efficiency measurement of the buck-type power optimizer of model 450W-P2 at static input voltage $U_{IN} = 32$ V relative to the input/output current ratio.

3 SENSITIVITY ANALYSIS

A high-resolution shading analysis tool for the PV system simulations has been developed at ZHAW IEFE. In detail, the simulated PV system, including shading objects are modelled in 3-dimensional space, based on existing systems such as is shown in Figure 4, and the shading situations are calculated for each position of the sun during the year with any temporal resolution [22]. Global horizontal irradiance data provided by MeteoSwiss, the Swiss meteorological service, are used and transposed to the PV plane for each time stamp using the transposition model by Ineichen and Perez [23, 24]. The shadings on the PV module plane are calculated in such a way that there was no rasterization of the module plane, resulting in the highest possible resolution. Furthermore, based on the shading situation at each time step, the reduction of direct irradiance is determined for each solar cell in each PV module. Finally, the reduced direct irradiance together with the diffuse light, the ambient temperature, the electrical configuration of the PV modules, with 60 standard 6 inch cells and 3 bypass diodes, in the string are used to calculate the IV curves.

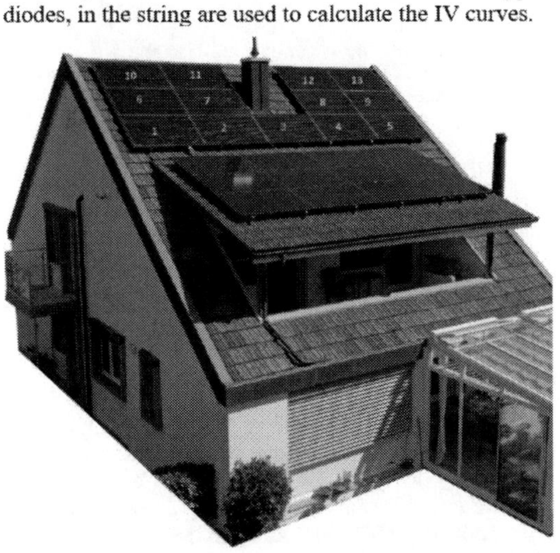

Figure 4: Single-family home with residential PV system and shading objects of type chimney and ventilation pipe. The 13kWp PV plant was implemented by PV installer in Switzerland. Reproduced with permission [22]. Copyright 2022, Alsona AG.

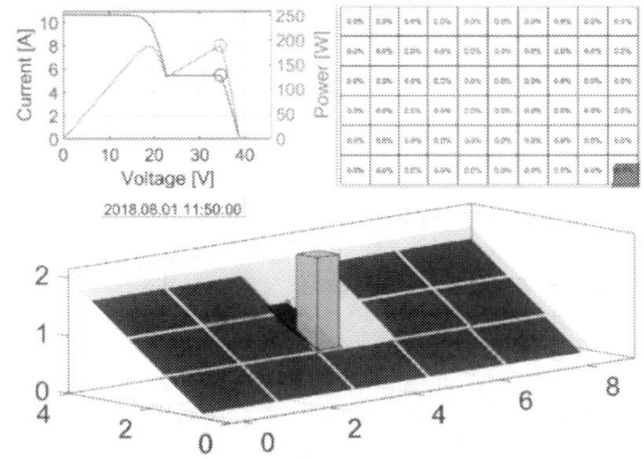

Figure 5: Reproduced PV system & simulated shading object (bottom), IV-curve (top-left) and shading (top-right) of Module 11, within the shading simulation MLPEShade.

The position of the shading object relative to the PV plant has a substantial impact on the annual energy yield of the plant, but also the potential gain of the MLPE system. Generally, the closer the object is placed to the PV modules and the more modules are partially shaded around the zenith of the days, and especially during summertime, the higher the annual yield gains of the MLPE system compared to the SINV system. However, by placing the modules further away from the shading object, the yield of the SINV is increased and, therefore, the MLPE yield gain decreases. The SINV might even perform better than the MLPE system in the annual yield analysis, if no relevant shading occurs at high daily irradiance situations.

In Figure 6, the annual MLPE yield gain is visualized with boxes, whereby their positions represent the position of the chimney,while their colours correspond to the magnitude of gain in percent. In Figure 12, the same PV system setup as in the previous section is visualized. For the simulations, the underlying system setup with SINV: SUN2000-3.68KTL-L1 ($\eta_{EURO} = 97.3$ %), and MLPE: SE3500H ($\eta_{EURO} = 98.8$ %) with P370, was chosen. According to Figure 6, the maximum MLPE yield gain resulted in 0.79 %, whereby the respective chimney position correlates with the chimney position on the actual roof top installation. Notably, this chimney position causes the greatest amount of shading, which is why the MLPE shows improved performance. On the other hand, if the distance between the PV modules on the left side of the roof and the chimney was greater, the MLPE gain could have been -0.2 %, meaning that the SINV system would perform better.

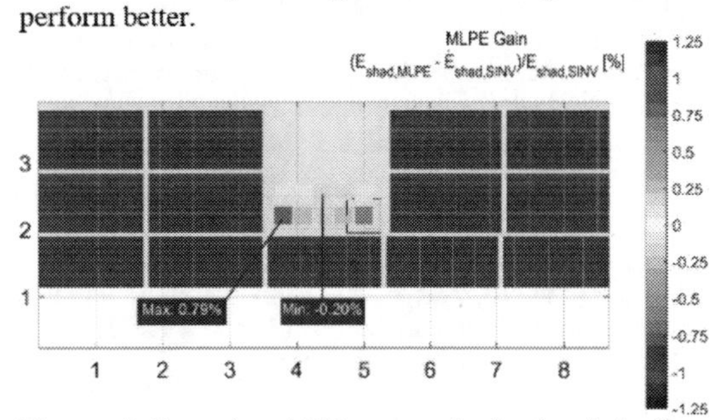

Figure 6: Reproduced PV system & simulated shading object (bottom), IV-curve (top-left) and shading (top-right) of Module 11, within the shading simulation MLPEShade.

4 SHADING ADAPTION EFFICIENCY

4.1 Proposed Concept

The described method was proposed as shading adaption efficiency (SAE), which is defined as the ratio of the output AC power to the maximum available DC power.[9] That DC power corresponds to the sum of all the individual PV modules k in the string, if all of them are operated in their individual absolute MPP. Therefore, this hypothetical DC power is identical to the output of a PV system, which is operated by 100 % efficient DC/DC converters and inverters. In the laboratory set-up described in Section 2, this ratio $\eta_{shad,a,n}$, will be calculated by the measured AC output power P_{AC} in a characteristic shading situation a, which is powered by specific IV characteristics of such partially shaded and unshaded PV modules in the string. According to the denominator of Equation 5, the DC powers of each PV module tracked at their individual MPP will be measured in the lab and aggregated to the total DC input power of all k PV modules. When testing SINV systems, the measured DC input power at the plugs of the inverter will always be smaller or equal to the term, $\sum_i^k P_{mod,i}$. Smaller values can be expected in situations, where the DC string current prevents that every PV module is operated in their absolute MPP.

$$\eta_{shad.a.n} = \frac{P_{AC.n}}{\sum_{i=1}^{k} P_{mod.i.n}} \tag{5}$$

Equation 5 holds for an individual point in time n. The goal is to determine the annual average $\eta_{shad,a,n}$, which would require a large amount of indoor measurements. In order to reduce the number of measurements, only a few typical shading moments in a year should be measured and weighted accordingly. In other words, the efficiencies of the limited set of operating points n, at a characteristic shading situation a is multiplied by their weighting factors w and added up, as given in Equation 6.

$$\eta_{shad.a} = \sum_{n=1}^{N} \eta_{shad.a.n} \cdot w_{shad.a.n} \tag{6}$$

The characteristic shading situations a will be identified by the binning of the annual system yields, then found at several moments n in the annual mapping and marked by the numbers therein. Meanwhile, the weighting factors are representations of the energy amount during the year at similar shading adaption efficiencies, which are different for MLPE and SINV. A similar approach is found by the established calculations of the average string inverter efficiency according to the standards, based on the irradiance binning at different locations, such as the European efficiency η_{EURO} or California inverter efficiency η_{CEC} [25, 26].

4.2 Derivation of Weighting Factors

For the 13-module PV system setup, which is presented in Section 3 (shown in Figure 4 and Figure 5), the annual simulation was performed. In detail, the performance model of the SUN2000-3.68KTL-L1 (η_{EURO} = 97:3 %) was chosen as SINV reference and the SE3500H (η_{EURO} = 98:8 %) with P370 Power Optimizer as MLPE system [27, 28, 29]. In order to identify appropriate weighting factors $w_{shad,a,n}$ for the SAE in Equation 6, the yield, of every simulated time step of the annual simulation of the SINV and MLPE system, is allocated into bars as shown in Figure 7. Within the mentioned figure, the results of the theoretical, 100% efficient MLPE system indicate higher yields for every power ratio $P_{MLPE\text{-}noLoss,Max} / P_{PV,rated}$, which shows the hypothetical benefit of the module-level tracking. However, for the real MLPE yield, the difference has shrunk significantly, due to the higher losses of the MLPE components compared to the SINV system. Accordingly, the real yield of the SINV system is higher than the real MLPE for a power ratio above 92 %.

This is owing to the fact that the SINV performs better in situations when little to no shading occurs, which is the case for the 13-module PV plant during noon, when the prevailing irradiance conditions offer the highest yields. The mentioned differences can be seen in more detail in Figure 8, where the surplus yield of the systems for every bin of Figure 7 are visualized. In detail, the cumulative difference of the 100% efficient systems results in 9.1 % and the effective difference in 0.8 %, which shows the higher overall performance of the MLPE system, but underlines the additional losses caused by the greater number of power electronic devices.

Now, for the definition of weighting factors $w_{shad,a,n}$ of the SAE, the number of bars must be reduced (e.g. to 6 bars), whereby their individual shares on $E_{MLPE\text{-}noLoss,Max,Annual}$ will represent the weighting values. Meanwhile, for every resulting bar, one strategically chosen operating point will be defined to represent a characteristic shading situation a with an efficiency $\eta_{shad,a,n}$ and, thereby exemplify the performance of the respective bars. As a consequence, the characteristic shading situations a, their weighting factors $w_{shad;a;n}$, and correlating operating points will be identified.

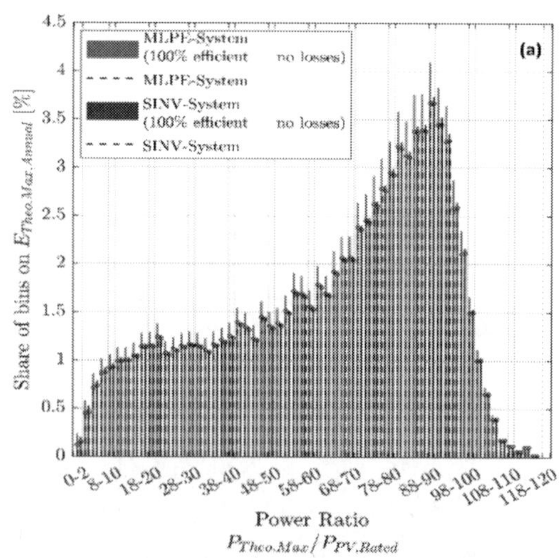

Figure 7: Annual yield of the SINV and MLPE system, once for the 100 % efficient and once for the effective yields after losses, are allocated in 2 %-bins. These bins represent relative values, based on the ratio of theoretical max. power of each time stamp to the rated power of the PV plant. The yield is shown as the share of each bin on the total, theoretical max. energy of the PV plant.

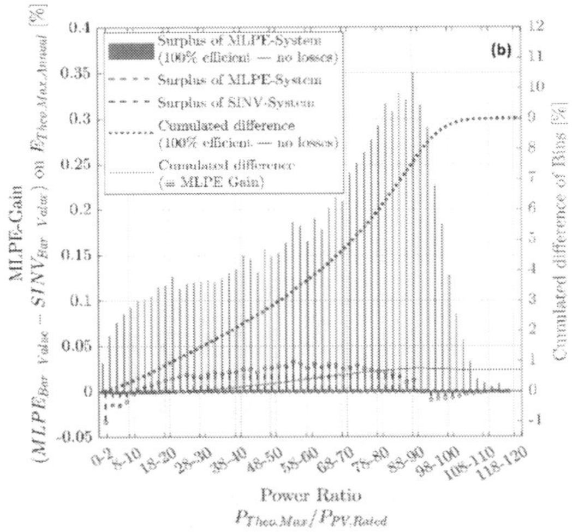

Figure 8: The yield differences of the SINV and MLPE bins of Figure 7 are visualized. Additionally, the cumulative sums of these differences are shown by lines and their values are based on the second y-axis on the right.

In Table 1, the results of the previously described reduction of bars and the definition of corresponding, characteristic shading moments n, are presented. In detail, the bars were reduced to six ranges (shown in Row 2), which shares on $E_{MLPE\text{-}noLoss,Max,Annual}$ resulted in the weightings $w_{shad,a,n}$ of Row 3 in Table 1. Within each of the six bars, a representative shading moment n was chosen, for which the respective $\eta_{shad,a,n}$ of the system-under-test shall be identified.

For the previously described 13-module PV system, which results are shown in Figure 7, the SAE calculation was performed. In further detail, the Equation 5 was applied for the shading moments n described in Table 1 and thus, by using Equation 6 with the defined $w_{shad,a,n}$, the SAE was identified. The resulting MLPE gain and in addition, the simulated results which represent the target values, are presented in Table 2. Although, the totals of the SAE calculation differ from its target values by 0.45 and 0.48 %, the difference of the MLPE gain results in only 0.04 %.

For the validation of the weightings $w_{shad,a,n}$ and shading moments n, the same PV plant was simulated with components that show a lower overall efficiency. In detail, the performance model of the SB3.6-1AV-41 inverter was used for the SINV system, and for the MLPE system, the SE3500 (non HD-Wave) inverter was utilized, while the power optimizer model remained the same [30]. Once again, the resulting SAEs in Table 3 show values, which are 0.44 or 0.47 % higher than the target values, i.e., the simulated annual SAE. However, the resulting MLPE gain nearly matches the target value with a difference of -0.02 %.

Consequently, the characteristic shading moments n are good indicators for the comparison of MLPE and SINV system components at different efficiency values. However, a future analysis of different characteristic shading moments might result in SAEs, which coincide more accurately with the target values. Nonetheless, for the purpose of demonstrating the SAE method with the mentioned example, the chosen values can be seen as sufficiently precise for now. Additionally, the MPPs of each PV module and every characteristic shading moment, n, are presented in Table 4 for the sake of completeness.

Table I: Description of the defined weighting factors wshad;a;n and characteristic shading moments n

Shading moments n:	1	2	3	4	5	6
Power ratio [%] $P_{MLPE-noLoss,Max}/P_{PV,rated}$	0-30	31-44	45-60	61-78	79-88	89-120
Weighting factors $w_{shad,a,n}$:	0.16	0.10	0.13	0.22	0.18	0.21

Table II: Previously defined weighting factors, operating points and corresponding simulated shading efficiencies. SINV: SUN2000-3.68KTL-L1 | MLPE: SE3500H + P370.

Operating Points:	1	2	3	4	5	6
Weighting factors:	0.16	0.10	0.13	0.22	0.18	0.21
MLPE SAE [%]:	96.60	97.13	97.77	97.49	96.96	96.65
SINV SAE [%]:	97.65	94.57	89.09	97.95	97.85	97.69
MLPE efficiency $\eta_{MLPE,Total}$ [%]:	96.60	97.13	97.77	97.49	96.96	96.65
SINV efficiency $\eta_{SINV,Total}$ [%]:	97.65	98.17	98.18	97.95	97.85	97.69

	Resulting SAE [%]	Resulting MLPE Gain [%]	Annual SAE [%] (5-min Sim.)	Annual [%] MLPE Gain (5-min Sim.)
MLPE SAE [%]:	**97.08**	**0.76**	96.63	0.80
SINV SAE [%]:	**96.34**		95.86	

Table III: Validation set for the defined weighting factors, operating points and simulated shading efficiencies. SINV: SB3.6-1AV-41 | MLPE: SE3500 + P370.

Operating Points:	1	2	3	4	5	6
Weighting factors:	0.16	0.10	0.13	0.22	0.18	0.21
MLPE SAE [%]:	94.96	95.51	95.33	94.74	94.41	94.59
SINV SAE [%]:	96.18	95.34	95.87	97.08	97.03	96.94

	Resulting SAE [%]	Resulting MLPE Gain [%]	Annual SAE [%] (5-min Sim.)	Annual [%] MLPE Gain (5-min Sim.)
MLPE SAE [%]:	**94.84**	**-1.79**	94.40	-1.77
SINV SAE [%]:	**96.57**		96.10	

Table IV: MPP values of each PV module for the characteristic shading moments n of the chimney case.

	1		2		3	
	I_{MPP} [A]	V_{MPP} [V]	I_{MPP} [A]	V_{MPP} [V]	I_{MPP} [A]	V_{MPP} [V]
Module 1 - 6, 8 - 10 and 12 - 13	2.06	35.37	4.19	35.37	5.70	35.37
Module 7	2.06	35.30	1.55	38.81	4.59	34.61
Module 11	2.06	35.37	4.18	23.40	5.07	34.74

	4		5		6	
	I_{MPP} [A]	V_{MPP} [V]	I_{MPP} [A]	V_{MPP} [V]	I_{MPP} [A]	V_{MPP} [V]
Module 1 - 13, (all PV-modules)	7.90	31.71	9.26	30.44	10.81	30.26

4.3 Resulting System Efficiencies

The corresponding efficiency results for both, the SINV and MLPE system are visualized as annual mappings in Figure 9 and 10. The colouring shows the shading adaption efficiencies of the mentioned systems for every day of a year and every moment of a day to identify the most beneficial performance of either SINV or MLPE. For the definition of the shading adaption efficiency, the value of the individual weighting factors, $w_{shad,a,1}$, $w_{shad,a,2}$ to $w_{shad,a,n}$, are found by aggregating similar operating points over the whole year, as described in Section 3.2.

For the shading situation *a*, the typical operating points are marked in the Figure 9. With the annual mapping of the SAE values, it is further possible to analyse the effects of shading on the performance of a system or to compare the two systems in detail and at any moment during the year.

It can be stated that the MLPE system (in Figure 9) is less impacted by the shading than the SINV (Figure 10). Although, the SINV performance is higher between 12:00 and 15:30, because there is no shading on the PV modules (as described in Section 3.2). During these times also high energy yields occur. The annual average shading adaption efficiency of the MLPE results in 96.6 % and for the SINV 95.9 %. Consequently, the MLPE can improve the performance by 0.7 % in this particular situation, where the chimney is directly positioned next to the PV system.

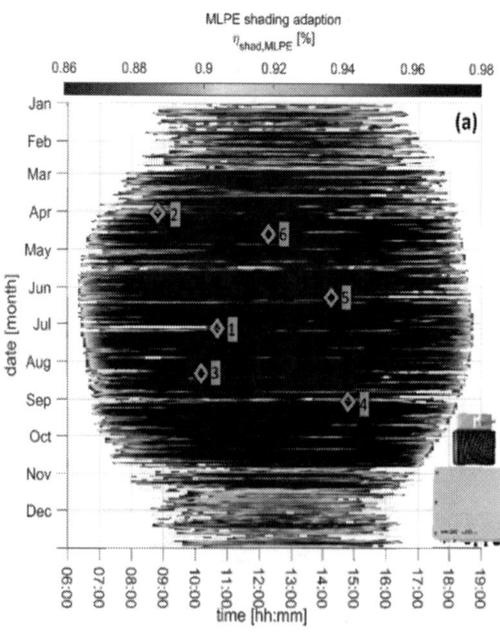

Figure 9: Momentary shading adaption efficiencies (SAE) for the MLPE as a mapping of days in the year and every timestep of the day. The efficiency values are based on the colorbar on top and colored correspondingly. The device images of the power optimizer and string inverter, whose performance models are used in the simulation, are shown next to the mappings.

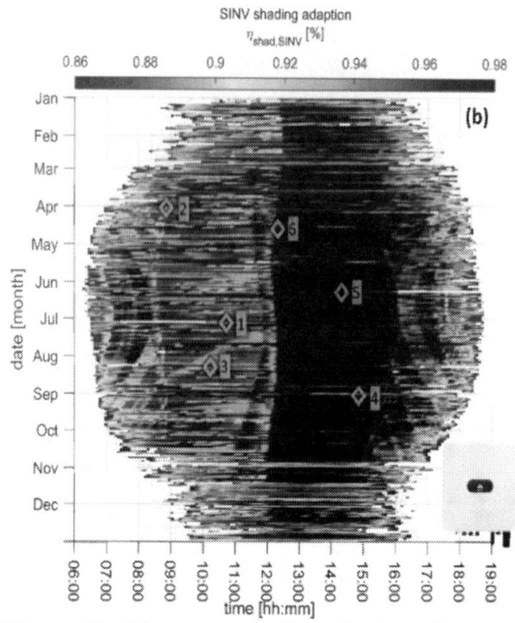

Figure 10: Momentary shading adaption efficiencies (SAE) for SINV system as a mapping of days in the year and every timestep of the day. The efficiency values are based on the colorbar on top and colored correspondingly. The image of the standard string inverter (SINV), whose performance models are used in the simulation, are shown next to the mappings.

5 DISCUSSION

Partial shading of PV systems has a significant impact on the annual performance. Additionally, the analyses of different plant configurations are time consuming. Up to now, there is no method available for installers to identify, which system components are best suited for certain specific situations they encounter. With the presented new approach, the shading adaption efficiency (SAE) can be used to assess the few most-common shading cases, which would need to be defined in future works as well as in standards. These cases will then stand as a benchmark for the comparison of power electronic components of string inverter (SINV) and MLPE systems (MLPE), which will help the decision-making of installers, especially, for the planning of residential roof-top PV plants. On the other hand, the method offers an approach for manufacturers to evaluate their system components in the benchmark situations by simply testing and measuring their components efficiency at a small number of standardized operating points, representing a few defined moments in a year, in order to assess the annual yield of the characteristic shading cases.

For a PV string consisting of 13 PV modules, an annual shading simulation for the SINV and MLPE system was conducted. In detail, this PV string exists in reality and is installed on a single-family house in Switzerland, whereby a chimney in the middle of the PV plant causes partial shading. Accordingly, it shows conditions in which MLPE devices could possibly increase the annual yield of PV system. However, the MLPE yield gain varies between -0.2 to 0.8% in the analysed case, based on the position of the chimney. Most notably, the further the shading object is located from the PV plant and the less shade occurs during noon, the greater the performance of the SINV compared to MLPE system. Generally, newer generations of components should be utilized, as they can offer at least 1% higher yields, regardless, if MLPE or SINV configuration is used. Furthermore, in the 13-module system, the SINV (SUN2000-3.68-KTL-L1) had a higher performance than the MLPE system at three chimney positions. The distances to the PV modules were the greatest in these positions, whereas the chimney still caused shading on the PV modules. Accordingly, in the mentioned cases, the additional losses of the power optimizers during the year were greater than the yield gain by their module-level tracking in times of shading, relative to the SINV. The mentioned losses might be reduced for an MLPE system, in which power optimizers are only installed at certain, most shaded modules, which is a system configuration that will be analysed in future works.

Finally, the methodology of the shading adaption efficiency is presented, and the annual performance of the previously mentioned PV system is visualized as a mapping of momentary SAE values. In summary, the performance of the MLPE system is less impacted by the shading, however the SINV system has a higher yield during 12:00 to 15:30, as there is less shading present. Given these points, the resulting annual average SAE arrives at 96.6 % for the MLPE and 95.9 % for the SINV, which shows that the MLPE system performs approximately 0.8 % better with the analysed system configuration.

ACKNOWLEDGEMENT

Within the IEC TC 82/WG 6, the shading adaption efficiency was presented [31]. To support the development of a future technical specification, benchmark cases for the SAE calculation need to be defined. Accordingly, the ZHAW is involved in the analysis of shading scenarios, to identify characteristic, benchmark shading situations, as a part of the collaboration by the International Energy Agency (IEA) PVPS Task 13 ST2.5 [32]. The MLPE research of the ZHAW is funded by the Swiss Federal Office of Energy, with Project Number: SI/502247-01 [15].

REFERENCES

[1] A. Gattlin, Why Top-Notch Solar Stock Enphase Is Outperforming Rivals Like SolarEdge, Investor's Business Daily, 2022, [Online: accessed 25.06.2022]. [2] SolarEdge Technologies Inc., Solaredge Fact Sheet, www.solaredge.com/sites/default/files/se-fact-sheet-na.pdf, [Online: accessed 15.05.2022].

[3] C. Deline, B. Marion, J. Granata, S. Gonzalez, A performance and economic analysis of distributed power electronics in photovoltaic systems, Technical report, National Renewable Energy Laboratory (NREL) and Sandia National Laboratories, 2011.

[4] C. Deline, S. MacAlpine, In Proceedings of the 2013 IEEE Energy Conversion Congress and Exposition. 2013 4801-4807.

[5] A. J. Hanson, C. Deline, S. M. MacAlpine, J. T. Stauth, C. R. Sullivan, IEEE Journal of Photovoltaics 2014, 4, 6 1618.

[6] K. Sinapis, C. Tzikas, G. Litjens, M. van den Donker, W. Folkerts, W. van Sark, A. Smets, Solar Energy 2016, 135 731.

[7] C. Deline, J. Meydbray, M. Donovan, Photovoltaic shading testbed for module-level power electronics: 2016 performance data update, Technical report, National Renewable Energy Laboratory (NREL) and PV Evolution Labs / DNV GL, 2016.

[8] ZHAW School of Engineering (ZHAW), Institute of Energy systems and Fluid-Engineering (IEFE), Homepage, www.zhaw.ch/en/engineering/institutes-centres/iefe/renewable-energies/photovoltaics/, [Online: accessed 18.05.2022].

[9] F. Baumgartner, R. Vogt, C. Allenspach, F. Carigiet, In Proceedings of the 38th European Photovoltaic Solar Energy Conference and Exhibition (EUPVSEC), Online. WIP Renewable Energies, Munich, Germany, 2021 650-654.

[10] C. Allenspach, V. G. de Echavarri Castro, S. Richter, C. Meier, F. Carigiet, F. Baumgartner, In Proceedings of the 37th European Photovoltaic Solar Energy Conference and Exhibition (EUPVSEC). WIP Renewable Energies, Munich, Germany, 2020 1188-1194.

[11] C. Allenspach, A. Bänziger, A. Schneider, F. Carigiet, F. Baumgartner, Performance comparison of a P370 power optimizer system and a string inverter system (poster), In 20th Swiss National Photovoltaic Conference, 2022.

[12] R. Bründlinger, N. Henze, G. Lauss, J. Liu, In Proceedings of the 26th European Photovoltaic Solar Energy Conference and Exhibition (EUPVSEC). WIP Renewable Energies, Munich, Germany, 2011, 3204-3211.

[13] SolarEdge Technologies Inc., SolarEdge Residential Offering for Installers, www.solaredge.com/sites/default/files/residential_catalogue_eng_sa.pdf; [Online; accessed 23:06:2022].

[14] Tigo Energy Inc., Tigo TS4-A-O - Case Study: Harvesting 36% Energy with Optimization, www.tigoenergy.com//product/ts4-a-o, [Online: accessed 23.06.2022].

[15] Swiss Federal Office of Energy (SFOE), EFFPVSHADE - Effizienzanalyse von dezentraler Photovoltaik Leistungselektronik bei Teilbeschattung [EFFPVSHADE - Analysis of efficiency of decentralized phototvoltaic power electronics under partial shading], SI/502247-01, 2021-2023.

[16] SolarEdge Technologies Inc., Datasheet - Power Optimizer: P300/P370/P404/P405/P500/P505, www.solaredge.com/sites/default/files/se-p-series-add-on-power-optimizer-datasheet.pdf, [Online: accessed 30.05.2022].

[17] I. Mehedi, Z. Salam, M. Ramli, V. Chin, H. Bassi, M. Rawa, M. Abdullah, Renewable and Sustainable Energy Reviews 2021, 146 111138.

[18] N. Dekker, M. Jansen, A. Burgers, M. Dörenkämper, R. Jonkman, R. van der Ven, E. Gramsbergen, In Proceedings of the 37th European Photovoltaic Solar Energy Conference and Exhibition (EUPVSEC). WIP Renewable Energies, Munich, Germany, 2020 945-951.

[19] Keysight Technologies, E4361A-J01 SAS, www.keysight.com/gb/en/products/dc-power-supplies/dc-power-solutions/e4360-series-modular-solar-array-simulators.html, [Online: accessed 02.06.2022].

[20] Newtons4th Ltd., PPA1500: Precision Power Analyzer, www.newtons4th.com/products/ppa1500-precision-power-analyzer/, [Online: 29.05.2022].

[21] M. Littwin, U. Jahn, F. Baumgartner, M. Green, W. van Sark, C. Allenspach, E. Bamberger, C. Biba, C. Deline, D. Gfeller, S. M. Goldroodbari, M. Köntges, C. Messner, U. Muntwyler, R. Neukomm, D. Riley, D. Rivola, J. S. Stein, M. Trommfsdor, Performance of New Photovoltaic System Designs, Technical Report Report IEA-PVPS T13-15:2021 (ISBN 978-3-907281-04-8), International Energy Agency (IEA) PVPS Task 13, 2021.

[22] Alsona AG, (PV installer and contractor for energy concepts), 13 kWp Ziegel EFH in Gontenschwil, www.alsona.ch, [Online: accessed 02.06.2022].

[23] Swiss Federal Office of Meteorology and Climatology (MeteoSwiss), IDAweb: Data portal of ground station measurements for education and research, www.meteoswiss.admin.ch/, [Online: accessed 18.06.2022].

[24] P. Ineichen, R. R. Perez, E. Maxwell, R. D. Seals, A. Zelenka, ASHRAE Transactions-Research 1992, 98 354.

[25] F. Baumgartner (author), N. Pearsall (editor), The Performance of Photovoltaic (PV) Systems, Chapter 5: Overall efficiency of grid connected photovoltaic inverters (references to EN 50530:2010+A1:2013), Elsevier, London, GB, 2017.

[26] W. Bower, C. M. Whitaker, W. Erdman, M. B. Behnke, M. Fitzgerald, Performance test protocol for evaluating inverters used in grid-connected photovoltaic systems, Technical report, Sandia National Laboratories, 2004.

[27] Huawei Technologies Co., Ltd, Datasheet: SUN2000-2/3/3.68/4/4.6/5/6KTL-L1, solar.huawei.com/en-GB/download?p=SUN2000-2-6KTL-L1.pdf, [Online; accessed 04.06.2022].

[28] Solaredge Technologies Ltd., Datasheet: Single-Phase Inverter SE3500H (HD-Wave), www.solaredge.com/sites/default/files/se-single-phase-HD-wave-inverter-setapp-datasheet.pdf, [Online; accessed 04.06.2022].

[29] SolarEdge Technologies Inc., Datasheet: Single-Phase Inverter SE3500H (HD-Wave), www.solaredge.com/sites/default/files/se-single-phase-16a-inverter-datasheet.pdf, [Online; accessed 05.06.2022].

[30] SMA Solar Technology AG, Datasheet: Sunny Boy SB3.6-1AV-41, files.sma.de/downloads/SB30-60-DS-en-52.pdf, [Online; accessed 04.06.2022].

[31] International Electrotechnical Commission (IEC), TC82: Solar photovoltaic energy systems - WG6: Balance-of-system components, https://www.iec.ch/dyn/www/f?p=103:14:605685340133361::::FSP_ORG_ID,FSP_LANG_ID:2749,25; [Online; accessed 09:06:2022].

[32] International Energy Agency (IEA) PVPS, Task 13: Performance, Operation and Reliability of Photovoltaic Systems, www.iea-pvps.org/research-tasks/performance-operation-and-reliability-of-photovoltaic-systems/, [Online; 08.06.2022].

HORIZONTAL SINGLE AXIS TRACKER IN WEATHERING STEEL, A SOLUTION WITH A LOW IMPACT ON THE LCOE AND LCA INDEX

Giuseppe Demofonti, Andrea Ricci, Antonio La Mura, Antonio Timidei, Alessandro Mercurio
Convert - Valmont (Roma, Italy)
contact e-mail: alessandro.mercurio@valmont.com

ABSTRACT

Specific aim of the research activity described in this paper was to improve the existing know-how and to generate new knowledge in the design of Horizontal Single Axis Tracker (in a 2P configuration) made in Weathering Steel for bifacial PV panels. Solar Trackers should be characterized in giving a relevant contribution to reduce to both the Levelized Cost of Energy (LCOE) and the Life Cycle Assessment (LCA) of large PV plants.

Specific research actions have been carried out focused on:

- The use of Weathering Steel for solar tracker fabrication, as a cheaper steel-coating solution to be used in aggressive corrosive environment (both air and soil).
- The experimental evaluation of wind loads to take into account for solar tracker design, in order to assure a safe and economical design of the structure.
- The Wireless Control & Powering Systems integrated to new generation of monitoring SCADA A.I., in order to reduce cables and time for commissioning and, at the same time, assure a high reliability of the monitoring.

All the research activities have been supported by an integrated experimental and analytical approach, characterized by a large scale experimental tests program performed on full scale mock-up.

This activity presented has been carried out in the framework of a European project H2020, GOPV (Global Optimization of integrated PhotoVoltaic system for low electricity cost).

Keywords: PV system, Solar Tracker, Energy Production, Single-axis Tracker, Weathering Steel, LCA, LCOE

1 INTRODUCTION

World Energy transition will be mostly driven by solar photovoltaic technology. Around 190 GW of new photovoltaic capacity is expected to be commissioned in 2022, which is 25% higher than 2021's [1].

Main case of International Energy Agency (IEA) report shows the yearly solar photovoltaic capacity additions will be constantly over 160GW in 2021-2026, reaching almost 200 GW in 2026. Utility scale solar PV provides, in the majority of countries, the lowest cost of adding new electricity capacity, especially in the actual context of increasing natural gas prices. These large projects represent the 60% of new yearly installation. [1]

In order to increase energy production from PV system, solar tracker technology is more and more deployed. As shown in the diagram of figure 1, IHS Markit has predicted that starting since 2019 solar trackers will represent more than one third of all ground installations; in terms of capacity trackers will exceed 150 GW during 2019 – 2023. That said, trackers are now being used more widely overseas, and in some markets their penetration has already reached nearly 50% of total installations.

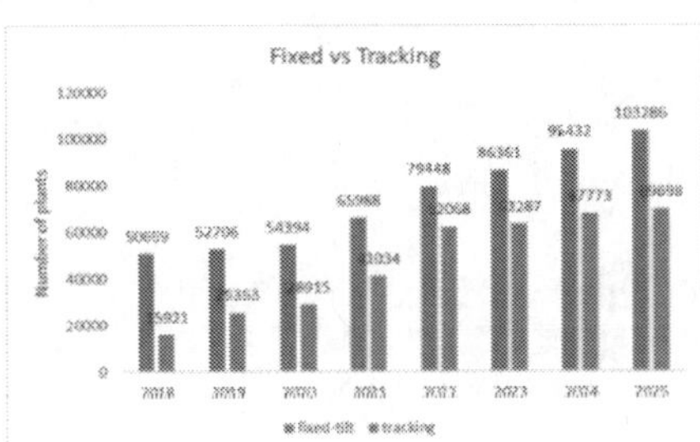

Figure 1 – Ground PV plants: in blue the fix-structure, in orange the trackers (one third of the total) – IHS Markit

Inside this scenario, the solar industry developed across the years two main key types of solar tracking systems: single-horizontal axis and dual-axis. The first, single horizontal rotational axis, moves the panels on one axis of movement, usually aligned with north and south: this means that the panels track the sun from east to west)as the sun rises and sets). The second, dual-axis tracker, allows panels to move on two axes, aligned both north-south and an east-west. In Figure 2 examples of horizontal single axis in tracker PV utility scale plants.

Figure 2 - Horizontal single axis tracker (1 and 2 modules) in PV utility scale plants

Single horizontal axis solar trackers provide significant advantages for renewable energy in terms of LCOE, in fact power output can be increased up to 30% (40% for dual axis tracker) [3], with a significantly lower cost increase compared to the two-axis solution. So, the deployment of single horizontal axis tracker for utility-scale PV plant is attracting the interest of several PV operators.

This paper describes the research activity carried out in the framework of a European project H2020, GOPV (Global Optimization of integrated PhotoVoltaic system for low electricity cost). [3] [4]

More in details the research activity performed by Convert – Valmont inside the GOPV project has been focused to improve the existing know-how, and also to generate new know-how, in the design of Horizontal Single Axis Tracker, for utility-scale PV plant optimized in terms of both Levelized Cost of Energy (LCOE) and Life Cycle Assessment (LCA) index. The technical approach followed, the most relevant experimental activities performed and the main results obtained are summarised in this paper.

2 APPROACH AND ACTIVITY

In order to achieve the above goals in terms of LCOE and LCA values, specific research actions have been carried out focused on:

- The use of Weathering Steel in the fabrication of tracker, as a cheaper steel-coating solution to use in aggressive corrosive environment, both for air and soil conditions. Solution with a lower environment impact.
- The experimental evaluation of wind loads to take into account in the tracker design, in order to assure a safe and economical design of the structure and in parallel to improve the existing international standards concerning the structural design of the tracker assuring also an optimization regarding the materials use,
- The develop of Wireless Control & Powering Systems integrated to new generation of monitoring SCADA A.I., in order to reduce cables and time for commissioning of PV plant and, at the same time, assure a high reliability of the monitoring

All the research activities have been supported by an integrated experimental and analytical approach, characterized by a large scale experimental tests program performed on full scale mock-up.

2.1 CORROSION RESISTANCE BASE MATERIAL / COATINGS

Solar Trackers are mainly fabricated in Hot Dip Galvanized (HDG) steel that has proven to be an excellent protection against corrosion and to provide service lifetime of 25 years, also in aggressive environments. Figure 3 shows HDG tracker structure, with its typical silver-grey color.

Nevertheless, HDG process is estimated to account about 20% - 25% of the cost of the tracker. Alternative materials should be explored to reduce costs, ensuring similar or higher mechanical and anticorrosion performances, and increase lifetime up to 35 years: Weathering Steel (W.S.) could be a technical and economical valid solution.

Figure 3 - HDG tracker structure, with the typical silver-grey colour.

It has been patented in 1933 by United States Steel Corporation, called *Cor-Ten,* as a low-alloy carbon steels with improved resistance to atmospheric corrosion on the time. In fact, the Weathering Steel in aggressive environments assure a self-protection from corrosion through the formation, on exposed surfaces, of a compacted layer of oxides (patina), obstructing oxygen diffusion. In general, the use of W.S. steels eliminates the galvanizing process (or other coating systems, e.g. painting) in the construction of the steel structure, with a significant effect on both LCOE and LCA.

Corrosion rate is estimated to be 4-5 times lower than traditional carbon steel, depending on the aggressiveness of environment and on the chemical composition of the steel. They are particularly indicated for atmospheres in categories C1, C2, C3 (ISO 9223 as reference). Moreover, changing its color from yellow to dark brown in the time made the W.S. very interesting structural steel for civil applications, so it has been used in many bridges, civil structures, guard rails, architectural buildings [5] [6]. Figure 4 shows typical applications of W.S. in the area of civil construction.

Figure 4- Typical applications of W.S. in the area of civil construction.

Nevertheless it was not used extensively in PV systems due to a lack of specific know-how dedicated to the design trackers for PV systems made in W.S., limit made more difficult to overcome due to the reduced thicknesses used in a tracker (2-4 mm) compared to those used in civil applications (>8 mm).

Starting from the basis of both the knowledge available in the literature and the international standards, we followed analytical procedures to easy classify the environment aggressiveness (atmosphere and soil) and to evaluate the sacrificial extra-thickness for tracker cross sections when weathering steel is used, in particular for this specific activity W.S. grade S235 ÷ S355, EN-10025, have been used.
A relevant improvement of the available know-how about the corrosion behavior of Weathering Steel in aggressive environment has been achieved through an out-door long term corrosion experimental test program on small specimens (three test-sites with different environmental conditions and more of two years of exposition) and specific laboratory analyses. Specific topics of this study were:

- Prediction of the corrosion rate of Weathering Steel in air and soil,
- Mechanical connections and bolts and possible use of hot galvanized bolts,
- Connection between Weathering Steel and Hot galvanized steel

In parallel, the monitoring of a tracker exposed in the environment of Tuscania (VT, Italy) allowed to collect knowledge about the transferability of data obtained on laboratory test specimens to real structures. Figure 5 shows the prototype in Weathering Steel tested in Tuscania.

Figure 5 - Tuscania, Italy: Horizontal Single Axis Tracker made in Weathering Steel

After one year of soil and air exposure, the 2P tracker and laboratory samples were analyzed. The same deep analysis has been repeated after about 2,5 years.
Results lead to the following indications/conclusions:

- Availability of validated analytical models for both evaluation of sacrificial extra-thickness for W.S. application and also an easy classification of the environment aggressiveness to use in the design regarding the lifetime of tracker.

- There is a full compatibility between W.S. components and hot-dip galvanized bolts to be used for structure and modules fixation, in figure 6 details of connection W.S. and hot-dip galvanized bolts are shown.

- No particular grounding issues occurred due to passivated connections, in particular no electric grounding issues due to passivation of connections has been detected. The experimental values shown a R-Earth bellow of the maximum accepted values of resistivity R-Earth ≤ 1 Ohm, indicated by standard.

- Passivated layer in W.S. can be measured by Coating Thickness US Gauge with successful, as direct measurements of coating thickness on metallographic sections confirmed.

- No trouble on the embedded part of pile, for medium aggressive soil, has been detected after approximately 2.5 years of burying; figure 7 shows a phase of the survey performed in Tuscania on the tracker piles.

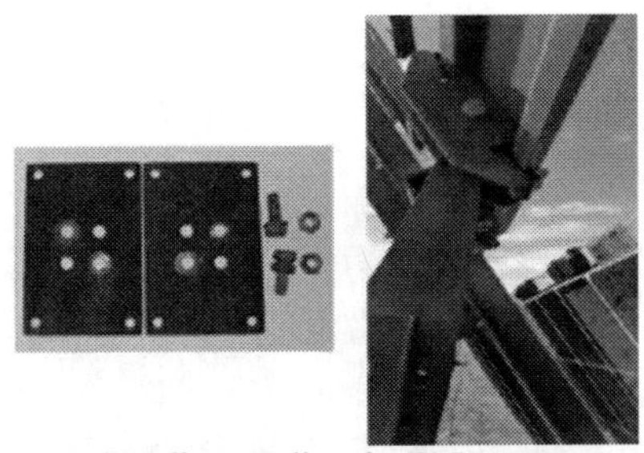

Figure 6 - Details regarding the W.S. components and hot-dip galvanized bolts

Figure 7 - A phase of the survey performed in Tuscania on the tracker piles

2.2 WIND STATIC AND DYNAMIC LOADS

Regarding the evaluation of wind loads effect on the strength and stability of the tracker, an integrated numerical-experimental approach has been developed and tested inside the GOPV project, with the final goal to evaluate the wind loads to use in the tracker design, in order to assure a safe and economical dimensioning of all tracker structural components and also its stability at

higher wind values.
Moreover, this was carried out to validate the CFD approach in order to use it as an analytical tool to perform sensitivity analyses, and so to be able to apply the wind tunnel results in scenarios differing from the one assumed in the testing.
Wind Tunnel (W.T.) tests has been performed considering in-scale GOPV tracker, and was carried out at the Polytechnic of Milan (POLIMI) [7]; On the other hand, the numerical CFD analysis has been carried out using the ANSYS code.
The Wind Tunnel test focused on three specific topics:

- Evaluation of wind pressure on single tracker vs wind speed (wind coefficients), for different tilting angle of tracker.
- Influence of the tracker position within the field layout in terms of wind coefficients (Internal rows vs External rows).
- Definition of wind conditions/tilting angle that may determine the instability of tracker, in terms of both Static Torsional instability (Divergence) and Dynamic Torsional Instability (Flutter/Galloping).

Figure 8 shows pictures of the tests lay-out, both to detect the static pressure load coefficients and to quantify the aerodynamic instability of tracker developed.

Figure 8 - Pictures of the tests lay-out inside the POLIMI Wind Tunnel Laboratory.

The W.T. results allow a more accurate plant design, distinguishing among internal and external rows. Internal rows are around 80 - 85% of the overall structures: the optimization of their design assure an important cost reduction. In particular has been estimated that inner rows suffer around 40% less pressure and torque reduction in stowing/safety position, and 50% less pressure and torque reduction in service positions. These experimental indications were also confirmed by CFD simulations performed using ANSYS code (figure 9).

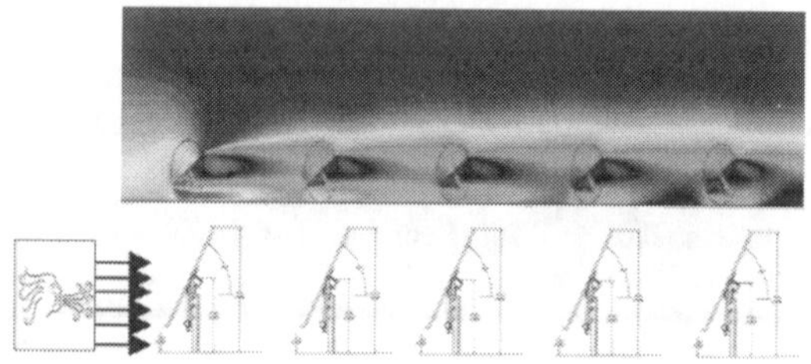

Figure 9 – CFD with Ansys software

Regarding the aeroelastic instability phenomena on the developed 2P tracker, the key results can be summarized as following:

- Torsional stability – Diverge: the phenomenon can born only in Stow position (0° +/-2°), with wind speed above 40 m/s
- Dynamic Instability Flutter: All other Service positions – stable up to 40 m/s
- Not Galloping in the same range of wind velocity

Values valid for a wind speed velocity referred to 10 minutes averaged values measured at 10m height above ground, to scale at gust factor can been used a multiplier of 1,46.

2.3 CONTROL & POWERING SYSTEMS

A new compact electronic control board and powering system for a performance large scale PV plant should satisfy the following general requirements, in order to achieve a redaction of LCOE:

- The new electronic device functioning in a "distributed inverters scenario" and with the possibility also to be powered directly by the inverter output.
- The new electronic device and powering system should assure a reduction of cables to install and consequently a decrease of commissioning time.
- The new monitoring and tracking should be based on a full new Centralized Tracker Wireless Control System.
- The dimension of these new electronic control boards/systems (and its switchboard/box) should be smaller of the existing one, in order to reduce shadowing on the rear side of modules.
- The Actuator – Motor device should ensure adequate values of both actuator force and stroke, in order to ensure the designed tracking operations.

In order to meet these requirements described above, and more generally to ensure an optimum cost/performance ratio of the system applicable in photovoltaic fields of different sizes, a new electronic control and powering system have been developed inside the GOPV project and tested in agreement with the EN-IEC-62817.
This system referred to as "Control & Power 100M" is integrated to a new Tracker Wireless Control System (SCADA A.I). It is based on a "Master - Slave" configuration, where a single "Master" board, controls and power a series of "Slave" boards of very reduced dimensions that are positioned inside each actuator.

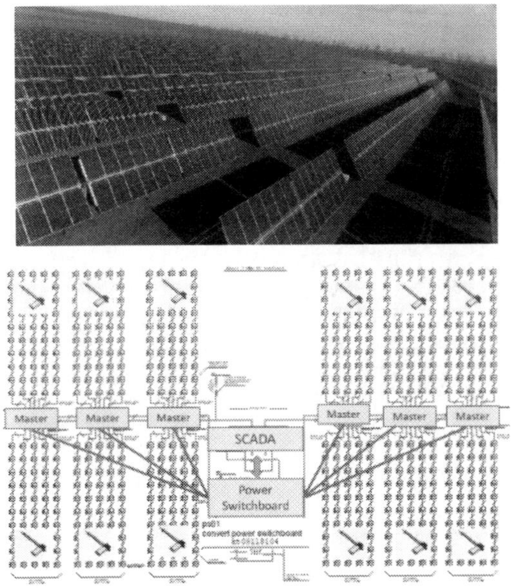

Figure 10 – A real application of 100M and its schematic configuration

The developed "Control & Power 100M" system can connect up to 100 Actuator – Motor device, with a reduced number of cables to install, with a consequently reduction commissioning time. The evaluation made show that technical/economic benefit increases with the plant size. In addition to the development of the "Control & Power 100M" system, a full new Centralized Tracker Wireless Control System, a has been developed: the SCADA A.I. In figure 10 a real application inside an utility scale PV plant of 100 M configuration is shown.

Finally figure 11 and 12 show Control System, SCADA A.I. and Driving Device components under testing at low and high service temperature (-30° and +85°C) inside a testing climatic chamber.

Figure 11 – Climatic chamber for testing

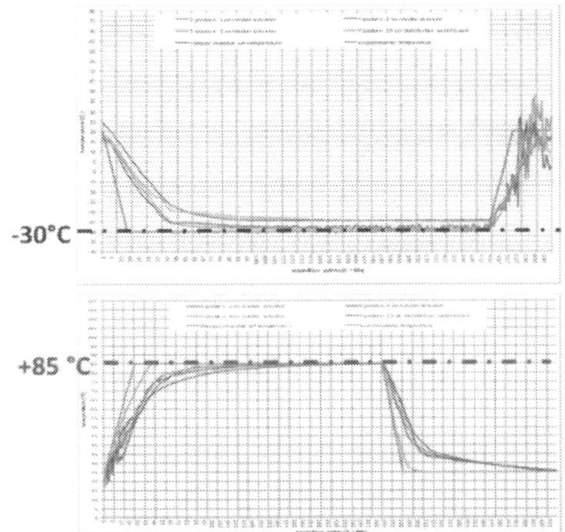

Figure 12 - Control System, SCADAA.I. and Driving Device components under testing at low and high service temperature, data record.

3 APPLICATION OF GOPV KNOW-HOW FOR NEW DEVELOPMENT

Results of the GOPV research projects gives an important know-how for new applications which are following the last trends in Photovoltaics, such as Agrophotovoltaic applications. Convert-Valmont has developed the first prototype of horizontal axis tracker for AgopPV plants, integrating fruit trees (citrus). This prototype has been made in cooperation with EF Solare Italia and Le Greenhouse for the Scalea Plant (south of Italy), figure 12.

This solution used all three topics developed and already presented. More in detail:

- Thanks to W.S. it has a lower cost steel-coating solution to use for aggressive corrosive environment (close to the see and environment with pesticides). It has also lower visual impact than standard steel structures, thanks to the brownish colour of the patina. This gives a positive feedback in term of acceptance (considering that visual impact is always a critical topic regarding the acceptability of a new PV plants).
- Thanks to experimental evaluation of wind loads it was possible to ensure a safe and economical design of the structure, considering its height and an important increase of static and dynamic effect of wind on structure.
- The Wireless Control & Powering Systems integrated to new generation of monitoring SCADA A.I., in order to reduce cables and time for commissioning and, at the same time, assure a high reliability of the monitoring. SCADA A.I. has been design to assure the monitoring and control also of agro-parameters .

Figure 12 - AgroPV structure, Scalea

4 CONCLUSIONS

The research activity performed inside the GOPV H2020 project and the obtained technical data have a key role in the optimization of the Technical- Economical design of tracker for large scale PV-Plant and in the reduction of both LCOE and LCA. These decreases could be summarized as follows:

The Use of Weathering Steel (W.S.) as a lower cost structural steel/coating:

- It has been confirmed for class of environment C1, C2, C3 and length of life 25 years; moreover, the W.S. solution can be used to design a tracker with a lifetime above 25 years.
- Allows to eliminate the HDG coating with a reduction of tracker cost of about 10% - 20% (with a redaction of LCOE) and in parallel a relevant redaction of CO_2 related to the tracker fabrication process. With a good influence on LCA of PV plant.

The new experimental Wind Loads to take into account in the tracker design:

- The improvement of tracker design (in comparison of existing international code/guideline) can be made using the Wind Tunnel tests results and CFD analysis, reducing the total weight of steel to use in PV utility scale plant of 10% - 20%, in relation of dimension/layout of plant and severity of site.
- Regarding the Aeroelastic instability phenomena, the results showed that the tracker developed inside GOPV project is stable: up to a wind speed of 40 m/s (144 km/h) in relation with the Static Torsional Stability - Divergence; in relation to the Dynamic Instability- Flatter the tracker is stable at all service positions up to 40 m/s (108 km/h); finally no Galloping phenomenon was not observed in the same wind speed range.

New Wireless Control & Powering Systems integrated to SCADA A.I.:

- Reduction of length/weight of cables and trenches of 10%÷60% is possible, in relation of the dimension/layout of plant. In parallel also time for commissioning can be reduced;
- A higher reliability in the monitoring is possible using the new SCADA A.I., integrated also with new PV plant configuration, as AgroPV.

All these tracker design optimizations that lead to a reduction in the amount of steel/materials used to manufacture the tracker and its electronic control, while maintaining high performance and reliability. Reduction which results in a positive effect on LCOE and LCA. Values.

5 REFERENCES

[1] Renewable Energy Market Update Outlook for 2022 and 2023– IEA[2]https://ihsmarkit.com/products/solar-intelligence-service-premium.html

[3] Grant Agreement number: 792059 — GOPV — H2020-LCE-2016-2017/H2020-LCE-2017-RES-IA

[4] https://www.gopvproject.eu/project/

[5] European design guide for the use of weathering steel in bridge construction N°143, 2nd Edition, 2021 Published by: ECCS – European Convention for Constructional Steelwork publications@steelconstruct.com

[6]https://www.promozioneacciaio.it/cms/it5537-acciaio-corten-un-materiale-per-tante-soluzioni.asp

[7] https://www.windtunnel.polimi.it/

ACKNOWLEDGMENTS

This paper is supported by European Union's Horizon 2020 research and innovation programme under grant agreement No 792059., project GOPV (Global Optimization Of Integrated Photovoltaic System For Low Electricity Cost).

LIFE EXPECTANCY OF PV INVERTERS AND OPTIMIZERS IN RESIDENTIAL PV SYSTEMS

Dr. Christof Bucher[1], Dr. Jasmin Wandel[2], David Joss[1],
Bern University of Applied Sciences / Department of Engineering and Computer Science
[1]Institute for Energy and Mobility Research / Laboratory for Photovoltaic Systems
[2]Institute for Optimisation and Data Analysis
Jlcoweg 1, 3400 Burgdorf, Switzerland, christof.bucher@bfh.ch

The average Time To Failure (TTF) of PV inverters and PV optimizers is investigated in this paper. The focus is on residential and small commercial systems. The data used in this paper includes 1195 PV systems consisting of 2121 inverters and 8542 optimizers. The data is obtained by means of an online survey sent to the system operators (343 systems) and by analysing system service databases of professional PV system installers and operators (852 systems). Since most of the systems examined are still in operation and have never failed so far, the results are presented using Kaplan-Meier survival curves. After 15 years, 34.3 percent of inverters show a first failure. The most important factors influencing the TTF are the installation location of the inverter (indoor installations have a lower TTF than outdoor installations), the manufacturer and the inverter topology (installations with optimizers have a lower TTF than installations without optimizers).
Keywords: Life expectancy of PV inverters, Time to Failure (TTF), survival curve, Kaplan-Meier estimator

1 INTRODUCTION

PV inverters are typically said to have a life expectancy of 15 years and must therefore be replaced once in the service lifetime of a typical PV system [1]. Accordingly, the warranties for inverters usually only cover about half the time that the performance warranties for PV modules are valid for. However, optimizer manufacturers usually provide longer warranties than inverter manufacturers.

The cost of maintaining and replacing inverters accounts for the largest share of the operating and maintenance costs (O&M) of a PV system. Therefore, it is advantageous for investors and operators of PV systems to have an accurate estimate of the expected lifetime of an inverter and to know the factors that influence its lifetime.

1.1 Study to assess life expectancy of PV inverters

In a study by the Bern University of Applied Sciences, the life expectancy of PV inverters is investigated. Due to various practical limitations such as a lack of information on inverter repairs, the study uses the time to energy-related failures (called TTF in this paper) as the relevant parameter.

1.2 Literature research

P. Hacke et al. [2] makes a detailed inventory of the PV industry in terms of reliability and safety. The publication assesses inverter reliability in relation to existing regulations and standards. The study shows that there are big differences in the information on the quality of inverters and that these can have an impact on O&M costs. The recommendation is made to collect financial data for O&M. To improve quality standards, it is recommended that design validation tests should be conducted under realistic environmental and operation conditions including end-use field conditions.

In Gatla [3], the lifetime of a single-phase 3kW inverter (600V/30A IGBT) is simulated with PLECS. The differences in the top ten PV sales countries are compared. This reveals that not only the maximum temperature at the site but also the temperature and irradiation cycles have an influence on the service life. At the system level, in Australia an average of 10% of the devices failed after 11 years, while this rate for England is theoretically only reached after 99 years (Germany 47 years).

An IEEE publication from Aalborg University [4] deals with the influence of PV module degradation on the lifetime of inverters. The publication assumes inverters to have a higher life expectancy if a lower PV power capacity is connected to the inverter. A system in Denmark is compared with a more degrading one in Arizona (Denmark: 0.15%/a, Arizona: 1%/a). The lifetime of the inverter in Arizona is underestimated by 54% (7 years).

In [5] from Aalborg University it is shown that DC oversizing can have an effect on the lifetime, depending on the location, while the effect is more serious at other locations. In Arizona, for example, the lifetime hardly changes due to the persistently high irradiation values, while in Denmark it deteriorates significantly with increasing oversizing. The main cause is the increasing thermal load.

According to Jordan [6], inverters are still the most frequently failing components of PV systems, accounting for 4–6 % of all cases evaluated. The failure frequency of residential PV systems is slightly lower than that of commercial or industrial systems, possibly due to more reliable inverters. On the other hand, the average production loss of large-scale industrial systems is much lower than that of commercial and residential PV systems. According to this study, the failure frequency of module inverters and power optimizers, which together constitute a device class, is underestimated.

Herz et al. [7] focus mainly on the detailed type of failure, but do not investigate correlations of specific situations or the overall probability of a failure occurring.

Häberlin et al. [8] analysed the inverter reliability of PV systems in Switzerland between 1992 and 2009. He noted that the average age of the inverters has stagnated at around 110 months. However, many of the inverters had been repaired in the first years after commissioning. The TTF definition used in this paper is therefore assumed to be lower than the life expectancy calculated by Häberlin. The discontinued data collection does not allow any direct conclusions to be drawn about current inverter technologies.

2 METHODOLOGY

2.1 General approach

The aim of this study is to quantify inverter failures in the field. The focus is on systems in residential buildings. The aim is not to specify the details of failures, but rather their year of occurrence and their relative frequency compared to inverters without failures.

In addition, local conditions such as sizing ratio (SR), installation location, type of PV system and others are evaluated to investigate their influence on the inverter lifetime.

2.2 Data collection

Data is gathered from three sources between March and May 2022:

1. By means of an online survey, mainly filled in by private owners / operators of PV systems (343 systems).
2. Data of the PV system portfolio of a professional PV system operator (83 systems).
3. Data of the PV system portfolio from professional PV system installers and operators (769 systems).

2.3 Data gross error detection and correction

In the context of this survey, a large part of the data was collected with the help of people who know the history of the PV systems well. These are, in particular, the system operators. However, they are often not familiar with the technical details. For this reason, the survey gave respondents the opportunity to add their own comments to the technical data. With the help of these comments, the data was then manually corrected where possible. The corrections made are described below.

2.3.1 Deletion of incomplete or unrealistic data-sets:

- Incomplete and unanalysable data records were deleted. If data-sets are incomplete but can still be evaluated (e.g. inverter manufacturer is missing) they were not deleted completely, but they were only included in the evaluations for which the missing information is not relevant. Therefore, not all investigations are based on the same number of data-sets.
- Unrealistic entries deleted (e.g. if the first malfunction occurred before the year of commissioning).

2.3.2 Creation of a consistent nomenclature

- Comma replaced with full stop.
- Obviously wrong units replaced / corrected.
- Spelling of products and companies standardised.

2.3.3 Correction of obvious errors

- Inconsistent data corrected or deleted (e.g. obvious factor 1000 for power values).
- Incorrect category assignments of inverters corrected. For example, the category "hybrid inverter" is not listed, but recorded as "string or multistring inverter".
- Uniform power information used for known products (e.g. AC power of inverters).
- Faults considered or not based on description. Two fictitious examples:
  - Fuse has blown, after switching back on the inverter was running again. This is not classified as a malfunction.
  - Inverter was making strange noises, so it was replaced. This was classified as a malfunction.

2.4 Survival curves

In order to assess the reliability of the inverters, the duration until the first energy-relevant fault occurs is considered. This duration is defined as Time to Failure (TTF).

The TTF is calculated and plotted using the concept of the nonparametric Kaplan-Meier estimator together with the corresponding Kaplan-Meier survival curve. The survival curve shows what percentage of the population is still alive after how many years. Applied to this study, this means what percentage of the inverters has still not had any yield-relevant faults after the specified number of years, while it adjusts for so-called censored data points. This means that the Kaplan-Meier estimator accounts for the fact that a lot of inverters did not show a failure up to the time point of investigation. However, those inverters are still included in the analyses with the given observation time used as lower bounds for the actual failure time sometime in the future. With the help of the survival curve method, various other information can be derived:

- The Kaplan-Meier estimator can be adjusted for different influencing factors, e.g. to the year of commissioning, the inverter manufacturer or the nominal power ratio.
- Statistical measures can be read from the survival curves. For example, it can be calculated what share of inverters has a first fault after five years. Or it can be seen what percentage of inverters reach their tenth year without any energy-relevant error.

The survival curve also takes into account how many inverters are still under observation in the corresponding year of operation ("samples at risk"). This means that, although the majority of the inverters have never had a malfunction, it can be estimated in which year of operation a certain percentage of the inverters will have their first expected malfunction. However, because only little data is available for old inverters, the uncertainty in the survival curve increases with increasing inverter life. The confidence interval is shown by the coloured area in the diagrams.

2.5 Avoidance of cross-correlations between dependent variables

In the survey, a set of system parameters (variables) is gathered for each inverter. Several of these variables cannot be assumed to be independent. For example, preliminary results have shown that both large inverters and inverters installed outside a building have a shorter TTF than small indoor inverters. In addition, small inverters are more likely to be installed indoors, while large inverters are predominantly installed outdoors. Comparing the TTF of inverters installed indoors and outdoors therefore shows results that do not reflect causality.

The qualitative dependency of variables is determined by means of mosaics. As shown in Figure 1, protected outdoor inverters are by far the biggest category for large inverters (> 20 kVA), whereas the share of protected outdoor inverters is a relatively small category for small inverters (< 20 kVA). Figure 2 shows, that manufacturer m1 is mainly represented with recent inverter installations, whereas old installations are dominated by manufacturer m5 and to some extent m4.

In order to exclude the influence of dependent variables, the method proposed in [9] is applied. According to Terry M Therneau et al., data is adjusted for several confounding factors by a (inverse probability) reweighting of the observed inverters before survival curves are computed. For example, it is known that the place of installation (indoor vs. outdoor) depends on the confounders power and manufacturer of the inverter. To adjust for these two confounders, the observations from the comparison indoor vs. outdoor are reweighted such that the two comparing groups are balanced with respect to the confounding variables power and manufacturer of the inverter.

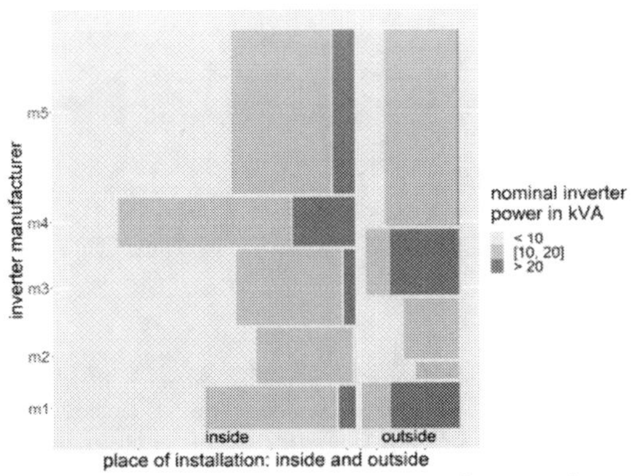

Figure 1: Mosaic showing inverter manufacturer, place of installation and inverter power.

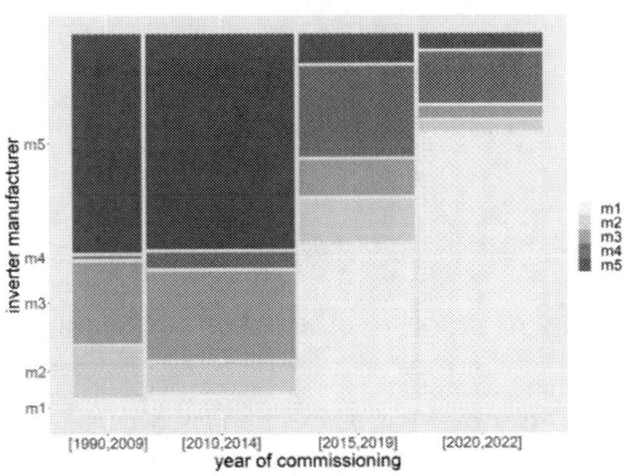

Figure 2: Mosaic showing inverter manufacturer, place of installation and inverter power.

3 RESULTS

The quantitative results of the survey and data gathering are presented using pie charts, histograms and other descriptive graphics. The TTF of the inverters are shown using Kaplan-Meier survival curves.

3.1 Description of the data

In the following figures (Figure 3 to Figure 10), the numbers of inverters are listed in different categories. Table I gives an overview of all data available in this paper. It shows the clear tendency, that data systems from professional owners have more inverters per system and higher power ratings than data from private owners and data from professional installers.

Table I: Data sources (abbreviations: sys. = system; inv. = inverter, priv. = private, prof. = professional)

Source of data	No. of sys.	Total power (kVA)	No. of inv.	No. of inv. per sys	Power per sys (kVA)	Power per inv. (kVA)
Priv. owner	343	6172	546	1.6	18	11.3
Prof. owner	83	7050	314	3.8	84.9	22.5
Prof. installer	769	13443	1261	1.6	17.5	10.7

The inverter manufacturers in this survey (Figure 3) are not representative of today's PV market. The manufacturers with most inverters analysed in this paper are Fronius, Huawei, Kostal, SMA, SolarEdge and Sputnik (alphabetic order, not corresponding to m1 to m5).

The inverter types are predominantly string and multi string inverters (Figure 4). Central inverters are underrepresented in a global context. In addition, several of the central inverters examined in this study are old, low-power models that are now usually replaced by transformerless multistring inverters.

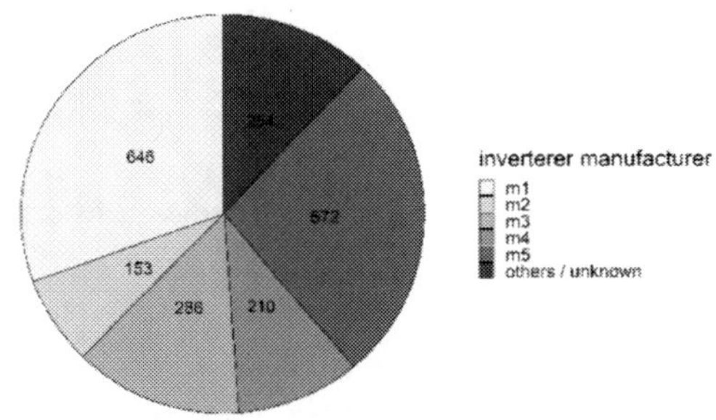

Figure 3: Number of inverters by manufacturer.

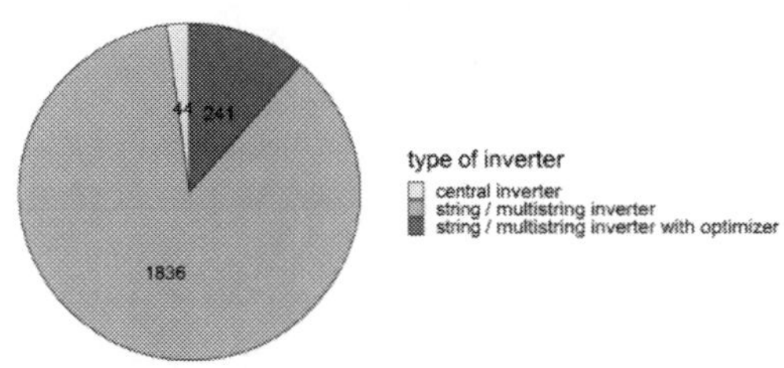

Figure 4: Number of inverters by type of inverter.

The oldest inverters investigated in this study are from the early 1990ies (**Figure 5**). However, most of the inverters were commissioned between 2008 and 2013 (first big installation boom in Switzerland due to the feed-in tariff introduced in 2009) and after 2017 (mainly driven by self-consumption).

As this paper is concerned with residential systems, most inverters have power ratings below 50 kVA (**Figure 6**). The biggest share of inverters has a power rating of 10 kVA to 15 kVA.

The distribution of Sizing Ratios SR (defined in this paper as AC power divided by DC power) is shown in **Figure 7**, peaking at roughly SR=0.9. Although it was assumed SR for new systems would change over time, the distribution stayed fairly uniform throughout the observed period.

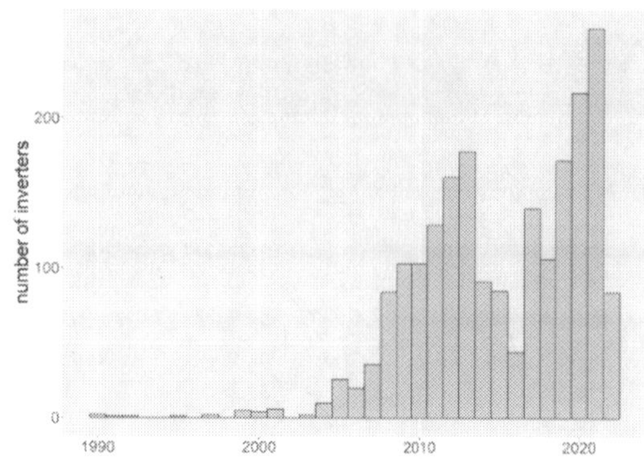

Figure 5: Number of inverters by commissioning year.

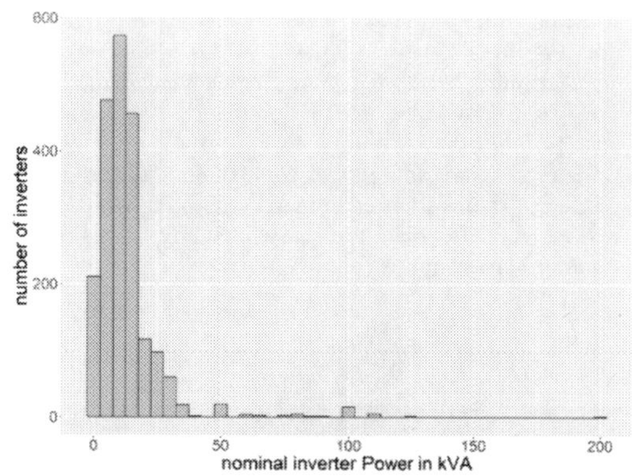

Figure 6: Number of inverters by nominal inverter power (AC power in kVA).

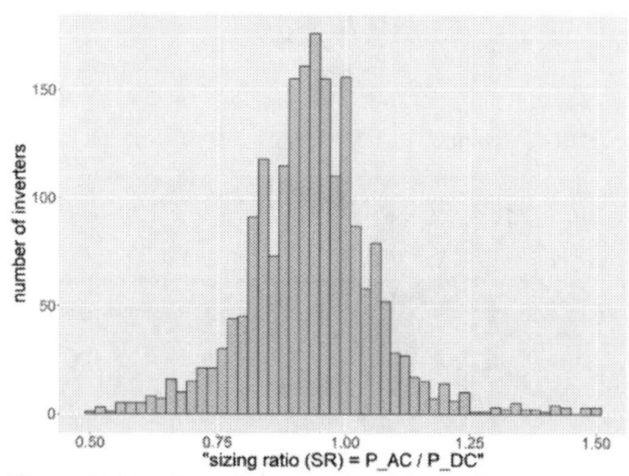

Figure 7: Number of inverters by sizing ratio.

The number of inverters by place of inverter installation is given in Figure 8. About half of the inverters are installed inside buildings. Only a minor share is installed outdoors, mostly protected from rain and direct sunshine. For a third of systems, the place of installation is unknown.

The tilt angles of the PV modules connected to the inverters are given in Figure 9. The type of the system (installation) is given in Figure 10. However, it was found that neither of these categories have a measurable impact on the TTF of the inverters.

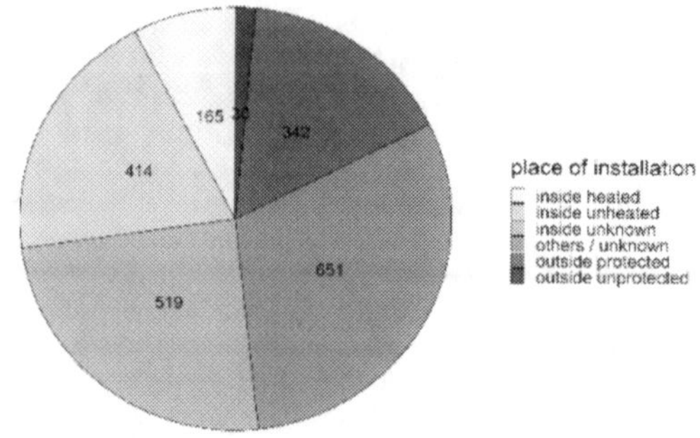

Figure 8: Number of inverters by place of installation.

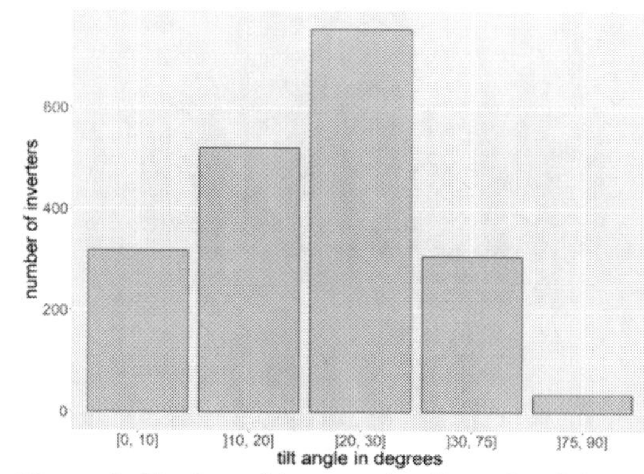

Figure 9: Number of inverters by tilt angle of the PV modules (typically corresponding to rooftop tilt angle).

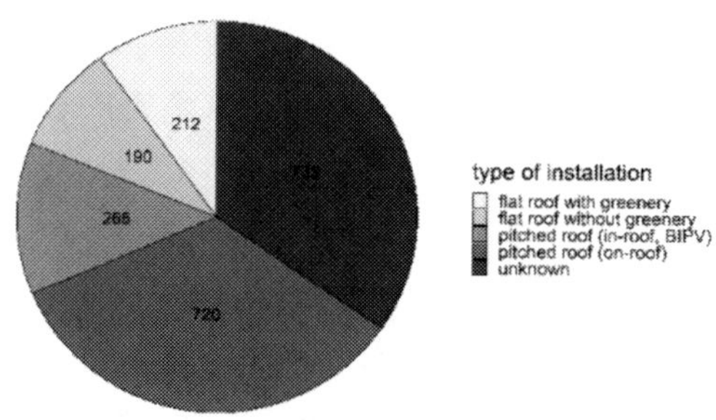

Figure 10: Number of inverters by type of installation.

3.2 Survival curves

The survival probability of all inverters is given in Figure 11. Because only a few inverters are older than 18 years, the survival curve is only plotted until year 18. In year 18, only 15 inverters were still in operation, resulting in a large confidence interval in the survival curve. After 18 years, the survival probability of all inverters is still 59.1% with a confidence interval of [52.1%, 67.0%].

Optimizers could only be observed during 8 years. Figure 12 shows the survival curve of inverters with and without optimizers. Throughout the observation time inverters with optimizers show roughly twice as many energy-relevant failures than inverters without optimizers.

For individual optimizers, the survival curve is relatively flat (Figure 13). However, as optimizers are normally used in larger quantities even for small systems, their TTF should rather be calculated for a whole fleet of optimizers than for a single device. Figure 14 shows the TTF after 5 and 10 years respectively for a given number of optimizers. Comparing the survival probability of a fleet of optimizers in Figure 14 with inverters without optimizers in Figure 12, a set of roughly 20 optimizers shows the same TTF after 5 and 10 years respectively as an inverter without optimizers. However, this comparison is only indicative, as optimizers cannot be operated without inverter.

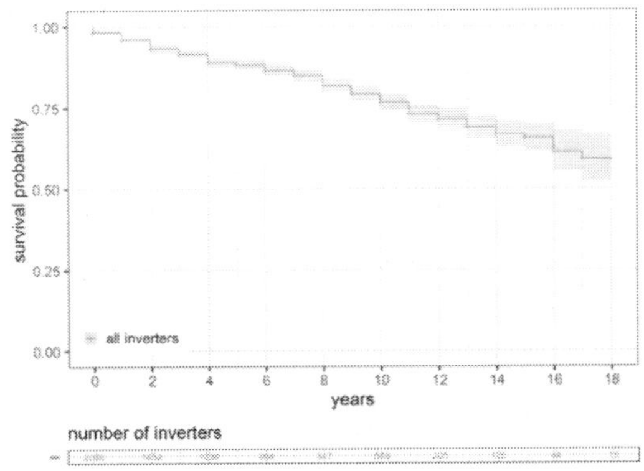

Figure 11: Survival probability of all inverters.

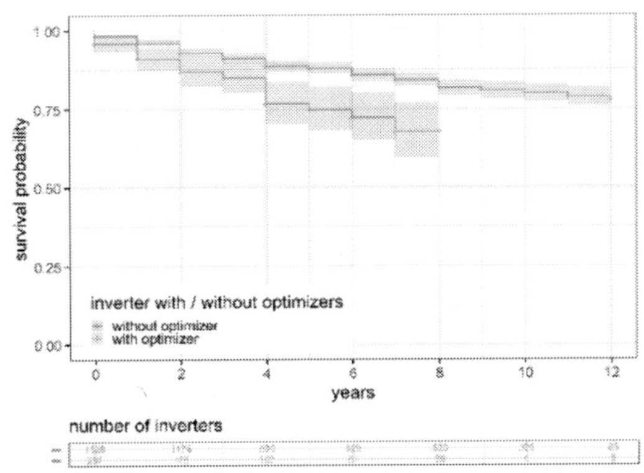

Figure 12: Survival probability for systems with and without optimizers.

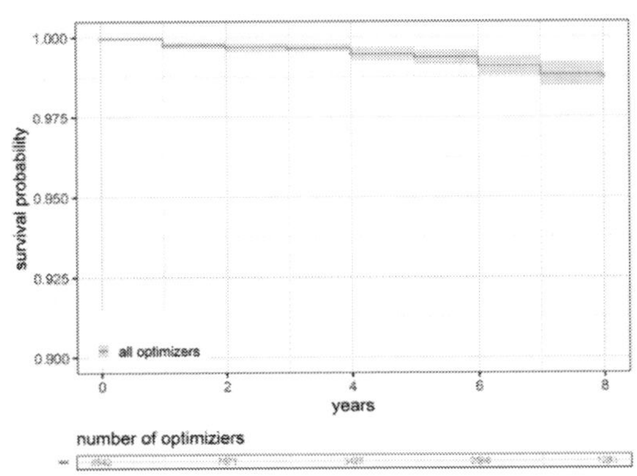

Figure 13: Survival probability of all optimizers (without inverter).

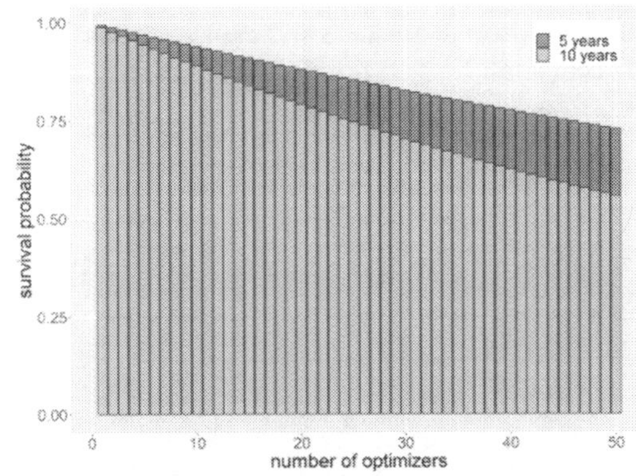

Figure 14: Survival probability after 5 and 10 years for a given number of optimizers (without inverter).

The survival curves of inverters split into three power categories are shown in Figure 15. A trend of more failures in large inverters can be observed. However, if these curves are adjusted for inverter manufacturer and place of installation (indoors/outdoors), this trend is not confirmed (Figure 16). This is most likely based on the fact that inverters installed outside a building have a lower TTF than inverters installed inside a building (Figure 17).

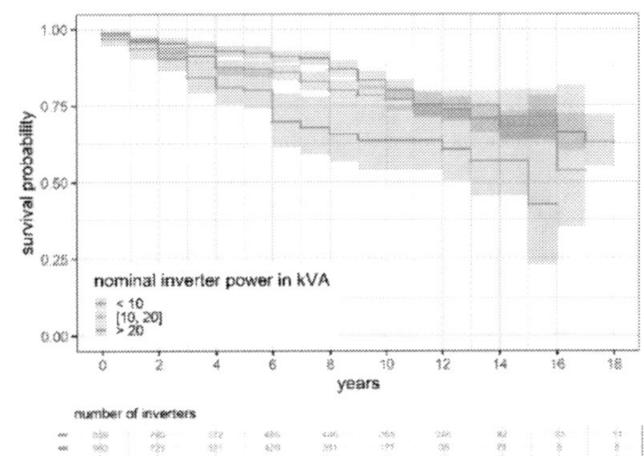

Figure 15: Survival probability for inverters of different power categories.

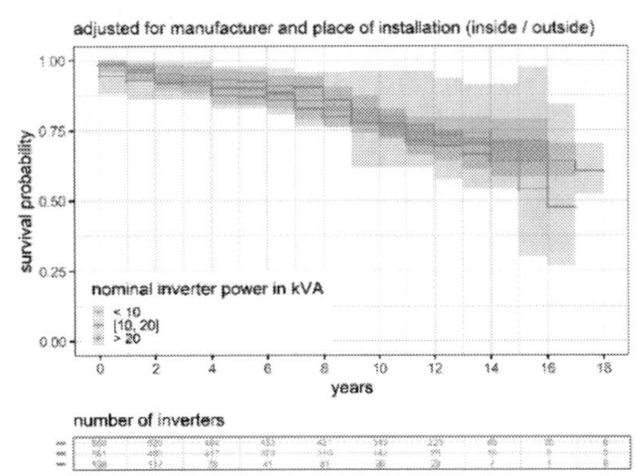

Figure 16: Survival probability for inverters of different power categories adjusted for manufacturer and place of installation.

Figure 17 shows the difference between the survival curves of inverters installed inside and outside a building. Because these variables are correlated to the inverter manufacturer and the nominal inverter power, an adjusted graph is shown in Figure 18. However, both figures show higher survival probabilities for systems installed inside a building. In the adjusted survival curve, this effect only becomes apparent after eight years.

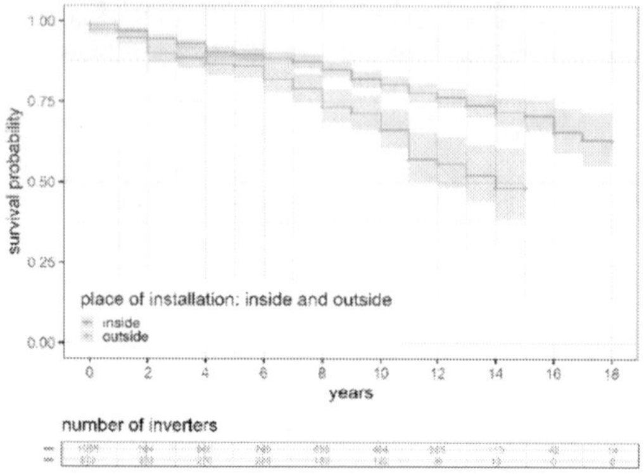

Figure 17: Survival probability of inverters installed inside and outside of a building.

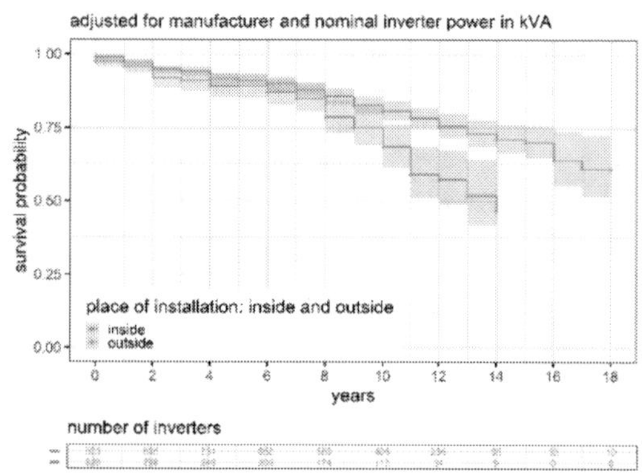

Figure 18: Survival probability of inverters installed inside and outside of a building, adjusted for manufacturer and inverter power.

Both inside and outside a building, inverters can face different situations. The survey asked whether installation rooms were heated or not (Figure 19). Outside the building, it was asked if the installation was protected against rain and direct sunlight (Figure 20). Both indoors and outdoors, stable conditions (heated room, protected installation site) seem to have a positive effect on the TTF of inverters. However, only little data is available to quantify these effects. The confidence intervals are overlapping in the outside case. Especially in this case, a clear conclusion therefore cannot be drawn from the available data.

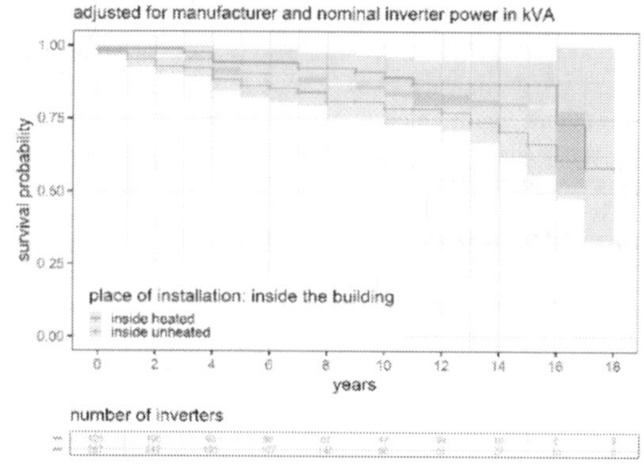

Figure 19: Survival probability of inverters installed inside the building.

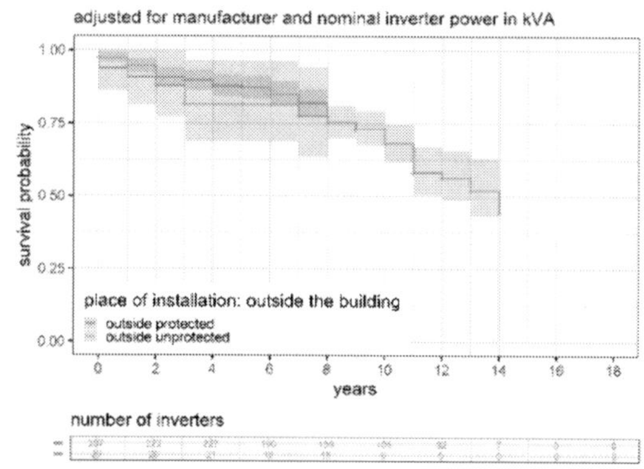

Figure 20: Survival probability of inverters installed outside the building.

No clear conclusion is possible from the survival curves of inverters installed with different sizing ratios (Figure 21). In the data-set, it is not the heavily loaded inverters with sizing ratios < 0.9 that have the lowest survival curve, but rather inverters with sizing ratios between 0.9 and 1.0. However, the confidence intervals overlap. This can indicate that too little data might be available to make a definite statement about the influence of the sizing ratio on the TTF. However, a clear conclusion is not possible.

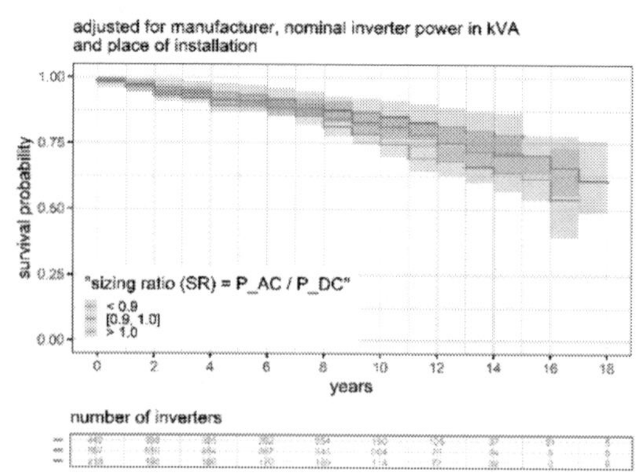

Figure 21: Survival probability of inverters with different sizing ratios.

The commissioning year has an influence on the survival probability of the inverters under investigation (Figure 22). Old inverters seem to have less energy-relevant failures in the beginning of their life span. The authors of this study cannot say whether this observation is biased by an effect not taken into account in this study, e.g. repair measures forgotten by the data providers.

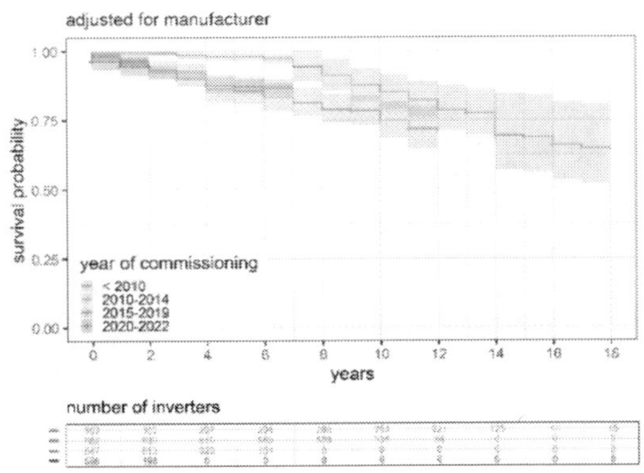

Figure 22: Survival probability of inverters with different commissioning years.

Finally, the survival probability of inverters from different manufacturers is shown in Figure 23. Although the confidence intervals of the curves are partly overlapping, the percentage of inverter failures observed varies by about a factor of three. The differences between the TTF varies thus remarkably between different manufacturers.

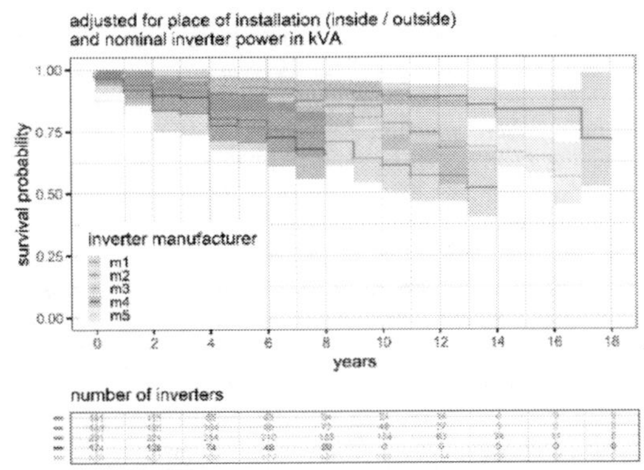

Figure 23: Survival probability of inverters from different manufacturers.

Several other questions have been asked in the survey to classify the PV systems, such as tilt angle of the roof or type of roof. However, the resulting data does not show relevant differences for such properties. Therefore, the corresponding survival curves are not shown in this paper.

4 DISCUSSION

4.1 Shortcomings of this study

Although the method chosen in this study allows the TTF of inverters to be quantified, there are several influences on the data that may negatively affect the results. Possible sources of error are discussed in the following sections.

4.2 Inverter as the system boundary

The relevant unit of this study is one inverter. For PV systems with optimizers, all optimizers are counted as part of one single inverter. Few systems in the scope of this study have many inverters with optimizers (>5); one case even had very many (>20) inverters with optimizers. In the case of the reported optimizer failures, it is not clear which inverter they were connected to. However, since the failures are assigned to one specific inverter in this study, it is unclear whether the reported defective optimizers are to be considered multiple failures on one inverter or many individual failures on several inverters. It was assumed that the optimizers with failures were randomly distributed within the PV systems. However, since the majority of the reported systems only involve one inverter, this error can be assumed to be insignificant.

4.3 Cause of failure unknown

The causes of the failures and defects are unknown. For example, it is not always possible to distinguish between a minor fault (such as the failure of a display) and a total loss (such as a full inverter breakdown including the failure of the display).

4.4 Rating of optimizer and inverter failures

The failure of an optimizer usually results in less yield loss than the failure of an inverter. However, in the survival curve (TTF), both failures are represented in the same way. If the failure of an optimizer triggers a maintenance service, the financial impact for the system owner is close to the financial impact of an inverter failure. However, if only the power of one module is lost and no maintenance is done, the impact of an optimizer failure is lower than that of an inverter failure. In the course of this study, the tendency was noted that individual defective optimizers were replaced in each case. Optimizers that are defective and have not been replaced were probably not detected within the scope of this study.

Under this consideration, however, some communication failures would also have to be evaluated as TTF as they can trigger maintenance. In this study, this is neglected, because the reason for communication errors is normally outside the PV system (e.g. poor mobile phone reception or security updates on the internet connection).

While inverters are usually repaired after the first fault occurs, even relatively young power optimizers with defects are not repaired but disposed of.

The number of optimizers resulting in the same TTF as one inverter is biased because the possible inverter failure is neglected in the case of optimizers. In reality, the TTF of inverters with optimizers would therefore be slightly lower than indicated in this study.

In general, the system comparison between plants with and without optimizers is critical because the monitoring and maintenance concept is different. Monitoring concepts with optimizers have the potential to detect more failures, which can trigger more maintenance activities, but also provide input for optimised maintenance concepts.

Inverter and optimizer technology has evolved greatly in the past. While the inverter industry is diverse and has decades of experience, optimizers are relatively new to the mass market and dominated by only a few manufacturers. Therefore, the learning curve of inverters can be assumed to be more advanced than the learning curve of optimizers. The results observed in this study might change with future inverter and optimizer generations.

4.5 Inverter replacement without failure

Especially in older systems, inverters are sometimes replaced pre-emptively without any immediate technical reason. In this study, this is counted as an inverter failure. This consideration is correct with regard to the economic calculation. However, it underestimates the life expectancy of the inverters.

4.6 Repair vs. end of life

It cannot be said without doubt whether an inverter failure indicates the end of the life of an inverter. Because installers and manufacturers often install second-hand replacement devices, even an inverter replacement does not necessarily mean the end of the inverter's service life. Due to this uncertainty, the TTF is defined as "time to first energy-relevant fault" in this study.

4.7 Non-representative data set

It can be assumed that the inverter brands and manufacturers that feature in the online survey are represented roughly proportionally to their market shares in German-speaking Switzerland. In the data from the institutional plant operators, however, individual products are naturally overrepresented. Because the inverters of different manufacturers have different TTFs, this leads to a distortion of the TTF of the entire inverter fleet.

It is not known whether one group of system operators felt particularly addressed by this survey. It is conceivable that system operators with defective systems would tend to fill out this survey more often than operators of systems without inverter defects. However, the comparison of the data from the online surveys with the data from institutional operators does not reveal a strong bias in this respect. On the other hand, it can be assumed that professional plant operators tend to pay more attention to quality and optimum operating conditions and therefore tend to have lower inverter failure rates.

Institutional system operators sometimes own systems that were predominantly built in certain time periods (for example, in the years 2008–2014 when Swiss feed-in tariffs, or FIT, were introduced). As a result, installations from these periods are overrepresented among these data providers. This leads to correlations in the statistics that have no causal link with the data providers.

4.8 Unknown initial failures

It is assumed that initial inverter failures are often not known to the building owner because they are either fixed by the installation companies or the inverters are replaced during installation. In the case of optimizers in particular, it must be assumed that initially defective devices were not reported in this study, as their replacement during plant commissioning causes only minor additional work and costs.

4.9 False information

It must be assumed that not all the information provided for this study was correct. The resulting errors cannot be quantified. However, from the personal comments made in the survey it can be assumed that most people correctly understood the questions. Obvious misunderstandings were corrected manually.

Certain errors are interpreted by the system operators as inverter errors, though fact they are not inverter errors. One example is earth faults in the module field, which lead to an error message on the inverter. As far as possible, these faults were masked as such and not evaluated within the scope of this study.

When optimizers are replaced in larger systems, it is not always clear how many optimizers were replaced. If no information is available, it is assumed that only one optimizer was replaced.

The data has shown some large (45 inverters, SR>1.5) and small (27 inverters, SR<0.5) sizing ratios. It is not known for sure if this data is realistic or not. However, it seems unrealistic and has therefore been deleted for the sizing ratio survival curve evaluation.

It was reported that many inverters were repaired in the years between 1992 and 2010 [8]. It is not clear if repairs of old inverters are always reported in this survey.

4.10 Misinterpretation due to correlated variables

As described in section 2.5, some survival curves are adjusted for possible confounders. However, not all possible influencing factors could be identified and eliminated. Therefore, it cannot be said for the graphs that the correlations found in the results correspond to the causes for the observed TTF.

5 FUTURE RESEARCH

The results of this study clearly show that the TTF of inverters depend on many different factors. It was possible to identify and quantify several of these factors. However, many questions remain open and many shortcomings could not be excluded (see chapter 4). Therefore, researchers are motivated to repeat this study and expand the data set to improve the quality and quantity of the results. The questionnaires should also be improved in order to reduce possible misinterpretations.

6 CONCLUSIONS

Data on inverter and power optimizer failures in PV systems were collected from PV system operators.

The inverters and optimizers in this study represent a broad, albeit not representative, cross-section of the inverters and optimizers used in PV systems on buildings in Switzerland and Europe. Inverters from the manufacturers Fronius, Sputnik and SMA are most widespread, while most optimizers are produced by SolarEdge. Compared to the European inverter market, the products from Sputnik (SolarMax) are probably overrepresented. In an international comparison, the European manufacturers are overrepresented.

The results across all inverters and power optimizers examined show that over 65% of inverters do not have a yield-relevant fault by their 15th year of operation.

Furthermore, the investigation shows that the TTF is dependent on various factors, in particular:

- Manufacturer: Different manufacturers have different TTF.
- Power: More powerful inverters have shorter TTFs; individually considered power optimizers have the longest TTFs. However, the overall reliability of a PV system increases significantly with the reduction in the number of power electronic components.
- Installation location: Outdoor power electronics have a shorter TTF than indoor power electronics.
- Topology: In PV systems with inverters with power optimizers, the first fault occurs earlier than in PV systems that do not have power optimizers.

Care should be taken not to overinterpret the results. Various factors play a role in the reliability of a PV system that are not taken into account in this study. It is also possible that certain influencing factors and correlations were not found within the scope of this study and were therefore wrongly neglected.

7 REFERENCES

[1] Peter Toggweiler, *Betriebskosten von Photovoltaikanlagen: Solarstromanlagen effizient betreiben.* [Online] Available: www.bundespublikationen.admin.ch.

[2] P. Hacke, S. Lokanath, P. Williams, A. Vasan, P. Sochor, G. TamizhMani, H. Shinohara, and S. Kurtz, "A status review of photovoltaic power conversion equipment reliability, safety, and quality assurance protocols," *Renewable and Sustainable Energy Reviews*, vol. 82, pp. 1097–1112, 2018.

[3] R. K. Gatla, S. S. Kshatri, P. Sridhar, D. S. Malleswararao, D. G. Kumar, A. S. Kumar, and J. Lu, "Impact of Mission Profile on Reliability of Grid-Connected Photovoltaic Inverter," *JESA*, vol. 55, no. 1, pp. 119–124, https://www.researchgate.net/profile/Ranjih-Gatla/publication/359465305_Impact_of_Mission_Profile_on_Reliability_of_Grid-Connected_Photovoltaic_Inverter/links/623f434a57084c718b6a0b85/Impact-of-Mission-Profile-on-Reliability-of-Grid-Connected-Photovoltaic-Inverter.pdf, 2022.

[4] A. Sangwongwanich, Y. Yang, D. Sera, and F. Blaabjerg, "Lifetime Evaluation of Grid-Connected PV Inverters Considering Panel Degradation Rates and Installation Sites," *IEEE Trans. Power Electron.*, vol. 33, no. 2, pp. 1225–1236, 2018.

[5] A. Sangwongwanich, Y. Yang, D. Sera, and F. Blaabjerg, "Impacts of PV array sizing on PV inverter lifetime and reliability," in *2017 IEEE Energy Conversion Congress and Exposition (ECCE)*, Cincinnati, OH, 102017, pp. 3830–3837.

[6] D. C. Jordan, B. Marion, C. Deline, T. Barnes, and M. Bolinger, "PV field reliability status—Analysis of 100 000 solar systems," *Prog Photovolt Res Appl*, vol. 28, no. 8, pp. 739–754, 2020.

[7] Magnus Herz et al., "Quantification of Technical Risks in PV Power Systems," Report IEA-PVPS T13-23:2021, IEA PVPS Task 13, 2021. [Online] Available: https://iea-pvps.org/wp-content/uploads/2021/11/Report-IEA%E2%80%93PVPS-T13-23_2021-Quantification-of-Technical-Risks-in-PV-Power-Systems_final.pdf.

[8] Prof. Dr. Heinrich Häberlin und Philipp Schärf, Ed., *Langzeitverhalten von PV-Anlagen über mehr als 15 Jahre*, 2010.

[9] Terry M Therneau, Cynthia S Crowson, Elizabeth J Atkinson, *Adjusted Survival Curves.* [Online] Available: https://cran.r-project.org/web/packages/survival/vignettes/adjcurve.pdf.

STUDY OF ABRASION RESISTANCE OF STRUCTURED GLASS

Cristina L. Pinto, Eugenia Zugasti, Jaione Bengoechea
National Renewable Energy Centre, CENER
C/Ciudad de la Innovación 7, 31621 Sarriguren (Navarra) Spain
Telephone: + 34 948 25 28 00
E-mail: cpinto@cener.com

ABSTRACT: In this work, the durability of a novel anti-reflective photovoltaic glass surface is going to be analysed. Subwavelength structures (SWSs) on glass surface have shown an enhanced broadband omnidirectional anti-reflective property, increasing the transmittance up to values higher than commercial PV glasses with anti-reflective coatings (ARC). Although some research has been done with respect to the anti-reflective and anti-soiling properties of these structured glasses, abrasion resistance measurements is still missing. To this aim, this study demonstrates the durability of these SWSs against dry abrasion. In this respect, abrasion produced by Nylon brushes with and without Arizona Test Dust A2 standard dust has been quantified. The reduction of the anti-reflectance porperty in function of brushing cycles (100, 200, 300 and 500) has been characterized. Additionally, the damage produced to the SWS has been visualized through SEM images.

Keywords: Glass, Subwavelength Structures, Abrasion Resistance, Anti-Reflectance

1 INTRODUCTION

For photovoltaic energy production solar light absorption in the active material is essential. However, several factors such as glass reflectance and soiling can decrease the amount of light arriving into the solar cell.

Reflectance is produced due to the refractive index mismatch between air and glass, where 4% of the light is reflected. To overcome this problem, nowadays, ARC are added to this first interface to increase the transmittance around 3% absolute [1]. Since, these ARCs must have a refractive index between glass' and air's, most of them are based on porous dielectric materials, such as silica (SiO_2), deposited by sol-gel [2]–[5]. However, the durability of these coatings is under development, in order to guarantee an average lifetime of at least 15 years [1].

On the other hand, soiling refers to the accumulation of dust, soil, organic materials, or other particles on PV module's front glass. When these dust particles settle on the glass, they cover the surface, reducing the amount of light that reaches the solar cell and decreasing the power output. In fact, in regions where the solar resource is more abundant, i. e. in the Sun Belt, normally the soiling rate also increases [6], [7]. Even though this soiling rate might depend on the location, environmental conditions and tilt angle of the module, in arid regions like deserts, the transmittance can be reduced up to 60% in a month [8]–[13]. Therefore, regular cleaning is mandatory to remove dust from the module surface. However, cleaning tools and air borne dust can scratch the front cover material, producing damage to the solar glass surfaces and to the ARC or even remove it completely from the glasses, decreasing the efficiency and the power generation of the plant.

To overcome these limitations, more resistant anti-reflective and anti-soiling surfaces are highly desired. With this aim, bioinspired graded refractive index surfaces have been recently developed on glass, which show an enhanced broadband omnidirectional anti-reflective property [14]. These structures on glass have the advantage of being robust, mechanical stable and durable due to their geometry morphology and homogeneity of materials, being quite durable against outdoor conditions [15]–[17]. Moreover, these structures are quite resistant to linear brush abrasion [18]. However, their durability against abrasion due to cleaning procedures is still missing.

For this reason, in this study, different abrasion tests are going to be performed to characterize the durability and resistance of random SWSs on glass. Two types of abrasive medium are going to be used: Nylon brushes and Nylon brushes with Arizona Test Dust A2. The characterization of the abrasion impact is going to be measured through spectral reflectance increase and SEM images.

2 MATERIALS AND METHODS

In this section, the materials and the methodology applied in the present work to quantify the damage produced by different types of abrasion materials will be presented.

2.1 Materials

Borofloat 33® glass samples (155 x 155 x 2 mm^3) have been structured with random SWSs by means of a single-step self-masking Reactive Ion Etching (RIE) process. SF_6 and Ar gasses have been used as active species to promote the chemical and physical reaction.

Two different abrasion materials have been used in this work.

- First, only a Nylon brush has been used. The brush block is 40 mm x 25 mm in area and 10 mm in thickness. The brush bristles consist of polyamide 612 (Nylon), which extend 18 mm from the brush block. The bristle tufts are staggered in a 5/4 pattern along the brush with a column diameter of 9 mm.
- Arizona (AZ) Test Dust A2 fine (defined in ISO 12103-1) has also been used as an abrasive medium. It is mostly composed of silica (69.0-77.0% of weight) and around 57% of particles are sized between 0.97 μm and 11 μm.

2.2 Abrasion procedure

In this work, the potential effect of abrasion due to a cleaning procedure was investigated on randomly structured glass. The procedure was set following the work carried by M. Zahid Khan [19] which is based on the standard IEC 62788-7-3 ED1[20]. A pressure of approximately 1500 Pa was applied to the glass by the brush and the test velocity of about 24 cm/s was defined on the abrasion machine (Figure 1). Different cycles were carried out (100, 200, 300 and 500) in the same glass sample, where one abrasion cycle consists of one straight forward and backward movement of the brush. For abrasion with AZ dust, 0.5 g of dust was deposited on glass sample before the abrasion (Figure 2), and it was re-deposited every 100 cycles.

As can be appreciated from Figure 3, glass samples were divided in different sections (guide-lines on non-structured surface) where the different test cycles were performed. The marked guide-lines in the glass are 7 cm long where the brush moves at constant velocity. The characterization was performed in this area.

After the abrasion and before the optical measurements, glass samples were cleaned in an ultra-sonic bath with water and PV cleaning soap for 30 minutes at 50 °C in order to remove soiling particles and contaminants from the surface.

Figure 1. General view of the abrasion test equipment.

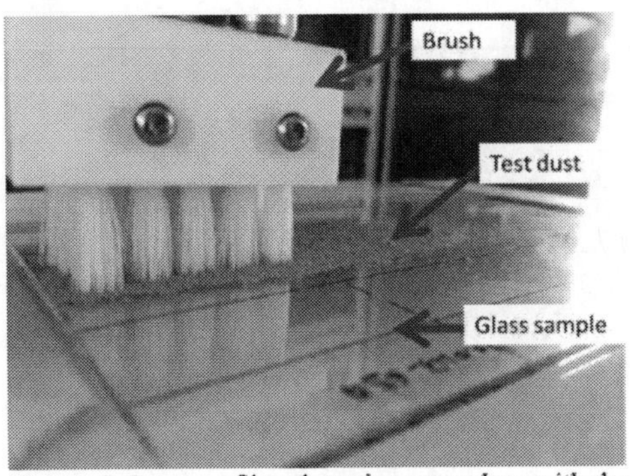

Figure 2. Images of brush testing procedure with dust.

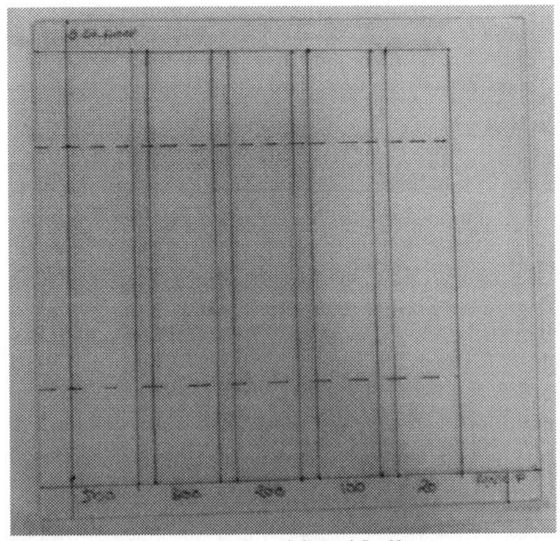

Figure 3. Glass sample with guide-lines.

2.3 Characterization equipment

Before and after the abrasion procedure, the spectral reflectance from 300 nm to 1600 nm was measured by means of a spectrophotometer (Instrument Systems) and integrated sphere.

In order to analyse the damage suffered by the nano-structures, Hitachi S-4700 Scanning Electron Microscope (SEM) was used to visualize the glass surface.

3 RESULTS

Glass is functionalized after structuring its surface by means of RIE process. It presents random nanocones in its surface (Figure 4a) which creates a graded refractive index showing an enhanced broadband anti-reflective property as can be shown in Figure 4b. According to the spectral reflectance measurements, structured glass presents less than 4.3% of reflectance averaged from 300 nm to 1100 nm, in comparison to flat glass, which presents 7.2% (absolute) reflectance averaged in the same spectral region.

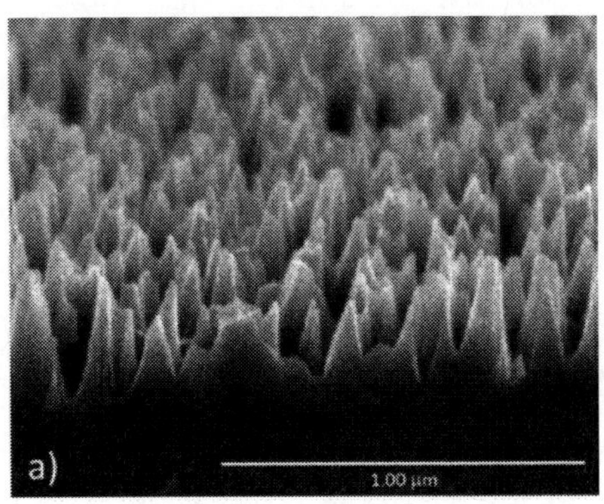

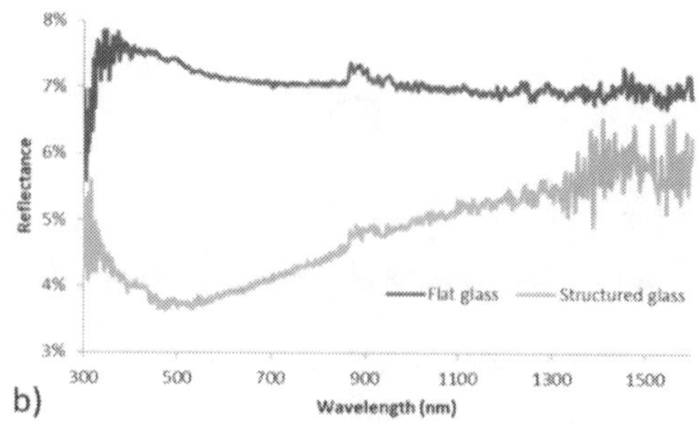

Figure 4. a) SEM images of the structured glass surface and b) spectral reflectance of flat and structured glass.

3.1 Abrasion Effect of Nylon Bristles without Dust

After the abrasion procedure performed by Nylon brushes, the spectral reflectance was measured obtaining very promising results. Figure 5a) shows the spectral reflectance of a flat glass, structured glass and structured glass after brush abrasion. The damage produced by the bristles is negligible, making any notable worsening in the anti-reflective property. Even after 500 cycles, the structured glass still presents a good optical behaviour, reflecting approximately 4.6% (absolute) in the spectral region from 300 nm to 1100 nm. Moreover, from SEM images (Figure 5b), it can be seen that the nano-structures present no appreciable damage. Based on these results, it can be concluded that these random SWS show a good resistance to Nylon brush abrasion.

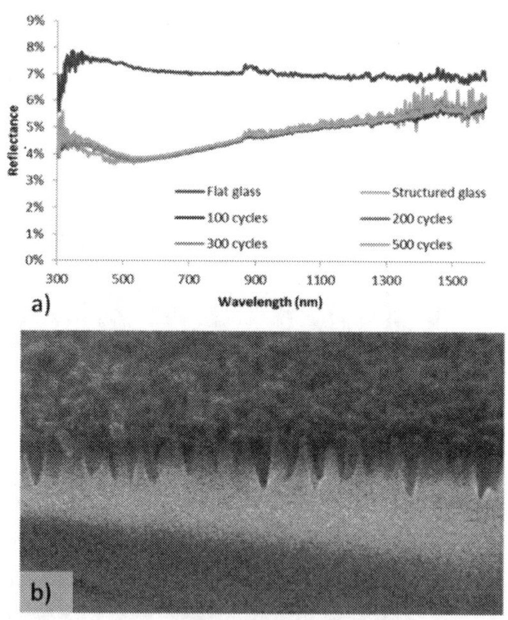

Figure 5. a) Spectral reflectance of tested glass with Nylon brush. b) SEM image of the sample after 500 cycles of abrasion test with Nylon brush.

3.2 Abrasion Effect of Nylon Bristles with Dust

After testing the abrasion resistance of glass samples without dust, Arizona Test Dust A2 was used to promote the sand abrasion test. In this case, 0.5 g of dust were deposited on the glass surface for every 100 cycles. Up to 500 cycles were performed, which will simulate the abrasion suffered in approximately 19 years of weekly cleaning frequency (Table 1).

Table 1. Approximation of cleaning damage scenarios [22]

	Cleaning frequency		
Test cycles	**Daily**	**Weekly**	**Monthly**
100	½ year	3.5-4 years	16-17 years
200	1 year	7-8 years	33 years
300	1.5 years	11-12 years	n. a.
400	2 years	15 years	n. a.
500	2.5-3 years	19 years	n. a.

After the test and the cleaning procedure, glass samples' spectral reflectance was measured. Figure 6a) shows the reflectance of a flat glass, structured glass and structured glass after several abrasion cycles. A worsening of 1% on the anti-reflective property can be observed with respect to the initial structured glass. As can be appreciated from this figure, the initial damage occurs early, after 100 cycles. Analysing the topography of the structured glass (Figure 6b) the abrasion caused by the dust and brush can be clearly seen, where most of the nano-structure were damaged. Nevertheless, the remaining structures still present enough anti-reflective property.

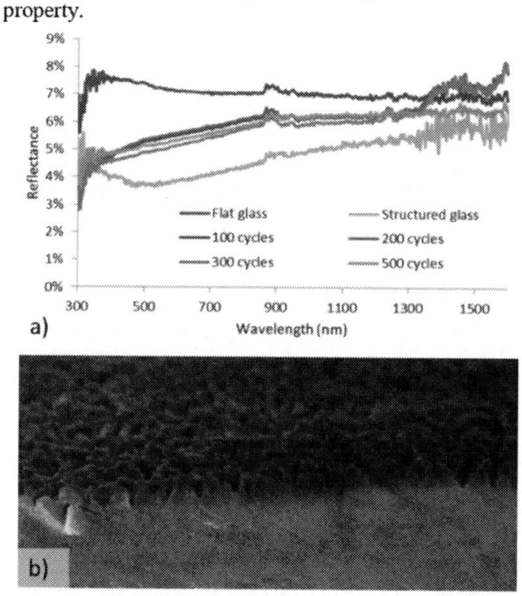

Figure 6. (a) Spectral reflectance of tested glass with AZ dust, (b) SEM image of the structured glass after 500 cycles of abrasion test with dust.

4 DISCUSSION

If we compare our results with the durability of commercial glasses, our structured glass present a reflectance of 5.6% averaged from 300 nm to 1600 nm. In comparison, commercial glasses, after suffering the same abrasion test [19],[22], their reflectance increases until almost 7.5%, averaged in the same spectral region for 500 abrasion cycles.

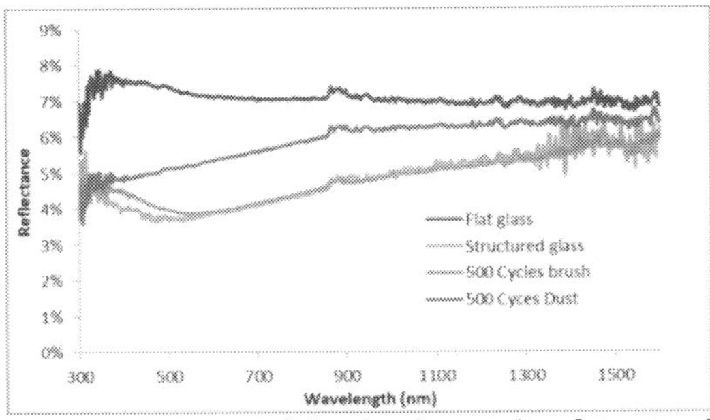

Figure 7. Summary of the abrasion results of structured glass, with and without abrasive dust.

5 CONCLUSIONS

Aiming towards a characterization of the resistance and durability of structured glasses, indoor abrasion tests have been performed. In this regard, 100, 200, 300 and 500 abrasive brush cycles have been applied to structured glass samples in order to simulate outdoor conditions related to cleaning procedures or sand storms. Nylon brushes with and without Arizona Test Dust A2 have been used as abrasive mediums following standard IEC 62788-7-3.

After tests, samples have been characterized by means of spectral reflectance increase. Interestingly, the tests show promising results, since after 500 cycles of Nylon brush abrasion, no noticeable decrease in the AR property was detected. Moreover, after 500 cycles of abrasion performed with dust particles, the structured glass still showed 5.6% of reflectance, representing a degradation of only 1% in the anti-reflective property. Besides, the status of the nano-structures has been verified by means of SEM images, showing a significant resistance.

Hence, the durability of random nano-structured glasses has been demonstrated in this work. It shows that randomly structured glasses can be used in PV industry in any region providing an enhanced anti-reflective and anti-soiling properties in addition to an increased resistance to abrasion.

ACKNOWLEDGEMENTS

Cristina Pinto gratefully acknowledges the Department of University, Innovation and Digital Transformation of the Government of Navarra for the grant for hiring doctoral students and doctoral students by companies, research centres, and technology centres: Industrial Doctorates 2020, with file number 0011-1408-2020-000003, received to carry out this study.

Our team want to thank Alicia Buceta for the cleaning process.

BIBLIOGRAPHY

[1] VDMA, "International Technology Roadmap for Photovoltaic (ITRPV) - 2020 Results," 2021. [Online]. Available: https://itrpv.vdma.org/documents/27094228/29066965/20210ITRPV/08ccda3a-585e-6a58-6afa-6c20e436cf41.

[2] D. B. Mahadik, R. V. Lakshmi, and H. C. Barshilia, "High performance single layer nano-porous antireflection coatings on glass by sol-gel process for solar energy applications," *Sol. Energy Mater. Sol. Cells*, vol. 140, pp. 61–68, Sep. 2015, doi: 10.1016/j.solmat.2015.03.023.

[3] D. Chen, "Anti-reflection (AR) coatings made by sol-gel processes: A review," *Sol. Energy Mater. Sol. Cells*, vol. 68, pp. 313–336, Mar. 2001, doi: 10.1016/S0927-0248(00)00365-2.

[4] H. K. Raut, V. A. Ganesh, A. S. Nair, and S. Ramakrishna, "Anti-reflective coatings: A critical, in-depth review," *Energy Environ. Sci.*, vol. 4, no. 10, pp. 3779–3804, 2011, doi: 10.1039/c1ee01297e.

[5] A. S. Sarkin, N. Ekren, and Ş. Sağlam, "A review of anti-reflection and self-cleaning coatings on photovoltaic panels," *Sol. Energy*, vol. 199, no. June 2019, pp. 63–73, 2020, doi: 10.1016/j.solener.2020.01.084.

[6] N. Cherupurakal, M. S. Mozumder, A. H. I. Mourad, and S. Lalwani, "Recent advances in superhydrophobic polymers for antireflective self-cleaning solar panels," *Renew. Sustain. Energy Rev.*, vol. 151, p. 111538, Nov. 2021, doi: 10.1016/J.RSER.2021.111538.

[7] H. Qasem, "Effect of accumulated dust on the performance of photovoltaic modules," Loughborough University, 2013.

[8] T. Sarver, A. Al-Qaraghuli, and L. L. Kazmerski, "A comprehensive review of the impact of dust on the use of solar energy: History, investigations, results, literature, and mitigation approaches," *Renew. Sustain. Energy Rev.*, vol. 22, pp. 698–733, 2013, doi: 10.1016/j.rser.2012.12.065.

[9] S. A. Sulaiman, A. K. Singh, M. M. M. Mokhtar, and M. A. Bou-Rabee, "Influence of dirt accumulation on performance of PV panels," *Energy Procedia*, vol. 50, pp. 50–56, 2014, doi: 10.1016/j.egypro.2014.06.006.

[10] F. Mejia, J. Kleissl, and J. L. Bosch, "The effect of dust on solar photovoltaic systems," *Energy Procedia*, vol. 49, pp. 2370–2376, 2014, doi: 10.1016/j.egypro.2014.03.251.

[11] M. Gostein, T. Duster, and C. Thuman, "Accurately measuring PV soiling losses with soiling station employing module power measurements," *2015 IEEE 42nd Photovolt. Spec. Conf. PVSC 2015*, no. November, pp. 3–7, 2015, doi: 10.1109/PVSC.2015.7355993.

[12] L. L. Kazmerski, M. Al Jardan, Y. Al Jnoobi, Y. Al Shaya, and J. J. John, "Ashes to ashes, dust to dust: Averting a potential showstopper for solar photovoltaics," *2014 IEEE 40th Photovolt. Spec. Conf. PVSC 2014*, pp. 187–192, 2014, doi: 10.1109/PVSC.2014.6925524.

[13] L. J. Simpson, S.-S. Jiang, C. Perkins, and H. Mountinho, "New Discoveries in Soiling of PV modules," 2018.

[14] C. L. Pinto, I. Cornago, A. Buceta, E. Zugasti, and J. Bengoechea, "Parametric analysis of random subwavelength structures with anti-reflective properties on glass applied to photovoltaics," *Sol. Energy Mater. Sol. Cells*, vol. 236, p. 111506, Mar. 2022, doi: 10.1016/J.SOLMAT.2021.111506.

[15] H. A. Macleod, *Thin-Film Optical Filters*, Third Edition. .

[16] A. Rao, R. Pillai, M. Mani, and P. Ramamurthy, "Influence of dust deposition on photovoltaic panel performance," *Energy Procedia*, vol. 54,

pp. 690–700, 2014, doi: 10.1016/j.egypro.2014.07.310.

[17] S. Chattopadhyay, Y. F. Huang, Y. J. Jen, A. Ganguly, K. H. Chen, and L. C. Chen, "Anti-reflecting and photonic nanostructures," *Mater. Sci. Eng. R Reports*, vol. 69, no. 1–3, pp. 1–35, 2010, doi: 10.1016/j.mser.2010.04.001.

[18] J. M. Newkirk *et al.*, "Artificial linear brush abrasion of coatings for photovoltaic module first-surfaces," *Sol. Energy Mater. Sol. Cells*, vol. 219, p. 110757, Jan. 2021, doi: 10.1016/J.SOLMAT.2020.110757.

[19] M. Z. Khan *et al.*, "Resilience of industrial PV module glass coatings to cleaning processes," *J. Renew. Sustain. Energy*, vol. 12, no. 5, 2020, doi: 10.1063/5.0024452.

[20] G. Kelly, "IEC 62788-7-3 ED1 Measurement procedures for materials used in photovoltaic modules," 2021.

[21] VDMA, *International Technology Roadmap for Photovoltaic*, 11th ed. 2020.

[22] K. Ilse, P.-T. Miclea, V. Naumann, and C. Hagendorf, "Cleaning resistance of glass coatings." Fraunhofer, 2018, [Online]. Available: https://www.fsolar.de/media/download/test-report-v403-2018-cleaning-resistance-of-glass-coatings.pdf.

PV MODULE SAND ABRASION TESTING

G. Mathiak, J.J. John, Nithin Sha, H. Alshanqiti, Afra S.P., Prashanth G. S.; Yogesh S., M. Weston, L. Pomares, V. Alberts
DEWA R&D Center, MBR Solar Park, Dubai (UAE),
Email: Gerhard.Mathiak@dewa.gov.ae

ABSTRACT: An increasing number of PV systems are being installed in deserts with hot-dry climates. Higher module temperatures, ultraviolet radiation, sandstorms and soiling by dust are the main challenges for special module design and module testing. In this work, the abrasion effect of blowing sand on glass samples and PV modules is investigated in the blowing sand test chamber at the MBR Solar Park in Dubai.

Currently there is no suitable standardized method for testing the abrasion resistance of PV modules and solar glass under sandstorm conditions. Qualification testing of PV modules for arid climate conditions must include a sand abrasion test and a dry-cleaning resistance test. Blowing sand tests were performed in a wind tunnel chamber according to IEC 60068-2-68Lc2, using different PV modules and glass samples with different coatings. The air temperature can be adjusted up to 70°C and wind speeds from 18 m/s to 30 m/s can be set. The sand concentration of 2.2 g/m³ is used for most PV module tests. The typical test duration for PV applications is 2 hours with normal exposure to the module vulnerable areas.

Coaxial microscopy, short-circuit current, and reflectance measurements were used to characterize PV modules with and without anti-reflective coating as well as glass samples before and after the sand abrasion tests.

Keywords: Qualification and testing, Sandstorm, Sand Twister, Sand abrasion, Blowing sand, Glass coating

1 INTRODUCTION

An important aspect for the use of photovoltaic modules and large PV systems in desert areas, beside the high intensity and irradiation, is the number of sunny days in per annum, which additionally leads to a large energy gain. This is a major advantage for the use of photovoltaic systems, as they are directly dependent on the levels of available solar energy. Another great advantage for desert regions is that the high space requirement of PV systems does not pose a problem. However, there are some environmental situations in desert areas that can have a negative effect on the efficiency of PV modules. In addition to dust and sand contamination (soiling), resistance to sand drifts and sandstorms, higher UV radiation, cleaning and higher temperatures are the main challenges for the PV module design and product qualification testing. Sandstorms can have a great influence on the glass surface of the modules and thus strongly influence their performance [1]. The kinetic energy of the sand particles accelerated by wind gusts and sandstorms is sufficient to cause damage to the glass surface of the modules.

In most cases, standardised ageing tests are also available for the evaluation of PV modules especially the front glasses at certain locations. To date, no accelerated ageing guideline has been developed for the degradation of PV modules due the impact of sandstorms, in order to realistically take into account performance losses caused by sand particle abrasion. In general, the type of stress on solid-state abrasion of materials is described by the properties of the erosive particles (size, shape, hardness and density) and the external conditions (impact velocity, angle of incidence and temperature). The numerous parameters that can occur simultaneously, which define the sandstorm erosion of materials in desert areas, show the complexity of this observation. The threat of erosion effects on front and rear side of PV modules leads to a growing expectation for a suitable test procedure. With the help of test procedures, plant operators could adapt and develop an optimized risk analysis for performance and cost calculations to the expected service life of PV modules.

Fig. 1: Photo of the Blowing sand test chamber at the DEWA R&D Center.

2 SAND STORMS IN UAE

Dubai regularly suffers from dust storms with limited visibility. These dust clouds consist of small particles (d<10μm) and come from the direction of Kuwait (Shamal winds) or Oman. Sand twister can also occur in the desert and they can kick up much larger particles. These sandstorms also have a strong impact on soiling and thus on the required cleaning frequency of the PV modules.

2.1 Sand and Dust Storm Simulation

Numerical weather prediction models are useful for characterizing dust source regions and ambient dust loads. This provides information on seasonal trends and informs site selection suitability for PV panels. Models give information on dust characteristics in areas where there are no observations or when remote sensing is not possible. Remote sensing of dust is difficult over desert surfaces due to high reflection or when clouds are present. The Weather Research and Forecast (WRF) Model is a numerical weather model with chemistry extensions to simulate dust emission and transport. A good example of the added value of model data is presented for a dust storm case that occurred from 13 to 15 August 2022. Dust was generated over Oman due to strong down drafts from a mature tropical low-pressure system. The dust was generated on 13 August around 15 UTC (19h00 local time) but was not visible in the satellite image due to cloud cover (Fig. 2a). However, dust is emitted in the model simulation, which is evident in the surface level PM10 concentration (Fig 2b). The dust was transported north across the Arabian Peninsula and reached the UAE on the 14th of August, where the model shows a good spatial alignment with the satellite dust detection (Fig. 2 c and d).

a.) MSG Dust RGB Composite (2022-08-13_15Z)

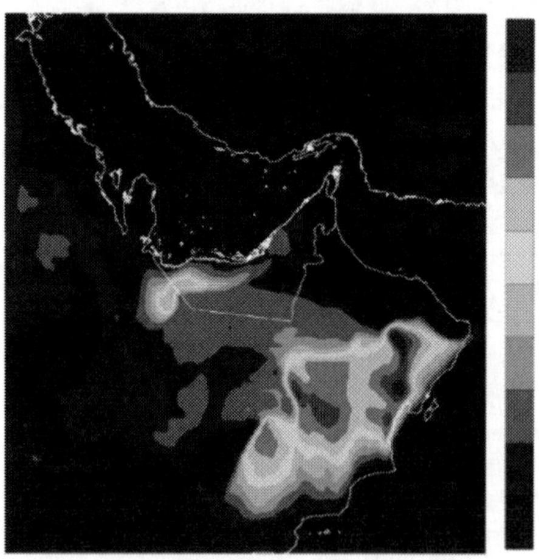

b.) WRF Surface Level PM10 (2022-08-13_15Z)

Fig. 2: a.) EUMETSAT Dust RGB composite product from MSG, a geostationary satellite. This product is used for night time scenes b.) Surface level PM10 (ug/m^3) from WRF-Chem

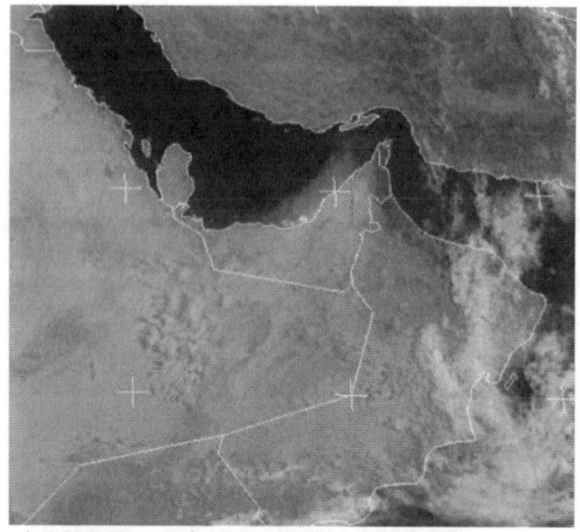

c.) MSG True Color Image (2022-08-14_06Z)

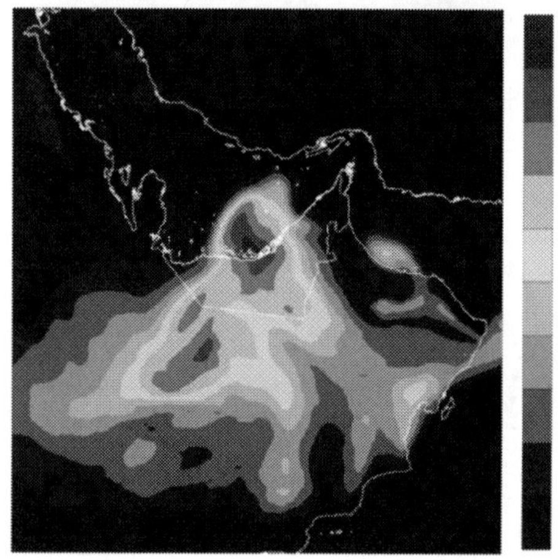

d.) WRF Surface Level PM10 (2022-08-14_06Z)

Fig. 2: c.) True color composite from MSG shows dust exiting the Peninsula over the Gulf. True color is used for day time scenes. d.) Surface level PM10 (ug/m^3) from WRF-Chem. Local time is UTC+4

2.2 Sand Twister and Dust Devils

Besides the sandstorms, there is another sand transport which can often be observed in solar parks in hot deserts. Sand twisters can arise when the ground is heated non-uniformly and if there is no strong wind. One twister can exist up to 1 hour and can lift sand up to several hundred meters high. The sand grains will fall down and can damage the modules.

Fig. 3a: Sand twister in the middle of the solar park.

Fig. 3b: Same sand twister, as in Fig. 3a, after leaving the solar park

3 Sand Abrasion Test Standards

For sand abrasion testing typically sand trickling tests and blowing sand tests are applied. In IEC 62788-7-3 [2], several test methods are described mainly for the investigation of glass samples without pass/fail criteria for PV modules.

3.1 Blowing sand

Blowing sand tests according to IEC 60068-2-68Lc2 [3] are performed in a closed wind tunnel chamber (see Fig. 4) with different PV modules and glass samples with different coatings. The air temperature can be adjusted between ambient and temperatures up to 70°C. During the test, wind speeds up to 30 m/s can be set. Winds with a speed lower as 18 m/s cannot cause the sand grains to become airborne. The sand concentration of 2.2 g/m³ is used for most PV module tests. The typical test duration for PV applications is 2 hours normal to the module vulnerable areas i.e., the front glass and backsheet. The tilt angle of the module in the DEWA sand test chamber is 30°.

The sand blowing test according to MIL STD 810G [4] was originally developed for environmental tests of military materials and components. The test method "510.5 Procedure II: Blowing Sand" uses as abrading sand silica sand with at least 95 % SiO_2 particles with grain diameters in the range 150µm to 850µm. The test is applied for materials likely to be used close to helicopters operating over unpaved surfaces. Typical test duration for PV applications is 1.5 hours per vulnerable area i.e., on the front glass and the backsheet.

Blowing Sand Test Parameters		
Maximum power of turbine	80	kW
Chamber Volume	40	m³
Test chamber height	1.00	m
Test chamber breadth	1.30	m
Test temperature	60	°C
PV module tilt angle	30	°
Sand concentration	2.2	g/m³
Humidity at test temperature	10	%
Wind speed	20	m/s
Test duration with blowing sand	4	hours
Consumption of sand (after cleaning of the chamber)	2.5	kg

Tab. 1: Test parameters of a typical Blowing sand test at DEWA R&D Center.

Fig. 4: View into the test chamber with a tilted framed PV Module, vertical glass sample, and sand catcher. One string was shaded by adhesive Al foil for measuring the effect of strong vibrations due to the high-speed wind in the chamber.

The chamber was prepared by measuring uniformity with a sand collector and glass samples at different positions. Sand concentration time sequences were conducted to minimize overshooting of the concentration and to optimize the test flow. Glass samples of different tilt angles were analyzed to verify the angle dependency of the effective test time:

(1) $t_{effective} = t_{test} \sin(\text{tilt angle})$.

In this unique chamber, it is possible to characterize the abrasion effect of a full-size PV module by the short-circuit current measured with a pulsed solar simulator. Due to the size of the exposition area of several other blowing sand abrasion test facilities, several tests have to stitch together to cover the full module.

Alternatively single cells in the module can be measured, either directly contacted through the backsheet foil or in a non-destructive way with partially shading of the cell under test [5], [6].

3.2 Sand trickling

The sand trickling tests according to DIN 52348 [7] and ASTM D968-05 [8] are usually used in the industry to investigate the abrasion resistance of glass and surface coatings. The DIN test setup consists of a funnel for the sand and a tube with two sieves to guide the free fall of the sand. The sieves homogenize the particle concentration over the exposition area. The final tube piece is 1.025 m high and has a diameter of 120 mm. The maximum speed of the free-falling sand particles is 4.4 m/s. Standardized natural silica sand (ASTM 20/30 sand) with 95 % SiO_2 and grain diameter of 0.5 mm to 0.7 mm is mainly used for this test. The ASTM-test setup has a thinner tube and no sieves.

The sand trickling test is simpler and faster than the blowing sand test. In the applied configuration the exposed area of the sand trickling test is smaller than the solar cell size.

An acceleration factor with regard to transmittance losses is the sand grain size distribution. A factor of 2 was determined for the change from the standard sand (d = 0.5-0.7mm) to the coarser sand (d = 0.7-0.9 mm) [9].

4 Results of Blowing Sand for Glass Samples

Several flat, low-iron glass samples (50mm x 50mm x 6mm) without anti-reflective coating were tested under different tilt angles and test durations. Small glass samples (d<10cm) cannot be thermally hardened so that the abrasion effect is different to tempered solar glasses of full-size PV modules.

The samples were tested at different tilt angles (30°, 60°, and 90°) for 2 hours.

4.1 Coaxial Microscopy

The effect of the tilt angle is significant (see Fig. 5). Small impacts and chipping (diameter about 100µm) can be seen at 30° and 60°tilt angle. At 90°, the complete surface is damaged.

A: 30° tilt angle

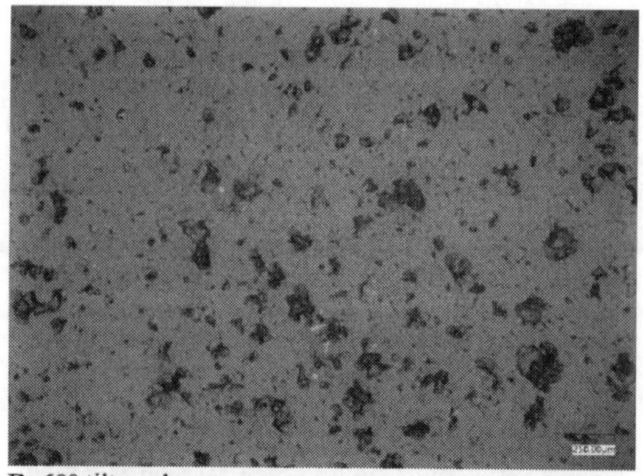

B: 60° tilt angle

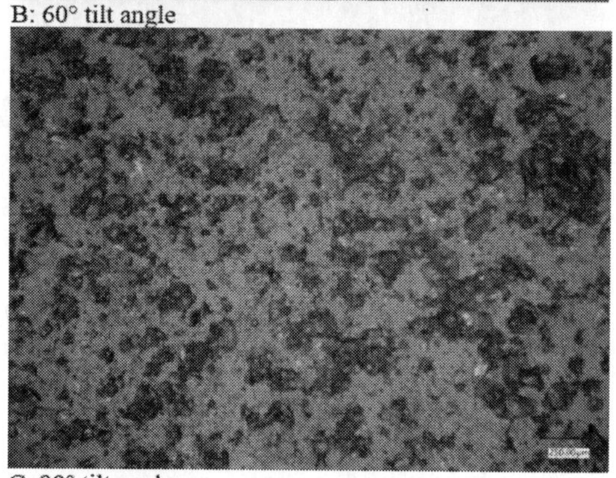

C: 90° tilt angle

Fig. 5: The glass samples without coating were tested at different tilt angles (30°, 60°, and 90°) for 2 hours. 100x magnification.

4.2 Reflectance

The hemispherical reflectance of the samples with different angles show an increase with the tilt angle (see Fig. 6) The reflectance of an untreated glass sample was about 8%. which is the reflectance of two glass surfaces

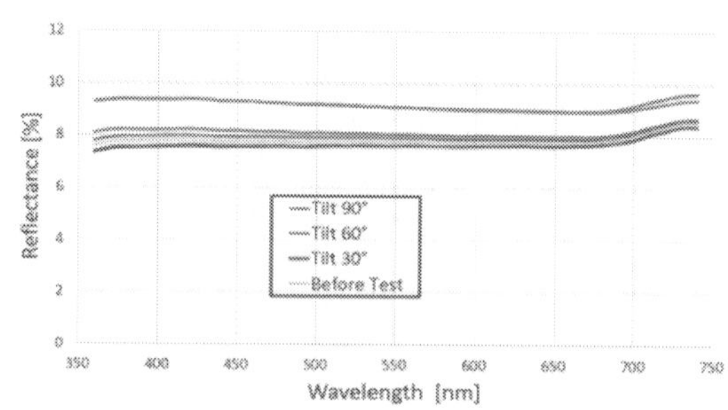

Fig. 6: The hemispherical reflectance of the samples was about 7.5% (30° tilt), 8% (60° tilt) and 9% (90° tilt). The reflectance of an untreated glass sample was about 7.8%.

Hemispherical reflectance measurements of glass samples with anti-reflective coating of different manufacturers can demonstrate a complete removal of the antireflective coating (see Fig. 4) up to 4% reflectance at undisturbed glass surface with n=1.5 [10]. To avoid the reflectance of the backside surface a light trap was used.

5 RESULTS OF BLOWING SAND FOR PV MODULES WITH ANTI-REFLECTIVE COATING

Two commercial PV modules with microstructured tempered glass and anti-reflective coating were tested at a tilt angle of 30°. The anti-reflective coating can be detected due to the different thicknesses in the hills and valleys of the microstructure. One module was covered with an Al-foil on a string to investigate the effects of module vibration due to high wind speeds. The other module was tested on uniformity by comparing the Isc value of the three strings.

5.1 Short-circuit Current

Isc measurements of the modules with and without shading by Al-foil showed a decrease in Isc with time and a small Isc loss of the covered string. The uncovered and heavily abraded middle string looks dull but has the highest hemispherical reflectance. The string-selected Isc measurements of the modules showed (see Tab 2a) and (see Tab. 2b)

String	Initial	2 hours		4hours	
Top	9.368	9.338	cover -0.32%		
Middle	9.403	9.299	-1.03%	9.227	-1.81%
Bottom	9.388	9.328	-0.63%		

Tab. 2a: Short-circuit currents and changes of the three strings of tested module with and without shading by Al foil. No effect due to vibrations in electroluminescence was detected.

String	Initial	4hours	
Top	9.509	9.415	-0.989%
Middle	9.607	9.366	-2.509%
Bottom	9.709	9.371	-3.481%

Tab. 2b: Short-circuit currents and changes of the three strings of tested module to check uniformity of abrasion effect on the three module strings. No effect due to vibrations in electroluminescence was detected.

5.2 Reflectance

The hemispherical reflectance of the samples with covered strings increases with test duration when the sample is uncovered.

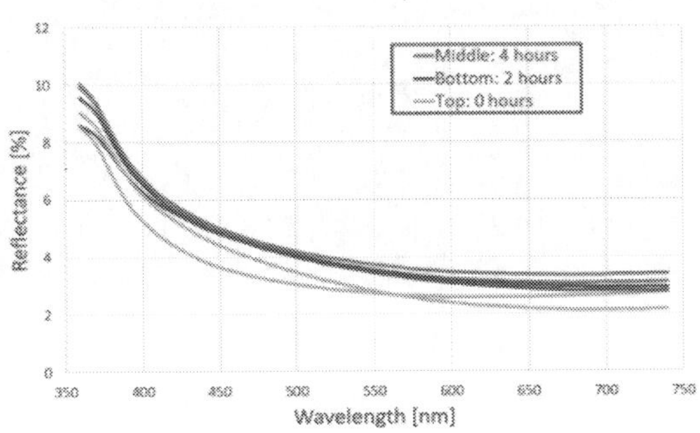

Fig. 7: Commercial c-Si module: Hemispherical reflectance on the cell after 0, 2, and 4 hours blowing sand test.

5.3 Coaxial Microscopy

The commercial module has a nanoporous silica coating on a patterned glass with hills and valleys. The thickness of the coating is different on the hills and the valleys, so that in the coaxial microscope the hills are brown and the valleys are blue (see Fig. 8).

Sand abrasion is caused by the formation of cracks, which develop almost parallel to the surface to split off. The lateral cracks are mainly responsible for the chipping of the glass material. Fig. 8 shows a microscopic image of a solar glass with ARC, which has been exposed to blowing sand. Here, the destructive effects of the sand impact on the glass surface and glass coating can be clearly seen.

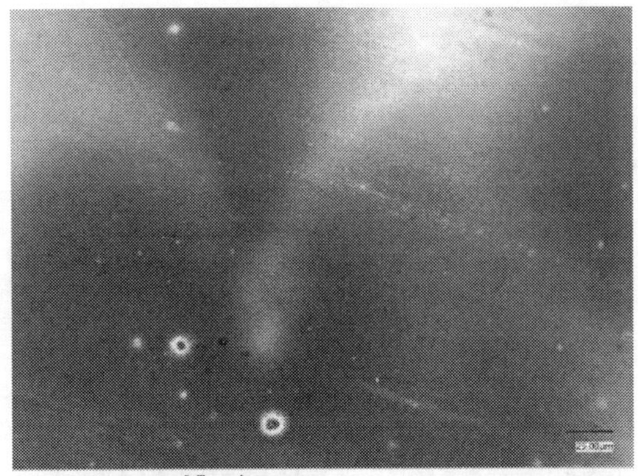

A: 1000x magnification

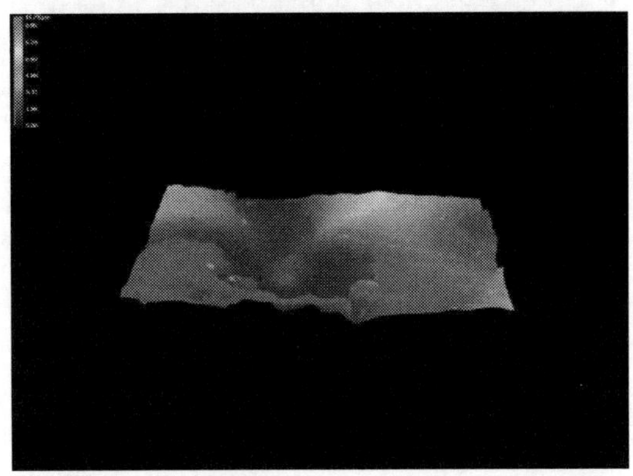

B: 3-dimensional image

Fig. 8: The thickness of the coating is different on the hills and the valleys, so that in the coaxial microscope the hills are brown and the valleys are blue

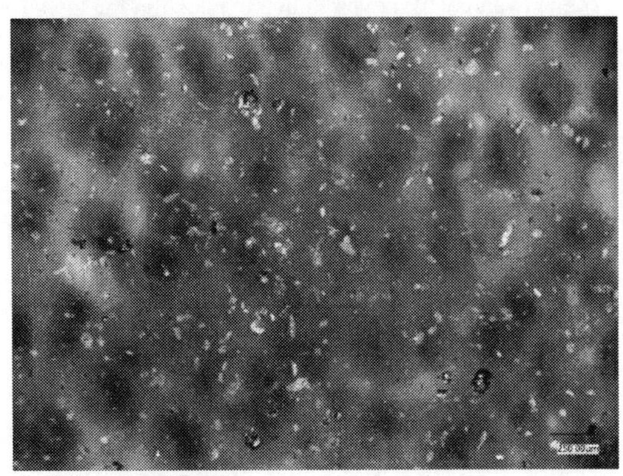

A: 100x magnification

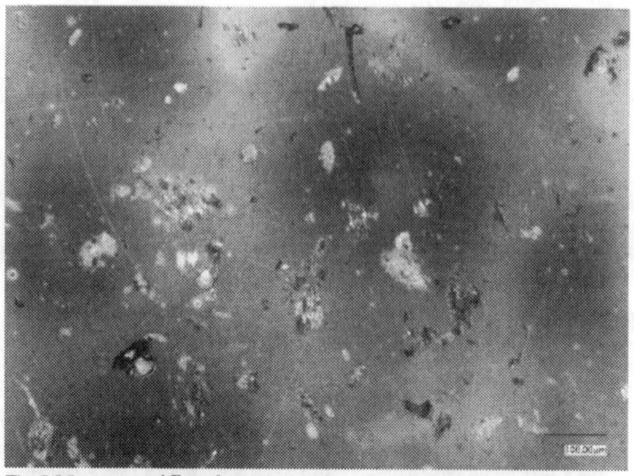

B: 300x magnification

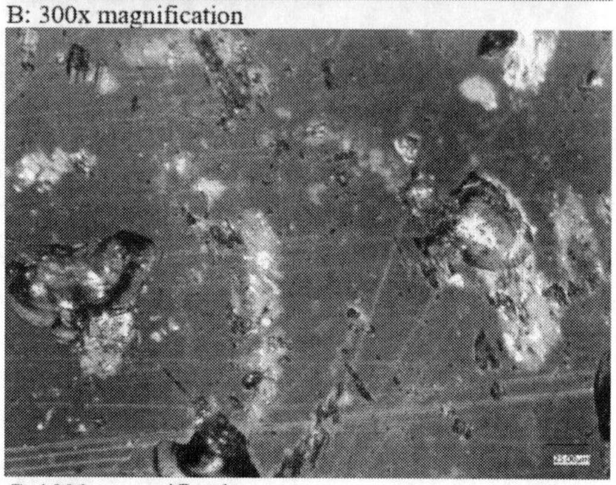

C: 1000x magnification

Fig. 9: Glass chipping and small impacts after a 4-hours blowing sand test on a structured glass with antireflective coating of a commercial module (30° tilt angle). Scratches caused by cleaning the module after the sand abrasion test.

6 Results of Blowing Sand for PV Modules without Anti-Reflective Coating

Two single cell modules without anti-reflective coating on the microstructured, tempered front glass were tested. Both modules were tested for four hours but the tilt angles were different: 25° and 90°.

6.1 *Short-Circuit Current*

The Isc value was measured before and after the test. The effect of the tilt angle was significant. Isc loss of samples mounted at 90° was more than 5 times higher than the samples mounted at 25° (See Tab. 3). Under the assumption of a linear effect and formula (1) a factor of 2 was expected.

Tilt angle	Initial	4hours	
25°	9.192	9.148	-0.478%
90°	9.191	8.949	-2.633%

Tab. 3: Short-circuit currents and changes of the of tested single cell modules at different tilt angles.

6.2 *Reflectance*

The hemispherical reflectance of the 90° sample was almost 6% and that of the 25° sample was almost 5%. The reference samples before the tests were displayed reflectance values of almost 4.5%. In the UV range the reflectance in comparison to the pure glass is increased.

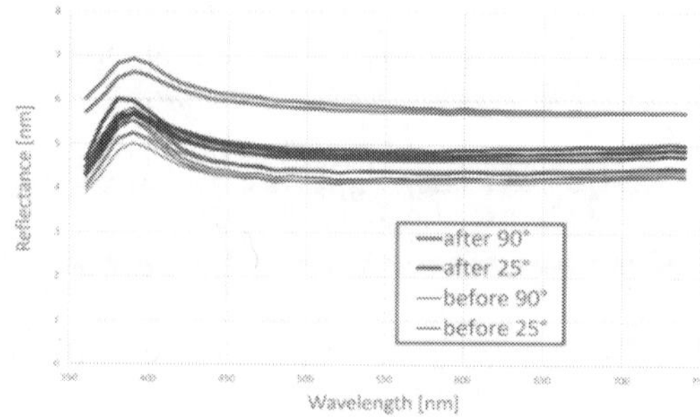

Fig. 10: Glass chipping and small impacts after a 4-hours blowing

6.3 *Coaxial Microscopy*

The hardened glass shows a valley structure (Fig. 11). At 25° tilt angle small impacts can be seen (Fig. 12). The abrasion effect is clearly to see at 90° tilt angle with strong chipping (Fig. 13).

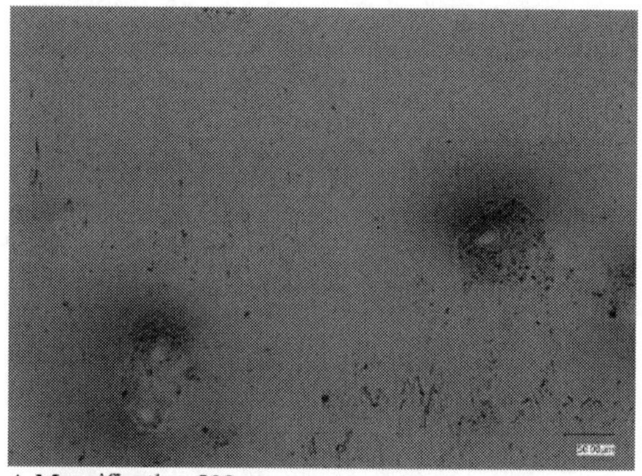

A Magnification 500x

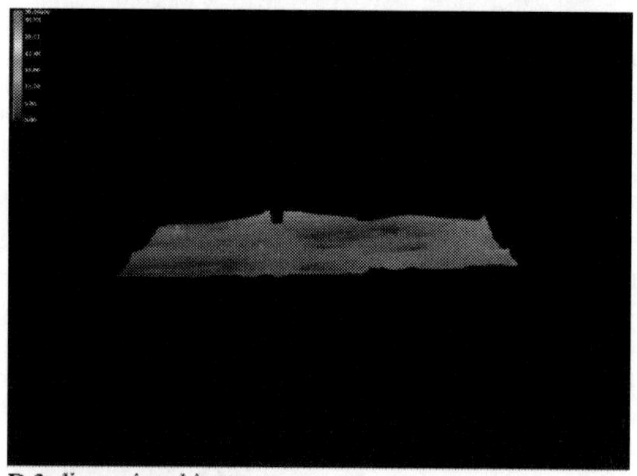

B 3-dimensional image

Fig. 11: Single cell module before testing.

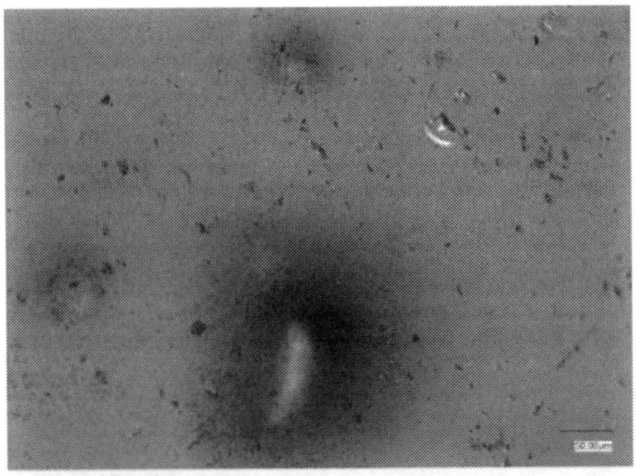

A Magnification 500x

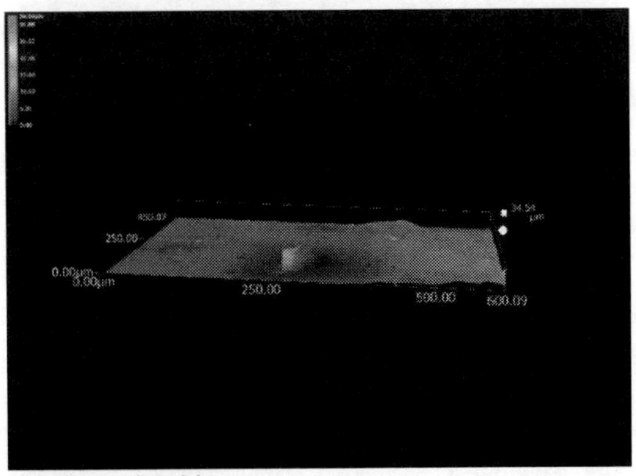

B 3-dimensional image

Fig. 12: Single cell module after a 4-hours blowing sand test with 25° tilt angle. Small impacts. Small bump.

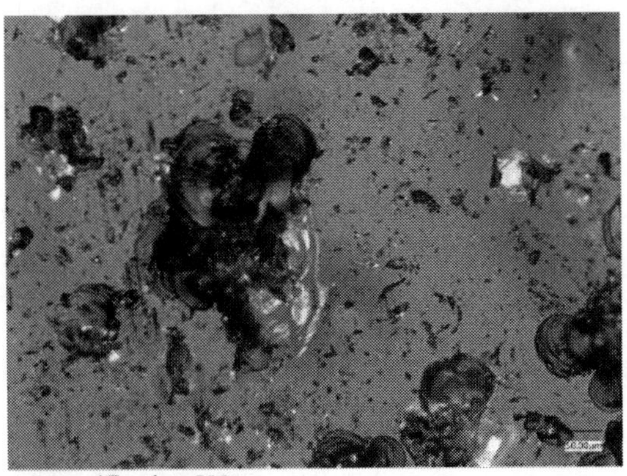

A Magnification 500x

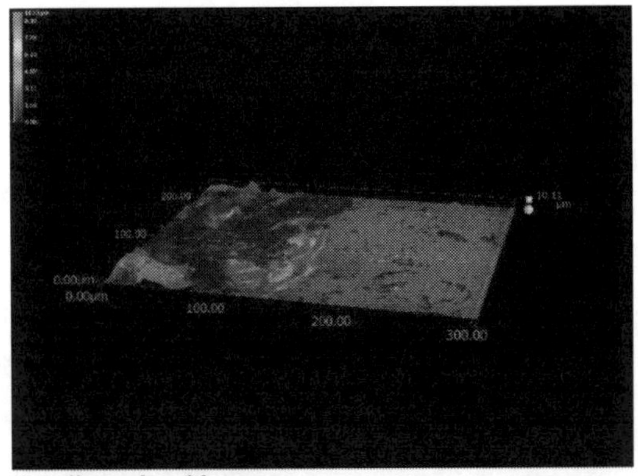

B 3-dimensional image

Fig. 13: Single cell module after a 4-hours blowing sand test with 90° tilt angle. Strong chipping with rainbow patterns.

7 Conclusions

7.1 *Status*

Using the unique blowing sand abrasion testing facility at DEWA R&D Center, PV modules and solar glass samples were tested according to IEC 62788-7-3 and IEC 60068-2-68 Lc2.

- The glass samples were analyzed by reflectance measurements and coaxial microscopy.
- The PV modules were analyzed by short-circuit current measurements, reflectance, and coaxial microscopy.

7.2 *Future Work*

- Experimental

The blowing sand test setup will be optimized to improve repeatability and reproducibility. Sand trickling experiments will also be carried out. Different sand types (grain size and chemical composition) will be tested. The toolbox for analysis is be extended to include other techniques (e.g.: EQE and roughness measurements).

- Standard

In a comprehensive standard test for arid climates, it is necessary to provide a sand abrasion test for PV modules. A new standard for the sand abrasion test of solar glass is currently published as a part of IEC 62788-7-3. In this standard, small glass samples are tested. Differences in the tempering of the glass samples and the full-size glass sheets can lead to differences in the test results. Verification measurements with full-size modules and an interlaboratory test with different glass samples will be performed. The goal is to develop a standard with pass/fail criteria for PV modules.

- Outdoor verification

Suitable test parameters for standard test procedures need to be derived from real weather data and reported damage to PV modules in the field. More data must be collected from PV modules in the field and from laboratory tests to define pass/fail-criteria on a scientific basis. This concerns the transmittance of the front glass with durable coating and the safety aspects for backsheet insulation. Sand twister (dust devils) are common in solar parks and can be a cause for accelerated sand abrasion in the field.

8 REFERENCES

[1] J. P. Thornton: The Effect of Sandstorms on PV Arrays and Components, NREL/TP-411-4787, 1992.

[2] IEC 62788-7-3:2022: Measurement procedures for materials used in photovoltaic modules - Part 7-3: Accelerated stress tests – Methods of abrasion of PV module external surfaces

[3] IEC 60068-2-68:1994 Environmental testing - Part 2-68: Tests - Test Lc: Dust and sand.

[4] MIL STD 810G: Department of Defense Test Method Standard: Environmental Engineering Considerations and Laboratory Tests, 2008. – Method 510.5, Proc. II: Blowing Sand.

[5] G. Mathiak, M. Hansen, M. Schweiger, L. Rimmelspacher, W. Herrmann, J. Althaus, F. Reil: „PV Module Test for Arid Climates including Sand Storm and Dust Testing" Proc. 32nd European Photovoltaic Solar Energy Conference and Exhibition, 2016, 5BO.11.4, 1568-1572.

[6] W. Herrmann, M. Schweiger, L. Rimmelspacher: "Solar simulator Measurement Procedures for Determination of the Angular Characteristic of PV Modules" Proc. 29th EU PVSEC 2014, 5CO.16.2, 2403-2406.

[7] DIN 52348: 1985:02: Testing of glass and plastics, abrasion test, sand trickling method.

[8] ASTM D 968-05:2010: Standard Test Methods for Abrasion Resistance of Organic Coatings by Falling Abrasive.

[9] G. Mathiak, N. Pfeiffer, L. Rimmelspacher, W. Herrmann, J. Althaus, F. Reil, C. Holze, A. Morlier: "PV Module Sand Abrasion Testing", Proc. 33rd EU PVSEC, Amsterdam, 2017, 5CO.5.2, 1383-1390.

[10] G. Mathiak, D. Grimm, L. Rimmelspacher, W. Herrmann, F. Reil, J. Althaus, A. Morlier: PV Module and Solar Glass Trickling Sand Proc.35th EU PVSEC, 2018, 5DO.7.2, 1092-1096.

AN EXPLORATORY STUDY ON THE ROLE OF SAND EROSION ON THE DEGRADATION OF WET INSULATION RESISTANCE OF PHOTOVOLTAIC MODULES IN A DESERT-LIKE ENVIRONMENT

Umang Desai[a,b], Aparna Singh[a,*]
a: Department of Metallurgical Engineering and Materials Science, Indian Institute of Technology Bombay, Mumbai–400076, India.
b: National Centre of Photovoltaics Research and Education, Indian Institute of Technology Bombay, Mumbai–400076, India.
*: Corresponding author (A. Singh: Aparna_s@iitb.ac.in)

ABSTRACT: Multi-layered polymeric backsheets in PV modules act as the first line of defense against both environmental and mechanical stressors. Continued exposure of the field stressors during the lifespan of the PV modules can promote degradation of its components, including the backsheets. Degradation of backsheets poses a life hazard since it increases the risk of electrification at high voltages. Desert-storm like conditions can potentially erode the outer layer of the backsheets thereby affecting the insulation capability of the backsheets. Therefore, in this work, we have prepared mini modules of size 5 cm X 5 cm having polyvinylidene fluoride (PVDF) based backsheet. These laminates have been subjected to erosion by sand particles for 30 minutes and 10 m/s velocity at an angle of attack of 30°. It is observed that the erosion of the outer layer of the backsheets affect the wet-insulation resistance of the mini-modules when backsheet surface is in wet condition. However, no significant changes after the sand erosion have been observed in the insulation resistance for the dry condition and when the glass is wet. The findings of this study give merit to the sand erosion process of the PV backsheets as an important field stressor.

Keywords: backsheets, sand erosion, insulation resistance, PV reliability

1 INTRODUCTION

Photovoltaic installations in arid regions like the deserts is economically attractive because of low cost of the land. However, the climatic conditions of the deserts are generally harsh that involve high ultraviolet radiation (UV) and diurnal temperature cycling. These factors can accelerate the degradation of the outer layer of the backsheets and may promote yellowing, cracking and complete separation at some spots in the modules[1]–[3]. Furthermore, the photovoltaic installations in the arid regions are also prone to erosion of the outer layer of backsheets during the desert-storm like conditions. Previous investigations, carried out on the backsheets, have unraveled that erosion of the outer layer of BS can degrade its volume resistivity because of sand erosion[4]. However, the effect of the sand erosion on the dry and wet insulation resistance of PV modules has not been reported yet in the literature.

Therefore, in this study we have explored the degradation of the dry and wet-insulation resistance of PV modules with polyvinylidene fluoride (PVDF) based backsheet due to erosion by the sand particles sourced from desert of Rajasthan, India. The backsheet samples have been subjected to erosion at 30° angle with a flow velocity of 10 m/s for 30 minutes to simulate desert storm like condition using a custom-made setup. It has been observed that dry-insulation resistance of the mini-modules due to erosion by desert sand particles is not affected. On the other hand, for the wet condition, the insulation resistance is only found to affect when the BS is wet.

2 MATERIALS USED

Mini-modules have been prepared having the dimensions of 5 cm X 5 cm and the architecture of glass/EVA/cell/EVA/BS. Annealed glass having 3.2 mm thickness has been procured from Gujarat Borosil Ltd., India, which have been cut into 5 cm X 5 cm pieces. Backsheet having polyvinylidene fluoride (PVDF) as the outer layer has been selected in this study owing to its popularity in the PV industry. Renewsys India Pvt. Ltd. provided industrial-grade EVA films. The laminates of glass/EVA/cell/EVA/BS have been fabricated using AWS laminator at 150°C under vacuum. Copper based adhesive tape has been used to emulate the frame of the mini-modules (Fig.1)

Fig. 1: Mini-module having architecture of glass/EVA/cell/EVA/BS and dimentions of 5 cm X 5 cm having copper based adhesive tape to emulate the frame.

The sand has been procured from the desert of Rajasthan, India and has been found to be angular (Fig. 2) by using the scanning electron microscopy (SEM). The mineralogical content has been determined through the energy dispersive spectroscopy (EDX) and showed the presence of silicates and calcites.

Fig.2: SEM image of the sand particles used for the sand erosion experiments[4]

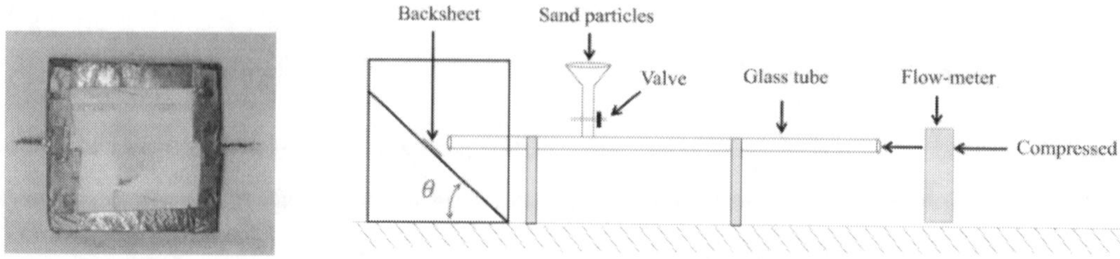

(a) (b)

Fig. 3: (a) backsheet of a mini-module after sand erosion, (b) Schematic of the set-up used for the erosion of the backsheets.

3 EXPERIMENTS

The pristine mini-module samples have been characterized for the dry and wet insulation resistance. These pristine mini-modules have been subsequently subjected to erosion by the particles of desert sand. The dry and wet insulation resistances have been measured for the samples post sand erosion. The details have been presented in this section.

3.1 Insulation resistance measurement

The insulation resistance test is done to determine the ability of the packaging materials to insulate the current carrying parts of the modules from the frame. The insulation test in this work is carried out at 500 DC voltage with electrification time of 60 seconds. The measurement involves short circuiting the two terminals of the cells and maintaining at positive terminal while the frame is maintained at the negative terminal (Fig. 4 a)

Furthermore, the wet-IR has also been measured for the mini-modules using a wet lint-free cloth. The wet cloth has first been put on the glass and BS side to measure the wet IR for the mini-modules. The voltage and electrification time have been kept as 500 V DC and 60 seconds, respectively.

3.2 Sand erosion experiments

The erosion process of PVDF based backsheets has been studied using desert sand particles of Rajasthan, India. The experiments have been carried out using a 10 mm inner diameter glass tube, which allows the flow of compressed air and has a provision to allow the sand particles to flow along the stream of the compressed air (as shown in Fig. 3). For all the erosion experiments, the backsheets were allowed to be impinged by sand particles for 30 min. The target surface of the backsheets was placed on a wooden plank at an inclination angle of 30° to the stream of compressed air. 30° is chosen as an angle of attack for erosion experiments since it emulates the condition of latitude-tilt type mounting for the PV modules installed Rajasthan, India.

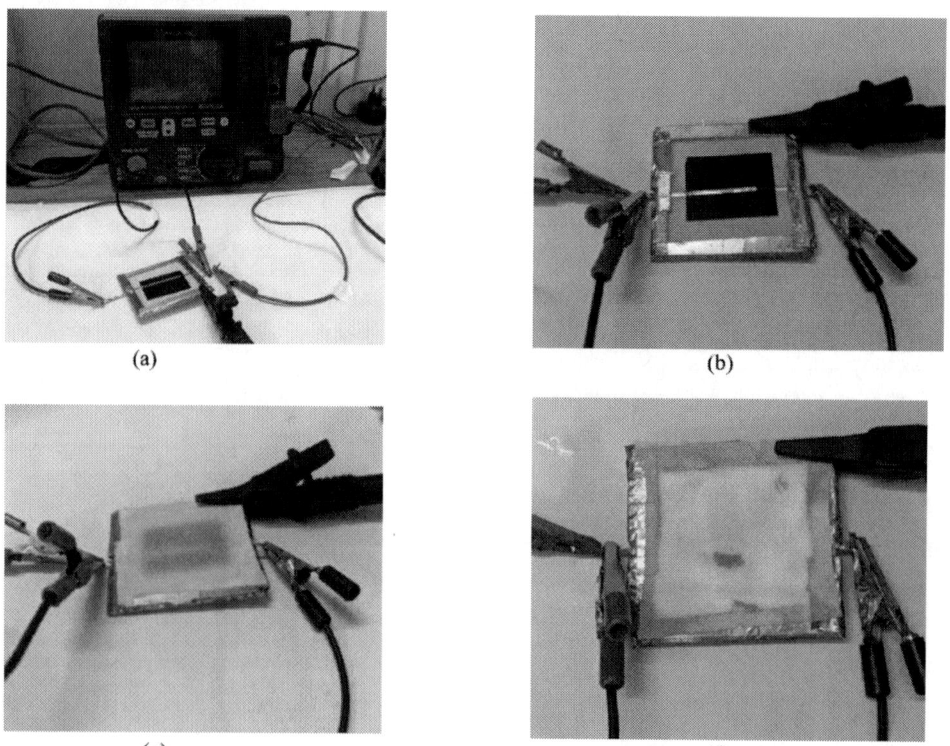

(a) (b) (c) (d)

Fig. 4: (a) testing set-up, (b) Dry insulation resistance measurement, (c) Insulation resistance measurement with glass wet, (d) Insulation resistance measurement with backsheet wet.

4 RESULTS

Normalized insulation resistance for mini-modules in the pristine and sand eroded condition has been presented in Fig. 5. Furthermore, the effect of wetting the glass and BS surfaces, on the normalized insulation resistance, using the lint-free cloth has also been presented in the Fig. 5.

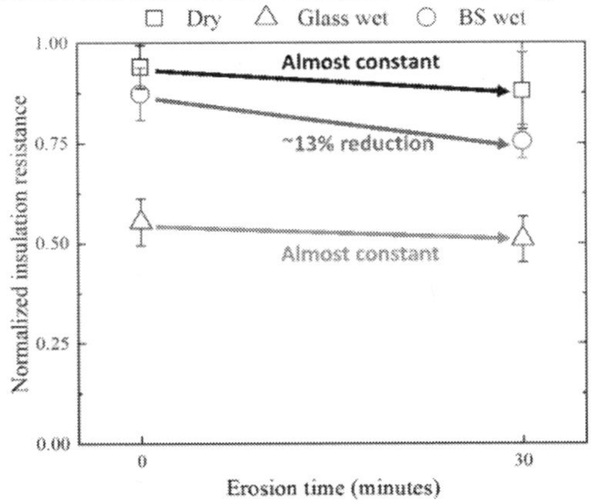

Fig. 5: Normalized insulation resistance for the pristine and sand eroded specimen in both dry and wet conditions.

For the pristine samples, it has been observed that the wet insulation resistance for glass and BS tends to reduce by 40% and 7% respectively when compared to the dry condition. The surface resistivity of glass diminishes when the glass is wet[5], which could be the reason for the reduction of the wet insulation resistance. Similarly, the loss of wet insulation resistance for the case of wet backsheet in the pristine condition could be because of its reduced surface resistivity.

Furthermore, it has been observed that the sand erosion process tends to reduce the wet insulation resistance for the case of wet BS. It has been shown recently that the volume resistivity tends to reduce when the outer surface of a backsheet is sand eroded[4], which could be the reason for the reduction in its wet insulation resistance. However, no significant changes have been observed for the case of either the dry condition or the wet glass.

5 Conclusions and outlook

This study is conducted to understand the effect of sand erosion on both the dry and wet insulation resistance of the mini-modules. We have tried finding the wet insulation resistance for either glass or backsheet being wetted using wet lint-free cloth.

It has been observed that the insulation resistance for the mini-modules tends to reduce when either the glass or the BS is wet for the pristine condition. However, maximum reduction in the wet condition is observed for the case of wet glass. Furthermore, upon the sand erosion process of 30 minutes has been found to reduce the wet insulation resistance only for the case of BS being wet. This could be because of reduction in its volume resistivity as the outer layer gets eroded due to sand particles. However, no significant changes have been observed for either the dry condition or when the glass is wet.

The results of the study presents interesting case of reduction in the wet insulation resistance of BS (i.e., subjecting the surface of BS with the wet lint-free cloth) which could be useful consideration for the BS manufactures.

6 References

[1] D. Chianese *et al.*, "Analysis of weathered c-Si PV modules," *Proceedings of the 3rd World Conference on Photovoltaic Energy Conversion*, vol. C, pp. 2922–2926, 2003.

[2] W. J. Gambogi *et al.*, "Backsheet and Module Durability and Performance and Comparison of Accelerated testing to long term fielded modules," *28th European Photovoltic Solar Energy Conference and Exhibition*, no. March, pp. 2846–2850, 2013, doi: 10.4229/28thEUPVSEC2013-4CO.10.2.

[3] N. Kim *et al.*, "Study on the degradation of different types of backsheets used in PV module under accelerated conditions," *Solar Energy Materials and Solar Cells*, vol. 120, no. PART B, pp. 543–548, 2014, doi: 10.1016/j.solmat.2013.09.036.

[4] U. Desai, S. Rathore, and A. Singh, "Exploration of degradation pathways resulting in reduction in volume resistivity in photovoltaic backsheets under sand erosion and aging in desert conditions," *Solar Energy*, vol. 236, pp. 860–869, Apr. 2022, doi: 10.1016/j.solener.2022.03.045.

[5] N. G. Dhere, N. S. Shiradkar, and E. Schneller, "Evolution of leakage current paths in MC-Si PV modules from leading manufacturers undergoing high-voltage bias testing," *IEEE J Photovolt*, vol. 4, no. 2, pp. 654–658, Mar. 2014, doi: 10.1109/JPHOTOV.2013.2294764.

Acknowledgement

The authors would like to acknowledge the support of the National Centre for Photovoltaic Research and Education (NCPRE) Phase-II (Grant No. 16MNRE002) funded by Ministry of New and Renewable Energy (MNRE), Government of India towards this work.

EFFECT OF DEGRADATION FACTORS ON SOLAR MODULE ENCAPSULANT MATERIALS

Hsin-Hsin Hsieh[1], Syh-Homg Chen[1], Wei-Lun Yang[2], Chiou-Chu Lai[2], Min-Tsung Kuan[2], Yi-Chun Liu[2]
[1] Center for Measurement Standards, Industrial Technology Research Institute, Hsinchu, 310, Taiwan ROC.
[2] Material and Chemical Research Laboratories, Industrial Technology Research Institute, Hsinchu, 310, Taiwan ROC.

ABSTRACT: Solar encapsulant material is considered to be used in PV module for its high chemical stability and optical performance. And Ethylene Vinyl Acetate (EVA) has been widely used in c-Si solar module in the past decades. However, reviewing the literature, PV degradation concerning the encapsulant materials is around 25 %, which is higher than the other module components. To approach its performance and reliability, the accelerated aging tests such as UV exposure, humidity, and thermal variation. In this research, a cyclic UV exposure/Thermal cycling test/Humidity Freeze sequential aging test cited from IEC 61215 is proposed to observe the power degradation in each step. According to the test results, it is detected that most encapsulant materials can sustain four aging cycles with a power loss less than 5 %. To study the variation in power, it is noted that a higher power drop during humidity freeze test (generally > 0.5 %), while a slight recovery in UV exposure. Humidity freeze test is a testing condition of high temperature (85 °C) and humidity (85 % RH), and the following UV test is an exposure chamber at 60 °C. Thus, it is suggested that the key factors degrading EVA are humidity and temperature (Humidity freeze testing condition), yet UV exposure is supposed to evaporate the humidity from module.
Keywords: Polymeric encapsulant; Temperature variation; Humidity freeze test

1 INTRODUCTION

Ethylene Vinyl Acetate (EVA) has been widely used in c-Si solar module in the past decades. Solar encapsulant material is considered to be used in PV module for its high chemical stability and optical performance. However, reviewing the literature shown in Fig. 1, PV degradation concerning the encapsulant materials is around 25 % (Delamination 5 % and Optical failure 20 % caused by EVA browning), which is higher than the other module components [1]. Recently, several encapsulants (POE, silicone etc.) have been developed to verify the replaceability under various application circumstance. Furthermore, issues of circular economy and reduction of carbon emission during manufacturing/recycle process is also highly discussed. Thus, a recyclable encapsulant has been carried out to investigate the availability for replacing the conventional encapsulant.

To approach its performance and reliability of the innovative encapsulant material, the recyclable PV modules were laminated with a massive production process of 150˚C for (12-15) min. And the accelerated aging tests such as UV exposure, humidity, and thermal variation, are applied. And an extended damping heat test is also considered for durability.

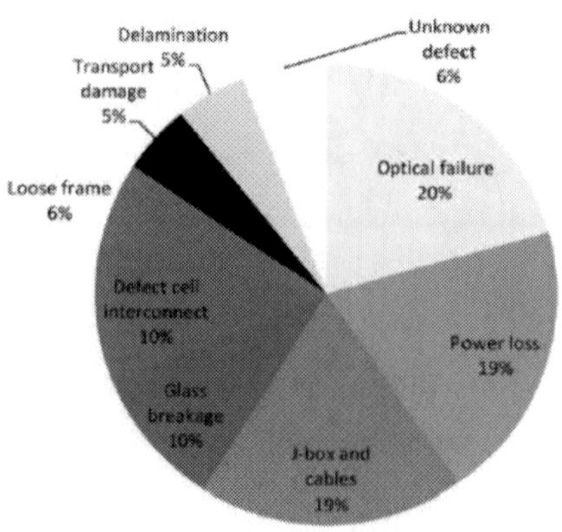

Figure 1: Failure modes and the corresponding rates of PV module in field application [1].

2 EXPERIMENTAL PROCEDURE

The testing modules are laminated in commercial size with an innovative recyclable encapsulant and half-cut solar cell in G1 type. The as-laminated solar modules were nominated 410 W of power output and also detected by electroluminescence (shown in Fig. 2). The testing items are deployed to characterize the performance and safety requirements based on IEC standards. The common IEC standards have been widely referred due to the testing scope covers a wide and general weathering condition. According to the standards, the performance and safety requirements for solar modules are referred to IEC 61215 and IEC 61730, respectively.

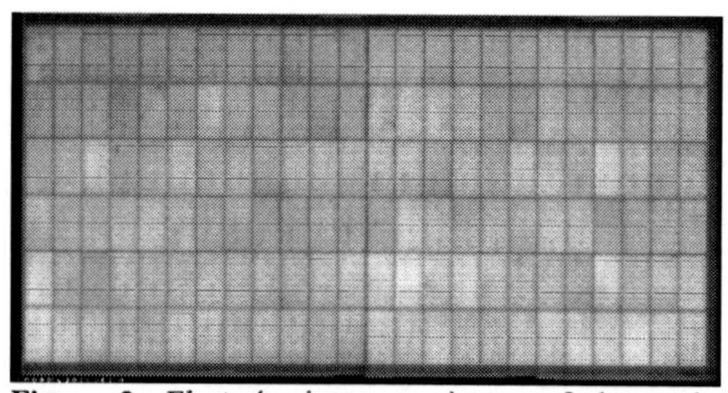

Figure 2: Electroluminescence image of the testing module.

To evaluate the module efficiently, a combined testing flow were deployed in Table 1. It is shown that the testing condition cover high temperature (and temperature change), high humidity, UV exposure, and their sequential tests. As shown in the table, sever sequential tests are to Reliability series test are referred from testing IEC standards, and the power measurement were recorded for research purpose.

On the other hand, an extended damping heat test 3000 hr is performed to inspect the degradation of IV. And the degradation factors are discussed by the following detailed electrical analyses.

Table 1: Combined testing flow of IEC 61215 and IEC 61730.

Testing Standards	Testing Items					
IEC 61215	Stability	Hot spot				
	Wet leak.	DH1000	Hail test			
IEC 61730	Wet leak.	Bypass diode				
	DH200	UV60	HF10	UV60	HF10	
	Cold	Dry	HF10	Cold	HF10	cut- BD
	Impulse Volt.					
	Module breakage					
IEC 61215+61730	UV15	IV test	UV15	TC50	HF10	Bypass diode IV test
	Wet leak.	IV test	Wet leak.	TC200	Bypass diode IV test	
	Wet leak.	DH1000	Mechan. Load			

3 RESULTS

3.1 Reliability test

After reliability test, the power of testing modules generally shows degradation < 5 %, which meets the requirement of IEC standards. The test with damping heat test generally has a higher degradation, yet the power recovered during the dried/high temperature condition.

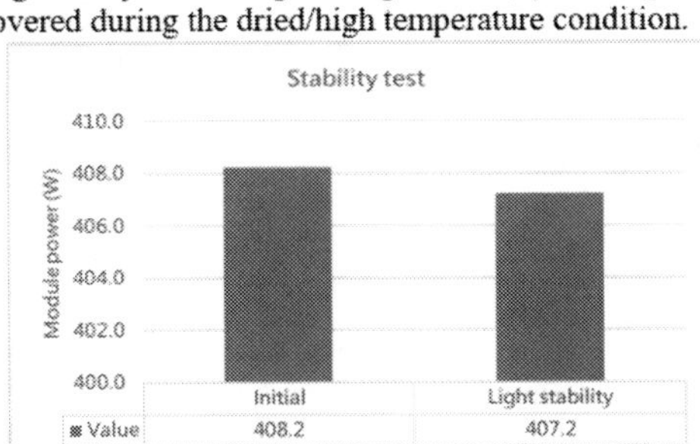

(a)

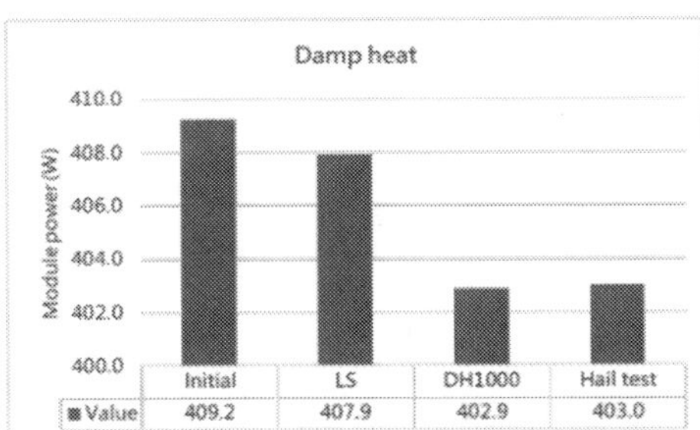

(b)

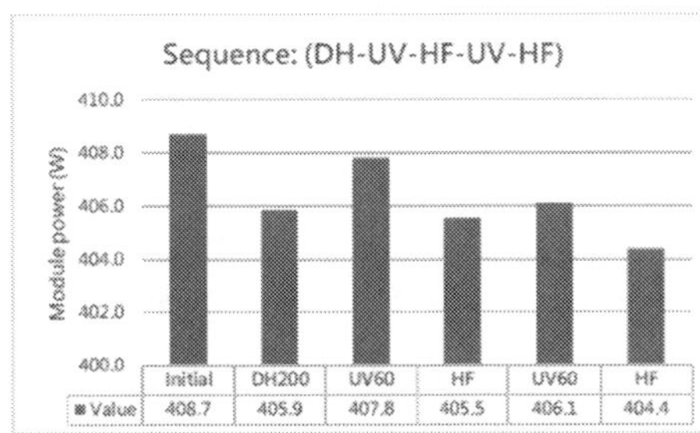

(c)

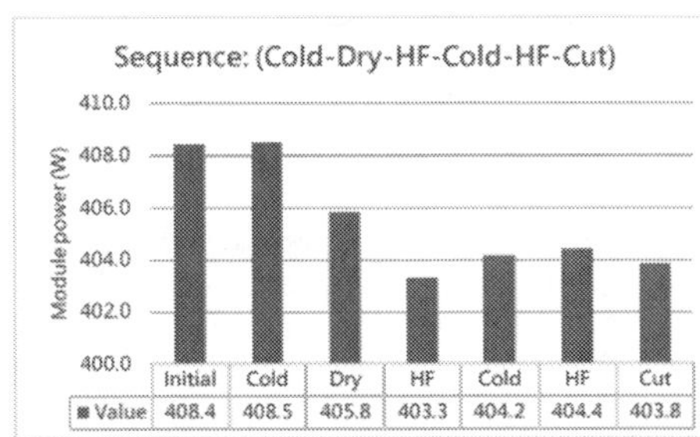

(d)

Figure 2: Reliability test of innovative PV modules.

3.2 Extended damping heat test

High temperature and high humidity are common failure factors for photovoltaic modules. In addition to the IEC International Electrotechnical Regulations, Taiwan's high-efficiency solar photovoltaic module technical specifications also pay special attention to their high temperature and high humidity resistance, and play an important role in testing projects. Therefore, in order to understand the product performance as a reference for improvement, 3000 hours of test evaluation and analysis of the humidity and heat test target of solar modules are considered in the current study. The power decay of the innovative solar module is 1.81 %, and the power decay of the commercial counterpart solar module is 1.43 %. Further analysis shows that the green module and the commercial material are detected degraded for reasons: the main reason for the decline of the green module is the Imp (maximum output current), and the module with commercial materials is the Vmp (maximum output voltage). It shows that the green module was observed degrades in photocurrent performance during the aging duration, and the voltage decline is mostly due to the solar cell or microcracks. However, the FF does not cause terminal cracking (increased series resistance) due to the aging test. Furthermore, a visual inspection for solar module was qualified without bubble generation or delamination, which can meet the requirements of international electrical regulations for solar module.

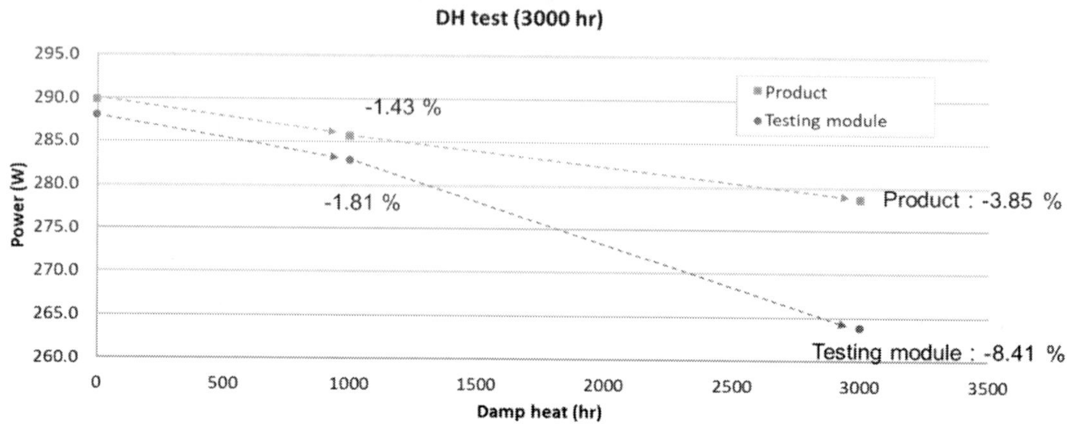

Fig. 3. Damping heat test for 3000 hr

4 CONCLUSIONS

The reliability and safety of a testing module laminated with an innovative PV encapsulant material is proposed in the research. Conclusions are addressed below:

(1) Power degradation of testing modules after IEC reliability generally shows < 5 %, indicating the performance can fulfill the requirement of IEC standards.

(2) In the further analyses, the test with damping heat test generally has a higher degradation, yet the power recovered during the dried/high temperature condition.

(3) For the extended damping heat test 3000 hr, the power losses of testing module and conventional product are similar in the 1000 hr, while a higher degradation (8.41 %) of testing module is detected due to the optical loss.

5 REFERENCES

[1] Review of Failures of Photovoltaic Modules, IEA-PVPS T13-01:2014.

LOW TEMPERATURE SOLAR CELL ENCAPSULATION
WITH NOVEL SILICONE ELASTOMER FOR BUILDING INTEGRATED PV

Guy Beaucarne[1], Mantas Zelba[2], Emmanuel Jadot[1], Jonathan Curon[1], Frédéric Gubbels[1], Valérie Hayez[1], Beatriz Sanabria Arenas[1], Gregory Chambard[1], Rimvydas Karoblis[2]
[1] Dow Corning Europe SA, Rue Jules Bordet Parc Industriel Zone C, B-7180 Seneffe, Belgium
[2] Via Solis, B. Brazdžionio str. 23, 48184 Kaunas, Lithuania

ABSTRACT: In this paper we introduce a new silicone solar cell encapsulant technology based on a two-part condensation cure chemistry, and implement with it an encapsulation process involving a dam and fill approach and in-lamination sealing. Bubble-free laminates have been fabricated both in a laboratory laminator and in a large area plate laminator. In order to achieve a transparent edge with water vapor barrier properties, a double edge materials procedure is introduced, where a transparent PIB is used as dam material, and a clear moisture cure sealant is backfilled into the groove between the glass plates and the transparent PIB. This technology is anticipated to be highly suited to BIPV applications, where module durability, hotspot resilience and better fire resistance are critical.

Keywords: Encapsulation, silicone, BIPV

1 INTRODUCTION

Policies tackling the climate and energy crises encourage incorporating photovoltaics in buildings to bring them closer to zero emission. Many buildings have limited roof space, and therefore integration of PV in façades is increasingly practiced and anticipated to become commonplace in the future. Building-Integrated Photovoltaics (BIPV) has existed for a long time, but uptake has been slow due to various challenges relating to technical aspects, excessive cost, and compliance issues to building codes. One by one hurdles are being overcome and BIPV projects are becoming more commonplace and affordable. Most façade BIPV modules are glass-glass laminates, which require an encapsulation process quite different from that for standard modules.

Conventional encapsulation materials based on ethylene vinyl acetate or polyolefin elastomers are not perfect fits for BIPV. Ideally a BIPV component would have a lifetime of 50 years or more, to match the lifetime of other components in a building skin. Resistance to high temperatures is desired, so that hot spots created during exceptional operating conditions do not lead to permanent laminate damage. Good fire resistance properties are also required. Another specific aspect of façade BIPV is that some laminates are very thick, much thicker than conventional PV laminates. These high mass panels require a lot of energy to be heated up to lamination temperature and involve a long process time because of the slow heating up and cooling down. A low temperature or even room temperature lamination process enables faster manufacturing and is therefore advantageous for BIPV applications.

Silicone encapsulation of solar cells is almost as old as photovoltaics itself. Early solar panels used silicone as encapsulant, and it is still the material of choice for space solar panels. The properties of silicone encapsulants in operating PV modules have been observed to degrade very little over long periods of time [1], resulting in modules showing lower degradation rate during field operation[2]. Also in extended accelerated ageing studies, exceptionally low degradations have been observed[3, 4].

In this work we introduce a new type of silicone solar cell encapsulant which enables lamination at temperatures down to room temperature, we describe the lamination process and show results at blank laminate and mini-module levels, after lamination and also after accelerated ageing.

2 NOVEL SILICONE SOLAR CELL ENCAPSULANT

The silicone encapsulant used in this work is DOWSIL™ 9955 Encapsulation and Lamination Silicone, which is part of a new generation of silicone elastomers. It is a two-component material, which requires mixing right before application. Some key properties of the material are listed in Table 1.

Table 1: Properties of silicone encapsulant DOWSIL™ 9955

Material property	Value
Before cure	
Viscosity	2800 mPas
After cure	
Transmittance	> 90% (5 mm slab, 450-760 nm)
Hardness	10 shore A
Elongation at break	190%
Flammability rating	UL94 HB

Unlike all other transparent silicone encapsulants, this new type of material is based on a condensation cure technology instead of a platinum-catalyst addition cure chemistry. This results in a more robust cure chemistry (fewer incompatibilities with other materials) and the development of strong adhesion on most common substrates (glass, aluminum, etc.) without the need to prime the surface prior to application.

Because of the intrinsic fire resistance properties (silicones get charred when exposed to a flame rather than burn and melt), it was possible to obtain a flammability rating under a small flame test, in contrast to all organic polymer-based encapsulants.

Laminating solar cells with a liquid silicone is very different from the conventional film-based processes but is not necessarily more complex. In this paper we explain a convenient process to achieve good encapsulation results.

3 ENCAPSULATION PROCESS USING SINGLE EDGE MATERIAL

Because the encapsulant is liquid, a dam is required

to confine the encapsulant during the process. This dam is useful for the fabrication sequence but can also provide additional functions, for instance vapor barrier properties.

3.1 General process

The schematic process sequence is given in Figure 1. A dam is first applied by applying a line of hotmelt edge material along the perimeter by extrusion. The silicone encapsulant is then dispensed on the glass, the solar cell or solar cell matrix placed, onto which some additional silicone is dispensed. The top glass is placed on top without pressing. In the vacuum chamber, air is first evacuated including the air present inside the sandwich structure. The evacuation times range between 5 and 15 min. Then the top glass is pressed down slightly while in the vacuum chamber, enough to compress the dam material a little and create a tight seal. Finally, the vacuum chamber is vented, and the atmospheric pressure compresses the whole structure to its equilibrium status. As silicone is incompressible, the final thickness of the region between the glass plates is determined by the amount of encapsulant used. A thickness of 0.8 mm as is typical for crystalline Si solar cells and conventional ribbon interconnection is typically targeted, but other thicknesses can be achieved by using less material if the solar cell string allows it, as for instance is the case for some thin-film solar cell technologies.

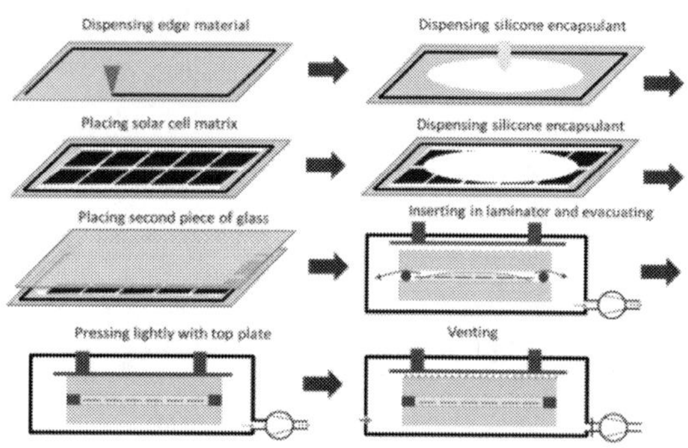

Figure 1: Suggested process with single edge material

Note that the silicone encapsulant is not yet cured after lamination and is still liquid. The encapsulant cures over the following hours but the module can already be moved. The long pot life of the encapsulant and liquid state right after encapsulation helps eliminate the remaining bubbles in the laminates, which is a common challenge in glass-glass lamination.

3.2 Small area laminate with silicone hotmelt as dam material

The application was first done on small area samples (20 x 20 cm^2) using a small vacuum chamber equipped with a top pressing plate. The glass was 4 mm float glass, the solar cell was a 156 mm multicrystalline Si solar cell, tabbed and attached with bussing ribbons. The dam material was a silicone hot melt product, DOWSIL™ 2400 Silicone Assembly Sealant. This type of material is solid as delivered, but becomes a viscous liquid when heated up. After application, if exposed to moisture in ambient air, chemical cross-linking occurs, resulting in a strong solid elastomer that will no longer melt when heated up, with excellent adhesion to glass. Silicone hotmelts come in different colors, but this one is perfectly transparent. The process with this edge material takes place completely at room temperature. Bubble-free laminates were repeatedly achieved. In figure 2 we show a picture of such mini-module.

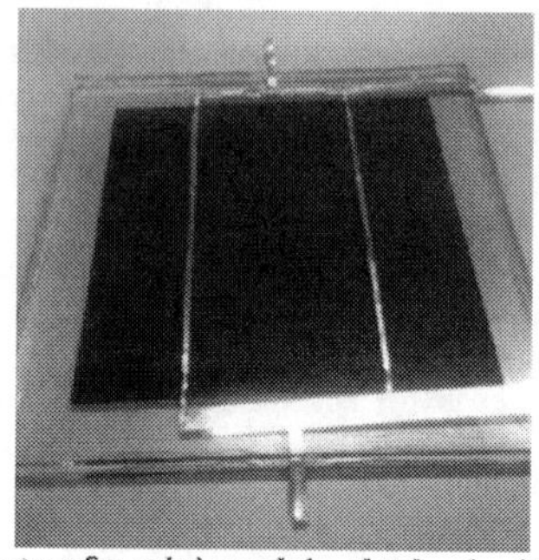

Figure 2: Photo of a mini-module obtained with silicone hotmelt as edge material

While silicone hot melt enables an easy process and can lead to visually pleasing laminates, it has the disadvantage that it does not provide a good water vapor barrier. Silicone hotmelts, like all silicones, have a comparatively high water vapor transmission rate (WVTR). Many solar cell technologies are sensitive to moisture, and therefore an effective vapor barrier at the edge of the glass-glass laminates is needed to pass the standard damp heat test at 85 °C and 85% relative humidity.

3.3 Larger area laminate with TPS as edge material

Experiments with a higher WVTR edge material were carried out in a larger area equipment. The edge material was a so-called thermoplastic spacer (TPS) material. TPS is a polyisobutylene (PIB)-based hotmelt material containing reinforcing fillers, desiccant particles and adhesion promoters. This type of material is used as an alternative to pre-shaped aluminum or silicone foam spacers in the fabrication of insulated glass, but also as edge material for thin-film PV modules. The equipment was a laminator from equipment manufacturer LISEC (model VPL-42/17 vacuum laminator). The glass was standard 3 mm extra-white solar glass and the solar cells were modern monocrystalline silicon PERC cells interconnected with a multi-busbar approach. As TPS requires a minimum temperature to soften sufficiently and develop adhesion, the process was carried out at 110°C.

Several mini-modules were made using the process described above. A picture of such a mini-module with TPS edge sealant is shown below.

Figure 3: Photo of a four cells mini-module with TPS edge

4 ENCAPSULATION PROCESS USING DOUBLE EDGE MATERIALS

4.1 Process

While the TPS edge seal provides the moisture barrier function required to pass damp heat module testing, it creates a wide opaque (black) area along the edge of the PV laminate, which is undesirable for BIPV applications. The silicone hot melt on the other hand provides a perfectly clear edge but is an insufficient water vapor barrier. To solve this the following process was developed. Instead of the TPS material, a transparent, unfilled PIB material is used as dam material. Pure PIB is a non-reactive material that slowly flows at room temperature when unconstrained. It is applied with a hot melt gun similarly to silicone hot melt or TPS. The line of transparent PIB is dispensed at a distance from the glass edge, in such a way that, after lamination, the outer surface of the PIB joint is a few mm from glass edge. Then, after the laminate has been taken out of the laminator, a transparent or translucent sealant is backfilled into the groove formed by the two glass panes and the dam to prevent the creep of the transparent PIB and provide anchoring of the edge joint. In our experiments we either used the translucent neutral sealant DOWSIL™ 791T Weatherproofing Sealant or a silicone hotmelt, for an even more transparent edge.

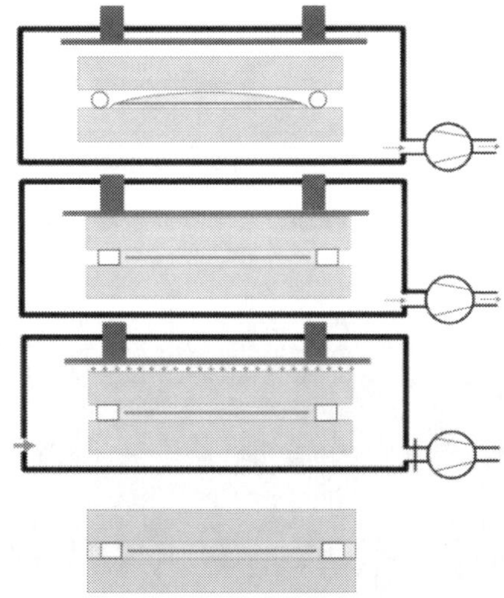

Figure 4: Schematic process with double edge materials. The last step (backfilling step) is carried out subsequently, outside the laminator

The back-filling process is commonly implemented in insulated glass unit manufacturing and can be automated. Figure 5 shows an example of manual backfilling.

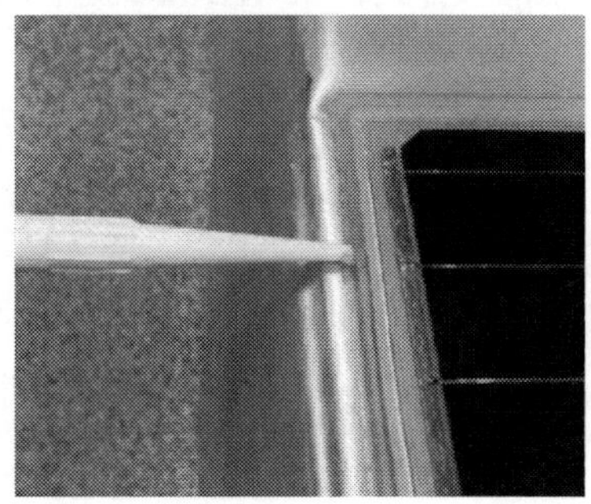

Figure 5: Backfilling process

Note that this fabrication sequence takes place completely at room temperature. Figure 6 shows a mini-module obtained through this process. For these tests, monocrystalline PERC solar cells with 4 busbars were used, in combination with 3 mm float glass pieces. The outer edge was obtained by backfilling with a translucent silicone sealant.

Figure 6: Picture of completed mini-module fabricated with the double edge materials process

4.2 Vapor barrier effectiveness

The double edge materials process was investigated with a comparative test. Two small laminates were produced with different edge structures. One only had silicone hot melt as edge sealant, and the second had the double edge structure transparent PIB / silicone sealant. For this experiment, no solar cell was inserted during lamination, and the backfilled silicone sealant in case of the double edge materials sample was a silicone hot melt.

Figure 7: Glass-glass laminates with the two types of transparent edge structures, prior to damp heat ageing (right: laminate with only silicone hot melt edge material, left: laminate with combination transparent PIB and silicone hotmelt back-filling)

Note that the transparent PIB does not contain any desiccant, in contrast with TPS material, and is therefore expected to be a less effective moisture barrier. However, because of the low WVTR of PIB itself, the relatively large width of the PIB seal and the small cross-section of the space between the two glass panes, it is expected that the PIB seal will provide a sufficient barrier. The laminates were put in a climate chamber for 1000 hours in 85 °C/ 85 % relative humidity conditions.

The silicone encapsulant has the property to show slight haziness when in presence of moisture. This property is used here to determine whether or not moisture has reached the inside of the laminate. Photos of the laminates after ageing are shown in Figure 8. A milky halo can be seen for the sample with only silicone hotmelt as edge sealant. On the contrary, no such halo can be seen for the sample featuring the double edge sealant structure.

a) b)

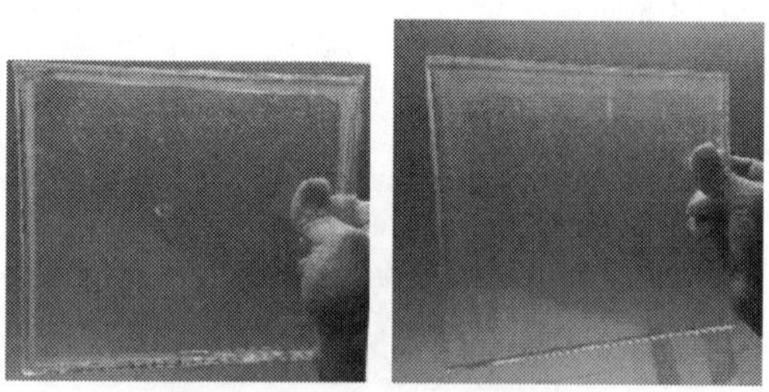

Figure 8: Photos of glass-glass test laminates after 4 weeks in 85 °C/ 85 % RH, a) with double seal transparent PIB/backfilled silicone hot melt, b) with only silicone hotmelt edge seal.

This absence of degradation of laminates with double edge materials in damp heat is also observed at mini-module level. Three mini-modules similar to the one shown in Figure 6 were measured with an illuminated IV system, placed in 85/85 damp heat conditions, and then measured again. Table 2 give the results after 475 hours of damp heat.

Table 2: Illuminated IV results of mini-modules with double edge materials before and after 475 hours of damp heat treatment (average of 3 mini-modules)

	Isc (mA)	Voc (mV)	FF (%)	Power (W)
Before damp heat	8992	666	72.1	4.32
After 475h damp heat	9017	663	73.2	4.38
Average change	+0.28%	-0.4%	+1.51%	+1.37%

As can clearly be seen there is no degradation after 475 hours of damp heat. The accelerated ageing procedure will be continued until 1000h.

4.3 UV resistance

In contrast with TPS material, pure PIB does not contain any pigment or filler. UV light can therefore reach not only the interface between glass and PIB, but also the bulk of the transparent PIB joint. To ascertain whether this could degrade the double materials edge structure, a laminate with encapsulated solar cell and double materials edge was exposed to UV (50 W between 300 and 400 nm) for 1000 h. Visual observation after ageing revealed no sign of degradation and discoloring.

5 DISCUSSION

Solar cell encapsulation at room temperature is not new. In fact early PV modules were made used silicone in a casting process, a process which is still used sometimes where laminating equipment is not available[5]. The process combining the in-lamination sealing and the special silicone encapsulant introduced here is however quite different. Process speed can be much higher as lamination and curing are decoupled. Moreover, no priming is needed, as the encapsulant provides strong and durable chemical adhesion, in contrast with addition cure transparent encapsulants.

This encapsulation process is not only applicable for BIPV, but potentially also for all PV applications where superior longevity and temperature resistance is desirable, such as desert modules or floating PV. The low temperature aspect of the lamination process is also of great interest for emerging solar cell technologies that cannot withstand typical lamination temperatures, such as perovskite-based solar cells.

6 CONCLUSION

Solar cell encapsulation with a transparent silicone elastomer has advantages for several PV applications, including BIPV. In this paper we introduced a new silicone solar cell encapsulant technology based on a two-part condensation cure chemistry, and implemented with it an encapsulation process involving a dam and fill approach and in-lamination sealing. Bubble-free laminates could be easily fabricated both in a laboratory laminator and in a large area plate laminator. In order to achieve a transparent edge with water vapor barrier properties, a double edge materials procedure was introduced, where a transparent PIB is used as dam material, and a clear moisture cure sealant was backfilled into the groove between the glass plates and the transparent PIB. This structure was shown to be effective in preventing moisture to enter the laminate.

7 REFERENCES

[1] G. Beaucarne, A. Dupont, D. Puthenmadom, N. Shephard, T. Sample, Material study of photovoltaic modules with silicone encapsulation after long-term outdoor exposure, Solar Energy Materials and Solar Cells, 230 (2021) 111298.

[2] A. Skoczek, T. Sample, E.D. Dunlop, The results of performance measurements of field-aged crystalline silicon photovoltaic modules, Progress in Photovoltaics, 17 (2009) 227-240.

[3] B. Ketola, A. Norris, The role of encapsulant moisture permeability in the durability of solar photovoltaic modules, 25th European Photovoltaic Solar Energy Conference, (2010).

[4] G. Beaucarne, S. Wang, X. Sun, Y. Wu, Y. Huang, N. Shephard, Back in the PV galaxy : The return of the silicone module, in: 32nd European Photovoltaic Solar Energy Conference, Munich, Germany, 2016.

[5] F. Dross, A. Labat, M.A. Perez Lopez, M.A. Perez Lopez, R. Raudez, A. Bruce, S. Kinne, R. Komp, Vacuum-free, cost-effective, developing-country-material-available solar cell encapsulation, Solar Energy materials and Solar Cells, 90 (2006) 2159-2166.

EARLY DETECTION OF POTENTIAL-INDUCED DEGRADATION VIA THE SHUNT RESISTANCE IN CIGS THIN FILM SOLAR MODULES

Lukas Gerstenberg, Rohith Krishnan Bala Krishnan, Viktor Wesselak
Institut für Regenerative Energietechnik, University of Applied Sciences Nordhausen
Weinberghof 4, 99734 Nordhausen, Germany
E-Mail: lukas.gerstenberg@hs-nordhausen.de

ABSTRACT: Additionally to a reduction of V_{OC}, CIGS Thin Fim Solar Modules also show shunting-type PID under negative bias stress. In this paper, a known method to monitor the progress of PID-s in c-Si modules via the shunt resistance is applied to CIGS TFSM. Measuring the dark IV curves of CIGS TFSM directly in the climatic chamber enables linking the change in shunt resistance to the occurring power loss. Results show that the shunt resistance decreases before a change in V_{OC} occurs, which makes the shunt resistance an appropriate PID indicator for CIGS TFSM.
Keywords: CIGS, degradation, shunt resistance

1 INTRODUCTION

The cumulative energy conversion during the operating time of photovoltaic (PV) modules should be as high as possible. With lifetimes of up to 20 to 25 and more years, PV modules experience numerous stress factors reducing their conversion efficiency. Climatic conditions, electrical topology such as string length and type of inverter can lead to degradation of the module efficiency. Especially modules in plants with high string voltages just below 1000 V and highly efficient transformer-less inverters are prone to degradation due to potential against ground (Potential-Induced Degradation (PID)). Particularly negative potential against ground causes degradation in all PV technologies including substrate Copper-Indium-Gallium-Diselenide (CIGS) Modules [1,9,10,11,12]. Typically, the potential difference between cell and frame causes positive sodium ions to migrate from the glass into the cells. Depending on the PV technology, different PID types and effects predominate. In crystalline Silicon (c-Si) modules, the positive sodium ions accumulate in stacking faults and cause shunts [1,2], this effect is called shunting–type PID or PID-s. The resulting reduction of the modules' total shunt resistance can be used to monitor the PID progress and detect PID in an early stage [2,3]. Detecting beginning PID in an early stage enables plant operators to take countermeasures as applying positive potential to the modules during non-operating hours and thereby increasing the modules lifetime and keeping the conversion efficiency as high as possible.

2 POTENTIAL-INDUCED DEGRADATION IN CIGS TFSM

The exact degradation mechanisms for substrate CIGS Thin Film Solar Modules (TFSM) are still unclear [4]. Recent publications explain several effects. CIGS modules having a substrate structure makes the back contact an easy entrance point for ions from the back glass. Lee *et al.* [5] reported that Na^+ ions migrate into the metal back contact under negative bias against ground. Having reached the active cell layer structure, the electric field provided by the potential against ground no longer forces the ions to move further. Once in the metal contact, the ions diffuse further into the CIGS absorber and CIGS/CdS interface due to concentration differences. Jahandardoost *et al.* reported increased recombination in the CIGS/CdS interface leading to a decreased open circuit voltage [6]. Similar to c-Si, CIGS cells show shunting-type PID [5,11] by accumulation of Na^+ ions in stacking faults. Usually multiple effects overlap, depending on the production process and material composition. For early detection of PID it is advisable to observe one easy-to-measure effect if it is clear that other degradation mechanisms happen later in the module lifetime.

3 EARLY DETECTION OF PID IN CIGS TFSM – EXPERIMENTAL SETUP

With shunting being the predominant PID effect in c-Si modules, correlating power loss and the change in shunt resistance is a promising monitoring solution. Florides *et al.* [2] have shown that a correlation between power loss and reduction in shunt resistance is possible for c-Si modules and have shown application of the monitoring process in the field [3]. Since CIGS TFSM show shunting as a PID effect, a similar approach of monitoring the shunt resistance under accelerated ageing was investigated.

Experimental Setup: Two identical CIGS TFSM M05 and M06 aged under accelerated conditions 85 °C and 85 % relative humidity (r.h.) and -1000 V bias connected to the cells against the grounded module edge, see Figure 1. All modules were mounted on typical aluminium profiles and held in aluminium mounting rails via four rubber clips. During the ageing process, the bias was removed every 12 hours to measure the dark IV curve directly in the climatic chamber (CC) under remaining high temperature and moisture. The climatic chamber test was interrupted every 60 hours to measure light IV curves under standard test conditions (STC) to determine module power P_{STC}. Before STC measurements, the CIGS modules underwent 8 hours of light soaking at one-sun illumination and 50 °C ± 5 K to stabilize metastable effects in the TFSM. The dark IV curve measurements enabled determination of the shunt resistance using equation (eq.) (1) between 0 V and 5 V. Smaller calculation intervals lead to high inaccuracy of the shunt resistance, higher calculation intervals underestimate the shunt resistance, since the linear fit with eq. (1) misses the IV curve. Even though a 15-bit analog to digital converter provides excellent resolution for the current measurement (around 6 µA), there is still a rather high variance when calculating the shunt resistance from the dark IV curves since the current flow near V=0 V is minute.

$$R_{Sh} = \frac{\Delta V_{SC}}{\Delta I_{SC}} \tag{1}$$

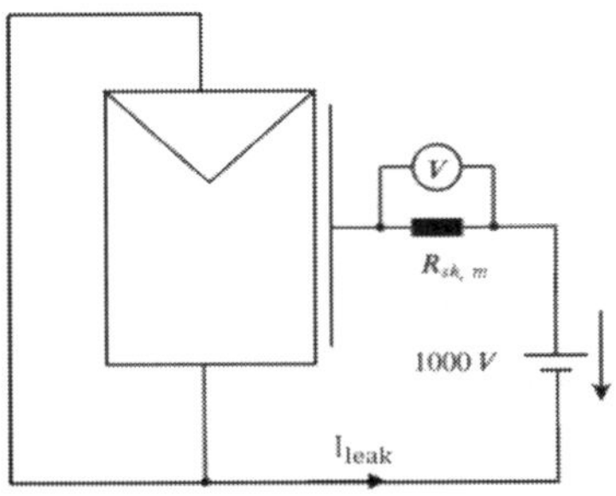

Figure 1: Circuit used for accelerated ageing in climatic chamber. Module is negatively biased at -1000 V against grounded mounting structure. Measurement resistance $R_{Sh, m}$ enables leakage current measurement

In a second experiment, the measurement resolution was increased to investigate the initial degradation phase with a new module Me01. The parameters were set to 65 °C, 85 % r.h. and -1000 V against grounded module edge. The reduced temperature resulted in slower ageing to enable detailed monitoring of the change in shunt resistance. Dark IV curve measurements took place every hour. The first intermediate STC measurement took place after five hours, after that the STC measurements followed every three hours. In both PID experiments, there was a reference module added to the climatic chamber without bias.
Note that the shunt resistance of PV modules is dependent on temperature. For this type of CIGS TFSM the temperature effect on the shunt resistance was investigated in detail in [7]. All shunt resistance values shown in Fig. 2 and Fig. 3 are for 85 °C. Shunt resistance values shown in Fig. 4 are for 65 °C.

4 RESULTS AND DISCUSSION

In the first accelerated ageing test at 85 °C both CIGS TFSM M05 and M06 showed drastic power loss already after 60 hours. Figure 2 shows the normalized power at STC (right y-axis) and the shunt resistance derived from dark IV curves (left y-axis) over the cumulated testing time for M05. The module M05 lost 22 percent of power (after 8 hours of light soaking) after the first 60 hours of PID testing. The power loss continues over the testing time and after 240 hours M05 shows 50 percent power loss. Along with the power loss, the shunt resistance decreased dramatically. The shunt resistance dropped from the staring value of 25.5 kΩ down to 0.93 kΩ within the first 24 hours of accelerated testing, see Fig. 2. Afterwards the shunt resistance decrease stabilized at around 0.11 kΩ. This stabilization is clearly visible in Fig. 3, showing the same data as Fig. 2 with decreased y-axis limits. The visible leaps in the shunt resistance course result from the intermediate measurements. The necessary pre-treatment of the CIGS TFSM prior to STC measurements lead to partial recovery of shunting-type PID. A similar observation was found in CIGS cells [8]. Module M06 showed similar behaviour concerning power loss and shunt resistance change. The power and shunt resistance in the reference module remained stable throughout the testing time, indicating that the negative potential against ground is the main degradation factor.

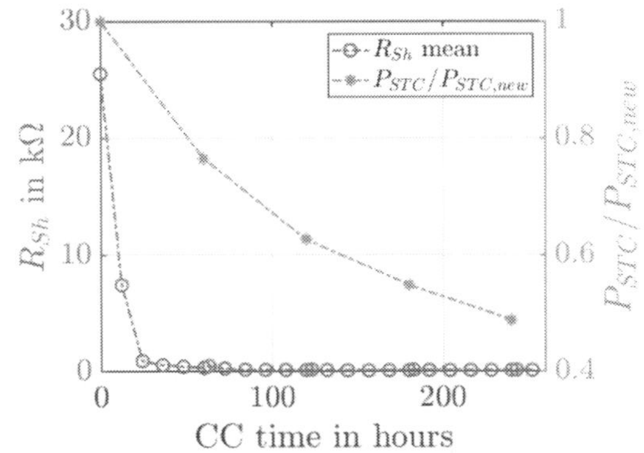

Figure 2: Shunt resistance in kΩ mean of five measurements (blue circles left axis) and normalized power (green stars right axis) of CIGS TFSM M05 during Bias Damp Heat Test at 85 °C, 85 % r.h. and cells at -1000 V against grounded module edge

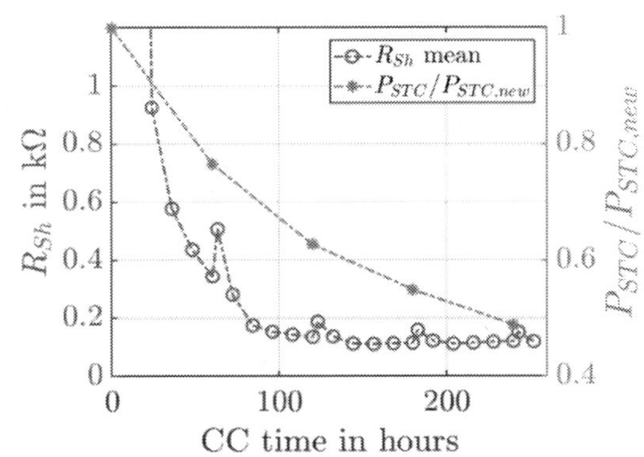

Figure 3: Shunt resistance in kΩ mean of five measurements (blue circles left axis) and normalized power (green stars right axis) of CIGS TFSM M05 during Bias Damp Heat Test at 85 °C, 85 % r.h. and cells at -1000 V against grounded module edge. Same data as in Fig. 2 with left y-axis limited to 1.2 kΩ

For this experiment, the shunt resistance decreases in three phases, compare with Fig. 3:
1) Initial drastic decrease (0 h-24 h)
2) Slowing continuous decrease (24 h-100 h)
3) Stabilization (100 h-240 h)
The initial drastic decrease and the stabilization are uncommon in other PV technologies showing shunting-type PID. The initial drastic decrease was not reported for c-Si modules, they rather show a continuous decrease simultaneously with power decrease until the module is short-circuited [2]. However, it has to be mentioned that the shunt resistance was measured every twelve hours and an initial plateau phase might be missed. The same accounts for the change in module power, since the first measurement occurred after 60 hours of testing, there is no information of how the initial change took place.

Additionally both modules M05 and M06 also showed reduction in V_{OC} and FF during this first test. Fig. 4 shows how V_{OC} and FF for Module M05 declined over the testing time. During the first 60 hours, V_{OC} dropped by ten percent and FF reduced by 20 percent while the shunt resistance reduced by 96 percent (Fig. 2). The shunt resistance gives a larger range of change in the same ageing time and thereby promises a higher potential to serve as a PID indicator and determine beginning degradation.

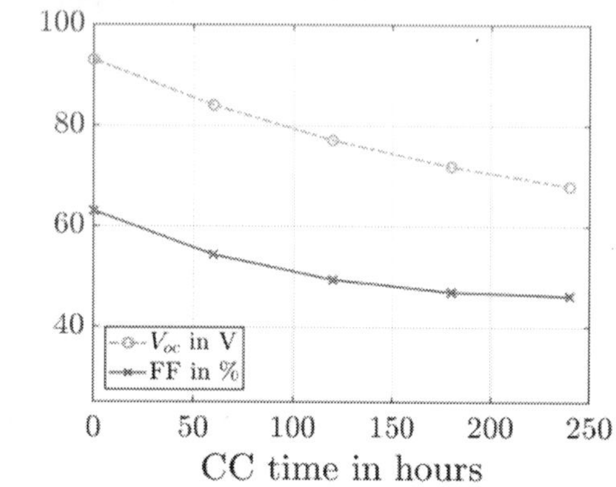

Figure 4: V_{OC} (green circles) and FF (blue crosses) of M05 CIGS TFSM during Bias Damp Heat Test at 85 °C, 85 % r.h. and cells at -1000 V against grounded module edge (CC= climatic chamber). Data taken from STC IV curves from intermediate measurements

The second accelerated PID experiment proceeded at a reduced temperature of 65 °C to slow down the ageing process. Additionally both dark and light IV curve measurements took place at higher frequency: Dark IV curve measurement every hour, light IV curve measurement every 3-5 hours. Fig. 5 shows the shunt resistance and the corresponding standard deviation (blue circles and error bars, left axis) and STC power (green crosses, right axis) over the testing time for module Me01. For the module power, there is only minor change visible. After ten hours of testing, the power loss was three percent, after 13 hours of testing the power loss was barely above five percent. The module power seems to transition from a stable plateau into a degradation phase.

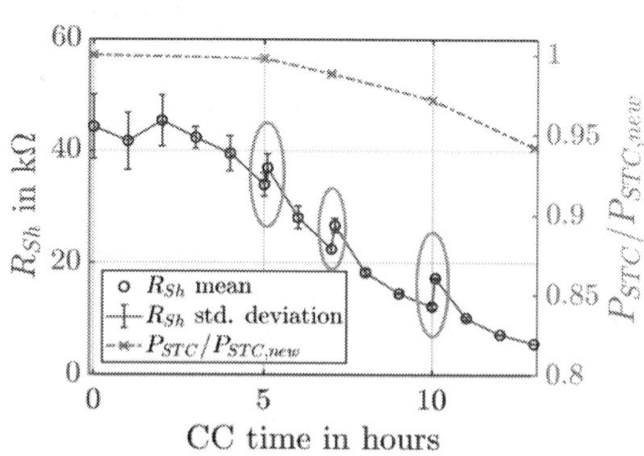

Figure 5: Shunt resistance in kΩ including standard deviations from five measurements each (blue circles left axis) and normalized power (green crosses right axis) of CIGS TFSM Me01 during Bias Damp Heat Test at 65 °C, 85 % r.h. and cells at -1000 V against grounded module edge

Opposite to the first test showing an initial drastic decrease for the shunt resistance, the measurements in this test captured a continuous decrease with an initial stable phase, see Fig. 5. Until around four hours of testing, the shunt resistance showed a stable value – considering the high standard deviation for high shunt resistance values. Between four and five hours, the shunt resistance starts to decrease. The first intermediate measurement at five hours lead to partial recovery of the shunt resistance, followed by continuous decrease, highlighted by the grey oval marks in Fig. 5. Instead of the planned three hours of ageing, the bias was only applied for two hours due to connection issues. The following hours of applied bias caused further decrease of the shunt resistance. The following intermediate measurement also lead to a partial recovery of the shunt resistance. After 13 hours of testing, the module power reduced by six percent and the shunt resistance reduced by 87 percent.

Module Me01 also showed slight decrease of V_{OC} during the first 13 hours of accelerated ageing at 65 °C. The V_{OC} decreased from 111.0 V to 108.7 V, see Table I. This drop-off of two percent is minor compared to the 87 percent reduction of the shunt resistance. This confirms the cautiously expressed observation from the first experiment that the decline of the shunt resistance is highly more significant than the decline of V_{OC}. Furthermore, the V_{OC} remained unchanged until the second intermediate measurement (110.7 V), while the shunt resistance already showed a decrease of 50 percent. For the adoption as PID indicator, the shunt resistance is the right choice for this type of CIGS TFSM.

Table I: V_{OC} and FF of Me01 from intermediate measurements at STC during Bias Damp Heat Test at 65 °C, 85 % r.h. and cells at -1000 V against grounded module edge

Time in h	0	5	7	10	13
V_{OC} in V	111.0	110.9	110.7	109.6	108.7
FF in %	74.1	73.8	74.0	72.8	70.8

To illustrate the relation between power loss and shunt resistance, Fig. 6 shows power loss of Me01 over the shunt resistance. The values shown are after light soaking.

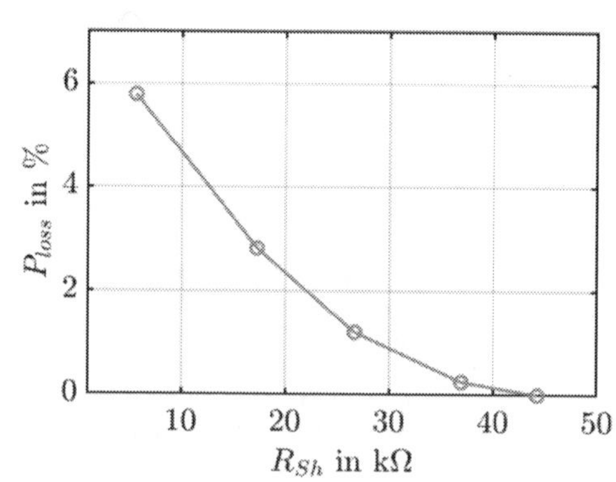

Figure 6: Power loss of Me01 over shunt resistance. Data shown are after light soaking.

Considering that V_{OC} and FF are only marginally affected until six percent of power loss, it becomes apparent that the power loss can be attributed to the reduction in shunt resistance. Further testing of Me01 is necessary to investigate the further progression of the change in module parameters. A revision of this test with several module will give insight about the repeatability and deviation between several CIGS TFSM.

5 SUMMARY AND OUTLOOK

Results from the first test at 85 °C and 85 % r.h. suggested that there is an initial strong decay of the shunt resistance right at the beginning of PID in CIGS TFSM. From the second test at reduced temperature and with increased measurement frequency it became apparent that the shunt resistance in CIGS TFSM does in fact decline continuously with a short plateau phase at the beginning.

In both tests, the intermediate measurements lead to partial recovery of the CIGS TFSM. Both power and shunt resistance are effected by the eight hours of light soaking prior to STC measurements. It is difficult to quantify by how much the power and shunt resistance would have declined more if STC power determination had been possible without light soaking. However, modules in the field are constantly light soaked during operation and only the stable power can be considered for analyzing the degradation.

In general, accelerated ageing tests in a dark climatic chamber at high temperature only barely resemble the actual environment a module experiences during its lifetime. In the field, high voltage stress is always accompanied by high irradiation, which also means that the modules operate in the maximum power point. Thus the modules operating voltage is present between the module leads. In the CC, modules typically operate in short circuit to apply the negative bias at both module terminals. This may effect ion movement in the active layers.

For adapting shunt resistance monitoring in the field, several points must be considered. From the data shown it seems possible to monitor V_{OC} and conclude occurring power loss from a decrease in V_{OC}. There are two main difficulties with this concept. It is possible to derive V_{OC} from dark IV curves by superposition into the fourth quadrant by I_{SC}. However, this technique requires measurement of the complete dark IV curve. This needs costly and powerful equipment to run the module under forward bias. The derived V_{OC} from dark IV curves will be rather inaccurate. Deriving V_{OC} using day light would require brief disruption of plant operation.

It is considerably simpler to apply small forward bias to the module during nighttime and acquire the shunt resistance for the module via the forward current. Considering temperature dependency, this promises to give an accurate indicator for progressing PID in CIGS TFSM. There might occur difficulties when measuring the dark IV curve for a module string. Here all shunt resistances of the individual modules are connected in series. Whether this technique is useable for module strings will be investigated in future work.

ACKNOWLEDGEMENT

The authors would like to thank the German Federal Ministry of Education and Research (BMBF) for funding the project "PV-FeldLab" within the funding program Forschung an Fachhochschulen under the support code FKZ 13FH600IB6.

REFERENCES

[1] Luo, W. et al.: Potential-induced degradation in photovoltaic module: a critical review. In: Energy & Environmental Science, Vol. 10, 2017

[2] Florides, M. et al.: Shunt Resistance Relation to Power Loss due to Potential Induced Degradation in Crystalline Photovoltaic Cells. In: IEEE 46th Photovoltaic Specialists Conference (PVSC) (2019), S. 1950-1954

[3] Florides, M. et al.: Early Detection of potential induced degradation in the field: testing a new methodology on silicon PV modules. In: 37th European Photovoltaic Solar Energy Conference and Exhibition, 2020

[4] Yilmaz, P. et al.: Potential induced degradation of CIGS PV systems: A literature review. In: Renewable and Sustainable Energy Reviews, Volume 154, 2022

[5] Lee, S. et al: Characterization of Potential-Induced Degradation and Recovery in CIGS Solar Cells. Energies 2021, 14, 4628.

[6] Jahandardoost, M. et al: Potential Induced Degradation and Recovery Effects in CdS/CIGS Solar Cells with Na and Rbf Treatments. In: IEEE 48th Photovoltaic Specialists Conference (PVSC) (2021), S. 1693-1696

[7] Gerstenberg, L. et al.: Investigations on the Temperature Dependency of the Shunt Resistance in CIGS Thin-Film Solar Modules using Dark IV-Curve Measurements. In: 38th European Photovoltaic Solar Energy Conference and Exhibition, 2021, p. 787-790

[8] Sakurai,, K. et al.: Rapid Recovery of CIGS Solar Cells from PID Stress with Light Soaking. In: IEEE 46th Photovoltaic Specialists Conference (PVSC) (2019), S. 2000-2002

[9] Boulhidja S, Mellit A, Voswinckel S, Lughi V, Ciocia A, Spertino F, Pavan AM. Experimental evidence of PID effect on CIGS photovoltaic modules. Energies 2020;13:537.

[10] Sakurai K, Tomita H, Schmitz D, Tokuda S, Ogawa K, Shibata H, Masuda A. Exploring suitable damp heat and potential induced degradation test procedures for Cu(In,Ga)(S,Se) photovoltaic modules. Japanese Journal of Applied Physics 2018; 57: 08RG02-1.

[11] Boulhidja S, Mellit A, Voswinckel S. Potential-induced degradation test on CIGS photovoltaic modules. Proceedings of 5.th International Conference on Electrical Engineering 2017: 1-4.

[12] Voswinckel S, Mikolajick T, Wesselak V. Influence of the active leakage current pathway on the potential induced degradation of CIGS thin film solar modules. Solar Energy 2020; 197: 455-461.

A SCREENING PROTOCOL TO ASSESS THE STABILITY OF THE INKS USED TO MASK METALLIC INTERCONNECTS IN BIPV MODULES

Alejandro Borja Block[1]*, Jordi Escarre Palou[2], Antonin Faes[1,2], Alessandro Virtuani[2], Christophe Ballif[1,2]

[1]École Polytechnique Fédérale de Lausanne (EPFL), Institute of Electrical and Micro Engineering (IEM), Photovoltaics and Thin Film Electronics Laboratory (PV-Lab), Rue de la Maladière 71b, 2000 Neuchâtel, Switzerland
[2]CSEM, Sustainable Energy Center, Jaquet-Droz 1, 2000 Neuchâtel, Switzerland

*e-mail: alejandro.borjablock@epfl.ch

ABSTRACT: Aesthetics plays a major role in the development of BIPV products. Several BIPV module manufacturers attempt to mask the reflective metallic interconnects to obtain uniformly colored modules. This is achieved using expensive manufacturing steps, such as colored stripes with manual positioning. Inks could be an alternative to perform this process automatically. However, inks used in BIPV products must be extremely stable and do not alter their visual appearance over time. In this work, we propose a screening protocol to assess the long-term stability of seven commercial black colored inks used to mask ribbons. We apply it to coupons mimicking the structure of conventional glass/glass (G/G) and glass/backsheet (G/Bs) structures encapsulated with EVA and POE, and to one-cell modules with ink-coated ribbons with the help of an in-house developed ribbon coating equipment. The protocol includes three different test sequences. Sequence A involves 250 h of DH (85°C, 85%RH) followed by to 60 kWh/m^2 of UV light exposure, sequence B the UV test only, and sequence C adhesion tests. The results show no degradation after the DH test and yellowing with unstable inks after the UV exposure. The electrical performance of modules coated with UV unstable inks, is only slightly affected (~-1%).

Keywords: Building Integrated PV (BIPV), aesthetics, reliability, metallic interconnects, ink

1 INTRODUCTION

Worldwide, the energy demand for buildings accounts for one-third of the total final energy consumption [1]. Building integrated photovoltaics (BIPV) will play an important part to minimize the energy consumption of buildings by generating energy on site.

Aesthetics plays a major role for building integrated photovoltaic (BIPV) products and projects. Architects and other stakeholders are often interested in solutions with unique design and customization properties, in which often the presence of a PV panel is partly or completely hidden. In many cases, BIPV modules are still considered a non-aesthetic add-on by building designers [2].

In recent years, several manufacturers of BIPV modules attempt at masking the highly reflective metallic interconnects to minimize inhomogeneous patterns of the solar modules. This is often achieved using expensive and inefficient manufacturing steps, such as applying colored stripes or tapes with manual positioning. Other solutions to modify the appearance of the BIPV modules is the use of digital ceramic printed glass in the front of the module, this solution does not cover fully the metallic interconnects and represents power losses of more than 60% depending on the opacity of the coating [3]. Inks could be used to mask the metallic ribbons in BIPV applications to make the modules more aesthetically without considerably impacting the power output. No specific protocols to screen the inks that are suitable for BIPV from the ones that are not exist. Furthermore, inks could prospectively be applied using highly automated processes reducing cost and takt time in production lines.

Understandably, inks used in BIPV products should be stable and do not alter their visual appearance over time. In this work, we propose a screening protocol to assess the long-term stability of inks used to mask ribbons and check the effect of unstable ink on coated one solar cell modules with the use of an in-house developed ribbon coating equipment.

2 EXPERIMENTAL

In this work, we have selected seven different black colored commercial inks (see table I) to coat the metallic interconnects used in PV modules. The details of the inks are listed in Table I. The ink dispensing technique and type are critical in analyzing the applicability of the inks and the coating process into PV manufacturing line.

Table I: Inks investigated in this work. Most of them are UV curable except ink #2.

INK #	Type	Dispensing technique
1	UV curable	Inkjet
2	Solvent based	Inkjet
3	UV curable	Inkjet
4	UV curable	Inkjet
5	UV curable	Inkjet
6	UV curable	Screen printing
7	UV curable	Screen printing

The ribbons are coated manually or according to the dispensing technique suggested by the ink manufacturer. Similarly, the ribbons on solar cells are coated with the help of an ink-compatible ribbon coating equipment. The samples are subjected to 3 main sequences to test stability and adhesion of the coatings based on common IEC standards and previous work [4]. It is critical to assess the stability of the coatings in order to understand their long-term stability to weathering agents and the adhesion to the ribbons in view of their industrial application. The characterization is done with image processing, reflection measurements of color palettes and in the case of the one solar cell mini modules, additionally, with electrical performance measurements.

2.1 Sample preparation

Once the metallic ribbons are coated, they are cut in 3 cm pieces. They are then laminated at approximately 150°C, depending on the recipe for the used encapsulant (EVA or POE) in usual PV configurations with glass-glass and glass-backsheet (see fig. 1). The final size of the samples is 5 cm by 5 cm using front glass and rear backsheet or glass.

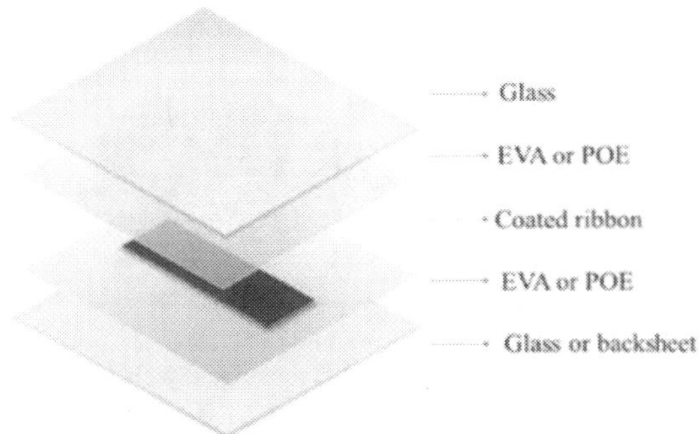

Figure 1: The sample configurations to test the stability of the coating have a glass/glass (G/G) and glass/ backsheet (G/Bs) structure and are made with either EVA or POE as encapsulant.

One-solar cell mini modules are prepared by first soldering the cell, subsequently coating the ribbons with the help of an in-house developed inkjet ribbon coating equipment (see fig. 2) that can coat ribbons even on top of the solar cells. Due to the compatibility of coating with the equipment this kind of sample was only manufactured with ink #3.

Figure 2: Solar cell with black coated ribbons (left) and a solar cell with uncoated ribbons (right). The high reflection of the uncoated ribbons can be observed.

The coated solar cell is encapsulated and laminated with front glass, EVA and a rear black backsheet. This is equally done with uncoated solar cell (see fig. 3).

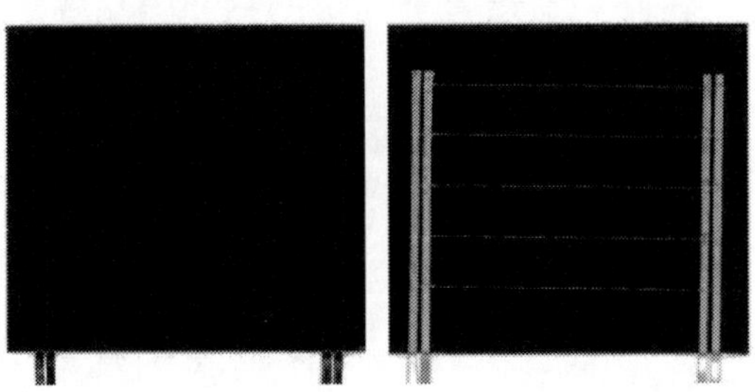

Figure 3: The solar cells (see fig. 2) are laminated into mini modules with EVA and a black backsheet. On the left the appearance is dark and uniform while on the right the high contrast of the metallic interconnects can be appreciated.

2.2 Testing sequences

The samples are subjected to 3 sequences in parallel (see fig.4).

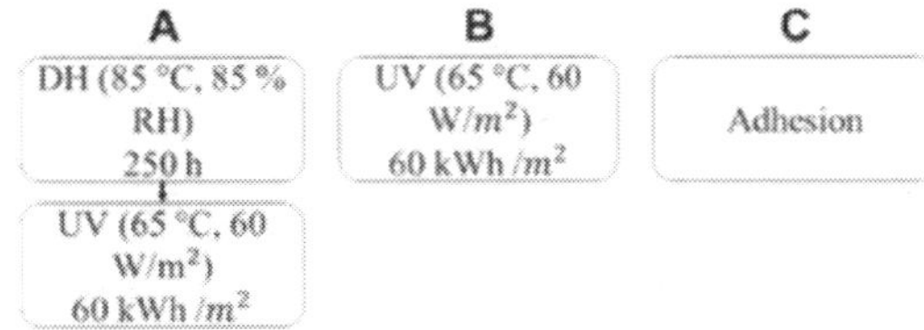

Figure 4: Sequences of the screening protocol. "A" starts with DH followed by UV to investigate the effect of humidity in the samples. "B" involves only UV with the same dose as "A" of 60 kWh/m^2. Sequence "C" tests the adhesion of the coating to the ribbon. All sequences are carried out in parallel.

The samples following sequence "A" are placed on a Weiss-Voetsch C 180, -40 chamber for damp heat (DH, at 85°C, 85% of relative humidity) according to IEC 61215-2 standard [5] for 250 h and posteriorly, such as sequence "B", inside a Q-Sun Xenon chamber (Q-Lab) chamber with a noon summer sunlight filter at the following testing conditions (see IEC 62788-7-2 [6]):

- Irradiance: 0.8 W/m^2at 340 nm
- Relative humidity: 20%
- UV dose (UV-A + UV-B): 60 W/m^2
- Chamber air temperature: 65°C
- Black panel temperature: 90°C

The samples are left in the chamber for a total of 1000 h equivalent to 60 kWh/m^2 which is approximately equivalent to 1 year of outdoor exposure in Central Europe.

In the case of sequence "C" the samples are coated ribbons not laminated of 6 to 8 cm length. The test follows the ASTM D3359 – 17 X cut tape test [7]. With the aid of a sharp tool an "X" of 4 cm lines is carved on the coated ink. As the ribbons are 8 cm by 0.5 cm there will be acute angles in the direction of the longest side of metallic interconnect (see fig. 5). The ribbons are fixed to a flat clean surface. Transparent pressure sensitive tape is cut in 8 cm pieces and glued manually with the help of pressure done by an eraser on top of the "X" avoiding bubble formation. After the tape is positioned, 1.5 min pass before pulling the tape in the direction of the acute angles of the "X" as close as 180° as possible. The result is classified on 5 levels depending on the amount of coating that detaches the metallic interconnect surface.

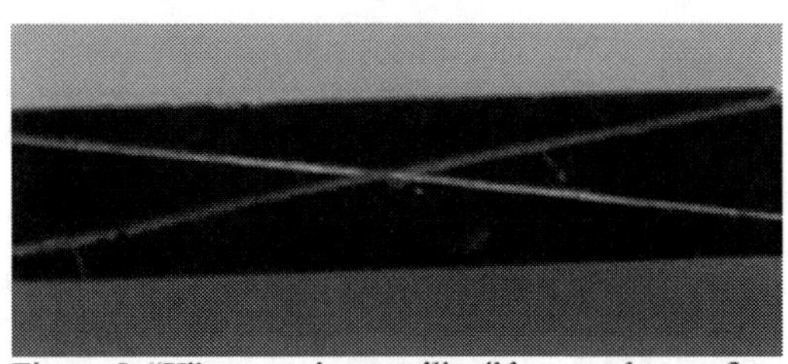

Figure 5: "X" cut on the metallic ribbon used to perform the adhesion test. The acute angles are observed on the lengthiest side of the ribbon, which is the pulled direction.

2.3 Characterization techniques

The measurements performed in this work aim to assess the effects of the coatings in common PV materials under accelerated aging tests. This effects can be aesthetical and electrical in the case of PV modules.

Image processing
Images of the samples are taken periodically with high illumination and resolution. A software was developed using image processing to compare the RGB color coordinates with a reference color, black in the case of this work RGB (0, 0, 0), and the color in our samples employing the CIEDE2000 standard [8]. The software allows to observe a relative change in color while the sample is degrading. A higher color change (ΔE) will be observed for less stable inks.

Reflectance of color palettes
Reflectance measurements directly on the samples is very challenging, therefore color palette such as "Reichsausschuß für Lieferbedingungen und Gütesicherung" (RAL) and "Natural Color System" (NCS) are used to measure the reflectance. The color of the sample is compared with the color of RAL or NCS charts and using a colorimeter X-Rite the reflectance is measured directly from the chart.

Electrical performance
The IV curve and electroluminescence are measured on the one solar cell mini modules to investigate the difference between coated PV module and uncoated one.

3 RESULTS

Out of the seven investigated inks four passed the experimental protocol (see table II). The inks that passed are UV curable and are dispensed by inkjet or screen printing.

Table II: Result of the protocol. Three inks failed the test due to color change or lack of adhesion to the metallic interconnects.

INK #	A	B	C	Failure cause
1	Pass	Pass	Pass	-
2	Pass	Pass	**Fail**	Adhesion
3	**Fail**	**Fail**	Pass	Yellowing
4	**Fail**	**Fail**	**Fail**	Adhesion Yellowing
5	Pass	Pass	Pass	-
6	Pass	Pass	Pass	-
7	Pass	Pass	Pass	-

Inks that failed the adhesion test (sequence C) were harder to manufacture because there was coating detachment during sample preparation and after the lamination process.

3.1 Stability

The stability of the coatings was analyzed with image processing and color palettes. With unstable inks a change in color was observed (see fig. 6)

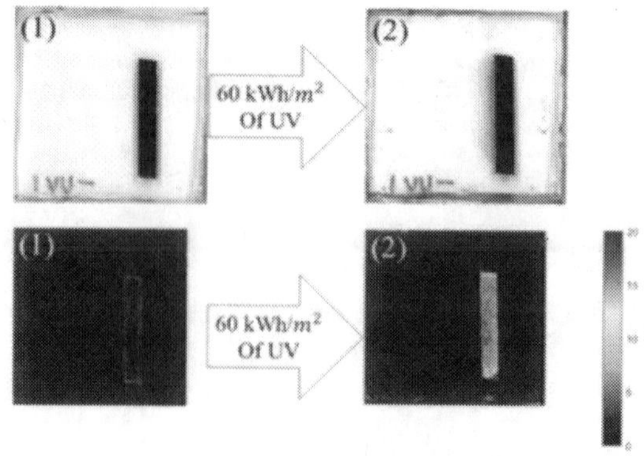

Figure 6: Input (top) and output (bottom) images of the software with G/G samples encapsulated with EVA using ink #3 (1) before and (2) after 60 kWh/$m^2$of UV light (sequence B). Components of the ink are causing change in color in the encapsulant and on the ribbon itself.

DH has not shown any effect on the investigated inks. By the end of the protocol sequence A and B were equivalent. There were no inks that passed sequence B and failed sequence A. The most stable bill of materials was G/G with POE as encapsulant while G/Bs with EVA produced the biggest color change (see fig. 7).

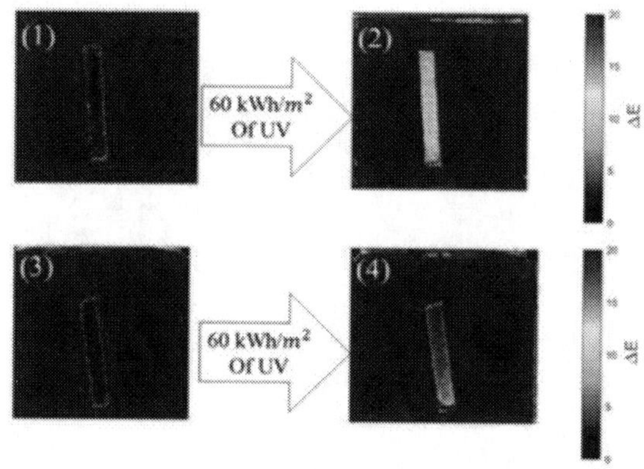

Figure 7: Software output images of (1) G/Bs with EVA and ink #3 sample before sequence B and (2) same sample after. (3) Software output image of G/G with POE and ink #3 sample before and (4) after sequence B. It is observed that (2) has a severe color change while (4) has a mild one in part of the metallic interconnect.

The reflectance measurements using color palettes (see fig. 8 and 9) clearly show a difference in the red side spectrum meaning that the black color is turning brownish after the UV exposure.

Figure 8: Comparison of samples with the RAL color palette, on the left side a one solar cell module (G/Bs and EVA) degraded after 180 kWh/m^2 of UV (3 times sequence B) and on the right side a reference PV module sample with the same ink #3 and materials. The procedure is done under same illumination.

The color palettes are needed because with current color characterization techniques (e.g. colorimeters and UV-Vis spectrometers) it is challenging to accurately measure the reflectance directly on these kind of samples, as several measurement artefacts were observed.

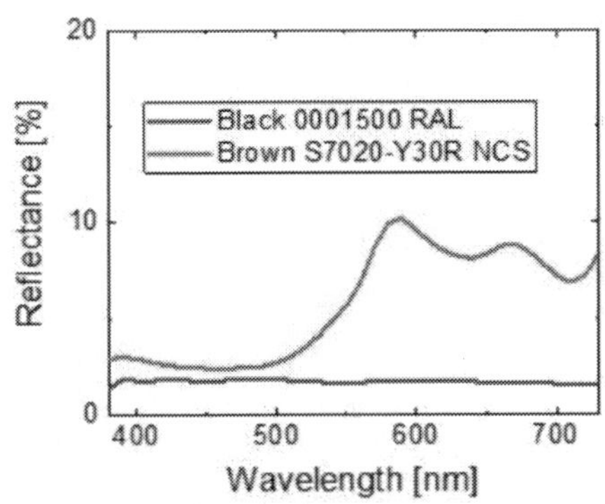

Figure 9: Reflectance measurement of the color palettes using an X-rite colorimeter. Color palettes are matched with the samples shown in fig. 8. Black is the reference module with same materials but has been stored in the darkness and brown is the degraded module which was degraded on 180 kWh/m^2 of UV light.

The encapsulant adjacent and on top of the metallic interconnects is becoming yellow or brown (see fig. 10). We speculate that the browning is due to the diffusion of ink components in the encapsulant. Further analysis is required to validate this hypothesis. In the region nearby the edges of the module the original black colored is maintained, the ink is stable, suggesting that oxygen may play a role to avoid the change in color.

Figure 10: On the left the reference module and on the right the degraded module after 180 kWh/$m^2$of UV light (3 times sequence B), both with ink #3.

We have encapsulated in typical module configurations (G/G and G/Bs) with EVA or POE 2-phenoxyethyl acrylate, which is the component with the highest weight in unstable ink #3. After 60 kWh/m^2 of UV exposure a clear yellowing has appeared, showing that this component is unstable under UV exposure with common PV materials (see fig. 11).

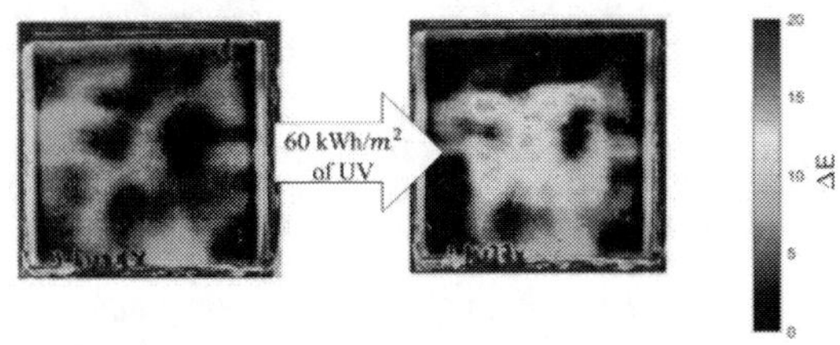

Figure 11: G/G coupon with EVA and 2-phenoxyethyl acrylate, main component on ink #3, before UV exposure (left) and after 60 kWh/m^2 of UV light (right). Clear change in color is observed because the sample is becoming yellow.

Aromatic or chromophoric functional groups may absorb light in the visible region [9], such as the conditions of sequence A and B, this may be the cause of the yellowing.

3.2 Electrical performance

We measured the IV curve to retrieve the power of the module before and after ageing comparing a ribbon coated one solar cell module with an uncoated one (see fig. 12).

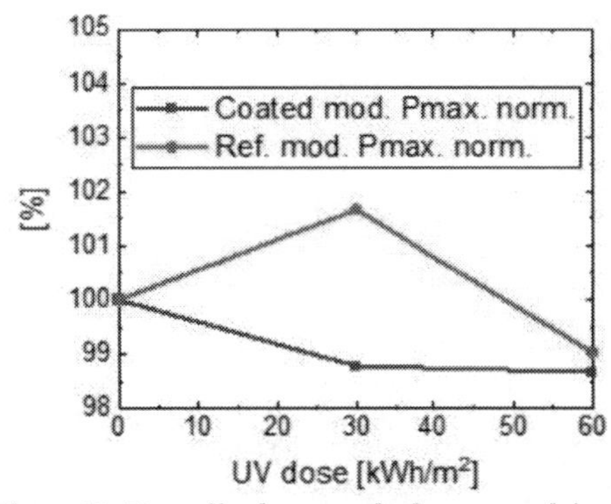

Figure 12: Normalized power of reference module without coated ribbons and coated module with ink #3 on UV exposure. The reproducibility of the measuring equipment is ±1% and uncertainty may be larger so the power loss in both cases is minimal.

The electrical performance measurements suggest that the investigated inks do not influence the power output of the modules. There will be reduction in short circuit current (Isc) due to shading losses if the coating is done on top of the solar cell, but after ageing the loss of power is not incrementing considerably even with the diffusion of ink components into the encapsulant. Besides the IV curve we performed EL measurements and did not show any degradation.

3.3 Adhesion

Most inks have passed the adhesion test. It highly depends on the curing of the coating, when it is not fully cured the expected result is that the ink will fail. Ink #2 is a clear example of an ink with poor adhesion to the metallic interconnects (see fig. 13).

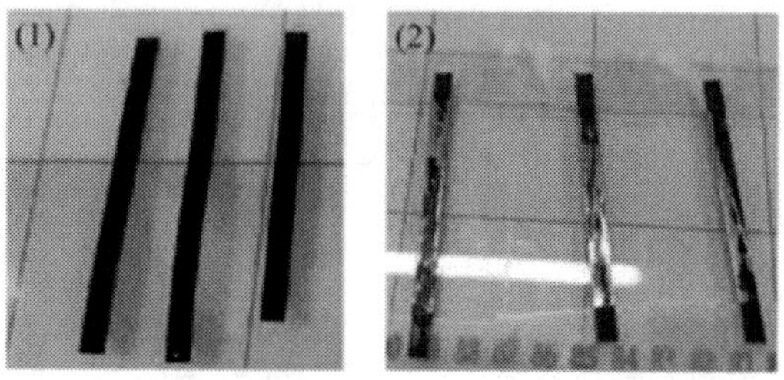

Figure 13: Metallic interconnects masked with ink #2 (1) before adhesion test and (2) after. The adhesion of this coating is poor and could cause detachment during PV module manufacturing in the production line.

Ink #2 is solvent based, it is not possible to improve the curing by increasing the UV dose, as it is possible with the other investigated inks. This is one of the drawbacks that solvent based inks have.

4 CONCLUSION

The purpose of this work is to investigate the stability and adhesion of seven commercially-available black inks to check if they are suitable to mask the metallic ribbons of PV modules, and to investigate the effect of unstable inks in PV modules. The G/G and G/Bs configurations with EVA and POE were exposed to a protocol which has three main sequences in parallel. Sequence A has 250 h of DH (85 °C, 85%RH) followed by 60 kWh/m^2 of UV light exposure (corresponding to approximately 1 years of operation in a Central European temperate climate), sequence B involving only the UV step, and sequence C related to adhesion testing. The protocol was designed based on already existing IEC standards. Additionally, the metallic ribbons of one solar cell modules were coated with unstable ink and aged under UV to investigate their electrical performance and aesthetics.

All investigated inks passed the DH test. Unstable inks (#3 and #4) produce a change in color from black to brown or yellow when exposed to UV light which is diffused into the encapsulant. The use of the investigated unstable inks would represent an aesthetical defect in the long term, but the electrical performance would remain almost constant (less than 1% power loss after 60 kWh/m^2 of UV exposure). The tested screen printed inks pass the protocol, but the implementation into production line would be challenging due to alignment of stringed cells. Among the seven investigated inks four would pass this protocol.

The proposed protocol sets only the minimum requirement that coatings for masking metallic interconnects should pass. To assess the stability in a longer term the accelerated ageing test should be extended. We are further investigating stability of coatings for BIPV applications and color characterizations techniques.

5 REFERENCES

[1] REN21, *Renewables 2022 Global Status Report*. 2022.

[2] C. Ballif, L.-E. Perret-Aebi, S. Lufkin, and E. Rey, "Integrated thinking for photovoltaics in buildings," *Nat. Energy*, vol. 3, no. 6, pp. 438–442, Jun. 2018, doi: 10.1038/s41560-018-0176-2.

[3] M. H. Saw, J. P. Singh, Y. Wang, K. E. Birgersson, and Y. S. Khoo, "Electrical Performance Study of Colored c-Si Building-Integrated PV Modules," *IEEE J. Photovoltaics*, vol. 10, no. 4, pp. 1027–1034, 2020, doi: 10.1109/JPHOTOV.2020.2981820.

[4] A. Borja Block, A. Virtuani, and C. Ballif, "STABILITY OF INKS USED FOR MASKING METALLIC INTERCONNECTS IN BIPV MODULES," in *38th European Photovoltaic Solar Energy Conference and Exhibition*, 2021, vol. 38, pp. 723–727.

[5] IEC, "IEC 61215-2: Terrestrial photovoltaic (PV) modules – Design qualification and type approval – Part 2: Test procedures," 2016.

[6] IEC, "IEC TS 62788-7-2: Measurement procedures for materials used in photovoltaic modules - Part 7-2: Environmental exposures - Accelerated weathering tests of polymeric materials," 2017.

[7] A. International, "Standard Test Methods for Measuring Adhesion by Tape Test 1." pp. 1–8, 2017, doi: 10.1520/D3359-09E02.2.

[8] G. Sharma, W. Wu, and E. N. Dalal, "The CIEDE2000 color-difference formula: Implementation notes, supplementary test data, and mathematical observations," *Color Res. Appl.*, vol. 30, no. 1, pp. 21–30, Feb. 2005, doi: 10.1002/col.20070.

[9] J. Segurola, N. S. Allen, M. Edge, A. McMahon, and S. Wilson, "Photoyellowing and discolouration of UV cured acrylated clear coatings systems: Influence of photoinitiator type," *Polym. Degrad. Stab.*, vol. 64, no. 1, pp. 39–48, 1999, doi: 10.1016/S0141-3910(98)00169-4.

6 ACKNOWLEDGMENTS

This project has received funding from the European Union's Horizon 2020 research and innovation programme under the Marie Skłodowska-Curie grant agreement No 754354 and has been funded in part by the European Commission (EC) under the H2020 Be-SMART (#818009). We gratefully acknowledge support from all PV-Lab team members.

ASSESSMENT OF ECA TO RIBBON INTERCONNECTION STABILITY BY CURRENT PATH AND POWER LOSS IMAGING

S. Großer[1*], M. Schak[1], T. Timofte[2], M. Turek[1]
[1] Fraunhofer Center for Silicon Photovoltaics CSP, Otto-Eissfeldt-Strasse 12, 06120 Halle (Saale), Germany
[2] International Solar Energy Research Center Konstanz e.V., Rudolf-Diesel-Strasse 15, 78467 Konstanz, Germany
*Corresponding author: stephan.grosser@csp.fraunhofer.de

ABSTRACT: Solar cells with back-junction technology exhibit the metallization at the rear side of the solar cell thus gaining from a shading free sunny side. Electrical interconnections between the solar cells and ribbons in a string are exposed to various stresses during manufacturing of the module. Additionally, they are stressed by environmental conditions during field operation. We present a study of electrical-conductive adhesive (ECA) joints for back-junction (BJ) solar cell technology focusing on advanced defect diagnostics of the joints during and after climate stress. To assess the stability and prevent failures, advanced diagnostic methods for joints are needed which provide fast non-destructive and spatial-resolved information on the lateral current flow. Commonly, inspection techniques rely on electroluminescence imaging (EL) or visual inspection (VI). However, they are not sufficient since the information depends on indirect impacts. In this work, a comprehensive application by means of the magnetic-field imaging (MFI) and Lock-In Thermography (LIT) will be presented and compared to standard imaging techniques. MFI allows imaging of the lateral current paths or local current interruptions even inside a laminate while LIT is related to thermal power losses. The benefits of the combined MFI/LIT method for advanced failure diagnostics will be shown on selected defect ECA-joints. The MFI/LIT application offers the opportunity to improve the assessment of ECA-joints aging due to the spatial resolution and fingerprint defect signatures like single defective joints and current-stressed junctions. This will improve the risk assessment of process and design related late defects of interconnections of BJ-solar cells and other technologies. Furthermore, due to the defect localization, advanced material analysis on microscopic scale is applicable by target preparation allowing the determination of the underlying failure mechanism. This approach supports the detailed assessment of ECA to ribbon interconnection and allows a deeper understanding of root causes in terms of degradation mechanisms of different materials and processes.
Keywords: Interconnection, Degradation, Ribbons, Characterization, Magnetic Field

1 INTRODUCTION

Back-junction solar cells benefit from a shading free sunny side since both contact polarities are on the rear side of the solar cell. To contact back-junction solar cells, ECAs are a very promising candidates which allow furthermore an ecologically friendly lead-free joint formation [1].

Electrical interconnections between cells and ribbons must resist various stresses [2, 3, 4]. Reliable joints between solar cells and ribbons are essential in solar module operation because the properties of the joints have a strong impact to the module performance. Degradation of joints due to aging leads to power losses. In this work, a method to study ECA joint degradation even at early stages is presented. Identification and imaging of joint degradation is important to localize aging of electrical-conductive adhesive (ECA). Magnetic Field Imaging (MFI) [5] as a versatile new characterization method and Lock-In Thermography (LIT) have been applied to detect lateral current paths and the power dissipation. In this paper, a correlative analysis of the non-destructive characterization methods will be presented to identify representative positions of degradation classes in modules. By means of microscopy, characteristic positions of ECA joints between solar cells and ribbons are investigated to determine the structural root cause for interconnection degradation, confirming the non-destructive approach and enable root cause analysis of the failure mode in detail.

2 EXPERIMENTAL

For evaluation of the rapid test methods half-cell minimodules with ZEBRA IBC solar cells in M2 format and 4 busbars were manufactured (Figure 1).

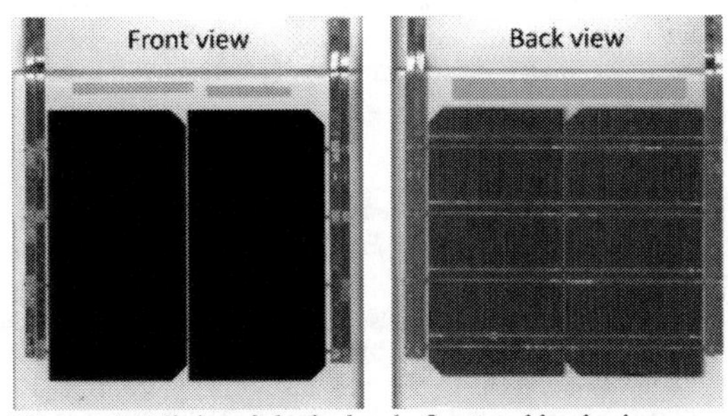

Figure 1: Minimodule design in front and back view.

A group of samples with soldered interconnections acts as a reference and an ECA interconnection process as the process to study. The ECA interconnection group has been divided into 3 sub-groups (ECA 1 - 3) differing in the used ribbon coating and process recipe. For ECA 1 and ECA 2 the interconnection consists of ECA and an Ag-coated Cu-ribbon and differs in the curing condition (fast / slow). The third ECA group has been fabricated by using a SnPb-coated Cu-ribbon with slow curing condition. Figure 2 shows the reported groups in tabular form. ECA interconnections to $Sn_{60}Pb_{40}$-coated ribbons can degrade under accelerated aging conditions. Therefore, the group ECA 3 is expected to exhibit failures which shall act as obvious test sites. The half-cell strings have been soldered to cross-connectors and encapsulated by EVA between uncoated float glass and a transparent back sheet. Each group consist of 3 samples for aging and 1 retained sample.

Group	Solder	ECA 1	ECA 2	ECA 3
Solar cell	ZEBRA M2 half cell (2x)			
Ribbon	Coated copper, 1.5mm x 0.2mm			
Method	Soldering	Electrically conductive adhesive (ECA)		
Ribbon coating	$Sn_{60}Pb_{40}$	Ag_{100}	Ag_{100}	$Sn_{60}Pb_{40}$
Recipe	320°C / 1s/BB	170°C / 4min	180°C / 2min	170°C / 4min
Encapsulation	Glass / EVA / transparent backsheet minimodule, with soldered cross-connectors			
Aging	TCT 200			
Quantity	3+1	3+1	3+1	3+1

Figure 2: Experiment design with 4 groups containing 3 different ECA and 1 solder interconnection processes.

Each group has been characterized with respect to their integral electrical performance (IV) and by locally resolved imaging techniques. EL has been measured with a LOANA tool by PV-tools. Thermal loss imaging has been performed by dark Lock-In Thermography using a InfraTec ImageIR 9400 with a resolution of 1280x1024px². For Magnetic Field Imaging (MFI) a measurement system equipped with an MFI-sensor from DENKweit has been applied during current feed. These characterization methods have been performed before and after accelerated aging by temperature cycling test with 200 cycles. Cross-sections for imaging the contact geometry between the ribbon and the solar cell have been prepared using gentle mechanical separation and metallographic preparation (grinding and polishing of resin embedded samples). Inspections of cross-sections have been performed by means of Scanning Electron Microscopy (SEM) using a HITACHI SU-70 system.

3 RESULTS

3.1 Electrical performance after temperature cycling

After TCT200 aging of all groups, exclusively the group ECA 3 (ECA/$Sn_{60}Pb_{40}$-coated Cu-ribbon) degraded very strongly in efficiency (Figure 3). The electrical degradation is dominated by the loss in the fill factor and related to aging of the interconnection (Figure 4). Stable electrical performance (η, I_{sc}, V_{oc} and FF) have been found for all other groups.

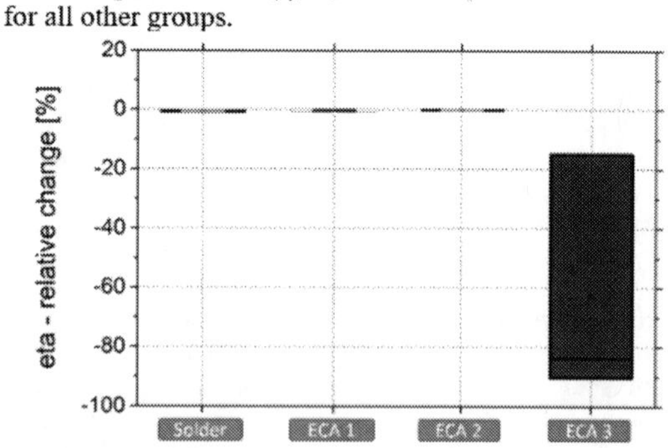

Figure 3: Boxplot of the relative change in efficiency after TCT 200.

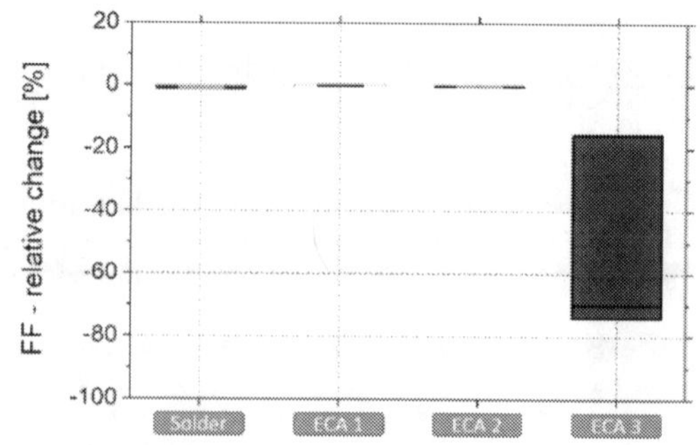

Figure 4: Boxplot of the relative change in FF after TCT200.

3.2 Contact structure of stable group

A reliable investigation of the microstructure depends in particular on an artifact-free or -minimized preparation. Preparation artifacts complicate the interpretation of observations. Therefore, a cross-section of a ECA interconnection without accelerated aging has been prepared and investigated regarding the occurrence of artifacts. In this paper, the SEM images exhibit the section of the copper ribbon in the upper half of the image and the solar cell in the lower half. Figure 5 a) shows the representative SEM image of the interconnection microstructure from group ECA 1 (retained sample without TCT aging). The ECA can be seen in between the busbar and the thin Ag ribbon coating. ECA's microstructure is represented by the (dark) non-conductive adhesive and the (bright) conductive Ag-filler. A widely form-closed join can be observed without artifacts or failures and confirms the gentle preparation suitable for detection of even small changes. Figure 5 b) present the observation of the contact microstructure (same group) after accelerated aging with TCT 200. In comparison to Figure 5 a) small cracks were found at the interface between the ECA and the Ag-coating of the ribbon. These cracks appear locally and extends in length over several micrometers whereas the crack width is well below 1 µm. Cavities within the ECA appear occasionally in ECAs independent from aging. Since the electrical measurements did not show evidence of electrical degradation, the observed cracks seem to be less severe. Nevertheless, the result indicates the decrease of the adhesion at the Ag-interface.

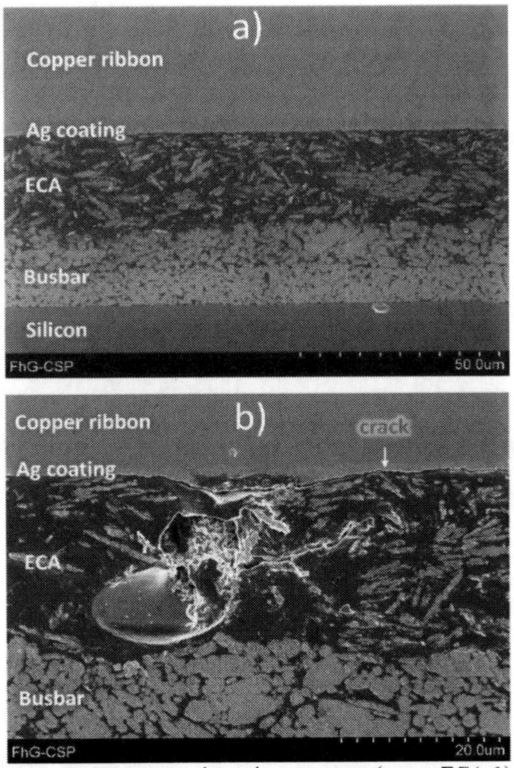

Figure 5: Interconnection microstructure (group ECA 1) of the a) initial status (retained sample) and b) after TCT 200. An arrow in b) indicates a thin crack at the interface between ECA and Ag-coating.

3.3 Localization of interconnection failures

Microstructure analysis allows detailed insights in the geometry of the contact. Nevertheless, a cross-section is a very local state on large interconnection system on a solar cell. A non-destructive approach is needed to screen contacts on large scales regarding contact issues and furthermore enable classification of different states.

Interconnection issues can result in thermal losses as a consequence of contact resistance increase. A negligible thermal loss is expected for infinite as well as low contact resistance. These states can be differentiated by observing the local current changes since an isolated contact position will not inject electrons in the busbar and the current in the ribbon keeps constant (Kirchhoff's circuit law). In this paper, we tested a correlative imaging approach to locally detect thermal losses by LIT technique and current paths by MFI-technique to locate and subsequent investigate the corresponding microstructure of the interconnection degradation. Figure 6 exhibits the collage of the results obtained by the employed imaging techniques including photography for identification of optical changes as well as electroluminescence (EL). The same module (group ECA 3) was measured before (Figure 6, left) and after (Figure 6, right) TCT 200. After TCT 200 the efficiency of the module dropped by -90%$_{rel.}$ to around 2 %.

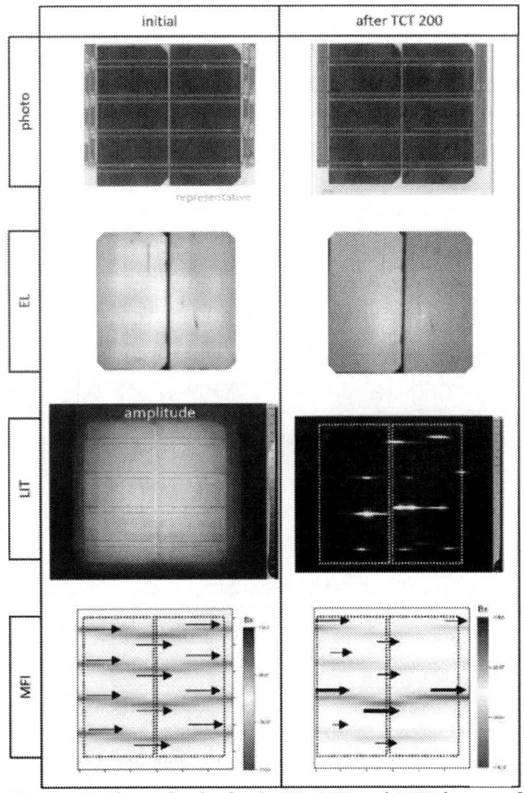

Figure 6: Photo (back view), EL, LIT and MFI image of the identical sample (ECA 3) initially and after TCT 200. Half-cell outlines drawn as a guide to the eye. MFI image represents the magnetic flux density caused by current parallel to the busbars. Implicated currents are drawn as arrows.

The initial measurements indicate a reasonable interconnection of all busbars. With focus on LIT and MFI, the LIT amplitude image shows a homogeneous thermal loss within the solar cell by recombination. The MFI image (magnetic flux density caused by current parallel to the busbars) can be interpretated as an equal distribution of the current densities within the 4 parallel ribbons. The intensity in the MFI image is the measured signal of B_x-component of the 3d-magnetic flux density. This magnetic flux density is induced by the local current. In the measurement setup, the intensity of the B_x-component correlates with the current density parallel to the busbars. Each of the 4 parallel (horizontal) ribbons conducts the comparable current from the cross connector into the cell. As the current decrease along the ribbon (connection to cross-connector) the nearby next ribbon (connection to the next half-cell) gathers the current. By comparison of all 4 parallel ribbons at the equivalent position (vertical line) the intensity of the B_x-component (current) is comparable and therefore indicate a homogeneous current feed. On the right-hand side one can observe the corresponding measurement after accelerated aging. The optical image shows 1 burn mark (red arrow) whereas the EL image reveals no additional locally resolved information. By contrast to EL, the lateral distribution of resistance losses in the interconnection can be observed by LIT. Thermal losses appear exclusively at positions of the interconnection within the solar cell.

Furthermore, the generated heat depends on the position within the module and differs strongly. A comparison of the MFI image after TCT with the initial state allows the assessment changes within the current distribution. Since the intensity of the signal correlates qualitatively with the current density one can deduce that the currents differ between parallel ribbons after TCT 200. Ribbons with high current densities can be found as well as ribbons with current densities below the detection limit. As a guide to the eye and for illustration black arrows are drawn in the image of Figure 6 symbolizing the deduced current densities. Combining the information of LIT and MFI allow to characterize the interconnection state more clearly than optical or EL imaging alone, as it allows to classify as well as localize different states of interconnection aging.

Two characteristic interconnections have been compared by means of microstructure diagnostics after target preparation. Figure 7 shows a selected position within the module of Figure 6. Considering the optical and current density information there is no evidence of differences. Whereas the LIT measurement clearly shows that power losses are increased in the lower ribbon (orange marked in Figure 7). The power loss is related to heating within the interconnection and not by conduction losses within the ribbon. To proof this hypothesis, the interconnection structure of this particular position has been investigated. Figure 8 a) represents the interconnection structure observed at the orange marked position in Figure 7 whereas Figure 8 b) shows the blue marked position. It can be observed that Figure 8 a) exhibits extended and wide delamination at the interface of the ECA to SnPb-coating. Please note, since the observed delamination is laterally extended, the formed cavities trap unavoidable debris from the preparation. Debris is an artifact and will be ignored for interpretation. Nevertheless, the wide delamination results in an interruption of the contact. On the left side of the delamination, the SnPb-coating is still connected to the ECA indicating that not the full interconnection was interrupted. One can assume that after TCT 200, the current from the busbar is expected to be conducted through a smaller interconnection area leading to enhanced power losses with the observed generated heat. By comparison to the nearby interconnection shown in Figure 8 b) the simplified interpretation can be confirmed. In cross-section view, the structure in Figure 8 b) shows a different characteristic with the ECA still being connected to the SnPb-coating. Thin cracks at the interface were also observed in this case similarly to the results shown in Figure 5 b) for the non-degraded group ECA 1.

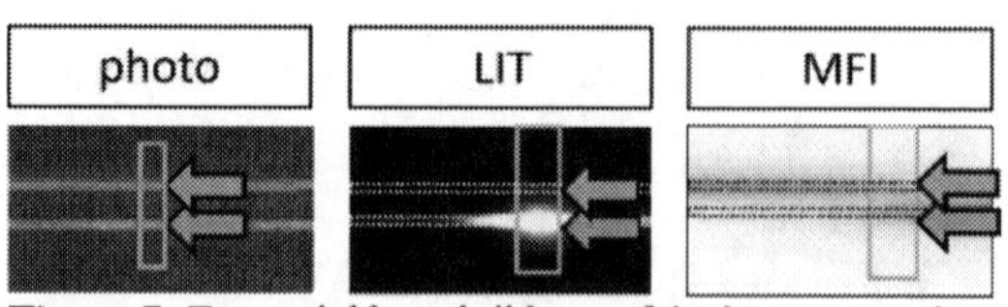

Figure 7: Two neighbored ribbons of the interconnection. Different thermal losses and but comparable current densities are deduced from the corresponding LIT and MFI information. Position of the ribbons have been indicated as a guide to the eye.

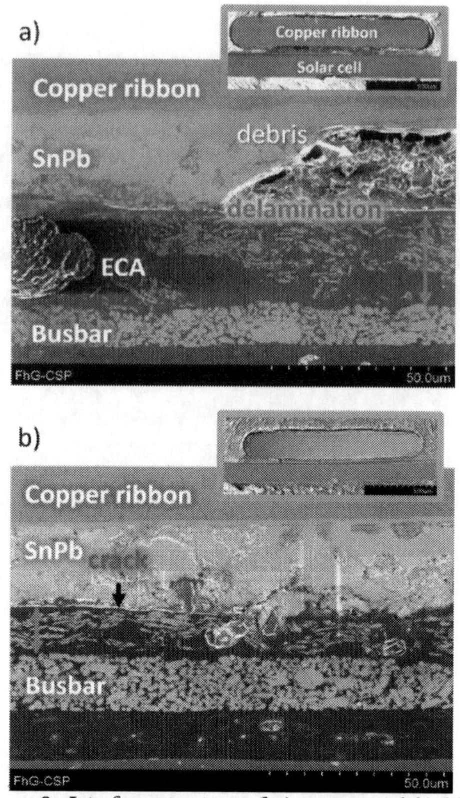

Figure 8: Interface structure of a) orange and b) blue marked position in Figure 7. Overview (inset) and detail view are shown. A thin crack in b) has marked by an arrow.

4 CONCLUSION

The assessment of ECA to ribbon interconnection stability has been investigated with focus on its microstructural properties. Based on a comparison of different interconnection materials and accelerated aging, the method of local assessment by lateral current path and power loss imaging could be demonstrated. Lock-In Thermography in combination with Magnetic Field Imaging have been applied to support locally resolved information about interconnection degradation between ribbons and the solar cell. Furthermore, due to the physically independent measuring method, current path and power loss information can be obtained. It has been confirmed by microstructure analysis of specific positions that the interpretation of LIT and MFI information allows further differentiation and classification of interconnection on larger scales.

The presented approach can improve the assessment of interconnection between ribbons and solar cells for new material combinations and support quality improvement due to degradation localization and clear correlation to the root causes found at the related contact microstructure.

5 ACKNOWLEDEMENTS

The authors like to thank Denise Ulm for preparation and microstructure analysis. Financial support from the Federal Ministry for Economic Affairs and Climate Action within the funded project “Zquadrat” (FKZ: 03EE1005) is gratefully acknowledged.

6 REFERENCES

[1] D. Tune et al., „The sun is rising on conductive adhesives“ Photovoltaics International 47 (2021) 57-77

[2] G. Oreski et al., „Reliability of Electrically Conductive Adhesives“ Photovoltaics international 42 (2019) 38-45

[3] T. Timofte et al., „Reliability Assessment for Industrial Soldered and Glued BJ-BC Solar Cell Interconnections“ 38th EUPVSEC (2021) 611-618

[4] T. Geipel et al., „Reliable interconnection of the front side grid fingers using silver reduced conductive adhesives “ Energy Procedia 55 (2014) 336 – 341

[5] D. Lausch et al., „Magnetic Field Imaging (MFI) of Solar Modules“ 35th EUPVSEC (2018) 1060-1064

ASSESSMENT OF INTERCONNECTION AGING IN SHINGLED CELL MODULES

S. Großer*, M. Schak, M. Pander, M. Turek, B. Jäckel
Fraunhofer Center for Silicon Photovoltaics CSP, Otto-Eissfeldt-Strasse 12, 06120 Halle (Saale), Germany
*Corresponding author: stephan.grosser@csp.fraunhofer.de

ABSTRACT: Shingled solar cells with Electrically Conductive Adhesive (ECA) interconnection represent a technological path to make very esthetic and efficient modules with lead free interconnection technology. The reliability assessment of the interconnection stability in shingled modules is one major aspect in the material qualification for selection of proper ECAs and process requirements. For shingle technology, commonly used electroluminescence (EL) image interpretation for string or module inspection gets more complex due to the lack of simple fingerprint-like clear features. The aim of this work is the evaluation of non-destructive proof, localization, and quantification of ECA failures in shingled solar cell modules. The Magnetic Field Imaging (MFI) technique has been applied to specifically detect defects in the interconnection. A simplified visualization scheme has been developed and successfully tested. A comparison between stressed shingled strings and their initial data allows the quantitative assessment of mechanical stress-related (bending and cycling) ECA-interconnection failures, their lateral distribution und occurrence frequency. With the application of the MFI-based imaging technique, it is feasible to detect and assess aged ECA interconnections as well as identify string-dependent inhomogeneities in aging. The applicability of the MFI-based approach has been successfully demonstrated and extends EL characterization with specific sensitivity to interconnection reliability assessment.
Keywords: Interconnection, Shingle, Stability, Characterization, Magnetic Field

1 INTRODUCTION

Shingled solar cells represent a technological path to make aesthetic, efficient and highly performant modules even under partial shading conditions [1]. Assessment of cell interconnections can be challenging due to complex EL patterns in example for material qualification [2, 3]. In contrast to ribbon or wire-based interconnections the shingled interconnections differ in the current flow. The busbar in shingle interconnection is perpendicular to the string direction and is used for interconnection to the next shingle cell. EL images for string or module inspection get more complex disallowing fingerprint-like clear features in the EL image.

ECA can play an important role in the PV interconnection technology but beside the cost, the reliability plays an important role [4, 5]. Methods to unambiguously identify the interconnection specific properties after stress tests or manufacturing are needed to characterize the reliability.

The aim of this work is the evaluation of an imaging technique for non-destructive proof, detection, and localization of ECA failures in shingled solar cell module. We employ Magnetic Field Imaging (MFI) technique [6] to detect and validate specifically current patterns caused by interconnection failures. We show that the application of MFI to embedded shingled solar cells allows the identification and localization of ECA aging and extends the commonly used EL technique. Shingle module development, ECA assessment and reliability test benefit from the presented approach. Sequential characterization and stress tests were performed to age the ECA-interconnection. Test specimen (small shingle modules) were characterized in terms of IV-performance and measured by EL as well as MFI. A comparison between initial and stressed shingles strings allows the assessment of stress related ECA-interconnection failures and their distribution.

2 EXPERIMENTAL

In this work, two comparable non-commercial shingle glass / back sheet modules with two parallel connected strings and 32 shingled solar cells have been tested, see Figure 1. Both modules have been characterized and stressed with the same conditions. The experimental approach is schematically shown in Figure 2.

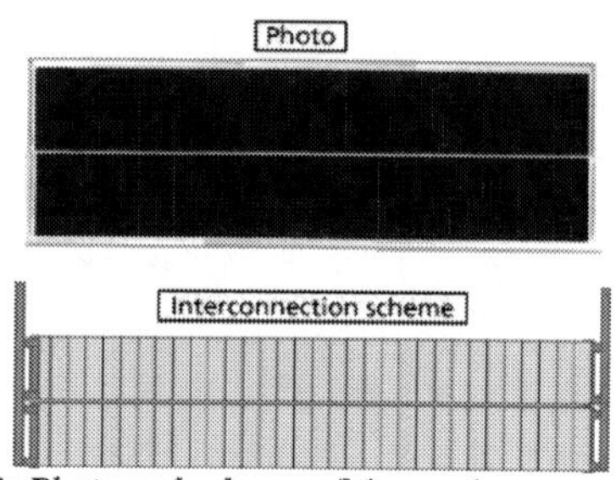

Figure 1: Photo and scheme of the used non-commercial shingle module design.

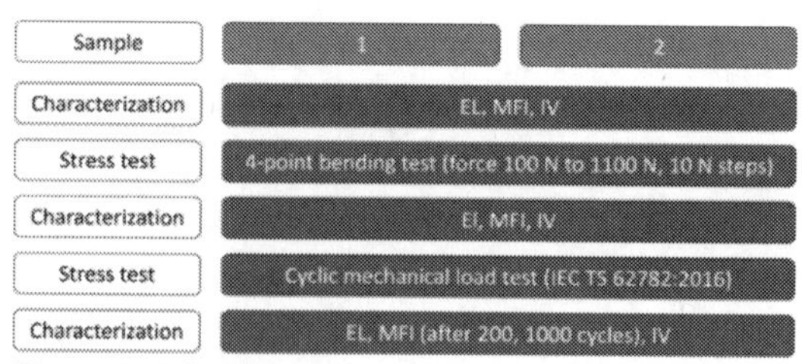

Figure 2: Schematic experiment plan with two stress tests and various characterization steps.

The four-point bending test setup with in-situ EL has been developed by Sander [7]. The assembly is installed in a Zwick Z400 universal testing machine. The distance between the load rollers is 400 mm and the support distance 840 mm. An EL camera is placed on the load rollers which enables in-situ EL. In the mechanical load test, the load is increased step by step in 10 N increments up to a maximum of 1100 N. An EL recording is made in each step. The same setup is used for cyclic mechanical loading. Here the cyclic load amplitude was chosen based on consideration made in Ref. [8] to represent the IEC TS 62782 cyclic load test conditions (1.000 Pa, 1000 cycles

[9]). Electroluminescence images are taken using a Greateyes Lumisolar Professional system with different current injection and exposure time settings (e.g. I = 3 A ~ I_{SC} and 1s). Current-voltage-curves are measured using a Berger Lichttechnik class AAA flasher under STC conditions. For Magnetic Field Imaging (MFI), a measurement system equipped with an MFI-sensor from DENKweit has been applied during current feed with a Keithley 2461 SMU. EL, MFI and IV characterizations have been performed before and after stress test.

3 PRELIMINARY CONSIDERATION

3.1 Implication of shingle cell contact irregularity to the lateral current path

A scheme of the current flow of three shingled cells as part of a horizontal string is shown in Figure 3. For simplification, only the presumed lateral (in-plane) current as well as the major (dominating) current flows are sketched in the scheme by arrows. Figure 3 a) represents the reference situation where the black-indicated perpendicular ECA interconnections are undamaged or disturbed. Horizontal lateral currents along the string dominate the current path. By introduction of an extended discontinuity like an ECA interconnection interruption, the dominating current paths are assumed to be disturbed like sketched in Figure 3 b). Additional perpendicular (along the shingle cell) current paths are expected to bypass the affected part of the interconnection. In case of the left shingle, the current flow points downwards (blue arrow) and it points upwards in the neighbored (middle) cell (blue arrow). Consequently, a disturbed ECA interconnection in a shingle string will affect generally two cells and not only a single cell.

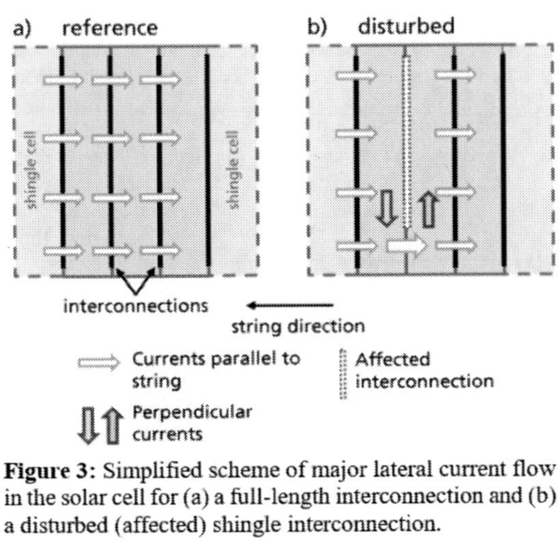

Figure 3: Simplified scheme of major lateral current flow in the solar cell for (a) a full-length interconnection and (b) a disturbed (affected) shingle interconnection.

3.2 Current path detection by MFI – instructive examples

Taking the considerations from the former subchapter into account, ECA interconnection failures in shingle strings can be observed if the current can be locally measured as well as their direction detected which we call current path. MFI technique delivers the opportunity to measure the magnetic flux density arising from electrical currents in conductors. The magnetic flux density has to be expressed in form of a vector. By separate detection of the magnetic flux density in the three directions (X, Y, Z), a simplified examination is feasible to detect in-plane current paths. The absolute value of B is for a wire proportional to the current I and inverse proportional to the distance. Therefore the MFI sensor must maintain a constant distance to the sample during measurement. Considering only the measured X and Y-components (2D planar expression of the magnetic flux density) and a suited orientation of the line sensor (along a current conduction wire direction), the current paths in an electrical device can be imaged and separated into parallel and perpendicular contributions. For illustration, instructive examples are presented in Figure 4. Two basic shapes of current conducting circuits have been measured by MFI. The "C"-shaped example shows clearly the separation of horizontal and perpendicular currents. The "split-up" shape exhibit additionally the current split in the lower half with reduced B_x intensity and the junction in the upper half according to Kirchhoff's circuit law.

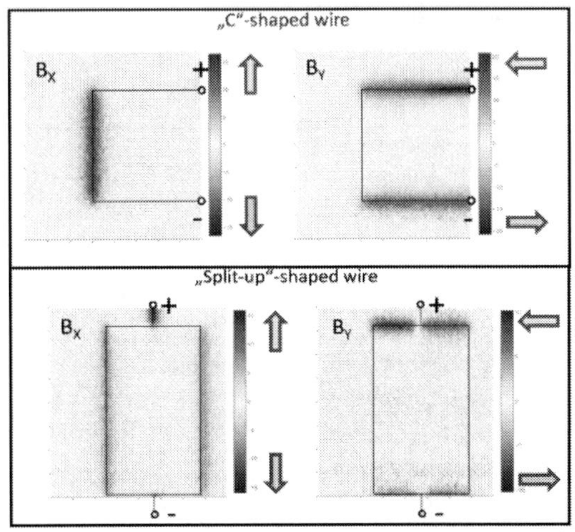

Figure 4: Image of B_x- and B_y-component for two types of electrical circuits. The position of the wire in the circuit is drawn as an overlay. The arrows indicate the deduced (physical) current direction.

4 RESULTS

4.1 Characterization of lateral currents in two shingled cells

To validate the lateral current flow, a simplified analysis approach has been validated. Module 1 in the initial state have been imaged with EL technique at low current (0.3 A, well below the I_{SC}) which is presented in part in Figure 5 a). Obviously, one shingle cell in the string is darker than its left and right neighboring cells. In serial connection, the same integral current flows through all cells. The voltage drop within the solar cell dominates exponentially the EL intensity ratio between two cells in a serial interconnection [10]. The lower intensity of the electroluminescence could be related to a different dark IV characteristic of the cell or to a failure in the interconnection. A measurement with the simplified MFI approach is presented in Figure 5 b) where the left image represents the perpendicular current and the right image the horizontal current contribution separately. The dashed box is a guide to the eye to allow local correlation of the cell pair. Noticeable perpendicular currents as well as a locally increased horizontal current at the interconnection edge are observable (colored arrows at the scale bare explain the responsible current direction). An implication of the lateral current paths in the existing shingle interconnection can be concluded taking the former considerations shown in Figure 3 and Figure 4 into account, see Figure 5 c). The detected perpendicular current pattern indicates clearly that the failure is located between the middle cell (dark in EL) and right cell (bright in EL) and exclude interconnection to the left cell.

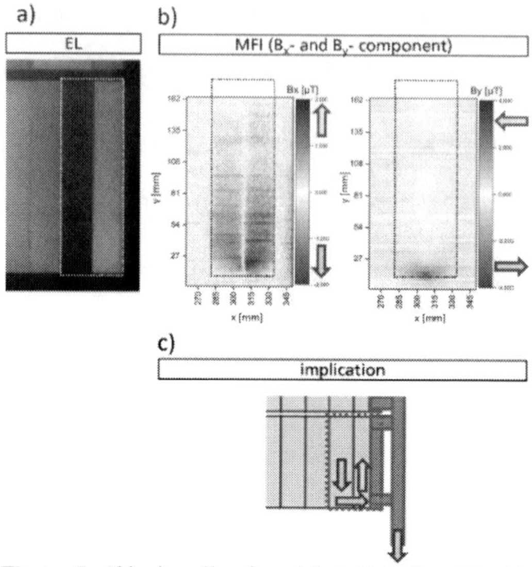

Figure 5: Shingle cells of module 1 (initial state) with different intensities in a) EL. Corresponding images of b) B_x- and B_y-component detected with MFI and c) implicated current paths. (EL, MFI recorded at I = 0.3 A)

4.2 MFI for module interconnection assessment

In the former subchapter, the application has been validated for one shingle cell pair of the module. The assessment has been extended to a shingle module with multiple shingle cell pairs and mechanical stress induced aging. This subchapter will propose a further simplification of the data evaluation to assess ECA interconnections in shingle modules.

As observed from the results shown before the value especially of the perpendicular (to the string direction) current paths is for shingled cells in most cases correlated with a failure. Figure 6 shows a collage of EL and MFI after 200 cycles and after 1000 cycles of the cyclic load test measured with comparable conditions. Both EL images seem to be comparable. This is in contrast to the MFI image which clearly shows areas with obviously larger values of perpendicular currents exemplary marked in MFI and EL images. Thus, the MFI-approach is supposed to be more sensitive to contact failures in shingle cell string than EL due to the detection of failure-related bypassed currents. Since MFI and EL are based on different physical mechanisms. MFI can detect large signals even at negligible EL intensity. Presumable good or less affected areas in the string can also be assessed by low MFI-intensity e.g., in the middle of the upper string of Figure 6.

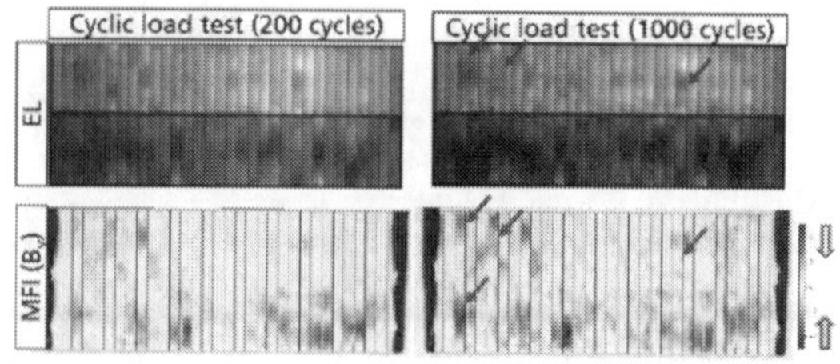

Figure 6: Module 1 after 200 and after 1000 cycles of the cyclic load test. a) EL and b) MFI (B_y-component) with overlay of a pattern indicating the cell outlines as guide to the eye. Red arrows exemplarily mark increasing values in B_y after 800 cycles.

One can see from Figure 7 a) that the MFI (B_y-component) of a module exhibits a large information depth. We propose a simplified approach to reduce the information further for a relative comparison of changes e.g. after stress or between comparable designs. Figure 7 a) exhibit the MFI (B_y-component) of module 1 after 1000 cycles of the cyclic load test. On the left and the right side, the current feed from the cross connectors is dominating. Using a pattern of the shingle cell outlines the size of the B_y-component was measured and transformed in a 3-color class representing the range of the B_y (see Figure 7 b)). This approach is a qualitative class definition and comparable only for comparable measurement conditions.

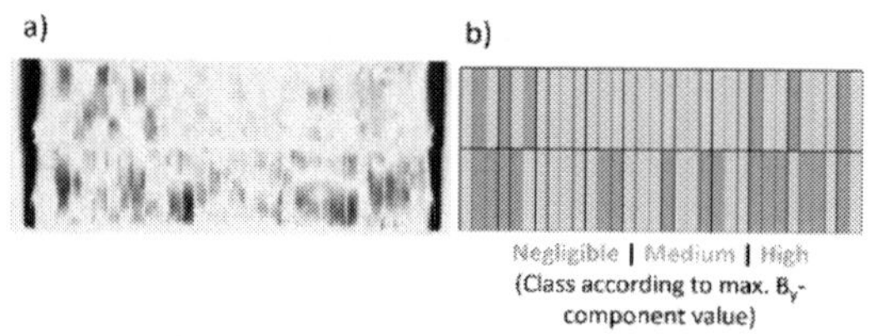

Figure 7: a) B_y-component raw image of module 1 after 1000 cycles of the cyclic load test and b) deduced simplified 3 classes scheme (< 2µT | 2-4µT | > 4µT).

This approach has been applied to three stages of experiment, see Figure 2. EL and the MFI-based interconnection assessment illustration are presented for module 1 in Figure 8. The initial state is shown in the top and the state after four-point bending test and after additional cyclic mechanical load is shown at the bottom, respectively. Initially, the lower string exhibits more failures. After four-point bending test, aging was found predominantly in the lower string (black arrow). The subsequently applied cyclic mechanical load test (after 1000 cycles) provoked changes in both strings (black arrows).

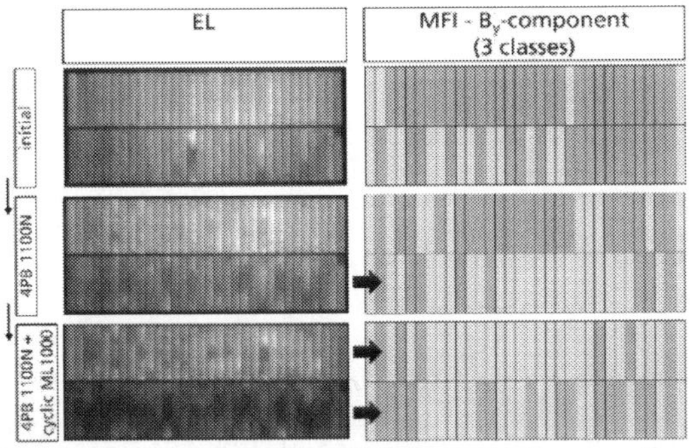

Figure 8: EL image (left) and MFI-based interconnection assessment illustration (right) obtained on module 1 at initial state, after four-point bending test and after additional cyclic mechanical load test. Arrows mark strings with larger changes.

The impact of the cyclic mechanical load test with 1000 cycles has been compared between the two tested shingle cell modules to cross-check the simplified approach. The comparison is shown in Figure 9 where the relative change of the frequency of the three groups has been plotted for both samples. A decrease of intact faultless interconnected cells by up to 80 % is caused by the cyclic mechanical load test in both modules after 1000 cycles. The number of strongly affected cells is increased by around 300 % after the four-point bending test since only a few cells were assigned to this class initially (see Figure 8). Therefore, the cyclic load test stressed the interconnections considerably in comparison to the four-point bending tests. The results for both modules are comparable indicating a similar process of manufacturing and that the simplified approach is applicable and reproducible under comparable test and measurement conditions.

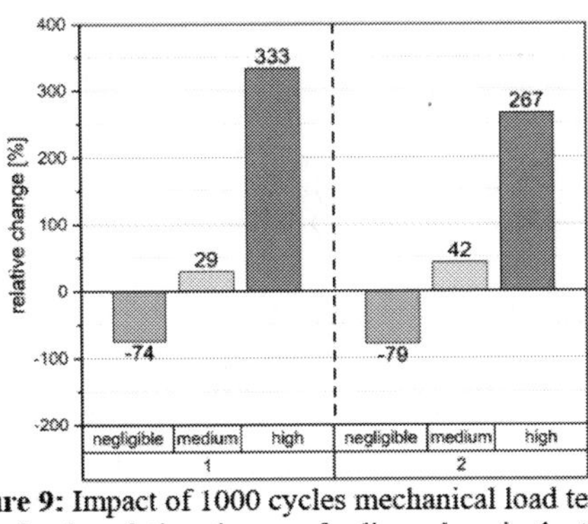

Figure 9: Impact of 1000 cycles mechanical load test given by the relative change of cell numbers in the three quality classes for module 1 and 2 (values rounded).

5 CONCLUSION

The stability of shingled solar cells with electrically conductive adhesive (ECA) interconnection has been investigated with a new approach. Mechanical stresses have been applied to provoke failures in the interconnection to test and evaluate a new and more precise assessment of the stability. Extending electroluminescence imaging, Magnetic Field Imaging (MFI) has been validated to provide additional information about precise failure sites as well as failure reasons. Due to the different physical principle of EL and MFI, the smart discrimination of directed currents allows the MFI technique to detect specifically currents in the shingle strings which can be attributed to failures in the string. We describe and show that currents perpendicular to the string directions give clear evidence of failures and occur in shingled cell pairs. Larger groups with perpendicular currents arise in case of multiple interconnection failures of cells in series. For this case an approach to simplify the evaluation has been proposed and successfully tested. Considering exclusively the one-dimensional magnetic flux density component related to perpendicular currents in the string allows the categorization based on the maximum value which is proportional to the maximum perpendicular current. This approach allows the quantification of interconnection aging and has been demonstrated for two different mechanical stress tests indicating different stress levels of these tests.

The presented approach improves the understanding and the quality of the assessment of interconnection aging in ECA of shingled solar cells. The information depth of the MFI technique enables distinct material analysis, allows quantification and deeper insight in failure mechanism, and qualification of new material combinations. However, simplification and comparability are needed for advanced interconnection stability assessment of shingle cells. This paper presents a proposal how future shingle cell module manufacturing and quality control can benefit in future from advanced measuring methods.

6 ACKNOWLEDEMENTS

Financial support from the Federal Ministry for Economic Affairs and Climate Action within the funded project "HOSSA" (FKZ: 03EE1014A) is gratefully acknowledged

7 REFERENCES

[1] N. Klasen et al. „Performance of shingled solar modules under partial shading" Prog. Photovolt. Res. Appl. (2021) 1–14.
[2] C. Carrière et al., „Towards shingling interconnection with SHJ solar cells" 37th EU-PVSEC (2020) 840-844
[3] I. Ullmann et al., „Investigation on the quality of adhesive joints of shingled solar cells by accelerated lifetime testing" 35th EU-PVSEC (2018) 1210-1213
[4] D. Tune et al., „The sun is rising on conductive adhesives" Photovoltaics International 47 (2021) 57-77
[5] G. Oreski et al., „Reliability of Electrically Conductive Adhesives" Photovoltaics international 42 (2019) 38-45
[6] D. Lausch et al., „Magnetic Field Imaging (MFI) of Solar Modules" 35th EU-PVSEC (2018) 1060-1064
[7] M. Sander et al., "Systematic investigation of cracks in encapsulated solar cells after mechanical loading" Solar Energy Materials and Solar Cells 111 (2013) 82-89
[8] M. Pander et al., "Lifetime estimation for solar cell interconnectors" 28th EU-PVSEC (2013) 2851 – 2857
[9] IEC TS 62782:2016, Photovoltaic (PV) modules - Cyclic (dynamic) mechanical load testing (March

2016)
[10] F. Frühauf and M. Turek "Quantification of electroluminescence measurements on modules" Energy Procedia 77 (2015) 63 – 68

Study of Hot Spots in Partially Shaded PV Modules with Large-Area Wafers

Haifeng Chu*, Constantin Klyk, Andreas Halm, Valentin D. Mihailetchi
International Solar Energy Research Center Konstanz e.V.
Rudolf-Diesel-Str. 15, 78467 Konstanz, Germany
*haifeng.chu@isc-konstanz.de

ABSTRACT: We study the temperature of state-of-the-art PERC modules under various partial shading conditions. Peak temperature of almost 150°C was measured when one of the half cell of an M6 PERC Module was shaded. For our simulated M6 module, the peak temperature is shown to be higher than 140°C when the shading percentage ranging from 15% to 35% (with respective to one cell in the module). Influence of various module layout (e.g. different wafer size from M0 to M12, different number of cells in a string) is studied by performing FEM simulation. Critical temperature of module materials has likely reached. However, several measures exist which can mitigate the high temperature issue of the shaded module, e.g., reducing number of cells in a module string.

Keywords: shading, hot spot, module

1 Introduction

PV modules degrade because of various reasons, of which hot spot is reported as the most important degradation mode for the PV systems that were installed in recent years (between 2007 and 2017) [1]. Bird droppings and shadow from surrounding objects are two common reasons of partial shading, which are also hard to avoid completely. The majority of PV modules uses high shunt resistance solar cells, and the worst shading case usually take places when 15% to 30% of one solar cell is shaded. Peak temperature of the partially shaded cell can exceed 150°C, leading to concern about reliability and safety [2–4]. The general public might not be well informed about this risk, while roof-top PV applications are expected to grow significantly in the following years.

In this work we study the temperature of state-of-the-art crystalline silicon modules under various partial shading conditions. We measured indoor *IV* characteristics of a PERC module consisting of 120 half-cell and three bypass diodes. Then we performed outdoor measurements to study the module temperature when one cell was partially shaded. The shaded module power and temperature are then simulated by using SPICE and finite element method, respectively. Furthermore, our simulation models are used to study the influence of shading area and module layout. Fortunately, several measures are already known to be able to mitigate the hot spot issue, which will be discussed at the end.

2 Indoor characterization

A commercially available PERC module is used for this experiment. The front side of the module is covered with tempered glass, whereas the rear side is covered with a white backsheet. It has a standard module layout: 120 half-cell (M6 wafer format, 9 busbars) and in total six strings (each with 20 half-cell connected in series). Three bypass diodes are included, where each of the bypass diodes is connected with two strings in parallel. A flasher is used to measure the *IV* curve when no shading presents, and the result is listed in Table I.

Table I: *IV* data of the unshaded PERC module under the standard test condition (1 Sun, 25 °C)

V_{oc} (V)	J_{sc} (A)	J_{mpp} (A)	FF (%)	P (W)	Eta (%)
40.775	10.630	10.046	78.8	341.6	19.0

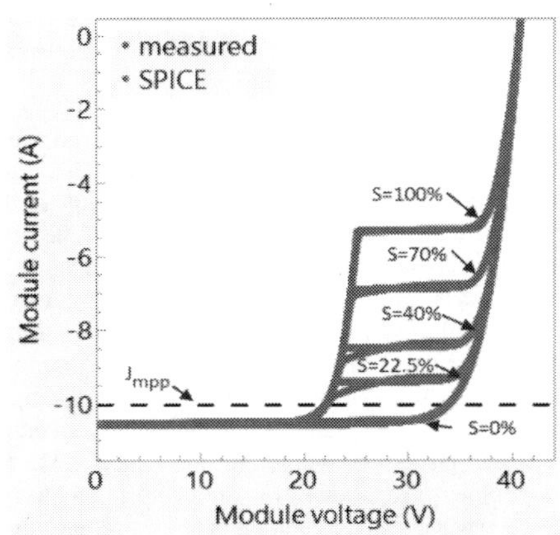

Figure 1: Measured and simulated *IV* curves when Cell A1 was shaded with different size of cardboard. The black dash line indicates the J_{mpp} of the unshaded module.

A part of the experimental work in this paper has followed the IEC hot spot test norm [5], e.g. the selection of cells to be shaded. One of the selected cell is the one with the highest shunt resistance (hereafter called Cell A1). We performed *IV* measurements when Cell A1 was shaded by using various size of cardboard (see results in Fig. 1). We also built a SPICE model for this PERC module; the basic procedure of the SPICE model is the same as that described in [6], and we have assumed reverse *IV* characteristic of Cell A1 is shown in Fig. 2. The simulation matches qualitatively well to the experimental data, especially at the operation points near J_{sc} and J_{mpp}, but a larger mismatch is observed for the voltage from 22 V to 36V.

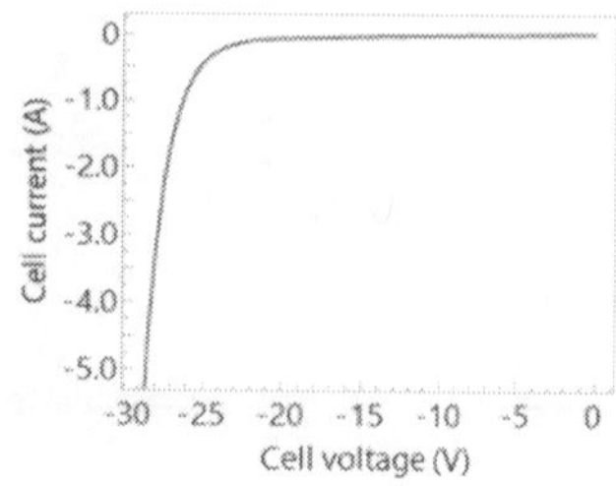

Figure 2: Assumed reverse *IV* curve of Cell A1 that is used for SPICE simulation.

3 Outdoor shading measurement

The PERC module was then installed outdoor (location: Constance, Germany). The module was south-facing and the inclined angle was 30° from the horizontal plane. One of the short edges of the module was placed directly on the ground, whereas the center of this module (total length of 1.76 m) is 0.44 m away from the ground. The module is under short circuit condition during the shading test.

Temperature was recorded by three thermocouples that were taped on the backsheet. An extra thermocouple was placed near the module to record ambient temperature. In-plane irradiance was measured by an ISET sensor (IKS Photovoltaik GmbH) that was installed nearby. Wind speed data was taken from the nearby weather station [7] , instead of on-site measurement.

Fig. 3 shows the change of module temperature after being shaded by a rectangular cardboard, and the shading percentage is 22.5% of the entire Cell A1. This is the worst (or very close to the worst) shading percentage for this cell. All three thermocouples were placed along the center line of the unshaded part of Cell A1. After 15 minutes the temperature increased to about 140°C and then it stabilized. Fig. 4 shows photographs of the module front and rear sides during the test. When the temperature of Cell A1 stabilized, an infrared image was taken by using an infrared camera, and the emissivity ε is scaled to fit the measured backsheet temperature (ε = 0.914). In this way the temperature shown in Fig. 4 (c) might be more representative for the temperature on the backsheet. One can observe that temperature of the unshaded part of Cell A1 appears to be quite uniform, whereas the shaded part of Cell A1 is about 30°C cooler than the unshaded part, but it is still about 50°C warmer than adjacent cells without any shade. This IR image suggests a uniform breakdown of the shaded cell, which is the ideal way. The shading test has been repeated with another seven solar cells and all showed a uniform breakdown behavior, whereas the peak temperature among all tests reached almost 150°C. This finding shows that for the state-of-the-art industrial solar cells, it seems that the breakdown takes place uniformly over the entire cell [4], instead of showing local hot spots as observed in the multicrystalline Si solar cell ten years ago [2]. More experimental evidence is needed to confirm whether this conclusion can be applied for all industrial modules now.

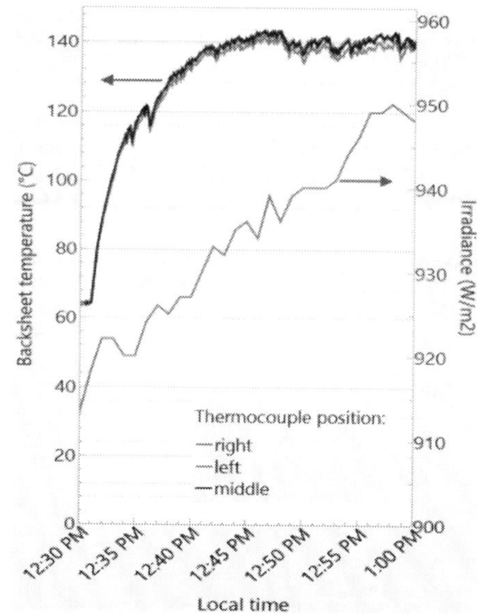

Figure 3: Measured backsheet temperature and irradiance. Shading percentage is 22.5%; ambient temperature and wind speed are 36.4°C and 10 km/h, respectively.

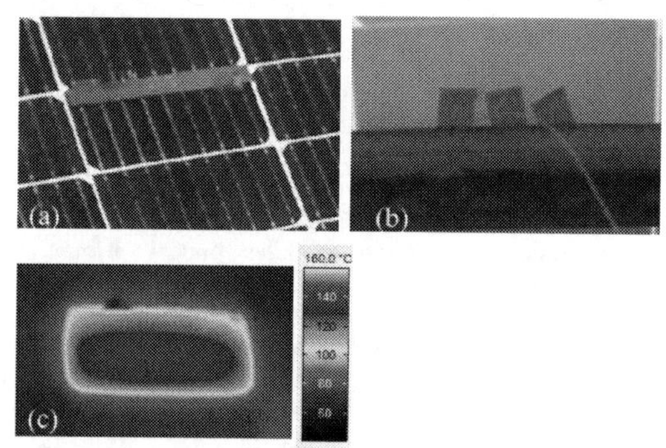

Figure 4: Photographs of the module's front (a) and rear (b) sides during the outdoor shading test. Center of the shaded cell is about 0.79 m above the ground. Distance between two adjacent thermocouples is about 28 mm. Infrared image taken from the front side of the module is shown (c).

4 Temperature simulation

4.1 Introduction to the simulation model

To save computational efforts, we limit our simulation domain to a 3 x 6 half-cell array. The CAD layout consists of Si solar cells (only Si, excluding surface passivating dielectric and metal grid), Cu ribbons, encapsulant, backsheet, and front glass. To simulate the steady-state temperature, heat conduction, radiation, and convection should be calculated. Figure 5 shows the basic flow chart for the module temperature simulation. We prepare the CAD layout and compute the mesh (about three million elements) with the software Salome [8], whereas the software ELMER [9] is used for temperature simulation by using finite element method. A plugin program is used to connect these two software [10]. All software mentioned here are open source.

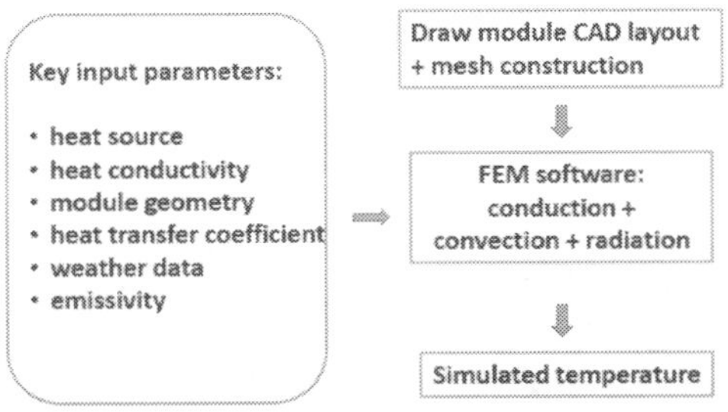

Figure 5: Flow chart of shaded module temperature simulation.

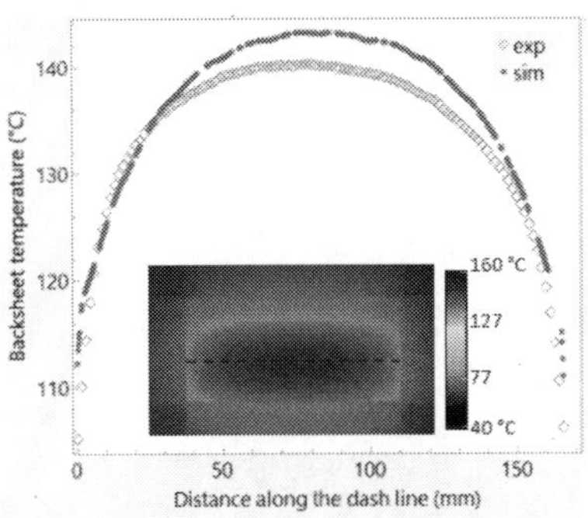

Figure 6: Simulated temperature distribution of Cell A1 (conditions are exactly the same as those in Fig. 4). The measured and simulated temperature along the black dash line is also shown.

For the calculation of heat sources, we first consider the background heat source because even without any shading, the short-circuited module is already much warmer than ambient temperature during the experiment shown in Fig. 3. For the glass, backsheet and encapsulant, we have selected the same heat source as the ones in Ref. [11], while we adjust the heat source for solar cells to fit the temperature that we measure when no shading exists. As for the shading case, an extra heat source is added to the unshaded part of Cell A1 because it consumes most of the power generated by all other unshaded cells from the same string. We found out that for the worst shading case the power consumed by the shaded cell approximately equals $0.97 \times (N\text{-}1) \times P_{unshaded\_cell}$, where N is the number of cells in a string.

Fig. 6 shows the result of FEM simulation for Cell A1, which matches to the experimental IR image shown in Fig. 4 (c). We have repeated the simulation for another two shading experiments. In one case the weather condition is different, while in the other case the shading percentage is changed to 40%. For both cases the absolute deviation of peak temperature between experiment and simulation is less than 3 °C. Note that many input parameters are not known precisely, such as the heat transfer coefficients, and therefore our simulation can only provide a rough estimation to learn how hot the shaded module can turn under a given condition.

4.2 Influence of shaded area and module layout

In this subsection, we assume a 120 half-cell module with a higher power, i.e. 360 W (Eta = 19.5%; M6 wafers), and we also assume that the module operates at a current that equals J_{mpp} of an unshaded module; as such the power generated by the entire PV system (e.g. a roof-top system consisting of 20 modules) is maximized. The weather data is set as follows: irradiance of 1000 W/m², wind speed of 10 km/h, and ambient temperature of 30 °C.

Fig. 7 shows the influence of changing shaded area. One can observe that when the shading percentage ranges from about 15% to 40%, the corresponding peak temperature on the backsheet is at a high level, reaching at least 90% of the temperature in the worst shading case.

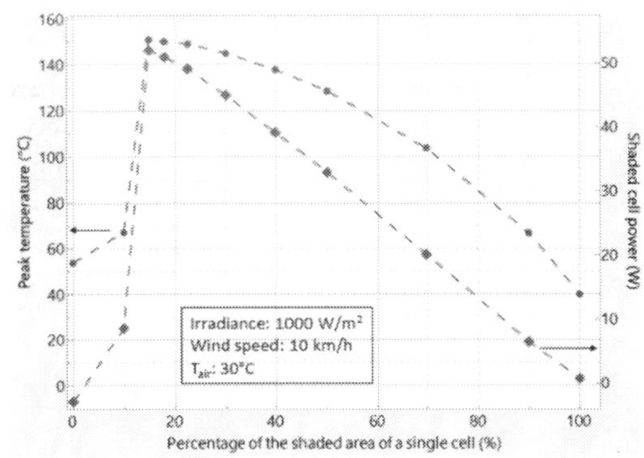

Figure 7: Simulated backsheet temperature and power consumed by the shaded cell with respective to shading percentage.

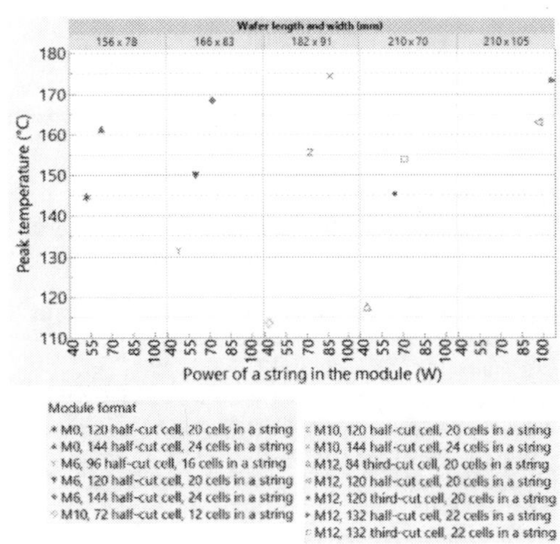

Figure 8: Simulated backsheet temperature for various module layout. Each module is assumed to consist of six strings and three bypass diodes. The percentage of shading area is assumed to be 22.17% with respective to a half/third cell.

We also simulate the temperature for various module layout (from M0 to M12). The simulated module power is scaled linearly according to the change of module area, and the reference point is the 360 W M6 module as in Fig. 7. From Fig. 8 we observe that for a given format, e.g., 120 half-cell module, using a larger wafer results in a hotter cell. But a shaded module with larger wafers can be designed to have a lower temperature than a shaded module with smaller wafers, when one reduces the number of cells in a module string. References

Furthermore, our simulation shows that the shaded module temperature increases with increasing cell efficiency, and it increases with increasing irradiance and ambient temperature.

5 Discussion and summary

The results shown in this paper are most relevant for a small/middle roof-top system (e.g. 20 modules).

Critical temperature of modules materials is probably reached for many state-of-the-art modules under partial shaded conditions, e.g. glass transition temperature of encapsulant (usually <140°C). Thus, reliability and safety need to be reexamined. For other cell technologies that have a high breakdown voltage, e.g., passivated contact and HJT, similar results to this paper are expected because the reverse bias behaviors of these cells are alike. This assumption needs to be experimentally verified.

Each of the following measures is known to be able to reduce the peak temperature: (1) reduce the number of cells in a module string, e.g. using 12 cells instead of 20 cells; (2) increase the number of bypass diodes [4]; (3) use an inverter that chooses the high-voltage local maximum power point [3].

References

[1] D. C. Jordan et al., "Photovoltaic failure and degradation modes," Prog. Photovolt: Res. Appl., vol. 25, no. 4, pp. 318–326, 2017.

[2] I. Geisemeyer et al., "Prediction of silicon PV module temperature for hot spots and worst case partial shading situations using spatially resolved lock-in thermography," Solar Energy Materials and Solar Cells, vol. 120, pp. 259–269, 2014.

[3] E. E. Bende et al., "Performance and safety aspects of PV modules under partial shading: a simulation study," in Proc. 29th Eur. Photovoltaic Sol. Energy Conf., 2014, 2546–2552.

[4] R. Witteck et al., "Three bypass diodes architecture at the limit," IEEE J. Photovoltaics, vol. 10, no. 6, pp. 1828–1838, 2020.

[5] Terrestrial Photovoltaic (PV) Modules—Design Qualification and Type Approval—Part 2: Test Procedures, IEC Standard 61215-2, 2019

[6] H. Chu et al., "Shade performance of a back-contact module assembled with cells featuring soft breakdown characteristic," Energy Procedia, vol. 92, 540–545, 2016.

[7] [Online] Available: https://meteostat.net/en/.

[8] SALOME. [Online] Available: www.salome-platform.org. Accessed on: Sep. 09 2022.

[9] ELMER. [Online] Available: https://www.csc.fi/web/elmer/. Accessed on: Sep. 09 2022.

[10] ElmerSalomeModule. [Online] Available: https://github.com/physici/ElmerSalomeModule. Accessed on: Sep. 09 2022.

[11] M. R. Vogt, p82, chap6, "Development of physical models for the simulation of optical properties of solar cell modules," Leibniz University Hannover, 2015.

CORROSION TESTING OF FPV MATERIALS IN OFF-SHORE

Myenggil Gang*[1], Dong-Chan Kim[1], Seong Dae Kim[1], Hyun-Jun Kim[2], Hyunsik Jo[2], Seungwook Shin[3], Chang-Sub Won * * [1]

[1]SCOTRA Corporation, R&D Center, Seoul, Korea, Rep. of South

[2]K-Water Research Institute, Water Energy & Infrastructure Research Center, Daejeon, Korea, Rep. of South

[3]Rural Research Institute, Korean Rural Community Corporation, Ansan-si, Korea, Rep. of South

ABSTRACT: In the present work, we demonstrate the case study and outcomes of the corrosion effect based on the Marine Research Floating photovoltaic (FPV) test-bed installed in 2021 on Saemangeum (seawater lake) in South Korea. We developed the floating PV system in collaboration with K-Water (water resource manager) to install marine solar power facilities. However, the corrosion control and preservation of floating PV structures are important factors in increasing durability in these regions. Accordingly, the possibility of installation in the corrosive environment was studied through corrosion exposure tests of floating PV structure materials. To study this, the corrosion grades of floating PV structure materials (metals) in marine environments were classified based on ISO 9223 standard. At the same time, the corrosion rate of standard specimens for the corrosive evaluation was determined based on ISO 9226 standard. After the test specimen was installed, the corrosion rate study was conducted every 3 months. The atmospheric corrosion of PosMAC [Ternary alloy coated steel (Zn-3%, Mg-2.5%, Al)], stainless steel, and aluminum was evaluated in Saemangeum (seawater lake) for 1 year. The specimen holder unit was composed of a material that did not affect the corrosion of the specimen. In addition, it was designed to minimize the occlusion of the specimen. The fabricated specimen holder unit was installed on a test-bed located in the sea. After the exposure test had been completed, the corrosion layers were removed by the chemical cleaning procedure according to metal type. Finally, the corrosion rate was evaluated by measuring the weight loss.

Keywords: Floating PV, Corrosion test, Off-shore

1. INTRODUCTION

1.1 Aim and approach

With the rapidly increasing demand for Floating Photovoltaics (FPV) systems, extensive R&D work needs to be performed for long-lasting FPV. These developments are needed to be ready to deploy FPV in various locations, including coastal cities, as a valuable energy source that can meet rapidly growing energy demand with fueled population growth. Though, for successful deployment in rough waters, FPV systems must be built with strong capabilities such as high resistance to ocean waves, saltwater corrosion, bio-fouling, and other challenges of the FPV system.

Most FPV platforms are designed for calm water environments like freshwater bodies or sheltered locations, while marine PV suffers from rough environmental conditions in the sea. Consequently, there is a growing demand for marine PV research.

A few Asian companies like SCOTRA take the lead in reliable product development to meet the increasing energy demand in potential marine application areas. The FPV systems need investigation and improvements to be ready for deployment in marine environments to withstand exposure to waves and saltwater corrosion.

Marine FPV can be integrated near-shore as the main power source in densely populated coastal areas. Test-bedding is needed to understand the functionality and operation of FPV structures with salt corrosion resistance, high humidity, strong waves, biofouling, and other severe marine conditions.

Recently, an evaluation of corrosion aggressivity was carried out in a harbor of Taiwan Isle.[1] They tested four metals in outdoor exposure: carbon steel, zinc, copper, and aluminum. Atmospheric corrosion of carbon steel, copper, galvanized steel, and aluminum was evaluated in a coastal industrial port zone in Cuba for 1 year. Samples were exposed at seven outdoor corrosion sites at different distances from the sea.[2]

Data on two and three outdoor exposure sites in Mexico and Cuba's port zones has been published.[3] A high corrosion rate is reported for those zones. Outdoor corrosivity classification reported ranged from C3 to C5 according to ISO 9223 standard. The salinity of the atmosphere plays a very important role in coastal sites.[4] Differences in corrosion mechanism have been reported depending on salinity level.

1.2 Scientific innovation and relevance

The floating PV structure is composed mainly of metallic materials. The marine environment increases the corrosiveness of FPV structure. The application of protective measures according to corrosivity categories assures durability of materials, as also an extension of maintenance and repair tasks.

Information on corrosivity is important for selecting FPV structure materials and protective systems. Most of the available information on the corrosivity of atmospheres in which the classification system and the methods of corrosivity are standardized.

This study presents information on the corrosivity of atmospheres for the proper selection of structure materials and anti-corrosive measures.

2. EXPERIMENTAL DETAILS

The corrosivity test units are designed for monitoring in outdoor environments. The corresponding test units are shown in Fig. 1. These were produced and used to monitor Saemangeum [seawater lake: salty conditions 20.0 [%] (35°50′28.9″ NL and 126°33′49.8″ WL)] from South Korea. The corrosivity test unit design used for acquiring experimental results follows the ISO 9223 standard.

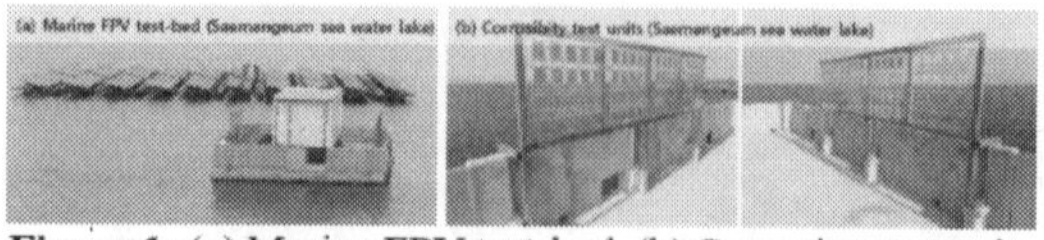

Figure 1: (a) Marine FPV test-bed, (b) Corrosion test units in Seamangeum seawater lake

Corrosivity test unit racks (Polycarbonate) for metal samples exposure were used. The specimen holder unit was composed of a material that did not affect the corrosion of the specimen. In addition, it was designed to minimize the occlusion of the specimen. The fabricated specimen holder unit was installed on a test-bed in the sea. (Figure 1).

Specimens of PosMAC [Ternary alloy coated steel (Zn-3%, Mg-2.5%, Al)], stainless steel, and aluminum were exposed for exposure periods of 3, 6, 9, and 12 months (Figure 1).

After the exposure test had been completed, the corrosion layers were removed by the chemical cleaning procedure according to metal type. To evaluate corrosivity, we measured the initial mass of samples and the chemically removed corrosion products to determine the corrosion rate of specimens according to established standard procedures (Figure 2 and Table 1).

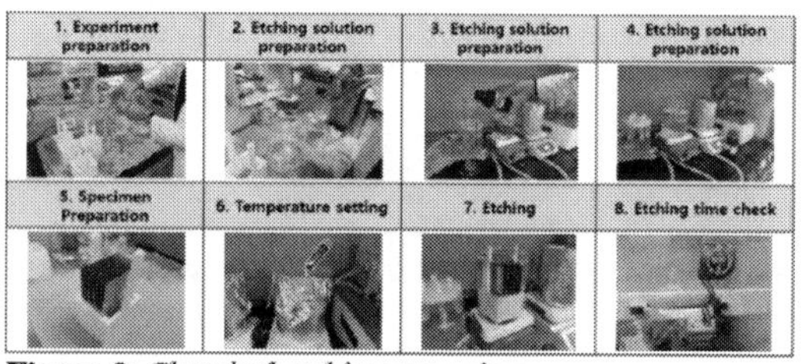

Figure 2: Chemical etching procedures

Table I: Etching conditions

Type of samples	Etching solutions	Etching conditions	
PosMAC300 PosMAC600 Pure Zn	CrO_3	1 min.	80 °C
SUS304 SUS316	HNO_3	60 min.	20 °C
Al5052 Al6061	H_3PO_4 CrO_3	10 min.	80 °C

All the specimen's size (10 cm x 15 cm) was kept equal and constant. The atmospheric corrosion rate was determined for different exposure periods, such as 3, 6, 9, and 12 months.

The samples were exposed quarterly for every material and exposure period. The average annual corrosion rate (12 months of exposure) was used to determine the corrosivity categories of the atmosphere for each metallic material.

3. RESULT AND DISCUSSION

Metallic materials exposed were the following:

- PosMAC [Ternary alloy coated steel (Zn-3%, Mg-2.5%, Al)]: PosMAC300 (Coating amount 300g/m^2, coating thickness ≈ 25 μm), PosMAC600 (Coating amount 600g/m^2, coating thickness ≈ 50 μm)
- Stainless steel: SUS304, SUS316
- Aluminum: Al5052, Al6061
- Pure Zinc: 98.5% purity minimum.

Chemical cleaning previous exposure of samples and procedure for removing corrosion products after exposure to determine corrosion rate (mass loss per area over time) were carried out according to established standard procedures.[6,7] The specimen thickness was considered equal and constant. The atmospheric corrosion rate was determined at different periods of exposure: 3, 6, 9, and 12 months. Three samples were exposed for every material and exposure period. The exposure period starts on Jun 2021 and up until May 2022 (Figure 3).

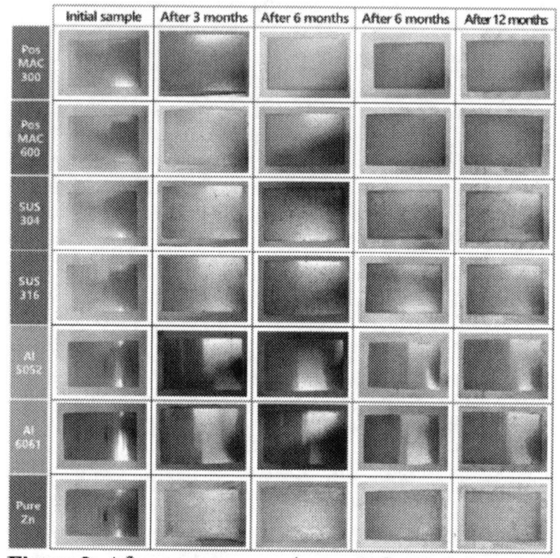

Figure 3: After exposure specimens (After 3, 6, 9, and 12 months exposure test)

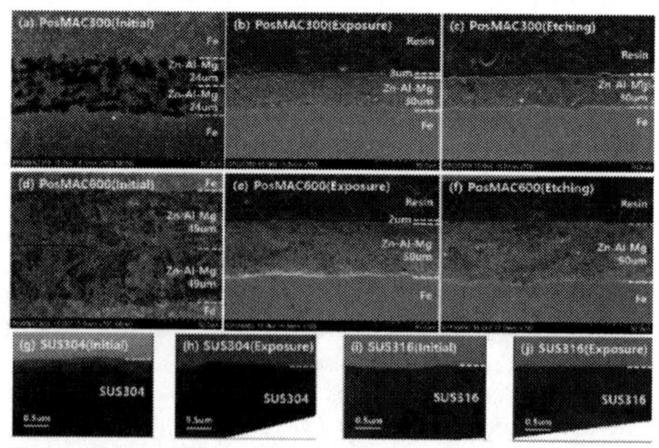

Figure 4: FE-SEM image for coating layer thickness (a) Initial PosMAC300, (b) Exposed PosMAC300 (c) Etched PosMAC300, (d) Initial PosMAC600, (e) Exposed PosMAC600, (f) Etched PosMAC600, (g) Initial SUS304, (h) Exposed SUS304, (i) Initial SUS316, (j) Exposed SUS316

The specimen's corrosion rate and coated layer thickness properties were studied using the weight change calculation and field emission scanning electron microscopy (FE-SEM). Figure 4 shows the cross-sectional field emission scanning electron microscopy (FE-SEM) images of PosMAC and Stainless steel. All of the samples show relatively smooth morphology. The thin layers at the top region with thicknesses of 2~3 μm are observed in exposed PosMAC samples. In particular, the thin layer was not observed, as confirmed by FE-SEM observation after the etching cleaning.

The corrosion grades of FPV materials in marine environments were classified based on ISO 9223 standards, and the corrosion rate of standard specimens for corrosion evaluation was determined based on ISO 9226 standards.[5,6]

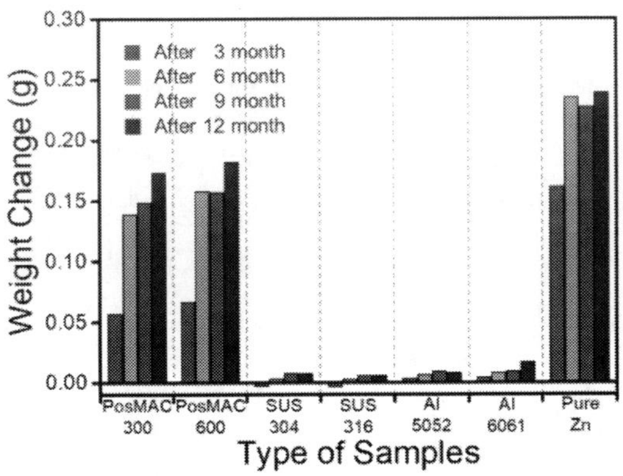

Figure 5: Specimens weight change graph for PosMAC, stainless steel, aluminum, and pure Zn for 3, 6, 9, and 12 months

The corrosivity of the exposure location is deduced from the corrosion rate, calculated from the loss of mass per unit area of standard specimens following the removal of corrosion products from the specimens after exposure periods of one year (Figure 5 and 6). Corrosion rates for the 1 year of exposure can be used to calculate corrosion rates for long-term exposures following ISO 9224 standards.[8]

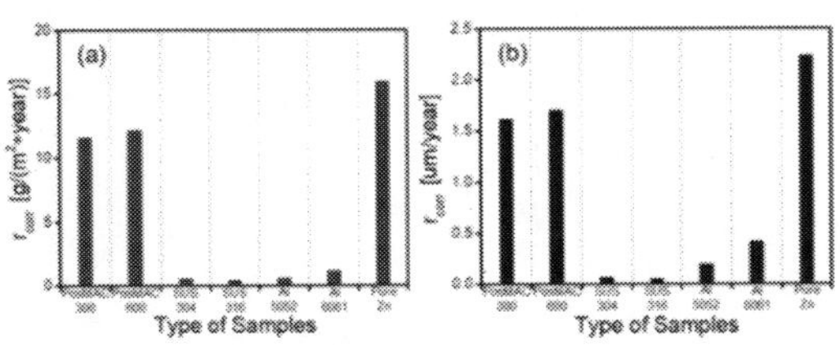

Figure 6: (a) Specimens weight change graph (After 12 months of exposure test)

The corrosion rate (r_{corr}) for each metal, expressed as a mass loss per unit area over time, is given by Equation (1):

$$r_{corr} = \frac{\Delta m}{A \times t} \quad (1)$$

Where,

r_{corr} is the corrosion rate, expressed in grams per square meter per year [g/(m^2 · year)];

Δm is the mass loss, expressed in grams (g);

A is the surface area, expressed in square meters (m^2);

t is the exposure time, expressed in years (a).

The corrosion rate can also be expressed as a thickness reduction rate given by Equation (2):

$$r'_{corr} = \frac{\Delta m}{A \times \rho \times t} \quad (2)$$

Where,

r'_{corr} is the corrosion rate, expressed in micrometers per year (μm/year);

ρ is the density of the metal (Fe: 7.98 g/cm^3; Zn: 7.14 g/cm^3; Al: 2.7 g/cm^3);

Δm is the mass loss, expressed in grams (g);

A is the surface area, expressed in square meters (m^2);

t is the exposure time, expressed in years (a).

The average annual corrosion rate (12 months of exposure) was used to determine the corrosivity categories of the atmosphere for each metallic material. The values obtained from the measurements (corrosion rates for the first year of exposure) are intended to be used as classification criteria for evaluating atmospheric corrosivity according to ISO 9223.[5]

4. CONCLUSIONS

Corrosivity categories of the atmosphere for PosMAC [Ternary alloy coated steel (Zn-3%, Mg-2.5%, Al)], stainless steel, and aluminum in the Saemangeum (seawater lake) site were determined according to ISO 9223, based on annual data of atmospheric corrosion rate. Prediction of corrosivity categories of the atmosphere was made according to ISO 9224 standards. The Saemangeum (seawater lake) environment shows atmospheric corrosivity of grade C4. Further, the extended prediction of corrosivity categories after exposure of 20 years reveals durability and useful life of the FPV structure can be used to build guaranteed metallic materials (PosMAC, stainless steel, and aluminum could be the metallic materials less affected in selected to build FPV system). Corrosion resistance, high strength (durability), structure weight, and material price must be considered when selecting offshore floating photovoltaic materials.

ACKNOWLEDGEMENT

This work was supported by the New and Renewable Energy Technology Program of the Korea Institute of Energy Technology Evaluation and Planning, granted financial resources by the Ministry of Trade, Industry and Energy, Republic of Korea under Grant 20183010014260.

REFERENCES

[1] C. M. Lo, Mater. Corros. 68 (2017) 1.

[2] Abel Castaneda, Mater. Crros. 1 (2018) 16.

[3] F. Corvo, Corros. Sci. 50 (2008) 220.

[4] M. Morcillo, Mater. Corros. 66 (2015) 882.

[5] ISO 9223, corrosion of metals and alloys. Corrosivity of atmospheres. Classification, determination, and estimation.

[6] ISO 9226, corrosion of metals and alloys. Corrosivity of atmospheres. Determination of corrosion rate of standard specimens for the evaluation of corrosivity.

[7] ISO 8407. Corrosion of metals and alloys. Corrosivity of atmospheres. Removal of corrosion products from corrosion test specimens.

[8] ISO 9224, corrosion of metals and alloys. Corrosivity of atmospheres. Guiding Values for the corrosivity categories.

Effects of Wind Load on the Mechanics of a PV Power Plant

[1,*]Pascal Romer, [1]Kishan Bharatbhai Pethani, [1]Andreas J. Beinert

[1] Fraunhofer Institute for Solar Energy Systems ISE,
Freiburg, Germany, Heidenhofstraße 2, 79110 Freiburg i.Br., Germany

*Corresponding author, e-mail to: pascal.romer@ise.fraunhofer.de

ABSTRACT: In contrast to the IEC 61215, in real life applications PV modules must withstand inhomogeneous load distributions, for example caused by wind. This work investigates the wind effects onto a PV power plant, containing ten rows with 40 modules each, using computational fluid dynamics simulations coupled to a mechanical finite element method model. The paper focuses on the impact of three factors on the mechanical stability of a PV power plant, namely: Module orientation, wind direction and module inclination angle. A crosswind scenario is found to be most critical. Furthermore, higher module inclination angles result in higher stresses. Finally, general thermomechanical rules are extracted allowing for a deeper understanding of the underlying effects, and therefore help to build more robust PV module installations in the future.

Keywords: FEM; CFD; Wind; Simulation; PV Module; Mechanics

1 INTRODUCTION

For photovoltaic (PV) modules it is mandatory to withstand a homogeneous mechanical load of at least 2400 Pa to pass certification according to IEC 61215-2:2021 [1]. However real-life application shows more often inhomogeneous load distributions on modules instead of homogeneous ones, e.g. due to wind loads. Such inhomogeneous loads can lead to significantly different states of stress compared to the homogeneous ones leading to fatal damage such as cell or glass breakage in the worst case [2].

There are many papers available that determine the wind load on modules using computational fluid dynamics (CFD) simulations or experiments [3–8]. However, none of these papers calculated the occurring stress states of wind loads passed onto a mechanical finite element method (FEM) simulation. In order to close the existing knowledge gap this paper simulates the wind load on a representative section of a PV power plant and determines the resulting stresses of the highest loaded PV module. Finally, the influence of module orientation, wind direction as well as module inclination angle is analyzed.

This is an abridged version of a paper currently in the review process [9].

2 METHODS

2.1 CFD simulations

In the CFD simulations, a representative section of a PV power plant, consisting of 10 rows of PV modules and the mounting rack is simulated. Each row consists of 40 modules with a 20×2 (portrait) or a 10×4 (landscape) arrangement. The influence of the wind direction and the module inclination angle is then examined using the portrait orientation. The simulated parameters are shown in Table I, where bold values correspond to the reference model.

Table I: Varied parameters within the CFD simulation, where bold values are taken for the reference.

Parameter	Values
Module orientation	Landscape, **portrait**
Module inclination angle [°]	0, 15, 30, **35**, 40, 45, 60, 70
Wind direction [°]	**0**, 15, 30, 45, ..., 180

The boundary conditions used in the CFD simulation are shown color-coded in Figure 1. Rigid walls with a vanishing wind speed are the floor (light orange) and the PV modules with the mounting rack (brown). The sidewalls (light blue) as well as the top surface (not shown) are simulated as open boundaries with vanishing stress. An outlet boundary condition with vanishing pressure was assumed for the back side (dark blue). Ultimately, an inlet with a specified wind speed from DIN EN 1991-1-4 [10] type 2, was assumed on the front side (green). The corresponding wind direction is indicated by the wind rose.

To obtain this wind profile, the base wind speed v_{Base} is processed according to:

$$v_{Wind}(z) = 1.3\, v_{Base} \qquad ; z \leq 4\,m$$

$$v_{Wind}(z) = 1.45\, v_{Base} \left(\frac{z}{10}\right)^{0.12} \qquad ; z \geq 4\,m$$

Based on the CFD simulations, the PV module with the highest wind pressure is identified and both the average and the maximum wind pressure on the front and rear are evaluated. The difference between the rear and the front pressure is referred to as the resulting pressure in the following.

Table II: Material properties of the reference PV module. *: provided by manufacturer, †: measured.

Layer	Material	Density [g/cm³]	Young's modulus [GPa]	Poisson's ratio [-]	CTE [$10^{-6}K^{-1}$]
Front glass	Soda-lime glass	2.5*	70*	0.2*	9*
Encapsulant	EVA	0.96 [15]	T-dep.†	0.4 [15]	T-dep.†
Solar cell	Cz-Silicon	2.329 [15]	Elasticity matrix [15]		T-dep. [13,14]
Backsheet	TPT	2.52 [15]	3.5 [15]	0.29 [15]	50.4 [15]
Frame	Aluminium	2.7 [16]	70 [16]	0.33 [16]	23 [16]
Frame-inlay	Rubber	0.067*	0.0074*	0.3*	769*

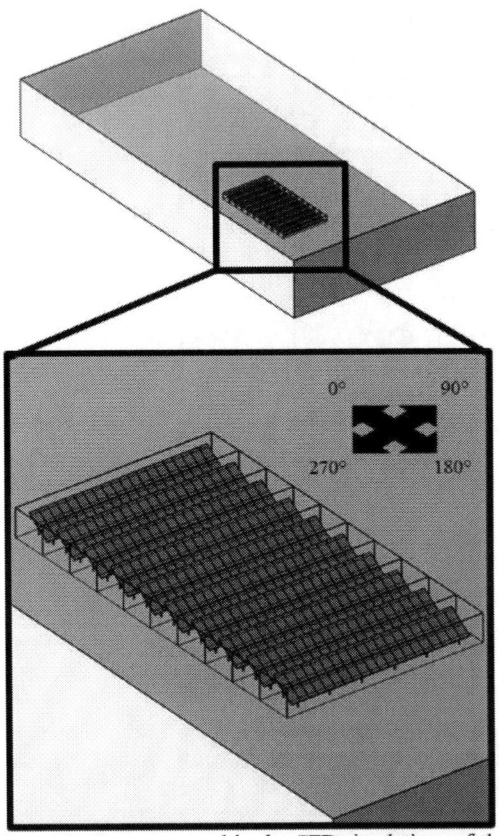

Figure 1: Geometry used in the CFD simulations of the reference portrait configuration. Orange and brown boundaries are simulated as solid walls, light blue boundaries as open boundaries, dark blue boundaries as outlets and the green boundary as an inlet. The wind rose depicts the wind direction.

2.2 Mechanical simulations

The mechanical FEM model consists of a 120 half-cell PV module with the overall size of 1773.1 x 1072.1 mm² and is based on previous works with neglected metallization and ribbons [11,12]. In contrast to the model mentioned, a complete PV module must be simulated in this work due to the inhomogeneous load distribution. Additionally, geometric nonlinearity is considered in this paper. Furthermore, the PV modules production process, consisting of soldering and lamination, is neglected in this work in order to study the pure wind load induced stress. The used material properties are summarized in Table II.

The wind loads calculated in the CFD simulation are set in the mechanical FEM simulation as surface loads on the front glass and the backsheet.

Due to the brittleness of glass and silicon, the evaluation of the mechanical FEM simulation focuses on the first principal stress in these two layers.

3 RESULTS AND DISCUSSION

With pure head wind, as it is simulated in the reference model, the first row of PV modules blocks the wind from the ones behind, depicted in Figure 3. Therefore, the modules with the highest resulting wind pressure are in the first row. Note that a negative resulting pressure corresponds to a load facing towards the rear of the module and vice versa.

3.1 Module orientation

In a first step a portrait orientation is compared to a landscape orientation. The wind pressure on the front of the module is shown in Figure 2. If this is offset against the wind pressure on the backside (not shown), the resulting maximum pressure is -780 Pa for the portrait orientation and -760 Pa for the landscape orientation. The resulting mean pressure is found to be -610 Pa for the portrait orientation and -640 Pa for the landscape orientation.

The implementation of those pressures in the mechanical FEM simulation leads to a marginal stress difference of 0.6 MPa in the solar cells and 0.3 MPa in the glass, shown in Figure 2 (middle and bottom).

To sum up:

1. Modules in landscape orientation face a marginally higher amplitude in the first principal stress (below 1 MPa) compared to those in portrait orientation.

3.2 Wind direction

For a fixed module inclination angle of 35°, the wind direction α_{wind} is varied between 0° (headwind) and 180° (tailwind) in increments of 15°. Both the average and the maximum resulting pressure (not shown due to the same trend than the average pressure) show a maximum amplitude in the suction pressure of -840 Pa and -1840 Pa respectively, at a wind direction of 45°. At a wind direction of 135° and 120°, the average and the maximum resulting pressure show a maximum in the amplitude of 950 Pa and 1920 Pa, respectively.

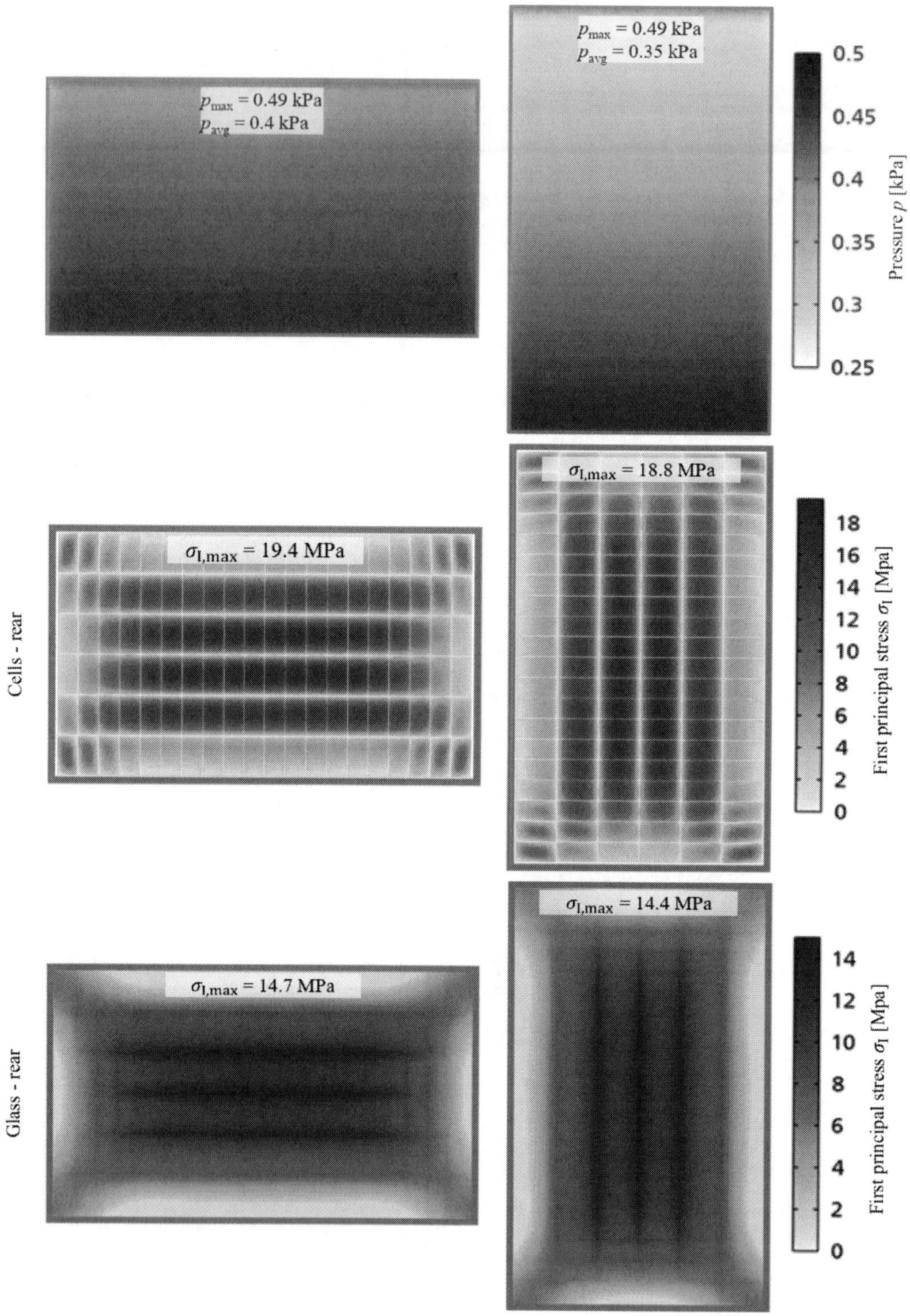

Figure 2: Front side pressure simulated by CFD (top) alongside the first principal stress in the cells (middle) and the frontglass (bottom) both simulated by mechanical FEM for a landscape(left) and portrait (right) orientation.

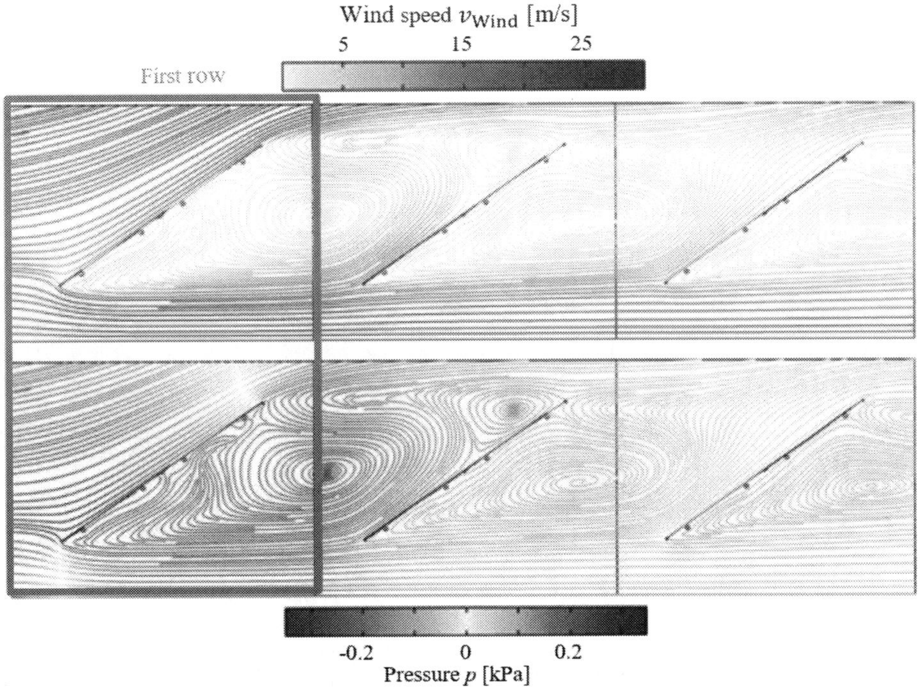

Figure 3: Cut plane through the center modules of the first three rows of PV modules with the wind speed at the top and the pressure at the bottom. In the back rows, the resulting pressure is significantly lower than in the front rows.

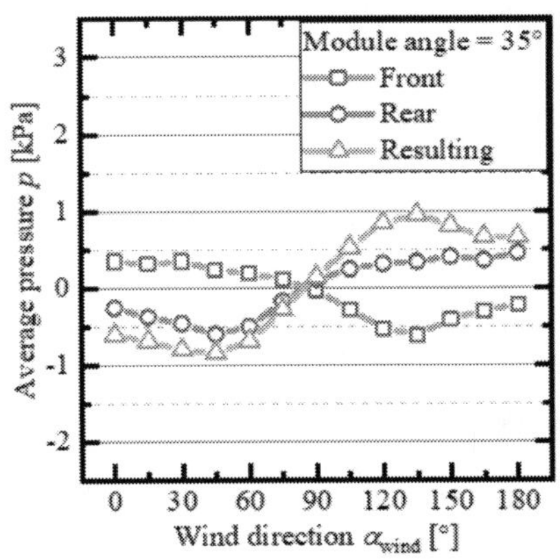

Figure 4: Average pressure on modules front (green) and rear side (blue) and the resulting pressure (orange) at wind directions between 0° and 180° with a module inclination angle of 35°.

Figure 5 shows the resulting first principal stress and the z-displacement for the different wind direction. Note that here and in the following, a negative z-displacement corresponds to a deformation in the direction of the modules rear site. As with the wind pressure, the z-displacement and the first principal stress also show peaks at 45° and 135°. Due to the position of the solar cells below the neutral axis within the PV module, the stress peak at 45°, corresponding to a downward deformation, is more pronounced with 23.2 MPa than the one at 135° with 14.7 MPa. Consequently, for the glass, with its top surface above the neutral axis, the peak at 135° is more pronounced with 22.1 MPa compared to the one at 45° with 17.9 MPa.

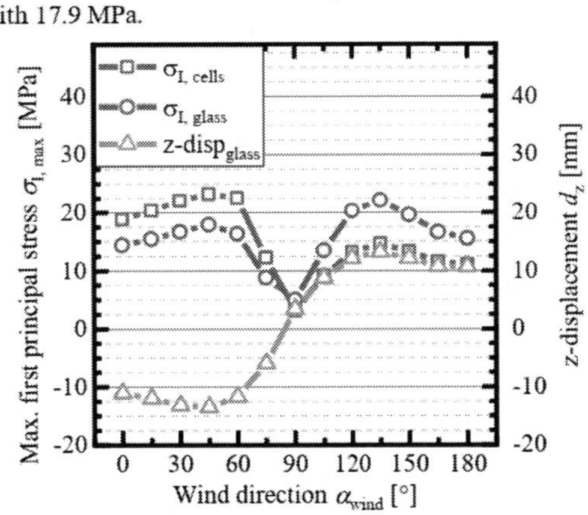

Figure 5: First principal stress in solar cells (green) and glass (dark blue) alongside the PV modules z-displacement (orange) for wind directions between 0° and 180°.

To sum up:

2. Cross wind (45° and 135°) is more critical than head (0°), tail (180°) and side (90°) wind.
3. In a glass-foil module, head wind scenarios are more critical for the solar cells, whereas tail wind scenarios are more critical for the front glass, due to the position relative to the neutral axis.

3.3 Module inclination angle

The module inclination angle β_{module} is varied between 0° and 75° for a wind direction of 45°, being the wind direction with the highest wind pressure. The average and the maximum resulting pressure are shown in Figure 6 and Figure 7, respectively. While the average resulting pressure increases monotonically with the module inclination angle up to -950 Pa at 75°, a peak with -1800 Pa occurs at a module inclination angle of 35° in the maximum resulting pressure. This peak is caused by vortexes arising on the modules rear side.

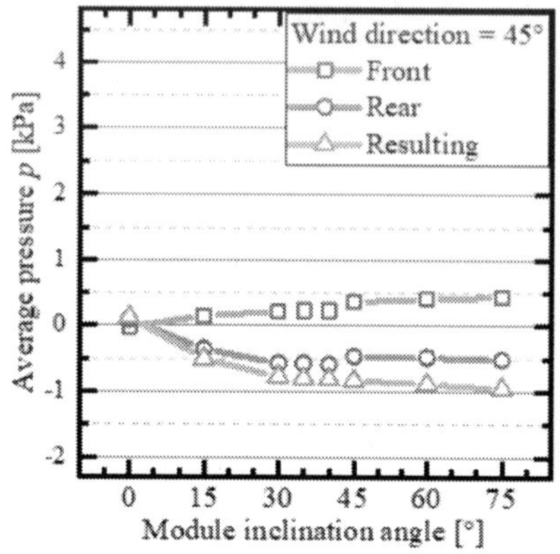

Figure 6: Average pressure on modules front (green) and rear side (blue) and the resulting pressure (orange) at module inclination angles between 0° and 75° with a wind direction of 45°.

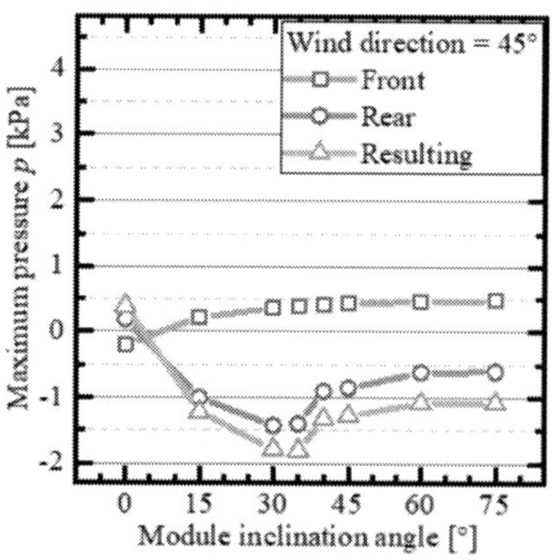

Figure 7: Maximum pressure on modules front (green) and rear side (blue) and the resulting pressure (orange) at module inclination angles between 0° and 75° with a wind direction of 45°.

The z-displacement and the maximum first principal stress in both solar cells and glass can be seen in Figure 8. The mechanical FEM simulation shows an increase of the z-displacements amplitude towards higher inclination angles. While the first principal stress in the cells and the glass, also follows the general trend of the average resulting pressure, additionally a local peak occurs at a module inclination angle of 35°. This is exactly the angle at which the maximum resulting pressure peak occurs.

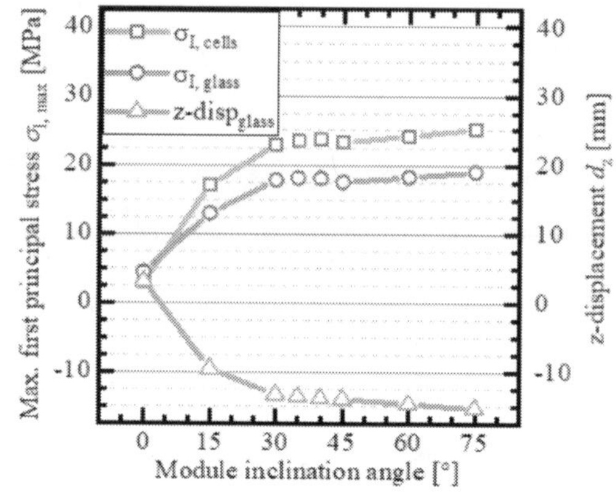

Figure 8: First principal stress in solar cells (green) and glass (dark blue) alongside the PV modules z-displacement (orange) for a module inclination angle between 0° and 75° with a wind direction of 45°.

To sum up:

4. Higher module inclination angles lead to a higher first principal stress.
5. The global trend in the first principal stress and the z-displacement is determined by the resulting average pressure.
6. Peaks in the resulting maximum pressure can cause local maxima in the first principal stress.

4 CONCLUSIONS

The influence of wind on the mechanics of solar modules is systematically examined. The influencing variables module orientation, wind direction and module inclination angle are considered. With a difference of less than 1 MPa, the influence of the module orientation is negligible. In contrast, the other two parameters, wind direction and module inclination angle, have a significant impact on the resulting first principal stress. The highest first principal stresses occur with 23.2 MPa in solar cells with a wind direction of 45° and 22.1 MPa in the front glass at a 135° direction of wind. However, it is worth to note, that all simulated stresses are not critical and correspond to a low probability of fracture. Looking at the module inclination angle, the first principal stress increases globally with increasing angle. However, a local peak develops at a module inclination angle of 35°. This is caused by turbulence at the rear of the module at this angle.

These general rules, regarding the influence of wind on the mechanics of PV modules, are derived:

1. Cross wind (45° and 135°) is more critical than head (0°), tail (180°) and side (90°) wind.
2. In a glass-foil module, head wind scenarios are more critical for the solar cells where tail wind scenarios are more critical for the front glass, due to the position relative to the neutral axis.
3. Higher module inclination angles lead to a higher first principal stress.
4. The global trend in the first principal stress and the z-displacement is determined by the resulting average pressure.
5. Peaks in the resulting maximum pressure can cause local maxima in the first principal stress.

REFERENCES

[1] International Electrotechnical Commission (IEC), IEC 61215-2:2021 Terrestrial photovoltaic (PV) modules – Design qualification and type approval: Part 2: Test Procedures 2, accessed 21 April 2021.

[2] P. Romer, A.J. Beinert, Effects of inhomogeneous snow load on the mechanics of a PV module, in: Proceedings of the 38th European Photovoltaic Solar Energy Conference and Exhibition (EU PVSEC), 2021, pp. 602–606.

[3] M. Shademan, R.M. Barron, R. Balachandar, H. Hangan, Numerical simulation of wind loading on ground-mounted solar panels at different flow configurations, Can. J. Civ. Eng. 41 (2014) 728–738. https://doi.org/10.1139/cjce-2013-0537.

[4] S.-T. Hsu, H.-H. Hsieh, Non-Uniform Mechanical Loads due to Wind Effect on Photovoltaic Module, in: Proceedings of the 33rd European Photovoltaic Solar Energy Conference and Exhibition (EU PVSEC), 2017, pp. 1698–1701.

[5] S.-T. Hsu, T.-C. Wu, Simulated Wind Action on Photovoltaic Module by Non-uniform Dynamic Mechanical Load and Mean Extended Wind Load, Enrgy Proced 130 (2017) 94–101.

[6] S.-T. Hsu, Standardization Work of Non-Uniform Wind Loads Test on PV Module, 37th European Photovoltaic Solar Energy Conference and Exhibition (2020) 894–899. https://doi.org/10.4229/EUPVSEC20202020-4CO.3.2.

[7] C.M. Jubayer, H. Hangan, Numerical simulation of wind effects on a stand-alone ground mounted photovoltaic (PV) system, Journal of Wind Engineering and Industrial Aerodynamics 134 (2014) 56–64. https://doi.org/10.1016/j.jweia.2014.08.008.

[8] C.M. Jubayer, H. Hangan, A numerical approach to the investigation of wind loading on an array of ground mounted solar photovoltaic (PV) panels, Journal of Wind Engineering and Industrial Aerodynamics 153 (2016) 60–70. https://doi.org/10.1016/j.jweia.2016.03.009.

[9] Pascal Romer, Kishan Bharatbhai Pethani, Andreas J. Beinert, Effect of Inhomogeneous Loads on the Mechanics of PV Modules, Progress in Photovoltaics: Research and Applications in review.

[10] Deutsches Institut für Normung e.V. (DIN), Eurocode 1: Einwirkungen auf Tragwerke - Teil 14: Allgemeine Einwirkungen, Windlasten: Eurocode 1: Einwirkungen auf Tragwerke - Teil 1-4: Allgemeine Einwirkungen, Windlasten, Beuth Verlag GmbH, 2005, p. 147.

[11] A.J. Beinert, P. Romer, M. Heinrich, M. Mittag, J. Aktaa, H. Neuhaus, The Effect of Cell and Module Dimensions on Thermomechanical Stress in PV Modules, IEEE J. Photovoltaics 10 (2020) 70–77. https://doi.org/10.1109/JPHOTOV.2019.2949875.

[12] P. Romer, G. Oreski, A.J. Beinert, D.H. Neuhaus, M. Mittag, More Realistic Consideration of Backsheets Coefficient of Thermal Expansion on Thermomechanics of PV Modules, in: Proceedings of the 37th European Photovoltaic Solar Energy Conference and Exhibition (EU PVSEC), 2020, pp. 772–776.

[13] R.B. Roberts, Thermal expansion reference data: silicon 300-850 K, Journal of Physics D: Applied Physics 14 (1981) L163-163.

[14] R.B. Roberts, Thermal expansion reference data: silicon 80-280K, Journal of Physics D: Applied Physics 15 (1982) L119-120.

[15] U. Eitner, S. Kajari-Schroeder, M. Koentges, H. Altenbach, Thermal stress and strain of solar cells in photovoltaic modules, in: H. Altenbach, V.A. Eremeyev (Eds.), Shell-like Structures: Non-classical Theories and Applications, Springer, Berlin/Heidelberg, 2011.

[16] W.M. Haynes (Ed.), CRC handbook of chemistry and physics, CRC Press, 2014.

LEAD-FREE PV MODULES: INDUSTRIAL REALIZATION AND EVALUATION OF ENVIRONMENTAL IMPACT

Paul Gebhardt*, Stephan Hoffmann, Timo Wenzel, Lorenz Friedrich, Sina Herceg,
Dilara Maria Subasi, Angela De Rose, Andreas Lorenz, Daniel Philipp
Fraunhofer-Institute for Solar Energy Systems ISE, Heidenhofstr. 2, 79110 Freiburg, Germany

ABSTRACT: With the large amount of photovoltaic (PV) modules required for a more renewable energy sector, strategies to avoid the toxic material lead, which is still present in most conventional modules today, becomes more and more relevant. This work presents the result of the government-funded research project "BermUDa", in which the industrial realization and environmental impact of lead-free modules was investigated. The herein chosen approach was to directly replace lead-containing module materials (namely metallization paste and solder cell connectors) with lead-free alternatives, while leaving the industrial production process as unchanged as possible. For this, a large number of pre-processed silicon wafers underwent cell (PERC) and module production, where the processability was tested and compared with conventional lead-containing materials. Subsequently, the performance of the products (cells and PV modules) was tested and compared in terms of electrical performance and reliability in power measurements and accelerated aging tests, respectively. The project thereby demonstrated the feasibility to work with lead-free technologies and adapt currently established industrial processes to produce completely lead-free PV modules that can compete with their lead-containing counterparts in terms of sustainability and costs (LCOE). The work also suggests focus on cell metallization pastes, driven by future market demand, to develop lead-free pastes that can achieve better fine line printing capabilities to allow greater flexibility when converting an existing production line to produce lead-free PV modules.
Keywords: Sustainable, Ribbons, LCA, Lead-free, Lead, Bismuth, Cell interconnection, Metallization

1. INTRODUCTION

For c-Si photovoltaic (PV) modules, lead is a common material contained in cell metallization pastes and as solder material in cell connectors: Our estimations yield that a conventional PV module typically contains around 8-10 g/m² of lead. Although PV modules are currently exempt from the European "Restriction of Hazardous Substances Directive" (RoHS[1]), a restriction of the use of lead is currently discussed on different regulatory levels, *e.g.*, as part of the EU-ecodesign guidelines.

Although the approach to replace lead with alternative materials while retaining conventional industry processes is not completely new, few sources for a study on the feasibility and sensibility, as well as the resulting environmental effects are available. It is, however, crucial for policy makers and module manufacturers alike to evaluate the current situation and determine, how easy a lead-free PV module production is realizable and which steps are necessary to react on a respective regulation in the future.

The herein chosen approach of the research project was to adapt the "state-of-the-art" module and cell manufacturing processes do make them lead-free. For this, the lead-containing materials in the screen printing and soldering process were replaced with lead-free alternatives to develop a "green" RoHS-compliant solar module. To ease the transition and minimize costs, solutions compatible with established industrial production lines are investigated

For this, 4500 pre-processed silicon wafers underwent cell production tests in the backend line of Fraunhofer ISE PV-TEC in which the screen printing properties of lead-free silver-based front side metallization pastes and the electrical properties of the resulting cells were investigated and compared to conventional lead-containing pastes. Subsequently, the produced cells underwent soldering tests on an industrial stringer, using lead-free ribbons. The applicable temperature window for soldering was determined as well as the mechanical stability of the joints.

Subsequently, the produced PV modules were tested and compared in terms of electrical performance and reliability by power measurements at STC and accelerated aging tests, respectively.

Furthermore, sustainability and economic aspects of lead-free materials were investigated in life-cycle-assessment (LCA) and a cost analysis, respectively. The LCA particularly focused on a comparison of the environmental impact of the lead-free materials and the different allocation methods for the co-products lead and bismuth. The LCA was supported by leaching experiments, which determined the heavy metal concentration leaching out of PV modules in a simulated worst-case end-of-life scenario.

2. METHODS

2.1. Cell Metallization

Throughout the project, around 4500 passivated emitter and rear solar cells (PERC) were processed in the backend line of Fraunhofer ISE PV-TEC. We used two types of industrially pre-fabricated p-type Czrochalski-grown silicon (Cz-Si) PERC precursors with rear side passivation and SiNx anti-reflection coating on the front side and a n-type emitter with a nominal sheet resistance of $R_{sh1} \approx 100\ \Omega/sq$ (type 1) and $R_{sh2} \approx 120\ \Omega/sq$ (type 2). The rear side was metallized with silver (Ag) pads and surrounding full area aluminum (Al) metallization on the fully automatic Asys XH-2 screen printing lines at Fraunhofer ISE PV-TEC. On the front side, an H-pattern grid consisting of 100 fingers and 5 busbars was applied, using commercially available lead-free and lead-containing Ag paste as a reference. Throughout the project, 9 different lead-containing and lead-free Ag and Al pastes for front and rear side metallization were tested.

2.2. Cell Interconnection and Reliability

For cell interconnection, an industrial stringer was used to operate at comparable condition with respect to an industrial cell production. The stringer used was the TT1800 of *teamtechnik* that uses IR light as primary heat source for the soldering process. We compared three different lead-free solder alloys with one lead-containing reference. The solder alloys were used as coatings of 0.9 mm × 0.22 mm copper ribbons, to interconnect solar cells with five busbars.

The herein presented solar cells with a lead containing cell metallization and a lead-free cell metallization were used (s. Section 2.1), as well as commercially available solar cells.

To find an optimal process window for interconnection of each solder alloy, the temperatures within the stringing process (IR radiation & hotplates) were varied. For each solder alloy, three adapted temperature profiles were used. Additionally, a fourth temperature profile, representing the standard soldering process for Sn60Pb40, was tested.

For each parameter set, 90° peel tests of the solder joints were performed on a peel testing machine of *Zwick Roell*. Due to the tapered busbar design of the front side, the measured values at the soldering pads were evaluated. Same for the rear side, only the maximum achieved peel force for each solder pad was used as value for the analysis.

In addition, 3-cell-strings were made to build small-scale modules to perform the temperature cycling test. A further cell was used to investigate the solder alloys in terms of optical behavior (effective width measurement), and microscopic examinations namely cross sections and imaging as well as SEM/EDX measurement to investigate the solder joint in detail. For the best interconnection parameters, full size modules (1700 mm × 1000 mm, 60 full cells each) were manufactured for accelerated aging tests according to IEC 61215-2:2016.

2.3. Sustainability and Cost Assessment

The study was modelled in the LCA software Umberto LCA+. For the life cycle inventory data, it was ensured that corresponding data from upstream processes are available in the Ecoinvent database. The scope of investigation was defined according to ISO 14040: The product system under investigation was an industrially manufactured lead-free Cz PERC module. The functional unit was –in accordance with PEFCR and PVPS– 1 kWh of electrical energy. The lifetime for a standard module was assumed to be 30 years with an annual power degradation of 0.7 %.

A cradle-to-grave analysis was performed to map the environmental impacts of the module from production through use to end-of-life. The focus of the actual analysis was primarily on the toxicity-related impact categories such as human toxicity, acidification and eutrophication. It has to be mentioned that especially the results from toxicity-related impact categories showed a higher uncertainty than for example the CO2 footprint. The USEtox model in particular is considered to be particularly reliable and was therefore selected for the analysis. For the LCA three different leaching scenarios were considered: a) groundwater contamination, b) agricultural soil contamination and c) contamination of soils used for other purposes.

The cost analysis was carried out using the Total Cost of Ownership (TCO) model developed at Fraunhofer ISE [2] and extended for the lead-free PERC modules to be developed in this work. Cross-cutting cost analyses were carried out across the value-added stages of cell, module and system. A monofacial p-type Cz M6 wafer 60-cell PERC module (reference) was compared with a lead-free PERC module (reference). with a lead-free PERC module.

3. RESULTS AND DISCUSSION

3.1. Cell Metallization

During several printing pre-tests, it was found that the selected lead-free paste did not perform equally well as the conventional lead-containing paste with respect to the printability of fine lines. While the lead-containing paste was able to print contact fingers down to a nominal finger width of $w_n = 33$ µm, the lead-free paste was limited to $w_n = 40$ µm (status 2019). Furthermore, depending on the fine line printing capability and contacting behavior of the metallization paste, the efficiency of the lead-free cells could be up to 0.3 $\%_{abs.}$ lower compared with the lead-containing reference.

It is assumed that the fine line printing performance of lead-containing silver pastes is generally on a higher level due to a substantially higher market demand and thus R&D activities for the optimization of lead-containing pastes. An increasing market demand regarding lead-free pastes will probably enable further paste development activities in order to achieve a comparable fine line printing capability of lead-free pastes.

After several pre-tests, an optimal configuration regarding pastes, screens and printing/firing process was selected to metallize the required large number of solar cells for interconnection and module integration.

3.2. Cell Interconnection and Reliability

A cross comparison of lead-free and lead-containing cell metallization with lead-free and lead-containing solder alloys was carried out at Fraunhofer ISE Module-TEC. We compared three different lead-free solder alloys (thus three different melting temperatures), namely Sn42Bi58, Sn60Bi40 and Ecosol, with the leaded reference Sn60Pb40.

The evaluation of the 90° peel tests showed that the weakest interface of the leaded cell group was always the interface between metallization paste and silicon wafer. The metallization sticked to the ribbon and was separated from the solar cell (s. Figure 1 a). This result correlates well to previous findings, published by other researcher [3–5], highlighting the necessity to optimize metallization and interconnection hand-in-hand. In comparison, for the commercial reference cell, a cohesive fracture within the metallization was observed. The lead-free cell metallization showed a mixed fracture pattern (s. Figure 1 b): cohesive within the metallization paste as well as adhesive between busbar and wafer, or even rupture of silicon pieces from the wafer.

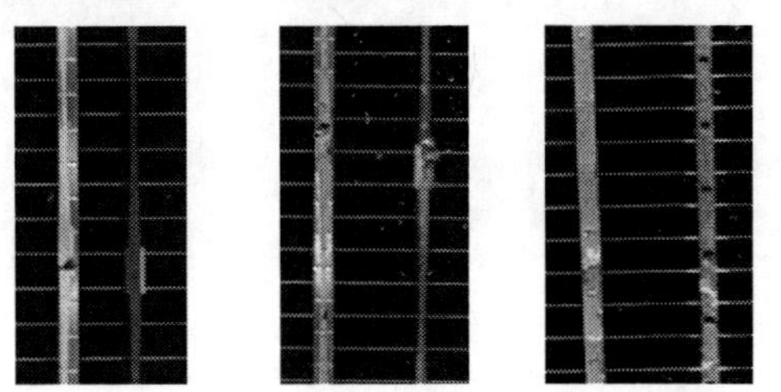

Figure 1: Top view images of the fracture pattern of leaded cell metallization (left), lead-free cell metallization (middle) and commercial reference cell (right).

These findings cannot be generalized for all combinations of metallization pastes and solder alloys, since only two different metallization pastes in combination with one type of precursor was tested. However, the most important finding is that lead-free metallization pastes can be contacted by soldering, resulting in a joint that is comparably stable as for established lead-containing materials.

The peel forces measured on commercial solar cells in combination with the different solder alloys are presented in Figure 2. Our results show that lead-free solder alloys have the potential to perform at the same level as a lead-containing solder alloy or even above. Regarding mechanical stability directly after soldering, lead-free solder alloys can easily replace lead-containing solder alloys. The achieved peel forces reached > 1 N/mm for a broad range of process parameters and remained the same as for a common lead-containing alloy.

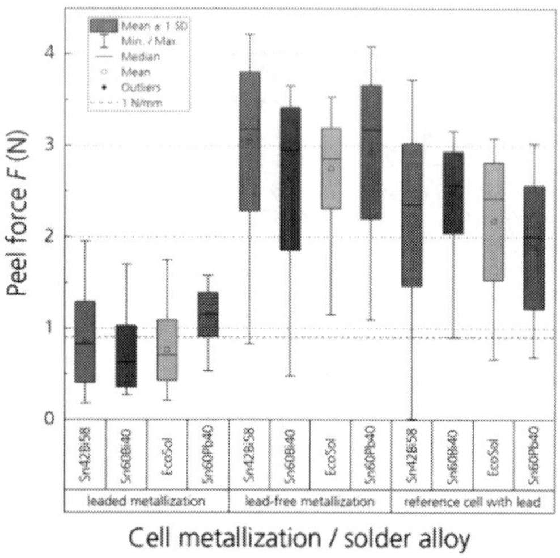

Figure 2: Measured 90° peel forces of the cell front side after soldering. For the group with lead-containing paste, the achieved peel forces were limited by the weak cell/metallization interface. The results show that lead-free pastes (middle) can operate on a comparable or even higher level to lead-containing pastes.

Some of the soldered strings were used to prepare metallographic cross sections for microscopy analysis. With SEM and EDX characterization, an intermetallic Ag_3Sn phase was observed (s. Figure 3 right), confirming a successful solder joint formation with SnBi solder.

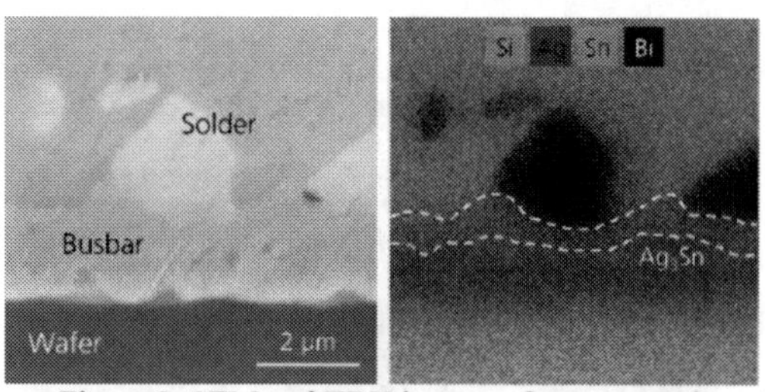

Figure 3: SEM and EDX images of a cross section of a lead-free solder joint. An Ag_3Sn phase is formed at the interface between solder and Ag metallization.

To further evaluate the technologies, reliability tests with the focus on thermal cycling were carried out on small-scale modules with a wide range of soldering conditions and material combinations, including lead-free and -containing metallization and solder, respectively, with a focus on resistance against thermomechanical stress (not shown). Based on these experiments 60-cell modules were produced, characterized in terms of *I-V*-parameters and electroluminescence imaging. The group achieving the best average value in terms of P_{MPP} was a lead-free group.

The reliability of the produced modules was demonstrated in accelerated aging tests according to IEC 61215. Full-size lead-free modules passed the criteria for damp heat, thermal cycling and combined UV/TC/HF tests with a good margin to the pass-fail criterion of -5 % relative P_{MPP} loss, which is typical for state of the art modules [6].

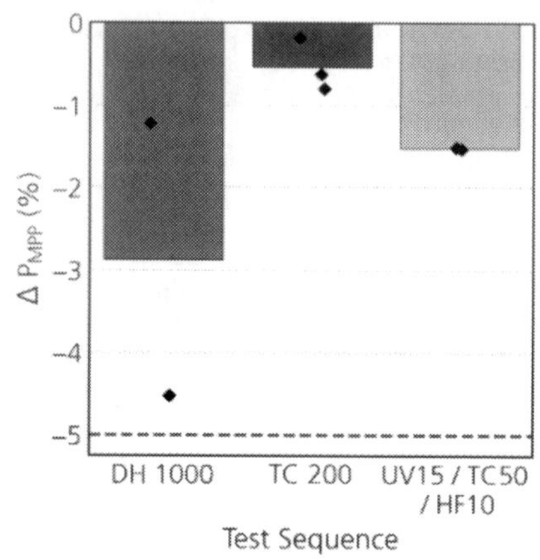

Figure 4: Power loss of lead-free modules after reliability tests according to IEC 61215-2:2016. The modules pass the tests with a good margin to the pass-fail criterion (-5 %).

3.3. Sustainability and Cost Assessment

A comparative LCA was performed for both module variants. For the analysis of the lead-free module, the lead, was replaced by bismuth in a mass ratio of one-to-one. Since no production or emission data for bismuth is available in the common LCA databases (*e.g.* GaBi or ecoinvent), it was modelled as a co-product of lead production according to literature [7, 8]. Different approaches for allocation (economic, physical) were used in a sensitivity analysis. The life cycle impact results for the indicators climate change, toxicity, eutrophication, acidification, ozone depletion and photochemical oxidation show a difference of < 1 % for cell as well as module production in all impact categories. Furthermore, literature suggests that the life cycle impact of bismuth may be worse in some categories than that of lead, since additional processing steps can be required for both production and recycling [7]. Since bismuth is often a by-product of lead mining, increased demand for bismuth could additionally increase the amount of lead mined.

However, our analysis also shows the toxicity of possible lead leaching into different soils, which may well be possible due to improper end-of-life storage in countries without a prober end-of-life treatment scheme in place [9]. Experimental leaching tests, into different solutions like simulated ground water or acidic rain, in simulated end-of-life scenarios were performed during the project, which showed a lower tendency of bismuth to leach out compared to lead. Additionally, less leaching was observed for modules after aging. Bismuth is also considered a critical, possibly rare resource for terawatt-scale PV expansion [10].

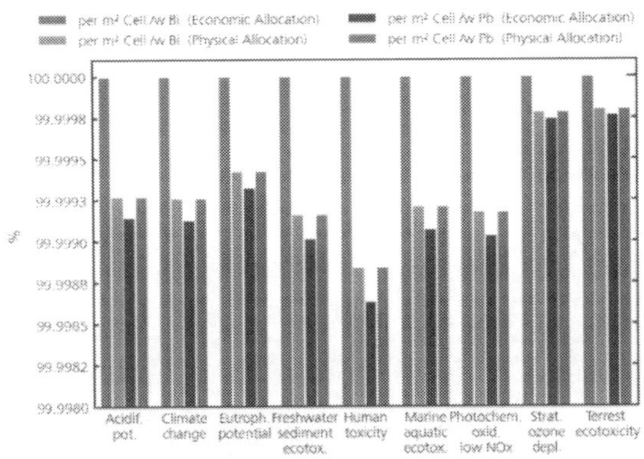

Figure 5: Life-cycle assessment of lead-containing and lead-free solar cells across different categories.

In conclusion, bismuth does not necessarily perform better as an alternative to lead in PV production, when considering the whole value chain. Nevertheless, lead, in contrast to Bismuth, has been shown to be toxic to humans and the environment when released into the environment.

In the Total Cost of Ownership (TCO) analysis, cross-cutting cost analyses were carried out across the value-added stages of cell, module, and system. Overall, the lead-free variant and the lead-contaning reference differ only slightly, *i.e.*, < 0.5 % for absolute module production costs and < 1 % for electricity production costs at Freiburg, even if the module efficiency of the lead-free variant is assumed < 0.3 $\%_{abs.}$ lower than the leaded reference.

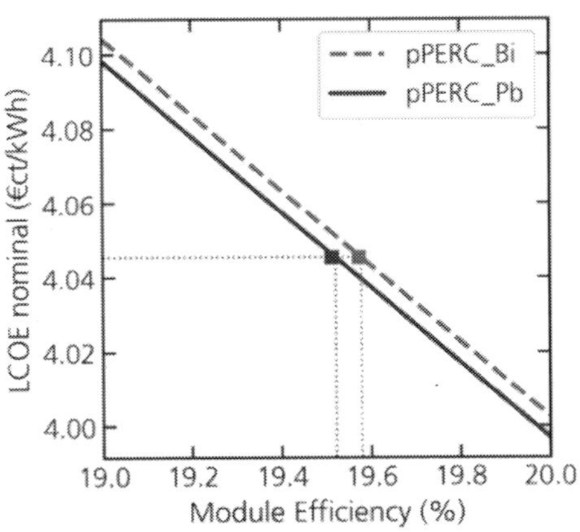

Figure 6: Sensitivity analysis of the influence of module efficiency on the LCOE at the Freiburg site

4. CONCLUSION AND OUTLOOK

This project investigated and evaluated the feasibility of producing lead-free PV modules using state-of-the-art production techniques such as screen printing and solder interconnection. To this end, nine different lead-containing and lead-free Ag and Al pastes for front and rear side metallization as well as four different solder alloys for cell interconnection were tested.

On cell level, the processability and printing properties of the lead-free metallization pastes required a relatively large finger thickness and showed slightly reduced efficiencies in some cases. An increasing market demand regarding lead-free pastes will probably enable further paste development activities in order to achieve a comparable fine line printing capability of lead-free pastes.

Concerning the cell interconnection, peel tests demonstrated that lead-free solder alloys can replace conventional alloys. Furthermore, they allow variations in soldering temperatures which can be beneficial for temperature-sensitive next-generation cell technologies like heterojunction [11, 12] or perovskite-Si-tandem [13].

As demonstrated on 60-cell PV modules, the power and reliability of lead-free modules was comparable and the modules passed critical accelerated aging tests.

The LCA showed that using bismuth is not necessarily better than lead in terms of environmental impacts. Further studies need to be done due to the availability of other lead-free soldering methods, which also do not contain bismuth. It is important to identify other soldering metals regarding the environmental aspect which can be than test in modules.

Besides the investigated technologies, there are already PV module products on the market, that are intrinsically lead-free, e.g. based on silicon heterojunction [14, 15] or IBC cells.

Furthermore, there exist new alternative metallization and interconnection concepts, e.g. plating [16] and conductive adhesives [17] respectively. These approaches are more complex to implement into an existing production line but are under continuous development and can be a viable alternative to achieve lead-free and sustainable PV modules. An in-depth comparison of different technological options is under development.

Acknowledgement: The authors acknowledge the funding from the German Federal Ministry for Economic Affairs and Energy (FKz.: 0324362 "BermUDa").

5. REFERENCES

[1] *Restriction of the use of certain hazardous substances, Directive 2011/65/EU: RoHS*, 2011. [Online]. Available: http://data.europa.eu/eli/dir/2011/65/oj

[2] S. Nold, Fraunhofer-Institut für Solare Energiesysteme -ISE-, Freiburg, Albert-Ludwigs-Universität Freiburg, Technische Fakultät, E. R. Weber, and S. W. Glunz, "Techno-ökonomische Bewertung neuer Produktionstechnologien entlang der Photovoltaik-Wertschöpfungskette. Modell zur Analyse der Total Cost of Ownership von Photovoltaik-Technologien," Stuttgart: Fraunhofer Verlag, Stuttgart.

[3] A. De Rose, D. Erath, T. Geipel, A. Kraft, and U. Eitner, "Low-temperature soldering for the interconnection of silicon heterojunction solar cells," in *Proceedings of the 33rd European Photovoltaic Solar Energy Conference and Exhibition (EU PVSEC)*, Amsterdam, Netherlands, 2017, pp. 710–714.

[4] D. Güldali and A. De Rose, "Material Joint Analysis of Lead-Free Interconnection Technologies for Silicon Photovoltaics," in *45th International Spring Seminar on Electronics Technology (ISSE)*, 2022. [Online]. Available: https://ieeexplore.ieee.org/document/9812798

[5] T. Geipel, D. Eberlein, and A. Kraft, "Lead-free solders for ribbon interconnection of crystalline silicon PERC solar cells with infrared soldering,"

in *AIP Conference Proceedings 2156*, Konstanz, 2019, p. 20015.
[6] P. Gebhardt, G. Mülhöfer, A. Roth, and D. Philipp, "Statistical analysis of 12 years of standardized accelerated aging in photovoltaic-module certification tests," *Prog. Photovolt: Res. Appl.*, 2021, doi: 10.1002/pip.3450.
[7] Umweltbundesamt, *ProBas Datensatz Wismut*. Projessorientierte Basisdaten für Umweltmanagement-Instrumente. [Online]. Available: http://www.probas.umweltbundesamt.de/php/prozessdetails.php?id={7FB2B6A5-43ED-4A2F-8866-F2E6A8126C68}
[8] T. Ibn-Mohammed *et al.*, "Life cycle assessment and environmental profile evaluation of lead-free piezoelectrics in comparison with lead zirconate titanate," *Journal of the European Ceramic Society*, vol. 38, no. 15, pp. 4922–4938, 2018, doi: 10.1016/j.jeurceramsoc.2018.06.044.
[9] P. Sinha and A. Wade, "Comment on "Long-term leaching of photovoltaic modules"," *Jpn. J. Appl. Phys.*, vol. 57, no. 1, p. 19101, 2018, doi: 10.7567/JJAP.57.019101.
[10] L. Bongartz, S. Shammugam, E. Gervais, and T. Schlegl, "Multidimensional criticality assessment of metal requirements for lithium-ion batteries in electric vehicles and stationary storage applications in Germany by 2050," *Journal of Cleaner Production*, vol. 292, p. 126056, 2021, doi: 10.1016/j.jclepro.2021.126056.
[11] B. A. Korevaar, J. A. Fronheiser, X. Zhang, L. M. Fedor, and T. R. Tolliver, "Influence of annealing on performance for hetero-junction a-Si/c-Si devices," in *Proceedings of the 23rd European Photovoltaic Solar Energy Conference and Exhibition*, Valencia, Spain, 2008, pp. 1859–1862.
[12] S. De Wolf, A. Descoeudres, Z. C. Holman, and C. Ballif, "High-efficiency Silicon Heterojunction Solar Cells: A Review," *Green*, vol. 2, no. 1, pp. 7–24, 2012, doi: 10.1515/green-2011-0018.
[13] M. de Bastiani, M. Babics, E. Aydin, A. S. Subbiah, L. Xu, and S. de Wolf, "All Set for Efficient and Reliable Perovskite/Silicon Tandem Photovoltaic Modules?," *Solar RRL*, vol. 6, no. 3, p. 2100493, 2022, doi: 10.1002/solr.202100493.
[14] A. Schneider, *Meyer Burger opens high tech module factory in Freiberg, Germany*, 2021. Accessed: Sep. 7 2022. [Online]. Available: https://www.meyerburger.com/de/newsroom/artikel/meyer-burger-eroeffnet-hightech-solarmodulfabrik-in-freiberg
[15] A. Schulze, *Another green check for the lead-free1 REC Alpha Pure: Certified as a Low Carbon Footprint solar panel*. Munich, 2022. [Online]. Available: https://www.recgroup.com/en/news/another-green-check-lead-free1-rec-alpha-pure-certified-low-carbon-footprint-solar-panel
[16] B. Grübel *et al.*, "Progress of plated metallization for industrial bifacial TOPCon silicon solar cells," *Prog. Photovolt: Res. Appl.*, 2021, doi: 10.1002/pip.3528.
[17] L. P. Bauermann *et al.*, "Qualification of conductive adhesives for photovoltaic application - accelerated ageing tests," *Enrgy Proced*, vol. 124, pp. 554–559, 2017, doi: 10.1016/j.egypro.2017.09.266.

EXPERIMENTAL STUDY OF NON-UNIFORM MECHANICAL LOADS AND DEFORMATION ON PHOTOVOLTAIC MODULE

Shu-Tsung Hsu*, Yean-San Long
Industrial Technology Research Institute (ITRI)
Rm. 609, Bldg. 51, No. 195, Sec. 4, Chung Hsing Rd., Hsin-Chu 31057, Taiwan
*E-mail: andersonhsu@itri.org.tw

ABSTRACT: In order to avoid photovoltaic (PV) systems installed in the outside might damage by typhoons, hurricanes and cyclones, consequently, recent years have been increasing interest in developing new mechanical loads technologies that provide guarantees to simulate the real wind effects. However, strong gusts always generated non-uniform wind-loads on PV system, with lot of dynamic forces from vibrate, torque and moment, and deformation. To date, the quality of anti- wind for PV modules typically use the mechanical load tests such as IEC 61215-2:2021 MQT 16 (static mechanic load, SML) and IEC TS 62782:2016 (dynamic mechanic load, DML). However, these documents may not fully fulfil the testing challenges induced by non-uniform wind-loads due to environmental condition like wind velocity (V), wind direction angle (β), and module tilt (α). In this paper, two kinds of designed cyclic loads (uniform, non-uniform) are used to evaluate the mechanical properties of twelve PV modules (mono, 72 cells, α = 25°) with two frames support. This work also considers the specific wind condition (V = 41.4 m/s, 50.9 m/s and 61.2 m/s; β = 0° and 180°), and follows two test process which include SML (case A & C) and DML (case B & D). The major self-developed equipment includes a non-uniform dynamic mechanical load (NUDML) system, a deformation measurement system (DMS) and thirty-two units of linear displacement sensor (LDS). The results successfully achieved a quantifiable deformation recording method for PV module under two types of mechanical load process (SML, DML) combining with two kinds of cyclic loads (uniform, non-uniform).
Keywords: Photovoltaic; Deformation; Non-uniform;

To date, the quality of anti- wind for PV modules typically use the mechanical load tests such as IEC 61215-2:2021 MQT 16 (static mechanic load, SML) and IEC TS 62782:2016 (dynamic mechanic load, DML). However, these documents may not fully fulfil the testing challenges induced by non-uniform wind-loads due to environmental condition like wind velocity (V), wind direction angle (β), and module tilt (α). In this paper, two kinds of designed cyclic loads (uniform, non-uniform) are used to evaluate the mechanical properties of twelve PV modules (mono, 72 cells, α = 25°) with two frames support. This work also considers the specific wind condition (V = 41.4 m/s, 50.9 m/s and 61.2 m/s; β = 0° and 180°), and follows two test process which includes SML (case A & C) and DML (case B & D). The major self-developed equipment includes a non-uniform dynamic mechanical load (NUDML) system, a deformation measurement system (DMS) and thirty-two units of linear displacement sensor (LDS).

1 TEST PLAN AND TEST EQUIPMENT

The installation of a PV module is of necessity concerned with the wind uplift produced by typhoons. Breakage of the glass covers of PV modules and frames is mainly due to the strong wind uplift and the associated large deflection. Therefore, an analytical model about the mean extended wind load [1] is essential for preventing such damage. Currently, the IEC standards [2-3] still lack of detailed documents to define wind test for PV module, especially considering the non-uniform wind loads due to moment effect. In this work, a self-developed NUDML system can simulate both the non-uniform loads (Mean Surface Pressure Pattern, MSPP [4]) and uniform loads, and use thirty-two LDSs to record the deformation data of test PV module.

1.1 Non-uniform loads

In general, the front wind induces a pressure load and the back wind induces a suction load (Fig. 1(a)). In Fig. 1(b), pressure coefficient C_p developed by computational fluid dynamics (CFD) using a test condition like a PV panel at fixed wind velocity (V) and tilt (α). This Cp stands for pressure load or suction load should have the near-symmetric trapezoidal distribution along the length (L) of module [1]. The value of ΔC_p (Eq 1) on suction load is little bigger than pressure load due to the clearance height of PV module from ground. Furthermore, when such air flow follows the independent Reynolds number and fluid similarity formula, then its C_p value always keeps the same no matter V is different. In order to simulate the uneven wind loads by mechanical load (ML) system, the designed non-uniform loads named MSPP@p_1-p_{18} which can be obtained by averaging the ΔC_p distribution on module into 18 zones in order to meet 18 pistons' forcing for ML test (Fig. 1(d)), and considering the required safety factor (SF). In Fig. 1(c), six types of MSPP@p_1-p_{18} are transferred from the wind pressure data of CFD [5] at wind velocity V_a = 20 m/s by Bernoulli equation (Eq 2, air density ρ = 1.1649 kg/m^3), and considering three wind velocities (V_b = 41.4 m/s, 50.9 m/s and 61.2 m/s) to stand for three scenarios of Beaufort Scale (BS13, BS15 and BS17), and two wind direction (β = 0°, 180°).

$$\Delta C_p = C_{p\,upper} - C_{p\,lower} \quad (1)$$

$$C_p = (P_a - P_\infty) / (0.5\rho V_a^2) = (P_b - P_\infty) / (0.5\rho V_b^2) \quad (2)$$

1.2 Uniform loads

In order to compare with non-uniform loads of MSPP, the value of according uniform loads in this work are defined by averaging MSPPs data (Figs. 1(c),1(e)).

1.3 Test flow and test condition

Below is the test flow for four cases (A-D) with two designed cyclic loads, and each case have three test modules for three wind velocity (TABLE I).

- Pre-test: EL and I-V test.

- Four cases of mechanical loads test.
  1) Case A: SML test [2] + designed cyclic loads (uniform).
  2) Case B: DML test [3] + designed cyclic loads (uniform).
  3) Case C: SML test + designed cyclic loads (non-uniform).
  4) Case D: DML test + designed cyclic loads (non-uniform).
- Post-test: EL and I-V test.

(a)

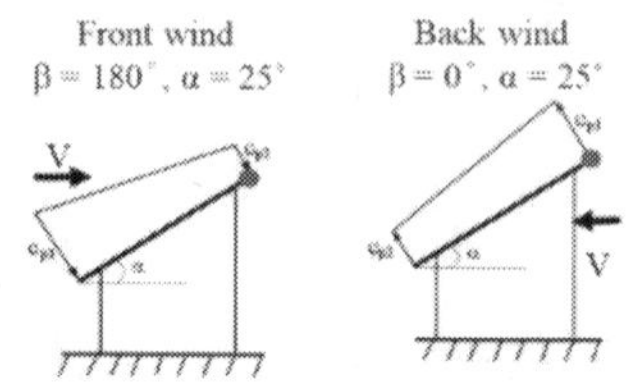

(b)

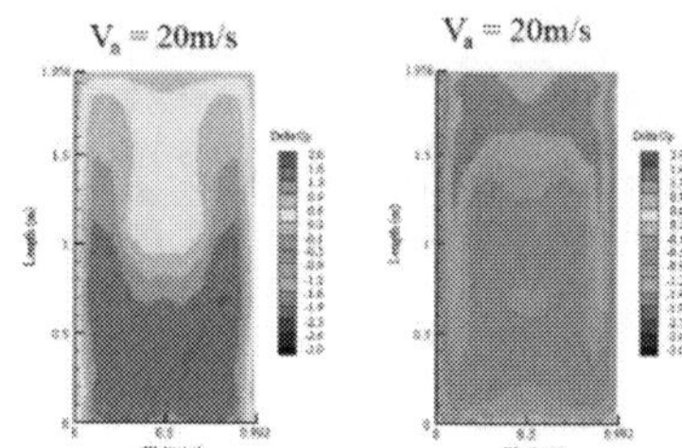

(c)

MSPP1@ V_b= 41.4 m/s		
1193	677	1193
1972	1076	1972
2568	1051	2568
3570	3045	3570
3447	4416	3447
3131	4119	3131

MSPP2@ V_b= 41.4 m/s		
-3686	-4579	-3686
-3334	-3587	-3334
-2411	-2192	-2411
-2275	-1586	-2275
-1982	-1608	-1982
-1115	-925	-1115

MSPP3@ V_b= 50.9 m/s		
1803	1023	1803
2981	1627	2981
3881	1589	3881
5397	4603	5397
5211	6674	5211
4733	6226	4733

MSPP4@ V_b= 50.9 m/s		
-5571	-6922	-5571
-5039	-5422	-5039
-3645	-3314	-3645
-3439	-2397	-3439
-2997	-2431	-2997
-1685	-1398	-1685

MSPP5@ V_b= 61.2 m/s		
2607	1479	2607
4309	2352	4309
5611	2297	5611
7802	6655	7802
7533	9649	7533
6842	9001	6842

MSPP6@ V_b= 61.2 m/s		
-8054	-10007	-8054
-7285	-7838	-7285
-5269	-4791	-5269
-4972	-3465	-4972
-4332	-3514	-4332
-2436	-2021	-2436

(d)

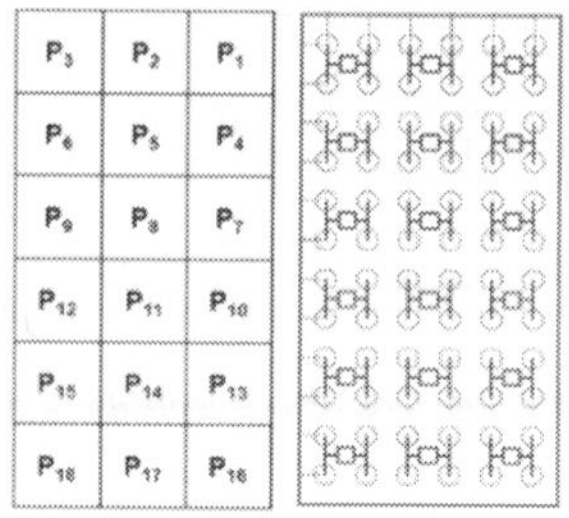

(e)

AVE (MSPP1) = +2565 Pa
AVE (MSPP2) = -2448 Pa
AVE (MSPP3) = +3876 Pa
AVE (MSPP4) = -3702 Pa
AVE (MSPP5) = +5601 Pa
AVE (MSPP6) = -5352 Pa

Figure 1: (a) front wind (β = 180°), back wind (β = 0°) and α = 25° (b) ΔC_p contour at V_a = 20 m/s (c) six MSPPs (unit: P_a) developed at three wind velocities and two wind direction (d) the diagram of MSPP@p_1~p_{18} and ML system with eighteen pistons (e) uniform loads. Note: ●: is the origin of a Cartesian coordinate system; AVE means average; "+" pressure load, "-" suction load.

TABLE I: Test condition (α = 25°, SF = 3)

designed loads		V=41.4 m/s	V=50.9 m/s	V=61.2 m/s
uniform load	pressure (β = 180°)	+2565 P_a	+3876 P_a	+5601 P_a
	suction (β = 0°)	-2448 P_a	-3702 P_a	-5352 P_a
module	case A	M2	M6	M10
	case B	M1	M5	M9
non-uniform load	pressure (β = 180°)	MSPP1	MSPP3	MSPP5
	suction (β = 0°)	MSPP2	MSPP4	MSPP6
module	case C	M4	M8	M12
	case D	M3	M7	M11

1-4 Test equipment

- NUDML system owns eighteen pistons with four suction cups (Fig. 2(a)).
- Pulsed solar simulator (BERGER): 1000 W/m², 25 °C, AM 1.5G (Fig. 2(b)).
- Electroluminescence (EL) system (Z4-PLUS) contains 4 high-resolution CCD cameras in NIR technique, biased current 8A (Fig. 2(c)).
- Module sample (196.8 cm x 99.2 cm x 4 cm) with two frames (4 cm x 4 cm x 110 cm) of support (Fig. 2(d)).
- KEITHLEY DAQ6510 data acquisition (Fig. 2(e)).
- Deformation measurement system and LDS under module (Fig. 2(f)).
- The allocation diagram of 32 (=4x8) LDSs on half module (Fig. 2(g)).
- Record the maximum and minimum deformation data for each LDS (Fig. 2(h)).

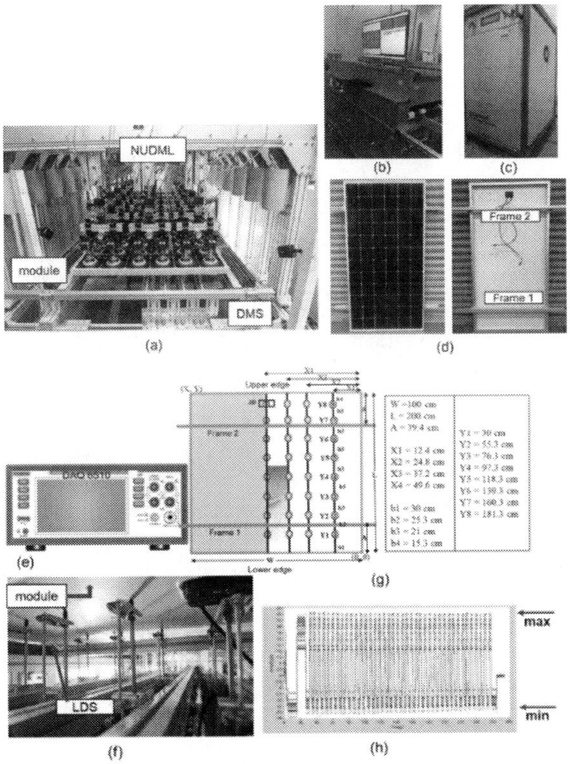

Figure 2: Test equipment (a) NUDML system (b) pulsed solar simulator (c) Electroluminescence (EL) system (d) module sample (e) KEITHLEY DAQ6510 data acquisition (f) deformation measurement system and LDS (g) the allocation diagram of 32 (=4x8) LDSs (h) record the maximum and minimum deformation data for each LDS. Note: L: module length; W: module width.

2 RESULTS AND DISCUSSION

In Fig. 3, the uniform load case, whether in the pressure or suction region, the maximum amount of deformation is concentrated at the center of the long side of the module (at Y = L/2). However, in the non-uniform load case, the maximum deformation in the pressure region occurs near the lower edge of the module (near frame 1), but in the pressure region occurs near the upper edge of the module (near frame 2). The two types of load (uniform and non-uniform) are similar in that the maximum deformation is larger in the suction region than in the pressure region, and increases with the increase of wind speed. In addition, the maximum deformation of the two regions (pressure and suction) increases gradually as the LDS approaches the center of the short side of the module (at X = W/2). In Fig. 4, the higher BS value has the higher wind speed, three modules (M7, M11, M12) due to non-uniform load have visible cracks, and two modules (M7, M11) of them due to DML case owns the maximum pressure happened at the lower edge then cause larger cracks and damage, but the maximum suction load happened at the upper edge is less damage. In addition, two modules (M9, M10) due to uniform load also have visible cracks, and module M9 due to DML case owns the greater damage happened near center of module. Upon performance, the P_{max} loss over 5 % is only happened on two modules (M9, M11) under BS17 for DML case, and the causes could be due to the visible cracks in their EL images.

3 CONCLUSIONS

The results successfully achieve a quantifiable deformation recording method for PV module under two types of mechanical load process (SML, DML) combing with two kinds of cyclic loads (uniform, non-uniform), especially for test module with two frame support.

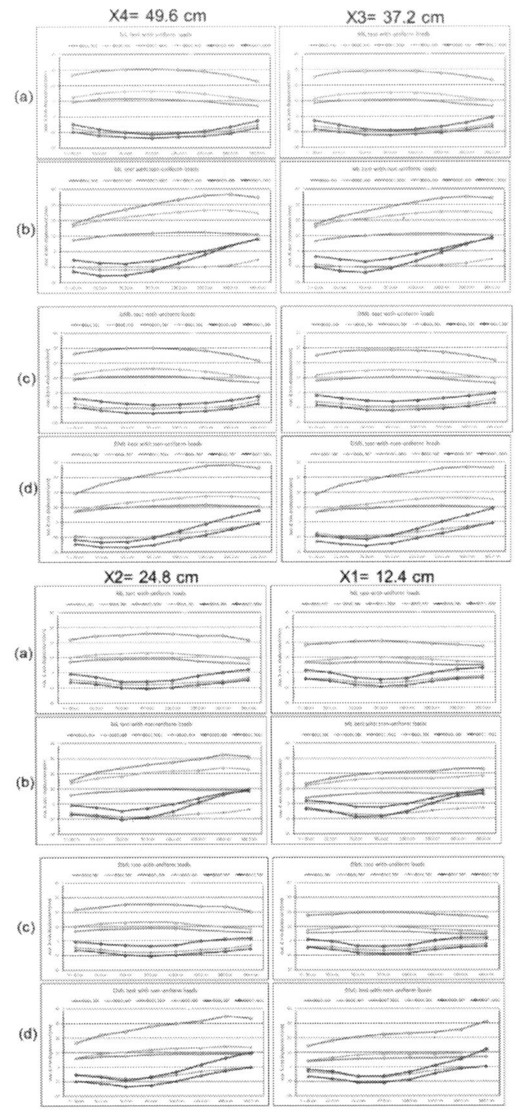

Figure 3: Four cases of deformation curves (X axil: Y1~Y8; Y axil: max. displacement) for both pressure and suction load under three velocities (a) case A (b) case C (c) case B (d) case D. Note: positive data means pull up; negative data means push down"; hollow symbol means suction load; filled symbol means pressure load.

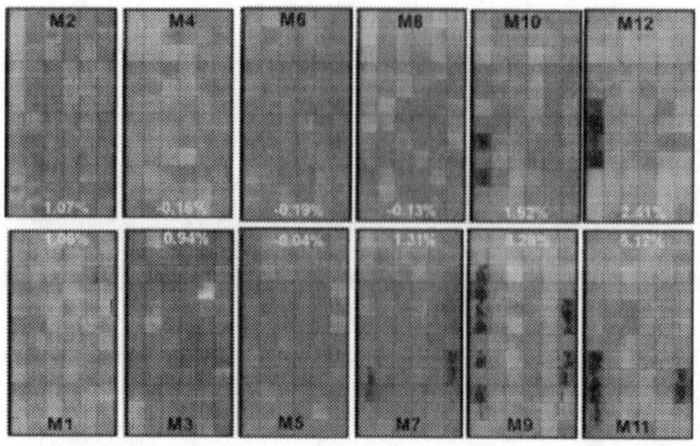

Figure 4: EL images and P_{max} loss (%) at post-test for twelve modules (M1-M12)

4 REFERENCES

[1] S. T. Hsu, T. C. Wu, Simulated Wind Action on Photovoltaic Module by Non-uniform Dynamic Mechanical Load and Mean Extended Wind Load, Energy Procedia (2017) pp. 94-101, DOI information: 10.1016/j.egypro.2017.09.401.

[2] IEC 61215-2:2021 MQT 16, Static Mechanical Load Test.

[3] IEC TS 62782:2016, Photovoltaic (PV) Modules – Cyclic (Dynamic) Mechanical Load Testing.

[4] S. T. Hsu *et al.*, Environmental Factors for Non-uniform Dynamic Mechanical Load Test due to Wind Actions on Photovoltaic Module, Energy Procedia (2018) pp. 50-57, DOI information: 10.1016/j.egypro.2018.09.008.

[5] K. C. Su, K. M. Chung, S. T. Hsu, Numerical Simulation of Wind Loads on Solar Panels, Modern. Physics Letters B, Vol. 32, Nos. 12 & 13 (2018) 1840009 (5 pages), DOI: 10.1142/S0217984918400092.

MEASURING AND MODELING THE RATE OF POWER LOSS IN N-PERT CELLS ASSOCIATED WITH PID-P AS A FUNCTION OF ENCAPSULANT RESISTIVITY AND IRRADIANCE

Brian M. Habersberger[1,*] and Peter Hacke[2]
1: Dow, Inc. 2: National Renewable Energy Laboratory
*230 Abner Jackson Parkway
HHD Building 1B228
Lake Jackson, TX 77566 USA

ABSTRACT: PID-p is commonly observed on the rear and front faces of bifacial PERC and n-PERT cells, respectively. While PID-p has been previously reported to be recoverable under light exposure as well as prevented or mitigated via the use of high-resistivity encapsulants, the relative rates of degradation and recovery processes, as well as their dependence on factors such as temperature, irradiance, and component resistivity are not well understood. Here, we expand on a recent report describing time-dependent PID measurements with simultaneous irradiance on minimodules comprising n-PERT cells and four encapsulants of varying resistivity. Additional physics are introduced with the goal of relating laboratory PID measurements to fielded module performance.
Keywords: Encapsulation, Degradation, n-type, PERT, Reliability

1 INTRODUCTION AND APPROACH

Under applied voltage, charge transport within a photovoltaic module can cause a variety of degradation phenomena collectively known as potential-induced degradation (PID). While multiple mechanisms that contribute to observed power loss during voltage stress have been identified, a single PID test protocol is generally used for module qualification. However, the various PID mechanisms, including shunting induced by sodium migration (PID-s), polarization of a dielectric layer in the cell (PID-p), and electrochemical corrosion (PID-c) may be subject to different acceleration factors and recovery phenomena, limiting the utility of module qualification protocols to accurately predict and screen field-relevant degradation.

While a variety of approaches to preventing or managing PID have been reported, the use of encapsulants such as polyolefin (POE) or ionomer instead of conventional ethylene-vinyl acetate copolymer (EVA) has been associated with prevention of both PID-s and PID-p.[1] In the case of PID-s, this is due to the impermeability of sodium ions in POE and ionomer relative to EVA, while for PID-p, it is due to the higher resistivity of POE and ionomer.[2,3]

Because PID-p is associated with polarization of a surface layer of a cell, the rate at which it occurs should be directly related to the flux of charge through the encapsulant, typically described as the volume resistivity or conductivity. Additionally, PID-p is reported to be recoverable via subsequent illumination, and in some cases preventable by simultaneous illumination during voltage stress.[4] Since illumination evidently depolarizes the cell, we may expect that the rate of depolarization is related to the irradiance. Thus, there are at least two critical factors determining the susceptibility of an encapsulated cell to PID-p: the rate of charge accumulation, primarily determined by the encapsulant, and the rate of depolarization, related to the irradiance.

To elucidate these factors, we have constructed minimodules using n-PERT cells and four encapsulants of varying resistivity (one EVA and three POEs). The minimodules have been subjected to voltage stress under varying levels of irradiance, from dark to 1000 W/m^2, with time-dependent monitoring of device performance. A rate-based model was developed relating the observed power loss to encapsulant resistivity and irradiance and is being expanded to include additional physics including silicon nitride photoconductivity and dielectric polarization of the encapsulant.

2 RESULTS AND DISCUSSION

The open-circuit voltage (V_{OC}) of minimodules was monitored during simultaneous voltage stress (-1000 V on the cell circuit) and illumination (Oriel filtered Xe solar simulator). The results of tests conducted with irradiance of 100 W/m^2 are shown in the top row of Figure 1. At this modest level of illumination, reduction in V_{OC} occurs rapidly for EVA-encapsulated cells, and minimal loss is observed with POE A, which has the highest resistivity (ρ). Cell encapsulated in POEs B and C, with intermediate resistivity, exhibit measurable loss, and those with POE C exhibit partial recovery after reaching a V_{OC} minimum.

As outlined above, power loss associated with PID-p is believed to be due to the accumulation of charge in a dielectric layer such as the silicon nitride. We have previously suggested a model describing this charge concentration in terms of the flux into the cell (primarily controlled by encapsulant resistivity) and out of the cell (presumably controlled by SiN_x photoconductivity), shown in Equation 1.[3]

$$\frac{dQ}{dt} = \frac{V}{\rho l} - kEQ \qquad \text{Eq. 1}$$

Here, Q is the charge concentration (C/cm^2), V is the applied voltage, ρ and l are the encapsulant resistivity and thickness, E is the irradiance, and k is an empirical constant. An empirical relation is additionally used to calculate the power loss associated with Q, obtained from time-dependent PID data reported by Yamaguchi et al.[6]

This very simple model is capable of qualitatively capturing many of the experimental observations, as shown in the bottom row of Figure 1. However, some observations, such as the partial recovery of minimodules encapsulated with POE C, are not represented in the model, and furthermore it is desirable to ground the empirical components of the model in physics, so we now seek to update this model through the incorporation of additional data and physical relationships. The second row

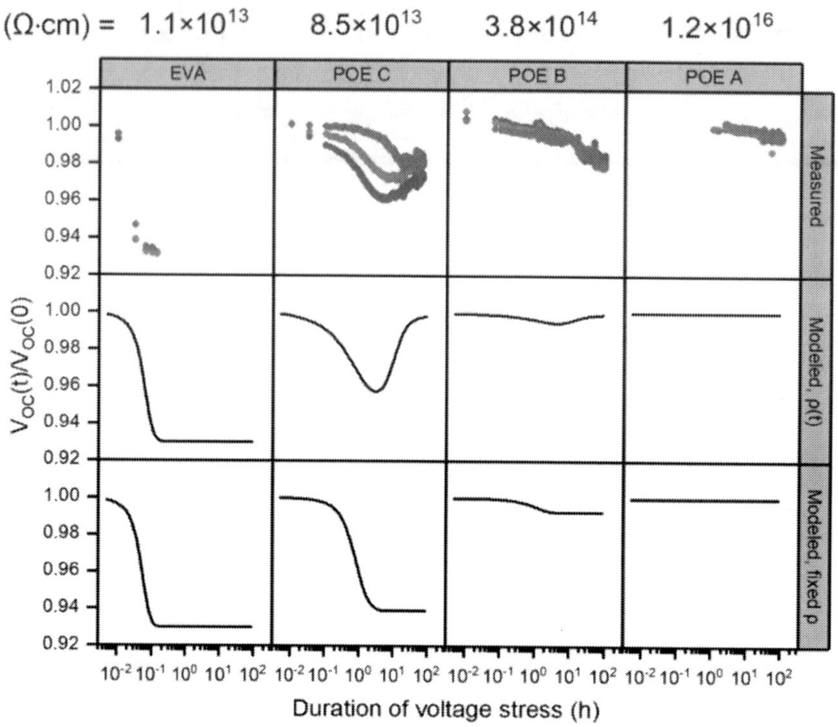

Figure 1. Top row: Normalized V_{OC} as measured on minimodules under -1000 V and 100 W/m^2 of irradiance. Different colored curves represent replicate specimens. Middle row: Predicted V_{OC} using Eq. 1 with time-dependent encapsulant resistivity. Bottom row: Predicted V_{OC} using Eq. 1 with fixed encapsulant resistivity.

of Figure 1 shows the results when the model is updated to include time-dependent charge flux through the encapsulant, represented through a time-dependent resistivity.

During resistivity measurements, typically a fixed time interval is chosen at which the current is recorded and used to calculate resistivity (for the values on Figure 1, a 10-minute interval was used). However, current usually continues to decline such that the longer the time interval, the higher resistivity value will be recorded. Instead of using such a fixed resistivity metric, our updated model now includes this physical behavior through the use of the time-dependent resistivity, $\rho(t)$, measured for each encapsulant. As shown in the second row of Figure 1, this allows the model to capture the recovery observed by minimodules containing POE C. While current flow through this material is initially sufficiently high to overcome the light-induced recovery process, over time it diminishes, allowing charge to drain and the device's power loss to recover, even while still under voltage stress.

Additional refinement of the model, incorporating known device and materials physics to describe charge drainage and the relation between charge accumulation and power loss, can be combined with climate conditions to potentially enable forecasting of power loss caused by PID-p.[7]

3 REFERENCES

[1] Luo, W.; Khoo, Y. S.; Hacke, P.; Naumann, V.; Lausch, D.; Harvey, S. P.; Singh, J. P.; Chai, J.; Wang, Y.; Aberle, A. G. Potential-induced degradation in photovoltaic modules: a critical review. Energy & Environmental Science 2017, 10 (1), 43-68.

[2] Habersberger, B. M.; Hacke, P.; Madenjian, L. S. In Evaluation of the PID-s susceptibility of modules encapsulated in materials of varying resistivity, 2018 IEEE 7th World Conference on Photovoltaic Energy Conversion (WCPEC)(A Joint Conference of 45th IEEE PVSC, 28th PVSEC & 34th EU PVSEC), 2018; IEEE: 2018; pp 3807-3809.

[3] Habersberger, B. M.; Hacke, P. Impact of illumination and encapsulant resistivity on polarization-type potential-induced degradation on n-PERT cells. Progress in Photovoltaics: Research and Applications 2021.

[4] Luo, W.; Hacke, P.; Hsian, S. M.; Wang, Y.; Aberle, A. G.; Ramakrishna, S.; Khoo, Y. S. Investigation of the Impact of Illumination on the Polarization-Type Potential-Induced Degradation of Crystalline Silicon Photovoltaic Modules. IEEE Journal of Photovoltaics 2018, 8 (5), 1168-1173.

[5] Yamaguchi, S.; Van Aken, B. B.; Masuda, A.; Ohdaira, K. Potential-Induced Degradation in High-Efficiency n-Type Crystalline-Silicon Photovoltaic Modules: A Literature Review. Solar RRL 2021, 5 (12), 2100708.

[6] Yamaguchi, S.; Nakamura, K.; Masuda, A.; Ohdaira, K. Rapid progression and subsequent saturation of polarization-type potential-induced degradation of n-type front-emitter crystalline-silicon photovoltaic

modules. Japanese Journal of Applied Physics 2018, 57 (12), 122301.
[7] Hacke, P.; Smith, R.; Kurtz, S.; Jordan, D.; Wohlgemuth, J. In Modeling current transfer from PV modules based on meteorological data, Photovoltaic Specialists Conference (PVSC), 2016 IEEE 43rd, 2016; IEEE: 2016; pp 1083-1089.

STUDY ON CORROSION REACTIONS AT THE REAR SIDE OF BIFACIAL PERC SOLAR CELLS

Yuanpang Hao[1], Xin Huang[1], Xiaocai Ke[1], He Wang[1], Hong Yang[1,*]
[1]School of Physics, Xi'an Jiaotong University, Xi'an 710049, People's Republic of China

ABSTRACT: In this paper, the corrosion reactions on the rear side of bifacial silicon PERC solar cells under different polarity system bias were investigated. It was found that there are various degrees of corrosion at the Al/Ag finger due to the difference in the metal electrode potential and polarity system voltage. A novel finding is that the phenomenon of broken grid was observed when the back of PERC silicon solar cells was under positive voltage bias. Then, SEM and EDS were performed to analyze the microstructure of the back of the bifacial PERC solar cells after corrosion. The corrosion reactions will cause serious performance degradation and have adverse effects on the reliability of bifacial PERC solar cells.
Keywords: Corrosion reactions, PERC, Microstructure, Degradation

1 INTRODUCTION

In the past few years, the long-term reliability and durability of photovoltaic modules in outdoor environments have been extensively studied [1-2]. It is commonly known that crystalline silicon photovoltaic modules are usually exposed to high humidity, high temperature, and high system voltage conditions, such an environmental can lead to the occurrence of corrosion reactions and the degradation of c-Si solar modules [3-5]. Moisture and gases can diffuse in from the breathable back-sheets or the edge of solar modules [5-6]. Some published papers have reported that once water vapor comes into contact with soda-lime glass, the corrosion reactions of glass would produce Na^+ ions, which is an important factor in the potential induced degradation of photovoltaic modules [7-10]. Peike et al. [11] reported that metal finger corrosion is one of the causes of degradation of solar cells. Bai et al. [12] have investigated the sliver finger on the front side of solar cells was corroded due to the presence of water under negative voltage bias. However, bifacial silicon passivated emitter and rear cells (PERCs) as a new type of solar cell have dominated the current photovoltaic industry in recent years. Because the back structure of bifacial silicon PERC solar cells is not equipped with a full-area metallization compared to aluminum back surface field solar cells, there is more extensive attention about the degradation caused by electrochemical corrosion reactions of bifacial PERC solar cells.

In this paper, some of the electrochemical corrosion reactions of the rear side of bifacial PERC solar cells under different polarity system bias in sodium chloride solution were investigated. Through simulation studies, we found that there are various degrees of corrosion at the Al/Ag finger due to the difference in degradation mechanisms under different polarity bias. In particular, the phenomenon of broken grid was observed when the back of PERC silicon solar cells was under positive voltage bias. Finally, the corrosion reactions of the back of the bifacial PERC solar cells were verified by using SEM and EDS mapping.

2 EXPERIMENTAL DETAILS

In this work, bifacial PERC solar cells were prepared by p-type single crystalline silicon wafers with dimensions of 182 mm×182 mm and thickness of 170±20 μm. Figure 1 shows the schematic structure of bifacial PERC solar cells. In Figure 1, bifacial PERC solar cell consists of the front silver contact, SiN_x: H ARC layer, n+ emitter, p-type Cz-silicon substrate, AlO_x passivation layer, SiN_x layer, the local Al Back Surface Field, rear Ag contact and rear Al contact. In this paper, the Al foil PID test method was used to simulate the polarity system bias of modules, which means 1500V voltage was applied between the bifacial PERC solar cell and Al foil. Sodium chloride solutions were prepared to corrode the rear side of bifacial silicon PERC solar cells in our lab. The microstructure characteristic of the rear side of corroded bifacial silicon PERC solar cells was analyzed by Scanning Electron Microscopy (SEM) and Energy Dispersive X-ray spectroscopy (EDS).

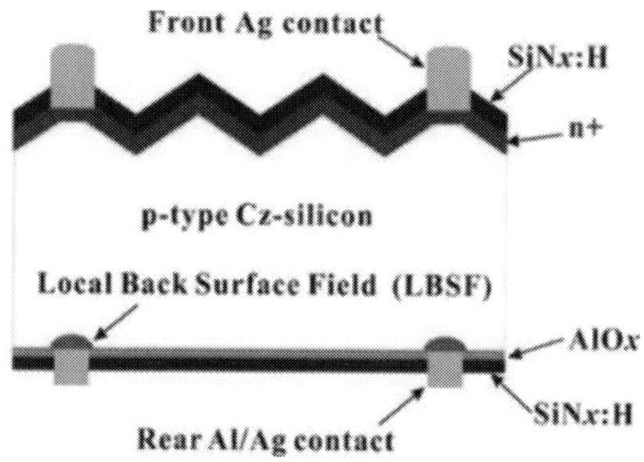

Figure 1: The schematic structure of bifacial PERC solar cells

3 RESULTS AND DISCUSSION

3.1 The electrochemical reactions on the rear side of bifacial PERC solar cells without bias

Figure 2 shows the electrochemical reaction diagram when the voltage applied between the back of PERC silicon solar cells and the Al foil is zero. In this situation, some galvanic corrosion reactions would happen on the rear side of bifacial PERC solar cells.

It is well known that the electrode potential of Ag/Ag^+ and Al/Al^{3+} are +0.799V and -1.622V, respectively. According to the difference of the electrode potential of different metals, the lower the metal potential, the easier the corrosion. Hence, The Al fingers act as anode under the action of electrolyte will occur oxidation reaction, which means Al fingers are firstly oxidized generating Al^{3+} ions, as expressed in Eq. (1). Due to the existing of sodium chloride solutions, we can clearly analyze that H_2O as the reduction regent is reduced producing H_2 and OH^-

ions near the sliver electrode, which can be expressed by Eq. (2). At the same time, Ag as reducing agent can accelerate the occurrence of water reduction reaction [13]. Owing to the presence of OH^- ions, Al^{3+} ions will combine with hydroxide ions to form sodium aluminum precipitation, as expressed in Eq. (3).

$$Al - 3e^- \longrightarrow Al^{3+} \quad (1)$$

$$2H_2O + 2e^- \xrightarrow{Ag} 2OH^- + H_2 \quad (2)$$

$$Al^{3+} + 3OH^- \longrightarrow Al(OH)_3 \quad (3)$$

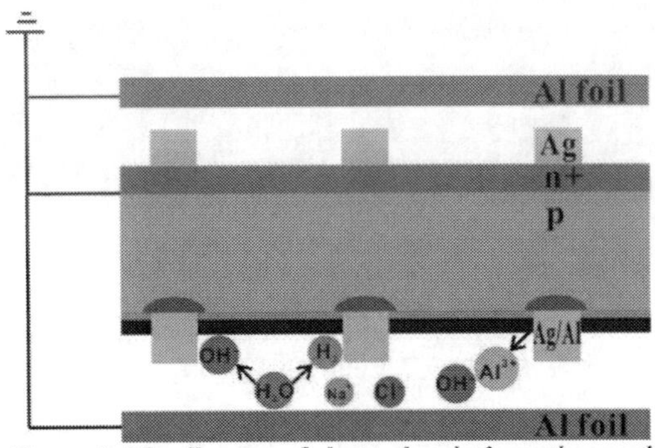

Figure 2: The diagram of electrochemical reaction on the back of bifacial PERC solar cells without bias.

3.2 The electrochemical reactions on the rear side of bifacial PERC solar cells under negative voltage bias

Figure 3 shows the diagram of electrochemical reaction on the back of PERC silicon solar cells under negative voltage bias. From Figure 3, the direction of electric field is from Al foil to the rear side of bifacial PERC solar cells. It is found that when the potential of the rear side of bifacial PERC solar cells to Al foil is negative, the rear side of solar cells act as cathode occurring reduction reactions. Due to the presence of sodium chloride solution, water is reduced under the action of sodium chloride electrolyte to form H_2 and OH^- ions, as given in Eq. (4), which indicates there is a basic environment on the rear side of bifacial PERC solar cells. In this situation, the Al fingers are corroded by sodium hydroxide.

$$2H_2O + 2e^- \xrightarrow{Ag} 2OH^- + H_2 \quad (4)$$

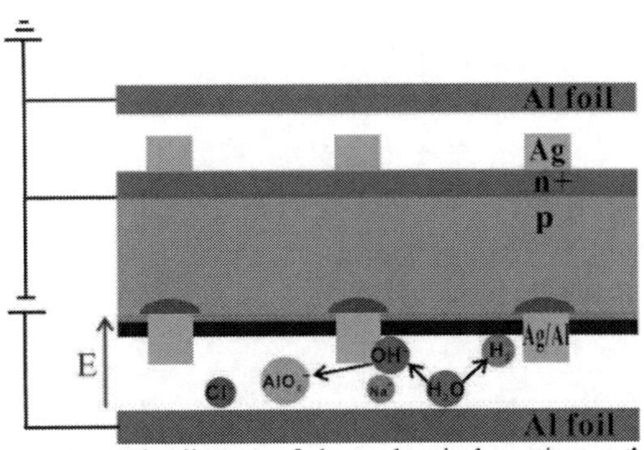

Figure 3: The diagram of electrochemical reaction on the back of bifacial PERC solar cells under negative voltage bias.

3.3 The electrochemical reactions on the rear side of bifacial PERC solar cells under positive voltage bias

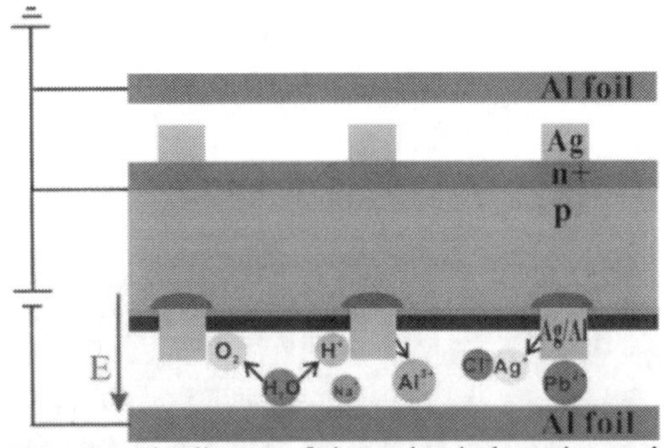

Figure 4: The diagram of electrochemical reaction on the back of bifacial PERC solar cells under positive voltage bias.

Figure 4 shows the diagram of electrochemical reaction on the back of PERC silicon solar cells under positive voltage bias. Hence, the rear side of bifacial PERC solar cells will sever as an anode occurring oxidation reactions.

As the previously description, the lower the electrode potential of metal, the easier the corrosion. Therefore, the Al fingers are preferentially corroded forming Al^{3+} ions, as shown in Eq. (5). It is worth noting that the phenomenon of broken grid was observed with the reaction proceeding. Moreover, Ag fingers also act as anode are oxidized generating Ag^+ ions on the rear side of bifacial PERC solar cells, as expressed in Eq. (6). The SEM and associated EDS mapping on the back of bifacial PERC solar cell as shown in Figure 5. The presence of Al and Ag confirmed the truth that the ions formed by the corrosion of Al and Ag fingers migrate to the surface of silicon nitride. The presence of elemental lead can be interpreted as corrosion of lead in the Ag/Al fingers, which can be expressed by Eq. (7). Meanwhile, at the anode, the water will be oxidized producing O_2 and H^+ ions, as given in Eq. (8). The production of H^+ ions will accelerate the metal fingers corrosion on the rear side of bifacial PERC solar cells.

$$Al - 3e^- \longrightarrow Al^{3+} \quad (5)$$

$$Ag - e^- \longrightarrow Ag^+ \quad (6)$$

$$Pb - 2e^- \longrightarrow Pb^{2+} \quad (7)$$

$$2H_2O - 4e^- \longrightarrow 4H^+ + O_2 \quad (8)$$

4 CONCLUSION

In this paper, the corrosion reactions on the rear side of bifacial silicon PERC solar cells under different bias voltages are investigated. The results indicate that there are different corrosion mechanisms on the rear side of bifacial PERC solar cells due to the difference of system bias voltage and metal electrode potential. It is found that the aluminum fingers are corroded under zero, negative and positive 1500V cell-to-Al foil bias. A novel discovery is that the phenomenon of broken grid was observed when the back of PERC silicon solar cells under positive voltage bias. The SEM and associated EDS mapping were performed on the back of bifacial PERC solar cells and confirmed the truth that the corrosion of Al and Ag fingers. The corrosion reactions will lead to serious performance degradation of

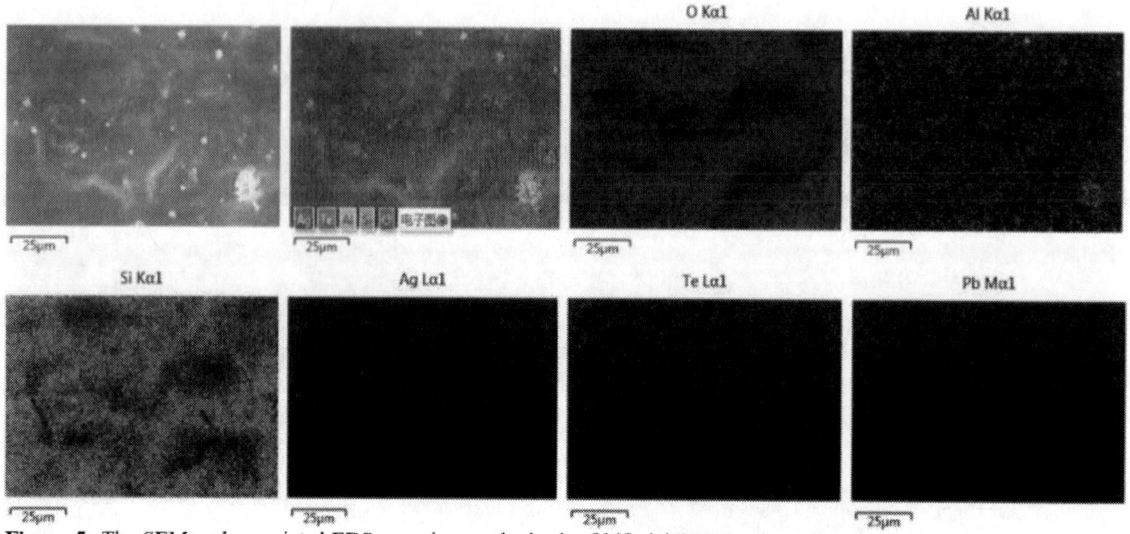

Figure 5: The SEM and associated EDS mapping on the back of bifacial PERC solar cell under positive voltage bias after corrosion.

bifacial PERC solar cells. The obtained results lay a foundation for further research on the corrosion mechanism of bifacial PERC solar cells.

5 ACKNOWLEDGEMENT

The authors would like to thank the support of the National Key R&D Program of China (Grant No. 2018YFB1500700). (*Corresponding author: Hong Yang, hongy126@126.com).

6 References

[1] H. Yang, W.S. He, H Wang, et al. Assessing power degradation and reliability of crystalline silicon solar modules with snail trails[J]. Solar Energy Materials and Solar Cells, vol.187, pp.61-68, 2018. https://doi.org/10.1016/j.solmat.2018.07.021

[2] P. Rajput, G.N. Tiwari, O.S. Sastry,B. Bora,V. Sharma. Degradation of mono-crystalline photovoltaic modules after 22 years of outdoor exposure in the composite climate of India[J]. Solar Energy, vol.135, pp.786-795, 2016. http://dx.doi.org/10.1016/j.solener.2016.06.047

[3] Q.Q. Bai, H.X. Tian, C.H. Nan, et al. Origin of ions for crystalline silicon solar module under high system voltage, 2021 IEEE 48th Photovoltaic Specialists Conference (PVSC), pp. 0252-0255, 2021. doi: 10.1109/PVSC43889.2021.9518659

[4] H.P. Xiong, C.H. Gan, X.B. Yang, et al. Corrosion behavior of crystalline silicon solar cells[J]. Microelectronics Reliability, vol.70, pp.49-58, 2017. http://dx.doi.org/10.1016/j.microrel.2017.01.006

[5] O.K. Segbefia, A.G. Imenes, T.O. Sætre. Moisture ingress in photovoltaic modules: A review[J]. Solar Energy, vol.224, pp.889-906, 2021. https://doi.org/10.1016/j.solener.2021.06.055

[6] M.D. Kempe. Modeling of rates of moisture ingress into photovoltaic modules[J]. Solar Energy Materials and Solar Cells, Vol.90, pp.2720-2738,2006. https://doi.org/10.1016/j.solmat.2006.04.002

[7] P.L. Hacke, M.D. Kempe, J. Wohlgemuth, et al. Potential-induced degradation-delamination mode in crystalline silicon modules[R]. National Renewable Energy Lab.(NREL), Golden, CO (United States), 2018.

[8] J. Carolus, J. A. Tsanakas, A. van der Heide, et al. Physics of potential-induced degradation in bifacial p-PERC solar cells[J]. Solar Energy Materials and Solar Cells, Vol. 200, 2019. https://doi.org/10.1016/j.solmat.2019.109950

[9] M.D. Kempe, G.J. Jorgensen, K.M. Terwilliger, et al. Acetic acid production and glass transition concerns with ethylene-vinylacetate used in photovoltaic devices[J]. Solar Energy Materials and Solar Cells, Vol. 91, pp.315-329, 2006. https://doi.org/10.1016/j.solmat.2006.10.009.

[10] A. Masuda, N. Uchiyama, Y. Hara, et al. Degradation by aceticacid for crystalline Si photovoltaic modules[J]. Japanese Journalof Applied Physics, Vol. 54, 2015.

[11] C. Peike, S. Hoffmann, P. Hülsmann et al. Origin of damp-heat induced cell degradation[J]. Solar Energy Materials and Solar Cells, Vol.116, pp.49-54, 2013. https://doi.org/10.1016/j.solmat.2013.03.022

[12] Q.Q. Bai, H. Yang, C.H. Nan, et al. Analysis of the electrochemical reactions and ions migration for crystalline silicon solar module under high system voltage[J]. Solar Energy, Vol. 225, pp.718-725, 2021. https://doi.org/10.1016/j.solener.2021.07.050

[13] J.C. Li, Y.C. Shen, P. Hacke, et al. Electrochemical mechanisms of leakage-current-enhanced delamination and corrosion in Si photovoltaic modules[J]. Solar Energy Materials and Solar Cells, Vol. 188, pp.273-279, 2018. https://doi.org/10.1016/j.solmat.2018.09.010

ANALYSIS OF ENCAPSULANT DEGRADATION IN DESERT EXPOSED PHOTOVOLTAIC MODULES WITH EXTERNAL QUANTUM EFFICIENCY AND UV-FLUORESCENCE

Ahmad Alheloo, Ali Almheiri, Sagarika Kumar, Baloji Adothu, J. J. John, Gerhard Mathiak, Vivian Alberts
Dubai Electricity and Water Authority (DEWA)
DEWA R&D Center, Dubai, UAE

ABSTRACT: In 2015, a comparative study of different PV modules was started at the Outdoor Test Field at the DEWA R&D Center at the MBR Solar Park in Dubai. Thirty different module types (multi- and monocrystalline silicon) were tested, out of which two were picked for this study. Non-destructive techniques such as ultraviolet fluorescence imaging, visual inspection, and module external quantum efficiency analysis were used to detect the encapsulant degradation occurring on the modules, where UV-F visualizes the discoloration while module EQE quantifies and pinpoints the effect and causes. Results show that all modules show browning both under visual light and through UV-F, which was correlated to a loss in EQE in the shorter wavelengths corresponding to the UV range and rear side of the module. Affected regions within the cell show a reduced EQE over the entire spectrum, indicating a multi-layered effect on the cell's packaging.

Keywords: PV module, encapsulant degradation, discoloration, EVA, quantum efficiency, UV-F, spectral response

1 INTRODUCTION

With the global rise in energy demand and environmental awareness in recent years, dependence on renewable energy for reliable energy production is quickly growing, of which solar energy constitutes a significant portion. This is especially true in the MENA region, where arid desert climates prevail, with a very high amount of solar irradiance. For instance, the city of Dubai aims to obtain 25% of its energy supply from solar energy by 2030 and aims to do so through the MBR Solar Park, which currently generates 2 GW by 2022. However, the arid desert climate of the region also brings about 2.5 times the amount of ultraviolet (UV) dosage that other climates have, greatly accentuating UV-related degradation affecting the encapsulation of the PV module [1,2].

While the probable effects of UV dosage on PV module layers are known, a deep analysis of the impact of such a high UV dosage on these desert-fielded has room for exploration. Further understanding the impact of UV encapsulant degradation on PV module performance and reliability is essential for the proper utilization of solar PV in desert climates, ensuring smooth and reliable operation for large-scale power generation. For this reason, this study aims to combine the reliable detectability of Ultraviolet Fluorescence (UV-F) imaging with the spatial and quantifiable nature of External Quantum Efficiency (EQE) and spectral response analysis in order to obtain further insight into encapsulant degradation causes and explore areas of correlation between the two methods.

2 LITERATURE REVIEW

The feasibility and success of utility-scale PV plants are highly dependent on the lifetime of the PV module and its reliability. In turn, the lifetime of a module is greatly influenced by the degradation modes affecting the module, including encapsulant degradation, potential induced degradation (PID), cracks and breakages, and more [3]. Most of these degradation mechanisms are more pronounced in the intense climate of a desert, due to high temperatures, humidity, dust, and UV dosage.

Packaging materials within the module play a significant role in the primary reduction of PV module performance [4]. Among them, the laminate's encapsulant is one of the most critical parts. The encapsulant acts as protection against a multitude of external and internal factors. However, it is prone to discoloration and delamination in hot climates [5,6].

The discoloration degradation occurring to the encapsulant, specifically the commonly used Ethylene-vinyl acetate (EVA), can take on two primary forms: browning and photobleaching. The former type occurs due to high temperatures combined with prolonged UV exposure, while the latter is caused by the permeation of oxygen at high temperatures [7]. Both discoloration forms cause a direct decrease in the amount of incident light transmitted through the solar cell, reducing short circuit current (I_{sc}) and maximum power (P_{max}) [6,8,9]. Hence, spectral response and EQE measurements, which quantify the solar cell's response to different light wavelengths within various layers in the cell, would potentially provide significant insight into the cell-level optical and electrical details of this reduced transmission due to encapsulant degradation. UV-F imaging is commonly used to visualize this discoloration in a simple yet effective way. However, using UV-F alone wouldn't provide enough information on the discoloration for an in-depth analysis.

Many studies within the literature perform UV-F on desert-fielded modules across the world[10]. However, other studies attempt to correlate the visualization of UV-F imaging with other testing means. For instance, Li et al. combines it with optical microscopy and X-ray Photoelectron Spectroscopy (XPS), linking discoloration with surface roughness and chemical bonds formation leading to increased series resistance [11]. Meanwhile, another study by Klemchuk et al. demonstrates that the degradation is caused by interactions between peroxides and additives in the EVA, finding a lack of chemical bond formation in their desert climate [12].

The present paper aims to present a means by which such discrepancies within the literature through the use of spectral response as a quantitative aspect that provides proper verifiable evidence that can potentially bridge gaps in the industry's understanding of EVA discoloration.

3 METHODOLOGY

A comparative analysis was carried out on two different desert-fielded modules (out of 30 samples), these modules were installed in the local Outdoor Testing Facility (OTF) within the Dubai Electricity and Water Authority's Research and Development Centre (DEWA R&D). The facility, shown in Fig. 1, is within the Mohammed Bin Rashid (MBR) Solar Park, making it a prime model for the climate that utility-scale modules are exposed to in the region. The modules were placed outdoors for seven years of exposure to the desert sun, at the optimum tilt angle for the region (25°).

Figure 1: The OTF within the MBR Solar Park

These two selected aged modules, shown in Fig. 2, are of different make and type. The first is multicrystalline (named Module A), and the second is monocrystalline (named Module B), specifically selected to compare both technologies. Moreover, two other modules, named Control_A and Control_B, are un-aged fresh modules of the same make as Module A and Module B, respectively. However, they are always kept indoors and away from sunlight, and as such, are used as a reference comparison for Module A and Module B, respectively.

The comparative analysis is "cross-characterization", where the two modules and their "Control" counterparts undergo specific analysis techniques, including Visual Inspection, UV-F (using a UV lamp), and module External Quantum Efficiency (EQE) analysis (using a customized setup). While the UV-F analysis aims to visually detect UV encapsulant degradation and its shape, the patterns are then analyzed through spatial module EQE using a module spectral response measurement device. This allows the quantification and in-depth understanding of the effects of encapsulant degradation on the module's performance.

Figure 2: The selected modules for the study.

3.1 Module EQE Measurement

For the simple and accurate measurement of spectral response, and hence EQE, the study utilizes the Bunkoukeiki CEP-M77 spectral response measurement device system. This system is able to apply bias light onto a PV module while concentrating monochromatic light on a 5cm x5cm area within one of the module's cells, allowing the measurement of SR and EQE for a particular area within the module.

4 RESULTS AND DISCUSSION

The investigation of degradation and its causes can hardly be effective when a single characterization tool is utilized, as such, this section aims to demonstrate the results of both methods used (UV-F and EQE) and shed light on any correlations observed.

4.1 Visual Inspection and UV-F Imaging

A common and suitable way to visualize the effects of UV dosage and degradation at the encapsulant layers on a PV module is through UV-F Imaging. As such, it was carried out on the modules in this study, with the prominent results shown in Fig. 3. Both module types demonstrated a frame-like bright pattern on all cells, possibly resulting from encapsulant degradation. Moreover, this pattern has been observed in all 30 field-aged modules, albeit with different geometries and shapes.

Figure 3: UV-F and Visual Imaging for the two selected modules.

However, all patterns can be categorized into two main types:

1. **Edge pattern:** Brightness is within the outline of the frame shape (Similar to Module A in Fig. 3.)
2. **Center pattern:** Brightness fills the frame shape entirely (Similar to Module B in Fig. 3.)

These repeated patterns can also be seen visually as browning on the cells. Moreover, some cells within the same module show swerves and gaps within the pattern, while other cells don't show any kinks. This is possibly due to a location-sensitive difference in UV irradiation across the cells in a single module. This difference is also noticeable across modules of the same make under different tilt angles, where 90° tilted modules showed very faint UV-F patterns when compared to 25° and 0° tilted modules, mainly due to the reduced incident direct UV on the module surface [13].

4.2 Module EQE Analysis

To further understand and quantify the observed encapsulant degradation on the two modules, module EQE was employed. Fig. 4 demonstrates the selected measurement areas, while Fig. 5 shows the resulting EQE measurements for Module A and B, respectively, as compared to their "Control" counterparts. Hence, an EQE reading was taken at both the center and edge of cells in each module, as shown in Fig. 4., with the aim of comparing EQE in pattern-affected areas within the cell to pattern-less areas.

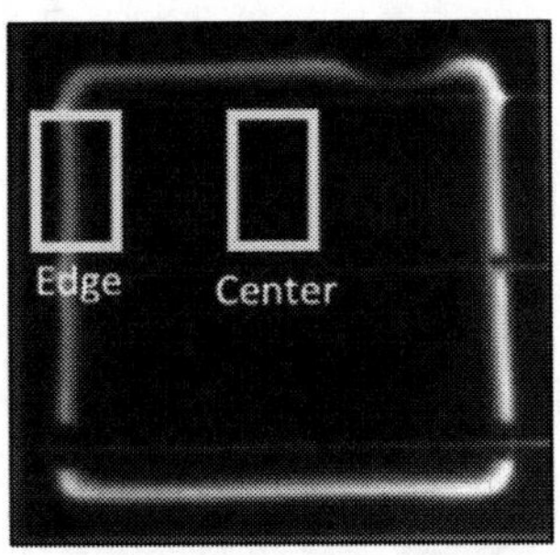

Figure 4: Selected areas for EQE testing within the selected modules.

Initially, it can be seen that there is a clear reduction in EQE between the aged modules and the unaged control modules in the UV range and beyond (100nm – 650nm). This reduction in EQE at lower wavelengths correlates with the encapsulant degradation seen in UV-F imaging. Moreover, since the drop occurs at lower wavelengths, corresponding to rear layers, it can be inferred that the UV-F is accompanied by the diffusion of UV-blocker from the rear side EVA foil.

When comparing the EQE of the same cell over an area with UV pattern and an area without such a pattern, it is vital to note that the area with a pattern (edge in Module A and center in Module B) exhibits lower EQE across the entire spectrum when compared to the pattern-less region (center in Module A and edge in Module B), indicating an effect across multiple layers within the cell. However, when comparing pattern and pattern-less regions of the control module, no difference can be noted between them, validating that this effect is due to the UV pattern seen in UV-F.

5 CONCLUSION

Thirty PV module types were investigated after seven years at an outdoor testing facility in desert conditions; among them, two types were chosen for this study which involves a combination of UV-F imaging and module EQE, aiming to correlate visual identification with quantifiable losses.

• A bright "frame-like" signature pattern was recognized for the identification of encapsulant degradation in the UV-F images. This effect was also detected as yellowing through visible light.
• Many different UV-F patterns were found for the two module types; they were categorized into two types: edge and center patterns. EQE measurements can be used to understand the probable causes for the patterns and to quantify their effect.
• Using spatial-resolved EQE, it was shown that this UV-F pattern is accompanied by a reduction in EQE values in lower wavelengths, including the UV range, most probably due to a diffusion of UV-blocker from the rear side EVA foil.

The combination of UV-F and EQE is an excellent method for determining the degradation mechanisms of PV modules affecting the encapsulant layer or caused by UV irradiation. This is especially true in hot desert climates where UV irradiance is exceptionally high, leading to more pronounced encapsulant degradation. Additionally, further in-depth implementation of this testing method on a larger scale would potentially provide valuable information on encapsulant degradation and its effects within the region.

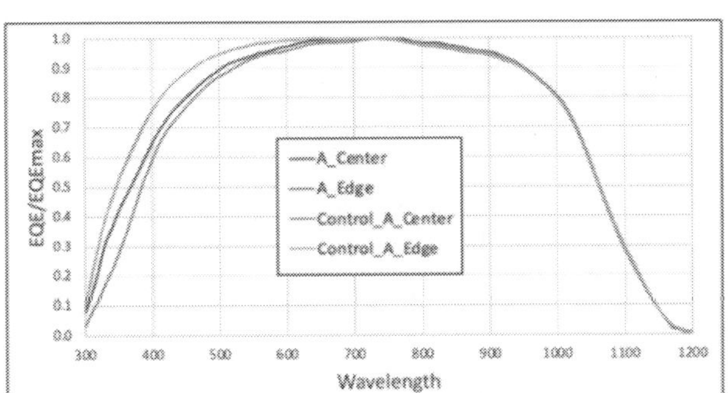

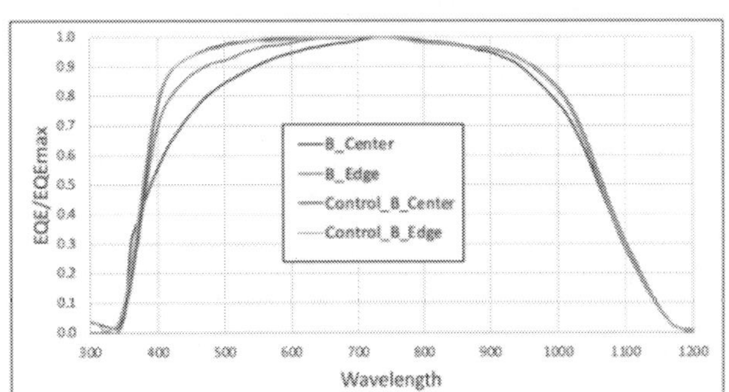

Figure 5: Various EQE measurements for the two modules

6 REFERENCES

[1] Cordero, R. R., Damiani, A., Jorquera, J., Sepúlveda, E., Caballero, M., Fernandez, S., ... & Labbe, F. (2018). Ultraviolet radiation in the Atacama Desert. *Antonie Van Leeuwenhoek, 111*(8), 1301-1313.

[2] Marzo, A., Ferrada, P., Beiza, F., Besson, P., Alonso Montesinos, J., Ballestrín, J., ... & Fuentealba, E. (2018). Standard or local solar spectrum? Implications for solar technologies studies in the Atacama desert. *Renewable Energy, 127*, 871-882.

[3] Köntges, M., Oreski, G., Jahn, U., Herz, M., Hacke, P., Weiss, K. A., ... & French, R. H. (2017). Assessment of photovoltaic module failures in the field. *Report IEA-PVPS T13-09*, 2017.

[4] Ndiaye, A., Kébé, C. M., Charki, A., Ndiaye, P. A., Sambou, V., & Kobi, A. (2014). Degradation evaluation of crystalline-silicon photovoltaic modules after a few operation years in a tropical environment. *Solar energy, 103*, 70-77.

[5] de Oliveira, M. C. C., Cardoso, A. S. A. D., Viana, M. M., & Lins, V. D. F. C. (2018). The causes and effects of degradation of encapsulant ethylene vinyl acetate copolymer (EVA) in crystalline silicon photovoltaic modules: A review. *Renewable and Sustainable Energy Reviews, 81*, 2299-2317.

[6] Chattopadhyay, S., Dubey, R., Kuthanazhi, V., John, J. J., Solanki, C. S., Kottantharayil, A., ... & Sastry, O. S. (2014). Visual degradation in field-aged crystalline silicon PV modules in India and correlation with electrical degradation. *IEEE Journal of photovoltaics, 4*(6), 1470-1476.

[7] Sinha, A., Gopalakrishna, H., Subramaniyan, A. B., Jain, D., Oh, J., Jordan, D., & TamizhMani, G. (2020). Prediction of climate-specific degradation rate for photovoltaic encapsulant discoloration. *IEEE Journal of Photovoltaics, 10*(4), 1093-1101.

[8] Pern, F. J. (1996). Factors that affect the EVA encapsulant discoloration rate upon accelerated exposure. *Solar energy materials and solar cells, 41*, 587-615.

[9] Bouguerra, S., Agroui, K., Kaaya, I., Bouraiou, A., Yaiche, M. R., & Mansour, D. E. (2021). Modeling the Effect of PV Module Orientation on the Encapsulant Browning Degradation Rate in Algeria Region. *IEEE Journal of Photovoltaics, 12*(1), 274-284.

[10] Correa-Puerta, J., Ferrada, P., Häberle, P., Díaz-Almeida, D., Sanz, A., Zubillaga, O., ... & del Campo, V. (2021). Comparing the effects of ultraviolet radiation on four different encapsulants for photovoltaic applications in the Atacama Desert. Solar Energy, 228, 625-635.

[11] Li, F., Buddha, V. S. P., Schneller, E. J., Iqbal, N., Colvin, D. J., Davis, K. O., & TamizhMani, G. (2021). Correlation of UV fluorescence images with performance loss of field-retrieved photovoltaic modules. *IEEE Journal of Photovoltaics, 11*(4), 926-935.

[12] Klemchuk, P., Ezrin, M., Lavigne, G., Holley, W., Galica, J., & Agro, S. (1997). Investigation of the degradation and stabilization of EVA-based encapsulant in field-aged solar energy modules. *Polymer degradation and stability, 55*(3), 347-365.

[13] Kumar, S., Alhamadani, H., Hassan, S., Alheloo, A., Hanifi, H., John, J. J., ... & Alberts, V. (2021, June). Comparative investigation and analysis of encapsulant degradation and glass abrasion in desert exposed photovoltaic modules. In *2021 IEEE 48th Photovoltaic Specialists Conference (PVSC)* (pp. 0793-0798). IEEE.

This is a short version of the paper submitted to the journal Progress in Photovoltaics.

It has been selected by the Committee of the WCPEC-8 for submission to Progress in Photovoltaics.

"LITTLE MAINTENANCE OVER 25 YEARS?" – A COMPREHENSIVE STATISTICAL ASSESSMENT OF 35,000 O&M TICKETS ACROSS 5 COUNTRIES AND THEIR TECHNO-ECONOMIC IMPACTS

S. Lindig[1], S. Gallmetzer[1], M. Herz[2], D. Moser[1], G. Oviedo-Hernandez[3], S. Southern[4], L. Moretti[5]

[1]Institute for Renewable Energy, EURAC Research, Viale Druso 1, 39100 Bolzano, Italy
[2]TÜV Rheinland Energy GmbH, Am Grauen Stein, 51105 Cologne, Germany
[3] BayWa r.e. Operation Services S.r.l., Via di Settebagni 390, 00139 Rome, Italy
[4]Statkraft Uk Ltd, United Kingdom
[5]Enel Green Power S.p.A., Rome, Italy
*Phone: +39 0471 055 715; E-mail: sascha.lindig@eurac.edu

ABSTRACT: In the framework of the H2020 project TRUST PV, steps towards an automatized, data-driven, decision support for the PV O&M sector have been investigated. Thereby, two important working areas are covered, the statistical as well as the economical evaluation of PV O&M failure tickets. A statistical evaluation is only possible if a common dictionary for ticket categorization is applied. Therefore, the TRUST-PV Risk matrix, a technical risk matrix covering all possible events within the operational phase of a conventional PV plant, has been created. Using the nomenclature of this matrix, we statistically evaluated more than 35,000 O&M tickets of over 100 PV. It has been shown that inverter and monitoring related tickets make more than 75% of all recorded tickets and there seems to be a bias towards automatically generated tickets and an underrepresentation of tickets related to components where a manual insertion is required. In the second part of this work, a concept to calculate the economic impact of O&M tickets has been developed based on information of individual tickets, PV system specific information as well as PV monitoring data.
Keywords: PV system performance, PV system maintenance, Data imputation, Monitoring

1 INTRODUCTION

The rapid decline of PV system prices experienced in recent years [1, 2] forces all market players to optimize their products and activities. Installation costs are mainly driven by advancements in PV related technology, a rapid PV deployment enabling economies of scales, and the application of installation standards through best practices. Installation costs are summarized under the capital expenditure (CAPEX), while operational costs can be described by the operational expenditure (OPEX). OPEX costs are usually calculated as a percentage share of the CAPEX costs. Thereby, decreasing installation costs put price pressure onto operation and maintenance (O&M) activities. Initially, O&M activities were dominated by manual labour and the expert knowledge of employees in the control room. Decreasing price margins require a different approach. In this context, the abundance of ticketing and performance data of PV systems offers a great opportunity to optimized O&M activities. Available data represent a complex source of information, which is continuously growing. Automation of data collection, processing and analysing are key tasks to exploit big data streams. Data preparation and standardization is a necessary tool to overcome this problem and is a core topic in this work.

A standardization framework of O&M failure tickets is presented through the development of the TRUST-PV [3] Risk Matrix. 37,328 O&M tickets, coming from 103 PV systems, have been aligned according to the risk matrix and statistically and economically evaluated. For the economic evaluation, the cost priority number (CPN) methodology [4], a cost-based failure modes & effect analysis (FMEA) method, has been taken as a basis and further refined. Initially, the CPN was developed as a scenario-based approach to provide a cost for a PV system O&M ticket describing a system issue. In this work, the CPN is the cost of an individual PV system ticket. The CPN combines the costs created to fix an issue in the field and costs related to the missing energy produced because of the issue. The application of the CPN methodology is only possible if, next to the failure tickets, PV performance data for the respective plants are available.

In this work, the TRUST-PV Risk Matrix is introduced, a statistical evaluation of the collected tickets is performed, and the concept of a tailored CPN approach together with its first application results is presented.

2 DATASETS

The datasets for this study were provided by three large O&M operators. O&M tickets of 103 different PV systems have been used. Five of these systems are installed in South Africa, while the rest is located across Europe, namely the UK, Italy, Spain and the Netherlands. The system sizes vary between 20 kWp and 476 MWp with a mean of 27 MWp and a median of 5 MWp. With 60% most plants are ground-mounted utility-scale systems with a fixed tilt between 15° and 30°. Another 16% are installed on flat roofs of industrial buildings with low tilt angles and the remaining ones are ground-mounted with single and dual axis trackers. Most of the provided ticketing and performance data are from the last four years, regardless of the lifetime of the plants, as they were later included into the portfolios of the O&M companies.

Performance data of 24 PV systems were provided in form of AC power data. The datasets have a resolution of one hour or higher and are stored at inverter level. This allows us to calculate detailed downtime costs, as done in Section 5, for individual inverters, a number of affected inverters and the whole plant, always depending on the component level on which an issue occurs. Data quality routines, as proposed by the IEA PVPS Task 13 [5], were applied to verify the goodness of the datasets and to identify and remove invalid data points. Datasets are rated based on the amount of outliers, missing data and the longest gap within the dataset. The data quality routines have been applied to hourly aggregated inverter time series. The available time series range from 11 months to four years. While the data show almost so signs of outliers, around 20% of data are missing in average. Single plants in Italy and South Africa have between 30% and 40% of missing data. With data gaps of up to 270 consecutive days, the high amount of missing data will inevitably lower the amount of tickets, for which the CPN can be computed.

3 TRUST-PV RISK MATRIX

While monitoring/performance data are usually stored as time series, in which case they are easier to handle, maintenance tickets, stored in the ticketing systems of different O&M providers, vary significantly in their data format, the amount and detail of information provided, language, and above all nomenclature. The lack of adequate classification and standardization in ticketing nomenclature is the main motivation for the TRUST-PV Risk Matrix development. It is a technical risk matrix covering all possible events within the operational phase of a conventional PV plant. The events include failures in PV systems, the appearance of degradation modes, defects, performance losses as well as inspection visits and routine maintenance activities.

In existing literature, the terms failure, degradation, defect, and performance loss are often used interchangeably although they have been defined previously for PV applications. Jordan et al. [6] as well as Lindig et al. [7] adopted the definition of the term failure for electrotechnical devices as ``the termination of the ability of an item to perform a required function" [8]. For PV modules it usually defines the end-of-life but for other components of a PV system, such as communication or inverter, a failure, resulting in the temporal inability to perform its intended purpose, can be fixed by resetting, repairing, or replacing the affected component. A degradation mode instead is an observable and irreversible variation in appearance, performance, and/or safety of a PV module. The appearance of a degradation mode represents a gradual change in the properties of a solar module while a failure is equivalent to the end-of-life of a module, either due to a defined threshold reached for the reduction in output power or if further operation of the component would create a safety hazard. Performance losses on the other hand include all performance impairing effects a PV system is experiencing, reversible and irreversible [9]. Finally, defects are defined by the International Energy Agencies (IEA) Photovoltaic Power Systems Programme (PVPS) as anything which triggers a deviation from normal operation and includes thereby failures, degradation, and performance losses into one parameter [10]. In the course of this work, we simply use the terms "issue" or "event" to describe the cause for the creation of a maintenance ticket.

The Risk Matrix aims to standardize appearing issues in the field. This is crucial when attempting to gain knowledge from a vast amount of maintenance tickets and for assessing the economic impact of the issues they correspond to. The TRUST-PV Risk Matrix is divided by component, subcomponent, and individual event. The components are shown in Figure 1. In summary, nine components, 62 subcomponents, and 341 events have been identified. Each component has thereby a number of subcomponents in order to account for events on a higher level and each event has a unique ID so that it can be easily identified with its correlating component and subcomponent. If the whole component is affected, the subcomponent is defined as "Entire component", where component is the respective component class. Target systems for the risk matrix are ground mounted systems with fixed tilt and tracking systems as well as standard building applied/integrated PV systems. The TRUST-PV Risk Matrix is freely available and can be downloaded under https://trust-pv.eu/reports/risk-matrix/. Next to the mandatory information on component, subcomponent and event, descriptive ticket information about the root cause of an appearing issue, the accountability, which detection method has been used and how the issue was resolved (solution) can be added to the standardized ticket scheme. Furthermore, an addition to the TRUST-PV Risk Matrix for floating PV is in preparation.

Not just failure/event related fields, but the entire ticket table structure should be prepared in a way so that tickets can be statistically as well as economically evaluated with the highest grade of automation. Therefore, a suggested ticket table structure is shown in TABLE III in the Appendix. Information are inserted using drop-down menus or are automatically generated to minimize human error when creating tickets.

The usage of a standardized taxonomy reduces alignment issues, provides a more efficient tool to exploit the ticketing system of an O&M company for statistical analysis of O&M operations and is a step forward towards an automated decision support system. Additionally, it gives field technicians a useful tool, as the workflow for entering ticketing information is completely standardized and can for example be implemented in the form of an app with drop-down menus, to easily identify and name issues encountered in the field.

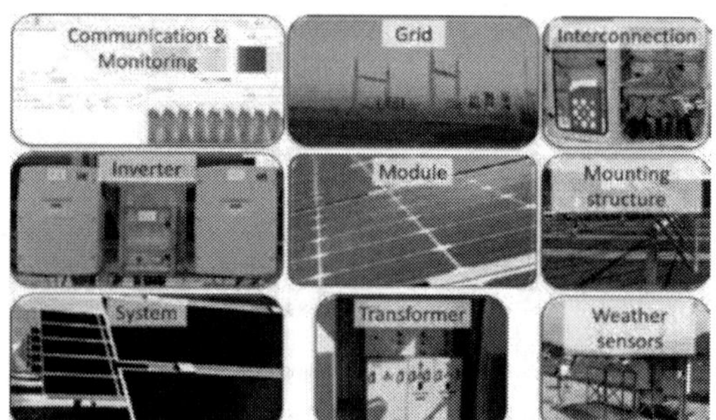

Figure 1: Components identified in the TRUST-PV Risk Matrix

4 STATISTICAL ASSESSMENT

In this study, 37,328 tickets have been collected from over 100 PV systems (see Section 2) and aligned according to the TRUST-PV Risk Matrix. The results of the statistical assessment are summarized in Figure 2 and Figure 3. The earliest tickets date back to 2010 while the newest have been collected in April of 2022. Most tickets were generated between 2018 and 2021 and the average time between the first and last ticket per plant is 2.5 years. Not all tickets result in performance related energy losses but are still important to be categorized and tracked such as "Site visit/inspection" events, which, if there is no need to momentarily disconnect a part of a PV plant, does not entail any energy loss.

Different techniques for aligning the tickets to the Risk Matrix have been applied. Inaccurately described issues, the usage of abbreviations and different languages, varying ticket table structures, inconsistent naming of the same component, and other issues have presented a particular difficulty while aligning the tickets and required an enormous amount of manual labour. Nevertheless, this activity was crucial to be able to create a standardized framework in which O&M tickets of different players, operating under different conditions, can be made comparable.

Figure 2 shows the distribution of the collected and aligned tickets among the components on the left and subcomponent on the right. In Figure 2b), the normalized share of tickets per component is divided into individual subcomponents. This is done within the component bars with graded colours. For each component, the most frequent subcomponent is named within the main bar. In the following, the main characteristics of component/subcomponent related statistics are summarized:

Component:

- 62.5% off all tickets are inverter related
- More than 75% of all tickets under categorized under “Inverter” and "Communication & Monitoring"

Subcomponent:

- 49.5% off all tickets are under the subcomponent "Entire inverter"
- 54 different subcomponents (of 62 in total in the Risk Matrix) are represented in collected tickets
- for seven components, over 70% of the tickets belong to one subcomponent
- components "Communication & Monitoring" and "Grid" have only one subcomponent
- component "Transformer": tickets are almost equally shared between subcomponents "Breaker" and "Entire transformer"
- component "Interconnection" is subdivided in a 4-2-2-1-1 distribution among the subcomponents "Potential equalization & Earthing", "AC connection and distribution boxes", "DC cabling", "AC cabling" and "String combiner boxes"

The summary above shows a high share of tickets are categorized under the two most affected components. This is because tickets categorized under these two components are usually automatically triggered, as their software is directly connected to the ticketing system in place. Furthermore, it seems that only a few subcomponents are affected frequently. On one hand, this can be traced back to typical and frequently appearing issues in the field. On the other, it is assumed that a number of issues in some of the plants have not been recorded and stored in the ticketing system. Especially issues which require manual input and which are not automated such as PV module or mounting system related tickets. For example, a site visit at one of the plants in Italy revealed many cases of glass breakages in frame-less glass-glass modules due to over-tightened clamps, requiring the subsequent exchange of the modules under warranty regulations. However, for these activities no tickets were found in the database. For this reason, it is assumed that issues on a higher level may be underrepresented in this study. Inverter related tickets, as well as tickets concerning the monitoring system, are expected to be realistically represented due to the automatic ticket generation mentioned earlier. This circumstance becomes apparent when examining the ticket distribution in Figure 2. Nevertheless, as this approach of ticket collection is widely spread in the industry and can thereby be considered as best practice, it is expected that the trends shown in this study are representative across the majority of ticketing systems and O&M providers.

Figure 3 shows the number of tickets per event/failure and to which components these events belong. The main event related statistics are summarized here:

Events:

- primary event per component accounts for more than 30%
- for five components, share is above 50%
- 90% of all tickets per component are categorized with less than 10 different events (exceptions are "Interconnection" with 13 events & "Inverter" with 12 events)

It is worth noting that 22% of all tickets are related to inverters which had turned off. Considering the 1,461 inverters in our ticketing database and an average ticket

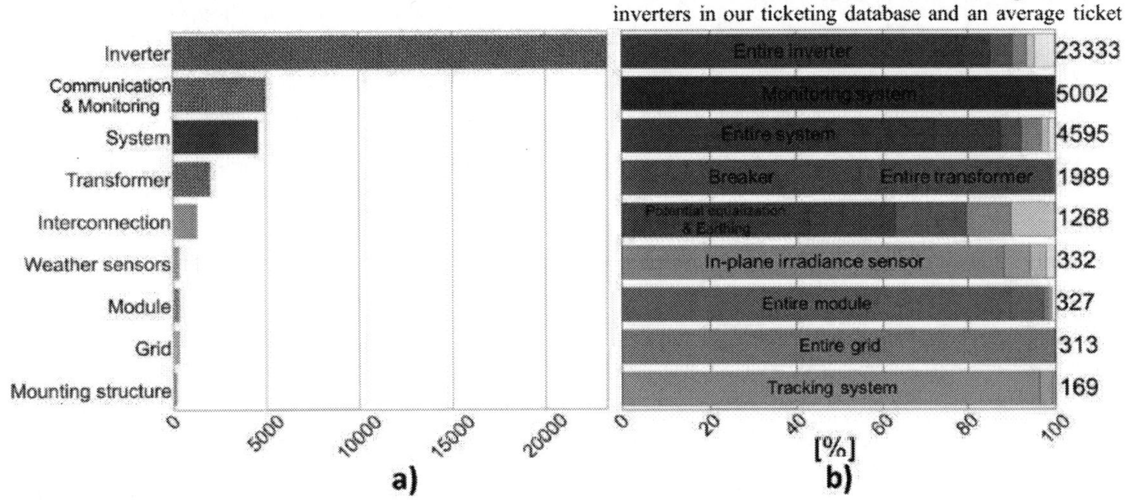

Figure 2: Distribution of tickets; a) by component; b) Normalized by subcomponent

acquisition time of 2.5 years, an average inverter turns off roughly 2.3 times a year. Furthermore, expanding this estimate to all inverter related tickets, 6.4 events are generated per inverter and per year. The 10 events with the highest individual proportions contribute to 70% of all collected tickets. With the top three events being "Off" (subcomponent "Entire inverter"), "Data communication /transmission error" and "Loss of generation", it is visible that a high number of issues in the field (43%) can be traced back to problems with software and data connection, which do not always require manual intervention, nor do they cause long-lasting problems, as these issues can often be dealt with remotely. The following two events are "Preventive maintenance" and "Site visit/inspection" (12.4%), which are pro-active activities to ensure reliable operation and high performance of PV plants. The real number of these events might even have a higher share because it is assumed that these activities are not thoroughly tracked in some of the investigated systems. Most of the remaining events shown in Figure 3 are the result of technical issues or a deviation from normal operation of some parts within the plant and oftentimes require manual intervention to fix the issue.

When only looking at the component "Module" it is interesting, and to a certain extent unexpected, to see that issues on a higher level and degradation modes were recorded very rarely. Common degradation modes such as corrosion or light-induced-degradation (LID) have been recorded once as well as other common issues such as hot-spots, defect bypass diodes, soiling or moisture ingress. Most of the investigated systems have been installed in the early 2010's, a probable reason as to why modern module issues such as light and elevated temperature-induced degradation (LeTID) are not reported. Nevertheless, based on statistics regarding the appearance of degradation from literature [10, 11, 6, 12], it can be assumed that far more performance impairing issues on PV module level have occurred than have been recorded. Nevertheless, due to their minor performance impact on plant level, it is possible that they have not been detected or treated due low priority.

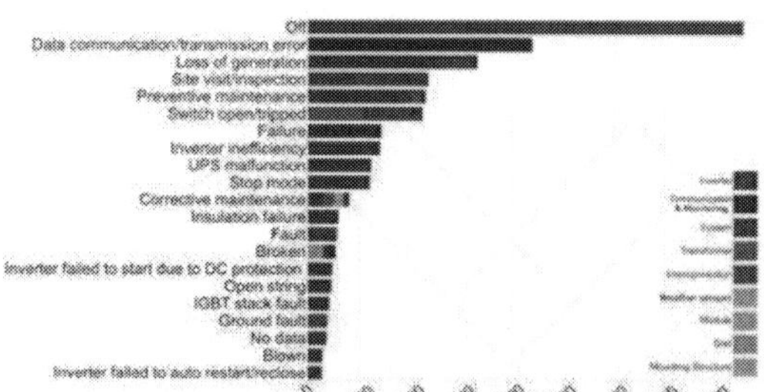

Figure 3: Categorized failures divided by Component with minimum threshold of 250 tickets

While a detailed event acquisition would be beneficial for a comprehensive failure analysis, it also raises the question whether it is always economically beneficial. It is possible that, with current O&M strategies, additional manual labour combined with sophisticated detection and prevention techniques to detect and overcome all performance impairing issues on PV plant level is more expensive than simply accepting a slightly lower PV system performance. To answer this question, the cost of PV plant events together with corrective measures has to be taken into acocunt and weighted against scenarios where no action is taken. The CPN, applied on individual tickets within the database in Section 5, is a suitable methodology to address this important question and a tool to improve current O&M decision making practices.

5 COST PRIORITY NUMBER METHODOLOGY

This method assigns a cost to an issue that appears in the field normalized over the plant's nominal capacity in €/kWp. The results can be used to get insights into various aspects. First, it determines the direct cost of lost energy due to the deviation from normal operation in case of an event. Second, the repair costs together with required detection costs are determined. The CPN is the sum of costs due to downtime and fixing costs due the failure/event appearance:

$$CPN\ [€/kWp/year] = C_{down} + C_{fix} \quad (1)$$

CPN related studies until today are still scenario-based. This work instead uses real monitoring data to evaluate the cost of appearing issues in the field with high certainty. The general calculation steps are shown in Figure 4. The sections 5.1 and 5.2 discuss the individual steps and the results of this study.

In previous CPN related work, the total duration of a ticket from opening to closing was divided into several periods, namely detection, response, and repair time [13, 14]. While this information would allow very interesting insights, the investigation of the collected tickets indicated that this information is usually not provided, and only a start (event/failure time) and an end time (resolution time) is recorded. Hence, we reduce the evaluation only to t_{down} and t_{fix}. Here, t_{down} describes the time period a ticket is open, and t_{fix} is the fixing time of the individual issue. The estimation of t_{fix} is further explained in Section 5.3.

In this study, historical O&M events that occurred in PV plants are evaluated. Thereby, the focus is on gaining a greater understanding of performance impairing issues in the field, and of the cost and effectiveness of performed corrective maintenance measures. This information is collected and stored in a "solution matrix", which, as it grows, will offer increasingly detailed information on the most cost-effective corrective maintenance measures to implement in case of future events and failures. Although in this work we apply the CPN to generate these statistical insights based on available historical ticketing data, the overall aim is to develop the CPN methodology in such a way that it can be implemented in the framework of a Decision Support System (DSS) in the future, where the CPN is used as a metric to select the most cost-effective maintenance actions. Additionally, the collected ticketing data and the solution matrix allow for a self-learning feedback loop, which further improves the CPN calculation, and the suggestions made by the DSS. Thereby, also site-specific parameters such as country or O&M subcontract dependent costs are considered.

Findings from these analyses are summarized in Section 5.3.

5.1. Downtime costs

Nowadays, downtime costs of PV plants are often calculated using a digital twin, although the definition of this term leaves room for interpretation, i.e. how it is created, what kind of models are used etc. Digital twin approaches in the O&M sector use plant metadata or data of the first months of operation together with weather data from a short measurement campaign or typical meteorological year (TMY) datasets from satellite data as input. In many of today's simpler digital twin approaches, the actual performance of the PV plant as well as performance loss mechanisms and degradation processes are not taken into account and therefore, the results are expected to have high uncertainty. If they are taken into account, then often as estimated values defined at the beginning of the lifetime of a PV plant without being updated during operation.

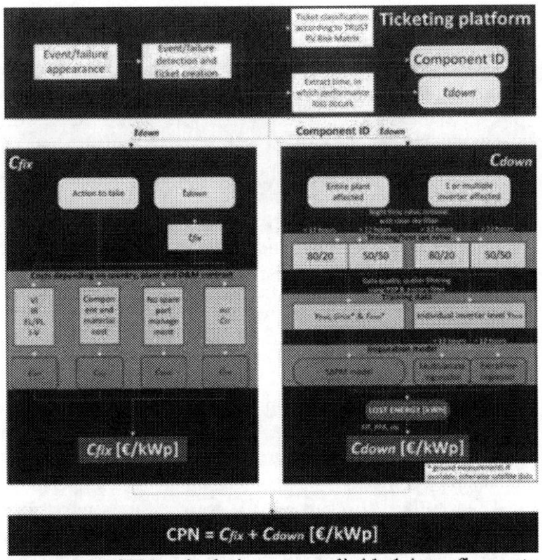

Figure 4: CPN calculation steps divided into fix costs (C_{fix}) and downtime costs (***C_down***); abbreviations - G_{POA}: plane-of-array irradiance, T_{amb}: ambient temperature, FIT: feed-in tariff, PPA: power-purchase agreement

In this work, a new methodology is presented using training data adjacent to the occurrence of a performance impairing event to better mimic the actual performance of the plant/component and thereby calculate values with higher accuracy. Of course, a failure/event might appear during a period a PV plant suffers from other performance reducing events such as high soiling, strong curtailment, or others. As we are only interested in the impact the individual issue, described by the respective O&M ticket, has on the performance, the effect of these other events are considered as "normal" performance and as such taken into account. In this section, this methodology is described and applied to the PV system data collected.

Methodology

The right side of the diagram in Figure 4 shows the workflow of calculating the downtime costs for a ticket generated in an O&M platform. The variables to consider when calculating C_{down} are time duration when the ticket is open, t_{down}, training/test set data ratio, which component is affected, predictor availability and the used data-filling, also called imputation, method. t_{down} as well as the component ID of the affected component, or components, are inputs coming from the individual ticket.

In a previous study [15], a matrix of well performing imputation techniques has been created in respect to the amount of missing data, training data availability and imputation model performance. If less than 12 hours are to be imputed, a training/test set ratio of 80/20 has been found to yield reliable results. If more than 12 hours must be recovered, a 50/50 ratio is sufficient without a major impact on model accuracy. Thereby, the data around the event are selected as training data. Data quality routines are applied to find the most suited training dataset, either before, after, or around the test set.

It has been shown that the most desirable training data scenario is the availability of neighbouring power signals, which could be close-by plants or inverters/strings in the same plant, which are usually available if one or multiple, but not all, inverters are affected by the event described in an O&M ticket [15]. Thereby, the non-affected power signals are set as predictor space.

In that case, multivariate linear regression (MLR) or ExtraTree regression [16] (ETR) is used as imputation method, depending on the period to be imputed. The models have been implemented with the Python library scikit-learn [17].

If the complete plant is down, the inverter signals of the plant before and after an event can be used to build a performance model. In this study, a simplified version of the Sandia PV array performance model (SAPM) [18, 19] for maximum power estimation was selected. The final yield of the complete PV plant is calculated to fit the model parameter using ground measured plane-of-array irradiance, G_{POA}, and ambient temperature, T_{amb}, or, if not available, satellite-based data [20]. In case of satellite-based data, the irradiance has to be decomposed [21] and transposed to the plane-of-array [22]. The SAPM model is described by:

$$P = G_{POA}(P_{nom} + k_1 \ln(G_{POA}) + k_2 \ln(G_{POA})^2 + T_{amb}(k_3 + k_4 \ln(G_{POA}) + k_5 \ln(G_{POA})^2) + k_6 T_{amb}{}^2) \quad (2)$$

Here, P_{nom} refers to the nominal power of the plant and k_1 to k_6 are empirical coefficients, which are fitted using non-linear least squares.

Before the application of the imputation models, the predictor space is prepared. First, night-time values are removed by applying a clear sky filter, implemented with pvlib [23]. Next, the test dataset is created based on the start and end time of the O&M ticket in question and the training set is selected based on the test set duration with a training/test set ratio of 80/20 or 50/50, as explained above. To verify the completeness and goodness of the available predictor time series, data quality routines are applied. Three different tests are applied. The completeness of the data is evaluated using a simple check for unexpected zeroes and NaN values. Predictors are removed which pass a threshold of 10% from the overall dataset. The other two tests look for outlier and trend behaviour in the datasets. Therefore, an average time series across all available predictors is calculated and used to compute the Pearson correlation [24] between the average time series and all individual predictors (called peer-2-peer filter), and the correlation of the *z*-scores of these parameters. The thresholds for predictor removal are set to 80% for the Pearson correlation and 60% for the *z*-score correlation. The thresholds have been defined experimentally. After the preparation of the predictor space, the aforementioned imputation methods are applied to simulate the energy, which would have been produced in absence of the event described in the O&M ticket. By subtracting this value with the actual produced energy during that time, the $Lost\ Energy$ in kWh is calculated.

Finally, C_{down} is computed by dividing the $Lost\ Energy$ with the nominal DC power of the plant and the result is multiplied with the agreed price per kWh. The price is defined by a feed-in tariff (FIT), a power-purchase agreement (PPA) or a leasing contract for each individual PV plant.

Results

As mentioned before, performance data of 24 PV systems were collected. Unfortunately, for many systems the available time period for performance data was shorter than the period for ticket collection. Furthermore, some tickets did not include information on start or/and end time of the recorded event. That is why from the initial 37,328 tickets, only for 20,242 tickets the downtime costs were attempted to be calculated. Another problem, which appeared more often than expected, are that some of the performance time series have many gaps. This is the case for 16 plants. The reason is probably that either the data transmission has failed or the data acquisition and storage routines are faulty. In any case, this instance together with the aforementioned fact that not all tickets describe a performance impairing event, reduced the number of calculated values to 4,680 tickets.

From these 4,680 tickets, 4,438 affect individual inverters, 101 multiple inverters and 141 affect the whole PV plant. The minimum time considered in this work is one hour and the ticket with the longest time period was recorded with 275 days, where an IGBT (insulated gate bipolar transistor) stack fault of one inverter was detected. All tickets have an average duration of 68.7 hours and a median value at 5.7 hours which means that the distribution is strongly right skewed towards smaller values. For all tickets, the yield loss [kWh/kWp], the overall energy loss [kWh], the monetary loss [€] and the normalized monetary loss (C_{down} in [€/kWp]) was calculated. With 0.2, the correlation between the yield loss and the time period the ticket was open is very small. That is because the impact of tickets on the overall plant output depends, among others, on the type and number of components affected, if the output is down to zero or only reduced due to the issue, the location of the plant or the time of the year. Apart from the yield loss, C_{down} depends on the price paid per kWh, which is fixed in the agreement for a particular plant in form of a FIT, a PPA or a leasing contract. Nevertheless, the yield loss and C_{down} are with 0.99 highly correlated as the vast majority (94%) of tickets are from PV systems in the same region, commissioned at a similar time and have therefore a very similar price per kWh. The systems under observation have been commissioned between 2009 and 2020, and fall under very different payment schemes, ranging from a fixed FIT of 0.26 €/kWh down to 0.04 €/kWh. C_{down}, calculated with the available tickets in this study, ranges from 0 €/kWp to 16.19 €/kWp with an average value of 0.027 €/kWp (translating to an average monetary loss of 1242.52€). Although these values may seem minor compared to the total value of a project, one has to keep in mind that C_{down} is normalized over the installed capacity of the plants, and for utility scale systems with several megawatts this can quickly translate to losses in the order of thousands or tens of thousands of €.

The difference in the price paid per produced kWh may also influence the maintenance strategy for a particular plant. The higher the FIT, the more money is lost in absolute terms in case of the same amount of lost power. The best course of action, purely from an economic point of view, is determined by leveraging the evolution of C_{down} and C_{fix}. After the detection of a performance impairing issue, repair costs depend, for example, on the response time and the availability of staff and material. If an issue does not have a great impact on performance, it might be economically cheaper to delay the resolution, for example if the issue appears on a weekend where usually a premium on labour cost is paid. The creation of a solution matrix, which is trained and optimized with maintenance tickets within a decision support system, can give suggestions for the best course of action. This approach is further discussed in the next section.

5.2. Fixing costs

The calculation of the economic impact of fixing or repairing an event in a PV plant was first introduced by the Solar Bankability project [4] as a purely theoretical concept. This study goes beyond the theoretical basis and a concept of how to estimate the fixing costs is developed by analysing the dataset introduced in Section 2. Similar to the estimation of downtime costs, the decision support system requires an automated calculation methodology of the costs arising due to repair in order to be integrated into an O&M platform.

Methodology

The evaluation of the collected O&M tickets showed that several provided a proposed action to take. The proposed actions have been collected into a solution matrix together with the time for repair and the respective cost parameters.

Once the action to take is known, it is possible to define the time required for repair starting from the parameter t_{down}. For an accurate determination of the repair time, it would be necessary that the maintenance tickets provide the exact period the site technician is actually spending in the field repairing the issue. Since this is not the case in the collected maintenance tickets, the time for fixing is estimated. While studying the tickets it became apparent that some tickets only entail the time for fixing while others cover the whole ticket duration from event/failure appearance (or detection) until issue resolution.

The separation of tickets is done by studying all available tickets with a duration of up to one day, corresponding to eight working hours. Like this, the tickets are divided into the ones covering the complete event duration and the ones only accounting for the time to fix.

The costs that arise for fixing an issue in the field were first defined by the Solar Bankability project [4] as the sum of the cost of detection and corrective maintenance, normalized over the total nominal power of the plant.

$$C_{fix} = \frac{(C_{det}+C_{rep}+C_{trans}+C_{lab})n_{fail}}{P_{nom}} \quad (3)$$

C_{det} are the costs for detection, C_{rep} the costs of repair, C_{trans} the costs of transport and C_{lab} the costs of labour. n_{fail} indicates the number of components that present the issue and P_{nom} is the nominal power of the plant.

All these cost parameters are necessary to correctly determine C_{fix} and therefore the CPN of the apparent issue. The focus in previous studies concerning the estimation of the economic impact of technical issues was mainly on the costs related to lost energy. The fixing costs, instead, have not yet been addressed on a real-data basis.

Therefore, an accurate and generalized estimation methodology of fixing costs is developed in this section.

The first parameter to consider is the cost of the detection technique employed, C_{det}. To define the cause of energy loss, sophisticated root-detection methods may be applied, which are usually only performed once a year or in case of root-cause detection of performance deviating events. These methods include visual inspection, current-voltage (I-V) curve tracing, infrared thermography (IR-T), electroluminescence (EL) or also photoluminescence (PL). All these techniques are subject to certain costs as described by Koester et al. [14].

The cost of repair, C_{rep}, represents the largest portion contributing to C_{fix} and includes the cost of repair as well as possible replacement of damaged or broken components. The ideal way to determine this cost is to know the type of components installed in the specific plant, and the corresponding market prices to estimate the cost of substituting the affected component. Since this method would require a detailed analysis of the specific plant, the repair costs used in this study rely on values provided by O&M providers active in the field.

Besides the costs of repair and detection, the fixing costs also encompass costs arising if a new component must be ordered and delivered to the site, C_{trans}. Finally, the last cost contributor are the labour costs C_{lab}.

$$C_{lab} = t_{fix} n_P C_P \tag{4}$$

As shown in equation 4, they are the product of the time to fix the issue, t_{fix}, the number of personnel required, n_P, and the hourly rate of the personnel, C_P. The latter parameter depends on the required level of expertise, as shown in TABLE I.

All these four cost parameters included within the fixing costs are highly dependent on country, plant and the respective O&M contract.

TABLE I: Hourly cost of the personnel depending on the working field; abbreviations - C_P: Cost of personnel per hour.

Personnel	C_P [€/h]
Engineer, Specialist	80
Technical staff	50
Unskilled personnel (manual worker)	30

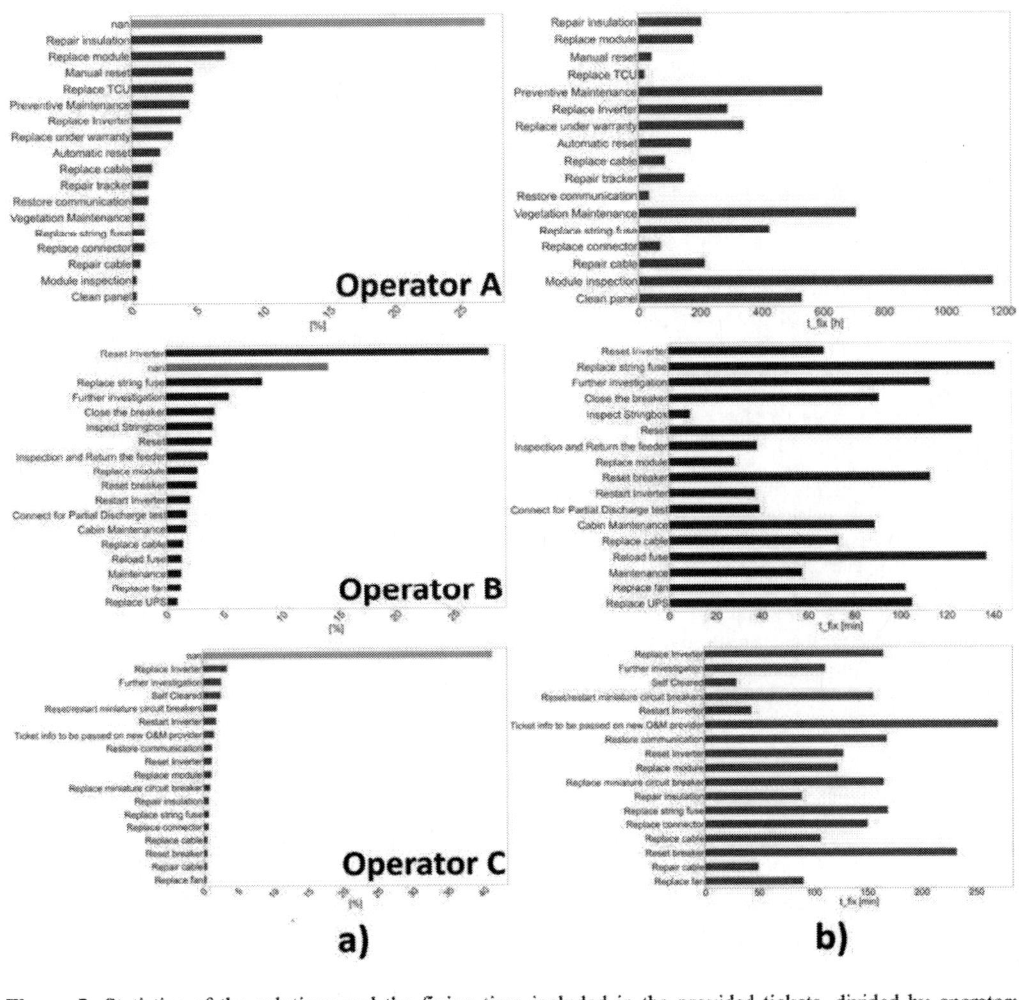

Figure 5: Statistics of the solutions and the fixing time included in the provided tickets, divided by operator; a) Percentage share of solutions in database; b) Average recorded time period of solutions (operator **A** in hours, **B** & **C** in minutes)

Results

For C_{fix} a comparison of tickets for three different O&M operators is performed. The results are shown in Figure 5, with operator **A** on top, **B** in the centre and **C** in the third row of the graph. The first column presents the solutions provided in the tickets and the second column shows an evaluation of the average time for the specific solution. Operator **B** delivered the biggest amount of information with more than 30,000 tickets, thus the conclusions are primarily based on its maintenance tickets.

By analysing the graphs of the first column in Figure 5, it is visible that all three providers present similar solutions. Depending on who must cover the costs of repair and if there is still warranty in place, the suggestion is to repair or replace the broken component, whether it is an inverter, a string fuse, a PV module or a cable. For roughly 30% of the tickets no solution, signed as "nan" in the graph, was provided. Thereby, operator **B** presented the highest share of solutions.

Instead, the second column shows the average time needed for fixing. Hereby, it is clearly visible that operator **A** has a different understanding of the time parameters compared to **B** and **C**. There is a large deviation for the values of t_{fix} with different orders of magnitude. That is why the time on the x-axis is given in hours in the first graph while it is provided in minutes for operators **B** and **C**. It can be expected that the time needed for the actual repair of most issues is usually less than one day or eight working hours. That is why we suspect that operator **A** considers within the time parameter also the time needed for response and procurement. Moreover, this operator may not have an efficient spare part management in place leading to long response periods in case a broken component has to be replaced. Instead, operator **B** and **C** present expected values for the different actions.

Further steps have to be taken to integrate a reliable automated C_{fix} calculation approach into a DSS platform. The fixing costs are comprised of labour, repair, detection, and transport costs. These values are dependent on various parameters which should be integrated into the metadata of the respective PV plant or be retrievable from an accessible database. If costs are subject to time dependent variations, a market price query should be implemented. A solution matrix, shown as an example in TABLE II, is an attempt to further simplify the cost calculation of a maintenance ticket, in particular for labour and repair costs. The matrix is presented as a list of all actions collected from the tickets. It associates the average time of fixing, an estimated number of personnel required, the hourly cost of individual personnel and the price of the component to be substituted to the respective solution. The last two parameters are highly situation specific and country and plant dependent. Therefore, a range of cost options should be considered.

TABLE II: Example idea of the solution matrix; abbreviations - t_{fix}: fixing time, n_p: number of personnel required, C_P: cost of personnel per hour, C_{rep}: repair cost (including possible component costs) normalized per kWp.

Solution	t_{fix} [h]	n_P [-]	C_P [€/h]	C_{rep} [€/kWp]
Replace Panel	1	2	30-80	150-300
Replace Cable	1.5	2	30-80	7
Replace Fuse	2.5	1	30-80	2
Replace Inverter	3	2	30-80	30-70
To be continued…				

In order to finalize the solution matrix, a high amount of input data are required. Only with standardized O&M tickets all information can be extracted in a cohesive manner and an insightful statistical and economical evaluation can be performed.

6 SUMMARY & OUTLOOK

In this study, more than 35,000 O&M tickets have been statistically evaluated and, based on the lessons learned, the integration of PV system data-streams into an automated decision support system (DSS) was investigated. For that purpose, the Cost Priority Number (CPN) methodology, a cost-based FMEA for PV systems, has been further refined to be used within day-to-day O&M activities. Data from three large O&M operators have been used to conduct this work.

The first important outcome of this work is the acknowledgement that, to properly exploit operational data of PV plants, a common dictionary/taxonomy is required to classify all the information extracted from different O&M tickets. A standardized structure for O&M ticket categorization is proposed where an event/failure in the field is defined by the affected component type, affected subcomponent, event/failure name, cause of the issue, accountable person, available resolution (if any), the ID of the affected component/s and the related time steps (start/detection/end/repair). Thereby, the TRUST-PV Risk Matrix categorizes issues in the field with a standardized set of parameters concerning component/subcomponent/ event. Overall, 341 events/failures have been identified affecting nine different components of a PV plant. In this work, historical tickets coming from the field have been aligned using the TRUST-PV Risk Matrix to provide a common nomenclature. The matrix will enable O&M operators to create tickets more efficiently, making them easier accessible and exploitable for a statistical and economical evaluation. Based on statistical insights, best-practices as well as shortcomings can be identified and overcome.

After the alignment of the collected O&M tickets, they have been statistically evaluated looking at the most appearing issues in the field and common practices of different operators. Thereby, inverter related tickets as well as the ones under the component "Communication & Monitoring" are with an overall share of 76% highly represented. Many tickets affecting other components must be included manually whereas tickets related to inverter and monitoring systems are automatically triggered because of automatic alarms. It is believed that this fact is a large contributor to the high share, especially of inverter related tickets. For most tickets, one primary subcomponent is affected, such as "Entire inverter" for the component "Inverter". This high share can be connected to frequently appearing issues in the field. The top ten events collected in this study account for 70% of all collected tickets.

We believe that module tickets, and tickets related to other "non-vital" components are highly underrepresented because of non-detection and a relatively low performance impact.

Another factor influencing the total number of tickets as well as their distribution is the collection strategy. The following related factors might affect the total number of tickets:

- Are tickets which affect multiple inverters stored multiple times, each accounting for one individual inverter?
- Are tickets stored repeatedly?
- Training of the field technicians varies
- Through higher grade of ticket acquisition automation for inverter and monitoring related issues "less" relevant or false alarms might be stored

It is difficult to say which approach is preferable in terms of cost effectiveness. While fully relying on an automated ticket acquisition system may save day-to-day O&M costs, it could also negatively affect the performance of PV systems by not detecting, and counteracting against, appearing issues on a higher level. This in turn may have grave long-term reliability effects once undetected issues affect other components or even pose a safety risk. The CPN methodology presented in this work is intended to be used to continuously train DSS with new information so that false alarms are gradually diminished by connecting ticketing data with monitoring data in real time.

In case of a system issue triggering a ticket, the CPN is the sum of costs related to downtime of the affected system part and the cost of detection and repair-/replacement required to resolve the issue. In this work, historical data have been studied, but the algorithms and methodologies developed are also applicable to live monitoring data. This will be highly valuable for immediately selecting the best course of action in case of an appearing issue in the field to keep the economic impact at a minimum. The gained knowledge through the evaluation of historical data is summarized into a solution matrix to evaluate all affecting parameters and automatically provide the best solution. A standalone attempt of such a solution matrix has been developed in this work and will be integrated, together with the TRUST-PV Risk Matrix, guidelines of data acquisition, and algorithms for the calculation of downtime costs, into a service management system.

ACKNOWLEDGMENTS

This project has received funding from the Horizon 2020 programme of the European Union under GA. No. 952957 – Project TRUST PV.

REFERENCES

[1] SolarPower Europe, "Global Market Outlook For Solar Power 2022-2066," SolarPower Europe, Brussels, 2022.

[2] S. Philpps and W. Warmuth, "Photovoltaics Report," Fraunhofer Institue for Solar Energy Systems, Freiburg, 2022.

[3] "H2020 TRUST PV Project," [Online]. Available: https://trust-pv.eu/. [Accessed 11 01 2021].

[4] "H2020 Solar Bankability Project," [Online]. Available: http://www.solarbankability.org/home.html. [Accessed 20 07 2021].

[5] French, R. H., Bruckman, L. S., Moser, D., Lindig, S., et al., "Assessment of Performance Loss Rate of PV Power Systems," IEA PVPS Task 13, 2021.

[6] D. C. Jordan, T. J. Silverman, J. Wohlgemuth, S. R. Kurtz and K. VanSant, "Photovoltaic failure and degradation modes," *Progress in photovoltaics, Research and application,* vol. 25, pp. 318-326, 2017.

[7] S. Lindig, I. Kaaya, K. Weiss, D. Moser and M. Topic, "Review of Statistical and Analytical Degradation Models for Photovoltaic Modules and Systems as Well as Related Improvements," *IEEE Journal of Photovoltaics,* vol. 8, no. 6, pp. 1773-1786, 2018.

[8] IEC60050-191, "International Electrotechnical Vocabulary: Chapter 191: Dependability and quality of service," International Electrotechnical Comission, Geneva, Switzerland, 1990.

[9] S. Lindig et al., "International collaboration framework for the calculation of performance loss rates: data quality, benchmarks and trends (towards a uniform methodology)," *Under Review,* 2020.

[10] M. Köntges et al., "Review of failures of photovoltaic modules - IEA PVPS T13-03:2014," IEA PVPS Task 13, 2014.

[11] M. Köntges, G. Oreski, M. Herz, U. Jahn, P. Hacke and K.-A. Weiß, "Assessment of Photovoltaic Module Failures in the Field," IEA PVPS Task 13, Subtask 3 Report IEA-PVPS T13-09:2017, 2017.

[12] M. Herz, G. Friesen, U. Jahn, M. Koentges, S. Lindig, D. Moser, S. Bordihn, G. Eder, U. Hupach, G. Oviedo Hernández, B. Herteleer, W. Herrmann, J. Lin and C. Schill, "Quantification of Technical Risks in PV Power Systems," IEA PVPS Task 13, 2021.

[13] G. Oviedo-Hernández, S. Lindig, D. Moser and P. Chiantore, "Optimization of the Cost Priority Number (CPN) Methodology to the Needs of a Large O&M Operator," in *36th EU PVSEC,* Marseille, 2019.

[14] L. Köster, S. Lindig, A. Louwen, A. Astigarraga, G. Manzolini and D. Moser, "Review of Photovoltaic Module Degradation, Field Inspection Techniques and Techno-Economic Assessment," *Renewable and Sustainable Energy Reviews,* vol. 165, 2022.

[15] S. Lindig, A. Louwen, M. Herz, J. Ascencio-Vásquez, D. Moser and M. Topic, "Performance Imputation Techniques for Assessing Costs of Technical Failures in PV Systems," in *38th European Photovoltaic Solar Energy Conference and Exhibition,* 2021.

[16] P. Geurts, D. Ernst and L. Wehenkel, "Extremely Randomized Trees," *Machine Learning,* vol. 63, pp. 3-42, 2006.

[17] F. Pedregosa, G. Varoquaux, A. Gramfort, V. Michel, B. Thirion, O. Grisel, M. Blondel, A.

Müller, J. Nothman, G. Louppe, P. Prettenhofer, R. Weiss, V. Dubourg, J. Vanderplas, A. Passos, B. M. Cournapeau, M. Perrot and E. Duchesnay, "Scikit-learn: Machine Learning in Python," *J. Mach. Learn. Res.*, vol. 12, p. 2825.2830, 2011.

[18] D. L. King, W. Boyson and J. Kratochvill, "Photovoltaic Array Performance Model," Sandia National Laboratories, Alberquerque, 2004.

[19] T.A. Huld et al., "A power-rating model for crystalline silicon PV modules.," *Solar Energy Mater. Solar Cells,* vol. 95, pp. 3359-3369, 2011.

[20] 3E, "3E's Satellite-based irradiation data," 2022. [Online]. Available: https://solardata.3e.eu/. [Accessed 01 03 2022].

[21] D. Erbs, S. Klein and J. Duffie, "Estimation of the diffuse radiation fraction for hourly, daily and monthly-average global radiation," *Solar Energy,* vol. 28, no. 4, pp. 293-302, 1982.

[22] P. Loutzenhiser, H. Manz, C. Felsmann, P. Strachan, T. Frank and G. Maxwell, "Empirical validation of models to compute solar irradiance on inclined surfaces for building energy simulation," *Solar Energy,* vol. 81, no. 2, pp. 254-267, 2007.

[23] W. Holmgren, C. Hansen and M. Mikofski, "Pvlib python: A python package for modeling solar energy systems," *J. Open Source Softw.,* vol. 3, no. 29, p. 884, 2018.

[24] J. Benesty et al., "Pearson correlation coefficient," in *Noise reduction in speech processing*, Springer, 2009, pp. 37-40.

[25] "CAMS: Surface solar radiation data," [Online]. Available: https://atmosphere.copernicus.eu. [Accessed 12 01 2021].

[26] Copernicus-Climate-Change-Service-ERA5, "Fifth generation of ECMWF atmospheric reanalyses of the global climate. Copernicus Climate Change Service Climate Data Store (CDS)," 2020. [Online]. Available: https://cds.climate.copernicus.eu/cdsapp#!/home . [Accessed 15 July 2020].

[27] D. Moser et al., "Identification of Technical Risks in the PV Value Chain and Quantification of the Economic Impact on the Business Model," *Progress in Photovoltaics: Research and Applications,* vol. 25, no. 7, pp. 592-604, 2017.

APPENDIX

TABLE III: Ticket structure integrating TRUST-PV Risk Matrix for automation of CPN calculation.

Field	Explanation	Mandatory/Optional	Additional information	Type
ticket_id	Unique ID for every ticket	Mandatory	Automatically generated number	Integer
plant_id	Which plant is affected	Mandatory	Automatically generated from metadata table	Integer
t_failure	When did issue appear	Mandatory	Manual input	Datetime
t_ackn	When was issue acknowledged/found	Mandatory	Usually ticket creation	Datetime
t_inter	When does intervention starts	Mandatory	Manual input	Datetime
t_res	When is issue resolved	Mandatory	Usually ticket closure	Datetime
failure_id	Unique failure ID*	Mandatory	Automatically inserted based on failure/event	Text field
Component	Which Component is affected*	Mandatory	Drop-down menu	Text field
Subcomponent	Which subcomponent of component is affected*	Mandatory	Drop-down menu	Text field
failure/event	Name of failure*	Mandatory	Drop-down menu	Text field
fail_desc	Description of failure*	Optional	Automatically inserted based on failure/event	Text field
Cause	What caused failure**	Optional	Drop-down menu	Text field
Accountability	Who is accountable**	Optional	Drop-down menu	Text field
Detection	How was failure detected**	Optional	Drop-down menu	Text field
Origin	Which phase of plant live did failure origin from**	Optional	Drop-down menu	Text field
Solution	Which solution was applied**	Optional	Drop-down menu	Text field
cab_affected	ID of affected cabin/subsystem (multiple possible)	Mandatory	Drop-down menu for 1 type of components and rest is automatically generated from hierarchy tree of PV plant component	Text field
trafo_affected	ID of affected transformer (multiple possible)	Mandatory		Text field
inv_affected	ID of infected inverter (multiple possible)	Mandatory		Text field
tracker_affected	ID of affected tracker (multiple possible)	Mandatory		Text field
string_affected	ID of affected string (multiple possible)	Mandatory		Text field
mod_affected	ID of affected module (multiple possible)	Mandatory		Text field
nr_cab_aff	Amount of affected cabins/subsystems	Mandatory	Sum cab_affected	Integer
nr_trafo_aff	Amount of affected transformers	Mandatory	Sum trafo_affected	Integer
nr_inv_aff	Amount of affected inverters	Mandatory	Sum inv_affected	Integer
nr_tracker_aff	Amount of affected trackers	Mandatory	Sum tracker_aff	Integer
nr_str_aff	Amount of affected strings	Mandatory	Sum string_affected	Integer
nr_mod_aff	Amount of affected modules	Mandatory	Sum mod_affected	Integer
Comments	Free field for additional information	Optional		Text field

*TRUST-PV Risk Matrix.
**Optional information from TRUST-PV Risk Matrix.

This is a short version of the paper submitted to the journal Solar RRL.

It has been selected by the Committee of the WCPEC-8 for submission to Solar RRL.

VIPV IRRADIATION MODELING AND VERIFICATION WITH TRIP DATA

Evgenii Sovetkin*, Jonas Noll, Neel Patel, Andreas Gerber, and Bart E. Pieters

IEK-5 Photovoltaics, Forschungszentrum Jülich
52425 Jülich, Germany

ABSTRACT: This work proposes a benchmark dataset for irradiance modeling for vehicle-integrated photovoltaics (VIPV). Our vehicle trip data consist of trips in the state of North Rhine-Westphalia (NRW) in Germany starting from March 2021, which amounts to more than 73 hours and a total distance of 3422 km. The sensor box is equipped with GPS, a magnetic compass, acoustic wind, and irradiance sensors and records at a rate of 0.58 Hz. The irradiance sensors are positioned on four sides of the vehicle: roof, left, right and rear. In addition to the data, the paper discusses a model that uses high-resolution topography (LIDAR) and low-resolution weather data to forecast the effective irradiation of modules mounted on a moving vehicle. The utility of our simulation approach is demonstrated by computing irradiation over long periods for various driving profiles and comparing results with the collected measurement data. Our model attains 98% accuracy of the irradiation computed over 8 months. Our data is published as a challenge, where everyone can propose and test their model for irradiance modeling.

Keywords: vehicle-integrated PV, geospatial data, measurement data

1 INTRODUCTION

Recently an increasing number of integrated photovoltaic (PV) applications appeared [14, 20, 6]. The local environment in such applications plays a significant role in PV modules' performance, as modules are no longer unobstructed from the sky dome. The vehicle-integrated PV (VIPV) is one such application. PV modules are installed on the car's roof, sides, hood, or trunk. The captured solar energy then contributes to vehicle motion and reduces the frequency of grid charging.

The main challenge in VIPV comes from the effects of the local environment on the modules' performance. Araki et al. [3, 4] model solar irradiance by a random distribution of shading objects and car orientation with the correction of the curved surface of the PV modules. Araki et al. [4] also perform year-long experimental measurements on a vehicle using multiple pyranometers. [13, 5] use the proprietary ArcGIS software to model irradiation for vehicles in Lisbon, Portugal.

Several projects aim to collect measurement data [16, 10]. Their solutions can be potentially scaled up to many users with different driving profiles, yielding a unique dataset and assisting in building accurate irradiance models.

Despite this progress, we argue that the question of the accurate irradiance and yield modeling for VIPV remains open. At the same time, precise irradiance modeling is essential in yield simulations for VIPV, allowing proper investment planning and facilitating global adaptation of the VIPV technology. Furthermore, simulations are crucial for designing different VIPV subsystems, such as inverters, cell topology, and bypass diodes.

We believe publicly available benchmark measurement data benefits irradiance/yield modeling problems. Such a dataset allows researchers to compare their modeling approaches robustly. Unfortunately, to our knowledge, no such dataset is available.

Therefore, this paper aims to achieve two goals. Firstly, we publish a large measurement dataset, [19, 18]. The data is published in a challenge [19], where a large portion is made public, and the rest is blinded and used to benchmark submitted models. Secondly, we highlight some of the irradiance/yield modeling problems in VIPV and propose our solution framework.

Our modeling approach is implemented in several open-source tools. The Simple Sky Dome Projector (SSDP) library [12] implements the irradiance modeling. The PV-Geographic Raster Image Processor (PV-GRIP) tool [17, 9] is a distributed-computing framework, and it addresses a technical challenge, as VIPV modeling requires intensive geospatial data processing.

The paper is organized as follows. First, we discuss our data logger setup in Section 2. There we provide the specification of our sensor box and describe the collected data and where one can find it. Then, Section 3 elaborates on our irradiance modeling approach and presents the simulation data. Lastly, Section 5 concludes the paper.

2 EXPERIMENTAL SECTION

Our experimental measurement setup consists of a sensor box mounted on a van. The vehicle performs

*Corresponding author: e-mail: e.sovetkin@fz-juelich.de

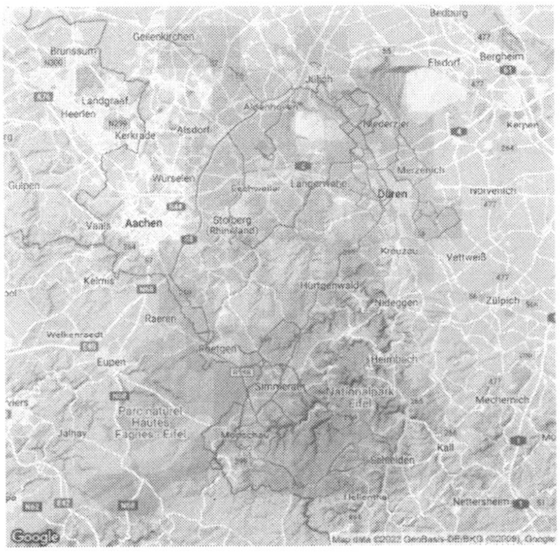

Figure 1: March 2021–October 2021 driving profile. Different colors indicate training (blue) and testing (red) regions. Blue and red points (curves) show the vehicle's movement. The semi-transparent blue and red rectangles show the grid generated by a geohash of length 5

regular trips in Germany's state of North Rhine-Westphalia (NRW) starting in March 2021. The travel time is more than 73 hours, 150000 observations, and a 3422 km covered distance. The vehicle travels through urban areas in Aachen, Jülich, and Düren, as well as areas with agricultural fields, forests, and the hilly regions in Eifel. The road types range from narrow roads and freeways. Figure 1 depicts the driving profile between March 2021 and October 2021.

Each observation consists of the outputs from GPS, a magnetic compass, and irradiance sensors. The average sensor read-out frequency is 0.58 Hz. The box is powered via the car battery; hence, it does not record when the engine is off. Therefore, our data contains travel data without observations while the vehicle is parked. Figure 2 shows a photograph of our sensor box setup.

Our vehicle covers 3422 km distance. The car takes several routes with a total of 330 km unique road surface. The total travel time equals 73 hours. Most trips start around 8 am and go in the south-west direction. The return trips happen between 1 and 3 pm in the northern direction. Such a regular travel pattern introduces an asymmetry in irradiation of the vehicle's left and right sides (see Table II in Section 4).

We publish the collected data in a challenge form [19, 18], where approximately 70% of the data is entirely open. The rest of the data irradiance sensor outputs (irradiance and temperature) are blinded and only used for evaluation. Publication of the data as a challenge allows it to be dynamic by adding additional stages when new data becomes available. It also allows other research groups to benchmark their irradiance models and compare their performance with other models. TableI summarises information on the provided fields and their units.

To evaluated the challenge honestly, we ensure that data in the training and evaluation part of the dataset do not correlate. We achieve this by placing training and evaluation data away from each other. We utilize geohash [21] to space the data, which splits our region into regular rectangular areas. Approximately 70% of those areas are made public. Figure 1 depicts those regions using blue- and red-colored rectangles.

3 IRRADIANCE MODELING

Our irradiance modeling uses low- and high-resolution topography and low-resolution weather data to compute effective irradiation on the surface of the moving vehicle. Our approach is implemented through the SSDP [12] library and PV-GRIP [9] distributed-computing framework.

The SSDP library [12] implements the Perez All-Weather Sky Model [11] to account for a non-isotropic diffuse light component. Furthermore, the SSDP tool implements several VIPV-specific simulation modes to model irradiance along a route, where the orientation and tilt of the incident plane are adapted to the course, road tilt, module orientation and height.

SSDP works with topography data represented either with a raster image (e.g., Digital Elevation Model) or an irregular mesh of points (unprocessed LIDAR point clouds). The library does not use any specific coordinate system. Hence we resample topography data in a metric coordinate system and feed it to the SSDP. Furthermore, it allows us to sample topography, take tilts in roads, and specify the offset and orientation of modules to model the module position correctly.

The SSDP irradiance model has its limitations. The topography is modeled as a surface and does not capture any 3d objects. For instance, in our model, trees cast more shadows than in reality. Furthermore, SSDP is not a ray-tracing library; hence the light reflections are modeled with significant simplifications. Nevertheless, such an approach dramatically improves the performance of simulations, allowing VIPV modeling over long routes.

The dynamic nature of a moving vehicle is the main challenge in the irradiance modeling that distinguishes VIPV from other PV applications. When a car moves, the environment and shading conditions are highly dynamic. Therefore, unlike the problem of irradiation modeling for building-integrated PV, VIPV requires intensive topographical LIDAR data

Figure 2: View on the sensor box from the rear-right side of the vehicle. The data logger is encapsulated in a weather-proof box and mounted with other sensors on aluminum profiles. The acoustic wind sensor is located right next to the right irradiance sensor. The rear sensor sees many reflections from the white roof of the vehicle

Table I: Description of fields

Field	Description
"gps time utc"	unix timestamp
"timestr"	time string in the format "%Y-%m-%d_%H:%M:%S" with UTC time zone (reformatted "gps time utc")
"latitude", "longitude"	gps coordinates
"gps altitude (m)", "gps speed (m/s)", "gps course (deg)", "gps climb (m/min)"	altitude, speed, bearing and climb rate
"gps EPX Estimated * Error"	gps estimated errors
"Magnetic Bearing [deg.], Magnetic Tilt [deg.]"	magnetic compass bearing and tilt
"module", "sensor_azimuth", "sensor_zenith"	side of the irradiance sensor ("module" attains values: "roof", "left", "right", "rear"). "sensor_azimuth" and "sensor_zenith" define orientation of the irradiance sensor. azimuth computed from magnetic compass, zenith 0 corresponds to the sky orientation
"irradiance_Wm2"	irradiance sensor irradiance and temperature readings
"GHI", "DHI", "Clear sky GHI", "Clear sky DHI"	copernicus irradiance values
"geohash"	region id which is used in training dataset split

processing. For example, our vehicle covers over 330 km of the unique road surface. Therefore, we must resample and prepare large volumes of data for the SSDP library's input. We developed a scalable distributed-computing framework implemented in the PV-GRIP tool to accelerate this process. For the architecture description we refer the reader to [17].

We need two kinds of geospatial data for our simulations: topographical data and atmospheric irradiation data. We use data from three sources for topography data: LIDAR data provided by the NRW state [1], Shuttle radar topography data (SRTM) [15], and Advanced Spaceborne Thermal Emission and Reflection Radiometer (ASTER) [2]. In addition, we resample data in UTM metric coordinate system to ensure that our shadows are simulated with the correct size.

The LIDAR data is provided in a "laz" format, and we resample the cloud points by computing different statistics for every region of size $30 \times 30 \mathrm{cm}^2/\mathrm{pixel}$. Among those statistics for us are interesting maximum and minimum statistics. Figures 3 demonstrate the so-called "hillshade" images of the sampled LIDAR data. The left image in Figure 3 is generated using the maximum value statistic that captures the shape of vegetation and buildings. The minimum value statistic (the right image, Figure 3) gives a view with buildings but without trees. The foliage is not visible in the minimum statistic plot since one of the reflections of the laser comes from the ground.

In addition to LIDAR, we use SRTM and ASTER data, the satellite-based topography data. These data sources have a courser resolution of $90 \times 90 \mathrm{m}^2/\mathrm{pixel}$ and $30 \times 30 \mathrm{m}^2/\mathrm{pixel}$ for SRTM and ASTER data, respectively. The satellite-based topography data does not capture any building or vegetation but provides a generic topography outlook.

In addition to the topography data, we use Copernicus [7], a satellite-based atmospheric data service. It provides GHI, DHI, and corresponding clear sky values with a spatial resolution of approximately 5 km and temporal resolution of about 10 minutes.

4 RESULTS

For simulations in this paper, we look at the model's performance implemented in the SSDP library using different topography sources: LIDAR, ASTER, and SRTM. These data sources are resampled with different resolutions. Furthermore, for each observation data point, the topography raster data is sampled so that a square of a specific size with the center at the observation point is wholly contained inside the sampled raster image. We call this square a minimal square. This sampling strategy ensures that at least the minimal-square-size worth of topography data is used to compute shading.

We sample the LIDAR data with $30 \times 30 \mathrm{cm}^2/\mathrm{pixel}$ resolution, maximum and minimum statistics, and a minimal square size of 200 m. Next, the ASTER data is sampled with a resolution of $10 \times 10 \mathrm{m}^2/\mathrm{pixel}$ and a minimal square size of 500 m. Lastly, the SRTM is sampled with $50 \times 50 \mathrm{m}^2/\mathrm{pixel}$ and a minimal square size of 2000 m.

Figure 4 shows simulated versus measured irradiance for each observation (left) and integrated over 1 minute (right). Unfortunately, many measurements do not agree with simulations (Figure 4, left). This can be attributed to two factors. Firstly, the limitation of the SSDP discussed in Section 3 result in imprecise modeling of geographical objects and their shading. Secondly, our model does not include any shading from cloud coverage. Nevertheless, the model yields correct results in a statistical sense: when we integrate irradiance over a short time (Figure 4, right), the model's errors cancel each other, and prediction improves. This improvement is illustrated in the right part of Figure 4, where the R^2 value increases from 0.37 (left) to 0.63.

Table II provides simulation results using different topography data sources. To integrate irradiance, we assume it to be constant between measurements and split all data into parts; there are no observations with a time difference of more than 30 seconds in every part of the data. Integrated parts are then summed up.

Using different topography data sources allows us to answer how various environmental features influence irradiation in general for VIPV. For example, comparing "LIDAR min" and "measured" concludes that vegetation decreases irradiation by 17%. Furthermore, vegetation and buildings reduce irradiation by 43% (compared with SRTM simulations). As shown in Figure 2, the rear sensor sees many reflections from the roof, resulting in the rear sensor measurement being more significant than the "LIDAR min" simulations. Generally, we expect "LIDAR max" to underestimate irradiation, as it gives more prominent shadows. Nevertheless, those simulations over all modules provide a relative error of 2%.

The simulations in Table II assume modules are located at 2m height (irradiance sensors are located on the van's roof). Simulations on the ground-level lead to a 10% reduction (for the roof's sensor) and 15% (for all sensors). Such a decrease emphasizes the importance of setting the correct elevation of the simulated modules.

Our simulations have been performed on eight heterogeneous nodes with 160 GB of RAM combined. Our dataset of 3422 km of the route takes around 4 hours to sample LIDAR data and perform irradiance simulations. Sampling and simulation for SRTM and ASTER data require approximately 30 minutes. We

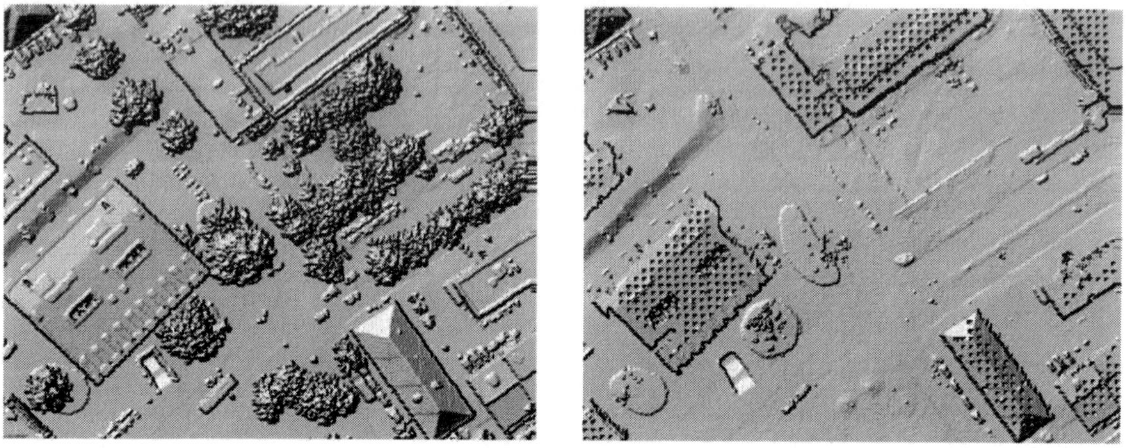

Figure 3: "Hillshade" visualization of terrain sample from LIDAR in the region of the "Super C" building in Aachen, sampled with the maximum statistic (left) and the minimum statistic (right). Resolution $30 \times 30\,\mathrm{cm}^2/\mathrm{pixel}$

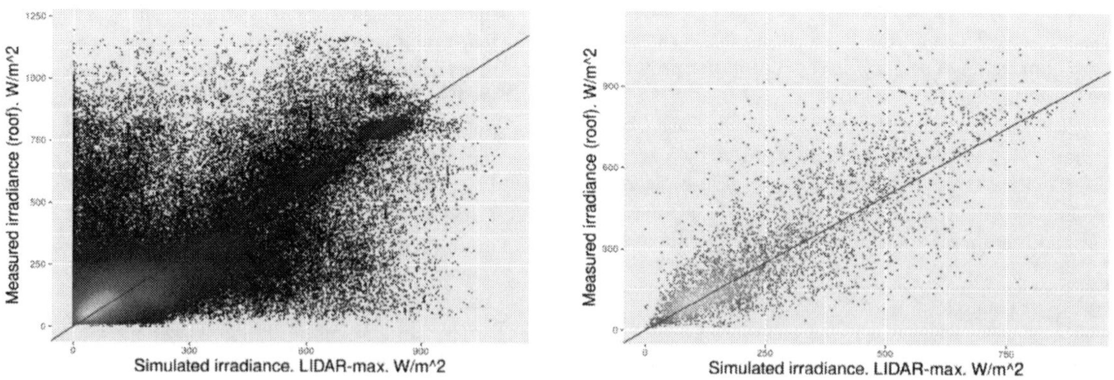

Figure 4: Simulated vs. measured irradiance for every observation (left) and integrated irradiance over 1 minute (right) for the roof sensor. Bright colors indicate a denser observation population. The decreased mean squared error and increased R^2 value (from 0.37 to 0.63) suggest that simulation errors reduce after integration

Table II: Irradiation (kWh/m^2) over 73 hours. Note that the rear sensor captures reflections from the white roof (see Figure 2)

	measured	LIDAR max	LIDAR min	ASTER	SRTM	GHI
top	21.85	19.71	23.76	27.30	27.74	27.95
left	12.00	13.07	15.94	18.84	19.80	
right	7.33	9.54	11.18	13.36	14.05	
rear	14.58	12.29	14.55	17.07	17.84	
total	55.77	54.62	65.42	76.56	79.43	

use 660 GB for raw data and 200 GB for sample data in terms of storage.

5 CONCLUSIONS

This paper introduced a public dataset containing VIPV measurement data. The dataset is published as a challenge [19, 18], where everyone may benchmark their irradiance model. Generally, VIPV irradiance modeling introduces technical challenges: analyzing the environment around a moving vehicle requires processing a large amount of geospatial data. We overcome this challenge by introducing a distributed-computing-based modeling tool PV-GRIP. Furthermore, irradiance modeling is achieved using our open-source C-library SSDP, which presents an efficient way to perform simulations over long routes. We remark, however, that our model is not perfect and can be gradually improved. Hence we invite others to benchmark their models on our dataset.

ACKNOWLEDGMENTS

The authors gratefully acknowledge the Helmholtz Data Federation (HDF) for funding this work by providing services and computing time on the HDF Cloud cluster at the Jülich Supercomputing Centre (JSC) [8]. This work is supported by the HGF project "Living Lab Energy Campus (LLEC)", as well as the "Street" project (Street — Einsatz von hocheffizienten Solarzellen in elektrisch betriebenen Nutzfahrzeugen, Förderkennzeichen: 0324275A). Thanks to the IBG-3 (FZJ) for providing the vehicle and collecting the data.

References

[1] *3d-measurement data in North Rhine-Westphalia*. Online; accessed 29.09.2022. URL: https://www.bezreg-koeln.nrw.de/brk_internet/geobasis/hoehenmodelle/3d-messdaten/index.html.

[2] *AIST, Japan Spacesystems, and US/Japan ASTER Science Team, 2009: ASTER Global Digital Elevation Model. NASA EOSDIS Land Processes DAAC, NASA, METI*. Online; accessed 29.09.2022. URL: https://cmr.earthdata.nasa.gov/search/concepts/C1711961296-LPCLOUD.html.

[3] K. Araki, K.-H. Lee, T. Masuda, Y. Hayakawa, N. Yamada, Y. Ota, and M. Yamaguchi. "Rough and Straightforward Estimation of the Mismatching Loss by Partial Shading of the PV Modules Installed on an Urban Area or Car-Roof". In: *2019 IEEE 46th Photovoltaic Specialists Conference (PVSC)*. IEEE. 2019, pp. 1218–1225.

[4] K. Araki, Y. Ota, and M. Yamaguchi. "Measurement and modeling of 3D solar irradiance for vehicle-integrated photovoltaic". In: *Applied Sciences* 10.3 (2020), p. 872.

[5] M. C. Brito, T. Santos, F. Moura, D. Pera, and J. Rocha. "Urban solar potential for vehicle integrated photovoltaics". In: *Transportation Research Part D: Transport and Environment* 94 (2021), p. 102810.

[6] B. Commault, T. Duigou, V. Maneval, J. Gaume, F. Chabuel, and E. Voroshazi. "Overview and perspectives for vehicle-integrated photovoltaics". In: *Applied Sciences* 11.24 (2021), p. 11598.

[7] *Copernicus Open Access Hub*. Online; accessed 29.09.2022. URL: https://scihub.copernicus.eu/.

[8] B. Hagemeier. "HDF Cloud–Helmholtz Data Federation Cloud Resources at the Jülich Supercomputing Centre". In: *Journal of large-scale research facilities* 5.FZJ-2020-00569 (2019), A137.

[9] *PV-GRIP: Photovoltaics Geographic Raster Image Processor*. Online; accessed 29.09.2022. URL: https://github.com/esovetkin/pvgrip.git.

[10] *PV2Go: solar potential for vehicles*. Online; accessed 29.09.2022. URL: https://pv2go.org/.

[11] R. Perez, R. Seals, and J. Michalsky. "All-weather model for sky luminance distribution—preliminary configuration and validation". In: *Solar energy* 50.3 (1993), pp. 235–245.

[12] *SSDP: Simple Sky Dome Projector*. Online; accessed 29.09.2022. URL: https://github.com/IEK-5/SSDP.

[13] T. Santos, N. Gomes, S Freire, M. Brito, L Santos, and J. Tenedório. "Applications of solar mapping in the urban environment". In: *Applied Geography* 51 (2014), pp. 48–57.

[14] A. K. Shukla, K Sudhakar, and P. Baredar. "Recent advancement in BIPV product technologies: A review". In: *Energy and Buildings* 140 (2017), pp. 188–195.

[15] *Shuttle radar topography data*. Online; accessed 29.09.2022. URL: https://dds.cr.usgs.gov/srtm/version2_1/.

[16] *SolarJinie VIPV evaluates the solar power generated from the bodywork of electric vehicles*. Online; accessed 29.09.2022. URL: https://www.enerjin.com/SolarJinie_VIPV_1p0_spec_sheet.pdf.

[17] E. Sovetkin, N. Patel, A. Gerber, and B. E. Pieters. "5BV.4.13 Open-source geospatial data service with an application in irradiance modelling for VIPV". In: *38th European Photovoltaic Solar Energy Conference and Exhibition*. 2021, pp. 1152–1160.

[18] *VIPV irradiance modelling challenge*. Online; accessed 29.09.2022. URL: http://doi.org/10.34730/3c8732aca9e3423aad57af2bba3a1dea.

[19] *VIPV irradiation modelling challenge*. Online; accessed 29.09.2022. URL: https://helmholtz-data-challenges.de/web/challenges/challenge-page/94/overview.

[20] S. R. Wadhawan and J. M. Pearce. "Power and energy potential of mass-scale photovoltaic noise barrier deployment: A case study for the US". In: *Renewable and Sustainable Energy Reviews* 80 (2017), pp. 125–132.

[21] *Wikipedia: Geohash*. Online; accessed 29.09.2022. URL: https://en.wikipedia.org/wiki/Geohash.

HIGH-VOLTAGE VEHICLE INTEGRATED PHOTOVOLTAIC SYSTEM CONCEPT AND DEMONSTRATOR TRUCK

Marc Andre Schüler[1], Christoph Kutter, Luis Eduardo Alanis[1], Lars Saroch[2], Christoph Barz[2], Jörg Bornwasser[1], Stefan Reichert[1], Holger Neuhaus[1], and Martin Heinrich[1]
[1]Fraunhofer Institute for Solar Energy Systems ISE,
[2]Fraunhofer Institute for Transportation and Infrastructure Systems IVI,
Heidenhofstrasse 2, 79110 Freiburg, Germany
Zeunerstrasse 38, 01069 Dresden, Germany

ABSTRACT: Certain commercial vehicles feature large roof areas that can be easily equipped with lightweight photovoltaic (PV) modules to harvest solar energy. Typically, heavy commercial vehicles (>12 t trucks and semi-trailer) can be fitted with an installed photovoltaic power of 3 kW_p to 7 kW_p with today's mainstream cell technologies integrated on the roof, and harvest significant amounts of energy. Therefore, a high voltage photovoltaic system can significantly reduce the amount and the cross-section area of the cables, and therefore cost and weight. This is why we propose a high voltage photovoltaic system concept for commercial application. The difference to conventional automotive high voltage systems, which are controlled by a central unit, is that within the photovoltaic array, high voltage is present whenever the PV modules are illuminated, and hence need to be considered under power at all times. To comply with safety standards and reduce the risks linked to high voltages (e.g. after an accident), we present a smart junction box with integrated disconnector that automatically separates the serial connection of the modules in order to lower the voltage into the range of safety extra-low voltage. The developed VIPV system is able to feed in solar energy during operation and in standby mode. We present a 15-t demonstrator truck, a Framo E-165, equipped with the abovementioned VIPV system and show yield calculations based on an energy yield model.
Keywords: DC-DC Converter, High-Voltage System, Vehicle-Integrated Photovoltaic (VIPV), Standalone PV

1 INTRODUCTION

Equipping vehicles with photovoltaic (PV) systems is seen as having a great potential for reducing CO_2 emissions in the transport sector, while also leading to further self-sufficiency from fossil fuels [1–4]. In the REPowerEU plan of the European Union, it is targeted to achieve a cross-sectoral expansion of PV to 320 GW by 2025 and 600 GW by 2030, as part of which Vehicle Integrated PV (VIPV) serves [4,5]. The Potential of vehicle-integrated photovoltaic (VIPV) on trucks has been estimated to 90 GW in the EU [4] and previous studies have looked at fuel savings of diesel-powered refrigeration trucks using irradiance measurements in the EU, USA [6] and calculations for Germany [7] and Australia [8]. Since then, cost of c-Si cells and modules have decreased further, so the implementation of solar cells on the roof of commercial trucks has become more economically attractive [6]. Additionally, with the accelerating transition from internal combustion engine drivetrains towards battery or fuel-cell electric drive trains, it is easier to utilize and store the solar energy harvested from VIPV within the vehicle. Recently, several parties (e.g. Sono Motors GmbH in cooperation with Chereau S.A.S.) are investigating VIPV Systems for diverse vehicle types, such as cooling trailers or heavy duty vehicles [9,10].
In this document we present the concept and introduce the several components to realize a photovoltaic system which is equipped on the roof of a demonstrator truck and feeds into the high-voltage battery of the vehicle, with an example given in Figure 1.

Figure 1: Demonstrator truck with integrated photovoltaic modules in the roof

2 SYSYTEM CONCEPT OF HIGH VOLTAGE INTEGRATION

This chapter introduces the relevant components to realize a high-voltage integration of the demonstrator's PV system. The system consists of a box-body integrated photovoltaic module array that feeds directly into the **high voltage battery** and is, to the authors' knowledge, the world's first integrated PV system for trucks operating fully in the high voltage infrastructure. Within the array, each module is dimensioned to comply with **IEC 61730** and Safety Extra Low Voltage (SELV) at all reasonable operation modes (up to 1000 W/m² at -20°C). Depending on the photovoltaic array size, system voltages can reach 400 - 800 V. Within the PV array, all modules are connected in series via their individual smart junction box with integrated connectors.

Superior to the smart junction box, the array is controlled with an automotive compliant SiC-based, transformer-less DC/DC converter. The DC/DC converter features a superfast Maximum Power Point (MPP) tracking technology that adjusts to the new MPP in under 1 ms to mitigate impacts of dynamic shading while driving. The DC/DC converter is connected to the vehicle control system as a subordinate via redundant communications and adjusts its power to the system parameters, only supplying the HV system when the feed-in release is granted. Therefore, the DC/DC converter is embedded in the vehicle's high voltage interlock circuit and can be galvanically separated from the high voltage DC link via manual service disconnect switches in case of emergency or maintenance. The VIPV system feeds into the high voltage (800 V) drive train battery via the DC link. In general, the system is designed as an extension of the electric truck, which enables an effortless adaptation to any make or model using the developed soft- and hardware interface.

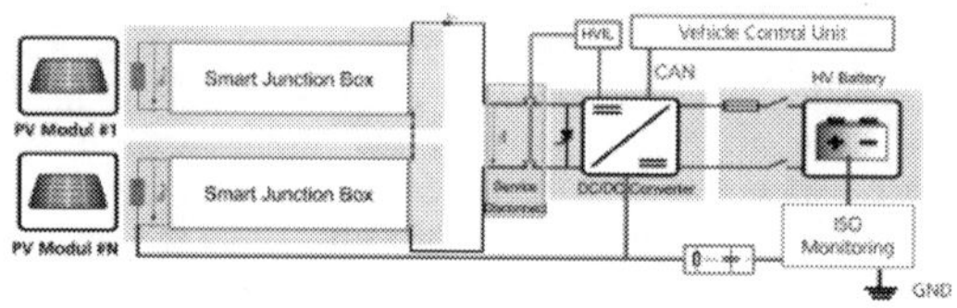

Figure 2: VIPV system concept developed in the project "Lade-PV"

2.1 Lightweight PV modules

PV modules for integration into vehicles must withstand harsh environmental conditions and are required to be lightweight to reduce load-related CO_2 emissions of the vehicle. Previous work presented some approaches to achieve the various constraints [4,11]. The PV modules used in the LadePV demonstrator truck, designed at Fraunhofer ISE and produced by SUNSET Energietechnik GmbH, feature a rated I_{SC} of 9.7 A and a V_{OC} of 28.56 V at Standard Testing Conditions (STC). The module configuration is shown in *Figure 3*.

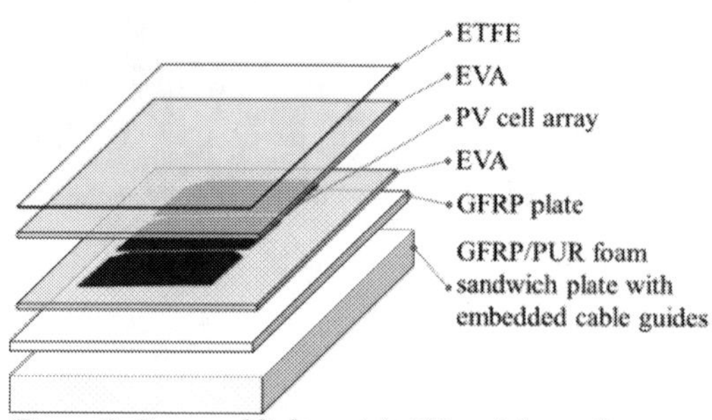

Figure 3: Proposed light-weight PV module configuration for integration on top of a GFRP/PUR foam sandwich structure [2,4]

2.2 Photovoltaic Array Featuring Smart Junction Box

For effectively conducting the harvested current to the DC/DC converter, all panels are connected in a series string, which allows to minimize the cross-section area of the cable. Furthermore, the installation has been designed to keep electromagnetic coupling between cables and modules at a minimum. A total of 14 Modules were attached on the roof of the demo truck, resulting in a maximum PV system voltage of 399.84 V at 25°C. The installed PV array can reach a power of 3.2 kW_p at STC, hence all components in the system are rated according to this value.

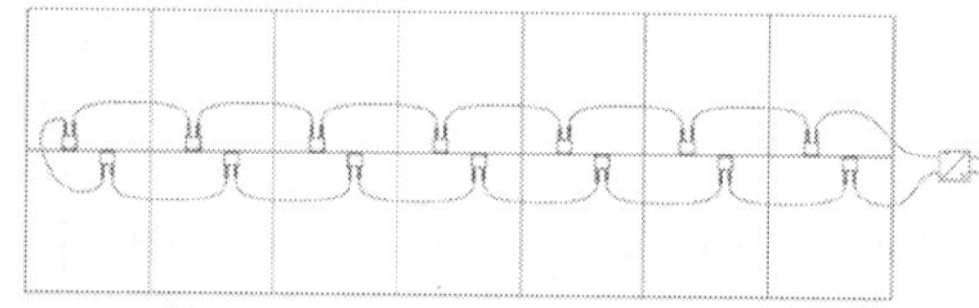

Figure 4: Schematic of wiring between PV modules and DC/DC converter

Open contact points could occur during an accident or because of cable damage, which might constitute a safety hazard, particularly if it happens where the full voltage of the string is tangible. This is not an acceptable scenario according to the ECE-R100 standard for electric vehicles, therefore it must be guaranteed that, for example, in the event of a cable defect, the string voltage is reduced to a non-hazardous level (<60 VDC) as quickly as possible [12]. Therefore, in the following states of the vehicle, the PV string has to switch into open state to ensure safe conditions:

- Separated contacts e.g. cable break, loosened plugs, faulty insulation
- Vehicle involved in an accident
- Maintenance of the vehicle
- HV circuit of the vehicle is not connected
- DC/DC converter is not active

To assure a safety level 2 (SELV) in such scenarios, a transistor-based smart switch circuit has been developed and implemented in every junction box of each PV module. In combination with the DC/DC converter, this smart junction box constantly monitors the parameters prevailing in the string and thus can separate and reconnect the series interconnection at module level accordingly via switches. In this way, a maximum of 60 VDC per single module is present at open contacts during the above-mentioned cases. Furthermore, the smart junction box features active bypass diodes to guarantee safe operating conditions of the PV modules when shading occurs.

2.3 DC/DC on board converter with fast MPPT

Among the PV components, a DC/DC converter was developed by M&P Motion Control and Power Electronics GmbH, that supplies the HV battery of an electrically powered truck with the harvested solar energy. This converter is connected between the photovoltaic module array (input) and the HV DC link of the truck (output). The hardware architecture of the converter is built on fast-switching, wide bandgap SiC MOSFETS in the power stage, allowing in- and output voltages of up to 900 V. This way a high efficiency of 98% is achieved at nominal power of 3.2 kW. Hence, the dimensions of the converter can be kept relatively small at 350x590x180mm without active cooling.

Besides transferring the energy, the converter adjusts the global operating point of the PV array by MPPT with the vehicle in both static and dynamic states. For this reason, an ultra-fast Incremental Conductance algorithm was developed to optimize energy harvest in dynamic mode by settling to the MPP of the PV array in 1ms. Communication to the truck's control unit is built via Controller Area Network (CAN) such that the converter acts as a slave via redundant communication. Hence, the DC/DC converter is integrated into the safety concept of the control unit and adjusts its power to the system parameters, and only supplies the HV system when the feed-in release is granted. Moreover, a connector for an external solar irradiance sensor is provided. Considering the information on actual irradiance, the PV system can ramp-up the vehicle's DC link in stationary state. Moreover, the converter is transformer-less, so the insulation monitoring device of the vehicle is able to detect faulty insulation in the PV circuit and therefore prevent hazardous failure states.

Figure 5: Automotive-compliant prototype of the developed DC/DC converter for LadePV

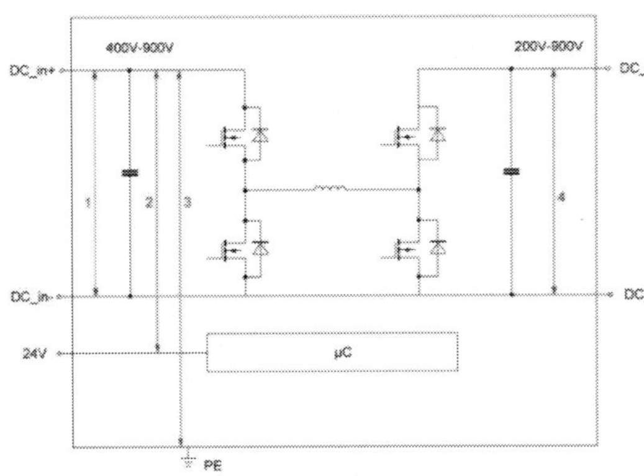

Figure 6: Schematic Overview of the developed DC/DC converter

Table I: Datasheet of the DC/DC on board converter

	Value	Unit
Nominal Power	3.2	kW
Max Input Voltage	900	V (DC)
Output Voltage	0-900	V (DC)
Efficiency (max.)	98	%
Current rating	10	A
Dimensions	350x590x180	mm
MPPT Settling	1	ms
Communication Interfaces	CAN, Ethernet	

3 SIMULATION BASED MODEL FOR HARVESTED ENERGY IN THE ELECTRIC TRUCK

3.1 Setup of a digital twin for energy prognosis

A digital twin for the demonstrator truck was created in IVIsion to allow an estimation of the energy harvested by the PV system. IVIsion is an in-house tool developed at Fraunhofer IVI for determining the driving dynamics and energy requirements of road and rail vehicles. The prognosis tool features a model for the relevant actuators and electrical loads connected to the on-board system. The solar radiation, which depends on the time of day, location, and direction of travel, is also modeled, making IVIsion particularly suitable for simulating VIPV.

Furthermore, the parametric basis of the model is extended by including conversion-efficiencies for the PV components such as PV modules, cables, and the DC/DC converter. Thereby, the modeling for the Framo E-165 includes the energy flow for major components of the drivetrain such as engine, gearing, power electronics, HV and LV system, but also auxiliary systems of the vehicle such as cooling and compressed air infrastructure [13].

The hourly-averaged annual global irradiation dataset for the location Freiburg (Breisgau) in Germany, acquired from the climate data center by the German Weather Service (DWD), was taken as input parameter for the irradiation perceived to the PV modules. For a more realistic approximation, the user profile, which considers the usage and operational patterns of the vehicle, was aligned with the logistic partner responsible for testing the demonstrator truck, resulting in a driving time between 05:00 and 10:00 on operational days. In the evening from 20:00 the truck is then recharged from the grid. An explanatory user profile taken from the simulation is shown in Figure 7, with an additional focus on the state of charge (SoC).

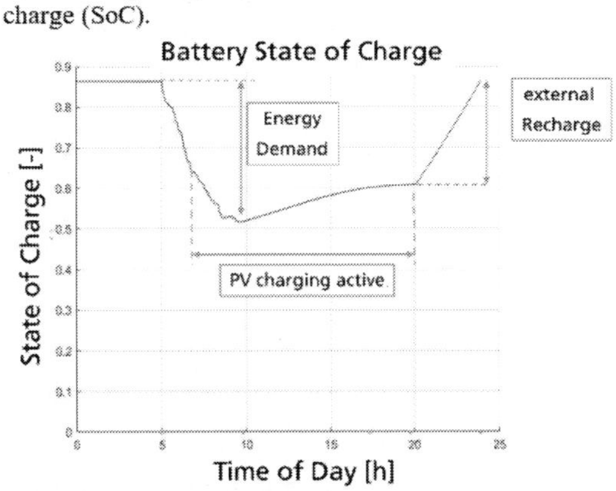

Figure 7: Simulated State of Charge for a given day in summer, taken form IVIsion

3.2 Results of the energy prognosis

Two simulation scenarios are set in IVIsion for every month considering the obtained user profile. One considers an electric truck based on the model for the Framo E-165 without integrated PV; in the second scenario the truck is equipped with a solar system as described in the previous chapter. Additionally, the energy demand of auxiliary actuators for stationary scenario is approximated to around 54 W in July, while in winter 19 W are considered. Considering the hourly-average irradiation data for the given location, a prognosed solar yield is calculated for every month in the second scenario. The results of the prognosed energy demand for the two scenarios and the solar yield are shown in Figure 9.

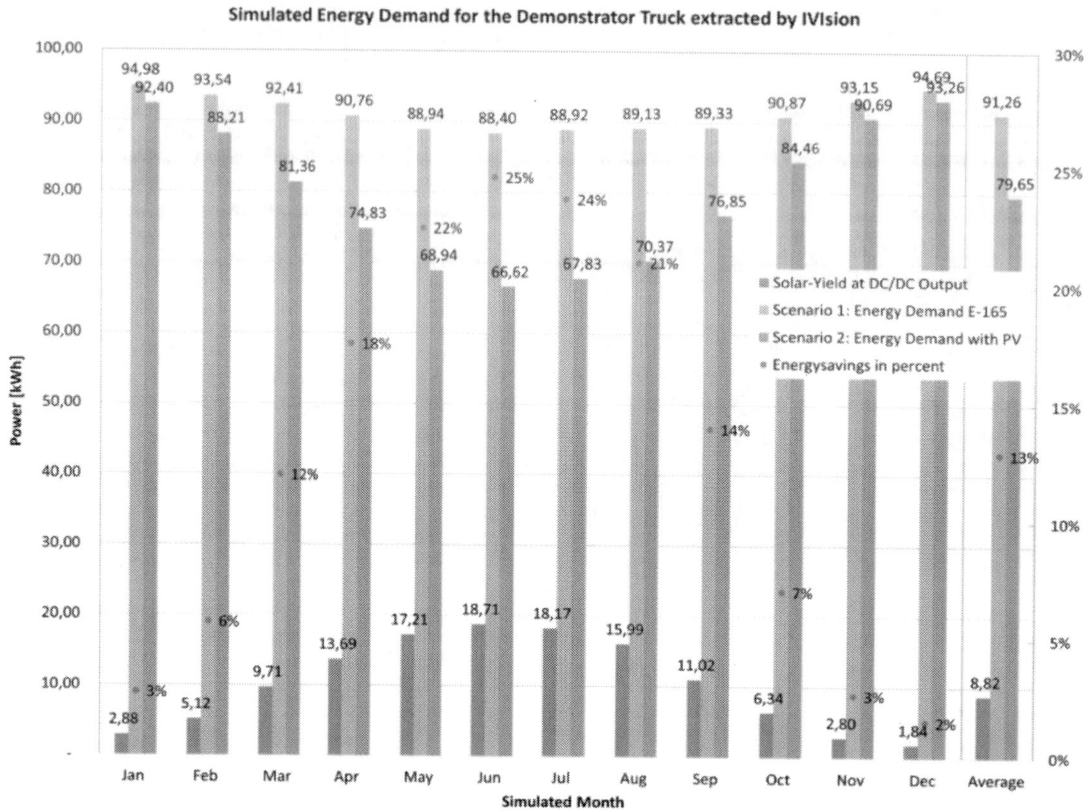

Figure 8: Simulated average solar yield in kWh (orange) and related energy savings in percent (blue) per day for every month. Based on the digital twin of the demonstrator in IVIsion and hourly-averaged irradiation data for Freiburg im Breisgau

As shown in Figure 9, comparing the energy demand of the simulation scenario for the truck without PV system and the demonstrator results in the latter expecting yearly energy savings at the charging station of 13% in average. In both scenarios, the boosted auxiliary energy demand while charging from the grid is included, a factor not considered in previous studies that focused mainly on the energy savings resulting from the additional solar range [14]. To further analyze the impact of this factor in the energy calculations, the model takes vacations and weekends into account, such that a total of 250 days of operation are considered, while the solar charging is simulated for the whole year.

Within the framework of the partner project PV2Go [15], five trucks have been equipped with irradiation sensors to create and validate a solar irradiation atlas for German traffic routes. As the shading profile of the assumed routes is yet unknown, the energy model neglects influence of shading to the vehicle but is likely to be extended in the future.

4 MONITORING OF THE DEMONSTRATOR

To collect data under real world conditions, the demonstrator truck was equipped with the described PV system in October 2021, and ever since operated by the electronic wholesaler Alexander Bürkle GmbH. Energy yield of the PV system and irradiation measurements are tracked in parallel by the datalogger STW TCG-4 with an integrated 4G antenna. In this way, the operation state of the truck can be monitored remotely, and a continuous validation of the energy prognosis is possible.

4.1 Outdoor exposure of the demonstrator:

Within the system ramp-up and operation period, different challenges had to be overcome to realize a smooth operation between the retrofitted PV system and the vehicle control unit. One of the major requirements from **ECE-R100** and **VDE C 0140-479-1:2007-05** concerning the insulation of all components in a vehicle, is to guarantee a critical minimum limit for the insulation of 100 Ω/V. To accomplish this, the Battery Management System (BMS) integrated isolation monitoring device tracks the electrical insulation R_{iso} continuously and forces the vehicle control unit to block driving permission if the measured value is below the limit. Furthermore, a warn signal is sent if the insulation drops below 500 Ω/V. As the maximum system voltage of the battery is 760 V in our demonstrator, the following limits are respectively programmed in the BMS for these safety conditions.

$$R_{iso_{min}} \geq 760\,V \cdot 100\,\Omega/V \geq \mathbf{76\ k\Omega}$$
$$R_{iso_{warn}} \geq 760\,V \cdot 500\,\Omega/V \geq \mathbf{380\ k\Omega}$$

Industrialized PV module requirements as per **DIN EN 61646; DIN IEC 61215** and **DIN VDE 0126-1-1** set a $R_{iso} > \mathbf{40\ M\Omega \cdot m^2}$ or a total minimum of at least $\mathbf{200\ k\Omega}$. However, as the temperature in the module rises during operation, and especially under the influence of moisture, the insulation might drop to as low as 90% of its initial value [16]. Since conventional insulation monitors measure the entire parallel impedance of a vehicle, the influence of moisture on the measured value is even stronger. More specifically, in our demonstrator setup a value of $125\ k\Omega$ is seen by the insulation monitoring system when water has pooled on the modules, even though the modules were measured individually in dry conditions and show an average value of $1.6\ M\Omega$. Furthermore, Wang et al. proposed a settling time of approximately 30 s before the first insulation measurement to avoid parasitic capacitive disturbances of PV modules on isolation monitoring devices [16].

In November 2021 the driving permission appeared to be locked by the insulation monitoring system. It was found through error diagnostics that one of the PV modules suffered a major fracture in one of the full cells, as shown in the electroluminescence (EL) image in Figure 11. The damage was determined to have been caused by accidental heavy mechanical impact while servicing the vehicle. This was determined to be the source of the loss of insulation and, therefore, the entire module was taken off the PV array. Similar faults have not been observed since.

Figure 9 Close-up of EL measurement to a damaged PV module on the solar roof taken by a daylight EL camera.

Aside from the effect of moisture on the modules, the temperature behavior when exposed to solar irradiation was also measured with an InfraTec VarioCAM infrared (IR) camera with a FPA microbolometer. A sample IR image is shown in Figure 12, which shows that the PV array's temperature behaves homogenously, and the measured temperature of the cells is approximately 64,4°C at the 08th of July around 1:30 pm. According to the weather station nearby, at the University of Freiburg, an irradiation intensity of 913 W/m² and a temperature of 19.8°C are present at this specific time.

Figure 10: Picture and sample IR image of a PV module of the demonstrator truck, which is exposed to solar irradiation on 8th July 2022 at 1:30 pm in Freiburg, Germany.

4.2 Energy measurements of the demonstrator in operation

First measurement data revealed that the behavior of the user, e.g. the charging-profile, needs to fit the assumptions of the energy prognosis. As can be inferred from the measurements in Figure 13 of the exemplified case from 28th of May, the electric truck is plugged to the charging station directly after its delivery service around 2:30 pm. It must be noted that, when the vehicles battery is above 60% of SoC, the control unit deregulates the energy transfer provided by the PV system, this SoC percentage has been arbitrary chosen by the vehicle manufacturer and will be adjusted at a later stage. Under these circumstances, the total harvested energy sums up to 10.23 kWh, whereas in the energy prognosis, the assumed user profile considers that the vehicle is only charged after sunset, which optimizes the transfer of available PV energy.

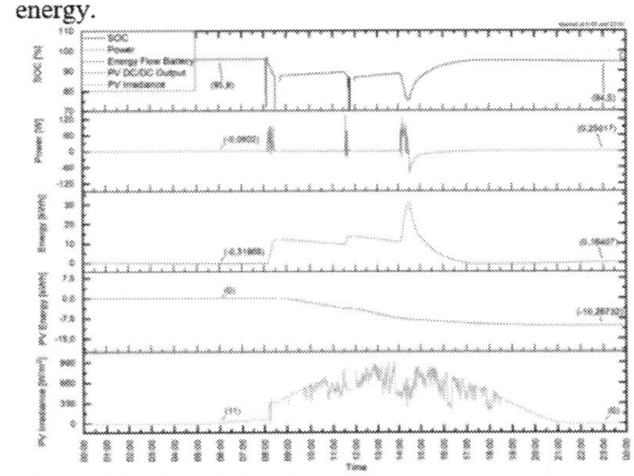

Figure 11: Plotted data for SOC, power-demand, energy flow from the battery, PV energy output at DC/DC for the sample case from 28th of May 2022

Figure 14 shows the measurements on the 19th of July, in which the usage profile aligns closely with the assumed profile in the simulation. The harvested PV energy on this day totaled 15.2 kWh. Throughout this specific day the vehicle was driven twice and consumed a total amount of 62.5 kWh for the trips. Accounting for the secondary consumers and the continuous energy consumption of the vehicle while in PV charge mode (e.g. HVIL safety system, monitoring), a total charge boost from SOC of 66.5% to 73.9% is achieved between 9:40 am and 7:00 pm. This results in a positive balance of 10.9 kWh that were charged into the battery. In this case, the solar coverage factor equals to 17.45% for this day. Accounting for the total energy transferred to the drive train versus the consumed energy for driving, a potential solar coverage factor of 24.3% was possible.

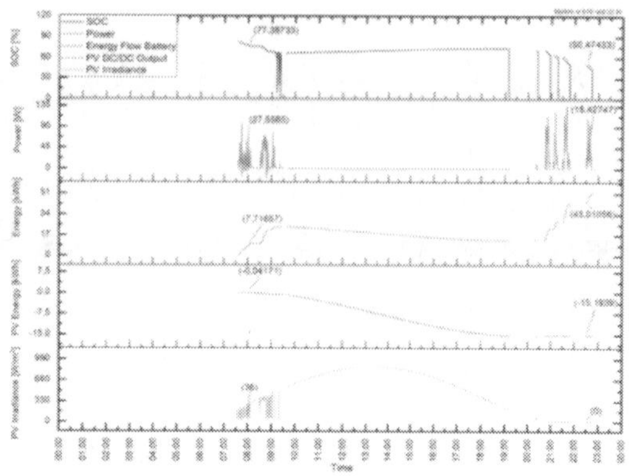

Figure 12: Plotted data for SOC, power-demand, energy flow from the battery, PV energy output at DC/DC for the sample case from 19th of July, 2022 with clear sky

Considering the area-under-curve (AUC) of the irradiation data, a total solar energy of 7.8 kWh/m² was available on the 19th of July. Under these circumstances, an active PV area of 13.34 m² would perceive approximately 104 kWh throughout the day. Considering factors for cell efficiency at 20%; a cell-to-module CTM loss of 98%; the DC/DC overall performance ratio of 98.5%; a MPPT tracking efficiency of 98%; and a temperature coefficient of -0.35%/°C with a cell temperature of around 65°C [4], a total energy harvest of 16.84 kWh is theoretically possible which is close to the simulated average for July of 18.17 kWh. The difference of 10% in the measured yield suggests that partial shading, dirt or other factors e.g. none optimal component behavior might have affected the actual power yield. Nonetheless, the chain efficiency from the PV array to the DC/DC output can be calculated around 73% for this scenario.

4.3 Validation of the energy prognosis model in stationary conditions (reference scenario)

To evaluate the installed system in parallel with the simulated energy model, a reference scenario in stationary conditions was set for the measurement of the energy yield. A parking lot was found, which excludes the effect of shading produced by surroundings (such as buildings or trees). With a relatively low SoC of around 28.9% as starting point, the vehicle was positioned to guarantee enough headroom for charging the battery with solar energy for a few days. During the observed period, beginning on Friday 01st of July around 11:20 am and ending on Friday 8th of July 1:00 pm, the PV system was able to harvest and deliver 150.68 kWh into the drivetrain of the vehicle, which translates to an average of 21.53 kWh/day. The measured amount of energy fed into the powertrain is in line with the predictions of the energy model for the month of July, which gives an average of 18.2 kWh/day transferred to the battery.

Within this period the SoC raised to 88.0 % after the seven days, which represents an increase in charge of 13.86 kWh/day. The charge that went into the battery as compared to the energy transferred into the drivetrain differs mainly because of the standby power consumption in the demonstrator, which is measured between 400 - 500 W while parking, which is also in line with other investigations [17,18]. In the energy prognosis, a mean standby consumption of 54 W is assumed for July. The actual power demand shows that this is an underestimation. As can be seen in Figure 15, the PV system was also able to fully charge within the following two days. Note that, within the reference week, the datalogger stopped operation from the 2nd of July, around 10:00 am, and began to operate again on the 6th of July around 16:25 pm. Nevertheless, the PV system operated as expected and the information of the supplied energy in between is taken from the SOC change, as well as the internal power meter of the DC/DC converter.

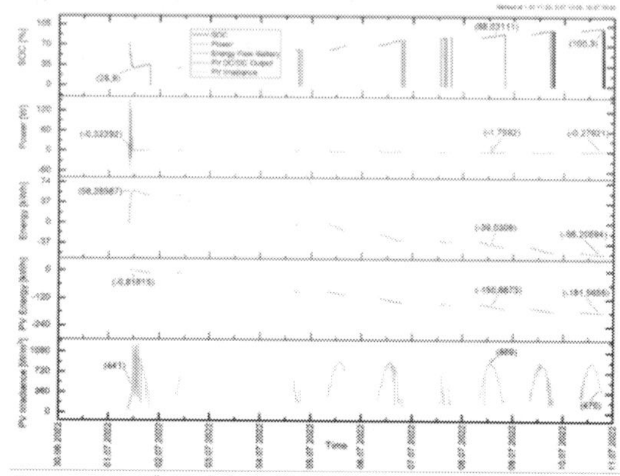

Figure 13: Plotted data for SOC, power-demand, energy flow from the battery, PV energy output at DC/DC for the static reference period in between 1st and 11th of July 2022

5 SUMMARY AND DISCUSSION

We presented a successfully and complete integration of the PV system into the vehicle's safety routine within our demonstrator. This approach of feeding into the HV system enhances the chain efficiency from PV module to the vehicle's battery, as well as shrinking the demand for various electrical converters. Another advantage is the high battery capacity available in electrified trucks, which avoids the expansion of the LV battery and therefore minimizes the additional weight and resources.

The reference measurement scenario exceeds and thereby validates the assumption on solar yield in the prognosed model by IVIsion. indicating the model holds for stationary and unshaded conditions.

Whereas, on the provided data of the 19th July, the approximated energy harvest overestimates the measured yield by 19.5 %. This is partly explained by the assumption made in simulation, which are based on average loss factors determined in laboratory environment compared to a specific day. Additionally, shading may also have been present during this day on the vehicle and even further, module losses such as the turned off module (see Fig. 11) and further losses due to degradation may also play a role. However, this last effect may be only very little since the stationary conditions over a week showed an overperformance of the harvested energy, compared to the average in simulation.

On the contrary, the relatively high standby losses result from the vehicle's concept, which is based on the conversion from a diesel to an electric vehicle. The auxiliary aggregates are intersected and are completely booted up by the control unit in the charging state. This indicates, there is still the need to optimize the system functions regarding PV charging, to achieve the prognosed solar coverage also in the battery during parking throughout the whole year.

Even though, the demonstrator vehicle was not

optimized in energy consumption and the prototype components can be optimized further, the implementation of direct feed into the high voltage intermediate circuit shows a promising chain efficiency of 73% for an exemplary day in summer. This result outperforms other approaches as for example to contribute the power to a low voltage buffer first [18].

Apart from the requirements on component level, we show that the evaluation of systems performance towards safety aspects should not be underrated in the design phase of VIPV systems. A careful adjustment of e.g. limits for isolation faults is necessary for a smooth but also safe operation. Another major learning and field for further studies is that the focus on system level is necessary, e.g. to optimize the status of continuous power consumption during standby. We showed in simulations and supported by experimental results that an average energy saving of 13% can be achieved if lower standby losses are in place.

6 ACKOWLEDGEMENT

We thank Torsten Peppel, Ralf Nöske and Steffen Margenberg, from M&P Motion Control and Power Electronics GmbH, Dr. Aswin Linsenmeyer from SUNSET Energietechnik GmbH, Stephan Möllinger from Alexander Bürkle GmbH and Rolf Huhn from TBV Kühlfahrzeuge GmbH for exceptional good collaboration within the project "Lade-PV". Finally, we would like to thank the German Ministry of Economic Affairs and Climate Action (FKZ 03EE1002A "Lade-PV") for their funding.

References

[1] A.J. Carr, E. van den Tillaart, A.R. Burgers, T. Köhler, B.K. Newman, Vehicle Integrated Photovoltaics - Evaluation of the Energy Yield Potential through Monitoring and Modelling, 5 pages / 37th European Photovoltaic Solar Energy Conference and Exhibition; 1701-1705 (2020).

[2] S. Neven-du Mont, M. Heinrich, A. Pfreundt, C. Kutter, A. Tummalieh, D.-H. Neuhaus, Energy Yield Modelling of 2D and 3D Curved Photovoltaic Modules, 6 pages / 37th European Photovoltaic Solar Energy Conference and Exhibition; 822-827 (2020).

[3] M. Heinrich, C. Kutter, F. Basler, M. Mittag, D. Eberlein, A. Schmid et al., Potential and challenges of vehicle integrated photovoltaics for passenger cars, in: EUPVSEC.

[4] C. Kutter, F. Basler, D.H. Neuhaus, M. Heinrich, J. Markert, L.E. Alanis, Integrated Lightweight, Glass-Free PV Module Technology for Box Bodies of Commercial Trucks, 37th EUPVSEC European PV Solar Energy Conference and Exhibition, Lisbon, Portugal, (2020) 1711–1718.

[5] THE EUROPEAN ECONOMIC AND SOCIAL COMMITTEE AND THE COMMITTEE OF THE REGIONS, EU Solar Energy Strategy, EU Solar Energy Strategy SWD(2022) 148 final 2022, https://eur-lex.europa.eu/legal-content/EN/TXT/?uri=COM%3A2022%3A221%3AFIN.

[6] U. Eitner, M. Ebert, T. Zech, C. Schmid, C.V. Schmid, A. Watts et al., Solar Potential on Commerical Trucks: Results of an Irradiance Measurement Campaign on 6 Trucks in Europe and USA, in: Proceedings of the 33rd European Photovoltaic Solar Energy Conference and Exhibition (EU PVSEC), Amsterdam, Netherlands, 2017, pp. 2147–2150.

[7] M. Kühnel, B. Hanke, S. Geißendörfer, K. von Maydell, C. Agert, Energy Forecast for Mobile Photovoltaic Systems with Focus on Trucks for Cooling Applications, 4 pages / 32nd European Photovoltaic Solar Energy Conference and Exhibition; 2539-2542 / 32nd European Photovoltaic Solar Energy Conference and Exhibition; 2539-2542 (2016).

[8] B. Elliston, M. Dennis, Feasibility of Solar-Assisted Refrigerator Transport in Australia, Canberra, 2009.

[9] IM Efficiency, Empower your truck with solar energy: SolarOnTop, https://imefficiency.com/solarontop.

[10] Sono Motors und CHEREAU, Sono Motors und CHEREAU unterzeichnen Vertrag und steigen in den Markt für solarbetriebene Kühlauflieger ein, https://sonomotors.com/de/press/press-releases/sono-motors-and-chereau-sign-contract/, [August 17, 2022].

[11] M. Yamaguchi, T. Masuda, K. Araki, D. Sato, K.-H. Lee, N. Kojima et al., Development of high-efficiency and low-cost solar cells for PV-powered vehicles application, Prog. Photovolt: Res. Appl. (2020).

[12] United Nations Economic Commission for Europe (UNECE), Regulation No 100 of the Economic Commission for Europe of the United Nations (UNECE) — Uniform provisions concerning the approval of vehicles with regard to specific requirements for the electric power train [2015/505]: UNECE R100, 2015.

[13] Christoph Kutter, Lars Saroch, Jörg Bornwasser, Luis Eduardo Alanis, Martin Heinrich, High-voltage Vehicle Integrated Photovoltaic Demonstrator Truck – System concept and energy flow modeling, 1st PV in Motion, 6th - 8th December 2021 (2021).

[14] C. Kutter, L.E. Alanis, D.H. Neuhaus, M. Heinrich, Yield Potential of Vehicle Integrated Photovoltaics on Commercial Trucks and Vans, 9 pages / 38th European Photovoltaic Solar Energy Conference and Exhibition; 1412-1420 (2021).

[15] A.K. Christian Schill, SOLARPOTENTIAL BEI AUTOFAHRTEN ERFORSCHEN: PV2Go: Ein Citizen Science Projekt, https://pv2go.org/.

[16] G. Wang, Y. Jiang, Y. Wei, H. Gong, J. Zhu, Study on Insulation Resistance Test of PV Module, 4 pages / 28th European Photovoltaic Solar Energy Conference and Exhibition; 3296-3299 (2013).

[17] ADAC e.V., Elektroauto und Ladeverluste: So können Sie Kosten vermeiden, [September 19, 2022], https://www.adac.de/rund-ums-fahrzeug/elektromobilitaet/laden/ladeverluste-elektroauto-studie/.

[18] R. Peibst, H. Fischer, M. Brunner, A. Schießl, S. Wöhe, R. Wecker et al., Demonstration of Feeding Vehicle-Integrated Photovoltaic-Converted Energy into the High-Voltage On-Board Network of Practical Light Commercial Vehicles for Range Extension, Sol. RRL 6 (2022) 2100516.

PV ROOFING FOR HIGH RANKING ROAD NETWORKS: TECHNICAL FEASIBILITY AND YIELD ESTIMATION

[1]Dr. Rita Ebner, [1]DI Christoph Mayr, [1]Dr. Marcus Rennhofer, [1]DI. Karl A. Berger, [2]Dr. Martin Heinrich, [2]Felix Basler, [2]Andreas J. Beinert, [2]Jonas D. Huyeng, [1]DI. Manfred Haider, [1]DI. Dr. Dominik Prammer, [1]DI. Dr. Alois Vorwagner, [3]Ing. Markus Fehringer, [3]Tobias Beck
[1]AIT Austrian Institute of Technology GmbH (AIT), Austria,
[2]Fraunhofer-Institut für Solare Energiesysteme ISE, Germany,
[3]Forster Industrietechnik GmbH, Austria

ABSTRACT: As renewable energies need to be extended massively, new concepts are necessary to prevent land conflicts with other uses. Such concepts should have a high generality to offer a swift expansion of renewables anywhere. Within the project *Photovoltaic Road Roofing Concept (PV-SÜD)*, a concept for the roofing of roadways with solar panels was investigated and its effects on the road infrastructure were analyzed. The technical boundary conditions and requirements resulting from this application area were determined with regard to the photovoltaic technology used, the possible energy generation and the required supporting structure. This was done in a generalized and multipliable approach differing from previous studies, which mainly aimed at a singular site or demonstrator project. The requirements were maximum flexibility in the road integration potential and that interferes as little as possible with traffic operation and which further guarantees an electric yield with minimized variation from site or orientation. The solution was selected in the form of a gable roof elevated in steel for road sections with a predominantly north-south course and in the form of an elevated pent roof for road sections with a predominantly east-west course in accordance with the optimum PV yield utilization. This roofing is to be realized as a steel frame construction with active PV elements in glass-glass technology.

KEYWORDS: Building Integrated PV (BIPV), Roofing Systems, PV System

1 INTRODUCTION

Energy generation by photovoltaics (PV) is one of the key technologies for increasing the share of renewable energy sources in primary energy production, which, however, requires the availability of corresponding areas. The high-ranking road network and the associated areas in Germany, Austria and Switzerland currently represent a largely untapped potential in this respect [1]. The technical feasibility of roofing sealed street areas has been demonstrated in principle. However, the economic viability and yield potential is so far often only given in special cases, such as the energy supply of rest areas and tunnel lighting. This study uses a more general approach and identifies a modular concept for the flexible use in the high-level road network, its energy yield potential, and further beneficial effects. The following goals were used as a guideline for the technical developments and analysis:

- Energy generation through photovoltaics with the help of suitable PV module technology
- Utilization of existing sealed street areas
- Flexible use in the high-level road network
- Flexible integration with minimum yield variety for different sites and geographical orientation
- Increasing the durability of the roadway by protecting against overheating, precipitation
- Positive effects on road surface properties

2 PV-SÜD CONCEPT

In the PV-SÜD project, the concept for such a photovoltaic road roofing concept was developed, taking into account the diverse requirements of the high-ranking road application area, both for the solar modules and for the supporting structure.

Supporting structure

The main goals were to develop a concept that could be easily multiplied (to allow for mass manufacturing of components), was using established concepts from other road constructions (to operate on a high technology readiness level) and provide a simple scheme for planers. Essential basic decisions in the design of the supporting concept were made with regard to structural safety and were chosen in agreement with needs of the end-user stakeholders in the D-A-CH region:

a) The canopy structure must be easily adaptable to the existing road layout.
b) The design of the basic geometry shall be optimized for power generation, aesthetics and road operation requirements.
c) The design in terms of geometry, length and cross-section shall be chosen in such a way that no rules and regulations for tunnel structures have to be mandatorily applied
d) The supporting structure shall be designed in such a way that no system failure can occur in the exceptional load case of fire or impact.

PV-system

The main fundamental decisions in the design for the PV-plant were examined with regard to safety, yield optimization and later scalability of the solution. The plant should meet the following requirements:

a) The PV solution should be scalable on highways with two and three lanes with and without shoulder and be duplicable, i.e. not require a site-specific special solution.
b) Regular commercially designed products should be preferred over specially designed solutions
c) The energy generation should be optimized and maximized with regard to the structural conditions; i.e., the entire covered area should be used by PV if possible, excluding openings for lighting or ventilation where necessary.
d) Energy generation should be as independent as possible from the direction of travel of the roadway (N-S, E-W, etc.) due to the type of construction and structural solution.
e) Facilities should be constructed to eliminate or minimize self-shading.

Further constraints related to PV-design as well as construction design were fire protection on high-ranked

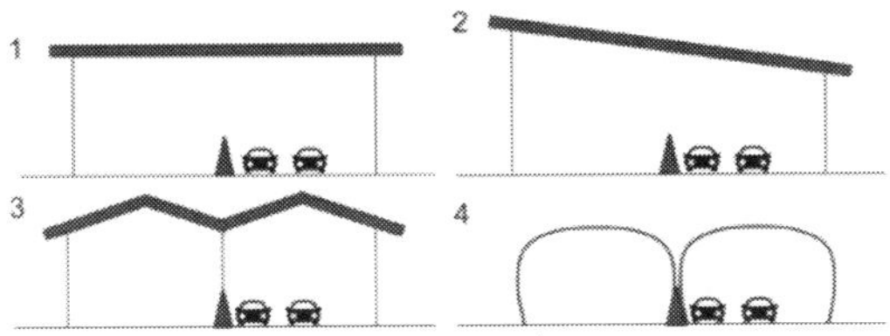

Fig. 1: The four variants considered in the discussion on the design of the roofing in the transverse direction: a completely flat roof (1), a monopitch roof inclined at an indeterminate angle (2), a gable roof inclined at an indeterminate angle (3) and a roofing with an arched cross-section (4).

roads, rain water drainage, ice formation protection, wind- and snow-load, load from traffic noise, low impact on maintenance work, safety and security in agreement with road network regulations (e.g. fire or accident impacting the construction).

Evaluation of different geometric shapes

In the design process, possible shapes were evaluated against common and often successfully implemented technical designs. Particular attention was paid to which designs satisfy as many of the set technical requirements as possible. The following Figure 1 shows four variants for possible designs. All of them were evaluated and also assessed in terms of their flexibility for different lane widths and PV module mounting types.

Depending on the east-west or north-south course of the traffic axis, a monopitch or gable roof is more suitable due to the energy yield. Flat roofs have a disadvantage in drainage, however, an energy-efficient elevation of photovoltaic modules could always be realized on them. Constructions with an arch shape are not suitable due to the energy yield and also the manufacturing costs.

The evaluation of different support structure forms in the longitudinal direction showed that a design with four supports and a cantilevered roof is the most suitable (variants see figure 2). It can be assumed that a structure with four columns will not collapse immediately in case of failure of one column. A certain distance between the supports has a disadvantage in terms of the number of foundations, but on the one hand there is the advantage of lower material consumption of the steel structure due to the reduced spans.

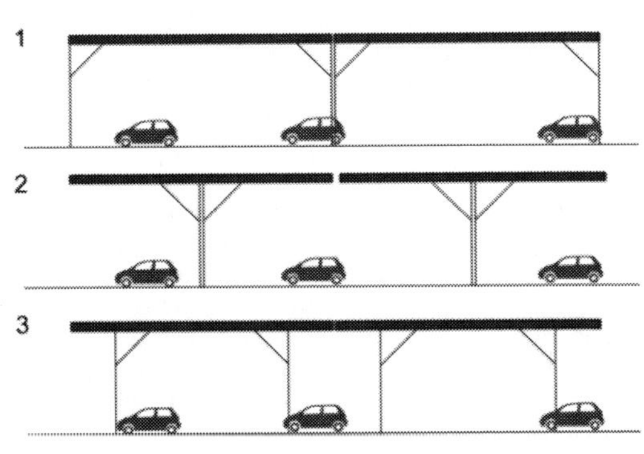

Figure 2: Three variants of roofing in the longitudinal direction : supports in the corners (1), supports in the center (2), cantilever with supports in the perimeter (3).

Solution for PV-SÜD

The result is a concept that provides a 10 m long base element consisting of a supporting structure made of steel, which are arranged either in gable roof or pent roof form, depending on the street orientation, which is used to mount overhead-certified silicon PV modules, with glass/glass configuration. A gable roof for road sections with a predominantly north-south course and a pent roof for road sections with a predominantly east-west course (see figures 3 and 4), were chosen. They offer a very flexible solution and do not negatively affect the traffic as well as a stability of the influence of orientation on energy yield.

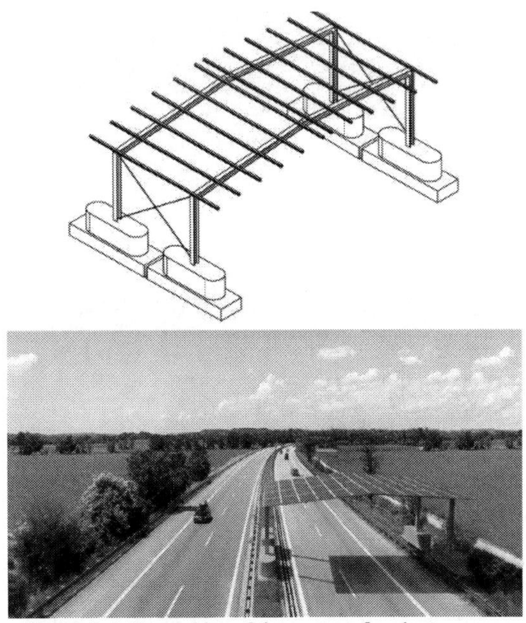

Figure 3: 3-D-graphics of the pent roof variant.

Most requirements can be met with (certified) components suitable for overhead installations, currently in a growing supply for e.g. BIPV applications. The performance of different PV module setups was investigated, by means of FEM simulation, in line with previous studies [2]. This concept was analyzed in detail with regard to the achievable solar yields as well as the expected side effects on the roadway and propagation.

3 RESULTS

3.1 YIELD SENSITIVITY OF PV-SÜD

Locations for a potential positioning of the PV-SÜD demonstrator on highways in Austria and Germany were analyzed. For this purpose, energy yield simulations were carried out for the different locations. If the road runs north-south, a gable roof with east/west (E/W) orientation of the roof surfaces and thus of the PV module arrays was selected. If the street runs from west to east, a monopitch roof was selected with the roof surfaces and thus the PV module fields oriented to the south (S). Table 1 gives the planning parameters for the simulation in PV-Sol.

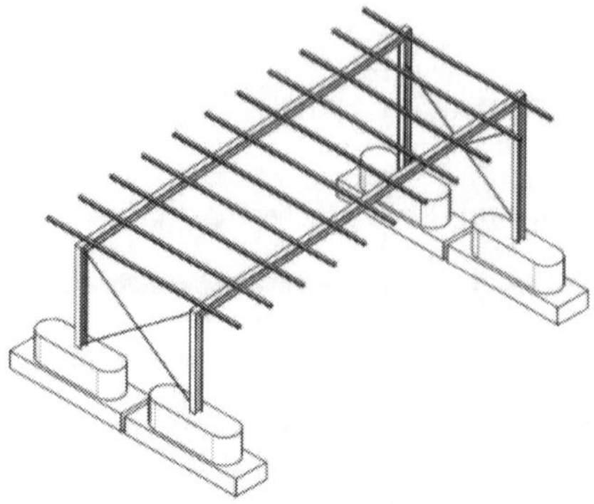

Figure 4: 3-D-graphics of the gable roof variant.

Component	Techn. data	unit
Structure	ca. 10 x 17	m, L x W
PV generator	11,88 x 16,8	m, L x W
PV area	199,6	m²
Nr. modules	120	#
Power p. module	315	Wp
PV nominal power	37,8	kWp
Inclination	5	°
Inverter, 3-phase	2 x 17,5	kW AC

Table1: Planning parameters for PV-Sol

The sites for the sensitivity study were the following in regions of Austria and Germany, distributed in various regions in the south as well as in the north, respectively.

a) ASFINAG Rastplatz Leobersdorf A2
coordinates: 47.910139, 16.208328

b) A1 Raststation St. Pölten SÜD
coordinates: 48.177632, 15.555647

c) A2 Klagenfurt am Wörthersee
coordinates: 46.648452, 14.253953

d) PWC-Anlage „Brandbühl" an der B 33 (Radolfzell am Bodensee, LK Konstanz)
coordinates: 47.770195, 8.965138

e) P+M Anlage Engen an der A 81 (Engen, LK Konstanz),
coordinates: 47.856474, 8.794141

f) T+R Anlage Hegau an der A 81 (Engen, LK Konstanz),
coordinates: 47.862292, 8.785754

g) Hamburg Seevetal, Rastplatz Sunderblick A1
coordinates: 53.377801, 9.951588

Table 2 gives the results for the yield calculation for all sites. Note, that they are scattered in two countries from the sea in the north to the south of the Alps. They further represent the two solutions in different orientations.

Table 2: Site and orientation sensitivity of yield.

Capcacity each 37,5 kW	Spec. yield [kWh/kWp]	AC-yield [kWh/Jahr]
Leobersdorf (E/W)	1 016	38 416
St. Pölten (E/W)	992	37 509
St. Pölten (S)	1 034	39 104
Klagenfurt (E/W)	1 132	42 815
Klagenfurt (S)	1 166	44 080
Hamburg (E/W)	836	31 612
Hamburg (S)	872	32 988
Brandbühl (E/W)	1 021	38 614
Brandbühl (S)	1 062	40 177
Engen (E/W)	1 014	38 348
Engen (S)	1 049	39 655
Hegau (E/W)	1 013	38 325
Hegau (S)	1 041	39 379

It can be seen from the table that two main influencing factors are geographical altitude and also orientation (design).

From latitude the results vary for about 25% for (S)-orientation from the maximum in Klagenfurt with 1166 kWh/kWp and year to the minimum in Hamburg with 872 kWh/kWp and year. The comparison of (E/W)-orientation gives almost the same result with a variation of 26% from Klagenfurt with 1132 kWh/kWp and year to Hamburg with 836 kWh/kWp and year. The (S)-orientation compared to (E/W)-orientation gives a variation per site of between 2.8% more yield for (S) than (E/W) in Hegau to 4.4% in Hamburg. Nevertheless, the variation in latitude has to be seen in a regional context also. Comparing sites only in Austria the total variation between all sites and orientations is in maximum only 15% (Klagenfurt (S) vs. St. Pölten (E/W)). In Germany the same calculation gives a maximum total variation of about 5% (Brandbühl (S) vs. Engen (E/W)) excluding Hamburg. The result for Hamburg would give 18% less and 22% less than Brandbühl (S) for the (S)-orientation and the (E/W)-orientation, respectively.

The yield values for one unit of PV-Süd (10 m roofing of road network) lie between 37,5 MWh / year (St. Pölten (E/W) and 44,0 MWh / year (Klagenfurt (S)) – except 31,6 MWh / year and 33,0 MWh / year for Hamburg (E/W) and Hamburg (S), respectively. The regional annual energy yield planning values for Austria would be between 37,5 MWh and 44,0 MWh with a mean value of 40,4 MWh. The regional annual energy yield planning values for Germany would be between 38,3 MWh and 40,2 MWh with a mean value of 39,1 MWh which is only 1,3 MWh or 3,2% less than the Austrian site evaluation. The total mean annual energy yield planning value of all sites (except Hamburg) is 39,7 MWh / year and 10 m unit.

Concluding the results show low variation in even wide regional variation of site and very low variation between different orientations. This proofs that the PV-Süd design is very robust against the impact of site and orientation for its final yield values. This finding makes the planning of energy output for several 10 m-units at different sites easier for the final utilization of the stakeholders hosting the road network.

3.2 RESULTS FOR DEMONSTRATION SITE

The site Hegau (A 81, Engen, LK Konstanz, Cooridnates: 47.82292, 8.785754) was chosen as the location for real demonstration because it has sufficient consumers and suitable access and exit routes.

The positioning of the system for the purposes of the calculation is shown in Figure 5. For both roof types (see Figures 6a and 6b) the energy yield calculation was performed over the course of the year. The monthly yield of the system in both configurations can be found in Figure 7. At the Hegau site, the PV system on the pent roof with 40° orientation to the west generates more energy in each month of the year than the variant with the gable roof (O/W), see Figure 7 and also Table 2 from above.

Different PV module technologies were compared at the Hegau site on a pent roof with south orientation. A 37,8 kWp system composed of monofacial modules was compared to a system of bifacial modules of the same nominal generator capacity. The yield predictions and system performance indicators for both variants are summarized in Table 1 and Figure 6. With the simulation, it could be shown that the bifacial PV modules only provide a small additional yield in this case. On the other hand, glass-glass modules provide a higher level of security in terms of overhead installations.

Figure 5: Position of the PV-SÜD demonstrator at the site Hegau

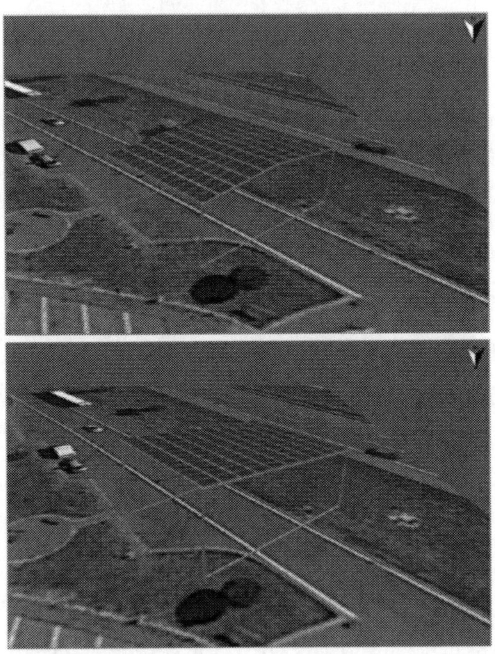
Figure 6: PV-SÜD demonstrator at the site Hegau: system variants: (a): Top: Roof type variant: pent roof, (b) - Roof type variant: gable roof.

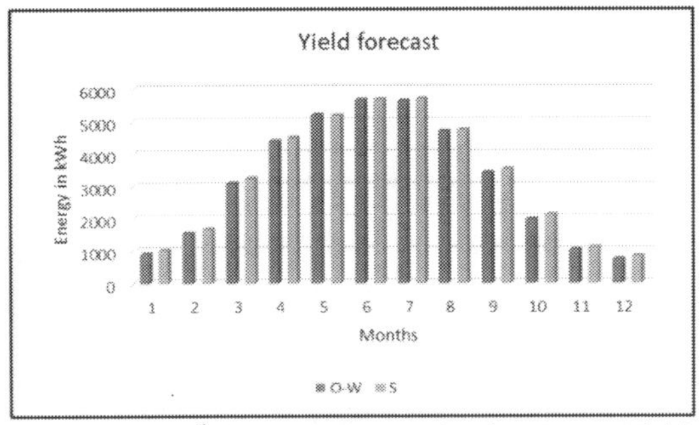

Figure 7: PV-SÜD demonstrator at the site Hegau: Yield forecast for O-W and S orientation

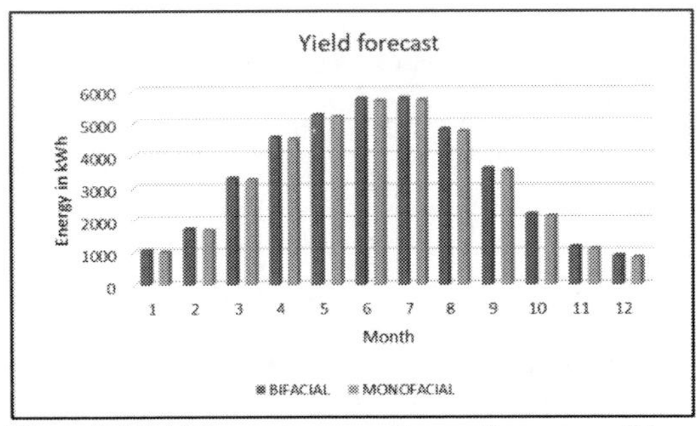

Figure 8: Yield forecast per month over the course of the year in comparison between monofacial and bifacial PV module

Table 3: Comparison of different module technologies

	Hegau (S) Monofacial	Hegau (S) Bifacial	Unit
PV-Generator Output	37.8	37.8	[kWp]
Generator Surface	199.6	201.8	[m²]
Number of modules	120	120	Quantity
Module	60 cells Module size: 1680 mm x 990 mm	60 cells Module size: 1682 mm x 1000 mm	
Spec. Annual Yield	1 041.17	1 058.89	[kWh/kWp]
Performance Ratio	88	89.8	[%]
Grid Feed -in	39 379	40 048	[kWh/Year]

3.3 ECONOMIC EFFICIENCY

The pure investment costs are only one part of the profitability calculation. The profitability also depends on how many years the system must be operated in order to compensate for the investment costs and ongoing operating costs through the electricity cost savings (or electricity feed-in).

By comparing the energy yield of the photovoltaic system and the type of use of the electricity (in this case, complete self-consumption), the profitability of the investment can be estimated.

The photovoltaic yield is one of the decisive factors for a possible return on investment in a PV roofing concept. The yield is offset by the acquisition costs, financing costs if applicable, and the costs for operation and maintenance of the system. The use or sale of the generated electricity is crucial. The profitability analysis was carried out for the Hegau (DEU) site using the simulation and calculation tool PV*SOL.

The sensitivities for different electricity purchase prices (0,12 €/kWh, 0.2 €/kWh and 0.3 €/kWh) and total investment of about 185.000 € (planning costs of 2021) are compared in Figure 9. Here, it becomes clear that the current electricity price is a significant factor influencing the economic viability of a PV system for captive use. E.g., assuming an electricity purchase price of 30 cents/kWh, a payback can be achieved within 16 - 18 years. It was assumed that the energy yield of the PV system is used entirely to cover own demand at the site and that there is no excess feed-in to the grid. Accordingly, the value of each generated kWh of the system is measured with the electricity purchase price at the site.

Economic efficiency for electro mobility consumption

As second consideration, 100% electricity consumption for electro-mobility was done based on certain assumptions. This would mean that the yield is used for charging of electro vehicle at a gas station or rest area. In a simplified approach the range of basic economic efficiency was evaluated by comparing electro mobility to the standard fossil mobility in the following way: First, the costs per kWh for gasoline vehicles with gasoline prizes between 0,95 €/l and 2,0 €/l was calculated. Second, the consumption of kWh per 100 km driving distance for a fossil vehicle and an electric vehicle was compared to evaluate the efficiency gain per kWh input energy for an electric engine. Values for gasoline engines between 4 l/100km consumption to 8 l/100 km consumption and a mean value for electric vehicles of about 20 kWh/100 km [3] were used.

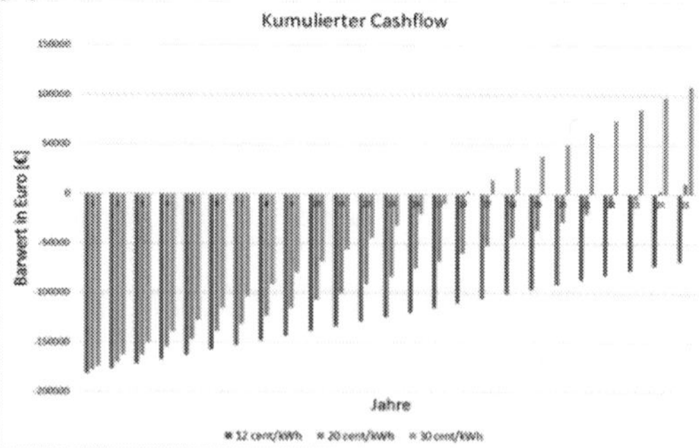

Figure 9: PV-SÜD demonstrator at the site Hegau: system profitability for one unit under assumption of different electricity consumption costs and 100% self-consumption.

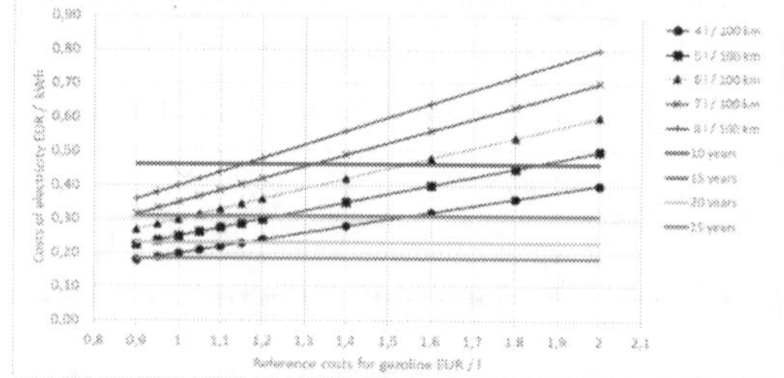

Figure 10: Cost limit of competitiveness for electro vehicles fed directly 100% by PV Süd electricity. Black lines give the electric reference costs to a gasoline engine of a certain consumption at a certain gasoline prize. Colored lines give the levelized costs of electricity for PV-Süd.

Third, the limit for costs of electricity produced from the PV-Süd unit was calculated in order to reach the same costs per kWh in electro-mobility as performed by the gasoline vehicle. The result in figure 10 shows costs for different consumption values of gasoline vehicles (l/100 km) and gasoline prizes per liter for fossil vehicles as black lines, respectively. The colored lines give the levelized costs of electricity of the PV-Süd unit, calculated for different values of lifetime (10 years to 25 years) and the planned installation costs of the demonstration site (planning costs of 2021). For all values above a colored line the gasoline driven vehicle is more expensive than the electric vehicle fed by 100% PV-Süd electricity. For values below a colored line the PV-Süd driven electric variant is not competitive to a gasoline driven vehicle.

The PV-Süd unit would be very competitive for electricity production for electro-mobility. If a lifetime of 25 years is assumed, only the 4l/100km gasoline vehicle with a prize of below 0,90 €/l is cheaper. Even with a very short lifetime of 10 years fossil vehicles with more than 7 l/100km and 8 l/100km are more expensive for gasoline prices above 1,35 €/l and 1,15 €/l, respectively.

Especially in areas of weak grid connection this might help to establish better and competitive charging capacities.

Consideration of the positive effects on the service life of the road surface:

For the Hegau site, there is a saving of 62 euros/year if it is assumed that the road surface is renewed every 23 years. A saving of 109 euros/year results if the pavement is renewed after 26 years.

Economic efficiency taking into account scaling effects:

The overall economic calculation, taking into account scaling effects, shows a low impact due to the side effects on the infrastructure and depends mainly on the investment costs and the applied electricity prices, assuming that the generated electricity is used for the road operators' own consumption [4].

4 CONCLUSION

The investigations have shown that the technical feasibility of a PV roofing in the high-ranking road network is basically given with currently available technology and can make a relevant, short term contribution to sustainable energy production in the related energy-sector.

The boundary conditions and challenges of a location have to be taken into account. With regard to the solar energy production of the roofing and the design of the structure, it was possible to find a robust concept that is not too dependent on specific locations and that also promises quite considerable solar energy yields on already sealed surfaces, due to the mostly horizontal orientation of the PV modules. With regard to durability and the condition of the road surface, the comparative studies indicate additional positive effects of the roofing.

Finally, it was shown within certain assumptions that the PV-Süd system is not only technically feasible, but that seen from a commercial point of view it also can be operated competitive to other sources of energy. The utilization may be for supplying resting areas as well as feeding electro-mobility.

5 REFERENCES

[1] H. Wirth et al., "Potenziale der Integrierten Photovoltaik in Deutschland," in 36. PV-Symposium. (https://www.pv-symposium.de)

[2] A. J. Beinert, P. Romer, M. Heinrich, M. Mittag, J. Aktaa and D. H. Neuhaus, "The Effect of Cell and Module Dimensions on Thermomechanical Stress in PV Modules," in IEEE Journal of Photovoltaics, vol. 10, no. 1, pp. 70-77, Jan. 2020, doi: 10.1109/JPHOTOV.2019.2949875

[3] The values were based on data sheet values of Nissan Leaf, Tesla Modell S, Skoda ENYAQ iV 60, Audi e-tron S, Hyundai Kona, which were reduced to 2/3 of battery capacity and then the mean value was taken.

[4] Weninger-Vycudil, A., Simanek, P., Haberl, P., Rohringer, T. (2009) *Straßenforschungsheft Nr. 584 Handbuch Pavement Management in Österreich 2009*, Bundesministerium für Verkehr, Innovation und Technologie BMVIT. Wien

Acknowledgement:
The research projects PV-SÜD-K and PV-SÜD-D are financed within the D-A-CH cooperation Transport Infrastructure Research 2019 by the Federal Ministry for Digital and Transport (Germany), the Federal Ministry for Climate Action, Environment, Energy, Mobility, Innovation and Technology (BMK) (Austria) and the Federal Roads Office ASTRA (Switzerland).

This is a short version of the paper submitted to the journal EPJ Photovoltaics.

It has been selected by the Committee of the WCPEC-8 for submission to EPJ Photovoltaics.

MULTI-OBJECTIVE OPTIMISATION OF AGRIVOLTAICS SYSTEMS

Bas B. Van Aken[1] , Sebastiaan Smeenk[2], Bernardo Maestrini[3], Eduardo Barros[2], Frank De Ruijter[3] and Ilkay Cesar[1]
[1]TNO Energy Transition, Solar Energy, PO box 15, 1755 ZG, Petten, the Netherlands
[2]TNO Energy Transition, Applied Geosciences, PO box 80015, 3508 TA, Utrecht, the Netherlands
[3]Wageningen University & Research, Agrosystems Research, PO box 16, Wageningen, the Netherlands

ABSTRACT: Agrivoltaic systems combine agricultural production with solar electricity generation to make better use of the available land. We calculated the time-dependent ground irradiance of an agrivoltaic system with BIGEYE as input for calculations of the agricultural yield for potato with the crop growth model SIMPLE. Land use efficiency, (bifacial) energy yield and crop yield are optimised using the multi-objective optimisation tool Everest. The SIMPLE crop growth simulation shows that in the Netherlands a reduction of irradiance results in a decrease in crop yield. Different locations, within the same agrivoltaic system, with the same annual irradiance can have a 10% difference in crop yield, depending on the distribution of light over the season, the pattern of crop soil coverage and light interception and occurrence of water stress. We will show that for the Netherlands, an agrivoltaic system can yield at the same time 70% of the single-use potato yield and 70% of the single-use kWh yield. For Southern Germany, with more rain in the crop season, these percentages are lower.
Keywords: agrivoltaics, integrated PV, crop modelling, optimization, energy yield

1 INTRODUCTION

The allocation of scarce land to agriculture, housing and renewable energy generation is a very topical and political issue and leads to demands for dual land use, like agrivoltaic systems [1]–[3]. At locations with high irradiance, crops can suffer from water and heat stress. In these cases, crops can be protected by providing shade. Going for dual-land use and harvesting both solar electricity and crops is then an obvious choice. However, in case the shading leads to a reduction in agricultural yield, the agrivoltaic systems have to be optimised taking into account different objectives to make the best use of the available land.

2 METHODS

We used BIGEYE [4] to calculate the time-dependent ground irradiance below and between the solar panels. These ground irradiance time-series were used as input for the crop growth model SIMPLE [5] to calculate potato yield. Land use efficiency, (bifacial) energy yield and crop yield are optimised using the multi-objective optimisation tool Everest [6], [7]. An APV system simulation with elevated panels on an overhead construction was created in TNO's PV simulation package BIGEYE.

2.1 BIGEYE – ground irradiance

BIGEYE is a state-of-the-art tool developed by TNO to simulate the performance of PV systems, with a focus on bifacial systems [8], [9]. BIGEYE implements a 3D view factor model to accurately handle the ground and other diffuse reflectors, for instance, the irradiance from the ground to the rear side of bifacial modules. These diffuse reflections contribute most of the light reaching the rear side of the modules, whereas they make up only 1–2% or even less of the total irradiance on the front. Therefore, in bifacial systems, elements of the support structure are more likely to cast shades on a light receiving surface of the module than in conventional systems. BIGEYE has flexibility in simulating such shades, and their impact on the IV curves (mismatch) of modules and strings. Amongst others, it takes into account the layout of the modules in blocks of cells protected by by-pass diodes [9].

BIGEYE deploys the Erbs model [10] to estimate diffuse horizontal irradiance (DHI) from global horizontal irradiance (GHI) when DHI is not available. BIGEYE uses the Perez model [11] to break down DHI into sky dome and circumsolar components, and to assess horizon brightening and darkening.

2.2 SIMPLE – crop growth

The crop growth model SIMPLE [5] is a generic crop growth model based on the radiation use efficiency and the harvest index principle using daily time steps. SIMPLE can simulate the growth of different crops – at the moment 14 crops have been parameterised – taking as input weather (global solar radiation, rainfall, temperature), management (planting date, cultivar, irrigation, harvest date) and soil physical properties. In the SIMPLE model, solar radiation influences crop growth in two ways. First, it enters in the equation to calculate daily produced biomass, which is simply the product of radiation, fraction of absorbed radiation, radiation use efficiency and abiotic factors (water, extreme temperatures, CO_2 concentration). Second, it influences potential evapotranspiration and thus the water stress reduction factor. In the model, water stress has a two-fold effect on crop growth: it reduces the radiation use efficiency that transforms absorbed solar radiation into new biomass, and it accelerates canopy senescence. The model can also be run without the activation of the water stress mode, thus assuming that the crop is receiving sufficient water.

For this study we simulated the growth of a rainfed potato crop in a sandy soil. Rainfall and maximum and minimum daily temperature were derived from Nasapower gridded historical weather data for the agrometeorology community from the year 2016 (an average year in terms of rainfall and radiation in the Netherlands). We used the cultivar parameters of the Russet Burbank variety and soil "Sand1", as parameterised by Zhao et al. [5]. We set the base temperature at 2°C, radiation use efficiency to 1.5, sowing date at 15 April and harvest date at 15 September, unless the model simulated that the crop reached maturity before.

2.3 Everest – multi-objective optimisation

Most real-world design and planning problems must also satisfy multiple objectives. Very often these objectives are in conflict with each other, which means that one objective must be decreased to increase the other objective. In such cases it is not trivial to find solutions which satisfy different objectives. In this study we used Everest, a multi-objective optimisation tool that uses the recently developed Stochastic Simplex Approximate Gradient (StoSAG) gradient-based optimization technique [6].

StoSAG is an approach which has shown to be powerful for optimization problems which need to consider uncertainty and need flexibility in terms of coupling with different simulators, in this case PV simulation with BIGEYE and crop yield simulation with SIMPLE. To approximate the gradient of the objective function with respect to the design variables subject to optimisation, an initial control point is chosen. Around this initial choice, the algorithm generates a set of perturbations by varying all the control variables simultaneously. The objective function values for each of these new points will be evaluated (in this work, by running the required energy and crop yield simulations). An approximate gradient is calculated as a linear regression through this set of perturbations and corresponding objective function values. The calculated approximate gradient provides the direction to update the control variables to improve the objective function.

One of the most popular approaches to resolve multi-objective optimization is the generation of a Pareto front for the different objectives. A Pareto front provides a decision maker with a range of solutions all of which are optimal depending on the relative importance between the different objectives. One of the most popular and often used multi-objective optimization techniques which can also be effectively used to generate a Pareto front is the weighted sum method.

Each combination of weights, results in an optimisation experiment which results in an optimal solution based on the chosen relative importance between the different objectives. Thus the user can generate a curve (range of solutions) which is representative of a Pareto curve thereby providing more robustness and choice to a decision maker.

The scientific focus of this study is not to present any newly developed optimization methods, but to showcase the application of the optimization framework based on the state-of-the-art StoSAG method for agrivoltaic system optimisation.

3 RESULTS

3.1 Ground irradiance and potato yield

We calculated the ground irradiance with BIGEYE for a model APV system for the same growing season as used in SIMPLE, 15 April till 15 September. The APV system consists of south-facing PV tables at 5 m above the ground. The tables, with two modules in portrait orientations, are spaced at a pitch of 19 m, yielding a coverage ratio of 21%, the position of the panels is indicated in Figure 1 by the black lines. The ground irradiance profile for south-facing panels, cumulative over the full growing season, shows narrow regions with lower irradiance, about 50-55% of the global horizontal irradiance in the open field. The remaining, wider regions have over 80% of the open field irradiance. On average, the ground irradiance over the full pitch is reduced by 13%. The spatial variation in ground irradiance leads to differences in crop growth. A homogeneous ground irradiance is more favourable for agriculture [12].

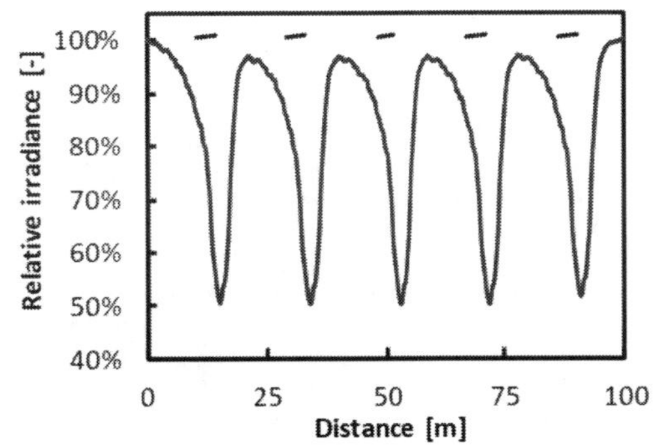

Figure 1: Light profiles for an elevated APV system as used in the calculations.

We coupled the ground irradiance output of BIGEYE to the SIMPLE crop growth model using daily irradiance values for the period 15 April till 15 September. Figure 2 shows the cumulative radiation and the predicted crop yield over a transect perpendicular to the panel rows, normalised to the irradiance and crop yield for an open field. At the top of the graph the location of the PV panels is indicated. Clearly, the curves for irradiance and crop yield follow similar trends, but the loss in irradiance just north of the panel rows is much higher than the maximum loss of crop yield of ~30%.

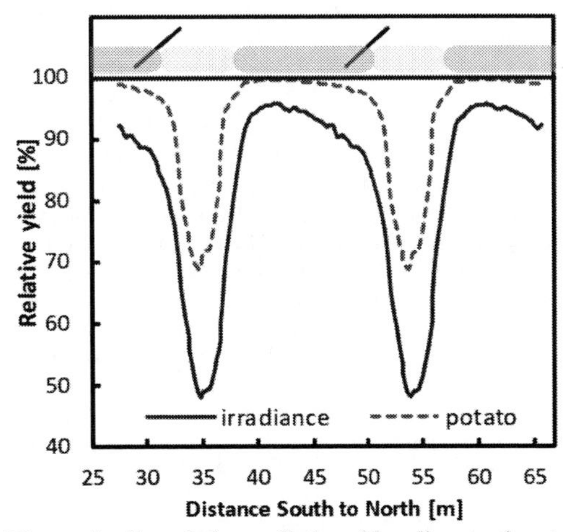

Figure 2: Cumulative radiation, blue line, and potato yield, dashed red line, as function of distance across the rows of PV panels. Both are normalised to their value for an open field situation. The sketch above the graph indicates the position of the PV tables and in which regions crop production is water limited (green) or light limited (yellow).

The SIMPLE model also returns the cumulative water stress per point on the transect. Combining the water stress data with the yield and irradiance profiles, we have identified that the parts of the field that receive relatively more light have also relatively higher water stress (indicated in green) and the parts receiving relatively less light have limited to no water stress (yellow). In the parts that are water limited a reduction in irradiance has a smaller impact because the crop cannot take full advantage of the irradiance. In the parts that are not subject to water stress, the reduction in irradiance available for photosynthesis is reflected in a directly proportional decrease of crop growth.

3.2 Irradiance timing effect

We show, in Figure 3, the direct relation between the irradiance and crop yield for a full pitch. The open symbol in the top right corner indicates the irradiance and crop yield for the open field. Again the (mostly) water limited (green) and mostly irradiance limited (yellow) regions are indicated in the graph.

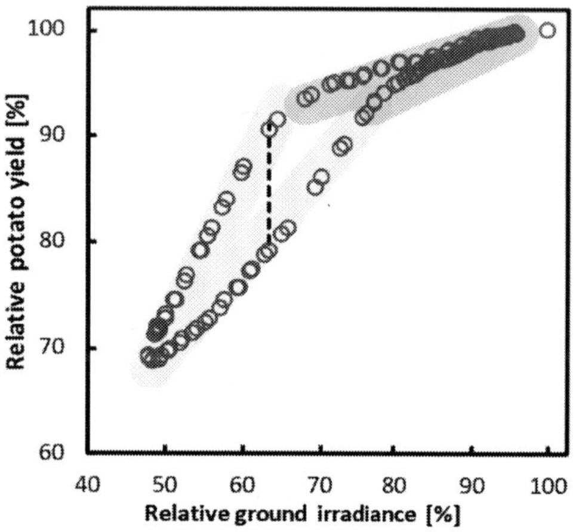

Figure 3: Relation between potato yield and cumulative radiation. The green and yellow bands indicate whether the crop yield is mostly limited by water (green) or light (yellow) availability.

The region that is mostly irradiance limited (yellow) shows two branches where equal amounts of cumulative irradiance over the crop season result in different crop yields. For instance, the two points with 63% relative ground irradiance, corresponding to 1550 MJ/ha/season, indicated with the dashed vertical line, correspond to two positions. One is located directly below the higher, north end of the PV table, x = 52 m. The crop simulation for this point gives the lower yield of 79% (~39 Mg/ha). The second point is located about 4 metres North of the first point, x = 56 m and has a crop yield of 90% (~44 Mg/ha). During the season, these points will receive different amounts of irradiance per day.

In Figure 4 we have plotted the progress of the cumulative irradiance for these two points from mid-April to mid-September to show the different distribution of irradiance per day over the growing season. Early in the season, the point x = 52 m, that is the point directly below the high end of the PV table, shows a faster increase of cumulative irradiance. The other point shows a faster increase in the months of June and the first half of July, around the summer solstice.

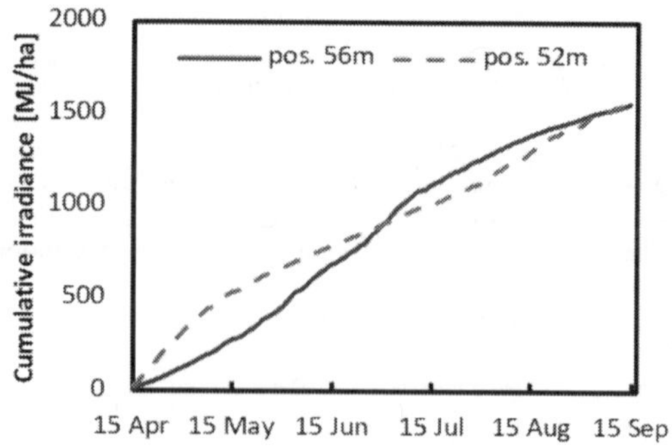

Figure 4: Progress of cumulative radiation for two positions with the same annual irradiance, as explained in the main text.

Around the summer solstice, the sun's elevation is highest and the shadow of the PV table will be projected closer to its position above the ground. Likewise, earlier and later in the season, the elevation is lower and the shadow will be further towards the North. This means that the point directly below the high end of the PV table will have more shadow around the summer solstice and the point 4 m further North will have more shadow earlier and later in the season. This is in good agreement with the calculated progress of the cumulative irradiances of these two points.

For the crop yield, one has to look at both the available irradiance and the ground coverage of the canopy. In crop growth models, like SIMPLE, that are based on radiation use efficiency, the newly produced biomass is directly proportional to the amount of intercepted radiation, thus daily crop growth is the product of irradiance by fraction of intercepted radiation, which depends largely on how much ground is covered by the canopy. Earlier in the season, there is little coverage by the canopy and a (large) part of the irradiance will reach the soil instead of being used for photosynthesis. In the middle of the season the canopy covers the ground fully. When the period of high irradiance coincides with that of high leaf coverage, this will be favourable for crop yield. The second point, x = 56 m, has the higher irradiance around the summer solstice and is projected to have a much higher yield of potatoes.

3.2 Optimisation of crop and energy yield

We are considering two objectives that are in competition for the available light: renewable electricity generation and photosynthesis for potato growth. We first establish the crop yield for a potato field without PV and the energy yield with lowest levelised cost of electricity, LCoE, for a solar farm, under conditions of sufficient ground irradiance to maintain soil quality [13]. For the other APV systems, the crop yield and the electricity yield will be normalised to these single-use solutions. Note that the LCoE-optimised PV system still yields 42% potato growth beneath the PV panels.

The APV system has been optimised for a range of weight factors changing the emphasis from potato growth to electricity generation. The tilt angle, height above the field and the pitch have been varied to optimise the weighted sum of objectives. Crop growth was calculated for the full area, and land loss due to the presence of the mounting structure was not yet taken into account in this optimisation study.

Figure 5 shows the pareto front for optimised APV system using meteorological data for Amsterdam, the Netherlands. The red, dotted line links the crop only yield with the point for the LCoE-optimised PV system, with a 42% potato yield. All solutions, resulting from the weighted sum optimisation by the Everest software package, are located above this line. This indicates that an optimised PV layout increases the land use efficiency compared to single-use optimised solar or potato farms. Land use efficiency values range from ~115% for an APV farm with ~90% potato yield and a small PV generation, to ~140% for an APV farm with 70% potato and 70% electricity yield.

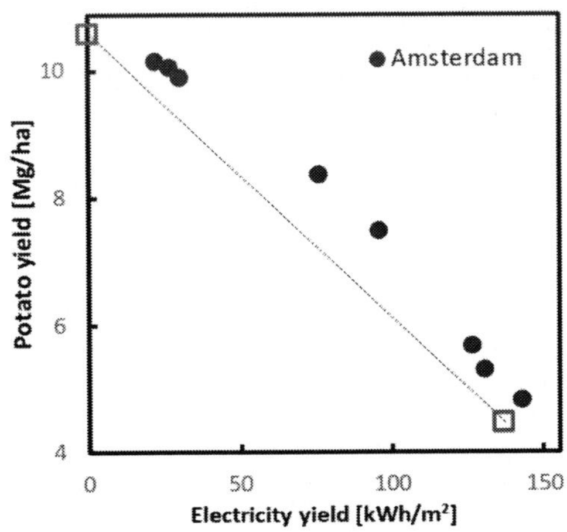

Figure 5: Pareto front for nine different combination of weighing factors for the APV system in the Netherlands. The open, red, square (top left) indicates the open field potato yield (dry matter); the other open symbol (bottom right) indicates the LCoE-optimised PV farm.

As an example, assume that there is a requirement for 50 ha of PV systems to be located on (former) potato fields. That could lead to a reduction in potato growing area of 50 ha. Applying the LCoE-optimised system with 137 kWh/m^2, instead of a regular land-based PV system, these 50 ha could still harvest the equivalent yield of 21 ha potato field, a reduction of 29 ha available arable land.

Applying one of the Everest-optimised APV systems, say the solution with 96 kWh/m^2 and 7.5 Mg/ha dry matter potato, one would need 70 ha of APV to achieve the kWh requirement. However, those 70 ha of optimised APV system also yield a potato crop equivalent to 49 ha, reducing the land available for crop growth only with 21 ha.

For the solutions with a very high crop yield and a kWh yield around 20% of the single-use system, the reduction in available land is even smaller, 13-15 ha, but one needs 200 to 300 ha for the APV systems to generate the required renewable electricity.

In Figure 6, the pareto fronts are plotted for the relative yields of electricity and potato harvest, comparing Amsterdam with Munich, Germany. Compared to the oceanic climate in Amsterdam, the conditions in Munich are on the border of humid continental with more sunshine and higher temperatures during the growing season but also more precipitation in the summer. This is reflected in the weather data used in the simulations. The absolute potato yield, under these model conditions, is 10% higher in Munich and the PV yield for the LCoE optimised system is 13% higher.

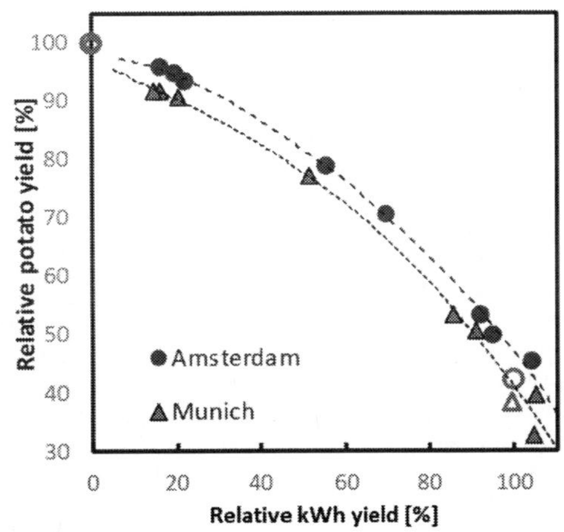

Figure 6: Pareto fronts for Amsterdam, blue circles, and Munich, green triangles. Open symbols indicate for both locations the LCoE-optimised PV yield and the potato only result. Dashed lines are guides to the eye.

Clearly, for both locations and the chosen crop there is no gain in the agricultural yield due to a reduction in light and reduction in heat and water stress by the PV tables. The absolute potato yields in Munich are higher compared to Amsterdam, but in Munich the relative yield reduction in the APV solutions is stronger than in Amsterdam. Comparing the two pareto fronts in Figure 6, the land use efficiency in Munich is about ~6% lower than in Amsterdam, for the simulated season.

The current comparison between Amsterdam and Munich is based on a single growing season, where radiation did not differ that much between the two locations, but rainfall did; also the same soil parameterisation was used for the crop simulation in both locations. In Munich the simulated crop yield is more sensitive to a reduction in irradiance caused by the introduction of PV above the field, as the crop yields are not limited by water stress. In contrast, in Amsterdam, the lower amount of rainfall in combination with the sandy soil leads to a situation where a reduction in irradiance by PV above the crops mostly alleviates the water stress without reducing the crop yield significantly.

This is confirmed by an analysis of more growing seasons. Rainfall during the growing season is always larger in Munich compared to Amsterdam, giving more water stress in Amsterdam and a relatively smaller impact of PV above the crops as the light interception by the PV panels alleviates the waters stress.

In the SIMPLE model, the crop yield is determined by both light and water availability. Water availability is the result of precipitation and irrigation, soil type (water holding capacity) and evapotranspiration (affected by irradiance, temperature, humidity and crop status). We note that the presently used SIMPLE model does not take into account effects of light saturation of the photosynthesis process or possible physiological adaptations to shade conditions such as changes in leaf thickness.

4 CONCLUSIONS

We simulated the effect of the shading of a PV system on the crop yield of a potato field. The crop growth simulation using the SIMPLE model, applying the ground irradiance time-series of BIGEYE as input, shows that in the Netherlands a reduction of ground irradiance results in a decrease in potato crop yield. For the locations with lowest shading the crop loss is small as water stress already limits the crop yield.

Detailed analysis of the time-dependent ground irradiance and crop yield, for locations with the same seasonal irradiance, shows that relatively more light around the solstice leads to a 10% higher crop yield, compared to having relatively more light in May and August. It is likely that a higher irradiance in the middle of the growing season coincides with the period that the fully developed leaf canopy will fully intercept the incoming radiation, whereas the higher irradiance earlier and later in the season coincide with less developed plants or plants that have already started senescence.

This APV system is optimised using the weighted sum method and optimisation tool Everest. For the given set of parameters, the APV system always has a smaller crop yield than a single-use agricultural field. But if we allow for a reduction in potato yield, we can get an energy yield gain that is much larger. The combined crop and energy production leads to a land use efficiency ratio of 115% to 140%.

ACKNOWLEDGEMENTS

This work was partially funded by the Dutch Ministry of Economic Affairs and Climate, in the MOOI project Sunbiose.

REFERENCES

[1] G. A. Barron-Gafford *et al.*, "Agrivoltaics provide mutual benefits across the food–energy–water nexus in drylands," *Nat. Sustain.*, vol. 2, no. 9, Art. no. 9, Sep. 2019, doi: 10.1038/s41893-019-0364-5.

[2] C. Dupraz, "Preface: AgriVoltaics2020 Conference Launching Agrivoltaics World-Wide," *AIP Conf. Proc.*, vol. 2361, no. 1, p. 010001, Jun. 2021, doi: 10.1063/12.0004866.

[3] M. Trommsdorff *et al.*, "Combining food and energy production: Design of an agrivoltaic system applied in arable and vegetable farming in Germany," *Renew. Sustain. Energy Rev.*, vol. 140, p. 110694, Apr. 2021, doi: 10.1016/j.rser.2020.110694.

[4] G. J. M. Janssen, B. B. Van Aken, A. J. Carr, and A. A. Mewe, "Outdoor performance of bifacial modules by measurements and modelling," *Energy Procedia*, vol. 77, pp. 364–373, 2015, doi: 10.1016/j.egypro.2015.07.051.

[5] C. Zhao *et al.*, "A SIMPLE crop model," *Eur. J. Agron.*, vol. 104, pp. 97–106, Mar. 2019, doi: 10.1016/j.eja.2019.01.009.

[6] R. M. Fonseca, B. Chen, J. D. Jansen, and A. Reynolds, "A Stochastic Simplex Approximate Gradient (StoSAG) for optimization under uncertainty," *Int. J. Numer. Methods Eng.*, vol. 109, no. 13, pp. 1756–1776, 2017, doi: 10.1002/nme.5342.

[7] E. G. D. Barros, B. B. Van Aken, A. R. Burgers, L. H. Slooff-Hoek, and R. M. Fonseca, "Multi-Objective optimization of solar park design under climatic uncertainty," *Sol. Energy*, vol. 231, pp. 958–969, Dec. 2021, doi: 10.1016/j.solener.2021.12.026.

[8] G. J. M. Janssen *et al.*, "How to Maximize the kWh/kWp Ratio: Simulations of Single-Axis Tracking in Bifacial Systems," in *Proceedings of the 35th European Photovoltaic Solar Energy Conference and Exhibition*, Nov. 2018, pp. 1573–1577. doi: 10.4229/35thEUPVSEC20182018-6BO.7.5.

[9] A. R. Burgers and B.B. Van Aken, "BIGEYE - simulation under shadow conditions," presented at the 6th workshop on bifacial PV, Amsterdam, Sep. 2019.

[10] D. G. Erbs, S. A. Klein, and J. A. Duffie, "Estimation of the diffuse radiation fraction for hourly, daily and monthly-average global radiation," *Sol. Energy*, vol. 28, no. 4, pp. 293–302, Jan. 1982, doi: 10.1016/0038-092X(82)90302-4.

[11] R. Perez, R. Seals, P. Ineichen, R. Stewart, and D. Menicucci, "A new simplified version of the perez diffuse irradiance model for tilted surfaces," *Sol. Energy*, vol. 39, no. 3, pp. 221–231, Jan. 1987, doi: 10.1016/S0038-092X(87)80031-2.

[12] M. Beck, G. Bopp, A. Goetzberger, T. Obergfell, C. Reise, and S. Schindele, "Combining PV and Food Crops to Agrophotovoltaic – Optimization of Orientation and Harvest," *27th Eur. Photovolt. Sol. Energy Conf. Exhib.*, pp. 4096–4100, Oct. 2012, doi: 10.4229/27thEUPVSEC2012-5AV.2.25.

[13] I. Cesar, B. B. Van Aken, L. Scholten, R. De Goede, and A. Schotman, "Design rules based on ground solar irradiation to maintain soil quality in solar parks," presented at the 8th World Conference on Photovoltaic Energy Conversion, Sep. 2022.

BIODIVERSITY AND ECOSYSTEM SERVICES IN SWEDISH SOLAR FARMS: CURRENT PRACTICE AND A PLANNING FRAMEWORK FOR NET GAIN

Michiel Van Noord*[1], Ida Pettersson[2], Tora Råberg[3], Karin Morell[3], Lars-Henrik Björnsson[1], Ursula Zinko[2], Sara Ghaem Sigarchian[1], Agnes Sandström[2]
[1] RISE Research Institutes of Sweden, Division Built Environment, Energy and Resources, Drottning Kristinas väg 61, Stockholm
[2] Ecogain AB, Östra Strandgatan 26A, Umeå
[3] RISE Research Institutes of Sweden, Division Bioeconomy and Health, Agriculture and Food, Beta 5, Scheelevägen 17, Lund

ABSTRACT: The Swedish solar farm market is expanding rapidly as ground-mounted solar plants are getting larger in size and number. This development inevitably leads to an increased land use change, which is the foremost cause of biodiversity loss. Along with a decrease in biodiversity, nature's supply of ecosystem services decreases, as well as chances to meet several (inter)national environmental and sustainable development goals.

This study presents results of the Eco-Sun project, developing practical guidelines for the planning, construction and management of solar farms with a net zero, or net positive, impact on biodiversity and ecosystem services. It shows that biodiversity measures are present but limited in Swedish solar farms, while agrivoltaic crop cultivation is limited to one single site. In order to address environmental challenges and increase multiple land use, a step-by-step process is presented for the inclusion of biodiversity, agricultural and non-agricultural ecosystem services into the planning and design of ground-mounted solar PV.
Keywords: Environmental Effect, Solar Landscaping, Sustainable, Agrivoltaics, Biodiversity and ecosystem services

1 INTRODUCTION

Land use change is the single largest cause of biodiversity and ecosystem service loss according to the Intergovernmental Science-Policy Platform on Biodiversity and Ecosystem Services [1]. Both the World Economic Forum [2] and the research behind the planetary boundaries [3] point to biodiversity loss as one of humanity´s greatest challenges. In Sweden, grazed pastures and mowed meadows but also forests are pointed out as the nature types with poorest conservation statuses, whereas butterflies and beetle species, partly dependent on the above nature types, are some of the species with the worst conservation statuses [4]. Improving (or at least not deteriorating) conditions for these species and nature types should be a priority for all land use changes including solar farms.

Biodiversity is a prerequisite for nature's ability to provide many ecosystem services. Ecosystem services are the products and services that nature freely delivers and that human beings and societies benefit of, such as food provisioning, climate and water regulation, photosynthesis or natural heritage. Through loss of biodiversity virtually all ecosystem services are affected negatively. The UN Convention on Biological Diversity is therefore proposed to adopt targets of "No Net Loss by 2030" for quantity and quality in biodiversity [5].

This study has been conducted within the Eco-Sun (Swedish: "Eko-Sol") project [6] that aims to develop a guidebook with hands-on planning process and guidance for development and management of solar farms with a net positive impact on biodiversity and ecosystem services. Earlier publications within the project are a compilation of current knowledge on possibilities and challenges for biodiversity and ecosystem services in solar farms [7] (Swedish only) and a report on the status and practices for Swedish ground-mounted PV-plants [8] (Swedish only). The latter reports that the total number of PV-plants in Sweden increased by more than 600% between 2017 and 2021, from just over 15,000 to more than 90,000. During the same period, the installed power increased by almost 690%, from 231 MWp to 1500 MWp [9]. For PV-plants with capacities of 1 MWp or greater, which are to a large extent ground-mounted, the growth rate was about 1500%, from 8MW in 2017 to about 122 MW in December 2021 [10]. The largest commissioned solar farms currently have a capacity of around 20 MWp, but farms in the order of magnitude of 500 MWp are currently being planned [11]. This development has triggered an increased awareness of associated risks and is leading to increasing requirements for environmental impact assessments from regional authorities. Also, a societal debate has emerged in Sweden, very similar to that in other countries, about the environmental and agricultural impact of solar farms through land use and the possibilities of multiple uses through agrivoltaics or biodiversity measures.

Some recent work has been done internationally on solar farm planning using the mitigation hierarchy framework, in parallel with our current work [12]. Guidance on Agrivoltaics have been developed amongst others by Fraunhofer ISE [13] and Solar Power Europe [14]. Focus for the present study has been to provide practical guidelines for the planning process of solar farms with no net loss of biodiversity or ecosystem services, i.e. "Net Gain solar farms". This work is, to our knowledge, the first in its kind for Scandinavian conditions. While some parts are specific for Sweden, much of the general planning process guidance is relevant regardless of the geographical location.

Furthermore, the case of Swedish solar farm development, currently evolving from niche market to commercial lift-off, can serve as an example for other countries in an earlier phase of market development.

2 SOLAR FARMS PRACTICES IN SWEDEN UNTIL 2021

Essences of the work published in [8] (Swedish only) are summarized in this section to illustrate the relevance of planning for Net Gain solar farms.

By combining inventories from the national green certificate system [15] and the solar industry association [16] a total of 27 larger solar farms (capacity >1 MWp) were identified in Sweden.

GIS-analyses were made to identify historical land use and assess whether any solar farms were located within or near valuable nature. These analyses were made using ArcGIS software and data from [17]–[19] for the identified larger solar farms as well as 18 out of 20 smaller solar farms (capacity 255 - 1000 kWp) that were listed in [20]. Two smaller farms were excluded as one was only in the planning phase and the other could not be located with sufficient accuracy.

Also, a survey was subjected to owners or developers associated with the 27 larger solar farms, resulting in 18 responses. The survey included questions on siting considerations, multiple land use and design of the PV-plants.

2.1 Siting of solar farms

The survey results showed that main siting considerations for solar farms >1 MWp were (in order of priority): costs for purchase/lease of land, proximity to electricity grid, acceptance by the local community, avoidance of valuable cropland, consideration of natural values, and the need for preparatory earthworks.

GIS analysis complemented with survey results showed that 65% of the PV-plants >1 MWp were built on arable land. Other "open" land use types, such as pasture or industrial land, were also relatively common as previous land use. Use of former woodland was uncommon (5%).

For the smaller solar farms, the use of arable land seems much less common. GIS-analysis found 35% of 20 investigated sites to be on arable land.

Five of all sites (regardless of size) are (partially) overlapping with protected natural areas; one with a nature reserve and three with larger areas of national interest for nature conservation. In addition, 14 sites were located within non-protected but designated natural areas. These designated natural areas are mainly coherent areas with a relatively high prevalence of valuable natural habitats, compared to surrounding lands, designated by the county authorities.

2.2 Multiple land use

When asked whether PV was combined with additional activities, all but 25% of survey respondents indicated some type of active biodiversity measures or agricultural activities, Fig. 1. When asked about details of the specific measure, the respondents indicated that sowing meadow flowers was the most common biodiversity invervention (included in 6 of 8 responses). The choice of livestock was sheep in all grazed parks and the choice of cultivated crop was silage.

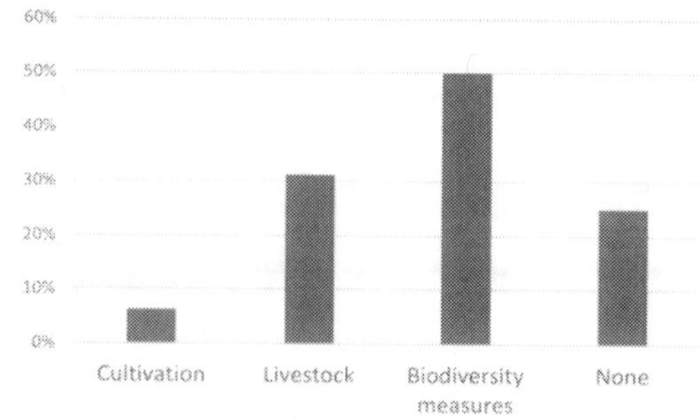

Figure 1: Survey responses to the question: "Is the PV system combined with any of the following?"

2.3 Discussion

Results regarding former land use for larger PV-plants, as compared to smaller ones could indicate that larger PV-plants generally are more likely to be (partially) located on land previously classified as arable land. More recently, after conduction of the survey, a number of cases have arisen where planned solar farms on arable land were prohibited by the county administrative board due to the planned use of cultivable land of the highest quality [21]. These cases have received considerable attention in media. Communication in press releases and social media by solar industry actors indicate that plant developers may be turning more to clear-cut woodlands than what is shown in the survey results. Also, a member of the project's reference group pointed out that their company's siting considerations nowadays gives high(er) priority to the risk of prohibition of plant establishment. This implies increased priority to avoidance of valuable cropland or land with high natural values.

Another relevant development is that (planned) project sizes are increasing rapidly, as described in the introduction section. To our understanding none of the ca 500 MWp PV-plants have passed through all mandatory spatial planning procedures yet, so actual establishment is still subject to uncertainties.

3 PROCESS FOR PLANNING OF NET GAIN SOLAR FARMS

As part in the creation of a Swedish guidebook for Net Gain solar farms, a suggested planning process was defined. This planning process was shaped by a combined effort of project members with expertise on the three main disciplines involved:

- Biodiversity, ecosystem services, and ecological restoration, represented by:
  - Ph.D. in Ecology with 18 years of working experience from research, government agencies and consultancy on a.o. ecological restoration and biodiversity surveys for city planning, mining, energy industries, etc.
  - M.Sc. in Physical Geography and Ecosystem Analysis with 5 years of experience in consultancy and applied research on biodiversity, ecological compensation, and environmental impact assessment for a.o. wind power, construction, and mining industries.
  - M.Sc. in Landscape Architecture plus Urban Design & International Planning. 5

years of working experience within landscape architecture and consultancy on biodiversity and environmental impact assessment.
    - M.Sc. in Environmental Science (nature conservation/conservation biology). 2.5 years of working experience in research and consultancy on biodiversity and social sustainability.
- Agriculture, represented by:
    - Ph.D. in Agronomy and M.Sc. in horticulture with 20 years of working experience from research in agricultural and urban ecosystems and biodiversity in the agroecosystem, work at the Board of Agriculture, international farm employee, and as a private consultant in sustainable gardening.
- PV applications, represented by:
    - M.Sc. in Applied Physics (Technology in Sustainable Development). 16 years of working experience as PV system planner/designer, project manager, consultant and researcher within solar PV applications and sustainability consultancy.
    - Ph.D. in Energy Technology. 6 years of working experience in consultancy, business development and research on PV and energy solutions.

A literature review was conducted to identify the possibilities and challenges for combining ground-mounted PV with measures to support biodiversity, with agricultural activities or (too some extent) with other ecosystem services [7] (Swedish only). The results of this literature review were combined with study visits at four solar farms and put into practice in theoretical exercises on how three of these solar farms could be enhanced.

3.1 Process for Net Gain

A top-level planning process on how to include the goal of Net Gain biodiversity and ecosystem services in regular solar farm planning, is defined as described below (see also Fig. 2).

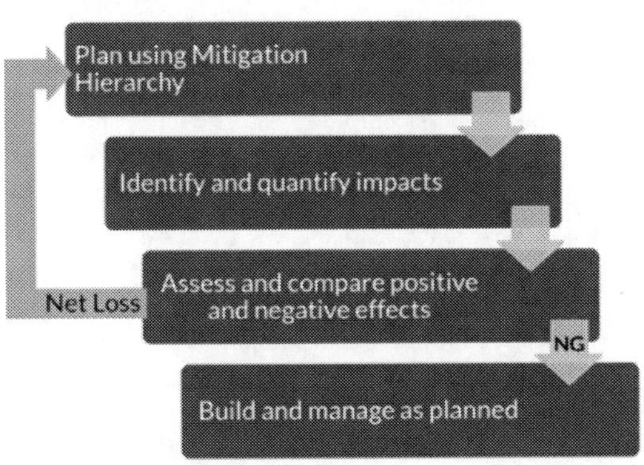

Figure 2: Top-level planning process for Net Gain biodiversity and ecosystem services. "NG" represents Net Gain.

A draft design of the ecovoltaic or agrivoltaic plant should be made to start with, using the principles of the Mitigation Hierarchy (Fig. 3). This implies that damage connected to the establishment of a solar farm should primarily be avoided to the largest extent possible, e.g. by avoiding sites with high natural (or agricultural) value or by conserving valuable trees. Secondly, damage should be minimized (e.g. by cautious planning and construction). Thirdly, final impacts should be restored as much as possible. And finally, any remaining damage should be compensated for to reach net zero impact. Additional compensational efforts can lead to Net Gain.

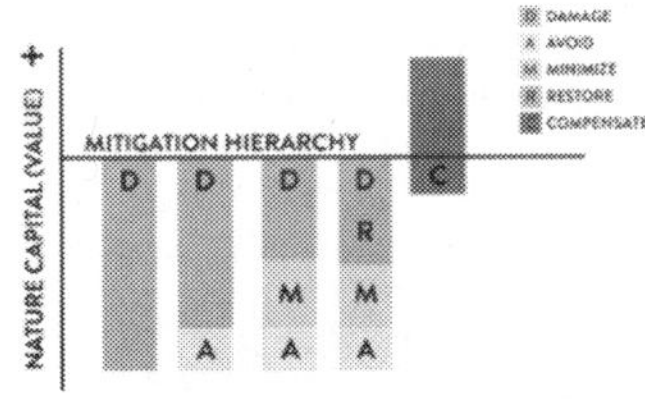

Figure 3: Mitigation Hierarchy as used for Biodiversity Net Gain. Adaptation of [22, Fig. 1]

Once a draft plant design is made, the impacts of the design should be assessed and quantified as good as possible. Quantification of biodiversity is a challenge, certain simplifications must be made, while the essence of biodiversity's complexity should not be lost. Several method development initiatives are ongoing. Some suggestions and initiatives are listed below, both for biodiversity and for ecosystem services.

- SS 199000: Swedish Standard for Biodiversity Surveys. Defines natural value classes that could be weighted with land areas in order to get a quantification. Not applicable for compensation assessment. [23] (Swedish only)
- Biodiversity Metric: British metric to measure biodiversity, including calculation tool. Developed for British conditions. [24]
- CLIMB (Changing Land Use Impact on Biodiversity): Ongoing project to value biodiversity in case of land use change. Method based on Biodiversity Metric (see above) and developed for Swedish conditions. [25]
- ESTER: Method to analyze ecosystem services before and after an intervention. This is a general level assessment, that might need additional detailed analyses. Developed by Boverket, the Swedish National Board of Housing, Building and Planning. [26]
- SPIES (Solar Park Impacts on Ecosystem Services): Decision support tool for comparison of the impact on ecosystem services of different measures in solar farms. Developed by Lancaster University. [27], [28].
- LER (Land Equivalent Ratio): A measure for the total yield of land productivity, comparing multiple land use (agrivoltaic) with single land use (i.e. separate agriculture and PV). [29]

The next step is to do an overall assessment of the impact on biodiversity and ecosystem services. As we consider the net impact, different effects should be compared and weighted. This requires some methodological choices, such as to what extent damage has to be restored or compensated by similar ecological structures (e.g. tree lines by tree lines in a different location) and when other structures can suffice.'

If the steps in the mitigation hierarchy have been followed thoroughly, and the final net assessment is considered net zero or net positive, Net Gain can be achieved and the plant is more likely to be approved by the authorities. Building and management need to fulfil the plan accordingly. Should the overall assessment result in a net negative value then the plant design should be adjusted and later re-assessed.

3.2 Plant planning and design process

In the previous section, the planning and design of the plant was described as a single process step, which is broken up into different steps described in this section. The main sub-processes are a pre-study phase and a planning and design phase, as shown in Fig. 4.

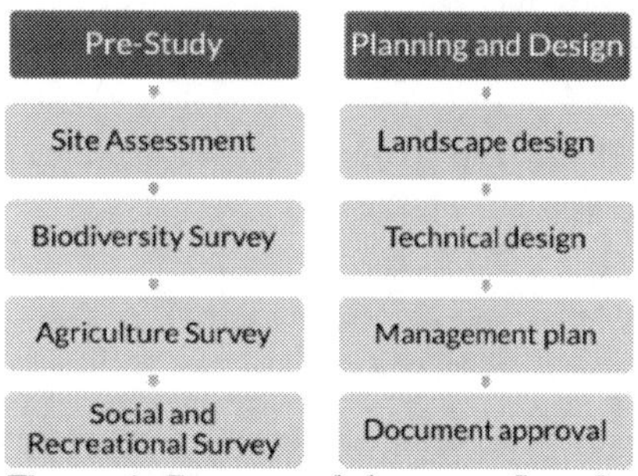

Figure 4: Recommended process flow for a PV-plant design process, with sub-processes in pre-study and planning and design phases

The pre-study phase starts with a site assessment aiming to do a first selection of the most suitable site(s). This can be performed in the form of a desktop study, typically involving analysis of GIS-data and national/regional/local authorities' plans and programs for spatial planning and nature conservation, as well as an initial site visit. This step typically includes a number of siting options and should identify whether sites (or their direct vicinity) contain high natural values, high agricultural values, and/or high recreational values, and/or if it is providing other important ecosystem services. If any of these is true, the risks (i.e. potential impacts and their likeliness) of a solar farm establishment should be considered. Based on these results the most suitable site alternative(s) can be identified, based on avoiding negative impacts. Of course this should be combined with typical factors valid for siting of conventional solar farms, that we will not go into in detail here.

From here on, more detailed analyses should be conducted into the different values on and surrounding the selected site:

- Biodiversity survey, aiming to exclude high natural value sites and identify suitable plants, animals, structures and biotopes relevant for the local biodiversity. This survey should assess the occurrence of both species and habitats, and optionally prevalence of invasive species or high-value trees, and detailed surveys on specific, protected habitats. When using Swedish Standard [23], we suggest that areas of natural value classes 1 and 2 should always be avoided, whereas classes 3 and 4 as well as unclassified areas should be assessed individually. A field inventory should be conducted by an ecologist. In addition to the field inventory, local nature interest groups can be a relevant source of information.
- Agricultural survey, aiming to identify and adapt to current agricultural activities on site and in the vicinity, and to identify potential crops (or livestock) suitable for the specific site. This can be achieved assessing soil maps, statistical data, cultivation zones, taking soil samples for analysis of nutrient levels and contamination, and by dialogue with local farmer(s) on e.g. cropping history and machine park. This survey should also include assessment of historical land use and whether this use makes the land unsuitable for food production.
- Social and recreational survey, with the objective to exclude high value sites and to identify how to adapt to or support local recreation or appearance. This can be done based on landscape analysis and stakeholder dialogues.

With the above surveys conducted a thorough basis is present for a final qualification of the site as suitable or non-suitable for Net Gain solar farms. Avoiding damage is a key question in that qualification, but this will go hand-in-hand with the potential to minimize damage and to what extent restoration seems reasonable.

The next phase for suitable sites is to plan for the design and maintenance of a future plant, starting with a landscape design. A major objective for this design is to conserve and support existing natural values, agricultural values or non-agricultural ecosystem services (i.e. minimizing, restoring and compensating damage). The use of a zonal design approach could be preferable, as some of the different activities or purposes of the site can be interfering. For example, crop cultivation requires soil with high nutrient levels whereas creation of meadows for biodiversity purposes benefits from low levels. Clear distinction between zones for the different purposes will make management easier and is expected to increase the quality of each purpose. Typical suggested zone types include:

- cultivation zones, where the annual crops can grow and where the farmer and agricultural machinery should be able to move more or less freely. Or where perennial crops for horticultural production can grow without the risk of shadowing the PV modules;
- installation zones, where the PV installation and balance of system (BoS) are located. These zones should be accessible for operating and service personnel, but only occasionally, so some interference with other zones is acceptable and usually also unavoidable;
- service zones, where the PV installation and BoS are located and where operating and service personnel need to move or access equipment on a more regular basis. Typically inverters and grid stations should be located in service zones;
- grazing zones where grazing animals can move freely and safely. These require some type of (moveable) fencing that might be integrated with the PV mounting structure;
- biodiversity zones where certain wildlife and plants can proliferate relatively undisturbed. This could also include wildlife passages and species selective

fencing or other barriers; and
- recreational zones where visitors can move safely and largely unrestricted on their own or at organised visits.

Keep in mind that (some) zones should be defined three-dimensionally, as for instance installation zones can be relevant over (PV modules) and below (cabling) cultivation zones in air and soil.

Next is the technical design of the PV installation. Some of the most important inputs to this is the needs of the plants or crops (from now on: plants) that will have to grow in between or below the PV structures. Therefore, an integrated plant selection and PV-design process is suggested, see Fig. 5. This process is typically most critical for agricultural crops, but should also be considered for biodiversity intended plants, be it maybe in a more comprehensive way.

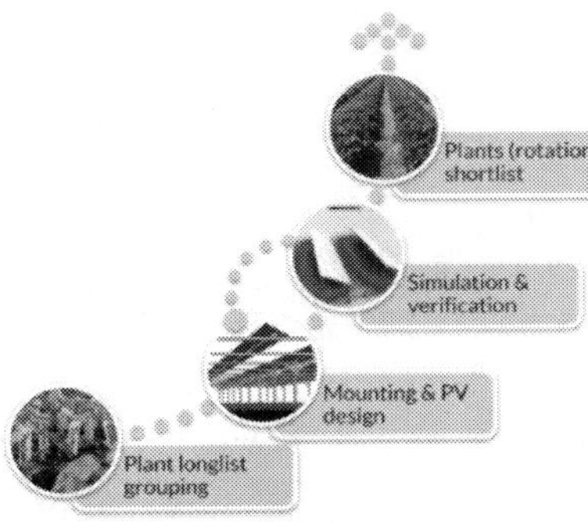

Figure 5: Suggested steps for integrated plant/crop and PV-installation design

Based on the different surveys in the pre-study phase a list of plants that are suitable for the current site can be created. Plants in this longlist may then be categorised according to light needs (daily light integral-levels from e.g. [30], [31]) and whether they have high or low sensitivity to, for instance, drought, extreme rain events, extreme temperatures, mould, etc. Note that these needs and sensitivities could change over time, e.g. night freezing protection might be relevant in early spring time, and crops like tomatoes require higher levels of light for the fruits to mature than for the plant to grow.

Now, PV mounting configurations and module types can be optimized for specific groups and maybe connected to specific site zones. In many cases irradiation on the crops is expected to be the main defining factor. Relevant design parameters are: system height (above ground level), PV array/row dimensions, distance between PV arrays/rows, module transparency, module tilt, module azimuth, fixed or dynamic (tracking) mounting. Also rainwater and/or irrigation management should be considered. If (locally) available machine parks for biodiversity management or agriculture are known, their sizes should also be taken into account in the design.

A draft design for a certain plant group should be verified using irradiation simulations and ideally also micro-climate simulations in combination with crop growth models. In general, a verification against preliminary irradiation criteria for maintaining proper soil health, proposed by [32] (Dutch only), should also be considered. Net Gain solar farms should aim at fulfilling all criteria, while considering that these are developed for conditions in the Netherlands and are still highly preliminary.

If the right conditions for certain plant groups can be verified with an economically feasible PV-design, a shortlist of plants can be compiled and, if relevant, crop rotations can be discussed with potentially interested farmers.

A final step in the design process, before approval of the designs and plans, is the creation of a management plan for biodiversity and/or agriculture. The right management is crucial for biodiversity measures to become effective and relevant management activities should be assessed to avoid possible conflicts with the current PV-design. Any identified conflicts should be addressed by either adapting the design or the management activity. Larger adaptations in the PV-design should ideally be avoided already in the previous steps as they might impact the optimization in relation to the plants and require additional design iterations.

To conclude, a final approval should be given to the plans and designs by the main stakeholders of the solar farm (e.g. developer, farmer, land owner). The approval process step will usually include an economic evaluation and feasibility assessment, including e.g. capital and operational expenditures for the solar farm and any agricultural activities, levelized cost of electricity, and return on investment.

4 DISCUSSION AND CONCLUSIONS

The presented work indicates that the market for solar farms in Sweden is rapidly expanding and that the exploitation of arable lands has been common practice. In combination with the critical status of biodiversity in Sweden and globally, a focus on solar farms with Net Gain for biodiversity and ecosystem services (including agriculture) is highly relevant.

Currently, few PV farms in Sweden are agrivoltaic cultivation sites and even though biodiversity interventions are much more common, they have mainly been limited to meadow flowers for pollinators and the effects of the interventions are typically not verified by long-term monitoring.

The authors have therefore proposed a process for Net Gain planning of PV farms that should support PV industry, agricultural sector and nature conservation stakeholders in developing and assessing agrivoltaic and ecovoltaic plants with net gain based definitions.

The proposed planning process is based on historical experience from relevant industries, yet still unverified in actual planning projects, which should be a focus for future research. Also, the assessment of net impacts of a planned site in the planning phase, as well as of an established site during the management stage, is a complex assignment and requires more work on tool and method development and verification.

Finally, while scientific and societal interest in ecovoltaics and agrivoltaics is high, many of the potential multiple land uses are still to be verified in general or, more specifically, for local conditions in Sweden or other countries. We are aware of and involved in some work in progress, but this is far from being sufficient.

5 REFERENCES

[1] IPBES, 'Global assessment report on biodiversity and ecosystem services of the Intergovernmental Science-Policy Platform on Biodiversity and Ecosystem Services.', IPBES, Bonn, 2019. [Online]. Available: https://doi.org/10.5281/zenodo.3831673

[2] World Economic Forum, 'The Global Risks Report 2020 - Insight Report, 15th Edition', World Economic Forum, 2020. [Online]. Available: http://www3.weforum.org/docs/WEF_Global_Risk_Report_2020.pdf

[3] W. Steffen *et al.*, 'Planetary boundaries: Guiding human development on a changing planet', *Science*, vol. 347, no. 6223, pp. 1259855–1259855, Feb. 2015, doi: 10.1126/science.1259855.

[4] SLU Artdatabanken, SLU Institutionen för akvatiska resurser (Aqua), and SLU Centrum för biologisk mångfald (CMB), 'Sveriges arter och naturtyper i EU:s art- och habitatdirektiv', Naturvårdsverket, 2020.

[5] CBD, Convention on Biological Diversity, 'Zero Draft Report', CBD, CBD/WG2020/2/L.1, Feb. 2020. [Online]. Available: https://www.cbd.int/doc/c/d9da/8747/30def58e30f47984ab032f91/wg2020-02-l1-en.pdf

[6] RISE, 'Planning of solar farms that benefit biodiversity and ecosystem services', *RISE*. https://www.ri.se/en/what-we-do/projects/planning-of-solar-farms-that-benefit-biodiversity-and-ecosystem-services (accessed Sep. 28, 2022).

[7] T. Råberg, M. van Noord, L.-H. Björnsson, I. Pettersson, and U. Zinko, 'Solcellsparker, biologisk mångfald och ekosystemtjänster: Påverkan och möjligheter för multifunktioner', 978-91-89385-93-1 (ISBN), 2021. Accessed: Dec. 22, 2021. [Online]. Available: http://urn.kb.se/resolve?urn=urn:nbn:se:ri:diva-57378

[8] L.-H. Björnsson, K. Morell, M. van Noord, and I. Pettersson, 'En kartläggning av solcellsparker i Sverige 2021', 978-91-89711-04-4 (ISBN), 2022. Accessed: May 24, 2022. [Online]. Available: http://urn.kb.se/resolve?urn=urn:nbn:se:ri:diva-59284

[9] Svensk Solenergi, 'Svensk solkraft', *Svensk Solenergi*. https://svensksolenergi.se/statistik/solkraft/ (accessed May 05, 2022).

[10] Energimyndigheten, 'Kraftig ökning av installerade solcellsanläggningar', Mar. 31, 2022. https://www.energimyndigheten.se/nyhetsarkiv/2022/kraftig-okning-av-installerade-solcellsanlaggningar/ (accessed May 05, 2022).

[11] J. Scully, 'Ilmatar to build 550MW solar PV project in Sweden', *PV Tech*, Sep. 12, 2022. https://www.pv-tech.org/ilmatar-to-build-550mw-solar-pv-project-in-sweden/ (accessed Sep. 27, 2022).

[12] L. Bennun *et al.*, *Mitigating biodiversity impacts associated with solar and wind energy development: guidelines for project developers*. IUCN, International Union for Conservation of Nature, 2021. doi: 10.2305/IUCN.CH.2021.04.en.

[13] 'Agrivoltaics: Opportunities for Agriculture and the Energy Transition - A Guideline for Germany', Fraunhofer Institute for Solar Energy Systems ISE, Freiburg, Oct. 2020.

[14] SolarPower Europe, 'Agrisolar Best Practices Guidelines Version 1.0.', Solar Power Europe, Brussels, May 2021.

[15] Energimyndigheten, 'Godkända anläggningar'. http://epi6.energimyndigheten.se/SharePoint/Eugen/Godk%C3%A4nda%20anl%C3%A4ggningar.xlsx (accessed Oct. 28, 2021).

[16] Svensk Solenergi, 'Anläggningar över 1 MW « Svensk Solenergi'. https://svensksolenergi.se/om-solenergi/anlaggningar/solcellsparker/ (accessed Oct. 28, 2021).

[17] Länsstyrelserna (County Administrative Boards), 'GeodataKatalogen', *Länsstyrelserna*. https://ext-geodatakatalog.lansstyrelsen.se/GeodataKatalogen/ (accessed Sep. 28, 2022).

[18] Naturvårdsverket (The Swedish Environmental Protection Agency), 'Miljödataportalen', *Naturvårdsverket*. https://miljodataportalen.naturvardsverket.se/miljodataportalen/ (accessed Sep. 28, 2022).

[19] Skogsstyrelsen (The Swedish Forest Agency), 'Ladda ner geodata', *Skogsstyrelsen*. https://www.skogsstyrelsen.se/sjalvservice/karttjanster/geodatatjanster/nerladdning-av-geodata/ (accessed Sep. 28, 2022).

[20] Svensk Solenergi, '☼ Sveriges största solcellsanläggningar – Google Mina kartor', *Google My Maps*, 2021. https://www.google.com/maps/d/viewer?mid=1rsRvEzEqJUG2yf0KhTx1sVGs50JWbEpS (accessed Aug. 30, 2021).

[21] TT, 'Stoppar bygget av Sveriges största solcellspark', *Ny Teknik*. https://www.nyteknik.se/energi/stoppar-bygget-av-sveriges-storsta-solcellspark-7032325 (accessed Sep. 29, 2022).

[22] Business and Biodiversity Offsets Programme (BBOP), 'Working for Biodiversity Net Gain. An Overview of the Business and Biodiversity Offsets Programme (BBOP) 2004–2018'. BBOP, 2018. [Online]. Available: https://www.forest-trends.org/bbop_pubs/overview2018

[23] SIS Swedish Standards Institute, 'Biodiversity survey - Implementation, assessment and reporting', SIS Swedish Standards Institute, Stockholm, SS 1999000, 2014.

[24] Stephen Panks *et al.*, 'Biodiversity metric 3.1: Auditing and accounting for biodiversity – User Guide', Natural England, York, Natural England Joint Publication JP039, 2022.

[25] Ecogain, 'CLIMB', *Ecogain*. https://www.ecogain.se/climb (accessed Sep. 09, 2022).

[26] Boverket, 'ESTER - verktyg för kartläggning av ekosystemtjänster', *Boverket*. https://www.boverket.se/sv/PBL-kunskapsbanken/teman/ekosystemtjanster/verktyg/ester/ (accessed Sep. 09, 2022).

[27] R. J. Randle-Boggis *et al.*, 'Realising co-benefits for natural capital and ecosystem services from solar parks: A co-developed, evidence-based approach', *Renew. Sustain. Energy Rev.*, vol. 125,

p. 109775, Jun. 2020, doi: 10.1016/j.rser.2020.109775.

[28] Lancaster University, 'SPIES Project'. https://www.lancaster.ac.uk/spies/ (accessed Sep. 09, 2022).

[29] C. Dupraz, H. Marrou, G. Talbot, L. Dufour, A. Nogier, and Y. Ferard, 'Combining solar photovoltaic panels and food crops for optimising land use: Towards new agrivoltaic schemes', *Renew. Energy Gener. Appl.*, vol. 36, no. 10, pp. 2725–2732, Oct. 2011, doi: 10.1016/j.renene.2011.03.005.

[30] P. Zambrano-Prado *et al.*, 'Assessment of the food-water-energy nexus suitability of rooftops. A methodological remote sensing approach in an urban Mediterranean area', *Sustain. Cities Soc.*, vol. 75, p. 103287, Dec. 2021, doi: 10.1016/j.scs.2021.103287.

[31] M. Cossu *et al.*, 'Agricultural sustainability estimation of the European photovoltaic greenhouses', *Eur. J. Agron.*, vol. 118, p. 126074, Aug. 2020, doi: 10.1016/j.eja.2020.126074.

[32] Kay Cesar, Bas van Aken, Luuk Scholten, Ron de Goede, and Alex Schotman, 'Nieuwe ontwerptoets verankert bodemkwaliteit in zonneparken', *Bodem*, vol. 2022, no. 2, pp. 34–36, Mar. 2022.

6 ACKNOWLEDGEMENTS

The authors like to acknowledge the reference group of the Eko-Sol project for sharing their insights and hands-on experience from industry and public authorities with us. This has been valuable input to the development of this work specifically and the project in general.

Additional thanks goes to the companies Energiengagemang, Svea Solar and Kalmar Energi for guiding us at their solar farms, as well as all companies that responded to the survey.

Furthermore, the authors thank the Swedish Energy Agency for financing the Eko-Sol project under project number P51007-1.

NOVEL MEASUREMENT CONCEPT FOR AGRIPVPLUS SYSTEMS - A TRIPLE USE APPROACH

Hugo Sánchez[1], Sebastian Dittmann[1], Santiago Tosello[1], Carlos Meza[1], Sandra Dullau[1], Maren Helen Meyer[1], Pascal Scholz[1], Sabine Tischew[1], Ralph Gottschalg[1,2]
[1]Hochschule Anhalt University of Applied Sciences, Bernburger Str. 55, 06366, Köthen, Germany
Email: hugo.sanchez@hs-anhalt.de
[2]Fraunhofer Center for Crystalline Silicon Photovoltaics CSP

ABSTRACT: The European Green Deal aims for Europe becomes the world's first climate-neutral continent by 2050. In this strategy, Photovoltaic (PV) is one of the key technologies to reduce significant greenhouse gas emissions. However, large ground-mounted PV power plants generate conflict between stakeholders interested in the same land. Farmers are worried about losing land to grow crops and thus their income. Conservationists are concerned about the dramatic loss of biodiversity, and energy producers want to build profitable PV power plants. Therefore, we must bring all stakeholders together to work out innovative and sustainable land-use strategies combining agriculture, biodiversity, and energy production.

Agrivoltaic (Agri-PV) involves the use of land both for agriculture and solar energy generation. Such dual systems have been developed in recent years. So far, however, their potential to contribute to biodiversity and ecosystem services restoration has been rarely discussed. Our demonstrator aims to investigate an innovative and sustainable triple-use approach, called the AgriPVplus system, that combines crop production, solar energy generation, and biodiversity measures.

Our measurement concept enables the investigation of energy yield measurements of different module technologies, different orientations to investigate the grid compatibility through vertical east-west installed bifacial PV modules, and the measurement of abiotic parameters such as photosynthetic active irradiance, soil temperature, and moisture to investigate the impact on the agriculture crops and biodiversity.

This paper will show a novel measurement concept to demonstrate the interfaces between energy production, crop yield, and nature conservation.

Keywords: agrivoltaics, vertical bifacial PV, biodiversity

1 INTRODUCTION

Food security and decarbonization require innovative schemes to fulfill the necessities of a growing world. To address these challenges, generating and studying new and innovative frameworks in the synergy of food and energy production is necessary. Agrivoltaics (Agri-PV) concepts were developed in the early '80s [1]. Nevertheless, most of the literature goes into the theoretical and simulation process to understand the integration capabilities between agriculture and Photovoltaic (PV) systems [2-4]. Only a few commercial integrations have been realized worldwide [5]. Most of them use a tilted racks system or tracking systems.

Land competition is a critical factor in the adoption of this technology. Landowners are not entirely convinced that agrivoltaics can be an affordable implementation for them [6]. It is not clear that productivity is enhanced with this technology. Farmers do not want to "affect" their activities by installing solar panels without having a clear benefit. One of the main challenges associated with this vision stems from the lack of planning, design, and evaluation tools for these technologies to gain the attention and confidence of farmers, stakeholders, and decision-makers [7-9]. The uncertain productivity enhancement is one of the main obstacles to developing more feasible applications.

On the other hand, producing renewable energy through ground-mounted PV systems requires large areas. Most designs are energy-centralized, putting behind the integration of agriculture. In this sense, developing new mechanisms to plan new PV systems is required [9-10]. Regarding biodiversity conservation within PV systems, only a few studies exist. That gives a valuable opportunity to study new concept schemes for affordable production systems. These studies should consider an interdisciplinary integration to reach a holistic, more critical approach to future applicability in society.

Therefore, an interdisciplinary team at Hochschule Anhalt University of Applied Sciences in Germany started by developing a new demonstration concept. Our demonstrator aims to investigate an innovative and sustainable triple-use approach, called the AgriPVplus system, that combines crop production, PV energy generation, and biodiversity measures. This team, including researchers for photovoltaic, agriculture, and biodiversity, defined the design and experiments to have a demonstrator that complines all three disciplines. This will provide a reliable model for scientific experimentation, testing new concepts, and sharing the results with the scientific community and civil society.

2 APPROACHE

Our study aims to present a new novel measurement concept for AgriPVplus systems, an innovative and sustainable triple-use approach that combines crop production, solar energy generation, and restoration of biodiversity

2.1 AgriPVplus

The novelty is that three different, but complementary areas are integrated. First, photovoltaics energy production has become a leader in the energy transition. The advances and adoption of the technology allow us to address the urgency of decarbonizing our energy. Second, food security is essential for human survival. Moreover, it is not only about producing it but also about doing it efficiently and sustainably. Third, and not less important, it is required to protect the resources available and ensure biodiversity conservation in our lands. Every of the previous three areas is not exclusive, but synergy options must be sought between the different components. Figure 1 shows a schematic concept for the triple-use approach called AgriPVPlus.

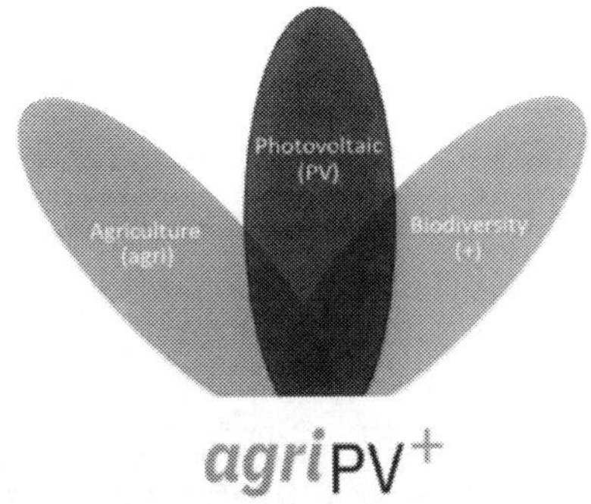

Figure 1: The triple-use approach of AgriPVPlus

The concept AgriPVplus is an innovative and sustainable land-use strategy combining agriculture, biodiversity, and energy production through Photovoltaic. The concept aims to optimize the resources and explore possibilities for different exciting agriPV systems, as shown in Figure 2.

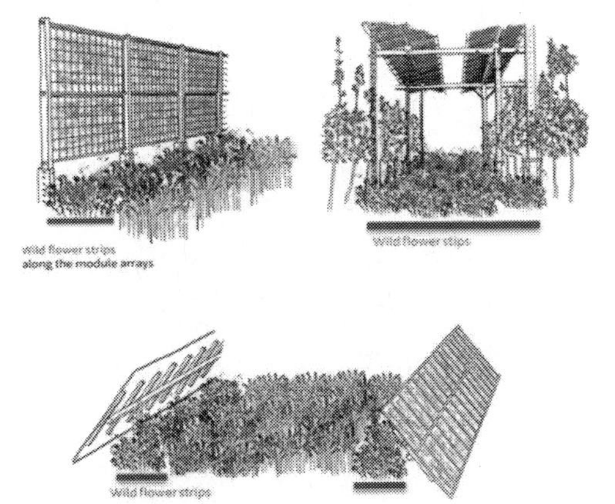

Figure 2: AgriPVPlus approach in different exciting AgriPV systems. Vertical mounted Bifacial PV modules (top left), elevated Roof (top right) and single axes tracker.

2.3 Anhalt's AgriPVplus Demonstrator

Hochschule Anhalt University of Applied Sciences developed an AgriPVplus demonstrator to investigate a comprehensive multifunctional land use system with scientists from agriculture, PV, and biodiversity. In specific for every area are:

Agriculture:
- Impact on soil erosion and humidity due to wind reduction and shading – input for crop modelling.
- Impact of natural pollinators and beneficial organisms.

Biodiversity:
- Increase of natural wildflowers and pollinators
- Investigation of connectivity between natural biotopes

Photovoltaic:
- Grid compatibility with vertical mounted bifacial PV modules
- Input for performance modelling approaches, e.g., albedo
- Comparison of different PV module technologies and installation methods

The AgriPVplus demonstrator is currently under construction at Hochschule Anhalt University of Applied Sciences in Bernburg, Germany, consisting of vertical east-west installed bifacial PV modules with a total power of 75 kWp and 12 kWp south-facing stem with different module technologies. The distance between arrays is 14m, allowing for the investigation of different kinds of crops (mainly dough resistance crops such as Millet, Soybean, and Durum Wheat). Wildflower strips are placed under the PV arrays (1m on each side) to investigate their influence in the crops and the Ecosystem services.

The demonstrator includes a high-precision monitoring system with a weather station, an enhanced irradiance and albedo measurement system [6,7], PV performance measurement on string and PV module level, and sensors for abiotic parameters. This allows us to investigate the impact on the PV energy yield and crop yield, e.g., less water evaporation due to the shading effect, less erosion due to lower wind speed, or higher PV energy yield due to seasonal variation of the albedo. Figure 3 shows an overview of the demonstrator.

Figure 3: (Right) Overview of the AgriPVplus demonstrator with vertical east-west installed bifacial PV modules. (left) 3D sketch of the demonstrator and south-facing reference system with 35° inclined PV modules.

The demonstrator will also be used as a source of high-quality data for energy prediction modelling for vertical PV systems and as a test facility for the intercomparison of different bifacial PV module technologies. As shown in figure 4 (colored modules), the demonstrator provides spaces for single module monitoring. The energy yield will be compared to 35° tilted south-facing mounted modules.

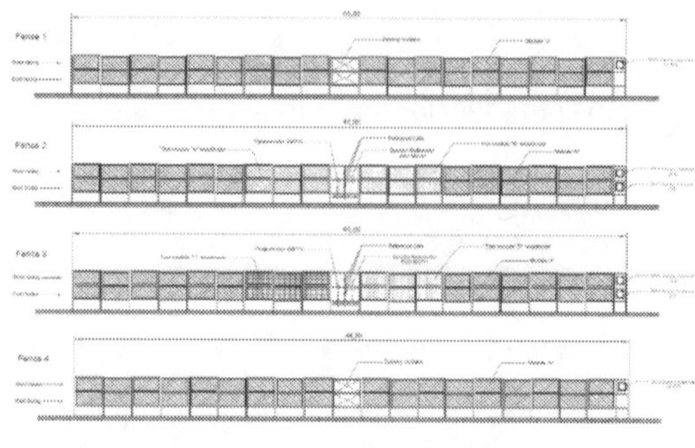

Figure 4: Overview of the AgriPVplus demonstrator. Gray modules are grid-connected with AC and DC monitoring; single PV module monitoring (colored modules) modules can be changed; irradiance, albedo, and spectral irradiance measurement within PV arrays.

2.3 Anhalt's AgriPVplus Monitoring system

The monitoring system includes the necessary equipment for PV systems and measurement systems for agriculture and biodiversity. Figure 5 and 6 shows the general diagram for the monitoring systems in the demonstrator.

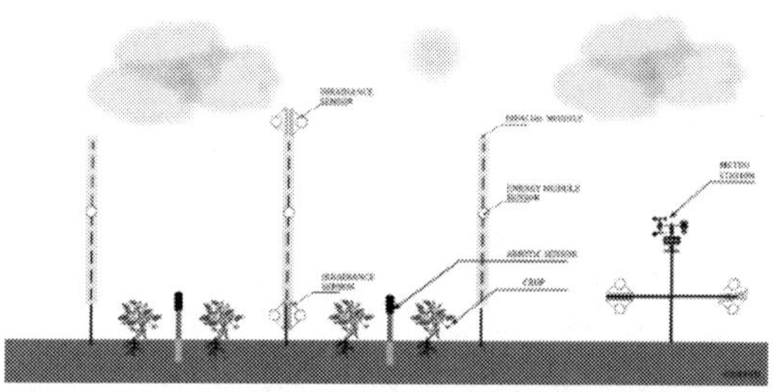

Figure 5: Concept of the AgriPVplus demonstrator, including a weather station with an enhanced albedo measurement system, and sensors for abiotic parameters such as photosynthetic active irradiance, soil temperature and moisture.

The AgriPVplus demonstrator is equipped with high-precision meteorological equipment, including irradiance and spectrum measurements, electrical variables (i.e., voltage, current, and power), and measurements for abiotic parameters for agriculture and biodiversity.

All the information is connected to the highly reliable data center at the university to further analysis and experimental validation. The proposed framework will provide the capability of interoperation between several communication protocols to get only one standardized data set.

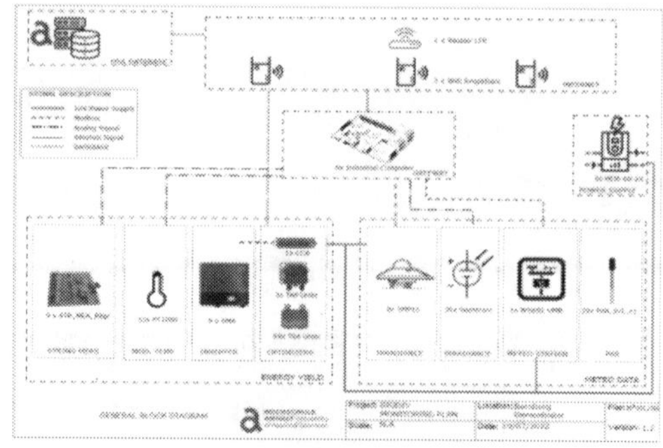

Figure 8: Monitoring Measurement Layout for the AgriPVPlus demonstrator.

A dedicated sensor for the abiotic parameters is developed to enhance crop behavior analysis. The sensor will provide the capability of creating Photosynthetically Active Radiation (PAR), soil temperature, and moisture maps to analyze the impact of the integration between PV systems and crops. Each sensor in the field will send data to the centralized server throughout the MQTT protocol. The current test shows that the network sampling measurement can be up 30 seconds, especially for the stabilization of the moisture sensor. Future publications will provide more results on these sensors. Figure 7 provides the general architecture of the MQTT network.

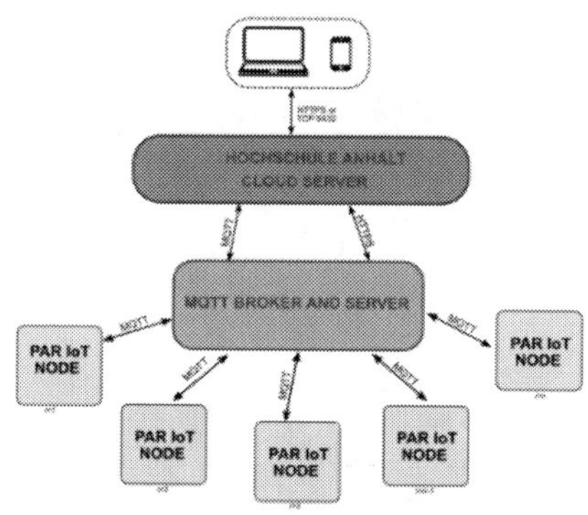

Figure 7: Monitoring Measurement Layout for the AgriPVPlus demonstrator.

3 CONCLUSIONS AND OUTLOOK

With the AgriPVplus system, the paper presents an innovative and sustainable land-use strategy that combines agriculture, biodiversity, and energy production. The novel measurement concept can contribute to the scientific discussion about a) the behavior of the albedo under several crops for vertical bifacial installations, b) the realistic and expectative energy yield in AgriPVplus systems, c) the interaction between several crops, biodiversity, and PV energy production. The final paper will present a detailed design of the AgriPVplus demonstrator and the first results of the first-year measurement campaign.

2 ACKNOWLEDGMENTS

This work is supported by the German Federal Ministry of Education and Research under the project "Biodiversity in solar parks - Innovative concepts and construction of demonstrators for a better compatibility of photovoltaic systems, nature conservation and agriculture (BIODIV-SOLAR)". Funding code: 13FH133KX0

3 REFERENCES

[1] Goetzberger A,ZastrowA. On the coexistence of solar-energy conversion and plant cultivation. International Journal of Solar Energy;1(1):55–69. (1982).

[2] Dinesh, H., Pearce, J.M.: The potential of agrivoltaic systems. Renewable and Sustainable Energy Reviews 54, 299–308 (2016).

[3] Hassanpour Adeh, E., Selker, J.S., Higgins, C.W.: Remarkable agrivoltaic influence on soil moisture, micrometeorology and water-use efficiency. PloS one 13(11), e0203256 (2018)

[4] Riaz, M.H., Imran, H., Butt, N.Z.: Optimization of pv array density for fixed tilt bifacial solar panels for efficient agrivoltaic systems. In: 2020 47th IEEE Photovoltaic Specialists Conference (PVSC). pp. 1349–1352); 2020).

[5] Weselek, A., Ehmann, A., Zikeli, S., Lewandowski, I., Schindele, S., & Högy, P. . Agrophotovoltaic systems: applications, challenges, and opportunities. A review. Agronomy for Sustainable Development, 39(4), 1-20. (2019)

[6] Piscaries, A.S.; Schelly, C.; Pearce, J.M. A First Investigation of Agriculture Sector Perspectives on the Opportunities and Barriers for Agrivoltaics. Agronomy 2020, 10, 1885.

[7] Agostini, A., M. Colauzzi, and S. Amaducci. "Innovative agrivoltaic systems to produce sustainable energy: An economic and environmental assessment." Applied Energy 281 (2021)

[8] Amaducci, Stefano, Xinyou Yin, and Michele Colauzzi. "Agrivoltaic systems to optimise land use for electric energy production." Applied energy 220 (2018): 545-561.

[9] R. Cameron, B. Cohen und S. Morrison, "An Approach to Enhance the Conservation-Compatibility of Solar Energy Development, "PLoS One, Bd. 7, Nr. 6, (2012).

[10] Zaplata, M.K.; Dullau, S.Applying Ecological SuccessionTheory to Birds in Solar Parks: AnApproach to Address Protection andPlanning. Land 2022,

[11] Dittmann, S., Sanchez, H., Burnham, L., Gottschalg, R., Oh, S.Y., Benlarabi, A., Figgis, B., Abdallah, A., Rodriguez, C., Rüther, R., et al.: Comparative analysis of albedo measurements (plan-of-array and horizontal at multiple sites worldwide. In: 36th European Photovoltaic Solar Energy Conference (EU PVSEC). pp. 1388–1393(2019).

[12] Sánchez, H., Meza, C., Dittmann, S., Gottschalg, R.: The impact of real albedo values on energy estimation for bifacial modules. In: 38th European Photovoltaic Solar Energy Conference (EU PVSEC). pp. 808–810 (2021).

RAINWATER MANAGEMENT IN AGRIVOLTAIC SYSTEMS
-
RESEARCH & DEVELOPMENT POTENTIAL

Johanna-Viktoria Rößner, johanna-viktoria.roessner@ise.fraunhofer.de
Fraunhofer Institute for Solar Energy Systems ISE
Heidenhofstraße 2, 79110 Freiburg im Breisgau, Germany

Agrivoltaics is a concept in which a piece of land is simultaneously used for both energy and food production by mounting photovoltaic modules at a certain height above (or in between strips of) agricultural land. A local and system-level incorporation of water management is imperative to the sustainable implementation of agrivoltaics. Water raining on the modules can be gathered and used for distinct purposes: groundwater recharge, crop irrigation, and cleaning and cooling of the PV modules. This research provides an initial overview of positive and negative impacts for each water use concept and outlines issues that should be taken into consideration and the potential for research and development. Various Managed Aquifer Recharge (MAR) technologies are a way to clean and store the water periodically in an underlying aquifer. Irrigation increases yield within the plant level and therefore increases the system's output. Thanks to the power supply generated by the PV modules, high-tech irrigation systems can be implemented in agrivoltaic systems; the special adaption of irrigation systems to agrivoltaics poses significant potential for research and development. Meanwhile, the necessity, i.e., profitability of cleaning and/ or cooling PV modules depends on local environmental and economic factors. Several cleaning techniques have been developed to mitigate soiling, ranging from manual cleaning to fully automatic cleaning systems. In agrivoltaic systems, the soiling risk can increase. Semi-automatic systems seem to have the greatest potential for agrivoltaics, because they can be used with farming equipment. Multiple cooling techniques have been developed to decrease cell temperature to increase power output, with some of them involving water. Water flowing over the module surface is a promising cooling technique for agrivoltaic applications. Attaching a perforated tube to the upper edge, the entire module can be covered in a thin film of water which cools very effectively (while also cleaning the surface). A closed-circuit system could be created involving the technical components used for rainwater harvesting. The economic feasibility of cooling panels in agrivoltaic systems needs to be investigated. In certain locations, rainwater-harvesting could also be relevant for ground-mounted PV systems.

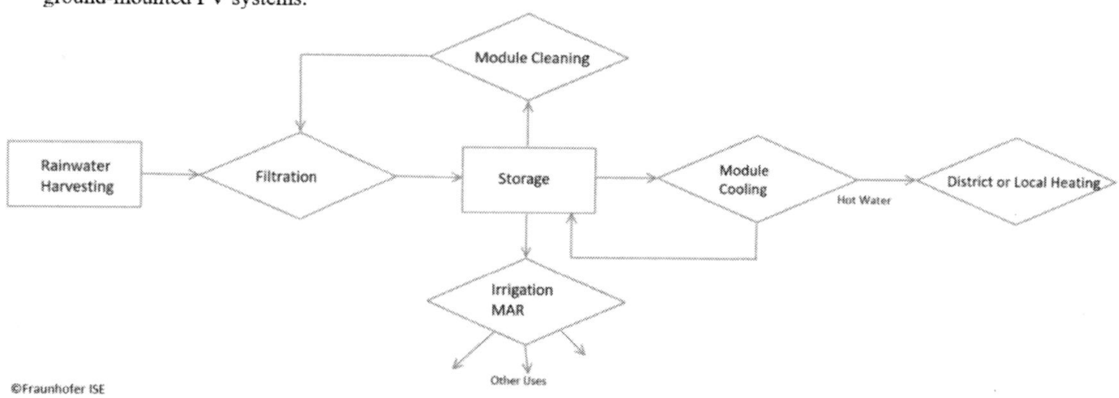

Figure 1: Flowchart of an integrated water management approach for agrivoltaic systems

1 SCIENTIFIC INNOVATION AND RELEVANCE

With elevated modules in agrivoltaic systems, rainwater collecting at the lower edge of the module ("dripping edge") and hitting the soil with a high speed can cause severe problems not just within the system: The force with which the water drips from the module is enough to damage plants and deform the soil (splash erosion) [1]. Also, surface runoff is increased, which can sweep seedlings away and on a larger scale, can aggravate flooding while also disturbing the local hydrologic balance and ecosystem. One Solution to this problem, which is currently in development, is attaching a disperser to the dripping edge to facilitate a more homogenous distribution of rainwater beneath the modules. Another approach is attaching a gutter to the lower edge of the module and "harvesting" the rainwater, which can then be used for several purposes. It is important to consider water management on a local and regional level as to implement the agrivoltaic system successfully. Under unfavorable circumstances, the agrivoltaic system could aggravate or create problems for local water management that need additional technical workarounds.

2 AIM AND APPROACH

The research presented here aims to provide an initial overview of positive and negative impacts for each water use concept, as well as agrivoltaics-specific issues that have to be taken into consideration. These act as a basis for pointing out research and development potential. For each application, literature was reviewed to get a general overview over different technologies and possible impacts. After that, the implementation into an agrivoltaic system was theorized. With regards to synergetic and adverse effects, the arising questions and ideas where then classified as research and (in some cases) development potential.

3 RESULTS

3.1 Managed Aquifer Recharge (MAR)

MAR involves infiltrating water through the vadose zone or directly into the aquifer whereby its quality is improved [2]. MAR can eliminate the need for storage devices and can be used to mitigate seasonal change in water availability. Several techniques are distinguished, depending on means of infiltration and extraction. The simplest method of infiltration involves filling infiltration trenches or basins with water, which then infiltrates through the vadose zone by means of gravitational force. This is an increasingly common way to ensure drainage in urban areas while reducing rainwater discharge to water treatment plants. The water is filtered while passing through the vadose zone; it's quality can improve immensely.

Other applications focus on a controllable storage of the water. Wells can be used to infiltrate water into an aquifer and extract it at a later point in time. In "Aquifer Storage and Recovery" (ASR), only one well is used for both infiltration and extraction. Depending on flow velocity in the aquifer, the water forms a bubble around the infiltration site. Only the outer layer of this bubble interacts with the aquifer medium and the native groundwater and therefore experiences a change in water quality. Therefore, ASR is only suitable if a change in water quality is not desired.

"Aquifer Storage, Transfer and Recovery" (ASTR) uses two wells: one for infiltration, another one for extraction. This causes a flux toward the extraction well, which pulls the entire infiltrated volume of water through the porous medium that makes up the aquifer. While traveling through the aquifer, the water interacts with the medium on a small-scale level, resulting in an improved water quality.

"Soil Aquifer Treatment" (SAT) was originally developed as a wastewater treatment technique. Water is infiltrated via infiltration basins and filtered while passing the vadose zone. An extraction well is placed somewhere in the vicinity of the infiltration site to regain the water. The extracted water is of suitable quality for irrigation. [2]

The suitability of a location for infiltration / MAR depends on an array of hydrogeological, hydrochemical and geo-chemical factors, such as the coefficient of permeability, hydraulic conductivity, the chemical makeup of the involved waters and that of the aquifer, presence and makeup of suspended solids, etc. [2]. Thus, MAR requires large scale considerations and technical effort, which is why it makes the most sense in conjunction with large AV systems, where a significant amount of water can be gathered and infiltrated. Coupling the two systems could eliminate the need for technical components for storage or intense cleaning of the water. To really be considered a part of the agrivoltaic system, the extracted water should at least partially be used to increase the output of the system (either through irrigation or cooling/cleaning the panels). As the infiltrated water can be used multi-purposely, MAR can be coupled with any other use of water described here, which would otherwise require storing water in tanks/ ponds or increase consumption of potable water. The infiltration and extraction wells should be somewhat close to the agrivoltaic system, otherwise long pipelines would be necessary to transport the water back and forth. This seems inefficient and would further separate the two systems. Finding a location that is suitable for both technologies might prove difficult, especially because some MAR-techniques require significant amounts of space. A main goal of agrivoltaics however, is mitigating land use conflicts by increasing land use efficiency, making it more likely to be implemented in regions suffering from land scarcity. Research needs to be conducted to define criteria for the suitability of locations for an integrated agrivoltaics-MAR-approach. The possibility of the water picking up contaminants while running over the modules' surface is also cause for consideration. Besides that, the precise technical execution and the economical feasibility should be investigated.

Outside of agrivoltaics, covering infiltration sites with overhead solar energy collectors to reduce evaporation and achieve dual land use seems promising. One more issue that persists (at least in Germany) is the legality of infiltrating gathered rainwater directly into aquifers. Current rainwater management practices need to be reconsidered to combat climate crisis impacts.

3.2 Irrigation

Irrigation increases yield within the plant level and therefore increases the system's output. This effect is especially interesting in conjunction with high value crops. In arid regions, the danger of salinization must be considered. The applied irrigation water is partially used by the plant, while the rest evaporates or seeps into the soil. Both the plants' uptake and the evaporation leave behind dissolved salts, causing the salts to accumulate in the soil and therefore in the seepage water. This can cause the permanent degradation of land. Salinated water seeping downward can lead to the salinization of the underlying aquifer. Measures taken to combat this include applying additional water to wash out the salts or using desalinated water for irrigation [6]. Reduced evapotranspiration in agrivoltaic systems may alleviate the risk of salinization. Thanks to the power supply generated by the modules, high-tech irrigation systems like overhead or drip irrigation can be implemented in agrivoltaic systems even in remote or undeveloped locations. The creation of a self-sufficient system that doesn't need additional water or power input seems within reach and could revolutionize agriculture.

There is a high potential for the development of agrivoltaics-adapted irrigation systems, e.g., using the modules' mounting structure for installation. Moreover, the power supply enables the operation of monitoring equipment and data processing units on site. Linking sensor data from within the field with plant models and weather data/forecasts could be used to create a highly efficient irrigation control system. Additionally, this data can be used to develop a custom PV-tracking algorithm to increase total system efficiency.

One more aspect to look into is water storage: irrigating with rainwater necessitates water storage. As storage tanks and ponds use up space and involve the risk of algae growth and other contamination, more elegant storage solutions should be developed.

3.3 Cleaning

Dirt on the surface of the modules reduces effective irradiance which in turn reduces power output [3]. Several techniques have been developed to mitigate this. They can all be placed on a spectrum ranging from manual cleaning in low-labor-cost regions to fully automatic cleaning systems in extremely exposed or isolated locations [3].

In agrivoltaic systems, the agricultural activity beneath the panels might increase the risk of soiling, while the elevated position of the modules and the plants underneath make it hard to reach them. Recently, the possibility of residuals from pesticides and fungicides depositing on the backside of the modules has been cause for concern. It is not yet clear whether this will require counter measures to maintain the bifacial gain.

Automatic cleaning systems are costly and thus only profitable in locations with very high exposure. Considering that in most locations, the exposure might only be intensified during certain seasons (e.g. harvesting season), semi-automatic systems seem to be a fitting, more economical alternative. This indicates significant potential for the development of semi-automatic systems that can be used with the farming equipment, e.g. brushes that can be attached to tractors. For systems with low clearance heights in areas with low labor costs, manual cleaning could also be an option.

3.4 Cooling

As power output decreases with rising cell temperature, cooling the modules is beneficial in some locations. Again, a range of techniques for mitigation has been developed; with respect to the topic at hand pertaining to rainwater, only the ones using water for cooling were investigated: photovoltaic/thermal (PV/T), water spraying and flowing water. From an energy efficiency point of view, PV/T systems would be the preferred option, because they increase the efficiency of solar energy conversion. But with the backside covered with tubes and insulation, PV/T modules are opaque. As the crops under the panels need light and the light distribution should be as even as possible, using PV/T modules in agrivoltaic systems would require special consideration. System parameters like elevation and row-to-row distance would have to be adjusted. An opaque backside would also eliminate the possibility of utilizing the amount of radiation that is reflected by the ground with bifacial modules. Due to the elevated position of the modules and greater row distances, this portion is more significant in agrivoltaic systems compared to ground-mounted PV systems. Further analysis might uncover agrivoltaic applications where PV/T can be implemented successfully.

A different possibility is attaching sprinklers to the modules to spray them with water. But as the module isn't fully covered with water, only a few spots are actually cooled while the rest of the modules surface isn't covered by water or isn't covered long enough to create a lasting effect. Additional efficiency losses are caused by wind or inexact orientation of sprinklers.

Out of the cooling techniques involving water, free flowing water seems to be the most promising for agrivoltaic applications. It involves attaching a perforated tube to the upper edge of the module. Depending on the flow rate, water covers the entire front of the module and cools it very effectively: Luboń et al. managed to increase power output by 20% [4]. With the gutter for rainwater harvesting already attached to the lower edge, the technical effort to create a closed circuit should be within reasonable limits. Including a heat exchanger could increase the energy output of the system. However, pumping the water through the system and compensating losses also requires energy. Whether the increase in power output can cover this energy expenditure is unclear and depends on module technology, irradiance, configuration of the water circuit and other factors. Rough calculations of different system configurations might be able to give an indication. Moreover, installing a water circuit would put additional load on the mounting structure, which is already one of the main economic concerns in agrivoltaics [5]. Additional technical components may also influence the light distribution beneath the panels.

The necessity, i.e., profitability of cleaning and/ or cooling PV modules is inherently dependent upon local environmental and economic factors, which is why these should be explored with special regard to agrivoltaics.

4 CONCLUSION

The above stated aspects need to be investigated to achieve a successful and sustainable implementation of not just water-utilizing applications but agrivoltaics in general. Agrivoltaics can be powerful tool in the Water-Energy-Food Nexus. Current research is largely focused on understanding the interplay of crop and PV-system. It is of great importance to consider water as an equally implicated sector, not only in research, but in the conceptualization and technical design of future agrivoltaic projects. There is great potential for agrivolatic systems to be an asset to water management and to address challenges even beyond the agricultural and energy sector. To achieve this goal, research projects and experiments must be conducted to examine each of the applications from a technical, agricultural and economical point of view.

5 REFERENCES

[1] Weselek, Axel, et al. 2019. Agrophotovoltaic systems: applications, challenges, and opportunities. A review. In: Agronomy for Sustainable Development. 39, 2019.

[2] Maliva, Robert G. 2020. Anthropogenic Aquifer Recharge. Cham: Springer Nature Switzerland AG, 2020.

[3] Gareis, Gabriel Hernán. 2021. Verschmutzung in der Photovoltaik auf einen Blick. Freiburg: Fraunhofer-Institut für Solare Energiesysteme ISE, 2021.

[4] Luboń, Wojciech; Pełka, Grzegorz; Janowski, Mirosław. 2020. Assessing the Impact of Water Cooling on PV Modules Efficiency. In: Energies. 13(10), 2020.

[5] Trommsdorff, Maximilian. 2016. An Economic Analysis of Agrophotovoltaics: Opportunities, Risk and Strategies towards a More Efficient Land Use. Constitutional Economics Network, Paper No. 03-2016. University of Freiburg.

[6] Ali, M. H. 2010. Fundamentals of Irrigation and On-farm Water Management Volume 1. New York : Springer, 2010.

EFFECTIVENESS OF SHORT GROUND-BASED MEASUREMENT CAMPAIGNS OF SURFACE ALBEDO

Vicente Lara-Fanego[a], Christian A. Gueymard[b], Tomas Cebecauer[a], and Marcel Suri[a]
[a] Solargis s.r.o., Bottova 2A, 811 09 Bratislava, Slovakia; vicente.lara-fanego@solargis.com
[b] Solar Consulting Services, Colebrook, NH 03576, USA

ABSTRACT: Bifacial PV technology is now becoming a key component of the solar energy industry. Several studies have investigated the quantification of bifacial gains as a function of surface albedo. Because of the latter's high spatiotemporal variability and lack of appropriate global data at high resolution, there is now a strong demand for the precise characterization of the local albedo at the design stage of many solar projects. A practical and inexpensive strategy to quantify the surface albedo and its spatiotemporal variability—as currently used in some real projects—consists in obtaining in-situ observations with a portable albedometer through short measurement campaigns of just a few days. Assuming that the mean deviation between these scarce observational measurements and those from moderate-resolution satellite-derived long-term time series of global albedo data is reasonably invariant over the entire long-term period covered by the latter, these satellite-based estimates can be corrected, using a sort of site-adaptation method, to ultimately confer the appropriate long-term variability to the albedo data. This work investigates the preliminary performance of such a strategy, focusing on the albedo's temporal variability. A database of albedo measurements recorded at 7 reference radiometric stations in the USA is used. Results show that very short-term campaigns of ground-based measurements and the ulterior satellite correction are not sufficient for the proper characterization of the surface albedo at any of the sites studied. Notwithstanding, it is found that a significant reduction in the systematic deviations between modeled and observed values is possible through the proposed site-adaptation technique. Depending on location, the latter can provide valid results with typically a few months of ground measurements, i.e., shorter than the 12-month period usually recommended for site adaptations of the solar resource.
Keywords: Albedo; Bifacial PV; Radiometry

1 INTRODUCTION

Proper consideration of the reflected irradiance is fundamental in many solar energy applications, e.g., [1][2]. This is particular true for the increasingly present bifacial photovoltaic (BPV) technology, which exploits the reflected irradiance incident on the modules' backsides. According to IEA estimates, it is anticipated that the share of BPV modules within the solar market will reach 50% by the end of 2023 [3]. Therefore, this technology is undoubtedly set to play an important role in the generation of solar electric power. In parallel, the energy production of BPV plants is significantly more sensitive to the magnitude and variations in local surface albedo than those based on conventional monofacial technology. For instance, it is currently considered that the dependence of the BPV energy gain on albedo is of the order of 30% per unit albedo [4]. Therefore, the proper characterization of the surface albedo is of key importance in order to reduce the uncertainty in the cost effectiveness of BPV projects.

The surface reflection process is determined by the bihemispherical reflectance, customarily called albedo (ρ). In broadband terms, the surface albedo is the ratio between the global reflected horizontal irradiance (RHI) and the global horizontal irradiance (GHI) incident on that surface, i.e., ρ = RHI/GHI. In nature, most of the land areas not covered by ice or snow have an albedo in the range ≈0.10–0.35. For decades, it has been customary to consider a fixed albedo of 0.2 for all calculations—a practice that is now called into question because it appears too simplistic. For instance, massive water bodies usually have a low albedo, typically well below 0.1, whereas regions covered with fresh snow have a very high albedo, which can exceed 0.85.

The problem here, however, is that albedo is not a truly invariant characteristic of the surface (whose physical properties often vary over time), but also depends on the atmospheric and illumination conditions of the moment. This combination of rapidly changing external drivers results in the albedo's remarkable spatiotemporal variability—a site-specific characteristic for the most part. For solar applications, the spatial variability of surface albedo has already been analyzed in the literature, which has highlighted various difficulties, e.g., [5].

In parallel, the present study is aimed at improving one's understanding of the *temporal* albedo variability, toward a refined characterization of this key variable. Specifically, the scientific question being addressed is whether such characterization can be satisfactorily achieved with *short* measurement campaigns—a practice that has started to become commonplace in recent years, because a long-term (12 months at least) high-quality measurement campaign with multiple radiometers is difficult, requires costly equipment and maintenance (particularly at remote sites), and increases the cost and time constraints of BPV projects.

Short albedo measurement campaigns are typically carried out with a portable albedometer during only a few days. In order to confer the appropriate long-term variability to the data, the possible systematic deviation (bias) of the long-term time series of satellite estimates is then empirically adjusted according to the recorded observational data. This approach is equivalent to a simple site-adaptation method of satellite-based surface albedo estimates. Such methods are routinely used in solar resource assessments to decrease the bias and uncertainty in satellite-derived irradiance estimates, using statistical methods and short-term in-situ ground measurement campaigns [6]. Since ground measurements are normally assumed to be the most accurate source of radiation data, the solar industry anticipates that this correction method provides a reliable characterization of the surface albedo too. In this context, the present work presents a first in-depth study to investigate the performance and limitations of such a strategy.

2 METHODOLOGY

The main experiment conducted here consists in simulating multiple campaigns of local measurements with a determined number of consecutive days. To match the resolution of the albedo retrievals obtained from space-borne MODIS observations, time series of daily albedo values during a 5-year period (2011–2015) are derived from the measured data at three NOAA-SURFRAD radiometric stations (Desert Rock [DRA], Table Mountain [TBL], and Fort Peck [FPK]) as well as four AmeriFlux stations (Walnut Gulch Kendall Grasslands [US-WKG], Metolius Mature Ponderosa Pine [US-Me2], Morgan Monroe State Forest [US-MMS], and Willow Creek [US-WCr]) in the Central and Western USA. These seven stations are selected here because they are considered *representative* of a larger, 500-m nominal MODIS pixel area [5]. Moreover, they cover a wide range of different climatic conditions. Their geographical distribution is shown in Figure 1.

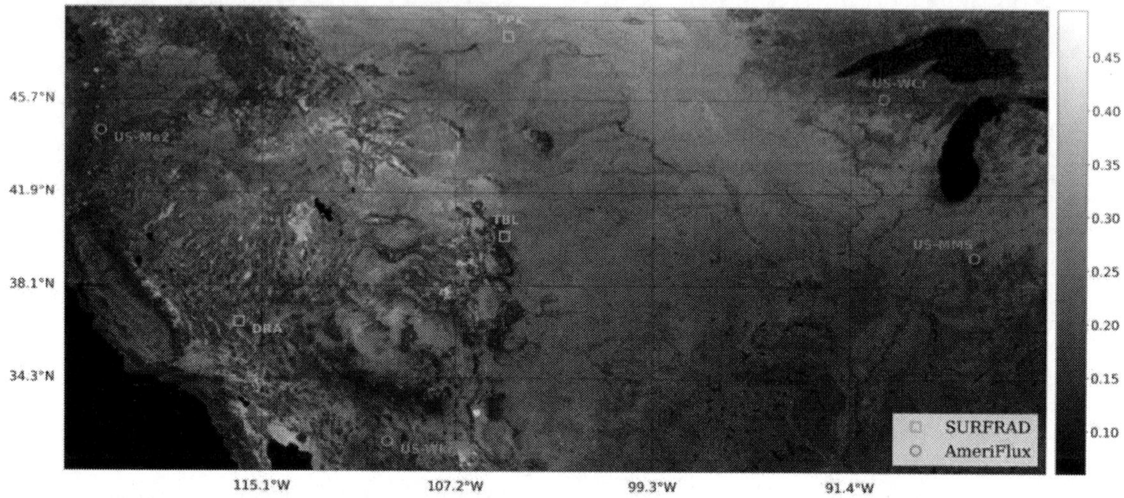

Figure 1: Location of the 7 radiometric stations under scrutiny, with a color code that defines their respective network. The background's grayscale indicates the mean annual albedo from the Solargis database.

Snow is known to introduce a strong and variable source of uncertainty, hence the albedo of ephemeral snow is filtered out here by excluding albedo values above 0.30. The satellite albedo estimates are obtained from the Solargis database at daily temporal resolution and 500-m spatial resolution[1]. The year 2015 is selected as the period when the fictitious measurement campaigns are supposed to have been conducted. The temporal duration of these campaigns is established in groups of consecutive days from a minimum of 1 day, increasing discontinuously to cover the whole year (up to 365 days, depending on the number of available valid days at each site). The campaign's initial date is selected randomly without repetition. The number of samples (number of measurement campaigns) is at most 200. For instance, for measurement campaigns of 8 days, a total of 200 distinct measurement campaigns is simulated. To that effect, 200 dates are randomly selected during 2015, each of them being the first of 8 consecutive days of measurements. To avoid border effects when the campaign would finish later than December 31, the remaining days to complete the campaign are taken starting January 1 of the same year 2015. Only valid pairs of observations and satellite data are considered. Finally, the satellite deviations (with respect to the ground-based data of the different measurement campaigns) are corrected for the long-term (5-year period), and the adjusted time-series thus obtained are then evaluated against the corresponding ground observations.

The resulting distributions of mean bias differences (MBD) are then analyzed. To characterize the method's value in practice and evaluate whether it is worth the effort to carry out a measurement campaign of a specific duration, it is assumed that the desirable improvement is to reduce the systematic deviation of the satellite estimate by at least half. Using a higher value, e.g., a 75% reduction in MBD, would not change the conclusions of this work with respect to the value of short-term measurement campaigns.

3 RESULTS

Figure 2 shows the results for the arid region of Desert Rock (DRA), Nevada, a case in principle favorable to the strategy under scrutiny thanks to the relatively low temporal variability of albedo under such climatic conditions and sparse vegetation. The bottom plot shows the time series of daily values of both the ground observations (gray line) and the satellite estimations (blue line) during 2015. In addition, two gray dashed horizontal lines represent the mean annual values of albedo of the ground measurements for 2015 and the long-term period, respectively. Remarkably, these values are almost equal at that site, thus denoting a very low interannnual variability at that site. Similarly, the dashed blue horizontal line marks the long-term mean value of the satellite estimates. Some days of sharp drops in albedo are observed in the ground measurements. These correspond to rain events, most likely (since water has a low albedo). Although those days have been taken into account in the random selection of dates for the campaigns, the realistic framework envisioned here calls for their omission in the analysis below for two essential reasons. First, it is highly

[1] https://solargis.com/docs/methodology/albedo

likely that the campaign would have been suspended under these particular and unusual conditions. Second, the deviations they produce are comparably extreme and could undermine the generality of the results. In contrast, two unexceptional cases of the customary 3-day measurement campaigns are highlighted with three red and green dots, respectively, for further discussion below. The red dots indicate situations when the ground-measured albedo is larger than the concomitant satellite estimate (i.e., the solid gray line is *above* the blue one in the bottom plot). Conversely, the green dots correspond to situations when the daily observations are lower than the satellite values (solid gray line *below* the blue one).

In all cases, the average albedo is calculated for each simulated campaign. The deviation relative to the average satellite-derived albedo is determined for the same dates. This is considered as the mean deviation (MBD) of the satellite with respect to the observations. This MBD is then used to correct the long-term time-series provided by the satellite. This corrected time-series is evaluated against the long-term values of the ground measurements, which results in a new MBD. The red and green horizontal dashed lines in the bottom plot of Figure 2 respectively represent this final MBD for each example campaign at DRA. This process is then repeated for all campaigns, and the MBD of the corrected satellite time-series is finally compared with that of the original (uncorrected) satellite time-series.

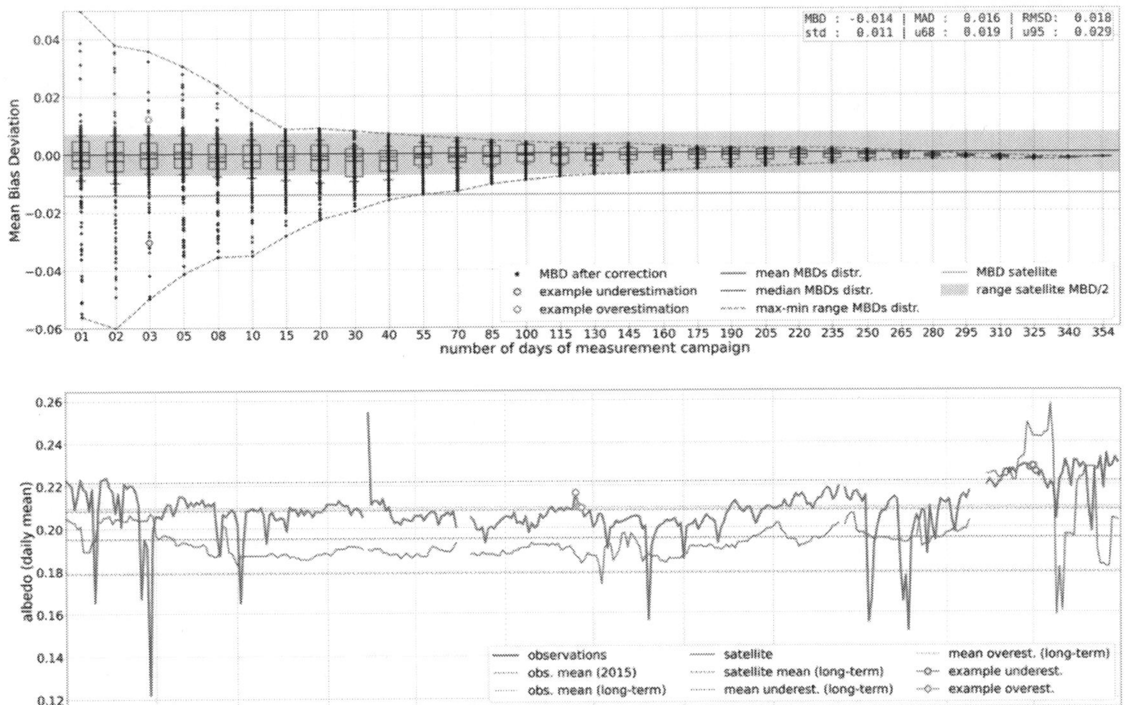

Figure 2: (Top) MBD distributions for fictitious measurement campaigns hypothetically conducted at Desert Rock (DRA), Nevada, as a function of their duration (blue dots). The boxplots of their distribution are also shown, including the mean and median values. The MBDs obtained for the measurement campaigns highlighted with red and green dots in the bottom plot appear as a red and green dot, respectively. The MBD corresponding to the uncorrected satellite data is shown (orange horizontal line), along with the range of MBD values that are lower or equal to half of that MBD. The MBD, MAD, RMSD, STD, and the uncertainty at 68% and 95% of confidence for the uncorrected satellite data with respect to the ground observations are also provided. (Bottom) Time-series of the daily albedo data at DRA; both the ground-based observations (solid gray line) and satellite estimations (solid blue line) are shown. Examples of 3-day in-situ campaigns are highlighted as red and green dots. Two gray dashed horizontal lines indicate the mean observed values for 2015 and the long-term value. The dashed red and green lines denote the mean long-term values after correction for the corresponding values of the campaigns highlighted as red dots and green dots, respectively.

The distributions of the daily MBDs resulting from the comparisons between the site-adapted satellite estimates and the 5-year ground observations are represented in the upper plot of Figure 2. The solid blue dots represent the MBDs of all the simulated measurement campaigns with different period durations (x-axis, uneven intervals). Negative MBDs indicate an underestimation of the corrected satellite-based albedo with respect to the ground observations. The gray zone in the upper plot indicates the range of values lower than, or equal to, half of the satellite's actual MBD (orange horizontal line). This is also indicated in the top-right text box along with other statistics. In particular, three common metrics are provided: standard deviation (STD), mean absolute deviation (MAD), and root-mean-square deviation (RMSD) of the original (uncorrected) satellite time-series. In addition, the uncertainty values at 68% and 95% of confidence, as obtained from the distribution of the residuals, are indicated. Boxplots, as well as the mean and median values, are also shown to fully describe the MBD distributions.

It is clear that the dispersion of results decreases rapidly when the measurement campaign period increases, as could be expected. In this example, all experiments produce results (blue dots) within the limits of the established criterion (gray area) when the ideal measurement campaigns are at least 145 days. It is emphasized that this number is only *indicative* because of the random selection of the starting day of the measurement campaigns, hence it might vary slightly depending on circumstances. Furthermore, this minimum number of days is found to also vary depending on location (not shown for brevity), although it is always relatively large (typically 3–4 months for the majority of the sites considered here), and considerably larger than the usual 2 or 3 days of very short-term measurement campaigns actually undertaken for many solar projects.

Most critically, there are many site-specific campaign situations at DRA for which the results would not attenuate the satellite systematic underestimation (orange horizontal line, negative MBD in Figure 2), or would even introduce a larger systematic error (in absolute terms). This enhanced bias can either increase the underestimation or rather create a positive systematic deviation, depending on site, climate, season, etc. In the upper plot of Figure 2, the resulting MBD of the two situations described just above are highlighted consistently with dots colored in either green (whenever the corrected satellite time-series underestimates the observed long-term values) or red (when it overestimates the observations) within their corresponding group of 3-day measurement campaigns. More precisely, the highlighted red dot in the upper plot is the MBD obtained from the measured albedo values highlighted also in red in the lower plot. Similarly, the green dots of the upper and lower plots are related in the same way.

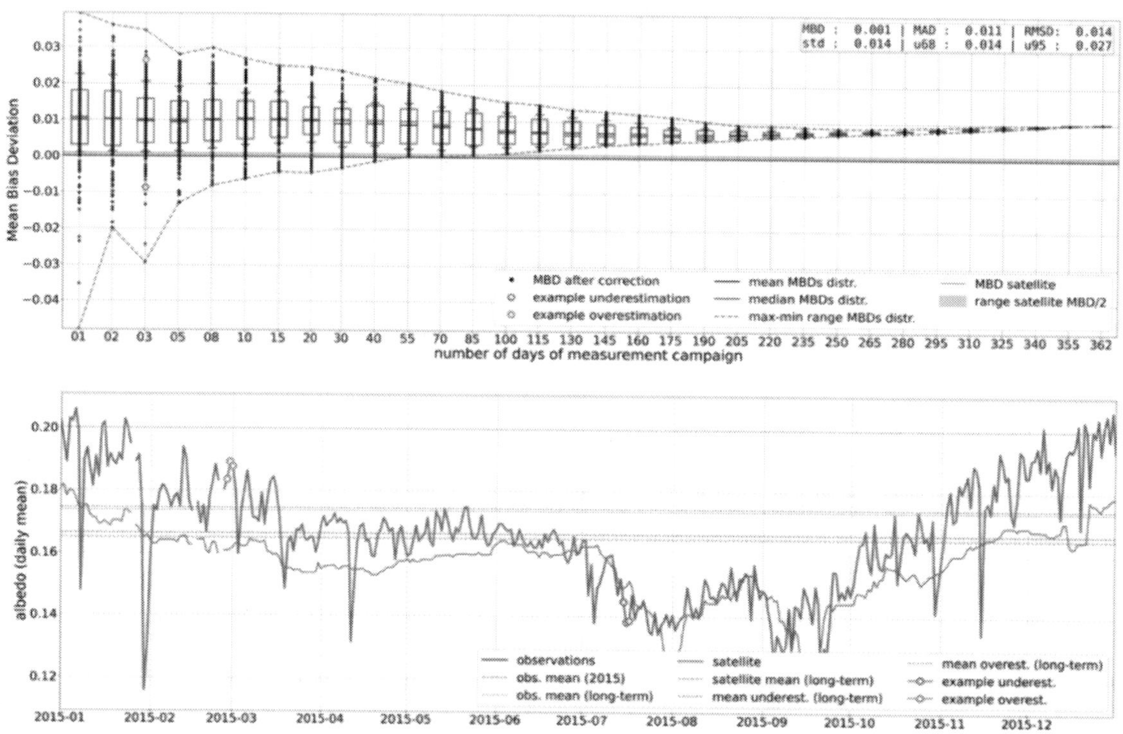

Figure 3: Same as Fig. 2, but for Walnut Gulch Kendall Grasslands (US-Wkg) in Arizona.

In the case of albedo overestimation (red dots), the fictitious measurement campaign for that example took place at the end of June. For that period, the systematic deviation is larger than the deviation between the original non-corrected satellite data and the long-term observational data. This results in an overcorrection of the satellite long-term time series with respect to the ground-based values, and thus in an albedo overestimation. The opposite situation occurs towards the end of November and beginning of December, when the satellite estimates are higher than those observed with the ground-based albedometer. In that case the deviation's sign is opposite to that of the actual mean deviation, thus erroneously indicating that the correction should be made in the opposite direction. This results in an increase in the magnitude of the albedo underestimation that is observed in the upper plot of Figure 2, and thus in an aggravation of the initial estimate, contrary to expectations.

Figure 3 presents the same information as Figure 2 but for the case of US-Wkg, located in a grassland area—hence with much more vegetation and a milder climate than DRA. This exemplifies the extreme adverse results that can occur with the strategy under scrutiny. For this location, the overall deviation of the original non-corrected satellite long-term estimation with respect to the observations is negligible (≈0). Therefore, the expected albedo correction factor is also 0. However, conducting a measurement campaign of 3 days, for example, like those highlighted again with red and green dots, would inevitably induce an error. This is because the deviations of the satellite with respect to the observations during those days are not 0, as can be seen in the bottom part of Figure 3. Consequently, an unwanted overcorrection (correspondingly negative or positive) occurs. In fact, as can be seen in the upper part of Figure 3, in almost no case can the method improve the original estimate of the uncorrected satellite in terms of bias, but rather almost always worsen it. This is a logical consequence of the fact that, in that case, the original satellite estimate is virtually unbiased. (In parallel, the deviations are relatively small in absolute terms.) For the best practice, an additional analysis would be required in this kind of situation to evaluate the benefit of any possible correction of the satellite data. In the particular case of US-Wkg (Figure 3), the station's albedo is characterized not only by seasonal variability but also by interannual variability. If bias corrections were to be devised there based on very long campaigns (close to 12 months), the variability situation would likely cause an overcorrection of the satellite data. A deeper analysis of this situation would be necessary, but is out of the scope of this work.

Finally, it is emphasized that the limit used as a criterion to evaluate the goodness of the method, i.e., to reduce the MBD of the non-corrected satellite albedo data by at least half, does not restrict the generality of the conclusions obtained from the results presented here. For instance, selecting a more permissive reduction of bias of only 25%—thus yielding an upper limit of 75% of the original MBD—implies that this criterion would always be reached with a shorter measurement campaign. Thus, considering DRA for example, the upper part of Figure 2 additionally indicates that the situation for which all cases have an MBD lower than that of the uncorrected satellite (horizontal orange line, or 100%) occurs for 70 days. Therefore, to achieve a reduction of 25% in the MBD of the uncorrected satellite data, the number of days would be necessarily ≥70, such as ≈85. Similar reasoning and observation can be made in the case of setting a more restrictive reduction of error of, e.g., 75%, which would lead to a lower final MBD of only 25% relatively to the raw MBD. In that case, more days would be needed to ensure that all cases meet the prescribed limit. Indeed, the upper plot of Figure 2 indicates that this number of days would increase to ≈190, i.e., slightly more than 6 months.

4 CONCLUSIONS

Currently, very short-term surface albedo measurement campaigns are being carried out for the characterization of this variable for future solar projects based on bifacial PV technology. The costs and complexity of the operation and maintenance of radiometric instrumentation over long periods of time are at the origin of this practical and economical solution. The essential question that was addressed here is whether this simple procedure actually constitutes a good approach.

In the real world, the local albedo is most often affected by high temporal variability, which introduces the risk that a short-term in-situ campaign, intended to characterize the long-term albedo value, might occur at the wrong time. This could eventually produce albedo estimates that are worse than those that are typically obtained from reliable satellite estimates, for instance. Such a situation would increase the uncertainty in the albedo data and therefore negatively affect the bankability of the solar project, which is the exact opposite of the expectation.

The objective of this work was to evaluate the actual potential benefit of in-situ albedo measurement campaigns by simulating their results under realistic conditions. To that effect, this study used five years of albedo data measured in the USA at seven high-quality radiometric stations, known to be located over relatively homogeneous terrain and thus representative of satellite albedo retrievals at ≈500-m resolution. For each single year, many fictitious campaigns were simulated with variable duration, after random selection of the starting day. The measured albedo values were compared with the corresponding satellite-based estimates. Their deviations relative to the ground measurements were used as correction factors to adjust (or "site adapt") the satellite data. The time series of the corrected albedo values were then compared with the corresponding ground-based measurements to assess the effectiveness of the correction method.

The results showed that there is a wide range of possible cases—a consequence of the temporal variability of albedo—that causes a wide dispersion in plausible results. That dispersion is larger for a shorter measurement campaign, which in many cases generates larger errors than those of the satellite retrievals.

Overall, it is found that the local albedo cannot be characterized by a small sample of observational data collected during only a few days, and that, moreover, such data cannot generally be considered usable for correcting satellite estimates.

Nevertheless, an appropriate site-adaptation technique—similar to those lasting ≈12 months in solar resource assessments—can be recommended to fully capture the seasonal variability. In effect, the results presented here do show that it is possible to significantly reduce the deviations and uncertainties of albedo estimates with limited periods of in-situ measurements. In all the cases analyzed the duration required to obtain a reliable correction is much longer than the 2- or 3-day campaigns that have become typical in the solar industry. Moreover, the optimum duration is found to largely depend on local conditions, even at the representative sites evaluated here. Stated differently, a correction based on very short-term measurement campaigns is applicable only to those particular sites where the deviation of the satellite-based data is invariant over time relative to ground observations. Consequently, it is obvious that even longer in-situ campaigns would be required at sites with heterogeneous terrain or surface conditions.

Finally, it is concluded that, contrary to expectations, the use of short-term measurement campaigns often leads to an increase in albedo uncertainty, which is precisely what is to be prevented to obtain a bankable solar system design.

6 REFERENCES

[1] C. A. Gueymard, V. Lara-Fanego, M. Sengupta, Y. Xie, Solar Energy 182 (2019) 194–212.

[2] B. Marion, Solar Energy 215 (2021) 321–327.

[3] G. Masson et al., Trends in Photovoltaic Applications 2021. Report IEA-PVPS T1-41 (2021).

[4] M. Aghaei, M. Korevaar, P. Babal, H. Ziar, Solar Radiation - Measurement, Modeling and Forecasting Techniques for Photovoltaic Solar Energy Applications, London: IntechOpen (2022).

[5] C. A. Gueymard, V. Lara-Fanego, M. Sengupta, A. Habte, Solar Energy 227 (2021) 655–673.

[6] J. Polo, S. Wilbert, J.A. Ruiz-Arias, R. Meyer, C. Gueymard, M. Súri, L. Martín, T. Mieslinger, P. Blanc, I. Grant, J. Boland, P. Ineichen, J. Remund, R. Escobar, A. Troccoli, M. Sengupta, K.P. Nielsen, D. Renne, N. Geuder, T. Cebecauer, Solar Energy 132 (2016) 25–37.

This is a short version of the paper submitted to the journal EPJ Photovoltaics.

It has been selected by the Committee of the WCPEC-8 for submission to EPJ Photovoltaics.

SIMULATION OF THE IRRADIANCE AND YIELD CALCULATION OF BIFACIAL PV SYSTEMS IN THE USA AND GERMANY BY COMBINING RAY TRACING AND VIEW FACTOR MODEL

Eva-Maria Grommes, Felix Schemann, Frederik Klag, Sebastian Nows, Ulf Blieske
Cologne Institute for Renewable Energy
University of Applied Sciences Cologne
Betzdorferstraße 2, 50679 Cologne, Germany.

ABSTRACT: Compared to conventional photovoltaic (PV), there are more influencing factors in bifacial photovoltaics to be considered to calculate incoming irradiance and energy yield. Accurate models are required to investigate the influences of the elevation, the albedo of the ground, the shading conditions between the PV rows and many other rear-side related factors. This paper combines the ray tracing (RT) and view factor (VF) models to calculate the irradiance with a subsequent electrical yield calculation using the one-diode model. To verify the results of the developed open-source simulation program BifacialSimu, accurate data from a plant in Golden, USA and a commercially operated plant in Germany are used. Comparing the actual data, it can be concluded that a combination of RT and VF models seems valid for more extended simulation periods of several months since the relative errors balance out. The RT-only simulation accurately reproduces the precise hourly radiation and electrical yield pattern. Still, a continuous positive deviation was found, which does not even out over long periods and is thus less accurate than the VF/RT combination. A simulation for a single month with RT can take several hours. Thus, the best simulation mode results according to user requirements.

Keywords: Photovoltaic, Bifacial, Solar Radiation, Software, Energy yield

1 INTRODUCTION

Bifacial photovoltaics are becoming increasingly important in the PV industry [1]. The behaviour of these modules is analyzed in numerous field studies. In addition, several commercial and freely available simulation programmes are already available on the market. Compared to monofacial PV, bifacial PV provides up to 30% more yield on the same area for fixed-angle generators. With tracked systems, up to 40%, additional yield can be achieved. As a result, PV systems with bifacial modules can achieve the lowest possible Levelized Cost of Electricity (LCOE). [2]

Detailed and accurate simulation programmes are necessary to consider all aspects of the irradiance and yield calculation during the planning of these systems [2]. In addition, daily yield forecasts play a significant role in the direct marketing of PV electricity, where incorrect forecasts can lead to losses. The elevation height, the albedo of the ground, and other properties significantly affect the rear irradiance and, thus, the additional yield of bifacial PV systems [3]. The contribution of this scientific publication is the combination of two different irradiance simulation methods. The simulation methods are the python libraries bifacial _radiance [4] of the National Renewable Energy Laboratory (NREL) and pvfactors [5] of SunPower Corporation. The two programmes are based on different radiation calculation models, one is the ray tracing model, and the other is the view factor model. For the validation of the simulation methods, actual radiation and yield data of two different plants are available. The plants are located in the USA and Germany. The actual data of these plants will be compared with the simulation data and analyzed for deviations.

2 STRUCTURE OF BIFACIALSIMU

To combine both simulation methods, the simulation proramme BifacialSimu [6] is used. The script with which the simulation is started is the main script. In the main script, all variables are declared necessary for the simulation. In addition, different simulation modes can be set for the irradiance simulation. Altogether there are three simulation modes, listed in Table 1 in the Appendix.

BifacialSimu is divided into three main sections: irradiance simulation, albedo simulation and electrical simulation. In every sector, a choice between different calculation methodologies is possible, as shown in Figure 1. This option gives the user the opportunity to adapt the simulation to their needs and to the individual project.

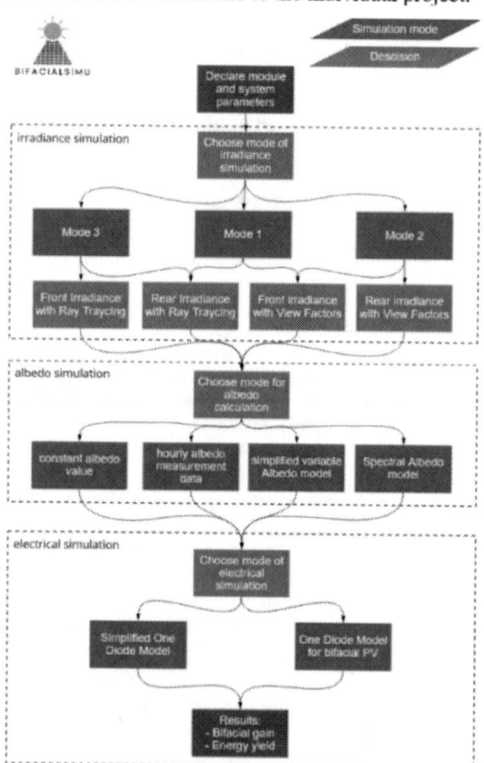

Figure 1: Simulation Procedure

The overall structure, containing the essential functions, is displayed in a Flowchart in Figure 5 in the Appendix.

3 SIMULATION

The procedure for running a simulation is briefly explained below. First, the simulation variables [6] must be entered via the main script or in the Graphical User Interface (GUI). The simulation mode, the orientation of the modules, height, electrical module parameters, and other essential parameters can be set via variables.

The weather data must be read in CSV format. Here the content must correspond to the TMY3 data format [7], it can be inserted manually or download from a TMY3 database. The global horizontal irradiance (GHI), diffuse horizontal irradiance (DHI), direct normal irradiance (DNI), ambient temperature and albedo must be included as hourly values for the period under consideration. If the hourly albedo is not available in the weather data, a constant albedo of the subsurface can be specified in the variable declaration. If the albedo is defined as a variable, it will be read from the weather file.

If all inputs are defined, the simulation can be started. The modules are divided into segments in both models to represent the inhomogeneity of the radiation, especially on the rearside of the modules. These segments divide the module on the Y-axis and are considered radiation sensors.

4 CALCULATION METHODOLOGY

The calculation can be logically divided into the irradiance calculation, the electrical simulation and the yield calculation. The irradiance calculation is conducted using the VF and RT models, while the electrical simulation uses the one-diode model. A simplified electrical simulation approach can be applied for missing rear module data. The energy yield calculation assumes an average module and is indicated per area.

4.1 Irradiance Calculation

The two libraries, bifacial_radiance [4] and pvfactors [5], are used for the irradiance simulation. bifacial_radiance creates a three-dimensional model of the hourly conditions in the year on the geometries entered. These three-dimensional models are then used to simulate the irradiance with ray tracing. pvfactors, on the other hand, simulate the irradiance via view factors and create a two-dimensional model. In pvfactors, the pvlib Python implementation of the Perez Model [8] is used [5]. For combining both computational methods within a simulation, some simplifications are used, which will be subsequently explained. While bifacial_radiance offers the possibility to specify geometries with gaps and spaces up to the cell level, it is not possible with pvfactors. Therefore, this function is not used in the programme.

Also, the simulation cannot consider more complex arrangements of modules with different orientations or different numbers of modules in each row and back shading caused by the mounting system. Both libraries require a sun position calculation and calculation of the module inclination angles [5]; [9]. To speed up the simulation, these calculations are performed only once via bifacial_radiance, which performs this calculation via pvlib [9]. The results of the calculations are then transferred to pvfactors in the form of a data frame.

The simulation mode cumulative sky of bifacial_radiance has limited use in this programme. The mode accumulates all sky states over the considered period for the irradiance simulation.

4.2 Electrical Calculation

The simulation of the electrical energy yield of the bifacial PV system is performed using the one-diode model. Bifacial values for fill factor FF_b, short-circuit current I_{SC_b}, and open-circuit voltage V_{OC_b} are first calculated using the module parameters from the datasheet of the modules. Then, the power P_{mpp} of the whole module is calculated by Equation 1 [2].

$$P_{mpp} = FF_b * V_{OC\_b} * I_{SC\_b} \quad (1)$$

4.3 Yield Calculation

The electrical energy yield of the PV system is calculated using electrical power. Since power values are calculated hourly, they can be equated with electrical energy. Each hour, the yield is summed up for an average module row by row. The area-specific electrical energy yield Y_b is used in the evaluation, which is the sum of the hourly energy yields Y_{bh} of the system and the area A_{module} of the solar module (Equation 2).

$$Y_b = \sum_{i=0}^{n} \frac{Y_{bh}}{A_{module}} \quad (2)$$

5 VALIDATION AND EVALUATION

In the following, the simulation results for both plants are validated using actual data and examined for deviations. Therefore, the measured front irradiation and the yields of two plants at two different locations are compared with the simulation results. There are no measured data for the back irradiance in both cases, which can not be compared directly with the simulation data.

NREL operates the plant in the USA at the site in Golden, USA [10]. Here, measurement data of the DC yield of module Y_b, albedo, irradiance onto the front side of the modules Irr_{front}, GHI, DNI, DHI and ambient temperature are available in an hourly resolution and high data quality. Data of "Row2" from the dataset is used as a reference. The module row consists of 19 modules, which are connected in series. This plant uses single-axis tracking. The modules are oriented to the south, and the rotation axis is at 1.5 m. In addition, data from a nearby weather station of the NREL Solar Radiation Research Laboratory (SRRL) are available [11]. Unrealistic measured data, such as negative radiation or missing values, are filtered before comparison. In case of missing values, values from the previous hour are used. The module parameters needed to simulate the electrical yield are taken from [12]. The rated power of the modules is approximately 350 Wp. The modules are poly-p-type PERC modules. A period of half a year between 01.10.2019 and 30.04.2020 is considered for simulation due to the completeness and quality of the measurement data.

Plant Germany is operated in Niedersachsen, Germany. Hourly AC yield data and irradiation on the front side of the modules are available here. The yield data are grid feed-in values, so inverter, interconnection, and line losses are already included. This must be considered when evaluating the data. Only percentage deviations between measured data and simulation data and the annual yield can be presented in this paper. The plant has a nominal output of 744.8 kWp. The modules used are 380 Wp modules STP380S-72/Pfh+ from the Suntech Power company. The modules use a fixed tilt. The radiation and weather data, GHI, DHI, DNI and the ambient temperature required for the simulation are obtained from the German

Weather Service (DWD) nearby. For simulation purposes, a test reference year (TRY) is used for the location of the plant [13]. The measured data are prepared to be read into the programme. A period of one year between 01.05.2020 and 30.04.2021 is considered.

5.1 Evaluation of the results of Plant USA

Figure 2 shows the relative deviation of the simulation results Irradiance on the front side (Irr_{front}) bifacial Energy yield (Y_b) from the actual data in each month. In addition, the difference between mode 1, mode 2 and mode 3 with constant albedo is shown in the graph.

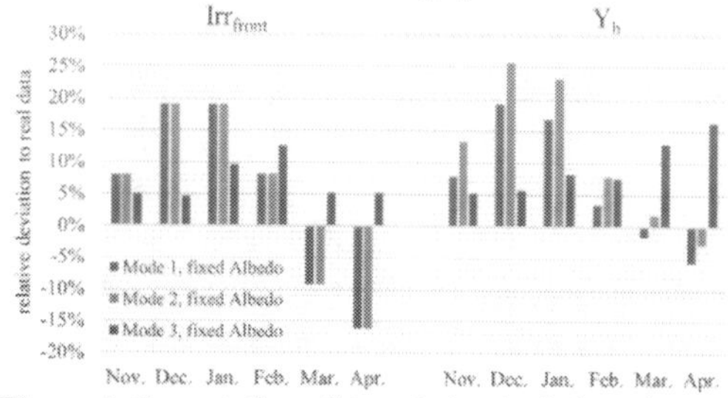

Figure 2: Presentation of the relative deviation of simulation data Nov 2019 – Apr 2020 to the measured data with a fixed albedo

The simulation results of mode 1 deviate from the measured data by 19% in December. Nevertheless, the deviations of mode 1 and 2 level out in the course of the measurement period. This is, however, not the case for mode 3.

Mode 3 simulates the actual course of the electrical energy yield of the plant qualitatively better. This can be seen in Figure 3. Here two selected daily courses of the simulated and measured electrical energy yield are shown. On the left side of the diagram, the 05.11.2019 and the right side, the 06.04.2020 are shown.

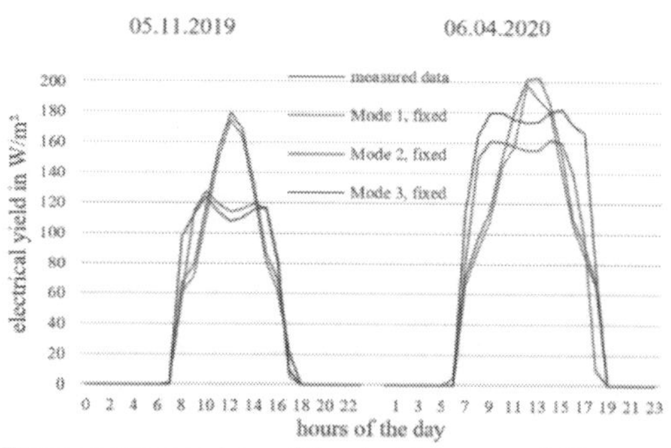

Figure 3: Comparison of the electric energy yield 05.11.2019 and 06.04.2020

On 05.11.2019, it can be seen that the energy yield simulated with modes 1 and 2 exceeds the actual yield between 10 am and 2 pm. The simulated yield has a peak of 170 W/m² there, which is not present in the actual data.

Mode 3 follows the measured yield closely and the deviations are minor. They can also be attributed to the lack of consideration of mismatching. At noon, the measured yield is 114 W/m² and the simulated 107 W/m² corresponds to a relative deviation of -6.5%.

On 06.04.2020, it can be seen that the electrical yield of the VF calculation is significantly below the actual data in the morning and evening hours and above it in the midday hours. This fits with the slightly negative deviation of mode 1 and mode 2 in April.

Overall, mode 3 represents the profile of actual electrical yields more closely. Mode 1 and mode 2 are strongly defined by the GHI course, producing midday peaks that are too high and morning and afternoon values that are too low.

The module array's total measured electrical energy yield between November and April is approximately 148 kWh/m². These are DC values, so inverter and cable losses are not yet included. Despite the significantly higher relative deviation of the simulation results from the measured values in mode 1, the irradiation on the front of the modules and the total electrical yield is closer to the measured data over the entire period under consideration. The deviation of the total electrical yield is 4.4% for mode 1, 8% for mode 2 and 10.4% for mode 3 (fixed albedo). Mode 1 is more accurate in the total consideration because negative and positive values balance each other out. This mechanism does not affect mode 2 as strongly and mode 3 not at all.

5.2 Evaluation of the results of Plant Germany

In the following, the accuracy of the simulation results for *plant Germany* is evaluated with the previously described actual data. Only the irradiance on the front side and the simulated DC yield are compared since these are the only real datasets available. The exact values for the deviations cannot be disclosed in this paper.

The results, in Figure 4, show that the front side irradiance values Irrf of mode 1 and mode 2 are identical since VF calculates the front side irradiance in both modes. The simulated values are only slightly below the actual measured data in this case. In mode 3, in which RT is used to calculate the radiation of the front side, the deviation from the measured data is higher, so the simulated values are even lower than modes 1 and 2.

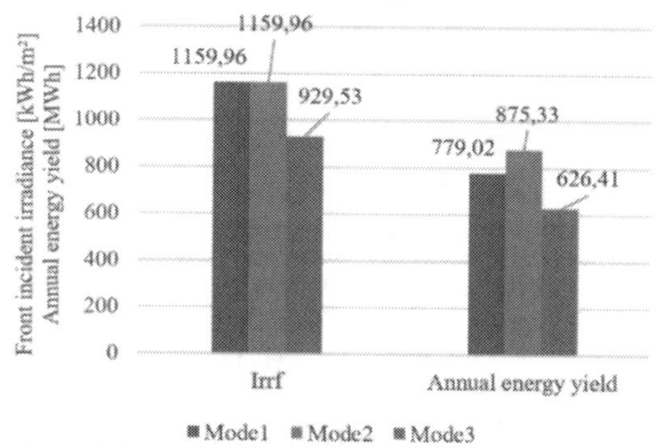

Figure 4: Front irradiance and annual energy yield für simulations of *plant Germany*

Since the weather data used for the simulation is a TRY of the observation period 1995-2012, certain deviations between simulation and actual data are expected. According to a weather map of the DWD, the global radiation of 2020 in the region can deviate from the TRY by up to 15% [14].

Comparing the electrical yield between the three simulation modes and the actual data gives similar results to the irradiance simulation. A medium deviation in the annual energy yield is obtained for the mode 1 and mode 2 simulations. Here, mode 2 is higher than the actual data, and mode 1 is lower. Mode 3, on the other hand, results in a significantly lower annual yield than the actual data.

Overall, mode 1 is closest to the measured energy yield.

6 ERROR CONSIDERATION AND CONCLUSION

In the simulations, loss mechanisms such as soiling, mismatching and interconnection losses are not considered. In addition, the entire module area is assumed to be the absorbing area, which is likely to be smaller in reality due to the edges of the modules, junction boxes and gaps between the cells. Also, shading effects from the mounting system on the rear side are not considered. These neglected effects should lead to a positive deviation from the presented simulation results. This is the case for the simulation with mode 3 of the *Plant USA*, where the relative deviations in irradiance and yield are positive every month. Therefore, this indicates that the RT simulation results are more in line with the actual data.

In theory, the simulation results of *Plant Germany* should also deviate positively overall. In addition to the effects already described here, DC yields from the simulations are compared with AC yields from the actual data and conversion losses from DC to AC voltage are not factored in the calculations.

In the simulations, static reflection losses of 3% on the front side and 5% on the backside are assumed. These angle-dependent losses would have to be considered for a more detailed analysis.

In summary, combining the two calculation models, VF and RT provides more accurate results for the cumulative electrical yield and the irradiance on the module surface for a half-year or year-round simulation. Positive and negative deviations in the combined VF/RT simulation tend to balance out, but this is not the case with the pure RT simulation. On the other hand, the RT simulation more clearly reproduces the actual course of the irradiation and thus of the electrical yield. Still, the permanently positive deviation does not balance out over more extended periods, resulting in a more significant deviation for several months. The pure view factor calculation was the least accurate in the day-by-day view and had higher deviations when viewed over a more extended period than the combination of VF and RT. As a result, combining the two calculation models is suitable for yield prediction over longer periods and thus can be applied in PV system design. However, the RT model should be superior when forecasting electrical yields for intraday energy trading. A further reduction in the deviations from measurement results can be expected if a 2D instead of a 3D VF model is included into the program.

7 REFERENCES

[1] J. S. Stein et al., 'Bifacial Photovoltaic Modules and Systems: Experience and Results from International Research and Pilot Applications: Report IEA-PVPS T13-14:2021', USA, Germany, Switzerland, Italy, Chile, Finland.

[2] R. Kopecek, Ed., *Bifacial Photovoltaics: Technology, applications and economics*. Institution of Engineering and Technology, 2018.

[3] U. A. Yusufoglu, T. M. Pletzer, L. J. Koduvelikulathu, C. Comparotto, R. Kopecek, and H. Kurz, 'Analysis of the Annual Performance of Bifacial Modules and Optimization Methods', *IEEE J. Photovoltaics*, vol. 5, no. 1, pp. 320–328, 2015, doi: 10.1109/JPHOTOV.2014.2364406.

[4] D. Berrian and J. Libal, 'A comparison of ray tracing and view factor simulations of locally resolved rear irradiance with the experimental values', *Prog. Photovoltaics Res. Appl.*, vol. 28, no. 6, pp. 609–620, 2020, doi: 10.1002/pip.3261.

[5] M. A. Anoma, D. Jacob, B. C. Bourne, J. A. Scholl, D. M. Riley, and C. W. Hansen, 'View Factor Model and Validation for Bifacial PV and Diffuse Shade on Single-Axis Trackers', pp. 1549–1554, 2018, doi: 10.1109/pvsc.2017.8366704.

[6] E.-M. Grommes and U. Blieske, 'BifacialSimu: Hollistic Simulation of large-scale Bifacial Photovoltaic Systems', *To be Publ. J. Open Source Softw.*, 2022.

[7] B. Marion, 'User's Guide for Albedo Data Sets', *National Renewable Energy Laboratory*, 2020. .

[8] R. Perez, R. Seals, P. Ineichen, R. Stewart, and D. Menicucci, 'A new simplified version of the perez diffuse irradiance model for tilted surfaces', *Sol. Energy*, vol. 39, no. 3, pp. 221–231, 1987, doi: 10.1016/S0038-092X(87)80031-2.

[9] S. A. Pelaez, C. Deline, P. Greenberg, J. S. Stein, and R. K. Kostuk, 'Model and Validation of Single-Axis Tracking with Bifacial PV', *IEEE J. Photovoltaics*, vol. 9, no. 3, pp. 715–721, 2019, doi: 10.1109/JPHOTOV.2019.2892872.

[10] NREL, 'NREL-bifacial-experimental-single-axis-tracking-field NREL Bifacial Experimental Single-Axis Tracking (BEST) field'. https://datahub.duramat.org/dataset/best-field-data.

[11] SRRL BMS Daily Data, 'NREL Measurement and Instrumentation Data Center'. https://midcdmz.nrel.gov/apps/day.pl?BMS.

[12] S. A. Pelaez *et al.*, 'Field-Array Benchmark of Commercial Bifacial PV Technologies with Publicly Available Data', *Conf. Rec. IEEE Photovolt. Spec. Conf.*, vol. 2020-June, pp. 1757–1759, 2020, doi: 10.1109/PVSC45281.2020.9300379.

[13] A. de Miguel and J. Bilbao, 'Test reference year generation from meteorological and simulated solar radiation data', *Sol. Energy*, vol. 78, no. 6, pp. 695–703, 2005, doi: https://doi.org/10.1016/j.solener.2004.09.015.

[14] DWD Deutscher Wetterdienst, 'Solar energy', 2022.https://www.dwd.de/EN/ourservices/solarenergy/solarenergy.html.

8 APPENDIX

Table I: Simulation Modes

Mode 1	VF model for irradiance simulation of the front side and RT model for irradiance simulation of the backside
Mode 2	VF model for front and back irradiance simulation
Mode 3	RT model for front and back irradiance simulation

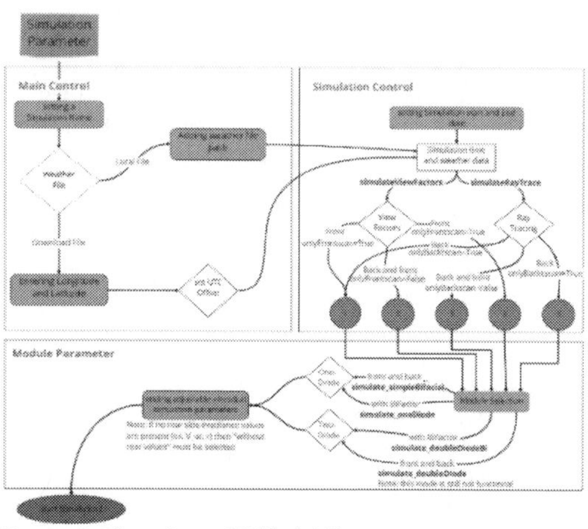

Figure 5: Flowchart of BifacialSimu

EVALUATION OF WRF-SOLAR CLOUD FORECAST USING THE NSRDB

Jaemo Yang[1], Manajit Sengupta[1], Yu Xie[1], Aron Habte[1], Pedro A. Jimenez[2], Ju-Hye Kim[2]

[1]National Renewable Energy Laboratory, Golden, CO, USA
[2]National Center for Atmospheric Research, Boulder, CO, USA

ABSTRACT: The cloud forecast is a crucial component in predicting solar irradiance from numerical weather prediction (NWP) models. Assessing cloud properties from NWP models requires significant work due to the need for high-quality data, spatial analysis covering the model's extent, and detailed analysis of the model's performance for different types of clouds. This study presents an evaluation of the Weather Research and Forecasting (WRF)-Solar cloud forecast using the National Solar Radiation Database (NSRDB). We propose an evaluation framework applied to a single model prediction as well as ensemble-based forecasts. Various cloud detection metrics are calculated for comparison with the satellite-derived data set. The mismatched clouds from WRF-Solar are quantified using nine cloud types classified by cloud top height and cloud optical depth. The results based on the WRF-Solar forecasts covering the entire United States for the full year of 2018 show mismatched cloud frequencies ranging from 8%–46% for thick and high-level (deep convective) to thin and low-level (cumulus) clouds.

1 AIM AND APPROACH

Accurately reproducing clouds in numerical weather prediction (NWP) models is important for predicting solar irradiance because clouds directly influence the solar radiation reaching the surface of the earth; however, cloud forecasts are not easy to evaluate because they require analysis covering a wide range of locations where different types of clouds occur. Due to their coverage at high spatial and temporal resolutions, satellite-derived cloud properties and solar radiation are good sources of data for assessing cloud forecasts if the satellite-based observations are of high enough accuracy to be used for NWP evaluations. The National Solar Radiation Database (NSRDB) [1], [2] has been demonstrated to be of high accuracy when compared to high-quality ground measurements and therefore it is deemed adequate for this evaluation. In this study, we present an evaluation of cloud forecasts simulated from the Weather Research and Forecasting-Solar ensemble prediction system (WRF-Solar EPS) [3]–[5] using the NSRDB. This work aims to investigate which types of clouds are difficult to forecast using WRF-Solar. In addition, we analyze the error and bias of the modeled cloud properties (e.g., cloud optical depth) for each cloud type (this will be included in the extended version of the abstract). The cloud forecasts covering the entire contiguous United States (CONUS) are used to analyze the ability of WRF-Solar EPS to represent cloud properties for various cloud types.

2 SCIENTIFIC INNOVATION AND RELEVANCE

2.1 Data

WRF-Solar EPS is a state-of-the-art ensemble model developed for solar energy applications. In this study, the model is employed to simulate day-ahead forecasts covering 2018. WRF-Solar EPS is configured with 9-km horizontal grid spacing (Fig. 1) and initialized at 6 UTC for each set of day-ahead forecasts (48 hours). The National Centers for Environmental Prediction Global Forecast System (0.25°×0.25°; 3-hour intervals) forecast data are used to provide the initial and boundary conditions for the model. The NSRDB is used to evaluate the cloud forecasts from WRF-Solar EPS. The native NSRDB data sets (2-km spatial resolution) are reprocessed to produce data for originally excluded parts over the ocean and then spatially interpolated to the 9-km WRF-Solar EPS model grid. The 9-km NSRDB data used in this work have been evaluated against surface observations by [6] and have been validated to be sufficient for use in the WRF-Solar model assessment.

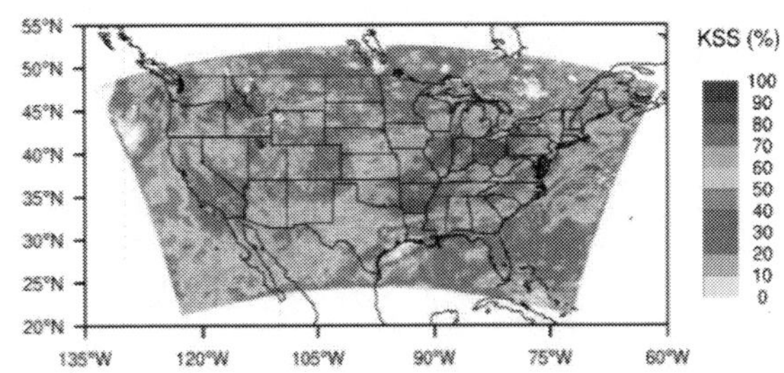

Figure 1. KSS (see Eq. (9)) calculated for the entire domain of WRF-Solar EPS.

2.2 Methodology

To analyze the cloud mask forecasts from WRF-Solar EPS, we build a methodology to filter the pixels from the NSRDB and WRF-Solar EPS data sets for clear and cloudy conditions. First, an absolute difference between all-sky global horizontal irradiance (GHI) and clear-sky GHI is calculated for both data sets as follows:

$$\Delta GHI = \left|GHI_{all-sky} - GHI_{clear-sky}\right| \quad (1)$$

Here, ΔGHI is calculated at each grid point and time step. Second, each pixel is filtered using the criteria for ΔGHI specified in Table 1. For the NSRDB data, if ΔGHI is less than 1.0 W/m^2, the pixel is treated as clear sky; otherwise, the pixel is regarded as cloudy sky. The method can be applied to forecasts comprising a single member as well as ensemble members. In the case of a single forecast, a method similar to that applied to the NSRDB can be used. For ensemble forecasts, an additional criterion is considered. To filter the ensemble data sets comprising ensemble members, if more than 50% of the ensemble members satisfy the case of $\Delta GHI \geq 1.0\ W/m^2$, that pixel is regarded as cloudy sky.

Table 1. Criteria for data-filtering the NSRDB (observation) and WRF-Solar (prediction) for clear/cloudy sky conditions

	NSRDB (observation)	**Prediction from a single member**	**Prediction from ensemble members**
Clear sky	$\Delta_{GHI} < 1.0$ W/m^2	$\Delta_{GHI} < 1.0$ W/m^2	> 50% of ens. members are: $\Delta_{GHI} < 1.0$ W/m^2
Cloudy sky	$\Delta_{GHI} \geq 1.0$ W/m^2	$\Delta_{GHI} \geq 1.0$ W/m^2	≥ 50% of ens. members are: $\Delta_{GHI} \geq 1.0$ W/m^2

To quantify the results, we calculate cloud detection metrics [7], [8] based on a 2 × 2 contingency table (Table 2), which represents the population of occurrences for each category "A" – "D". An important analysis in this work is to confirm the model's capability to predict different cloud types classified by cloud top height (CTH) and cloud optical depth (COD). To enable this analysis, we filter the occurrences in Table 2 by CTH and COD obtained from the NSRDB.

Table 2. Contingency table for NSRDB (observation) and WRF-Solar EPS (prediction)

Prediction / Observation	**WRF-Solar EPS**		
NSRDB	Scenario	Clear sky	Cloudy sky
	Clear sky	A	B
	Cloudy Sky	C	D

The frequency of clouds and seven additional metrics—including probability of detection in clear sky (PODCLR), probability of detection in cloudy sky (PODCLD), false alarm rate in clear sky (FARCLR), false alarm rate in cloudy sky (FARCLD), hit rate, Kuiper's skill score (KSS), and mismatached cloud frequency (MCF)—are used to assess the accuracy of clouds produced by WRF-Solar EPS [Eqs. (2)–(10)].

$$FOC_{NSRDB} = \frac{C+D}{A+B+C+D} \times 100\% \tag{2}$$

$$FOC_{WRF-Solar} = \frac{B+D}{A+B+C+D} \times 100\% \tag{3}$$

$$PODCLR = \frac{A}{A+B} \times 100\% \tag{4}$$

$$PODCLD = \frac{D}{C+D} \times 100\% \tag{5}$$

$$FARCLR = \frac{C}{A+C} \times 100\% \tag{6}$$

$$FARCLD = \frac{B}{B+D} \times 100\% \tag{7}$$

$$Hit\ rate = \frac{A+D}{A+B+C+D} \times 100\% \tag{8}$$

$$(where\ 0 \leq HR \leq 100\%)$$

$$KSS = \frac{A \cdot D - B \cdot C}{(A+B) \cdot (C+D)} \times 100\% \tag{9}$$

$$(where\ -100\% \leq KSS \leq 100\%)$$

$$MCF = \frac{C}{C+D} \times 100\% \tag{10}$$

3 RESULTS

We quantify the overall performance of WRF-Solar EPS in terms of forecasting skills for nine cloud types. MCF is calculated and used as an adequate metric to estimate how many cloudy occurrences are missed by the model. Figure 2 shows the averaged MCF across all seasons for CONUS in 2018. The MCF scores for WRF-Solar EPS show that the ensemble model performs better for high-level and thick clouds than for low-level and thin clouds. For three CTH levels, WRF-Solar EPS exhibits scores ranging from 8%–27%, 16%–33%, and 19%–46% for high- (7 km < CTH), middle- (3 km < CTH ≤ 7km), and low-level (CTH ≤ 3 km) clouds. In terms of MCF classified in three COD ranges, the model shows MCFs ranging from 8%–19%, 13%–34%, and 27%–46% for thick (20.0 < COD), middle-thickness (3.0 < COD ≤ 20.0), and thin (COD ≤ 3.0) clouds.

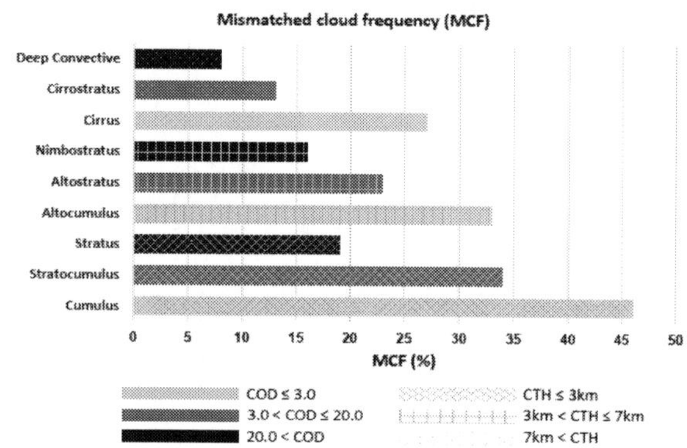

Figure 2. Averaged MCF of WRF-Solar EPS calculated over CONUS for 2018.

4 CONCLUSIONS

In this study, we present an evaluation approach for NWP cloud forecasts using satellite information from the NSRDB. Day-ahead cloud mask forecasts are simulated from WRF-Solar EPS on a 9-km grid. To assess the performance of WRF-Solar EPS in forecasting clouds over the ocean as well as for CONUS, the NSRDB data sets are newly processed and aggregated to the 9-km model grid. We quantify the model's cloud mask forecast by using cloud detection metrics and MCFs classified in different COD and CTH. Preliminary results indicate that WRF-Solar EPS provides accurate forecasts for thick (MCF: 8%–19%) and high-level (MCF: 8%–27%) clouds, whereas thin (MCF: 27%–46%) and low-level (MCF: 19%–46%) clouds cause difficulties for the model in predicting cloud masks. This analysis produces useful information about types of clouds that are not adequately reproduced by the model. This information eventually enables model developers with information that can lead to further model improvements. An evaluation of WRF-Solar models with increased spatial resolution, different settings in the physics options, and long-term simulations will be further considered in a future study.

ACKNOWLEDGMENTS

This work was authored in part by the National Renewable Energy Laboratory, operated by Alliance for Sustainable Energy, LLC, for the U.S. Department of Energy (DOE) under Contract No. DE-AC36-08GO28308. Funding provided by the U.S. Department of Energy Office of Energy Efficiency and Renewable Energy Solar Energy Technologies Office. The views expressed in the article do not necessarily represent the views of the DOE or the U.S. Government. The U.S. Government retains and the publisher, by accepting the article for publication, acknowledges that the U.S. Government retains a nonexclusive, paid-up, irrevocable, worldwide license to publish or reproduce the published form of this work, or allow others to do so, for U.S. Government purposes.

This material is also based upon work supported by the National Center for Atmospheric Research, which is a major facility supported by the National Science Foundation under Cooperative Agreement No. 1852977.

REFERENCES

[1] Sengupta, M., Xie, Y., Lopez, A., Habte, A., Maclaurin, G. and Shelby, J., 2018: The national solar radiation data base (NSRDB). *Renewable and Sustainable Energy Reviews*, 89, 51–60.

[2] Habte, A., Sengupta, M., Gueymard, C., Golnas, A. and Xie, Y., 2020: Long-term spatial and temporal solar resource variability over America using the NSRDB version 3 (1998–2017). *Renewable and Sustainable Energy Reviews*, 134, 110285.

[3] Yang, J., Kim, J.H., Jimenez, P. A., Sengupta, M., Dudhia, J., Xie, Y., Golnas, A. and Giering, R., 2021: An efficient method to identify uncertainties of WRF-Solar variables in forecasting solar irradiance using a tangent linear sensitivity analysis. *Solar Energy*, 220, 509–522.

[4] Kim, J. H., Munoz, P. A. J., Sengupta, M., Yang, J., Dudhia, J., Alessandrini, S. and Xie, Y., 2021: The WRF-Solar Ensemble Prediction System to provide solar irradiance probabilistic forecasts. *IEEE Journal of Photovoltaics*, 12(1), 141–144.

[5] Sengupta, M., Jimenez, P., Yang, J., Kim, J. H. and Xie, Y., 2021: New Developments in Ensemble-based Probabilistic Forecasting of Solar Radiation: The WRF-Solar Ensemble Prediction System (No. EMS2021-347). Copernicus Meetings.

[6] Jiménez, P. A., Yang, J., Kim, J. H., Sengupta, M. and Dudhia, J., 2022: Assessing the WRF-Solar model performance using satellite-derived irradiance from the National Solar Radiation Database. *Journal of Applied Meteorology and Climatology*, 61(2), 129–142.

[7] Karlsson, K. G. and Johansson, E., 2013: On the optimal method for evaluating cloud products from passive satellite imagery using CALIPSO-CALIOP data: Example investigating the CM SAF CLARA-A1 dataset. Atmospheric Measurement Techniques, 6(5), 1271–1286.

[8] Yang, J., Sengupta, M., Jiménez, P.A., Kim, J. H. and Xie, Y., 2022: Evaluating WRF-Solar EPS cloud mask forecast using the NSRDB. *Solar Energy*, 243, 348–360.

IMPROVED BLENDING OF PV POWER FORECASTS IN CASE OF MEASUREMENTS WITH LIMITED RELIABILITY

Wiebke Herzberg[1], Nicolas Holland[1], Tobias Zech[1], Jefferson Bor[1], and Elke Lorenz[1]
[1]Fraunhofer Institute for Solar Energy Systems ISE, Heidenhofstr. 2, 79110 Freiburg, Germany

ABSTRACT: For forecasting the power output of a photovoltaic (PV) power plant, solar irradiance forecasts are an essential input. Forecasts generated from different sources and models such as satellite data, numerical weather models, or irradiance measurements, perform differently depending on the forecast horizon. An optimized forecast for each horizon can be derived by combining several different source forecasts via a machine learning (ML) model to create a resulting 'blended' forecast. The training of an ML blending model requires power generation data of the considered PV power plant as a target variable. Typically, power measurements from the plant are used for this purpose. However, poor quality or limited availability of power measurements will lead to low quality blended forecasts as a result. In this work we consider intra-day forecasts, obtained from blending numerical weather predictions with satellite-derived forecasts, for a large PV plant with a capacity of 1 GW. The plant is frequently subject to curtailment by grid operators, which limits the reliability of its power measurements. This poses a significant challenge for the optimization of the blending model, resulting in a notable underestimation of the power forecasted by the model in high-power situations. We present an approach to improve forecast blending by using satellite derived power as a replacement for measurements in model training. We evaluate the effect of this replacement for two different ML blending models: a Huber linear regressor and a neural network. The performance of the different models is characterized by employing commonly used metrics such as RMSE and BIAS, as well as a distribution-oriented evaluation framework.
Keywords: PV Forecasting, Machine Learning, Satellite

1 INTRODUCTION

Reliable forecasting of PV power production is an important tool for a variety of applications: power grid operators require feed-in forecasts for grid regulation, power plants with integrated storage capabilities utilize production forecasts to optimize their battery usage and energy traders need forecasts when selling energy produced by PV plants. For forecasting the power output of any PV power plant, solar irradiance forecasts are an essential input. Irradiance forecasts can be obtained from different sources and models, such as satellite data, numerical weather models, or irradiance measurements. Each source, and the specific forecast type generated from it, will vary in performance depending on the considered forecast horizon. For example, satellite-based forecasts are viable for short-term horizons of up to several hours, while predictions from numerical weather models are viable for up to several days.

To obtain an optimized forecast for a given location and plant, different input forecasts can be combined through a weighting scheme, resulting in an optimal forecast for each forecast horizon (see, e.g., [1] and references therein). We refer to this process as forecast blending. The blending can be accomplished by a variety of different machine learning models, which may be trained using PV measurements from the power plant. We can express this problem as

$$f(x) = y \quad (1)$$

with a forecast blending model f, power measurements y, and x representing all input features of the model. Depending on the model, input features can consist of irradiance forecasts from different sources, PV power forecasts obtained by converting irradiance forecasts to power through application of a PV simulation, or other ancillary features such as e.g., solar position.

Training of the model is performed on a dataset containing features and corresponding measurements $\{(x_t, y_t)\}_{t\in T}$ spanning a certain time range T. The quality of the forecasts produced by the forecast model very much depends on the quality of the available power measurements used in training. If the quality of the measurements is poor, or their availability is limited, the quality of the resulting blended forecast will be poor as a consequence.

One important issue causing challenging quality and availability of measurements are curtailment events. Curtailment occurs when e.g. grid operators send a signal to artificially reduce the production of a PV plant in order to ensure grid stability. Curtailment will result in PV power measurements which do not reflect the power that could potentially be generated by the plant under the given irradiance conditions. We denote with $\hat{y}$ the actual production of the power plant, which can be different from the potential production y of the plant i.e., for some instances it holds true that $\hat{y} < y$ due to curtailment. While for the applications we mentioned it is usually necessary to forecast the potential output y of the plant, often only the actual production including curtailment $\hat{y}$ is measured and therefore available as a reference for model training. If a power plant is frequently subject to curtailment, the resulting effect on the measurements can cause significant issues in the optimization process of the blending models. In previous works [2] we investigated how different forecast sources can be best combined without the presence of curtailment and in following works we explored ways to handle curtailment by detecting it using a curtailment detection model [3] and adapting model training to the presence of curtailment by tweaking loss functions used in model training [4].

In this work we present an approach to improve forecast blending by introducing satellite derived power as the target variable in model training, thereby replacing the PV power measurements. We investigate this approach for intra-day forecasts up to horizons of 4 hours that are generated for a power plant with 1 GW installed capacity in China. The forecasts are obtained by blending a satellite-based forecast with numerical weather predictions. For the blending model we explore two options, a linear Huber model [5], which possesses a built-in resilience against outliers and a neural network. We evaluate the different models following verification techniques recently illustrated by Yang et. al. [6].

The remainder of the paper is structured as following: Section 2 discusses the measurement data, curtailment and the challenges it introduces to the measurements, as well as the dataset of input forecasts used in this work. Section 3 describes our approach to improve the generated blended forecasts, including the satellite data used as replacement in model training and the blending models that were employed. Section 4 compares our results obtained from different models and training data. Section 5 summarizes and concludes.

2 DATASETS

2.1 PV power measurements

The power plant for which we evaluate our approach is a PV plant of 1 GW installed capacity in the Qinghai province of China. Measurements of produced power are available, but the plant is subject to frequent curtailment by grid operators which limits the reliability of the measurements. We work with measurements of the total power plant output resampled to 15-minute mean values. Our dataset includes data from March 2020 to March 2021.

2.2 Curtailment

Curtailment represents the artificial reduction of the power production of a PV plant at times where less power feed-in to the grid is required. The actual information of when curtailment is being applied at the plant and to what extent the power production is curtailed is often not available. In our previous work [3] we developed a curtailment detection method d which aims to predict whether a datapoint $(x_t, \hat{y}_t)$ is subject to curtailment, i.e.,

$$d(\hat{y}) = \begin{cases} 1, & if\ \hat{y} = y \\ 0, & else \end{cases} \quad (2)$$

The curtailment detection method is based on a comparison of measurements and Active Generation Control (AGC). An example of a day with detected curtailment is shown in Figure 1. The forecast model training is enhanced by replacing the training dataset with a curtailment filtered version $\{(x_t, \hat{y}_t) | d(\hat{y}) = 1\}_{t \in T}$. While we expect some improvement from the application of a curtailment filter to training data of a blending model before the model optimization process, since the number of uncorrelated data points $(x_t, \hat{y}_t)$ will be reduced, it will also introduce new challenges. Applying a curtailment filter will result in many missing data points, especially for periods with potentially large power generation (see Figure 1). Also, due to imperfections in filter performance, the filtered data can still contain curtailed data points as outliers, reducing the quality of the remaining data.

2.3 Model input

To create a blended forecast which is optimized for each horizon, we combine input forecasts from different sources and additional solar position features via a machine learning (ML) model. In this work, we use numerical weather predictions (NWP) from the European Centre for Medium-Range Weather Forecasts (ECMWF),

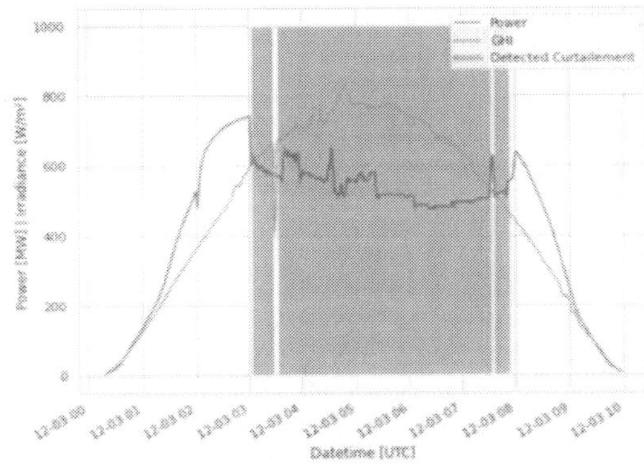

Figure 1: PV power and global horizontal irradiance (GHI) measurements for a day with mostly clear-sky conditions. The highlighted area is labeled as curtailed by the curtailment detection model.

and cloud motion vector forecasts from images of the geostationary Himawari satellite, which is operated by the Japan Meteorological Agency.

The NWP data with an original resolution of 1 hour is upsampled to time steps of 15 minutes by interpolation of the clear-sky index. The satellite-based forecasts are directly created in 15 minute steps by applying the heliosat method [7] and cloud motion vectors [8] on satellite images. We make use of the deepflow algorithm [9] to compute cloud motion vector fields, using the implementation from the OpenCV library [10]. Both the resampling of NWP irradiance as well as the heliosat method require modelled clear-sky irradiance, for which we use the clear-sky model by Dumortier [11].

To convert these irradiance forecasts to forecasts of PV power, we apply a PV simulation adapted for the specific plant. The simulation consists of several steps: First, global horizontal irradiance (GHI) values are converted into plane of array irradiance using the DIRINT separation model [12] and the transposition model by Perez et. al. [13]. For both models we rely on the implementations in PVLIB [14]. The subsequent steps to derive PV power from plane of array irradiance are implemented inhouse by a parametric model-chain [15]. The simulation also requires an ambient temperature input for which we employ the ECMWF NWP temperature forecasts. A more detailed description of the source forecasts and the PV simulation is available in [4].

2.4 Training and testing set

For the evaluation of the performance of the different blending models, we divide the entire available dataset into a training and a testing set. The models are optimized on the training data, and evaluation is carried out on the testing data. We divided the data along the base times (i.e. forecast creation times), and randomly sampled 67% of forecasts into the training set. The remaining 33% of the forecasts are used for testing.

Training and evaluation for all models is performed on data where curtailment has been removed by applying the filtering method described in Section 2.2.

3 APPROACH AND MODELS

3.1 Blending models

For the blending models we explore two different

options, a linear Huber regressor and a Neural Network (NN). The models generate intra-day forecasts with a time resolution of 15 minutes. New forecasts are also generated every 15 minutes. Here, we evaluate the forecasts for horizons up to 4 hours.

3.1.1 Huber regressor

The Huber regressor [5] is a linear regression model with a built-in resilience against outliers, which can be tuned via a hyperparameter. For samples where the deviation between model and measurement falls below a certain threshold, the squared loss is optimized, whereas for samples with deviations above the threshold, the absolute loss is optimized. Thereby, less weight is attributed to samples constituting large deviations i.e., outliers. In this work, the scikit-learn [16] implementation of the Huber model was used and the corresponding hyperparameter ε was kept fix at a value of 1.05.

Three input features were used for this model: the PV simulated values of both, the satellite-derived irradiance forecast and NWP irradiance, and the cosine of the solar zenith angle. A separate model was trained for each forecast horizon. In the training for each horizon, data from a window of a total of 5 horizons around the target horizon were included, to avoid overfitting and to generate a smooth output forecast.

3.1.2 Neural network

In order to compare the linear Huber model with a non-linear model we chose to train a neural network with 2 hidden layers and a 100 neurons per layer. For the implementation we rely on the multi-layer perceptron [17] implemented in scikit-learn [16]. The network was trained using stochastic gradient decent [18] to minimize MSE (see Section 3.3) and the input features for the model were PV simulation values of NWP and satellite-based predictions and the solar zenith angle. In addition to those features we added the forecast horizon as another input feature, allowing the model to make predictions for several horizons in one forward pass of the model.

3.2 Replacement of training data

To avoid model training being affected by curtailment, or potentially other circumstances causing challenging quality and availability of measurements, we introduce a satellite-derived target variable as a replacement for the measurements in model training. The new target variable is obtained from converting satellite-derived irradiance values of our real-time processing chain (forecast horizon of 0 minutes) to PV power by means of a PV simulation. We refer to this as 'nowcast' values. This change in the target variable can be expressed as replacing $\hat{y}$ with simulated nowcast values $s(x_{sat})$. One advantage is that these simulated values are not subject to curtailment, meaning they do not artificially underestimate the true output potential of the power plant. They could also be used to mitigate data gaps which arise from removing curtailed datapoints, this is however not investigated in this work. Possible disadvantages include s being merely an approximation of the potential power plant output $s(x_{sat}) \approx y$.

Figure 2 shows power values from the satellite-derived nowcast against power measurements where curtailed data were removed by applying the curtailment detection method outlined in Section 2.2. The measurements show a reasonable agreement with the nowcast, suggesting that its use as a replacement in model training is worth to investigate. For our evaluation, the blending models were trained once on the actual measurements and once on the satellite-derived nowcast.

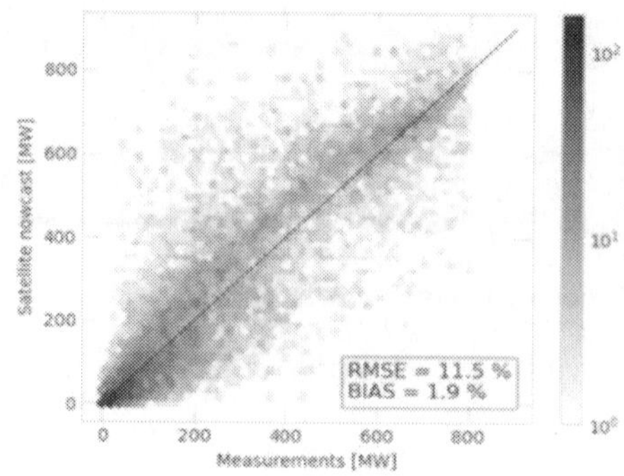

Figure 2: PV power derived from a satellite-based irradiance nowcast against measured power. Only data points for which no curtailment was detected are displayed. Relative RMSE and BIAS are calculated with respect to installed capacity of 1 GW using daylight values only.

3.3 Evaluation metrics

A common metric used in the evaluation of PV power forecasts is the root-mean-square error (RMSE), which is the root of the mean-square error,

$$MSE = \frac{1}{N}\left(\sum_t (f(x_t) - \hat{y}_t)^2\right). \quad (3)$$

The RMSE can be complemented by considering the BIAS

$$BIAS = \frac{1}{N}\left(\sum_t f(x_t) - \hat{y}_t\right). \quad (4)$$

In addition, to obtain a more in-depth understanding of forecast model performance, a distribution-oriented approach to evaluation is of great value. In this framework, data comprising forecasts generated by the blending models and corresponding measured power can be expressed as a 2-dimensional distribution $p(f,\hat{y})$. This joint distribution $p(f,\hat{y})$ can be related to marginal and conditional distributions via two Murphy-Winkler factorizations [19]

$$p(f,\hat{y}) = p(\hat{y}|f)\, p(f) \quad (5)$$

$$p(f,\hat{y}) = p(f|\hat{y})\, p(\hat{y}) \quad (6)$$

called *calibration-refinement* and *likelihood-base rate* factorization respectively. Following these factorizations, the MSE can also be decomposed in two distinct ways (as presented e.g., in [20], [6])

$$MSE = V(\hat{y}) + E_f[f - E(\hat{y}|f)]^2 - E_f[E(\hat{y}|f) - E(\hat{y})]^2 \quad (7)$$

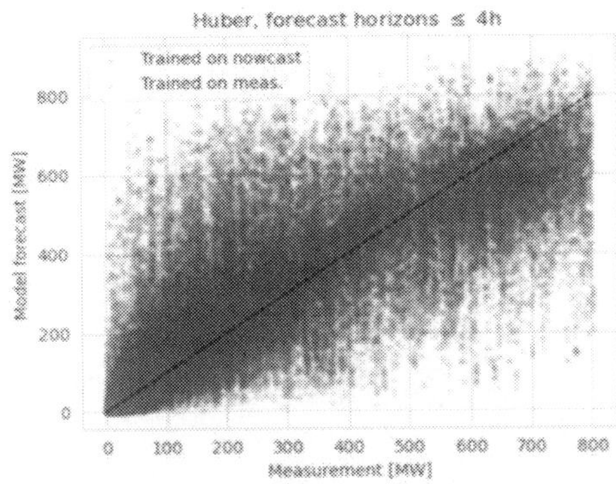

Figure 3: Scatter plot of forecasts from the Huber model for horizons up to 4 hours, trained on measurements (blue) or satellite-derived nowcast (red).

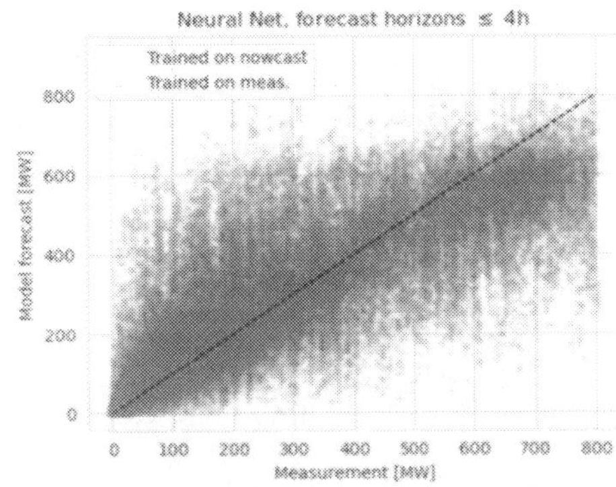

Figure 4: Scatter plot of forecasts from the NN model for horizons up to 4 hours, trained on measurements (cyan) or satellite-derived nowcast (orange).

$$MSE = V(f) + E_{\hat{y}}[\hat{y} - E(f|\hat{y})]^2 - E_{\hat{y}}[E(f|\hat{y}) - E(f)]^2 \quad (8)$$

with E indicating the expectation value. In the first decomposition (Eq. (7)), $V(\hat{y})$ denotes the variance of the measurements, while the second and third term can be referred to as *type-1 conditional bias* and *resolution* respectively. In the second decomposition (Eq. (8)), $V(f)$ denotes the variance of the forecast values, while the second and third term can be referred to as *type-2 conditional bias* and *discrimination,* respectively. Note that $V(\hat{y})$ is a fixed property of the dataset, while $V(f)$ is subject to the optimization. When evaluating forecast models, the conditional biases are to be minimized while the resolution and discrimination components are to be maximized.

4 RESULTS

Evaluations shown in this section were all carried out on the testing set.

4.1 Model comparison using RMSE and BIAS

We begin the comparisons by looking into the difference between models trained on power measurements $\hat{y}$ and satellite derived power nowcast values $s(x_{sat})$. Figure 3 and 4 show scatter plots of forecasted vs. measured PV power for the Huber model and NN model respectively. For both ML models, we find that the model trained on measurements has a strong tendency to underestimate large measurements between 600 and 800 MW, which can be explained by those situations being underrepresented in the dataset as well as undetected curtailed values still being present during training. This lowers the correlation between input forecasts and the target variable (the measurements) which in turn tends to pull down the forecasted values towards the mean. Training on the nowcast visibly alleviates this underestimation, the improvement being slightly more noticeable for the Huber model.

However, forecasts from the nowcast-trained models exhibit a higher RMSE and BIAS when evaluated against the measurements, as presented in Figure 5. In part this is to be expected given the relation between measurements and satellite nowcast shown in Figure 2. We note though, that the increase in both metrics is much stronger for the Huber model than for the NN.

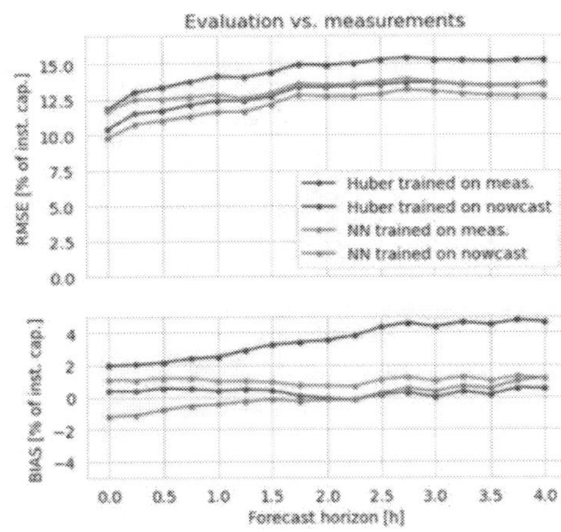

Figure 5: RMSE and BIAS of forecasts generated by different models in dependence of forecast horizon. Values are normalized with respect to installed capacity of 1 GW.

4.2 Model comparison using conditional distributions

In order to further investigate the difference of the measurement-trained and nowcast-trained models we divide the forecasted values into bins and compute conditional distributions following the Murphy Winkler factorizations (Eq. (5) and (6)). In Figure 6, the panels on the left side show conditional distributions binned along the measured and forecasted values for an example horizon of 2 hours for the Huber model. The panels on the right side of Figure 6 show the same for the neural network. We find that the underestimation in the high-power range of 600 to 800 MW of the measurement-trained models (blue and cyan distributions) is very prominent: The distributions in this range are barely present for the forecast binning (Figure 6, upper panels) and the distributions clearly underestimate the measurements for the measurement binning (Figure 6, lower panels). The nowcast-trained models (red and orange distributions) alleviate this underestimation but

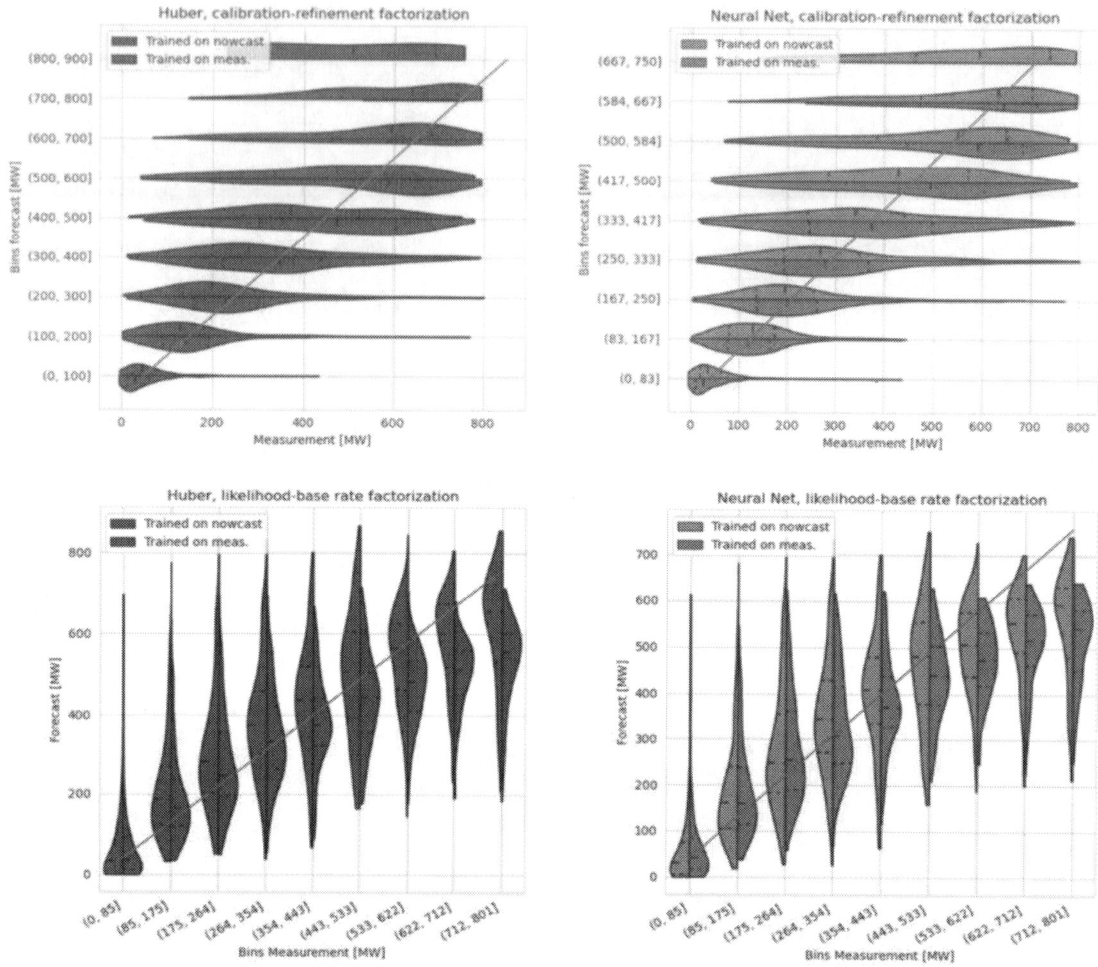

Figure 6: Comparison of conditional distributions for Huber models (left panels) and NN models (right panels) trained on either measurements or nowcast, for an example forecast horizon of 2h. Upper panels: distributions conditioned on forecasts $p(\hat{y}|f)$. Lower panels: distributions conditioned on measurements $p(f|\hat{y})$.

also introduce an overestimation in the low to intermediate power region of 100 to 300 MW, which is particularly strong for the Huber model. Apart from that, it is interesting to note that the measurement distributions for the upper most forecast bins, which are only visible for the nowcast-trained models (see Figure 6, upper panels) tend towards a uniform distribution. This highlights again that the measurements have almost no correlation to forecast values in that region.

4.3 Comparison using MSE factorization

Lastly, we compare the different models with respect to the components of the MSE factorizations given in Eq. (7) and Eq. (8). The upper left panel of Figure 7 shows the three components of Eq. (7), the variance of measurements, the type-1 conditional bias, and the resolution, for all four investigated models and with increasing forecast horizon. The respective MSE components are of similar magnitude for all four models, with the contributions from variance and resolution being much larger than the type-1 conditional bias.

We observe that training on the nowcast (red and orange lines) tends to decrease the resolution and increase the type-1 bias for both the Huber and NN model, meaning both MSE components become slightly worse when evaluated against measurements. This results in the overall increased RMSE as seen in Figure 5.

The lower left panel of Figure 7 shows the three components of the MSE decomposition given in Eq. (8), the variance of the forecasts, type-2 conditional bias, and discrimination, for the investigated models and with increasing forecast horizon. We can see more prominent differences between the models, particularly in the variance of forecast values and the discrimination. It is evident that training on the nowcast lowers the type-2 conditional bias for both the Huber and NN model and additionally increases the discrimination, which are in general favorable qualities. The lower type-2 conditional bias reflects forecast distributions which are better centered on the diagonal (see Figure 6, lower panels) and a higher discrimination means forecasts will differ more strongly for different measurement situations. Taken together, this means we obtain a better mapping of the forecasts to the potential range of power measurements. However, the significantly larger variance of the forecasts outweighs the favorable trends in type-2 conditional bias

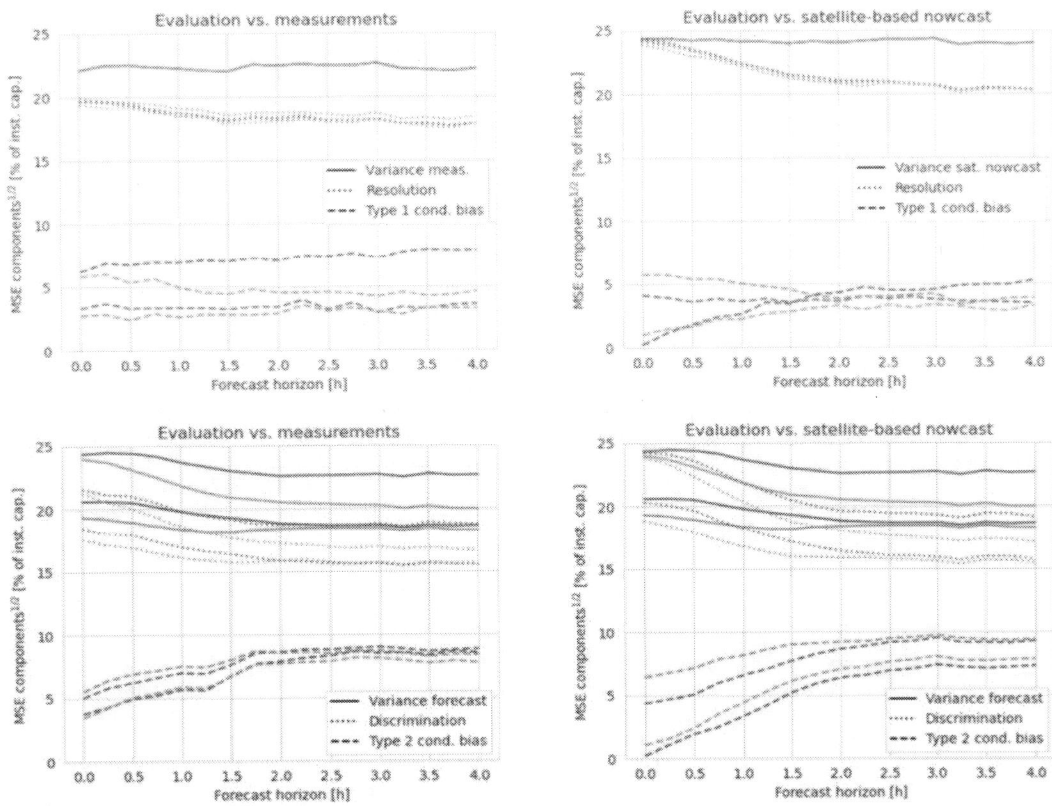

Figure 7: Square root of the components of the MSE factorization given in Eq. (7) (upper panels) and Eq. (8) (lower panels) normalized with respect to installed capacity for measurement-trained Huber (blue) and NN (cyan) models, as well as nowcast-trained Huber (red) and NN (orange) models in dependence of forecast horizon. Left panels: Calculation with respect to measurements. Right panels: Calculation with respect to satellite-derived nowcast.

and discrimination and leads to the overall increased RMSE for both nowcast-trained models, when evaluated against measurements, as displayed in Figure 5.

It should be noted here that J. Moskaitis points out in [20] that a comparison of discrimination between two models is only fair if their variances are similar, because forecast variance and discrimination are related. In their work they therefore split the MSE only into two terms, the conditional bias and a so-called shape term, which comprises both variance and discrimination. Here, following Yang et al [6] we chose to present the full decomposition into three terms to allow a more detailed discussion.

4.4 Complementary evaluation with respect to the satellite-derived nowcast

Due to the limited reliability of the power measurements, an evaluation of the forecasts based solely on these measurements is of course inherently problematic. Therefore, we complement the measurement-based evaluation with an evaluation against the satellite-based nowcast which is free of curtailment and was used to replace the measurements in model training. In this evaluation a lower RMSE of the nowcast-trained models compared to the measurement-trained models can be expected, since the evaluation is carried out with respect to the target variable used in training. Figure 8 shows the RMSE and BIAS for all models with respect to the satellite-derived nowcast. Indeed, we see an improved RMSE and BIAS of the nowcast-trained models, except for horizons >1.5 hours for the Huber model, where an increased BIAS manifests, leading also to an increased RMSE (this can occur since the Huber model does not optimize MSE in training). For a horizon of 0 minutes, we can observe the Huber model having an RMSE of zero, which is due to the fact that the nowcast values are an input feature for the models and that the Huber model was trained separately for each horizon. The NN does not drop to an RMSE of zero since several horizons of the model are trained together in one pass.

The upper right panel of Figure 7 shows the MSE components of Eq. (7) calculated with respect to the satellite-derived nowcast. There is an overall higher variance of the nowcast values compared to the variance of measurements, which is accompanied by a higher resolution for all four models. As opposed to the evaluation against measurements in the upper left panel, we see here that the type-1 conditional bias is improved for the nowcast-trained models, apart from the Huber model for horizons larger than 1.5 hours, exhibiting an increased type-1 bias. Since the difference between the variance and resolution is smaller or similar here than in the evaluation against measurements this gives together with the improved type-1 bias the overall smaller RMSE for the nowcast-trained models compared to their measurement-trained counterparts (see Figure 8). The

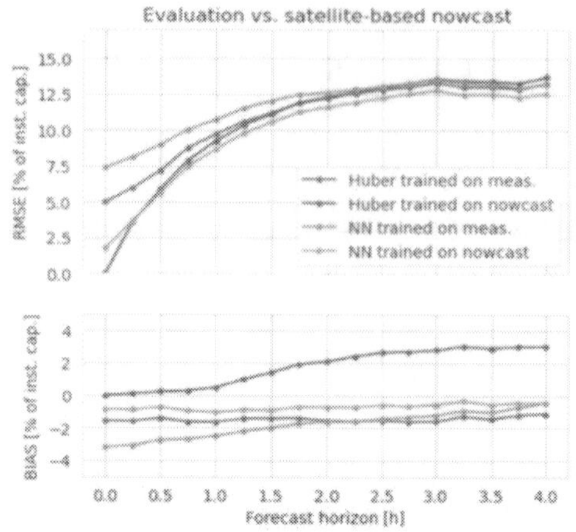

Figure 8: RMSE and BIAS of forecasts generated by different models in dependence of forecast horizon, calculated with respect to the satellite-derived nowcast. Values are normalized with respect to installed capacity of 1 GW.

exception being again the later horizons of the Huber model, where the type-1 bias was not improved.

In the lower right panel of Figure 7 the components of the MSE decomposition given in Eq. (8) are shown. The improvement of the discrimination and conditional type-2 bias components of the nowcast-trained models compared to the measurement-trained models, which were already discernable in the measurement-based evaluation (lower left panel), are again present, and here outweigh the increased variance of the nowcast-trained models, leading to the overall smaller RMSE (see Figure 8). These results emphasize, although to some degree expected, the improvements obtained from the nowcast-trained models.

5 SUMMARY AND CONCLUSION

For a dataset with measurements of limited reliability, we explored the replacement of measurements by a satellite derived PV power nowcast in model training to improve the results of forecast blending. Two types of blending models were investigated, a linear Huber model and a neural network. For both models we compared results after training the model on measurements and on a satellite-derived nowcast by considering standard metrics such as RMSE and BIAS and a distribution-oriented evaluation framework.

Due to the lack of reliable ground truth data, it is difficult to assess which model has the best performance. Evaluating a model against the variable it was trained on is generally expected to give a better RMSE (assuming training is based on MSE optimization). Nowcast-trained models for short horizons have the additional advantage of having an input (the satellite-derived forecast) that is highly correlated to their target variable, allowing their RMSE to reach even lower values when calculated with respect to the nowcast. When evaluating against measurements, the observed higher RMSE of the nowcast-trained models compared to the measurement-trained models is therefore expected and based on the differences between measurements and nowcast. First there is the inherent difference between satellite data and ground-based measurements, next there is the difference between a PV power simulation and actual power measurements. These two sources of error would be present in any model trained on satellite and evaluated against measurement data. The third component in our case are the effects of curtailment, which further lower the correlation between measurements and satellite-derived nowcast. Because, even though a model for curtailment detection was applied, some undetected curtailed values remain in the data set. In turn, when evaluating against the nowcast, as expected the nowcast-trained models exhibit a lower RMSE than their measurement-trained counterparts (an exception being larger horizons of the Huber model, which can be explained by Huber models not optimizing the MSE). Therefore, from these results alone we cannot find a definitive indicator of forecast quality.

In general, conventional comparisons of summary error metrics like RMSE or BIAS are not suitable to highlight different aspects of forecast quality. In our case they do not convey the fact that large power values are underrepresented in the forecasts. Therefore, we investigated the forecasts with a distribution-oriented approach, where different attributes of forecasts can be distinguished. We observe that the nowcast-trained models have lower type-2 conditional biases and larger discriminations, in both the evaluation against measurements and against the nowcast. This corresponds to a better mapping of forecasts to the range of power values and reflects the visible alleviation of the underestimation in the high-power region. When evaluated against measurements, this partial improvement in conditional type-2 bias and discrimination is mitigated by a strongly increased variance, leading to an overall higher RMSE of the nowcast-trained models. When evaluated against the satellite-derived nowcast, the drawbacks of disproportionally large forecast variances disappear (except for horizons >1.5 hours of the Huber model), while the improvements in discrimination and type-2 conditional bias become more pronounced, thus leading to an overall lower RMSE of the nowcast-trained models compared to the measurement-trained ones.

Though we cannot claim that evaluation with MSE decompositions do directly lead to "best-performing" model recommendations, the detailed evaluations give additional insight into model performance. The results presented in this work suggest it worthwhile to explore the use of satellite based nowcast-trained models. These models have shown to overcome the underestimation for large PV power values that occur when the measurement data used in training is of limited reliability. Which kind of model (e.g., NN, Huber or other options) is most suitable may depend on the individual use-case and input data. In the work presented here, the Huber model alleviates the underestimation in the high-power range of 600 to 800 MW more strongly than the NN, but this comes at the cost of also introducing a stronger overestimation in low to intermediate power values around 100 to 300 MW. The neural network (which is MSE optimized) produces more balanced results and with its nonlinear nature additionally might have more flexibility in adaption to the data. It exhibits the lowest RMSE in the evaluation based on the nowcast for almost all horizons.

Further enhancements in the generated forecasts can be expected from a fine tuning of the models or improving the satellite-derived PV power data. The latter could be achieved for example by improving the underlying satellite method and PV simulation, or by adaption of the satellite-derived data to measurements through postprocessing (e.g., bias or trend removal). The more accurate we can infer PV power from satellite data, the more accurate will the models be when compared to actual measurements. Furthermore, using satellite based nowcasts for model training holds the potential to mitigate data gaps that arise from removing curtailed datapoints, which is expected to be an additional advantage.

6 REFERENCES

[1] Sengupta, M., Habte, A., Wilbert, S., Gueymard, C., Remund, J. (2021). Best Practices Handbook for the Collection and Use of Solar Resource Data for Solar Energy Applications: Third Edition.

[2] Holland, N., et al. (2019). Solar and PV forecasting for large PV power plants using numerical weather models, satellite data and ground measurements, 2019 IEEE 46th Photovoltaic Specialists Conference.

[3] Holland, N., et al. (2020). Trainable curtailment detection for the enhancement of PV power forecasting, 2020 PVSEC-30 & GPVC.

[4] Holland, N., et al. (2021). Combination of physics based simulation and Machine Learning for PV power forecasting of large power plants, Proceedings of the 38th European Photovoltaic Solar Energy Conference and Exhibition.

[5] Huber, P. J. (1981). Robust Statistics, John Wiley and Sons, New York.

[6] Yang, D., et al. (2020). Verification of deterministic solar forecasts, Solar Energy, Vol. 210, Pages 20-37.

[7] Hammer, A., Heinemann, D., Hoyer, C., Kuhlemann, R., Lorenz, E., Müller, R., Beyer, H.G. (2003). Solar energy assessment using remote sensing technologies, Remote Sensing of Environment, 86(3): 423–32.

[8] Kühnert, J., Lorenz, E., Heinemann, D. (2013). Satellite-Based Irradiance and Power Forecasting for the German Energy Market, Solar Energy Forecasting and Resource Assessment, Elsevier.

[9] Weinzaepfel, P., Revaud, J., Harchaoui, Z., Schmid, C. (2013). Deepflow: Large displacement optical flow with deep matching, ICCV – IEEE International Conference on Computer Vision, pages 1385–1392.

[10] Bradski, G. (2000). The opencv library, Dr. Dobb's Journal of Software Tools.

[11] Dumortier, D. (1995). Modelling global and diffuse horizontal irradiances under cloudless skies with different turbidities, Daylight II, jou2-ct92- 0144, final report vol. 2. Tech. rep., CNRSENTPE.

[12] Perez, R., Ineichen, P., Maxwell, E., Seals, R., Zelenka, A. (1992). Dynamic Global-to-Direct Irradiance Conversion Models. ASHRAE Transactions-Research Series, pp. 354-369.

[13] Perez, R., Ineichen, P., Seals, R., Michalsky, J., Stewart, R. (1990). Modeling daylight availability and irradiance components from direct and global irradiance. Solar Energy 44 (5), 271-289.

[14] Holmgren, W.F., Hansen, C.W., Mikofski, M.A. (2018). pvlib python: a python package for modeling solar energy systems, Journal of Open Source Software, 3(29), 884.

[15] Müller, B., et al. (2015). Yield predictions for photovoltaic power plants: empirical validation, recent advances and remaining uncertainties, Progress in Photovoltaics: Research and Applications: published online

[16] Pedregosa, F., et al. (2011). Scikit-learn: Machine Learning in Python, JMLR 12, pp. 2825-2830.

[17] Hinton, G.E. (1989). Connectionist learning procedures, Artificial intelligence 40.1, 185-234.

[18] Kingma, D., and Ba, J. (2014). Adam: A method for stochastic optimization, arXiv preprint arXiv:1412.6980.

[19] Murphy, A. H., & Winkler, R. L. (1987). A General Framework for Forecast Verification, Monthly Weather Review, 115(7), 1330-1338.

[20] Moskaitis, J.R. (2008). A Case Study of Deterministic Forecast Verification: Tropical Cyclone Intensity, Weather and Forecasting, 23(6), 1195-1220.

This is a short version of the paper submitted to the journal Solar RRL.

It has been selected by the Committee of the WCPEC-8 for submission to Solar RRL.

SINGLE-SITE FORECASTS FOR 130 PV SYSTEMS AT DSO LEVEL, USING A HYBRID-PHYSICAL APPROACH, TO ENABLE FUTURE SMART-GRID OPERATION

D. Koster, D. Fiorelli,, Christian Braun
Luxembourg Institute of Science and Technology (LIST), Environmental Research and Development (ERIN),
5 Avenue des Hautes-Fourneaux, L4362 Esch-sur-Alzette, Luxembourg,

P. Bruneau
Luxembourg Institute of Science and Technology (LIST), IT for Innovative Services (ITIS),
5 Avenue des Hautes-Fourneaux, L4362 Esch-sur-Alzette, Luxembourg

ABSTRACT: With rising shares of photovoltaic (PV) generated electricity in our power grids, PV power forecasting is rapidly gaining importance for energy suppliers and grid operators. But specifically the distribution system operators (DSO) will require detailed forecasts at higher spatial resolutions, to manage grid congestions and avoid voltage band violations, which are local phenomena. In view of future smart grid operation, we propose a forecasting scheme using numerical weather prediction (NWP) data to feed a PV performance model and compute single-site power forecasts. The intra-day outcomes of this rather conventional forecasting scheme are combined with a supervised-learning model, based on recurrent neural networks (type: long- short-term memory, LSTM) which is fed by the 15 minutes measurement data of the two previous days for each PV system. This approach accounts for different local effects, not covered by the NWP-based physical model, such as partial shading or snow cover. Further, it will mitigate impacts of "imperfect" data on side of the DSOs, used to parametrize the PV performance model. This hybrid-physical approach has been applied to about 130 PV systems, validated over a two years period and yielded a normalized root mean square error of 8.3% and 8.7% in 2020 and 2021.

Keywords: solar power forecasting, photovoltaics, recurrent neural networks, numerical weather prediction model, long short-term memory

1 AIM AND APPROACH

1.1 Demand for forecasting (for the future DSO)

Fluctuating renewable energy resources, such as wind and solar power, become the backbone of our energy system, specifically of our electricity system. Energy suppliers, transmission system operators and balance responsible parties have growing demands on accurate photovoltaic (PV) power forecasts on aggregated levels of different granularity. The forecast requirements for distribution system operators (DSO) will become much more "detailed" in the future, specifically on the spatial scale. Grid integration of high shares of fluctuating renewables might cause grid congestions and voltage issues and will require flexibility provision. The operation of flexible grid assets requires detailed forecasts at high spatial resolutions. The approach described in this paper thus aims at single site forecasts instead of a regionalized approach. On regionalized level the overall performance of PV power forecasting benefits from a smoothing effect, as described by [1]. Single site forecasts do not profit from this smoothing effect and are also much more vulnerable to "imperfect" data utilized to parametrize a PV performance model. Simplified models, used e.g. since DSOs do not have detailed information on each PV system, do not represent well the individual system. Although simplified models could still result in decent forecast performance on aggregated time frames [2], there are some common simplifications that have impact on accuracy of the forecasted production curve: for example that only one orientation or inclination per system is used. Besides these challenges, physics-based forecasting models will not account for malfunction of the system or local effects on the irradiance, such as partial shading or snow cover, which could be an argument for data-driven methods.

2.2 Approach

The forecast scheme described here is using a numerical weather prediction model (NWP) feeding a PV performance model [2] to compute bottom up single site forecasts of roughly 130 PV systems in the distribution grid of a small DSO in Luxembourg. Currently, the NWP forecasts stem from European Centre for Medium ranged Weather Forecasts (ECMWF). The NWP forecasts (horizontal irradiance, ambient temperature), actually on hourly time resolution, are interpolated to get 15 min values, to be in line with the measurements. The forecast is computed and issued once per day in the early morning, as soon as the NWP data is available, and covers 72h ahead, starting from midnight of the same day. The 72h single-site forecasts are split in three forecast horizons: Intra-day (0-24h ahead), day-ahead (24-48h) and 2 days-ahead (48-72h). The overall scheme of the forecasting model, representing the part presented in this publication and further described below, is depicted in Figure 1.

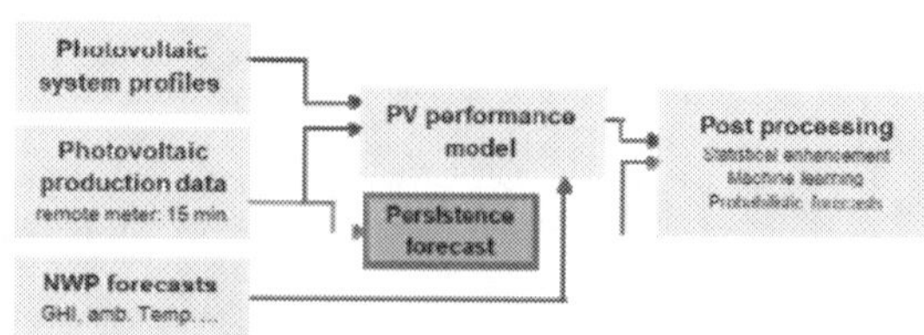

Figure 1 – overall scheme of the forecast approach

The irradiance forecasts of NWP models, specifically the model of ECMWF, delivers the best available performance up to date [3]. But besides the uncertainties that are inherent to forecasts generally, there are local effects which might reduce the irradiance that reaches the modules surface, in practice. Two simple reasons can be partial / temporal shading effects or snow cover. On technical level, the accuracy of a PV performance model has its limitations in the availability of data, accuracy of

data, and that there might be technical malfunctions which cannot be represented in the model. But since these factors influence the performance of the PV system, those aspects are inherently represented by the measurement time series data of the PV systems. Hence, a suitable approach is to combine a physical PV performance model, fed by predictions from a NWP model, with a machine learning approach to correct for the influence of such effects described above.
The forecasting approach presented in this paper uses a supervised learning algorithm, based on recurrent neural networks (RNN) of the type Long Short-term Memory (LSTM). The algorithm, known as DeepAR in the literature, is specifically developed to forecast time series of similar nature, by training a single model [4][5]– a characteristic which fits the forecasting of PV production profiles of several PV system within a region for a given, common period. The model has been trained for 130 PV systems, within the same region, based on a 2-years period dataset (split to 60% used for training the model, 20% for validation, 20% for evaluation). In its current form, the DeepAR model has been used to forecast intra-day time series for a full day for each PV system, computed in the early morning. The model works in an auto-regressive fashion by encoding measurement time series of the two previous days, and recursively forecasting intra-day steps by reinjecting forecasts as input. Besides, it allows to input additional covariates in view to increase the forecasting capacity of the model. For example, time-related features (e.g. hour in the day, day in the month, season) can be used, as well as custom features. In the context of this paper, we use these time features as covariates, but also the NWP / PV performance model-based power forecasts as custom covariates, thus implementing the projected hybrid-physical approach. Further, time-independent parameters ("static features") are fed into the algorithm, which are characteristics of each PV system. DeepAR models the series of expected PV power for a given system, and uses the covariates to obtained improved forecasts.
Regarding the application of this hybrid-physical approach by a DSO for an active distribution grid operation, the approach has the advantage, that besides the smart meters of the PV systems, no additional sensors or other hardware are required. Only smart meter measurements, NWP data and the technical information on the PV systems are necessary.

2 THE USE CASE

The PV power forecasting model in the CombiCast project is developed in collaboration with Electris (a brand of Hoffmann Frères Energie et Bois s.à r.l.), an energy supplier and grid operator (DSO) in the region of Mersch, Luxembourg. Currently, about 150 PV system are connected in the distribution grid of Electris, which are ranging from 1,4 kWp to almost 250 kWp, with an average of 17 kWp nominal power. For about 130 of these systems, 15 minutes resolved measurement values are available.

2 SCIENTIFIC INNOVATION AND RELEVANCE

The combination of different types of forecasting models is described in the literature and methods joining a statistical method with a physical model, such as in our case, are called hybrid-physical approach [6]. But the examples in the literature differ substantially in which part of the forecasting is assigned to the statistical method or the physical model: Some try to use statistical methods in time domains where NWP models are weak(er), such as intra-hour or few hours ahead. Others compute PV power directly, via statistical, data-driven models but the representation of the physical model seems to be rather limited to very specific aspects. Our approach differs from those and focuses on the PV performance model, fed by NWP forecasts and uses a data-driven approach to correct for e.g. shortcomings in the parametrization of the performance model, due to lacking degree of details on the individual systems, and to improve on local effects on the in-plane irradiance (such as shading and snow cover). *(State-of-the-art is much more detailed in the full paper version to be published open access in Solar RRL journal.)*

3 RESULTS

3.1 NWP-data and PV performance model based forecast results

The performance of the purely NWP-based forecast using the PV performance model (before feeding it into the DeepAR model) has been evaluated by common performance metrics, such as root mean square error (RMSE), mean bias error (MBE) and skill score (s – based on RMSE and in reference to a persistence forecast). Only daylight values were uses and the results are normalized to nominal power.
The overall forecast performance, as indicated in Table I is already on a good level. The MBE indicated a tendency to overestimation, but in a very moderate manner. The skill score, also for the individual forecasts, is for almost all the systems positive. The nRMSE for all 130 system can be seen in Figure 2 and indicates few outliers. For some of these outliers, the DeepAR shall be able to improve the performance substantially.

Table I: forecast performance metrics averaged over 2 years for 130 PV systems and the three forecast horizons

	Intra-day (0h – 24h)	*Day-ahead (24h – 48h)*	*2-days-ahead (48h – 72h)*
nRMSE 2020	9.9 %	10.1 %	10.5 %
nRMSE 2021	10.3 %	10.9 %	11.1 %
nMBE 2020	0.6 %	0.5 %	0.9 %
nMBE 2021	1.2 %	1.4 %	1.2 %
s (2020)	29.6 %	32.6 %	-
s (2021)	33.8 %	35.5 %	-

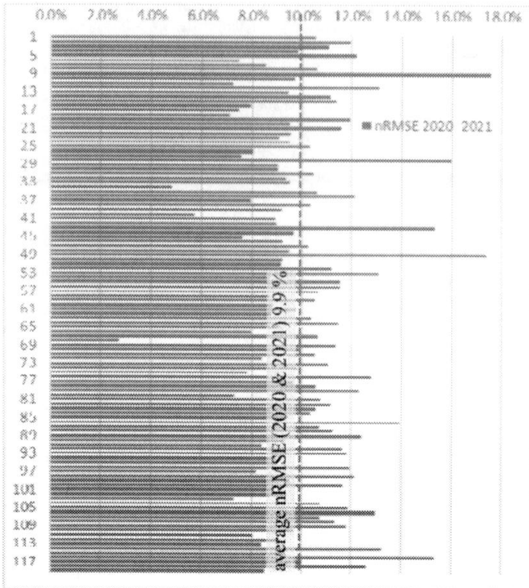

Figure 2: nRMSE of the intra-day forecast, over the years 2020 & 2021 for all PV systems

In order to identify the source of higher nRMSE for some systems, we performed a visual check of all systems. We analyzed the forecast- and production curves and their alignment, checking the system data (nominal power, location and orientations) as well as observing aerial photos of each system. For some of the less well aligned curves, we have been able to identify potential reasons:

a) *System with versatile orientations*: Due to the simplified information on the individual PV systems, as mentioned to be common for DSOs, TSOs and utility companies [7], our data set only contained a single orientation and inclination per system. This is an important parameter in the PV performance model. Visual checks identified 9 systems where PV systems consisted of sub-systems of different orientations, which degrades forecast performance.
b) *Partial, temporal shading*: Based on the production curves and the aerial photos, 5 PV systems could be identified to be partially shaded, e.g. by other buildings or trees, which affected the forecast performance during specific seasons or time of day.
c) *Temperature / limited inverter capacity*: one PV system does not show a "bell-shaped" production curve during clear-sky days – the curve seems to be cut. This could be indicating to temperature problems or nominal capacity limits of the inverter.

Besides those effects harming the forecast performance for specific systems during certain periods, some short periods of "snow cover" could be found, which degraded the overall forecast performance for the whole system portfolio.

3.2 Forecast improvements, based on DeepAR

The improvements of the NWP-based ID forecast by the data-driven post processing, using the DeepAR algorithm, are in the following analysed on an overall level, across the whole portfolio of PV systems. In the sub-chapter, the improvements of the forecast for specific systems, with known forecast deficits, will be addressed shortly (*more details in the full paper published open access in the Solar RRL journal*).

The overall nRMSE for all PV systems, estimated for the ML-enhanced ID forecast in 2020 and 2021 is 8.3%, respectively 8.7%, which represents a reduction of the nRMSE by 1.6% absolut, in 2020 and 2021 (see Table 2). The already low nMBE changed to -0.6% in 2020 and 0.2% in 2021, representing an improvement at least in 2021. The overall skill score substantially increased up to 44% and 47.5%, as depicted in the table below.

Table II: Comparison of forecast metrics for NWP-based forecast and the machine learning- enhanced model

	Intra-day (0h – 24h) NWP based forecast	*Intra-day (0h – 24h) ML enhanced forecast*
nRMSE 2020	9.9 %	8.3 %
nRMSE 2021	10.3 %	8.7 %
nMBE 2020	0.6 %	-0.6 %
nMBE 2021	1.2 %	0.2 %
s (2020)	29.6 %	44.0 %
s (2021)	33.8 %	47.5 %

But more than the improvements on the average over all PV systems, the project team was targeting the improvements for those systems that could not be matched very well with the NWP-based forecast. The ML-based model was able to improve specifically on this temporal alignment of the curves, which results in the improvements shown in Figure 3: It can be observed that the forecast for all PV systems improved, on the yearly level, but specifically the forecast for the systems showing the weakest skill score improved the most.

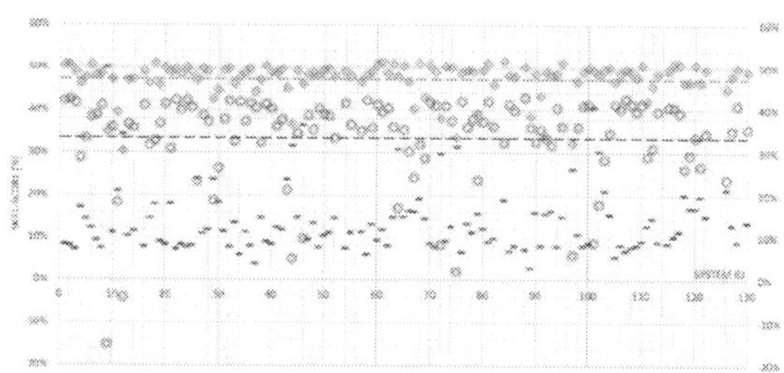

Figure 3: – Skill score of NWP-driven forecast 2021 (blue), compared to the Skill-score of the ML-enhanced forecast (orange), both on left axis, and the improvement of the skill-score (grey lines, referring to right axis)

3.3 Improvements of the forecast for specific systems, with known forecast deficits

Among the systems for which the NWP-based / PV-performance model forecast was least performant were many of the systems which were classified to have specific issues:

a) versatile orientations not reflected by the system information of the DSO
b) partial / temporal shading
c) temperature problems or inverter limitations

Hence, for these systems, the ML-enhance forecast generally improved significantly the forecast metrics. Exemplarily, the Figure 4 depicts the problem of forecast mismatch in case of versatile orientations not covered by the model and the respective improvements based on the DeepAR model.

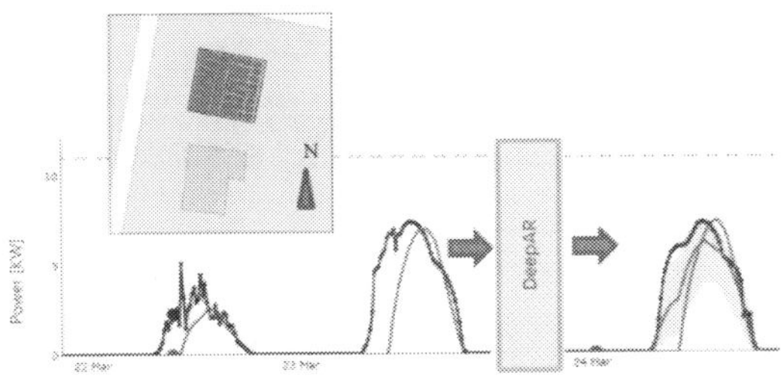

Figure 4: Illustration of system no.44 (top) and the measured production curve (blue) over the NWP-based forecast (red) for three days, depicted in the right graph, incl. the ML-enhanced forecast on the third day (green)

For this specific system, the simplification of the real situation to a uni-directional system caused high systemic deviations of the forecast during the morning hours. The averaged MBE in March 2020 for the morning hours between 8:00 and 9:00 lies around 3 kW. Such systemic deviation can be attenuated by the ML-enhance forecast and resulted in the overall improvement of the skill score from s= 5.1% (NWP-based) to s=36,7% (ML-enhanced), in case of this specific system (system no. 44).

Above described example is an outstanding result, since the forecast for this system was extremely affected by the simplifications of the orientation. The overall improvements for the classified systems are the following:

- *Multiple orientations*: the reduction of systemic forecast deviation during certain time of day has been reduced and resulted in an improved skill score rising from 27.9% in that category to 43.7% – the forecast and productions curves were much better aligned.
- *Partial / temporary shading*: also in this category, the skill score improved, but curves were less well aligned. But the impact on skill score was still high, improving from s=22.2% to 42.8% on average over all system in that category.

Also temporary limited effects, such as snow cover, were corrected for, but the data set did not provide much evidence for it, since snow cover over several days, was scarce in the data. Since the ML model uses measurements of the two previous days as covariates, it takes up to two days until such effects are fully incorporated, and reflected in the forecast.

4 CONCLUSIONS

In this paper we presented a hybrid-physical approach for solar power forecasts, targeted to the application by DSO, in view a grid management and operation at high PV shares. Thus, aiming at single-site forecasts on ID and DA level, we proposed a NWP-driven PV performance model, as the basis of the model. In a second step, we proofed the ability of a data-driven approach, using a RNN model of the LSTM type, to mitigate the temporarily appearing systemic deviations due to either local effects, such as shading, or lacking data quality especially on the orientation of the systems.

In a first step, the performance of the NWP-driven PV performance model based forecasts have been evaluated by comparing to measurements for two full years (2020/2021). The model showed small bias towards slight overestimations, with a nMBE of 0.6% and 1.2% in 2020 and 2021. The nRMSE around 10% can be considered a good result, compared to other studies, as well as the RMSE-based skill score of 29.6% and 33.8% for both years, averaged over all systems.

The forecast data has, in a second step, been introduced as covariates to a LSTM-model, which forecasts the PV power output by considering the two previous days production time series. Overall, the ML-approach succeeded in a further reduction of nRMSE, to 8.3% and 8.7% (in 2020 and 2021). The average skill score over all PV systems (in 2021) improved from 33.8% to 47.5%.

Concluding, it can be stated that the benefits of the presented approach lie within its simplicity and that no additional devices or sensors are necessary, while working exclusively with data available to the future DSOs. The improvements of the forecast suffering from local effects such as shading and more complex orientations of the PV systems, is specifically beneficial for small scale, roof top PV systems in urban areas, where such effect are very common. Further, the effect of local shading on the forecast quality is more severe in regions with high shares of direct irradiance, here this forecast model could be even more beneficial. This also holds true for regions with longer periods or more frequent periods of snow cover. Furthermore, the presented approach, using the LSTM-type model DeepAR, provides a probabilistic forecast, which could be a clear benefit, but needs further investigation on the quality of the given confidence intervals.

5 ACKNOWLEDGEMENTS

This work was supported by the Luxembourg National Research Fund (FNR) in framework of the FNR BRIDGES Project "CombiCast" (BRIDGES18/IS/12705349/Combi-Cast). Furthermore, we would like to thank our partner Electris (a brand of Hoffmann Frères Energie et Bois s.à r.l.), for their trust, the very supportive partnership throughout the whole project duration and their contribution to the common project, financially as well as in terms of man-power and data.

6 REFERENCES

[1] M. Lave, J. Kleissl, and E. Arias-Castro, "High-frequency irradiance fluctuations and geographic smoothing," Sol. Energy, vol. 86, no. 8, pp. 2190–2199, Aug. 2012.

[2] D. Koster, F. Minette, C. Braun, and O. O'nagy, "Short-term and regionalized photovoltaic power forecasting, enhanced by reference systems, on the example of Luxembourg," Renew. Energy, vol. 132, pp. 455–470, 2019.

[3] R. Perez et al., "Comparison of numerical weather prediction solar irradiance forecasts in the US, Canada and Europe," Sol. Energy, vol. 94, pp. 305–326, 2013.

[4] K. Cho, B. van Merriënboer, D. Bahdanau, and Y. Bengio, "On the properties of neural machine translation: Encoder–decoder approaches," in Proceedings of SSST 2014 - 8th Workshop on Syntax, Semantics and Structure in Statistical Translation, 2014, pp. 103–111.

[5] I. Amazon Web Services, "aws - DeepAR Forecasting Algorithm," 2022. [Online]. Available: https://docs.aws.amazon.com/sagemaker/latest/dg/deepar.html.

[6] J. Antonanzas, N. Osorio, R. Escobar, R. Urraca, F. J. Martinez-de-Pison, and F. Antonanzas-Torres, "Review of photovoltaic power forecasting," Sol. Energy, vol. 136, pp. 78–111, 2016.

[7] IEA-PVPS, "Data Model for PV Systems," 2020.

DESIGN OF "HOTSPOT-FREE" BIPV MODULES – PERANAKAN PV

Min Hsian SAW[a,b], Srinath NALLURI[a], Mauro PRAVETTONI[a]
[a]Solar Energy Research Institute of Singapore (SERIS), National University of Singapore
[b]Department of Mechanical Engineering, Faculty of Engineering, National University of Singapore

[a]7 Engineering Drive 1, #06-01 Block E3A, Singapore 117574
[b]9 Engineering Drive 1, #07-08 Block EA, Singapore 117575

ABSTRACT: Coloured PV modules are nowadays gaining interest and becoming more prevalent, with their potential for architecturally harmonious integration in buildings. However, with the printing of different colours onto the module front surface, the series-interconnected solar cells are subjected to non-uniform shading: these colourful modules suffer from mismatch losses and are susceptible to developing hot-spots. In this paper, we present an innovative colourful PV module technology with mitigated mismatch loss and no hot-spot formation risk. It is curated based on two novel design principles: (i) selection of colours (two and above) and tuning of printing parameters for each colour such that the different colours have the same amount of optical transmittance, i.e. equivalent short-circuit current; and (ii) applying a repetition design across the PV module, i.e. all solar cells have the same pattern coverage area/ratio. As a result, all solar cells in the PV module are designed to receive uniform optical irradiance. Highlighting the connection with the local environment and traditions, we demonstrate our innovative module design technology by developing colourful and patterned PV modules that are inspired by the Peranakan culture uniquely to Singapore and Malaysia, coined as "Peranakan PV". This unique BIPV solution combines both tradition and innovation in a reliable and beautiful PV product that is well suited for integration into the façades (building-integrated PV, BIPV) of heritage as well as modern buildings. The design principles have been filed with the Registry of Patents Singapore in October 2021 as "Design principles for multi-coloured and patterned photovoltaic (PV) modules" (SG Non-Provisional Application No. 10202111422S).

Keywords: Building Integrated PV (BIPV), colourful and patterned PV, hot-spot, mismatch, module

1 INTRODUCTION

Building integrated solar photovoltaics (BIPV) has immense potential in transforming buildings into power generators and shift towards sustainability. With their potential for architecturally harmonious integration in buildings, coloured PV modules are nowadays gaining interest and becoming more prevalent. At present, the state-of-the-art are monochromatic PV modules as a number of challenges have been identified for more colourful designs, in particular non-uniform shading induced by the different colours. The series-interconnected solar cells experience different illumination conditions, i.e. more solar energy hits the solar cells of the PV module in one area than in other areas. This potentially causes two issues: (1) short-circuit current and power losses due to cell-to-cell mismatch; and (2) risk of "hot-spots" development during operation of the PV module. Hot-spots phenomena is undesirable as it could lead to long-term and non-irreversible destruction to the PV modules [1]. In addition, hot-spot has been found to be one main factor associated with fire in PV systems [2].

In view of the abovementioned, we propose two novel design principles to create hot-spot free colourful and patterned PV modules with no current mismatch: (1) selection of appropriate colours (two and above) and tuning of printing parameters for each colour such that the different colour areas have the same amount of optical transmittance, i.e. short-circuit current; and (2) applying repetition design/pattern across the PV module, i.e. all solar cells in the module are covered with the same pattern, such that they are shaded with same coverage area/ratio.

Tailoring multi-coloured PV modules with the proposed design principles, it circumvents the concerns of mismatch and hot-spots formation and thus lifts the constraint of using only monochromatic PV modules for building applications which is by far the mainstream. Furthermore, an assortment of patterns can be used jointly with combination of colours, offering architects and designers high design flexibilities. Such multi-coloured PV modules do not only generate energy, but also retain the buildings' appearance or even add aesthetical value to the buildings. In addition, by adopting our innovative technology, the risk of hot-spot formation is diminished, and this is particularly crucial for integration of solar PV into buildings where fire safety is of utmost importance. Simultaneously enriching the visual appearance of coloured PV modules and address the associated technical challenges, this opens a path to wider adoption of BIPV.

2 DESIGN OF EXPERIMENTS

2.1 Design principles

In this study, digital printing technology on glass was adopted to create coloured PV modules. To allow for a uniform light transmittance to the solar cells, the first design principle relates to colour science: the printing parameters of different colours were carefully tuned to allow them to have the same amount of optical transmittance (and translated to short-circuit current), avoiding cell-to-cell mismatches. Taking red and green prints as example, as shown in Figure 1, we see that the convolution of the absolute spectral responsivity with the spectral irradiance of the reference AM1.5G spectrum, $\int SR(\lambda) * E_{AM1.5G}(\lambda)\, d\lambda$, is different for the two colours. Compared to the red print, the green print has lower spectral current generation between 575-700nm; but shows higher generation between 400-575nm. Yet, when we integrate the spectral current generation from 300-1200nm (the responsivity range of c-Si solar cells), the STC short-circuit current generated by both colour prints are the same: 20.49 mA/cm^2 and 20.21 mA/cm^2 for the green and red prints, respectively.

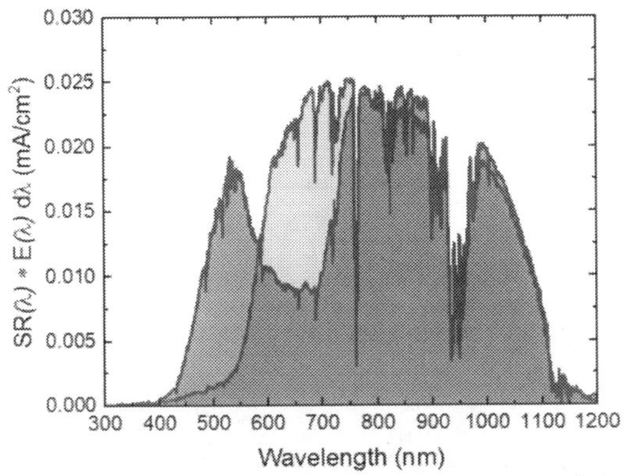

Figure 1. Spectrally resolved current generation of the green and red prints in the wavelength range between 300-1200nm.

The second design principle relates to creating the same amount of pattern coverage area for each solar cell in the PV module, i.e. by applying repeated pattern: a pattern is made to cover the area of a single solar cell, and this design is reused throughout the entire PV module. In terms of aesthetics, repetition design brings a sense of unity, consistency, and cohesiveness to the viewers. Adopting both design principles, all solar cells within the PV module are subjected to uniform shading, thereby mitigating cell-to-cell mismatch loss and the risk of hot-spot development.

2.2 Demonstrating the design principles

Highlighting the connection with the local environment and traditions, we demonstrate the proposed design principles on two colourful PV modules with repeated patterns that are inspired by the Peranakan culture, a hybridization of ancient Chinese culture with the local cultures of the Malay Peninsula, which is now Singapore and part of Malaysia [4]. Colourful decorative tiles with floral and geometrical motifs are commonly used in the Peranakans homes to adorn the façades. Figure 2 shows the two Peranakan tile designs: (a) blue-white Peranakan PV—these two colours symbolises peace, purity, and serenity and are commonly used in the porcelain and dinnerware of Peranakan families; (b) Peranakan PV with red and green— these bright colours are often seen in the objects of the daily life, ornaments and fashion of the Peranakan people. The designs were first created in an imaging software, then printed onto glass with the digital printing technology and subsequently fabricated into 60-cell PV modules with our in-house pilot production line. These modules are coined as "Peranakan PV". A reference module with the same cell-layout but without printing on the glass (transparent) were made for comparative study. To highlight the importance of our proposed design principles and demonstrate contradistinction, apart from the Peranakan PV modules, the results of a patterned PV module, coined as "Singapore-landscape module", is also presented: this module was designed with little attention paid to non-uniform shading and hot-spot reliability concern.

Figure 2. top: blue-white Peranakan PV; bottom: red-green Peranakan PV.

2.3 Module characterisations

Following, the modules were characterized for their electrical performance under Standard Test Conditions (STC) in our laboratory which is ISO/IEC 17025:2017 accredited for electrical testing of PV modules (LA-2012-0526-E-1). In addition, to validate the mitigation of hot-spot, we also conducted outdoor hot-spot endurance test and map the surface temperature distribution of solar cells in the modules with infrared thermography (IR) [5]. The outdoor measurements circumvent the limitation of spatial uniformity and allow uniform irradiance onto the test modules. The hot-spot test was conducted with the modules mounted on a solar tracker (normal incidence angle to the module surface) and on a day with clear sky conditions, with an average measured irradiance of 950W/m^2 (Figure 3).

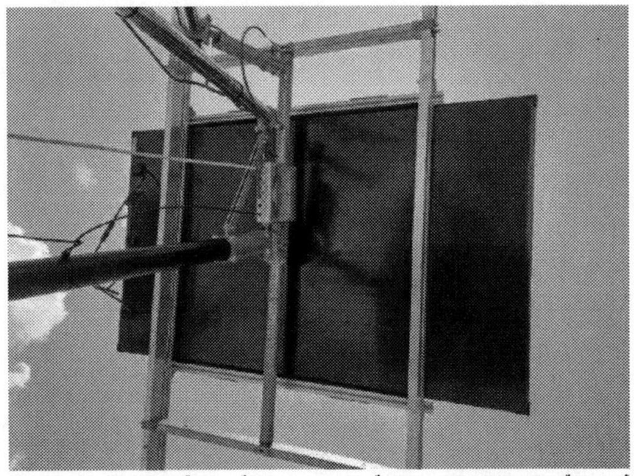

Figure 3. Outdoor hot-spot endurance test conducted under clear sky conditions, with an average irradiance of 950W/m^2.

3 RESULTS

Figure 4 shows the I-V and P-V curves of the Singapore-landscape module (solid lines), comparing against its reference module (dash lines, made with transparent glass with no printing) with the same cell-layout. As compared to its reference module, the coloured module shows a significant drop in short-circuit current, while the module voltage is minimally affected. In addition, it also suffers current and power loss caused by the cell-to-cell mismatch, as clearly indicated by the step in the I-V and P-V curves.

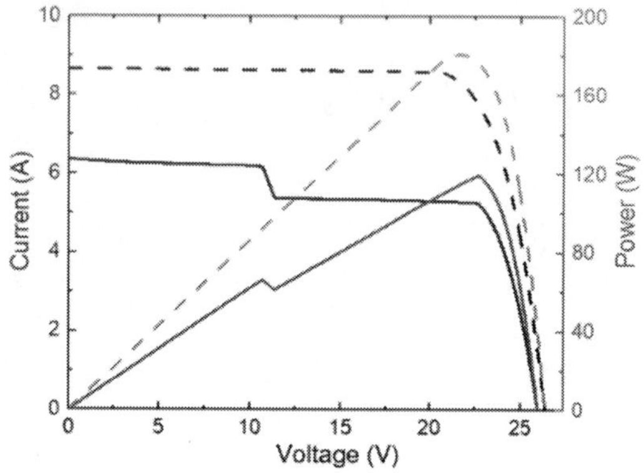

Figure 4. I-V and P-V curves of the Singapore-landscape module. The dash lines represent the I-V and P-V curves of a reference module made with the same cell-layout.

Figure 5 depicts the measured I-V and P-V curves of the Peranakan modules (solid lines) and their reference module with the same cell-layout (dash lines). The module power is dependent on the amount of shading and operating conditions. Compared to the reference module, the power loss of the Peranakan PV modules is dominantly attributed to the loss in module current that arises from the reflection and absorption losses induced by the colour printing, while the module voltage is minimally affected. Essentially, there is no mismatch effect observed for the Peranakan PV modules, implying uniform light transmittance to the underlying solar cells.

(a) Blue-white Peranakan PV

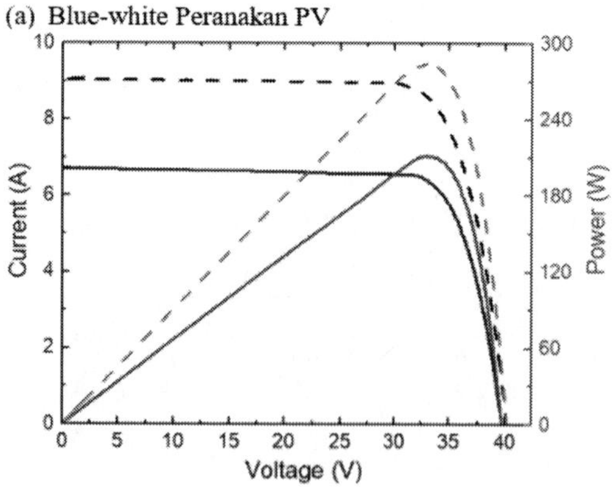

(b) Red-green Peranakan PV

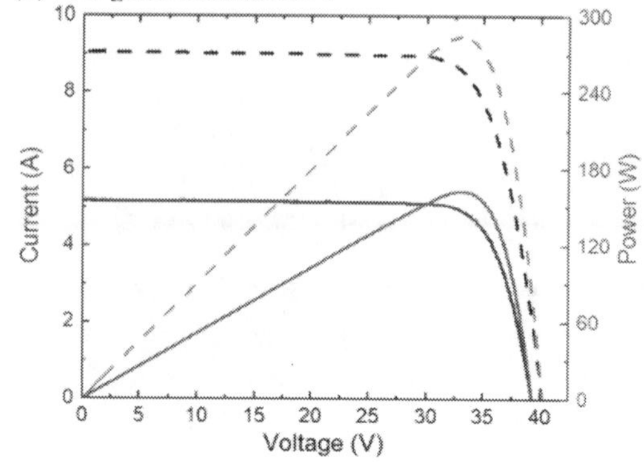

Figure 5. I-V and P-V curves of (a) the blue-white Peranakan PV; and (b) the red-green Peranakan PV. The dash lines represent the I-V and P-V curves of a reference module made with the same cell-layout.

Figure 6 illustrates the IR images of the modules. The pictures on the left show the front view of the modules: for the Peranakan modules in (a) and (b), they are reflected vertically to match the perspective of the IR images; while for the Singapore-landscape module, the image is reflected horizontally. As seen from the IR images, two solar cells in Singapore-landscape PV (indicated by the arrows in the corresponding picture) exhibit temperatures exceeding 100°C: this is due to the larger shading of these two cells compared to other cells connected in the same string. This suggests a potential of localised heating and hot-spot development for them. In contrast, despite also having colour and pattern printings on the front glass, the solar cells in the Peranakan PV modules show fairly homogeneous temperature distribution measured under the same exposure conditions: they range from 50°C to 70°C.

From both I-V and IR results, it is implied that the Peranakan PV modules are least susceptible to mismatch and hot-spot, proving the feasibility of the proposed design principles. The comparison with the Singapore-landscape module also highlights the significance of this work: simply decorating the PV modules with beautiful graphics, patterns or images with little attention paid to non-uniform shading could have detrimental impacts on the module functionality and safety.

4 CONCLUSIONS

We have successfully demonstrated the validity of our proposed design principles in creating colourful and patterned PV modules which are free from current mismatch and hot-spot: they are based on our colour science algorithm and the adoption of repeated patterns. With this innovative technology, we simultaneously enrich the visual appearance of coloured PV modules for building applications and address the associated technical challenges. This eliminates the constraint of using monochromatic PV modules which are by far the mainstream for BIPV applications, allowing for higher design flexibilities and better integration into both heritage and modern buildings. Moving forward, we further synthesize the design principle based on colour science into "pixelation-based" design schemes: this work will be presented in the MRS Fall 2022 conference.

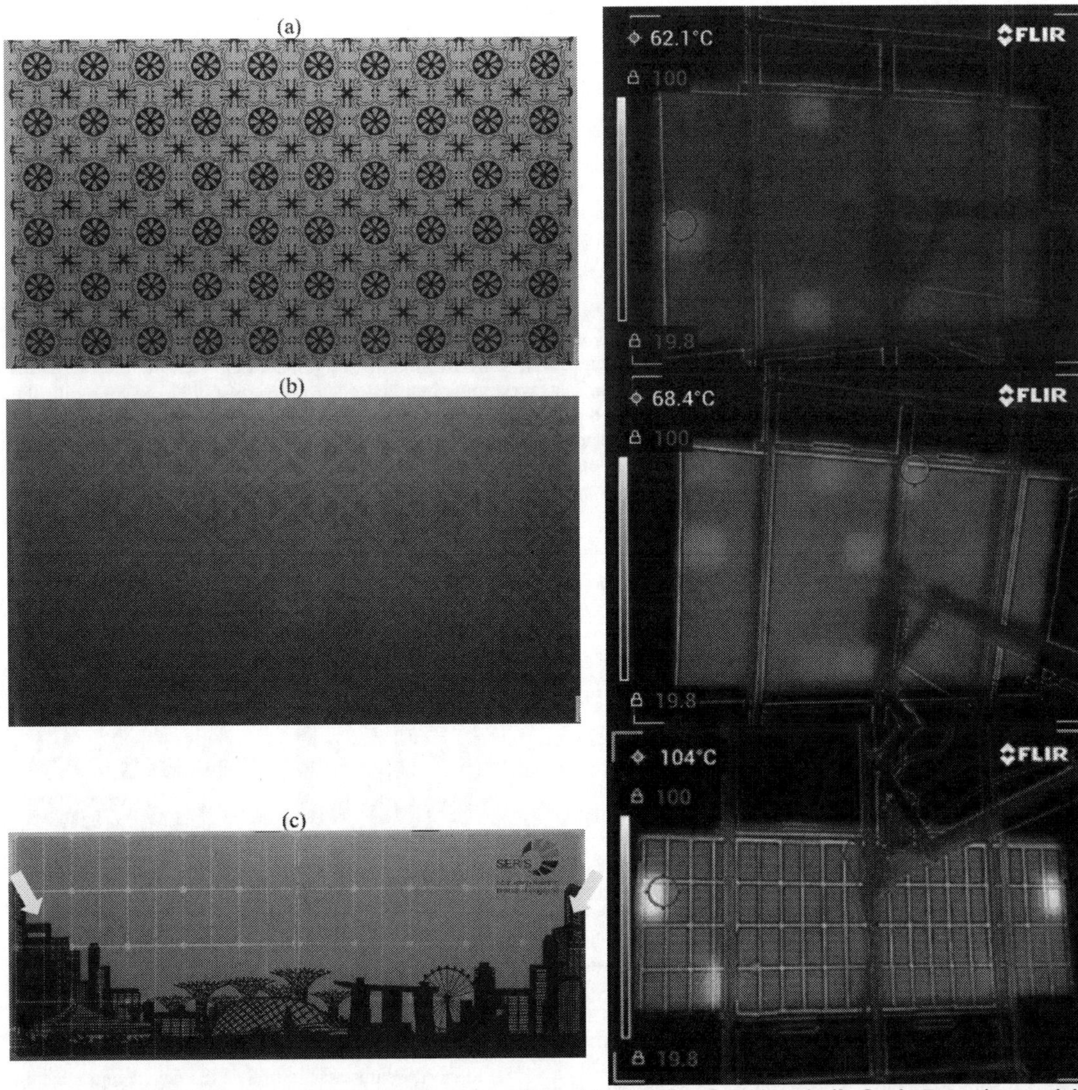

Figure 6. Infrared (IR) thermography (right) and corresponding pictures (left, reflected vertically for the Peranakan modules, and reflected horizontally for the Singapore-landscape module), to represent the same perspective of the IR thermographies of the Peranakan modules and Singapore-landscape module, respectively. The IR images were captured from the outdoor hot-spot endurance testing under the same exposure conditions: the arrow indicates the hot-spot cells in the Singapore-landscape module.

5 ACKNOWLEDGMENT

SERIS is a research institute at the National University of Singapore (NUS). SERIS is supported by NUS, the National Research Foundation Singapore (NRF), the Energy Market Authority of Singapore (EMA) and the Singapore Economic Development Board (EDB). The authors would like to thank Leow Shin Woei, Gavin Raharjo, Jack Garcia Villanueva and Abdul Razak Bin Saman for their scientific and technical supports.

6 REFERENCES

[1] Jordan, D. C., Silverman, T. J., Wohlgemuth, J. H., Kurtz, S. R., & VanSant, K. T. (2017). Photovoltaic failure and degradation modes. Progress in Photovoltaics: Research and Applications, 25(4), 318-326.

[2] Dhere, N. G., & Shiradkar, N. S. (2012). Fire hazard and other safety concerns of photovoltaic systems. Journal of Photonics for Energy, 2(1), 022006.

[3] Saw, M. H., Singh, J. P., Wang, Y., Birgersson, K. E., & Khoo, Y. S. (2020). Electrical Performance Study of Colored c-Si Building-Integrated PV Modules. IEEE Journal of Photovoltaics, 10(4), 1027-1034.

[4] Lee, S. K. (2008). The Peranakan Baba Nyonya culture: resurgence or disappearance?. Sari (ATMA), 26, 161-170.

[5] IEC 61215-2:2021: Terrestrial photovoltaic (PV) modules - Design qualification and type approval - Part 2: Test procedures

BIFACIAL PV BALCONY AND TERRACE FENCES

Roland Valckenborg[1], Simona Villa[1], Bas B. Van Aken[2] and Teun Burgers[2]
[1] TNO – Energy Transition, High Tech Campus 21, 5656 AE Eindhoven, The Netherlands
2 TNO – Energy Transition, Westerduinweg 3, 1755 LE Petten, The Netherlands
Corresponding author: Roland Valckenborg, email: roland.valckenborg@tno.nl

ABSTRACT: We have developed a bifacial PV fence that can be used both on balconies and terraces. For aesthetical reasons the PV fence mimics balusters, by alternating vertical columns of PV cells and (partially) transparent areas. As a result the solar cells cover about 60% of the bifacial fence. We present the full year performance of such a vertical bifacial BIPV-system, by experimental validation of two prototypes. These results are supported by simulation studies, in particular quantifying the shading behaviour, both of fence on fence, but also of the intrinsic horizontal and vertical construction elements of the balcony or terrace fences. One prototype is a west-facing balcony, about one meter from the façade. The bifacial modules on this balcony produce 624 kWh/kWp which corresponds to a $PR_{bifacial}$=88%. The other prototype is a square-shaped roof terrace fence, consisting of three modules on each side. We present a full analysis of the annual yield for all four sides. The east, west and south side have a specific yield in-line with the performance of a perfectly oriented monofacial module in the Netherlands. The BIGEYE modelling of this system results in a very good agreement between modelled and measured kWh generation per month. Also, we give a detailed insight into which horizontal or vertical construction element causes the most shading losses for each module. The vertical posts on the inside of the northern fence create the largest shading loss, followed by the horizontal toprails above the south, west and east fence and the vertical posts inside of the west and east side. These shading losses are incurred mostly by the north fence (9%) and least by the south (3%).

Keywords: Building Integrated PV (BIPV), Bifacial, Energy Performance, Façade, Modelling, Shading

1 INTRODUCTION

1.1 The market of balconies and terrace fence

Making existing homes more sustainable in densely populated, urbanized areas with many high-rise buildings is very challenging: there is little roof surface available for the installation of conventional PV-modules. And on flat roofs PV is competing with installation equipment (e.g. outside units of heat pumps, air heat recovery systems, installations for window & façade cleaning), with rain-retaining, temperature-reducing green roofs or with roof terraces for leisure. However, for these buildings, the fences on balconies and around terraces are available to explore for PV electricity production.

The European Near Zero Energy Building (NZEB) directive has led to laws in all the member states. For the Netherlands - where we developed this product - the implementation of this law [1] has led to three requirements for new buildings. We aim to fulfil the third requirement that imposes a minimum amount of renewable energy that needs to be produced on the building itself. Especially for high-rise buildings with a small roof-to-façade ratio, this requirement pushes architects to design PV-active façade elements, and makes them interested to use every available m² of the façade, hence also potentially the balconies in that façade.

1.2 Product development

The Dutch consortium, of knowledge institute TNO and companies Balkonhekken BV, Durin BV, and Rebor BV, has developed a bifacial PV fence that can be used both on balconies and terraces. For aesthetical reasons the PV fence mimics balusters, by alternating vertical columns of PV cells with (partially) transparent areas.

The aim of this contribution is to show the full year performance of such a vertical bifacial BIPV-system, by experimental validation and simulations of two prototypes. The application of bifacial PV-cells instead of monofacial ones in a terrace fence is a no-brainer, because the sun's position varies throughout the day and seasons. As a result, the two sides of an, e.g. west- and east-facing balcony fence, receive both a significant amount of light For the application in balconies, there is clearly a front-side in the same orientation as the façade and a back-side 'looking back to the building'. The building will limit the amount of back-side irradiance, and only a small amount of sky diffuse irradiance and light that is reflected from other parts of the façade on the back side of the PV-balcony will contribute to the total performance. On the other hand, the added price of a bifacial PV-cell in comparison to a standard monofacial PV-cell is declining rapidly. It was decided to use exactly the same cells and other component to assemble exactly the same PV-modules for the balcony as for the terrace fence application. The same layout will make logistics in the factory much easier, hence will lead to less installation mistakes and a lower cost price.

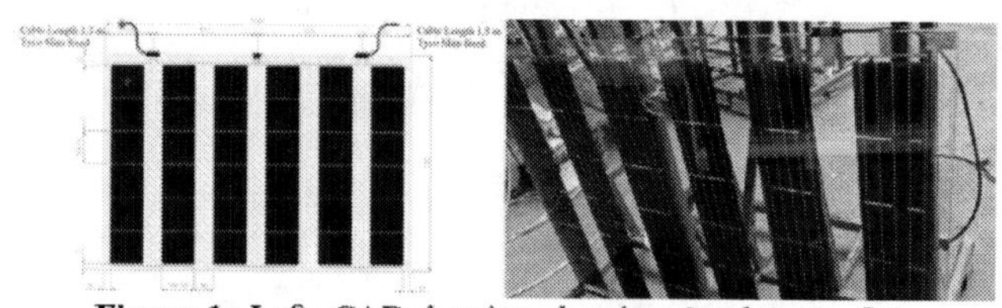

Figure 1: Left: CAD-drawing showing 6 columns of 6 cells with 3 sidewing junction boxes. Right: first assembled prototype.

The PV-modules were assembled by third party Ertex from 36 n-type TOPCON bifacial cells with a bifaciality factor of ~80%. The indoor IV-flash data shows that the electrical parameters of all modules show little variation. The most important of them, i.e. the rated power, P_{mpp}, is on average 170.9 Wp, with a standard deviation of just 0.9 Wp (and min and max values of: 169.7 Wp and 173.0 Wp respectively), which is within the measurement accuracy of the IV-flasher and measurement equipment in the outdoor testing. The other relevant electrical parameters are: $V_{mpp} = 19.1$ V, $I_{mpp} = 8.9$ A, $V_{oc} = 23.9$ V, $I_{sc} = 9.4$ A and a fill factor of 75.7%.

The safety requirements for balconies in the Netherlands are prescribed by EN 12600 [2], in which the mechanical strength is tested with a heavy bag falling into the product. Based on this test, it was decided to use a laminate with two times 5 mm glass. Regarding the size, the maximum open area between the floor and the lowest rail of the balcony needs to be 100 mm, and the minimum height of the toprail needs to be 1000 mm (1200 mm when installed on the 4th floor or higher).

1.3 Measurement setups

At the Solar Outdoor research facility SolarBEAT [3] (Eindhoven, The Netherlands) two measurement setups were built, as shown in Figure 2. For the PV balcony three prototypes have been installed in a balcony in the west façade of a high-rise building. For the PV terrace fence 12 prototypes have been installed on the roof of the building. These are installed in a square with 3 PV modules for each orientation. Please note that always the front side of the bifacial module has been oriented to the outside of the terrace. Therefore, also the north side of the terrace has a north facing front side, for research purposes, not for optimal electricity production.

Figure 2: Picture of the two demonstrators at outdoor research facility SolarBEAT. Left: PV balcony. Right: PV terrace fence.

The measurement period was August 2021 up to and including July 2022. for all results and simulations presented in this paper. The overall data completeness of the one year outdoor testing is around 97%. This value is so close to 100% that it was decided to not interpolate the few missing hours, because such an interpolation procedure is mostly prone to hidden effects.

For the PV balcony setup, the electrical DC data was collected directly from the SolarEdge power optimizer M2640. The accuracy of this data as reported by SolarEdge is accurate enough for the purpose of this setup. For the PV terrace fence setup, however, we want to analyse the complicated effects of various types of shading. Therefore each module is electrically connected to an IV-tracer with MPPT-tracking in between the IV-sweeps. It was found that one IV-sweep for each module every 5 minutes gives enough information on all the shading types. For DC energy yield measurements, the P_{mpp} was measured with the same equipment but a faster time resolution of 1 minute. The used EKO Instruments PV blocks [4] equipment measures all 12 PV modules simultaneously (so not sweeping).

1.4 Modelling with BIGEYE

TNO's BIGEYE software package simulates the output of standard and specialty PV systems. The output during a year, approximated in 1-hour steps, has been shown to be within 1% accurate compared to the full-year output, varying in real time. Details of the BIGEYE software package have been published elsewhere [5,6]. BIGEYE was developed when bifacial and semi-transparent PV modules could not be simulated with existing software packages. As, for many situations, the major part of the modules' rear irradiance is due to ground-reflected light, so-called albedo, BIGEYE calculates the irradiance distribution on the ground in great detail, taking into account the distribution of shadow and beam light, the semi-transparency of the PV modules and the partial blocking of the sky, indirect/diffuse shadow, by surrounding PV tables. Likewise, the irradiance and shadow distribution on both sides of the semi-transparent modules is calculated. And the irradiance per cell for each timestep is determined. The resulting maximum power point, V_{mpp} and I_{mpp}, per module or string of modules is then either approximated by applying the Rodrigo model or by generating the full I-V curve of the string or module.

2 PV BALCONY

Figure 3: Three prototypes labelled B1, B2, B3. On the two pillars in between (highlighted with orange rectangle) backward facing photodiodes measure $G_{meas,back}$

2.1 Full Year Measurement Results

The DC sum of energy over this year is 106.6 kWh per module. The building industry is used to quantify everything related to balconies per running meter, with unit m^1. Hence, this gives 71 kWh/m^1. In comparison with other applications one should realize that the height of the balcony is 1.0 m. Therefore expressing the result per area will give the same numerical value. Expressed in specific yield the full year result is: 624 kWh/kWp. The frontside irradiation was measured to be 598 kWh/m^2. The backside irradiation with respect to front side is 35% for the photodiode between module B1 and B2, and 28% for the photodiode between module B1 and B2. This difference can be understood from the heterogenous light distribution caused by the different building materials in the back of the balcony, like windows, frames, bricks, colored glass (see Figure 3). For the $PR_{bifacial}$ calculation [7] only one value of $G_{meas,back}$ can be used. Therefore we take the averaged value of the two measurements. The final result is $PR_{bifacial} = 88\%$, which is comparable to a good installed and well ventilated BIPV-module.

2.2 Comparison PV balcony with (colored) PV Façade element

A PV-active façade element will nearly always be a colored (and sometimes also textured) PV-module with non-standard dimensions. Pelle et.al. [8] presents a recent market survey for such products. And also an internal TNO survey for specifically the Netherlands was done. Both studies show products in a power range of between 110 and 150 Wp/m^2, with a median value around 130 Wp/m^2. If we assume a PR=88% and calculate the needed power rating of such a virtual (monofacial) PV-active façade module, to give the same full year energy yield of one m^2 balcony (71 kWh/m^2), then we find R=135 Wp/m^2. That implies that to comply with the Dutch NZEB-law, the architect can choose between this PV balcony or a colored PV facade module (of 135 Wp/m^2) to get the same full year energy yield per m^2. When assuming the same ratio between backwards and frontside irradiation, this conclusion is valid for all orientations. Moreover, this conclusion will roughly hold for any place in the world, by assuming that the precise trajectory of the sun through the sky will only give second order deviations in backwards irradiation on a balcony.

3 PV TERRACE FENCES

Figure 4: In total 12 prototypes with horizontal toprails and vertical pillars in between the modules (see orange rectangles).

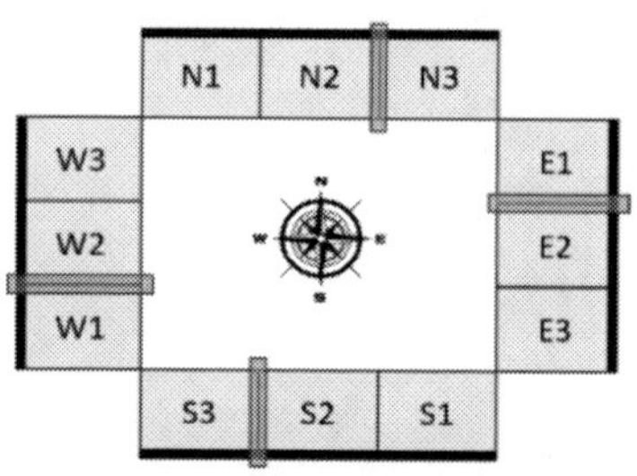

Figure 5: Top view with PV-module (yellow) labelling. Black bars shows front side orientation. Orange rectangles denote the pillars that contain photodiodes, for each pillar 3 pieces in height, and both facing back and front. Hence a total of 4x3x2 = 24 photodiodes are installed.

It will be shown later that the middle module of each orientation is least affected by shading (but not completely shade free). Therefore the results of the middle modules (nr.2 in each orientation) are presented first; see Figure 6. One of the most striking results is the fact that the east (E2) and west (W2) module are benefitting most from the long daytime in the months April till August. Secondly, March 2022 was an exceptionally high irradiance month, of which the south (S2) module benefitted the most.

This monthly energy yield per PV-module can be presented to the wishes of the user, to be either kWh/m^1, kWh/m^2 or kWh/kWp. Because these graphs look similar to Figure 6 but only with a different y-axis, they have not been displayed in view of space. The overall yearly energy results are shows in Figure 7 below.

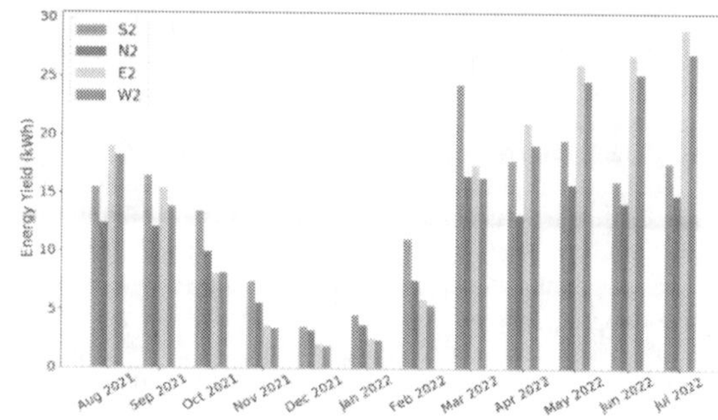

Figure 6: The Energy yield per module (nr.2 of each orientation) in a monthly bar plot.

	E [kWh]	Y [kWh/m1]	[kWh/kWp]
S2	167	111	978
N2	129	86	753
E2	177	118	1035
W2	166	110	969

Figure 7: The total full year energy yield (nr.2 of each orientation) per module, per running meter, and per specific power.

Although the comparison with a 'normal' pitched roof mounted PV-system is a bit strange, because of the completely different application area, it does show a remarkable result. For this specific location, PVGIS [9] gives 1026 kWh/kWp for a south oriented roof top installation under perfect tilt angle (35° tilt). From Figure 7 it can be seen that the bifacial PV terrace fences produce roughly the same amount of energy. Only the north oriented fence is performing significantly less, however in a commercial installation, one would turn around the PV-module between the vertical pillars and change it into a south facing installation.

Like explained in the previous chapter from monthly energy yield together with $G_{meas,back}$ and $G_{meas,back}$ the $PR_{bifacial}$ is calculated and shown in same barplot style.

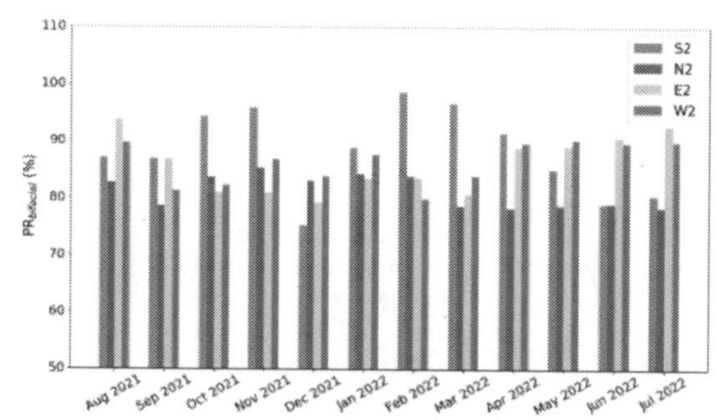

Figure 8: The $PR_{bifacial}$ per module (nr.2 of each orientation) in a monthly bar plot.

It is difficult to explain why the $PR_{bifacial}$ can be on the order of 10% higher or lower for specific modules for specific months. Taking the measurement accuracy into consideration, these differences are significant and not related to measurement accuracy. For the lower performance of S2 in December, we found out that this is caused by the measurement cabin. Checking all webcam images (full year round, every 10 minute) we see that only in the situation of lowest solar elevation, the measurement cabinet is shading only the south oriented modules. The other significant deviations are also cause by shading, which will be investigated in the next chapter. Analyzing full year irradiances and yield gives the following bifacial PR-values: S2=89%, W2=86%, N2=77%, and E2=87%. Only the north oriented module is significantly lower than the other orientations which are roughly between 85% and 90%. Like concluded for the PV balcony, this level of (DC) PR is expected for a good installed and well ventilated BIPV-module.

4 SHADING

4.1 Shading Identification (Qualitative)

Visual inspection of the demonstrator setup shows a huge variety of different types of shading. Moreover, throughout the seasons, other types become more prominent and others less or disappearing. Pictures have been taken to identify at least six types of shading, as shown in Figure 9.

Figure 9: Pictures showing the various types of shading. 1: far-away building, 2: nearby measurement house, 3: module-to-module from opposite row, 4: module-to-module onto adjacent module, 5: construction element horizontal toprail, 6: construction element vertical pillar.

We exclude the far-away horizon shading and nearby objects shading in the further analysis, because the effect is limited. Moreover, these types of shade will be very site dependent, whereas the construction shading is present in all installations (although also depending on the precise layout and number of terrace fences of each customer). The overall effect of the far-away shading will most probably be judged by the craftsmanship of the e-installer, or by a commercial PV system software (like PVSYST) that is used when making a quote (with yield prediction) for a specific customer.

4.2 Shading on one typical sunny day

Focusing now on the construction elements themselves, a very sunny day suits best to pinpoint the shading caused by these. Figure 10 shows the irradiance on the 8th September 2021 for all 24 photodiodes.

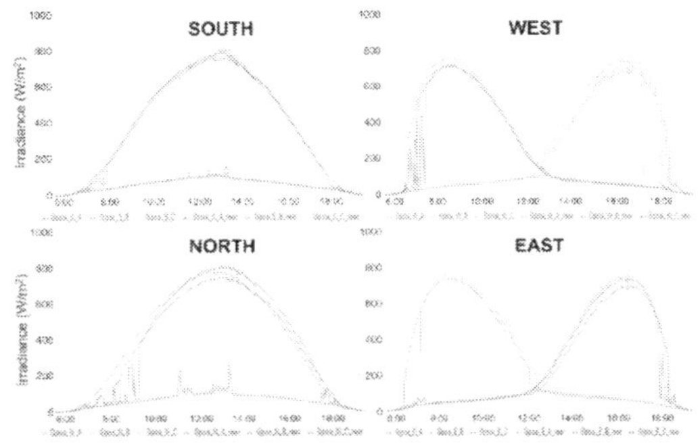

Figure 10: For each orientation, the 3 frontside and 3 backside photodiodes are plotted as a function of time of the day (8th September 2021).

All curves show a smooth profile with a shape that is completely matching expectations; the small drops and spikes are caused respectively by shading and very short glare of scattered irradiance on nearby objects. IV-curves of the west oriented module W1 are plotted in Figure 11(b) for 4 instructive moments in the morning of this day. In Figure 11(a) from top to bottom I_{mpp}, V_{mpp}, and P_{mpp} are plotted. The vertical dotted lines indicate the moment at which the IV-curve are taken.

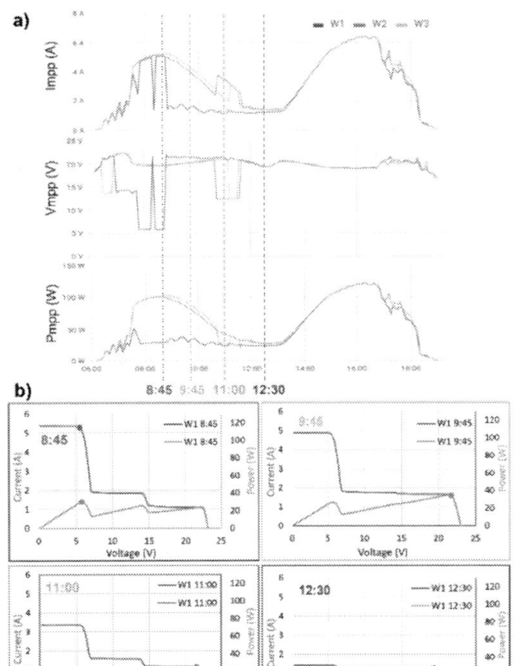

Figure 11: Module W1 for 8th September 2021. a) from top to bottom: I_{mpp}, V_{mpp}, and P_{mpp}. Vertical dotted lines in blue, cyan, pink and red refer to the times 8:45, 9:45, 11:00, and 12:30 of which the IV/PV-curves are plotted in b).

The activation of the bypass diodes can be clearly seen in the IV-curves. Together with the camera-images we determined that first the shading from module S3 will hit W1 and results in: mpp @ high I, low V, see blue dot in middle-top plot of 8:45. Thereafter the same shading gives mpp @ high V, low I, see orange dot in right top plot of 9:45. On about the same moment, also the self-shading from the construction appears. Finally after around 14:30 all shade is out of module W1.

The above procedure is so tedious that one cannot quantify the overall losses for a full year by running through all days of the year. This is our main motivator to model the demonstrator setup.

5 MODELLING AND VALIDATION

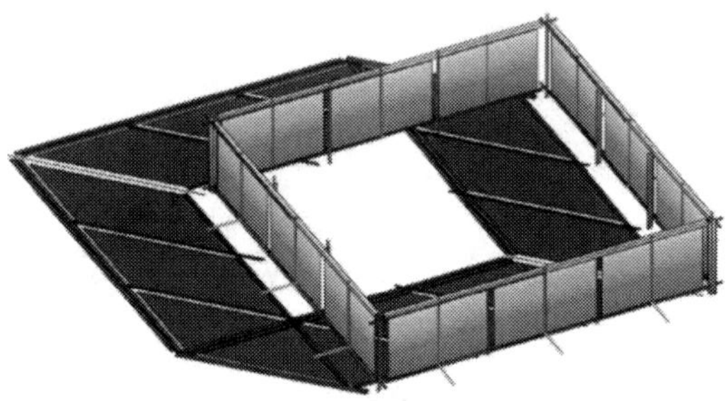

Figure 12: Screenshot of irradiance distribution (from dark blue = 0 W/m^2 till yellow = 800 W/m^2) on the PV fence demonstrators as composed virtually in BIGEYE. All construction elements that give shade (like vertical pillars and horizontal toprails) have been included.

5.1 Validation

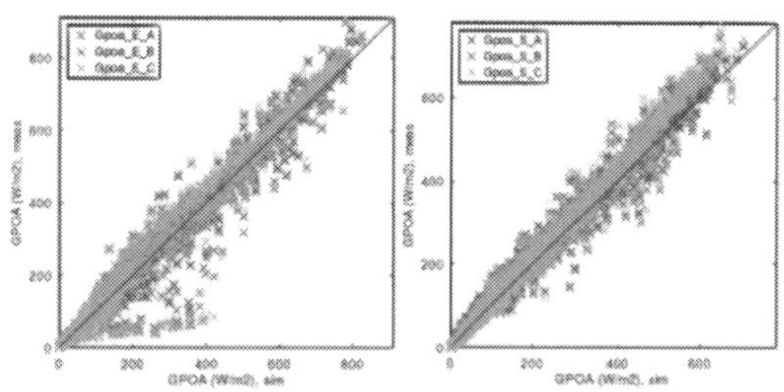

Figure 13: Measured versus simulated G_{POA} for orientation east (left) and south (right). '_A' denotes the upper PD, '_B' the middle PD, and '_C' the lower PD. Time period is last two weeks of July 2021.

The measurements and model give 24 values for irradiance per minute. This gives a huge amount of data. For Figure 13, we zoom into two weeks of data at the start of the measurement period and for only two orientations, east and south (in view of space). One can see that the correlation between model and measurements are quite good. For north side orientation (not depicted) this correlation is a bit lower, which we attribute to the more difficult shading situation.

Apart from looking to scatter plots of measurement versus model, one can also look to summed energy yield.

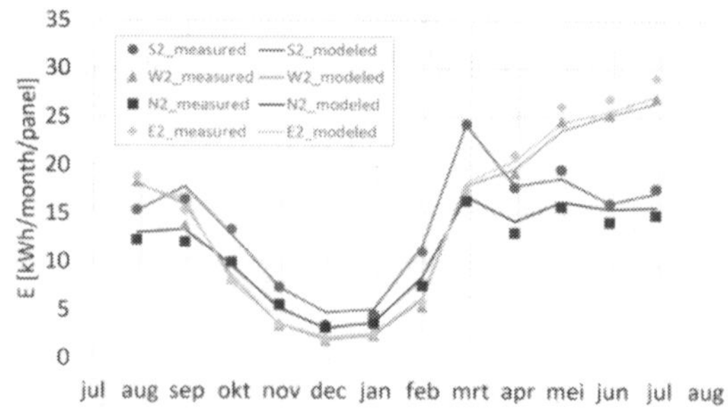

Figure 14: Monthly sum for the middle (nr.2) module of each orientation. Markers are measurements. Solid line drawn between modeled results (for clarity no marker).

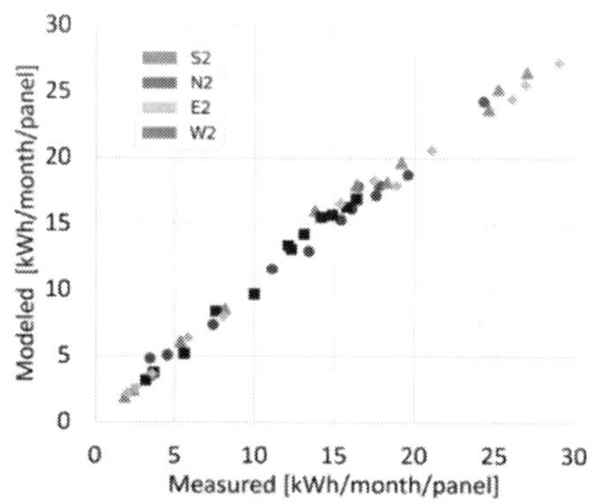

	Full year in [kWh/module]			
	measured	modeled	Δ [kWh]	Δ [%]
S2	167.2	169.0	1.8	1%
N2	128.8	135.2	6.5	5%
E2	177.0	173.3	-3.7	-2%
W2	165.7	169.6	3.9	2%

Figure 15: Top: Measured versus modelled monthly sum. Bottom: yearly sum measured versus modeled. Difference expressed both in absolute [kWh] and in [%] of yearly yield.

The timeseries graph of Figure 14 gives a nice impression of the good quality of the BIGEYE modelling. For a more quantitative validation, one can also study the scatter plot of modelled versus measured monthly energy sum, see Figure 15, and the yearly sum presented in the table below the figure. We see a slight underestimation of the model for the east and west oriented modules for the months April till July. The reason for this is somehow speculative because we are searching for small deviations. The far-away horizon shading of the direct light and the hemisphere coverage of the diffuse part of the irradiation might be the cause for this, because in the model these far away objects are not included. Most remarkable is the quite accurate prediction of the model for the very high performing S2-module in the high irradiance month March 2021.

5.2 Shading Losses Quantification

With a validated model, we have a tool to study the quantitative effect on full year energy yield for each type of shade. Please note that we will limit ourselves to the self-shading types originating from the construction elements, as explained in the previous paragraph.

Focusing on the self-shading, there are still many construction elements that can be toggled ON/OFF in the simulation. A full setup of all possible options is impossible to model. Therefore we have chosen to start with a model that has no construction element at all; the PV-modules are hanging freely in space. We define this as 'no shading'. Then we add the element of which we have the impression that it is a main contributor, and run the full year model for all PV-modules. The relative energy yield with respect to the unshaded is shown below in Figure 16. Thereafter we add the second-most contributing element, run the model, and analyze again. This procedure is repeated so to fill the table in Figure 16, until we have a model that has all the construction elements added with dimensions exactly equal to the demonstrator,

	N_1	N_2	N_3	E_1	E_2	E_3	S_1	S_2	S_3	W_1	W_2	W_3
no shading	100%	100%	100%	100%	100%	100%	100%	100%	100%	100%	100%	100%
N pillars	96%	94%	97%	100%	100%	100%	100%	100%	100%	100%	100%	99%
NW, NE pillars	96%	94%	96%	100%	100%	100%	100%	100%	100%	100%	100%	99%
next W and E pillars	96%	94%	96%	97%	100%	100%	100%	100%	100%	100%	97%	99%
first W and E pillars	95%	94%	96%	97%	97%	100%	100%	100%	100%	100%	97%	97%
SW, SE pillars	95%	94%	96%	97%	97%	99%	100%	100%	100%	99%	97%	97%
S pillars	95%	94%	96%	97%	97%	99%	99%	99%	99%	99%	97%	97%
N toprail	95%	93%	95%	96%	97%	99%	99%	99%	99%	99%	97%	96%
W toprail	92%	92%	95%	96%	97%	98%	99%	99%	98%	97%	96%	94%
S toprail	92%	92%	95%	96%	96%	96%	98%	97%	97%	95%	95%	94%
E toprail	92%	92%	91%	95%	95%	95%	97%	97%	97%	95%	95%	94%

Figure 16: To the left, the name of the construction element added. Top row shows PV-module label (including same color coding as in previous Figures). Thereafter energy yield is color coded shown relative to the 'no shading' situation of the first row of results.

Another way of expressing the results is by looking to the specific effect of each shading element in terms of energy loss, as reported in Figure 17. For that insight we notice that an absolute scale of [kWh/module] is better distinguishing than a relative scale.

	N_1	N_2	N_3	E_1	E_2	E_3	S_1	S_2	S_3	W_1	W_2	W_3
N pillars	5.24	8.77	4.29	0.59	0.06	0.01	0.03	0.03	0.03	0.02	0.11	1.27
NW, NE pillars	0.54	0.10	0.69	0.24	0.04	0.01	0.01	0.04	0.04	0.02	0.04	0.21
next W and E pillars	0.09	0.20	0.52	4.49	0.60	0.01	0.04	0.06	0.16	0.37	4.33	0.02
first W and E pillars	0.40	0.18	0.16	0.08	4.22	0.38	0.17	0.04	0.09	0.05	0.39	4.50
SW, SE pillars	0.01	0.04	0.02	0.01	0.05	0.63	0.46	0.07	0.32	0.63	0.05	0.01
S pillars	0.05	0.01	0.04	0.02	0.23	0.94	0.88	1.63	0.98	0.82	0.16	0.03
N toprail	0.73	0.86	0.59	1.01	0.28	0.07	0.19	0.26	0.26	0.08	0.23	1.30
W toprail	3.73	0.93	0.22	0.58	0.87	0.89	0.13	0.27	0.75	2.13	2.44	2.43
S toprail	0.32	0.19	0.13	0.26	0.96	4.14	2.07	2.34	2.22	3.98	0.95	0.20
E toprail	0.23	0.94	4.64	2.24	2.25	2.01	0.78	0.30	0.14	0.36	0.61	0.61

Figure 17: To the left, the name of the construction element added. Values are [kWh/module] losses due to the specific shading (not cumulative) for one full year.

5.3 Modelling Full Terrace Demonstrator Setup

After implementing all the construction elements in BIGEYE, one could ask themselves which construction element is causing the most shading losses. The product developers should put most effort in reducing the size of this construction element, if at all possible. Please note that changing the construction element leads to a specialized version which will have a much smaller market volume and hence gives a financial penalty to the total system costs. Nevertheless, it is for sure of academic interest to find out which construction element is causing how much overall shading.

Figure 18: Left: the absolute loss of energy on all modules caused by a specific construction element [kWh/element]. Right: only adopting the coloring in a visualization with sizes of pillars and toprails that are more closely resembling the actual demonstrator.

Finally, the question can be raised about the total self-shading losses (caused by all construction elements in the demonstrator) for each PV-module, see Figure 19.

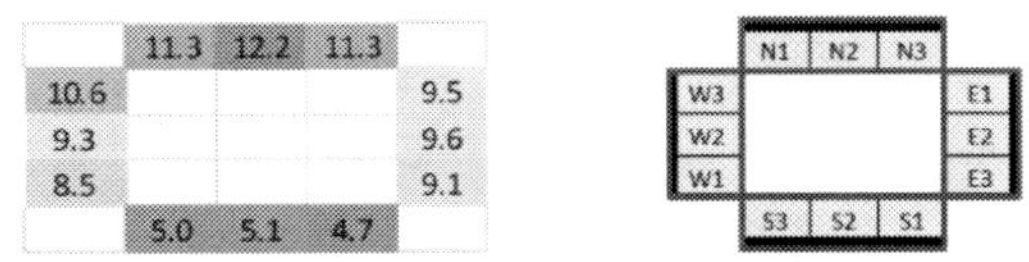

Figure 19: Left: the absolute loss of energy for one module [kWh/module] caused by the shading of all construction elements. Right: PV-labelling as reminder.

In conclusion, at first hand the shading losses in this PV terrace demonstrator seem to be very severe when looking to the IV-curves and the visual appearance of the shade looking live at the system. However, apparently people have the tendency to focus on the moments of the year in which the shading is prominent. Whereas there are many more 'dull' minutes and hours of the year in which there are hardly shading losses. With the help of the validated model in BIGEYE we have seen that the worst shaded module, i.e. N2, has a full year round energy loss of 12 kWh, which corresponds to a relative loss of 9%. The least shaded module, i.e. S2, has a full year round energy loss of 5 kWh, which corresponds to a relative loss of only 3%.

6 CONCLUSIONS

6.1 Building Requirements, Safety and Aesthetics

With the more and more stringent energy requirements for renewed and newly built buildings, every squared meters of the building needs to be exploited for PV energy production. From here, the motivation of this work to investigate in depth a BIPV application which was not yet well-studied: PV balconies and terrace fences. With a consortium of partners from both the building and PV side, as well as research institute, we developed a bifacial PV fence that can be used both on balconies and terraces. A commercial product was produced, meeting both building, safety and aesthetical requirements. Two demonstrators, a PV-balcony and a roof PV-terrace, were installed and their performance was tested outdoor for one full year. Modelling work was performed to better understand specific issues, such as the irradiance distribution and shading losses.

6.2 Performance Balcony and Terrace Fences

Over a full year, a PV balcony element installed on the west façade of the SolarBEAT building yielded 71 kWh/m^1, where the unit is intended as energy produced per running meter (in line with building industry metrics used when referring to balconies). It was found that on average over a year, the backside irradiation with respect to front side is around 30%. The bifacial performance ratio for the west side balcony modules resulted to be $PR_{bifacial}$ = 88%, comparable to a good installed and well ventilated BIPV-module. Interestingly, it was found that, per m^2-façade, the energy yield produced by the semitransparent bifacial balcony element is exactly comparable to a (monofacial) colored BIPV façade module as commercially available nowadays, i.e. with rated power in the order of 100-150 Wp/m^2.

The PV terrace also showed very good results. The south, east and west modules all yielded comparable amounts, in the order of 1000 kWh/kWp, which is also comparable to a standard monofacial south-oriented 35°-tilted PV system. The north facing elements underperformed slightly due to the more severe shading. The yearly $PR_{bifacial}$ ranged from 77 to 90%, depending on the orientation.

6.3 Shading

For vertical bifacial PV application like this one, shading plays a crucial role. Therefore, an in-depth analysis was carried out both with measurements (IV tracing) and simulations. We were able to identify and distinguish between different shading types; in particular, we focused on the construction shading casted by the vertical/horizontal support element of the balcony itself. On sunny days and especially around noon, the toprails and pillars inevitably cast a shade on the PV balcony elements, but with a good cells design and the use of bypass diodes the modules are still able to find their mpp and limit the losses. With the help of the validated model in BIGEYE we saw that on the full year the shading losses range from 9% for the worst shaded module (north-oriented) to 3% for the least shaded (south-oriented).

6.4 Modelling

For accurately modelling irradiance, yield and shading effects we built and validated a model using the TNO-developed software BIGEYE. When looking at yearly energy yield, the model presented an accuracy of around 1-2% (up to 5% for the north-facing module, due to the more frequent and severe shading conditions which are difficult to accurately reproduced). Thanks to the model, it was possible to investigate and quantify, in terms of energy loss, the effect of each single shading element on each module of the demonstrator. The obtained results are of interest not only for academic purposes, but also for commercial interest, as they could be used as guidelines to improve the design of the product.

All in all, the bifacial PV fence to be used in balconies or terraces proved to be an extremely interesting BIPV product. Thanks to the good performance and the energy gains resulting from the bifaciality of the element, this application can significantly contribute to meet the energy requirements of newly built buildings.

7 ACKNOWLEDGEMENTS

This project is supported by a Topsector Energy subsidy provided by the Dutch ministry of Economic Affairs (TGOM120009, BOUWDIT)

8 REFERENCES

[1] https://www.rvo.nl/onderwerpen/wetten-en-regels-gebouwen/beng/indicatoren
[2] EN 12600 'Glass in building - Pendulum test - Impact test method and classification for flat glass'
Energy Conference and Exhibition, pp. 1573-1577, 2018.
[3] R.M.E. Valckenborg, A. de Vries, W. Folkerts, G.P.J. Verbong, 'The BIPV research facility SolarBEAT in the Netherlands',.6AV4.25, EU-PVSEC (2014), pp. 3707-3713
[4] https://www.pvblocks.com/
[5] Burgers A.R., et al. "BIGEYE: Accurate energy yield prediction of bifacial PV systems", BifiPV Workshop, Denver, USA, 10-11 September 2018.
[6] Janssen G.J.M., et al. "How to Maximize the kWh/kWp Ratio: Simulations of Single-Axis Tracking in Bifacial Systems", 35th European Photovoltaic Solar
[7] NEN-EN-IEC 61724-1: 2021 on PV system performance – Monitoring
[8] Martina Pelle et.al, 'Coloured BIPV Technologies: Methodological and Experimental Assessment for Architecturally Sensitive Areas', Energies 2020, 13, 4506; doi:10.3390/en13174506
[9] PVGIS, https://re.jrc.ec.europa.eu/pvg_tools/en/

MOST DEMANDING STANDARD COMBINATION TO SUPPORT BIPV DEVELOPMENT AND SUSTAINABILITY – FIRE SAFETY

Simon Boddaert[1], Fabio Parolini[2], Pierluigi Bonomo[2], Muhammad Awais[3], Claudio Pirota[3], José Jimenez[4], Jose M. Vega de Seoane[5], Daniel Valecia-Caballero[5]

[1] CSTB, 290 Route des Lucioles, FR 06904, Sophia Antipolis, France
[2] SUPSI-ISAAC, Campus Mendrisio, CH 6952, Canobbio, Switzerland
[3] PIZ Srl, Via dei Molini 22, IT 23013, Cosio Valtellino, Italy
[4] ONYX SOLAR, C/ Río Cea 1, 46, ES 05004, Ávila, Spain
[5] TECNALIA, Mikeletegi Pasealekua, 2, ES -20009 Donastia - San Sebastian, Spain

ABSTRACT: The growing development of renewable energies including photovoltaics [1] and increasing request for reducing the energy consumption of the building stock has furthered the photovoltaic integration into architecture (BIPV) in more and more different solutions to finally become components of construction. Therefore, the BIPV components should not be evaluated separately as a constructive element or as a simply photovoltaic product, but as an active set element, able to produce energy, which supports a constructive role and different additional functionalities. These products are complex to be evaluated according to traditional standards based on specific sequences or procedures to fulfil exclusively the building requirements. A combined approach is rarely proposed to evaluate the BIPV components in real operating conditions. In the European project BIPVBOOST [2], aspects related to the construction and photovoltaic sector were considered and missing gap are analysed in the project report D5.1 [3]. A new approach based on performance was proposed that could define a set of new test procedures for the qualification of BIPV products considering them as active building products. The objective is to determine whether the proposed methodology is representative (validation) and whether the electrical effects observed are significant and require the definition of new standards. This work takes into consideration of the photovoltaic cell technologies used as well as the mode of implementation in the building.
Keywords: Building integrated PV, Façade, Performance, Fire, Safety

1 FOREWORD

In the development of the BIPVBOOST project, a most demanding procedure dealing with the fire reaction of BIPV components/systems, studying the influence of BIPV specific features (high operating temperature, cablings, electronic devices, flammable mass, etc.) on fire safety has been studied. The EN 50583-1 [4] and EN 50583-2 [5] standards made the first step defining the properties and the applicable regulatory framework for photovoltaic modules used as construction products. Nevertheless, specific and representative evaluations remain few. To illustrate this investigation in the definition of a new test procedure (NTP), it has begun with the resistance and reaction fire requirements relating to the field of construction. In the EU context, according to the Regulation no. 305/2011, named CPR [6]. Accordingly, product is covered by a harmonized CPR standard, the fire reaction tests must be carried out following the indications given by EN 13501-1 [7] and EN 13501- 5 [8] (standard for roofs exposed to an external fire). In order to test the BIPV product according to the construction requirements, the following standards were also considered EN 11925-2 [9] the requirements of façades EN 13823, [10] and roofing products B_{roof} - CEN TS 1187 [11] as the main elements dealing with BIPV components. BIPV modules are made with an active part inside able to generate energy and so the IEC standards shall be considered also as IEC 61215 [12] [13] and IEC 61730 [14] [15]. Nowadays the tests on BIPV elements are carried out without an electric charge as if they are no active product. However, the presence of an electric load subjected to a fire attack or stress (under operating conditions) can generate additional damage due to electric arcs generation, resulting in the deterioration of initial fire capabilities. The energy density can be significant and can thus contribute to a notable acceleration of the thermal effects linked to fire. The possible acceleration of fire behaviour under an electrical load is the aim of this work, in addition to developing the most significant test procedure which uses the new combined performance-based approach.

2 CONTEXT AND AIM

As previously described, BIPV products are elements that, in addition to performing the constructive function for which they were developed, produce electricity supplied by a solar source. As they are construction products, they must be evaluated under the construction product regulations while maintaining similar performance to non-active components that perform the same architectural function.

However, compliance with the requirements of construction products alone cannot be sufficient to describe the behavior of BIPV modules as they contain an electrical active part capable of producing energy. This electrical part, in fact, must be verified according to the standards of photovoltaic electrical products.

The international standard IEC 61730-2 [15] in the assessment of fire behavior, express that when a PV product is used on the building refers national and local regulations are needed. "Fire resistance requirements for a PV module intended for building applications are defined in local or national building codes".

The same reference is provided in the information annex, the ANNEX B: "PV modules mounted in or on buildings, in general, shall fulfil national building and construction regulations and requirements".

The Direction of the International Standard is therefore to consider the PV on the building as a construction product that must follow the building regulations about fire behavior.

The presence of the active parts requires the verification of the photovoltaic part in innovative building products (BIPV components) at the same time as the verification of the building part. It is, therefore, necessary to take a complete and exhaustive approach that defines test procedures that adapt to the products considering both the construction and electrical aspects at the same time.

However, nowadays, there are still several missing gaps between the requirements for the electrical part and the requirements for construction products, creating a barrier for the market [16], so, new testing procedures have been developed to consider both electrical and building aspects, well describing, and fitting the BIPV products.

3 THE APPROACH

Starting point of this work was to fill the gap between building and photovoltaic codes and define the best tests combination as possible to support and assess correctly innovative BIPV solution. Many works are currently being carried out by IEA PVPS program (T15) and TC 82 revision activity or Join working Group (JWG11) with cooperation with BIPVBOOST project members, to improve characterization conditions, dedicated to active innovative components. Based on this work in progress, we developed and validated in the framework of BIPVBOOST project a new test procedure on challenging BIPV solutions (EN 50583-1 category C) by merging fire test related to building envelop elements (EN 13823 – SBI) validated with ignitability tests (EN 11925-2) with the most relevant electrical condition to define the most realistic operating conditions. Electrical operating conditions are defined by the cell type technologies (C-Si), as well manufacturer datasheets and electrical standard requirements (EN 61215, EN 61730). This multi-normative approach made it possible to define the maximum electrical conditions to be applied to test these new construction equipment as well as possible. Dedicated equipment and necessary setup to then apply these conditions have been defined and tested to ensure the representativeness of the solution.

4 NEW TEST PROCEDURE METHODOLOGY

This New Test Procedure (NTP) methodology aims to detect possible deviation or defects brought by the active part based on a combination of building and electrical standards as a first step forward. Tests are carried out to identify reference performance result of façade BIPV solution first, then the same BIPV solution are tested several times including an electrical load. The load definition is based on modules electric specifications to define the maximum running conditions in normal conditions (according to EN-61215 requirements) and to run the test in the most demanding conditions. Results obtain allow to define the level of influence induced by the electric load on fire results.

4.1 NTP procedure

The purpose of the test procedure is to identify a possible negative effect under adverse test conditions or to apply the most realistic test conditions considering effect of electric generation under fire attack taking part of LVD advises. The criterion for passing the test is defined by the initial (reference) performance result compared with the combined effect results according to the worth test conditions, regarding the operating conditions and setup in accordance with the procedure explained in Figure 1.

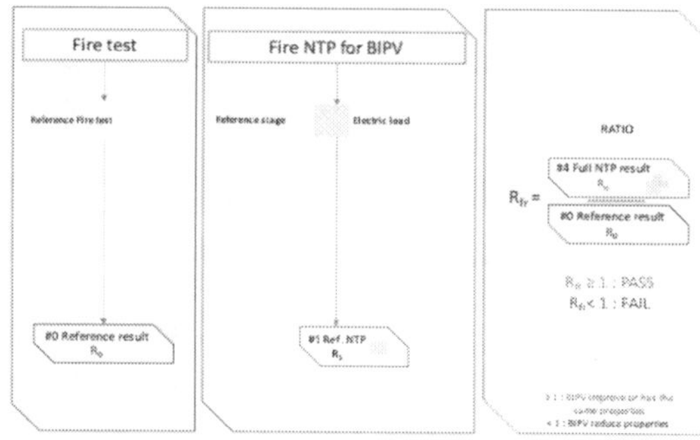

Figure 1: Fire Ratio $\mathbf{R_{fr}}$ determination to assess BIPV exclusively with electrical load under NTP.

If R_{fr} is higher than 1 the test is passed and the BIPV solution does not bring negative impact even if electric load is applied. Somehow, we could hypothesize that the NTP demontrates a positive results on final ratio. Conversely, if R_{fr} value is lower than 1, that means that NTP affect significantly BIPV test results and is very sensitive to an electrical load. As an example, for façade BIPV solution, if initial test reaches "B-s3, d0" classification, the same BIPV solution under these more severe conditions have to reach an equal or higher SBI classification to pass the test. This result demonstates that BIPV solution does not bring any negative contribution according to the fire test point of view.

4.2 Testing setup

To be the most representative in any new test procedure, the reference standard setup (here EN 13823 – SBI for façade systems) is strictly respected thanks to the active support of manufacturers involved in the project (PIZ, ONYX, SCHWEIZER, FLISOM) to manufacture tailored modules that meet the standard requirements. All BIPV elements fit with the required size of elements to test in the most critical fire conditions.

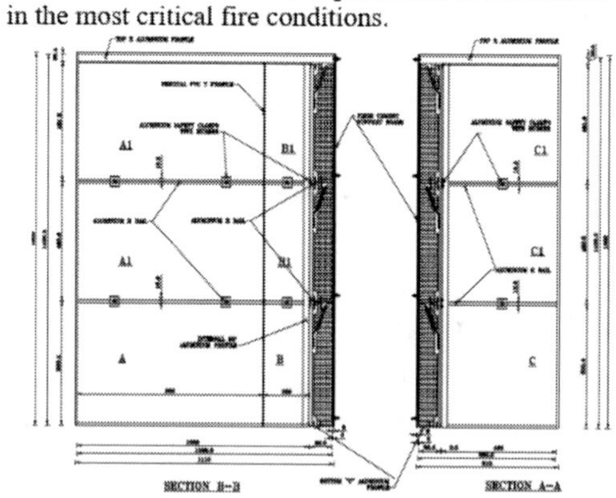

Figure 2: Left-side view (long wing/wall) and right-side view (short wing/wall) of test bench with module size.

Tailored modules are then installed to match perfectly with requirements and all modules are electrically connected in order to represent a real BIPV field (Figure 3). According to the test process shown in the Figure 1, the test is carried out with BIPV façade elements installed in accordance with the Figure 3 setup and the results obtained make it possible to define the reference value $\mathbf{R_0}$. $\mathbf{R_0}$ is the average value of at least three tests. Subsequently, the NTP tests are then carried out with the commissioning of the electrical load.

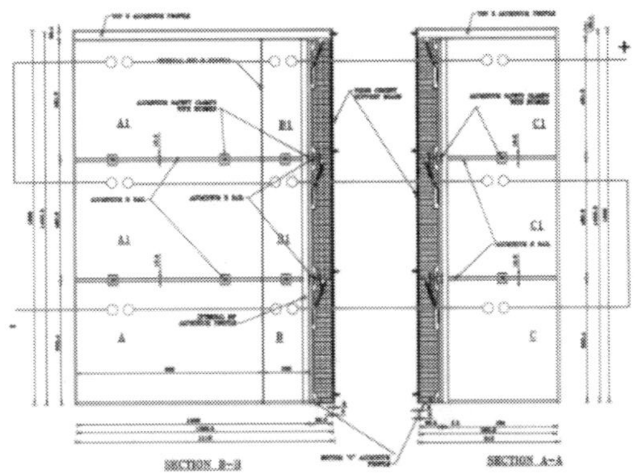

Figure 3: Module's arrangement and electrical connections.

Figure 4: Arrangement of modules made in accordance with the standard EN 13823.

4.3 Electric load definition for the NTP

Based on this specific approach to integer the electric generation during the fire test, we can apply a voltage (U) and current (I) values specifically to the manufacturer specifications and adapt the effect of electrical part accordingly.

The electric load used is a PV field Generator (REGATRON PV SERIES) allowing to set the electric conditions according to two operating modes, switching from I and P to U and P mode imposed. The starting mode used is I and P values imposed according to PV datasheets to define the maximum current value to be applied during the fire test. The Figure 4 shows the setup used to apply an electric load during the fire test.

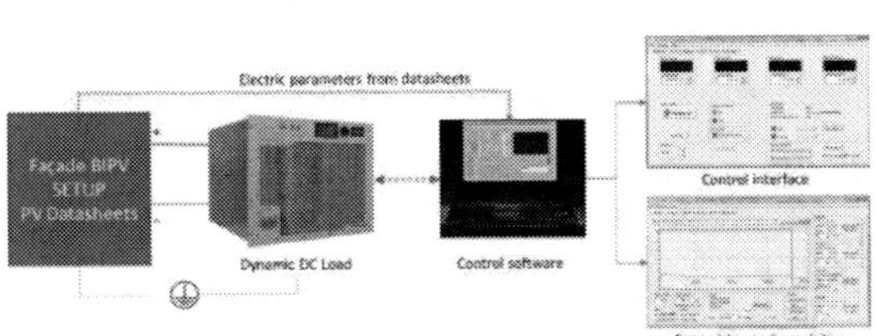

Figure 5: Details of the complete experimental setup of the NTP to impose an electrical load during fire tests.

The I value is set to the maximum current in short circuit according to the manufacturer prescription. The maximum power value is fixed according to the maximum voltage allowed according to IEC 61730; the common value is 1000 V DC. According to this operating mode, Pmax is fixed by the value of I_{sc} multiplicated by the maximum voltage value.

Considering an open circuit due to destruction of bus bars by the fire attack or due to an electric arc, the maximum voltage value remains the same, 1000V.

4.4 Test execution

Afterwards, the DC generator is connected to the PV field plugs in reverse mode (++ and --) to overide the bypass diodes. Before starting the fire test, an initial electric test is performed in order to validate the electric connection and check the perfect operating conditions. The set current value must be displayed on the DC generator, and Power mode have to be switched on. A multimeter is used to check that the current value imposed is correct.

Main aims are to apply an electrical load to force the arc generation in case of busbar or ribbons degradation due to fire attack and to detect a possible current failure. Current failure expresses bus bar destruction through combustion of combustible part, or electrical arc generation in the electrical circuit. In case of appearance of current default, that reach a null value, a nominal voltage value is imposed by the DC PV Generator equivalent to the global V_{oc} values of modules.

Before to run the NTP test The DC PV generator is preliminary connected to a BIPV module sample, in order to test the connection quality and to validate the setup. We run a first load test to validate the process. The corresponding electric load is applied on PV array 1 mn before to start the fire test to stabilize the voltage value according to the room temperature and to check possible current leakages.

After validation of this step, the DC PV Generator is then connected to the BIPV field to be fire tested with the application of the electrical load. Only BIPV elements and cables are located in the room test, only cables pass-through room wall. DC PV GENERATOR is located outside the test room, on a movable bench to be closed as possible of the tested mockup and protected from thermal conditions.

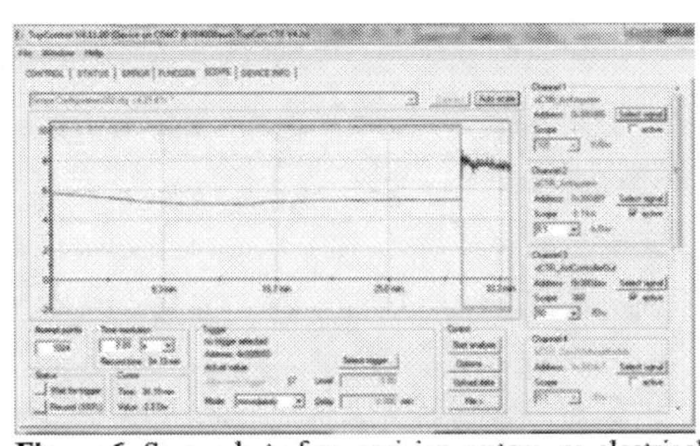

Figure 6: Screenshot of supervision system, no electrical breakage observed before the forced breakdown.

4.5 Test duration

The test duration is often depending on the thermal energy generation during the test, a major or time required by the standard. This test duration must be adapted according to the BIPV element to be assessed and the most relevant standard for performing the test. For façade BIPV solution, the EN 13823 impose a maximum time of 20 mn, or it observes a divergence in combustion energy generation that reflects runaway and divergent propagation. in this specific case an emergency stop is performed during the test.

5 RESULTS

Tests under electrical load are also performed carried out on three times on similar mockups to have representative results. Thus, the reference results and the results with the electric charge allow to have a general validation of the procedure. $\mathbf{R_1}$ is the average value of at least three tests.

Figure 7: NTP running on BIPV façade system and results after test.

5.1 Fire NTP exploitation of result.

As explained in the NTP proposed as part of this work, the impact of the electrical load on the fire tests is expressed by the $\mathbf{R_{fr}}$ ratio in order to determine if the impact is negative, expressed by an acceleration of the fire kinetics

$$R_{fr} = \frac{R_1}{R_0}$$

Concerning the classification criteria, they can be expressed for the SBI test with several representative parameters. Either the overall classification of type fire tests or the characteristic parameters of the SBI test with, for example, THR(600), FIGRA or TSP(600).

The absence of recorded events can also be an element of classification and can be used as a $\mathbf{R_{fr}}$ comparison ratio.

Table I: SBI classification results for standard test and NTP test with electric load, and standard deviation.

	Av.(R_0)	Av.$_{load}$.(R_1)	Std.dev.
FIGRA(0.2) W/s	37.7	38.3	0.4
THR(600) M	5.1	4.95	0.1
SMOGRA cm2/s2	2.2	2.2	0
TSP(600) m2	32	28.3	-2,1

All these parameters are presented in the Table I with average values obtained, to demonstrate deviation between test carried out without electric load and with electric load. On every parameter, we have observed no significant deviations on results, the deviations observed are well below the measurement errors and above all the admissible range of the results. These detailed results make it possible to define on which elements the fire test with electric load can be influenced and which are the components which present a sensitivity to the electric load.

In terms of general results, the potential classification of the components tested can also be used to have an overall analysis. The Table II shows the overall results of the test campaigns and deviation and trends observed for this case study.

Table II: SBI overall results between tests and trends observed.

Potential classification	Av.	Av.$_{load}$	Trend
Class	A2/B	A2/B	=
Smoke production	s1	s1	=
Flaming droplet/particules	d0	d0	=

Without and with an electric load applied, tests are carried out and we observed a similar trend in results with the same fire potential classification on the average values with reaches B-s1, d0 in both cases. A Glass/Glass BIPV module with C-Si cells, integrated in a façade solution does not significantly change classification et does not have a degradation effect.

5.2 Electric analysis

If the analysis of the criteria is done on the fire parameters, it is also important to analyze the electrical results obtained to determine whether the active part of the facade components can withstand the test conditions. Lessons learned with the monitoring solution demonstrate the absence of any electrical discontinuity despite the direct busbars and ribbons fire exposition during the tests.

We do not observe any electric arc generation or any electric breaks. Moreover, no acceleration of the fire kinetics due to the presence of electricity has been identified with abnormal voltage variation during the test as showed on Figure 8.

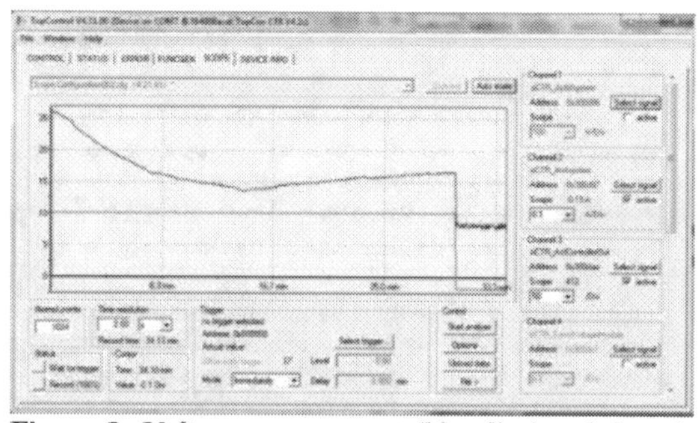

Figure 8: Voltage measurement (blue line) and forced current (green line) during SBI test. Active break at 30mn when the test is over.

Voltage variation is due to temperature increase during the test operation but never the voltage value reach Voc, that express an open circuit due to an electric break. In both cases (without and with electric load) we observe an absence of modification of behavior on the fire results

5.3 NTP validation on SBI test

Obtained results demonstrate feasibility of this NTP but also the method validation allowing to demonstrate that a combined test could be carried out and allow to apply in reverse mode an electric load that allows to force energy to go throw the BIPV components to represent a more severe and more realistic operating conditions than current standard tests contributing analysis supporting the BIPV progress.

Many other observations and results obtained with these test campaigns could be highlighted over the absence of electrical discontinuity or circuit breaks and no electric arcs detected. First observation underlines no acceleration of the fire kinetics due to the presence of electricity that results in the absence of modification of behavior on the results of the fire tests.

Secondly, we could analyze the use on the NTP on test times duration. This procedure could be implemented in the fire laboratory under the supervision of qualified staff, at the same time as the fire test operators. The implementation of an electrical circuit did not generate additional time for carrying out the tests, only the preliminary assembly time is impacted to ensure good electrical connections and good electrical continuity of the BIPV components. The electrical connection part represents less than 10% additional time, trained operators can significantly reduce this time, with the use of suitable equipment. 3-5% extra time is reasonable.

Thirdly, we have carried out a short investigation on the effect of the NTP on delays and overall cost. Even if this investigation is a research work carried out with the CSTB fire Lab team, to carry out the combined tests also makes it possible to avoid logistics costs between different laboratories, a great decrease in the number of samples or spares elements and avoid a too large quality discrepancies of samples/prototypes due to storage conditions, travels and handling issues that could occur.

Last but not the least, dual feedback on fire and electric load conducted at the same time allows to the manufacturers to improve or to change components as quick as they can to fix any weak points observed after the NTP procedure and save many mouths of step-by-step upgrades. This work also supports optimization manufacturing and support manufacturers in their validation certification process.

All these arguments, collected from results and observation make it possible to make cost reductions and to reach the market in much shorter times which is a crucial element for manufactures, but also for the complete BIPV value chain (planners, architects, insurance) and could help to support BIPV policy based on explicit results.

Next step of this work is to extend the NTP to other photovoltaic components (sensitive layers) and different integration categories to reach a full process validation on other BIPV systems

6 METHODOLOGY EXTENSION AND VALIDATION

The same methodology is applied on other façade BIPV solution with thin films modules (Figure 9) and on roof solution with corresponding standard (CEN TS 1187). Identical test series are carried out with at least three tries by BIPV solution to reach more significant mean values.

We observed that for CIGS solution results are bad, with a slight improvement on smoke generation but not significant in front of other parameters. These kinds of elements do not meet requirements for building application (Table III) with a E-s2, d2 classification, but the method is still valid and can be applied and provide useful results.

Figure 9: NTP running on BIPV façade system with CIGS modules and results after test.

We can conclude that electric load does not provide any negative effects on well burning solution and no electric deviation are observer during the complete test duration.

Table III: SBI classification results for standard test and NTP test with electric load, and standard deviation.

Potential classification	Av.	Av.$_{load}$	Trend
Class	E	E	=
Smoke production	s2	s1	^
Flaming droplet/particules	d2	d2	=

Concerning roof systems, the electric load is applied with the same methodology and results are expressed accordingly CEN TS 1187 parameters and the BRoof T3 classification and significant test observations like fire penetration, flame spread and flaming droplets.

Figure 10: NTP running on CIGS BIPV Roof system and results after test.

In both cases, flame spread never reaches to top part of the mockups with a well self-extinguish process. Neither electric arcs nor breaks are observed. The BRoof T3 classification is missed only because of flaming droplets observed on the bottom part of the mockup.

In terms of general results, the potential classification of the components tested can also be used to have an overall analysis. The Table III shows the overall results of the test campaigns and deviation and trends observed for this case study.

7 CONCLUSION AND PERSPECTIVES

7.1 Main results and conclusion

All investigations performed during this work in the framework of BIPVBOOST project allow to validate the first step of the NTP on two category families (A and C) from EN 50583 standard.

If we note a strong result dependance with the mounting specification of BIPV solution, preliminary results of an electric load demonstrated an unsignificant influence on the spread of fire and from a main aspect do not affect fire classification of those solutions.

These most demanding test conditions demonstrate a feasible combination of standard from building requirements and electrotechnical standard mandatory for building application. Identification of missing gap and NTP suggestion are strongly supporting BIPV development by bringing more confidence in the use of PV in the building environment.

7.2 Perspectives

Next steps will be to combine accelerated aging effects on fire results with an electric load. With the combined efforts of projects partners, new investigations had been carried out to promote NTP on mechanical, electricity and energy aspects. The full combination of all these parameters aim to bring the most demanding conditions to test BIPV solution to prove that the BIPV solutions are safe during their implementation but also after a long lifespan. We expect that R_{fr} ratio taht currently express exclusively the contribution of the electric load in a possible deviation in the fire test results could integer other NTP in its final contribution tackling oll categories of EN 50583-2.

Further works must be carried out to finalize and fully validate this NTP with the development of more appropriate devices specifically developed for fire tests. On the other and the integration of other NTP is another challenge.

A large request focus on the capability to develop a module per module monitoring system to have a per module investigation and analysis solution. Objective is to detect possible default on the most solicited module during the fire test and use them to improve the learning curves.

One last suggestion could be the add specific arc sensor used in each junction box for example or between two module connections to discriminate the most affected modules by the combination of fire attack and electric load when NTP are running. Identification and location of weak point or weak elements or compounds could strongly support manufacturers components selection and tune specifically the BIPV element to meet the requirements of every BIPV projects. We could also expect component optimization in cases of over quality to reduce cost of manufacturing.

Challenges to overcome on standardization and parallel works are in progress to stress out the normalization work to support development of new procedures to fully integer these NTP in next standards. This work aims to bring more confidence in BIPV and more support from public policy and insurance to support the growth of BIPV and promote efficient solution to fight the global warming.

8 ACKNOWLEDGEMENTS

This project has received funding from the European Union's Horizon 2020 research and innovation programme under grant agreement No 817991. Authors thank the massive contribution of all project parters and specifically manufacturers for their specifics efforts making this work possible.

9 REFERENCES

[1] E. Commission, «A European Green Deal,» [Online]. Available: https://ec.europa.eu/info/strategy/priorities-2019-2024/european-green-deal/delivering-european-green-deal_en.

[2] H. 2020. BIPVBOOST, *European project - This project has received funding from the European Union's Horizon 2020 research and innovation programme under grant agreement No 817991,* 2019.

[3] BIPVBOOST - Horizon 2020. www.bipvboost.eu

[4] *EN 50583-1 Photovoltaics in buildings - Part 1: BIPV modules,* 2016.

[5] *EN 50583-2 Photovoltaics in buildings - Part 2: BIPV systems,* 2016.

[6] European commission, *Regulation (EU) N° 305/2011 of the European Parliament and of the Council of March 9th 2011 - Construction Products Regulation, CPR 305/2011,* 2011

[7] *EN 13501-1 Fire classification of construction products and building elements - Part 1: Classification using data from reaction to fire tests,* 2018.

[8] *EN 13501-5:2016 Fire classification of construction products and building elements - Part 5: Classification using data from external fire exposure to roofs tests,* 2016

[9] *EN ISO 11925-2:2020 Reaction to fire tests - Ignitability of products subjected to direct impingement of flame - Part 2: Single-flame source test (ISO 11925-2:2020),* 2020

[10] *EN 13823:2020 Reaction to fire tests for building products - Building products excluding floorings exposed to the thermal attack by a single burning item,* 2020.

[11] *CEN/TS 1187:2012 Test methods for external fire exposure to roofs,* 2012

[12] *IEC 61215-1 Terrestrial photovoltaic (PV) modules – Design qualification and type approval – Part 1: Test requirements,* 2021.

[13] *IEC 61215-2 Terrestrial photovoltaic (PV) modules – Design qualification and type approval – Part 2: Test procedures,* 2021.

[14] *IEC 61730-1 Photovoltaic (PV) module safety qualification - Part 1: Requirements for construction,* 2016..

[15] *IEC 61730-2 Photovoltaic (PV) module safety qualification – Part 2: Requirements for testing,* 2016..

[16] F. Parolini, P. Bonomo, S. Boddaert et Al. , «Performance assessment of bipv systems: research on bipv characterization methods,» in *37th European Photovoltaic Solar Energy Conference and Exhibition,* 2020.

THE ART AND SCIENCE OF INTEGRATING PHOTOVOLTAICS IN THE BUILT ENVIRONMENT: CASE STUDY OF THE UFSC SOLAR ENERGY AND GREEN H_2 LABORATORY

Isadora Pauli Custódio[1], Ísis Portolan dos Santos[2], Ricardo Rüther[1]
[1]Solar Energy Research Laboratory Fotovoltaica/UFSC (www.fotovoltaica.ufsc.br)
Universidade Federal de Santa Catarina (UFSC), Caixa Postal 476, 88040-900, Florianópolis – SC, Brazil
Tel.: +55 (48) 3721-4598; E-mails: isadora.custodio@posgrad.ufsc.br; ricardo.ruther@ufsc.br
[2]Landscaping and Architecture Laboratory PARQ/UFSM
Universidade Federal de Santa Maria (UFSM), 97105-900, Santa Maria – RS, Brazil
Tel.: +55 (55) 3220-8304; E-mail: isis.santos@ufsm.br

ABSTRACT: The progressive use of photovoltaic (PV) modules as construction materials is changing the shape of the built environment. Architectural integration quality can be achieved by virtue of the available technical information, but to explore the technology's full potential, non-technical features must also be considered. Architects and engineers need to get a better insight of how the application of PV modules can be interpreted from theories of architectural perception. This work proposes a method to identify how the perception of PV systems influences architectural spatial experience, that is, how they affect people's movement in spaces. It complements the investigations that were proposed so far, as architectural criticism should consider both constructions and spaces. The new green H_2 laboratory of the Solar Energy Research Laboratory Fotovoltaica/UFSC in Florianópolis/Brazil is presented as a practical example of the theory and practice of integrating solar photovoltaics in the built environment, addressing simultaneously the onsite-generation of electricity for the building, and the production of green H_2 directly from solar electricity and water electrolysis. The work means to serve as inspiration, expanding the use of PV modules as a way of architectural expression and narrowing the gap between the design and the PV communities.
Keywords: Building Integrated PV (BIPV), Solar Architecture, Green Hydrogen, Aesthetics evaluation, Interpretations of architecture.

1 INTRODUCTION

The desire of implementing active and passive strategies for harvesting solar energy in buildings has increased the use of photovoltaic (PV) modules as construction elements, making building envelopes more complex and multifunctional [1]. This is an elegant and efficient way of installing a solar generator, since the building skin doubles as the supporting structure for PV modules, and the building's electrical installation doubles as the electrical interface between the solar generator and the public electricity grid. Therefore, PV modules can produce energy and compose the architectural envelope at the same time [2]. This way, the cost of typical construction elements can be offset, and new aesthetical features are added to the environments [3].

In the European context, plans for the transformation of the building sector into nearly zero-energy buildings (NZEBs) are being drawn [4]. Being PV a requisite for achieving the net zero target [5], it becomes clear that the global cumulative capacity of Building-integrated Photovoltaic (BIPV) systems has been increasing. This growth will radically change the forms of buildings and cities [6], hence it is necessary that architects and designers explore new architectural expression possibilities [7]. The interest in achieving "architectural integration quality" [8] within a solar expansion scenario should come from all the involved stakeholders, and the final design of the integration should achieve an optimal compromise among all relevant factors [16], [17].

To explore the full potential of the PV technology, scientists and architects must work together and understand each other's languages [18]. Architects and designers must improve their knowledge about PV using all the available technical information. However, non-technical issues should also be considered [19]. Architects and engineers need to get a better insight of how PV integration can be understood from theories of architectural perception. PV modules must be used as a way of architectural expression; that is, cognitively, the technology should be handled as any other material of the building envelope [20].

A review on the literature on the aesthetics of PV building integration demonstrated that there is a desire to achieve objectivity on the creation of a method or on the definition of criteria that should be considered to design, judge, or describe PV architecture [3], [5]–[8], [11]–[13], [15], [17]–[41]. In the present work, the methods shown by these investigations are considered the "traditional interpretations of PV architecture". These contributions transmit a connection between PV and architecture. Still, PV also needs to be related to the core of the architectural subject to obtain its highest integration capability.

Architecture, painting, and sculpture have aesthetic values and are involved with the beauty that pleases human eyes. In paintings and sculptures, man remains outside just as an observer. Therefore, the description of their function, technique, plastic, and pictorial data is enough to make their value understood. Differently, architecture includes man. It is a functional art which defines interior and exterior spaces where men can live and be at. Its forms and surfaces are built not to be merely observed; architecture allows men to experience its spaces. Therefore, purely determining its visible features does not define this art; a spatial interpretation is also necessary. Through words it is possible to help others feel the spatial experience, that is, the pleasure provided by architecture [42], [43].

Existing investigations, or the traditional criticism of PV architecture, are based on the visible features of the constructions. Even if important, these characteristics alone are unsuccessful in making the whole value of PV architecture recognized. Criticism ignores space because it is a void that comes by itself, or a denial of what is solid. However, the Gestalt theory regards the notion of perception and interpretation of elements as a whole, and not only from the parts in isolation [44]. Since architecture is composed both by solid matters and voids which rely on each other, the criticism of PV architecture should consider both constructions and spaces to make its whole value understood.

This work proposes a method for the spatial interpretation as to provide, together with the traditional interpretations, a full understanding of PV architecture. The idea is to go beyond tradition, to complement the investigations that were proposed so far. Associating traditional and spatial interpretations helps to understand the aesthetic pleasure of PV architecture. The establishment of a good critique makes the public understand that aesthetical emotion can be found in monuments, historical constructions, and buildings of renown architects, but also in the common buildings that people pass by every day [43]. Furthermore, the method is illustrated by the interpretation of a case study in spatial terms. The new green H_2 laboratory of the Solar Energy Research Laboratory Fotovoltaica/UFSC in Florianópolis/Brazil is presented as a practical example of the theory and practice of integrating solar photovoltaics in the built environment, addressing simultaneously the onsite-generation of electricity for the building, and the production of green H_2 directly from solar electricity and water electrolysis. The results should serve as inspiration to the use PV modules as a way of architectural expression, improving the acceptance of architects and designers on the use of PV in buildings, as well as to contribute to the definition of PV architecture. Since successful PV building integration is a consequence of optimum interdisciplinary teamwork [1], results should also give engineers and architects a better insight of how PV integration can be interpreted from theories of architectural perception, bringing the PV and the design communities together.

2 CONCEPTUAL REVIEWS

2.1 The traditional interpretation of architecture

The traditional interpretation of architecture relates to content (political, philosophical-religious, scientific, social-economic, materialist, and technical interpretations), physiopsychological, and formalist features.

Regarding content features, the political interpretation relates architecture with political events; the philosophical-religious interpretation relates architecture with philosophical-religious historical events; the scientific interpretation relates architecture with mathematical and geometric conceptions; the social-economic interpretation establishes analogies between architecture and economic systems and social institutions; the materialist interpretation relates architecture to geographic and geological conditions of places, or to archaeological investigations, or even to the morphology of humans and nature; and the technical interpretation pursues art and technique together [43]. Political, philosophical-religious, scientific, and social-economic interpretations were not found in existing investigations on PV architecture. The materialist interpretation was found in 4 studies about the artificiality/naturality of the rigid pattern of PV plants in relation to the natural setting of landscapes [38]–[41], and the technical interpretation was found in 20 investigations about the mounting structures, energy performance, and economic analysis of PV architecture [1], [3], [6]–[8], [16]–[20], [22], [23], [30], [31], [33], [34], [37], [38], [40], [41].

The physiopsychological category relates architecture and humans, reducing art to a science where architecture and its forms transcribe states of mind or human expressions [43]. It is highly influenced by the sensitivity, mentality, education, environment, and state of mind of the spectator. Nine examples from the literature on PV architecture follow this approach [1], [7], [15], [17], [22], [25], [26], [32], [39]. They regard the pleasantness, impressiveness, motivational effect, degree of satisfaction, level of excitation, and feeling of familiarity of the observers concerning PV architecture.

Lastly, the formalist category relates the architectural composition with a series of qualities, such as unity, symmetry, balance, emphasis, contrast, proportion, scale, expression or character, truth, property, urbanity, and style [43]. A review of the literature revealed 33 examples on PV architecture that employ this interpretation [1], [3], [5]–[8], [11]–[13], [17]–[35], [37]–[41]. They debate about the geometry of the PV systems (field's size, shape, and position), their materiality (modules' colors, textures, and reflexivity), and modular pattern (PV modules plus mounting system).

Even if the traditional interpretations are essential to understand architectural aspects of PV integration, they alone are not enough to talk about PV architecture. Being space the protagonist of the architectural subject [43], it is what provides the aesthetic experience of PV building integration. Accordingly, spaces must also be interpreted.

2.2 The spatial interpretation of architecture

As tradition, the aesthetics of PV building integration is interpreted mostly in technical and formalist ways. As only space can provide the specific value of architecture, an openness on the critic subject is necessary to make the movement of change to the spatial interpretation, releasing from the tradition that is enforced [45]. The imaginary is essential in this practice, as it gives us the ability to see in a thing what it is not, or to see it differently from what it is [46]; it helps us to "see" the void of space. Attributes of architecture such as scales, proportions, colors, textures, natural light, and contrasting effects [42] provide a way to transcribe the imaginary as reality [46].

The spatial interpretation has to do with the movement of humans in a given space [43]. Since childhood, a person perceives forms and surfaces as hard or soft, heavy or light, smooth or rough, and even tense or loose [42]. These senses lead to different feelings of the spaces, as each architectural element, form, or surface, will provide particular effects over human activities, behavior, and communication [48]. The way humans notice contrasting effects, colors, scales and proportions, rhythm, textures, natural light, and sounds influences the movement in space and consequently its interpretation [42].

The existing investigations on PV architecture help us to see the concrete matter of this art, lacking the comprehension of the spatial experience which is provided by the buildings and their integrated PV systems. As stated, the spatial interpretation has to do with the path of people who experience architectural spaces, which are influenced by constructions. Therefore, the traditional interpretations are the base for the spatial interpretation [43]. Since architecture is not mechanics, but an art, there is not a fixed list of categories to be assessed [47]; each case has different evaluation needs. The analysis will lead to the understanding of how the PV systems influence people's paths, that is, how their visual perception affects the architectural spatial experience. While the cognitive vocabulary proposed by Scognamiglio [20] describes how the PV systems are visually perceived, the spatial interpretation relates to how this perception influences spatial experience; so, they are complementary approaches.

Comprehension and interpretation come from the sensitivity, mentality, education, environment, and state of mind of the observer. The social imaginary is a set of images and ideas that are associated regarding something, which is molded and built over time, depending on experienced situations [46]. The spatial interpretation could depend directly on the critic subject, as his social imaginary can have great influence on it. This work, however, proposes the spatial interpretation of PV architecture as a morphological analysis of a human body experiencing a space. This way, the influence of the social imaginary of the critical subject is softened [49].

3 GENERAL METHODOLOGY FOR THE SPATIAL INTERPRETATION OF PV ARCHITECTURE

This assessment contemplates the interpretation of PV in the built environment, considering buildings and urban infrastructures that enclose spaces for human activities.

Architectural spatial analysis is very subjective. However, for teaching purposes, it is important to establish directions [49]. Therefore, a guide was created for the spatial interpretation of PV architecture. In urban areas, there are spaces of movement (the streets) and spaces of experience (sidewalks, squares, and allotments). Also, distances of 100-25 m from the architecture that will be experienced and interpreted have little influence over the human senses; distances of 25-7 m have more influence as details can be better recognized; and from 7-0 m all human senses are used, and all details observed. Furthermore, humans find it easy to look down and to the sides, but the upper field of view is limited. Therefore, upper floors of buildings can sometimes only be seen from certain distances. Finally, humans can perceive objects better when standing or walking (0-5 km/h); running or riding a bike (10-12 km/h) lowers perception; and driving a car (50-100 km/h) lowers it even more [48]. Therefore, the spatial interpretations should be done from spaces of experience, respecting the distance range of 25-7 m and from 7-0 m in relation to the building that will be analyzed. Since PV modules are frequently used in rooftops, it is possible that in some cases they will influence the spatial experience only from distances over 7 m. Lastly, interpretations must be carried out considering that the spectator is either standing still or walking, as this improves the spatial perception. Briefly, the guides for the spatial interpretation of PV architecture are:

- The analysis should be done from spaces of experience (sidewalks, squares, and allotments)
- Distance ranges of 25-7 m and 7-0 m from the buildings should be considered
- The spectator must be standing still or walking (0-5 km/h)

4 RESULTS AND DISCUSSIONS

4.1 Description of the case study

A case study was assessed to illustrate the methodology. The Solar Energy Research Laboratory at Universidade Federal de Santa Catarina (Fotovoltaica/UFSC Laboratory (www.fotovoltaica.ufsc.br) is located at Sapiens Parque (www.sapiensparque.sc.gov.br) technology park in Florianópolis (27°S, 48°W) – Brazil. The Laboratory integrates solar photovoltaics in the built environment, addressing simultaneously the onsite-generation of electricity for the building, and the production of green H_2 directly from solar electricity and water electrolysis. The Laboratory has two buildings (building A and building B), a third building that is under construction (building C), a solar carport, and an electric bus (eBus) charging station and bus stop. All these constructions have integrated PV systems which have been analyzed considering the spatial interpretation of architecture. Fig. 1 shows an aerial view of the laboratory buildings, as well as a 3D model of the buildings and the morphology of their PV systems. A utility-scale PV power plant of 96 kWp and some other ground-mounted PV systems are also present, but they are not included in the analysis as they do not comprehend spaces for human activity.

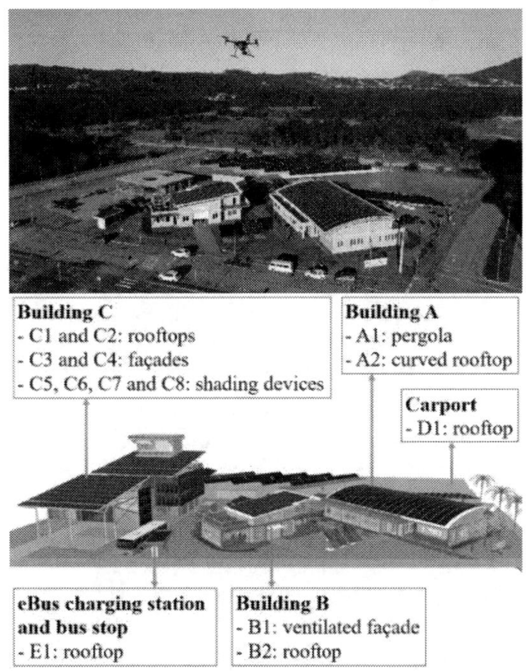

Figure 1: Top (a) Aerial view of the Fotovoltaica/UFSC Laboratory buildings; bottom (b) 3D model of the buildings and the morphology of their PV systems.

Building A, building B, and the solar carport were inaugurated in 2015. In 2017, the eBus charging station and bus stop was built, and building C is currently under construction. The first buildings were conceivable mainly through support from the Ministério da Ciência e Tecnologia (currently the MCTIC) and ANEEL's R&D Project PE-0403-0027/2011, also counting on donations and technical assistance from other institutions and companies. Building C (under construction) is an object of study of ANEEL's R&D Project PD 0403-0042/2016 and of a more recent Research Grant Agreement 74/2021 brought by the German Technical Cooperation Agency GIZ. It will house a large bank of Li-ion electrochemical batteries for the storage of PV-generated energy and produce green H_2 with PV in house. Considering standard conditions (1 $Nm^3 H_{2(g)}$ = 1 m^3 $H_{2(g)}$ at 0°C and 1 atm = 0.08988 kg $H_{2(g)}$ = 1.2699 L $H_{2(l)}$), the PV rooftops of building C will be able to produce up to 27.3 Nm^3/h, that is, 2.5 kg/h of green H_2. The laboratory aims to disseminate knowledge about the application of the PV technology in urban areas, through the development of research and education programs. Its researchers work for the widespread adoption and research on renewable energies, mostly solar photovoltaics, making the laboratory a reference center in Brazil. The PV systems of the laboratory contribute for the production of green H_2 directly from solar electricity and water electrolysis, and can fully supply the energy consumption of the buildings and recharge the eBus after each of its five daily round trips to the main university campus [50], [51]. The laboratory is a Positive Energy Building (PEB), so in months with surplus of PV energy, the energy credit fed to the utility's grid is consumed by the main campus of the university. Furthermore, the architectural concept of the buildings is based on the idea of a living laboratory, where all civil installations are apparent for experimental and teaching purposes. This makes the buildings look like a machine, or an equipment. The architectural design is focused on the shape and orientation of the PV rooftops, and on the use of sustainable materials, bioclimatic strategies, and energy efficiency measures. The buildings are made of exposed brick walls and Aluminium Composite Panels (ACP), metallic pillars and beams, PVC frames, and reforested wood decks and pergolas. The PV modules are integrated as rooftops, façades, and shading devices, which in some places act as sealing materials, that favor the internal thermal comfort especially due to the shade provided by the modules. Other employed strategies of sustainability are crossed ventilation, green roofs, white double-modular metal tiles, use of rainwater and reuse of greywater, greywater treatment, water mirrors, grass block pavers, native vegetation, use of natural lighting and LED artificial lighting with independent drive circuits, and other native and low environmental impact materials.

The buildings of the laboratory have a variety of PV technologies: thin-film cadmium-telluride (CdTe), multi-crystalline silicon (m-Si), thin-film microcrystalline-silicon (µc-Si), mono-crystalline silicon (c-Si), and thin-film copper-indium-gallium-diselenide (CIGS). Currently, m-Si is the one with the highest installed capacity (68.6 kWp), but the rooftops of building C will add 192.7 kWp of c-Si. Table I presents some technical features of the PV systems integrated to the buildings. Systems A2, B1, B2, C5, C6, C7, and C8, are installed above the building surfaces, following their tilt and orientation, and acting as shading elements that reduce the buildings thermal loads; systems A1, C1, C2, C3, C4, and D1 act as sealing materials and are also used as comfort strategies; and system E1 is installed above a horizontal rooftop, following its orientation but not its tilt angle, also acting as a shading element that reduces the buildings thermal loads; all systems have non-optimal tilts and/or orientations (for the city of Florianópolis, Brazil, the optimal tilt is 27° and the optimal orientation is N).

Table I: Technical features of the PV systems integrated to the Fotovoltaica/UFSC Laboratory buildings.

PV system	PV technology	Installed capacity (kWp)	Azimuthal deviation (°)	Tilt angle (°)
A1	CdTe	4.4	90	0
A2	m-Si	66.1	0/0/180	17/5/4
B1	m-Si	2.5	90	90
B2	µc-Si	13.5	0	6
C1	c-Si	72.2	320	10
C2	c-Si	64.2	320	10
C3	c-Si	15.0	50	90
C4	c-Si	10.7	230	90
C5	c-Si	4.5	320	90
C6	c-Si	10.8	230	90
C7	c-Si	4.5	140	90
C8	c-Si	10.8	50	90
D1	CIGS	13.4	270	5
E1	CdTe	2.4	36	3
Total		295		

4.2 Spatial interpretation of the case study

The spatial interpretation consists in analysis from two distance ranges, as established by the spatial interpretation general methodology. Fig. 2 presents the area concerned by the distance ranges of 25-7 m (in blue) and 7-0 m (in red) from the buildings. The first analysis (25-7 m) considers the adjacent streets, from where a neighboring building (a software company named Softplan) can also be seen, while the second analysis (7-0 m) concerns the plot of land of the laboratory and its sidewalks.

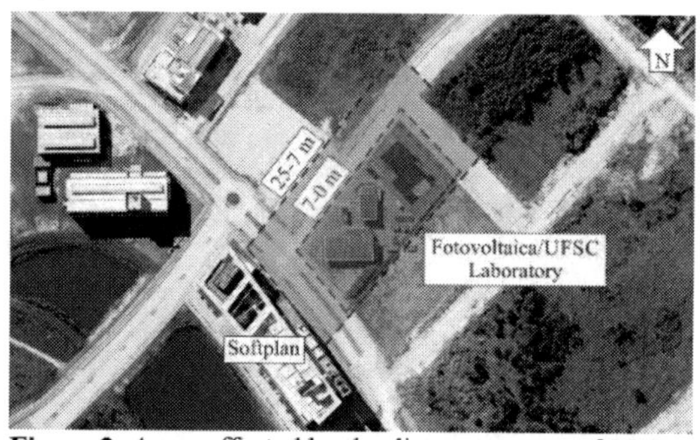

Figure 2: Areas affected by the distance ranges of 25-7 m (in blue) and 7-0 m (in red) from the Fotovoltaica/UFSC Laboratory buildings.

Fig. 3 presents the spatial interpretation for a pedestrian in the distance range of 25-7 m. Someone that walks by Sapiens Parque technology park's main street (dashed black arrow) sees the buildings of the laboratory and its neighbor, the Softplan software company. However, while Softplan's building is huge and has an entrance that is elevated from the street level, the dimensions of the Fotovoltaica/UFSC Laboratory are more in line with the human body scale and the entire building is at street level. Moreover, the laboratory has exposed brick walls, wood decks and pergolas, and its plot is covered by grass and growing palm trees, which are all familiar materials, while Softplan has mainly cold and more industrialized materials such as concrete, metal, and glass, with no vegetation in its surroundings. These characteristics make the laboratory more alluring than the Softplan buildings and persuade pedestrians to go in the direction of its main entrance (red arrow). Besides this access, there is also going to be a pedestrian's entry in building C (blue arrow), which will be an open area with two huge PV rooftops (C1 and C2). The two large PV surfaces will be highlighted from the rest of the buildings due to their dimensions. These features tend to generate curiosity and persuade pedestrians to go look at the structure. The modules will be oriented to the street, hence someone standing in the dashed black arrow is going to be able to see the PV generators from the technology park entrance, all the way up to the buildings. Moreover, the concave shape of this entrance feels welcoming and, since PV generators C2, C3, and C4 using bifacial PV modules will double as the rooftop's and façades' sealing material, a person will be able to appreciate them both from the inside and the outside of the building. The desire to enter this concavity will be emphasized by neighboring vegetations that compose a large plan, or a background to the building, which will stop us from going further on the street. Finally, the vehicles access (green arrow) is also located at a strategic place, from where people can see PV system D1, which is highlighted by its orientation. The views from the black, blue, and green arrows are illustrated in Table II.

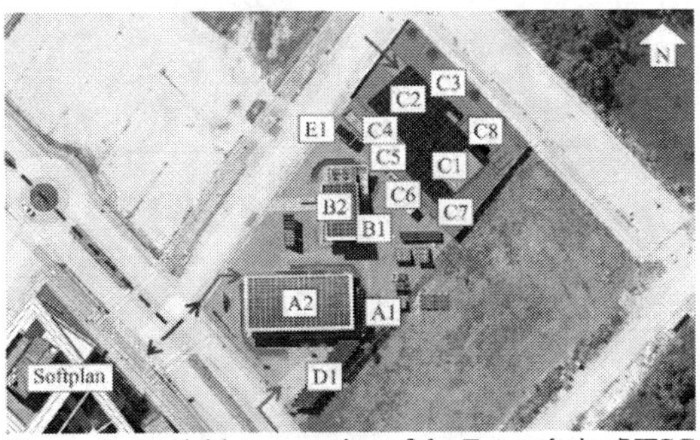

Figure 3: Spatial interpretation of the Fotovoltaica/UFSC Laboratory buildings from the distance range of 25-7 m.

Table II: Illustration of the spatial interpretation of the Fotovoltaica/UFSC Laboratory buildings from the distance range of 25-7 m.

View angle	Picture/image
Black arrow	
Black arrow	
Black arrow	
Blue arrow	
Green arrow	

Fig. 4 and Fig. 5 present the spatial interpretation for a pedestrian in the distance range of 7-0 m. The laboratory has four convex, blue water mirrors that embrace buildings A and B. While concave shapes are welcoming, convex shapes rs lead to repulsive movements. These movements, illustrated by red arrows in Fig. 4, take people away from the buildings to the spots named P1, P2, and P3, from where the PV systems can be better and naturally perceived, that is, from where the human upper field of view limitation is softened. From P1 it is possible to see PV systems A2, C1, C2, C4, C5, and C6; from P2 it is possible to see PV systems A1, A2, B1, C1, C2, C4, and C6; and from P3 it is possible to see PV system D1. All the visualizations are emphasized by the orientation and tilt angle of the PV systems. Furthermore, the blue water mirrors are the elements that visually predominate from pedestrian viewing level. The blue color contrasts with the orange color of the exposed brick walls, since they are complementary colors. Therefore, when an observer goes away from the buildings to the strategic points of view P1, P2, and P3, everything in blue is highlighted, so both the water mirrors and the blue PV systems become clearer. This is emphasized on days of clear sky. Also, one of the water mirrors has a water fountain (Fig. 5). This element draws attention due to its sound and its appearance, and, from that point of view, it is possible to better visualize PV systems A2, B1, C1, C2, C4, and C6. Table III illustrates the views from P1, P2, P3, and from the water fountain.

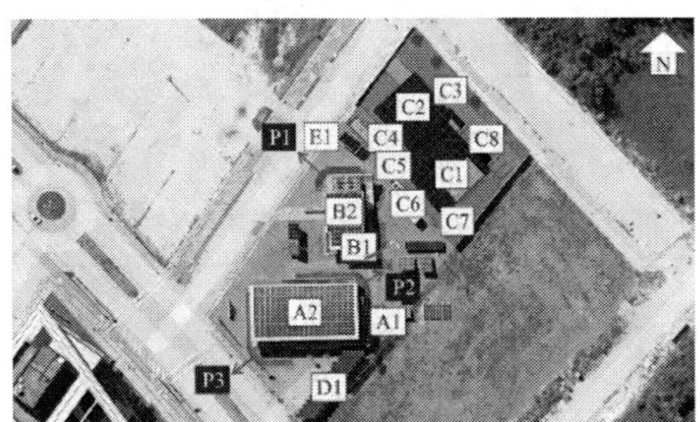

Figure 4: Spatial interpretation of the Fotovoltaica/UFSC Laboratory buildings from the distance range of 7-0 m, considering the water mirrors: displacement from the water mirrors to strategic points of view.

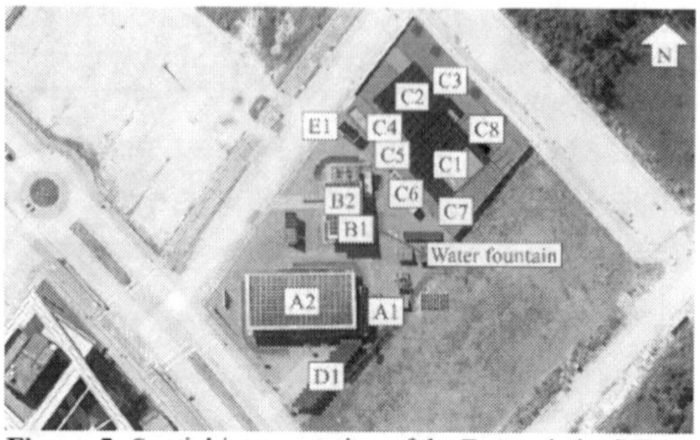

Figure 5: Spatial interpretation of the Fotovoltaica/UFSC Laboratory buildings from the distance range of 7-0 m, considering the water fountain.

Table III: Illustration of the spatial interpretation of the Fotovoltaica/UFSC Laboratory buildings from the distance range of 7-0 m.

View angle	Picture/image
P1	
P1	
P2	
P2	
P2	
P2	
P3	
Water fountain	
Water fountain	
Water fountain	

The spatial interpretation is infinite, as all the visual characteristics of the buildings and their surroundings can influence the paths of people that experience the spaces. This work has described the main characteristics of the spatial experience provided by the UFSC Solar Energy and Green H_2 Laboratory. Its main purpose is being a showcase for the dissemination of PV building integration. Therefore, its spatial experience, or the movement provided by its architectural features, leads pedestrians to places where the visualization of the PV building integrations is highlighted. The moment of invitation to enter the outside area of the laboratory is provided by features of its surroundings, but as the spectator arrives, he/she is invited to go away, so the upper view limitation is not an obstacle to contemplate the PV systems. Therefore, the integrated PV systems are the protagonists of the spatial experience, as they lead the paths of people in the architectural space.

4.3 Definition of PV architecture

From the spatial analysis of the Fotovoltaica/UFSC Laboratory, the definition of PV architecture as a whole becomes clearer. Considering the traditional and spatial interpretations, PV architecture can be defined as building coherent PV integration, that is, as functionality, technology and construction, but also as art, beauty and emotion; music, poetry and symbol; and history, culture and society.

5 CONCLUSIONS

With the progressive use of PV in buildings, modules are becoming more visible and changing the forms of built environments. Investigating PV architecture is essential for achieving architectural integration quality and preserve urban and architectural contexts. To explore the full potential of the PV technology, architects and designers need to improve their technical knowledge. However, non-technical issues must also be considered. Architects and engineers need to get a better insight of how PV modules can be interpreted from theories of architectural perception. Methods presented by existing investigations, namely the tradition of PV architecture, are fixed on the visible surfaces of the constructions. However, the essence of architecture is the space, which is a void, or a denial of what is solid. Since architecture is arranged both by constructions and spaces that rely on one another, architectural criticism should regard both, to provide a full understanding of the subject.

This work aimed to give some insight on the art and science of integrating PV modules on buildings, and to study how it influences the experience of architectural spaces. It intended to complement the investigations that were proposed so far. The concept of the architectural subject was delineated on its essence; the notion of the traditional interpretations of PV architecture was clarified; the concept of spatial interpretation was explained; and a guide for the spatial interpretation of PV architecture was presented. While the traditional interpretations have to do with visual features, the spatial interpretation has to do with the movement of people in spaces. The interpretation of spaces defined by PV building integration helps to understand the role of the systems in the experience of the architectural space. The method was illustrated by the interpretation of a case study, the UFSC Solar Energy and Green H_2 Laboratory in Brazil, in spatial terms. Its new green H_2 laboratory is presented as a practical example of the theory and practice of integrating solar photovoltaics in the built environment, addressing simultaneously the onsite-generation of electricity for the building, and the production of green H_2 directly from solar electricity and water electrolysis. The spatial interpretation demonstrated that the integrated PV systems lead the spatial experience. It is expected that this work offers engineers and architects a better insight of how PV integration can be interpreted from theories of architectural perception and serve as inspiration, expanding the acceptance of the use of PV in buildings as a way of architectural expression. The demonstration of the art and science of integrating PV in the built environment can narrow the gap that exists between the design and the PV communities, improving PV integration quality to preserve urban and architectural contexts, as well as contributing to the definition of PV architecture.

ACKNOWLEDGMENTS

This work was financially supported by Coordination for the Improvement of Higher Education Personnel – CAPES. The authors acknowledge with thanks the financial support of the Brazilian Ministry of Science and Technology – MCTIC, the Brazilian National Scientific Development Council – CNPq, and the Brazilian Electricity Regulatory Agency – ANEEL through its R&D Program, who enabled the Fotovoltaica/UFSC Laboratory buildings and research equipment which led to this work. The authors also acknowledge with thanks the financial support of the German Technical Cooperation Agency GIZ, through its Research Grant Agreement, which funded the construction, infrastructure and equipment of the new green H_2 laboratory presented in this paper.

REFERENCES

[1] I. Hagemann, "Architectural considerations for building-integrated photovoltaics," *Progress in Photovoltaics: Research and Applications*, vol. 4, pp. 247–258, 1996, doi: 10.1002/(SICI)1099-159X(199607/08)4:4<247::AID-PIP135>3.0.CO;2-E.

[2] R. Rüther, *Edifícios solares fotovoltaicos: o potencial da geração solar fotovoltaica integrada a edificações urbanas e interligada à rede elétrica pública no Brasil*, 1st ed. Florianópolis, 2004.

[3] J. Cronemberger, M. A. Corpas, I. Cerón, E. Caamaño-Martín, and S. V. Sánchez, "BIPV technology application: Highlighting advances, tendencies and solutions through Solar Decathlon Europe houses," *Energy Build*, vol. 83, pp. 44–56, 2014, doi: 10.1016/j.enbuild.2014.03.079.

[4] EPBD recast, "Directive 2010/31/EU of the European Parliament and of the Council of 19 May 2010 on the energy performance of buildings (recast)," 2010. [Online]. Available: https://eur-lex.europa.eu/legal-content/EN/TXT/?uri=CELEX%3A02010L0031-20210101

[5] A. Scognamiglio and F. Garde, "Photovoltaics' architectural and landscape design options for Net Zero Energy Buildings, towards Net Zero Energy Communities: spatial features and outdoor thermal comfort related considerations," *Progress in Photovoltaics: Research and Applications*, vol. 24, pp. 477–495, 2016, doi: 10.1002/pip.2563.

[6] A. Scognamiglio and H. N. Røstvik, "Photovoltaics and zero energy buildings: a new opportunity and challenge for design," *Progress in Photovoltaics: Research and Applications*, vol. 21, no. 6, pp. 1319–1336, 2012, doi: 10.1002/pip.2286.

[7] P. de Berardinis and P. Bonomo, "Towards a system for evaluation of the project of PV integration in buildings. BIPV.tool," in *27th European Photovoltaic Solar Energy Conference and Exhibition*, 2012, pp. 4183–4187. doi: 10.4229/27thEUPVSEC2012-5BV.1.5.

[8] IEA SHC, "Task 41 A.2 - Solar Energy Systems in Architecture: integration criteria and guidelines," 2012. [Online]. Available: https://task41.iea-shc.org/Data/Sites/1/publications/T41DA2-

Solar-Energy-Systems-in-Architecture-28March2013.pdf
[9] M. Pelle, E. Lucchi, L. Maturi, A. Astigarraga, and F. Causone, "Coloured BIPV Technologies: Methodological and Experimental Assessment for Architecturally Sensitive Areas," *Energies (Basel)*, vol. 13, no. 17, p. 21, 2020, doi: 10.3390/en13174506.
[10] F. Rosa, "Building-Integrated Photovoltaics (BIPV) in Historical Buildings: Opportunities and Constraints," *Energies (Basel)*, vol. 13, no. 14, p. 28, 2020, doi: 10.3390/en13143628.
[11] R. Xu, S. Wittkopf, and C. Roeske, "Quantitative evaluation of BIPV visual impact in building retrofits using saliency models," *Energies (Basel)*, vol. 10, no. 5, 2017, doi: 10.3390/en10050668.
[12] R. Xu and S. Wittkopf, "Visual assessment of BIPV retrofit design proposals for selected historical buildings using the saliency map method," *Journal of Facade Design and Engineering*, vol. 2, no. 3–4, pp. 235–254, 2015, doi: 10.3233/fde-150022.
[13] P. Florio, C. Roecker, and M. C. Munari Probst, "Urban acceptability of solar installations: LESO-QSV GRID, a software tool to support municipalities," in *CISBAT 2015 International Conference "Future Buildings and Districts - Sustainability from Nano to Urban Scale,"* 2015, pp. 981–986. doi: 10.13140/RG.2.1.2017.0728.
[14] Becquerel Institute, "BIPV market and stakeholder analysis 2019," 2019. [Online]. Available: https://bipvboost.eu/public-reports/
[15] N. Sánchez-Pantoja, R. Vidal, and M. C. Pastor, "Aesthetic perception of photovoltaic integration within new proposals for ecological architecture," *Sustain Cities Soc*, vol. 39, pp. 203–214, 2018, doi: 10.1016/j.scs.2018.02.027.
[16] I. Custódio, T. Quevedo, A. P. Melo, and R. Rüther, "A holistic approach for assessing architectural integration quality of solar photovoltaic rooftops and shading devices," *Solar Energy*, vol. 237, pp. 432–446, May 2022, doi: 10.1016/j.solener.2022.02.019.
[17] V. Kosoric, S. Wittkopf, and Y. Huang, "Testing a design methodology for building integration of photovoltaics (PV) using a PV demonstration site in Singapore," *Archit Sci Rev*, vol. 54, no. 3, pp. 192–205, 2011, doi: 10.1080/00038628.2011.590052.
[18] A. Scognamiglio and C. Privato, "Starting points for a new cultural vision of BIPV," in *23rd European Photovoltaic Solar Energy Conference*, 2008, pp. 3222–3233. doi: 10.4229/23rdEUPVSEC2008-5BP.1.5.
[19] I. Hagemann, "Examples of successful architectural integration of PV: Germany," *Progress in Photovoltaics: Research and Applications*, vol. 12, no. 6, pp. 461–470, 2004, doi: 10.1002/pip.561.
[20] A. Scognamiglio, "A trans-disciplinary vocabulary for assessing the visual performance of BIPV," *Sustainability*, vol. 13, no. 10, p. 38, 2021, doi: 10.3390/su13105500.
[21] A. S. Bahaj, P. A. B. James, and M. F. Jentsch, "Photovoltaics: added value of architectural integration," *Proceedings of the Institution of Civil Engineers - Energy*, vol. 160, no. 2, pp. 59–69, 2007, doi: 10.1680/ener.2007.160.2.59.
[22] P. Bonomo, P. de Berardinis, and F. Frontini, "Analysis of BIPV case-studies through a multicriteria evaluation tool," in *28th European Photovoltaic Solar Energy Conference and Exhibition*, 2013, pp. 4373–4379. doi: 10.4229/28thEUPVSEC2013-5CV.7.12.
[23] W. Celadyn and P. Filipek, "Investigation of the Effective Use of Photovoltaic Modules in Architecture," *Buildings*, vol. 10, no. 145, p. 20, 2020, doi: 10.3390/buildings10090145.
[24] I. Custódio, C. Zomer, and R. Rüther, "A worldwide approach to the LESO-QSV method for assessing the visual impacts of solar systems in urban environments," *Solar Energy*, vol. 212, no. July, pp. 178–190, 2020, doi: 10.1016/j.solener.2020.10.067.
[25] IEA PVPS, "Task 7 - Photovoltaic power systems in the built environment," 1997.
[26] H. Kaan and T. Reijenga, "Photovoltaics in an architectural context," *Progress in Photovoltaics: Research and Applications*, vol. 12, no. 6, pp. 395–408, 2004, doi: 10.1002/pip.554.
[27] M. C. Munari Probst and C. Roecker, "Solar Energy Promotion & Urban Context Protection: Leso-QSV (Quality-Site-Visibility) Method," 2015.
[28] M. C. Munari Probst and C. Roecker, "Criteria and policies to master the visual impact of solar systems in urban environments: The LESO-QSV method," *Solar Energy*, vol. 184, no. September 2018, pp. 672–687, 2019, doi: 10.1016/j.solener.2019.03.031.
[29] M. C. Munari Probst and C. Roecker, "Urban acceptability of building integrated solar systems: LESO-QSV approach," in *30th ISES Biennial Solar World Congress 2011, SWC 2011*, 2011, vol. 6, pp. 4359–4367. doi: 10.18086/swc.2011.27.10.
[30] C. Polo Lopez, F. Frontini, P. Bonomo, and A. Scognamiglio, "PV and façade systems for the building skin. Analysis of design effectiveness and technological features," in *29th European Photovoltaic Solar Energy Conference and Exhibition*, 2014, pp. 3613–3618. doi: 10.4229/EUPVSEC20142014-6DO.7.3.
[31] I. P. dos Santos, "Desenvolvimento de Ferramenta de Apoio à Decisão em Projetos de Integração Solar Fotovoltaica à Arquitetura," Doctoral thesis, Federal University of Santa Catarina, 2013. [Online]. Available: https://fotovoltaica.ufsc.br/Teses/Tese_Isis_Portolan_dos_Santos.pdf
[32] T. Schoen, D. Prasad, D. Ruoss, P. Eiffert, and H. Sørensen, "Status Report of Task 7 of the EIA PV Power Systems Program," in *18th European Photovoltaic Solar Energy Conference and Exhibition*, 2000, pp. 1–4.
[33] A. Tablada, V. Kosorić, H. Huang, S. S. Y. Lau, and V. Shabunko, "Architectural quality of the productive façades integrating photovoltaic and vertical farming systems: Survey among experts in Singapore," *Frontiers of Architectural Research*, vol. 9, no. 2, pp. 301–318, 2020, doi: 10.1016/j.foar.2019.12.005.

[34] G. Becker, F. Flade, R. Krippner, B. Schiebelsberger, and W. Weber, "High Quality Solutions of Building-Integrated Photovoltaics (BIPV) – Results of the World Wide Competition in 2017," in *35th European Photovoltaic Solar Energy Conference and Exhibition*, 2018, pp. 1828–1832. doi: 10.4229/35thEUPVSEC20182018-6BV.1.42.

[35] M. Devetaković *et al.*, "Photovoltaics on landmark buildings with distinctive geometries," *Applied Sciences*, vol. 10, no. 19, p. 22, 2020, doi: 10.3390/app10196696.

[36] I. Hagemann, "PV in buildings - the influence of PV on the design and planning process of a building," *Renew Energy*, vol. 8, no. 1–4, pp. 467–470, 1996, doi: 10.1016/0960-1481(96)88900-2.

[37] C. Peng, Y. Huang, and Z. Wu, "Building-integrated photovoltaics (BIPV) in architectural design in China," *Energy Build*, vol. 43, no. 12, pp. 3592–3598, 2011, doi: 10.1016/j.enbuild.2011.09.032.

[38] A. Scognamiglio, "'Photovoltaic landscapes': Design and assessment. A critical review for a new transdisciplinary design vision," *Renewable and Sustainable Energy Reviews*, vol. 55, pp. 629–661, 2016, doi: 10.1016/j.rser.2015.10.072.

[39] A. del C. Torres-Sibille, V. A. Cloquell-Ballester, V. A. Cloquell-Ballester, and M. Á. Artacho Ramírez, "Aesthetic impact assessment of solar power plants: An objective and a subjective approach," *Renewable and Sustainable Energy Reviews*, vol. 13, no. 5, pp. 986–999, 2009, doi: 10.1016/j.rser.2008.03.012.

[40] IEA SHC, "Task 51 C.3 - Lessons Learnt from Case Studies of Solar Energy in Urban Planning," 2018. doi: 10.18777/ieashc-task51-2018-0003.

[41] G. Lobaccaro *et al.*, "A cross-country perspective on solar energy in urban planning: Lessons learned from international case studies," *Renewable and Sustainable Energy Reviews*, vol. 108, pp. 209–237, 2019, doi: 10.1016/j.rser.2019.03.041.

[42] S. E. Rasmussen, *Arquitetura Vivenciada*, 3rd ed. São José dos Campos, 2015.

[43] B. Zevi, *Saber Ver a Arquitetura*, 6th ed. São Paulo, 2009.

[44] S. Vezzani, B. F. M. Marino, and E. Giora, "An early history of the Gestalt factors of organisation," *Perception*, vol. 41, pp. 148–167, 2012, doi: 10.1068/p7122.

[45] H.-G. Gadamer, *Verdade e Método I: Traços fundamentais de uma hermenêutica filosófica*, 15th ed. Petrópolis, RJ; Bragança Paulista, SP, 2015.

[46] C. Castoriadis, *A Instituição Imaginária da Sociedade*. Rio de Janeiro, RJ, 1982.

[47] G. Scott, *The Architecture of Humanism: a Study in the History of Taste*. Florence: Houghton Mifflin Harcourt, 1914.

[48] J. Gehl, *Cidades para Pessoas*, 3rd ed. São Paulo, 2015.

[49] G. Kohlsdorf and M. E. Kohlsdorf, *Ensaio Sobre o Desempenho Morfológico dos Lugares*. Brasília: FRBH, 2017.

[50] C. Zomer, I. Custódio, A. Antoniolli, and R. Rüther, "Performance assessment of partially shaded building-integrated photovoltaic (BIPV) systems in a positive-energy solar energy laboratory building: Architecture perspectives," *Solar Energy*, vol. 211, no. September, pp. 879–896, 2020, doi: 10.1016/j.solener.2020.10.026.

[51] C. Zomer *et al.*, "Energy balance and performance assessment of PV systems installed at a positive-energy building (PEB) solar energy research centre," *Solar Energy*, vol. 212, no. November, pp. 258–274, 2020, doi: 10.1016/j.solener.2020.10.080.

LIGHT REFLECTION ANALYSIS OF PV MODULES: COMPARISON TO BUILDING FACADES AND ASSESSING THE POSSIBILITY OF GLARE

Janina Moereke[1], Peter Borowski[1], Stefan Grünsteidl[1], Jörg Palm[1], Subarna Babu Sapkota[2] and Thomas Dalibor[1]
[1]Avancis GmbH, Otto-Hahn-Ring 6, 81739 München, Germany
[2]Avancis GmbH, Solarstraße 3, 04860 Torgau, Germany
Phone: +49(0) 89 219620 442, Email: janina.moereke@avancis.de

ABSTRACT: Light reflection measurements have been performed on several PV modules as well as a range of passive façade materials. The measurements have been performed in specular direction according to DIN EN ISO 2813 and resulted in the confirmation of very matt appearance for special façade PV modules compared to commonly used building façades. BRDF analysis of these façade PV modules further showed the absence of a specular reflection peak compared to glossy samples and established maximum reflection to occur at reflection angles above 80° to the normal of the module and in-plane with light incidence. Luminance estimations suggest that levels of absolute glare are only reached with both maximum Sun irradiation as experienced on Earth and geometries of maximum reflection. Since these geometries and irradiances occur for an observer of a south-facing façade BIPV installation only in the morning and evening hours, the risk of glare from the analysed façade PV modules is considered negligible.

Keywords: Building Integrated PV (BIPV), Façade, Solar Architecture, Glare, CIGS

1 INTRODUCTION – GLOSS AND GLARE IN BIPV

As photovoltaic (PV) modules are being integrated into urban structures such as building façades, the importance of the visual appearance of the modules increases and the question of gloss or how strongly light is reflected from PV modules has become relevant not only regarding yield, but also regarding potential risks and disturbances in urban environments through glare. Especially for environments such as airports, glare assessments are part of safety hazard prevention [1]. In urban environments the oversight of assessing how a structure may reflect sunlight has made headlines in the past, even mentioning sun rays being reflected at such strength that surrounding buildings and parked cars had been damaged [2][3][4].

Glare describes the phenomenon when the human eye is no longer able to adjust to the brightness of the incoming light, in the case considered here, the light reflected from the PV module's front glass. Among other measures, front glass design options aiming to increase transmittance and avoid reflection have been explored [5][6]. As a result, a variety of PV modules is available that differ in their material properties, especially regarding their visual appearance. However, the assumption that light is reflected as strongly from all PV modules as from other glazing is still common when assessing the risk of glare and represents an obstacle in encouraging the use of PV modules by architects and planners [7][8]. As a consequence, solutions for reducing the risk of glare often address only the orientation of the modules [9]. However, since a diversity of technologies and design options exists for building-integrated PV (BIPV) installations, a more rigorous analysis of the reflection behaviour of PV modules is of interest.

A previous study has assessed light reflection focusing on the specular case, for which classifications for glossiness exist from other industries [10]. Others have included a more detailed analysis of light reflection by performing experiments across a range of angles and recognized that there is currently no universally accepted evaluation method for the risk of glare from PV modules [11]. The present study goes a step further by combining the two methods and putting the results into context in the building industry: making use of the speed and ease of gloss measurements to compare PV modules with commonly used non-PV façade materials, but also providing thorough information regarding light reflection from a highly matt PV surface using BRDF (bidirectional reflectance distribution function) measurements. Together with actual measurements of glare on a BIPV façade, concerns of disturbing or dangerous levels of light reflection when using PV modules in urban environments are addressed, particularly in comparison to conventional non-active building materials and façades.

2 EXPERIMENTAL METHODS

Two methods were used to assess light reflection from PV modules. Specular reflection, i.e. when the angle of reflection equals the angle of incidence, was compared for different materials, while scattered reflection was analysed for our SKALA BIPV module. SKALA is a thin-film CIGS PV module available in different colours. It has been designed specifically for the use in façade installations in urban environments.

2.1 Gloss measurements

The way in which light is reflected from a material or a surface results in a visual impression that is described across different industries using terms such as “glossy”, “semi-glossy” or “matt”. While these terms are often used without a clear definition of how much and in which direction light is reflected, a standard has been created in the coating and varnish industry [12] relating these terms of visual impression to measurements of light reflection. The norm describes a standard procedure defining measurement and incidence angles. Comparable results are achieved with this measurement procedure and enable the assessment of different materials according to their glossiness. Gloss measurements are performed only in specular direction, i.e. at measurement configurations for which the angle of incidence equals the angle of reflection (see an illustration of the measurement setup in Figure 1).

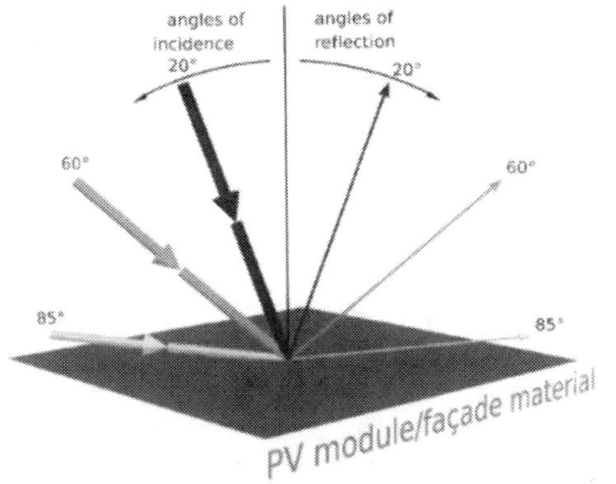

Figure 1. Angles of incidence and reflectance used in gloss measurements. Only specular reflection, where the angle of incidence equals the angle of reflection is assessed in this measurement setup.

Specular reflection is measured at three distinct angles: 20°, 60° and 85°. Results are given in gloss units (GU). The gloss units scaling is defined using light reflection from a perfect mirror and defining the equivalent gloss values to be 2000 GU, 1000 GU and 199 GU at 20°, 60° and 85°, respectively, for the measurement equipment used in this study. The measured gloss value of a material can therefore be converted back into percent reflected light using the conversion factors listed in Table I. Since these values are based on the reflection from a standard material as defined in [13] rather than a manufacturer specific definition, an independent publication derives factors differing in the decimal places [14]. For this study the factors provided by the equipment manufacturer have been used since the measurement equipment was calibrated as such by the manufacturer.

Table I: Conversion from gloss units to percent reflection according to the equipment manufacturer and [14]

Measurement angle	% reflection (acc. to equipment manufacturer)	% reflection (acc. to [14])
20°	GU(20°)/20	GU(20°)/20.376
60°	GU(60°)/10	GU(60°)/9.994
85°	GU(85°)/1.99	GU(85°)/1.674

Table II: Gloss measurement angle according to [12]

Gloss value at 60°	Measurement angle according to [12]
<10 GU	85°
10-70 GU	60°
>70 GU	20°

While the measurements were performed at three different angles, only one value will be quoted as the sample's gloss value, depending on the value measured at 60° and according to the standard (see Table II). For very matt samples (<10GU at measurement angle of 60°), a measurement angle of 85° is used, while for highly glossy samples (>70GU at a measurement angle of 60°) the measurement angle of 20° is most appropriate.

Measurements were performed using the glossmeter Rhopoint IQ-S. This handheld equipment measures light reflection in accordance with the previously mentioned standard [12] and converts percent reflection automatically into GU as defined in Table I (middle column). Measurement results were available in GU only. The manufacturer states that results of repeat measurements may vary by ±0.2% while the results are reproducible to ±0.5%. In the current study, repeatability was found to be even lower at around ±0.1% for measurements in the lab. Samples measured with the Rhopoint IQ-S included not only different PV modules, but also façade materials as well as existing building façades in Munich and Torgau, Germany. Each sample was measured between 3 to 5 times on different positions and the average taken. A measurement position dependent variation of up to ±0.3 GU was found for the matt appearing PV modules, while measurement variations of up to ±4 GU were found for semi-glossy PV modules.

2.2 Bidirectional reflectance distribution function (BRDF)

While an assessment of the glossiness of a surface can provide an indication of risk of glare in specular direction, it cannot provide information on the intensity of reflected light outside the specular direction. The measurement of the bidirectional reflectance distribution function (BRDF) provides a method to assess reflection into all directions other than, but also including the specular direction [15]. With this method, light is directed at a particular angle onto the sample and reflection measured at a range of reflection angles using a gonioreflectometer. This equipment allows the precise positioning of both light source and detector. Measurements presented in this study were performed in two independent labs. The sample under investigation was the BIPV module SKALA Green from the manufacturer AVANCIS. The BRDF measurement was performed in the first lab using a 75W Hamamatsu Xenon lamp as light source. Its unpolarized broadband spectrum is directed onto the sample and the reflected light detected by the spectrometer from Instrument Systems, CAS 140CT. The sensitivity of the detector ranges from 380nm to 1100nm. Reflected light is

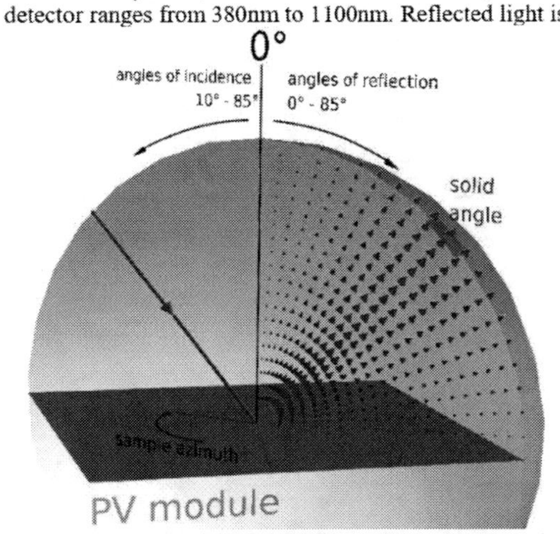

Figure 2. Geometry for 2D BRDF measurements: angles of incidence ranging from 10 to 85° while reflection is measured at angles ranging from 0 to 85° (for lab 1). The measurement is performed in plane with the light incidence.

measured spectrally resolved within this range of wavelengths as radiance through the solid angle, i.e. W/m²sr. Incident light is given as the irradiance in W/m². BRDF results are defined as the ratio of scattered radiance to irradiance. Consequently, BRDF results are quoted as unit per solid angle, i.e. sr^{-1}. The second lab applied the same principle, but used a monochromatic light source at a wavelength of 638nm. In this setup, small format samples of AVANCIS SKALA Green, Grey and Black were measured, showing different appearances of glossiness.

In the present study, incident angles are varied from 10° to 85° in 5° steps, while scattered light is detected at angles ranging from 0° to 85° in 1° steps in plane with light incidence for the first lab and from -85° to 85° with the same step size in the second lab (see Figure 2 for an illustration of the measurement setup). The setup therefore allows measurement of reflection not only in specular direction (i.e. where incident angle equals reflection angle), but also in scattering directions. The direction into which light is reflected most strongly can thus be determined.

2.3 Luminance and glare

In a scientific report concerning glare [16] the German commission on radiation protection defines absolute glare to occur when the incoming light source is so strong that the human eye can no longer adapt to this luminance and the person is blinded. For glare assessments a threshold of 1.6×10^5 cd/m² is typically applied as the critical luminance at which this occurs. Luminance levels of 10^4 cd/m² can produce similar effects under certain luminance contrasts. Other countries, while being equally concerned with the phenomenon of glare, have not included a threshold value of luminance for glare [17]. The appropriateness of available criteria for the assessment for glare have been questioned by a previous study [18]. However, since the threshold value reported in [16] is used frequently in glare assessments as well as the cited publications on glare, it will also serve as a reference point for this study.

Luminance is defined as the radiation per area in the visible wavelengths considering the sensitivity of the human eye as well as the solid angle over which the radiation is measured. It is given in units of candela/m² which are equivalent to the luminous flux (in lumen/m²) per steradian.

The sensitivity of the human eye is expressed by the spectral luminous efficiency function, V(λ). This shifts slightly depending on the surrounding brightness. For daylight vision the photopic spectral luminous efficiency function should be used (see Figure 3).

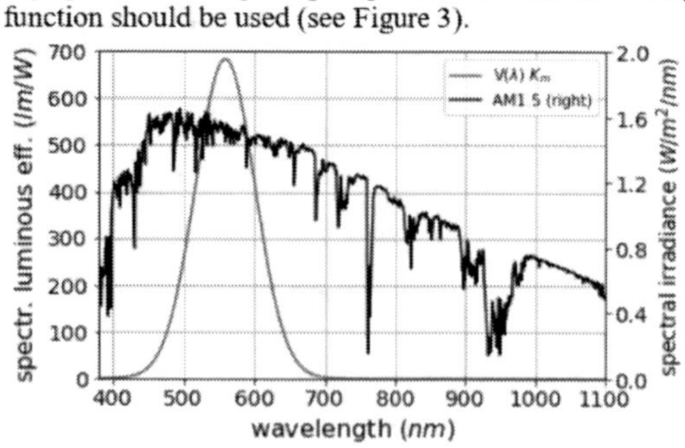

Figure 3. AM1.5 spectrum (in black) and the photopic spectral luminous efficiency function V(λ) (in blue) used to convert radiometric units, W/m², into photometric units, cd/m².

Irradiance of the Sun, on the other hand, is generally given in units of W/m² (see Figure 3 for the AM1.5 spectrum of 1000 W/m² [19]). In radiometry, the whole range of optical electromagnetic radiation is considered as opposed to photometry which only considers the spectrum in the visible range.

In order to be able to judge whether reflected radiation may cause glare, a conversion from radiometric to photometric units is required. Conversion into luminous flux Φ_V, i.e. into lm/m², is performed using Equation (1):

$$\Phi_V = K_m \int_\lambda \Phi_{e,\lambda}(\lambda) V(\lambda) d\lambda \qquad (1)$$

where K_m is the maximum luminous efficiency for photopic vision (683 lm/W at 555nm), $\Phi_{e,\lambda}(\lambda)$ is the irradiance in W/m²/nm and $V(\lambda)$ is the unitless photopic spectral luminous efficiency function (Figure 3) [20].

Dividing by the solid angle of the light source results then in luminance in cd/m². For the AM1.5 spectrum and using the solid angle subtended by the Sun on the sky as seen from Earth, this conversion leads to a luminance of the Sun of 1.6×10^9 cd/m² during midday on a clear day. Accepting the threshold for glare as stated earlier, only surfaces with a reflection of less than 0.01% would prevent glare at these conditions.

3 RESULTS

3.1 Specular reflection through gloss measurements

Measured gloss values of various PV modules, conventional façade materials as well as of existing building façades are shown in Figure 4. Results are expressed in gloss units, or GU, a scaling system related to the percentage of light reflected as described in Table I. The angle at which the measurement was performed for each material was chosen according to the definition in Table II.

According to [12], materials with a gloss value below 10 GU are described as "matt", while materials with a gloss value above 70 GU are considered glossy. Materials with gloss values in between these boundaries are considered "semi-glossy". The majority of passive façade materials and façades measured falls into the "semi-glossy" category. Conventional façade materials (yellow in Figure 4) cover the whole range of glossiness with a few materials exhibiting matt properties, while others reflecting light strongly enough that they are categorised as glossy. Roof-top PV modules from different suppliers (orange in Figure 4) all fall into the category of semi-glossy. Our product line AVANCIS SKALA Colour (in green in Figure 4) falls entirely into the "matt" category exhibiting very low gloss values below 10 GU, often even below 5 GU. A notable exception is AVANCIS SKALA Black with reflection properties comparable to conventional PV modules. AVANCIS SKALA Colour represent one of the least glossy materials in this study. Only among the measured building façades (shown in blue in Figure 4) can materials be found that reflect even matter than these PV modules, that have been particularly designed for the installation in building façades. Stone façades of prominent old buildings in Munich resulted in very low values. It should be noted, however, that results of historic buildings are likely associated with a large error as placing the measurement equipment onto the macroscopically very rough surface proved to be rather

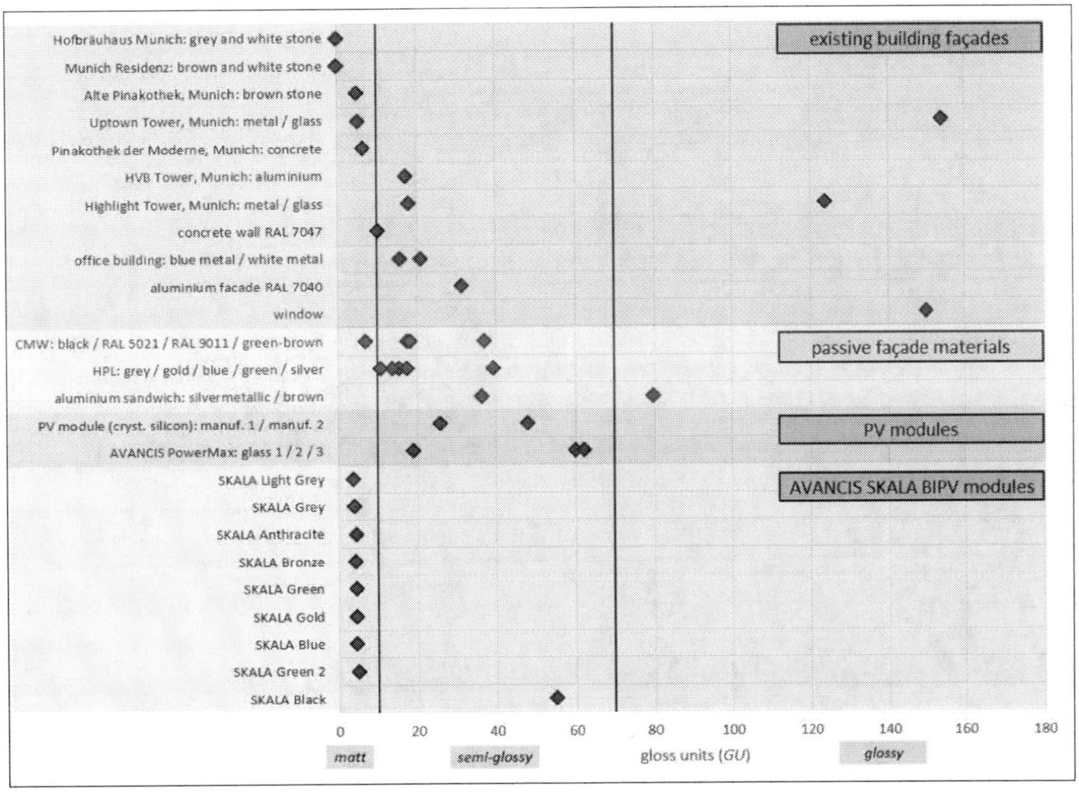

Figure 4. Gloss values of PV modules and façade materials as well as already existing building façades. CWM: compressed mineral wool, HPL: high pressure laminate. See subsection 2.1 for details on how data was acquired as well as an estimate of the measurement error of these values.

difficult. Diffuse light entering or leaving the measurement cavity of the equipment may have affected the measurement in these cases.

Interestingly, glass façades of existing buildings have the highest gloss values in this study. Even roof-top PV modules, for which glare assessments are often demanded by concerned stakeholders, do not reach these levels of light reflection in specular direction. Previous studies echo these findings by pointing out that anti-reflective coatings are typically applied to PV module glass surfaces, thus significantly reducing light reflection [21].

Repeating gloss measurements on the very matt BIPV modules SKALA (in green in Figure 4) after three months of outdoor exposure, heavy soiling and extensive cleaning did not show any change in gloss values.

Repeating these measurements again on SKALA modules in wet conditions showed that reflection strongly increases for comparably matt PV modules when the surface is wetted, while this effect was not as strong for more glossy modules or materials. However, once the modules were irradiated with sunlight, they dried within minutes, making these wet conditions practically irrelevant for vertical BIPV installations since sunshine and rain usually do not appear together.

Judging from the measured gloss values, as presented in Figure 4, an increased concern regarding disturbing glare from PV installations in an urban environment appears questionable. Since the light reflection properties of market-typical passive cladding materials lie in the range of those of PV modules, there seems to be a bias assuming an increased risk of glare from PV modules that is, however, not reflected in industry-approved light reflection measurements. If concern is justified, it should rather focus on conventional glass façades. Further, the measured values clearly dissent the still-common practice by some entities conducting glare assessments assuming reflective properties of window glass for PV modules.

Assuming the validity of the threshold for glare as stated in subsection 2.3 as well as reliable measurement values of the applied gloss meter for very low reflection values, none of the considered passive façade materials is free of disturbing glare effects and even concrete and stone façades of existing buildings would pose a potential risk of glare.

3.2 Scattered reflection through BRDF

Light reflection is often described using the 3-component-model which simplifies light reflection into a specular, a haze and a diffuse or Lambertian component [22]. In this model, the strongest intensity is reached in specular direction through the specular component. Less intense light is reflected in the haze component which widens the specular peak. Finally, the diffuse component describes light reflected in all directions at a rather low intensity. As gloss measurements only measure the specular component, a low gloss value may indicate, that a different component in the model is dominating. It is therefore hypothesized that matt samples, such as the BIPV modules SKALA, scatter light at a larger range of angles with a comparatively low intensity compared to glossy samples, such as window glass. Photographs of special façade PV modules as well as roof-top PV modules

Figure 5. Pictures of various BIPV modules AVANCIS SKALA with incident sunlight at small angle to normal direction (top left) and at large angle to the normal (top right). As a comparison, roof-top PV modules AVANCIS PowerMax with incident sunlight at large angle to normal (bottom).

illustrate the visual appearance of matt modules (top) and semi-glossy modules (bottom) (see Figure 5).

Exploring this further, BRDF results extend on gloss measurements by measuring the scattered radiance in the plane of incident light (see Figure *2* in the previous section). Figure *6* illustrates the difference in reflection behaviour for glossy and matt samples. The results show a measurement at an incident angle of 60° at a wavelength of 638nm performed in the second lab described in the experimental setup. In this case, reflection was not only measured in the quadrant opposite to light incidence such as illustrated in Figure *2*, but also in the quadrant of incident light. Negative angles on the x-axis refer to angles in the latter quadrant. Three almost identical curves with an almost non-distinguishable peak in specular direction are shown in grey, green and blue. By comparison, a fourth curve in black shows a prominent peak with its maximum in specular direction, i.e. at a measurement angle of roughly 60°. Photographs of the illuminated samples (below graph in Figure *6*) confirm that the sample with the strong specular peak has a semi-glossy appearance and represents a small sample of a roof-top PV module, while the curves with a rather smooth gradient correspond to samples equivalent to BIPV modules AVANCIS SKALA Colour. As hypothesized from gloss measurements, these results confirm that a semi-glossy sample mainly reflects

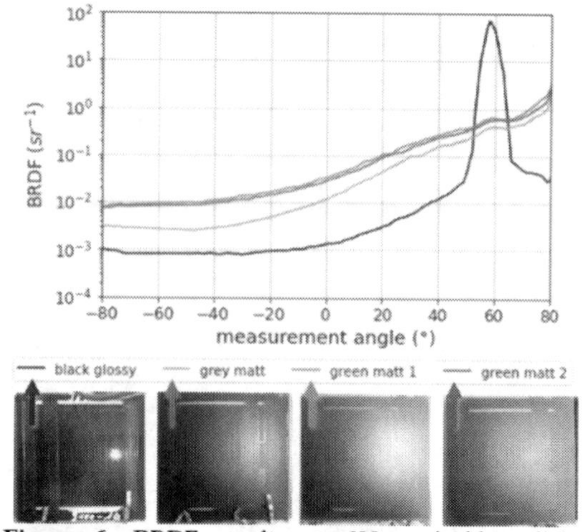

Figure 6. BRDF results at 638nm incident light wavelength and 60° incident angle for four different samples. The black curve represents the measurement on a module stack equivalent to the PV module AVANCIS PowerMax and the three other curves the measurement on modules stacks equivalent to the BIPV modules AVANCIS SKALA Colour (Grey, Green 1 and Green 2).

light into specular direction (hence the strong peak for the curve in black) while matt samples scatter the light over a range of angles (hence the lack of a clear peak for grey, green and blue curves). In fact, the graph shows, that the matt samples reflect most strongly at the largest angles. The maximum reflection in this case does not occur in specular direction, but at angles above 80°. The maximum values of the matt samples are, however, still lower by two orders of magnitude compared to the semi-glossy module. For any of the curves, backscattered light, i.e. at negative measurement angles, is of rather low intensity and the analysis in the following will therefore concentrate on the reflection measured at positive measurement angles.

For further tests, a commercially available BIPV module AVANCIS SKALA Green (corresponding to 'green matt 1' in Figure *6*) was investigated with BRDF in the first lab as described in the experimental setup. In Figure *7*, BRDF results are shown for different illumination angles, from 10° to 80° in 10° steps. These results confirm that reflection is stronger at large angles. The combination of large incident and large measurement angle produced the strongest light reflection (see grey

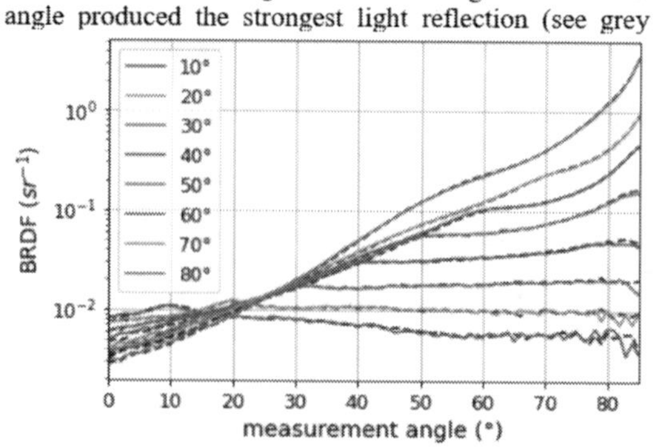

Figure 7. BRDF results for AVANCIS SKALA Green for two different sample orientations at incident angles ranging from 10° to 80° in 10° steps. Dashed lines represent the same measurements with the sample azimuth turned by 90° (see Figure 2).

curve in Figure 7). Starting from an incident angle of 40°, the maximum is pushed to measurement angles above 80°. Increasing the incident angle increases the slope of the curve and the absolute value of the maximum. Additionally, Figure 7 shows two sets of BRDF results for the same sample, but with different sample azimuths, i.e. with the sample rotated around the normal direction (see Figure 2). Results are identical for both sample orientations. An isotropic module surface can be assumed. In a vertical installation, turning the module orientation around the normal direction of the module would not change reflection behaviour.

4 ESTIMATING REFLECTED LUMINANCE

Gloss measurements have shown that our BIPV modules SKALA Colour reflect light rather weakly compared to other materials and PV modules. It is therefore expected that the risk of glare for these modules remains rather low. BRDF measurements have shown that the most severe reflection from these very matt PV modules occurs at very large observation angles to the normal, particularly when the sun rays hit the module at large incident angles. It is therefore expected that the risk of glare may only occur at very specific situations and times of the day.

Using the standard spectrum AM1.5 [19] of 1000 W/m^2 corresponding to the maximum irradiance by the Sun during the day and performing a conversion into luminance based on Equation (1) and multiplying by BRDF results to get reflected intensity, luminance levels at or above 10^5 cd/m^2 were reached at incident and reflection angles above 70° only. However, further converting these calculations into percentage reflection and comparing to gloss measurements, revealed an order of magnitude discrepancy between BRDF and gloss values. A similar discrepancy was observed for BRDF measurements conducted at two independent labs on the same type of PV module: The BRDF curves showed very good agreement qualitatively but an observed order-of-magnitude discrepancy in the actual measurement values could not be explained. It was therefore concluded that a quantitative interpretation of BRDF is not appropriate, but that the results give information regarding the relative distribution of light reflection over a range of angles only. In this regard, the curves of both labs matched showing maximum reflection at large incident and reflection angles for the PV module SKALA Green.

Plotting BRDF in specular direction spectrally resolved (see Figure 8) further confirms increasing reflection with increasing angles, but also shows how the visual impression changes at large angles. While the colour impression is clearly visible at smaller angles with the normal to the module, showing a peak at wavelengths just above 500nm, (i.e. green colour) where the human eye is most sensitive (compare V(λ) in Figure 3), at large angles the curve flattens, reflecting wavelengths almost equally across the visible spectrum and at a stronger intensity. An observer will therefore experience white light at an intensity an order of magnitude stronger at this angle (compare visual impression in Figure 5, top panel). Gloss results suggest that up to 2% of incident light is reflected at these large angles. According to the definitions of luminance and glare given in section 2.3 this would result in critical levels of glare at and near the Sun's maximum irradiation, thus at high elevation, when large

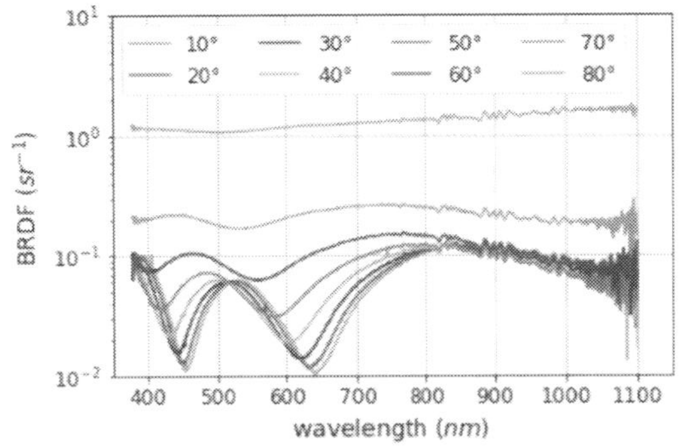

Figure 8. Spectrally resolved BRDF results in specular direction, i.e. where the angle of incidence corresponds to the angle of reflection.

reflection angles in-plane with light incidence are directed to the ground and not into the eye of a passer-by.

In fact, measuring luminance directly during a sunny day in August on a south-facing vertical installation of the BIPV modules AVANCIS SKALA, one was not able to detect critical luminance levels at or above 10^5 cd/m^2. Even positioning the equipment at a large observation angle near 80° in plane with the Sun at maximum elevation (65° in Munich) did not result in these critical levels. For a module installation at eye level of an observer either looking at the installation in the normal direction of the modules or at angles above 80°, the maximum luminance levels detected reached values of the order 10^3 cd/m^2.

5 CONCLUSIONS

Since gloss measurements showed that conventional passive façade materials as well as window glass tend to reflect light even stronger than special façade PV modules, an unnecessary discrepancy seems to exists in how appropriate a material is considered for façades, depending on whether they are PV modules or conventional passive materials. The BIPV modules AVANCIS SKALA Colour were among the least reflective materials measured in this study. Closer investigation of light reflection from this kind of modules with BRDF measurements showed that maximum reflection occurs indeed at large measurement and incident angles. This combination of large incident and reflection angles would become relevant to an observer of a south-facing façade installation only in the morning and the evening hours when the Sun's elevation is relatively low. In this scenario the field-of-view of the observer would also include the Sun itself, so that the risk of glare from reflection is considered negligible. In fact, luminance measurements for an observer at a large angle to the normal of a vertical south-facing BIPV installation with AVANCIS SKALA Colour detected luminance levels of only up to 10^3 cd/m^2 throughout a clear summer day in Munich, Germany, not reaching critical levels for glare.

The study showed further that material properties of different glass surfaces (window glass and PV front glass of different module types and manufacturers) can affect light reflection significantly, so that basing an analysis of light reflection from PV modules on window glass properties will significantly and unnecessarily overestimate the risk of glare.

7 REFERENCES

[1] "Technical Guidance for Evaluating Selected Solar Technologies on Airports", Federal Aviation Administration, Version 1.1, April 2018

[2] https://www.nationalgeographic.com/history/article/130904-walkie-talkie-building-london-melts-sunlight-physics-science

[3] https://www.theguardian.com/artanddesign/shortcuts/2013/sep/03/walkie-talkie-death-ray-buildings-heat

[4] A. Kandt, R. Romero, "Siting solar photovoltaics at airports", NREL, June 2014. https://www.nrel.gov/docs/fy14osti/62304.pdf

[5] T. E. Kuhn et al., "Review of technological design options for building integrated photovoltaics (BIPV)", Energy & Buildings 231 (2021) 1 10381

[6] M. Schiavoni et al., "Anti-glare effect of deeply textured cover glasses for solar modules for installations close to airports and heliports", Proc. of the 28th EuPVSEC (2013), p.3069-3072

[7] C. Xiang et al, "Facade Integrated Photovoltaic, state of the art of Experimental Methodology", IOP Conf. Ser.: Earth Environ. Sci. 352 (2019) 012062

[8] M. Herz et al., "Determining of reflection and glare caused by reflected sunlight on PV modules surfaces at PV systems", Proc. of the 28th EuPVSEC (2013), p.3607-3609.

[9] J. Zehndorfer, "Teure Folgekosten vermeiden", Photovoltaik 10/2018.

[10] R. Trattnig et al., "Smart Glass Coatings for Innovative BIPV Solutions", Sustainability 2021, 13, 12775

[11] H. R. Wilson et al., "Evaluating BIPV modules with respect to textural appearance, glare, colour and electricity yield" 36. PV-Symposium 2021, https://publica.fraunhofer.de/entities/publication/25247c7b-d9c2-40ed-9f1d-5fd1dab87fff/details

[12] DIN EN ISO 2813: "Paints and varnishes - Determination of gloss value at 20°, 60° and 85°"

[13] ASTM D 2457 – 03: "Standard Test Method for Specular Gloss of Plastic Films and Solid Plastics"

[14] A.R. Hanson, "Good Practice Guide for the Measurement of Gloss", Nat. Meas. Good Practice Guide No. 94, NPL, 2006

[15] Fred E. Nicodemus et al., "Geometrical considerations and nomenclature for reflectance." (1977).

[16] Strahlenschutzkommission (SSK), "Blendung durch natürliche und neue künstliche Lichtquellen und ihre Gefahren", February 2006, report by German commission on radiation protection

[17] E. Hjorth, F. Wochele, „Glare from photovoltaic systems", Master thesis, Lund University, EEBD - 15 /17

[18] M. Battaglia et al., "Quantification of glare from sunlight reflected on solar installations", Proc. of the 32nd EuPVSEC, 2016, 2503-2507.

[19] IEC 60904-3: "Photovoltaic devices – Part 3: Measurement principles for terrestrial photovoltaic solar devices with reference spectral irradiance data"

[20] Bureau International des Poids et Mesures (BIPM) "Principles governing Photometry", 05/2019 2nd edition

[21] https://www.nrel.gov/state-local-tribal/blog/posts/research-and-analysis-demonstrate-the-lack-of-impacts-of-glare-from-photovoltaic-modules.html

[22] A. Prokhorov, "Effective emissivities of isothermal blackbody cavities calculated by the Monte Carlo method using the three-component bidirectional reflectance distribution function model", Applied Optics 2012.

GLASS DIGITAL PRINTER CHARACTERIZATION FOR OPTIMIZATION OF BIPV AESTHETIC VS PERFORMANCE BALANCE

Benjamin RIEDEL, Philippe THONY, Ioannis TSANAKAS, Christophe MENEZO
CEA-INES, Université Savoie Mont-Blanc
50 avenue du Lac Léman, 73170 Le Bourget-du-Lac / benjamin.riedel@univ-smb.fr

ABSTRACT: This paper shows experimental studies and theoretical calculations of colorimetry as well as solar efficiency of glass digital printing. The main goal is to optimize the aesthetics and performance balance of building integrated photovoltaics customized (BIPV) with printed glass. The primary inks used by the printer are characterized. Color and power production measurements of PV module that used printed glass are then compare with theory. The influence of the ink thickness is also evaluated from measured spectra. Despite an important difference between theoretical results and measurements, a ranking of best transmitters' ink is observed. Therefore, the measured values can help to predict the power production and the final color when drawing a future BIPV design. Decreasing the ink thickness allows increasing the power production in the same time of decreasing the color saturation. The ranking of best transmitters ink permit to elaborate an optimization method for design drawing, when the control of ink thicknesses permit to match the expected color.
Keywords : BIPV, Optical Losses, Performance, Aesthetic, Digital Printing

1 INTRODUCTION

The deployment of BIPV draws increasing interest from stakeholders in the buildings sector, especially in new buildings or renovation projects. The appearing aesthetics limitations is among the key barriers towards the market uptake of BIPV and their acceptance/adoption by architects and end-users, especially for façade integration. To answer these issues and improve artistic freedom, it is necessary to develop new strategies that permit to change the aspect of a BIPV module while minimizing losses in its power output (thus its energy yield throughout its service lifetime).

Several technologies already exist, like the multilayer optical filters, which permit to obtain saturated colors minimizing significantly the power losses [1]. Others techniques refer e.g. to the use of colored components, such as colored glass, encapsulant or solar cells [2]. However, these technologies limit the aspect at saturated colors or strongly limit the photogenerated current, thus the module's electrical performance. Thanks to the infinite possibilities of computer drawing, the digital printing on glass with ceramic inks can help to control the balance between aesthetics and performance [3]. With this technique, "drawing" approaches, based on dot patterns that deceive the human eye appear to be promising; in fact, allowing to choose the covering rate and, as a consequence, to control the balance between the saturation and the power losses. An additional optimization strategy consists to create a color with side-by-side dots of different primary colors as in the case of pixels-based color reproduction on computer screens. In this case, the light will cross only one ink instead of successive several ones. Yet, in order to evaluate the aspect and the power of a colored BIPV module as a function of the input drawing, the behavior of the printer and its inks characteristics needs to be studied.

2 METHOD

2.1 Glass digital printing process

To use the digital printer, an input image must be drawn, taking into account the different colors and patterns that will be studied. After a conversion into pdf format, the printer software realizes an assignment of different mixes (Figure 1) of the six primary inks to the corresponding pixels. During the printing process, each primary ink is deposited one after the others, providing a superposition of several coatings. Finally, the glass is tempered and the ceramic inks become fixed at its surface for a long time (more than 50 years).

Input Color	RGB	Blue	Red	Yellow	Green	White	Black
	167,103,38	0%	71%	95%	0%	17%	44%

Figure 1 : Exemple of an input color and its corresponding mixes of printer primary inks

2.2 PV module fabrication and flash testing

Once the printed glasses are ready, colored PV modules are laminated with the components indicated in Figure 2. A black ionomer is used as back encapsulant to minimize the contrast with the solar cells and make it non-perceptible to the human eye.

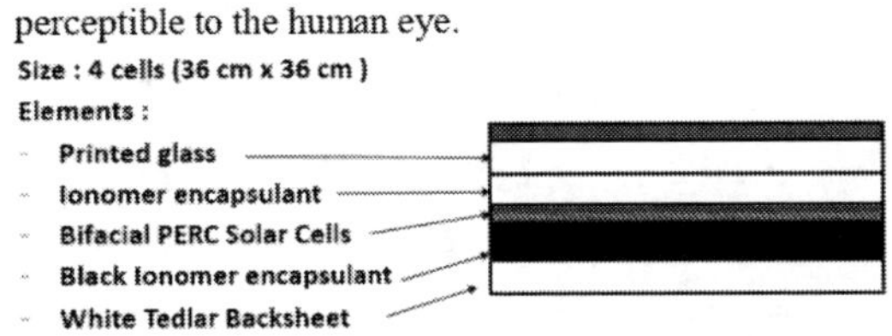

Figure 2 : Elements used for the PV module fabrication

Type of drawing	Reference	Full Covered	Ordered Raster	Pixelized Raster
Pmax (Wc)	18,15	3,87	14,92	4,37
Power losses (%)	0 %	79 %	18 %	76%
Coating density	0 %	100 %	25,3%	75,6%
Module aspect				

Figure 3 : Power losses of colored PV modules associated to different type of drawing

After the lamination, the colored PV module undergoes IV characterization at standard test conditions (STC), by means of a AAA+ solar simulator, at CEA INES test facilities. This method allows testing the impact of different drawing techniques (like rastering and pixelization) on the electrical performance. The results for three different drawings are shown in the Figure 3.

In qualitative terms, the results confirm that the pixelization technique works well for creating colors with additive synthesis.(Figure 3). Nevertheless, every printed dot is a superposition of several primary inks coatings, inducing successive absorptions of light and important power losses. In order to improve the performance optimization, the printing must be a pixelization with dots that contain only one primary inks. The control of each color thickness allows controlling its intensity and matching the expected color in the additive synthesis, at the same time of increasing or decreasing the light transmittance. As such, an optimization method must be elaborated, dealing with the primary inks colors, transmittances and impacts of their thicknesses variations. Thus, each primary ink should be characterized in terms of colorimetry and solar transmittance and the thickness impact must be evaluated. Glasses that were printed with the primary inks were spectrally measured thanks to a spectrophotometer and a colorimeter before the PV modules' lamination stage (Figure 4). The blue ink is so transparent that the silver interconnections are clearly visible, so the manufacturer usually prints a white layer below.

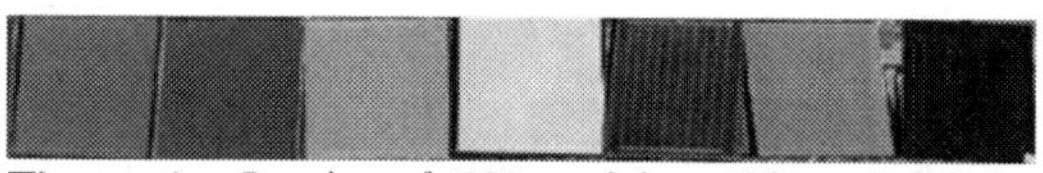

Figure 4 : Laminated PV modules with each of the primary inks (from left to right : blue, red, yellow, white, blue, white superposed with blue, reference).

2.3 Theoretical calculations of colors based on reflectance spectra

Thanks to the printed glasses reflectance spectra R(λ), a corresponding theoretical color can be calculated according to the CIE 1931 standard [4]. Considering the usual D65 illuminant (Figure 5) as the source S(λ) and the color matching functions of the human eye as the receptor, colorimetric coordinates XYZ can be calculated integrating the curves between 380 and 780 nm (visible range) :

$$X = \frac{1}{N}\int_{380}^{780} S(\lambda).R(\lambda).\bar{x}(\lambda)d\lambda \quad Y = \frac{1}{N}\int_{380}^{780} S(\lambda).R(\lambda).\bar{y}(\lambda)d\lambda$$

$$Z = \frac{1}{N}\int_{380}^{780} S(\lambda).R(\lambda).\bar{z}(\lambda)d\lambda \quad N = \int_{380}^{780} S(\lambda).E(\lambda).\bar{y}(\lambda)d\lambda$$

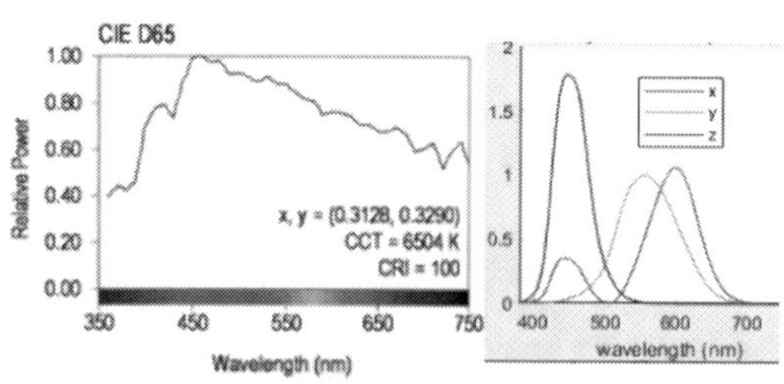

Figure 5 : CIE Standard D65 illuminant spectrum (left) [7] and human eye color matching functions (right)

The XYZ coordinates are then converted into the HSV (Hue, Saturation, Value) color space. The hue correspond to the tint, represented by an angle along a wheel of the rainbow colours, in a range between 0 and 360. The saturation corresponds to the distance from the grey while the value correspond to the distance from the black. Both are in the range between 0 and 1.

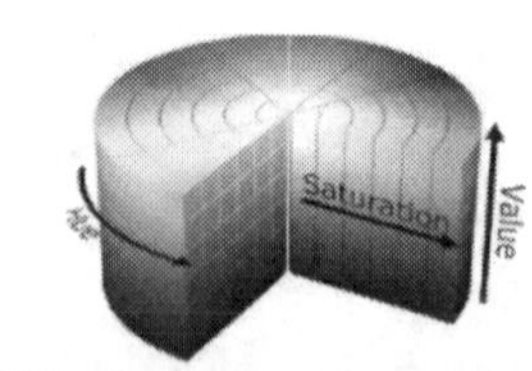

Figure 6 : HSV color space representation [5]

2.4 Theoretical calculations of power losses based on transmittance spectra

A theoretical relative solar efficiency is evaluated calculating the ratio between short-circuit currents of the printed module and the reference: $\eta_{ink} = \frac{I_{sc,ink}}{I_{sc,reference}}$

These theoretical short-circuit currents are calculated by integrating the AMG1.5 solar spectrum $S(\lambda)$ and the solar cell spectral response $SR(\lambda)$ with transmittances of glass $T_g(\lambda)$ and encapsulant $T_e(\lambda)$ (Figure 7) :

$$I_{sc,ink} = \int_{380}^{1200} S(\lambda).T_{g,i}(\lambda).T_e(\lambda)SR(\lambda)d\lambda$$

$$I_{sc,reference} = \int_{380}^{1200} S(\lambda).T_g(\lambda).T_e(\lambda)SR(\lambda)d\lambda$$

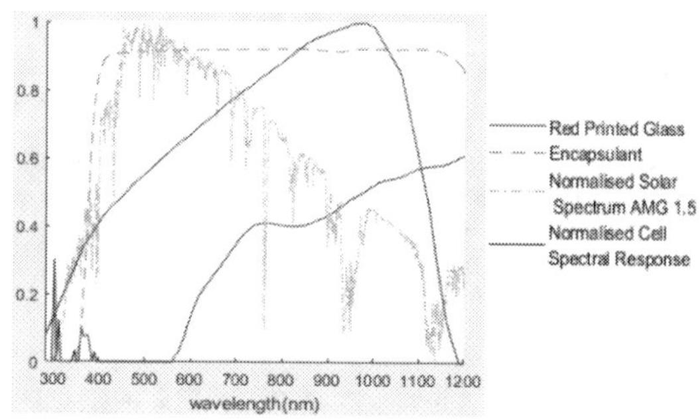

Figure 7 : Spectra used in solar efficiency calculation

3 RESULTS AND DISCUSSION

3.1 Colorimetry

Calculation gives colors that are visually darker than the colors obtained thanks to the measurements (Figure 8). Regarding the HSV coordinates, the value V is always lower in case of the calculations, explaining the darker side of the color. On the other hand, the saturation S is always higher than in the measurements.

Primary Colour	Red	Yellow	Green	Blue
Mesured HSV	3,46,47	34,62,74	138,40,45	211,51,46
Calculated HSV	7,55,18	35,72,44	151,64,19	220,73,16

Figure 8 : Colorimetric results from measurements and calculations (case of 40 microns thick inks)

These differences can be explained by the black and white points references, that are always concidered using a colorimeter, but have to be chosen in the case of the calculations. Thus, it seems that the black point reference of the colorimeter is lighter than the absolute black considered in theoretical calculations. The fact that glasses are transparent also increases the comparison difficulty, as the color depends on the background on which it is printed. Finally, a difference of Hue H that is always slightly higher in calculation is noted (between one to nine degrees more in the rainbow wheel angle, depending on the color, corresponding to a shift to the left in reflectance spectra).

3.2 Relative power efficiency

The results of relative power efficiency allow to rank the best inks in terms of light transmission. The same tendency – in qualitative terms – is observed for both cases of measurements (Figure 9) and calculations (Figure 10). Such ranking could be used to dictate an "order of preference", to take into account for designing and producing colored BIPV modules. However, in quantitative terms, there are important deviations between theory and measurements. In particular, actual measurements generally give rather better efficiency values compared to the calculated ones, i.e. up to 40% higher for the case of the white ink and up to 20% higher for the red ink. These differences can be explained by the influence of scattering effects of the light that could be induced by the printed coating. It indeed might cause irregular (non-uniform) incident light on the solar cells, while the spectral response measurements were realized under normal incident light conditions. Other scientists have also observed already such issues [6].

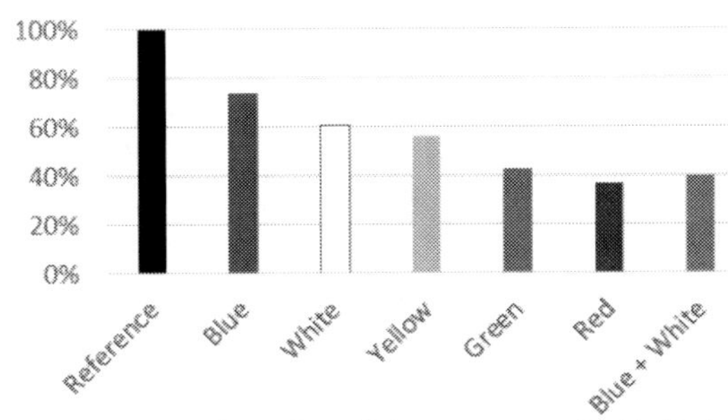

Figure 9 : Relative solar efficiency measured with flash-test.

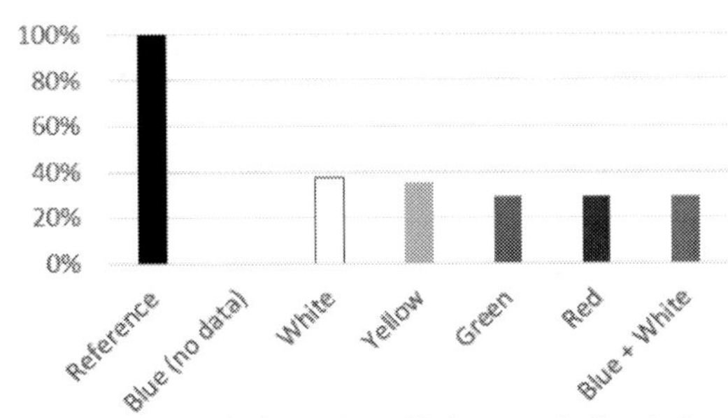

Figure 10 : Relative solar efficiency calculated from spectra.

3.3 Influence of ink thickness

Two different thicknesses are compared in terms of color and solar efficiency, i.e. 20 µm and 40 µm. For each ink studied the colorimetric results show that when the thickness increases, the saturation and value coordinates increase too. However, the differences are not the same, but it depends on the color. Indeed, there is only an increase of a few % of saturation concerning the red and green inks while the yellow and blue ink result in an increase of 20% and 40% respectively. In the case of the value coordinates, there is a small increase of a few percent for the red ink while there is no change concerning the green ink. The increase is higher for the yellow, blue and white inks (about 15% more).

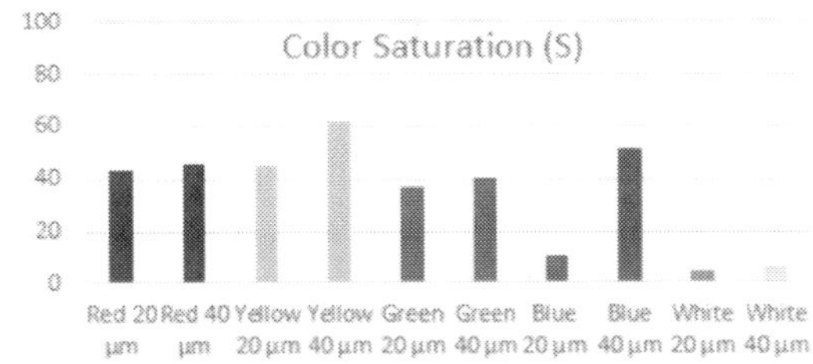

Figure 11 : Colorimetric coordinate S in function of ink color and thickness

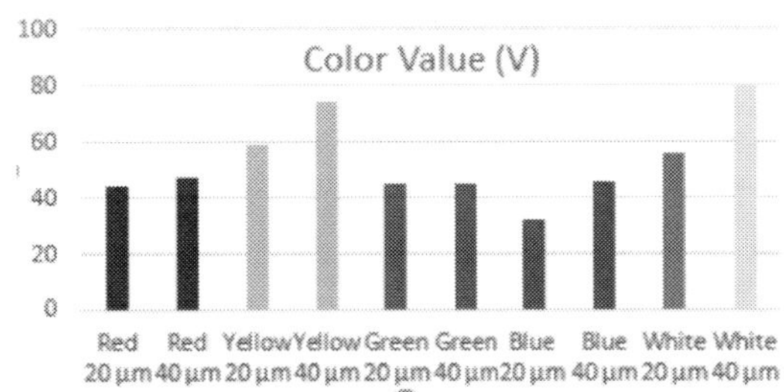

Figure 12 : Colorimetric coordinate Value V in function of ink color and thickness

Concerning the power losses, it is already known that the absorbance of a given coating increase with the thickness (Beer-Lambert law). The theoretical calculations of relative solar efficiency show this tendency, with different value of decrease depending of the color (figure 13). Indeed, when the ink thickness is divided by 2, the solar efficiency of the red ink is doubled whereas that of the yellow ink increase of 25%.

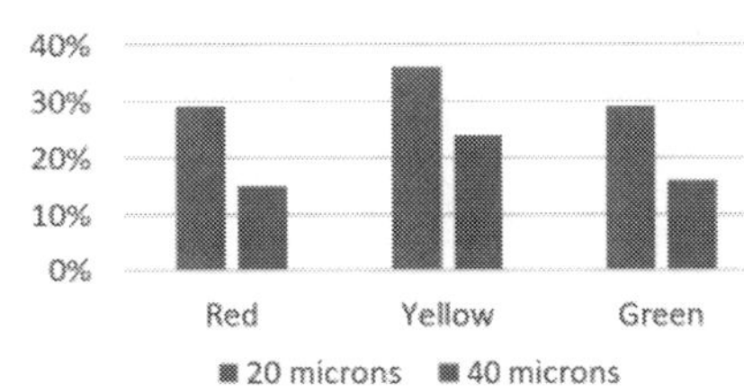

Figure 13 : Theoretical relative solar efficiency in function of ink thickness

4 CONCLUSION

The results obtained with measurements are more reliable than the theoretical calculations in terms of power production but also in the case of colorimetry. This is certainly linked to the involvement of more complex radiative phenomena that we must better understand. Therefore, the measured values can help to predict the power production and the final color when drawing a future BIPV design. Decreasing the ink transmittance allows increasing the power production in the same time of decreasing the color saturation. The ranking of best transmitters ink permit to elaborate an optimization method for design drawing, when the control of ink thicknesses permit to match the targeted color.

5 REFERENCES

[1] B. Bläsi, "Morpho butterfly inspired coloured BIPV modules," 2017.

[2] C. Kutter, "Decorated building_integrated photovoltaic modules : Power losses, color appearance and cost analysis," 2018.

[3] L. Slooff, "An architectural approach for improving aesthetics of PV," 2017.

[4] A. Broadbend, Colorimetry, Methods, 2017.

[5] "Wikipédia.org," [Online]. Available: https://en.wikipedia.org/wiki/HSL_and_HSV.

[6] C. Kutter, "Light Attenuation Model to Predict Nominal Power of Modules with Light-Scattering Ceramic Printed Front Glasses," 2018.

[7] "Wikipédia.org," [Online]. Available: https://en.wikipedia.org/wiki/Illuminant_D65.

LED MODULE WITH A MINIMUM SHADING LOSS FOR MEDIA FAÇADE ON A BUILDING INTEGRATED PHOTOVOLTAIC MODULE

Dong-Youn Shin[a,*], Woo Gyun Shin[b], Gi-Hwan Kang[b]
[a)] Department of Nanotechnology Engineering, Pukyong National University,
45 Yongso-ro, Nam-gu, Busan, 48513, Republic of Korea
[b)] Photovoltaic Research Department, Korea Institute of Energy Research,
152 Gajeong-ro, Yuseong-gu, Daejeon, 34129, Republic of Korea

ABSTRACT: The anthropogenic climate changes render zero-energy buildings mandatory to suppress the emission of CO_2 but the monotonous design of conventional photovoltaic modules has not been welcomed by architects. In order to improve the public acceptance of conventional photovoltaic modules, photovoltaic modules have begun decorated with LEDs or even used as the background of LED modules, acting as media façade with an information function. However, the lifetime mismatch between photovoltaic and LED modules has demanded replaceable LED modules with a minimum shading loss. In this study, an in-house developed LED module is presented, which is designed to be assembled and replaced like LEGO bricks when it is malfunctional. The grid-type design of an LED module also dwindles the shading loss of a photovoltaic module. By applying a reflective coating on the grid walls of an LED module, the sunlight at an incident angle will be re-captured by a photovoltaic module, preserving the photoconversion efficiency of a photovoltaic module.
Keywords: Building Integrated PV, Media Façade, LED Module

1 INTRODUCTION

The adverse outcomes of anthropogenic climate changes have become apparent and the more determined measures all through daily lives have just begun to be taken for urgently reducing the emission of CO_2. The concept of a zero-energy building where the amount of energy to consume is produced by itself using a building integrated photovoltaic module (BIPV) has been deemed as one possible measure [1,2]. However, the poor aesthetic value of a conventional photovoltaic module renders it not welcomed by an architect. In order to endow functional and aesthetic values on a conventional photovoltaic module, an LED module with a minimum shading loss is in-house developed to act as media façade on a BIPV and its impact on the photoconversion efficiency of a BIPV is discussed herein.

2 APPROACH

A conventional LED module for media façade is generally made with transparent glasses or films where LEDs are powered using fluorine tin oxide based transparent electrodes [3]. However, the lifetime of LEDs in the range of 42000 and 60000 hours [4] is incomparably shorter than the lifetime of a photovoltaic module in the range of 25 and 30 years and hence an LED module is likely to be malfunctioned in the middle of the service of a BIPV. An LED module is required to be not only transparent for minimizing the shading loss of a photovoltaic module but also replaceable when malfunctioned.

Such requirements can be fulfilled using a grid-type LED module which can be assembled on a BIPV like LEGO bricks, as shown in Fig. 1(a). If an LED module is malfunctioned, then only the broken one is replaceable. This grid-type design also allows more sunlight to pass through with a minimal shading loss, compared to a conventional plate-type LED module, as can be seen in Fig. 1(b).

Another innovative feature of an in-house developed LED module enables a minimal shading loss even at an incident angle of sunlight. As can be seen in Fig. 1(c), the grid walls of an LED module are coated with an aluminum tape for reflecting incident sunlight. By re-guiding sunlight toward a photovoltaic module, the shading loss of a photovoltaic module by an LED module itself can be minimized.

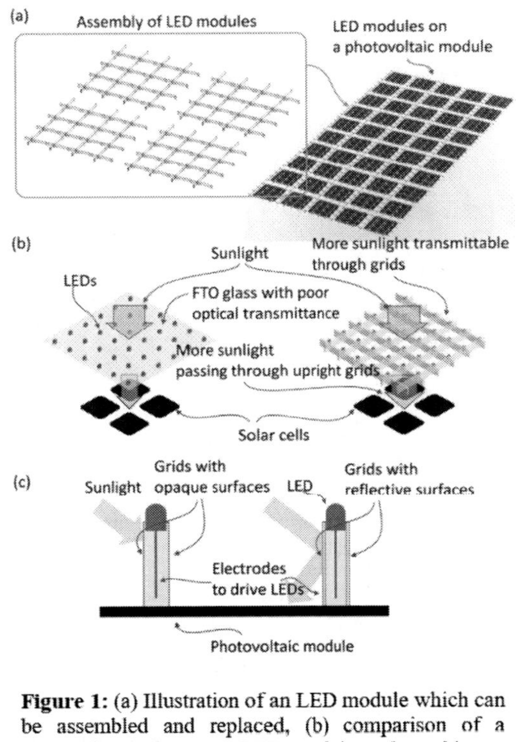

Figure 1: (a) Illustration of an LED module which can be assembled and replaced, (b) comparison of a conventional plate-type LED module and a grid-type LED module, and (c) influence of the surface conditions of an LED module on the collection of sunlight.

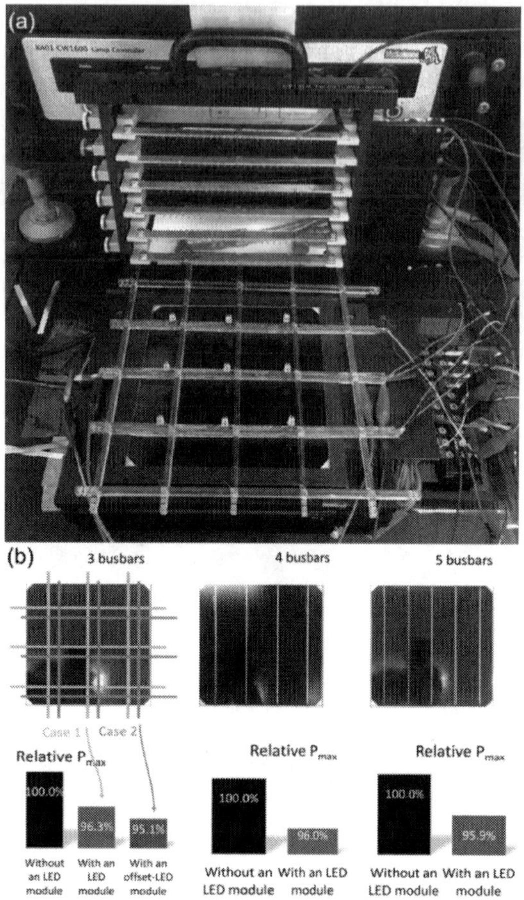

Figure 2: (a) Photoconversion efficiency measurements with an LED module placed on a photovoltaic module and (b) impact of an LED module on the photoconversion efficiencies of photovoltaic modules with busbars from 3 to 5.

3 RESULTS AND DISCUSSION

Photoconversion efficiency measurements were conducted, as can be seen in Fig. 2(a) where an LED module was placed on a photovoltaic module. As shown in Fig. 2(b), the shading loss by an LED module which was placed on a photovoltaic module with 3 busbars was as low as -3.7% especially when the grids of an LED module were located directly on the busbars of a photovoltaic module. Even when the grids of an LED module were not located directly on the busbars of a photovoltaic module, its shading loss was still as low as -4.9%. The shading losses by an LED module on the photovoltaic modules with 4 and 5 busbars were -4.0% and -4.1%, respectively.

No significant shading loss was also observed even under solar irradiation at a steep solar incident angle when the grid walls are reflective, as illustrated in Fig. 3(a). The grid walls with a reflective coating preserved at least 85.6% of the maximum peak power with respect to a reference photovoltaic module even at a solar incident angle of 60°. On the other hand, an LED module with the opaque grid walls decreased the maximum peak power down to 70.6% at a solar incident angle of 60°.

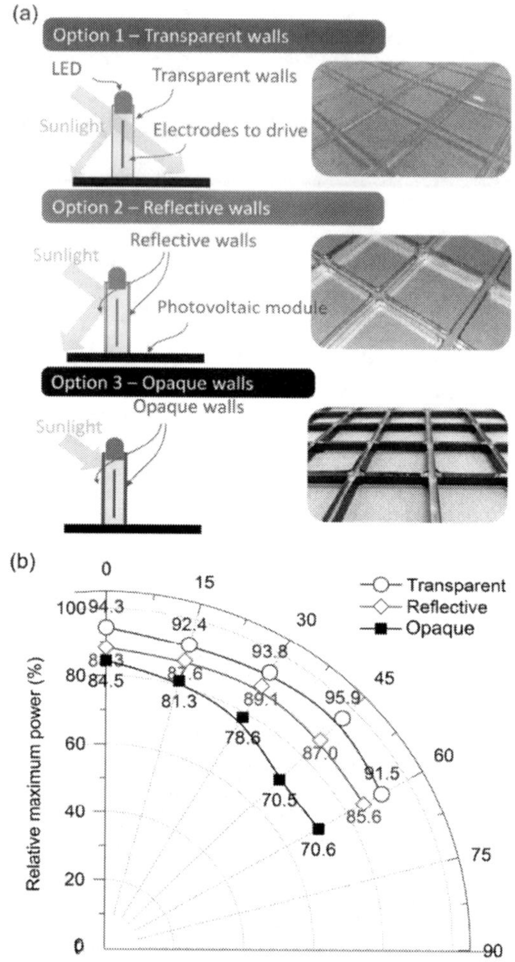

Figure 3: (a) Different surface conditions of the grid walls and (b) impact of the grid wall surface condition of an LED module on the photoconversion efficiency of a photovoltaic module.

Figure 4 shows an LED module on a photovoltaic module where 9 LEDs were all turned on. Its brightness was measured using a spectroradiometer (CS-1000, Konica Minolta, Inc., Tokyo, Japan). At the distance of 1 m, the brightness of an LED module was too high to be measurable at the driving voltage of 6 V. Therefore, the driving voltage was reduced down to 3 V and the measured brightness of an LED module was 9038 cd/m^2. At the distance of 7 m, the measured brightness of an LED module with the drving voltage of 6 V was 3286 cd/m^2.

4 CONCLUSIONS

In summary, a grid-type LED display is presented here, which not only endows functional and aesthetic values on a BIPV but also minimizes the shading loss of a BIPV. Considering that photovoltaic modules with commercially available transparent film-type LED modules exhibited a shading loss of -22.59%, the shading loss with a grid-type LED display presented in this study might be regarded as a record low value to the author's best knowledge.

The unique trait of a grid-type LED module is that its shading loss could be minimized with the reflective grid walls because the reflective grid walls redirect sunlight to a photovoltaic module. The measured shading loss of an LED module with the reflective grid walls was at most -14.4% even at a steep solar incident angle of 60°. Compared to the shading loss of -29.4% with the opaque grid walls at the same solar incident angle, the developed grid-type LED module with the reflective grid walls has a considerably smaller shading loss. Consequently, the grid-type LED module would be widely applied in zero energy buildings by increasing the public acceptance of BIPVs.

5 ACKNOWLEDGEMENTS

This work was supported by the Basic Science Research Program through the National Research Foundation of Korea (NRF), which is funded by the Ministry of Education, Science, and Technology (Grant No. 2021R1I1A2059733).

6 REFERENCES

[1] P. Heinstein, C. Ballif, and L. E. Perret-Aebi, Building Integrated Photovoltaics (BIPV): Review, Potentials, Barriers and Myths. Green, 2013;3(2):125-156.

[2] A. J. Marszal, P. Heiselberg, J. S. Bourrelle, E. Musall, K. Voss, I. Sartori, and A. Napolitano, Zero Energy Building – A review of definitions and calculation methodologies. Energy and buildings, 2011;43(4):971-979.

[3] A. Menéndez, A. Martínez, A. Santos, B. Ruiz, K. Moritz, I. Klein, J. Díaz, A. R. Lagunas, T. Sauermann, and D. Gómez, A multifunctional ETFE module for sustainable façade lighting: Design, manufacturing and monitoring, Energy and Buildings, 2018;161:10-21.

[4] H. Zhang, Reliability and lifetime prediction of LED drivers, 14th China International Forum on Solid State Lighting: International Forum on Wide Bandgap Semiconductors China (SSLChina: IFWS), Beijing, China, 2017.

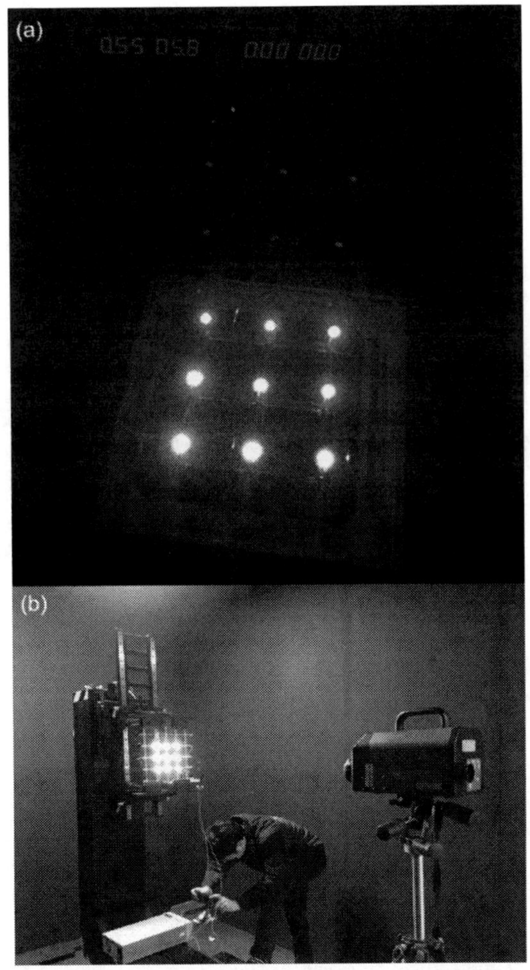

Figure 4: (a) LED module on a photovoltaic module where 9 LEDs were all turned on and (b) brightness measurement set-up of an LED module using a spectroradiometer.

NEW ARTIFICIAL INTELLIGENCE METHODS FOR PV ANNUAL DEGRADATION RATE ESTIMATION

Benoît Braisaz[1], Anton Sanaa[1], Oume Lgheit Rhazi[1], Marine Bila[2], Ibrahima Tounkara[3], Axel Becker[3]
[1] EDF R&D, TREE – E38 - PV Team, avenue des Renardières 77250 Moret sur Loing-Orvanne, France
[2] EDF Renewables North America, 15445 Innovation Dr, San Diego, CA 92128, United-States
[3] EDF Renewables, 100 esplanade du Général de Gaulle, 92932 Paris La Défense Cedex, France
Phone : +33 (0)1 60 73 62 69, e-mail : benoit.braisaz@edf.fr

ABSTRACT: Artificial Intelligence methods are overwhelming every R&D area where data are used, including PV module performances. Here we propose to test new Artificial Intelligence (AI) methods in order to estimate the degradation rate of PV Modules. Among these AI methods, we implemented Long Short Term Memory (LSTM) neural network. In the learning step, known degradation rates were simulated in a **digital power plant** and by using meteorological data set of an EDF Renewables outdoor test facility (almost 4 years of data, PV module production & meteorological conditions).
Then, these generated data were used to teach AI systems. In the test step, the AI is expected to assess ou find the degradation rate by analyzing a 3 years-long dataset. In addition, we developed a method for estimating electrical parameters (Isc, nds, Rs) and another method for saving the state of plant in order to replay it later with different meteorological data. The results are promising with the last method and show the performance and efficiency of AI when estimating degradation rate.
These methods have many interests, but one of the most important is to combine PV model and the exact field meteorological dataset to simulate various degradation amplitudes and shapes in order to feed an AI. By this way and knowing the meteorological conditions, AI can learn and propose the expected PV behavior (every shape is normalized) and interpolate the degradation rate corresponding to the field values of PV module production.
A final comparison is done with data from EDF R's test zone in US by comparing the estimated AI degradation rate and the STC (Standard Test Conditions) flashes of modules carried out recently.
Keywords: Photovoltaic, Annual Degradation, Energy Performance, Modelling, Artificial Intelligence

1 CONTEXT

Artificial Intelligence proved many times its efficiency to estimate and predict parameters in various area. In the photovoltaic area, regression methods are used in ([1],[2]) for estimating PV Performance under outdoor operating conditions. Long short-term memory network (LSTM [3]) have been efficiently applied in [4] for photovoltaic array fault diagnosis in decentralized MPPT systems. Deep IV curve analysis have been also successfully applied in [5] and [6] for PV diagnosis and faults identification. In order to detect faults in real-time on massive number of PV operational plants, machine learning methods (k-nearest-neighbor techniques on plant with not enough historical data for the regression) were used in [7]. Real time monitoring method in [8] associates low cost sensors and Neural Networks for PV monitoring. Aerial Infrared Thermography images can feed Convolutional Neural Networks (CNNs) algorithms for identifying disconnected substrings, hot spots, and disconnected strings by data image analysis ([9]).
Deep convolutional generative adversarial network (DA-DCGAN) has been implemented in [10] for DC series arc fault diagnosis.

To sum up, many AI use cases are continuously emerging and testing these AI methods remains an active R&D area in photovoltaic.

Like all the conventional methods described in [11] or [12], Artificial Intelligence has often to face with sensor uncertainties for estimating instant PV values from instant input data. A strong learning phasis gives usually better results. If the output target is a global parameter (rather than an instant one like PR or power) the results can be more robust. Indeed, degradation rate and electrical parameters vary slowly during the learning phasis. Consequently, if the input dataset is large enough, AI methods can extract accurate information of a global parameter even in the case of noisy data.

The proposed use case here is to estimate the degradation rate of PV plants by using AI algorithm, meteorological conditions and measured output power over 3 years. AI algorithms are able to calculate the expected power depending on the weather from a large range of data.

Three methods and its results will be described in the next chapters:

- Direct Degradation Rate Estimation by application of LSTM algorithm [3] from Pytorch Python libraries
- Degradation Rate Estimation by electrical parameter estimation and application of Catboost algorithm [13]
- Degradation Rate Estimation by saving neural network state, which consists of succession of LSTM, CNN, and Dense neural networks under Keras Tensorflow libraries

2 AVAILABLE DATA AND HYPOTHESIS

In its Test Zone in California (Figure 1), EDF Renewables monitors several small PV Systems and weather stations. The available data period is 2017-2021 (less than 4 years). 3 full years dataset have been used and consists of two parts : meteorological dataset (POA Irradiance, Temperature, etc.) and electrical dataset (Pdc, Pac, current and voltage etc.). The initial time step of data acquisition is 10 seconds (one minute averaged) and each small PVSystem is made of a 10 modules string.

Figure 1 : Photo of the test zone in California

Among these small PV Systems, four systems have been tested which will be named as : INV6A, INV8A, INV6B and INV3D. Some PV modules of theses 4 systems were flashed by CFV Labs and we observed degradations (see Table I). The last row is the annual degradation rate and gives an indication of what degradation rate the algorithms should guess.

Table I: Comparison 2021 Flashes versus 2016 Flashes

	INV6A	INV8A	INV6B	INV3D (same as INV6B)
Comparison between 2021 versus 2016 flashes (%)	0,54	-5,92	-6,39	-6,39
Annual degradation rate (%/year)	**0,11**	**-1,18**	**-1,28**	**-1,28**

The meteorological dataset (Figure 2) will be used for electrical data generation (Voltage and current) via Digital Power Plant for learning and test phasis, and the electrical dataset for validation phasis of the AI algorithms.

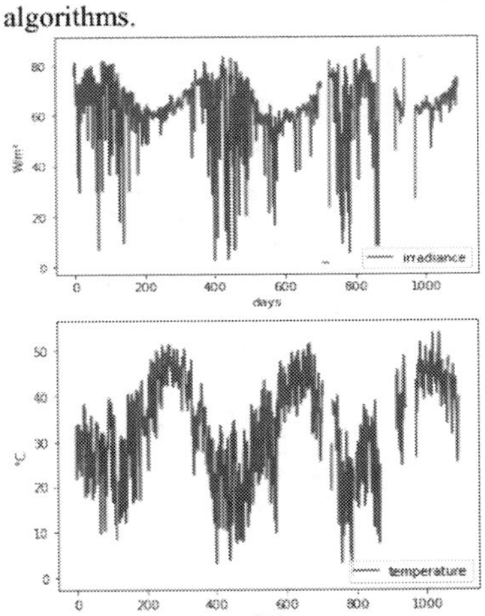

Figure 2 : Example of temperature and irradiance with data gaps and noise

The data is filtered :

- Every power is normalized between -1 and 1 (with a min max scaler)
- Time chosen between 10:00-16:00
- Hourly average
- 80% of the dataset is used for learning

The developed code uses Python and its Pytorch & Keras Libraries.

3 METHODS

3.1 Digital Power Plant PVSYSTEM for data generation

In order to generate the data in the learning phasis, a Python code has been developed which uses an advanced one-diode model, also used in the EDF PV modelling tool PVNOV® and similar to .PAN file structure (PVSYST).

Several equivalent electric models exist for representing a PN-junction. The current default one used in the tool is a single-diode model composed of a current source, an anti-parallel diode, a series resistance and a shunt resistance, with corrective terms for the dependency of the model to low irradiance and temperature. The single-diode can be represented by the electric scheme Figure 3.

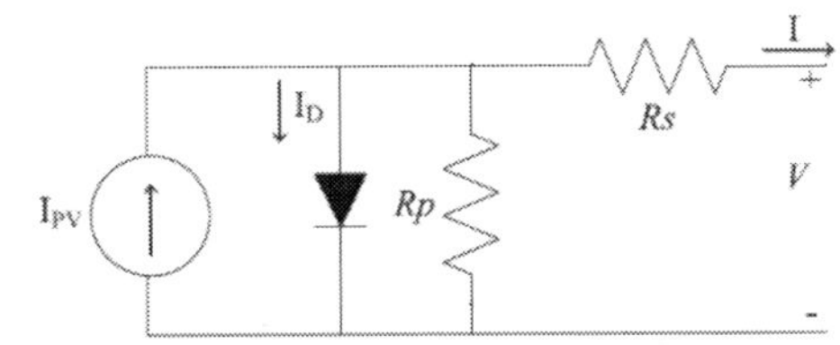

Figure 3 : One-diode equivalent scheme

The current is calculated by the well-known one-diode Shockley equations :

$$I = I_{Ph} - I_0 \cdot [e^{\frac{V+I \cdot R_S}{n \cdot V_T}} - 1] - \frac{V + I \cdot R_s}{R_{sh}}$$

and the saturation current I0 estimated by the following equation :

$$I_0 = C_s\,(T)^{m_s(T)} \cdot S_{cell} \cdot \exp\left[\frac{qE_g}{n_{Ds}kT}\right]$$

Temperature correction :

A dependency with the temperature of the exponent m_s in the saturation current is added (coming from an improved 2-diodes model published in [14], [15]) :

$$m_s(T) = \alpha_{m_s}(T - T_{STC}) + \beta_{m_s}$$

And considering a dependency with the temperature of the series resistance ([16]):

$$Rs = Rs_{STC}(1 + \mu_{Rs}(T - T_{STC}))$$

Irradiance correction :

In low light conditions, relative efficiency of PV modules is affected. The behavior given by the default parameters may give better or worse relative efficiencies than the current ones – particularly when irradiance level is under 400 W/m². Consequently, the following corrections were implemented:

Exponential dependence of Rshunt to irradiance according to [17]:

$$R_{sh} = R_{sh,STC} + [R_{sh,0} - R_{sh,STC}] \cdot \exp\left(-k \cdot \frac{E}{1000}\right)$$

Where RshSTC is the STC shunt resistance, E is the irradiance received by the cell (W.m⁻²), Rsh0 and k are two coefficients to adjust.

Linear and quadratic dependence of the saturation coefficient with irradiance, according to [18]:

$$C_s = (a_{cs2}\ (1000 - E) + a_{cs}\ )\,(1000 - E) + C_{s,STC})$$

Where Cs_STC is the saturation coefficient at STC and a_{cs} and a_{cs2} are coefficients. All these coefficients are to be adjusted.

Nomenclature :

- Iph : the photocurrent (A)
- α is the coefficient temperature of the short-circuit current I_{sc} (about 0.04 %/°C)
- E is the irradiance received by the cell ($W.m^{-2}$)
- T is the cell temperature (in K)
- Rs is the serie resistance (in Ohm m^2)
- Rsh is the shunt resistance (in Ohm m^2)
- C_s is the saturation coefficient
- E_g is the energy gap (silicium, 1.12 eV)
- n is the ideality factor of the diode
- kb is the Boltzmann's constant (1.3854 x 10^{-23} $J.K^{-1}$)
- q is the elementary charge (1.6021 x 10^{-19} C)

3.2 Method 1 : Direct Degradation Rate Estimation

The principle of the first method is to estimate the degradation rate by using as inputs the measured DC power at the entrance of the inverter.

The diagram in Figure 4 summarizes the method: the first step is the data generation. Dataset is generated by running Digital Power Plant (DPP PVSYSTEM) with the meteorological dataset of the analyzed plant (in our case 3 years). Many simulations are run with a linear degradation rate (%/year) between -5%/year and +5%/year.

The used data is a daily average of filtered data between 10:00 and 16:00.

In a second step, this dataset is used for teaching an AI in order to estimate the degradation rate. The step 3 is the data validation on measured inverter data.

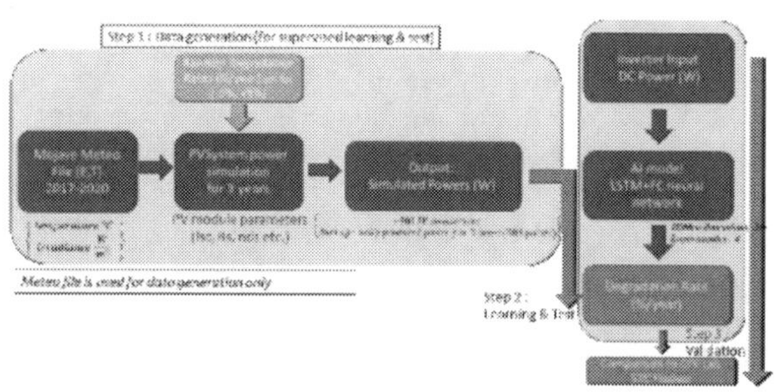

Figure 4 : Method 1, Direct Degradation Rate Estimation

The AI part is made of a succession of LSTM (Long short-term memory [3]) and FC (Fully Connected) networks implemented with Pytorch.

The input vector size is the produced power (3 years): 365 days*3 points. The output is the degradation rate (%/year).

Many simulations (~1h simulation time) were necessary to find the most appropriate hyperparameters for the neural network LSTM structure. This structure consists of a LSTM and a fully connected neural network.

- Structure with the best results :
  - Test duration : 20% (80% used for learning)
  - Epoch number : 3000
  - Hidden dimensions : 20
  - Layer number : 4
  - Dataset : 365*3 (days)
  - Learning rate : 0,0001

3.3 Method 2 : Degradation Rate Estimation by electrical parameter estimation

The second method goes beyond the first one : electrical parameters are first estimated on a monthly basis and then their evolution is used for degradation rate estimation.

The diagram Figure 5 exhibits the method :

- Step 1 : Data generation
  - Many temporal current and voltage datasets are generated with random Isc, nds and Rs and the meteorological dataset in the test zone
- Step 2 : Supervised learning for parameter estimations
  - A CatBoost algorithm learns to guess Isc, nds, and Rs from hourly data on one month (7*30 days = 210 points)
- Step 3 : parameter estimations from data inverter
  - This algorithm is run on DC inverter data on Month 1 and then on Month 36
- Step 4 : Degradation rate
  - Degradation rate is calculated by comparing Pmpp at 1000 W/m² and 25°C from Month 1 and 36 parameters.

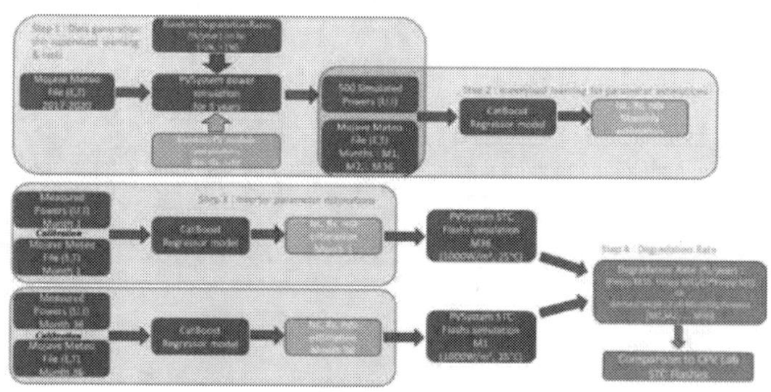

Figure 5 : Method 2, Degradation Rate Estimation by electrical parameter regressive estimation

3.4 Method 3 : Degradation Rate Estimation by saving neural network state

The last method saves the state of a plant at the first months and replays it several years later with different meteorological conditions. Comparison between these results and the measured powers should give the degradation.

The diagram Figure 6 describes the method :

- Step 1 : Data generation for the learning phasis
  - Many temporal power data are generated with different degradation rates
- Step 2 : Supervised learning for power estimation :
  - A LSTM and CNN neural network learns

to estimate powers from Month 1 data (7*30 days = 210 points) of a given PV system

- Step 3 : Month 1 network save
- Step 4 : Simulation of Month 1 network with Month 36 meteorological dataset

- The Month 1 network is saved and then is rerun on Month 36 weather data of the same given system

Degradation rate is then calculated either with the deviation between measured Month 36 powers and Month 1 network powers applied on Month 36 meteorological data or by comparing powers at 1000W/m² and 25° C of month 1 and 36 networks.

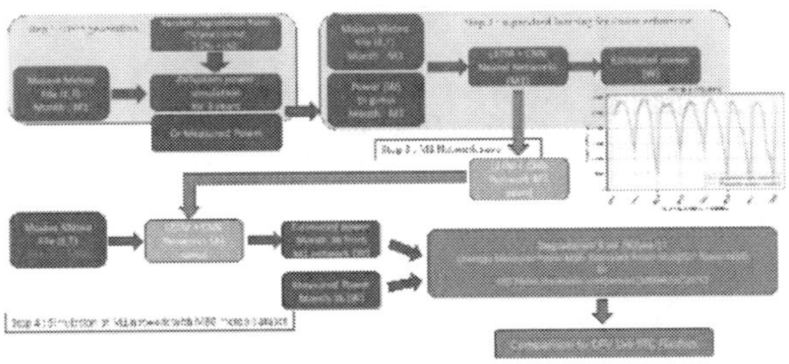

Figure 6 : Method 3 : Degradation Rate Estimation by saving neural network state

4 RESULTS ON THE TEST PHASIS

4.1 Method 1 : Direct Degradation Rate Estimation

The test phasis uses generated data with known degradation rate in order to evaluate if the algorithm is able to estimate degradation rate or not. In Figure 7, we concluded that the algorithm can efficiency estimate degradation rate from simulated data.

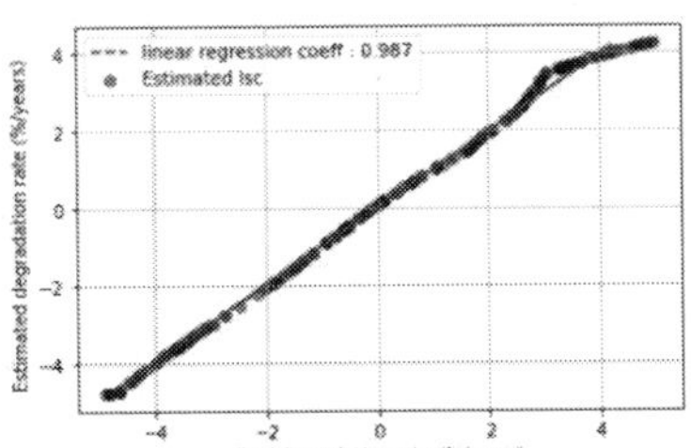

Figure 7 : Estimated degradation rate versus simulated one

4.2 Method 2 : Degradation Rate Estimation by electrical parameter estimation

The goal of the test phasis on the method 2 is to estimate electrical parameters (Isc, nds and Rs). Figure 8 shows the estimated parameter (Isc, nds and Rs) function of the true values inserted in PVSystem. The Catboost algorithm can estimate electrical parameters with no noise, Rs is harder to estimate but its impact on the output power is lower than Isc.

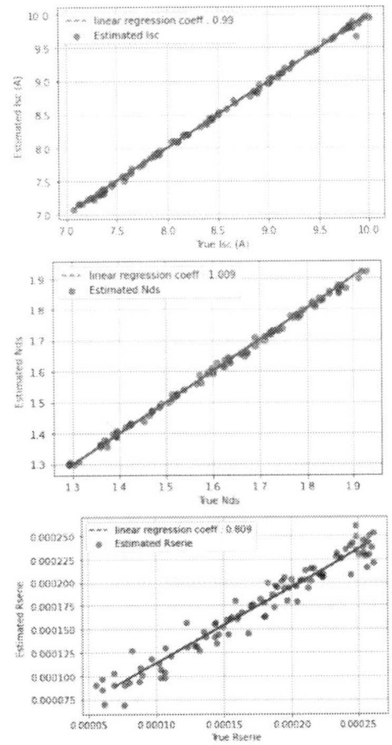

Figure 8 : (Isc, nds and Rs) estimated and true values

4.3 Method 3 : Degradation Rate Estimation by saving neural network state

After running the Month 1 network with Month 12 meteorological data, the deviation between PVsystem powers at Month 12 and predictions at Month 12 with Month 1 network (Figure 9) should be the degradation over 11 months.

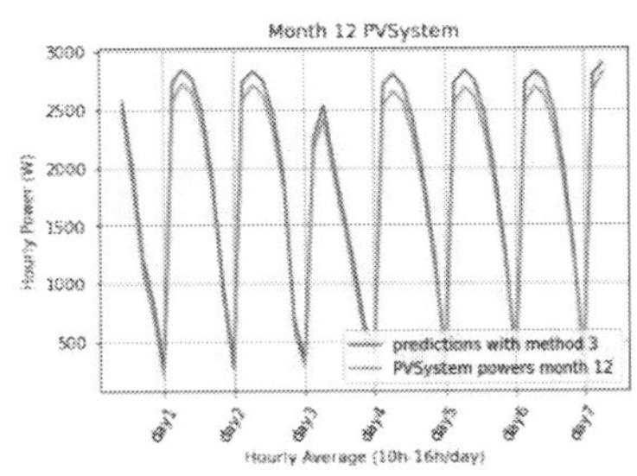

Figure 9 : Simulated power on Months12

The algorithm can give estimations of the degradation rate which are summarized in Figure 10.

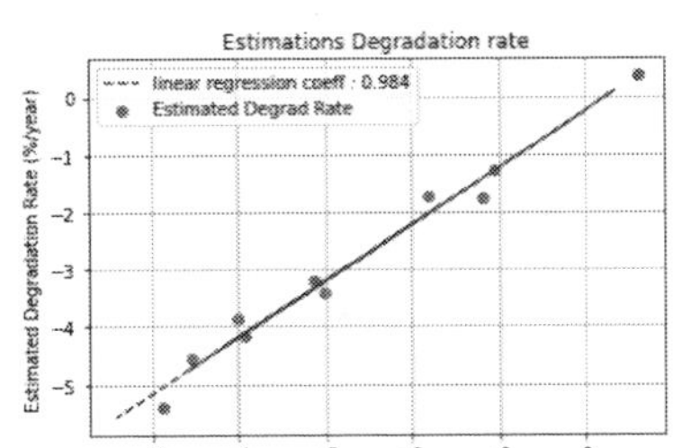

Figure 10 : Estimation degradation rate

5 RESULTS ON THE VALIDATION PHASIS

5.1 Method 1 : Direct Degradation Rate Estimation

In the validation phasis, we compare the estimated degradation rate obtained from the method 1 and the CFV flashes of few PV modules.

Unfortunately, results were not as those expected, we have observed a strong deviation between degradation rate with flash and the one from Method 1 (Figure 11). There are several reasons behind this noticeable deviation. Firstly, our model learnt data without any noise contrary to inverter inputs. Consequently, each similar weather inputs give the exact same output. AI can theoretically learn from very few points the simulated learning data set and give the exact degradation rate. But when it comes to inverter data, even the same meteorological conditions can give different current and voltage response due to uncertainties on sensors, environmental conditions etc. Hence, AI can strongly be mistaken on low degradation rate in case of noisy data.

In addition, installed modules are supposed to be unknown (We don't know all the exact module parameters of the analysed PV Plant). Consequently, modules used during the learning phasis were not the same as those on the inverters. Electrical parameters are not the same and unknown. If PV modules behave differently at low irradiances or higher temperature, it will trigger errors in the algorithm. That's why, we developed methods which estimate electrical parameters.

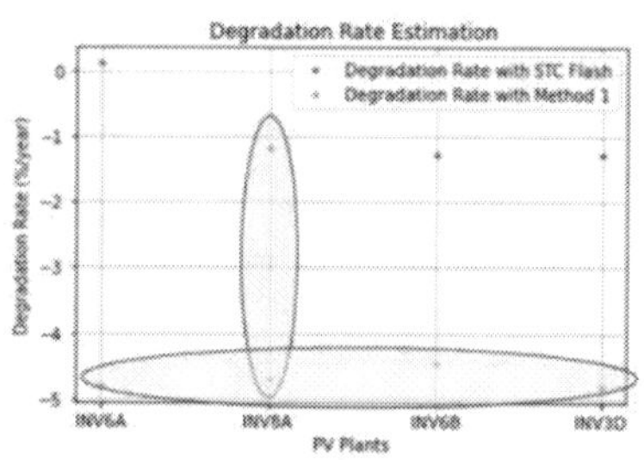

Figure 11 : Results on Method 1 : comparison between flashes and Method 1

5.2 Method 2 : Degradation Rate Estimation by electrical parameter estimation

In this method, as explained before, the first step is to estimate the electrical parameters (Isc, Nds, Rs) by using the data of the first month and then the last month of the analyzed period. It is possible to calculate 2 powers with these two parameters set (first month and last month) and then estimate the degradation loss.

On the Figure 12, one can see the deviation between flashes and method 2 results. The curve tendency is caught and the magnitude order is the same but many reasons can explain the constant deviation :

- Uncertainties of measurements: outdoor and indoor
- Flash done only at STC conditions contrary to the outdoor data used
- Validity of one diode model
- Further electrical parameters are not taken into account (cs, shunt resistance etc.)

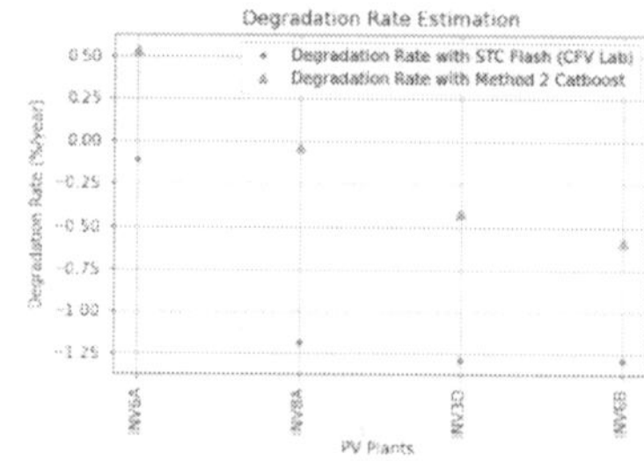

Figure 12 : Results on Method 2 : comparison between flashs and Method 2

5.3 Method 3 : Degradation Rate Estimation by saving neural network state

The objective of the neural network method 3 is to save the state of a plant at a specific month by learning output power and meteorological conditions. Then, the network is run on new weather after several years and estimated power is compared to actual power.

The neural network simulates the expected power at M3 from M1 data (Figure 13). The first result shows the ability of the network to efficiently estimate the PV production of the next months after a learning phasis.

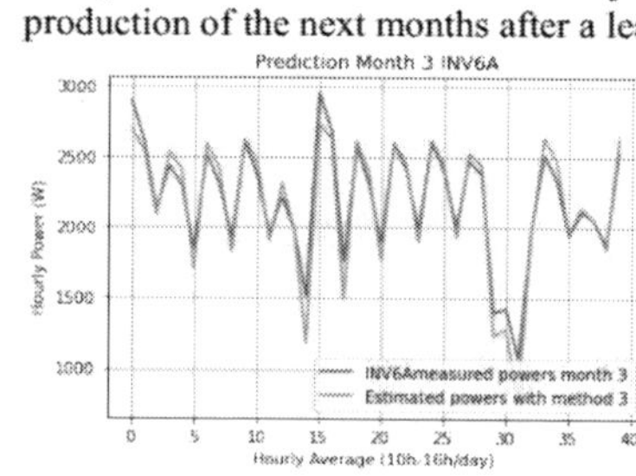

Figure 13 : Prediction of Month 3 INV6A production from Month 1 data learning

The deviation between degradation rate calculated from STC Flash (CFV lab) and Method 3 is under 0.5% (Figure 14). The method 3 seems interesting in order to estimate degradation rate versus STC flashes. It is very important to keep in mind that even if the deviation remains high, the CFV flashes estimate (with uncertainties) only the module degradation and not the degradation of the whole system (mismatch, inverter, MPPT, soiling etc.). The degradation rate overestimation can be explainable by this way.

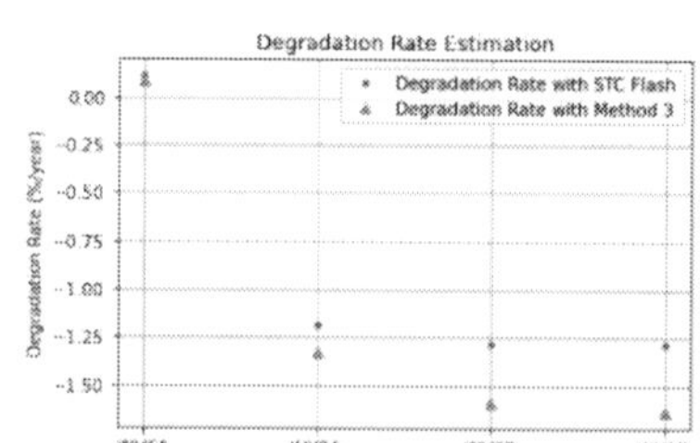

Figure 14 : Results on Method 3 : comparison between flashs and Method 3

6 DISCUSSIONS AND PERSPECTIVES

Three AI innovative methods were tested in this study for degradation rate estimation. Each of them has pros and cons and could be strongly improved thanks to many other data and tunning hyperparameters.

The use of neural networks for degradation rate estimation and its comparison with field data (more than 3 years) and STC flashes data promotes the method 3 demonstration. Results are promising even if comparison to flashes may not be the most appropriate as a PV System production degrades not only at the clean PV module level. That's probably why, Method 3 overestimated module degradation. Once the learning phase is done, the algorithm is very fast. We don't need any PV model for running the method, only weather and inverter data are necessary.

On the other hand, by learning on many simulated datasets (with different amplitude and degradation trends) in the same environment as the observed PV System, AI methods could have a strong advantage versus conventional methods. AI uses all the available information for degradation rate estimation : data (sensors & powers) and models.

Theoretically less than two years could be sufficient for degradation rate estimation because AI learn the normal shape of day (knowing the weather of the day, Day N is a sunny day, day N+1 is cloudy…).

The next steps are clearly to confirm these first results on other outdoor data and to make more hyperparameter improvements.

7 REFERENCES

[1] A. K. Tossa *et al.*, « Artificial intelligence technique for estimating PV modules performance ratio under outdoor operating conditions », *J. Renew. Sustain. Energy*, vol. 10, n° 5, p. 053505, sept. 2018, doi: 10.1063/1.5042217.

[2] A. Saberian, H. Hizam, M. A. M. Radzi, M. Z. A. Ab Kadir, et M. Mirzaei, « Modelling and Prediction of Photovoltaic Power Output Using Artificial Neural Networks », *Int. J. Photoenergy*, vol. 2014, p. 1-10, 2014, doi: 10.1155/2014/469701.

[3] S. Hochreiter et J. Schmidhuber, « Long Short-Term Memory », *Neural Comput.*, vol. 9, n° 8, p. 1735-1780, nov. 1997, doi: 10.1162/neco.1997.9.8.1735.

[4] A. Y. Appiah, X. Zhang, B. B. K. Ayawli, et F. Kyeremeh, « Long Short-Term Memory Networks Based Automatic Feature Extraction for Photovoltaic Array Fault Diagnosis », *IEEE Access*, vol. 7, p. 30089-30101, 2019, doi: 10.1109/ACCESS.2019.2902949.

[5] R. A. Guejia Burbano, G. Petrone, et P. Manganiello, « Early Detection of Photovoltaic Panel Degradation through Artificial Neural Network », *Appl. Sci.*, vol. 11, n° 19, p. 8943, sept. 2021, doi: 10.3390/app11198943.

[6] M. Laurino, M. Piliougine, et G. Spagnuolo, « Artificial neural network based photovoltaic module diagnosis by current–voltage curve classification », *Sol. Energy*, vol. 236, p. 383-392, avr. 2022, doi: 10.1016/j.solener.2022.02.039.

[7] C.-C. Hsu, J.-L. Li, et Y.-S. Chen, « Real-time fault detection in massive multiarray PV plnats based on machine learning techniques », p. 4.

[8] S. Samara et E. Natsheh, « Intelligent Real-Time Photovoltaic Panel Monitoring System Using Artificial Neural Networks », *IEEE Access*, vol. 7, p. 50287-50299, 2019, doi: 10.1109/ACCESS.2019.2911250.

[9] Oliveira & Al., « Automatic fault detection of photovoltaic array by convolutional neural networks during aerial infrared thermography.pdf ». PVSEC 2019.

[10] S. Lu, T. Sirojan, B. T. Phung, D. Zhang, et E. Ambikairajah, « DA-DCGAN: An Effective Methodology for DC Series Arc Fault Diagnosis in Photovoltaic Systems », *IEEE Access*, vol. 7, p. 45831-45840, 2019, doi: 10.1109/ACCESS.2019.2909267.

[11] A. Phinikarides, N. Kindyni, G. Makrides, et G. E. Georghiou, « Review of photovoltaic degradation rate methodologies », *Renew. Sustain. Energy Rev.*, vol. 40, p. 143-152, déc. 2014, doi: 10.1016/j.rser.2014.07.155.

[12] R. H. French, L. S. Bruckman, D. Moser Lindig, EURAC, van Iseghem, Mike, ISI, Fraunhofer, Stein, Joshua S., et M. Richter Herz, Magnus, van Sark, Wilfried, Baumgartner, Franz, Integration of Photovoltaic Solar Energy, Energy and Resources, *Assessment of Performance Loss Rate of PV Power Systems*. IEA-PVPS, 2021.

[13] L. Prokhorenkova, G. Gusev, A. Vorobev, A. V. Dorogush, et A. Gulin, « CatBoost: unbiased boosting with categorical features », p. 11.

[14] B. Braisaz, P. Dupeyrat, A. Lindsay, et K. Radouane, « An advanced model of PV power plants based on Modelica ® », *PVSEC 2013*.

[15] P. Dupeyrat, B. Braisaz, M. Schumann, A. Lindsay, J. Herth, et D. Boublil, « Improved Method for the Extraction of PV Module Parameters », *PVSEC 2013*, p. 5.

[16] K. Lee, « Improving the PV Module Single-Diode Model Accuracy with Temperature Dependence of the Series Resistance », 2017.

[17] Mermoud, « Performance assessment of a simulation model for PV modules of any available technology ». PVSEC 2010.

[18] S. Elies, « Improved Two-Diode Model for More Detailed Simulation of I-U-Curves for Solar Cells and Modules », *PVSEC 2009*.

PRECISE ON-SITE POWER ANALYSIS OF PHOTOVOLTAIC ARRAYS BY SELF-REFERENCE ALGORITHM

Martin Scheler[1], Darwin Daume[1], Dharmik Sojitra[1], Tina Neumeyer[1],
Simon Steinbach[1], Tamara Beck[1], Achim Schulze[2], and Bernd Hüttl[1*]
[1]Coburg University of Applied Sciences, Dept. of Electrical Engineering and Computer Sciences, Coburg, Germany
[2]Rosenheim Technical University of Applied Sciences, Dept. of Appl. Natural and Social Sciences, Rosenheim, Germany
[*]E-mail: bernd.huettl@hs-coburg.de

ABSTRACT: To detect degradation of photovoltaic systems at an early stage, precise performance determinations are essential. Current-voltage measurements on single modules in indoor labs under well-defined conditions are state-of-the-art. However, this method causes a logistical and economic effort. On-site measurements reduce the effort but are prone to error because of poorly determinable test conditions: The detected temperature on the backside of modules as well as the irradiance detected by pyranometer in tilted module plane usually differs from the actual operating conditions. In consequence, the accuracy of measured current-voltage characteristics is poor. A precise on-site power determination is achieved by a self-reference algorithm. Improvements by self-referencing are not achieved by correcting the raw electrical data, but by replacing the measured test conditions (temperature and irradiance) with effective data. These effective values correspond better to the actual module irradiance and temperature values. Furthermore, precise power determination of photovoltaic arrays requires to identify and exclude temporarily deformed current-voltage curves from consideration, as it can arise from non-uniform irradiation within array like shading. In this paper, the application of a digital curve filter is reported and performance determinations by a self-reference algorithm are demonstrated on a photovoltaic array. We review and discuss the measurement concept in terms of its ease of use.

Keywords: Environmental Effect, Evaluation, Experimental Methods, Performance, PV Array

1 INTRODUCTION

On-site current-voltage (I-V) measurements of photovoltaic arrays are a state-of-the-art method for determining the performance of photovoltaic (PV) systems. Such analyzes are performed in accordance with the IEC 61829 standard [1]. They require defined conditions of irradiance and module temperature to allow the determination of power at standard test conditions (STC). Analyzes at STC can be performed more precisely – only on single modules – in standardized indoor labs according to the IEC 60904-1 standard [2].

However, I-V outdoor analyzes of PV arrays offer some advantages in comparison to indoor characterizations: Outdoor analyzes do not need time and logistics for module transport to an indoor laboratory. The electrical performance is monitored under natural operating conditions and effects such as light soaking or light-induced degradation are considered [3]. In addition, I-V measurements can be performed slowly (within 2 to 10 seconds) and transient errors [4] can be avoided.

On the other hand, on-site electrical measurements are prone to error because of poorly determinable test conditions. The detected temperature on the backside of modules (T_{mod}) usually differs from operating temperature of module cells. Also, there are impacts of wind cooling, sun heating and thermal capacities resulting in temperature deviations of temperature sensors and pn-junctions. Furthermore, the irradiance detected by pyranometer in tilted module plane (G_{mod}) is mostly different to that reaching the module's cells. One reason is the different angular dependence of modules which distinguishes from cosine function of pyranometers. Moreover, the aperture may be different for modules and sensors. And at last, the spectra of natural sunlight may differ from the air mass global standard spectrum AM1.5g [5] in the PV-relevant spectral range due to variable water, ozone, or aerosol concentrations.

In consequence, the accuracy of measured I-V characteristics is poor, as shown for example for measurement campaigns on a single PV module: Fig. 1a presents noisy curves of normalized power at maximum power point (P_{mpp}) versus measured irradiance (G_{mod}) for three temperatures (T_{mod}). The P_{mpp} is normalized to the nominal power at STC.

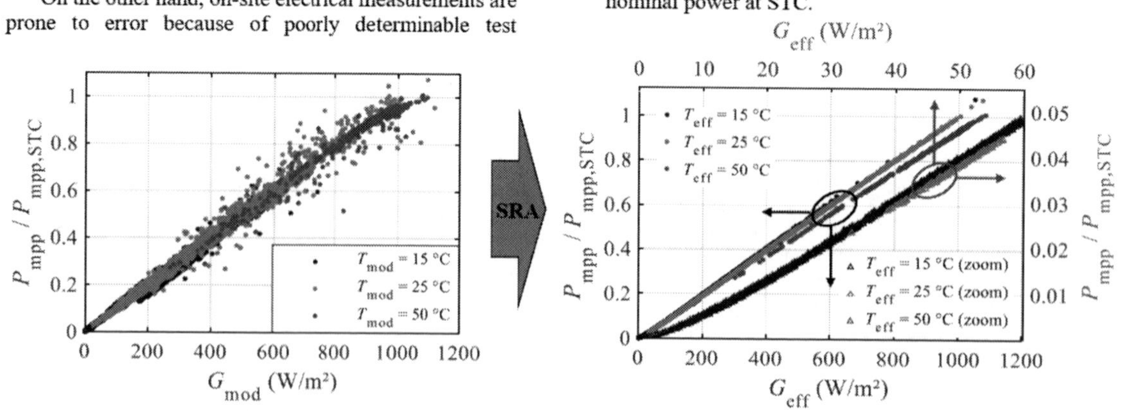

Figure 1: Normalized P_{mpp} of a single PV module vs. irradiance. **a.** Measured irradiance (G_{mod}) for three backside temperatures (T_{mod}: 25 °C, 50 °C and 75 °C in intervals of $\Delta T_{mod} = \pm 2$ °C). **b.** Effective irradiance (G_{eff}) for three effective temperatures (T_{eff}: 25 °C, 50 °C and 75 °C) and processing interval widths of $\Delta G_{mod} = \pm 10$ W/m² and $\Delta T_{mod} = \pm 2$ °C.

Uncertainties in outdoor measurements are reduced by applying the self-reference algorithm (SRA), as shown in fig. 1b. Improvements by SRA are not achieved by correcting the raw electrical data, but by replacing the measured test conditions (G_{mod} and T_{mod}) with effective data (G_{eff} and T_{eff}). These effective values correspond better to the actual module irradiance and temperature than the measured values. The effective data are determined by self-referencing to the measured short-circuit current (I_{sc}) or open-circuit voltage (V_{oc}).

According to the first axiom of SRA, V_{oc} data are used to determine T_{eff} by means of the linear relationship between temperature and open-circuit voltage [6, 7]. It should be noted that linearity is valid only for a certain irradiance G_{mod} or a small interval ΔG_{mod} around G_{mod}. In conclusion, measured values V_{oc} can be converted into temperature values (T_{eff}) separately for each irradiance interval. Moreover, according to the second axiom of SRA, the effective irradiances G_{eff} can be determined by measured I_{sc} due to linearity between both data [6, 7]. If the electrical parameters like P_{mpp} are allocated to the effective data (G_{eff} and T_{eff}), smooth characteristics are obtained, as shown for example in fig. 1b. For application of the SRA and precise determination of module's performance the recording of 10,000 to 100,000 I-V curves is required.

Thus far, the SRA concept has been successfully applied to the electrical characterization of single modules in special outdoor laboratories [6, 7]. This allows accurate P_{mpp} determination of single modules for different temperatures and irradiances. In this paper, we investigate and demonstrate the applicability of SRA for precise PV array power measurements.

2 APPLICATION OF THE SRA TO PV ARRAYS

Locally distributed operating conditions, for example due to shading, can cause inhomogeneities (non-uniformities) of temperature and irradiance within arrays. On-site array measurements therefore require an alternative approach to consider the temperature and irradiance ($T_{mod,i}$ and $G_{mod,i}$) of each module individually. This can be done by measuring the operating conditions of each module with independently equipped temperature and irradiance sensors in synchronization with the I-V curve of the array.

In accordance with the first axiom of the SRA, the effective temperature T_{eff} is now calculated by using the average temperature value T_{avg} of all temperature sensors [8]:

$$T_{avg} \propto V_{OC,array} \, . \qquad \textbf{(1)}$$

The linearity between the short circuit current and irradiance, as described in the second axiom, applies only under homogeneous irradiance conditions ($G_{mod,1} = G_{mod,2} = \ldots = G_{hom}$) [8]:

$$G_{hom} \propto I_{SC,array} \, . \qquad \textbf{(2)}$$

Individual irradiance sensors for each module of the array can be used to detect such inhomogeneities in test PV fields, but in commercial applications, this approach is cost intensive. In addition, irradiance sensors are subject to failure and soiling, especially in large sensor networks. An alternative to this approach is to analyze the synchronously measured I-V curves because inhomogeneities, especially large shadings, deform the shape of the curve. These deformations could be detected visually, however machine or digital filtering is more efficient. Therefore, such digital curve filter is introduced. Moreover, further improvements of the SRA are performed to reduce the required number of I-V curves drastically, e.g., to 1,000 records.

3 EXPERIMENTAL SET-UP AND DIGITAL I-V CURVE FILTER

3.1 Experimental

For fundamental understanding, the influence of temperature and irradiance inhomogeneities on the I-V curves is investigated using a PV array with 12 modules (mc-Si modules equipped with bypass diodes). Each module of array is equipped with temperature ($T_{mod,i}$) and irradiance sensor ($G_{mod,i}$). The temperature inhomogeneity is defined as

$$\Delta T_{inhom} = \frac{\max(T_{mod,i}) - \min(T_{mod,i})}{T_{avg}} \, , \qquad \textbf{(3)}$$

where the temperature is given in Kelvin. The irradiance inhomogeneity is defined as:

$$\Delta G_{inhom} = \frac{\max(G_{mod,i}) - \min(G_{mod,i})}{G_{avg}} \, , \qquad \textbf{(4)}$$

where G_{avg} is the average irradiance of the modules. The I-V curves of the array are recorded under various temperature and irradiance conditions over a period of 36 days in summer. The data base includes 2013 measurements (records). The temperature and irradiance values were found to be in the range of 20 to 65 °C and 200 to 1200 W/m², respectively. In addition, large temperature or irradiance inhomogeneities are achieved by artificial heating, cooling or artificially shading with transparent foils, respectively, of a single module.

For digital filtering and evaluating the I-V curve shape, the measured I-V curves of the array are normalized to the short-circuit current and the open-circuit voltage. The normalized I-V curve of a single module is used as the non-deformed reference curve for the array as shown in fig. 2 and fig. 3. The digital filter classifies the measured I-V curves binary in deformed and non-deformed curves as described in more detail in the following chapter. Fig. 2 and fig. 3 represent the impact of temperature or irradiance inhomogeneities on the I-V curve shape, respectively. The records are sorted in the figures by ascending temperature and irradiance inhomogeneity. It is found that the temperature inhomogeneity in the array has no influence on the curve shape as shown in fig. 2. On the other hand, irradiance inhomogeneities can affect the curve shape significantly (particular at inhomogeneities > 10 %) as shown in fig. 3. Hence, the shape of the I-V curve provides information about the irradiance homogeneity of the PV

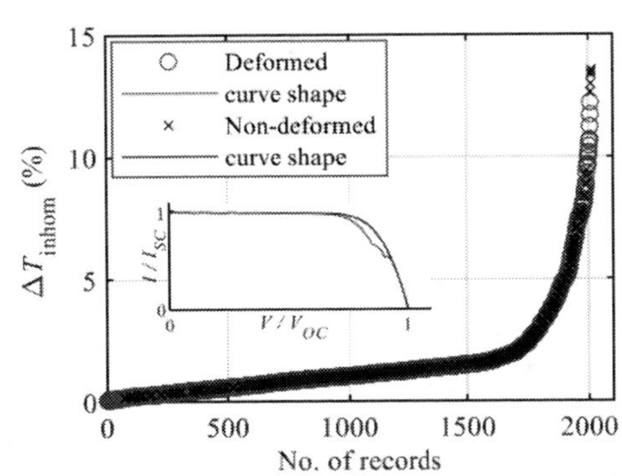

Figure 2: Influence of the temperature inhomogeneity on the I-V curve shape of the PV array.

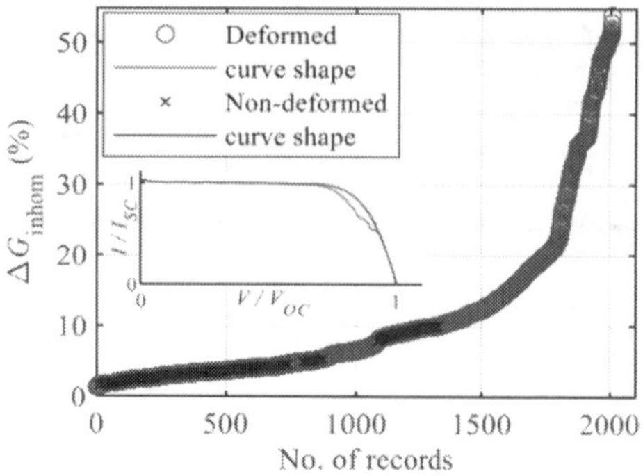

Figure 3: Influence of the irradiance inhomogeneity on the I-V curve shape of the PV array.

array. Non-deformed curves at low irradiance inhomogeneities (< 10 %) could have been caused by local shading, which is not detected by the pyranometer.

3.2 Filter algorithm

The digital filter must identify automatically only temporarily deformed I-V curves which are caused by irradiance inhomogeneities. Permanently deformed curves which may result from different types of degradation should therefore not be identified. The identified deformed I-V curves are excluded from further processing by means of the SRA.

A flow chart of the filter algorithm is shown in fig. 4 and it is described in the following. The normalized I-V curves are used for processing. The slopes (or derivatives) of the normalized curves ($\partial I/\partial V \cdot V_{OC}/I_{SC}$) at normalized I_{SC} and V_{OC} are calculated by linear regression. The interval width of the regression is as large as the I-V curve shows linear behavior. The fill factor of each curve is also calculated. Fig. 5 illustrates the determined parameters on a non-deformed and two deformed I-V curves. The regression intervals and the fill factor depend on the I-V curve shape.

In the first step, all records with too high slopes at normalized I_{SC} are excluded. These curves are assumed as "deformed". The threshold value indicating "too high" slopes is defined as followed. The slopes at normalized I_{SC} of the records are sorted in ascending order. For example, the reference slope value $S_{5\%}$ is assigned as the slope value at 5 % of the data base size (considering a data base size of 100, the 5 % value would be at position 5). The maximum tolerable slope value S_{max} (threshold value) is defined as a multiple of the reference slope value:

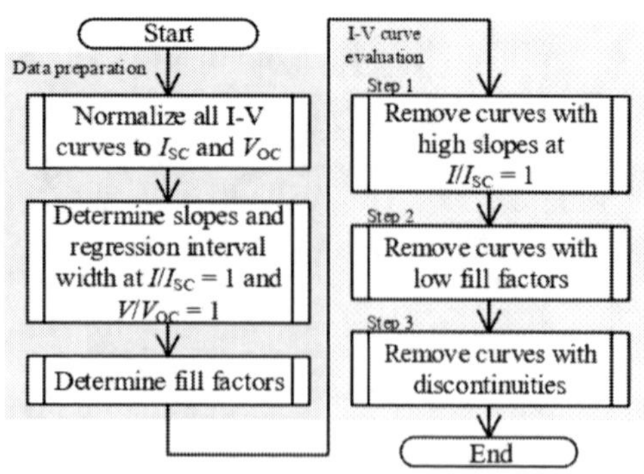

Figure 4: Flow chart of the filter algorithm.

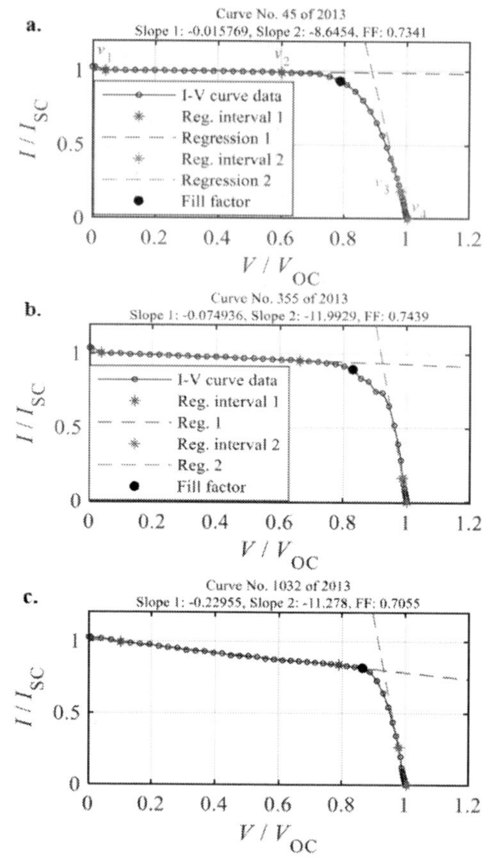

Figure 5: Determined input parameters. **a.** non-deformed I-V curve. **b.** and **c.** deformed curves.

$$S_{max} = \varepsilon_1 \cdot S_{5\%} \,. \tag{5}$$

For example, a value of $\varepsilon_1 = 2$ is suitable, however, depending on the PV array and the measured temperature and irradiance conditions, an adjustment of ε_1 might be necessary. All records with slopes higher than S_{max} are excluded. Since S_{max} is based on measured I-V curves, the digital filtering of I-V curves of differently degraded PV arrays is possible. However, not all temporarily deformed curves are recognized by step 1 of the filter algorithm, hence the steps 2 and 3 are applied according to fig. 4.

In step 2, deformed I-V curves with low fill factors are identified and sorted out. Fill factors of I-V curves depend on the temperature (T_{avg}) and irradiation (G_{hom}) too, which must be considered by filter step 2. Different fill factor thresholds are determined for some groups of comparable temperatures and irradiances.

In step 3 of the algorithm, records are identified whose I-V curves show discontinuities or steps given by irradiance inhomogeneities. The derivative of the I-V curve in the interval v_2 to v_3 (indicated in fig. 5a) is evaluated. I-V curves with significant peaks in the derivative are sorted out. The almost linear behavior of the I-V curve (intervals v_1 to v_2, and v_3 to v_4) is not evaluated.

3.3 Filter results

The filter is applied on the same data base with 2013 records as described above. The results are summarized in tab. I. The first column shows the current filter step according to fig. 4. The second column shows the number of remaining records after the filter step. The third column

Table I: Application of the digital I-V curve filter

Filter step[1]	Remaining records	Filtered records	Curve ratio (%)
Start	2013	0	33.5
Step 1	841	1172	80.1
Step 2	714	1299	94.4
Step 3	699	1314	96.4

[1]according to fig. 4

summarizes the filtered records which contain deformed I-V curves. The last column shows the ratio of non-deformed curves (in sum 699, determined by visual evaluation) to remaining records. The ratio becomes 100 %, if all remaining records contain non-deformed I-V curves. After applying the digital filter, in this example, about 96 % of the remaining records in the database contain non-deformed I-V curves.

4 EVALUATION OF THE SRA ON PV ARRAYS

The self-reference algorithm is applied to the same data base of I-V curves generated by the PV array (12 mc-Si modules) as described in section 3. Note that the SRA is applied to the filtered database using only the 699 non-deformed I-V curves (as pointed out in tab. I) recorded under almost uniform irradiation conditions. To determinate T_{eff}, we use T_{avg} and for G_{eff}, the short circuit current of $I_{SC,STC} = 9.91$ A, measured indoor under STC, is used as a calibration value. The temperature coefficient for short-circuit current α is taken from the datasheet. Moreover, the mathematical concept of SRA was improved in comparison to ref. [7] so that a data base with less than 1,000 records are sufficient for precise calculations. These improvements will be reported in an additional paper elsewhere.

For comparison, the 12 modules were electrically tested, and power rated by means of a flasher in a standard test laboratory. The measurements resulted in a maximum module power of $P_{mpp,mod} = 314.7 \pm 0.5$ W at STC. Therefore, a power of $P_{mpp} = 3776 \pm 6$ W is expected for the whole array at 25 °C and 1000 W/m². The deviation of SRA- and flasher-determined STC-power is about 0.1 %. The SRA results are presented in tab. II as a power matrix according to IEC 61853-1 standard [9] with additional temperature and irradiance values. As expected, the power increases with increasing irradiance and decreases with increasing temperature. The density of measuring points at low irradiances is lower than for high irradiances and this is even enhanced for filtered data. For high irradiances, the results can be assumed to be more accurate.

5 SUMMARY

The self-reference algorithm is applied to PV arrays.

Table II: Calculated power matrix

P_{mpp} / W	15 °C	25 °C	45 °C	50 °C	75 °C
100 W/m²	227	223	213	211	202
200 W/m²	643	635	590	573	513
400 W/m²	1475	1432	1381	1375	1240
500 W/m²	1891	1824	1722	1704	1655
600 W/m²	2308	2215	2064	2033	1922
800 W/m²	3140	2998	2747	2691	2456
1000 W/m²	3973	3781	3430	3350	2990
1100 W/m²	4389	4172	3772	3679	3257

Temperature and irradiance inhomogeneities can occur in a PV array due to its spatial distribution. Large irradiance inhomogeneities affect the characteristic I-V curve of the array and thus lead to a mismatch in calculating the maximum power point. A digital I-V curve filter has been presented to detect irradiance inhomogeneities based on the I-V curves shape. By comparing measured I-V curves to others, only temporal inhomogeneities are recognized and excluded from evaluation. The use of the filter substitutes to measure the irradiance inhomogeneity of the array and therefore only one pyranometer is necessary for plausibility and calibration purposes. The application of the SRA in combination with filtered data source allows a precise determination of the array performance under different conditions. The SRA is suitable for on-site measurements on PV arrays. Adjustments are being made to the SRA to increase the precision for small data sizes.

ACKNOWLEDGEMENT

The authors gratitude the German Federal Ministry of Education and Research (BMBF) under funding no. FKZ 13FH600IA6 for their financial support on this work.

REFERENCES

[1] IEC 61829: Photovoltaic (PV) array – On-site measurement of current-voltage characteristics, 2015.

[2] IEC 60904-1: Photovoltaic devices – Part 1: Measurement of PV current-voltage characteristics, 2020.

[3] M. Gostein and L. Dunn, "Light soaking effects on photovoltaic modules: Overview and literature review," in *35th IEEE Photovoltaic Specialists Conference (PVSC)*, 2011, pp. 3126-3131.

[4] T. Roth, D. Wichmann, K. Meyer, and M. Orlob, "In-depth analysis of transient errors of inline IV measurements," in *Energy Procedia 8*, 2011, pp. 82-87.

[5] IEC 60904-3: Photovoltaic devices – Part 3: Measurement principles for terrestrial photovoltaic (PV) solar devices with reference spectral irradiance data, 2019.

[6] H.-D. Mohring, D. Stellbogen, D. Geyer, and P. Lechner, "Comparative assessment of different thin film technologies from outdoor I-V characteristics," in *26th European Photovoltaic Solar Energy Conference (EUPVSEC)*, 2011, pp. 3625-3628

[7] B. Hüttl, L. Gottschalk, S. Schneider, D. Pflaum, and A. Schulze, "Accurate performance rating of photovoltaic modules under outdoor test conditions," in *Solar Energy*, Vol. 177, 2019, pp. 737-745.

[8] D. Daume, M. Scheler, T. Neumeyer, T. Beck, S. Steinbach, D. Sojitra, A. Schulze, and B. Hüttl, "Leistungsbestimmungen von PV-Strings unter Freifeldbedingungen," in *5. Regenerative Energietechnik Konferenz (RET.Con)*, 2022, pp. 136-143 (in German).

[9] IEC 61853-1: Photovoltaic (PV) module performance testing and energy rating – Part 1: Irradiance and temperature performance measurements and power rating, 2011.

EFFECT OF AVAILABILITY AND QUALITY OF DATA ON THE DETECTION OF DEFECTS UTILIZING ARTIFICIAL NEURAL NETWORKS IN PV SYSTEM'S MONITORING DATA

David Dassler[1,2,*], Stephanie Malik[1,2], Ralph Gottschalg[1,2,3], Matthias Ebert[1,2]

[1] Fraunhofer Institute for Microstructure of Materials and Systems IMWS, Walter-Hülse-Straße 1, 06120 Halle, Germany

[2] Fraunhofer-Center for Silicon-Photovoltaics CSP, Otto-Eißfeldt-Straße 12, 06120 Halle, Germany

[3] University of Applied Sciences, Bernburger Straße 55, 06366 Köthen, Germany

[*] Corresponding author: phone +49 345 5589-5214, e-mail David.Dassler@csp.fraunhofer.de

ABSTRACT: For analyzing high volumes of operational data from grid-connected PV systems, meanwhile, algorithms help to detect exceptional deviations from normal behavior. Machine learning methods in particular are used to overcome complex relationships within the data. One category are artificial neural networks, which are utilized to map the specific system performance as a function of environmental parameters. This paper investigates the influence of the data used for training on the detectability of yield deviations. Based on four different training scenarios the effect of missing data, the selection of irradiance sensor data, the data resolution as well as the monitoring resolution are analyzed in detail. Recommendation in respect to spatial irradiance sensor coverage and data aggregation will be given. This paper will explain the approach of defect detection, the data modification as well the impact of training quality on the detectability.

Keywords: Artificial Neural Networks, PV System, Monitoring, Machine Learning, Data Quality

1 INTRODUCTION

Monitoring systems of grid-connected photovoltaic (PV) plants record the behavior of various electrical and meteorological parameters over time, which can be used to remotely assess the functionality and efficiency of the plant. In addition, the collected data can provide important information about emerging defects or technical problems [1]. Defects and disruptions in the operation of the system can cause energy yield losses. A significant proportion of grid-connected PV systems do not deliver the expected energy yield due to technical or operational problems. Empirical studies on PV systems, e.g. [2, 3], have shown that energy yield losses are mainly related to modules or inverters that do not meet the expected warranty conditions. It is in the interest of an operator to detect these quickly and to fix them in an economically reasonable way.

To extract relevant information from monitoring's big data efficiently, a variety of solutions already exist. Among others, they utilize physical [4] or empirical models [5] to determine e.g., yield losses, string deviations or technical failures. For empirical, machine learning (ML) approaches have become very popular in the last decade. Some of the most widely used methods [6] consider artificial neural networks (ANN), which provide good results due to their universal and robust application. However, studies [7, 8] show that the current challenge in the application of ML lies in the insufficiently studied relationship between process complexity, available data quantity as well as quality and mathematical complexity.

Nevertheless, the number of publications on this topic shows that the application of ANN is promising, but an independent and general approach is not easy to determine. For this reason, this paper aims to demonstrate the limitations of using ANN in the field of defect detection. Here, defect detection means the procedure of detecting and localizing potential defects in PV systems. One common approach here is the comparison between the monitored system performance (actual behavior) and the simulated performance (predicted behavior).

It is not intended to present an additional new performance model, but to show how data and ANN training processes can be approached and applied. The work focuses on the application of ANN on grid-connected large systems. It formulates recommendations to better understand the modeling and application phase of big data from PV systems and to prepare upcoming monitoring strategies for ML implementation.

Therefore, this work investigates the dependencies of several factors of the data used on the quality of ANN training and of the detectability of deviations in large data. In addition to the careful choice of training parameters, the impact of the sensor quantity and distance from modeled modules as well as the data resolution and the data integrity.

2 METHODS

This paper continues preliminary investigations [9, 10, 11] on the applicability of ANN for defect detection in PV systems on a multi-MW sized PV plant. In the following the approach will be described.

2.1 Impact analysis

In a comprehensive study of varying training parameters, resolution and availability of sensor data, the impact on the detectability of defects was determined. To analysis this, based on a predefined ANN and under reproducible conditions only one aspect within the used data was changed. The accuracy of the trained ANN model was compared. In Table I these modeling scenarios are summarized. In the following, the motivation and approach each of these modeling scenarios will be described.

Table I: Overview on the several performed modeling scenarios

Scenario	Parameter changed
A	Distance of available irradiance sensors
B	Replacement weather with external data
C	Data resolution: 15 min, 60 min, or 1 day
D	Monitoring resolution: inverter vs. string data

Scenario A: Monitoring systems mainly aim to record the system behavior. Due to a good cost-benefit ratio during the construction phase, only important sensors are installed. Based on the example of irradiance sensors, this analysis helps to show the advantage of considerable coverage of sensors and measurement quality in the field of defect detection and the use of ANN.

Scenario B: One major performance indicator of monitoring systems is data availability. This can be affected due to missing data, signal issues or defective sensors. An approach to improve data availability is the utilization of external (e.g., local weather services) or computed data. In this scenario external data from PVGIS [12] for irradiance and temperature estimation model [13] are used to obtain replacement data. This scenario compares the detectability using original and external data.

Comparison C: Depending on the monitoring equipment and the data platform data is stored differently. Since data handling is time and cost dependent, data can be stored after certain time as aggregated mean values. Therefore, a question appears which data resolution is necessary for defect detection.

Comparison D: Defects can have a different impact on the performance of the string or inverter. Single problems such as glass breakage, hot spots or partial delamination may only affect the string in which they occurred. Wider problems such as soiling, inverter failures, or module degradation usually affect more than one string. In this context, defect detection modeling was performed on both inverter (central inverter) data and associated string data.

2.2 Data

The monitoring data of a multi-MWp sized plant in Germany with more than 10 years of operation was used. During operation, considerable problems and defects have occurred from time to time. These were investigated in detail in previous work [9, 10] and identified within the monitoring data. For instance, known defects are inverter failures, module degradation, cell and glass breakages and string failures.

The data provided from a commercial monitoring system are irradiation data, module, and ambient temperature as well as numerous inverter and string data (e.g., DC current, voltage, and power) have been taken for the analysis. Data from three pyranometer (in plane of array) and module temperature sensors (module rear side) are used, distributed over the entire plant. The minimal distance between the sensors is around 650 m.

Five years of weather and performance data (on the DC side) were used, available as minute data. In addition, further parameters are obtained or calculated: sun position angles (sun azimuth and sun elevation) as well as the angle of incidence (AoI).

2.2 Artificial Neural Networks

Artificial neural networks (ANN) are mathematical systems from the field of supervised machine learning methods. ANN can be applied for different types of modeling problems, whereas here ANN are used to determine plant behavior within a regression model. The model proposes the estimation of the system performance to be determined only based on meteorological parameters. The subsequent detection of potential defects (significant deviations) is based on the comparison between measured and estimated performance. This approach assumes that the data used for training does not show deviations which want to detect within the analysis.

The modeling considers on multilayer perceptron feedforward networks, for simplicity, in the following termed as ANN. A feedforward ANN is composed of different layers, which themselves consist of different numbers of nodes (so-called neurons). Each individual neuron is completely connected with all neurons of the adjacent layers. The connections between the neurons (so-called weights) take over indirectly the function of storing the structural dependency between given input and output information. Furthermore, an activation function within a neuron transforms the incoming signal into the final output signal.

The weights of an untrained ANN are predefined systematically or randomly. To adapt the weights to the data, a training process must be carried out using an initial data set (training data). During training, the weights are iteratively adjusted by an optimization algorithm until a termination criterion is reached. It is important that the training data already contains all the information needed for later use.

Training and modeling of the ANN models were done completely with Keras library 2.9.0 in Python 3.10.

The ANN designed consist of multiple hidden layers, each with a different number of neurons. On the input side, environmental data are used: irradiance POA, module and ambient temperatures as well as the sun position angles.

In preliminary work [10, 11], these hyperparameters were determined for the selected monitoring data set to obtain an accurate initial model with less training loss. The training period contains two full years.

2.3 Detection of Significant Deviations

Defects and disruptions in PV systems can appear differently, e.g. [2, 15], and have different technology-, material-, and on-site dependencies [16]. As soon as a defect or disruption appears and influences the electrical behavior, it may be possible that it can be captured by a monitoring system. The approach presenting find statistically significant deviations between actual and predicted behavior. These deviations are indicating possible problems with the system.

After completion of a training session the training and model quality was determined. For this purpose, for each target parameter y the root-mean-square error (RMSE, see Equation 1) between predicted and observed values are calculated (N as the number of data points considered during training):

$$\mathrm{RMSE} = \sqrt{\frac{1}{N}\sum_{i=1}^{N}\left(y_{pred}(i) - y_{obs}(i)\right)^2} \tag{1}$$

The detection of significant deviation the trained models were applied and tested on an independent period, which was not used during training.

$$\text{daily nMSE} = \frac{\sum_{i\in d}\left(y_{pred}(i) - y_{obs}(i)\right)^2}{N_d \sum_{i\in d} y_{obs}(i)} \quad (2)$$

The daily normalized mean squared error (daily nMSE, see Eq. 2) shows the average error per day d normalized with corresponding observed values (with N_d as the number of values for day d).

The range of acceptance (Eq. 3) of parameter y was determined as a statistical estimation based on the training accuracy. Values of daily nMSE outside this range will be declared as significant deviations.

$$\mu_y^T \pm k\,\sigma_y^T \quad (3)$$

where $k = 2$, $\varepsilon(i) = y_{pred}(i) - y_{obs}(i)$ and

$$\mu_y^T = \frac{1}{N^T}\sum_{i=1}^{N^T}\varepsilon(i), \sigma_y^T = \sqrt{\frac{1}{N^T-1}\sum_{i=1}^{N^T}\left(\varepsilon(i) - \mu_y^T\right)^2}$$

3 RESULTS

To identify the influence of each combination A - D the respective data was modeled and evaluated. Figure 1 shows exemplarily the daily nMSE and range of acceptance for each electrical parameter. The appearance of the errors and ranges depicted varies between the parameters, this leads to various points in time when the errors exceed the range of acceptance.

It is important to mention, that both the daily nMSE and the range of acceptance depend on the simulated values y_{pred}. This means, on the one hand the accuracy of the model generated during training affects the magnitude of daily nMSE and on the other hand the width of the acceptance range.

3.1 Data Suitability

In the following the training and testing results of the different combinations are presented.

Scenario A: Data quality for irradiance sensors

Figure 2 shows the influence on model accuracy (for DC current) using different pyranometers within the PV plant. The sensors are installed across the plant. The entire PV system was trained for each inverter using one of these pyranometer data for the irradiance input. The DC current RMSE increases linearly with distance between sensor used and the respective inverter area.

Assuming a linear dependency between distance and its RMSE, an RMSE increase of 10 A/km can be estimated. Furthermore, in respect to the geographical size of the entire PV system, using only one sensor or an averaged irradiance value will worsen the model accuracy.

In addition, the influence on the detectability will be demonstrated. Figure 3 shows exemplarily the testing results for DC current data of one inverter trained using different pyranometer data. At least one exceeds in March can be observed for all pyranometers, even the magnitude varies. of this varies due to training. However, a further exceed can only be seen for using one pyranometer (Pyranometer 1), which has a low distance to the inverter area used.

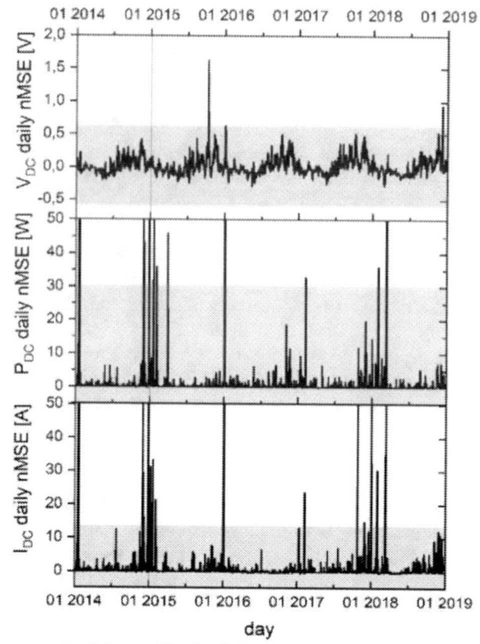

Figure 1: Normalized daily mean error of inverter monitoring data (DC voltage, DC power, DC current); green: acceptance range, outside of which errors are declared as significant deviations

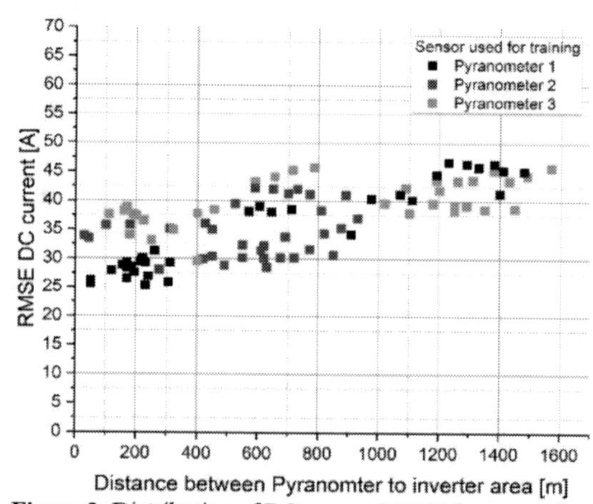

Figure 2: Distribution of DC current RMSE for test period in respect of used pyranometer during training.

Scenario B: Replacement with external data

To identify the impact of using external data, the training results (RMSE) of this model must be evaluated first. The RMSE increases for DC current and power by a factor of four and for DC voltage by a factor of two. However, this variation depends strongly on the external data source and further preprocessing. In the following, the focus is more the qualitative than on absolute differences.

In Figure 4 the daily nMSE is depicted. Because of higher training error, the model with foreign data shows a wider acceptance range and higher peaks. Despite this, these are for the most part still within the acceptance range. For this reason, only 50% of the exceeds are detected.

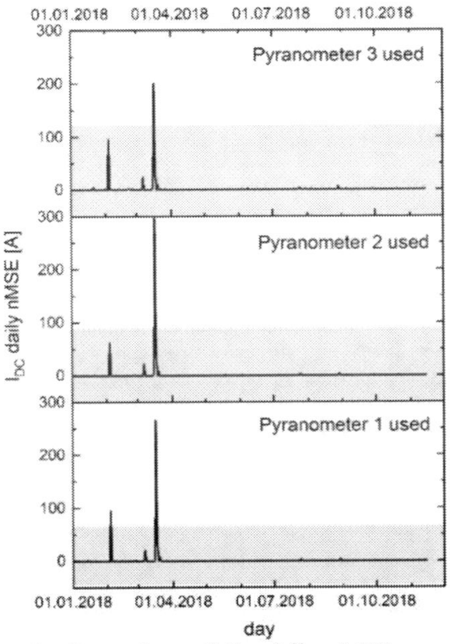

Figure 3: Comparison of the daily nMSE error and acceptance ranges for one inverter area using three different pyranometer data for ANN training; for this inverter Pyranometer 1 has lowest distance, followed by Pyranometer 2 and Pyranometer 3 with largest distance

Scenario C: Data Resolution

ANN training was performed for three different data time resolution: 15 min, 1 hour, 1 day. Through this aggregation the amount of data is reduced, and the performance data is averaged. This leads to an improvement in model accuracy for aggregated data for all target parameters. For instance, DC current RMSE decreases from 37.1 A (15 min) to 19.5 A (1d). This means, that averaging of data increases training accuracy, because there is less interference between meteorological and electrical parameters.

Furthermore, Figure 5 shows the daily nMSE of DC current. Also here, the acceptance range decreases in accordance with the decreasing data resolution (15 min → 1 day). Nevertheless, this reduction is counterproductive, since at the same time the daily nMSE increases significantly. This has the consequence of raising the number of significant deviations (potential defects).

Scenario D: Monitoring Resolution

The last scenario depends mainly on the on-site available monitoring equipment. An independent training for inverter and string data took place, with 15 min resolution. Figure 6 compares the daily nMSE over three years between inverter and string data. The acceptance range between for multiple strings are determined based on the lowest and highest acceptance range levels.

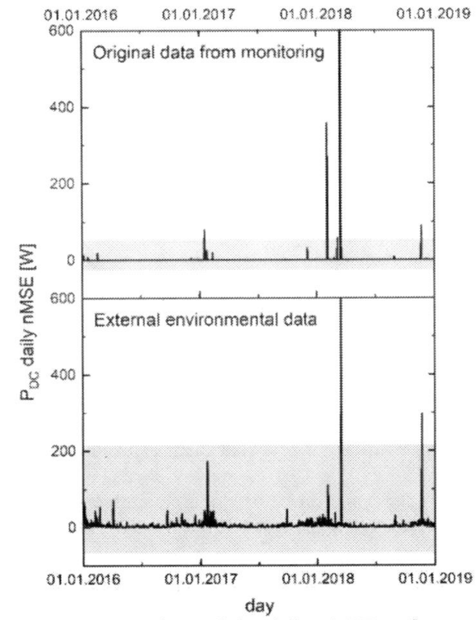

Figure 4: Comparison of the daily nMSE and range of acceptance of DC power between two trained model, one with original irradiance and module temperature data and one with externally determined data.

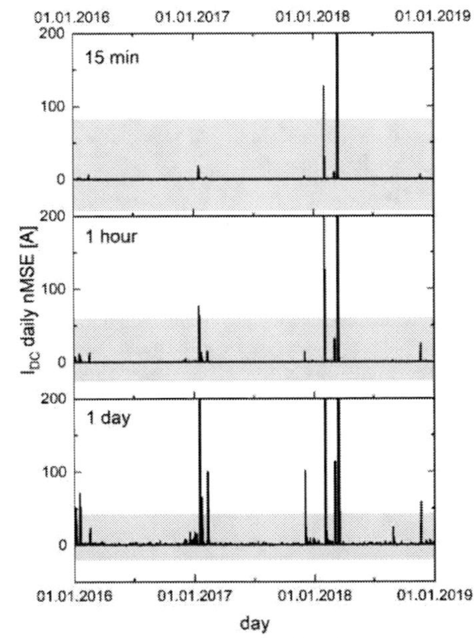

Figure 5: Comparison of the daily nMSE and range of acceptance of DC current between two trained model, one with original irradiance and module temperature data and one with externally determined data.

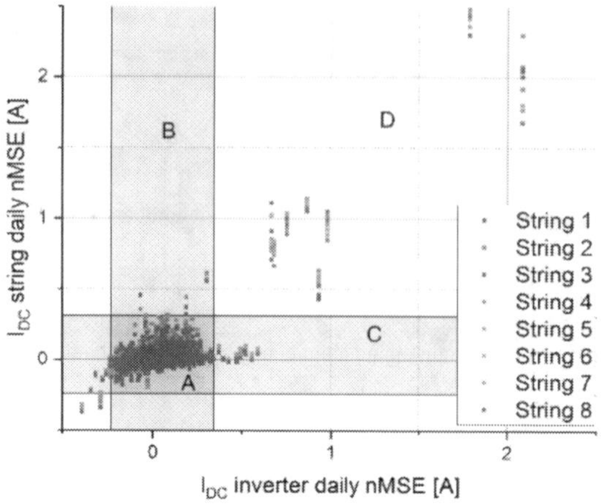

Figure 6: Daily comparison of the errors (nMSE) of an inverter with eight associated strings from one of the related combiner boxes, exemplary for DC current; green: acceptance range, errors outside are possible deviations for the respective component:

Area A: pairs of errors here are okay
Area B: points inside show a significant high error (potential problem) at string level, but not at inverter level
Area C: points here indicate a problem at the inverter, but not at string level
Area D: points here show a problem which affects string as well as inverter

This comparison shows a linear tendency between inverter and string errors. However, in all areas A, B, C, and D nMSE pairs are present. Inverter-string pairs – especially in area B and D – demonstrate that there are days with deviations (potential defects) they are only detectable using string data.

4 SUMMARY AND DISCUSSION

A parameter study was conducted to better understand the influence of different input parameters for the utilization of artificial neural network in respect of defect detection. Based on training results the daily nMSE and the acceptance range can be determined for application purposes. Both quantities directly based on the training quality.

Comparison within the scenario A to D have demonstrated that a change of input data within the training step has an undeniable effect on the detectability of defects.

The role of the choice of irradiance sensor was investigated. Closer sensors result in an increase of the model quality. Otherwise, this decreases linearly with increasing distance of the sensor to the modules to be examined. It can be concluded that large plants should have a large number of proper maintained sensors. Simply averaging the irradiance values will reduce the model quality. Depending on the plant size, a maximum distance between two irradiance sensors of 500 m is recommended.

In the second step the question of whether synthetic or third-party data sets can be used in the case of missing or problematic data was investigated. Replacing the data strongly depends on the data source and the way of further computation (e.g., fitting). The here used external data has caused a deterioration of the detectability of deviations due to larger training errors. It is not recommended to use only non-system data.

Furthermore, the impact of data resolution was evaluated. Although aggregated values show less variation and can therefore be trained better, there are larger evaluation errors when applying. This also increases the number of (suspected) deviations observed. However, a reduction of resolution primarily leads to a high number of reports, including false reports.

Finally, it is advisable to use multiple electrical quantities for the defect detection. As seen, for instance in Figure 1, system's deviations can appear only for voltage but not for current or power, or vice versa. Similarly, the extended training of string data can help, to identify problems in strings, they don't show up within the daily nMSE of the corresponding inverter. Further analysis of these deviations' patterns helps to identify problems faster.

These recommendations only can be applied as long the system is maintained properly, equipped with enough sensors and string-wise monitored. It is necessary that in future systems are monitored in more detail to ensure the applicability of machine learning methods on PV system data.

Furthermore, it is important to check the correct origin of the data for training and to remove outliers to provide a good model quality. The choice of this data used for training plays a significant role how good defects can be detected in the monitoring data.

5 Acknowledgment

This work was partially funded by the German Federal Bank Investitionsbank Sachsen-Anhalt, the German Federal State Saxony-Anhalt and the European Regional Development Fund (EFRE) in the project Mon-KI (No. 2004/00009) and by the Federal Ministry for Economic Affairs and Climate Action (BMWK) within the project "PVOptDigital" (project no. 03EE1107A).

6 References

[1] IEA PVPS Task 13, Subtask 2, Analytical Monitoring of Grid-connected Photovoltaic Systems, Good Practices for Monitoring and Performance Analysis, Report IEA-PVPS T13-03: 2014, March 2014, ISBN 978-3-906042-18-3.

[2] M. Halwachs, L. Neumaier, N. Vollert, L. Maul, S. Dimitriadis, Y. Voronko, G.C. Eder, A. Omazic, W. Mühleisen, Ch. Hirschl, M. Schwark, K.A. Berger, R. Ebner, Statistical evaluation of PV system performance and failure data among different climate zones, DOI: https://doi.org/10.1016/j.renene.2019.02.135.

[3] D. C. Jordan, S. R. Kurtz, Photovoltaic Degradation Rates—an Analytical Review, Prog. Photovolt: Res. Appl., 21: 12-29. doi:10.1002/pip.1182.

[4] M. J. Mayerab, G. Gróftt, Extensive comparison of physical models for photovoltaic power forecasting, Applied Energy, 2021, https://doi.org/10.1016/j.apenergy.2020.116239.

[5] S. Randsom, J. Sutterlueti. A Systematic Comparison of 12 Empirical Models Used for Energy Yield Predictions vs PV Technology, 33rd European Photovoltaic Solar Energy Conference

and Exhibition, 2017, DOI: 10.4229/EUPVSEC20172017-5CO.7.6.
[6] S. Rodrigues, H. G. Ramos, and F. Morgado-Dias, "Machine Learning in PV Fault Detection, Diagnostics and Prognostics: A Review," 2017 IEEE 44th Photovoltaic Specialist Conference (PVSC), 2017, pp. 3178-3183, doi:10.1109/PVSC.2017.8366581.
[7] D. Weichert, P. Link, A. Stoll, S. Rüping, S. Ihlenfeldt, S. Wrobel, A review of machine learning for the optimization of production processes, The International Journal of Advanced Manufacturing Technology, 2019, DOI: https://doi.org/10.1007/s00170-019-03988-5.
[8] J. Ferrero Bermejo, J.F. Gómez Fernández, F. Olivencia Polo, A. Crespo Márquez, A Review of the Use of Artificial Neural Network Models for Energy and Reliability Prediction. A Study of the Solar PV, Hydraulic and Wind Energy Sources, Appl. Sci., 2019, DOI:10.3390/app9091844.
[9] G. D. Rupakula, D. Dassler, S. Malik, M. Ebert, R. Schmidt, Automatic Fault Detection and Classification in PV Systems by the Application of Machine Learning Algorithms, 38th European Photovoltaic Solar Energy Conference and Exhibition, 2021, DOI: 10.4229/EUPVSEC20212021-5DO.1.2.
[10] D. Dassler, S. B. Kuppanna, S. Malik, R. Schmidt, M. Ebert, Training and Evaluation for Yield-Driven Detection of Losses in PV Systems Utilizing Artificial Neural Networks, IEEE PSSC 47 2020
[11] D. Dassler, S. Malik, S. B. Kuppanna, B. Jaeckel, M. Ebert, Innovative Approach for Yield Evaluation of PV Systems Utilizing Machine Learning Methods, IEEE PVSC 46 2019.
[12] PHOTOVOLTAIC GEOGRAPHICAL INFORMATION SYSTEM, PVGIS Version 5.2, https://re.jrc.ec.europa.eu/pvg_tools/en/
[13] J. A. Davies and D. C. McKay, "Evaluation of selected models for estimating solar radiation on horizontal surfaces," Solar Energy, vol. 43, pp. 153-168, 1989.
[14] J.A. Kratochvil, W.E. Boyson, D.L. King, Photovoltaic Array Performance Model; Albuquerque, N.M., Livermore, C.A., Eds.; MDPI: Basel, Switzerland, 2004; Volume 8.
[15] IEA PVPS Task 13, Performance, Operation and Reliability of Photovoltaic Systems, Quantification of Technical Risks in PV Power Systems, Report IEA-PVPS T13-23:2021, October 2021, ISBN 978-3-907281-11-6.
[16] M. Aghaei, A. Fairbrother, A. Gok, S. Ahmad, S. Kazim, K. Lobato, G. Oreski, A. Reinders, J. Schmitz, M. Theelen, P. Yilmaz, J. Kettle, Review of degradation and failure phenomena in photovoltaic modules, Renewable and Sustainable Energy Reviews 159, 2022, DOI: https://doi.org/10.1016/j.rser.2022.112160.

This is a short version of the paper submitted to the journal Progress in Photovoltaics.

It has been selected by the Committee of the WCPEC-8 for submission to Progress in Photovoltaics.

MULTI-CLIMATE STUDY OF CONDENSATION AND SOILING OF PV MODULES

Benjamin Figgis[1], Marilia Braga[2], Aline K. V. de Oliveira[2], Ricardo Rüther[2], Muhammad Zahid Khan[3,8], Carlos D. Rodríguez-Gallegos[4], Thomas Reindl[4], Aaron Sahm[5], William Snyder[6], Laurie Burnham[6], Fabian Wolferstetter[7], Fernanda Norde Santos[7], Sebastian Dittmann[8,3], Ralph Gottschalg[8,3], Veronica Bermudez[1], Brahim Aissa[1], Giovanni Scabbia[1]

1. Qatar Environment & Energy Research Institute, Hamad bin Khalifa University, Doha, Qatar
2. Universidade Federal de Santa Catarina (UFSC), Florianópolis, Brazil
3. Fraunhofer Center for Silicon Photovoltaics CSP, Halle (Saale), Germany
4. Solar Energy Research Institute of Singapore, National University of Singapore, Singapore
5. Center for Energy Research, University of Nevada, Las Vegas, USA
6. Sandia National Laboratories, Albuquerque, USA
7. German Aerospace Center (DLR), Institute of Solar Research, Almeria, Spain
8. Anhalt University of Applied Sciences, Koethen, Germany

ABSTRACT: This study investigated variation in condensation (dew) and its effect on PV soiling, on a global level. Qualitatively, it is well known that condensation plays several roles in PV soiling: capillary adhesion and cementation cause dust particles to stick to the surface of modules, while run-off of dew droplets can clean them. However because outdoor condensation sensors are rare, the quantitative link between condensation and soiling is less known. Here, we developed a low-cost sensor to measure condensation outdoors and deployed it at seven PV test sites worldwide, spanning different climates. It was found that the condensation data helped explain differences in soiling rate between geographical locations (sites with higher average condensation had higher average soiling rates), however it did not explain differences in soiling rate at a given location over time (there was negligible correlation between daily soiling rate and daily condensation, at each location). Regarding performance of the sensor itself, it produced stable and expected measurements of the daily maximum condensation, however at four sites the daily minimum values suggested gradual contamination or drift of the sensor. It was concluded that condensation does help explain differences in PV soiling rates between different locations, but it cannot be used (in a simplistic way, at least) to predict daily variation each location's soiling rate.

Keywords: photovoltaic, soiling, cleaning, condensation, dew, modeling

1 INTRODUCTION

Soiling greatly degrades the performance of PV systems, especially in desert climates where rain is scarce. Condensation strongly affects PV soiling via three main mechanisms [1][2][3]: (i) Capillary adhesion, where dew droplets trap dust particles and adhere them to the module surface; (ii) cementation, where soluble material in the dust dissolves in the condensation and, after drying, forms a strong material bridge between dust particles and the module surface; and (iii) run-off of droplets when a sufficient quantity of condensation is reached, which partially cleans the module. Notably, the third mechanism counteracts the first two, which makes the link between condensation and soiling complicated and non-obvious.

There have been anecdotal reports that dry, inland areas experience much less PV soiling than coastal areas, which are exposed to humidity from the sea. However the quantitative relationship between condensation and PV soiling rate has not been widely studied in the field. A main reason is that PV sites rarely have condensation sensors — they are not a standard instrument in weather stations, and stand-alone condensation sensors tend to be specialized scientific-grade devices that are expensive and not commonly available.

The Photovoltaic Collaborative to Advance Multi-climate Performance and Energy Research (PV CAMPER) is an international collaboration of research institutes that conducts PV field studies [4] (datahub.duramat.org/project/about/pvcamper). In 2019 the group started a project to develop a low-cost condensation sensor and deploy it worldwide. The goals of the project were to: (i) evaluate performance of the low-cost sensor, (ii) describe variation in PV condensation in different climates, and (iii) determine whether condensation data predicted differences in PV soiling rates between locations, and over time at a given location.

Qatar Environment & Energy Research Institute (QEERI), as the project lead, developed the low-cost condensation sensor based on an existing industrial electrical component, mounted on a glass coupon. Ten devices were built and deployed to project participants worldwide, including PV CAMPER members and other institutes. Participants collected condensation, PV soiling rate and various meteorological ("meteo") data. This paper used data collected from seven locations (others had not yet acquired sufficient data or had faults).

2 METHODS

2.1 Condensation Sensor Development

QEERI had previously purchased and installed a scientific-grade condensation sensor (CiSmst Dewing Sensor BTF 11344) at its Outdoor Test Facility (OTF) in Doha, Qatar, and used it to study the effect of condensation on PV soiling at that particular location [5]. As the PV CAMPER group includes PV test sites around the world, it was an ideal platform for a global follow-up study of cross-climate differences in PV condensation and soiling. The study was further expanded though the participation of additional institutes. It was decided to develop and use a novel low-cost condensation sensor, because the scientific-grade (CiSmst) sensor previously used by

QEERI was expensive and no longer in production. It served as a high-fidelity reference device for evaluating the low-cost sensor developed herein.

The low-cost condensation sensor was developed by QEERI using an off-the-shelf industrial electrical component (part number HDS10). The product's datasheet describes it as a switching-mode element whose resistance changes at high relative humidity. The 7 × 5 mm sensor was mounted with heat-conductive ceramic paste to a 100 × 100 × 3 mm glass coupon (Figure 1). The glass coupon served to approximate the sky-facing, soiling-exposed, glass surface of a PV module. A microcontroller was connected to supply power to the sensor and record its output signal. The output signal was then scaled from 0 (dry sensor) to 1 (sensor saturated with water). This dimensionless parameter is labelled "Dew" in the results figures.

Figure 1: An example of the low-cost condensation sensor developed for this project. It comprised an HDS10 moisture-sensitive component attached to a glass coupon, resting on a box with a power supply and data logger.

The response of the low-cost condensation sensor was compared to the scientific-grade sensor, and also to the amount of dew predicted by environmental conditions (namely, the dew-point and module surface temperatures). The scientific-grade device was attached to a full-size, glass/backsheet PV module. The device was supplied factory-calibrated, and its output had dimensions µg/mm^2.

In principle condensation will commence when the module surface temperature T_S falls below the dew-point temperature T_{DP}, i.e. $T_S - T_{DP} < 0$, although in practice condensation has been observed to start even when T_S is slightly above T_{DP} [3]. The minimum of $T_S - T_{DP}$ each night was calculated and named ΔT_{min}, dimension °C.

The maximum values of the two condensation sensors, and ΔT_{min}, were recorded for seven consecutive nights at the OTF in Doha. The results are shown in Figure 2, with the nights arranged in order of increasing condensation (decreasing ΔT_{min}). The response of the low-cost condensation sensor "qualitatively matched" the trend in ΔT_{min}, i.e. as the module surface temperature approached the dew point (on different nights), the sensor values increased. Also, the low-cost sensor had a roughly similar response as the scientific-grade device, except for night #4. On that night, the ΔT_{min} parameter predicted a greater amount of condensation (relative to night #6), which was indeed measured by the low-cost sensor but not by the scientific-grade one.

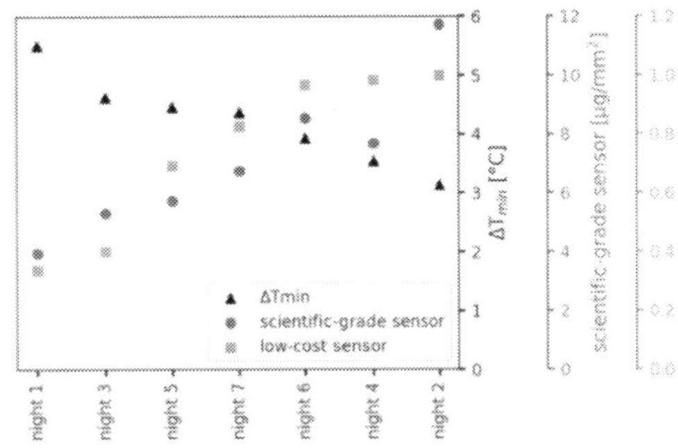

Figure 2: Comparison of three condensation metrics over seven nights: Maximum value from the scientific-grade condensation sensor (red markers); maximum value from the developed low-cost condensation sensor (orange markers); minimum difference between PV module and dew-point temperatures, ΔT_{min}. Nights ordered by decreasing ΔT_{min}

Overall, this preliminary study demonstrated the "proof of concept" of the low-cost sensor. In particular, the direction of its response (increasing or decreasing) matched expectations from the dew-point and surface temperatures, while the magnitude of its response approximated a scientific-grade condensation sensor. This provided sufficient confidence to use the low-cost sensor for the multi-site, global study.

2.2 Sensor Deployment

Next, we built ten identical low-cost condensation sensors with the design specified in the previous section, and distributed them to participating research institutes around the world, spanning a variety of climates. All participants collected condensation, soiling rate and rain data (some also collected additional meteo data which was not used in this study). This paper used data from seven sites; the other three have not yet collected a sufficient length of data or experienced technical faults. The locations of the sites are shown in Figure 3. The seven sites whose data were used in the study were: Albuquerque (USA), Al Kharsaah (Qatar), Doha (Qatar), Florianopolis (Brazil), Halle (Germany), Las Vegas (USA), and Singapore.

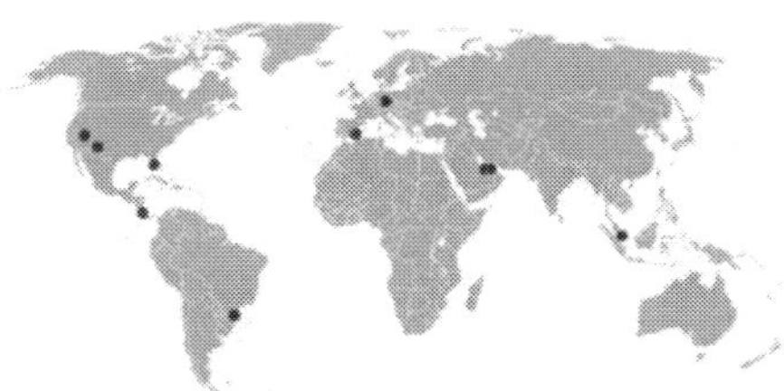

Figure 3: Locations of participating test sites in the project. Red markers: Sites providing data used in this study. Grey markers: Sites not yet able to provide data

Using their existing instruments and methods, some sites measured soiling continuously (e.g. every 1 minute) while others measured it once per day. To standardize the data for analysis, continuous soiling datasets were converted to daily rates. To compare the soiling data to condensation data, it was also necessary to also convert the condensation data from 1-minute recordings (inherent to the instrument) to daily values. Data exploration revealed that daily extremes of condensation (minimum and maximum each day) were the most useful parameters; the average amount of condensation each day had less physical relevance.

In preparing the data, rain days were excluded from each location's dataset. This decision was made for two reasons: Rain and dew are distinct physical phenomena which we did not want to confound (the aim was to investigate the effect of condensation specifically), and rain tends to clean PV modules whereas condensation is usually expected to worsen their soiling. Excluding rain days particularly reduced the size of the Singapore dataset (only 53 days out of 237 were rain-free).

3 RESULTS AND DISCUSSION

The daily condensation ("Dew") range throughout the study for each site is shown in Figure 4. The locations are ordered top-to-bottom by their approximate condensation intensity, with Las Vegas having the least dew and Florianopolis the most. Plotting the raw dew values over time (Figure 4.a) shows that condensation occurred frequently at all locations except Las Vegas (where it was rarely observed) and Albuquerque (where it was sporadic). This is consistent with an earlier literature review [3] which reported that condensation on PV modules is common, due to radiative cooling at night.

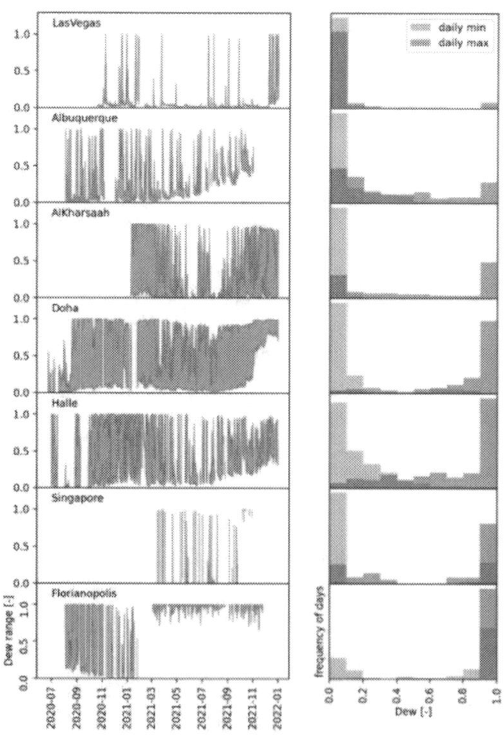

Figure 4: Condensation data from each location. (a, left) Daily condensation range (minimum to maximum) at each site for the duration of the study. (b, right) Histograms of the daily minimum (orange) and maximum (blue) condensation level, overlapping histograms appear grey. Rain days are excluded. Dew values are scaled such that 0 = dry sensor, 1 = sensor saturated with water.

Histograms of the same data (Figure 4.b) further show that condensation has a fairly "binary" distribution — at most sites, the condensation reached saturation each night (the blue bars are clustered at Dew ≈ 1). Only in Albuquerque and Halle did the condensation often peak at a sub-saturation levels (blue bars < 1). At most sites the condensation sensor completely dried out almost every day (orange bars clustered Dew ≈ 0); only in Florianopolis did the sensor often remain wet throughout the entire day.

However, there is evidence of drift or degradation of the condensation sensors at some locations in Figure 4.a, which casts doubt on some of the data. In the Albuquerque dataset, for example, the bottom end of the daily dew range starting increasing around March 2021, whereas prior to that it was consistently around zero. It is unlikely that this was due to changing climate conditions. It is speculated that the sensor gradually become contaminated or its electrical characteristics (resistance as a function of wetness) started to drift. Increasing minimum dew values are also discernable in the Doha, Halle and Florianopolis plots. In Florianopolis there was a large change in the sensor response after a faulty memory card was replaced in March 2021. Overall the trends in Figure 4.a suggest that the instruments' measurement of daily maximum condensation remained stable, but the daily minimum values could become unreliable after months of exposure. In the remainder of this report, only maximum daily dew measurements are used.

Next, the average soiling rate at each site was compared to the average of its daily maximum dew values (Figure 5). It is seen that the driest locations (Las Vegas and Albuquerque) had the lowest soiling rates, both 0.02 %/day. Overall there was an approximate trend of increasing soiling with increasing condensation, with a major outlier being Al Kharsaah in Qatar. Al Kharsaah was an active construction site for a utility-scale PV plant during the study period, with extensive earth works in progress, which may have caused its very high soiling rate (0.63 %/day). The occurrence of low soiling rates at the two dry inland deserts (Las Vegas and Albuquerque), coupled with high soiling rates in the more humid coastal deserts of Qatar (Doha and Al Kharsaah), strongly suggests that — for arid climates at least — more condensation corresponds to greater soiling.

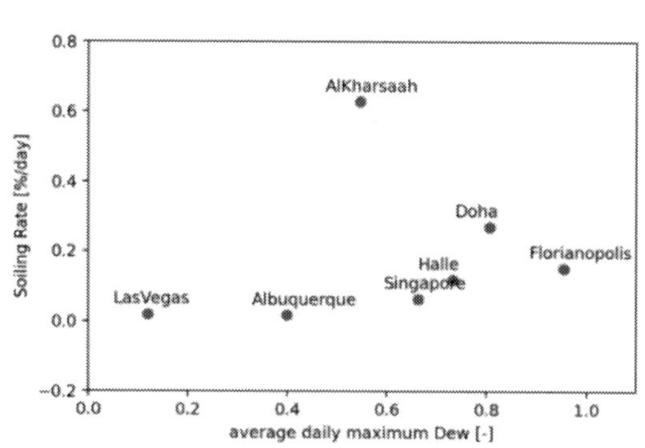

Figure 5: Average daily soiling rate vs. average daily maximum dew level, for each location

Finally, we examined whether condensation could predict soiling rate variation over time at each location. The correlation of daily soiling rate to daily maximum dew, for each site, is shown in Figure 6, labelled with the Pearson correlation coefficient R. Overall, there was low and sometimes negative correlation between the two variables, indicating that dew is not a strong predictor of variation in the daily soiling rate. This finding was consistent with the mentioned soiling-modeling exercise at the Doha site specifically, which found that it was difficult to predict the daily soiling rate from environmental conditions, including condensation [5].

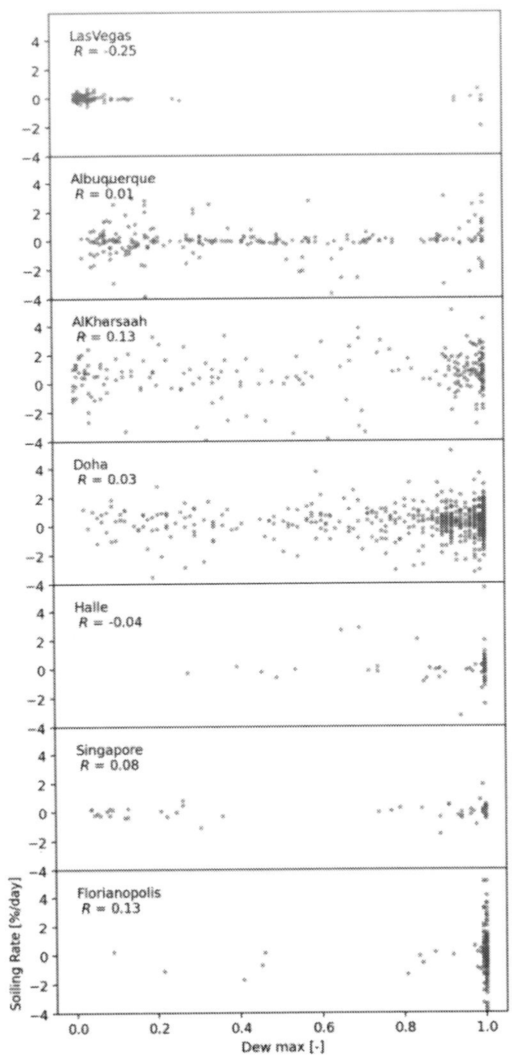

Figure 6: Daily soiling rate vs. maximum daily dew for each location. Each marker represents one day. R values are Pearson correlation coefficients

On face value it is unexpected that condensation is predictive of soiling differences between locations, but not predictive of soiling differences at given location between days. Three possible explanations for this discrepancy are considered:

(i) Lab experiments by Ilse et al [2] have shown that dust becomes cemented to glass when the dew point is only just reached, i.e. cementation does not require copious amounts of condensation, and in fact very wet glass may (conversely) shed water droplets and clean the sample. Yet, the condensation data used in this study ranged from dry to saturation (Dew = 0 to 1), well beyond the range in which the physical phenomenon of interest may have occurred (Dew = 0 to slightly above 0). This allows the possibility that humid locations experience greater soiling than very dry locations, while the severity of soiling each day would depend on whether dew-forming conditions were just reached, rather than the quantity of dew that then accumulated.

(ii) Soiling is known to be a complex phenomenon that is influenced by many environmental factors, making it difficult to predict on short timescales [6]. The location comparison (Figure 5) used long-term data averages, whereas the same-site analysis (Figure 6) used one-day data granularity, which may account for the higher correlation seen in the former.

(iii) If a variable does not vary much, it is unlikely to be a good predictor of any phenomenon. Therefore another theoretical explanation for the results is greater variability in the average condensation between locations, than in the daily condensation at a given location. However common experience tells us that humidity and condensation do vary greatly throughout the year at a given place, so this may not be the strongest explanation.

4 CONCLUSIONS

This study investigated cross-climate differences in condensation, and the ability to predict soiling rates from condensation data. To measure condensation at multiple field sites globally, a low-cost sensor was developed and found to have a similar response to a scientific-grade condensation sensor during testing. Ten units of the sensor were built and provided to study participants worldwide; usable data was obtained from seven locations. At some locations the condensation data showed signs of contamination or drift of the sensor after several months, at the low (dry) end of the measurement range. Overall the study indicated the developed sensor is a useful and low-cost tool for measuring condensation in the field, but its data should be checked for possible drift over time.

The two inland arid locations (Las Vegas and Albuquerque) rarely experienced condensation, however all other sites frequently experienced dew. This was consistent with previous reports that condensation on PV modules is common due to radiative cooling at night. Generally, the results indicate that locations with more intense condensation (specifically, higher average of daily maximum dew) have greater soiling rates. Notably, the two mentioned inland arid locations had very low soiling rates, despite being surrounded by deserts.

Although condensation was linked to differences in long-term soiling rates between locations, it was not predictive of daily differences in the soiling rate at a given location. A possible reason is that soiling is a complex phenomenon that has much random fluctuation on the short (1-day) timescales analyzed here. Another is that dust becomes cemented to glass when dew conditions are only just reached, whereas the condensation data used in this study ranged from dry to fully saturated. Thus, condensation data might still be useful for predicting soiling rate variation over time, but using more other characterization or modeling approaches.

5 ACKNOWLEDGEMENTS

SERIS is a research institute at the National University of Singapore (NUS). SERIS is supported by NUS, the National Research Foundation Singapore (NRF), the Energy Market Authority of Singapore (EMA) and the Singapore Economic Development Board (EDB).

6 REFERENCES

[1] K. K. Ilse et al., "Comprehensive analysis of soiling and cementation processes on PV modules in Qatar," Sol. Energy Mater. Sol. Cells, vol. 186, no. June, pp. 309–323, 2018, doi: 10.1016/j.solmat.2018.06.051.

[2] K. Ilse, B. Figgis, M. Z. Khan, C. Hagendorf, and V. Naumann, "Dew as a Detrimental Influencing Factor for Soiling of PV Modules," IEEE J. Photovoltaics, vol. 9, no. 1, pp. 287–294, 2018, doi: 10.1109/jphotov.2018.2882649.

[3] B. Figgis et al., "Investigation of factors affecting condensation on soiled PV modules," Sol. Energy, vol. 159, no. October 2017, pp. 488–500, 2018, doi: 10.1016/j.solener.2017.10.089.

[4] L. Burnham et al., "Photovoltaic Collaborative to Advance Multi-Climate Performance and Energy Research (PV CAMPER)," 36th Eur. Photovolt. Sol. Energy Conf. Exhib., no. September, p. 1, 2019, doi: 10.13140/RG.2.2.11460.78720.

[5] B. Figgis, G. Scabbia, and B. Aissa, "Condensation as a predictor of PV soiling," Sol. Energy, vol. 238, no. December 2021, pp. 30–38, 2022, doi: 10.1016/j.solener.2022.04.025.

[6] B. Figgis, B. Guo, W. Javed, S. Ahzi, and Y. Rémond, "Dominant environmental parameters for dust deposition and resuspension in desert climates," Aerosol Sci. Technol., vol. 52, no. 7, pp. 788–798, 2018, doi: 10.1080/02786826.2018.1462473.

A FAST SOFTWARE CHECK FOR PV SYSTEMS

Markus Rinio[1], Ulf Enarsson[1], Clifford Hansen[2]
[1] Karlstad University, 65188 Karlstad, Sweden
[2] Sandia National Laboratories, Albuquerque, New Mexico 87185, USA

ABSTRACT: After installation of a PV system, the question arises if it works correctly. Comparing the yield with some prediction gives a hint if the system is OK, but takes a long time and depends on the weather during that period. Also the effect of shading by trees or buildings is hard to estimate and often not included in the prediction. We present a solution based on the evaluation of the power curve of the PV system from a sunny day. The peak power of the PV system is obtained by simulating this power curve using the peak power as a fit parameter. The simulation uses the public library pvlib python including the clear sky model of Ineichen and Perez, and a database for the air turbidity. The temperature of the PV modules is calculated from air temperature and wind speed. The software with its graphical user interface for Microsoft Windows™ is freely available. The simulation of irradiance and module temperature is compared to experimental data. As an example, the peak power of some newly installed PV systems was evaluated. Furthermore, a degradation analysis of the peak power over a 10-year period of a test system is performed.
Keywords: Software, Monitoring, PV system, Simulation, Degradation

1 INTRODUCTION

The amount of installed PV systems world-wide is rapidly growing. Quality control is therefore of increasing importance. Smaller systems like those mounted on single house roofs are typically not equipped with sensors for irradiance or module temperature. This makes it harder to directly check their performance. The typical solution is to compare the energy yield from a certain period of time, for instance a year, with the prediction from the installer or via the photovoltaic geographical information system PVGIS or the PVWatts calculator by NREL [1], [2]. However, such a yield depends on varying irradiance and air temperature during the year. Moreover, periods of shading by nearby objects like trees or buildings are typically not included in the yield prediction.

A much faster solution is the comparison of the power curve of a single sunny day with a simulation for precisely that day. This makes it possible to exclude effects of shading. Temperature effects on the system performance can be included with data from nearby weather stations. The approach allows for faster responses in case of malfunctions.

2 METHOD

The AC power curve can be downloaded from most inverters as an ASCII file containing the system power as a function of time. The rated system AC power is usually given for standard test conditions (STC) - that is, illumination with 1000 W/m² light intensity with the atmospheric mass 1.5 (AM1.5) spectrum perpendicular to the surface of the PV modules, and 25 °C module temperature. Since these conditions are practically never reached under real-life conditions, a calculation of the power under different module temperatures and irradiances is needed.

There exist models for the calculation of the clear sky radiation incident on an inclined surface [3]–[5].The free software presented here simulates the real-life conditions for a given place and time on the earth using the clear sky model of Ineichen and Perez [4]. The simulation includes a database of the average atmospheric turbidity, interpolated for every geographic location and time of the year.

The system AC power at STC is used in this software as a variable parameter to fit the simulation to the measured data. Here, STC serves as a convenient reference condition, because models that translate irradiance and temperature to power are formulated relative to ratings at STC.

The free software consists of a graphical user interface with import filters for different measurement data file formats [6], [7]. The simulation is performed by an integrated part based on pvlib python [8]. The main parameters needed for the calculation is the position and angle of the PV modules as well as the day and time zone. Air temperature and wind speed can be given as constant values, but are recommended to be imported as time-dependent parameters over the day from nearby weather stations.

From these parameters and the simulated irradiance, the module temperature is simulated for different module types and mounting situations [9], [10]. The lowlight behavior of the modules as well as the inverter efficiency can be imported as functions of the light intensity and inverter load, respectively. Cable losses and soiling are neglected so far but could be included in the inverter efficiency, if needed. Diffuse light reflection from the ground can be taken into account via an assumed albedo. If a nearby weather station provides data of the global horizontal irradiance (GHI), such data can be used to correct the clear sky model. The calculation of the global tilted irradiance (GTI) in plane of the modules as well as of the module temperature can be bypassed if direct sensor data is available.

To evaluate the simulations, measurements of the GTI and the module temperature were performed using a newly calibrated irradiance detector (type Si-RS485TC-T-Tm-MB from Ingenieurbüro Mencke & Tegtmeyer GmbH, 31789 Hameln, Germany) based on a multicrystalline solar cell. Effects of the detector temperature were corrected with an in-built temperature sensor. The temperature of the first test PV system was measured with an external temperature sensor from this detector that was glued to the back of one of the PV modules. The nominal precision of the irradiance measurement is about 3 %.

3 RESULTS AND DISCUSSION

3.1 Test of the module temperature simulation

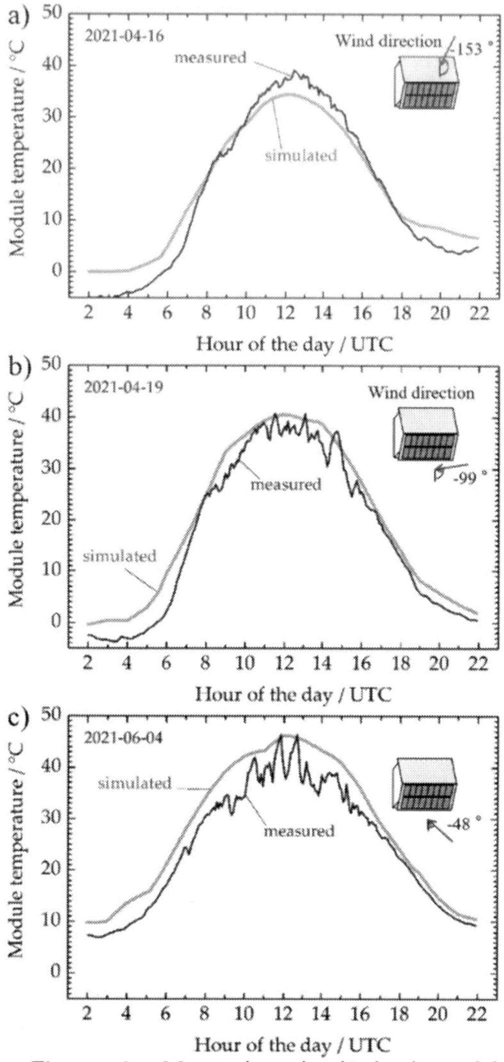

Figure 1: Measured and simulated module temperature on a roof mounted test system during three different sunny days in spring 2021. a) Wind direction from behind the roof leading to the PV modules being in wind shadow. b) Wind direction nearly parallel to the modules. c) Wind direction diagonal from the front towards the PV modules [7].

On a first roof mounted test system in Karlstad, Sweden, the temperature of the PV modules was measured during three different days. The air temperature and wind speed was obtained from a weather station in about 12 km distance. Then the module temperature was simulated with pvlib python and compared to the measured temperatures.

The results together with the average wind direction relative to the PV modules (also obtained from the weather station) are shown in figure 1. When the wind was blowing from behind the roof leaving the PV modules in wind shadow, the measured module temperature was slightly higher than the simulated one. The opposite behavior is seen when the wind blew diagonal from the front.

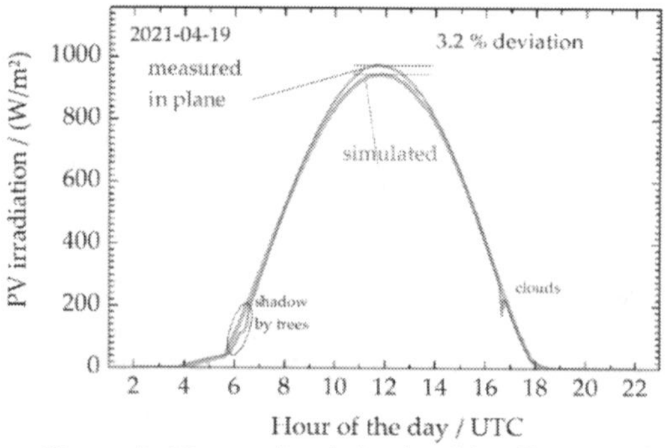

Figure 2: Measured and simulated irradiance on the roof mounted system during one day in spring 2021 [7]

The best match was obtained when the wind blew nearly parallel to the modules, which represents a kind of average situation. More details about this experiment are given in the previous publication [7]. The typical temperature deviations of about 5 °C would lead to an inaccuracy of the peak performance of about 2 to 3 %.

3.2 Irradiation simulation

For our first test system, the simulated GTI was compared to the measured one. For the three days of spring analyzed in the previous chapter, the measured GTI was between 1 and 3 percent higher than the simulated one. An example is shown in figure 2.

A somewhat larger deviation of about 3 to 6 percent was obtained when comparing the simulated GHI with the measured one by a local weather station [7]. The software presented here has an option to correct this deviation, if such measurement data from a nearby weather station is available.

3.3 Evaluation of three test systems in Mora, Sweden

Three newly installed PV systems located in Mora, Sweden are evaluated with our software [11]. The simulated PV system power over a sunny day (2022-04-20) together with the measured power is shown in figure 3. The curves show how periods of shading can be excluded "by-eye". This is done by modifying the peak power setting in the simulation until the simulated and the measured curves match each other for time periods without shading. System A exhibits an extreme case of shading with more than half of the day during this time of the year. Yet, it is still possible to precisely adjust the simulation to the measurement for the remaining hours of the day. The temperature of the PV modules was simulated using air temperature and wind speed during the day from a nearby weather station.

The systems also had sensors to measure the module temperature directly. The system peak performance was evaluated using the module temperature simulation and the direct module temperature measurement. The result is shown in table I. Systems A and C comply with their rated peak power within a marginal of 3 %, which is better than the estimated accuracy of this evaluation. When using the measured module temperature, system B seems to have a performance problem delivering only 92 % of the rated peak power. Yet, an analysis revealed an unusual low measured module temperature, not rising as expected when the modules were under direct sunlight. Since the peak power was 97 % of its rated value when using the

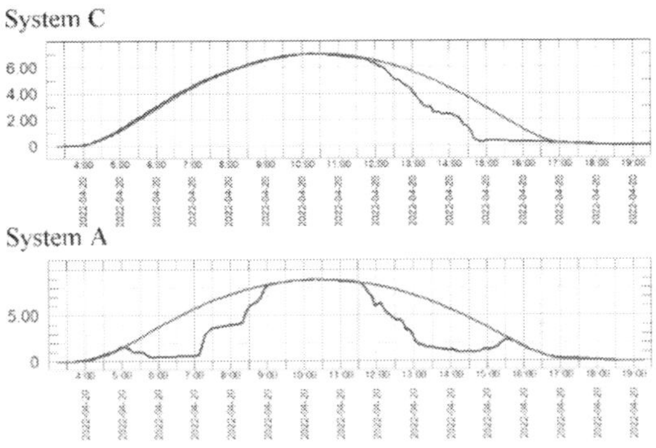

Figure 3: Simulated (red) and measured (blue) PV system power over a sunny day for two newly installed systems (A and C) in Mora, Sweden. The simulation was done with the clear sky model and calculated module temperatures from air temperature and wind speed. Deviations of these curves are solely due to shading by nearby trees.

Table I: Quotient of measured peak power and rated peak power. The simulated module temperature is calculated from air temperature, wind speed, and simulated irradiance.

System	Peak power / rated peak power	
	using simulated module temperature	using measured module temperature
A	97 %	98 %
B	97 %	**92 %**
C	97 %	100 %

simulated module temperature, we suspect that the temperature sensor was possibly detached from its module. An investigation of this question was not finished until the submission of this paper.

3.4 Long-term degradation analysis

We evaluated the long-term degradation of a test PV installation located at Glava Energy Center, Glava, Sweden [11]. Over a 10-year time period between 2012 and 2022, about six sunny days were identified for each year and evaluated. The system was set up about three years before 2012, and should therefore have finalized an initial stronger degradation during the first year. It should also have collected some initial dust before our investigation period started. The producer guaranteed an annual degradation of the peak power of not more than 0.7 %.

Figure 4 shows the measured peak power as a function of time. The evaluation is based on the Ineichen/Perez clear sky model. The module temperature was calculated from the simulated irradiance in combination with the air temperature and wind speed measurements. This weather data was taken from an on-site station. In a few cases where data points were missing from the on-site weather station, wind speed- and air temperature data from a weather station at 15 km distance was used.

The peak power values shown in figure 4 show a degradation of 0.22 % per year. The data points exhibit also a certain scatter with deviations of up to 7 % mainly due to seasonal variations. In a second step, all data corresponding to months from April to August were marked with filled symbols. These data points now lie all

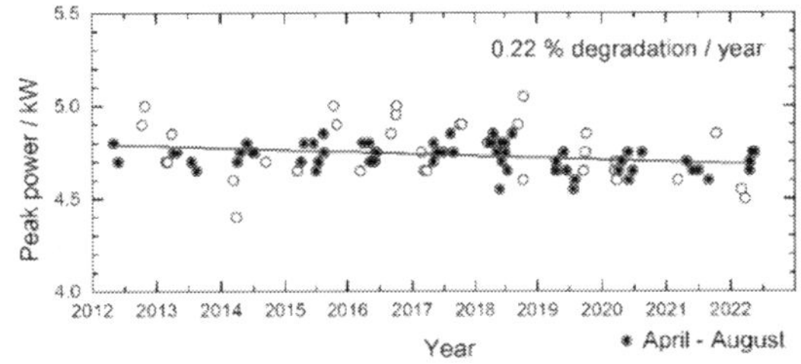

Figure 4: Peak power of the test system at Glava Energy Center over a 10-year period. The filled symbols correspond to dates from April to August. The red line is a linear fit to all data points showing a degradation of 0.22 % per year.

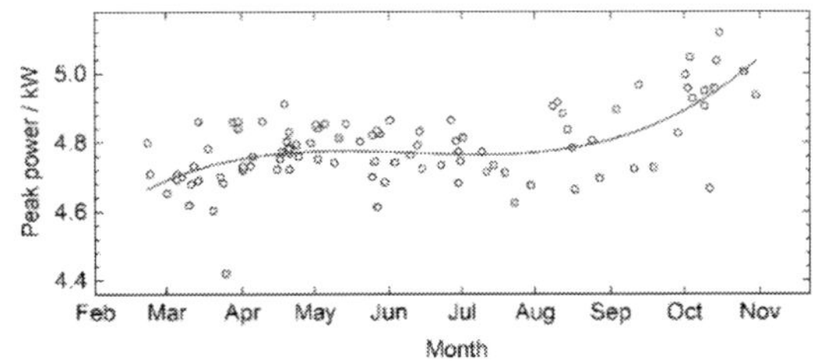

Figure 5: Peak power values of the test system at Glava Energy Center from a 10-year period after degradation correction all plotted into one year. The red line is a 3rd order polynomial fit to the data points.

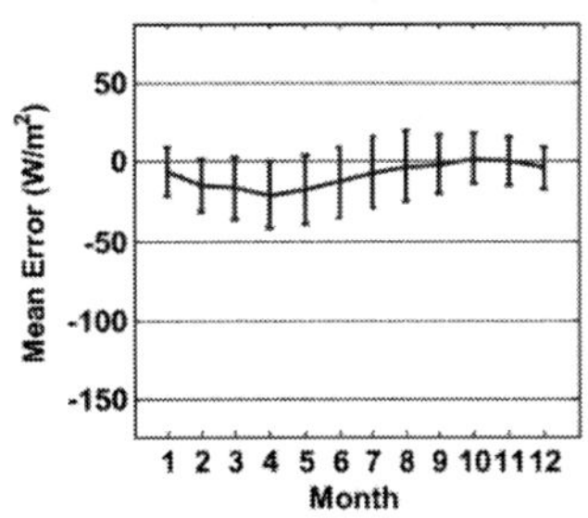

Figure 6: Seasonal deviation of the calculated irradiance of the Ineichen clear sky model from measured values averaged over 30 locations and some hundred clear sky days per location in the USA [12].

in a range of ± 3 % around the red line.

To get a clearer image of the seasonal effect, the degradation was first removed from the data points by adding the degradation of 0.22 % of the starting power to the measured power values per year. The points were then all plotted into one year. The result is shown in figure 5.

An earlier analysis of nine different clear sky models by Reno, et al. for 30 locations in the USA over a number of years and hundreds of sunny days per location confirmed some seasonal variations in these models [12]. The model of Ineichen used for our software was among those having the least seasonal variations. However, seasonal deviations about 2 - 3 % have been seen also for this model as depicted in figure 6. Such variations could be different for other locations like Sweden.

In our case, the red line in figure 5 has up to 5 % larger values in November compared to June. It might therefore be recommended to only use data from April to August with that model in Sweden.

4 CONCLUSION

A pure software solution for a fast measurement of the peak power of a PV system, as it would be measured under standard test conditions, is presented. The technique is based on fitting the simulated power curve (power as a function of time) to the measured power curve. The simulation of the module temperature using air temperature and wind speed had a good agreement with the measurements within about 5 °C.

The ability of our software to simulate the module temperature could even be useful when having temperature sensors mounted to the PV system. As revealed for the installation in Mora, faulty temperature measurements could possibly be detected.

In general, hourly based radiation data is also available from PVGIS [1], where even the angle of the modules can be taken into account. Unfortunately, no actual data for the previous years was available, e. g. for Sweden, when writing this paper.

When evaluating sunny days between April and August in Sweden, the measured global tilted irradiation matches the simulated one with about 3 % accuracy. If measurement data of the global horizontal irradiation is available, the simulated irradiation can be corrected. Such measurement data can be taken from nearby weather stations, most likely on places with nearly perfect view towards the horizon. Generally, it has to be mentioned that the simulated global tilted irradiation (GTI), with or without such a GHI correction, might be a little higher than the one at the place of the PV system: The diffuse component of the GTI constituted about 12 % of the GTI in our cases. If the PV system is surrounded by sky-shading objects like trees, the real GTI could be a few percent smaller. This would lead to an underestimation of the peak power of the PV system.

The power curves in figure 3 demonstrate that is easy to exclude periods of shading "by-eye", which is seen as a major advantage of this software.

The long-term degradation analysis of the test system at Glava Energy Center, Sweden, revealed an excellent stability with only 0.22 % power degradation per year.

An estimation about the total accuracy of our approach is not easy, but it seems that cases where the measured peak power is lower than 90 % of the rated peak power indicate a problem of the system. The advantage of this approach is that owners of PV systems could react faster to fix the problem.

ACKNOLEDGEMENTS

This work was funded by Region Värmland, Sweden and Tillväxtverket, Sweden, under project number 20201237, and continued within the project SOLVE supported by the Swedish Energy Agency under project number 52693-1, and Region Värmland under project number Nyps 20304629. The authors would like to thank Thomas Carlund, Sveriges meteorologiska och hydrologiska institut (SMHI), for discussions about correct radiation measurements and the supply of additional weather data, and Daniel Gustafsson and Tobias Pulls at Karlstad University for help with legal questions regarding the software publication.

REFERENCES

[1] European Commission, 'Photovoltaic Geographical Information System (PVGIS)', 2021. https://ec.europa.eu/jrc/en/pvgis (accessed Jun. 24, 2021).

[2] National Renewable Energy Laboratory (NREL), 'PVWatts Calculator', *PVWatts Calculator*, 2021. https://pvwatts.nrel.gov/ (accessed Jun. 24, 2021).

[3] J. E. HAY, 'Calculation of the Solar Radiation Incident on Inclined Surfaces', *Proceedings first Canadian Solar Radiation Data Workshop, Toronto, Ontario, Canada 1978*, 1978, Accessed: Oct. 08, 2021. [Online]. Available: https://ci.nii.ac.jp/naid/10025764105/

[4] P. Ineichen and R. Perez, 'A new airmass independent formulation for the Linke turbidity coefficient', *Solar Energy*, vol. 73, no. 3, pp. 151–157, 2002.

[5] M. A. Atwater and J. T. Ball, 'A numerical solar radiation model based on standard meteorological observations', *Solar Energy*, vol. 21, no. 3, pp. 163–170, Jan. 1978, doi: 10.1016/0038-092X(78)90018-X.

[6] M. Rinio, 'PVcheck Software', 2021. https://pvcheck.hotell.kau.se/

[7] M. Rinio, 'PVcheck—A Software to Check Your Photovoltaic System', *Energies*, vol. 14, no. 20, Art. no. 20, Jan. 2021, doi: 10.3390/en14206757.

[8] W. F. Holmgren, C. W. Hansen, and M. A. Mikofski, 'pvlib python: A python package for modeling solar energy systems', *Journal of Open Source Software*, vol. 3, no. 29, p. 884, 2018.

[9] D. L. King, W. E. Boyson, and J. A. Kratochvil, 'Photovoltaic array performance model. Report SAND2004-3535', *Photovoltaic System R&D Department, Sandia National Laboratories*, 2004.

[10] 'pvlib.temperature.sapm_cell — pvlib python 0.9.0+0.g518cc35.dirty documentation'. https://pvlib-python.readthedocs.io/en/stable/generated/pvlib.temperature.sapm_cell.html (accessed Oct. 08, 2021).

[11] U. Enarsson, *An Evaluation of the Condition and long term Degradation of Photovoltaic Installations using PVcheck*. 2022. Accessed: Sep. 22, 2022. [Online]. Available: http://urn.kb.se/resolve?urn=urn:nbn:se:kau:diva-91225

[12] M. Reno, C. Hansen, and J. Stein, 'Global horizontal irradiance clear sky models : implementation and analysis', Jan. 2014, doi: 10.2172/1039404.

DEVELOPMENT AND EVALUATION OF AN AUTOMATIC FAILURE DETECTION SYSTEM FOR O&M OF PV PORTFOLIOS

Eduardo A. Sarquis Filho[1,2], Bernd Kollosch[3], Björn Müller[1], Nicolas Holland[4], Klaus Kiefer[4], Christian Reise[4], Paulo J. Costa Branco[2]

[1] Enmova GmbH, Basler Str. 115, 79115 Freiburg, Germany, e.sarquis@enmova.de, b.mueller@enmova.de

[2] IDMEC, Instituto Superior Técnico, Universidade de Lisboa, Av. Rovisco Pais, 1049-001 Lisboa, Portugal, pbranco@tecnico.ulisboa.pt

[3] Pohlen Solar GmbH, Am Pannhaus 2-10, 52511 Geilenkirchen, Germany

[4] Fraunhofer Institute for Solar Energy Systems ISE, Heidenhofstrasse 2, 79110 Freiburg, Germany, klaus.kiefer@ise.fraunhofer.de

ABSTRACT: Tools that help automate operation and maintenance (O&M) processes are fundamental to expanding the ability of service providers to meet the demand of the growing number of PV systems. Within maintenance processes, the daily routine of online monitoring is a repetitive and time-consuming activity that relies heavily on manual analysis by specialized personnel. Automating this routine is challenging and requires a reliable automatic fault detection (AFD) system. This study proposes an AFD system consisting of nine algorithms primarily grounded in experts' knowledge that analyze different aspects of the PV system for abnormal behavior. The algorithms were individually developed, tested, and validated with the help of 5 years of data, including tickets created by manual monitoring, from 80 rooftop PV systems. The complete AFD system was then tested by monitoring 170 systems for one month, where it correctly detected 99% of the problems and issued less than 12% of false alerts. It also identified problems (13.2% of the total of actual faults) that passed unnoticed by conventional monitoring. The proposed AFD system provides a promising solution for automatizing part of the online monitoring routine and helps operators expand their capacity to monitor more systems.

Keywords: PV System, Monitoring

1 INTRODUCTION

Tools that help automate maintenance processes are key to expanding the capacity of O&M service providers to meet the demand of the growing number of PV systems. Within maintenance processes, the daily online monitoring routine is a repetitive and time-consuming activity that requires specialized personnel to analyze hundreds (even thousands) of performance charts, one by one, creating tickets for the abnormalities found.

The main challenge to automating this routine is the need for automatic fault detection (AFD) algorithms capable of reliably monitoring all critical aspects of the PV system [1]. In our experience, AFD systems are often not used in practice due to the high number of alerts. Too many false alerts or unnoticed major failures are unacceptable and undermine the reliability of an automatic monitoring system. They end up deactivated and only human fault detection is used.

Several studies and procedures for AFD in PV systems are available in the literature [2, 3, 4, 5], but there is still little experience with these procedures with real-time field data. Therefore, there is no consensus on the best approach. This is mainly due to the lack of a common basis for comparing the different methods, as well as a reference of actual faults to evaluate the effectiveness of fault detection, as noted by a recently published IEA PVPS Task 13 report on the topic [6].

This study proposes an innovative approach to address this challenge and create an AFD system adopting not only one but a set of nine AFD algorithms grounded in expert knowledge, assisted by statistical and machine learning methods, validated with field data, and designed to be flexible and adaptable to the characteristics of PV systems and O&M service providers.

The algorithms were individually developed, tested and validated with the help of 5 years of data from 80 rooftop PV systems. The performance of the algorithms was measured by comparing the automatic alerts with the tickets created using conventional monitoring, where the alerts are based on human observation.

The performance of the complete AFD system was tested in monitoring 170 systems for one month. False alerts were analyzed in details in order to refine the performance indicators and identify limitations of the proposed system.

This work is part of the OptOM research project (see acknowledgements), within which an integrated, cost-optimized operational management concept for the economic lifetime of PV power plants will be developed and implemented. The study complements previous studies and fault detection [7] on long-term performance loss [8] in the pursuit of a cost-optimized maintenance plan to correct faults and reduce power losses.

2 ONLINE MONITORING OF PV SYSTEMS

According to recent best practices guidelines for the O&M of PV systems, the "operations team of the O&M service provider is responsible for continuously monitoring and supervising the solar PV power plant conditions and its performance" [9]. This is usually done remotely in plant operations centers where the monitoring team performs its data analysis, detects faults, and creates "tickets" for the maintenance service team.

The monitoring team counts on monitoring software systems to assist their analysis. Still, in the authors' experience, the automatic alert features within these systems are often not used due to the high number of alerts. Basic alerts are typically related to the breach of predefined ranges or thresholds. Without further investigation, they convey little information. The lack of time to investigate a large number of alerts leads to the deactivation of the automatic alerts. Failure detection is done solely by observing graphs and indicators.

3 AUTOMATIC FAULT DETECTION SYSTEM

The development of an automatic fault detection (AFD) system to assist in monitoring a portfolio of photovoltaic systems should provide a reliable daily evaluation of the system status. This means that ideally, no fault should go unnoticed, and the tolerance for false alerts is very low. Previous studies suggested that the combination of AFD algorithms potentially increases overall sensitivity [7]. Thus, the AFD system proposed here considers a set of nine AFD algorithms that analyze different aspects of the PV system in search of abnormal behavior.

The development of these AFD algorithms focused on automating the daily monitoring routine, and was inspired primarily by expert knowledge but also by literature research and O&M contract requirements. Here is a brief description of the nine AFD algorithms developed:

- **Communication Fault (COMM)**: Identify large data gaps in the data sent by the inverters. Detects loss of communication with the PV system or its inverters.
- **Inverter Outage (IOUT)**: Identify days when the inverter produced very little or no power, but significant solar irradiance was available. Detects when the inverter is completely out of service.
- **Open Circuit Condition (OPC)**: Identify the occurrence of inverters in open circuit condition, i.e., zero (or almost zero) AC power output and input of DC voltage above normal. Detects the deactivation (idle state) of inverters (temporary or permanent).
- **Inverter Relative Underperformance (IRU)**: Compares the daily specific yield of inverters connected to arrays with the same orientation. Detects sections of the PV system operating with very different efficiencies (daily).
- **DC Current Comparison (DCC)**: Compares the DC current of inverters connected to arrays with same orientation and identifies deviations from the median behavior. Detects sections of the PV system operating with different efficiencies (every instant).
- **Inverter Late Wake Up (LWU)**: Compares the starting time of inverters connected to arrays with the same orientation. Detects sections of the PV system sensitive to morning dew due to insulation problems.
- **PR Sudden Drop (PRSD)**: Compares the daily PR with the average daily PR from the last months. Detects any problem that affects the PV system performance suddenly (daily).
- **High Specific Energy Loss (HSEL)**: Measures the difference between actual and simulated daily specific energy yields for each inverter. Detects any problem that causes significant energy loss compared to simulation (daily).
- **High-Performance Loss (HPL)**: Measures the relative difference between actual and simulated daily specific energy yield for each inverter. Detects any problem that causes significant performance loss compared to simulation (daily).

Each AFD algorithm was designed so that its sensitivity can be adjusted by setting a detection threshold. This flexibility is key for two reasons:

(a) Allow algorithms to be adapted to the singularities of the individual PV system under monitoring;

(b) Allow adjusting the fault detection sensitivity for the whole portfolio of PV systems, or parts of it, according to the characteristics of the O&M service provider or its clients and contracts (responsiveness, contractual requirements, available resources, logistics, etc.).

The AFD algorithms work on the data collected by a conventional monitoring system, which measures power, voltage, and current at the inverters' input and output, as well as the irradiance at the solar PV array's plane of incidence.

It is of fundamental importance that the automated fault detection system is adaptable to the monitored systems and adjustable to the level of sensitivity desired by the O&M service provider. To this end, a preparation phase dedicated to modeling the system and adjusting the fault detection thresholds should precede the use of the system in the daily routine. Figure 1 depicts the complete flowchart of the proposed AFD system.

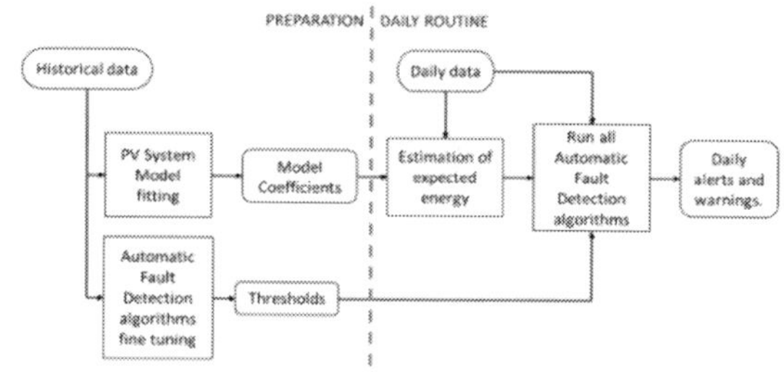

Figure 1: Flowchart of the proposed AFD system.

4 TESTING METHODOLOGY

For the development, test and validation of the AFD algorithms, we used 5 years of monitoring data and daily maintenance tickets from a portfolio of 80 rooftop PV systems. The fault alerts are also calculated daily for comparison with the daily maintenance tickets. The occurrence of an alert on a day with a maintenance ticket is a True Positive (TP), i.e., a correct fault detection. True Negatives (TN) are days with no alerts and no tickets, that is, days of normal operation. A False Positive (FP) happens if an alert occurs on a day with no ticket, which is a false alert. Conversely, a False Negative (FN) is when no alert occurs on days with tickets. Those are undetected faults. The sums of the occurrences in these classes are used to calculate sensitivity and specificity using the following equations.

$$\text{Sensitivity} = \sum TP \,/\, (\sum TP + \sum FN) \quad (1)$$
$$\text{Specificity} = \sum TN \,/\, (\sum TN + \sum FP) \quad (2)$$

Sensitivity measures how many existing faults (assumed a maintenance ticket is equal to an actual fault, see below) were correctly detected. Specificity measures how many of the negatives are truly negative. Additionally, the average number of false alerts per year per system is calculated for a more practical understanding of the specificity of the methods.

Not all tickets in the maintenance records are associated with problems that significantly affect system performance. The daily specific energy loss (SE_{loss}) indicates the relevance of the maintenance ticket in terms of system performance. The calculation of SE_{loss} is defined by

$$SE_{loss} = (E_{exp} - E_{out}) / P_{nom.} \quad (3)$$

where P_{nom} is the nominal power of the PV system, E_{out} is the energy output obtained from the integrated power output over time, and E_{exp} is the expected energy according to a prediction model.

A sensitivity weighted by the ticket relevance can be calculated using (4), where TP_{rel} is the sum of the SE_{loss} of all true positive alerts, and FN_{rel} is the sum of the SE_{loss} of all false negative alerts.

$$\text{Weighted Sensitivity} = TP_{rel} / (TP_{rel} + FN_{rel}) \quad (4)$$

The detection thresholds can adjust the sensitivity of the AFD algorithms individually. An increase in the level of sensitivity generally leads to a decrease in specificity, i.e., an increase in the occurrence of false alerts. A large number of false alerts is not tolerable in the monitoring of PV systems, so achieving high specificity is critical for an AFD algorithm. In this study, the thresholds were manually set to maximize sensitivity while keeping specificity above 0.95 whenever possible.

To evaluate the performance of algorithms combined, we merge all alerts before calculating the performance indicators. To measure the value added by each algorithm, we used the best first search forward selection strategy. In this strategy, we start with an empty set of algorithms and add one algorithm per round. The order in which the algorithms are added is defined according to the performance increment obtained. In each round, the algorithm that brings the most performance increase to the AFD system is selected.

Maintenance tickets are a record of the plant's history and the best available reference for measuring the quality of AFD algorithms. However, they are not a perfect reference as they reflect the actions of monitoring personnel and are subject to human error. To reduce the influence of these errors on the evaluation of the proposed AFD system, an additional and more detailed test was performed with a small and more recent set of monitoring data. The dates with FP or FN alerts are further investigated to verify if they are indeed false alerts or if they are unrecorded faults correctly detected by the AFD system. For this test, 170 PV systems were monitored for one month, from October 1, 2021 to November 4, 2021, summing up to 5491 days.

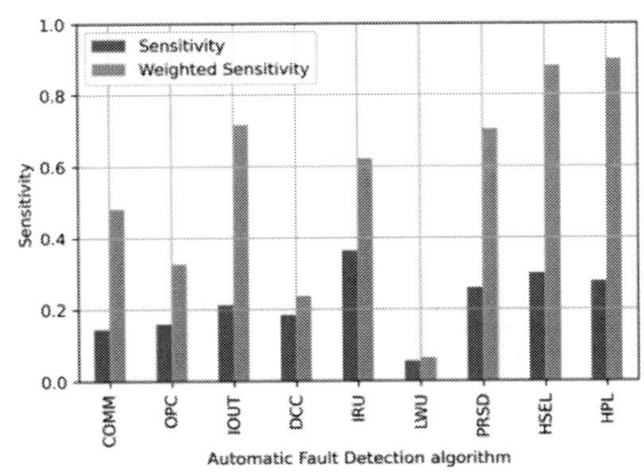

Figure 2: Sensitivity of AFD algorithms.

5 RESULTS AND DISCUSSION

The individual performance of the AFD algorithms showed sensitivity values ranging from 0.055 to 0.363, which means that in the best case, the algorithms could detect 36.3% of the tickets. However, the weighted sensitivity range reached up to 0.90, which indicates that the algorithms can detect up to 90% of the specific energy losses. See the summary of all individual performances in Table I and a graph of the sensitivities in Figure 2.

The combination of AFD algorithms resulted in higher sensitivity and weighted sensitivity, 0.547 and 0.985, respectively. This means that together the algorithms can detect 54.7% of the tickets and 98.5% of the energy losses. As expected, the specificity of the combination is lower than that of the algorithms in isolation. It is down to 0.81, which translates to an average of 46 false alerts per year per system. This number of false alerts is somewhat high, but could be due to real failures not recorded in the tickets. The next test brings more clarity about the actual percentage of false alerts.

Table I: Summary of AFD algorithms individual performances.

Algorithm	Sens.	W.Sens.	Spec.	FP/year
COMM	0.145	0.481	0.972	5.216
IOUT	0.212	0.715	0.973	4.110
OPC	0.159	0.326	0.970	6.061
DCC	0.185	0.236	**0.975**	3.956
IRU	**0.363**	0.620	0.882	**27.404**
LWU	0.055	0.064	0.970	9.011
PRSD	0.258	0.703	0.954	8.296
HSEL	0.299	0.880	0.932	15.841
HPL	0.277	**0.900**	0.954	8.688

Table II: Summary of the performance of the combined AFD algorithms.

Round	AFD added	Sens.	W.Sens.	Spec.	FP/year
1	IRU	0.363	0.620	0.882	27.40
2	PRSD	0.446	0.887	0.861	31.81
3	HSEL	0.502	0.966	0.836	39.12
4	COMM	0.521	0.975	0.828	40.82
5	OPC	0.539	0.979	0.818	43.55
6	HPL	0.542	0.984	0.818	43.60
7	LWU	0.544	0.984	0.812	45.36
8	DCC	0.546	0.984	0.810	46.06
9	IOUT	0.547	0.985	0.810	46.08

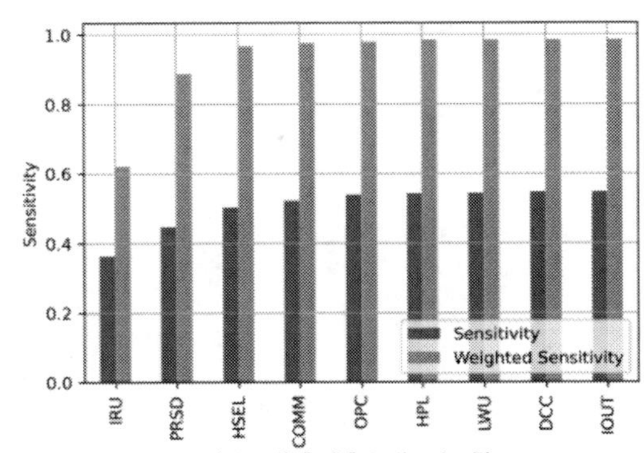

Figure 3: Sensitivity of AFD algorithms combined.

In the last test, the evaluation of the 5491 days of operation from 170 PV systems resulted in 2047 alerts from the AFD system, while the conventional routine created 1153 tickets. When comparing these tickets, we observed that 1105 tickets referred to common events, 942 tickets were apparently false alerts, and the automated system did not report 48 problems. See Table IV. The resulting sensitivity and specificity are 0.958 and 0.782, respectively. These values are consistent with the previous tests with historical data.

Table IV: Confusion matrix that compares the conventional and automatic alerts prior to detailed analysis.

		Automatic Alerts	
		P	N
Conventional Alerts	P	1105	48
	N	942	3396

Detailed analysis of false alerts revealed that of the 942 apparent false alerts, 705 are real problems not included in the manual tickets:

- 351 are problems caused by shading;
- 25 are problems caused by soiling;
- 90 are minor events considered to have little or no practical relevance (some happened during low irradiance); and
- 239 are actual problems not identified in the conventional monitoring routine.

Although none of the algorithm specializes in detecting shading or soiling, it is interesting to note that the AFD system detected the impacts caused by these problems on the PV system.

On the other side, just 48 problems passed unnoticed by the AFD system, from which only 17 were serious problems. Figure 4 presents a summary of this comparison and Table V presents the confusion matrix resulting from the comparison after detailed analysis.

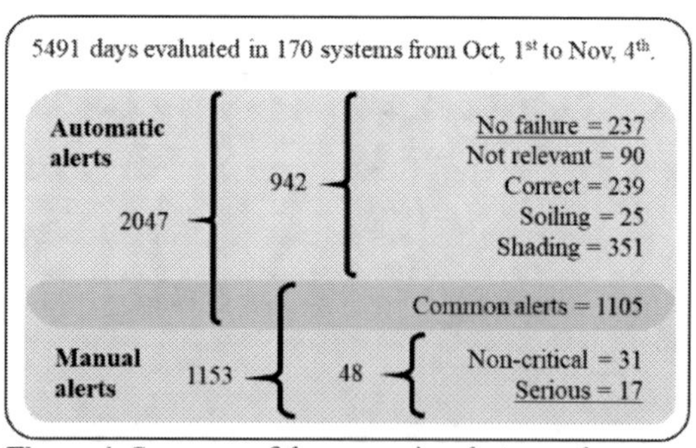

Figure 4: Summary of the comparison between the tickets created by the conventional monitoring and the proposed automated system.

Table V: Confusion matrix comparing the conventional and automatic alerts after the detailed analysis.

		Automatic Alerts	
		P	N
Conventiona l Alerts	P	1810	17
	N	237	3427

From the confusion matrix, the resulting sensitivity is 99.07%, i.e., more than 99% of the problems observed in the period were correctly detected by the automated system. The new specificity is now 0.935, an increase of 19.4% from the initial value. This indicates that a significant part of the false alerts is indeed fault events and the AFD algorithms' performance is probably better than observed in the previous tests. The system's precision, which is the number of false alerts (FP) in relation to all alerts (TP+FP), is 88.42%, i.e., less than 12% of the automatic tickets are false alerts.

It is important to note that 13.2% of the actual problems went unnoticed by the conventional monitoring routine, a result that concretely reveals the limitations of conventional monitoring tickets.

6 FINAL CONSIDERATIONS AND OUTLOOK

The proposed AFD system combining different AFD algorithms provided a promising solution for automating part of the online monitoring routine. In a practical test of monitoring 170 systems for one month, it achieved a sensitivity level of 99% with fewer than 12% of false alerts. It also detected problems (13.2% of the total actual faults) that passed unnoticed by the conventional monitoring routine.

The limitations on the conventional monitoring tickets are probably due to the limited time dedicated to each system (due to high-cost pressure from the O&M contracts) and its strong dependency on repetitive manual tasks, something naturally prone to human errors. The proposed solution will help the operator focus on the systems showing abnormal/suspicious behavior, leading to better-performing PV systems. Furthermore, it will expand the operator's capacity to monitor more systems.

The study showed that expert knowledge provides a solid foundation for developing high-performance AFD algorithms. Future studies will focus on further improving these algorithms with data-driven approaches and artificial intelligence.

7 ACKNOWLEDGEMENTS

The authors thank all co-workers at Fraunhofer ISE, Pohlen Solar, Enmova, and IDMEC for supporting this work. This work was partly supported by FCT, through IDMEC, under LAETA, project UIDB/50022/2020. This work was partly financed by the Federal Ministry for Economic Affairs and Energy of Germany (BMWi) within the framework of the research program "Innovations for the energy transition" under contract number 03EE1058.

8 REFERENCES

[1] A. Louwen, D. Miorandi, C. Torrero, F. Venturini, and D. Moser. "*Fault Detection in Operation and Maintenance of PV Systems*." In Proceedings of the 38th European Photovoltaic Solar Energy Conference and Exhibition, pp. 967-972, 2021.

[2] S. R. Madeti, S. N. Singh, A comprehensive study on different types of faults and detection techniques for solar photovoltaic system, Solar Energy 158 (2017) 161–185. doi:10.1016/j.solener.2017.08.069.

[3] A. Mellit, G. M. Tina, S. A. Kalogirou, Fault detection and diagnosis methods for photovoltaic systems: A review, Renewable and Sustainable Energy Reviews 91 (2018) 1–17. doi:10.1016/j.rser.2018.03.062.
[4] D. S. Pillai, F. Blaabjerg, N. Rajasekar, A Comparative Evaluation of Advanced Fault Detection Approaches for PV Systems, IEEE Journal of Photovoltaics 9 (2) (2019) 513–527, conference Name: IEEE Journal of Photovoltaics. doi:10.1109/JPHOTOV.2019.2892189.
[5] M. N. Akram, S. Lotfifard, Modeling and Health Monitoring of DC Side of Photovoltaic Array, IEEE Transactions on Sustainable Energy 6 (4) (2015) 1245–1253, conference Name: IEEE Transactions on Sustainable Energy. doi:10.1109/TSTE.2015.2425791.
[6] M. Green et al., "*The Use of Advanced Algorithms in PV Failure Monitoring,*" tech. rep., Report IEA-PVPS T13-19, 2021.
[7] E. A. Sarquis Filho, K. Kiefer, N. Holland, B. Kollosch, B. Müller, and PJ Costa Branco. "*Analysis of Automatic Fault Detection Methods for Commercially Operated PV Power Plants*." In Proceedings of the 38th European Photovoltaic Solar Energy Conference and Exhibition, pp. 961-966, 2021.
[8] K. Kiefer, B. Farnung, B. Müller, K. Reinartz, I. Rauschen, and C. Klünter, "Degradation in PV Power Plants: Theory and Practice," in *36th European Photovoltaic Solar Energy Conference and Exhibition*, Oct. 2019, pp. 1331–1335.
[9] A. Ara, R. Gottschalg and B. Clarke. "*O&M Best Practice Guidelines Version 5.0.*" tech. rep., SolarPower Europe, 2021.

ROBOPV: AN AERIAL ROBOTS' EMBEDDED CODE FOR INTELLIGENT MONITORING AND FAULT DETECTION OF LARGE-SCALE PV PLANTS USING DEEP NEURAL NETWORK MODELS

A. M. Moradi Sizkouhi[1,2], S.M. Esmailifar[2], M. Karimkhani[2], M. Aghaei[3,4*]
[1]Department of Electrical and Computer Engineering, Concordia University, Montreal, QC H3G 1M8, Canada
[2]School of Mechanical, Aerospace and Marine Engineering, Amirkabir University of Technology, 15875-4413 Tehran, Iran
[3]Department of Ocean Operations and Civil Engineering, Norwegian University of Science and Technology (NTNU), 6009 Ålesund, Norway
[4]Department of Sustainable Systems Engineering (INATECH), University of Freiburg, 79110 Freiburg, Germany

*mohammadreza.aghaei@ntnu.no; aghaei@studysolar.uni-freiburg.de

ABSTRACT: In this paper, a novel embedded software package, called RoboPV, is introduced for autonomous aerial monitoring of large-scale photovoltaic (PV) plants. RoboPV enables aerial robots to automatically perform aerial monitoring of PV plants, from optimal trajectory planning to real-time fault detection using deep neural network models. This software package is made of four systematically integrated parts: boundary detection, path planning, dynamic processing, and fault detection. The first requirement of efficient path planning is to identify the PV plant boundary closed curve. This requirement is prepared by an encoder-decoder deep learning architecture which is trained by labeled real flight aerial images of PV plants. Subsequently, a novel path planning algorithm is conducted by RoboPV to design an optimal flight path with full coverage of whole regions of the PV plant. Another deep neural network model has been also trained for real-time fault detection in PV plants through PV faults labeled dataset. Using this highly precise trained neural network, aerial images are analyzed in real-time during the flight and when a probable fault is detected, RoboPV commands the aerial robot to maneuver and concentrate more on the detected fault. In this study, several decision-making and maneuver algorithms are developed for various real-world flight conditions to improve the performance of RoboPV during an autonomous aerial inspection. RoboPV is a modular processing library that can be installed on any micro-computer processor with a low computational burden. Moreover, supporting the MAVLink communication protocol enables RoboPV to connect with an intelligent Pixhawk flight autopilot and navigate a wide range of multi-rotors. To demonstrate the performance of RoboPV, a six degrees of freedom dynamic model of a multi-rotor has been developed in a Simulink environment, and processor in the loop (PIL) tests have been performed for aerial monitoring of two different real megawatt-scale PV plants.

Keywords: Photovoltaic (PV) power plant, Autonomous aerial monitoring, Aerial Robots, Encoder-decoder architecture, autonomous fault detection, RoboPV

1 INTRODUCTION

The aerial inspection provides detailed information about Photovoltaic (PV) systems' conditions in a reliable and faster way [1]. In traditional inspection, human operators have played a significant role in the monitoring of PV systems [2]. However, aerial inspection techniques facilitate the faults diagnosis process for PV plants, especially for large-scale ones [3]. The failures and defects can be detected on PV modules by different sensors and special instruments. In this approach, the conventional methods have been replaced with artificial intelligence (AI)-based techniques and high-precision multi-rotors in periodic inspection missions of PV plants [4]. In a semi-autonomous inspection, the multi-rotor flies on an optimal path created by a Ground Control Station (GCS). In this method, flight data are received during the flight, but video streams transmitted to the GCS are not analyzed in real time. In this method, the multi-rotor cannot perform any maneuver to improve the accuracy of the aerial monitoring mission, and offline data analysis is still one of the disadvantages of this method.

An autonomous system for aerial monitoring of PV plants has not yet been developed, with desired capabilities, namely, bi-directional communication with a wide range of different multi-rotor configurations, accurate detection of failures in PV arrays with higher accuracy via deep learning models, guidance and control the multi-rotor in flight and plan a new trajectory. In one of our previous works [5], an automated control system has been developed using a designed multi-rotor and a GCS platform to provide accurate information on PV plants' operating conditions. In another recent study, an encoder-decoder architecture of CNNs has been developed for accurately detecting the potential defects on PV modules using aerial images collected through several experimental flights by multi-rotor [6].

Here, we have developed and designed a novel PV plant monitoring software platform, called "RoboPV", for autonomous inspection of PV plants, along with the project development and based on our previous research studies [4, 5, 7-10]. The proposed processing platform enables different multi-rotors to automatically monitor large-scale PV plants and analyze PV systems' performance and conditions. RoboPV is an intelligent data processing platform which can be installed on an onboard microprocessor, e.g., RaspberryPi 4, and interacts with the 3DR Pixhawk autopilot to command different multi-rotors.

RoboPV employs improved deep learning algorithms to enhance the performance of aerial inspection compared to conventional and semi-autonomous methods. An improved encoder-decoder architecture is trained with a designed Fully Convolutional Network (FCN) as a backbone to detect the PV plant's boundary points accurately. Furthermore, for automatic fault detection, the real-time collected images are analyzed by RoboPV to localize potential faults on PV modules. For this purpose, RoboPV uses an accurate trained image segmentation model to predict the exact position of any type of failure on PV modules with pixel-level accuracy.

This paper is organized as follows: Section 2 introduces RoboPV and provides detailed information about its construction and how it processes flight data and executes decision-making during the aerial inspection mission. Section 3 presents the dynamic modelling of the multi-rotor used in the simulation verification and also the results and performance verifications of RoboPV in autonomous aerial monitoring of two PV plants are discussed. Finally, Section 4 draws some conclusions with recommendations for future research.

2 ROBOPV SOFTWARE PACKAGE

Using RoboPV, any multi-rotor can be upgraded with different intelligence levels so that entire strings of PV plants can be inspected automatically, from path planning and image collection to online image processing for failure detection, flight data analysis, and decision-making for remedial action. An automated and integrated embedded software package is proposed for RoboPV, which includes four modules: boundary detection, path planning, fault detection, and dynamic processing. As can be observed in Fig. 1, these four units perform the whole procedure of PV plant monitoring in four phases:

1. Boundary Detection: RoboPV extracts the boundary area of the PV plant in a pre-flight step using an encoder-decoder network based on an FCN.
2. Path Planning: Using the path planning algorithm, an optimal flight path is generated with the shortest total distance to cover the entire PV plant. Depending on the multi-rotor's flight endurance and maneuverability, this path is designed.
3. Dynamic Processing: The flight computer of the multi-rotor interacts with the onboard microprocessor of RoboPV during the aerial inspection. RoboPV monitors all flight data to make decisions based on defined algorithms. Whenever any PV module failures are detected during the mission, RoboPV sends a control signal to Pixhawk to maneuver the multi-rotor on the possibly faulty region for further investigation. Dynamic processing examines the multi-rotor's ability to complete the mission after each maneuvering process. For the remainder of the mission, an optimal path is replanned if necessary.
4. Fault Detection: An onboard microprocessor processes the collected video stream of PV modules online to extract feature maps of target defects and failures on the PV modules for accurate position location.

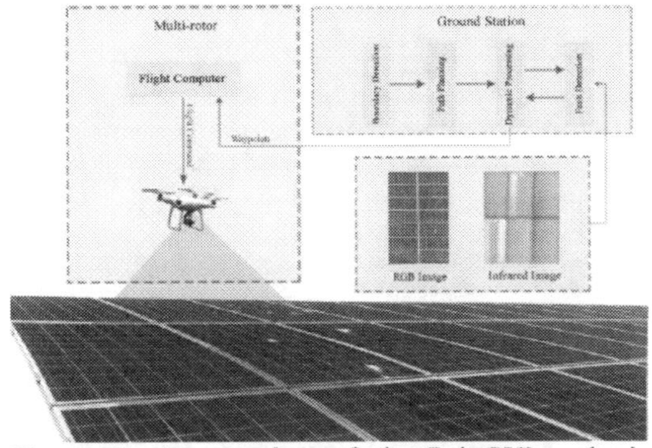

Figure 1: An overview of the RoboPV's role in autonomous aerial monitoring of a PV plant.

2.1 Boundary Detection

As an encoder part of RoboPV, a modified FCN backbone is used to determine the pixel-level boundary of a PV plant. With this technique, each pixel is labelled to determine whether it belongs to the PV string's aerial image. The structure of the proposed network is depicted in Fig. 2.

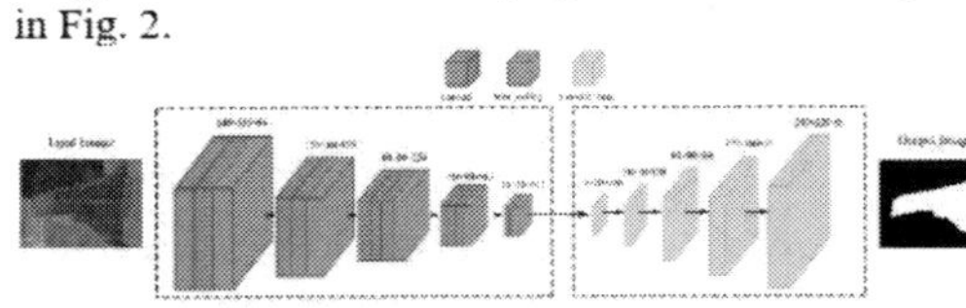

Figure 2: The architecture of the encoder-decoder network that is used for boundary detection.

To develop an accurate image segmentation model, a dataset of aerial images of large-scale PV plants, entitled "Amir", was developed elsewhere [11]. There are 3584 aerial images of PV plants from twelve countries in Amir. Through convolutional operations, the encoder calculates low-resolution feature maps from aerial images of PV plants and their masks. To reconstruct images with extracted PV plants, the features are imported into the decoder part of the network and up-sampling is performed. For training and evaluating the model, 80% of the total images are randomly used for the training step, and the remaining 20% are used for examining the trained model's performance in predicting unknown data. Training and testing processes have an accuracy rate of 97.61% and 96.99%, respectively [11]. As illustrated in Fig. 3, the trained network's prediction is accurate enough in boundary detection on three given PV plants.

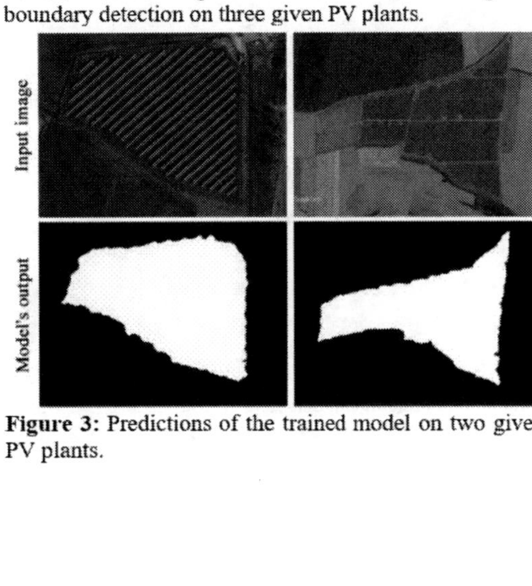

Figure 3: Predictions of the trained model on two given PV plants.

2.2 Path Planning

To send waypoints to an aerial robot, an accurate set of waypoints must be calculated. A precise estimation of the PV plant's boundary is required for the path planning algorithm. The best path for the multi-rotor is one that enables it to inspect all PV modules in the shortest distance without creating blind spots in the camera's field of view, especially near the border regions .During path planning, the main objective is to generate the shortest possible path on a PV plant to minimize aerial inspection time. Due to 90° field of view (FoV) of the aerial camera used in this experiment, the multi-rotor covers a 30 × 30 m2 area on the ground by flying at a fixed 15m altitude. The final calculated flight trajectories are shown with red lines on these PV plants in Fig. 4. Using the generated optimal path, the multi-rotors will cover all areas of the PV plant, specifically the corners, based on the camera's field of view.

Figure 4: The outputs of the path planning on three sample PV plants.

2.3 Dynamic Processing

RoboPV transmits the waypoints calculated by the path planning to Pixhawk through an application programming interface (API) established for MAVLink. As a client, RoboPV communicates with Pixhawk via APIs. The dynamic processing of PV plants involves three main subsystems, each with a defined role. The schematic of the dynamic processing is shown in Fig. 5. As the multi-rotor flies, the dynamic processing analyzes its flight data to make fast decisions. As illustrated in Fig. 5, this unit also interacts with other components of RoboPV and the Pixhawk flight controller. The dynamic processing is therefore responsible for executing the maneuver action, including a set of actions for precise inspection of the target region, after investigating the fault detection. The decision-making subsystem of the dynamic processing is called up after each monitoring maneuver to analyze the flight condition and make fast decisions. This step determines whether the mission can be completed by the multi-rotor on its current path.

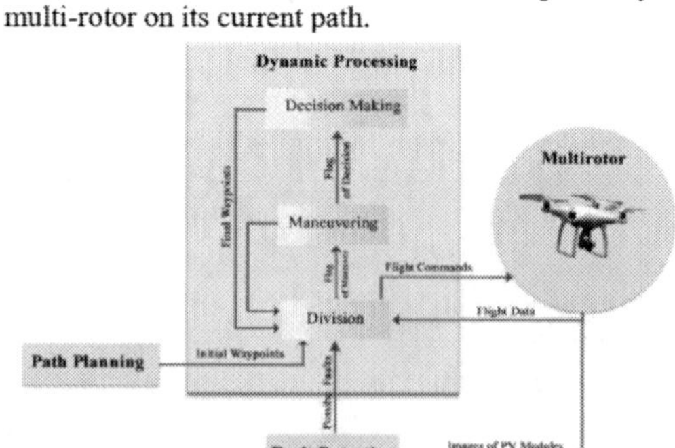

Figure 5: The architecture of the Dynamic Processing unit.

Upon determining the possibility of damage to the PV modules, the maneuver command is issued. A maneuver action is defined as a change in multi-rotor's states for the purpose of examining the target area in a PV string carefully. When an aerial inspection of a PV plant is conducted, the following situations can be considered. In the first step, the damage is specified, and the multi-rotor is beginning to maneuver. In order to get closer to the PV modules affected, the multi-rotor reduces its altitude. In this phase, the multi-rotor hovers at a new altitude to analyze the region precisely. Once the PV module failure has been detected, the multi-rotor returns to its previous flight altitude and waits for new commands from RoboPV.

A significant part of autonomous aerial monitoring is decision-making. It determines whether the multi-rotor can continue its mission on its current path after every maneuver. An increase in flight altitude is used to increase the multi-rotor's field of view on the ground when designing a new trajectory. Using the MAVLink communication protocol, decision-making sends the waypoints to the Pixhawk autopilot if the new altitude is feasible.

2.4 Fault Detection

After the multi-rotor begins flying over the PV plant, the fault detection unit begins analyzing the aerial images of the PV modules. For image collection, different cameras are available with different frames per second rates. By contrast, the multi-rotor flies at a constant 5 meter per second cruise velocity and moves 5 m forward per second. As a result, while the fault detection unit analyzes the frames, the multi-rotor has moved 5 meters since the previous time step. The displacement of the multi-rotor during the image processing by fault detection on the microprocessor has no effect on reducing the accuracy of the RoboPV in the automatic fault detection, and it is sufficient overlap (20 × 30 m^2) between the frames recorded in the previous and current time step. An aerial RGB image is first imported into a PV module extractor. Using some image processing techniques, the module extractor recognizes modules.

Based on the VGG16 modified model, feature maps are extracted from images and masks first. According to the dimensions of the input image to the model, the FCN structure adjusts the first layer dimensions, called the input layer, as x_i. Input images have dimensions of 640 pixels wide, 480 pixels high, and 3 pixels deep. A convolutional layer is used later in the encoder unit to increase the dimension of each layer. Image and mask two-dimensional matrices are entered into the network as inputs. Next, the features are imported into the decoder part of the network for up-sampling to reconstruct images with bird droppings. Enhanced drop segmentation is achieved by enhancing the decoder part of the proposed encoder-decoder network.

The trained network's output is a binary image where pixels predicted to belong to the multi-rotor's dropping have 255 values and the rest have 0. Several well-known performance measures were used to evaluate the network performance, including the Dice (F1) score, Jaccard (J) or Intersection-Over-Union (IoU) score and pixel accuracy. In segmentation problems, the Jaccard score is commonly used. This segmentation index is calculated by dividing the overlap region between the predicted segmentation object and the annotated mask by the union area. The neural network is trained in two stages. A down-sampling operation of the images is performed first to calculate low-resolution feature maps; a second up-sampling operation is performed to reconstruct the image with the desired region of the potential bird's drops.

A base learning rate of 0.0005 is used for network training. Training and validation loss rates and accuracy rates are reported by the system at each stage.

To prepare extracted modules for training, some preprocessing operations are required. A bird drop examination is conducted on each PV module to determine if there are bird drops present. There are 23 epochs in the training. A batch size of 32 is selected, and images are shuffled to produce diverse images in each batch. Binary cross entropy is used as a loss function. The encoder-decoder model is then given extracted modules of PV strings for testing its performance. The outputs are shown in Fig. 6. Using training data that remained unseen during training, the trained model can accurately determine the location of the multi-rotor's dropping on PV modules with an average accuracy of 93.33%.

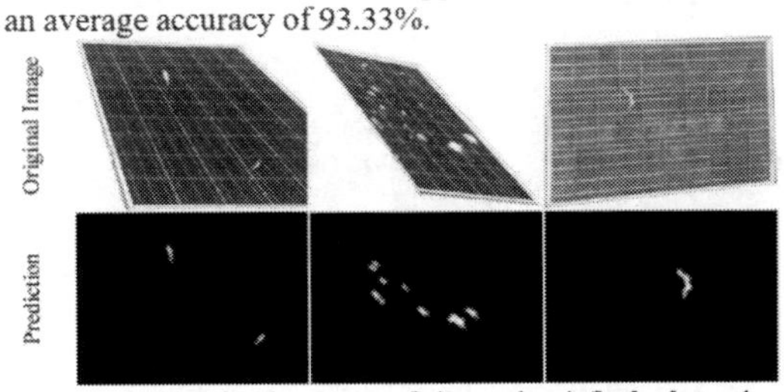

Figure 6: The final output of the trained fault detection model for different affected PV modules by birds' drops.

Table I: The performance measurements of the FCN with defined metrics.

Activations	Pixel Accuracy	IoU Score	F1 Score
RELU	0.93	0.42	0.63
TanH	0.90	0.41	0.61
Sigmoid	0.95	0.46	0.62

3 SIMULATION AND RESULTS

For model-based design of control system, first requirement is the mathematical model of the plant which is a multi-rotor, in this paper. Multi-rotors and especially quad-rotors are a highly non-linear, under-actuated, and intrinsically unstable system which are used as a test case of control system design in several researches [12]. The flight control system has several functions, which includes vehicle stabilization and position tracking. In our laboratory, based on the performance of the Pixhawk 4 Holybro, it is selected for control the aerial vehicles in different operational modes. Pixhawk control system possesses of three separate cascade PID based controllers as:

1. Altitude controller: Control and change the multi-rotor's altitude by increasing or decreasing all motor angular velocity simultaneously during flight.
2. Attitude controller: Changing multi-rotor's Euler angles by generating pitch, roll and yaw moments.
3. Position Controller: The desired roll and pitch angles are computed and fed into the attitude controller to track generated path via multi-rotor.

3.1 MATHEMATICAL MODELING

To demonstrate the performance of RoboPV, a six degrees of freedom dynamic model of a multi-rotor is developed in a SIMULINK environment with a defined aerial monitoring mission on two different real large scale PV plants. The PIXHAWK flight control and SIMULINK environments in the PIL test were connected via MAVLink API. The six degrees of freedom dynamics model consist of rotational and translational equations of motion of a multi-rotor is derived based on the Euler-Newton formulas [13] as show in Fig. 7.

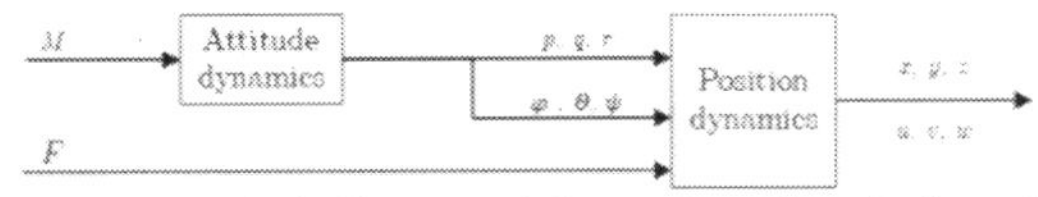

Figure 7: Block diagram of the quadrotor's attitude and positions dynamics.

3.2 RESULTS OF MONITORING

In order to show the accuracy of RoboPV's performance in an autonomous aerial monitoring of a PV plant, it is necessary to simulate the real-world conditions in the loop of Pixhawk autopilot processor as PIL tests. For this purpose, as mentioned in previous sections, firstly, the boundary detection unit extracts the PV plants' boundary points using a trained deep neural network. Then, these points enter the path planning unit as inputs. The trajectory planning unit, by considering various parameters such as the length and width of the PV plant, the dimensions of the panels, the FoV of the camera, and the velocity and altitude of the multi-rotor, designs an optimal path to cover the whole PV plant with minimum path length.

Fig. 8 show the PIL simulation trajectories of two mentioned PV Plant. It is obvious that the multi-rotor starts to fly from take-off point along the preplanned paths. It keeps flying until faces with a potential fault. After the fault detection unit has determined the potential faults on the PV modules, the dynamic processing unit commands the multi-rotor to reduce its velocity and stop at the fault detected point. Later on, the multi-rotor decreases its altitude from 15m to 5m, and hovers over the point for 60s, while sends the video stream to RoboPV in real-time for detailed investigation. At this time, the fault detection investigates the area more precisely. This procedure can be seen in some points in Fig. 8 where the multi-rotor has reduced its altitude. At the end of 60s, the multi-rotor returns to the prior altitude automatically and waits for the next command from the RoboPV.

when the multirotor performs fault detection maneuvers, it is possible that the multirotor cannot accomplish the preplanned trajectory due to its low state of charge (SOC). In this situation, the decision-making unit of RoboPv replans the trajectory with higher altitude. The replaned trajectory with higher altitude results in larger field of view with lower accuracy but shorter path length. The shorter trajectory guarantees accomplishment of the mission but maybe with lower performance. It can be seen in Fig. 9, that the multirotor has climbed to the relative hight of 18 meters about second 310. However, this phenomenon didn't occur in the second mission (Fig. 10), because of its sufficient SOC.

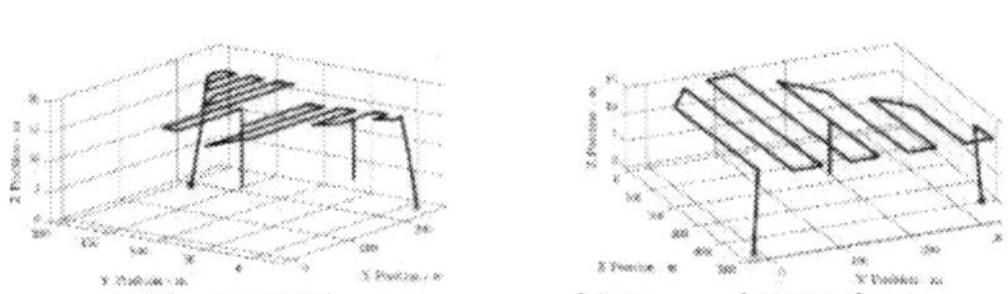

a) First PV Plant b) Second PV Plant

Figure 8: The simulated aerial inspections over PV plants. (Red dot: Take-off area, Green dot: Landing area).

The detailed of multi-rotor states during simulation and consequently fault detection of two mentioned PV Plant presented in Fig. 9 and 10.

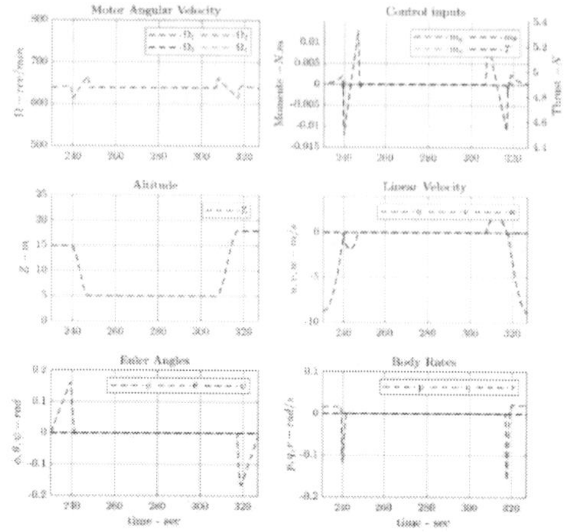

Figure 9: Multi-rotor states during the first PV plant monitoring.

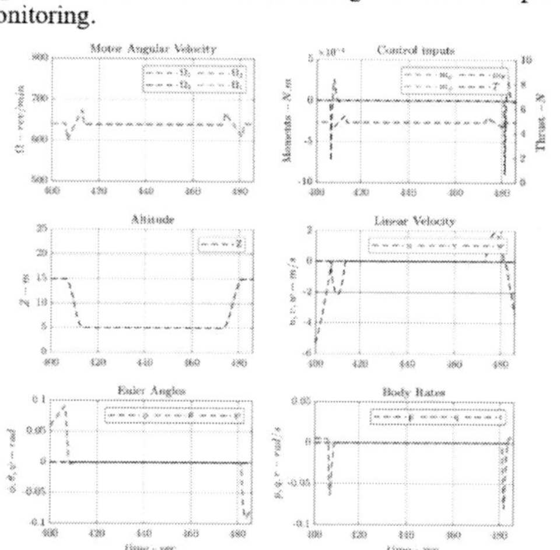

Figure 10: Multi-rotor states during the second PV plant monitoring.

4 CONCLUSIONS

In this study, an embedded software package, called RoboPV, was introduced. The main purpose of RoboPV is aerial robots' management in large scale PV plants monitoring. In order to validate RoboPV's performance in an aerial inspection, comprehensive processor in the loop (PIL) tests are conducted. To perform these PIL simulations, Pixhawk 4 Holybro autopilot is employed, as a processor, to control and navigate the multi-rotor on the desired trajectories calculated by RoboPV. RoboPV functions including path planning, dynamic processing as well as path following guidance and the multi-rotor dynamic model are simulated in real-time, in the computer environment. The results prove that RoboPV can execute the autonomous aerial inspection with an overall accuracy of 93% for large-scale PV plants. The detailed information about RoboPV have been provided elsewhere [14].

5 REFERENCES

[1] Grimaccia, F., Leva, S., Dolara, A. and Aghaei, M., 2017. Survey on PV modules' common faults after an O&M flight extensive campaign over different plants in Italy. IEEE Journal of Photovoltaics, 7(3), pp.810-816.

[2] Kandilli C. Performance analysis of a novel concentrating photovoltaic combined system. Energy conversion and management. 2013 Mar 1;67:186-96.

[3] Al-Housani, M., Bicer, Y. and Koç, M., 2019. Experimental investigations on PV cleaning of large-scale solar power plants in desert climates: Comparison of cleaning techniques for drone retrofitting. Energy Conversion and Management, 185, pp.800-815.

[4] Grimaccia, F. A. M. M. M. E. A., et al. "Planning for PV plant performance monitoring by means of unmanned aerial systems (UAS)." International Journal of Energy and Environmental Engineering 6.1 (2015): 47-54.

[5] Aghaei, M., Grimaccia, F., Gonano, C.A. and Leva, S., 2015. Innovative automated control system for PV fields inspection and remote control. IEEE Transactions on Industrial Electronics, 62(11), pp.7287-7296.

[6] Moradi, A. M., M. Aghaei, and S. M. Esmailifar. "A deep convolutional encoder-decoder architecture for autonomous fault detection of PV plants using multi-copters solar energy." Sol Energy (2021): 217-228.

[7] Eskandari, A., Milimonfared, J. and Aghaei, M., 2020. Fault detection and classification for photovoltaic systems based on hierarchical classification and machine learning technique. IEEE Transactions on Industrial Electronics, 68(12), pp.12750-12759.

[8] Eskandari, A., Milimonfared, J., Aghaei, M. and Reinders, A.H., 2020. Autonomous monitoring of line-to-line faults in photovoltaic systems by feature selection and parameter optimization of support vector machine using genetic algorithms. Applied Sciences, 10(16), p.5527.

[9] Sizkouhi, A.M.M., Esmailifar, S.M., Aghaei, M., De Oliveira, A.K.V. and Rüther, R., 2019, June. Autonomous path planning by unmanned aerial vehicle (UAV) for precise monitoring of large-scale PV plants. In 2019 IEEE 46th Photovoltaic Specialists Conference (PVSC) (pp. 1398-1402). IEEE.

[10] Eskandari, A., Milimonfared, J. and Aghaei, M., 2020. Line-line fault detection and classification for photovoltaic systems using ensemble learning model based on IV characteristics. Solar Energy, 211, pp.354-365.

[11] Sizkouhi, A.M.M., Aghaei, M., Esmailifar, S.M., Mohammadi, M.R. and Grimaccia, F., 2020. Automatic boundary extraction of large-scale photovoltaic plants using a fully convolutional network on aerial imagery. IEEE Journal of Photovoltaics, 10(4), pp.1061-1067.

[12] Zhang, X., Li, X., Wang, K. and Lu, Y., 2014, October. A survey of modelling and identification of quadrotor robot. In Abstract and Applied Analysis (Vol. 2014). Hindawi.

[13] Najm, A.A. and Ibraheem, I.K., 2019. Nonlinear PID controller design for a 6-DOF UAV quadrotor system. Engineering Science and Technology, an International Journal, 22(4), pp.1087-1097.

[14] A.M. Moradi Sizkouhi, S.M. Esmailifar, M. Aghaei, M. Karimkhani, RoboPV: An integrated software package for autonomous aerial monitoring of large scale PV plants, Energy Convers. Manag. 254 (2022) 115217.

PHOTOVOLTAICS MODULES AND SYSTEMS MONITORING WITHIN HIGHLITE H2020 PROJECT

Antonin Faes[1*], Renaud Langou[1], Gizem Nogay[1], Samuel Harrison[2], Hervé Colin[2], Carolyn Carriere[2], Vincent Barth[2], Florian Buchholz[3], Andreas Halm[3], Constantin Klik[3], Loic Tous[4], Patrick Choulat[4], Jonathan Govaerts[4], Ismail Kaaya[4], Hugo Quest[5,6], Andrew Fairbrother[5], Alessandro Vituani[5], Andrea Ingenito[5], Nicolas Wyrsch[5], Frank Feldmann[7], Dirk Rainwand[7], Max Mittag[7], Martin Heinrich[7], Felix Haase[8], Matevz Bokalic[9], Kristijan Brecl[9], Marko Topic[9], Josco Kester[10], Stefan Wendlant[11], Marco Galiazzo[12], Alessandro Voltan[12], Giuseppe Galbiati[13], Liesbeth Theunissen[13], Frank Torregrossa[14], Frederik Bamberg[15], Michael Grimm[15], Julius Denafas[16], Tadas Radavicius[16], Povilas Lukinskas[17], Jarno Kaakkunen[17], Tuukka Savisalo[17], Thomas Regrettier[18].

[1] CSEM, Rue Jaquet-Droz 1, 2002 Neuchâtel, Switzerland, [2] CEA-INES, 50 Avenue du Lac Léman, 73370 Le Bourget-du-Lac, France, [3] ISC Konstanz, Rudolf-Diesel-Straße 15, 78467 Konstanz, Germany, [4] imec, Kapeldreef 75, 3001 Leuven, Belgium, [5] EPFL, Route Cantonale, 1015 Lausanne, Switzerland, [6] 3S Swiss Solar Solutions, Schörenstrasse 39, 3645 Thun, Switzerland, [7] Fraunhofer ISE, Heidenhofstraße 2, 79110 Freiburg im Breisgau, Germany, [8] ISFH, Am Ohrberg 1, 31860 Emmerthal, Germany, [9] UL-FE, Trzaska 25, 1000 Ljubljana, Slovenia, [10] TNO, Westerduinweg 3, 1755 LE Petten, the Netherlands [11] PI-Berlin, Wrangelstraße 100, 10997 Berlin, Germany, [12] AMAT Italia SRL, Via Postumia Ovest, 244, 31048 Olmi TV, Italy, [13] HENKEL Belgium NV, Nijverheidsstraat 7, 2260 Westerlo, Belgium, [14] IBS, Rue Gaston IMBERT prolongée, ZI Peynier Rousset, 13790 Peynier, France, [15] 3D-Micromac AG, Technologie-Campus 8, 09126 Chemnitz, Germany, [16] Solitek, Mokslininkų g. 6A, Vilnius 08412, Lithuania, [17] Valoe Oyj, Insinöörinkatu 5, FI-50150 Mikkeli, Finland, [18] Voltec Solar, 1 Rue des Prés, 67190 Dinsheim-sur-Bruche, France

*Corresponding author, work phone: +41 32 720 5633, e-mail: Antonin.faes@csem.ch

ABSTRACT: In this work we give an overview of the module monitoring results achieved within the H2020 funded project called HighLite. A dedicated online monitoring platform has been developed for the project to group and treat the data from all partners with homogeneous procedures. Monitoring of advanced prototype modules developed in the project and based on various cells and interconnection approaches, like silicon-heterojunction (SHJ)-shingle, TopCon-shingle, cut interdigital back-contact cells (cut-IBC) are reported. We describe key learnings from monitoring modules for various applications including building-applied PV (BAPV) and building-integrated PV (BIPV). Annual energy yield up to 1900 kWh/kWp is expected in Cadarache (France) based on HighLite PV technology. Temperature corrected performance ratios are used for a ranking of the new technologies against the reference PERC modules.

Keywords: Building Applied PV (BAPV), Building Integrated PV (BIPV), Outdoor monitoring, Energy yield, Performance ratio.

1. INTRODUCTION

The photovoltaic (PV) industry in Europe has been under strong pressure over the past decade with low cost and high performing modules based on p-type PERC solar cells.

The HighLite project focuses on bringing 2 n-type c-Si technologies, taking profit of the strength of the project partners [1]. The first technology is based on shingle assembly of n-type heterojunction (SHJ) cells. The second one is based on industrial n-type interdigitated back-contact (IBC) cells cut in half or smaller. Shingle SHJ assembly enables the manufacturing of high-performance lead-free modules (improving sustainability) while modules with IBC cut-cells are interesting for applications demanding high power density combined with improved aesthetics.

The main objectives of the HighLite project partners are to: first, develop innovative manufacturing solutions and processes, then develop innovative PV equipment and materials, also implementing novel layers and processes in existing pilot-production lines (cells and modules), and finally demonstrate the superior performance of the optimum integration achieved with the developed technologies.

2. MONITORING IN H2020 HIGHLITE

In work-package 6 (WP6) on "Outdoor demonstrators", monitoring activities on photovoltaics (PV) modules aim to demonstrate the advantage of the module technologies developed within HighLite compared to reference commercial modules from the market.

For this purpose, monitoring of PV systems are taking place at the sites of 8 partners as shown in Fig. 1. The sites are located from North to South, in Vilnius (Lithuania), Genk (Belgium), Berlin (Germany), Konstanz (Germany), Neuchatel (Switzerland), Ljubljana (Slovenia) and Cadarache (France). In WP6 two types of installation are monitored: The first one is dedicated to Buildings Applied PV systems (BAPV) and the second one to Buildings Integrated PV systems (BIPV).

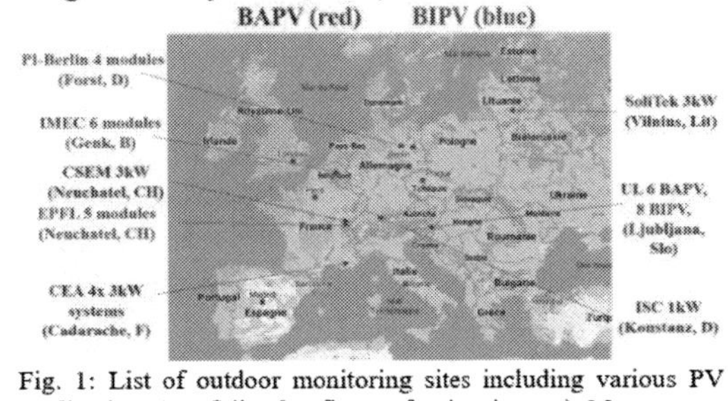

Fig. 1: List of outdoor monitoring sites including various PV applications (see following figures for the pictures). Map source: Google. (n.d.).

A. Standard for Monitoring

The IEC 61724-1 standard on "Photovoltaic system performance – Part1 [1]: monitoring" is dedicated to PV systems producing electricity where the minimum PV installation size to be monitored ranges from 0 up to 5 MWp. In HighLite, the monitoring is performed on single PV modules and, , on systems up to 3-4 kW_p in CEA-

INES' facility. One effect of having small systems is that most of the performance parameters are given easily by direct current (DC) without the loss induced by the inverter; as we will compare benchmark reference modules in the same configuration, the comparison is acceptable. One advantage of characterizing single modules is the possibility to record current-voltage curves (IV-curves) at a time interval of 3 min while the rest of the time the module is at maximal power point (mpp) with a mpp tracker, which enables to extract short circuit current (I_{sc}), open-circuit voltage (V_{oc}), the fill-factor (*FF*) as well as compare the power at maximal power point from the IV-curve (P_{max}) to the P_{mpp} during tracking.

To compare PV module performance during monitoring, the performance ratio of the array (PR'_{bi}) uses the sum over the time-interval (τ_k) of the module DC-power (P_{mp}) normalized by the indoor STC rated power (P_0), this is divided by the front side irradiance (G_i) normalized by the reference STC irradiance ($G_{i,ref}$) as shown in equation 1 [1]:

$$PR'_{bi} = \frac{\left(\sum_k P_{mp,k} \cdot \tau_k\right)}{\left(\sum_k \frac{\left(C_{k,25°C} \cdot P_0\right) \cdot G_{i,k} \cdot BIF_k \cdot \tau_k}{G_{i,ref}}\right)}$$
Equation 1

$$C_{k,25°C} = 1 + \gamma \cdot \left(T_{mod,k} - 25°C\right)$$
Equation 2

$$BIF_k = \left(1 + \varphi_{max} \cdot \frac{G_{rear,k}}{G_{i,k}}\right)$$
Equation 3

Equation 2 gives the temperature adjustment factor ($C_{k,25°C}$) linked to the power temperature coefficient (γ) and the module temperature (T_{mod}). The third equation gives the bifacial irradiance factor (*BIF*) in function of the bifacial factor (φ_{max}) and rear side irradiance (G_{rear}). For the standard performance ratio (*PR*) *C*=1 and *BIF*=1 and for the temperature corrected performance ratio (*PR'*): *BIF*=1.

B. List of Modules

The list of modules was initially defined in the proposal and refined during the development phase within WP5. For T6.1 about BAPV, five types of modules are monitored: (a) PERC half-cells (benchmark from Voltec Tarka 120) (b) SHJ shingled bifi (for bifacial), (c) IBC half-cells + ribbons bifi (d) PERC shingled (e) TopCon shingled bifi. For T6.2 about BIPV, three types of modules are monitored: (a) BIPV PERC black (benchmark from 3S-MS-II) (b) BIPV PERC Terracotta (Benchmark from 3S-MS-II) (c) BIPV IBC cut-cells lightweight (d) BIPV SHJ shingled black (e) BIPV SHJ shingled terracotta. For each type of module, one extra module noted "ref" was sent for specific measurements, in particular: stabilization, evaluation, and temperature coefficient extraction.

Fig. 2: BAPV site in University of Ljubljana (UL) in Slovenia (azimuth: south, tilt: 30°, individual module monitoring including IV-curve)

Fig. 3: BAPV site at CEA in Cadarache in France (azimuth: south, tilt: 30°, albedo: 0.3, 5 to 10 modules string-level monitoring)

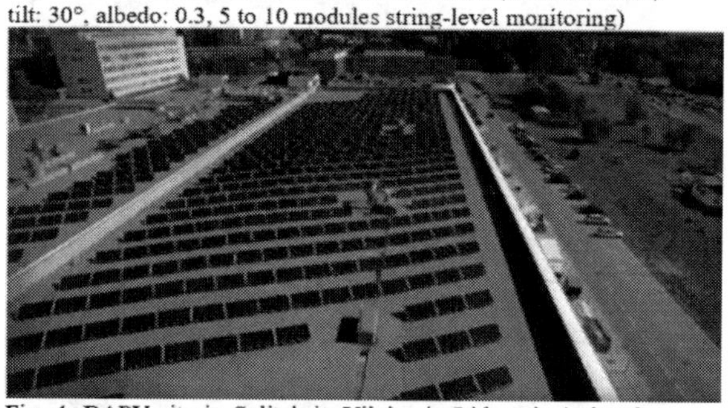

Fig. 4: BAPV site in Solitek in Vilnius in Lithuania (azimuth: south, albedo: bitumen roof, 10 modules string-level monitoring)

Fig. 5: BAPV site at ISC in Hohentengen in Germany (azimuth: south, tilt: 20°, albedo: grass).

Fig. 6: BIPV site at IMEC in Genk in Belgium (azimuth: south and east, tilt: 90°).

Fig. 7: BIPV site at PI-Berlin in Forst in Germany (azimuth: east/south, tilt: 90°).

Fig. 8: BIPV site in UL in Slovenia (azimuth: south, tilt: 90° and 20°, individual module monitoring including IV-curve)

Fig. 9: BIPV and BAPV site for EPFL and CSEM in Neuchatel in Switzerland (azimuth: south, tilt: 17°, individual module monitoring including IV-curve).

The structure of HighLite project and WP6 on monitoring is shown in Fig. 10. Important relations are made with WP5 on module development and fabrication, as all the module delivery apart from the reference modules comes from WP5. The link with WP7 is also very important as this WP provides the characterization used to calculate the various performance parameters. In particular, stabilization tests and procedures based on IEC 61215 [2] for each module family from HighLite, thermal coefficients and reference light current-voltage (LIV) curve and parameters are performed in WP7. WP3 and WP4 delivered the cell and string materials for module fabrication within WP5. WP8receives the monitoring data from WP6 to be included in the various calculations of Levelized Cost of Electricity (LCOE), life-cycle assessment and CO_2 impact of the various technologies. Each WP6 partner owning a monitoring site measured the received module(s) after stabilization, installed them and has been monitoring the module outdoor performance and meteorological data. Data are checked and transferred to the common HighLite monitoring website. This website developed by CSEM includes all information on the sites, modules and strings. Indoor characterization (LIV, EL, pictures) and time series for module performance, temperature and meteorological data are also transferred to the monitoring website. A strong advantage is that all the data are treated the same way and the performance indicators are calculated identically.

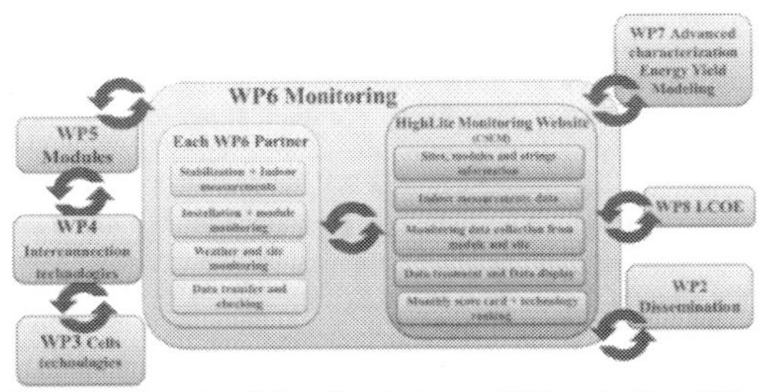

Fig. 10: List of collaboration between WP6 and other WPs, procedure at each partner site and data for the HighLite monitoring website developed by CSEM.

3. RESULTS AND DISCUSSIONS

3.1 Performance ratio analysis

The summary of the performance ratio corrected by temperature (*PR'*) of nearly all modules monitored in the project is presented in Fig. 11. Bifacial modules are showing PR' above 1 only at two monitoring sites which could be due to the different mounting situations. In case of CSEM/EPFL monitoring site, the elevation is only about 0.15 m, which strongly limits the gain from the backside irradiance (see Fig. 9). For the UL monitoring site, the elevation is higher (see Fig. 2) but the mounting system has bars shadowing the backside of the modules.

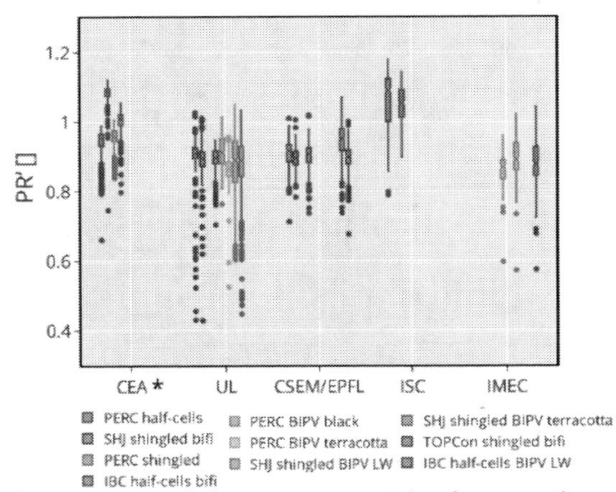

Fig. 11: Summary of all temperature corrected performance ratio

(PR') for the modules monitored within HighLite with daily period aggregation. *P_{mpp} instead of P_{max}.

For the CEA monitoring site in Cadarache, the PR' from mono-facial PERC cells in modules gives a value around 0.95 by changing the module design to 1/6 shingled PERC interconnection no major gain is observed in PR' (see Fig. 12). A gain is observed for the bifacial modules with respect to other cell technologies. IBC half-cut cells modules from Valoe show a gain of 6.3% PR' compared to the PERC half-cell over the same period. Whereas, the SHJ-shingle bifi modules show a gain of 14.7%. The reason can be a higher bifaciality for the SHJ-shingle modules versus IBC modules with 0.9 versus 0.69, respectively. Another reason could be the exact location of the modules in the monitoring site. The SHJ-shingled modules are placed at the upper row in the Fig. 3 (at 1.7 m from ground) whereas the IBC-shingle modules are placed in the lower row (at 0.7 m from ground). With a higher elevation the rear side irradiance is higher and the shadowing from the surrounding is reduced. The Irradiance-Temperature map of Performance Ratio (PR) shown in Fig. 13 gives some interesting information on the behavior of the different module strings. For example, the PERC half-cells and PERC shingled show low PR at low temperature. This is most probably due to shadowing in the morning when the modules are cold and the sun is at a low angle. IBC half-cut bifi module strings show limitations at low irradiance <100 W/m^2 at all temperatures.

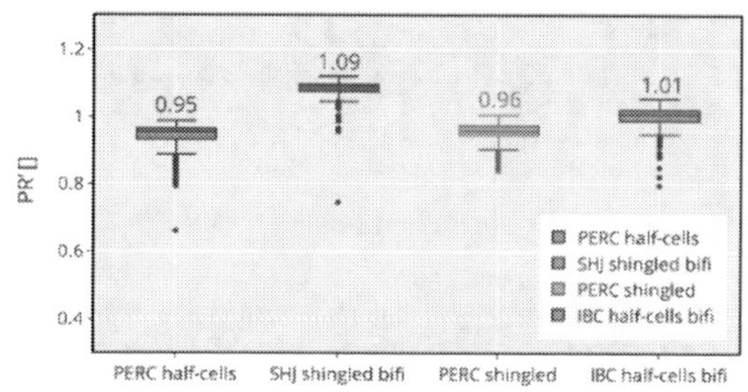

Fig. 12: Temperature corrected performance ratio (PR') for the module strings monitored at CEA Cadarache with daily period aggregation.

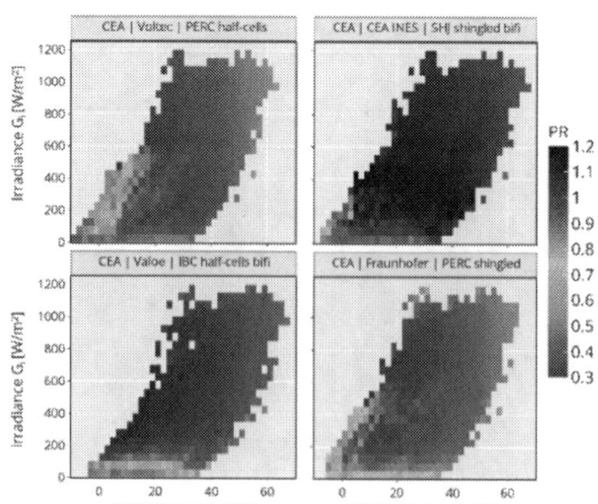

Fig. 13: Irradiance-Temperature map of Performance ratio (PR) for the module strings monitored at CEA Cadarache with 10 min period aggregation.

3.2 Outdoor temperature coefficient extraction

By filtering the irradiance and sorting the monitoring maximal power of the modules at various temperatures, the temperature coefficient of power (γ) can be extracted using a linear regression (see Fig. 14) [4]. During outdoor monitoring the STC conditions of 25 °C and 1000 W/m^2 are met only in few cases. The values of γ are nearly constant with irradiance for both shingle SHJ and PERC half-cell modules as shown in Fig. 15. The outdoor temperature coefficient for the power (γ) is close to the indoor measurement values from WP7 (see Table 1).

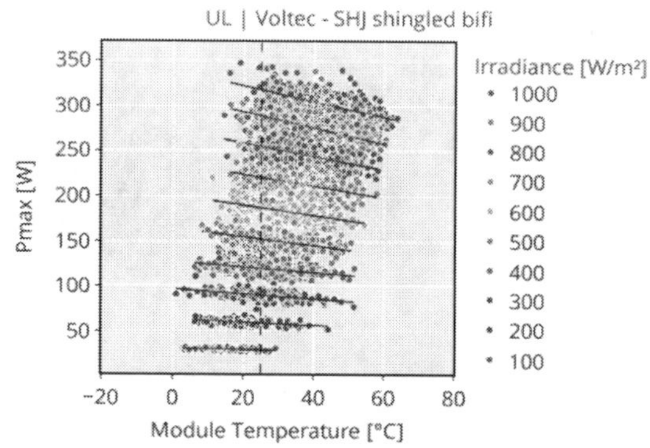

Fig. 14: P_{max} versus temperature at all irradiances for the Voltec SHJ shingled module monitored at UL with 10 min aggregation.

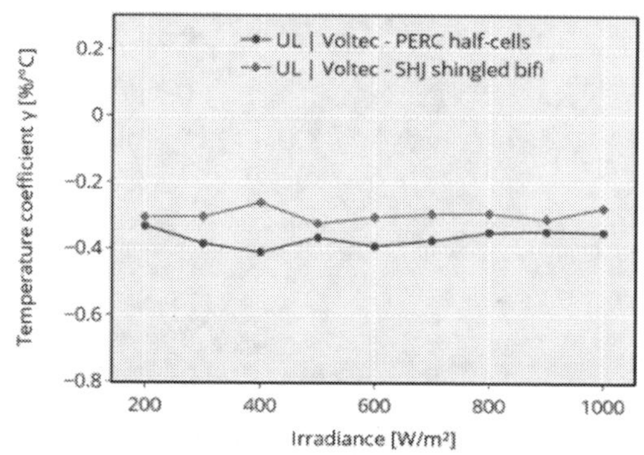

Fig. 15: P_{max} temperature coefficient (γ) versus irradiance for the Voltec SHJ shingled and PERC half-cell benchmark modules monitored at UL.

Table 1: P_{max} temperature coefficient (γ) measured in indoor conditions and computed from outdoor measurements at 1000 W/m^2 for the Voltec SHJ shingled and PERC half-cell benchmark modules.

Module	Indoor γ [%/°C]	Computed γ [%/°C]
Voltec Tarka 120	-0.3403	-0.35
Voltec SHJ bifi	-0.30	-0.28

3.3 Energy yield (EY) calculation

The energy yield can be calculated over each month as shown in Fig. 16 for PERC half-cell systems in 2021 and 2022 and for SHJ-shingled systems in 2022. The differences on the PERC system from year to year are quite significative in particular in July and August. For the SHJ-shingled system, the energy yield is measured only over 7 months, and for the remaining 5 months Direct-Diffuse Power Rating model (DDPR) has been used within HighLite-WP7 to simulate the monthly energy production [5].

Table 2 gives the annual EY of the PERC half-cell system over 2021 and from September 2021 to August 2022. The increase of EY for this system in 2022 is principally due to an increase in irradiation. The SHJ-shingled system shows an EY of more than 1900 kWh/kWp in Cadarache. This value is relatively high and will have to be validated with measurement until the end of the year, however it is realistic, knowing that the yearly global tilted irradiation in Cadarache at 30° is 1932 kWh/m^2 based on PVGIS- SARAH 2.

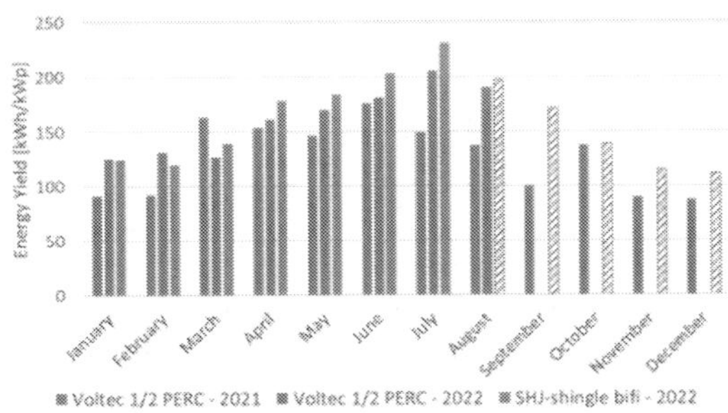

Fig. 16: Monthly Energy Yield (*EY*) for the module strings monitored at CEA Cadarache. The dashed bars are simulated data.

Table 2: Annual Energy Yield for PV systems installed at CEA Cadarache (see Fig. 3 and Fig. 16).

Modules	Voltec 1/2 PERC	Voltec 1/2 PERC	SHJ shingled bifi
Dates	1.1.-31.12.2021	1.9.2021-31.8.2022	1.1.-31.12.2022 (with simulation)
Annual Energy Yield (kWh/kWp)	**1524.2**	**1704.4**	**1912.0**

3.4 Data Quality Check

This section is proposing a methodology focusing on PV monitoring data check and how to point out monitoring issues. On the morning and evening, *PR'* can be strongly influenced by the position of the reference cells (or pyranometer) as shown in Fig. 17. From the 8th to the 17th of May a first reference cell was used and the *PR'* over time shows square shape and *PR'* versus irradiance shows low values at low irradiance (possibly partial shadowing of the string at sunrise and/or sunset). By changing the reference cell (after 19th of May), the shape of the PR' at sunrise and sunset changes drastically (possibly due to an under-estimation of the irradiance by the new reference cell), showing spikes of *PR'* in the morning and evening. This effect is well seen on the graph of *PR'* versus irradiance (Fig. 17). As shown by the grey square in the graphs PR' vs. irradiance (top figures in Fig. 17), a threshold filter removes data with irradiance <50 W/m^2, so that the influence of the position of the reference cell is not be observed in the monthly *PR'* value. The period of aggregation is important to extract *PR'* values, less than 10 min aggregation might lead to inaccuracy on *PR'*.

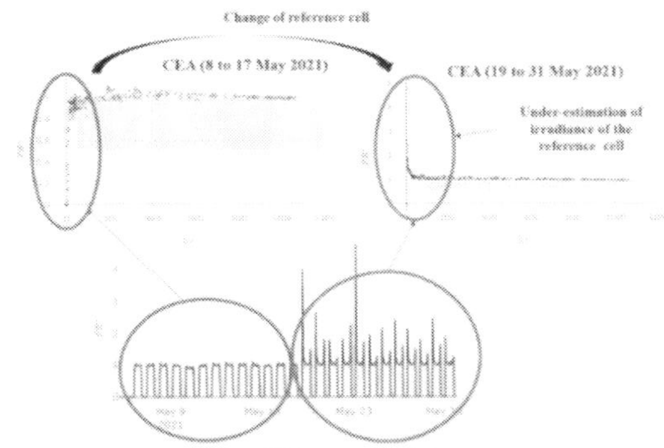

Fig. 17: Top graphs: *PR'* corrected by temperature versus irradiance for the string of 10 benchmark modules Tarka-120 measured in Cadarache, bottom graph: *PR'* corrected by temperature between 8th and 31st of May 2021 (30 min aggregation period).

Preliminary results on monitoring of the reference commercial modules tested within HighLite H2020 project are shown in Fig. 18. BIPV terracotta MS-II from 3S Swiss Solar Solutions modules show a stable array performance ratio (*PR'*) over the year, maybe with reduced values during winter (possibly due to partial shadowing of the modules). BAPV glass-back-sheet Tarka-120 modules from Voltec also show stable PR' over the first year. After the first year of monitoring the modules were unmounted and remeasured indoor, the results are shown in Fig. 19. The indoor performances are stable after one year of outdoor monitoring with less than 1% variation of 4 parameters. The variation in power output of the module is between -0.3 and -0.9% for both reference technologies.

All equipment of the PV system can have an influence on the performance of the system, such as shown in Fig. 20 where the maximal power point tracker (MPPT) has been changed on a module. We clearly see an effect on the *PR'* corrected by temperature for the irradiance between 50 and 200 W/m^2.

The final project deliverable will report on the monitoring of the new technologies developed in HighLite H2020 project compared to reference modules during their first year of operation.

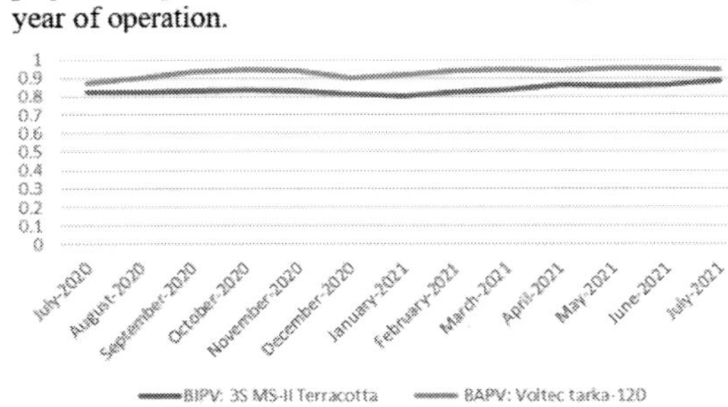

Fig. 18: Performance ratio of the array (*PR'*) corrected by temperature for the benchmark module technologies based on PERC (monthly aggregation for *PR'*).

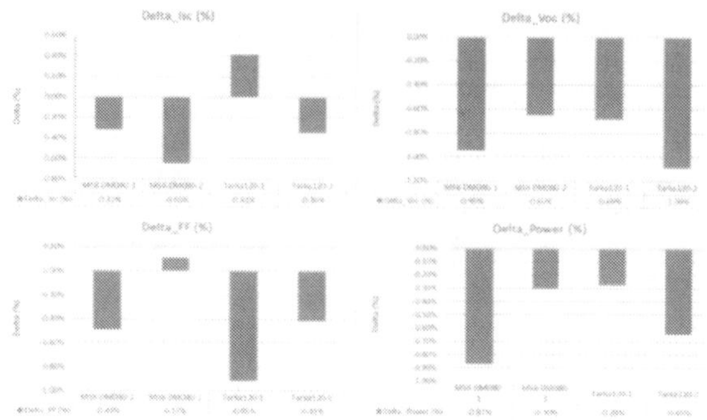

Fig. 19: Module indoor performance variation after one-year of outdoor monitoring of corresponding modules shown in Fig. 18.

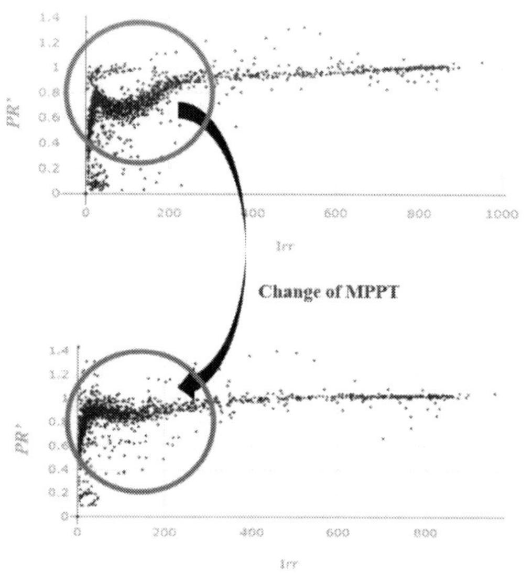

Fig. 20: *PR'* corrected by temperature versus irradiance for the benchmark module Tarka-120 (top) with defective MPPT (top) and with new MPPT (bottom) (30 min aggregation period).

4 CONCLUSION

With about 60 modules of 10 different technologies monitored over 7 sites, the volume and variety of data to analyse is important.

The major results of this initial PV module monitoring phase within H2020 project HighLite are the following:

- Bifacial technologies can increase the performance ratio corrected by temperature (*PR'*) of the PV system (up to 1.09) if the modules are installed correctly, i.e. 1 meter above the ground no backside shadowing, no surrounding shadowing.
- Temperature coefficients of power can be extracted during outdoor monitoring and correspond to the indoor coefficients.
- Irradiance-Temperature maps of Performance ratio (*PR)* can detect shadowing problems of the PV systems.
- Energy yield above 1900 kWh/kWp is expected for the HighLite modules installed in Cadarache.

5 REFERENCES

[1] L. Tous *et al.* "Overview of key results achieved in H2020 HighLite project helping to raise the EU PV industries' competitiveness", Submitted to Progress in Photovoltaics, 2022.

[2] IEC 61724-1:2021 Photovoltaic (PV) system performance – Part1: Monitoring; International Standard, Publication date: 2017-03-03, Edition 1.0, Solar photovoltaic energy systems, ICS 27.160 - Solar energy engineering, Stability date 2021, number of Pages 118.

[3] IEC 61215-1:2016 Terrestrial photovoltaic (PV) modules - Design qualification and type approval - Part 1: Test requirements; International Standard, Publication date: 2016-03-09, Edition 1.0, Solar photovoltaic energy systems, ICS 27.160 - Solar energy engineering, Stability date 2021, number of Pages 37

[4] H. Quest *et al.* "DECOUPLING PERFORMANCE GAINS OF SILICON HETERO-JUNCTION BIFACIAL MODULES", this proceeding.

[5] B. Kim, K. Brecl, M. Topic, "A new PV module performance model based on separation of diffuse and direct light", Sol. Energy 113, 212–220, 2015.

5 ACKNOWLEDGMENTS

This project has received funding from the European Union's Horizon 2020 Programme for research, technological development and demonstration under Grant Agreement no. 857793. We gratefully acknowledge support from all H2020 Highlite project members.

This is a short version of the paper submitted to the journal Progress in Photovoltaics.

It has been selected by the Committee of the WCPEC-8 for submission to Progress in Photovoltaics.

IDENTIFY, ANALYSE, MITIGATE – QUANTIFICATION OF TECHNICAL RISKS IN PV POWER SYSTEMS

Magnus Herz*[1], Gabi Friesen[2], Ulrike Jahn[3], Marc Koentges[4], Sascha Lindig[5], David Moser[5]
[1]TÜV Rheinland, Cologne, Germany, Magnus.Herz@de.tuv.com
[2]SUPSI, Switzerland, [3]VDE Renewables, Germany,
[4]ISFH, Germany, [5]Eurac Research, Italy

ABSTRACT: Technical risks are important criteria to consider when investing in new and existing PV installations. Quantitative knowledge of these risks is one of the key factors for the different stakeholders, such as asset managers, banks or project developers, to make reliable business decisions before and during the operation of their PV assets. Within the IEA PVPS Task 13 Expert Group, we aim to increase the knowledge on methodologies to assess technical risks and mitigation measures in terms of economic impact and effectiveness. The developed outline provides a reproducible and transparent technique to manage the complexity of risk analysis and processing in order to establish a common practice for professional risk assessment.
Semi-quantitative and quantitative methodologies are introduced to assess technical risks in PV power systems and provide examples of common technical risks described and rated in the new created PV failure fact sheets (PVFS). Besides the PVFS based on expert knowledge and expert opinion, an update on the statistics of the PV failure degradation survey is given. With the knowledge acquired and data collected, the risk and cost-benefit analysis is demonstrated in a case study that shows methods for prioritizing decisions from an economic perspective and provided important results for risk managing strategies.
Keywords: PV System, Risk Analysis, Failure Rate

1 INTRODUCTION

PV risk analysis serves to identify and reduce the risks associated with investments in PV projects. The key challenge in reacting to failures or avoiding them at a reasonable cost is the ability to quantify and manage the various risks. There are several interpretations of the concept of risk, but in general risk can be defined as the probability of failure multiplied by the consequences of its failure. The common approach in evaluating technical risks is to apply a classical Failure Mode and Effects Analysis (FMEA) [1]. It is widely used in the automotive, aerospace, and electronics industries to identify, rank, and mitigate potential failures. Root causes and impact of a failure can be analysed. The disadvantage of this approach is that the risk is evaluated in a qualitative way and cannot provide a framework for the calculation of the economic impact. Thus, a cost-based FMEA was proposed in 1993 [2] and enhanced in 2003 [1]. Several applications of cost-based FMEA can be found in the literature [3], often related to automotive or wind energy [4]. In 2017, a cost-based FMEA was presented within the Solar Bankability Project [5] [6] as a first attempt to implement a cost-based FMEA to the PV sector. The metric Cost Priority Number (CPN) was applied as one Key Performance Indicator (KPI) for the risk assessment of PV investments. In [7], the CPN method was further developed with the focus on the needs of large O&M operators. In the 2020 launched H2020 project TRUST-PV [8], the improved Cost Priority Number methodology is used to calculate the cost and thereby financial impact of individual PV system issues described in O&M tickets, which were provided by several major O&M companies and asset managers across Europe [9].

In this context, another widely used quantification method is the Reliability, Availability and Maintainability (RAM) analysis. The RAM analysis aims to identify any significant performance losses and then recommend improvements to the maintenance strategy. In IEC TS 63265 – "Reliability practices for the operation of photovoltaic power systems", coordinated by Roger Hill, with the foreseen publication in 2022, another toolkit will be provided of many methods of how different stakeholders can demonstrate the effective of reliability increasing measures from technical and economic point of view.

The aim of this work is to increase the knowledge of methodologies to assess technical risks and mitigation measures and to investigate the most important risks by collecting case studies and updating the database with the acquired information.

2 COMMON PRACTICE

According to the Project Management Body of Knowledge (PMBOK) guide, a set of standard terminology and guidelines for project management [12], "Risk quantification is a process to evaluate identified risks to produce data that can be used in deciding a response to corresponding risks". This implies that the first step is to identify the technical risks and subsequently determine the probability of occurrence and the impact on the energy yield. Previous works within IEA PVPS Task 13 [10] [11] , Moser et al. [5] and the PV failures fact sheet in Chapter 3.1 have identified and described the most common technical failures that could impact the performance of a PV power plant. In addition to failures, there are also other technical risks during operation caused by varying performance loss rates as analysed in [13] [14]. How to respond to these risks with preventive or corrective actions is discussed by Jahn et al. in [15] and [16]. In the following, these evaluation processes are classified into semi-quantitative and quantitative methods with a focus on photovoltaics. This chapter gives an insight into common methods used, how technical risks in PV plants can be evaluated and minimised, and provides recommendations for best practices.

The semi-quantitative methods use human problem-solving strategies, based on expert knowledge and expert opinion. The best ways to use such a knowledge-based method is to conduct on- or offline workshops where experts can discuss and consequently assign values to the risks identified. They can prioritise the identified risks using a pre-defined rating scale. Risks will be scored based on their probability or likelihood of occurrence and their impact.

2.1 FMEA

One typical approach is a classic Failure Modes and Effects Analysis [17]. In the FMEA, each identified risk is evaluated for its Severity (S), Occurrence (O) and Detectability (D), which describe the impact on performance, the probability of appearance and how easy it is to detect the risk. Each of these evaluation parameters is usually rated with numbers on a scale of 1 to 10.

$$RPN = S \cdot O \cdot D \quad (1)$$

With the resulting Risk Priority Number (RPN), the evaluated risk can be ranked and compared with other risks. Figure 1 gives an example of FMEA rating of PV module failures. The disadvantage of this approach is that further usage, i.e. within a financial model, is limited [5].

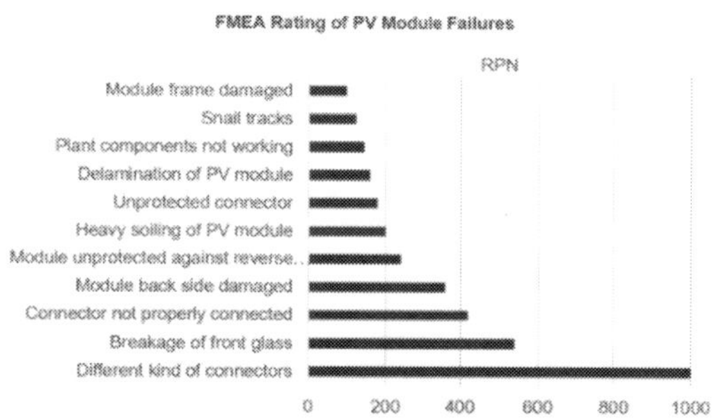

Figure 1: Example of rating of PV module failures based on classic FMEA. The rating of the technical risks was based on the statistics of failure reports from TÜV Rheinland.

2.2 CPN

Quantitative Methods involve assessing the probability and impact of risks using numerically based techniques, such as simulation and fault tree analysis. The results provide information about the effects of the identified risks and represent a given reality in the form of a numerical value that can be utilized in economic and financial models for quantitative decision making.

For PV systems the CPN enables accurate economic quantification of PV degradation modes and other performance impairing effects of operating PV plants. It therefore has enabled risk assessments of investments in PV power plant projects [5]. The CPN methodology used assessed the economic impact of PV projects based on factors such as performance loss and downtime. Thereby, a cost-based Failure Mode and Effect Analysis methodology for the PV sector has been developed in form of the CPN. In its initial form, it was developed using theoretical scenarios to calculate extreme values for the CPN metric, expressed in €/kWp/year. Thereby, all phases of a PV power plant's life cycle (from product testing to decommissioning) have been included. The methodology helps to identify and classify technical risks and their economic impact by assigning a cost metric that, based on collected statistics, supports preventive and corrective measures, which would then lower the impact of failures on the availability and performance of a PV plant. Thereby, it was possible to create a database which gives indicators of failure appearance likeliness and severity. Such results could then be used to improve O&M activities.

2.3 RAM

Technical risks and the reliability of a component are complements of each other, as long as they cover the same sample space. In this context another widely used quantification method is the Reliability, Availability and Maintainability analysis. RAM analysis aims to identify any significant performance losses and then recommends improvements to the maintenance strategy. In this bottom-up approach a Reliability Block Diagram (RBD) or the Fault Tree Analysis (FTA) is recommended to determine the effects of the failure of individual components .

In RAM modelling, the reliability term R is defined as the probability that a system or component performs adequately within a given time.

$$R(t) = \int_t^{\infty} f(t)dt \quad (2)$$

The probability density function PDF of failures f(t) with increasing lifetime is expressed by an exponential, normal, Weibull or lognormal distribution. Weibull distributions are applicable to a broad range of failure modes and mechanisms. The normal distribution is preferred for items that have a wear out mechanism such as bearing or motors. Derived from [18], the best-fit PDFs for the different components are shown in Table 1.

Table 1: Best-fit PDFs for the components of a PV plant adapted from [18].

Component	PDF
PV modules	Exponential
Bypass diode	Weibull
DC switch	Lognormal
AC switch	Weibull
AC circuit breaker	Weibull
Connector	Exponential
Inverter	Lognormal

The failure rate λ is the frequency of component failure. The mean time to failure (MTTF) of a component defines the expected life of non-repairable items.

$$\lambda = \frac{f(t)}{R(t)} \quad (3)$$

$$MTTF = \int_0^{\infty} R(t)dt \quad (4)$$

Availability (A) is defined as the percentage of time that the plant was successfully operating. A is the MTTF divided by the total operating time and can be calculated with MTTF and Mean Down time (MDT), as follows:

$$A = \frac{MTTF}{MTTF + MDT} \quad (5)$$

3 RISK DATABASE

According to the PMBOK guide [12], the Risk

Database (RDB) is the central repository for all information regarding the identified risks. In terms of technical risks, the RDB provides the range of affected components, the description with causes and consequences, failure rates, the probability of occurrence, the impact on KPIs and the recommended control and mitigation actions. It should be updated and maintained as a growing data hub through all phases of the project. In this chapter, we present a systematic approach to identify the main technical risks, define the most important risk parameters and collect these failure, loss and occurrence data.

3.1 PV Failure Fact Sheets (PVFS)

The PV failure fact sheets (PVFS) summarise some of the most important aspects of single failures. The target audience of the PVFS are PV planners, installers, investors, independent experts and insurance companies, and anyone interested in a brief description of failures with examples, an estimation of risks and suggestions of how to intervene or prevent these failures. Front delamination (Figure 5) is taken as one example of the 30 created PVFS. The complete list with all PVFS are publicly available in the Annex of Herz et al. [19].

The format of the PVFS is based on the failure description presented within the H2020 Solar Bankability project [6]. A rating system for the estimation of the severity of a failure is used here which simplifies the approach proposed within the IEA PVPS Task 13 [10] by implementing the rating system proposed by the Sinclairs [20]. The correlation between the different failures is highlighted in the text by using bold characters. Each PVFS is structured into one to three pages. The first page is a descriptive page, whereas the remaining pages contain examples composed of a picture, a legend and an estimation about its severity. The first page is structured as follows:

Component

The PV system components are divided into:

(1) PV module (including junction box)
(2) Cables and interconnectors (at module, string and combiner box level)
(3) Mounting (structure, clamps and screws)
(4) Inverter

Defect

Short name describing the failure/defect.

Appearance

Description of how the defect looks like.

Detection

Description of methods which can be used to detect the failure. Detection methods in brackets lists secondary methods, which do not detect the failure with absolute certainty or which can be used in addition to other methods. The following abbreviations are used:

Origin

Description of the failure and its main causes and origin (1. Material and production, 2. Transport and installation, 3. Operation and maintenance).

Impact

Description of the impact on the safety, performance and reliability of the component and system and its severity. For every failure, a range of possible ratings is given, one for the safety and one for the performance.

A failure is defined as a safety failure when it endangers somebody who is applying or working with PV modules or simply passing the PV modules. Three categories are defined in Figure 2.

Safety category	Description
(grey hexagon)	Failure has no effect on safety.
f e m	Failure may cause a fire (f), electrical shock (e) or a physical danger (m) if a follow-up failure and/or a second failure occurs.
f e m	Failure can directly cause a fire (f), electrical shock (e) or a physical danger (m).

Figure 2: Safety category

A failure is defined as a performance failure when it impacts the performance and/or reliability of a system. Five categories are defined in Figure 3. They go from 1 (low severity) to 5 (high severity).

Performance category	Description
1 — 5	The defect has no direct effect on performance.
1 — 2 — 5	The defect has a minor impact on performance.
1 — 3 — 5	The defect has a moderate impact on performance.
1 — 4 — 5	The defect has a high impact on performance.
1 — 5	The defect has a catastrophic impact on performance.

Figure 3: Performance category

For each category, the expected loss is estimated on the component level and if no mitigation measure is implemented. It can range from no power degradation (0%) over power degradation below detection limit (<2-3%), power degradation within warranty (<0.7-1%/year) and power degradation out warranty (>0.7-1%/year) to catastrophic power degradation (>3%/year).

Mitigation

Description of the corrective actions to be done on a short and medium term when detecting a failure and preventive actions to be implemented to avoid the failure from the beginning. Preventive actions are separated into recommended actions, representing the minimum requirement for small residential systems and optional actions for large scale systems.

The general rule for intervention in case of a failure is: All components with a direct safety risk or a performance severity of 5, highlighted in red, should be replaced or repaired. Regular inspections should be performed to monitor the status of the not replaced or repaired components.

Component Defect	**Module** **Front delamination**		**PVFS 1-3**
Appearance	Any local separation of the layers between (i) the front glass and the encapsulant or (ii) the cell and the encapsulant, visible as bubbles or as bright, milky area/s. It may appear continuous or in spots. The position and size of the delamination or bubble depends on the origin and progress of the failure.		
Detection	VI, (INS)		
Origin	The adhesion between the glass, encapsulant, active layers, and back layers can be compromised for many reasons. Typically, it is caused by the manufacturing process (e.g. poor cross linking of EVA, too short lamination times, too high pressure in the laminator, contaminations, improper cleaning of the glass, incompatibility of EVA with soldering flux, inadequate storage of the raw material) or environmental factors (e.g. thermal stresses, external mechanical stresses, UV). Delamination is generally followed by moisture ingress and **corrosion**. It is therefore more frequent and severe under hot and humid conditions.		
	Production	Installation	Operation
Impact	Delamination or bubbles do not automatically pose a safety issue, but they can result in **reduced insulation** of the component and increased safety risk when they form a continuous path between electric circuit and the edge due to possible water ingress. Moisture in the module will decrease performance due to an increase of series resistance, affect long term reliability and in some cases also the structural integrity of the module. Moreover, delamination at interfaces in the optical path will result in additional optical reflection and subsequent decrease in current. This can be the origin of current mismatch. If the mismatch is significant, it will trigger the bypass diode and cause further power loss. The inverter might also shut down due to leakage current's leading to a further performance loss. Manufacturing related delamination issues often affects a relevant percentage of modules within the same production batch and consequentially has a big impact on system performance.		
	Safety:	Performance:	
Mitigation	Corrective actions	Preventive actions (recommended)	Preventive actions (optional)
	Modules with a direct safety risk or a severity of 5 should be replaced. Regular inspections should be done to monitor the status of the not replaced modules. In case of individual module testing all modules which failed the insulation and/or wet-leakage test should be replaced.	Check validity of IEC 61215 certification and BOM, ground fault detection by inverter or other devices at all time.	Extended testing (e.g. damp heat), pre-shipment inspections (e.g. cross linking level of EVA) regular visual system inspections.

Figure 4: First page of PVFS example with general information

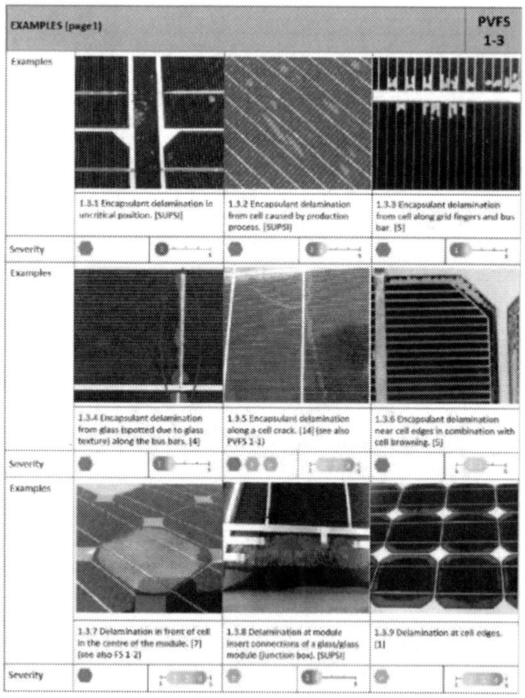

Figure 5: Remaining pages of a PVFS contain examples composed of a picture, a legend and an estimation about its severity.

3.2 PV Failure Degradation Sheets (PVDS)

Besides the PVFS collection we provide an update on the statistical risk data of the PV Failure Degradation Sheet (PVDS) survey developed in Koentges et al. [11]. It requires a large amount of measured input data but it is able to generate statistical data on degradation rates and power loss of PV systems based on failure types. Due to the high requirements on the PVDS much less input data can be collected. In the following, we introduce the collected data, the way of analysing the data and the analysis results.

The failure data is collected in an excel sheet which is sent to system owner, experts, installer or manufacturer. Some data is also collected by scientific publications or an Australian internet survey. The survey structure is first presented in the IEA PVPS TASK 13 report "Assessment of Photovoltaic Module Failures in the Field" [11]. The plain survey and the survey explanation can be downloaded here [21].

Compared to the first presented survey structure in [11], we added two new failure categories for PV modules: LID/LeTID degradation and potential induced delamination [22]. Since the last failure data evaluation [11], we added 76 new complete PVDSs to the PV system data collection. The data collection consists of 226 PVDSs in total.

Figure 6 shows the frequency distribution for PV module failures with an impact on the power generation of the PV systems. The distribution is split into failures which lead to a degradation and suddenly occurring failures. Most reports on failures with power loss are given in the first 10 years of operating time. This is to be expected as it is often too expensive to repair PV systems older than 10 years. Therefore, no detailed analysis is made. The main results of the last report "Assessment of Photovoltaic Module Failures in the Field" remain true. PID effects, cell cracks and defective bypass diode failures seem to dominate the failure statistic in the first seven years of operation. This dominance now becomes even more pronounced in comparison with the statistics presented in [11]. Additionally, the failure type "burn marks" has been detected more frequently. For sudden events, also shown in Figure 6, the failure glass breakage and dust soiling fully dominate the failure statistic.

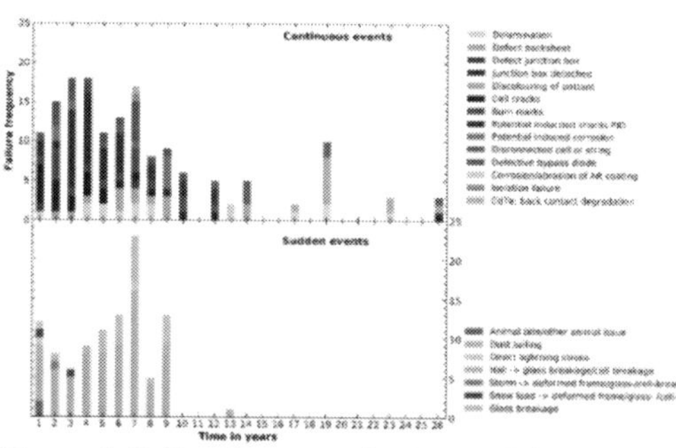

Figure 6: Failure frequency for PV module defects with an impact on the system power. The upper graph is showing PV module failure frequency with a slow degradation over time and the lower graph failure frequency for sudden events.

Figure 7 shows the power loss impact of sudden events on PV system performance. Documented glass breakage events lead in temperate climates to a loss of 1% to 2% of a system's power, with one exception in the dataset. These events seem to occur everywhere but appear to be not so severe for the whole system. Dust soiling appears everywhere except for tropical climates. In temperate climates, the impact is at a maximum 7% of the total system power whereas up to 15% power loss occurs in dry climates and over 25% for continental climates. As expected, the deformed PV module frame due to snow load occurs only in the continental and polar climate.

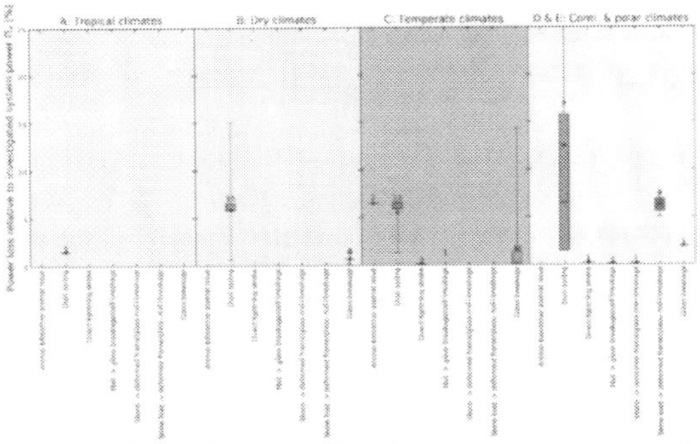

Figure 7: Power loss of sudden failure events on the total power of the PV system.

4 CASE STUDY

Having the methods presented and data collected, the following chapter demonstrates a risk and cost-benefit analysis using a case study that show techniques for prioritising decisions from an economic perspective and provide important results for risk managing strategies.

4.1 Risk Analysis

Risk analysis enables users with statistical and reliability data to develop and run scenarios in which PV performance and costs are affected by components that can fail.

How the risk quantification method can be also applied in practice is demonstrated using a 10 MW PV plant with PID affected PV modules. The assumptions in Table 3, derived from an operating PV plant, serve as input for this case study. Not considered are financial parameter such as depreciation, interest or taxes.

Table 2: Metadata of investigated PV plant

Parameter	
Risk	Potential induced degradation (PID)
Detection time	4 years
Response time	1 year
Repair time	1 year
Size of plant	10 MW
Module tilted/ orientation	20°, 0° south oriented, 2 portraits
Type of installation	Free Field Installation
Base frame	Fixed Installation
Modules	40000 x 250 Wp
Inverters	20 x 500 kVA
PPA	0.25 €/kWh
CAPEX	20 Mio€
OPEX	50 k€/a
Inverter nominal power	500 kW

Taking the behaviour of the identified root cause into account, the potential future Performance Loss Rate (PLR) is expected to increase further with an expected saturation of 50%. After this value is reached, the PLR is expected to stagnate at a constant level of 0.7% per year. This prediction of performance development for 20 years of operation is shown together with the exceedance probability P10 and P90 for a confidence level of 68.2% in Figure 12. Taking CAPEX, OPEX and annual revenues into account, the project's financial profit after 20 years of operation is 48% below original expectations for the defined scenario without mitigating actions.

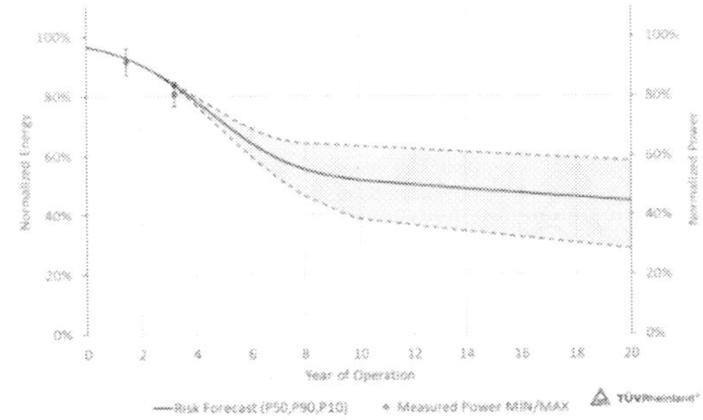

Figure 8: Energy forecast of No-Mitigation Scenario.

4.2 Cost-Benefit Analysis

The CPN methodology allows the estimation of the economic impact of failures on the LCOE and on business models of PV projects and has been developed not only to determine the economic impact of technical risks, but also to be able to assess the effectiveness of mitigation measures. Specific failures have to be examined in order to draw recommendations on how to mitigate the economic impact for, e.g. soiling, or potential induced degradation (PID). Some failures can be prevented or mitigated through specific actions at different project phases (e.g. for PID); other events (e.g. soiling) can be prevented or mitigated through a more generic action. For example, the monitoring of performance or visual inspection can be considered as generic mitigation measures that can have a positive impact on the reduction of the CPN of many failures. In practice, it is important to understand how mitigation measures can be considered as a whole to be able to calculate their impact and thus assess their effectiveness.

The cost-benefit analysis is also a tool to determine whether the benefit of one option will justify its costs. It can identify the best mitigation options from an economical point of view. The analysis continues the case study. Three mitigation scenarios are defined:

- **No-Mitigation** option without intervening into the current plant operation
- **Mitigation option 1 - PID Box**: Installing PID-boxes and allowing the performance of the PV modules to recover to a certain level
- **Mitigation option 2 - PID Box & partial repowering**: Installing PID-boxes and replacing very low performing PV modules by high-power-modules.

The expected annual production of energy yields for the three scenarios is illustrated in Figure 9. After the mitigation measures were applied in year 5 of operation, the energy yields show a steep rise. The expected PV plant output after 20 years of operation is calculated at 45% rated energy output for the no-mitigation scenario and at 84% and 91% for mitigation options 1 and 2, respectively.

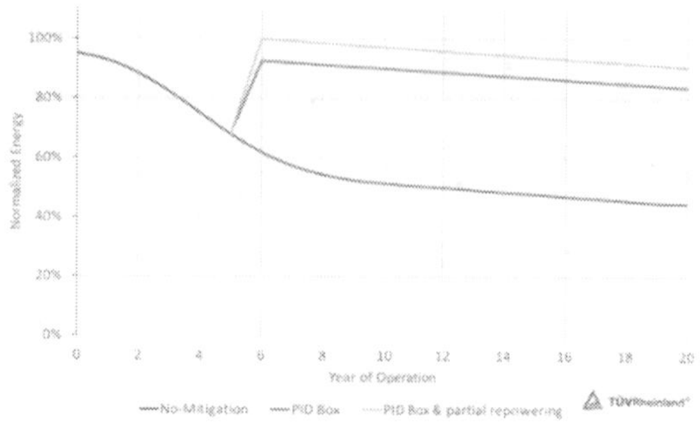

Figure 9: 20-year forecast for three mitigation scenarios; the repowering is carried out with a higher module power class.

Table 3: Costs of mitigation scenarios.

	No-Mitigation	PID Box	PID Box & partial repowering
Cost [k€]	15	238	3233

The cost-benefit analysis also takes the associated costs of the available options into account, as described in Table 3. The costs are derived from a real project and include basic O&M costs for the no-mitigation scenario and the costs of additional detection actions and equipment for mitigation options 1 and 2 when installing PID-boxes or replacing PV modules. The impact on the annual cash flow is demonstrated in Figure 10. In the reference scenario, the monetary yield of the PV project after 20 years is expected to be around 225% of the CAPEX (dashed line). If no mitigation measures are taken, the lowest result of around 115% of CAPEX is forecasted. Mitigation options 1 and 2 result in 6.0%, respectively 4.6% below expectations, which both represent successful projects results. It can be concluded that both mitigation options should be considered and taken as a solution compared to non-action. However, the additional investments in year 5 of operation for option 2 are with a factor of 8 significantly higher

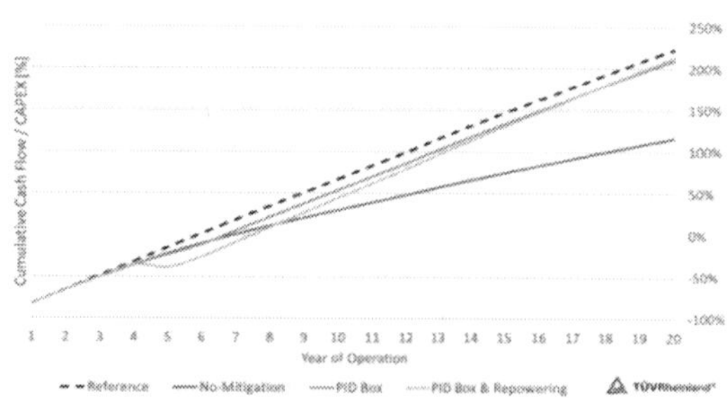

Figure 10: Annual cumulative cash flow of the mitigation scenarios with CPN and loss of revenue after 5 and 10 year of operation if no action is taken.

5 CONCLUSIONS

We introduced semi-quantitative and quantitative methodologies to assess technical risks in PV power systems and provided 30 examples of common technical risks described and rated in the new created PV failure fact sheets [19]. Besides the PVFSs based on expert knowledge and expert opinion, an update on the statistics of the PV failure degradation survey, developed in Koentges et al. [11], was given. With the knowledge acquired and data collected, the risk and cost-benefit analysis was demonstrated in one case study that showed methods for prioritising decisions from an economic perspective and provided important results for risk managing strategies.

However, providing an overview of quantification methods, we draw the conclusion that more standardisation is required. Risk definitions are not fully structured and event databases (solar logbooks) are not harmonised. Data analysis would benefit from the use of a standardised language and metadata formats. The development of an automated and therefore time-efficient solution for extracting key parameters from maintenance tickets is required to gain statistical insights from a large number of PV plants. Also, the development of a software tool for field technicians is recommended that would allow the precise and error-free recording of standardised parameters for the calculation of the O&M contractors KPIs necessary for an efficient implementation of the methodology [7]. In summary, the O&M field practices must certainly move away from a manual input of tickets in text format and adopt a more standardised approach where human intervention is limited.

All things considered, we believe that data-driven evaluation of techno-economic performance indicators is a significant key to take decision support on LCOE to the next level.

6 ACKNOWLEDGEMENTS

This work has received funding from the Horizon 2020 programme of the European Union under contract number 952957.

This work is supported by the German Federal Ministry for Economic Affairs and Energy (BMWi) under contract no. 0324304A, 0324304B and 0324304C.

This work is supported by the Swiss Federal Office of Energy (SFOE) under the contract SI/501788-01 and SH/8100038-02-02-02.

7 REFERENCES

[1] C. M. Spencer, S. J. Rhee, Cost based failure modes and effects analysis (FMEA) for systems of accelerator magnets, in Pro-ceedings of the 2003 Particle Accelerator Conference, 2003, vol. 4, pp. 2177–2179 vol.4, doi: 10.1109/PAC.2003.1289057.

[2] W. Gilchrist, Modelling Failure Modes and Effects Analysis. Int J Qual Reliab Manag, 1993, 10:16–23. doi: 10.1108/02656719310040105.

[3] A. Hassan, I. Dayarian, A. Siadat, J.-Y. Dantan, Cost-based FMEA and ABC concepts for manufacturing process plan evaluation, in 2008 IEEE Con-ference on Cybernetics and Intelligent Systems, 2008, pp. 197–202, doi: 10.1109/ICCIS.2008.4670795.

[4] M. Shafiee, F. Dinmohammadi, [Accessed: 09-Jun-2021], An FMEA-Based Risk Assessment Approach for Wind Turbine Systems: A Comparative Study of Onshore and Offshore, Energies, vol. 7, no. 2, pp. 619–642, Feb. 2014, doi: 10.3390/en7020619. [Online]. Available: https://www.mdpi.com/1996-1073/7/2/619.

[5] D. Moser, M. D. Buono, U. Jahn, M. Herz, M. Richter, and K. D. Brabandere, "Identification of technical risks in the photovoltaic value chain and

quantification of the economic impact,", Progress in Photovoltaics: Research and Applications, vol. 25, no. 7, pp. 592–604, 2017, doi: 10.1002/pip.2857.

[6] "Solar Bankability project website," accessed: 2020-04-28. [Online]. Available: http://www.solarbankability.org/home.html.

[7] G. Oviedo Hernández, S. Lindig, D. Moser, and P. V. Chiantore, "Optimization of the Cost Priority Number (CPN) Methodology to the Needs of a Large O&M Operator," in 36th EUPVSEC Proceedings, Sep. 2019, pp. 1613 - 1617.

[8] "H2020 TRUST PV Project," [Online]. Available: https://trust-pv.eu/. [Accessed 11 01 2021]..

[9] Lindig, S., Gallmetzer, S., Herz, M., Louwen, A., Koumpli, E., Enriquez Paez, P. S., Moser, D., "Towards the development of an optimized Decision Support System: a comprehensive Statistical and Economical Assessment of over 35,000 O&M tickets", Under Review 2022.

[10] M. Köntges, S. Kurtz, C. Packard, U. Jahn, K. A. Berger, K. Kato, T. Friesen, H. Liu, M. Van Iseghem, J. Wohlgemuth, D. Miller, M. Kempe, P. Hacke, F. Reil, N. Bogdanski, W. Herrmann, C. Buerhop Lutz and G. Friesen, Review of Failures of Photovoltaic Modules, Report IEA-PVPS T13-01: ISBN 978-3-906042-16-9, 2014.

[11] M. Köntges, G. Oreski, U. Jahn, M. Herz, P. Hacke, K. A. Weiss, G. Razongles, M. Paggi, D. Parlevliet, T. Tanahashi and R. H. French, Assessment of Photovoltaic Module Failures in the Field, Report IEA-PVPS T13-09: ISBN 978-3-906042-54-1, 2017.

[12] Project Management Institute. (2017). A guide to the Project Management Body of Knowledge (PMBOK guide) (6th ed.). Project Management Institute.

[13] C. Schill, D. Parlevliet, A. Anderson, B. Stridh, L. Burnham, C. Baldus-Jeursen, L. Micheli, E. Urrejola, E. Whitney and G. Mathiak, Soiling Losses – Impact on the Performance of Photovoltaic Power Plants, Report IEA-PVPS T13-21: ISBN 978-3-907281-09-3, 2021.

[14] R. H. French, L. S. Bruckman, D. Moser, S. Lindig, M. v. Iseghem, B. Müller, J. S. Stein, M. Richter, M. Herz, W. v. Sark, F. Baumgartner, Assessment of Performance Loss Rate of PV Power Systems, Report IEA-PVPS T13-22, ISBN 978-3-907281-10-9, 2020.

[15] U. Jahn, M. Herz, D. Moser, G. Belluardo and M. Richter, Managing technical risks in PV investments: How to quantify the impact of risk mitigation measures for different PV project phases?, Prog Photovolt Res Appl. 2018; 26: 597– 607. https://doi.org/10.1002/pip.2970.

[16] U. Jahn, B. Herteleer, C. Tjengdrawira, I. Tsanakas, M. Richter, G. Dickeson, A. Astigarraga, T. Tanahashi, F. Valencia, M. Green, A. Anderson, B. Stridh, A. R. Lagunas Alonso, Y. Sangpongsanont, Guidelines for Operation and Maintenance of PV Power Plants in Different Climates, Report IEA-PVPS T13-25, ISBN 978-3-907281-13-0, 2021.

[17] D. H Stamatis, Failure Mode and Effect Analysis: FMEA from Theory to Execution. Quality Press, 2003.

[18] A. Sayed, M. El-Shimy, M. El-Metwally, M. Elshahed, Reliability, Availability and Maintainability Analysis for Grid-Connected Solar Photovoltaic Systems, Energies 2019, 12, 1213. https://doi.org/10.3390/en12071213.

[19] M. Herz, G. Friesen, U. Jahn, M. Köntges, S. Lindig, D. Moser, IEA PVPS Task 13: Performance, Operation and Reliability of Photovoltaic Systems - Quantification of Technical Risks in PV Power systems. International Energy Agency; 2021..

[20] K. Sinclair, M. Sinclair, "Silicon solar module visual inspection guide: Catalogue of Defects to be used as a Screening Tool", https://www.engineeringforchange.org/wp-content/uploads/2017/09/Solar-PV-Product-Visual-Inspection-Guide.pdf.

[21] https://iea-pvps.org/research-tasks/performance-operation-and-reliability-of-photovoltaic-systems/documents/ follow link to "PV Failure Fact Sheets", last access Aug 2021.

[22] P. Hacke, M.D. Kempe, J.H. Wohlgemuth, J.C. Li, and Y. Shen: Potential-Induced Degradation-Delamination Mode in Crystalline Silicon Modules, Presented at the 2016 Workshop on Crystalline Silicon

This is a short version of the paper submitted to the journal EPJ Photovoltaics.

It has been selected by the Committee of the WCPEC-8 for submission to EPJ Photovoltaics.

TECHNICAL CONSIDERATIONS RESULTING FROM PHOTOVOLTAIC MODULE HEATING DURING ELECTROLUMINESCENCE INSPECTION

Lukas Koester[a,b,*], Emanuel Vallarella[c], Atse Louwen[a], Sascha Lindig[a], David Moser[a]
[a]Institute for Renewable Energy, EURAC Research, Viale Druso 1, Bolzano, 39100, Italy
[b]Dipartimento di Energia, Politecnico di Milano, Via Lambruschini 4, 20156 Milano, MI, Italy
[c]Huawei Nuremberg Research Center, Power Conversion Technology Laboratory, Nürnberg, Germany
[*]Corresponding author, e-mail: lukas.koester@eurac.edu

ABSTRACT: Electroluminescence imaging experienced uprising use in the last years. In future photovoltaic plants the time-consuming connection of power supplies to photovoltaic module strings could be overcome by use of inverters with bi-directional functionality, allowing backpowering of connected module strings directly. During backpowering, the limits of the power supply must be considered. As temperature influences the open-circuit voltage, it is a crucial parameter to remain in these limits. The EL inspection itself impacts the solar cells temperature, as they warm up during the process, resulting in a shift of open circuit voltage. This work investigates the heating of a hetero junction photovoltaic module during electroluminescence inspection. The saturation temperature is estimated by energy balance calculations and experimentally verified to be around 20°C, with resulting voltage drop above 2 V. The importance of this is not only to verify that the measurement is within the limits of the power supply but also to consider changes of the recorded luminescence intensity due to the temperature, shown to have a decrease of up to 4 %.

1 INTRODUCTION

Numerous methods are available to inspect photovoltaic (PV) power plants during their operational phase, one of which is electroluminescence (EL) inspection. EL works through forward biasing the PV solar cells p-n junctions by injecting a forward current into the module [1], commonly described under the synonym backpowering. This forces majority carrier to cross the junction and therefore results in increased recombination processes with emission of luminescence radiation from the solar cells [2]. The luminescence signal is small compared to sun irradiation, wherefore EL is commonly performed during the night, although daylight EL is possible as well [3]. The imaging speed of EL increased strongly resulting in decreased costs. While the technical specification (TS) by the international electrotechnical commission (IEC) 60904-13 [1] describes EL inspections of single PV modules in laboratory conditions, a new TS is currently under preparation which will deal with EL measurements of online PV strings in utility scale [4]. A key factor impacting the speed of EL measurements is the procedure of applying the necessary power to the PV modules. Commercially available power supplies can supply sufficient power to forward-bias up to single strings of PV modules, while prototypes exist that are able to perform automated switching between up to 100 strings, which allows unmanned aerial vehicle (UAV) mounted inspections of up to 500 kWp, without moving the power supply [5]. Nevertheless, the extra hardware needs installation and reconfiguration of the cabling before every EL inspection. In future plants the time-consuming installation part of power supplies could be overcome by using inverters with bi-directional functionality, allowing backpowering of connected module strings directly. Nowadays, existing inverters can be upgraded to perform backpowering for EL [6].

During EL inspection, the cell temperature is increased due to the supplied power. This results in a decrease of the cells open-circuit voltage (V_{OC}). Especially in the case for EL-ready inverters, these considerations need to be taken into account in the planning phase of a PV plant already. Thus, to identify reliably the best conditions for EL imaging on large PV strings, the heating process must be understood and considered fully. While the heating of PV modules during normal operation under sun irradiation is described well [7]–[11], the heating process of cells during EL inspection is not. Another factor is the stability of the recorded EL signal: The intensity of luminescence radiation depends exponentially on temperature and voltage. A cell temperature increase will result in a reduction of luminescence signal intensity [12]. Although a small effect, it needs to be investigated and quantified for future EL inspections.

A power supply unit (PSU), such as standard power supplies or EL-ready inverters, comes with a specific operational range, usually defined by voltage range and a maximum current. The latest TS for EL defines the amount of current to feed into a PV module for reliable EL inspection (i.e. I_{SC} and $0.1 \times I_{SC}$) [1]. The TS recommends to have a power supply with sufficient voltage to provide I_{SC}, which should be at V_{OC} level, but can be significantly higher [1]. Considering that these currents are mandatory for proper EL inspection and the specific operational range of PSUs, it minimizes the range of how many modules are allowed per string.

To address these questions, here, the temperature increase of PV modules during EL inspection is investigated from a physics-based point of view. For an understanding of the processes, the investigation is approached for single modules of different technologies. These results are then discussed in the context of PV module strings. First, the saturation temperature during backpowering is derived by energy balance calculations and then examined by experimental temperature measurements during EL. Continuous EL imaging allows to investigate the impact heating has on the EL signal intensity.

This work builds up as follows: Section 2 describes the theoretical calculation of the maximum temperature and introduces the theory behind the luminescence intensity. Further, the setup of the experiments is described. In Section 3, the results of calculations and experiments are shown. These results are then discussed in Section 4. Finally, Section 5 summarizes the work and gives an outlook on the future of EL inspections.

2 METHODS, MATERIALS AND THEORETICAL MODEL

2.1 Theoretical background

The process of PV module heating under normal operation conditions is well known. It depends mainly on environmental conditions like solar irradiance, ambient temperature and wind speed [13]. While the heating is mainly caused by the sun irradiation, the cooling process strongly depends on the wind speed. The cell temperature affects V_{OC} and I_{SC}, so that an increase in PV module temperature results in reduced maximum power [13], typically around 0.3-0.4 %/K, depending on the PV module technology. Therefore, PV module data sheets are usually providing the temperature coefficients for short circuit current, open circuit voltage and maximum power (α_{Isc}, β_{Voc} and γ_{Pm} respectively).

The heating process during EL inspection instead is independent of the solar irradiance, as performed by night (daylight EL is not considered here). Therefore, heating depends on the supplied power P_{in}, while ambient temperature and wind speed still have a cooling effect on the PV module, which leads to a saturation temperature T_{sat} when in equilibrium. In forward bias, a forward current is pushed through the p-n junction. Thus, the power is dissipated and heats the solar cells [14]. Further, the forward current enhances the transition of majority carriers through the p-n junction which results in enhanced recombination, including radiative recombination and photon emission. These photons, emitted in the near infrared regime, are the luminescence signal, and can be captured by an Indium Gallium Arsenide (InGaAs) camera for EL imaging. The amount of photons emitted by the solar cell is given by the EL signal intensity Φ_{EL}, which is correlated to the cell voltage V_{cell} and temperature T_{cell} by [15]

$$\Phi_{\text{EL}} = EQE \cdot \Phi_{\text{BB}}(\exp\left(\frac{qV_{\text{cell}}}{k_B T_{\text{cell}}}\right) - 1) \tag{1}$$

which further depends on the external photovoltaic quantum efficiency EQE, the black body radiation Φ_{BB} and elementary charge q. For the present investigation, the exponential impact of voltage and temperature will be of interest.

2.2 PV Module under investigation

For the experimental part of our investigation of the heating effect, a bi-facial PV module with hetero junction technology (HJT) cells, is chosen. It represents one of the main PV technologies which is expected to be installed in the next decade [16]. The electrical parameters of the selected module are given in Table 1.

Table 1: Data sheet electrical parameters of HJT module under investigation

Properties	HJT Module
Bi-facial	Yes
P_{nom} / W	360
V_{OC} / V	52.6
I_{SC} / A	8.9
V_{MP} / V	42.8
I_{MP} / A	8.4
α_{Isc} / %/K	0.014
β_{Voc} / %/K	-0.285
γ_{Pm} / %/K	-0.327

2.3 Energy balance principle

The calculation of balance between fed and emitted energy allows to estimate the final cell temperature. The fed energy is given by the supplied power P_{in} consisting of current I_{in} and voltage V_{in} provided by the power supply:

$$P_{\text{in}} = V_{\text{in}} I_{\text{in}} \tag{2}$$

The solar cells cool down until an equilibrium is reached between input power and ambient conditions. The cooling procedure can be described by effects of conduction (q_c), convection (q_{cv}) and radiation (q_{rad}) as shown in Equations (3), (4) and (5), respectively. All parameters are explained in Table 2.

$$q_c = \frac{k}{s} A(T_{\text{cell}} - T_{\text{env}}) \tag{3}$$

$$q_{cv} = hA(T_{\text{cell}} - T_{\text{env}}) \tag{4}$$

$$q_{rad} = \sigma A\left(\epsilon T_{\text{cell}}^4 - \alpha T_{\text{env}}^4\right) \tag{5}$$

The energy emitted due to conduction must be considered for each surface and material layer (encapsulation, glass on both sides) individually. Convection describes the energy emitted due to cooling on one surface. As the calculation performed will be compared to actual temperatures measured in the laboratory, the wind speed is considered 0 m/s on both sides. The energy emitted due to radiation is given by the Stefan-Boltzmann law of an object in a cooler environment.

As the energy is emitted on both sides of the cell equally (due to the shape of a solar cell, the heat distribution is simplified to be one dimensional), all described emission processes must be considered for both sides of the PV modules and their potentially different layers. Adapting Rim et al. [17], the whole energy balance can therefore be expressed as

$$\frac{T_{cell} - T_{env}}{2R_{c,EVA} + 2R_{c,g} + 2R_{cv,g}} + \sigma A(\epsilon T_{cell}^4 - \alpha T_{env}^4) = I_{in} V_{in} \tag{6}$$

With the thermal resistance of conduction $R_{c,glass} = L/Ak_{glass}$ and $R_{c,EVA} = L/Ak_{EVA}$ for glass and encapsulant and the thermal resistance of convection $R_{cv} = 1/hA$ for glass on both sides, respectively. All parameters are explained and listed in Table 2. Solving Equation (6) for T_{cell} gives the maximum temperature the cell reaches in equilibrium of supplied power versus ambient temperature and wind speed.

2.4 Experimental Setup

The experimental approach is conducted in a laboratory with the setup schematized in Figure 2. The ambient temperature was kept at 25 °C (± 1 °C), monitored by an external thermometer placed one meter away from the module. The power to backpower the module is provided by a *TTi* power supply. The parameters (current, voltage, power) are measured every 0.25 seconds by a *Fluke* multimeter. The module temperature is measured every second by a *Lutron* thermometer with thermocouples in four positions depicted in Figure 1. In front of the PV module, two cameras are placed. A *FLIR* IR camera to survey possible formation of hot spots and a *Raptor* camera with InGaAs sensor to capture the EL images, both imaging every 15 seconds. The power supply is started with the module specific current (I_{SC} and $0.1 \times I_{SC}$, one by one) and the voltage range is chosen with 30 % above V_{OC} to ensure no current limitation. For the whole heating and cooling time, all data is logged.

Table 2: Description of used parameters for energy balance calculations in Equation 6 to achieve saturation temperature

Variable	Meaning	Unit	Value
T_{cell}	solar cell temperature	K	
T_{env}	environmental temperature	K	298.15
h	Convective coefficient	W/m²/K	6 [18]
k_{EVA}, k_g, k_{bs}	Thermal conductivity of encapsulant, glass and backsheet	W/m/K	0.34 [19], 0.8 [20], 0.25 [21]
L_{EVA}	encapsulant thickness	mm	0.45 [20]
L_g	glass thickness	mm	2
L_{bs}	backglass thickness	mm	2
σ	Stefan-Boltzmann constant	W/m²/K⁴	$5.67 \cdot 10^{-8}$
A	solar cell area	mm²	24370
ϵ_g	emissivity of glass	-	0.9 [22]
ϵ_{bs}	emissivity of backsheet	-	0.83 [21]
α	absorptivity of solar cell	-	0.93 [23]
t	time		

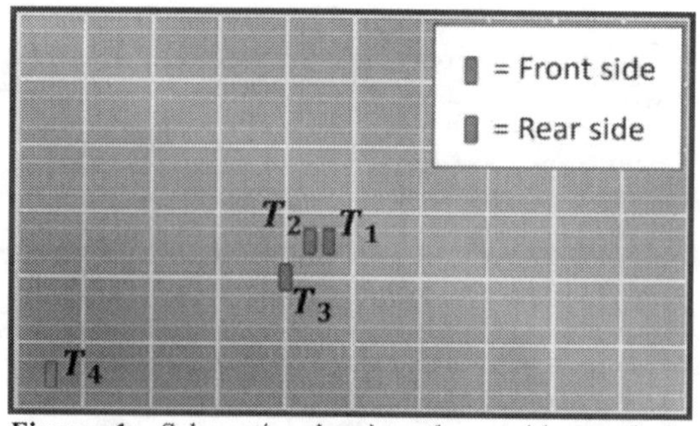

Figure 1: Schematic showing the positions of the thermocouples to survey the module temperature. Including the used measurement point names, T1 to T4. T1: front side above a center cell; T2: rear side above the center cell; T3: rear side in cell gap; T4: rear side above corner cell.

The temperature will be measured on the module surfaces, while the calculations are done for the solar cells themselves. The offset between cell and backsheet is described for normal module operation at 1000 W/m² to be between 2 °C and 3 °C [24]. Therefore, a deviation between measurement and calculation is expected.

3 RESULTS

3.1 Calculation

In Table 3, the cell saturation temperatures calculated for the HJT module following Equation (6) are given, as maximum temperature $T_{sat,calc}$ and in comparison to ambient temperature of T_{amb} = 25 °C as $\Delta T_{sat,calc}$ =

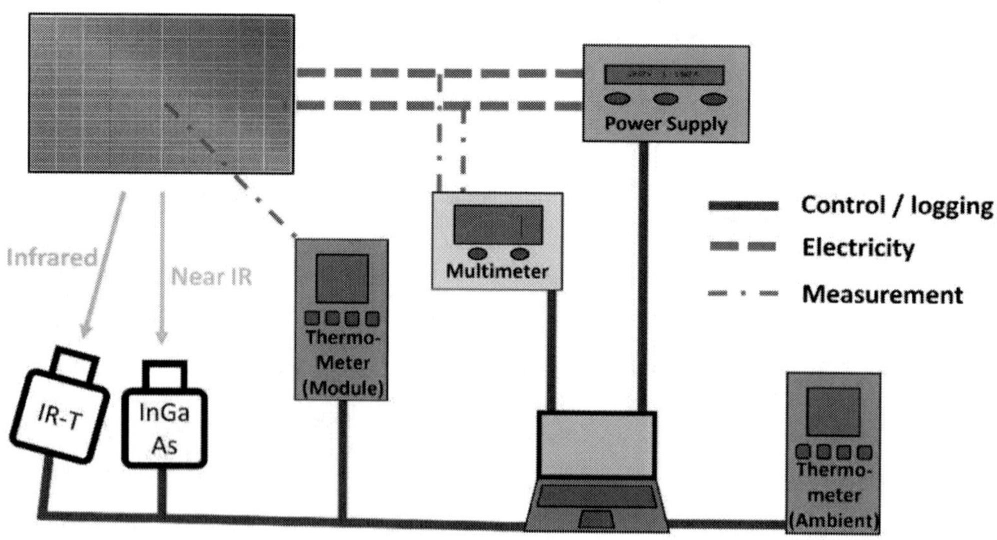

Figure 2: Schematic of the experimental setup to survey the heating of a PV module during EL inspection.

$T_{sat,calc} - T_{amb}$. The temperature increase is calculated to be 20.1 °C.

3.2 Experimental results

The measurements of $0.1 \times I_{SC}$ resulted in temperature increases of below 2 °C. This alteration is too small to allow deeper inspections with the used setup. Therefore, all results and discussion are given for the I_{SC} inspection. From the measurements, the following plots are depicted: For the whole measurement cycle of each PV module, the supplied current, voltage and front-cell temperature as well as ambient temperature are shown in Figure 3 (a). A detailed inspection of the different measured module temperatures is given in Figure 3 (b), together with ambient temperature and the input voltage for a better orientation. Finally, the results of the EL inspections are shown by the overall EL intensities of the module in Figure 3 (c). No hot spots are present during the measurement, as shown by the IR-T images, an example image is shown for the HJT module in Figure 4.

Table 3: Results of the experimental measurements of heating of the HJT module during EL.

Symbol	Explanation	Value	Unit
$T_{sat,calc}$	From eq. 6	45.1	°C
$\Delta T_{sat,calc}$	From eq. 6	20.1	°C
$T_{1,max}$	Front, center cell	42.3	°C
$\Delta T_{1,max}$	Front, center cell	17.1	°C
$T_{2,max}$	Rear, center cell	42.4	°C
$\Delta T_{2,max}$	Rear, center cell	17.2	°C
$T_{3,max}$	Rear, cell gap	40.5	°C
$\Delta T_{3,max}$	Rear, cell gap	15.3	°C
$T_{4,max}$	Rear, corner cell	39.1	°C
$\Delta T_{4,max}$	Rear, corner cell	14	°C
V_{start}	PSU on	55.5	V
V_{end}	PSU off	53.3	V
ΔV	$V_{start} - V_{end}$	-2.2	V
β_V	Measured average	-0.26	%/°C
$\sigma_{\beta,dev}$	Standard deviation	±0.02	%/°C
t_{heat}	Heating time	2576	s
t_{cool}	Cooling time	2674	s
$\Delta\Phi_{EL}$	EL intensity drop	3.93	%
$\Delta\Phi_{EL,cells}$	$\Delta\Phi_{EL,center} - \Delta\Phi_{EL,corner}$	-1.34	%

The voltages at begin (V_{start}) and end (V_{end}) of supplying power are given, as well as their difference (ΔV). For the measured range, the temperature coefficient of the voltage compared to the front-cell temperature is calculated with an average of σ_V with a standard deviation $\sigma_{\beta,dev}$. The heating velocity is given by the time until saturation temperature is reached (t_{heat}) is calculated, defined as the time from PSU start until the temperature is stable, meaning variation in the following 10 minutes is below 0.5 °C. Also, the cooling time (t_{cool}) is presented, defined as time from PSU-off until the module temperature is within 1 °C of the ambient temperature. All results are listed in Table 3.

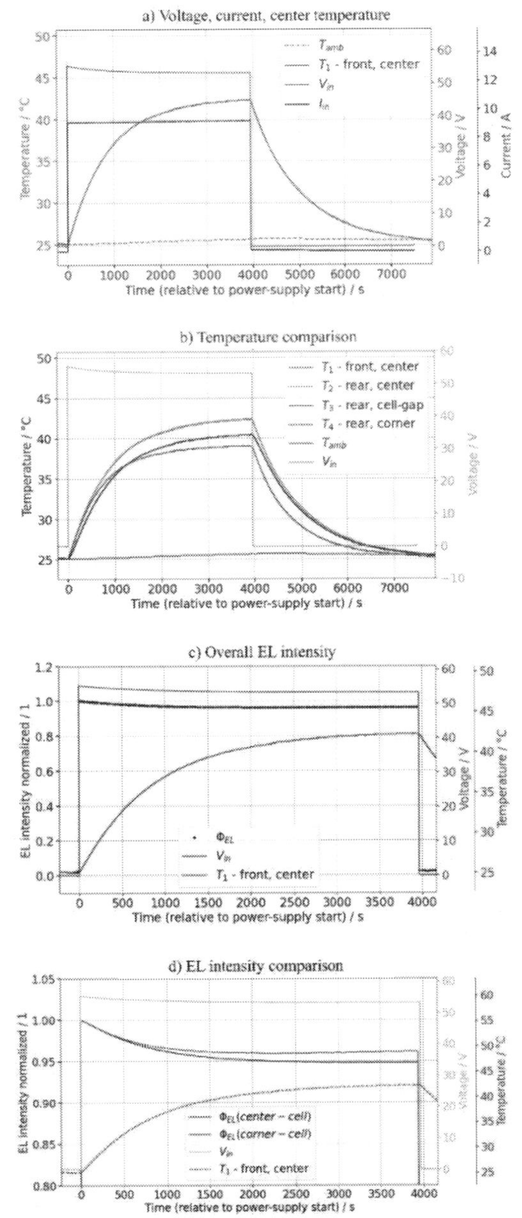

Figure 3: Plots of experimental results. a), Measured Voltage, current, centered cell temperature and ambient temperature over the whole measurement cycle. b), Comparison of all temperatures measured at the four points depicted in Figure 1. c), The evolution of the electroluminescence signal during the measurement, in comparison to the voltage necessary to provide short circuit current and the temperature measured on the module front. d), Comparison of the two electroluminescence signals measured for different cells in the same module: One in the center and the other in the corner.

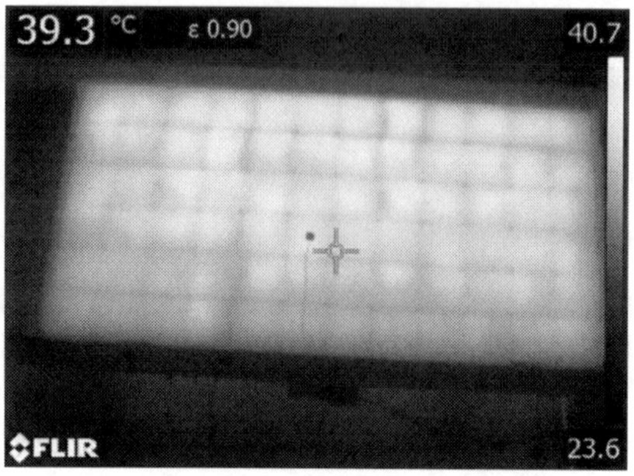

Figure 4: Infrared thermography image of the HJT module under reverse biasing with short-circuit current. No formation of hot spots visible at the temperature measurement points.

The EL intensity is taken as average over the whole module and then normalized to the maximum intensity of the respective module. Next to the plot in Figure 3 (c), the total EL-intensity drop $\Delta\Phi_{EL}$ is given. Further, the normalized average intensity of a center cell and a corner cell of the same PV module are measured over the whole heating process and compared to each other in Figure 3 (d). The maximum difference between these EL intensities of the two cells are given by $\Delta\Phi_{EL,cells}$. Both intensities are given in Table 3.

4 DISCUSSION AND CONCLUSION

4.1 Temperatures

As expected, a difference between calculated and measured temperature can be seen: the calculated one ($T_{sat,calc}$) is between 3 °C higher than the measured one ($T_{1,max}$). Due to the same configuration of front and backside of the HJT module (glass-glass), no difference between the rear side (T_2) and front side (T_1) temperatures can be seen in Figure 3 (b). The offset between calculated and measured temperature can be explained by the fact that the calculated temperature was done for the cells, while the temperature measured was done on the module surface. Thus, a cooler measured temperature is expected. Further, the thermocouples are covered solely by an aluminum tape. Therefore, the surrounding temperature has a cooling effect on the measurement. Furthermore, some material parameters (material emissivity, cell absorptivity) are generalized, and possible technology-specific differences are not considered. In real life conditions, the wind speed would have another effect on cooling of the PV module, which might be different for front and rear side, depending on installation. This effect played no role in the laboratory measurements.

The effects of lateral distribution of thermal energy through the modules can be seen in curves T_3 and T_4 of Figure 3 (b). Although the corner cell heats up as quick as the center cell in the beginning, it is only partially surrounded by other cells and ends in a much cooler equilibrium compared to the center cell. The center cell instead is surrounded by other cells, thus heat sources, and therefore has lower lateral distribution of heat. This results in a 3°C difference for corner compared to center cell. T_3 instead shows the delay of lateral heat distribution, as it increases slowly in the beginning (longer distance to the actual heat source), but in the end results in a saturation temperature between the module temperatures measured above cells in the center and at the corner. The same effects can be seen in a reversed manner during the cooling process.

t_{heat} and t_{cool} show that heating and cooling of a module takes about the same time. Considering EL inspections performed by UAVs, the flight-over of single strings takes maximum a few minutes. Therefore, if the backpowering is planned precisely, the module temperatures will stay below half of their saturation temperatures (about 10 minutes of backpowering). Ground based inspection of one string might take longer. Enough time for the modules to get close to saturation temperature. To maintain the backpowering time as short as possible is advisable, also to minimize the energy fed into the system.

4.2 Voltage

While the cell temperature increases, the module voltage (necessary to maintain supply of short circuit current) drops. The resulting thermal coefficient for the voltage (β_V, cf. Table Table 3) are comparable with the one given for V_{OC} by the data sheet ($\beta_{V_{OC}}$ cf. Table 1) with a standard deviation of 0.02 %.

4.3 EL intensity

The EL intensity behavior shown in Figure 3 (c) shows the decrease. In total, the signal intensity is reduced by 4 %, following the exponential voltage drop.
The temperature dependency of the EL signal comes also into place within a single module. This is shown by comparison of the EL signal intensity of a cell in the center with a corner one. The intensity developments over time of both cells, each normalized to one, can be seen in Figure 3 (d). The corner cell has a slightly lower EL intensity variation of 1.3 percentage points and therefore maintains a stronger signal compared to the center cell.

5 SUMMARY AND OUTLOOK

In this paper, the heating effect of backpowering PV modules during EL inspections is investigated, theoretically and experimentally. An energy balance calculation allows to estimate the saturation cell temperatures under a forward current of I_{SC}. The expected temperature increase is in the range of 20 °C, which could be verified with experimental measurements. Calculation of the heating time shows that the heating plays a minor role for fast EL imaging by use of UAVs. Nevertheless, the resulting voltage drop due to temperature might be low in a single module, but in a module string, it can reduce the string voltage by as much as the voltage of one whole PV module. Therefore, planning of a PV plant with EL-ready inverters must be done carefully to ensure the voltage ranges of the inverters are not reached.
For EL inspections, a drop in EL intensity of up to 4 % due to the backpowering induced heating could be observed. Although the generated EL signal is already small, the 4 % decline in intensity makes no practical difference for qualitative analyses done in outdoor conditions. The same is the case for the variation in EL intensity through the PV

module, which can be 30 % between center and corner cell. But both described phenomena should be considered for quantitative EL measurements in laboratory conditions.

Next to the described HJT module, two module types of different technology (interdigitated back contact and passivated emitter and rear contact) are investigated as well. These results and the discussion of technology dependent differences will be published in a dedicated journal article, which is under review during the time of this publication. That article will also present the *EL window*, a simulation of dark I-V curves in correlation to the EL-ready inverter ranges, as a practical tool to plan a PV plant layout with respect to the EL functionality. Further, the applicability of the described heating effect for snow melting purposes is discussed.

For future work, the usage of EL-ready inverter will be tested in a real-life application. Although still a prototype, if its functionality is proven and it becomes a standard in future PV installations, it will make EL inspections much faster. The time saved due to cease of the time intensive PSU installations will be a game changer as it can drive the outstanding detection possibilities of EL also financially in the range of IR-T inspections.

6 FUNDING

This work has received funding from the Horizon 2020 programme of the European Union under GA. No. 952957 – Project TRUST PV.

7 REFERENCES

[1] *IEC TS 60904-13 - Photovoltaic Devices - Part 13: Electroluminescence of Photovoltaic Modules.* Geneva, CH: International Electrotechnical Commission (IEC), 2018.

[2] D. Kiliani, "Luminescence Imaging Techniques for Silicon Photovoltaics," Universität Konstanz, Konstanz, 2013.

[3] W. Herrmann *et al.*, *Qualification of Photovoltaic (PV) Power Plants using Mobile Test Equipment: International Energy Agency Photovoltaic Power Systems Programme: IEA PVPS Task 13: Report IEA-PVPS T13-24:2021.* International Energy Agency, Photovoltaic Power Systems Programme, 2021.

[4] *IEC TS 62446-4 ED1 Photovoltaic (PV) systems - Requirements for testing, documentation and maintenance - Part 4: Electroluminescence Measurement of Photovoltaic Arrays.* Geneva, CH: International Electrotechnical Commission (IEC), Under Development.

[5] S. Koch, T. Weber, C. Sobottka, A. Fladung, P. Clemens, and J. Berghold, "Outdoor Electroluminescence Imaging Of Crystalline Photovoltaic Modules: Comparative Study Between Manual Ground-Level Inspections And Drone-Based Aerial Surveys," 2016, pp. 1736–1740. doi: 10.4229/EUPVSEC20162016-5DO.12.2.

[6] J. Ballestín-Fuertes *et al.*, "Novel Utility-Scale Photovoltaic Plant Electroluminescence Maintenance Technique by Means of Bidirectional Power Inverter Controller," *Applied Sciences*, vol. 10, no. 9, p. 3084, Apr. 2020, doi: 10.3390/app10093084.

[7] A. Louwen, A. C. de Waal, R. E. I. Schropp, A. P. C. Faaij, and W. G. J. H. M. van Sark, "Comprehensive characterisation and analysis of PV module performance under real operating conditions," *Progress in Photovoltaics: Research and Applications*, vol. 25, no. 3, pp. 218–232, 2017, doi: https://doi.org/10.1002/pip.2848.

[8] S. Lindig, A. Louwen, D. Moser, and M. Topic, "Outdoor PV System Monitoring—Input Data Quality, Data Imputation and Filtering Approaches," *Energies*, vol. 13, no. 19, 2020, doi: 10.3390/en13195099.

[9] S. Bouchakour, D. V. Caballero, A. Luna, E. R. Medina, K. B. El Amin, and P. R. Cortés, "Monitoring, modelling and simulation of bifacial PV modules over normal and high albedos," in *2020 9th International Conference on Renewable Energy Research and Application (ICRERA)*, 2020, pp. 252–256. doi: 10.1109/ICRERA49962.2020.9242869.

[10] G. M. Tina, F. B. Scavo, and A. Gagliano, "Multilayer Thermal Model for Evaluating the Performances of Monofacial and Bifacial Photovoltaic Modules," *IEEE Journal of Photovoltaics*, vol. 10, no. 4, pp. 1035–1043, 2020, doi: 10.1109/JPHOTOV.2020.2982117.

[11] B. G. Bhang, W. Lee, G. G. Kim, J. H. Choi, S. Y. Park, and H.-K. Ahn, "Power Performance of Bifacial c-Si PV Modules With Different Shading Ratios," *IEEE J. Photovoltaics*, vol. 9, no. 5, pp. 1413–1420, Sep. 2019, doi: 10.1109/JPHOTOV.2019.2928461.

[12] L. Koester, A. Astigarraga, S. Lindig, and D. Moser, "Development Of Daylight Photoluminescence Technique For Photovoltaic Modules And Investigation Of Temperature Dependency," in *37th EU-PVSEC*, Online, Sep. 2020, p. 6. doi: 10.4229/EUPVSEC20202020-4CO.4.2.

[13] E. Skoplaki and J. A. Palyvos, "On the temperature dependence of photovoltaic module electrical performance: A review of efficiency/power correlations," *Solar Energy*, vol. 83, no. 5, pp. 614–624, May 2009, doi: 10.1016/j.solener.2008.10.008.

[14] T. Ghanbari, "Permanent partial shading detection for protection of photovoltaic panels against hot spotting," *IET Renewable Power Generation*, vol. 11, no. 1, pp. 123–131, 2017, doi: https://doi.org/10.1049/iet-rpg.2016.0294.

[15] U. Rau, "Reciprocity relation between photovoltaic quantum efficiency and electroluminescent emission of solar cells," *Phys. Rev. B*, vol. 76, no. 8, p. 085303, Aug. 2007, doi: 10.1103/PhysRevB.76.085303.

[16] ITRPV, "International Technology Roadmap for Photovoltaic (ITRPV), 12th Edition," VDMA Photovoltaic Equipment, Frankfurt am Main, Germany, 2021. [Online]. Available: https://itrpv.vdma.org/

[17] S. Rim, M. Baldrias, and M. Morse, "Temperature of solar cells in reverse bias: Theory and applications," in *2010 35th IEEE Photovoltaic Specialists Conference*, Honolulu, HI, USA, Jun. 2010, pp. 002698–002700. doi: 10.1109/PVSC.2010.5616916.

[18] L. Zhu, A. Raman, K. X. Wang, M. A. Anoma, and S. Fan, "Radiative cooling of solar cells," *Optica*, vol. 1, no. 1, p. 32, Jul. 2014, doi: 10.1364/OPTICA.1.000032.

[19] "MakeItFrom Database." [Online]. Available: https://www.makeitfrom.com/material-properties/10-Percent-Vinyl-Acetate-EVA

[20] A. Pfreundt, A. J. Beinert, D. Yucebas, L. V. Mesquita, L. Pitta-Bauermann, and M. Mittag, "Post-Processing Thickness Variation of PV Module Materials and its Impact on Temperature, Mechanical Stress and Power," in *36th EU-PVSEC*, Marseille, France, 2019, p. 6. doi: 10.4229/EUPVSEC20192019-4CO.4.3.

[21] J. Oh *et al.*, "Reduction of PV Module Temperature Using Thermally Conductive Backsheets," *IEEE J. Photovoltaics*, vol. 8, no. 5, pp. 1160–1167, Sep. 2018, doi: 10.1109/JPHOTOV.2018.2841511.

[22] A. Riverola *et al.*, "Mid-infrared emissivity of crystalline silicon solar cells," *Solar Energy Materials and Solar Cells*, vol. 174, pp. 607–615, Jan. 2018, doi: 10.1016/j.solmat.2017.10.002.

[23] R. Santbergen and R. J. C. van Zolingen, "The absorption factor of crystalline silicon PV cells: A numerical and experimental study," *Solar Energy Materials and Solar Cells*, vol. 92, no. 4, pp. 432–444, Apr. 2008, doi: 10.1016/j.solmat.2007.10.005.

[24] J. A. Kratochvil, W. E. Boyson, and D. L. King, "Photovoltaic array performance model.," *Sandia Report*, Aug. 2004, doi: 10.2172/919131.

[25] M. Villalva, J. Gazoli, and E. Filho, "Modeling and circuit-based simulation of photovoltaic arrays," in *2009 Brazilian Power Electronics Conference, COBEP2009*, Nov. 2009, vol. 14, pp. 1244–1254. doi: 10.1109/COBEP.2009.5347680.

OUTDOOR PHOTOLUMINESCENCE IMAGING OF PV MODULES BASED ON STRING INVERTER'S IV CURVE SWEEP

Marija Vuković, Kristian Hovde Liland, Ulf Geir Indahl, Marko Jakovljević, Andreas Svarstad Flø, Espen Olsen, Ingunn Burud
Faculty of Science and Technology, Norwegian University of Life Sciences,
Drøbakveien 31, 1430 Ås, Norway
marija.vukovic@nmbu.no

ABSTRACT: Photoluminescence imaging of field-deployed silicon photovoltaic modules with sun's excitation has so far been conducted with a lock-in technique, which implies connecting to the modules during imaging and switching their operating point between two predetermined conditions. As opposed to this approach, the present study investigates aspects of acquiring images which cover a range of operating points on the IV curve. This can be achieved by conducting imaging during IV curve scans, which are performed by string inverters for the purpose of optimal power production. Such imaging method has the potential to be completely contactless with no other equipment necessary apart from the imaging apparatus and no need to connect to the imaged modules. This can be achieved if the images can be obtained without knowing when an IV curve scan takes place by processing images in real-time and in that way identifying when the photoluminescence signal appears due to an IV curve sweep. The present study investigates the possibility to use an image selection algorithm based on the Pearson Correlation Coefficient.
Keywords: photoluminescence, Pearson correlation coefficient, inverter, IV curve sweep, monitoring, degradation

1 INTRODUCTION

Compared to other outdoor imaging characterisation techniques, photoluminescence (PL) images of silicon photovoltaic (PV) modules give more information than thermal infrared images and the same level of detail as electroluminescence (EL) images. At the same time, PL imaging has the potential to be faster than EL imaging because it requires no electrical connection with the PV modules [1].

When performed in controlled laboratory settings, PL imaging of silicon solar cells and wafers is a contactless imaging technique which detects the radiative band-to-band recombination within the semiconductor material. There is in essence no need to connect to the sample because one uses light as excitation source and the only other necessary piece of equipment is the imaging device [2]. However, when the PL imaging is conducted outdoors with sunlight excitation, both the operating point of the PV modules and the sun's reflection affect the imaging procedure. This leads to the fact that PL imaging in such conditions ceases to be a contactless imaging method and is based on interference with the PV system and its production [3-9].

A PL imaging procedure which does not require control of the PV system's operating point and therefore has the potential for high efficiency and throughput is based on data collection during the IV curve sweeps performed by the string inverters [10]. IV curve sweeps are built-in functionalities of string inverters for the purpose of global maximum power point (MPP) search and monitoring. They can take place on regular basis, as often as every 10 minutes. The approach described in [10] has been applied with the current measurements, which means that it has been necessary to connect to the PV system in order to collect them.

However, developing the method further and conducting image acquisition without impacting the PV system implies collecting images without measuring the current, i.e., without knowing how many minutes ago the last IV curve scan took place and how long time it will take until the next one. Since the image acquisition time can be as low as 6-ms, the imaging apparatus records 10 000 images for one minute. This results in an immense amount of data being generated and a considerable storing capacity is necessary [10]. It is therefore highly desirable to have a quick processing algorithm in place which would enable detection of the current fall based on the sudden appearance of the PL signal on the images. A promising approach is to use the Pearson Correlation Coefficient (PCC) for image processing and selection in real-time.

2 EXPERIMENTAL

The equipment used in this study is a Raptor Photonics Owl 640S with an InGaAs detector, optical range 900 nm – 1700 nm and 640x512 resolution together with an optical band-pass filter by Edmund Optics with the wavelength range 1100-1150 nm and optical density ≥4.0. The three imaged bifacial PERC modules (JA Solar JAM72D20) were connected in a string to an inverter Fronius Primo 3.0. Images of reflectance and photoluminescence emitted from the string together with current measurements were acquired as described in [10].

The image set used in the present study comprises 500 images which span the whole IV curve scan. This is a 3D image matrix with the dimension 640x512x500. One of those images is shown in Fig. 1(a) and the average pixel values of the image series are shown in purple in Fig. 1(b). The current measurements obtained simultaneously with the images are shown in black. With the falling current during the IV curve scan, the PL signal increases, as is expected.

The algorithm usually applied during processing of PL images is subtraction of images with low PL signal from the images with high PL signal [3,4,6,9,10], which results in a difference photoluminescence image, ΔPL. For the data set described above, this requires selection of two image sub-sets with a sufficiently high difference in the PL signal [10]. This is time consuming when processing images in real-time. An alternative approach for this purpose would be obtaining a PCC image based on correlating every pixel vector in the spatial dimension of an image stack with the average pixel values for the whole image stack shown in Fig. 1(b). It is expected that the pixels representing the modules will have high correlation with the average values, while those representing the background (surroundings) will have low correlation. The validity of the PCC image will be compared with a ΔPL image.

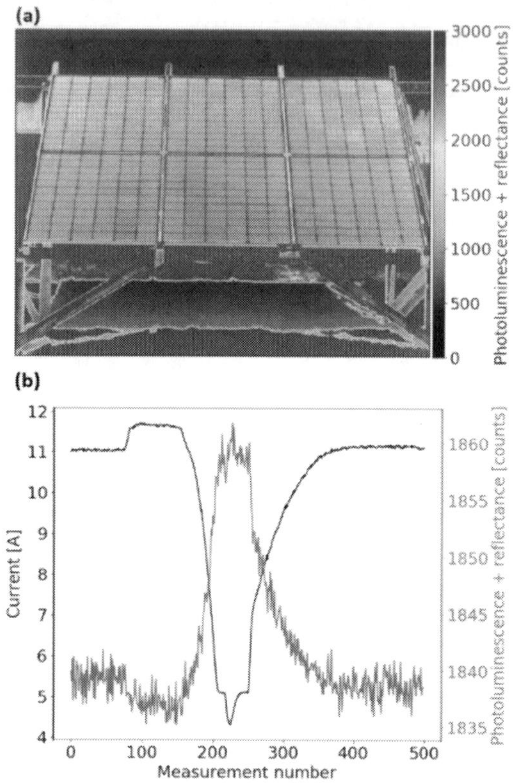

Figure 1: One of the 500 images in the timeseries acquired during an IV curve sweep showing a bifacial module string (a) and the average pixel count of the 500 images overlaid with the current measurement (b).

3 RESULTS AND DISCUSSION

The PCC image calculated between every pixel vector in the image stack and the average pixel values of the whole stack is shown in Fig. 2(a). The correlation image corresponds well with the ΔPL values shown in Fig. 2(b). Red colour represents high values, while blue colour represents low values in both images. The cells in all three modules seem to be intact, without any serious series resistance problems. This conclusion can be drawn from both the ΔPL image and the correlation image.

The correlation image also corresponds well with the ΔPL image in the sense that the upper parts of the two modules on the sides have higher values than the lower parts of the same modules. This is in accordance with the fact that the modules are bifacial, and the upper parts are elevated from the ground resulting in a higher gain.

Looking at the background pixel values in Fig. 2(b), it is visible that some of them are not eliminated and that they are higher than the ΔPL values. These values are the difference in reflection between two sub-sets chosen among the 500 images to calculate the ΔPL image. They should not have been visible if the irradiation during image acquisition had been constant. This is an indication that the variation in reflection during image acquisition leaves traces in the ΔPL images. At the same time, the background values in Fig. 2(a) are not in the same range as the correlation values, but are considerably lower, indicating that the PCC might be more robust to variation in irradiation. Further investigation has shown that the background correlation values are lower than 0.6.

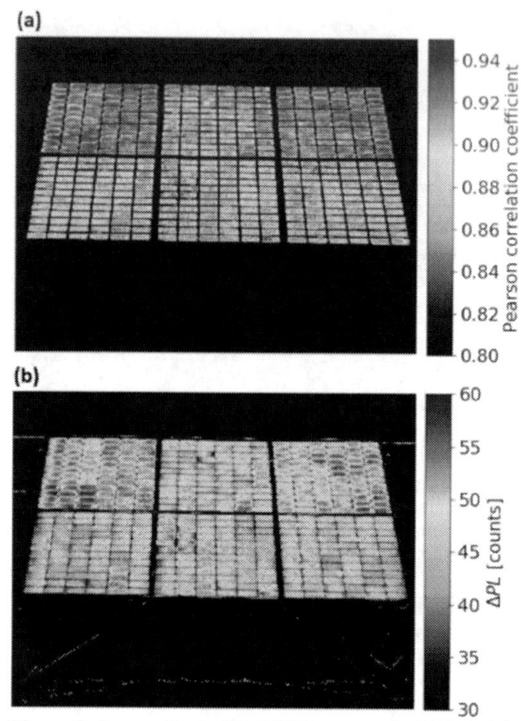

Figure 2: Image displaying PCC in every pixel (a) and the ΔPL image obtained with the established subtraction method (b).

The way the PCC approach can be used for image selection in real-time is to read in a set of images as they are being acquired and to calculate the PCC. If no current fall takes place during acquisition of one image set, the PCC is low across both the module pixels and the background pixels. That image set is discarded, and a new recorded set of images is used to calculate a new PCC image. This is done until a PCC image is obtained where the correlation values representing the modules are considerably higher than those representing the background.

This approach depends on the fact that the PL signal can be detected from the modules, meaning that the modules produce the PL signal when the current drops. The question poses itself what kind of results the approach can give with faulty modules producing a weak PL signal. Another relevant question is how big part of the images needs to consist of the module pixels as opposed to the background pixels so that the increase in the PL signal is clearly visible in the average pixel values as shown in Fig. 1(b). These questions will be addressed in future studies.

4 CONCLUSIONS

Photoluminescence images acquired during an IV

curve sweep performed by a string inverter cover the whole range of the current fall and rise. The image collection procedure itself results in a vast number of acquired images and image processing in real-time is desirable in order to save storage capacity. An algorithm based on PCC is promising in order to achieve this goal. For its calculation, it suffices to use one image set instead of having to identify two sets with sufficient difference in the PL signal between them for calculation of a ΔPL image. The amount of information obtainable from correlation images is comparable to ΔPL images at the same time as the PCC seems to be more robust to irradiation variation during image acquisition.

ACKNOWLEDGEMENTS

This work was performed within The Norwegian Research Center for Sustainable Solar Cell Technology (FME SUSOLTECH, project number 257639/E20). The center is co sponsored by the Research Council of Norway and its research and industry partners.

REFERENCES

[1] W. Herrmann, G. Eder, B. Farnung, G. Friesen, M. Koentges, B. Kubicek, O. Kunz, H. Liu, D. Parlevliet, I. Tsanakas, J. Vedde, Report IEA-PVPS T13-24:2021.
[2] T. Mehl, Energy Procedia 92 (2016).
[3] T. J. Silverman, M. G. Deceglie, K. VanSant, S. Johnston, I. Repins, Proceedings 44th Photovoltaic Specialist Conference (2017).
[4] M. Guada, A. Moreton, S. Rodriguez-Conde, L. A. Sanchez, M. Martinez, M. A. Gonzalez, J. Jimenez, L. Perez, V. Parra, O. Martinez, Energy Science & Engineering 8 (2020).
[5] L. Stoicescu, M. Reuter, J. H. Werner, Proceedings 29th European Photovoltaic Solar Energy Conference and Exhibition (2014).
[6] R. Bhoopathy, O. Kunz, M. Juhl, T. Trupke, Z. Hameiri, Prog Photovolt Res Appl 26 (2018).
[7] O. Kunz, G. Rey, M. K. Juhl, T. Trupke, Proceedings 48th Photovoltaic Specialists Conference (2021).
[8] L. Koester, A. Astigarraga, S. Lindig, A. Louwen, M. Antinori, D. Moser, G. Manzolini, Proceedings 38th European Photovoltaic Solar Energy Conference and Exhibition (2021).
[9] M. Vukovic, I. E. Høiaas, M. Jakovljevic, A. S. Flø, E. Olsen, I. Burud, Prog Photovolt Res Appl 30 (2022).
[10] M. Vukovic, M. Jakovljevic, A. S. Flø, E. Olsen, I. Burud, Appl Phys Lett, 120 (2022).

EVALUATION OF SNOW REMOVAL METHODS FOR ROOFTOP PHOTOVOLTAICS

*Alexander Granlund, Mattias Lindh, Tommy Vikberg, Anna Malou Petersson
RISE Research Institutes of Sweden AB, Industrigatan 1, 941 38 Piteå, Sweden

*Corresponding Author: Tel. +46 10 516 61 84, alexander.granlund@ri.se

ABSTRACT: Avoiding snow on photovoltaic (PV) installations is motivated for two reasons: to decrease power losses from shading, or to decrease mechanical loads to avoid damage to the PV-installation and the underlying construction. We experimentally investigated the effectiveness and suitability of four different snow removal methods at our facility in the north of Sweden (Piteå, 65°N), throughout three winters. The layout of a PV installation and the underlying roof, together with meteorological conditions and snow characteristics, impact which methods are best suited for snow removal. A simple roof rake with a rectangular toolhead works well when the snowpack is compact and not too thick, whereas a roof rake with a slide works better when the snow is dry and packed. Neither the investigated passive hydrophobic surface coatings, nor the active forward bias electrical heating methods induced shedding of the accumulated snowpack in our experiments without additional intervention. At our test facility in Piteå, the roof rake with a slide was the most effective and user-friendly snow removal. Despite maximum snow loads of approximately 1 kPa, far below the modules' rating, cell damage was observed for both snow removal groups (except for the slide roof rake group) and the control group.
Keywords: Durability, Snow, Evaluation, Electroluminescence, Rooftop

1 INTRODUCTION

In regions that regularly experience sub-zero temperatures and snowfall, avoiding snow on photovoltaic (PV) installations is typically motivated for two reasons: to decrease power losses from shading, or to decrease mechanical loads to avoid damage to the PV-installation and the underlying construction.

Both snow loads and snow removal might contribute to an accelerated degradation in PV-modules. Cells might crack under heavy loads or careless snow removal, frames might loosen across many thaw-freeze cycles, and modules might delaminate. Studies have indicated that a single exposure to temperatures below -20°C can significantly increase a glass/backsheet-module's susceptibility to crack formation under relatively low mechanical loads [1]–[3]. As a PV-module has an expected technological lifespan of 25 years minimum, initial small cracks are likely to grow larger across multiple winters and snow-load-cycles. If they are a common occurrence in cold climates, their prevention further motivates snow removal, even at low snow loads. The most commonly used standard for mechanical load testing, IEC 61215 [4], is based on laboratory conditions and does not consider low temperatures, inclined modules, or in-homogenous loads that are to be expected for real world applications. IEC 62938 tests in-homogenous loads, but also at room temperature [5].

To avoid snow on PV installations, the PV installation can either be constructed to avoid any snow accumulation, or snow accumulating on the installation can be removed. There is however no clear consensus of how snow should be removed, if at all. Some methods might cause more harm than they prevent by exerting additional stress or coming into direct contact with the modules; some methods could put the user at risk, for example, if performed at height; and some methods might be inaccessible due to high costs or limited availability.

Snow removal is a problem that largely remains unexplored when it comes to side-by-side tests. This study aims to evaluate the performance of different methods and investigate whether they cause or prevent damage to the PV-modules in real-world conditions.

1.1 Background

In the literature, often proposed methods for removing snow includes snow and ice repelling coatings [6]–[8] and electrical heating [9]–[12]; whereas, based on user testimonies, what is typically used on-site at residential installations is an assortment of commonly available tools such as shovels, brooms, or roof rakes that mechanically removes snow.

Many coatings based on different technologies have been tested in literature [6]–[8]. Due to limited performance gain or low mechanical durability, many of these coatings are not yet ready for market.

Forward bias electrical heating applies an electrical current in the opposite direction of what the module typically delivers. This results in heat generation due to resistive heating. In literature, there are predominantly two different outcomes from using this method. Either the snow is melted away completely, or the meltwater lowers the friction between the snowpack and module sufficiently for the modules to shed the snow entirely. The latter requires less energy and is generally considered the desired outcome [9]–[10].

The use of mechanical snow removal tools is generally discouraged in literature as they risk to permanently damage the PV-modules [12], [13]. Some commercially available tools might minimize these risks and could offer an attractive solution to the problem for residential installations.

In a pre-study, different snow removal methods were compiled from literature as well as through interviews with early adopters of solar PV in northern Sweden. In addition, commercially available methods were identified. These methods were evaluated by their expected performance regarding availability, initial cost, operational cost, labour intensity, risk of module damage, and risk of injury [14]. This evaluation suggested that two types of roof rakes, forward bias electrical heating, and hydrophobic surface coatings were promising for further evaluation in field.

2 METHOD

The test facility, presented in Figure 1, is located in Piteå, Sweden, at 65°N; where the snow depth on average peaks at 70 cm each winter and roofs are built to withstand a snow load of 3 kPa [15], [16]. The test facility consists of a 12 m × 5 m construction with a 14° single pitch corrugated steel roof. A low inclination was chosen to promote snow accumulation and represent an installation where snow removal might be necessary. The roof was fitted with 20 framed monocrystalline glass/backsheet 60-cell PV-modules (*Longi LR6-60BP 305M*) divided into five equal groups mounted on a rail and clamp system (*Schletter Solo* and *Rapid*). The modules had a mechanical load rating of 5.4 kPa according to *IEC 61215*. Four of the groups were designated snow removal methods, whereas the centre group was the control. Cable connections were accessible from within the facility, enabling on-site electroluminescent (EL) imaging. The modules were stored at open circuit as mechanical degradation was the focus of this study. Time-lapse images were recorded with a surveillance camera throughout the second and third winter (*Reolink Go PT*).

Figure 1. The test facility in november 2021. Module groups from left to right: heated, coated, control, roof rake with slide, and rectangular roof rake.

2.1 Snow removal methods

For this study forward bias electrical heating, hydrophobic surface coatings and two types of roof rakes were tested, based on the pre-study described in Section 1.1.

The first coating tested (*Nanoflex VP20*) was a hydrophobic coating claiming to be suitable for PV, although not specifically for snow repulsion [17]. After the first winter these modules were exchanged to another set of modules (*SoliTek Standard M.60-B-310,* with a mechanical load rating of 3.6 kPa) that had been treated with a different coating developed in the project *Super PV*. This coating had indicated an improved snow performance for modules at higher inclination than what is examined in this study [18].

The forward bias electrical heating employed a DC power supply (*EA-PS 9200-25 T*) to deliver 1200 W to the four 305 W modules connected in series. Heating was performed up to 8 hours.

For this study two different toolheads were tested. One was designed to be pushed up from the roof eave through the snowpack and cut out blocks of snow. The blocks would then slide down a plastic tarpaulin off the roof. The steel toolhead was mounted on plastic wheels to keep some distance between the it and the underlying surface. The other rake featured a simple rectangular toolhead in hard plastic designed to pull and scrape down snow off the roof. One long edge was fitted with less abrasive cell foam. The toolhead must be lifted above the snow and lowered with sufficient force to grab it, before pulling it off the roof. Both roof rakes are presented in Figure 2.

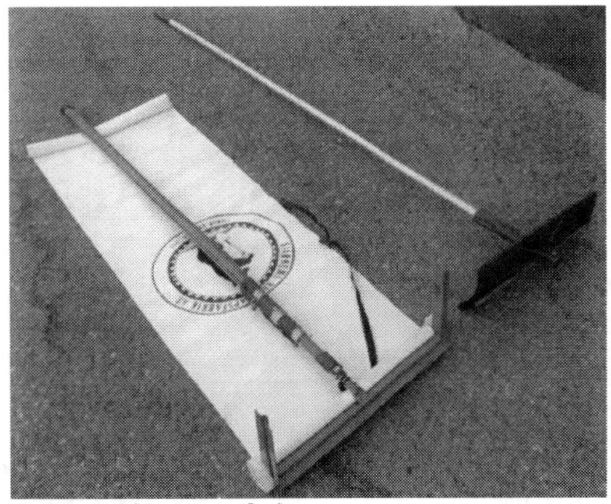

Figure 2. The two roof rakes.

As the coated and control groups required no active snow removal their snow covers were left untouched during the winter seasons. The active snow removal groups (roof rakes and electrical heating) would at the project's start be cleared from snow once they had accumulated more than 10 cm, unless more snow was expected within the coming days. This was done regardless of other external conditions to encompass a variety of ambient temperatures, snow characteristics and times of year. As testing proceeded, at what conditions to perform snow removal was adapted based on preliminary results. E.g., it was quite quickly found that electrical heating was not feasible at very low temperatures. Prior to each snow removal event, the entire test facility was photographed by drone or from the ground. Snow depth at the eave below each module group was measured and snow density measurements would be made on the ground depending on the availability of untouched snow. The snow was characterised (e.g., dry, wet, packed, loose, crusty), and after snow removal the facility was photographed again. The subjective relative effort required (low, medium, high) to perform the active snow removal methods, respectively, was also estimated.

2.2 Module inspection

Visual inspections and EL imaging were performed between each winter. Visual inspections looked for module damage such as damaged glass, frames, or contacts, as well as damage to the mounting system and the facility in general. EL imaging was used to identify cracks or otherwise defective and non-active areas of the solar cells.

Initial EL imaging was performed indoors, prior to installing the modules. Subsequent EL imaging was performed without removing the modules from the roof installation to minimize risks related to handling the modules. A DC power supply (*EA-PS 9200-25 T*) was connected to one module at a time and delivered a maximum of 45 V or 4.5 A. An InGaAs-sensor camera (*WiDy SenS 640V-ST*) with a 1 100 nm high-pass filter was used for imaging together with an 8 mm short wave infra red (SWIR) lens (*VS Technology*), 1.5 mm of spacer rings were used to increase sharpness. Exposure time typically varied from 2 – 10 ms. Imaging was performed at dusk or after sunset, when ambient noise was minimal. Whole modules were photographed and close-ups were taken of noticeable defects.

3 RESULTS AND DISCUSSIONS

The first winter, of 2019–2020, received atypically little snow fall and the temperature often fluctuated above and below freezing, almost no substantial snow accumulation occurred on the test facility. This resulted in a single snow removal event. The winter of 2020–2021 instead featured atypically heavy snow fall and the final winter of 2021–2022 was more ordinary, and the temperature was consistently below freezing. This encompasses a large variance in snow accumulation and conditions for the timeframe of the study. Most snow removal occurred during the latter two winters.

The single-pitch roof typically accumulated snow unevenly. Less snow gathered along the top and sides, meaning that the thickest part of the snowpack typically was slightly above the eave at the centre of the roof. An example of how the snowpack thickness varied along the height of the roof is presented in Figure 3.

Figure 3. Blue, snowpack inclination; red, roof inclination.

3.1 Hydrophobic coatings

The coatings tested in this study did not indicate any decrease in snow accumulation or increase in snow shedding performance compared to the control group. The second set of coated modules performed worse in terms of snow accumulation and shedding, which was likely not due to the coating itself but rather the sharp edges of the frame-edge where it meets the glass, compared to the Longi modules. For the SoliTek modules, that edge was a sharp right angle protruding 2 mm from the glass surface, for the Longi modules the edge also protruded 2 mm but featured a smooth curved edge. Figure 4a shows an occurrence when the frame likely inhibited snow shedding and Figure 4b the end of one winter where the coated modules (*SoliTek)* showed no distinct difference from the control modules.

It is possible that better performance could be discerned with a higher inclination or frameless modules. The Super PV-coating has shown improved snow performance for higher inclination modules that were less susceptible to edge effects [18].

3.2 Electrical heating

Forward bias electrical heating would consistently initiate melting but was by itself insufficient for inducing snow shedding. Since this test limited the use time to 8

Figure 4. Coated SoliTek module group marked with red. a) Module frames likely preventing snow shedding. b) Snow still covering parts of the coated and control groups at the end of winter.

hours, unless all snow was removed, what meltwater remained in the snowpack would refreeze. To isolate factors that might have prevented shedding, other than the low inclination, the snow on the eave below the modules was removed as seen in Figure 5. This was done on multiple occasions, and in this case, after 6 hours of heating, the snowpack still showed no signs of shedding.

Figure 5. Snow unable to shed after 6 hours of heating with no snow on the roof eave.

On two occassions the snow cover was also cut along the side edges of the module group, which enabled snow shedding as seen in Figure 6. The ambient temperature was on this occasion above melting. Thus, shedding was found to be possible under certain circumstances.

Forward bias electrical heating, although theoretically possible, does not seem like a practical or economical solution at this point for the residential market; as it requires additional hardware such as a large DC power supply, and a more involved electrical installation. Efficient use also likely requires the entire snow layer to be removed in one go, as to prevent otherwise unnecessary thaw-freeze cycles. It is likely beneficial to use the method in conjunction with mild weather, as to direct more energy towards melting rather than heating. A net benefit from electrical snow removal would require it to either increase the energy yield to a point that surpasses the energy used for heating, negate the costs or labor required to remove the snow by other means, or prevent costly damage to the installation.

Further testing should be done on larger module groups and regard electricity use (and possible subsequent production gain), both for complete melting and for enabling snow shedding. To make the method accessible for the residential market, it would also be of interest to explore implementations that removes the need for an external DC power supply to deliver the required power, for example an adapted hybrid inverter.

Figure 6. March 2nd, 2022, the ambient temperature ranged between +4–7°C. Left, snowpack separated from anchor points prior to heating. Right, successful snow shedding after heating.

3.3 Roof rakes

Both types of tested roof rakes could often remove most of the accumulated snow from their respective group. The roof rake with a slide did not completely remove snow due to the gap between modules and toolhead formed by the plastic wheels. The wheels and general design did however make it easy to use in most cases, but certain types of snow proved problematic. This included when snow was loose and grainy, very icy, or wet and compact. These conditions made it problematic to get the toolhead underneath the snow cover, especially with a fully extended telescopic handle, and very icy crusts made it difficult to cut through the snow. The rectangular roof rake also struggled in certain conditions. Dry and wind-packed snow often adhered especially well to the glass surface of the modules, which increased the required effort for snow removal markedly, especially with a fully extended handle, see Figure 7 for an example of performance in such conditions. Both methods struggled with snow that had experienced multiple thaw-freeze cycles and had icy crusts. The roof rake with a slide removed the snow without exerting significant stress on the modules. On the other hand, the long handle in combination with the weight at the toolhead of the rectangular roof rake made it difficult to handle and increased the risk of damaging the modules. In one case, the tool head was hence dropped above a module, resulting in a cracked cell wafer, see Section 3.5.

Figure 8 presents the experienced effort required for snow removal at different snow densities. At high snow densities, when the snow is very wet, the rectangular roof rake required less effort since the snowpack more easily slipped downwards at these conditions, whereas the slide roof rake was difficult to force through the dense snowpack. At lower snow densities, the slide roof rake in general required less effort.

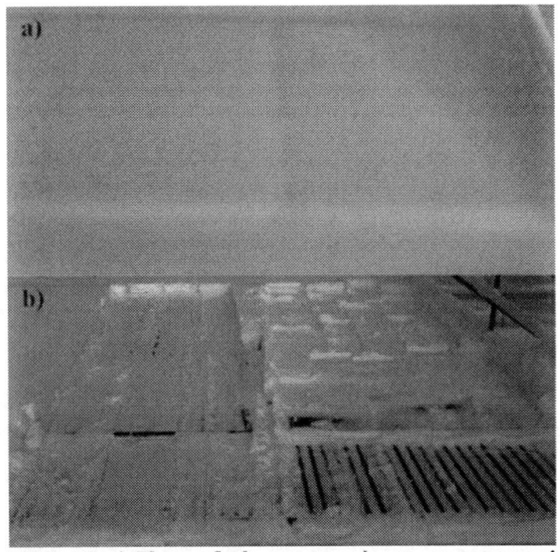

Figure 7. a) The roof rake groups prior to snow removal on December 15th 2021 and b) after snow removal. Left group, roof rake with slide, and right group, rectangular roof rake. Snow was dry and wind packed.

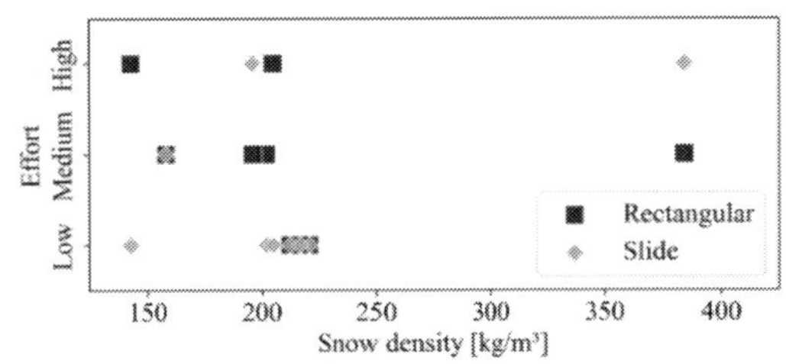

Figure 8. The effort required to remove snow of different densities with the two roof rakes.

The test facility proved to be especially well suited for these ground-operated methods due to the low roof eave. The use of telescopic tools is possibly only feasible for single-story roofs as longer handles are difficult to operate, thus increasing both risks and required effort. The underlying principles behind each method can however still be applied to other types of installations and tools but would require risk assessments and feasibility studies.

Removing snow completely with these methods proved difficult. The adhesion of cold snow to the modules' glass is significant. Complete snow removal was most easily achieved during mild and sunny days as snow which was wet from melting was far easier to remove, especially with the rectangular rake. However, leaving a few cm of snow can be beneficial since it reduces the risk of damaging the module during the final snow removal, and the thin cover was found to come off during sunny days. Leaving just a small part of the black modules bare would accelerate snow melting during sunny days as their surface would absorb more light than the surrounding snow and therefore heat up. Performing this kind of mechanical snow removal in conjunction with a snowmelt period could therefore have two benefits: avoiding crust formation and accelerating ambient melting.

Due to the risks of mechanical snow removal and the

labour intensity of complete snow removal, it might be preferential to focus on reducing loads, not removing them entirely.

3.4 Reasons for snow remaining on PV installation

The snowpack seemed to remain on the PV-installation for several different reasons. Snow on the roof eave below the modules might hinder snow shedding, as seen on the control module group in Figure 4b, likely due to more friction on the roof than the modules. Figure 5 indicates that this alone cannot explain why shedding did not occur for thicker snowpacks. The relatively small module groups as compared to typical PV-installations gives a proportionally large circumference relative to the area. This means that edge effects from frames, rails and neighbouring snow – that the snowpack can cling to - might have a larger than typical impact when the snowpack is thick enough to drape over the module groups' edges. Throughout thaw-freeze cycles these effects will likely grow more prominent. Figure 6 shows that once the snowpack was separated from these points of attachment, snow shedding could occur for the heated group. Lastly the frame edges also affect snow remaining on the PV installation through obstruction, as seen in Figure 4a, where the sharp edges of the SoliTek modules' frames are believed to prevent snow shedding.

This signifies the importance to consider the reasons for snow to remain on the PV installation, and how they affect performance, when designing and constructing a PV installation in a snow rich climate.

The increase in ambient melting from a partly exposed solar cell surface, as observed for the roof rakes, should be generally applicable regardless of what method is used. Thaw events might hence assist in snow removal, but subsequent refreezing is likely to further cling a remaining snowpack to the modules. It might therefore be less labour-intensive to remove snow prior to or in conjunction with a thaw event.

3.5 Crack formation

Modules on the lower row and not directly next to the roof edge would be subject to the largest snow depth and load. The largest snow load, estimated from snow depth and density, occurred on February 25th, 2021, and measured approximately 1 kPa near the heated, coated and control groups' lower modules. As snow had routinely been cleared from the roof rake groups, they did not exhibit the same load.

As previously mentioned, snow removal with the rectangular roof rake caused an entirely cracked cell. No further cracks were detected on any of the modules belonging to the roof rake groups. The modules of the lower row of the other groups did however all have cracks. They show a slight tendency of following an "X-pattern", which is associated with load related cracks [19]. The cracks imaged with EL in May 2022, their module group, and their relative position superimposed on a 60-cell module is presented in Figure 9 along an outline of the X-pattern. Cells in the coated modules *(SoliTek)* show crack formation with similar origin points for multiple cells, implying that these might be manufacturing defects that have remained undetected until they propagated to a noticable degree. As mentioned, these modules are rated for a load of 3.6 kPa compared to 5.4 kPa for the other modules *(Longi)*, which might also explain why more cells from these modules are cracked.

Figure 9. Superposition of cracked cells on a 60-cell module. Orange, surface coated (*Solitek*); red, electrical heating; light blue, control; dark blue, rectangular roof rake. Dashed lines represents the typical *X-pattern* for cracks formed by mechanical loads.

None of the cracked cells, except for the cell damaged by the rectangular roof rake, give rise to dark isolated areas, and might hence not significantly affect the modules' energy output at this point. The cracks might however be likely to grow larger throughout subsequent winters and mechanical load cycles.

It is notable that the heaviest measured snow load of approximately 1 kPa is much lower than what the modules were rated for. Most cracks were detected after the winter that this snow load was observed. Other studies have indicated that glass/backsheet modules are prone to crack formation at seemingly low mechanical loads when exposed to sub -20°C temperatures [1–3]. It is possible that this, plausibly in conjuction with prior defects from manufacturing, have led to the crack formation.

One might expect that it would be sufficient if the modules were rated for mechanical loads similar to or greater than the general construction requirements where they will be mounted. Although we cannot judge in every case for sure why cells have cracked, these results indicate that such is not necessarily the case, and emphasizes the disparity between conditions for certification tests and the real world. We believe that in the long-term, certification procedures should include realistically low temperatures to better reflect real-world conditions. One way to circumnavigate this issue in the short term is to design PV-installations in such a way that they do not accumulate significant snow loads, for example by mitigating attachment and obstruction of the snowpack as mush as possible or opting for high inclination, with smaller snow depth and higher tendency of natural shedding.

We see a need for development of modules designed specifically for cold conditions, realistic module test procedures, and careful monitoring of the snow loads during operation and maintenance of a PV installation in snow rich locations.

4 CONCLUSIONS

The layout of a PV installation and the underlying roof, together with meteorological conditions and snow characteristics, impact which methods are best suited for snow removal. A simple roof rake with a rectangular toolhead works well when the snowpack is compact and not too thick, whereas a roof rake with a slide works better when the snow is dry and packed. Neither the investigated passive hydrophobic surface coatings, nor the active forward bias electrical heating methods induced shedding of the accumulated snowpack in our experiments without additional intervention. At our test facility in Piteå, the roof rake with a slide was the most effective and user-friendly snow removal method. Despite maximum snow loads of approximately 1 kPa, far below the modules' rating, cell damage was observed for the groups designated a snow removal method (except for the slide roof rake group) as well as the control group.

5 ACKNOWLEDGMENTS

This study was performed as part of the project *Impact and management of snow loads for roof-mounted PV installations*, funded by the Swedish Energy Agency, Länsförsäkringar's Research Fund, Lindbäcks Bygg, and PiteEnergi.

6 REFERENCES

[1] E. J. Schneller, H. Seigneur, J. Lincoln, and A. M. Gabor, "The Impact of Cold Temperature Exposure in Mechanical Durability Testing of PV Modules," in IEEE 46th Photovoltaic Specialists Conference, Jun. 2019, pp. 1521–1524. doi: 10.1109/PVSC40753.2019.8980533.

[2] M. W. Rowell, S. G. Daroczi, D. W. J. Harwood, and A. M. Gabor, "The Effect of Laminate Construction and Temperature Cycling on the Fracture Strength and Performance of Encapsulated Solar Cells," in 2018 IEEE 7th World Conference on Photovoltaic Energy Conversion (WCPEC) (A Joint Conference of 45th IEEE PVSC, 28th PVSEC 34th EU PVSEC), Jun. 2018, pp. 3927–3931. doi: 10.1109/PVSC.2018.8547978.

[3] H. Seigneur et al., "Microcrack Formation in Silicon Solar Cells during Cold Temperatures," in 2019 IEEE 46th Photovoltaic Specialists Conference (PVSC), Jun. 2019, vol. 2, pp. 1–6. doi: 10.1109/PVSC40753.2019.9198968.

[4] International Electrotechnical Commission, Terrestrial photovoltaic (PV) modules - Design qualification and type approval - Part 2: Test procedures (IEC 61215-2:2016), vol. 2017.

[5] International Electrotechnical Commission, Photovoltaic (PV) modules – Non-uniform snow load testing (IEC 62938), 1.0., vol. 2020.

[6] R. W. Andrews, A. Pollard, and J. M. Pearce, "A new method to determine the effects of hydrodynamic surface coatings on the snow shedding effectiveness of solar photovoltaic modules," Solar Energy Materials and Solar Cells, vol. 113, pp. 71–78, Jun. 2013, doi: 10.1016/j.solmat.2013.01.032.

[7] P.-O. Andersson, B. P. Jelle, and Z. Zhang, "Passive Snow Repulsion: A State-of-the-art Review Illuminating Research Gaps and Possibilities," Energy Procedia, vol. 132, pp. 423–428, Oct. 2017, doi: 10.1016/j.egypro.2017.09.650.

[8] E. Andenæs, B. P. Jelle, K. Ramlo, T. Kolås, J. Selj, and S. E. Foss, "The influence of snow and ice coverage on the energy generation from photovoltaic solar cells," Solar Energy, vol. 159, pp. 318–328, Jan. 2018, doi: 10.1016/j.solener.2017.10.078.

[9] A. Rahmatmand, S. J. Harrison, and P. H. Oosthuizen, "An experimental investigation of snow removal from photovoltaic solar panels by electrical heating," Solar Energy, vol. 171, pp. 811–826, Sep. 2018, doi: 10.1016/j.solener.2018.07.015.

[10] A. Weiss and H. Weiss, "Photovoltaic cell electrical heating system for removing snow on panel including verification," presented at the 5th International Conference on Renewable Energy Research and Applications, Nov. 2016. doi: 10.1109/ICRERA.2016.7884484.

[11] I. Frimannslund and T. Thiis, "A feasibility study of photovoltaic snow mitigation systems for flat roofs," Czasopismo Techniczne, vol. 2019, no. Volume 7, Art. no. Volume 7, Jul. 2019, doi: 10.4467/2353737XCT.19.073.10724.

[12] B. B. Aarseth et al., "Mitigating Snow on Rooftop PV Systems for Higher Energy Yield and Safer Roofs," 2018. doi: 10.4229/35THEUPVSEC20182018-6CO.3.5.

[13] H. A. Walker, "Best Practices for Operation and Maintenance of Photovoltaic and Energy Storage Systems; 3rd Edition," NREL/TP-7A40-73822, 1489002, Dec. 2018. doi: 10.2172/1489002.

[14] A. Granlund, and A.M. Petersson, 2021, Provning av metoder för snöborttagning vid takmonterade solelanläggningar. RISE ETC Report 2021-23, 2021.

[15] SMHI, "Normalt största snödjup under vintern, medelvärde." https://www.smhi.se/data/meteorologi/sno/normalt-storsta-snodjup-under-vintern-medelvarde-1.7931 (accessed Sep. 21, 2022).

[16] Boverket, "Karta med snölastzoner," Nov. 13, 2019. https://www.boverket.se/sv/byggande/regler-for-byggande/om-boverkets-konstruktionsregler-eks/sa-har-anvander-du-eks/karta-med-snolastzoner/ (accessed May 28, 2020).

[17] NewPro, "NewPro NANOFLEX H9 / VP 12 / VP 16 / VP 20." https://www.newpro.de/en/newpro-nanoflex.html (accessed Sep. 21, 2022).

[18] Super PV, "Advanced functional nanocoatings to improve PV modules performance - Blog." https://www.superpv.eu/blog/Advanced-functional-nanocoatings-to-improve-PV-modules-performance/ (accessed Sep. 21, 2022).

[19] M. Köntges, S. Kurtz, C. Packard, U. Jahn, K. A. Berger, and K. Kato, Performance and reliability of photovoltaic systems: subtask 3.2: Review of failures of photovoltaic modules: IEA PVPS task 13: external final report IEA-PVPS. Sankt Ursen: International Energy Agency, Photovoltaic Power Systems Programme, 2014.

WIND EFFECTS ON PV-STRUCTURES: WHEN AND HOW TO DESIGN FOR WIND

Francesco Dorigatti, Matthew T.L. Browne, Chiara Pozzuoli1, Zachary J. Taylor
RWDI
Via Thaon Di Revel, 21, 20159 Milano, Italy/ 600 Southgate Drive, Guelph, ON N1G 4P6 Canada
Email: francesco.dorigatti@rwdi.com

ABSTRACT: Wind design is an important aspect for structural efficiency to help minimize cost and maximize the long-term reliability of any PV installation. Over the last several years, prescriptive load methodologies have arisen for the static wind loads of rooftop PV systems and ground-mounted PV systems. However, these prescriptive load methodologies assume that the PV structure does not move in the wind. Movement of flexible structures in the wind can lead to additional loads or in some cases can lead to wind-induced divergent motion causing structural failures. These failures have led to substantial economic losses due to downtime of the PV installations affected and to significant costs for retrofits and improvements to existing sites.
Keywords: Wind loading, PV systems, Reliability, Stability, Experimental Methods

1 INTRODUCTION

One of the challenges facing building code committees is the definition of what makes a solar structure flexible or not. The ASCE 7 standard [1] defines a flexible structure based on the natural frequency threshold of 1 Hz. This criterion was developed primarily with buildings in mind and has been shown to be inadequate for smaller structures such as solar PV installations. Experience from full-scale installations and evidence from wind tunnel testing on scale models clearly show that ground mounted single-axis trackers are flexible enough to move due to wind even though most have frequencies beyond 1 Hz, to a point where these structures have had well documented structural failures due to wind. Despite the related economic losses, it is from the analysis and retrospective investigations of these failures, as well as from the increasing amount of wind tunnel testing on scale models carried out to support the solar system design, that an improved understanding of the interaction between solar structures and wind has been gained. This knowledge can now be leveraged to strive towards more robust and reliable structural designs.

Since the existing codes and standards do not properly capture the flexibility of PV structures, this study synthesizes extensive wind tunnel and analytical research that has been carried out and is currently ongoing to establish new guidelines that define when a PV structure should be considered susceptible to wind and what are the methodologies to use in order to analyze and design it for wind.

2 SCIENTIFIC INNOVATION AND RELEVANCE

There are two main concerns related to the interaction between solar structures and wind: (i) aerodynamic instability, and (ii) dynamic wind loading. The approach to assess the aeroelastic susceptibility of solar structures is to first assess the likelihood for aerodynamic instability, which has the potential to cause divergent motion of the structure and, in turn, structural failure, and then investigate the dynamic wind loading, which provides the information needed for structural design, once the risk of instability has been ruled out.

2.1 Aerodynamic instability

Divergent aerodynamic instabilities causing large amplitude rotational motions have been observed primarily in ground-mounted single-axis solar trackers, which tend to present a relatively low torsional stiffness by design. For these single-axis trackers there are two broad categories of aerodynamic instability: (i) stiffness-driven, and (ii) damping-driven.

Although the large amplitude oscillations of stiffness-driven and damping-driven instability appear similar, measurements of aerodynamic stiffness and damping undertaken through wind tunnel testing have clarified the differences of the underlying phenomena that cause each type of instability. Wind tunnel tests on full aeroelastic models (e.g., Figure 1) are currently undertaken to establish the critical wind speeds for the onset of aerodynamic instabilities.

Figure 1: Full aeroelastic scale model in the wind tunnel

The stiffness-driven instability consists of a loss of structural stiffness in the tracker system caused by 'negative' aerodynamic stiffness being introduced by the wind action. As the negative stiffness contribution introduced by the wind increases with the wind speed, at elevated wind speed its magnitude becomes large enough to eventually overcome the inherent structural stiffness of the system, which causes significant motions. Stiffness-driven instability has been observed and well documented through wind tunnel testing on aeroelastic scale models. It often occurs at low tilt angles (e.g., < 20°) and, due to non-linearities in the combined fluid-structure system, the divergent motions that it causes are often observed to lead to limit cycle oscillations.

For greater tilt angles (e.g., > 20°) the aerodynamic instability is observed to be driven by changes in damping. At elevated wind speeds the aerodynamics of the flow around photovoltaic modules is observed to create negative aerodynamic damping. Once the negative aerodynamic damping exceeds the inherent structural damping of the system, then the system becomes negatively damped. When this condition is reached, the tracker system absorbs more energy from the wind flow than what it can dissipate internally, which leads to excessive dynamic rotations.

2.2 Dynamic wind loading

Wind loading is generally composed of three parts: the mean wind load, the fluctuating background load, and the inertial wind loads. In the solar industry the mean and background wind loads are referred to as the "static" wind loads since they do not depend on the motion of the structure. However, the fluctuating background wind loads can cause a structure to move which induces additional inertial wind loads that must be resisted by the structure. These inertial wind loads are referred to as the "dynamic" wind loads.

There is currently an open question about what threshold or criterion should define when dynamic loads may be important. As previously discussed, the 1 Hz threshold used for buildings [1] has been proven to be not appropriate for solar structures. The commentary to the National Building Code of Canada [2] suggests that 10 Hz may be a suitable threshold above which dynamic loads are not likely to be important.

Dynamic wind loads are complex and therefore difficult to account for through prescriptive methods. Wind tunnel testing on rigid models instrumented for pressure measurements is the most accurate method to determine dynamic wind loads. Alternatively, an analytical approach assessed in this study to derive the inertial wind loads, (i.e., dynamic wind loads) is based on first principles of wind loading and atmospheric turbulence and relies on a relatively straightforward equation that can be computed with readily available structural properties and design criteria. It can be shown that the ratio of peak inertial forces, $\hat{F}_i$, to the total force, $\hat{F}_t = \hat{F}_s + \hat{F}_i$, (where $\hat{F}_s$ is the static force) can be approximated through

$$\frac{\hat{F}_i}{\hat{F}_t} = 0.3 \cdot \left(\frac{3\,m}{L_c}\right)^{\frac{1}{2}} \zeta^{-\frac{1}{2}} \bar{U}^{\frac{5}{3}} xL_u^{-\frac{1}{3}} f_n^{-\frac{1}{3}} \left(\frac{1}{\bar{U}^{\frac{4}{3}} + (2 f_n L_c)^{\frac{4}{3}}}\right) \quad (1)$$

In Eq. (1), ζ is the structural damping ratio to critical, $\bar{U}$ is the mean wind speed at structure height, xL_u is the longitudinal integral length scale, f_n is the natural frequency for the mode of interest, and L_c is the chord length of the PV module. It should be noted that a dependance of the peak inertial forces on the aerodynamic admittance function is implicit in this formulation, which assumes that the dynamic loads collapse reasonably well when scaled by the reduced frequency $f_n L_c/\bar{U}$.

3 RESULTS AND DISCUSSION

3.1 Aerodynamic instability

Results from several different aeroelastic model tests carried out at RWDI have been anonymized and the critical reduced gust speeds are shown in Figure 2. The critical reduced gust speed is defined as the peak structurally averaged gust speed U divided by the natural frequency f_n (typically taken for the first torsional mode of vibration) and chord length L_c. The envelope of the data is shown in the gray shaded region in Figure 2 while the blue shaded region shows one standard deviation either side of the mean. To explain the relatively large scatter of this dataset, it should be noted that there are significant differences in terms of chord, tracker elevation, structural damping, and overall architecture of the ground-mounted single-axis systems from which the data was derived.

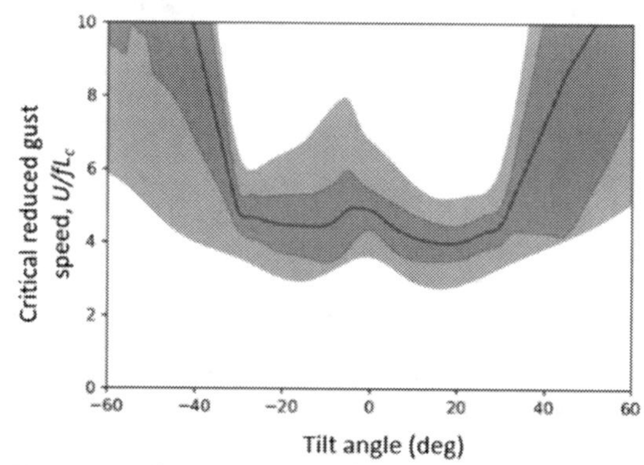

Figure 2: Critical wind speeds from several aeroelastic model wind tunnel studies

Based on these results, a conservative lower bound can be provided such that aerodynamic instability could occur when

$$U > 2.5 L_c f_n. \quad (2)$$

The peak structurally averaged gust speed (U) is not straightforward to calculate; however, it can be approximately related to the mean wind speed ($\bar{U}$) by $U = 1.4\bar{U}$ such that the criterion becomes

$$\bar{U} > 1.8 L_c f_n \quad (3)$$

This criterion for aerodynamic stability assumes relatively low inherent damping and is conservative. The intent of the criterion is to flag at which wind speed a particular PV structure may start to become susceptible to aerodynamic.

3.2 Dynamic Wind Loading

Several dynamic wind load measurements obtained through wind tunnel testing of rigid models instrumented for pressure [3] of a wide range of different ground-mounted solar structures are compared to predictions obtained from Eq. (1) in Figure 3. The agreement between the upper bound of the experimental data and the analytical results from Eq (1) confirms that this equation can be used conservatively to predict the dynamic wind load contribution on a solar system depending on its structural and geometrical parameters and on the characteristics of the atmospheric wind. Based on that prediction, a criterion is suggested that considers not significant for design any dynamic wind loading less than approximately 5% of the total loading, i.e., $\hat{F}_i/\hat{F}_t < 0.05$. However, it is recommended that wind engineering guidance be sought if the suggested criteria are not met.

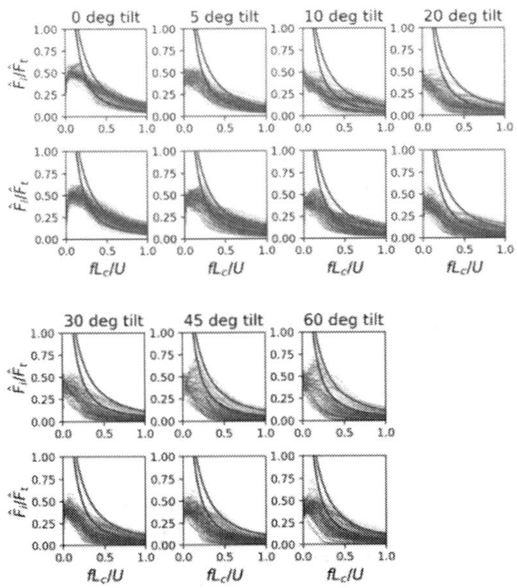

Figure 3: Evaluation of Eq. (1) compared to data points as measured from wind tunnel tests. Structural damping is taken as 1% of critical in each case. Top and bottom rows represent perimeter and central zones, respectively; solid lines show Eq. (1) while dots represent wind tunnel data points; blue is for $L_c \sim 4$ m, red for $L_c \sim 2$ m

6 REFERENCES

[1] ASCE/SEI, ASCE 7-22: Minimum Design Loads and Associated Criteria for Buildings and Other Structures (2021).

[2] NRC, Structural Commentaries: User's Guide - NBC 2015: Part 4 of Division B (2017).

[3] Browne, Z. J. Taylor, S. Li, and S. Gamble. A wind load design method for ground-mounted multi-row solar arrays based on a compilation of wind tunnel experiments, J. Wind Eng. Ind. Aerodyn., vol. 205, no. July, p. 104294 (2020).

BIFACIAL PV CANOPY SYSTEM IN HIGH LATITUDE, MODEL DEVELOPMENT AND VALIDATION WITH THE FIRST MONTHS OF MONITORING DATA

Samuli Ranta*[1], Hugo Huerta Medina[1], Sami Jouttijärvi[2], Aleksi Heinonen[1], Anton Driesse[3]

[1]New Energy group, Turku University of Applied Sciences (TUAS), Joukahaisenkatu 7, 20520 Turku, Finland,
*principal author: samuli.ranta@turkuamk.fi
[2]Department of Mechanical and Materials Engineering, University of Turku, Vesilinnantie 5, 20500 Turku, Finland
[3]PV Performance Labs, Emmy-Noether-Str. 2, Freiburg, Germany

A seaside canopy with an integrated bifacial photovoltaic system (PV) was set up for a residential district in the city of Naantali, Finland. This system is part of the PV demonstration built for the Naantali Housing Fair in 2022 and integrates a bifacial 100 kWp photovoltaic power plant in the residential district. A modelling-based simulation tool was earlier developed to predict PV performance of bifacial systems in Nordic conditions for vertical configurations. The model was validated by comparing the simulation results with performance data recorded from East/West vertical bifacial arrays at separate locations. The aim of this study is to find the best tools to model canopy systems in case, especially the incident irradiance calculation for lower tilt angles, high clearance, and high latitude configurations to accurately forecast the energy yield. A more advanced ray-tracing tool Radiance was used for the study. The aim was to define the minimum level of complexity needed for a reliable model.

Keywords: Building Integration, Modeling, Ray Tracing, Bifacial PV, System Performance

1 INTRODUCTION

Bifacial solar modules are versatile and suitable for different configurations as they can capture sunlight on both the front and rear sides of the module. Many studies have been made about the benefit that they can have in vertical East-West installations [1,2]. Other uses, such as solar canopy integration in Nordic conditions have been less studied even though bifacial PV offers good potential to increase PV energy production in densely populated areas. Due to the partly see-through glass-glass structure, bifacial modules are especially appealing for this kind of dual-use, where PV plant is integrated into the urban infrastructure. Bifacial modules provide shade, protection from the weather, and aesthetic pleasure for the users of the canopies while the land use of the PV plant is minimized.

In an urban environment, carports are another potential use case for bifacial PV. They have similar advantages as canopies, like high ground clearance and relatively low ground coverage ratio, which are beneficial for bifacial energy gain. However, commercial simulation software products are not well suited to simulate such a system and can greatly underestimate the potential of the bifacial gain. This study aimed to model and simulate bifacial gain by modelling a canopy with *Radiance* [3] and *Bifacial Radiance* [4] and compare the results for the field-measured data. Then the complexity of the model was gradually reduced, to learn what are the minimum requirements for a reliable model. These results can be in future used to develop even more simplified and easier-to-use tools, based on open-source tools.

Figure 1. Drone picture of the Bifacial PV canopies in Naantali Finland (TUAS), the canopy under study is indicated by the green arrow.

2 AIM AND SYSTEM DESCRIPTION

The aim of this study was the modelling of energy yield from bifacial solar canopies at high latitudes and furthermore the optimization and validation of the methodology by comparing its results against field monitoring data. The starting point of the study was the design of the bifacial solar canopies using commercial software for planning the PV system. In addition, a single solar canopy has been modelled using the Turku University of Applied Sciences PV Performance Modeling tool (TUAS-PVPM) [5].

Figure 2. Two strings of the PV system are monitored by a high-accuracy data acquisition system (DAQ).

The system under study is a 100 kWp bifacial PV plant integrated into seaside canopies (Fig. 1) at the newbuild residential district in the city of Naantali, Finland (60° N 22° E). This system consists of 6 canopies, each having 36 to 60 bifacial 350 Wp modules. These modules are frameless glass-glass construction, and they have 10 % transparency thanks to cell spacing. Modules are attached to the sub-structure by a watertight profile system, and they have a dual purpose as roofing elements. Canopy frames are constructed of wood and the ground clearance of the arrays is 3 meters on the front side and 4.75 meters on the rear side of the canopy. The understructure of canopies is covered with a wooden deck and gravel the further back. The tilt angle of the modules is 15° and the azimuth angle of the canopy roofs varies between 135° and 190°.

The canopy under this study is monitored at the string level by a data acquisition (DAQ) (Fig. 2). Two of the three strings in the canopy are monitored with a high-resolution 24-bit analogue to digital converter, simultaneously string voltage and current are sampled with a one-second resolution. The site is also equipped with an irradiation station which has a Class A ventilated pyranometer for the global tilt horizontal irradiance (GTI), six silicon monitoring cells, one for the front-side plane of array (POA) and five for the back of arrays irradiance (Fig. 3); this configuration allows Irradiance Ratio (IR) calculation (monitoring cell irradiance / POA irradiance). A weather transmitter for all other relevant weather variables and an Internet Protocol (IP) camera facing towards the studied canopy was included as well. IP camera was proven to be very useful when analyzing situations in which modelled and measured data had the highest deviation.

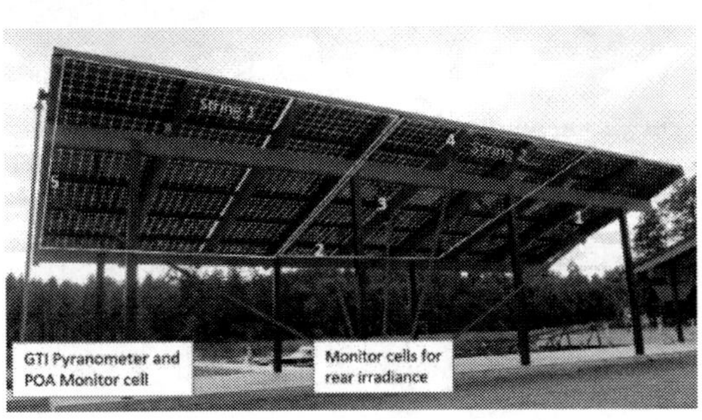

Figure 3. GTI, POA, back irradiance sensors and monitored strings.

3 MODELING DESCRIPTION

The PVPM under development is based on the PVlib toolbox, with the aim of a simplified model based on irradiance translation calculations performed on both sides of the PV array. However, since the tool was focused on vertical configurations and the canopies addressed here are tilted surfaces with high clearance, there were significant differences in irradiation modelling. Tilt angle issues can be addressed by using a view factor model [6,7] which has been coupled to PVPM, but the shading effects by the supporting structure are still to be tackled. Therefore, more advanced modelling tools were selected to understand better what the most critical elements of the model are. Radiance software was chosen as the principal optical modelling package as it is one of the most known for simulating light environments and its advantages have been demonstrated within the PV sector in several publications [8,9].

For the electro-thermal modelling, which is responsible for converting incident irradiation to electricity output, the differences are expected to be smaller, allowing better compatibility with models that are developed and optimized with data from East-West oriented systems.

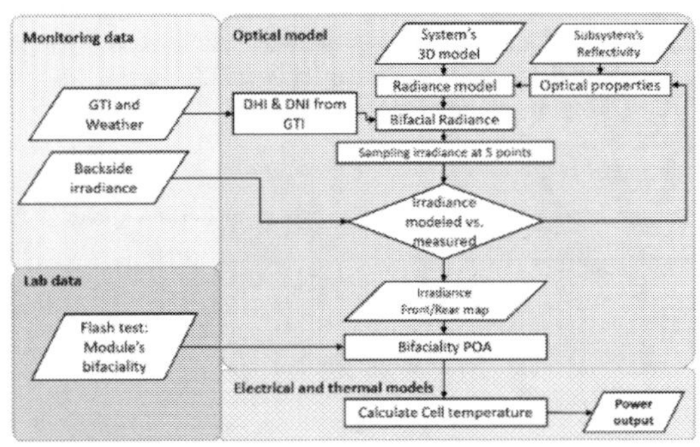

Figure 4. The methodology used during the study.

A methodology, illustrated in Figure 4, is the basis of this study to understand the challenges in developing a simplified model, especially when there are several elements affecting the light conditions, mainly at the rear of the PV array. In this study, besides the supporting structure and several wooden beams, the different surfaces at the front, rear and below the system, like water, gravel, and the dock all at different elevations, must be considered.

An advanced optical model was built starting with the 3D modelling of the canopy using Siemens NX 3D design software. The model has the objective to be as close to reality as possible yet still simple enough to allow efficient data processing. PV modules up to a cell level, supporting structure, monitoring cells and all different ground surfaces nearby the canopy, as well as part of the water surface, were included in the scene. This 3D model was then exported to the lightening simulator Radiance. In Radiance, the optical properties for each of the elements present in the optical model from now on just referred to as the model, were added based on the measurements taken with a spectrophotometer. These values were just a starting point or an approximation for the colour and reflectance of the materials. A 3D view from the canopy system modelled with Radiance is shown in Figure 5.

Figure 5. Bifacial PV canopy modelled with Radiance.

To model the sunlight conditions, two months of measured irradiance were used. Since only Global Tilt Irradiance (GTI) is measured, the Diffuse Horizontal (DHI) and Direct Normal Irradiance (DNI) had to be calculated. To do it, two different methods, clear sky and B. Marion [10] were used to evaluate their results when comparing them with GTI-measured data. Clear sky results proved to be close enough to GTI data (Fig. 6), on the other hand, results from the Marion method had limitations in estimating irradiance components for low irradiance conditions. The calculated DNI and DHI were then used to generate the sky conditions within Radiance.

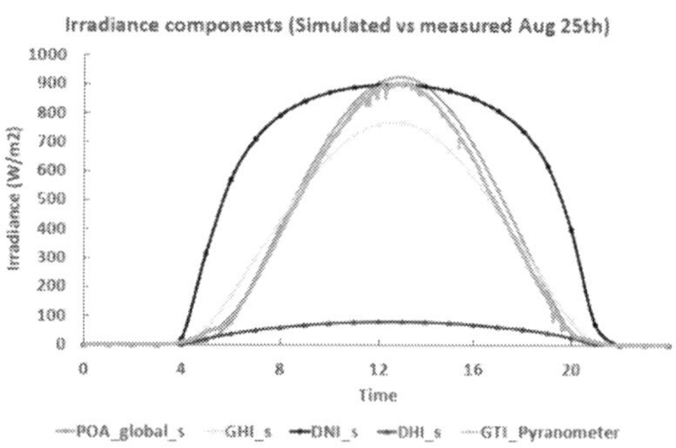

Figure 6. Modelled irradiance vs. measured POA Irradiance, August 25th 2022.

Bifacial Radiance software, a tool which works as a wrapper between Radiance and Python PVLIB [11] is used as the central process for tuning the model by adjusting the material properties, mainly the reflectance of the large surfaces below the PV array. The tuning process consisted of sampling the rear irradiance on five reference points (representing the monitoring cells) included in the optical model and comparing those values with the measured ones until the materials properties (colour, reflectance and roughness) are adjusted to such that the deviation between simulated and measured is acceptable. Table 1 shows the deviation evolution of the simulation results from measured data. In addition, Figure 7 shows the irradiance match achieved for sunny days after slightly tuning the model within a few iterations.

Monitor Cell	Deviation (%)				
	Round 1	Round 2	Round 3	Round 4	Round 5
1	9	1	-6	-9	-11
2	24	10	-4	-3	-4
3	26	15	7	3	-1
4	27	26	23	16	2
5	39	26	16	10	5

Table 1. Simulation deviation from measured data after tuning material's properties on each run.

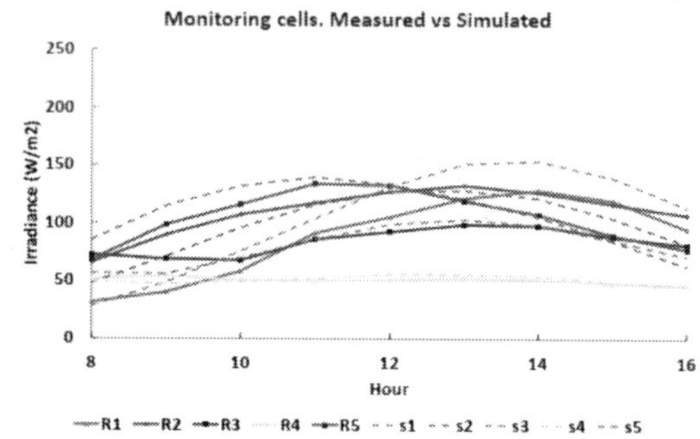

Figure 7. Monitoring cells. Measured (R) values compared with simulated (s) ones, August 25th 2022.

From Figure 7 can be seen that the simulated data is close to measured one from hour 10:00 up to 15:00 during the day. However, the early and late hours had higher deviations. Further analysis from IP camera photos of the system showed that during those hours of the day, shadows cast by the other canopies and some other elements like hills and trees in the surroundings are affecting the incident light on the sensors. This affects the most on monitor cell number 1, especially in the afternoon (Figure 8).

Figure 8. IP camera picture of shading elements affecting measured data, 19:01, August 25th, 2022. The monitored canopy is the one in the foreground of the figure.

Once the model simulates light conditions close to the measured data, the model is ready to simulate the incident irradiance on the PV array and hence the Irradiance Ratio (IR) calculation and Irradiance maps (front & rear) can be made. To have an initial idea about the behaviour of the expected results from simulations, the Irradiance ratio from measured data was calculated, and the ratio's trend followed during the day is shown in Figure 9. Two clear days are presented, June 24th and August 25th; on both, the ratio presents higher variability at the beginning and the end of the day, reaching ratios as high as 1.0 early in the morning when the sun is rising from Nort-East. The ratio in both cases stabilizes around 0.13 when the sun is higher than 30 degrees altitude. Additionally, the trend tends to increase late in the evening but does not present symmetry due to the orientation of the system as well as the shadings affecting readings during those hours (Figure 9).

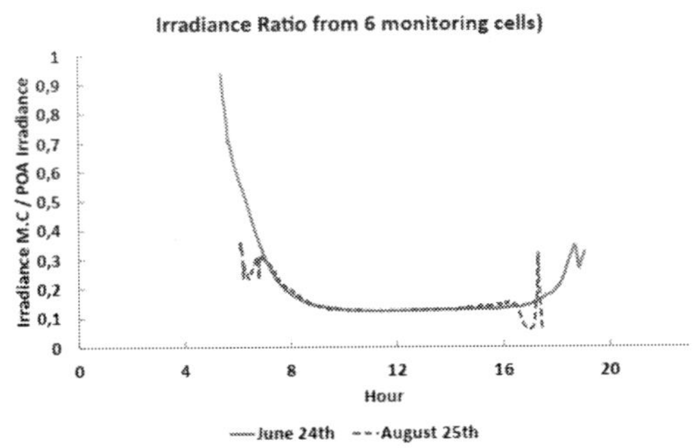

Figure 9. Irradiance Ratio (measured) from monitor system, second resolution data averaged to 1 minute.

4 MODELLING VALIDATION

To validate the modelling, the next step in the study was to compare the Irradiance Ratio from the model to the measured values seen in Figure 9. Bifacial-Radiance libraries facilitate sampling the incident irradiance on any surface within the model. It is well known that ray-tracing models are computationally expensive, especially if there is a high number of sampling points within the scene, therefore, to evaluate the influence on the irradiance ratio calculation, initial simulations considered two sampling scenarios, sampling irradiance only at 5 points (modelled monitoring cells) as well as sampling every single cell within the PV array (2880 points). Results are shown in Figure 10.

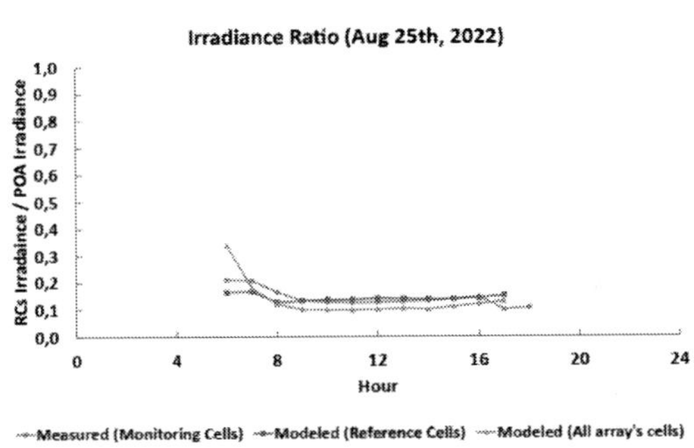

Figure 10. Modelled irradiance vs. measured POA Irradiance, August 25th 2022.

Results from the evaluation, demonstrate the accuracy of the model while comparing measured versus modelled values. When sampling under the same conditions (5 points) simulation results equals 0.14 for the IR (blue curve), compared to 0.13 from measured data (green curve). Early and late irradiance ratios are not shown as they are more affected by the objects in the surroundings, which are not yet part of the 3D model. Additionally, the irradiance ratio calculated from sampling with a higher resolution is presented in the same Figure 10. Results show that the irradiance ratio (orange curve) goes down to 0.1 for the same time-lapse, which means that the shading effects are better estimated.

An additional simulation with the same high sampling condition was run to evaluate the effects of the structures below the PV array (supporting profiles and wooden beams) by deactivating all of them below except one (vertical poles). The irradiance maps with a resolution to cell level were created for visual evaluation, Figure 7. The maps generated show two scenarios at three different hours of the day (10:00, 12:00 and 14:00), on the left side the simple one including 8 vertical poles and, on the right, the detailed one including all the present substructures.

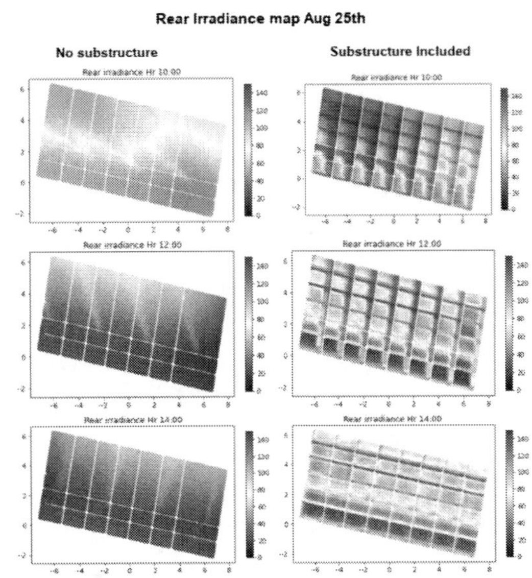

Figure 7. Rear irradiance maps, simple and detailed models, when sampling at PV cell level.

From the irradiance maps, it is evident that the distribution of the light is highly affected by the structure supporting the array, the ratio estimated without including all the elements in the model was 0.19 compared to 0.1 as it was seen in Figure 10.

As was mentioned earlier, during the planning of the project, the canopies were modelled with two other software, a commercial one and the TUAS-PVPM. Both underperform in IR estimation, and in the black-box one underestimates and the second one overestimates the IR. This raises the question of what are the boundaries to simplify a model, easy to use and fast enough but still accurate in results.

5 SUMMARY AND CONCLUSIONS

The model provided promising results in estimating the incident irradiance as can be seen for example in Figure 10. A longer dataset with varying weather conditions is however needed for proper model validation; especially to evaluate the performance under snowy conditions. Oversimplification of the model does tend to overestimate the IR and hence lowers the usability of the model. The next target is to find the best compromise between complexity and accuracy.

The horizon profile of the location was not part of the model which did create a higher deviation, especially for the afternoon hours as there are shading objects on the west side of the system. A horizon profile will be added for the 3D model in the future. Another important development point is sampling on systems where the azimuth is other than its equivalent in cartesian coordinates. The model samples correctly only with the systems facing directly to the South but for the system under study, which has 191° azimuth, a labour-intensive manual correction was required. Adding the azimuth correction will greatly improve the usability of this workflow.

Finally, it can be concluded that once the model is in future optimized, it can be used as an aid for decision-making when bifacial solar canopy or carport structures are designed. However, for yield prediction alone it is overcomplicated, and a simpler model should be sufficient.

6 ACKNOWLEDGEMENTS

The work has been carried out under the New Energy research group at TUAS Finland and the Department of Mechanical and Materials Engineering, University of Turku Finland. Special thanks to TUAS student assistants Jori Koskinen, Kimmo Paldanius, Eelis Niemi, and staff members Miikka Säteri and Maximilien Marc for all the background support. The project has received funding from the European Union's Horizon 2020 research and innovation programme under Grant Agreements n° 957751 (RESPONSE) and n° 957769 (TIGON).

7 REFERENCES

[1] Stein, Joshua S.; Reise, Christian; Bonilla Castro, Johanna; Friesen, Gabi; Maugeri, Giosuè; Urrejola, Elías; Ranta, Samuli 2021: Bifacial Photovoltaic Modules and Systems: Experience and Results from International Research and Pilot Applications. IEA PVPS.]

[2] Joshua S. Stein, Laurie Burnham, Matthew Lave. One Year Performance Results for the Prism Solar Installation at the New Mexico Regional Test Center: Field Data from February 15, 2016 - February 14, 2017

[3] Ward, Gregory J., "The RADIANCE Lighting Simulation and Rendering System," Computer Graphics (Proceedings of '94 SIGGRAPH conference), July 1994, pp. 459-72

[4] Ayala Pelaez and Deline, (2020). Bifacial_radiance: a python package for modeling bifacial solar photovoltaic systems. Journal of Open-Source Software, 5(50), 1865, https://doi.org/10.21105/joss.01865

[5] S. Ranta, H. Huerta, O. Huhtala, A. Heinonen, J. S. Stein. "Modeling Tool Validation for the Yield Prediction of Bifacial East-West Vertical PV System in Nordic Conditions". EU PVSEC (2020)

[6] Anoma, M., Jacob, D., Bourne, B.C., Scholl, J.A., Riley, D.M. and Hansen, C.W., 2017. View Factor Model and Validation for Bifacial PV and Diffuse Shade on Single-Axis Trackers. In 44th IEEE Photovoltaic Specialist Conference.

[7] B. Marion *et al.*, "A Practical Irradiance Model for Bifacial PV Modules," *2017 IEEE 44th Photovoltaic Specialist Conference (PVSC)*, 2017, pp. 1537-1542, doi: 10.1109/PVSC.2017.8366263.

[8] Ayala Pelaez, Deline C, MacAlpine M, Marion B, Stein J, Kostuk K. Comparison of Bifacial Solar Irradiance Model Predictions with Field Validation. IEEE J Photovoltaics. 2019; 9(1):82-87.

[9] Ayala Pelaez S, Deline C, Greenberg P, Stein JS, Kostuk RK. Model and validation of single-axis tracking with bifacial PV. IEEE J Photovoltaics. 2019;9(3):715-
721. https://ieeexplore.ieee.org/document/8644027

[10] Bill Marion, A model for deriving the direct normal and diffuse horizontal irradiance from the global tilted irradiance, Solar Energy, Volume 122, 2015, Pages 1037-1046, ISSN 0038-092X, https://doi.org/10.1016/j.solener.2015.10.024. (https://www.sciencedirect.com/science/article/pii/S0038092X15005757)

[11] William F. Holmgren, Clifford W. Hansen, and Mark A. Mikofski. "pvlib python: a python package for modeling solar energy systems." Journal of Open-Source Software, 3(29), 884, (2018). https://doi.org/10.21105/joss.00884

A NEW METHOD TO ANALYSE THE SOILING IN MULTI-JUNCTION III-V BASED CPV MODULES

Gianluca Timò, Nicola Castagnetti, Alessandro Minuto, Edoardo Celi e Stefano Rizzi
RSE
Strada Torre della Razza, le Mose, 29100 Piacenza, Italy

ABSTRACT: The side-by-side (SBS) method allows evaluating the concentrator photovoltaic (CPV) performance degradation due to soiling by monitoring the power or current ratio between a module under test and a reference module that must always be kept clean. For monitoring the performance of the soiling in CPV power plants, the SBS method introduces cost issues as a continuous cleaning of the reference module must be guaranteed by manual or automatic means. With this contribution, a new data filtering method is proposed that allows evaluating the photovoltaic performance degradation due to soiling in multi-junction based CPV modules without the need of using a reference module. The new method is based on the measurement of the electrical parameters of the CPV module along with the direct irradiance data in the spectral bands of the top and middle sub-cells and on the subsequent application of a proper filtering procedure. For a validation of the new methodology, a ten-month testing campaign has been carried out and the results obtained by the data filtering method have been compared with those obtained by the SBS one.
Keywords: III-V semiconductors, Multijunction Solar Cell, Concentrators, Characterization, Monitoring, Soiling

1 INTRODUCTION

The performance of a CPV module based on Multijunction (MJ) solar cells is influenced by several factors, namely: i) current mismatch among the MJ solar cells, ii) temperature, iii) possible solar cell degradation, iv) direct normal irradiance (DNI), v) solar spectrum composition (current mismatch among sub-cells) and vi) soiling and its spectral dependence, again effecting the current balance among sub-cells [1],[2].

Soiling can be monitored if all the above effects can be isolated from the soiling itself. Temperature and DNI effects can be respectively isolated by filtering the data and considering the I_{sc}/DNI ratio, where I_{sc} is the short circuit current of the CPV module. Solar cell degradation can be excluded by monitoring the dark I-V curve (in real CPV plants it could performed by inverters able to measure the dark I-V curve during the night). Eventually, spectral effects can be isolated by applying the SBS method, which requires keeping a reference module always clean [3].

However, for monitoring the effect of the soiling in CPV power plants, the SBS method introduces cost issues as a continuous cleaning of the reference module must be guaranteed by manual or automatic means.

With this contribution we present a new method called "data filtering method" that, assuming a uniform soiling distribution over the lenses, allows eliminating the spectral effects with a simpler soiling monitoring, which does not require side-by-side measurements.

The new method is based on the measurement of the I_{sc} of the CPV module along with the direct irradiance data in the spectral bands of the top and middle sub-cells and on the subsequent application of a proper filtering procedure.

2 DATA FILTERING METHOD

2.1 Theory assumptions

In CPV module based on two terminals triple-junction (TJ) InGaP-InGaAs-Ge solar cells, the bottom Ge sub-cell never limits the I_{sc} of the whole device, therefore, the following assumption is considered:

$$I_{sc} = min\{I_{sc}^{top}, I_{sc}^{mid}\} \quad (1)$$

Where: I_{sc}^{top} is the short circuit current produced by the top cell, I_{sc}^{mid} is the short circuit current produced by the middle cell.

Equation (1) tell us that since the sub-cells are connected in series the short circuit of the TJ device can be limited either by the top-cell or by the middle-cell.

Furthermore, if suitable spectral bands are chosen (see Fig.1) such that DNI_{top} and DNI_{mid} are the normal solar irradiance in the wavelength region where respectively the top-cell and middle-cell are spectral sensitive, the following assumption can be introduced for the limiting cell (with the integral extended to the full DNI spectral range as shown in Fig.1):

$$I_{sc} = I_{sc}^{i} = \int DNI(\lambda)SR^{i}(\lambda)d\lambda = k_i DNI_i \quad (2)$$

Where $SR^i(\lambda)$ is the spectral responsivity and i is the junction index (i=1,2; for top and middle cell respectively). Equation (2) simply expresses a linearity relationship between the short circuit of the limiting cell and the solar irradiance in its spectral band. This means that, when the middle-cell is limiting, the following relationship holds:

$$\frac{I_{sc}}{DNI_{mid}} = k_{mid} \quad (3)$$

While, when the top-cell is limiting, we get:

$$\frac{I_{sc}}{DNI_{top}} = k_{top} \quad (4)$$

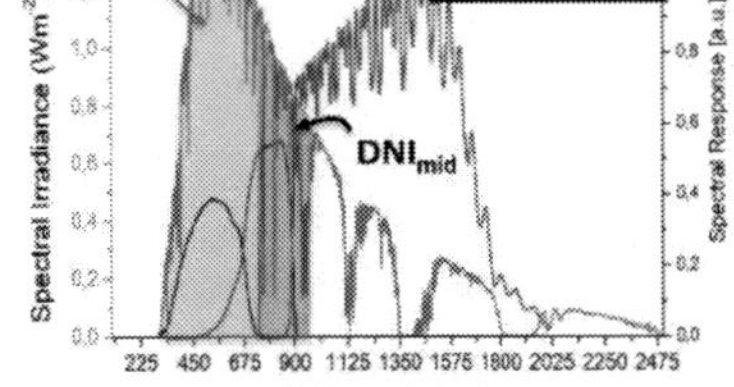

Figure 1: Spectral response of the TJ InGaP-InGaAs-Ge cell, spectral irradiance and selection of the DNI_{top} and DNI_{mid} spectral bands. The DNI_{top} spectral range is between 350-680 nm, while DNI_{mid} spectral range is between 680-950 nm.

Since the change in the sun spectrum composition can be described by the ratio $\frac{DNI_{top}}{DNI_{mid}}$, if we plot I_{sc}/DNI_i vs DNI_{top}/DNI_{mid}, we should find a portion of the graph in which either Equation (3) or the Equation (4) is fulfilled. That is, we can identify I_{sc}/DNI_i data that are not spectrum dependent, therefore, the variation of I_{sc}/DNI_i along the time can be analyzed to study the soiling (assuming that the temperature effects have already been isolated).

2.2 Experimental

The study of the soling has been carried out by considering two CPV modules based on InGaP/InGaAs/Ge triple junction solar cells, with silicon on glass Fresnel lenses (see Fig.2), having acceptance angle of ±0.5°. For validation purpose, a reference CPV module was always kept clean, while a second one, the test module, was left free to accumulate soiling. During the ten-month testing campaign, the electrical parameter of the CPV modules as well as the direct normal irradiance were measured. A pre-filtering procedure has been applied by selecting the data fulfilling the following conditions: i) DNI > 500 W/m^2, ii) DNI/GNI > 0.8 (where GNI is the global normal irradiance), iii) solar tracking error < 0.2°.

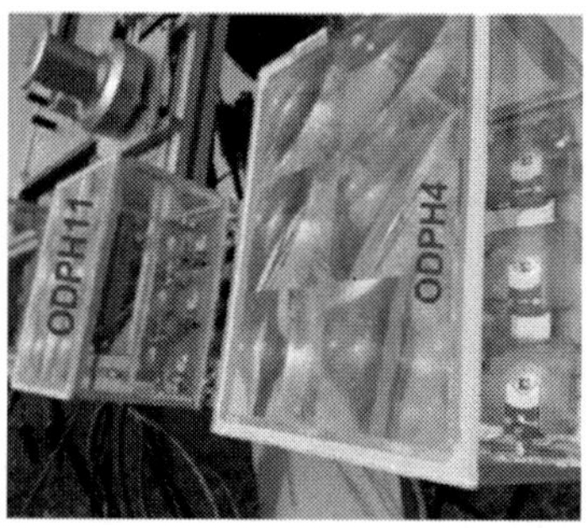

Figure 2: CPV modules considered in the soiling study.

2.3 Study of the I_{sc}/DNI_i vs DNI_{top}/DNI_{mid} plot

In Fig.3, a I_{sc}/DNI_i vs DNI_{top}/DNI_{mid} plot is shown.

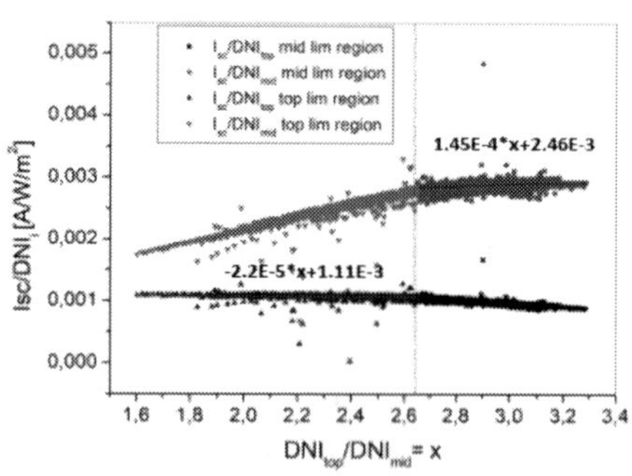

Figure 3: I_{sc}/DNI_i vs DNI_{top}/DNI_{mid} plot. A linear data fitting and related equations are shown in two different portions of the plot. Data have been acquired on the reference CPV module.

In the data region of the plot shown in Fig.3, where the DNI_{top}/DNI_{mid} ratio is low (on the left side of the vertical line), the TJ is limited by the top-cell; therefore, we expect to find a constant I_{sc}/DNI_{top} ratio. The data fitting shows an almost constant I_{sc}/DNI_{top} ratio with a very weak dependence on spectrum composition. Always referring to Fig.3, on the middle-cell limiting side of the plot, that is, for the higher values of the DNI_{top}/DNI_{mid} ratio, (i.e., on the right side of the vertical line) we find again a data region where the I_{sc}/DNI_{mid} is almost constant. In this case, a slightly more pronounced dependence on the spectrum composition can be pointed out.

To understand this dependence, it can be assumed, without losing generality, that the increase in the DNI_{top}/DNI_{mid} ratio (x) is obtained by increasing DNI_{top} and keeping DNI_{mid} constant. In the middle-cell limiting side of the plot, hence, the I_{sc}/DNI_{mid} should be constant, because as much as we increase x, the middle cell current and the DNI_{mid} do not change. However, when x increases, at the same time, the extra current produced by the top cell with respect to the middle cell increases as well. The top cell excess current is partially dissipated by radiative recombination (luminescence), which can be absorbed by the middle cell. Therefore, there is a photon coupling effect between the top and middle cell which induces a small increase in the I_{sc} of the middle cell as much as x increases. This means that equation (3) must be corrected by considering the photon coupling:

$$\left(\frac{I_{sc}}{DNI_{mid}}\right)_{mid\ cell\ lim\ region} = k_{mid} + L_{1,2} \cdot x \tag{5}$$

Where $L_{1,2}$ represents the photon coupling coefficient. Continuing the study of the I_{sc}/DNI_i vs DNI_{top}/DNI_{mid} plot, is straightforward to get the direct proportionality of the I_{sc}/DNI_{mid} and the invers proportionality of the I_{sc}/DNI_{top} with respect to the spectrum composition, respectively in the top-cell and in the middle-cell limiting side of the plot, that is:

$$\left(\frac{I_{sc}}{DNI_{mid}}\right)_{top\ cell\ lim\ region} = k_{top} \cdot x \tag{6}$$

$$\left(\frac{I_{sc}}{DNI_{top}}\right)_{mid\ cell\ lim\ region} = \frac{k_{mid} + L_{1,2} \cdot x}{x} \tag{7}$$

In the top-cell limiting side of the plot, by increasing the DNI_{top}/DNI_{mid} ratio, necessarily the I_{sc} of the TJ device (i.e. the I_{sc} of the top cell) proportionally increases, while in the middle-cell limiting side of the plot, the I_{sc} of the TJ device (i.e. the I_{sc} of the middle cell), remains almost constant, as there is a weak increase in the middle cell current owing to the photon coupling effect, however, DNI_{top} increases much more than I_{sc}, therefore I_{sc}/DNI_{top} proportionally decreases with x.

Eventually, to understand the reason related to the weak dependence of I_{sc}/DNI_{top} on the spectrum composition, in the top-cell limiting side of the plot, it is worthwhile to point out that Equation (2) requires a suitable selection of the DNI_{top} and DNI_{mid} spectral bands: the DNI_i spectral band should be overlapped to the spectral responsivity of the limiting cell. However, since the exact spectral responsivity wavelength range of the top and middle cell was not known a priori, in order to draw the plot reported in Fig.3, a "reasonable" selection of the DNI_i spectral bands was accomplished. Actually, these bands could not be well overlapped with the spectral responsivity of the cells, therefore, the validity of Equation (2) should be assessed by choosing different DNI_i spectral bands.

We therefore assume to change the wavelength range

from the exact one, DNI_{mid}, to another one, $DNI_{mid}{}^{(1)}$, by considering the following equations:

$$DNI_{mid}{}^{(1)} = \alpha^{(1)} \cdot DNI_{mid} \quad (8)$$

$$DNI_{mid}{}^{(1)} + DNI_{top}{}^{(1)} = DNI_{mid} + DNI_{top} \quad (9)$$

$$\frac{DNI_{top}{}^{(1)}}{DNI_{mid}{}^{(1)}} = x^{(1)} \quad (10)$$

Where $\alpha^{(1)}$ can be > 1 or < 1. Equation (9) tells us that while changing the single spectral range of DNI_{mid} and DNI_{top}, we keep the total DNI spectral range unchanged, therefore, if we choose a narrow wavelength range for DNI_{mid} we must consider a wider wavelength range for DNI_{top}. It can be shown that Equation (4) and Equation (5) are straightforward replaced by:

$$\left(\frac{I_{sc}}{DNI_{mid}{}^{(1)}}\right)_{mid\ cell\ lim\ region} = \frac{k_{mid}+L_{1,2}\cdot x^{(1)}}{\alpha^{(1)}} \quad (11)$$

$$\left(\frac{I_{sc}}{DNI_{top}{}^{(1)}}\right)_{top\ cell\ lim\ region} = k_{top}\left(1+\frac{1}{x^{(1)}}\frac{\alpha^{(1)}-1}{\alpha^{(1)}}\right) \quad (12)$$

Equation (12) shows that indeed the weak dependence of I_{sc}/DNI_{top} on the spectrum composition, in the top-cell limiting side of the plot, can be due to an inadequate selection of the DNI_i spectral bands. If $\alpha^{(1)} > 1$, the slope of I_{sc}/DNI_{top} is negative, as shown in Fig.3. If the value of $\alpha^{(1)}$ is decreasing, the slope becomes less and less negative and then can become positive as soon as $\alpha^{(1)}$ becomes <1.

In conclusion, by filtering the data of the I_{sc}/DNI_i vs DNI_{top}/DNI_{mid} plot, for example, by analyzing I_{sc}/DNI_{mid} ratio along the time, in the wavelength region where the middle cell is limiting, we can neglect the spectrum dependence with a good approximation and study the soiling effect.

3 PRELIMINARY ASSESSMENT OF THE DATA FILTERING METHOD

Equations (11) and (12) allow carrying out a preliminary assessment of the data filtering method. Let's select two different DNI_i band, such that:

$$DNI_{mid}{}^{(1)} = \alpha^{(1)} \cdot DNI_{mid} \quad (13)$$

$$DNI_{mid}{}^{(2)} = \alpha^{(2)} \cdot DNI_{mid} \quad (14)$$

With $\alpha^{(1)} > 1, \alpha^{(2)} < \alpha^{(1)}$. From Equation (11) it holds:

$$\left(\frac{I_{sc}}{DNI_{mid}{}^{(2)}}\right)_{mid\ cell\ lim\ region} > \left(\frac{I_{sc}}{DNI_{mid}{}^{(1)}}\right)_{mid\ cell\ lim\ region} \quad (15)$$

While from Equation (12) we can write:

$\left[\left(\frac{I_{sc}}{DNI_{top}{}^{(2)}}\right)_{top\ cell\ lim\ region}\right]_{slope}$ will be less negative than

$$\left[\left(\frac{I_{sc}}{DNI_{top}{}^{(1)}}\right)_{top\ cell\ lim\ region}\right]_{slope} \quad (16)$$

The validity of the above condition has been assessed in Fig.4 and Fig.5 on the test CPV module.

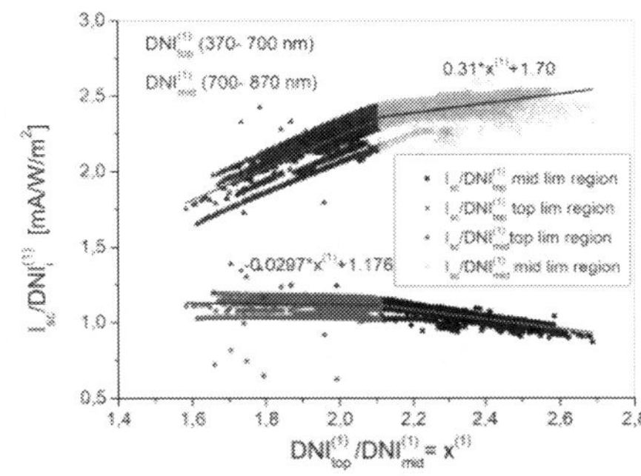

Figure 4. I_{sc}/DNI_i vs DNI_{top}/DNI_{mid} plot. The DNI_{top} and DNI_{mid} wavelength bands, as well as the linear fitting equations are shown.

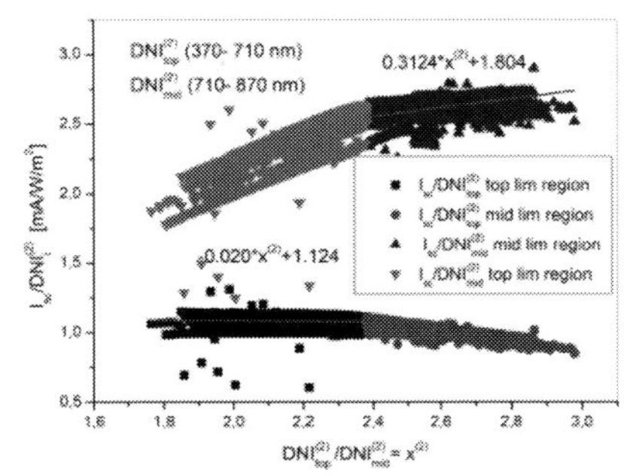

Figure 5. I_{sc}/DNI_i vs DNI_{top}/DNI_{mid} plot. The DNI_{top} and DNI_{mid} wavelength bands, as well as the linear fitting equations are shown.

In Fig.4, I_{sc}/DNI_{top}, in the top -cell limiting side of the plot, shows a negative slope, therefore, we can assume that $\alpha^{(1)} > 1$. In Fig.5, the DNI_i spectral bands have been changed; i.e the DNI_{top} band is wider while the DNI_{mid} band is narrower and therefore $\alpha^{(2)} < \alpha^{(1)}$. Accordingly, by looking at the fitting equations, the conditions (15) and (16) are fulfilled. It is possible to assess that the conditions (15) and (16) are also fulfilled by selecting $\alpha^{(3)} < 1$.

By comparing the plots of Fig.4 and Fig.5 with the plot of Fig.3, it easy to point out the higher data dispersion present in the formers, owing to the soiling.

4 TEMPERATURE EFFECT AND MEASUREMENT UNCERTAINTY

As reported in the introduction, soiling can be monitored if temperature effects are isolated from the soiling itself and the instrumental uncertainty is much lower than the soiling effect.

We have analyzed the temperature effect by monitoring the lens temperature and the I_{sc}/DNI_{mid} ratio of the reference module (see Fig.6). The data have been prefiltered as previously described in section 2.2, acquired during the ten-month testing campaign and the I_{sc}/DNI_{mid} ratio daily averaged.

Excluding the soling, the I_{sc}/DNI_{mid} ratio includes all the uncertainty introduced by temperature variation, and by the measurement of the electrical parameters and DNI.

The data are randomly distributed along the time. This behavior can be explained by assuming that while the I_{sc} has a positive temperature coefficient, the lens optical efficiency decreases with temperature, furthermore, the I_{sc}/DNI_{mid}, in particular, suffers of the instrumental uncertainty of the spectral irradiance measurement. Hence, if the data are not filtered by temperature, we can consider a global uncertainty band within which soling cannot be distinguished.

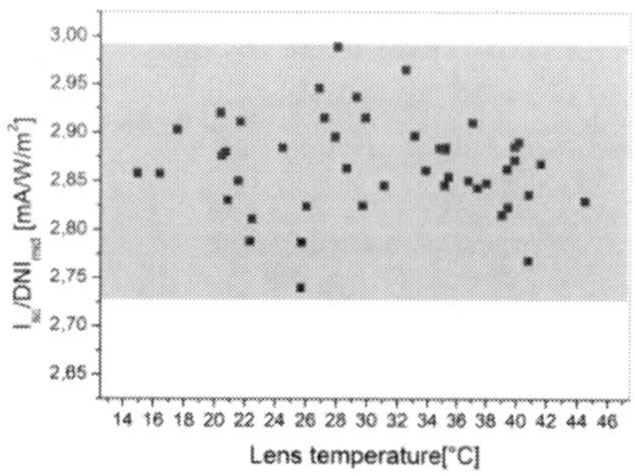

Figure 6. I_{sc}/DNI_{mid} vs lens temperature. The DNI_{mid} data, once acquired, have been prefiltered and the daily averaged. The band within which I_{sc}/DNI_{mid} varies is highlighted in light blue.

5 SOILING ANALYSIS WITH THE DATA FILTERING METHOD

By filtering the I_{sc}/DNI_{mid} data, i.e., by selecting to monitor I_{sc}/DNI_{mid} ratio when the middle cell is limiting, we could analyze the soiling effect on the CPV module (see Fig.7).

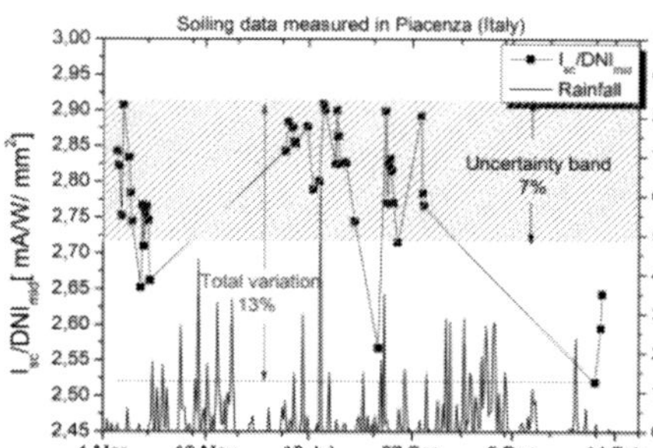

Figure 7. Filtered I_{sc}/DNI_{mid} data and rainfall data registered in the ten-month testing campaign. The uncertainty band, introduced in section 4, is also shown.

We can point out that as long as the I_{sc}/DNI_{mid} ratio remains mostly within the uncertainty band, it can recover its initial value, thanks to the rainfall natural cleaning. Therefore, it does not seem a real disadvantage if the soiling cannot be distinguished within this band.

Along the time, the soiling increases and a recovery process is no longer complete. During the whole test period, a 13% total variation in the I_{sc}/DNI_{mid} ratio due to soiling has been measured.

6 FILTERING METHOD VALIDATION

To validate the proposed new methodology for analyzing the soiling on CPV module based on MJ solar cells, the soiling results obtained by the data filtering method have been compared with those obtained by the SBS method. Since in the SBS method, the soiling is analyzed by the variation of the $I_{sc}(dirty)/I_{sc}(clean)$ ratio, where $I_{sc}(dirty)$ is the short circuit current of the test CPV module, while $I_{sc}(clean)$ is the short circuit current of the reference CPV module, the normalized data have been compared (see Fig.8).

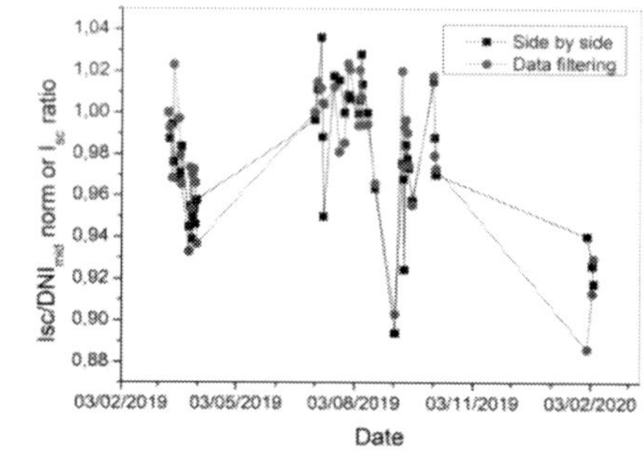

Figure 8. Normalized I_{sc}/DNI_{mid} data obtained with the data filtering method and normalized I_{sc} data obtained with the SBS method.

Fig.8 shows a good agreement between the soiling data obtained with the data filtering method and those obtained with the SBS method, thus validating the new proposed methodology.

7 CONCLUSIONS

Assuming a uniform soiling distribution over the lenses, the data filtering method allows simplifying the soiling monitoring of CPV modules based on MJ solar cells by analyzing the variation of the daily I_{sc}/DNI_i ratio when the top cell or the middle cell are limiting, thus avoiding the use of a reference module to be always keep clean. While the I_{sc}/DNI_{top} ratio can be made spectrum independent by a suitable choice of the DNI_i bands, the I_{sc}/DNI_{mid} ratio maintains a weak dependence on the spectrum composition due to photo-coupling effects. The data filtering method is less accurate than the SBS method, since it includes the uncertainty introduced by the DNI measurement; however, its accuracy can be improved by using isotype solar cells instead of the spectroradiometer.

Besides, it's worthwhile to point out that if the reference module has a heavier degradation than the test module, the SBS method could not recognize the soiling effects. As presented, the data filtering method can be implemented in the algorithm of next generation inverters for identifying threshold values to be used for planning the CPV module cleaning. With respect to the SBS method, the data filtering method is expected to be cheaper as it replaces a reference module and its continuous cleaning with a spectroradiometer or isotype sensors.

The data filtering method can also be applied in case of not unform soiling distribution, by monitoring the P_{max}/DNI_i ratio instead of the I_{sc}/DNI_i ratio.

7 REFERENCES

[1] G. Siefer, F. Almonacid, A. J. Garcı, and E. F. Ferna, Sol. Energy, vol. 92, pp. 221–229, 2013.
[2] P. D. Burton, B. H. King, and D. Riley, Sol. Energy, vol. 112, pp. 469–474, 2015.
[3] J. Montes-Romero et al., AIP Conf. Proc., vol. 2149, August 2019.

8 AKNOWLEDGMENT

The authors gratefully acknowledge the Ministry of Economic Development and the Italian Electrical System for funding this work in compliance with the Decree of April 16th, 2018. Ioffe Institute of Saint-Petersburg is acknowledged for CPV modules manufacturing.

OPTICAL DESIGN AND OPTICS MANUFACTURING OF A HIGHLY INTEGRATED MICRO-CPV FOR SPACE APPLICATIONS

Anderson BERMUDEZ-GARCIA[1], Philippe VOARINO[1], Olivier RACCURT[2]
[1] Univ. Grenoble Alpes, CEA, Liten, Campus Ines, 73375 Le Bourget du Lac, France
[2]Univ. Grenoble Alpes, CEA, Liten, DEHT, SAMA, F-38000 Grenoble, France
Corresponding author: philippe.voarino@cea.fr

ABSTRACT: This work presents the first optical developments of a novel concept of micro-Concentrators PhotoVoltaic (CPV), based on a highly integrated miniaturized mirror. This approach is performed with the objective of significantly reducing the recurring cost of the standard PhotoVoltaic Assembly (PVA) made with classical Coverglass Interconnected Cells (CIC) for space applications. Optics have been optimized for < 1 mm² micro solar cells. Free forms optics, primary and secondary elements have been studied and an acceptance angle along the y-axis of 8.2° has been obtained. A parabolic shape has then been chosen for proof-of-concept testing. The concentration ratio is 26X and the first estimate of optical efficiency is of 82 % without any antireflective optimization. For the experimental part, honeycomb structural assemblies, optical molds and optics have been realized. The fabrication process is described, and the first characterizations are achieved.

Keywords: devices, concentrators, space.

1 INTRODUCTION

Photovoltaic developments for space are booming due to the enormous needs of constellations for telecommunications applications driven by the New Space race. In addition to this need, space agencies want to send human beings on the Moon and Mars. Finding a way to produce power for these harsh environments also becomes a challenge. Moreover, missions launched into Deep Space remain essential for scientific developments and knowledge. For all these applications, photovoltaic technology remains a widely studied solution. Micro-CPV could be a way to produce more energy using less III-V materials with the best Cell to Module Ratio (CTM), which is the ratio between the module efficiency on the cell efficiency, compared to classical CPV [1]. Some CPV systems have already been launched into space, and reflectors are the most deployed optical system: AstroEdgeTM array, Orbital-ATK CellSaver [2], DSS FACT (Functional Advanced Concentrator Technology), spacecraft and Hughes 702 designs [3]. In terms of space CPV heritage, these systems have been shown to use less mass and are particularly suitable for addressing missions with high radiation environments, as well as having great potential to improve PhotoVoltaic Assembly (PVA) efficiency and reduce cost [4]. These systems have an optical efficiency ~ 90 %. Then, two recent micro-CPV systems have been developed since 2016 at Air Force Resarch Laboratory [5] and since 2019 at Penn University [6]. Both systems have a concentration ratio below 30X, used micro-cells with a size of less than 200x200 μm², no structure support considered and have a theoretical (experimental) optical efficiency close to 80 % (respectively ~ 60 %). Therefore, the optical efficiency must be improved while maintaining a robust mechanical structure.

To characterize a CPV system, the CTM allows to evaluate the capacity to integrate photovoltaic solar cells into a system. Another critical metric is the Acceptance Angle (AA), defined as the angle that allows more than 90% of the incident solar flux to be collected. This parameter is closely related to the sun-tracker and the geometrical concentration ratio. The larger the acceptance angle, the lower the concentration factor. The last value to characterize is the distribution of the optical flux over the solar cells surface, called PAR (Peak Average Ratio), which is the maximum irradiance divided by the average irradiance integrated on the solar cell area [7].

The proposed patented solution is based on the idea of integrating optics into a honeycomb structure used for planar PVA. The integration of optics into a honeycomb structure can bring several benefits (e.g. structural and thermal) without increasing the overall mass of the PVA. The underlying principle is as follows: sunlight passes through the protective glass and is reflected by the mirrors located into the honeycomb structure. Reflected light is then focused onto the surface of Multi-Junction Solar Cells (MJSC), which are localized on the backside of the coverglass. This mirror network is then known as Primary Optical Element (POE). A Secondary Optical Element (SOE) can be placed on the upper surface of the MJSC to increase the AA of the system and homogenize the flux. Figure 1 presents the system for an area of 40x40 mm² (half a space solar cell) and a thickness close to 3 mm, using 30 interconnected micro solar cells. The aim of the work is to present the optical design studied and chosen for this highly integrated micro-CPV system, and to evaluate theoretically the electrical power as a function of the MJSC subcells.

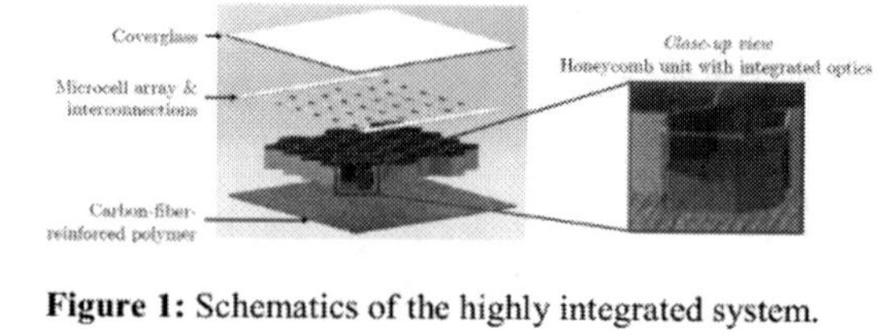

Figure 1: Schematics of the highly integrated system.

2 OPTICAL DESIGNS

To study which will be the most suitable shape to concentrate the light onto the solar cells, two reflectors are studied: one based on the parabolic shape and the second based on an optimized free-form. These two POEs will then be associated with a SOE. Optics will be evaluated in terms of PAR and AA.

2.1 configuration

First, optical simulations were carried out using the commercial software TracePro for Monte Carlo ray-tracing. For a preliminary study, to be independent of the hexagonal honeycomb shape, the modeling was performed with radial optical symmetry. A standard honeycomb with a medium size cell of 3/16" (4.8 mm) was chosen, defining an optics diameter slightly inferior at 4.6 mm. The solar cell has a diameter of 0.9 mm giving a geometrical concentration ratio of 26X. The solar source is extracted from the AM0 solar spectrum (1366 W/m²), an incidence half angle of 0.27° (ASTM E-490). To save modeling time, only three wavelengths (0.3, 0.555 and 1.7 µm) with a solar weight of 0.17, 0.74, 0.01, respectively, have been considered.

To model the system the characteristics of the materials used are:

- a coverglass with no scattering and non-waviness simulated by a refractive index of Schott's N-BK7 and a thickness of 500 µm (see Figure 2);
- a silver reflector embedded into the honeycomb with the refractive index and absorption coefficient defined in the Figure 2;
- a circular MJSC of 900 µm diameter and 200 µm thickness with measured EQE (External Quantum Efficiency), shown in Figure 3. Giving a current density for the top subcell of 17.18 mA/cm², the middle subcell of 14.9 mA/cm², and the bottom subcell of 27.22 mA.cm² for AM0.

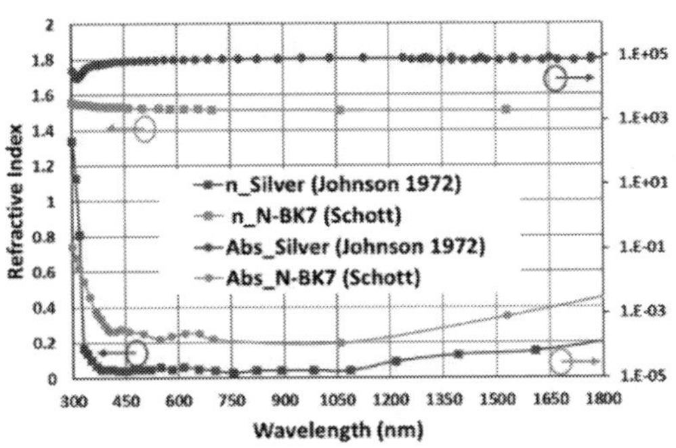

Figure 2: Refractive index (square shape) and absorption coefficient (round shape) of the silvered parabolic reflector (red) and of the coverglass (blue).

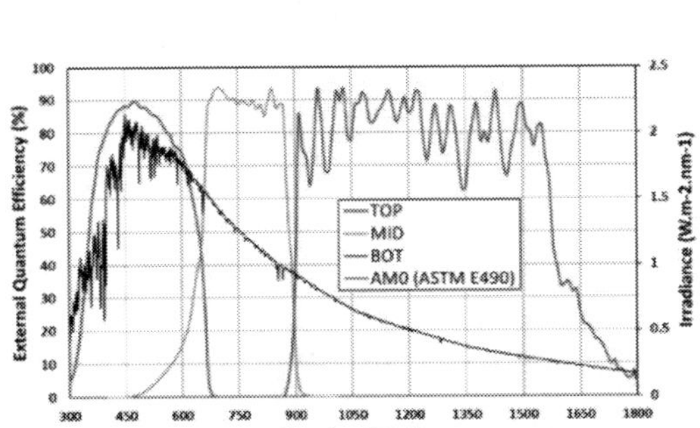

Figure 3: External Quantum Efficiency of the MJSC and AM0 ASTM E490 spectrum versus the wavelength from 300 to 1800 nm used in the modelling.

2.2 Parabolic Shape

First modelling was carried out with a parabolic shape defined by the following equation (1) with a focal length of 2.25 mm:

$$z = \frac{x^2}{4.f} \quad \text{(Eq. 1)}$$

Figure 4 presents a schematic view of the parabolic mirror with the definition of the reference axis. If the parabolic mirror is located at the focal, then all the flux will be focalized in a point at the center of the solar cell, and unfortunately providing a higher PAR. By defocusing the parabole, for instance, by sweeping along the z-axis, it is possible to decrease the PAR by spreading the focal spot (see Figure 5). Thus, the PAR decreases from 1342 to 32.2 by shifting the solar cell -0.35 mm along Oz.

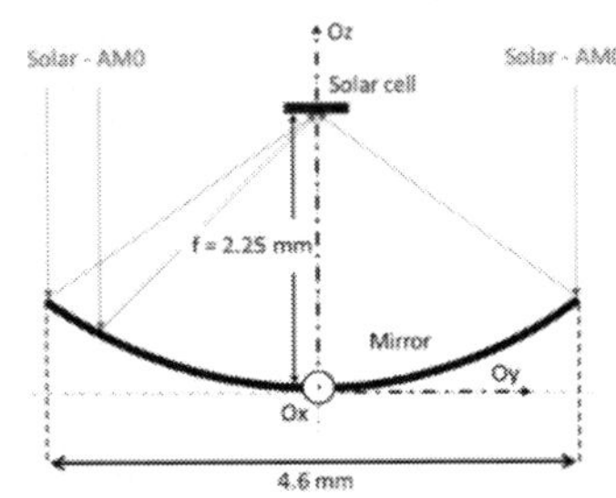

Figure 4: Schema of the parabolic reflector with the definition of the axis.

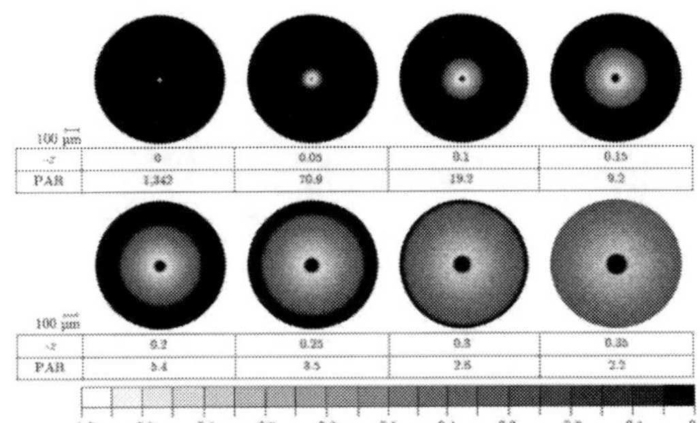

Figure 5: Normalized to peak irradiance maps for different detector positions along the Oz axis and its associated PAR.

Reasonably and arbitrarily, 60% of the receiver surface was chosen to be a limit. So, the solar cell is moved from -0.225 mm of the focus point. Results are shown in Figure 6 (a). A PAR of 4.4 is calculated from the irradiance map, and AA is extracted from curves of the Figure 7 considering the normalized irradiance at 90%. AAy = 4° presents the results with a rotation angle around Oy.

2.3 Free-form shape

By modeling a free-form shape, the objective is to find an optical design with the best uniformity (low PAR) over the receiver while maintaining a high acceptance angle. For this purpose, two objective functions have been established, the first one is to maximize the incident flux on the cell, and the second one is to verify that the flux profile is over 60 % for the receiver surface, distributed as homogeneously as possible. The previous parabolic surface with a focal of 2.25 mm and the solar cell located at -0.225 mm from the focus point are the staring data. As the symmetry is radial, the parabolic surface could be defined as a parabolic curve where some variables points are defined. The position of these points can move along Σz (800 μm at most). By increasing the number of variable points, the optimization becomes more time-consuming, but a smoother surface is obtained. The Trace-Pro optimization process uses the Nelder-Mead method which seek to minimize continuous functions in a multidimensional space [8]. The results are presented in Figure 6 (b), and show that peaks are more numerous than previously, with a high center point which is not the case for the parabolic reflector (see Figure 6 (a)). A PAR of 6.7 is calculated from the irradiance map, and in Figure 7 an AAy = 3° is obtained.

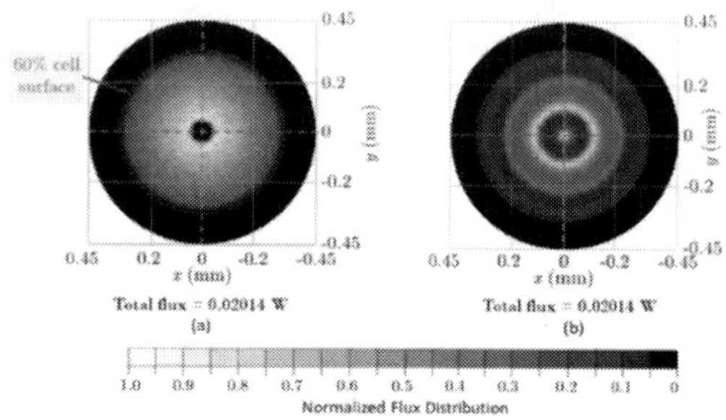

Figure 6: Irradiance maps normalized to the peak with a solar cell located at -0.225 mm of the focal point: (a) Parabolic Reflector and (b) Free Form reflector.

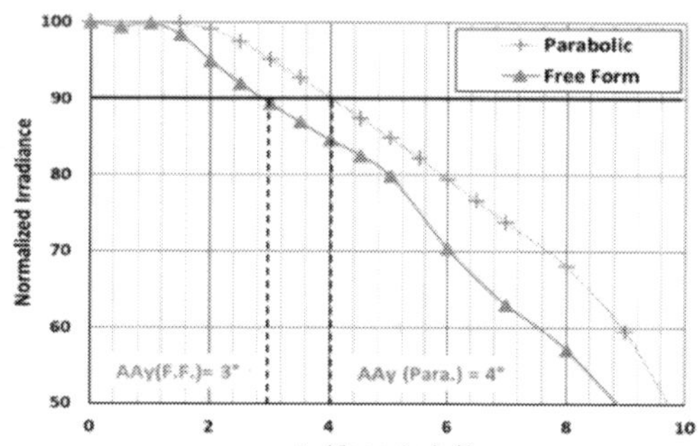

Figure 7: Normalized irradiance in function of the incidence angle around the Oy axis (AAy) for the parabolic reflector and the free-form reflector.

2.4 With secondary optical elements

By using SOE, the aim is to improve the flux distribution on the receiver and increasing the AA. By excluding the SILO (for SIngLe Optical surface), which remains a good solution and have already be addressed [9], two refractive SOEs with different shapes are studied. The material used for modeling is a low Fe Glass [10]. The first optics is a hemisphere or half-sphere and the second is an optimized free-form, both of which have been sized with the same radius as the solar cell (0.45 mm), so that they do not cast a shadow. To define the free-form SOE, three solar sources light simultaneously the optical system with angles of 0°, 5° and 10° around the y-axis. The task of the optimizer is to maximize the total received flux. To begin with, a dome shape was defined as shown in Figure 8 Model #1. The red line in the center of the SOE presents the radial symmetry of the optical system. Relative variables were used, one-dimensional for the central point A (Σz = 0.2 mm) located at 0.5 mm, and two-dimensional for B and C (Σz = 0.2 mm and a Σy = 0.08 mm). Subsequently, following an iterative process, the mobility range of the control points (or optimization space) is reduced to about a half, to refine the solution presented in Model #2 on the Figure 8.

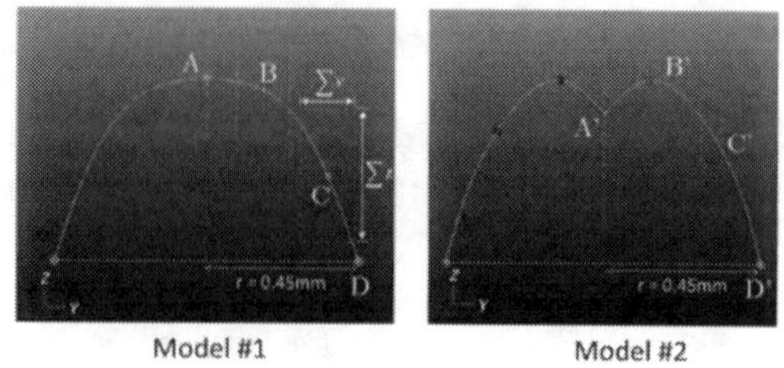

Figure 8: Optimization models of the free-form SOE. Model #1 presents the initial model. Model #2 presents the final profile of the free-form SOE.

By associating these two SOE to the free-form POE, the AA increases to reach 8.2° for the radial free-form (Figure 9). The PAR calculated from the irradiance map is worse for the half-sphere (PAR = 12.5) and improves for the radial free-form (PAR = 5.6).

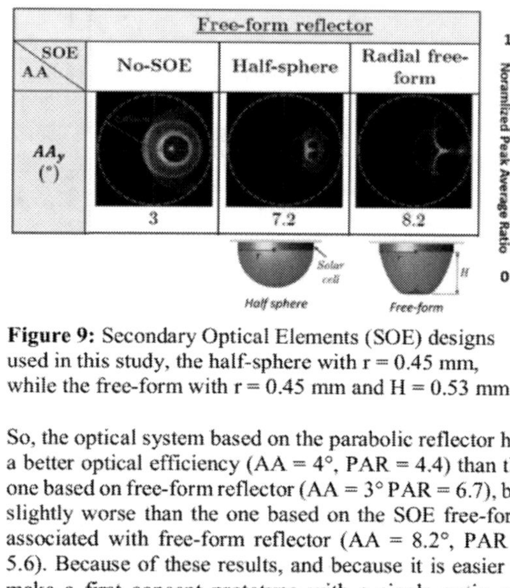

Figure 9: Secondary Optical Elements (SOE) designs used in this study, the half-sphere with r = 0.45 mm, while the free-form with r = 0.45 mm and H = 0.53 mm.

So, the optical system based on the parabolic reflector has a better optical efficiency (AA = 4°, PAR = 4.4) than the one based on free-form reflector (AA = 3° PAR = 6.7), but slightly worse than the one based on the SOE free-form associated with free-form reflector (AA = 8.2°, PAR = 5.6). Because of these results, and because it is easier to make a first concept prototype with a single optic, we decided to prototype only the parabolic reflector without any SOE at first.

To have a better spectral understanding, EQE with the full spectrum extracted from AM0 ASTM E-490 (from 300 nm to 1800 nm with a step of 10 nm), has been used with the optical model to estimate the short circuit current of the system with the parabolic reflector. The current density map of the middle subcell is presented on Figure 10 as example. Thus, the simulation results give the following current for each subcell of the MJSC: the top subcell (2.35 mA) the middle subcell (2.15 mA) and the bottom subcell (3.95 mA). The middle subcell is the limiting subcell, and the current distribution is similar for the three maps.

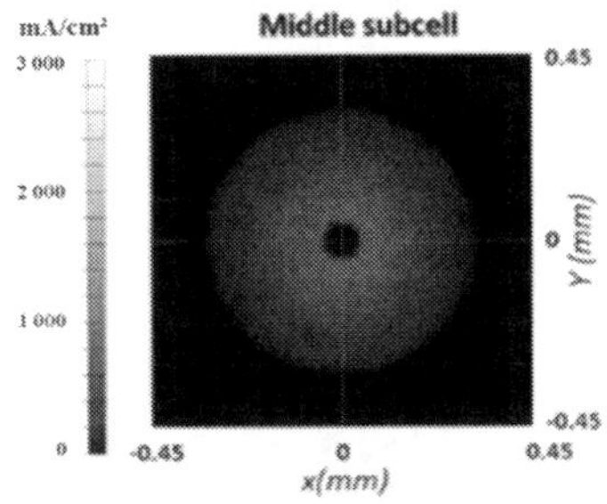

Figure 10: Current density map for middle subcell of the MJSC used in the optical modeling.

3 MANUFACTURING PROCESS

Apart from the manufacture of the mold, which was subcontracted, all the steps in the manufacture of the optics were carried out in the laboratory. The first part describes how the honeycomb samples incorporating the optics are manufactured. The second part details the molding process using the mold designed in the laboratory. The proof-of-concept developed is the same size as a standard space solar half-cell, i.e. 40x40 mm², and will include 30 concentrating optics.

3.1 First step: honeycomb manufacturing

For prototyping mini-modules of about 40x40 mm², samples of 200x200 mm² were made first. These samples are composed of a honeycomb core (CR-III® 5.7 3/16 5052) which is bonded with a CRFP (TeXtreme 86 PW HS40 WD/20:4320:43-1000 fibres), prepreg by means of an epoxy films (Hexbond 312UL-100). With the exception of the honeycomb core, the materials used are not space qualified, an epoxy adhesive film was used instead of e.g. M18 (for space), since the latter is 15 times more expensive.

After being laminated, the 200x200 mm² samples must be cut to generate 40x40 mm² mini-modules. For this purpose, the protocol shown in Figure 11 was followed: (a) first, a protective layer was placed on the top of the module; (b) the mini-modules were sliced using a diamond wire cutter with water cooling method, typically used to cut silicon wafers; (c) image showing the 5 transverse and longitudinal cuts to form 25 mini-modules from the 200x200 mm² samples; (d) these were then cleaned using an ultrasonic bath for 20 min, in sweep mode, using iso-propanol to remove any residue from the cutting process; (e) and finally dried in an oven, at 60°C for 3 min to evaporate the remaining potential liquids.

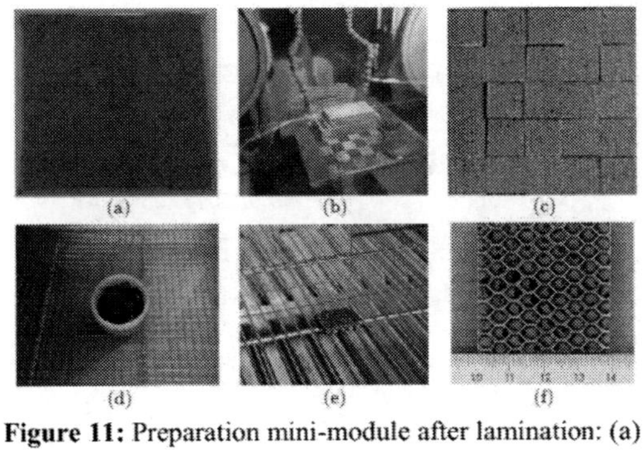

Figure 11: Preparation mini-module after lamination: (a) Protective tape on; (b) Cutting process of the laminated sample into mini-modules; (c) 25 mini-modules cut; (d) Cleaning of a module with ultrasound and iso-propanol; (e) Drying process; (f) Mini-module laminated, cleaned and dried.

3.2 Second step: mold manufacturing

In order to design a mold that allows the formation of parabolic surfaces inside a honeycomb network (i.e. a laminated composite structure), it is necessary to design cavities in the mold that have the same shape of the network to allow the parabolic shape to enter inside, as shown in Figure 11 (f). In addition, it is necessary to take into account a mechanical tolerance between each honeycomb-cell to make a smooth insertion, and to consider possible small variations in the shape of the honeycomb core. On the other hand, the depth of these cavities will determine the height rim of the parabola. Then, to ensure that the system is hermetic during polymer injection, a bottom clamping plate was designed to allow the composite structure to be attached to the mold using four mounting socket cap screws (a) (b) (c) and (d). In addition, the mold has a rectangular cavity that houses the polymer during molding (backside of the mold), which has a height of 8 mm to contain the material when it is subjected to vacuum (i.e., which increases its volume) to extract potential bubbles. On the other hand, eight demolding holes distributed around the network of injection cavities have been designed to demold the prototype. A roughness of the parabola Ra < 150 nm has been specified. Figure 12 shows the real mold used in our study.

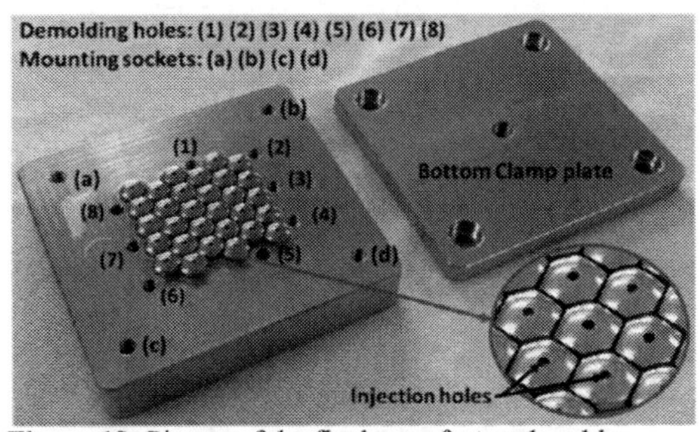

Figure 12: Picture of the final manufactured mold, showing a close-up view of 7 units.

3.3 Third step: optics manufacturing

Then, the process proposed for molding optics using polymers can be divided into five stages: (1.) It starts by preparing the surfaces involved, the mold is cleaned using an ultrasonic and iso-propanol bath and dried with compressed air. While the composite stack is pre-treated to clean and improve the adhesion on surfaces using a plasma treatment (N_2) during 3 min. and a primer coat (Dowsil™ 1200 OS) is then applied over the CFRP and the inner aluminum honeycomb walls; the latter is allowed to dry for 30-40 min. at room temperature. (2.) Once the primer is dry, the polymer is injected into a confined geometry that will shape the optics; (3.) Once the molding material has been introduced, the assembly is placed in a vacuum chamber at 0.01 mbars, for at least 20 min or until all the bubbles present have been removed. (4.) The assembly is then heated in an oven at 150°C for 10 min. (5.) Finally, the mold is extracted using compressed air and, if necessary, extraction screws. Then, optics are set

into an electron beam deposition chamber to coat the polymer with silver reflective layer to get a parabolic mirror.
The first result is presented on Figure 13. As optics are tiny (< 10 mm), the quality of the surface and the shape is a key point for such micro-CPV systems. So experimental shapes of the optics will be compared with the mold and with the theoretical one.

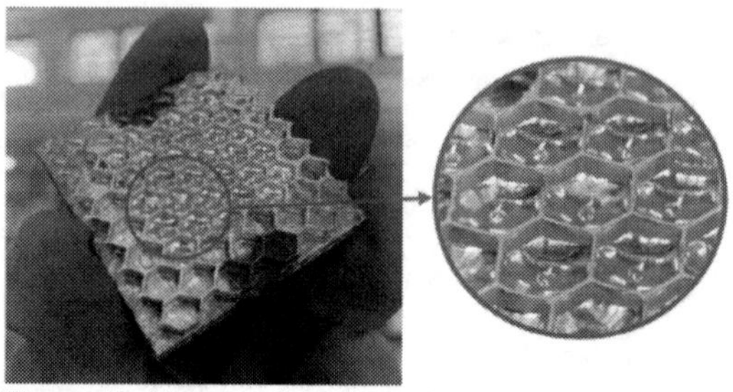

Figure 13: Pictures of the first mini-module with reflective silver coating.

4 OPTICAL CHARACTERIZATIONS

The method used to characterize the manufactured mirrors is based on optical microscopy. The first step consists of acquiring the topographic surface using a 3D optical microscope. A matting powder is deposited on the surface of the optics to enable the acquisition and determination of the image of the reflector. Once the image has been acquired, the surface is sampled as points and then approximated using quadratic shape recognition. The model used determines the closest quadratic shape using the least square method. Once the quadratic shape is found, a Poisson algorithm is used to divide the surface into 20,000 facets (TracePro takes no more than 25,000 facets). Once this surface is entered into the optical model, ray-tracing allows us to estimate the optical efficiency of the reflector. It should be noted that this method does not take into account the possible roughness of the molds and optics, the objective being to evaluate mainly the shape errors. For more details, and to see the validation of the method with a perfect ball, it is necessary to refer to the following thesis [11]. The solar cell serving as a detector is placed at the focal point. The first results of an optic and the corresponding mold are shown in Figure 14.

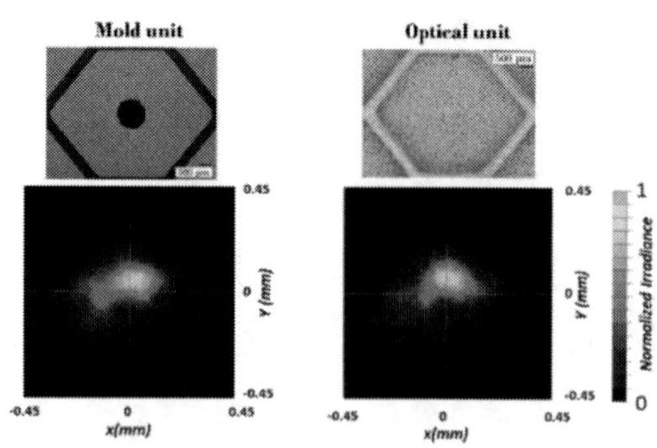

Figure 14: Pictures of a mold unit and the associated optics with the matting layers, and normalized irradiance maps of the spot at the focal point.

The optical efficiency of the mold is less than 13% compared to the optical efficiency of the specified and ideal dish. Once the optic is made, another 7% of the optical flux is lost. A detailed analysis of the mold and the characterization method is planned in order to determine the exact causes of these shape losses.

5 CONCLUSION

A new concept based on a miniaturized mirror highly integrated inside a honeycomb core sandwich panel has been proposed, with the objective of increasing the specific and volumetric power of a micro-CPV. While providing a monolithic structure that avoids deployment in space, between the optical network and the cells, decreasing the associated risk and complexity.

Furthermore, to establish an adapted reflector system for micro-CPV in space, in terms of angular tolerances and flux uniformity, a comparative study of two idealized systems, a parabola with different defocusing planes and an optimized free-form was performed.
The parabolic reflector, therefore, proved to be an efficient solution in terms of flux homogeneity, as well as in terms of manufacturing simplicity. For this reason, the latter design was chosen to validate the concept of a highly integrated reflector in a honeycomb. It presents a theoretical optical efficiency up to 88% (considering an ARC on both sides of the coverglass), i.e. an improvement of up to 8% over the state of the art micro-CPV for space.

To move from a theoretical concept to a validated prototype with industrial potential, a molding method has been proposed. For characterizing the surface and shape defects of an optical surface (at millimeter scale) and thus estimating its optical efficiency, a characterization method has been developed, which is based on topographic measurements by optical microscopy combined with ray-tracing simulations.

ACKNOWLEDGEMENTS
This work was supported by Institut Carnot Energies du Futur.

REFERENCES

[1] A. Ritou, P. Voarino, et O. Raccurt, « Does micro-scaling of CPV modules improve efficiency? A cell-to-module performance analysis », *Solar Energy*, vol. 173, p. 789-803, oct. 2018, doi: 10.1016/j.solener.2018.07.074.
[2] M. Eskenazi, « Design, analysis testing of the CellSaver concentrator for spacecraft solar arrays », in *Conference Record of the Twenty-Ninth IEEE Photovoltaic Specialists Conference, 2002.*, mai 2002, p. 935-938. doi: 10.1109/PVSC.2002.1190734.
[3] C. J. Gelderloos *et al.*, « Characterization testing of Hughes 702 solar array », in *Conference Record of the Twenty-Eighth IEEE Photovoltaic Specialists Conference - 2000 (Cat. No.00CH37036)*, Anchorage, AK, USA, 2000, p. 972-975. doi: 10.1109/PVSC.2000.916048.
[4] A. Bermudez-Garcia, P. Voarino, et O. Raccurt, « Environments, needs and opportunities for future space photovoltaic power generation: A review », *Applied Energy*, p. 34, 2021.

[5] C. J. Ruud, J. S. Price, B. Fisher, B. Wang, et N. C. Giebink, « Lightweight Monolithic Microcell CPV for Space », in *2018 IEEE 7th World Conference on Photovoltaic Energy Conversion (WCPEC) (A Joint Conference of 45th IEEE PVSC, 28th PVSEC 34th EU PVSEC)*, juin 2018, p. 3535-3538. doi: 10.1109/PVSC.2018.8548170.

[6] M. P. Lumb *et al.*, « Design, modeling, and experimental results for CPV arrays built using heterogeneously integrated III-V micro-cells », in *Physics, Simulation, and Photonic Engineering of Photovoltaic Devices IX*, San Francisco, United States, mars 2020, p. 3. doi: 10.1117/12.2549037.

[7] R. Herrero *et al.*, « Indoor Characterization of Multi-Junction Solar Cells Under Non Uniform Light Patterns », Freiburg, (Germany), 2010, p. 36-38. doi: 10.1063/1.3509226.

[8] J. A. Nelder et R. Mead, « A Simplex Method for Function Minimization », *The Computer Journal*, vol. 7, n° 4, p. 308-313, janv. 1965, doi: 10.1093/comjnl/7.4.308.

[9] S. Askins, M. Victoria, R. Herrero, C. Domínguez, I. Antón, et G. Sala, « Hybrid dome with total internal reflector as a secondary optical element for CPV », Freiburg, Germany, 2016, p. 050002. doi: 10.1063/1.4962084.

[10] K. R. McIntosh, J. N. Cotsell, J. S. Cumpston, A. W. Norris, N. E. Powell, et B. M. Ketola, « An optical comparison of silicone and EVA encapsulants for conventional silicon PV modules: A ray-tracing study », in *2009 34th IEEE Photovoltaic Specialists Conference (PVSC)*, Philadelphia, PA, USA, juin 2009, p. 000544-000549. doi: 10.1109/PVSC.2009.5411624.

[11] A. Bermudez-Garcia, « Highly integrated photovoltaïc micro-concentrator for space applications », Université Grenoble Alpes, Le Bourget du Lac, 2022.

ANALYSIS OF POWER GRIDS WITH HIGH-PENETRATION OF PHOTOVOLTAICS: THE CASE OF HONDURAS

Gustavo Gómez-Ramírez
Costa Rica Institute of Technology, Costa Rica

Carlos Meza
Anhalt University of Applied Sciences, Germany

Gonzalo Mora-Jiménez
Costa Rica Institute of Electricity, Costa Rica

Isaac Luévano-Reyes
ETAP, Mexico
Luis García-Santander
University of Concepción, Chile

J. Rodrigo Rojas
National University, Costa Rica

ABSTRACT: In 2020, Honduras was the country with the highest penetration of solar energy in its electricity matrix. Moreover, the Honduran power system is interconnected with other countries in the Central American region, with a common regional market. On the other hand, the Honduran electricity system has a high impedance and low inertia. The conditions described above make this country an interesting case study to analyse the behaviour of a weak power grid with a high penetration of solar photovoltaic energy. This paper briefly discusses these conditions and analyses an event where a blackout occurs in Nicaragua due to the shutdown of several PV plants in Honduras. This paper also explores a solution that avoids the blackout by integrating distributed storage. The analysis and proposed solution are performed using a model of the Central American electricity grid developed using ETAP software.

1 INTRODUCTION

Honduras is a low-income country that has struggled to develop its energy sector. Honduras highly depended on fossil fuels for electricity generation during the last half of the 20th century and the first decade of the 2000s. Moreover, Honduras imports all the fuel fuels and has no refinery capacity. This situation made the electricity prices highly dependent on international oil prices. Due probably, to the fragile economic situation of the country, there was very little investment in large power projects that used local energy resources [1].

Two elements have made it possible for Honduras to improve its energy situation. Namely, the consolidation of the Central American power system and the development of photovoltaics projects.

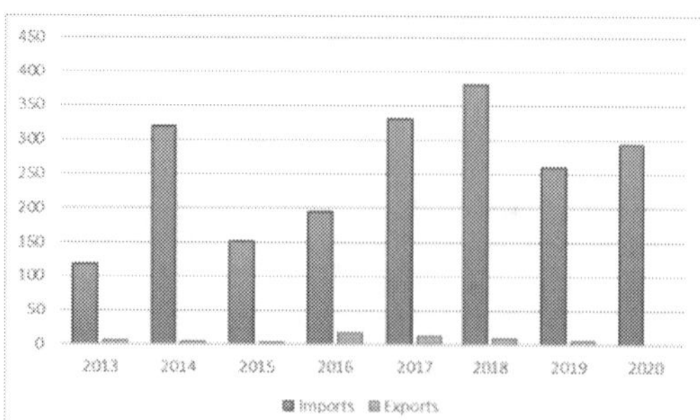

Figure 1: Historical power imports and exports according to [3].

The Honduran local power system is connected to the Central American power system. This is a dedicated transmission infrastructure with an extension of 17289.23 km and connects the power systems of the region's countries [2]. A Central American Regional Market allows countries to buy and sell energy among each other. The Central American Power infrastructure enables Honduras to get power from other countries that produce energy from cheaper, price-stable, and cleaner sources. As shown in Figure 1, Honduras is mainly an energy importer.

From the second decade of the 2000s, Honduras has been able to develop several photovoltaic projects. Several factors have contributed to the fast development of PV projects in Honduras, such as the relatively low initial investment of PV compared to other energy projects (e.g.Hydro), the intrinsic modularity of PV projects, the low prices of PV technology and the high-irradiance present in the region. As a result, in 2015, Honduras became the second biggest solar market in Latin America. Additionally, in 2017 Honduras became the first country with 10% PV in its electricity market. Furthermore, according to data provided by the IE PVPS in 2020 [4], Honduras is the country with the highest PV penetration, see Figure 2.

Figure 2: Countries with highest PV penetration according to [4].

Nevertheless, the Honduran power system faces important challenges given the intrinsic variability in the power production of PV plants and the limitations of the Honduran transmission and distribution infrastructure. The Honduran power system has a significant share of intermittent power, i.e., wind and photovoltaic energy. Photovoltaic power generation comprises 12.7 % of the contribution to electricity demand, but wind and photovoltaic together represent more than 20%. Moreover, Honduras Power System presents high technical losses in the grid, around 32%, [3]. Power generators are concentrated in one part of the grid. The distribution and transmission grid cannot support the generation-load un-balance due to losses in case of significant disturbances. Therefore, the conditions mentioned above can originate significant problems for Honduras's power system and its interconnected neighbours, as was the case on June 9th, 2021.

2 CASE OF STUDY OF CONTINGENCIES IN THE HONDURAN POWER SYSTEM

On June 9th, 2021, a series of events disconnected several PV power plants in Honduras, resulting in a 4-hour blackout in the neighbouring country of Nicaragua. These events are shown in Figure 3 according to [5].

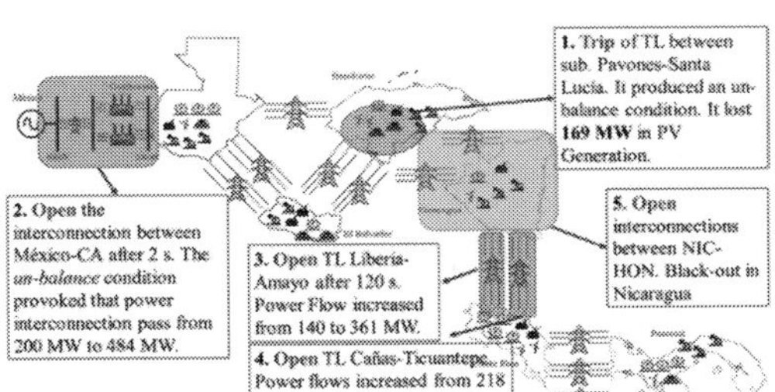

Figure 3: Events that happened on June 9th due to a contingency in the Honduran Power System.

To reproduce this event, we have developed a model of the Central American power systems. This model considers all the power system's main elements, as seen in Figure 4. Previous validation of the modelled power system has been done using the steady-state power flow values for specific cases measured in the actual power system. Also, as seen in Figure 5, we could simulate the same chain of events that caused the blackout in Nicaragua.

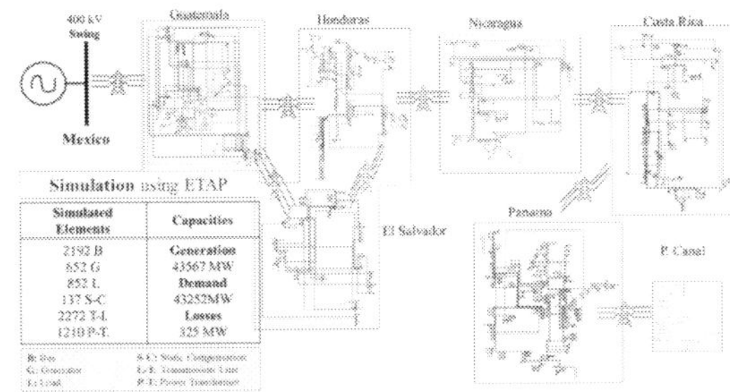

Figure 4: Screen capture of the model of the Central American Power Network using ETAP ®.

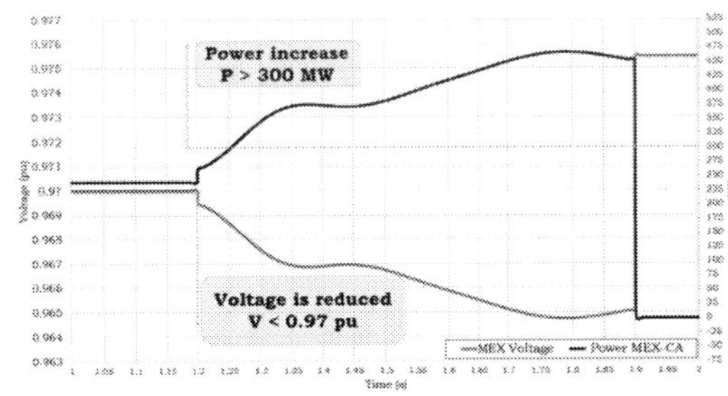

Figure 5: Simulation results of the events that happened on June 9th due to a contingency in the Honduran Power System using the model in Figure 4.

3 DISTRIBUTED STORAGE AS A SOLUTION FOR DEALING WITH CONTINGENCIES IN THE HONDURAN POWER SYSTEMS

As a first solution, we integrated storage at different points in the Honduran network to avoid the events that led to the blackout in Nicaragua.

Eight storage elements of 20 MW and 135 MWh were modelled, as shown in Figure 6. They were placed following an iterative exploratory process at different points in the network that, according to the simulation, showed the most significant impact on the bus bar's voltage levels.

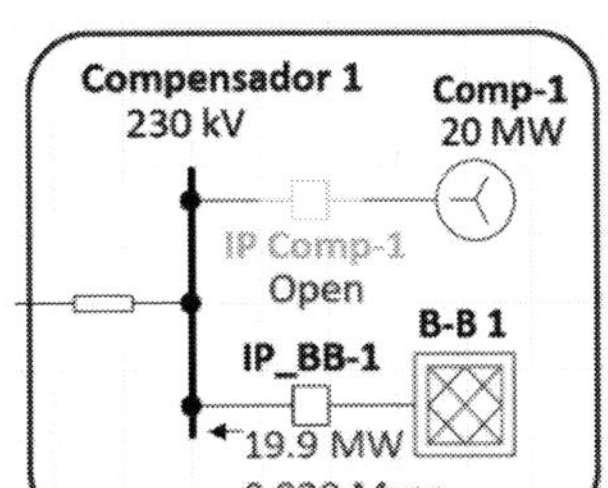

Figure 6: Model of the storage units.

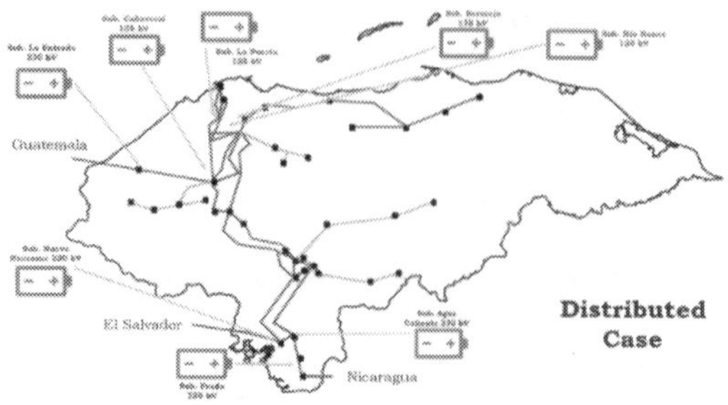

Figure 7: Location of the storage units.

4 SIMULATION RESULTS

The simulation results are shown in Fig C. Notice that even though there is a voltage drop below the normal operation limits, its duration is not long enough to activate the protection.

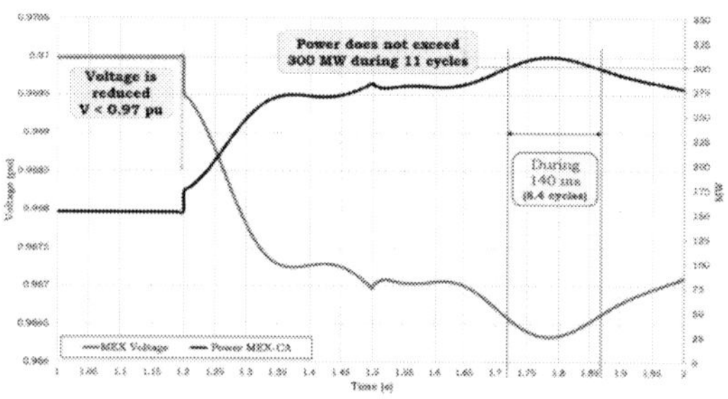

Figure 8: Simulation results of the events that happened on June 9th due to a contingency in the Honduran Power System using distributed storage.

5 CONCLUSIONS AND FUTURE WORK

Honduras is a fascinating study case for PV high-penetration weak power grids connected in a multinational power transmission system.

The Central American (CA) power network model has been developed and validated. This model allows emulating contingencies in the CA power system due to PV generation.

Distributed storage is a possible solution to support PV's high penetration in weak power grids.

In future work, we will identify high-probable contingencies in PV in the CA power network. Also, we will explore different criteria and methodologies to locate storage units in the CA power network and evaluate the impact of "smart" inverters' contribution to stabilising the CA power network.

6 REFERENCES

[1] Meza, C. (2014). A review on the Central America electrical energy scenario. Renewable and Sustainable Energy Reviews, 33, 566-577.

[2] EOR (2019). Tomo i: Planificación de largo plazo de la generación y la transmisión regional correspondiente al período 2019-2028.

[3] CEPAL(2021) Estadísticas del subsector eléctrico de los países del Sistema de la Integración Centroamericana (SICA), 2019 y avances a 2020.

[4] IEA (2022). Photovoltaic power system technology: Trends in Photovoltaic Applications 2022.

[5] EOR (2021). Informe del evento ocurrido en el sistema eléctrico reigional (SER) el 09 de junio del 2021, a las 11:03 horas.

A MODEL CORRECTING THE EFFECT OF SUB-HOURLY IRRADIANCE FLUCTUATIONS ON OVERLOAD CLIPPING LOSSES IN HOURLY SIMULATIONS

Adrien Villoz, Bruno Wittmer, André Mermoud, Michele Oliosi, Agnes Bridel-Bertomeu
PVsyst SA
Route de la Maison-Carrée 30 - 1242 Satigny – Switzerland, support@pvsyst.com

ABSTRACT: Simulations of PV systems that proceed in hourly steps can underestimate overload losses arising from sub-hourly irradiance fluctuations. The aim of this study is to estimate and address these additional overload clipping losses to improve the accuracy of the simulations in PVsyst.

The current trend for designing PV systems is to increase the DC/AC ratio. This trend is mainly due to PV modules plummeting in price and PV plants being constrained by grid limitations. The increase in DC/AC ratio allows PV systems to maximize production for the same AC capacity, by having the inverters operating at full output power for more hours during the day. When the inverter is operating close to maximal power, it can happen that irradiance spikes drive it into clipping conditions. This behavior is not captured by a simulation that proceeds in hourly steps, which leads to a systematic underestimation of the clipping losses (1-5% or more over a year).

The purpose of this study is to find a way to estimate from sub-hourly irradiance fluctuations a correction that can be applied in a general way to the clipping losses of an hourly simulation. In a first step, an analysis of hourly average irradiation data and sub-hourly data was done, to identify a way to calculate the behavior of overload clipping losses when applying a threshold. The analysis then went on using the same irradiation data for complete simulations in four climates and covering multiple cases of orientations of PV modules and DC/AC ratios.

The goal of the study, to improve simulations in PVsyst that are done in hourly steps, was achieved by using coefficients extracted from the sub-hourly data, applied on a simple geometrical model. This model corrects the error on overload losses significantly and thereby reduces the average difference of minute- and hourly-based AC generation below 1% for each DC/AC ratio.
Keywords: Clipping, Modelling, Simulation, System Performance

1 MOTIVATION

Sub-hourly clipping losses are a well-known phenomenon. If the simulation of the PV performance is performed in hourly steps, then the calculations will be based on hourly average values. If the hourly average irradiance value is close to the clipping level, the irradiance fluctuations will cause clipping losses that are missed in the hourly simulation. Several studies suggest that this effect may amount in up to 5% overestimates of the yearly generation for high DC/AC ratios, and a few percent for more regular systems. A non-exhaustive list of such studies can be found in [1, 2, 3, 4, 5, 6, 7, 8, 9, 10].

To visualize how this comes about, we take the irradiance evolution in one-minute steps for one hour and compare it to its mean hourly value and the level at which clipping sets in.

We have four distinct cases, which are depicted in Fig. 1. The first two on the top, are when the clipping level is above the highest or below the lowest irradiance value within the hour. Then the hourly average will not miss any additional clipping, since we are always either in non-clipping mode or permanently clipping, making the hourly average equivalent to the minute level treatment.

The third case on the bottom left is when the clipping level is in between the highest and the mean irradiance. The hourly treatment will detect no clipping at all, which means that the orange area above the clipping level will be missed.

Finally, if the clipping level is in between the lowest and the mean value, we get to a situation that is harder to grasp. The hourly treatment will detect clipping and remove all the area in between the mean and the clipping level. This will miss entirely the area above the mean, but it will contain also a part of the non-clipping portion towards the lower irradiance values. This second part will never fully compensate the missed part above the mean, since in total the areas above and below the mean are the same. Therefore, also this situation leads to under-estimated clipping losses.

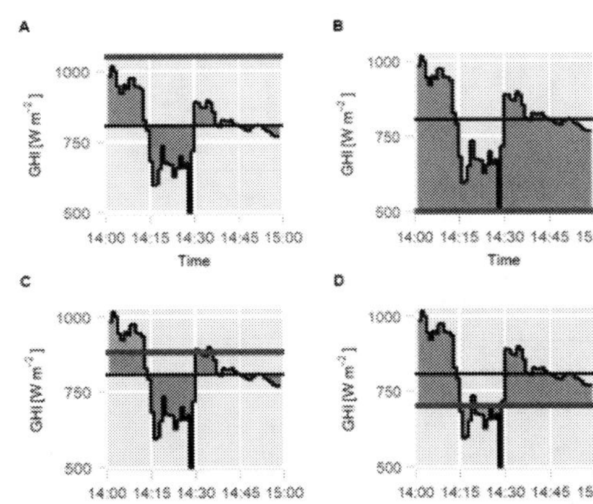

Figure 1: Irradiance evolution within one hour. On the left (cases A and C) the clipping threshold is above the hourly average, on the right (B and D) it is below. In the top row (A and B) the hourly averaging will capture correctly the clipping losses, while in the lower row (C and D) the clipping losses will be underestimated by an amount proportional to the orange surfaces.

In conclusion we can say that the clipping losses that are calculated with the help of hourly averages will either be correct, or underestimate the clipping losses, they will never overestimate them. The missed clipping losses are highest in hours where the clipping threshold coincides with the average irradiance value. As result, the hourly simulations will be biased towards a higher PV generation. The exact amount will depend on the specific weather as well as on the DC/AC ratio of the system. Studies have shown that this bias can reach a few percent, which is a non-negligible amount, and that calls for a correction. This is the subject of the present study.

2 APPROACH FOR THE SUB-HOURLY CLIPPING CORRECTION

One way to handle the sub-hourly irradiance fluctuations would be to move to a minute-level simulation. However, this can become extremely time consuming for large PV systems or PV systems with complex shading situations. Other approaches aim at finding corrections for hourly simulation values, based on the DC/AC ratio and general features of the weather files used in the simulations, like weather variability or climate category [4, 5, 11, 12, 13, 14, 15, 16, 17].

In this study, we choose an approach that is a combination of both, and which is schematically shown in Fig. 2. In a first step, the minute-level weather data is analyzed in order to extract coefficients that are stored together with the hourly values. These coefficients will allow to correct for the sub-hourly fluctuations at the stage when the clipping losses in the inverter are calculated.

The key idea is based on a reordering and linearization of the irradiance values and will be explained in the following.

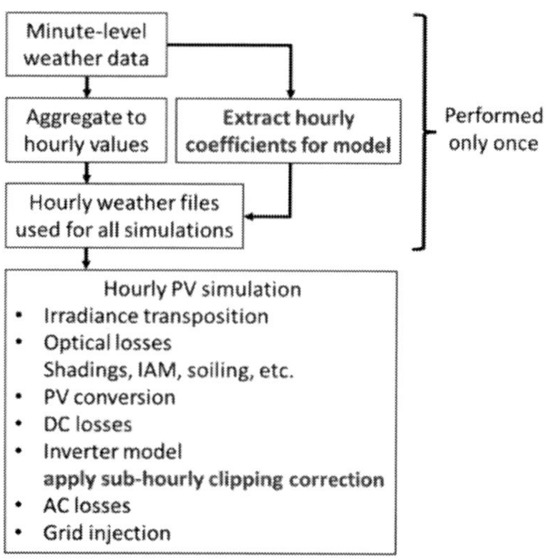

Figure 2: Approach for correcting the sub-hourly clipping losses. When aggregating the minute-level weather data to hourly values, additional coefficients are stored together with the hourly values, and which are used in the inverter modeling to estimate the sub-hourly clipping losses.

2.1 Reordering irradiance values

The performance of a PV system is in first order a linear function of the effective irradiance reaching the PV modules. Other conditions like temperature or spectral corrections affect the output only in second order. The PV system reacts almost instantaneously to the irradiance level, which, in a minute-level simulation, would in principle allow us to reorder the minute-values within each hour, without affecting the simulation result. Therefore, we can sort the minute irradiance values in ascending order and get a simplified irradiance evolution over one hour. This is shown in Fig. 3.

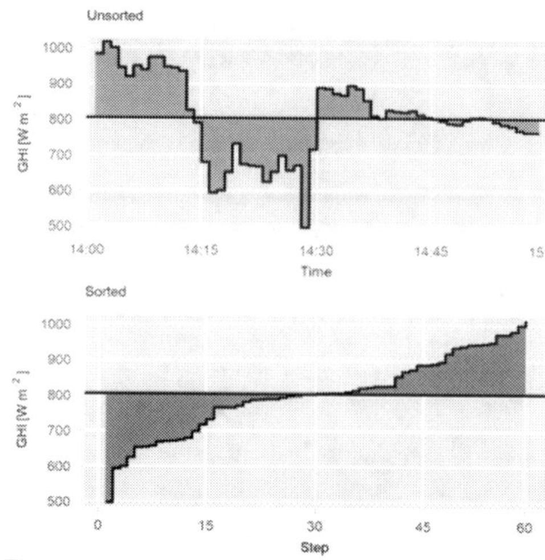

Figure 3: Re-ordering the minute values within an hour does not affect the simulation results. Sorting the values in ascending irradiance values allows to describe the irradiance value distribution with a simple approximation.

The reordering simplifies considerably the shape of the irradiance evolution, which can now be described with a few parameters only. This will allow to reconstruct this irradiance evolution at the stage where the clipping of the inverters is treated. In this way the amount of irradiance that will lead to clipping can be estimated allowing to correct the hourly clipping losses.

2.2 Linearized irradiance evolution

We approximate the ordered irradiance evolution by two stretches of straight lines, as shown in figure 4. The lines are limited by the lowest and highest irradiance and touch at the point where the irradiance evolution crosses the mean value. This can be described by a total of three parameters.

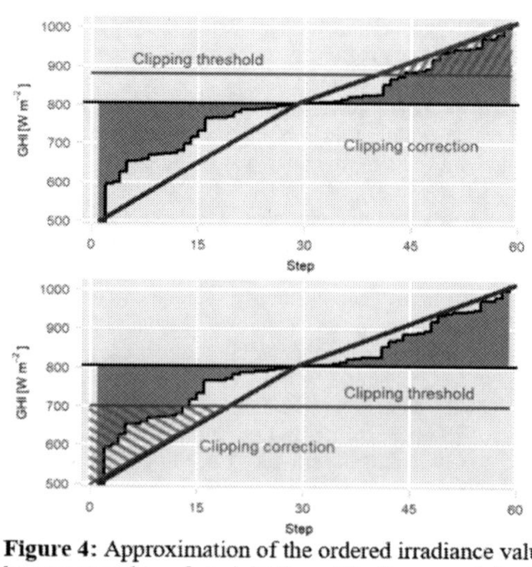

Figure 4: Approximation of the ordered irradiance values by two stretches of straight lines. The lines are delimited by minimal and maximal irradiance, and touch at the point where the values cross the mean. The clipping correction is approximated by the area of the textured triangle.

In the example shown in Fig. 4, one can see that this approximation is not always perfect. The values above the mean are quite well approximated by the triangle, whereas the values below the mean are only grossly described by the triangle.

We chose to start with this simple approach, but it is possible to use other parametrizations with more parameters in order to optimize the fitting to the minute-level data.

2.3 From horizontal irradiance to AC power

The minute level input data contains the irradiance on a horizontal plane. In our approach we derive the additional coefficients from this data, and store them together with the hourly irradiance averages, in order to be able to reproduce the irradiance distribution within the hour, when we perform the hourly simulation steps.

This horizontal irradiance is subject to several transformations before the final effective irradiance on the PV modules is obtained. These include the transposition to the plane of incidence, the incidence angle losses, shading losses and soiling losses. After this, follows the PV conversion, the losses in the DC part of the system and the inverter model, before the AC power on which the clipping will be applied is known. Since not all these effects are linear, the shape of the horizontal irradiance evolution will suffer modifications, as shown in Fig. 5. In our approach we neglected these non-linearities and converted the linear description at the irradiance level directly to AC power by applying the ratio of the mean values. This is a simplification that can easily be improved in further sophistications of the model.

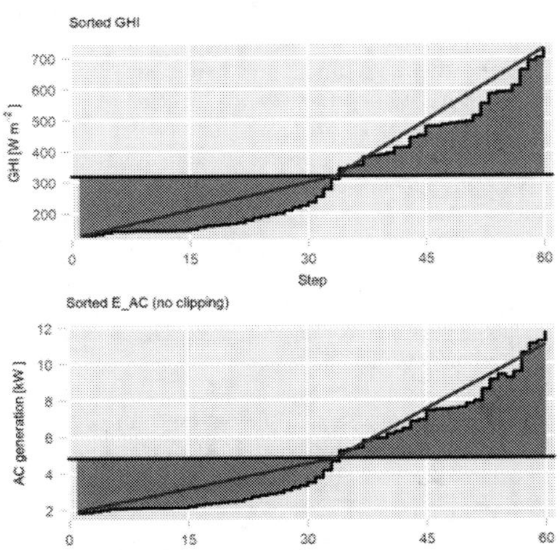

Figure 5: When going from global horizontal irradiance to generated AC power, the shape and ordering of the minute-level values changes. The simple linear approach uses just a linear rescaling of the two straight lines.

3 SIMULATIONS WITH ONE-MINUTE DATA

In order to check the effectiveness of the clipping loss correction, we performed simulations of a PV system in one-minute steps. Since the PVsyst simulation that was used in this study only allows to simulate in hourly steps, a special handling was necessary to obtain the one-minute simulation data.

The minute-level weather data can be sliced into 60 sets, one for each minute in the hour, as indicated in Fig. 6. Introducing a corresponding time shift to each of the slices, it is possible to run the hourly simulation on them and re-assemble the results in order to get effectively the results of a one-minute level simulation.

Since the hourly simulation uses a stationary thermal model, these results would correspond to the hypothesis of PV modules without any thermal inertia. This is an extreme case that will bias the results. The true temperature of the PV modules will not follow instantaneously the irradiance fluctuations but have a time constant in the order of 10 minutes. Therefore, we performed the set of simulations with the PV module temperature fixed during every hour at the value obtained by the hourly simulation using the averaged weather input data. This represents the case where the PV modules have a rather high thermal inertia. We chose these results as reference to which we compare the hour-level simulations, with and without the clipping loss correction.

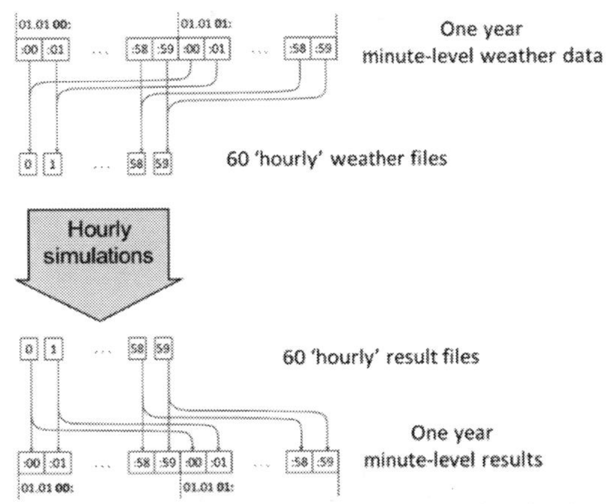

Figure 6: Decomposition of one-minute weather data into 60 'hourly' files. Each of the files is used as input for one hourly simulation and the results are re-assembled to give an effective one-minute simulation.

3.1 Data considered

The minute simulation approach was applied on four weather datasets, and we considered several PV system scenarios.

The data was obtained from four stations in the U.S. listed by the NREL Measurement and Instrumentation Data Center (MIDC): the NREL SRRL Baseline Measurement System [18], in Golden, Colorado; the University of Oregon [19], in Eugene, Oregon; the Natural Energy Laboratory of Hawaii Authority [20], in Kailua-Kona, Hawaii; and the University of Louisiana at Lafayette [21], in Lafayette, Louisiana. The four listed sites belong to four different climates under the Köppen climate classification, respectively, semi-arid continental (BSk), warm-summer Mediterranean (Csb), tropical semi-arid (BSh), and humid subtropical (Cfa). Temporally, the data subset used is that of year 2020.

As PV system scenarios, we considered four different DC/AC ratios and seventeen orientations. The DC/AC ratios are, 1, 1.33, 1.67, and 2, and are obtained using a 9kW inverter AC nominal power, and a variable number of strings in parallel, for systems amounting to 9, 12, 15, and 18 kWp DC. The orientations cover azimuths among [-135°, -90°, -45°, 0°, 45°, 90°, 135°], with the south as azimuth 0° and east at -90°, and tilts among [0°, 30°, 60°,

90°] with additional conditions: tilt 90° combined only to full E and full W azimuths, and tilt 0° computed only once. When accumulating multiple scenarios or climates, we have given the same weight to all cases.

4 HOURLY RESULT COMPARISON

Our correction modifies the resulting AC power at the level of the hourly steps of a simulation. As explained in section 2, the missing clipping loss at each hourly step is estimated based on the three coefficients extracted from the irradiance minute-data. In order to corroborate the model, the first step is therefore to investigate whether the correction effectively improves the hourly level clipping losses.

In Fig. 7, we compare the hourly clipping losses to the sum of the clipping losses found in the minute simulations for the corresponding hours. The points represent only the hours that have non-negligible clipping in the minute simulation. The data used for the plot covers only one site (Lafayette) and one orientation (South 30° Tilt), but the results are representative of all other cases. In the plot at the top, the clipping losses from the hourly simulation are not corrected, while the values of plot at the bottom have been corrected using our model.

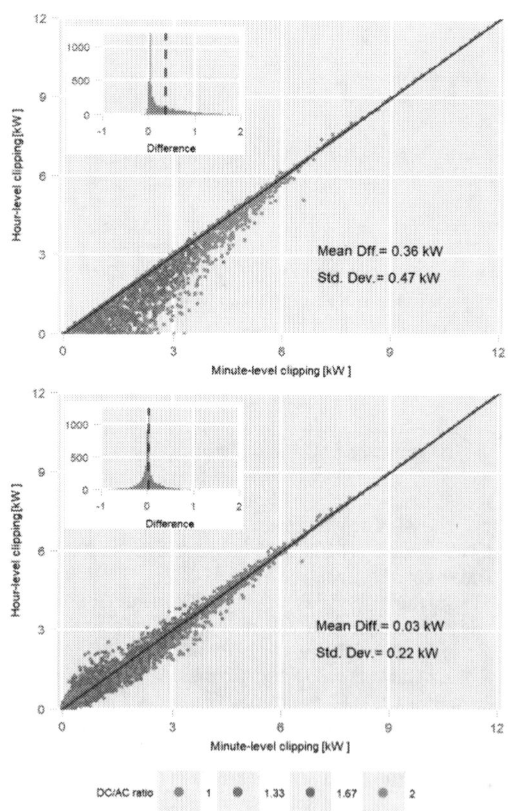

Figure 7: Comparison of the clipping losses of the hourly simulation with the hourly sums of the minute simulation for a 9kWAC PV system and four different DC/AC ratios. The first plot uses values without correction, while the second plot uses the corrected values.

As expected from the discussion in section 1, the non-corrected hourly data clipping is always smaller than the minute-data clipping. In other words, the distribution of the clipping losses from the hourly simulation exhibits a non-negligible bias towards lower values. The mean difference is 0.36 kW, to be put into perspective with the fixed system AC nominal power, 9 kW. The standard deviation is 0.47 kW. The clipping losses correction brings all points back towards the diagonal. Using corrected hourly clipping losses, the mean difference is 0.03 kW, and the standard deviation is 0.22 kW. The model therefore improves both the accuracy and the precision of the hourly simulation which thus becomes more consistent with the minute-level simulation.

5 IMPACT ON YEARLY RESULTS

In the previous section, we have shown that both the spread and bias of the distribution of the clipping loss errors for each hourly step did significantly improve with our model. In this section we will look at the impact of the correction on the yearly AC power output.

In Fig. 8, we show a histogram of the distribution of the difference in AC generation between hourly simulation and minute simulation. All cases under study have been considered, each count thus representing the yearly simulation for a given site, DC/AC ratio, and orientation. The lower side (negative counts) of the histogram denotes the results without correction, while the upper side (positive counts) denotes the results with the clipping correction. For both sides, we show the different DC/AC ratios with different colors, to highlight that the improvement spans all DC/AC ratios.

In Fig. 9, the same data is presented in a different fashion: the data is split according to different DC/AC ratios, and the distributions are represented using violin plots, with the non-corrected and corrected results shown side-by-side and in different colors (red and blue respectively). The central points denote the median of the distribution, and the error bars show the 25% and 75% quantiles. This second plot allows us to better quantify the improvement that is achieved by the clipping correction.

It can be noted again that both the spread and the bias of the distributions are improved. The mean and the standard deviation for the different DC/AC ratios are presented in Table I.

Table I: Mean and standard deviation of the distribution of the errors among the yearly AC generation results, comparing the hourly simulation with an without correction, to the minute simulation.

AC power error	Mean [%]		Standard Deviation [%]	
	Uncorr.	Corr.	Uncorr.	Corr.
Global	-2.61	-0.77	1.5	0.72
DC/AC = 1	-1.04	-0.92	0.73	0.72
DC/AC = 1.33	-2.1	-0.61	0.84	0.64
DC/AC = 1.67	-3.34	-0.64	1.09	0.67
DC/AC = 2	-3.96	-0.88	1.24	0.81

The improvement in the AC generation result distribution is negligible for the DC/AC ratio 1 but is important for the higher DC/AC ratios, with the most improvement for the highest ratio. Still there remains an overall bias around 0.6%, which is not removed by the correction. As explained before, the differences in AC generation also include other sources of bias. Our preliminary studies point towards a bias introduced by the transposition from the horizontal to the tilted plane.

In summary, we can say that the model captures most of the error, as the AC generation bias can be reduced below the 1% level for all DC/AC ratios. The correction of the remaining bias will be the subject of a subsequent study.

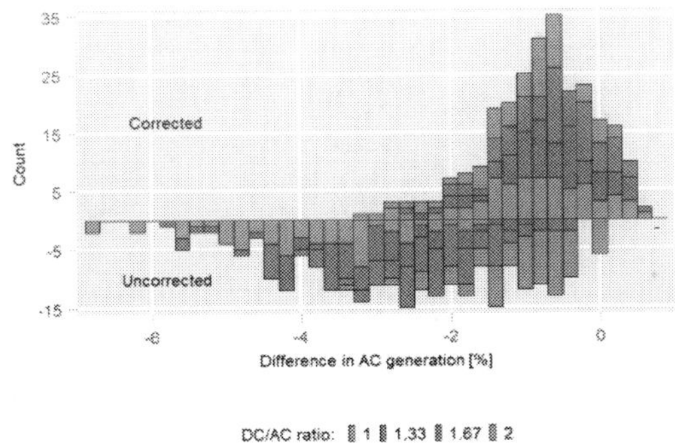

Figure 8: Histogram of the distribution of errors comparing the yearly AC generation from the hourly simulations with that of the minute simulations. All cases (site, orientation, and DC/AC ratio) have been considered. The distribution of errors without correction is shown as negative counts, the one with correction as positive counts. Different DC/AC ratios are shown in different colors.

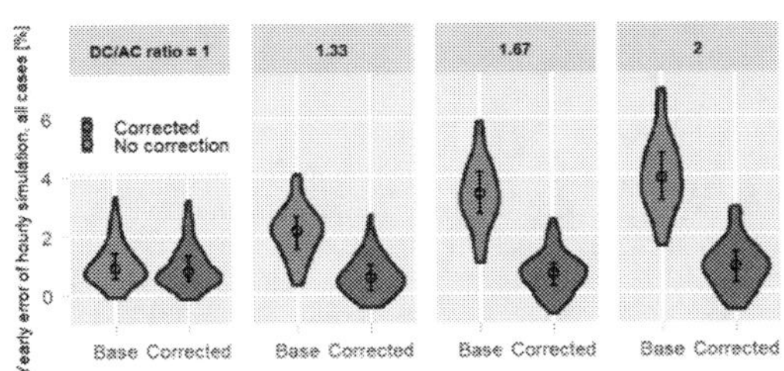

Figure 9: Violin plot of the distribution of errors, comparing comparing the yearly AC generation from the hourly simulations with that of the minute simulations. The different DC/AC ratios have been separated and corrected vs. uncorrected results are shown side by side. Central points denote the median of the distribution and and the error bars show the 25% and 75% quantiles.

6 CONCLUSION

When simulating PV systems with an hourly resolution, some effects with a shorter time scale may be missed, potentially causing a bias in the simulation results. One of the most important higher resolution effects that may cause such a bias is the clipping losses due to sub-hourly clipping, which generally leads to overestimates in the simulated generation of PV systems. Several studies have found that this effect may lead to up to about 5% overestimates of the yearly generation for high DC/AC ratios, and a few percent for more regular systems.

Since performing higher-resolution simulation comes at a greater cost in terms of performance, we propose an approach to correct the hourly resolution clipping by means of few coefficients extracted from minute-resolution irradiance data. We utilize the near linearity of the PV generation in terms of irradiance to reorder the minute-data without affecting the results significantly. This allows us to estimate the distribution of irradiance values within each hourly step using a simple linear approximation with three coefficients, and ultimately model the missed clipping loss using this linear estimate.

We applied the model to four sites with minute-level data and scenarios spanning different orientations and DC/AC ratios. To analyze the results, we compared both the uncorrected and corrected results to a minute-resolution simulation. The results show that our approach corrects both the spread and the bias of the hourly clipping loss distribution, by a factor of two and ten respectively. We also compared the yearly results, showing that the bias error on the AC generation is on average reduced to the sub-percent level.

Further investigation suggests that the remaining bias may be related to effects other than the missing sub-hourly clipping. Such other effects have been shown to exist [22, 23, 24], for example the shadings and transposition evaluation. A bias comparable to the remaining error at AC generation is already observed after application of the transposition model. Correcting this will need further study. In parallel to this, our model still has room for improvement. Furthermore, we would like to validate the model for tracking systems as well as more climates and weather data sources.

The model presented in this publication will be made available in a future PVsyst version and be improved further.

7 REFERENCES

[1] K. S. Anderson, W. B. Hobbs, W. F. Holmgren, K. R. Perry, M. A. Mikofski and R. A. Kharait, "The Effect Of Inverter Loading Ratio On Energy Estimate Bias," in *2019 46th IEEE Photovoltaic Specialists Conference (PVSC)*, 2019.

[2] B. Burger and R. Rüther, "Inverter sizing of grid-connected photovoltaic systems in the light of local solar resource distribution characteristics and temperature," *Solar Energy,* vol. 80, p. 32–45, January 2006.

[3] H. Capdevila and T. Gerstmaier, "Impact of Temporal and Spatial Discretization on the Agreement between Measured and Modeled Performance of Large Photovoltaic Power Plants," *31st European Photovoltaic Solar Energy Conference and Exhibition; 2150-2156,* 2015.

[4] D. Cormode, N. Croft, R. Hamilton and S. Kottmer, "A method for error compensation of modeled annual energy production estimates introduced by intra-hour irradiance variability at PV power plants with a high DC to AC ratio," in *2019 IEEE 46th Photovoltaic Specialists Conference (PVSC)*, 2019.

[5] R. Kharait, S. Raju, A. Parikh, M. A. Mikofski and J. Newmiller, "Energy Yield and Clipping Loss Corrections for Hourly Inputs in Climates with Solar Variability," in *2020 47th IEEE Photovoltaic Specialists Conference (PVSC)*, 2020.

[6] J. Luoma, J. Kleissl and K. Murray, "Optimal inverter sizing considering cloud enhancement,"

Solar Energy, vol. 86, p. 421–429, January 2012.

[7] A. Parikh, K. Perry, K. Anderson, W. B. Hobbs, R. Kharait and M. A. Mikofski, "Validation of Subhourly Clipping Loss Error Corrections," in *2021 IEEE 48th Photovoltaic Specialists Conference (PVSC)*, 2021.

[8] I. Meyer, A. Salles, J. Tellez Mejia, V. Vorasitchai, S. Chitiva Bocanegra and N. Cherdsanguan, "Optimisation of DC/AC Ratio in Different Climatic Zones: A Technical Analysis on a Sub-Hourly Level," *37th European Photovoltaic Solar Energy Conference and Exhibition; 1610-1615,* 2020.

[9] J. Good and J. X. Johnson, "Impact of inverter loading ratio on solar photovoltaic system performance," *Applied Energy,* vol. 177, p. 475–486, September 2016.

[10] R. Darawali, L. D. Tafur and A. Benson, "Advanced Solar PV Energy Pre-Construction Losses: Losses due to uneven terrain, tracker wind stowing, and sub-hourly clipping error," in *Sandia PV Performance Modeling Collaborative*, 2022.

[11] J. O. Allen, W. B. Hobbs and M. Bolen, "Classification Method to Predict the Effect of Short-Term Inverter Saturation on PV Performance Modeling," in *Sandia PV Performance Modeling Collaborative*, 2022.

[12] K. Anderson and K. Perry, "Estimating Subhourly Inverter Clipping Loss From Satellite-Derived Irradiance Data," in *2020 47th IEEE Photovoltaic Specialists Conference (PVSC)*, 2020.

[13] D. A. Bowersox and S. M. MacAlpine, "Predicting Subhourly Clipping Losses for Utility-Scale PV Systems," in *2021 IEEE 48th Photovoltaic Specialists Conference (PVSC)*, 2021.

[14] K. Bradford, R. Walker, D. Moon and M. Ibanez, "A Regression Model to Correct for Intra-Hourly Irradiance Variability Bias in Solar Energy Models," in *2020 47th IEEE Photovoltaic Specialists Conference (PVSC)*, 2020.

[15] M. Ghiz, "DNV's machine learning model compensates for errors introducedby hourly PV modeling," in *Sandia PV Performance Modeling Collaborative*, 2022.

[16] A. Mueller and E. Greeno, "Estimating subhourly clipping from hourly data: a simplified approach," in *Sandia PV Performance Modeling Collaborative*, 2022.

[17] T. Townsend, K. J. Sauer, B. Littmann and B. Grenko, "THE TRIPLE-C METHOD FOR CORRECTLY SIMULATING PV CLIPPING LOSS," in *Sandia PV Performance Modeling Collaborative*, 2022.

[18] T. Stoffel and A. Andreas, *NREL Solar Radiation Research Laboratory (SRRL): Baseline Measurement System (BMS); Golden, Colorado (Data),* Not Available, 1981.

[19] F. Vignola and A. Andreas, *University of Oregon: GPS-based Precipitable Water Vapor (PWV),* NREL-DATA (National Renewable Energy Laboratory - Data (NREL-DATA), Golden, CO (United States)), 2013.

[20] K. Olson and A. Andreas, *Natural Energy Laboratory of Hawaii Authority (NELHA): Hawaii Ocean Science & Technology Park; Kailua-Kona, Hawaii,* NREL-DATA (National Renewable Energy Laboratory - Data (NREL-DATA), Golden, CO (United States)), 2012.

[21] NREL, *University of Louisiana at Lafayette,* NREL-DATA (National Renewable Energy Laboratory - Data (NREL-DATA), Golden, CO (United States)).

[22] S. Ransome and P. Funtan, "Why hourly averaged measurement data is insufficient to model PV system performance accurately," *20th European Photovoltaic Solar Energy Conference,* January 2005.

[23] A. Neubert, M. Hamer, P. Rainey and M. A. Mikofski, "The Impact of Tracking Algorithms and Time Resolution on Energy Yield Modelling of Single Axis Tracker Systems," *38th European Photovoltaic Solar Energy Conference and Exhibition; 1290-1293,* 2021.

[24] C. W. Hansen, J. S. Stein and D. Riley, "Effect of time scale on analysis of PV system performance," *SANDIA Report,* 2012.

GaN BASED MODULE-INTEGRATED, HIGH-EFFICIENCY DC/DC OPTIMIZER FOR MAXIMIZING THE YIELD OF THE LARGE PHOTOVOLTAIC POWER PLANT

Van Sang Nguyen*, Stephane Catellani*, Anthony Bier*, Tony Delaplagne*, Sylvain Lespinats*
Patrick Merhej**, Carlotta Greco**, Fabrizio Bizzarri**

*Univ. Grenoble Alpes, CEA, Liten, INES, 73375 Le Bourget du Lac, France
**Enel Green Power SpA, 1 Via Giosuè Carducci, Milano 20123, Italy

ABSTRACT: In a photovoltaic (PV) power plant, due to the low-voltage standard regulation, the maximum open-circuit voltage of a string of PV module cannot exceed 1500 VDC. This work deals with the design, characterization and optimization of PV modules integrating a 400 W DC-DC converter (optimizer). These PV modules with integrated power electronics will be implemented in a 1500 V PV string. These optimizers will allow regulating the output voltage of each PV module to a value inferior to the open-circuit voltage. Consequently, we can increase the number of PV modules connected in series in a single string and raise the power of the string. In this work, two compact demo-boards of a 400 W hardware are developed for complete characterizations in different systems configurations using Silicon based transistors for the first one and Gallium Nitride HEMT (high electron mobility transistor) for the second one; respectively. Finally, this works shows a comparison of the electrical behaviour of both hardware solutions.

Keywords: Bifacial, DC-DC-Converter, Energy Performance, Power Optimizer, PV Module

1 INTRODUCTION

Thanks to its high performance, the wide-band-gap (WBG) semiconductors such as Gallium Nitride (GaN) and Silicon Carbide (SiC) devices become the key components for the efficient conversion systems connected to renewable energy sources [3-4], in particular, for photovoltaic installations where the high efficiency is a primordial requirement. Going along with the emergence of the bifacial PV module with a higher power, the integrations of the power electronics such as the micro-inverter and the DC/DC optimizer become one of the mainstreams at this moment [5-6], using WBG devices as GaN devices in these situations brings a huge potential for new performant generation of PV power plants.

This work aims to increase the installed power of a 1500 V string within a large PV power plant. The main idea is the integration of the GaN based power electronics elements inside the junction box to step down the output voltage of each PV module before connecting in series within a string.

By using the closed-loop regulation of DC-DC optimizer, the output voltage of the string is always under 1500 V_{DC} and the output power is increased thanks to the higher number of the installed modules in a single string. In addition, with the local MPPT (maximum power point tracking) it is possible to extract the maximum available power from each PV module independently from the electrical conditions of others PV modules belonging to same string.

Finally, real-time measurements of current and voltage of each "smart PV module" are available and collected by a central unit. These data could be used to simplify the management of the power plant for real-time diagnostic identifying the position of under-performed PV modules using same hardware.

Experimental results show two demo-boards, which are developed at laboratory level using Si & GaN transistors. Electrical characterization & efficiency comparison are fully presented in this article.

2 SMART PV STRING

2.1 More powerful 1500 V_{DC} PV string

The approach re-determines the maximum number of PV modules per string. With the conventional approach, the maximum voltage of a string (1500 V_{DC} in this case) and the open-circuit voltage (V_{OC}) of a single module give the maximum number of PV module, N_{S1}. The open circuit voltage of the last generation such as the bifacial PV module (as 3S Dual 72N) is close to 50V, and then the maximum number of a string can be calculated as equation (1):

$$N_{S1} = \frac{1500\ V_{DC}}{V_{OC}} = \frac{1500\ V_{DC}}{50\ V_{DC}} \approx 30 \qquad (1)$$

With this new approach, each PV module is integrating a smart electronics board (S.E.B); this S.E.B is a compact high efficiency synchronous buck converter as Figure 1 with multi-functions: voltage regulation, MPPT, and diagnostics.

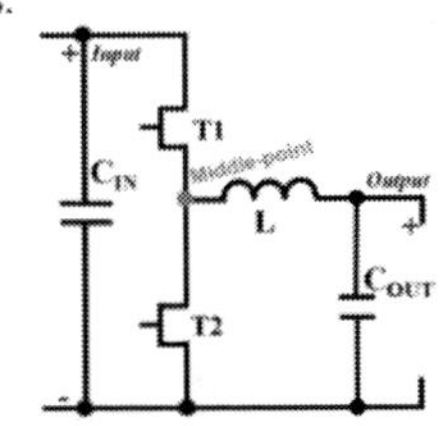

Figure 1: Topology of S.E.B, a synchronous buck DC/DC converter

Figure 2 shows a smart PV string where the S.E.B is integrated at each PV module to manage its output voltage.

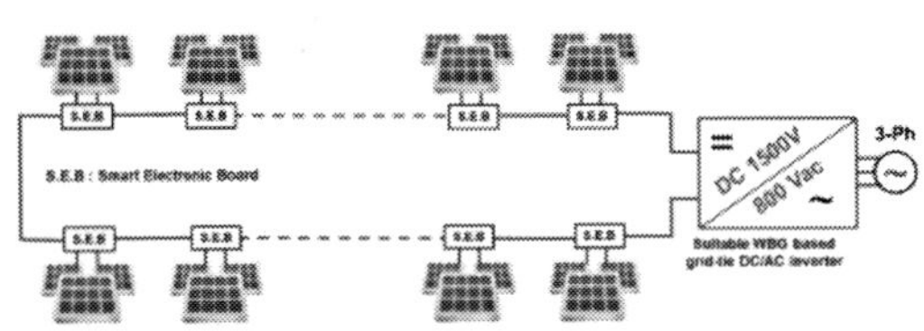

Figure 2: Configuration of a 20 kW of a smart string with the integration of S.E.B

The maximum voltage of a string and the output voltage of each S.E.B ($V_{OUTPUT\text{-}DC}$) give the maximum number of PV module, N_{S2}. Since the S.E.B is a step-down topology, the output voltage of each "smart module" is always smaller than the working voltage of the module. However, to optimize the energy conversion's efficiency of each S.E.B, this output voltage should be not too small; for example, a value twice smaller than the working voltage of the module, is considered in equation (2)

$$N_{S2} = \frac{1500\ V_{DC}}{V_{Output-DCDC}} = \frac{1500\ V_{DC}}{25\ V_{DC}} \approx 60 \qquad (2)$$

As the number of PV modules with S.E.B connected in series is doubled thanks to this approach, the string is twice more powerful than a conventional string without S.E.B. Consequently, it will be possible to reduce the quantity of electrical components in a PV power plant as well to raise the power density of the DC-AC converter units. To make this solution more relevant in term of performance, the optimizations of S.E.B are carefully taken into account with the criteria of the performance, the compactness, the reliability and the cost. In this configuration, as in Figure 2, smart string operates with a suitable grid-tie inverter.

Without the DC/DC optimizers, when one or more solar modules are under-performing, they impose a general reduction in the current over the entire string due to the series connection. Maximum power points are decreased not only in the under-performing modules but also on the modules; this degrades the power of entire string. In the similar scenario, DC/DC optimizers with their closed-loop control rebalance the output voltage of "smart modules" in series connection keeping each module at its proper MPP.

On another side, the integrations of electronics and power electronics to substitute traditional junction box boost the digitalization of the PV power plant and could simplify the maintenance through artificial intelligence.

3 Si & GaN BASED S.E.B DESIGNS

In the first step, the Silicon based S.E.B is built up to evaluate and validate the full-custom controller as reported in Figure 3.

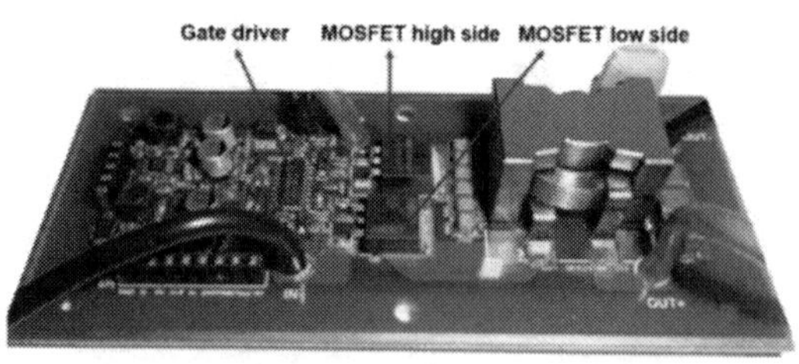

Figure 3: Si based S.E.B of 400W with the smart functions (120 x 55 x 22 mm)

In this design, the currents and voltages of S.E.B are collected in order to monitor the function state of the power electronics part and the electrical output behaviour of PV module.

In a hard switching topology, as the synchronous Buck converter of the PV optimizer, high-speed GaN transistors already implemented in several designs [7-8] are used. Decreasing the rise time and the fall time could reduce the switching losses directly.

With similar controller and auxiliary circuitry, the power part of Si based S.E.B has been modified. Based on the consideration of the ON-state resistance values of the 100V GaN transistors compare to the 100V Si MOSFETs, an IC GaN (an 100V GaN HEMT with an integrated gate driver) and a second 100V GaN HEMT replace the 100V Si MOSFETs and their gate driver as the GaN design in Figure 4. The integration of the gate driver simplifies and ensure the design steps of the optimizer where the GaN devices are well known as extreme high-speed devices.

In both Si based and GaN based DC/DC optimizers, measurements of the input/output currents and voltages, the control signal of the power elements, the checkpoint of the auxiliary power supplies can be obtained for demo-board analysis and diagnostics evaluation at module level.

Figure 4: New design of a GaN based S.E.B (94 x 55 x 22 mm)

Figure 5 shows the main waveforms of the GaN based S.E.B where we can see the output voltage, the current of the inductance and the voltage on the middle point of the synchronous buck converter.

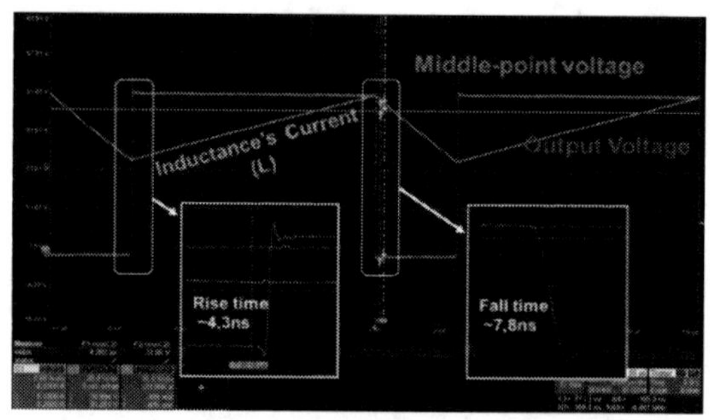

Figure 5: GaN based S.E.B, main waveforms

Thanks to the high speed of GaN devices, the rise time and the fall time of the voltage on the middle point are 4.3 ns and 7.8 ns, respectively. Consequently, the hard switching losses of the transistors of this buck converter decrease. In the case of Si based S.E.B, these rise time and the fall time of the Si based transistors are above 20 ns.

Beside the GaN transistors, the output inductance plays an important role on the global loss distribution of S.E.B. Figure 6 shows the characterized incremental inductance value as the function of the applied time-voltage-integral under DPG10 choke tester; the value of the inductance begins to collapse at 300 µVs. The output inductor should not be used for a higher applied impulse voltage; otherwise, the performance of the systems would be decreased with the collapsed value of the inductance.

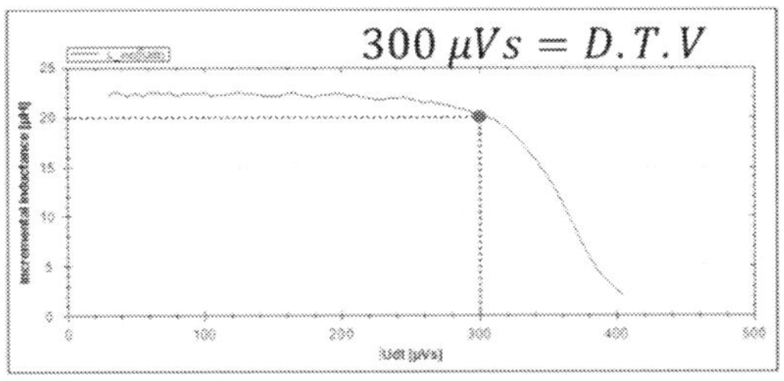

Figure 6: Inductance's value under choke tester

This value corresponds to: $300\ \mu Vs = D \times T \times V$

Where, D is the duty cycle of the Buck converter, this value is equal to the output voltage divided by the input voltage. T is the maximum period of the applied pulse. V is the applied voltage on the inductance.

The minimum switching frequency is driven as reported in equation (3):

$$f_{min} = \frac{1}{T} = \frac{D \times V}{300\ \mu Vs} = \frac{\frac{32\ V}{44\ V} \times 44\ V}{300\ \mu Vs} \approx 106\ kHz \qquad (3)$$

Theoretically, the efficiency of the S.E.B could be kept increasing by a slower working frequency, however, the current output inductance needs to be replaced if the S.E.B performs with a switching frequency lower than about 100 kHz.

In Figure 7, the comparison of the efficiencies of the Si based and GaN based 400 W S.E.B are shown. The highest efficiency of Si based S.E.B is 98% where the Si MOSFETs switch at 135 kHz. With a similar switching frequency, GaN based S.E.B increases the efficiency to 99% with identical passive power components.

The working frequency has been varied to modify the switching losses of the GaN transistors in order to find out the most performant working-point of the S.E.B. At the working frequency of 100 kHz, the highest efficiency of the experiment is 99.1% with a CEC efficiency of 99%.

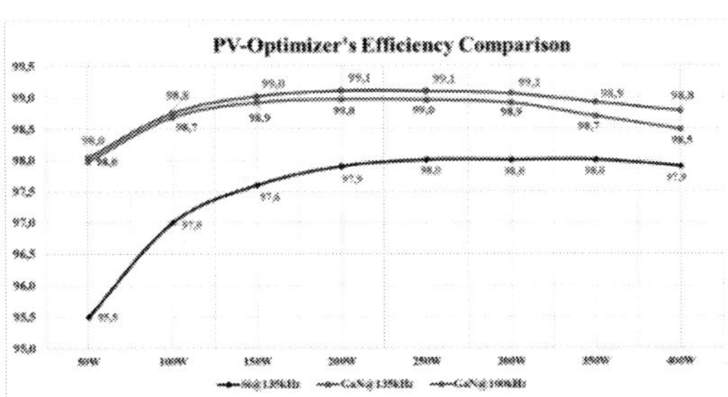

Figure 7: Efficiencies of Si based S.E.B and GaN based S.E.B

Beyond the efficiency, it is possible to modify other parameters of the solar power system by reconfiguring the frequency of the GaN devices. For instance, the working frequency of GaN devices could be increased, keeping under control the amount of losses respect to Si-based solution, it is possible to raise the reliability of the solution by reducing the quantity of passive components such as the capacitors.

In order to estimate the losses due to mismatches, we simulated a set of PV modules with the same IV curves (based on measurements) except for short-circuit current (I_{SC}) and open circuit voltage (V_{OC}) where difference follow a normal distribution. The quantity of mismatch tends to increase with the coefficient of variation (CV) and with the number of module in the string. Mismatches would decrease the production of the system if plugged on a unique MPP optimizer. These losses can be taken into account for comparison with the proposed solution. In the test system, the level of mismatch is of course very low. For measured CV, the relative Pmax value is consequently close to 1 and the losses due to mismatches are negligible (fig. 8, blue curve). However, we can observe that the efficiency quickly decreases with CV. In a likely case, CV could reach higher values (10, 25 or 50 typically). We estimate that CVx50 could be a credible in a large PV plant. In such a case, the losses due to mismatches would be clearly higher than S.E.B losses.

As we can see the theoretical estimation in Figure 8, using the Si based S.E.B could compensate the mismatch of CVx50 thanks to its efficiency of 98%. GaN based S.E.B with the efficiency up to 99.1% could mitigate the mismatch; bring it very close to the ideal case and eventually, optimizes the yield of the PV plant.

It must be emphasize that our estimation of mismatch losses is probably under-estimated. Indeed, the variation of modules specifications in real life may be higher; the shape of IV curves may show other variations than I_{SC} and V_{OC}; shading, dust, temperature variation, defaults, aging, etc.. may occur.

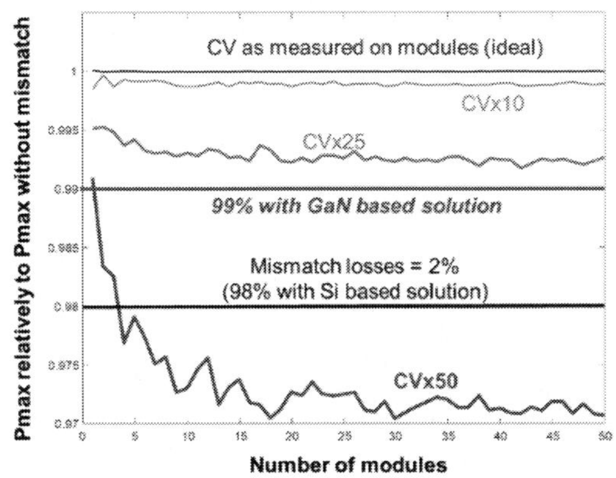

Figure 8: Impact on the theoretical estimation of mismatch losses from an IV curve with different S.E.B

4 SYSTEMS IMPLEMENTATION

Following up the designs of different versions of S.E.B, step-by-step implementations are used for the demonstration of this approach. In laboratory, the first systems implement Si based S.E.Bs in order to evaluate and validate the full-custom controller in a small-scale number of modules. With a stand-alone characterization, the output voltage of Si based S.E.B is fixed at 32 V where the MPP voltage of the module is 44 V, this S.E.B reaches its optimized performances.

Figure 9 shows a group of four bifacial PV modules in series connection through four Si based S.E.B. On this demonstration, 36 bifacial modules are implemented and an electronics load is used at string output level. In the next configuration, a suitable grid-tie DC/AC inverter, able to regulate its input voltage up to $1500\ V_{DC}$, will be implemented.

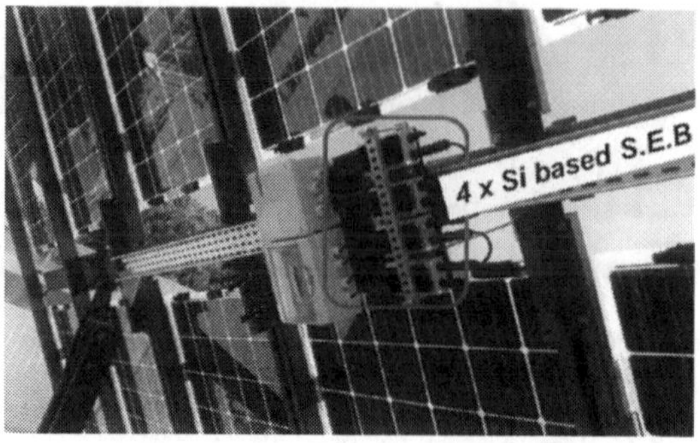

Figure 9: Si based S.E.Bs with four bifacial PV modules

Figure 10 shows the voltage, the current, the power and the irradiance along a day of a demonstration. In this experiment, an unexpected mechanical issue of the tracking occurs as shown on the power and the irradiance curves. However, the strings output power in the range of 10.8 kW follows the irradiance and even glides the fluctuations of the irradiance.

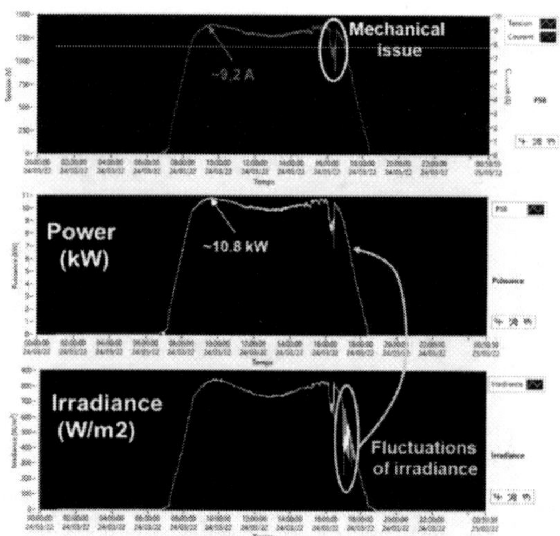

Figure 10: Si based S.E.B, its measured efficiency and the demonstration with 4 bifacial PV modules

Ongoing activities aim to increase the number of modules in a smart string to at least 50 modules with GaN based S.E.B. Another topic also gives the possibility to implement high voltage GaN devices in a suitable grid-tie DC/AC inverter for the configuration of this smart string [9]. Finally, a power performance analysis will be evaluated between two power plants: one arranged with by GaN based S.E.B and GaN based DC/AC, and the other one will be a conventional installation.

5 CONCLUSIONS

This work presents a new concept of PV string with the integration of GaN based power electronics in order to regulate the output voltage, to maximize the extracted power from PV modules and to effect the electrical diagnosis at module level.

By using the integrated power electronics, the length of a string is increased due to the possibility to raise the number of the PV modules connected in series. The harvested energy of the PV plant is increased thank to the fact that every PV modules will work at its proper MPP and the efficiency, maintenance activity will be optimized due to the localization of under-performed PV modules.

High-speed GaN devices plays an important role for improving the parameters of the PV power plant such as the efficiency and the reliability; it could be a starting point to enable the use of optimizers in commercial or in industrial applications.

ACKNOWLEDGMENT

This work was realized in the collaborations between CEA-INES and ENEL Green Power.

This work was also supported by the French National Program "Programme d'Investissements d'Avenir – INES.2S" under Grant ANR-10-IEED-0014-01.

REFERENCES

[1] L. Scarpa, G. Chicco, F. Spertino, P-M. Tumino, M. Nunnari "Technical Solutions and Standards Upgrade for Photovoltaic Systems Operated Over 1500 Vdc" 2018 IEEE 4th International Forum on Research and Technology for Society and Industry (RTSI), 2018

[2] J. He, A. Sangwongwanich, Y. Yang, F. Iannuzzo "Performance Comparison of PV Inverter Systems Considering System Voltage Ratings and Installation Sites" 2021 IEEE Applied Power Electronics Conference and Exposition (APEC), 2021

[3] Yole report on Power SiC 2020: Materials, Devices and Applications, Yole 2020

[4] Yole report GaN power 2021: epitaxy, devices, applications and technology trends, Yole 2021

[5] VDMA "International technology roadmap for photovoltaic ITRPV", 10th edition, March 2019

[6] V-S. Nguyen, S. Catellani, A. Bier, J. Martin, H. Zara, and J. Aimé, "A compact high-efficiency GaN based 400W solar micro inverter in ZVS operation," PCIM Europe digital days 2020, Jul. 2020

[7] P. Chen, T. Liu, Y. Cheng, H. Wu, and J. Zhu"A GaN-based High Power Density Power Optimizer for Solar-Powered Aircraft Applications" 2021 IEEE Workshop on Wide Bandgap Power Devices and Applications in Asia (WiPDA Asia), Aug 2021

[8] "A GaN-Based High-Efficiency Solar Optimizer with Reduced Number of Power Devices"

[9] V-S. Nguyen, A. Bier, H. Es-seghier, U. Soupremanien, G. Delette, S. Catellani "A 20 kW, 3-level flying capacitor 1500 V inverter with characterized GaN devices for grid-tie applications" EPE ECCE Europe 2022, Sept 2022

DATA-DRIVEN PV SYSTEM ENGINEERING: COMPREHENSIVE MODELLING AND SIMULATION OF A GRID-CONNECTED THREE-PHASE INVERTER

Valeriya Titova[1], Leon Lahrmann[1], Andre Avila Alves[1], Frerk Haase[1], Zouhair Khadiri-Yazami[2], Martin Lapke[1]
[1]Faculty for Engineering and Computer Science, Hamburg University of Applied Sciences, Berliner Tor 5, 20099 Hamburg, Germany
[2]SMA Solar Technology, Sonnenallee 1, 34266 Niestetal, Germany

ABSTRACT: In a data-driven modelling approach, the intelligent integration of data between physical and virtual entities of a photovoltaic (PV) system can facilitate innovation, control and optimization of the entire PV power plant. In an inverter, the insulated-gate bipolar transistor (IGBT) power electronics are used to distribute and convert the direct current from the PV module to alternating current. Due to the demand for performance improvements, the need to model these IGBT stacks is constantly increasing [1, 2]. Consequently, to reduce thermal resistance and improve the reliability of power electronics, new data points such as thermal loading and internal humidity are increasingly moving into the scientific focus [3]. In this contribution, we pursue two approaches towards anomaly detection in the operation of a PV plant. A comparison-based approach, analyzing the operation-induced data of an inverter to a physically motivated inverter model, and a failure-motivated approach using numerical modelling methods to reveal the hidden characteristics for latter data analysis. We present a MATLAB/Simulink model of a three-phase grid-connected inverter. We demonstrate the responsiveness of our model by changing the polarity in the active and reactive current control. This model shall serve as an idealized digital twin of the inverter to aid in anomaly detection within measured inverter data. Furthermore, we present initial ideas on a failure-motivated modelling approach to improve our model's accuracy and enable optimized predictive maintenance algorithms. The use of hidden characteristics, such as IGBT junction temperature and power module interior humidity can serve as markers for failure detection in the operation of a PV system. Therefore, we have analyzed initial power loss calculations of an IGBT, which serve as a basis for thermal loss simulations to predict overheating of the switch components within the inverter. We examine the correlation between modelled and measured temperatures of the IGBT stacks. In addition, we propose a humidity model to prevent humidity-induced failures in power electronics.

1 INTRODUCTION

To optimize current predictive maintenance approaches (e.g. to detect anomalies in data records of solar inverters for failure prediction of these power electronic devices), this study aims to provide model-based improvements by developing a data-driven model of an IGBT-based inverter and comparing the modelled IGBT temperature stress to that of physical PV inverters in operation.

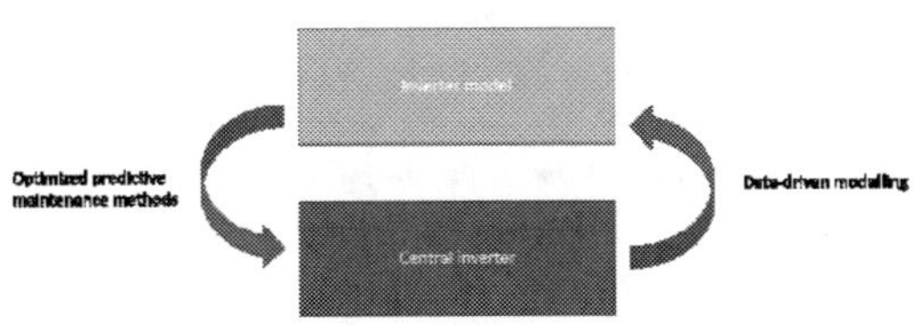

Figure 1: Workflow establishing a digital twin of a grid-connected three-phase inverter.

We pursue two approaches towards anomaly detection in the operation of a PV plant. First, we present a comparison-based approach, where the aim is to analyze the quality of data generated by a physics-motivated inverter model in relation to measured real-world data. Figure 1 shows an establishing process of a digital twin of a grid-connected three-phase inverter. In this study, we examine MATLAB/Simulink modelling of a grid-connected three-phase inverter. This enables the implementation of various modern predictive maintenance strategies to reduce the operation and maintenance costs and increase the performance of the PV system components and thus of the entire PV system.

With the second approach, we aim to establish failure models using analytical and numerical modelling methods to reveal hidden impact of unapparent indicators, such as junction temperature and humidity content in an inverter.

The analysis of the recorded anonymized data of a central inverter shows a correlation between modelled and measured IGBT temperatures of an inverter. Including the influence of humidity in the analysis might enable the identification of the electronic components aging and a reduction of failures during operation.

2 PHYSICALLY MOTIVATED MODELLING OF AN INVERTER

Our model aims for detection of thermally induced anomalies and early detection and prevention of PV inverter failures. To investigate the effects of inverter internal temperature on possible loss mechanisms, a PV system with detailed inverter features is modelled in Simulink using Simscape components.

Figure 2 shows a schematic representation of a three-phase inverter model developed in the program MATLAB/Simulink. The model consists of the grid-connected inverter and six regulation blocks to control (proportional-integral (PI)-controller) this three-phase inverter in the dq reference-frame: phase locked loop (PLL) block, voltage dq transformation, current dq transformation, two blocks for setting the active and reactive current reference, respectively, and the retransformation to the natural reference frame in order to generate the pulse-width modulation (PWM) signal.

The input of the modelled three-phase inverter consists of a direct voltage source and can later be replaced with a PV array. The modelled inverter is connected to the low voltage (460 V) AC test grid.

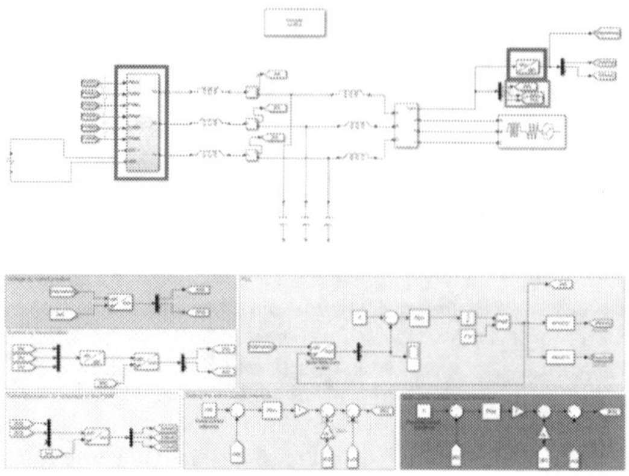

Figure 2: Top: A simplified Simulink model of a three-phase grid-connected inverter in MATLAB/Simulink. A box with the red frame indicates a subsystem, which includes the IGBT bridge. The output of the voltage phase measurement (green box) is used by the Clarke-transformation (blue box) and the PLL-subsystem to synchronize the inverter to the grid. Bottom: Regulation blocks for inverter control.

Figure 3 illustrates the performance of the grid-connected inverter model by controlling the energy flow. Figure 3A shows the three voltage phases of the grid. Figure 3B displays the three current outputs generated by the inverter. After a short settling time, which corresponds to the settling time of the PI controller, the signals of current and voltage are in phase. Black vertical lines indicate timestamps for orientation and simple comparison of both signal outputs. The first black vertical line at a time of 0.05 seconds shows that the two blue sine curves of current and voltage are in phase. Beginning with a time of about 0.08 seconds, the sign of the active current controller changes from - to + (triggered by the input of the user). After a short settling time, the blue sine curves of current and voltage are 180° out of phase, shown by the second black vertical line, clearly demonstrating the controllability of our model.

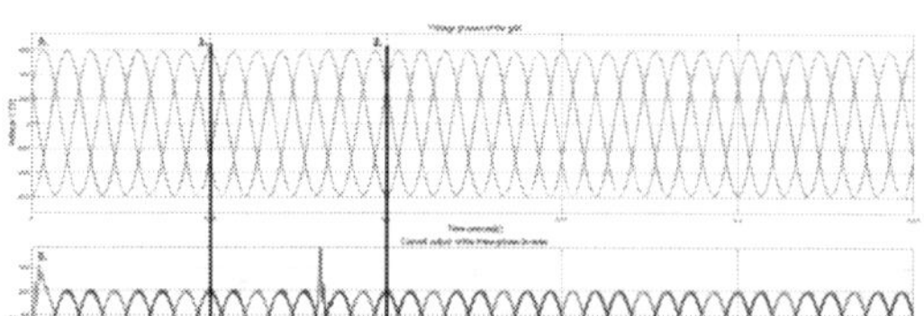

Figure 3: Output signal of a three-phase grid-connected inverter: A (top): voltage, B (bottom): current.

Calibrating the model based on measured values of real devices (temperature, current-voltage measurements) installed on the anonymized large-scale photovoltaic power plants is currently our focus to further specify and evaluate our inverter model. By using recorded data over several years, we can classify anomalies by comparing yield predictions with actual measurements. Challenges due to timescale difference of the model to the large observation scale are under investigation. The evaluation of deviations to expected inverter behavior are ongoing.

3 FAILURE MOTIVATED MODELLING APPROACH

The fact that power electronics act as a heat source, metals inside an inverter act as a temperature reservoir and polymers/silicones in the interior of power electronic modules act as a moisture reservoir gives a lot of potential for a failure-motivated approach. The internal temperature and internal humidity in the inverter are two of the most common causes of corrosion of the components and aging of the power electronics. The relationship between the ambient conditions such as temperatures and humidity inside the inverters brings us to the setup of the transfer functions. Since power electronics modules are the source of deterioration of the overall system, sensor data, e.g. the temperature at the module adaptor and the ambient relative humidity, provide input for a failure-motivated analysis. Our goal is to use inverter modelling to investigate the effects of temperature recorded by sensors in a central inverter on system dynamics and power loss in the AC area.

3.1 Temperature-driven modelling

We start our modelling approach by analyzing the existing dataset by means of implementing an equivalent thermal circuit model, commonly used [4].

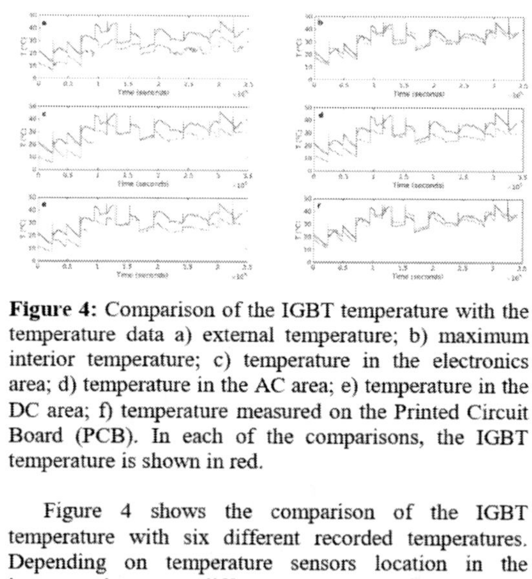

Figure 4: Comparison of the IGBT temperature with the temperature data a) external temperature; b) maximum interior temperature; c) temperature in the electronics area; d) temperature in the AC area; e) temperature in the DC area; f) temperature measured on the Printed Circuit Board (PCB). In each of the comparisons, the IGBT temperature is shown in red.

Figure 4 shows the comparison of the IGBT temperature with six different recorded temperatures. Depending on temperature sensors location in the inverter, there are different levels of offsets in the temperature readings. However, it is noticeable that the basic course of the curves is almost identical. From the comparison of the IGBT temperature and the external temperature T_{EXT}, a constant offset of about 10°C is determined (see Fig. 4 a).

In order to model the relationship between the power losses and the temperature measured in the insulated-gate bipolar transistor (IGBT), we model the IGBT temperature as the sum of the external temperature T_{EXT} added with a constant and a matched fraction of the power dissipation P_{LOSS}:

$$T_{IGBT} = (T_{EXT}+C_1)+(P_{LOSS} \cdot C_2) \qquad (3.1)$$

The coefficients C_1 and C_2 are determined by matching the IGBT and external temperature. The coefficient C_2 represents the thermal resistance R_{IGBT_EXT}, whereby C_1 reflects a temperature offset.

The matching is based on the complete dataset from one inverter for eight randomly selected non-contiguous days. For the modelling, equation (3.1) is supplemented with the coefficients C_1=10 and C_2=0.65. The value of C_2=0.65 is obtained by fitting to the peak values, testing the range between 0.5 and 0.75.

Figure 5 shows the comparison of the modelled IGBT temperature with the measured one. The calculated temperature is composed of the external temperature corrected by the coefficient $C_1 = 10$ and the power dissipation multiplied by a constant correction factor $C_2 = 0.65$.

The temperature rises to the similar maximum. The modelled temperature's increases as well as fall is much steeper than the measured one. This is due to the material properties of the installed electronic components and the transient behavior of the temperature. These assumptions have to be investigated further.

By adjusting for extraneous influences, the determination of power dissipation in the semiconductors (IGBT) can thus be made more precisely, since the modelling of the IGBT temperature is not oriented to the measured values recorded by sensors, but to output current variables. In addition, the junction temperature of the power modules as an important parameter can be analyzed in a similar fashion and will be modelled in the future.

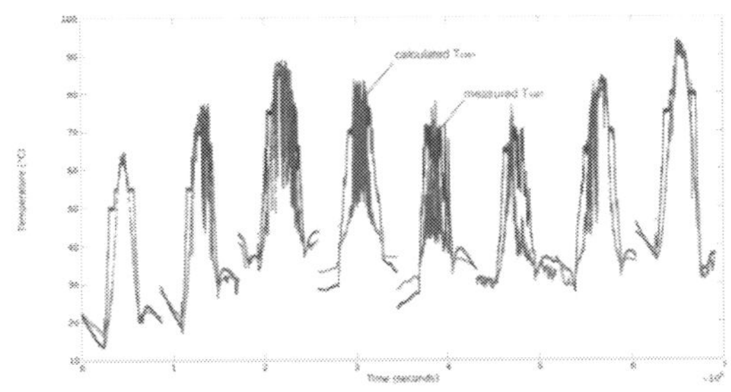

Figure 5: The IGBT temperature calculated using equation (3.1) (red line) compared with the actually measured IGBT temperature (blue line).

3.2 Humidity modelling

A diffusion model can be used to reveal the influence of external humidity on power electronics. Based on advanced numerical methods, the commercial software COMSOL Multiphysics offers an opportunity to model humidity transfer in the power electronics module inside a central inverter, and to extend it to consider temperature influences by means of the *finite element method* (FEM).

In this study, we reproduce an experiment described in [5] in a simulative way. Figure 6 shows numerical results of humidity distribution of an experimental setup consisting of a beaker with an evaporated area of A=3.63 cm^2 (xy-plane) filled with silicone (specific polymer with humidity absorbing properties) of the thickness of d_{gel}=33 mm (z-plane). The beaker is enclosed with an empty volume of V_{air}=13.6 cm^3 inside a glass tube, where a humidity sensor and a temperature sensor are placed. Such an experimental setup provides a one-dimensional moisture flow through the area A and enables the simulation of the water vapor distribution in the given empty volume. The experiment takes place under defined conditions in the humidity chamber with 10% relative humidity and a temperature of 40°C. It can be seen that moisture is diffusing into the beaker.

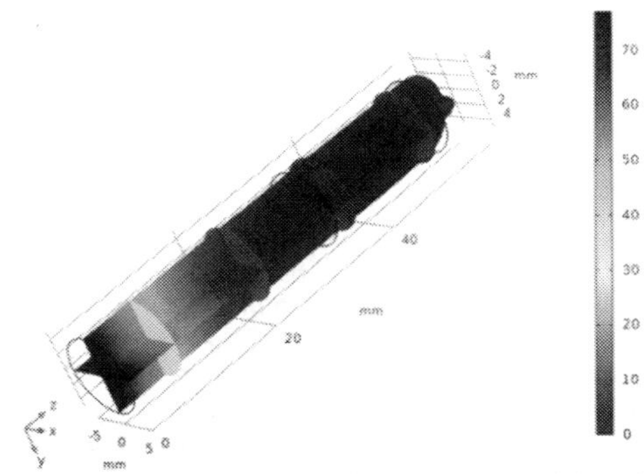

Figure 6: Time-dependent diffusion model using axisymmetric simulation in COMSOL Multiphysics. The distribution of the relative humidity in percent is plotted after a time of 2 h.

Figure 7 shows the moisture content (green line) at the position of the humidity sensor in the glass tube as a function of time. The red line describes the humidity content in the chamber of approx. 40 g/m^3 throughout the experiment, which matched the humidity level defined in the climate chamber in the experiment.

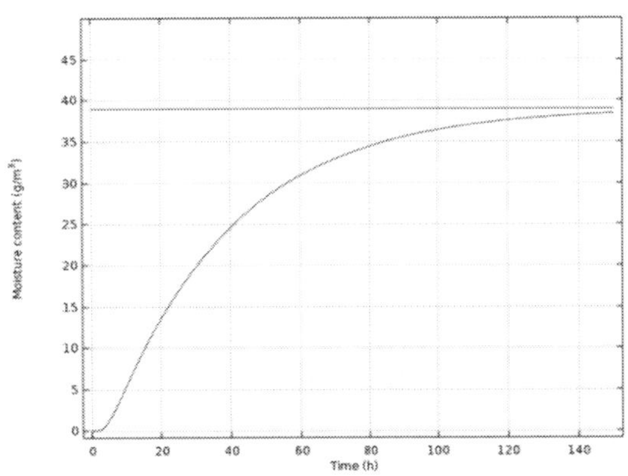

Figure 7: Moisture content in the setup of Fig. 6 as a function of time. The humidity sensor signal (green line) is rising according to a multi-exponential function to the humidity content in the chamber of approx. 40 g/m^3.

As the next step, the experiment of Bayerer et al. [5] is to be extended to power modules and the transfer function from the external humidity to the moisture content in the modules is to be determined.

4 CONCLUSIONS

We demonstrated a Simulink inverter model as solid base for improving data-driven anomaly detection and/or enhanced preventive maintenance approaches. With the second approach, we performed failure-motivated modelling by using recorded dataset of the real PV power plant. We showed the correlation between measured and calculated IGBT temperature. For further investigations, the model we have developed for determining the IGBT temperature is passed on to the project collaborators with a focus on data analysis and machine learning. By eliminating extraneous influences, the estimation of power loss in the semiconductors can be made more precisely, as the modelling of the IGBT temperature is not oriented to the measured values recorded by sensors, but to output current variables.

Furthermore, we showed an excellent verification of the experimental results of Bayerer et al. [5] on the moisture content in an adsorbing material by numerical modelling with COMSOL Multiphysics. The results demonstrate that the humidity transport in power modules can be well described by means of numerical simulations and thus take into account the influence of humidity on the lifetime of power electronic devices, and on the performance of the inverter. Additional measurements are currently under way, aiming the verification of physical models with sensor data installed in an operating PV plant with a known geographic location and humidity measurements on module level in defined environments by means of a climate chamber.

5 ACKNOWLEDGMENTS

This work was funded by the German Federal Ministry for Economic Affairs and Climate Action within the research project "Digital-Twin-Solar" (contract no. 03EI6024F).

6. REFERENCES

[1] K. Ma, N. He, M. Liserre, F. Blaabjerg, Frequency-domain thermal modeling and characterization of power semiconductor devices, *IEEE Transactions on* Power Electronics, 31(10), 7183-93 (2015).

[2] K. Ma, and F. Blaabjerg, Multi-timescale modelling for the loading behaviours of electronics converter, *Proceedings of the IEEE Energy Conversion Congress and Exposition* (ECCE, 2015), pp. 5749-56.

[3] P. Hacke, S.Lokanath, P. Williams, A. Vasan, P. Sochor, G. Tamizhmani, H. Shinohara, S. Kurtz, *A status review of photovoltaic power conversion equipment reliability, safety, and quality assurance protocols*, Renewable and Sustainable Energy Reviews (2018)

[4] M. Maerz, P. Nance, *Thermal Modeling of Power Electronic Systems*, Infineon Technologies AG Munich, PCIM Europe Mag, 2 (2000).

[5] R. Bayerer, M. Lassmann, S. Kremp, Transient hydro-thermal-response of power modules in inverters – mission profiling for climate and power loading, *Proceedings of the 8th International Conference on Integrated Power Electronics Systems* (CIPS, 2014), pp. 1-8.

MICROINVERTER PV SYSTEMS: NEW EFFICIENCY RANKINGS AND FORMULA FOR ENERGY YIELD ASSESSMENT FOR ANY PV PANEL SIZE AT DIFFERENT MICROINVERTER TYPES

Stefan Krauter, Jörg Bendfeld
Paderborn University, EET-NEK, Pohlweg 55, D-33098 Paderborn, Germany, E-mail: Stefan.Krauter@upb.de

ABSTRACT: The market for microinverters is growing, especially in Europe. Driven by rising electricity prices, many small photovoltaic energy systems are being installed. Since 2014, microinverters have been studied indoors and outdoors at Paderborn University. In the indoor lab, conversion efficiencies as a function of load have been measured with high accuracy and ranked according to Euro- & CEC weightings: The latest rankings of 2022 are included in this paper. In the outdoor lab, energy yields have been measured using identical and calibrated crystalline silicon PV modules. Until 2020, measurements were carried out initially with 215W_p modules. Because of increasing module power nowadays, 360 W_p modules are now being used. These extremes have been investigated: An older module with 215 W_p and a new module offering 360 W_p (both with 60 cells). Appling the low-power modules, the challenge for the microinverters was during weak-light conditions, for high-power modules, some inverters temporarily reached their power limits, so output was reduced. A method to determine actual energy yield has been developed, using a linear equation and a reference configuration: So yield for any module & inverter configuration can be determined by just two coefficients.
Keywords: Microinverter, System Performance, Energy Rating, Yield

1 INTRODUCTION

Microinverters are inverters that are connected mostly to a single PV module (occasionally to two modules, as indicated in the tables; few are available for four modules, but these are not considered here), so each module–inverter combination acts as an independent power plant. The microinverter consists of a maximum power point tracker (MPPT), the DC-AC inverter, and an islanding protection unit (see e.g., [1]). For higher power requirements, several module-inverter combinations are interconnected in parallel on the AC output side. This configuration offers various advantages: Easier planning and installation, easy up- and downscaling of a plant, including extensions or repair that could be carried out even during power plant operation. Logistics is simplified. Effect of shadowing is very limited, and due to low system voltages, potential induced degradation (PID) does not occur. An excellent overview of the development and the advantages of microinverters has been compiled by H. Oldenkamp [2]. However, costs of power plants based on micro inverters are about 10–20% higher. Some of the inverters cannot be operated by themselves and require a control unit (often combined with a remote shutdown option and a monitoring system), or a protective device for grid interfacing (depending on national regulations), thus adding extra costs. Also, conversion efficiency may not be as high as for central inverters. Due to smart master–slave concepts centralized solutions with multiple but relatively large inverters may offer higher yields under weak light conditions. [3] is giving a performance comparison of a microinverter, a power-optimizer, and a central inverter.

2 MEASUREMENTS

2.1 Indoor tests for conversion efficiency measurements

Due to the reproducible test conditions in the indoor lab, the inverters have been examined individually with predefined and controlled input data. Input has been a PV module simulator with data being set corresponding to the modules used in the outdoor test. The main output data being recorded is the delivered AC power of the inverters. Besides input power, output is also a function of input voltage. If input voltage is getting too low, the inverters even stop operating. The following examinations are based on the possible range of input data (including voltage) given the specific PV module also used for the outdoor investigation

Peak efficiency is often reached close to the maximum load of the inverter. Peak efficiency (often promoted in data sheets) is not a helpful value since most of the time the inverters operate in the range of 20% to 40% of their rated power – at least under non-arid conditions. Consequently, an adequately weighted efficiency is a more adequate value to rate conversion devices. One type of weighted efficiency is the so-called "European Efficiency" η_{Euro} which it is calculated according to:

$$\eta_{Euro} = 0.03 \cdot \eta_{5\%} + 0.06 \cdot \eta_{10\%} + 0.13 \cdot \eta_{20\%} + 0.1 \cdot \eta_{30\%} + 0.48 \cdot \eta_{50\%} + 0.2 \cdot \eta_{100\%} \quad (1)$$

The other is the "CEC efficiency" by the California Energy Commission (CEC). CEC efficiency is computed as an average value of DC–AC conversion efficiencies at six pre-defined relative output values between 10% and 100% of its rated power (with an emphasis on higher irradiance levels) is determined by:

$$\eta_{CEC} = 0.04 \cdot \eta_{10\%} + 0.05 \cdot \eta_{20\%} + 0.12 \cdot \eta_{30\%} + 0.21 \cdot \eta_{50\%} + 0.53 \cdot \eta_{75\%} + 0.05 \cdot \eta_{100\%} \quad (2)$$

For the "European Efficiency", weighting factors for high relative power values are lower.

The output power values used for the inverters (adjusted by controlling the DC input current) are continuously increased in 1024 steps from 0 to maximum. Each step takes eight seconds while the measurement duration is 500 ms. Figure 1 shows an example for the measuring procedure.

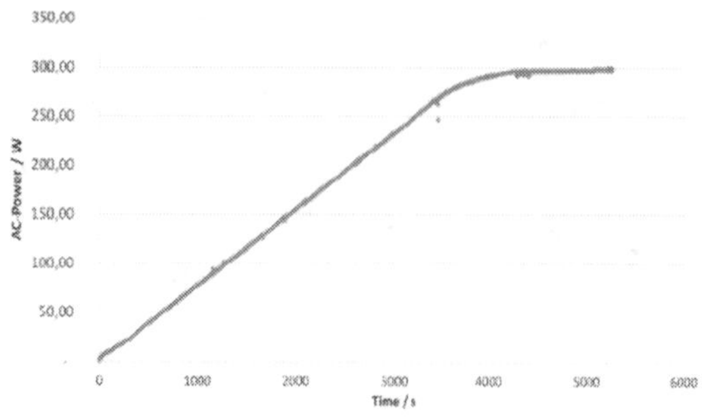

Figure 1: Example for measured AC power output (in Watt) as a function of measurement duration (in seconds) for linear increasing DC input current.

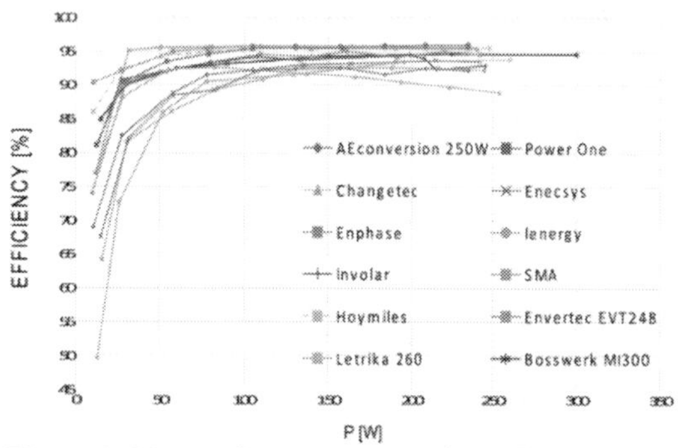

Figure 2: Measured DC-AC conversion efficiencies as a function of power output, for twelve microinverters with single PV module inputs (updated for 2022).

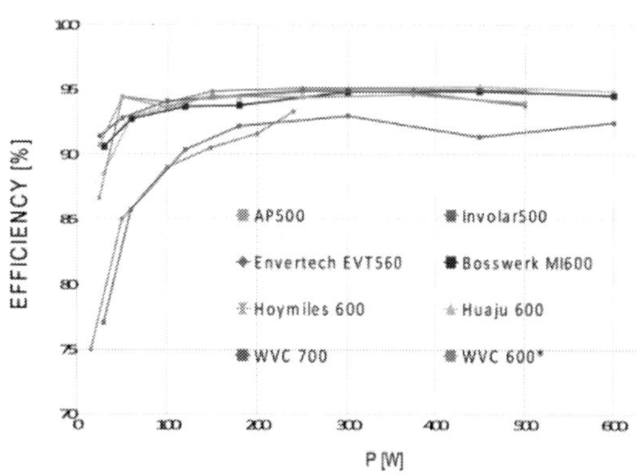

Figure 3: Measured DC-AC conversion efficiencies as a function of power output, for eight microinverters with two PV module inputs, including the latest models of Hoymiles, Huaju, Bosswerk.

The measured DC-AC conversion efficiencies of all inverters are shown in Fig. 2 and Fig. 3. Based on those measurements, the European (EU) efficiencies and the CEC efficiencies for the micro inverters have been calculated according to (1) and (2), eleven micro-inverters are designed for single modules, eight inverters have inputs for two PV modules: Involar MAC 500, APS YC 500, Envertech EVT-560, WVC 600, WVC 700, Hoymiles MI 600, Huaju HY 600, and Bosswerk Mi600.

Comment: WVC 600* stopped operating at a measured power of 250 W. After a test run at higher temperatures, the inverter failed constantly. Since the documentation of WVC 600 and WVC 700 inverter has been extremely poor, its rated power has been assumed. For this reason, WVC 700 is shown first with the assumed rated power of 600 W, then with 700 W. The maximum measured power of the WVC 700 inverter was 600 W only.

The ranking considering the European (EU) conversion efficiency is shown in Table I.

Comments: Envertech EVT 560 and PowerOne/ABB Micro-0.25-i have the same conversion efficiency, thus sharing rank number 5, so do Involar MAC 500 and Bosswerk Mi600, thus sharing rank 8 as well as Aptronic INV 250-45 and Enecsys SMI-S-240W, thus sharing rank 16.

Table I: Ranking of all tested microinverters by "European Conversion Efficiency", according to (1).

Rank No.	Manufacturer Model, Type	European Efficiency	Relative eff. of #1
1	SMA Sunnyboy 240	95.4%	100.0%
2	Enphase M 215	95.2%	99.8%
3	Hoymiles MI 500	95.0%	99.5%
4	Hoymiles MI 600	94.7%	99.3%
5	Envertech EVT-560	94.6%	99.2%
5	PowerOne/ABB Micro-0.25-i	94.6%	99.2%
7	Huaju HY 600*	94.5%	99.0%
8	Involar MAC 500	94.3%	98.8%
8	Bosswerk Mi600*	94.3%	98.8%
10	APSystems YC 500	94.1%	98.6%
11	Bosswerk MI300*	93.5%	98.0%
12	Envertech EVT-248	93.2%	97.7%
13	Involar MAC 250	92.7%	97.2%
14	WVC 700 (at 600W)	91.6%	96.0%
15	Changetech ELV 300-25	90.9%	95.3%
16	Aptronic INV 250-45	90.4%	94.7%
16	Enecsys SMI-S-240W	90.4%	94.7%
18	Ienergy GT 260	89.9%	94.3%
19	Letrika 260	88.7%	93.0%
20	WVC 700 (at 700W)	73.3%	76.8%
21	WVC 600 (failed)	0.0%	0.0%

* new in 2022

Table II shows the same, but with the CEC-efficiency formula (2) applied. Comment: Envertech EVT-560, Involar MAC 500, and Bosswerk Mi600 have the same conversion efficiency, thus sharing rank 7.

Table II: Ranking of all microinverters by "CEC Efficiency", calculated according to (2).

Rank No.	Manufacturer Model, Type	CEC-Efficiency	Relative eff. of #1
1	Enphase M 215	95.6%	100.0%
2	PowerOne/ABB 0.25-i	95.5%	99.9%
3	Hoymiles MI 500	95.4%	99.8%

4	SMA Sunnyboy 240	95.1%	99.5%
5	Hoymiles MI 600	95.0%	99.4%
6	Huaju HY 600*	94.9%	99.3%
7	Envertech ENV-560	94.6%	99.0%
7	Involar MAC 500	94.6%	99.0%
7	Bosswerk Mi600*	94.6%	99.0%
10	APSystems YC 500	94.5%	98.9%
11	Bosswerk Mi300*	94.1%	98.5%
12	Envertech EVT-248	94.1%	98.4%
13	Involar MAC 250	93.9%	98.2%
14	Enecsys SMI-S-240W	92.0%	96.3%
15	WVC 700 (at 600 W)	91.6%	95.9%
16	Letrika 260	91.5%	95.8%
17	Ienergy GT 260	91.5%	95.7%
18	AEconversion 250	91.2%	95.5%
19	Changetech ELV 300-25	90.9%	95.1%
20	WVC 700 (at 700W)	87.5%	91.6%
21	WVC 600 (failed)	0.0%	0.0%

* new in 2022

2.2 Thermal issues

In 2021/22, we started also to investigate thermal behavior of newer inverters: Some of the inverters failed when operated close to their nominal power (WVC 600, 700), independent of temperature (see Table I and II). Others (e.g., the Huaju HY 600) reduce their power output at 60°C, but operate normally at 25°C (see Fig. 4), but after cooling down, the inverters operated normally, without any damage. Possibly, that overheating issue was due a relatively compact inverter size.

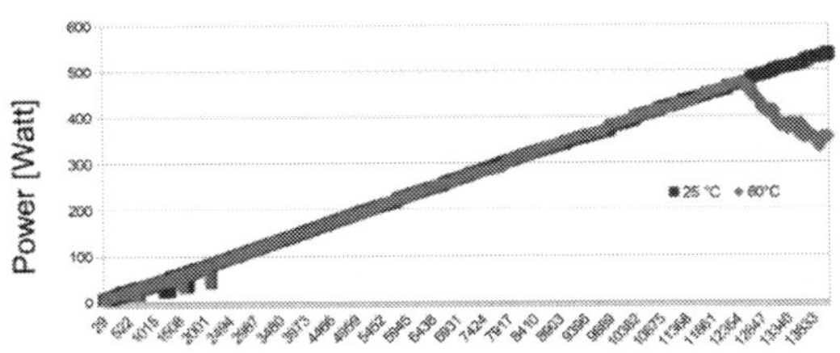

Figure 4: AC power output of the Huaju HY 600 inverter at 25°C (blue line) and at 60°C (red line) ambient temperature (heat chamber in the laboratory).

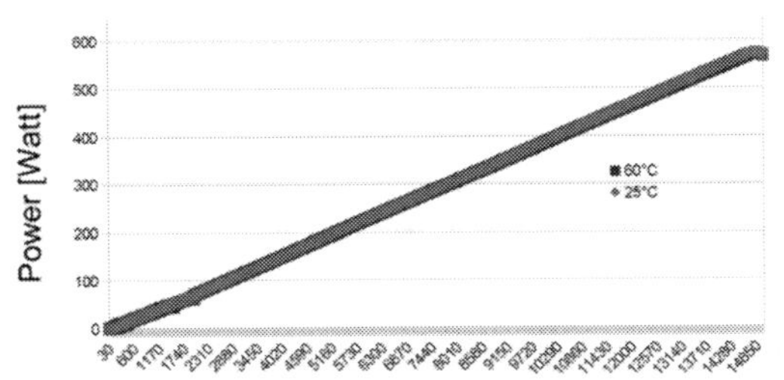

Figure 5: AC power output of inverters at 25°C (blue line) and at 60°C (red line) ambient temperature of Bosswerk Mi600 and Hoymiles HM 600.

Other inverters did not show any power reduction (e.g., Bosswerk Mi600, Mi300, and Hoymiles HM600), even at elevated temperatures (see Fig. 5).

2.3 Outdoor measurements for yield assessments

The new configuration for the tests, using ten 360 W_p modules (lower row, from left), is shown in Figure 6. Modules have been manufactured by Solarwatt®, the power output at STC of each module has been measured in the factory in Dresden (Germany). Additionally, one module has been sent for a precision measurement to the testing laboratory ISFH in Hameln (Germany). It turned out that the factory measurements have been very accurate (362 W_p vs. 359.34 W_p ±3% at ISFH in July 2021).

Besides the effects already observed with the 215 W_p modules, such as distinct conversion efficiencies at different irradiance levels, speed of MPPT algorithms, minimum thresholds for initiating operation; additionally, temporal saturation effects are observed at some inverters with the new 360 W_p modules applied.

Figure 6: Configuration of PV modules of PV outdoor laboratory for electrical energy yield comparison of microinverters using eight equal, calibrated PV modules (of 360 W_p each) as inputs.

The resulting electrical energy yields during the course of a day for the different microinverters and module configurations are shown in Figure 7 for a daily course and in Table III (see [4]) over a longer period of time for the 215 W_p modules. To some extent, the above-mentioned effects can be observed.

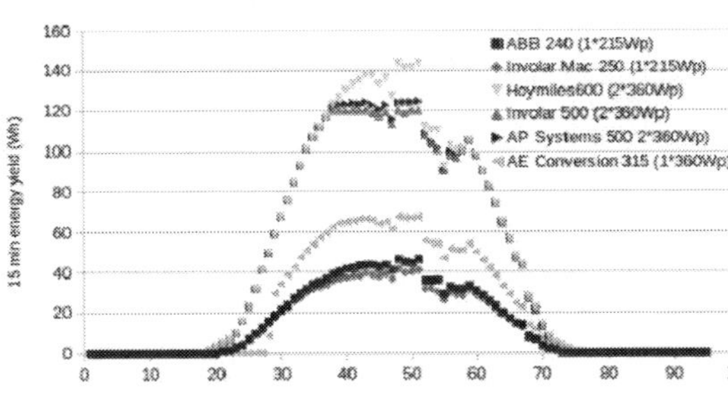

Figure 7: Example of electrical energy yield measurements (during an interval of 15 minutes) of different inverters and 2 different PV module sizes during a mostly clear day (some clouds in the afternoon).

Table III: Former ranking of microinverters by relative energy yield measured, using the 215 W_p modules (1 or 2 of them, as number of inputs), referenced to former #1 (Power One/Aurora/ABB, type Micro-0.25-i) [4].

Rank No.	Manufacturer Model, Type	Relative yield vs. ABB
1	Involar MAC 500	100.7 %
2	Power One/Aurora/ABB Micro-0.25-i	100.0 %
3	APSystems YC 500	99.3 %
4	Hoymiles MI 600	97.4 %
5	SMA Sunnyboy 240	95.2 %
5	Enphase M215	95.2 %
7	Involar MAC 250	94.2 %
8	Envertech EVT 300	94.0 %
9	WVC 700 (at 600 W)	91.7 %
10	AEconversion/Aptronic INV 250-45	92.5 %
11	Envertech EVT 248	92.1 %
12	Ienergy, GT 260	91.5 %
13	Enecsys SMI-S-240W	88.7 %
14	Hoymiles MI 250	78.4 %
15	Changetec ELV 300-25	75.6 %

While the different types of effects make it quite cumbersome to predict an energy yield for a certain configuration at a certain location, a more consumer-friendly yield-predicting method has been elaborated by performing some yield data analysis.

Each microinverter has been directly connected to a calibrated electrical energy meter with a S_0-interface. To secure an accurate yield measurement, the calibrated electrical energy meters are replaced on a regular base with new freshly calibrated ones. All S_0-interfaces have been connected to a server-based data acquisition system.

3 UNIVERSAL YIELD ASSESSMENT

To ease the characterization of a specific combination of PV module & microinverter, a linear equation has been applied to a well investigated reference characteristics of a very good inverter without issues for low irradiance, MPPT, and saturation. The inverter chosen as a reference has been the Enphase M 215, which ranked #1 at the CEC-efficiency rankings, see [4].

Plotting a function of the actual yield (y) over the reference yield (x), that function would be $y = a\,x + b$ with the trivial coefficients $a = 1$ and $b = 0$ for the reference configuration (Enphase M 215 with the Q-cells 215 W_p module). Figure 8 shows the original configuration with the inverters for single modules and the 215 W_p modules attached.

The coefficients of the different inverters for the relative yield equation $y = ax + b$ have been elaborated in Table IV: It can be observed that for low daily yields Involar MAC 250 is performing a little bit better than the reference, so b is above 0, for high yields its performance is decreasing (relative to the reference), so a is above 1. For the Envertech EVT 300 the characteristics is vice versa: Performance at low yields is worse than the reference, so b is negative; relative performance is increasing towards high reference yields, so steepness of curve is higher, resulting in an $a > 1$.

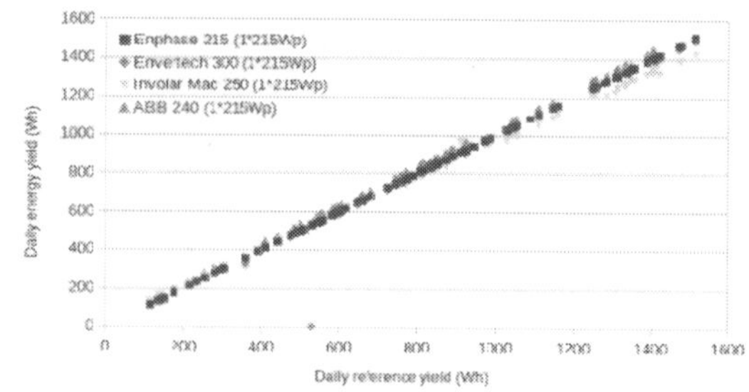

Figure 8: Electrical energy yields of different inverters for single modules with a 215 W_p module attached. Daily reference yield (x-axis) is the energy yield (AC) achieved by an Enphase M215 inverter with a single 215 W_p module applied.

Table IV: Coefficients for relative daily yield $y = a\,x + b$ (referenced to Enphase M 215, all with <u>single</u> 215 W_p modules), yield is given in AC electrical energy

Manufacturer	Type	a	b (Wh)
Involar	MAC 500	0.923	+43.35
Power One /Aurora/ABB	Micro-0.25-i	1.011	+25.90
Envertech	EVT 300	1.020	-33.45
Enphase	M 215	1.000	±0.00
Bosswerk*	Mi 300	0.969	+4.58

*new in 2022

Figure 9 shows the characteristics of different microinverters that can serve two modules, either with two 215 W_p (older measurements) or two 360 W_p modules (latest measurements). Table V shows the corresponding coefficients a (for "steepness") and b (for "offset") of the relative daily yield curve.

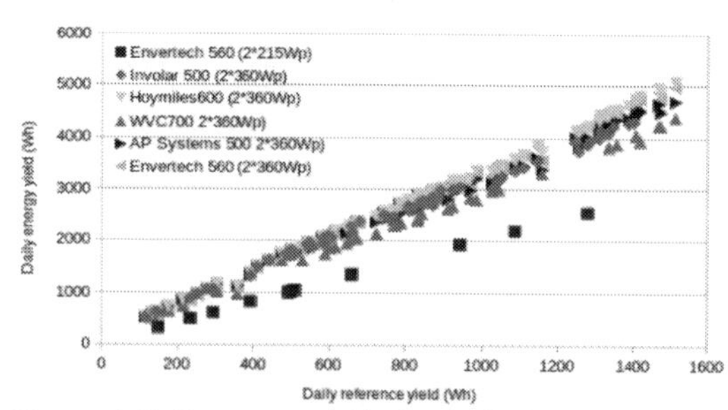

Figure 9: Daily energy yields (AC) of different inverters for two modules with two 215 W_p or 360 W_p modules attached. Reference yield (x-axis) is the yield achieved by an Enphase M215 with a single 215 W_p module applied.

The coefficients of determination R^2 for all regressions of the measurement values to determine the coefficients a and b have been in the vicinity of 0.99 or above.

Table V: Coefficients for relative daily yield $y = a\,x + b$ for microinverters serving two modules, either 215 W_p or 360 W_p types (referred to Enphase M215 with a 215 W_p module), yield is given in AC electrical energy

Manu-facturer	Type (module power)	a	b (Wh)
Envertech	EVT 560 (2 x 215 W_p)	1.983	+37.80
Envertech	EVT 560 (2 x 360 W_p)	3.227	+109.97
Hoymiles*	MI 600 (2 x 360 W_p)	3.189	+168.32
Involar	MAC 500 (2 x 360 W_p)	2.889	+180.70
AP Systems	YC 500 (2 x 360 W_p)	2.953	+254.77
WVC	WVC 700 (2 x 360 W_p)	2.750	+172.39
Bosswerk*	Mi 600 (2 x 360 W_p)	3.122	+112.17
Huaju*	HY 600 (2 x 360 W_p)	3.141	+153.90

* new or updated (Hoymiles MI 600) in 2022

4 CONCLUSION AND OUTLOOK

The use of a reference configuration together with the two coefficients of a linear equation enables a simple method to describe quite accurately the daily yield performance of any microinverter in combination with any PV module, even with under- or oversized ones. While prices of PV modules are decreasing at a higher pace than prices for microinverters, we will see more configurations with oversized modules and more saturated microinverters more often in the future. This underlines the necessity of a method (e.g., as described) to extrapolate energy yield.

5 REFERENCES

[1] W-F. Lai, S-M. Chen, T-J. Liang, K-W. Lee, A. Ioinovici: "Design and Implementation of Grid Connection Photovoltaic Inverter", Proceedings IEEE Energy Conversion Congress and Exposition (2012) 2426.

[2] H. Oldenkamp, I.J. de Jong: "The return of the AC-module inverter", Proceedings 24th European Photovoltaic Solar Energy Conference (2009) 3101.

[3] D. Stellbogen, P. Lechner, M. Senger: "Field and laboratory performance characterisation of microinverter and Power optimizer systems", Proceedings 32nd European Photovoltaic Solar Energy Conference (2016) 1654.

[4] S. Krauter, J. Bendfeld: "Micro-Inverters: An Update and Comparison of Conversion Efficiencies and Energy Yields". Proceedings of 37th European Photovoltaic Solar Energy Conference (2020) 935.

DEVELOPMENT FROM A SMALL TO LARGE SCALE LUMINESCENT DOWNSHIFT LAYER FABRICATION ON SILICON SOLAR CELLS

Cass, B.*, Chandra, S., Ortega, A.& McCormack, S. J.
Solar Energy Applications Group, Dept of Civil, Structural and Environmental Engineering, School of Engineering, Trinity College Dublin, College Green, Dublin 2, Ireland
Email: bicass@tcd.ie

ABSTRACT:

This paper present the issues that have been faced during the scaling up fabrication of a Luminescent Downshifting layer (LDS) for large-scale compound parabolic photovoltaic system-based requirement. The development process in fabricating large-scale LDS layers changes significantly through the upscaling process.

The requirement of these LDS layers remain the same from being applied on a 2 x 2cm PV cell onto two different 12.5 x 12.5 cm PV cell systems. During this process, the characteristics which influence the LDS' efficiency on the PV cell were altered to remain giving the best possible output. From the upscaling process, it will be shown that various issues arise for the larger cell but can be fixed with a change in fabrication processes.

Keywords: Solar Cell Efficiencies, Photovoltaic, Building Integration, Fabrication, Sustainable

1. Introduction

Luminescent Downshifting (LDS) improves the performance of solar cells by shifting the range of wavelength that can be absorbed more efficiently by different types of photovoltaic[1]. The LDS layer is a thin film polymer [2] layer that can shift the wavelength to a longer wavelength depending on the composition of the LDS layer. Figure 1. Below demonstrates how an LDS layer can affect the interaction of light photons and the solar module substrate underneath.

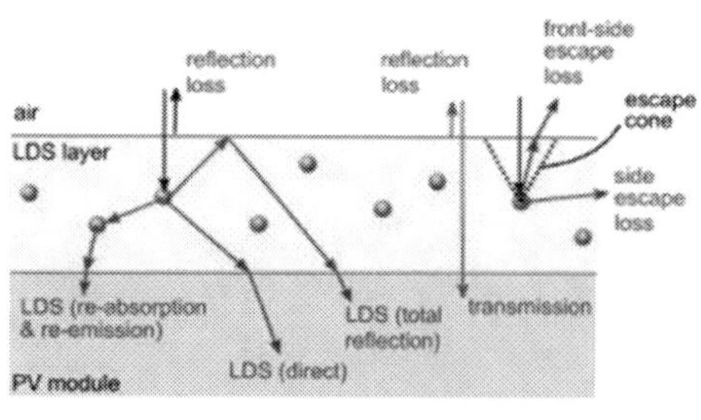

Figure 1. Illustration of an LDS Layer applied to a PV[2]

The large-scale fabrication of an LDS layer based system had not been produced of this magnitude so far. With the layering process being done individually in small batches to try and produce repeatable quality and consistency. With the end goal of producing multiple systems that have a higher efficiency in power production and can be integrated into both domestic & commercial settings.

2. Methodology

The LDS used in this project was composed of a chloroform solution mixed with a polymer of Poly[methyl methacrylate] (PMMA) as the base solution; the downshifting occurs with the addition of a luminescent dye. In this case, it was methyl orange supplied from Sigma Aldrich. The LDS solution was made using a specific recipe, this contained 14g (10ml) chloroform, 4g of PMMA & 0.2wt% of Methyl Orange dye.[2] These layers were applied using a spin coater system to apply at both 2x2cm & 12.5x12.5cm sizes. The physical characteristics of the layer include a uniform layer (as uniform as possible through multiple fabrications across numerous batches), a thickness of 20μm and a physical adhesion (no peeling across the cell once dried). Any variation to these characteristics can disrupt the effectiveness of the LDS layer[3]

The LDS layers were made in batches. The purpose of this was to make the upscaling process easier and more viable without having to alter more areas that can effect the characteristics of the LDS during the fabrication process. To maintain the uniformity of the layers on larger cells, minimisation of changing variables was required, which included the environment in the LDS solution process and pre-polymerization . To limit any effects from the air (pre-polymerization before application of LDS), a nitrogen-rich glove box was used for the preparation and mixing of the polymer solution for LDs layers. . The thickness of the layers was maintained by applying 3 ml and 10ml of LDS layer solution to the small and large cells, respectively. The average spin speed required for an even distribution of LDS was 1000 rpm using varying speed ramps and time lags. 3 minutes was required for a full LDS application. Figures 2.1-2.3 show the different stages of the application of the LDS layers on various cells and types of cells.

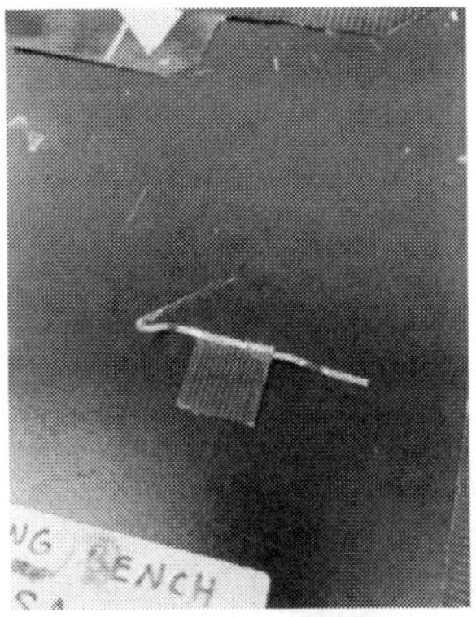

Figure 2.1 Progression of LDS on 2 x 2 cm PV cells

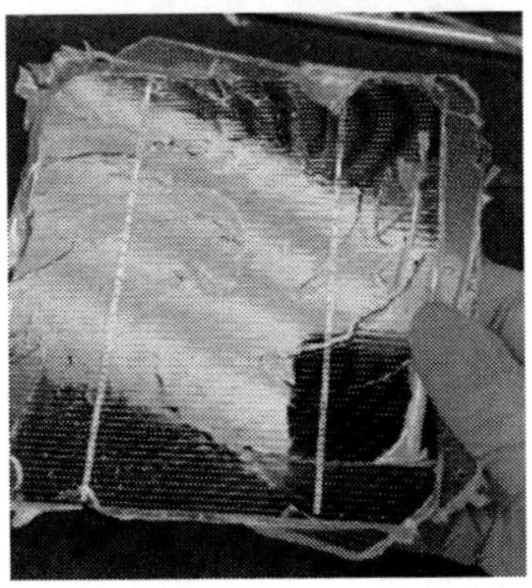

Figure 2.2. Progression of LDS 12.5x12.5cm (front contact) PV cells

Figure 2.3. Progression of LDS 12.5x12.5cm (back contact)cells

When scaling up from the 2 x 2cm to 12.5x12.5cm PV cells, different variations to the previous application were required. For the 2 x 2cm PV cells, a ratio of 10 ml of chloroform to 3.6g of PMMA was used. When applied to the larger cells, however, the LDS thickness was less than the desired 20 μm thickness. To counteract this, the addition of 0.4g of PMMA was added and brought the thickness up to the required size.

As seen in figure 2.2 while an uneven application, air gaps and peeling is also occurring. This in part was due to the polymer mixture being processed in air environment , to try and fix this issue, the use of an inert environment was used. The inert environment allowed for a cleaner environment for the mixing of the chloroform, PMMA & dye solution. It was found to massively reduce the any air gaps/bubbles that would appear into the LDS mixture. While also reducing any impurities entering into the mixture such as dust or dirt as the LDS solution would be moved from the inert environment in a closed container. For this work, a nitrogen-filled glove box was used. This allowed the PMMA polymer to be fully dissolved into the chloroform solution without oxygen getting trapped within the mixture and a pre-polymerisation hardening of LDS solution.

With the introduction of an inert environment, it was found that an elongated mixture time was also required, whereas previously ,smaller batches of 15/20 ml chloroform base only required 30-40 mins, the larger batches of a 30 ml base required between 45-60 mins for full dispersion of the PMMA polymer. In part, this was due to less evaporation occurring in the nitrogen-rich glove box compared to a room temperature laboratory setting.

These batches consisted of 30 ml chloroform, 12g of PMMA and a 0.2wt% of methyl orange dye. The purpose of facilitating the upscaling to 12.5x12.5 cm PV cell. These batches were able to produce 3 LDS layers of 12.5x12.5cm consistently. Trying to producelarger batch sizes would result in the reduction in quality of the LDS layer (impurities in the layer due to mixing not being fully complete), but also a very large requirement of time allocated to mixing (30ml chloroform based batch would take 50 – 60 mins, a 40 ml chloroform based batch would take 90-100mins).

3. Innovation

The purpose of the large-scale LDS layered cells was to integrated into a Compound Parabolic Concentration (CPC) system as part of the IDEAS Project within Horizon 2020. These CPC systems use 32 solar cells across 4 rows with 8 cells connected in series and will be integrated into a domestic dwelling or commercial structure by being placed on the facades or roof (depending on most optimal light exposure) of the buildings. The purpose of the IDEAS project is to develop a self-sufficient building using renewable sources, including solar technology, heat pump implementation, phase change material incorporation with each source of energy, all while being maintained and managed through automated and fully integrated system. For the systems shown below in figure 3., there was 4 systems with 32 LDS layered cells required. The results of which will be discussed in a further section.

Figure 3. CPC system with LDS layered cells installed at University of Ferrara, Ferrara, Italy

4. Results

The successful assembly of the 4 CPC systems was sent and installed in Ferrara. It was required 128 LDS layered cells, with an estimated assembly time of between 32 – 42 hours of lab fabrication, only considering the mixing process. Itdoes not include the fabrication time of the LDS, nor does it include any defunked/broken cells occurring during the fabrication process.

A Promising outcome of this fabrication process, even though it was completed by manually, was the maintenance of the overall quality and consistency of the LDS layers. As the LDS were produced in batches of 3 layers at a time, there was a large possibility for issues to arise given the many variables occurring during the production process. An important aspect of this consistency was achieved using an inert environment, in this case, a nitrogen-filled glove box reduced the environmental effects on the LDS solution fabrication.

While large systems of PV panels have been successfully assembled and implemented before, an attempt had not been made to integrate LDS layered PV cells into a large-scale system, with the influential inclusion of a CPC benefit. LDS had previously been produced on much smaller scales of single cells of various sizes but not fabricated and produced in such quantities (128 cells in the system, with 12 LDS cells as a backup in case of damage).

Figure 4. Damage sustained by the PV cell in CPC systems.

Unfortunately, some of the LDS layered cells sustained damage at some point after assembly in the CPC system and before/during the installation process. This damage may be attributed to the fact that once the LDS layer has been applied onto the cell, the cell itself can become more brittle because of the spin coater process, which in some cases can lead to small hairline cracks that only develop or show when further pressure is applied. It should be noted, however, that when a hairline crack appears, the LDS layered cell can still function completely, only when further pressure is applied will a brake occur, causing loss of functionality. As shown in Figure 4 above.

The most important part of this upscaling is to ensure that there is power production occurring from the CPC systems that include the LDS cells. While also showing an improvement over a regular PV panel.

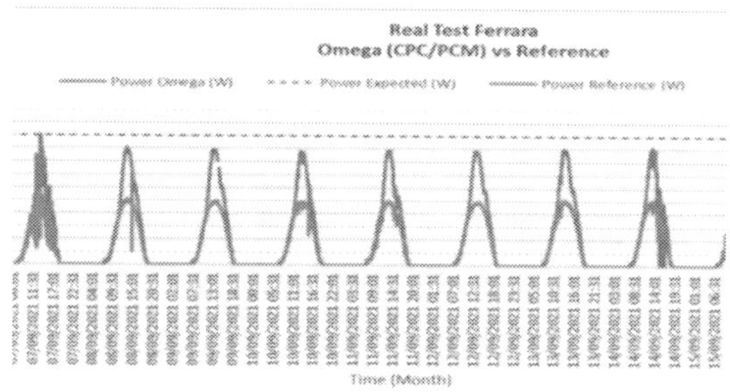

Figure 5. Graph of comparison between PV panels & LDS CPC system. Where the X-axis is time & Y-axis is Power (Watts). The expected power is 100W as denoted by the dotted line.

Figure 5 above demonstrates that the LDS filled CPC system can out preform a system that is just PV silicon-based cells. With the aim of producing 100W per system being achieved and, in some cases, also exceed given beneficial circumstances.

5. Conclusion:

The upscaling of the LDS layer from a 2x2cm to 12.5x12.5cm silicon-based cell has shown to be a successful transition, while it can be difficult to maintain a consistent quality for these LDS layers, mostly due to the fact these layers are applied at a time using a single spin coater rather than in a more commercial or industrial setting. It is possible to produce working LDS based systems that can out preform equivalent system of the same size/quantity of blank silicon-based cells. Further work must be conducted into the durability and maintenance of these LDS cells once layered, as the damage that can occur within the systems could reduce the overall efficiency of each system. While again trying to further improve the production time required and/or the quantity of LDS layers that are being produced from any given batch, ideally a minimum of 5 LDS layer per batch within that previously mentioned time frame of 60 mins of mixing of the LDS solution. It leads onto the question of how can we further improve both the LDS layers to be more user friendly and reliable, and the process in which the production of these LDS can be more streamlined and more commercially viable.

6. Acknowledgements

The authors would like to acknowledge the European Research Council grant H2020 IDEAS (815271) and support funding from Science Foundation Ireland (SFI) and insights from the PEARL PV COST Action Network.

References:

[1] Klampaftis, E., Ross, D., McIntosh, K. R. & Richards, B. S. 2009. Enhancing the performance of solar cells via luminescent down-shifting of the incident spectrum: A review. Solar Energy Materials and Solar Cells 93, 1182-1194

[2] R. Rothemund,Optical modelling of the external quantum efficiency of solar cells with luminescent down-shifting layers,Solar Energy Materials and Solar Cells,Volume 120, Part B,2014,Pages 616-621,

[3] Cass, B., Chandra, S., Sethi, A., & McCormack, S. J. (2022). Characterisation of Luminescent Downshifting Layer Effects on Silicon Solar Cells. 871–874. https://doi.org/10.1007/978-3-030-76221-6_97

CRITICAL ISSUES AND NOVEL CONCEPTS OF COOLING BIPVT AND ITS INTEGRATION WITH LOW-TEMPERATURE HEATING

George Aspetakis, Qian Wang
Department of Civil and Architectural Engineering, KTH Royal Institute of Technology
Teknikringen 78B, Stockholm, Sweden

ABSTRACT: The efficiency of photovoltaic (PV) modules is inversely correlated with their operating temperature. Extensive research has shown that cooling PV systems to maintain a stable and low surface temperature enhances PV efficiency, power generation and prolongs the life span of the modules by limiting the degradation caused by thermal fatigue. This applies particularly for Building Integrated Photovoltaic-Thermal (BIPVT) systems where the extracted heat can be potentially recovered by the rest part of building energy services. To implement this, BIPVT systems can be integrated with low temperature heating systems, such as existing heating and ventilation installations, running on low temperature level, or integrated with heat pumps. The effective cooling of such systems can be instrumental for meeting nearly Zero Energy Building (nZEB) standards, by enhancing their overall conversion efficiency, while meeting both the electrical and thermal demand of buildings. In this study, key challenges related to this topic are outlined and crucial parameters for thermal management concepts are presented. Practical aspects of the design and integration processes are explored as well, alongside the cost-effectiveness, sustainability and aesthetics of said systems. Meeting such criteria plays a vital role for the penetration of the next generation of BIPVT to the market, yet current literature does not address these implementations comprehensively. In view of these characteristics, new BIPVT cooling concepts are proposed along with methods for simulating their thermal behavior utilizing numerical solutions, i.e., computational fluid dynamics (CFD).
Keywords: BIPVT, cooling, solar air heater

Nomenclature		Symbols	
		A	Area (m^2)
BIPVT	Building Integrated PVT	Cp	Specific heat of air (J/Kg K)
PV	Photovoltaic	Dh	Hydraulic Diameter (m)
PVT	Photovoltaic-Thermal	f	Friction Factor
TEF	Thermal Enhancement Factor	h	Convective heat transfer Coefficient (W/m^{2}K)
VG	Vortex Generator	L	Length (m)
		$\dot{m}$	Mass flow rate (kg/s)
		Nu	Nusselt number
		Pr	Prandtl number
		Re	Reynolds number
		T	Temperature (K)
		U	Free flow Velocity (m/s)
		$\dot{V}$	Volumetric flow rate (m^3/s)
		ΔP	Pressure drop (Pa)
		v	Kinematic viscosity (J.s/kg)
		ρ	Density (kg/ m^3)

1 INTRODUCTION

Only a fraction of the total incident radiation on a photovoltaic (PV) panel is converted directly to electrical energy. The rest is either transmitted to the surroundings or absorbed by the PVs material, in the form of thermal energy [1]. This results in an increase of PV modules temperature. Studies have shown that this causes an almost linear drop in efficiency of up to 0.5 %/C° and attribute this to electrical resistance losses in the cells [2]. Additionally, the effective lifecycle of the modules is reduced due to thermal fatigue. Therefore, cooling PV systems is beneficial in multiple aspects: e.g., by enhancing the generation efficiency and prolonging their lifespan. PVT/BIPVT systems generate electrical and thermal energy simultaneously, by utilizing various forms of cooling.

Previous studies have explored different cooling configurations, some particularly focusing on using air as the cooling medium [3]. Although great interest is detected, no significant breakthroughs have been made lately, with no sound solutions penetrating the market [4]. This is partly due to the insufficient increase in either electrical or thermal energy generation. Solar air heaters have shown potential to be suitable for dealing with this challenge, given that the physical conditions, heat transfer mechanism and overall purpose are of very similar nature. Thus, in this study, heat transfer enhancement designs of solar air heaters, with proven performance, are discussed and their suitability for BIPVT systems is explored. Further improvements and novel concepts compatible with such systems are suggested.

2 NUMERICAL DESCRIPTION

Assuming an adiabatic backplate, the convective heat transfer between the PV surface and the stream of air is:

$$Q = hA\Delta T = \dot{m}C_p(T_{out} - T_{in}) \tag{1}$$

In the case of utilizing ambient air as the cooling medium, it is not sensible to cool the stream beforehand (therefore increasing ΔT) or increasing the mass flow $\dot{m}$ significantly, since those operations require active power consumption. In contrast, increasing the heat transfer coefficient h and heat transfer area A only increases the load of the fan, which requires energy, albeit significantly less that the former methods. Nusselt number is widely used in the investigations which indicates the ratio of convective over conductive heat transfer :

$$Nu = \frac{hD_H}{k} \tag{2}$$

D_H is the hydraulic diameter and k the conductivity of air. The friction factor of the channel plays a crucial role as well, since it is directly responsible for the pressure drop that the fan blower must provide.

$$f = \frac{\Delta P}{\frac{1}{2}\left(\frac{L}{D_H}\right)\rho U^2} \tag{3}$$

The goal here is to improve Nu with minimum friction penalty. To effectively compare methods with a single metric, studies in similar applications use the thermal enhancement factor TEF (or thermohydraulic parameter) as an indicator, which combines the heat transfer improvement and subsequent friction increase. A definition was proposed by Webb and Kim [5] in which it assumed that the pumping power of the channels are equal:

$$P_0 = P \rightarrow \left(\dot{V}\Delta P\right)_0 = \dot{V}\Delta P \rightarrow (fRe^3)_0 = fRe^3 \tag{4}$$

Where values with the subscript 0 indicate a smooth channel and those without the configuration/geometry analyzed.

With $Re = \frac{UD_H}{\nu}$ being the Reynolds number which describes the turbulent nature of the flow.

Both sides of Eq. 4 are multiplied by $1/f_{0,Re}$ which is the friction factor of the smooth duct but for the Reynolds number of the modified one.

$$Re_0^3 \frac{f_0}{f_{0,Re}} = Re^3 \frac{f}{f_{0,Re}} \tag{5}$$

The Blasius equation for friction in smooth ducts $f_0 = \frac{C1}{Re^{m1}}$ is used to get:

$$\left(\frac{Re_0}{Re}\right)^3 \left(\frac{C1/Re_0^{m1}}{C1/Re^{m1}}\right) = \left(\frac{f}{f_0}\right)_{Re} \rightarrow \left(\frac{Re_0}{Re}\right)^{3-m1} = \left(\frac{f}{f_0}\right)_{Re} \tag{6}$$

TEF is defined as : $TEF = \frac{Nu}{Nu_0}$ for equal pumping power. The Dittus-Boelter equation can be then used to predict the Nu in smooth ducts: $Nu_0 = CRe_0^{m2}Pr^n$. Inserting it to the former equation gives:

$$TEF = \frac{Nu}{CRe_0^{m2}Pr^n} = \frac{Nu}{CRe_0^{m2}Pr^n(Re^{m2}/Re^{m2})} = \frac{\frac{Nu}{Nu_0}}{\frac{Re_0^{m2}}{Re^{m2}}}$$

Combining with Eq. 6 gives:

$$TEF = \left(\frac{Nu}{Nu_0}\right)_{Re} \Bigg/ \left(\frac{f}{f_0}\right)_{Re}^{\frac{m2}{3-m1}}$$

A simpler definition is commonly found in existing literature:

$$TEF = \frac{\frac{Nu}{Nu_0}}{\left(\frac{f}{f_0}\right)^{0.33}}$$

Most studies validate their mathematical models or CFD analyses with the known Dittus-Boelter equation $Nu = 0.023Re^{0.8}Pr^{0.4}$ (for heating of fluid) or the Gnielinski equation:

$$Nu = \frac{\left(\frac{f}{8}\right)(Re - 1000)Pr}{1 + 12.7\left(\frac{f}{8}\right)^{0.5}(Pr^{0.66} - 1)}$$

3 HEAT ENHANCEMENT DESIGNS

3.1. Principles of heat enhancement designs

Solar air heating studies primarily address the limitation of utilizing air as a cooling medium, specifically the much lower thermal capacity of air in comparison to water. However, water shows disadvantages for certain applications, leading to investigations into the efficient heating of air through solar energy. Among the heat enhancement mechanisms possible, emphasis has been given to mostly two aspects: the disruption of the thermal boundary layer and the increase of turbulence of the flow [6]. Turbulent flow displays higher convection heat transfer rates due to its tumultuous movement which distributes heat further. In applications of forced convection above surfaces or in piping, surfaces exert friction on the particles adjacent, impending their movement and transmission of heat. In such manner, a boundary layer and an equivalent thermal one are formed, hindering efficient heat transfer. Different strategies have been reported that attempt to disturb this layer at short intervals and obtain the thermal characteristics of the free flow. Rashidi et al. [7] reviewed the prospective utilization of inserts in solar thermal energy systems, which covered twisted tapes, wired coils, vortex generators and baffles. Various numerical and experimental studies were evaluated, giving insight into the specifics of each concept and recommendations for future research was given. However, it was reported that hardly any studies performed economic assessments of these systems, which is a vital factor for their wider adoption as a viable solution.

Given the development, selected high-performing methods and designs are presented in further detail. Given the infancy of this technology, the terminology used is still hazy. A straightforward and comprehensible designation was given by Promvonge and Skullong [8] who define baffles and winglets as extremely thin ribs. They distinguish the former from the latter when their height is equal or more than 25% of the height of the channel and their angle of attack is equal or more than 30°.

In summary, solar air heating technology has seen significant developments, which could potentially cover the gap found in PVT/BIPVT performance enhancement, which is reflected in this study.

3.2 Development of Baffles

Baffles are surfaces that direct the airflow with the primary aim of interrupting the thermal and laminar boundary layer. Shapes vary from rectangular, triangular to trapezoid profiles. Starting from the edge of the baffle, the flow separates, creates recirculation zones and impinges back on the surface. This, in combination with secondary flows formed, results in better mixing of the fluid and subsequently in improved heat transfer, though with an increased pressure drop. Combinations of factors like height, width, and position need to be tested to procure optimum thermohydraulic behavior. The key parameters can be summarized as blockage ratio, pitch ratio and angle of attack. Baffles can be further subdivided according to their shapes and arrangements. Some early studied adopted shapes in terms of such baffles, such as : Z-shaped [9], U-shaped [10] and cuboid-shaped [11].

The sub-category of baffles which continuously outperform other designs, by achieving higher and consistent TEF values, is the V-shaped baffle and its multiple derivations. Researchers have detected its potential and dedicate efforts to develop its performance further. V-shaped designs break the main flow at regular intervals and create secondary flow jets, which progress from the leading edge up to the trailing one, eventually boosting heat transfer. Kumar et al. [12] investigated a discretized broken V-shape baffle arrangement. The idea with the utilization of discretized baffles instead of continuous ones, is to direct and join the main and secondary flows through these gaps. This results in the latter accelerating the former, energizing its slowed-down boundary layer. Optimal values of parameters like the position and size of the gaps were identified. Promvonge and Skullong [8], motivated by perforated designs, aimed for a configuration that would decrease the pressure drop penalty as well as introduce impingement jets right behind the baffles that cool the adjacent surface. It was made possible with the forming of flaps from square shaped cuts on the baffles. Air flows unobstructed though the cavity and is directed by the flap towards the heated surface, creating a jet that impinges on it. Tests delivered promising results and specified optimal flap angles and pitch ratios. Eiamsa-ard et al. [13] further developed this concept by employing semi-circular flaps (or hinges) instead, with the intend of reducing dead zones behind the baffle.

From the channel design perspective, Tamna et al. [14] performed one of the few studies exploring staggered an in-line configurations of baffles on both sides of the air channel. It was shown that an arrangement on a single wall (that of the absorber) provides higher TEF values in comparison to a double sided configuration. This difference was attributed to the lower blockage ratio of the single arrangement. Should a double arrangement be chosen, the in-line version achieved only slightly better results. Fawaz et al. [15] analyzed an in-line arrangement in a numerical simulation and found out that TEF values increase as the blockage ratio is decreased. Furthermore, fully developed periodic flow behavior was detected a few characteristic lengths in.

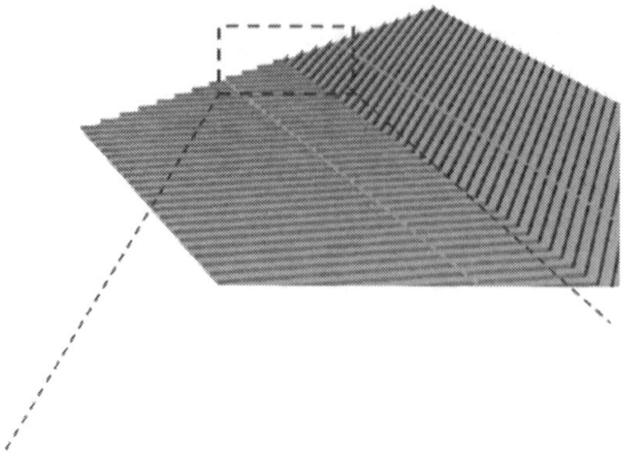

Figure 1: Example of a discrete V-shaped baffle with two gaps.

3.3 Vortex Generators

An alternative critical design element, Vortex generators (VGs), as the name suggests, are meant to create vortices in the main flow, resulting in increased turbulence and mixing. As described previously, turbulent flows transfer heat more effectively because they prevent thermal and mass layers from forming. Similarly to baffles, vortex generators with various shapes have been investigated, such as rectangular [16], trapezoidal [17], delta shaped, curved [18] or even truncated half conical [19]. Increase of the pressure drop due to the blockage of the flow is always present. VGs are commonly protrusions on surfaces, that initially disturb the boundary layer and the wake created from their trailing edges hinder it from forming again.

Sawhney et al. [20] inspected the performance of wavy delta winglets. The aim was for the wavy shape to create pairs of counter rotating vortices, which interacted with each other, superimposing them towards one direction. Thus, the effect on the surface heat transfer is multiplied. To fully take advantage of this concept, critical pitch ratios were investigated, so as to strengthen the vortices at regular intervals before they break down. Some studies incorporated perforated designs as well. In this case, openings are created in the VGs, through which the flow enters. A lower pressure drop is achieved due to the overall smaller blockage. Baissi et al. [21] investigated the thermal and friction profiles of perforated and non-perforated curved delta shaped vortex generators. All perforated cases showed reduced friction rates and improved TEF values. Skullong et al. [22] tested perforated rectangular and trapezoidal winglets and compared different hole diameters on those configurations. The study was further developed for the combination of grooves on the surface with VGs [23]. Zhao et al. [24] explored the combination of VGs and baffles by placing delta winglet VGs in front of V-shaped continuous ribs. It was reported that longitudinal vortices are able to alter the direction of the back flow produced by such ribs with a positive effect on heat transfer. Optimization of nine parameters, related to delta shaped winglet design, for flow-up and flow-down arrangements, was performed by Dezan et al. [25]. It was concluded that the most impactful parameters were chord and angle of the first row, followed by their equivalents in the third and first rows. Flow-up configurations displayed marginally improved results only with low Reynolds numbers. Optimized setups tended to generate and maintain longitudinal vortices along the main flow direction and close to the heated surface, while their corner vortices were of reduced size and dissipated rather quickly.

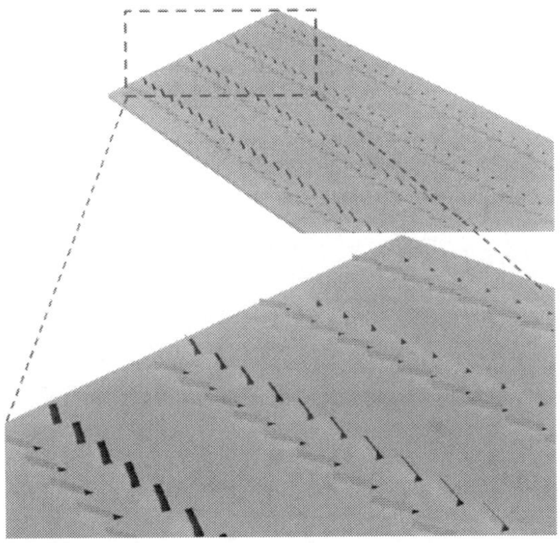

Figure 2: Example of rectangular vortex generators.

3.4 Impingement jets and ribs

To combat the pressure drop increase caused by the addition of inserts in solar air heaters as introduced in Section 3.2 and 3.3, researchers have attempted to minimize their size considerably. One strategy is to focus mainly on the disruption of the thermal boundary layer through minute extrusions, called usually ribs, and artificial roughness on the heated surface. The result is separation and reattachment of the flow that enhance heat transfer. Such configurations can potentially be combined with the technique of impingement flow cooling. Impingement jets strike the selected surface, enhancing the heat transfer coefficient locally and maintain a slim thermal boundary layer.

A selection of rib profiles were analyzed numerically by Ngo and Phu [26], regarding their heat transfer and friction factor. The highest thermohydraulic performance was achieved by the hyperbolic ribs. The smoother profile of the hyperbolic ribs resulted in smaller zones of ineffective convection, particularly upstream. Deo et al. [27] applied the broken V-shape to their rib arrangement and complemented it with staggered rib units that disrupted the flow between the gaps. Optimal rib geometry parameters were identified, which lead to higher TEF values. Multiple studies investigated the arc shape for ribs with broken, staggered and multi-arrangements [28–30].

Nayak and Singh [31] examined the effect of channel spacing and mass flow rates of a non-conventional cross-flow jet plate solar air heater. Longer spacing and larger mass flow rates were favored. It was determined that, even though the friction factor did not change considerably in the proposed configuration, its thermal behavior was improved and was therefore deemed more fit for solar air heating than the typical setup. N. Kumar et al. [32] constructed conical ring obstacles for enhancing the cooling ability of impingement jets. Due to the longer swirl flow path and the greater mixing rate caused by the rings, the heat transfer was improved.

Table I: Studies investigating heat transfer enhancements in solar air heaters

Study	Type	Max TEF	Critical parameter
Promvonge and Skullong (2021) [8]	Flapped V-shaped baffle	2.5	Pitch, flap angle
Sriromreun et al. (2012) [9]	Z-shape baffle	2.98	Height, Pitch
Bopche and Tandale (2009) [10]	U-shape baffle	1.97	Pitch
Pasra et al. (2021) [11]	Cuboid baffle	3.28	Height
Eiamsa-ard et al. (2022) [13]	V-shaped w/ circular hinges	2.6	Hinge angle
Tamna et al. (2014) [14]	Multi V-shape baffle	1.83	Pitch
Fawaz et al. (2018) [15]	V-shape baffle	0.78	Blockage ratio
Depaiwa et al. (2010) [16]	Rectangular winglet VGs	1.17	Re number
Xiao et al. (2020) [17]	Trapezoidal VGs	1.89	Upwind arrangement
Zhou and Ye (2012) [18]	Curved VGs	1.6	Attack angle
Bezbaruah et al. (2021) [19]	Truncated half conical VGs	2.04	Re number
Sawhney et al. (2017) [20]	Wavy delta-shaped VGs	2.09	Pitch
Baissi et al. (2020) [21]	Perforated VGs	2.26	Pitch
Skullong et al. (2018) [22]	Perforated VGs	2.01	Hole diameter
Skullong et al. (2017) [23]	Grooves + VGs	2.12	Groove Pitch
Zhao et al. (2021) [24]	Ribs + VGs	1.4	Shape of ribs
Ngo and Phu (2020) [26]	Conic ribs	1.56	Conic constant
Deo et al. (2016) [27]	V-shaped, staggered ribs	2.45	Rib height

4 PVT AND BIPVT APPLICATIONS

Currently, studies regarding Air-cooled PVTs mainly aimed to increase the heat transfer through the addition of fins, which increase the total heat exchange area. Unsurprisingly, sub-optimal arrangements cause significant pressure drop, thereby reducing the efficiency of the system. So far, studies omitted this impact with limited evidence on the performance and do not investigate its effectiveness by having such a high pumping power demand. This can be illustrated in the study of Shrivastava et al. [33] where multiple fin and baffle combinations for a PVT system were tested, regarding their thermal performance, but the overall cost-effectiveness was not reported. PVT/BIPVT research primarily focused on heat transfer related to water, Phase Change materials and nanofluids as shown by the comprehensive review on PVT of Chandrasekar et al. [3]. Those that did employ air as the cooling medium mostly investigated regular air channels and the addition of fins but did not explore the impact of inserts found in solar air heater studies.

The aforementioned heat transfer enhancement strategies indicate the proven performance increase for solar air heaters. It is sensible to extend their conclusions to PVT/BIPVT collectors too, based on the similarity of these technologies. There is strong potential for improving their efficiency, by applying the further adapted techniques, with the aim of maximizing the cooling of the PV modules and the generation of thermal energy. Operational and differences do exist though that will determine which methods are most appropriate for these applications.

From the module level, one critical impactful decisive parameter is to maintain a uniform temperature reduction, specifically avoid large temperature gradients along the PV surface. This causes the cells to operate at different maximum power points, reducing their electrical efficiency and lifespan. Thus, a form of cooling is required that would avert hot and cold zones from forming or at least minimize them. Flapped or perforated baffles are able to counter that aspect, by reducing the total area of the dead transfer zones behind the installed baffle arrangement. In that sense, the combination of both cooling mechanisms, the increased rate of vortices and the presence of impingement jets can provide spatially even heat extraction.

It should be noted that in most solar air heater applications, the surface of the absorber could be modified by various manufacturing processes, in order to attain the desired morphology that would enhance heat transfer. For instance, ribs and protrusions can be either be added or manufactured directly on the absorber material. In the case of PVT, it is the photovoltaic panel area that is heated by radiation. The backsheet of this is usually made of tedlar material and is hard to process. Future designs should take this difference of materials into consideration.

For the case of BIPVT, key challenges lie in the weight of the systems. Installations in the residential sector require low weight solutions, given the risk of damage on building envelopes. Heavy weight materials, such as metals, shows limitations for practical applications. Polymers, on the other hand, are a lightweight solution suitable for such applications. 3D additive manufacturing has progressed rapidly the past years, now being able to construct complex geometries with greater ease and in a shorter timeframe. Metals, polymers and composites are presently employed. Vortex generators in general require less material than baffles, thus are more fit for instances where low weight is critical.

In the end, it is of outmost importance for prospective designs to keep friction levels low, thus reducing pumping power demand. Although different methods might have similar TEF values, options with less severe pressure drop ought to provide better results, even if other solutions perform better thermally.

5 CONCLUSIONS

This study presents the prospective enhancements to the performance of PVT/BIPVT systems, with a thorough summary of key operational aspects in the development of solar air heater technology. The study concluded that:

- A parallel between solar air heaters and Air-based PVT technologies was drawn, emphasizing that methods employed for enhancing heat transfer in the former show potential for the latter as well.
- A thorough description of heat transfer improvement techniques that are apt for PVT and BIPVT systems was compiled based on literature review, mainly focusing on baffles, vortex generators, impingement jets and ribs
- Characteristics specific for such applications were described as well, with suggestions on how to counter them using novel approaches such as 3D printing or the utilization of polymers instead of metals.
- The study also pointed the challenges of optimizing pressure drop increase with any heat transfer enhancement method, to discover those solutions with minimum friction, in order for the system to be cost-effective.

The authors gratefully acknowledge the financial support from the Swedish Energy Agency (grant number: 52488-1)

6 REFERENCES

[1] J.K. Kaldellis, M. Kapsali, K.A. Kavadias, Renewable Energy 66 (2014) 612–624.
[2] E. Skoplaki, J.A. Palyvos, Solar Energy 83 (2009) 614–624.
[3] M. Chandrasekar, T. Senthilkumar, Journal of Cleaner Production 322 (2021) 128997.
[4] E.D. Rounis, A. Athienitis, T. Stathopoulos, Renewable Energy 163 (2021) 1729–1753.
[5] R.L. Webb, E.R.G. Eckert, International Journal of Heat and Mass Transfer 15 (1972) 1647–1658.
[6] Y.D. Khimsuriya, D.K. Patel, Z. Said, H. Panchal, M.M. Jaber, L. Natrayan, V. Patel, A.S. El-Shafay, Applied Thermal Engineering 214 (2022) 118817.
[7] S. Rashidi, M.H. Kashefi, F. Hormozi, Solar Energy 171 (2018) 929–952.
[8] P. Promvonge, S. Skullong, International Journal of Heat and Mass Transfer 172 (2021) 121220.
[9] P. Sriromreun, C. Thianpong, P. Promvonge, International Communications in Heat and Mass Transfer 39 (2012) 945–952.
[10] S.B. Bopche, M.S. Tandale, International Journal of Heat and Mass Transfer 52 (2009) 2834–2848.
[11] H. Parsa, M. Saffar-Avval, M.R. Hajmohammadi, International Journal of Mechanical Sciences 205 (2021) 106607.
[12] R. Kumar, R. Chauhan, M. Sethi, A. Kumar, Experimental Thermal and Fluid Science 81 (2017) 56–75.
[13] S. Eiamsa-ard, A. Suksangpanomrung, P. Promthaisong, International Journal of Thermal Sciences 177 (2022) 107577.
[14] S. Tamna, S. Skullong, C. Thianpong, P. Promvonge, Solar Energy 110 (2014) 720–735.
[15] H.E. Fawaz, M.T.S. Badawy, M.F.A. Rabbo, A. Elfeky, Alexandria Engineering Journal 57 (2018) 633–642.

[16] N. Depaiwa, T. Chompookham, P. Promvonge, Proceedings of the International Conference on Energy and Sustainable Development: Issues and Strategies, ESD 2010 (2010).
[17] H. Xiao, Z. Dong, Z. Liu, W. Liu, Applied Thermal Engineering 179 (2020) 115484.
[18] G. Zhou, Q. Ye, Applied Thermal Engineering 37 (2012) 241–248.
[19] P.J. Bezbaruah, R.S. Das, B.K. Sarkar, Renewable Energy 180 (2021) 109–131.
[20] J.S. Sawhney, R. Maithani, S. Chamoli, Applied Thermal Engineering 117 (2017) 740–751.
[21] M.T. Baissi, A. Brima, K. Aoues, R. Khanniche, N. Moummi, Applied Thermal Engineering 165 (2020) 113563.
[22] S. Skullong, P. Promthaisong, P. Promvonge, C. Thianpong, M. Pimsarn, Solar Energy 170 (2018) 1101–1117.
[23] S. Skullong, P. Promvonge, C. Thianpong, N. Jayranaiwachira, M. Pimsarn, Applied Thermal Engineering 122 (2017) 268–284.
[24] Z. Zhao, L. Luo, D. Qiu, Z. Wang, B. Sundén, Energy 224 (2021) 119944.
[25] D.J. Dezan, A.D. Rocha, W.G. Ferreira, Applied Energy 263 (2020) 114556.
[26] T.T. Ngo, N.M. Phu, Journal of Thermal Analysis and Calorimetry 139 (2020) 3235–3246.
[27] N.S. Deo, S. Chander, J.S. Saini, Renewable Energy 91 (2016) 484–500.
[28] V.S. Hans, R.S. Gill, S. Singh, Experimental Thermal and Fluid Science 80 (2017) 77–89.
[29] R.S. Gill, V.S. Hans, R.P. Singh, Applied Thermal Engineering 191 (2021) 116871.
[30] I. Singh, S. Singh, Solar Energy 162 (2018) 442–453.
[31] R.K. Nayak, S.N. Singh, Solar Energy 137 (2016) 434–440.
[32] N. Kumar, A. Kumar, R. Maithani, Thermal Science and Engineering Progress 17 (2020) 100493.
[33] A. Shrivastava, J.P.A. Jose, Y.D. Borole, R. Saravanakumar, M. Sharifpur, H. Harasi, R.K.A. Razak, A. Afzal, Sustainable Energy Technologies and Assessments 49 (2022) 101782.

THERMAL INFLUENCES OF ROOFTOP PHOTOVOLTAIC PANELS ON BUILDING PERFORMANCE: A META-ANALYSIS

Delia D'Agostino [a], Danny Parker [b]

[a] European Commission, Joint Research Centre (JRC), Ispra (VA), Italy
[b] Florida Solar Energy Center, University of Central Florida, USA

ABSTRACT

Solar electric photovoltaic (PV) panels are becoming common worldwide and are increasingly integrated as a building material. However, an advantage of conventional rooftop stand-off PV arrays on rack-mounting with an air space below the panels, comes from roof shading of direct sun and modification to roof surface long-wave radiation to the sky. These influences affect thermal transmittance and, consequently, cooling and heating loads of the roof structure underneath. Thermal impacts of PV stand-off array on roof heat transfer performance are complex depending on array shadow, spacing, tilt, wind as well as roof view-factor to the sky and infrared heat exchange at night. Arrays also strongly influence thermal long-wave radiation exchange to the sky for both cooling and heating. When arrays cover much of buildings' roof surfaces, these impacts begin to exert a larger influence on roof/ceiling related heat gains and building thermal performance. This can be particularly important for existing buildings with lower levels of roof insulation. This paper summarizes a meta-analysis the existing research into this influence as well as EnergyPlus simulation analysis conducted by the authors for specific examples. Our evaluation is bolstered by infrared thermography illustrating the phenomenon as well as experimental temperature data taken from actual roof assemblies. Simulation analysis is used to examine array shading impacts and interaction with exiting roof insulation levels. Potentials for existing buildings are discussed as well as how these influences might become more important with increased summer heat waves associated with climate change. Comfort impacts are simulated for residential buildings without mechanical cooling systems.

Keywords: Roof insulation; energy efficiency; photovoltaic systems; residential building retrofit; building energy simulations.

1 INTRODUCTION

In the view of being the first climate-neutral continent by 2050, renewable energy will play a key role in Europe [1]. The need to electrify energy demand to facilitate greenhouse gas (GHG) emission reductions, and reduce climate change warming potentials, makes it important to locate new available areas for renewable energy production and particularly PV [2].
The influence of PV array shading of roofs has received limited attention within the scientific community compared to other aspects related to renewable production, efficiency, climate. However, empirical evidence of how array shading impacts to end-use cooling are compelling.
We assemble relevant studies and sources for the first time. For instance, an average 26% reduction in submetered and measured space cooling was observed in after rooftop PV panels were installed atop Japanese houses [3]. Furthermore, a 38% reduction in roof heat gain was seen during the cooling season in San Diego, California [4]. In Phoenix, Arizona, a large decrease in roof-related space cooling was observed [5] both in measurement and simulation. A 17.6% decrease to space cooling in Western Greece was measured and simulated [6]. In this paper, we detail each study briefly, provide full references for existing work and summarize key points of agreement or difference.

Decreased solar irradiance to the roof and building below is important in winter, potentially increasing heating through lost roof solar absorption. However, as we detail in the paper, reduced long-wave re-radiance from covered roofs provides some compensating effect for heating since winter solar irradiance is lower and nights are longer with colder effective sky temperatures.

2 METHODS

We examine the influence of stand-off PV arrays on building roof thermal performance by conducting a detailed meta-analysis of existing studies, both empirical and simulation based. We also add to this body of knowledge by conducting thermography and measurement of our own of buildings with PV systems. Finally, we conduct detailed simulation analysis for typical buildings in Milan examining how the array shading influence interacts with existing roof insulation levels (Figure 1). Both EnergyPlus and TRNSYS hourly simulations are used for the evaluation.

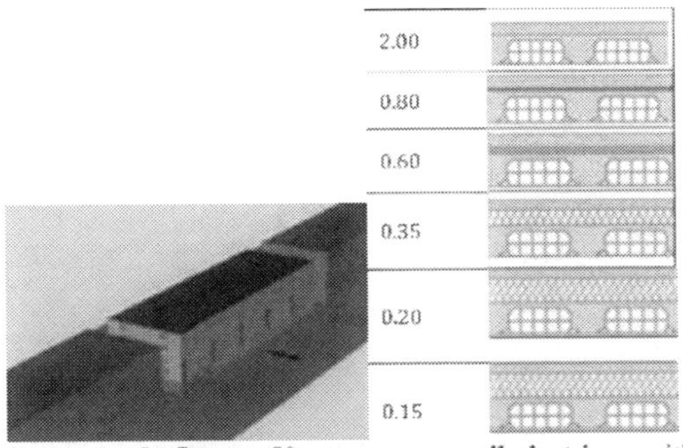

Figure 1. 5 floors, 50 apartments, all-electric with efficient heat pumps (COP=3.5 heating, 4.1 cooling).

We aim to investigate the thermal influence of stand-off PV arrays on buildings and its impact on thermal transmittance, cooling and heating loads to:

- reduce energy consumption, GHG emissions linked to buildings;
- increase comfort and renewable production, electrification;
- consider climate change and extreme events.

We address the specific challenges of modelling the influence of stand off arrays within energy simulations. The largest influence is the solar shading influence of stand-off array. However, changes to roof long-wave irradiance to day and night skies are also determinant for capturing effects as the altered roof view factor to the sky, and to a lesser extent the altered roof surface convective heat transfer.

3 RESULTS

Mirroring empirical data, simulation suggests the PV array shading and alteration to long wave radiation may be important, particularly for poorly insulated roof assemblies. For instance, in one early evaluation, a 39% reduction in cooling during summer was obtained both in simulations and measurements of covered roof assemblies [7].

With increasing installation of larger rooftop PV arrays, the fraction of exposed roofs that are covered increases significantly and thus the impact of PV shading and interruption of roof night sky radation.

In a detailed hourly simulation analysis by the authors in Milan (Italy) [2][8], both shading and infrared exchange with the sky influences were addressed. Impacts were small save for uninsulated or poorly insulated roof assemblies. This is accomplished within the EnergyPlus simulation by assuming that the PV array covered portion of the roof has a 90% interruption of incident solar radiation compared to the exposed sections. However, the long-wave emittance of this covered section is also altered from 0.92 for fully exposed sections vs. 0.10 for the covered segments. If the array vertical spacing from the roof is close, there can also be an insulating effect from the air below the stand-off array. However, this can vary considerably depending on array spacing and wind conditions. Close spacing may also lead to secondary heating of the roof below from the elevated temperature of the PV arrays. Available sources indicated this a weak effect.

We found large impacts to cooling from covering PV arrays during summer. Our simulation results of winter influences in Milan were similar to those of [9].

With simulations, we examined the air temperatures within the top floor apartment flats with the cooling system disabled on the hottest summer day. We examined predicted performance both with and without insulation and a shading cover from an expansive rooftop PV system (Figure 2).

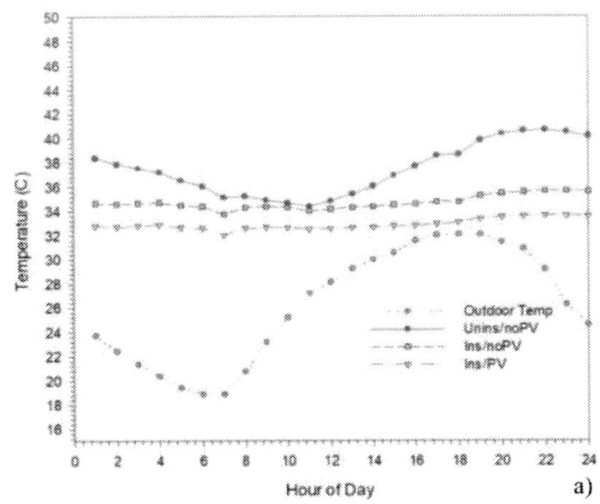

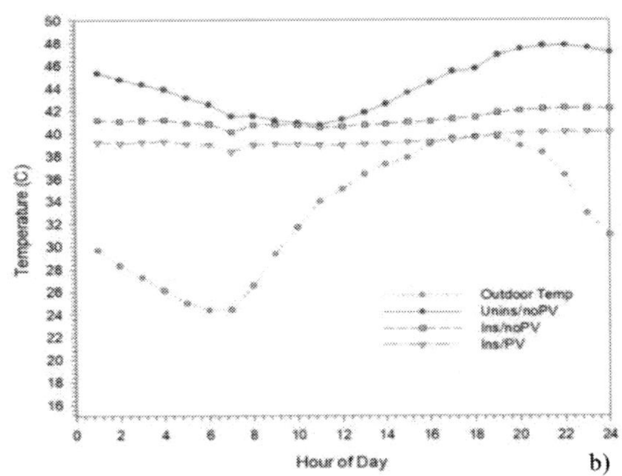

Figure 2. Comparison of simulated top floor apartment air temperatures adjacent to roof on summer peak day with and without roof shading from PV arrays and insulation (Unins/Ins). Blue dotted line and symbols are the outdoor temperatures on 18 July on the 2018 TMY (typical meteorological year) Milan weather file.

Fig. 2b shows the same date for a weather file morphed to the 90% IPCC condition in 2060.

Shading from the rooftop PV system appeared effective in controlling excess heat accumulation in the top floor dwellings, particularly when matched with insulation (Figure 3). On 18 July, the hottest day on the Milan weather file, the elevated indoor temperatures in simulated inside flat 47 next to the uninsulated roof were much more moderate with the PV system overhead. During evening hours around 10 pm on this very hot day, the peak inside air temperature in the uncooled and uninsulated flat was predicted at 40.7 °C against 35.6 °C with insulation added (Figure 3).

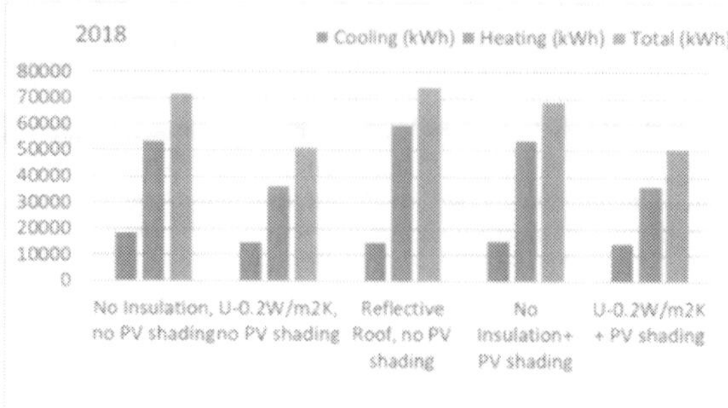

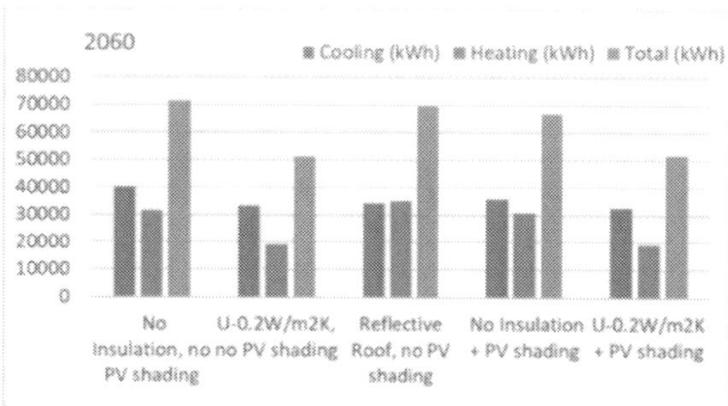

Figure 3: Predicted energy performance with and without PV shading using 2018 and 2060 Extreme Weather.

However, providing PV shading over the top of the roof in the apartment complex with insulation was predicted to drop the inside flat temperature without cooling to 33.6 °C, only slightly above the outdoor temperature. While not comfortable, it indicates a significant improvement in interior comfort especially to vulnerable occupants. As seen with the much hotter extreme weather for 2060 (peak outdoor temperature is 40 °C), the differences in heat accumulation could easily impact occupant mortality as the differences in insulation and shading represent a 7 °C difference in peak room temperatures without mechanical cooling. Predicted energy savings are shown in Table 1 and Table 2.

2018	Cooling savings (%)	Heating savings (%)	Heat/cool energy savings (%)
No Insulation, no PV shading	0.0	0.0	0.0
U-0.2W/m², no PV shading	19.1	31.9	28.7
Reflective Roof, no PV shading	19.4	-11.5	-3.6
No Insulation + PV shading	16.5	-4.0	3.9
U-0.2W/m²K + PV shading	21.1	31.2	28.9

Table 1: Predicted energy savings in 2018.

2060	Cooling savings (%)	Heating savings (%)	Heat/cool energy savings (%)
No Insulation, no PV shading	0.0	0.0	0.0
U-0.2W/m², no PV shading	16.9	39.5	28.7
Reflective Roof, no PV shading	14.6	-12.0	2.6
No Insulation + PV shading	11.0	1.8	6.6
U-0.2W/m²K + PV shading	18.4	31.2	27.5

Table 2: Predicted energy savings in 2060.

We found that heating impact to insulated assemblies are near adiabatic with effectively no influence. The largest influence for either cooling or heating was found in uninsulated roofs. From simulations, we anticipate some geographic differences in PV panel-roof interaction since high wintertime solar gains may create modest increases to heating from PV panel shadowing as shown by [6]. However, making allowance for variations due to climate and roof constructions, our simulation results line up well with both previously cited empirical studies. In summary:

- Insulation cuts heating by 32%, cooling by 19% (uninsulated apartments).
- PV can provide >46% annual energy needs.
- PV arrays shading can reduce overheating, cooling by 17% (uninsulated case, 3% insulated apartments).

We provide further empirical evidence of the phenomenon from thermography (Figure 4).

Figure 4. Visible and IR image illustrating impact of shading from photovoltaics on thermal heat gain to brown tile roof. Colour is proportional to temperature, with white and red hottest, and black and blue coolest.

The shading impact of a South-facing stand-off PV array is evident from such an assessment: the roof surface is hotter outside the PV array shadow and much lower underneath.

Roof insulation retrofit with added PV can also have large potential impacts on comfort and safety in existing apartment buildings during summer heat waves. Many flats in older apartment buildings with low levels of insulation in Europe lack cooling systems. With increased summertime heat waves expected from climate-related warming, the top floor apartments adjacent to the roof deck under full sun can experience much higher levels of heat accumulation during summer heat waves that can endanger the health of occupants, particularly the elderly [10][11].

Even in flats with cooling systems, there is the possibility of equipment failure. Also with increased heatwaves in the future, there are increased chances that utilities may experience "brownouts" where power for cooling and or heating systems are temporarily unavailable [12]. Under full sun, the upper floor apartments can become exceedingy hot and are notoriously difficult to ventilate with limited cross ventilation [13][14].

4 CONCLUSIONS

The meta-analysis of existing sources as well as those produced within our own evaluation shows that shading from stand-off PV arrays have measurable impacts on building thermal performance. The impacts vary with roof coverage and also inversely with roof insulation, in all cases:

- improving building envelope is a key to controlling thermal losses and cutting space heating and cooling loads;
- insulation and stand-off PV arrays reduce overheating in upper-floor apartments, improving occupant health and safety;
- shading from stand-off PV arrays impacts building thermal performance, especially during summer heat waves and under in future climate conditions.

Installation of PV on existing older roofs with poor thermal properties that cannot be easily insulated may be important for helping with summer overheating of spaces immediately underneath the roof ceiling. Comfort implications for buildings without mechanical cooling systems are predicted within the simulation analysis.
Practically, we suggest that insulating and installing PV on existing buildings that can be refurbished at the same time may be a winning combination for improving building thermal performance. Finally, we further show evidence in simulation analysis using morphed future weather data for Milan, that shading from rooftop PV arrays will become more important during summer heat wave events associated with anticipated future climate change.

ACKNOWLEDGMENTS

The views expressed are purely those of the authors and do not represent any official position of the European Commission or the University of Central Florida. The authors wish to thank Christian Thiel and Maurizio Bavetta (JRC) and James Fenton (FSEC/UCF) for supporting this research.

REFERENCES

[1] D. D'Agostino, D. Parker, I. Epifani, D. Crawley, L. Lawrie, How will future climate impact the design and performance of nearly zero energy buildings (NZEBs)?, Energy, Volume 240, 2022.

[2] D'Agostino Delia, Parker Danny, Melià Paco, Giovanni Dotelli, Optimizing photovoltaic electric generation and roof insulation in existing residential buildings, Energy and Buildings, Volume 255, 2022.

[3] Tsurusaki T., Tanaka A., Nakagami H., Evaluation of Photovoltaic Power Generation Systems in Residential Homes in Japan: A Partnership Program of Utility and Consumers' Cooperative, 2000, Proceeding of the ACEEE 2000 Summer Study on Energy Efficiency in Buildings, Vol.1, 355-366, American council for an Energy Efficient Economy, Pacific Grove, CA.

[4] A. Dominguez, J. Kleissl, J.C. Luvall, Effects of solar photovoltaic panels on roof heat transfer, Sol. Energy. 85 (2011) 2244-2255. https://doi.org/10.1016/j.solener.2011.06.010

[5] W.A. Miller, E. Brown, R.J. Livezey, Building-integrated photovoltaics for low-slope commercial roofs, (2005). doi: 10.1115/1.1877514.

[6] V. Kapsalis, D. Karamanis, On the effect of roof added photovoltaics on building's energy demand, Energy Build. 108 (2015) 195-204. https://doi.org/10.1016/j.enbuild.2015.09.016

[7] Peng, C., Yang, J., The Effect of Photovoltaic Panels on the Rooftop Temperature in the EnergyPlus Simulation Environment, International Journal of Photoenergy, 2016, http://dx.doi.org/10.1155/2016/9020567

[8] D'Agostino, D., Parker, D., Melià, P., Dotelli, G., Data on roof renovation and photovoltaic energy production including energy storage in existing residential buildings , Data in Brief 41,107874

[9] A. Dominguez, J. Kleissl, J.C. Luvall, Effects of solar photovoltaic panels on roof heat transfer, Sol. Energy. 85 (2011) 2244-2255. https://doi.org/10.1016/j.solener.2011.06.010.

[10] Conti S, Meli P, Minelli G, Solimini R, Toccaceli V, Vichi M, Beltrano C, Perini L. Epidemiologic study of mortality during the Summer 2003 heat wave in Italy. Environ Res. 2005 Jul;98(3):390-9.

[11] Kovats, R.S., Kristie, L.E, 2006, "Heat Waves and Public Health in Europe," European Journal of Public Health, Vol. 16 (6), (2006), 592-599.

[12] Quinn, A., Kinney, P and Shaman, J. "Predictors of summertime heat index levels in New York City apartments," Indoor Air, 2017, Jul: 27(4) 840-851

[13] D'Agostino, D., Tzeiranaki, S.T., Zangheri, P., Bertoldi, P., Assessing Nearly Zero Energy Buildings (NZEBs) development in Europe, Energy Strategy Reviews 36,100680, 10.1016/j.esr.2021.100680

[14] D'Agostino, D., Parker, D., Melià, P., Environmental and economic data on energy efficiency measures for residential buildings, Data in Brief 28,104905, 10.1016/j.dib.2019.104905

GLASS BIPV FOR COMMERCIAL BUILDING INTEGRATION IN CYPRUS

Maria Hadjipanayi[1], Yerasimos Yerasimou[1], Matthew S.H. Norton[1], George Elias Georghiou[1], Eirini Chrysovalanto Miltiadous[2], Annita Karantoni[2], Fanos Karantonis[2], Rafaela Nikolaidou[3], Eliza Loucaidou[3]
[1]University of Cyprus, Department of Electrical Engineering, 1 Panepistimiou Avenue, 2109, Aglantzia, Nicosia, Cyprus, +357-22892287, hadjipanayi.maria@ucy.ac.cy
[2]K-Energy, 6 Pentadaktylou Street, Palaiometocho, 2682, Nicosia, Cyprus
[3]Deloitte, Innovation and Entrepreneurship Centre, 1 Ayias Zonis and Thessalonikis Street, 3025 Limassol, Cyprus

ABSTRACT: The work reported involved the development of a glass PV system for the purpose of integrating it into a commercial building façade in Cyprus. Firstly, the potential for building integrated photovoltaics (BIPV) in Cyprus was analysed considering the different type of buildings available and their respective needs, available space, and envelope retrofitting possibilities together with possible specific applications of BIPV. Moreover, the solar resource of Cyprus, the expected annual energy yields for different PV technologies at vertical tilt and different façade orientations were analysed to derive the best-case scenarios for BIPV systems in specific building types. The BIPV solution was installed on the façade of a commercial building in Nicosia, Cyprus and after initial tests it is currently fully operational. The pilot BIPV installation is also being monitored in terms of PV generation, module temperature, irradiance resource and environmental parameters. Overall, this work has enabled a first-time hands-on experience for a Cypriot SME to be involved in launching a new product in the local solar PV market with the prospect of being fully certified after the end of this project.
Keywords: Photovoltaic, Building Integrated PV (BIPV)

1 INTRODUCTION

Building-integrated photovoltaics (BIPV) are considered the optimal method for installing effectively and seamlessly renewable energy sources in urban built-up areas. Their widespread use is expected to be the backbone for materialising the European targets on nearly-zero energy buildings (NZEB). Despite the potential for BIPV to accelerate the decarbonisation of EU's building stock, the technology is still experiencing a slow up-take and the BIPV market is yet to meet its growth expectations. However, as building architecture trends nowadays show emphasis on large openings and fully-glazed façades, particularly in commercial buildings, there is enormous untapped potential and opportunity for glass BIPV installations to contribute to reducing significantly building energy consumption. In fact, the biggest market for glass BIPV installation in the near future is expected to be the façades [1].

To this end, the Impact-BIPV project [2] was conceived and implemented in order to study practically the BIPV façade potential for buildings in Cyprus. The project focused on the development of a glass PV system and its pilot installation on a commercial building façade in Cyprus as a BIPV element. An equally important aspect of this project was to support a local SME in creating capabilities in research and innovation, which is still a low performing metric in Cyprus, particularly towards addressing the lack of local expertise in the BIPV market, design, installation and operation. The project essentially envisaged to achieve the following:

- the development and testing of a full cost-effective BIPV solution customized for the conditions of Cyprus considering local building stock architectural trends;
- to integrate the developed solution as small-scale demonstration for the first time in Cyprus on a selected existing building;
- to create technical skills required in the local industry to grow and sustain this new market in Cyprus, and
- to attract interest in energy efficiency from other local building professionals as well as public authorities and local policymakers.

In this paper we discuss the implementation of this project, the challenges encountered, and some preliminary results from the pilot BIPV operation.

2 BIPV POTENTIAL IN CYPRUS

Initially, the potential for BIPV applications in Cyprus was analysed in terms of the available (and new) building type stock. The majority of buildings in Cyprus (approximately 67%) are residential [3]. Buildings that can be classified as residential may include a single-family house, duplex, a flat that is part of an apartment block and terraced family house. Non-residential buildings include offices, commercial and industrial buildings. Typical examples of the different types of buildings available in Cyprus are presented in Table I.

The various features of such typical buildings, such as the indoor area, window ratio and energy performance as well as parameters for both new and existing buildings are presented for each building type in Table I [4]. This exercise was useful in identifying the different type of buildings that can be considered when it comes to possible specific applications.

In general, the average energy performance of the buildings depends on many parameters, such as the type of glazing, the efficiency class of the electrical loads installed within the building, and the insulation properties of the building's surfaces. Nonetheless, as shown in Table I, even new buildings that are built using advanced materials with high thermal insulation properties and are equipped with energy class A electrical appliances have substantial energy requirements. Therefore, there is a need for onsite renewable energy, in order to reduce the reliance on fossil fuel, on which Cyprus is highly dependent [5]. As large buildings have the largest window ratio – also observed in Table I – they offer the greatest opportunity for BIPV integration into the building windows. Moreover, the largest wall surface for all the example office and retail buildings faces the South, which is also important for integrating ventilated façades in the building envelope.

Table I: Comparison of building types [4]

Building type	Surface-to-Volume ratio	Window ratio (%)	Area (m^2)	Average energy (kWh/m^2)
		Existing		
Single family house	0.36	8.82	194	666
Single family house	0.36	12	195	321
Apartment block	0.17	14.83	519	402.5
Apartment block	0.11	14.04	826	439
Office building	0.33	21.19	1448	537
Retail building	0.29	17.81	412	822
		New		
Single family house	0.33	13.6	176	479
One-bedroom apartment	0.34	21.85	46	521/363/664*
Two-bedroom apartment	0.34	12.65	88	507/351/635*
Three-bedroom apartment	0.35	16.28	103	497/354/617*
Office building	0.29	27.2	2515	402.7

** Note: As 1st, 2nd or 3rd floor.*

The solar resource of Cyprus, the expected annual energy yields for different PV technologies at vertical tilt, and different façade orientations were analysed in order to derive the best-case scenarios for BIPV systems in specific building types. In Cyprus the annual solar irradiation can reach 2000 kWh/m² [6] therefore, the maximum generated annual energy for monocrystalline silicon is around 400 kWh/m², whereas for thin-film is around 300 kWh/m², by assuming cell efficiencies of 20% and 15%, respectively and that the PVs are positioned optimally. For different surfaces of buildings, different percentage of the incident solar radiation is accepted. For south-facing vertical surfaces, the maximum solar irradiation is 70% of the maximum available, whereas for east and west-facing vertical surface the maximum available solar irradiation is 50%. Therefore, installing monocrystalline silicon PV modules on a south-facing vertical wall would yield a total of 280 kWh/m²/year while on an east and west-facing vertical wall the PV module would yield 200 kWh/m²/year. Similarly, a thin-film PV module on a south-facing vertical wall would yield a total of 210 kWh/m²/year while on an east and west-facing vertical wall the PV module would yield 150 kWh/m²/year [6]. These are the maximum energy values for each building surface for the given cell efficiencies and by assuming that no shading exists. Overall, after studying the case of certain typical buildings in Cyprus with their respective surface areas in the different orientations it was found that for both residential and office buildings, BIPV vertical façades combined with roof-mounted applications resulted in the best energy-demand coverage scenarios (80% for a residential reference building and 50% for an office reference building, whereby all surfaces were assumed to be covered by non-transparent and transparent BIPV).

3 DESIGN AND ASSEMBLY OF BIPV SOLUTION

Originally, the main aim of the project was to develop a solar PV glass from scratch, which in turn would be integrated within an aluminium system for BIPV application. The possibility of procuring photovoltaic cells has been explored to embed them in glazing in collaboration with glazing manufacturers in Cyprus. Various parameters needed to be considered to design the solar glass such as module size and shape, covering glass, glass quality, strength, structure, coating, colour, cell background or reverse side of module, semi-transparency, arrangement of solar cells in the module, interconnections, etc. The options in Cyprus in terms of components, infrastructure and technical expertise in manufacturing PV glazing for building integration were limited and the risk of failure was high so it was deemed more appropriate to acquire PV cells already embedded in double-glazing that could meet all the above-mentioned requirements and utilize this to develop a complete and functional BIPV solution specifically for façades and balconies. The ultimate aim is for the final BIPV solution, after being piloted in an existing building in Cyprus, to be ready for official certification so as to fully comply with PV regulations and building safety (e.g. EN 50583 'Photovoltaics in Buildings', as well as national building and renovation regulations).

3.1 Criteria for PV glass selection

The main parameters that were considered when choosing the PV glass to use were:

- accordance with the architectural requirements of the building design particularly paying attention to glass transparency and colour as well as the solar factor (e.g. low-e lamination)
- sensitivity to high temperatures which is particularly important for high insolation countries like Cyprus and for BIPV façades where there is the issue of low ventilation for the BIPV
- accordance to building regulations for NZEB in Cyprus whereby the maximum average coefficient of thermal permeability of the window (including aluminium frames) must not exceed the value of 2.25 W/m²K.
- Electrical compliance requirements including IEC safety qualifications (in terms of BIPV module and system safety, BS EN 50583-1 & 2: 2016) and BS 7671 wiring regulations
- Elimination of thermal bridging effect in the glazing.

3.2 Characteristics of chosen PV glass

Taking into consideration all the requirements described above, the technology selected for the PV Glass to be assembled into a façade component was the amorphous silicon (a-Si). In particular, two units of a-Si technology with high transparency (30%), good thermal insulation (1.6 W/m²K), peak power 28 Wp/m², and dimensions 1000mm x 1000mm were acquired from Onyx Solar [7]. Prior to installation, the glass PV modules were subjected to indoor tests for validating their performance. Their output performance under standard test conditions (STC) was investigated using a class AAA solar simulator (IEC 61215, Ed. 2, Procedure 10.2) and was found to be in line with the nominal expected performance. The modules also passed the requirements of the IEC standards for insulation tests, exhibiting resistance values well in excel of the minimum requirement.

3.3 Junction box positioning and PV frame

Next followed the development of a solution for the integration of the junction box in the metal frames of the BIPV solution. Ideally, the junction box would need to run within the aluminium profile allowing both good ventilation and protection against moisture. In this respect, two options were considered: positioning the junction box (a) on the edge of the PV glass (edge junction box), or (b) behind the PV glass (rear junction box). Due to space limitations in the aluminium profile, the latter was chosen, which also enabled the easier connection of the different modules together (see Figure 1).

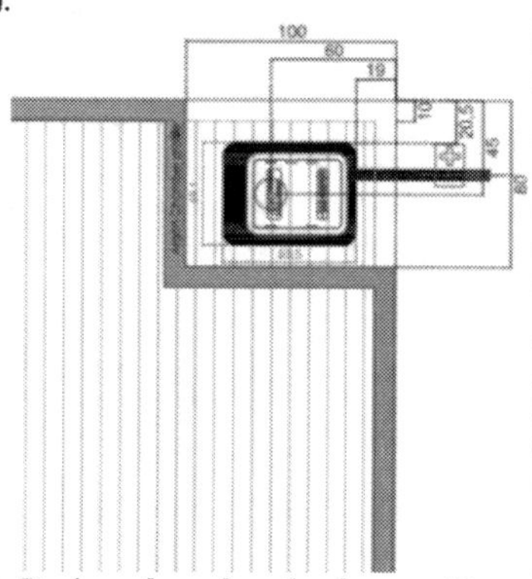

Figure 1: Design of rear junction box position.

For the PV frame selection, the following requirements were taken into consideration: waterproofing of the cable routes and connection points, compliance with electrical wiring safety regulations (BS 7671, 17th Edition), as well as accessibility for operation and maintenance. To this end, the selected PV glass was framed using a building system called 'Envelop 3D' invented by RABEL SYSTEMS [8]. Essentially this technology allows the creation of multiple thermally broken 3D envelopes with integrating components (e.g. shading devices, railing, louvres, etc.) and could easily adapt to the project's requirements for a custom frame enabling safe cable routing and cable connection accessibility (see technical characteristics in Table II).

Table II: Technical specifications of Envelope 3D [8]

Technical Specifications	**Value**
Post width	35/42/71 mm
Frame height	57 mm
Sash height	6500 mm
Glazing possibilities	32-59 mm
Thermal break width	30/53 mm
Sealing type	Gasket
Performances	**Value**
Fabrication thermal transmittance EN ISO 10077-2 (Uw)	0.6-1.4 W/(m²K)

Based on the 3D envelope method, a tailor-made aluminium frame was developed and the PV glass was sealed appropriately and assembled into a window and a balcony glazing solution (Figure 2).

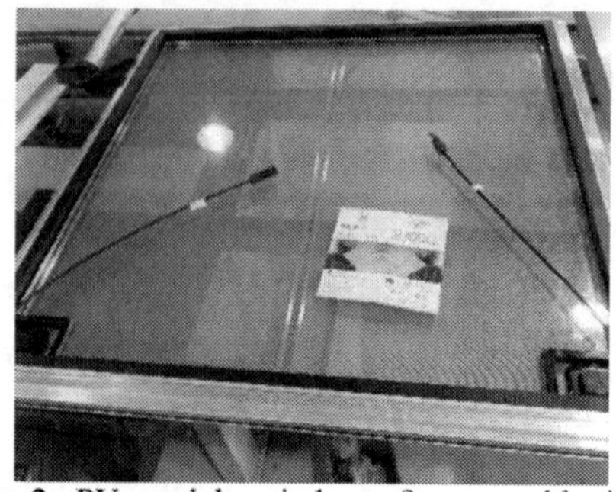

Figure 2: PV module window after assembly into an aluminium frame.

4 BENCHMARKING STUDY

4.1 Simulation approach

As part of the performance evaluation of the BIPV solution developed, the system was benchmarked existing PV technologies to give a clearer understanding of how the new system compares with existing PV solutions for the built environment. For this analysis, the annual energy yield of the developed system was simulated using the pvlib python tool [9] for the installation site in Cyprus. This tool takes as inputs a typical meteorological year (TMY) obtained from the PVGIS solar irradiation database [10] as well as the laboratory-measured performance characteristics of the system. In addition, the same analysis is performed using commercially available alternative PV panels, including: monocrystalline silicon panels (glass/polymer), multi-crystalline silicon panels (glass/polymer), and thin-film PV panels (glass/polymer). Each of these were simulated using two temperature models: (1) A close-mount arrangement, where the panels are close to, but not sealed against, a rear support structure, (2) An insulated back, where there is no free convection at the rear of the module. For the three generic types listed above, three representative modules were selected for the simulation from the California Energy Commission (CEC)'s module database, each with an active area of ~1m². For the Impact-BIPV modules, input parameters to describe their performance were compiled from various sources summarized in Table III.

Table III: BIPV module parameters compiled for the purpose of performance benchmarking simulations.

Parameter	**Source**
Temperature coefficients for current and voltage	Manufacturer data sheet
STC power, current, and voltage	Measured under STC laboratory conditions (IEC 61216 ed.2) at Univ. Cyprus
Power output at PVUSA test conditions	Calculated from data sheet and outdoor measurements
Normally operating cell temperature	Regression analysis of in-situ outdoor measurements
Modules series and shunt resistances	Calculated from measurement data using NREL's SAM software

The simulation was performed using the pvlib software. Spectral effects were ignored due to the lack of detailed spectral response data for the different panels, and therefore the uncertainties that this analysis would introduce were avoided. Furthermore, detailed site data with a horizon profile and ground albedo measurements were not introduced to the simulation, so the energy yields calculated are likely to be an overestimation for the particular installation site. Nonetheless, the results can be expected to give a good representation of the relative technology performances for the purposes of benchmarking.

4.2 Energy yield results

The annual energy yield of the four systems, each comprising the same inverter and two modules of area $1m^{-2}$ were simulated using the pvlib software. The modules were oriented at an angle of 90 degrees to the horizontal and an azimuth angle of 206 degrees. The results for each system under two different mounting scenarios are shown in Figure 1.

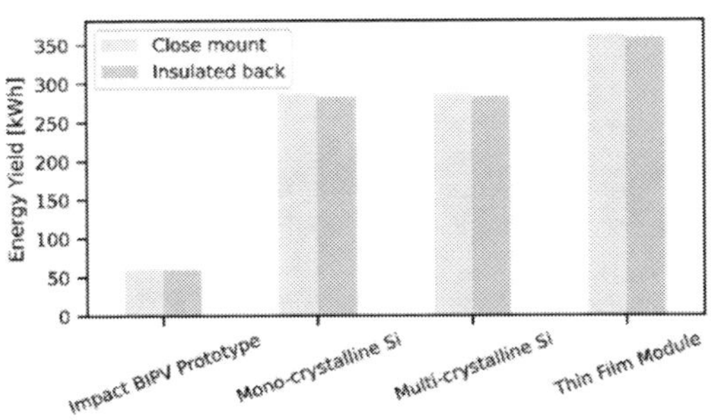

Figure 1: Simulated energy yields for the four PV technologies at the pilot building in Nicosia, Cyprus, oriented in the same manner as the planned installation.

The comparatively low energy yield of the prototype is attributed to the high level of transparency of the module, as it is also intended to allow the passage of light into an interior. The superior performance of the thin-film system is attributed to the better performance of amorphous silicon under the low light conditions that are prevalent in vertical façade mounting arrangements. Finally, the different mounting scenarios were seen to have only a minor effect on the overall expected energy yield.

4.3 Performance Ratio (PR) results

In addition to energy yield studies, the Performance Ratio (PR) of the four PV module case studies was investigated. The PVGIS database was used to obtain the global solar irradiance resource in the plane of the chosen installation location. This was combined with the nameplate performance efficiencies of each module to arrive at an anticipated energy yield under 'optimal' conditions. The pvlib energy yield simulation considers all the losses typically encountered in PV systems, such as temperature and angle-of-incidence effects, apart from the factors mentioned in section 4.1.

The ratio of the values for each simulated scenario are plotted in Figure 2, and show that the Impact-BIPV prototype can be expected to outperform all the other technologies, in terms of PR. This is attributed to the better low light performance and comparatively low operating temperature of these modules.

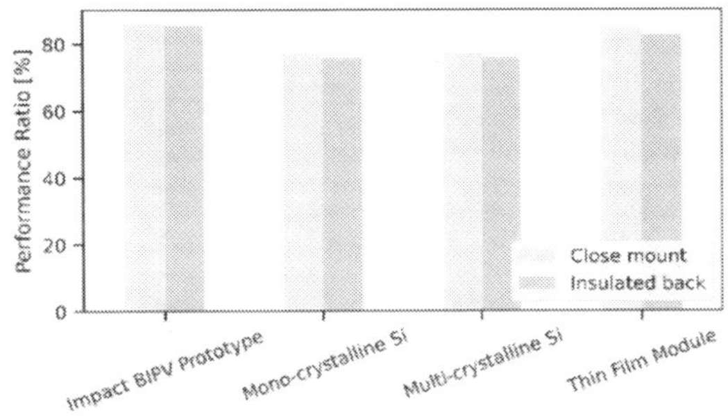

Figure 2: Comparison of the annual performance ratios for the four PV systems simulated.

5 PILOT INSTALLATION

5.1 Selection of pilot site

Cyprus experiences amongst the highest solar irradiation in Europe, across the whole island (see Figure 5). For this reason, the selection of the installation site in the country was mainly based on criteria related to the status of the building and not the renewable source level. The selected building for the pilot integration of our BIPV solutions is a commercial building located in the southern suburbs of Nicosia. The selection of the building that is going to accommodate the ImpAct-BIPV prototypes was made based on the following criteria:

- The location and orientation of the building is favourable as there are no shading problems from the surroundings.
- The structure of the building will be convenient to monitor, control and demonstrate the prototypes.

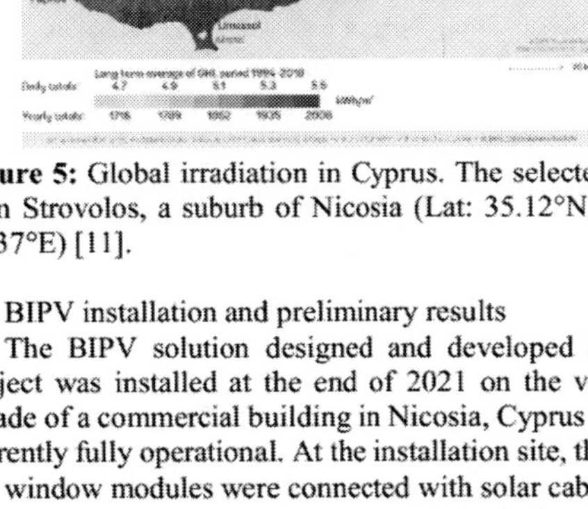

Figure 5: Global irradiation in Cyprus. The selected site is in Strovolos, a suburb of Nicosia (Lat: 35.12°N, Lon: 33.37°E) [11].

5.2 BIPV installation and preliminary results

The BIPV solution designed and developed in the project was installed at the end of 2021 on the vertical façade of a commercial building in Nicosia, Cyprus and is currently fully operational. At the installation site, the two PV window modules were connected with solar cable in a series circuit connection. The initial challenges of achieving onsite weather-proofing, thermal insulation, effective positioning of the junction box and relevant cabling, and enabling access for future maintenance and repairs have been overcome.

The assembled solutions were incorporated with a data collection setup including a custom-built MPPT controller

(MPPT DC/DC converter) for recording the electrical measurements and sensors for recording temperature readings and irradiance. A temperature sensor was attached to the rear side of each module and the irradiance sensor mounted in the same plane as the modules. The MPPT controller was installed in a shady and sufficiently ventilated space, to ensure its optimal operation so that external environment conditions do not affect its efficiency. An electronic load is used to dissipate the generated power. The MPPT converter was integrated to a datalogger for collection and storing of the data from all the sensors connected to the BIPV pilot. For the purpose of data collection, a GSM module is used and the data is transferred to an (Internet of Things) IoT data cloud for later analysis.

The pilot BIPV is currently being monitored (after the initial tests and troubleshooting) in terms of PV generation, module temperature, irradiance levels and environmental parameters on site. From the preliminary measurements, the BIPV modules indicated initial power generation (together) of approximately 50W under full sunlight.

The performance of the modules during the first months of operation showed a reduction over time which is largely attributed to the change of the angle-of-incidence as we move from winter to summer time in Cyprus and the sun gets higher in the sky. An overview of the daily instantaneous power measurements from the two modules is shown in Figure 6.

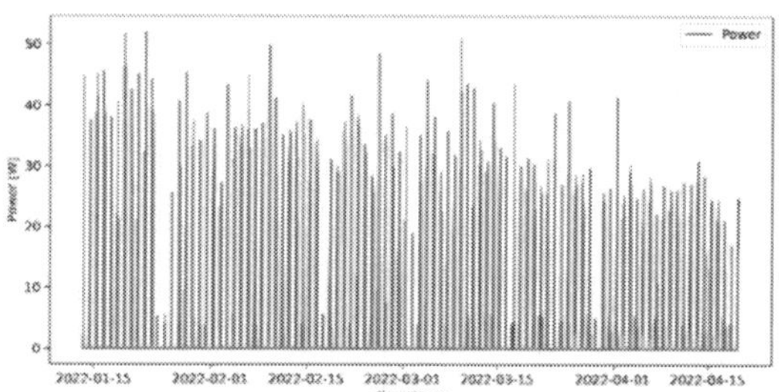

Figure 6: Instantaneous power measurements from the installed BIPV modules recorded over the period January -April 2022.

The performance efficiency of the installed modules was studied further against the angle of incidence and irradiance levels. Figure 7 shows the average performance efficiency binned by angle of incidence and irradiance. From this we can learn that above 600 W/m^2 the system performance is relatively constant at roughly 2% power conversion efficiency (PCE). When the sun is behind the panel, i.e. angles greater than 90° the performance is quenched. At angles below 80°, the system exhibits improved performance, except between 200-400 W/m^2 of irradiance when a dip in performance occurs. Further studies are needed to understand the origin of this dip. Below 200 W/m^2 there's an increase in performance, possibly indicating that this thin-film PV material behaves better than the silicon reference sensor under highly diffuse / cloudy conditions.

Monitoring and data analysis will continue for at least six more months beyond the end of the project in order to extract a good first insight into the power performance and possible degradation of the final pilot solution developed as well as help identify any maintenance issues regarding mechanical and electrical parts of the installed system.

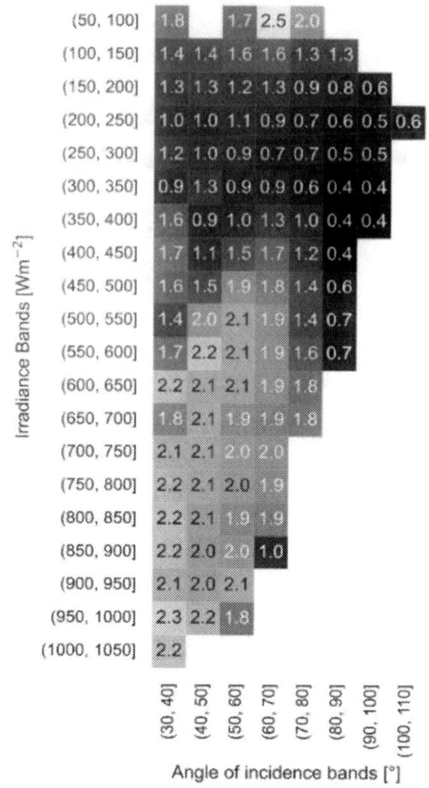

Figure 7: Performance efficiency (in %) against angle of incidence and irradiance.

6 CONCLUSIONS

Overall, this work resulted in new and important learning curves for a Cypriot SME, particularly concerning first-time hands-on experience in developing, commissioning, and testing in real ambient and operational conditions a new BIPV solution targeting the Cypriot solar PV and building market. Essentially, the project successfully brought together a local solar PV company and a local glazing company, with no prior experience in BIPV, to integrate effectively solar glazing components in high-performance aluminium profiles for local façade, window, and balcony applications.

A number of challenges have been overcome regarding appropriate glazing embedding for the glass modules, appropriate handling/positioning of electrical components of the façade (junction box, cables), proper installation and fixation in building, etc.

The qualification and safety tests performed showed operation at values close to the nameplate values, and certainly within the ±3% confidence interval of the laboratory measurements. As regards the benchmarking study, although it showed the anticipated energy yield of the prototype BIPV system to be low in comparison with other PV technologies, this is an inherent aspect of technologies that must balance the competing demands of capturing light for generating electricity and allowing light into an interior to provide pleasant and free illumination for the building occupants, not to mention the impact of window designs on interior temperatures through solar gain and insulation. An encouraging result of the benchmarking exercise is that the prototype can be expected to exhibit a very good performance ratio, maximising the benefit of the included PV material. This is attributed to the comparatively low operating temperatures of the modules as well as the fact that it is based on a thin-film technology that operates well in the low-light conditions that can frequently occur on building façades.

Finally, the pilot system will continue to be investigated (e.g. in terms of low light performance, shading impact, temperature response, degradation, long-term performance in time, etc.) before pursuing an official full quality certification by accredited bodies.

7 ACKNOWLEDGEMENTS

The project ImpAct-BIPV (title: 'Innovative Design for Improved Application of Glass BIPVs on Buildings') is co-financed by the European Regional Development Fund and the Republic of Cyprus through the Research and Innovation Foundation (Grant ID: ENTERPRISES/0618/0052).

8 REFERENCES

[1] PVSites Horizon 2020 project (Grant ID: 691768), 2016. BIPV market and stakeholder analysis and needs (Project report from website: www.pvsites.eu, Accessed 12/08/2022).

[2] The ImpAct-BIPV project at the website https://www.k-energy.com.cy/impact-bipv/

[3] Deloitte, Cyprus Real Estate Market Outlook Real Estate Advisory Services June 2016.

[4] Calculation for Setting the Minimum Energy Performance Requirements at Cost Optimal Levels According to Article 5 of the Directive 2010/31/EE for the Energy Performance of Buildings (recast), April 2013.

[5] T. Mesimeris et. al, Cyprus' Draft Integrated National Energy and Climate Plan for the period 2021-2030, European Commission 2019.

[6] JRC Photovoltaic Geographical Information System (PVGIS), European Union, 2001-2022.

[7] "Onyx Solar," https://www.onyxsolar.com/, Accessed: 12/08/2022

[8] "RABEL SYSTEMS," https://rabel.com.cy/aluminium-systems/rabel-envelop-3d-building-system, Accessed: 12/08/2022

[9] William F. Holmgren, Clifford W. Hansen, and Mark A. Mikofski. pvlib python: a python package for modeling solar energy systems. Journal of Open Source Software, 3(29):884, 2018.

[10] Huld, T., Müller, R. and Gambardella, A., 2012. "A new solar radiation database for estimating PV performance in Europe and Africa". Solar Energy, 86, 1803-1815.

[11] 2020 The World Bank, Source: Global Solar Atlas 2.0, Solar resource data: Solargis

INSPIRING BIPV IN UAE, A JURY EXPERIENCE IN SOLAR DECATHLON MIDDLE EAST (SDME 2021) CONTEST

Cristina S. Polo López[1], Francesco Frontini[1], Edwin Rodriguez-Ubinas[2]
[1]SUPSI, University of Applied Sciences and Arts of Southern Switzerland, Mendrisio, Switzerland Via Flora Ruchat-Roncati 15, CH-6850 Mendrisio, T +41 (0)58 666 63 14, cristina.polo@supsi.ch
[1] SUPSI, University of Applied Sciences and Arts of Southern Switzerland, Mendrisio, Switzerland Via Flora Ruchat-Roncati 15, CH-6850 Mendrisio, T +41 (0)58 666 63 20, francesco.frontini@supsi.ch
[2] Dubai Electricity and Water Authority, DEWA R&D Centre, Dubai, United Arab Emirates, edwin.ubinas@dewa.gov.ae

ABSTRACT: Traditional photovoltaic applications dominate the PV market in the United Arab Emirates (UAE), with only a few examples of Building Integrated Photovoltaics (BIPV). However, the UAE movement toward more sustainable development is bringing interest to innovative and better integrated renewable energy solutions for buildings. The Dubai Electricity and Water Authority organised the Solar Decathlon Middle East (SDME) to raise social awareness of sustainability and clean energy technologies. This competition challenges university students to design, build, and test sustainable Zero Energy Houses. SDME juries conduct a comprehensive evaluation of the houses through its ten contests. Furthermore, the SDME BIPV special award recognises outstanding Building Integrated Photovoltaics systems. Researchers from SUPSI were on the juries of this award on the SDME 2021. The present study summarises their experiences and findings, focusing on the top four projects and including the analysis of the architectural integration, societal messages, local culture relevance, multifunctionality, and energy production.
Keywords: Building Integrated; BIPV; Solar Architecture; NZEB; Education and Training

1 INTRODUCTION

Worldwide sustainability policies recognise Building Integrated Photovoltaics (BIPV) as one of the most effective ways to exploit the huge potential of the decentralised generation of renewable energy in the built environment [1]. The EU and many countries around the world have set demanding goals related to nearly zero or zero energy buildings. These objectives can be achieved by replacing conventional surfaces of building roofs and façades with active PV claddings; it is possible to achieve these goals [2, 3]. The United Arab Emirates (UAE) construction sector has been growing steadily over the last decades at a speedy pace, favouring the housing and tourism sector [4]. The UAE construction industry is expected to expand by 4.4% in 2022 and register an annual average growth of 3.9% from 2022 to 2025. Hosting the Expo 2020 event (1st October 2021 - 31st March 2022) provided positive support to the economy compared to the levels of 2020 [5]. In October 2021, the UAE became the first country in the Gulf Cooperation Council (GCC) region to commit to carbon neutrality, announcing a target toward net zero by 2050 [6]. To achieve this target, the government plans to invest AED600 billion (US$163.4 billion) in renewable energy over the next three decades.

The Dubai Electricity and Water Authority DEWA (http://www.dewa.gov.ae/en) is a key enabler of Dubai's energy transition to net zero by 2050. In 2014, it started Mohammed bin Rashid Al Maktoum Solar Park, the largest single-site solar park in the world. With this project and for several years now, the Middle East once again confirms the growth and importance of commissioning large solar projects and launching further phases of their renewable and solar energy programmes. In the UAE, as well as in other countries in the region, many projects are in the pipeline and launching their first major projects [7]. Shams Dubai launched two years later in 2016, is the DEWA's first smart initiative to connect solar energy to buildings. The initiative encourages households and building owners to install PV panels to generate electricity and connect them to DEWA's grid. Over 11,000 new residential units are currently being constructed across the country as part of the Sheikh Zayed Housing Programme [8]. Other related initiatives included the R&D Centre and the Innovation Centre, equipped with the latest renewable and clean energy technologies to raise awareness of sustainability while enhancing national capabilities and increasing competitiveness [9].

DEWA and the US Department of Energy signed an agreement to collaborate on developing the Solar Decathlon Middle East (SDME). The SDME challenges university students to design, build, and operate sustainable zero-energy houses. Its 1st edition was held in 2018, and despite the global pandemic, DEWA successfully organised the second edition of the competition in 2021. Both editions occurred at the Mohammed bin Rashid Al Maktoum Solar Park, integrating the unique local and regional characteristics of Dubai, United Arab Emirates.

1.1 Solar Decathlon Competition

Solar Decathlon (https://www.solardecathlon.gov/) competitions started 20th years ago in the USA with the aim to prepare the next generation of building professionals to design and build high-performance, low-carbon buildings powered by renewables, mainly solar. Since the first competition in 2002, the Solar Decathlon has expanded internationally to include six additional worldwide competitions: Solar Decathlon Africa (started in 2016), Solar Decathlon China (started in 2013), Solar Decathlon Europe (initiated in 2007), Solar Decathlon India (launched in 2020), Solar Decathlon Latin America and Caribbean (initiated in 2015), and Solar Decathlon Middle East (first edition in 2018). For approximately two years, multidisciplinary teams worked together in their projects' design, construction, and operation, first in their country and then in the competition's host country. Teams present their designs to expert jurors and have the opportunity to enhance their professional network, make connections with industry professionals, share ideas with students from around the world, and explain their projects to visitors. Only fully operational houses enter the final phase of the competition. In that phase, the performance of the houses and teams are evaluated according to different KPIs set by the Solar Decathlon rules.

Solar Decathlon's challenges regarding sustainability

include many aspects, such as the use of renewable energy, rational consumption (energy, materials, resources, food, etc.), and raising social awareness about clean energy and environmental responsibility.

1.2 SDME 2021 Competition

Solar Decathlon Middle East (SDME 2021) [10] was held in Dubai (UAE), organised under the patronage of His Highness Sheikh Hamdan bin Mohammed bin Rashid Al Maktoum, Crown Prince of Dubai and Chairman of the Dubai Executive Council.

The SDME embraces the goal of developing and promoting ideas, capacities, and technologies that can be implemented for the benefit of the inhabitants responding to the Middle East's cultural, climatic, and social contexts. Although, since the early 1970s, architectural styles in Dubai have changed significantly (see Figure 1), SDME proposals show a balance between the energy benefits of traditional architecture's passive design strategies and contemporary architecture's innovation and advanced technologies.

Figure 1: Dubai City, contrasting traditional architecture (in the old town) and modern architecture. Photo: © Polo C. (SUPSI).

The SDME Rules encourage the teams to develop competition prototyping houses that meet the sustainability triple challenge: energy, environment, and society. They also established that the prototype houses must have a net usable area ranging from 70 m^2 to 120 m^2, organised for living and working. All proposals focused on addressing issues and needs for sustainable living in this region, where high temperatures, high humidity, and dust conditions challenge people's daily lives most of the year. The teams must assemble their highly energy-efficient solar-powered homes in a "Solar Village" (Solar Hai), the competition and exhibition site (Figure 2). Students (decathletes) and their houses are evaluated in ten different areas: 1. Architecture; 2. Engineering and construction; 3. Energy efficiency; 4. Sustainability; 5. Communication; 6. Innovation; 7. Energy management; 8. Comfort conditions; 9. House functioning and 10. Mobility (sustainable transportation).

Five groups of juries assess the first six categories. In contrast, the points for the last three come from the houses' continuous monitoring, test results, and the completion of the tasks required by the competition. In addition to the competition's overall and ten contest prizes, SDME grants three special awards BIPV (Building Integrated Photovoltaics), Interior Design, and Smart Solutions (developed by the students).

Figure 2: The "Dubai Solar Hai" at Mohammed bin Rashid Al Maktoum Solar Park. Source: SDME Organization.

The final phase of the competition comprises three periods: Assembly, Competition/Exhibition, and Disassembly. The assembly period lasts fifteen days and includes the setup of the houses, the rules compliance check, the building inspections, and the final connection to the electrical grid. The second period is a mixed one. During its fifteen days, the houses are opened to the public on specific hours since the monitoring, and competition contests evaluation also takes place during this period. On these days, the houses are fully operational, teams receive the visit of the juries, and award ceremonies are held. After the closing ceremony, start the disassembly. The houses are disassembled and transported to their permanent location.

Successful participating houses present a mix of sustainable solutions, good architectural practices, efficient technologies, and solar systems to balance their consumption and help them to achieve the Net Zero Energy NZEB level.

2 AIM AND APPROACH

This paper aims to present some latest results on Building Integrated Photovoltaics (BIPV) in highly efficient buildings, using as case studies the SDME 2021 houses. After introducing the SDME BIPV Special Award, the authors summarise the findings on the four better-rated houses by the BIPV jurors. The analysis of the case studies includes architectural integration, societal messages, local culture relevance, multifunctionality, and energy production.

2.1 BIPV Contest in SDME 2021

Since its first edition, the SDME competition has included a special prize to recognise outstanding BIPV solutions. The BIPV juries comprehensively analyse the presented solutions and assess how well the photovoltaics elements are incorporated or blended within the house's envelope (roof or facades) or their exterior components. DEWA invited two researchers from the University of Applied Sciences and Arts of Southern Switzerland (SUPSI) to participate as jury members for Building Integration Photovoltaics Special in the Solar Decathlon Middle East (SDME) 2021 competition. The SUPSI Institute of Applied Sustainability to the Built Environment (ISAAC) hosts the Swiss BIPV Competence Centre, one of the first institutes globally operating in the BIPV field. Since 2005, this centre has been involved in EU and international research activities. Researchers from SUPSI University work on the education, dissemination, and promotion of solar buildings, shifting the attention from technology to architecture. Furthermore, aiming to encourage architects, clients, and the new generations, they created the solar architecture website (www.solarchitecture.ch). From there, they showcase real examples of today's best practices, demonstrating the feasibility and the quality of solar buildings, as well as the BIPV benefits in terms of aesthetics, construction technology, and sustainability.

SDME 2021 was one of the parallel activities of the Dubai Expo 2020. That offered the participating universities and their partners the opportunity to visit one of the most important global events of the year and the world's biggest stage for innovations and solar architecture (Figure 3).

Figure 3: Dubai Expo 2020. BIPV solar integration in the Sustainability Pavilion, Grimshaw Architects. Photo: © Polo C. (SUPSI).

During Dubai Expo 2020, SDME teams had the opportunity to exhibit their futuristic vision of locally adapted and culture-integrated solar-powered houses. In addition, the SDME 2021 and the participating teams shared Dubai Expo values "*Connecting Minds and Connecting the Future*" and showed the future of building sustainably. As a result, hundreds of young minds became connected to ideate a future building sector characterised by the responsible use of natural resources, clean energy utilisation, and human-centric approach implementation (exploiting the latest technologies to improve people's health and well-being).

3 SHOWCASING OF SELECTED AWARDED PROJECTS – SDME 2021 BIPV CONTEST

SDME 2021 was an excellent platform for testing and displaying materials, components, equipment, and systems developed or improved by the participant universities or their research and industrial partners. The teams adapted Middle Eastern traditional concepts and looked for new solutions to respond to the SDME 2021 challenges. The result was houses with high efficiency, smart energy management, low embodied energy, clean energy production, water conservation, and occupant comfort. They also made efforts to integrate their photovoltaic systems into the architecture of their projects.

Eight teams representing twelve universities worldwide (US, UK, China, Bahrain, and UAE) participated in the final phase of the second edition of the SDME. These teams complied with all the competition requirements and overcame the added challenges of the COVID-19 pandemic. For the present analysis, the authors selected as case studies the four projects judged worthy of being awarded and recognised on the BIPV Special Award of the Solar Decathlon Middle East 2021. As Table 1 shows, these projects were developed by Team Sharek (University of Sharjah, UAE), Team Harmony (British University, UAE), Team ScutxCSCEC (South China University of Technology), and Team GO-Smart (University of Bahrain). They all reveal the multidisciplinary approach intrinsic to the realisation of solar architecture with photovoltaics. Besides, they successfully performed in all contests during the competition, as shown in Table 1 and Figure 4.

Table I: BIPV Special Award and overall rankings (SDME 2021).

Team	BIPV contest Ranking	SDME 2021 Ranking	SDME 2021 Score
SHAREK	1st place	2nd place	834.85
HARMONY	2nd place	5th place	608.97
SCUTxCSCEC	3rd place	1st place	934.34
GO-SMART	Honorable Mention	3rd place	785.41

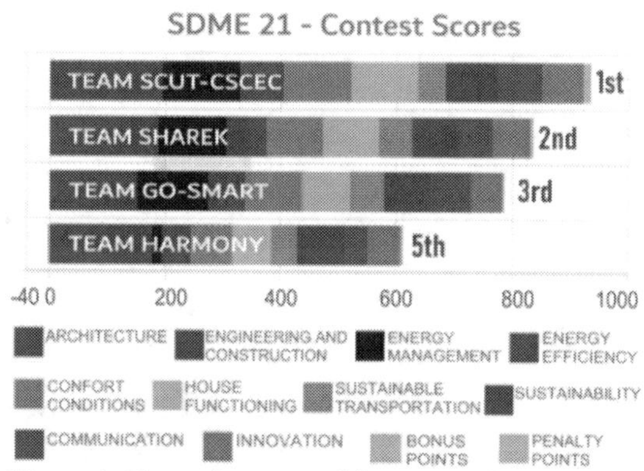

Figure 4: The performance of the case studies on the ten official contests and their overall ranking. Data source: SDME Organization.

3.1 Team SHAREK, University of Sharjak, (UOS), UAE

Sharek is a unique iconic project that uses BIPV as a fundamental part of the design and architecture, specially conceived for Dubai's hot climate and severe environment. The dynamic and circular envelope with BIPV collaborates in a smart way with the building system creating a harmonised construction that never appears the same, continuously changing during the day. Furthermore, this team used the PV modules as an essential part of the building itself, integrating them into roof and façade elements (Figure 5).

Figure 5: Team Sharek's house [11]. Source ©Photo: Polo C., Frontini F. (SUPSI).

The team integrated an innovative custom BIPV solution (2.4 kWp) into the circular facade (Figure 6) with a triple purpose: shade the south courtyard, produce energy and communicate the name of the team (printed on the protective glass). These vertical modules are mounted in a mobile structure that enables them to follow the sun and reach an electricity production of 3.9 MWh/y.

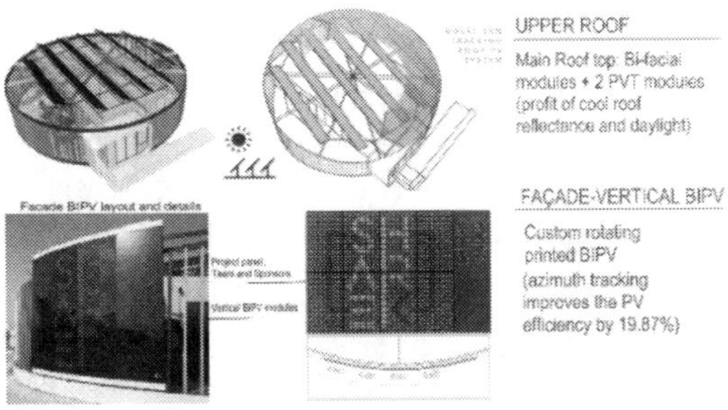

Figure 6: Team Sharek's house details. Source: SDME Organization.

Over the roof, the team installed twenty-two high-efficiency bi-facial PV modules with a tilt angle of 30° on a rotating aluminium structure (11.8 kWp). These modules are benefitted from cool roof reflectance since they can capture solar energy on both sides. The support structure is well designed to meet the module colours and re-call the vertical louvers of the façade. The rotating system tracks the sun's movement according to the irradiation level, optimising the modules' energy production. This rotating PV solution contributes to the passive cooling strategy and creates pleasant shadows in the garden during the day contributing to the multifunctionality of the BIPV installation (Figure 7). The azimuth tracking improves the modules' generation by 19.87% and enables them to produce 26.5 MWh/y.

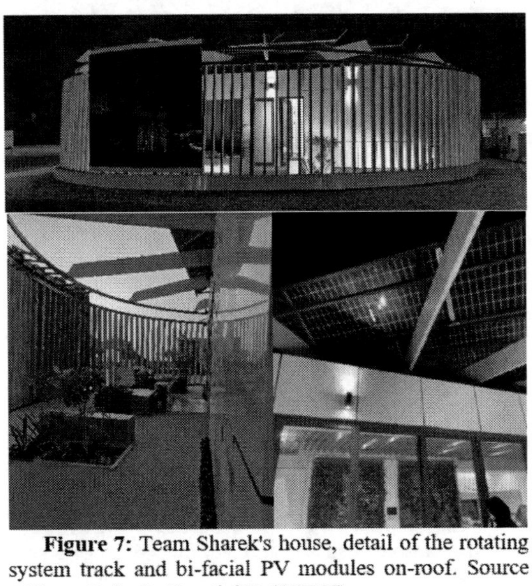

Figure 7: Team Sharek's house, detail of the rotating system track and bi-facial PV modules on-roof. Source ©Photo: Polo C., Frontini F. (SUPSI).

Sharek team of the University of Sharjah (UAE) took the second award at the SDME 2021 competition by well-balancing their house architecture, comfort conditions, energy management, and efficiency (Figure 8).

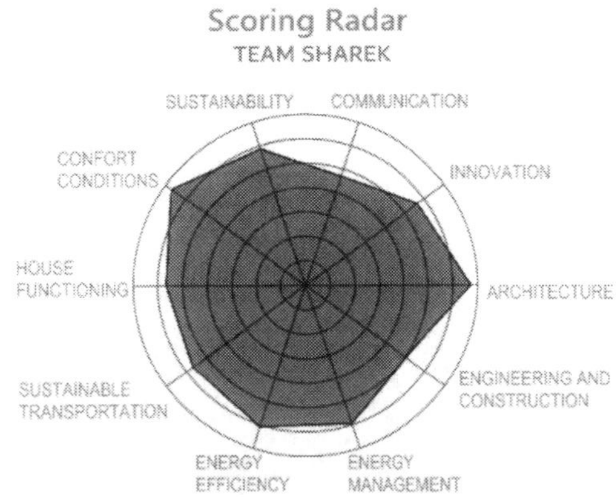

Figure 8: Team Sharek performance in the competition (2nd BIPV Special Award and 2nd SDME overall ranking). Source: SDME Organization.

3.2 Team HARMONY, The British University in Dubai, UAE

Harmony House has an external shell that serves two functions: to avoid indoor overheating (protecting the thermal envelope from solar radiation) and to be the support structure for the energy generation system (Figures 9 and 10). The photovoltaic modules are integrated into the external shell's pitched roof (facing south) and its west and east facades. The roof system (21° tilted South, 9.7 kWp, 17.5 MWh/y) is made from eighteen frameless glass-glass modules to reduce dust deposition and guarantee easier cleaning. The roof PV system is fully ventilated thanks to the house's "modern barajeel" which works as a traditional ventilation system and reduces the thermal losses of the photovoltaic modules.

Figure 9: Team Harmony's house [12]. Source © Photo: Polo C., Frontini F. (SUPSI).

The uniqueness of the Harmony project lies in the west and east facades (52 active solar sheets, 3.3 kWp) that use a novel BIPV technology (Cosmiq Art by Kameleon Solar). This technology allows the creation of multi-coloured designs on the PV modules. The team exploited this innovation to communicate messages linked to sustainability. A local artist contributed to the final design, and the result was BIPV facades that can be considered a piece of art that promotes local calligraphy as a cultural value. This façade system was properly designed, providing appropriate ventilation and fastening the panels with a simple but effective system. Moreover, the BIPV facades act as a second skin creating a comfortable overshadowed corridor around the house.

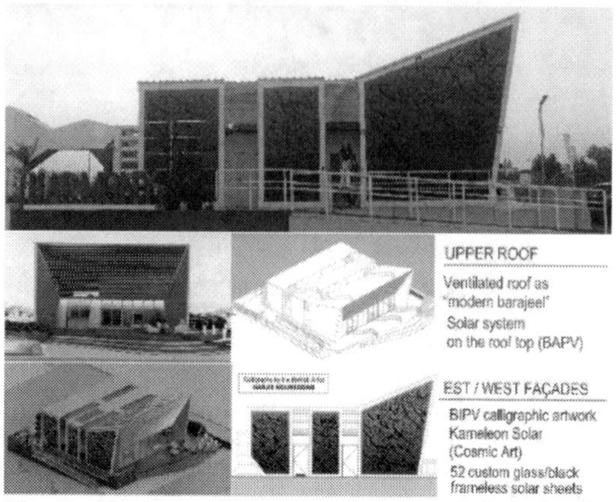

Figure 10: Team Harmony's house technical details. Source©: Courtesy of Team Harmony.

Team Harmony took fifth place at the SDME 2021 and second place at the BIPV contest (Figure 11).

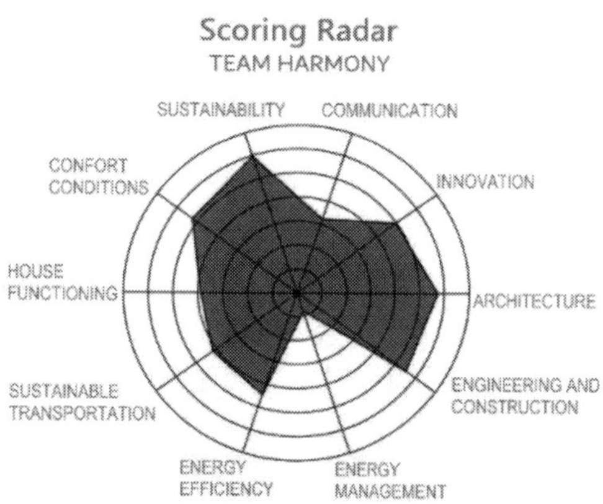

Figure 11: Team Harmony performance in the competition (2nd BIPV Special Award and 5th SDME overall ranking). Source: SDME Organization.

3.3 Team SCUTxCSCEC, South China University of Technology, China

This house was very well designed and built and reveals the great effort of the team to achieve the competition goal: to develop a sustainable construction that maximises energy efficiency and solar supply.

This team integrated forty-six PV modules into a canopy over the house roof and the exterior screen wall. These two systems are used simultaneously as shading and power generation components, helping the X-House to achieve zero energy balance in the challenging Middle East climate (Figure 12).

One of the distinctive elements of this house is the canopy that "flies" over its roof. This canopy is perfectly integrated with the building energy system and the interesting cooling system. The team ideated this cooling system inspired by traditional Middle Eastern cooling towers and wind catchers. It has the double function of redirecting the cool air to the house's central patio and rejecting the high-temperature winds. Under the canopy, four movable and electronically activated plates with an X arrangement redirect the wind, and operable shutters control the airflow to or from the patio.

Team SCUTxCSCEC installed a reflective-radiative film (Radi-Cool) on the house's roof with 91% solar reflectance. The film's high reflectivity reduces the solar heat gains of the house. It also increases the energy production of the bi-facial PV modules integrated into the canopy, providing 30% more radiation. The high-efficiency bi-facial PV modules use crystalline silicon PV cells of the PANDA n-type. The double-sided photovoltaic modules can achieve an annual energy production of approximately 27 MWh (2'6811 kWh), which is four times the annual consumption of the building. During the competition period in Dubai, it generated 1.8 times the energy consumption in the late autumn (10 days).

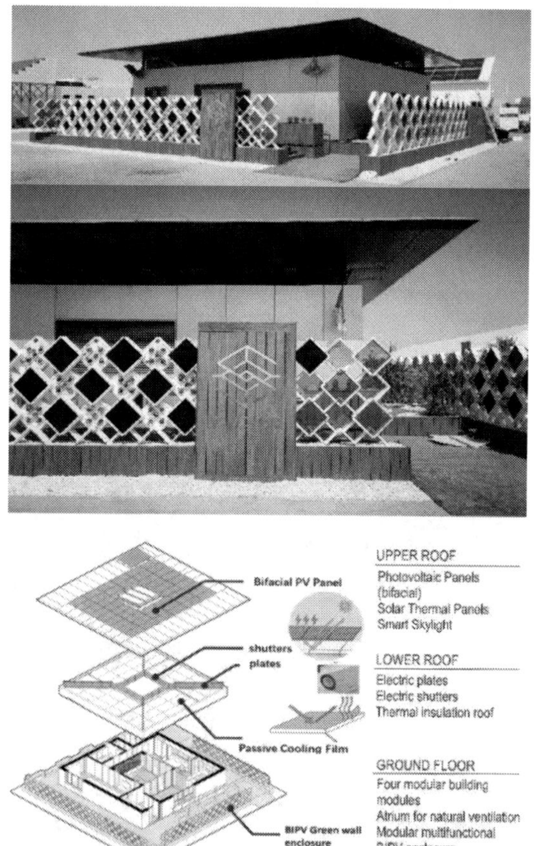

Figure 12: Team SCUTxCSCEC's house [13]. Source © Photo: Polo C., Frontini F. (SUPSI); ©Technical details: Team SCUT x CSCEC.

The team named the exterior screen wall "BIPV green wall enclosure". This wall is made of modular components of 400 x 400 mm size for a total of 384, with multifunctional purposes (Figure 13). The main frame of these components is prefabricated by UHPC (Ultra-High-Performance Concrete) as structure and to be modularly assembled, while users can customise the inner panel. The enclosure design is achieved through white 3D printed and easy-to-assemble frames that integrate two different BIPV technologies: black crystalline solar cells for the opaque element and semi-transparent CdTe thin-film cells for the transparent one with 70% light transmittance. The BIPV elements supply enough energy (5V each) to power the appliances to improve outdoor comfort. The design is well integrated with the house's architecture, contributing to pleasant ventilation of the outdoor space and providing the appropriate privacy for the house occupants. The internal area of this multifunctional enclosure can be customised and combined with green elements or connected to various DC appliances to ensure a variety of uses and applications powered by solar (e.g., fan modules, light-emitting modules, mobile phone charging modules, planting modules that can improve the comfort of the courtyard).

Figure 13: Detail of the vertical BIPV multifunctional enclosure, X-House. Source © Photo: Polo C., Frontini F. (SUPSI); ©Technical details: Team SCUT x CSCEC.

Team SCUT x CSCEC was the competition's overall winner, ranking first in nine of the ten official contests at the Solar Decathlon Middle East 2021 with the X-House project (Figure 14).

Figure 14: Team SCUTxCSCEC performance in the competition (3rd BIPV Special Award and 1st SDME overall ranking). Source: SDME Organization.

3.4 Team GO-SMART, University of Bahrain (UOB), Bahrain

The photovoltaic system integrated into the GO-Smart house is made of 4 arrays of e thirty-six solar panels of mono crystalline type (14.4 kWp and total energy production of about 21 MWh/y) facing south, supported on tilted structures installed over the roof (Figures 15 and 16). The PV modules shade the house from Dubai's intense solar radiation. The BIPV jury appreciated the simplicity of its concept, which is flexible, cost-effective, and highly replicable. This BIPV roof can be easily installed on almost any flat roof, as the one presented by the Go-Smart team. The system can be self-installed by the owners of new or existing houses. Its design facilitates the proper ventilation of the PV modules, thanks to the wood lamellas and the space under the solar modules that catch the wind, cooling down the system, increasing its efficiency, and reducing the house overheating. This space could also be used as part of a garden terrace for other purposes, such as drying or protecting goods from the sun.

Figure 15: Team GO-SMART's house [14]. Source © Photo: Polo C., Frontini F (SUPSI).

Figure 16: Team GO-SMART's house details. Source © Photo: Polo C., Frontini F.

Team GO-SMART, University of Bahrain, took third place at SDME 2021 competition and received an Honorable Mention from the BIPV Jury. Their house redefines the way of living in the Middle East with a contemporary revival of the traditional courtyard house. It combined traditional architecture and passive design strategies with smart technologies, renewable energies, and a flexible design that meets the standards for a sustainable lifestyle (Figure 17).

Figure 17: Team GO-SMART performance in the competition (Honorable Mention BIPV Special Award and 3rd SDME overall ranking). Source: SDME Organization.

4 RESULTS AND CONCLUSIONS

Until now UAE PV market was focused on conventional photovoltaic applications, and integration into buildings is not common. However, UAE Government programmes encourage distributed renewable resources generation and promote the installation of PV panels in buildings to diversify the energy mix and to reach carbon neutrality and net zero targets by 2050. As highlighted in the introduction, strong growth is predicted for Middle Eastern solar PV. For this reason, in 2016, Dubai government launched the Shams Dubai programme to encourage building owners to install solar PV panels, promoting awareness of the benefits of solar energy for the country's sustainable development. Following this mission, in 2021, DEWA organised the international competition Solar Decathlon Middle East (SDME). Two editions of SDME have been organised, SDME 2018 and SDME 2021. As a result, thousands of participating students, student visitors and visitors have benefitted from the education and social awareness of sustainability, Zero Energy Buildings, and the use of renewable energy. Regarding renewables, SDME special emphasises the seamless integration of solar systems in the building envelope and exterior elements, recognising the outstanding PV integration with the BIPV Special Award.

A total of eight teams representing 12 universities around the world have participated in the 2nd SDME to design, build and operate smart, sustainable homes that use solar energy and take into consideration the region's climate." SDME 2021 showcased some iconic projects that use BIPV as a fundamental part of the design and architecture, specially designed for hot climates and harsh environments such as Dubai. The teams integrated the photovoltaic modules into south-facing pitched roofs, facades, and exterior screen walls.

The participating teams adopted innovative BIPV techniques to communicate messages linked to sustainability. Solar energy provides electricity and, simultaneously, creates a strong relationship with the landscape and its immediate surroundings. Drawing inspiration from traditional local architecture to introduce active and passive conditioning techniques that reduce the energy load of dwellings, BIPV solar systems are successfully integrated into modern and traditional architectural design.

The four architectural projects awarded by the BIPV jury aim to challenge the way we build our homes, pushing the boundaries of what is possible by seeking out alternative construction methods that save energy, cost, and time with flexible, cost-effective, and replicable solutions. Table II summarises the main information on the BIPV systems of these four houses' prototypes.

Table II: Summary of BIPV systems used by the SDME awarded teams (BIPV Special Award).

	BIPV Installed capacity			Annual energy production
	Roof kWp	Façade kWp	Total kWp	MWh
SHAREK	11.77	2.40	14.17	26.49[1]
HARMONY	9.70	3.30	13.00	17.50[2]
SCUTxCSCEC	18.63	n.a.	-	26.80[3]
GO-SMART	14.40	-	14.40	21.00[4]

Annual production estimation: [1]BIPV façade (3.87 MWh) and bi-facial BIPV roof (22.62 MWh); [2]Energy estimation roof BIPV system; [3]Only bi-facial BIPV modules installed on the roof (405 Wp each, 19.9% efficiency); [4]36 BIPV on-roof modules (400Wp each).

4.1 Scientific innovation and relevance

Through design, architecture, and solar technology, SDME 2021 reflects on the identity and concept of new residential buildings, highlighting the importance of education on environmental issues. Following the EXPO 2020 Dubai tagline "*Connecting Minds, Creating the Future*", Building-integrated photovoltaic (BIPV) panels and solar energy can be a resource in communicating sustainability and raising awareness in new generations for more energy efficiency, contributing to decarbonisation of the built environment, showing the path to follow. The event held in parallel with EXPO Dubai 2020 [15] facilitated public and citizen participation and a broader audience to learn about the latest developments in terms of energy efficiency and the integration of solar renewables and BIPV technologies in housing to achieve the new energy and societal challenges prospected for the country and whole Middle Est region. For the first time in this edition, the importance of BIPV has been recognised by paying attention to the integration of solar in the house concept. Inspiring projects presented to this edition of the contest and synergies with BIPV experts revealed important potential follow-ups for future academic projects and research activities for mutual join collaboration and long-term partnerships that could give a new impulse to the BIPV market in the next years.

Despite the fact that today there is no regulatory framework established in the country for BIPV to be used as a part of the building envelope, BIPV has a huge potential for implementation in buildings in the future UAE. Architectural projects assessed by the BIPV jury proved the technological attractiveness of BIPV in the solutions presented.

5 ACKNOWLEDGE AND REFERENCES

5.1 Acknowledge

Authors from the University of Applied Sciences and Arts of Southern Switzerland (SUPSI) would like to thank the Dubai Electricity and Water Authority (DEWA) and the Dubai Supreme Council of Energy for the invitation to take part as the jury for the BIPV special award, within the SDME 2021 competition, all the DEWA team who have collaborated in the organisation (Tawaddod Alkindi, Sarah Abdulmajeed Alzarouni, and Habiba Ebrahim Ahmed Mohamed) and Mr. Claudio Montero, Energy Endeavour Foundation.

Furthermore, special thanks are given to all the team members of selected and awarded projects. Especially to Mr. Wenhao Zhang from the South China University of Technology, project manager of the X-House, and to Ms. Nesrine Oussadou, Project Manager and Site Operations Coordinator of Harmony Home, for the valuable information about the projects and for collaborating in the dissemination and transfer of knowledge through the website www.solarchitecture.ch.

5.2 References

[1] V. D'Ambrosio, M. Losasso, E. Tersigni, Towards the Energy Transition of the Building Stock with BIPV: Innovations, Gaps and Potential Steps for a Widespread Use of Multifunctional PV Components in the Building Envelope. *Sustainability* **2021**, *13*, 12609. https://doi.org/10.3390/su132212609

[2] European Commission. *EU Solar Energy Strategy* and *REPowerEU plan*; SWD(2022). European Commission: Brussels, Belgium, 2022. COM/2022/230 final.

[3] P. Corti, P. Bonomo, A. M Ramachandran and A. Asok (2022). *Indian BIPV Report 2022: Status and Roadmap. Status Report 2022*. SUPSI-NIIST. https://solarchitecture.ch/wp-content/uploads/2022/08/BIPV-Report-India_v01.pdf (accessed 25.08.22)

[4] United Arab Emirates Construction Industry Databook Series - Market Size & Forecast by Value and Volume (area and units) across 40+ Market Segments in Residential, Commercial, Industrial, Institutional and Infrastructure Construction, Q1 2022 Update, Report February 2022, pag. 170.

[5] United Arab Emirates (UAE) Construction Market Size, Trends and Forecasts by Sector - Commercial, Industrial, Infrastructure, Energy and Utilities, Institutional and Residential Market Analysis, 2022-2026. Report, 44 Pages, April 2022, GlobalData. ID: 5610510.

[6] UAE Energy Strategy 2050 (2017), https://u.ae/en/about-the-uae/strategies-initiatives-and-awards/federal-governments-strategies-and-plans/uae-energy-strategy-2050. (accessed 25.08.22)

[7] MENA Solar and Renewable Energy Report (2021). 20 pages. MESIA Middle Est Solar Industry Association in collaboration with Middle Est Energy Dubai. https://www.middleeast-energy.com/content/dam/Informa/Middle-East-Electricity/middle-east-energy-2021/reports-mee/MEE_Mesia_Report_v3.pdf (accessed 25.08.22)

[8] UAE's Ministry of Energy and Infrastructure (2021). Press release October 2021

[9] Dubai Electricity and Water Authority DEWA (https://www.dewa.gov.ae/en) and DEWA Research & Development (R&D) Centre, https://www.dewa.gov.ae/en/about-us/strategic-initiatives/mbr-solar-park/research-and-development-centre (accessed 25.08.22)

[10] Solar Decathlon Middle East SDME, Instagram: https://www.instagram.com/sdmedubai/?hl=en (accessed 01.09.22)

[11] Team Sharek webpage: https://www.sharjah.ac.ae/en/Research/Partnerships/SHARE2020/pages/default.aspx
Instagram: https://www.instagram.com/teamsharjah2020/ (accessed 01.09.22)

[12] Team Harmony BIPV insights: https://solarchitecture.ch/sdme-2021-team-harmony/ (accessed 25.08.22)

[13] Team SCUTxCSCEC, X HOUSE Virtual Reality www.jsvry.com/p/aVnOva0XWk; BIPV insights: https://solarchitecture.ch/sdme-2021-team-x-house (accessed 01.09.22)

[14] Team Go-Smart webpage: https://sdmeteam.uob.edu.bh/ (accessed 01.09.22)

[15] Expo 2020 Dubai, https://www.expo2020dubai.com/ (accessed 25.08.22)

EXPERIENCE AND RESULTS FROM THE SOLARISE LIVING LABS

Bert Herteleer[1], Cas Lavaert[1], Bert Uytterhaegen[1], Anastasios Kladas[1],
Jan Cappelle[1], Victor Becerra[2], Ahmed Rachid[3]
[1] KU Leuven Research Group ELECTA Ghent, Gebroeders De Smetstraat 1, 9000 Ghent, Belgium
[2] University of Portsmouth
[3] Université de Picardie Jules Verne, UPJV
bert.herteleer@kuleuven.be

ABSTRACT: This paper describes the Living Labs developed and completed by the three academic partners (UPJV, KU Leuven and University of Portsmouth) in the EU Interreg 2 Seas project SOLARISE. The Living Labs demonstrate a variety of technologies, with all three having PV and battery storage, and UPJV and KU Leuven demonstrating thermal energy storage solutions. While usage and visits of the Living Labs have been hampered by the covid pandemic, lessons about the development of the Living Labs, such as regulatory delays even for limited PV power installations are discussed. Historical monitored data from the Living Labs will be made available for download via the partner websites and the SOLARISE project website, www.interregsolarise.eu.
Keywords: living lab, performance, data

INTRODUCTION

The SOLARISE project co-funded by the European Interreg 2 Seas programme aims to boost the adoption of solar energy in the 2 Seas region, covering North France, England, Belgium and The Netherlands. A key metric of the complete SOLARISE project is enabling CO_2 emission savings through project As part of the SOLARISE project and to support the adoption of solar technologies, the three academic partners (Université de Picardie Jules Verne (UPJV, France), KU Leuven Technology Campus Ghent (KUL, Belgium) and University of Portsmouth (UoP, England)) have developed and installed Living Labs on their premises. Living labs are defined as user-centered, open innovation ecosystems based on a systematic user co-creation approach integrating research and innovation processes in real-life settings. The key stakeholders of a university-based living lab are academic staff, researchers, students, and external partners. The Living Labs had the following design goals:

- To demonstrate solar photovoltaic technology changes and inspire curiosity;
- To enable teaching activities to occur, aimed at varying degrees of sophistication;
- To leverage and illustrate the link between energy consumption and generation;
- To demonstrate the design considerations and challenges in using PV and PVT systems in combination with storage; and,
- To illustrate the challenges of managing power and energy flows on a power-constrained connection and,
- To contribute to CO_2 savings over the system lifetimes.

The three academic partner Living Labs use and installed similar photovoltaic technologies, with local accents that illustrate local enablers and challenges. All three living labs had to deal with space constraints to install the PV arrays, resulting in occasionally energetically sub-optimal locations compared to fully shade-free environments, yet the solutions developed result in specific benefits for visits and onsite activities.

RESULTS

All three Living Labs were completed in the course of 2021. The KU Leuven Living Lab [1] (Figure 1) has an active awning visible from street level and a rooftop PVT/PV array, demonstrating a range of PV module types and aesthetics, with agrivoltaic (semi-transparent, with larger cell spacing), bifacial, coloured and "standard" crystalline silicon modules. All modules are equipped with optimisers. For the active awning, currents and voltages are measured per module with dedicated custom signal conditioning boards. The active awning has an educational switching matrix, which allows users to disconnect the modules from the respective optimisers and connect these to the inverter directly, thus allowing experiments illustrating array mismatch or shading losses to be done. Historical data of the Living Lab is available at https://solarise-kuleuven.one/. Data can be downloaded at 15 s to 5 min resolution.

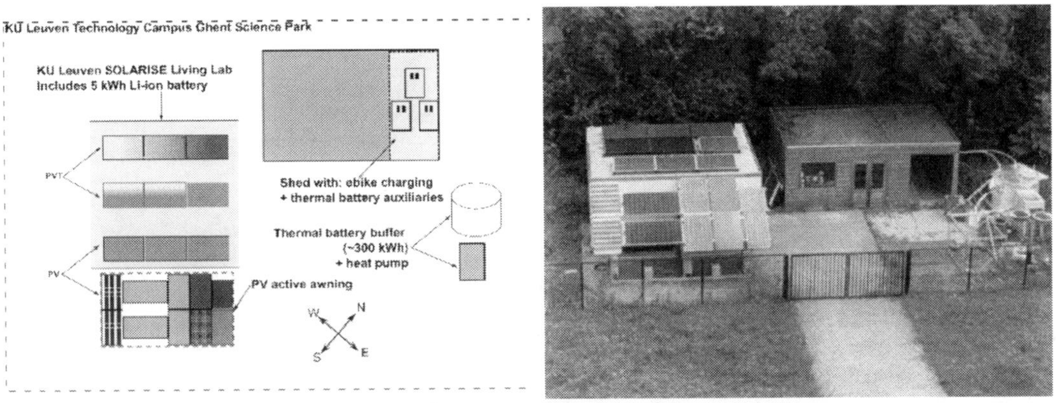

Figure 1: Bird's eye view of the KU Leuven Living Lab, schematic left and photograph right.

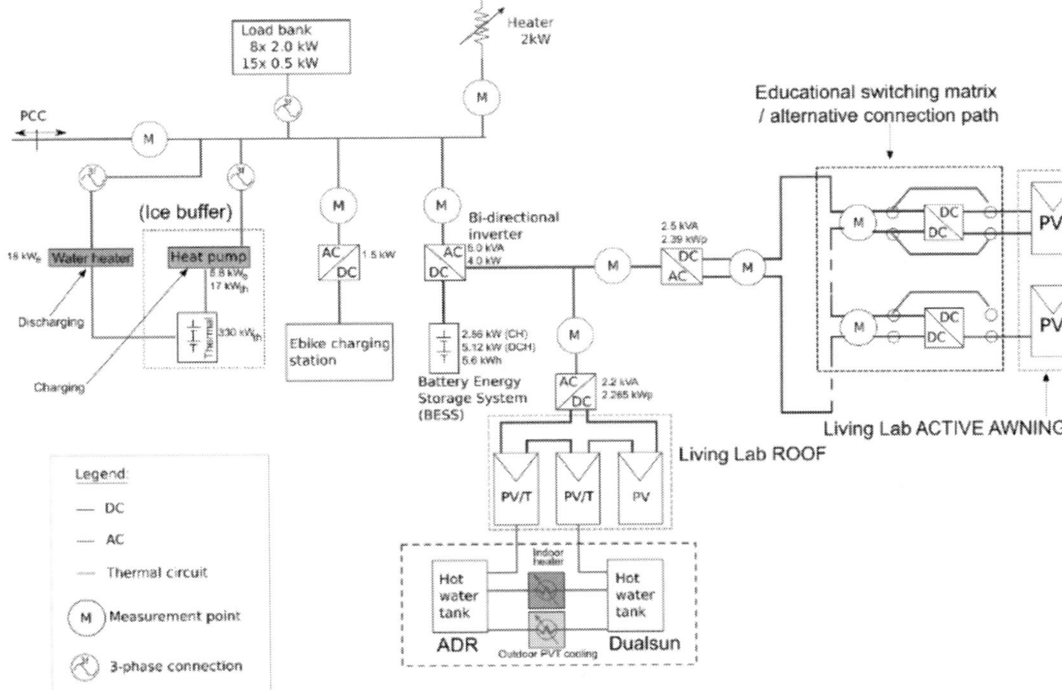

Figure 2: High-level schematic of the KUL Living Lab

The UPJV Living Lab "SOLLAB" [2], shown in Figure 3 is located in Amiens city centre, hosting a research team working on electrical energy topics. To reduce the risk and the need for systematic access to the roof, a 3D camera with raise/lower functionality is employed, allowing visitors and researchers to remotely view the PV technologies employed. The SOLLAB SCADA system was developed in Labview, which controls the various components.

Figure 3: The UPJV Living Lab building Productique *and SOLLAB components on the roof*

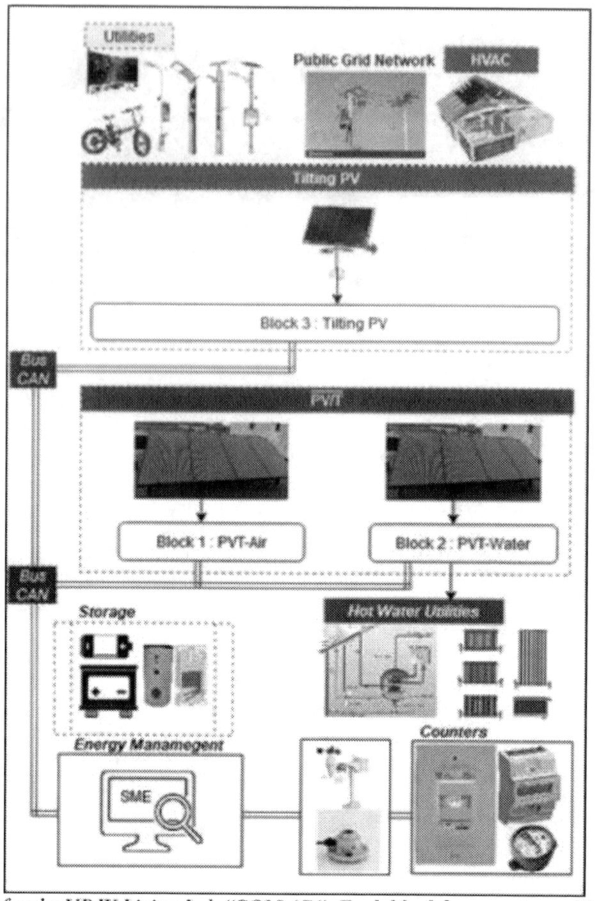

Figure 4: Block diagram for the UPJV Living Lab "SOLLAB". Each block has a universal board based on an ESP32 microcontroller that enables communication with IoT (Internet of Things) modules.

The University of Portsmouth Living Lab is spread over two buildings at the UoP campus: the Port Eco-House (PEH) and the Future Technology Centre (FTC). These enable research on smart grids, building management and energy management systems. The PEH schematic is given in Figure 6; the FTC schematic is similar [3].

Figure 5: The University of Portsmouth Living Lab locations: the Port-Eco House (left) with its ground-mounted PV array, and the Future Technology Centre (right)

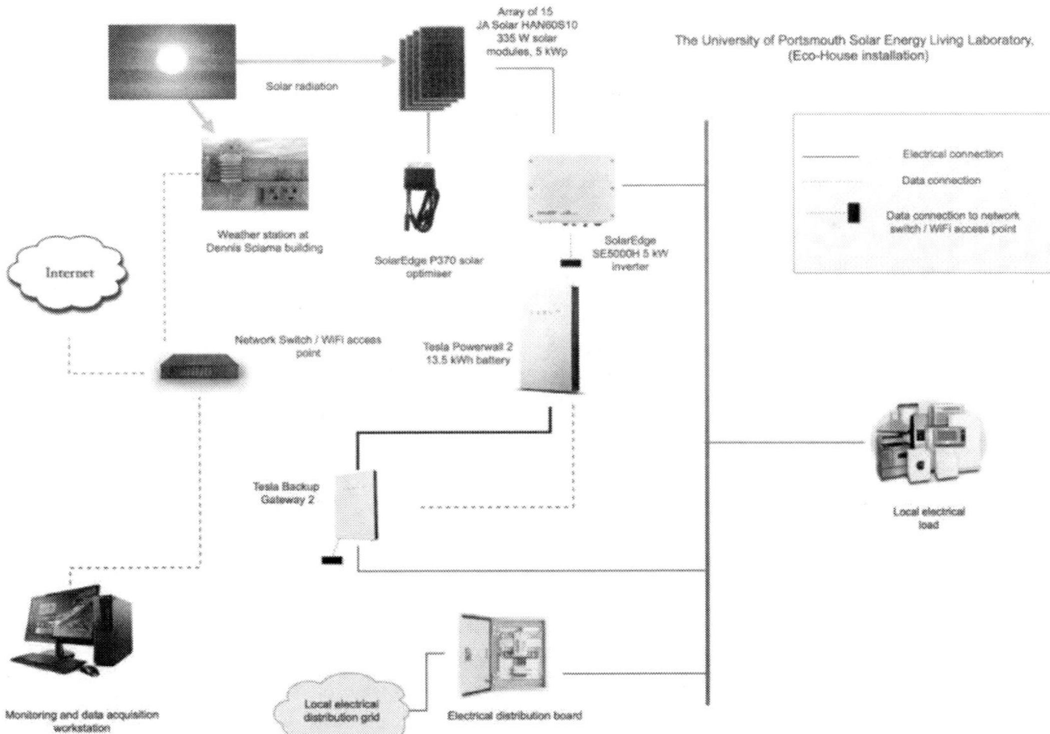

Figure 6: UoP Port-Eco House data and energy flow schematic

Figure 7 and Figure 8 illustrate the operation of the KUL and UoP Living Labs, where the KUL Living Lab sees significant shading from a nearby building on campus in autumn-winter-spring (the blue shaded area in the middle pane). The UoP ground-mounted array at the Port Eco-House also sees significant shading throughout the year due to its location on campus: ground-mounted between buildings. For both Living Labs, the PV systems appear to be performing better than the PVsyst models, although the KUL Living Lab PVsyst model is particularly challenging to implement, given the large number of different PV modules used in both arrays.

Table 1 summarises the main technologies employed and characteristics of the three Living Labs

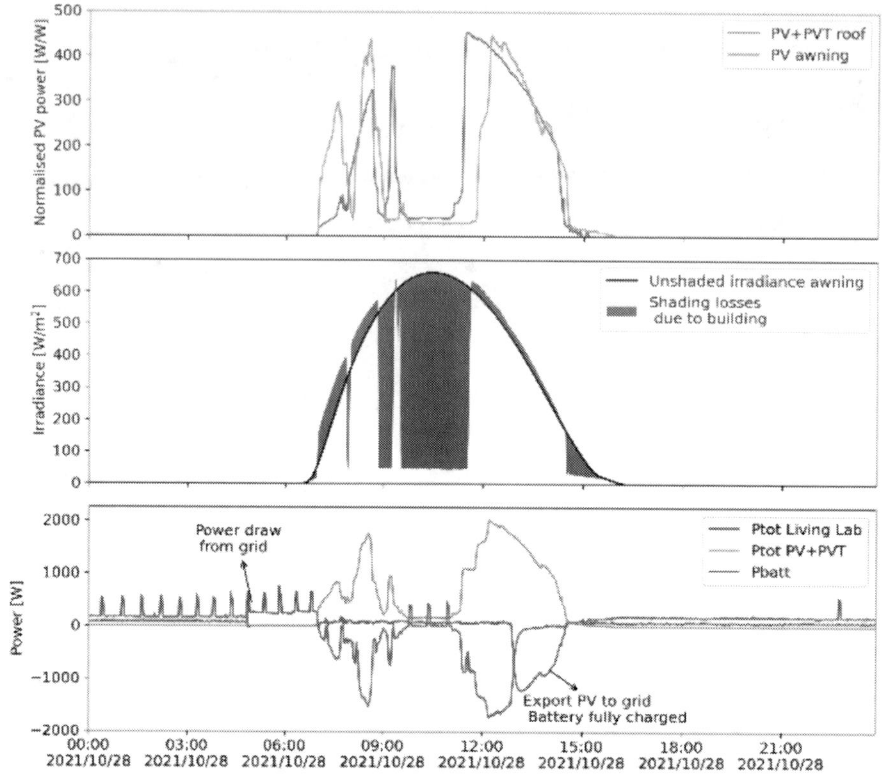

Figure 7: Measured data at the KUL Living for a single day. The KUL Living Lab sees significant shading (blue shaded area in the middle pane) on this clear-sky day.

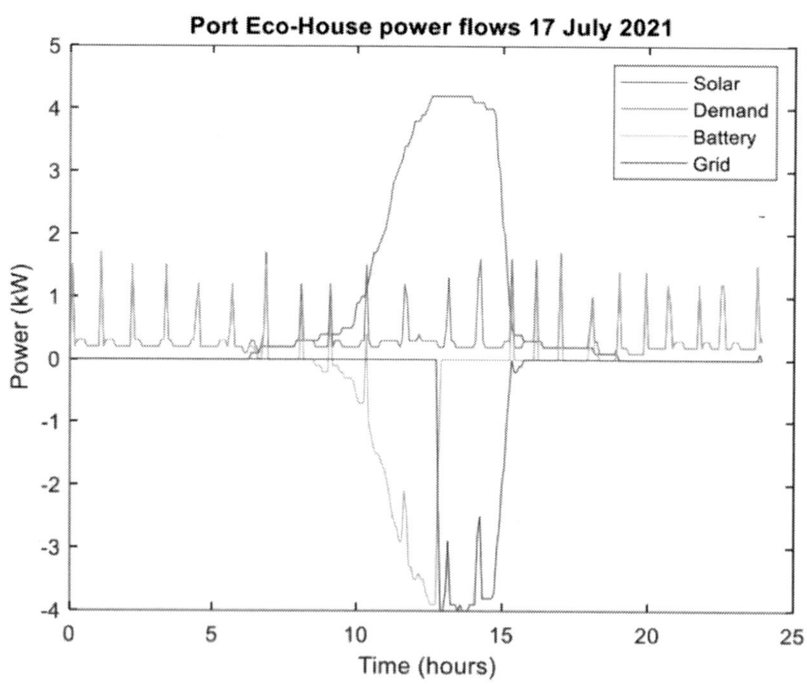

Figure 8: Measured data at the UoP Living Lab (right) for a single day.

Table 1: Overview of SOLARISE Living Lab key technologies and characteristics

	UPJV	KU Leuven	UoP
Location	Amiens, France	Ghent, Belgium	Portsmouth, England
Completed	December 2021	June 2021	May 2021
PV systems	PVT-Air 1.2 kWp elec + 2.4 kWth PVT-Water 1.2 kWp elec + 2.4 kWth	• Rooftop PV & PVT 2.27 kWp / 2.2 kVA @15°, 137° (SE) • Active awning PV 2.4 kWp / 2.5 kVA @38°, 137° (SE)	• Ground-mounted PV 5.025 kWp / 5 kVA (Port-Eco House) • Existing 25 kW array at Future Technology Centre building
Optimisers	Yes – per 2 modules	Yes – individual	Yes – individual
PVT systems	PVT-Air, PVT-Water	2 independent PVT-water systems	None
Electrical energy storage	4.8 kWh lead-acid BESS 1.536 kWh Li-ion BESS	5.76 kWh, 5 kVA Li-ion BESS	• 13.5 kWh, 5 kVA Li-ion BESS at PEH • 40.5 kWh, 15 kVA Li-ion BESS, at FTC building
Thermal energy storage	421 L hot water tank	2 separate hot water tanks (197 L; 156 L) 330 kW_{th} thermal battery (ice buffer)	None
Extra technologies	• 2-axis tracker for PV module outdoor I-V curve measurements • 3D camera with raise/lower function for remote visits • Weather station Vantage Pro 2	• Charging station for 3 ebikes • Living Lab EMS and power control at point of common coupling (PCC) with campus grid • Educational switching matrix on active awning to illustrate mismatch and shading losses	• PEH: instrumented 3-bedroom for energy efficiency and building performance research • Weather data available from Dennis Sciama Building at UoP campus
Monitoring	Labview	Siemens PLC and python on dedicated server, SolarEdge	Tesla, SolarEdge
Time resolution	1 min	15 s	5 min
Data link	https://bit.ly/3BuvWUU	https://solarise-kuleuven.one/	www.interregsolarise.eu

A key lesson from the development period is that the Living Labs encountered is a lengthy regulatory delay, as permission needed to be obtained from local planning authorities and/or grid operators, despite the PV and battery power levels being a fraction of the campus loads. These regulatory bottlenecks will need to be addressed to facilitate the rapid uptake of PV technologies to decarbonize grids.

CONCLUSIONS

The SOLARISE Living Labs have been built by the three academic partners (UPJV, KUL and UoP) and completed in a time of flux due to the covid pandemic. Visits to the Living Labs have been hampered compared to pre-covid expectations due to local and national restrictions, with information and videos about the Living Labs available online, including training videos at https://www.interregsolarise.eu/training_videos/. Historical monitored data from the Living Labs is available for download via the partner websites and the SOLARISE project website, www.interregsolarise.eu. Despite the small array sizes and limited electrical energy storage at the Living Labs, lengthy regulatory delays were experienced. A clear, streamlined application and approval process for PV, battery and PV + battery systems is recommended to reduce barriers to deployment and lower system costs.

ACKNOWLEDGEMENTS

The three Living Labs of KU Leuven Technology Campus Ghent, Université de Picardie Jules Verne (UPJV), and University of Portsmouth were developed and funded by the SOLARISE project. The SOLARISE project received funding from the Interreg 2 Seas programme 2014-2020 co-funded by the European Regional Development Fund under subsidy contract N° 2S04-004. KU Leuven Technology Campus Ghent received additional co-funding by the Province of East Flanders.

REFERENCES

[1] T. Motoasca, "Development of an Academic Living-Lab as Sociotechnical Imaginary to Facilitate the Uptake of Solar Technologies in the 2Seas Region," in *36th EU PVSEC*, Marseille, 2019.

[2] A. Rachid, "A Solar PV/T Living Laboratory as a Cyber-Physical System," in *36th EU PVSEC*, Marseille, 2019.

[3] V. Becerra, "OUTPUT 5.3 Living Lab Pilot (UoP)," 2022. [Online]. Available: https://www.interregsolarise.eu/downloads-o5-3/.

BIPV ENVELOPE DESIGN OPTIMIZATION AT CONCEPTUAL DESIGN STAGE

Samarasinghalage Tharushi Imalka, Pabasara Wijeratne, Rebecca Yang, Ron Wakefield
RMIT University, Solar Energy Application Lab, School of Property Construction and Project Management, Melbourne, Australia

ABSTRACT: The design of BIPV envelopes involves a large set of envelope-related parameters, PV-related parameters, and conflicting performance criteria. Both building functionality and the energy system are influenced by BIPVs. Therefore, BIPV envelope design has become a complex process. Optimization is an impressive way to support BIPV façade design and is rarely used in early building design. Consideration of building envelope-related parameters and product-related parameters simultaneously for different BIPV application areas has not been fully realized due to the complexity of the parameterization of variables. Therefore, this study utilizes a genetic algorithm (GA) based multi-objective optimization (MOO) framework to identify optimal alternative BIPV envelope design options at the conceptual design stage which maximizes life cycle energy (LCE) generation and minimizes life cycle cost (LCC) simultaneously considering multiple BIPV products for different BIPV application areas. Six BIPV application areas including, curtain wall, window, double-skin façade, balcony, parapet, and shading devices were tested using the utilized framework. The case study results indicated that the utilized optimization framework could be a useful tool for designers in the conceptual design phase of a BIPV façade. In addition, the optimization process provided an insight into the design space as well as trade-off patterns between multiple design objectives.

Keywords: Building Integrated Photovoltaics (BIPV), optimization, Life cycle energy (LCE)

1 INTRODUCTION

At present, building attached solar photovoltaic (BAPV) applications are dominating Australia's solar market while the contribution of Building-integrated photovoltaics (BIPV) which are specified as PV modules that can be integrated into the surfaces of the building (roof or facade) replacing conventional building materials, is very limited [1]. A major reason for the lack of application of BIPV systems is there are information gaps within and across the building and PV industry sectors concerning BIPV application areas, BIPV products, and energy and cost-effective project solutions.

Building Integrated Photovoltaics (BIPVs) are of top interest with their potential to meet the energy demands of the world in near future and rapidly developing technology. It is widely emphasized that most crucial design decisions concerning sustainability need to be made in the early design stages and further, building performance analysis needs to be merged into the design process to assess the dependencies of performance criteria on the form, material, and technical systems. At present, still, a small share of the PV market represents BIPV systems due to stakeholders' belief that BIPV integrations are more expensive than conventional building material installations. However, choosing better BIPV products and designing better building layouts will prove the incorrectness of the above belief.

Among the challenges faced throughout BIPV design and planning, deciding best performing module types and optimal module configurations are crucial. Furthermore, it is a difficult task to find better design alternatives satisfying several conflicting criteria, especially, economical and energy performance. Although the application of BIPV into building envelopes will contribute to the Net Zero building target, sustainable building envelop design with BIPV is very complicated even in the conceptual design phase. Accordingly, rigorous attention must be paid to building envelope design at the conceptual stage of a building design [2]. Moreover, a more comprehensive approach at the conceptual design stage is needed to explore how to use different BIPV modules to increase the performance of a BIPV system. The major challenges in BIPV design are the low efficiency and high PV cost [3], and accordingly building envelope design should be optimized to best utilize PV modules. Building Integrated Photovoltaics (BIPV) modules integrated into building envelopes are used in three major application areas namely, roofs, facades, and external devices. There are several BIPV application type subcategories under these major categories such as skylight, roof sheet, rainscreen, curtain wall, double-skin facades, windows, balcony, parapet, canopy, and shading devices [4, 5]. Thus, For BIPV envelopes to play these multi-functional roles in different applications, numerous factors should be considered However, past studies have not considered different PV Modules as a decision variable in multi-objective BIPV design optimization models and BIPV application types have not been fully considered inside the multi-objective building envelope optimization process. Therefore, this paper utilizes a genetic algorithm-based multi-objective optimization framework to identify optimal alternative BIPV envelope design solutions for different BIPV application areas. Six case studies related to six different BIPV application areas (curtain wall, window, double-skin façade, balcony, parapet, and shading devices) were simulated using the optimization framework and a different number of optimal alternative design solutions were identified for different application areas. These optimal solutions will be further compared and discussed based on the performance criteria and the BIPV application area.

Being informed by the identified design solutions, it is expected that more building professionals in Australia will consider using optimization frameworks as a viable option to plan the BIPV envelope design, thereby increasing the BIPV usage in Australia.

2 METHODOLOGY

This paper utilizes a multi-objective optimization framework (Figure 1) to identify and compare optimal

BIPV envelope design options for different BIPV application areas considering PV tilt angle, D/L ratio (Distance-to-length), PV placement, and PV product type as design variables to maximize LCE and minimize LCC in the conceptual design stage.

Figure 1: Optimization framework adopted for the investigation
Source: Samarasinghalage, et al. [6]

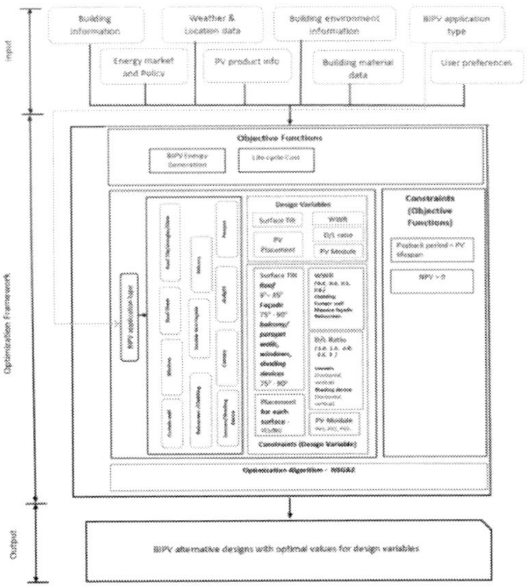

Six application types were selected to apply BIPV namely, curtain wall, window, double-skin façade, balcony, parapet, and shading devices. Both LCC and LCE were selected as objective functions for the optimization process.
Design variables selected for each application type are listed in table I.

Table I: Design variables

	BIPV Products	PV tilt angle	PV placement	D/L ratio
curtain wall	32	✓	✓	
window	8		✓	
double-skin façade	16		✓	
balcony	4		✓	
Parapet	16		✓	
shading device	32	✓	✓	✓

Table II: Building models for each application type

Application type	Building model
curtain wall shading devices/parapet	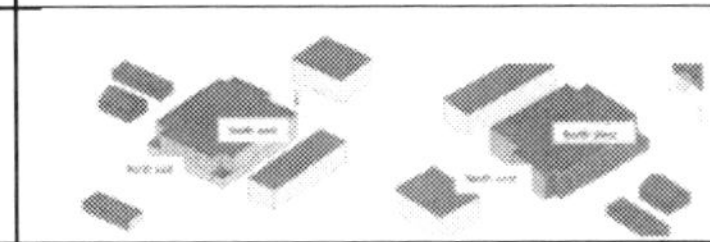
window	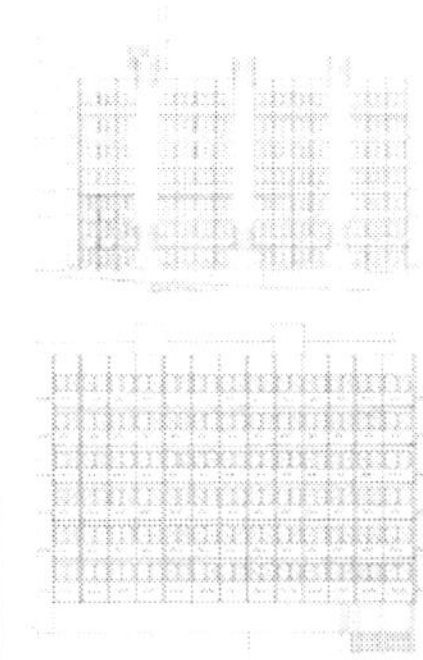 north south
double-skin façade	 East West North
balcony	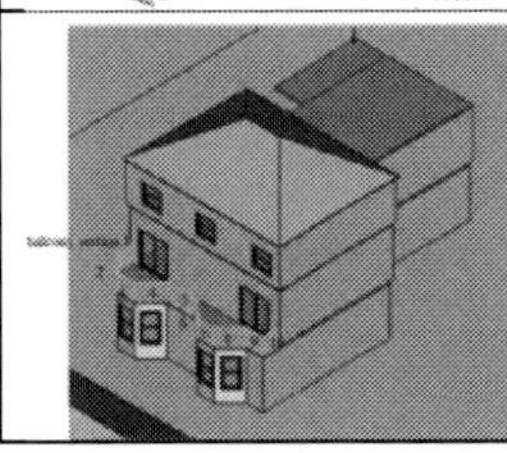

The initial population and number of iterations were selected between 30-50 and 40-100 respectively. These are selected based on the convergence of the results.

3 RESULTS

3.1 Curtain wall

Figure 2 illustrates the Pareto-optimal solutions and the LCC and LCE values of the other design alternatives generated throughout the optimization process. Final design solutions are far different from the initial population values and only three design solutions are outputted as optimal solutions. Table III lists the parameter values of the design alternatives and their corresponding LCC and LCE values. According to the optimal results, a tilt angle of around 90 degrees for the south facade and an angle of around 50 degrees for the north façade are recommended. Only the first alternative places PVs on both facades and it has the highest LCE and LCC. Compared to the other two designs, design 3 has a high payback period of 23.369 years and therefore, designers could go with the first two options if a much lower payback period is preferred. Likewise, the proposed framework suggests the Pareto optimal solution along with their crucial parameter values such as capital cost, first-year energy generation, NPV, payback and total PV area so that building professionals could select the designs based on any preferred parameter.

Figure 2: Pareto optimal results curtain wall BIPV

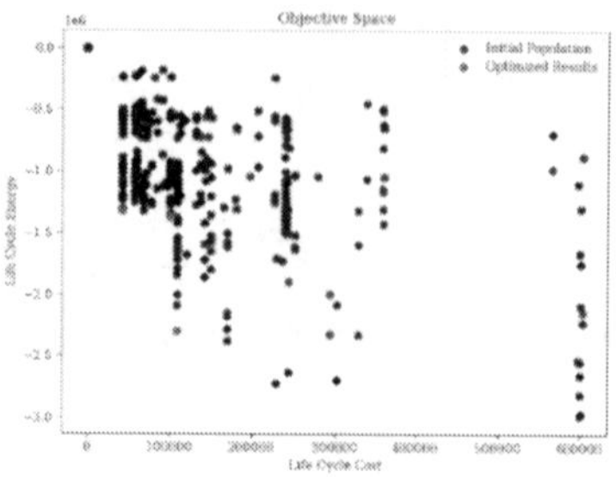

Table III: Optimal design alternatives curtain wall

	ALT 1	ALT 2	ALT 3
Southeast tilt	80	90	90
Southeast PV placement	1	0	0
Southeast No of PVs	266	0	0
Northeast tilt	55	40	40
Northeast PV placement	1	1	1
Northeast No of PVs	175	175	448
PV Type	109	109	904
PV Life Span	25	25	25
Life cycle cost (AUD)	89098.9	35356.71	83768.8
Life Cycle Energy	2303426	1322701	1369337
Payback Period	10.4455	5.3406	23.369
NPV	48294.21	48139.69	2323.05
Total PV area	780.0849	309.5575	322.56

3.2 Window

Table IV: Optimal design alternatives window

	ALT 1	ALT 2	ALT 3	ALT 4
North PV placement	YES	NO	YES	YES
North no of PVs	168	0	168	168
South PV placement	YES	YES	YES	YES
South no of PVs	156	156	156	156
PV Type	706	701	702	701
PV Life Span	25	25	25	25
life cycle cost (AUD)	57527.29	26010.75	57527.29	54022.33
Life Cycle Energy	799656	485137	799656	778331
Payback Period	22.0724	12.1303	22.0724	20.3925
NPV	3039.62	10303.45	3039.62	4929.47
Capital Cost	45432.3	20514.18	45432.3	42606.37
LCOE	0.087225	0.065076	0.087225	0.084244
First Year Energy Generation	33947.07	20595.08	33947.07	33041.82
System Size	32.4	14.82	32.4	30.78
Total No of modules	324	156	324	324
Total PV area	305.8767	147.274	305.8767	305.8767

3.3 Double-skin façade

The double-skin façade is a special type of façade application that consists of two layers usually glass where air flows through the intermediate cavity. The major difference between other BIPV facades with double-skin façades is the difference in alternative materials. Only two options have resulted for the double-skin façade BIPV application. According to the generated optimal results in table V, only design 1 places BIPV in both east and north facades and it shows the highest life cycle energy generation. In terms of payback periods, all the designs have payback periods between 3 and 6 years, which are significantly lower values compared to other BIPV application areas. This is because there is a high alternative material cost involved for double-skin façade design. For instance, an Aluminum frame with double glaze costs between 990- 1175 AUD per square meter. Alternative design 2 has lower payback, lower cost and lower energy generation than design 1.

Table V: Pareto optimal results double skin façade

	ALT 1	ALT 2
bld1_facade1 Southeast tilt	90	90
bld1_facade1 Southeast PV placement	1	0
bld1_facade1 Southeast No of PVs	270	0
bld1_facade2 Northeast tilt	90	90
bld1_facade2 Northeast PV placement	1	1
bld1_facade2 Northeast No of PVs	180	180
PV Type	503	503
PV Life Span	30	30
Module Weight	34.5	34.5
LCC (AUD)	79327.44	31730.97
LCE (kW)	2786840	1464443
Payback Period	5.5005	3.6259
NPV	72890.71	54884.12
Capital Cost	48872.92	19549.17
LCOE	0.040883	0.03112
First Year Energy Generation	102665.4	53949.17
System Size	126	50.4
Total No of modules	450	180
Total PV area	765.45	306.18

Figure 3: Design variable variations double skin façade results

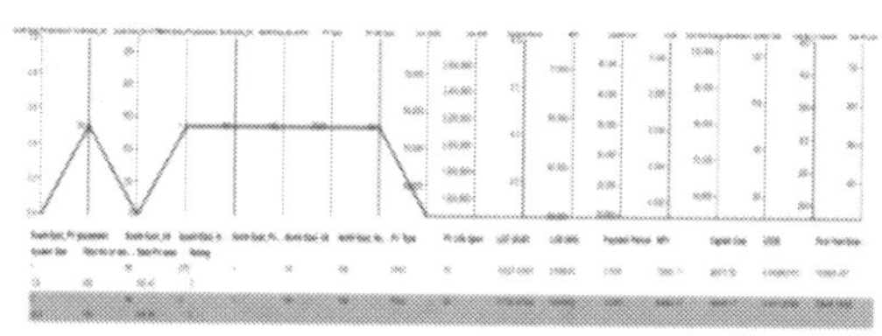

3.4 Balcony

The alternative material for the balcony is glass and therefore, the alternative material price is significantly higher. No optimal results were identified for the balcony BIPV because all the alternative designs generated through the optimization model violated the payback and NPV constraints.

3.5 Parapet

Three alternative designs have been identified where only design 1 places PVs on both southeast and northeast sides. Out of the 16 products utilized, only two products were indicated in the optimal solutions. Thus, this framework enables the building professional to filter better BIPV products out of a large set of products. Payback periods are considerably lower here as BIPV system sizes are lower. According to table VI, designs 1 and 3 has the same payback periods while design 3 has a bit higher payback. Alternative design 2 has the highest LCE as well as the highest LCC.

Table VI: Pareto optimal results parapet BIPV

	ALT 1	ALT 2	ALT 3
parapet Southeast PV placement	0	1	0
parapet Northeast placement	1	1	1
PV Type	1102	1101	1101
PV Life Span	25	25	25
Module Weight	17	17	17
LCC (AUD)	3587.57454	8180.505893	3241.332524
LCE (kW)	343903	659377	291547
Payback Period (Years)	0.6148	0.71	0.6148
NPV	21825	38973.43	18302.46
Capital Cost (AUD)	1103.74	2361.62	935.74
LCOE	0.01866709	0.022481863	0.020146562
First Year Energy Generation (kW)	13822.4562	26502.30093	11718.14193
System Size	12.6	29.68	11.76
Total No of modules	84	212	84

3.6 Shading device

Figure 5 illustrates the Pareto optimal results identified through the optimization framework. There are 19 alternative designs generated for the shading devices application. Among the alternative designs, almost all the designs suggest placing PVs on the northeast façade. The Southeast tilt angle is 0 or 15 degrees while the northeast façade tilt angles are 15, 30, 45 or 60 degrees. Alternative design 2 has the lowest payback period of 6.81 years while alternative design 1 has the highest life cycle energy

Figure 4: Pareto-optimal results shading device BIPV

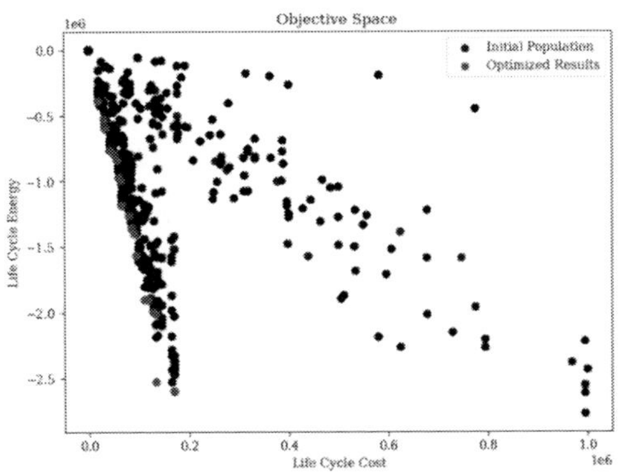

Figure 5: Design variable variations shading device results

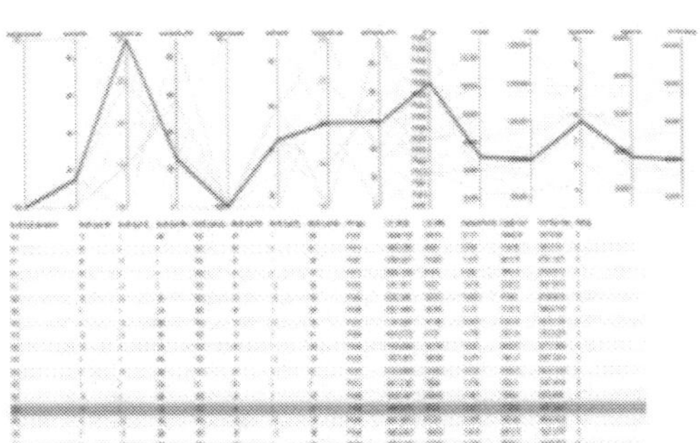

4 CONCLUSION

Six BIPV application areas were optimized using the utilized optimization framework and a different number of design solutions were achieved for different application areas. These design alternatives included the optimal values of their design variables which optimized life cycle energy and life cycle cost simultaneously. The framework enabled the use of multiple design variables having different ranges inside the optimization process. If the optimization process is carried out manually, multiple values for a design variable may significantly increase the cases to be optimized and would increase the effort and the time outstandingly.

Curtain wall, shading device, and parapet BIPV were simulated using the same building model and according to the results, parapet application has significantly better payback periods and lower LCE. Results indicate that different BIPV application areas act differently in a BIPV system design scenario and thus, BIPV designers should be watchful about the different design options while carrying on a BIPV system design. The framework helps select the best BIPV application area for a given building model. Thus, application-level optimal solution comparison which is very helpful in taking early BIPV design decisions is also possible in the proposed framework. For shading devices BIPV application 19 alternative designs were identified while parapet BIPV application case study did not output any optimal design solution. This is because the generated designs for parapet application does not satisfy the constraints (payback period less than PV lifespan and Net Present Values is positive) added in the optimization model. The Pareto optimal solutions generated by the framework do not suggest one best alternative design, but multiple design options so that designers could select any design based on their priorities and constraints. As all the generated designs are non-dominated solutions based on both life cycle energy (LCE) and life cycle cost (LCC), building professionals can use any of the generated designs for their decision making. For instance, if energy generation is preferred over the cost, users could select a high energy generation design solution from the set of all the alternative design options. If a lower payback is preferred, designers could go with the design with the lowest payback period. The utilized framework is useful in optimizing cost and energy for any preferred BIPV application area. Thus, application-level optimal solution comparison which is very helpful in taking early BIPV design decisions is also achieved.

5. ACKNOWLEDGEMENTS

This study is supported by the 'BIPV Enabler' research project funded by the Commercialization of R&D Funding Initiative Pilot at the Australia Renewable Energy Agency, and the EOF scheme at RMIT University. It is also supported by the industry partners in this research project. The authors appreciate the comments and suggestions of reviewers.

6 REFERENCES

[1] ARENA, "Solar energy," https://arena.gov.au/what-is-renewable-energy/. 2021.

[2] J. P. Basbagill, F. L. Flager, and M. Lepech, "A multi-objective feedback approach for evaluating sequential conceptual building design decisions," *Automation in Construction*, vol. 45, pp. 136-150, 2014.

[3] Y. Lu, R. Chang, V. Shabunko, and A. T. L. Yee, "The implementation of building-integrated photovoltaics in Singapore: drivers versus barriers," *Energy*, vol. 168, pp. 400-408, 2019.

[4] A. Henemann, "BIPV: Built-in solar energy," *Renewable Energy Focus,* vol. 9, no. 6, pp. 14-19, 2008.
[5] M. Pagliaro, R. Ciriminna, and G. Palmisano, "BIPV: merging the photovoltaic with the construction industry," *Progress in photovoltaics: Research and applications,* vol. 18, no. 1, pp. 61-72, 2010.
[6] T. I. Samarasinghalage, W. P. U. Wijeratne, R. J. Yang, and R. Wakefield, "A multi-objective optimization framework for building-integrated PV envelope design balancing energy and cost," *Journal of Cleaner Production,* vol. 342, p. 130930, 2022.

TECHNICAL ASPECTS FOR ROAD INTEGRATED PHOTOVOLTAICS TOWARDS A MORE SUSTAINABLE MOBILITY SECTOR

Jonas D. Huyeng, Jacob Forster, Felix Basler, Pascal Romer, Andreas J. Beinert, Christian Schill, Martin Heinrich, Holger Neuhaus, Harry Wirth
Fraunhofer Institute for Solar Energy Systems ISE
Heidenhofstr. 2, 79110 Freiburg, Germany

ABSTRACT: Photovoltaic modules can be integrated into the mobility infrastructure, to realize a dual use of already sealed land. While the theoretical potential is very large, actual technical solutions are unlikely to be installed in every road, railway and bike path. In this work, we evaluate three different integration categories: Into the traffic area, *e.g.*, PV-Pavement, beside the traffic route, *e.g.*, PV-Noise barriers, and above the traffic area, *e.g.*, PV-Roofing. We discuss four quantitative technical aspects, the Specific Yield (kWh/kW$_p$), Specific Costs (€/kW$_p$), Non-PV Cost Share (%), and Integration Capacity (kW$_p$/rm) and two qualitative aspects, the availability of specific Site Requirements and the PV Module Complexity. For each aspect, we assign a rating to the three categories and discuss the advantages and disadvantages of the current technical solutions. We find that PV upgrade solutions for existing noise barriers have a high technical feasibility, while PV-Roofing has the highest practical potential. Solutions for PV-Pavements are currently mostly relevant for niche applications, but all approaches can contribute in different places to the generation of sustainable energy in already sealed areas.
Keywords: Road Integrated PV (RIPV), Module Integration, Dual-Use Infrastructure

1 INTRODUCTION

The mobility sector has enormous need to reduce the associated greenhouse gas emissions (s. [1] and updates). With the current shift towards e-mobility it is also important to supply enough sustainable energy [2]. With the addition of photovoltaics (PV), this energy can be harvested on the existing transportation infrastructure, without the need to seal additional areas.

We refer to this dual use of already sealed land in the present infrastructure or associated areas of the mobility sector traditionally as "Road Integrated Photovoltaics" (RIPV). More generally, we include other transport as railways, bike or foot paths in this category and address applications beyond the substitution of pavement.

The theoretical potential for such applications is enormous. In Germany for example, around 5 % of the land area are is dedicated to transport (18,000 km^2), according to the Federal Statistical Office [3]. While certainly not all of it can be combined with PV, a significant amount of nominal PV capacity (GW$_p$) could be installed here. Previous estimations of the theoretical potential for RIPV in Germany have identified up to 660 GW$_p$ on motorways, national roads, regional roads, and railways [4], due to an available area of 3,300 km^2. To further the discussion, one should go from a theoretical assessment (*i.e.*, multiplying the available area by a common assumption of 200 W/m^2) to a technical evaluation of applications and available technology and ultimately a practical-economical potential. This work describes several technical aspects, as they are found today, to take into consideration.

To utilize the sealed land in the mobility infrastructure, different approaches for RIPV have been developed by several groups. The most direct integration might be the replacement of pavement or railbed by adopted PV modules. Studies have been performed in the US [5], the Netherlands [6] and France [7] for example. Another approach is utilizing associated infrastructure, such as noise barriers. Here, the first studies have been started already in 1989 [8] and are continued until today [9]. To circumvent the mechanical load on the PV modules and utilizing the technology of carports and other related applications, lately roofings with PV modules have been suggested as dual-use installation on roadways or associated areas [10, 11].

For a more general discussion, we categorize these into integration into the traffic area, *e.g.*, PV-Pavement, integration beside the traffic route, *e.g.*, PV-Noise barriers (PVNB), and integration above the traffic area, *e.g.*, PV-Roofing (s. illustrations in Fig. 1).

2 APPROACH

In this work, we look into the three categories with five different applications and compare them in six categories. To evaluate the examples in these categories, we use published data from literature, expert opinions from public or private sources, and internal project results from our

Figure 1: Illustrations of the three RIPV categories: integration into the traffic area (top), integration beside the traffic route (middle) and integration above the traffic area (bottom). Copyright: Fraunhofer ISE.

ongoing research. Where current data is not conclusive, we discuss it in the text.

To compare the different examples, we rank their described advantages and disadvantages in each category on a one (largest disadvantage) to five (highest advantage) scale. The results of this rating are displayed in Fig. 2.

In the following, we first describe the five examples and then discuss them under each of the six categories.

3 INTEGRATION EXAMPLES FOR RIPV

3.1 PV-Pavement

For the integration into the traffic area, we chose the common and widely discussed example of PV-Pavement, where specialized PV modules with highly durable front sheet are used as the road surface (*cf.* [12]).

Fundamentally, PV-Pavement needs to cope with harsh conditions from vehicle loads (including braking action from heavy trucks), corrosion from winter service and soiling, temperature cycling (without rear ventilation) and high safety requirements (both electrical and mechanical). On the other hand, the installations are relatively simple, and the theoretical potential is quite high.

Several specialized products have been developed and are being tested (*e.g.*, [5–7, 13, 14]). Most of them use thicker front sheets to address the increased requirements, which unfortunately reduces the solar yield. In addition, shading from traffic has to be taken into account.

As previous studies have shown, the currently available products are not yet ready for a broad deployment on all traffic areas. Therefore, we refer to the current state of the art and consider applications that seem suitable and the associated solar yield. These are for example, the integration into motorway shoulders or other less frequently used areas such as bike or foot paths.

3.2 PV-Noise barrier (PVNB)

For the integration beside the traffic route, we chose PV-Noise barriers (PVNB). As mentioned above, developments for PVNB have started more than thirty years ago already [8]. Today however, several parameters have significantly changed. For example, the efficiency and costs of a PV module has improved a lot, which makes the application more feasible than before.

An easy integration is the mounting of PV modules on the ridge of existing barriers, referred to as "PVNB (Upgrade)". Here, the noise barrier remains with its previous functionality and the generated energy can repay the costs of installation etc. An illustration is shown in Fig. 3. Several products have been recently submitted to an innovation tender in Austria [15]. Of course, the statics of the noise barrier have to be taken into account.

Figure 3: Illustration of an upgraded noise barrier, referred to as "PVNB (Upgrade)". The modules are installed on the ridge of existing barriers.
Copyright: Fraunhofer ISE.

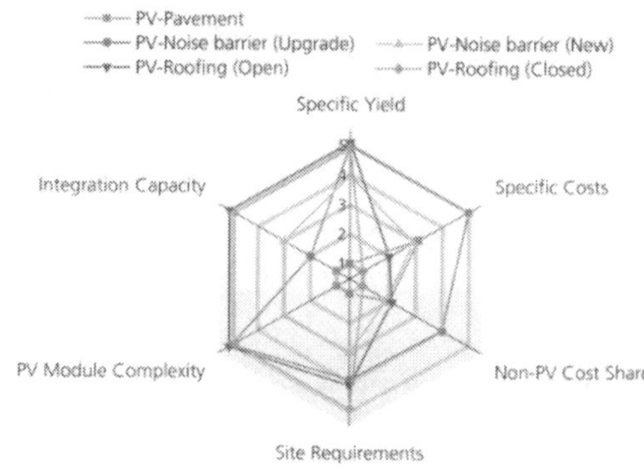

Figure 2: Combined rating of RIPV applications, derived in this work. The examples and ratings are explained in the text.

An alternative is the replacement of the noise barrier elements inside the wall (as illustrated in Fig. 1, middle). From our experience and due to the involved efforts, we expect that this is more likely to be considered for new walls, which is why we refer to it as "PVNB (New)" in this work.

Another integration level is the addition of PV modules on the street-averted side. As technical requirements are similar to those of standard PV installations, mostly regulatory questions remain. As these are very dependent on local legislation, we do not discuss these here in further detail. However, the high theoretical and technical potential for this application, especially for south facing walls, should not be forgotten.

3.3 PV-Roofing

For the integration above the traffic area, we consider PV-Roofing on motorways, as is currently investigated [10, 16]. While certainly having to fulfill the highest safety standards due to the overhead installation, it also has the highest theoretical potential, considering the vast road network.

In the near term, one should expect that PV-Roofing will be mostly done for shorter tracks close to consumers. This could be roadhouses or nearby residents, but also road infrastructure such as tunnels. A major challenge is the balance between safety standards and costs. While the supporting structure should be as simple as possible, it also has to withstand extreme events, such as collision of a truck.

Another issue with the roofing of a road is the noise propagation. As sound cannot escape here, the noise level in direct vicinity to the installation is increased. We therefore consider an open and closed installation, where the latter would be complemented by a common noise barrier at both sides, to contain and absorb road noise.

PV-Roofing is also a promising candidate for integrated PV in other areas, such as car ports, parking spaces, bike and foot paths or other urban areas. Some states in Germany have already started to extend the PV obligation from new buildings to public parking lots [17]. These new market segments might lead to synergistic advancements of necessary components.

4 RATING OF RIPV EXAMPLES

4.1 Specific Yield

The specific yield refers to the amount of generated power over one year (kWh), compared to the nominal installed capacity (kW_P).

This is obviously very dependent on the site of the installation and the orientation of the PV modules. Also, the expected performance ratio (effective vs. nominal efficiency) and systematic effects such as shading or soiling from the traffic areas need to be taken into account.

We calculated several different systems and locations using PV*SOL and compared the relative results, to derive some general rule of thumb for an expectable specific yield.

The ideal installation conditions for central Europe are commonly known to be slightly tilted towards the south, to get the most direct insolation. With this one can harvest around 1000 kWh/kW_P per year (still depending on the exact location). With current PV module technology for roof top installations with 200 W/m^2, this amounts to 200 kWh/m^2.

However, horizontal installations can regularly harvest 85% - 95% of energy in the same places, as the sun passes over them during the day.

PV-Pavement, PVNB (Upgrade) and the PV-Roofings are all installed horizontally or only at slight tilt for self-cleaning effects. Only for PVNB (New) are the modules installed in the vertical wall. Therefore, also the yield is dependent on its cardinal direction, ranging from 50 % - 70 % compared to the ideal installation. However, if the vertical installation is bifacial, even with a bifaciality factor of only 0.7, the yield can be up to 100% for an ideal East-West orientation, with the additional benefit of a shifted generation profile (peaks in the morning and afternoon).

When evaluating PV-Pavement, it should also be considered that it does not have the desired back ventilation, will be systematically shaded by traffic, and more severely soiled over the year. While some bike paths have performed with less than 5 % of their expected yield in some studies [18], others have been able to provide a yield of around 73 kWh/m^2 [19], which was later increased to 93 kWh/m^2. Compared to a mounted installation this is less than 47 %.

On the one hand, one can certainly assume further improvements as the technology matures and more companies are engaging in this. On the other hand, the fundamental limitations of shading and soiling cannot be eliminated. We estimate that today an optimized installation might be able to reach around 50% compared to a mounted installation.

Concluding our analysis, we give an expected specific yield range (± 30 %) for the five considered examples in Table I.

We assign the highest rating to the three horizontal installations, PVNB (Upgrade), PV-Roofing (Open) and PV-Roofing (Closed). To account for the high variation with cardinal direction described for PVNB (New), we deduct one point. We assign the lowest rating (highest disadvantage) to PV-Pavement, which has a systematically lower expected yield, even assuming further improvements of demonstrated technology.

Table I: Expected specific yields (kWh/kW_P) with an uncertainty of ± 30 % for central Germany and derived rating (1-5).

	Specific Yield (kWh/kW_P)	Rating
PV-Pavement	450	●○○○○
PVNB (Upgrade)	900	●●●●●
PVNB (New)	800	●●●●○
PV-Roofing (Open)	900	●●●●●
PV-Roofing (Closed)	900	●●●●●

4.2 Specific Costs and Non-PV Cost Share

With the specific costs (€/kW_P) of the different application examples, we include all estimated costs for PV- and "non-PV"-components together and divide by the estimated nominal installable capacity.

However, these integrated applications fulfill a dual use and some require higher safety regulations in the mounting structure. To differentiate the costs of the adapted PV modules and the remaining costs, we also give the resulting "Non-PV Cost Share", for each example.

We consulted several experts, *e.g.*, in the field of civil engineering, to discuss the costs of mounting structures and other "non-PV" components. We also consider the additional costs for adapted module materials and necessary engineering work, estimated from current research projects. Where possible, we also refer to price information given by suppliers of current (RIPV) products and compare these to our estimations. As most of these data is confidential, we can only discuss a qualitative comparison.

The upgrade option for PVNB can be realized at comparably low costs, as standard PV modules from mass manufacturing can be mounted on the barrier ridge. Still the price is likely more than double that of a standard rooftop installation, as the mounting structure must be more durable and reliable. Some intelligent engineering is necessary, to design a system that does not impede the structural integrity of the underlying noise barrier. From the present data, we estimate that the non-PV cost share will be around 55% in the end.

A fully integrated PVNB will likely be two to three times more expensive than the ridge top option, due to the higher material consumption and more complex PV modules. On the other hand, such a product also replaces part of the noise barrier, whose value could be deducted from the costs. From current commercial products [15] and our own designs, we derive a non-PV cost share of around 75%. Those PVNB however are mostly sound-insulating. For sound-absorbing barriers, more complex elements have to be built, which allow for less PV area. Therefore,

Table II: Estimated overall specific costs (€/kW_P), here relative, for different applications, and derived rating (1-5).

	Specific Costs (a.u.)	Rating
PV-Pavement	2 ×	●●●○○
PVNB (Upgrade)	1 ×	●●●●●
PVNB (New)	3 ×	●●●○○
PV-Roofing (Open)	4 ×	●●○○○
PV-Roofing (Closed)	5 ×	●○○○○

the non-PV cost share will be significantly higher for (highly) absorbent barriers, with a lower solar yield. Unfortunately, current legislation in Germany often promotes highly absorbent sound barriers.

The prices for PV-Pavement can be estimated somewhere in between both PVNB options, although the current prices vary significantly between suppliers [20]. Evaluating the different concepts, we estimate that a high non-PV cost share of around 70% is necessary for the save integration of adopted modules into the traffic area .

As part of the pavement is replaced, one could consider deducting a certain amount from the costs, similar to the discussion of noise barriers. However, it has to be demonstrated that PV-Pavement can achieve the same lifetime as standard paving, or additional replacement and maintenance costs must be added.

The PV-Roofing has the highest specific costs in our current estimation, as no standard mounting constructions are available and as we are considering motorway installation, with the highest safety level.

On the other hand, we assume that reliable glass-glass PV modules with an overhead certificate should be sufficient. Those are already available on the mass market at a slight premium compared to standard products. We therefore expect the non-PV cost share to be between 85% - 90%, as a steel and concrete installation is probably necessary on motorways [16]. For the closed installation, the specific costs and non-PV cost share is higher, due to the additional noise barriers. We estimate that compared to the PVNB (Upgrade), the specific costs will be three to four times as high. However, this can be significantly reduced for applications, where a less massive mounting structure is sufficient, *e.g.*, bike paths.

Concluding this analysis, we give the relative expected specific costs and the non-PV cost share in Table II and Table III.

Table III: Estimated non-PV cost share (%) and derived rating (1-5).

	Non-PV Cost (%)	Rating
PV-Pavement	70	●●○○○
PVNB (Upgrade)	50	●●●●○
PVNB (New)	75	●●○○○
PV-Roofing (Open)	85	●●○○○
PV-Roofing (Closed)	90	●○○○○

We assign the highest ratings to PVNB (Upgrade), where we still see room for improvement in the non-PV cost share, but already a feasible specific cost range.

The considered PV-Roofing of motorways has a low rating mostly due to the high non-PV cost share, which needs to be reduced by new and improved mounting structures. The currently resulting specific costs are not attractive for a mass deployment and need to be improved.

For PV-Pavement and PVNB (New), we see high specific costs and assign a rating of three points. For the non-PV cost share, we deduct an additional rating point, to account for the fact that the dual use in these cases is only preserved by a high investment in the overall installation, making the RIPV application less attractive. As discussed, one could however argue that in both cases, part of the costs should not be attributed to the PV-installation, but to the replaced infrastructure.

4.3 Site requirements

For site requirements, we specifically acknowledge requirements of the considered applications compared to the theoretical potential, taking into account commonly available sites in the mobility infrastructure. Our rating is listed in Table IV.

Table IV: Rating for specific Site Requirements (1-5) for the considered applications, compared to the overall (theoretically) available mobility infrastructure.

	Site Requirements	Rating
PV-Pavement		●○○○○
PVNB (Upgrade)		●●●●○
PVNB (New)		●●●○○
PV-Roofing (Open)		●●●●○
PV-Roofing (Closed)		●●●●●

Currently, most PV-Pavement solutions have demonstrated significant short comings in the realized yield and durability. This still leaves several applications in slow traffic areas, bike or foot paths, as companies are currently focusing. This is a significant reduction compared to the theoretical potential, and we therefore assign the lowest rating.

Similarly, the case of vertical integrated PV in noise barriers (PVNB (New)), currently has less practical applications, due to the aforementioned requirements of highly absorbent barriers. However, we think that the dual use of PVNB will be acknowledged more strongly in the future, and therefore assign the rating of three.

The closed PV-Roofing can be practically installed in almost any location, as it does not interfere with the infrastructure and residents. There might be cases, where the roofing might be unwanted by travelers (*e.g.*, panoramic views), but we do not consider these cases to be significant and assign the highest rating.

For the noise barrier upgrade and the open roofing, we deduct one point. The open PV-Roofing might cause an unwanted increase of noise in populated areas, which reduces its applicability compared to the theoretical potential. The ridge top PV mounting requires an existing noise barrier with sufficient structural reserves. Where noise barriers are already present, this should be given in many cases, but not all.

4.4 PV Module Complexity

With the category of PV Module Complexity, we acknowledge the applicability of existing PV module products, development potential and fundamental constraints identified for the applications. Our rating is listed in Table V.

For the PVNB (Upgrade) and both PV-Roofing applications, currently available PV modules can be used. The PV modules are also available from mass manufacturing and profit from the ongoing improvement of the industry. This is a strong advantage, and they all therefore receive our highest rating in this category.

For the vertical PVNB, different approaches exist. Most of those use adopted PV modules, which are however still close to standard fabrication. Also, certain synergies are assumed with the growing sector of Building

Table V: Rating for derived PV module Complexity (1-5) for the considered applications, with 1 indicating a low rating (high complexity) and 5 a high rating (low complexity).

PV Module Complexity	Rating
PV-Pavement	●○○○○
PVNB (Upgrade)	●●●●●
PVNB (New)	●●●○○
PV-Roofing (Open)	●●●●●
PV-Roofing (Closed)	●●●●●

Integrated PV (BIPV). We therefore give the rating of three points to this application. It should be noted that this does not apply to PV modules for highly absorbent solutions, which are likely more complex.

The PV-Pavement solution uses a highly adopted PV module concept, including several different materials and additional process steps. We therefore assign this application the lowest rating, indicating the highest comparative complexity.

4.5 Integration Capacity

We estimate the available installation capacity per running meter of a road (W_p/rm), as an indication for the fraction of technical potential compared to the theoretical potential, which considers the available area as a whole.

As an example, we consider a two lane motorway including a shoulder (emergency lane) and noise barrier and compare what installation capacity could be realized with the different applications today. We again use these estimations to rank the examples. This rating is listed in Table VI.

Table VI: Estimated Integration Capacity per running meter (W_p/rm) and derived rating (1-5).

	Integration Capacity (W_p/rm)	Rating
PV-Pavement	240	●○○○○
PVNB (Upgrade)	340	●●○○○
PVNB (New)	370	●●●○○
PV-Roofing (Open)	3400	●●●●●
PV-Roofing (Closed)	3400	●●●●●

Current solutions for PV-Pavement have a lower specific capacity (W_p/m^2) than standard PV modules, to account for the additional requirements. Also, the existing solutions cannot be integrated in the whole roadbed. For our analysis, we assume a specific capacity of 150 W_p/m^2, derived from some product data sheets, and consider only the less used motorway shoulder with a width of about 1.6 m. This leads us to an Integration Capacity of 240 W_p/rm. This is the lowest value of our examples.

For the PVNB, we assume a standard PV module with 200 W_p/m^2, to be mounted on the ridge, and an adopted module with 185 W_p/m^2 for the vertical installation. We assume that for an optimistic mounting, we can fit one standard sized module (1.7 m × 1.0 m) per running meter, resulting in a maximum Integration Capacity of 340 W_p/rm. For the vertical installation, we assume a mean barrier height of 4 m, whereas the PV installation is only implemented above 2 m from the ground, due to soiling (dirt, snow, …) and potential stone impacts. With this, we arrive at 370 W_p/rm. One might consider that the height of noise barriers varies and that it is more likely that higher barriers are equipped with PV. We appoint a rating of two and three accordingly to these examples.

All other examples are dwarfed by the enormous potential of PV-Roofing. Considering a two lane street with emergency lane, a width of about 17 m can be assumed. Using standard PV modules with about 200 W_p/m^2, this results in 3400 W_p/rm. This is of course highly depending on the width of the street or rather the available area. In any way, PV-Roofing clearly has the highest Integration Capacity and is ranked highest for both application examples.

5 CONCLUSION

In this work, we have derived a qualitative comparison for three levels of RIPV integration. Considering the current state of the art, the installation of PV modules on the ridge of existing noise barriers as an Upgrade received the highest rating (mean: 4.2). While the solar capacity that can be installed with this approach is limited, it should already be a very attractive option for the mobility sector.

A very high solar capacity can be realized by PV-Roofing. However, the specific costs for such installations are dominated by the mounting structure if it is to be installed above a motorway. A couple of additional benefits could be considered to justify the high costs of the mounting structure, such as a lower degradation of the pavement, less winter service or environmental hazards for vehicles or noise insulation.

A very attractive option might be the extension of tunnel portals, where the PV installation can be used to adapt the lighting from bright daylight to the darkness of the tunnel with semi-transparent PV modules and the generated power can be directly used for lighting and ventilation, one of the major power consumptions in road infrastructure.

Another attractive option may be private entryways or road sections of distribution companies. Here, a roof may protect vehicles and personnel from rainfall and winter service may not be necessary anymore.

Solutions for PV-Pavement are less attractive (mean rating: 1.5), given the current limitations, also described in several studies. It is however possible that such installations remain to be seen in niche applications with controllable conditions such as bike paths or walkways with less requirements on the module pavement.

Another application with significant potential is vertical PV-installations to be used in PV-Noise barriers. While current legislation is limiting the applicability, costs can likely be further reduced when a sizeable market is opened. Currently, several countries are looking into this, e.g., Austria [21] and Switzerland [22].

6 ACKNOWLEDGEMENTS

The authors acknowledge funding by the German Federal Ministry for Economic Affairs and Climate Action under grant 03EE1062A ("PVwins"). In addition, funding has been received from the research projects "PV-SÜD-K" and "PV-SÜD-D" within the D-A-CH cooperation Transport Infrastructure Research 2019 by the Federal Ministry for Digital and Transport (Germany), the Federal Ministry for Climate Action, Environment, Energy, Mobility, Innovation and Technology (BMK) (Austria) and the Federal Roads Office ASTRA (Switzerland).

7 REFERENCES

[1] European Environment Agency, *Greenhouse gas emissions in Europe: A retrospective trend analysis for the period 1990 - 2008*. Literaturverz. S. 118 - 119. Luxembourg: Publ. Off, 2011. [Online]. Available: http://publications.europa.eu/en/publication-detail/-/publication/525c3315-c592-4f63-8a64-e8c58f12d3d2

[2] "Why the automotive future is electric," *McKinsey & Company*, 07 Sep., 2021. https://www.mckinsey.com/industries/automotive-and-assembly/our-insights/why-the-automotive-future-is-electric (accessed: Sep. 12 2022).

[3] Statistisches Bundesamt, *Bodenfläche insgesamt nach Nutzungsarten in Deutschland*. [Online]. Available: https://www.destatis.de/DE/Themen/Branchen-Unternehmen/Landwirtschaft-Forstwirtschaft-Fischerei/Flaechennutzung/Tabellen/bodenflaeche-insgesamt.html (accessed: Sep. 12 2022).

[4] H. Wirth *et al.*, "Potenziale der Integrierten Photovoltaik in Deutschland," in *36. PV-Symposium*.

[5] *Solar Roadways*. [Online]. Available: https://solarroadways.com/ (accessed: Sep. 12 2022).

[6] *SolaRoad*. [Online]. Available: https://www.solaroad.nl/ (accessed: Sep. 12 2022).

[7] *Wattway by colas*. [Online]. Available: https://www.wattwaybycolas.com/ (accessed: Sep. 12 2022).

[8] T. Nordmann and L. Clavadetscher, "PV on noise barriers," *Prog. Photovolt: Res. Appl.*, vol. 12, no. 6, pp. 485–495, 2004, doi: 10.1002/pip.566.

[9] *PVwins – Development of Wall-Integrated PV Elements for Noise Protection - Fraunhofer ISE*. [Online]. Available: https://www.ise.fraunhofer.de/en/research-projects/pvwins.html (accessed: Sep. 12 2022).

[10] *PV-SÜD-K*. [Online]. Available: https://projekte.ffg.at/projekt/3725909 (accessed: Sep. 12 2022).

[11] *PV-Süd – PV Roofing Over Roads - Fraunhofer ISE*. [Online]. Available: https://www.ise.fraunhofer.de/en/research-projects/pv-sued.html (accessed: Sep. 12 2022).

[12] H. Hu, D. Vizzari, X. Zha, and R. Roberts, "Solar pavements: A critical review," *Renewable and Sustainable Energy Reviews*, vol. 152, p. 111712, 2021, doi: 10.1016/j.rser.2021.111712.

[13] *PLATIO SOLAR*. [Online]. Available: https://platiosolar.com/ (accessed: Sep. 13 2022).

[14] *Solmove – Photovoltaik für horizontale Flächen*. [Online]. Available: https://www.solmove.com/ (accessed: Sep. 13 2022).

[15] Innovationspartnerschaft, *Die Autobahn als Sonnenkraftwerk: Mit der Lärmschutzwand Strom produzieren*. [Online]. Available: https://www.ioeb-innovationsplattform.at/challenges/detail/die-autobahn-als-sonnenkraftwerk-mit-der-laermschutzwand-strom-produzieren/ (accessed: Sep. 13 2022).

[16] R. Ebner *et al.*, "PV Roofing for High Ranking Road Networks: Technical Feasibility and Yield Estimations," *WCPEC-8*, to be published.

[17] PV Magazine, *Baden-Württemberg fördert 2 Pilotprojekte für Parkplatz-Photovoltaik mit 300.000 Euro*. [Online]. Available: https://www.pv-magazine.de/2021/04/16/baden-wuerttemberg-foerdert-2-pilotprojekte-fuer-parkplatz-photovoltaik-mit-300-000-euro/ (accessed: Sep. 13 2022).

[18] PV Magazine, *France's solar road: International solar experts give their analysis*. [Online]. Available: https://www.pv-magazine.com/2016/11/30/frances-solar-roadc2a0international-solar-experts-give-their-analysis_100027057/ (accessed: Sep. 13 2022).

[19] A. Shekhar *et al.*, "Harvesting Roadway Solar Energy—Performance of the Installed Infrastructure Integrated PV Bike Path," *IEEE Journal of Photovoltaics*, vol. 8, no. 4, pp. 1066–1073, 2018, doi: 10.1109/JPHOTOV.2018.2820998.

[20] H. Hu, X. Zha, Z. Li, and R. Lv, "Preparation and performance study of solar pavement panel based on transparent Resin-Concrete," *Sustainable Energy Technologies and Assessments*, vol. 52, p. 102169, 2022, doi: 10.1016/j.seta.2022.102169.

[21] *Die Autobahn als Sonnenkraftwerk: Pilotprojekt PV-Anlagen auf Lärmschutzwänden*. [Online]. Available: https://a3bau.at/die-autobahn-als-sonnenkraftwerk-pilotprojekt-pv-anlagen-auf-laermschutzwaenden (accessed: Sep. 13 2022).

[22] PV Magazine, *Schweiz stellt Flächen an Nationalstraßen kostenlos für Photovoltaik zur Verfügung*. [Online]. Available: https://www.pv-magazine.de/2022/08/18/schweiz-stellt-flaechen-an-nationalstrassen-kostenlos-fuer-photovoltaik-zur-verfuegung/ (accessed: Sep. 13 2022).

DESIGN AND CHARACTERIZATION OF SEMITRANSPARENT PATTERNED PV MODULES FOR GREENHOUSE APPLICATIONS

Carlos Toledo[1,2], Alessandra Scognamiglio[1]*, and Francesco Frontini[3]
[1] ENEA Centro Ricerche Portici, Energy Technologies Department, Photovoltaics and Smart Devices Division, Innovative Devices Lab, largo Enrico Fermi 1, 80055 Portici (NA), Italy. *Corresponding author: alessandra.scognamiglio@enea.it
[2] Technical University of Cartagena (UPCT), C/ Doctor Fleming, s/n. - 30202 – Cartagena, Spain.
[3] SUPSI, University of Applied Sciences and Arts of Southern Switzerland, Trevano CP 105, Canobbio CH-6952, Switzerland.

ABSTRACT: Continuous progress in photovoltaic (PV) technology has provided additional possibilities for integration in buildings meeting aesthetical requirements for innovative architecture designs while producing electricity directly from the sunlight to be utilized to mitigate its energy demand. Among the more promising fields to transfer the multifunctional character of the PV panels, greenhouse plant production is one of the most potentially application able to combine food and energy production on the same land unit. This paper presents the design, development and characterization of an innovative glass-glass semitransparent patterned PV modules where a careful approach in terms of integration of PV technology in the landscape is considered. By the scanning of a specific image which mimics with the surrounding landscape, using pixelation technique in the design, and ink-jet process for manufacturing of the outer glass layer, the first prototypes have been developed. These modules have been implemented in the renovation of a commercial-size greenhouse located in an historical environment, the Botanical Garden in Portici (Italy), and tested under steady state conditions in order to characterize electrical performance as well as global transmittance of the patterned glass. Both experiences serve as the basis for improving new designs, advancing towards PV-agri solutions which find the best compromise between functional characteristics and aesthetic ones.

Keywords: PV module, Design, Characterization, PV greenhouse, Solar Landscaping.

1 BACKGROUND and CASE STUDY

In agreement between ENEA and the Museum of Agricultural Sciences (MUSA centre – Naples, Italy), it was decided to develop a restoration project reinforcing the actual structure and replacing existing glass in one of the roof's southwest eaves with PV glass technology. It is stressed the importance of an aesthetic module design due to the historic context where the greenhouse is located: the botanical garden of Portici (40°48'47.0" N, 14°20'37.8" E), a historic garden dating back to the XVIII century. Figure 1 shows the greenhouse's condition prior to intervention. The southwest eave (azimuth -45°) has an area 104 m^2 (19.3 m length x 5.4 m width) and its slope is 20°.

Figure 1: Greenhouse at the botanical garden of Portici, Naples (Italy). Status before intervention: Southwest eave.

2 DESIGN and IMPLEMENTATION

Both aesthetical, given the presence of tree vegetation, and technological feasibility to ensure photosynthetic light transmission, were factors considered when developing the PV module design.

From the aesthetical point of view, the PV module is designed to harmonize the artificial appearance of the silicon-based PV module (characterized by orthogonal grids from the blue PV cells) with the organic aspect of the vegetation. Regarding technical feasibility, each PV module must ensure a degree of opacity of no more than 30% according to the scientific literature to guarantee sufficient light transmission so that the species cultivated in the greenhouse to photosynthesize [1].

Based on the *Boogie-Woogie* patent [2], pixelation technique is used in the design of the PV glass. It consists of a graphic process of generating an architectural surface that passes through the scanning of a specific image, its reading in terms of pixels, and the transformation of the obtained map into an architectural surface texture. The external layer (the outside glass) of the PV module is therefore conceived in such a way that it would show a decorative pattern, which would fit into the texture of lights and shadows generated by the foliage of the trees. Figure 2 shows the reference image used for the design of the external glass (Figure 2-a) and the layers conceived for the construction of the PV module (Figure 2-b).

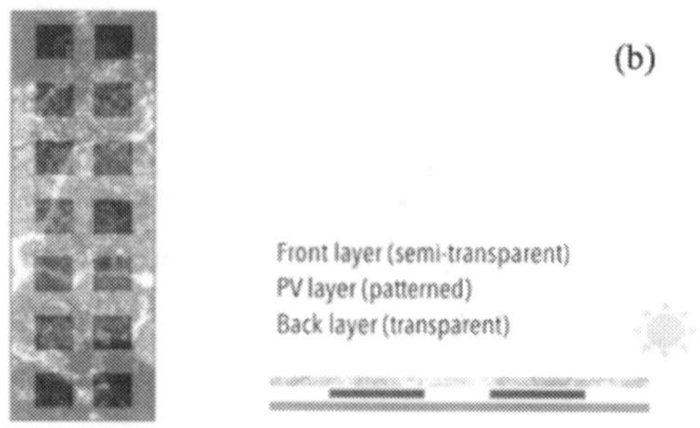

Figure 2: Design phase: pixelation and layering of the customized PV module. (a) Reference image (left, photo: courtesy of Mario Laporta) and Pixelation (right) (b) Design of the semitransparent patterned PV module.

The image was scaled to fit the dimensions of the roof and divided into 6 rectangular zones than can be broken down into 5 submodules with the size of the glass components to be replaced (0.6 m x 1.5 m). Hence, 5 different components offer the possibility to achieve a high degree of variability in the coverage image of the surface.

A research collaboration with *Sunage* (CH) [3] was established for the production of the custom designed PV glass. *Sunage* has a production process that allows the outer layer of the glass to be ink-jet printed starting from a given design (See Figure 3-a). 5 different typologies of module were produced with sizes 60 cm x 180 cm. 90 PV modules in total to cover the area of the roof (3 rows and 30 modules per row) which contains 14 crystalline silicon (c-Si) cells per module (Figure 3-b view from inside, Figure 3-c view from outside). The color selected was blue to not penalize significantly the PV production since photocurrent density loss for c-Si technology is lowest at wavelengths between 420 nm and 640 nm [4].

Figure 3: Construction phase: customized PV module and implementation on the greenhouse roof. (a) Modified front glass through ink-ket printing with a pattern designed to mimic the historical environment. (b) Semitransparent patterned PV module integrated into the structure of the greenhouse. (c) Integrated glass-glass PV module into the roof. View from outside.

3 LABORATORY CHARACTERIZATION

Because the degree of customization is very high, the optical properties of the colored-patterned front layer will affect the power conversion efficiency, as well as the quantity (percentage transmittance) and quality (spectrum) of solar radiation that reaches the plant. Within a research collaboration with SUPSI (CH) [5], the characterization of one prototype was carried out to define its electrical behavior (photovoltaic) and its degree of transmittance (photosynthesis and daylight transmission).

The sample tested is the previous prototype to the installed on the roof. It has the same size although there is a slight variation on the distance and number of PV cells (see Figure 4-a) for each module.

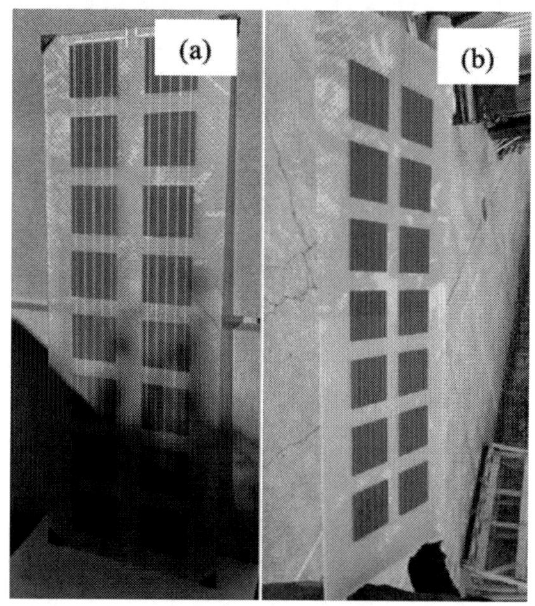

Figure 4: Customized PV glass prototypes under study. (a) Prototype to test at SUPSI: 180 cm x 60 cm, 16 cells. (b) Customized PV glass installed into the roof structure: 180 cm x 60 cm, 14 cells.

Mechanical and electrical features are specified in Table I.

Table I: Mechanical and electrical features of the PV module under test

Mechanical features	
Module length	180 cm
Module height	60 cm
Number of cells in pararell	1 cell
Number of cells in series	16 cells
Single cell area	249.64 cm^2
Electrical features	
Maximum power (P_{max})	65 W
Open circuit voltage (V_{oc})	10.94 V
Short circuit current (I_{sc})	8.09 A
Maximum power voltage (V_{mpp})	8.42 V
Maximum power current (I_{mpp})	7.82 A

Considering the PV module installed on the roof has 14 cells instead of 16. The estimated nominal power of each module is about 56 W, this means the installed capacity is about 5 kW_p (56 W x 90 modules). That is 48 W_p/m^2. With an active area of 31.4 m^2 (249.64 cm^2 x 14 cells/module x 90 modules), which means about 30% of the surface covered with opaque PV cells.

3.1 Electrical performance

Electrical performance was measured using PASAN 3B under standard test conditions (STC). The values obtained are shown in Table II.

Table II: Electrical performance measurement at STC

Test results	
Maximum power (P_{max})	66.12 W
Open circuit voltage (V_{oc})	10.81 V
Short circuit current (I_{sc})	7.82 A
Maximum power voltage (V_{mpp})	8.80 V
Maximum power current (I_{mpp})	7.51 A
Fill factor (FF)	78.3 %
Module efficiency	6.1 %

The I-V curve at STC is shown in Figure 5.

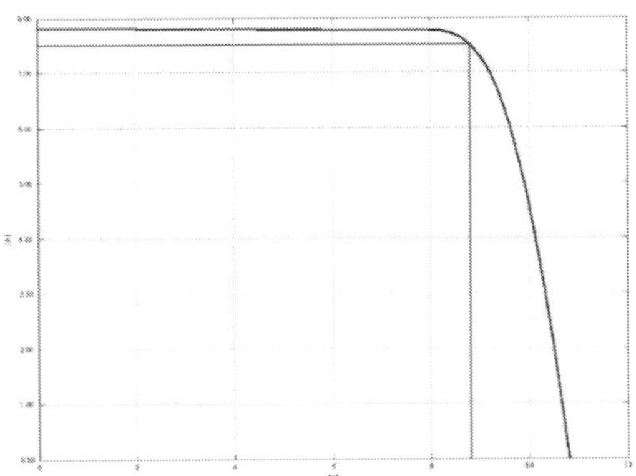

Figure 5: I-V at STC for the prototype tested at SUPSI PVlab.

3.2 Transmittance

The front glass modification affects the glass transmittance and light spectrum that passes through the greenhouse. In order to deepen synergies between photovoltaics and photosynthesis designs, this must be quantified in terms of quantity (percentage of solar radiation) and quality (reflection or absorption of the solar spectrum).

In order to characterize and evaluate the customized PV glass in terms of its possible impact on crop response, the global transparency factor (TRF) of the module can be then estimated for wavelengths in the range of 400 – 700 nm based on the approach proposed by Fernández et al [6].

Since the prototype color is not homogeneous over the entire surface of the glass as explained above, different zones have been considered: PV cells (opaque) and semitransparent patterned glass with different densities, the global transparency of the PV module can be achieved by a combination of different areas. For the semitransparent-patterned zones, it is built a matrix establishing a distance among reference points as shows in Figure 6.

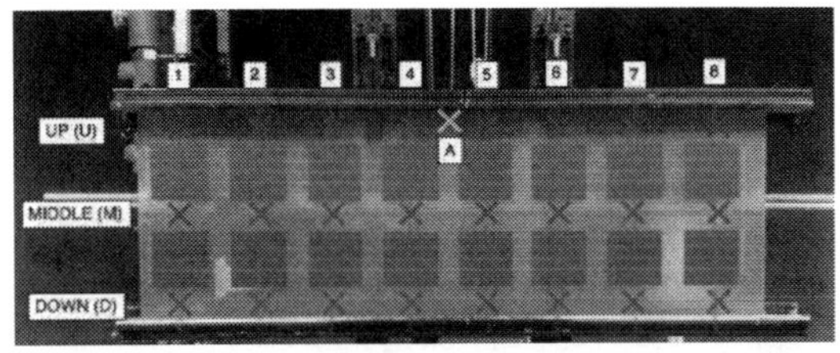

Figure 6: The PV module's semitransparent-patterned regions are represented by the reference point matrix. 24 points spread over three zones: up (U), middle (M), and down (D) using the PV cells as reference for the distance. One additional point (A) is tested as representation of a pattern area with low level of saturation.

The global TRF can be estimated the by an average value of the glass's transmittance ($\tau_{glass,avg}$), and then with the expression:

$$TFR = [\tau_{glass,avg} \cdot (1 - A_{active}/A_{module}) + (\tau_{PV} \cdot (A_{active}/A_{module}))]$$

where τ is transmittance and A is area, so A_{active} is the solar cell area and A_{module} means total module area. The second part of the expression (marked in red), the transmittance of the PV area (τ_{PV}) is set equal to 0, as c-Si cells are opaque.

The measurements were performed using AAA class Pasan Flasher as light source and fiber-optic spectrometer (Avantes) to measure percentage of transmittance. To limit the area of each reference point, a mask (~15 cm^2) was used during the tests to avoid light reflections and interreflections.

As example, Figure 7 shows the measurement results for two points located at the first column of the matrix with values of transmittance of 54% and 45% in the photosynthetic active radiation (PAR) region.

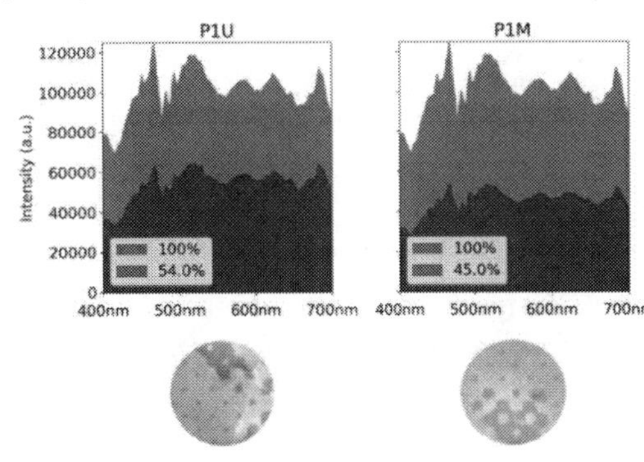

Figure 7: Percentage of transmittance in the PAR region in two points of the matrix. Left: area above column 1 (P1U). Right: middle area of column 1 (P1M). In grey, reference spectrum (100% light transmission). In blue, intensity captured by the spectrometer when is located behind the matrix's point.

Results of light transmittance for all the points of the matrix together with the additional point "A" are shown in Figure 8.

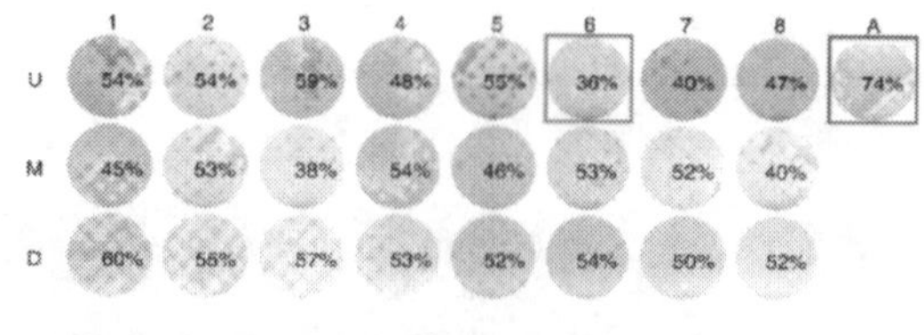

Figure 8: Percentage of transmittance in the PAR region for all the matrix's point and the additional point "A".

As shown Figure 8, the range of transmittance varies from 36% to 74%, with an average of 51%. Hence, a TRF of 0.3 is obtained.

4 CONCLUSIONS

This paper presents the first results of a glass-glass semitransparent patterned PV module design, with the aim to mimic with the surrounding environment, so matching the technical requirements and the visual ones. The first prototype was realized and installed and in the roof of a commercial-size greenhouse located in the botanical Garden in the Royal Palace of Portici (XVIII century). Similar prototype was tested in laboratory to perform electrical characterization and light transmittance. Depends on the density of the pattern, transmittance varies from 36% to 74%, which means a global transparency factor of 0.3 which might limit the crop yield since, according to the literature, a minimum value of 0.7 is recommended. These first results allowing for the improving designs in order to achieve an optimal balance of operational and aesthetic aspects.

Future work includes the monitoring of the PV performance, microclimatic conditions of the indoor space and crop response of different species within the PV greenhouse, as well as the optimization of new designs, including color in the patterns.

REFERENCES:

[1] C. Toledo and A. Scognamiglio, "Agrivoltaic Systems Design and Assessment: A Critical Review, and a Descriptive Model towards a Sustainable Landscape Vision (Three-Dimensional Agrivoltaic Patterns)," *Sustainability*, vol. 13, no. 12, 2021, doi: 10.3390/su13126871.

[2] A. Scognamiglio, G. Graditi, F. Pascarella, and C. Privato, "Boogie Woogie, BIPV module. Patent RM2005A000060," 2005.

[3] Sunage, "Design, development and production of photovoltaic modules." https://sunage.ch/en/ (accessed Jan. 19, 2022).

[4] G. Peharz and A. Ulm, "Quantifying the influence of colors on the performance of c-Si photovoltaic devices," *Renewable Energy*, vol. 129, pp. 299–308, Dec. 2018, doi: 10.1016/j.renene.2018.05.068.

[5] SUPSI, "SUPSI PVLab laboratory," 2022. https://www.supsi.ch/isaac_en/servizi/SUPSI-PVLab.html (accessed Sep. 09, 2022).

[6] E. F. Fernández *et al.*, "Global energy assessment of the potential of photovoltaics for greenhouse farming," *Applied Energy*, vol. 309, p. 118474, Mar. 2022, doi: 10.1016/j.apenergy.2021.118474.

Acknowledgments:

This work was supported by the Italian Ministry of Economic Development in the framework of the Operating Agreement with ENEA for Research on the Electric System. Carlos Toledo is grateful for postdoctoral fellowship 21227/PD/19. Fundación Séneca. Región de Murcia (Spain). The authors want to acknowledge Sunage SA for providing the glass and the samples as well as SUPSI PVlab team (Flavio Valoti and Moreno Ronchi) for their support with the measurements.

VALIDATION OF A BIFACIAL SIMULATION SOFTWARE AGAINST MEASURED ENERGY YIELD FROM A UTILITY-SCALE VERTICAL FIXED-TILT AGROPHOTOVOLTAIC PLANT

*Oliver Pfeiffer[1], Christopher Enrico Marks[1], Jorit Hanneforth[1], Eva-Maria Grommes[1], Ulf Blieske[1]
[1]University of Applied Science Cologne
Cologne, Germany
oliver.pfeiffer@smail.th-koeln.de, christopher_enrico.marks@smail.th-koeln.de, jorit.hanneforth@smail.th-koeln.de, eva-maria.grommes@th-koeln.de, ulf.blieske@th-koeln.de

ABSTRACT: This paper handles a validation study involving the open source bifacial photovoltaic (PV) simulation tool *BifacialSimu*. The tool is based on the existing python libraries *pvlib, bifacial_radiance* and *pvfactors* and offers to apply different simulation modes (*view factor* and *ray tracing*) for front and back side irradiance calculations. Measured PV energy yield (2021 and 2022) from a 4.1 MWp vertical fixed tilt agro-PV plant in Germany was used as reference data. In addition, onsite albedo measurements were conducted over a two month period in 2022. Over the measuring period, the albedo averages out to 0.19 and its daily characteristics are relatively coherent to the findings in literature. A recorded mowing cycle of the grass between the PV rows increases the median albedo by one percentage point. An overall yield comparison between *BifacialSimu* and the measured yield results in a mean deviation of -6.4 % in 2021 and -14.4 % in 2022 for the considered months. It can be said that *BifacialSimu* performs better at times when the share of diffuse irradiance is high, leading to more precise yield prediction in winter months. A similar assessment of *PVsyst* shows opposing behavior.
Keywords: Bifacial, Evaluation, Large Grid-connected PV systems, Ray Tracing, Simulation, ViewFactor

1 INTRODUCTION

The bifacial photovoltaic (PV) and agrophotovoltaic (APV) technologies have gained increasing attention worldwide in recent years, so there are various approaches to combine these. The potential of APV to resolve future challenges like climate change, water shortages and increasing demand for electrical energy and food is promising, as shown in studies by Fraunhofer ISE [1]. The motivation for integrating photovoltaics in the agriculture sector is a tradeoff in land use between the agricultural and the energy industry. Therefore, the mounting systems of APV plants are differentiated into high elevated, axis tracking or vertical systems [2]. Vertical fixed-tilt bifacial plants have advantages in load-shifting to encounter the south-facing plants, soiling and ground coverage through the modules over tilted south-facing panels [3]. Precise yield predictions of large-scale bifacial PV plants are also required to allow an economic operation of the plant.

A tool for bifacial PV simulation using *view factor* and *ray tracing* simulation methods is the python based *BifacialSimu*, which has not been extensively validated yet [4]. Hence this work aims to offer a validation study in the field of utility-scale vertical fixed-tilt APV plants applying the mentioned simulation methods. Therefore, the measured yield data of a 4.1 MWp PV plant in Aasen (Donaueschingen), south Germany is used. The irradiance input data for the simulation program is estimated by measurements of the global horizontal irradiance and diffuse horizontal irradiance of a nearby weather station [5]. The direct normal irradiance is calculated over the trivial geometric relation for a horizontal surface [6]. Albedo values from the literature for undergrounds with similar surface conditions have been used for previous simulations and have been compared to the results using on-site albedo measurements [7].

2 METHODOLOGY

2.1 Description of *BifacialSimu*

This simulation tool *BifacialSimu* combines the python libraries of *bifacial_radiance*, developed by National Renewable Energy Laboratory (NREL), and *pvfactors* developed by SunPower Corporation. Because of this it is given to use the two different radiation calculation models *view factor* and *ray tracing*. The view factor model uses "isotropic scattering of reflected rays, allowing for calculation of irradiance by integration". On the other hand, the ray tracing model uses "multipath reflection and absorption of individual rays entering a scene". This is beneficial if complex simulation is necessary because of the including shading and finite-system edge effects.[8]

For a better understanding of the program a short explanation of the simulation execution is conducted. After running the *GUI.py* file, the graphical user interface opens in a separate window. Firstly, there is the option to define the simulation parameters. To define how the panels are mounted, there is a check mark for single axis tracking. Then the definition of the surface tilt, the modules height and dimensions and plant size by number of modules in each direction are necessary. Other parameters to fill in are the surface reflectivity of the PV rows, the ground coverage ratio, and the albedo. There is the option given to use an average measured albedo, an hourly measured albedo, an hourly spectral albedo or an hourly variable albedo.

In the submenu of the main control a weather file can be read in. The format must be in TMY3 format. In the submenu *simulation control,* there is the possibility to choose between the radiation calculation models. There are five options given. Moreover, the simulation period and the UTC must be defined. In the last submenu *module parameter* there are important parameters of the module listed, which can be found in the data sheets of the modules.

The results of the simulation are given in various graphs and CSV files. For example, a graph shows the bifacial power output over the time and the value of bifacial gain in percentage. Moreover, the absolute irradiance over the simulation period is displayed, where the front and back irradiance are considered separately.

2.2 Description of the PV plant in Aasen

The plant in Aasen has been built by *Next2Sun Technology GmbH* and is now operated by *Solverde Bürgerkraftwerke Energiegenossenschaft*. The 11,000 modules ordered in 550 strings containing 20 modules each are installed with an azimuth of 90 or rather 270 degrees with a fixed tilt of 90 degrees. The rated power of the N-PERT modules (Jolywood JW-D72N Series) is specified with 380 Wp and 385 Wp (two different modules are used).

The measured DC yield data is available in an hourly resolution [8].and high data quality [9]. The weather and irradiance data for the simulation input are given by the database of the German Weather Service (DWD) from a nearby weather station.

2.3 Data procurement and processing

2.3.1 Measured yield data

The yield data of the PV plant in Aasen is provided string wise (DC) for the year 2021, for the year 2022 only the overall AC yield of the whole plant is available. Hence the 2021 data contains quite a lot of data gaps due to tracker communication issues, strings with missing data were filtered. Moreover, only strings that are not located at the edge of the plant are regarded, as *BifacialSimu* does not consider the different shadowing conditions of the outer modules. All in all, 120 strings are considered for 2021. For 2022 all 550 strings are regarded. The yield is given in cumulated energy per hour and needs to be converted to an hourly power yield in W/m^2 by subtracting the yield of the preceding hour and divide it by the related module area.

2.3.2 Weather and Irradiance data

Since the PV plant in Aasen has a commercial and not a research background, there is no scientific on-site irradiance measurement system. Thus, the irradiance data, needed for the simulation in *BifacialSimu* was taken from the climate data center from the German weather service (DWD) who runs a weather station 15 km from Aasen. Therefore, the irradiance and temperature conditions may slightly deviate from what is present in Aasen. The irradiance data (GHI and DHI) is available as a ten-minute sum in J/cm^2. To derive the mandatory format for *BifacialSimu*, the values from thirty minutes prior and twenty minutes after a full hour were summed up and transformed into W/m^2, representing the full hour value.
The given weather files do not contain the direct normal irradiance (DNI), which is a necessary input variable for the simulation. Therefore, the DNI was calculated with using the *pvlib.irradiance.dni* function contained in the python library *pvlib*. Finally, the adjusted weather file could be used for the simulation with *BifacialSimu*.

2.4 Assumptions and simplifications

Due to missing or insufficient data, the following assumptions have been made to be able to compare simulated and measured yield.

Table I: Problems and assumptions in data usage

Problem	Solution / Assumption
Discontinuous yield data of some strings	Only inverter strings with valid measurement data are considered
Missing yield data for the months May and June 2021	These two months are excepted from the consideration
BifacialSimu is only able to simulate rectangularly shaped plant layouts	The simulated layout contains one string per row; All strings except the first and last row receive equal irradiance; From the measured data, only strings in the middle of the plant layout plus one first and one last row string were considered
Upper row modules are east-facing and lower row modules are west-facing; *BifacialSimu* is only able to simulate commonly orientated modules	Only upper-row modules are considered in 2021, as one string only contains commonly orientated modules
Missing on-site albedo measurements for most of the time	Monthly albedo measurements of a comparable location are used [7]; part-time on-site albedo measurements have been done and compared
Incomplete irradiance data for the months February, March, October and December in 2021	These four months are excepted from the consideration (overall, four months of 2021 are remaining)
Measured yield data for 2022 is only available for the whole plant	DC yield is calculated by an assumed inverter efficiency of 0.97, different shadowing conditions and module orientations are neglected
Measured yield data for 2022 does not differ between east and west orientation of the modules	It is assumed that the overall energy yield is not affected significantly due to a leveling out of the occurring yield deviation between morning and evening
Missing on-site irradiance measurements	Irradiance data of a nearby weather station is used
DNI data is not measured	DNI is calculated from DHI and GHI [6]
Different resolution of measured yield and irradiance data	Irradiance data have been hourly averaged as described in 2.3.2

2.5 Albedo measurement in Aasen

Especially in vertical PV plants the ground albedo has a high influence on the rear and front side gain of bifacial modules. To evaluate the impact of the albedo on the simulation results, on-site albedo measurements using two pyranometers from *Kipp and Zonen* (CM11 and CM21) have been made.

For the measurement the two pyranometers were installed as an extension of a module row at its top end. The installations height was three meters and the distance from the row was one meter. To minimize the yield altering effect of reflection from exposed metal, the installation gear was wrapped with black tape. The adjustment of the orientation of the level was done with preinstalled leveling units in the pyranometers. The data was stored in an SD card installed in the *Kipp and Zonen logbox_SE*.

In six second intervals, a voltage value was stored, that needed to be transformed in a corresponding irradiation value via pyranometer specific sensitivity factors. The irradiation values were then averaged over an hour (from 30 minutes before and 20 minutes after a full hour). The measurements were done from June 14th to August 4th. On July 5th the grass between the PV rows was mown.

3 VALIDATION AND EVALUATION

3.1 Influence of measured albedo data

The Albedo measurements were carried out from June 14th to August 4th, including a mowing cycle of the grass that grows between the PV rows.

In Fig. 1, the average measured albedo is shown for selected hours of the day over the whole measuring period, averaging out at 0.19.

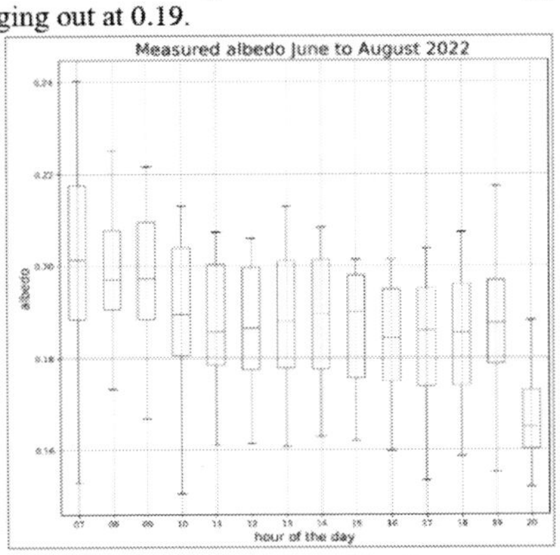

Figure 1: Hourly measured albedo from June 14th to August 4th 2022

The characteristics in the first half of the day are coherent to the findings in the literature, as median measured albedo decreases until noon [7]. In the second half of the day however, the curve recovers slightly, drops again in the early evening, and recovers again until sundown. On contrary, the mentioned report forecasts a steady increase of the hourly albedo after noon [7].

Fig. 2 shows the influence of the albedo on the simulated energy yield on a representative summer day. A literature value of 0.247 (constant over the whole day) is compared with the measured albedo. Since the average of the measured Albedo is lower than the literature value, the simulated energy yield using the literature value exceeds the measurement-based simulation.

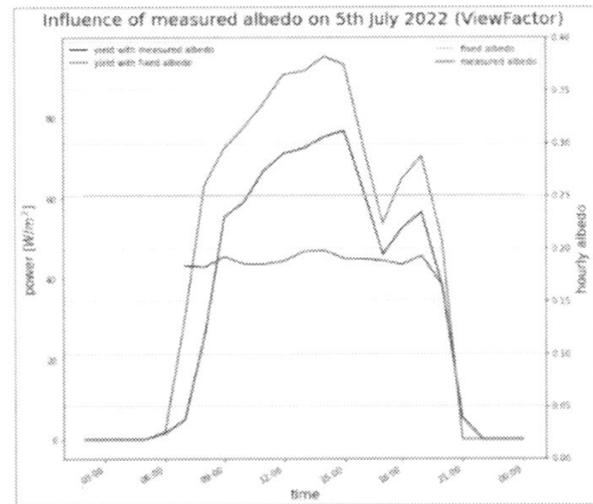

Figure 2: Influence of albedo on simulated energy yield

In Fig. 3 the influence of the mowing can be seen. After July 5th, the day of mowing, the median of the daily albedo increases by about one percentage point.

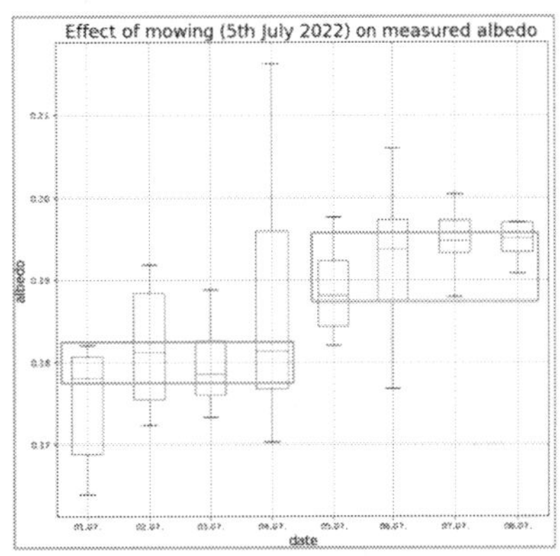

Figure 3: Influence of grass height on albedo

3.2 Validation regarding long term simulation

The yield forecast precision of the simulation tool against real yield data needs to be examined. Therefore, the simulation results are compared to the given yield data of the PV plant in Aasen and to the simulation results of the established simulation program *PVsyst*.

Fig. 4 compares the measured energy yield in Wh/m² to the simulation results of *BifacialSimu* and *PVsyst* in selected months of 2021 and 2022. Since *PVsyst* is a *view factor* based modelling tool, the *view factor* simulation mode was used in *BifacialSimu*.

In 2021 the deviation of the simulation results with *BifacialSimu* against the measured data varies between -11 % and +10 % for the monthly consideration, with an average of -6.2 % for 2021. The lowest deviation can be recognized in March 2021. The deviation of *PVsyst* against the measured data varies between -43 % and +1 %, with an average of -14.3 %.

In 2022 the results with *BifacialSimu* show the lowest deviation in January with -1.5 %. Between February and June, the discrepancy is around -8 % and -11 %. The highest deviation occurs in July being -16 %. In 2022 the measured yield with an overall average of -9.6 % exceeds the simulated in every month, so other than in 2021 the monthly deviations do not balance out over time. The

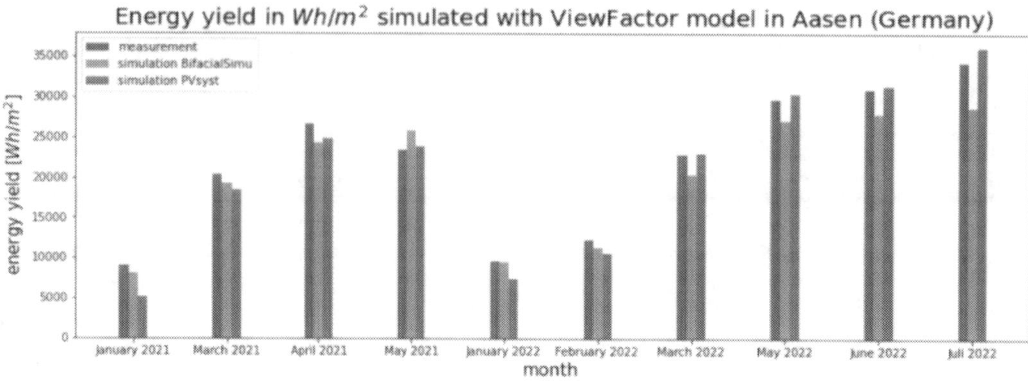

Figure 4: Monthly yield comparison between measurement, *PVsyst* and *BifacialSimu*

simulation with *PVsyst* shows a contrary behavior in winter and summer time. In January and February, the simulation deviates by -23 % and -13 % against the measured data. However, the deviation in the months March till July is particularly low, with a maximum of +5 %. The overall average for 2022 is -4.51 %.

This consideration shows that *BifacialSimu* performs better in winter months and *PVsyst* in summer months. This behavior can be explained by a higher share of diffuse irradiance in winter and a higher share of direct irradiance in summer months. Section 3.3 discusses this topic in more detail, through an examination of the characteristic curve over the course of a day.

3.3 Validation of the characteristic yield curve

Subsequently, the daily characteristic yield curves for days in January 2021 and June 2022 are discussed. Assumptions have been made to explain the deviation of the measured and the simulated energy yield.

Fig. 5 shows the energy yield for a partly sunny day in June 2022 with a higher share of direct normal irradiance in the morning and an increase of diffuse irradiance starting at noon time. There is a slight time deviation between the peak values of the measured and simulated yield. The overall characteristic of the yield curve is estimated suitable for the first half of the day whereas the *view factor* simulation estimates the height of the first peak slightly better than the *ray tracing* simulation mode. The underestimation of the yield in the afternoon with both simulation modes is noticeable. This is assumed to be explained by an underestimation of direct irradiance on the backside energy yield within *BifacialSimu*.

Another aspect that can be pointed out is that the yield values are approximated much better on days with a higher share of diffuse irradiance. This can be seen in Fig. 6, where the curves of the measured and the simulated data are more closely aligned, on an overcast day in January 2021. The differences between both models seem to be lower when the share of the diffuse irradiance is especially high, as it is shown on an overcast day.

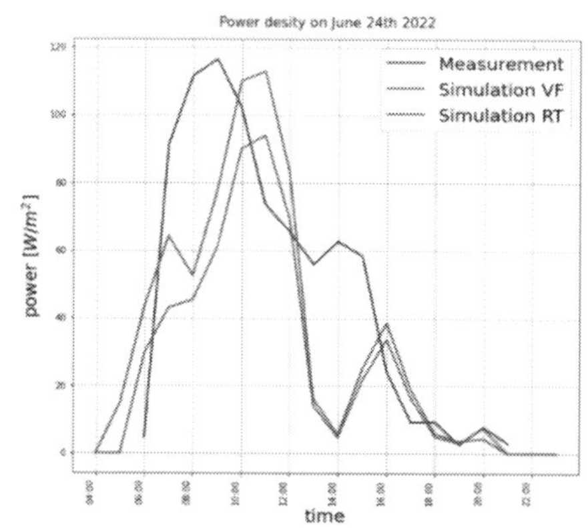

Figure 5: Measured and simulated yield on a partly sunny day in June 2022

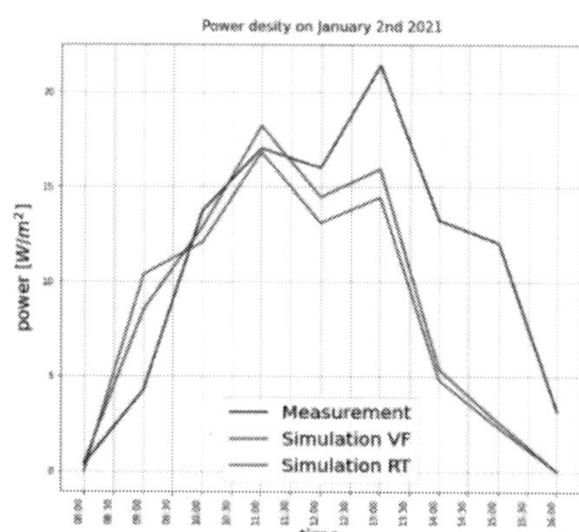

Figure 6: Measured and simulated yield on an overcast day in January 2021

4 DISCUSSIONS

Even though the simulation results of *BifacialSimu* are plausible and the deviations are placed in the same order of magnitude than the ones of an established simulation tool (*PVsyst*), differences especially in hourly simulations occur and need to be considered. The deviations may have two main reasons, the input data of the simulation and the simulation itself.

Regarding the input data, possible error sources are the

different orientation of the upper and lower row modules. This is important, because the yield difference between back and front side depends on the bifaciality factor. Since the yield data was available string wise in the year 2021, only east facing modules are considered to handle this problem. In 2022 no differentiation could be made, because the yield data was only available for the whole plant. For the 2022 data, an additional impact is that only AC yield data is available while DC yield is simulated. A significant reason for deviations, especially influencing hourly simulations is the missing on-site irradiance data, hence GHI and DHI data from a weather station at a distance of 15 km is used and DNI is calculated therefrom. A slight deviation may be attributed to the simplification of the plant layout in the simulation as well as the averaging of the irradiance data.

As *BifacialSimu* performs better in winter than in summer months, it is assumable that diffuse irradiance conditions have a positive impact on the simulation accuracy. This assumption is supported when considering the daily characteristic curves. Overall, a high share of diffuse radiance contributes to better simulation results. On the other hand, an underestimation of direct irradiance on the back side yield can be observed.

5 CONCLUSION

Taking all aspects into consideration, it can be concluded that *BifacialSimu* is a suitable tool to estimate the overall energy yield of vertical fixed tilt bifacial PV plants. The accuracy for hourly simulation strongly depends on the irradiance conditions, leading to more accurate results in times with a high share of diffuse irradiance. The impact of direct irradiance conditions to the back side yield simulation has to be further studied to keep improving *BifacialSimu's* performance.

6 ACKNOWLEDGMENT

We want to thank Peter Bendix (*Next2Sun Technology GmbH*) for the cooperation and for providing measured yield data of the PV plant in Donaueschingen-Aasen. Furthermore, we want to thank Rudolf Gecke[1] for enabling the on-site albedo measurements.

7 REFERENCES

[1] M. ,. e. a. Trommsdorff, "Agri-Photovoltaik: Chance für Landwirtschaft und Energiewende - Ein Leitfaden für Deutschland," Fraunhofer ISE, October 2020. [Online]. Available: https://www.ise.fraunhofer.de/content/dam/ise/de/documents/publications/studies/APV-Leitfaden.pdf. [Accessed 1st February 2022].

[2] M. G. M. F. J. Scharf, "Agri-Photovoltaik - Stand und offene Fragen," Technologie- und Förderzentrum im Kompetenzzentrum für Nachwachsende Rohstoffe (TFZ), Straubing, May 2021

[3] e. a. R. Younas, "Agrivoltaic Farm Design: Vertical Bifacial vs. Tilted Monofacial Photovoltaic Panels," Lahore, Pakistan, 2019.

[4] E. Grommes, U. Blieske, „BifacialSimu:Holistic Simulation of large-scale Bifacial Photovoltaic Systems," To be published in: Journal of Open Source Software, 2022

[5] Deutscher Wetterdienst (DWD), "Climate Data Center - Deutscher Wetterdienst," [Online]. Available: https://cdc.dwd.de/portal/. [Accessed 30th January 2022].

[6] J. Brownson, "Solar energy conversion systems," Oxford, Elsevier's Science and Technology, 2014, p. p. 214.

[7] B. Marion: „Albedo Data for Bifacial PV Systems Update", NREL, Golden, USA 2020

[8] Pelaez er al., (2020). Bifacial_radiance: a python package for modeling bifacial solar photovoltaic systems. Journal of Open Source Software, 5(50), 1865 [Online] doi.org/10.21105/joss.01865

[9] Next2Sun GmbH, "Hourly Yield Data of Bürgersolarpark Donaueschingen-Aasen," 2021

AGRIVOLTAICS AS A TECHNOLOGICAL OPTION FOR ECOSYSTEM RECOVERY: A CASE STUDY FOR THE MAR MENOR COASTAL LAGOON IN SOUTH EAST SPAIN

Carlos Toledo*[1,2], Adela Ramos[2], Lucía Serrano[2,3], Antonio Urbina[4]
[1]Italian National Agency for New Technologies, Energy and Sustainable Economic Development (ENEA), Centro Ricerche Portici, Portici, Italy.
[2]Technical University of Cartagena (UPCT), Dep. of Applied Physics, Cartagena, Spain.
[3]Rey Juan Carlos University (URJC), Department of Computer Science and Statistics, Móstoles (Madrid), Spain.
[4]Institute of Advanced Materials and Mathematics (INAMAT2), Department of Sciences, Pamplona, Spain.
*Corresponding author: carlos.toledo@upct.es

ABSTRACT: The potential contribution of photovoltaic systems to the recovery of the damaged Mar Menor ecosystem in south-east Spain has been evaluated. The Mar Menor coastal lagoon is threatened by strong eutrophication of its water due to several factors, of which nitrogen load from agriculture is pointed out as the main cause of ecosystem degradation. The introduction of photovoltaic systems in the watershed area of Mar Menor will reduce the land use devoted to agriculture and livestock and thus reduce the entrance of contaminants into the lagoon. Life Cycle Assessment (LCA) and Geographical Information System (GIS) methodologies have been combined to provide a model in which agrivoltaics is introduced either by substituting crops or by sharing the land with the crops. The results show that using a small percentage of each parcel to set a PV system, 745 tons of nitrogen can be saved as fertilizer while generating 1928 GWh/year of electricity production on the study area. It has been shown that agrivoltaic systems can be used for ecosystem restoration in agricultural areas where intensive production has led to important land and water contamination problems.

Keywords: Photovoltaic, Sustainable, LCA, Agrivoltaics, GIS.

1 INTRODUCTION

In recent decades, intensive agriculture and urban development have triggered an environmental crisis in the Mar Menor (Southeast of Spain), the largest saltwater lagoon in Europe [1]. The land surrounding the lagoon has undergone significant changes to the agricultural model, increasing irrigated lands for an intensive farming practice (between 1988 and 2009, intensive irrigation increased by more than 140% of occupied land), thus contaminating the soil and the groundwater with nitrates and phosphates from fertilizers that leak into the lagoon. On the other hand, Murcia Region is one of the regions where the solar resource is superior to that of most EU countries (1800-2200 kWh/m^2/y [1]). This is a clear example of where utility-scale photovoltaic (PV) systems increase the pressure on agricultural activities and threaten to trigger land-use competition. In this context, PV systems integrated in agricultural land and where sunlight and land are shared but agricultural yields are prioritized, also known as agri-PV or agrivoltaics (APV), provide a feasible solution for farmers to contribute to climate resilience while also generating extra revenue through the production of electricity [2]. Furthermore, in the case study of the Mar Menor watershed basin, the inclusion of PV systems in the area will reduce the use of fertilizers and fungicides and therefore contributes to ecosystem restoration. Agrivoltaics is currently at the top of the European agricultural policy agenda as a primary driver of the transition to a more sustainable agricultural sector and several countries are developing policies to regulate this activity [3, 4].

2 METHODOLOGY

The combination of photovoltaic power modelling with Life Cycle Assessment (LCA) and Geographical Information Systems (GIS) provides a tool to evaluate the benefits of introducing agrivoltaic systems in a heavily exploited agricultural area. The need to reduce waste emissions (specially nitrates) from industrial agricultural systems together with the reduction of water consumption offers a good case study to propose agrivoltaics as an environmental restoration tool which may contribute to the recovery of the Mar Menor coastal lagoon and to improve the sustainability of food production. Details about the used methodologies are provided in the following subsections:

2.1 Photovoltaic system design and power rating

Photovoltaic system design and modelling have been carried out for three kinds of systems to be installed in the watershed basin of Mar Menor. First, a partial coverage of existing greenhouses and floating systems in small irrigation ponds (no modification of crop´s area); secondly, the installation of conventional PV systems on the ground (thus reducing crop´s area) and finally, an agrivoltaic approach in which the land is shared by the crops and the PV systems. According to the literature, a 30% PV cover ratio on greenhouse roofs is compatible with a wide range of crops [5]. Therefore, in this scenario, it is assumed that there are no losses in crop production and the PV energy generated is estimated with a PV cover ratio of 30% on south-facing roof eaves. In the second scenario, each parcel is assumed to devote a certain percentage of its land to PV production. To prioritize agriculture productivity, this proportion is set at 20%, resulting in 80% agricultural production with a 20% reduction in fertilizer, water, and agrochemical use due to the implementation of a PV system; the model enables researchers to choose other land use percentage. The electricity generated in this scenario is calculated using a conventional land use energy intensity of 2 ha/MW$_p$. Finally, an elevated agrivoltaics system using 100% of each parcel area is under consideration. In this scenario, the criteria are a fixed system with a clear height of 2.10 m, optimal slope and azimuth angles, and 70% of sunlight at ground level. The PV systems have been designed and modelled for commercial crystalline silicon (c-Si) technology using PVGIS JRC database and system performance tool.

2.2 Life Cycle Assessment (LCA)

LCA methodology following the ISO standards 14040 and 14044 has been applied to study the environmental impacts of the agriculture production and the PV systems manufacture and operation by several research groups [6]–[8]. For each LCA approach, a different functional unit (FU) was used to compile the inventory of crops and PV systems, with a FU as 1 kg of the seven main crops (artichoke, broccoli, lettuce, melon, lemon, mandarin and orange trees) and a FU of 1 kWh_{AC} for solar electricity produced by the PV systems. Life cycle impact assessment has been carried out using *SimaPro* software and *Ecoinvent* database. This study's LCA results are reported by Martin-Gorriz et al. [9]. The fruit and vegetable production were studied from cradle to gate life cycle phases. On top of standard LCA, particular focus has been devoted to calculating nitrogen load from agriculture into Mar Menor via eutrophication processes and reducing this load by replacing of crops with PV systems. This factor has been pointed out as the leading cause of ecosystem degradation of Mar Menor [2]. A balance of energy inputs and outputs of the combined PV system electricity generation and agricultural cumulative energy demand (CED) has been carried out.

2.3 Geographical Information System

GIS has been the tool used to calculate land use for each crop (according to the official registry) and to simulate the share of land that will be occupied by the proposed PV systems that have been modelled. The GIS maps enabled the possibility of a detailed simulation of agricultural land occupation by PV systems for every crop in each cultivated plot, identified by geographical coordinates and cadastral codes. The GIS software used for the analysis was *ArcGIS* from the ESRI family (version 10.2.1). The primary source of geographical crop data is the SIOSE database (Information System on Land Occupation in Spain [10]) and for solar radiation is the Photovoltaic Geographical Information System (PVGIS) of the Joint Research Centre (JCR), European Commission [1]. The cadastral information for the GIS crop base layer is extracted from the Cadastral shapefile map (ETRS89/UTM zone 30). The layer is split into different parcels and sub-parcels. The former makes a distinction on the owner and the latter, within the same parcel, on the crop type.

3 RESULTS AND DISCUSSION

The main innovation of the research work is the combination of three methodologies to provide clear recommendations to policymakers and stakeholders to solve an urgent ecosystem degradation problem. The results are based on simple models that can be further refined but provide a clear indication of short and middle-term impacts reducing nitrogen load into Mar Menor coastal lagoon thanks to the introduction of PV systems in its watershed area.

3.1 Available land for photovoltaics

In the first scenario the PV systems are integrated into greenhouses with a 30% coverage of the greenhouse roof and thus sharing the space with pepper production (in this case, the production area is kept, and no reduction of fertilizers or agrochemicals is considered), additional calculations have been carried out for the potential electricity production of floating systems installed on existing irrigation ponds. The second scenario of substituting a percentage of farming land use for photovoltaic generators directly reduces chemicals discharged into Mar Menor Lagoon, i.e., fertilizers, fungicides, insecticides and herbicides. In all scenarios, the PV installation complies with the legal framework and preservation of natural areas protected by Red Natura 2000 (shown in red in Figure 1). In some areas (yellow in Figure 1), restrictions apply but do not eliminate the possibility of PV distributed system installation (the main restriction comes from the necessary preservation of waterways and "high quality" landscape, according to the Spanish ministry of ecological transition).

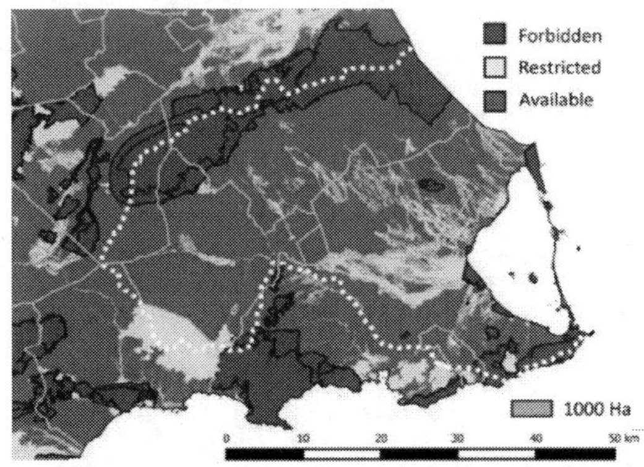

Figure 1. Mar Menor in South East Spain (Murcia Region), the watershed area affecting Mar Menor is indicated with the white dotted line and regulations for PV installation are classified in three groups: red areas are forbidden (Natura 2000 protected areas), yellow are restricted to special conditions and green are available. Grey rectangle shows an area of 1000 ha, which approximately can accommodate 500 MW_p PV systems [11].

Table I shows the total crop yield per hectare, parcel area, and irrigation use for the most representative crops, as well as the energy generated annually in the second scenario where the total area that can be used for PV ground systems if PV modules replace 20% of the seven considered crops is 3788 hectares. As an example of the obtained results that are calculated in detail for the geographical location of each crop by using GIS tools, the aggregated solar electricity production can be up to 1655 kWh per kW_p (for an average irradiation of 2118 kWh/m^2 per year), thus leading to a total electricity production of 1592 GWh/year for c-Si cells with PR = 0.86, considering thermal losses, and a ground requirement ratio (GRR) = 4 (nominal aggregated capacity of all systems: 962 MW_p). While keeping the technical parameters of the PV systems constant, a very large range of solar electricity production is obtained according to the percentage of land used. Within the second case study, a simple option has been selected to be presented in this communication: 20% land use in all parcels, independently of the kind of crop and the size of the parcel. More refined selection of land has been modelled, by chosing different percentages according to the kind of crop (for example, depending on the amount of fertilizers required, or the profitability of each crop. Stakeholders may decide the final objective: reduce environmental impacts on Mar Menor and maximize combined profits from crops and PV.

Table I. Total crop yield per hectare, parcel area, and irrigation for crops grown in the study case zone. The production of PV energy corresponds to the second scenario.

	Yield (ton/ha-year)	Surface (ha)	Water (m^3)	PV (GWh/year)
Artichoke	12.5	1966	9285	165.3
Broccoli	28.6	1883	4090	158.3
Lettuce	53.0	3291	3816	276.7
Melon	43.0	3400	4943	285.9
Lemon	29.5	4342	4463	365.1
Mandarin	22.5	1892	4850	159.1
Orange	20.9	2165	4463	182.0

The third scenario assumes that 100% of the parcel area is covered by PV panels with a power density that allows 70% of the sunlight to reach the ground. As shown in Figure 2, this is performed by simulating the projection of annual shadows and determining the ground cover ratio (GCR) value. A GCR of 0.35 is estimated from simulations. That corresponds to a land use energy intensity of 5 ha/MW$_p$. Under this configuration, the potential installed capacity is 3788 MW$_p$. Based on the PVGIS energy production method (14% system losses), that means an annual electricity production of 6666 GWh/year, about four times that of scenario 2. However, since there is no evidence in the literature of the impact of the PV modules' shadows on their use, this scenario does not consider fertilizer savings. As a result, but also because agrivoltaics is proposed as a method for lagoon ecological recovery, the impact of fertilizer savings is focused on scenario two.

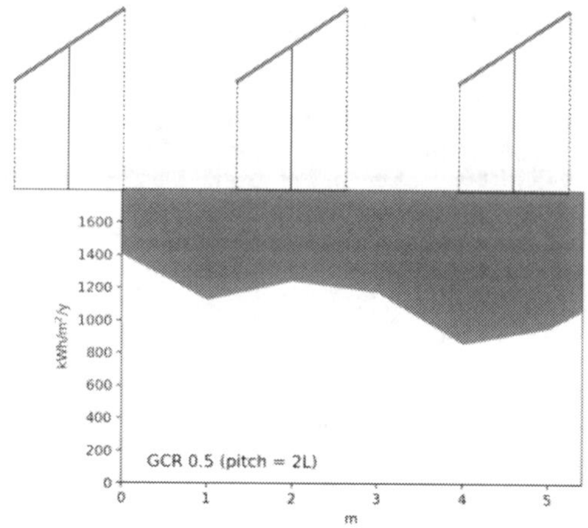

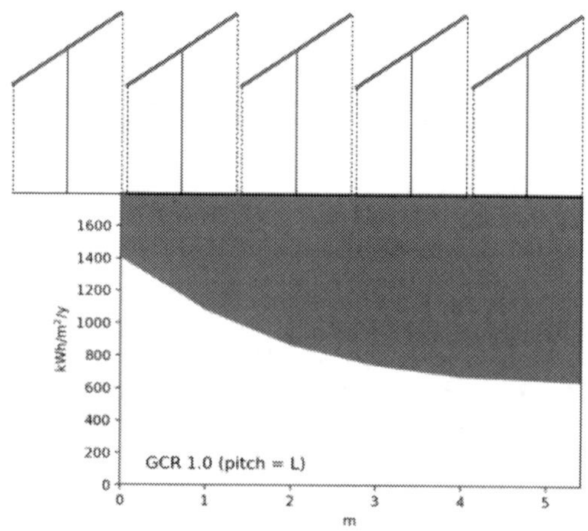

Figure 2. Simulations to determine the GCR value that allows 70% of sunlight to reach ground level. Optimal angles of inclination and slope are assumed, as well as a free height of 2.1 m and a length of the PV module (L) of 1.3 m.

3.2 Use of fertilizers, agro-chemicals and water consumption

The farming developed in the area under study requires the use of fertilizers. The fertilizers considered in this study are nitrogen (N), phosphorous pentoxide (P_2O_5), and potassium oxide (K_2O), whose amounts per hectare were obtained from official local reports [12]. Nevertheless, the production of fertilizers has one of the most detrimental impacts on the environment, joined with the irrigation and fieldwork operations [9], compared to other stages during the life cycle of the agriculture activity.

Water consumption is one of the main concerns in southeast Spain, since the area and surroundings have suffered from droughts during the last decays [13]. Besides, groundwater is being overexploited by wells, pipes and desalination plants, many of which are neither authorized [14].

The inventory of fertilizer and agrochemical use for the most representative crops in the study area is shown in Table II. A detailed calculation for nitrogen load reduction when 20% of the land is devoted to PV ground systems (second scenario) indicates that it could lead to reductions of up to 744 metric poured tons of nitrogen per year.

Table II. Total amount of fertilisers and agrochemicals inventory per cycle and hectare in the agricultural area affecting the Mar Menor coastal lagoon (calculated with data from [9] and [15]).

	Fertilizers (ton/year)			Agro-chemicals (ton/year)	
	N	P_2O_2	K_2O	Fungicides/ insecticides	Herbicides
Artichoke	489.5	194.6	336.2	25.5	7.9
Broccoli	357.8	120.5	220.3	15.0	5.6
Lettuce	694.4	154.7	500.2	26.3	23.0
Melon	561.0	197.2	584.8	30.6	17.0
Lemon	790.2	156.3	573.1	39.1	17.4
Mandarin	404.9	81.3	179.7	24.6	11.3
Orange	426.5	86.6	270.6	15.1	10.8

3.3 Land use, PV installed capacity and annual energy production. GIS approach.

The previous sections' results can be specified by parcel, providing a clearer picture of the agrivoltaic potential in each scenario. Figure 3 shows the installed potential and energy generated in each parcel for the study area and using scenario 2 as a reference and with the value choice option of 20% land use on each parcel for PV ground mounted systems.

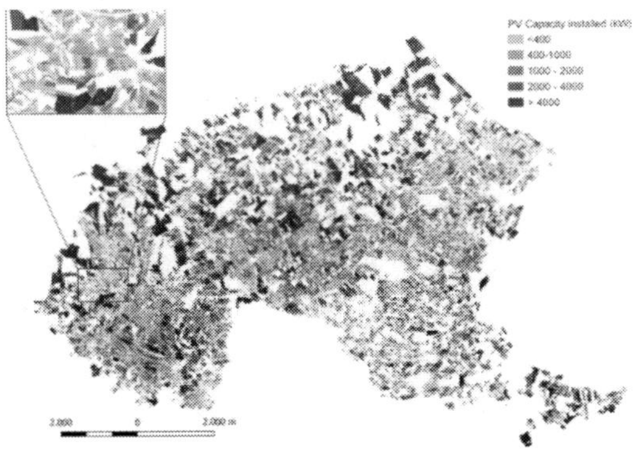

(a)

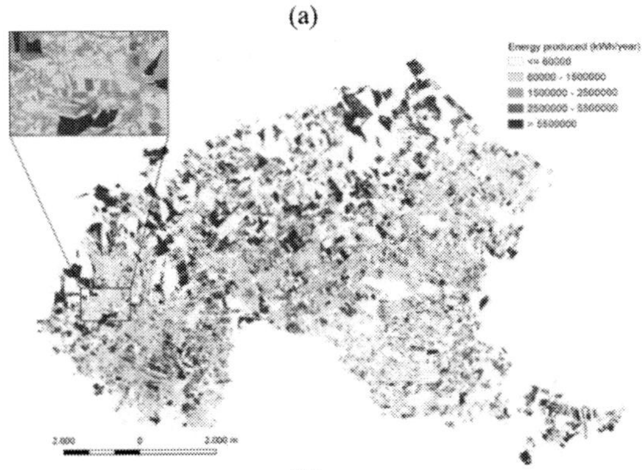

(b)

Figure 3. (a) PV installed capacity and (b) energy production in scenario 2 with a 20% land use for PV.

4 CONCLUSIONS

Ground mounted and greenhouse roof-top PV systems in rural areas as well as agrivoltaic systems are presented as valuable tools for contributing to the recovery of a coastal area by diversifying farmer income and alleviating the pressure of using fertilizers toward a more sustainable business model. A small percentage of each parcel (20%) can be used to install a PV system allowing for a reduction of at least 745 tons of nitrogen as fertilizer. A calculation of solar electricity production with ground mounted and greenhouse roof-top PV systems provides an initial estimate of 1928 GWh per year (scenario 1+2). If agrivoltaic systems are included, higher electricity production can be obtained although in this case, the use of agrochemicals is not reduced. The results are relevant to bringing the energy transition to the primary sector, which mainly contributes to GHG emissions. The developed model provides a tool to combine several value-choice parameters and to provide different solutions to stakeholders and politicians, which ultimately should take the decisions in the energy transition.

Acknowledgements:

This work is supported by the Spanish Agencia Estatal de Investigación MCIN/AEI/10.13039/501100011033, grant PID2019-104272RB-C55 and MCIN/AEI/10.13039/501100011033/ grant PID2021-126605NB-I00 both including FEDER funds; C. T. is grateful for his postdoctoral fellowship "Margarita Salas" as part of the Spanish University System requalification program funded by the Ministry of Universities with EU Next Generation funds.

REFERENCES:

[1] European Commission, "Photovoltaic Geographical Information System (PVGIS v5.2)," 2020. http://re.jrc.ec.europa.eu/pvg_tools/en/tools.html (accessed Nov. 29, 2021).

[2] C. Puertes, I. Bautista, A. Lidón, and F. Francés, Agricultural Systems, 188 (2021) 103029.

[3] Solar Power Europe, "Agri-PV: how solar enables the clean energy transition in rural areas," (2020).

[4] Chatzipanagi A., Taylor N., Thiel C., Jaeger-Waldau A., Dunlop E., and Kenny R., "Agri-photovoltaics (Agri-PV): how multi-land use can help deliver sustainable energy and food," European Commission, Scientific analysis or review JRC129225,(2022).

[5] C. Toledo and A. Scognamiglio, Sustainability, 13, (2021) 12.

[6] J. Damásio et al., Aquatic Toxicology, 87, (2008) 4.

[7] D. Tilman, K. G. Cassman, P. A. Matson, R. Naylor, and S. Polasky, Nature, 418, (2002) 671.

[8] J. A. Campillo, M. Albentosa, N. J. Valdés, R. Moreno-González, and V. M. León, Aquatic Toxicology, 142 (2013) 365.

[9] B. Martin-Gorriz, B. Gallego-Elvira, V. Martínez-Alvarez, and J. F. Maestre-Valero, Journal of Cleaner Production, 265(2020) 121656.

[10] Instituto Geográfico Nacional., "Sistema de Información sobre Ocupación del Suelo de España - SIOSE." https://www.siose.es/ (accessed Sep. 22, 2022).

[11] Sistema de información Territorial de la Región de Murcia "Mapa de Adecuación Territorial para la Generación de Energía Solar Fotovoltaica." https://sitmurcia.carm.es/ (accessed Sep. 22, 2022).

[12] Ministerio de and Agricultura, Alimentación y Medio Ambiente (MAGRAMA), "Resultados técnico-económicos de Cultivos Hortícolas 2014," (2015).

[13] J. A. Redondo-Orts, M. A. Sáez-García, and M. I. López-Ortiz, Water, 14 (2022) 14.

[14] I. Katerina, "Agroecologies for the Stateless (master thesis)," Delft University of Technology, 2022.

[15] Comunidad Autónoma de la Región de Murcia (CARM), "Estadística Agraria Regional 2020-2021," 2020. https://www.carm.es/ (accessed Sep. 22, 2022).

A SUSTAINABLE AGRIVOLTAIC EXPERIENCE IN A DRY MEDITERRANEAN AREA: CO-EXISTENCE OF AGRICULTURAL ACTIVITIES IN UTILITY SCALE PLANTS OF EGP FOR MULTIPLE LAND USE

Massimo Fagnano[1], Miriam Di Blasi[2], Maria Genovese[2], Gian Maria Baldi[3], Michele Falce[3],
[1]University of Naples Federico II, fagnano@unina.it, via Università 100, Portici (Naples), Italy,
[2]Enel Green Power, Innovation, miriam.diblasi@enel.com, maria.genovese@enel.com, via A. Pisano 120, Pisa, Italy,
[3]Novamont, gianmaria.baldi@novamont.com, michele.falce@novamont.com, via G. Fauser 8, Novara, Italy

ABSTRACT: The results reported in the document are related to one of the agrivoltaic (APV) tests in progress in Kourtesi photovoltaic (PV) plant in Greece, as part of the wide experimental program launched by Enel Green Power (EGP) in January 2021. Aromatic species have been implemented in three different areas: in the corridors between two modules rows, underneath the modules and in a control area, free from panels. In addition, in unused marginal areas specific crops have been implemented, enabling new business models, with a circular bioeconomy approach, delivering new win-win solutions for the management of the PV plant and nearby farmers. Proximity and remote sensing networks allow to collect data about all the crops, as a valid support tool to schedule irrigation and fertilization activities. Sustainable agricultural practices have been also implemented, as the use of Riterra®, a biodegradable in soil mulching film made of Mater-Bi®, a family of completely biodegradable and compostable bioplastics produced by Novamont. Preliminary results for the aromatic species have shown no differences in crops growth between the corridor and the control area. Crops growth monitoring will continue up to 2023, to assess the results and the best sustainable agronomic practices for improving the crops growth.

Keywords: agrivoltaic, sustainable solar park, sustainable agricultural practices, circular economy, agriculture 4.0.

1 INTRODUCTION

EGP launched a massive agrivoltaic program in January 2021, involving nine site demonstration tests distributed among Spain (5 sites), Greece (2 sites) and Italy (2 sites). The program has been developed with a holistic approach, for standard photovoltaic (PV) plants, both in brownfield and greenfield sites, to improve their environmental/ecological sustainability and social acceptability. The program results in a paradigm shift from industrial solar plant to sustainable solar parks, consisting of two main pilars:

1) Solar inclusivity, that stands for multi-purpose land use (used for crops, vegetation, grazing) and for a collaborative multistakeholder approach, with the engagement of local communities;

2) Solar diversity, that means biodiversity integration everywhere, through the wildlife habitat preservation, as well as ecosystem services improvement.

Different climate areas, PV standard technologies (fixed structures or trackers, equipped with mono or bifacial modules) and eight layouts (different sizes of free corridors from 3 m to 8 m) have been considered in the experimental program. In each agrivoltaic (APV) tests tailor made agro-zoological solutions have been implemented according to a common scientific protocol, consisting of: 1) a testing area ranging from 3 up to 6 ha, 2) a five step methodology, shown in Figure 1 (site visit, soil characterization, design, implementation and results evaluation), 3) the crops implementation in three areas (in the corridors between two modules rows, underneath the modules and in a control area reproducing the open field demonstration) and 4) low height crops selection, ranging from aromatic species, food, medicinal crops and flowers mix.

The experimentation is enabling a massive data collection, through a proximity sensors network. These monitoring sensors will allow to enable sustainable and more effective agricultural practices to support the markets competitiveness of the agricultural partners.

Figure 1. The common methodology considered in the demonstration tests

2 MATERIALS AND METHODS

2.1 Kourtesi PV plant

The results reported in this study are related to the agrivoltaic test ongoing in EGP Kourtesi PV plant, located near Kourtesi, Municipality of Kyllini, Prefecture of Elis (western Greece). The area has a Mediterranean climate with mild winter and hot-dry summers. Minimum temperatures are observed in January and February, but on average the temperature never drops below 10°C. The maximum temperatures are reached in July and August, on average about 26°C.

2.2 Kourtesi APV test

The APV test in Kourtesi PV plant has been designed and implemented in order to provide a holistic vision, covering different aspects. The activities have been carried out with the support of Novamont SpA and University of Naples Federico II.

Tailor made solutions have been designed in compliance with the specificity of the area, through surveys of morphological characteristics, adaptability, and market demands. All the implemented crops have enabled the definition of even more sustainable practices for the management and the operation of the plant, by guaranteeing a soil regeneration in terms of organic matter and soil porosity.

The main features of the selected APV solutions for Kourtesi test are reported below:

- Local market interest and relevance for the area;
- Sustainable agricultural practices;
- Circular bioeconomy approach;

- Monitoring through proximity and remote sensing, based on the agriculture 4.0.

The crops have been implemented in different areas. Aromatic species have been implemented in the PV plant in three zones (Figure 2 and Figure 3): in the corridors between two modules rows, underneath the modules and in a control area, free from panels and reproducing the open field cultivation. The selected crops are thyme (*Thymus vulgaris* L.*)*, oregano (*Origanum vulgare* L.), rosemary (*Rosmarinus officinalis* L.) and mountain tea (*Sideritis raeseri* L.).

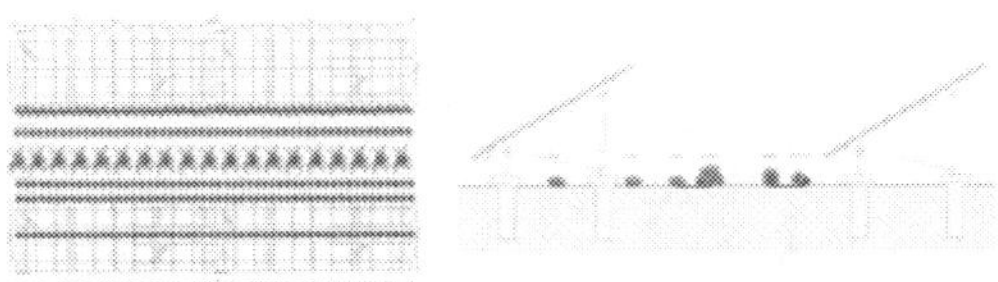

Figure 2 - Design of the APV tests in the corridors between two panels rows (left) and underneath the panels (right)

Figure 3 - Crops implementation: in the corridors (left) and in the control area (right)

Within the APV test, specific solutions aimed at valorizing the unused marginal areas have been also implemented. These solutions have been selected to guarantee soil regeneration and, at the same time, to create value for local communities, enabling the development of new business models, with a circular bioeconomy approach. The crops that have been implemented in the marginal areas are bermuda grass (*Cynodon dactylon* L)., cardoon (*Cynara cardunculus* L. *var. altilis*) and safflower (*Chartamus tinctorius* L.). For the cardoon cultivation the Trinaseed variety selected and implemented by Novamont, for developing innovative agricultural value chains, has been used. These crops promote a model of sustainable agriculture based on the cascading use of renewable raw materials and on the development of bioproducts. The adoption of these crops enables the development of new business models, with a bio-economy circular approach, also representing an interesting option for farmers and can contribute to increase soil organic carbon stock [1].

Bermuda grass has been selected to increase the soil erosion resistance due to sheep grazing [2], that is a well-established practice in Kourtesi PV plant, increasing at the same time local biodiversity. and creating a win-win solution. Sheep farmers takes benefits from grazing their animals on fields with high nutritional vegetation value, in fact the bermuda grass can be an effective way to improve forage quality and reduce the need for additional supplementation for livestock [3].

Cardoon and safflower (Figure 4) are two oilseed dry crops which fit well with Mediterranean climatic conditions and enable the development of a business models based on a circular economy approach (Figure 5). They are low input management species [4 - 5], and their cultivation offers to the agricultural industry and farmers an opportunity to grow on abandoned and uncultivated soils, regenerating them [6-7]. Thanks to their features, cardoon and safflower can provide oil and biomass that meet the requirements of biorefinery for the production of low environmental impact bioproducts [7].

Figure 4 - Cardoon (left) and safflower (right)

From their seeds an oil can be extracted, that can be used as raw materials for biorefineries (production of bioplastics, bioherbicides, bio lubricants, cosmetics ingredients), while the solid residual can be used as protein panel to be integrated in the animal's ration [8]. In Kourtesi demo site the cardoon and safflower seeds will be used to obtain protein panel (non-OGM) to be integrated in the sheep feeding ration, partially substituting soybean meal. If compared to soybean meal, they have advantage in terms of environmental impact and crops resilience [6]. The oil will be used by Novamont for the production of biopolymers, which in turn will be used to produce the Mater-Bi® biodegradable in soil mulching film used for aromatic plants cultivation in the APV test, thus demonstrating that even a marginal area can contribute to generate value. From this scheme is evident as an appropriate development of dry crops and the use of innovative transformation technologies could create examples of bioeconomy and green chemistry (from renewable sources).

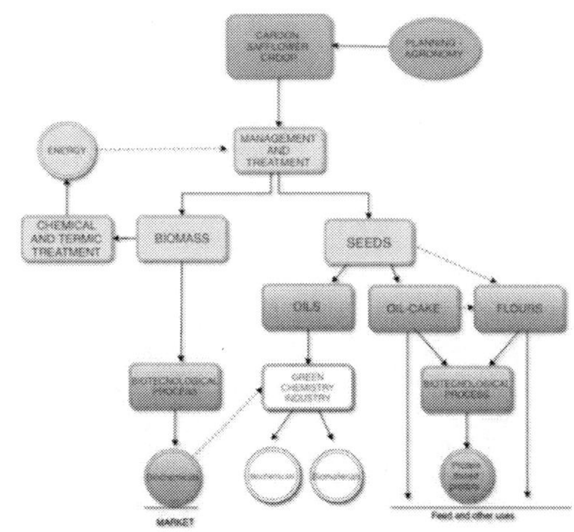

Figure 5 - Example of bio-economy and green chemistry model through the cultivation of cardoon and safflower in marginal areas

As part of the sustainable agricultural practices, the Riterra® biodegradable in soil mulching film made of Mater-Bi®, a family of completely biodegradable and compostable bioplastics produced by Novamont, has also been implemented in the crops cultivation (Figure 6).

Figure 6 - Mater-Bi® biodegradable mulching film implemented in Kourtesi APV tests

By minimizing the impact on the environment and saving time and resources in managing the end of life of mulched crops, biodegradable Mater-Bi® mulching film provides an agronomically and environmentally efficient alternative to traditional mulch film.

The biodegradable film can contribute to environmental protection by eliminating the problems of post-harvest collecting and disposing plastic film [9].

The agriculture 4.0 concept has also been included in Kourtesi APV test, by monitoring the crops growth through a sensors network, made up of proximity and remote sensing. This network has allowed to control and schedule irrigation and fertilization activities remotely and represents a further opportunity for creating value for local farmers, by enhancing their skills, that can increase their competitiveness on the market.

3 RESULTS AND DISCUSSIONS

3.1 Crops monitoring

The monitoring of the crops' growth has been carried out through proximity and remote sensing.

Regarding proximity sensors, soil water potential, soil temperature, climate conditions underneath the panels and in the corridors have been implemented.

For remote sensing, processing of satellite images to develop vegetation indexes, as NDVI (Normalized Difference Vegetation Index), MSAVI (Modified Soil-Adjusted Vegetation Index) and the Management Unit Zones (MUZ), have been considered, to evaluate and quantify the agricultural inputs (water and fertilizer) based on plant development and on the spatial variability of the soil characteristics (Figure 7).

Figure 7 - MSAVI image (left) and MUZ (right) on PV Kourtesi plant

Site-specific applications of agricultural inputs can, indeed, be implemented by dividing a field into smaller management zones that are more homogeneous in properties of interest rather than by considering the field as a whole [10].

A management zone is defined as "a portion of a field that expresses a homogeneous combination of yield-limiting factors for which a single rate of a specific crop input is appropriate" [11]. In order to define homogeneous zones Novamont staff used the cluster analysis algorithms K-means [12].

The vegetation cover was measured by taking a photo from 1,5 m above of each sampling point and image elaboration has been made with ImageJ software (Figure 8). Images elaboration included image scaling, sampling area measurement (cm^2), and vegetated area identification (through color analysis) and measurement (cm^2). Sampling area was about 1.000 cm^2. The vegetation cover percentage was then calculated as the ratio between vegetated area and sampling area multiplied by 100.

The crops height has been estimated through measurements collected in June 2022.

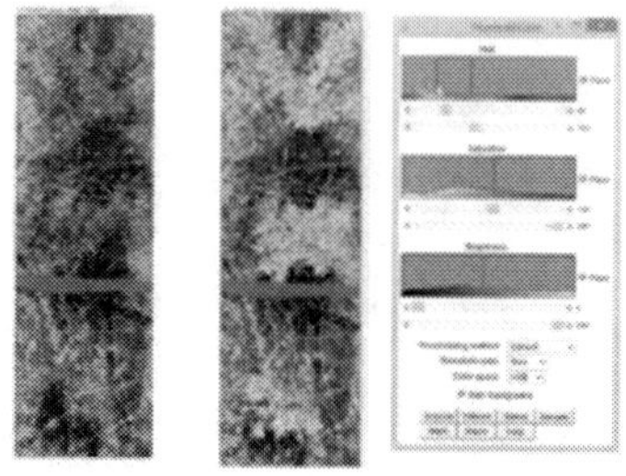

Figure 8 - Image elaboration: rosemary cover evaluation

3.2 Results

Despite the 2022 characterized by an exceptionally arid spring-summer, survival of plantlets of aromatic plants was very good (>90%) and engraftment of bermudagrass and safflower seeds was regular, while cardoon crop showed a very slow germination rate. Sampling made in June reported that biomass cover and plant height were not significantly different between the corridors of the testing area, and the ones measured in the control area and, on average, these parameters were higher in the corridors (Figure 9 and Figure 10):

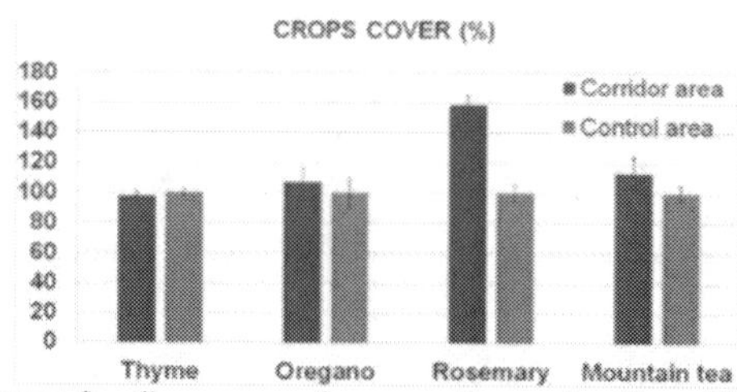

Figure 9 – Crops cover for aromatic species (June 2022)

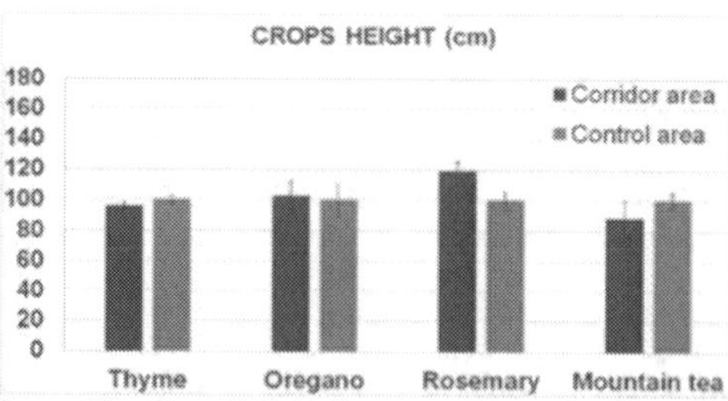

Figure 10 – Crops height for aromatic species (June 2022)

In July 2022, the first oregano harvest was made with the aim to inhibit the apical growth and promote lateral branching (Figure 11).

Figure 11 - From left to right: oregano before the harvest, fresh biomass and oregano dried and chopped

4 CONCLUSIONS

The first data collected in spring-summer 2022, allowed to draw some preliminary conclusions for Kourtesi dry and semi-arid environment.

For aromatic species, there are no differences in crops growth between the corridor and the control area and, on average, crops growth is slightly higher in the corridors, due to the shading that has a beneficial effect on the crops.

Bermuda grass meadow grows well also in aridity conditions allowing to extend the grazing period until the month of June.

Safflower has shown a good adaptability to marginal soils of the area and an excellent drought resistance.

Monitoring of crops growth will continue up to 2023, to assess the results and the best sustainable agronomic practices for improving the crops growth as well as the dosing of fertilizers and of water for irrigation.

The data provided by the sensors network will be combined with the offline measurements on the crops and an assessment of the coexistence of the crops and the PV modules will be carried out, in terms of agricultural yield and quality and energy production.

5 REFERENCES

[1] D'Avino L., Di Bene C., Farina R. Razza. Introduction of Cardoon (*Cynara cardunculus* L.) in a Rainfed of Rotation to Improve Soil Organic Carbon Stock in Marginal Lands. *Agronomy* 2020, 10(7), 946.

[2] Chen F., Zhang J., Zhang M., Wang J. Effect of *Cynodon dactylon* community on the conservation and reinforcement of riparian shallow soil in the Three Gorges Reservoir area. *Ecological processes* 4, Article number: 3 (2015).

[3] Hendricks T. J., Tucker J. J., Hancock D., Mullenix M. K., Baxter L.L., Stewart Jr. R. l., Segers J. R., Bernard J. K. Forage accumulation and nutritive value of bermudagrass and alfalfa-bermudagrass mixtures when harvested for baleage. *Crop Science* 2020, 60 (6).

[4] Ottaiano L., Di Mola I., Impagliazzo A., Cozzolino E., Masucci F., Mori M., Fagnano M. Yields and quality of biomass and grain In *Cynara cardunculus* L. grown in southern Italy, as affected by genotype and environmental conditions. *Italian Journal of Agronomy* 2017, volume 12:954.

[5] Farneselli M., Falcinelli B., Pannacci E., Paolitti L., Boggia A., Rocchi L. Circular agricultural system in safflower: agronomic practices and oil extraction. Environmental impact.Evoluzione dei sistemi agronomici in risposta alle sfide globali, Udine 15-17 settembre 2021. Società Italiana di Agronomia, 50° Convegno Nazionale

[6] Nannucci L., Mariottini F., Parrini S., Sirtori F., Bozzi R., Falce M., Aquilani C., Confessore A., Cannas A., Brajon G. Cardoon Meala s Alternative Protein Source to Soybean Meal for Limousine Bulls Fattening Period: Effects on Growth Performances and Meat Quality Traits. *Animals* 2021, 11, 3383.

[7] www.novamont.com

[8] Fedi C. Il cardo, nuova chance per le terre marginali. *L'informatore Agrario* 17-18/2019

[9] Guerrini S., Yan C., Malkinconico M., Mormile P. Agronomical Overview of Mulch Film System. *Polymers for Agri-Food Application* pp 241-264.

[10] Zhang N., Wang M., Wang N. Precision agriculture – a worldwide overview. *Computers and Electronics in Agricolture.* Volume 36, Issue 2-3, 2022, Pages 113-132.

[11]Doerge T. Defining management zones for precision farming. *Crop Insights* 1998, 8-21.

[12]Gavioli A., Godoy de Suaza E., Bazzi C. L., Schenatto K., Betzek N. M. Identification of management zones in precision agriculture: An evaluation of alternative cluster analysis methods. *Biosystems Engineering* 181 (2019) 86-102.

SURVEY OF THE FLOATING PHOTOVOLTAIC POTENTIAL OF PORTUGAL

Dorivaldo Duarte [a,b], Luís Fialho [a,b], Manuel Collares-Pereira [a,b], Pedro Horta [a,b]
[a] Renewable Energies Chair, University of Évora. Pólo da Mitra da Universidade de Évora, Edifício Ario Lobo de Azevedo, 7000-083 Nossa Senhora da Tourega, Portugal
[b] Institute of Earth Sciences, University of Évora, Rua Romão Ramalho, 7000-671 Évora, Portugal

ABSTRACT

Portugal is characterized by a high solar potential with annual global horizontal solar irradiation exceeding 1800 kWh/m2 [1]. Due to this radiation availability and the sharp reduction in costs that has occurred in the last decade, solar photovoltaic has shown a market growth and an increasing penetration in the Portuguese national electricity generation system. The application of floating photovoltaic (FPV) technology brings the advantage of using an otherwise potentially unused area and can present a yield increase due to its installation over water, at the same time providing a reduction in water evaporation [2]. This study provides a comprehensive analysis of the potential for FPV installation in mainland Portugal, using water body data by the Portuguese Environment Agency (APA). This analysis results in a mapping of the potential areas for FPV applications in the national territory, thus establishing a relationship between the availability of solar radiation and the geographical distribution of water bodies in the different regions of the country. This work intends to determine if the FPV potential in Portugal can contribute to the installed capacity target as defined in the Portuguese National Energy and Climate Plan 2030 (PNEC 2030), which defines the goal of 7GW of solar photovoltaic systems in 2030 [3].

Keywords: Floating Photovoltaic, Water, Solar Energy, Photovoltaic System.

1 INTRODUCTION

Floating photovoltaic systems represent a technical innovation, the first FPV systems date back to 2007 [5]. However, several gaps were identified in the state of art regarding this technological application, and this work intends to point out results to some of them, within the Portuguese context (e.g., territory, regulation, policies). No study was found on the potential for application and growth of this technology in Portuguese territory [6], and given the current context, this work aims at adding knowledge to the existing state of the art. The Portuguese government has recently launched the first floating photovoltaic public tender process, which presents a portfolio of 263 MW for new installed capacity for this technology, in seven dams [7]. This public tender marks the expansion of FPV technology in Portugal and is aligned with the policy defined in the National Energy and Climate Plan 2030 [8] and international decarbonization targets [9].
This work presents the analysis of the Portuguese potential for FPV systems, considering the total area available and establishing a correlation not only with the solar resource as well as the transmission infrastructure and coexisting assets, generating outcomes of both the technical and economic potential.

2 AIM AND METHODOLOGY

The main objective of this work is to survey the FPV potential of continental Portugal. For this purpose, a very broad approach was used to collect all the information concerning water bodies at national level, from dams, lakes, wetlands, near onshore water bodies, collecting georeferenced data on GIS files. After this collection, the next step consisted in the application of some boundary conditions, namely the exclusion of all the area crossing the boundaries of the Portuguese administrative regions, the exclusion also of the near onshore water bodies because there is no clear and evident study of the degradation curve in higher salinity water conditions and for this reason the study is restricted to onshore water bodies, and finally the last condition imposed was the exclusion of wetlands, since after a thorough analysis of these areas it can be concluded that they are intended for agricultural purposes which are an impeditive factor for FPV installations. With this analysis methodology and the application of these restrictions, a first preliminary result was obtained regarding the available potential area for FPV systems to be deployed and considering that the water surface area calculated is based at full storage level of the water bodies.
In order to closely align this study with the objectives of the Portuguese National Energy and Climate Plan 2030 (PNEC 2030). The PNEC 2030 defines a goal to install a total of 7GW of solar photovoltaic capacity in Portugal, in order to achieve the Portuguese decarbonization targets.
The methodology used is shown in Figure 1. After this broad and extensive data collection and exclusion criteria application, it was decided to applied a 85% reduction in the total net available area, in order to simulate further exclusion and selection criteria not considered at this stage of the work. Selection criteria will be addressed in future stages of this work and will allow the fine tuning of the end-results. Selection criteria will include application of some restrictions of both technical and environmental nature, in particular hybridisation or connection with existing assets such as hydroelectric dams, wind farms, centralized energy storage systems, proximity to available energy injection points of the

national transmission and distribution grid, protected areas, reserved areas for recreational navigation or fishing and water sports. Another shortcoming identified in the present results concerns additional fine tuning according to different scenarios related to the surface limits of water bodies, due to shadowing, the bathymetry of dams and lakes, and impacts on water availability (water storage levels) due to climate changes.

A 1 to 1 relationship between area and installed power was used, i.e. 1MW of installed FPV power for an area (water surface) of 1 ha [11].

The next step of the methodology used was to use the SISIFO [4] software simulation tool to simulate the PV annual generation yields for each region, considering the final available areas. A standard tilt of 15° was used for the FPV modules, and South facing.

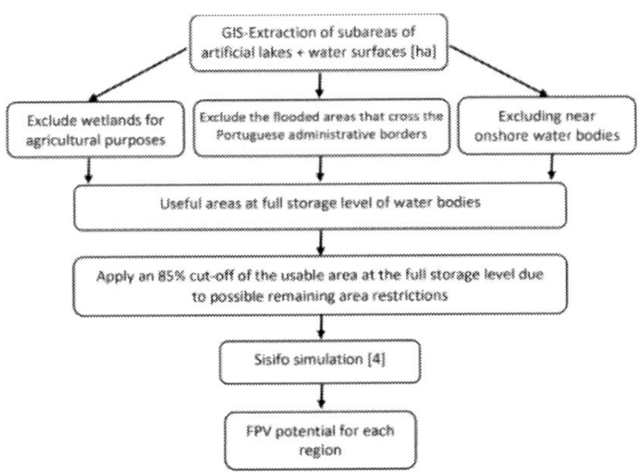

Figure 1: Methodology simplified Flowchart

3 RESULTS ANALYSIS

3.1 Distribution of installed FPV power in Europe

As first approach, the European FPV context and installed capacity is being monitored and updated in a internal database, enabling further in depth analysis and evaluation of the market and national policies evolution. The currently the installed FPV capacity distribution in Europe is represented in Figure 2. The Portuguese total installed capacity is about 3% of the total European installed power [10], currently being led by France with a total of 17MW, the largest installed FPV capacity in Europe, followed by Albania and Belgium. The total European installed capacity regarding FPV systems is 47.2MW, however several FPV investments can be identified through public data records, at different development stages (investment, engineering, installation, etc.), that will upscale this technology in the next years.

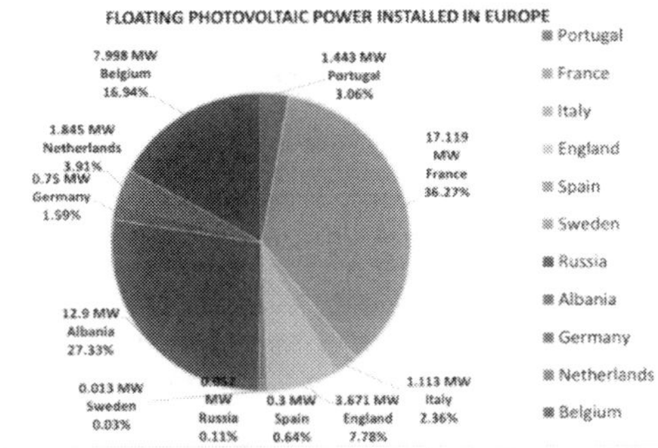

Figure 2: FPV installed power capacity in Europe [10]

3.2 Characterization of the Portuguese FPV potential

Concerning the Portuguese mainland territory, official administrative regions were considered to assess the potential FPV distribution scenarios. Mainland Portugal is then divided into five distinct regions, namely the Northern region (Norte), the Central region (Centro), Lisboa e Vale do Tejo, Alentejo (large region in the south) and Algarve (most southern region).

For each region a careful and detailed analysis of the existing water bodies was made, resulting in the illustrated potential, as shown in the next figure.

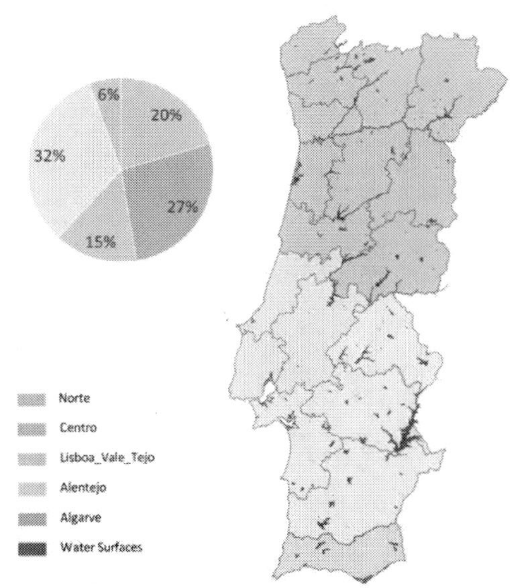

Figure 3: Portugal map with the distribution of available water bodies and surfaces

As shown in the graph of Figure 3, the Alentejo region leads with the largest available area for floating photovoltaic deployment with 32% of the total national available area, mainly due to Alqueva dam and artificial lake, one of the largest artificial water reservoirs in Europe. The Center region (in orange in the Figure) represents 27% and the 3rd largest available area is in the region of Lisboa e Vale do Tejo with 15%.

However, apart from water surface area availability, also the solar resource should be considered in order to assess the FPV potential for energy generation.

In order to assess this, solar PV simulations were conducted with the SISIFO simulation tool, for the entire potential water surface area in each region. A correlation of 1MWp per hectare was used, considering the average area usage per installed FPV power. The PV simulation was performed considering the centroid location (coordinates) for each region. No FPV simulation software is publicly available and with full model validation, so a standard PV simulation software was used, considering that this approach introduces some uncertainty in the end results.
The results for this correlation between available area and the solar radiation, are shown in figure 4.

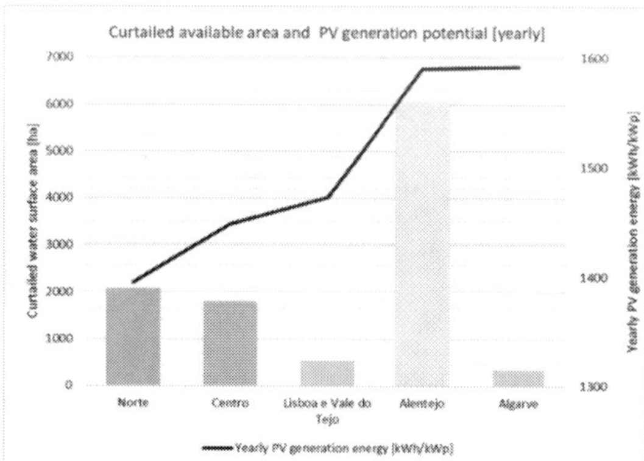

Figure 4: Curtailed available area [ha] (*columns*) vs yearly PV generation yield per region [kWh/kW_p] (*line*)

From the results presented, it can be concluded that Alentejo and Algarve are the regions with the highest solar radiation availability, justified by the higher annual PV yield [kWh/kW_p]. These two regions present very similar values of solar resource potential, and correlating it with water surface availability, the Alentejo region stands out, combining a large available area for FPV installations with one of the highest solar resource potential.
Table 1 presents further detailled results from these simulations, per region, as well as a ratio between the available water surface area and the region total area.

Table 1. FPV annual yields vs area per region

	Values obtained by curtailment 85% of the area at full storage level				
Region	**Water surface area [ha]**	**Maximum PV power [MW]**	**Yearly PV generation energy [kWh/kWp]**	**PV Generator Energy [GWh/year]**	**Ratio of curtailed areas [%]**
Norte	2076.76	2076.76	1394.66	2896.38	0.11
Centro	1804.86	1804.86	1447.96	2613.37	0.08
Lisboa e Vale do Tejo	537.09	537.09	1472.12	790.67	0.04
Alentejo	6071.61	6071.61	1589.46	9650.58	0.21
Algarve	345.42	345.42	1591.51	549.73	0.07
Total national	10835.75	10835.75	7495.71	16500.7	-

At a national level, the calculated maximum FPV potential for installed capacity is 10.8 GW, with a corresponding yearly energy generation of 16500.7 GWh. The Alentejo region presents the major contribution for this result, with a potential for the installation of 6.0 GW of FPV systems, representing 0.21% of the total region area. This FPV installed capacity will generate about 9650 GWh/year, this value being higher than the sum of the solar energy generation of all the other regions. On the other hand, despite having a very large territorial area, its ratio between area available water surface and total region area is 0.21%, pointing that Alentejo presents itself as a promising region for the installation of FPV systems.
The potential installed capacity of Alentejo and Algarve regions surpass the national target (PNEC 2030) of 7 GW.

4 CONCLUSIONS

This work intends to carry out the first analysis of the Portuguese potential for the installation of floating photovoltaic systems technology. The results obtained in this preliminary analysis allow to conclude that the potential FPV capacity exceeds the national target of 7 GW, as defined in the PNEC 2030 for the photovoltaic energy sector.
The analysis of the results at a regional level point to the Alentejo region as having the greatest potential, in terms of both water surface and solar resources.
Despite the use of a conservative approach, applying a curtailment of 85% of the total available area at the full storage level, according to the simulation results, the FPV national capacity potential is 10.8 GW.
The software tool (SISIFO) does not allow the accurate simulation of a FPV system and, without a validated simulation model at the moment, it was, nevertheless, used. The accuracy of the results would be improved with a FPV simulation software/model, taking into account, for instance, the cooling effect due to the aquatic environment or suitable losses scenario for FPV systems (soiling, misalignment, electric isolation failures, etc.).

5. NEXT STEPS

Several shortcomings were identified during this work, that will be addressed in the next steps, in order to reduce the results associated error:

- Update the results with a simulation model developed and validated specifically for floating photovoltaic systems;
- Collect all the shapefiles referring to the bathymetry of water bodies, in order to be able to calculate the available area for any water storage level;
- Collect and apply diverse existing restrictions for each water body, such as nature reservations, hydro-generation reserved areas, water sports or fishing areas.
- Analyze scenarios taking into account the proximity of strategic co-existing assets such as grid connection points, power lines and transformer stations, or other renewable energy power plants;

ACKNOWLEDGMENTS

The authors would like to thank the support of this work, partly funded by European Union's H2020 programme as part of the SolaQua project (Accessible, reliable, and affordable solar irrigation for Europe and beyond), Grant agreement ID 952789.

7 REFERENCES

[1]"GLOBAL SOLAR ATLAS," 26 01 2022. [Online]. Available: https://globalsolaratlas.info/map?c=39.391918,-4.134201,5&s=39.774871,-8.230036&m=site.

[2] R. Gonzalez Sanchez, I. Kougias, M. Moner-Girona, F. Fahl and A. J€ager-Waldau, "Assessment of floating solar photovoltaics potential in existing hydropower reservoirs in Africa," ELSEVIER, 2020.

[3] PNEC 2030 "PLANO NACIONAL ENERGIA E CLIMA 2021-2030"

[4] IES-UPM, "SISIFO – PV simulation framework", available at: www.sisifo.info

[5] M. Acharya and S. Devraj, "Floating Solar Photovoltaic (FSPV): A Third Pillar to Solar PV Sector?," 2019.

[6] M. Rúben Rodrigues Sobral, Avaliação do potencial fotovoltaico flutuante em Portugal, Lisboa: UNIVERSIDADE DE LISBOA, 2018.

[7] J. d. negócios, "Primeiro leilão solar em barragens terá 263 MW," [Online]. Available: https://www.jornaldenegocios.pt/empresas/energia/detalhe/primeiro-leilao-solar-em-barragens-tera-262-mw. [Accessed 02 2022].

[8] Ministério do Ambiente e Transição Energética, "Plano Nacional Integrado Energia e Clima 2021-2030, PNEC 2030," 2019.

[9] U. Nations, "The Paris Agreement," [Online]. Available: https://unfccc.int/process-and-meetings/the-paris-agreement/the-paris-agreement. [Accessed 26 05 2020].

[10] R. Alexandre Rodriguez Reb, Estudo do potencial de sistemas fotovoltaicos flutuantes em massas de água doce e marinhas, Évora: Universidade de Évora, 2021.

[11] Rádio Renascença [Online] Available:EDP inaugura "maior central solar flutuante" em albufeira na Europa - Renascença (sapo.pt)

ON THE PERFORMANCE OF PHOTOVOLTAICS ON CAR ROOFS: CASE STUDY UNDER DIFFERENT STATIC SCENARIOS

Javier Macías, Rebeca Herrero, Rubén Núñez, Ignacio Antón
Instituto de Energía Solar, Universidad Politécnica de Madrid
Madrid, Spain. javier.macias@ies.upm.es

ABSTRACT: Vehicle Integrated Photovoltaics presents particular characteristics such as the curvature of the surface or the vehicle motion and requires the development of specific modelling to evaluate the solar resource on roads, streets or parking lots. This paper describes a methodology for the calculation of the different components of the solar resource on photovoltaic surfaces integrated in vehicles. This methodology is based on LiDAR (Light Detection and Ranging) technology which provides a 3D point cloud representing the positions of objects in urban landscapes including buildings, ground or vegetation. These 3D maps provide the necessary information for the surface reconstruction of obstacles, particularly buildings. Then, ray-tracing methods are applied to compute shadow maps on the photovoltaic surface at any location and specific time of the year, providing a binary information (shade/no-shade) to calculate the effective Direct Normal Irradiance. In addition, the Sky View Field observed from the vehicle is determined and used for the calculation of the diffuse component of irradiance. Following the description of the methodology, an example case is presented where the solar resource is calculated in the different car parks around the Instituto de Energía Solar of Universidad Politécnica de Madrid.
Keywords: LiDAR, Sky View Field, Irradiance Components.

1 INTRODUCTION

The transport sector is one of the most polluting sectors with around 30% of greenhouse gas emissions, of which around 73.4% come from road [1]. Therefore, the European Green Deal aims to reduce these emissions to 55% in 2030 and 100% in 2035, compared to 2021 [2].

To achieve these goals, new energy sources such as renewable energies and electric vehicles (EVs) are called to lead the change. In fact, EV sales reached another record high in 2021 and already account for 9% of the world's vehicle fleet (17% of sales in Europe), despite the current materials and supply crisis [3].

However, this quick deployment of the fleet poses new challenges that have yet to be overcome. Among them, the range of EVs is still much shorter than that of conventional internal combustion cars, the number of charging points is still reduced, charging time are still long and the energy necessary to power these vehicles will represent a significant increase in the global energy mix. In addition, the rising cost of energy worldwide due to geopolitical issues makes it even more expensive to power the vehicle.

Therefore, Vehicle Integrated Photovoltaics (VIPV) technology is an optimal solution to partially mitigate the problem with the aim of increasing the autonomy of the vehicle by installing photovoltaic (PV) cells on the vehicle's surfaces.

In the past, VIPV was considered as a power source for secondary systems, but most car manufacturers now consider it as a solution for vehicle propulsion.

However, the conditions involved in this technology (curved surfaces, vehicle movement or shading effects) make it necessary to develop new tools to study its feasibility.

In particular, the study of the solar resource on flat surfaces has already been extensively studied before. However, there are few tools that study the solar resource on the PV surface of a vehicle, considering its movement. This motion will cause dynamic shading effects on the PV surface caused by the surrounding obstacles. This paper describes a methodology for estimating the effective solar resource on these surfaces including shading conditions.

The first step is the calculation of the solar irradiance components Direct Normal Irradiance (DNI) and Diffuse Horizontal Irradiance (DHI) on the PV surfaces, using conventional methods and without considering shadows.

Subsequently, a ray-tracing algorithm has been developed based on LiDAR technology for the simulation of the shadow maps on the PV surface caused by the surrounding obstacles. This shadow maps, together with the DNI estimated before, allow to determine the effective direct component of the irradiance on each point of the vehicle surface. Additionally, the diffuse component of the irradiance is corrected through the estimation of the Sky View Field (SVF) from the vehicle.

This methodology provides set of irradiance maps that can be elaborated for a particular place, a specific route or a whole city, with different time resolutions, from a particular moment of the year to yearly average values. These maps provide the necessary input info to calculate the impact of the partial shading in the performance of VIPV at static (parked) or dynamic (on route) conditions. A case study will be presented where the effective solar resource is determined for three parking lots in the Instituto de Energía Solar of Universidad Politécnica de Madrid (IES – UPM).

2 IRRADIANCE MODELLING

The study of irradiance modelling has been extensively studied before, especially for flat modules. There are several ways for calculating irradiance on a horizontal surface:

2.1 Clear Sky Models:

These models estimate the terrestrial solar radiation under a cloudless sky as a function of the solar elevation angle, site altitude, aerosol concentration, water vapor and various atmospheric conditions. Ineichen, Haurwitz or Bird are examples of these models [4].

2.2 Calculation from Databases:

Another way to estimate the irradiance on the PV surface is by using databases with real data series measured chronologically. An example of this is the Typical Meteorological Year (TMY).

The TMY is a set of meteorological data with data values for every hour in a year for a given geographical location. The data are selected from hourly data in a longer time period (normally 10 years or more) [5].

These methodologies provide an estimation of the irradiance in the horizontal plane at a specific location. However, the conditions defining VIPV technology differ from those for flat modules.

In particular, the curvature of the vehicle surfaces means that the irradiance is not homogeneous at each point. The car roof, hood or doors are useful areas for the installation of PV cells.

In addition, the movement of the vehicle along its trajectory will also imply the solar resource to be different at each moment. This movement will also cause the shadow pattern on its photovoltaic surface to be dynamic.

All these variables will influence the solar resource and subsequent PV generation considerably. Depending on the PV technology of the module (cell size and number of cells, interconnection between cells or use of bypass diodes), the power output will vary, especially under partial shading conditions, as has been discussed in other papers [6].

3 CASE STUDY: SOLAR RESOURCE IN THE SURROUNDINGS OF IES – UPM

The Instituto de Energía Solar of the Universidad Politécnica de Madrid (IES - UPM) has been selected as a case study, as it is an office building with three adjacent car parks north, east and south, with 60, 80 and 60 parking spots, respectively, as presented in Figure 1:

Figure 1: IES - UPM and adjacent car parks.

For the estimation of the irradiance, we have used the typical values provided by the TMY from the Photovoltaic Geographical Information Sistem (PVGIS) and selected a clear day in June (12/06) as it has suitable irradiance values. We also defined the roof of the vehicle as the PV surface to be studied.

In this case, a geometric surface of double curvature with a projected area of 1.6 metres long and 1.3 metres wide has been considered to simulate the roof (Figure 2). The curvature on the long axis is 2 m and on the short axis is 2.95 m. It is considered that double curvature surfaces are considered a good approximation of real car roofs.

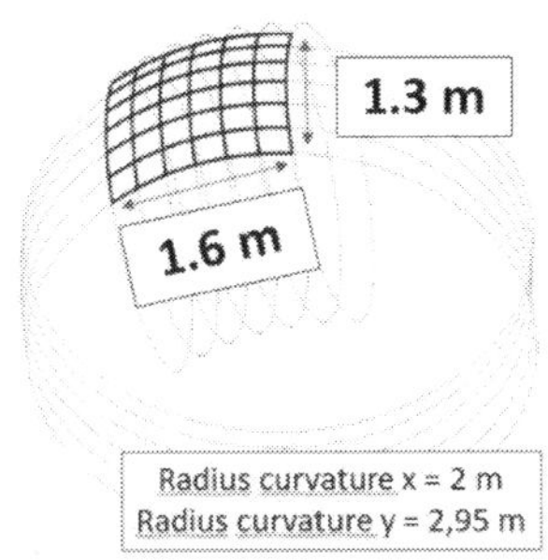

Figure 2: Car-Roof defined as a double curvature surface.

4 EFFECTIVE IRRADIANCE ON THE CURVED SURFACES OF AN EV

A Python-based tool has been developed to calculate the solar resource and the energy generation for the PV surface of the car roof [7].

This tool has three different modules. The first one is a *'Solar Resource'* module based on the open source library PVLIB [8] to estimate the solar resource on the geometric surface along its trajectory without the presence of obstacles that produce shading. This tool supports both surfaces defined from CAD files and geometrically defined surfaces.

The second module (*'Shading Module'*) uses LiDAR technology for city mapping and uses ray-tracing methods for the estimation of shadow maps on the PV surface.

The third module (*'PV Generation* Module') consists of a power generation module with which, from the estimated solar resource, we can calculate the power output for different configurations. However, in this paper we will focus on the first two modules of the tool that allow us to estimate the solar resource.

4.1 Solar Resource modelling (no shadows):

As an example, the study of the solar resource on the car roof of a vehicle located in a parking spot (latitude = 40.453160, longitude = -3.7266616, elevation = 0°, azimuth = 0°) from the car park – East is carried out.

For this position, the TMY provides the horizontal components of irradiance: Global Horizontal Irradiance (GHI), Direct Normal Irradiance (DNI) and Diffuse Horizontal Irradiance (DHI), as well as the meteorological conditions (ambient temperature or wind speed). The hourly values of the irradiance on the horizontal plane are shown in Figure 3 for the parking spot selected:

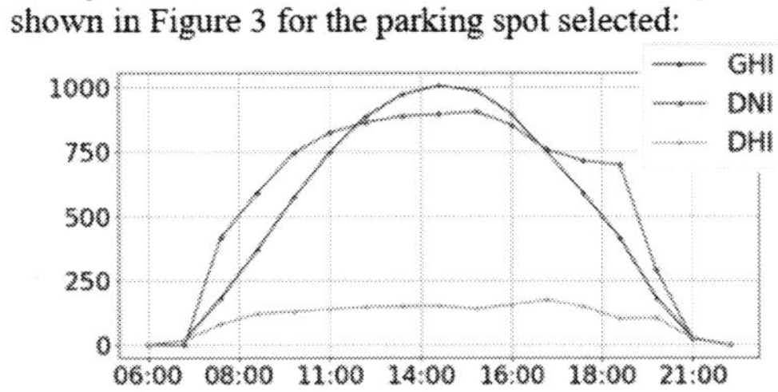

Figure 3: Irradiance values [W/m^2] from TMY for the parking spot selected (12/06).

These values are useful if we consider a horizontal plane without any elevation. However, to calculate the effective irradiance (POA Global) on the car roof it is necessary to know the direct (POA Direct) and diffuse (POA Diffuse) components on each point of the surface.

For this purpose, the car roof is divided into infinitesimal planes and the irradiance on each of them is estimated. We define POA as the Plane Of the Array of each infinitesimal.

To calculate the direct component, we start from the DNI component and correct it geometrically with the angle of incidence (AOI) of the solar vector at each point of the surface (Figure 4):

$$POA\ Direct = DNI * \cos\left(AOI(x,y)\right)$$

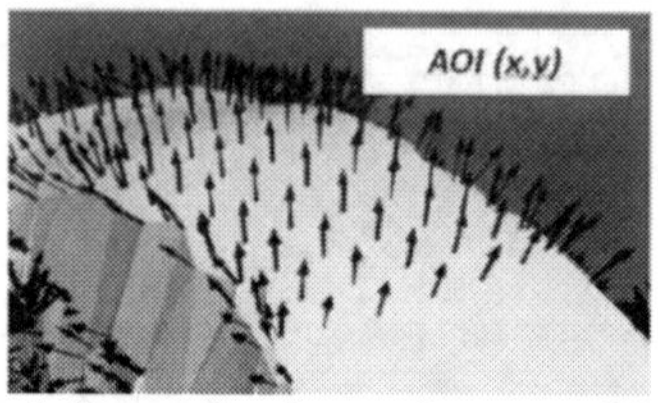

Figure 4: Car-roof and normal vectors of the different infinitesimals.

The diffuse component of the irradiance is estimated by correcting the DHI by the Sky View Field (SVF) [9].

$$POA\ Diffuse = DHI * \text{SVF}$$

The Sky View Field is the percentage of sky visible in a hemispheric photo. In order to calculate the SVF in the different positions of the car park, a measurement campaign was carried out in which 86 images were taken in different positions of the north, south and east car parks.

Subsequently, by processing the images, the percentage of sky visible at each position was obtained. As an example, the sky image and the processed image at the case study position are shown (Figure 5):

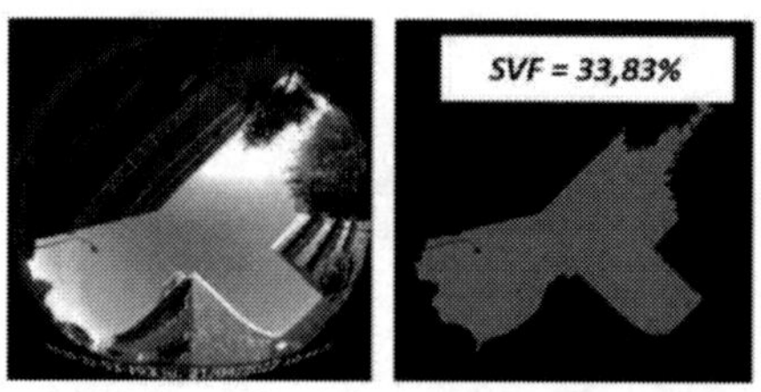

Figure 5: On the left: hemispheric image on the parking spot of study. On the right: Processed image and SVF estimated.

Once the direct and diffuse components have been obtained, the global component (POA Global) at each point of the curved surface can be calculated (without considering shadows of the urban landscape) as the sum of both.

$$POA\ Global = POA\ Direct + POA\ Diffuse$$

4.2 Solar Resource Modelling including shadows:

The shadows cast by the urban landscape on any point of the surface will cause the direct component of the irradiance (POA Direct) to decrease, so the POA Global will be corrected according to the shadow pattern. In this case the POA Direct component is considered zero if the vehicle is under shadows.

Using the *Shading Module*, the shadow maps for any position and time of day are calculated. This module is based on LiDAR data available at the National Centre for Geographic Information (CNIG). Specifically, a point cloud is obtained from the surroundings of the IES – UPM. A LiDAR Point Cloud consists of a large number of points collected through the LiDAR sensor that represents a 3D shape of the objects.

However, the density of points offered by LiDAR technology is not high and it does not provide information about vertical surfaces (i.e., building façades), so a reconstruction of the surfaces has been carried out using a 3D mesh (Figure 6):

Figure 6: LiDAR Point Cloud (left) and 3D Mesh (right).

Once we have the urban landscape surface simulated, ray tracing methods are performed to estimate the shadow maps on the car-roof throughout the day (Figure 7):

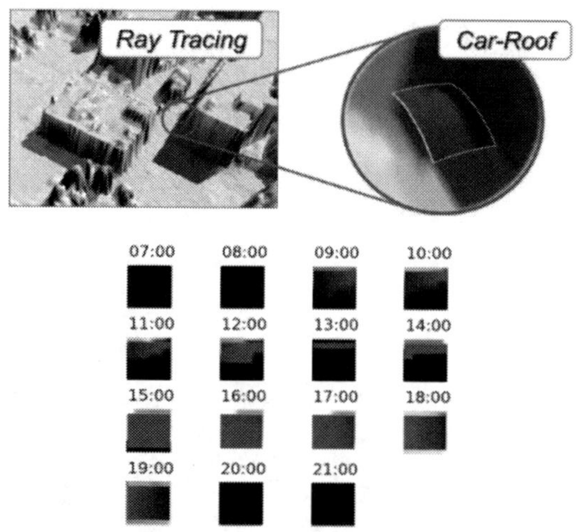

Figure 7: Ray-tracing method to estimate the shadow maps (up) and hourly shadow maps on the car-roof (12/06) (down).

From these shadow maps, the direct component of the irradiance is corrected, obtaining the irradiance maps on the curved surface (POA Global), as presented in Figure 8:

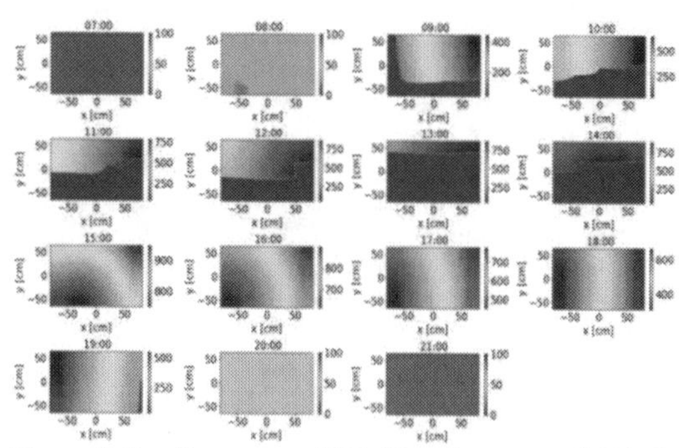

Figure 8: Irradiance maps [W/m^2] on the car-roof (12/06).

Finally, the curves for each of the irradiance components are shown in mean value (Figure 9). The dark blue curve shows the average on the surface of the estimated POA Global without shadows, and the dotted red curve shows the average on the surface of the POA Global including shadows. The differences in these curves come from the shadows and from the variation of the SVF.

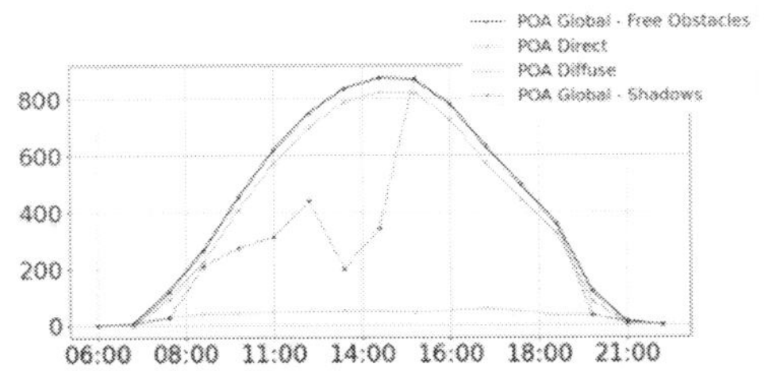

Figure 9: Hourly Solar Resource (average) [W/m^2] on the car-roof (12/06).

This methodology for calculating the POA Global including shadows has been used for 86 different car park positions. In the different spots, the shadow pattern will be different depending on the surrounding urban landscape (which will affect the direct component) and the SVF will also vary (affecting the Diffuse component).

Once the POA Global including shadows has been obtained at each of the car park positions, a linear interpolation has been carried out to estimate the irradiation (kWh) at any point of the surroundings of IES – UPM for the typical day selected (12/06). The heat map of the irradiation is shown in Figure 10:

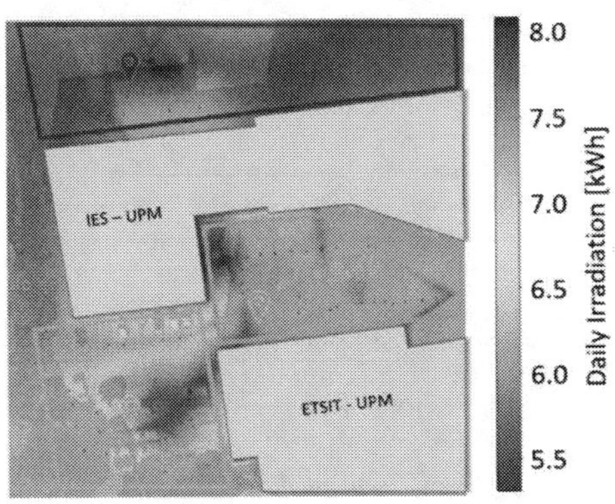

Figure 10: Irradiation [kWh] on the different car parks around the IES – UPM for the day of study (12/06).

5 CONCLUSIONS

In this work, a methodology for the estimation of the solar resource on the roof of an EV is proposed.

Curvature, EV motion or shading effects significantly affect the solar resource reaching the surface. In particular, the value of the direct component of the irradiance is highly dependent on shading patterns and the value of the diffuse component depends on the SVF.

The GHI, DNI and DHI values are obtained from the TMY on the study position. The direct component of the irradiance at the different infinitesimals of the curved surface is calculated with the AOI.

However, these values do not include the shading effects. Therefore, the *Shading Module* is used to know the pattern of shadows on the PV surface and to correct the direct component of the irradiance.

The diffuse component is estimated from the DHI and the SVF, obtained in a measurement campaign in 86 different positions in the different car parks of the IES - UPM.

Finally, an irradiation heat map is obtained by linear interpolation of the global irradiance (mean value on the car roof surface) at the different parking spots. The irradiation at any point in the surroundings of the office building is calculated for the day of study.

To sum up, the methodology proposed allow us to estimate the solar resource on the different surfaces of an EV for a particular place considering shading scenarios, but also can be applied for trajectories.

6 ACKNOWLEDGEMENTS

This work has been supported by the project DETECT-PV (reference PID2021-128853OB-I00) financed by the Fondo Europeo de Desarrollo Regional (FEDER) and the Spanish Ministerio de Ciencia e Innovación and Agencia Estatal de Investigación. We acknowledge to open-source libraries PVLIB, PyVista and PVGis Database for their contribution to this project.

7 REFERENCES

[1] «Transport emissions». https://ec.europa.eu/clima/eu-action/transport-emissions_en.

[2] «European Climate Law». https://ec.europa.eu/clima/eu-action/european-green-deal/european-climate-law_en.

[3] «Global Electric Vehicle Outlook 2022», p. 221, 2022.

[4] J. Stein, C. Hansen, and M. Reno, «Global horizontal irradiance clear sky models: implementation and analysis.», SAND2012-2389, 1039404, mar. 2012. doi: 10.2172/1039404.

[5] «TMY generator». https://joint-research-centre.ec.europa.eu/pvgis-photovoltaic-geographical-information-system/pvgis-tools/tmy-generator_en (accedido 25 de septiembre de 2022).

[6] J. Macías, R. Herrero, R. Nunez, and I. Anton, «On the effect of cell interconnection in Vehicle Integrated Photovoltaics: modelling energy under different scenarios», in *2021 IEEE 48th Photovoltaic Specialists Conference (PVSC)*, Fort Lauderdale, FL, USA, jun. 2021, pp. 1336-1339. doi: 10.1109/PVSC43889.2021.9518935.

[7] J. Macías, R. Herrero, I. Antón, and R. Núñez, «EVALUATION OF THE SOLAR RESOURCE AND ENERGY GENERATION IN VEHICLE INTEGRATED PHOTOVOLTAICS», *in 2020 37th EU PVSEC* p. 4.

[8] W. F. Holmgren, C. W. Hansen, and M. A. Mikofski, «pvlib python: a python package for modeling solar energy systems», *J. Open Source Softw.*, vol. 3, n.° 29, p. 884, sep. 2018, doi: 10.21105/joss.00884.

[9] I. García, M. de Blas, B. Hernández, C. Sáenz, and J. L. Torres, «Diffuse irradiance on tilted planes in urban environments: Evaluation of models modified with sky and circumsolar view factors», *Renew. Energy*, vol. 180, pp. 1194-1209, dic. 2021, doi: 10.1016/j.renene.2021.08.042.

3D-PRINTED WALKABLE SOLAR TILES

M. Canino*[1], S. Fasahat[1, 2], L. Maserati[2], V. Boldrini[1], E. Centurioni[1],
L. Ferlauto[2], R. Rizzoli[1], C. Sangiorgi[3], D. Cavalcoli[2]
[1] CNR-IMM - Consiglio Nazionale delle Ricerche, Istituto per la Microelettronica e Microsistemi, Bologna (Italy)
[2] Physics and Astronomy Dept., University of Bologna, Bologna (Italy)
[3] Department of Civil, Chemical, Environmental and Materials Engineering, University of Bologna, Bologna (Italy)

ABSTRACT. This work reports on the realization and characterization of 100 cm^2 walkable solar tiles for infrastructure-integrated PV (I2PV) made by combining a solar cell with a 3D-printed walkable tile, both made up of commercial products. With further engineering, the resulting panels should be integrated in pavements and bike lanes. Transparent tiles are characterized. To determine material inhomogeneities, a map of the transmittance in the UV-Vis is made. The transmittance is correlated with the J_{SC} of a small crystalline silicon solar cell (3.6 cm^2 in area) mechanically stacked in different positions under the tile, as well as with photocurrent spectra. The resistance of the printing resin to plasma, thermal, UV, and mechanical stress is analysed by UV-Vis transmittance and Fourier-transform infrared (FTIR) spectroscopy. The material compatibility with PV applications proves to be promising, while routes for future process development are identified.
Keywords: Building Integrated Photovoltaics (BIPV), optical properties, polymer film, 3D-printing, reliability

1 INTRODUCTION

Installations based on PV cells can be used in different urban structures like roads and buildings. This is useful to have the combination of electricity generating with other features of urban structures materials. One of the main factors that should be considered in the future application of solar photovoltaic is Infrastructure-integrated PV (I2PV). In the frame of I2PV, several solutions are explored to integrate PV panels in transport infrastructures such as railways [1,2], highways [3,4], and bike lanes [5]. The development of walkable solar cells is of interest both at commercial [6,7] and at research [8] level. Furthermore, 3D-printing of solar panels is an emerging trend in solar cell fabrication [9].

The idea of combining solar cells with a transparent 3D-printed coating, could thus be exploited in a future manufacturing line based on the same technology to fabricate both the solar panel and the coating.

This works reports on the fabrication and characterization of 100 cm^2 walkable solar tiles for I2PV, made by combining a solar cell with a 3D printed walkable tile. With further engineering, the resulting panels should be integrated in pavements and bike lanes.

Transparent tiles such as the one shown in Figure 1 are characterized. To determine tile material inhomogeneity, a map of the transmittance in the UV-Vis is made. The transmittance is correlated with the J_{SC} and the photocurrent spectra of a small crystalline silicon solar cell (3.6 cm^2 area), mechanically stacked in different positions under the tile. The reliability of the printed material under illumination, thermal, and mechanical aging is investigated.

2 EXPERIMENTAL

2.1 3D-printing

A 10 x10 cm^2 sample tile was printed with a Form 3 SLA printer using a Clear transparent resin marketed by Formlabs. The tile, shown in Figure 1, is made of a continuous layer, approximately 8 mm thick, partially covered with half-truncated octahedrons, 10 mm in width and 5 mm in height. The planar mutual distance between the rows of octahedrons is approximately 12.5 mm. This type of surface texture was conceived to target skid resistance in wet or icy conditions, without compromising the tile transparency. For the measurements, three regions have been identified: top, faces, and bottom of the half-truncated octahedrons. Regions corresponding to half-truncated octahedron top are about 13 mm thick, while half-truncated octahedron bottom, corresponding to the base transparent layer, is 8 mm thick. A matrix of measuring positions has been identified through letters and numbers as shown in Figure 1.

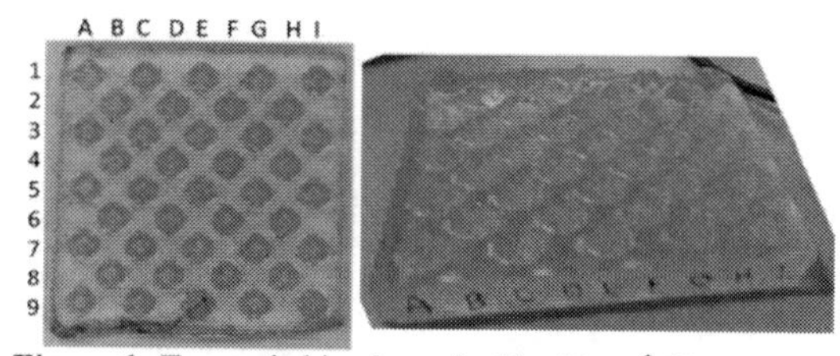

Figure 1: Top and side view of a 10 x10 cm^2 tile.

A thinner flat foil of resin with thickness ranging between 0.5 mm and 2 mm was also 3D printed, in order to characterize the Clear resin optically. No polishing was applied to the material. The foil was also cut into smaller pieces to perform aging experiments.

2.2 Characterization

Transmittance spectra of the tile were acquired in the range 250-1100 nm with a fiber optics spectrophotometer by Ocean Optics and coupled with an integrating sphere. The Transmittance spectra of the pieces of the thin foil were acquired through a Lambda 950 UV/Vis spectrophotometer by PerkinElmer, with an integrating sphere. Fourier-Transform Infrared Spectroscopy (FTIR) measurements were acquired in Attenuated Total Reflectance (ATR) mode by an Agilent Cary 630 FTIR with a diamond tip.
IV characteristics under illumination were acquired under an AM1.5G spectrum with no temperature control.

Photocurrent spectra were acquired using a Czerny-Turner monochromator by Sciencetech, with diffraction gratings covering the spectral range UV-vis-NIR equipped with QTH and Xe-lamps. The system is coupled with lock-in amplifier allowing for the detection of very low photocurrent signal, down to 10 fA.

2.2 Aging treatments

A 3D-printed foil with thickness ranging between 0.5

and 2 mm was put under plasma, thermal, UV, and mechanical stress in order to reproduce reasonable aging conditions. Aging tests were performed on pieces of the same foil, and each treatment was performed on two pieces: a treatment-specific piece and one that underwent all the aging treatments in sequence. The results here reported compare the aged samples with a piece of as-printed material which was kept as reference. It underwent no ageing, except the time of about six month that passed from its deposition to completion of the characterization.

The oxygen plasma ageing was achieved by exposing the samples to a 24.7% O2 + 75.3% Ar plasma for 15 minutes using model 1020 plasma cleaner by Fischione. For the thermal stress, two situations were considered: 70 °C for 3 hours in air and 90 °C for 85 hours in vacuum. The ramp up took approximately 1h while the ramp down proceeded with the furnace thermal inertia. The UV aging took place under solar simulator with AM1.5G spectrum at 1.1 sun intensity for 10 h. The sample temperature was not controlled during this treatment. The mechanical stress test consisted in creep recovery and dynamic force experiments. in a dynamic mechanical analyzer model Q800 DMA. In the creep recovery test a 18 N force was applied. In the dynamic force test, two rates were used: 50 N/min and 100 N/min.

3 RESULTS AND DISCUSSION

3.1 Characterization of the solar tile

In the top view shown in Figure 1, left side, darker regions correspond to regions with lower transmittance, while bright regions correspond to regions with higher transmittance. As expected, half-truncated octahedron tops are darker due to their higher thickness. The faces of the half-truncated octahedrons are darker as well due to their slope.

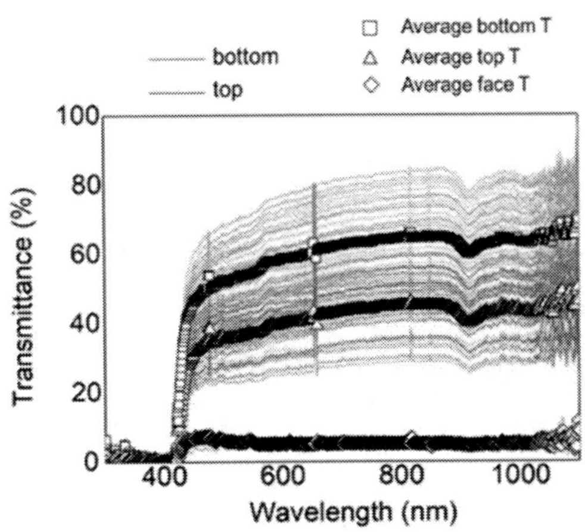

Figure 2: Transmittance spectra of all the half-truncated octahedra tops (pink) and bottoms (green). Symbols represent average transmittance spectra of the three different features: tops, bottoms, and faces of the half-truncated octahedra.

Transmittance spectra taken in different positions, reported in Figure 2, show a high dispersion in transmittance data due to material and shape imperfections. The sharp increase in transmittance at 400 nm makes the material suitable for photovoltaic application.

Transmittance values at 800 nm are reported in the map of Figure 3. This wavelength value is representative since the transmittance values reach almost constant values between 600 nm and 1000 nm.

T@800 (%)	A	B	C	D	E	F	G	H	I
1	50	76	51	79	51	73	40	59	57
2	80	50	79	45	65	51	52	35	49
3	52	81	45	83	56	74	39	68	42
4	81	59	83	48	78	59	40	36	61
5	55	76	55	74	66	66	50	60	30
6	78	54	72	45	56	51	62	47	54
7	45	69	46	55	28	36	34	54	43
8	69	35	49	30	51	33	47	34	54
9	40	62	36	59	4	64	43	46	40

Figure 3: Map of the transmittance at 800 nm measured in different positions. The pink values refer to measurements on the top of the half-truncated octahedra and the green values refer to measurements taken at the bottom of the half-truncated octahedra.

Figure 4 shows the I-V characteristics under illumination taken on the bare cell and on the cell covered by the tile. The ratio between the two curves is about 74%.

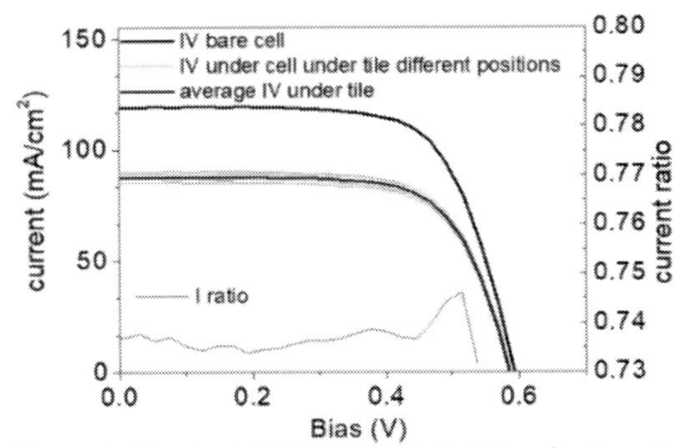

Figure 4: Illuminated IV curves of a 2×1.8 cm^2 solar cell either bare (black) or covered by different regions of the tile (cyan curves). The average is the blue curve. The right axis reports the current ratio between the bare and stacked cell (red curve).

The photocurrent of a solar cell bare either bare or stacked with the tile is reported in Figure 5. The red curve is the ratio between the photocurrent in cell stacked under the tile with respect to the bare cell. The average ratio over the spectrum is about 65%, while the ratio between 780 nm and 820 nm is (67±4)%

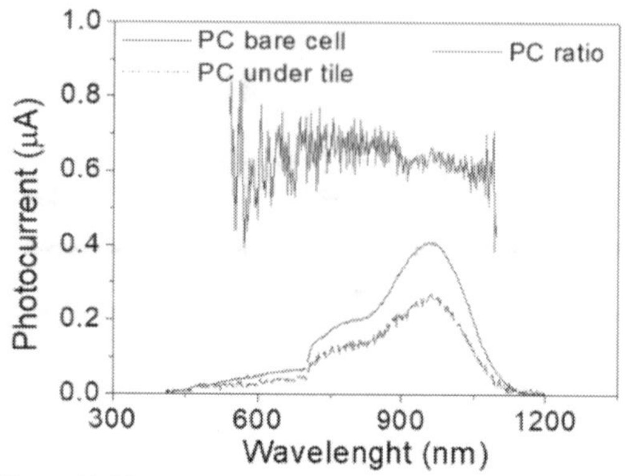

Figure 5: Photocurrent of the bare cell (black) and of the tile/cell stack (blue). The red curve represents the photocurrent ratio of the tile/cell stack to the bare cell.

3.2 Aging test

Aging tests were performed on different pieces of a thin foil of printed resin. The graph in Figure 6 shows the average transmittance of the as received pieces and the average transmittance of the pieces after aging, both single and cumulative. The aging treatments did not substantially modify the transmittance in the wavelength range above 400 nm, with a decrease in transmittance <1% after aging. The decision to plot the average values is due to the different thickness of the pieces. The reported graphs don't show the dispersions of the curves before and after ageing. The standard deviation of the transmittance at 800 nm is 1.6% in the case of the as received samples and 2.3% for the aged samples. The change in the onset below 300 nm needs further verification. However, it occurs in a wavelength range that is not useful for the final device since the transmittance of the tile, that is much thicker, is zero (see Figure 2).

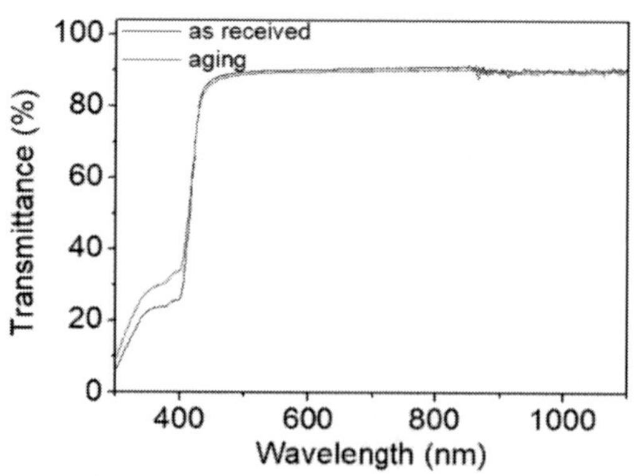

Figure 6: Average UV-Vis Transmittance spectra of the foils before (black) and after aging (red).

Figure 7 shows the Mid Infrared Transmittance spectra of the as received sample and of the sample that underwent cumulative aging. The curves reported in this graph are the average of measurements acquired on different spots of the same piece in order to avoid artifacts induced by imperfections in the material. The spectra do not show any substantial difference in the material chemical composition.

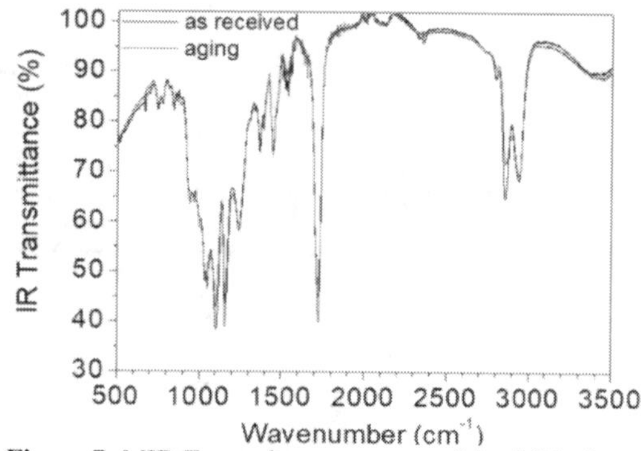

Figure 7: MIR Transmittance spectra of the foil before (black) and after aging (red).

This work represents a preliminary effort to identify the most relevant advantages and issues in the production of 3D-printed solar tiles.

The printed material proved to be highly resistant to the performed aging stress (Figures 6 and 7). Its transmittance is almost constant T in the range 400-1000 nm (Figure 6), which makes it suitable for a stack with single-junction solar cells. The choice of silicon wafer technology or multicrystalline silicon or perovskite thin film solar cells needs to be explored depending on the tile-cell mechanical stacking technologies that will be applied. The transmittance of the resulting tile (Figure 2) is however lower and not constant in the 400-600 nm range due to the higher sample thickness. A high dispersion of transmittance values as a function of the investigated spot is also observed. These two considerations indicate that the tile design needs optimization, which is also possible given the mechanical resistance demonstrated by the material. A further source of dispersion in transmittance values has been identified in shape and material imperfections, such as bubbles and scaffolding residues. This observation highlights the need to improve not only the printing process but also the distribution of scaffoldings in the design of the 3D printing element. In principle, this issue can be solved by other types of 3D printing methods or up-scaling to industrial moulding of transparent resin paving tiles.

The differences in average transmittance measured by the UV-Vis spectrophotometer and estimated by IV curves under illumination and by photocurrent, depend on the different illuminated area, i.e. about 1mm^2 for spectrophotometer, 3.6 cm^2 for illuminated IV curves, and 4 cm^2 for photocurrent. The different Transmittance values between measured by illuminating small spots or larger regions indicates that light scattering occurs in the material and also point out the need for optimization in the tile design.

4 CONCLUSIONS

This work, based on commercial materials, allowed for the assessment of preliminary routes for development of production processes of walkable 3D-printed transparent solar tiles. Among them we can mention: the optimization of the printing process (deposition, detachment and post-printing treatments); the creation of a texture design yielding more homogeneous transmittance values without jeopardizing the skid resistance and other functional properties typical of outdoor walking surfaces; the identification of suitable solar cell technologies.

Dr. Andrea Barbieri and Dr. Elisa Bandini (CNR ISOF) are acknowledged for their assistance in FTIR and UV-Vis-IR measurements; Dr. Marta Tessarolo (UNIBO) is acknowledged for mechanical stress measurements.

4.1 References

[1] https://www.pv-magazine-india.com/2021/12/23/solar-for-railways/?utm_source=USA+%7C+Newsletter&utm_campaign=3711e3fe36-RSS_EMAIL_CAMPAIGN&utm_medium=email&utm_term=0_80e0d17bb8-3711e3fe36-159283698

[2] https://www.pv-magazine-india.com/2021/09/01/up-to-28-of-railway-traction-load-could-be-viably-run-with-direct-solar-feed/

[3] https://www.pv-magazine.com/2021/09/06/hybrid-wind-solar-for-highways/?utm_source=USA+%7C+Newsletter&utm_campaign=d93b113bef-RSS_EMAIL_CAMPAIGN&utm_medium=email&utm_term=0_80e0d17bb8-d93b113bef-159283698

[4] https://www.solaroad.nl/about/

[5] https://apnews.com/article/technology-europe-business-environment-and-nature-9f9a06fe6fbe9f4c6c5b8edfb274f43d

[6] https://www.inventsrl.it/en/invent-floor-the-walkable-photovoltaic-tile/

[7] https://energyglass.gruppostg.com/it/elementi-fotovoltaici.html

[8] H. Hu, D. Vizzari, X. Zha, R. Roberts, Solar pavements: A critical review, Renewable and Sustainable Energy Reviews 152 (2021) 111712.

[9] Adel El-Shahat, 3D Printing Makes Solar Power Less Expensive, Gives Wider Range, Natural Gas Electricity, (2017) 21962.

HOT-SPOT ENDURANCE TEST FOR VEHICLE INTEGRATED PHOTOVOLTAICS

K Naga Bhavya Jyothi, Ruchita Korgaonkar, Narendra Shiradkar
Indian Institute of Technology Bombay, Mumbai, Maharastra, India, 400076
bhavyajenny18@gmail.com, naren@ee.iitb.ac.in

ABSTRACT: This paper explains the hot-spot endurance test for vehicle integrated photovoltaics (VIPV). The qualification test identifies the reliability issues due to the hot-spots which is most commonly observed in the field deployed PV modules. As per the standard IEC 61215-2:2021, a module must be incident with 800 to 1000W/m^2 light intensity. In this case all the cells would be incident with the same irradiance and the test is valid as the module is placed on a flat surface. However, for VIPV, flexible modules are placed on vehicle roof that has curved layout due to which the irradiance incident on the cells would vary. Thus, there is a need to validate if the worst-case hot-spot condition for a module placed flat works for curved surface. PSpice simulations are performed considering all the cells in the module having high shunt resistance. The shading for worst-case hot-spot condition is simulated for flat and curved module. The obtained conditions are validated experimentally with a 72-cell flexible module, whose results show that for the module placed flat has only one worst-case hot-spot condition for high shunt resistance cells. Whereas, for a module on curved surface, the worst-case shading conditions vary with the light intensity on the cells indicating the dominant effect of differing incident illumination.
Keywords: Solar cell, hot-spot, solar module, vehicle integrated photovoltaics, partial shading

1 INTRODUCTION

The emerging demand for Vehicle-Integrated Photovoltaics(VIPV) has attracted research towards its reliability. VIPV differs in the performance compared to the conventional modules because of the curved layout of the vehicle roof on which solar panels would be mounted. Apparently for VIPV, the irradiance levels would be different on the cells due to the curvature of the car roof. This would result in mismatch in the current generation development. Current mismatch sometimes leads to hot-spots that heats up the cells resulting in the back sheet damage, desolder etc. Therefore, there is a need to understand the worst-case hot-spot condition for further analysis.

Basically, hot-spot endurance test for a flat module doesn't have difference in irradiance received as the entire solar panel surface is incident uniformly. But this needs to be examined for a VIPV case as the curved module doesn't receive irradiation uniformly. As per IEC 61215 standard [1] the worst-case hot-spot condition is obtained based on the amount by which a cell must be partially shaded such that the current in the cell would be maximum power point current [1-2]. Under module short circuit condition and bypass diode turned on point, the partially shaded cell would be in reverse bias region and would dissipate maximum power leading to worst-case hot-spot condition[4]. This amount of partial shading calculated needs to be verified for a module placed flat and on a curved surface.

This paper first explains the worst-case hot-spot condition for series connected modules. Secondly, PSpice simulations are performed for a 72-cell module to understand the hot-spot condition for a flat module and VIPV. Then, the obtained simulation results are validated experimentally with a 72-cell flexible module.

2 SHADING FOR WORST-CASE HOT-SPOT CONDITION

IEC 61215 conducts hot spot endurance test by shading cells to find low and high shunt resistance cells. In this paper, to analyse the procedure to be followed for VIPV, the cells are considered to have high shunt resistance. Only the high shunt resistance cells are considered to understand the effect of the curvature solely. The high shunt resistance cells would be shaded at different levels as per standard. Basically this is to find how much a cell, namely target cell in the module must be shaded such that the current in the target cell string is equivalent to maximum power point current(I_{mp}) of the module. This leads to maximum power dissipation on the target cell, which would damage the target cell affecting the performance of the module. This is a non-destructive method mentioned in the standard to find the worst-case shading for hot-spot.

Theoretically, current developed by a shaded cell(I_{shaded}) is directly proportional to the intensity of the light incident on the cell(L_{cell}) and the unshaded fractional area of the cell(A_{fu}) as shown in equation (1). k is a proportionality constant, $A_{unshaded}$ is the amount of unshaded area of the cell and A_{Total} is the total area of the cell.

$$I_{shaded} = k\, L_{cell}\, A_{fu} = k\, L_{cell} \frac{A_{unshaded}}{A_{Total}} \quad (1)$$

The cell short circuit current under no partial shading i.e., A_{fu} =1 and receiving 1000W/m^2 irradiance can be written as

$$I_{sc(1000)} = k\,(1000) \quad (2)$$

From equation (1) and (2),

$$A_{fu} = \frac{I_{shaded}}{I_{sc(1000)}}\, x \frac{1000}{L_{cell}} \quad (3)$$

If the cell is shaded in such a way that the shaded current(I_{shaded}) is maximum power point current(I_{mp}), then equation (3) can be written as,

$$A_{fu} = \frac{I_{mp}}{I_{sc(1000)}}\, x \frac{1000}{L_{cell}} \quad (4)$$

This equation (4) sets the shading level for the target cell to be at worst-case hot-spot condition. This equation is analysed further with Pspice simulations.

3 SIMULATION RESULTS

A 72-cell module with three bypass diodes is developed in PSpice to validate equation (4). The cells are given 1000W/m^2 and are considered identical with high shunt resistance to ease the analysis regarding the shading condition for worst-case hot-spot. As stated earlier for hot-spot testing, the target cell must be shaded to produce I_{mp} current. IV and PV characteristics of the model were simulated and shown in Fig. 1 to find I_{mp} . The short circuit current of this module with 1000W/m^2 uniform irradiance is 8.63912A. The peak power current from Fig. 1 is 8.059A.

$$\frac{I_{mp}}{I_{sc(1000)}}\, x\, \frac{1000}{L_{cell}} = \frac{8.059}{8.639}\, x\, \frac{1000}{1000} = 0.9328 \qquad (5)$$

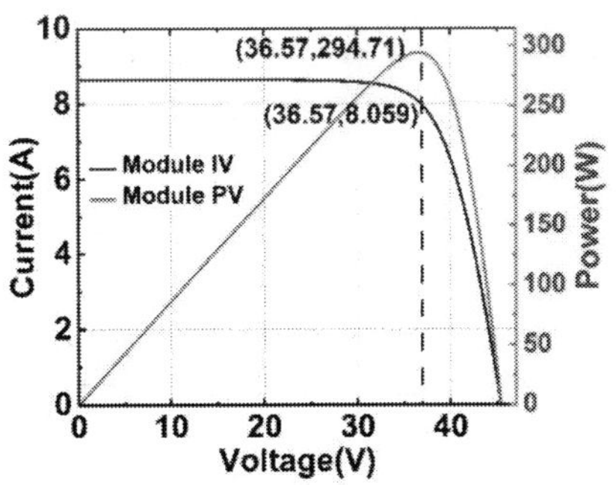

Fig. 1. Simulated module IV and PV for a flat surface module with no partial shading showing the required I_{mp} to be maintained for worst-case hot-spot.

From equation (4) and (5) the amount of unshaded portion of the cells turns out to be 93.28% of the total cell area. This can be accurately modeled in PSpice by considering the unshaded portion of the target cell area proportional to the

amount of light incident on the target cell for Imp. It can be clearly understood that the ratio, 0.9328 commensurate to 93.28% of 1000W/m2 light incident on the target cell. Therefore, 932.8W/m2 intensity would be given to the target cell in PSpice and the simulations are studied considering that the module is short circuited for hot-spot testing. Under this condition, target cell gets reverse biased by other cells in the string with -12.04V across it. The current is 8.08A which deviates from the expected peak current i.e., 8.059A by nearly 20mA due to the leakage current at reverse voltage of 12.04V which dissipates 97W on the target cell. This high-power dissipation would lead to physical or electrical damage of the module. This shows that equation (1) gives the worst-case result for a flat module.

Now, the primary goal of this research work is to verify the applicability of equation (4) for VIPV hot-spot endurance test. Firstly, to anticipate the effect of curved roof, we considered approximate dimensions of an AudiR8 car roof. This data was given as input to a python code developed to track the car roof surface and foresee the module shape. Fig. 2 depicts the layout of the module on car roof. The curvature on car front and back side is neglected for earlier stage of analysis due to large radius which is nearly flat.

In our approach, we first considered a 1.92m x 0.96m module of 72 cells and 3 bypass diodes which fits perfectly on the car roof. The size of each cell is 16cm x 16cm. Due to the curved base, the irradiance is different at various positions on the cell itself. Therefore, effective irradiance on each cell needs to be assessed based on the irradiance received at various points on the cell. For modeling this, each cell is divided into small elements of 1cmx1cm. The normal irradiation on each element is calculated assuming that the sun is in overhead position. The effective normal irradiation on each cell based on the configuration of 256 elements is computed and it varies from 967W/m2 to 998W/m2 from edges to middle due to symmetrical roof as shown in Fig. 2.

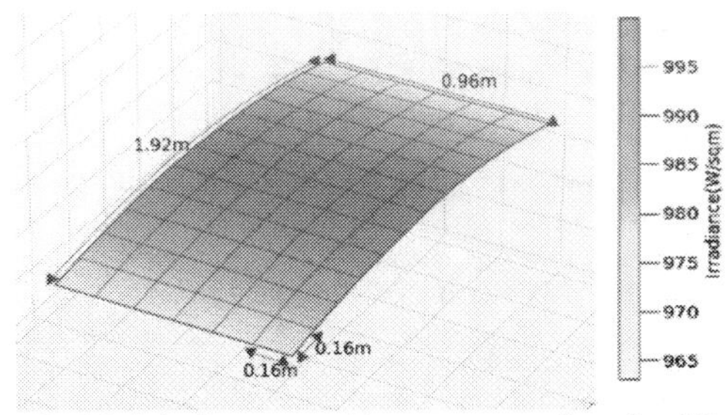

Fig. 2. Module layout tracking a car roof surface[3] generated in python and effective normal irradiance(W/m2) on each cell is enumerated for PSpice simulations

The acquired irradiance values are then given as input to a PSpice model developed for this 72-cell module as stated earlier. In PSpice, cells can be forcibly shaded to perceive hot-spot susceptibility for a VIPV system.

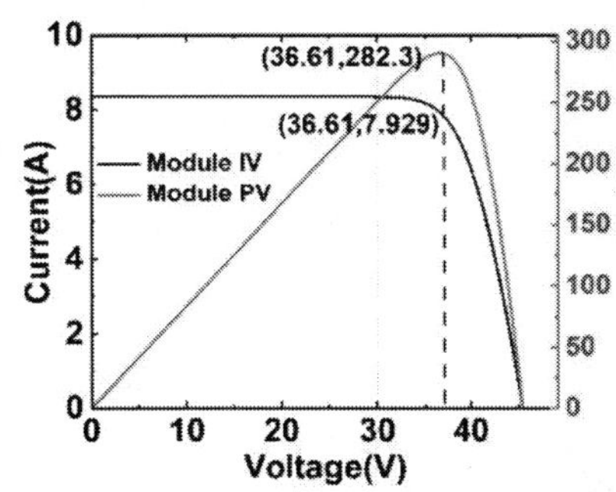

Fig. 3. Simulated module IV and PV for a curved surface module with no partial shading with I_{mp} mentioned for worst-case hot-spot condition.

To validate the hot-spot test for VIPV, the same process followed for the conventional module is repeated again. From Fig. 3, peak power current is 7.929A and short circuit current is 8.359A. As per the irradiance on all the 12 cells in the strings, 967W/m2 is the least with 998W/m^2 maximum. Due to the presence of the 967W/m2 irradiation in all the strings, this would limit the module short circuit current to the current generated for 967W/m^2.

For a cell receiving low irradiance,

$$\frac{A_u}{A_T} = \frac{I_{mp}}{I_{sc\_1000}}\, x\, \frac{1000}{Intensity_{cell}} = \frac{7.929}{8.64}\, x\, \frac{1000}{967} = 94.9\%$$ unshaded cell area

Thus, 5.1% partial shading is the shading required for the cell receiving low irradiance.

Now, considering cell receiving high irradiance,

$\frac{A_u}{A_T} = \frac{I_{mp}}{I_{sc\_1000}} x \frac{1000}{L_{cell}} \frac{7.929}{8.64} x \frac{1000}{967}$=91.67% unshaded cell area.

Thus, 8.33% partial shading for the 998W/m^2 cell. Comparing both the shading level, nearly 3% shading level difference can be seen under the condition that all cell have high shunt resistance. This difference would vary with the curvature of the car roof and the angle of incidence of light.

Also, the power dissipated was nearly 95.97W on the target cell which is less compared to flat module(97W). In view of finding this, we also observed that the cells receiving 967W/m2 intensity in the other strings are reverse biased by 2.2V due to comparatively less irradiance than other cells in the respective strings. However, this leads to 18.84W power dissipation which is still less than the maximum.

Finally, the simulation results show that for identical cells with same high shunt resistance in VIPV require different shading level. The level of this can be found using equation (4) with respect to irradiance on the cell. Also, the shading level obtained in case of flat module also differs for VIPV, indicating the significance of an additional procedure for VIPV. This is also validated experimentally in the next section.

3 EXPERIMENT RESULTS

Fig. 4. EL image of the 72-cell flexible module taken before the experiment. The target cell considered for shading during the experiment is highlighted

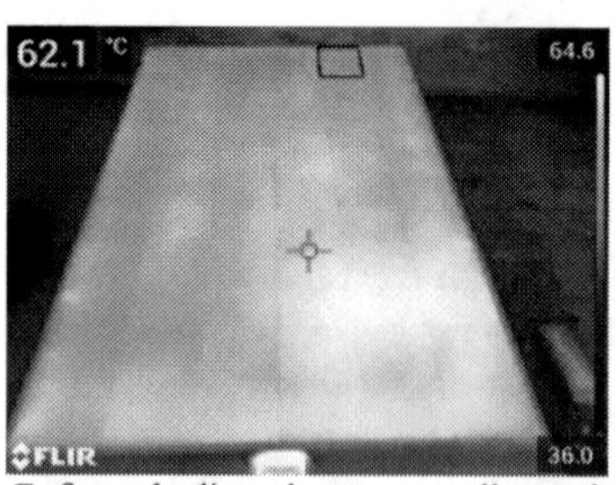

Fig. 5. Before shading the target cell, module is short circuited and thermal image is clicked to check if the cell develops hot-spot without shading. No hot-spot developed by target cell.

A 72-cell 360W flexible module is considered for experiment purpose. Before performing experiment, EL image was taken as shown in Fig. 4. A good cell without any cracks is considered for the experiment as highlighted in Fig. 4. The hot-spot experiment is performed under natural sunlight. The procedure followed in section 2 for a flat module is repeated by placing the module on ground. Fig. 5. shows the short-circuited module highlighting the target cell. It is checked that the cell doesn't develop any hot-spot under short circuited condition without shading. IV of the module is taken and the I_{mp} is obtained from Fig. 6 and included in equation (4)

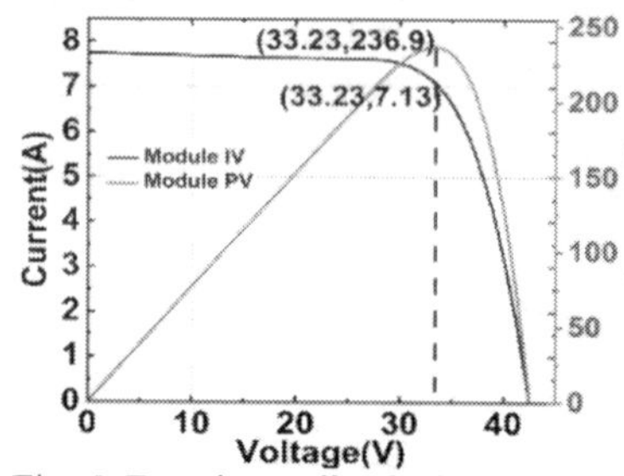

Fig. 6. Experimentally obtained module IV and PV for a flat surface module with no partial shading

$$\frac{A_u}{A_T} = \frac{I_{mp}}{I_{sc\_1000}} x \frac{1000}{L_{cell}} = \frac{7.13}{9.67} x \frac{1000}{791.6} = 0.93 \quad (5)$$

Thus, 93% of 252cm^2 cell unshaded area generates I_{mp} current. Therefore, 17 cm^2 cell area shaded with silver tape for worst-case hot-spot. After shading the cell by 17cm^2 area, hot-spot is developed as shown in Fig. 7. Cell shaded develops temperature from 68^{0}C to 90^{0}C in less than a minute. Successfully, the shading level calculated in equation (5) develops hot-spot in the target cell.

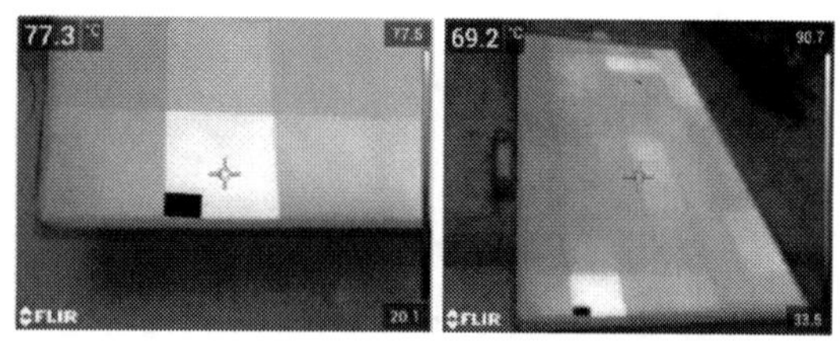

Fig. 7. The shaded cell develops hot-spot under 17cm^2 cell shaded area.

To validate equation (4)for VIPV, flexible module is placed on a curved surface, a structure developed to replicate a car roof as shown in Fig. 8. Now, before shading the target cell, module is short circuited to check if the target cell develop hot-spot before shading.

Fig. 8. Flexible module placed on the structure developed to give a car roof shape

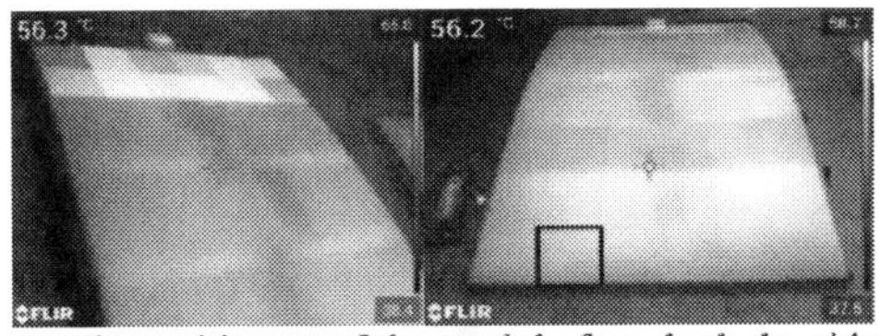

Fig. 9. Thermal image of the module from both the sides. Cells near bypass diodes show high temperature as the light incident on them is comparatively less than other cells in the module. Highlighted cell is the target cell receiving high irradiance.

Fig. 9. shows that the target cell doesn't develop any hot-spot. But some cells have shown hot-spots. These are the cells receiving relatively less irradiance than the other cells. This could be the reason for higher temperature of these cells. This confirms that some of the cells would be in reverse region due to difference in irradiation. The shading level for these cells at high temperature would be different from the cells receiving higher intensity of light.The worst-case shading level for low irradiance and high irradiance receiving cells is obtained from IV and PV characteristics of the module shown in Fig. 10,

$$\frac{A_u}{A_T} = \frac{I_{mp}}{I_{sc\_1000}}\, x \,\frac{1000}{L_{cell}} = \frac{7.11}{7.46} = 0.953 \quad (6)$$

$$\frac{A_u}{A_T} = \frac{I_{mp}}{I_{sc\_1000}}\, x \,\frac{1000}{L_{cell}} = \frac{7.11}{9.67}\, x \,\frac{1000}{1030} = 0.914 \quad (7)$$

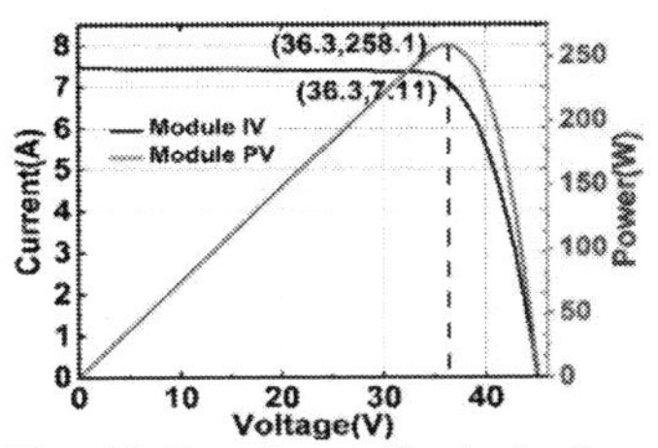

Fig. 10. Experimentally obtained module IV and PV for a curved surface module with no partial shading

From equation (6) and equation (7), 95.3% and 91.4% are unshaded levels. Thus, 10.12cm^2 and 21.6cm^2 cell area shadings are considered for worst-case hot-spot. In equation (6), 7.46A is the short circuit current of the VIPV. Since, this is the output current when the cells receiving less intensity of light limit the module, Isc_1000x(L_{cell}/1000) will be VIPV short circuit current.

Fig. 11 shows hot-spot being developed by the target cell for 21.6cm^2, but not for 10.12cm^2. Thus, it is evident that the high irradiance cell develops hot-spot and this depends on the irradiance received.

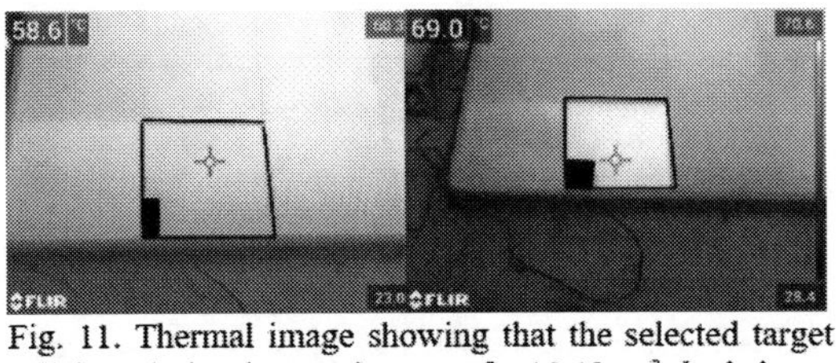

Fig. 11. Thermal image showing that the selected target cell doesn't develop any hot-spot for 10.12cm^2 shaded area and hot-spot for 21.6cm^2 shaded area

Therefore, the experiment results confirm that the cells worst-case shading levels are different based on irradiance received as observed in simulation work. This also shows that the shaded area level is different from flat module condition.

7 CONCLUSIONS

The worst-case hot-spot shading procedure in IEC 61215 considers high and low shunt resistance cells for hot-spot endurance test. In this paper, a worst-case shading condition is considered for high shunt resistance cells based on irradiance and area of fraction shaded to generate maximum power point current. In case of high shunt resistance cells, current would be nearly constant in the current limiting region of a shaded cell and also when the bypass diode turns on. PSpice simulations are performed for a module with high shunt resistance cells to find the worst-case shading for hot-spot when the module is kept flat and on curved surface. Results show that for the same high shunt resistance cells, the worst-case shading levels vary for flat and curved surface cases. In case of curved surface i.e., VIPV, the shading conditions are different due to the difference in the incident light over the curved layout of car roof. This is validated experimentally by placing a flexible module on a curved surface and repeating the test for low and high irradiance receiving cells. The worst-case shading conditions are observed to be different. The shading condition obtained for the cell incident with low irradiance doesn't even develop a hot-spot when tested on the cell with high irradiance. This confirms that the conditions for the hot-spot vary in VIPV even for the module with cells having high shunt resistance due to the varying incident irradiance. Also, the shading conditions of VIPV doesn't match with flat module case. Thus, underlining the fact that the hot-spot endurance test followed for flat module can't be applied exactly for VIPV. The angle of incidence of light and the curvature of car roof plays a significant role during the test. A global worst-case scenario for hot-spot shading can be obtained by varying the radius of curvature of the curved surface. This global worst-case can be used as a limit for the cells to operate safely without leading to destructive effects.

ACKNOWLEDGEMENT

This work is supported by National Centre for Photovoltaic Research and Education (NCPRE) funded by Ministry of New and Renewable Energy (MNRE), Government of India through the Project No. 31/09/2015-16/PVSE-R&D dated 15th June 2016

REFERENCES

[1] IEC 61215, "Terrestrial photovoltaic (PV) modules – Design qualification and type approval – Part 2: Test procedures," 2021.

[2] A. Y. Liu, S. L. Lai, and J. Yeh, "Discussion of a Failure Hot-Spot Endurance Testing Case for CIGS Thin-Film Photovoltaic Module," *Energy Power Eng.*, vol. 05, no. 04, pp. 189–193, 2013, doi: 10.4236/epe.2013.54b036.

[3] https://free3d.com/3d-model/audi-r8-14024.html

[4] A. Y. Liu, S. L. Lai, and J. Yeh, "Discussion of a Failure Hot-Spot Endurance Testing Case for CIGS Thin-Film Photovoltaic Module," Energy Power Eng., vol. 05, no. 04, pp. 189–193, 2013, doi: 10.4236/epe.2013.54b036.

DAY-AHEAD FORECASTING OF SOLAR IRRADIANCE: KNN-BASED ENSEMBLES

Yiannis Kamarianakis[1], Yannis Pantazis[1], Evangelia Kalligiannaki[1], Konstantinos Kotsovos[2], Issam Gereige[2], Marwan Abdullah[2], Athanasios Tzavaras[3], Theodoros Katsaounis[1,4]

[1]FORTH, IACM, Heraklion, Greece, [2]ARAMCO, Renewable Energy, Carbon Management Division, Thuwal, Saudi Arabia, [3]KAUST, CEMSE, Thuwal, Saudi Arabia,
[4]University of Crete, Dept. of Mathematics & Applied Mathematics, Heraklion, Greece

ABSTRACT: This work presents a novel, outlier-robust forecasting scheme, which is derived by combining two specifications. Specifically, a recently proposed time series KNN (TSKNN) ensemble procedure is combined with spline-based daily profiles (SDP), estimated via median regression. An experiment that mimics real-world implementation of an online system is performed, to evaluate day-ahead solar irradiance forecasts of the proposed method, dubbed $TSKKN_{RQ}$, versus TSKNN, SDP and the persistence model. Alternative specifications are evaluated with a new relative error metric which is outlier-robust. The analyzed measurements are collected at KAUST, Thuwal, Saudi Arabia for a full year, with 10-minute granularity.

1 INTRODUCTION

Solar plant power production is uncertain, mainly due to the stochastic formation and movement of clouds. Accurate forecasts of photovoltaic (PV) output are essential to Distribution System Operators [1], as they assist efficient solar energy trading and management of electricity grids [2]. This work evaluates a recently proposed, computationally-light time series KNN-regression scheme (TSKNN) [3], for day-ahead solar irradiance (SI) forecasting. SI is an essential predictor of photovoltaic (PV) energy output.

Time series KNN schemes (TSKNN) are straightforward to implement in online applications as they are based solely on historical data: next-day meteorological predictions are not required as additional input. Of course, advanced predictive models (e.g. penalized regressions, random forests, neural networks; [4] that use meteorological information, should dominate such schemes by a significant margin, to justify their substantially increased cost, both computational and in terms of the price to acquire continuously updated meteorological forecasts. Online forecasting systems for PV and wind-farm energy outputs are expected to employ such "cheap" specifications while in "safe-mode" operation, when for instance, access to day-ahead meteorological forecasts has been interrupted.

Outlying SI measurements may occur, due to extreme weather or malfunctioning equipment. This work emphasizes outlier-robust estimation methods and accuracy metrics, in contrast to the vast majority of published research. To the best our knowledge this is the first application of TSKNN to forecast SI; furthermore, it is the first study that constructs a forecast-combination-scheme, which aims at improving TSKNN performance.

2 DATA

SI measurements (W/m^2) are collected at KAUST, Thuwal, Saudi Arabia: a challenging location for solar panels, which often need to operate in the presence of dust and/or at temperatures far beyond the Standard Test Conditions (STC). Specifically, historical observations are collected at the New Energy Oasis (NEO) test field near the Red Sea coast (22.30 N, 39.10 E), during 2016 [364 consecutive days (52 weeks), with starting (ending) date, 01 January 2016 (29 December 2016)]. For each day, an observation is recorded every 10 minutes, starting at 8:00 and ending at 17:00 (earlier than 8 AM and later than 5 PM, SI levels are low, with negligible resulting PV energy yields; hence these measurements are omitted). Thus, the maximum number of daily observations, at times $t_n > 0, n = 1, \dots, T$, is $T = 55$. Data cleaning and outlier-removal have been applied to eliminate statistical artifacts (measurements the lie outside a range of reasonable values in W/m^2).

3 METHODOLOGY

3.1 TSKNN

The analyses presented herein are based on the methodology proposed in [3], which is implemented in R via the `tsfknn` package. Interestingly, [3] proposed an ensemble scheme that, instead of choosing the number of nearest neighbors k (typically a tuning parameter), it averages forecasts corresponding to $k = 3$, 5, 7. This is the default choice in `tsfknn` package, which provided very satisfactory levels of accuracy in the 111 time series of the NN3 competition [6]. Our preliminary analyses confirmed the above-mentioned finding; all results depicted in the next sections are based on forecast combinations for k = 3, 5, 7. In the application, TSKNN implementation is based on the default values regarding a) the distance metric (Euclidean is widely adopted), b) multi-step ahead forecasting strategy and c) the time-lags used as input variables. Regarding a) and b), preliminary investigations suggest that the default settings in `tsfknn` package provide superior levels of accuracy, whereas regarding c), we follow the recommendation in [3] for periodic time series and set the number of time lags equal to the number of time-periods per day. The effect of the length of the training period, N_{TSKNN} is examined in the forecasting experiment, for N_{TSKNN} = 1,...,5 weeks.

3.2 Robust smooth daily SI profiles

A typical day-ahead forecasting benchmark is based on average daily profiles (DP). A significant advantage of DP estimators is that uncertainty quantification (forecasting intervals) is straightforward, via quantile regression [7]. The simplest method to calculate such profiles, utilizes time-specific dummy variables [8]. Such specifications are straightforward to implement using conventional least-squares-based estimators, or outlier-robust procedures [e.g. weighted least squares, least absolute deviations, a.k.a. median regression]. Unfortunately, the resulting number of unknown parameters is large in our case (55 coefficients) and requires an extensive historical period of available measurements to produce, reasonable, smooth daily profiles.

A second class of DP estimators is based on sine and cosine functions [8], which induce smooth variation into periodic models. Again, the resulting statistical models are straightforward to estimate with conventional or robust procedures. Their main disadvantage, is that model building (selection of statistically significant sine/cosine frequencies) is computationally demanding. On the other hand, our experiments indicated very similar levels of accuracy to the spline-based estimators that are presented next; hence performance of such periodic models will not be discussed further.

The specifications described thus far correspond to parametric models. An alternative, non-parametric modeling strategy, which is fast to compute without compromising accuracy, uses regression splines [9]. The procedure implemented here (dubbed SDP, for smooth daily profiles), fitted piece-wise cubic polynomials with 15 knots. Knots are breakpoints in the third derivative (the number of knots was chosen in a preliminary cross-validation experiment) arranged at equally-spaced time instants. The function `bs` in the `R` package `splines` was used to construct B-spline basis expansions.

Daily profiles can be again derived using least squares estimation or robust alternatives. An additional robust estimation procedure that wass evaluated in this case, is likelihood-based and adopts the heavy-tailed *t* distribution for the residuals (`R` package `mgcv`).

Similar to TSKNN, the effect of the length of the training period, N_{SDP} is examined in the forecasting experiment, for $N_{SDP} = 1,\ldots,8$ weeks. The forecasting experiments that follow, focus on median-regression-based SDPs, which according to our experiments dominate alternatives for the adopted accuracy criterion that is discussed in Section 3.4. Furthermore, forecasting intervals are straightforward to compute by switching to quantiles other than the median.

3.3 Robust TSKNN

TSKNN is expected to perform well when environmental conditions are unstable: for instance, in a simplified situation with two daily regimes, namely clear sky and sandstorm, when a series of days with clear skies are followed by days with sandstorms, TSKNN will produce a forecast that is based on historical information from the most recent regime. On the other hand, spline-based average daily profiles are expected to perform well in time periods which are stable, since the profiles are computed using a substantially larger sample size (number of days) relative to TSKNN. Furthermore, robustly estimated (e.g. via median regression) smooth profiles essentially neglect measurements that correspond to extreme environmental conditions and exploit the vast majority data.

This work proposes a robust TSKNN variant, dubbed $TSKKN_{RO}$. $TSKKN_{RO}$ is motivated by shrinkage estimators, which are widely applied in regression problems via the LASSO [10]. Essentially, given that TSKNN exploits a relatively small number of days from the training data to compute forecasts for daily profiles, $TSKKN_{RO}$ ``shrinks" TSKNN towards a smooth, outlier-robust SDP profile, derived from a significantly larger sample. An interesting research question, relates to the mixing weights in $TSKKN_{RO}$; should they be kept fixed or do different periods within a year require different weights? Obviously, a scheme based on varying weights is expected to cope with periods with different characteristics throughout a year; unstable periods are expected to correspond to relatively large weights for TSKNN relative to SDP. In general, an estimator that combines TSKNN with robust SDP can also be viewed as a classic forecast combination scheme. In light of the forecast combination puzzle [11], the application that follows examines a simple scheme with equal weights, which does not require any additional tuning.

3.4 Benchmark Models and Accuracy Metrics

The persistence model (PRS) is a typical benchmark in renewable-energy-related forecasting experiments. This naive scheme implies that current conditions remain unaltered and that a future daily profile will be very close to the one most recently observed. Hence observations that correspond to the most recent day, are used to forecast unobserved SI profiles, for one or more days ahead.

An accuracy metric that is widely used in engineering is the mean absolute percentage error (MAPE). As shown, among others in [5], MAPE systematically favors methodologies that produce low forecasts. Specifically, MAPE regression can be viewed as a weighted median regression with the observed measurements taking the role of weights; hence the lowest measurements are more influential and the predictive specification is pulled towards them. To overcome the above-mentioned shortcomings, [5] proposed an alternative relative error metric, namely *LnQ*, which is formulated as follows

$$LnQ = \frac{1}{N}\sum_{t=1}^{N}\left[ln\left(\frac{\widehat{Y(t)}}{Y(t)}\right)\right]^2 \quad (1)$$

with $Y(t)$ denoting observed and $\widehat{Y(t)}$ forecasted values.

Inspired by *LnQ*, the accuracy criterion that is reported in our forecasting experiments, is an outlier-robust' variant, dubbed LnQ_R

$$LnQ_R = \frac{1}{N}\sum_{t=1}^{N}\left|ln\left(\frac{\widehat{Y(t)}}{Y(t)}\right)\right| \quad (2)$$

It is worth highlighting that *LnQ* complies with a multiplicative error model, which is appropriate for heteroscedastic processes [5]. *LnQ*-optimal regression estimates via least squares, coefficients of a linear model on log-transformed responses, whereas LnQ_R -optimal regression uses median regression instead.

4 RESULTS AND DISCUSSION

Figures 1 and 2 depict static SDP variants where differences between a) parametric versus nonparametric specifications and b) conventional versus outlier-robust estimation procedures are illustrated. The forecasting experiment that is discussed next evaluates four day-ahead forecasting methodologies in terms of their LnQ_R –performance: a) the naïve PRS, b) TSKNN, c) robust-SDP that combines smoothing splines with median regression and d) $TSKKN_{Ro}$, which does not require additional tuning as it corresponds to simple averaging of TSKNN and SDP-based daily SI forecasts.

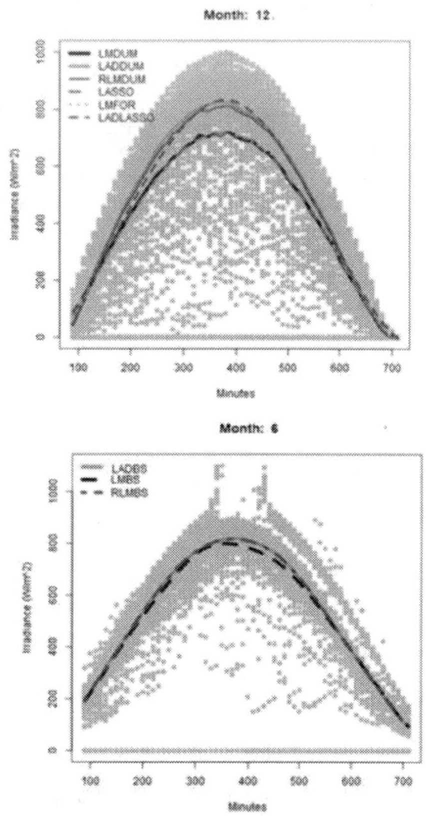

Figure 1: Irradiance SDP for December (top) based on 6 parametric procedures (three dummy-variable based and three based on periodic functions). The effect of extremely low measurements on outlier-sensitive least-squares-based daily profiles is clearly observed: ls-based SDPs lie substantially below their robust variants. SI SDP for June based on 3 non-parametric procedures (bottom); robust (LADBS, RLMBS) SDPs are very close to outlier-sensitive LMBS in this case. This example illustration uses more data relative to the ones utilized in our forecasting experiments.

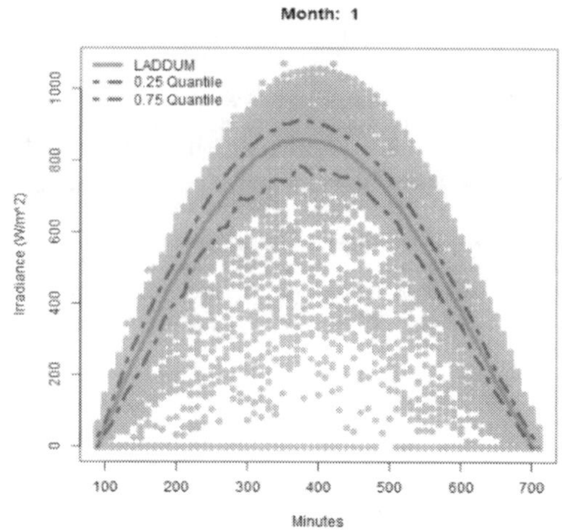

Figure 2: Irradiance SDP for January, based on median regression on dummy variables; estimated forecasting intervals correspond to 0.25 and 0.75 quantiles. This example illustration uses more data relative to the ones utilized in our forecasting experiments. Still, the curve that corresponds to the lower forecasting boundary is not smooth enough.

The forecasting experiment comprises $D = 76$ testing days (Figure 3) dispersed in the training data and mimics a realistic online implementation: sometime during each testing day $d = 1,\dots,D$, a forecast is computed for the SI profiles of the next day. Results are summarized in Table I; a subset of observed and forecasted values is shown in Figure 4. Figure 5 depicts SDP forecasts for the last day of the experiment, coupled with quantile-regression-derived intervals that quantify uncertainty on point forecasts. In accordance with prior expectations, uncertainty increases towards mid-day.

Table I: LnQ_R performance with increasing amount of training data for the four alternative, day-ahead forecasting specifications.

# of tr. weeks	PRS	TSKNN	SDP	$TSKNN_{R0}$
1	0.1323	0.2030	0.1117	0.1351
2	0.1323	0.1527	0.1114	0.1180
3	0.1323	0.1432	0.1052	0.1104
4	0.1323	0.1413	0.1070	0.1105
5	0.1323	0.1409	0.1092	0.1118

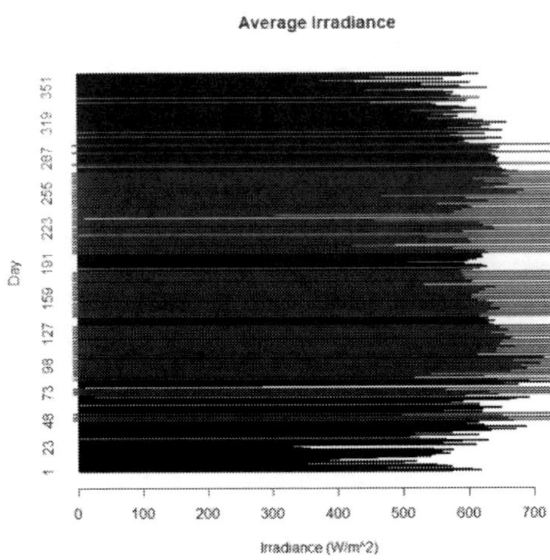

Figure 3: Average levels of daily solar irradiance across the examined period. The testing days of the forecasting experiment are highlighted in red.

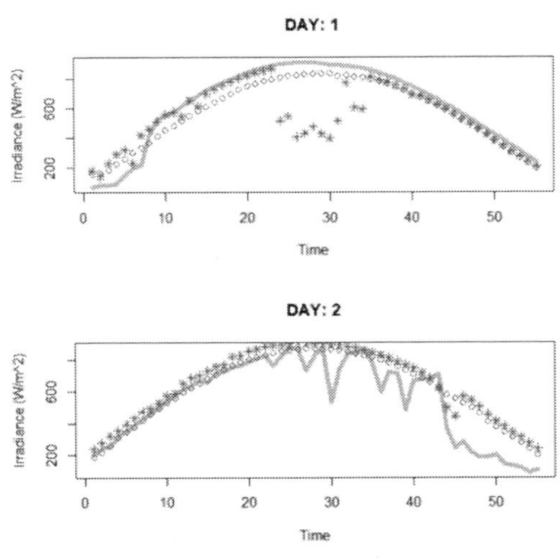

Figure 4: Observed SI values (grey line) versus PRS (green stars) and $TSKKN_{R0}$ (red circles) forecasts, for the first 2 days of the forecasting experiment.

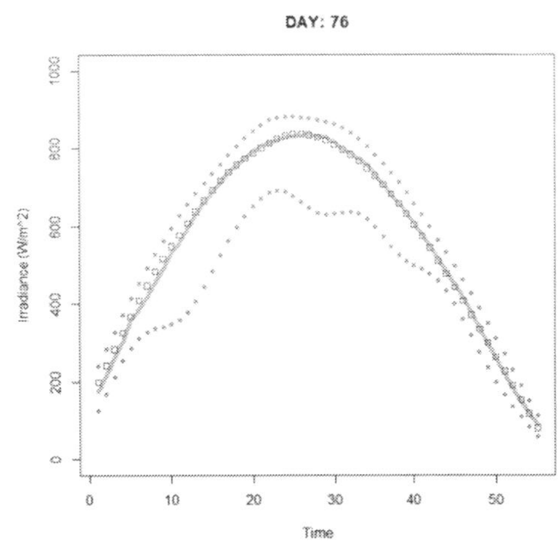

Figure 5: Observed SI values (grey line) versus SDP forecasts (brown squares) and SDP forecasting intervals (0.1 and 0.9 quantiles) shown as green points.

A striking result in Table I is the clear dominance of SDP relative to the recently proposed TSKNN scheme. SDP performs well even with a small training dataset. On the other hand, TSKNN performance improves very fast initially, when the available historical data increase from one to two weeks and slowly afterwards, as the number of training weeks increases from three to five. It is surprising that TSKNN is less accurate relative to the persistence model. On the other hand, the proposed combined scheme, $TSKKN_{R0}$ performs significantly better relative to TSKNN, with performance close to SDP.

5 CONCLUSIONS

This work evaluated an easy-to-implement recently proposed, times-series KNN, ensemble scheme for day-ahead SI forecasting. Although its performance is not satisfactory, especially versus spline-based smooth daily profiles estimated with median-regression, our investigation suggests that a novel scheme, which shrinks TSKNN towards SDP may save practitioners from extremely low levels of accuracy. The new scheme, namely $TSKKN_{R0}$ does not require additional tuning as it weighs equally TSKNN and SDP. A research question that is worth examining is related to the forecast combination puzzle: is it worth trying to optimize the forecast combination scheme?

6 ACKNOWLEDGMENTS

The authors acknowledge the support of KAUST and Saudi Aramco R&D Center - Carbon Management Division for their financial support in developing this work. This work was partially supported by grant #OSR–2020-4433.02 from KAUST and Saudi Aramco.

REFERENCES

[1] J. Antonanzas, et al., Solar Energy (2017) **158**, 140-146.
[2] EPIA (2012) Available online: http://pvtrin.eu/assets/media/PDF/Publications/other_publications/263.pdf.
[3] F. Martínez, et al., Artificial Intelligence Review (2019) **52**, 2019-2037.
[4] A. Nespoli et al., Energies (2019) **12**, 1621. https://doi.org/10.3390/en12091621
[5] C. Tofallis, Journal of the Operational Research Society (2015) **66**, 1352-1362.
[6] S.F. Crone at al., International Journal of Forecasting (2011) **27**, 635-660.
[7] R. Koenker, Quantile Regression, Cambridge University Press (2005).
[8] P.S.P. Cowpertwait and A.V. Metcalfe, Introductory Time Series with R, Springer (2009).
[9] S.N. Wood, Generalized Additive Models: An Introduction with R, Chapman and Hall/CRC (2006).
[10] T. Hastie et al., The elements of Statistical Learning: data mining, inference and prediction. Springer (2009).
[11] G. Claeskens et al., International Journal of Forecasting (2016) **32**, 754-762.

IMPROVED ADAPTIVE NEURO-FUZZY INFERENCE MODEL FOR PHOTOVOLTAIC POWER FORECAST

Mustapha Habib, Annika Gram and Qian Wang

Division of Building Technology and Design, KTH Royal Institute of Technology, Department of Civil and Architectural Engineering, 11428 Stockholm, Sweden

Abstract-Photovoltaic (PV) systems are recently the most used sustainable energy source to fit with the energy demand growth. Generally, batteries, as storage systems, are installed along with PV modules. When it comes to an optimal power management of PV/battery hybrid systems, the uncertain and intermittent behavior of PV power production can provoke some challenges, with which, the real-time operation of the hybrid system can be degraded, therefore, PV power forecast is highly needed. Data-driven models are became nowadays very efficient methods to build regression models for the purpose of PV power forecast. In this paper, Adaptive Neuro-Fuzzy Inference (ANFIS) is chosen as a data-driven technique, to build up forecasting models. Standard ANFIS, which uses only weather data, cannot avoid the confusing scenarios like PV modules covered by the snow in clear-sky days. This work proposes an improved ANFIS model taking historical generated power into account. The developed model is validated on a real case, using the PV system of the institute of energy system technologies in Offenburg. When adding the average of produced power of the last 72 hours as additional input, the model was able to follow the rapid changes in weather conditions and overcome the unexceptional situations like the problem of snow on the PV modules.

Keywords-photovoltaic power forecast, adaptive neuro-fuzzy inference, photovoltaic/battery hybrid systems management.

I. INTRODUCTION

In the last few years, there has been a growing interest on weather forecast in many applications. This task is highly recommended when optimal energy management is targeted for systems integrating renewable resources. When it comes to islanded photovoltaic (PV)-based microgrids (MG), PV power forecast is a fundamental task [1-3], surplus PV generation can occur overvoltage at the connection points even with optimal design of PV systems. In case of connected storage system, PV system can cause frequent fast and deep charging/discharging scenarios, which may decrease the batteries life-time. Having enough information about the future PV production can make it possible avoiding all the mentioned challenges. Generally, PV power is predictable when enough information about the insolation and temperature is available, this data is usually provided thanks to many online services available on internet.

PV generation forecast methods can be broadly classified into four approaches: statistical approach, Artificial Intelligence (AI) approach, physical approach and hybrid approach [4]. Statistical approaches are based on data-driven formulation using historical measured data to forecast solar time series [5]. AI approaches utilize advanced AI techniques, such as Artificial Neural Networks (ANNs), to construct solar forecasters, which can be also classified into the category of the statistical approach [6]. Physical models are based on numerical weather prediction or satellite images that predict solar irradiance and PV generation [7], [8]. Finally, hybrid approaches are combination of the three afore-mentioned methods [9]. In practice, different forecasting approaches are preferred depending on different scales of prediction horizons to meet the requirements of the decision-making process [10].

In the case of PV plant, the power generation mechanism depends on several factors like the number and model of solar cells; number of installed modules; positioning and inclination of modules; instantaneous values of solar irradiation and temperature. Unlike other factors, insolation and temperature are not constants, they are varying and fluctuating all the time in an uncertain behavior, which makes it a challenging for any forecast method to follow up the real produced power.

Different prediction horizons will correspond to the specific needs of decision-making activities in a MG as follows [11]:

1. Short-term (between 24 and 72 hours ahead): Such forecasts are crucial for different decision-making problems involved in the electricity market and power system operation, including economic dispatch, unit commitment, etc.
2. Medium-term (up to one week ahead): Medium-term forecasting would be useful for e.g., maintenance scheduling of PV plants, conventional power plants, transformers, and transmission lines.
3. Long-term (up to months to years): Long-term prediction/estimation can be applied for long-term solar energy assessment and PV plant planning.

Introduced by Jang in 1993 [12], Adaptive Neuro-Fuzzy Inference (ANFIS), belongs to AI approach, are among the candidates for PV power forecast development. This marriage of learning capability of neural network and knowledge representation ability of fuzzy logic has given birth to fuzzy neural networks. As a result, the drawback of neural network black box inability to explain decision (lack of transparency), and weakness of learning in fuzzy logic have been conquered. In this paper, and for a particular PV system, different ANFIS models with different parameters are built up using two main inputs: real-time information of solar irradiation and temperature.

In cold zones, after snowing periods, PV modules could stay covered by the snow for a long time even during the followed sunny days, which makes it confusing for standard ANFIS models, that are based solely on insolation/temperature information, to predict the generated PV power. To overcome this challenge, this paper suggests the usage of an improved ANFIS regression model by providing enough information about historical changes in power production. In this fact, additional input is to be added which is the average of produced power for the last few hours. As a validation case, real dataset for a 6.3 KWp PV system, installed in the institute of energy system technologies (INES) in Offenburg, is used (Fig. 1). The general characteristics of INES PV system is listed in Table 1. First, a standard ANFIS model, using only insolation and temperature as inputs, is developed and simulated. Then, the same simulation scenario will be executed for an ANFIS model integrating an additional input, which is the average of produced power during the last 72 hours.

Table 1. INES PV power system parameters

	Parameter	Value
PV power system	Number of modules	27
	Inclination	9 x 35° 18 x 30°
	Alignment	180° south
	Power	6.3 kWp
PV module	Power at mpp Pmpp	240 Wp
	Voltage at mpp Vmpp	30.0 V
	Current at mpp Impp	8.1 A
	Open circuit voltage Voc	37.4 V
	Short circuit current Isc	8.6 A
	Temperature coefficient	-0.46 %/K
	model	Bosch solar module c-Si M 60

Fig. 1. INES 6.3 kWP PV generator

The rest of the paper is organized as follow: in Section II, more explanation about the structure of a default ANFIS model is presented; in Section III, ANFIS standard forecast model is explained with a simulation study; while Section IV will be the stage for simulating the improved ANFIS model.

II. ADAPTIVE NEURO-FUZZY INFERENCE

The success of ANFIS can be attributed to the robustness of results it provides. ANFIS has as highly generalization capability as neural networks and other machine learning techniques [13-15]. It is based on Takagi-Sugeno-Kang model, or simply Sugeno fuzzy model, proposed by [16] where a rule R_k can be represented as:

$$R_k\text{: IF } \mu_{A_i}(x) \text{ AND } \mu_{B_i}(y) \text{ THEN } f = p_k x + q_k y + r_k \quad (1)$$

Where k is the number of rules, A_i and B_i are n fuzzy membership functions of any shape i.e., gaussian, triangular, trapezoidal, etc., denoted by μ in the antecedent part of the rule R_k, and p_k, q_k, r_k are the linear parameters of consequent part of the kth rule. The parameters of membership functions (antecedent or premise parameters) and consequent part of the rule (consequent parameters) are tuned during the training process.

In this article, ANFIS five-layers architecture comprises of two types of nodes: fixed and adaptable (Fig. 2). The nodes in membership function layer and consequent layer are tunable, the rest of the nodes are fixed. In Layer 1, the node i is a membership function i.e., triangle, trapezoidal, or gaussian, etc. For example, if μ_{A_1}, μ_{A_2} and μ_{B_1}, μ_{B_2} are the membership functions of gaussian shape with two parameters center (c) and width (σ). Layer 2 calculates the firing strength of a rule via product Π operation. Layer 3 is normalized firing strength of a rule from previous layer. In Layer 4, each node represents consequent part of fuzzy rule. The linear coefficients of rule consequent are trainable. Nodes in Layer 5 perform defuzzification of consequent part of rules by summing outputs of all the rules. Further detail on computation performed in ANFIS can be found in the related paper [16].

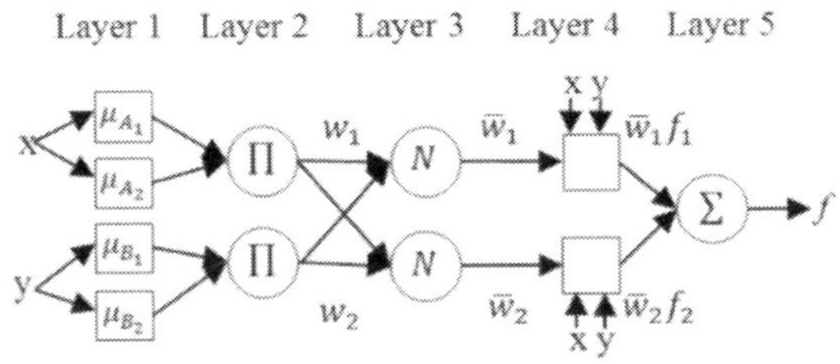

Fig. 2. ANFIS architecture

ANFIS learns by tuning all its tunable parameters (c, σ and p_k, q_k, r_k) in order to map input to the desired output with minimum error. The default ANFIS learning algorithm employs gradient descent (GD) for tuning membership functions and least square estimation (LSE) for training the consequent parameters. it is a two pass learning process as presented in Table 2.

The two pass learning algorithm tunes consequent parameters by LSE in forward pass and while back-propagating error back to first layer, it updates membership functions using GD. Since, GD is influenced by back propagation (BP) algorithm of ANN, which has the drawback to be likely trapped in local minima. On the other hand, the convergence of gradient method is also very slow; depending on initial parameter values.

Table 2. Two-Pass ANFIS Learning Algorithm

	Forward Pass	Backward Pass
Antecedent Parameters	Fixed	GD
Consequent Parameters	LSE	Fixed
Signals	Node Output	Error Signals

III. ANFIS FORECAST MODEL

In this section, standard Sugeno fuzzy-bazed model, that uses solar irradiation and temperature as inputs, is studied and validated. In the context of higher supervisory layer for INES experimental MG, the devlopped model is to be implemented on a high performance hardware solution like industrial computer, this later will then communicate with the Programmable Logic Controller (PLC), that manages the whole MG prototype, in the framework of a smart energy management. In INES experimental MG, weather data is accessible thanks to online service called *Deutscher Wetterdienst*. It provides up to 3-day ahead with one-hour resolution.

Bellow, the ANFIS training process is explained, in which, four months dataset for Offenburg region, from Setpember 2016 to January 2017, is downloaded from the local database. For testing, two weeks from February 2017 selected, these weeks are characteriszed by a rapid switching between snowing and sunning days. Figs 3, 4 and 5, and Table 3 show the characteristics of the chosen training data. As a development platform, MATLAB version R2014a, with Neuro-Fuzzy Designer toolbox is used on Intel® Core (TM)2 Duo CPU. The training parameters are chosen with default values.

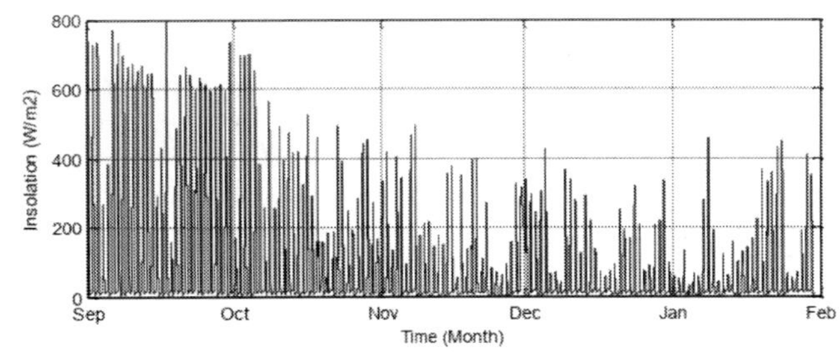

Fig. 3. Hourly insolation in the training dataset.

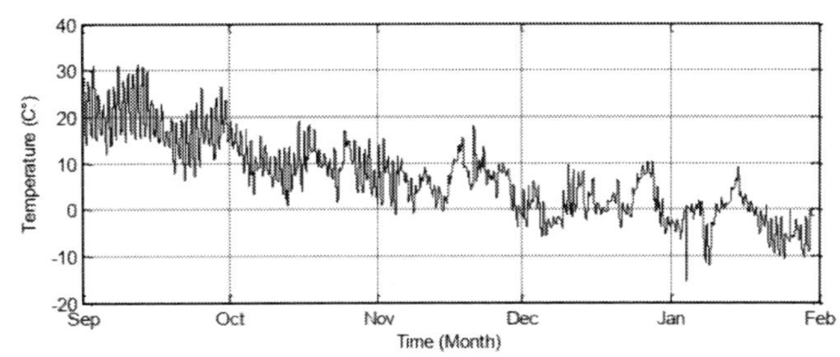

Fig. 4. Hourly temperature in the training dataset

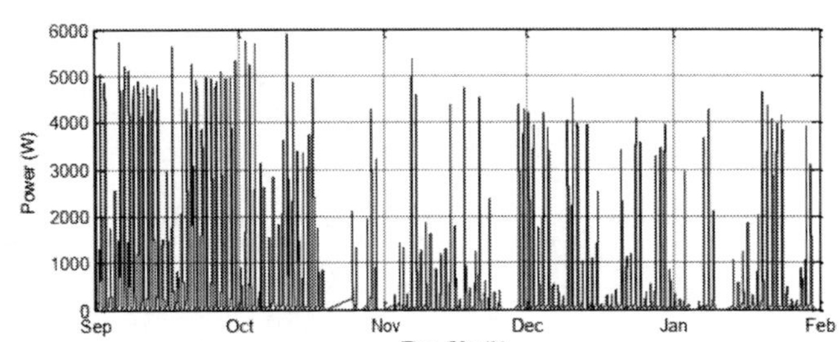

Fig. 5. Hourly produced PV power in the training dataset.

Table 3. Main characteristics of dataset

Characteristics	Power (W)	Insolation (W/m^2)	Temperature (C°)
Mean value	500	76	6
Min value	0	0	-15
Max value	5904	798	31
Time step	1 Hour		
Data length	3574 Hours		

Case 1:

The first training case will be for a Sugeno fuzzy model with three triangle membership functions for each input variable. The training result is shown in Fig 6. The evaluation of the generated model performance using the testing data is presented in Fig 7.

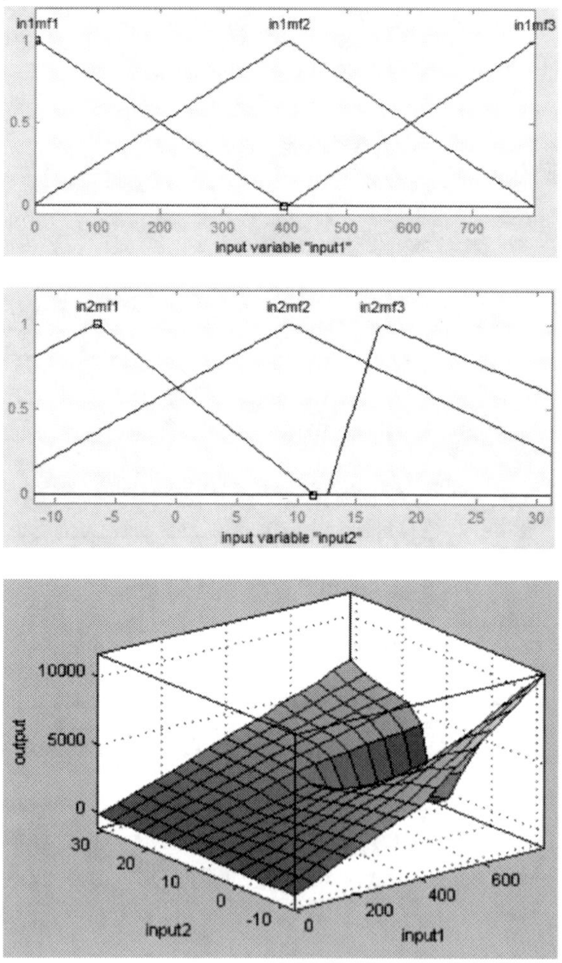

Fig. 6. Fuzzy surface map and membership functions (case 1), in which, "input1" is the insolation, "input2" is the temperature and "output" is the PV power

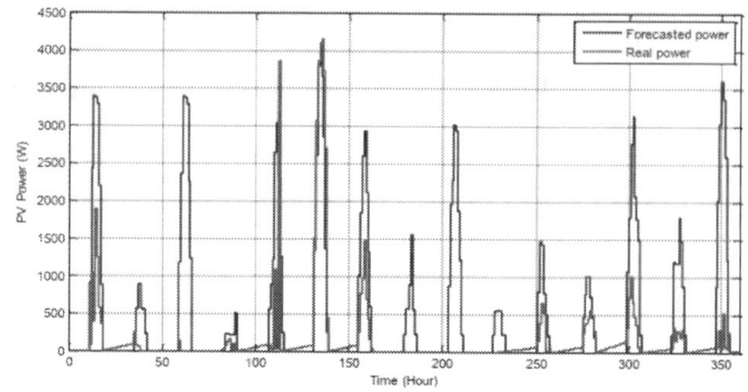

Fig. 7. Sugeno fuzzy model prediction performance (case 1)

Case 2:

In this case, a Sugeno fuzzy model is built using trapezoidal membership functions, two for insolation and three for temperature. The training result is shown in Fig 8. The evaluation of the generated model using the testing data is presented in Fig 9.

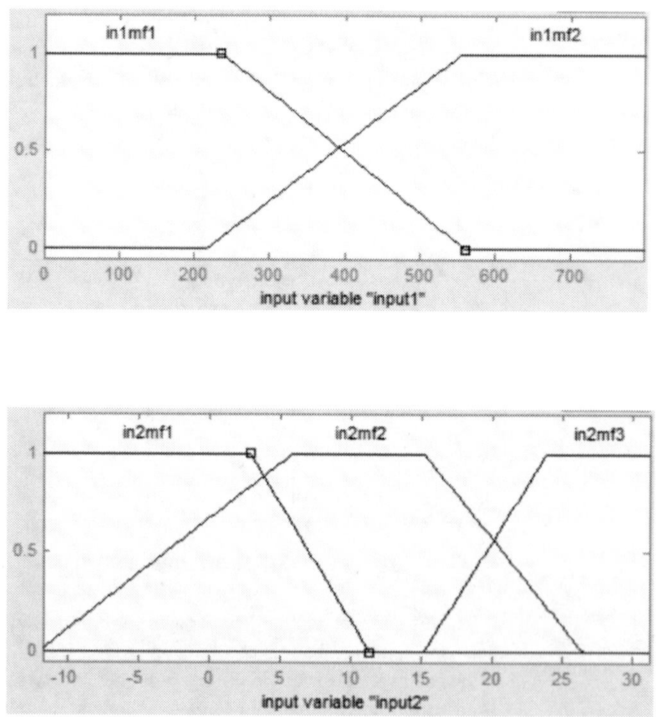

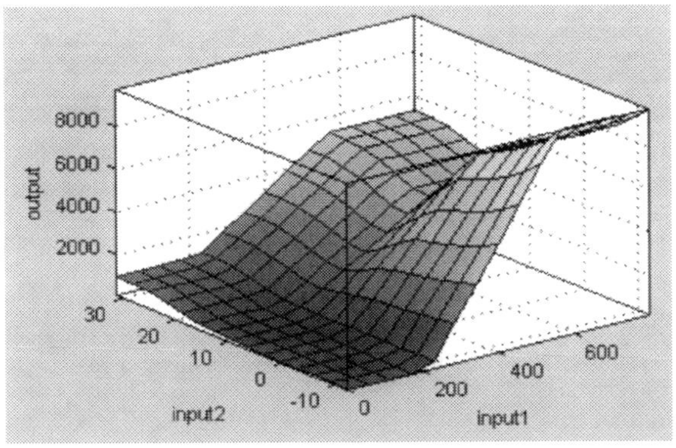

Fig. 8. Fuzzy surface map and membership functions (case 2), in which, "input1" is the insolation, "input2" is the temperature and "output" is the PV power

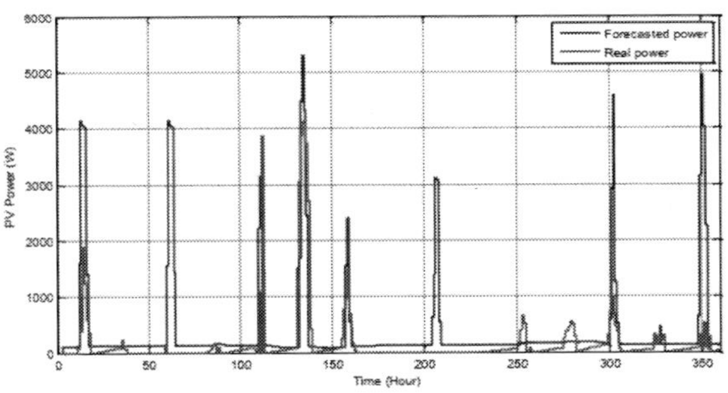

Fig. 9. Sugeno fuzzy model prediction performance

(case 2)

Case 3:

In this case, a Sugeno fuzzy model with three Generalized bell-shape membership functions for each input variable. The training result is shown in Fig 10. The evaluation of the generated model using the testing data is presented in Fig 11.

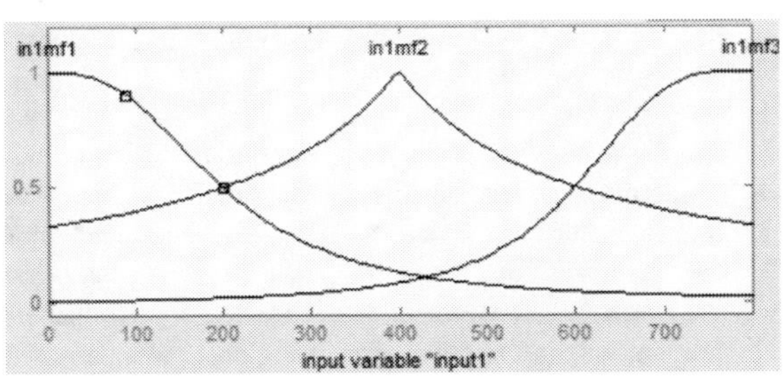

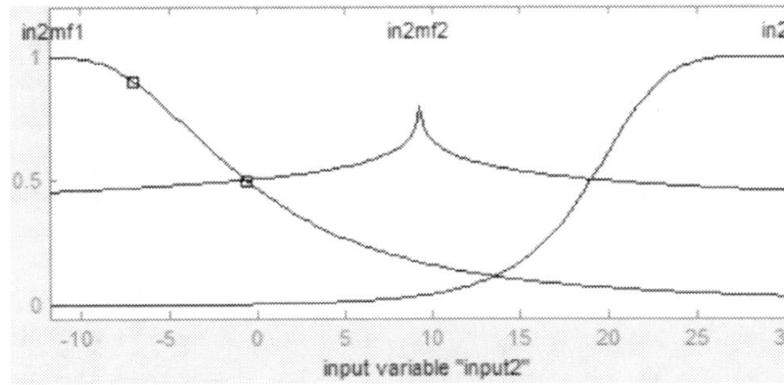

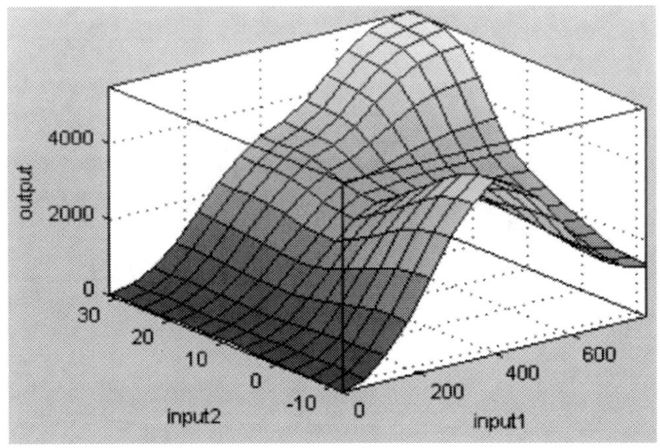

Fig. 10. Fuzzy surface map and membership functions (case 3), in which, "input1" is the insolation, "input2" is the temperature and "output" is the PV power

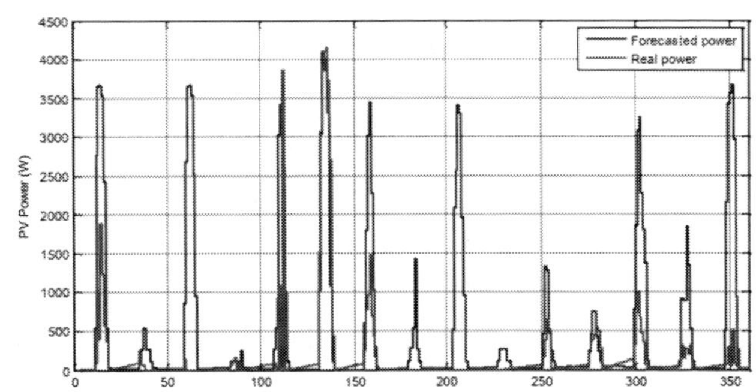

Fig. 11. Sugeno fuzzy model prediction

performance (case 3)

Comments:

Through previous results, it is clearly visible that the rapid alternance between snowing and sunny times makes a big confusion for all developed Sugeno fuzzy models. Even with a high accurate online weather forecast service at INES, the short cloudy periods (terms of hours) cannot be efficiently predicted. In addition, during snowing days (Fig. 12), the snow can keep covering PV modules during the subsequent clear-sky days, provoking another challenge for standard ANFIS models to find out the real produced power, which is translated to big prediction errors visible in Fig. 7, Fig. 9 and Fig. 11, i.e., the snow layer on PV modules makes the solar cells not accessible for the solar irradiation.

Fig. 12. INES PV modules during a snowing day

IV. IMPROVED ANFIS

In this section, we will investigate the possibility to improve the developed standard ANFIS models to prevent the effect of the alternance snowing/sunny days. As mentioned previously, this improvement is based on adding historical power production evolution as a new input to the Sugeno fuzzy model. The reason behind this technique is to decrease or eliminate the impact of weather fluctuations on the performance of forecasting models. For this purpose, the average of historical produced power, for different horizons, is calculated. (Figs 13, 14 and 15).

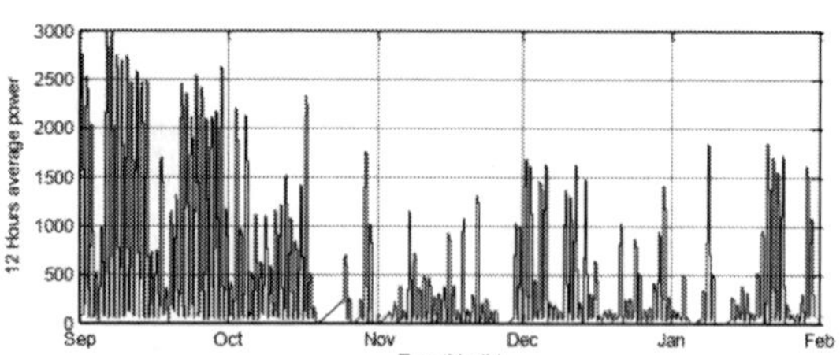

Fig. 13. Average power production for the last 12 hours for every hourly value of training data.

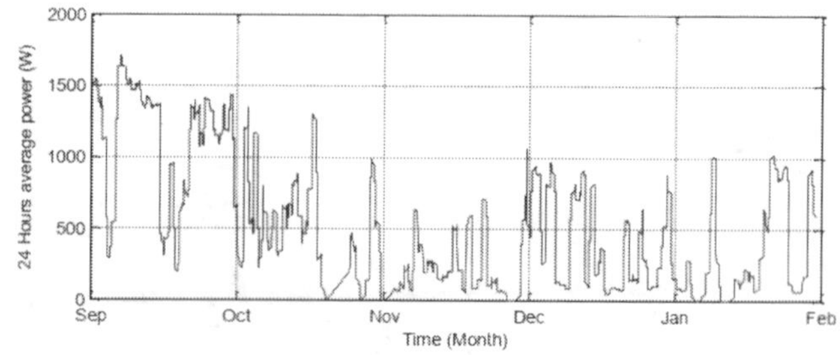

Fig. 14. Average power production for the last 24 hours for every hourly value of training data.

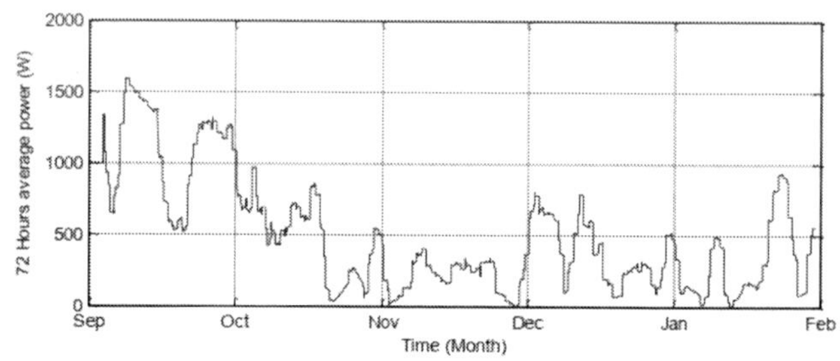

Fig. 15. Average power production for the last 72 hours for every hourly value of training data

With the average of produced power for the last 12 hours as additional input, ANFIS model can have an information about the weather changes curve that has effect even in the current day. This is important when we aim to compensate the error caused by very short fluctuations on the insolation level due to the clouds movement. ANFIS will not track, with exactitude, the power to be produced, rather it will compensate the global forecast error caused by these fluctuations for one day scenario. This technique is suitable in case of daily energy management of PV/batteries hybrid system.

The radical climatic changes between snowing and sunny days has a slower dynamic compared to the insolation fluctuations. The ANFIS model can, with the average of produced power during the last 24 hours as additional input, improve the standard model in order to not rely solely on the online weather forecast service. This later usually cannot predict such rapid daily changes in weather conditions.

The average of produced power during the last 72 hours can give information about the weather trend in a general way and the state of the PV modules in particular. Adding it as an additional input to the ANFIS model, can provide significant improvement on the prediction performance as it offers information on the global weather trends regardless of the short-term weather fluctuations.

Case 4:

In this case, the training process will be carried out on a Sugeno fuzzy model with only two Generalized bell-shape membership functions for each input. The membership functions correspond to the trainined improved ANFIS model are shown in Fig 16. The evaluation of the created model using the testing data is presented in Fig 17.

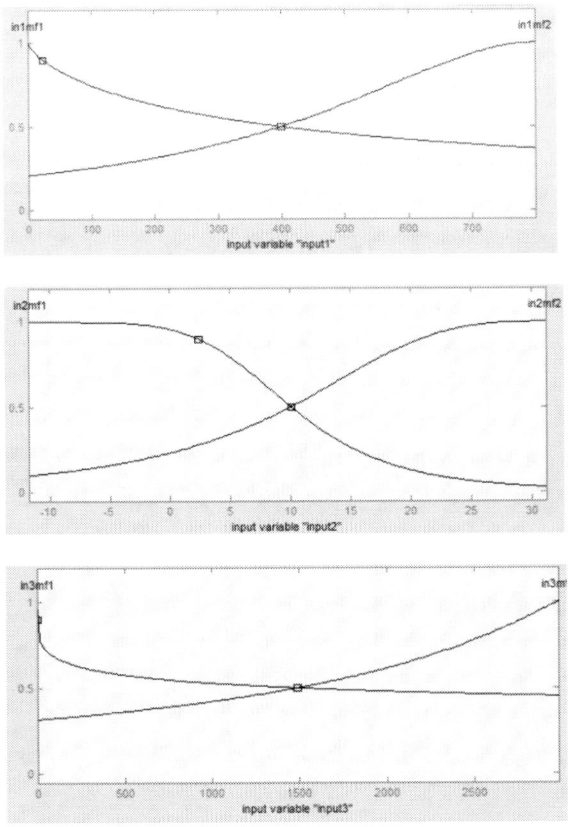

Fig. 16. Membership functions (case 4), in which, "input1" is the insolation, "input2" is the temperature, "input3" is the produced power everage of the last 72 hours, and "output" is the PV power

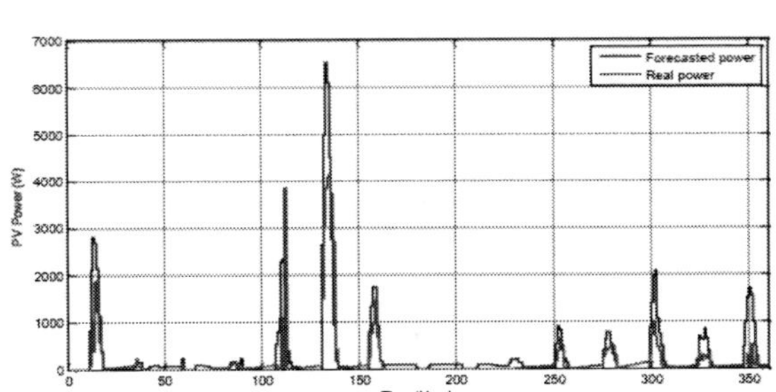

Fig. 17. Sugeno fuzzy model prediction performance (case 4)

Comments:

Providing historical data, as additional input, improves significantly the forecasting accuracy as shown in Fig. 17. With information about the power production trend during the last 72 hours, along with real-time insolation and temperature data, the model is capable of decreasing the effect of fast changes in weather conditions. This can correct partially the problem of the inaccessibility of solar irradiation to the snow-covered PV modules.

CONCLUSION

In this paper, PV power forecast using improved ANFIS model is investigated. Standard models, which are based on the use of online services for weather data, cannot overcome the challenge of insolation fluctuations and the rapid changes in the general weather conditions. Adding the historical PV production power to the standard model, as an additional information to the weather data, can enhance significantly the ANFIS model performance, and even, detect critical situations like the effect of snows on PV modules during clear-sky days.

As a future work, the improved ANFIS model performance can be also evaluated in the context of PV models degradation impact. As it is well known, the operation degradation of PV cells can be accelerated under some specific conditions, mainly weather-related. The proposed ANFIS model can be tested in this regard using different historical data lengths. Obviously, the historical data in this case is significantly longer compared to the studied on in this article. The effect of adding different horizons can be also evaluated to find out the optimal input combination.

Acknowledgement

This work is partially financed by Swedish Energy Agency under the grant number: 52488-1.

Reference

[1] H. Kanchev, D. Lu, F. Colas, V. Lazarov, and B. Francois, "Energy Management and Operational Planning of a Microgrid With a PV-Based Active Generator for Smart Grid Applications," IEEE Trans. Ind. Electronics, Vol. 58, No. 10, Oct. 2011, pp. 4583-4592.

[2] A. Molderink, V. Bakker, M. G. C. Bosman, J. L. Hurink, and G. J. M. Smit, "Management and Control of Domestic Smart Grid Technology," IEEE Trans. Smart Grid , Vol. 1, No. 2, September 2010, pp. 109-119.

[3] Mohamed E . El-Hawary,'' The Smart Grid—State of the art and Future Trends'', Electric Power Components and Systems, 42:3-4, 239-250, Feb 2014.

[4] Can Wan, Jian Zhao,Yonghua Song, Zhao Xu, Jin Lin, and Zechun Hu,'' Photovoltaic and Solar Power Forecasting for Smart Grid Energy Management'', CSEE JOURNAL OF POWER AND ENERGY SYSTEMS, Vol. 1, NO. 4, December 2015

[5] P. Bacher, H. Madsen, and H. A. Nielsen, "Online short-term solar power forecasting," Solar Energy, Vol. 83, No. 10, pp. 1772–1783, Oct. 2009.

[6] A. Sfetsos and A. H. Coonick, "Univariate and multivariate forecasting of hourly solar radiation with artificial intelligence techniques,"Solar Energy, Vol. 68, No. 2, pp. 169–178, Feb. 2000

[7] R. Perez, E. Lorenz, S. Pelland, M. Beauharnois, G. Van Knowe, K. Hemker Jr,et al., "Comparison of numerical weather prediction solar irradiance forecasts in the US, Canada and Europe," Solar Energy,vol. 94, no. 0, pp. 305–326, Aug. 2013.

[8] E. Geraldi, F. Romano, and E. Ricciardelli, "An advanced model for the estimation of the surface solar irradiance under all atmospheric conditions using MSG/SEVIRI data,"IEEE Transactions on Geoscience and Remote Sensing, vol. 50, no. 8, pp. 2934–2953, Aug. 2012

[9] S. I. Nanou, A. G. Papakonstantinou, and S. A. Papathanassiou, "A generic model of two-stage grid-connected PV systems with primary frequency response and inertia emulation," Electric Power Systems Research,vol. 127, no. 0, pp. 186–196, Oct. 2015.

[10] R. Perez, S. Kivalov, J. Schlemmer, K. Hemker Jr, D. Renn´e, and T. E. Hoff, "Validation of short and medium term operational solar radiation forecasts in the US," Solar Energy, vol. 84, no. 12, pp. 2161–2172, Dec. 2010.

[11] F. Ueckerdt, R. Brecha, and G. Luderer, "Analyzing major challenges of wind and solar variability in power systems," Renewable Energy, vol. 81, no. 0, pp. 1–10, Sep. 2015.

[12] J-SR Jang. Anfiss: adaptive-network-based fuzzy inference system. IEEE transactions on systems, man, and cybernetics, 23(3):665-685, 1993.

[13] Samarjit Kar, Sujit Das, and Pijush Kanti Ghosh. Applications of neuro fuzzy systems: A brief review and future outline. Applied Soft Computing, 15:243-259, 2014.

[14] Massimo Panella. A hierarchical procedure for the synthesis of anfis networks. Advances in Fuzzy Systems, 2012:20, 2012.

[15] Hossein Ali Zamani, Shahin Rafiee-Taghanaki, Masoud Karimi, Milad Arabloo, and Abbas Dadashi. Implementing anfis for prediction of reservoir oil solution gas-oil ratio. Journal of Natural Gas Science and Engineering, 25:325-334, 2015.

[16] Michio Sugeno and Kazuo Tanaka. Successive identification of a fuzzy model and its applications to prediction of a complex system. Fuzzy sets and systems, 42(3):315-334, 1991.

EVALUATION OF SOLAR RADIATION POTENTIAL AT THE SHOOTING POINT BY IDENTIFYING THE SKY AREA OF SKY IMAGE USING U-NET

Akito Sago and Yuzuru Ueda
Tokyo University of Science
Niijuku, Tokyo, 168-0072, Japan, 4321528@ed.tus.ac.jp

ABSTRACT: In recent years, electric vehicles (EVs) have become increasingly popular because of becoming carbon neutral. The use of photovoltaics (PVs) in EVs is expected to increase their cruising range and enable them to use clean energy. However, it is difficult to estimate the amount of solar radiation reduced by the srounding obstructions and gain by reflective objects such as buildings, windows, etc. We use U-Net to identify areas in Fisheye image of the sky, estimate the amount of solar radiation based on the degree of shielding and reflection. As an accuracy verification, we use this model to compare the estimated amount of solar radiation for a day with the actual measured amount.
Keywords: Solar Radiation, Evaluation, PV System

1 INTRODUCTION

In 2020, countries all over the world have set a goal of virtually zero greenhouse gas emissions based on the Paris Agreement's goal of a decarbonized society. To achieve this goal, the transportation sector, which is heavily dependent on fossil fuels, is reviewed its energy and promoted the sale of EVs. With the promotion of EVs, attention has been focused on the advantages of increased cruising range and clean power generation by installing PVs. However, in residential areas, there are many obstacles such as buildings and roadside trees, and those affect the amount of solar irradiation by casting shadow and/or reflect the sunlight. Therefore, the amount of electricity generated differs greatly depending on the point and time of day, making it difficult to estimate the amount of solar radiation. This paper proposes a method to estimate the amount of solar radiation by automatically identifying obstacles and sky areas in the sky image at the shooting point.

2 METHOD

We identify obstacles from sky images taken in a residential area and estimate the amount of solar radiation at the shooting location based on the typical solar radiation data in METPV-11 [1].

2.1 Identifying sky area in sky images

We used semantic segmentation to identify obstacles and sky areas in the sky image. Semantic segmentation is a deep learning algorithm that recognizes the pixel attributes in the image from the surrounding pixels and labels them. In this study, we used U-Net [2] for image segmentation by transfer learning. We prepared a dataset of 333 sky images and manually labeled them, and pre-trained the dataset with U-Net. Figure 1(a) shows one of sky images in dataset. Figure 1(b) shows one of manually labeled sky images.

The sky image of a residential area in Tokyo is shown in Figure 2(a), and the segmentation result is shown in Figure 2(b). There are four semantic class: Building, Sky, Window, and Ground, because they have significant characteristics of reflection and shielding.

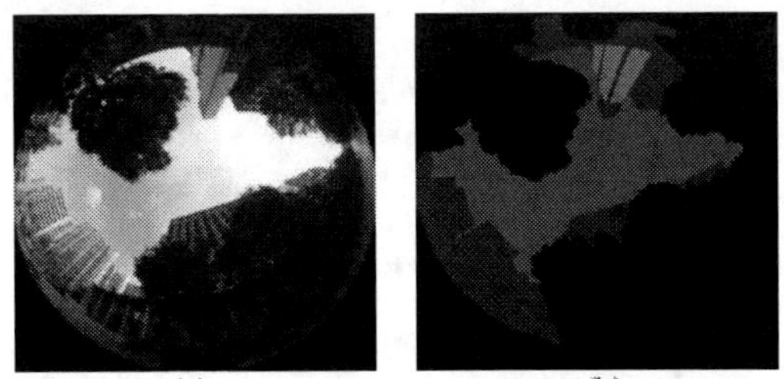

(a) (b)

Figure 1: Example of sky images from dataset. (a) a sky image we shot. (b) a manually labeled sky images.

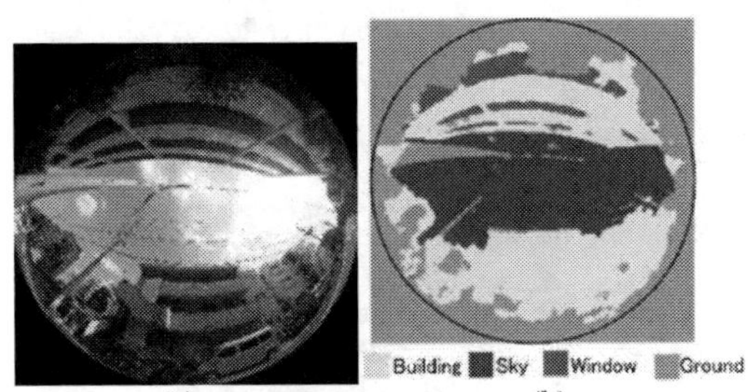

(a) (b)

Figure 2:(a) a sky image we shot. (b) a manually labeled sky image.

2.2 Calculation of solar orbit

The amount of solar radiation is affected by the condition whether the sun is hidden by buildings, etc. It is necessary to determine the position of the sun at each time. Equation (1) shows the solar altitude angle a. Equation (2) shows the solar zenith angle Ψ.

$$a = \arcsin(\sin\varphi \ \sin\delta + \cos\varphi \cos\delta \cos t) \quad (1)$$

$$\Psi = \arctan\{\sin\varphi \sin\delta \sin t / \sin\varphi \cos a - \sin\delta\} \quad (2)$$

Where δ is the solar latitude, φ is the latitude, and t is the hour angle.

2.3 Solar Radiation Correction

The global horizontal radiation, which represents the solar radiation obtained from the horizontal surface, is divided into two major components at the observation point: direct solar radiation that arrives directly from the sun, and diffuse solar radiation that is obtained through clouds and other atmospheric scattering. Since these two components are affected differently by obstacles, it is necessary to divide the measured global horizontal solar radiation into two components and take each into account. The procedure for correcting the solar radiation is as follows.

-Estimate direct solar radiation H_b and diffuse radiation H_d by using method of [3] of separating direct and diffuse insolation on global horizontal surface radiation H_g from METPV-11.
- Convert the sky image of the segmentation result obtained in Section 2.1 to the orthogonal image, calculate the sky view factor *SVF*, which is the ratio of the sky area excluding buildings, etc., and apply the sky coverage as a correction to H_d.
- From the sun's orbit obtained in Section 2.2, calculate the shielding factor *SF*, which indicates the degree to which the sun is obscured, and from the shielding and the position of the sun, calculate the reflection factor *RefFa*[4], which indicates the degree of reflection. Equation (3) shows the reflection factor *RefFa* and Figure (3) shows the reflection of the sky image.

$$RefFa = \sum_{i=1}^{N_{ray}} Hit_{o,i}\, Hit_{s,o,i}\, d\Omega_i \cos\alpha_i\, \rho_{o,i} \cos\alpha_{s,o,i}\, SVF \quad (3)$$

Where $Hit_{o,i}$ is the function of whether ray hits an obstacle [0 1], $Hit_{s,o,i}$ is the function whether the obstacle receives direct solar radiation [0 1], $d\Omega_i$ is solid angle of the ray, α_i is the angle between the ray and observation point, $\rho_{o,i}$ is Albedo value of the obstacle and $\alpha_{s,o,i}$ is the angle between the obstacle and the sun.

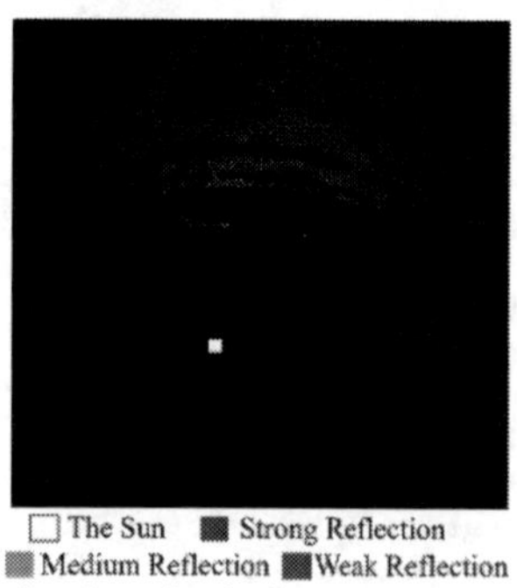

Figure 3: Reflection of a sky image at a certain time.

From the sum of the above corrected H_d, H_b, and the newly added reflected solar radiation H_r, we obtain the estimated solar radiation H_g' at the shooting point. Equaton (4) shows H_g' and Equation (5) shows H_r.

$$H_g' = H_b SF + H_d \mathrm{SVF} + H_r \quad (4)$$

$$H_r = H_b\, RefFa \quad (5)$$

3 DAILY RADIATION POTENTIAL EVALUATION

Figure 4 shows the sky image taken in September 15 with the accompanying solar orbit, and Figure 5 shows the comparison of the solar radiation before correction (black solid line) and after correction (thin bars) obtained from METPV-11 for one hour.

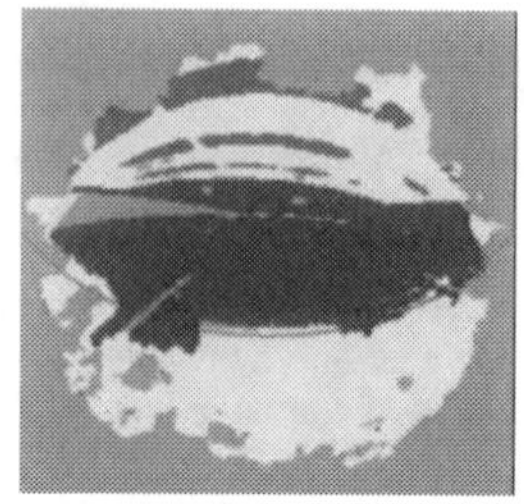

—the sun appearing, — the sun hidden

Figure 4: Segmentation with solar orbit

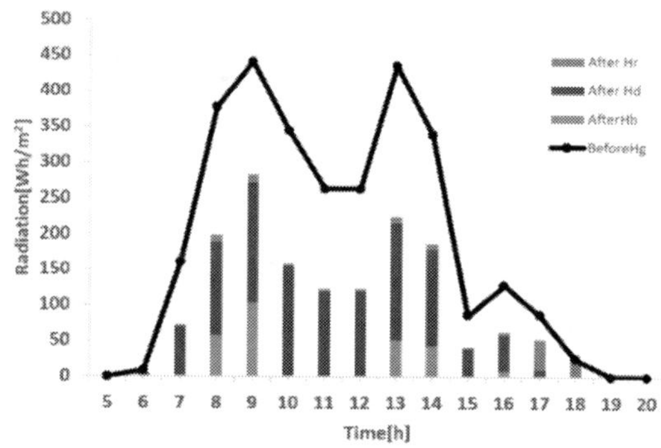

Figure 5: Segmentation with solar orbit

On September 15, the sun is hidden during the daytime hours and appears in the morning and evening. Figure 5 shows the hourly average solar radiation between 6:00 to 18:00, when the sun is out. During the daytime hours, when the sun is hidden (10:00~11:00), the direct solar radiation indicated by the orange bar cannot be obtained, and since the area is surrounded by buildings, the estimated solar radiation is low even during the hours when the sun is out. We can see that it is possible to estimate the amount of solar radiation for the shooting point from the segmentation.

4 ACCURACY VERIFICATION AND DISCUSSION

To check the accuracy of the solar radiation estimation, we measured the solar irradiation from the images taken on the rooftop of the research building at Tokyo University of Science. 30 minutes data was combined into one set, and the estimated values were compared with the measured values. In estimating the solar irradiation, corrections were made based on reference data of the measured values at the point with sun exposure. The sky image used is shown in Figure 6(a), and the calculated orbit of the sun is shown in Figure 6(b), which is the segmentation image. A comparison of the solar radiation intensity before correction, the estimated value after correction, and the measured value is shown in Figure 6.

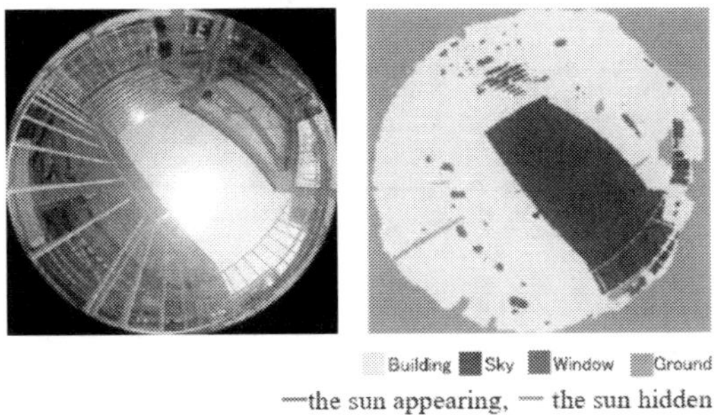

(a) (b)

Figure 6: Shooting point (a) The sky image. (b) Result of image segmentation.

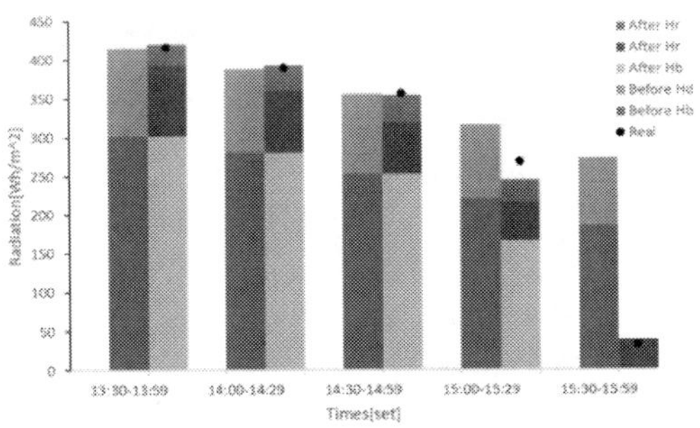

Figure 7: Comparison of segmentation

Table I: Integrated value of radation

	Radiation
Reference value [kW/m²]	1.75
Estimate value [kW/m²]	1.45
Measured value [kW/m²]	1.46

Table II: The error rates for each set

	Error Rates [%]
Set 1	1.7
Set 2	1.3
Set 3	0.1
Set 4	8.2
Set 5	25.4

$$Error\ Rate = \frac{|Reference\ Data - Estimate\ Value|}{Reference\ Data} \quad (6)$$

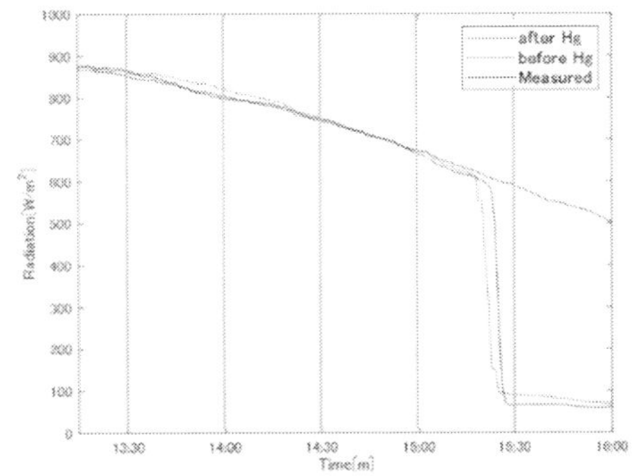

Figure 8: Comparison of solar radiation per minute.

The three integrated values are shown in Table I and error rates for each set are shown in Table II. Furthermore, the error rates are shown in Equation (6).

In sets 13:00~14:59, the sun was always out. In set 15:00~15:29, the sun was hidden in the middle of the day. In set 15:30~15:59, the sun was always hidden. Sets 13:00~14:59 could capture the increase in solar radiation due to reflections from buildings, so the error rate was exceedingly small, and the estimation was done with good accuracy. Set 15:00~15:29, on the other hand, had a larger error rate than the others, and its accuracy was slightly worse. This may be because the time when the sun hides the buildings was different by minutes. Figure 8 shows deviation of sunset time.

However, Table I shows that the estimation method is useful because it is close to the measurement based on the integrated values of solar radiation before correction.

5 CONCLUSION

In this paper, we were able to estimate the solar irradiance at the shooting point by identifying obstacles and sky from the sky image, combining it with the solar orbit, and applying the direct and scattered irradiance corrections estimated from the horizontal all-sky irradiance. Since the reflected solar radiation was assumed to be completely scattered, we plan to propose a better model that takes into account the characteristics of reflection by buildings and other objects in detail as a future task.

REFERENCE

[1] NEDO, Database on solar radiation, Web:https://www.nedo.go.jp/library/nissharyou.html

[2] Philipp Fischer, Thomas Brox. U-Net: Convolutional Networks for Biomedical Image Segmentation, MICCAI, Springer, LNCS, Vol.9351: 234-241, (2015)

[3] the Solar Energy Society of Japan, New Solar Energy Utilization Handbook, p13,p25-27, 2020

[4] NREL, Solar Radiation Reflected from Exterior Surfaces, Web:https://www.energyplus.net/sites/default/files/docs/site_v8.3.0/EngineeringReference/06-SolarRadiation/index. html

IMPROVED MODELS FOR NOWCASTING SUNSHINE NUMBER

Marius Paulescu[1], Nicoleta Stefu[1], Ciprian Dughir[2], Andreea Sabadus[3]
[1]Faculty of Physics, West University of Timisoara, V. Parvan 4, 300223 Timisoara, Romania
[2]Electronics, Telecommunications and Information Technology Faculty, Politehnica University of Timisoara, V Parvan 2, 300223 Timisoara, Romania
[3]Institute for Advanced Environmental Research, West University of Timisoara, V Parvan 4, 300223 Timisoara, Romania
*Corresponding author: marius.paulescu@e-uvt.ro

ABSTRACT: This study deals with the performance in intra-hour forecasting of the relative position of the Sun and clouds. Sunshine number SSN, a binary quantifier showing if the Sun shines or not, is used as a measure for the Sun position with respect to clouds. The quality of SSN forecasts issued by statistical, (ARIMA modeling of various radiometric time-series) and physics-based models (sky-imagery analysis) is assessed. Irrespective the forecasting procedure, the overall results show a low percentage of SSN forecasts misclassified. Total misclassification increases with increasing the forecast horizon and with the increase in variability of the state-of-the-sky. At the transition from a state to another, the different statistical models experience a different inertia. The level of accuracy in the SSN forecasting by sky-imagery is strongly dependent on the procedure applied for cloud displacement estimation.

Keywords: solar irradiance; clouds; forecasting

1 INTRODUCTION

The smart-grid concept is more and more present in the management of power grids with high penetration of the photovoltaic (PV) sources. A key task for providing smartness to the grid is the accurate forecasting of PV power output [1]. A smart-grid management requires intra-hour forecasts of PV power, the technical constraints being related to the start-up time of conventional power sources connected as balancing instruments.

The performance of a PV plant is intimately related to the presence of the in-plane solar irradiance. Consequently, the relative position of the Sun and clouds becomes decisive for accurate nowcasting of solar irradiance and PV power.

This study aims an increasing in the accuracy of intra-hour PV power forecasting. Specifically, the study is focused on nowcasting the sunshine number SSN [2], a binary quantifier for the relative position of the Sun and clouds, showing whether the Sun shines or not. Accurate forecast of SSN ensures high-performance in PV power forecasting [3]. Three procedures for nowcasting SSN are evaluated: (1) Statistical extrapolation of the SSN time series. (2) Nowcasting the solar irradiance components and subsequent use of the WMO sunshine criterion for SSN. (3) Nowcasting SSN based on direct observation of the sky using a sky imager. The models are briefly described in Sec. 2 while their specific implementation is acknowledged in Sec. 4.1, followed by the presentation of the test results (Sec 4.2).

2 FORECASTING MODELS

2.1 Predictor variables

The proposed statistical forecasting models operate on the SSN time series and the clearness index k_t time series. The two physical quantities are defined next:

SSN is defined as a random binary variable [2]:

$$SSN_t = \begin{cases} 0 & \text{if the Sun is covered by clouds at time } t \\ 1 & \text{otherwise} \end{cases} \quad (1)$$

A series of *SSN* values can be derived from radiometric measurements on basis of the WMO sunshine criterion [4]: the Sun is shining at time t if the direct-normal solar irradiance $G_{dn,t}$ exceeds 120 W/m^2, i.e.:

$$SSN_t = \begin{cases} 1 & \text{if} \quad G_{dn,t} > 120\ \text{W/m}^2 \\ 0 & \text{otherwise} \end{cases} \quad (2)$$

The average value of SSN over a given period Δt equals the relative sunshine over Δt.

The atmospheric clearness index [5] is defined as the ratio of the global solar irradiance G_t measured at ground level and its counterpart $G_{ext,t}$ estimated at the top of the atmosphere:

$$k_t = G_t / G_{ext,t} \quad (3)$$

The clearness index isolates the stochastic component in a solar irradiance time series. The traditional approach in nowcasting solar irradiance is based on the extrapolation of the clearness index time series (e.g. [6]).

2.2 Forecasting models

Two different in nature approach for nowcasting SSN are assessed: (1) A statistical one considering the class of Autoregressive Integrated Moving Average ARIMA models [7], implemented in three variants differentiated by the time series on which the model operates and (2) A physics-based one, processing information about the state-of-the-sky captured by an all-sky imager [8]. On short, the following forecasting models were tested:

PS - ARIMA modeling of the SSN time series, which can be viewed as a reference approach.

PC - ARIMA modeling of the diffuse and global clearness index time series followed by the evaluation of SSN using the WMO sunshine criterion.

PN - ARIMA modeling of direct-normal clearness index time series followed by the SSN evaluation

PA - All-sky imager-based SSN forecasting. On brief, this approach consists of the following steps: defining a polar grid attached to the Sun, cloud fraction evaluation on every element of the grid, applying the cross-correlation analysis for cloud displacement vector, cloud fraction estimation on the Sun position at the forecasting horizon and issuing the SSN forecast [9].

Figure 1: Photo of the experimental setup FORPV.

3 DATABASES

This study was conducted with data collected on the Solar Platform of the West University of Timisoara (45°46'N, 21°25'E, 85m asl) [10]. It includes the radiometric station (class A pyranometers ISO 9060:2018), a weather station and four PV experimental setups. The station is fully monitored. All electrical, radiometric, and meteorological sensors are read simultaneously at equal time intervals of 4 seconds.

Data collected from the experimental setup FORPV (Fig. 1) are used in this study. FORPV consists of a 540 Wp PV system, feeding a resistive constant load, pyranometers for global, diffuse and in-plane solar irradiances and a Delta-T Devices WS-GP2 weather station. An all-sky imager EKO Instruments ASI-16 monitors the state-of-the-sky at 1-minute sampling.

Several distinct sets of data were used to build and test the forecasting models. The first dataset, FIT, contains global, diffuse and direct-normal solar irradiance data series. It was developed for fitting the ARIMA models. The second dataset, TEST, was developed aiming to verify the SSN forecasting models. The tests were performed against data collected in four days (25 to 28 June 2022) characterized by a moderate to high variability in the state-of-the-sky. Figure 2 shows the variation of global solar irradiance and sunshine number recorded during these days. The third dataset, SKY, consists of all-sky images recorded during the test days. It includes the raw fisheye images and the processed images with the FindClouds Trinity software provided by the ASI-16 manufacturer (Fig. 3). Elimination of objects on the edges of the view horizon, projection from the hemisphere to a 2D disk ("defishing"), converting the image into binary values for each pixel, etc., are all performed during the processing stage.

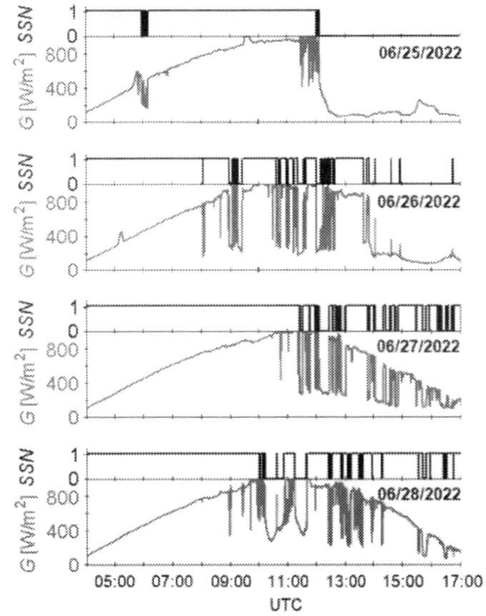

Figure 2: Time variation of global solar irradiance G and sunshine number SSN during the test days.

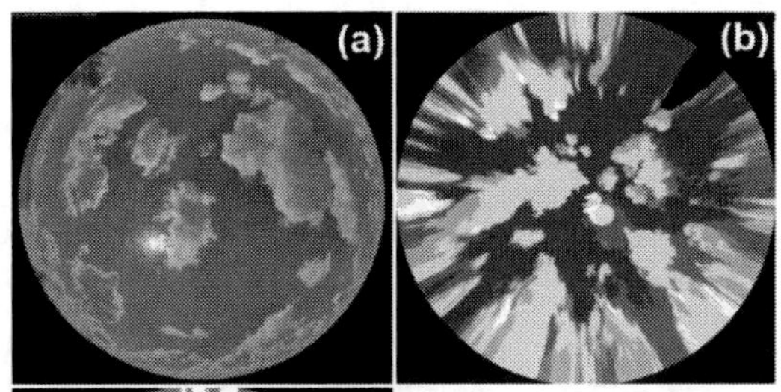

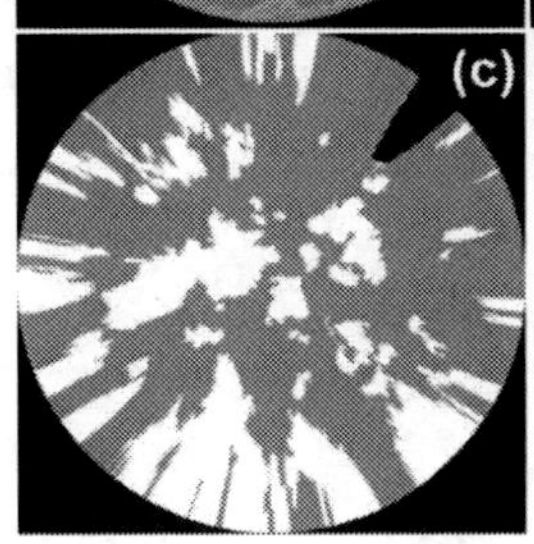

Figure 3. Original and processed images of the sky: (a) Fisheye picture taken by ASI-16 on 06/26/2022 at 10:59 UTC; (b) The defished, mirrored image with cloud detected. The Sun position (yellow disk) is superimposed; (c) The binary image.

4 RESULTS AND DISCUSSIONS

4.1 Models implementation

The ARIMA models were selected on the basis of parsimony principle, i.e. keeping a minimum number of coefficients (see e.g. [3] for details). The algorithm for SSN forecasting based on sky-imagery follows up-to a point, the procedure introduced in [8]. The algorithm is described in [9].

4.2 Models assessment

The performance in forecasting a binary variable like SSN can be properly investigated as a classification problem. The results are organized as Table 1, where p_{00}, p_{01}, p_{10}, p_{11} represent the probabilities of occurrence of the corresponding events.

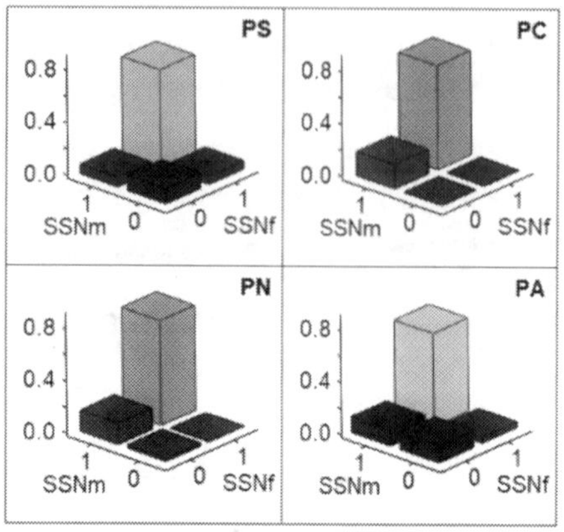

Figure 4: Misclassification of the SSN forecasts in all the test days. Forecast horizon = 5 minutes.

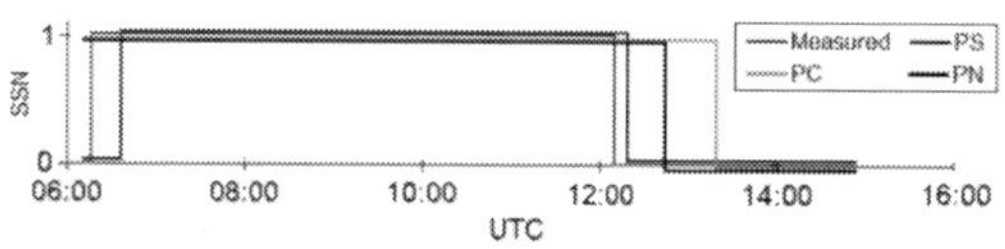

Figure 5: Time variation of measured and forecasted SSN on 25/06/2022.

From these estimates, the total misclassification $TM = p01 + p10$, an attribute relevant for the model quality, can be computed.

Figure 4 shows the misclassification of the SSN forecasts at a forecast horizon FH = 5 min, considering data recorded in all the test days. At a first sight all the models demonstrate a low percentage of misclassified forecasts. The PS model achieves the best performance (TM = 0.136) followed by PA (TM = 0.168). Definitely, the total misclassification increases with increasing the forecast horizon (Table 2). Likewise, the increase in the variability of the state-of-the-sky leads to the increase of misclassification. This is visible in Table 3 which presents the total misclassification of the SSN forecasts broken down by the test days. Looking at Fig. 2 it can be seen that the lowest variability in the state-of-the-sky was on 06/25 (the lowest TM = 0.067 was achieved by PA) while the largest variability was on 06/26 (the lowest TM = 0.300 was achieved by PS).

The tested statistical models are mostly differentiated by the persistence characteristic, i.e. the tendency of a model to preserve the current state in the future. This represents a major limitation of the accuracy in nowcasting PV power. Persistence characteristics of the tested models are illustrated in Fig. 5, which displays the variation in time of the measured and forecasted SSN values on 25/06/2022. The transition from SSN = 1 to SSN = 0 is performed by the statistical models PS, PC, and PN with different delays. Extrapolation in the future of the SSN time series (PS) shows the lowest inertia. Although many efforts were devoted for increasing the accuracy of nowcasting SSN (e.g. [11]), further research are required for mitigation the effect of persistence in SSN prediction. A substantial advance in the fight against the persistence effects may be obtained by looking deeper to the solar radiative transfer and bringing into the statistical forecasting algorithms more insight from clouds dynamics.

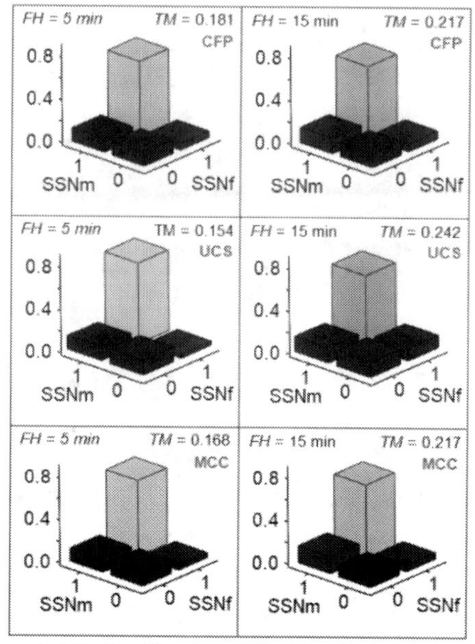

Figure 6: Misclassification of the SSN forecasts in all the test days. Two forecast horizons FH are considered: 5 and 15 minutes. The PA procedure is applied with different assumption for cloud displacement: CFP - Cloud field persistence, UCS - Uniform cloud speed and MCC - Maximum cross-correlation analysis

Table 1. Probabilities of different combinations of true and modeled outcomes

		Model-based classification	
		SSN = 1	SSN = 0
Measurement-based classification (truth)	SSN = 1	$p11$	$p10$
	SSN = 0	$p01$	$p00$

Table 2: Total misclassification of the SSN forecasts for all test days as function of the forecast horizon FH.

FH	5 min	15 min
PS	0.136	0.238
PC	0.193	0.238
PN	0.171	0.214
PA	0.168	0.217

Table 3: Total misclassification of the SSN forecasts for the test days at FH = 15 min.

Day	06/25	06/26	06/27	06/28
PS	0.100	0.300	0.233	0.200
PC	0.266	0.333	0.200	0.166
PN	0.133	0.333	0.200	0.166
PA	0.067	0.367	0.233	0.200

All-sky imagery analysis is a good way to reduce persistence. Our preliminary results shows that different procedures for cloud displacement estimation led to different levels of accuracy in the SSN forecasting. This is illustrated in Fig. 6, which shows the misclassification of the SSN forecasts in all the test days, at two forecast horizons. The PA procedure was applied with different assumption for cloud displacement: CFP - Cloud field persistence, UCS - Uniform cloud speed and MCC - Maximum cross-correlation analysis. For a forecast horizon $FH = 5$ min the lowest value of the total misclassification ($TM = 0.154$) was achieved by the UCS procedure while for $FH = 15$ min the lowest value of the total misclassification ($TM = 0.217$) was achieved by MCC.

5 CONCLUSIONS

This study evaluated the performance in intra-hour forecasting of the relative position of the Sun and clouds. The focus was on nowcasting sunshine number at two forecast horizons: 5 and 15 minutes. Statistical and physics-based models were tested. None of the SSN forecasting models can be ranked the best. At forecast horizons of minutes, the statistical models perform with reasonable accuracy. At larger forecast horizons the all-sky imager-based model becomes high-competitive. Although many efforts were devoted to increase the accuracy in nowcasting sunshine number, there is still room enough for mitigation the effect of persistence.

The accurate forecast of sunshine number is a key task for high-quality nowcasting PV power. For example, the PV2-state model [3], based on the sunshine number forecasts, proved a high-performance in nowcasting PV power. When sunshine number is correctly predicted, the PV power forecasts trace with accuracy the measurements. Forecasting SSN based on all-sky imagery analysis is expected to generate a strong increase in overall performance of the two-state model, aiming beyond the state-of-the-art.

6 ACKNOWLEDGEMENTS

The authors thank Dr. Robert Blaga for support and the useful discussions. This work was supported by a grant of the Romanian Ministry of Education and Research, CCCDI-UEFISCDI, Project No. PN-III-P2-2.1-PED-2019-3942.

7 REFERENCES

[1] C. Wan, J. Zhao, Y. Song, Z. Xu, J. Lin, Z. Hu. CSEE J. Power Energy Syst. 1(4) (2015) 38.
[2] V. Badescu, M. Paulescu. Atmos. Res. 101 (2011) 194.
[3] M. Paulescu, N. Stefu, C. Dughir, A. Sabadus, D. Calinoiu, V. Badescu, Renew. Energy 195 (2022) 322.
[4] World Meteorological Organization WMO-No.8/2008 https://community.wmo.int/activity-areas/imop/wmo-no_8, 2022 (accessed August 2022).
[5] B.Y.H. Liu, R.C. Jordan. Sol. Energy 4 (1960) 1.
[6] M. Paulescu, E. Paulescu. Renew. Energy 143 (2019) 985.
[7] G.E.P. Box, G.M. Jenkins, Time series analysis. Forecasting and Control. Holden-Day (1970).
[8] S. Quesada-Ruiz, Y. Chu, J. Tovar-Pescador, H.T.C. Pedro, C.F.M. Coimbra. Sol. Energy102 (2014) 267.
[9] Blaga R, Dughir C. Sunshine number nowcasting based on all-sky images (2022) To be published by the AIP Conference Proceedings.
[10] Solar Platform of the West University of Timisoara, Romania. http://solar.physics.uvt.ro/srms/ (accessed August 2022).
[11] M. Brabec, M. Paulescu, V. Badescu. Sol. Energy 101 (2014) 272.

A COMPUTATIONALLY LIGHT DATA-DRIVEN ALTERNATIVE TO CLOUD-MOTION PREDICTION FOR PV FORECASTING

Rafael E. Carrillo[1], Pierre-Jean Alet[1], Stephan Müller[2] and Jan Remund[2]
[1]CSEM, Sustainable Energy Center, Rue Jacquet-Droz 1, 2002 Neuchâtel, Switzerland
Emails: rafael.carrillo@csem.ch, pierre-jean.alet@csem.ch
[2]Meteotest, Fabriktrasse 14, 3012 Bern, Switzerland
Emails: stefan.muller@meteotest.ch, jan.remund@meteotest.ch

ABSTRACT: In this paper we compare a state-of-the-art cloud-motion tracking solution against a data-driven solution for photovoltaic (PV) power production to provide insights into their performance drivers. The solutions are MeteoTest's CloudMove, based on cloud-motion tracking from satellite images, and CSEM's data-driven solution based on graph machine learning forecasting models that use only past PV production data from a network of PV systems. The study compared the two approaches in terms of accuracy in different scenarios for a forecasting horizon of up to six hours ahead, with a resolution of 15 minutes, on a dataset of 21 days and 18 locations in Switzerland. The days and locations were selected as a representative sample of the whole range of different conditions in terms of seasons, weather conditions, terrain, and distance to other instrumented sites. Over the whole benchmarking set, CSEM's data-driven technique yields a lower normalized root-mean-square error than the cloud-motion tracking method for forecasting horizons above one hour, though the error spread is larger with CSEM's data-driven solution. Regarding computational load, the data-driven methods can accelerate the computation of forecasts by a factor 100, after an initial training, thus offering a viable alternative to satellite-based cloud-tracking methods.
Keywords: Solar radiation, forecasting, photovoltaic, machine learning.

1 INTRODUCTION

Operating power systems with large amounts of renewables requires predicting photovoltaic (PV) power production with fine temporal and spatial resolution [1]. Since the high variability of solar irradiance mainly comes from cloud movement and its stochastic blocking of sunlight, state-of-the-art methods have focused on predicting cloud dynamics using ground-based cameras, satellite images, and numerical weather predictions (NWP). The main problem with these approaches is the low resolution (temporal or/and spatial) of the input data and thus of the resulting forecasts. Precise numerical weather forecasts, based on satellite images and weather models, provide high spatial and temporal resolution but at the expense of a high computational processing load and storage.

One of the services that Meteotest offers is CloudMove [2], a nowcasting service for irradiance and PV power production. CloudMove uses weather models and cloud positions from satellite images to propagate the cloud's movement in the future to forecast the radiation for up to six hours ahead. Additionally online ground data (solar radiation or PV production) is used to correct the forecasts. CloudMove yields state-of-the-art accuracy for predictions in the six hours ahead horizon.

Recent approaches have proposed data-driven forecast methods that use multi-site spatio-temporal historical data for PV forecasting without requiring NWP or cloud movement information [3]. The premise is that using measurements from ground sensing networks increase the accuracy and resolution of short-term forecasts.

CSEM has recently developed graph machine learning methods for multi-site PV forecasting that rely entirely on production data and exploit the intuition that PV systems indeed provide a dense network of simple weather stations [4], [5]. The underlying idea is that temporal correlations between the outputs of PV systems are linked with their spatial relationship, thus providing enough information to accurately forecast power production (or irradiance) at localized stations. These methods can accelerate the computation of forecasts by a factor 100, after an initial training, while maintaining the high temporal and spatial resolution of short-term forecasts.

In this paper, we compare CSEM's data driven solution against CloudMove in different scenarios, for a forecast horizon of up to six hours. The analysis includes comparisons in different seasons and weather conditions. A representative set of 18 locations and 21 days were selected to cover the whole range of possible conditions in Switzerland, in terms of weather, terrain, and distance to other instrumented sites, to evaluate the accuracy of both approaches. By comparing the two state-of-the-art methods in identical conditions, this study provides insights into their respective performance drivers and limitations.

2 METHODOLOGY

In this section we briefly present the two methods used in the study: Meteotest's CloudMove forecasting solution and CSEM's graph-based forecasting model. We also present the dataset and the scenarios considered in the study.

2.1 CloudMove

CloudMove is the state-of-the-art solution from Meteotest for short-term solar irradiance forecasting [2]. CloudMove predicts the cloud movement based on weather models and cloud positions from satellite images. The most recent clouds (clear sky index) retrieved from satellite images are propagated based on wind vectors from weather models. Additionally, online local ground data (solar radiation or PV production) is used to correct the forecasts.

CloudMove produces forecasts for the next 6 hours with a time resolution of 15 minutes and is updated every 15 minutes. Computational time is in the order of seven minutes for the whole area once the satellite image is available. The system can forecast the radiation for any site in Europe, the Middle East, Africa, India, North and South America.

2.2 Graph-based forecasting model

CSEM's data-driven solution is a multi-site PV forecasting method which relies entirely on production data and exploits the intuition that PV systems can be used as a network of virtual weather stations. The solution is based on graph neural networks (GNN), which can model jointly the spatial and temporal relations of PV production data. The GNN model used in this paper is the Graph-Convolutional Long Short Term Memory (GCLSTM) model [5]. The LSTM is a recurrent neural network used to model the temporal correlations of the signal while the graph convolutional layers model the spatial correlations. The model is based on an encoder-decoder LSTM architecture, see Figure 1. The encoder GCLSTM is used to encode the past data (similar to a Kalman filter) and the decoder uses this encoded information to make the prediction over the entire horizon. Additionally, the decoder also uses the computed clear sky irradiance for all locations in the forecasting horizon (six hours ahead) as input. The GCLSTM model is trained using a year of representative data. Once the model is trained, it uses observed PV production data from the past 3 hours to forecast the next 6 hours with a resolution of 15 minutes.

At evaluation time, CSEM's data-driven solution generates the forecasts about 100 times as fast as the method based on satellite imaging though its computational complexity depends on the number of nodes instead on the area covered. Its main computational burden is in the training phase, that can be done a few times per year offline, thus alleviating the computational load in real-time forecasts. Also, the amount of data that the proposed solution needs to store to produce the forecasts is minimal, as it just needs to store the learned machine learning model.

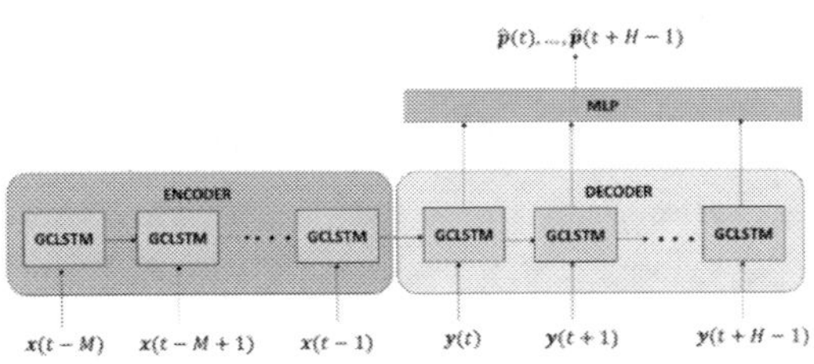

Figure 1. Encoder-decoder architecture of the graph-based forecasting model.

2.3 Evaluation set-up

A representative set of 18 locations and 21 days were selected to cover the range of possible conditions in Switzerland (weather, terrain, distance to other instrumented sites). The study includes comparisons in different seasons with at least five representative weather days per season. The data-driven model uses data from 304 PV stations scattered over Switzerland for both training and evaluation. Figure 2 shows the locations of the stations used for the evaluation as well as the stations used as input for the data-driven model. One additional criterion to select the sites is the density of PV stations around selected sites. The training was performed using data from the whole 2016 and the evaluation was performed in 2017 on the selected 21 days. The metric used in the evaluation is the normalized root-mean-square error (NRMSE), normalized by the peak production value over the year. Nighttime values were excluded from the error computations.

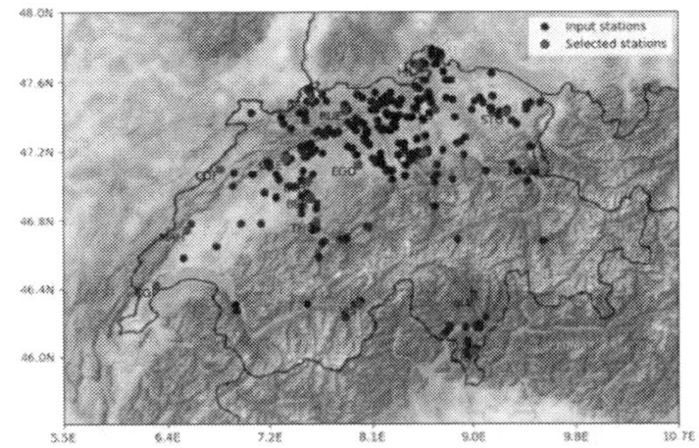

Figure 2. Locations of the selected stations for the evaluation (in red) and locations of the stations used as input data to the GCLSTM model (in blue).

3 RESULTS

3.2 Overall results

We start by comparing the accuracy of the two methods over the whole dataset, i.e., all locations and days. The NRMSE of a smart persistence model was also included as benchmark. In all figures, solid lines show the median error over sites and the shaded areas show the interquartile distance. Figure 3 shows the NRMSE evolution over the forecasting horizon of 6 hours ahead with 15 minutes steps. Over the whole set, CSEM's data-driven technique yields a lower NRMSE than the cloud-motion tracking method for forecasting horizons above one hour. The error spread is larger with GCLSTM than with CloudMove, with some peripheral, lakeside locations the most challenging to predict. In contrast, CloudMove errors spreads almost uniform over the whole country, except for some Alpine sites where precise local data is not available to correct the forecasts. The advantage of CSEM's data-driven technique becomes clearer for forecasting horizons above three hours, when its upper quartile of error values gets lower than the lower quartile with the cloud-motion tracking.

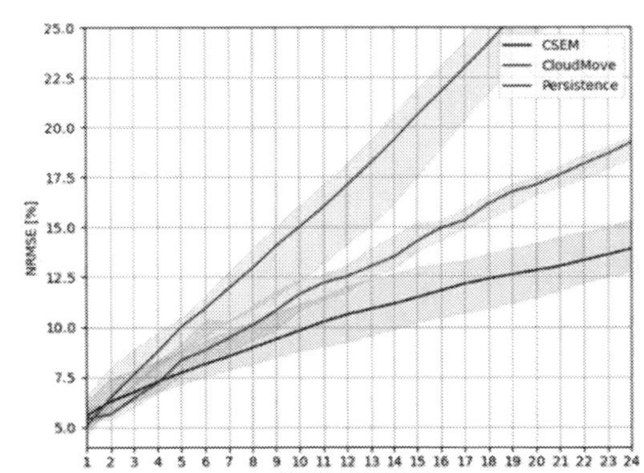

Figure 3. NRMSE evolution over the forecasting horizon of 6 hours ahead in steps of 15 minutes.

The next set of results show the dependency of the different methods' accuracy on the time of the day, both at the time when the forecast is computed (prediction time) and the desired time of the forecast (target prediction time). Figure 5 and Figure 6 show the NRMSE per prediction time and target prediction time, respectively, of the day (hourly) for the three methods. With all considered

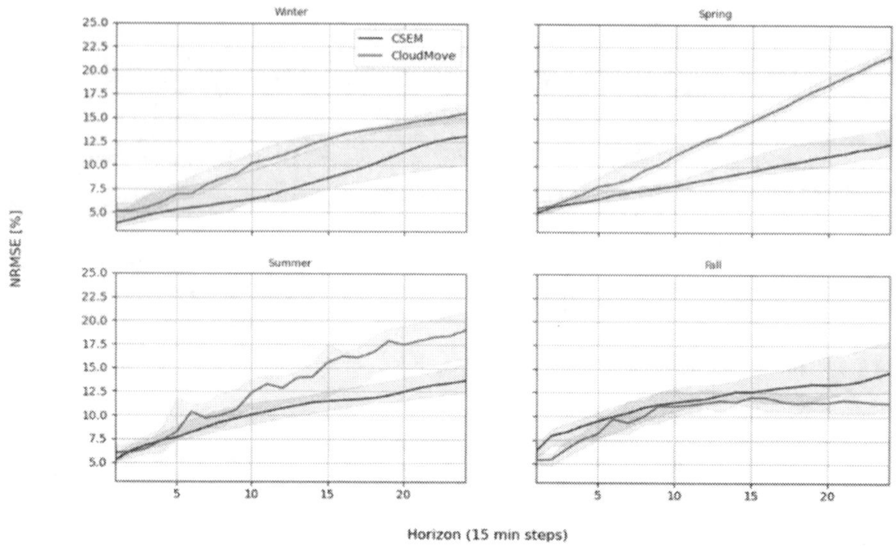

Figure 4. NRMSE evolution over the six hours ahead forecasting horizon for the different seasons of the year.

forecasting techniques, the largest errors are obtained for forecasts calculated in the early morning hours, when there is still scarce information on the day's weather. In terms of target prediction time, the peak error corresponds to the solar noon. It is sharpest with the persistence forecasting method, and flattest with CSEM's technique.

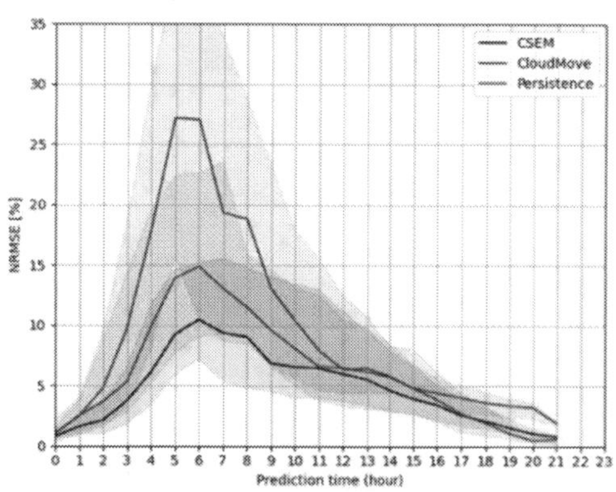

Figure 5. NRMSE per prediction time (hourly) of the day.

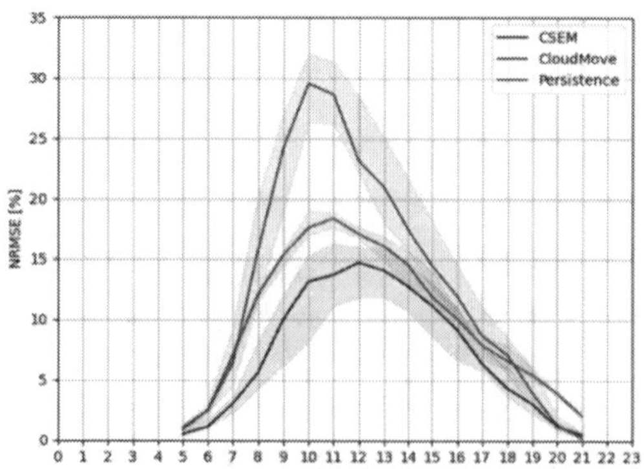

Figure 6. NRMSE per target prediction time (hourly) of the day.

3.3 Seasonal dependency

Next, we present results on how dependent the forecast's accuracy is on the different seasons. Figure 4 shows the error evolution over the forecasting horizon for all seasons. CloudMove shows larger errors in winter and summer than CSEM's GCLSTM. Particularly, CloudMove shows a significantly larger slope in the error evolution in spring (see next paragraph for an explanation). GCLSTM yields larger errors in fall. Also, GCLSTM yields a larger error spread in general, except in summer.

Figure 7 shows the NRMSE vs prediction time of the day (hourly) for all seasons. CloudMove yields larger errors for winter and summer and significantly larger errors in spring. Specially, CloudMove has large errors (above 15%) for forecast computed in the early morning hours during spring. This explains the large errors of CloudMove for spring in general. Also, CloudMove has a larger error spread in winter. In summer, GCLSTM yields a large error spread around the solar noon due to the large peak production in the season.

4 CONCLUSSIONS

We have presented the results of the comparison of a cloud-motion tracking method, Meteotest's CloudMove, and a data-driven approach based on past PV production data, CSEM's GCLSTM. The study shed light on the different factors that drive the performance of the two methods. Over the whole benchmarking set, CSEM's data-driven technique yields a lower NRMSE than CloudMove for forecasting horizons above one hour, with a similar error for the first hour.

The error spread is larger with CSEM's data-driven solution with some border and lakeside locations the most challenging to forecast. The error is low for sites with a high density of PV stations but high for sites with low density and at the border of the graph. CloudMove achieves a lower error in fall but CSEM's solution yields lower errors (and spread) during summer when the production is the highest. Also, for both methods, forecasts computed in early morning hours have the largest errors due to the scarcity of weather's information at that time.

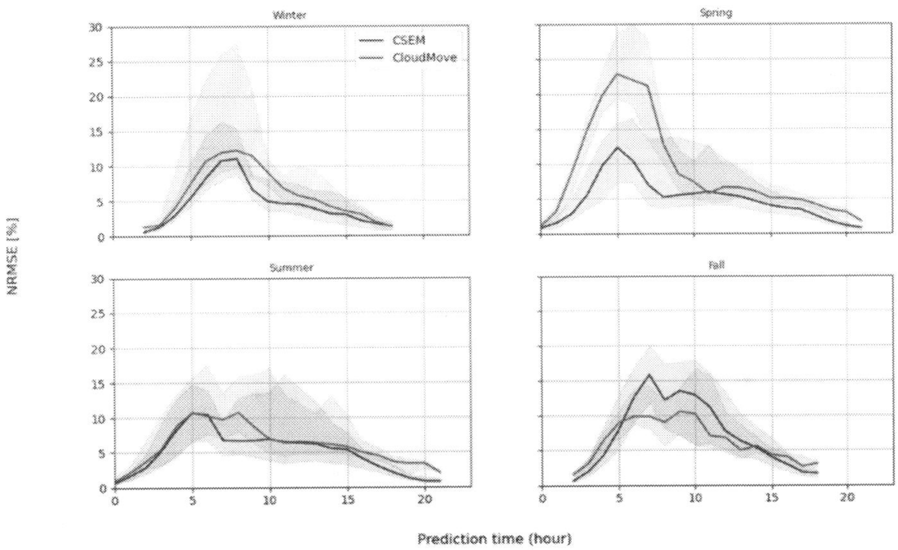

Figure 7. NRMSE per prediction time of the day (hourly) for all seasons.

5 ACKNOWLEDGMENTS

This research was co-financed by Innosuisse – Swiss Innovation Agency – project 57765.1 IP-EE: DIGERATI.

6 REFERENCES

[1] P.-J. Alet *et al.*, 'Forecasting and Observability: Critical Technologies for System Operations with High PV Penetration', EUPVSEC, 2016.

[2] S. C. Müller and J. Remund, 'Satellite Based Shortest Term Solar Energy Forecast System for Entire Europe for the Next Hours', EUPVSEC, Amsterdam, Sep. 2014.

[3] L. Benavides Cesar, R. Amaro e Silva, M. Á. Manso Callejo, and C.-I. Cira, 'Review on Spatio-Temporal Solar Forecasting Methods Driven by In Situ Measurements or Their Combination with Satellite and Numerical Weather Prediction (NWP) Estimates', *Energies*, vol. 15, no. 12, Art. no. 12, Jan. 2022.

[4] R. E. Carrillo, M. Leblanc, B. Schubnel, R. Langou, C. Topfel, and P.-J. Alet, 'High-Resolution PV Forecasting from Imperfect Data: A Graph-Based Solution', *Energies*, vol. 13, no. 21, Art. no. 21, Nov. 2020.

[5] J. Simeunović, B. Schubnel, P.-J. Alet, and R. E. Carrillo, 'Spatio-Temporal Graph Neural Networks for Multi-Site PV Power Forecasting', *IEEE Transactions on Sustainable Energy*, vol. 13, no. 2, pp. 1210–1220, 2022.

PV VERY SHORT-TERM POWER FORECASTING METHOD BASED ON CLOUD MONITORING DATA

João Esteves[1], Rui Pestana[2], Yang Cao[1], Nuno Pinho da Silva[1], Zheng Wang[3]

[1]Centro de Investigação em Energia REN-State Grid, S.A., Rua Cidade de Goa, nº4-B, 2685-038 Sacavém, Portugal, +351 210011387, joao.esteves@rdnester.com
[2]Rede Elétrica Nacional, Av. Estados Unidos da América, 55, 1749-061 Lisboa
[3]EPRI – Renewable Energy Research Center Department, Beijing, China

ABSTRACT: In the wake of climate change, more attention is being focused on green energy production. Renewable energy resources play an important role in the goal of reducing greenhouse gas emissions as they allow the reduction of fossil fuel use. One method to produce energy in a sustainable way is to make use of photovoltaic (PV) panels, which convert the solar energy directly into electricity. Even though the deployment of solar energy production contributes to the decarbonisation of the energy system, it also has its obstacles. The power that can be generated by PV units is proportional to the amount of irradiation reaching the PV panels, which introduces high levels of uncertainty and volatility. In this work, a PV power forecast method based on cloud monitoring data is presented to tackle the very short-term forecast horizon (up to two hours-ahead). The developed methodology uses an all-sky imager (ASI) installed in a solar PV power plant located in the South of Portugal.

Keywords: solar forecast, all-sky imager, cloudiness, image processing

1 INTRODUCTION

In recent years, the solar power installed capacity increased significantly all over the world. This trend is going to continue and to be intensified in the next decades in the scope of achieving a clean and green energy transition. In Fig. 1, it is shown the evolution of the solar power in Portugal.

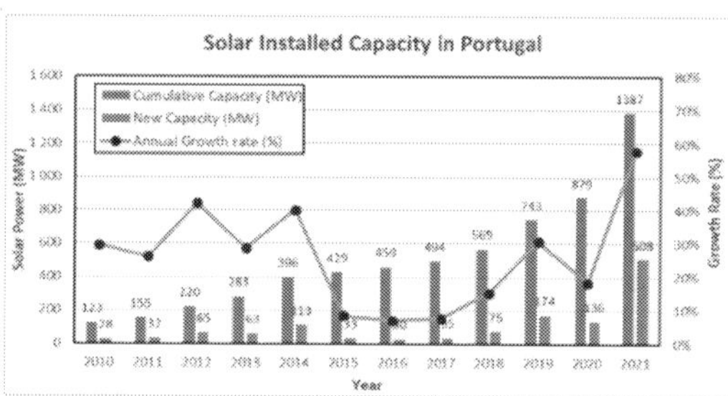

Figure 1: Solar Power in Portugal [1]

Given the volatility and uncertainty of the solar renewable energy source (RES), there is a need for a continuous improvement of forecast tools that could tackle different time horizons. On the one hand, from a network operator perspective, the priority is given to larger forecast horizons (from day-ahead forecasts to weeks-ahead forecasts). On the other hand, from a solar power plant owner or aggregator perspective, the priority is given to shorter forecast horizons such as day-ahead and even hours-ahead. This is especially important in order to allow a more efficient and effective participation in the day-ahead and intraday markets.

One of the factors that contributes to the above mentioned is the presence of clouds in the sky, which affects the available ground irradiation level. This leads to variations in the output power of solar PV systems. Therefore, forecast tools are being developed and used in a wide range of applications. Numeric Weather Prediction (NWP) models are being used, especially by utilities, to have an overview of solar generation in a range of several days ahead. When aiming for shorter forecast horizons, different approaches are used such as persistence methods, physical models, satellite images or ground station data.

In this work, a PV power forecast method based on cloud monitoring data is presented to tackle the very short-term forecast horizon (up to two hours-ahead). The developed methodology uses an all-sky imager (ASI) installed in a solar PV power plant located in the South of Portugal.

In a first stage, the sky images are processed and analyzed in order to have the computational ability to distinguish cloud from sky pixels. It is also used a developed index to distinguish thin clouds from thick clouds. Thus, providing a better understanding of the solar irradiation being blocked and not reaching the ground. These images are used together with a satellite image of the PV solar power plant.

In the second stage, based on the Sun position (using the zenith angle), the clouds shadow projection in the ground is estimated. Thus, providing an estimated area of the solar PV power plant panels covered by the shadows (coverage). The solar power forecast is calculated using a theoretical clear sky radiation formulation (latter converted to power) and the estimated PV power plant panels coverage.

In the final stage, using NWP forecasts (wind speed, wind direction and cloudiness at different atmosphere levels) the cloud shadow movement is forecasted up to 2 hours-ahead. Thus, for each available image (in this work it was used a 15 minutes time granularity) the power forecasts for the next two hours are provided.

2 METHODOLOGY

The methodology presented in this paper analyzes ASI images to provide PV power forecasts p to 2 hours-ahead. This is achieved using three major processes: i) Cloud Recognition Method, ii) Cloud Shadow Projection and, iii) Cloud Shadow Movement. Furthermore, it is also calculated the theoretical clear sky power as the base to apply the reduction caused by the shadow effect. In this chapter, the methodology is going to be presented and explained.

2.1 Cloud Recognition Method

The **Cloud Recognition Method** enables to process and transform the ASI images and performs a pixel analysis to calculate a cloud index. The cloud index represents the amount of clouds present in each image. The range of processes and transformations are fully detailed in [2]. I Fig. 2 it is shown an example of the visual result of the Cloud Recognition Method, applied to three different ASI images, which allow the calculation of the cloud index.

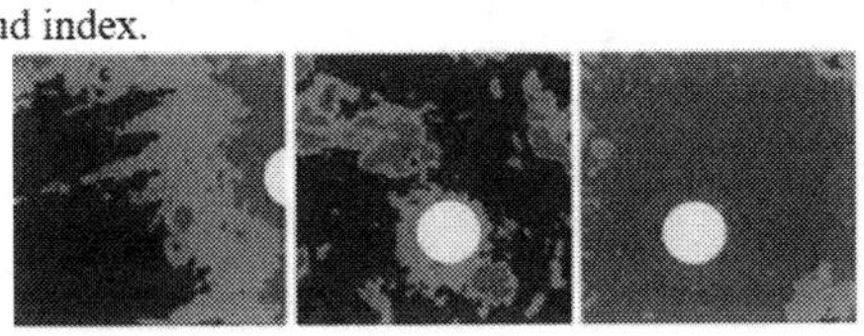

Figure 2: ASI image analysis

In this methodology, the percentage of thin clouds over all the clouds detected in the ASI image it is used, in a threshold pixel analysis, formulated as in (1).

$$cloudThin = \frac{Thin\ Cloud}{Thick\ Cloud + Thin\ Cloud} \quad (1)$$

2.2 Cloud Shadow Projection

In order to have an estimation of the percentage of the solar site covered by the shadows from the clouds present in the sky, it is necessary to project those shadows in the ground. To perform such task two things are necessary: i) the height of the clouds and, ii) a satellite image with the layout of the solar site.

In the case of **the clouds height**, an estimation method based in NWP forecasts. The estimation method uses Cloudiness forecasts at different atmosphere levels (50 meters to 12000 meters) to retrieve the most likely height of the clouds at a certain time stamp.

In the case of the **satellite image with the layout of the solar site**, it was used the image shown in Fig. 3. In the computation process, only the PV panels were considered as the area of the solar site.

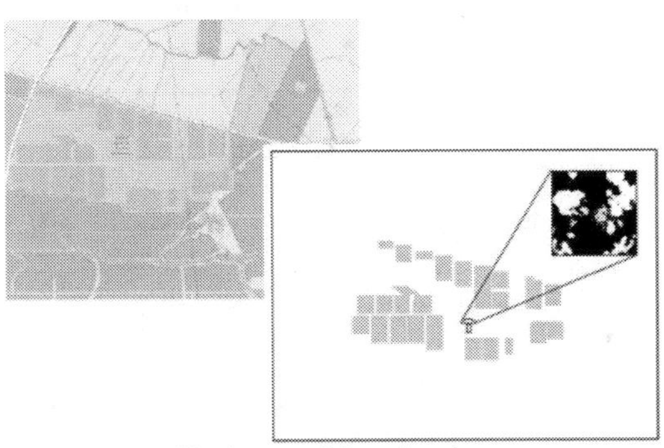

Figure 3: Satellite image of the solar site

In order to project the shadow created by a cloud in the satellite layout of the solar site, the Solar position was used in terms of zenith angle:

$$\theta_Z = \cos^{-1}(\sin\varphi \cdot \sin\delta + \cos\varphi \cdot \cos\delta \cdot \cos\omega) \quad (2)$$

In (2), δ is the equator angle and ω is the hour angle:

$$\delta = 23.45° \cdot \sin\left[2\pi \cdot \frac{284 + dayth}{365}\right] \quad (3)$$

$$\omega = (dayhour - 12) \cdot 15° \quad (4)$$

In (3), *dayth* represents the day number of the year, ranging from 1 (January 1st) to 365 (December 31st). In (4), *dayhour* is a representation of the time affected by the longitude zone of the ASI site:

$$dayhour = hour + \frac{minutes}{60} + \frac{Longitude * 4}{60} \quad (5)$$

With the knowledge of the solar **zenith angle** and the **cloud height**, for a specific time point, each pixel of the shadow can be derived from the pixels from the ASI image clouds. If one considers $(\boldsymbol{p_i}, \boldsymbol{p_j})$ as the cloud pixel position in the ASI binary image and, the $(\boldsymbol{L_x}, \boldsymbol{L_y})$ as the shadow pixel position in the layout image, the following equations can be formulated [3]:

$$(L_x, L_y) = f(p_i, p_j) \quad (6)$$

$$L_x = (p_j - sunloc_y) \times \tan(\theta_Z) \times \text{CloudHeight} \quad (7)$$

$$L_y = (p_i - sunloc_x) \times \tan(\theta_Z) \times \text{CloudHeight} \quad (8)$$

In (6) and (7), $(sunloc_x, sunloc_y)$ represent the pixel location of the Sun in the ASI image [1].

It is worth mention how the solar zenith angle and cloud height impacts the shadow. On the one hand, the zenith angle, which varies over the day, defines the direction in which the shadow is being projected in the ground. On the other hand, the cloud height defines how far this projection reaches in the ground. An example of a cloud shadow projection is shown in Fig. 4.

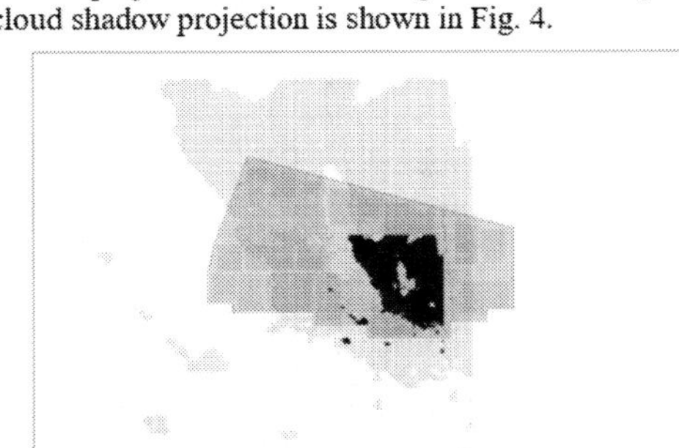

Figure 4: Shadow projection example

2.3 Cloud Shadow Movement

In order to estimate the movement of the clouds, and respective shadow projection, NWP wind speed and direction forecasts are used. In this case, the atmosphere level of the wind speed and wind direction forecasts is selected to be the closest to the cloud height.

The clouds movement is calculated based on the creation of the cloud motion vector (CMV):

$$\boldsymbol{V} = (u, v) \quad (9)$$

In (9), the CMV is composed by two wind vectors:

$$u = ws \times \cos(wd) \quad (10)$$

$$v = ws \times \sin(wd) \quad (11)$$

In (10) and (11), *ws* represents the wind speed and *wd* represents the wind direction.

Reaching this stage, it is possible to describe the methodology that allows to provide 2 hours ahead solar power forecasts. For a specific time point, denominated as base point (**BP**), the wind speed and wind direction is retrieved from the NWP forecast files. If one considers *tx* as the time increment (in minutes) in relation to the time point of **BP** image (*bpp*), the wind speed for *tx* can be defined as follows:

$$ws' = \left(\frac{60}{1000}\right) \times ws \times tx \quad (12)$$

The corrected wind speed for *tx* in (12) is given in kilometers by *tx* minutes. With this, the CMV wind vectors for the time point *tx* can be redefined as follows:

$$u = ws' \times \cos(wd) \quad (13)$$

$$v = ws' \times \sin(wd) \quad (14)$$

Finally, the new image for tx time point BP_{bpp+tx} is created by applying the CMV to shadow pixels of the **BP** image, this is formulated as follows:

$$BP_{bpp+tx}[i,j] = BP_{bpp}[i+u, j+v] \quad (15)$$

In Fig. 5 it is shown an example of the application of the cloud shadow movement for 2 hours-ahead (tx as 15 minutes).

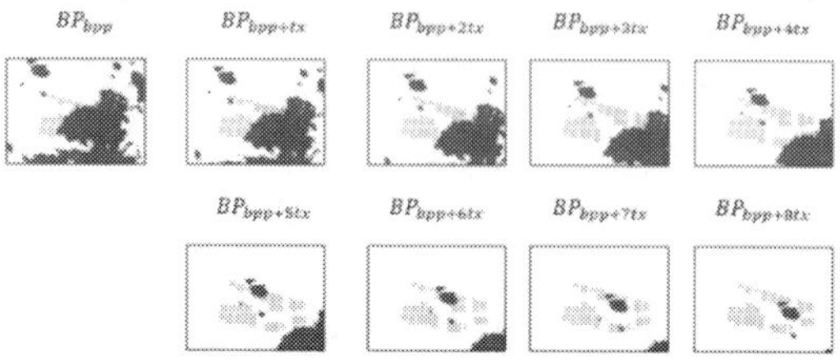

Figure 5: Shadow movement example

For each image shown in Fig. 5, the solar power plant coverage (***solarCOV***) is calculated (value between 0 and 1, being 1 full shadow coverage).

2.4 Theoretical Clear Sky

The theoretical clear sky irradiance is calculated, being converted to power using the algorithm fully detail in [4]. If one considers $I_0 = 1366\frac{W}{m^2} \pm 7$ as all the extra-terrestrial irradiance, the direct normal irradiance (in W/m^2) can be formulated as follows:

$$I_d = I_0 \times e^{\frac{-m \times T_L}{0.9 \times m + 9.4}} \quad (16)$$

In (16), m the equivalent atmospheric mass and T_L is the atmospheric perturbation factor given by:

$$m = \frac{1}{\sin \alpha_s} \times (1 - 0.1 \times z) \quad (17)$$

$$T_L = 2.6 + 1.6 \times \beta_A(\pm 0.3) \quad (18)$$

In (18), β_A is a factor equal to 0.0045. In (17), z is the solar zenith angle defined in (2), and α_s is the solar altitude:

$$\alpha_s = \sin^{-1}(\sin(\varphi)\sin(\delta) + \cos(\varphi)\cos(\delta)\cos(\omega)) \quad (19)$$

In (19), φ is the latitude of the site, δ is the equator angle and, ω is the hour angle.

For the diffuse irradiance in W/m^2:

$$I_{df} = \frac{I_0}{25} \times \sqrt{\sin(\alpha_s)} \times \left(T_L - 0.5 - \sqrt{\sin(\alpha_s)}\right) \quad (20)$$

Finally, to obtain the total clear sky irradiance (in W/m^2), the following formulation applies:

$$I_{cs} = I_d + I_{df} \quad (21)$$

The Theoretical clear sky power is calculated as follow:

$$P_{CS} = EFIC \times NMOD \times AREA \times 10^{-06} \times \cos(\theta_i) \quad (22)$$

In (22), θ_i represents the incidence angle. This angle depends on the type of PV panels (fixed, double axis Sun tracking, and so on). The different types of Sun tracking devices and the respective incidence angles are deeply detailed in [4]. In the case of fixed solar panels, the incidence angle is given by:

$$\theta_i = \cos^{-1}[\sin(\alpha_s)\cos(INC) - \cos(\alpha_s)\sin(INC)\cos(\gamma_s)] \quad (23)$$

In (23), γ_s is the solar azimuth angle and is given as follows:

$$\gamma_s = \cos^{-1}\left(\frac{\sin(\delta)\cos(\varphi) - \cos(\delta)\sin(\varphi)\cos(\omega)}{\cos(\alpha_s)}\right) \quad (24)$$

2.4 Power Forecast

The power forecast for each image is given as follows:

$$pred = P_{CS} \times \boldsymbol{solarCOV} \quad (25)$$

3 CASE STUDY

3.1 Solar site

The case study is focused on a solar PV site located in the South of Portugal, see Fig. 6. The ASI is located in the command center building of the solar site. In Table I, one can observe the solar site technical characteristics.

Figure 6: Case study solar site

Table I: Solar site characteristics

Efficiency	10%
Number of Solar Panels	59500
Area of the PV panel	1.5 m^2
Type of Sun Tracking	Fixed
Panels Inclination	16°
Installed Capacity	12 MW

3.2 NWP power forecasts to compare

In order to test the performance of the proposed methodology a case study was elaborated in which NWP power forecasts were used to have a comparison reference. The month of September 2021 was used to develop the performance test. In this particular case, two different NWP power forecast providers are used, the NWP models are received in a daily basis at 07h00m UTC (named ***NWP1***) and at 10h00m UTC (named ***NWP2***). To have a clear understanding of the performance of the developed tool (named for the purpose of this case study ***pred***) when compared to the NWP's forecasts, the initial two hours (due to the short-term purpose of the tool) of each model are used to compute the normalized root mean squared error (NRMSE). Thus, when analyzing NWP1, one is looking at the forecast horizon between 07h00m UTC and 11h00m UTC and, when analyzing NWP2, one is looking at the forecast horizon between 10h00m UTC and 14h00m UTC. For the proposed methodology, the last image available before the start of the forecast horizon is used.

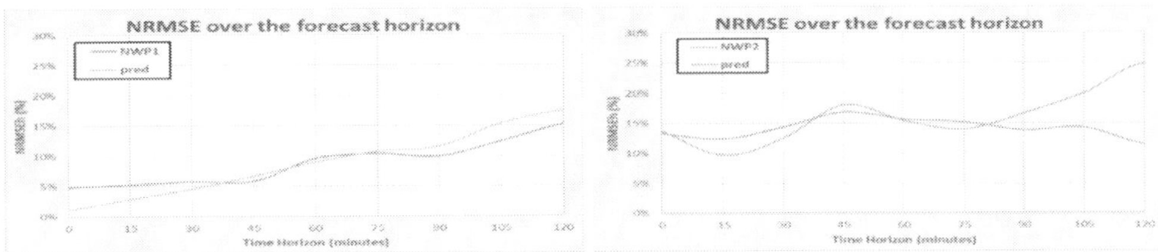

Figure 7: September 2021 analysis.

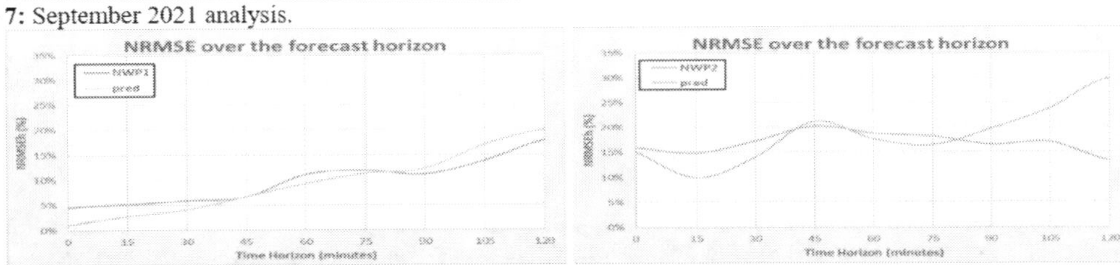

Figure 8: September 2021 cloudy days analysis.

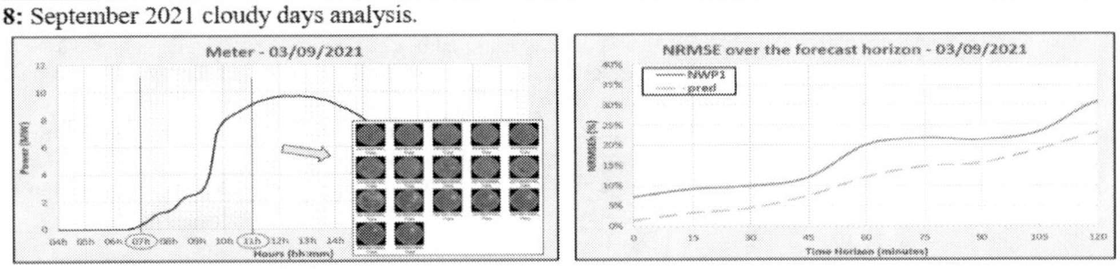

Figure 9: 3rd September 2021 analysis.

4 RESULTS AND DISCUSSION

4.1 All days of September 2021

In Fig. 7 it is depicted the results of the NRMSE over the 2-hours forecast horizon considering all days in September 2021. It is clear that the proposed ASI local approach can have some advantage in the initial period of the forecast horizon when compared with the NWP's forecasts. As the time horizon advances this advantage starts to dissipate.

4.2 Cloudy days of September 2021

In Fig. 8 it is depicted the results of the NRMSE over the forecast horizon considering only the cloudy days in September 2021. In this case it is more evident that the proposed methodology has advantage in the initial period (initial 80 minutes) of the forecast horizon when compared with the NWP's forecasts. The local approach with real time information of the sky conditions can result in a better forecast when analyzing a range of 1 hour-ahead.

4.3 Single Day Analysis

In Fig. 9 it is depicted a single day, 3rd September 2021, to better illustrate the advantage of the proposed methodology (when compared with NWP1, as an example). This day faced heavy clouds during the morning, which dissipated completely during the afternoon. The NWP models were not able to predict this volatile phenomenon as the proposed methodology was able to reproduce better the sky conditions in the generated forecasts.

5 CONCLUSIONS AND FUTURE WORK

The proposed methodology relies on the use of an ASI to provide power forecasts to a solar PV power plant in the range of 2 hours-ahead. In these preliminary results, the methodology have shown that could benefit from the observation of the sky to provide power forecasts. The performance test, comparing with NWP power forecasts, shown that the proposed methodology have the potential to provide better results in the range of 1 hour-ahead. The results achieved so far also indicate that, this is especially true when analyzing cloudy days, which have higher probability of being miss-predicted using NWP models.

6 ACKNOWLEDGEMENT

This work was supported by CEPRI (Chinese Electric Power Research Institute), in the scope of the R&D NESTER (Centro de Investigação em Energia REN-SGCC, S.A.) project 'Renewable Energy Power Forecasting Model Research and Prediction Results Analysis'.

7 REFERENCES

[1] REN – Redes Energéticas Nacionais S.A., "Thecnical Data", available at: https://www.ren.pt/en-GB/media/publicacoes

[2] João Esteves, Yang Cao, Nuno Pinho da Silva, Rui Pestana, "Identification of clouds using an all-sky imager", 2021 IEEE Madrid PowerTech, 28 June – 02 July 2021.

[3] J. J. Simpson and J. R. Stitt, "A Procedure for the Detection and Removal of Cloud Shadow from AVHRR Data over Land," IEEE TRANSACTIONS ON GEOSCIENCE AND REMOTE SENSING, vol. 36, no. 3, May 1998.

[4] J. Esteves, R. Pastor, N. Pinho da Silva, R. Pestana and Z. Chen, "Forecasting PV/CPV at National Level - Portuguese Experience," IET Renewable Power Generation, vol. 13, no. 5, 2019.

HYBRID NUMERICAL ENSEMBLE METHOD AND COMPARATIVE STUDY FOR SOLAR IRRADIANCE FORECASTING

Jayesh Thaker [1], * and Robert Höller [2]
[1] Department of Physics, Carl von Ossietzky Universität Oldenburg, 26129 Oldenburg, Germany
[2] School of Engineering, University of Applied Sciences Upper Austria, Stelzhamerstrasse 23, 4600 Wels, Austria; robert.hoeller@fh-wels.at
* Correspondence: jayesh.thaker@uni-oldenburg.de; Tel.: +49 15168 797550

ABSTRACT: Renewable energy sources like solar energy and wind power, which are not available on-demand must be stored or used instantly. The solar irradiance is strongly dependent on unpredictable atmospheric conditions. Thus, the performance and the accuracy of solar forecasting model is highly influenced by the complex and stochastic atmospheric conditions. The forecasting challenges of solar energy augment with increasing deployment of photovoltaic (PV) system into electrical grids. As the global horizontal irradiance (GHI) is the most important key input in PV power prediction systems, increasing effort is spent on research on forecasting of GHI as a basis for corresponding power forecasts.

The proposed approach in this paper is based on forecast of the global numerical weather prediction (NWP) model Global Forecast System (GFS). The proposed model uses day-type classification benchmark based on hourly vectors generated using clear sky index and hybrid numerical formulation. A novel GHI prediction system using input data from an GFS and two levels ensemble is presented and evaluated by providing day ahead forecasts with an hourly time resolution. The methodology applies various machine learning (ML) and recurrent neural network (RNN) models along with a parametric model as the basis to build an ensemble model.

The focus of this paper is the comparison of the capability and performance of developed models and build a hybrid ensemble model for producing reliable day-ahead forecast horizon and an hourly resolution of GHI. The detailed study and analysis of probabilistic time series forecast is presented. The reliability and the sharpness of the forecast models is evaluated by using various skill scores of probabilistic forecasts.

Results show that the hourly irradiance classification approach reduces the forecasting error by a considerable margin and the optimized hybrid ensemble model outperforms all other models. These findings also emphasize that all the models perform better than GFS.

Keywords: Solar irradiance forecasting; NWP; MOS; probabilistic forecast; machine learning; deep learning, ensemble models; weather classification; clear sky index, global horizontal irradiance

1 INTRODUCTION

Due to the global warming caused by anthropogenic green-house gas emissions, solar energy, in particular PV, is becoming a significant component of the world's energy mix and its share is anticipated to increase in the future.

The forecasting of solar energy involves several complex issues due to its highly stochastic nature, which creates difficulties for power grid operation in planning and operation [1,2,3]. The accurate forecasting of solar irradiance is required to mitigate the challenges to the operation of power grids due to inherent variability of the solar energy [4].

The forecast models are usually developed depending on the prediction horizon, the purpose of the predictions and applications. Several methods are being explored in the field to this aim, which could be grouped as: NWP, all-sky imagers (ASI), satellite imaging, statistical, and machine learning (ML). They could also be further categorized based on their characteristics, such as whether a method considers spatial-temporal features, whether it is deterministic or probabilistic, whether it considers exogenous inputs or just its data features, whether it is hybrid, i.e., using two or more approaches simultaneously, or ensemble, which uses the outputs of various models, etc.

For intra-hour and intra-day forecasts, ML, and artificial neural network (ANN) techniques have proven to be the most effective. The analysis of cloud and motion vectors using all sky imagers (ASI), or satellite imagery are usually coupled with radiation transfer models to produce reliable forecasts up to 1- 6 h ahead. A generalize short-term forecasting model of solar irradiance without local telemetry using satellite data is proposed in [5] whereas [6] proposes intra-hour irradiance forecast by cloud height measurement using network of all-sky imagers.

However, for day-ahead predictions, the evolution of atmospheric condition is highly uncorrelated with the present condition and therefore, statistical methods solely are insufficient to accurately forecast solar irradiance [7]. NWP are specially designed to resolve complex atmospheric phenomenon, and it has been shown that they are likely to improve solar irradiance prediction for day-ahead forecasting [7,8].

However, even state-of-the-art NWP models remain relatively inaccurate despite recent improvements in NWP forecasts of surface solar irradiance in Weather Research Forecast (WRF-Solar) model [9,10]. This is primarily due to propagation of numerical errors and approximations caused by unknown initial boundary conditions, highly unpredictable atmospheric phenomenon coupled with complex local orography. Nevertheless, these errors may be systematic and often exhibit patterns that can be learned which is demonstrated in this paper. Therefore, post-processing techniques are frequently used in conjunction with NWP to predict and rectify future bias by learning from prior NWP forecast errors. NWP post-processing termed as Model Output Statistics (MOS) [11] determines statistical relationship between a target variable and a set of predictors taken from the output of a numerical weather model. In the recent literatures many authors have leveraged the advances in statistical learning to propose improved NWP post-processing models, particularly in the field of solar irradiance forecasting.

A plethora of statistical learning methods have been proposed and shown to outperform linear regression. They include, for example, gradient boost, nearest neighbors, regression splines, random forest, and polynomial regression. However, although the MOS framework allows for many predictors to be used as input to a regression algorithm, most of the MOS algorithms proposed in the solar forecasting literature only use a reduced number of standard variables e.g., temperature, zenith angle and some cloud-coverage. The ability of high dimensional MOS input set to enhance NWP forecasts needs be further investigated because NWP can be adjusted to generate a richer and more useful set of predictors. The recent article [12] provides not only benchmarking for statistical learning for NWP MOS but also excellent chronology of the publications in this field.

This paper aims to present a novel hybrid MOS system to predict day-ahead global horizontal irradiance (GHI) with hourly resolution using exogenous inputs from GFS. The motivation for the development of such a system is to contribute towards applications that reliably forecast solar irradiance in a flexible and robust manner using open-source GFS data.

The paper confirms that all the individual models perform statistically much better compared to GFS. The detailed probabilistic forecast analysis is presented which demonstrates that the two-level hybrid ensemble model presented in this study outperforms all the models both in deterministic and probabilistic forecasts.

The structure of this article is as follows: evaluation metrices used for deterministic and probabilistic forecasts in Section 2.1, theoretical background in Section 2.2 and data pre-processing with a focus on the clustering of data and benchmark definition of day-type in Section 2.3. We discuss models applied in section 3 and results in Section 4.

2 THEORY, EVALUTATION METRICES & DATA

2.1 Evaluation Metrices

The three most widely used metrices for point forecast accuracy are the mean bias error (MBE), the mean absolute error (MAE) and the root mean square error (RMSE) [13]. The normalized MAE (nMAE) and RMSE (nRMSE) are calculated by dividing the respective metrices by the average value of the observed parameter (measured GHI in this case) under consideration.

A detailed overview of the error metrices for deterministic and probabilistic forecast is presented by the authors [14] which are applied in this paper as well. Additionally, this paper also presents following metrices for probabilistic forecasts.

2.1.1 Prediction interval coverage probability

The reliability of a forecast can also be estimated based on the probability of its central prediction intervals (PIs). The prediction interval coverage probability (PICP), also known as PI confidence level, is the probability that the measured target values lie within the PI of the model [15]. PICP is derived by Equation (1):

$$PICP = \frac{1}{n}\sum_{i=1}^{n} c_i \qquad (1)$$

where c_i is binary (0 and 1). If measured target (y) is within the PI range then $c_i = 1$; otherwise, $c_i = 0$. Normally, a higher PICP has a lower false rate, and the model is under-confident and vice versa. Ideally, PICP should be close to 1 [15].

2.1.2 Prediction interval normalized average width

The average width of the PIs for different nominal coverage (α) is a common measure of the sharpness and measures the wide degree of PI. PINAW is the average of PI widths normalized by the range of testing targets and is defined by Equation (2) [15]:

$$PINAW = \frac{1}{nr}\sum_{i=1}^{n}(U_i - L_i) \qquad (2)$$

where $r = y_{max} - y_{min}$. As the models produce forecasts for different time of the day and the hourly average value of GHI varies, the PI width must be normalized to account for the hourly variation of the irradiance. Ideally, a smaller PINAW and a larger PICP are desirable for greater reliability [15].

2.2 Theoretical Background

GHI (W/m^2) is the total amount of shortwave radiation received by a horizontal surface and can be expressed in terms of the Diffuse Horizontal Irradiance (DHI) and the Direct Normal Irradiance (DNI) as:

$$y(t) = GHI = DHI + DNI \times cos(\theta_z) \qquad (3)$$

where θ_z is the solar zenith angle of the sun. $\cos(\theta_z)$ is used in all models as one of the key predictors in training and testing, as it strongly influences the GHI.

The clear sky index is the ratio of global, $y(t)$ to clear sky global irradiance.

$$K_t = \frac{y(t)}{I_{cs}(t)} \qquad (4)$$

The **persistence model** is used as the baseline model as per the standard practice. If $\hat{y}(t+h)$ denotes the forecast value, the persistence model can be expressed as: $\hat{y}(t+h) = y(t)$.

Smart persistence model is typically used for solar irradiance forecast as it also accounts for the expected variation based on the clear-sky irradiance with time. It is defined by the following equation:

$$\hat{y}(t+h) = y(t).\frac{I_{cs}(t+h)}{I_{cs}(t)} = K_t . I_{cs}(t+h) \qquad (5)$$

2.3 Database Clustering

The ground measurement of GHI in this analysis is taken from the Kipp & Zonen (CM11) installed at the University of Applied Sciences Upper Austria (FHOOE). The university building is in Wels, Austria (latitude 48.17°N, longitude 14.03°E, elevation 317 m). The granularity of measured GHI is 10 minutes which is averaged into hourly data. The exogenous NWP data – forecast GHI (W/m^2), temperature (^{0}C), and relative humidity used in this study are from GFS, along with K_t and $\cos(\theta_z)$.

The ground measured data for 363 days in 2018 along with GFS forecast data are used to train various models. The trained models are assessed for forecasting day-ahead from 01-01-2019 till 24-06-2019 covering mainly the winter and spring seasons where maximum variability is typically observed in central Europe.

The modelling is based on classifying data into six clusters depending on the value of clear sky index (CSI), K_t. Subsequently, we defined a benchmark for the type of day based on the daily average value of K_t. It must be noted that that the maximum value of K_t is truncated to 1.0 while calculating daily average failing which it will give erroneous higher average K_t resulting in wrong classification of the day type. The criteria for the classification for different hourly and day type clusters is illustrated in the following tables.

Table I: Hour type classification based on CSI

Class	Type	CSI Range
HT1	Overcast	$0 \leq K_t < 0.3$
HT2	Highly cloudy	$0.3 \leq K_t < 0.5$
HT3	Cloudy	$0.5 \leq K_t < 0.7$
HT4	Mostly Clear	$0.7 \leq K_t < 0.9$
HT5	Clear	$0.9 \leq K_t \leq 1.0$
HT6	Sub-hour	$K_t > 1.0$

Table II: Day type classification based on CSI

Class	Type	CSI Range
DT1	Overcast	$0 \leq K_t < 0.3$
DT2	Highly cloudy	$0.3 \leq K_t < 0.5$
DT3	Cloudy	$0.5 \leq K_t < 0.7$
DT4	Mostly Clear	$0.7 \leq K_t < 0.9$
DT5	Clear	$K_t \geq 0.9$

The above five defined forecast day types based on the value of forecast clear sky index value are used as one of the key predictors for all the models. This benchmarking clearly defines a pattern of GFS forecast GHI versus the ground measured actual GHI as illustrated in Table III.

Table III: Distribution of actual vs. forecast day type

	Actual day type	Number of forecast day types				
		DT1	**DT2**	**DT3**	**DT4**	**DT5**
A- 2018 Training with zenith angle correction	DT1	14	19	10	10	9
	DT2	5	8	14	23	9
	DT3		4	12	30	17
	DT4			2	23	60
	DT5				6	88
B- 2018 Training w/o zenith angle correction	DT1	8	11	5	10	
	DT2	5	12	24	27	5
	DT3	1	2	14	51	4
	DT4			3	40	48
	DT5				9	84
C- 2019 Testing with zenith angle correction	DT1	12	10	7	1	2
	DT2		5	11	13	3
	DT3		2	9	16	10
	DT4			3	10	24
	DT5				5	50

Two important observations are deduced from this classification analysis – (i) Almost all actual clear sky days are forecasted as clear sky days by GFS. (ii) It is evident that GFS is not very efficient in detecting and forecasting clouds as it forecasts mostly clear sky days and cloudy days as clear sky days which is the main source of forecast error. This is substantiated by the statistical analysis which shows that GFS has a positive bias of 58.75 and 52.20 for the data for 2018 and 2019, respectively.

2.3 Zenith angle correction for measured data

Direct radiation falling on a flat horizontal surface originating from a point source with a defined zenith position will have an intensity value proportional to the cosine of the zenith angle of incidence as defined in Equation (3). In this study the GHI data are filtered out for sub-hour for $\theta_z > 80°$ angle of incidence with respect to 1000 W/m^2 irradiance at normal incidence (0°) as per the technical specification of the pyranometer [16]. By comparing A and B in Table III. it is observed that this data pre-processing step is particularly important failing which it results in erroneous classification of the hour and day types adversely affecting the output of the model.

2.4 Predictors feature analysis and selection

Month, day of the month, hour of the day, year, season are added as additional feature values as they define the seasonality of GHI. GHI, temperature, and relative humidity values are scaled, whilst all other variables are employed as feature values (binary 0 or 1) in machine learning and linear regression models for training and testing.

Multiple Fourier series in the form of $\sin(2\pi)$ and $\cos(2\pi)$ are added for each season of the year, hour of the day and week of the year as additional feature values for training and testing ML and RNN models [14].

As CSI removes inherent seasonal trends from the GHI measurement due to the position of the sun, it is used for all ML and LSTM models. Model forecast GHI is calculated by the following equation:

$$GHI_{model} = GHI_{GFS} \times \frac{K_{t,model}}{K_{t,GFS}} \qquad (6)$$

Feature selection is an extremely important pre-step for any machine learning model to identify important features for the model. Redundant features not only result in higher computational cost but also adversely affect the forecast accuracy. Random forest, extreme gradient boost and correlation analysis are used to select the important predictors.

3 MODELS

In this study, various ML and RNN (Long Short-Term Memory - LSTM) models have been applied, analyzed, and compared to allow higher degrees of model complexity. The focus of this paper is on improving the forecast accuracy of GFS through the combination of different models based on day type classification generated using hourly CSI vectors. The flow chart of algorithm used for modelling is illustrated in Figure 1.

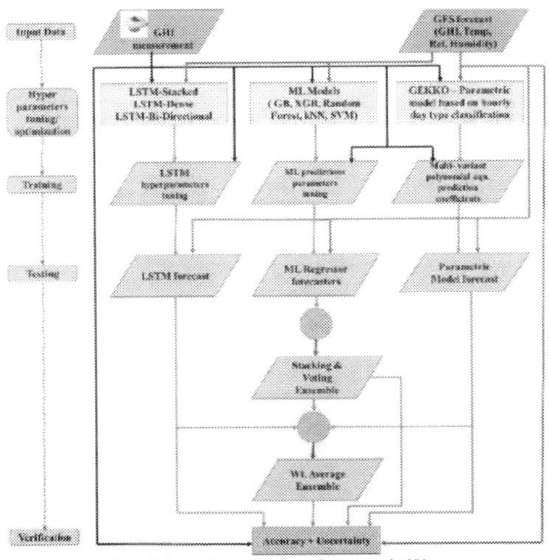

Figure 1: Algorithm flow chart for modelling

The decision to use these techniques over the many other ML methods is based on a comparison of the models to find the best fit and the possibility of merging them to form ensemble models to increase forecast accuracy. Furthermore, the probabilistic forecast of the best performing modes is evaluated as a measure of uncertainty and the resulting quantile forecasts evaluated by different probabilistic forecasting skill scores.

A detailed overview of theory and description of Random Forest (RF), Gradient Boosting (GB), Extreme Gradient Boosting (XGB) and Gekko which are used for modelling in this study, is presented by authors [14]. The other models used in this study are:

3.1 Support Vector Machine (SVM)

Support Vector Machines (SVM) have been recently developed in the framework of statistical learning theory (SLT) [17,18]. SVMs are supervised learning models with associated learning algorithms that analyze data for classification and regression analysis and are widely applied to several applications.

A detailed overview of theory and algorithm of SVM for classification and regression is presented in [19,20].

We applied the most commonly used non-linear kernel space called radial basis function (rbf) for our study while the other options are linear, poly, sigmoid and precomputed.

3.2 k-Nearest Neighbors (kNN)

The kNN is a non-parametric supervised machine learning algorithm which is easy, user-friendly that may be used to solve both classification and regression problems by storing all available informations during the training phase and predict the numerical target based on a similarity measure (e.g., distance functions). kNN is therefore, also known as a lazy learning algorithm as training primarily consists of just storing the data. The kNN algorithms locate the k closest or most similar neighbors of a query point and predicts by computing the class label (classification) or continuous target (regression). The basic concept of kNN is that it approximates the target function $f(x) = y$ locally rather than globally for each forecast [21].

The basic structure of k nearest neighbor algorithm is presented in [21].

The kNN is one of the most widely used algorithms in time series forecasting due to its simplicity and intuitiveness. It assumes that that future time series sequences will resemble the sequence that has emerged in the past [22].

3.3 Long Short-Term Memory (LSTM)

Recurrent neural networks (RNN) based on the idea of sequential information with loops within themselves allow information to persist. The RNN has a memory component, where the current output is a function of current input as well as a previous step. However, the major drawback of RNN is in its failure to detect long term dependency due to its inability to remember long-range information because of the vanishing gradient problems [23].

LSTM is a variant of RNN, proposed by Hochreiter and Schmidhuber [24] which is specially designed to memorize very long-term temporal dependencies through memory cells containing several types of gates to address the vanishing gradient problem of RNN. Additionally, LSTM can learn and handle nonlinearity effectively.

The memory blocks that make up the LSTM architecture and the step-by-step description of the gates and the mathematical equations associated with them are presented in [25].

There are many variants of LSTM model based on the topologies of the layers such as the simplest Vanilla or Stacked with multiple hidden layers on top of each other or Bidirectional where the sequence is presented forward and backward to two separate LSTM network connected to the same output layer [26].

It is noteworthy that the input data are split into batches of windows while training LSTM model A model is called a **stateless** model if the batches are independent of each other. Alternatively, the model is called a **stateful** model if batch to batch dependency is considered [26].

It can be observed in the literature that there is growing interest in applying LSTM for solar irradiance forecasting. However, most of the research is in short-term intra hour forecasting [27,28,29,30,31,32] mostly using ASI as an exogenous input. We experimented with three variants of LSTM for our study namely, stacked, bidirectional and deep (consisting of two dense layers instead of simple LSTM layers) for day ahead forecasting.

3.4 GEKKO

Gekko is a parametric model used in this study to solve fourth degree polynomial equation [14]. GHI is used instead of CSI for modelling as this is independent of seasonality. The algorithm shown in Figure 2 is used to train the Gekko model to solve the equation of the form:

$$\hat{y} = a_1x_1 + a_2x_1^2 + a_3x_1^3 + a_4x_1^4 + b_1x_2 + b_2x_2^2 + b_3x_2^3 + b_4x_2^4 + c_1x_3 + c_2x_3^2 + c_3x_3^3 + c_4x_3^4 + d_1x_4 + dx_4^2 + d_3x_4^3 + d_4x_4^4 + e_1x_5 + e_2x_5^2 + e_3x_5^3 + e_4x_5^4 \quad (7)$$

where: $\hat{y}$ = Model forecast GHI
x_1= GFS GHI
x_2= GFS temperature
x_3= GFS relative humidity
x_4= $\cos(\theta_z)$
x_5= GFS clear-sky index

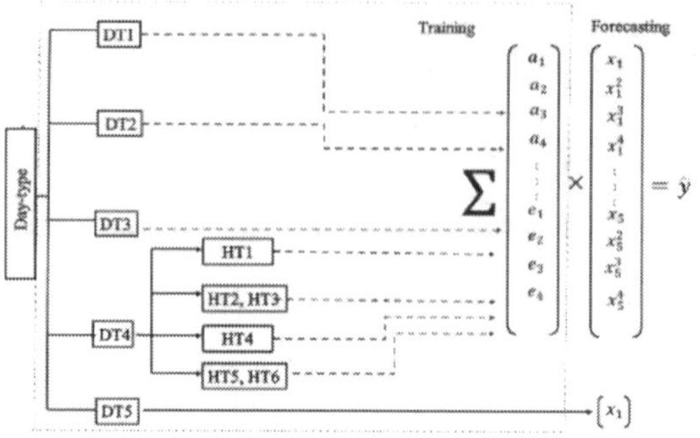

Figure 2 Schematic of algorithm applied to GEKKO

3.5 Ensemble models

Ensemble learning is the process of strategically generating and combining various experts or classifiers of different models to solve a particular computational intelligence problem [33]. Ensemble methodoly has become a significant examplar after a study found that forecasts made by a combination of a set of classifiers are offen more accurate than made by single individual classifier [34].

The ensemble learning techniques can be broadly classifed into following three types based on the type of the models used, the data sampling (with or without replacement etc.) and the decision function (meta, voting, average) [33]:

Stacking The underlying idea of stacking (also known as stacking generalization) is to minimize the generalization error of various generalizers (i.e. ML models). This is most commonly performed in two stages: (i) a *Meta* model is generated which is made up of the predictions of a set of ML models (i.e. weak learners) through the k-fold cross validation technique. (ii) The Meta-Model is then trained using a second ML model (also referred to as the "final estimator" or "final learner") [33] .

Bagging is derived from the combination of abbreviation of *B*ootstrap and *Agg*regat*ing*. It uses bootstrap sampling to generate data subset for training the base learners. The decisions of individual classifiers are then aggregated by simple majority vote [33].

Voting ensemble is thus an extension of bagging where base classifiers can be homogenous or heterogenous. The evaluation of any voting ensemble consisting of number of ML models is peformed by two methods: (i) hard mode in which the the models with "the most votes" is the best combination, and (ii) soft mode considers the probabilites generated by each model which are weight averaged. The combination which generates the highest weighted average is the forecaster.

Boosting As the name suggest, boosting algorithms covert weak learners to strong learners. The underlying concept is similar to bagging, however, in boosting the resampling is strategically boosted to provide the most informative data for each consecutive classifer while creating an ensemble of classifiers. Three weak classifiers are produced by each iteration of boosting; the first classifier, C1, is trained using a randomly selected portion of the training data. Given C1, the most instructive subset is the training data subset for the second classifier, C2. In particular, only half of the training data used to train C2 is correctly classified by C1, and the other half is incorrectly labeled. Instances on which the first two classifiers disagree are used to train the third classifier, C3. A three-way majority vote is used to combine the three classifiers [33].

In this study, we applied stacking and voting algorithm to build first level ensemble model using ML models, viz RF, XGB, GB, kNN and SVM.

Weighted Average Ensemble Model In the weighted average ensemble, the weight of each model performance is first evaluated for by using inverse value of MAE [14]. The MAE of six models is considered for this second and final level of ensemble viz. stacking, voting (outputs from the first level ML ensemble), Gekko, stacking LSTM, dense LSTM, and bidirectional LSTM.

The forecast of weighted average ensemble is then calculated on hourly basis from the forecast and respective weight of each model.

4. RESULTS & DISCUSSION

4.1 Deterministic Forecast

One of the key observations of this study is that this day type classification clearly shows that GFS forecasts clear sky day (DT5) quite well as shown in Table III. The nMAE for GFS forecast of GHI for DT5 for 2018 is 16.8% and for the test period in 2019 is 15.5%. The same drops to 6.7% and 6.9% respectively when the forecast DT5 is actual DT5 for roughly 90 days in a year.

We, therefore, forced all model to forecast values of GFS GHI for DT5 events as none of the models could improve on the forecast errors better than that of GFS.

Table IV shows the nMAE nRMSE and their respective skill scores of all the models. All the models show significant improvement over GFS. Further the advantage of applying ensemble technique is clearly illustrated as Stacking ensemble outperforms all its constituent ML models in nMAE skill score whereas hybrid weighted average ensemble proves to be the best performing model both in nMAE and nRMSE skill score.

Table IV: Metrices of deterministic forecasts of all models

Model	nMAE	nRMSE	Skill Score	
			nMAE	nRMSE
Wt. Ensemble	**21.11**	**31.75**	**46.25**	**43.83**
LSTM-Stack	21.57	32.22	45.07	43.00
Stacking	21.60	32.68	44.99	42.18
RF	21.61	32.38	44.97	42.71
Dense	21.70	32.59	44.74	42.35
XG-Boost	21.71	32.62	44.71	42.29
Voting	21.75	32.60	44.62	42.33
GB	21.81	32.74	44.82	42.08
SVM	21.85	33.32	44.37	41.06
Bidir LSTM	21.87	33.38	44.31	40.95
kNN Bagging	21.91	33.08	44.22	41.48
kNN	21.92	33.07	44.19	41.49
Light GBM	22.10	33.27	43.72	41.15
Gekko	22.20	33.81	43.48	41.18
GFS	25.64	38.26	34.72	32.31
Persistent	39.28	56.53		

In the subsequent discussion, we present further analysis of the weighted average ensemble model and its constituent models and comparison with GFS for different day types, hour types and hours of the day.

Weighted average ensemble model outperforms all other models for DT1, DT2 and DT4 in both nMAE and nRMSE metrices as illustrated Table V. However, dense LSTM seems to perform better for DT3. It can also be inferred that overcast days have the much lower variation of clear sky index as nRMSE for DT1 is lower compared with DT2, DT3 and DT4. Clear sky days DT5 has the least variation as anticipated.

All the models show excellent improvement by 12% or more over GFS for DT1, DT2, DT3 and DT4. Weighted average ensemble shows improvement by almost 35% for highly cloudy day type (DT2) over GFS which is significant considering high variability under such cloudy conditions.

Table V: nMAE and nRMSE for different day types

Model	nMAE				
	DT1	DT2	DT3	DT4	DT5
Wt_Ensem	**29.34**	**41.68**	35.91	**31.13**	14.80
Bidir LSTM	31.40	43.22	39.13	33.08	14.80
Stacking	32.05	42.81	39.05	31.78	14.80
Voting	36.47	44.21	39.23	31.89	14.80
Stack LSTM	37.33	46.48	37.17	31.47	14.80
Gekko	38.13	48.16	39.38	33.24	14.80
Dense LSTM	38.69	45.76	**35.84**	32.78	14.80
GFS	39.08	64.04	50.48	42.62	14.80
			nRMSE		
Wt_Ensem	**38.64**	**63.67**	55.17	**41.32**	**22.97**
Bidir LSTM	41.22	67.79	61.43	44.86	22.97
Stacking	41.96	63.06	59.32	43.46	22.97
Voting	47.94	63.42	59.31	42.93	22.97
Gekko	49.82	72.41	61.26	45.94	22.97
Dense LSTM	50.36	68.60	**53.35**	44.62	22.97
Stack LSTM	51.36	69.00	55.08	42.35	22.97
GFS	51.68	91.93	68.20	57.69	22.97

Table VI shows the nMAE and nRMSE scores for the models for different hour types with the best score highlighted in bold. This clearly illustrates the importance of ensemble model as each constituent model perform better than other depending. It is important to note that all the models show significant improvement in sub-hour variation compared to GFS with Bidirectional LSTM performing the best in HT6.

Table VI: nMAE and nRMSE for different hour types

Model	nMAE					
	HT1	HT2	HT3	HT4	HT5	HT6
Wt_Ensem	40.77	40.17	38.45	28.06	16.07	22.70
Stacking	43.89	41.44	39.93	29.32	16.15	23.20
Bidir LSTM	44.29	45.59	40.24	28.28	16.56	23.08
Dense LSTM	45.39	43.83	38.91	27.81	16.59	22.94
Gekko	45.64	39.23	43.74	29.79	16.56	23.35
Voting	45.75	42.53	40.73	29.39	16.15	22.97
Stack LSTM	45.82	45.64	39.67	27.59	16.27	22.97
GFS	48.59	50.56	51.49	36.63	18.52	25.89
			nRMSE			
Wt_Ensem	71.35	67.88	60.68	38.14	23.64	33.64
Voting	71.58	69.03	62.74	40.09	24.11	34.26
Stacking	72.80	66.90	62.46	40.87	24.13	34.26
Dense LSTM	73.17	71.12	59.68	38.46	24.62	33.70
Stack LSTM	75.05	75.40	62.36	37.55	23.84	33.69
Gekko	78.04	60.52	68.33	41.82	24.94	34.69
Bidir LSTM	78.68	80.05	68.22	39.24	24.21	33.61
GFS	84.67	77.26	68.37	51.12	28.04	38.53

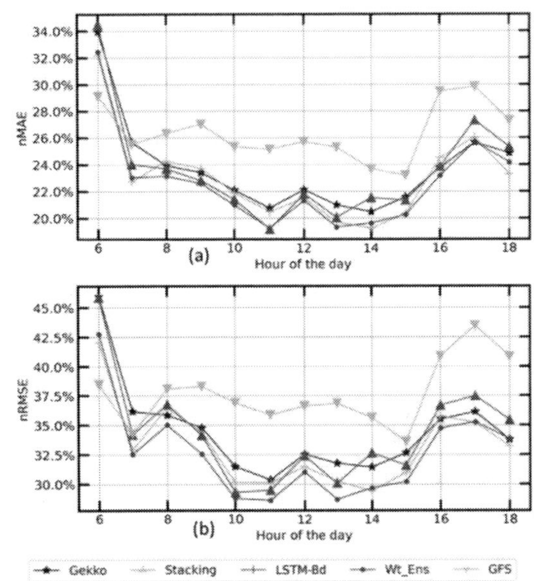

Figure 3 Variation of (a) nMAE and (b) nRMSE at different hours of the day

The hourly variation in Figure 3 shows that GFS seems to perform better at the first morning hour (6:00). This is due to the fact there are limited data available for the modelling at this sub-hour especially after applying zenith angle correction in data pre-processing. Consequently, the performance of other models drops for extremely low GHI at this hour of the day. All the models perform excellently better than GFS at other hours of the day. It is also noteworthy that nRMSE for weighted average ensemble model remains constant 28-30% during peak hours from 10:00 – 15:00 confirming that the model is stable for peak hours.

4.2 Probabilistic Forecast

The uncertainty is bound to play a critical role in forecasting of solar irradiance because of the associated uncertainty in meteorological parameters due to stochastic nature of weather conditions. Probabilistic forecasting plays a crucial role in assessing prediction uncertainty by providing a complete predictive distribution.

For each time step, eleven quantiles [0.05, 0.1, 0.2, 0.3, 0.4, 0.5, 0.6, 0.7, 0.8 ,0.9, 0.95] are used to generate prediction intervals and analyze models for probabilistic forecasts. The prediction intervals express the uncertainty associated with each point forecast.

Prediction interval for each time step for eleven quantiles of each model is calculated.

Figure 4 shows that PICP for $\alpha \leq 0$ is well calibrated for all the models and close to ideal values. However, it drops for higher α values suggeting that the models are overconfident. It is observed that the deviation of GFS from 1:1 line is much steeper indicating its poor quality and extremly high over confidence in forecasting high quantiles.

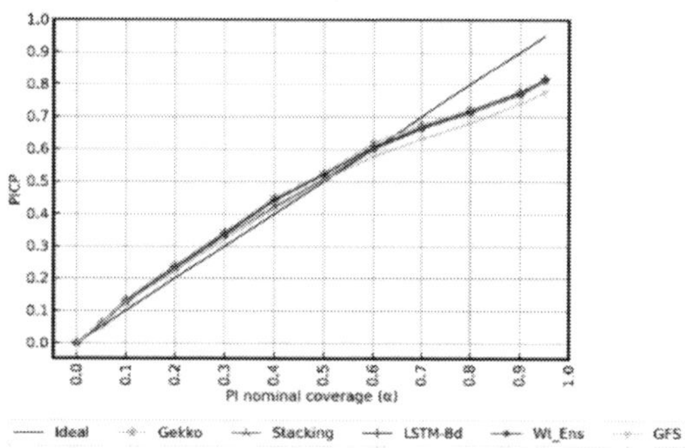

Figure 4 PICP as a function of nominal coverage, α

PINAW is a measure of the sharpness of a model and is directly correlated with PICP. An increase in PICP will broaden the PINAW and likewise a wider PINAW has a better PICP value. However, a wider PINAW will fail to detect false rate. Therefore, a smaller PINAW with larger PICP are ideal to construct prediction intervals [15].

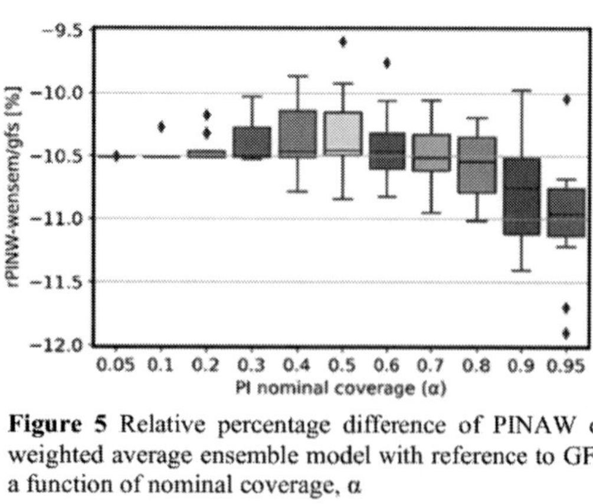

Figure 5 Relative percentage difference of PINAW of weighted average ensemble model with reference to GFS a function of nominal coverage, α

We, therefore, evaluated relative percentage difference of PINAW of each model with reference to that of GFS to quantify the sharpness of prediction interval width. Higher the negative percentage difference in the width, better the sharpness. Weighted average ensemble model outperforms all other models with narrow PINAW and higer PICP. It is also noticed that PINAW of all models gets narrower compared to that of GFS for higher α. Figure 5 illustrates this observation for weighted average ensemble model while other models also have a similar pattern.

Figure 6 shows the variation of nCRPS with type of the day. Weighted average ensemble performs the best and its nCRPS is almost 20% better compared with GFS for DT2 and DT3 and 10% for DT1 and DT4.

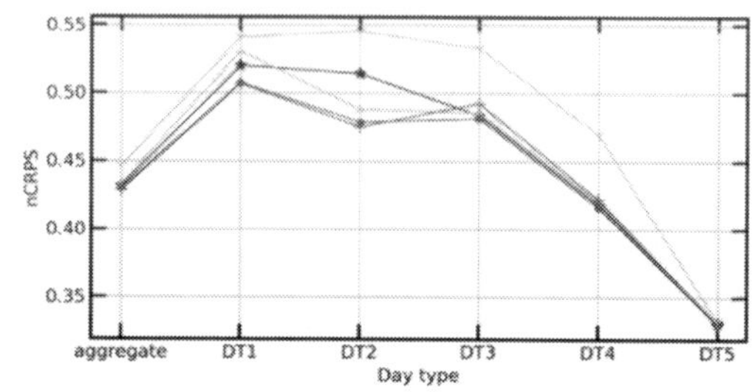

Figure 6 Variation of nCRPS with type of the day

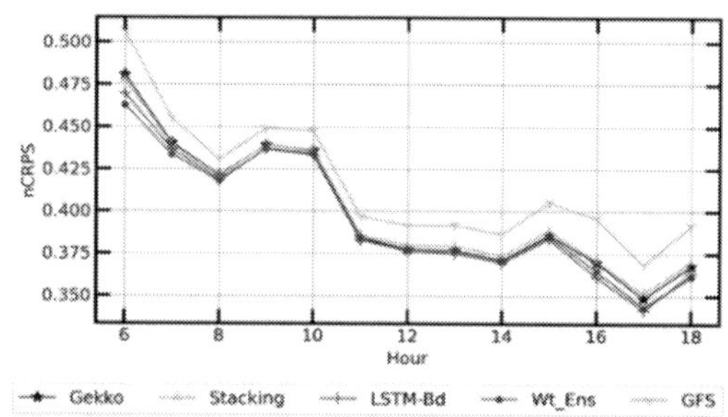

Figure 7 Variation of nCRPS with hour of the day

All the models perform better compared to GFS at each hour of the day as illustrated in Figure 7.

Brier Score (BS) is typically used for discrete variable whereas GHI is a continuous variable. However, we applied BS to compare the performance of the models to predict two rare events and one normal event as described below:

- Rare event (RE1) where GHI is lower than its 10% percentile
- Rare event (RE2) where GHI is higher than its 90% percentile
- Normal event (NOR) where GHI is between 10% and 90% percentile

The threshold value is kept at 0.5 of maximum GHI value for each hour within the cluster BS evaluation as the GHI changes at each hour.

As illustrated in Figure 8. GFS appears to be more skillful for RE1 in predicting extremely low value GHI less than 100 W/m^2. This further validates earlier observations of GFS performing better in sub-hour in PIT histogram.

Overall, weighted average ensembles proves to be the most skillful for NOR and RE2 followed by its constituent models Gekko, stacking and bidirectional LSTM.

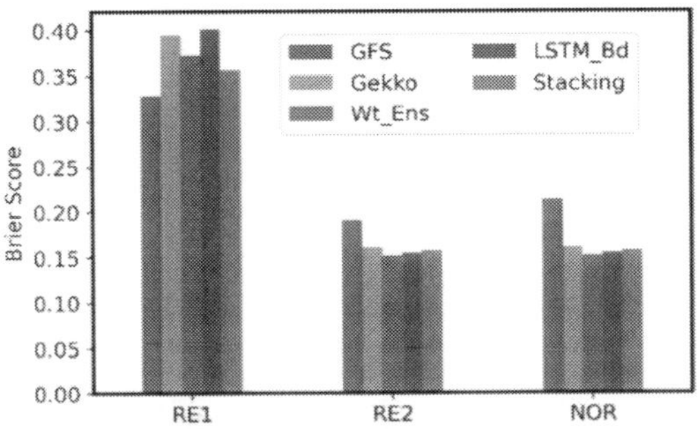

Figure 8 Brier scores for three events

Figure 9 (a,b,c) show the same for the three events. Overall weighted average ensemble is the most skillful from 7:00 hr. onward till sunset. BS is minimum during peak hours between 9:00 to 16:00 for NOR confirming that all the models are very skillful and predict forecast with high degree of certainty. Bidirectional LSTM also proves to be excellent performer during peak hours except at 12:00 hr. for NOR and RE2 events.

GFS performs better for RE1 event at most of the hours of the day as anticipated by observations from other metrices.

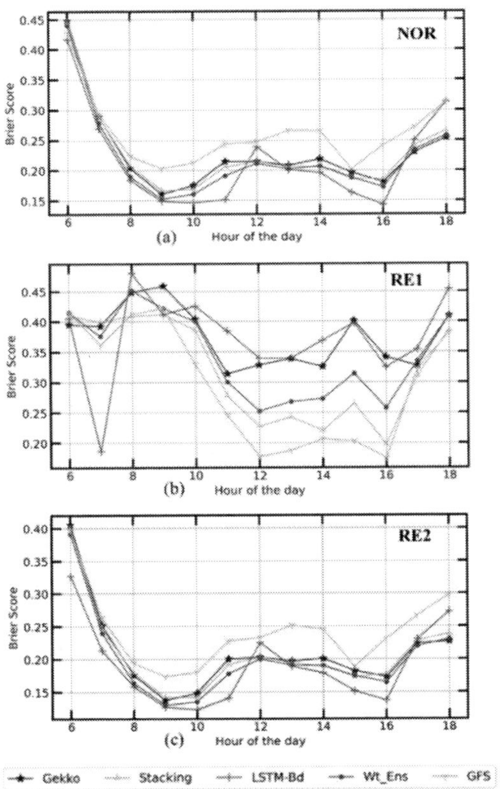

Figure 9 Variation of Brier Score with hour of the day for three events: (a) NOR, (b) RE1, (c) RE2.

4 CONCLUSIONS AND OUTLOOK

In this paper, we present a state-of-the-art two-level hybrid ensemble model for day-ahead irradiance forecasting with hourly resolution. We developed a novel methodology of classification which forms the basis of this study. This paper presents a distinct pattern of GFS forecasting and its major areas of forecasting errors.

Several machine learning models are applied and combined to build stacking and voting regression ensemble models. These along with Gekko parametric model and LSTM are combined to build weighted average ensemble model which proved to be the best performing model yielding aggregate nMAE and nRMSE score of 21% and 31.7% respectively during the highly variable seasons of winter and spring. The results do not include summer during which the model performance is expected to increase substantially. All the models are evaluated for probabilistic forecasts and the quality and certainty of forecasts is compared with that of GFS. The results show that the models developed are highly skillful and sharp to forecast with high degree of certainty.

This study demonstrates the potential of applying day type classification based on forecast irradiance in combination with ML and RNN techniques to build multilevel hybrid model to improve forecasting of solar irradiance to next level. Further improvement can be explored by applying adaptive boosting (AdaBoost) hybrid model to all ML models and even exploring Gated Recurrent Unit (GRU) LSTM and by experimenting with ECMWF and WRF day-ahead NWP forecast as exogenous input data.

5 REFERENCES

[1] Eltigani, D., Masri, S., 2015. Challenges of integrating renewable energy sources to smart grids: A review. Renew. Sustain. Energy Rev. 52, 770–780.

[2] Stram, B.N., 2016. Key challenges to expanding renewable energy. Energy Policy 96, 728–734.

[3] Souvik Sen and Sourav Ganguly. Opportunities, barriers, and issues with renewable energy development – A discussion. Renewable and Sustainable Energy Reviews, 69:1170–1181, 2017. ISSN 18790690. doi: 10.1016/j.rser.2016.09.137

[4] Perez, R., Lauret, P., Perez, M., David, M., Hoff, T.E., Kivalov, S., 2018. Solar resource variability. In: Perez, R. (Ed.), Wind Field and Solar Radiation Characterization and Forecasting: A Numerical Approach for Complex Terrain. Springer International Publishing, Cham, pp. 149–170.

[5] Lago, Jesus; Brabandere, Karel de; Ridder, Fjo de; Schutter, Bart de (2018): Short-term forecasting of solar irradiance without local telemetry: A generalized model using satellite data. In Solar Energy 173, pp. 566–577. DOI: 10.1016/j.solener.2018.07.050.

[6] Niklas Benedikt Blum, Bijan Nouri1, Stefan Wilbert, Thomas Schmidt, Ontje Lünsdorf, Jonas Stührenberg, Detlev Heinemann, Andreas Kazantzidis, and Robert Pitz-Paal: Atmos. Meas. Tech., 14, 5199–5224, 2021, DOI: doi.org/10.5194/amt-14-5199-2021

[7] Diagne, M., David, M., Lauret, P., Boland, J., Schmutz, N., 2013. Review of solar irradiance forecasting methods and a proposition for small-scale insular grids. Renew. Sustain. Energy Rev. 27, 65–76.

[8] Lorenz, E., Hurka, J., Heinemann, D., Beyer, H.G., 2009. Irradiance forecasting for the power prediction of grid-connected photovoltaic systems. IEEE J. Sel. Top. Appl. Earth Obs. Remote Sens. 2 (1), 2

[9] Jiménez, P.A., Alessandrini, S., Haupt, S.E., Deng, A., Kosovic, B., Lee, J.A., Delle Monache, L., 2016a. The role of unresolved clouds on short-range global horizontal irradiance predictability. Mon. Weather Rev. 144 (9), 3099–3107.

[10] Jiménez, P.A., Hacker, J.P., Dudhia, J., Haupt, S.E., Ruiz-Arias, J.A., Gueymard, C.A., Thompson, G., Eidhammer, T., Deng, A., 2016b. WRF-solar: Description and clearsky assessment of an augmented NWP model for solar power prediction. Bull. Am. Meteorol. Soc. 97 (7), 1249–1264.

[11] Glahn, H.R., Lowry, D.a., 1972. The use of model output statistics (MOS) in objective weather forecasting. J. Appl. Meteorol. 11 (8), 1203–1211.

[12] Verbois, Hadrien; Saint-Drenan, Yves-Marie; Thiery, Alexandre; Blanc, Philippe (2022): Statistical learning for NWP post-processing: A benchmark for solar irradiance forecasting.

[13] Jie Zhang; Bri-Mathias Hodge; Anthony Florita: NREL; Siyuan Lu; Hendrik F. Hamann: IBM TJ Watson Research Center; Venkat Banunarayanan: U.S. Department of Energy. Metrics for Evaluating the Accuracy of Solar Power Forecasting

[14] Thaker J., Höller R. A Comparative Study of Time Series Forecasting of Solar Energy Based on Irradiance Classification. Energies 2022, 15(8), 2837; https://doi.org/10.3390/en15082837

[15] Pang, Jingyue; Liu, Datong; Peng, Yu; Peng, Xiyuan (2018): Optimize the Coverage Probability of Prediction Interval for Anomaly Detection of Sensor-Based Monitoring Series. In Sensors (Basel, Switzerland) 18 (4). DOI: 10.3390/s18040967.

[16] https://www.kippzonen.com/Knowledge-Center/FAQ?item=28

[17] Cortes C. and Vapnik V. 1995, "Support vector networks", Machine Learning, 20:1—25

[18] Vapnik V. 1998, "Statistical Learning Theory", Wiley, New York

[19] Evgeniou, Theodoros; Pontil, Massimiliano (2001): Support Vector Machines: Theory and Applications. In G. Goos, J. Hartmanis, J. van Leeuwen, Georgios Paliouras, Vangelis Karkaletsis, Constantine D. Spyropoulos (Eds.): Machine Learning and Its Applications, vol. 2049. Berlin, Heidelberg: Springer Berlin Heidelberg (Lecture Notes in Computer Science), pp. 249–257.

[20] Fletcher Tristan, Support Vector Machines Explained. https://www.csd.uwo.ca/~xling/cs860/papers/SVM_Explained.pdf

[21] Sebastian Raschka, Machine Learning Lecture Notes. https://sebastianraschka.com/pdf/lecture-notes/stat479fs18/02_knn_notes.pdf

[22] Yakowitz, S.: Nearest-neighbour methods for time series analysis. Journal of Time Series Analysis 8(2), 235–247 (1987)

[23] Hochreiter S (1998) The vanishing gradient problem during learning recurrent neural nets and problem solutions. Int J Uncertainty Fuzziness Knowl Based Syst 6(02):107–116. https://doi.org/10.1142/S0218488598000094

[24] Hochreiter S, Schmidhuber J (1997) Long short-term memory. Neural Comput 9(8):1735–1780. https://doi.org/10.1162/neco. 1997.9.8.1735

[25] https://colah.github.io/posts/2015-08-Understanding-LSTMs/

[26] Staudemeyer, Ralf C.; Morris, Eric Rothstein (2019): Understanding LSTM -- a tutorial into Long Short-Term Memory Recurrent Neural Networks. Available online at http://arxiv.org/pdf/1909.09586v1

[27] Benali L, Notton G, Fouilloy A, Voyant C, Dizene R (2019) Solar radiation forecasting using artificial neural network and random forest methods: application to normal beam, horizontal diffuse and global components. Renew Energy 132:871–884. https://doi.org/10.1016/j.renene.2018.08.044

[28] Caballero R, Zarzalejo LF, Otero Á, Piñuel L, Wilbert S (2018) Short term cloud nowcasting for a solar power plant based on irradiance historical data. J Comput Sci Technol. https://doi.org/10.24215/16666038.18.e21

[29] Caldas M, Alonso-Suárez R (2019) Very short-term solar irradiance forecast using all-sky imaging and real-time irradiance measurements. Renew Energy 143:1643–1658 https://doi.org/10.1016/j.renene.2019.05.069

[30] Feng C, Cui M, Lee M, Zhang J, Hodge BM, Lu S, Hamann HF (2017) Short-term global horizontal irradiance forecasting based on sky imaging and pattern recognition. In: 2017 IEEE Power & Energy Society General Meeting, IEEE, pp 1–5. https://doi.org/10.1109/PESGM.2017.8274480

[31] Huang X, Zhang C, Li Q, Tai Y, Gao B, Shi J (2020) A comparison of hour-ahead solar irradiance forecasting models based on lstm network. Math Probl Eng. https://doi.org/10.1155/2020/4251517

[32] Li G, Wang H, Zhang S, Xin J, Liu H (2019) Recurrent neural networks based photovoltaic power forecasting approach. Energies 12(13):2538. https://doi.org/10.3390/en12132538

[33] http://www.scholarpedia.org/article/Ensemble_learning

[34] Hansen L.K., Salamon P (1990) Neural network ensembles. IEEE Transactions on Pattern Analysis and Machine Intelligence(Volume: 12, Issue: 10). https://doi.org/10.1109/34.58871

HOW DOES THE USE OF SATELLITE-DERIVED INSOLATION DATA IMPACT THE ACCURACY OF PERFORMANCE RATIO ESTIMATES?

Ebrar Özkalay[1,2*], Alessandro Virtuani[2,3], Andrew Fairbrother[2], Artur Skoczek[4], Gabi Friesen[1], Christophe Ballif[2,3]
1 – University of Applied Sciences and Arts of Southern Switzerland (SUPSI-PVLab), Via Flora Ruchat-Roncati 15, 6850 Mendrisio, Switzerland
2 – École Polytechnique Fédérale de Lausanne (EPFL), Institute of Electrical and Micro Engineering (IEM), Photovoltaics and Thin-Film Electronics Laboratory, Rue de la Maladière 71b, 2002 Neuchâtel, Switzerland
3 – CSEM PV-Center, Rue de Jaquet-Droz 1, 2002 Neuchâtel, Switzerland
4 – Solargis s.r.o., Bottova 2A, 811 09 Bratislava, Slovakia
Email: ebrar.ozkalay@supsi.ch

ABSTRACT: PV systems' short- and long-term performances are often analyzed using satellite-derived insolation data. Despite the fact that higher uncertainties are associated with satellite-derived insolation data, no studies have been conducted to quantify the impact on the assessment of the PV plant's performance ratio (PR), used to judge the overall health status of a plant. In order to evaluate the performance of a PV system accurately, the uncertainty contribution needs to be carefully computed. In this work, we quantify: (1) the accuracy of two satellite-derived insolation data-set (open-source and paid service) with respect to the irradiance ground measurement obtained by properly maintained sensors; and (2) their effect on the uncertainty of the plant's PR computed using satellite-derived insolation with different temporal resolutions. The uncertainty of PR computed using insolation data from a well-maintained and regularly recalibrated pyranometer computed as 3.6% (k=2), and it assumed constant for all temporal resolutions (daily, monthly and yearly) as a conservative assumption. The statistical evaluation shows that monthly and yearly PR values, which have an uncertainty of less than 5%, can be obtained using satellite-derived insolation data without having to deal with regular calibration and maintenance of an on-site pyranometer. We showed that depending on the quality of satellite-derived data and temporal resolution, uncertainty of PR differs. The PR computed using paid-service satellite-derived data has approximately a 1.7% lower uncertainty (k=2) than the PR computed using open-source satellite-derived data for the temporal resolutions considered (day and month), with the exception of the yearly resolution (for which this value is 0.6% lower).

Keywords: Performance Ratio (PR), ground-based insolation data, satellite-derived insolation data, uncertainty

1 INTRODUCTION

Photovoltaic (PV) module manufacturers, plant owners, and operation and maintenance (O&M) companies all need to analyze the performance of the PV systems accurately. The performance ratio (PR) is a widely accepted metric used to evaluate the performance of PV systems. As a first approximation, this is the ratio between the system's energy yield (kWh) (over a given temporal interval) and the amount of solar resource received over the same interval.

Usually, irradiance sensors to measure received solar resources are not widely used in residential (or commercial/industrial) installations, whereas for larger systems, these sensors are frequently reported to lack maintenance or regular recalibrations. Data from local meteorological ground stations are valuable and accurate since the instruments are usually well-maintained, but they are sparsely distributed and hence not a suitable data source for many PV systems. In such cases, satellite-derived insolation data can be a valuable alternative for evaluating the performance of a PV system. However, the uncertainty of satellite-derived insolation data can be much higher than that of ground-based data recorded with well-maintained sensors. As a result, the performance metric PR, computed using satellite-derived data and evaluating the PV system's performance, is less accurate. It is, however, known that when longer integration times are used to determine insolation (kWh/m^2) (e.g. hour, day, month, and year), the discrepancy between ground-based and satellite-derived insolation data is considerably reduced, leading to a reduction in uncertainty.

This work quantifies the accuracy of satellite-derived insolation of two sources (open-source and paid service) and understands the impact of satellite-derived insolation data on the accuracy of PR estimates. In order to do that, ground-measured insolation data and PR time series computed using ground-mounted sensors are used as benchmarks. In parallel, possible ways to reduce uncertainty are discussed, such as the quality of satellite-derived data and temporal resolution.

2 EXPERIMENTAL

2.1 Ground-based and Satellite-derived Insolation Data

Table 1 shows details of the ground-based monitoring system and satellite-derived insolation data-sets. Ground-based global horizontal irradiance (GHI) and the plane of array irradiance (G_{POA}) (20° tilt and 4° azimuth from the South) were monitored every minute for over three years (between December 2015 and May 2019) by two secondary standard pyranometers mounted at SUPSI's premises (Canobbio, Switzerland, 46.02°N and 8.91°E).

The CAMS satellite-derived insolation data are freely available (open-source) and have GHI, diffuse horizontal irradiance (DHI) and direct normal irradiance (DNI) components with 1-minute intervals for the same time period (between December 2015 and May 2019) [1]. Since G_{POA} data is unavailable, it was calculated using Perez's model in PVlib [2]. The second data-set is a paid service with propriety data processing. GHI, DHI, DNI and G_{POA} components were retrieved with a 5-minute frequency for the same period (between December 2015 and May 2019) from Solargis [3]. For the location analyzed in this study, both Solargis and CAMS mostly use the same input parameters from the satellite but use different models to derive insolation (e.g. clear sky irradiance, cloud index, all-sky irradiance, etc.) [3, 4].

Table I: Ground-based and satellite-derived irradiance monitoring systems (UC: uncertainty, k: coverage factor).

	Ground	CAMS	Solargis
Instrument	Pyranometer CM11 (UC = 2.7%, k=2)	Satellite (open-source)	Satellite (paid-service)
Time-step	1 minute	1 minute	5 minute
GHI	✓	✓	✓
DHI	–	✓	✓
DNI	–	✓	✓
G_{POA}	✓	– (PVlib *Perez* [2])	✓

The single PV module's long-term performance was recorded for over three years between December 2015 and May 2019 at SUPSI's premises (Canobbio, Switzerland, 46.02°N and 8.91°E). The module was installed at a tilt angle of 20° and 4° azimuth from the South and monitored using a maximum power point (MPP) tracker (MPPT3000) especially developed by SUPSI [5]. The hardware combines IV tracing performed every 5-minute, whereas the module is otherwise operated at its maximum power using an MPP tracker.

3 RESULTS AND DISCUSSION

3.1 Uncertainty of Satellite-derived Insolation Data

As a first step, ground-measured irradiance values between 50 W/m^2 and 1250 W/m^2 were removed to exclude anomalies and erroneous measurements. In addition, due to mountains on the west and the east sides, the data outside of our quality intervals, which are early morning and late afternoon data, were removed.

To provide insight into the satellite-derived data relative to ground-based measurement, statistical metrics, including mean bias error (MBE) and root mean square error (RMSE) for intended temporal resolutions, were calculated using the following formulas.

$$\text{MBE}\,[\%] = \frac{\frac{1}{N}\sum_{t=1}^{N}(G_{sat_t} - G_{gr_t})}{\frac{1}{N}\sum_{t=1}^{N} G_{gr_t}} \times 100$$

$$\text{RMSE}\,[\%] = \frac{\sqrt{\frac{1}{N}\sum_{t=1}^{N}(G_{sat_t} - G_{gr_t})^2}}{\sqrt{\frac{1}{N}\sum_{t=1}^{N} G_{gr_t}^{\;2}}} \times 100$$

where N represents the total number of measurements, G_{sat} is satellite-derived GHI or G_{POA}, and G_{gr} is ground-based GHI or G_{POA}.

As shown in Figure 1, Solargis GHI and G_{POA} data have better precision (lower RMSE) and accuracy (lower MBE) for all temporal intervals compared to the CAMS data. The precision of both satellite-derived data (RMSE) improves with longer temporal intervals for both GHI and G_{POA}. However, the accuracy (MBE) is independent of the temporal resolution.

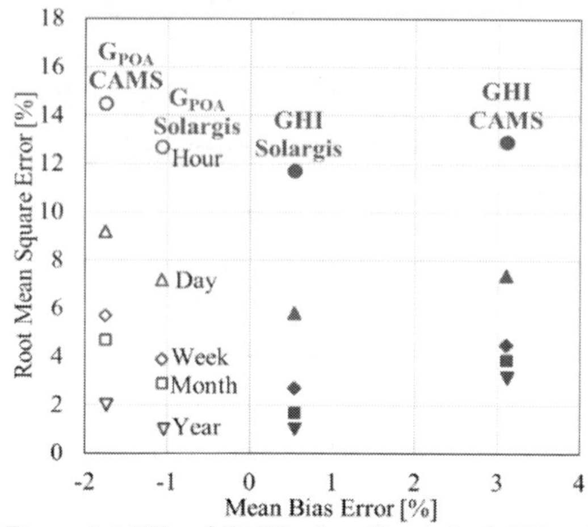

Figure 1: MBE and RMSE of satellite-derived GHI and G_{POA} data from CAMS and Solargis with respect to ground-measured data using well-maintained thermopile pyranometers (Circle: Hour, Triangle: Day, Tilted square: Week, Square: Month and Reverse triangle: Year).

The overall uncertainty of satellite ($UC_{Satellite}$) data was calculated using the approach in the report of IEA PVPS Task 16 [6]. The method uses MBE and RMSE of satellite-derived insolation data and the ground irradiance measurement uncertainty (UC_{Pyra}) to compute the uncertainty of satellite-derived insolation. This method assumes a Gaussian distribution and provides 95% confidence (coverage factor, k=2). Considering the pyranometer specifications, measurement conditions, calibration procedure, secondary calibration, maintenance and environmental conditions, the uncertainty of the pyranometer is 2.7% (k=2) for the monitoring period [7]. As suggested by IEA PVPS Task 16 [6], we assumed that the uncertainty of the pyranometer is constant at different temporal resolutions despite this being a conservative assumption. In addition, some random errors cancel out for larger time intervals. The following formula was used to calculate $UC_{Satellite}$ [6].

$$UC_{Satellite}[\%] = k \times \pm\sqrt{\left(\frac{UC_{Pyra}}{k}\right)^2 + \left(\frac{MBE}{k}\right)^2 + \left(\frac{RMSE}{k}\right)^2}$$

Figure 2 shows the $UC_{Satellite}$ of G_{POA} for CAMS and Solargis data (relative to ground-based measurement) with different temporal resolutions. The uncertainties of G_{POA} decrease significantly with longer temporal intervals as some errors average out, as seen in the decrease of RMSE with longer temporal intervals. The uncertainties of Solargis G_{POA} are below 5% for weekly, monthly and yearly resolutions, whereas only yearly CAMS G_{POA} has uncertainty below 5%. Solargis G_{POA} has, on average, 1.8% lower uncertainty than CAMS G_{POA} for all temporal resolutions, except for the yearly temporal resolution, which is 0.7% lower. In this analysis, we refrain from commenting on the results of the yearly resolution to avoid possible erroneous results since only 3.5 years of data are available.

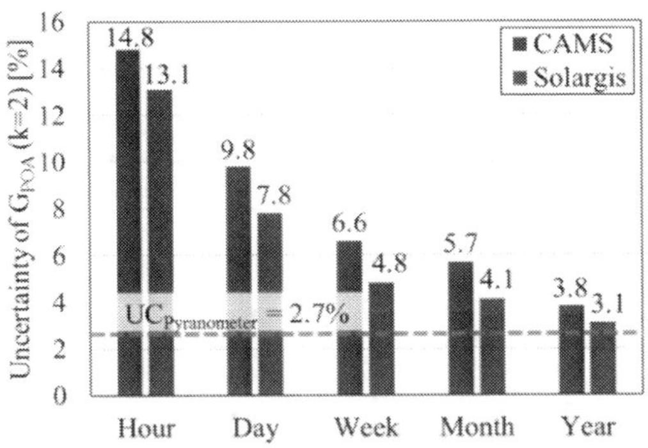

Figure 2: Uncertainty (k=2) of G_{POA} for CAMS and Solargis satellite-derived data for different temporal resolutions. The green dashed line shows the uncertainty (k=2) of ground-based insolation measured by pyranometers.

3.2 Uncertainty of Performance Ratio

The overall uncertainty of PR (UC_{PR}) was calculated by propagating the uncertainties of G_{POA}, the standard test condition (STC) power measurement by Pasan IIIb at SUPSI PV-Lab (1.6%, k=2, spectrum corrected) and the outdoor power measurement by maximum power point tracker (MPPT3000) (1.74%, k=2, including uncertainty of tracker [5]).

Table II shows the uncertainties of PR computed using the Solargis ($PR_{Solargis}$) and CAMS (PR_{CAMS}) satellite-derived and ground-based (PR_{Pyra}) insolation data-sets for different temporal resolutions (day, month and year). Since all the uncertainty contributions to the uncertainty of PR_{Pyra} are independent of temporal resolution, the uncertainty of PR_{Pyra} is constant at different temporal resolutions. As expected, since the uncertainty of satellite-derived insolation decreases with longer temporal intervals, the uncertainty of PR also decreases. $PR_{Solargis}$ has an approximately 1.7% lower uncertainty than for PR_{CAMS} for daily and monthly temporal resolutions, and for the yearly temporal resolution, this discrepancy is lowered to 0.6%.

Table II: Uncertainties (k=2) of PRs computed using ground-based and satellite-derived CAMS and Solargis G_{POA} data.

	Uncertainty of PR (k=2)		
Temporal res. / Insolation data	**Day**	**Month**	**Year**
Ground	3.6%		
CAMS	10.1%	6.2%	4.5%
Solargis	8.2%	4.7%	3.9%

The uncertainties of monthly and yearly PR computed using Solargis are 1.1% and 0.3% higher than the best case (3.6%, which is the uncertainty of PR_{Pyra}), respectively. Depending on the purpose of use, different time resolutions and satellite-derived data quality can be preferred for analysis using PR as a performance index. For example, if only monthly or annual PR is the matter to check the general status of a PV system, PR values, which have an uncertainty of less than 5%, can be obtained using satellite-derived insolation data without having to deal with regular calibration and maintenance of an on-site pyranometer. However, there is a trade-off since the accuracy and precision drop compared to the best case.

4 CONCLUSION

In general, having reliable data from on-site radiometric sensors would lead to lower uncertainties of irradiance and PR. Unfortunately, these sensors are missing, or their data are not entirely reliable since they are not well-maintained and have no regular recalibrations. In this work, we computed PR's uncertainty using ground-based and satellite-derived insolation data from different sources (CAMS and Solargis) and for different temporal resolutions. Solargis satellite-derived G_{POA} insolation data (paid service) has about 1.8% lower uncertainty than CAMS (open-source) data, except for the yearly temporal resolution (for which it is 0.7% lower). This lower uncertainty is reflected in the uncertainty estimation of the PR: daily and monthly $PR_{Solargis}$ have about 1.7% lower uncertainty than PR_{CAMS}, except for the yearly temporal resolution (0.6% lower). The uncertainty of PR decreases significantly with longer temporal intervals as some uncertainty sources of satellite-derived insolation cancel each other out.

We showed that it is possible to obtain monthly and yearly PR with less than 5% uncertainty using satellite-derived insolation data. The monthly and yearly $PR_{Solargis}$'s uncertainties are only 1.1% and 0.3% higher than the uncertainty of the PR_{Pyra} (3.6%), respectively. Depending on the purpose of use and requirements, one of the appropriate satellite-derived data sources (open-source or paid service) or ground-based measurement options can be preferred.

5 ACKNOWLEDGEMENT

The project called "Reliability of PV Systems integrated into the built environment" (REBIPV) is supported by the Swiss National Science Foundation under COST IZCOZ0_182967 and the Swiss Federal Office of Energy. In addition, the COST Action PEARL PV (CA16235) provided additional (collaborative) support. Finally, the authors would like to thank SUPSI-PVLab (University of Applied Sciences and Arts of Southern Switzerland), the BIPV-Innovative Building Skin Team researchers, and all the staff members who contributed to supplying the relevant data for this work.

6 REFERENCES

[1] M. Lefèvre, A. Oumbe, P. Blanc, B. Espinar, B. Gschwind, Z. Qu, L. Wald, M. Schroedter-Homscheidt, C. Hoyer-Klick, A. Arola, A. Benedetti, J. W. Kaiser and J.-J. Morcrette, "McClear: a new model estimating downwelling solar radiation at ground level in clear-sky conditions," *Atmos. Meas. Tech.*, vol. 6, p. 2403–2418, 2013.

[2] W. F. Holmgren, C. W. Hansen and M. A. Mikofski, "PVlib python: a python package for modeling solar energy systems," *Journal of Open Source Software*, 2018.

[3] Solargis, "GIS data layers of Global Horizontal Irradiation, Air Temperature and Photovoltaic Electricity Potential," 2017.

[4] Z. Qu, A. Oumbe, P. Blanc, B. Espinar, G. Gesell, B. Gschwind, L. Klüser, M. Lefèvre, L. Saboret, M. Schroedter-Homscheidt and L. Wald, "Fast radiative transfer parameterization for assessing the surface solar irradiance: The Heliosat-4 method," *Meteorologische Zeitschrift,* vol. 26, pp. 33-57, 2017.

[5] D. Chianese, E. Burà and A. Realini, "MPPT3000 multifunction testing device for PV modules," in *23rd European Photovoltaic Solar Energy Conference and Exhibition*, Valencia, 2008.

[6] IEA PVPS Task 16, "Best Practices Handbook for the Collection and Use of Solar Resource Data for Solar Energy Applications: Third Edition," 2021.

[7] D. Dominé, A. Jagomägi, A. G. d. Montgareuil, G. Friesen, E. Mõttus, H.-D.-. Mohring, D. Stellbogen, T. Betts, R. Gottschalg, T. Zdanowicz, M. Prorok, F. Fabero, D. Faiman and W. Herrmann, "Uncertainties of PV Module - long-term outdoor testing," in *25th European Photovoltaic Solar Energy Conference and Exhibition*, Valencia, 2010.

DAY-AHEAD PHOTOVOLTAIC PRODUCTION FORECASTING USING XGBOOST AND POST-CALIBRATION MODELING

Spyros Theocharides[1], Javier Lopez Lorente[1], George Makrides[1] and George E. Georghiou[1]
[1] FOSS Research Centre for Sustainable Energy, Photovoltaic Technology Laboratory, Department of Electrical and Computer Engineering, University of Cyprus, Nicosia, 1678, Cyprus

ABSTRACT: This work proposes a methodology for accurate day-ahead photovoltaic (PV) production forecasting models that utilized machine learning techniques for forecasting and clustering. Specifically, the core element of the proposed method is a power output predictive model (trained in an extreme supervised learning regime) that incorporates an output post-processing step for calibrating the forecasts against cluster-based irradiance profiles. The initial forecasting model was developed using an extreme gradient boosting (XGBoost) algorithm, while the clustering was performed using the TSClust algorithm and the post-calibration using extreme learning machines (ELM). The TSClust indicated 5 types of clusters that were used to assign the type of cluster to each daily solar irradiance profile. The ELM used as input parameters the type of cluster and the output of the XGBoost model (power forecast) to improve the quality and accuracy of the PV production forecasting. The proposed methodology demonstrated absolute improvements of up to 2% compared to forecasting without post-calibration.

Keywords: clustering, forecasting, machine learning, photovoltaic, solar irradiance.

1 INTRODUCTION

The high penetration of grid-connected photovoltaic (PV) systems introduces new challenges to the management and stability of electricity grids. Balancing the demand with an increasing share of PV production is becoming a very challenging task due to the inherently variable weather-dependent production and requires grid management services to ensure availability, stability and energy quality. In this respect, accurate PV production forecasting is a proven method to alleviate the resource uncertainty posed by large shares of PV systems through active grid management. In particular, accurate day-ahead PV production forecasting can assist utilities and plant operators in the direction of energy management through optimal operation, dispatchability and operation mode planning.

Day-ahead PV production forecasting is a feature that has been recently added to the electricity system operation with the most commonly adopted forecasting approaches based on time series analysis and the development of parametric models [1–3]. These approaches require PV system technical characteristics (capacity, temperature coefficients, installation information, etc.) to simulate the behaviour of the system and since accurate knowledge of system characteristics and actual performance is not always available, this has direct implications on their ability to forecast. In this sense, the forecasting accuracy of parametric models is driven by the performance of the model component parameters.

In the scope of refining the forecasting accuracy of the power produced by PV systems, adaptive data-driven machine learning techniques have been gaining ground. Machine learning models that do not require knowledge of the physical parameters but only construct relationships between input and output variables based on data-driven approaches have already been developed and presented in the literature [4–9]. However, a primary objective is to develop a methodology that imports the forecasting accuracy beyond the state-of-the-art pursuing in parallel minimal computational and data complexity. In addition, there is yet no complete universal forecasting model and methodology to ensure accurate forecasts according to the technology and location.

The scope of this work is to present a holistic methodology for constructing accurately performing day-ahead PV production forecasting models that leverage novel machine learning techniques both for supervised learning predictions and cluster profiling. In particular, the core element of the proposed method is a power output predictive model (trained in an extreme supervised learning regime) that incorporates an output post-processing step for calibrating the forecasts against cluster-based irradiance profiles.

The approach followed includes the development of a data-driven machine learning model based on the Extreme Gradient Boosting (XGBoost) algorithm. Specifically, the predictive model was optimised according to the input parameters used for training and the duration of the dataset. In addition, a clustering assessment on historic daily solar irradiance patterns was performed in the scope of calibrating the output of the forecasting model to the obtained profiles using an extreme learning machine (ELM). The performance of the day-ahead forecasting model was evaluated on actual measurements of a test-bench PV system installed at the outdoor test facility (OTF) of the University of Cyprus (UCY), Cyprus, and against predefined performance metrics.

2 EXPERIMENTAL SETUP

2.1 OTF of UCY

Historic performance and weather measurements were acquired over a year from a 1 kWp test-bench grid-connected PV system installed at the OTF of UCY in Nicosia, Cyprus (Koppen's classification BSh; hot semi-arid). The performance of the system and environmental weather conditions were monitored and continuously recorded at a resolution of 30 minutes according to the requirements set by the IEC 61724 [10]. The meteorological measurements included the incident global irradiance (G_{I}), wind direction (W_{a}), wind speed (W_{s}) and ambient temperature (T_{amb}). The PV system operational measurements included the maximum power current (I_{mp}), voltage (V_{mp}) and power (P_{mp}), as measured at the output of the PV system (DC side) [11,12].

2.2 Weather Data

The solar irradiance data were derived from the Weather Research and Forecasting (WRF) Model, which is a mesoscale NWP model designed for atmospheric research and operational forecasting applications [13]. In addition, the Copernicus Atmosphere Monitoring Service (CAMS) all-sky radiation service provided solar radiation data at the investigated location. Time series data of global horizontal irradiance (GHI) observations at a 15-minute temporal resolution were obtained from the Copernicus portal.

3 METHODOLOGY

3.1 Approach

This work aims to provide a holistic method for developing advanced day-ahead PV production forecasting models that are based on state-of-the-art machine learning principles and enhanced with output post-calibration techniques. The approach followed includes the development of a data-driven power output predictive model based on the XGBoost ensemble algorithm [14]. Specifically, the predictive model was constructed by applying the XGBoost regressor to the historical data-sets of a test-bench PV system (train set of yearly evaluation period), installed at the OTF of UCY, under an extreme supervised learning regime. In this context, the hyperparameters of the constructed predictive model were optimized by selecting the most significant input features. The methodology further entailed the clustering of historic solar irradiance measurements using the TSClust technique [15], in an attempt to obtain calibration factors for the forecasted power output according to the day-ahead irradiance predictions using an ELM. The performance of the day-ahead PV production forecasting model was evaluated based on acquired historic measurements (test set of yearly evaluation period) and predefined metrics. The overall methodology is illustrated in Figure 1.

3.2 XGBoost

XGBoost is an enhanced distributed gradient boosting (GB) framework developed to be very efficient and flexible. XGBoost provides a parallel tree boosting (also known as GBDT, GBM) that solves regression and classification problems [14].

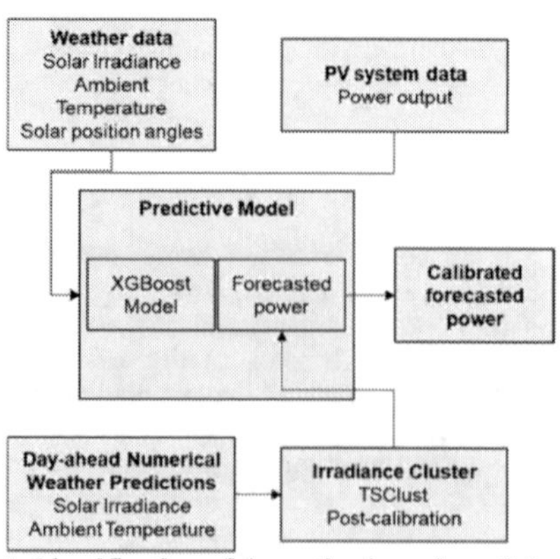

Figure 1: Identification of the optimal number of clusters using the Clustree function.

3.3 TSclust

TSclust is a type of partitional and fuzzy clustering procedure that uses a custom implementation to assign the residuals into the fitting clusters [15]. The measure of clustering evaluation can be found in (1):

$$Sim(G, C) = \frac{1}{k}\sum_{i=1}^{k} max_{1\leq j\leq k} Sim(G_i, C_j) \qquad (1)$$

where,

$$Sim(G_i, C_j) = \frac{2|G_i \cap C_j|}{|G_i| + |C_j|} \qquad (2)$$

where |.| indicating the cardinality of the parameters in the set. This measure has been used for comparing different clustering methodologies [16,17].

3.4 Post-calibration correction

The post-calibration stage was implemented to improve the forecasted power from the XGBoost model, which was a combination of the TSClust clustering, to assign each day to a specific cluster, and an ELM [18], where its parameters do not need tuning. The input parameters of the ELM are the clustered irradiance and the output of the XGBoost model (forecasted power).

3.5 Performance Evaluation

The performance evaluation of the proposed day-ahead PV production forecasting model is based on several performance metrics applied to the test set evaluation period: (i) the mean absolute percentage error (MAPE); (ii) the nRMSE, which is the relative root mean square error (RMSE) normalised to the nominal capacity of the PV system (P_{nominal}); and (iii) the mean bias error (MBE) which is the methodical error or the overall bias error. These are defined by equations (4) to (6), respectively.

$$\text{MAPE} = \frac{100}{n} \cdot \sum_{i=1}^{n} \left| \frac{y_{\text{observed,i}} - y_{\text{forecasted,i}}}{y_{\text{actual,i}}} \right| \qquad (4)$$

$$\text{nRMSE} = \frac{100}{P_{\text{nominal}}} \sqrt{\frac{1}{n} \cdot \sum_{i=1}^{n} (y_{\text{observed,i}} - y_{\text{forecasted,i}})^2} \qquad (5)$$

$$\text{MBE} = \frac{100}{n} \times \sum_{i}^{n} \frac{y_{\text{observed,i}} - y_{\text{forecasted,i}}}{y_{\text{actual,i}}} \qquad (6)$$

4 RESULTS

4.1 Performance Evaluation of Ground vs Satellite data

The performance evaluation metrics for the location of the ground measurements and satellite data demonstrated an overall nRMSE of 10.1% and 10.6% respectively, while the MBE demonstrates that there is not a significant bias within the data since is lower than 5%. Additionally, Table 1, demonstrates more detailed analysis of the performance evaluation metrics providing evidence that improvements can be made by applying the proposed post-calibration method.

Table 1: Performance of the solar irradiance forecasts compared to the ground- and satellite-based measurements for the test-bench PV system site, over the evaluation period.

Ground Measurements		
MAPE (%)	nRMSE (%)	MBE (%)
8.45	10.10	2.08
Satellite Data		
MAPE (%)	nRMSE (%)	MBE (%)
9.82	10.60	-3.51
Ground vs. Satellite		
MAPE (%)	nRMSE (%)	MBE (%)
0.95	1.49	-0.09

4.2 Solar Irradiance Clustering

The clustering analysis was applied to the average daily solar irradiance profiles and the number of clusters was determined using the Clustree function [19], which enables the identification of the optimal number of clusters using a tree visualization. The location of the largest single leaf of the tree, which is also preceded by a single leaf, indicates the optimal number of clusters, Figure 2 demonstrates the results from the Clustree function. For this study, 5 solar irradiance clusters were identified. The 5 daily solar irradiance profile clusters identified from this analysis were divided as follows: (a) C1: Clear sky day, (b) C2: high-medium solar irradiance day, (c) C3: medium solar irradiance day, (d) C4: medium-low solar irradiance day and (e) C5: low solar irradiance day. The characteristic (averaged) daily solar irradiance profile of each cluster is illustrated in Figure 3.

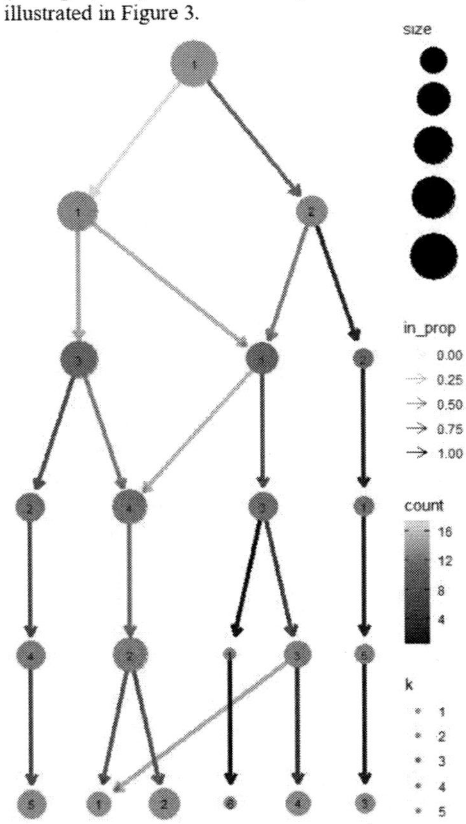

Figure 2: Identification of the optimal number of clusters using the Clustree function.

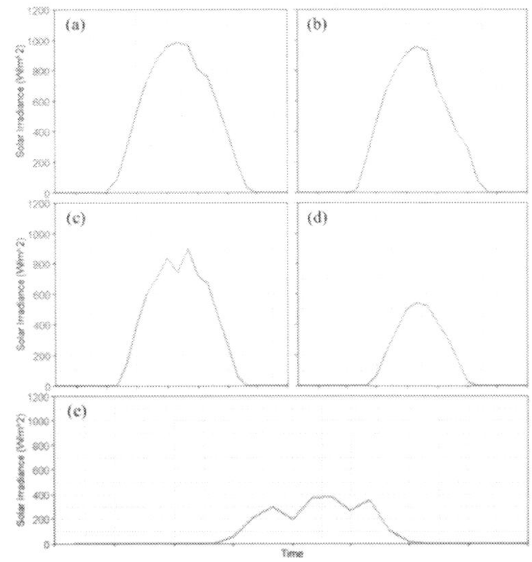

Figure 3: Daily solar irradiance profile clusters of: (a) C1, (b) C2, (c) C3, (d) C4 and (e) C5. The clusters are labeled progressively from clear-sky day to low solar irradiance day.

4.3 Post-calibration methodology

As already mentioned the ELM model does not require any kind of tuning or optimization to elevate its performance. Therefore, the ELM model can be developed using only the training and testing stages. Table 2 summarizes the number of days assigned to each cluster and the number of days used as training and testing for the ELM model.

Table 2: Number of days assigned to each cluster and the number of days used as training and testing by the ELM.

Cluster	Days		
	Total	**Training**	**Testing**
C1	143	100	43
C2	100	70	30
C3	64	45	19
C4	31	22	9
C5	27	19	8

4.4 Performance evaluation

The scatter plot in Figure 4 demonstrates the forecasted and actual residuals of the XGBoost model and the forecasting acquired by the post-calibrating methodology. The data distribution in figure shows a more concentrated (to the fitting line) residuals (Figure 4a) compared to the single XGBoost model (Figure 4b), which demonstrated more scattered residuals. This indicates that the proposed methodology accurately improved the quality and accuracy of the forecasted power output, helping reduce any systematic bias inherent to the single XGBoost model.

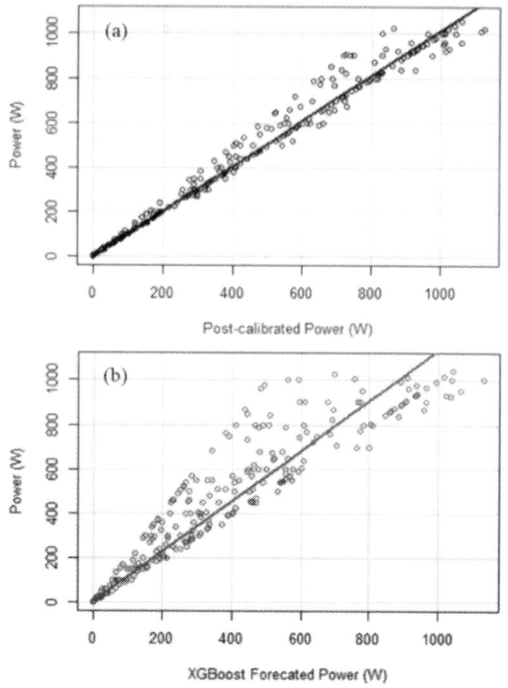

Figure 4: Scatter plot forecasted against the actual power residuals for (a) the post-calibrating model and (b) the single XGBoost model.

Additionally, to exemplify and compare the behaviour of the proposed post-calibration against the single XGBoost and the observations Taylor diagram, which associates the heuristic distance between the proposed methodologies against the respective actual data [20]. The standard deviation and cantered root mean squared (RMS) difference results for each of the two models, are exhibited depicted in Figure 5. The figure indicates that the post-calibrating model (the model with the less distance from the observations) demonstrated the most accurate behaviour compared with the single XGBoost model.

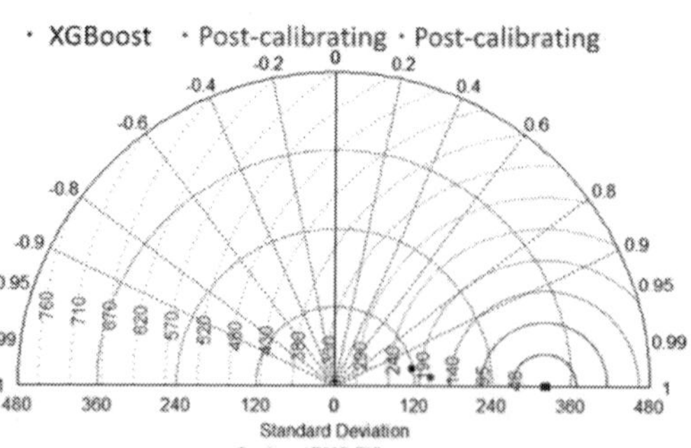

Figure 5: Visual representation of the best-performing model Taylor diagram. The blue and red dots represent the single XGBoost and post-calibrating methodology respectively, while the green rectangular represents the observations.

In addition, Table 3 summarises the performance metrics of the XGBoost and the proposed post-calibrating methodology against the predefined performance metrics. The table illustrates that the post-calibrating accomplished a *MAPE*, *nRMSE* and *MBE* of 4.51%, 5.98% and 1.37%, respectively, when applied to the test set evaluation dataset and outperformed both the model without post-processing correction and the PM (baseline model).

Table 3: Performance evaluation metrics for the XGBoost and post-calibrating methodologies.

Metrics	Methods	
	Post-calibrating methodology	**XGBoost Model**
MAPE (%)	4.51	5.17
nRMSE (%)	5.98	7.69
MBE (%)	1.37	-1.82

Finally, Figure 6 demonstrates a visual representation of the actual against the forecasted power provided by the post-calibration methodology to verify whether the proposed methodology captures the behaviour of the PV system or not. It can be observed how the forecast was close to the actual readings with some slight underestimations during the central hours of the exemplified days. It is however important to mention that post-calibrating model is overall more prone to overestimations as illustrated by the positive *MBE* compared to the single XGBoost model, which presented a negative (underestimation) *MBE*.

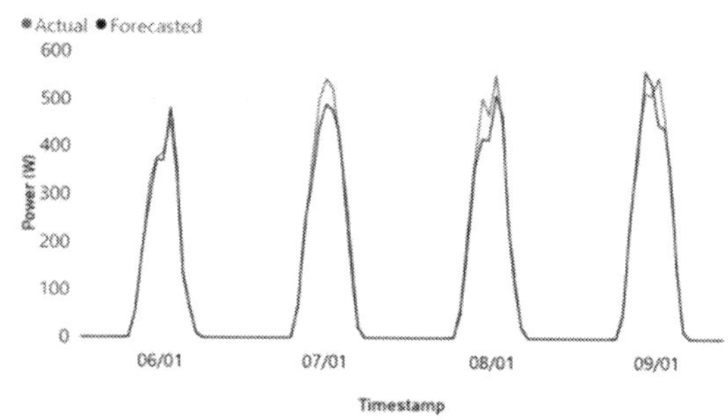

5 CONCLUSIONS

The high shares of grid-connected photovoltaic (PV) systems and the intermittent nature of solar generation introduce new reliability challenges to electricity grids. In particular, the inherent balancing operations of the demand to the supply require new grid planning tools that consider the weather-dependent variability of solar generation. To that end, performance improvements of day-ahead PV production forecasts can be achieved by leveraging state-of-the-art machine learning principles and alleviating the uncertainty of solar irradiance predictions through analytical post-processing.

The scope of this work was to present a holistic methodology for constructing accurately performing day-ahead PV production forecasting models that leverage novel machine learning techniques both for supervised learning predictions and cluster profiling. In particular, the core element of the proposed method is a power output predictive model incorporating an output post-processing step for calibrating the forecasts against cluster-based irradiance profiles.

The approach followed includes the development of a data-driven machine learning model based on the Extreme Gradient Boosting (XGBoost) ensemble algorithm. Specifically, the predictive model was optimised according to the input parameters used for training and the duration of the dataset. In addition, a clustering assessment on historic daily solar irradiance patterns was performed in the scope of calibrating the output of the forecasting model to the obtained profiles and the post-calibration was achieved using an extreme learning machine (ELM) model. The performance of the day-ahead forecasting model was evaluated on actual measurements of a test-bench PV system installed at the outdoor test facility (OTF) of the University of Cyprus (UCY), Cyprus, and against predefined performance metrics.

The solar irradiance forecasts were calculated using a conventional Weather Research Forecasting (WRF) model and were benchmarked against historic ground measurements and satellite data, which exhibited errors of around 10%. Thus, providing evidence that significant improvements can be achieved by calibrating the power output predictions. The application of the clustering following the TSclust technique to the yearly historic irradiance measurements resulted in 5 identified solar irradiance daily profiles.

Finally, the results demonstrated that the application of the proposed forecasting solution provided accurate forecasts with a daily nRMSE of 5.98% and MAPE at 4.51%. That indicates that the application of post-calibration presents an absolute improvement of up to 2% for all types of days.

ACKNOWLEDGEMENTS

This work was co-funded by the European Regional Development Fund and the Republic of Cyprus through the Cyprus Research and Innovation Foundation (RIF) in the framework of the project "ELECTRA" with protocol number: INTEGRATED/0918/0071.

REFERENCES

[1] S. Pelland, J. Remund, J. Kleissl, T. Oozeki, K. De Brabandere, Photovoltaic and Solar Forecasting: State of the Art, Int. Energy Agency Photovolt. Power Syst. Program. Rep. IEA PVPS T14. (2013) 1–40. https://doi.org/978-3-906042-13-8.

[2] E. Lorenz, T. Scheidsteger, J. Hurka, D. Heinemann, C. Kurz, Regional PV power prediction for improved grid integration, Prog. Photovoltaics Res. Appl. 19 (2011) 757–771. https://doi.org/10.1002/pip.1033.

[3] S. Pelland, G. Galanis, G. Kallos, Solar and photovoltaic forecasting through post-processing of the Global Environmental Multiscale numerical weather prediction model, Prog. Photovoltaics Res. Appl. 21 (2013) 284–296. https://doi.org/10.1002/pip.1180.

[4] I. Ashraf, a Chandra, Artificial neural network based models for forecasting electricity generation of grid connected solar PV power plant, Int. J. Glob. Energy Issues. 21 (2004) 119–130. http://inderscience.metapress.com/content/N0R76HDTKC19GWVH.

[5] D. Caputo, F. Grimaccia, M. Mussetta, R.E. Zich, Photovoltaic plants predictive model by means of ANN trained by a hybrid evolutionary algorithm, in: Proc. Int. Jt. Conf. Neural Networks, 2010. https://doi.org/10.1109/IJCNN.2010.5596782.

[6] S.A. Kalogirou, Applications of artificial neural-networks for energy systems, Appl. Energy. 67 (2000) 17–35. https://doi.org/10.1016/S0306-2619(00)00005-2.

[7] U.K. Das, K.S. Tey, M. Seyedmahmoudian, M.Y. Idna Idris, S. Mekhilef, B. Horan, A. Stojcevski, SVR-based model to forecast PV power generation under differentweather conditions, Energies. 10 (2017) 1–17. https://doi.org/10.3390/en10070876.

[8] M.P. Almeida, O. Perpiñán, L. Narvarte, PV power forecast using a nonparametric PV model, Sol. Energy. (2015). https://doi.org/10.1016/j.solener.2015.03.006.

[9] S. Theocharides, G. Makrides, A. Livera, M. Theristis, P. Kaimakis, G.E. Georghiou, Day-ahead photovoltaic power production forecasting methodology based on machine learning and statistical post-processing, Appl. Energy. 268 (2020) 115023. https://doi.org/10.1016/j.apenergy.2020.115023.

[10] M. Richter, J. Kalisch, T. Schmidt, E. Lorenz, K. De Brabandere, Best Practice Guide On Uncertainty in PV Modelling, Public Rep. Perform. Plus WP2 Deliv. 2.4. (2015) 1–37.

[11] S. Theocharides, M. Theristis, G. Makrides, M. Kynigos, C. Spanias, G.E. Georghiou, Comparative Analysis of Machine Learning Models for Day-Ahead Photovoltaic Power Production Forecasting, Energies. 14 (2021) 1081. https://doi.org/10.3390/en14041081.

[12] G. Makrides, B. Zinsser, M. Norton, G.E. Georghiou, M. Schubert, J.H. Werner, Potential of photovoltaic systems in countries with high solar irradiation, Renew. Sustain. Energy Rev. 14 (2010) 754–762. https://doi.org/10.1016/j.rser.2009.07.021.

[13] J.G. Powers, J.B. Klemp, W.C. Skamarock, C.A. Davis, J. Dudhia, D.O. Gill, J.L. Coen, D.J. Gochis, R. Ahmadov, S.E. Peckham, G.A. Grell, J. Michalakes, S. Trahan, S.G. Benjamin, C.R. Alexander, G.J. Dimego, W. Wang, C.S. Schwartz, G.S. Romine, Z. Liu, C. Snyder, F. Chen, M.J. Barlage, W. Yu, M.G. Duda, The weather research and forecasting model: Overview, system efforts, and future directions, Bull. Am. Meteorol. Soc. 98 (2017). https://doi.org/10.1175/BAMS-D-15-00308.1.

[14] T. Chen, C. Guestrin, XGBoost: A scalable tree boosting system, in: Proc. ACM SIGKDD Int. Conf. Knowl. Discov. Data Min., 2016. https://doi.org/10.1145/2939672.2939785.

[15] P. Montero, J. Vilar, TSclust: An R Package for Time Series Clustering, JSS J. Stat. Softw. 62 (2014) 1–43. https://doi.org/10.18637/jss.v062.i01.

[16] K. Kalpakis, D. Gada, V. Puttagunta, Distance measures for effective clustering of ARIMA time-series, in: Proc. - IEEE Int. Conf. Data Mining, ICDM, 2001. https://doi.org/10.1109/icdm.2001.989529.

[17] J.M. Vilar, J.A. Vilar, S. Pértega, Classifying

Time Series Data: A Nonparametric Approach, J. Classif. 26 (2009). https://doi.org/10.1007/s00357-009-9030-3.
[18] G.-B. Huang, Q.-Y. Zhu, C.-K. Siew, Extreme learning machine: Theory and applications, Neurocomputing. 70 (2006) 489–501. https://doi.org/10.1016/j.neucom.2005.12.126.
[19] L. Zappia, A. Oshlack, Clustering trees: a visualization for evaluating clusterings at multiple resolutions, Gigascience. 7 (2018). https://doi.org/10.1093/gigascience/giy083.
[20] K.E. Taylor, Summarizing multiple aspects of model performance in a single diagram, J. Geophys. Res. Atmos. 106 (2001) 7183–7192. https://doi.org/10.1029/2000JD900719.

EVALUATION OF IRRADIANCE DECOMPOSITION AND TRANSPOSITION MODELS FOR A NORDIC CLIMATE

Dina C. Martinsen[1,2], Mari B. Øgaard[2], Magnus M. Nygård[2], Marie S. Wiig[2], Marisa Di Sabatino[1], Erik S. Marstein[2], Heine N. Riise[2*]
[1]Norwegian University of Science and Technology, Department of Materials Science and Engineering, Sem Sælands vei 12, 7034 Trondheim, Norway
[2]Institute for Energy Technology, Department for Solar Power Systems, Instituttveien 18, 2007 Kjeller, Norway
* Corresponding author: E-mail: heine.riise@ife.no, Tel.: +47 948 688 28

ABSTRACT: Cost reductions and technological advancements continue to increase profitability of photovoltaics (PV) also in the northern parts of the world. In the planning of PV systems, estimation of accumulated energy generation throughout the system lifetime is essential for estimating levelized cost of electricity (LCOE) and the financial return for the system. Robust estimations of energy output require high quality solar resource data and accurate modeling of the irradiance in the plane of the PV modules. However, existing transposition and decomposition models for estimating the irradiance in tilted planes are not widely validated for locations at high latitude. In this work, the performance of a selection of models for decomposition and transposition of irradiance was evaluated at three locations in Norway. The impact of cloud conditions and season on model performance was also investigated. The results show large variation in model performance for the three studied locations, and the model performance was found to be affected by both season and cloud cover. The decomposition models were found to yield larger uncertainties compared to the transposition models. For the decomposition models, the rRMSD ranged from 17.9% to 71.4%, and for transposition models the range was from 6.2% to 12.6%.
Keywords: Solar irradiance, Irradiance modeling, Nordic Climate

1 INTRODUCTION

Cost reductions and technological advancements continue to increase profitability of photovoltaics (PV) also in the northern parts of the world. In the Nordic countries, the interest in PV technology has increased drastically in the last years. In Norway, the annual domestic PV energy generation is expected to reach 4 TWh by 2030 considering moderate growth rates [1]. Accumulated energy generation throughout the system lifetime is an essential parameter for estimating the levelized cost of electricty (LCOE) and the financial return for PV installations. Hence, making accurate energy yield assessments for PV systems is a key prerequisite to determine if a PV project is bankable [2]. Accurate energy yield assessments of PV power plants require high quality solar resource data and accurate modeling of the irradiance in the plane of the PV modules.

Most PV modeling approaches use the horizontal irradiance (*GHI*) as input parameter for the performance modeling of systems. As PV modules are commonly tilted in order to maximize sunlight collection, it is a necessary step in PV modeling to estimate the irradiance in the tilted plane of array (*POA*) of the PV modules. *POA* irradiance modeling using *GHI* as the only irradiance input measurement involves two steps: decompositon [3,4] and transposition [5,6] of the irradiance. In the decomposition step, direct normal irradiance (*DNI*) and diffuse horizontal irradiance (*DHI*) are estimated from the *GHI* measurements. In the transposition step, *DNI* and *DHI* are transposed to the plane of the tilted module array. The total *POA* irradiance is then calculated from the estimated diffuse, direct and reflected in plane irradiance.

Numerous models for decomposition and transposition exist, but there is no combination of decomposition and transposition models that is widely accepted as a standard. The performance of the models has been compared in several studies. Over 140 different decomposition models were assessed in a comprehensive study carried out by Gueymard and Ruiz-Arias in 2016 [7]. Validation of transposition models has also been performed for numerous locations and for a wide range of models [3,4,8,9]. In general, different combinations of decomposition and transposition models yield varying results in different locations. An explanation for this, is that most of the models are not based on theoretical relations, but are developed on site specific empirical observations over a limited time span. Thus, assessment of the performance of the different models at different irradiation and climate conditions is necessary to understand which model combination that will yield accurate irradiance modeling in different regions [4,7].

Irradiance model performance assessments demand measurements of *GHI*, *DNI*, *DHI* and *POA* irradiance as well as albedo, temperature and humidity measurements. It is essential to ensure a long time series, preferably several years, with high temporal resolution. However, suboptimal inputs are not uncommon in the studies presented in the literature as the site of interest often lack either measurements of albedo or some of the irradiance components. In addition, many of the evaluation studies only consider locations in a limited subset of climate zones, and hence, excludes influence of variability in location specific parameters like cloud conditions, precipitation, snow cover, temperatures and solar elevation.

Consequently, there is a need for further validations of irradiance models, for a broad range of operating conditions. There is especially a lack of irradiance model validation for sites located in Nordic climates and latitudes. Within the Nordics, there is a wide range of latitudes with large variations in climate. The climates include milder climates in coastal regions, continental subarctic climates in inland areas, and arctic conditions at the highest latitudes [10]. In general, the irradiance conditions are characterized by low solar elevation, significant seasonal variations and a large fraction of diffuse light. These conditions are significantly different from the conditions at lower latitudes where most models are developed and mosts exsisting irradiance model performance assessments are performed.

One exception is a study by Gueymard et al. [3] that looks into the performance of transposition models for snow-free and snow-covered ground conditions as well as summer conditions, both for overcast and clear skies in Golden, Colorado USA (39.75°N, 105.23°W) with Köppen-Geiger (KG) [11] climate zone *Dfb*. *Dfb* is the same KG climate zone as for the lcoations studied in this work. The results indicate that the overall best performing models, regardless of cloud conditions, are the Gueymard model and the Klucher model [3]. For clear skies, the Perez model gave the best results. However, upon division of the dataset into subsets of season and albedo effects, after applying decomposition models to estimate *DNI* and *DHI* instead of using measured values, the Klucher, Perez and Gueymard models no longer yield the best results. Using the common assumption of a constant albedo value of 0.2, the Reindl and the Hay models perform best while for snow covered winter days the Klucher model perform best. The results indicate that when *POA* is simulated using *GHI* as the sole input, the dominating contribution to the overall uncertainty in modeled *POA* irradiance is introduced by the decomposition model.

The aim of the present work is to reduce the uncertainty in modeled yield of PV power plants in Nordic conditions through improving the accuracy of irradiance modeling. Irradiance measurements carried out at three locations in southeastern Norway are used to investigate the accuracy of selected decomposition and transposition models, as well as combinations of models. Morover, the impact of cloud cover and seasonal changes on modeling accuracy is investigated and discussed.

2 METHODOLOGY

2.1 Irradiance datasets

Measurements carried out at three locations in southeastern Norway are analyzed in this work. The three sites are: (1) the PV module outdoor test-site at IFE Kjeller, (2) the field station for bioclimatic studies (BIOKLIM), at NMBU in Ås, and (3) the outdoor PV lab J5 at the Grimstad campus of the University of Agder (UiA). The geographical location of each of the three locations is illustrated in Figure 1, and an overview of the site specifics and analysis periods used is given in Table I. As can be seen from Figure 1, the three sites are situated in the same KG climate zone, i.e. *Dfb*.

The site at Kjeller is equipped with a pyranometer measuring *GHI*, two pyranometers mounted at a 45° tilt angle measuring *POA*, and a Kipp&Zonen SHP1 pyrheliometer mounted on a SOLYS2 sun-tracking device measuring *DNI*. All pyranometers are of type Kipp&Zonen SMP10 equipped with CVF4 ventilation units. The site in Ås is equipped with one Kipp&Zonen CMA11 pyranometer measuring *GHI*, one Kipp&Zonen pyranometer mounted at a 42° tilt angle measuring *POA* and a Kipp&Zonen CMP11 pyranometer mounted with a shadow ring measuring *DHI*. The site in Grimstad is equipped with one pyranometer measuring *GHI*, one pyranometer mounted at a 45° tilt angle measuring *POA* and a Kipp&Zonen SHP1 pyrheliometer mounted on a SOLYS2 sun-tracking device measuring *DNI*. The pyranometers are of type Kipp&Zonen CMP11 equipped with CVF4 ventilation units. All three sites are also equipped with a weather monitoring station, and measurements of relative humidity and ambient temperature are used in this work. For the dataset from Grimstad, measurements of relative humidity are obtained from the Norwegian center for climate services, measured at a station located 4 km from the site [12].

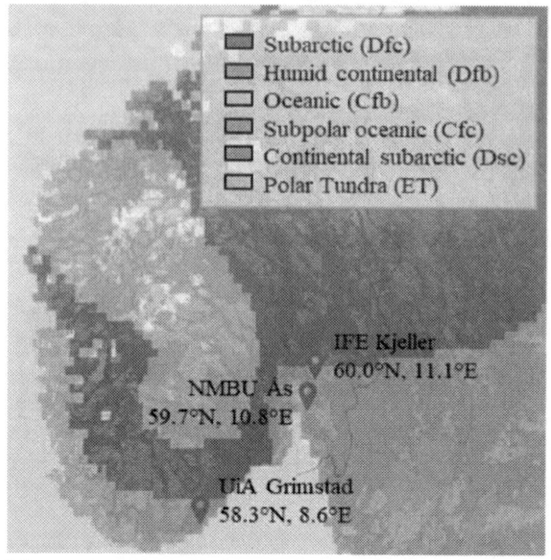

Figure 1: Map of the geographical locations of the collected irradiance and meteorological measurements analyzed in this work with coordinates and KG climate zone [11] indicated.

Table I: Irradiance components measured, altitude and length of analysis period for analyzed datasets.

Site	Period	Altitude (m)	Measurements
Kjeller	10/21 - 07/22	126	*GHI*, *DNI*, *POA* (45°)
Ås	07/18 - 12/21	93	*GHI*, *DHI*, *POA* (42°)
Grimstad	01/21 - 12/21	55	*GHI*, *DNI*, *DHI*, *POA* (45°)

2.2 Data processing

To eliminate invalid data points, initial filtering was applied to the *GHI* data prior to modeling. The initial filtering involved a solar elevation angle filter in combination with a low irradiance filter, as in [13]. Solar elevation angles were restricted to values above 5° to ensure only daytime values were included as input for the modeling. Irradiance values below 50 W/m^2 in the *GHI* data series were also filtered out to exclude time periods with shading from surrounding buildings during the winter months and higher errors in measuring equipment at low irradiance conditions. To remove unrealistically high *GHI* values, measurements larger than 1.1 times simulated clear sky *GHI* was filtered out. Lastly, periods where measured *POA* was larger than zero but the measured *GHI* was zero, was removed. This situation may occur when snow covers horizontal sensors, but has slipped of sensors mounted at a tilt angle. After filtering, there are 8754, 40923 and 22705 remaining daytime data points with five-minute frequency in the three datasets from Kjeller, Ås and Grimstad, respectively.

Many of the evaluated decomposition and transposition models are developed for datasets with one-hour resolution. However, for energy yield modeling of PV systems, having as high-resolution data as possible is beneficial for accuracy. This has led to higher resolution data series beeing applied to the existing irradiance models, and validation studies using higher temporal resolution data have been performed for several locations [5]. The model accuracy in this work is calculated for the five-minute median of the measured irradiance data. The pvlib python toolbox v9.1 [14] was used in the processing of the data.

2.3 Irradiance modelling

The decomposition models evaluated in this work estimates *DNI* and *DHI* using *GHI* as irradiance input. Some models also rely on additional predictors, like air mass, ambient temperature, and relative humidity. Most of the models use the empirical correlation between the diffuse fraction, *Kd*, and the clearness index, *Kt*, to estimate *DNI* and *DHI* components [5]. The two fractions *Kd* and *Kt* are defined by Equation (1) and (2), respectively.

$$Kd = \frac{DHI}{GHI} \quad (1)$$

$$Kt = \frac{GHI}{G_{\text{ext}}} \quad (2)$$

The models commonly estimate *Kd*, by correlations to *Kt*, the extraterrestrial irradiance (G_{ext}), and *GHI*, in order to subsequently estimate *DNI* using the relation between DNI, GHI and DHI given in Equation (3), by inserting the estimated parameter *Kd* as shown in Equation (4). In Equation (3) and (4), θ_Z is the sun zenith angle.

$$DNI = \frac{GHI - DHI}{cos\,(\theta_Z)} \quad (3)$$

$$DNI = \frac{(1 - Kd)GHI}{cos\,(\theta_Z)} \quad (4)$$

Transposition models commonly estimate the *POA* irradiance using the three irradiance components *GHI*, *DNI* and *DHI*, and information on the tilt and orientation of the PV array. The aim of the models is to estimate the contributions from the direct, diffuse and reflected components of the *POA* irradiance. Transposition models of varying complexity exist. The simplest models consider the diffuse radiation to be uniformly distributed over the sky dome. The simplest models also assume that reflections from the ground has the same characteristics as diffuse light [15]. More complex models account for increased irradiance in the cicumsolar region and close to the horizon [6].

The selection of models included in this analysis is based on a combination of prior assessments of irradiance model performance in literature and the selection of available models in commonly used modeling software like PVsyst, pvlib and PV*SOL. The models included are listed in Table II and Table III, along with predictors and the abbreviation assigned for each model. All decomposition models use *GHI* as a predictor and all transposition models use *GHI, DNI* and *DHI* as predictors. Only the additional predictors to *DNI*, *DHI* and *GHI* are tabulated.

2.4 Model performance analysis

The decomposition models were evaluated by comparing the output to measured *DNI* irradiance. For the dataset from Ås, measurements of *DNI* were not available, and *DNI* was calculated using Equation (3) and the measurements of *GHI* and *DHI*. The performance of the transposition models was investigated by using the measured irradiance components *GHI*, *DNI* and *DHI* from each of the sites as input. If the measured DNI and DHI value was not available, Equation (3) was used to calculate the value. The output of the model was compared to measured POA irradiance. The accuracy of the individual models and model combinations were evaluated both in terms of relative root mean squared deviation (rRMSD) and the relative mean biased deviation (rMBD). Both metrics are given as relative to the mean measured value. The rRMSD represents an unbiased magnitude of the model deviations, giving more weight to larger errors as each error effect is squared. The rMBD represents the model bias.

Table II: Evaluated decomposition models.

Decomposition models	Abbr.	Predictors[1]
Erbs [16]	E	*Kt*
Reindl 1 [17]	R1	*Kt*
Reindl 2 [17]	R2	*Kt*, θ_{el}
Orgill and Hollands [18]	OH	*Kt*
Maxwell (Disc) [19]	Dc	*Kt*, θ_Z, G_{ext}, *AM*
Skartveit [20]	S	*Kt*, θ_{el}
Yao 2 [21]	Y	*Kt*
Hollands 2 [22]	H2	*Kt*, ρ
Engerer 2 [23,24]	En2	*Kt*, θ_Z, *t*, *AST*, *V*, GHI_c
Perez et al. (Dirint) [25]	Dt	*Kt*, θ_Z, *t*, *V*, GHI_c, *AM*, T_{dp}

[1]Predictor symbols: *Kt* – clearness index, θ_{el} - Elevation angle, θ_z - zenith angle, G_{ext} - Extraterrestrial irradiance, *AM* - air mass, ρ – albedo, *t* - time of day, *AST* - solar time, *V* - variability index, GHI_c - clear-sky global horizontal irradiance, T_{dp} - temperature dew point.

Table III: Evaluated transposition models.

Transposition models	Abbr.	Predictors[2]
Isotropic [15]	I	β, ρ
Klucher [26]	Kl	β, ρ, θ, θ_Z
Hay and Davies [27]	HD	β, ρ, θ, θ_Z, G_{ext}
Reindl [17]	R	β, ρ, θ, θ_Z, G_{ext}
King [14]	Ki	β, θ_Z
Perez [28]	P	β, ρ, θ, θ_Z, G_{ext}, *AM*

[2]Predictor symbols: ; β - tilt angle, ρ - albedo, θ - incidence angle, θ_z - zenith angle, G_{ext} - Extraterrestrial irradiance, *AM* - air mass.

In the evaluation of the irradiance models, three different aspects were investigated. First, the overall best performing models for the sites were identified. Second, the impact of cloud conditions/clearness index (*Kt*) on the model performance was evaluated. Third, the combined impact of cloud conditions and season on model performance was determined. In the evaluation of the impact of cloud conditions on the model performance, the dataset was sectioned by *Kt* into three cloud condition categories. The use of *Kt* to classify irradiance measurements by clear, partially overcast and overcast cloud conditions has been reported in the literature with a variety of different partition boundaries for the three categories [16,18,29–31]. Most studies use a Kt in the range of 0.2 to 0.35 as the upper partition boundary of the overcast category, and a Kt in the range of 0.6 to 0.75 as the lower partition boundary for clear conditions. There are multiple reasons for this variation in partition boundaries used. One is that there are no unique set of partition boundaries that will represent the three categories well for all locations, as Kt depends on local climate conditions and the applied irradiance measurement methods [32]. Another reason is that the transition between the three cloud categories is continuous, and it is difficult to objectively determine where the boundary should be set. In this work, the choice of partition boundaries is based on a combination of reported partition boundaries from literature, the partition boundaries used by common modeling software in the industry, e.g. Homer Pro, and the frequency distribution of Kt from the studied locations. The partition boundaries used are defined in Table IV. The influence on the results of adjusting the partition boundaries is not investigated in this work.

Table IV: Kt partition boundaries used to determine different cloud conditions.

Cloud conditions	Kt range
Clear	$Kt > 0.6$
Partly overcast	$0.3 < Kt < 0.6$
Overcast	$Kt < 0.3$

In the evaluation of the impact of season on the modeling performance, the dataset was further sectioned into the four seasons in addition to the sectioning by Kt, giving a total of twelve categories. The sectioning of the dataset by season was simply performed by dividing the dataset by seasons in terms of date ranges. The date ranges are specified in Table V.

Table V: Season ranges for subdivision of datasets

Season	Date ranges [DD/MM]
Winter	01/12 – 28/02 (29/02 leap years)
Spring	01/03 – 31/05
Summer	01/06 – 31/08
Autumn	01/09 – 30/11

3 RESULTS

3.1 Decomposition model validation

Figure 2 shows the accuracy of the decomposition models in terms of rRMSD and the absolute value of the rMBD for the three studied locations. There is large variation in the accuracy of the decomposition models included in the analysis for all the three locations. The overall calculated model fit in terms of rRMSD range from 17.9% to 71.4%. However, there are several models that yield good results, and there is no model that is significantly better than others. Additionally, the overall best performing decomposition model is not the same for the three locations. For each of the three locations, the overall best performing decomposition models in terms of both rRMSD and rMBD are listed in Table VI. Dirint is found to be a good choice in terms of rRMSD for all three locations. In terms of rMBD, the models Reindl1, Orgill and Hollands and Disc yields good results for the three locations Ås, Grimstad and Kjeller, respectively.

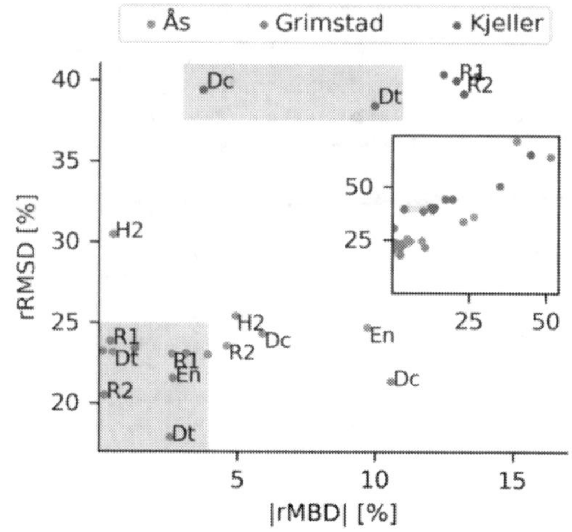

Figure 2: Decomposition model accuracy in terms of rRMSD as a function of the absolute rMBD value for the three studied sites. Labeled models in the shaded area are the overall best decomposition models for the sites. Selected values are indicated by the model labels provided in Table II.

Table VI: Overall best performing decomposition models for the three studied sites in terms of both rRMSD and rMBD.

Site	Models	rRMSD [%]	rMBD [%]
Ås	Dirint	23.2	0.5
	Reindl1	23.0	-2.6
Grimstad	Dirint	17.9	2.5
	Orgill and Hollands	23.2	-0.1
Kjeller	Dirint	38.4	-10.0
	Disc	39.4	-3.7

In Figure 3, the model accuracy in terms of rRMSD for all the included decomposition models in the analysis is plotted for each of the three locations upon division of the dataset into the three cloud condition categories *clear*, *partly overcast* and *overcast* defined in Table IV. The figure shows a general trend of higher rRMSD and rMBD values and lower accuracy for overcast cloud conditions for all decomposition models for all three locations. The model bias results in terms of rMBD are shown in Figure 4. From the figure, it is clear that the models tend to strongly underpredict irradiance for overcast cloud conditions. It is interesting that the only decomposition model with a positive bias for the overcast cloud condition is the Skartveit decomposition model. This model was developed based on a dataset from Bergen in Norway, a location with large degree of overcast weather. Moreover, the Skartveit model achieve the largest rMBD for the clear sky category in contrast to all the other decomposition models. In addition, the Skartveit model shows the largest variations in accuracy both in terms of rRMSD and rMBD between the three locations. Again, we observe that several of the best models have similar performance. The best results, both in terms of rRMSD and rMBD, for the overcast cloud cover category are achieved by the Orgill and Hollands decomposition model for all three locations. For clear and partly overcast cloud conditions, the best results are achieved by the Dirint decomposition model, the same model found to yield the overall best results for all three locations in terms of rRMSD.

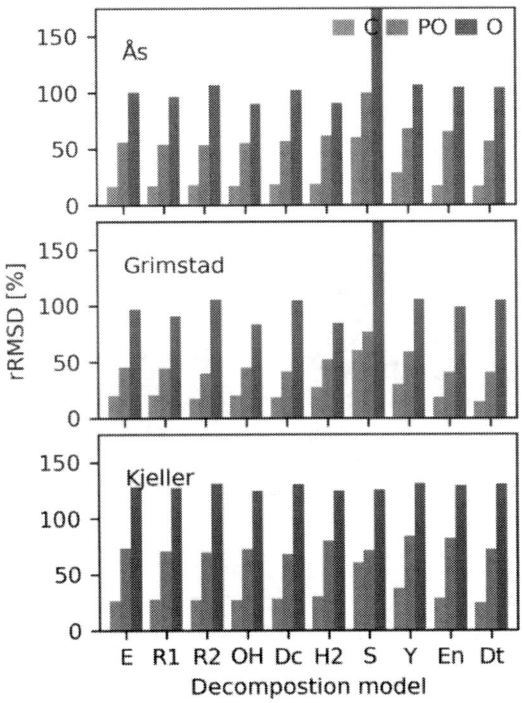

Figure 3: Model accuracy in terms of rRMSD for all the studied transposition models, for the three studied locations, and for different cloud conditions: clear (C), partly overcast (PO) and overcast (O) (defined in Table IV). The models are indicated by the labels provided in Table II.

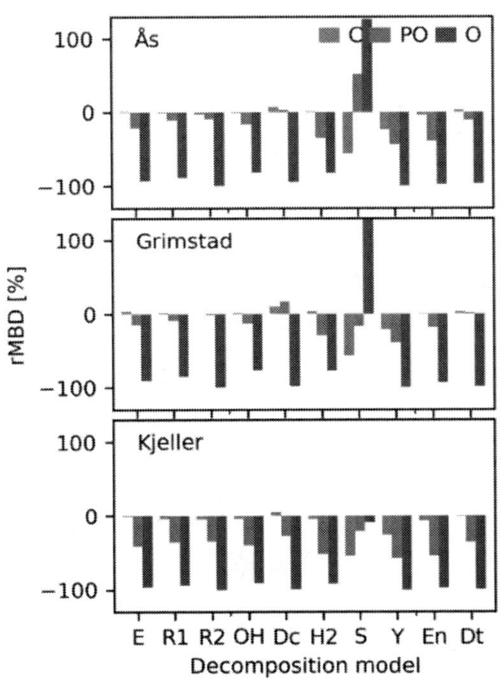

Figure 4: Model bias in terms of rMBD for all the studied transposition models, for the three studied locations, and for different cloud conditions: clear (C), partly overcast (PO) and overcast (O) (defined in Table IV). The models are indicated by the labels provided in Table II.

Figure 5 shows the combined impact of season and cloud conditions on modeling accuracy. It was observed large variation in model performance between the twelve categories of season and cloud conditions for all three locations. The overcast cloud condition category represents the major source of error for any season, but especially for winter and autumn months. The best results are achieved for the clear cloud cover category for any season.

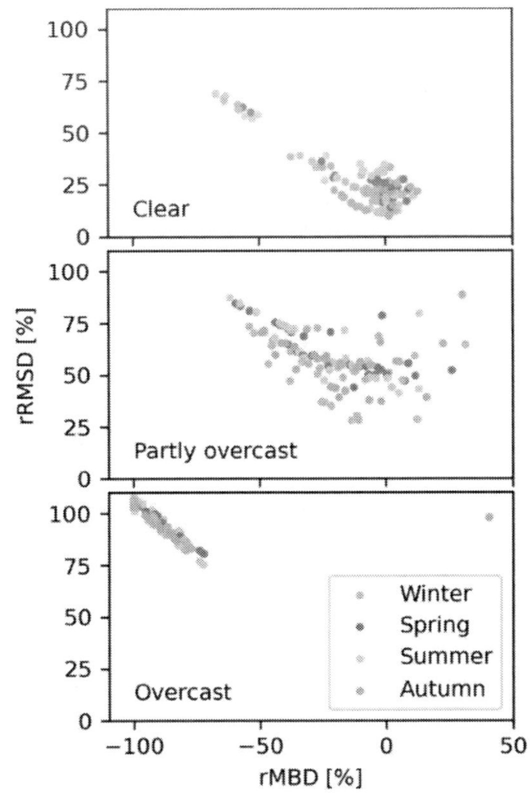

Figure 5: The combined impact of season and cloud conditions on model accuracy for all evaluated decompostion models, for the three studied sites.

3.2 Transposition model validation

Figure 6 compares the absolute rMBD values and rRMSD for the studied transposition models for the three studied locations. The figure clearly shows that there is large variation in performance of transposition models, but the variation is smaller than for the decomposition models. That the main part of the error in *POA* irradiance modeling is expected to originate from the decomposition step is in line with the observations made by Gueymard et al. [3]. For each of the three locations, the overall best performing decomposition models in terms of both rRMSD and rMBD are listed in Table VII. The rRMSD range from 6.2% to 12.6%. For all locations, the lowest rRMSD is achieved with the Perez transposition model (P), while the Klucher transposition model (Kl) seems to be a good choice to achieve low rMBD values. However, it should be noted that the exact values differ between the locations.

Figure 7 shows the model bias results in terms of rMBD for the three locations upon division of the dataset into the three categories *clear*, *partly overcast* and *overcast* cloud conditions (Table IV). For all three locations, all the transposition models achieve the highest uncertainty for the *overcast* category. All transposition models tend to underpredict *POA* irradiance for Grimstad. However, this is not the case for Ås and Kjeller where several of the models overpredict the irradiance for either all or some of the cloud cover categories. The results in terms of rRMSD for different cloud conditions are presented in Figure 8. The results in Figure 8 indicate that the transposition model Perez still is the optimal choice for the three locations Ås, Grimstad and Kjeller in terms of rRMSD.

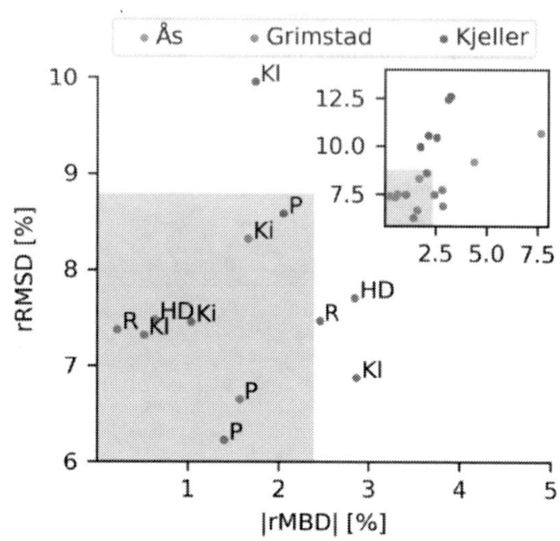

Figure 6: Transposition model accuracy in terms of rRMSD and absolute rMBD for the three studied locations. Labeled models in the shaded area are the overall best transposition models for the sites. Selected values are indicated by the model labels provided in Table II.

Table VII: Overall best performing transposition models for the three locations Ås, Grimstad and Kjeller in terms of both rRMSD and rMBD.

Location	Models	rRMSD [%]	rMBD [%]
Ås	Perez	6.6	1.5
	Reindl	7.4	-0.2
Grimstad	Perez	6.2	-1.4
	King	7.5	-1.0
Kjeller	Perez	8.6	2.0
	Klucher	10.0	1.7

Figure 9 shows the combined impact of season and cloud conditions on the performance of the transposition models, it was observed that there is a large spread in model performance in the twelve season and cloud conditions categories for the three locations, but still less variation than for the decomposition models. Like observed for the decomposition models, the *overcast* category represents the major source of error for any season, especially for winter and autumn months. The best results are achieved for the *clear* category, with clear conditions during winter and autumn achieving slightly better results compared to spring and summer. The dataset from Kjeller shows the least consistent results with both the worst and the best results in the analysis in terms of both rRMSD and rMBD.

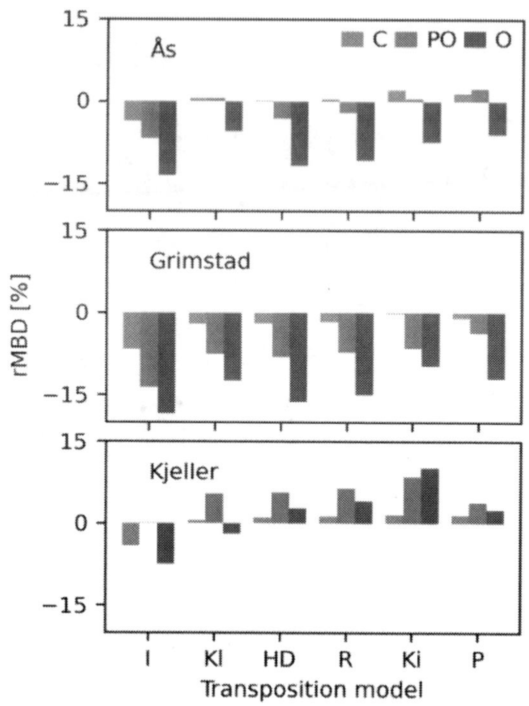

Figure 7: Model bias in terms of rMBD for all the studied transposition models, for the three studied locations, and for different cloud conditions: clear (C), partly overcast (PO) and overcast (O) (defined in Table IV). The models are indicated by the labels provided in Table III.

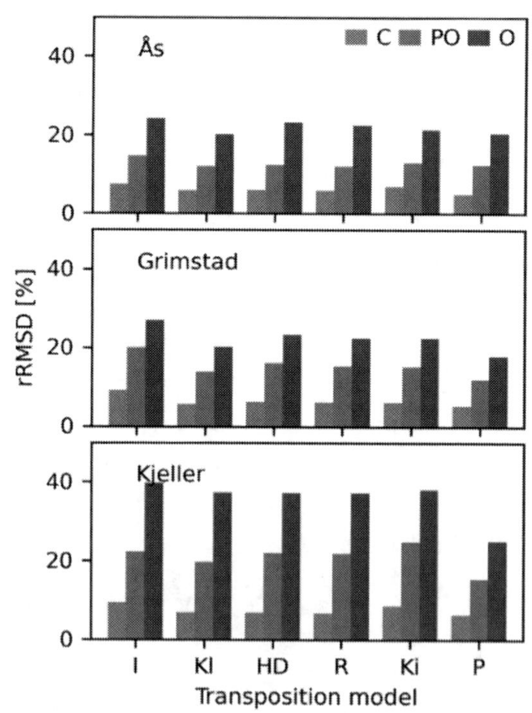

Figure 8: Model fit in terms of rRMSD for all the studied transposition models, for the three studied locations, and for different cloud conditions: clear (C), partly overcast (PO) and overcast (O) (defined in Table IV). The models are indicated by the labels provided in Table III.

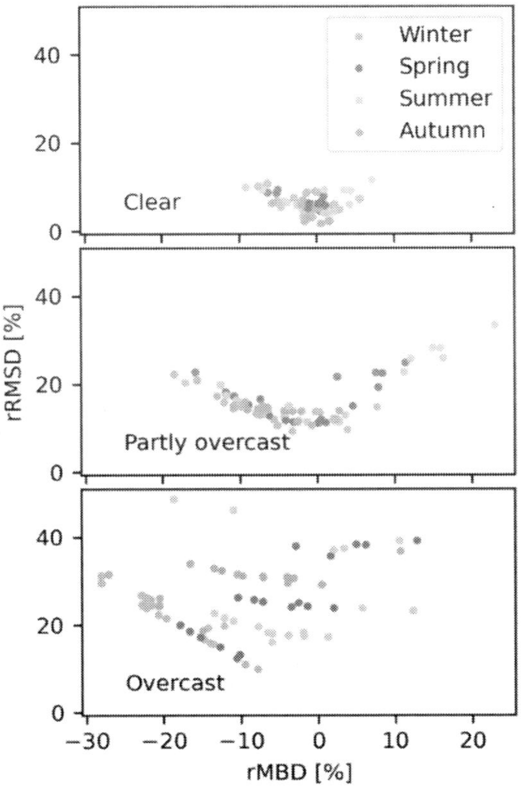

Figure 9: The combined impact of season and cloud conditions on model accuracy for all evaluated transpostion models, for the three studied sites.

3.3 Model combination validation

For each of the three sites, an assessment of combinations of decomposition and transposition models with measured *GHI* as the sole irradiance input were performed in order to identify the model combination yielding the overall highest accuracy in modeled *POA* irradiance. Figure 10 shows the results for the three locations. For each of the three locations, the overall best performing combination of decomposition and transposition models in terms of rRMSD and rMBD are listed in Table VIII. The Dirint decomposition model is found to be the optimal choice in terms of rRMSD for all three locations but combined with varying transposition model for each location. In terms of rMBD, the optimal combination is different for all the three locations. The optimal transposition model Perez found by assessing the accuracy of transposition models alone is only found to yield the best result in combination with a decomposition model for one of the locations, Grimstad, only in terms of rRMSD. A possible explanation for this, is that the transposition models that yields best result in this investigation, have errors that outweigh the errors in the decomposition models. This can also explain why the combined error when decomposition and transposition modeling is combined give higher accuracy than the decomposition step alone.

For both the decomposition and transposition models it is interesting that the model performance results have significant variation between the three studied datasets. The three measurement sites are located in similar climate and at similar latitudes, and large variations in not expected. One potential explanation for this result, is that the relatively small variations in weather, albedo and latitude between the sites gives significant impact on model performance. Another explanation is that there might be variation in the quality of the measurements. An issue with the DNI measurement could for example explain why the performance of the different models are lower for the Kjeller site when evaluated individually, but when model combinations are tested with GHI as the only input, the performance is similar as for the two other sites.

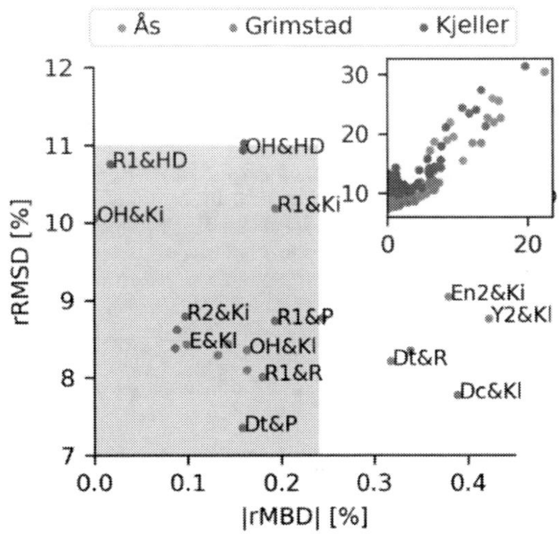

Figure 10: Performance of model combinations of decomposition and transposition models in terms of rRMSD as a function of absolute rMBD for the three locations Ås, Grimstad and Kjeller. Labeled models in the shaded area are the overall best model combinations for the sites. Selected values are indicated by the model labels provided in Table II and Table III.

Table VIII: The overall best performing combinations of decomposition and transposition models for the three locations Ås, Grimstad and Kjeller in terms or rRMSD and rMBD.

Location	Models	rRMSD [%]	rMBD [%]
Ås	R1&R E&Kl	8.0 8.4	-0.2 0.1
Grimstad	Dt&P OH&Ki	7.4 10.1	-0.2 0.0
Kjeller	Dt&HD R1&HD	9.9 10.8	1.3 0.0

4 CONCLUSION

In this work the performance of a selection of decomposition models and transposition models was investigated on datasets collected at three locations in Norway. The impact of cloud conditions and season on model performance was also investigated. The variation in model performance for the three studied locations was found to be large. Moreover, the model performance was found to be impacted by both season and the cloud conditions. The decomposition models were found to be less accurate compared to the transposition models. In the estimation of *POA* irradiance using only *GHI* as input, the decomposition models were thus determined to be the major source of uncertainty.

We did not identify one model that had significantly better performance than the rest. Several models had similarly high performance. By individually assessing the transposition models, the Perez model and Klucher model was found to yield the overall lowest uncertainty for all three locations in terms of rRMSD and rMBD, respectively. However, when applied in combination with any of the included decomposition models in this analysis, the Perez and Klucher models did not achieve the best results. By assessing combinations of decomposition and transposition models, the Dirint decomposition model yielded the lowest uncertainty in modeled irradiance for all three locations in terms of rRMSD but combined with a different transposition model for all locations. In terms of rMBD, no common optimal model combination was found. In terms of rRMSD and rMBD, the accuracy of the decomposition and transposition models is influenced by the season and degree of cloud cover. Most of the tested decomposition and transposition models seem to strongly underpredict irradiance for overcast cloud conditions. This is especially clear during winter and autumn for all three locations. One exception is the Skartveit decomposition model, as the only model found to overpredict irradiance for overcast cloud conditions. Our results show that cloud cover and season have a large impact on accuracy of irradiance modelling, indicating that optimal model choice for a specific region, depend on the local climate and irradiance conditions.

5 REFERENCES

[1] E. Stensrud Marstein, T. Kopstad Berentsen, L.T. Kilde, I. Strand, A. Ekern, S.E. Foss, J. Selj, et al., Veikart for den norske solkraftbransjen mot 2030, FME Susoltech and Solenergiklyngen Report. (2020).

[2] R. Kopecek, J. Libal, Bifacial photovoltaics 2021: Status, opportunities and challenges, Energies. 14 (2021).

[3] C.A. Gueymard, From global horizontal to global tilted irradiance: How accurate are solar energy engineering predictions in practice?, in: American Solar Energy Society - SOLAR 2008, 2008.

[4] M. Lave, W. Hayes, A. Pohl, C.W.Hansen, Evaluation of global horizontal irradiance to plane-of-array irradiance models at locations across the United States, IEEE J Photovolt. 5 (2015).

[5] J.A. Ruiz-Arias, C.A. Gueymard, Solar resource for high-concentrator photovoltaic applications, Green Energy and Technology. 190 (2015).

[6] S.A. Kalogirou, Solar Energy Engineering: Processes and Systems: Second Edition, 2014.

[7] C.A. Gueymard, J.A. Ruiz-Arias, Extensive worldwide validation and climate sensitivity analysis of direct irradiance predictions from 1-min global irradiance, Solar Energy. 128 (2016).

[8] P.G. Loutzenhiser, H. Manz, C. Felsmann, P.A. Strachan, T. Frank, G.M. Maxwell, Empirical validation of models to compute solar irradiance on inclined surfaces for building energy simulation, Solar Energy. 81 (2007).

[9] D. Yang, Solar radiation on inclined surfaces: Corrections and benchmarks, Solar Energy. 136 (2016).

[10] G. Ketzler, W. Römer, A.A. Beylich, The Climate of Norway., In: Beylich, A.A. (Eds) Landscapes and Landforms of Norway. World Geomorphological Landscapes. Springer, Cham. (2021).

[11] M.C. Peel, B.L. Finlayson, T.A. McMahon, Updated world map of the Köppen-Geiger climate classification, Hydrol Earth Syst Sci. 11 (2007).

[12] Norsk Klimaservicesenter, Seklima Observasjoner og værstatistikk, (2020). https://seklima.met.no/observations

[13] M.B. Øgaard, H.N. Riise, H. Haug, S. Sartori, H. Selj, Photovoltaic system monitoring for high latitude locations, Solar Energy. 207 (2020).

[14] W.F. Holmgren, C.W. Hansen, M.A. Mikofski, pvlib python: A python package for modeling solar energy systems, J Open Source Softw. 3 (2018).

[15] B.Y.H. Liu, R.C. Jordan, The interrelationship and characteristic distribution of direct, diffuse and total solar radiation, Solar Energy. 4 (1960).

[16] D.G. Erbs, S.A. Klein, J.A. Duffie, Estimation of the diffuse radiation fraction for hourly, daily and monthly-average global radiation, Solar Energy. 28 (1982).

[17] D.T. Reindl, W.A. Beckman, J.A. Duffie, Diffuse fraction correlations, Solar Energy. 45 (1990).

[18] J.F. Orgill, K.G.T. Hollands, Correlation equation for hourly diffuse radiation on a horizontal surface, Solar Energy. 19 (1977).

[19] E.L. Maxwell, A quasi-physical model for converting hourly Global Horizontal to Direct Normal Insolation, Report number: SERI/TR-215-3087. (1987).

[20] A. Skartveit, J.A. Olseth, A model for the diffuse fraction of hourly global radiation, Solar Energy. 38 (1987).

[21] W. Yao, Z. Li, Y. Lu, F. Jiang, C. Li, New models for separating hourly diffuse and direct components of global solar radiation, in: Lecture Notes in Electrical Engineering, 2014.

[22] K.G.T. Hollands, S.J. Crha, An improved model for diffuse radiation: Correction for atmospheric back-scattering, Solar Energy. 38 (1987).

[23] N.A. Engerer, Minute resolution estimates of the diffuse fraction of global irradiance for southeastern Australia, Solar Energy. 116 (2015).

[24] J.M. Bright, N.A. Engerer, Engerer2: Global re-parameterisation, update, and validation of an irradiance separation model at different temporal resolutions, Journal of Renewable and Sustainable Energy. 11 (2019).

[25] R.R. Perez, P. Ineichen, E.L. Maxwell, R.D. Seals, A. Zelenka, Dynamic global-to-direct irradiance conversion models, in: ASHRAE Trans, 1992.

[26] T.M. Klucher, Evaluation of models to predict insolation on tilted surfaces, Solar Energy. 23 (1979).

[27] J.E. Hay, Calculation of monthly mean solar radiation for horizontal and inclined surfaces, Solar Energy. 23 (1979).

[28] R. Perez, P. Ineichen, R. Seals, J. Michalsky, R. Stewart, Modeling daylight availability and irradiance components from direct and global irradiance, Solar Energy. (1990).

[29] A.I. Kudish, A. Ianetz, Analysis of daily clearness index, global and beam radiation for Beer Sheva, Israel: Partition according to day type and statistical analysis, Energy Convers Manag. 37 (1996).

[30] M. Iqbal, Prediction of hourly diffuse solar radiation

from measured hourly global radiation on a horizontal surface, Solar Energy. 24 (1980).
[31] M. Chikh, A. Mahrane, M. Haddadi, Modeling the diffuse part of the global solar radiation in Algeria, in: Energy Procedia (2012).
[32] H.F. Assunção, J.F. Escobedo, A.P. Oliveira, A new algorithm to estimate sky condition based on 5 minutes-averaged values of clearness index and relative optical air mass, Theor Appl Climatol. 90 (2007).

IMPROVING SITE SHADING ASSESSMENT FROM DIGITAL SURFACE MODELS WITH ONSITE LIDAR MEASUREMENTS

Daniel Alvarez Mira, Martin Bartholomäus*, Sebastian Poessl, Thøger Kari Hass, Peter Behrensdorff Poulsen, Gisele A. dos Reis Benatto, Sergiu V. Spataru
Technical University of Denmark (DTU), Department of Electrical and Photonics Engineering
Frederiksborgvej 399, 4000 Roskilde, Denmark

ABSTRACT: Product integrated photovoltaic (PIPV) systems, can be severely impacted by shading in the surroundings as their collector area is typically small. To ensure the proper functioning of these devices, methodologies to estimate solar irradiance and near shading accurately and efficiently in built up locations are needed. Digital Surface Models (DSM) have proven to be a good alternative to estimate solar irradiance on building surfaces and urban areas. However, high-accuracy DSM are scarce, with limited availability and tend to show the maximum elevation of the obstructions elements disregarding their real shape. We propose a methodology with onsite LiDAR measurements to improve the accuracy to which the solar irradiance can be estimated in urban environments. This methodology produces horizon matrices which can reduce the estimated irradiance's RMSE by 25-40% compared to fisheye sky imaging methodologies, and by 45-75% to the ones obtained with low-resolution DSMs. However, at locations with low shading profiles the differences are negligible.
Keywords: Shading, Solar Radiation, Stand-alone PV Systems, Modelling, Digital Surface Models (DSM), LiDAR.

1 INTRODUCTION

Solar photovoltaic (PV) systems rely on the solar resource to generate electricity. The assessment of the available solar resource at the design and feasibility phase at any potential installation location is a crucial step. In the case of product integrated photovoltaic (PIPV) systems, where the collector area is small, the impact of shading created by local topography is more significant than conventional rooftop systems. These devices are usually deployed in urban environments, where light obstructing elements tend to be more abundant.

In addition, the potential deployment locations of PIPV devices can be challenging if a large number of locations are considered, and therefore, the use of methodologies to estimate solar irradiance accurately and efficiently in built up locations is needed.

Historically, fisheye sky imaging has been used to produce a representation of the potentially shading elements in the surroundings and estimate solar irradiance as shown in [1]–[3]. They are based on the extraction of a horizon line, also known as skyline, to characterize the surroundings and determine the amount of solar irradiance blocked. However, this method presents some drawbacks as the dependency on atmospheric conditions to properly obtain the horizon line.

Another alternative is the use of Digital Surface Models (DSMs). DSMs are geospatial digitized maps representing the objects' height that include the Earth's surface elevation, as well as natural and man-made objects (such as buildings and vegetation) [4], [5]. The most common applications of DSMs are urban/architectural planning and navigation systems. However, its use to estimate solar irradiance on building surfaces and urban areas is beginning to grow for indoor climate and photovoltaic potential estimation purposes, as in [6], [7].

DSMs are usually created using Light Detection and Ranging (LiDAR) or stereo photogrammetry [8]. Although these methodologies can achieve high accuracies in their models, these are sometimes outdated (not including new buildings or accounting for trees growth) and their availability is limited [9]. Moreover, these DSMs merely show the maximum height of the elements captured disregarding their real shape of the objects in the surroundings.

During this work, we have developed a methodology to combine DSM with onsite LiDAR measurements to improve the accuracy to which the solar irradiance and near shading can be estimated in urban environments. This methodology has been tested at different locations and its effects analyzed for different DSM resolutions.

2 METHODOLOGY

2.1 LiDAR measurements

Light Detection and Ranging (LiDAR) is a remote sensing technique that uses laser light to measure the distance from objects to the sensor. This distance is acquired by measuring the time since the signal is emitted until its reflection returns to the sensor. The measurements are stored as point cloud files and can be transformed into DSMs [10].

LiDAR sensors can be used in a wide range of different situations but are broadly categorize as airborne or terrestrial based on its acquisition mode. Aerial or airborne LiDAR allows to create accurate and dense topographic data which makes it suitable for covering large areas such as cities or regions. On the other side, terrestrial LiDAR sensors exist that can scan the environment from the ground, covering regions and details that airborne sensors cannot access. Therefore, the combination of both can provide complete details of the region of interest [10], [11].

In this work, to accurately characterize the surroundings, we have used Robosenses' "RS-Bpearl" LiDAR sensor (Figure 1). It features a 90° vertical and 360° horizontal field of view, a range of up to 100m and an accuracy of ±3 cm, making it suitable for its use to characterize the obstruction elements at a specific location.

When acquiring LiDAR data at a specific location, we recorded data snapshots at different heights by adjusting the extension of the center column on the tripod to which the sensor was mounted, in order to increase the vertical resolution and density of the data even more. These snapshots contain xyz data points that were merged while accounting for the height difference between measurements.

Figure 1: LiDAR measurements setup. Including the Robosenses' "RS-Bpearl" LiDAR sensor using for the obstruction characterization of the surroundings

2.2 The use of DSMs for shading modelling

DSMs represent height information of the ground surface, vegetation such as trees, as well as man-made features such as buildings. This information can be used to model the shading losses caused by such elements, and therefore estimate plane of array irradiances on any surface contained in the model [12], [13].

DSMs allow to create horizon lines similar to those acquired from fisheye sky imaging which can be used to estimate shaded irradiance as explained in [14]. However, the results are highly influenced by the DSM resolution which is dependent on the acquisition method and data treatment.

As presented in [9], publicly available DSM exist both on the local and worldwide level, and present a wide range of resolutions from half a meter cells, up to 30 or 90 meters. High-accuracy DSM are rarely publicly available and only countries such as Denmark or Sweden offer DSMs with resolution of 1 meter or better.

For this work, the publicly available DSM covering the whole surface area of Denmark (DK-DEM) was used. The DK-DEM was generated from airborne LiDAR measurements, and cyclically updates a fifth of the country's land in five-year periods since 2018 [15], [16]. The DK-DEM is stored in raster maps with a resolution of 0.4 by 0.4 meters and can be downloaded at the Danish Agency for Data Supply and Efficiency (SDFE) website [17]. Figure 2 shows the DK-DEM coverage and the tile containing the relevant DSM for this project.

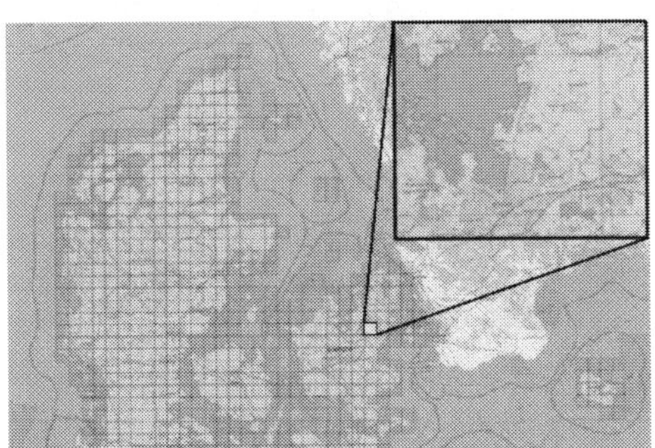

Figure 2: DK-DEM coverage area and tile containing the DSM of the studied area

2.3 Enhanced DSM with onsite LiDAR measurements

Accurate solar irradiance modelling in urban environments is crucial to ensure the proper performance of PIPV systems, which have small light collection surfaces and can be severely impacted by nearby shading, even from small obstructions, that might not be captured by low resolution or airborne LiDAR DSMs.

The methodology introduced bellow proposes a solution to increase the DSM-based shading estimation accuracy by combining low resolution DSMs with onsite LiDAR measurements which is relevant for solar applications deployed in built-up environments.

For PIPV applications, due to their small active surface area, it is important that close-by elements are accurately represented as they can generate deep shadows on the device's surfaces. On the contrary, the effects of far distant object is homogeneous and therefore, they can be more vaguely represented. Additionally, for obstructions with irregular shapes (e.g. trees), local LiDAR measurements can be used to increase the accuracy of the DSM.

To that end, our method proposes to remove DSM data in a 20 meter radius at the studied locations and replace it with onsite LiDAR measurements (Figure 3).

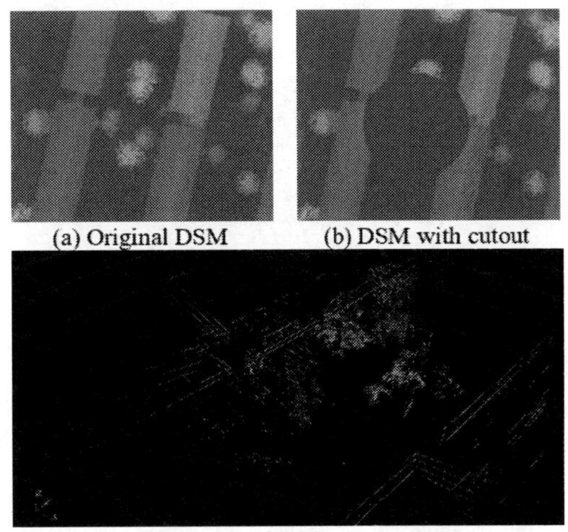

(a) Original DSM (b) DSM with cutout

(c) Onsite raw LiDAR measurements

Figure 3: DSM enhancement with onsite LiDAR measurements

In our method, the DSM data in a circle of 20m radius is replaced with a horizontal plane of ground height at that location, using the open source geoprocessing tool QGIS. Hereafter, a horizon line is obtained with the grass r.horizon algorithm. After obtaining onsite LiDAR measurements at the site, the data is projected onto a binary azimuth by elevation matrix referred to as horizon matrix. The matrix is binned at one degree resolution. If any LiDAR data point falls within a certain bin, it is counted as obstruction and given the value 0, whereas bins without LiDAR data are counted as unobstructed and given the value 1 (see Figure 4b). The LiDAR horizon matrix and DSM horizon line from the same location are then combined into a matrix by addition, counting all bins at or below the horizon line as obstructed. Finally, a kernel algorithm is employed to fill in gaps originated from the LiDAR sensor's vertical resolution (2.8°).

From airborne DSM only maximum elevations can be obtained. This is how horizon lines (HL) are created and means that concave or irregular shapes cannot be capture. However, with the proposed methodology, by combining DSMs with onsite LiDAR measurements we can produce horizon matrices (HM) which show a more realistic representation of the obstruction elements. Figure 4.d shows a comparison between the HL and SM at one studied location.

(a) Panoramic image of the studied location

(b) Horizon matrix created from local LiDAR measurements

(c) Horizon matrix after merging with DSM cutout

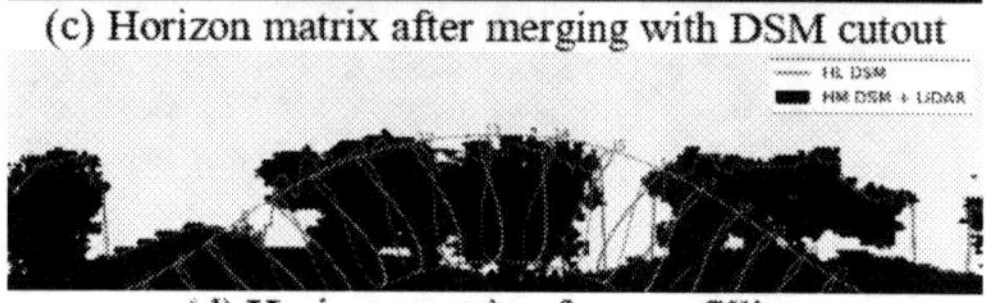

(d) Horizon matrix after gap-filling

Figure 4: Onsite LiDAR measurements combination with DSM and showing the yearly sun path at the site.

By reproducing more accurately irregular shapes such as trees or light poles we can obtain more accurate results as presented in Section 3.

2.4 Experimental validation

The aim of this project is to create a methodology to enhance DSM resolution with onsite LiDAR measurements and evaluate its accuracy in estimating shading and solar irradiance modelling in urban environments.

Irradiance measurements are taken onsite to evaluate the proposed methodology. The unshaded solar irradiance components: global horizontal irradiance (GHI), direct beam irradiance (DNI), and diffuse horizontal irradiance (DHI); are measured onsite by class A radiometers (ISO 9060:2018).

These irradiance measurements are used to model shaded irradiance at different built-up locations by means of a publicly available DSM (DK-DEM), as well as the combination of that DSM and onsite LiDAR measurements to increase the resolution of the potential light obstructions in the DSM such as trees. The modelled shaded irradiance is later compared to ground solar irradiance measurements.

As previously mentioned, the DK-DEM has an extraordinarily high accuracy. Therefore, to properly evaluate the effects of combining low-resolution DSMs with onsite LiDAR measurements we have created synthetic DSMs with lower resolutions from the DK-DEM as explained in [9]. We have chosen 3.2 and 12.8 meters resolutions as they are close to the publicly available worldwide DSMs with 2.5 and 12 meters resolutions.

Study cases:

Three different locations (Figure 5) were studied at DTU Risø campus, Roskilde, Denmark. These locations simulate the kind of environment where PIPV could be installed in cities. The first location (Figure 5.a) presents a low shading profile as it could be the case in some squares where the buildings are located far away. The second one (Figure 5.b), shows a street canyon with tall buildings on both sides of the street. Finally, location 3 (Figure 5.c) resembles to an environment that could be found on a park.

(a) Location 1

(b) Location 2

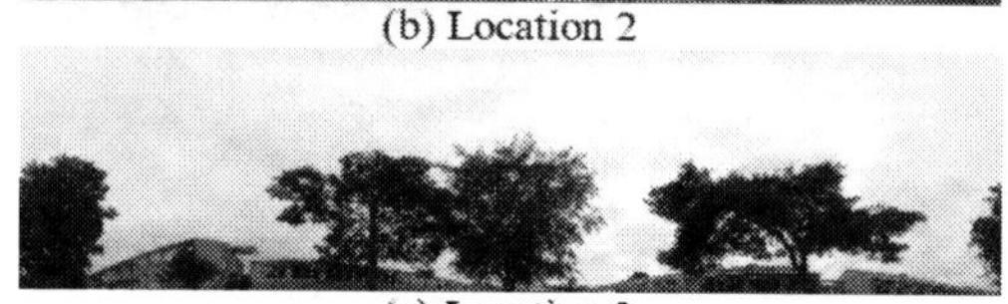

(c) Location 3

Figure 5. Panoramic images of the three studied locations.

Class C (ISO 9060:2018) pyranometers were installed – either permanently or temporally – at each of these locations to measure shaded irradiance. Onsite LiDAR measurements were taken at each of them as well with the set up shown in Figure 1. Those measurements were used to enhance the resolution of the different DSMs.

For each of the studied locations, 7 different inputs have been used for the surroundings characterization and therefore, 7 shaded irradiances have been calculated. The analyzed inputs are the following:

1. A horizon line obtained from a fisheye image of the sky hemisphere – *Fisheye based*.
2. A horizon line obtained from the synthetic DSM with a 12.8 m resolution generated from the DK-DEM – *DSM 12.8m*.
3. A horizon matrix obtained from onsite LiDAR measurements combined with the 12.8 m resolution synthetic DSM – *DSM 12.8m + LiDAR*.
4. A horizon line obtained from the original DK-DEM with a 0.4 m resolution – *DSM*.
5. A horizon matrix obtained from onsite LiDAR measurements combined with the DK-DEM – *DSM + LiDAR*.
6. A horizon line obtained from the synthetic DSM with a 3.2 m resolution generated from the DK-DEM – *DSM 3.2m*.
7. A horizon matrix obtained from onsite LiDAR measurements combined with the 3.2 m resolution synthetic DSM – *DSM 3.2m + LiDAR*.

Since the main objective of this work is to evaluate the increased accuracy in the irradiance modelling in urban environments by using the proposed methodology over low-resolution DSMs, we have focused our analysis in the first three cases. Furthermore, independently of the DSM resolution used as an input to our method, all resulted horizon matrices (cases 3, 5 and 7) are equivalent with negligible differences and therefore, for simplicity and readability of the results, only the ones from the worst case scenario (case 3) are considered further. However, the results of all seven cases are briefly compared at the end of the following section.

3 RESULTS AND DISCUSSION

The shaded irradiance for each of the analyzed locations has been estimated from 15-minute averages of the unshaded solar irradiance components, and assuming the isotropic distribution of the diffuse irradiance.

As the results are greatly influenced by the nature of the shading profile at each location, to extract more relevant conclusions, the analyses are performed in each location independently and presented below separately.

In Figure 6 the HL used for the analysis are plotted over the HM used as well for each of the locations.

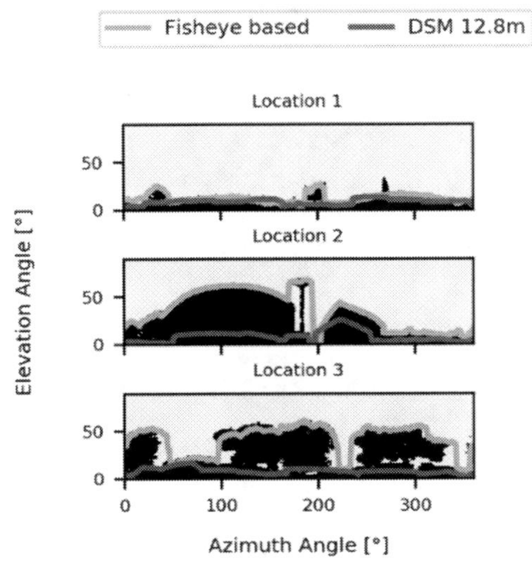

Figure 6: Surroundings characterization inputs used at each location.

3.1 Location 1 analysis

In this section the results for a location with a low shading profile (location 1) are presented. We have taken irradiance measurements at this location for a year and compared them to our predicted irradiance with the different inputs. After quality assessment and filtering, over 14.000 15-minute averaged data points are available.

As shown in Figure 5.a, this location presents a low shading profile with almost no obstruction elements along the sun path. As a result, the accuracy of the shaded irradiance prediction is very similar for all investigated methods with the exception of the 12.8 m resolution DSM which shows a slightly greater overestimation as seen in Table I and Figure 7.

Table I: Statistical errors to measured shaded irradiance at location 1

Location 1	*Fisheye*	*DSM 12.8m*	*DSM 12.8m + LiDAR*
RMSE [W/m²]	18	25	19
MBE [W/m²]	5	12	6

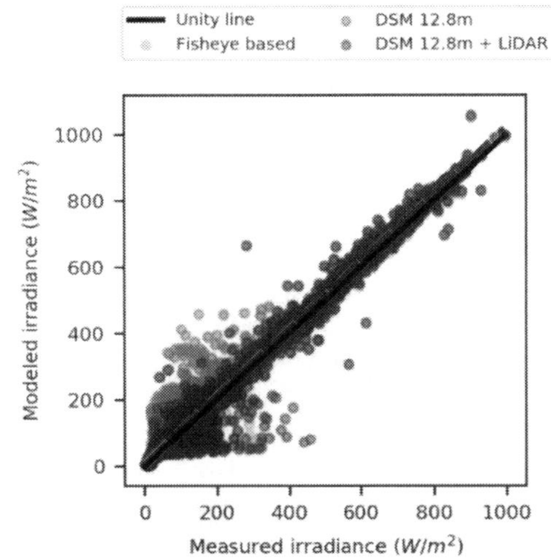

Figure 7: Modelled plane of array irradiance (shaded irradiance) over measured irradiance at location 1

This slight overestimation for the 12.8 m DSM in comparison to the other methods is due to its inability to represent objects will smaller longitudinal dimensions than its cell size. In this particular case, both trees in the location aren't capture by the DSM, and therefore unconsidered in the solar irradiance modelling.

3.2 Location 2 analysis

Location 2 (Figure 5.b) shows a greater shading profile than the previous analyzed location. The results in here show that for locations with greater obstruction elements, the choice of the surroundings characterization method makes a big difference as shown in Table II and **Error! Reference source not found.**.

Table II: Statistical errors to measured shaded irradiance at location 2

Location 2	*Fisheye*	*DSM 12.8m*	*DSM 12.8m + LiDAR*
RMSE [W/m²]	75	185	43
MBE [W/m²]	-21	106	7

When using horizon lines for the shaded irradiance modelling, obstruction elements with non-convex shapes (e.g. street light pole) are misrepresented (see Figure 6 – Location 2). It assumes everything under the head of the pole is an obstruction element as HLs only consider maximum elevation. This leads to big underestimations for two reasons.

Firstly, when the sun position is within the area wrongly considered as obstructed by the HL the direct irradiance is blocked in the model and it should not (see Figure 9).

Secondly, the Sky View Factor (SVF) which determines the amount of diffuse irradiance to be reduced, is underestimated. This leads to consider smaller fraction of the diffuse irradiance reaching the location than it does in reality.

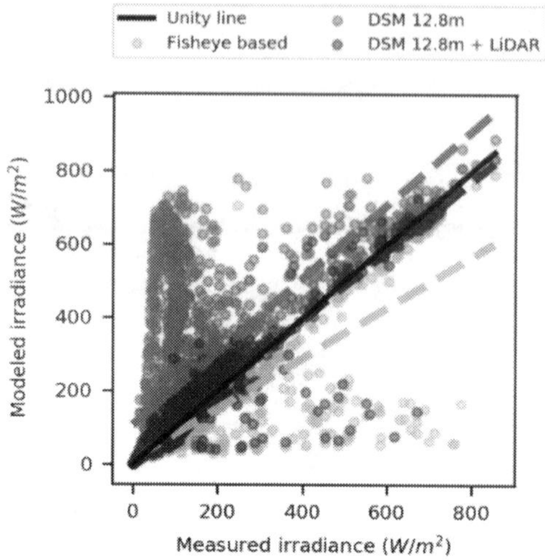

Figure 8: Modelled plane of array irradiance (shaded irradiance) over measured irradiance at location 2. The discontinuous lines represent the regression line of each of the methods

Therefore, when using the high-accuracy HM obtained from the onsite LiDAR measurements, the objects are better represented and the RMSE is reduced by more than 40% and75 % to the fisheye and the low-resolution DSMs methods respectively. Our proposed methodology reduces the bias error as well compared to the other methods as seen in Table II: Statistical errors to measured shaded irradiance at location 2.

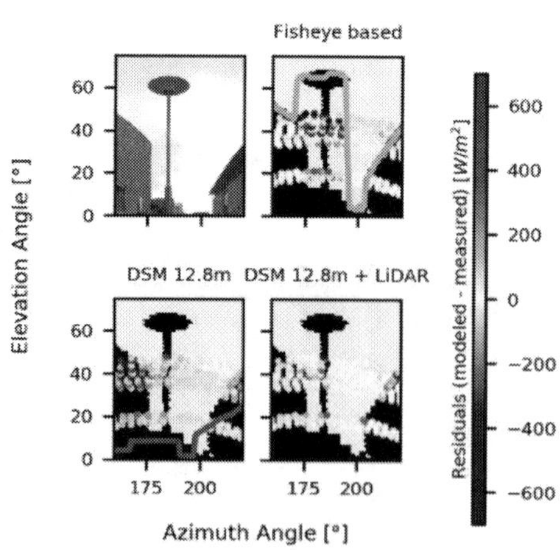

Figure 9: Residuals of each of the studied methodologies as a function of solar position plotted over the horizon matrix obtained with our method for a specific sky region in location 2

3.3 Location 3 analysis

Finally, we analyzed location 3 (Figure 5.c) which shows a great shading profile produced by vegetation. As shown in Table III and **Error! Reference source not found.**, the results show that all models perform poorly in this particular location.

Table III: Statistical errors to measured shaded irradiance at location 3

Location 3	*Fisheye*	*DSM 12.8m*	*DSM 12.8m + LiDAR*
RMSE [W/m²]	113	155	84
MBE [W/m²]	-63	101	-25

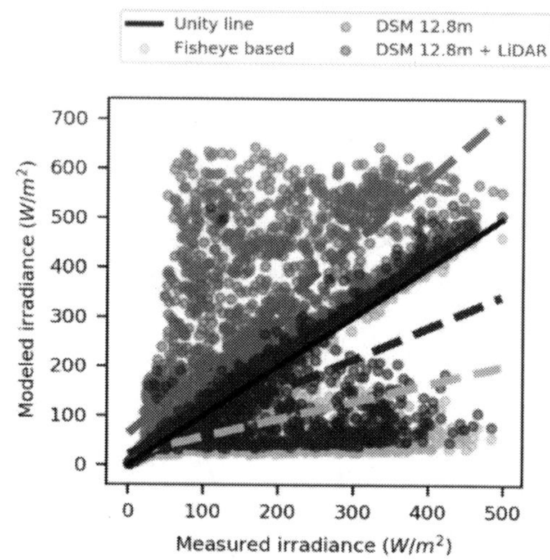

Figure 10: Modelled plane of array irradiance (shaded irradiance) over measured irradiance at location 3. The discontinuous lines represent the regression line of each of the methods

In urban locations where the vegetation is abundant (e.g. parks or green areas), the modelling of the shaded irradiance cannot be reduced to binary variables. Trees branches and leaves do not fully block the incident irradiance, but some light passes through. This can be seen in Figure 11, where every data point where the Sun is behind a tree underestimates solar irradiance.

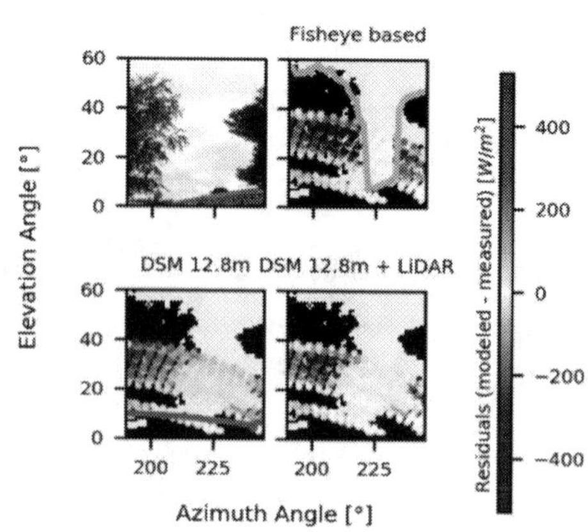

Figure 11: Residuals of each of the studied methodologies as a function of solar position plotted over the horizon

matrix obtained with our method for a specific sky region in location 3 covered by trees

Further research needs to be done towards more accurate ways of modelling solar irradiance reduction due to trees. It is a complex challenge as not all trees have the same constitution which can also vary seasonally.

However, as onsite LiDAR measurements produce point clouds, it is believed that the density of this ones can be used to categorize the kind of tree and its effects on the solar irradiance.

3.4 Errors comparison of all studied inputs at highly shaded locations

Finally, we present the RMBE and MBE of the locations with high shading profiles combined (locations 2 and 3) for all the analyzed cases (see Figure 12).

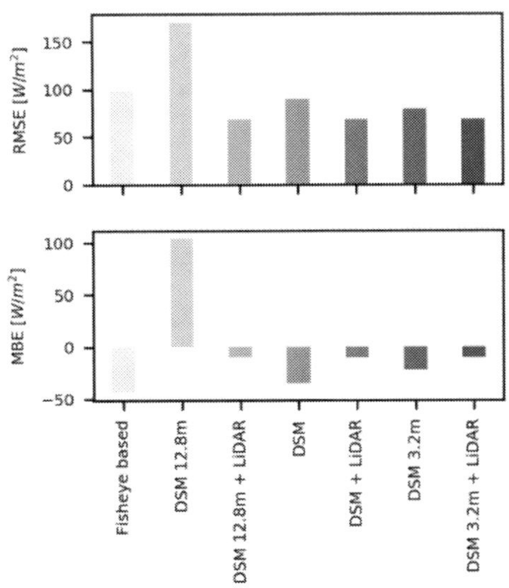

Figure 12: Statistical errors to measured shaded irradiance at both analyzed locations with high shading profiles combined

The results show that all cases where DSMs have been combined with onsite LiDAR measurements perform equally disregarding the resolution of the DSM. Furthermore, the results prove that our proposed methodology improve the results in all studied cases, even in comparison to one of the most accurate DSM available (DSM with 0.4 m resolution), due to its ability to represent irregular objects as street light poles or trees more accurate.

The results obtained with the 3.2 m resolution DSM seem to be more accurate than those obtained with the 0.4 m resolution one. However, this is believed to be caused by the random effects of disregarding some shapes as the street pole in the case of location 2 and the width of the trees in location 3.

4 SUMMARY AND CONCLUSIONS

A new methodology has been introduced to enhance the accuracy to which low-resolution DSMs can represent urban environments for irradiance modelling purposes on buildup locations.

Although it does not improve results in location with low shading profiles, it has proven to reduce both short-term and long-term errors noticeably at location with medium/high shading profiles.

The main driver of this improvement is that this new methodology allows to create horizon matrixes which represent better irregular or non-convex shapes such as trees or light poles.

The characterization of trees represent a big challenge as they do not fully block solar irradiance. However, onsite LiDAR measurements could help to create more accurate methodologies of modelling trees based on the point density of their measurements.

The proposed methodology can improve further. The cutout radius has been chosen arbitrarily but can be adjusted to the distance to the main shading elements. In the future, a dynamic acquisition system for the LiDAR measurements will allow to do better cutouts and improve the characterization of greater areas (e.g. city squares or full parks/green areas).

REFERENCES

[1] R. Carrasco-hernandez, "Calculation of patterns of solar radiation within urban geometries," 2015.

[2] A. Calcabrini, H. Ziar, O. Isabella, and M. Zeman, "A simplified skyline-based method for estimating the annual solar energy potential in urban environments," *Nat. Energy*, vol. 4, no. 3, pp. 206–215, 2019, doi: 10.1038/s41560-018-0318-6.

[3] X. Li and C. Ratti, "Mapping the spatio-temporal distribution of solar radiation within street canyons of Boston using Google Street View panoramas and building height model," *Landsc. Urban Plan.*, vol. 191, no. June 2017, 2019, doi: 10.1016/j.landurbplan.2018.07.011.

[4] Geospatial Agency Innoter, "Digital surface model." https://innoter.com/en/products/spatial-data/dsm-generation/ (accessed Jan. 11, 2022).

[5] I. Geodetics, "DEM, DSM & DTM: Digital Elevation Model – Why It's Important." https://geodetics.com/dem-dsm-dtm-digital-elevation-models/ (accessed Dec. 26, 2021).

[6] M. C. Brito, S. Freitas, S. Guimarães, C. Catita, and P. Redweik, "The importance of facades for the solar PV potential of a Mediterranean city using LiDAR data," *Renew. Energy*, vol. 111, no. October, pp. 85–94, 2017, doi: 10.1016/j.renene.2017.03.085.

[7] R. Compagnon, "Solar and daylight availability in the urban fabric," *Energy Build.*, vol. 36, no. 4, pp. 321–328, 2004, doi: 10.1016/j.enbuild.2004.01.009.

[8] UP42 GmbH, "Everything you need to know about Digital Elevation Models (DEMs), Digital Surface Models (DSMs), and Digital Terrain Models (DTMs)," 2021. https://up42.com/blog/tech/everything-you-need-to-know-about-digital-elevation-models-dem-digital (accessed Jan. 16, 2022).

[9] D. Alvarez Mira, M. Bartholom, S. Poessl, P. B. Poulsen, and S. V Spataru, "Comparing the Accuracy of Horizon Shade Modelling Based on

Digital Surface Models Versus Fisheye Sky Imaging," *2022 IEEE 49th Photovolt. Spec. Conf.*, 2022.

[10] J. Sreevalsan-Nair, "LiDAR BT - Encyclopedia of Mathematical Geosciences," B. S. Daya Sagar, Q. Cheng, J. McKinley, and F. Agterberg, Eds. Cham: Springer International Publishing, 2020, pp. 1–4.

[11] B. Lohani and S. Ghosh, "Airborne LiDAR Technology: A Review of Data Collection and Processing Systems," *Proc. Natl. Acad. Sci. India Sect. A - Phys. Sci.*, vol. 87, no. 4, pp. 567–579, 2017, doi: 10.1007/s40010-017-0435-9.

[12] D. Bredemeier, C. Schinke, T. Gewohn, H. Wagner-Mohnsen, R. Niepelt, and R. Brendel, "Fast Evaluation of Rooftop and Façade PV Potentials Using Backward Ray Tracing and Machine Learning," *Conf. Rec. IEEE Photovolt. Spec. Conf.*, pp. 294–299, 2021, doi: 10.1109/PVSC43889.2021.9518660.

[13] D. Lingfors, J. M. Bright, N. A. Engerer, J. Ahlberg, S. Killinger, and J. Widén, "Comparing the capability of low- and high-resolution LiDAR data with application to solar resource assessment, roof type classification and shading analysis," *Appl. Energy*, vol. 205, no. September, pp. 1216–1230, 2017, doi: 10.1016/j.apenergy.2017.08.045.

[14] D. A. Mira, M. Bartholomaus, P. B. Poulsen, and S. V. Spataru, "Accuracy Evaluation of Horizon Shading Estimation Based on Fisheye Sky Imaging," *Conf. Rec. IEEE Photovolt. Spec. Conf.*, pp. 2052–2059, 2021, doi: 10.1109/PVSC43889.2021.9519063.

[15] Dataforsyningen, "Danmarks Højdemodel - Overflade." https://dataforsyningen.dk/data/928#description (accessed Sep. 13, 2022).

[16] Dataforsyningen, "DHM product specification v1.0.0," pp. 1–21, 2020, [Online]. Available: https://dataforsyningen.dk/asset/PDF/produkt_dokumentation/dhm-prodspec-v1.0.0.pdf.

[17] Dataforsyningen; Styrelsen for Dataforsyning og Effektivisering, "Kortforsyningen download," Oct. 27, 2020. https://download.kortforsyningen.dk/forside (accessed Sep. 13, 2022).

IMPACT OF DATA QUALITY ON THE PERFORMANCE EVALUATION OF LARGE-SCALE PV PLANTS

Giuliano L. Martins[1,2]*, Elena Koumpli[1], Sarah Southern[1], Anastasia Panoui[1] and Jan Muller[1]
[1]Statkraft UK Ltd, 22 Bishopsgate, London, UK
[2]Anhalt University of Applied Sciences, Köthen, Germany
*Corresponding Author: giuliano.luchettamartins@statkraft.com

ABSTRACT: Quality assurance is essential to achieve reliable performance analysis on large-scale PV monitoring data. In this paper, two novel methods are demonstrated for the quality control (QC) of irradiance data. QCs are applied on four large-scale PV power plants (>30MW_{peak}), both tracker and fixed tilt types. Faults distribution and energy impact are shown to depend highly on-site type and location, with tracker stows and deviations by nearby reflections and dew being the most common issues for tracker and fixed tilt sites, respectively. A comparative analysis between raw and QC Performance Ratio (PR) is performed. The impact of data quality can be major, on both performance analysis and PR calculations. Daily PR deviations can go up to -12.5%, strongly affecting the performance analysis of the power plant, carried out on a daily basis. Significant monthly deviations of up to -3.3% are also found, with the potential of causing contractual liabilities and affecting stakeholders' expectations.

Keywords: data quality, performance ratio, irradiance, solar resource

1 INTRODUCTION

The most established key performance metric in the Solar industry is the Performance Ratio (PR), which is actual to the expected power generation of a photovoltaic (PV) plant. PR is commonly assessed on a monthly and annual basis and compared against a threshold guarantee.

Performance metrics calculations and their guaranteed targets are normally defined by contractual agreements, amongst various stakeholders such as Operation & Maintenance (O&M), Asset Managers, and Owners. Negative deviations cause significant concerns as to the root of potential under-performance and can lead to large penalties for the stakeholders involved, thus accurate evaluation of PR becomes paramount.

The main reason for PR calculation inaccuracy is the poor quality of sensor data, an increasing challenge for large-scale (>30 MW_{peak}), and more complex (tracking, bifacial, mixed topography) PV plants. Reliable monitoring data is required for the accurate evaluation of PV system performance [1]. In addition, historical key performance indicators (KPI) calculations can assist in the optimization of power plants and generation forecasting. Irradiance is the main and one of the most complex environmental variables involved in this calculation, due to the many factors that can affect its measurement (position of sensor, shading, soiling etc.) and as such, requires accurate quality assessment.

Quality control (QC) of irradiance measurements is vital to guarantee the quality of KPI calculations. Baseline Surface Radiation Network (BSRN) [2] standards are one of the most sophisticated methods presented in literature [3]. In [4], Lorenz et al. highlight the lack of control methods for tilted irradiance, even though it is the most relevant measurement in the scope of PR; they proposed improvements in the quality assessment of such variables. This remark becomes even more relevant within the general increase of 1-axis tracking PV power plants.

This work presents the application of the best performing QC methods from literature and outlines a step-by-step methodology to mitigate the impact of data quality on global plane-of-array irradiance (GPOAI) measured by pyranometers, based on common challenges in real large-scale PV plants. Apart from validating the methods against real scenarios, the goal is to also demonstrate the impact on PR, by correcting the anomalous raw data. A simplistic, but efficient, gap-filling method is applied, eliminating potential bias from missing data [5].

2 MATERIALS & METHODS

This section comprises detailed information about the methods used to identify anomalous data, and the order of priority they are applied. The methodology is then applied on the PR analysis.

The dataset has a 15-minute timestep resolution and all data points are filtered for daylight values above of 50 W/m². For the data quality routines (DQR) implemented here, the following calculations are required: a) the variability index (VI) as presented in [7], calculated for every third part of the day, to avoid smoothing results due to the lower time resolution.

$$VI = \frac{\sum_{k=2}^{n} \sqrt{(GPOAI_k - GPOAI_{k-1})^2 + \Delta t^2}}{\sum_{k=2}^{n} \sqrt{(CS_k - CS_{k-1})^2 + \Delta t^2}} \quad \text{Eq.1}$$

b) the clearness index (kt), defined as the ratio between measured irradiance and the calculated clear sky irradiance and c) the day type classification based on Iqbal [8]. To calculate the clearness index the Ineichen and Perez clear sky (CS) model is used [6].

2.1 Quality control methods

This section describes the analytical steps of the DQR applied (QC1 to QC6). The following steps are applied daily, where "j" refers to the analyzed or reference data.

QC1 – Adapted BSRN for tilted irradiance sensors as can be found in [4].

QC 2.1 – Sensor comparison (all sensors available in same orientation) (when $VI_{max} < 3$)

$$\frac{GPOAI_j - GPOAI_i}{GPOAI_i} > \pm 0.30; 0.50; for\ (\theta_z < 75°; \theta_z > 75°) \quad \text{Eq.2}$$

Where $GPOAI_i$ refers to the GPOAI median of all available sensor measurements and θz refers to the solar zenith angle.

QC 2.2 – Sensor comparison (sensor nearby)

$$\frac{GPOAI_j - GPOAI_i}{GPOAI_i} > \pm 0.10; 0.20; for\ (\theta_z < 75°; \theta_z > 75°) \quad \text{Eq.3}$$

Where $GPOAI_i$ refers to the GPOAI measurement of the nearby sensor.

QC 3 – Clear sky threshold (when $VI_{max} < 3$)

$$\frac{GPOAI_j - GPOAI_{CS}}{GPOAI_{CS}} > \pm 0.08; 0.12; for\ (\theta_z < 75°; \theta_z > 75°) \quad \text{Eq.4}$$

QC 4 – Shading identification, adapted from [4]. When day type is clear and $VI_{max} \leq 1.5$.

$$(kt_{Sat} > 0.75)\&(kt_j < 0.75); for\ \theta_z < 75° \quad \text{Eq.5}$$

Where sat: satellite data.

QC 5 – Comparison with power profile curve

If both conditions below apply, the flag is removed.

$$\left(\frac{Power_{j\,Norm} - GPOAI_{j\,Norm}}{GPOAI_{j\,Norm}}\right) > \pm 0.10; for\ \theta_z < 90° \quad \text{Eq.6}$$

$$\left(\frac{[Power_{j\,Norm}]' - [GPOAI_{j\,Norm}]'}{[GPOAI_{j\,Norm}]'}\right) > \pm 0.80; for\ \theta_z < 90° \quad \text{Eq.7}$$

Where []' represents the derivative, and Norm refers to Normalized curve, defined in Eq.8.

$$X_{Norm} = \left(\frac{X - X_{Min}}{X_{Max} - X_{Min}}\right) \quad \text{Eq.8}$$

QC 6 – Tracker stows identification, mainly based on tracker logs.

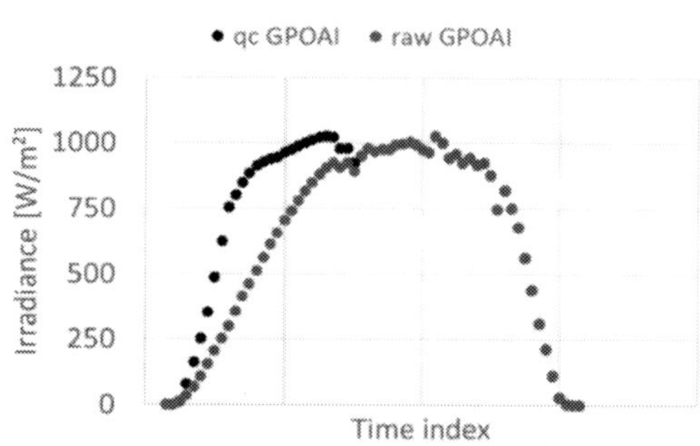

Figure 1: Quality controlled and raw data illustrating an occurrence of tracker stows.

2.2 Data quality procedure

The methods are organized in such order that the more relevant QCs and root causes (e.g. shading) will overwrite the less descriptive ones in the subsequent steps. Figure 2 shows how QCs are organized. On the left column, the QCs include comparative equipment and physical boundaries. On the right-column, the QCs use satellite data or mathematical models. To further improve their efficiency a comparison with the (normalized) power profile curve is added as the final step in the series. The latter overwrites the former, in case of a flag in the same timestamp.

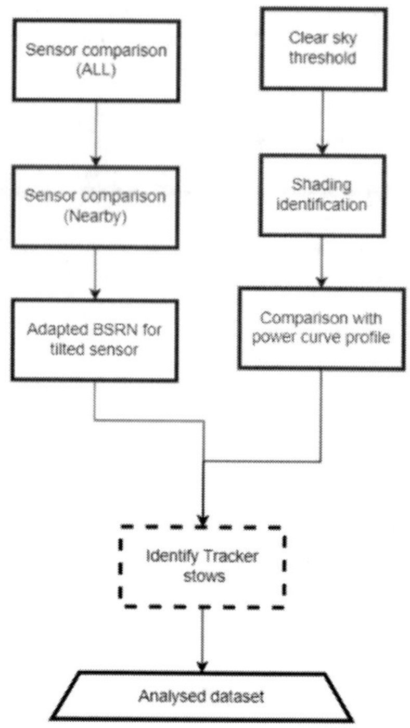

Figure 2: Flowchart illustrating algorithm steps for irradiance data quality control.

Flags from the clear sky threshold are automatically considered to be caused by nearby reflection or dew. The detection of dew formation on sensors is based on [9].

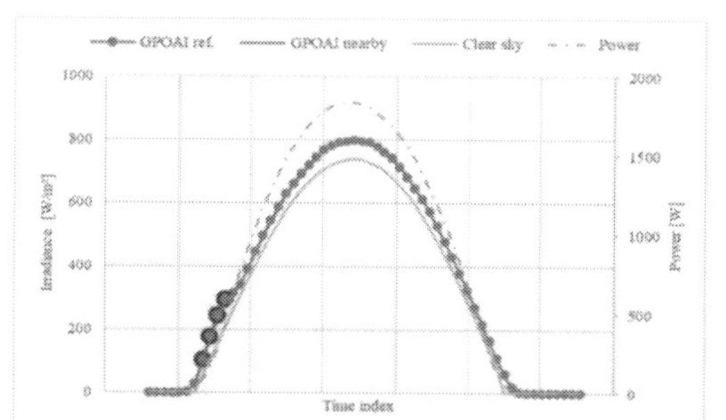

Figure 3: Example of effective application of clear sky threshold (QC3), illustrating reference and nearby sensor and clear sky irradiance, as well as power output.

Figure 3 demonstrates the effectiveness of QC3. Literature methods are not able to identify the anomalous data, since data lies under physical limits and nearby sensors follow the same pattern as the reference one. Using the power profile on top of QC3 demonstrates that PV system is not observing the enhanced pattern in the early morning, thus it refers to anomalous irradiance data. In this specific case, the root cause was identified as dew.

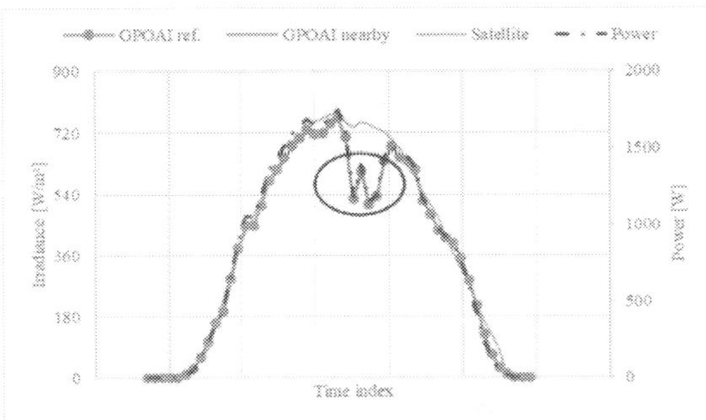

Figure 4: Example of mitigation of anomalous data using power profile (QC5), illustrating reference and nearby sensor and satellite irradiance, as well as power output.

In addition, circled data in Figure 4 was previously flagged by QC4; however, un-flagged by QC5, since power profile data alligns correctly with the irradiance data, therefore illustrating the importance of the novel methods in a robust data quality procedure for irradiance data.

Figures 3 and 4 demonstrate how the two novel methods (QC2 and QC5) are useful in flagging anomalous data, and reducing false positives. Therefore, improving the robustness of the DQR.

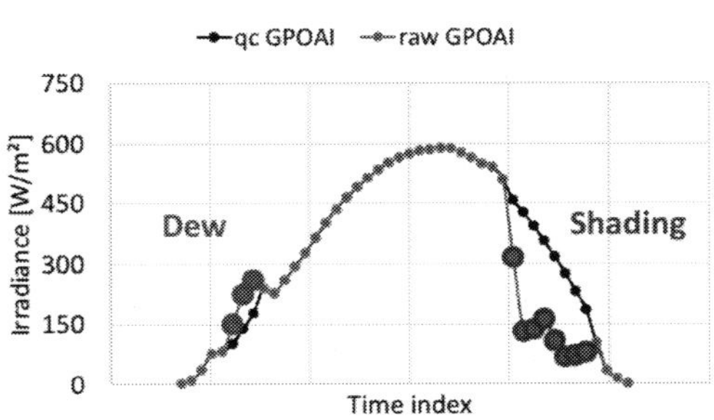

Figure 5: Quality controlled and raw data after applying the proposed algorithm.

Furthermore, as irradiance data goes through the quality control routines, anomalous data points are flagged. Flagged data are removed and replaced with gaps. The gaps are then filled by transposing global horizontal irradiance, based on Perez model [10], measured by a horizontal pyranometer, if available on site or satellite data if not. For a tracking system, important tracker characteristics (tracker rotation angles, axis tilt, ground cover ratio etc.) are required as additional inputs to this calculation.

2.3 Performance ratio analysis

Performance ratio, as by IEC-61724-1, is defined as:

$$PR = \frac{E_{actual} \cdot G_{STC}}{P_{STC} \cdot H} \qquad \text{Eq.9}$$

Where E_{actual} and H are total generation and insolation respectively; G_{STC}, P_{STC} are irradiance, rated power at standard testing conditions (STC).

When PR is calculated using raw GPOAI data, it is defined as rawPR. While it is calculated using treated data, it is defined as qcPR.

The variation is defined as PR variation:

$$varPR = \frac{qcPR - rawPR}{rawPR} \qquad \text{Eq.10}$$

2.4 Dataset

Four different large-scale PV plants are analysed, each > $30MW_{peak}$. There are two tracker and two fixed tilt sites.

Table 1: Description of site types and dataset lengths

Site	Type	Dataset (years)
T1	Tracker	1.5
T2	Tracker	1.5
F1	Fixed tilt	2.2
F2	Fixed tilt	1.3

3 RESULTS & DISCUSSION

This section comprises the main results and considerations of the analyzed dataset on a daily and monthly basis. It starts with an overview of the impact of data quality on PR using the entire dataset, followed by how these "faults" are distributed across all systems

Table 2: Descriptive results per site, with PR variation for the entire dataset.

Site	Dataset (years)	Clear sky days	Days with flags	varPR
T1	1.49	63%	19%	-0.40%
T2	1.48	64%	18%	-0.18%
F1	2.23	27%	8%	-0.01%
F2	1.28	30%	5%	0.01%

From Table 2 it is clear that flagging anomalous data depends highly on clear sky days, which is expected by how the clearness and variability index thresholds are defined in QCs.

Fixed tilt sites present minor deviations in PR, for all datasets while tracker sites present more significant deviations. This can be explained due to the impact of tracker stows in irradiance data, in addition, none of the "F" sites has major shading or soiling problems, thus their faults have minor energy impact, which are generally related to dew and punctual soiling situations. Nonetheless, this consideration will be further discussed in the following sub-sections.

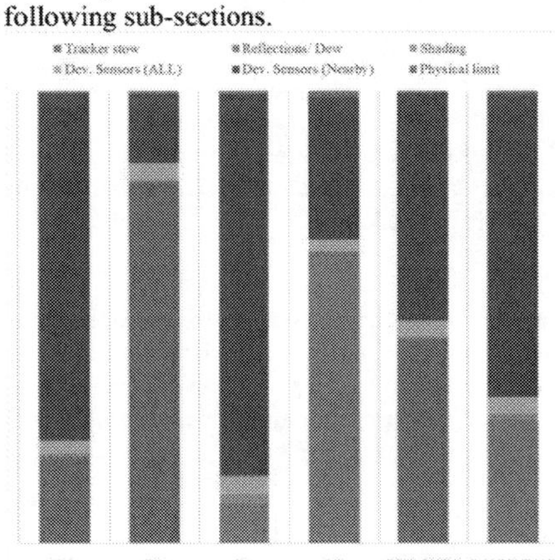

Figure 6: Distribution of faults per site, and overall site type.

Figure 6 illustrates how faults are distributed per site, and per plant type. Tracker sites are majorly impacted by stows, the analysis considers only stows by technical issues, and not from standard behaviour (for example for protection from string winds); the energy impact depends on irradiance intensity and duration of the event. The second most recurrent fault is the deviation between sensors, this QC highlights issues that cannot be easily tagged due to satellite and modelled data accuracy limitations, for example: soiling by bird droplets, early/late shading in a particular sensor and minor deviations in sensor installed angles. These anomalous data have low impact in energy contribution, and/or can be readily solved once flagged to the relevant team.

The remaining QCs are useful to identify major installation problems, shading/reflection by nearby objects, that are prone to happen in plant early operations months.

3.1 Effect on daily analysis

Table 3 summarizes the results for each site. The impact of QC on data can reach up to -12.5% and -5.5% on tracker and fixed tilt sites, respectively.

Moreover, such magnitude in deviation, can potentially mask performance analysis in PV power plants, where underperforming strings could go unnoticed in such cases. In addition, the standard deviation ranges between 2% - 1% overall, which is a significant deviation, especially in large-scale PV plants, which translates to several MWh of energy loss.

Table 3: Overall statistical information on daily performance ratio variation per site in %, and per site type.

	Daily - varPR (%)					
	T1	T2	F1	F2	Tracker	Fixed tilt
Mean	-1.8	-0.9	-0.1	0.2	-1.4	0.0
Median	-0.8	-0.1	-0.1	0.2	-0.4	0.0
Max.	2.5	2.9	5.2	1.6	2.9	5.2
Min.	-12.5	-11.1	-5.5	-0.3	-12.5	-5.5
Std. Dev	2.5	2.1	1.3	0.4	2.3	1.1

The scatterplot, Figure 7, illustrates the daily deviation between raw and QC. Tracker sites outliers are impacted by stows (not related to strong wind effects). For fixed tilt sites, PR deviations refer to compared data to other sensors on site, root cause varies, for example: shading caused by bird droplets, early dew on the pyranometer's dome, nearby reflections from structures and soiling.

Nevertheless, daily variations are significant. Irradiance is the core variable for many models to analyze equipment performance. Thus, quick maintenance response from O&M can be impacted due to misleading results, caused by anomalous data.

Data quality issues might be seasonal such as dew, shading and soiling; however, installation-related problems might occur (leveling, installation at a shaded location, etc.), therefore, early identification of anomalous data and root cause assignation is essential to mitigate PR impact, and not propagating issues further on.

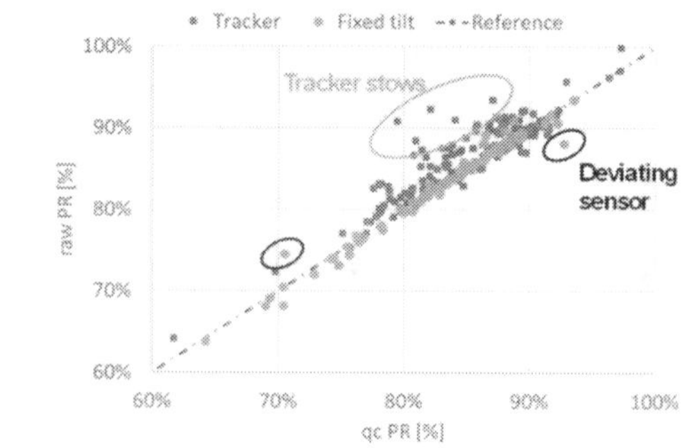

Figure 7: Scatterplot relating raw and quality controlled daily PR, highlighting main fault conditions.

3.2 Effect on monthly analysis

The Figure 8 shows the varPR and cumulative faults per month, for each site.

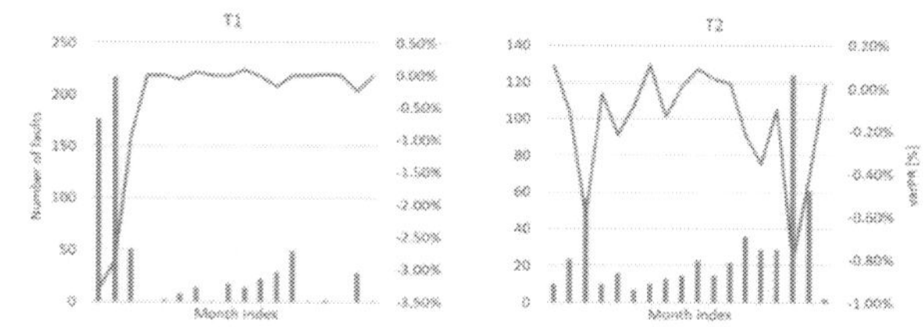

Figure 8: Amount of faults and performance variation per month, for sites T1 (left) and T2 (right).

T1 has a high varPR impact on early site operation, caused by initial technical problems in tracker operation, which affected the plant majorly, moreover, problems were mainly solved and as consequence the effect in PR was mitigated. For T2, tracker stows were punctual events, even though proportionally more recurrent, thus varPR was not as high magnitude as in T1.

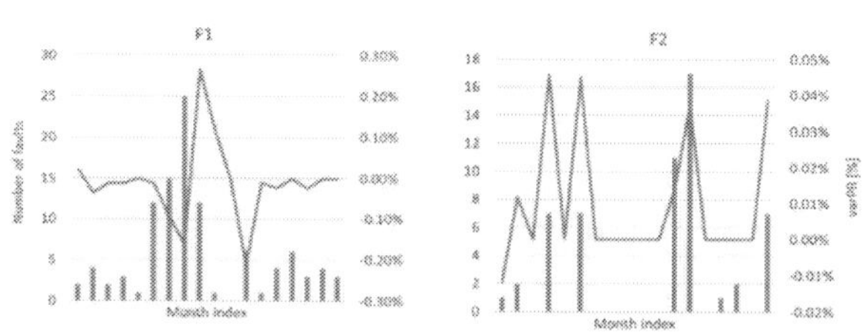

Figure 9: Amount of faults and performance variation per month, for sites F1 and F2.

In the case of fixed tilt sites, monthly number of faults is shown to be seasonal, due to deviating from its pair sensors. Root cause ranges from dew, soiling, and nearby reflections. Faults are not propagated throughout weeks, which supports the fact of it being punctual problems, caused by idiosyncratic conditions on specific days, and removed by sensor cleaning. Nonetheless, the event has an impact on irradiance and must be treated to not bias analysis.

4 CONCLUSIONS

This paper presents the implementation of the best performing literature-based quality checks combined with two novel data quality routines into a step-by-step process with well-defined and prioritized steps, that accurately identifies and mitigates anomalous data, facilitating decision-making and solution on-site.

Distribution of faults are shown to vary per site type, and location. Tracker stows have the highest proportion and impact for tracker sites among all detected flags. On the other hand, sensor intercomparison is the main tool for identifying anomalous data on fixed tilt sites.

PR analysis between raw and QC data show daily PR deviations up to -12.5% and -5.5% for tracker and fixed tilt sites, respectively, which shows that if irradiance data are left untreated, they can bias performance analysis routines significantly and especially for tracker sites. These deviations reduce, on monthly, to -3.3% and -0.27% for tracker and fixed tilt sites, respectively.

Deviations in daily analysis have major impact on performance analysis, since they can mask equipment failure and malfunction. Furthermore, monthly deviations can mislead stakeholders' expectations. Finally, yearly datasets still present deviations in PR of up to -0.4% for tracker sites. Thus, potentially compromising energy prediction, acceptance test results and long-term performance analysis.

Novel methods improved data quality for irradiance sensors. Existent faults on site depends heavily on type (tracker/ fixed tilt). Quality controlled data is vital, as PR deviations can be high, even for larger datasets; potentially propagating errors if root causes are not flagged and solved.

5 REFERENCES

[1] International Electrotechnical Commission. IEC 61724-1:2021 Photovoltaic system performance - Part 1: Monitoring n.d.

[2] Ohmura A, Gilgen H, Hegner H, Müller G, Wild M, Dutton EG, et al. Baseline Surface Radiation Network (BSRN/WCRP): New Precision Radiometry for Climate Research. Bull Am Meteorol Soc 1998;79:2115–36. https://doi.org/10.1175/1520-0477(1998)079<2115:BSRNBW>2.0.CO;2.

[3] Gueymard CA. A review of validation methodologies and statistical performance indicators for modeled solar radiation data: Towards a better bankability of solar projects. Renew Sustain Energy Rev 2014;39:1024–34. https://doi.org/10.1016/j.rser.2014.07.117.

[4] Lorenz E, Guthke P, Dittmann A, Holland N, Herzberg W, Karalus S, et al. High resolution measurement network of global horizontal and tilted solar irradiance in southern Germany with a new quality control scheme. Sol Energy 2022;231:593–606. https://doi.org/10.1016/j.solener.2021.11.023.

[5] Livera A, Theristis M, Koumpli E, Theocharides S, Makrides G, Sutterlueti J, et al. Data processing and quality verification for improved photovoltaic performance and reliability analytics. Prog Photovoltaics Res Appl 2021;29:143–58. https://doi.org/10.1002/pip.3349.

[6] Ineichen P, Perez R. A new airmass independent formulation for the Linke turbidity coefficient. Sol Energy 2002;73:151–7. https://doi.org/10.1016/S0038-092X(02)00045-2.

[7] Stein JS, Hansen CW, Reno MJ. The variability index: A new and novel metric for quantifying irradiance and pv output variability. World Renew Energy Forum, WREF 2012, Incl World Renew Energy Congr XII Color Renew Energy Soc Annu Conf 2012;4:2764–70.

[8] Iqbal M. An Introduction to Solar Radiation. ACADEMIC PRESS CANADA; 1983.

[9] Geuder N, Wolfertstetter F, Wilbert S, Schüler D, Affolter R, Kraas B, et al. Screening and Flagging of Solar Irradiation and Ancillary Meteorological Data. Energy Procedia 2015;69:1989–98. https://doi.org/10.1016/j.egypro.2015.03.205.

[10] Perez R, Ineichen P, Seals R, Michalsky J, Stewart R. Modeling daylight availability and irradiance components from direct and global irradiance. Sol Energy 1990;44:271–89. https://doi.org/10.1016/0038-092X(90)90055-H.

6 ACKNOWLEDGEMENTS

The authors acknowledge funding from the Horizon 2020 programme of the European Union under GA. No. 952957-Project TRUST PV. Would also like to acknowledge Statkraft Ltd. And ENCAVIS for the site data.

ESTIMATING GLOBAL HORIZONTAL IRRADIANCE AT THE URBAN LEVEL:
A SENSITIVITY ANALYSIS USING DIFFERENT DIGITAL SURFACE MODELS

Rodrigo Amaro e Silva, Philippe Blanc
O.I.E. Centre Observation, Impacts, Energy, MINES Paris - PSL Research University
06904 Sophia Antipolis, France
rodrigo.amaro_e_silva@minesparis.psl.eu

ABSTRACT: Digital Surface Models are key for integrating urban-level shadowing in PV modelling. Despite the increasing number of such products and their diversity in terms of acquisition method and post-processing, current state-of-the-art research on solar resource and PV modelling in urban environments consider only a single DSM and disregard the uncertainty associated to the choice of such product. Here, seven different DSMs, with different levels of detail and accuracy were used to calculate the incident global horizontal irradiance with 15-min resolution for the city of Nantes, France. First, an exploratory analysis across three DSMs is done to illustrate the visual and practical differences between products. The root-mean-squared deviation between DSM pairs was also quantified regarding the surface altitude (0.6 to 3 m) and incident solar irradiance (32 to 70%, relative to the annual mean daytime value). Lastly, the uncertainty for annual irradiance is estimated to be 1.0-24.9% for rooftops and 5.8-44.8% for ground level surfaces, while it is highlighted that uncertainty can be considerably higher for given short time periods with clear-sky conditions.
Keywords: Digital Surface Model, Shading, Solar radiation, PV, Modelling.

1 INTRODUCTION

Distributed photovoltaics (PV) is expected to grow exponentially in urban areas in order to achieve local net-zero goals [1], with the concept of integrated-PV starting to expand beyond buildings [2] to, for example, parking infrastructures [3] and PV-coupled electric vehicles [4].

This will imply a growing need for shadow modelling at the urban level, where PV is prone to shadowing periods due to the rich and complex morphology of such areas (e.g., buildings with different shapes and sizes, trees) [5]. Urban shadowing can not only hamper the PV annual yield but also introduce considerable power ramps [5], reducing revenues while increasing integration costs. Thus, shadow modelling will become an ever more common module for various tools, e.g., solar cadasters, PV self-consumption simulators, or solar forecasting algorithms. And the cornerstone for urban shadow modelling is having a three-dimensional and digital representation of the surface topography, i.e., a Digital Surface Model (DSM).

DSMs can be acquired from different sources (spaceborne or airborne photogrammetry, airborne LiDAR) and with different levels of detail, spatial resolution, and 3D uncertainties. And while there is already a considerable amount of literature exploring DSMs for solar studies at the urban level [6,7], most tend to explore a single DSM, since this is often a not so easily accessible asset. This work aims to perform a sensitivity assessment of incident irradiance to the uncertainty of the DSM – since DSMs are difficult to validate in a comprehensive manner – and infer, even if only for a single case study, the degree of uncertainty associated to the choice of a DSM. The proposal of an inter-comparison of different DSMs for a single region becomes more pertinent when it is considered that there is no true perfect reference against which DSM models can be validated.

2 METHODS AND DATA

To achieve this, a 500 x 500 m2 area in the city of Nantes, France, is considered, for which seven different DSMs from different sources and methodologies were compiled (c.f. Table 1 for more details).

Here, 1000 pairs of (latitude, longitude) coordinates were randomly sampled inside our region of interest. For each location it was calculated: the corresponding surface height; the local horizon (i.e., the corresponding highest angular elevation of the blocking surrounding as a function of the azimuth, around the point of interest), and a timeseries of 15-min global horizontal irradiance (GHI) over one year. The latter is obtained from integrating satellite-derived irradiance estimates (namely, from the HelioClim3 product [8,9]). The shadowing impact is modelled by: i) decomposing the GHI in its various components (i.e., direct, diffuse circumsolar, diffuse) using the Ruiz-Arias model [10]; ii) a binary shadowing factor is applied to direct and diffuse circumsolar, where irradiance is unaffected/nullified if the sun elevation, at a given location and given time/sun azimuth, is higher than the corresponding horizon value; iii) the remaining diffuse irradiance was assumed isotropic and, thus, was mitigated according to the fraction of the sky dome obstructed by the horizon profile.

It was also possible to differentiate rooftop and ground-level cases: the former are those whose coordinates correspond to the inside of the building polygons defined in the building cadaster of DSM A; the latter were identified as the coordinates where the absolute difference in altitude between the DSM C (which presents the highest level of detail while being sharp) and the DSM A one (which only describes buildings) was smaller than 0.6 m (an empirically defined threshold).

Finally, the obtained values were cross compared.

Table I: Description of DSMs used in this work. All DSMs are downscaled to a resolution of 10 cm. IGN stands for Institut national de l'information géographique et forestière.

DSM ID	Provider	Source	Description
A	IGN	Building cadaster coupled with DEM.	Often designated as shoebox model. The ground level is defined by a Digital Elevation Model and from a building cadaster with contour polygons and one height value per building, each polygon is vertically extruded assuming cuboid format (i.e., flat surfaces).
B B.1	IGN	B to C.3: DSMs acquired from airborne stereo photogrammetry.	Despite being from the same provider, B and B.1 were taken at different times and B has coarser native horizontal resolution (which depends on the altitude of the plane for each acquisition).
C C.1 C.2 C.3	Google		C.1 is an artificially degraded version of C (i.e., coarser horizontal resolution from different Levels of Detail (LoDs) of the underlying 3D triangulation model) which then is downscaled back to original resolution using a bilinear interpolation; C.2 and C.3 are natively coarser versions of C (by a factor of 2 and 4, respectively), which undergo the same downscaling process.

3 RESULTS

3.1 Exploratory visual analysis and discussion of inter-DSM variability

To better illustrate the potential variability between DSMs, three very different cases (A, B, C) were selected out of the complete set of seven DSMs.

The first one is based on a simple shoebox model, using the RGE ALTI a Digital Elevation Model (DEM) with 1-m resolution [11] is fused with the BD TOPO building cadaster [12], where buildings are described using contour polygons and a single height value. This is an oversimplistic approach (shoebox model), as the polygon describing the contour of a given building is simply extruded vertically according to the height value while assuming a flat rooftop (resulting in sharp building outlines from a top view, c.f. Figure 1, top). It is important to note that here the rooftop geometry, including local slopes and superstructures, is disregarded, as well as non-building structures (e.g., trees). However, it contains accurate information which is not often found in DSMs by default, namely the footprint of the buildings and their maximum height.

The second and third DSMs (Figure 1, middle and bottom) both show a more complete description of the urban topography, although the second DSM shows some blurriness: they correspond to the same Ground Sampling Distance (10 cm) but their effective spatial resolutions are significantly different. This can be the result from a coarser native resolution of aerial photography and/or a lower number of images acquired at the same type for the stereoscopic photogrammetric process. For example, the size of the windows used for the correlation process maybe different and then induce different blurring effect in the DSM retrieval.

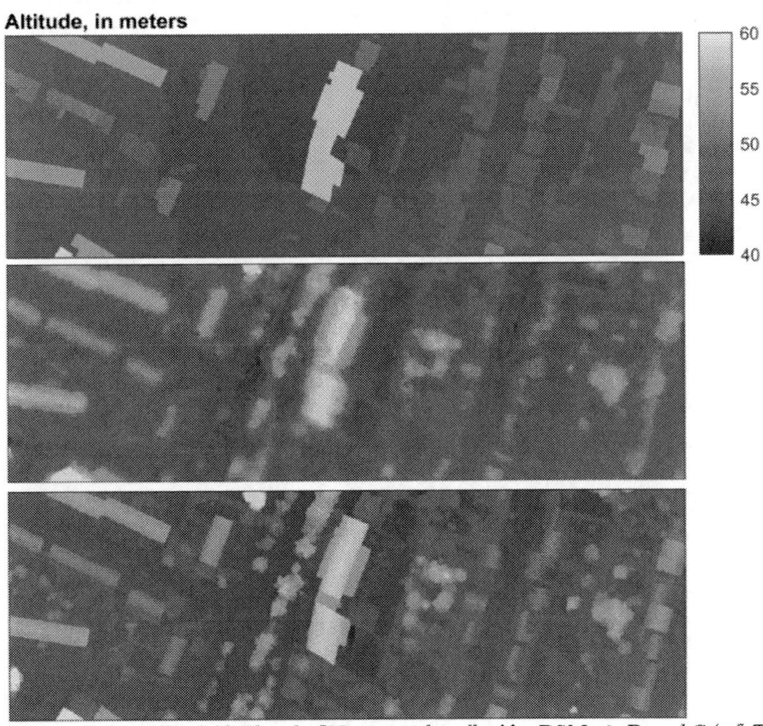

Figure 1: A small neighborhood of Nantes as described by DSMs A, B, and C (c.f. Table 1), with considerably degrees of detail and sharpness (c.f. text for more details).

3.2 Difference in DSM altitude values

The variability in the altitude estimates across the DSMs was evaluated by calculating a pair-wise root mean square deviation (RMSD), ranging between 0.5 and 3 m (Figure 2). While the DSMs C to C.4, which come from the same provider, show to be quite similar (with low RMSD), this same cluster and all the other DSMs show higher RMSD values. Although not shown here, it was found that there is a bias (i.e., a systematic deviation) when DSM #3 is compared to the other DSMs which plays a great role in its associated high RMSD. This is to be expected since all non-building structures are not described in this DSM, assuming instead the ground-level altitude.

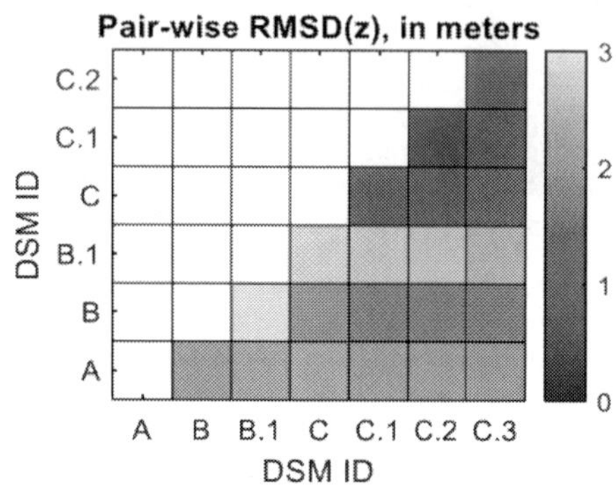

Figure 2: RMSD for altitude estimates, considering different DSM pairs.

3.3 Difference in DSM GHI values

A similar analysis was done for the resulting GHI timeseries, considering only daytime 15-min values. Figure 3 shows that the RMSD on the estimated shadow-impacted GHI using two different DSMs can easily surpass 50%. The highest values are found when comparing DSMs A, B, and B1 with DSM C. For DSMs C to C.3, coming from the same provider just with different LoDs and post-processing, it can also be seen that the difference between them increases with the difference in native resolution (e.g., $|RMSE_C - RMSE_{C.3}| > |RMSE_C - RMSE_{C.1}|$), but also that the worse the resolution of reference, the lower the difference with the further degraded versions (e.g., $|RMSE_C - RMSE_{C.2}| > |RMSE_{C.2} - RMSE_{C.3}|$).

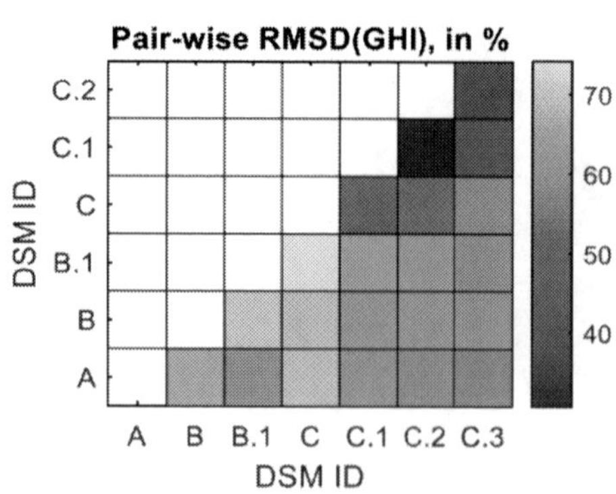

Figure 3: RMSD for daytime GHI values, considering different DSM pairs.

As a complementary analysis, the total annual solar resource was calculated for each location and each DSM. The uncertainty around these is of 1.0 to 24.9% for rooftop locations, whereas for ground level is 5.8 to 44.8%. These were calculated as the 10th and 90th percentile of deviations normalized by the mean annual solar resource. This higher uncertainty at the ground level is to be expected, since is naturally exposed to higher horizon profiles.

3.4 Brief analysis on higher temporal resolution results

To complement the previous overall analysis, the focus was placed on a small subset of data, choosing two illustrative winter days (one clear and the other cloudy) and on three DSMs (A, B, C).

The inferred horizon profiles A and C show considerable differences when compared to DSM B (Figure 4), which was shown to have the highest level of detail with minimal signs of blurriness (Figure 1, bottom). DSM A greatly underestimates the degree of obstruction around the location of interest, while the DSM C, despite being more similar, seems to frequently overestimate the horizon profile.

Figure 5 shows the impact from the horizon profile on the estimated GHI impacted by shadow. On a clear day, DSM A disregards the shadowing impact for most of the day, while the difference between DSM B and C is significantly lower. In clear-sky conditions, a considerable share of GHI is direct irradiance and both DSMs result in an identical direct obstruction (i.e., for both cases, the sun trajectory is always below each DSM horizon profile, c.f. Figure 4). For a cloudy day, where diffuse irradiance is predominant, all DSMs show a relatively similar GHI profile, which decreases non-linearly with the integral of the horizon profile.

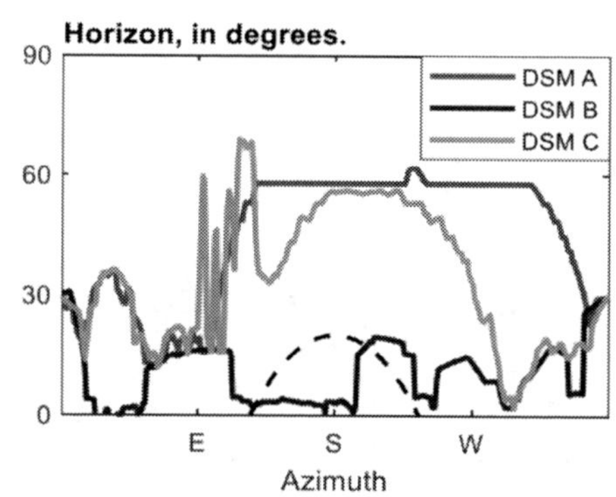

Figure 4: Inferred horizon profiles for three DSMs. The dashed dark line indicates the sun trajectory for a given winter day.

4 CONCLUSIONS

Here it is shown that the inferred horizon profiles and, therefore, the shadow-impacted global horizontal irradiance (GHI) are highly sensitive to the choice of DSM as an input for the horizon calculations. This raises awareness for the need that end-users understand the quality of the DSM they are using.

It was also calculated the uncertainty of the total annual solar resource, due to the availability of multiple DSMs. This uncertainty was inferred for rooftops and the ground level as of 1-24.9% and 5.8-44-8%, respectively,

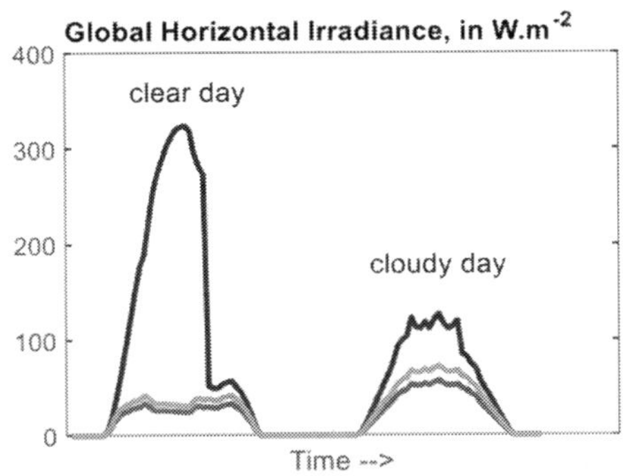

Figure 5: Illustrative two winter days with different weather conditions, and the resulting shadow impacted GHI considering three different DSMs,

with the ground level showing greater uncertainty. This is to be expected, since at the ground level there is a natural exposition to higher horizon profiles.

Finally, it was also shown that when looking at shorter time periods instead of overall RMSD values, it is possible to find moments of the day where the difference in inferred GHI for different DSMs can vary drastically. This is also dependent on the weather condition, since that impacts the contribution of direct and diffuse components to the GHI (with direct being considerably more sensitive to the horizon profile).

Future work will naturally involve going beyond the impacts on only GHI for rooftop structures. It is also important to evaluate the impact of the DSM on the inferred geometry (i.e., tilt and orientation), as well as the irradiance estimate for the inferred planed of array (i.e., global tilted irradiance – GTI).

5 ACKNOWLEDGEMENTS

This research was partially completed under the framework of the Horizon 2020 project e-shape (grant agreement 820852). It is also important to state that this work was only made possible thanks to the contribution of the two DSM providers: the Institut national de l'information géographique et forestière (IGN) from France and Google from the United States of America.

6 REFERENCES

[1] International Energy Agency (IEA), Empowering Cities for a Net Zero Future: Unlocking resilient, smart, sustainable urban energy systems (2021).

[2] M. Panagiotidou, M.C. Brito, K. Hamza, J.J. Jasieniak, J. Zhoue, Prospects of photovoltaic rooftops, walls and windows at a city to building scale, Solar Energy (2021).

[2] A. Małek, Ł. Wojciechowski, Carports as an element of energy security for electric vehicles, IOP Conference Series: Materials Science and Engineering (2022).

[4] M.C. Brito, T. Santos, F. Moura, D. Pera, J. Rocha, Urban solar potential for vehicle integrated photovoltaics, Transportation Research Part D: Transport and Environment (2021).

[5] P. Moraitis, B. Kausika; N. Nortier, W. van Sark, Urban Environment and Solar PV Performance: The Case of the Netherlands, Energies (2018).

[6] S. Freitas, C. Catita, P. Redweik, M.C. Brito, Modelling solar potential in the urban environment: state-of-the-art review. Ren. Sust. Energy Rev. (2015).

[7] E. Fakhraian, M.A. Forment, F.V. Dalmau, A. Nameni, M.J.C. Guerrero, Determination of the urban rooftop photovoltaic potential: A state of the art, Energy Reports (2021).

[8] SoDa – Solar Radiation Data, HelioClim-3 (HC3) Overview, available online: http://www.soda-pro.com/help/helioclim/helioclim-3-overview

[9] P. Blanc, B. Gschwind, M. Lefèvre, L. Wald, The HelioClim Project: Surface Solar Irradiance Data for Climate Applications, Remote Sensing (2011).

[10] J.A. Ruiz-Arias, H. Alsamamra, J. Tovar-Pescador, D. Pozo-Vázquez, Proposal of a regressive model for the hourly diffuse solar radiation under all sky conditions, Energy Conversion and Manag. (2009).

[11] IGN, RGE ALTI, https://geoservices.ign.fr/rgealti.

[12] IGN, BD TOPO, https://geoservices.ign.fr/documentation/donnees/vecteur/bdtopo.

SHORT-TERM SOLAR IRRADIANCE FORECASTING BASED ON THE INTEGRATION OF NUMERIC WEATHER PREDICTION AND DEEP LEARNING

Shirzadi Navid [1], Nasiri Fuzhan [1], Menon Ramanunni [1], Eicker Ursula [1]
Concordia University
[1]Gina Cody School of Engineering and Computer Science, Concordia University, 1455 Boulevard de Maisonneuve, Montréal, Canada

ABSTRACT: Accurate short-term solar irradiance forecasts would prove valuable for optimal management and effective operational control of microgrids integrated with the conventional power grid. Because of the intermittent characteristics of renewable energy resources, predicting global solar irradiance is a challenging task. The current forecasting models mainly use historical data, while the unexpected behavior of other environmental indicators, such as cloud movement, is not considered. Therefore, over a short period of time, the present-day forecast results are found to be unreliable, especially for control purposes requiring higher accuracy. To tackle this issue, the article proposes an integrated model based on the hybridization of the numeric weather prediction (NWP) model along with long short-term memory (LSTM) in the day-ahead (24 hours) horizon. Downward shortwave and longwave solar radiation features are selected from the NWP forecast output to be integrated with the LSTM model. The results show that hybridizing models could significantly improve the forecasting accuracy in overcast and scattered climate conditions by reducing the root mean squared error of the single LSTM model's prediction results from 197.19 and 192.81 to 62.96 and 72.01, respectively.

1 INTRODUCTION

To mitigate the environmental pollutions and emissions caused by fossil fuels (and their contributions to climate change), renewable resources are receiving an increasing attention globally. One of the promising renewable resources is solar energy. Based on the International Renewable Energy Agency report, about 709 GW of solar photovoltaic (PV) systems installed in 2020 shows about a 21% rise in just one year (compared to 2019 capacity). However, with the dispersion of distributed renewable generations (such as small and large PV plants) more and more inside the cities, the development of smart grids will be essential for energy management purposes [1]. Therefore, to keep the power grid in balance and for proper and precise energy management, the prediction of the available renewable resources is necessary [2]. Accurate solar irradiance forecasting is beneficial for power management purposes and can increase the system's reliability. However, precise forecasting is challenging due to the solar resource's high volatility and intermittent characteristics.

Various method has been used to forecast solar irradiance in previous studies. These methods could be classified into two main categories: numeric weather prediction (NWP), statistical, and learning times series methods. Persistence predictions as the simplest time series models that consider the forecast of the irradiance model in the next prediction horizon equal to the current irradiance at the surface [3]. Persistence models are typically used as a naïve predictor, and the forecasting performance of the other robust methods is compared with them [4]. While statistical models use historical data to predict future solar irradiance, NWP models forecast the current and future states of the atmosphere and surface based on duplicating the physical phenomenon. Moreover, recently, different machine learning techniques have been widely used for solar irradiance forecasting based on learning the pattern of historical data.

A comparison of different time series solar radiation forecasts has been done by Reikard [5] for different time resolutions (5, 15, 30, and 60 minutes). The result of his work shows that the autoregressive integrated moving average (ARIMA) model can obtain better accuracy in most of the forecast horizons (1, 2, 3, and 4 hours) compared with other statistical and learning models such as neural network (NN). In another study, Mellit et al. [6] used a multilayer perceptron (MLP) technique to predict day-ahead solar irradiance by using temperature as an independent variable. Their forecasting results showed a better correlation coefficient for sunny days (98%) and less for cloudy days (95%). In [7], the authors a multistage neural network instead of a single-stage neural network for daily global horizontal irradiance forecasting. Their result shows that their proposed model could reduce the mean bias error (MBE) from about 30% (single-stage NN) to about 20%. However, in all mentioned statistical and learning models, the impact of unexpected cloud movement could not be captured from the historical data. Especially on cloudy days, the inaccuracy of these models will rise.

As advanced machine learning methods, deep learning algorithms are also employed to better deal with the dependencies in historical data. Sun et al. [8] used a convolutional neural network (CNN) to forecast PV output using sky images. In another study, Qing et al. [9] investigated the ability of the LSTM method (using meteorological features) for the day ahead prediction with hourly resolutions. Moreover, the hybridized models consist of CNN, and LSTM models developed recently for solar radiation forecasting [10,11]. Other types of hybrid machine learning models were also developed for solar irradiance forecasting to improve efficiency, and a complete study has been done in [12] on them.

On the other hand, the NWP models that integrate hydrodynamic equations with numeric methods are mainly used with an extended forecasting horizon (up to15 days) [13]. Although the NWP model could be a good choice for longer-term forecasts, it could result in lower accuracy when it comes to short and very short-term forecasting. Moreover, it is considerably impacted by its initial conditions.

Although several studies have been carried out for solar irradiance forecasting, there is still a need to development of methods that can more accurately capture unexpected solar irradiance behavior. This paper aims to provide a methodology for precise solar irradiance forecasting in a short-term (day ahead) horizon that could contribute to better smart grid and microgrid energy management. The novelty of the proposed approach rests in the integration of deep learning and the NWP model for better forecasting results. In this sense, the main objectives of this study could be summarized as follows:

- Development and application of an LSTM model as a time-series learning method for dealing with long-term dependencies
- Comparison of deep learning forecasting results with NWP solar irradiance forecasting results as single models for three days with the different climatic situations.
- Proposing a hybrid model integrating the results of LSTM and NWP models to better deal with unexpected trends in solar irradiance that could not be captured using historical data.

The rest of this paper is organized into three sections. Section 2 represents the related methodology of data preprocessing, NWP results, and the LSTM model. Section 3 describes the Implementation, while in section 4 the comparison between each model and the final proposed model will be made. Finally, a conclusion providing a summary of the research is discussed in section 5.

2 METHODOLOGY

The proposed approach for solar irradiance forecasting will be explained in this section. Based on the overview in Fig. 1, the overall methodology could be explained in 4 main steps. In step one, the preprocessing of the historical data is carried out, consisting of normalization, outlier detection, and dealing with missing values. Then the prediction results of downward longwave and shortwave irradiance are extracted from the NWP simulation model and preprocessed before being employed as predictors in the second step. Then in the third step, the LSTM is developed based on the 3-dimension input, including historical solar irradiance, downward longwave, and shortwave solar irradiance. In the last step, hyperparameter optimization is conducted to find the optimum value for the LSTM parameter.

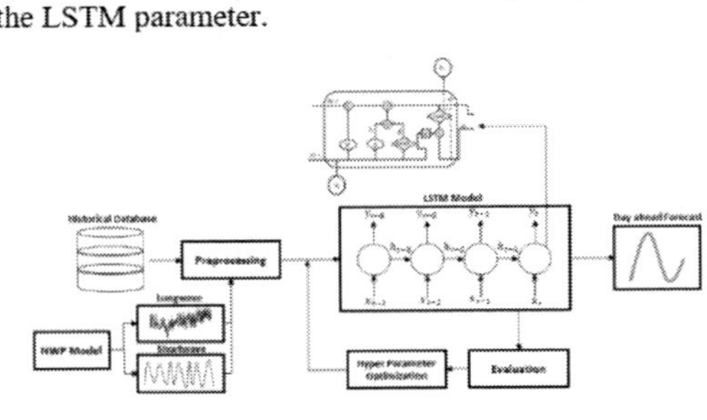

Figure 1: The overview of the proposed model

2.1 Preprocessing

In the first step, the missing values of both historical solar irradiance data and long and shortwave irradiance (NWP results) are detected and replaced with the average of the next and previous values.

Then to normalize the input data, the Min-Max Scaler method is employed to scale the input data:

$$X_N = \frac{X - X_{\min}}{X_{\max} - X_{\min}} \quad (1)$$

With quantile-based methods, the outliers of the predictors and target variables are detected and removed from the data set. Since the number of outliers is not considerable, the deletion does not have a significant impact on the dataset.

2.2 LSTM

Recurrent neural network (RNN) is one of the deep neural networks developed for time series forecasting [14]. The recurrent feedback connections make RNN a proper algorithm for dealing with sequential data [15]. However, inappropriate dealing with data with long-term dependencies could be considered one of the weaknesses of RNN [16]. LSTM method, as one of the types of RNN, was developed to solve the long-term dependencies challenge [17,18]. An LSTM cell includes several layers such as forget layer, update layer and regeneration layer that distinguishes it from the RNN's cell that only contains a tang hyperbolic function. The complete methodology of the LSTM method can be found in [19].

2.3 Hybrid Model

The structure of the hybrid model is the same as the single LSTM model. However, the input dimension of the hybrid model is 3, while the single LSTM model only employs solar irradiance historical data. To hybridize the NWP model with the LSTM model, first, the prediction results of the NWP model are extracted and then scaled up along with the target variable and then fed to the LSTM model as the predictor.

2.4 Evaluation Metrics

For estimating the performance of the forecasting models, root mean square error (RMSE) and mean absolute error (MAE) are employed. RMSE and MAE could be calculated via equations (2) and (3), respectively.

$$RMSE = \sqrt{\sum_{i=1}^{n} \frac{(\bar{y}_i - y_i)^2}{n}} \quad (2)$$

$$MAE = \frac{\sum_{i=1}^{n} |\bar{y}_i - y_i|}{n} \quad (3)$$

where y is the observed value and $\bar{y}$ is the predicted value.

3 INPUT DATA

The city of Montreal is selected as the case study. Montreal is located in the Southern part of the province of Quebec, Canada, with latitude and longitude coordinates of 45 N and -73 E degrees, respectively [20]. Montreal's solar irradiance historical data is extracted from the NASA data access port [21] in hourly resolution from Jan 2021 until the end of May 2021. Furthermore, the downward shortwave and longwave solar irradiance are extracted from the NWP model's result for the same period (from Jan 2021 until the end of May 2021). The overall trend and seasonality of the longwave and shortwave irradiance are illustrated in Fig. 2. There is a clear increasing trend in both shortwave and longwave irradiance from winter to summer, along with the observed solar irradiance. This rise is more evident in shortwave irradiance.

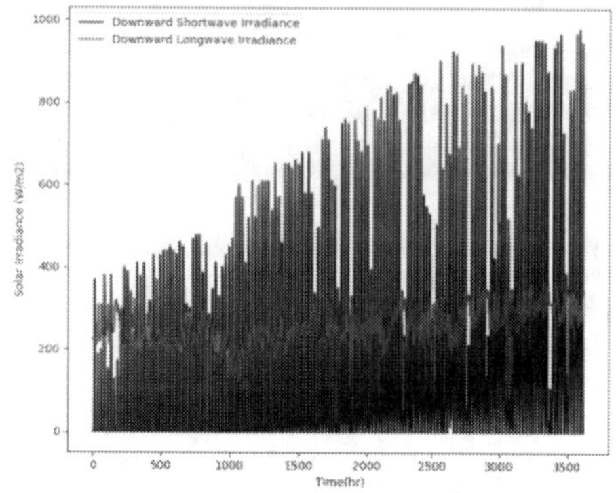

Figure 2: Downward longwave and shortwave solar irradiance

4 IMPLEMENTATIONS

The LSTM model is developed using the Keras library with a Tensorflow backend. A grid search optimization method is carried out to find the optimum values for LSTM parameters, and the final results are shown in Table 1. Three lookback periods are considered, and the lookback period equal to 24 showed the fastest convergence (because of the smaller window size) and less error in the final prediction.

Table I: LSTM parameter selection

Parameters	Values/Types	Selected Parameters
Lookback period	{24,48,72}	24
No. hidden layers	4	-
No. neurons per hidden layer	60	-
Activation function	Sigmoid	-
Optimizer types	{Adam, RMSprop}	Adam
Batch Size	{1,32,64,128}	32
No. of epochs	{50,100,150}	100

Increasing the number of epochs of the neural network results in further reduction of errors with the training data, as shown in Fig. 3; however, above 100 epochs, the overfitting tends to cause a reduction of the accuracy of the model in forecasting the test data set.

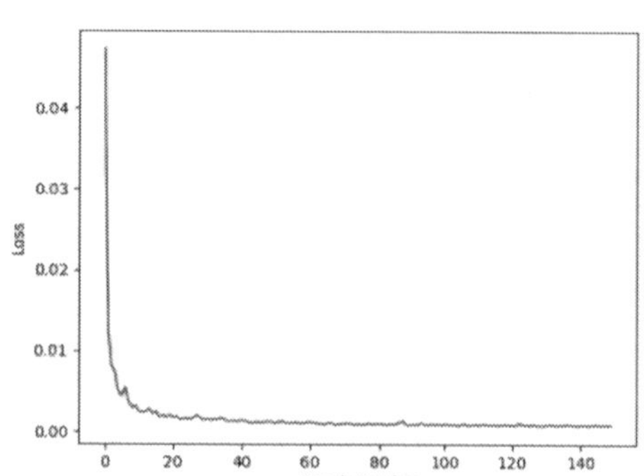

Figure 3: Integrated model convergence plot

5 RESULTS AND DISCUSSION

The confusion matrix for predictors and target variables is shown by a heatmap in Fig. 4. The heatmap shows a strong correlation between downward shortwave solar irradiance (NWP-Forecast-SW) and the observed irradiance. Although, as Fig 4. shows, there is a poor correlation between the observed irradiance and the downward longwave irradiance (NWP-Forecast-LW), the predictor is still considered as an input since it could positively impact the results.

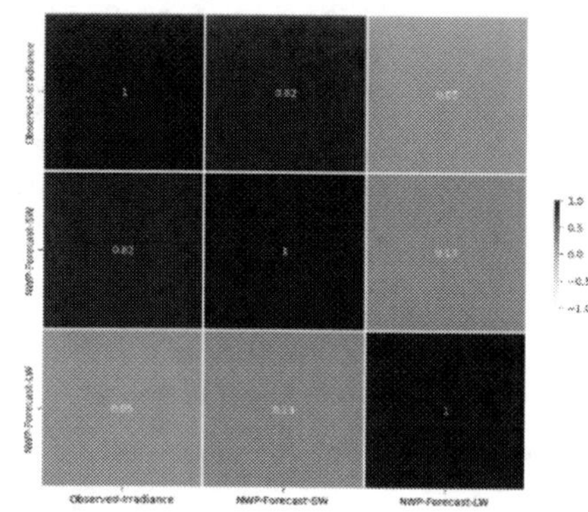

Figure 4: Correlation heatmap between independent and target variables

The forecasting results for all models: single LSTM, NWP, and the proposed hybrid model are shown in Fig. 5 to 7, respectively. Three different days in May are selected to test the model. The days are chosen based on the clearness index of the sky and the sky cloud coverage. Although there are about six different cloud coverage conditions based on the Oktas unit [22], in this study, three scenarios are considered for sky conditions: clear sky, scattered (partly cloudy), and overcast (cloudy). The days are selected in a way to challenge the prediction ability of different models, especially the single LSTM model that only uses historical data. On May 6, with a clear sky situation, the forecasting result (Fig. 5) shows an inaccurate forecast for all models; however, the proposed integrated model forecasting result has a better fit with the observed values compared with other single models. Moreover, since the NWP shortwave prediction is shifted to the right and misses the peak, it also caused a small shift in the hybrid model and decreased its accuracy.

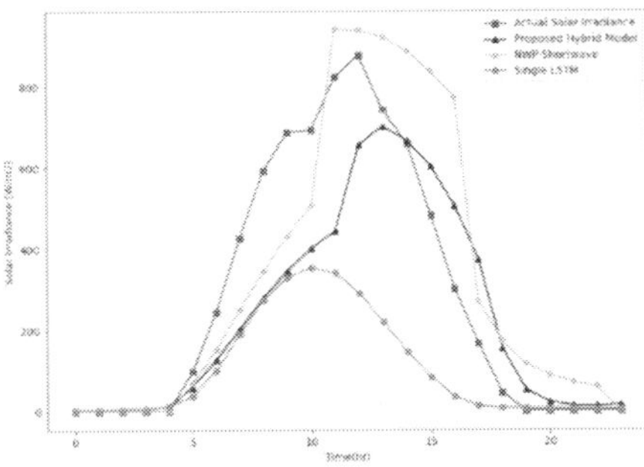

Figure 5: Forecasting results for a Clear Sky day (May 7)

On the other hand, on May 11, the sky is scattered, and the single LSTM model cannot capture the unexpected movement of the clouds from the historical data; it predicts a pattern based on the historical trend while it is far from the actual trend. However, during this time, the NWP prediction could contribute to the hybrid model by providing more accurate forecasts. As shown in Fig. 6, the hybrid model forecasted values have less error than single LSTM and NWP models.

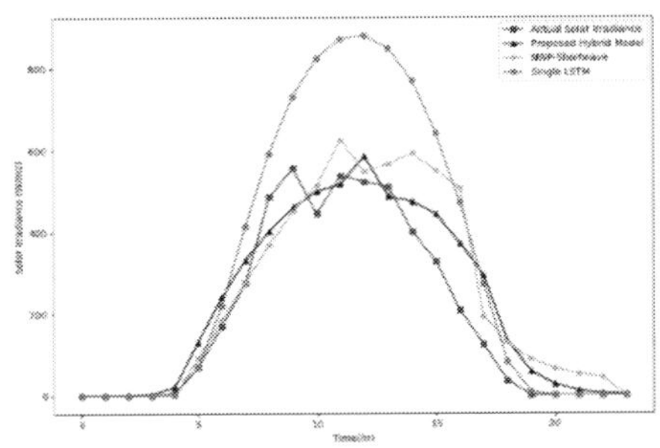

Figure 6: Forecasting results for a scattered day (May 11)

The effectiveness of the proposed method is shown even more significantly in the case of the sky overcast on May 4 (in contrast to the clear sky on previous days). The day-ahead forecast has shown that the proposed model can deal with the existence of clouds and adequately predict solar irradiance, especially in peak hours.

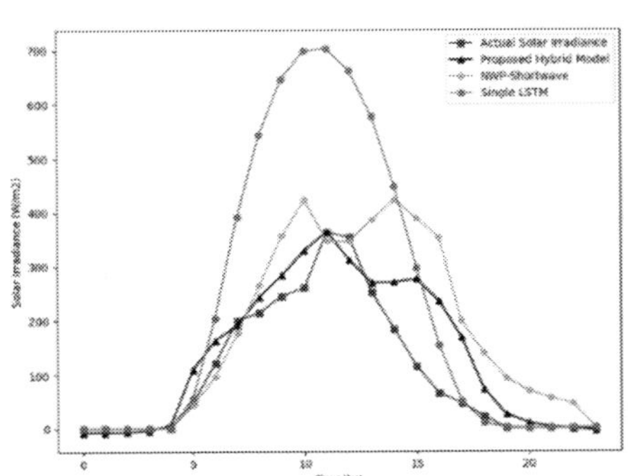

Figure 7: Forecasting results for an overcast day (May 4)

Table 2 shows the day ahead solar irradiance forecasting results (hourly resolution) of different models for three discussed days. The evaluation metrics show a significant improvement in the results of the single LSTM model when it is hybridized with the NWP model in scattered and overcast scenarios (presents of the clouds in the sky).

Table II: Models' evaluation results

Model	RMSE	MAE
May 4 (Overcast)		
Single LSTM	197.19	124.43
NWP-Shortwave	117	77
Proposed Model	62.96 (±0.10)	40.60
May 11 (Scattered)		
Single LSTM	192.81	126.91
NWP-Shortwave	101.48	67.76
Proposed Model	72.01	53
May 6 (Clear-Sky)		
Single LSTM	272.64	185.82
NWP-Shortwave	171.60	123.26
Proposed Model	167.13	114.43

6 CONCLUSIONS

This article proposes a deep learning-based integrated model for day ahead hourly solar irradiance forecasting. The long short-term memory (LSTM) method is employed as the basic structure of the hybridized model to deal with long-term dependencies in the historical data with 24 hours lookback period. Then the numeric weather prediction results were adopted as a predictor of the LSTM model to consider the cloud movement and other physical indicators that could impact the solar irradiance received on the earth's surface. Three different forecasting scenarios (Clear-sky, scattered, and overcast) were considered.

The results show that the proposed hybrid model could significantly improve the forecasting accuracy in the presence of clouds in the sky (on scattered and overcast days). On clear sky days, the inaccuracy is high in all models; however, the proposed model can still make stronger predictions compared with the other two single models. The effect of the forecasting horizon on the final accuracy of the model should be investigated in future studies.

REFERENCES

[1] Cannizzaro D, Aliberti A, Bottaccioli L, Macii E, Acquaviva A, Patti E. Solar radiation forecasting based on convolutional neural network and ensemble learning. Expert Syst Appl 2021;181:115167. https://doi.org/10.1016/j.eswa.2021.115167.

[2] Wan C, Zhao J, Song Y, Xu Z, Lin J, Hu Z. Photovoltaic and solar power forecasting for smart grid energy management. CSEE J Power Energy Syst 2016;1:38–46. https://doi.org/10.17775/cseejpes.2015.00046.

[3] Kumler A, Xie Y, Zhang Y. A New Approach for Short-Term Solar Radiation Forecasting Using the Estimation of Cloud Fraction and Cloud Albedo 2018. https://doi.org/10.2172/1476449.

[4] Das S. Short term forecasting of solar radiation and power output of 89.6kWp solar PV power plant. Mater

Today Proc 2019;39:1959–69. https://doi.org/10.1016/j.matpr.2020.08.449.
[5] Reikard G. Predicting solar radiation at high resolutions: A comparison of time series forecasts. Sol Energy 2009;83:342–9. https://doi.org/10.1016/j.solener.2008.08.007.
[6] Mellit A, Pavan AM. A 24-h forecast of solar irradiance using artificial neural network: Application for performance prediction of a grid-connected PV plant at Trieste, Italy. Sol Energy 2010;84:807–21. https://doi.org/10.1016/j.solener.2010.02.006.
[7] Kemmoku Y, Orita S, Nakagawa S, Sakakibara T. Daily insolation forecasting using a multistage neural network. Sol Energy 1999;66:193–9. https://doi.org/10.1016/S0038-092X(99)00017-1.
[8] Paquin F, Rivnay J, Salleo A, Stingelin N, Silva C. Multi-phase semicrystalline microstructures drive exciton dissociation in neat plastic semiconductors. J Mater Chem C 2015;3:10715–22. https://doi.org/10.1039/b000000x.
[9] Qing X, Niu Y. Hourly day-ahead solar irradiance prediction using weather forecasts by LSTM. Energy 2018;148:461–8. https://doi.org/10.1016/j.energy.2018.01.177.
[10] Ghimire S, Deo RC, Raj N, Mi J. Deep solar radiation forecasting with convolutional neural network and long short-term memory network algorithms. Appl Energy 2019;253:113541. https://doi.org/10.1016/j.apenergy.2019.113541.
[11] Yan K, Shen H, Wang L, Zhou H, Xu M, Mo Y. Short-term solar irradiance forecasting based on a hybrid deep learning methodology. Inf 2020;11:1–13. https://doi.org/10.3390/info11010032.
[12] Guermoui M, Melgani F, Gairaa K, Mekhalfi ML. A comprehensive review of hybrid models for solar radiation forecasting. J Clean Prod 2020;258. https://doi.org/10.1016/j.jclepro.2020.120357.
[13] Lorenz E, Heinemann D. Prediction of solar irradiance and photovoltaic power. vol. 1. Elsevier Ltd.; 2012. https://doi.org/10.1016/B978-0-08-087872-0.00114-1.
[14] Jozefowicz R, Zaremba W, Sutskever I. An empirical exploration of Recurrent Network architectures. 32nd Int Conf Mach Learn ICML 2015 2015;3:2332–40.
[15] Koutník J, Greff K, Gomez F, Schmidhuber J. A clockwork RNN. 31st Int Conf Mach Learn ICML 2014 2014;5:3881–9.
[16] Greff K, Srivastava RK, Koutnik J, Steunebrink BR, Schmidhuber J. LSTM: A Search Space Odyssey. IEEE Trans Neural Networks Learn Syst 2017;28:2222–32. https://doi.org/10.1109/TNNLS.2016.2582924.
[17] Hochreiter S. Untersuchungen zu dynamischen neuronalen Netzen. Master's Thesis, Inst Für Inform Tech Univ Munchen 1991:1–71.
[18] Yoshua Bengio, Patrice Simard, Paolo Frasconi. Learning Long-term Dependencies with Gradient Descent is Difficult. IEEE Trans Neural Netw 1994;5:157.
[19] Shirzadi N, Nasiri F, El-Bayeh C, Eicker U. Optimal dispatching of renewable energy-based urban microgrids using a deep learning approach for electrical load and wind power forecasting. Int J Energy Res 2021:1–16. https://doi.org/10.1002/er.7374.
[20] Where is Montreal, Quebec, Canada on Map Lat Long Coordinates n.d. https://www.latlong.net/place/montreal-quebec-canada-27653.html (accessed May 30, 2021).
[21] NASA POWER | Prediction Of Worldwide Energy Resources n.d. https://power.larc.nasa.gov/ (accessed February 16, 2022).
[22] Stull R. Practical Meteorology: An Algebra-based Survey of Atmospheric Science. 2017.

CLOUD FLOW CENTRING IN SKY AND SATELLITE IMAGES FOR DEEP SOLAR FORECASTING

Quentin Paletta[1, 2 *], Guillaume Arbod[2], Joan Lasenby[1]

[1] University of Cambridge, Engineering Department, Trumpington St, Cambridge CB2 1PZ, UK
[2] Engie, Lab CRIGEN, 4 Rue Joséphine Baker, 93240 Stains, France

* Corresponding author: qp208@cam.ac.uk

ABSTRACT: In solar energy, most of the short-term spatiotemporal variability can be attributed to the displacement of clouds over solar facilities. To accelerate the expansion of this variable source of energy, sky cameras and satellites have been used as complementary perspectives on the cloud cover to better anticipate its impact on the local energy generation. Recently, machine learning (ML) algorithms have shown considerable potential at modelling cloud movements from observations (e.g., high forecast skill, quantifiable uncertainty, improved interpretability). In addition to optimising the network architecture, data-centric methods have been used to facilitate model learning. For instance, representing the scene with polar coordinates centred on the region of interest (the sun in sky images or a solar site in satellite observations) significantly improves predictions. In this paper, we propose to further augment this method by consistently centring the polar representation on the incoming flow of clouds using an optical flow algorithm. Although flow centring images might benefit predictions in particular contexts such as low data availability settings, the method is hindered by the breaking up of some critical spatial features and the difficulty in generating accurate optical flow in highly changing cloud aspect conditions. In these conditions, increasing the temporal resolution of observations would facilitate the calculation of cloud motion vectors. In addition, predicting the clear-sky index instead of the absolute irradiance level could facilitate the integration of daily and yearly solar variability into the modelling when the estimation of the sun position is hindered by flow centring.

Keywords: Solar forecasting, Computer vision, Deep learning, Optical flow, Sky images, Satellite observations.

1 INTRODUCTION

Unlike conventional energy sources (e.g. gas, coal, nuclear), renewable energies such as wind and solar are inherently variable at multiple timescales, thus more difficult to integrate into an energy mix [6]. In solar energy for instance, most of the short-term intermittency is caused by cloud cover changes which can induce large local power shifts. As most of the incoming solar flux directly originates from the sun (direct versus diffuse solar irradiance), the power output of a solar farm significantly decreases when it is occluded by a cloud shadow. Anticipating these events without delay would benefit a range of mitigation solutions such as energy storage management, hybrid power plant optimisation, smart grids or energy trading, hence, making solar energy cheaper and more reliable [21].

Beside statistical methods, a promising approach to forecasting the spatiotemporal variability of solar irradiance resulting from the cloud cover dynamics is based on the analysis of sequences of ground-taken sky images [11] or satellite images [7]. Traditionally, this computer vision task is addressed by breaking down the problem into the following [17, 19]: cloud segmentation based on pixel values [10], cloud localisation [2, 12], cloud tracking [18], cloud physical properties quantification [13] and irradiance map estimation [2]. More recently, data-driven algorithms have benefited from increasing data availability. Specifically, convolutional neural networks can be trained to extract relevant spatiotemporal features from a sequence of past cloud observations to improve short-term solar forecasting [20, 22].

In addition to novel deep learning architectures [4, 9], data centric approaches have been shown to improve predictions from both sky and satellite images. In particular, the approximate rotational invariance around the region of interest (the sun in a sky image or a solar farm in a satellite image) of this computer vision task can be leveraged by randomly rotating the scene during training (data augmentation) or representing the image with polar coordinates (data transformation) which facilitate learning with convolutional architectures (Figure 1). An example of the latter method, SPIN (Simplifying Polar Invariance for Neural networks), was shown to provide the most accurate predictions from sky and satellite images while enabling faster training [14].

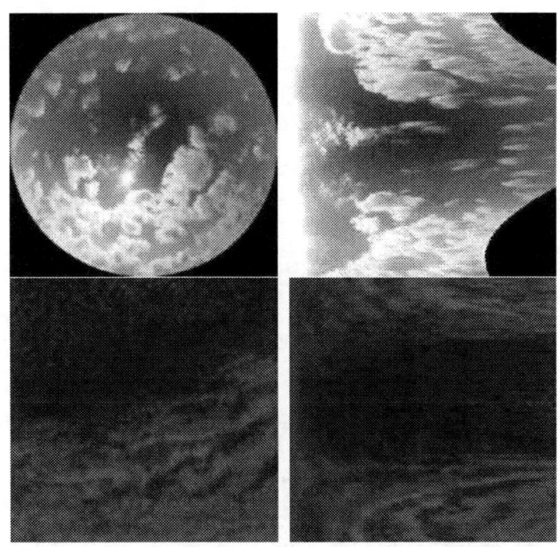

Figure 1: From top to bottom: images of the cloud cover taken by a hemispherical camera and a satellite (source: SIRTA [5], EUMETSAT [3]). From left to right: raw images and corresponding polar representations (SPIN).

Contributions To further facilitate the extraction of spatiotemporal features by convolutional architectures, we propose to augment the SPIN method by vertically centring the polar representation on the incoming flow of

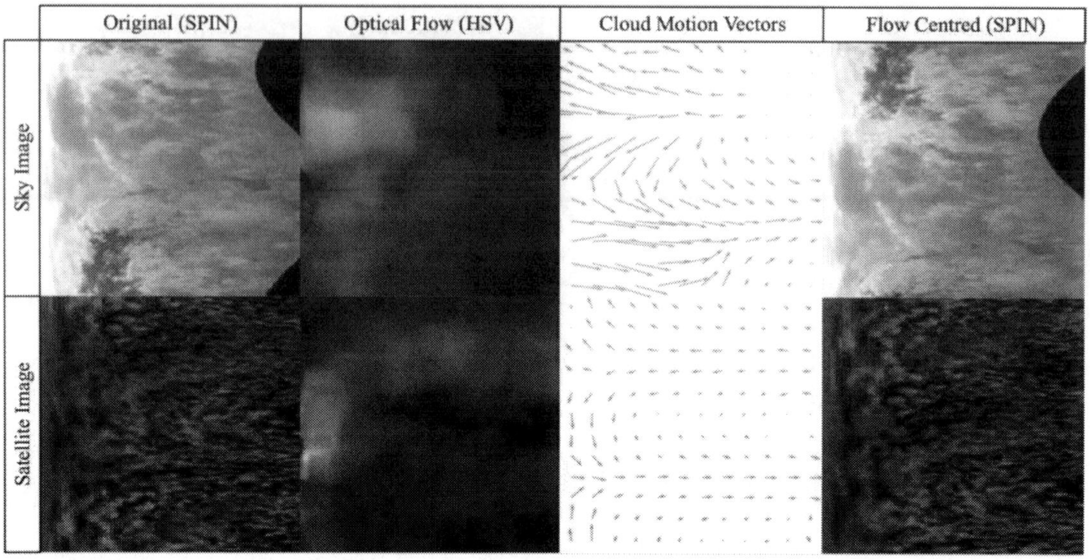

Figure 2: Illustration of the flow centring method with sky and satellite images. The principal incoming cloud flow is derived from the cloud motion vectors. Colours (HSV) represent directions and magnitudes.

clouds (Flow centring: FC) [1]. The principal cloud flow can be derived from the cloud motion vectors (CMVs) determined by an optical flow algorithm. This new transformation standardises the scene by consistently showing critical clouds in the same region of the image while avoiding the arbitrary cropping of these clouds by the horizontal image border. We show with two solar forecasting tasks (from sky or from satellite images) that this method has the potential to improve predictions at different time scales and in different training contexts.

2 APPROACH

2.1 METHODOLOGY

At each time step, an optical flow algorithm (Gunnar-Farneback) is applied to determine the $I \times J$ CMV map from the last two frames of the past sequence (Figure 2). The continuous ranked probability score (CRPS) is used to measure the dissimilarity between the discrete probability distribution f_i of cloud direction θ of each row i of the cloud direction map ($\theta_{i,j} : [1{:}I] \times [1{:}J] \rightarrow [0{:}2\pi[$) with the target distribution function f_{target} corresponding to clouds going towards the centre of the coordinate system (horizontal to the left: $f_{target} = \delta_{\pi}$) [8]. In practice, the cloud direction cumulative distribution function F_i (Equation 1) of each row i is compared row-wise to the target cumulative distribution function F_{target} (Equation 2). The image is then translated vertically such that the row $\hat{\imath}$ corresponding to the best match (i.e. lowest CRPS) is centred (see Equation 3).

$$F_i(\theta) = \int_{-\infty}^{\theta} f_i(x)dx = \int_{-\infty}^{\theta} \frac{1}{J}\sum_{j=1}^{J} \delta_{\theta_{i,j}}(x)\,dx = \frac{1}{J}\sum_{j=1}^{J} \mathbf{1}_{\theta \geq \theta_{i,j}} \quad (1)$$

$$F_{target}(\theta) = \int_{-\infty}^{\theta} f_{target}(x)dx = \int_{-\infty}^{\theta} \delta_{\pi}(x)dx = \mathbf{1}_{\theta \geq \pi} \quad (2)$$

$$\hat{\imath} = \underset{0 \leq i \leq I}{\operatorname{argmin}} \int_{0}^{2\pi} \left[F_{target}(\theta) - F_i(\theta)\right]^2 d\theta \quad (3)$$

To improve visual consistency between the frames of a sequence, all the frames in the sequence are translated by the same number of rows.

2.2 Dataset

The proposed FC method is tested on both sky images and satellite observations. The sky image dataset originates from SIRTA's lab in France (48.713° N; 2.208° E) [5]. The temporal resolution of images ranges from 1-min in 2017 to 2-min in 2018 and 2019. The corresponding irradiance values are measured on site by a ventilated pyranometer (Kipp and Zonen CM22) and reported as a per minute average. Sensors are cleaned from 3 to 5 times a week and are regularly calibrated in Davos by the World Radiation Center.

The satellite observation dataset was collected over the same period (2017 to 2019) with a 5-min temporal resolution [3]. Original images cover a 2.2×2.2° region centred on SIRTA's lab. From the pixel statistics we are able to derive the ground albedo corresponding to cloud-free sky conditions, as well as the maximum observed solar radiation reflected by clouds at a given time of the

	Sky images (SIRTA [5])	**Satellite observations (EUMETSAT [3])**
Training set	2017 & 2018 - 10k or 70k samples	2017 & 2018 - 10k or 180k samples
Validation set	even days from 2019 - 30k samples	even days from 2019 - 10k samples
Test set	odd days from 2019 - samples 30k	odd days from 2019 - 10k samples

Table 1: Training, validation, and test set characteristics.

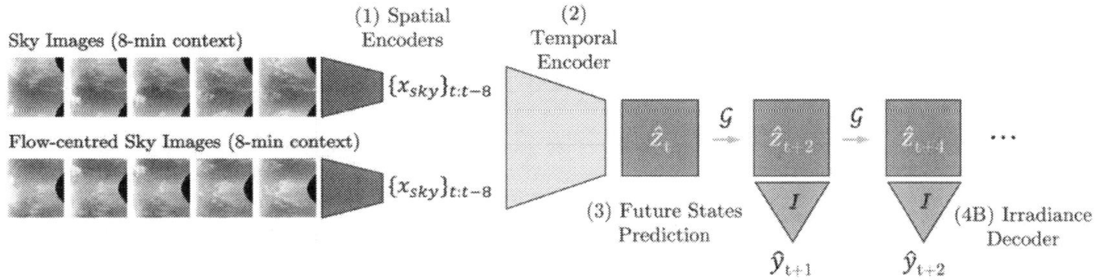

Figure 3: The model takes one or two sequences of past sky images as input (e.g. SPIN and FC SPIN). The spatial encoder is made of ResNet modules. The temporal encoder uses 3D convolutions to extract spatiotemporal features, while the Future States Prediction module is a spatial Gated Recurrent Unit layer. Future states are decoded into irradiance levels with 2D CNNs and fully connected layers. More details on the neural network can be found in [15].

day. Combining both results in a cloud index map ranging from 0 (no cloud) to 1 (thick cloud). This effective cloud albedo is less dependent on diurnal changes than the original satellite image. More details on this data pre-processing steps can be found in [14]. To match the temporal resolution of satellite images, irradiance measurements are averaged over the 5-min interval separating two satellite observations.

3 RESULTS

The different data-centric approaches are compared based on the performance of an irradiance forecasting deep learning model called ECLIPSE [15] from sky images or row satellite images together with their concomitant cloud index map as an additional channel. In a previous work, ECLIPSE was adapted to model cloud dynamics from two types of observation (sky and satellite images) [16]. Similarly here, the model is designed to take one or two sources of input corresponding to four different combinations of input categories: SPIN, FC SPIN, SPIN + SPIN or FC SPIN + SPIN (Figure 3). It is composed of two parallel spatial encoders, a temporal encoder, and a future states prediction module. Each future state is then passed through a regression head to predict the corresponding irradiance level. In addition to the root mean square error (RMSE), the forecast skill (FS) is reported. This metric quantifies the model performance relative to a baseline mode, the smart persistence model, which takes into account the diurnal changes of the extra-terrestrial irradiance. The training, validation and test sets are defined as detailed in Table 1.

Figures 4 and 5 together with Tables 2 to 5 compare the different data-centric approaches based on the performance of the deep learning model on two datasets (sky images and satellite observations) and with different sizes of the training set (large and small). In Tables 2-5, the rows corresponding to models trained on a single input (SPIN or FC SPIN) are separated from those trained on two inputs (SPIN + SPIN or FC SPIN + SPIN) because the resulting differences in terms of model complexity hinder proper comparisons. Reported results are average over three trainings with random initialisations.

The results show that FC does not seem to offer a consistent FS gain and even penalises the model in some configurations. For instance, in the large training set configuration (180k samples for sky images and 70k for satellite images), results are conflicting as FC SPIN + SPIN seems to outperform the baseline (SPIN + SPIN) on all three horizons for satellite images (Figure 5b and Table 5) whereas it performs consistently worse for sky images (Figure 4b and Table 3). A reason for that could be one of the drawbacks of the FC method: it partially or completely breaks some spatial features. For example, the vertical position of the black area corresponding to the frame in sky images (see Figure 1, top row) depends on the principal cloud motion with FC. However, the vertical position of the black area together with its horizontal extent, were shown to correspond to the second and third principal components of the learnt spatiotemporal representation of the sequence of past sky images [14].

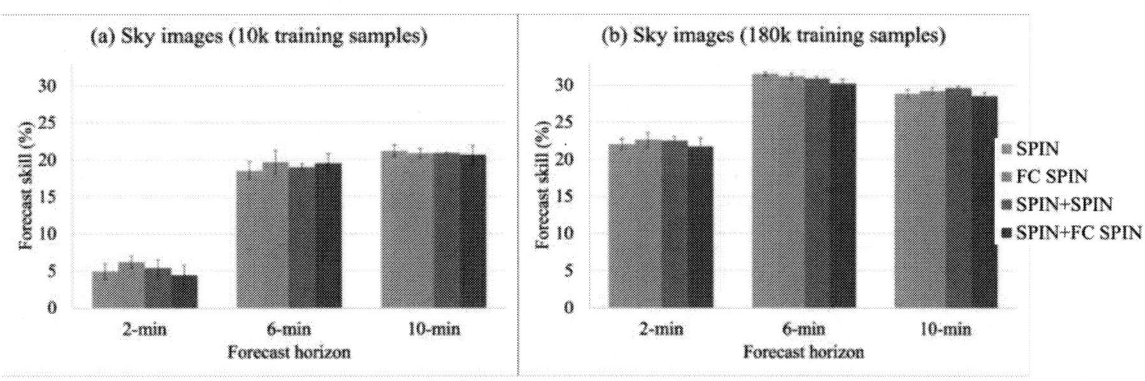

Figure 4: Comparison of the different types of sky image inputs based on ECLIPSE's forecast skill relative to the smart persistence model. The model was trained on (a) small (10k samples) and (b) large (180k samples) training sets. See Tables 2 and 3 for detailed results.

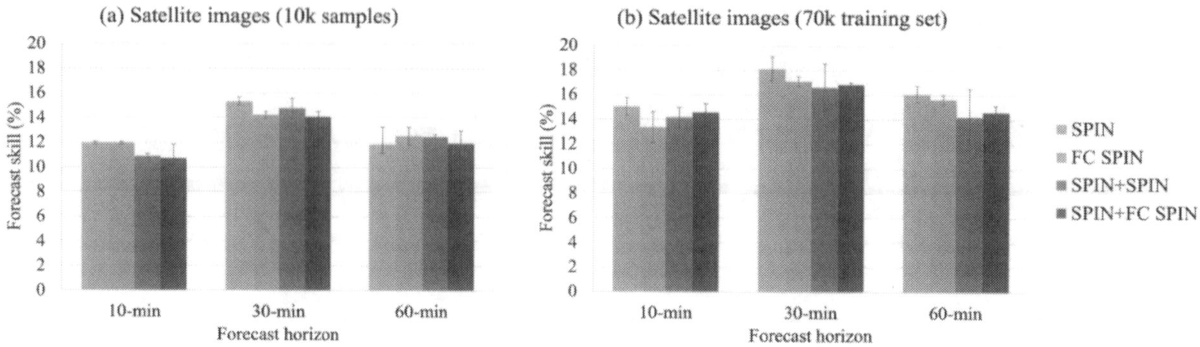

Figure 5: Comparison of the different types of satellite image inputs based on ECLIPSE's forecast skill relative to the smart persistence model. The model was trained on (a) small (10k samples) and (b) large (70k samples) training sets. See Tables 4 and 5 for detailed results.

Indeed, the position of the sun in the image can be used to determine the concomitant clear-sky irradiance level. Therefore, translating the whole image vertically based on the direction of clouds disentangles the position of the sun from the vertical position of the black area. Despite an improved principal flow reading facilitated by FC, losing that informative spatial feature penalises the model performance. In the two inputs configuration (SPIN + SPIN or SPIN + FC SPIN), allocating a portion of its parameters to a less informative source of data (FC SPIN) might lower the model performance if its overall complexity is too low.

An analogous analysis with satellite images might reveal the existence of similarly critical spatial features penalising DL models trained on cloud flow centring. For instance, the principal direction of the cloud cover in raw satellite images might correlate with the time of the day or the time of the year.

For sky images, centring the polar representation on the flow of incoming clouds provides a consistent FS gain on the 2-min ahead predictions for both small and large training sets (Figure 4). In addition, FC SPIN improves by 1.1% FS over the baseline (SPIN) for the 6-min ahead forecast with the small training set. Even though FC SPIN is on average comparably worse than the original SPIN for satellite data (Figure 5), the difference narrows when the training set is smaller. This suggests that guiding cloud motion modelling with FC could benefit smaller training sets more. In other words, the model improves its flow reading with more data.

4 CONCLUSION

The proposed approach aims at facilitating vision-based short-term irradiance forecasting with deep learning by standardising sky or satellite images based on the principal flow of clouds. Images unwrapped in polar coordinates are centred on the angular direction corresponding to incoming clouds. Although no consistent improvement was observed, the proposed flow centring method could provide performance gains in some contexts such as low data availability. However, despite facilitating critical cloud perception, centring the image on the flow of incoming clouds suffers from a few limitations such as a difficulty to estimate optical flow in some situations (e.g. highly changing cloud appearance). In addition, the partial or complete break of some of critical spatial features (e.g. vertical position of the black frame in sky images) appears to penalise the DL model. In that case, predicting the clear-sky index instead of the absolute irradiance level could facilitate the integration of daily and yearly solar variability into the modelling. Additionally, it seems that, given enough data, models would recognise the principal flow of clouds on their own. Applying the same technique to raw (not polar coordinates) sky or satellite images by rotating them based on the current cloud flow might provide similar results.

We hope that future research will enhance the method to provide more reliable estimation of the principal cloud flow. In particular, optical flow applied to sky images is hindered by the stochastic nature of clouds and by the dependence of their appearance on their position in the sky relative to the camera. Lowering the temporal resolution of observations would amplify visual similarities between subsequent timesteps, hence could facilitate CMV estimation and FC. Otherwise, robust statistics could be used to estimate principal flow more reliably, for instance by taking the median of the sequence of past principal directions. In addition, all motion vectors contribute equally to the estimation of the principal flow in the present method, regardless of their magnitude or of the presence of clouds. Although the benefit of the method in cloud-free conditions for both satellite and sky images will remain limited, weighting motion directions accordingly in the final principal flow estimate would improve the technique.

5 ACKNOWLEDGMENT

The authors acknowledge SIRTA and EUMETSAT for providing the data used in this study. This research was supported by ENGIE Lab CRIGEN, EPSRC (EP/R513180/1) and the University of Cambridge.

6 REFERENCES

[1] Bernecker, D., Riess, C., Angelopoulou, E., Hornegger, J., 2014. Continuous short-term irradiance forecasts using sky images. Sol. Energy 110, 303-315, https://doi.org/10.1016/j.solener.2014.09.005

[2] Blanc, P., Massip, P., Kazantzidis, A., Tzoumanikas, P., Kuhn, P., Wilbert, S., Schüler, D., Prahl, C., 2017. Short-term forecasting of high resolution local DNI maps with multiple fish-eye cameras in stereoscopic mode. AIP Conf. Proc. 1850. https://doi.org/10.1063/1.4984512

[3] EUMETSAT (Organization), 1991. EUMETSAT : the European Organisation for Meteorological Satellites. EUMETSAT, [1991] ©1991, Darmstadt-Eberstadt, Germany.

[4] Gao, H., Liu, M., 2022. Short-Term Solar Irradiance Prediction From Sky Images With a Clear Sky Model. Presented at the Proceedings of the IEEE/CVF Winter Conference on Applications of Computer Vision, pp. 2475–2483.

[5] Haeffelin, M., 2005. SIRTA, a ground-based atmospheric observatory for cloud and aerosol research. Geophysicae 23, 253–275.

[6] Impram, S., Varbak Nese, S., Oral, B., 2020. Challenges of renewable energy penetration on power system flexibility: A survey. Energy Strategy Rev. 31, 100539. https://doi.org/10.1016/j.esr.2020.100539

[7] Kurzrock, F., Cros, S., Ming, F.C., Otkin, J.A., Hutt, A., Linguet, L., Lajoie, G., Potthast, R., 2018. A Review of the Use of Geostationary Satellite Observations in Regional-Scale Models for Short-term Cloud Forecasting. Meteorol. Z. 277–298. https://doi.org/10.1127/metz/2018/0904

[8] Lauret, P., David, M., Pinson, P., 2019. Verification of solar irradiance probabilistic forecasts. Sol. Energy 194, 254–271. https://doi.org/10.1016/j.solener.2019.10.041

[9] Le Guen, V., Thome, N., 2020. A Deep Physical Model for Solar Irradiance Forecasting with Fisheye Images, in: 2020 IEEE/CVF Conference on Computer Vision and Pattern Recognition Workshops (CVPRW). pp. 2685–2688. https://doi.org/10.1109/CVPRW50498.2020.00323

[10] Li, Q., Lu, W., Yang, J., 2011. A Hybrid Thresholding Algorithm for Cloud Detection on Ground-Based Color Images. J. Atmospheric Ocean. Technol. 28, 1286–1296. https://doi.org/10.1175/JTECH-D-11-00009.1

[11] Lin, F., Zhang, Y., Wang, J., 2022. Recent advances in intra-hour solar forecasting: A review of ground-based sky image methods. Int. J. Forecast. https://doi.org/10.1016/j.ijforecast.2021.11.002

[12] Nouri, B., Kuhn, P., Wilbert, S., Prahl, C., Pitz-Paal, R., Blanc, P., Schmidt, T., Yasser, Z., Santigosa, L.R., Heineman, D., 2018. Nowcasting of DNI maps for the solar field based on voxel carving and individual 3D cloud objects from all sky images. AIP Conf. Proc. 2033, 190011. https://doi.org/10.1063/1.5067196

[13] Nouri, B., Wilbert, S., Segura, L., Kuhn, P., Hanrieder, N., Kazantzidis, A., Schmidt, T., Zarzalejo, L., Blanc, P., Pitz-Paal, R., 2019. Determination of cloud transmittance for all sky imager based solar nowcasting. Sol. Energy 181, 251–263. https://doi.org/10.1016/j.solener.2019.02.004

[14] Paletta, Q., Hu, A., Arbod, G., Blanc, P., Lasenby, J., 2021a. SPIN: Simplifying Polar Invariance for Neural networks Application to vision-based irradiance forecasting, in: 2022 IEEE/CVF Conference on Computer Vision and Pattern Recognition Workshops (CVPRW). pp. 5182–5191. https://doi.org/10.1109/CVPRW56347.2022.00566

[15] Paletta, Q., Hu, A., Arbod, G., Lasenby, J., 2021b. ECLIPSE : Envisioning Cloud Induced Perturbations in Solar Energy. Applied Energy

[16] Paletta, Q., Arbod, G., Lasenby, J., 2021b. Omnivision forecasting: combining satellite observations with sky images for improved intra-hour solar energy predictions. ArXiv.

[17] Peng, Z., Yu, D., Huang, D., Heiser, J., Yoo, S., Kalb, P., 2015. 3D cloud detection and tracking system for solar forecast using multiple sky imagers. Sol. Energy 118, 496–519. https://doi.org/10.1016/j.solener.2015.05.037

[18] Quesada-Ruiz, S., Chu, Y., Tovar-Pescador, J., Pedro, H.T.C., Coimbra, C.F.M., 2014. Cloud-tracking methodology for intra-hour DNI forecasting. Sol. Energy 102, 267–275. https://doi.org/10.1016/j.solener.2014.01.030

[19] Sawant, M., Shende, M.K., Feijóo-Lorenzo, A.E., Bokde, N.D., 2021. The State-of-the-Art Progress in Cloud Detection, Identification, and Tracking Approaches: A Systematic Review. Energies 14, 8119. https://doi.org/10.3390/en14238119

[20] Srivastava, S., Lessmann, S., 2018. A comparative study of LSTM neural networks in forecasting day-ahead global horizontal irradiance with satellite data. Sol. Energy 162, 232–247. https://doi.org/10.1016/j.solener.2018.01.005

[21] Yang, D., Li, W., Yagli, G.M., Srinivasan, D., 2021. Operational solar forecasting for grid integration: Standards, challenges, and outlook. Sol. Energy 224, 930–937. https://doi.org/10.1016/j.solener.2021.04.002

[22] Zhang, J., Verschae, R., Nobuhara, S., Lalonde, J.-F., 2018. Deep photovoltaic nowcasting. Sol. Energy 176, 267–276. https://doi.org/10.1016/j.solener.2018.10.024

6 APPENDIX

Sky Images	RMSE (+ Forecast skill relative to the smart persistence model)		
	2-min	6-min	10-min
SPIN	88.7 W/m² (+4.9%)	105.0 W/m² (+18.6%)	**113.1 W/m² (+21.2%)**
FC SPIN	**87.5 W/m² (+6.2%)**	**103.6 W/m² (+19.7%)**	113.6 W/m² (+20.9%)
SPIN + SPIN	**88.2 W/m² (+5.4%)**	104.4 W/m² (+19.0%)	**113.6 W/m² (+20.9%)**
SPIN + FC SPIN	89.1 W/m² (+4.5%)	**103.7 W/m² (+19.6%)**	113.9 W/m² (+20.7%)

Table 2: Evaluation of the method on solar forecasting from sky images (SIRTA dataset – training set: **10k samples**). Results are averaged over three trainings with random initialisation.

Sky Images	RMSE (+ Forecast skill relative to the smart persistence model)		
	2-min	6-min	10-min
SPIN	72.7 W/m² (+22.1%)	**88.3 W/m² (+31.5%)**	102.2 W/m² (+28.8%)
FC SPIN	**72.1 W/m² (+22.7%)**	88.8 W/m² (+31.2%)	**101.6 W/m² (+29.2%)**
SPIN + SPIN	**72.3 W/m² (+22.5%)**	**89.1 W/m² (+30.9%)**	**101.1 W/m² (+29.6%)**
SPIN + FC SPIN	73.0 W/m² (+21.8%)	90.0 W/m² (+30.2%)	102.6 W/m² (+28.5%)

Table 3: Evaluation of the method on solar forecasting from sky images (SIRTA dataset – training set: **180k samples**). Results are averaged over three trainings with random initialisation.

Satellite Images	RMSE (+ Forecast skill relative to the smart persistence model)		
	10-min	30-min	60-min
SPIN	**104.8 W/m² (+12.0%)**	**120.8 W/m² (+15.3%)**	134.8 W/m² (+11.9%)
FC SPIN	105.0 W/m² (+11.9%)	122.3 W/m² (+14.2%)	**113.7 W/m² (+12.5%)**
SPIN + SPIN	**106.2 W/m² (+10.9%)**	**121.5 W/m² (+14.8%)**	**133.8 W/m² (+12.5%)**
SPIN + FC SPIN	106.5 W/m² (+10.7%)	122.5 W/m² (+14.1%)	134.6 W/m² (+11.9%)

Table 4: Evaluation of the method on solar forecasting from satellite observations (EUMETSAT dataset – training set: **10k samples**). Irradiance measurements are averaged over a 5-min time window (same as the temporal resolution of satellite images).

Satellite Images	RMSE (+ Forecast skill relative to the smart persistence model)		
	10-min	30-min	60-min
SPIN	**101.3 W/m² (+15.0%)**	**116.8 W/m² (+18.1%)**	**128.3 W/m² (+16.1%)**
FC SPIN	103.3 W/m² (+13.3%)	118.2 W/m² (+17.1%)	128.9 W/m² (+15.6%)
SPIN + SPIN	102.4 W/m² (+14.2%)	118.9 W/m² (+16.6%)	127.3 W/m² (+14.2%)
SPIN + FC SPIN	**101.9 W/m² (+14.6%)**	**118.6 W/m² (+16.8%)**	**130.5 W/m² (+14.6%)**

Table 5: Evaluation of the method on solar forecasting from satellite observations (EUMETSAT dataset – training set: **70k samples**). Irradiance measurements are averaged over a 5-min time window (same as the temporal resolution of satellite images).

FUTURE WEATHER DATA ANALYSIS FOR SOLAR HEATING AND COOLING

Jeremias Schmidli, Daniel Carbonell, Christof Biba
SPF Institute for Solar Technology, OST Eastern Switzerland University of Applied Sciences
CH-8640 Rapperswil, Switzerland

ABSTRACT: To tackle the challenges of global warming, efficient design of renewable energy systems is needed. To achieve this goal there is a need of accurate simulations considering uncertainties such as future solar resources. Within the present paper, some of the impacts of future weather scenarios on solar heating and cooling are evaluated. Transient simulations are carried out for different future scenarios to evaluate the impact of the weather data on the energetic efficiency of a PV heat pump system delivering heating, cooling, domestic hot water, and electricity to a multi-family residential building. The system efficiency of different future scenarios is assessed based on the grid purchase ratio. The results show that the overall grid purchase ratio is likely to decrease in the future for the location of Zurich Switzerland, which is mainly caused by an increase in cooling demand during summer due to rising temperatures that come together with high irradiation.
Keywords: Solar Radiation, System Performance, Solar Heating and Cooling, Solar Resources, Climate Scenarios

1 INTRODUCTION

With global warming and a European energy crisis, it is now more important than ever to get away from CO_2 intensive energy. The building sector in particular has one of the largest potentials for a reduction of non-renewable energy [1]. One of the most popular and efficient renewable heating systems are photovoltaic (PV) powered air source heat pump (ASHP) systems [2]. Clearly, the energetic efficiency of such a system depends on the climatic conditions and with climate change, not only temperature will change, but also other factors such as precipitation, wind, or irradiance, as is stated in the IPCC report on climate change [3]. The technical report on climate scenarios for Switzerland (CH2018) [4] evaluates different climatic scenarios. They include the Representative Concentration Pathway (RCP) scenarios 2.6, 4.5, and 8.5, where the numbers stand for the radiative forcing in W/m^2 and correspond to a low, medium, and high greenhouse gas concentration in the future, respectively.

Irradiance is particularly important for the performance of a PV ASHP system, as it is its main driver. There are some studies on the impact of climate change on solar radiation [5,6], but different models lead to quite different projections. Thus, there is a need to assess the impact of solar radiation changes in particular, and climate change in general, on the energetic performance of solar heating and cooling.

The selection of a weather dataset for simulations has a large impact on the simulation results [7–12] and previous work has shown that the weather changes of the future could be beneficial for solar heating and cooling [12]. However, a detailed evaluation of future solar heating and cooling that takes different data sources into account does – to the authors' knowledge – not yet exist.

This paper thus focuses on different data sources, models, and scenarios, and assesses how large their influence is on the energetic performance of domestic solar heating and cooling with a PV powered ASHP system. In Section 2, the different evaluated weather data sets are presented. Section 3 gives an overview of the simulation methodology and Section 4 shows general results of analyzing the data sources as well as the simulation results. Section 5 then summarizes the findings and draws a conclusion.

2 WEATHER DATA SETS

The following sources were used within this paper:

CH2018 Climate Scenarios [4]:
These scenarios contain 68 different datasets for regional and global climate models. They contain the three RCP scenarios 2.6, 4.5 and 8.5, and are available in daily resolution for the years 1980-2100.

Climate Scenarios Indoor Climate (IC) [13]:
These datasets are based on the CH2018 Scenarios and are intended to be used for building design. They contain design reference years (DRY), that represent a normal year as well as extreme years, labeled *1 in 10 hot summers* (1in10). The dataset includes a total of 6 subsets, consisting of each a DRY and a 1in10 set for the RCP8.5 scenario in both 2035 and 2060 as well as for the RCP2.6 scenario in 2060.

Meteonorm (MN) [14]:
A tool to get weather data for every location in the world. It contains datasets of the past, average datasets of the present, as well as datasets including future scenarios.

From these sources, the following five sets were chosen for the location of Zurich (weather station Fluntern/SMA) in Switzerland:

- The year 2060 was selected for all 68 CH2018 scenarios. This set is named **60S** in the following.
- For all 68 scenarios from the CH2018 set, the average irradiance over 30 years, i.e. from 2045 to 2075 was evaluated and compared. Three of the RCP8.5 scenarios were then selected: the one with the lowest average irradiance, the one with the highest average irradiance and the one where the average irradiance is closest to the IC DRY RCP8.5 2060 data. Those three sets of 30 years are referenced in the following as **30Y**.
- All *Climate Scenarios Indoor Climate* together make up the **IC** dataset.
- Based on the IC DRY RCP8.5 2060 dataset, the irradiance was artificially altered by multiplying the whole timeseries with a constant between 0.9 and 1.1 in order to get the influence of only the irradiance, independent of the temperature. This parametrically changed dataset is called **PAR**.

- From Meteonorm, a reference dataset for current weather, as well as the three RCP scenarios 2.6, 4.5 and 8.5 for the year 2060 were selected for the **MN** dataset.

An overview of the evaluated weather datasets is given in the following table:

Table I: Weather dataset overview.

Short-name	Source	Scenarios	Years	Set Size
60Y	CH2018 Climate Scenarios	RCP2.6 RCP4.5 RCP8.5	2060	68
30S	CH2018 Climate Scenarios	RCP8.5	2045-2075	90
IC	Climate Scenarios Indoor Climate	RCP2.6 RCP4.5 RCP8.5	2035 2060	6
PAR	IC RCP8.5 2060	-	2060	13
MN	Meteonorm	Reference RCP2.6 RCP4.5 RCP8.5	Presence 2060	4

When analyzing the data on a seasonal level, the seasons are defined as follows:

- Winter: December, January, February (DJF)
- Spring: March, April, May (MAM)
- Summer: June, July, August (JJA)
- Autumn: September, October, November (SON)

2.1 Data Processing

Weather data from the CH2018 scenarios is only available in a daily resolution. As the dynamic simulations need at least an hourly resolution for the input data, the data was resampled to hourly using Meteonorm.

3 SIMULATION METHODOLOGY

Dynamic simulations are carried out with the open source module pytrnsys [15] that uses the commercial software TRNSYS 18 [16]. The simulations are done in order to assess the influence of the different weather data files on the efficiency of a multi-family residential building with PV and an air source heat pump.

3.1 Simulation Environment

The simulated system contains the following components:

- Multi-family building
- PV panels
- Air source heat pump (ASHP)
- Two sensible heat storages, one for space heating/cooling and one for domestic hot water
- Control system

The multifamily building [2] has six apartments and a reference total yearly heat demand of 60.8 MWh/a for the MN reference weather file. The domestic hot water (DHW) demand is constant for all simulations and equals 17.2 MWh/a. The space heating (SH) demand changes depending on the weather file. For the reference MN case it equals 43.6 MWh/a, the corresponding cooling (SC) demand equals 5.8 MWh/a.

The household (HH) electricity demand is constant at 16.2 MWh/a. The PV field is sized to 12.2 kWp which corresponds to 10 Wp per m^2 energy reference area as suggested by the joint model cantonal provisions in the energy sector (MukEn [17]). The PV panels are 190 Wp monocrystalline modules. They are oriented towards south with an inclination of 15°. A 20 kW reversible air source heat pump was used for both heating and cooling. The DHW and SH storages have a sizes of 2 m^3 and 1 m^3, respectively.

3.2 Key Performance Indicator

In order to compare the different systems and weather files, the grid purchase ratio (R_{net}) was chosen as key performance indicator. The R_{net} expresses the ratio of the electricity taken from the grid ($E_{grid\ purchase}$) to the total amount of energy used by the system (E_{use}) including heating, domestic hot water, cooling and electricity demands. Thus, the aim is to achieve a low R_{net}.

$$R_{net} = \frac{E_{grid\ purchase}}{E_{use}} \quad 3.1$$

Where:

$$E_{use} = E_{HH} + Q_{DHW} + Q_{SH} + Q_{SC} \quad 3.2$$

E_{HH}:	*Household electricity*
Q_{DHW}:	*Domestic hot water demand*
Q_{SH}:	*Space heating demand*
Q_{SC}:	*Space cooling demand*

4 RESULTS AND DISCUSSION

In this section, first some general evaluations of the weather datasets are given in subsection 4.1. Then, the simulation results are presented in subsection 4.2. All evaluations are done for the location Zurich-SMA.

4.1 Weather Data Analysis

To see the overall development of the different CH2018 climate models and pathways over time, the yearly average global horizontal irradiance and ambient temperature of all 68 scenarios are given in Figure 1 with a corresponding linear fit. The lines are smoothed with a five-year gaussian filter. For the irradiance, a slight general decrease over time can be seen, though with the large variance, a general increase in irradiance is also possible. For the temperature, on the other hand, it is more clear: all scenarios show a significant increase in temperature.

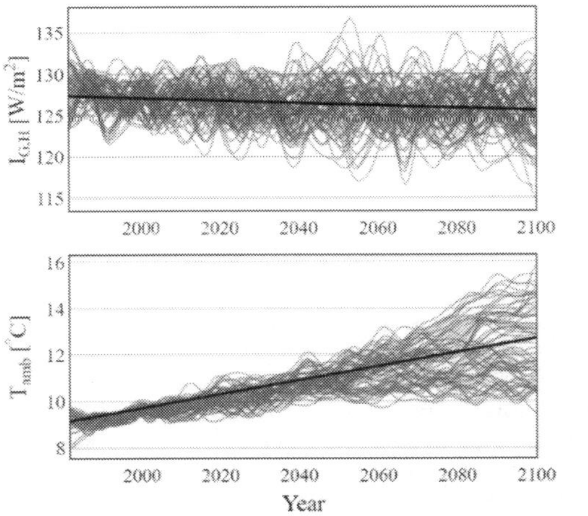

Figure 1: Global horizontal irradiance (top) and ambient temperature (bottom) development for all different scenarios in Zurich SMA. Blue lines are smoothed with a five-year gaussian filter, the black line shows a linear fit of the data.

The yearly averages of the different RCP scenario from the CH2018 data are depicted using a 20-year gaussian smoother in Figure 2. On the top, it can be seen that the slight decrease in irradiance is stronger for the more extreme scenarios (RCP8.5) but still quite small with a change of less than 2 W/m² over 120 years. The ambient temperature changes (bottom) are more relevant. For the RCP2.6, 4.5 and 8.5 scenarios, an increase of 1 °C, 1.9 °C and 3.8 °C can be seen, respectively.

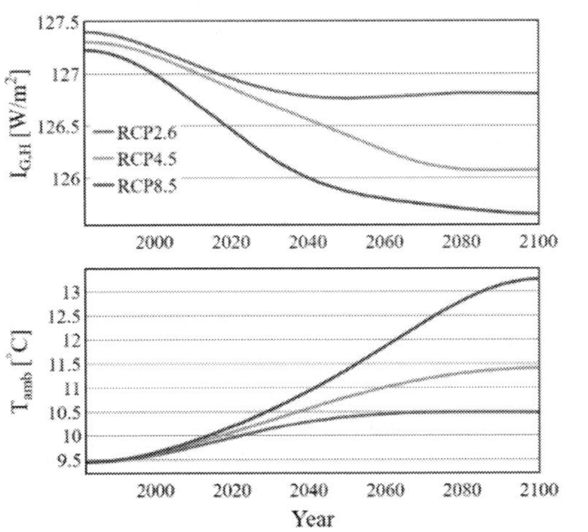

Figure 2: Average global horizontal irradiance (top) and outside temperature (bottom) development for the three evaluated RCP scenarios, smoothed with a 20-year gaussian filter.

A seasonal comparison of the different data sources are given in Figures 3 and 4 for the RCP8.5 scenario. The lines show the seasonal development of the CH2018 scenario, whereas the icons represent the MN and the IC datasets for the year 2060. In Figure 3, which shows the global horizontal irradiance development, it can be seen that the irradiance is slightly decreasing for winter and spring, whereas in summer there is a slight increasing trend. A comparison with the other datasets show that the MN data has generally larger irradiance values, whereas the IC data corresponds better to the CH2018 development, though slightly on the low side.

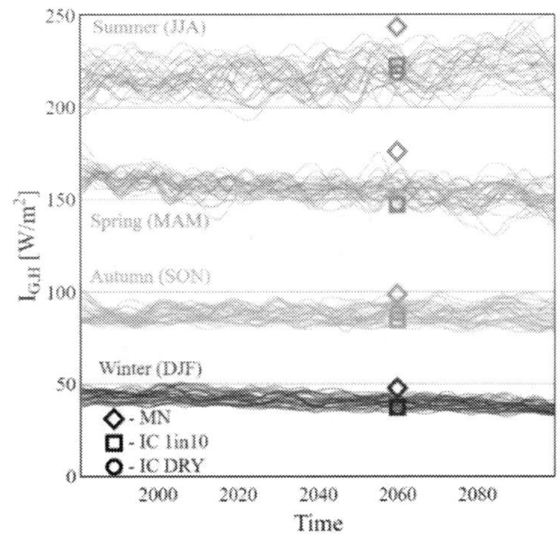

Figure 3: Seasonal global horizontal irradiance development over time for all RCP8.5 datasets, lines represent the values from the CH2018 scenarios and are smoothed with a five-year gaussian filter.

The temperature development, shown in Figure 4, shows an increase for all seasons. The MN, as well as the IC datasets correspond well to the projections, except during summer, where the MN dataset and the IC 1in10 are significantly above the average projection from the CH2018 scenarios. For the IC 1in10, this is expected, as it shows an extreme scenario. For the MN case, the different data on which it is based on can be taken as explanation.

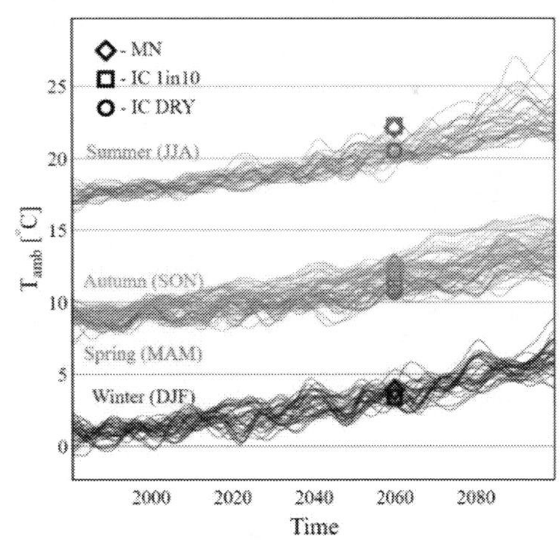

Figure 4: Temperature developement over time for all RCP8.5 datasets, lines represent the values from the CH2018 scenarios and are smoothed with a five-year gaussian filter.

4.2 Simulation Results

The grid purchase ratio of all simulation results are shown in Figure 5, sorted by dataset and RCP scenario. The higher RCP scenarios tend to lead to smaller R_{net} values, reducing thus the need of purchasing electricity from the grid. Furthermore, it can be seen that the CH2018 scenarios (60S and 30Y) lead to a large variety in grid purchase ratio (from 0.303 to 0.363, which corresponds to a difference of about 20%). This variability is a lot larger than the variability from the different scenarios within one dataset (IC or MN) or when parametrically changing only the irradiance of a dataset (PAR). For the MN dataset, which contains also the reference in blue, a difference of up to 6% in R_{net} is caused by climate change in the year 2060.

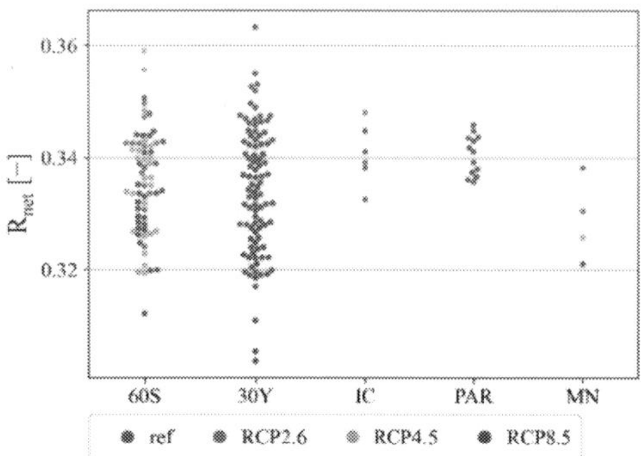

Figure 5: Swarmplot of the resulting grid purchase ratios for the analyzed datasets.

The dependence of the grid purchase ratio on ambient temperature and global horizontal irradiance is given in the scatterplot in Figure 6. Clearly R_{net} is decreasing both with increasing temperature and irradiance.

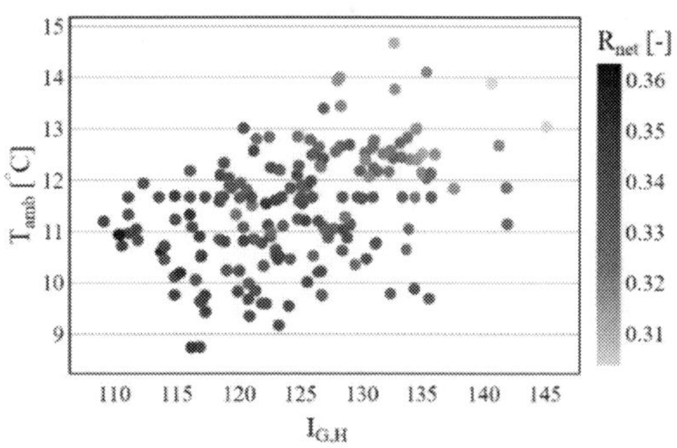

Figure 6: Scatterplot of all simulation results, grid purchase ratio as a function of ambient temperature and global horizontal irradiance.

To further analyze this dependency, the individual correlations are shown for each season in Figures 7 and 8.

The dependence on irradiance is depicted in Figure 7 where it can be observed that R_{net} decreases with increasing $I_{G,H}$. This can be explained by an increase in PV production. The r-value of the corresponding linear fits is highest in summer (-0.86), followed by winter (-0.71), autumn (-0.53) and spring (-0.41).

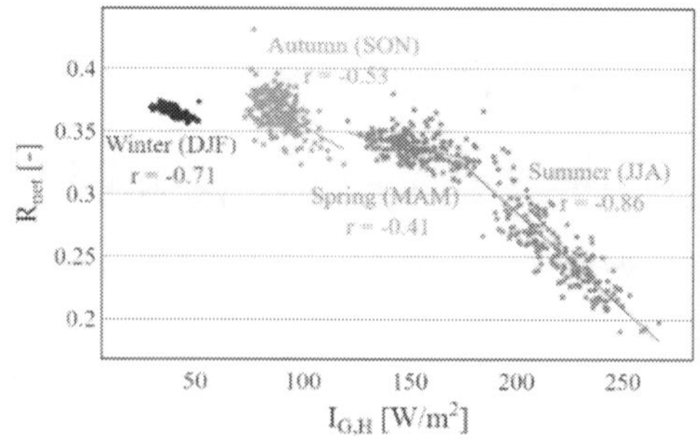

Figure 7: Seasonal dependence of grid purchase ratio on global horizontal irradiance.

The ambient temperature dependence is shown in ***Figure 8***. Only in summer, a clear correlation between R_{net} and T_{amb} (r-value of -0.94) is observed. There is a small positive correlation in spring (r = 0.26), but for autumn and winter, there appears to be no significant correlation (r-values -0.09 and -0.07, respectively).

During summer, when there is cooling needed, higher irradiance comes with higher temperatures. Therefore, the demand for cooling increases in times when there is a lot of irradiance available for PV electricity production. The high correlation between R_{net} and $I_{G,H}$ in winter is best explained by the general low total irradiance in winter. Most irradiation in winter can be used by the system, as the demand is comparably high. The low dependence on temperature in autumn, winter and spring can be explained with temperatures not strongly correlating with irradiance: e.g. a sunny and clear day in winter may come with very cold temperatures, whereas a clouded day without much irradiance may be significantly warmer.

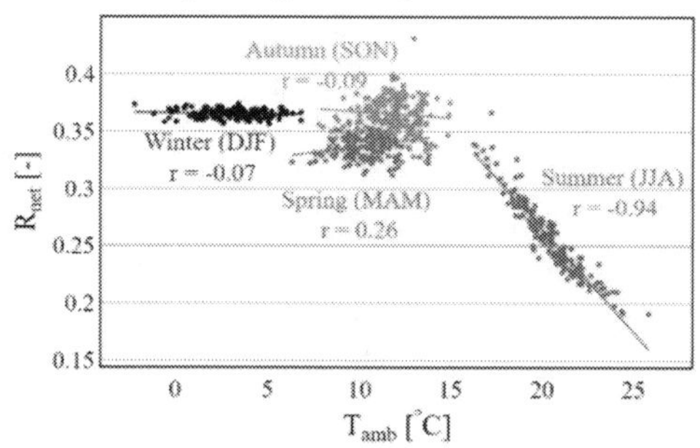

Figure 8: Seasonal dependence of grid purchase ratio on ambient temperature.

Finally, the development of R_{net} over time is depicted in Figure 9 as a function of the season. It can be seen, that R_{net} is decreasing overall and particularly in summer. There is no significant change in autumn and winter and in spring a slight increase can be seen. The decrease of R_{net} in summer can be explained mainly by the increasing temperature, as discussed before. Vice versa, the increasing R_{net} in spring is likely due to the increasing temperature and decreasing irradiation. The overall decrease R_{net} is thus driven by the summer period.

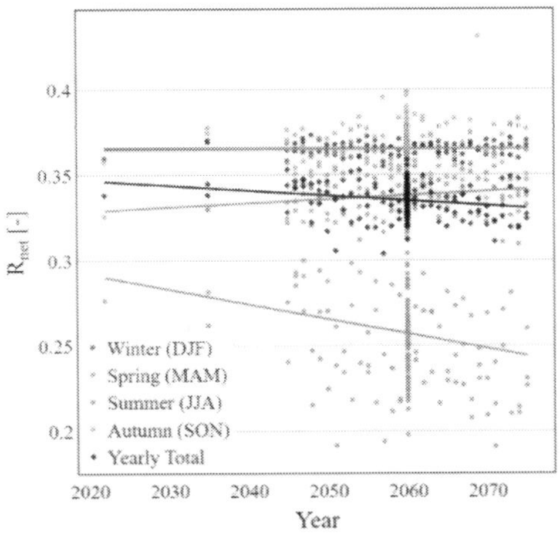

Figure 9: Seasonal dependence of grid purchase ratio on simulation year.

5 CONCLUSIONS AND OUTLOOK

Dynamic simulations in TRNSYS were used to assess the influence of different future weather data files on a PV driven air source heat pump system for domestic hot water, space heating, and cooling, of a multifamily building in Zurich, Switzerland.

The results show increasing temperatures for all seasons and RCP scenarios, while the irradiation is likely to decrease in winter and spring and increase in summer. With these climatic changes, the grid purchase ratio can be expected to decrease overall and during summer, while there may be a small increase in spring.

Further research will be done on how these results influence system costs as well as the need for storage.

6 ACKNOWLEDGEMENTS

This research was supported by the Swiss Federal Office of Energy (SFOE) in the frame of the project SolResHC and by OST Eastern Switzerland University of Applied Sciences. The results of these investigations are a contribution to the IEA PVPS Task 16.

7 REFERENCES

[1] Nejat P, Jomehzadeh F, Taheri MM, Gohari M, Abd. Majid MZ. A global review of energy consumption, CO2 emissions and policy in the residential sector (with an overview of the top ten CO2 emitting countries). Renewable and Sustainable Energy Reviews 2015;43:843–62.

[2] Wang X, Xia L, Bales C, Zhang X, Copertaro B, Pan S, et al. A systematic review of recent air source heat pump (ASHP) systems assisted by solar thermal, photovoltaic and photovoltaic/thermal sources. Renewable Energy 2020;146:2472–87.

[3] Masson-Delmotte V, Zhai P, Pirani A, Connors SL, Péan C, Berger S, et al., editors. Climate Change 2021: The Physical Science Basis. Contribution of Working Group I to the Sixth Assessment Report of the Intergovernmental Panel on Climate Change. Cambridge, United Kingdom and New York, NY, USA: Cambridge University Press; 2021.

[4] CH2018. CH2018 - Climate scenarios for Switzerland, Technical Report. National Centre for Climate Services, Zurich; 2018.

[5] Ruosteenoja K, Räisänen P. Seasonal Changes in Solar Radiation and Relative Humidity in Europe in Response to Global Warming. Journal of Climate 2013;26:2467–81.

[6] Bartók B, Wild M, Folini D, Lüthi D, Kotlarski S, Schär C, et al. Projected changes in surface solar radiation in CMIP5 global climate models and in EURO-CORDEX regional climate models for Europe. Clim Dyn 2017;49:2665–83.

[7] Carbonell D, Schmidli J, Philippen D, Haller M. Solar-ice systems for multi-family buildings: hydraulics and weather data analysis. Clima 2019, Bucharest, Rumänien: 2019.

[8] Oxizidis S, Dudek AV, Aquilina N. Typical Weather Years and the Effect of Urban Microclimate on the Energy Behaviour of Buildings and HVAC Systems. Advances in Building Energy Research 2007;1:89–103. https://doi.org/10.1080/17512549.2007.9687270.

[9] Bhandari M, Shrestha S, New J. Evaluation of weather datasets for building energy simulation. Energy and Buildings 2012;49:109–18.

[10] Schmidli J, Philippen D, Carbonell D, Haller M. Vergleich des Einflusses unterschiedlicher Wetterdatensätze auf die Systemeffizienz von Solar-Eis-Systemen. Symposium Solarthermie und Innovative Wärmesysteme 2019, Bad Staffelstein: Conexio GmbH; 2019.

[11] Schmidli J, Carbonell D, Haller M, Biba C. Weather Data Influence on a Photovoltaic Driven Heat Pump System for Net-Zero Energy Multi-Family Buildings. 37th European Photovoltaic Solar Energy Conference and Exhibition 2020:1906–10.

[12] Schmidli J, Bosshard I, Battaglia M, Haller M, Carbonell D. SolSimCC - Einfluss von Klimawandel und Nutzerverhalten auf das Kosten/Nutzenverhältnis solarer Energieerzeugung am Gebäude. Rapperswil: SPF Institute for Solar Technology; 2020.

[13] Climate scenarios indoor climate - MeteoSwiss n.d. https://www.meteoswiss.admin.ch/home/climate/swiss-climate-in-detail/climate-scenarios-indoor-climate.html (accessed September 9, 2022).

[14] Remund J, Kunz S. METEONORM: Global meteorological database for solar energy and applied climatology. Meteotest; 1997.

[15] pytrnsys — pytrnsys documentation https://pytrnsys.readthedocs.io (accessed September 18, 2022).

[16] Klein et al SA. TRNSYS 18: A Transient System Simulation Program, Solar Energy Laboratory, University of Wisconsin, Madison, USA 2017.

[17] Konferenz kantonaler Energiedirektoren, editor. Mustervorschriften der Kantone im Energiebereich (MuKEn 2014) 2018.

OPERATION ANALYSIS OF CONCENTRATED PHOTOVOLATIC/THERMAL FOR HEATING CRUDE OIL

Duolong Gao[1,2], Wu Long[1,2], Qiang Liu[1,2], Jing Guo[1,2], Wei Cui[1,2], He Wang[3], Hong Yang[3,*]
[1]SINOPEC Northwest Company of China Petroleum and Chemical Corporation, Urumqi 830011, China
[2]Key Laboratory of Enhanced Oil Recovery in Carbonate Fractured-vuggy Reservoirs, 830011, China
[3]School of Physics, Xi'an Jiaotong University, Xi'an 710049, China

ABSTRACT: Because the flowability of crude oil is worse at ambient temperature, it is usually heated to about 50 °C during transportation. It is important to develop an effective method heating crude oil. In this paper, an operation analysis of concentrated photovoltaic/thermal (CPV/T) for heating crude oil in Urumqi is carried out for the first time. The viability of CPV/T for heating crude oil was evaluated. It is found that the CPV/T system has a better capacity utilization because it can generate electricity and heat simultaneously during full-year operation, particularly in summer.
Keywords: CPV/T system, heating crude oil, solar energy

1 INTRODUCTION

Because the crude oil is highly viscous at ambient temperature, it needs to be heated and insulated to keep a certain flowability during transportation. At present, the crude oil in the tank is kept warm by a direct electric heater in most cases. The interest in using solar energy for heating crude oil is increasing more and more because it is clean and limitless. A. A. Badran et al. ever studied conventional flat-plate collector and trickle solar collector for heating fuel oil [1][2]. Li et al. reported application of photovoltaic/thermal system for crude oil heating [3]. But these systems operate under one sun, and cannot provide enough energy to meet crude oil heating in cloudy and rainy weather, particularly in severe cold area. In order to obtain high temperature hot water and enough energy, the CPV/T was investigated by other authors. T. P. Otanicar et al. conducted parametric studies of CPV/T systems with spectral filtering and high temperature photovoltaic [4]. I. Ceylan et al. gave out a detailed analysis of CPV/T solar air heater system with thermal energy storage in winter season application [5]. J. I. Rosell et al. simulated a low concentrating photovoltaic/thermal system [6]. PV systems directly produce electricity from photons but as the temperature of cell increases the efficiency decreases [7]. Many researchers have studied the temperature dependence of solar cells[8-10]. It is desirable to develop CPV/T systems which can provide more days for heating crude oil in a year. But almost no one reported the possibility of heating crude oil using low concentration CPV/T systems. In this paper, the electrical and thermal yield of CPV/T system were investigated according to months in Urumqi. The energy demand for heating crude oil also was analyzed according to months in Urumqi. It was found that the CPV/T could provide a better chioce for heating crude oil in a year, particularly in severe cold area, compared to conventional flat-plate collector and evacuated tube collector. When the energy demand for heating crude oil is less in summer, the electrical yield produced by CPV/T can be connected to grid. When the energy demand for heating crude oil is insufficient in winter, the electrical yield produced by CPV/T can be used to heat crude oil.

2 EXPERIMENTAL DETAILS

The CPV/T system used for heating crude oil is installed in Xinjiang, China (Latitude: 43.78 N; Longitude: 87.62 E). Fig. 1 shows the scene of low concentration CPV/T for heating crude oil. Fig. 1(a) shows the oil extraction plant. Fig. 1(b) shows the CPV/T collector. Fig. 1(c) shows the customized crystalline silicon solar module. Fig. 1(d) shows the inverter. Fig. 1(e) and (f) show the measured the receiver inlet and outlet temperature. The environmental temperature, receiver inlet and outlet temperatures were measured by Pt100. The direct normal irradiance (DNI), wind speed, the flow rate of pump and electrical power output were recorded at regular intervals.

Figure 1: The scene of low concentration CPV/T for heating crude oil. (a) oil extraction plant, (b) CPV/T, (c) low-concentrated crystalline silicon solar module, (d) inverter, (e) inlet temperature of water tank, (f) outlet temperature of water tank.

3 RESULTS AND DISCUSSION

Figure 2(a) gives out the average daily direct normal irradiance (DNI) according to months in Urumqi. It is found that the average daily DNI ranges from 3.905 kWh/m^2 to 6.607 kWh/m^2 according to months in Urumqi. August is the month when average daily DNI had the highest value, 6.607 kWh/m^2 and this value decreased to 3.905 kWh/m^2 in January.

Figure 2(b) shows the average daily environmental temperature according to months in Urumqi. It is found that the average daily environmental temperature ranges from -5.1 °C to 27.8 °C according to months in Urumqi. Augest is the month with the highest average daily temperature of

27.8 °C. This value decreases to -5.1 °C in December, which is the month with the lowest average daily temperature. Compared Figure 2 (a) with (b), we found that conventional flat-plate collector and evacuated tube collector cannot provide more days for heating crude oil compared to CPV/T system from the cost-effective viewpoint.

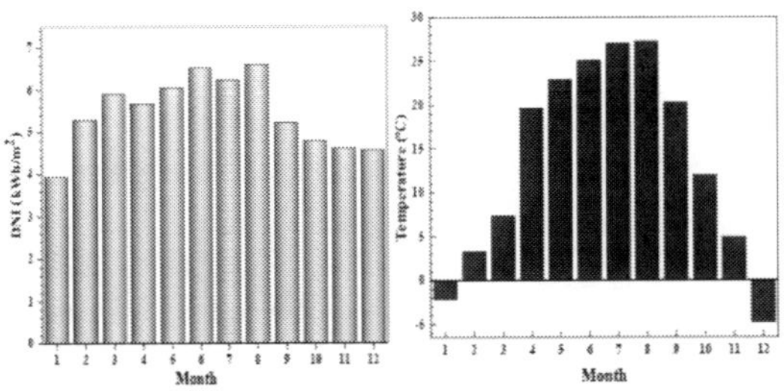

(a) (b)

Figure 2: Average daily direct normal irradiance and environmental temperature according to months in Urumqi.

The average daily effective thermal energy and electric energy of CPV/T system are given in the following equations.

$$Q_{termal}=\int H(t)\eta_{termal}(H,T_a,T_{inlet},T_{outlet})dt \quad (1)$$

$$Q_{electric}=\int H(t)\eta_{electric}(H,T_m)dt \quad (2)$$

Where H(t) is the direct normal irradiance, T_a is the ambient temperature, T_m is the receiver temperature, T_{inlet} is the inlet temperature of receiver, T_{outlet} is the outlet temperature of receiver, η_{termal} is the thermal efficiency of CPV/T system. $\eta_{electric}$ is the electric efficiency of CPV/T system, Q_{termal} is the thermal energy of CPV/T system, $Q_{electric}$ is the electric energy of CPV/T system.

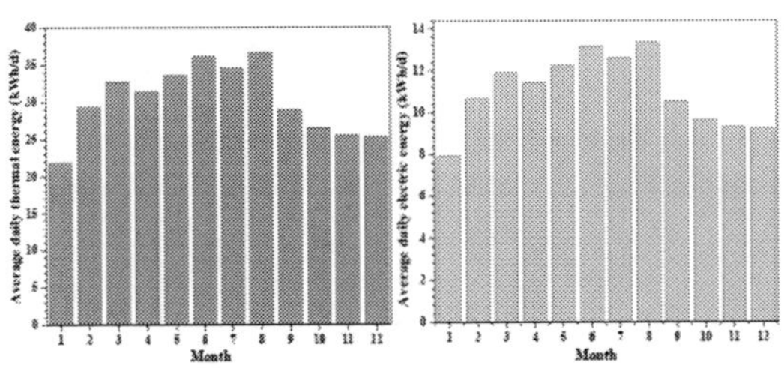

(a) (b)

Figure 3: Average daily thermal energy (a) and electric energy (b) according to months in Urumqi.

Figure 3(a) shows the average daily thermal energy produced by CPV/T system according to months in Urumqi. It is found that the average daily thermal energy produced by CPV/T system ranges from 21.7 kWh/d to 35.9 kWh/d. Augest is the month with the highest average daily thermal energy of 35.9 kWh/d. This value decreases to 21.7 kWh/d in January, which is the month with the lowest average daily thermal energy. Figure 3(b) lists the average daily electrical energy produced by CPV/T system according to months in Urumqi. It is found that the average daily electrical energy produced by CPV/T system ranges from 7.89 kWh/d to 13.19 kWh/d. Aguest is the month with the highest average daily electrical energy of 13.19 kWh/d. This value decreases to 7.89 kWh/d in January, which is the month with the lowest average daily electrical energy

Figure 4 (a) shows the average daily inlet temperature and needed temperature rise of crude oil according to months in Urumqi. It is found that the average daily inlet temperature according to months in Urumqi ranges from 33.2 °C to 41.8 °C. Aguest is the month with the highest average daily inlet temperature of 41.8 °C. This value decreases to 33.2 °C in December, which is the month with the lowest average inlet temperature of crude oil. And we found that conventional flat-plate collector and evacuated tube collector cannot provide more days for heating crude oil compared to CPV/T system from the cost-effective viewpoint. Setting that the outlet temperature of crude oil is 55 °C. The thermal energy demand of heating crude oil is expressed as:

$$Q_{crude\ oil}=m_{oil}c_p\Delta T=m_{oil}c_p(T_{setting}-T_{inlet}) \quad (3)$$

where m_{oil} is the crude oil mass flow rate (kg/h), c_p is the specific heat of crude oil, c_p=2.582 kJ/(kg·°C), T_{seting} is the designed outlet temperature, $T_{setting}$=55 °C, T_{inlet} is the crude oil inlet temperature.

Figure 4 (b) shows the average daily thermal energy demand for heating crude oil according to months in Urumqi. It is found that the average daily thermal energy demand for heating crude oil ranges from 385.7 kWh/d to 519.2 kWh/d according to months in Urumqi, which correspond to 1.5 Ton/h of crude oil mass flow.

According to the above analysis, it is found that many sets of CPV/T are needed so as to meet the average daily thermal energy demand for heating crude oil in the oil extraction plant. Because solar energy is unbalanced in summer and winter, it must balance cost and sets when hybrid energy is used for heating crude oil.

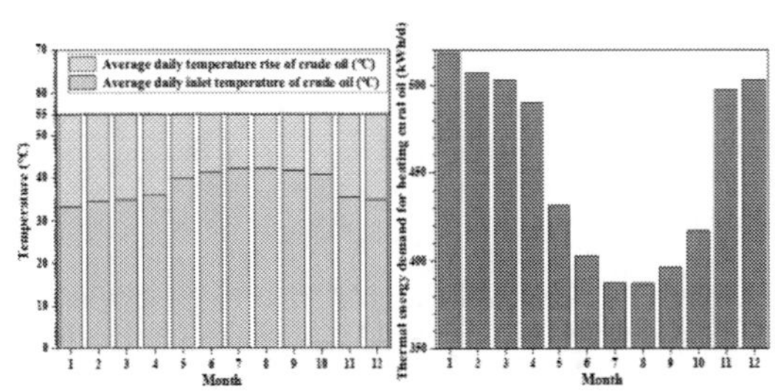

(a) (b)

Figure 4: Average daily inlet temperature and needed temperature rise of crude oil (a) and thermal energy demand for heating crude oil according to months in Urumqi.

4 CONCLUSION

In this paper, an operation analysis of CPV/T for heating crude oil in Urumqi is carried out for the first time. The electrical and thermal yield of CPV/T system were investigated. The viability of CPV/T for heating crude oil was evaluated. Because solar energy is unbalanced in summer and winter, it must balance cost and sets when hybrid energy is used for heating crude oil. It is found that the CPV/T system has a better capacity utilization because it can be grid-connected during full-year operation, particularly in summer

ACKNOWEDGEMENT

The authors would like to thank the support of China Petroleum and Chemical Corporation (34400007-21-ZC0613-0004). (*Corresponding author: Hong Yang, hongy126@126.com).

REFERENCES

[1] A. A. Badran, B. A. Jubran, Fuel oil heating by a trickle solar collector, Energy Conversion and Management 42 (2001) 1637-1645.
[2] Badran, A.A., Hamdan, M.A., 1998. Utilization of solar energy for heating fuel oil. Energy Conv. Manag. 39(1-2), 105-111.
[3] Wuping Li, et al. Application of photovoltaic/thermal technology for heating crude oil, Energy Conservation in Petroleum & Petrochemical Industry 11(3) (2021) 13-16.
[4] T. P. Otanicar, et al. Envisioning advanced solar electricity generation: Parametric studies of CPV/T systems with spectral filtering and high temperature PV, Applied Energy 140 (2015) 224–233.
[5] I. Ceylan A detailed analysis of CPV/T solar air heater system with thermal energy storage: A novel winter season application. Journal of Building Engineering 42 (2021) 103097.
[6] J. I.Rosell, et al. Design and simulation of a low concentrating photovoltaic/thermal system. Energy Conversion and Management 46 (2005) 3034–3046.
[7] Fan, J.C.C., 1986. THEORETICAL TEMPERATURE-DEPENDENCE OF SOLAR-CELL PARAMETERS. Solar Cells 17(2-3), 309-315.
[8] Cheng, X.L., Zhou, T.Z., Feng, S.S., Wang, H., Yang, H., Mao, S.C., Yu, H.C., Ieee, 2019. Comparative Study on Temperature Coefficients of Al-BSF Solar Cells and PERC Solar Cells, IEEE 46th Photovoltaic Specialists Conference (PVSC). Chicago, IL, pp. 304-307.
[9] Wang, H., Cheng, X., Yang, H., 2020. Temperature Coefficients and Operating Temperature Verification for Passivated Emitter and Rear Cell Bifacial Silicon Solar Module. IEEE Journal of Photovoltaics 10(3), 729-739.
[10] Bai, Q., Nan, C., Zhou, L., Yang, Y., Mao, S., Han, H., Yang, H., Wang, H., 2021. Effect of temperature on ideality factor for PERC monofacial crystalline silicon solar module, 2021 IEEE 48th Photovoltaic Specialists Conference (PVSC). pp. 0244-0247.

CLOUD-BASED DECISION SUPPORT SYSTEM FOR OPERATION AND MAINTENANCE IN PHOTOVOLTAIC SYSTEMS

Andreas Livera[1], Jose G. Franquelo[2], Ruben Gonzalez Bernal[2] and George E. Georghiou[1]
[1] FOSS Research Centre for Sustainable Energy, Photovoltaic Technology Laboratory, Department of Electrical and Computer Engineering, University of Cyprus, Nicosia, 1678, Cyprus
[2] Isotrol SA, Isaac Newton 3, 41092 Seville, Spain

ABSTRACT: A decision support system (DSS) is proposed in this work for providing recommendations of actionable decisions to resolve underperformance issues in photovoltaic (PV) systems. The proposed cloud-based DSS incorporates functionalities for data quality, PV system modeling, problem detection, energy loss breakdown, incidents criticality evaluation and operation and maintenance (O&M) recommendations. It was benchmarked experimentally using historical field data from a 10.09 MW_p PV power plant in the Mediterranean region. The results demonstrated the DSS's capabilities for automated detection of failures and performance losses and categorization into different fault causes. Over the 1-year evaluation period, the detected underperformance incidents accounted for 2100 MWh (or 14.30% of the total energy production) of lost energy. Given the DSS recommendations, approximately 7% of lost energy production could be recovered by performing field mitigation activities (e.g., corrective actions).
Keywords: decision support system, energy loss analysis, fault diagnostics, monitoring, operation and maintenance, performance, photovoltaic, renewable asset management.

1 INTRODUCTION

Over the last decade, the industry and research has shifted on the operating and maintaining aspects of photovoltaic (PV) systems [1]. In this domain, operation and maintenance (O&M) and monitoring strategies are important for maximizing the output power of PV systems while also minimizing the loss of energy production and operating costs [2].

Currently, intelligent monitoring systems integrate health-state, digital twin and fault diagnostic tools to ensure optimum PV performance by assessing in real-time the operating state of PV assets and detecting failures remotely [3]. Features such as energy loss breakdown, incidents criticality evaluation and quantification of faults economic impact should also be incorporated into the monitoring systems to provide recommendations of maintenance actions and to intelligently resolve the power reducing issues that occur during the PV system operation [1]. Such features are also necessary for optimizing/scheduling field O&M activities, thus lowering operating expenses and increasing PV energy yield.

A recent industry benchmark study indicated that the global PV industry could be losing $14.50 billion each year by 2024 if not executing an O&M strategy [4]. Bearing this in mind, focus is shed on technical assets managers who look for operational efficiency of PV power plants. Performance monitoring, early detection and prediction of failures leads to O&M recommendations for corrective, preventive and/or predictive actions. A cloud-based solution for PV O&M cost reduction and revenue optimization is thus developed in this paper; this is a short version of the paper submitted to a journal [5]. The proposed cloud-based decision support system (DSS) incorporates automated functionalities for data analysis, PV system modelling, early fault detection, criticality evaluation and O&M recommendations, thus allowing plant operators to take actions to maximize energy production and to reduce operational costs. The DSS was validated using historical measurements obtained from a PV power plant of 10.09 MW_p installed in the Mediterranean region.

2 METHODOLOGY

2.1 Approach

The proposed solution for PV assets diagnosis and maintenance operates entirely on the available field measurements. The DSS incorporates algorithms for data cleansing and quality, early fault detection, and it provides visualization features, O&M recommendations, prioritisation of field actions and follow up (see Fig. 1).

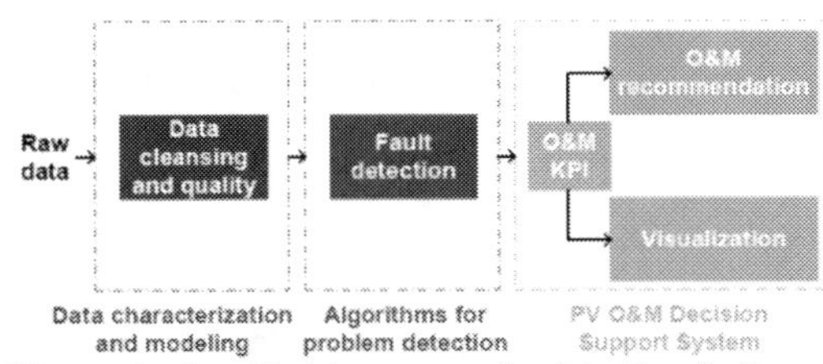

Figure 1: Overall data process for detecting faults and providing O&M recommendations by the DSS.

Initially, an Extract, Transform and Load (ETL) method is used to extract the raw data from supervisory control and data acquisition (SCADA) systems like Isotrol's Bluence [6] and then upload/store them to an analytical scalable database (Google Cloud Platform in this case).

Then, data aggregation and pre-processing techniques are applied to provide high fidelity data for further analysis (i.e., the application of fault detection algorithms). Statistical, comparative and machine learning (ML) algorithms are subsequently executed to detect faults and inefficiencies in PV systems. The results of this analysis are then used for the generation of O&M recommendations, visualization of results and provision of insightful information. The O&M recommendations (e.g., corrective actions to be implemented) are derived based on key performance indicators (KPI) and financial metrics that can quantify the power and economic impact of underperformance incidents (i.e., failures, losses, problems, and systemic issues). The provided information for resolving the detected incidents are then translated to mitigation actions (to be performed by plant owners/operators) to minimize the energy and economic losses. Finally, a criticality value (based on the impact in form of energy losses) is used to schedule and prioritize the field O&M actions.

2.2 Decision support system architecture

The proposed DSS consists of four modules (see Fig. 2): 1. Data quality analysis, 2. Out-of-normality analysis, 3. Inverter and solar field analysis and 4. Energy loss analysis and plant status. More details for each module are provided in the following subsections.

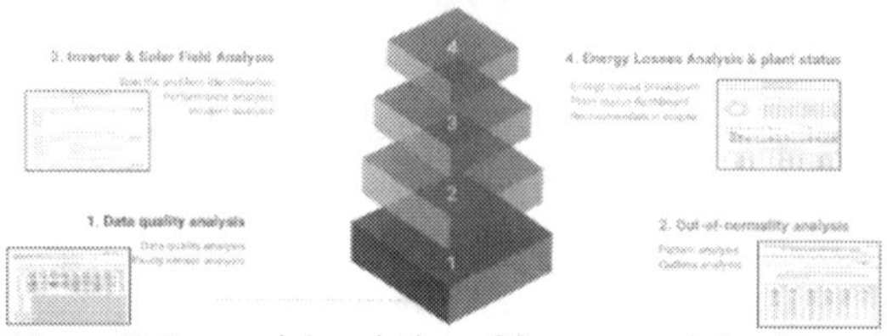

Figure 2: Layered description of the proposed cloud based DSS architecture for O&M in PV systems.

2.3 Data quality analysis

The data cleansing algorithms (i.e., the AC-DC cleansing, constant pattern detection, night-time anomalies detection, isolated outliers' detection, irradiance-power curve cleansing, etc.) are initially applied to the available field measurements to: a) include daylight measurements only (e.g., irradiance values between 20 W/m^2 and 1300 W/m^2), b) identify invalid data (missing and erroneous values, outliers, etc.), sensors faulty operation (e.g., measurement problems) and failures as frozen data, synchronization, communication, consistency, and performance issues (e.g., communication errors, storage problems, etc.), c) correct or filter out detected invalid values using data inference and deletion techniques and d) prepare the data for the execution of problem identification algorithms. More details about the data quality methodology are provided by Livera et al. [7].

The main data issues (which are common and the most difficult to solve) include communication errors with the plant, inverter and/or a sensor, saturation (when the sensor has a limit value within the operating range of the variable it measures), frozen (when the sensor remains constant within the precision range with which the measurement is taken or stored), inverter & stringbox coherence (the DC current, voltage and power are compared between the input of the inverter and the aggregation at the output of its solar field) and tracker blocking in the cells of the meteorological station (the measured irradiance describes a daily curve analogous to a blockage of trackers in the solar field).

Visualization of the results is provided in the form of time related plots (x-t plots), scatter plots of two parameters (x-y plots) and heatmaps.

Predictive models are also used to predict meteorological (e.g., global horizontal and diffuse irradiance) and PV electrical (e.g., DC current, voltage and power, active and AC output power) characteristics [8]. Detailed description of the utilized models and worldwide validation is given in [8,9].

2.4 Out-of-normality analysis

The objective of this step is to detect underperformance incidents and equipment malfunctions from the pattern and outlier analyses. The pattern analysis is used to detect PV electrical and thermal behaviour changes as well as to determine the normality state. For this purpose, the PV electrical and thermal characteristics are simulated using ML techniques (e.g., neural networks) and then compared with the real data [10]. Outliers (different from the outlying observations detected by applying statistical techniques such as sigma and boxplot rule methods [11] in module 1) are detected by comparing recorded measurements of similar elements (e.g., similar subarrays and strings) or from nearby sensors. Visualization of the results is provided in the form of heatmaps and time series plots.

2.5 Inverter and solar field analysis

Inverter failures (e.g., overheating, clipping, etc.) [12], solar field problems (e.g., soiling, bypass diode failures, etc.) [11,13] and PV plant faults (due to blocked tracker, string disconnection and deficient thermic behaviour) are detected in this module. The energy loss for each fault type is also estimated from the area under the curve (AUC) [1].

Events of inverter's underperformance are identified either by comparing the output of similar size rating inverters or by comparing the measured against the predicted inverter production [7]. Inverter shutdown failures or manual stops are detected when the inverter's production is near zero during hours above the generator's Minimum Irradiance Threshold (MIT) [14]. Such failures can also be identified by combining signal (measured or prediction) and alarm analysis at the inverter level. For late awakening, overheating and clipping problems the methodology is described in [5].

For solar field problem detection, the detailed methodology for detecting and classifying different types of commonly occurred PV faults is described in [11,13]. Specific devices (e.g., DustIQ by Kipp & Zonen), soiling stations and two-sensor systems (at PV cell or module level), where one of the sensors is cleaned regularly (manually or automatically), are used to estimate the soiling losses [15]. In case of unavailability of dedicated soiling sensors/station, systems or O&M procedures, the soiling losses are estimated using the method proposed in [16]. Other solar field problems (e.g., performance below best behaviour) are detected from the deviation between the actual behaviour of the devices with respect to the manufacturer's specifications.

Visualization of the results is provided in the form of heatmaps, time series and scatter plots.

2.6 Energy loss analysis and plant status

Details regarding the plant status, energy losses breakdown, incidents criticality evaluation (based on the Failure Mode, Effects and Criticality Analysis method [17]), alarms and O&M recommendations are provided in this module.

The performed energy loss analysis focuses on investigating the impact of problems and faults on the PV energy production. The detailed methodology is given in [5]. In general, the results of module 3 (Inverter and Solar Field Analysis) are used as inputs to a recommendation engine responsible for transforming underperformance events into actions to be conducted by the O&M team (see Fig. 3). A criticality value along with a text format are used as inputs to the engine for the generation of O&M recommendations. The list of text-based database of suggestions and actions allow the transformation of detected failures into readable actions or information provided to O&M personnel for the test PV plant.

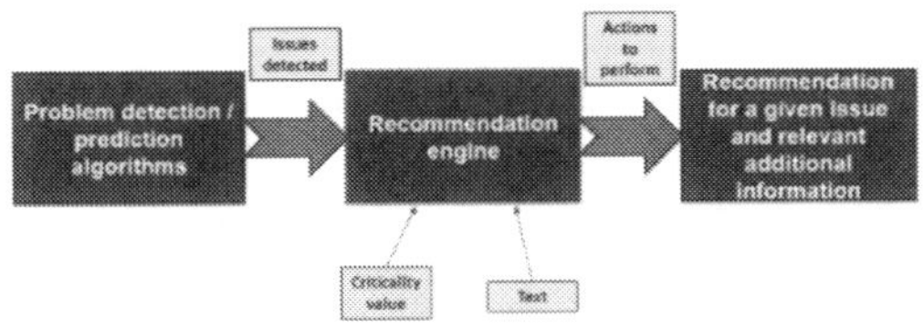

Figure 3: Process from problem detection to generation of recommendation.

For scheduling the field maintenance activities, O&M KPI are defined as depicted in Fig. 4.

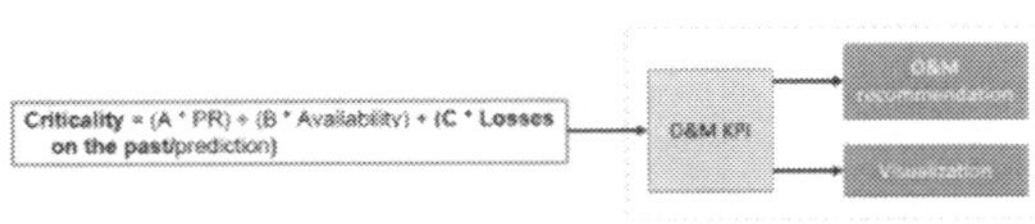

Figure 4: Definition of KPI for the generation of recommendations.

The provided O&M recommendations are finally sorted using the criticality as the key element and the results are stored in the data warehouse.

2.7 Cloud-based platform

The final DSS architecture was implemented by ISOTROL (i.e., a Spanish company focused on engineering, software and control services, that is currently managing more than 100 GW of installed electrical power worldwide), using tools provided by Google Cloud's environment. A centralized platform was developed by selecting the Google Cloud Platform (GCP) as the provider of the infrastructure.

The final solution was integrated to the GCP in a report-format, that divides all the accessible information into seven different categories: 1. Energy loss and plant status, 2. Incidents in inverters, 3. Incidents in the solar field, 4. Out-of-normality and patterns analysis, 5. Outliers, 6. Alarm analysis and 7. Sensors and data quality.

2.8 Validation

The proposed DSS was experimentally validated using a 1-year (January to December 2021) measurement data from a 10.09 MW_p PV power plant (currently administrated by Tegnatia) installed in the Mediterranean region (Köppen-Geiger-Photovoltaic climate classification DH; temperate climate with high irradiation) [19]. The grid-connected PV plant comprises of 38344 poly-crystalline silicon PV modules (36064 X 260 W_p JINKO + 2280 X 315 W_p TRINA), 10 inverters (9 X 1000 kW + 1 X 630 kW ABB) and 68 stringboxes. The system is installed in an open-field mounting arrangement and at a fixed tilt angle of 25°.

The system is connected to a data-acquisition platform, which is used for the monitoring and storage of measurement data based on the IEC 61724-1 [20]. The data are stored as 10-minute averages. The meteorological measurements include the in-plane irradiance, wind speed, module and ambient temperature. The PV system's operational measurements include the string current, inverter DC current, voltage and power, active power and the output AC power.

3 RESULTS

3.1 Data quality analysis

The data quality methodology was initially applied on the available field measurements (Module 1). The data quality over time (i.e., the behaviour of data in daily time series) is depicted in Fig. 5. It can be seen that data issues mainly took place in March-April.

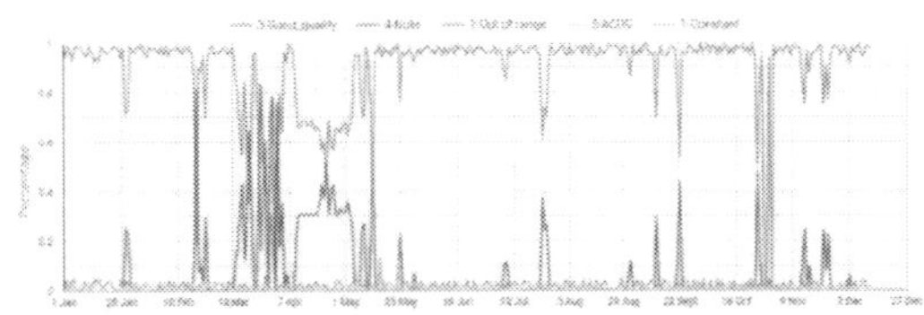

Figure 5: Data quality analysis over the evaluation period (January - December 2021) for the test PV plant.

Communication errors and sensor problems (i.e., coherence, saturation and frozen) were also detected during the evaluation period (see Fig. 6). The issues mainly occurred in February, March, May and September - December.

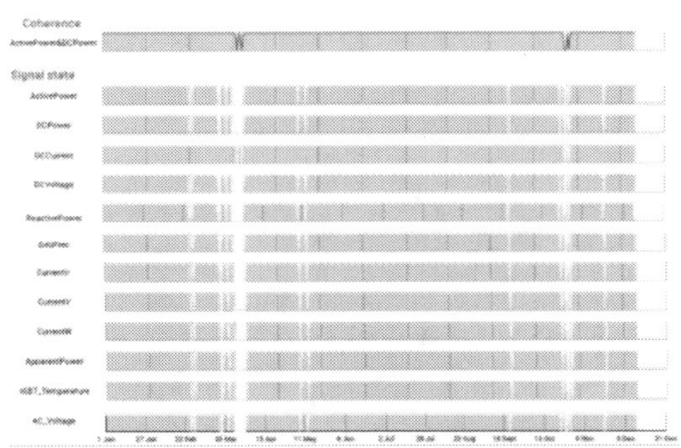

Figure 6: Time series by sensor at the inverter level over the evaluation period (January - December 2021).

The results of the weather signals analysis indicated that the weather sensors were working properly, except in March, November and December (see Fig. 7).

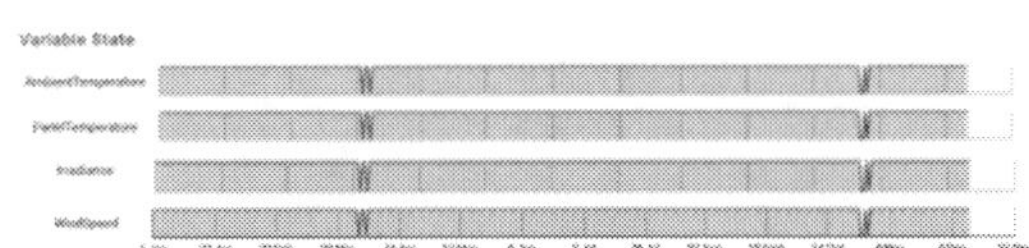

Figure 7: Weather sensor time series over the evaluation period (January - December 2021).

After the application of the data quality algorithms, predictive models were applied on the cleansed data to simulate the PV performance at the inverter level (see Fig. 8). The models performed well for predicting the PV electrical signals, exhibiting good agreement with the measured data.

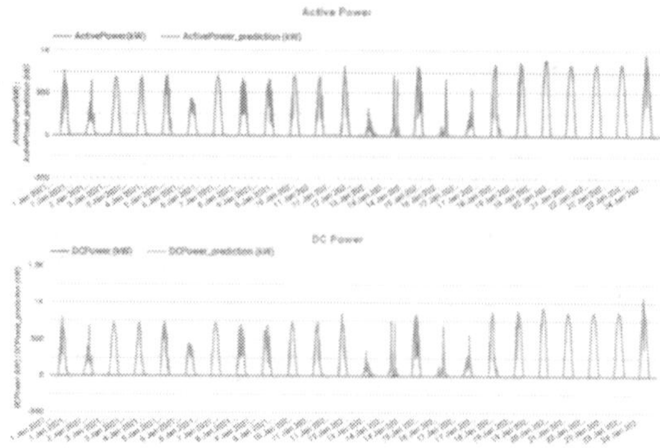

Figure 8: Measured and predicted Active and DC Power over a 3-weeks period in January 2021.

3.2 Out-of-normality analysis

The cleansed data were then analysed by Module 2. Pattern and outlier analyses were thus performed to identify underperformance incidents and equipment malfunctions.

The generated inverters out-of-normality heatmap is depicted in Fig. 9. The heatmap shows the most significant variables (i.e., Active and DC Power, DC voltage and insulated-gate bipolar transistor [IGBT] temperature) at the inverter level. The results indicated the normal health-state grade of inverters (and electrical signals) as shown in Fig. 9.

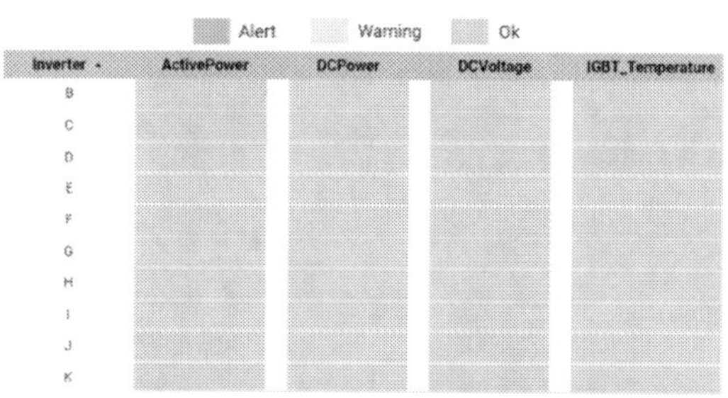

Figure 9: Out-of-normality heatmap (patterns) for the most significant variables at the inverter level. The green colour indicates the state of normality ("Ok"), the yellow colour indicates slight differences ("Warning"), while the red colour indicates significant differences ("ALERT").

We can also detect in which days anomalies occurred in the signals at the inverter level. Null days (coloured in white in Fig. 10) correspond to communication errors or to days in which inconsistencies were detected between inverter sensors with their solar field. Such errors and inconsistencies mainly occurred in March, May, July, September, October and December (see Fig. 10).

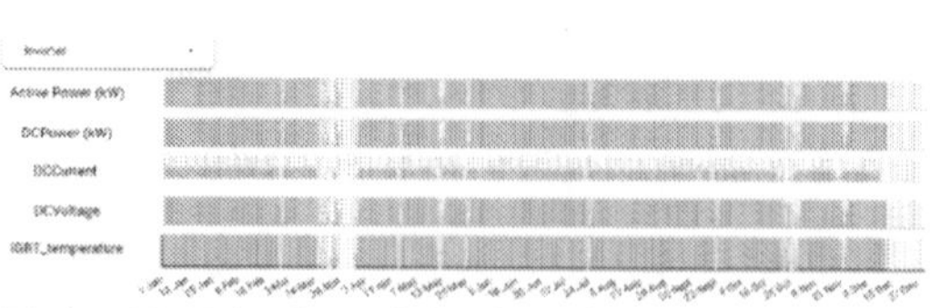

Figure 10: Behaviour of signals on the inverter timeline.

The analysis of anomalies by outliers then statistically compared the signals of similar elements (i.e., inverters and strings of the same capacity) and detected several occurrences of string disconnection and inverter low performance. Examples from the signal comparison between different strings and inverters is shown in Fig. 11, indicating string disconnections and the fault operation of inverter G over the first four days of January 2021.

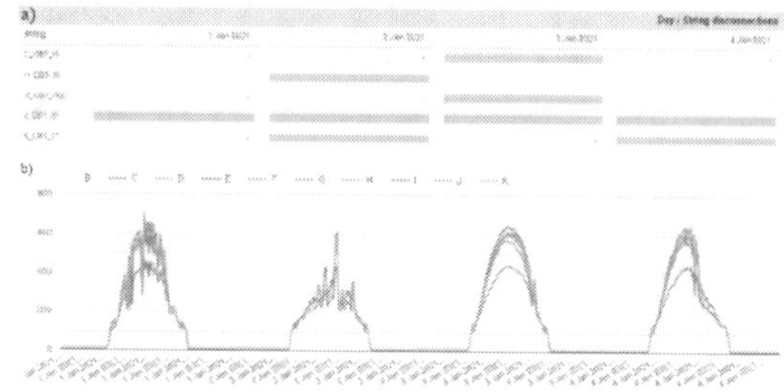

Figure 11: Signal comparison between a) strings and b) inverters during the first four days of January 2021.

3.3 Inverter and solar field analysis

Different algorithms were executed in this step to identify inverter failures (such as inverter shutdown failures as shown in Fig. 12) and solar field problems.

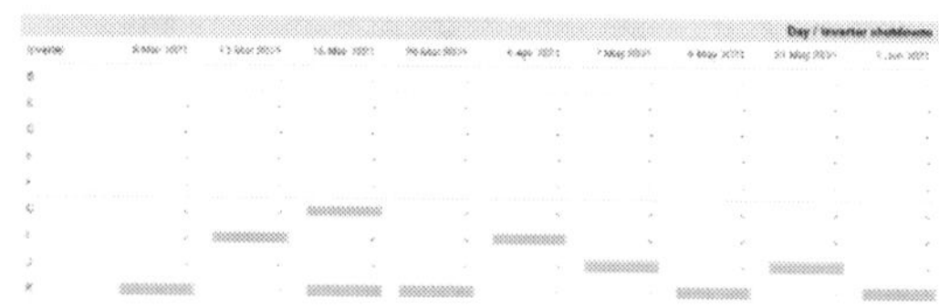

Figure 12: Detected inverter shutdown failures over the period from March to June 2021.

As part of the inverter analysis, a heat map depicting the efficiency of each inverter by relating the actual Active to the DC power and considering only datapoints similar to STC was used (see Fig. 13), with inverter F exhibited the lowest efficiency of 94.78%.

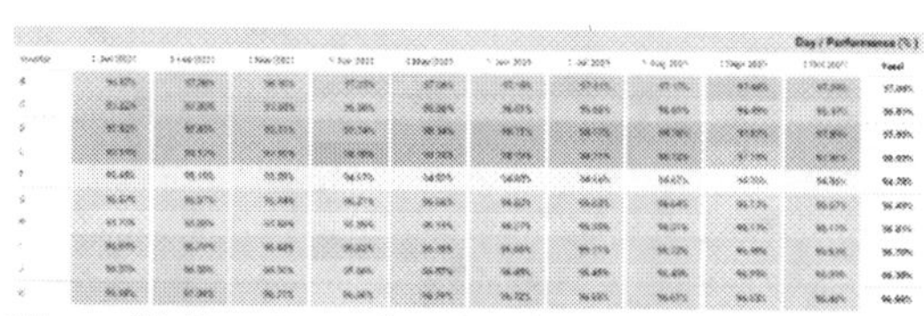

Figure 13: Inverter performance heat map at STC over the evaluation period.

Similarly, a heat map of solar field performance under optimal conditions was constructed (see Fig. 14) that shows the solar field performance at the inverter level and the deviation between the actual behavior of the devices with respect to the manufacturer's specifications.

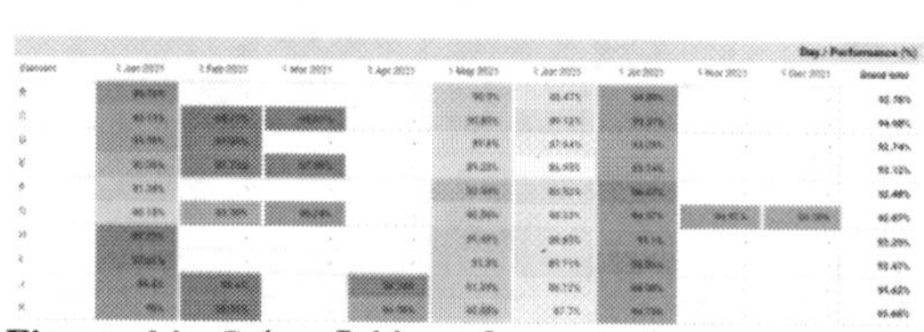

Figure 14: Solar field performance heat map under optimal conditions over the evaluation period.

The performed inverter and solar field analysis identified different loss incidents, including disconnected strings, inverter related failures, low performance and other solar field problems. As an example, the detected incidents during the first six months period in 2021 are summarized in Fig. 15, where the bar graph groups all the detected losses at the inverter level and shows the respective lost energy (in MWh) for each fault type over the period from 1 January to 30 June 2021.

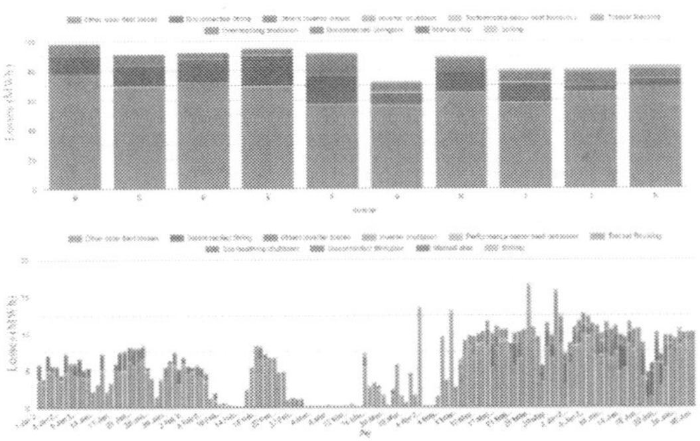

Figure 15: Detected incidents, amount and type of energy losses at the inverter level over the period from 1 January to 30 June 2021.

3.4 Energy loss analysis and plant status

Finally at Module 4, the energy loss analysis was performed. Over the 1-year evaluation period, the test PV plant produced 12586 MWh (85.70%), while the detected incidents (i.e., inverter failures, PV module faults and string disconnections) accounted for 2100 MWh (14.30%) of lost energy (see Fig. 16).

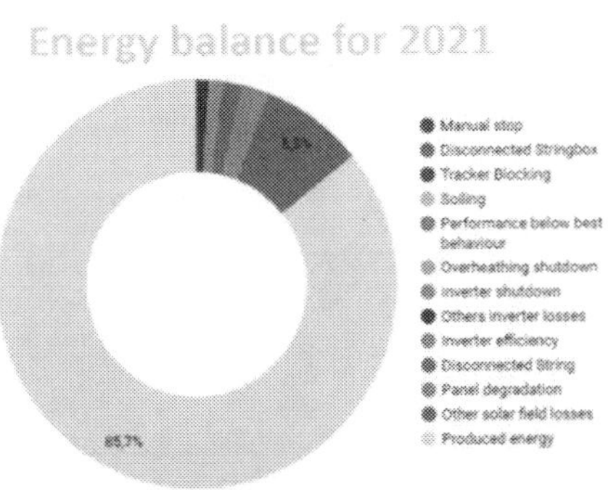

Group	Location	Type of losses	Losses (MWh)	Percentage
Total theoretical energy			14686	100%
Produced Energy			[illegible]	[illegible]
Nominal losses	Inverter	Inverter efficiency	167	1.14%
	Solar field	Panel degradation	313	2.13%
[illegible]	Inverter	Other performance inverter	[illegible]	[illegible]
	Solar field	Performance below best behaviour	0.4	0%
		Other performance solar field	1226	8.34%
Incidences	Inverter	Shutdowns	[illegible]	[illegible]
		Manual stop	0	0%
		Overheating	0.5	0%
		Curtailment temp	0	0%
	Solar field	Stringbox disconnections	0	0%
		String disconnections	238	1.62%
		Soiling	0	0%
		Tracker blocking	0	0%

Figure 16: Energy loss breakdown for the test PV plant over the evaluation period (January to December 2021).

Based on the Failure Mode, Effects and Criticality Analysis (FMECA) approach and plant's O&M KPI, a criticality value was then assigned to each of the detected incidents. The DSS engine then provided a list of automatically generated recommendations for resolving the underperformance incidents (i.e., performance losses and failures) and data issues and errors (see Fig. 17).

Figure 17: Screenshot depicting some of the O&M recommendations generated by the DSS engine along with the criticality value assignation for the detected incidents.

3.5 Estimation of recoverable energy

Given the DSS recommendations for resolving the detected issues, approximately 7% of the lost energy production per year could be recovered by performing corrective actions. The recoverable energy was calculated using the resolution times specified in [21], while the response time was defined as stated in the contract made between the PV plant owner and the O&M company [22].

4 CONCLUSION

In this work, a DSS architecture, that operates on field measurements, was developed for enhancing the PV performance. The DSS was validated in a real environment that manages the complete process of acquisition and standardization of plant data on a cloud platform, validation, and cleaning of the received information, identification of failures, generation of recommendations for maintenance, prioritization of actions and follow-up.

The obtained results showed the effectiveness of the incorporated DSS functionalities for data cleansing and the system's capabilities for automatically detecting PV failures, categorizing the incidents into different loss causes and proposing mitigation actions. Over the 1-year evaluation period, the detected incidents accounted for 2100 MWh (or 14.30% of the total energy production) of lost energy. Given the DSS recommendations for resolving the detected issues, approximately 7% of the lost energy production could be recovered by performing corrective actions.

Finally, the results (extracted directly from the cloud-based platform, Bluence Asset Performance Diagnosis, developed by Isotrol) provide evidence of the importance of DSS platforms (capable of generating specific O&M actions) for improved PV production, optimized revenue and reduced O&M costs. Insights are also provided to the community regarding the procedure of optimizing/scheduling the field O&M activities in PV systems. Future work will focus on incorporating additional preventive and predictive functionalities to the developed cloud-based DSS solution for PV assets diagnosis and maintenance.

5 ACKNOWLEDGEMENTS

This work was funded by the 1C4PV project. Project 1C4PV is supported under the umbrella of SOLAR-ERA.NET Cofund by the Centro para el Desarrollo Tecnológico Industrial (CDTI), the Scientific and

Technological Research Council of Turkey and the Research and Innovation Foundation (RIF) of Cyprus. SOLAR-ERA.NET is supported by the European Commission within the EU Framework Programme for Research and Innovation HORIZON 2020 (Cofund ERA-NET Action, N° 691664). TEGNATIA ENERJI ÜRETIM SAN. VE TIC. is kindly acknowledged for providing the field data of the test PV plant.

6 REFERENCES

[1] A. Livera, M. Theristis, L. Micheli, E.F. Fernández, J.S. Stein, G.E. Georghiou, Operation and maintenance decision support system for photovoltaic systems, IEEE Access. 10 (2022) 42481–42496. https://doi.org/10.1109/ACCESS.2022.3168140.

[2] A. Livera, M. Theristis, G. Makrides, G.E. Georghiou, Recent advances in failure diagnosis techniques based on performance data analysis for grid-connected photovoltaic systems, Renew. Energy. 133 (2019) 126–143. https://doi.org/10.1016/j.renene.2018.09.101.

[3] A. Livera, M. Theristis, G. Makrides, G.E. Georghiou, Intelligent cloud-based monitoring and control digital twin for photovoltaic power plants, in: 49th IEEE Photovolt. Spec. Conf., 2022.

[4] Raycatch, Solar Asset Optimization, Industry Benchmark Study, (2021).

[5] A. Livera, G. Tziolis, F. Jose G., G. Ruben Gonzalez, Bernal George E., Cloud-based platform for photovoltaic assets diagnosis and maintenance, Energies [Submitted]. (2022).

[6] Isotrol, Bluence Asset Performance Diagnosis by Isotrol, (2018). https://bluence.isotrol.com/ (accessed August 10, 2022).

[7] A. Livera, M. Theristis, E. Koumpli, S. Theocharides, G. Makrides, J. Sutterlueti, J.S. Stein, G.E. Georghiou, Data processing and quality verification for improved photovoltaic performance and reliability analytics, Prog. Photovoltaics Res. Appl. 29 (2021) 143– 158. https://doi.org/10.1002/pip.3349.

[8] A. Phinikarides, C. Shimitra, R. Bourgeon, I. Koumparou, G. Makrides, G.E. Georghiou, Development of a novel web application for automatic photovoltaic system performance analysis and fault identification, 44th IEEE Photovolt. Spec. Conf. (PVSC),4th Photovolt. Spec. Conf. PVSC 2017. (2017) 2298–2300. https://doi.org/10.1109/PVSC.2017.8366379.

[9] C.A. Gueymard, J.A. Ruiz-Arias, Extensive worldwide validation and climate sensitivity analysis of direct irradiance predictions from 1-min global irradiance, Sol. Energy. 128 (2016) 1–30. https://doi.org/10.1016/j.solener.2015.10.010.

[10] A. Livera, M. Theristis, G. Makrides, S. Ransome, J. Sutterlueti, G.E. Georghiou, Optimal development of location and technology independent machine learning photovoltaic performance predictive models, in: 46th IEEE Photovolt. Spec. Conf., 2019: pp. 1270–1275. https://doi.org/10.1109/PVSC40753.2019.8980474.

[11] M. Theristis, A. Livera, C.B. Jones, G. Makrides, G.E. Georghiou, J.S. Stein, Nonlinear Photovoltaic Degradation Rates: Modeling and Comparison Against Conventional Methods, IEEE J. Photovoltaics. 10 (2020) 1112–1118. https://doi.org/10.1109/JPHOTOV.2020.2992432.

[12] A. Livera, M. Theristis, G. Makrides, J. Sutterlueti, G.E. Georghiou, Advanced diagnostic approach of failures for grid-connected photovoltaic (PV) systems, in: 35th Eur. Photovolt. Sol. Energy Conf. (EU PVSEC), 2018: pp. 1548–1553. https://doi.org/10.4229/35thEUPVSEC20182018-6BO.6.5.

[13] L. Micheli, M. Theristis, A. Livera, J.S. Stein, G.E. Georghiou, M. Muller, F. Almonacid, E.F. Fernández, Improved PV Soiling Extraction through the Detection of Cleanings and Change Points, IEEE J. Photovoltaics. 11 (2021) 519–526. https://doi.org/10.1109/JPHOTOV.2020.3043104.

[14] A. Livera, M. Florides, M. Theristis, G. Makrides, G.E. Georghiou, Failure diagnosis of short- and open-circuit fault conditions in PV systems, in: 45th IEEE Photovolt. Spec. Conf., 2018: pp. 0739–0744. https://doi.org/10.1109/PVSC.2018.8548161.

[15] K. Ilse, L. Micheli, B.W. Figgis, K. Lange, D. Daßler, H. Hanifi, F. Wolfertstetter, V. Naumann, C. Hagendorf, R. Gottschalg, J. Bagdahn, Techno-Economic Assessment of Soiling Losses and Mitigation Strategies for Solar Power Generation, Joule. 3 (2019) 2303–2321. https://doi.org/10.1016/j.joule.2019.08.019.

[16] M.G. Deceglie, M. Muller, Z. Defreitas, S. Kurtz, A scalable method for extracting soiling rates from PV production data, 43rd IEEE Photovolt. Spec. Conf. 2016-Novem (2016) 2061–2065. https://doi.org/10.1109/PVSC.2016.7749992.

[17] O.N. Sarr, F.I. Barro, O.A. Niasse, F. Dia, N. Mbengue, B. Ba, C. Sene, Analysis of Failure Modes Effect and Criticality Analysis (FMECA): A Stand-Alone Photovoltaic System, Sci. J. Energy Eng. 5 (2017) 40–47. https://doi.org/10.11648/j.sjee.20170502.11.

[18] Girdharee Saran, Top 7 Advantages of Using Google Cloud Hosting, (2020). https://www.whizlabs.com/blog/google-cloud-hosting/ (accessed July 12, 2022).

[19] J. Ascencio-Vásquez, K. Brecl, M. Topič, Methodology of Köppen-Geiger-Photovoltaic climate classification and implications to worldwide mapping of PV system performance, Sol. Energy. 191 (2019) 672–685. https://doi.org/10.1016/j.solener.2019.08.072.

[20] International Electrotechnical Commission (IEC), Photovoltaic system performance - Part 1: Monitoring, IEC 61724-1:2017, (2017).

[21] D. Moser, M. Del Buono, U. Jahn, M. Herz, M. Richter, K. De Brabandere, Identification of technical risks in the photovoltaic value chain and quantification of the economic impact, Prog. Photovoltaics Res. Appl. 25 (2017) 592–604. https://doi.org/10.1002/pip.2857.

[22] SolarPower Europe [SPE], Operation & Maintenance Best Practice Guidelines / Version 4.0, 2019.

SOILING DETECTION ON PHOTOVOLTAIC MODULES USING HYPERSPECTRAL IMAGING AND MACHINE LEARNING ALGORITHMS

Marko Turek
Fraunhofer Center for Silicon Photovoltaics CSP
Otto-Eißfeldt-Str. 12, D-06120 Halle (Saale), Germany

ABSTRACT: The automated remote detection of the state of PV power plants plays an increasing role for cost-optimized plant monitoring. Ideally, it yields additional information explaining observed power yield fluctuations by going beyond the data provided by the electrical inverter without requiring any direct access to the power plant locations. To this end, optical and electro-optical inspection techniques play an important role in the monitoring and maintenance of PV power plants. In our work, we present a proof-of-concept based on lab-samples showing that the combination of hyper-spectral imaging techniques together with machine learning approaches can detect and classify soiling on glass surfaces. We demonstrate a workflow including a) the acquisition of the high-dimensional spectral data, where each pixel of the image carries an entire reflection spectrum, b) the dimensionality reduction of the extended spectral data-set using principal component analysis, and c) a comparison of supervised classification and unsupervised clustering algorithms. As a result, we find that our workflow can identify surface-regions of PV modules with increased soiling. Applying a numerical cross-validation, we find recall and precision metrics well above 90% for the supervised classifiers. Although the unsupervised clustering approaches are naturally showing a lower predictive power, they nevertheless can be used to distinguish soiling-free surfaces from surfaces covered with soiling.
Keywords: Energy Performance, Monitoring, Optical Properties, Plant Contol

1 INTRODUCTION

The development of automated optical or electro-optical inspection methods as a further source of information in addition to the electrical power data of PV power plants is a current and very active research topic. Mostly, it includes the development of optical inspection systems, luminescence imaging, and infra-red imaging [1, 2, 3]. The acquisition of the data is often either performed by an operator with a mobile, ground-based imaging system or by aerial inspection, e.g. the usage of drones. An imaging-based monitoring aims at providing more detailed information on the state of a PV power plant and ideally results in defining specific maintenance actions. In particular, a reliable soiling detection and classification can be the basis for optimized cleaning schedules in order to achieve a cost optimized solar power generation [4]. This topic is also being addressed by machine learning approaches applied to converter or sensor data. [5, 6]

While hyperspectral remote sensing is a well-established tool in acquiring digital images of earth materials or ecosystems it has not yet been widely applied to the monitoring of PV power parks. In Refs. [7, 8], approaches have been described to detect photovoltaic solar panels using hyperspectral imaging and to provide an estimate for their solar power generation. Generally, there are several data sources for hyperspectral images available, e.g. satellite data. An analysis of Google Earth Engine images for the detection of soiling has been presented in Ref. [9]. Another approach has been based on an HySpex hyperspectral sensor which is an aircraft-borne imaging spectrometer [10]. Also, the spatial resolution of these data sources being in the range of few to several meters is sufficient to detect large scale PV power plants. However, there are no established PV-specific data analysis algorithms which aim at detecting soiling and classifying soiling yet. Also, the relevant wavelength range that is needed for a reliable detection of soiling has not been studied systematically so far. With our results, we present a proof-of-concept study that hyperspectral imaging in combination with automated classification algorithms is capable to detect soiling and distinguish soiled PV module surfaces from clean ones. Our innovative approach is designed in a way that it can be transferred to outdoor data, e.g. by compensating for overall intensity changes and focusing on spectral finger-prints to detect soiling. On the samples considered, a benchmark of the performance of the various supervised or unsupervised algorithms is presented. We consider this work as being of great relevance for the further development of remote sensing as a cost-efficient approach to monitor large-scale PV power plants.

With our work, we are aiming at the establishment of optical imaging techniques using spectral data as a further valuable source of information. In particular, hyperspectral imaging gathers spectral and position resolved data in one measurement run yielding complete images for a large number of wavelengths. In our case, we employ a VNIR camera covering wavelengths between 400 nm and 1000 nm. Thus, hyperspectral imaging results in a massively extended data set as it combines spatial and spectral information. On the one side, this makes it a very powerful technique to detect material changes or surface modifications, such as soiling. On the other hand, more advanced machine learning algorithms are required to process and analyze this hugely increased data set. Therefore, we further aim at providing and analyzing several clustering or classification algorithms where the relevant information is extracted by an automated machine learning approach. We present a comparison of the applicability and predictive power of several well-known supervised classification algorithms such as the random forest approach, support vector machines, or the multi-layer perceptron model. Also, we investigate to which extend unsupervised clustering algorithms, such as k-means or mean-shift can be applied instead of supervised classification. We show that both, unsupervised clustering and supervised classification leads to a reliable soling identification.

2 EXPERIMENTAL APPROACH AND DATA ANALYSIS

2.1 Samples and data acquisition

For our experiments, we have prepared mini-modules

made of representative PV-materials. We investigate organic and inorganic soiling by preparing corresponding lab-samples. In case of the inorganic soiling, a soiling chamber is used in which various environmental conditions such as the type of dust, the temperature, the humidity or even illumination can be modified to achieve realistic soiling conditions. For the organic soiling, we have prepared a second mini-module with various degrees of algae coverage, see Figure 1.

The hyperspectral imaging is performed using a VNIR spectral imaging systems operated in reflection mode. While the spectral resolution of the camera is in the range of several nm for wavelengths between 400 nm and 1000 nm, the spatial resolution is in the sub-mm-range. As the data sets are getting very memory-intensive, we have analyzed only certain representative parts of the entire mini-modules.

Figure 1: (left) Sample 1: Mini-module with inorganic soiling (dust) on the left third, no soiling on the middle third, and organic soiling (algae) on the right third. (right) Sample 2: Mini-module with no soiling on the leftmost quarter, low organic soiling (algae) on the second-left quarter, medium organic soling organic soiling on the third quarter and strong soling on the rightmost quarter.

2.2 Data preparation and analysis

In a first step, all data reflectance data are captured together with a list of wavelengths by the camera. Then, our algorithm applies a scaling of the entire data set based on two additional measurements of a white reference with known reflectance properties and a black background, respectively. Then, regions of interest (ROIs) are identified and extracted from the images. In a first data-preprocessing step, we perform a L1-normalization correcting for intensity differences. Also, a min-max feature scaling which regularizes the various wavelength values is applied such that wavelengths with lower intensities are obtaining equal weights as high-intensity wavelengths for the subsequent training or clustering algorithms.

In a second data-preprocessing step, we perform a dimensionality reduction. Each pixel is represented by a vector in a high-dimensional vector-space where the dimensions are given by the set of wavelength values. In our case, we have captured 196 different wavelengths (i.e. spectral bands) implying that all pixels are represented as vectors in a 196-dimensional vector space. This high-dimensional data-set is not suitable for most of the standard clustering or classification algorithms. Therefor, we have reduced the dimensionality to values between three and five by applying a principal component analysis which preserves the vector components with the largest variation in the data.

In a third step, we apply either a classification algorithm (support vector machines, random forest, multi-layer perceptron model) or clustering algorithms (k-means, mean-shift) on the reduced data set. In case of the classification, the training phase is performed on smaller sub-regions which are characteristic for the classes to be identified, see Figure 2. We have defined the six classes "Dust_BS" (backsheet area covered with inorganic dust), "Dust_Cell" (cell area covered with inorganic dust), "Clean_BS" (backsheet area without any soiling coverage), "Clean_cell" (cell area without any soling coverage), "Algae_BS" (backsheet area covered with organic algae soiling), and "Algae_Cell" (cell area covered with organic algae soiling). All pixels in these marked ROIs, see Figure 2, are labeled with the according class-label even though some of the pixels in these ROIs might not be affected by e.g. algae-soiling as the soiling is not entirely uniform. Nevertheless, the marked ROIs are considered as representative sample sets for the six different classes.

Figure 2: Top-part of sample 1 with six different ROIs marked representing six different classes which serve as a training data set for the classification algorithms.

All algorithms have been implemented in Python using the Scikit-Learn library. The calculations have been done on a state-of-the-art notebook without any specific hardware components. Typically, a single classification or clustering takes around few seconds to minutes for our data set.

3 RESULTS

In our analysis, we focus on two different case studies. In the first case study, we investigate to which extent a clean surface can be distinguished from a soiled surface and whether the organic soiling can be distinguished from the inorganic soiling. In the second case study, we analyze sample 2 which exhibits various degrees of organic soiling coverage. Here, we evaluate how sensitive our approach is in order to detect even lower degrees of soiling.

3.1 Case study one: Classification of different soiling types

The six selected ROIs in Figure 2, show very distinct reflectance spectra, see Figure 3. As expected, the area of the cell without any soiling shows the lowest reflectance. The two ROIs covered with organic algae soiling show a clear clorophyll-signal represented by an increased reflectivity around 550 nm (green) and reduced reflectivity in the wavelength range below and above these values. In particular, a clear reduction in reflectivity around 680 nm can be observed for the algae covered areas. Finally, the two backsheet areas without soiling or with inorganic dust soiling show the highest reflectance values. Also, all backsheet ROIs exhibit another specific spectral signature around 940 nm, where a second distinct reduction of reflectance can be found.

The second data-preprocessing step consisting of the dimensionality reduction using PCA can be presented by a false-color RGB image, see Figure 4. The PCA algorithm performs a linear transformation of the high-dimensional vector space given by the spectral data in such a way that the first PCA components represent the highest variance in the data set. Already this step clearly shows the differences between the ROIs with organic soling (red) and the clean ROIs (green) or ROIs with inorganic soiling (blue).

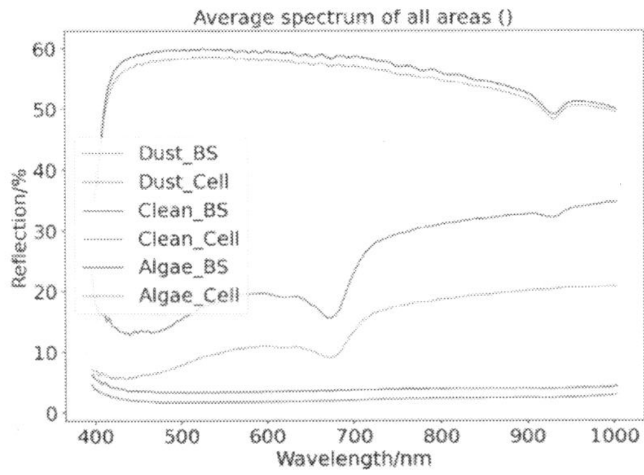

Figure 3: Average reflectance of the six ROIs defined in Figure 2.

The reduced data set obtained from the PCA is then analyzed using various clustering algorithms. In Figure 5, the results of a mean-shift clustering and a k-means clustering are shown as a color overlay. Each pixel is associated to a certain cluster depending on its spectral properties, i.e. pixels with similar spectra are grouped into one cluster. These algorithms can be applied without any further input regarding the data which means that these procedures are unsupervised. As a result, it is found that the algae coverage is clearly identified as one cluster while the areas covered with inorganic dust are associated with another cluster as well as the clean areas.

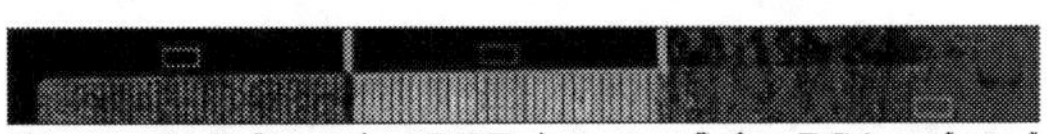

Figure 4: False-color RGB image of the PCA-reduced data set where the first three PCA-components are represented by the red, green and blue channels.

Figure 5: (top) Color overlay of a mean-shift clustering with six clusters. (bottom) Color overlay of a k-means clustering with seven clusters. All pixels associated to the algae cluster are colored in green.

Figure 6: (top) Color overlay of a random forest classification with six defined classes. (bottom) Color overlay of a support vector machine classification with six defined classes.

Finally, we have implemented several supervised classification algorithms. In this case, one first has to perform a training of the models using a labeled data set. We have chosen the marked ROIs in Figure 2 as representative training sets for the six different classes distinguishing cell and backsheet as well as clean and soiled areas. Then, these trained models are applied to the entire data set yielding a class for each pixel, both inside or outside the ROIs used for training. These classification results are represented by a color overlay where each pixel is presented in a certain color depending on its class association, see Figure 6.

Both classification approaches, random forest (RFO) and support vector machine (SVM), yield very similar results. For the three cell regions, the precision values of the random forest approach are 89%, 92%, and 82% for the inorganic dust, clean cell, and algae-covered cell, respectively. The corresponding recall values are 92%, 89%, and 81%, respectively. In comparison, we find for the support vector machine model precision values of 91%, 93%, and 85% as well as recall values of 89%, 89%, and 83%, respectively. Here, the precision value describes ratio between the correctly associated pixel of one class (true positive) to all pixels associated to this class by the model (sum of true positive and false positive). On the other hand, the recall value describes the number of all correctly classified pixels (true positive) relative to all pixels that are in this class (sum of true positive and true negative). Ideally, both indicators are as large as possible meaning that all pixels of one class are classified correctly (false negatives → 0) while no pixels of another class are predicted by the model to be in the same class (false positives → 0).

3.2 Case study two: Evaluation of various degrees of soiling coverage

Our second analysis investigates the impact of various degrees of algae coverage on the outcome of the investigated models. Here, we have applied both clustering and classification models obtained from sample 1 to another sample 2. Sample 2, see Figure 1 (right), has been prepared with four different parts: a clean part at the left, a part with heavy algae soiling at the right and low as well as medium soiling in between.

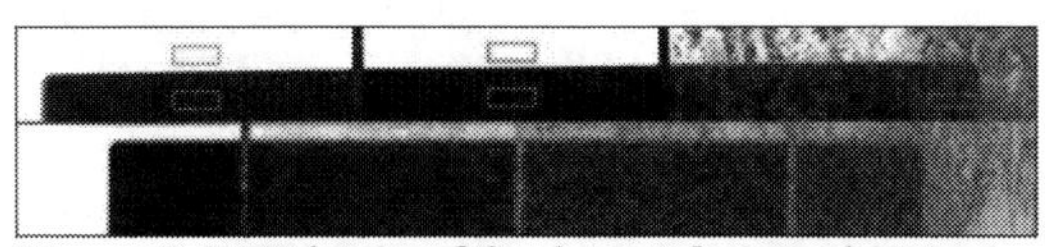

Figure 7: RGB-image of the data set for sample 1 (upper part of the image) and sample 2 (lower part of the image). The marked ROIs in sample 1 have been used as training data sets.

Thus, the training of the classification models has been performed on the marked ROIs shown in Figure 7 and then applied to the entire sample 1 (training sample) and sample 2 (test sample). In this case study, we have also added a multi-layer perceptron (MLP) model with 2 hidden layers of size 6 and 4, respectively, to our set of investigated models.

The clustering of the pixels for both samples once more identifies the algae-cluster reliably, also for the second sample. In all three algae-covered quarters, the color overlay in Figure 8 (top) shows that the pixels have been associated to the algae-cluster (color-coded in green). On the other hand, the support vector machine classification algorithms trained on sample 1 correctly classify pixel of the area with strong and medium algae-soiling on sample 2 while the area with low soiling coverage of sample 2 is partially classified as covered with inorganic dust. The clean part of sample 2 is correctly classified as clean. The application of the multi-layer perceptron model yields slightly different results. In particular for the part with low soiling coverage, there are noticeable differences between the two trained models. The multi-layer perceptron model classifies a slightly larger number of pixels as algae which is closer to the actual coverage, see Figure 1 (right).

Figure 8: (top) Color overlay of a mean-shift clustering applied to sample 1 (upper part of image) and sample 2 (lower part of image). (center) Color overlay of a support vector machine classification trained on marked ROIS in sample 1 and applied to both sample 1 (upper part of image) and sample 2 (lower part of image). (bottom) Same as (center) for multi-layer perceptron network.

3 DISCUSSION

In our experiments, we have investigated two different aspects of soiling detection using hyperspectral imaging. In a first case study, we compare an inorganic dust soiling with an organic soiling by algae-coverage. We find that our approach of data preprocessing by normalization and feature scaling combined with a dimensionality reduction using a PCA leads to a suitable low-dimensional data set for the application of clustering or classification algorithms. In case of the clustering, we have compared different types of well-established algorithms, i.e. mean-shift clustering and k-means clustering. In either case, the hyperspectral data set can be automatically grouped into several clusters corresponding to the different soiling types. In particular, the organic soiling is clearly localized on the sample. Hence, this approach is in particular useful for the detection and separation of PV module surfaces that are in a different state regarding their soiling properties. For example, this can be a valuable information for an optimized cleaning routine, if only certain parts of modules or PV power plants are affected by soiling. However, the unsupervised clustering approach does not allow to draw direct conclusions about the type of soiling.

To distinguish and classify different types of soiling, supervised algorithms which include a training phase on labeled data can be employed. In our case of organic and inorganic soiling, we can clearly distinguish between these two types and separate the soling-covered parts of the module from the clean parts. Hence, applying this approach to the remote surveillance of PV modules can give useful information on the different cleaning strategies, e.g. regarding the usage of cleaning agents, or different mitigation strategies, e.g. the application of chemicals preventing organic soiling. Our results show that a trained model based on one sample can be successfully applied to another sample. However, further improvements to the data analysis are required in order to detect also minor soiling coverage. One possible approach to be investigated in further studies would be to combine the unsupervised clustering with a subsequent supervised classification. In our tests, it became apparent that the clustering can separate the several groups of pixels corresponding to the various soiling types and degrees. Hence, this step could be included into the data pre-processing before the actual training and classification takes places.

4 SUMMARY AND OUTLOOK

In summary, we developed an imaging and data analysis approach based on hyperspectral imaging data that allows for a separation of regions with soiling from regions without soiling on PV modules. Also, it can distinguish different types of soiling such as inorganic dust and organic algae films. As our approach relies on machine learning classification or clustering algorithms, it can be applied with a large degree of automation. Hence, it is suitable to be applied to remote sensing data as for example provided by satellites. This would allow to gather valuable additional information about a PV power plant's state going beyond the electrical power output data available from the inverters. Thus, it can be used to trigger cleaning cycles, identify affected modules and provide information on the type of cleaning procedure needed.

ACKNOWLEDGMENTS

The author gratefully acknowledges support with sample preparation from G. Willers, support with measurements from S. Eiternick, and financial support from the Federal Ministry for Economic Affairs and Climate Action within the "PVOptDigital" (FKZ 03EE1107A) project.

REFERENCES

[1] I. Høiaas, K. Grujic, A. G. Imenes, I. Burud, E. Olsen, N. Belbachir. *Inspection and condition monitoring of large-scale photovoltaic power plants: A review of imaging technologies.* Renewable and Sustainable Energy Reviews 161, p112353 (2022)

[2] C. Buerhop, L. Bommes1, J. Schlipf, T. Pickel, A. Fladung, I. M. Peters. *Infrared imaging of photovoltaic modules: a review of the state of the art and future challenges facing gigawatt photovoltaic power stations.* Prog. Energy 4, p042010 (2022)

[3] W. Mühleisen, C. Hirschl, G. Brantegger, L. Neumaier, M. Spielberger, H. Sonnleitner, B. Kubicek, G. Ujvari, R. Ebner, M. Schwark, G.C. Eder, Y. Voronko, K. Knobl, L. Stoicescu. *Scientific and economic comparison of outdoor characterization methods for photovoltaic power plants.* Renewable Energy 134, p321 (2019).

[4] K. Ilse, L. Micheli, B. W. Figgis, K. Lange, D. Daßler, H. Hanifi, F. Wolfertstetter, V. Naumann, C. Hagendorf, R. Gottschalg, J. Bagdahn. *Techno-economic assessment of soiling losses and mitigation strategies for solar power generation.* Joule 3, p2303 (2019)

[5] M. Heinrich, S. Meunier, A. Samé, L. Quéval, A. Darga, L. Oukhellou, B. Multon. *Detection of cleaning*

interventions on photovoltaic modules with machine learning. Applied Energy 263, p114642 (2020)

[6] David Dassler, Shreyas Bangalore Kuppanna, Stephanie Malik, Ralf Schmidt, Matthias Ebert. *Training and Evaluation for Yield-Driven Detection of Losses in PV Systems Utilizing Artificial Neural Networks*. Proceedings of 47th IEEE Photovoltaic Specialists Conference (PVSC), p1903 (2020)

[7] D. Czirjak. *Detecting photovoltaic solar panels using hyperspectral imagery and estimating solar power production*. J. Appl. Remote Sens. 11, p026007 (2017).

[8] M. S. Karoui, F. Z. Benhalouche, Y. Deville, K. Djerriri, X. Briottet, A. Le Bris. *Detection and area estimation for photovoltaic panels in urban hyperspectral remote sensing data by an original NMF-based unmixing method*. Proceedings of the IEEE International Symposium on Geoscience and Remote Sensing (IGARSS). p1640 (2018) DOI 10.1109 / IGARSS. 2018. 8518204.

[9] H. Supe, R. Avtar, D. Singh, A. Gupta, A. P. Yunus, J. Dou, A. A. Ravankar, G. Mohan, S. K. Chapagain, V. Sharma, C. K. Singh, O. Tutubalina, and A. Kharrazi. *Google Earth Engine for the Detection of Soiling on Photovoltaic Solar Panels in Arid Environments*. Remote Sens. 12, p1466 (2020)

[10] C. Ji, M. Bachmann, T. Esch, H. Feilhauer, U. Heiden, W. Heldens, A. Hueni, T. Lakes, A. Metz-Marconcini, M. Schroedter-Homscheidt, S. Weyand, J. Zeidler. *Solar photovoltaic module detection using laboratory and airborne imaging spectroscopy data*. Remote Sensing of Environment 266, p112692, (2021).

THE INFLUENCE OF SYSTEM LAYOUT AND OPERATING CONDITIONS ON PV SYSTEM DEGRADATION

Aline Kirsten Vidal de Oliveira*[1], Claudia Buerhop-Lutz[2], Ian Marius Peters[2], Ricardo Rüther[1]
[1]Universidade Federal de Santa Catarina, Florianópolis, Brazil
E-mail: alinekvo@gmail.com, Tel.: +55 (48) 99668-9926
[2]Helmholtz Institut Erlangen Nürnberg, D-91058 Erlangen, Germany

ABSTRACT: This paper compares performance data of two installed PV systems with the same type of multicrystalline, p-type silicon PV module in two different climatic and installation conditions. The analysis was carried out on a 17.64kWp portion of a roof-top PV plant installed in 2013 in the Lisbon area, in Portugal (38.7°N; 9.3°W) and on a ground mounted 2.35 kWp system installed in 2016 at the Fotovoltaica-UFSC Solar Energy Research Laboratory in Florianópolis -SC (27.4°S, 48.4°W). The modules of both plants are installed with a tilt of 20° in a region close to the coast. The analysis showed that, overall, the Brazilian system presented higher performance rations (PR) than the system installed in Portugal (around 20% more than Portugal's PR). The curtailments that the Portugal's system suffers in high-irradiance months, the soiling and the high temperatures in these periods are among the possible explanations for the higher system losses. The data analysis also showed the effects of shadowing on Brazilian's system and Potential Induced Degradation (PID) that affected the system installed in Portugal. Although the Brazilian site is characterized by a hot and humid climate, it did not show the same degradation patterns, due to the lower voltages that the modules in Brazil are subjected to, because it is a system with one string with 10 modules (around 301 VMPP), while the Portugal system has strings with 24 modules in series (around 727.2 VMPP). The results showed that the degradation behavior of a PV installation is determined by several factors, including environmental conditions and especially the layout of the installation. These findings lead to a more accurate understanding of different climatic and layout related degradation mechanisms of PV modules, therefore having a significant contribution for the design of reliable PV systems.
Keywords: Degradation, System Performance, Multicrystalline Silicon.

1 INTRODUCTION

With the advancement of PV energy around the world, the assessment of systems performance is important to forecast long-term energy production accurately in different conditions and to understand the different degradation mechanisms of modules in different climates. Different studies in the literature try to draw conclusions on the environmental factors that influence PV system performance and degradation around the world [1], [2]. Delamination, Potential Induced Degradation (PID), and diode or junction box problems are some examples of PV degradation types that relate to climate factors like high temperatures and humidity [3], [4]. On the other hand, PV systems installation characteristics like the mounting structure, the number of modules in a string and the grounding and inverter topologies have showed to influence the degradation rates of the modules, because they can aggravate problems like PID and backsheet adhesion loss [5]–[7].

There are many variables that influence this analysis, and therefore, the comparison of similar PV modules under different weather and installation conditions is important to isolate the causes for different performances and degradation rates.

This paper presents a comparison of two different PV systems with the same multicrystalline silicon PV module in two different climatic and installation conditions. The goal of this analysis is to determine the impact of different layout and operating conditions on the PV systems' performance.

2 METHODS

This paper compares performance data of two installed PV systems with the same type of multicrystalline, p-type silicon PV module in two different climatic and installation conditions. The study analyses the different environmental conditions of the test sites (e.g., ambient temperature, irradiance, precipitation, installation conditions) and their impact on PV systems performance.

System 1 is a 17.64 kWp portion of a PV plant installed in 2013 in the Lisbon area, in Portugal [8]. The location is characterized by a Csa climate according to the Köppen-Geiger classification [9], meaning a temperate, hot dry-summer climate. As it is close to the Atlantic Ocean, the plant weather conditions also include high wind and moderate amounts of airborne seawater mist [8]. The PV plant is installed over several buildings, but this paper will mostly focus on the performance analysis of three strings of the system (as highlighted in Fig. 1), which are installed with an azimuth of 172° and a tilt of 20° on an office building next to a factory. The strings have no influence due to shading but are subjected to excessive heat from machines and chimneys, which release particles that can cause soiling. The performance and environmental data of the system was obtained from a Sunny Webbox with Sensorbox (SMA Solar Technology AG, Germany) in a temporal resolution of 5 min. Irradiance data was also validated with data from a weather station Davis Vantage Pro2 installed in the site.

Figure 1: Aerial image of the Portuguese PV system analysed in this study, in the Lisbon region, Portugal (39°N, 9°W).

System 2 was installed in 2016 at the Fotovoltaica-UFSC Solar Energy Research Laboratory in Florianópolis - SC and is a ground mounted 2.35 kWp system. The system has an azimuth of 0° and a tilt of 20° [10]. According to the Köppen-Geiger classification [9], the location has a humid subtropical climate (Cfa), meaning hot and humid summers and mild to cold winters. The system is also close to the sea (circa 800 m), present minimum soiling and it is affected by shading during certain times of the day and year from nearby buildings and a wind turbine, as is shown in Fig: 2. The electrical data of the system were acquired from the inverters by a Campbell CR6 datalogger with a sampling interval of 10 s and stored as 1-min averages.

The electrical energy parameters on the direct current (DC) side of the systems were used in the data analyses, as well as environmental data (plane of array irradiance, ambient temperature, wind speed and others) from a weather station close to each site. Some characteristics of both systems are presented in Table I. The data were also filtered for outliers and filling methods were applied when necessary. For comparison of performance, the fourth year of operation of each PV system was used (2017 for Portugal and 2020 for Brazil).

Figure 2: Aerial image of the Brazilian PV system analysed in this study, which is part of the outdoor PV testing rig at the Fotovoltaica-UFSC Solar Energy Research Laboratory in Florianópolis, Brazil (27°S, 48°W).

Table I. PV systems and test sites specifications.

Location	Portugal	Brazil
Installation year	2013	2016
Köppen-Geiger Climate Classification	Csa	Cfa
Latitude, Longitude	38.7°N, 9.3°W	27.4°S, 48.4°W
Tilt Angle	20°	20°
Evaluation Period	2014 - 2021	2017 - 2021
Temporal resolution	5 min	1 min
Modules in series	Up to 24	10
Module power (Wp)	245	235
Installed capacity analysed (kWp)	17.64	2.35
String V_{oc} at STC (V)	890.4	368.0
String V_{mpp} at STC (V)	727.2	301.0

3 RESULTS

Each test site presented some conditions that affected the results. For the Brazilian system, the shadow of nearby objects influenced the power generation, reducing the voltage of the string at the beginning and at the end of the day. Fig. 3 shows the voltage profile of this PV system through the year of 2020. The voltage value is normalized by the number of modules in the string to facilitate the analysis. The image shows that the shadows affect the system performance in different times of the day through the year. For cloudy days, the effect on the voltage is not significant. An example of effect of the shading in the voltage of the string is shown in Fig. 4, which presents the voltage profile (normalized by the number of modules in the string) for the day 13/12/2020, a clear day with maximum irradiance of 993.7 W and maximum ambient temperature of 29.0 °C.

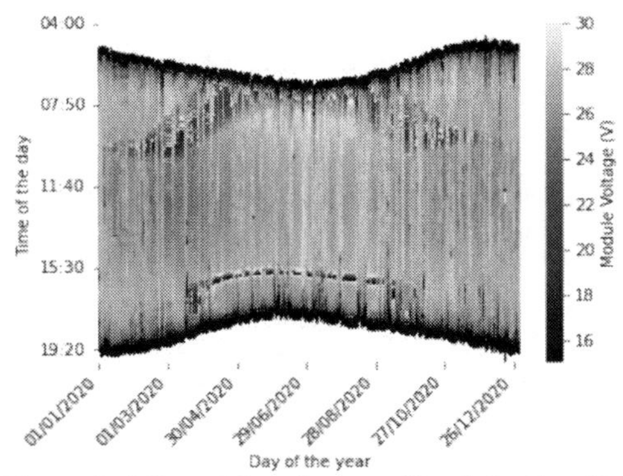

Figure 3: PV module voltage profile of the Brazilian PV system through the year, highlighting the impact of nearby shadings that affect the system in the morning and in the afternoon.

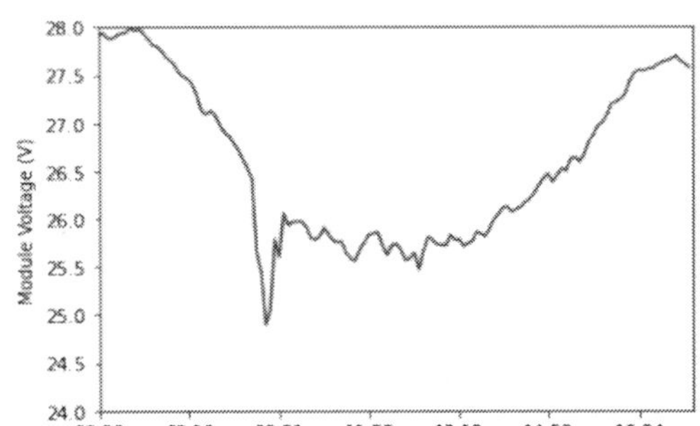

Figure 4: PV module voltage profile of the Brazilian PV system for the day 13/12/2020, highlighting the impact of nearby shadings that affect the system in the morning and in the afternoon.

Overall, the Brazilian system presented higher performance rations (PR) than the system installed in Portugal (around 20% more than Portugal's PR), as can be seen in Fig. 5, that also shows the average ambient temperature of each site for the third year of operation of each system. The blue bars show that for the Portugal system, the lower monthly temperatures and irradiances are correlated to the months with higher values of PR. Beside the high temperatures, another factor that affects performance during summer months (June to September) is the undersizing of the inverter and curtailments that are common in this system for this period. The region is also characterized by a dry summer, what increases the effect of soiling on the PV production.

For the system in Brazil, because of a weather station communication problem in March 2020, not enough valid data points were stored, and ambient temperature in this month was discarded. Still, the graph shows that the inverse correlation between ambient temperature and PR are not so strong, meaning that despite the high temperatures, other environmental factors increase PR during summer months (December to March) in this site.

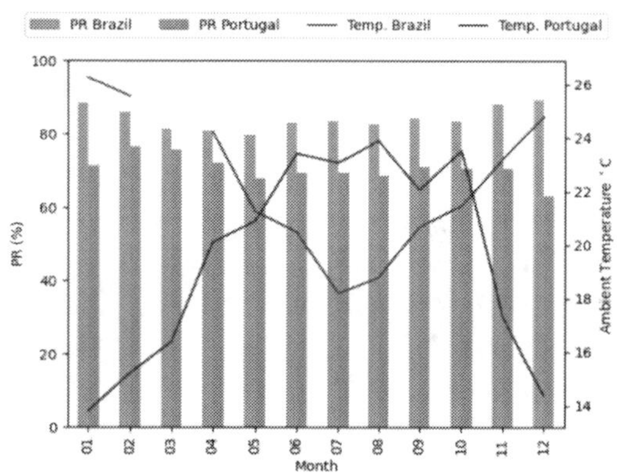

Figure 5: Estimated monthly PR (colour bars) and average monthly ambient temperature (lines) of the PV systems under study during their fourth year of operation. Data on the Brazilian system are shown in shades of red and on the Portugal system are shown in shades of blue.

Another possible explanation for the lower PR rates of the Portugal system is the PID presented by the system, that was already reported in [8]. A decrease of string voltage and consequently produced power through the years of operation caused by this degradation was observed. Fig. 6 show the comparison of normalized power of 3 strings of the PV plant distributed according to the instant tilted irradiance value in 2014 and 2020, showing the degradation that occurred between these two years. In the graph, the power was normalized by the installed PV capacity and the data was filtered for the period between 11 AM and 13 PM to facilitate the analysis. The mean operational voltage of the strings had a decrease of 8% and the annual PR of the system had a reduction of 9% between 2014 and 2021. Meanwhile, the Brazilian system did not present visible operational voltage reduction. Fig. 7 shows the decrease of string voltage (normalized by the number of the modules in the string) through the years of operation, another evidence of the PID degradation.

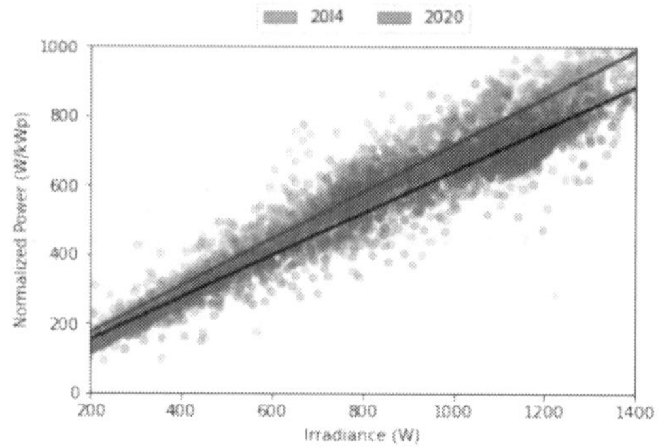

Figure 6: Comparison of PV module normalized power (Produced power over installed capacity) for different irradiance levels of the Portuguese PV system through the years of 2014 and 2020.

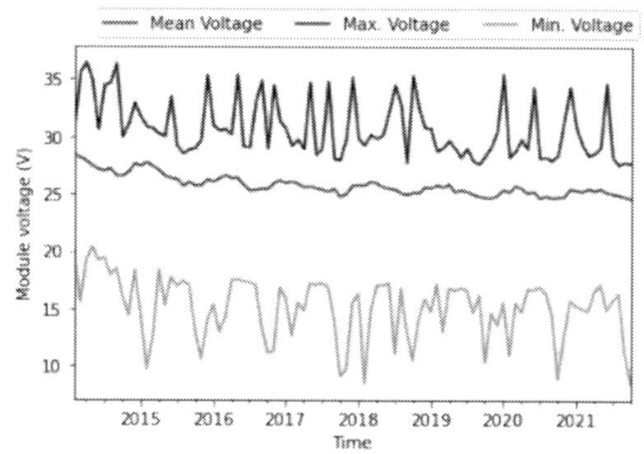

Figure 7: PV module voltage (daily maximum, minimum and average) of the Portuguese PV system through years of 2014 and 2020.

The PID degradation was also identified through a electroluminescence inspection held in 2020, that showed that several modules were affected. The number of cells affected in each module and the number of modules affected in each string varied according to the location of the modules in the building roof, which affected the temperature and soiling conditions to which they are exposed. The number of cells affected by PID in each string of the PV system (according to the electroluminescence images) was counted and Fig. 8 presents a correlation between affected cells in the string and the evolution of the mean module voltage through the years. In the graph, the string voltage was divided by the number of modules in the string and the data was filtered for the period between 11 AM and 13 PM to facilitate the analysis. It is evident that the strings with more cells damaged were directed correlated to a large reduction in module voltage, except for a few outliers. Fig. 7 and Fig. 8 show that the degradation affected the performance of the strings already from the first year of the plant operation, in different intensities through the plant. The degradation was mostly stabilized after the third year of operation.

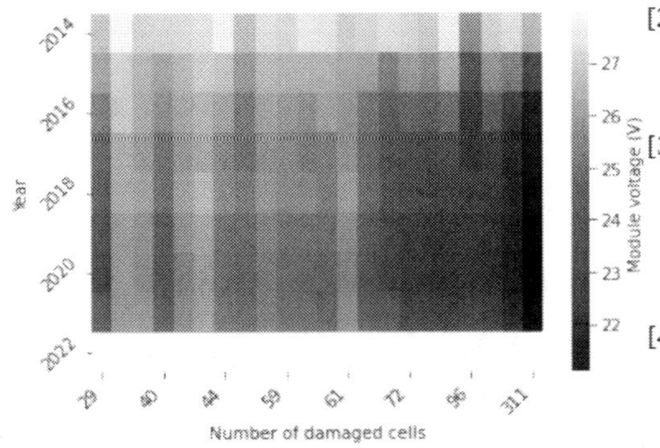

Figure *8:* Mean module voltage evolution during the years of the PV system operation of several strings of the Portugal system. The strings are placed in order of the number of damaged cells detected through an electroluminescence inspection held in 2020.

The same degradation has not yet been observed in the Brazilian modules, what suggests that the climatic and environmental differences (e.g., the layout of the systems) between the two sites are related to this type of degradation. The PR of the system had a reduction of only 2% between 2017 and 2021. One of these differences that explain the reduced degradation is the size of the strings of each site that can be seen in Tab. 1, since PID occurs in modules under high voltage, like is the case of the Portugal system [6].

4 CONCLUSIONS

This paper compared the performance of two two different PV systems with the same multicrystalline silicon PV module in two different climatic and installation conditions to determine the impact of these factors on the PV systems' performance.

The analysis showed that, overall, the Brazilian system presented higher performance rations (PR) than the system installed in Portugal, even with the high impact of shadows (around 20% more than Portugal's PR). One of the reasons for the difference of performance was curtailments and the PID that affected Portugal's system. The data analysis provided some evidences for the occurrence of this degradation type and demonstrated its effects on the plant performance.

Our results show that the degradation behavior of a PV installation is determined by several factors, including environmental conditions and especially the layout of the installation. Understanding the correlations between these factors and PV performance can improve PV systems design to avoid the major factors that contribute to the development of specific degradation types and increase long-term reliability.

5 REFERENCES

[1] T. Karin, C. B. Jones, and A. Jain, "Photovoltaic Degradation Climate Zones," in 2019 IEEE 46th Photovoltaic Specialists Conference (PVSC), Jun. 2019, no. 1, pp. 0687–0694. doi: 10.1109/PVSC40753.2019.8980831.

[2] E. L. Meyer and E. E. Van Dyk, "Assessing the reliability and degradation of photovoltaic module performance parameters," *IEEE Trans Reliab*, vol. 53, no. 1, pp. 83–92, 2004, doi: 10.1109/TR.2004.824831.

[3] D. C. Jordan, T. J. Silverman, J. H. Wohlgemuth, S. R. Kurtz, and K. T. Vansant, "Photovoltaic failure and degradation modes," *PROGRESS IN PHOTOVOLTAICS: RESEARCH AND APPLICATIONS*, no. January, pp. 318–326, 2017, doi: 10.1002/pip.

[4] A. Ndiaye, A. Charki, A. Kobi, C. M. F. Kébé, P. A. Ndiaye, and V. Sambou, "Degradations of silicon photovoltaic modules: A literature review," *Solar Energy*, vol. 96, pp. 140–151, 2013, doi: 10.1016/j.solener.2013.07.005.

[5] D. C. Jordan, C. Deline, M. Deceglie, T. J. Silverman, and W. Luo, "PV Degradation – Mounting & Temperature," in *2019 IEEE 46th Photovoltaic Specialists Conference (PVSC)*, Jun. 2019, no. c, pp. 0673–0679. doi: 10.1109/PVSC40753.2019.8980767.

[6] J. Hattendorf *et al.*, "Potential Induced Degradation in Mono-Crystalline Sillicon Based Modules: an Acceleration Model," in *27th European Photovoltaic Solar Energy Conference and Exhibition*, 2012, pp. 3405–3410.

[7] M. Köntges *et al.*, *Review of Failures of Photovoltaic Modules*. 2014. doi: 978-3-906042-16-9.

[8] C. Buerhop *et al.*, "Analysis of inhomogeneous local distribution of Potential induced degradation at a rooftop photovoltaic installation," *IET Renewable Power Generation*, vol. 11, no. 10, pp. 1253–1260, 2017, doi: 10.1049/iet-rpg.2017.0105.

[9] C. M. Peel, B. L. Finlayson, and T. A. McMahon, "Updated world map of the Köppen-Geiger climate classification," *Hydrol Earth Syst Sci*, vol. 11, pp. 1633–1644, 2007, doi: https://doi.org/10.5194/hess-11-1633-2007.

[10] A. M. Pires, M. Braga, and R. Rüther, "Performance assessment of bare and anti-reflective coated CdTe photovoltaic systems in comparison to multicrystalline Si in Brazil," *Progress in Photovoltaics: Research and Applications*, vol. 29, no. 10, pp. 1105–1124, 2021, doi: 10.1002/pip.3446.

EXPERIMENTAL OBSERVATIONS ABOUT THE CLOUD ENHANCEMENT PHENOMENON ON PV STRINGS

Kari Lappalainen and Seppo Valkealahti
Tampere University, Electrical Engineering
P.O. Box 692, FI-33101 Tampere, Finland
E-mail: kari.lappalainen@tuni.fi, seppo.valkealahti@tuni.fi

ABSTRACT: Due to the cloud enhancement (CE) phenomenon, irradiance on partly cloudy days can considerably exceed the expected clear sky irradiance. Although CE is a well-known phenomenon, its effects on the operation of photovoltaic (PV) power systems have not been studied comprehensively based on electrical measurements. In this paper, the characteristics of CE events are analysed based on irradiance measurements and by examining the operation of three PV strings based on current–voltage curves measured under CE conditions. The highest measured irradiances were about 1450 W/m^2 while the theoretical clear sky maximum irradiance is about 900 W/m^2. It means that the highest measured irradiance was about 1.7 times the expected clear sky irradiance. Correspondingly, the highest measured powers exceeded the highest theoretical clear sky powers. Although the measured maximum power point (MPP) powers exceeded the nominal standard test conditions MPP power, the MPP voltages were typically lower than the nominal MPP voltage. The reason for this is that during the CE events, temperature was typically much higher than in standard test conditions. The experimental results of this study show that CE events do not cause problematic increase of PV system operating voltages and confirm the findings of earlier simulation studies.
Keywords: Cloud Enhancement, Overpower, Plant Control, PV Array, Voltage Fluctuation

1 INTRODUCTION

Power production of photovoltaic (PV) generators is dependent on the irradiance incident on the PV cells of the generator. Solar radiation is highly variable due to shadings caused by moving clouds [1]. The nominal electrical values of PV modules are usually defined under the standard test conditions (STC), which are not common in practise. In many areas of the world, the global irradiance from clear sky around solar noon can be higher than the STC irradiance of 1000 W/m^2 [2]. Furthermore, on partly cloudy days, irradiance incident on the PV modules can considerably exceed theoretical clear sky irradiance as a consequence of a phenomenon called cloud enhancement (CE) [3] or overirradiance [4]. Clear sky irradiances exceeding the STC irradiance are typically not an issue, as the exceedance is less than 10% in most parts of the world and, correspondingly, PV power losses due to high module temperatures are typically greater than 10% compared to operation at the STC temperature of 25 °C. However, under CE conditions, irradiance can be much higher and PV cell temperatures can be lower leading to higher PV powers. For example, in [5], up to 30% higher power output of PV modules compared to the STC were reported.

Although CE is a well-known phenomenon [6], its effects on the operation of PV systems have been studied only in few articles: the operation of individual PV modules under CE conditions was studied in [5]; the operation of a PV plant under a single CE situation was studied in [7]; and inverter sizing and energy losses due to inverter saturation under CE conditions were studied in [8]. In [3], [9], the effects of CE on the operation of PV generators were extensively studied using operating condition measurements as inputs for simulations. However, comprehensive studies of the operation of PV generators under CE conditions based on actual electrical measurements have not been presented earlier.

In this paper, experimental observations about the CE phenomenon on PV strings are presented. Characteristics of CE events are analysed based on irradiance measurements and the operation of three PV module strings under CE conditions is studied based on measured current–voltage (*I–U*) curves of the strings. The global maximum power point (GMPP) characteristics of the strings are studied comprehensively. The GMPP characteristics of PV strings have previously been studied experimentally in [10], [11]. However, those studies were not focused on CE conditions, but instead analysed measurements of whole days. The aim of this experimental study is to investigate the effects of CE phenomenon on the operation of PV strings based on *I–U* curve measurements. The study is novel, because no such equally comprehensive studies based on actual electrical measurements have been presented earlier.

2 EXPERIMENTAL DATA AND METHODS

The experimental data consists of irradiance measurements of six summer months (June–August 2021 and 2022) and of 70 days of full-day *I–U* curve measurements from the PV research plant of Tampere University [12]. The layout of the studied PV strings, consisting of 6 (String 4), 17 (String 1) and 23 (String 1&4) series-connected PV modules, is presented in Fig. 1 and their details are compiled in Table I. The analysed measurements of String 1 (13 days) were performed on 14–19 June and 25–31 July, 2022. The measurements of String 4 (22 days) were performed on 1–2 June and 12–17 August, 2021 and on 11–17 July and 1–7 August, 2022. The measurements of String 1&4 (35 days) were performed on 3–9 June and 5–11 August, 2021 and on 27–30 June, 1–10 July and 8–14 August, 2022. Daily measurement period was from 8:00 to 18:00 (UTC+2). An *I–U* curve tracer, where IGBTs act as a variable load, was used to measure an *I–U* curve once a second. Hence, in total, 2520000 measured *I–U* curves were analysed. The irradiance incident on nine PV modules of the studied strings was measured by photodiode-based SP Lite2 pyranometers mounted at the same 45° tilt angle as the PV modules. Back-sheet temperatures of those modules were measured by Pt100 temperature sensors. Moreover, global horizontal

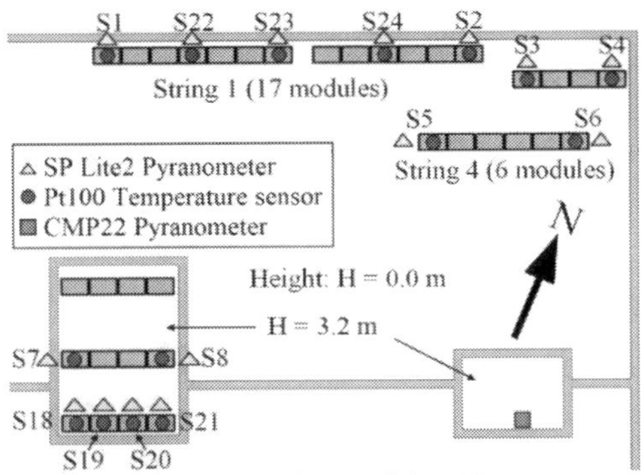

Figure 1: Partial layout scheme of the PV power research plant of Tampere University.

Table I: Details of the studied PV strings.

	String 1	String 4	String 1&4
Number of PV modules	17	6	23
$U_{MPP,STC}$ (V)	439	155	593
$P_{MPP,STC}$ (W)	3230	1140	4370

irradiance (GHI) was measured by a CMP22 pyranometer. All irradiance and temperature measurements were performed with a sampling frequency of 10 Hz. Locations of the irradiance and temperature sensors are presented in Fig. 1. Average irradiances and temperatures of the strings were calculated as the averages of all the sensors of the string.

Each measured *I–U* curve involves 4000 measured *I–U* value pairs. The following procedure was used to pre-process the *I–U* curves. First, the measurement points having identical voltage value were replaced with a single point by averaging their current values. After that, clearly abnormal measurement points were removed. A measurement point was removed if its power differed from the previous and following measurement point by more than 1.3 times the mean change of power between adjacent measurement points in its vicinity (previous and next 9 points). Finally, the measured current and voltage values were smoothed separately using smooth.m function in MATLAB.

A CE event is typically defined by comparing measured irradiance with expected clear sky irradiance [13, 14] or with a static irradiance limit. In that way, a CE event is defined to start when the measured irradiance exceeds the threshold and to end when the irradiance decreases below it. From the PV system operation point of view, the most reasonable choice for the static threshold is the STC irradiance. Another static threshold used is the solar constant [13, 15]. In this study, a continuous range of irradiance values starting from the STC irradiance of 1000 W/m^2 was used similarly than in [3, 16] and all events where the measured irradiance exceeded the irradiance limit were identified. That makes the obtained results more generalizable than use of a single static irradiance threshold. CE events were identified in measured powers by a similar approach using a continuous range of power values starting from the nominal STC powers (see Table I) of the strings. An overpower event starts when the measured power exceeds the threshold and ends when the power decreases below it.

3 EXPERIMENTAL RESULTS AND DISCUSSION

3.1 CE events identified in irradiance measurements

Fig. 2 presents the shares of the studied days as a function of an irradiance limit, where the irradiance caused by the CE phenomenon for one PV module, GHI, and the strings exceeded that irradiance limit. CE events measured by sensor S24 exceeded 1000 W/m^2 in around 80% of the days and 1200 W/m^2 in around 40% of the days. CE events were more infrequent for GHI and the PV strings due to a suboptimal tilt angle and a larger land area, respectively. CE events were more infrequent for String 4 than for Strings 1 and 1&4, although String 4 is the shortest studied string, since String 4 is located closer to a nearby built structure (see Fig. 1) and thus is more frequently shaded than Strings 1 and 1&4.

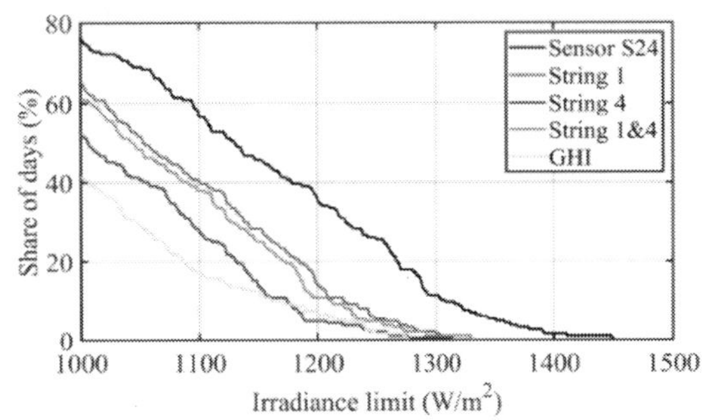

Figure 2: Shares of days with CE irradiance higher than the irradiance limit.

The numbers of identified CE events for the PV module, GHI, and the strings are presented in Fig. 3 as a function of the irradiance limit. There were about 20000 CE events for sensor S24 which exceeded 1000 W/m^2. For the PV strings the corresponding number was from 3300 to 5200. The highest irradiance measured by sensor S24 was 1449 W/m^2. It is worth noting that the highest theoretical clear sky irradiance in the Tampere region is just over 900 W/m^2. Thus, the highest measured irradiance was about 1.7 times the expected clear sky irradiance. The highest measured irradiances for the strings were over 1300 W/m^2. The highest measured irradiances are compiled in Table II.

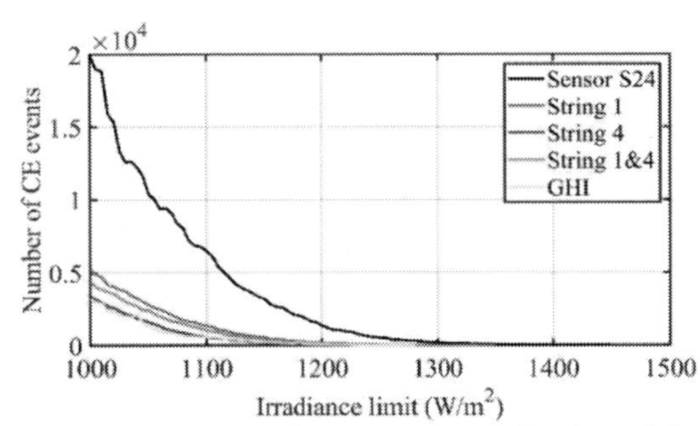

Figure 3: Numbers of CE events as a function of the irradiance limit.

Fig. 4 presents the cumulative and maximum durations of the CE events as a function of the irradiance limit. The durations decreased sharply with the increasing irradiance limit. The total durations of the CE events exceeding 1000 W/m^2 was 71 hours, i.e., 1.6% of the time,

Table II: The highest measured irradiances.

Sensor/String	Highest irradiance (W/m^2)	Day
S23	1399	July 29, 2022
S24	1449	June 4, 2022
String 1	1348	July 29, 2022
String 4	1315	June 4, 2022
String 1&4	1332	July 29, 2022
GHI	1276	July 15, 2022

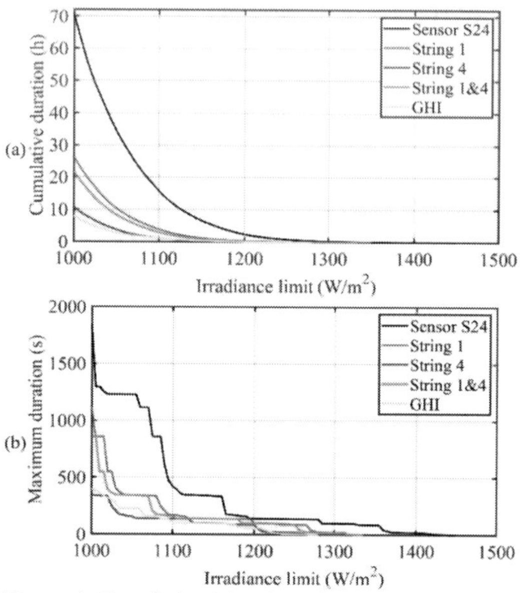

Figure 4: Cumulative (a) and maximum (b) durations of the identified CE events exceeding the irradiance limit.

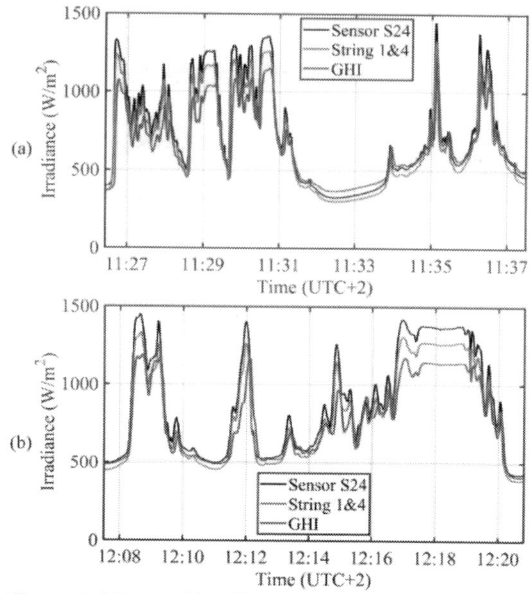

Figure 5: Measured irradiances on June 4 (a) and July 29 (b), 2022.

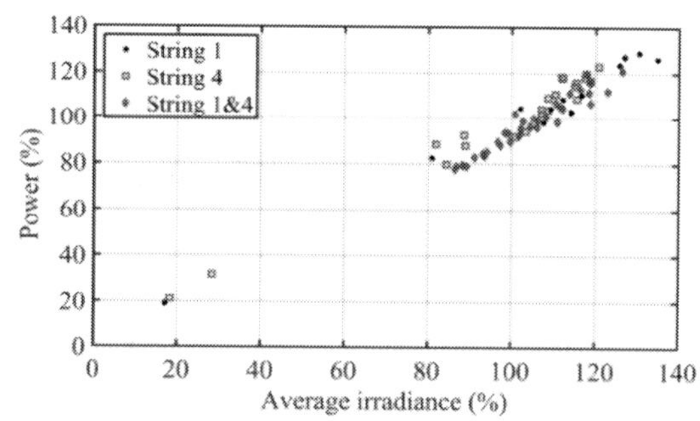

Figure 6: Scatter plot between the daily highest measured power and average irradiance for the studied PV strings.

for sensor S24. For the strings, the total time when the irradiance was over 1000 W/m^2 was from 11 (String 4) to 26 (String 1) hours. The maximum durations of the CE events exceeding 1000 W/m^2 for sensor S24 and String 1&4 were around 31 and 14 minutes, respectively. The average durations of these CE events were 12.8 s for sensor S24 and 17.8 s for String 1&4.

Fig. 5 presents two examples of measured CE events. In both cases, the peak irradiances of sensor S24 and String 1&4 were almost 1450 W/m^2 and 1350 W/m^2, respectively. During the shading period around 11:33 in Fig. 5 (a) the area of the PV string was shaded and the measured GHI was higher than the measured irradiances on PV modules, since the GHI sensor was exposed to a larger share of diffuse radiation. When the shading moved away, the measured on-plane irradiances exceeded the measured GHI.

3.2 CE events identified in *I–U* curve measurements

Fig. 6 presents a scatter plot between the daily highest measured power and average irradiance for the studied PV strings. The highest measured powers were 129%, 123% and 121%, with respect to the nominal STC power, for Strings 1, 4 and 1&4, respectively. The highest measured power and irradiance values are well in accord with each other. The main reason for the differences is that the operating temperature of the PV modules varies daily and is typically higher than the STC temperature of 25 °C. Increase of temperature decreases the maximum power of PV modules explaining why on most days the maximum measured irradiance was slightly higher than the maximum measured PV power.

The maximum durations of the overpower events are presented in Fig. 7 as a function of the power limit. The longest durations of the events exceeding the nominal STC power are 161, 145 and 234 seconds for Strings 1, 4 and 1&4, respectively. For String 1, overpower events exceeding 1.1 times the nominal STC power and lasting over two minutes were measured. The maximum durations in Fig. 7 are a bit shorter than was reported for a 32 kW PV array based on simulations in [9] being consistent with each other.

Fig. 8 presents the relative cumulative frequencies of the measured GMPP powers and voltages for the studied PV strings during the identified overpower events, i.e., when the GMPP power exceeded the nominal MPP power. The shapes of the distributions are roughly similar for Strings 4 and 1&4 while somewhat higher powers and voltages exist for String 1. The GMPP voltages of the strings were most of the time from 80% to 100% of the nominal STC value. The GMPP voltages of the same strings were studied in [11] during whole days (from 8:00 to 18:00) and over 100% GMPP voltages were found to be much more common than in this study. The rarity of higher

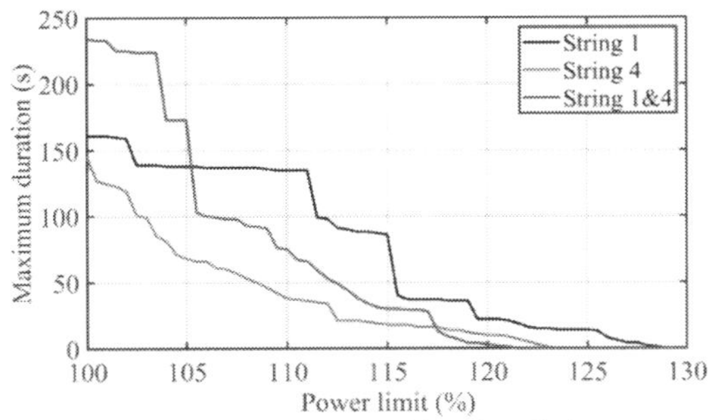

Figure 7: Maximum durations of the identified overpower events exceeding the power limit.

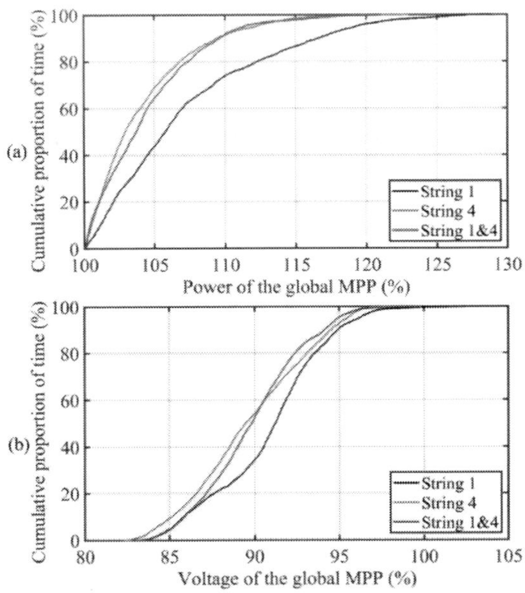

Figure 8: Relative cumulative frequencies of the GMPP power (a) and voltage (b) for the studied PV strings during the identified overpower events. The powers and voltages are with respect to the nominal MPP values.

voltage values results from the typical operating conditions of the studied PV strings under CE events: temperature was typically higher than in STC, and thus the GMPP voltages were lower than the nominal MPP voltage. The average back-sheet temperatures of the strings during the overpower events were around 30 °C. In warmer locations, even higher operating temperatures are expected. Thus, the results show that increase of PV generator operating voltages due to CE is not a problem.

Figs. 9 and 10 illustrate the behaviour of the GMPP power and voltage during CE events. The measured GMPP powers follow quite accurately the measured average irradiances of the PV strings even for the highest measured values. The measured GMPP voltage is quite steady during the CE events. Variations in the voltage exist mainly during sharp changes in the average irradiances. In these situations, the strings are partially shaded, which might lead to existence of multiple MPPs and changes in the GMPP voltage [17]. The power-voltage curves in Fig. 9 (b) measured during the CE event show that the PV string operates under nearly homogeneous irradiance conditions. The bottom curve in Fig. 9 (b) is measured during a sharp change in the average irradiance of String 1 and several local MPPs (LMPP) exist in the power–voltage curve.

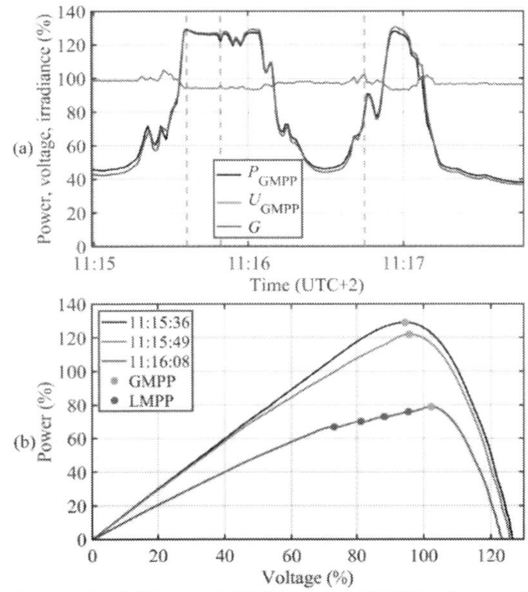

Figure 9: (a) Measured GMPP power, GMPP voltage and average irradiance of String 1 on June 16, 2022. (b) Three power–voltage curves of String 1 measured during the period presented in (a). Vertical dashed lines in (a) indicate the moments when the curves in (b) were measured.

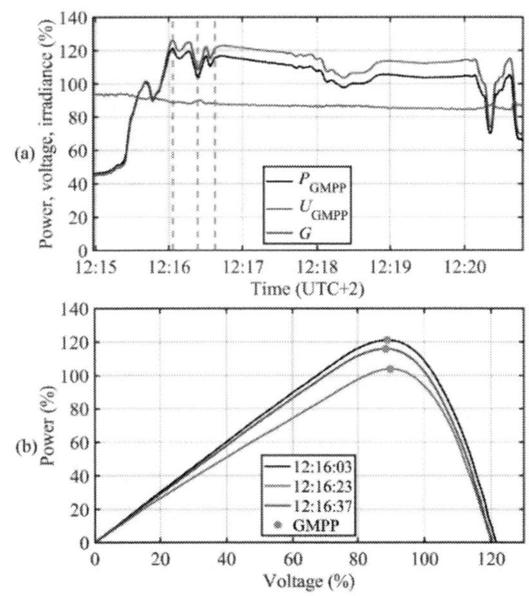

Figure 10: (a) Measured GMPP power, GMPP voltage and average irradiance of String 1&4 on July 7, 2022. (b) Three power–voltage curves of String 1&4 measured during the period presented in (a). Vertical dashed lines in (a) indicate the moments when the curves in (b) were measured.

Fig. 10 shows an example of a strong and long overpower event for String 1&4. The measured power stays one and a half minutes above 110%, with respect to the nominal power, and mainly above 100% for over four minutes. The measured power-voltage curves demonstrate that the PV strings operate under nearly homogeneous irradiance conditions during the CE event.

4 CONCLUSIONS

This paper presented an experimental study of the cloud enhancement phenomenon on PV strings. The characteristics of CE events were analysed based on irradiance measurements and the operation of three PV strings was studied under CE conditions based on measured *I–U* curves of the strings. In addition to the characteristics of overpower events, the GMPP characteristics of the strings were studied.

The highest measured irradiance was about 1.7 times the expected clear sky irradiance. Correspondingly, it was shown that PV string power can considerably exceed the highest theoretical clear sky power. Although the measured GMPP powers exceeded the nominal MPP power in STC, the GMPP voltages were typically lower than the nominal MPP voltage due to high operating temperatures. Thus, increase of PV generator operating voltages due to CE events is not a problem. In warmer locations, operating temperatures are expected to be even higher. The experimental results presented in this paper confirmed the findings of earlier simulation studies.

ACKNOWLEDGMENTS

We acknowledge the financial support from Business Finland for the research reported in this paper.

REFERENCES

[1] K. Lappalainen, S. Valkealahti, Solar Energy 138 (2016) 47. http://dx.doi.org/10.1016/j.solener.2016.09.008.
[2] V. Badescu, Solar Energy 61 (1997) 251. https://doi.org/10.1016/S0038-092X(97)00057-1.
[3] K. Lappalainen, J. Kleissl, Journal of Renewable and Sustainable Energy 12 (2020) 043502. https://doi.org/10.1063/5.0007550.
[4] G.H. Yordanov, O.-M. Midtgård, T.O. Saetre, H.K. Nielsen, L.E. Norum, IEEE Journal of Photovoltaics 3 (2013) 271. https://doi.org/10.1109/JPHOTOV.2012.2213581.
[5] M. Zehner, T. Weigl, M. Hartmann, S. Thaler, O. Schrank, M. Czakalla, B. Mayer, T.R. Betts, R. Gottschalg, K. Behrens, G. König-Langlo, B. Giesler, G. Becker, O. Mayer, Proceedings 26th European Photovoltaic Solar Energy Conference, Vol. I (2011) 3935. https://doi.org/10.4229/26thEUPVSEC2011-5AO.6.3.
[6] C.A. Gueymard, Solar Energy 153 (2017) 755. http://dx.doi.org/10.1016/j.solener.2017.05.004.
[7] T. Weigl, L. Nagl, J. Weizenbeck, M. Zehner, M. Augel, P. Öchsner, B. Giesler, G. Becker, O. Mayer, T.R. Betts, R. Gottschalg, Proceedings 27th European Photovoltaic Solar Energy Conference, Vol. I (2012), 3801. https://doi.org/10.4229/27thEUPVSEC2012-5CO.7.6.
[8] J. Luoma, J. Kleissl, K. Murray, Solar Energy 86 (2012) 421. https://doi.org/10.1016/j.solener.2011.10.012.
[9] M. Järvelä, S. Valkealahti, Energies 13 (2020) 2185. https://doi.org/10.3390/en13092185.
[10] K. Lappalainen, S. Valkealahti, Proceedings 38th European Photovoltaic Solar Energy Conference, Vol. I (2021), 1264. https://doi.org/10.4229/EUPVSEC20212021-5CV.2.39.
[11] K. Lappalainen, S. Valkealahti, EPJ Photovoltaics 13 (2022) 4. https://doi.org/10.1051/epjpv/2022001.
[12] D. Torres Lobera, A. Mäki, J. Huusari, K. Lappalainen, T. Suntio, S. Valkealahti, International Journal of Photoenergy 2013 (2013) 837310. https://doi.org/10.1155/2013/837310.
[13] J. Zhang, K. Watanabe, J. Yoshino, T. Kobayashi, Y. Hishikawa, T. Doi, Japanese Journal of Applied Physics 57 (2018) 08RG11. https://doi.org/10.7567/JJAP.57.08RG11.
[14] G.H. Yordanov, T.O. Saetre, O.-M. Midtgård, Solar Energy 115 (2015) 68. https://doi.org/10.1016/j.solener.2015.02.020.
[15] G.L. Martins, S.L. Mantelli, R. Rüther, Solar Energy 231 (2022) 47. https://doi.org/10.1016/j.solener.2021.11.050.
[16] M. Järvelä, K. Lappalainen, S. Valkealahti, Solar Energy 196 (2020) 137. https://doi.org/10.1016/j.solener.2019.11.090.
[17] K. Lappalainen, S. Valkealahti, Applied Energy 301 (2021) 117436. https://doi.org/10.1016/j.apenergy.2021.117436.

ONE YEAR PERFORMANCE ASSESSMENT OF SLICON HETERO-JUNCTION SOLAR MODULES ON HORIZONTAL SINGLE AXIS TRACKER

Maugeri Giosué*[1], Asier Sanz Martinez[2], Eneko Setien[2], Lorenzo Todaro[3], Alexandre Mignonac[4], Giuseppe Demofonti[5], Salvatore Guastella[1], Alessandro Virtuani[6], Stéphane Guillerez[4]

[1]RSE SpA - RSE SpA, via R. Rubattino 54, 20134 Milan, Italy
[2]TECNALIA, Basque Research and Technology Alliance (BRTA), Derio, Spain; [3]Enel Green Power, Catania, Italy;
[4]CEA Cadarache, France;
[5]Convert Italia Spa, Roma, Italy, [6]EPFL, Neuchâtel, Switzerland

ABSTRACT: The deployment of single-axis tracker bifacial PV systems is attracting the interest of several PV operators particularly targeting the utility-scale market segment. The main driver is the lowest Levelized cost of electricity (LCOE) achievable by this solution due to the increased energy yield. This paper presents the results of one full year of monitoring data from a pilot PV system realized in the framework of the H2020 European project, GOPV. The PV pilot plant is installed in Cadarache (France) and consists of bifacial half-cell silicon heterojunction (HJT) PV modules combined with Horizontal Single Axis Trackers (HSAT), east-west sun tracking.
This work compares the monthly gain of the energy yield of the bifacial HJT technology, installed both on a tilted fixed structure and on a HSAT tracker, with a PERC monofacial technology installed on tilted fixed structure, which serves as a reference system.
The results highlight that, compared to the benchmark technology, the energy yield of the bifacial HJT on a fixed structure can achieve an average of 15% gain on annual basis, and the HJT bifacial on HSAT can exceed 35% during the summer months. This study also includes insights from the experience gathered in one year of activity on the influence of wind parameters on tracking yield.
Keywords: PV system, HJT, bifacial gain, single-axis tracker, PV monitoring

1 INTRODUCTION

In recent decades, photovoltaic (PV) technology has proved to be one of the most reliable renewable energy sources contributing to the widespread use of crystalline silicon (c-Si) photovoltaic modules in various installations. Thanks to the technological improvements, the efficiency of the solar cells has progressively increased, and new efficient technologies have rapidly taken the place of the previous ones; let's think for example how the PERC technology (passivated emitter and rear-contact) has progressively replaced the traditional AlBSF (aluminium back surface field) cell, or other improvements such as the introduction of half-cells or the adoption of larger cell sizes.
In more recent times, the bifacial technology - capable of capturing solar radiation from both surfaces (upper and lower) - is rapidly eroding the market share of monofacial technologies and it is expected that within the next 10 years it will cover a share of the global market by more than 50%, while about 20% of the single-sided module market will use bifacial cells. [1]
The main advantage of this technology consists in the greater energy obtainable for the same surface area, quantifiable with the "Bifacial Gain" index, corresponding to the additional fraction of the total energy that a bifacial photovoltaic system can provide compared to a monofacial system of the same orientation and size.
The Bifacial Gain can be further expanded if other *factors* affecting the energy performance of the bifacial modules are considered in the design phase [1], such as the albedo, the type of mounting structure (fixed or single-axis tracker), the elevation of the modules, the Ground Coverage Ratio –(GCR), the space between the PV modules on the structure and, finally, the climatic zone of the installation site.
The energy gain also depends on the bifaciality of the PV module itself, i.e., the ratio between the nominal power of the rear side of the module and that of the front side.
Typically, the bifaciality ratio is in the range of 70% to 85% for most bifacial modules currently on the market; with silicon heterojunction technology - HJT - it is possible to have up to 85% of bifacial ratio and a potential of 100% that can be reached with careful optimization. [1]
The bifacial technology, coupled with single-axis tracking structures, can increase the energy yield of PV systems, compared to a monofacial installation on fixed tilted structures. A recent study has in fact indicated the bifacial single-axis trackers is the solution with the lowest LCOE (Levelized cost of electricity) for the vast majority of potential PV sites on the planet [2]; this result, combined with the progressive decline in the prices of single-axis trackers, makes this solution suitable for the utility- scale PV systems.
Although the bifacial technology combined with tracker is already largely used in utility scale PV installations, there is scientific interest in the study of models able to estimate, with high precision, the producibility of the PV installation, taking into account the *factors* that mainly affect the energy yield of this solution. Moreover, the PV operator are also interested in finding solutions to increase the producibility of bifacial PV system, optimize the O&M measures, reducing at the same time the LCOE costs.
This paper aims to illustrate the results of one full year of performance assessment of Heterojunction (HJT) bifacial technology combined with Horizontals Solar Axis Tracker (HSAT), with respect to a standard monofacial module (PERC) mounted on a fixed tilted structure that is considered as reference in this analysis. The pilot PV system has been realized in the framework of a H2020European project, GOPV (Global Optimization of integrated PhotoVoltaic system for low electricity cost).

2 METHODS

2.1 Cadarache Solar PV system description

The PV solar system is located in Cadarache (France,

43°N) and consists of a fixed tilted structure and three horizontal single-axis trackers (HSAT).
The fixed tilted structure is south oriented while the three trackers are aligned along the north-south axis.
Both types of structure host respectively two PV technologies: the monofacial PERC and the half-cell silicon heterojunction (Bfc-HJT), the latter made within the GOPV project.
The **Table 1** summarizes the nominal power on the DC side, for each PV plant.

Table 1 PV system installed in the Cadarache pilot and the respective Nominal Power of the DC string

	Fixed		HSAT	
N. Modules	8		8	
N. Strings	1		1	
Layout	P		2P	
Technology	HJT Bfc	PERC Mfc	HJT Bfc	PERC Mfc
P_{STC}*[kW]*	2,95	2,56	2,97	2,56
Tilt	30°	30°	+/-55°	+/-55°
Elevation [m]	1	1	2,3 (*)	2,3 (*)
Albedo [avg]	0,25			
Soil Material	Gray gravel			

P – Portrait; 2P – two module in Portrait;
Bfc – Bifacial; Mfc – Monofacial
(*) average height of the tracker

Figure 1 shows the modules layout of fixed tilted structure and of horizontal single-axis trackers.

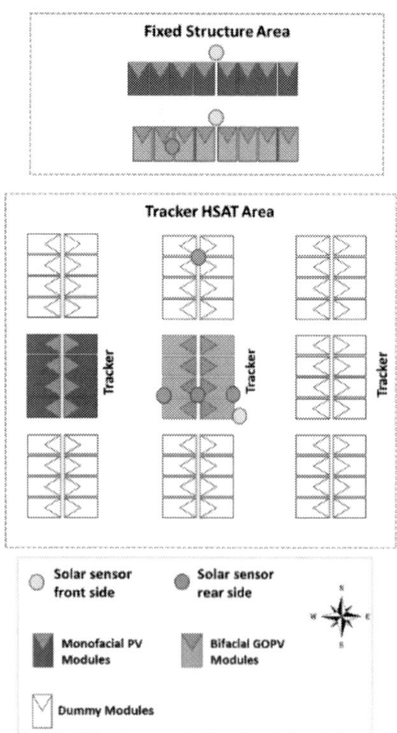

Figure 1 Modules layout of fixed tilted structure and horizontal single-axis trackers

The fixed test bench is equipped with one front side and one rear side solar sensors, as well as a sensor temperature respectively installed for each PV technology.
The tracker test bench is equipped with two front side and four rear side solar sensors, as well as a temperature respectively installed for each PV technology.

2.2 Energy Yield comparison

A comparative analysis is performed between bifacial HJT technology and monofacial PERC installed on the same fixed structure. Then a second comparison is performed between bifacial HJT on a tracker and the monofacial PERC on fixed structure (reference technology).
PV data have been collected from March 2020 to March 2022 for PV system on Fixed structure and from November 2020 to March 2022 for the PV system on tracker. All data are aggregated on a monthly basis.
In 2021, problems with the data collection occurred on all sets installed on trackers (from 10/06/2021 to 15/06/2021). For a reliable comparison analysis of the different PV systems installed, overall data (all data collected in the test period) and filtered data (removing invalid periods) are used for data analysis reported in this section. No main issues occurred in data collection of PV systems installed on Fixed.
All the evaluations have been performed considering the same time period of analysis and synchronizing all datasets: the corresponding records with invalid data were deleted from all datasets in order to avoid the underestimation or overestimation of the columns of variables unaffected by invalid values and errors.
Since, the bifacial and monofacial PV strings converge into a single inverter, the AC-side energy generation for each technology is not available. For this reason, all data and results reported in this section relate to DC data for all systems.

2.3 Performance Metrics

Two performance indicators are used for this analysis: the *Energy Yield* and the *Bifacial Gain*.
The Energy Yield represents the net DC energy output (E_{DC}, [kWh])from the PV string during a period (a month), normalized to the DC rated power in STC (P_{STC}, [kWp]) and it is calculated in accordance with IEC 61724-1 [3] as:

$$Y_{a,month} = \frac{E_{DC}}{P_{STC}}$$

The Bifacial Gain is defined as the additional fraction of the total energy that a bifacial PV system will produce compared to a monofacial system with the same orientation and inclination [1]it is calculated as:

$$Bg = \frac{(Y_b - Y_m)}{Y_m}$$

where Y_b and Y_m are respectively the Energy yield of bifacial and monofacial systems

3 RESULTS

3.1 Backside irradiance assessment

Figure 2 and **Figure 3** show the monthly irradiance respectively for the tilted fixed structures and the tracker structures. In both graphs, the orange bar represents the rear side irradiance contribution.
The months of February and April 2021 are particularly characterized by cloudy skies and rains, while on July 21, a failure lasting two days occurred in the front side solar

sensor of the fixed structure; this explains the lower radiation compared to previous months. (**Figure 2**)
In general, for both structures, it can be observed that the amount of rear-side radiation is less during the winter period due to fewer sunny days with clear skies, lower radiation, and lower sun elevation (winter solstice) that generate longer shadows that cover the ground. During the summer months, the rear-side radiation tends to increase.

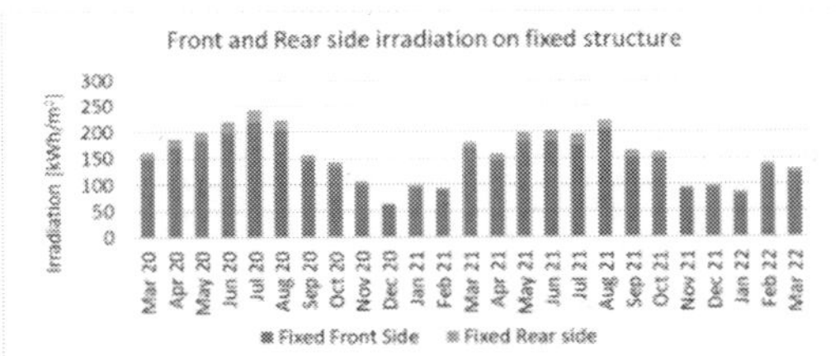

Figure 2 Front and Rear side irradiation on fixed structure

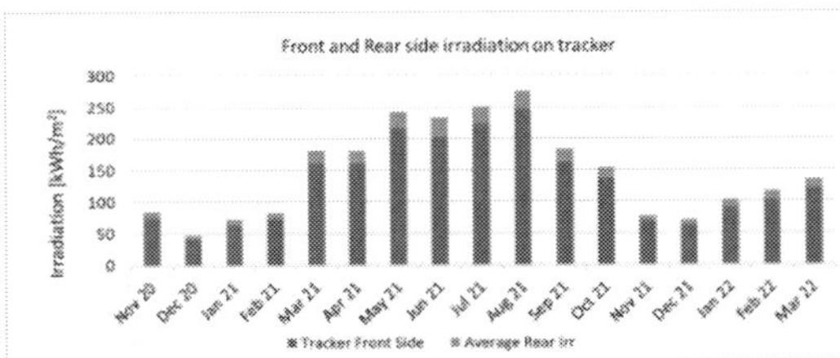

Figure 3 Front and Rear side irradiation on tracker.

Figure 4, shows the rear-side irradiation measured in the tracker where four solar sensors are installed. It can be noted that during the summer months the rear-side inhomogeneity becomes more evident between the middle and edge of the structure. The dotted red line represents the average value of the four solar sensors.

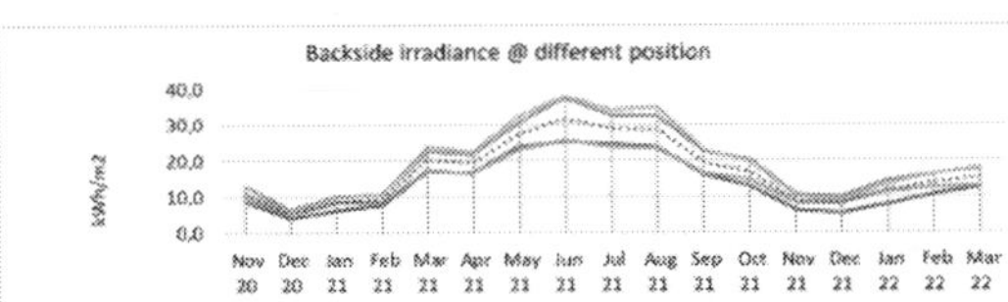

	Nov 20	Dec 20	Jan 21	Feb 21	Mar 21	Apr 21	May 21	Jun 21	Jul 21	Aug 21	Sep 21	Oct 21	Nov 21	Dec 21	Jan 22	Feb 22	Mar 22
Middle Rear Cell 3	8,8	4,1	6,3	7,5	16,9	16,5	23,9	25,2	24,6	23,6	16,0	12,8	6,3	5,4	7,8	10,5	12,5
West Rear Cell 4	13,1	6,4	10,2	10,8	22,5	21,6	30,4	37,2	32,3	32,2	22,5	20,3	10,5	10,0	14,4	16,2	17,9
Middle Rear Cell 5	9,4	5,3	8,3	8,4	17,4	16,3	23,1	25,6	23,9	23,3	16,3	14,6	8,2	8,1	11,3	12,5	13,0
East Real Cell 6	11,2	5,9	9,7	10,8	23,9	22,5	32,2	38,0	34,0	34,7	22,8	19,7	9,9	9,0	13,3	16,1	17,3
Average Rear Irr	10,6	5,4	8,6	9,4	20,2	19,3	27,4	31,5	28,7	28,4	19,4	16,9	8,7	8,1	11,7	13,8	15,2

Figure 4 Rear radiation distribution measured with four solar sensors installed on the tracker.

For the period April 21 – March 22, one full year for both installations, the cumulative rear and front irradiance captured from both structures has been acquired. For this analysis, only the measurements of the solar sensor positioned in the middle of the rear side have been used. According to the obtained results, it can be concluded that the backside contribution is higher on the tracker than on the fixed bench during the period. In fact, the rear-side irradiance contribution is 10.4% for the tracker and 8.4% for the fixed structure (see **Figure 5**). Additionally, for the same period the effective irradiation (rear + front) on the tracking bench is 6,72% higher compared to fixed one.

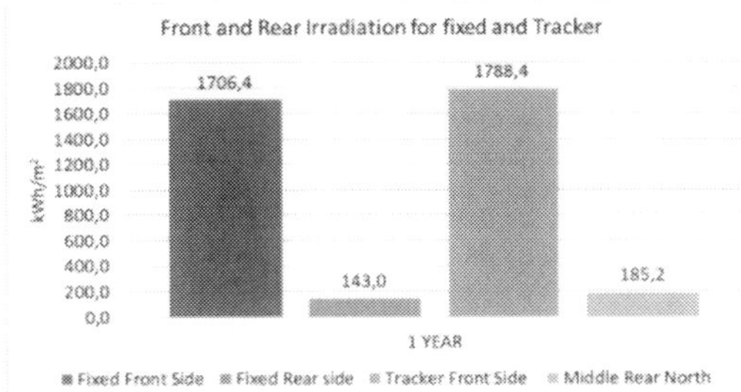

Figure 5 The cumulative rear and front irradiance for fixed and tracking structures.

3.2 Energy Yield assessment

3.2.1 Fixed structure: Bifacial vs. Monofacial

The monthly Energy Yield for fixed tilted structure is shown in **Figure 6.**

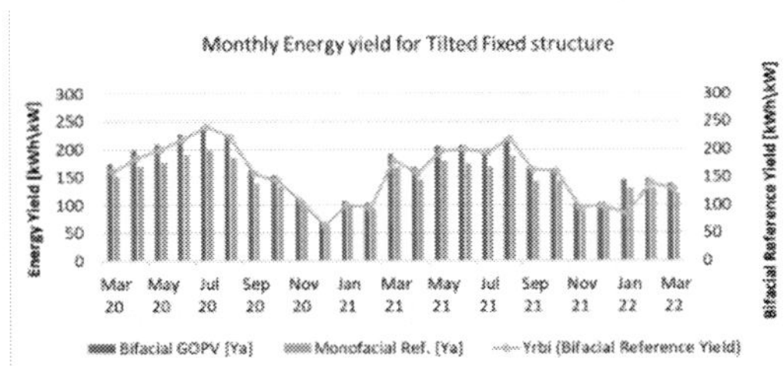

Figure 6 Bifacial HJT and monofacial PERC Energy Yield for a fixed tilted structure. The yellow solid line represents the Bifacial Reference Yield

The bifacial technology shows a better performance with respect to the monofacial reference for the whole testing period. The yellow curve of the graph represents the Bifacial Reference Yield [Y_{rbi}], calculated according to the recent standard IEC 61724-1:2021 with the irradiation measured in the plane of Array (PoA) of the fixed PV plants. The overall trend of the Bifacial Energy Yield follows the Bifacial Reference Yield.
For the fixed systems, the Bifacial Gain of GOPV bifacial with respect to the Monofacial reference technology for the period March 20 to March 22, is on average +15,1%, with the highest values obtained in summer months, toping at +21,2% in July 2020. See figure **Figure 7**

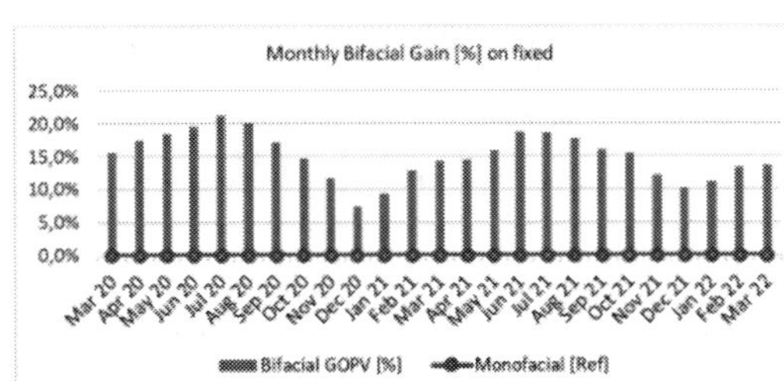

Figure 7 Monthly bifacial gain for the Bfc-HJT technology compared to the reference PERC-monofacial modules on fixed structures.

3.2.2 Tracker structure: Bifacial vs. Fixed Monofacial

Figure 8 shows the monthly energy yield gain evolution of the bifacial HJT on tracker, compared to the monofacial

PERC on tilted fixed structure, for the period November 2020 – March 2022. In general, from November to February – i.e. in the winter period - the gain is negative as the angular declination of the sun is unfavorable with respect to the horizontal plane in which the modules are located on the tracker system, while the solution on a fixed structure, with a 30 °-tilt, is less disadvantaged. The energy yield gain, starts to be positive from March to October, increasing gradually up to 35% in the summer months of June, July and August.

In general, a growing trend is observed starting from December, with the increase in the angular declination of the sun, which from the winter solstice (21 December) grows to reach its maximum value on the summer solstice (21 June). The higher gains in summer months (up to 35%), highly compensate the losses of the winter months, justifying the adoption of the tracker technology.

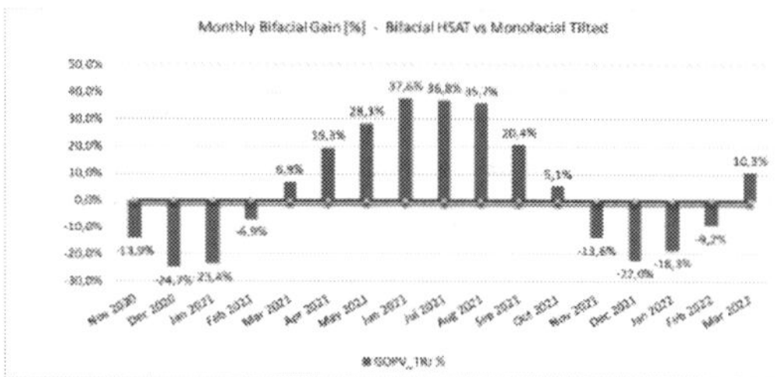

Figure 8 Evolution of the monthly bifacial gain (expressed in%) of the Bifacial tracking system compared to the Monofacial fixed structure.

It should be noted that, positive gain of tracked bifacial vs fixed bifacial is slightly delayed and starts in April.

The energy gain obtainable by the tracker has been assessed comparing the two bifacial HJT installed on fixed and tracker. As better explained in paragraph 3.2, the value of the gain of the tracker can be compromised if a threshold for the wind stow position is set on the tracker relatively low compared to the characteristics of the installation site. Especially during the winter period when the declination of the sun is more unfavorable with respect to the horizontal plan this could compromise the expected energy yield gain. Moreover, during the months of December and November some partial shading of neighboring buildings affected the energy production of the tracker.

For this reason, this assessment considers 12 clear skies days appropriately selected throughout the first year of monitoring (one day for each month), during which the production for the trackers was not affected by the stowage position for wind levels above the cutoff.

Figure 9 illustrate the energy yield of the tracker, comparing the two bifacial HJT modules installed on the fixed solution and on the tracker. From the obtained results, at Cadarache latitude it is possible to achieve a tracker energy yield gain of about 5,9 % (vs. expected ~7% of the PV model based on SAM Software, and verified with Radians ray tracing software).

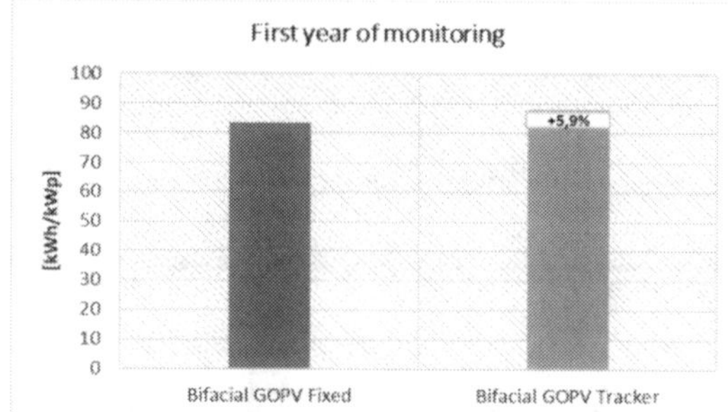

Figure 9 Energy yield gain comparing bifacial HJT installed on a tracker and on a fixed structure, performed considering one representative day per month during the first year of monitoring.

With the same methodology the gain of the GOPV solution (HJT bifacial + Tracker) respect to reference solution (monofacial PERC on tilted) has been calculated, giving a gain of 22% on annual basis.

3.3 Discussion on Tracker Operative Condition

During the monitoring period, the Bifacial tracking systems were set up with a lower wind limit than the one recommended by the tracker manufacturer, for local safety reasons due to imperfect module clamping system. This led to a lower gain than expected, since the trackers went too frequently in safety position. The wind limit recommended by the manufacturer was applied starting from the end of July 2021.

Figure 10 shows a typical day, in which for a few minutes the wind speed exceeds the limit value set for the trackers, bringing them in safety position. Net of two isolated cases (circles in orange), the power curve of the horizontal solar axis tracker system (in blue) is relatively broader than the power curve of the fixed system, thanks to the contribution of solar tracking. **Figure 11** shows another case of a windy day, with wind speed exceeding the safety limit starting from 10:30 and lasting all the day. In this case, the power curve, starting at 10:30, takes the typical shape of a fixed system. The additional power contribution is only that linked to the bifacial gain.

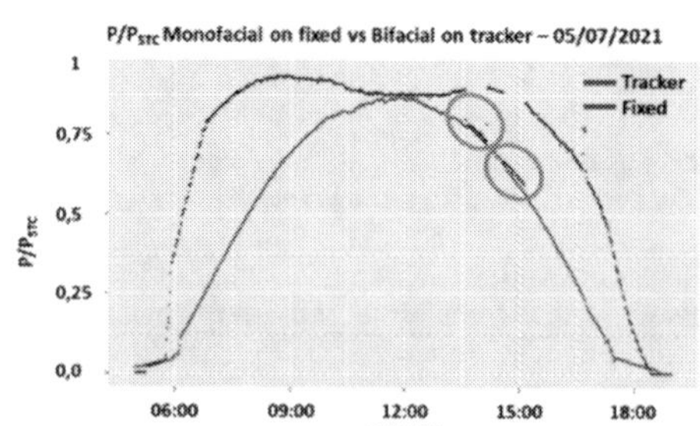

Figure 10 Power curve of a day, in which for a few minutes the wind speed exceeds the limit value set on the trackers

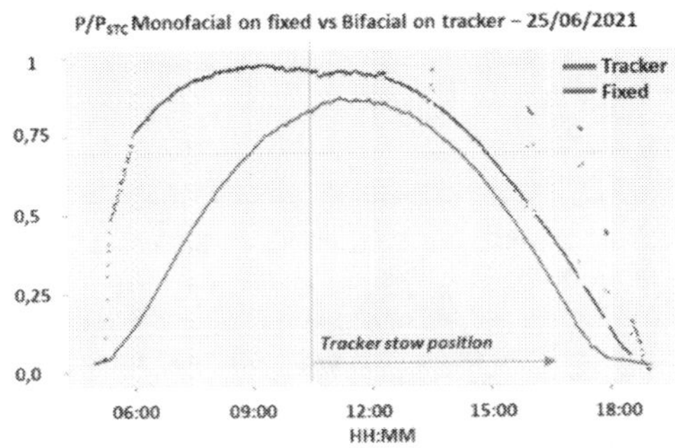

Figure 11 Power curve of a day, in which the wind speed exceeds the safety limit value set on the tracking systems starting from about 10:30; it takes the typical shape of a fixed system

4 CONCLUSION

This work illustrates the effectiveness of the GOPV solution consisting of half-cell HJT bifacial technology installed on Horizontal single axis tracker, compared to the reference solution consisting of PERC monofacial technology installed on fixed tilted structure. The analysis has been carried out with in the European GOPV project, using an **experimental PV pilot plant** installed in CEA's Cadarache, solar platform (France, 43 °,N).

The results highlights that on fixed tilted mounting structure, bifacial HJT technology improves the yearly Array Energy Yield by +15% compared to a reference monofacial PERC. Energy gain can even reach 20% in summer period..

The global GOPV solution (HJT bifacial technology + Horizontal solar axis tracker), outperforms the reference system (monofacial PERC fixed tilted) by more than 22% on an annual basis.

Therefore, the positive gains recorded from March 21 to October 21 (8 months) exceed, in absolute value, the losses of the winter period (4 months from November 2020 to February 2021). In fact, during the summer period, in the months of June, July and August, this gain have exceed the 35% of the monthly Array Energy Yield of the reference system.

Finally, when analyzing the tracker gain alone on energy yield, a +5,9% (vs. expected ~7% of the PV model) was observed compared to the same bifacial GOPV technology installed on a fixed structure. In conclusion, the results obtained in this experimental PV system demonstrated the effectiveness of combining the two technologies, Horizontal Single Axis tracker and Heterojunction silicon bifacial modules. Although the area constituted by the single-axis trackers worked for most of the annual period under disadvantageous conditions, let's think about the low wind speed limit set on trackers due to experimental constraints of the new GOPV modules, the results obtained hint at the possibility of obtaining greater energy gain, compared to the monofacial technology on fixed, in nominal condition without any extra constrain.

REFERENCES

[1] J. S. Stein, C. Reise, J. B. Castro, G. Friesen, G. Maugeri e E. Urrejola, Bifacial Photovoltaic Modules and Systems: Experience and Results from International Research and Pilot Applications, IEA PVPS Task 13 Performance, Operation and Reliability of Photovoltaic Systems, 2020.

[2] C. D. Rodriguez-Gallegos e H. Liu, «Global Techno-Economic Performance of Bifacial and Tracking Photovoltaic Systems,» *Joule,* pp. 1514-1541, 2020.

[3] *CEI EN IEC 61724-1 Photovoltaic system performance - Part 1: Monitoring,* 2021.

[4] «International Technology Roadmap for photovoltaic (ITRPV),» VDMA, 2020.

ACKNOWLEDGMENTS

This paper is supported by European Union's Horizon 2020 research and innovation programme under grant agreement No 792059., project GoPV (Global Optimization Of Integrated Photovoltaic System For Low Electricity Cost).

A FACILITY TEST TO GENERATE DATA FROM REAL PV SYSTEMS AFFECTED BY FAULTS AND VALIDATE FAULT DETECTION AND DIAGNOSIS ALGORITHMS

Maugeri Giosué*, Rossetti Andrea, Urciuoli Vittorio, Bombelli Davide, Salvatore Guastella
RSE SpA - RSE SpA, via R. Rubattino 54, 20134 Milan, Italy

ABSTRACT: The use of Machine Learning (ML) techniques in various application domains--in computer safety systems or within industrial processes--has proven benefits from an economic and efficiency standpoint. With these methods applied to datasets from monitoring systems, in real-time, it is possible to reduce the time to recognize faults, restore normal operation quickly, and avoid serious or permanent malfunctions, which would lead to longer downtime and higher repair costs. In the photovoltaic (PV) sector, the automatic fault detection through algorithms employing ML approaches can provide important clues to maintenance operators for subsequent troubleshooting phase, enabling them to optimize repair time and costs. However, developing models for the detection and diagnosis of faults is far from trivial and numerous challenges must be faced. One of the most difficulties for the development, validation and optimization of Failure Detection and Diagnosis (FDD) algorithms, lies in accessing the historical production datasets of PV plants in which there are enough failure events for the training and tests of ML algorithms. Furthermore, the FDD algorithms need to be ready in handling data that may be affected by communication errors, uncalibrated sensors, or other factors that can affect the PV performances (shadings, grid interruptions, ecc).
Within this work we are presenting an experimental Test Facility for the simulation of faults in PV systems, to generate large datasets from real PV systems that are affected by faults. A characterization of the main fault conditions that can occur in photovoltaic systems is presented, illustrating how the electrical parameter evolves in the specific fault. The real-field PV data generated by the RSE PV Facility Test allows having historical PV data with several separated failure events appropriately labeled for training and testing of ML algorithms.

Keywords: PV system, Fault Detection and Diagnosis, PV monitoring

1 INTRODUCTION

Nowadays photovoltaic (PV) technology plays a central role in the energy transition. In the first quarter of 2022, it was confirmed that the installed cumulative PV capacity globally reached the milestone of 1 TW.
The rapid growth of PV technology is also linked to its economic competitiveness, compared to other energy sources, allowing thus to be implemented independently of subsidy programs. In confirmation of this, in the last decade, there has been a fall of 77% in the cost of electricity from utility - scale solar photovoltaic (PV), allowing to reach competitiveness of LCOE compared to the conventional power plant.
Even if, new PV plants installation allows achieve the global environmental goals, it is necessary also that the PV plants (existent and new installation) maintain their reliability, energy performance over their lifetime, through measures that do not aggravate the Levelized Cost of Energy (LCOE). Among the factors that make it possible to keep a low LCOE, there is the optimization of Operation and Maintenance (O&M) measures. Recent studies have shown that an optimized O&M activity can allow a reduction in LCOE of up to 50% as it lowers the maintenance costs of the PV system (OPEX), ensures the maintenance of performance levels and contributes to increase the lifetime of the PV system. To this end, innovations and improvement, are becoming increasingly important because they help O&M services to keep up with market requirements. In fact, the automatic fault detection and diagnosis algorithms can provide important suggestions to Operation and Maintenance (O&M) operators, allowing them to optimize maintenance times and repair costs.
However, developing models for algorithms able to detect and diagnose the faults is far from trivial and several challenges have to be faced. One of the challenge for the development, validation and optimization of Failure Detection and Diagnosis (FDD) models, relies in accessing the historical production datasets of PV plants in which there are enough failure events for the training and tests of ML algorithms. Furthermore, the FDD algorithms need to be ready in handling data that may be affected by communication errors, uncalibrated sensors, or other factors that can affect the PV performances (shadings, grid interruptions, ecc).
Within this work we present an experimental setup realized at RSE PV Laboratory, named RSE PV Failure Test Facility, for the simulation of faults in photovoltaic systems. A characterization of the main fault conditions that can occur in photovoltaic systems is presented, illustrating how the electrical parameter evolves in the specific fault condition and thus providing specifications on data processing to make the fault condition easily detectable by the fault detection algorithms.

2 DESCRIPTION OF THE EXPERIMENTAL SETUP

The RSE PV Failure Test Facility allows the simulation of faults that in literature are classified as PV failures that mainly affect the PV performances and maintenance costs. [1]
It consists of three main sections, each designated to specific fault conditions that can be replicated on small-scale PV systems equipped with suitable data acquisition systems (Figure 1)
The facility sections are:

- *Inverter failure facility section:* allows simulating failures typically attributable to the inverter (e.g., ground fault, low insulation resistance)
- *Module\string failure facility section:* allows to simulate failures typically attributable to the PV modules and string (e.g., parasitic resistance, diode breakdown, hotspot)
- *Facility soiling section:* allows to simulate the presence of dust/dirt, even inhomogeneous, on the surface of the modules; the objective is to analyze, as the electrical and installation configuration of the PV plants varies, the

electrical effects of the phenomenon and the power losses generated due to mismatch on the PV strings, attributable to inhomogeneous soiling.

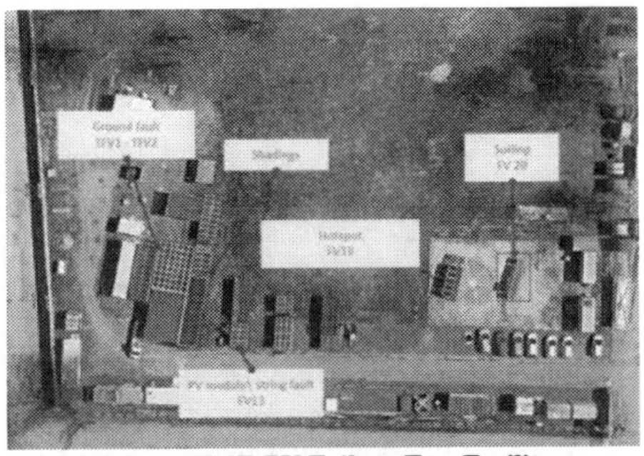

Figure 1 RSE PV Failure Test Facility

Table 1 summarizes the characteristics of the PV plants of the RSE PV Failure Test Facility and the respective type of reproducible failures.

Table 1 Characteristics of the PV plants and the respective type of reproducible failures.

	TFV1	TFV2	FV13	FV19	FV20
Reproducible Faults	Ground Fault Shadings Line-to-Line Fault	Ground Fault Shadings Line-to-Line Fault	Bypass Diode (SC; OC) Shadings Series Resistance	Hotspot	Soiling
Numbers of Strings	2	2	1	1	2
Module per Strings	24	24	8	6	1
String Voltage [V]	454	430	228	187,92	29,90
String Current [A]	8,19	8,17	7,72	8,35	7,86
Nominal Power [kW]	7,44	6,96	1,72	1,56	0,235 (x2)*
DETAILS DESCRIPTION					
SC: short circuit OC: open circuit			*(x2): the installation consists of two PV module connected to micro inverter		

The PV plant's parameters are acquired by dedicated monitoring systems and archived in a database. The generated large datasets will be linked to a public repository, organized by fault categories in order to better facilitate the study of identification and fault diagnostics algorithms based on Machine Learning techniques, for the PV sector.

The data collection and transmission scheme is schematically indicated in Figure 2.

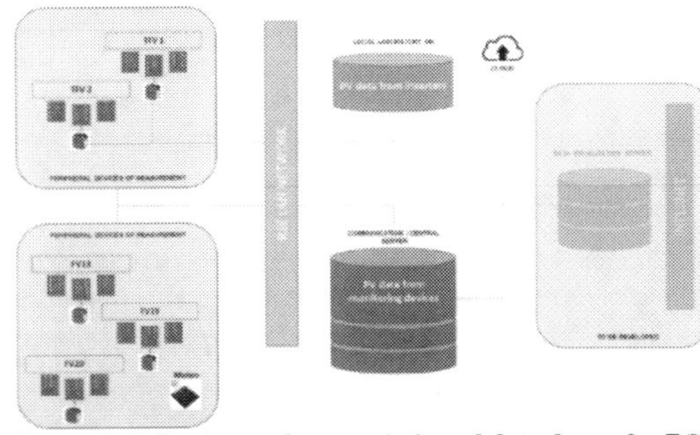

Figure 2 Collection and transmission of data from the RSE PV Failure Test Facility

3 RESULT OF REPLICATED FAULT IN THE RSE PV FAILURES TEST FACILITY

3.1 Low Insulation Resistance and Ground Faults

A low insulation resistance can be attributed to deterioration of the silicon of the junction box of one or more modules in the string, which causes water or moisture infiltration, or to the presence of moisture inside field switchboards and string parallel switchboards, or to the deterioration of the electrical insulation of cables; when the insulation resistance is lower than a preset value (about 1MOhm), the inverter generates the "Riso Low" alarm; depending on the logic adopted by the inverter, in some cases this alarm may cause disconnection from the grid [2]. The persistence of this fault can be also the cause of Ground Faults. It occurs as a result of current leakage to ground, both DC and AC side, which can occur inside the inverter and in the DC section of the plant. Damage in cable insulation or by corrosion and water infiltration within field switchboards in between 2 or more points of the string can be the cause also of Line-to-line (LL) faults. Block diodes between strings electrically connected in parallel are generally used to protect strings from voltage mismatch.

However, if an accidental connection to ground is created between two points in the same string, the effect is a decrease in string voltage and, in absence Block Diode, a mismatch among PV string that generate an overcurrent in the faulty string, which could be high enough to damage PV modules and conductors.

In Figure 3 is illustrated the string voltage observed in the three days of failure where a **low insulation resistance** was applied in the PV system TFV1. The figure shows how the reduction of insulation resistance to a value of 300 kΩ does not cause any instantaneous action by the inverter until the end day. The day after, the inverter, carrying out the check that precedes the connection to the grid, detects a Riso value <1MΩ and consequently remains in a Ground fault error status. In fact, during the second day, despite a good level of irradiation, the string voltage has a value close to open circuit voltage, meaning that the system is not in operation. On the third day, an operator eliminates the fault condition artificially applied to the string.

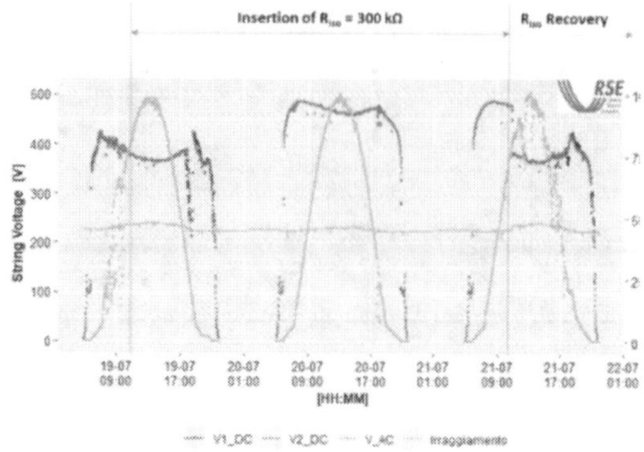

Figure 3 PV plant TFV1, string voltage observed in the three days of failure (blue and red curves); irradiation (yellow curve) and AC mains voltage value (orange curve) corresponding to the test days

The **Ground fault** has characteristics that make it more easily detectable by remote monitoring since R_{iso} around zero causes high values of leakage currents. The Figure 4 shows the instant in which the fault is caused by the RSE operator (red circle) and the effect that this fault has on the string voltage. It is possible to observe that, following the fault, the string voltage reaches values close to the open circuit voltage. The fault condition is recovered, and the inverter returns to be operative.

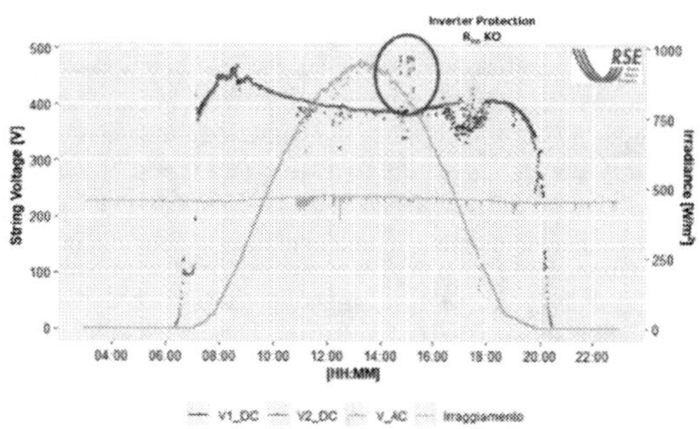

Figure 4 PV plant TFV1, a ground fault in the PV string

The **Line to line** faults result in string voltage mismatch and power loss. As the Figure 5 illustrate, the voltage of string 1 experiences a voltage drop proportional to the voltage of the PV modules underlying the fault (6 PV modules with a voltage of 19 V each).

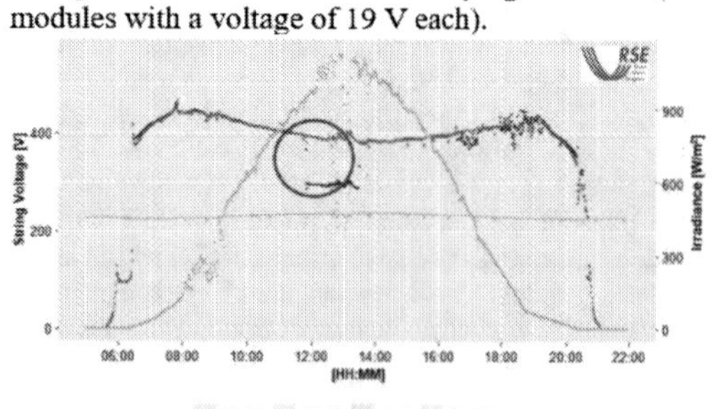

Figure 5 Line to Line fault: DC and AC voltage observed during the failure, while yellow curve represents the irradiation

Table 2 resume the fault characteristics of the Low resistance insulation and Ground Fault.

Table 2 Fault characteristics of Riso Low and Ground fault and related effects on the string parameters available from the monitoring system

Fault	Fault characteristics	Parameters behaviour
Rsio Low	Generally, the failure occurs in the early day, specially with low irradiation or on rainy days condition; the string voltage (Vdc) is close to the open circuit voltage (Voc).	$V_{dc} \approx V_{oc}$ $V_{ac} \approx V_{nom}$ $I_{dc} = 0$ $I_{ac} = 0$ for Irr ∈ [100; 300] W/m^2
Ground Fault and Line to line	Three cases can occur: 1. the string voltage reaches values close to the open circuit voltage while the grid voltage remains at the nominal grid levels even if slight drops occur due to the lack of supply of the inverter. 2. the string voltage goes to values close to the open circuit voltage and the grid voltage to zero due to the intervention of the protection on the AC side 3. the string voltage undergoes a drop proportional to the voltage of the single PV module	**Case 1:** $V_{dc} \approx V_{oc}$ $V_{ac} \approx V_{nom}$ $I_{dc} = 0$ $I_{ac} = 0$ for Irr ∈ [100; 300] W/m^2 **Case 2:** $V_{dc} \approx V_{oc}$ $V_{ac} = 0$ $I_{dc} = 0$ $I_{ac} = 0$ **Case 3:** $V_{dc} = V_{dc,max} - V_{dc,\,mod}*N$ with N The number of module under fault $V_{ac} > 0$ $I_{dc} > 0$ $I_{ac} > 0$

3.2 Increase in module/string series resistance:

An increase in series resistance (Rs) is often caused by the presence of module cell degradation phenomena, such as corrosion of the upper and lower metal contacts; in a PV string, the value of Rs is higher because the degradation phenomena of each module are added to the resistances of the contacts interposed along the current path [3]

The **increase of series resistance** in the PV string, can be identified by analyzing the parameter R_{imp} calculated as

$$R_{imp} = \frac{V_{oc\_stc} - V_{mp\_stc}}{I\_cor_{stc}}$$

where the V_{oc_stc} is the string open circuit voltage provided by the PV module manufacturer, while the V_{mp_stc} is the string voltage of the maximum power point measured and corrected under STC conditions.

Figure 6 shows the values of R_{imp} as the series resistance varies and for current values included in the range [5 A, 8 A]. corresponds to a respective working condition to which a series resistance has been applied.

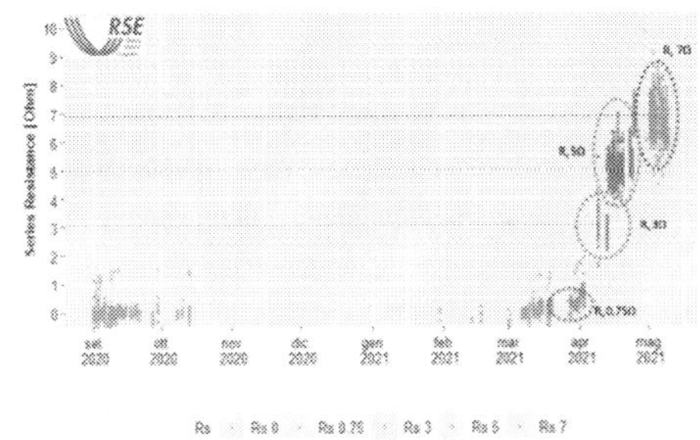

Figure 6 - PV plant FV13, values of Rimp as the series resistance varies, calculated on the data acquired by the plant

Table 3 resume the fault characteristics of the increase of Series Resistance in PV string

Table 3 Fault characteristics of Increase of Series Resistance in PV string fault and related effects on the string parameters available from the monitoring system

Fault	Fault characteristics	Parameters behaviour
Increase of Series Resistance in PV string	The phenomenon is detectable at high irradiation values (> 900W / m^2), when greater voltage drop across the resistors are more evident.	R_{imp}progressive increase

3.3 Bypass diode failure

The bypass diode is, in general located inside the junction box of each PV module, connected in anti-parallel to a series of PV cells, in the direction of the direct current; it has the function of allowing the module itself to bypass faulty or shaded cells, and thus prevent localized heating that over time could damage a PV module by, for example, the appearance of hot-spots; depending on the type of module and especially on the number of cells in series in the module, it is therefore possible to find one or more bypass diodes within a module; generally this component is subjected to continuous thermal and electrical stresses, and over time this can cause the diode inside the module junction box to fail [4]

Failure on bypass diodes can occur in two ways: failure of the diode in open circuit and failure of the diode in short circuit. As regards the short-circuit fault, it can be identified by analysing the string voltage (V_{dc}) of the PV string, in open circuit. This filter avoids to misclassificate the fault with the series resistance since the effect on string voltage during the operative condition is the same.

Figure 7 shows the V_{dc} value, corrected in temperature, measured in instant when string current was equal to zero.

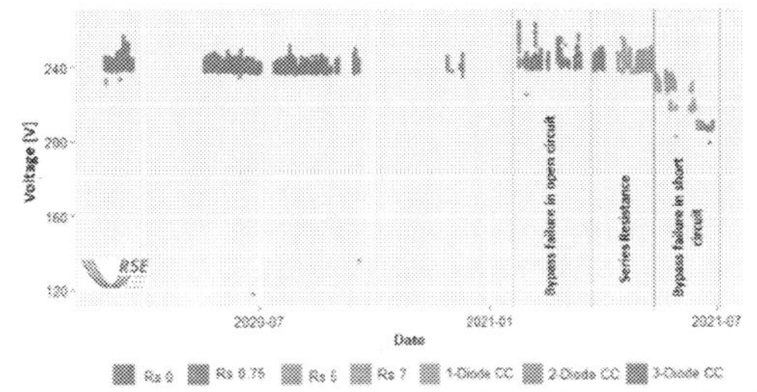

Figure 7 - PV plant FV13, trend of the string voltage in the FV13 PV system, corrected in temperature, obtained for irradiance value lower than 200 W/m2, and string current equal to zero

Table 4 resume the fault characteristics of the Short-circuited diode in PV string

Table 4 Fault characteristics of short-circuited diode in PV string fault and related effects on the string parameters available from the monitoring system

Fault	Fault characteristics	Parameters behaviour
Short-circuited diode	Generally, the fault is best identified in the early hours of the day, before the inverter enters the supply state. The string voltage (Vdc) during the string open circuit is reduced in proportion to the cells underlying each faulty diode.	Reduced Vdc, corrected in temperature, when Idc = 0

3.4 Soiling

The phenomenon of soiling is one of the environmental factors that influence the performance of photovoltaic systems, reducing their producibility. It is generated by the deposit of dirt, dust and contaminants (organic or inorganic) on the surface of the photovoltaic modules which causes a decrease in the solar radiation that reaches the photovoltaic cells. As consequence, this phenomenon affects the value of the current generated by the photovoltaic module, while that of the open circuit voltage remains unchanged.

To accelerate the soiling characterization tests, and to have an analytical correlation between the amount of dust present on the surface of a photovoltaic module and the respective decay of energy performance, RSE carried out the characterization of the soiling through an artificial fouling a PV module (PV1), comparing it with the performance of a second PV module (PV2), kept clean.

In the case of homogeneous soiling on the surface of the modules, a negative correlation was observed on the maximum power, as the amount of dust distributed on the surface of the module increased.

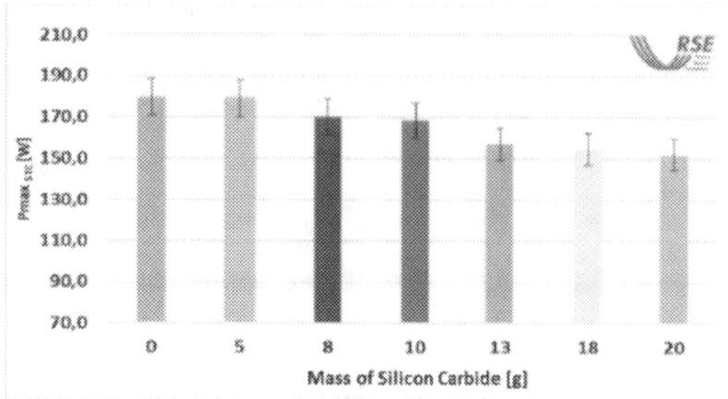

Figure 8 Homogeneous soiling: maximum power measured on module PV1 (dirty) as the silicon carbide mass varies

However, in the case of inhomogeneous soiling, being the characteristic of the voltage-current curve generally characterized by the presence of the so-called 'knees', generated by the activation of the bypass diodes present in the PV module, there is not the same correlation observed for homogeneous soiling. It has been observed that a better correlation can be seen by the short circuit current of the module.

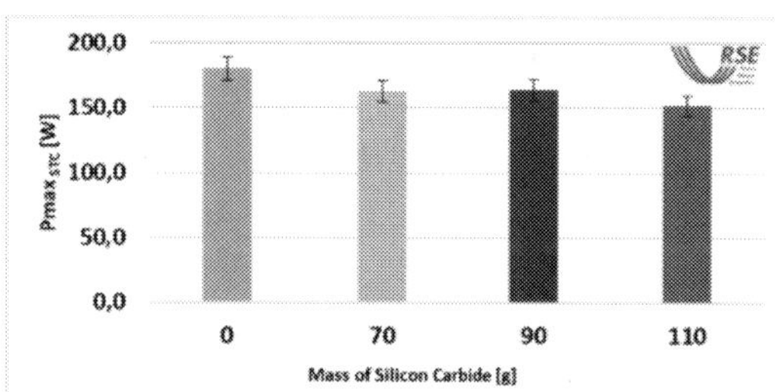

Figure 9 Inhomogeneous soiling: maximum power measured on module PV1 (dirty) as the silicon carbide mass varies

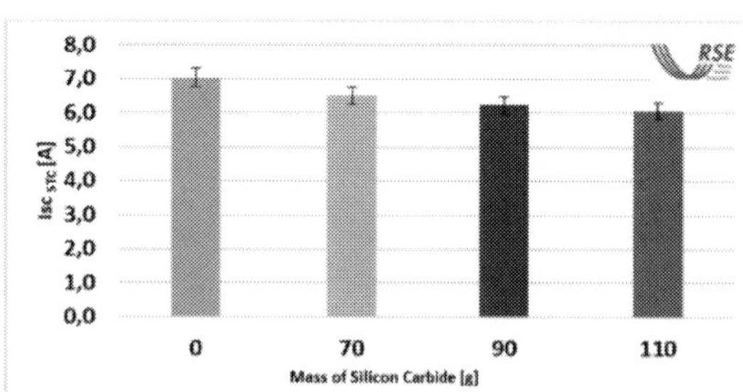

Figure 10 Inhomogeneous soiling: Short-circuit current measured on module 1 (dirty) as the silicon carbide mass varies

Table 5 resume the fault characteristics of soiling phenomena.

Table 5 Fault characteristics soiling in PV module and related effects on the string parameters available from the monitoring system

Fault	Fault characteristics	Parameters behaviour
Soiling	For homogeneous soiling the phenomenon is detectable observing the progressive reduction of the maximum power over the specific monitored period or if available from a progressive reduction of short circuit current. For inhomogeneous soiling, a progressive reduction of short circuit current	P_{max} progressive decrease Isc progressive decrease

4 CONCLUSIONS

In this work the RSE PV Failure Test Facility has been presented and result of faulty test on PV plants aiming at characterizing the electrical parameter under faulty condition has been presented.

From the characterization activity it emerged that the low insulation resistance faults can, be recognized by observing the behavior of the inverter in the early hours of the day, at irradiation levels between 100 and 300 W / m^2, in which the humidity level can be higher and therefore facilitate the phenomena. low insulation of the PV string, in the case of already compromised insulation. The ground fault attributable to a R_{iso} equal to zero, has characteristics that make it more easily recognizable. It can occur regardless of weather conditions as it is mainly linked to insulation in the cables or connectors which, deteriorating, causes significant leakage currents to earth, leading to evident effect on string and grid voltages.

A rapid increase in the string series resistance attributable to particularly aggressive corrosion phenomena, can be identified by analyzing the ΔVrs and R_{imp} parameters, taking into account that the analysis of ΔVrs must be carried out for high current values, that is close to the nominal values of the string.

Shorted bypass diode can be identified by analyzing the string voltage (Vdc) in open circuit, in the early hours of the day. Finally, the soiling effect have been analyzed, monitoring the maximum power of the module in dirty condition. As regard for homogeneous soiling it has been observed a negative correlation on the maximum power, as the amount of dust distributed on the surface of the module increased. For inhomogeneous soiling, the negative correlation is better observed in short-circuit current as the amount of dust of the module varies.

These results also allow us to state that the RSE PV Failure Test Facility has the potentiality to generate large datasets from real PV systems artificially affected by faults, allowing to collect historical PV data with several separated failure events appropriately labeled for training and testing of ML algorithms. By the RSE PV Failure Test Facility datasets, it is possible to make the Fault detection algorithms able in managing the real PV environment (shading, data affected by communication errors, sensors not calibrated, different module technology, or network outages), increasing their reliability in the task of fault detection and diagnosis prediction. The RSE PV Failure

Test Facility can offer large datasets of labeled PV failures which in general can take several years of PV production to collect all possible failure modes. Furthermore, it allows to demonstrate the applicability of the techniques developed to real photovoltaic systems and could be used as a test bench to validate ML algorithms in different photovoltaic plants affected by faults.
For this purpose, the large dataset with labeled fault conditions will be linked to a free-accessible repository, organized by fault categories, in order to better facilitate researchers interested in the study of identification and fault diagnostics algorithms based on Machine Learning techniques.

5 REFERENCES

[1] G. Maugeri, V. Barone and D. Bombelli, "Tecniche diagnostiche avanzate per impianti FV: database guasti e progetto facility sperimentale," RdS, 2019.

[2] R. Araneo and M. Mitolo, "On the Insulation Resistance in High-Power Free-Field Grid-Connected Photovoltaic Plants," 2019.

[3] G. Spagnuolo and H. Kalliojärvi-Viljakainen, "Identification of series resistance from the measured pv panel electrical characteristics," in *36th European Photovoltaic Solar Energy Conference and Exhibition*, 2019.

[4] W. G. S. e. S. W. Ko, "Origin of Bypass Diode Fault in c-Si Photovoltaic Modules: Leakage Current under High Surrounding Temperature," *Energies,* vol. 2416, p. 11, 2018.

[5] C. D. Rodriguez-Gallegos and H. Liu, "Global Techno-Economic Performance of Bifacial and Tracking Photovoltaic Systems," *Joule,* pp. 1514-1541, 2020.

[6] J. S. Stein, C. Reise, J. B. Castro, G. Friesen, G. Maugeri and E. Urrejola, Bifacial Photovoltaic Modules and Systems: Experience and Results from International Research and Pilot Applications, IEA PVPS Task 13 Performance, Operation and Reliability of Photovoltaic Systems, 2020.

[7] "International Technology Roadmap for photovoltaic (ITRPV)," VDMA, 2020.

[8] *CEI EN IEC 61724-1 Photovoltaic system performance - Part 1: Monitoring,* 2021.

[9] G. Maugeri and D. Bombelli, "Tecniche di diagnostica avanzata per impianti FV: caratterizzazione dei guasti e sviluppo di algoritmi FDD," Rapporto RdS prot. 21010992, RSE, Milano, 2021.

ACKNOWLEDGMENTS

This work has been financed by the Research Fund for the Italian Electrical System under the Contract Agreement between RSE S.p.A. and the Ministry of Economic Development - General Directorate for the Electricity Market, Renewable Energy and Energy Efficiency, Nuclear Energy in compliance with the Decree of April 16th, 2018".

PERFORMANCE EVALUATION AND OPERATIONAL FAULTS ANALYSIS FOR PV SYSTEM PROTECTION AND SAFETY STANDARD IMPROVEMENT

Tarikua Mekashaw Zenebe*[1], Lemu Kebede[2], Ole-Morten Midtgård[1], Steve Völler[1]
[1]Norwegian University of Science and Technology, Department of Electric Power Engineering, Norway,
[2]Private Researcher, Germany
*Corresponding author: tarikua.zenebe@ntnu.no

ABSTRACT: With increased interest in utilizing PV (photovoltaic) systems, several studies are recently focusing on ensuring their reliability and safety. Contrary to the conventional power system, research showed that existing PV systems protection and safety standards have not fully addressed PV systems protection and safety challenges. This paper aims to find gaps in PV system protection and safety standards by evaluating the performance of the existing PV system and operational faults through qualitative methods and literature review. The qualitative approach included interviewing solar experts with six to twenty-two years of experience in the PV sector in Africa and Nordic countries. As per the interviewees, the PV system works as desired if the existing standards and rules are applied appropriately. However, the literature review has shown that faults like hotspots, degradation, and arc faults are difficult to detect during regular operation. Therefore, an accurate system is needed to follow up and undertake detailed performance analyses. Besides, in the protection and safety standards designing a way of gathering detailed information that enables the PV system operator or owners to know the exact status of the PV system shall be a mandate. Data-driven methods and regular infrared imaging could fill the gap.
Keywords: PV system fault, PV protection and safety standards, Fire

1 INTRODUCTION

Renewable energy sources like solar PV (photovoltaic) have been growing steadily to meet the need for combating climatic change while ensuring energy access for all. Besides, the continuous cost reduction in PV panels and their system components make them preferable and competent to conventional energy sources [1]. With this increase in the utilization of PV systems, ensuring the system's reliability and safety is significantly more critical than before. In this regard, protection devices play a significant role. Nevertheless, due to the varying operational range of the PV system because of its non-linear behaviors and its dependance on environmental condition, it is still difficult to select the proper protective device. The analysis in [2] showed the protective devices' failure to detect line-to-line and ground faults during high impedance faults and low irradiance as well as in MPPT-based PV systems, even though the design followed the existing PV system protection standards. Furthermore, some faults, such as short circuits and series arcs, might show similar behavior, and correctly identifying them is a challenge with the existing protection devices [3]. In addition, faults that cannot be identified can may remain undiscovered and circulated for a long time, resulting in severe problems. Moreover, other challenges include random faults such as parallel arc faults and a lack of specified detection techniques in PV system standards [2-4].

In addition to what has been researched, with the increasing number of installed rooftop PV systems, recently, the report of fires caused by PV systems in the UK, USA, Netherland, and Japan is increasing. It is forecasted to keep on rising in the future [5]. As shown in [3,4,6], DC (direct current) arc faults and localized heating (caused by hotspots and degradation) are the main causes of fire in PV system. Leakage current due to an unidentified fault could also cause a fire in the PV system. Identifying such faults in time is challenging as they need very detailed data on the system. Therefore, looking into reliability and how to meet the required safety in PV systems is essential.

This research aims to use the results of the performance evaluation and operational faults analysis of installed PV systems to investigate protection and safety standards gaps so that they can address future system protection and safety challenges. In this study, quantitative methods and literature reviews have been used. Data have been gathered through a survey by interviewing solar experts with six to twenty-two years of experience in designing, installing, operating, and maintaining PV systems in Africa and Norway. The study focuses on small-scale standalone PV systems.

The findings show that with the current information, it is impossible to say 1) the PV system is reliable and safe within its lifetime and 2) the existing standards are enough to keep it reliable and safe. Due to the lack of necessary information in the studied PV system, a detailed analysis and evaluation of the errors could not be carried out. For systems installed in African countries, it is not common to do a detailed performance analysis unless the power production is entirely interrupted. In this scenario, slow degradation of the PV module, hotspot formation, and arc fault that led to severe conditions like fire cannot be identified in time. Even if there is better follow-up and preventive maintenance in PV systems installed in Norway, the level of detail is still not enough. Therefore, standards need to require methods for installed PV systems to get information about the system's reliability and safe operation. Data-driven methods and regular infrared imaging of PV modules could be used to identify hotspots, degradation, etc.

The paper is organized into five sections. Section two presents the methods followed. Then, section three summarizes the literature review and interview results, while section four discusses the results and the paper's limitations in detail. Finally, the study concludes in section five by highlighting the main findings.

2 METHODS

The information gathered from various literatures and interviews is the base for the results of this paper. The interview is conducted by first preparing a pre-planned survey. However, during the interview, the experts were free to talk about their experience even if it was not listed on the questionary. In this way, it was able to collect more essential points. The information in this paper from the interviewee entirely depends on their experience, not linked to the company they are working for. Relevant information about the seven interviewees (I1-. I7) and their field of experience is provided in Table I

Table I: Information about interviewee

	Country	Field of experience	Years of experience
In1,2	Ethiopia	D, Ins, O&M	20
In3	Kenya	D, Ins, O&M	16
In4	Niger	Ins, O&M	8
In5	Norway	D, Ins, O&M	10
In6	Uganda	Ins, O&M	6
In7	Uganda	D, Ins, O&M	22

Note: In: Interviewee D: Design, Ins: Installation, O&M: Operation and Maintenance

We have performed seven interviews—one from Norway and the others from African countries, including Ethiopia, Kenya, Niger, and Uganda. The interviewees have experience from 6 to 22 years in designing, installing, operating, and maintaining PV systems. In addition, they have been working with PV systems ranging from 30Wp to 1MWp. The PV systems in Africa are installed for rural community hospitals, schools, and private houses, and for urban community private institutions which want reliable power supply due to frequent grid power interruption. In Norway, the interviewee has experience in grid-connected PV systems without batteries, and the customers are private commercial building owners.

3 RESULTS

Hereinafter, the result of both the literature review and the interview will be presented.

3.1 Literature review

3.1.1. PV system protection and safety standards

There are a number of international and national standards related to the PV system protection and safety. NEC, UL, NRTL, and NFPA are in the USA [7]. In Europe, there are more than 28 standards related to PV system protection. The standards include guidelines to select overcurrent protection, lightning and surge voltage protection, and earth fault protection. Besides, the standards contain the general requirement for power converters used in PV systems. Moreover, they consist of requirements for testing, documentation, and maintenance of PV modules and plants [8]. In African countries, in addition to the national standards, they adapt international standards from USA or Europe, or both. Table 2 summarizes some of the typical international and national standards established to minimize and mitigate failures and thus ensure safety in PV systems.

Table II: International and national PV system protection and safety standards

	List of standards	Comment
IEC [9]	62109-1:2010	Safety of power converters for use in photovoltaic power systems – Part 1: General requirements
	62109 2:2011	Safety of power converters for use in photovoltaic power systems: Part 2: Particular requirements for inverters
	62109-3:2022	Safety of power converters for use in photovoltaic power systems : Part 3: Particular requirements for electronic devices in combination with photovoltaic elements *NB*: the electronics equipment include DC-DC or DC-AC power convertors, active diodes, protection, control, monitoring, or Communication devices
NEC [10]	690:2014	Ground fault, overcurrent, and DC arc fault circuit protection required
UL	1741[11]	Requirement for inverters, converters, controllers and interconnection system equipment for use with distributed energy resources
	2579[12]	fuses for PV system

PV system standards in Africa

Based on the available data on the internet, we have tried to get PV system standard information for the country listed in Table I.

Kenya

Kenya has drafted a regulation that clearly states which international standards must be followed. Based on [13], in Kenya a license by an accredited institution is mandatory to work in the PV sector. Also, they need to renew their licenses after a certain period. This is an excellent approach to encourage professionals to evaluate their competence with the new technology to fulfill the requirement of the regulatory body—the same works for companies and their employees. Even if the draft regulation states the PV standards for making the work simple, it would have been efficient and clear for the working force in the sector to prepare a separate document that comprises at least the mandatory requirements and direct the reader to the given standards for further understanding. Moreover, even if the draft regulation was in 2020, most of the standards listed are old.

Ugand and Niger

Based on our internet search, we could not find national standards for PV systems installed in Uganda. However, the interviewee told us they follow international and national standards, especially from the UK. Similarly, to Niger.

Ethiopia

So far, there is no well-organized document or body

on which the PV system's designer, installer, and operators can rely. For those systems where NGOs are involved, they try to apply international standards, but it entirely depends on their will. But related to quality assurance, recently, in June 2021, a recommendation document was prepared to ensure the development and implementation of high-quality solar equipment with a capacity of less than 300W [14].

PV systems standards in Norway:

Norway specifically follows NEK400:2018, which mainly refers to international, continental, and national standards like IEC. NEK400 (2018) [15] is linked to low voltage electrical installation in Norway: it is a way to meet the expectation of authorities that check electrical safety. In addition, NEK is developing a standard related to PV system testing, documentation, and maintenance due to the increase in PV system installation and understanding of the need for a separate standard that is a line with the Norwegian context. The standard is called NEK 446:2022. It consists of international standards such as IEC 62446-1:2016 and 2018, IEC 62446-2:2020, and IEC 62446-3:2017. It provides a way of securing the PV system throughout its lifetime by suggesting proper testing, documenting, and maintenance (including preventive, corrective, and performance-related maintenance requirements) [16]. Furthermore, a new addition in NEK 400:2022 requiring more protection on the DC switch to prevent fire spread is in place [17].

3.1.2. PV system faults leading to fire in the world

The visible consequences of faults in PV systems are power disruption and fire. However, the root cause for those consequences might originate long before it is identified. Unless there exist detailed analysis and continuous follow-up of the system, small changes in power production and degradation might not be seen on time. Faults that lead to fire are severe and result in the complete destruction of the PV system and all nearby building, especially if it is a rooftop PV system. So, what is known about PV faults and fire currently?

Extensive works have been done to collect PV system-caused fire data and their root cause to study the need for standard amendment by the British National Solar Center [6]. This study found that PV system fires are either caused by a design error, installation error, product defect, or/and insufficient operation and maintenance. Besides, in [6], based on the 2013 report, it is shown that most PV standards are related to the PV module and that there relatively less focus on other systems parts like connectors and additional protection.

Recently several reports have been seen regarding fires related to the PV system. Table III presents a summary of the latest news about fires in PV systems in well-known countries and the identified causes based on available reports in the news and publication. It can be clearly seen that DC connectors and arc faults are the main cause of the fire. The root cause can be related to all life cycle stages of the PV system: design, installation, operation, and maintenance. Several studies [3,18,19] have shown that arc faults are the most dangerous one but also hard to detect.

Table III: Recently reported PV system caused fire incidents

Country	Reported incident of damage caused by PV	Common cause; proposed solution
USA	155 solar system fire in 2015-2018 [20]	PV module; regular infrared (IR) imaging
Netherla-nds	29 PV-related fires in 2018 and 2019, 37 fires in the first ten months of 2020 [21]	70% of the fires were caused by faulty connectors for the fault from 2018-2019; proper and quality conector installation
Australia	20% increase in fires related to solar panels since last year, [22]	>50% caused by dc isolators; use dc isolator inside inverter
Japan	172 fire accidents occurred from 2008 to 2017 [23]	Mounting material; inspection

PV standards and arc fault protection

Looking at the standards in Europe, there is no mandate (but only recommendation) on installing arc fault detection and protection devices in PV systems [19,24,25]. In USA, it is mandatory to install an arc protection device. Nevertheless, as it is shown in [25], there is still a problem in suggesting the right methodology. Thus, it is important to follow the right and efficient methods.

3.1.3. PV faults leading to fire in Africa

When we come to the African countries (Ethiopia,, Kenya, Niger, and Uganda), it is difficult to find well-documented data about fire causes, especially those related to PV systems. The main reason is that most PV systems are in rural areas where no dedicated institution is responsible in case a fire happens. Nevertheless, looking at pictures of rooftop PV systems published by several organizations and companies, some assumption can be taken.

1. Looking at rural communities and small villages in Ethiopia, Kenya, Uganda, and Niger, the roofs of the houses are made of steel and wood.
2. By putting the PV panel directly on the roof, as shown in fig. 3, the efficiency of the PV panel will decrease due to cell temperature increase as it does not have enough ventilation on the rear side,
3. Suppose any potential fault that led to fire happens at the module or cable. In that case, the roof is very suitable for the fire to spread all over the area, and it is incredibly difficult to contain the fire at its origin. Moreover, if the houses are connected, as shown in fig. 3, the fire will further go to other places and destroy the settlement without well-equipped firefighters.

Figure 1: The mounting structures of PV system in rular areas in Africa. (Source: left picture [26], right picture [27])

3.2 Interview results

3.2.1. PV system operational faults causes in Africa

The interview results show that the root cause of the operational faults and low performance of PV systems in Africa could be linked to the different phases of the PV systems: Design, installation (construction), operation, and maintenance. Lack of adequate training of the designer, installer, and consumer takes major responsibility for the system malfunction. In addition, the quality of products, lack of funding for equipment replacement, and environmental influence have contributed to not using the PV system until the end of its lifetime.

Design phase: Error in this phase cause premature failure in the PV system. Improper system component and protection device selection is the major problem. Some of the errors during this phase include:

- Improper inverter size, charge controller rating, surge arrester rating, and battery fuse selection.
- Designing the PV system without fulfilling the standards.
- Selecting low-quality products.

Installation phase: Not following the design is the major problem during this phase. The error in this stage include:

- Discrepancy between design and installation in following the right panel and battery configuration. This has resulted in early-stage system failure, fire results from a short circuit in some scenarios.

Operational phase: During this phase, some errors during the early stage will contribute to the system's failure. In addition, customer behavior contributes a lot to the PV system failure in this phase. Those includes:

- Adding additional load than the system is designed for, especially connecting tempering loads like hair dryers.
- Decrease in output power due to hot spot (only one interviewee mentioned this).
- Over-discharging of battery due to sizing problem or change in consumer behavior. For instance, the PV system was designed considering the TV will only be used for 4 hours, but then the consumer changes its behavior over time and might use it for more than 4 hours.

Maintenance: Not getting proper and timely maintenance is one of the major problems that the interviewee raised. Most consumers are only interested in maintenance during failures instead of paying for regular preventive maintenance. On the one hand, the reasons are economical and lack of awareness of the advantage of preventative and routine maintenance. On the other hand, it is also related to the trust they have in the company to listen to what they advise them. Moreover, this is linked to the type of project design. The majority of the project, especially in Ethiopia, are NGO funded. And once the NGO commission the project, the consumer might not have the money and the know-how to get the preventive maintenance with a company. Or in the case of community schools and hospitals, the personnel who get training to maintain and run the system might no longer work there and did not correctly teach others before leaving. This can be mitigated by giving basic training to all as much as possible or preparing a manual that most people can understand. Problems related to maintenance include.

- Dust accumulation in the inverter and PV panel.
- Replace the wrong load type (bulb) and other components after the end of life or failure.
- Lack of proper maintenance in the battery.

Others: Other problems are related to environmental effects as the technologies are not adapted for Africa's environmental condition. This includes:

- Battery shorter life due to hot temperature and high humidity.
- Corrosion in inverter electronics part and cables

It is worth mentioning that corrosion is one of the causes of arcing, which is again the main cause of fire [6].

Concerning the standards applied in the PV system, even if some of the interviewees are aware of the international standards, as there is no strong regulatory body to demand this in each and every house in the rural area especially, the knowledge they have is limited.

Regarding protection devices, the interviewee primarily used fuses, circuit breakers, ground fault protection devices, and surge and lightning arrestors.

As it was mentioned, due to economical reasons, the inability to replace some components like batteries and inverters after their lifetime is one of the significant problems. Recently companies will give a performance warranty of 25-30 years of lifetime with some degradation for a PV module [28]. For the PV system to function fully, other components must last for the same period. For instance, the inverter has a warranty lifetime of about 5-25 years, depending on the type of inverter: central inverters have five years, string inverters have ten years, and microinverter have 25 years [29]. PV cable can last as long as the PV system, and usually, the company gives a warranty lasting 25-30 years of life [30]. Depending on how they are used, the lead acid batteries can last more than five years; however, it is one of the PV components that need replacement at least once during the entire lifetime of the PV system. Usually, lead acid batteries have about five years warranty [31]. Thus, they require replacement cost and create awareness while the purchaser sells and install such system instead of merely saying PV system work for 25-30 years is very important. The replacement cost must be told to the customer in an understandable way, for instance, as shown in Fig. 2.

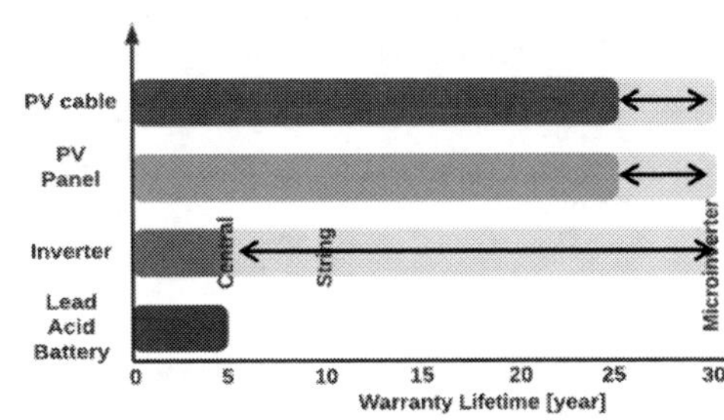

Figure 2: PV systems components versus their warranty lifetime.

3.2.2. PV system operational faults causes in Norway

Below the main challenges related to PV system operational fault in the case of Norway is presented. The results are based on PV systems that have been working for a maximum of ten years long.

According to the interviewee, electrical faults count

for less than 20% of the problems in rough estimation. The main issues are related to the mechanical challenges due to the harsh weather conditions. Power optimizer and inverter fan failure are the main problems so far faced. The reason for power optimizer failure is not known by the interviewee. It is worth mentioning here that the panel and battery degradation problem is less due to the cold weather. Not often, but hotspots in the panel during commission were identified. This could be related to transportation and manufacturing defects. Related to the mechanical fault, the PV module frame and glass breakage is the main problem which even might cause fire. There is no remote detection mechanism for such failure other than visual inspection by humans or drones.
The diagram in Fig. 3 summarizes the result of the interview.

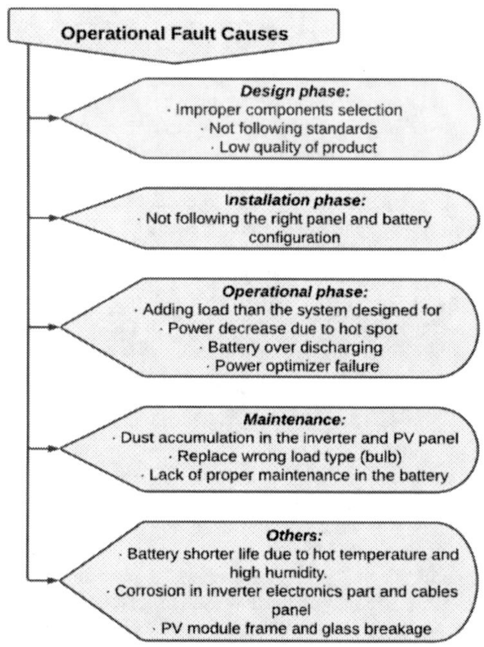

Figure 3: Summery of the causes of PV systems operational faults, as per the data collected from interviewee.

As we can notice, the challenges in the PV system installed in Norway are quite different from others and also less. The reason for not having some problems as in Africa, apart from having different weather conditions are listed below:

- Following national and international standards and rules.
- Make the design as simple as possible. Reduce the number of components as much as possible. In this way point of failure can be decreased.
- Remove the over-voltage protection instead; use the protection system built-in components like an inverter. But here, the level of protection accuracy by the component compared to a separate protection device near the fault point needs to be analyzed.
- Reduce connector, and avoid using fuse by limiting the number of solar PV panels per parallel to two, etc.
- Use higher quality equipment, including tiny components like connectors.
- Buy components and tiny parts like connectors from a similar company.
- Be strict with the design and installation stages and follow continuous maintenance during operation.
- Use a double isolated system and avoid using ground fault protection devices as they can be a source of failure.

Even if arc fault is mentioned as the major problem in the literature, based on the interview result, no DC arc fault detection systems are installed in the PV systems.

4 DISCUSSIONS

Comparing the interview results in African countries and Norway, apart from having a well-trained workforce and a strong regulatory body, the climatic condition has created an advantage related to the problem of battery and PV panel degradation. On the other hand, the mechanical challenge is a disadvantage and requires technological advancement in PV panel frame design and glass manufacturing.

Most of the problems that affect the performance and safety of the PV system in Africa can be alleviated by proper education and awareness creation in the community. But some need cooperation from other developed countries and NGOs working in the sector. Some of the proposed solutions are listed below.

- Strong regulatory body that checks if standards and rules are followed and if a quality product in the market is met. The other advantage of a dedicated regulatory body is that it will follow the updated standard and mandate companies to apply that. Especially for PV systems, as it is showing a vast increase in installation, new standards to make it more reliable and safer will be in place. Therefore, it is difficult to keep track unless there is a dedicated regulatory body. As shown in the result section, most faults are avoidable if strict measures are taken at the different stages of the PV system (design, installation, and operation).
- Revise the education sector to have qualified technicians. Having qualified engineers and technicians will play a significant role in mitigating the problem. As the engineers get experienced, they can select an efficient but simple and easy design. Furthermore, by including PV education in the primary educational system, the ability of the community to keep running the system will drastically improve.
- Preparing a simple manual so the consumer can make trouble shouting and provide routine services by themselves.
- For those helped by NGOs, a project design that grants continuous system functioning has to be placed.
- Create awareness in the community to buy quality products with more extended warranties rather than cheap and with less quality. A solid commitment to control the market to have a quality product is also needed from the government. In this regard, it is an appreciable

move by Ethiopia to draft a guideline that guarantees qualified PV products in the market.

- The world has to collaborate to combat climatic change. International funds that help the community keep the system running correctly shall be given. It can be a way of helping the rural people in terms of a loan, for example, for those who need replacement of some PV system components.
- To carry out more research in adapting the technology for a warmer climate.

When we ask about the mandatory protection devices, it is only fuses, and circuit breakers are mentioned. However, as we have seen, arc fault is the major cause of fire in PV systems. Thus, a device that can protect the arc and prevent fire from spreading has to be mandatory. Moreover, as shown in Table III, even if our interview result has not revealed this, from the literature it can be see that the majority of the fire causes are the disconnector. Therefore, in the standard, the quality and the installation of the connector have to get significant attention and emphasized during commissioning.

The interview result has indicated that there is a lack of evidence in Norway and Africa to say the existing standards are well enough to protect and maintain PV system reliability and safety in a longer term of operation. Because 1) with the existing monitoring techniques and level of detail, we have not seen a major fault in the PV panel; it does not mean fault will not occur in the longer term. 2) As shown in research, hotspot and arcing cannot be detected right away and are only visible when they cause a severe problem and require detailed analysis for early identification. 3) As indicated in some reports and can be deducted from the interview result, especially for PV systems in Africa, most systems installed are not continuously monitored. They do not have their status data after installation. This means we have a lack of data. Not only is this risk of fire due to PV systems increasingly being reported than before. Therefore, requiring additional, detailed, and consistent measurement of data and monitoring in the PV standards and codes, at least for those PV systems that can justify the extra cost, is necessary. One solution could be using online and data-driven methods such as machine learning techniques as an additional fault detection system and regular infrared imagining/ infrared thermography of PV panels [16].

The other problem the literature review has revealed is standards' availability, simplicity, and affordability. It was very difficult to get existing standards of PV systems in a structured way. We can imagine how difficult it will be for installers and people with little internet access and so on. Moreover, it is expensive to buy as there are multiple standards related to PV system design, installation, operation, and maintenance. Therefore, there is a need to work on the availability and simplicity of standards. In addition, as technology is growing, there might be a lot of amendments to the standards. There should be a way to disseminate the information and make it available for everyone if the PV market is needed to grow while keeping the good practices and ensuring the system's reliability, durability, and safety.

From a safety perspective, the mounting structure of PV systems in rural area need to get great attention. Here it is extremely important to train the local community on how to protect themselves and their property (like buildings) in case a fire emerges from the PV system. Those who do not have continuous follow-up service with the installer companies can be trained, and a manual that can be easily understandable by the community can be prepared. This might be something that needs to be planned from the NGO or government regulator side and make sure if this is fulfilled during commissioning, especially for community-based PV systems for hospitals or schools in the rural area. In general, proofing PV system reliability and safety is a key to its success and boost in the market because this will build its reputation and put trust in the community. On the contrary, a bad reputation in the community might lead to resistance to the technology and lead them to go back to using kerosene in the rural area in Africa. Therefore, strong measures to keep standards in place and work to solve the existing gap are required.

Looking at national standards, it seems some countries like Norway consider PV systems like other as other electrical systems for a long time, and the national standard is only for grid-connected PV systems. But considering its growth standards related to stand-alone PV systems and more standards dedicated to PV systems shall be developed considering the unique environmental and other country-specific conditions.

4.1 Limitation of the paper and future work

Due to the limited number of interviewees, which is seven, all the conclusions taken might need more study. The information from the interviewees about standards was limited, and the countries do not have recorded PV system fire data. In this regard, there is a lot to do in recording and categories fires or damages caused by PV systems and having dedicated and proper documentation. Therefore, looking at the criticality of the topic, the authors would like to give more time and continue this paper study for further investigation.

5 CONCLUSIONS

From the literature review, we have found out that arc fault is causing a severe problem in the PV system and leading to fire. Therefore, arc fault protection must be mandatory and included in international and national standards. From the interview for the PV system in Africa, the causes for operational faults are related to the error at different phases of design, installation, operation, and maintenance. In general, during the early operation of the PV system, faults are due to manufacturing defects, design, or/and installation problems. However, the lack of continuous follow-up and performance evaluation of the PV system makes it challenging to notice hidden faults as early as possible. Having qualified manpower and an educated community can alleviate most of the problems. But in addition, a strong regulatory body that follows the fulfillment and update standards are required in the countries. While in Norway, the major problem is related to mechanical faults due to the harsh environmental condition.

In general the PV systems installed, including in Norway, will not able us to do a detailed investigation that will indicate the status of the PV system concerning a hotspot, degradation, and arc fault. Therefore there is a need to revise existing protection and safety standards to know the status of the PV system as detailed as possible. This means, at least for those with cost-justifiable projects, there shall be a means of collecting detailed data. One suggested approach for hotspot detection is to have regular infrared imaging. Data drive methods could also solve the problem.

6 ACKNOLEDGMENT

We want to give special thanks to the people and institutions willing for the interview and for sharing their experience and their precious time. We would also like to thank the Department of Electric Power Engineering at NTNU for funding this project.

7 REFERENCES

[1] Detollenaere, A., et al, Snapshot of Global PV Markets 2020, PVPS Task 1 Strategic PV Analysis and Outreach (2020).

[2] Pillai, Dhanup S., and Rajasekar Natarajan, A compatibility analysis on NEC, IEC, and UL standards for protection against line–line and line–ground faults in PV arrays, IEEE Journal of Photovoltaics 9.3 (2019) 864-871.

[3] Artale, Giovanni, et al, Characterization of DC series arc faults in PV systems based on current low frequency spectral analysis, Measurement 182 (2021) 109770.

[4] Falvo, M. C., and S. Capparella, Safety issues in PV systems: Design choices for a secure fault detection and for preventing fire risk, Case Studies in Fire Safety 3 (2015) 1-16.

[5] PV Magazine USA (2021, July 2), https://pv-magazine-usa.com/2021/07/02/pv-fires-estimated-costs-and-the-road-to-better-data/. Accessed: 22.09.22

[6] Blazey, S, Fire and Solar PV Systems–Investigations and Evidence (2017).

[7] Campbell, C., The relationship between NEC, UL, NRTL, NFPA 79 (2019), https://righthandindustrial.com/the-relationship-between-nec-ul-nrtl-nfpa-79-listed-and-ahj-in-industrial-electrical-installations/. /. Accessed: 19.09.22

[8] European Standards, https://www.en-standard.eu/search/?q=PV+system+protection. Accessed: 23.09.22

[9] European Standards, IEC 62109, https://www.en-standard.eu/search/?q=iec+62109. Accessed: 25.09.22

[10] National Electrical Code Tips, Article 690, Solar Photovoltaic Systems, Part 2, http://www.codebookcity.com/codearticles/nec/necarticle690-2.htm. Accessed: 26.09.22

[11] UL Standard |, UL 1741. https://www.shopulstandards.com/ProductDetail.aspx?productId=UL1741. Accessed: 23.09.22

[12] UL 2579, Fuses for Photovoltaic systems, https://en.sekorm.com/doc/3102908.html. Accessed: 23.09.22

[13] Kenya, The Energy Act 2019, The Draft Energy (Solar Photovoltaic Systems) Regulations (2020)

[14] Foreign, Commonwealth and Development Office (FCDO), Africa Clean Energy Technical Assistance Facility, Tetra Tech International Development, Ethiopian Stand-Alone Solar Standards: Guidance For Adoption And Implementation (2021).

[15] Norsk Elektroteknisk Komite (NEK), Nek 400:2018 (2022), https: //www.nek.no/produkter/nek-400/. Accessed: 22.09.22

[16] Norsk Elektroteknisk Komite (NEK), Nek 446:2022, https://www.nek.no/produkter/nek-4462022/. Accessed: 22.09.22

[17] Strobl, Christian, and Peter Meckler, Arc faults in photovoltaic systems, Proceedings of the 56th IEEE Holm Conference on Electrical Contacts, IEEE (2010).

[18] Patrick Schahl Arc Fault Detection in PV Inverters and How Plant Operators Can Reduce Electrical Fire Threats, Delta Electronics (Netherlands) BV (2021).

[19] SolarEdge, Application Note Inverter Arc Detection in SolarEdge Systems (2020).

[20] PV Magazine USA, Solar system fires are on the rise in the U.S. (2020, April 22). https://pv-magazine-usa.com/2020/04/22/solar-system-fires-are-on-the-rise/. Accessed: 20.09.22

[21] PV Magazine International., How to protect rooftop PV systems from fire risk, https://www.pv-magazine.com/2021/03/23/how-to-protect-rooftop-pv-systems-from-fire-risk/. Accessed: 26.09.22

[22] PV Magazine Australia, Australian rooftop PV safety standards under fire in new report. https://www.pv-magazine-australia.com/2020/11/19/australian-rooftop-pv-safety-standards-under-fire-in-new-report/. Accessed: 19.09.22

[23] PV Magazine International, Japanese government warns against fire risk from rooftop PV, https://www.pv-magazine.com/2019/01/30/japanese-government-warns-against-fire-risk-from-rooftop-pv/. Accessed: 25.09.22

[24] IEC 60947-9-2 (2021)

[25] Norsk Elektroteknisk Komite (NEK). Nek 400:2022, https://www.nek.no/nye-krav-til-dc-bryter-i-nek-4002022-skal-forhindre-brann/. Accessed: 22.09.22

[26] Baurzhan, S., and Jenkins,G.P. Off-grid solar PV: Is it an affordable or appropriate solution for rural electrification in Sub-Saharan African countries? , Renewable and Sustainable Energy Reviews 60 (2016) 1405-1418.

[27] Free Images: house, rooftop, renewable, outdoor structure Free stock photos, https://pxhere.com/en/photo/1346014. Accessed: 23.09.22

[28] Prieto-Castrillo, et al, Warranty assessment of photovoltaic modules based on a degradation probabilistic model., Progress in Photovoltaics: Research and Applications 28.12 (2020): 1308-1321.

[29] PV Magazine USA, Inverter quality and warranties more important to owners, https://pv-magazine-usa.com/2017/08/07/inverter-quality-and-warranties-more-important-to-owners/. Accessed: 23.09.22

[30] Nassau National Cable , What Makes Photovoltaic Wire and Cable Different from Normal Cables? PV Wire vs. USE 2, https://nassaunationalcable.com/blogs/blog/what-makes-photovoltaic-wire-and-cable-different-from-normal-cables. Accessed: 25.09.22

[31] Solar Optimum, Comparing Solar Battery Storage Warranties, https://solaroptimum.com/blog/2021/03/19/comparing-solar-battery-storage-warranties/_(2021). Accessed: 25.09.22

IMPACT OF LARGE FORMAT SOLAR PANELS ON THE BALANCE OF SYSTEM COSTS OF PHOTOVOLTAIC POWER PLANTS

Djaber Berrian, Johannes Linder
Belectric Holding GmbH, Wadenbrunner Str. 10, 97509 Kolitzheim

ABSTRACT: New large-size photovoltaic (PV) modules with a power rating exceeding 500Wp have been introduced to the PV market recently. These large format PV panels, known as M10 made of 182mm cell-size, and M12 made of 210mm cell-size, are expected to dominate the PV module market share in the coming years due to their high-power output. We benchmarked - in terms of the balance of system (BoS) costs - the new format PV modules to the current dominated PV module size (M6) in the market, which is made of 166mm cell size. Based on an existing large-scale utility PV plant, we modeled the power, energy yield as well as the involved costs for the three PV module formats mentioned earlier. Both large formats PV modules have shown a significant BoS costs saving up to 6% for approx. 50MWp PV project in comparison to the current mainstream PV modules. Interestingly, we did not find a noticeable difference in terms of BoS costs between M10 and M12 sizes, although the M12 PV module size usually has a higher power rating than M10 PV modules.
Keywords: Cost reduction , Bifacial, Modeling, PV System, PV power plant, PV module

1 INTRODUCTION

For many decades improving cell efficiency was the main driver for achieving a low Levelized cost of electricity (LCoE), this has resulted in a steady increase of solar panel power output of about 5W to 10W yearly[1][2]. However, since 2016, new wafer formats have been introduced which drove rapidly the size and power rating of PV modules. From 2016 to 2021, the power rating of PV modules was increasing from 40W to 50W per year [3], as of this year commercial PV modules can deliver more than 600 Watt-peak. The new format PV modules, which are known as M10 and M12 (or G12) corresponds respectively the wafer or cell size of 182mm and 210mm, have disrupted the value chain - especially the compatibility of new size PV modules with standard PV inverters - and made the PV system designers confused on what technology to consider for better overall performance and cost-effective design of PV systems.

The impact of larger size based solar cells (M10, M12) on module power and efficiency in terms of cell-to-module loss (CTM) has been previously investigated, where it has been found that the power increase with larger solar cells while the electrical loss of interconnection is higher with larger cell sizes [4]. When it comes to large-size PV modules (M10 and M12), we find that different PV module manufacturers claim different opinions about what size to dominate the PV market in the coming years. Some rely only on the potential of M10 PV module size as it combines the high power and mechanical durability, while other PV module suppliers are determined to manufacture only M12 as this allows to achieve much high power per piece of PV module. Both groups of PV suppliers mentioned earlier agree on a significant cost reduction at the system level when using big-size PV modules, however, the effect of large-size PV modules (M10 and M12) on the balance-of-system (BoS) costs is hardly quantified in the literature. A similar analysis about the impact of PV module size on the BOS cost was made by the national renewable energy laboratory (NREL), which focus on other size of PV modules not specifically the comparison of M10 and M12 format [5]. The BoS cost of PV systems is defined as cost per installed capacity (€/kWp), all cost elements – excluding the PV modules costs - in a PV systems project which can be influenced by the PV module design and format are considered as BoS costs. The evaluation of BoS costs can be challenging as several sources of cost (such as inverters & MV stations, e-BoS cost, substructure cost, PV modules transportation, substructure transportation, construction & installation cost substructure, construction & installation cost modules, project development costs, project management, SCADA, facilities... etc.) have to be determined or usually are just estimated. Belectric has several years of experience in project development of PV plants as well as in engineering, procurement, and construction (EPC) of more than 4GWp of PV power plants worldwide. This has allowed Belectric to have a stack of real figures about different costs associated with PV projects, as well as to set up cost models by mapping cost relationships between different cost elements of utility-scale power plant projects. Moreover, Belectric has a dedicated team who works on procurement for different types of PV systems. In this work, we investigate the impact of large-size PV modules (M10 and M12) on the performance and the balance of system costs of utility-scale PV power plants with different installed capacities. We also quantify the BoS costs of large-size PV modules to the mainstream PV module size (M6) based on a concrete German PV power plant.

2 METHODOLOGY AND KEY MATERIALS

2.1 Large format PV modules

In this study three module formats were considered, M6 format - made on 166mm cell size - which currently dominates the global market share of PV modules [6], while 182 mm (M10) and 210 mm (M12) cell formats are new to the PV market with a huge potential of taking over M6 PV modules market share, especially in case of utility-scale PV systems [6]. The main difference between M6, M10, and M12 PV modules is the cell size used to make these solar panels, the cell size and the number of cells define the PV module dimensions. Table I summarizes the cell size, the number of cells per module, as well as the mechanical dimensions and power rating of the PV modules explored in this work.

2.2 Energy yield modelling

The methodology used to determine the impact of

Table I Properties summary of different PV modules formats used for the analysis

PV Module	Power (W)	Efficiency (%)	Technology	Voc (V)	Isc (A)	Size (m)	Area (m2)	Weight (kg)
M6 (166-72c)	445	19.56	p-type PERC, half cut, MBB	49.45	11.380	2.148x1.060x35	2.276	28.2
M10 (182-72c)	544	21.02	p-type PERC, half cut, MBB	49.50	13.850	2.284x1.133x35	2.588	32.3
M12 (210-60c)	595	21.02	p-type PERC, half cut, MBB	41.50	18.360	2.172x1.303x40	2.830	35.3

large format PV modules on performance and cost is based on three phases. The first phase aims to determine the specific energy yield for many scenarios of PV systems. For a given project site, we use the meteorological data (irradiance and temperature) as input to PVsyst 7.4.9 software and run a system design optimization through energy yield modeling for three formats PV modules, M6, M10, and M12 considering the PAN file of each PV module technology. The system design optimization is based on a series of different geometrical configurations of PV systems, which are generated based on a batch simulation that combines a range of tilt angles, row spacing values, DC/AC ratios, a stack of PV modules (2 portraits, 3 portraits, or 4 portraits), as well as the system type (fixed tilt, tracking, or east-west). Since the share of bifacial modules is increasing consistently every year [6] and the use of bifacial modules for PV power plants have the potential of reducing the LCoE [7-8] through high yield and low-price gap to monofacial modules, we also considered them along with the monofacial PV modules in our system design optimization. As a result, the total number of design configurations amounts to more than 5000, that are run in the batch mode of PVsyst 7.4.9.

2.3 Estimation of installed capacity per plot size

In the second phase, the AutoCAD-based solar design software 'PVcase' is used to determine the installed capacity in terms of megawatt peak for each PV module technology within the plot size of the site. PVcase is a fast-growing tool in the solar industry thanks to its accurate and efficient way of designing a PV power plant for a given plot size. PVcase tool allows to allocate the PV modules including the substructure (PV module supporting tables) within the site of interest, other PV components such inverters, MV stations, and cables can be also easily drawn in the plot size of the site, so the amount of installed PV capacity for a given PV module dimensions and orientation is determined. The surrounding objects which can cast shading on the power plant such as trees, overhead power lines, communication poles... etc., can be also introduced to estimate the loss due to shading. Knowing the amount of megawatt peak per the plot size accurately is crucial for the determination of the capital expenditure (Capex), as well as the sales of electricity for PV system projects. At Belectric, we have a dedicated surveying team in charge of this task, the installed capacity per plot size can change from one PV module to another depending on the PV module efficiency, size, and orientation.

2.4 PV system cost estimation

Belectric has developed a cash flow model based on the Levelized cost of electricity in order to determine the capital expenditure (Capex) and operating expense (Opex) of PV power plants, as well as the sales of electricity for a given specific case assumptions and electricity price offtake (or incentives) in €/kWh.

Several cost sources are taken into account in the LOCE tool, such as substructure costs, PV modules, installation and transportation, electrical balance of system (eBoS), medium power station (MV stations), operating and maintenance, in addition to financial and tax expenditure. The tool also allows for the calculation of the net present value (NPV) of each configuration of the PV power plant for the entire project lifetime which is in a range of 25-30 years. For instance, if we have simulated several configurations as mentioned in section 'Energy yield modelling', then the LOCE tool determines the NPV of all these configurations. In the end, the winning PV system configuration will correspond to the highest NPV of the project. The NPV formula is simplified and defied in the equation below, where S_t and C_t are the sales of electricity and total expenditure at year t respectively, and the discount rate is d.

$$NPV\ (€) = \sum_{t=1}^{n} \frac{S_t}{(1+d)^t} - \sum_{t=1}^{n} \frac{C_t}{(1+d)^t}$$

3 RESULTS AND DISCUSSION

3.1 Site description and constraints

The suggested site of the PV power plant is in Deubach village in southwest Germany. The coordinates of the location are 49.5638° N, 9.7746° E. The PV system will be mounted on a dry-grass field surrounded by green grass from a cultivating area in the south and northeast of the plot as shown in Fig. 1.

According to the surveying results, two plot areas are available for the PV plant, 10.76 ha area in the north-west and an area of 27.28 ha in the south-east of the location respectively, which makes a total plot area of 38.04 ha. These two plots are separated by high voltage power lines crossing from east to west as shown in Fig. 1. For safety reasons, the space under the high voltage power lines cannot be used for the PV plant, hence the area is left empty. To find out how many megawatts peak (MWp) can be the installed in Dubach, the PVcase software was used for an exemplary configuration made of four modules-in-portrait, tilted by 15°, oriented full south with

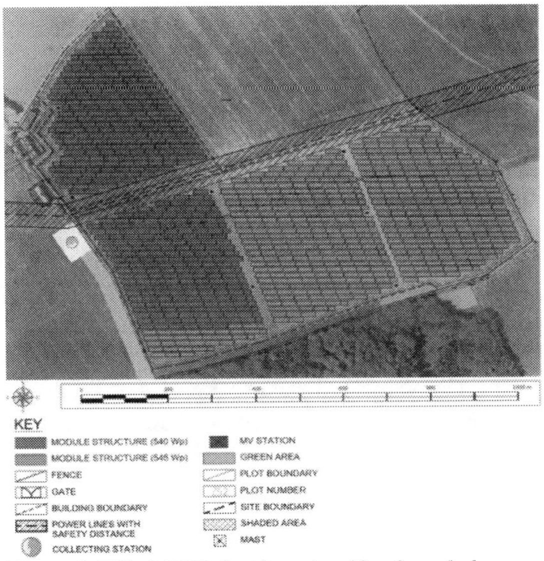

Figure 1: Bifacial PV plant layout and key boundaries

a row pitch of 11.81 m and a bifacial PV module of 544Wp average power (M10 format) have been chosen for that. As a result, and as it is shown in Fig. 1, the two plot sizes of Deubach can accommodate circa 49 MWp in the case of M10 PV modules, excluding all restricted areas for PV panels. The PVcase was also used for M6 and M12 PV module sizes following the same procedure as described above to find out the number of PV modules per plot size, hence the achieved capacity. The results are shown in Table II.

Table II: Installed PV capacity (MWp) for different PV module sizes

PV Module format	PV Module count	Achieved capacity (MWp)	Additional MWp to M6 format
M6 (166-72c)	104070	46.311	-
M10 (182-72c)	90220	49.080	+ 2.769
M12 (210-60c)	82689	49.200	+ 2.889

3.2 Solar resource assessment

The energy production of a power plant is directly linked to the weather and irradiance data. Therefore, the energy yield of a PV system depends on the site latitude, the season, the time of day, the orientation, and the inclination of solar panels from the ground. PVsyst is a worldwide known and bankable software for yield modeling of monofacial and bifacial PV systems, it was also validated for bifacial yield prediction [9-10]. The software has been used for this work, A meteorological file - includes irradiance and weather information - is essential to run PVsyst simulations. For this purpose, we used the solar resource platform called 'Solargis', the platform provides historical, recent, and forecasted meteorological data for almost any location on Earth. Solargis meteorological files known as 'typical metrological year' (TMY), are generated on satellite-based models [11]. A TMY P50 file is considered in this study which is usually sufficient for many financial institutes. The P50 detonates that the probability (50%) of the solar resource in the TMY being exceeded throughout the lifetime of the PV project. Due to bankability reasons, some financial institutes require a higher probability of exceedance of the solar resource in TMY files which can be up to 90% (P90), this is usually considered a conservative estimate, thus usually the P90 values are lower than P50. Another important parameter that impacts the yield considerably is the albedo (the ground reflection of light) of the site, as bifacial PV modules were also planned for this project, knowing the albedo of the location is required to run bifacial PV simulations in PVsyst. Based on the project surveying of the site, the albedo is assumed to be 20% and constant over the year. Table III summarizes the weather and irradiance data of the Deubach site.

Table III: Weather and irradiance data summary

Parameter	Yearly values
Yearly accumulated GHI (kWh/m2)	1135
Yearly accumulated DHI (kWh/m2)	581
Yearly accumulated BHI (kWh/m2)	554
Average ambient temperature (C)	8.9
Average wind speed (m/s)	6 m/s at height 100 meters

The following are the acronyms used in the table.
GHI: global horizontal irradiance
DHI: diffuse horizontal irradiance
BHI: beam horizontal irradiance

3.3 PV system-specific energy yield assessment

The specific energy yield at the AC level is modeled by PVsyst 7.4.9 for different types of PV system configurations and module sizes as described in section 2.2. We have benchmarked different bifacial PV module formats to two configurations, four-in-portrait fixed tilt, south oriented, with a tilt angle of 15°, and a ground coverage ratio (GCR) of 76%. And one-in-portrait horizontal single-axis tracker with a GCR of 42%. String inverters were used for this analysis with a DC/AC ratio of 1.2. The bifacial modules selected for this comparison have all the same bifaciality factor (70%) and are measured according to IEC 60904-1-2. Table IV shows the modeled energy yield for different PV module formats in comparison to M6 for two configurations: fixed tilt and tracking. As it can be seen, both M10 and M12 have higher specific energy yield than M6. This is mainly due to better low irradiance performance and lower temperature coefficient of large size PV modules (M10 and M12) than M6. Due to the higher operating current of M12 PV modules (more ohmic losses) than M10, a lower specific yield is achieved for those PV modules than M10 format which runs at a very similar current to M6.

Table IV: Specific energy yield for fixed tilt and tracking bifacial PV systems

	Fixed tilt configuration		Tracker configuration	
PV module format	Specific yearly energy yield (kWh/kWp)	Relative deviation to M6 (%)	Specific yearly energy yield (kWh/kWp)	Relative deviation to M6 (%)
M6 (166-72c)	1086.2	-	1326.0	-
M10 (182-72c)	1091.0	+0.44%	1330.04	+0.33%
M12 (210-60c)	1088.4	+0.20%	1328.6	+0.19%

Table V: NPV cash flow of different bifacial PV systems

Substructure	Yield (kWh/kWp/a)	Capex deviation (€)	Opex deviation (€)	Sale of electricity deviation (€)	NPV deviation (€)
FIX South, 4P, 15°, bifacial	1084,8	0	0	0	0
FIX South, 3P, 20°, bifacial	1098,6	-1.283.739 €	-276.405 €	-1.996.004 €	**-435.860 €**
SAT, 1P, bifacial	1323,2	-10.603.735 €	-1.768.385 €	-18.537.965 €	**-6.165.845 €**

Based only on the specific yield optimization, we observe the system design with a tracker has the highest specific yield, however, we will show in the next section that the most profitable system design - considering the yield as well as the associated costs (Capex, Opex, and sales of electricity) - is a fixed tilt system.

3.4 Techno-economic assessment

3.4.1 NPV assessment

This part is conducted based on the Belectric way for cost modeling of power plants as mentioned in section 2.4, in addition to its strong network with PV module, inverter, and equipment suppliers. The LOCE tool of Belectric was used with the latest prices for different components of PV systems for the Deubach power plant. For the cash flow analysis, we use the following case assumptions, an electricity offtake for the first 10 years (phase 1) is 6 c€/kWp, from year 11th (phase 2) to year 30th, the electricity offtake is assumed 4.8 c€/kWp. The discount rate for phases one and two are 3.25% and 4.5% respectively. We calculated the NPV of all simulated configurations (more than 5000 scenarios), Table V summarizes three design configurations based on NPV, i.e., the best tracker design configuration, and the 3P and 4P design configuration as well. As it can be seen, although the tracker has the highest energy yield and the lowest Capex and Opex among other PV system designs, however, in terms of sales of electricity, the tracker has the lowest outcomes which in turn makes its NPV value deviate a lot from the best system design (Fix south, 4P as a reference in Table V). This can be explained by the fact that, due to the very high electricity offtake (the market price of 2022 in Europe) and the available plot area (36 hectares), 1P solar trackers cannot achieve installed capacity as big as fixed tilt 4P or 3P. However, it is worth mentioning, that the tracker is still a profitable system design for this project (NPV>0), but not the most profitable. It also observed that the best system designs based on NPV ranking are all in combination with bifacial modules, we hardly see a system design (either, fix tilt or tracking) very well ranked in combination with monofacial PV modules. We found two reasons behind the bifacial PV module dominance in system design, one is obviously the bifacial gain (improved yield) in comparison to monofacial PV modules, second and with much more impact is the price gap between monofacial and bifacial PV modules which continues to squeeze down, as of the current time of this work the price gap is 3€/kWp. For the analysis of the impact of large format PV modules on BoS costs, we will only deem the most profitable system design i.e., fix south, 4P, 15°, with bifacial PV modules.

3.4.2 BoS cost assessment

The bar chart in Fig. 2 shows the breakdown of BoS costs of approx. 50MWp PV project in Germany considering a scenario of a four-in-portrait fixed tilt configuration with bifacial PV modules of different sizes. As can be seen, both PV module formats M10 and M12 have overall lower BoS costs than the M6 PV module format (-5.3% and -5.1% respectively). Based on the cost modeling, we found out there are several cost elements that contribute to the low BoS costs of large-size solar panels (M10, M12) in comparison to M6. Large PV modules mainly save on PV module count, i.e., fewer PV modules are needed to achieve the required capacity. As result, the bar chart shows the main cost reduction is associated with the PV module transportation (purple

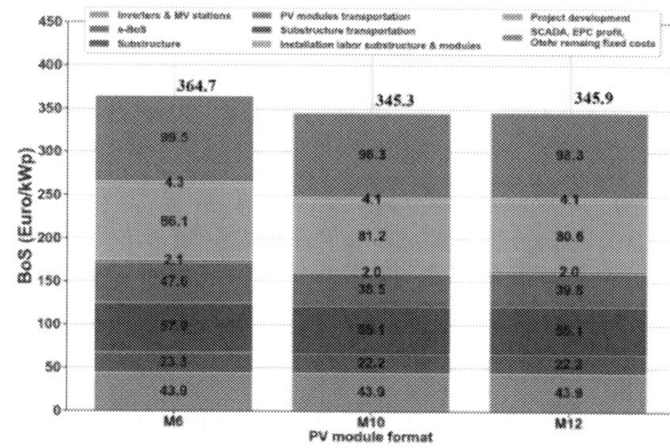

Figure 2: Balance of system cost (BoS) breakdown of the large-scale bifacial power plant for different PV module formats

zone), followed by costs related to labor work for installation of PV modules and substructure (gold zone). Other, costs that are also decreased for large-size PV modules, are the substructure costs and the electrical balance of system costs (e-BoS). This is explained by the fact that large PV modules allow for a smaller number of strings per the plot size to fulfill the needed power capacity in comparison to the standard PV module size (M6), which as a result leads to a smaller number of tables substructures and reduced electrical balance of system. Due to the high installed power capacity of large solar panels, fixed costs - such as project development - are also reduced in terms of €/kWp. Another important outcome of this study is the BoS cost difference between M10 and M12, several PV investors are wondering what large format PV module would save most in terms of BoS costs. The BoS saving due to large PV modules (M10 and M12) is shown in Fig. 3 for the fixed tilt four-in-portrait mounting, both solar panel sizes were found to save approx. 5% of BoS costs in comparison to the M6 PV module, with a very slight tendency of M10 to save more BoS costs than M12 for this specific layout. As a result, in terms of PV module selection between M10 and M12 for a given PV power plant, other factors are more important to consider, such as PV module price, yield, the compatibility of the PV module with other PV system components (inverters, substructure … etc.) as well as the reliability of large size PV modules throughout the PV project lifetime.

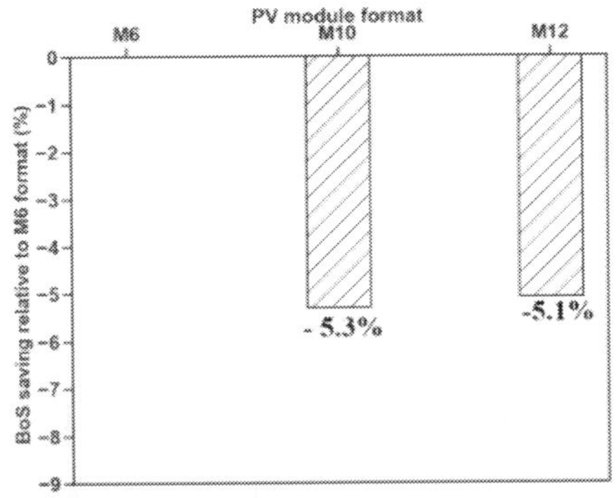

Figure 3: Balance of system cost (BoS) saving of large format PV modules (M10 and M12) in comparison to M6 for the large-scale bifacial power plant

3.4.3 BoS cost saving sensitivity analysis

To find out how robust are the findings from the previous section, we run a sensitivity analysis on three factors that were found to be the main reason for BoS cost saving for large format PV modules, which are the installation costs (PV modules and substructure), the PV module transportation costs, as well as the project volume (MWp). In our analysis, we assume PV modules are shipped from China to the project site. The PV module transportation costs are directly linked to the shipping price of PV module containers as well as the maximum power capacity that can be shipped inside the container for a given module type (M6, M10, or M12). Fig. 4 shows the BoS cost savings as a function of PV module container freight costs for M10 and M12. As it can be seen, the BoS cost saving depends strongly on these costs for both M10 and M12. BoS cost saving is very high (above 5%) when the PV module freight costs are above 12,000€/container, this coincides usually with solar projects in Europe. Whereas solar projects with PV module freight costs below 8000€/container have very low BoS cost saving (below 4.5%), this corresponds - for example - to solar projects in India or China. Another interesting fact about BoS cost saving in terms of M10 versus M12 is the dependency on the solar project site. M10 PV modules tend to save more BoS costs than M12 when a solar project is in Europe (or with very high PV module freight costs), while for solar projects in China (or with very low PV module freight costs), M12 has a lead in terms of BoS cost saving. Interestingly, for solar projects in India (or medium PV module freight costs), M10 and M12 can save the same amount of BoS costs, no preference is given to either PV module for better BoS saving.

Large format PV modules (M10 or M12), due to their increased size and weight come with additional installation costs in comparison to the M6 format. In the previous section to find out the BoS cost saving (Fig. 2), we assumed the installation costs of large-size PV modules to be 5.3% higher than M6. This figure was based on the Belectric way of installing of large format PV modules in the field, but new machineries are expected to be used in the future to install large format PV modules which might change the assumed additional installation costs of large format PV modules. To find out how this parameter could influence the BoS cost saving

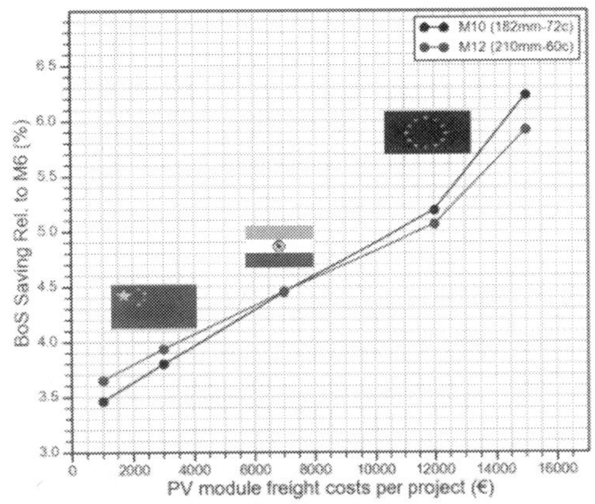

Figure 4: Balance of system (BoS) cost saving sensitivity to PV module freight costs

concluded in the previous section, we did BoS cost saving sensitivity on the additional installation costs of M10 and M12 which is shown in Fig. 5. Obviously, the BoS cost saving drops for both M10 and M12 for the solar project in Deubach when higher installation costs are assumed. Because these additional installation costs compromise the advantage of large format PV modules (low number of modules to achieve the needed MWp capacity) seen previously in the BoS cost breakdown bar chart. If a smaller number of modules require high installation costs than M6, then the advantage of the smaller number of modules with large format PV modules is compromised. However, it is important to mention that the rate of drop in BoS cost saving is very low compared to the rate of increase in additional PV module installation costs.

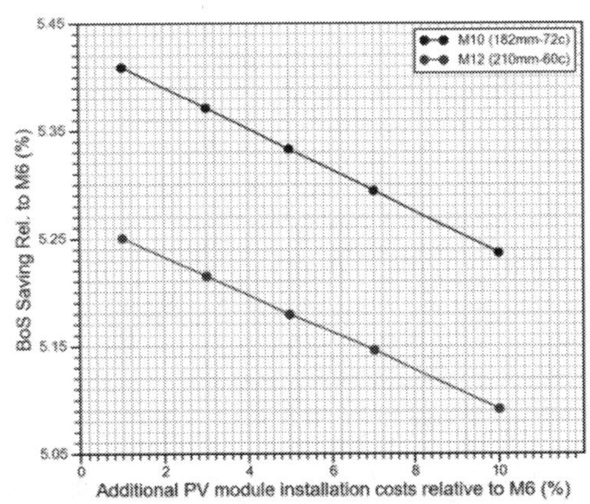

Figure 5: Balance of system (BoS) cost saving sensitivity to the additional PV module installation costs of large format PV module relative to M6

The sensitivity of the BoS cost saving relative to M6 in terms of PV project size is shown in Fig. 6 for M10 and M12, considering three cases of PV module freight costs of 3000€, 7000€, and 13000€, which reflect three regions China, India, and Europe, respectively. The BoS cost saving depends on the PV project size (MWp), the difference in BoS cost saving between M10 and M12 is almost negligible for project sizes below 20MWp. Whereas increasing the PV project size (MWp) the BoS cost saving relative to M6 drops, this behavior is seen for the three cases (or regions). This observation can be explained by the fact that when increasing the project size (MWp) only fixed costs (€) are impacted, but specific costs (€/kWp) remain the same. Hence, the fixed costs are becoming very small in terms (€/kWp) for big power plants, as well as the absolute difference in fixed costs (€/kWp) between M6 and M10/M12. Whereas, for small size power plants the fixed costs become significantly high in terms (€/kWp), hence the absolute difference in fixed costs (€/kWp) between M6 and M10/12 is big. When the BoS cost saving relative to M6 is calculated, the small size power plants have higher BoS cost saving due to the high absolute difference in fixed costs between M6 and M10 or M12 than the case of large power plants. We also see the decline rate in BoS costs saving from small size power plant (<50MWp) to large PV power plants (>50MWp) depends on the PV module freight costs, high freight costs does not imply a high drop in BoS cost saving between small and large scale power plants, because the impact of very high PV module freight costs dominate in the total BoS costs over the impact of large scale power plants on the fixed costs.

The results shown in Fig. 6 confirm the observations from Fig. 4, M10 and M12 PV modules can have different or the same BoS saving relative to M6 depending on the region (or PV module freight costs), on some occasions M10 can save more than M12, and vice versa.

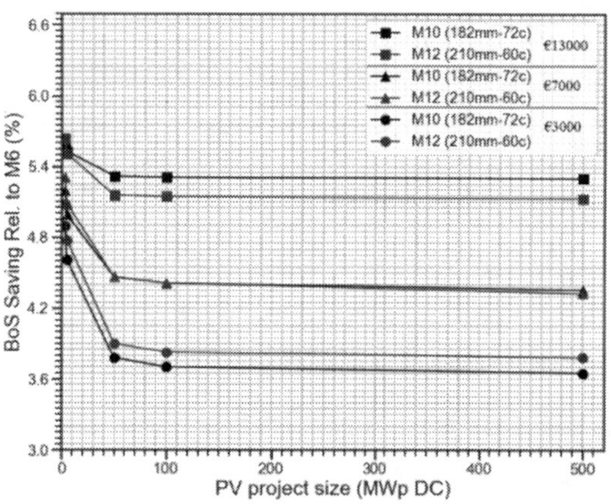

Figure 6: Balance of system cost (BoS) saving sensitivity to the PV project size for different locations, i.e., different PV module freight costs

4 CONCLUSIONS

Large format PV modules (M10 or M12) can reduce significantly the BoS costs of PV power plants by up to 6% in comparison to the mainstream PV module format (M6). PV module transportation costs, as well as labor costs for the installation of PV modules and substructures, are the main reason for the BoS cost saving in large-size PV modules. The BoS cost saving depends strongly on the project size as well as on the PV module freight costs or the PV project site. The difference in BoS cost saving between M10 and M12 is almost negligible for project sizes below 20MWp. In terms of PV module selection between M10 and M12 for a given PV power plant, we conclude that the choice of either size of PV module will be project-specific and other factors related to reliability are more important to consider when selecting between M10 and M12. It has been found also that the BoS cost saving drops for both M10 and M12 for the solar project in Deubach when higher installation costs are assumed, however, the rate of drop in BoS cost saving is very low compared to the rate of increase in additional PV module installation costs.

5 References

[1] Hermle, M., Feldmann, F., Bivour, M., Goldschmidt, J.C. and Glunz, S.W., 2020. Passivating contacts and tandem concepts: Approaches for the highest silicon-based solar cell efficiencies. Applied Physics Reviews, 7(2), p.021305.

[2] Wirth, H., Photovoltaic Modules/ Photovoltaic Module Technology, 2nd ed. Berlin: De Gruyter, 2020

[3]Lau, H., "Larger Wafer-based Solar Cells and Modules: Solar Industry Trends to Design Modules with Higher Power Ratings and Lower BOS Needs," Nov. 18, 2019

[4] Mittag, M., Pfreundt, A. and Shahid, J., Impact of Solar Cell Dimensions on Module Power, Efficiency and Cell to Module Losses

5] Horowitz, K.A., Fu, R., Silverman, T., Woodhouse, M., Sun, X. and Alam, M.A., 2017. An analysis of the cost and performance of photovoltaic systems as a function of module area (No. NREL/TP-6A20-67006). National Renewable Energy Lab. (NREL), Golden, CO (United States).

[6] ITRPV, International Technology Roadmap for Photovoltaic, 2022, https://www.itrpv.net (accessed: June 2022).

[7] Libal, J., D. Berrian, and R. Kopecek. "Overview: energy yield simulations and calculation of LCOE for bifacial PV systems." Bifi PV Work (2017).

[8] Rodríguez-Gallegos, C. D., Liu, H., Gandhi, O., Singh, J. P., Krishnamurthy, V., Kumar, A., ... & Peters, I. M. (2020). Global techno-economic performance of bifacial and tracking photovoltaic systems. Joule, 4(7), 1514-1541.

[9] Nussbaumer, H., Janssen, G., Berrian, D., Wittmer, B., Klenk, M., Baumann, T., ... & Mermoud, A. (2020). Accuracy of simulated data for bifacial systems with varying tilt angles and share of diffuse radiation. Solar Energy, 197, 6-21.

[10] Berrian, D., Libal, J., Klenk, M., Nussbaumer, H., & Kopecek, R. (2019). Performance of bifacial PV arrays with fixed tilt and horizontal single-axis tracking: comparison of simulated and measured data. IEEE Journal of Photovoltaics, 9(6), 1583-1589.

[11] Cebecauer, T. and Suri, M., 2015. Typical meteorological year data: SolarGIS approach. Energy Procedia, 69, pp.1958-1969

COMPARISON OF SOILING MEASUREMENT APPROACHES IN SOLAR PV PLANTS

Paula Rodriguez[1], Hector Aristizabal[2], Jose Muñoz[3], Shilpee Sinha[4]
[1]Mott MacDonald Colombia SAS, Calle 161 # 54-87, Bogotá, Colombia, paula.rodriguez@mottmac.com
[2]Mott MacDonald Colombia SAS, hector.aristizabal@mottmac.com
[3]Mott MacDonald Ltd Spain, jose.munoz@mottmac.com
[4]Mott MacDonald Ltd, shilpee.sinha@mottmac.com

ABSTRACT:
Soiling is a key system loss that could be minimised during the operational phase of a Solar PV system to improve its performance. Therefore, accurately measuring, monitoring and analysing soiling data at PV plants, especially in arid, semi-arid and air polluted environments, gains importance.
In this study we aim to understand how different variables such as ambient and PV module temperature, rainfall, relative humidity and solar angle, impact the soiling measurements obtained by the calibrated cells (CC) and optical sensors (OS) of two PV plants located in Mexico. We analyse how the output data from each of these two measurement approaches should be treated and analysed according to the variables that impact such measurement.
The analysis performed shows no correlation between the selected variables and the Soiling Ratio (SR) obtained with the CC and OS. The dispersion of SR obtained with the CC was significatively higher than the OS, which seems to be mainly due to the variation in the irradiance measured by the CC through the day, and thus, filtering the data by irradiance reduces the difference of the SR measured with the two different devices.
Keywords: Soiling, Monitoring, Shading, Energy Performance, Measurement Method, Data Analysis

1 INTRODUCTION

Over the last decades, PV modules' efficiency has increased considerably, and manufacturers continue expending effort to increase it. In parallel, PV system developers try to maximise energy production by reducing the energy losses associated with the PV System performance through proper equipment selection, design considerations, installation methods and Operation & Maintenance (O&M) activities.

One of the key system losses that could be minimised during the operational phase of a solar PV system is soiling. Soiling is defined as the accumulation of snow[1], dirt, dust (generally defined as a Particulate Matter (PM), less than 500 µm in diameter [1]), tree leaves, pollen, and bird droppings on PV modules, which leads to shading. As it is highly dependent on the environmental conditions, soiling differs considerably from one location to another, having a higher impact in arid, semi-arid and air polluted environments. Moreover, system design aspects such as PV modules tilt angles also impact the soiling levels. A recent study estimated an impact of 3% - 4% loss of the global energy yield in 2018 due the soiling [2].

Soiling on PV modules can lead to two types of shading: soft shading, that occurs when soiling particles reduce the overall irradiance that reaches the PV cells surface, negatively impacting the generated current but with no effect on the module voltage; and hard shading, which blocks the irradiance at the module's surface, hence affecting both the current and voltage of the PV module. Soft shading can cause current imbalance between PV modules within a strings and hard shading voltage mismatch between strings; both affecting performance loss associated with the inverter's Maximum Power Point Tracking (MPPT). In addition to the impact on the irradiance that reaches the PV cell surface and on the system performance, soiling can also generate hot spots on the PV modules, that could potentially lead to permanent damage (i.e., burnt cell, glass breakage, etc.).

Depending on the environmental conditions of the site, soiling could have a significant impact on the performance of a PV plant, reducing the energy generation and increasing the O&M costs and thus, increasing the Levelized Cost of Electricity (LCOE).

Two physical principles to measure soiling are defined in the IEC 61724-1:2021 [3] (Photovoltaic system performance. Part 1 Monitoring): (1) Calibrated Cells: comparing the measurements of a pair of PV reference cells, one of which is routinely cleaned (the "clean cell") and the other is let to soil naturally (the "soiled cell") and (2) Optical Sensors: using optical principles, to determine the effect of reflection or transmission of light in a surface due to soiling.

In general, Calibrated Cells (CC) and optical Sensors (OS) are very compact and easy to install. Calibrated Cells have low CAPEX; however, human error may affect the measurement, i.e., the soiled cell could be cleaned by mistake instead of the clean cell, leading to erroneous data. On the other hand, Optical Sensors have a higher CAPEX but do not suffer of human error as in the Calibrated Cells. Additionally, measurements using Calibrated cells are mainly affected by temperature and long-term degradation; while measurements using Optical Sensors are mainly affected by moisture and precipitation, as the water content in air affects the light transmission measurement. Further, measurements with the Optical Sensors are also influenced by the colour of the dust as this impacts the measured transmission loss. There is a wide dust variation around the globe in terms of colour and mineral composition [4].

The actual soiling impact on PV modules may be different to the soling measured by the Optical Sensors as those measurement methods do not consider the effect of PV module coatings (anti soiling or anti-reflective coatings) and frames. Moreover, the soiling impact on PV modules caused by humidity and dew point is different to the impact in the soiling measurement instruments. When humidity and dew point are large enough, PV modules

[1] In this study we have not considered the effect of snow.

could get naturally cleaned (depending on the amount of water, PV module tilt, type of glass surface); otherwise, dust cementation could be enhanced (dust particles could stick to the surface of the PV module and to one another) - soiling cementation depends on the dust composition. Soiling deposition rates are also affected by ambient conditions, including wind velocity, wind direction and seasonal variations [1].

Energy loss caused by soiling is also influenced by the daily irradiance variation due to the variation in solar angle incidence on PV module surface through the day. Measured soiling losses tend to be higher during early morning and late afternoon periods, however during these periods the energy generated by the PV plant is minimum.

With the aim to understand how different variables such as temperature, rainfall, moisture and solar angle amongst others impact on the two different soiling measurement approaches, as well as to identify how the output data from each measurement approach should be treated and analysed to obtain an accurate and reliable measurement of a PV system soiling loss; we have analysed the data obtained from Calibrated Cells and Optical Sensors installed in two PV plants in Mexico.

2 METHODOLOGY

We have used the three (3) year 5-minute data collected from the Calibrated Cells and Optical Sensors installed in two (2) PV plants located in Yucatan (Plant 1) and Jalisco (Plant 2) States in Mexico. The data analysed in this study was obtained from the pair of Calibrated Cells and Optical Sensors installed in the tracking structures of the PV plants - at the same plane of the PV modules.

By using programming tools, we have treated (cleaning and filtering) and analysed the data obtained from each of the soiling measurement methods as well as the key meteorological data: (1) ambient temperature - Tamb, (2) Rainfall - Rain, (3) relative humidity - RH, as well as (4) PV module temperature - Tmod. In this study, the difference in the measured values between the Optical Sensors and Calibrated Cells as well as the impact of the five aforementioned variables in the soiling measurements is analysed.

2.1 Data definition

The Soiling Ratio (SR) is defined in the IEC 61724-1:2021 as the ratio of the actual power output of the PV array under given soiling conditions to the power that would be expected if the PV array was clean and free of soiling; and the Soiling loss (SL) as the power loss due to soiling (SL = 1 - SR). Thus, the SR of a clean panel is 100% and for a soiled panel it tends to 0 %.

In this study we compare the recorded Soiling Ratio (SR) data from the Optical Sensors and the calculated SR for the Calibrated Cells by using the measured irradiance as defined below.

$$SR_{CC} = \frac{\text{Irradiance measured in the soiled cell}}{\text{Irradiance measured in the clean cell}} * 100\%$$

2.2 Data cleaning (Raw Data - Cleaned Data)

Data outliers for the measured variables: Tamb, RH, Tmod, Irradiance of the cleaned calibrated cell (Iclean), Irradiance of the soiled calibrated cell (Idirty), Soiling ratio of the optical sensors (SRos); as well as for the calculated SRcc, were removed.

For the purpose of this study, we have defined data outliers as the values that are outside the range of values summarised in Table I.

Table I: Range of accepted values

Variable	Min value	Max value
Ambient Temperature (°C)	0	40
Relative Humidity (%)	0	100
PV module Temperature (°C)	0	90
Irradiance (W/m2)	0	1,400[2]
Soiling Ratio (%)	50[3]	100

2.3 Data treating (Cleaned Data - Filtered Data)

We have analysed three filtering approaches for the soiling ratio:

(1) G filter - removing irradiance effect: In order to reduce the impact of the environmental stability, such as irradiance and temperature caused by cloud influence in the soiling measurements, the SR points in which the irradiance at the plane of the PV modules was below 800 W/m2, were removed.

(2) Tdew filter - removing dew temperature effect: the SR values in which the PV temperature is not 2°C above the dew temperature, were removed, as per the recommendation from the manufacturer of the optical sensors [6].

(3) G-Tdew filter - removing irradiance and dew temperature effect: applying G filter (1) and Tdew filter (2).

2.4 SR comparison

The daily Soiling Ratio with the calibrated cells and the optical sensors for each plant were compared.

The daily SR values were obtained by averaging the 5-minute data for each of the following datasets:

(1) Raw Data – data as taken from the SCADA system

(2) Clean Data – data after removing outliers (section 2.2)

(3) Filtered Data – data treated (section 2.3)

(3.1) G filter

(3.2) Tdew filter

(3.3) G filter - Tdew filter

Moreover, in order to analyse the impact of solar angle incidence on PV modules, three different datasets based on the time of the day (A) Daytime – from sunrise to sunset, (B) Around Noon –± 4h around solar noon and (C) Noon – ± 2h around solar noon, for each of the five datasets above (1, 2, 3.1, 3.2 and 3.3) were created. A total of 15 datasets were analysed (i.e. 1A: Raw Data – Daytime, 3.1 B: G filter – Around Noon, etc).

Furthermore, the difference between the daily Soiling Ratio obtained with the calibrated cells and the optical sensors is defined as:

$$dSR = \frac{SR_{CC} - SR_{OS}}{SR_{OS}} * 100\%$$

The mean and standard deviation of the dSR between the nine (9) filtered datasets (i.e 3.1A, 3.2A, 3.3A, 3.1B, etc), were compared.

[2] Calibrated Cell manufacturer datasheet [5]

[3] Applicable only for the optical sensor, as per the manufacturer datasheet [6]

The dSR for both Plants was plotted against the daily values of the key variables (Tamb, Rainfall, RH and Tmod) with the aim to understand if there is a correlation between the key variables and the dSR.

3 RESULTS

3.1 Soiling Ratio – CC vs. OS

Figure 1 presents the SR average (line) and standard deviation (shaded area) obtained with the CC and OS during the analysed period for each of the datasets for both PV plants.

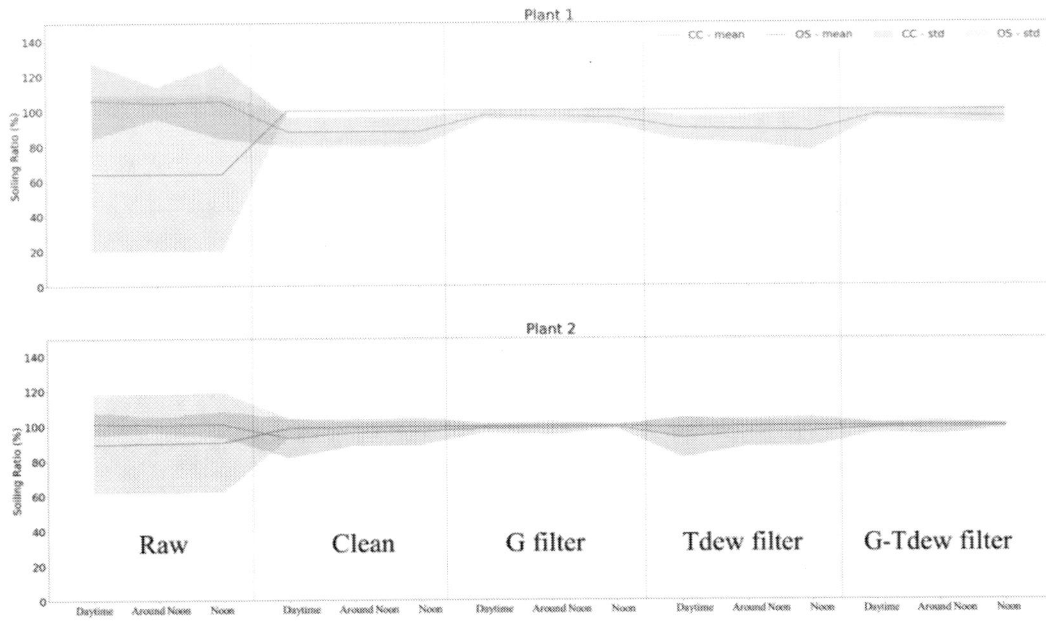

Figure 1: Daily Soiling Ratio – CC Vs OS

In graph above, it is observed that the dispersion of the data obtained from the calibrated cells is, in all datasets (except for 1A, 1B and 1C), higher than the data measured with the optical sensors in both PV plants. Moreover, it is observed that such dispersion, as well as the difference between the average SR obtained with the CC and OS is reduced as the data is cleaned and filtered.

As an example, Figure 2 presents the hourly SR (average of the 5-minute SR) for one day for the 2A dataset; in which it is observed the variability of the SR obtained with the CC through the day compared to the SR from the OS.

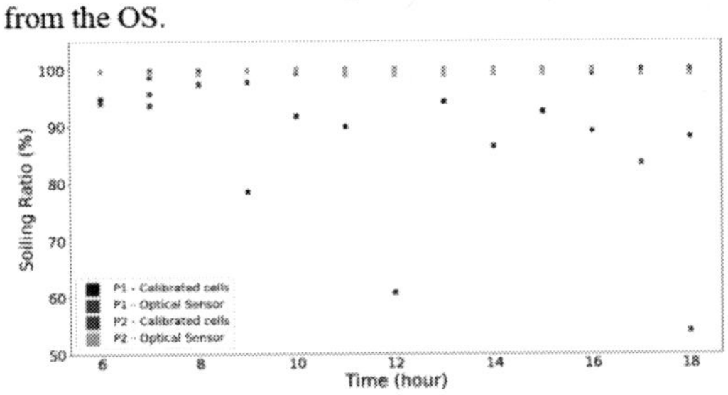

Figure 2: Hourly SR (CC V OS) – 2A

Figure 3 presents the hourly SR for the 3.1A dataset (G filter) for the same day as presented in Figure 2; in which it is observed that the hourly variability of the SRcc is removed by filtering the data with the irradiance (Figure 2).

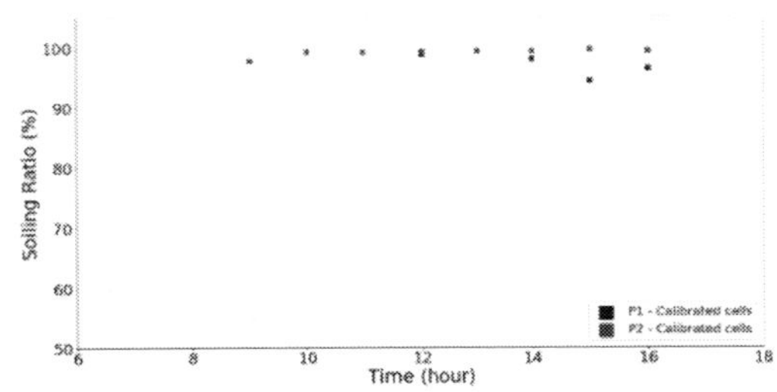

Figure 3: Hourly SR (CC V OS) – 3.1A

3.2 Difference in Soiling Ratio (dSR)

Figure 4 summarises the mean and standard deviation of the dSR for each of the three clean datasets and the nine filtered datasets; from which can be observed that the difference in the SR for both plants is higher in the Tdew filter dataset (3.2), compared to the G filter (3.1) and G – Tdew filter (3.3) datasets. Furthermore, it is observed that filtering the data with the irradiance (scenarios 3.1 and 3.3) has a positive effect on the data dispersion and difference in the SR values obtained by each of the measurements, as anticipated in section 3.1.

Since both PV plants have tracking systems, we would have expected the SR in the Daytime, Around Noon and Noon datasets to be similar, as the solar angle is optimised. However, the dispersion of the dSR in all datasets analysed (except for scenario 2) is higher as the data approaches to the Noon for Plant 1 and lower for Plant 2. As observed in Figure 4; the maximum difference between the SR from the CC and the OS in these datasets was 0.23% (3.2C scenario, Plant 1).

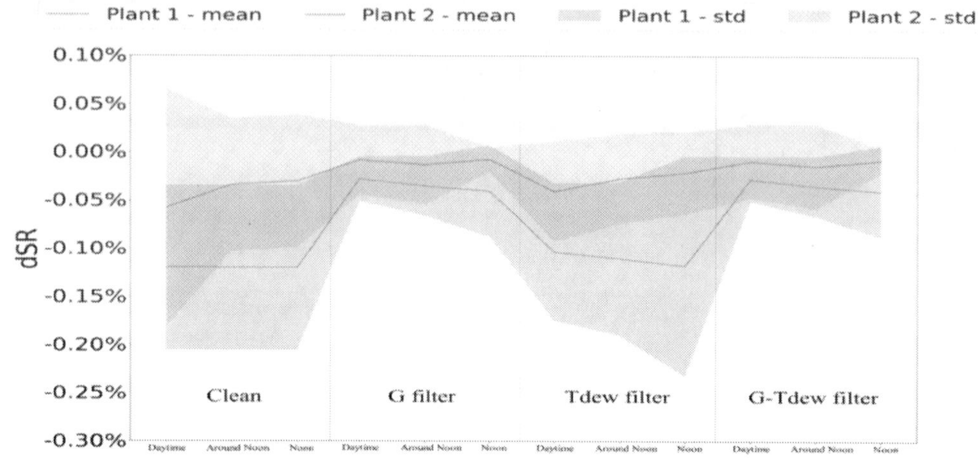

Figure 4: Difference (CC vs. OS) – dSR

3.3 SR correlation with the key variables

The difference of the SR obtained with the CC and OS (dSR) against each of the key variables in the Daytime G filter (3.1A) and the Tdew filter (3.2A) datasets for both PV plants (P1 and P2) was compared.

In Figure 5, it is observed that the average dSR as well as the dispersion, when applied the G filter is lower than when applying the Tdew filter, which indicates that the SR obtained from the OS and the CC are more aligned in the G filter scenario. As observed in Figure 5, no correlation between the dSR and each of the key variables was observed.

Thus, it is not possible to conclude that higher ambient, PV module temperatures or relative Humidity will lead to a higher difference in the SRos and SRcc.

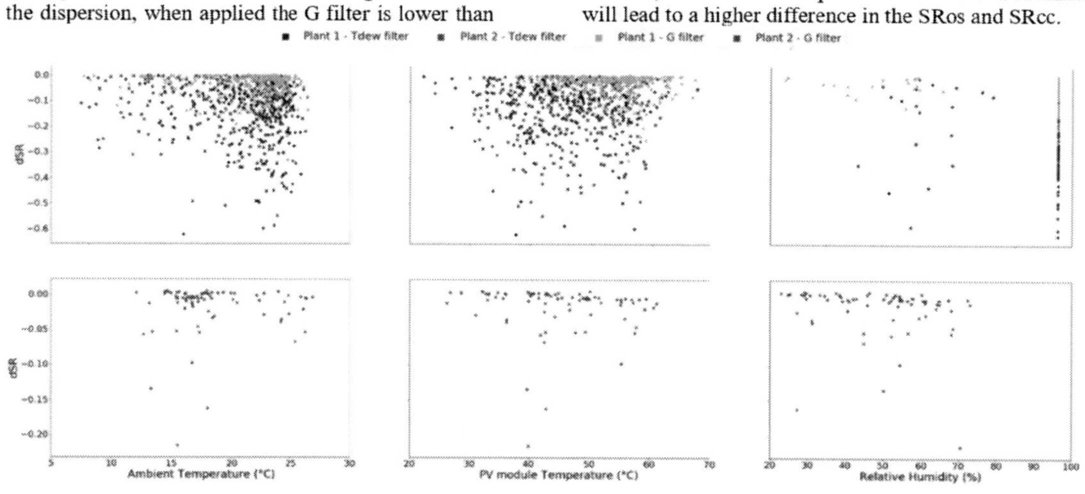

Figure 5: Average dSR Vs the key variables (* A negative dSR value indicates that the SR obtained from the OS is higher than the SR obtained with the CC).

The rainfall data obtained from the SCADA system of both plants presented significant number of gaps and erroneous values, which prevented the analysis of such variable on the Soiling Ratio for this study.

4 CONCLUSIONS

The dispersion of the SR from the calibrated cells is significantly higher than the data measured with the optical sensors in all the datasets analysed.

The SR obtained from the calibrated cells is influenced by the hourly variations of solar irradiance through the day, which is not the case for the soiling ratio obtained by the optical sensors. Such variability is reduced when filtering the data with a certain value of irradiance (800W/m2 in this study).

The removal of the dew temperature effect does not seem to have an impact on the OS measurements; the difference between the SR obtained with the CC and the OS was lower when such filter was not applied.

The difference in SRcc and SRos after filtering the data varied between 0.05% to 0.23% when analysing different periods of the day (Daytime, Around Noon and Noon).

No correlation between the ambient temperature, PV module temperature and relative humidity on the Soiling Ratio was observed. The results obtained do not present enough evidence to conclude that higher ambient temperature, higher PV module temperatures or relative humidity will lead to a higher difference in the SRos and SRcc.

The results obtained in this study should not be extended to other PV plants as it limits to the analysis of two PV plants located in Mexico and thus it might not be representative for other PV plants in different regions.

5 REFERENCES

[1] Mina Mesbahi. Solar Plaza. White Paper Soiling: The Science and Solutions. Available at: www.solarassetmanagementeu.com

[2] Klemens Ilse, Leonardo Micheli, Benjamin W. Figgis, Katja Lange, David Daßler, Hamed Hanifi, Fabian Wolfertstetter, Volker Naumann, Christian Hagendorf, Ralph Gottschalg, and Joorg Bagdahn3. (2019). Techno-Economic Assessment of Soiling Losses and Mitigation Strategies for Solar Power Generation. Joule pp. 2303–2321.

[3] IEC 61724-1:2021 Photovoltaic system performance. Part 1 Monitoring.

[4] Thurston Charles, Korevaar Marc, van Velsen Donald and Lee Clibe. Solar Module Soiling Whitepaper. Kipp&zonen. Available at: https://info.otthydromet.com/1928_KZ_CNT_c-met_DustIQ-Soiling-Whitepaper-all_EN_19-Landing-Page.html

[5] Manual de usuario Celula MET. Atersa. Available at: https://atersa.shop/app/uploads/2019/06/MU-38-S%20(Manual%20MET).pdf

[6] DustIQ Manual. Kipp&zonen. Available at: https://www.kippzonen.com/Product/419/DustIQ-Soiling-Monitoring-System#.YxjKP3bMKUk

MONITORING OF PHOTOVOLTAIC SOILING AND ASSESSMENT OF SOILING LOSSES CHARACTERIZATION STRATEGIES

Vardanyan R.R., Badalyan N.K.
National Polytechnic University of Armenia
105 Teryan Str. Yerevan 0009, Armenia

ABSTRACT: The accumulation of dust and other environmental contaminants on the photovoltaic (PV) modules, also known as PV module soiling, is a significant environmental influence reducing the output of the PV power plant. This problem is substantial for countries with low yearly precipitations level and dusty climates. In this paper, the influence of soiling on the performance of PV modules during a year in Yerevan (Armenia) is investigated. Two PV modules were tested, one of which is regularly cleaned with the use of a specially developed automated cleaning system, while the other is left to the soil naturally. The short circuit current, output power, and efficiency losses of two PV modules were measured. The measured data shows that the characterization of soiling losses by measuring the power losses is beneficiary since the output power is the main operational parameter of the PV module. The measurement of shirt circuit current is simpler, but it is not recommended use due to the dependence of short circuit current on the homogeneity of soiling. Meantime, if necessary, the measurement of the efficiency of a soiled PV module can be realized for indication of the amount of efficiency reduction due to the soiling.
Keywords: photovoltaic, soiling, power, efficiency, cleaning.

1 INTRODUCTION

The performance of photovoltaic (PV) modules is influenced by environmental factors such as solar irradiation and ambient temperature. Following irradiance and temperature, the accumulation of dust, dirt, pollen, and other environmental contaminants on PV modules, also known as PV module soiling, is the third most important environmental factor in determining the output of the PV power plant.

The module soiling results in a reduction of solar irradiance reaching the solar cells of the PV module and therefore reducing power generation. Soiling losses depend on meteorological parameters such as rain, temperature, wind, humidity, and cloud cover. The most important meteorological factor is the average frequency of rain since rain can wash soiling off of solar PV modules.

Soiling losses vary greatly from region to region. The rate at which soiling deposits depend on geographical factors such as proximity to deserts, agriculture, industry, and roads, as these are likely to be sources of airborne particles. If a location is close to a source of airborne particles, the risk of soiling losses is high.

Power reductions greater than 50% have been reported in the literature because of soiling [1, 2]. Average annual energy losses due to soiling are typically in the range of 3 – 6% [3]. For optimal tilt angle of PV modules (approximately 35^0 in Belgium), the annual power loss between 3 – 4% has been measured [4].

Thus, soiling can be a severe issue for PV systems worldwide, becoming even more concerning because of the rapid PV market expansion. Therefore the soiling monitoring strategy is an important issue [5].

There are different approaches to determining soiling losses. The level of soiling can be expressed with the soiling ratio (SR) defined in the International Standard IEC-61724-1 for PV systems monitoring [6 - 8]. This standard proposes that two PV modules are used, where one is left to accumulate soil, and the other is held clean. The soiling ratio is estimated by the ratio of the power output of the soiled device to its expected power output if it was clean. The expected power output is calculated using calibration values and the measured short-circuit current of the clean device. This setup is also referred to as a "soiling measurement station", or just a "soiling station". In accordance with the standard, an alternative metric is the soiling loss (SL), which is defined as SL = 1 – SR.

There is another approach for quantifying the soiling losses, based on the determination of the efficiency of the PV module [9]. There are other soiling monitoring strategies also, based on analysis of the structure of dust by scanning electron microscope and by the weight and color of soil [10].

In this paper, the effect of soiling on the performance of PV modules in Yerevan (Armenia) conditions during a year has been investigated. The short circuit current loss, output power loss, and efficiency loss due to the soiling are measured. Two identical PV modules are tested. One of the PV modules is cleaned regularly by a specially developed automated cleaning system, and the other is left to the soil naturally. The experiments are conducted for about a month in spring, summer, autumn 2021, and winter 2021 and 2022, providing the monitoring for one year. The influence of soiling on short circuit current, power, and efficiency of PV module is analyzed and the recommendations on soiling monitoring strategy are suggested.

2 EXPERIMENTAL SETUP AND METHODOLOGY

The experimental investigation of the effect of soiling on the performance of PV modules was conducted in accordance with International Standard IEC-61724-1. Two identical PV modules are tested in outdoor conditions at the research laboratory for "PV and semiconductor devices" at the National Polytechnic University of Armenia. The PV modules are made of 60 polycrystalline solar cells with a maximum power of 275 W and are produced in Armenia.

The PV modules are placed nearby to guarantee the same environmental influences on the modules. The modules are tilted at 30°, which is considered approximately optimal for the Yerevan location (Fig. 1).

One PV module (left) was cleaned periodically before the measurements and the other (right) module was not cleaned at all and used as the reference module.

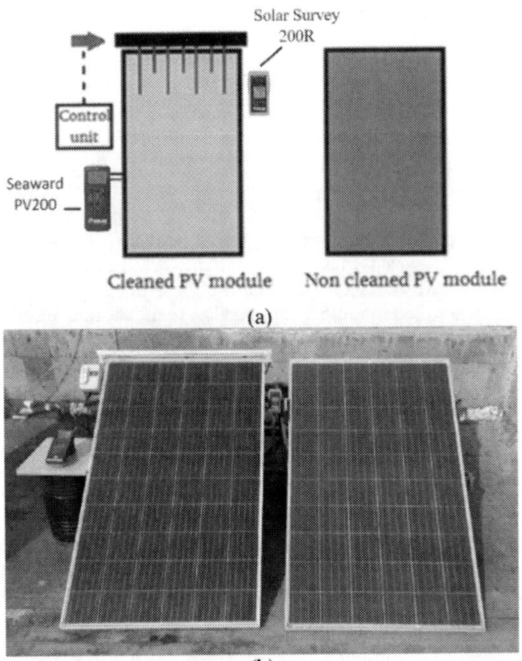

Figure 1. The structure of the experimental setup for measuring the soiling effect (a) and the tested PV modules with an automated cleaning system (on the left) and the Seaward measuring system (b)

To measure the parameters of the cleaned PV module a special automated cleaning system was developed. It was operating with the spray of water. Each day, before measurements, automatically, the water was supplied by the control unit to the tube installed above the PV module (Fig. 1). The water was sprayed out from the thin holes of the tube, thus cleaning the PV module.

The short circuit current and power of PV modules in accordance with standard IEC-61724-1 were measured. Besides these measurements, the efficiencies of the modules were measured as well. All measurements were conducted with the use of the SEAWARD PV200 measuring system. The intensity of the solar radiation, the temperature of PV modules, and the ambient temperature, were measured using the Solar Survey 200R, which was attached to the solar panel (Fig. 1). The measured data was uploaded and analyzed with the use of the PVMobile application.

The experiments were conducted in accordance with the following sequence.

- Each day at 13:30 one of the PV modules was automatically cleaned with the use of water spray during 10 minutes.
- After 15 minutes the temperatures of cleaned and soiled modules were equal, and the characteristics of the I-V curve, temperatures of cleaned and soiled modules, as well as the ambient temperature and the solar radiation intensity were measured.

3 RESULTS AND DISCUSSION

The short circuit currents (I_{SC}), maximum powers of the PV modules (P_m), and the efficiencies (η) are measured during the spring, summer autumn 2021, and winter of 2021-2022 (Fig. 2).

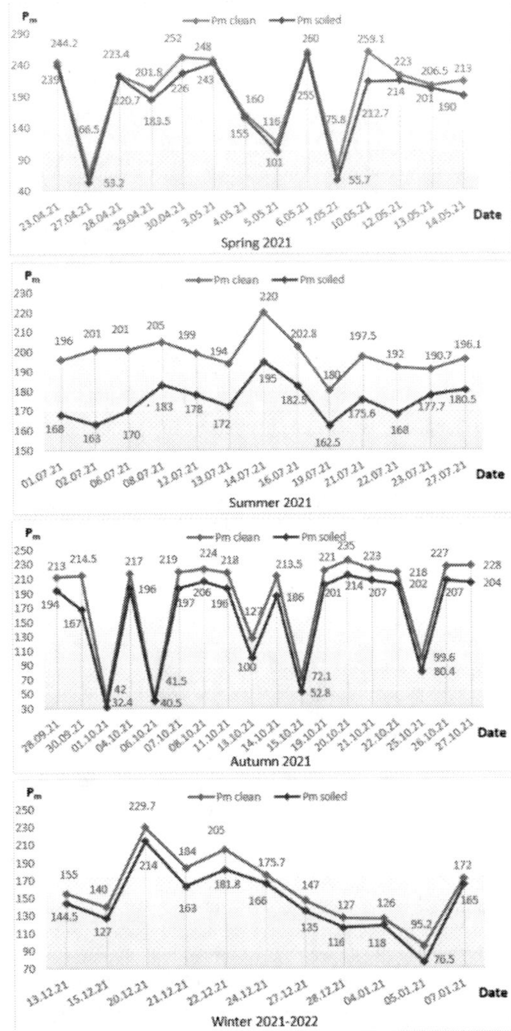

Figure 2. Output powers of cleaned and soiled PV modules during 2021 and 2022

It can be seen from Fig. 2 that the power losses due to the soiling are significant.

The soiling ratios (SR) for short circuit current I_{SC} and maximum power P_m are calculated with the use of the following formulas:

$$SR\,(I_{SC}) = \frac{I_m\ \text{soiled}}{I_m\ \text{clean}},$$

$$SR\,(P_m) = \frac{P_m\ \text{soiled}}{P_m\ \text{clean}}.$$

The losses due to the soiling are calculated as

$$SL = [1 - (SR)] \times 100\%.$$

The average values for soiling ratios of Isc and Pm as well as for soiling losses are calculated for each season in

2021 and 2022 (winter). The annual average values are calculated also. The obtained results are presented in Table 1.

Table I: Meausered average parameters

Measurement dates		Pr. (mm)	T_{amb} (°C)	Soiling Ratio (SR)		Soiling Loss (SL) x100 (%)	
Start	End			I_{sc}	P_m	I_{sc}	P_m
Spring 2021		89.4	28.1	0.96	0.91	4	9
23/04/21	14/05/21						
Summer 2021		90.5	36.8	0.89	0.88	11	12
01/07/21	27/07/21						
Autumn 2021		46.2	24.3	0.96	0.87	4	13
28/09/21	27/10/21						
Winter 2021-2022		29	8.1	0.96	0.91	4	9
13/12/21	07/01/21						
Annual average		**63.7**	**24.3**	**0.94**	**0.89**	**6**	**11**

It can be seen from Table 1 that the annual average power loss in our experimental site (in Yerevan, Armenia) is 11%. It is high in comparison with the typical loss of 3 – 6% [3] and 3 – 4% [4]. This comparatively high value of PV power losses can be described by the operation of a construction site located close to the research laboratory.

The annual average soiling losses determined by the short circuit current Isc (6%) and by maximum power Pm (11%) of PV modules are different. This difference is significant for the spring, autumn and winter periods (Fig.3).

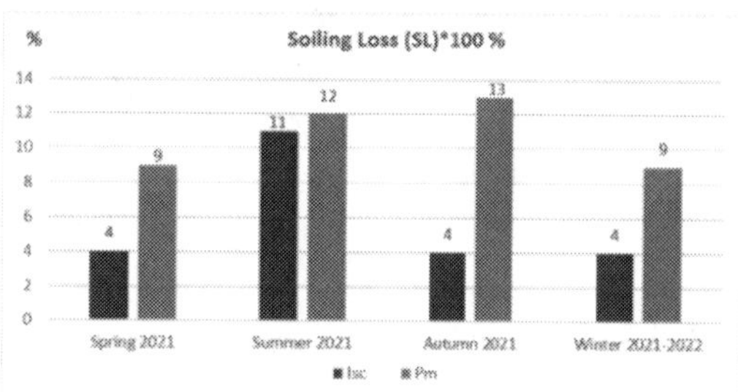

Figure 3. Soiling losses determined by I_{SC} and P_m of PV modules

The big difference between annual average soiling losses of Isc and Pm can be described by the non-uniformity of soiling on the surface of the PV module, causing the big changes in Isc. Therefore, the characterization of soiling losses by measurements of power losses is beneficiary, since this is the main important operational parameter of the PV module.

The efficiencies of the cleaned and non-cleaned PV modules are measured during spring 2021, summer 2021, autumn 2021, and winter 2020-2022 also. The efficiency of a PV module is determined as the fraction of maximum power P_m and incident power of solar radiation:

$$\eta = \frac{P_m}{P_{in}} = \frac{P_m}{AP_{sun}} = \frac{I_m P_m}{AP_{sun}},$$

where P_{in} is the incident power of solar radiation (W), A is the surface area of PV modules (m^2), P_{sun} is the solar radiant power per unit area (W/m^2), I_m is the maximum current (A) and V_m (V) is the maximum voltage at the maximum power point.

It follows from the presented formula, that the ratio of efficiencies of the cleaned and non-cleaned PV modules is the ratio of maximum powers of the modules. Note, that as the result of measurements the average values of the soiling ratio of 0.89% and soiling losses of 11% are obtained by the use of measured efficiencies, which are equal to the values, obtained using the maximum power measurements presented in Table.

Thus, the characterization of the soiling effect by determining of soiling ratio and soiling losses using the efficiency measurements, as it is done in publication [9], is similar to using maximum power measurements. Therefore, the assessment of the soiling losses is reasonable to conduct by measurements of maximum power losses, while the measurement of the efficiency of a soiled PV module can indicate the efficiency reduction of the soiled module in comparison to the cleaned module.

4 CONCLUSIONS

- The annual average power soiling losses of PV modules in Yerevan (Armenia) during 2021 and winter 2022 are investigated. It is shown, that the power losses are comparatively high (11%) due to the operation of a construction site located close to the research laboratory.
- There is a big difference between soiling losses determined by the measurements of short circuit currents and output powers of cleaned and non-cleaned PV modules. This big deviation is described by the strong dependence of the short circuit current on the non-uniformity of the dust on the surface of the PV module.
- The characterization of soiling losses by measuring the power losses is beneficiary since the output power is the main operational parameter of the PV module.
- The measurement of short circuit current is simpler, but it is not recommended use due to the dependence of short circuit current on the homogeneity of soiling.
- The measurements of power losses are simpler in comparison with the measurements of the efficiency of PV modules and are more beneficiary. If necessary, the measurement of the efficiency of a soiled PV module can be realized for indication of the amount of efficiency reduction due to the soiling.
- The assessment of soiling losses by optical characterization or by analysis of the structure of dust, weight, and color of soil is difficult and more complicated in comparison with the method based on measuring the power losses of PV modules.

Acknowledgment

This work was supported by the Science Committee of the Ministry of Education, Science, Culture and Sports of the Republic of Armenia, in the frame of the research project 21DP-2B005.

5 REFERENCES

[1] Sarver T, Al-Qaraghuli A, Kazmerski L.L. A comprehensive review of the impact of dust on the use of solar energy: History, investigations, results,

literature, and mitigation approaches. Renew. Sustain. Energy Rev. 22, 2013, p. 698–733.

[2] Costa SCS, Diniz ASAC, Kazmerski LL. Dust and soiling issues and impacts relating to solar energy systems: Literature review update for 2012–2015. Renew. Sustain. Energy Rev. 63, 2016, p. 33–61.

[3] S. Canada. Quality assurance: Impact of Soiling on Utility-Scale PV System Performance. SolarPro Magazine, no 6.3, May 2013, p. 14 -20.

[4] R. Appels, B. Muthirayan, et al. The Effect of Dust Deposition on Photovoltaic Modules. Conference Paper of the IEEE Photovoltaic Specialists Conference. June 2012.

[5] J. Bessa, L. Micheli, F. Almonacid, E. Fernandes. Review. Monitoring photovoltaic soiling: assessment, challenges, and perspectives of current and potential strategies. iScience 24, 102165, March 19, 2021, p. 1-37.

[6] M. Gostein, T. Duster, C. Thuman. Accurately Measuring PV Soiling Losses With Soiling Station Employing Module Power Measurements. Conference Paper June 2015.

[7] M. Gostein, B. Littmann, J. Caron, L. Dunn. Comparing PV Power Plant Soiling Measurements Extracted from PV Module Irradiance and Power Measurements. Proceedings of the 39th IEEE Photovoltaic Specialists Conference, Tampa, FL, June 2013.

[8] G.P. Smestand et all. Modeling photovoltaic soiling losses through optical characterization. Sci. Rep. 2020. Published online 2020. https://www.ncbi.nlm.nih.gov/pmc/articles/PMC6952442/

[9] Y. Andrea, T. Pogrebnaya, B. Kichonge. Effect of Industrial Dust Deposition on Photovoltaic Module Performance: Experimental Measurements on the Tropical Region. Hindawi, International Journal of Photoenergy. Volume 2019, p. 1-10.

[10] A. Hussain, A. Batra, R. Pachauri. An experimental study on the effect of dust on power loss in a solar photovoltaic module. Renewables 2017, p. 1-13.

1 MW PV SYSTEM AT MUNICH TRADE FAIR CENTRE - 25 YEARS OF SUCCESSFUL OPERATION

Gerd Becker, Fabian Flade, Bodo Giesler, Bruno Schiebelsberger, Walter Weber
Bavarian Association for the Promotion of Solar Energy, Friedrich List Strasse 88, 81377 München, Germany
Phone: +49 (0) 89/278134-28, info@sev-bayern.de, www.sev-bayern.de

ABSTRACT: A 1 MW photovoltaic rooftop system was built in 1997 on northern six roofs of the Munich Trade Fair Centre. At the time of its construction, it was the world's largest roof-mounted PV system - with an installed rated output of the modules of 1016 kW, see Figure 1. In November 2022, the system will be in operation for 25 years, the average specific yield is 1032 kWh/kW until the end of 2021. The system consists of 7812 modules (thickness of the solar cells: 280 µm to 300 µm) of 130 W each and a central inverter. Extensive evaluations were carried out annually, a lot of operational experience was gained. Thus, measures for an excellent long-term behaviour are regular reports (logfile, monthly / quarterly reports), functional checks e.g. efficiency, MPP control, insulation resistance) and yearly (or often) on-site inspections.
Keywords: PV System, Lifetime, Monitoring

1 GENERAL INFORMATION / FINAL YIELD - VERSUS PERFORMANCE RATIO

The photovoltaic rooftop system - installed on the roofs of the Munich Trade Fair Centre in 1997 by Bayernwerk, Siemens and Stadtwerke München SWM with funding from the federal and state governments - has a nominal output of the modules of 1016 kW. In November 2022 it will be in operation for 25 years.

Excellent operational results have been obtained, as Figure 1 illustrates. In the years 1999 to 2021 (1998: startup) an average annual specific yield of 1023 kWh/kW was achieved, in a range from 938 kWh/kW to 1160 kWh/kW. Performance Ratio PR was from 72% to over 80%, see **Figure 1**. It can be mentioned, that in this year from January to August 2022 an excellent final yield of 916 kWh/kW has been achieved.

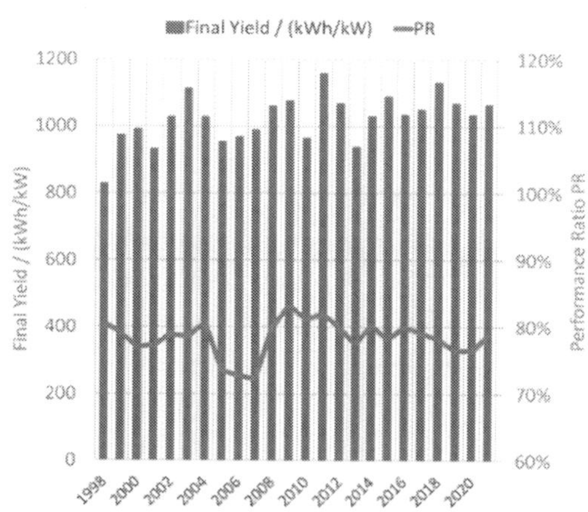

Figure 1: Final Yield and PR

2 GOAL OF THIS CONTRIBUTION

This paper intends to report operating results and malfunctions - and conclude what can be learned from them. Which measures are necessary to operate a big PV system for 25 years and more?

3 CONSTRUCTION OF THE PV PLANT

7812 modules - **Figure 2** - facing south – each with a nominal output of 130 W - are mounted with a tilt angle of 28° on six hall roofs, where they are well ventilated.

Each 130 W module (V_{MPP} = 20.4 V; I_{MPP} = 6.35 A) consists of 84 monocrystalline silicon cells, the thickness of the pre-sorted solar cells is 280 µm to 300 µm. 21 of the frameless solar modules are connected in series to form a string. The DC voltage is therefore at 400 V level. 6 strings are connected to each other in a generator connection box, their output cables in coupling boxes. The outputs of 6 coupling boxes form a sub-generator. The electrical energy generated by the 12 sub-generators is fed to the centrally located inverter (master-slave-slave operation) in an (originally) air-conditioned room.

Figure 2: Modules and Central Inverter

DC copper cables with cross-sections up to 400 mm² and lengths up to 500 m are used to connect the sub-generators with the inverter. They are adjusted to the length so that the ratio of length and cross-section is approximately constant.

DC switches were used in the original installation to separate the inverter and PV arrays. As of 2007, these have been taken away. PV energy is now fed directly into the 20 kV grid via a transformer.

A monitoring system records measurement data such as currents, voltages, power, global radiation, temperatures every minute.

Maintenance and operational management with an annual inspection are carried out by Stadtwerke München SWM in cooperation with the manufacturer of the inverter.

A logfile including all events is kept. A comprehensive evaluation of the operation is carried out annually – with an operational report.

4 OPERATIONAL EXPERIENCE

4.1 PV Modules

Only very rarely modules had to be replaced due to glass breakage, a total of 5 times. Modules with delamination (see **Figure 3 left**) were discovered on several hall roofs, as well as few modules with a browning.

Figure 3: Delamination (left) and browning (right)

Most of the modules with browning (**Figure 3 right**) were discovered on one roof. Some of them showed only a very slight, other modules a medium to strong brown discoloration.

In the string current measurements carried out as part of the annual inspections, no difference to the non-discolored modules was found. Slight module contamination (moss growth) on the lower retaining clips was found. The rough costs for the module cleaning would have been approx. 1500 €/roof, in the end no cleaning was carried out.

4.2 Inverter with monitoring system

From 2005 up to and including 2007, the plant showed poor specific yield and PR values. In 2005 it was 952 kWh/kW / 73.5%, in the year 2006 only 969 kWh/kW/ 72.8% and in 2007 only 988 kWh/kW at 72.2%.

Correspondingly, the logfiles for 2005 and 2006 should be mentioned, which identified the generator circuit breakers as the reasons of outages.

From March 16, 2007, a modernization of the inverter, also known as a "retrofit", was carried out. The DC distribution was modified, the power units were completely replaced, and the direct current breakers (**Figure 4 left**) were taken away. A new monitoring computer with the WINCC operating system was installed.

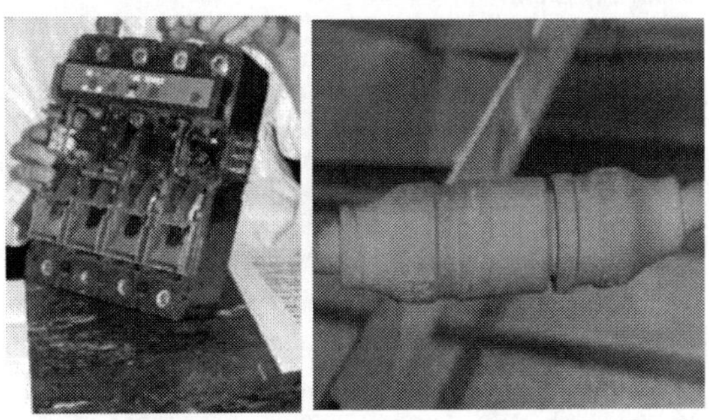

Figure 4: DC Breaker(left) and plug-in connectors (right)

On April 17, 2007, the "retrofitted inverter" went into operation. From now on, measured values were available in minute intervals and, if required, even higher resolution. The maximum power was now >1.25 MVA, so that power limitation no longer occurred.

A monitor with a continuous display of the operating data and error messages was installed in a control room of the grid operator SWM.

4.3 Other components - connectors, cables, boxes

The logfile and the inspection list problems in each year of operation. The most important in the 25 years of operation are:

- Due to moisture, a total of 66 generator junction box covers were replaced and an additional drain hole was drilled. These measures have been successful, the boxes have all been dry since then.
- Rubbish was found on the hall roofs.
- The air conditioner in the inverter room was not working properly.
- The beam claws used to attach the modules were checked for strength. Critical cases were identified, the beam claws were replaced (**Figure 5 left**).
- Cable bushings were defective (Figure 4 right).
- A defective module was detected by measuring the string current; the bypass diode in a junction box had burned. The typical charring of the terminal block may be due to insufficient contact after a long period of permanent overheating, see **Figure 5 on the right**.

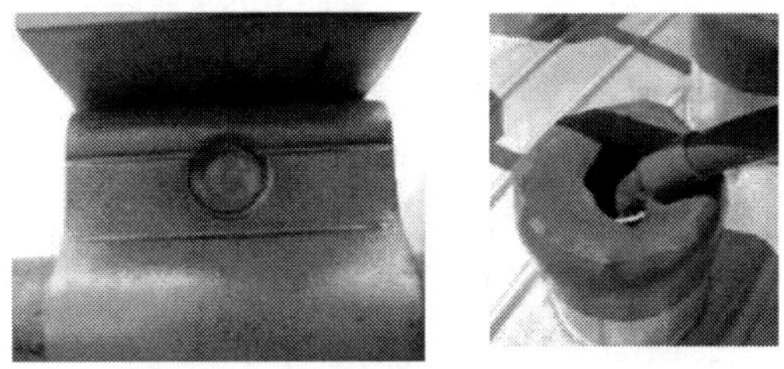

Figure 5: Beam claws (left) and cable bushings (right)

- A layer of corrosion was found on the busbars of the generator connection boxes.
- Various mechanical damages of the cable attachments were found. A close inspection showed that no cables were damaged. The damage was probably caused by a person "climbing on top".
- Cable bushings were damaged.
- There was normal dirt on the modules from bird droppings, pollen, etc. on all roofs. However, only one module with a "corroded" surface was discovered. The cause is not entirely clear, but it could be bird droppings.
- An open string was found by measuring the string currents. Many plug-in connectors of the module cabling on the roof sides were slightly open (**Figure 4, right**).

5 FUNCTIONAL CHECKS

Extensive functional tests were carried out with the help of the measurement data supplied by the WINCC operating system, but also with the help of external measurements.

5.1 Efficiency curve of the inverter

Every year the efficiency curve of the inverter in master-slave-slave operation was evaluated on the base of minutely recorded values, see **Figure 6**. For August 2020 e.g. more than 23.000 values have been applied.

The best efficiencies at high utilization are over 96% in July and November some values even exceed 97%. The maximum powers driven on the direct current side are over 1000 kW in the summer month and just under 800 kW in November due to the lower global radiation.

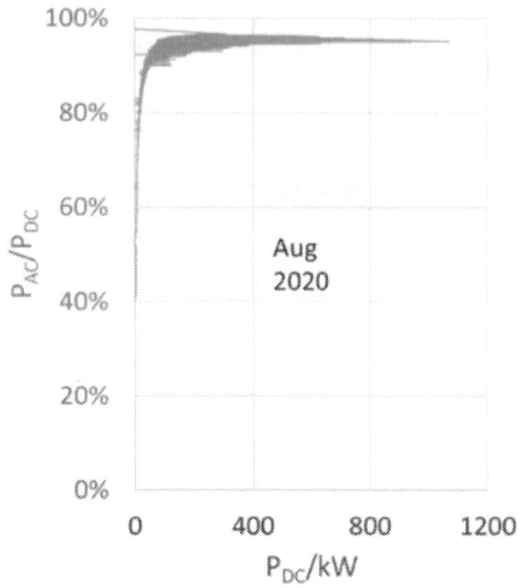

Figure 6: Efficiency curve in August 2020

5.2 MPP control of the inverter

The inverter must always adapt to the maximum power point of the solar generator, the MPP point (Maximum Power Point). The position of this MPP essentially depends on global radiation and module temperature. It should always be within the permitted input voltage range of 320 to 460 V for the inverter so that the PV generator can deliver maximum power.

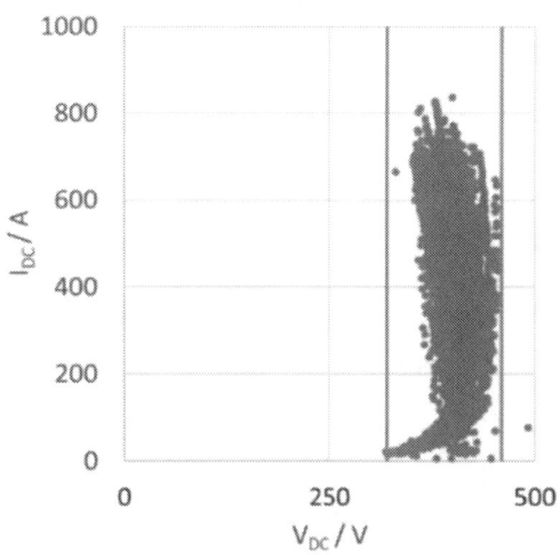

Figure 7: MPP control of the inverter in Aug 2020

A monthly graphic - **Figure 7** - in the DC current / DC voltage I_{DC}/V_{DC} characteristics field (e.g. August 2020) indicates that the PV generator is well adapted to the inverter. It can be seen in that the MPP control works well.

5.3 Insulation resistance of the DC cables

The behaviour of the insulation resistance of the DC cables at the end of Sep 2020 can be seen in Figure 8, there is no problem. A warning would appear at an insulation resistance of approximately 70 kOhms, and the maximum insulation resistances are always below 5000 kΩ.

In the rainy months of April to November, the limit value for the warning was briefly undershot. For a better analysis,

The insulation resistance of the DC cables from the modules to the central inverter are monitored and recorded as a minute mean value, see Figure 7.

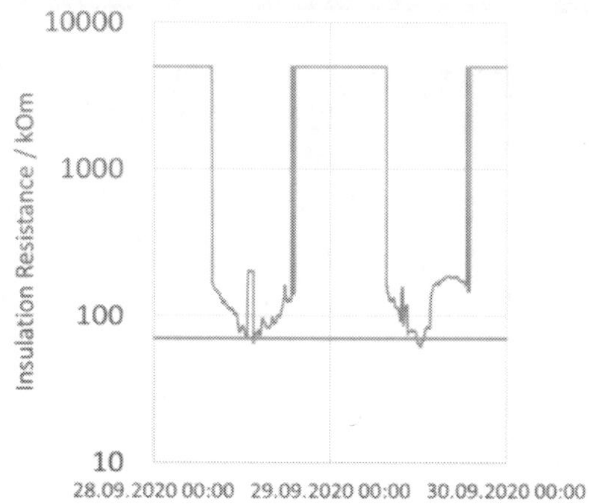

Figure 8: Insulation resistance at the end of Sep 2020

When there is no energy fed into the grid, the insulation resistance is 5000 kOhms, with PV energy production it is below 100 kOhms.

5.4 Thermographic investigations

A thermographic investigation - **Figure 9** - of the complete PV generator on all hall roofs of the Trade Fair Centre was carried out in 2012 and 2014. The facility was found to be in fair condition for its age.

Figure 9: Symbolic picture for the thermographic investigation

However: Significant heating of 30 module junction boxes was discovered. 7 modules showed heated solder points between the cross-connectors and the outlets to the junction box. A total of around 40 modules showed reactions to dirt-related shading of the cells directly above the lower terminals. In addition, on one day of the measurements, it was found that a total of 6 module strings and thus 16.4 kW were in open circuit operation.

5.5 Module measurements

The installed modules have been in operation since November 1997. In the years 2005, 2006, 2012 and 2017, measurements of the "performance at STC" and other investigations were carried out at two different test laboratories. **Table 1** shows the number of modules measured, the average power measured and the deviations from the nominal power of 130 W.

Table 1: Results of measuring modules

PV Testlab, parameters, year		2005	2006	2012	2017
Testlab 1	Number of modules	6	3	12	No measure-ment
	Mean output power / W	125,9	125,7	122,7	
	ΔP	-3,2%	-3,3%	-5,6%	
Testlab 2	Number of modules	No measure-ment	9	9	12
	Mean output power / W		122,3	121,6	121,9
	ΔP		-5,9%	-6,4%	-6,2%

In summary, all extensive measurements show that the modules have currently experienced a reduction of more than 6% compared to the nominal output. This effect cannot be seen when comparing the annual yields, the influence of other parameters such as global radiation and temperature is too great.

6 PARTICULAR FAULTS /

Two failures - caused by poor communication - should be mentioned.

First failure: At the beginning of May 2008, an external company installed a horn and switched off the PV system with the monitoring computer/server. Because of this server shutdown, the system could not send a message. The external company did not inform anyone, after a few days the switch-off was accidentally discovered, the loss was around 24 MWh.

The reason for the second system failure in June 2015 was a switch-on routine. The circuit breaker of the transformer for feeding into the medium-voltage level had switched off incorrectly every day after shutting down the system to zero power and then switched on again about 1 minute later. This behavior should be fixed on Friday, June 19, 2015 by changing the software remotely. The following steps led to the malfunction.

- On Friday, June 19, 2015, the software was adjusted via remote maintenance in such a way that it should no longer be switched on again after approx. 1 minute. However, during this reprogramming the software routine for reconnection was "corrupted" so that the next morning on Saturday 20 June 2015 and Sunday 21 June 2015 the transformer was not switched on.
- The managing SWM did not receive a message because "only the switch-on command was missing", which the system did not interpret as a fault.
- On Monday, June 22, 2015, the software supplier recognized the error and repaired it, and the transformer was switched on at around 10:20 am.

Figure 10 shows the behaviour of global radiation and power into grid.

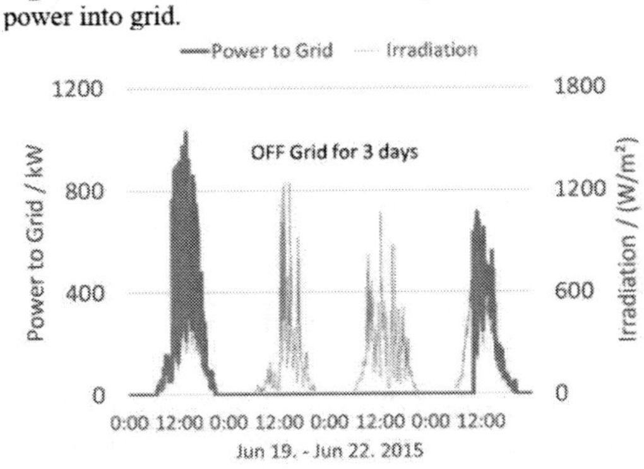

Figure 10: Global radiation and power into grid during failure

From June 20, 2015 to June 22, 2015 at around 10:20 a.m. there was no grid feed-in because the transformer was not switched on. There was an energy loss of 6 MWh.

7 LIGHTNING RODS

It should be mentioned here, that lightning rods had been installed. No outage due to a thunderstorm has been noted.

8 RESERVE MODULES

In total until now , about 120 PV modules have been exchanged. In the first years of operation, there were roughly 60 reserve modules available. Later, PV modules from other defective strings were used to replace defective modules.

In 2010 there were still 36 reserve modules available. In 2013, after defective modules were exchanged in November, only 5 spare modules left. From 2015 new modules were no longer available. Required modules were taken from PV strings on other halls, the installed power decreased .

9 SUMMARY

In about 25 years of operation, specific yields between 947 kWh/kW and 1131 kWh/kW were achieved, with operating quality values between 72.2% and 82.0%. These are good results for the Munich location.

The high availability of the modules could, among other reasons, result from the thickness of the solar cells with 280 µm to 300 µm, [1] makes the statement that "crystalline silicon PV modules" with cell thicknesses of much less than 200 µm are undoubtedly increasingly susceptible to any type of mechanical stress.

A weak point was the originally installed inverter - consisting of three inverters in master-slave-slave operation - with DC switches to separate the inverter from the PV generators.

Numerous "normal faults" occurred: glass breakage, delamination, junction boxes were damp, etc. They were rectified quickly and often preventively. But a lack of communication also led to shutdowns lasting several days.

The functional tests (efficiency, carried out with the recorded measured values show excellent behavior even after 25 years of operation. The efficiency of the inverter, the insulation resistances of the DC cables and the MPP control are all operating in the green.

10 WHAT CAN WE LEARN?

Very important for a good long-term behaviour are continuous monitoring, a well-kept logfile, a regular annual report and regular inspections, as well as additional investigations such as measuring the modules or thermographic recordings and quick troubleshooting.

Functional Checks are only necessary if the operational behaviour shows bad performance.

11 REFERENCES

[1] Cunow, E.; The development of cell thickness reduction of crystalline solar cells in PV modules and its impacts on large PV power plants; 33rd European Photovoltaic Solar Energy Conference and Exhibition; Conference Proceedings; PP 1941 – 1944

MITIGATING THE INFLUENCE OF CLIPPING VARIABILITY ON PERFORMANCE RATIO CALCULATIONS OF PV SYSTEMS

Elena Koumpli[1*], James C. Blakesley[2], George Koutsourakis[2], Christopher West[1] and Jan Muller[1]
[1]Statkraft UK Ltd, 22 Bishopsgate, London, United Kingdom
[2]National Physical Laboratory (NPL), Hampton Road, Teddington, Middlesex, TW11 0LW, United Kingdom
* Corresponding Author: Elena.Koumpli@statkraft.co.uk

ABSTRACT: The most established applied guarantee for solar photovoltaic (PV) system performance assessment is the calculated performance ratio (PR). PR values are usually reported monthly and annually and are compared to modelled PR. Negative deviations cause significant concerns as to the root of the observed underperformance and can lead to heavy financial penalties. One of the reasons of PR calculation variability is due to the uncertainty of oversizing losses of PV systems. Oversizing loss can be exaggerated by actual weather conditions resulting in excess "clipping" for a given reporting period that negatively impacts the PR, if not corrected. In this work, the influence of weather induced clipping variability on PR calculations is presented and a methodology is demonstrated to reduce such variability when calculating PR. The sensitivity of such metric is also tested for varying levels of under-performance for simulated systems with high DC:AC of 1.6. The clipping corrected PR is found to "correct" calculated annual PRs under variable weather datasets amongst 6 tested locations by an average of 7%, with a maximum of 20% for systems in very sunny locations and with high DC:AC ratios.

Keywords: performance ratio, oversizing, clipping, performance assessment

1 INTRODUCTION

In the rapidly growing PV systems industry, many critical decisions are based on a very few system-level performance metrics. One of the most commonly used metrics for PV system performance assessment is the performance ratio (PR), which is the ratio of the actual to the expected power generation of a PV array or system. Accurate PR calculation is crucial for the future or running finances of a PV system, as this is a common metric that most contracts in the PV system industry use for PV system evaluation. This means that any inaccuracies in the calculation of PR can lead to considerable damages for PV system companies, either they are the developers, owners or operators of the PV system. The latest standard IEC-61724 defines the procedure for accurate PR calculations [1]. Corrections for the effects of temperature on PR calculations are described in the same document, as it has been demonstrated that this is a crucial step for accurate PR assessment [2].

Additional PR calculation uncertainties which are currently not consistently considered are due to oversizing losses of PV systems. Oversizing occurs when a PV array has a rated DC power larger than its inverter's rated AC output power. This is a common practice in the PV systems industry, leading to inverter clipping on sunny days, in which the AC output is limited to a fixed value resulting in a generation "loss". Average sizing ratios (DC:AC ratios) can vary, depending on different geographical regions and local grid regulations. Cases of sizing ratios higher than 1.6 are becoming more and more frequent.

Clipping losses are taken into account during the design process of a PV system, based on historical weather data for a site. Nevertheless, clipping can still affect PR calculations, depending on the actual weather conditions over the specific period that PR is reported. For instance, a higher than average number of sunny days during the reporting period (especially for short-term performance testing such as provisional acceptance) results in excess unexpected clipping losses and this will result in a lower PR value, that would not express the real PR of the system. PR is currently not corrected for such clipping variability due to unexpectedly good (or bad) weather. In addition, the time resolution of meteorological data used for modelling PR and of data loggers used for measuring PR also affect performance metrics when clipping occurs [3].

In this work, cases where clipping affects PR calculations are presented. A simple methodology is demonstrated, to mitigate for such uncertainties and generate a more accurate PR value for oversized systems under variable weather systems.

2 METHODOLOGY

2.1 Formulae

The methodology presented in this paper, is based on a simple, yet accurate evaluation of the "lost" generation due to clipping, based on the temperature-corrected version of PR, calculated at a 15-min resolution, which is the minimum temporal resolution required in most Engineering, Procurement, and Construction (EPC) and Operation and Maintenance (O&M) contracts. The formula is based on the temperature corrected PR as in IEC 61724-1:2021 eq. (24):

$$PR_k = P_k \cdot G_{STC} / (P_{DC_{STC}} \cdot G_{POA,k} \cdot \mu_{STC,k}) \qquad (1)$$

Where P_k and $G_{POA,k}$ are power output and irradiance respectively at a given timepoint, k, $\mu_{STC,k}$ is the temperature correction factor (as defined in [2]), G_{STC}, $P_{DC_{STC}}$ are irradiance and DC rated power at standard testing conditions (STC) respectively.

The clipping-corrected PR metric, PR_i^*, for each reporting interval (normally a year), i, is then calculated using the following formula:

$$PR_i^* = \sum_{k,\ year\ i} PR'_k G_{poa_k} \mu_{STC_k} \Delta t_k \Big/ \sum_{k, year\ i} G_{poa_k} \mu_{STC_k} \Delta t_k \qquad (2)$$

Where PR'_k is the STC-corrected PR evaluated for interval k having a duration of Δt_k. Where clipping is detected PR'_k is evaluated as the *average of the value of* PR'_k *in the hour before* $PR'_{k_{before}}$ *and the hour after clipping,* $PR'_{k_{after}}$.

Eq. (1) can be re-written to calculate the total lost generation E_{VEP} (Virtual Energy Produced) during clipping over period i using PR'_k:

$$E_{VEP} = \sum_{k,\ i} \left(\frac{P_{DC_{STC}} PR'_k G_{poa_k} \mu_{STC_k}}{G_{STC}} - P_{AC_k}\right) \Delta t_k \qquad (3)$$

Where P_{AC_k} is the AC power output of the system. The estimation of lost generation can be applied at either site- or at inverter-level, depending on the level of granular analysis required. The methodology can therefore be used to also estimate the revenue loss that results from clipping at inverter lever, aligning with other fault detection and loss classification methodologies also applied at inverter level. To detect clipping several methodologies can be used, some of which are presented in [4].

2.2 Validation cases

To validate this method a simple system with a single inverter was simulated using real data of irradiance and temperature and with sizing ratio (SR) = 1.15 which is a very common case. The module and inverter characteristics of the system used correspond to the "default" system presented in Table I. The method was used to estimate clipping losses and calculate daily PRs using PVLib [5] to simulate the system. As can be observed in Figure 1, the proposed methodology can reconstruct the clipped data with good accuracy, when both clipping and temperature corrections are applied.

When the full power curve is reconstructed for the clipping intervals using (3), the estimated total generation during clipping is very close to the modelled case (where SR = 1.0), with a difference of 0.08%. If the temperature correction factor is omitted for simplicity, then this difference increases to 0.26%, as expected.

The PR correction methodology was also applied on a real system with sizing ratio of approximately 1.6, for a month's worth of real weather data. As can be observed in Figure 2, the deviation between the different PR calculation methods is considerable on days with higher clipping losses. These are high irradiance days, which provide most of the generation of a PV system and will have a significant impact in monthly or annual PR calculations especially on months with high clipping losses.

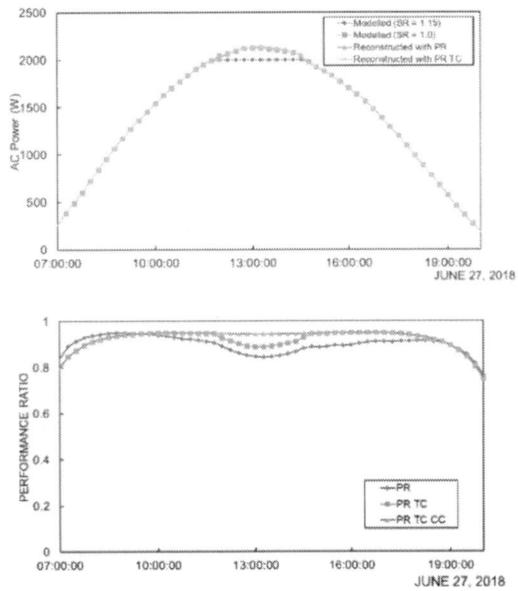

Figure 1. Modelled and reconstructed power curves for normal and oversized simulated power output *(top)*. Daily PR, temperature-corrected PR (PR TC) and clipping-corrected PR (PR TC CC) *(bottom)*. SR = 1.0 corresponds to the "unclipped" power curve.

3 SENSITIVITY RESULTS WITH WEATHER AND UNDER-PERFORMANCE

In this section, a system model was implemented to analyse the effect of weather variability and the impact of different types of common system faults on the reproducibility and sensitivity of the new performance metric. The study assumes that the model reproduces perfectly real behaviour and does not include the effects of model error, random effects in system behaviour, measurement uncertainty and degradation.

A system model, comprising a set number of arrays (each array comprising 1 inverter of 8 parallel strings, M_p and 24 modules in series, M_s, each) was implemented in PVLib [6] for the sensitivity analyses that were carried out in the following sections. A default array was defined

Table I. Definition of the arrays simulated and the value of each used parameter

Label	Tilt (°)	Azimuth (°)	Module name	Inverter name	Ms	Mp	Array losses						
							Failed modules	Failed strings	Soiling	DC mismatch	DC resistance	Inverter derate	Inverter overcliping
default	20	180	Canadian Solar	Huawei	24	8	0	0	0.02	0.01	0.01	0	0
perfect	20	180	CS6X-300M	SUN2000 36KTL US	24	8	0	0	0	0	0	0	0
1 string fail	20	180			24	8	0	1	0.02	0.01	0.01	0	0
1 module fail	20	180			24	8	1	0	0.02	0.01	0.01	0	0
2 module fail	20	180			24	8	2	0	0.02	0.01	0.01	0	0
derated inverter	20	180			24	8	0	0	0.02	0.01	0.01	0.05	0
overclipping 5 %	20	180			24	8	0	0	0.02	0.01	0.01	0	0.05
overclipping 10 %	20	180			24	8	0	0	0.02	0.01	0.01	0	0.1
5 % more soiling	20	180			24	8	0	0	0.07	0.01	0.01	0	0
mismatch	20	180			24	8	0	0	0.02	0.06	0.01	0	0
wiring 1	20	180			24	8	0	0	0.02	0.01	0.15	0	0
wiring 2	20	180			24	8	0	0	0.02	0.01	0.6	0	0
default 1.47	20	180			22	8	0	0	0.02	0.01	0.01	0	0
default 1.33	20	180			20	8	0	0	0.02	0.01	0.01	0	0
default 1.2	20	180			18	8	0	0	0.02	0.01	0.01	0	0
default 1.00	20	180			15	8	0	0	0.02	0.01	0.01	0	0
default 1.00 limited 1.6	20	180			15	8	0	0	0.02	0.01	0.01	0	0.375

which was used to define 10 additional model with added faults (module, string failures etc.) and/or other losses (soiling, mismatch etc.) and to also define systems of different sizing ratios. These array definitions are described in Table I. A perfect array model was also created, which assumes an exceptional system with no additional losses or other faults.

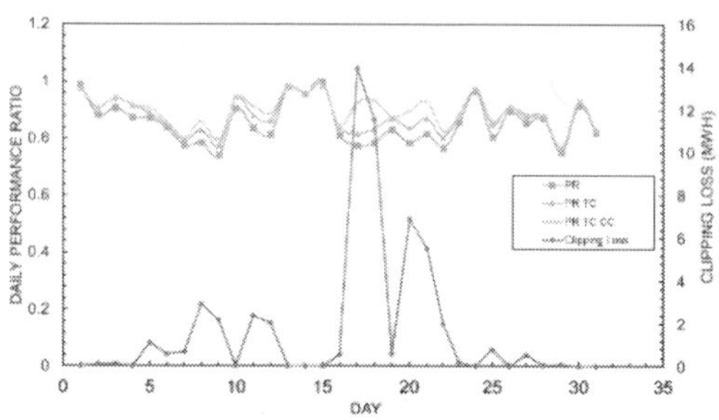

PR monthly	PR TC monthly	PR TC CC monthly
82.8%	86.1%	88.5%

Figure 2. Daily and monthly PR, temperature-corrected PR (PR TC), clipping- corrected PR (PR TC CC) and total clipping loss for a real site with SR = 1.6.

For the simulation, 1-min weather data were used from 6 different BSRN (Baseline Solar Radiation Network [6]) locations where the data are acquired from stations located above natural surfaces, and where good temperature and irradiance data are available for over 10 years. The selected stations are presented in Table II.

Table II. The BSRN stations used for the simulations.

Label	Name	Location	Latitude (°N)	Longitude (°E)	Duration of good data (years)
PAY	Payerne	Switzerland	46.815	6.944	15.1
LIN	Lindenberg	Germany	52.210	14.122	18.1
E13	Southern Great Plains	Oklahoma, US	36.605	-97.485	18.1
CAB	Cabauw	Netherlands	51.971	4.927	16.6
CAR	Carpentras	France	44.083	5.059	18.0
IZA	Izaña	Tenerife, Canary Islands	28.309	-16.499	11.8

For each simulation, 5 performance ratio calculations are performed, uncorrected PR as in (1), STC temperature-corrected PR, uncorrected PR with excluded clipping points, STC corrected PR with *excluded clipping points* and the new clipping corrected PR.

3.1 PR clipping-corrected variability with weather

In this section, each simulation was carried out annually for the default array in Table I and for all available years per location, for each of the 4 annual PR metrics and for increasing sizing ratios. The average PR per location and the standard deviation are shown in Figure 3 for 4 PR metrics.

The uncorrected PR presents a significant dependence on location for the same system, which is largely removed by the STC correction. However, a downward trend is apparent for both metrics as the sizing ratio increases, where there is a strong reduction in the measured PR as more clipping is present in those cases. For the case where clipping data are excluded from the PR calculation this dependence on the SR is much reduced, however for sunnier locations such the IZA case, the trend is still apparent. This is due to the fact that in sunnier locations and for systems with high SRs the clipping loss is much higher and at much lower irradiance, which will cause a bias on the PR calculation. The clipping corrected PR shows *significantly reduced variability* with location, year and SR compared to the other metrics with a median PR correction of 7% amongst all locations and SR =1.6 and almost 20% for the system in IZA.

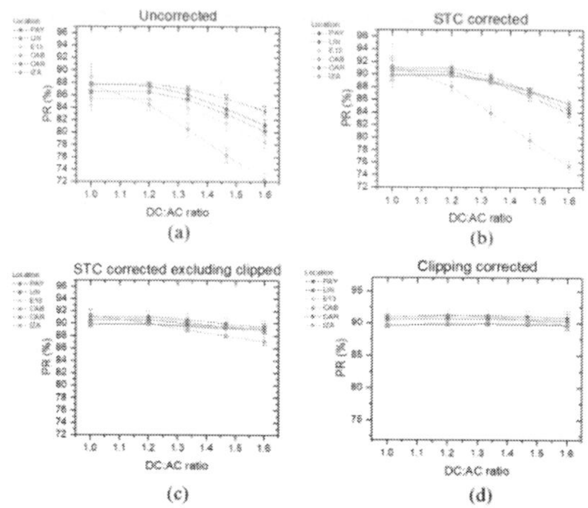

Figure 3. Average and standard deviation of annual PR values determined for nominally performing arrays as a function of DC:AC ratio for the default array in 6 different locations. Each graph shows the results of a different PR metric on the same simulated results. Error bars indicate 1 standard deviation, representing interannual variability in PR.

This is further demonstrated using the RMSE of annual PR compared to the "true" PR, where a single calculation is performed using all the available years, i per location, j.

$$RMSE(PR_{annual} - PR_{true}) = \sqrt{\frac{\sum_{i,j}\left(PR_{i,j} - PR_{true,j}\right)^2}{(n_i - 1)n_j}} \quad (4)$$

As shown in Table III, when SR = 1, the RMSE is higher for the uncorrected PR than for the corrected PRs. As the SR increases the RMSE of the uncorrected PR and STC corrected PR increase significantly as expected. The clipping corrected PR and PR metrics excluding clipped data do not change significantly as the SR increases with the clipping corrected PR showing the lowest, almost unvaried RMSE, for all SR cases.

Table III. RMSE of annual PR compared to "true" PR for all available years for the default array with different DC:AC ratios

DC:AC ratio	$RMSE(PR_{annual} - PR_{true})$				
	Clipping corrected	Uncorrected	STC corrected	Uncorrected excl. clipping	STC corrected excl. clipping
1.60	0.16 %	0.73 %	0.54 %	0.45 %	0.23 %
1.47	0.16 %	0.65 %	0.46 %	0.43 %	0.21 %
1.33	0.16 %	0.55 %	0.35 %	0.41 %	0.19 %
1.20	0.16 %	0.45 %	0.23 %	0.40 %	0.16 %
1.00	0.21 %	0.39 %	0.15 %	0.39 %	0.15 %

3.2 PR clipping-corrected sensitivity with faults

This study investigates the ability of different PR metrics to identify arrays that are underperforming, as an ideal metric should. For this, various arrays with different level of (under-)performance are simulated. These systems are defined in Table I. As previously, the "default" array is taken to be the expected performance of an array, with a normal level of soiling and balance-of-system losses. Each of the other simulated arrays has the same nominal DC and AC capacity and a DC:AC ratio of 1.6, but with varying loss factors. The "perfect" array is exceptional in that is has better performance than the default array. Each of these arrays was simulated in the 6 BSRN locations and generated the PR metrics for 20 one-year intervals using each of the PR methods.

Two ratios were used to evaluate the efficiency of the PR metrics with faulty systems: a) the "true energy performance index" (*EPI*) that is defined as the ratio of the energy generated for the given array over the entire weather dataset for the specific location divided by the energy generated by the default array over the same period and location. This ratio is a measure of how much energy an array will produce over a long period of time relative to an identical array that is performing as expected and b) the "annual PR_a/ expected PR_e", which is the ratio of the annual PR of a given array divided by the "true" PR of the default array. As there are multiple years in each weather dataset, 20 values of this ratio for each array and location were generated. The objective of this study is to understand how annual weather variations affect the ability of these PR metrics to discriminate between good and bad arrays, i.e., how the possible "randomness" of weather could wrongly determine the classification of the array. The results for two cases, uncorrected PR and clipping corrected PR are shown in Figure 4.

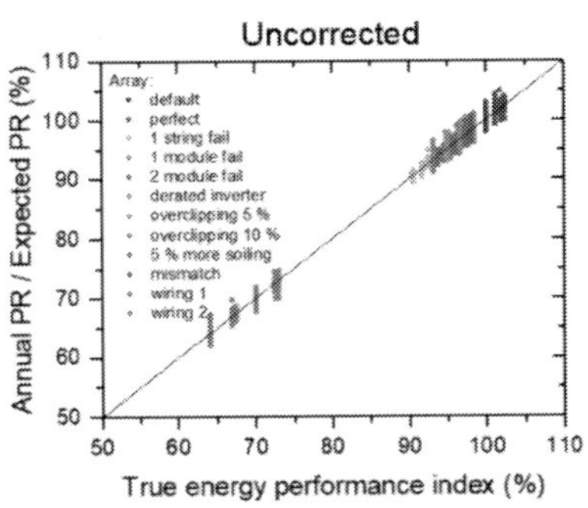

(a)

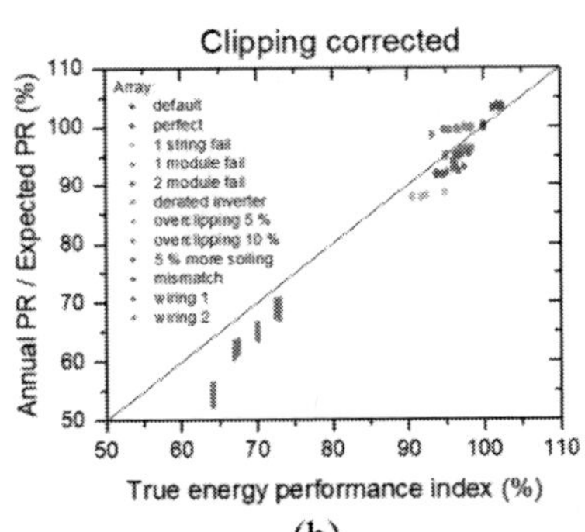

(b)

Figure 4. Scatter plot of ratio of annual / expected PR versus true performance index for differently performing arrays simulated in six different locations. Different point colours indicate arrays with different performance issues where SR = 1.6.

An ideal PR metric would be expected to have all points sitting neatly on a straight line, which would represent a threshold for identifying arrays with good and critical performance against passing a performance test. For example, points above the line could indicate a case of an array passing a performance test ("higher" PR_a/PR_e) despite having some performance issues ("lower" EPI). On the other hand, points below this line could indicate a case of an array failing a test ("lower" PR_a/PR_e) despite being of acceptable quality ("higher" EPI).

The sensitivity results are also summarised in Table IV where the RMSE of the difference between "annual PR_a/ expected PR_e" and the true EPI (energy performance index) is calculated for various types of performance losses and for each performance metric, where RMSE is calculated in the same manner as in Eq.(4).

Table IV. Root mean squared error of difference between ratio of PR_a to PR_e and the true EPI for different classes of array issues and PR metrics. SR = 1.6 in all cases.

Array group	$RMSE\left(\frac{PR_{annual}}{PR_{expected}} - EPI_{true}\right)$				
	Clipping corrected	Uncorrected	STC corrected	Uncorrected excl. clipping	STC corrected excl. clipping
All	2.91 %	0.92 %	0.66 %	3.29 %	2.54 %
Module, string, wiring and mismatch issues	3.54 %	0.87 %	0.62 %	3.96 %	2.96 %
5% more soiling	1.83 %	0.84 %	0.59 %	2.04 %	1.70 %
Derated inverter	0.16 %	0.90 %	0.65 %	0.49 %	0.25 %
Overclipping inverter	3.22 %	1.02 %	0.76 %	3.75 %	3.15 %

In this study, PR metrics that treat clipping either by excluding or reconstructing the curve show very similar behaviour as the one shown in Figure 4b. It is observed that these PR metrics are insensitive to the "overclipping" losses where the inverter operates under derating mode (due to high temperature). In the case of the PR clipping-corrected method this trend can be reduced if the clipping detection method implemented is able to automatically identify cases like that and exclude this loss from the reconstructed curve.

In the case of DC losses, these losses are relatively less important for a high DC:AC ratio arrays. However, correcting for clipping increases the weighting given to these losses and overemphasises their importance. On the other hand, using the uncorrected PR, these losses could go unnoticed, evolving into more critical performance issues in time. Thus, in terms of fault detection the clipping-corrected PR is a more efficient metric than the uncorrected (or STC corrected) PR.

Conversely, unexpected losses that could occur only during clipping are completely excluded from all PR metrics that treat clipping (either reconstructing or excluding entirely), however the impact of those will not be visible in any of the other metrics either and therefore it cannot really be quantified.

5 CONCLUSIONS

The widespread adoption of a clipping corrected PR metric by the industry can lead to more accurate PR assessment for PV systems, more straightforward financial risks for companies in the sector and increase in confidence of PV generation by consumers.

A new PR formula is suggested, based on the 61724-1 temperature corrected PR and a simple methodology for estimating clipping loss for oversized PV systems. The methodology has been implemented on a monofacial, fixed PV system in Netherlands with a high DC:AC ratio of 1.6, improving the monthly PR by 2.4%. The new PR metric was evaluated for its weather variability and its sensitivity to various system faults by using BSRN data from 6 locations and nearly 20 years' worth of data per location and simulating arrays of varying levels of under-performance.

The variability of the new metric is found to be significantly reduced when used on top of weather correction showing a median correction of 7% amongst all systems with SR=1.6, significantly reducing the uncertainty from weather inputs. This enables more confidence when setting the PR guarantees in contracts for large scale PV systems where a typical meteorological year (TMY) is used, and which might differ significantly from the actual weather data during performance evaluation.

The sensitivity of the new metric was also tested under 11 different faulty situations and is found to increase particularly for DC string and module faults, overemphasizing their impact on annual PR but also making it a useful metric in terms of identifying and correcting faults that could otherwise be largely masked by clipping.

ACKNOWLEDGEMENTS

Statkraft acknowledges funding from the Horizon 2020 programme of the European Union under GA. No. 952957–Project TRUST PV. NPL acknowledges funding by the UK Government's Department for Business, Energy and Industrial Strategy (BEIS) through the M4R programme and the UK's National Measurement System programmes.

REFERENCES

[1] International Electrotechnical Commitee, "IEC 61724-1: Photovoltaic system performance - Part 1: Monitoring," 2021.

[2] T. Dierauf, A. Growitz, S. Kurtz, and C. Hansen, "Weather-Corrected Performance Ratio," 2013.

[3] R. Kharait, S. Raju, A. Parikh, M. A. Mikofski, and J. Newmiller, "Energy Yield and Clipping Loss Corrections for Hourly Inputs in Climates with Solar Variability," in 2020 47th IEEE Photovoltaic Specialists Conference (PVSC), Jun. 2020, vol. 2020-June, pp. 1330–1334, doi: 10.1109/PVSC45281.2020.9300911.

[4] Perry, K., Muller, M. and Anderson, K., "Performance Comparison of Clipping Detection Techniques in AC Power Time Series". 2021 IEEE 48th Photovoltaic Specialists Conference (PVSC).

[5] William F. Holmgren, Clifford W. Hansen, and Mark A. Mikofski. "pvlib python: a python package for modeling solar energy systems." Journal of Open Source Software, 3(29), 884, (2018). https://doi.org/10.21105/joss.00884

[6] Driemel, A., et al., Baseline Surface Radiation Network (BSRN): structure and data description (1992–2017). Earth Syst. Sci. Data, 2018. 10(3): p. 1491-1501

CLEANING OPERATIONS PLANNING INDUCED BY MONITORING AND MODELLING OF SOILING ON PHOTOVOLTAIC MODULES

Thomas Carrière[1], Jean Catry[1,2], Alexandre Boilley[1]
[1]SOLAÏS, 55 Allée Pierre Ziller, 06 560 Sophia Antipolis, France
[2]MINES ParisTech, CS10207, 06904 Sophia Antipolis, France

ABSTRACT: Photovoltaic (PV) power plants are subject to soiling due to cementation or deposit of particles on the panels. This soiling turn losses of power generation which in turns causes non-negligible financial losses for the owner of the plant. PV plants can be cleaned leading to an improvement in power generation. However, it is generally unclear whether the cleaning of the plant will have a positive financial impact considering the cost of cleaning the panels and the probability that natural rain might clean the panels anyway. In this paper, we consider the use of DustIQ data to implement decision-making strategies related to soiling. The DustIQ is a device developed by Kipp&Zonnen to directly measure soiling losses. Although its measurements are robust, their raw value are difficult to use as the data requires some amount of pre-processing and in situ calibration. We propose a methodology to calibrate the DustIQ data based on measurements that are available on all monitored PV plants, and then propose a decision-making model taking into account local weather conditions to decide whether a PV plant should be cleaned or not at a given time.
Keywords: Monitoring, Performance, Deposition, Cleaning, Decision-making

1 INTRODUCTION

The soiling of photovoltaic (PV) modules accounts for a significant part of the losses in many PV power plants, up to 50% in extreme environments subject to dust storms. In relatively little dusty environments, such as southern Europe, this loss is usually lower than 10% [1], yet it is high enough to include it in financial models and to monitor it throughout the lifetime of the plant.

Today, most of the yield models assume an average transmission loss value of a few percentage points in Europe for monthly soiling values evaluated on a very shallow basis, while operators often rely on arbitrary cleaning periods. However, soiling is difficult to measure accurately, while being strongly dependent on local environment parameters, i.e., rain intensity and frequency, and the structure and composition of dust. Some models use large-scale dust composition or turbidity of atmosphere [1], but it appears much more relevant to use local measuring devices to record irradiance transmission loss since the effect of soiling is very location dependent.

Our study aims at analyzing local soiling data collected by a DustIQ [2] device developed by Kipp&Zonen,to build realistic soiling data-based models to help power plant developers or operators. Many studies have used local measurements of the transmission loss with optical measuring devices or short-circuit intensity and voltage monitoring [3][4], however they require to jointly monitor the characteristics of a soiled and a regularly cleaned module. On the other hand, the DustIQ allows automatic on-site collection of data, and therefore enabling the monitoring of multiple potentially remote sites. The DustIQ must be calibrated twice: first at the factory with Arizona Test Dust and then on-site with local dust, to account for its properties.

The contributions of this paper are the following. First, this paper shows effective preprocessing methods of the DustIQ data to get a meaningful signal. Although the literature provides some insight on how to filter DustIQ data, we could not find comprehensive comparisons of the different proposals.

The second contribution relates to the calibration of DustIQ data to the actual measured PV power losses caused by soiling. Indeed, K&Z calibration procedure must be performed under specific conditions which are quite hard to achieve: the Sun's incidence angle must be the smallest possible and the irradiance over 900-1000 W/m², which are only found in Europe around solar noon in summer. Moreover, transmission loss must be higher than 5%, which implies letting soiling accumulate on the device naturally for quite a long time before calibration. As a result, actual DustIQ devices are often poorly calibrated, rendering their data more challenging to analyze. We propose an affine re-calibration procedure for the collected data to correct these effects.

Our final scientific contribution in this paper is a proposed model that allows the simulation of soiling scenarios depending on a few parameters that can be derived from calibrated DustIQ data. In our studies, soiling follows quite a linear trend with time, and rainfall cleaning events of variable amplitudes regularly occur. From the DustIQ data, we could identify the mean daily increase in soiling, the frequency of cleaning events (e.g., rain, and the typical magnitude of these events). The cleaning events are then modelled by a Bernoulli trial for each day, and the amplitude of each cleaning event is modeled by an Exponential distribution independent from the Bernoulli trial. This simple model allows to easily calculate the expected levels of soiling of the panels at any time, based on real data.

The remaining of the paper is structured as follows: Section 2 describes the filtering of the data, while section 3 shows its recalibration. Finally, section 4 shows how we can use the recalibrated data as a decision-making tool to decide whether a plant should be cleaned or not.

2 FILTERING OF THE DUSTIQ DATA

To perform the analysis of this paper, we used data collected from a solar PV plant in southern France on a period of roughly one year between March 2021 and March 2022. We measured PV power generation, Global Horizontal Irradiance (GHI) using a pyranometer, ambient temperature and soiling loss using the DustIQ.

Figure 1 shows some raw DustIQ data over a few days. Measurements in the early morning appear faulty, and show an unexpectedly large variability given the relatively slow rate of increase of the dust on the panels.

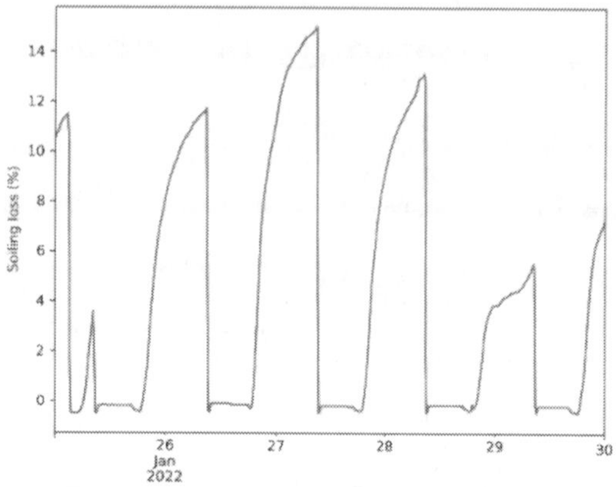

Figure 1: Raw DustIQ data on a short period

There are some standards provided by Kipp&Zonen (K&Z) to preprocess DustIQ data. It is recommended to only consider measurements when the DustIQ is dry, requiring either local weather measurements (rarely available) or the use of numerical reanalysis for dew temperature and rain which are not precise enough and do not allow real-time pre-processing.

Besides, the DustIQ monitors the soiling on the sensor from normal irradiance, while the actual panels do not receive normal irradiance. Depending on the angle of incidence of the Sun rays, the transmission loss due to soiling can be mitigated when the light is globally scattered towards the panels for low incident angles. These behaviors are very difficult to predict without heavy analysis of the local dust sources and temporal variability. The authors in [5] proposed an incident angle modifier on the transmission loss, however it is based on empirical data fittings and hence is difficult to generalize for other devices.

In this paper, we compare different preprocessing approaches from modeling of the weather effects to simple down-sampling of the data, to identify the most robust way to use DustIQ data. We considered the following filters.

K&Z recommendations: ambient temperature higher than the dew temperature + 5 C° to avoid water on the sensors, discard consecutive measurements with a difference of more than 5%.

Numerical filters: Daily mean, median and most frequent value of the measurements. The goal of these filters is to take care of the large variability of the measurements.

Sun position filters: measurement when the Sun is at its highest or at noon. The goal of these filters is to take care of the variability of the measurements by downsampling, while keeping a value where a) The Sun is close to the position it had at the time of the local calibration, which makes the transfer function to power loss of the device valid and b) the high elevation of the Sun greatly lowers the risk of having dew on the sensor.

These different filters were applied on DustIQ data from a measurement campaign on a PV power plant located in Southern France. They are then visually compared to see whether a meaningful signal can be extracted from the raw signal. Figure 2 shows the different filtered data.

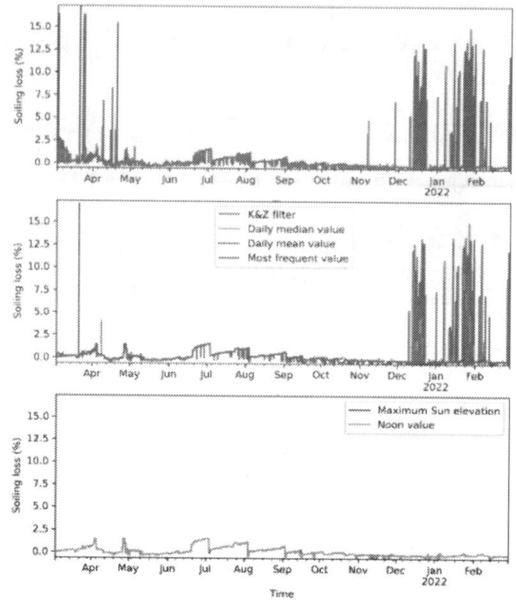

Figure 2: Application of different filtering processes to the DustIQ data; Top graph: raw data. Middle graph: Numerical filters. Bottom graph: Sun position filters.

It is quite clear that the K&Z recommendations and the numerical filters are not able to filter the dew data in the morning, since the spikes caused by dew are still visible on the figures. The filter for most frequent value removes most of this variability, however it still shows more sensitivity to noise than the filter of maximum Sun elevation, which is thus the preferred way to preprocess the data. Besides, to apply this filter very easily, we can use a filter for values measured at noon as it is a good proxy for the time of maximal Sun elevation at our location.

3 DUSTIQ DATA CALIBRATION

Although the raw DustIQ data can be improved by applying the relevant filter, it remains clear that the scale of the measurements is not right despite the local calibration. On Figure 1, measurements are regularly negative, which is a significant hint that the calibration is not right.

To help recalibrate the model, we propose to study the variations of the ratio between the power generated by the plant and the one simulated with a PV power model that does not take soiling into account, R_{soil}, with respect to the soiling loss measured by the DustIQ. If the DustIQ data has value, we should see a pattern between the DustIQ data and the ratio. If not, these data are independent and thus DustIQ data cannot be used to measure the soiling of the plant. The power model is taken from [6]:

$$P_{sim} = PR * S_{PV} G_{mod}\left(\eta_{STC} + \beta \log(G_{mod})\right)\left(1 - \alpha\left(T_{cell} - 25^{\circ}\right)\right)$$

$$R_{soil} = P_{real}/P_{sim}$$

Where P_{sim} is the simulated power, S_{PV} the surface of the modules, η_{STC} the efficiency of the module in standard conditions, G_{mod} the incoming Global Tilted Irradiance (GTI), T_{cell} the temperature of the modules, α the sensibility of the modules to temperature and β a parameter to be fitted. The GTI is derived from the Global Horizontal Irradiance (GHI) measured by a pyranometer, using a decomposition model. The cell temperature T_{cell} is derived from the ambient temperature measured in situ, using the Sandia model [7].

The raw efficiency of the model is displayed on Figure 3. The accuracy seems quite good, although the dispersion is quite large mainly because of the conversion from horizontal to plane-of-array irradiance.

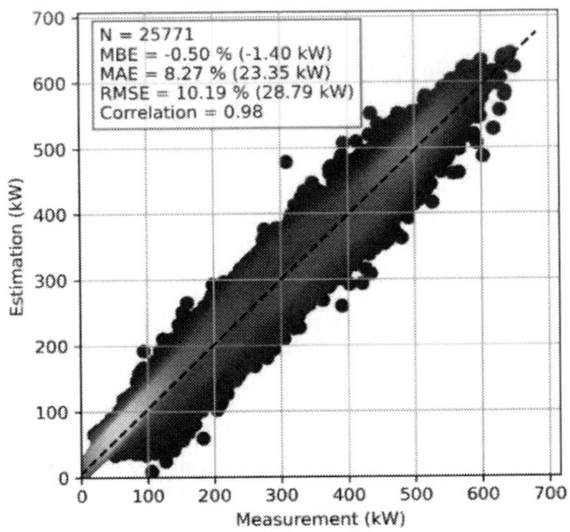

Figure 3: Performance of the model without soiling

Once this model has been applied, we can plot the ratio R_{soil} between real and simulated power with respect to DustIQ data. A time-series also showing precipitation can be found on Figure 4.

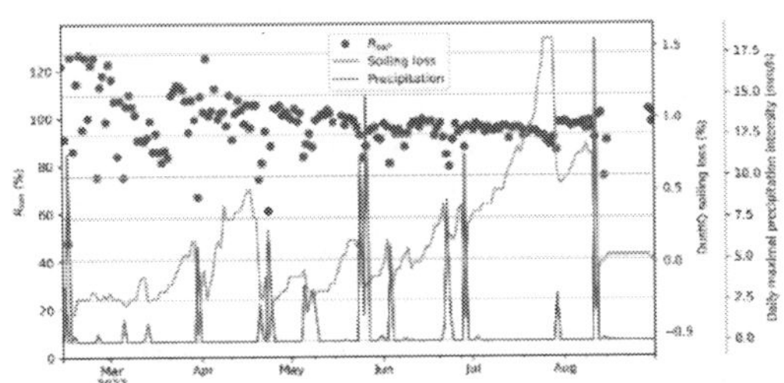

Figure 4: Ratio between real and simulated power (blue dots), DustIQ measurements (orange line) and precipitation (green line)

Whenever there is rain, the model overestimates the plant performance, resulting in lower R_{soil} values. This might be caused by a bad estimation of the module temperature from ambient temperature, or optical effects caused by the water on the module. In any case, it is quite clear that the cause of bad performance on these days is caused by rain and not soiling, and so these data points should be discarded in our analysis.

The dust level accumulates over several days. Whenever there is rain, the DustIQ signal lowers, which makes sense as it suggests the sensors are cleaned by the rainfall. This increases the trust we can have in the measured signal, as it corresponds to the intuition that dust accumulates every day via cementation or deposit but might be cleaned by natural rain. The amount of cleaning does not seem dependent on the amount of rain. This makes sense, as rain can clean the sensors but also bring dust when it evaporates, and many other factors might affect the cleaning effectiveness of rain such as the size of the droplets or the wind.

Finally, we can see from Figure 3 that the proposed power model tends to underestimate production when there is low irradiation i.e., in the 100-300 kW range, such as cloudy weather. To further isolate the effect of soiling, we only kept data points corresponding to clear-sky days.

On Figure 5, we plot the filtered values of R_{soil} against the measured DustIQ signal.

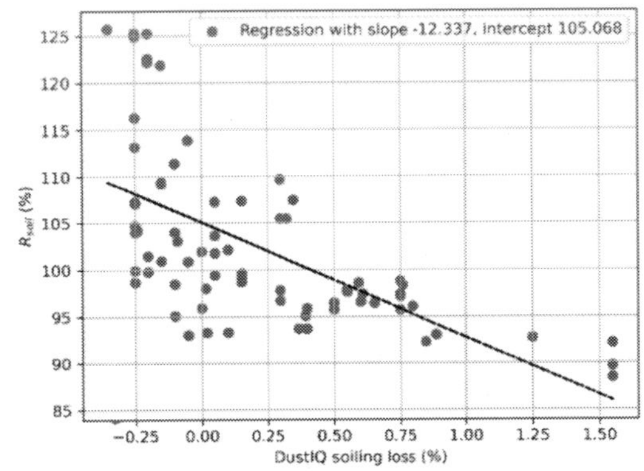

Figure 5: Figure 4: Scatterplot between R_{soil} and DustIQ

From Figure 5, it seems that there is a trend between R_{soil} and the DustIQ signal, that we can roughly approximate with a linear relationship. To account for this, we propose to update the power model as follows:

$$P'_{sim} = PR * S_{PV} G'_{mod}\left(\eta_{STC} + \beta log(G'_{mod})\right)\left(1 - \alpha\left(T_{cell} - 25^{\circ}\right)\right)$$

$$G'_{mod} = \rho G_{mod}, \rho = \alpha_{SL} SL + \beta_{SL}$$

With α_{SL} and β_{SL} are coefficients that perform an affine calibration of the soiling loss SL measured by the DustIQ. Using the updated model, we obtained the performance showed on Figure 6.

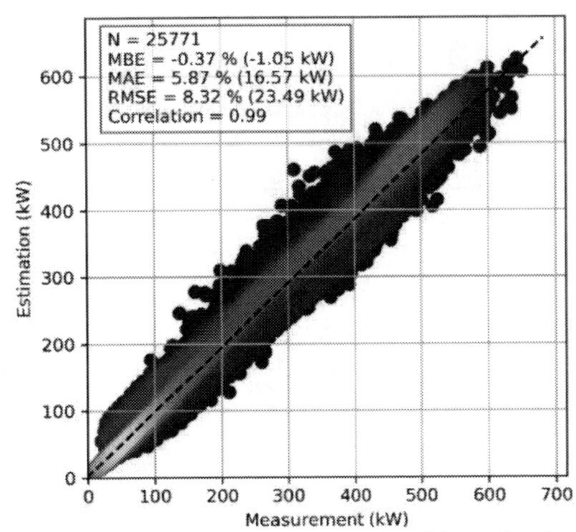

Figure 6: Performance improvement of the updated

power model

The fit of the data is much better, with an improvement of 0.13% on MBE, 2,4% on MAE and 1,87 on RMSE. Based on this, we can conclude that the calibrated measurements are much closer to the real soiling losses.

4 APPLICATION ON DECISION MAKING FOR PANEL CLEANING

Once that we are confident in the soiling loss values reported by the DustIQ, we can then try to implement decision making using it. One of the most common questions that PV asset owners have regarding soiling is whether they should clean the modules or not given the amount of soiling. We propose to derive the gain of the cleaning operation as follows:

$$G = \left[\sum_{t=0}^{+\infty} \pi\left(PV_{cleaned,t} - PV_{soiled,t}\right)\right] - CC$$

Where G is the financial gain in euros, CC the cost of the cleaning operation in euros, π the average price at which the electricity is sold, and $PV_{cleaned,t}$ and $PV_{soiled,t}$ are the generation of the cleaned and soiled plant at time t respectively.

To approximate this gain, we propose to neglect the term $\beta log(G_{mod}) \ll \eta_{STC}$. In which case, we have:

$$PV_{cleaned} - PV_{soiled} = (\rho_{cleaned} - \rho_{soiled})PR * S_{PV}\eta_{STC}\left(1 - \alpha\left(T_{cell} - 25^{\circ}\right)\right)$$

Where $\rho_{cleaned}, \rho_{soiled}$ are the calibrated soiling in the cleaned and soiled case at time t respectively. We propose the following model to estimate the soiling at day d:

$$\rho_{d+1} = \rho_d + D - R.Cl$$

With D the increase of soiling, R a binary variable for rain events and Cl the amount of cleaning provided by the rain event. We propose to model D by a constant value, R as a Bernoulli trial which parameter is estimated by the empirical precipitation probability, and Cl as a random variable following an exponential distribution, whose parameter is estimated from the data. The values are then clipped to the interval [0-1].

This model is quite basic, as both the increase in soiling and the amount of cleaning are unconditional from weather data, while it heavily depends on it. However, investigating this relationship would require much more weather measurements and is thus not considered in this work.

Figure 7 shows a few trajectories generated using this model. While they are distinguishable from the real signal, they still reproduce some properties of the original signal, such as the autocorrelation and the variability, while allowing the generation of diverse plausible trajectories.

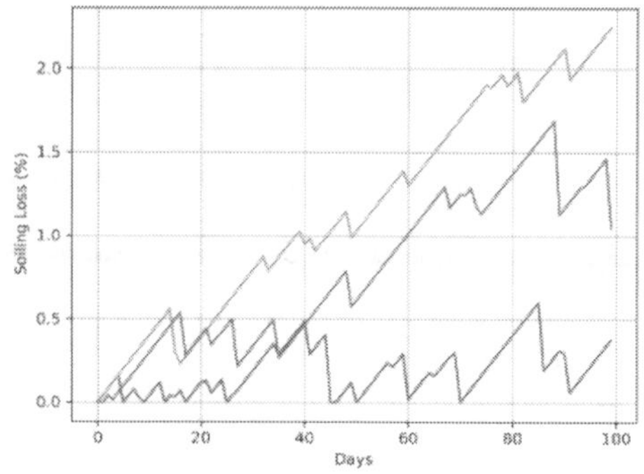

Figure 7: Generated trajectories for the soiling loss. Scenarios vary depending on the random results of the Bernoulli trials and cleaning events effectiveness.

With this model, the expected soiling loss is:

$$E(\rho_{d+1}) = E(\rho_d) + D - p_R/\lambda$$

With p_R and λ the parameters of the Bernoulli and exponential distribution. Thus, the expected value of the soiling either increases or decreases each day depending on these parameters. Whenever it reaches 1 (increasing case) or 0 (decreasing case), it does not change anymore. Thus, the infinite sum from the equation that derives the gain can be changed by a finite sum until the day for which both the cleaned and the soiled plant reaches the ceiling. After this point, the initially cleaned and soiled plants are in the same soiling state and thus there will be no more economical gain.

The number of days N_d to reach the ceiling is:

$$N_d = \left[\frac{1-\rho_0}{D - p_R/\lambda}\right] \text{ if } D - \frac{p_R}{\lambda} > 0$$

$$N_d = \left[\frac{-\rho_0}{D - p_R/\lambda}\right] \text{ if } D - \frac{p_R}{\lambda} < 0$$

Where [] is the integer part operator. Noting $N_d' = N_d * (timesteps\ per\ day)$ the number of timesteps that are needed to reach the number of days N_d, the expected gain of cleaning operation is thus:

$$E(G) = \left[\sum_{t=0}^{N_d'} \pi\left(E\left(PV_{cleaned,t}\right) - E\left(PV_{soiled,t}\right)\right)\right] - CC$$

Assuming that the prices are independent from the PV power generation. Replacing the expected values, we obtain:

$$E(G) = -CC + \pi PR.S_{PV}\eta_{STC} * \sum_{t=0}^{N_d'} \left(1 - \alpha(T_{cell} - 25^{\circ})\right)\left(E\left(\rho_{cleaned,t}\right) - E\left(\rho_{soiled,t}\right)\right)$$

And so

$$E(G) = -CC + \pi PR.S_{PV}\eta_{STC} * \sum_{t=0}^{N_d'} \left(1 - \alpha(T_{cell} - 25^{\circ})\right)\left(\rho_{cleaned,0} - \rho_{soiled,0}\right)$$

Assuming the same cell operating temperature for each time T_{cell}:

$$E(G) = -CC + \pi PR.S_{PV}\eta_{STC}N_d' * \left(1 - \alpha(T_{cell} - 25^{\circ})\right)\left(\rho_{cleaned,0} - \rho_{soiled,0}\right)$$

And finally, we can decide whether the plant should

be cleaned or not by comparing the expected gain with 0.

5 CONCLUSIONS

In this work, we proposed a complete method to implement decision-making on whether to clean a plant or not based on DustIQ data. This data is very useful, but it requires some pre-processing and calibration before it can be used.

The calibration procedure requires measurements of ambient temperature, GHI and PV power generation which are almost always available for existing power plants. Further work could investigate whether the use of irradiance probes is possible, considering that they are much more affected by soiling than ventilated spheric pyranometers, and thus might call into question the methodology we proposed here.

We also proposed a simple model allowing decision-making on whether a plant should be cleaned or not depending on its amount of soiling, the probability of rain and the typical amount of cleaning of the PV panels after a rain. This model sets a baseline, but it is still basic in its characterization of soiling evolution. Further work should condition the amount of dust increase and cleaning based on weather parameters such as Aerosol Optical Depth (AOD), wind speed or dew. This would allow generating more plausible soiling trajectories and better estimate the financial gain obtained from cleaning a PV power plant.

6 REFERENCES

[1] Supe, H., Avtar, R., Singh, D., Gupta, A., Yunus, A. P., Dou, J., ... & Kharrazi, A. (2020). Google earth engine for the detection of soiling on photovoltaic solar panels in arid environments. *Remote Sensing*, *12*(9), 1466.

[2] DustIQ Instruction Manual v2105-2, Kipp&Zonen, May, 19, 2021

[3] Javed, W., Guo, B., Figgis, B., Pomares, L. M., & Aïssa, B. (2020). Multi-year field assessment of seasonal variability of photovoltaic soiling and environmental factors in a desert environment. *Solar Energy*, *211*, 1392-1402.

[4] Figgis, B. (2018). *Investigation of PV soiling and condensation in desert environments via outdoor microscopy* (Doctoral dissertation, Strasbourg).

[5] Nepal, P., Korevaar, M., Ziar, H., Isabella, O., & Zeman, M. (2018). Accurate soiling ratio determination with incident angle modifier for PV modules. *IEEE Journal of Photovoltaics*, *9*(1), 295-301.

[6] Williams, S. R., Betts, T. R., Helf, T., Gottschalg, R., Beyer, H. G., & Infield, D. G. (2003, May). Modelling long-term module performance based on realistic reporting conditions with consideration to spectral effects. In *3rd World Conference on Photovoltaic Energy Conversion, 2003. Proceedings of* (Vol. 2, pp. 1908-1911). IEEE.

[7] Fuentes, M. K. (1985). A simplified thermal model of photovoltaic modules. *Sandia National Laboratories Report, SAND85-0330*.

DATA ANALYTICS AND THE PERFORMANCE RATIO OF PV PLANT WITH BIFACIAL SOLAR MONOCRYSTALLINE MODULES

Luis Mota Rodrigues, Lightsource bp, 33 Holborn, EC1N 2HT, London, UK, +442080494448, luis.rodrigues@lightsourcebp.com
Chris Buckland, Lightsource bp, 33 Holborn, EC1N 2HT, London, UK, +442036677966, chris.buckland@lightsourcebp.com

ABSTRACT: Bifacial module will remain the central component of the Solar PV industry independently of the introduction of new cell technologies for the next years where unprecedent expansion and investments on this increasing critical sector will happen. Understanding the performance of bifacial modules at plant level has several advantages: it can help to qualify best manufacturers and technology; it can lead to a better system optimization and better matching between modelling and "data from the field". We believed Data Analytics should has already been present across PV industry some time ago, but nowadays it looks even more pertinent to provide business analysts, techno-structural executives, and corporate operational units with in-house actionable knowledge about key performance indicators, from individual component assessment up to integrated performance. This paper is focus as example of the potential of Data Analytics at practical level of how a Solar Developer can use it to extract more value from utility scale bifacial PV systems in distinct fronts. In this study the focus was restricted to Performance Ratio [PR] analysis.

1 BIFACIAL MODULE LONGEVITY

Silicon crystalline bifacial module can still be considered a recent product in the Solar PV industry. It received a fast and widespread adoption from most players of the PV industry due to the additional (uplift) power generated from the rear side of the module taking advantage of the environmental albedo from the surroundings. Nowadays this type of product is being extensively deployed at utility scale installations across the globe becoming the dominant module technology. Currently mainstream products are based in p-type PERC/PERT cell technology and bifacial module is seen as a promising platform to be used as well in the forthcoming PV cell technology developments presently under intensive R&D development aiming volume manufacturing in the next 2-3 years. Practical examples of the continuing reuse of bifacial module technology are the novel n-type TopCon and n-type Heterojunction (HJT) products - and quite probably the bifacial platform will stay for multijunction based module expected for the end of current decade, like Perovskite along HJT layers in tandem structures.

2 DATA ANALYTICS AS BUSINESS INTELLIGENCE

The concept of Data Analytics (DA) in this study is as a systemic method of examining data sets to find trends and draw conclusions about the embedded information. DA can be divided into quantitative and qualitative analysis. Since PV generates huge numerical data sets in this work the two types are present to quantify variables and to interpret relationships.

From a developer perspective DA could be seen as a first step to approach a self-service Business Intelligence culture. Extracting the valuable knowledge hidden in the large datasets of operational data saved and promote job enrichment letting internal collaborators develop and run their own queries and build reports themselves.

In this study the programming tool used was R, from R Core Team (2021) R a language and environment for statistical computing. R Foundation for Statistical Computing, Vienna, Austria.

3 THE IMPACT OF THE BIFACIAL GAIN IN THE PERFORMANCE RATIO (PR)

Although welcomed by most of the PV players Lightsource bp [LSbp] understands there are relevant details which are not well understood neither discussed, from a practical and modelling standpoint, on the factors affecting the behavior of the performance ratio [PR] of a PV plant. PR is a common and recurrent metric used by Solar PV Asset Management units and Solar PV Asset companies, as well others relevant stakeholders across all PV Chain Value for contractual and operational purposes. But data generated by scada systems has a much wider potential besides tailor-made tools developed by external Data Analytics/IT companies since they can feed valuable know-how for multiple internal stakeholders not just one as usually it happens. As examples: data manipulated by DA helps to track down the evolution of the plant performance and availability after construction and during EPC management until FAT, it can periodically make the assessment of the PV plant condition to oversight operational activity at O&M level, and DA also contributes to gather relevant feed-back to support decision making and fine tune LCOE analysis for Procurement and Engineering teams to compare different technologies, to confirm accuracy of component specifications, to qualify products and to evaluate manufacturers of the key components – solar PV modules, power conversion systems and tracker systems. The principal motivation behind this work.is to show how DA is a pertinent approach to create added value to the PV plant owners and solar developers.

Due to the impact of the bifacial modules and relevance of the PR as a key metric we believe deeper understanding regarding PR is need around 3 pillars: 1) sensitivity across irradiance levels and impact in the energy generation, 2) role of the albedo and its simultaneous effect with irradiance, 3) understanding the time-related behaviour of PR, like seasonal, daily, and hourly behaviour.

3.1 Site chosen for this study

All dataset manipulated in this study is from a ground

mounted fixed tilted plant with bifacial modules. Besides simplicity, analysing performance with bifacial modules is more complex with trackers since tracked systems favour sun facing front side of the panels making less visible the albedo role. There is an additional practical reason, ground mounted bifacial plants remain a huge business segment across several regions of the global PV market as of today and surely will stay so in the forthcoming years.

The field data presented throughout this work was collected remotely via scada from an operative PV plant located in UK, Cambridge region, roughly 120 kms north of London energized in 2020. The total power DC installed is approximately 5 MW. It uses bifacial modules with individual rated power of 375W, connected to string inverters with individual nameplate power of 105kVA and a ground mounted fixed structure with a 25° tilt in which clearance from table to soil varies from 1 metre up to 3 metres, respectively from ground to lower and ground to top of the table. Modules were installed in 4 rows in landscape configuration. All data in this study comprise datasets collected with 15 min and 1 min interval for a complete year, from 1st January 2021 until 31st December 2021.

3.2 PR calculation

In this work PR was computed as instantaneous PR. It uses the value of irradiation from the irradiation sensor (Pyranometer) and summation of Output Active Power of all the inverters to infer PR value of the plant as expressed below.

$$\frac{(AC\ Active\ Power\ Exported\,[kVA] * 1000\,[W.m-2])}{(DC\ capacity[kWp] * Solar\ Irradiation[W.m-2])}$$

3.3 Albedo and bifacial modules

Definition of albedo throughout this work is the following: as the fraction of solar radiation which is reflected from the ground and ground cover, like gravel or grass, and indexed to the visible and infrared light of the solar spectrum.

Today is relatively easy to collect real and reliable data from equipment installed in a solar PV plant. Noteworthy to refer is of essence the need of ensuring correct calibration procedure, proper installation and regular check of positioning and condition of the pyranometers, extensive to other metering devices, to ensure quality on the collected numerical data.

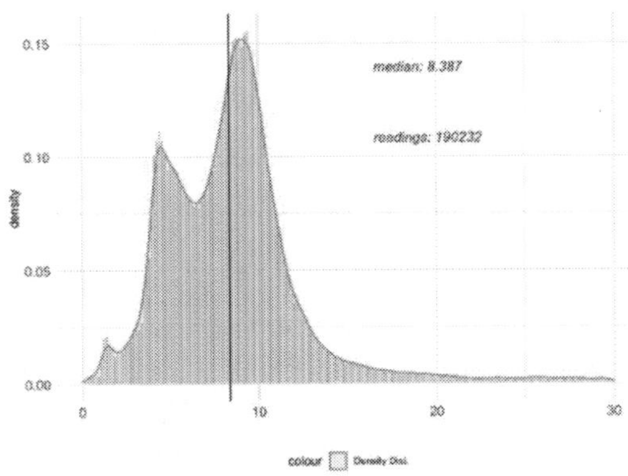

Figure 1: Albedo histogram as percentage of the reflected inclined irradiance over the total irradiation, Albedometer was on the Plane of Array – POA. Median value (8.387% ±0.123) %.

Above one can see the histogram of albedo during one year with datapoints collected each 1 minute. We can add albedo histograms may show minor or major differences for each location in the side. Each albedometer in the site although showing a common pattern collects its unique surroundings, its own signature indexed to the specific sub-location in the plant. This suggest how sensible could be the modelling of a bifacial plant to include all environmental and surrounding variables. As proven by the white sheets experience explained in following section.

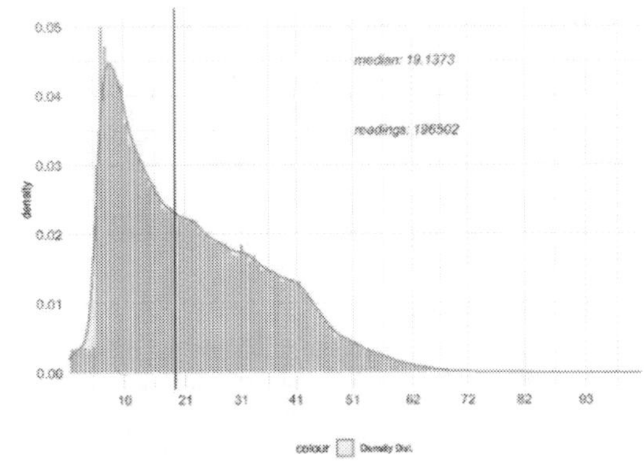

Figure 2: Albedo histogram of the reflected POA irradiance 1minute dataset collected in 2021. Median value (19.137 ±1.276) W.m^{-2}.

Figure 2 represents the distribution of the Albedo values in W m^{-2}. Median of the dataset is around (19.137 ±1.276) W.m^{-2} and mostly of readings are below 70 W.m^{-2} at POA.

As important as the quantitative values is the simultaneousness it occurs with POA irradiance. DA helps us to capture this matching based in 1min dataset for 1 complete year.

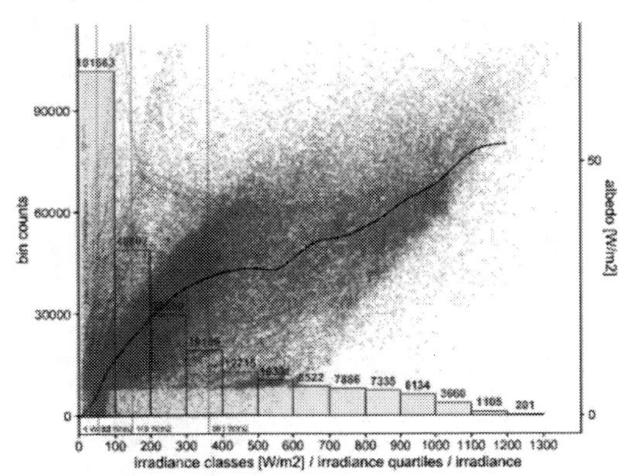

Figure 3: Albedo POA W.m^{-2} values (red), 1-minute datapoints, on top of the irradiance classes to pair irradiance at POA and albedo at same moment in time (same minute) during a complete year - 2021. The smooth line is computed using cubic splines showing as expected a positive correlation between POA irradiance and POA Albedo values.

To understand the impact of this additional irradiation it was constructed the histogram below in which both direct POA irradiance and POA reflected values were added and placed on top of the single irradiance POA histogram.

Irradiance values are represented now in the y axis as log10 to help visualising the difference. Somehow surprisingly the larger deltas are in the high irradiance classes suggesting it is at high irradiance levels that occurs the decisive contribution of bifacial modules for the uplift comparing to the monofacial modules.

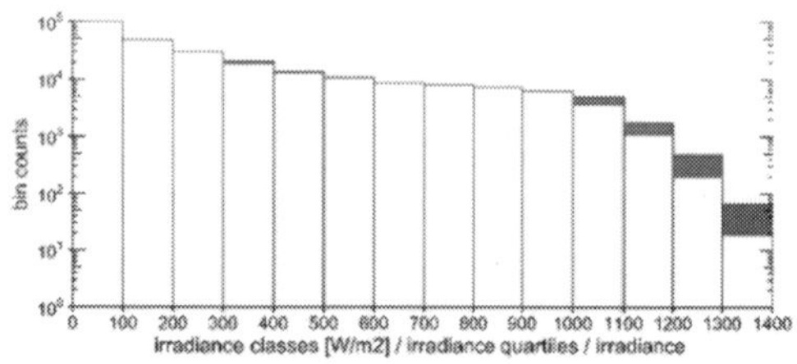

Figure 4: Histograms of 1) sum of irradiance POA plus albedo POA values in green and 2) Irradiance POA values in white.

3.4 The White Sheet Experiment

An experiment was done in this ground mounted and fixed tilted plant equipped with string inverters. The area underneath tables where modules allocated to seven string inverters was covered with white polymer foils ("white sheets") as shown in the fig.5. This installation was done in middle of winter season after PV plant was stable and in regular operating for some months.

Figure 5: Picture taken from field experiment with polymer-based white sheets covering ground area underneath all module tables connected to 7 out of 39 string inverters, all with same rated power.

Comparing the field data from the previous 15 days prior installation of the white sheets with the following 15 days it was clearly observed a change of the PR ranking as shown in fig.6. It was observed a global jump in PR with an average increment of (2.813±0.102) % in the set of inverters with white sheets underneath. This delta suffered minor variations in the following months proofing the increment was consistent and linked to the positive contribution of the increased albedo generated by the white sheets, independently of seasonal changes with irradiance levels and increasing sunny hours from Winter to Spring.

Figure 6: Classification of individual string inverters by boxplot of indexed PR values. 15 days data before and 15 days data after installation of white sheets it is clear the bias induced by the new material in the ~+2.8% uplift in performance. The 7 selected inverters had an initial PR value spread in the plant rank as seen at the left.

Besides the utility of using DA to quick infer and quantify the impact in the PR. This simple experiment also show how significant is the role of the surrounding – albedo – in the performance of a plant with bifacial modules and rises questions:

- How extensive can be the manipulation of the surroundings of a PV plant with bifacial modules to increase even more the uplift?
- Can extensive modification in the surroundings, instead of increment in fact reduces the complexity of modelling bifacial plants?

Related to this subject in the next figure a simple comparison between simulated values and field results taking into consideration the particularities of this bifacial plant, design inputs, including the technical specifications information from the module and inverter manufacturers. Without surprise PR value computed from generated data and yearly energy generation were higher (respectively ~7.2% and 5.8%) than the estimation for P50 case. The blue dots are the dataset as provided by the Scada system whereas the other were extracted from simulation done with PVsyst and Meteosat database. Fitting was done with cubic splines.

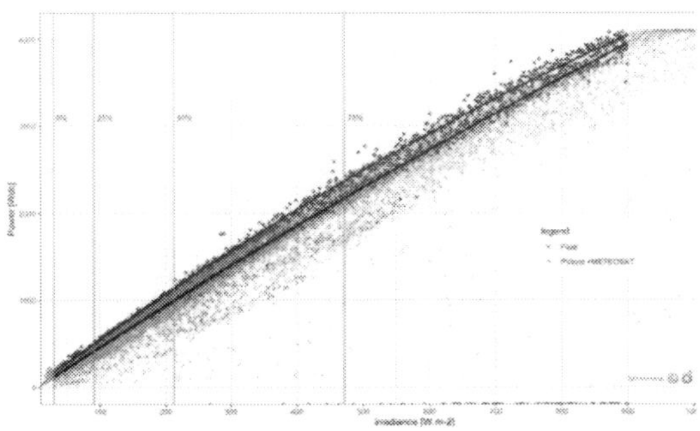

Figure 7: 8760-hour data extracted from simulation (fitting with dark line) compared with equivalent 8760-hour field data (fitting with blue line) after aggregating 15 min data to its respective hour.

This example shows that presently it looks quite reasonable Solar Developers use in-house developed DA tools to quantify bifacial projects. This know-how could be especially suitable for repowering projects, swapping monofacial by bifacial modules, since usually there is plenty of historical data to be computed as baseline reference.

4 DATA ANALYTICS, PERFORMANCE RATIO AND ENERGY GENERATION

We believe during present decade there will be an increasing need for general use and development of the DA area inside PV players to understand the mechanisms affecting Solar PV performance and the decision-making process in distinct situations:

- Interface between PV and specific environment of the site location is one of them and it is evolving with installation of utility scale Solar PV plants in increasing challenging places such water reservoirs, sea, fisheries, rooftops, BIPV.
- All new technologies in cell, modules, inverters, power electronics and introduction of BESS co-located with PV are in expansion in quite different geographies and under strong R&D shortening product development cycle raising the need to compare technologies and manufacturers also in shorter periods
- Systems are turning more complex and technology denser like introduction of BESS co-located with PV and use of hybrid systems to produce green hydrogen/ammonia.
- The investment scale is also evolving, soon +1GWp plants will be considered, possibly with pre-assembly structures, automated installation and with MVDC equipment instead of the usual MVAC.
- Finally, manufacturers do not have same position in the market, neither same quality of the components (including bifaciality gain) and engineering content. DA can be decisive in the selection of best: component mix, technology and manufacturer

Following examples will try to demonstrate how valuable quantitative and interpretative information can be obtained using simple DA in current situations.

4.1 Irradiance and Performance Ratio

4.1.1 Variation of PR per irradiance classes and variation of PR by irradiance quartile.

Below one can observe the profile of 1min datapoints showing relationship between PR and POA irradiance. PR by quartile indicates higher PR happened in the second quartile between [53 – 148] W.m^{-2}. Important to mention median of irradiance values >4 W.m^{-2} in the plant during 2021 was 148 W.m^{-2}.

Inside the 4th quartile it was separated median PR when inverter not clipping [363 – 820] W.m^{-2} and the media PR in all 4th quartile [363 – maximum irradiance] W.m^{-2}. As expected, within 4th quartile median PR incorporating the clipped values is lower than the unclipped one. This will be explained in the fig.8. Turning clear in this UK site, with fixed tilted installation, more than 50% of the POA irradiance readings lays below 200W.m^{-2} but in the other hand only generates roughly 10% of the energy while a share larger of 60% occurs in the last quartile (last 25%).

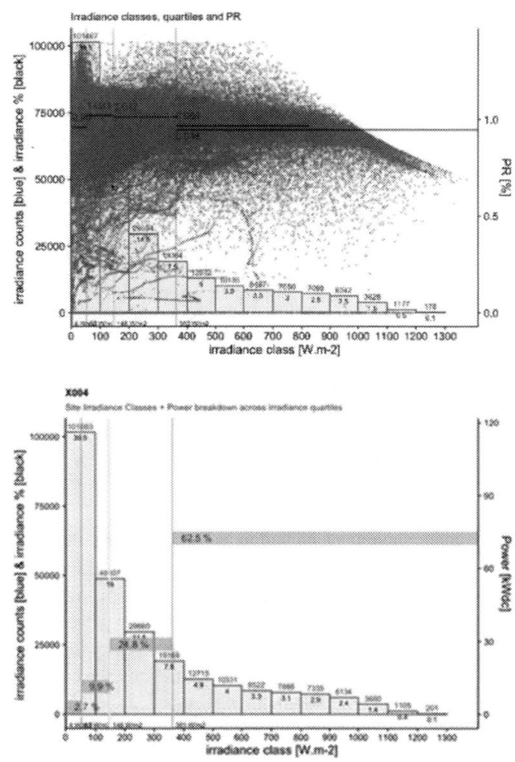

Figure 7: Comparison between PR and energy generation by irradiance quartile. Around 62.5% of the energy exported by the plant in 2021 was generated in the 4th

quartile, equivalent of 25% all irradiance, ranging from 361 W.m^{-2} up to ~1250 W.m^{-2} (maximum read),

The 1 min dataset allows to be seen clipping zone and the inverter forbidden area on top right side. The variation of values next to the clipping is partially explained in the following chart – fig.8 related to the effect of ambient temperature and operation of the components.

4.1.2 Single day analysis

The use of 1-minute datasets brings richer information. In the next chart with 1-minute data during a clear sky day for simplicity, we exposed principal observations that worth to be presented.

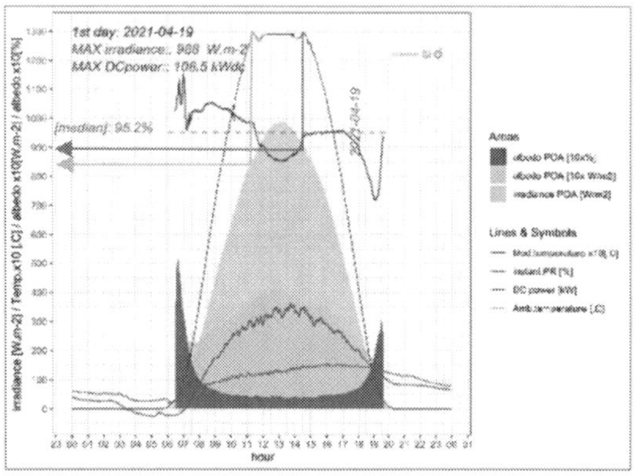

Figure 8: Clear day profile showing PR behavior, inverter output DC power, irradiance area and albedo area in % to the incident irradiation and in W.m-2. 1min data interval. Median of the instantaneous PR for the day was (95.2 ±0.8) %. Albedo values reached ~40 W.m-2 around 1pm.

It can be seen how instantaneous PR value is evolving during the day reflecting the interaction and dependency between environment, equipment and the resulting energy generated.

- Earlier in the morning module low temperature, due dependency as expressed by thermal coefficients, increases module wattage, and coupled together with high relative albedo (blue area) PR starts quite high and then continuously diminishes with increasing direct light share and increasing module temperature.
- When clipping occurs, as expected, PR lowers even more reflecting the loss derived by the cap imposed by the clipping of the inverter when irradiance is high.
- Nevertheless, due to 1min resolution we can see the irradiance values when enter and leaves the clipping zone are not the same. Clipping starts with lower irradiance values than when it stops clipping. That partially explains the jitter values around the inverter forbidden zone as shown in top right area of fig.7 (top).
- Resulting from the temperature effect it can be observed DC power curve is not aligned with irradiance area. Irradiance area is slightly shifted rightwards. This means, for clear days or when similar irradiation happens, energy production is larger in the first half of the sunny hours (morning) than in the last half (afternoon). Obs: Solar plant is facing south.
- Albedo as % to the direct irradiance showed peak rates at the dawn and at sunset. This is a quite distinctive from monofacial plant and changes PR values at low irradiance levels and subsequently changes overall plant PR since this is a place where median irradiance is typical low ~ 150 W.m-2

4.1.3 Temporal analysis

Temporal analysis helps to see profile according to simulated data and allow to investigate profile in distinct time related situations only possible with DA tools. The first one is the variation of the PR in all year aggregating 1min data to 1 hours to match the kind of data as given by the simulation software, it can see how stable the PR of the bifacial module is.

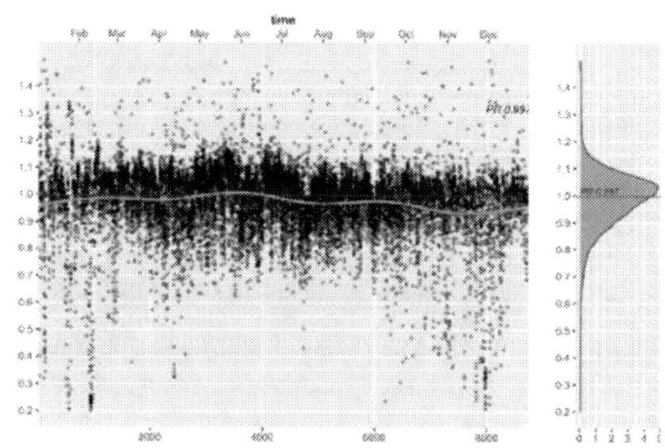

Figure 9: 1min data aggregated to plot 8760-hour dataset with field data to allow direct comparison with 8760-hour theoretical dataset (from modelling software).

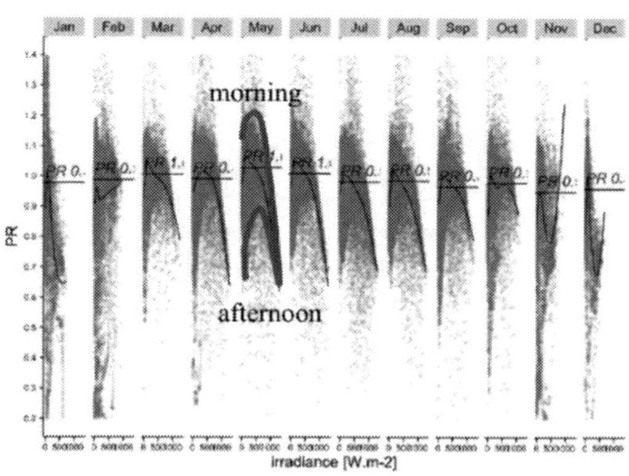

Figure 10: Monthly PR profile and median per month. Fitting with cubic splines. Indicated path followed by instantaneous PR in the morning and afternoon,

There is a common pattern in the PR profile between April and September associated to longer hours with irradiance when most of the energy is generated. May is the month with the highest PR value. This facet view also makes clear complementing what is observed in the previous plot. During the morning PR is systematic higher than in afternoon.

In the next figure, fig.11, we plot the variation of the albedo POA with irradiance POA by hour and by months showing a systemic relationship between irradiance and albedo. More pronounced during early and sunset hours of day than with month.

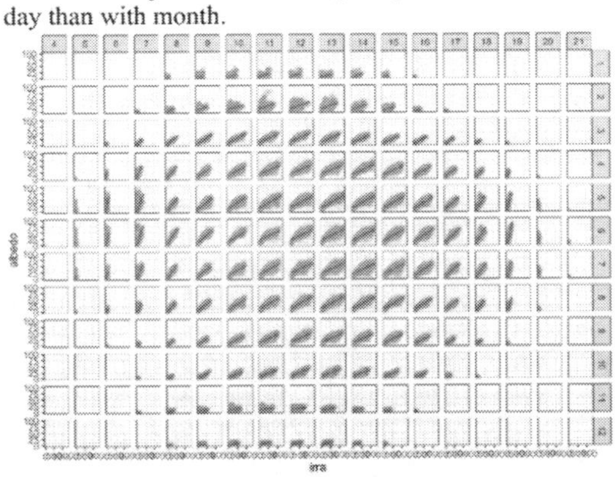

Figure 11: Monthly and hourly changes of the albedo with irradiance level during 2021 (both in POA). Fitting with linear regression. Albedo sensitivity to hour of day is higher than to the month of the year.

4.1.4 comparison of PR between monofacial and bifacial Solar PV plants

Data analytics helped us to quick compare performances across technologies for plants located very closed to each other to reduce uncertainty based in the location. Below the comparison of the PR profile with POA irradiance for 2021 year.

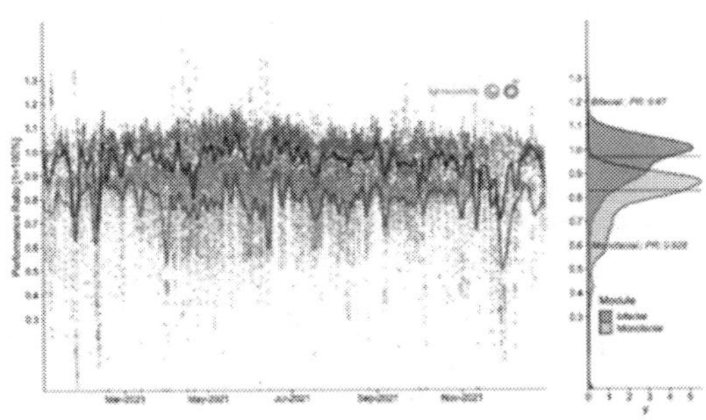

Figure 12: Comparison between monofacial and bifacial plants located in same vicinity with similar MWp installed. 15min datasets for both plants. PR density distribution represented on the right side.

The comparison can go further down to the hour level. As clear seen in the following 2 figures, bifacial plant has a consistent higher PR value than the monofacial one overall, per month, per hour and at low irradiance level. This multiple quantitative and qualitative analysis could be easily done to compare modules, cell technologies, inverters, design variables, manufacturers etc.

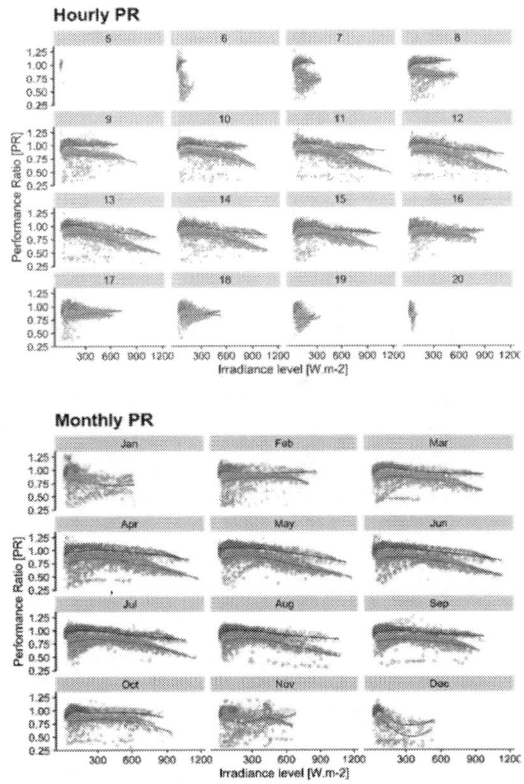

Figure 13: Comparison of different PV plants, close to each other, same year, through multiple analysis of behavior of the PR with irradiance in different temporal views.

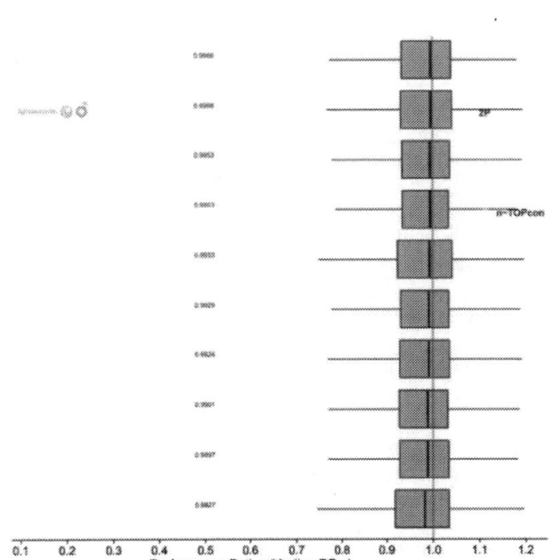

Figure 14: Through regular monitoring of PR from string inverters with different setups it is possible to carry out simple and robust side-by-side evaluations of components and configurations to optimize yield and reduce LCOE.

Here a particular table configuration: 2 in portrait – 2P and a specific bifacial n-TOPcon module was assessed against the baseline plant design: p-perc bifacial modules mounted in landscape configuration. It can be seen performance for 2P is comparable with 4L, 4 modules in landscape, contributing to validate 2P which has a lower CAPEX cost. For the n-TOPcon comparison the selected bifacial n-TOPcon module surprisingly was not outperforming bifacial p-perc (Obs: probably related to the BOM of the n-TOPcon module not to cell technology).

Below we present PR variation for 1min dataset for the best performance string inverter for 2021 coupled to the ambient temperature as measured by the plant weather station.

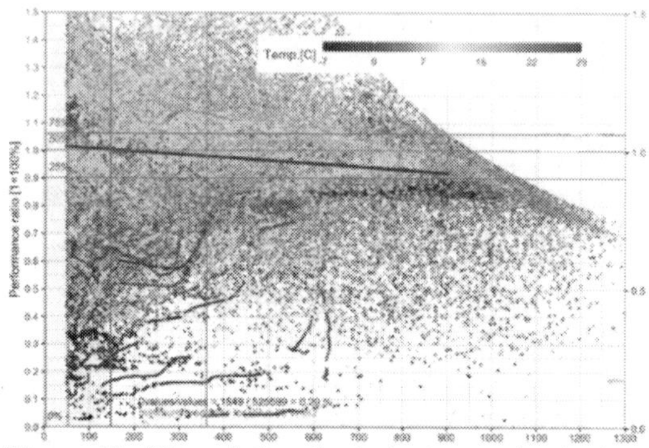

Figure 15: PR variation for 1min dataset for the best performance string inverter for 2021 coupled to the ambient temperature. POA Irradiance values higher than 50W.m^{-2}

It can be seen PR lowering when irradiance is high and ambient temperature is high as well. It can be seen also the interrow shading for a few days during the cold winter days through the blue dots.

5 CONCLUSIONS

Development of Data Analytics allow companies with PV assets to extract valid information to help decision making in various independent and relevant fronts: O&M, Procurement, Engineering (Modelling and LCOE analysis) and Asset Management.

Moreover, considering how challenging is to model environmental parameters alongside with products with increasing complexity and novel and more complicated systems, for solar developers and solar asset operators we cannot foresee better approach than the use of DA tools for an integrated optimization of systems and improved efficiency to drive competitiveness within the solar PV industry. Not only because the difficulty for IT external companies to capture and cope with the pace for all different pillars but mainly because developing in-house these business intelligence skills will become core for the massive evolution in scale and deployment of integrated renewable systems expected for the next decades.

Below the main topics raised in this study:

5.1 Sub-hour data and PR

The need to adjust modelling to reality requires control over sub-hour data not only to tackle what is currently an issue of sub-hour clipping (easily capture using 1,5, 15 minutes data), but also to investigate, as shown in fig.8 how clipping occurs, it may not start and end at same irradiance level and as seen in the same fig.8 together with fig.10 instantaneous PR is quite different along the day and it is reduced during clipping.

5.2 Assessing performance of key PV components with PR

It was present how feasible and convincing could be usage of DA to compare performance of components as modules with different cell technology (monofacial and n-TOPcon) and tables like ground mounted structures with landscape and portrait configuration. This allows reliable comparison to be done across different vendors and module technology even using equivalent power class and cell technology.

5.3 Albedo and bifacial PV modelling

It worth also to mention modelling bifacial modules requires better understanding how to integrate all irradiance related to the surroundings and the performance asymmetry during a single as shown in a single day in fig.8. Fig.7 represents not limitation of modelling tools, as seen in N.Riedel-Lyngskaer et al. [1], modelling of bifacial is improving but it requires data with higher resolution and with quality. As extensively shown through temporal analysis, fig.9, 10 and 11, PR varies greatly throughout the year and during the day.

5.4 PR and Energy generation

There is the need to understand there are some limitations to extrapolate PR as energy generation predictor. Higher PR values at low irradiance has not same magnitude and positive benefit as higher PR at high irradiation level. This implies that bifaciality gain should be monitored at full span of the irradiation level and possible according use case. And when computing PR at high irradiation it should not include datapoints from the clipped zone due to the natural masking from the clipping is counted as a loss which is not true.

ACKNOWLEDGEMENTS

I would like to thank all support, guidance and discussion maintained with Chris Buckland, head of the Innovation team at Lightsource bp. To Zach Davis by the work created, planned, developed, and supervised in the field with the white sheets experiment. I would like also to thank the corporate support in the time, total support and motivation received from David Garcia, head of Lightsource bp Central Engineering Unit.

6. References

[1] N. Riedel-Lyngskaer, D. Berrian, Daniel Alvarez Mira, Alexander Aguilar Protti,, Applied Sciences 10(8487) (2020).

[2] Janssen, G.J.M., et al., Outdoor Performance of Bifacial Modules by Measurements and Modelling. Energy Procedia, 2015. 77: p. 364-373.

[3] M. Chiodetti, “Bifacial PV plants: performance model development and optimization of their configuration,” M.S. Thesis, KTH Royal Institute of Technology, Stockholm, Sweden, 2015.

AN AUTOMATED FLEXIBLE DATA-DRIVEN ENSEMBLE APPROACH FOR ESTIMATING PV MODULE DEGRADATION AND COMPARING AGAINST WARRANTY LEVELS

Julián Ascencio-Vásquez*[1], Haohui Liu[2], Zoe Defreitas[1]
Envision Digital, Redwood City, United States
Envision Digital International, Singapore

ABSTRACT: Degradation is one of the critical topics in PV performance assessments. It affects the lifetime yield and bankability and can always signal underperformance events or trigger a PV module warranty claim. To support this issue, holistic Technical Asset Management tools are needed for most PV stakeholders to keep track of the optimal performance of PV systems. One example is the EnOS™ Advanced Analytics for Solar, which enabled operators, asset managers, and others to get highly accurate and detailed energy loss calculations based on loss breakdown tools. This tool also includes sections addressing the PV module degradation estimations and DC health in the context of degradation and module health. This article focuses on steps beyond typical degradation assessments, including a robust approach to computing degradation rates of operating PV systems using diverse combinations of statistical methods, filtering strategies and irradiance data sources, followed by the introduction of the first steps toward a data-driven PV module warranty claim approach. A case study of an entire operating system is included proving the robustness and highlighting the need for those further steps for PV module degradation assessment.

Keywords: PV Performance, Analytics. Degradation, Asset Management, Warranty, Monitoring

1 INTRODUCTION

The deployment of Photovoltaic (PV) systems keeps expanding, and the installed capacity globally has already overpassed the 1TW [1], a massive milestone for the industry since several years ago. This achievement confirms that PV technologies are and will continue to be a primordial component in electricity grids.

Besides the incredible and fast deployment, the expected lifetime of a PV module and, consequently, the lifetime of a PV system is already reaching 30 years as a typical value for long-term yield assessments (LTYA) and financial calculations [2]. More ambitious plans are trying to ensure a lifetime of 50 years [3]. However, to keep up those ambitious figures, the degradation of the components will play a critical role. High degradation rates in PV modules will impact the lifetime yield and project bankability, or they can always signal underperformance events, which, depending on their severity, can trigger a PV module warranty claim and, in the worst case, the decommissioning of the site before expected.

The research community has published comprehensive pipelines to estimate the degradation rates of PV modules and PV systems, including the challenges and uncertainties in accurately estimating degradation rates using operational data have been published [4]. In commercial applications, the Solar Centre of Excellence (COE) team of Envision Digital has been putting efforts to include robust degradation and DC Health algorithms in the EnOS™ Advanced Advanced Analytics for Solar [5], which enabled operators, asset managers, and others to get highly accurate and detailed energy loss calculations based on loss breakdown tools.

To be able to trigger some actions to improve the yield under scenarios of high degradation, the estimated degradation rates have to be compared to a baseline. A common case is to compare the estimations with the design assumptions or typical degradation rates (i.e. 0.5%/year). However, comparison to component warranty curves is a promising alternative to moving beyond degradation assessments.

The PV Module warranty curves are evolving due to technological improvements. The baseline of degradation warranty curves assumes a two-step model with 90% performance until year ten followed by 80% performance until year 20 or 25. In recent years, PV module manufacturers started considering lower degradation rates, moving to a linear model. This model considers an initial performance decrease after year 1 (from -3% to -1%), followed by a steady rate (from -0.6%/year to -0.2%/year) until years 20, 25 or 30.

The PV industry keeps increasing the maturity level; however, experience in PV module warranty remains scarce. This conference article introduces enhanced techniques for PV degradation assessment, followed by developing the first steps towards eventual data-driven PV module warranty claims using an analytics platform.

First, we propose a flexible degradation calculation tool, including the selection of the degradation model, irradiance source data, and filtering techniques. This innovation is highly relevant for O&M teams, asset managers and others to have trustful degradation rate estimations considering the high number of uncertainties presented in such calculations. Then, the second item is highly innovative since we propose the first steps towards data-driven PV module manufacturing warranty claims triggered by the results displayed on technical asset management tools. This item's relevance can be considered extremely high since early detection of possible underperforming PV modules could reflect substantial monetary savings and yield optimizations.

2 CONTEXT ON HOLISTIC TECHNICAL ASSET MANAGEMENT TOOLS

EnOS™ Advanced Analytics for Solar by Envision Digital is an industry-leading performance analytics app which is an exemplary illustration of the Technical Asset Management concept [5]. First, it does tight quality control of data, checking data quality and sensor health. This app calculates more advanced KPIs such as weather-corrected PR and analyzes inverter efficiency, degradation status, and downtime statistics with reliable input. Moreover, it dissects the performance data and employs sophisticated machinery of interconnected algorithms to disentangle performance losses into over 15 major loss categories.

Such categorization will allow the tool to isolate the PV plant's normal behaviour from a temporal issue, such as soiling or inverter downtime. This powerful characteristic will be crucial for a PV module degradation assessment. Figure 1 illustrates an example of the user interface of EnOS™ Advanced Analytics product, showing the loss breakdown summary.

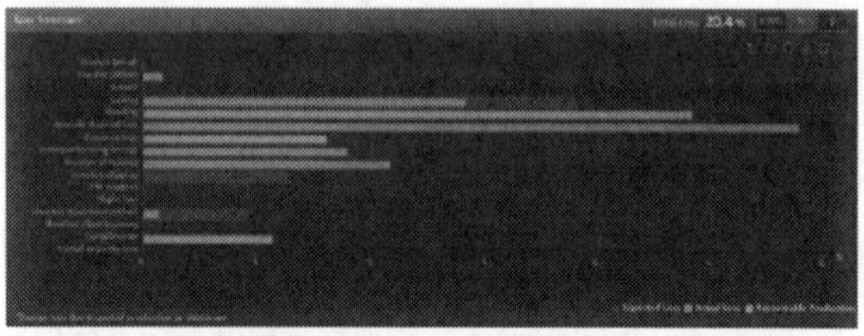

Figure 1: EnOS™ Advanced Analytics for Solar breaks down performance losses into refined categories according to root causes or mechanisms; doing so helps provide an in-depth understanding of asset performance and actionable insights.

3 FIRST STEPS TOWARDS EVENTUAL DATA-DRIVEN PV MODULE WARRANTY CLAIMS

An important aspect currently not covered by any technical asset management tool is correlating the degradation assessments with PV module manufacturing warranty items. This paper goes beyond state-of-the-art, establishing the first steps and definitions towards eventual data-driven PV module warranty claims.

The accurate estimation of degradation rates will enable us to cover a very delicate topic noted as warranty claims. This section shows the first of an innovative data-driven PV module warranty assessment actively running on a technical asset management tool like EnOS™ Advanced Analytics for Solar.

Hereby, we propose the first steps towards eventual data-driven PV module warranty claims, starting by first allowing the stakeholders to visualize on their technical asset management tool the scenarios and probabilities of being under the position of a warranty claim triggered by an unexpected very high degradation rate estimation.

3.1 Enhanced Techniques for PV degradation assessment

This section presents an enhanced approach to calculating degradation rates under different scenarios considering statistical methods, filtering techniques, and different irradiance data sources.

Such items are now available options for the users, and are listed below:

- **Degradation Methods**: Year on Year, Regression, Ensemble.
- **Irradiance Source**: Sensor, Modelled clear sky.
- **Filtering approach**: None, Normal status, Clear sky, Ensemble.

One aspect of those scenarios is the use of statistical methods to estimate the annual degradation rate. Year-on-Year (YoY) method to calculate degradation rates has been a widely used technique in the entire PV industry [6]; however, a set of different statistical methods have been published and benchmarked to get the most accurate degradation rate calculations [7]. Due to the task's difficulty, the main actors on the topic agree that an ensemble method could be a more reliable result. Following that logic, a linear regression model with seasonal decomposition and ensemble method is integrated as a valid alternative.

Among the available irradiance data sources, the usage of modelled clear sky irradiance, following the procedure standardized by NREL [8], is available, with the option to use measured sensor data. The first will be recommended when the certainty of measured data can not be confirmed (e.g., due to miscalibration or soiling).

Then, the filtering techniques give plenty of freedom to simulate several scenarios. One case is the usage of none filters, allowing to work on proper data considering all operating conditions. Another option is the "Clear sky only", which excludes overcast instances. In addition, taking advantage of the automatic recognition of operating events per timestamp, it is possible to compute degradation considering only "Normal status" instances, neglecting clipping, downtime, curtailment, and shading, among others. Finally, an ensemble filtering technique is available, considering all previous options. Figure 2 shows the front end of the web application.

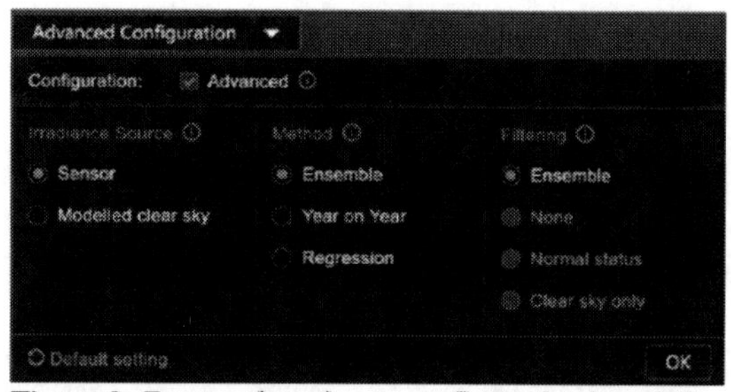

Figure 2: Front-end section to configure the degradation assessment [5].

3.2 Designing PV Module warranty curves

Warranty claims are a very delicate subject during the lifetime of PV projects. This item can heavily impact asset owners' bankability and financial figures and affect the PV module manufacturers' credibility. Warranty claims are usually filed after extended on-field and indoor tests on many PV modules.

Some predictions on the expected degradation rates used in the PV module warranty levels have been published by the International Technology Roadmap for Photovoltaic (ITRPV) [8], and illustrated in Figure 3. It is expected that the degradation rate under warranty will be under 0.5%/year for 30 operating year by 2031, and if this is true, the lifetime yield of PV systems will considerably increase, and the LCOE decrease substantially.

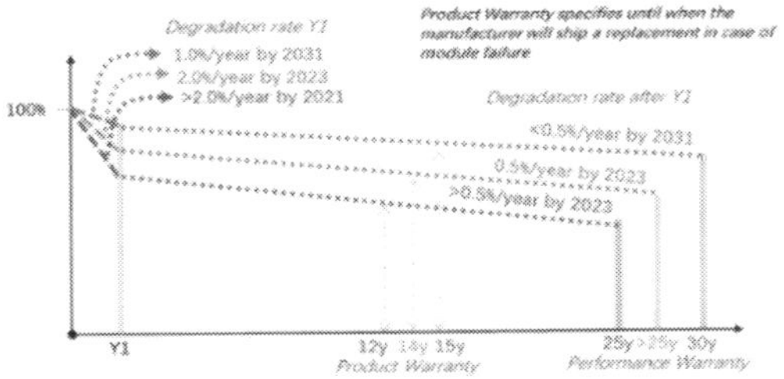

Figure 3: Temporal extrapolation of the ITRPV scenarios considering degradation at year one and later and years of product and performance warranty. Values are taken from [8].

In practice, the methodology proposed here is flexible to design two types of degradation warranty levels:

- Baseline warranty level: defined as a two steps curve with specific degradation rates and year of changepoint;
- Realistic warranty level: it follows the modern manufacturing warranty curves, considering linear degradation from year one until the year of performance warranty mentioned on datasheets.

Both types are illustrated in Figure 4.

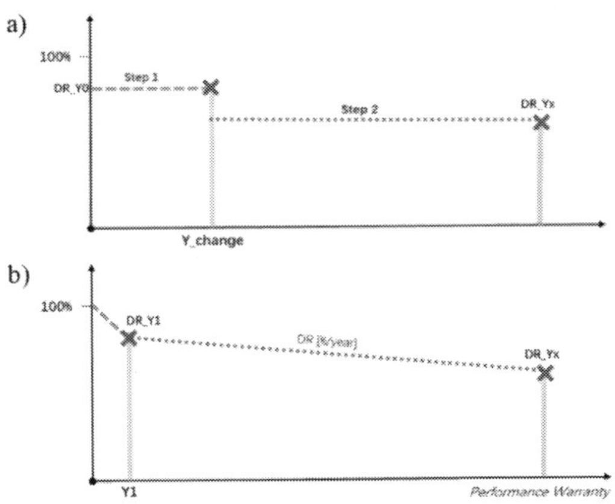

Figure 4: (a) Baseline warranty level; (b) Realistic warranty level.

4 RESULTS

As a quick overview of expected results, Figure 5 shows the overlap of a statistical degradation indicator calculated from operational data with PV module warranty curves. The use of both signals (degradation KPI and degradation warranty) will be crucial to trigger underperformance events, on-field performance tests, and always undesired PV module warranty claims. The extended linear model of the degradation KPI could be an interesting sign of a possible warranty claim.

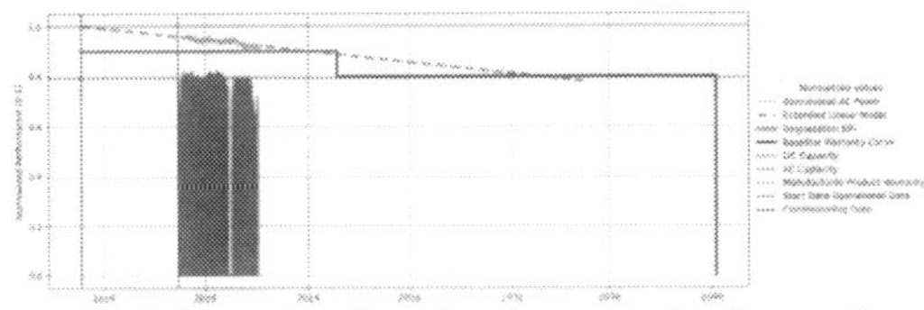

Figure 5: Example of overlapping normalized operational power, estimated PV module degradation KPI, and the PV module warranty curves.

A real case where a PV system is analysed using EnOS™ Advanced Advanced Analytics for Solar is shown below. The analysed system is operating in a Moderate Climate based on the KGPV Climate Classification [9], it was commissioned in 2015, and available operational data starts only from 2018.

Figure 6 shows one front-end section where the average degradation rate per inverter is displayed together with the confidence intervals. The settings for this scenario include an Ensemble statistical method with Ensemble filtering techniques and the usage of a measured irradiance sensor. Among inverters, it is shown that several inverters present a higher degradation rate at around -3%/year, while the average gets to -0.9%/year.

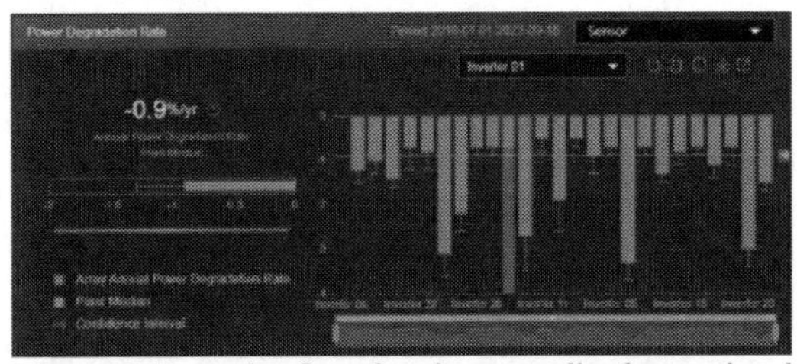

Figure 6: Example of overlapping normalized operational power, estimated PV module degradation KPI, and the PV module warranty curves.

In Figure 7, looking closer at the time-series of a single inverter with a degradation rate above average (> -0.9%/year), it is possible to observe a linear behaviour, considered normal, during the four years of recording data. This figure also serves as an example of extra features within this tool: even though operational data from commissioning until 2018 is unavailable, the algorithm can infer the daily power rating. This ability enable the tool to compare a particular PV module warranty curve with the inferred daily power rating and real degradation curves. After seven years of operation, the algorithm computes a power rating above 90% performance for this inverter, and further actions should not be triggered.

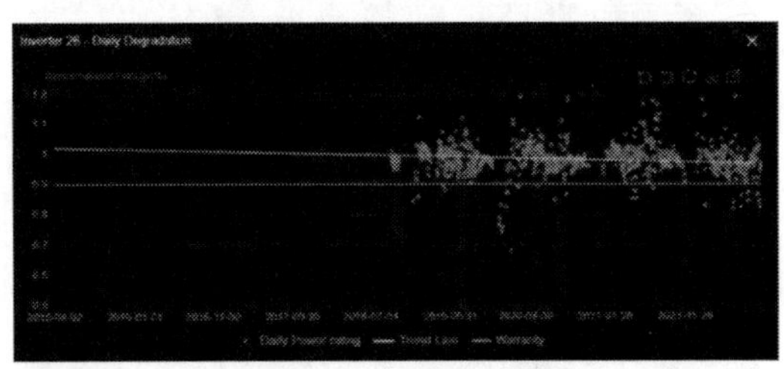

Figure 7: Front-end view of a degradation assessment for a single inverter from commissioning until now with "normal" behaviour.

On the other hand, shown in Figure 8, one of the underperforming inverters shows a normal behaviour for years 2018-2019, but a stronger degradation rate (i.e., abnormal behavior) from year 2020. This pattern is repeated in all inverters with a degradation rate lower than -1.5%/year.

This scenario is a good example for comparing the real degradation curves with the PV module warranty curves. When the desired pattern is not represented with the operation data and inferred KPIs, the operators and asset managers should consider further investigations into that sector of the PV plant.

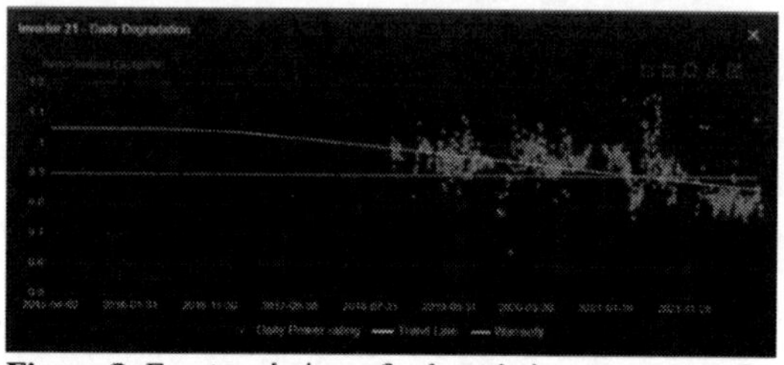

Figure 8: Front-end view of a degradation assessment for a single inverter from commissioning until now with "abnormal" behaviour.

5 CONCLUSIONS

This conference article covers operational aspects of PV module degradation and reliability. The need to accurately compute the degradation rates of operating PV systems is essential to identify early failures, optimize the yield, or trigger PV module warranty claims when the underperformance is severe.

The methodologies and solutions included in the EnOS™ Advanced Analytics for Solar by Envision Digital are presented to tackle this sensitive subject. First, computing accurate degradation rates include options regarding the statistical method, filtering technique and irradiance data sources. Those options give the analyst a robust and comprehensive way to adapt under diverse circumstances. The second aspect presented was the design and overlap of PV module warranty curves onto the real degradation curves, which the operator can freely configure. This innovative approach allows an in-depth assessment of PV module degradation and will support the triggering of corrective actions.

The presented work shows the first steps for moving into a data-driven PV module warranty claim framework supported by robust statistical methodologies.

6 REFERENCES

[1] PV Magazine, "Humans have installed 1 terawatt of solar capacity, generated over 1 petawatt of solar electricity in 2021," 2022. [Online]. Available: https://pv-magazine-usa.com/2022/03/14/humans-install-1-terawatt-of-solar-capacity-generate-over-1-petawatt-of-solar-electricity-in-2021/

[2] IEA PVPS Task 13, "Uncertainty in Yield Assessments and PV LCOE," Nov. 2020.

[3] V. Poulek, D. S. Strebkov, I. S. Persic, and M. Libra, "Towards 50years lifetime of PV panels laminated with silicone gel technology," *Sol. Energy*, vol. 86, no. 10, pp. 3103–3108, 2012, doi: https://doi.org/10.1016/j.solener.2012.07.013.

[4] S. Lindig *et al.*, "International collaboration framework for the calculation of performance loss rates: Data quality, benchmarks, and trends (towards a uniform methodology)," *Prog. Photovolt. Res. Appl.*, vol. 29, no. 6, pp. 573–602, 2021, doi: https://doi.org/10.1002/pip.3397.

[5] Envision Digital, "A Holistic approach to Technical Asset Management: Improving and optimizing solar PV assets," 2021. [Online]. Available: https://www.envision-digital.com/smart-solar/

[6] D. C. Jordan, M. G. Deceglie, and S. R. Kurtz, "PV degradation methodology comparison — A basis for a standard," in *2016 IEEE 43rd Photovoltaic Specialists Conference (PVSC)*, 2016, pp. 0273–0278. doi: 10.1109/PVSC.2016.7749593.

[7] M. Theristis *et al.*, "Comparative Analysis of Change-Point Techniques for Nonlinear Photovoltaic Performance Degradation Rate Estimations," *IEEE J. Photovolt.*, vol. 11, no. 6, pp. 1511–1518, 2021, doi: 10.1109/JPHOTOV.2021.3112037.

[8] ITRPV, "International Technology Roadmap for Photovoltaic, 2020 Results, 12. Edition.," Apr. 2021.

[9] J. Ascencio-Vásquez, K. Brecl, and M. Topič, "Methodology of Köppen-Geiger-Photovoltaic climate classification and implications to worldwide mapping of PV system performance," *Sol. Energy*, vol. 191, pp. 672–685, 2019, doi: https://doi.org/10.1016/j.solener.2019.08.072.

POSITIONING A UAV FOR INFRARED INSPECTION OF SEVERAL ROWS OF SOLAR PANELS SIMULTANEOUSLY

Tor Atle Solend, Jonas Moen, Anders Rødningsby
University of Oslo / Department of Technology Systems
Gunnar Randers vei 19, 2007 Kjeller, Norway

When photovoltaic (PV) solar panels in a PV power plant deteriorate after some time of operation, a thermographic inspection with unmanned aerial vehicles (UAVs) can be necessary. Often, faulty panels, or cells within panels, radiate heat that is visible with infrared (IR) cameras. When performing a thermographic inspection of solar panel faults from a UAV, the correct choice of position for the UAV depends on several factors. The camera's pixel resolution and field of view (FOV) set a distance threshold between the panel and the UAV according to a minimum resolution requirement. This threshold is also dependent on the angle between the panel and the UAV. Because of the vast size of some PV power plants, inspecting several rows of panels simultaneously is often appropriate. Latitude-dependent row-to-row distances limit the number of rows which fit inside the FOV which leads to site-dependent strategies. This paper analyzes several key factors, concerning the choice of position for a thermographic inspection of a solar PV power plant with a UAV.

Keywords: UAV, positioning, thermography, IR, solar photovoltaics, operation and maintenance, O&M

1 INTRODUCTION

Infrared (IR) thermography of solar photovoltaic (PV) power plants with unmanned aerial vehicles (UAVs) is a common method for inspecting the health of the PV panels [1]. The overview from airbourne sensors is in many cases superior to ground-based inspection. However, the UAV should be positioned such that the image quality is acceptable, and pointing the camera in the right direction such that all the panels of interest are contained within the boundaries of the field of view (FOV). The FOV of the camera is the angle at which the camera's sensor can record image data and is determined by the chosen lens and focal point, as illustrated in Figure 1. The number of pixels in the horizontal and vertical direction on the IR sensor in the camera, combined with FOV, determines the image resolution of the solar panels from a given distance.

The heat signatures of certain faults are identifiable with an infrared (IR) camera from a relatively long distance, but only classifiable at a shorter distance when the resolution of the IR camera is high enough. Resolution thresholds for classifications are proposed in [2]. This means that the maximum distance threshold between the UAV and solar panels, considering a resolution criterion, can be adjusted by changing lenses [3].

When PV power plants are large [4], the amount of data captured from UAV thermography missions is substantial. IR images will accumulate much space on hard drives, making processing cumbersome, and thus highlighting the importance of capturing data of the right quality. The chosen position of the UAV with the correct Line of Sight (LOS), should maximize the capture of PV panels within a quality boundary, and thus also minimize the amount of useless data, e.g. images of the ground.

For increased data acquisition efficiency, it can be favorable to inspect more than one row of PV panels simultaneously. Then there should be an optimal position for the UAV concerning the quality of the data and the number of rows inside the FOV. The optimum should maximize the number of rows within the boundaries of the minimum resolution criterion and account for noise and errors affecting both quality and acquisition efficiency.

While there have been many efforts to find the most efficient path for covering an entire PV power plant [5] [6], they usually assume that the image quality is acceptable. However, little focus has been given to the choice of the position of the UAV and the image resolution's dependency on the properties of the onboard camera, the inclination angle of the panels, and the number of rows inside the FOV. Figure 1 illustrates how the distance and angle transform the panels onto the image plane.

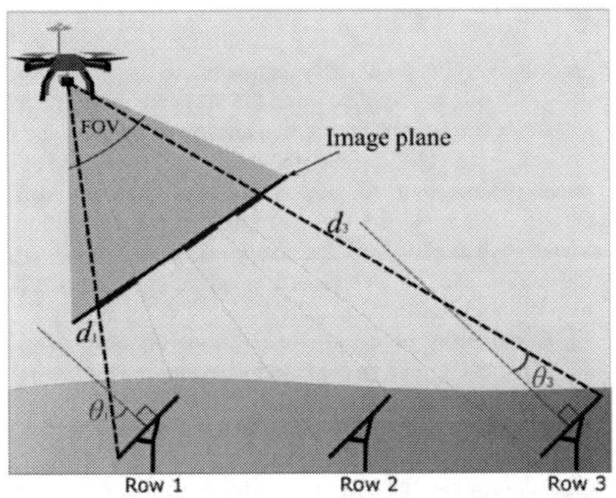

Figure 1: The footprints of the tilted panels, here represented as one-dimensional blue lines on the image plane, are transformed from the three-dimensional world onto the two-dimensional image plane. Both distance d and viewing angle θ affect the resolution.

Because of the limited flight time of a UAV, the downtime of changing batteries can be substantial for large PV power plants. In addition, the period of the day where IR thermography of PV panels is viable is limited, e.g. by a minimum of 500-700W/m^2 [7] of incident irradiance. Thus, it is vital to have a robust system that balances image resolution with the amount of useful data in the FOV optimized for efficient data acquisition. The magnitude of the different variables affecting the data quality and acquisition efficiency is however not clear. This paper gives an analysis of some of the crucial decision variables considering the choice of a UAV's position for IR PV power plant inspection of two or more rows of solar panels simultaneously.

A novel approach to find the maximum distance position of a UAV inspecting two or more rows of solar panels simultaneously with a fixed camera/lens setup is presented in this paper. The maximum will incorporate the contribution from the θ's, θ_1 and θ_3 in Figure 1, and distances, d_1 and d_3, to the two extremities of the nearest and furthest panels relative to the camera's position. In addition, the angular accuracy of the IR camera concerning emissivity as a function of θ is incorporated into the analysis.

2 BACKGROUND

When inspecting solar panels with an IR camera, it is often preferable to align the LOS with the normal line emanating from the panel in focus to avoid angular effects in the images [8]. However, if the UAV is inspecting two or more rows of panels simultaneously, the LOS should point somewhere in the middle. Thus, it is not possible to position the UAV normally to all rows unless the UAV is at an infinite distance, where all the panels appear to be of equal size in the image. In practice, IR cameras often have a non-changeable lens with a fixed FOV and thus have a maximum distance concerning pixel resolution to the panels, either for identification or classification of IR faults.

The footprint of each panel in the image will have non-equal sizes, or lengths as in the two-dimensional illustration in Figure 1, according to the angle θ between the UAV/camera and the panels. Depending on the task, it might be preferable to avoid too much difference in the size of the projected panel rows in the same image. Especially when considering an automated machine vision algorithm trying to classify hotspots etc., it would make the process less complex when the sizes are similar.

θ depends on the inclination angle of the panel and the position of the UAV. θ is different at the lower edge of the panel compared to the top. Thus, the resolution differs across the panel and the point where the resolution is lowest will decide how far away, and/or at which angle, the UAV can be positioned to comply with a given resolution criterion.

Considering a scenario where two or more rows of solar panels are inspected simultaneously, the extreme points on the outermost rows in the image will have the lowest resolution and define the boundaries for positioning the UAV. According to Figure 1, the locations of the points are at the lower or upper edge of row one, and either the higher or lower edge of row three. If the UAV's position is below the normal line emanating from row three, the upper edge of the row is the point with the lowest resolution. If the position is above the line, the lowest edge is the lowest and will serve as the resolution constraint.

The most common IR cameras for UAVs are long-wave IR cameras recording radiation in around 8-15 µm wavelengths. The front glass covering the solar panels is opaque in this part of the spectrum, leaving the actual solar cells behind the glass invisible in the images. Thus, the heat radiation from the glass, and its emissivity, determines what the IR camera reads. Angular temperature accuracy for an IR camera, its ability to determine the temperature of an equally hot pixel from different angles, depends partially on the directional emissivity (DE) of the objects in view. DE is the ratio of the intensity of radiation emitted by the surface, in a specified direction and wavelength, to a blackbody's radiation intensity with the same temperature and wavelength [9] and can is calculated with (1).

$$\varepsilon_{emitter}(\lambda,\theta,\phi,T) = \frac{I_{emitter}(\lambda,\theta,\phi,T)}{I_{blackbody}(\lambda,T)} \quad (1)$$

where $\varepsilon_{emitter}$ is the DE in the angle θ from the surface normal, with azimuth ϕ, for the wavelength λ and temperature T with emitted radiation intensity I from the emitter compared to a blackbody. Note that the blackbody is independent of the viewing angle. The DE can appear as light intensity differences of equally tempered objects in an IR image, as shown in Figure 2.

Figure 2: Two rows of solar panels in Arvika, Sweden, inspected by an IR camera pointing straight down. The difference in view angle between the two rows results in a difference in temperature accuracy, represented as light intensity in the image.

From a quality perspective, the angular temperature accuracy deterioration can make it difficult to differentiate between hotspots and other faults emitting heat, and measurement inaccuracy. In Figure 2 the hotspot visible on the top row appears equally bright as much of the bottom row where there are no apparent faults. This is not necessarily difficult to tell apart from a human but might cause problems for a machine vision algorithm. Thus, it should be considered when choosing the UAV's position. It must be noted that the difference in light intensity in Figure 2 might have different causes, e.g. reflection from the sun or reflected heat from other sources in the environment. An increase in θ also increases the reflection of incident radiation, and to avoid this it is advised to keep $\theta \in [-60°, 60°]$ [10].

The DE is difficult to model analytically because of its dependence on the surface structure. The front glass of solar panels is indeed a real material, as opposed to theoretical surfaces such as blackbodies, and emits radiation nonuniformly with regards to angle to the surface normal. A numerically fitted curve from measurements can be closer to the actual DE than a purely theoretical function [11]. Numerically fitting data with (2) follows measured DE [12] from solar panel front glass closely.

$$\varepsilon_e(\theta) = a\theta^b + c \quad (2)$$

where a and b are numerically fitted constants, and c is DE at the normal viewing angle.

In Figure 2, the distance from the camera to the rows is approximately equal, but θ is not. At low altitudes, the effect from θ is large, compared to high altitudes. However, if the camera is equipped with an unchangeable lens, then the possibility to vary the altitude is constrained by the resolution criterion. Theoretically, one could mount a very narrow FOV lens to the camera and fly at an extreme altitude to minimize the θ's contribution. A narrow FOV will introduce other sources of error. Imperfections in the lens might cause artifacts along the edges of the image and the LOS will be far more susceptible to angular vibrations in both the UAV and the camera gimbal. In addition, camera gimbal accuracy will drift with time [13].

3 METHOD

Two supposed locations in Oslo (Norway), and Rome (Italy), for solar PV power plants with south-facing fixed-tilt solar PV panels and several rows, are analyzed. The solar altitude profiles differ between latitudes causing the optimal fixed inclination angles of solar panels to be different. Here, it is implied that the panels are facing southward, with optimal inclination angles according to [14]. When inspecting more than one row, the row-to-row distance determines how many rows which will fit inside the camera's FOV. This distance is here calculated from the latitude-dependent solar altitude β in the morning at the winter solstice and panel inclination α, as illustrated in Figure 3.

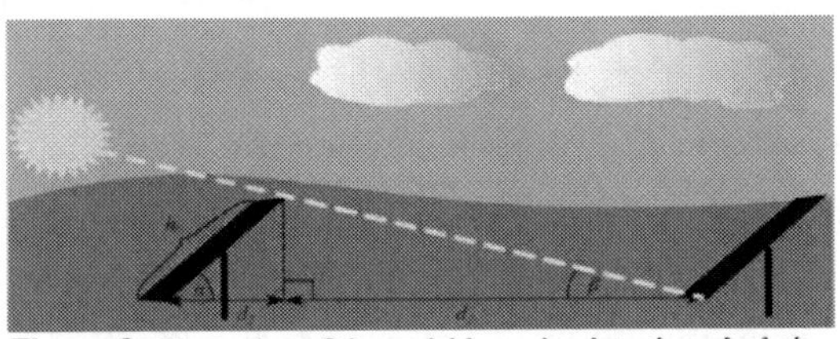

Figure 3: Illustration of the variables going into the calculation of row-to-row distance.

β is chosen at 10:00 am on the winter solstice [15]. Then the shadowing from one row to the next is minimal throughout the year. The row-to-row distance and total ground span for a given number of rows n_r can then be calculated with (3) and (4).

$$d_1 + d_2 = \cos(\alpha) \cdot h_p + \frac{\sin(\alpha) \cdot h_p}{\tan(\beta)} \quad (3)$$

$$Total\ span = (d_1 + d_2) \cdot (n_r - 1) + d_2 \quad (4)$$

Table 1 sums up the calculated values for the two sites.

Table 1: Site-specific PV power plant layout variables

	β	α	$d_1 + d_2$[m]	n_r	Total span [m]
Oslo	4	45	17.31	2	18.44
Rome	19	35	3.98	3	9.26

The position at the maximum distance to all points on the panels in view, given a resolution criterion, lies at the intersection of two perfect circles emanating from the two points which are furthest away from the camera on the UAV. The circles represent the maximum distance corrected for θ. The circles are calculated by using (5):

$$d_i = d_{max} \cdot \cos(\theta_i) \quad (5)$$

where d_i is the angle θ_i corrected maximum distance, d_{max} is the maximum distance at the normal viewing angle.

To incorporate IR temperature accuracy deterioration as a function of angle into the maximum angle corrected distance boundary lines, (2) is factored in to (5) as shown in (6). This IR temperature accuracy deterioration cannot be considered an absolute constraint but indicates the magnitude of the loss of quality as a function of θ.

$$d_{R,i} = d_{max} \cdot \cos(\theta_{R,i}) \cdot (a \cdot |\theta|_{R,i}^{b} + c), \quad (6)$$
$$\theta \in [-90°, 90°]$$

For this equation to make sense, the angular deterioration should equal one at the normal viewing angle, but the DE is always less than one. Consequently, the DE values from [12] translated to equal one when $\theta_{R,i}$ is equal to 0 as shown in Figure 4.

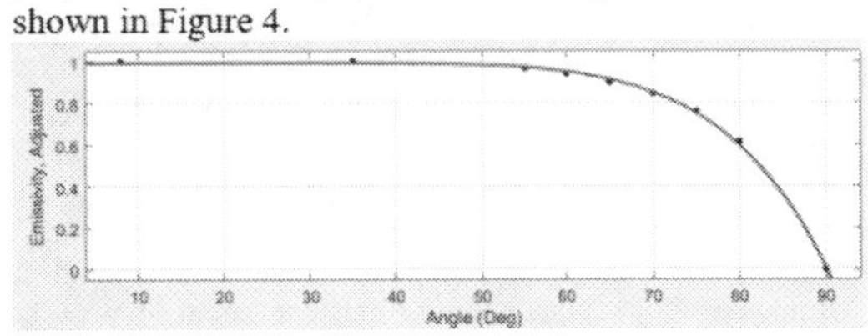

Figure 4: The emissivity of PV panel cover glass adjusted to equal one at $\theta = 0$. The curve is fitted numerically to the data.

The coefficients from the curve fitting are in Table 2.

Table 2: Numerically computed coefficients used in (6).

a	b	c	R^2
$-9.056e^{-16}$	7.694	0.9934	0.9988

4 RESULTS

The maximum distance circles calculated with (5) are plotted in Figure 5 for the supposed Rome PV power plant case. Here, d_{max} is set to equal 30m, which would result in a horizontal image resolution of approximately 2.7 cm/pixel with an IR camera with 640x512 pixel resolution and a horizontal FOV of 32°. This camera setup is used for all cases throughout this paper.

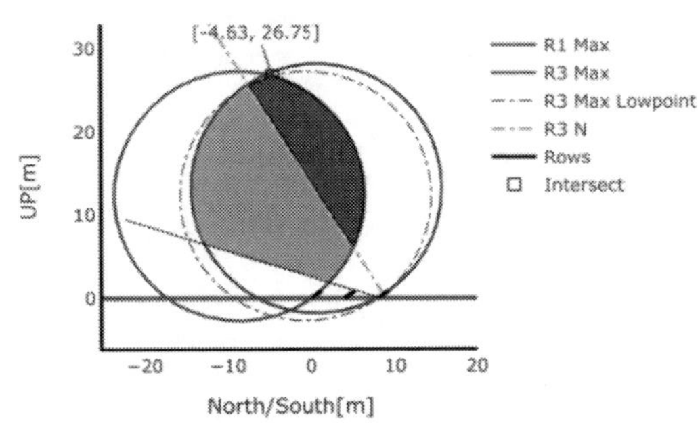

Figure 5: Angle-corrected maximum distance circles in supposed PV power plant in Rome. The dark green area is where the lowest point on row three is the resolution constraint, the light green is for the top point. The colored area is a union of three circles where the resolution criterion is met. The black line limits the area to where all rows are completely visible.

The figure shows that the intersection of the two circles emanating from the outermost edges of the panel rows in view is above the yellow stippled normal line from row three. This means that the lower edge of row three has a lower resolution than the upper. Thus, above this normal line, the stippled red circle is the constraint.

Figure 6 shows the incorporation of angular temperature deterioration (6) onto the maximum distance traces. The stippled traces indicate a small influence when theta approaches ±90°. The trace from row three is from the lower edge.

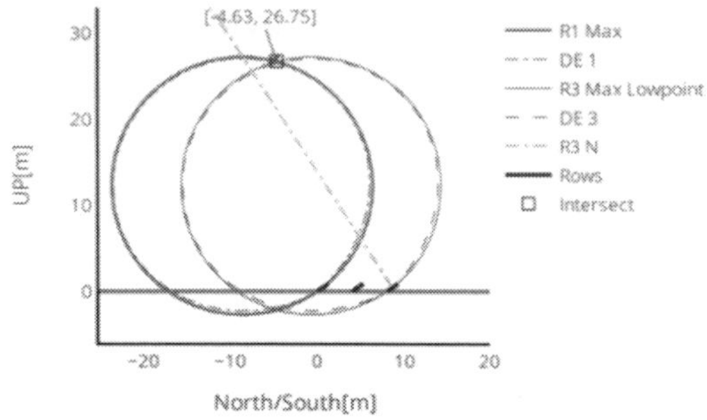

Figure 6: The stippled circles are the DE-modified versions of the solid line circles. A slight decrease in width happens when θ is large, but else hardly noticeable.

Figure 7 shows the corresponding situation in the supposed Oslo case, without DE temperature accuracy deterioration. The steeper inclination angle and row-to-row distance make it only possible to inspect two rows simultaneously with this setup.

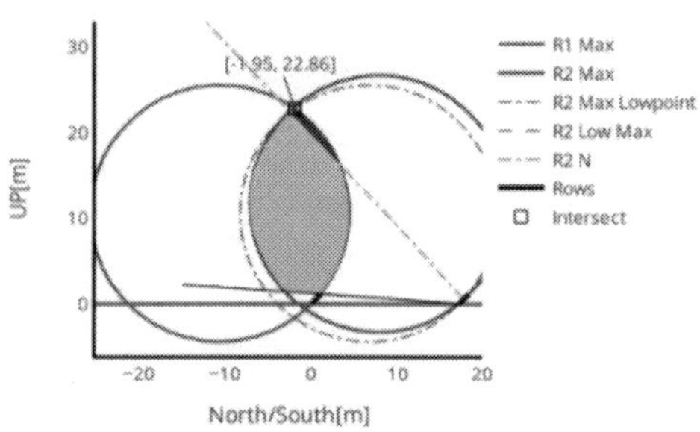

Figure 7: In Oslo, the intersection point is just barely above the normal line of the top point, resulting in a small area where the lower point of row three is the lowest resolution (dark green area).

Here the intersections are located close to the normal from the furthest row, and the area above the line is very small compared to Rome. The stippled circular trace is the constraint above the line, and the solid trace is the constraint below it.

The intersection point locations are presented in Table 3:

Table 3: Intersection points measured from the low edge of row one.

	North/South [m]	Up [m]
Rome	-4.63	26.75
Oslo	-2.11	22.97

5 DISCUSSION

The best or optimal position for the UAV must lie within the union of the maximum distance circles if the resolution criterion is a constraint. The optimum should incorporate the factors which affect the quality of the data and efficiency of the data acquisition. The most robust position against LOS and position errors should be near the intersection of the maximum distance circles. Here, the resolution criterion is met and the field of view should encompass all rows of interest. At the intersection, the buffer on each side of the outermost panels in the field of view is at its largest. Traveling downwards along the trace of row three maximum, or row two in Oslo, the buffer decreases. Figure 9 illustrates this where the grey triangle resembling the FOV encompasses all rows completely, while the identical triangle rotated and translated to a lower point on the maximum distance trace from row three loses sight of the lower part of row one.

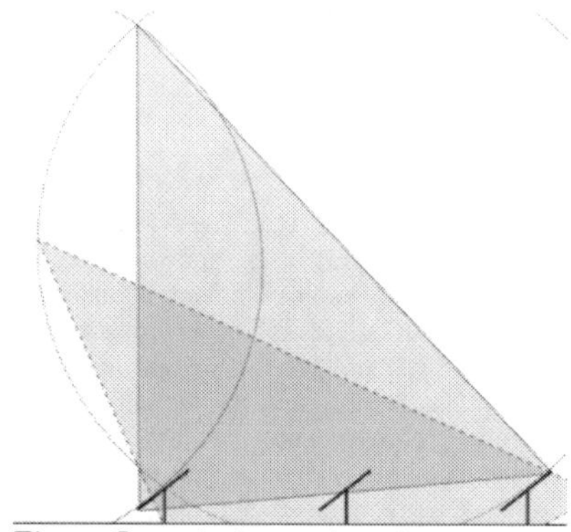

Figure 8: At the intersection, all rows are completely in view (grey area). By traveling lower, parts of either row one or three fall out of view (turquoise area).

This model does not incorporate position errors. Global navigation satellite system (GNSS) receivers on UAVs will estimate the position with some uncertainty, and inertial sensors will drift with time causing further uncertainty. To compensate for this, regarding the maximization of resolution within the criterion, the UAV cannot stay exactly at the intersection point.

However, higher resolution is achieved by positioning the UAV closer to the panels. Thus, it is a matter of preference which position is optimal for the PV power plant being inspected, concerning image resolution or other variables such as robustness.

The temperature accuracy deterioration as a function of θ has been shown to have a minor impact on image quality compared to θ's effect on resolution ($\cos(\theta)$). The difference in magnitude between the two effects is shown in Figure 9.

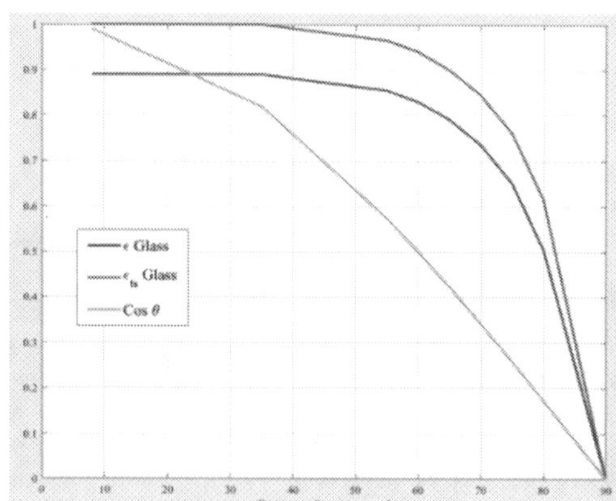

Figure 9: A comparison of $\cos(\theta)$ (yellow line) versus emissivity(θ) (blue line). The red line approximates the blue line translated upward to equal one at $\theta = 0$.

6 CONCLUSIONS AND FURTHER WORK

The operating area for a UAV inspecting more than one row of solar panels has been established for a minimum resolution criterion. The position of the intersection between two maximum distance circles from the lowest resolution points on the outermost panel rows is shown to be robust against angular errors on the LOS. However, position errors from global navigation satellite systems (GNSS) and the UAV's inertial navigation system (INS) are not included in the analysis. The magnitude of these errors depends on the UAV and the gimbal's onboard sensors and systems. The errors will affect the optimal position and add complexity to the problem. A simplified case is studied in [3] but should be studied further by incorporating θ and LOS errors.

The DE might not be the most impactful effect on the image quality regarding sharp angles. Reflection from the sun or heat radiation from other sources in the environment might be something to be aware of and should be studied.

An optimal position should incorporate the most impactful variables affecting the data quality and acquisition efficiency. The optimization is however difficult to solve analytically. Emissivity, and other angular radiation effects, as a function of angle, are difficult to model and add complexity. An evolutionary optimization algorithm or other numerical solution techniques can be viable methods and should be studied.

7 REFERENCES

[1] S. Gallardo-Saavedra, L. Hernández-Callejo, and O. Duque-Perez, "Technological review of the instrumentation used in aerial thermographic inspection of photovoltaic plants," *Renew. Sustain. Energy Rev.*, vol. 93, pp. 566–579, Oct. 2018, doi: 10.1016/j.rser.2018.05.027.

[2] U. Jahn, M. Herz, M. Köntges, D. Parlevliet, M. Paggi, and I. Tsanakas, *Review on infrared and electroluminescence imaging for PV field applications: International Energy Agency Photovoltaic Power Systems Programme: IEA PVPS Task 13, Subtask 3.3: report IEA-PVPS T13-12:2018*. Paris: International Energy Agency, 2018.

[3] T. A. Solend, H. Jonas Fossum Moen, and A. Rødningsby, "Modelling the impact of UAV navigation errors on infrared PV inspection data quality and efficiency," in *2021 IEEE 48th Photovoltaic Specialists Conference (PVSC)*, Jun. 2021, pp. 0991–0996. doi: 10.1109/PVSC43889.2021.9518810.

[4] "The world's largest solar parks," *pv magazine International*. https://www.pv-magazine.com/2021/09/23/the-worlds-largest-solar-parks/ (accessed Dec. 10, 2021).

[5] A. Pérez-González, N. Benítez-Montoya, Á. Jaramillo-Duque, and J. B. Cano-Quintero, "Coverage Path Planning with Semantic Segmentation for UAV in PV Plants," *Appl. Sci.*, vol. 11, no. 24, Art. no. 24, Jan. 2021, doi: 10.3390/app112412093.

[6] X. Luo *et al.*, "Optimal path planning for UAV based inspection system of large-scale photovoltaic farm," in *2017 Chinese Automation Congress (CAC)*, Oct. 2017, pp. 4495–4500. doi: 10.1109/CAC.2017.8243572.

[7] H. Glavaš, M. Vukobratović, M. Primorac, and D. Muštran, "Infrared thermography in inspection of photovoltaic panels," in *2017 International Conference on Smart Systems and Technologies (SST)*, Oct. 2017, pp. 63–68. doi: 10.1109/SST.2017.8188671.

[8] S. Vergura, "Correct Settings of a Joint Unmanned Aerial Vehicle and Infrared Camera System for the Detection of Faulty Photovoltaic Modules," *IEEE J. Photovolt.*, vol. 11, no. 1, pp. 124–130, Jan. 2021, doi: 10.1109/JPHOTOV.2020.3033777.

[9] Y. A. Çengel, A. J. Ghajar, and M. Kanoglu, *Heat and mass transfer: Fundamentals and applications*, 4th ed. New York: McGraw-Hill, 2011.

[10] G. Álvarez-Tey, R. Jiménez-Castañeda, and J. Carpio, "Analysis of the configuration and the location of thermographic equipment for the inspection in photovoltaic systems," *Infrared Phys. Technol.*, vol. 87, pp. 40–46, Dec. 2017, doi: 10.1016/j.infrared.2017.09.022.

[11] T. J. Warren, N. E. Bowles, K. D. Hanna, and J. L. Bandfield, "Modeling the Angular Dependence of Emissivity of Randomly Rough Surfaces," *J. Geophys. Res. Planets*, vol. 124, no. 2, pp. 585–601, Feb. 2019, doi: 10.1029/2018JE005840.

[12] I. Subedi, T. J. Silverman, M. G. Deceglie, and N. J. Podraza, "Emissivity of solar cell cover glass calculated from infrared reflectance measurements," *Sol. Energy Mater. Sol. Cells*, vol. 190, pp. 98–102, Feb. 2019, doi: 10.1016/j.solmat.2018.09.027.

[13] M. Okulski and M. Ławryńczuk, "Development of a High-Efficiency Pitch/Roll Inertial Measurement Unit Based on a Low-Cost Accelerometer and Gyroscope Sensors," in *2019 24th International Conference on Methods and Models in Automation and Robotics (MMAR)*, Aug. 2019, pp. 657–662. doi: 10.1109/MMAR.2019.8864628.

[14] "JRC Photovoltaic Geographical Information System (PVGIS) - European Commission." https://re.jrc.ec.europa.eu/pvg_tools/en/ (accessed Mar. 18, 2022).

[15] University of Oregon, "UO SRML: Sun chart program." http://solardat.uoregon.edu/SunChartProgram.html (accessed Sep. 13, 2022).

EFFECTS OF THE SOIL AND WASHING CYCLE ON THE PERFORMANCE OF A PV-ARRAY

[1]Yasuhiro Matsumoto, [1]Mario G. Moheyer, [1]Marco A. Ramos, [2]Edgar R. Sandoval,
[3]Nun Pitalúa, [1]José A. Urbano and [1]René Asomoza

[1]Electrical Engineering Department, CINVESTAV-IPN Mexico City, Mexico
[2]TecNM/Tecnológico de Estudios Superiores de Cuautitlán Izcalli, State of Mexico, Mexico
[3]Industrial Engineering Department, Sonora University, State of Sonora, Mexico

ABSTRACT: In the interval of 2017 to 2020, the performance ratio (PR) of the 60 kWp photovoltaic system (PVS) located on the northern side of Mexico City was measured. Due to the high altitude located at 2,200 meters above sea level, the PV modules' dirt was notorious; the condensation cycle of humidity and low temperatures starting in November facilitates the adherence of dust. Even though, in a new year and up to spring seasons, the natural rainfall washes the soiled PV array. Then, the cleaned PV-modules showed from 10 to 24% of increment in energy generation, i.e. the average PR increased from 66 % to 78 %. The obtained year-based average PR from 2017 to 2020 was 76.8% and in any circumstances, the soiling affects the PV system lowering drastically its PR. One of the major influences on the PV electric generation performance was PV module operating temperature, however, the PV-module soiling contributed notoriously reducing its PR. The total produced energy by the PV system during 4-years was 376,680 kWh, i.e., an average of 258 kWh a day.
Keywords: System Performance, Optical Losses, Environmental Effect, Evaluation

1 INTRODUCTION

Solar photovoltaic systems (PVS) have become one of the cheapest and environmentally friendly alternative energy sources in several countries. The PV system is silent, safe, and reliable with a low-maintenance cost without any on-site pollutant emissions. The utility grid-tied PVS is increasing rapidly in the world and the estimated global PV market grew to over 100 GW annually in the last three years. The global renewable energy introduction in 2040 as a "new policy scenario", the total volume forecast is 6,504 GW, including 2,540 GW of solar PV power [1].

In Mexico, up to 2010, 85% of the installed PV capacities were off-grid and the number of grid-tied systems grew fastest after 2007 when the grid connection was allowed. In regards to the Distributed Generation in Mexico, it is classified as any system with a capacity below 500 kW. The National Association of Solar Energy (ANES) reported approximately 21,600 interconnection permissions for distributed solar in 2015. In March 2017, the Energy Regulatory Commission (CRE) approved a regulation that allows net metering for distributed energy generation. Furthermore, the Mexican Secretary of Energy settled on a minimum participation of clean energies in electricity generation of 25 % for the year 2018, 30 % for 2021, and 35 % by 2024 [2], but also, carbon emission reductions of 50% below 2000 levels by 2050 was introduced in 2012 [3].

However, up to 2019, the total installed PV capacity has achieved only 4,426 MW, because, since march 2018, officially canceled its 4th long-term energy auction. This follows on the heels of a new government that favors fossil fuels for central power generation being now difficult to achieve objectives by using PV [4, 5].

Despite the national situation in their public policies, and the scarce academic research participation, several academic institutions in the Mexican Republic has continued to carry out different activities to maintain and promote solar cells and the use of fovotoltaic systems. In this study, were evaluated the performance of a 60kw PV system, taking into consideration that the PV module and ambient temperatures, wind velocity, and solar irradiations as the main parameters for the PV system performance evaluation (See Fig. 1).

Figure 1: PV-module arrays at the angle of 20° from the horizontal. The modules are connected as 12S and 4P, 5×12 kW arrays.

Now, the soiling of PV modules is a major problem in dry climate environments, where there is much dust and little rain. This is a representative case of Mexico City in the winter seasons. Soiling normally causes a decrement of a PV-specific yield (kWh/kWp) by roughly 12 %/month, but its rate in a desert environment could be from ~0.1% and up to 1 % per day [6]. Even though, without rain, the soiling losses can become large and have even exceeded 70%. Further, there are currently no simple and effective technologies to combat PV soiling, and the largest PV plants in several regions are still cleaned manually.

Cleaning machines and robots have been introduced to the market but are not yet widely adopted. The issue is especially pressing because the largest deployment of PV is planned in dry and dusty locations such as the Middle East, China, and India [7].

As for the background, typical PR in the 1980s was about 50~75% [8], 70~80% in the '90s [9, 10], and nowadays, above 80% [11]. In this work, it was demonstrated how the accumulated dirt on a PV array affects the PR by reducing notably the amount of irradiation reaching the solar cells. The accumulated soil is cleaned in different seasons by means of natural rainfall and analyzed these weather influences on PV-array performance. It evaluated the generated electricity from a 60kw PV system, comparing previous and after rainfall on a soiled PV array in terms of performance ratio.

2 SYSTEM DESCRIPTION

The PV module arrays are installed on the fifth-floor building roof and fixed on aluminum framed structures oriented 30° East-faced from the geographical South. The PV-module arrays are electrically subdivided into five sections. Each one of the five arrays is connected to its corresponding inverter; Fronius model IG Plus V11.4-3 DELTA with the capacity of 11.4 kW/each. The solar irradiance was measured using a pyranometer from Kipp & Zonen model cmp3 installed in the same PV-module plane of array (POA) with an angle of 20 degrees from the horizontal as Mexico City´s latitude.

2.1 Monitoring and performance concepts

Data monitoring is one of the important requirements for the PVS. Without accurate data monitoring, the PVS performance ratio (PR) can not be obtained with precision. Effective data monitoring not only helps to identify system performance but also helps to resolve possible troubles [12]. The Fronius inverter system integrates all the monitored data every 5 minutes and it is logged, and the solar irradiance was monitored by pyranometers and using a Campbell CR300 data-logger, every minute. IEC standard 61724 performance parameters have been used to define the overall system performance with respect to energy production, solar resource, and the overall effect of system losses [13].

PR is the ratio between the actual yield (i.e. annual production of electricity delivered at AC) and the ideal yield which is defined as:

$$\mathbf{PR} = \frac{Real\ Yield\ AC}{Ideal\ Yield\ DC}$$

Under normal PV system operating conditions, the measured data contains deviations caused by malfunctions such as string defects, shadings, modules, or inverter failures that influence the measured performance of a PV system.

One of the unavoidable energy losses is the temperature coefficient. In our case, the monocrystalline silicon solar cell has about -0.47/°C, which means, about a half percentage of power loss for every degree °C increase in temperature. It means that, if the PV module operates at the temperature of 60 °C, the temperature difference with the standard test condition (STC) is 35 °C, and the power loss might be around 16%. By the way, the temperature of the PV module detected in Mexico City frequently reaches around 60 °C and up to 70 °C during a clear sky at mid-day, and the PVS performance is greatly reduced [14].

3 PV SYSTEM PERFORMANCE, DISCUSSION, AND RESULTS

First, a performance ratio PR from January to December 2019 is shown in Fig. 2. The daily solar irradiation on a monthly average basis is displayed and the produced average daily energy for the corresponding month is given in kWh/day. The PR changes from its lower values in the winter season (December to February) to higher values from June to August. Despite the summer season in Mexico City, the effects of the rainy period (June to Sept.) the effective irradiance-time is less than that of the springtime of February to April. The best PR was in July, with 83.5%, and the worst was in December, with 69.9%.

The obtained average PR during 2019 was 77.1 % with an average daily irradiance of 350 kWh and an average produced energy of 270kWh.

Fig 3 shows the PR comparison between the entire months of January and August. As can see, the measured solar irradiation for both months is very similar in its magnitude. Even though, the generated energy has a difference which is noted in the PR. i.e., the PR in January is 71.1%, while in August is about 82.4%. To understand these PR differences, Fig. 4 is shown to consider PV module temperatures. As is shown, the corresponding PV module temperature for January is 33.2 °C, while in August is 38.2 °C. It means the average temperature during the daytime in August has of about 5°C greater temperature compared to January. Also, it is shown in Fig. 5 and Fig. 6 are ambient temperature and wind velocities, respectively, for the considered months. As can see, the

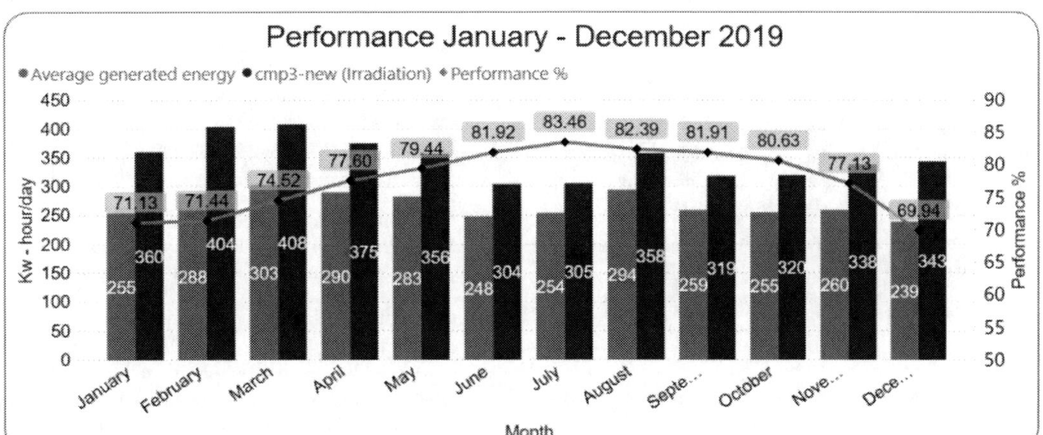

Fig. 2 Performance ratio (PR) of the 60 kW PVS monitored from January to December, 2019. It is displayed a month basis PR. The Solar irradiation and the generated energies are displayed in kWh/day, based on month average.

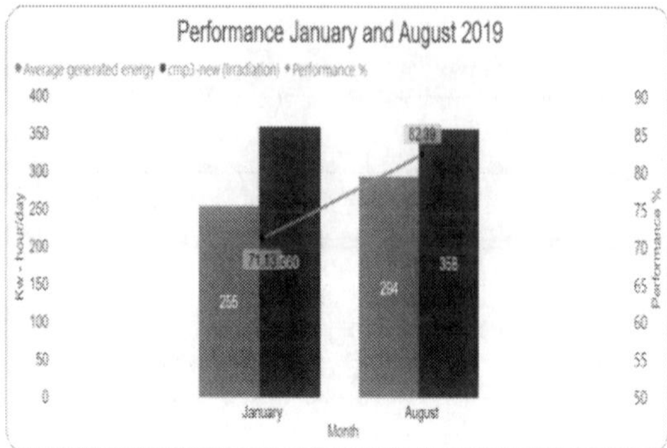

Fig. 3. Performance ratio PR of January and August, 2019. Also its average Solar irradiance and the generated energy form the PVS.

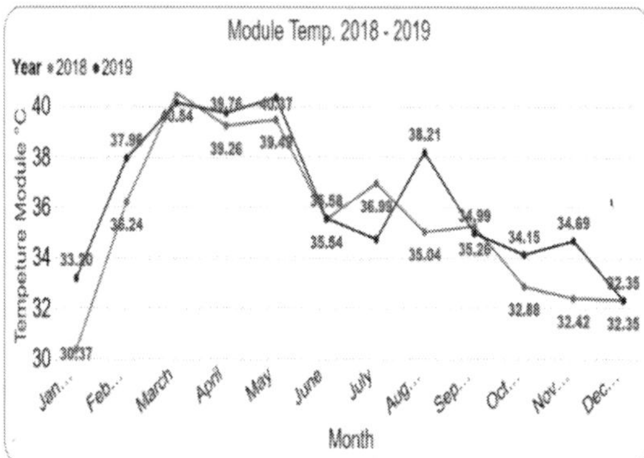

Fig. 4. Average PV module temperature during a day (7:00 to 19:00 hrs, which is the PVS operating time) for each of the months of 2018 and 2019.

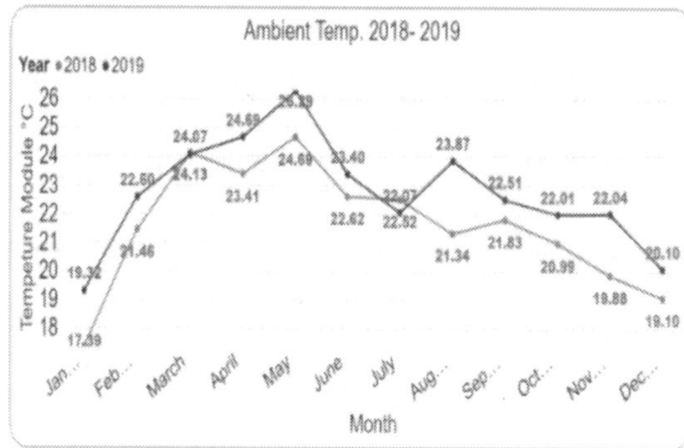

Fig. 5. Month based average ambient temperature for 2018 and 2019 in a day-time (PV operating time)

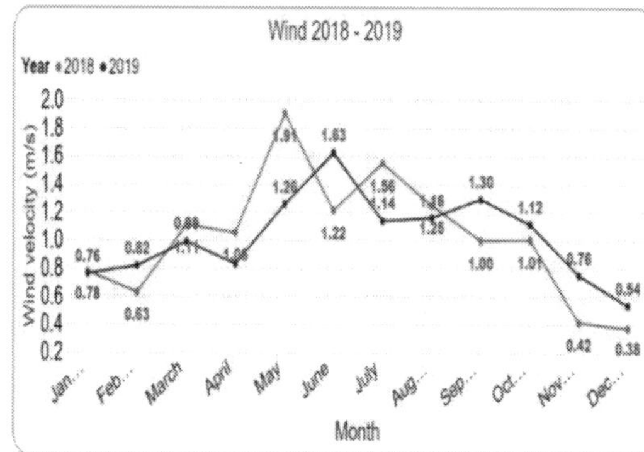

Fig. 6. Wind velocity based on daily and monthly averaged values during the daytime.

ambient temperature is also higher in August 2019 compared to January, and the wind velocity is greater in August with 1.25 m/s, compared to January with 0.76 m/s. It means, that despite higher wind velocity in August, the PV module temperature remains at a higher temperature compared to January.

There is a notorious PR difference between January and August as was shown in Fig. 3, which is not related to the solar cell power temperature coefficient, but on the contrary, the PV module temperature is greater in August than that in January.

So, the possible PR difference could be due to the external perturbation, as is the case of accumulated soil which resulted in less energy production in January. After analyzing PR differences between January and August 2019, the following section, continues to review more PR from 2017 to 2020 where was visualized the lower PR caused by the dust accumulation in the PV modules, i.e. the PV array got dirty due to the natural atmospheric soiling starting in the months of November to December, which is the dry season. Noteworthy that in Mexico City with a height of 2,200 meters above sea level, the soiling of the PV modules in the winter period is notorious; this is because of the condensation cycle of humidity and low temperatures which facilitates the adherence of dust [15, 16]. The PR of PVS decreases in the mentioned winter season as was shown in Fig. 2, which is sure, because of the soiling effects. In 2017, the first rainfall occurred in March, and the average PR was calculated within the three days previous and, three days after the rainfall.

Figure 7 shows the PR for the first days of March 2017. Where as a result of the previous accumulation of dust in the PV module, it showed a poor PR (see 1st to 3rd of March), however, after the first rainfall, which occurred on the night of March 3rd, the dirty PV array was cleaned, showing a significant average PR increase (March 4th to 6th) of approximately 23.6%; from 65.9 to 81.5% on average.

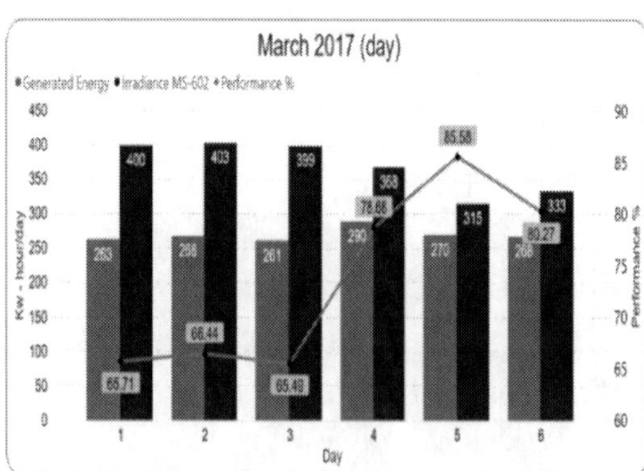

Fig. 7 Performance ratio obtained from a produced energy vs solar irradiation from March 1st to 6th, 2017. On the night of March 3rd, rain fell that dragged the dust accumulated in the PV array. The three days average PR increased by 23.6%; from 65.9 to 81.5%.

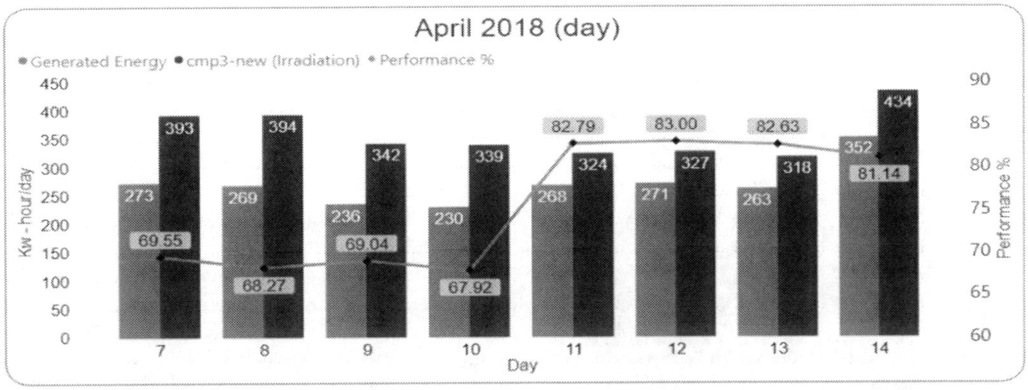

Fig. 8 Performance ratio obtained from a produced energy vs solar irradiation from April 7th to 14th, 2018. During the night of April 10 rain fell which washed up taking away the accumulated dust on the PV array. The average PR gain of 20%, from 68.7% to 82.4%.

Table I. Performance ratio of the photovoltaic system before and after the rains, during the corresponding month/year. Each of the averages of the data collected on those days is also indicated.

Performance Ratio (%)	**March 2017**	**April 2018**	**Jan. 2019**	**Jan. 2020**	**Averages**
Soiled PV array	65.9	68.7	68.2	62.3	66.3
Washed PV array	81.5	82.4	75.2	73.8	78.2
PR-gain (%) after rainfall	***+23.6***	***+19.9***	***+10.0***	***+18.5***	***+18.0***

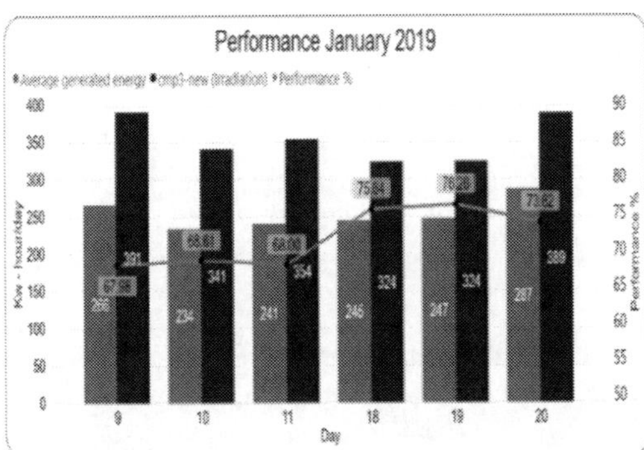

Fig. 9 Performance ratio obtained from January 9th to 20th, 2019. January 12 to 17, was rainy with cloudy days washing away the accumulated soil on the PV array. The PR increment was about 10%, from 68.2% to 75.2%.

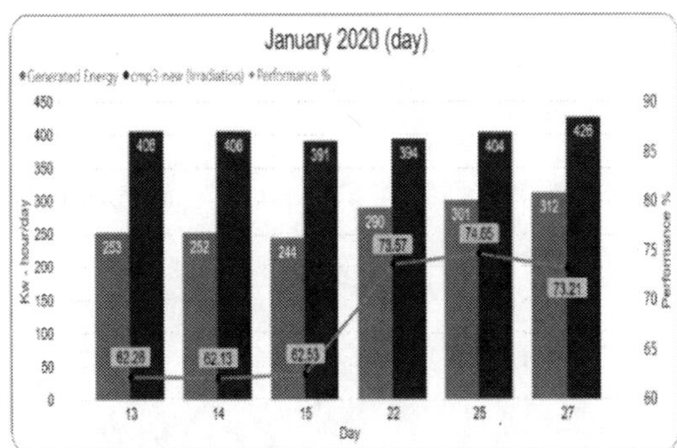

Fig. 10 Performance ratio obtained from January 13th to 27th, 2020. On January 18th, and some more days after, rain fell to wash away the accumulated dust on the PV array. The three-day average PR increase was 18.5%, from 62.3% to 73.8%.

Figure 8 shows the measured PR behavior in April 2018, where poor PR values were calculated for April 7 to 10th (four continuous days) corresponding to 68.7%, then the PV array was cleaned by means of rainfall, in accordance to CONAGUA (National Commission for Water Resources) [17] indicating corresponding precipitation on April 10th night. It can be seen that the PR increased to 82.4% (April 11 to 14th). This means an average PR increment of about 20 %, due to the cleaning effects. In addition, Fig. 9 and Fig. 10 show the increased average PR corresponding to January 2019 and January 2020, respectively. The rainy season around Mexico City starts in May - June and lasts until September - October every year. But from January, it may rain sparsely or intermittently.

Fig. 11 shows the dirt on the PV modules and the accumulated dust before and after the rains in January 2020. After the removal of dirt due to the rains, the average PR increases substantially as described in Table I. This Table I summarizes the average PR for the dirty and cleaned PV array, before and after the rain, respectively, and during the considered months of the year (Fig. 7 through Fig. 10). The PR gain is at least 10% and up to 23.6%, with dirty PV panels averaging 66.3%, but increasing to 78.2% after the rain falls. Fig. 12 shows the daily averaged/monthly-based PV module temperature, the ambient temperature, and the wind velocity during 2019. From the given PV module temperatures, it is possible to elucidate, that the detected

Fig. 11. One of the aspects of a soiled and clean PV-module array; previous (top) and after (bottom) the rain in January 2020.

lower PR in January was mainly due to PV module soiling. This assumption comes from observing the PV-module temperatures during the starting weeks of 2019, which is lower (33.2 °C) compared to August (38.2 °C), i.e. the influence of temperature coefficient in silicon solar cells, at a higher temperature of PV modules affects lowering the generated power (reduces its PR), but in spite of higher temperature in August, the obtained PR is higher than in January, as was demonstrated in this work.

4 SUMMARIES

The grid-connected PV system was monitored and evaluated mainly the performance variations during 48 consecutive months from 2017 to 2020.

The obtained average PR during 2019 was 77.1 % with an average daily irradiance of 350 kWh and an average produced power of 270kWh. The best PR occurred in July, with 83.5%, and the worst with 69.9% was in December.

It was focused on a different specific period during March 2017, April 2018, January 2019, and January 2020, to identify the origin of a poor PR. In these mentioned periods, it was found that PR increments from 10 to 24% with an average of 18% after natural rainfall.

The average PR over four years of the PV system was 76.8% and in any circumstances, the soiling affects the PV system lowering notoriously its PR. The daily average energy generated by the 60kW photovoltaic system was 258 kWh and the accumulated energy from 2017 to 2020, and during four consecutive years was 376,680 kWh. The PV-module soiling contributed notoriously to a PR reduction, but the natural cleaning processes as rainfall helps reduce losses due to dirt accumulation on the PV array.

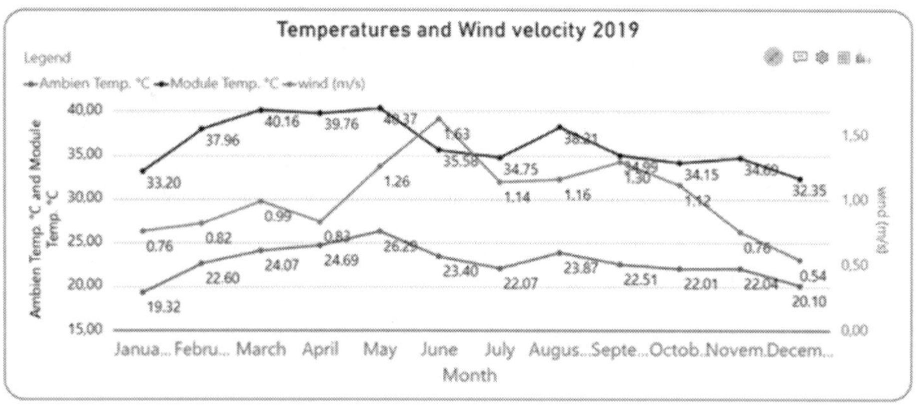

Fig. 12. Monthly-averaged PV module ambient temperatures, and wind velocity for 2019. The average values of all the parameters were considered taking into account PV system working hours of 7:00 am to 7:00 pm, and the detected data were logged every 5 minutes.

5 ACKNOWLEDGEMENTS

We are very pleased to have Mr. Miguel Angel Luna who has facilitated the measurements and their analysis.

References

[1] RTS Japan 2018. PV activities in Japan and global PV highlights. Volume 28, No. 11, (332)

[2] The low of energy transition LEY DE TRANSICIÓN ENERGÉTICA. Nueva Ley DOF 24-12-2015

[3].https://www.senado.gob.mx/64/gaceta_del_senado/documento/111987, Gaseta del Senado Sept 14, 2020. Gaceta:LXIV/3PPO-10/111987

[4] https://en.wikipedia.org/wiki/Solar_power_in_Mexico Solar Power in Mexico.

[5] Mexico's Solar Power Capacity Grew 32% In 6 Months With Its Cumulative Installed Capacity Exceeding 4 GW In June 2019: Mexican Solar Energy Association Asolmex, https://taiyangnews.info/markets/mexicos-solar-pv-capacity-exceeds-4-gw/

[6] Solar Energy, vol 107, sept. (2014), Arash Sayyah, Mark N. Horenstein, Malay K. Mazumder, Energy yield loss caused by dust deposition on photovoltaic panels.

[7] Benjamin Figgis. Investigation of PV soiling and condensation in desert environments via outdoor microscopy. Mechanics of materials [physics.class-ph]. Université de Strasbourg, 2018. English. ffNNT : 2018STRAD006ff. Pp22-26

[8] Decker B, Jahn U., Performance of 170 grid-connected PV plants in Northern Germany, analysis of yields and optimization potentials. Solar Energy April-June, 1997; 59(4–6) 127–133. Selected Proceeding of ISES 1995: Solar World Congress. Part IV. DOI: 10.1016/S0038-092X(96)00132-6

[9] Ulrike Jahn and Wolfgang Nasse, Operational Performance of Grid-connected PV Systems on Buildings in Germany, Prog. Photovolt: Res. Appl. 2004; 12:441–448 (DOI: 10.1002/pip.550)

[10] Reinders A.H.M.E., van Dijk VAP, Wiemken E, Turkenburg W.C., Technical and economic analysis of grid-connected PV systems by means of simulation. Progress in Photovoltaics: Research and Applications 1999; 7(1): 71–82.

[11] Ueda Y, K Kurokawa, K Kitamura, M Yokota, K Akanuma, H Sugihara. Performance analysis of various system configurations on grid-connected residential PV systems. Solar Energy Materials & Solar Cells 2009; 93: 945–949.

[12] Southern Energy, 2018. https://southern-energy.com/

[13] Marion, B. Adelstein, J. Boyle, K. et.al. 2005, NREL/CP-520-37358 "Performance Parameters for Grid-Connected PV Systems", 31-IEEE PVSC, Lake Buena Vista, Florida

[14] J.A. Urbano, et al. ISES Solar World Congress, "One-year 60 kWp photovoltaic system energy performance at CINVESTAV, Mexico City", Energy Procedia, Vol. 57, (2014) 217-225.

[15] Investigation of factors affecting condensation on soiled PV modules, Ben Figgis et al., January 2018, Solar Energy 159:488-500, DOI:10.1016/j.solener.2017.10.089

[16] Wu-zhi Yuan, Zhixing Liao, Kui He, Si-Min Huang, Mechanism of photovoltaic module soiling in the presence of dew, Solar Energy Materials and Solar Cells 248 (2022) 111962

[17].https://smn.conagua.gob.mx/es/climatologia/temperaturas-y-lluvias/resumenes-mensuales-de-temperaturas-y-lluvias (found on Feb. 2022)

DETAILED PERFORMANCE METRIC ESTIMATION OF SOLAR MODULES FROM TIME-SERIES MONITORING OF SOLAR POWER SYSTEMS

Julián Ascencio-Vásquez*[1], Yang Li[2], Haohui Liu[3], Kang Jian[2], Haoran Chang[2], Biyun Liu[2], Lu Zhao[3]
[1]Envision Digital, Redwood City, United States
[2]Envision Digital, Shanghai, China
[3]Envision Digital International, Singapore

ABSTRACT: Having reliable monitoring and analytics tools is the first step in tracking and evaluating the health status and yield of the components and entire solar PV plants. However, only robust algorithms and strong analytics can unleash the full potential to trigger cost-effective corrective actions enabling the highest revenue possible for a Photovoltaic (PV) system. To keep such products on top of the state-of-the-art, the Solar Centro of Excellence (COE) of Envision Digital has dedicated efforts to expand the current DC health status features, proposing and developing methods for detailed performance assessment by inferring and evaluating the Current-Voltage characteristics (i.e., *I-V* curves) of PV modules extracted from maximum power point (MPP) monitoring time-series. This conference paper uses analytical and physical methods to extract intrinsic PV module data, such as series and parallel resistances. Based on the single-diode model, those inherent data are then linked to certain reversible and irreversible performance losses like soiling and degradation. The proposed workflow will bring new technical aspects and metrics to complement the solar O&M and asset management sector.

Keywords: PV Performance, Analytics, IV Curves, Asset Management, O&M, Monitoring

1 INTRODUCTION

Solar photovoltaics became one of the mainstream energy sources for the world to achieve carbon neutrality and fight against climate change [1]. As solar penetration expands, operation and maintenance (O&M) become increasingly critical to ensure the full utilization of the installed solar capacity. However, some key performance metrics for O&M traditionally can only be measured from on-site inspections [2], [3]. While on-site inspections will be necessary to evaluate with very high accuracy the status of specific components, operators would rather reduce the number of annual interventions to the minimum since those activities are usually costly and complicated in terms of logistics when operating remote sites with difficult access. Moreover, such tests involve additional and dedicated equipment, and for certain, very particular weather conditions have to happen; This fact is one of the motivations behind O&M and asset managers to count on smart monitoring systems together with dedicated advanced analytics tools [4].

Metrics and key performance indicators (KPIs) that can represent the performance and health status of solar systems and subcomponents are essential and vital for the asset management of solar systems as well as the O&M activities carried out in the field [5]. For example, module cleaning can be scheduled based on the estimation of the soiling level, which is more efficient and cost-effective than regular cleaning. The degradation rate of modules can also determine the replacement, repowering and warranty claim. In addition to O&M, these metrics can be used in short and long-term power forecasts. More accurate estimates with detailed system metrics can provide better wholesale market bidding and reasonable long-term planning [6]. Finally, the benchmark of such metrics can reveal the optimal combination of technologies, installation types and geographical locations, which can be feedback to the system design or even the design and manufacturing of solar cells and modules [7].

There have been a variety of methods and equipment for data acquisition and analytics in the asset management of photovoltaic systems. As normally required by grid regulators, time-series data of voltage, current, and power of key components and weather stations are collected and reported through the SCADA system. However, to gain an in-depth insight into the module performance, imaging technologies, including infrared imaging, electroluminescence and photoluminescence, and *I-V* curve measurements, are still recommended [8].

The *I-V* curves are an option that can bring insightful knowledge about the PV system's state and performance; however, some limitations exist in collecting and processing them, making this a difficult task for typical monitoring and analytics tools. First attempts of an analytics product dedicated to *I-V* curves have been presented in [9], [10]; the benefits and potentials of the algorithms have the only drawback that input data needs to be measured on-site, sometimes forcing the disconnection of strings, interruption of normal operation and external hardware are often required.

Researchers recently proposed methods to estimate degradation [11] and loss mode classification [12] from synthetic *I-V* characteristics of modules together with time series monitoring data. In this paper, we proposes an improved *I-V* reconstruction method based on time-series monitoring data with more metrics extracted from physical and fitting methods.

2 PV PERFORMANCE AND *I-V* CURVES

It is well-known that changes in *I-V* curve parameters indicate certain failure modes in the PV modules. Table I shows a simplification of certain failures and how they are reflected in *I-V* curves [13]; the complete table information can be found in [10].

Also, due to PV module intrinsic properties, the *I-V* curves will vary their parameters and shape depending on the operating weather conditions. In Figure 1, this effect is shown by the characteristics of a PV Module operating at different irradiance-temperature conditions.

Table I: Simplified relation among the failure modes and the shapes of *I-V* curves based on information from [10], [13].

	I_{SC} Short-Circuit Current	V_{OC} Open-Circuit Voltage	R_{OC} Shunt Resistance	R_{SC} Series Resistance	Change in slope	Inflexion points
Failure						
Discolouration Yellowing	X			X		X
Corrosion	X		X	X	X	X
Delamination	X			X		X
PID		X				

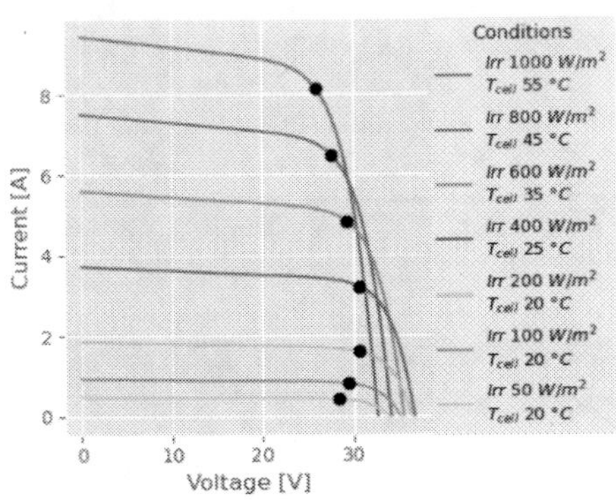

Figure 1: Examples of *I-V* curves operating under different weather conditions.

A health-status assessment of PV modules will consider "time" as a crucial factor. For example, the assessment of an irreversible performance loss such as PV module degradation will need to consider the analysis of specific Performance Metrics over a long period (at least two years of data) due to the very small variations (e.g., typically considered as 0.5%/year). On the other hand, a soiling assessment would provide a daily metric like the Soiling Rate (e.g., 0 to 0.15%/day depending on external factors and mounting type); thus, the calculation of a soiling rate will require a shorter time series and metrics with at least a daily resolution.

Using the above concepts, in this conference paper, we introduce the detailed performance metrics from *I-V* curves, where the changes in series resistances and saturation current can be used under different window periods to assess reversible and irreversible performance losses, as well as evaluate possible failure modes of the PV modules under operation.

3 METHODOLOGY AND DATASETS

3.1 An approach to infer *I-V* curves from MPP data

The following methodology, shown in Figure 2, gives a step-by-step approach to assessing failures from MPP monitoring data inferring variations in intrinsic PV module characteristics.

I. As field conditions (e.g., irradiance and temperature) change over time, the operating point of a module or a string varies accordingly, controlled by the maximum power point tracking (MPPT) of inverters. The relationship between field conditions and pairs of MPP Voltage (V_{MPP}) and MPP Current (I_{MPP}) can be analyzed with a physical model to extract the detailed metrics.

II. A data quality stage needs to be considered in all monitoring tools to remove outliers and unphysical metrics products for communication and sensor issues [14].

III. V_{MPP} and I_{MPP}, measured at the operating conditions of that moment, must be corrected, trying to get a normalization state at the Standard Test Conditions (STC) [15].

IV. Besides the monitoring data, the parameters of an initial *I-V* curve need to be extracted. In this workflow, algorithms embedded in the PVLIB python package are used [16]. Using the De Soto model [17], five parameters of the single diode equation are calculated; then, using the Lambert W transformation [18], the entire *I-V* curve is computed. This step will give the initial PV module characteristics.

V. Weather-corrected V_{MPP} and I_{MPP} are used to fit *I-V* curves over them by finding the optimal among different PV module parameters.

VI. Finally, to assess failures and trigger further actions, the results from step V are compared to the ones of step IV. Also, since this novel approach considers time-based data, comparing results within step V is possible.

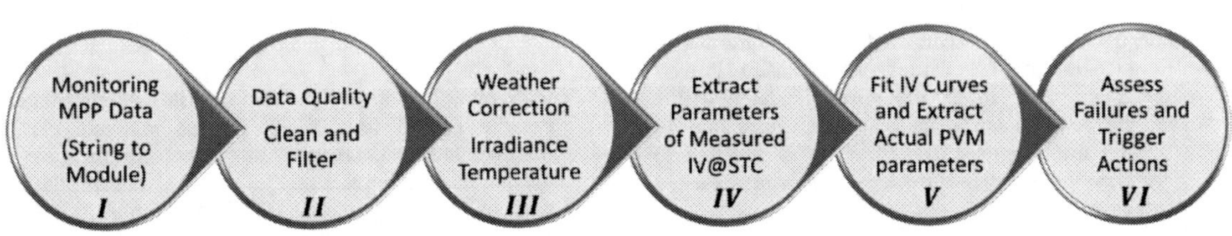

Figure 2: Workflow from monitoring data through *I-V* curve reconstruction and metrics estimation towards corrective actions.

3.2 Data used in this study

In this study, we analyse the DC health of a PV system operating Kyocera 295W PV modules in a tropical climate. The O&M operator has provided initial *I-V* curves of an installed PV module.

Monitoring data with the 1-minute resolution has been provided, including the plan-of-array (POA) irradiance, back-of-module temperature, maximum power point current and voltage of a string in a site during the first week of June, and the first week of September 2021 were collected. Additional information provided by the O&M operator confirms the decrease in power due to soiling, which serves as a hypothesis for the current study.

3 RESULTS

Following the pipeline introduced in the previous section, the collected MPP monitoring data were first filtered to avoid abnormal operation conditions, like clipping where the DC power reaches the rated AC power of the inverter or curtailment. Figure 3 shows the remaining data in blue and the data filtered out in red.

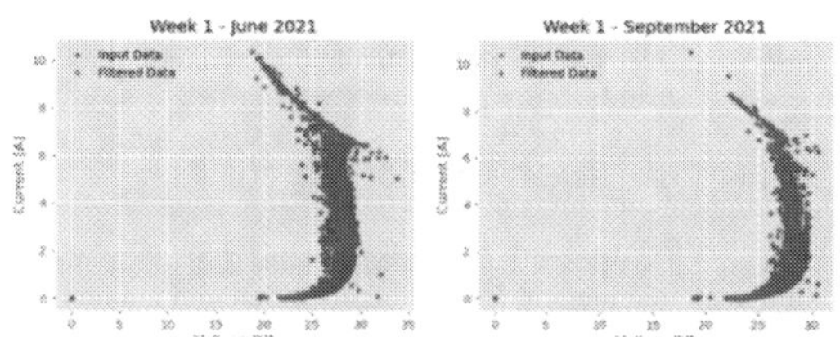

Figure 3: Input and filtered MPP monitoring data for the first week of June 2021 and the first week of September 2021.

The measured *I-V* curve at STC is processed to extract the initial *I-V* parameters, and the computational fit provides five parameters shown in Table II. The measured and fitted *I-V* curves are presented in Figure 4.

Table II: Five *I-V* curve parameters fitted from a measured *I-V* curve.

Parameters	Unit	Fitted Value
Photocurrent	A	9.293
Saturation Current	A	1.572 e-8
Series Resistance	Ω	0.135
Shunt Resistance	Ω	38.326
(n x Ns x V_{th})*	V	1.907

* Product of the Thermal Voltage V_{th}, Diode Ideality Factor "n", and the number of series cell "Ns".

To better use the *I-V* curve data at STC, the MPP monitoring data should be corrected, aiming to normalize the point cloud of the Current-Voltage and Power-Voltage pairs. The input monitoring data and the weather-corrected data are shown in Figure 5. By overlapping the measured *I-V* curve, it is possible to observe how the weather-corrected monitoring data get grouped close to the MPP.

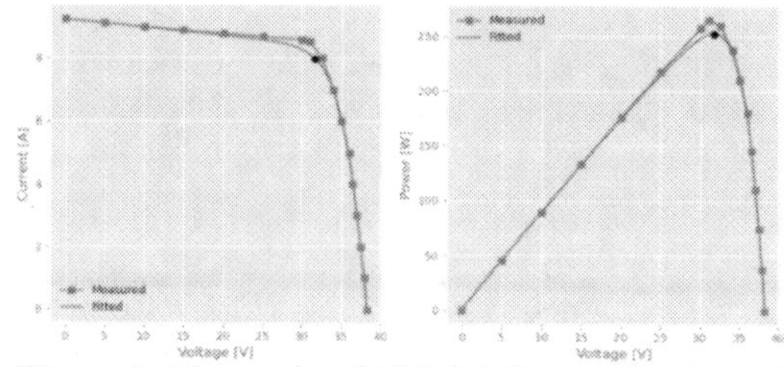

Figure 4: Measured and Fitted *I-V* curve in terms of Current (left-side) and Power (right-side) versus Voltage at STC.

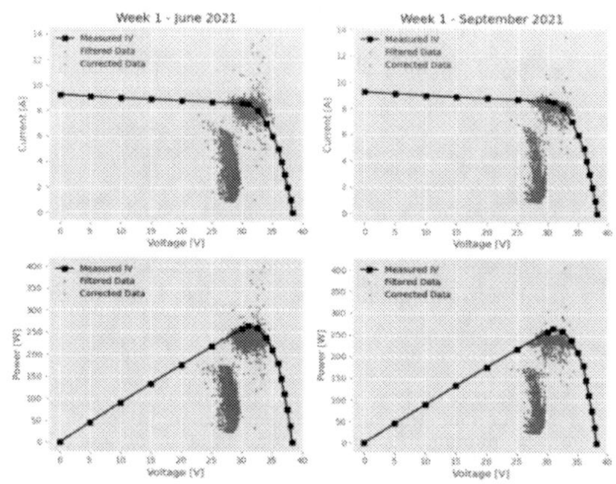

Figure 5: Filtered, weather-corrected data and measured I-V curve for the two periods evaluated.

After weather correction, extracting *I-V* curve parameters from the MPP monitoring data is applied at each timestamp. Since the studied period is very short, only the series resistance and photocurrent have been optimized in this study. New and detailed performance metrics can be analysed by comparing the mean values between both periods. In this study, the photocurrent presents a drop of 2.7% and the series resistances an increase of 6.04%. Figure 6 illustrates the initial and degraded I-V curves.

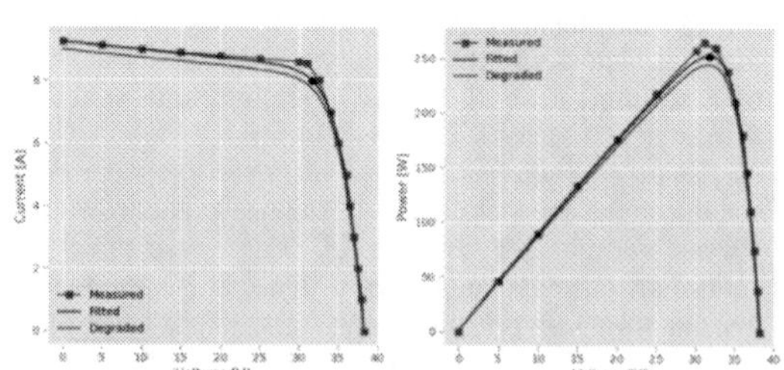

Figure 6: Measured, fitted and degraded *I-V* curves following the proposed workflow.

In this demonstration, the possible actions triggered can be related to soiling. The soiling rate can be calculated as the decrease of fitted photocurrent since it is the parameter representing the optical absorption of a module. As soil accumulates on the module glass and blocks irradiance transmission, the I_L decreases consequently. During this 90-day period, a soiling rate of 0.03% per day is estimated, which is a reasonable value.

The result is robustly calculated in these two weeks, including a large proportion of overcast days. Following this preliminary study, the estimated soiling rate will be compared with soiling sensor measurements to validate the result further.

4 CONCLUSIONS

Performance metrics, like performance ratio or performance index, are commonly used by many stakeholders in the PV industry, enabling the understanding and supporting the decisions along the value chain. This conference paper presents detailed performance metrics for PV module health status to enhance such actions.

The proposed workflow has been applied in a short period (i.e., 90-day window) of monitoring data, resulting in detailed performance metrics showing decreases in the photocurrent and increases in the series resistances, representing a soiling event as confirmed by the O&M provider. The same approach could be applied for more extended data periods aiming for more profound degradation assessments expecting to overachieve current DC health status tools.

This method can be used in predictive maintenance, better power forecasting and geographic-specific system design in future work.

5 REFERENCES

[1] IEA, "World Energy Outlook 2021," Paris, 2021. [Online]. Available: https://www.iea.org/reports/world-energy-outlook-2021

[2] IEA PVPS Task 13, "Qualification of Photovoltaic (PV) Power Plants using Mobile Test Equipment," Apr. 2021.

[3] IEA PVPS Task 13, Subtask 3.3, "Review on Infrared and Electroluminescence Imaging for PV Field Applications," Mar. 2018.

[4] National Renewable Energy Laboratory, Sandia National Laboratory, SunSpec Alliance, and and the SunShot National Laboratory Multiyear Partnership (SuNLaMP), "PV O&M Best Practices Working Group. 2018. Best Practices for Operation and Maintenance of Photovoltaic and Energy Storage Systems; 3rd Edition.," National Renewable Energy Laboratory., Golden, CO. [Online]. Available: https://www.nrel.gov/docs/fy19osti/73822.pdf.

[5] Envision Digital, "A Holistic approach to Technical Asset Management: Improving and optimizing solar PV assets," 2021. [Online]. Available: https://www.envision-digital.com/smart-solar/

[6] kWh Analytics, "Solar Risk Assessment: 2022," 2022. [Online]. Available: https://www.kwhanalytics.com/solar-risk-assessment

[7] J. Ascencio-Vásquez, K. Brecl, and M. Topic, "Geographical Approach for Weather Risk Identification and PV Performance Assessment," presented at the 37th European Photovoltaic Solar Energy Conference and Exhibition.

[8] L. Koester, S. Lindig, A. Louwen, A. Astigarraga, G. Manzolini, and D. Moser, "Review of photovoltaic module degradation, field inspection techniques and techno-economic assessment," *Renew. Sustain. Energy Rev.*, vol. 165, p. 112616, 2022, doi: https://doi.org/10.1016/j.rser.2022.112616.

[9] H. Liu, M. Arora, K. Jian, and L. Zhao, "Automatic IV Curve Diagnosis with Deep Learning," in *2021 IEEE 48th Photovoltaic Specialists Conference (PVSC)*, 2021, pp. 2242–2246. doi: 10.1109/PVSC43889.2021.9519033.

[10] IEA PVPS Task 13, Subtask 3.2, "Review of Failures of Photovoltaic Modules," Mar. 2014.

[11] X. Sun, R. V. K. Chavali, and M. A. Alam, "Real-time monitoring and diagnosis of photovoltaic system degradation only using maximum power point—the Suns-Vmp method," *Prog. Photovolt. Res. Appl.*, vol. 27, no. 1, pp. 55–66, Jan. 2019.

[12] M. Wang *et al.*, "Analytic Isc–Voc Method and Power Loss Modes From Outdoor Time-Series I–V Curves," *IEEE J. Photovolt.*, vol. 10, no. 5, pp. 1379–1388, 2020, doi: 10.1109/JPHOTOV.2020.2993100.

[13] J. Ascencio-Vásquez, "Modeliranje In Globalno Vrednotenje Učinkovitosti Delovanja Ter Staranja Fotonapetostnih Modulov In Sistemov," University of Ljubljana, 2021. [Online]. Available: https://repozitorij.uni-lj.si/IzpisGradiva.php?lang=eng&id=131079

[14] S. Lindig *et al.*, "International collaboration framework for the calculation of performance loss rates: Data quality, benchmarks, and trends (towards a uniform methodology)," *Prog. Photovolt. Res. Appl.*, vol. 29, no. 6, pp. 573–602, 2021, doi: https://doi.org/10.1002/pip.3397.

[15] International Electrotechnical Commission, "IEC 60891:2021 : Photovoltaic devices - Procedures for temperature and irradiance corrections to measured I-V characteristics." Oct. 27, 2021. [Online]. Available: https://webstore.iec.ch/publication/61766

[16] W. F. Holmgren, C. W. Hansen, and M. A. Mikofski, "pvlib python: a python package for modeling solar energy systems," *J. Open Source Softw.*, vol. 3, no. 29, p. 884, 2018, doi: 10.21105/joss.00884.

[17] W. D. Soto, S. A. Klein, and W. A. Beckman, "Improvement and validation of a model for photovoltaic array performance," *Sol. Energy*, vol. 80, no. 1, pp. 78–88, 2006, doi: https://doi.org/10.1016/j.solener.2005.06.010.

[18] A. Jain and A. Kapoor, "Exact analytical solutions of the parameters of real solar cells using Lambert W-function," *Sol. Energy Mater. Sol. Cells*, vol. 81, no. 2, pp. 269–277, 2004, doi: https://doi.org/10.1016/j.solmat.2003.11.018.

UTILIZING SYSTEM EFFICIENCY EVALUATIONS TO DETERMINE DC OUTPUT OF PV SYSTEMS

Bengt Jäckel*, Matthias Pander, David Daßler, Stephanie Malik, Jens Fröbel
*Bengt.jaeckel@csp.fraunhofer.de
Otto-Eissfeldt-Strasse 12
06120 Halle (Saale), Germany

ABSTRACT: A novel approach to investigate the state of health (SoH) of PV systems is presented. It utilizes simple monitoring data from the past minimum three years of the PV system. The data must include energy yield of the PV system, irradiance data and for good filtering logfiles to understand missing or flowed data sets. The available data should be at least on daily basis for trend charting of several years. The PV system itself consist of PV modules, cabling, and other balance of system components (BOS) including inverters. All those impact the herein used system efficiency SE evaluation. The level of detail obviously also depends on the number of individual measurement points (e.g. string, combiner box or inverter level) and can range from a few kWp to MWp.

By applying a linear regression fit into monthly or annual data a robust system efficiency SE can be determined even if some data outliers exist. Depending on scope of work even different day profiles can be separated e.g. for clear and sunny days vs days low light levels. However, the main goal here is to determine current and past DC output power of the PV modules from monitoring data while applying some assumption such as the system consist of always the same BOS components, same orientation, and homogeneous degradation of modules. Those, at least partly, can be easily validated by flash test or electroluminescence (e.g. for testing for PID).

The determination of DC output power requires reference points. Those can lay in the past for instance from system commissioning or / and can be from today (fielded modules are remeasured in laboratory). From that a reference data set for a particular system can be created with known DC power. Against this system a sister system nearby e.g. from a different technology or otherwise poorly performing system can be evaluated to detected possible system efficiency losses. This, for instance, can be applied to a c-Si system with a thin film sister system nearby (same irradiance profile) with very similar BOS (same non-PV module related losses), same orientation and known DC rating.

Results of four different PV technologies will be exemplarily presented and evaluated by the introduced methodology within the paper. The applied methodology predicts current DC ratings of the PV modules precisely and can determine degradation rates.

Keywords: energy yield, prediction, degradation

1 INTRODUCTION

Plant owners and O&M (Operation & Maintenance) providers benefit from proper operation of their PV systems. Unscheduled or too many maintenance interventions due to defects or shading lead to yield losses and increased operating costs, which have a negative impact on the profitability of the PV system. Automated analysis of system information, using monitoring or imaging inspection methods, can minimize such unnecessary effects and thus forms an important part of increasing yield and reducing operating costs.

Monitoring systems record the behavior of system components and environmental conditions over time in high resolution. These sometimes huge data sets provide important information about the state of health (SoH) of the system. Current analysis software uses energy yield and performance ratio (PR) [1][2][3] as standard for plant evaluation. However, new very efficient methods are needed for different kinds of system related questions such as relative and absolute degradation rates, relative and absolute power output values for instance for large scale claims. A routine to quickly and reliable determine relevant information values is needed that is robust to disturbances and yield-reducing local effects.

Newer approaches are based on machine learning and stochastic methods [4]. The focus to utilize such methods is on the possibility to screen large data sets and test different parameters, while keeping the statistical relevance in mind.

The general challenge is the lack or the difficult to assess system operation information such as cleaning, fixing smaller items (fuses, broken modules) or external events such as grid outages.

The procedure described here tries to filter out such events and only focus on the global SoH and how efficient the PV system can generate power. The aim is to determine SoH data and degradation rate, if applicable, and compare that to indoor data, or alternatively validate that with literature information [5][6][7].

2 APPROACH

The general approach is given in Figure 1. The monitoring data must provide certain inputs such as at least daily energy yield data of the system part under investigation and cumulative irradiance for the same time periods. The monitoring data should at least contain 3-5 years of detailed information to trace any kind of seasonal variations. The provided data should be at least daily energy yield data (EYA) and cumulative irradiance data (Int) for the same period.

In a first step the provided data must be filtered against outliers or otherwise faulty data. Logfiles of outages, system maintenance etc. are very useful to clean the given data. Furthermore, easy to detect false data points are energy yield data without irradiance data or irradiance data without details about the produced energy. However, there are also data points in between. Such data needs to be either filtered out or attributed to certain SoH conditions.

From the cleaned data EYA vs Int plots are generated for defined time intervals (e.g. monthly). Furthermore, clustering of the data can be done to look for different available cumulative irradiance levels (e.g. low, medium and high daily cumulative irradiance levels). From such plots system efficiency (SE) is calculated by linear regression fits for the predefined time interval.

The SE (slope of data) represents the system's capability to generate electrical energy from a given irradiance, called system efficiency SE. A higher slope represents a better capability of the system to generate electricity. Obviously higher efficient modules / other BOS will increase the slope, however this can be considered in normalization steps to compare systems correctly.

Monthly data is a good compromise between daily weather fluctuations and still be able to nicely track seasonal variations (general trend, change of performance of a particular cell technology)

The last step of data preparation and treatment is normalization to either a given start point such as begin of operation of the system or a point in time where additional information exist such as after a hand-over where extra data was collected such as remeasurement of modules, string measurement or IR-imaging. Alternatively, the values can all be normalized to e.g. the rated name plate rating of the system or laboratory reference measurements for the given module type used in the system under investigation (STC, low-light data, size of module (efficiency).

Figure 1: Schematic workflow for data evaluation and system efficiency SE calculations

3 EXEMPLARILY DATA SET

For this study of dataset consisting of more then six years of data was used. Four different modules were used, D1 and D2 are thin film based, D3 and D4 are multi-crystalline Silicon standard modules. The unfiltered, raw dataset is given in Figure 2.

3.1 Filtering

The data set was checked for false data points as described before. An example is given for non-traced EYA data in Figure 3. All data points for D4 fall onto the x-axis (only irradiance data).

Figure 4 displays a full dataset for the year 2020. The monthly changes in cumulative irradiance are clearly visible and are in that way expected.

This data set also illustrates the challenge for filtering data to certain cumulative irradiance levels. Obviously in wintertime there is limited high irradiance data, whereas in the summer it's the opposite. This requires finally a careful precheck prior the linear regression fitting, that sufficient data points in a selected time period exist.

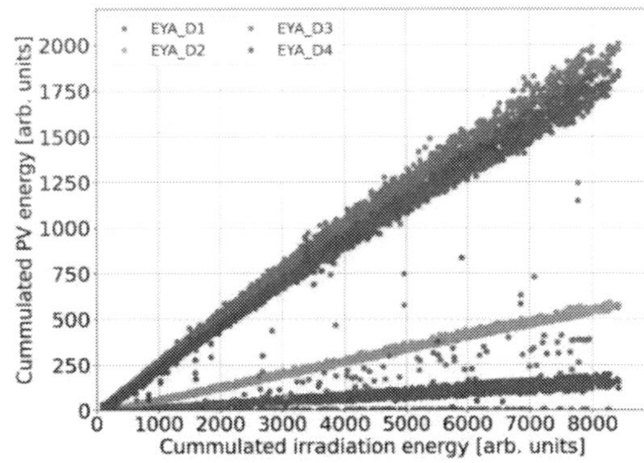

Figure 2: Overview plot of all unfiltered data of four modules with quite different efficiencies

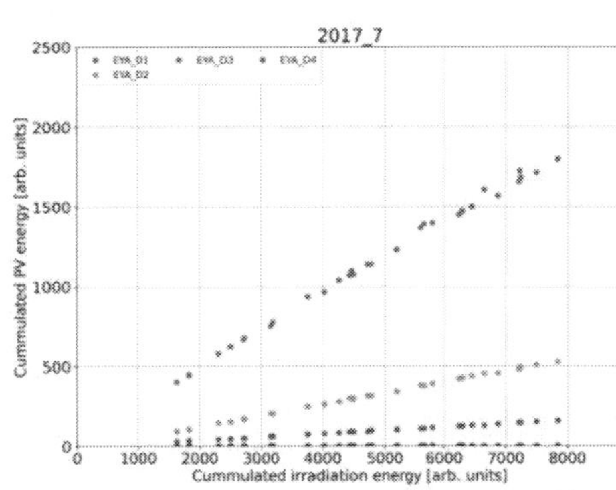

Figure 3: Example of a monthly data set with false data of D4 (all data-points sit on x-axis)

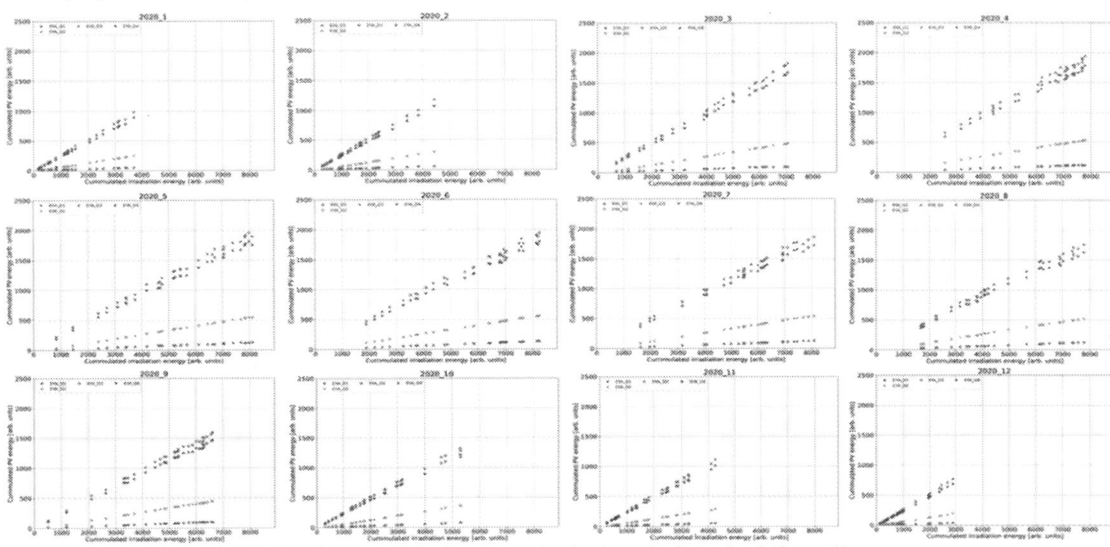

Figure 4: Exemplarily detail plots for year 2020 for a more in-depth analysis to check for outliers

3.2 Linear regression

After filtering the data, the monthly data linear regression fitting is done. The raw data result is given in Figure 5. As expected from module type and data the SEs of D3 and D4 are significantly higher compared to D1 and D2. Seasonal variations of SE are immediately visible and are mainly due to the impact of temperature. Filtered out periods that are extended show up as missing points such as for D4 in 2017-2018.

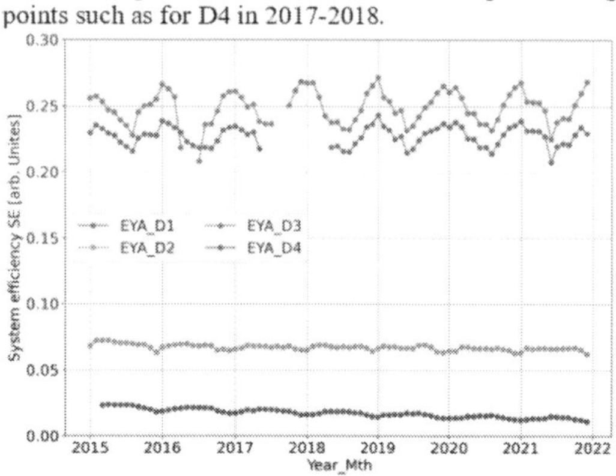

Figure 5: Non-normalized SEs for the four different modules. Missing points are due to filtering of "false" data.

The next step is to normalize the data and decouple the information from size of module/size of power plant and make the data more comparable. This is done to a point in the beginning of 2015 in Figure 6.

As the dataset used herein is extended and covers more than six years an additional reasonable linear fit is added to the graph. This fit guides the eye to show long term changes that is equivalent to the degradation of the system, not only necessarily to the module. However, most visible degradation with such shape will be mainly due to changes of the module, as inverter or fuse failures will more lead to "false" data and will be, depending on filter procedure, removed by the applied filters.

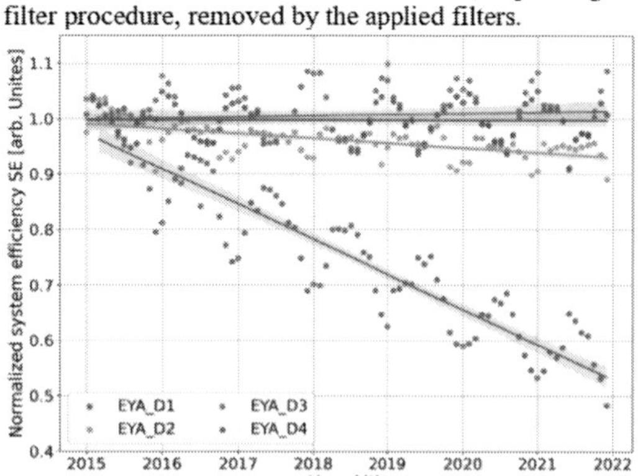

Figure 6: Normalized SEs to a datapoint in spring 2015 and a linear fit to guide for degradation

3.3 Comparison to Indoor STC data

Table 1 summarizes the determined relative losses from the regression fits from Figure 6. Module D1 shows with 7% a very high degradation rate, whereas D3 and D4 almost show no change and with "some good" will even a slight increase over the years (D3).

Table 1: Regression fit (RF) data

	D1	D2	D3	D4
Start in 2015	1.0	1.0	1.0	1.0
Value in 2021	0.53	0.94	1.014	0.99
Loss	0.47	0.06	0.014	0.01
Degradation rate [%/a]	-7.0	-0.9	+0.2	-0.1

All modules were also characterized indoors several times within the past years. The measured values for STC and low light conditions are given in Table 2 and Table 3. Degradation rates were determined and are given in all tables in the last row and are also marked as normalized data in Figure 7.

Table 2: STC data (1000W/m^2)

	D1	D2	D3	D4
Start in 2015 [W]	23.0	74.7	257.6	240.9
Value in 2021 [W]	10.1	71.6	258.7	238.4
Loss [W]	12.9	-3.1	1.1	-2.5
Degradation rate [%/a]	-8.3	-0.6	0.06	-0.15

Table 3: Low light data (200W/m^2)

	D1	D2	D3	D4
Start in 2015 [W]	4.3	12.2	52.0	47.0
Value in 2021 [W]	2.7	11.4	51.8	46.3
Loss [W]	-1.6	-0.8	-0.2	-0.7
Degradation rate [%/a]	-5.5	-1.0%	-0.06	-0.2

The indicated error bars for STC values are typical measurement uncertainty MU values for the different technologies where D1 is +/-10% as it had a very bad reproducibility in indoor measurements, D2 is +/-5% and typical for thin film. D3/D4 are crystalline silicon and have an attributed MU of 2.5%.

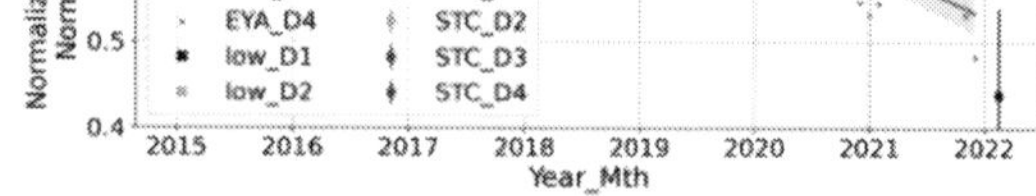

Figure 7: Added markers to indicate reference STC indoor data with estimated indoor measurement uncertainties (bars)

4 SUMMARY AND DISCUSSION

The overall accuracy between the determined loss by the system efficiency SE regression fits and the indoor data is quite good. For the quite stable technologies, here for D3/D4, the determined degradation rates are basically zero and SE and flash test data agree almost perfectly.

Thin film technology D2 agrees also quite good. Per SE a rate of ~0.9%/a was determined, however, indoor a range from 0.6%/a and 1.0%/a for STC and low light conditions, respectively. As the used SE method takes all level of irradiance into account it is reasonable that the SE determined value falls in between.

Within the same interpretation the strongly degrading technology D1 can be understood. However, the STC value seems to underestimate, the low light value to overestimate the degradation. It must be noted that reproducibility of power measurement was poor due to contacting, quite low currents, and a final fill factor of ~30%. Most degradation of that particular module came from FF-loss. Due to this the measurement uncertainty was also determined to be significant higher indoors compared to the other module types.

Further optimization of the SE algorithm can be done, however it will require more data pretreatment. Analyzing detailed irradiance bins, such as low, medium and high daily cumulative irradiance levels can be used. As introduced in the Approach section this can lead to the removal of larger periods in time. To compensate for that hourly data can be used for example. Further improvements can be made by further clustering for temperature intervals. Any extra filter, however, will add complexity and require further procedures to check data consistency. The presented SE data evaluation is simple and is based on averaging over larger periods, knowing that some details get lost, but the bigger picture is maintained. However, it is fast and robust as only a few data filters (see above) were applied.

To summarize, the SE approach is a fast and quite accurate determination procedure for degradation rate without extensive indoor testing. It requires a data set of at least 3-5 years of at least daily monitoring data. The evaluation can be used to recalculate approximate nameplate ratings of modules at the beginning of system operation (reverse extrapolation) on the base of today reference data (full data set from start of operation till today needed!). From the trend charting it can be further utilized to forecast upcoming performance and degradation rates and may help system operation providers to easily support decision making for e.g. re-powering of older systems.

5 LITERATURE

[1] IEA PVPS Task 13, Subtask 2, Analytical Monitoring of Grid-connected Photovoltaic Systems, Good Practices for Monitoring and Performance Analysis, Report IEA-PVPS T13-03: 2014, March 2014, ISBN 978-3-906042-18-3; https://iea-pvps.org/key-topics/analytical-monitoring-of-pv-systems-final/

[2] Y. Ueda et al, "Performance degradation analyses of different kinds of PV technologies in Hokuto mega solar project", EuPVsec 2011

[3] W. Luo et al, Analysis of the Long-Term Performance Degradation of Crystalline Silicon Photovoltaic Modules in Tropical Climates, IEEE Journal of Photovoltaics

[4] D. Dassler, "Innovative Methoden zur Ertrags-auswertung von PV-Systemen unter Anwendung von Verfahren des maschinellen Lernens", PV Symposium Bad Staffelstein, 2019

[5] D.C. Jordan, S.R. Kurtz, „Photovoltaic Degradation Rates - an Analytical Review" Prog. Photovolt: Res. Appl. 2013; 21:12–29

[6] D.C. Jordan et al, Compendium of photovoltaic degradation rates, Prog. Photovolt: Res. Appl. 2016; 24:978-989, doi: 10.1002/pip.2744

[7] K. Kiefer, B. Farnung, B. Müller, Degradation in PV Power Plants, Proceedings 36th EUPVSEC, 2018

PV-HYBRID SYSTEM DATA VISUALISATION RECOMMENDATIONS

Michael Müller
Office for Renewable Energy Systems (OFRES)
D-89081 Ulm, Germany

ABSTRACT: This work presents a standardised data visualisation form for battery based PV-Hybrid off grid systems. It is applicable for lead-acid and lithium-ion battery based systems. The data visualisation displays the battery data and provides a fast and efficient overview on the system behaviour and battery condition. With the help of a standardised data format and visualisation, it is very easy to compare different systems. It is also possible to compare different years of operation of one system and to identify changes, increases in the load demand or loss of performance. This helps to improve the system, the energy management and to increase the lifetime of the battery. All these aspects can reduce the operating costs of the system. The visualisation summarizes the profile of the battery voltage, current and daily amp-hour usage and the corresponding minimum and maximum values as well as calculating mean values and standard deviations for lead-acid and lithium-ion based battery PV Hybrid off-grid systems. Battery usage diagrams picture the condition of the battery and provide an in-depth look at the functionality of the state of charge algorithm. All battery data are displayed as raw data so the visualisation can be used for lead-acid and lithium batteries. Several statistically relevant histogram functions show a detailed picture of the usage of the system. This allows us to evaluate the system status.

1 INTRODUCTION

1.1 General situation

The first PV off-grid systems with diesel generators as backup power sources (PV-Hybrid Systems) were built in the 1980s and the technology has now seen more than 40 years of technological development. Nonetheless not all technical issues are solved yet. Despite this, a stable market exists, and many PV Hybrid systems are installed world-wide supported by advanced products and trained installers. The long-term operation of the systems however, has proved to be especially difficult. Often the remoteness of the installations makes spare-part supply complicated and the maintenance costly. Furthermore, the involvement of many different institutions, companies and governmental agencies has resulted in complicated processes which need to be followed in order to make decisions. Hence we often see situations where the systems are running and supplying energy, while the information on the performance and efficiency of the operating systems is mostly unavailable.

1.1 Purpose of the work

The key proposal of this work is to define an easy to use and standardised data visualisation for battery based PV-Hybrid systems. With the help of a suitable tool chain, big data sets of many PV-Hybrid systems can be visualised and analysed within a short period of time. This permits a fast review of the performance of the systems and allows to compare different systems. This standardisation and the possibility to compare systems allows for the identification of weak points of the operating and management strategy of single systems. This again enables the operators to improve the efficiency of the systems. At the same time, fuel costs can be reduced.

2 ENERGY AND PERFORMANCE ANALYSES

2.1 System Performance Analyses

The scope of the energy analyses proposed is to provide key figures describing the performance of the system. It is possible to perform the analyses with the help of pure in-out energy data which can be easily gained with the help of simple AC kWh counters. No detailed battery data or profile data is required. No Ah-counting and no energy throughput data for the battery are taken into account at this stage. The central aspect remains to characterise the system with the help of extremely little input data.

2.2 Measurement points

The following parameters need to be recorded.

FC : Fuel Consumption. The measurement unit typically is in liter [l]. This parameter describes the overall amount of all types of fossil fuels which had to be put into the system within the analysing period.

FG : Fossil Generation. This parameter measures the overall energy contribution from the generation side of all controllable fossil generators in the system. It is measured in [kWh] and should be the output energy coming from the generators. It will reflect the AC output side. For this counter it does not matter whether the generators supply load directly or provide an input to the PV-Hybrid System power converters.

RG : Renewable Generation. The energy contribution of all different renewable power sources used in the system is counted here. The parameter reflects the energy sum of all power generators. In the case of mixed generators the individual energy contribution is summed up. This overall sum can consist of Photovoltaic generators, wind generators, water turbines or other renewable power sources used in the system. The unit of the parameter is [kWh].

LC : Load Consumption. There is a need to count the overall energy consumption of the system. This is a pure load counter which does not take into account whether the power comes from power inverters, generators or others. It simply counts the overall energy consumption of all loads. The unit is [kWh].

2.2 Energy analyses key factors

The following key figures can be easily calculated based on the above recorded energy data:

- GFC : Genset Fuel Consumption. It is measured in [l/kWh] and calculated by :

GFC[l/kWh] = FC[l] / FG[kWh]

Input parameters are the Fuel Consumption in litres and the overall amount of Fossil Generation in [kWh]. The key figure Genset Fuel Consumption shall be as low as possible and not only depends on the used type of generators but also on the control strategy of the whole system. Many generators show relatively high consumption at low loads while the efficiency at high loads is rather good. To optimise the system, it is recommended to supervise the GFC key figure and eventually to change the genset control strategy according to the consumption profile of the specific genset. The genset shall only run at times in which it can operate in a high fuel efficiency operating point.

- REC : Renewable Energy Contribution. This is measured in [%] and calculated to

REC[%] = RG[kWh] / (RG[kWh] + FG[kWh])

The REC key figure is an important and easy to calculate figure. It shows the fraction of renewable energy sources in the system. It reflects the main sources to supply the load. This figure can also be used to optimise the energy management strategy of the system to increase renewable energy contribution and to reduce the fuel consumption. The REC figure shall be optimized to be as high as possible. In systems without a fossil generators it is 100%.

- HSEP : PV Hybrid System Efficiency Performance. The figure is measured in [%] and calculated by :

HSEP[%] = LC[kWh] / (RG[kWh] + FG[kWh])

This key figure shows the efficiency of the whole system and reflects the losses during energy storage within the battery. Clearly this parameter shall be as high as possible but cannot exceed 100%. The overall system efficiency can be optimised by reducing the load during night time and by increasing the renewable energy contribution. Also, the applied energy management strategy of the fossil generators has significant influence on this parameter. The efficiency of the generator depends on the load consumption. This means it is possible to increase the efficiency of the whole system if the type of generator is adapted to the load and managed according to efficiency rules. The aging of the battery also has a negative effect to the overall system efficiency as the battery losses increase by the number of used cycles.

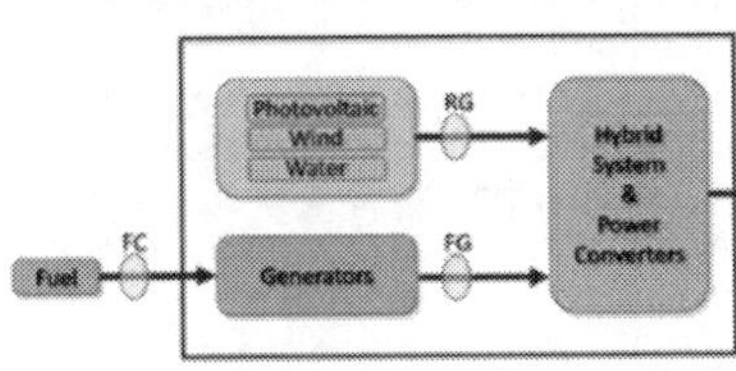

Figure 1: Measurement points to calculate key figures for Energy analyses.

3 DATA FORMAT

The central idea of the approach in this work is to compare the operation data of different systems, locations and the performance development over lifetime of battery based off-grid PV-Hybrid Systems. Therefore a central data format has been defined. All system monitoring data must be converted into this data format before the data can be process by the toolchain. The data format contains a minimum data set in order to allow the visualisation of all systems. The focus of the data is the battery. All values are meant to be battery relevant data.

Negative current values have been defined to be battery discharging currents. The decimal separator must be a point.

Date, Time U_batt[V] I_batt[A] T_batt[°C] SOC_batt[%] system status

Here a sample data set can be seen:

04.11.2016,17:16:51 24.600 -11.400 19.60 0.3200 3
04.11.2016,17:17:51 24.600 -11.500 19.50 0.3200 3
04.11.2016,17:18:51 24.600 -11.100 19.40 0.3200 3
04.11.2016,17:19:50 24.600 -11.300 19.30 0.3200 3

All values are separated by one blank space. First the time stamp in the given format followed by the voltage of the battery in V. The battery current in A followed by the temperature in °C. The fourth value is the State of Charge of the battery in % [0..1]. Finally a parameter describing any system status must be added. It must be there even if it does not contain data. It could be a fixed value over the whole data set. The time between two data rows can be defined individually, it is not necessary to keep this constant over the whole data set.

4 SINGLE PARAMETER VISUALISATION

4.1 General information

For each of the parameters in the data set a one page single parameter visualisation was defined. This applies to the battery voltage, battery current, temperature and the State of Charge of the battery. This single parameter visualisation focuses on displaying statistical information of the parameter in order to provide a fast overview over the provided time period.

Figure 2 shows an example of this standardised single parameter data visualisation. In this example the battery temperature was chosen. The system is a PV off-grid system in Germany which supplies a cooling system.

4.2 Type of displayed graphs

The standardised single parameter one page visualisation consists of several graphs.

The first section is to display the profile of the parameter without any additional analyses. An example can be seen in figure 2. It provides an overview on the parameter values over the given data period.

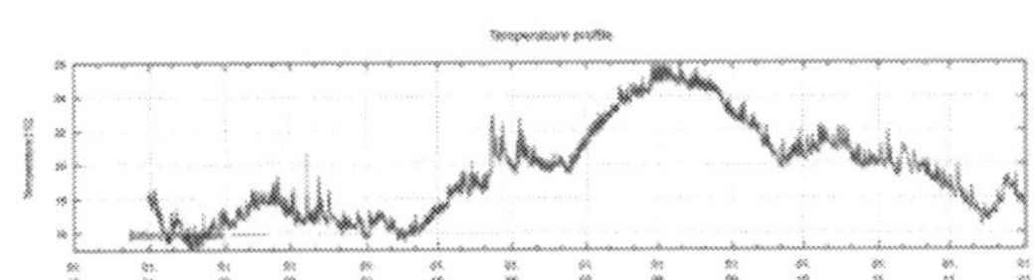

Figure 2: Profile of the parameter over time

The second part is to display the frequency distribution of the parameter. The analyses divides the whole range of data into sections. For each section the frequency of occurrence of the value is calculated. Finally a gauss fit is calculated to be displayed additionally to the graph. Figure 3 shows an example visualisation.

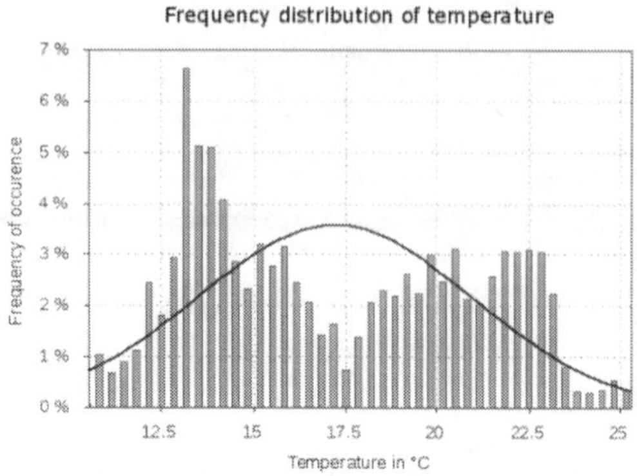

Figure 3: Frequency distribution of the parameter

Figure 4 is a graph which displays the frequency distribution of the mathematical derivation of the parameter. It is again combined with a gauss fit to the graph. It provides information on the fluctuation and changes within the parameter value.

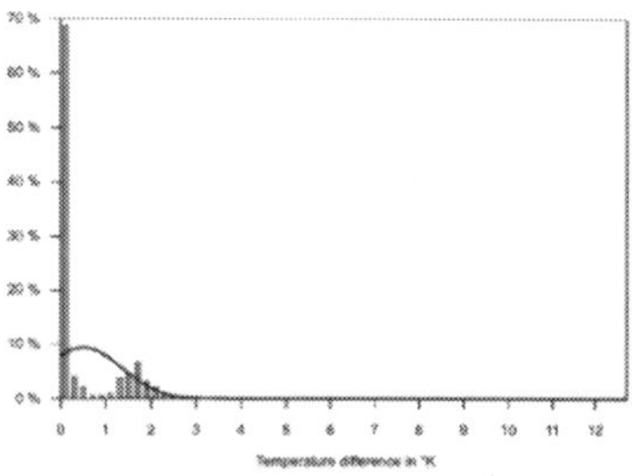

Figure 4: Frequency distribution of the derivation of the parameter

A good overview on what happened within the data set period shows the heat map of the parameter in figure 5. The color shows the value of the parameter while the x- and y-axis display the timestamp of the dataset. The x-axis show the day, while the y-axis display the hour within the day. The graph provides a good overview on the parameter value over the whole data set. This also applied for big data sets.

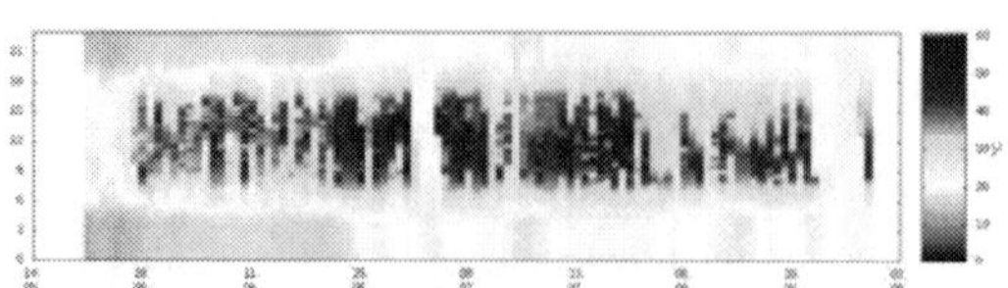

Figure 5: Heat map of the parameter

Finally in figure 6 the last graph displays the profile of the mathematical derivation of the parameter and is the base for the calculation of the frequency distribution. It provides information on the fluctuation over time.

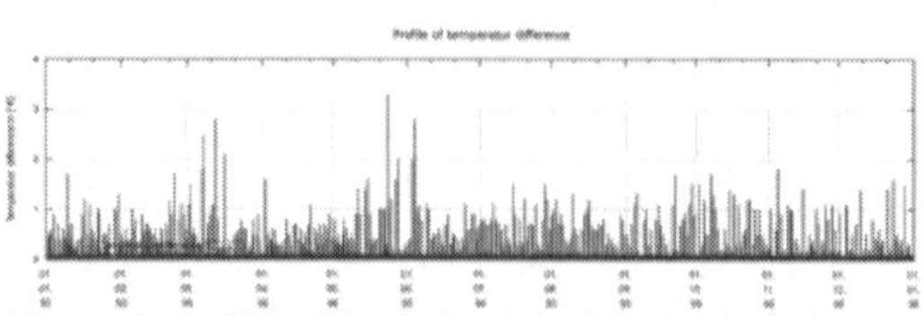

Figure 6: Profile of the derivation of the parameter

5 BATTERY PARAMETER VISUALISATION

5.1 General information

The main standardised graph displays the battery parameter in order to allow a fast visualization of the system. As the battery is the focus it can be seen that systems with a dominant battery structure are the application goal of the visualisation. The size of the system does not limit the application of this toolchain. Also the type of battery applied is flexible. This visualisation is applicable to lead acid batteries as well as to lithium ion batteries and other types as well. The visualisation is even applicable to grid connected systems dominated by a battery. It summarizes most important battery information.

5.2 Type of displayed graphs

The standardised battery data visualisation is a comprehensive graphical output.

The first graph can be seen in figure 7. It is a profile of the battery treatment. For each day the absolute maximum battery voltage is displayed in blue. The absolute minimum battery voltage for each day is red, while the average battery voltage is a black dot. The profile shows the development of this min/max/average values over whole data set over time.

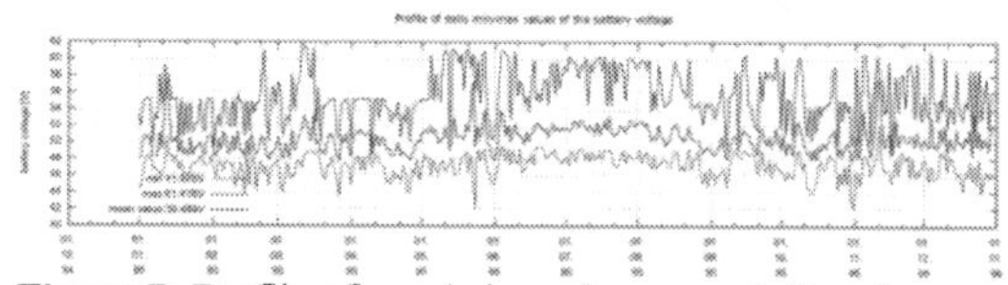

Figure 7: Profile of max/min und average daily values

The same calculation can be seen in figure 8. It displays the min/min/averag profile of the battery current in the next section.

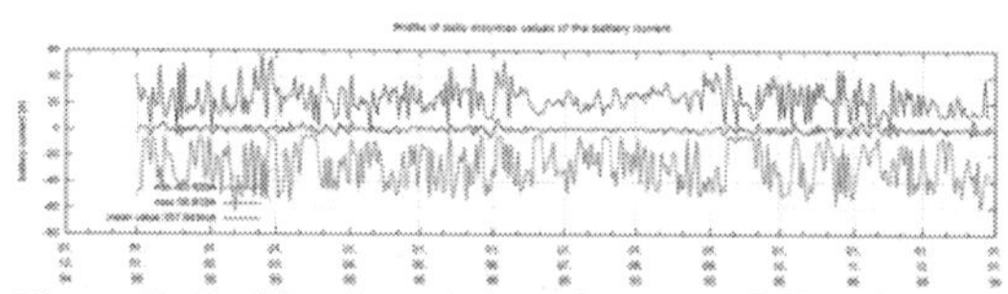

Figure 8: Profile of max/min and average daily values

Below that graph the daily energy balance is visualized. It can be seen in figure 9. For each day the overall Ah charging the battery is counted and displayed as a green bar. Also the overall Ah discharged from the battery is calculated and displayed as a red bar. The solar or wind current which might supply the load directly without flowing through the battery is not displayed as it cannot be determined from the provided data. Additionally the difference between charging Ah and discharging Ah is calculated and displayed as a black dot. As long as the black line is within the green area the energy balance is positive, once it falls into the red area the energy balance is negative. Over a long period of time the system can only survive with a positive energy balance. This graph also allows to look into the design of the system to see whether the sizes of the components fit to the load demand.

The lower part of the battery data visualisation displays several statistical information. That is especially helpful to analyse the treatment of the battery within the system.

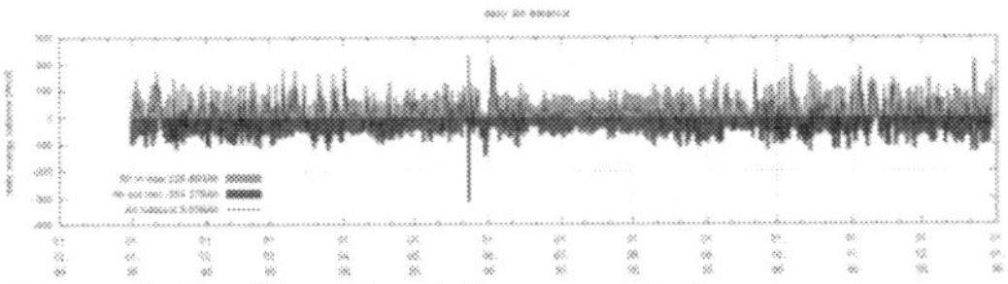

Figure 9: Profile of the daily energy balance

The left graph of figure 10 displays the battery current over the battery voltage. Each value of the dataset is one dot in the graph. The higher the dot density the more often the battery operates under that voltage-current constellation. In this graph all applied end of charge voltage levels should be seen as well as the deep discharge protection value. It will show the typical charging and discharging currents.

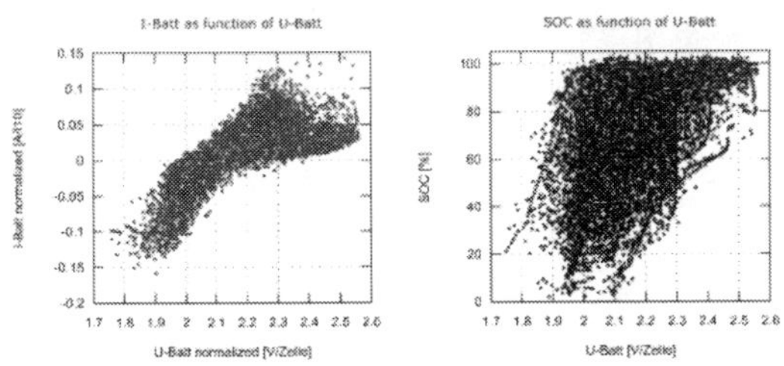

Figure 10: Current and SOC over battery voltage

The same type of diagram displays the State of Charge (SOC) of the battery over the voltage of the battery. This graph brings up interesting information on the status of the system, the SOC of the battery and the battery itself. Typically the battery charging and discharging hysteresis should be visible in that diagram. In the given example the hysteresis is more a block with strong rising and falling edges. That indicates a very week battery which already reached the end of life.

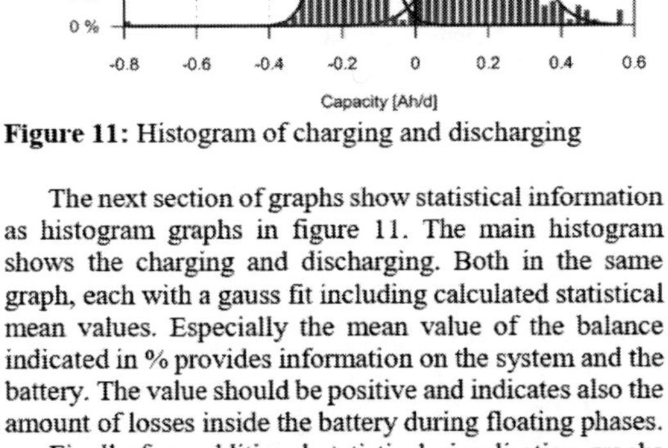

Figure 11: Histogram of charging and discharging

The next section of graphs show statistical information as histogram graphs in figure 11. The main histogram shows the charging and discharging. Both in the same graph, each with a gauss fit including calculated statistical mean values. Especially the mean value of the balance indicated in % provides information on the system and the battery. The value should be positive and indicates also the amount of losses inside the battery during floating phases.

Finally four additional statistical visualisation graphs provide information about the usage and treatment of the system. A histogram including a gauss fit is displayed for each of the following parameters:

- Battery voltage (figure 12, left)
- Battery current (figure 12, right)
- State of Charge of the battery (figure 13, left)
- Temperature of the battery (figure 13, right)

The calculated mean values are written within the figure and show the peak of the gauss fit.

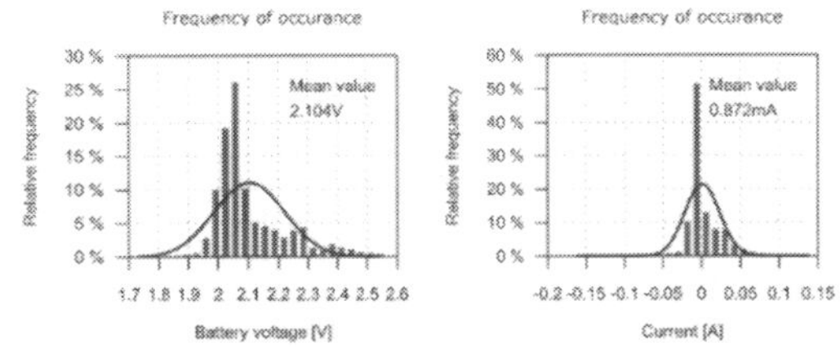

Figure 12: Histogram of battery voltage and current

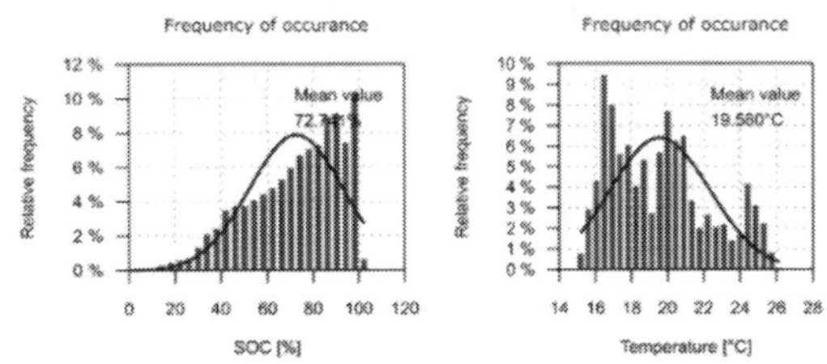

Figure 13: Histogram of State of Charge and Temperature

7 SUMMARY

The battery is a key part of many small and medium sized systems and is the focus of the data visualisation. Therefore, it is essential to have all relevant battery data available.

The main goal of this approach is to provide a proposal for a fast, easy to use and standardised monitoring data visualisation of PV-hybrid systems within an extreme wide power range. Small systems with several watt output power as well as systems in the mega-watt power range can be analysed with the help of the given visualisation. The ability to compare different types of systems makes the analyses usable for a huge variety of systems. Therefore, all parameters are normalized accordingly.

The approach focusses on visualisation of the battery performance as the most costly component of PV-Hybrid systems and cannot be applied to battery-less systems. The goal is to optimize the battery set points on the basis of the results of the visualization, in order to increase the battery life time and to optimize the usage of the battery.

The proposed approach does not reflect all possible analyses. Depending on the topology of the system, a much more detailed analysis may be needed to provide further information. The idea of the general visualisation is to have a good overview on the performance of the system and to visualise the systems data in order to have a fast impression of the usage and the efficiency of the system. Comparability and standardisation with the help of normalised data are important aspects and provide the foundation of this data visualisation.

The described tool chain exists and was the basis for this document. Anyone who is interested in performing the analyses can contact the author under mm@ofres.org. If the data are provided in the described data format (Chapter 3) the visualisation service is free of costs. Data converting services can be offered accordingly.

8 REFERENCES

The following excerpt of relevant literature shall provide an overview here:

[1] Gogla Pico-PV Report, The emerging market for pico-scale solar PV system in SubSaharan Africa: from donor-supported niches towards market based rural electrification. Ivan Nygaard, Ulrich Elmer-Hansen, Thomas Hebo-Larsen. ISBN 978-87-93458-09-3

[2] Off-grid Solar Market Trends Report 2018, International Finance Corporation, Washington D.C. 20433 under cooperation with Dalberg Advisors, Lighting Global, World Bank Group Gogle and Esmap.

[3] Survey of national and international standards, guidelines & QA procedures for stand-alone PV Systems, 2nd edition 2004. Alison Wilshaw, Jonathan Bates, Rolf Oldach IEA PVPS. 30.4.2004 https://www.iea-pvps.org

9 IMPRESSUM

This work has been realized within the Task 9 and 18 of the International Energy Agency (IEA) under the PVPS program. Please visit www.iea-pvps.org for more information.

The work is available under ISBN 978-3-907281-33-8: PV-Hybrid System Data Visualisation Recommendations as official publication.

ISBN 978-3-907281-33-8

9 783907 281338

Figure 14: ISBN bar code of publication

To use the data visualisation service free of costs please contact the author under mm@ofres.org

The work was partially realized and funded by the German Ministry of Economic Affairs and Energy (BMWi) under FKZ0324062.

Michael Müller, Office for Renewable Energy Systems (OFRES), D-89081 Ulm, Germany.

Figure 15: QR-code for data analysing services

SOILING ANALYSIS FOR PV POWER PLANTS OPERATED IN DESERT CLIMATES

W. Herrmann[1], T. Shahin[2]
[1]TÜV Rheinland Solar GmbH, Am Grauen Stein, 51105 Cologne, Germany
Email: werner.herrmann@de.tuv.com
[2]King Abdullah University of Science (KAUST), 23955 Thuwal, Kingdom of Saudi Arabia

ABSTRACT: The accumulation of dust on PV modules is one of the main causes of energy yield loss of PV power plants operated in desert climates. In this paper, we present analytical methods for assessing PV soiling impacts and a methodology for establishing a site-specific soiling metric. A generalized method for calculating the cost-optimized cleaning interval as a function of the average daily soiling rate is discussed. The experimental data for our study were obtained from a soiling station at KAUST University in Thuwal, Saudi Arabia. Soiling monitoring was performed with encapsulated c-Si reference cells (mini-module design) that were subjected to different cleaning intervals (daily, weekly and monthly). Our database included more than 3 years high-resolution data with a recording interval of 30 seconds.
Keywords: PV Array, Soiling Losses, Economic Analysis

1 INTRODUCTION

In the long-term operation of PV power plants, soiling of the PV module surface can lead to significant energy yield losses. The accumulation of dust means a lower radiation transmission through the front glass, resulting in lower effective irradiance reaching the solar cells in the PV module. The continuous process of dust accumulation and dust removal by wind or rain leads to variations in dust soiling during the year.

PV soiling losses are highly dependent on the site characteristics such as the climatic conditions or the surrounding environment (i.e. type of vegetation) and on the glazing characteristics of the PV module. The combination of these factors determines the speed at which soiling accumulates, but also the speed at which dust is removed by gravity, wind or rain. Figure 1 provides an overview of the factors that influence the extent of soiling of PV glazing [1].

Previous work has shown that the transmission loss due to dust soiling is not constant throughout the day, but depends on the incident angle (AoI) of the sunlight [2] [3]. Angular transmittance losses are minimal at solar noon and rise in the morning and afternoon with increasing value of AoI. An overview of the effects of soiling loss on the performance of PV power plants can be found in [4].

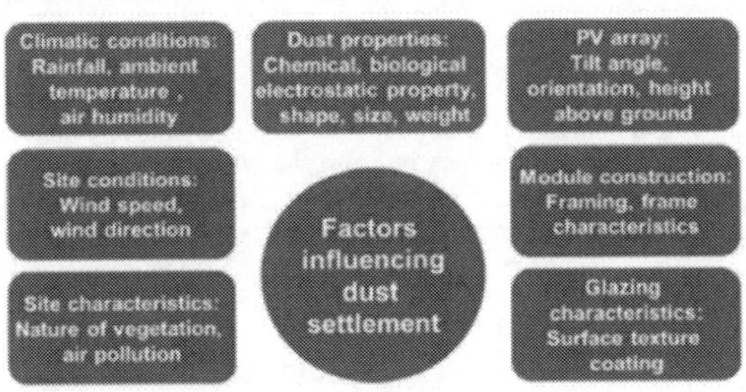

Figure 1: Overview of factors influencing the dust settlement on PV glazing

Locations in desert climates are characterized by high dust accumulation rates, which may show annual fluctuations, and infrequent precipitation events. This allows the application of simple statistical methods to calculate site-specific cleaning schedules with the aim to optimize the energy yield performance of PV power plants. In this paper, we use historical measurement data from a soiling station in Saudi-Arabia to develop such an application-oriented approach. This requires only two input parameters: a) the average daily soiling rate, b) the average costs for a single cleaning process, expressed as a percentage of the annual energy sales.

2 TEST LOCATION

The experimental data for our study were obtained from a soiling station at KAUST University in Thuwal (Saudi Arabia). Figure 2 and Table I give information on the test site.

Figure 2: TÜV Rheinland energy yield test installation at KAUST University, Thuwal (Saudi Arabia)

Table I: Information on TÜV Rheinland PV energy yield test installation at KAUST

Location	King Abdullah University of Science and Technology (KAUST) Thuwal, Saudi Arabia
Geographical position	22.3° N / 39.1°E 3 m above sea level
Inclination angle of test rack	25°
Surroundings	University ground (gravel covered), coastal environment (Read Sea)
Operation of test site	From March 2015

3 SOILING MONITORING

3.2 Soiling sensor

Although soiling sensors are commercially available [4], we have used c-Si mini-modules with a 3 x 3 cell layout in a glass/EVA/backsheet package. The shorted cell in the center acts as an irradiance sensor, while the cells at the edges are dummy cells to ensure a homogeneous dust soiling across the cell in the center. The short-circuit current of the center cell is measured via the voltage drop across a precision shunt resistor, and the cell temperature is measured with an encapsulated Pt100 temperature sensor. General information on soiling monitoring can be found in [5] [6].

The effective irradiance reaching the reference cell is calculated according to the following formula

$$\frac{I_{SC}/\{1+\alpha\cdot(T_{CELL}-25°C)\}}{I_{SC,STC}}\times 1000\frac{W}{m^2} \quad (1)$$

where α is the relative temperature coefficient of short-circuit current, I_{SC} the measured short circuit current and $I_{SC,STC}$ the calibrated short- circuit current at standard test conditions (STC). We have chosen a standard (patterned) PV glass without coating as the front cover of the mini-modules.

Figure 3 shows a view of the test set-up. It illustrates the side-by-side measurement of irradiance with two mini-modules, one of which is manually cleaned daily (clean reference cell), while the other is manually cleaned once a week (soiled reference cell). In addition, a third mini-module is used, which is cleaned monthly. In this way, we were able to study the effects of the momentary degree of soiling on the further dust accumulation process.

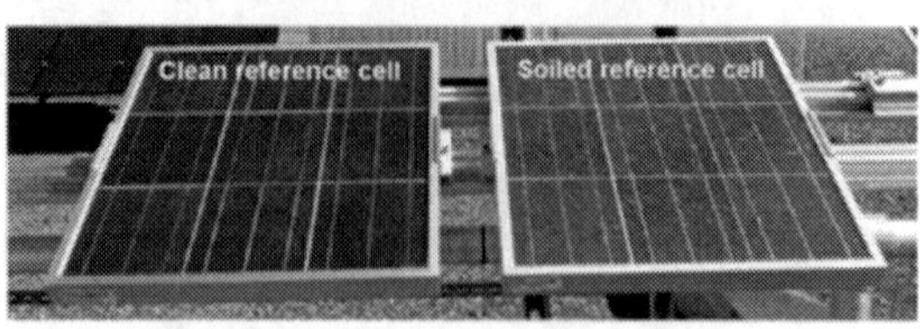

Figure 3: Soiling monitoring with c-Si reference cells

It should be noted that this type of monitoring does not take into account the PV soiling losses caused by a non-uniform soiling of PV modules due to wind effects, gravity or the PV module frame. Similarly, it does not take into account that local variations in soiling levels may occur in PV power plants covering a large terrain [7]. Uncertainties may also arise from the fact that the construction of the soiling sensor used may not be identical to the PV modules used in the field. This applies in particular to the surface properties of the front glass.

3.2 Soiling Ratio (SR)

Our database on dust soiling of PV glazing at the Thuwal site comprises more than 3 years high-resolution data (30s data recording interval). The most important parameter for assessing site specific PV soiling is the Soiling Ratio (SR), which is calculated from the instantaneous irradiance measurements of the clean and the soiled reference cell (RC). It equals to 1 if both cells are clean. Such reset is observed after manual cleaning of sensors or after rain events.

$$\text{SR} = \frac{Solar\ irradiance: soiled\ RC}{Solar\ irradiance: clean\ RC} \quad (2)$$

Figure 4 shows the time series of SR values in 2017 as an example. The sawtooth pattern is a result of manual cleanings of the "soiled" reference cell, which cause SR to be reset to 1. The time course of data also shows that the cleaning schedule of the "soiled" reference cell was not always followed. However, for the majority of the weekly cleaning cycles, it can be seen that SR is in the range 0.95 at the end of the week.

The diagram also reveals that the high-resolution SR data are subject to a considerable daily scatter. This scatter disappears if just data around solar noon are filtered.

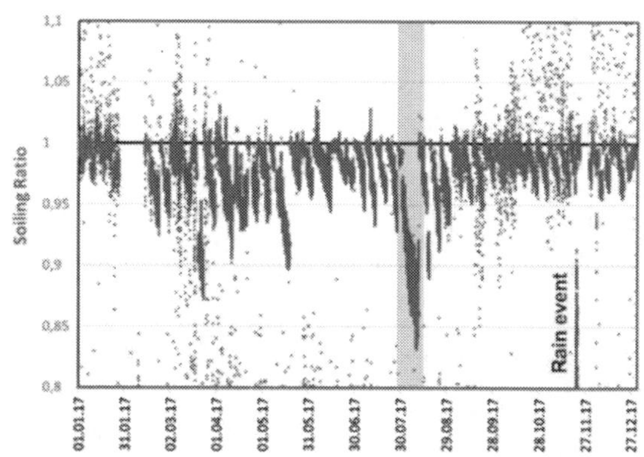

Figure 4: Soiling ratio (SR) at KAUST in the year 2017. The jumps in the data indicate the weekly cleaning of the "soiled" sensor. During the entire year, only one rain event was observed (peak of red data points). The green marked period is shown in Figure 5 with higher resolution.

4 DATA ANALYTICS FOR PV SOILING

4.1 Angle of incidence effects

To better understand the scatter of SR data points, Figure 5 shows a two-week period with continuous dust accumulation. The diagram shows that the daily SR curve is W-shaped [6]. Starting from the maximum value at noon, SR decreases in the morning and afternoon and rises again in the early morning and late afternoon due effects caused by diffuse irradiance. The transmission loss due to dust soiling is thus made up of two components:

a) A component for low incident angles of direct sunlight, which is proportional to the amount of accumulated dust. In Figure 5 this dust accumulation line is shown in red.
b) A variable angular component that correlates with day-time and the incident angle of sunlight. Figure 5 illustrates that the angular component increases as the dust deposition increases.

The angle-dependent part of the PV soiling loss is represented by the difference between the annual averages of Soiling Ratio for two conditions:

a) Calculation of $\overline{SR}_{YEAR}$ without any data filtering
b) Calculation of $\overline{SR}_{YEAR,Noon}$ taking into account only the data around noon time

For weekly cleaning intervals, this analysis showed that the angle-dependent contribution to the annual PV soiling loss is significant. For the 2017 data with mainly 1-week cleaning intervals, it is in the range of 20%.

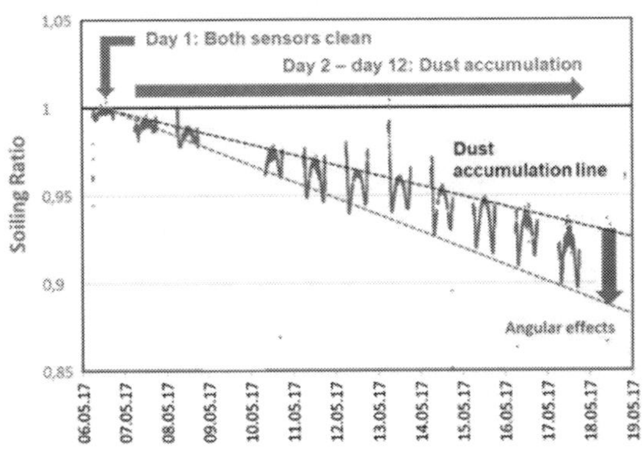

Figure 5: Development of the soiling ratio as a function of dust accumulation during a 2 weeks period

4.2 Daily soling rate

Further analyzes of dust soiling makes it necessary to introduce additional parameters.

a) Daily average soiling ratio $\overline{SR}_{DAY}$

$$\overline{SR}_{DAY} = \frac{\text{Daily insolated solar radiation measured with soiled reference cell}}{\text{Daily insolated solar radiation measured with clean reference cell}} \quad (3)$$

b) Daily soiling rate DSR

This parameter is defined as the day-to-day change of ratio $\overline{SR}_{DAY}$

$$DSR_N = (\overline{SR}_{DAY,N} - \overline{SR}_{DAY,N-1}) \times 100\% \quad (4)$$

where N is the current day and N-1 is the previous day. DSR_N is expressed as a percentage per day. It is negative for dust accumulation and positive for dust removal (e.g. caused by wind, gravity, dew formation or rain).

4.3 Soiling metric

A soiling metric should make it possible to compare the soiling conditions at different locations. Previous analyses have shown that the frequency distribution of the daily soiling rates (DSR_N) is the most suitable method for this purpose [8]. As an example, Figure 6 shows the soiling characteristic for Thuwal in 2017. Note that data points have been removed from days when cleaning of the "soiled" sensor took place and for weekends, when no cleaning of the "clean" sensor took place.

Figure 6 reveals that the DSR frequency distribution is bell-shaped and that it can be divided into two areas:

a) DSR <0: dust accumulation and
b) DSR >0: dust removal.

Besides statistical information such as median, annual mean and standard deviation, the diagram also contains information on extreme weather events (outliers) such as sandstorms (DSR <-2%) or rainfall. The most important parameter is the average daily soiling rate $\overline{DSR}$, which is well approximated by the peak of the distribution. Note that this indicator may not include days with sandstorm and rain events. Table 2 gives an overview of the annual soiling conditions at KAUST test site from 2016 to 2018. The data show a little variability, indicating that an annual mean value of –0.53%/day can be assumed for dust soiling studies.

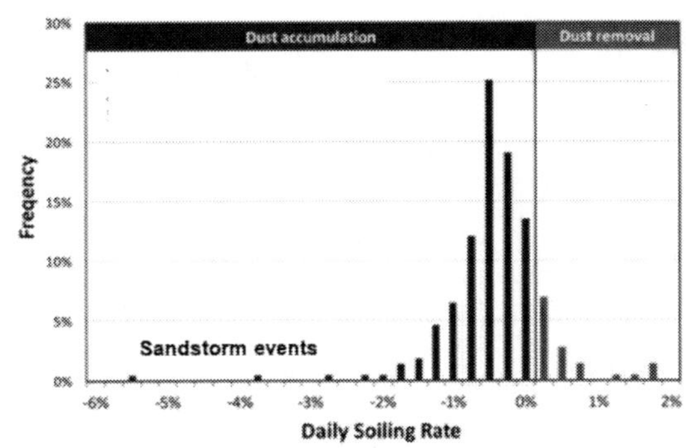

Figure 6: Frequency distribution of average daily soiling ratio (DSR_N) at KAUST test site in Thuwal in 2017

Table II: Variability of soiling parameters in Thuwal

	DSR Average	**DSR Standard deviation**
2016	-0.54%/day	0.51%
2017	-0.53%/day	0.47%
2018	-0.52%/day	0.58%
Average	**-0.53%/day**	**0.52%**

The soiling characteristics of different sites will mainly differ in the position of the maximum of the distribution and the width, expressed by the standard deviation. The diagram is therefore a suitable tool for comparing the soiling conditions at different sites.

Figure 7 shows that the daily soling rate DSR is subject to seasonal fluctuations. In Thuwal, the monthly averages range between 0.4%/day to 0.6%/day. In the next chapters, it will be shown that it makes no difference for the calculation of annual PV energy yield losses whether a constant annual mean or temporal fluctuations of DSR are used.

5 ANNUAL PV SOILING LOSS

The development of cleaning strategies for PV power plants is based on the time series of the daily soiling rate DSR and the daily solar insolation.

We used historical DSR data from 2016 to 2018 to investigate the impact of the number of periodical cleanings per year on the annual PV soiling loss. Dust accumulation was simulated by adding the daily soiling rates and the integral soiling rate was reset to zero at the end of the period. As an example, Figure 8 shows the typical saw-tooth pattern of the integral soiling rate during 2017, calculated for a 4-week cleaning schedule. Weighting of the integral soiling rates with historical daily solar radiation data (measured with the "clean" sensor) resulted in an annual PV soiling loss of -7.25%.

Figure 9 shows the calculated annual PV soiling loss for variable cleaning intervals. The data show that for cleaning intervals up to 4 weeks, the experimentally determined annual PV soiling loss is well approximated (deviation <±1%) by a theoretical annual PV soiling loss calculated for an assumed constant $\overline{DSR}$=-0,53%/day. This assumption leads to a regular sawtooth pattern, for which the annual soiling loss can be modelled by the following equation:

$$Annual\ PV\ soiling\ loss = \frac{365}{2 \cdot N_C} \cdot \overline{DSR} \qquad (5)$$

N_C is the number of cleanings per year. Equation 5 can be used to calculate cleaning schedules for a certain threshold of annual PV soiling loss. For example, –5% annual PV soiling loss and $\overline{DSR}$= –0.53%/day will require 18 cleaning operations per year, resulting in a three weeks cleaning schedule.

Our calculations have shown that for cleaning intervals up to 6 weeks the difference between the two methods is less than ±1%. This means that a good estimate for annual PV soiling loss can be achieved just by knowing $\overline{DSR}$.

6 OPTIMAL CLEANING SCHEDULE

As shown in the previous section, the annual PV soiling loss for a desert site can be described by a single parameter, the average daily soling rate $\overline{DSR}$. From this, a simple approach for optimum cleaning schedules can be derived by relating the costs of cleaning to the (projected) annual energy sales. It should be noted that cleaning costs for PV power plants are very site-specific and depend on many factors, e.g. the costs for investments, maintenance, electricity (robotic dry cleaning) or water consumption, transport and labor (manual wet clean).

Once a cleaning method has been chosen, the costs for a single cleaning operation (unit costs) can be assumed constant, so that the annual cleaning costs add up linearly. The cost-optimal cleaning schedule is then given by the intersection of two curves:

a) Annual PV soling loss as a function of the number of cleanings per year (equation 5)
b) Annual cleaning costs related to annual energy sales as a function of the number of cleanings per year N_C

$$Annual\ cleaning\ costs = N_C \cdot UCC \qquad (6)$$

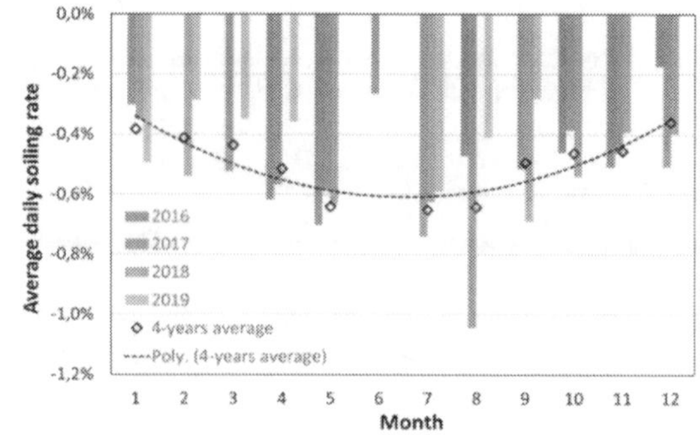

Figure 7: Seasonal variability of daily soiling rates at Thuwal, Saudi Arabia. The dashed line represents the annual average line.

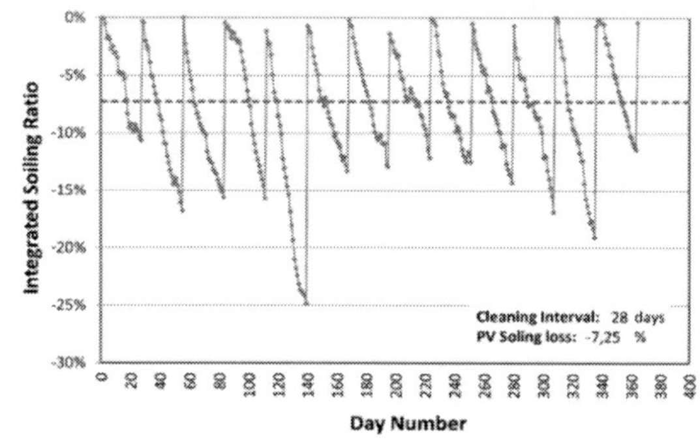

Figure 8: Integrated soiling ratio for a 4 weeks cleaning schedule. PV soiling loss resulted from weighting with the time series of daily solar insolation.

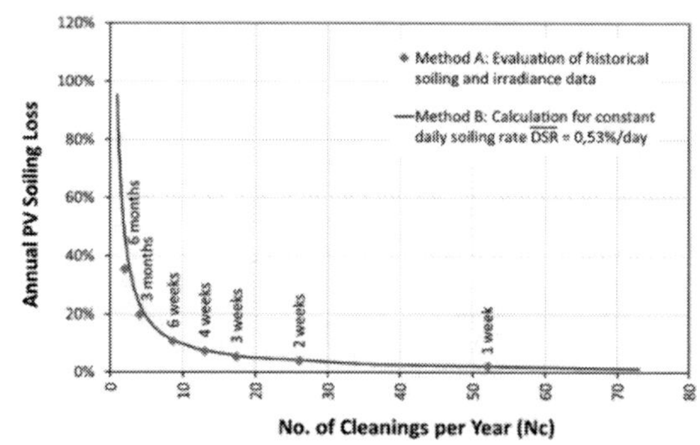

Figure 9: Annual PV soiling loss for variable periodical cleaning intervals

where UCC are unit costs for a single cleaning operation expressed as percentage of the annual energy sales resulting for the case without dust soiling.

Figure 10 shows the result of our parameter study, carried out for variable average daily soling rate $\overline{DSR}$ (red curves: -0.25%/day to -1%/day) and variable unit costs for cleaning UCC (blue lines: 0.25% to 1%). The optimal number of annual cleanings $N_{C.OPT}$ results from the condition that the annual PV soiling loss is equal to the annual cost for cleaning.

$$\frac{365}{2 \cdot N_{C,OPT}} \cdot \overline{DSR} = -N_{C,OPT} \cdot UCC \quad (7)$$

$$N_{C,OPT} = 13.51 \cdot \sqrt{-\overline{DSR}/UCC} \quad (8)$$

Example: A typical average daily soiling rate $\overline{DSR}$ = –0.5%/day and unit costs for cleaning UCC=0.5% will result in optimal 14 cleanings per year, which corresponds to cleaning intervals of 26 days. The cleaning schedule results in an annual PV revenue loss (energy sales) of –6.5% due to dust soiling (equation 5).

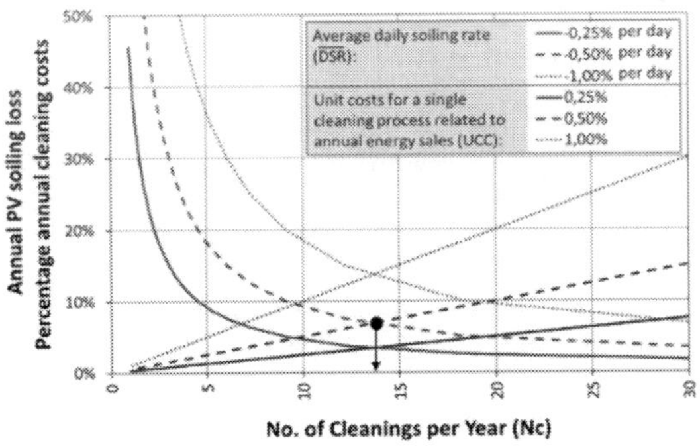

Figure 10: Normalized diagram to find the optimal number of cleanings per year ($N_{C,OPT}$) as a function of the average daily soiling rate ($\overline{DSR}$) and the unit costs for a single cleaning (UCC). The percentage annual cleaning costs are related to the annual energy sales.

7 SUMMARY

Desert climates in the MENA region are characterized by few precipitation events and a little seasonal variation of daily solar insolation. This makes it possible to base cleaning strategies for optimizing the energy yield of PV power plants solely on the statistical evaluation of daily soiling rates (DSR). The distribution of the experimental DSR values is bell-shaped and it is suitable for comparing soiling conditions at different slocations. The peak of the distribution represents the average daily soiling rate $\overline{DSR}$ with good approximation.

PV soiling losses are caused by transmission losses of incident light on the glass surface. We have found that incident angle effects account for a significant proportion of annual PV dust soiling loss. With a weekly cleaning schedule, the share of incident angle losses at the KAUST test site ($\overline{DSR}$ = –0.53%) adds up to –20%.

The optimal cleaning schedule can be derived from the average daily soiling rate $\overline{DSR}$ and the unit costs for a single cleaning operation (UCC), expressed as percentage of annual energy sales. With a typical $\overline{DSR}$= –0.5% for desert sites, at least 10 cleanings per year are required to keep annual PV soiling loss better –10%.

Uncertainties: In this study, the results for annual PV soiling loss were not corrected for the incidence angle component as function of the soling ratio. In addition, soiling effects for periods longer than 7 days were not fully taken into account. For example, a higher dust removal may be observed for high soiling ratios. Such effects will affect the shape of the soiling characteristic and the parameter $\overline{DSR}$. More precise dust soiling simulations may also require consideration of seasonal variations in daily soiling rates.

REFERENCES

[1] Report IEA-PVPS T8-01:2015, Energy from the Desert: Very Large Scale PV Power Plants for Shifting to Renewable Energy Future

[2] J. Zorrilla-Casanova et al.: Analysis of dust losses in photovoltaic modules, World Renewable Energy Congress, Linköping, Sweden (2011)

[3] W. Herrmann, M. Schweiger, G. Tamizhmani, B. Shisler, Kamalaksha CS, Soiling and self-cleaning of PV modules under the weather conditions of two locations in Arizona and South-East India, IEEE-PVSC42 (2015)

[4] Report IEA-PVPS T13-21:2022, Soiling Losses Impact on the Performance of Photovoltaic Power Plants

[5] M. Gostein, J. R. Caron, B. Littmann, "Measuring soiling losses at utility-scale PV power plants, IEEE-PVSC40 (2014)

[6] IEC 61724-1, Photovoltaic system performance – Part 1: Monitoring

[7] M. Gostein, K. Passow, M. G. Deceglie and L. Micheli: Local Variability in PV Soiling Rate, IEEE-PVSC45 (2018)

[8] Report IEA-PVPS T13-09:2017, Assessment of Photovoltaic Module Failures in the Field

EVALUATION OF LOCAL CONDITIONS AND THEIR IMPACT ON BIFACIAL PV PERFORMANCE AT HIGH LATITUDE

*Alexander Granlund, Anna Malou Petersson, Josefin Sundström, Jimmy Narvesjö, Mattias Lindh
RISE Research Institutes of Sweden, Industrigatan 1, 941 38 Piteå, Sweden

*Corresponding Author: Tel. +46 10 516 61 84, alexander.granlund@ri.se

ABSTRACT: Different conditions such as module orientation, ground albedo, shading and latitude are known to affect the performance of bifacial photovoltaic modules. We evaluate bifacial performance for one year at a site located at 65°N through comparison of measured and simulated front and back side plane-of-array irradiation. Each investigated module has a different azimuth, tilt, and exposure to shading from the surroundings. Local shading is found to severely impact the energy yield of the site in general, and individual modules to a varying degree depending on their location and orientation. Proper shading analysis appears to be required in the planning phase of a bifacial photovoltaic installation to accurately calculate the expected energy yield. The bifacial gain of the modules with azimuths in the east–west sector is found to span a range from 16 % to approximately the bifaciality factor, depending on the orientation. To fully utilize the potential of bifacial photovoltaics, this variability also needs to be carefully considered when planning and building bifacial photovoltaic installations.
Keywords: Bifacial, Shading, Simulation

1 INTRODUCTION

Bifacial photovoltaic (PV) modules utilize light from both front and back sides, which can make them particularly well suited for high latitudes. In the winter, snow increases the albedo of the ground giving enhanced reflections, and during summer, both sides can receive direct irradiance due to the wide range of solar azimuths. Both front and back side performance throughout a year must be taken into consideration to maximize the energy output of PV sites with bifacial modules. However, the methodology for rating nominal power output of bifacial modules remains to be standardized across the industry, and manufacturers use different approaches. One proposed set of bifacial standard test conditions consists of a front side irradiance of 1000 W/m² (equivalent to regular monofacial standard test conditions) and an additional back side irradiance of 135 W/m², which is meant to emulate a 37°-tilted module 1 m above ground with a ground albedo of 0.21 [1], [2]. The bifaciality factor (ϕ) is a measure of how efficient the back side is relative to the front side and lies between 0 (monofacial) and 1. Multiplication of ϕ with a back side irradiance of 135 W/m² indicates that the bifacial gain of the module, i.e. the increased energy yield from back side irradiance, is a constant smaller than 13.5 %. In real-world implementations, module azimuth and tilt can vary, and local shading, front and back side irradiance, and ground albedo are all subject to change over time, which will impact the bifacial gain.

Previous real-world studies have shown that the bifacial gain of south facing modules can be at least 10–27 % at high latitudes [3], [4], and that vertical east-west bifacial modules have similar annual energy output as south-facing latitude-tilt bifacial modules [3]. Vertical east-west oriented bifacial modules represent an extreme example of bifacial gain since they can receive similar irradiance on both sides and thus feature a bifacial gain close to their bifaciality factor.

1.1 Background

The field test site is a bifacial PV installation called Solvåg, located in Piteå, Sweden, at 65°N. It features 117 modules at different azimuths and tilts, see Figure 1. This unconventional PV-installation has previously proven to be difficult to analyse due to highly varied orientations, internal shading between neighbouring modules, and uneven shading from adjacent trees to the east and west. To the south there are also some tall buildings (not shown). Previous work found that a 35° tilted and south-west oriented module had the highest measured annual energy output, but simulations using PVsyst that did not take shading into account suggested that southward azimuth and a tilt of 60–70° should maximize the annual energy yield at this site [5], [6]. The discrepancy between simulation results and measurements motivates a more in-depth study of the local surroundings and its impact on PV and bifacial performance.

Figure 1. A photograph providing an overview of the PV installation Solvåg in Piteå, Sweden.

1.2 Aim

This study aims to explore two problems using ray-tracing lighting simulations in combination with real-world measurements. First, to what extent the surroundings and its shading affect on-site PV performance for different module orientations. Second, to explore bifacial performance for high latitude conditions and put it in the context of recently proposed bifacial standard test conditions.

2 METHOD

2.1 Field measurements

The data sheet of the modules (Prism Solar, Bi60-368) at the site specifies a bifaciality factor of 0.9 and a bifacial power rating of 368 W when subject to 300 W/m^2 back side irradiance. The modules were connected to an optimizer-system (P350 and SE25K, SolarEdge), allowing for module-level performance monitoring. Three of the modules were selected based on azimuth, tilt, location, and previous performance; see Table I for module orientation and Figure 2a for the position on site. In June of 2021, these modules were fitted with reference solar cells (SI-I-420TC and SI-V-10TC, IMT Solar) to measure plane-of-array (POA) irradiance on both their front and back sides. The global horizontal irradiance (GHI) was measured on site with a heated and ventilated class A pyranometer (SMP10, Kipp & Zonen). Both POA and GHI data were logged with gateways connected to the inverter and optimizer system (SE1000-CCG-G, SolarEdge).

The period of interest for this study is one year, from July 2021–June 2022. Due to a cable breakdown in the front POA-sensor for the 60°-tilted module, irradiance data is missing from December 3rd to April 4th. The missing data is left out and, importantly, the analyses presented in this paper are largely unaffected because of the comparatively low irradiation during that time of the year.

Table I. Detailed orientation with tilt and azimuth of the selected modules equipped with POA irradiance sensors.

Tilt	[°]	35	60	90
Azimuth*	[°]	206	217	271

*0° being north, 90° east, 180° south and 270° west

Unlike the procedure for measuring bifacial gain outlined in the standard IEC 60904-1-2 [7], the POA sensors at our site were carefully mounted in a non-interfering way slightly above the modules. This was done to enable long-term measurements that minimized shading of modules and production losses. The IEC 60904-1-2 standard requires multiple measurement points on the module surface to ensure uniform irradiance for short term analyses, which is unfeasible for long term studies of a commercial PV installation. The POA irradiance measurements can therefore be expected to differ slightly from the actual module irradiance.

2.2 Simulations

Analyses of irradiance and shading was performed with ray-tracing lighting simulations using the open-source Python package bifacial radiance. It was developed by the National Renewable Energy Laboratory (NREL) [8] and is based on Radiance [9]. The simulations were based on hourly GHI, direct normal irradiance (DNI) and diffuse horizontal irradiance (DHI) data from STRÅNG, provided by the Swedish Meteorological and Hydrological Institute (SMHI) [10]. Irradiance data was modelled using a Cumulative Sky approach, where data for extended periods of time is cumulatively modelled with the Perez-model on a single hemispherical sky divided into 145 patches according to a Tregenza discretization scheme [11–13]. This makes simulations less computationally intensive by condensing multiple single hour cases into one single cumulative case, at the cost of some precision.

The simulation domain featured 3D-representations of the wooden deck and the module mounting system at the site. The general 3D-model was provided by the architect of the PV installation [14] and subsequently adjusted using Blender [15] to better represent the scale and position of the wooden deck below the modules, the surrounding trees, and nearby buildings. The resulting 3D-model was converted from a generic .obj-file to a radiance-compatible .rad-file. See Figure 2a-b to compare the site and resulting full 3D-model. In unshaded simulations, neighbouring modules and surrounding trees and buildings were removed from the 3D-model; only the wooden deck, a single PV module under study and its mounting-system were included. Figure 2c shows a rendering of such a 3D-model in snowy winter conditions.

The spectral reflectance properties of the model materials were specified either using standard materials in the bifacial radiance software, or from the Spectral Materials Database [16]. For periods where the ground was covered with snow (Nov–Apr) the ground and wooden deck had their material properties changed to represent snow. Due to the varying albedo, the simulations for a full year were divided into three periods: start of year with snow (Jan–Apr), middle of year without snow (May–Oct), and end of year with snow (Nov–Dec). The albedo differences between fresh and aged snow was not taken into account.

The position and azimuth of each PV module was measured with accurate geo-positioning and verified using maps from Lantmäteriet, the Swedish agency responsible for national land surveying. Module and mounting system tilts were measured with a digital inclinometer (GAM 270 MFL, Bosch). A small horizontal PV module was added in the same location as the on-site pyranometer to verify the simulated GHI.

2.3 Data processing

Bifacial radiance provides a user specified set of point values of irradiance at the front and back sides of a PV module. The median value of 9 vertically spaced points along the centre of the modules were used in this study. Irradiance levels can also be visualized with false colour renderings. To quantify the total *useful* irradiance (G_U) for a bifacial module, the simulated and measured back side irradiance is multiplied with the bifaciality factor:

$$G_U = G_F + \phi \times G_B, \tag{1}$$

where G_F and G_B are the front and back irradiances, respectively. The bifacial gain is calculated according to

$$BG = \frac{\phi \times G_B}{G_F}. \tag{2}$$

The measured annual energy output (kWh/year) of a module is to a first approximation proportional to the annual irradiance (kWh/m^2/year) impinging on its surface. To facilitate our analysis, the annual useful irradiance is therefore converted to a simplified annual energy (E_{irr}) by multiplying it with the area (A) and efficiency (η) of the module, according to

$$E_{irr} = G_U \times A \times \eta. \tag{3}$$

The relative deviation between measured and simulated energy can then be given by

$$RD = \left|\frac{E_{irr} - E_{mea}}{E_{mea}}\right|, \qquad (4)$$

where E_{mea} is the measured annual energy.

Figure 2. a) Photograph of Solvåg. The left red circle marks the approximate location of the selected 35° and 60° modules, and the 90° module is positioned in the right circle. b) Rendering of the full 3D-model including surroundings and modules. c) Rendering of a 3D-model for an unshaded simulation of a single module during winter; all surroundings are removed, and both the ground and the wooden deck are covered in snow.

3 RESULTS AND DISCUSSIONS

A false color rendering of annual irradiation at the site is presented in Figure 3; the PV modules were excluded to show which areas on the ground are most exposed to shading, and the albedo was set constant for the entire year. From this figure it is evident that the trees, both to the east and to the west of the installation, are expected to significantly impact the total energy yield of the site. Note that Figure 3 does not present the irradiation on the actual module faces, but that the shading conditions are adverse for the positions of many of the east and west oriented modules.

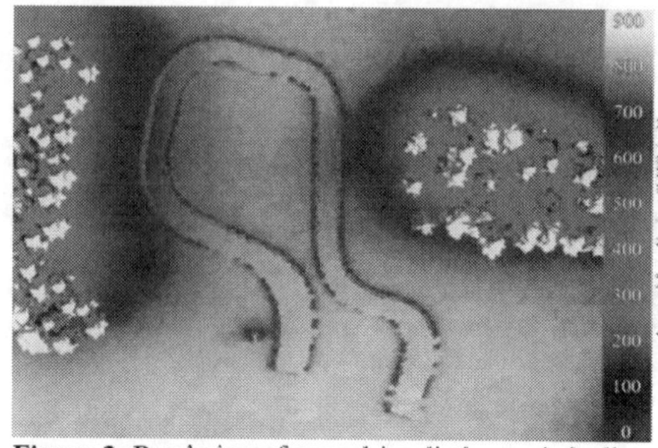

Figure 3. Rendering of annual irradiation and shading at the site without modules and mounting systems.

3.1 Simulated irradiance

The cumulative sky of the whole year for the site is presented in Figure 4 as reference for further discussions. For the actual simulations, shorter time periods were used, as outlined in the Section 2.1. Note that east/west is mirrored to what is typical for cardinal directions due to the ground-to-sky perspective.

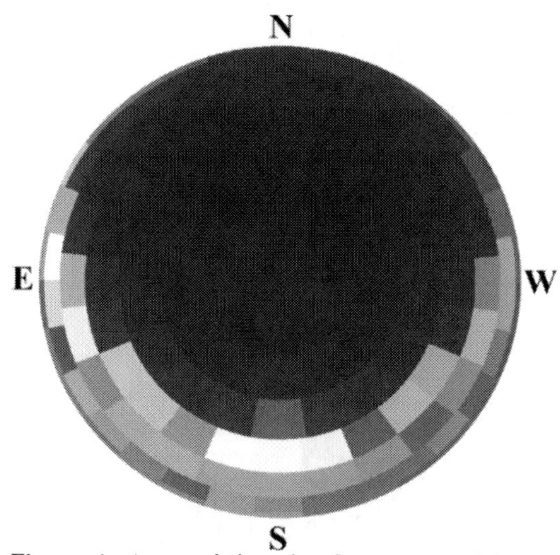

Figure 4. A cumulative sky for a year of hourly irradiance data. Bright areas indicate where the sun is most often positioned throughout the year and where radiance is high. The centre of the circle corresponds to the zenith and the edges to the horizon.

A comparison between GHI from measurements, STRÅNG-data, and simulated irradiance at the pyranometer position are presented in Table II.

The overall correspondence between measurements and simulation results are decent at the annual level, with differences < 4 % and based on this we deem the accuracy of the simulations in to be sufficient for our purpose. There are small deviations between the STRÅNG data (simulation input) and unshaded simulations (output), which ideally should be equal. This indicates that the processing of irradiance data into a sky model and back to irradiance introduces errors. A seasonal trend appears where the simulations overestimate the irradiation during the Nov-Dec period and underestimate it during the May-Oct period. We tentatively attribute these discrepancies to the limited spatial discretization resolution of the cumulative sky, shown in Figure 4.

Table II. On-site measured GHI, model data from STRÅNG (used as input to the bifacial radiance simulations) and the simulated irradiation at the pyranometer position with and without shading, for different periods of the year.

Period	Annual GHI [kWh/m²]			
	Measured	STRÅNG	Shaded	Unshaded
Jan-Apr	199	185	198	204
May-Oct	711	707	679	703
Nov-Dec	0.08	0.04	5	6
Total	910	892	882	913

A low solar elevation during the winter could be represented by a patch covering a larger elevation. Similarly, the low resolution could lead to an underestimation of the solar elevation during the summer.

The module irradiance from measurements and simulations of three selected modules equipped with POA sensors are presented in Table III.

Table III. The annual cumulative front, back and useful irradiance from the three different cases and the corresponding calculated bifacial gain (*BG*).

Case	Tilt	G_F	G_B	G_U	*BG*
	°	kWh/m²			%
Measured	35	1075	281	1328	24
	60	838*	324	1130*	35*
	90	689	568	1200	74
Shaded	35	969	212	1160	20
	60	886	288	1145	29
	90	485	506	940	94
Unshaded	35	1120	364	1448	29
	60	1068	482	1502	41
	90	659	853	1427	117

*Front sensor data missing from December 3rd until April 4th.

The shaded simulated front side irradiance was typically lower than the measured (except for the 60° module with data loss). We attribute this to both to the position of the POA sensors above the simulated modules, which reduces the amount of shading both from far-away objects and neighbouring modules, and to the evaluation method of distributed irradiance on the comparably large modules in the simulations. It is intriguing how the bifacial gain of the vertical west-facing module for the shaded and unshaded simulations reach 94 and 117 %, respectively, and thus exceed both the measured bifacial gain at 74 % and the bifaciality factor of 0.9. Assuming similar morning and afternoon irradiance, the front and back sides of the module should get the same average irradiation. Reassuringly, Figure 4 shows that the eastern part of the sky is slightly brighter during the analyzed year, which could explain the high bifacial gain. Presumably, this could in turn be caused by typical differences in morning and afternoon sky conditions. We expect the measured and the shaded simulation bifacial gain to be smaller than that of the unshaded simulation given the larger distance from the vertical module to the trees to the west than to the east. The trend in the data is in line with this argument although the bifacial gain of the shaded simulation does not fully reach that of the measured. Possibly this stems from the difficulty of properly capture non-uniform shading from the trees, or slight misplacements of the trees in the 3D-model.

While the simulated irradiance is largely in line with measurements and expectations, there is room for improvement. Higher spatial resolution of the cumulative sky would more accurately position the sun on the sky and lead to higher simulation accuracy at large, especially through improved shading analysis in locations where this is prominent. However, since the position of the sun also depends on time, a cumulative sky model with high spatial resolution requires input irradiance data with correspondingly high time resolution to yield credible results. Computationally heavy, temporally resolved, simulations (without a cumulative sky) could also be performed for comparison.

3.2 Local performance

The individual module performance in terms of measured energy output is presented in Figure 5a. North facing modules are uncommon and inherently inefficient in the northern hemisphere, which holds true also at this site. Modules oriented to the east and west produced approximately 240–285 kWh per year and appear comparatively insensitive to the tilt. The highest energy yields are found for the southwest-facing modules with tilt between 10 and 60°. The measured energy output of the 35°, 60° and 90° modules were 332, 309 and 238 kWh, respectively.

In Figure 5b-c we present the relative deviation in percent between the measured energy and the simulated simplified annual energy (E_{irr}, defined in Eq. 3) for the shaded and unshaded cases; note the different scales of the color bar. The shaded simulation outperforms the unshaded case for all modules at the site. Adding shading from 3D-modelled surroundings improves the average relative deviation at this site from 46 % to 10 %, and seemingly eliminate the baseline deviation present in the unshaded case (the smallest relative deviations are about 15 % for the unshaded simulations but close to 0 % for the shaded simulations). The modules with highest measured energy output feature the smallest deviations, whereas the largest deviations are seen for the modules facing north and for high tilts. As expected, the improvement is biggest for high tilt modules in the parts of the site that are most exposed to shading, which is close to the trees. For example, the relative energy deviation for a vertical east-facing module located next to the eastern trees is reduced by 71 % units, from 82 % to 11 %.

Clearly the modules at the site are heavily affected by shading, but more importantly, these results point to that comparatively accurate estimates of energy yield for bifacial PV modules can be inferred from general irradiance data and a detailed 3D-model using open-source software, also for quite demanding shading and irradiance conditions.

The 25 best performing modules from each case, as well as their position on site, is presented in Figure 6. The best performing modules from the measured and shaded simulation cases are located within the same sections of the site but do not overlap completely. These sections also seem to suffer the least from shading in Figure 3, apart from a few east-oriented modules on the western part of the site. The simulations tend to overestimate the performance of high tilt and/or east-west oriented modules compared to the field measurements. For simulations without shading, the overestimation of high tilt east-west oriented modules increases, as shown in Figure 5c.

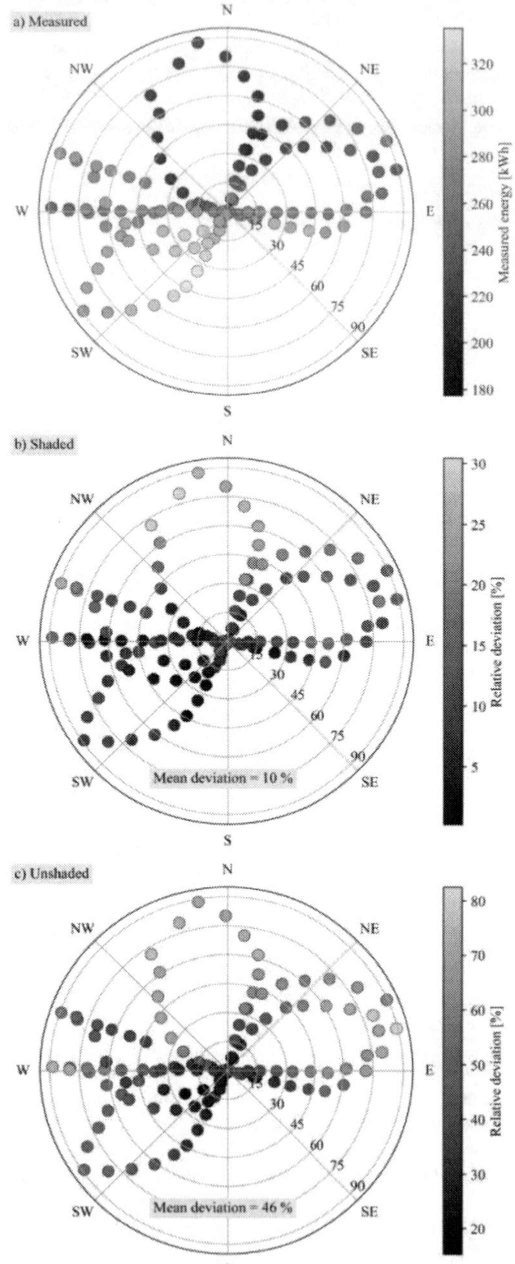

Figure 5. Polar plots of each module as a function of azimuth and tilt (radius). a) Measured annual energy. b-c) The modulus of the relative deviation of the simulated annual energy, E_{irr}, for the shaded and unshaded simulations compared to measured energy output. Only the shaded simulations had negative deviations, all > -5 %. Note the different scales in the color bar.

The unshaded simulations are accordingly unable match the position of the 25 best performing modules to the section where the shaded and measured cases reside, see Figure 6. Sites with less shading are needed for high tilt east-west modules perform at their potential, as implied from Figure 5c and 6.

Both simulations correctly position some of the best 25 modules in the south/southwest/west facing section of modules near the bottom left part of Figure 6. The gap of two modules in the middle of this section, where only unshaded results are represented, indicates that they probably are primarily shaded by the neighbouring modules.

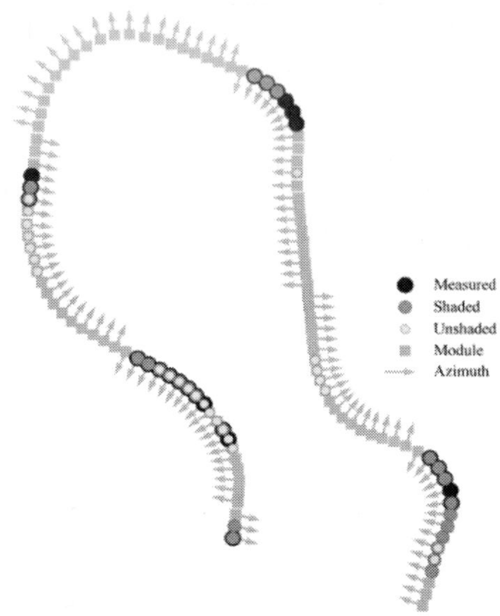

Figure 6. The position of the 25 best performing modules from each case and the azimuth of all modules.

4.3 Bifacial Performance

The bifacial gain from the shaded simulations is presented as a heatmap in Figure 7. The bins contain up to four modules of the same tilt (specified ±2.5°) and azimuth (specified ±22.5°), but sometimes different shading conditions. For bins with multiple modules, the minimum value of the bifacial gain is presented, as a conservative measure.

The bifacial gain of the east-facing modules tends to be lower than for the west-facing modules. As discussed in Section 3.1 this can be an effect of a higher irradiance in the morning compared to the afternoon, it can be an artefact caused by the low-resolution cumulative sky model, or a combination thereof.

The proposed standard for bifacial standard test conditions for power rating indicates a constant bifacial gain of maximum 13.5 %. The datasheet of the modules at Solvåg on the other hand specifies a back side irradiance of 300 W/m², which multiplied with the bifaciality factor would give a constant bifacial gain of 27 %. This difference alone is an example that highlights the need for a standard that harmonizes test conditions for bifacial PV power rating across the industry, and makes it easier to compare different bifacial modules. However, the vast range of bifacial gains presented for different orientations in Figure 7 emphasize that the bifacial gain cannot be expected to be constant, but rather to vary significantly depending on module orientation and local surroundings that cast shade. The orientation distribution in Figure 7 is unfortunately incomplete, with the south to south-east sectors being near empty. Nevertheless, it is notable that the *smallest* simulated bifacial gain we attain at this site is 16 %.

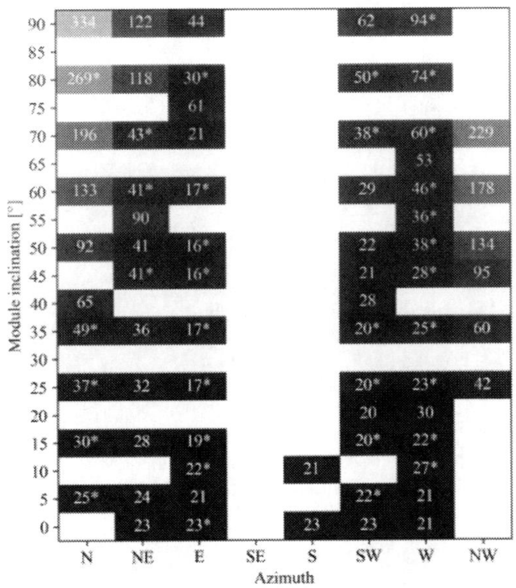

Figure 7. Bifacial gain in % from the shaded simulations as a function of module tilt and azimuth. A * signifies that it is the minimum value of multiple modules.

A large bifacial gain strengthens the case for employing bifacial instead of monofacial PV technology. But the facilitation of well performing future bifacial PV installations does not only need standardized bifacial test conditions—because the bifacial gain is so strongly dependent on the conditions and circumstances easy-to-use tools should be made accessible that can adapt the standardized rated bifacial module power to an expected power and energy production at any specified site, module orientation, latitude, or climate.

5 CONCLUSIONS

Shading causes significant general production losses at the Solvåg site. The shading impact on individual modules varies with position and proximity to shading objects but also with the module orientation. Vertical east-west facing bifacial modules are most severely affected, which is a consequence of the PV-array layout and the high latitude with low average solar elevation. We find that accurate and reliable predictions of the energy yield of bifacial PV at a site with complicated irradiation conditions can be acquired if shading is properly considered.

The bifacial gain in the azimuth sector east-to-west was found to range from 16 % up towards the bifaciality factor of the modules. This testifies to that real-world bifacial gain can deviate significantly from what currently proposed bifacial standard test conditions imply.

6 ACKNOWLEDGMENTS

This study was performed as part of the Arctic Solar 2.0 project, co-financed by the European Regional Development Fund, Region Norrbotten, Piteå Kommun, and PiteEnergi. We would like to thank Warm in the Winter for providing the original 3D-model of the site.

7 REFERENCES

[1] J. Stein, C. Reise, G. Friesen, G. Maugeri, E. Urrejola, and S. Ranta, *"Bifacial Photovoltaic Modules and Systems: Experience and Results from International Research and Pilot Applications.,"* SAND2021-4835R, 1779379, 695675, Apr. 2021. doi: 10.2172/1779379.

[2] C. Deline, S. MacAlpine, B. Marion, F. Toor, A. Asgharzadeh Shishavan, and J. Stein, *"Assessment of Bifacial Photovoltaic Module Power Rating Methodologies—Inside and Out,"* IEEE Journal of Photovoltaics, vol. PP, pp. 1–6, Jan. 2017, doi: 10.1109/JPHOTOV.2017.2650565.

[3] C. Pike, E. Whitney, M. Wilber, and J. S. Stein, *"Field Performance of South-Facing and East-West Facing Bifacial Modules in the Arctic,"* Energies, vol. 14, no. 4, Art. no. 4, Jan. 2021, doi: 10.3390/en14041210.

[4] J. S. Stein, D. Riley, M. Lave, C. Hansen, C. Deline, and F. Toor, *"Outdoor Field Performance from Bifacial Photovoltaic Modules and Systems,"* in 2017 IEEE 44th Photovoltaic Specialist Conference (PVSC), Jun. 2017, pp. 3184–3189. doi: 10.1109/PVSC.2017.8366042.

[5] M. Lindh, M. Svedjeholm, A. Granlund, J. Petersson, and A. M. Petersson, *"Handbok för nordlig solel."* RISE Energy Technology Center, 2020.

[6] A. Granlund, J. Narvesjö, and A. M. Petersson, *"The Influence of Module Tilt on Snow Shadowing of Frameless Bifacial Modules,"* presented at the 36th European Photovoltaic Solar Energy Conference and Exhibition, 2019. doi: 10.4229/EUPVSEC20192019-5CV.4.36.

[7] International Electrotechnical Commission, *"IEC TS 60904-1-2: Measurement of current-voltage characteristics of bifacial photovoltaic (PV) devices,"* vol. 2019.

[8] S. Ayala Pelaez and C. Deline, *"bifacial_radiance: a python package for modeling bifacial solar photovoltaic systems,"* Journal of Open Source Software, vol. 5, p. 1865, Jun. 2020, doi: 10.21105/joss.01865.

[9] G. Ward Larson and R. Shakespeare, *"Rendering with Radiance: The Art and Science of Lighting Visualization."* San Fransisco: Morgan Kaufmann Publishers, Inc., 1998.

[10] Swedish Meteorological and Hydrological Institute, Swedish Radiation Safety Authority, and Swedish Environmental Agency, *"STRÅNG - a mesoscale model for solar radiation."* http://strang.smhi.se/

[11] D. Robinson and A. Stone, "Irradiation modelling made simple: the cumulative sky approach and its applications," p. 5, 2004.

[12] R. Perez, R. Seals, and J. Michalsky, *"All-weather model for sky luminance distribution—Preliminary configuration and validation,"* Solar Energy, vol. 50, no. 3, pp. 235–245, Mar. 1993, doi: 10.1016/0038-092X(93)90017-I.

[13] P. R. Tregenza, *"Subdivision of the sky hemisphere for luminance measurements,"* Lighting Research & Technology, vol. 19, no. 1, pp. 13–14, Mar. 1987, doi: 10.1177/096032718701900103.

[14] Warm in the Winter, *"Warm in the Winter - Arkitektkontoret med fokus på stadsbyggnad."* https://www.warminthewinter.se/

[15] Blender Online Community, *"Blender - a 3D modelling and rendering Package."* Blender Foundation. [Online]. Available: http://www.blender.org

[16] J. A. Jakubiec, *"Data-Driven Selection of Typical Opaque Material Reflectances for Lighting Simulation,"* LEUKOS, pp. 1–14, Aug. 2022, doi: 10.1080/15502724.2022.2100788.

OPTIMAL DESIGN AND INSTALLATION OF SOLAR HOME SYSTEMS FOR BOTTOM-UP GRIDS: COMPARATIVE CASE STUDIES FROM KENYA AND NORWAY

Ida Fuchs*, Anja Myhre Waitz, Kari Thorset Lervik, Steve Völler, Ole-Morten Midtgård,
Jayaprakash Rajasekharan
*ida.fuchs@ntnu.no, +47 73559775
Norwegian University of Science and Technology / Department of Electric Power Engineering
Trondheim, Norway

ABSTRACT: In rural electrification and off-grid areas a new concept of providing energy access arises: Swarm electrification. The concept is based on energy sharing between individual solar homes systems (SHS). The aim of this study is to analyze how different designs of SHS and different azimuth & tilt angles of PV panels influence the energy sharing with the goal of enabling swarm electrification. The paper both looks at this effect for a case close to the equator and a different case in the northern hemisphere. The two case studies are from Kenya and from Norway. Different types of sharing scenarios were analyzed for two households in each case study. The findings show that the benefit of energy sharing highly depends on the sharing scenarios. Sharing only surplus energy showed very little potential of improvement. Sharing both surplus energy and battery capacity can improve the household with the initially smaller installed PV capacity but decreases the served load ratio for the household with the initially higher PV capacity. The results show that a deviation from the PV azimuth angle with maximum yield, to get a more diverse PV generation (i.e., all panels South oriented vs East or West), show no significant benefits in the sharing scenarios in the simulated cases.
Keywords: Stand-alone PV Systems, Energy Sharing, Rural Electrification, Developing Countries

1 INTRODUCTION

Decentralization and digitalization have provided bottom-up grids a momentum to prosper. Electricity generation for individual households has never been so attainable, as with nowadays solar home systems (SHS). A SHS consists of solar PV panels, battery units and power electronics. These systems have conquered off-grid areas and provided electricity where national and regional electrification attempts failed, mainly due to large distances or high costs. This led to a new emerging concept that focuses on organically grown bottom-up grids called "Swarm electrification", that is based on energy sharing between households in a community [1]. Swarm electrification describes a step-by-step electrification process with four phases:

- Phase 1 – Stand-alone SHS
- Phase 2 – Interconnected SHS
- Phase 3 – SHS connected to local grids
- Phase 4 – Local grids connected to national grids

The design and installation criteria of SHS that enable energy sharing and thereby foster swarm electrification, is subject of this study. When planning SHS in existing residential buildings, the choice where to set up the system is often limited to the existing roof or facade. In many cases the azimuth and tilt of the roof is a given parameter. If the choice is free, the common strategy is to maximize the yield of the PV system. This means choosing the optimal azimuth and tilt for a given location. Lately, new approaches have been published where the aim is to optimize the SHS in a way to maximize self-consumption rather than the yield, resulting in modified azimuth & tilt angles [2], [3]. In this study, the focus is on off-grid SHS and analyze different types of energy sharing between two residential households under different azimuth and tilt constellations.

The research question is: How do design (PV and battery sizes) and installation (PV orientation, system interconnection) of SHS influence energy sharing and thus the growing of a bottom-up grid?

The rest of the paper is organized in the following way: Section 2 provides background information and key data on the case studies. Section 3 explains the analyzed scenarios for the simulation. Section 4 introduces the method and tools used for modelling and simulation. Section 5 presents the results and finally the conclusions are made in section 6.

2 CASE STUDIES

The optimal azimuth and tilt for a PV system is mostly dependent on its geographical location, but also on local conditions like shading from buildings & trees or snow coverage. For instance, locations close to the equator have a rather flat optimal tilt, while locations far from equator have a higher tilt. Therefore, in Northern regions the most beneficial installation of PV panels might be on the façade. Additionally, at locations far from equator the seasonality (summer/winter) will have a larger impact on the SHS design and performance.

To cover these various aspects two entirely different case studies are chosen in this study: One case close to the equator and another case in the northern hemisphere rather far from the equator. The two case studies are from Kenya and from Norway.

2.1 Background Kenyan case

The progress on the UN Sustainable Development Goal (SDG) 7 – Affordable and Clean Energy – has been measured continuously. The latest tracking in the "Energy Progress Report 2020" [4] states that the 2030 goal for SDG 7 will not be achieved, if the progress continues at the current pace. Clearly, there is a need for new knowledge, tools, and methods to accelerate clean and affordable energy access. One solution might be PV and battery technologies, as they play a major role in electrification of emerging and developing countries. The fact that PV technology can be scaled down to the smallest possible system, makes it superior compared to other electricity generation technologies in many cases of the rural electrification framework. A recent report by the

Norwegian Solar Energy Cluster shows that decentral PV and battery systems are both faster commissioned and cheaper in investment & operation, compared to ordinary grid extensions [5]. As a consequence, individual SHS have had a rapid success, and installation rates are increasing in rural electrification. Therefore, a rural village in East-Africa in the county of Kilifi, Kenya, was chosen as a case study. It is one of the poorest counties in the country, with 70 % of the population living under the poverty line and having a high unemployment rate. A primary school called Eco Moyo Education Center is the case studies main subject. Eco Moyo is a charity project that provides education to reduce poverty in line with the SDGs [6]. The school wants to keep a green & renewable profile and wants to preserve the surrounding environment as much as possible. The school area consists of several individual buildings with different purposes, such as classroom buildings, learning lab, teachers housing, visitor huts, etc. This makes the case ideal for the purpose of this research study. For this study, two of the residential buildings at the school compound were chosen as subjects for the analyses. The input data for the modelling of the load was retrieved during a field visit in February and March 2022 [7].

2.2 Background Norwegian case

Since the first case study is located very close to the equator, a second case far away from the equator was chosen for comparison reasons. The effect of azimuth and tilt on the yield of a PV system is highly depended on the latitude. Local energy communities (LEC) have recently got more attention in Norway. Distributed and local energy generation is one of the main characteristics of LECs [8]. However, LECs are often connected to an existing distribution grid, and the local energy generation is additional. In [9] a list of different LEC projects across Europe is given, which also includes two projects in Norway. Energy sharing is subject of interest in such LECs. For reasons of comparability, the Norwegian case study will have similar characteristics (appliances, load, system design) as the case study in Kenya, i.e. an off-grid setting is chosen, which is not the most common form of LECs.

Norway has a large amount of leisure cottages in the mountains, at the coast or often remote areas. These cottages have become subject of discussion in terms of sustainability, since the standard in the cottages has increased from simple wooden cottages with a fireplace to full equipped modern houses, including electricity supply the whole year, even when not in use (heating and stand-by appliances). To meet recent attempts to turn around the trend and lower the footprint of such cottages, one idea is to use only off-grid SHS for necessary appliances (excluding heating), and then maximize the access to electricity by energy sharing and bottom-up grids. This would reduce the costs of electrifying remote cottage areas and avoid unnecessary electric grid infrastructure in the nature. Therefore, the case of two off-grid cottages with SHS was chosen and the possibility of energy sharing between the cottages is analyzed. The different azimuth scenarios are studied and the influence on the profitability of energy sharing is analyzed.

2.3 Key data for both cases

In both cases the electricity demand has many similarities, but also some major differences. Table I and II summarizes the framework and key data for the electricity demand. In both cases the same appliances are modelled, and the input data (appliance rating, time of use, ect.) can be retrieved from [7]. However, the Norwegian mountain cottages are only used on weekends (including Fridays), while the Kenyan rural residential house is used as permanent residence. This results in a similar daily demand (considering only the active days for the Norwegian case), but a very different annual demand.

Table I: Framework for demand modelling

	Usage	Appliances
Kenyan rural residential house	all days whole year	lights, phones, laptops, TV, radio, speaker, fridge
Norwegian mountain cottage	random weekends	lights, phones, laptops, TV, radio, speaker, fridge

Table II: Key data for demand modelling

	Annual demand [kWh/a]	Daily demand [kWh/d]
Kenyan rural residential house	1000	2.75
Norwegian mountain cottage	360	2.3 (active days)

Due to the very different location of the two case studies, the typical and optimal angles for azimuth and tilt are quite different. For the Kenyan case, the typical tilt is 10° and rather flat. Panels are faced towards North since the location is South of the equator. However, with such a flat tilt the azimuth of the panels has a low influence on the PV yield. For the Norwegian mountain cottage case, the situation is different in off-grid SHS settings. The panels are often mounted vertical at the façade with a tilt of 90° and the optimal azimuth is facing South. There are two reasons for that: Firstly, to achieve an increased production during the time of low sun in the winter half year, and secondly, to avoid snow cover. The disadvantage is a lower production during summer, especially midday when the sun stands high. However, East/West constellations can have benefits especially during summer months, as the morning and evening production follows the typical morning and evening demand, avoiding mid-day overproduction. Table III summarizes the used angles for the SHS, additional to the PV size as peak power capacity and battery size as energy storage capacity. The maximum charging and discharging power for the battery is set to 50 % of the energy storage capacity. However, mind that the batteries in simulated SHS are quite large compared to the PV size, which means that the maximum charging capacity is seldom limiting the charging, but rather the maximum PV generation.

Table III: Key data for PV yield calculation

	PV sizes [kW]	PV tilt	PV azi-muth	PV moun-ting site	Battery sizes [kWh]
Kenyan rural residential house	0.15-0.9	10°	North, West/ East	Roof	1-3
Norwegian mountain cottage	0.15-0.9	90°	South, West/ East	Facade	1-3

Figure 1 shows the technical setup for the modelled SHS. The inverter efficiency is set to 90 %. The roundtrip efficiency for charging is set to 90 %.

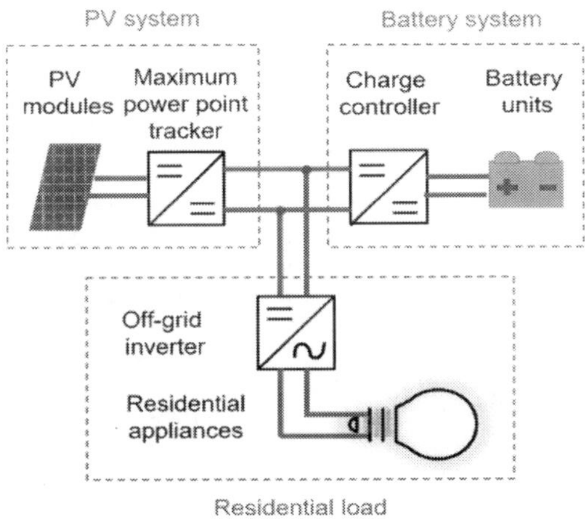

Figure 1: Technical setup for the simulated SHS.

3 SCENARIOS

To analyze how different PV panel orientations influence energy sharing for two households in Kenya and two cottages in Norway, this study looks at two scenarios. One scenario is where optimal yield at the given location is assumed, the second scenario is a constellation where one household faces to the West and the other household faces to the East. Reference cases with no energy sharing are simulated for comparison and to evaluate the improvement. Figure 2 visualizes the simulated scenarios with two households (HH1 and HH2). These scenarios are applied at both the Kenyan and Norwegian case.

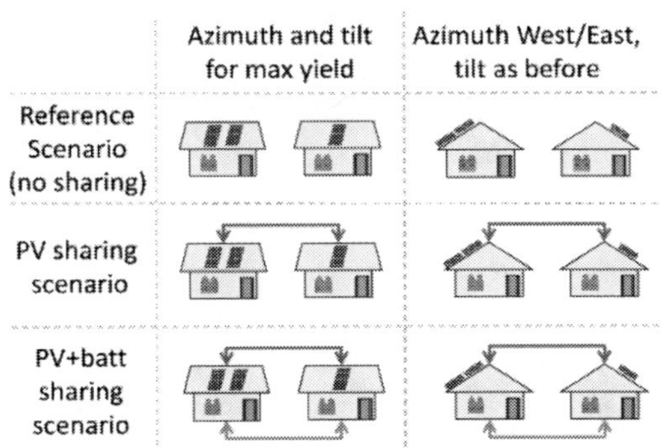

Figure 2: Simulated scenarios for HH1 and HH2 with different types of sharing and PV azimuth.

4 METHOD

The modelling and simulations are done in Python. The time series have a resolution of 15 minutes. A total year was simulated to capture seasonal effects. Several open-source tools were used for the study:

- **RAMP** – bottom-up stochastic model for generating multi-energy load profiles [10]
- **PVLib/PVGIS** – tool for simulating the performance of PV systems [11]
- **Prosumpy** – toolkit for the simulation of self-consumption in SHS [12]

The flow chart (see Figure 3) shows the structure of the simulations. The method applies both for the Kenyan and Norwegian case and each case is run separately.

Firstly, the stochastic load profile is computed with RAMP based on input data from field work. Secondly, the PV generation is calculated with the help of PVLib. The tool uses PVGIS to access meteorological data for a given location. For this study, the data from 2015 is used. Thirdly, the dispatch of the SHS is calculated for one year with Prosumpy, for various sizes of installed PV and battery capacity. Here, the installed PV capacity for HH1 is twice as high as for HH2 for each of the simulations, while the battery capacity is equal in both households.

During the dispatch of the reference system, the surplus energies for the base cases are calculated for HH1 and HH2. For the PV sharing scenario a new dispatch is calculated for HH1, adding the surplus energy from HH2 to HH1 if available. In the same way, the dispatch adds surplus energy from HH1 to HH2. The method gives an estimate of improvement potential which is at least as high as calculated, but could be higher, if a more advance method is used. However, for the given cases the method delivers good enough results in accuracy.

For the PV+batt sharing scenario it is assumed that the household share their systems as if it was one system. In that scenario PV sizes and battery sizes are added into one system, and the total demand consists of both the demand of HH1 and HH2.

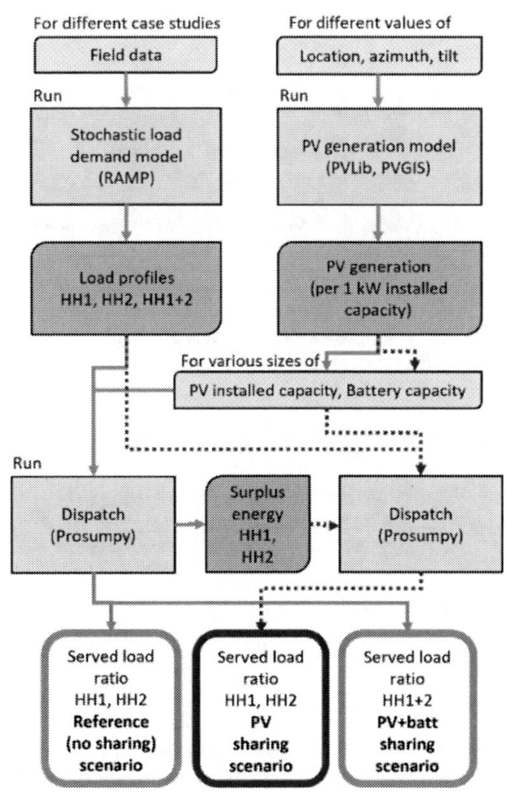

Figure 3: Flow chart of methodology and simulation.

To evaluate the results, the served load ratio (SLR) is calculated for each scenario, which is served load over a given time period divided by total demand over the same given time period. The time period in the simulated cases is one year. The served load is defined by the fraction of the load that can be supplied by the SHS. The remaining fraction that cannot be supplied, is usually called unmet load.

The SLR is calculated for the reference cases and for the PV sharing scenario for each household. For the PV+batt sharing scenario the SLR is calculated for the total system including both households. Since the households have similar demands, which only vary stochastically, it can be assumed that this total SLR is the same as the individual households would experience.

5 RESULTS

In this section of the paper the results from the simulations with the method introduced in section 4 are presented and discussed, for both the Kenyan and the Norwegian case introduced in section 2 and the scenarios introduced in section 3.

Figure 4 shows the simulated demand for both cases and for the two households, HH1 and HH2. Both households in both cases have similar demands. They vary only stochastically. This was chosen to minimize the impact of different demand on energy sharing, and thus better examine the effect of the system design and different angles. Further, the figures show clearly that the demand in the Kenyan case is continuous for all days, while in the Norwegian case one can see the 52 weekend demand peaks vs. no demand during weekdays.

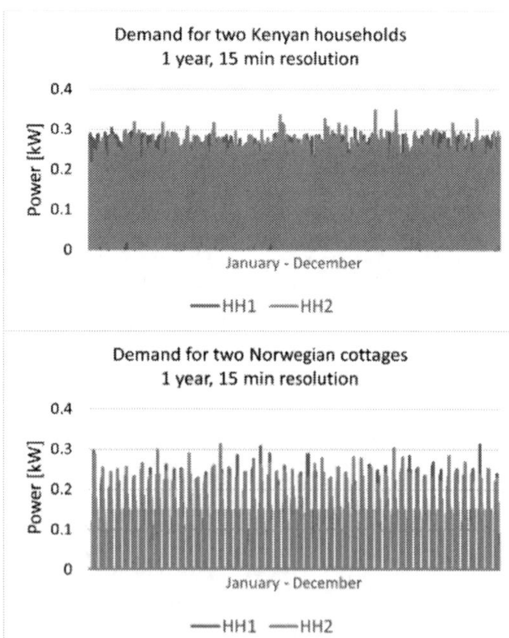

Figure 4: Demand for the Kenyan and Norwegian case for one year.

Figure 5 shows the PV generation profiles for a whole year for the Kenyan and Norwegian case for 1 kW installed PV capacity. For the Kenyan case, both the North, East and West azimuth direction of the panel is plotted. Since the tilt of the panels is only 10°, the influence of azimuth is small. During months May-August the North direction is slightly better, while East or West is better from November to February. In between those months the PV generation is the same for both. Table IV shows the PV yield for 1 kW installed capacity for the different angles for the simulated cases.

For the Norwegian case, where seasonality is much stronger and the tilt is 90°, the situation is very different (see Figure 5). The PV generation of the South direction is highest during spring and autumn months. During winter there are only few sun hours, and the sun stands very low. The low position is good for the 90° tilt; however, the low sun hours result in a low generation. During summer there are a high number of sun hours, however during midday the sun stands too high which is not optimal for a vertical panel. This leads to a lower generation in summer. For a West or East direction of the panel, the overall production during winter gets worse, but slightly better during summer. However, those directions have a lower yield in total compared to the South direction (see Table IV).

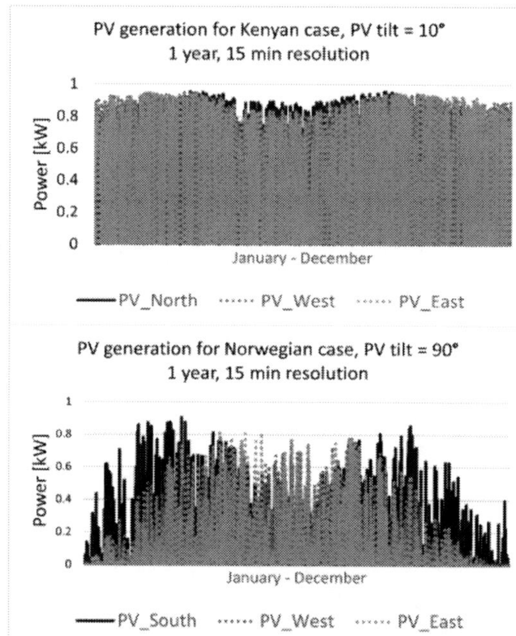

Figure 5: PV generation for 1 kW installed capacity for both Kenyan and Norwegian case for one year.

Table IV: Annual PV yield for 1 kW installed capacity for all simulated angles

	PV annual yield [kWh] (North, South)	PV annual yield [kWh] (West)	PV annual yield [kWh] (East)
Kenyan case	1948	1982	1899
Norwegian case	782	500	495

Figure 6 shows the power and energy flows for 5 typical days in January for the Kenyan reference case. It can be seen that HH1 meets its load and has surplus energy, while HH2 has unmet load each day during evening and night. The surplus energy of HH1 could charge the battery of HH2 and thus prolong the time until the battery gets empty. HH2 has no surplus energy during these days, that could benefit HH1.

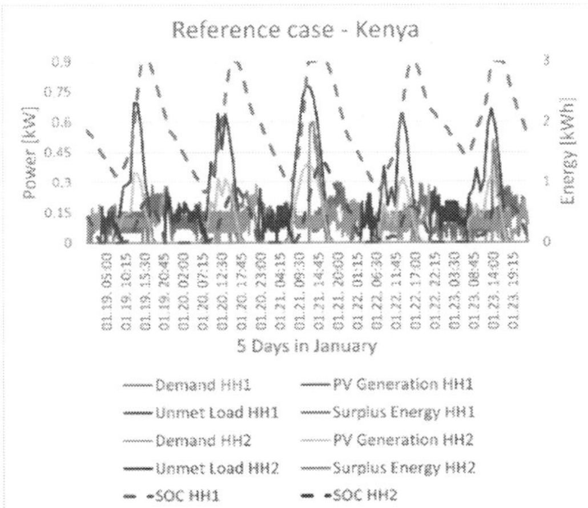

Figure 6: Power and energy flows for 5 days for reference case Kenya. The battery's state of charge (SOC) is plotted on the right axis, the rest is on the left axis.

Figure 7 shows the power and energy flows for the Norwegian reference case for 5 days that include a weekend, meaning the plot shows Thursday to Monday, while the active weekend days are Friday to Sunday. It can be seen that cottage HH2 empties its battery during early morning hours on Sunday, the third cottage day, while HH1 still has enough energy. Additionally, HH1 has surplus energy on the first cottage day, Friday. This energy could be used to fill the battery of HH2, such that it might last longer on the third day. It can also be seen that both cottages have surplus during the day before cottage active days. However, both batteries are full then, so any sharing will not benefit the other part on that day.

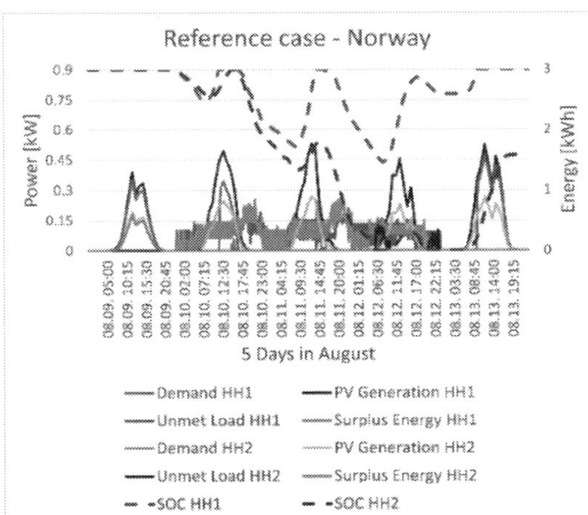

Figure 7: Power and energy flows for 5 days for reference case Norway. The battery's state of charge (SOC) is plotted on the right axis, the rest is on the left axis.

Figure 8 shows the SLR for the Kenyan case for the different PV-battery size combinations (e.g. "0.3/1" stands for 300 W solar power and 1 kWh battery capacity). The upper two plots are HH1 and HH2 for the reference case and for both the sharing scenarios with an azimuth to the North. In case of surplus energy sharing (PV sharing scenario) HH1 sees no difference, but HH2 improves in the cases with high PV size and available battery capacity. In the case of both PV and battery sharing, the SLR for HH1 decreases while it increases for HH2. In the scenario where HH1 has an azimuth to the West and HH2 to the East, there is no significant difference to the sharing scenarios compared to the North direction. The difference lays in the range of 1-2 percentage points.

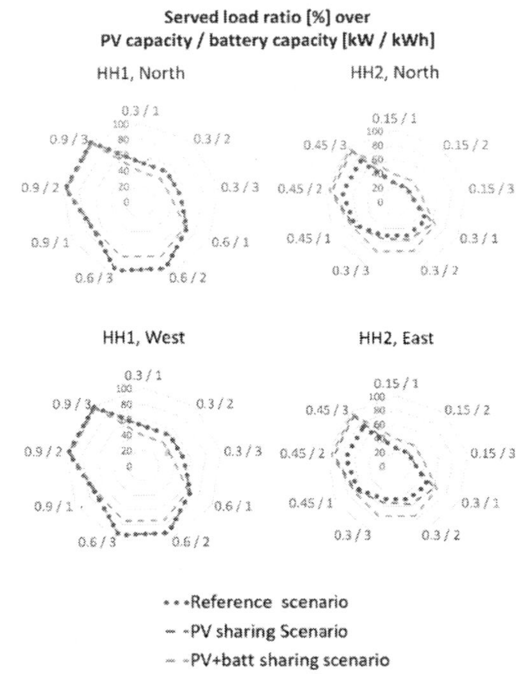

Figure 8: SLR in [%] plotted over the PV and battery size combinations for the Kenyan case for both PV angle scenarios.

Figure 9 shows the SLR for the Norwegian case. The upper two plots are HH1 and HH2 for the reference case and for both sharing scenarios with an azimuth to the South. The tilt is 90°, mounted vertical to the façade, which means the different azimuth has a larger effect in the Norwegian case than in the Kenyan case. Additionally, the different number of sun hours and position of the sun in the Norwegian case affect the results for the scenario with the panel constellation of West for HH1 and East for HH2. There is a large difference between the total PV generation between South constellation and East or West generation, which highly influences the amount of available energy for the cottages. The direction to the South gives the best results with the highest SLR for both cottages. The sharing scenarios behave similar as in the Kenyan case, as the PV sharing improves the SLR for HH2 only slightly, while the PV+batt sharing scenario improves it for HH2 but decreases the SLR for HH1. The main reason for that is that both demands are similar in size and do only have stochastic differences. If the loads had larger differences, for example the cottages would be used on different weekends, the positive effect of both sharing scenarios would be larger.

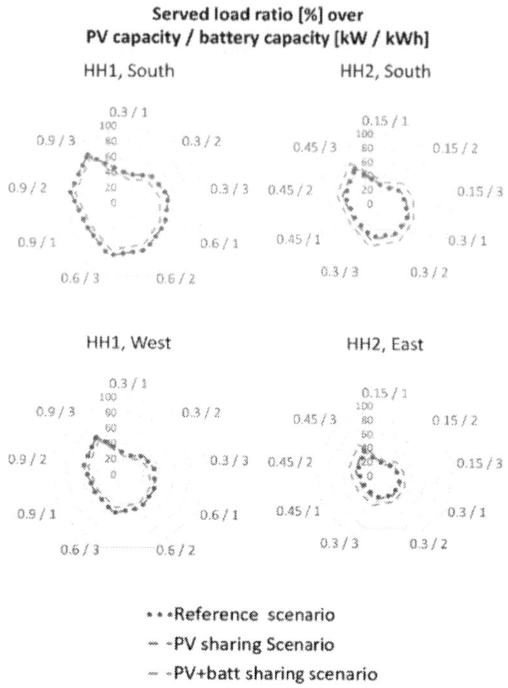

Figure 9: SLR in [%] plotted over the PV and battery size combinations for the Norwegian case for both PV angle scenarios.

6 CONCLUSIONS

Different PV and battery sizes, different azimuth and different types of sharing scenarios were analyzed for off-grid SHS in both Kenya and Norway. The following conclusions can be made:

- The PV sharing scenario has lower potential for improvement in both analyzed cases and for both azimuth constellations, than the PV+batt sharing scenario.
- The PV+batt sharing scenario can improve the SLR of household HH2 (with the lower initially PV size) with up to around 20 percentage points. However, at the same time, it can decrease the SLR for household HH1 with up to 10 percentage points.
- The different azimuth constellations have no significant influence on the energy sharing behavior in the simulated cases. The potential for improvement is slightly higher for the West/East scenario since the households start with a lower initially SLR.
- In the Kenyan case the total energy yield from the SHS varies little between the different azimuth scenarios, while in the Norwegian case the variation is significant. Even with energy sharing in the West/East azimuth scenario in the Norwegian case, the cottages do not reach the SLR level that they would have without energy sharing in the South azimuth scenario.

Based on the two case studies' results and the above made conclusions it is possible to evaluate which constellations of energy sharing and azimuth directions are more advantageous, and which are less beneficial for the involved participants.

Further work will look into how different and flexible loads could increase the sharing potential, how sharing between more than two households influences the potential of improvement, and how different modelling approaches influence the accuracy of the results.

7 REFERENCES

[1] S. Groh, D. Philipp, B. E. Lasch, and H. Kirchhoff, "Swarm electrification - suggesting a paradigm change through building microgrids bottom-up", Proceedings of 2014 3rd International Conference on the Developments in Renewable Energy Technology, ICDRET 2014, pp. 1–2, 2014.

[2] S. Wendlandt et al., "The Energy System Defined by PV: Balancing Generation and Load via Adequate Plant Design and Load Adaption", 31st Eur. Photovolt. Sol. Energy Conf. Exhib., pp. 2477–2481, 2015.

[3] S. Krauter, D. Rustemovic, and A. Khatibi, "Reduction of required Storage Capacities for a 100 % Renewable Energy Supply in Germany, if new PV systems are installed with East-West tracking systems at increased elevation angles", 37th Eur. Photovolt. Sol. Energy Conf., vol. 243, p. 118466, 2020.

[4] IEA, International Renewable Energy Agency, United Nations Statistics Division, The World Bank, and World Health Organization, "Tracking SDG 7: The Energy Progress Report 2020", World Bank, p. 176, 2019, https://trackingsdg7.esmap.org/.

[5] "Cheaper, faster, cleaner - Speeding up distributed solar solutions to meet development and climate goals", Nor. Sol. Energy Clust., April 2019.

[6] https://www.ecomoyo.com

[7] A. Waitz, K. Lervik, "Techno-Economic Analysis of Rural Microgrid at Eco Moyo Education Centre in Kenya", Master's thesis in Energy and Environmental Engineering, NTNU, June 2022.

[8] R. Rana, K. Berg, M. Z. Degefa, and M. Löschenbrand, "Modelling and Simulation Approaches for Local Energy Community Integrated Distribution Networks", IEEE Access, 2022. doi: 10.1109/ACCESS.2022.3140237

[9] A. Hernandez-Matheus, M. Löschenbrand, K. Berg, I. Fuchs, et al. *(in press)* "A systematic review of machine learning techniques related to local energy communities". Renewable and Sustainable Energy Reviews, 2022.

[10] F. Lombardi, S. Balderrama, S. Quoilin, E. Colombo, "Generating high-resolution multi-energy load profiles for remote areas with an open-source stochastic model", Energy, vol. 177, pp. 433–444, 2019, doi: 10.1016/j.energy.2019.04.097.

[11] W. F. Holmgren, C. W. Hansen, and M. A. Mikofski, "PVlib Python: a Python Package for Modeling Solar Energy Systems", Journal of Open-Source Software, vol. 3, no. 29, p. 884, 2018.

[12] S. Quoilin, K. Kavvadias, A. Mercier, I. Pappone, A. Zucker, "Quantifying self-consumption linked to solar home battery systems: Statistical analysis and economic assessment", Applied Energy, Elsevier, 2016, 182, pp. 58-67.

OPPORTUNITIES AND LIMITATION FOR PV-SYSTEMS WITH A DELTA EAST/WEST CONFIGURATION AT HIGHER LATITUDES AND SITES WITH LOW CLEARNESS INDEX

H.G. Beyer*, M. Zehner#
*University of the Faroe Islands
Torshavn, Faroe Islands
#Technische Hochschule Rosenheim
Germany

ABSTRACT: With the aim of a more balanced production pattern PV systems are installed in increasing number with in a "delta" configuration with module facing east and west with a certain inclination angle to the horizontal. The production patten of these systems should show a less "peaked" appearance around noon and improved generation in the morning and evening hours as compared to systems with south facing modules with the costs of a reduced overall energy gain. As both, the positive effect of a more balanced production and the reduced overall gain are expected to depend on location (by location specific incidence angles) and clearness index (by various mix of diffuse and direct irradiance additionally modifying the effective irradiance) the respective is expected to show a site dependence.
Keywords: System Performance, Solar Radiation, Modelling

1 INTRODUCTION

In recent years, an increasing number of PV-systems is installed in a Δ-configuration, i.e. with modules facing east and west with a certain tilt against the horizontal. One of the motivations for the use of such systems is the option to achieve a more levelized generation pattern. Some characteristics of systems of this type are e.g. discussed by [1,2].

Motivated by the installation of a respective PV-system on the Faroe Islands (~62° degrees North), the opportunities and limitations of systems in Δ-configuration are inspected for a site at a higher latitude and exposing a heigh cloudiness (annual average cloud cover ratio ~0.3. Besides comparing the yield of Δ-configurated system with the yield of south oriented systems, the variation of the yield of the system with the tilt against the horizontal of the east and west oriented modules is analyzed together with the variability of these characteristics with latitude and cloudiness of various sites inspected.

Figure 1:.250kWstc PV-System in east/west Δ-configuration on the island Suðuroy (Faerø Islansfs), view looking south. Image: SEV, www.sev.fo .

2 A PV-System ON THE FAROE ISLANDS

Figure 1 shoes the system, installed on the Island Suðuroy in the village Sumba, located close to te southernmost point of the archipelago. See map in Figure 2.

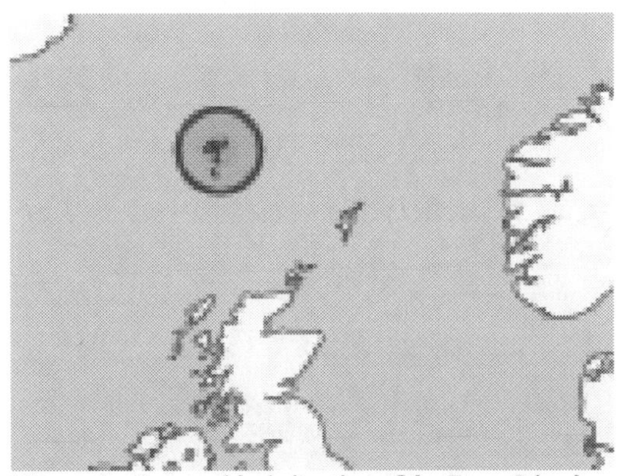

Figure 2: Geographical situation of the Faroe Islands.

The System is composed of 5 sections, each consisting of pairs of eastward and westward facing module rows. There are separate MPP-trackers for the eastward and westward facing rows feeding one common DC/AC inverter per section.

For comparison, a smaller system with south oriented modules at 30° is operated. Operational and meteorological data (global and diffuse horizontal irradiance, air temperature) are logged with hourly time resolution.

3 OPERATIONAL DATA OF THE PV-SYSTEM "SUMBA"

A data analysis reveals a mostly stable and homogeneous operation of the inverters, for both, both the DC/DC and the DC/AC steps. (see figures. 3a-d). The apparently simple linear response of power to horizontal irradiance, can - in general - be reflected well by simulation as shown in figures 3a-3e.

For winter month (January, November, December) the relation of power data and irradiance apperas mor aleatoric, as effecteted by shadowing by terrain and buildup structures,

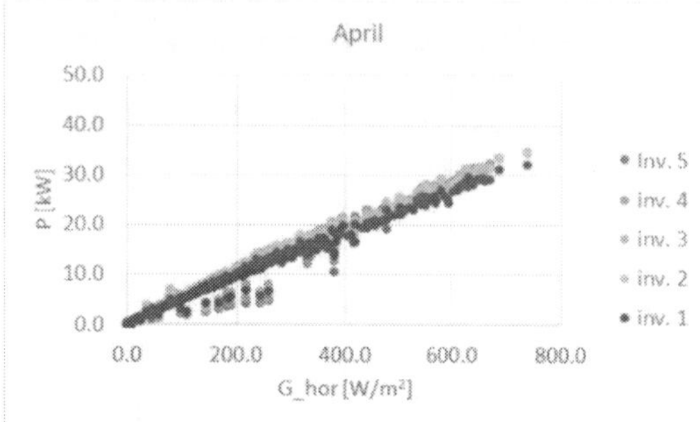

Figure 3a: Hourly power versus versus irradiance for the inverters of the 5 sub-system of the Sumba plant for April 2021 (Inverter 1 is connected to a lower number of module). The recognizable malfunctions stem from two days.

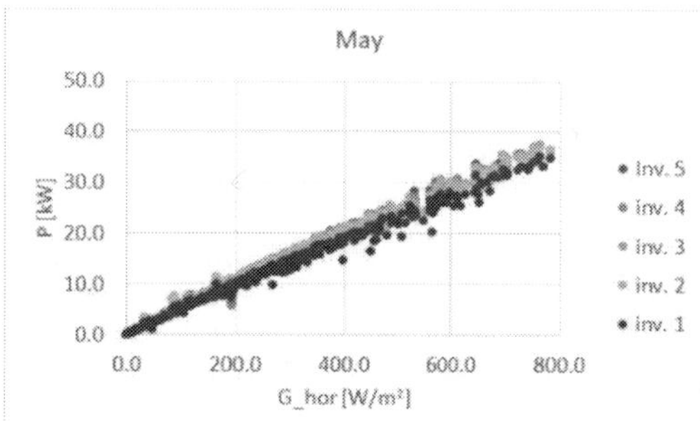

Figure 3b: As fig. 3a, but for the Month of May 2022.

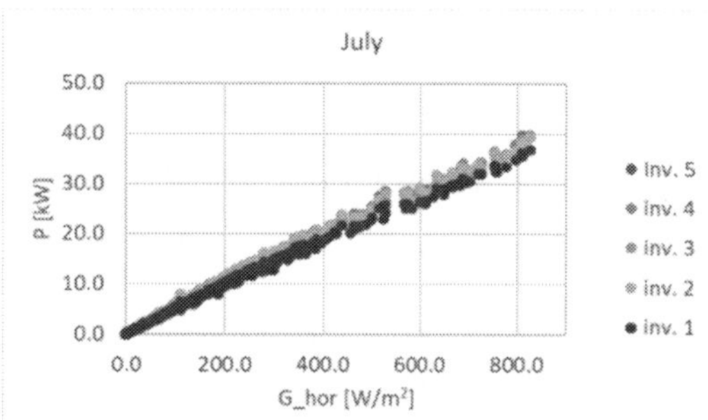

Figure 3c: As fig. 3a, but for the Month of July 2022.

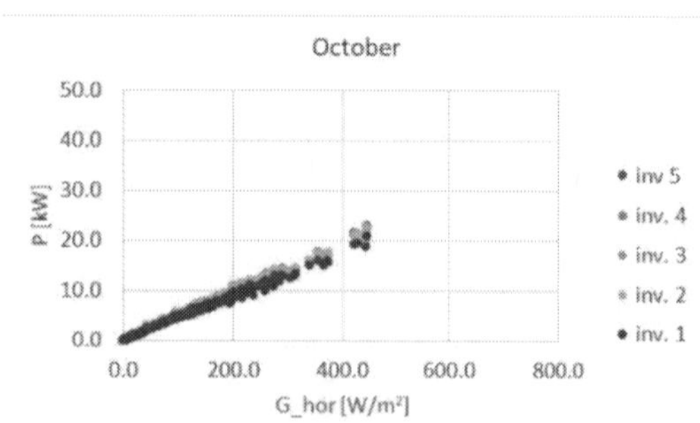

Figure 3d: As fig. 3a, for the Month of October 2022.

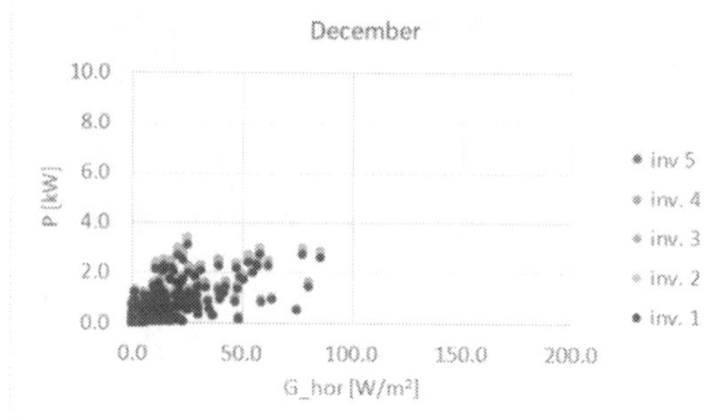

Figure 3e: As fig. 3a, for the Month of December 2022.

Based on these data, a first comparison of the performance of Δ- and south oriented systems can be done. For this the power values of a section of the Δ-system is compared to the power of the south-oriented system, after scaling according to the installed capacity (see figure 4). For this example, refering to a clear day, carefull inspection reveals some benefits for the Δ-system in the morning and the afternoon hours.

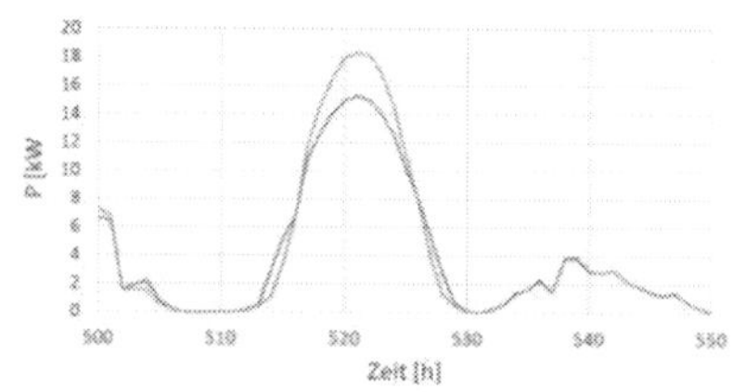

Figure 4: Sections of measured time series (data from June 2021) of the power output of a Δ-configured (blue,) and a southfacing (red, optimal tilt) system (scaled to same installed capcity).

4 MODELLING OF THE POWER OUTPUT OF THE SYSTEM IN SUMBA

Data from part system in Sumba are used for the set-up of a simulation model for its power output in dependence of of horizontal irradiance and ambient temperature.

For this, irradiance data effective for the east and west oriented modules are derived from the data on horizontal irradiance. For this a scheme given in [4] is adapted.

Figure 5 shows respectively simulated power for the east and west oriented modules and the resulting power for the whole system referring to the month og June. Modell parameters used are adapted according to measurement from the month of May. The almost linear relation of power tom horizontal irradiance apparent in the measurement is clearly reproduced by the modelled data.

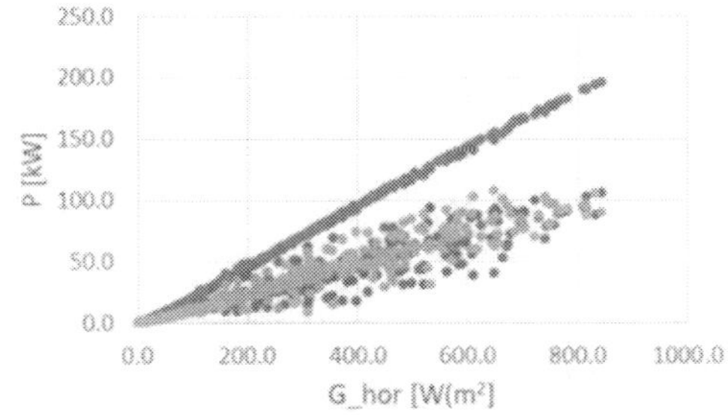

Fig.5: Simulated power data for the inspected PV-System versus the horizontal irradiance. Based on data for the irradiance on the east- and west-facing modules, derived from the horizontal irradiance, the respective power output values are modeled (green and red). The complete system power output (le) is gained by addition of the east and west values)

In the following section, the modelling scheme is used to assess the performance of Δ-systems in a more general way – including the analysis for other locations in Europe and the variation of details of the Δ-configuration (tilt of the east/west modules = on the basis of more extended meteorologic time series. The required meteorological data are extracted from the PVGIS Server [3]. Data from the SARAH2 database for the years 2005-2015 are consideted,

5 MODELLING LONG TERM PERFORMANCE OF Δ-CONFIGURED SYSTEMS AT 4 SITES IN EUROPE

The power time series for Δ-systems are formed by the addition of the series of power for east and west orientated system, as directly extractable from the PVGIS server. The time series of the power for the whole systems are analysed according to the expected multi-year yield, the variability (multi-year standard deviation) and the expected yield from the edge of day hours (defined here as the first and last 8 hours of the day). With respect to the Δ-systems titt angles of the east and west facing modules of 10, 20 and 30 degrees are considered. For comparison, south facing systems are assumed to are installed with site specific optimal tilt.

Besides the site Sumba (37.1 N_6.7W), sites in in Finland (61.5N_21.7O), Ireland (55.3N_6.7W) and Spain (37.1N_6.7W) are consider (see Tabel 1.)..

Table 1: The sites under inspection. Besides theit geographical coordinates, the mean diffuse fraction of the irradiance is given for charcterisation.

Coordinats	H_dif/H_glob
Spain 37.1 _-6.7	0.30
Ireland 55.3_-6.7	0.59
Faroe Islands 61.5_-6.7	0.63
Finland 61.5_21.7	0.46

For sorting of the results, figure 6 give the yields of horizontally installed systems, formalized by the yield of the system in Spain.

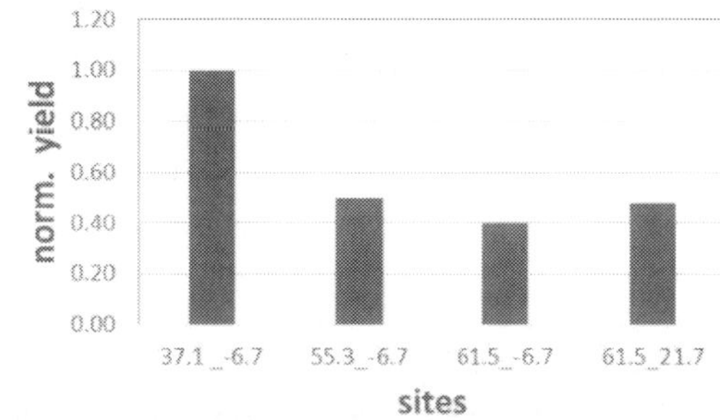

Figure 6: Tields of horizontally imstalled PV-systems, normalized by the yield of the system in Spain.

In the following charts, the yield values are normalized by the yield of a horizontally installed system at he site under inspection.

th

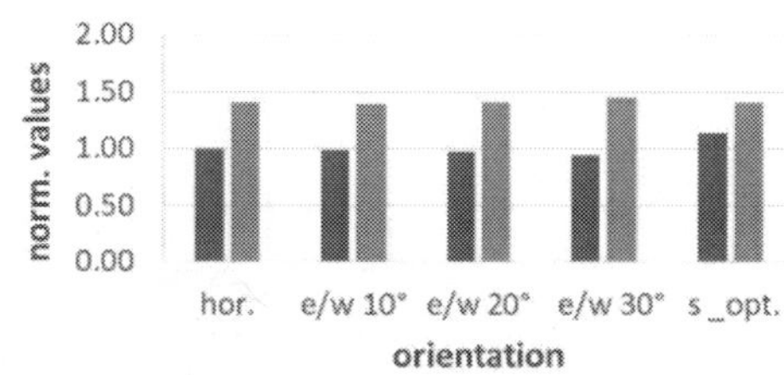

Figure 7a: The yield (normalized to the yield of a horizontally installed system, blue) and the ratio of the standard deviation to the average of the power time series for te site in Spain. Under inspection: systems installed horizontally (hor.) in Δ-configuration with various tilt to the horizontal (e/w X) and south oriented with tilt for optimal energy gain (s_opt).

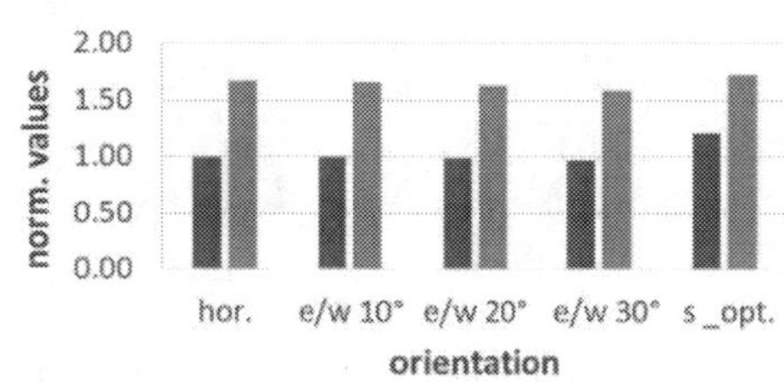

Figure 7b: As fig. 7a, but for the site in Ireland.

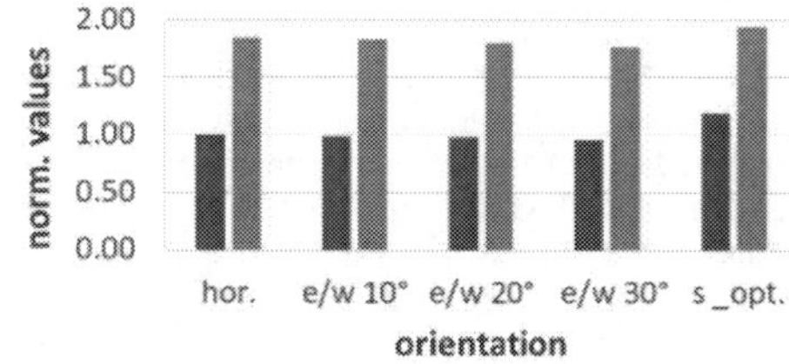

Figure 7c: As fig. 7a, but for the site on the Faroe Islands

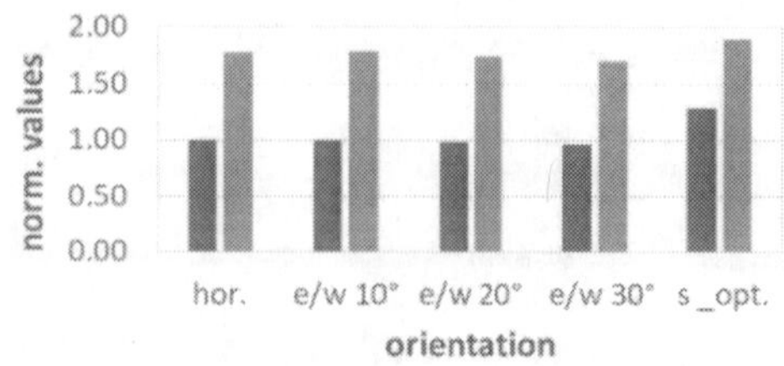

Figure 7d: As fig. 7a, but for the site in Finland

For all sites inspected, the yield data show a common pattern when varying the module orientation. The Δ-Systems show a decreased yield as compared to horizontally installed systems (and more pronounced lower yields than south oriented systems) by reduced relative fluctuations (decreased quotient of standard deviation and mean power)

The inspection of the yield values of the edge of day hours shows, as expected, a higher fraction of the yield in those hours to the overall yield for Δ-systems.
(see figure.6).

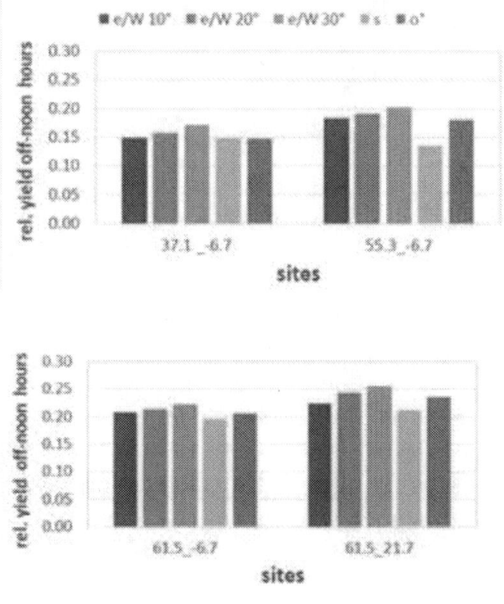

Figure 8: Fraction of the yield from edge of day hours (first and last 8 hours of the day, according to time references given by PV-GIS) for the 4 sites inspected (ref.: coordinates.

This measure for a smoothing increases for all systems types inspected with the latitude of the site South oriented systems show the lowest values here for all sites The differences in the faction of yield for the sites on the Faroe Island and in Finland- showing same latitude reflects the influence of the higher diffuse fraction of the irradiance on the Faroe Islands,

6 SUMMARY

PV Systems in Δ-configuration are expected to show benefits with respect to reduced variability and increased contribution of morning and afternoon hours to the overall system yield especially for sites at higher latitudes. The tendency to a higher diffuse fraction of the irradiance typical fir European sites at higher latitude is counteracting here,

In general, the decision on the installation of Δ-systems is made easier when information on an hour of day specific value of the generated power could be provided (see [2]).

ACKNOWLEDGEMENT

Thanks are due to the network operator of the Faroe Islands SEV for providing data,

REFERENCES

[1] N. Kreifels, S. Killinger, J. Mayer, R. Hollinger, Ch. Wittwer, Effekte Ost-/Westausgerichteter Solarstromanlagen, 30. Symposium Photovoltaische Solarenergie (2015)
[2] R.H. Almeida, L. Narvarte, E. Lorenzo, PV arrays with delta structures for constant
irradiance daily profiles, Solar Energy 171, 23-30 (2018)
[3] PVGIS https://re.jrc.ec.europa.eu/pvg_tools/en/ , besucht 06.05.2022
[4] a. Drews., A.C. de Keizer, H.G. Beyer, E. Lorenz, J. Betcke, D. Heinemann, W.G.J.H.M. van Sark, E. Wiemken, W. Heydenreich, S. Stettler, P. Toggweiler, S. Bofinger, G. Heilscher, M. Schneider, Monitoring and remote failure detection of grid-connected PV systems based on satellite observations, Solar Energy, 81, 548-564 (2007)

OPTIMAL DESIGN OF A SEMI GRID-CONNECTED PV SYSTEM FOR A SITE IN LWAK, KENYA USING HOMER

Josephine Nakato Kakande[a,b], Godiana Hagile Philipo[a,c], and Stefan Krauter[a]
[a] Electrical Energy Technology – Sustainable Energy Concepts
Faculty of Computer Science, Electrical Engineering and Mathematics, Paderborn University
Pohlweg 55, D–33098, Paderborn, Germany
[b] Department of Electrical and Computer Engineering, CEDAT, Makerere University
P.O.Box 7062, Kampala, Uganda
[c] Department of Materials and Energy Science and Engineering
The Nelson Mandela African Institution of Science and Technology, P.O.Box 447, Arusha, Tanzania
Email: jkakande@mail.uni-paderborn.de
Email: philipog@campus.uni-paderborn.de
Email: stefan.krauter@uni-paderborn.de

ABSTRACT: Extensive increases in electricity demand plus numerous power outages encountered in developing countries has led to the deployment of costly, polluting, and unreliable diesel generators to ensure continuity of power. This paper proposes a robust grid-interactive hybrid (PV/Battery-Storage/Diesel-Genset) power system for a site in Lwak, Kenya, the primary objective being the minimization of generator usage and grid cost, and also increasing reliability of supply. To model the entire system, a commercial software, HOMER Pro Version 3.14 is used to evaluate the performance of each component and perform sensitivity analysis to optimize the system. Based on measured load demand, energy potentials, and the related cost, various design configurations are assessed, and the optimal design is selected. The proposed system is selected to most optimally supply the demand with minimal Net Present Cost (NPC), Levelised Cost of Energy (LCOE), emissions, and generator fuel consumption. Demand Side Management (DSM) is also modelled to demonstrate how peak load and storage requirements can be reduced by maximizing usage of the solar production and utilizing excess energy immediately. The designed microgrid systems with individual connections to the main grid will later be interconnected to each other to form resilient and stable interactive power systems.
Keywords: Microgrid, DSM, Grid-Connected, Battery Storage and Control, Photovoltaic

1 INTRODUCTION

Optimal design of Photovoltaic energy systems is often constrained by an inaccurate knowledge or understanding of the demand profile for the site or community where the system will be installed. Estimated demand values are often used in sizing and designing such systems, which results in over- or undersized systems. This paper discusses the design of a grid connected hybrid solar PV system with storage and generator for a setup in Lwak, Kenya. The contribution of this paper lies on selected design criteria, that are chosen with the goal of interconnection of individual modular microgrid systems and incorporation of demand side management measures. Two schemes to implement demand side management (DSM) are presented and compared with the aim of increasing the reliability of the power supply to the site, lowering storage requirements and maximising the use of the generated renewable energy.

Loads can be classified as being deferrable and non-deferrable. DSM involves a combination of generation management, load control and storage management. DSM aims to align electricity supply with demand and minimize energy usage so as to avoid problems such as overloading of the system, over discharge of battery storage (BS), outages, as well as shortened lifetime of system equipment. DSM is possible based on load classification [1].

The need for DSM is more pronounced in microgrids because oversizing the microgrid system at the design stage results in underutilization of the system and hence a higher levelized cost of energy (LCOE) [2]. Common demand side schemes include energy efficiency, energy conservation, load shifting and peak shaving. Load shifting means customers are incentivized to change their demand from peak to off-peak hours, and thus the maximum or peak demand is reduced [3]. Peak shaving means some loads are dropped in order to limit the maximum power consumed to some specified limit. This helps consumers to reduce on or avoid peak demand charges.

The load management measures that can be deployed include banning of appliances that consume higher power than appropriate for the microgrid, use of circuit breakers and remote meters with relays that can be programmed to interrupt supply to or turn off certain deferrable appliances [4]. Price-based DSM utilizes aspects like time-of-use (TOU) pricing, critical peak pricing (CPP) and real time pricing, that elicit better responsiveness to supply-demand constraints than fixed tariffs [3].

HOMER (Hybrid Optimization of Multiple Energy Resources) Pro is commonly used for designing of hybrid microgrid and grid-connected systems [5][6]. Its simulation and optimization capabilities provide economic and technical insights into the performance and feasibility of energy systems based on local weather conditions and power consumption patterns. The version used in this paper is HOMER Pro Version 3.14.

In this paper a solar PV system with storage is designed for a facility currently served by the grid and with a generator as backup power. The generator sizing is larger than the current load and as such it operates with less than optimal efficiency when it runs. Generators operate at different levels of efficiency depending on the loading, and when they are very lightly loaded this has adverse effects such as a poorer fuel consumption rate and the reduced lifetime of the generator. The system to be installed at a convent in Lwak will be used for experimentation on interconnected microgrids. As such three single phase systems to supply the loads will be set up, able to operate independently but then capable of being connected such that they jointly supply the load. This paper focuses on the demand management aspects of ensuring the system operates optimally and with extended life of the system components. This will act as a benchmark for sustainable and resilient interactive microgrids in Africa.

2 CASE STUDY

2.1 Site Description

The Convent in Lwak, Kenya, East Africa is located at 0°8.4' S, 32° 21.4' E. The site, shown in Fig. 1, is located close to Lake Victoria.

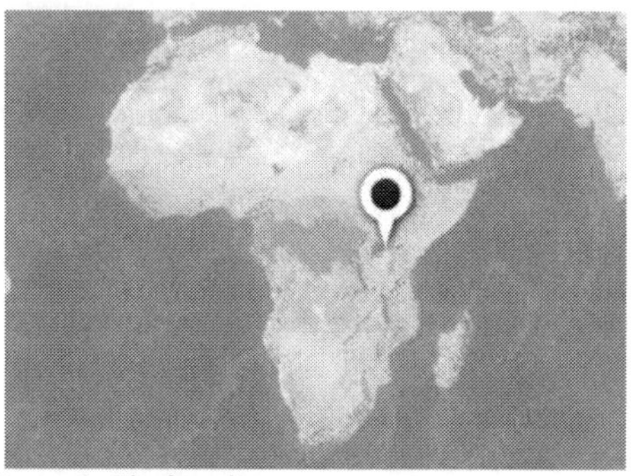

Figure 1: Site location

The location is currently supplied by grid power with a generator to act as backup energy source for electricity supply during grid outages. In the convent, a 40 kW diesel genset is installed.

2.2 Solar Resources

The NASA solar energy dataset was used to obtain monthly solar insolation. Lwak has an average annual insolation of 6.25 kWh/m^2/day, with peak daily insolation of 6.78 kWh/m^2/day during the month of February and a minimum of 5.81 kWh/m^2/day in July. The insolation profile is shown in Fig. 2. During the rainy months of June and July the insolation is lowest.

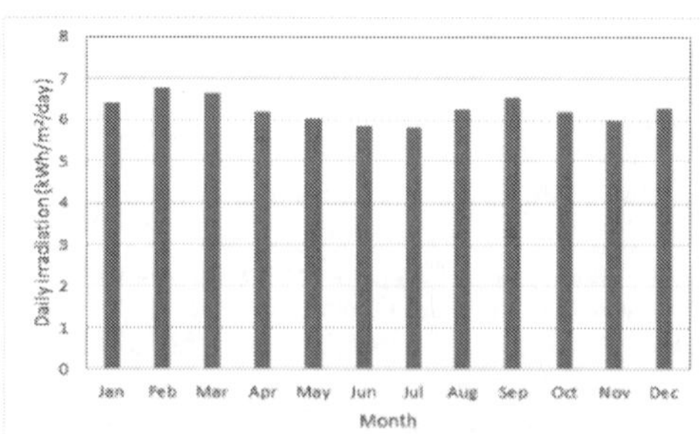

Figure 2: Monthly Average Solar Global Horizontal Irradiance (GHI) data (Source: NASA data downloaded in HOMER Pro)

2.3 Weather

A sensor that measures irradiance (W/m^2), wind speed (m/s) and ambient temperature (degrees Celsius) was installed at the location in Lwak. Fig. 3, Fig. 4 and Fig. 5 show the location's irradiation, wind speed and ambient temperature daily profiles for the dates 25th March to 22nd April 2022.

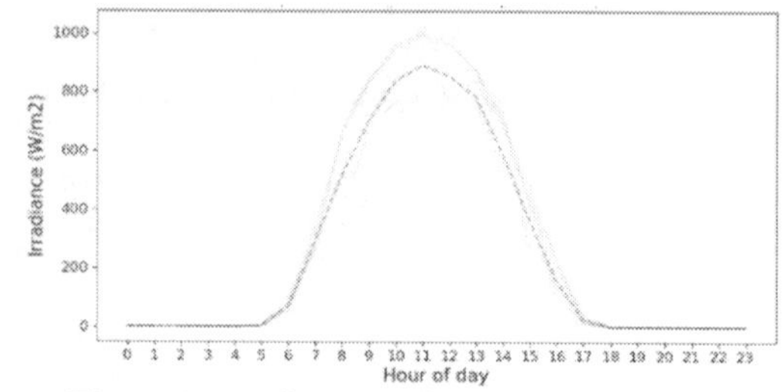

Figure 3: Irradiance profile at Lwak for 25 March 2022 to 22 April 2022 (dashed line represents average values)

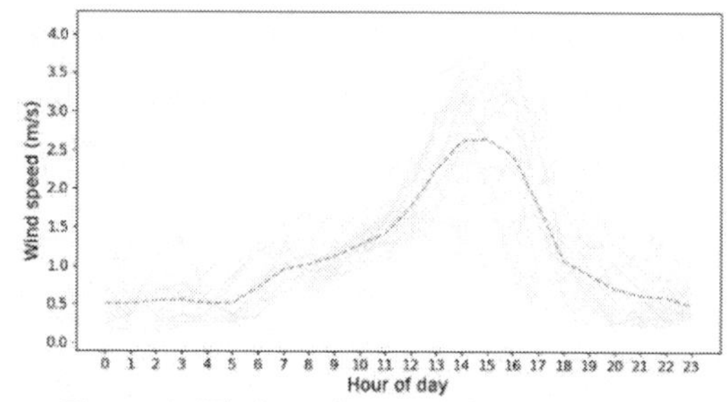

Figure 4: Wind speed at Lwak for 25 March 2022 to 22 April 2022 (dashed line represents average values)

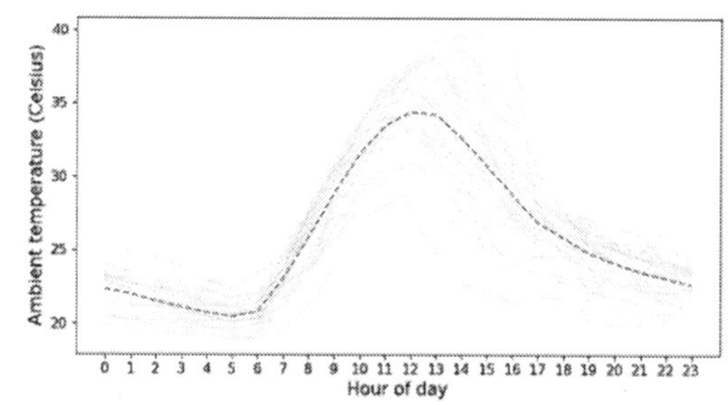

Figure 5: Ambient temperature at Lwak for 25 March 2022 to 22 April 2022 (dashed line represents average values)

The average irradiance is highest between 10 am and 2 pm. In general, increased temperatures reduce PV production. The average wind speeds are highest between 2 pm and 5 pm, which will be beneficial for lowering ambient and module temperatures during the high temperature periods as shown in Fig. 5.

Irradiance, ambient temperature, module temperature and wind speed affect PV production to varying extents. Knowledge of the irradiance, temperature and wind data will be useful at a later stage, for prediction and forecasting applications as well as energy system optimization.

2.4 Consumption Profile

VectoGraphs Power Quality measurement devices were used to measure power consumption at the site for the period 23rd March 2022 to 26th March 2022. The maximum power measured in that time was 8.2 kW. The average daily load profile is depicted in Fig. 6. It displays a bimodal distribution. It can be seen that demand peaks between 6 pm and 8 pm when residents return for the night. A smaller rise in consumption is also visible between 9 am and 10 am when some high consumption activities are being carried out, possibly meal preparation. The average daily consumption measured is 2.33 kW.

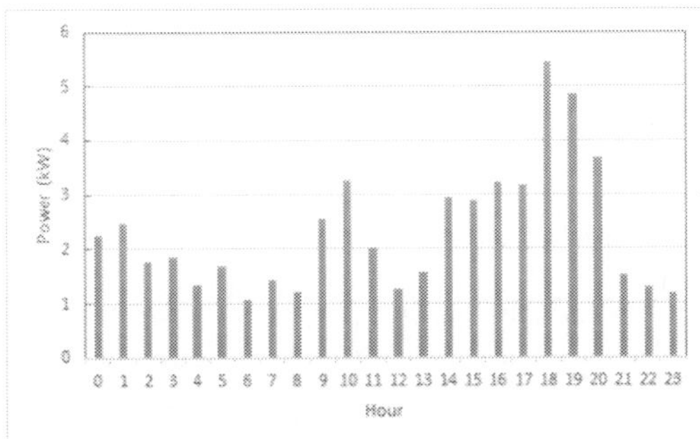

Figure 6: Average hourly load profile for Lwak site for 23rd March to 26th March 2022

It was estimated that the average grid outage duration per customer per year in Kenya in 2016 was 216.3 hours [7]. To estimate the reliability of the grid supply to adequately incorporate the grid uptime conditions into the simulation, generator run-time data from a school situated in the same location as the Convent was used. This outage data was then incorporated into the HOMER Pro simulation.

2.5 Appliances description

The site has a number of electrical consumption devices as indicated in Table I. Approximate ratings and estimated operation durations are specified therein [8] [9]. Deferrable or shiftable loads are those for which the start time is not important, while non-deferrable loads are those which should operate at a given time.

Table I: Load characteristics

Device	Rating (kW)	Quantity	Total (kW)	Duration (hrs)	Deferrable (D) or Non-deferrable (N)
TV	0.2	2	0.4	3	D
Radio	0.015	1	0.015	3	D
Water dispenser	0.53	2	1.06	6	D
Fridge	0.25	4	1	6.5	N
Freezer	0.12	2	0.24	6.5	N
Electric cooker	2	1	2	3	D
Electric Iron	2	2	4	2	D
Kettle	2	2	4	0.5	D

Typically, refrigerators and freezers remain switched on (though they operate with cyclic intermittency and as such do not draw power constantly), they can provide some demand flexibility by adjusting the power setpoint to a lower one [10]. Load shifting is also possible by switching off the appliance for some hours and then turning it on at other times. For this work, we consider the fridges and freezers to be non-deferrable loads. Additional loads such as lighting are not included in Table I.

2.6 DSM measures

From the load profile in Fig. 6, it can be seen that 6-8 pm are the peak hours, while between 11 am and 1 pm, which are high irradiance hours, the average load is relatively low. Load shifting and peak clipping are thus applied as described below.

For peak clipping, if the hourly load from 6 to 8 pm exceeds the peak value during the morning hours (which in this case is 10 am), then the load is limited to the peak morning value of 3.238 kW. Thus, the power requirements are lowered during the evening period, which in turn alleviates the demands on the storage system [11]. Rather than the average power [12], the peak morning value is selected here as the clipping point to prevent excessive load shedding at a time when the solar production is adequate or even more than enough to cover the load.

For load shifting, during the peak periods the power is limited to the average power, in this case 2.33 kW, and the additional loads are shifted to the low demand and high irradiation period of 11 am to 1 pm. The total load of the fridges and freezers running concurrently is about 1.24 kW so that leaves sufficient reserve for some other loads to run between 6 and 8 pm. Here it is assumed for simplicity that the amount by which the evening load is lowered is assigned equally to the hours to which it is rescheduled.

3 SYSTEM DESIGN

The system components are described in this section. A hybrid power system comprising of grid, PV, battery storage system and generator is simulated using HOMER Pro to optimize the capacity and costs of the setup. The schematic layout of the system is shown in Fig. 7.

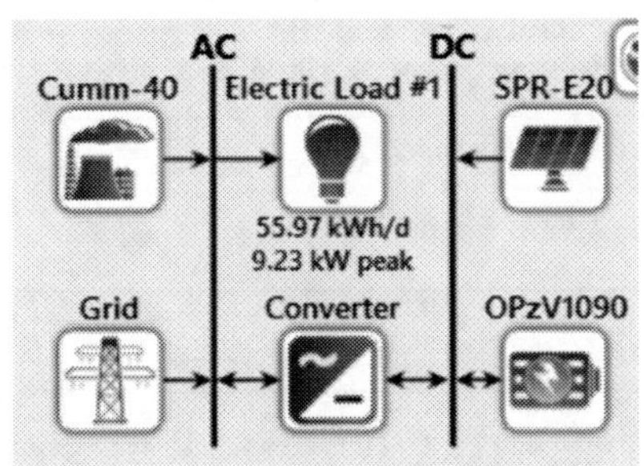

Figure 7: System schematic of Lwak convent

For the PV system the SunPower E20-327 modules are chosen as they have comparable specifications to the SPR-320E-WHT-D used for studies of solar PV systems in Kenya, Zimbabwe and other countries [13][14]. These monocrystalline Silicon 327 W_p PV modules have an efficiency of 20.1%. Capital and replacement costs are both set to US$ 3,300 / kW_p and US$ 10 / kW_p for O&M costs.

The HOMER Generic System converter that acts as both rectifier and inverter, with a lifetime of 15 years and 95% efficiency, is used in the simulation. Capital and replacement costs are both set to US$ 300.

For storage, Sunlight 2V7 RES OPzV 1090 is modelled, with capital and replacement costs of US$ 300 and US$ 10 for O&M. Lithium-ion storage is not used due to cost implications as well as the increasing trend of battery manufacturers and suppliers preferring to serve mobile applications or very large scale plants rather than stationary applications. The 120 kWh system consists of three strings each of 48V DC since the system to be installed will require integration of 3 microgrids each able to operate in isolation but also able to be interconnected.

A 40kW diesel generator (DG) with the efficiency characteristics of the Cummins 40 kW genset is used in HOMER. As there is already a generator on site, the initial capital costs for the generator are set to zero. Replacement cost is US$ 15,000 and the fuel price is US$ 1/Litre.

The Load Following (LF) dispatch scheme is used in the simulation. With LF control the generator is only able to supply the load, but not to charge the batteries. This is in contrast to HOMER's Cycle Charging (CC) scheme that operates the generator at full power and uses the surplus power that exceeds the load requirements to charge the storage.

For the grid a US$ 0.21/kWh price is set. Grid feed-in is not included as the Kenya regulatory framework does not allow net-metering currently, so excess energy cannot be fed into the grid.

The discount rate, inflation rate and project lifetime are set to 7%, 7.5% and 25 years respectively. The annual capacity shortage is set to 0%.

4 DISCUSSION OF RESULTS

4.1 Without DSM

Without DSM measures and using the load profile in Fig. 6, four scenarios are simulated. The first scenario (S1) is for HOMER Optimizer to select the PV system and battery size. The second scenario (S2) is when the battery voltage is set to 48 V and HOMER provides the optimal PV size, while in the third scenario (S3) the PV system size is further stipulated as 16.5 kW. Scenario 4 (S4) is the business as usual (BAU) scenario where the only power sources are the grid and generator. The resultant key optimal technical capacities and economic values are summarised in Table II.

Table II: Results without DSM

Scenario	PV (kW)	Converter (kW)	LCOE (US$/ kWh)	NPC (US$)	EE (%)	RF (%)
S1	12	4.96	0.169	91,559	6.1	91.4
S2	12.3	6.27	0.174	94,589	4.2	94.9
S3	16.5	6.13	0.192	104,360	25.8	98.8
S4	0	0	0.248	134,595	0	0

The renewable fraction (RF), which shows how much of the consumed energy is from PV, is always above 90% for the hybrid PV setups.

The excess electricity (EE) is the electricity produced per year that remains unused. For scenario S2, EE is only 1,015 kWh/yr, while with the 16.5 kWp setup it is 8,080 kWh/yr. This indicates that there is potential to use the excess electricity for other applications such as cooling, water heating or water purification.

NPC and LCOE are lower as expected with PV than for the BAU grid-connected scenario.

4.2 With Peak clipping and Load shifting

With peak clipping applied, the new load profile is shown in Figure 8, while the load shifted profile is shown in Figure 9.

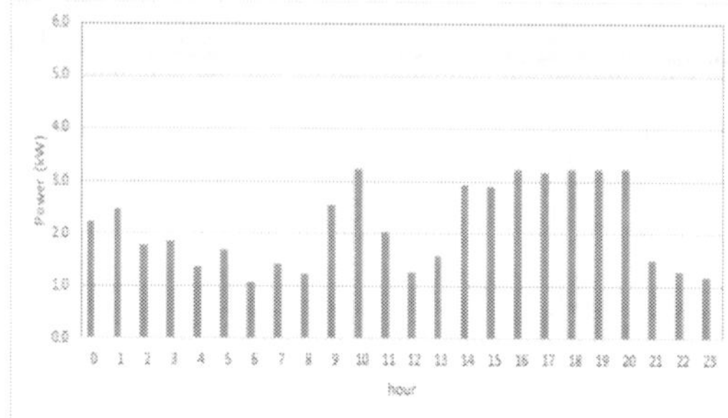

Figure 8: Profile after peak clipping (PC)

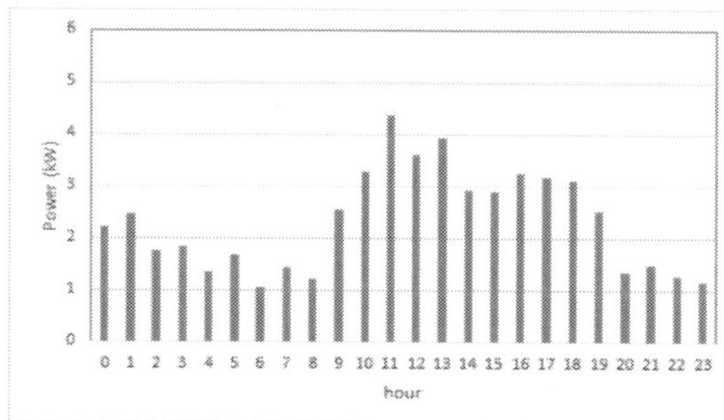

Figure 9: Profile after load shifting (LS)

Figure 8 shows that after the load clipping, the evening peak consumption visible in Figure 6 is gone and there is an almost constant profile from 2:00 pm to 8:00 pm.

Figure 9 for load shifting shows that there is decreased consumption from 6 pm to 8 pm relative to the profile without DSM given in Figure 6. The peak load hours are now 11 am to 1 pm.

Peak clipping lowers the daily energy consumption by 8% to 51.69 kWh while the peak demand is reduced to 5.53 kW. For load shifting the daily consumption remails unchanged at 55.97 kWh. With the random day-to-day and timestep variabilities of 10% and 20% respectively in HOMER, the PC and LS simulated peak demands are 5.53 kW and 7.6 kW as shown in Fig. 10 and Fig. 11.

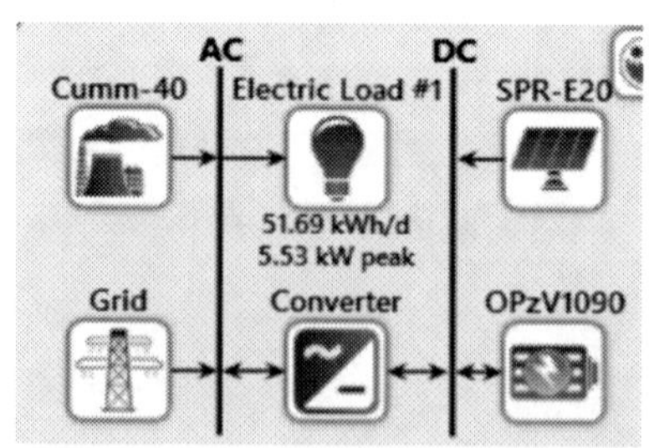

Figure 10: Peak clipping schematic in HOMER

In HOMER, the load clipping and load shifting renewable fractions are 96.8% and 97.5 % respectively when the PV size is determined by HOMER (scenario S2). Generator production for peak clipping and load shifting are 0.14% and 0.18% respectively for scenario S2. The system performance is as summarized in Table III, where BS (battery storage) refers to the number of batteries used. Each 2V battery has a capacity of 1.67 kWh.

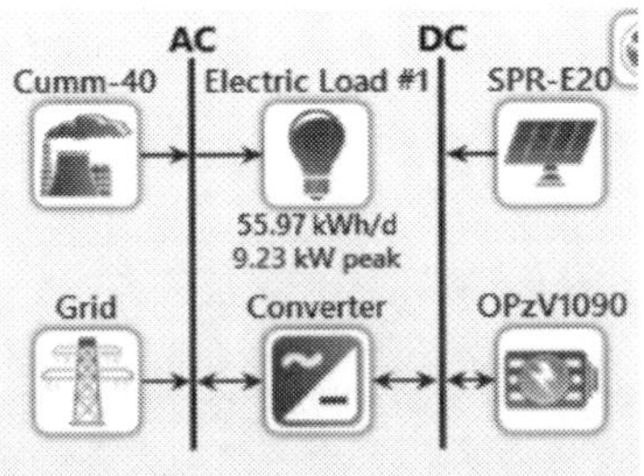

Figure 11: Load shifting schematic in HOMER

Table III: Peak Clipping (PC) and Load Shifting (LS) results

		PV (kW)	Converter (kW)	BS	EE (%)	RF (%)
S1	PC	11	4.55	45	5.1	91.9
	LS	12.4	5.18	44	7.9	94.0
S2	PC	11.7	5.30	72	5.8	96.8
	LS	12.8	6.56	72	7.2	97.5
S3	PC	16.5	4.88	72	31.3	99.7
	LS	16.5	5.75	72	26.6	99.5
S4	PC	0	0	0	0	0
	LS	0	0	0	0	0

When the PV size is stipulated as 16.5 kW (S3), the load shifting system obtains 1.96% of its energy from the grid while the peak clipping has 0.14% grid input, demonstrating the almost complete reliance on solar production.

A comparison of the no DSM case with the two DSM cases shows that excess energy and renewable fraction are always greater when DSM is applied.

The increase in excess energy for Scenario 3 relative to S1 and S2 is to be expected as additional PV capacity results in additional overall energy production. This means that in addition to the cost savings (see Table IV) from minimal consumption of grid power and less generator usage, rather than PV curtailment there is a clear opportunity to use the power not needed by the load for e.g. thermal applications.

Table IV: Economic results for Peak Clipping (PC) and Load Shifting (LS)

		PV (kW)	Converter (kW)	BS	LCOE (US$/ kWh)	NPC (US$)
S1	PC	11	4.55	45	0.161	80,831
	LS	12.4	5.18	44	0.148	80,410
S2	PC	11.7	5.30	72	0.169	84,476
	LS	12.8	6.56	72	0.158	86,012
S3	PC	16.5	4.88	72	0.194	97,355
	LS	16.5	5.75	72	0.175	95,243
S4	PC	0	0	0	0.251	125,681
	LS	0	0	0	0.248	134,723

5 CONCLUSION AND NEXT STEPS

In this paper, the incorporation of peak clipping and load shifting were evaluated as means of rightly sizing of a hybrid PV system using HOMER Pro, with the aim of achieving sustainable operation of the site's hybrid PV system. Both DSM schemes showed improvements in the fraction of renewable energy used relative to the case without DSM, as well as increased excess energy. Scenario 3 (S3) where PV size was set to 16.5 kW was the most optimal, yielding a moderate LCOE but a large amount of excess energy that can be applied towards thermal storage or other productive uses.

The next steps will include use of load data and weather data for the site to carry out optimization processes and assess the dynamic operation of the demand management algorithms. Incorporating forecasting methods to enable real time load and generation management so as to formulate the most appropriate dispatch strategies will be carried out. Investigation into optimal battery management strategies that extend battery lifetime and costs will be carried out as part of the evaluation of interconnected and resilient microgrids.

6 ACKNOWLEDGEMENTS

The authors are thankful to the German Federal Ministry of Education and Research (BMBF) for funding this project within the compound assignment "CLIENT II–Art-D Grids: Africa Research and Teaching Platform for Development –Sustainable Modular Grids for Grid Stability" Project: 03SF0607B.

7 REFERENCES

[1] Y. Zhang, T. Huang, and E. F. Bompard, "Big data analytics in smart grids: a review," Energy Informatics, vol. 1, no. 1, pp. 1–24, 2018, doi: 10.1186/s42162-018-0007-5.

[2] N. Narayan et al., "Stochastic load profile construction for the multi-tier framework for household electricity access using off-grid DC appliances," Energy Efficiency, pp. 197–215, 2018.

[3] H. P. Pourbabak, T. C. Chen, B. Z. Zhang, and W. S. Su, "Control and energy management system in microgrids," Clean Energy Microgrids, pp. 109–133, 2017, doi: 10.1049/pbpo090e_ch3.

[4] M. Buevich, D. Schnitzer, T. Escalada, A. Jacquiau-Chamski, and A. Rowe, "Fine-grained remote monitoring, control and pre-paid electrical service in rural microgrids," IPSN 2014 - Proc. 13th Int. Symp. Inf. Process. Sens. Networks (Part CPS Week), no. 1, pp. 1–11, 2014, doi: 10.1109/IPSN.2014.6846736.

[5] Restrepo, D., Restrepo-Cuestas, B. and Trejos. A., Microgrid analysis using HOMER: a case study. DYNA, vol. 85 (2018) 207. https://doi.org/10.15446/dyna.v85n207.69375

[6] K. Tamilarasu, C.R. Sathiasamuel, J.D.N. Joseph, R. Madurai Elavarasan, L. Mihet-Popa, Reinforced Demand Side Management for Educational Institution with Incorporation of User's Comfort. Energies, vol. 14 (2021) 2855. https://doi.org/10.3390/en14102855.

[7] J. Taneja, "Measuring Electricity Reliability in Kenya," Energy Policy, vol. 12 (2016).

[8] Subhasis Panda, Sarthak Mohanty, Pravat Kumar Rout, Binod Kumar Sahu, Mohit Bajaj, Hossam M. Zawbaa, Salah Kamel, Residential Demand Side Management model, optimization and future perspective: A review, Energy Reports, Vol. 8, 2022, https://doi.org/10.1016/j.egyr.2022.02.300.

[9] Olaniyan, Kayode & Mclellan, Benjamin & Ogata, Seiichi & Tezuka, Tetsuo. (2018). Estimating Residential Electricity Consumption in Nigeria to Support Energy Transitions. Sustainability. 10. 1440. 10.3390/su10051440.

[10] Zhengyi Luo, Jinqing Peng, Jingyu Cao, Rongxin Yin, Bin Zou, Yutong Tan, Jinyue Yan, Demand Flexibility of Residential Buildings: Definitions, Flexible Loads, and Quantification Methods, Engineering,2022. https://doi.org/10.1016/j.eng.2022.01.010.

[11] Dharani, R.; Balasubramonian, M.; Babu, T.S.; Nastasi, B. Load Shifting and Peak Clipping for Reducing Energy Consumption in an Indian University Campus. Energies 2021, 14, 558. https://doi.org/10.3390/en14030558.

[12] Philipo, G.H.; Kakande, J.N.; Krauter, S. Neural Network-Based Demand-Side Management in a Stand-Alone Solar PV-Battery Microgrid Using Load-Shifting and Peak-Clipping. Energies 2022, 15, 5215. https://doi.org/10.3390/en15145215.

[13] Samu, G. Poyrazoglu, and M. Fahrioglu, "The Potential and Economic Analysis of Grid-connected Solar PV Power in Kenya," Proc. - 2019 IEEE 1st Glob. Power, Energy Commun. Conf. GPECOM 2019, pp. 298–301, 2019, doi: 10.1109/GPECOM.2019.8778467.

[14] Uwineza, H. G. Kim, and C. K. Kim, "Feasibilty study of integrating the renewable energy system in Popova Island using the Monte Carlo model and HOMER," Energy Strateg. Rev., vol. 33, p. 100607, 2021, doi: 10.1016/j.esr.2020.100607.

SOLAR PHOTOVOLTAIC SYSTEMS FOR HIGH POWER PUMPING: A USE CASE

Márcia Horta [a,b], Luís Fialho [a,b], Carlos Marques [c], Pedro Horta [a,b]
[a] Renewable Energies Chair, University of Évora. Pólo da Mitra da Universidade de Évora, Edifício Ario Lobo de Azevedo, 7000-083 Nossa Senhora da Tourega, Portugal
[b] Institute of Earth Sciences, University of Évora, Rua Romão Ramalho, 7000-671 Évora, Portugal
[c] Associação de Beneficiários do Roxo, EN 383, Montes Velhos 7600-411, São João de Negrilhos – Aljustrel, Portugal

ABSTRACT: Solar water pumping combines solar photovoltaic (PV) technologies, hydraulic engineering, and high-efficiency technology to provide water management techniques and optimise irrigated farming. Within the context of the SolaQua project, a solar photovoltaic system design for water pumping was developed and implemented in Montes Velhos Pumping Station, located in Aljustrel, Portugal. This pumping station has 2.813 MW of total installed pumping power and a maximum flow rate of 3.08 m^3/s. The designed PV system has 320 kW_p installed on 1-axis North-South horizontal sun-trackers in a hybridised configuration with the electricity grid. This work aims to analyse the real operation data from that hydro-agricultural system from 2019 to May 2022. It is intended to analyse data regarding the electrical conversion efficiency of the pumping station, to calculate the annual performance ratio (PR) and other equally relevant variables for the pumping station and solar photovoltaic plants, such as self-consumption rates and annual savings. The designed system allowed the cost reduction of the pressurised pumped water and increased the share of renewable energy (RE), presenting a cost-effective alternative to the previous energy sources, diesel generators or electric grid for water pumping.

Keywords: Solar Water Pumping, Water Pumping, Pumping Station, PV System, Pressurised Pumped Water.

1 INTRODUCTION

Agricultural irrigation in Southern Europe is a high consuming activity: of electricity (24,000 GWh/year) and of water (70,000 Hm^3/year). In the last years, the electricity price has increased across the region (in Spain 1250% from 2008 to 2013 for the Irrigators Communities, in Portugal 226% and in Italy 32%), reducing the competitiveness of agriculture business models [1]. Futhermore, with the recent european crisi context, eletricity gross market prices have been increasing at faster rates.

Most of the current agricultural pumping systems are powered from the national electric grid or diesel generators. However, both the increase in the price of this conventional electricity and the trend to phase-out fossil fuels, has boosted the look for alternatives to power these systems. Solar photovoltaic irrigation/pumping systems appear as a sustainable solution due to the decrease of the PV modules prices in international market, as well as the removal of technological barriers that limited the power of PV pumping systems [4].

This work aims to analyse real data obtained from the hydro-agricultural system of Roxo (Portugal) over the period of 2019 to 2022, mainly from a high-power photovoltaic pumping setup. Solar pumping is based on a combination of photovoltaic (PV) technology, hydraulic engineering, and high-efficiency water management techniques to optimize irrigated farming. The designed system allowed the cost reduction of the pressurized pumped water as well as an increase of the share of renewable energy (RE), presenting a cost-effective alternative to the previous energy sources.

2 CONTEXT OF THE ROXO HYDROAGRICULTURAL DEVELOPMENT

The Roxo Beneficiaries Association (AB Roxo) is the entity responsible for managing the Roxo hydro-agricultural operation since 1968. This system was designed around 80 years ago, in the first half of the 1930's, and has allowed the conversion of thousands of hectares from rain-fed to irrigated land in an area around the Roxo stream and between the city of Beja and the towns of Ferreira do Alentejo and Aljustrel, in Portugal. The main infrastructures of this system, managed by this association are:

- The Roxo Dam;
- A gravity and a pressure water distribution network;
- The Montes Velhos pumping station, associated to the irrigation blocks with pressurized water distribution (Montes Velhos and Aljustrel);
- A drainage network, with a total length of 71.91 km [2].

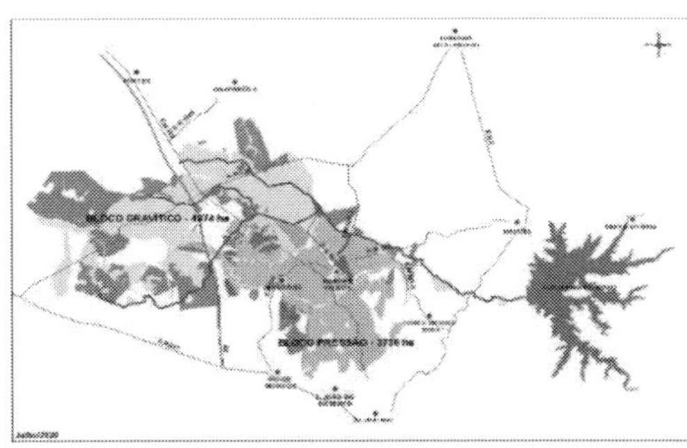

Figure 1: Water distribution network from the Roxo dam

3 PV GENERATION SYSTEM FOR SELF-CONSUMPTION AT THE MONTES VELHOS PUMPING STATION

In order to develop sustainable strategies and policies, the AB Roxo decided to invest in renewable energy sources with the construction of a photovoltaic power plant to generate the electricity needed to operate the pumps at the Montes Velhos pumping station, in order to reduce the total operational costs [3]. Thus, the Roxo photovoltaic power plant was designed with a total installed power of 320 kWp, on a setup with solar tracking on a horizontal north-south axis, allowing it to properly meet the seasonal and daily electrical needs of the pumping station, improving the key performance indicators, that would not be achieved with a standard south facing fixed tilt PV system [2]. This pumping station has 2.813 MW of total installed pumping power and a maximum flow rate of 3.08 m^3/s.

Figure 2: Montes Velhos Pumping Station and its PV plant, Portugal

3.1 Operation of the Pumping Station and Photovoltaic Plant

The next figure shows the simplified scheme for power flows of the hybrid configuration: pumping station, the photovoltaic plant and the public electric grid as well.

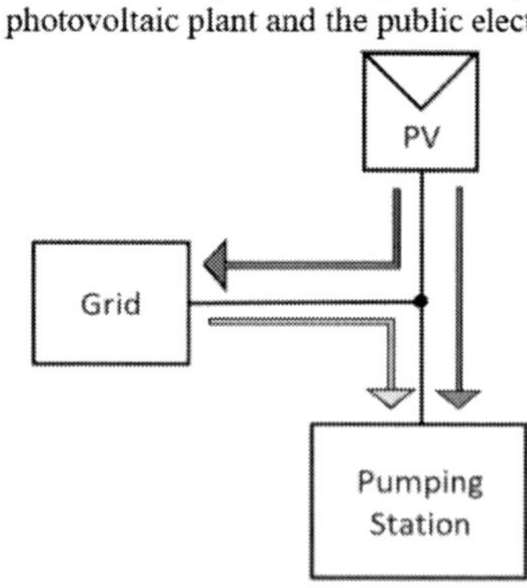

Figure 3: Simplifed scheme of power flow for the current setup at the Roxo Pumping Station

Part of the photovoltaic generation is self-consumed by the pumping station to meet its energy needs, represented in the scheme by a green arrow. The remaining PV electricity generated is injected into the grid and sold to EDP (market electricity trader), represented with the blue arrow. When the PV generation is not enough (or not available) to supply the pumping station demand, AB Roxo buys the necessary power from ENDESA (market electricity trader), injecting from the grid, represented by the yellow arrow.

This purchase/sale of electricity to ENDESA and EDP, respectively, is made possible under the decree-law no. 15/2022 [4], currently in force in Portugal.

4 RESULTS ANALYSIS

4.1 Irradiation in the Plane of the PV Modules

With the TMY available on the PVGIS website [5], a preliminary data quality check was done regarding the values collected by the pyranometer installed in this photovoltaic plant, concluding that are closely aligned in accordance with the typical seasonality values and monthly irradiation values.

The real data collected showed that July 2020 presented the highest monthly solar irradiation total so far. The higher difference of irradiation values summer-Winter is due to the horizontal tilt of the PV system and is an part of the goals for the customized design.

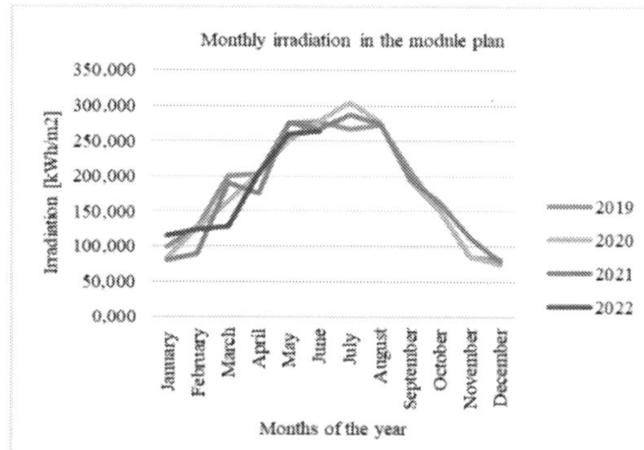

Figure 4: Monthly total Irradiation in the PV Modules plan

4.2 Annual PR

The concept of PR, *performance Ratio*, is used as a quality index with regard to the connection of photovoltaic systems to the grid and is also used to monitor the operating quality of the installation in the long term. This quality index is directly related to the efficiency of converting solar irradiation into electrical energy, and for this situation it was obtained using the following expression:

$$PR_{monthly} = \frac{E_{AC,monthly}}{E_{DC,monthly\ module\ output}}$$

Where, monthly AC Energy corresponds to the energy measured at the output of the photovoltaic installation's electricity meter and monthly DC energy corresponds to the power measured at the output of the photovoltaic modules.

The monthly DC energy at the output of the photovoltaic modules was obtained using the following expressions [5]:

$$\sum E_{DC\,(med,PV),monthly} = G_{monthly}{}^{*} \times A_{active,total} \times \eta_{module}$$

Where, the first term corresponds to the monthly irradiation and the following terms correspond to the total active area of the installation and the efficiency of the modules installed, respectively.

To calculate the total active area of this photovoltaic installation, it was necessary to consider the specification sheet of the modules installed [6], from which the following was calculated:

$$A_{active,total} = n^{\underline{o}}\ cells \times length \times width \times n^{\underline{o}}\ modules \leftrightarrow$$

$$A_{active,total} = 72 \times (0{,}156) \times (0{,}156) \times 1000$$

$$\eta_{module} = 16{,}5\ \%$$

When this photovoltaic plant was installed, a minimum annual PR was defined in its specifications (Technical Specifications, part of the installers contract documents) and it was required to be above 85%. All the calculated PR values, monthly and annual, should be higher than that figure, otherwise specific correctives measures have to take place by the installer in order to achieve the defined minimum annual PR.

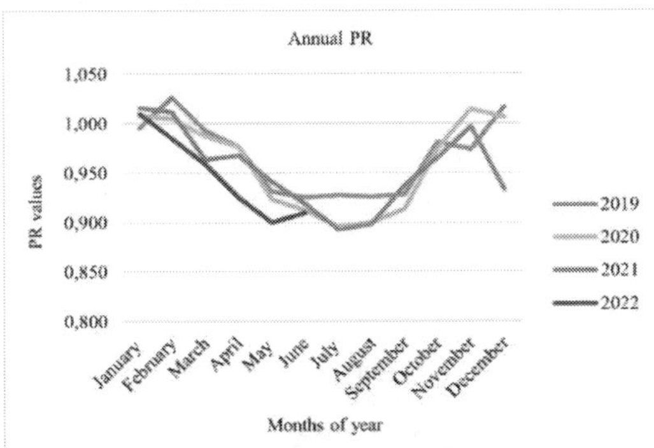

Figure 5: Annual PR – calculated from real data (2019 to May 2022)

The previous figure shows a clear seasonality regarding monthly PR values, with the majority of the months being above 90%. The PR seasonality is often related to the operating temperature and humidity for all equipments, cabling, soiling, etc. The precision of this PR values is clearly affected by the lack of proper O&M and periodic cleaning of pyranometers and PV modules, presenting several values above 100%. Recommendations were issued to the plant owner in order to avoid these issues in future operation. Missing O&M records also do not allow to perform an in depth analysis of these outliers.

4.3 Self-Consumption-Rate

The self-consumption rate is related to the ratio of electricity that is self-consumed according to the pumping station's energy needs. The self-consumption rate depends on the energy that is produced and that is consumed locally. In general, due to the characteristics of the irrigation profile in Mediterranean countries, in summer months the self-consumption rate achieved is lower, increasing the grid energy consumption. In particular, in years with less rainfall or during drought periods, the irrigation needs become higher and, in consequence, the pumping station electricity needs increases.

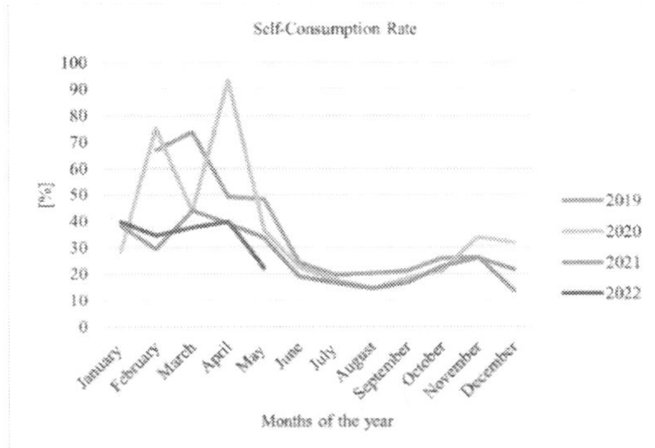

Figure 6: Self-Consumption Rate

4.4 Generation *Vs.* Consumption

Although the design of the PV system (horizontal tilt) produces the largest difference between summer and winter energy generation values, when compared with standard PV systems design, the seasonality effect of this application is critical for the improvement of technical-economic indicators. While the production is higher in summer, it is not enough to meet the energy needs of the pumping station, as can be seen (Figure 7) by the large difference between solar generation and consumption in summer. An carefully designed expansion of the existing PV plant is needed to further reduce the existing gap during summer energy needs and solar generation, without creating a large solar energy surplus in the winter period (economically less interesting due to lower tariffs).

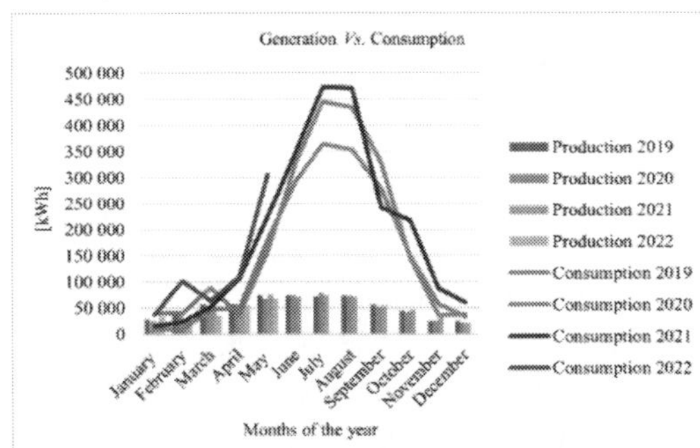

Figure 7: Generation Vs. Consumption

The generated solar energy in this period has enabled to avoid the emmissions of 809.8 tons of CO_2, considering a total production of 2.030 GWh and a average specific emmissions factor (ENDESA) of 398.88 g/kWh of CO_2.

4.5 Savings

Given that AB Roxo must meet their business model goals, this renewable energy project must accomplish important savings regarding the pumping energy cost, and, in consequence, pressurized water cost to be sold to the end-users (farmers, irrigators). Savings in the pressurized water cost (€/m^3) are reflected directly in the increased sustainability of the farmers business model. The current trend of improvement of hydraulic and irrigation efficiency often causes an increase of energy consumption (from gravity irrigation systems to pressurized precision irrigation systems), turning the electricity cost a major driver in the modern farming model.

Due to the night period (or cloudy periods) pumping needs, a alternative energy source will always be needed (or energy storage). This hybrid configuration enables important savings and guarantees energy supply safety.

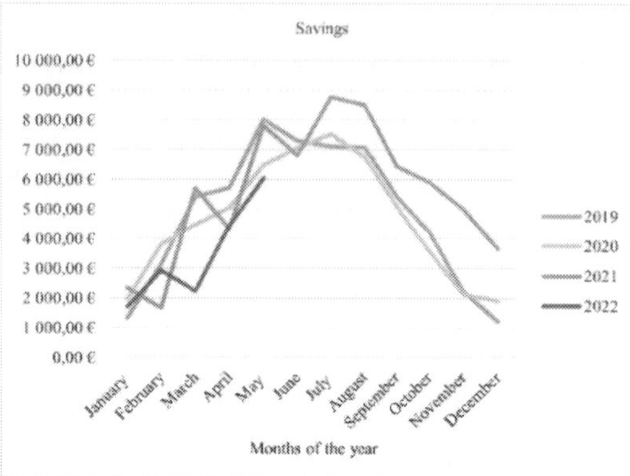

Figure 8: Monthly savings (self-consumed eletricity)

Total savings for the analysed period (2019 to May 2022) are 198 365€ (59.5% of CAPEX), an average of 56 675€ saved per year.

4.6 Surplus electricity injected into the grid

Energy generation surplus can be injected into grid and provides an additional cashflow for AB Roxo, when sold to an electricity trader. Furthermore, the project economic indicators can be improved trough selling the winter surplus energy at a higher tariff, and made possible by the current decree-law in force, e.g., with the creation of an self-consumption energy community with existing neighbors (peer-to-peer energy trading).

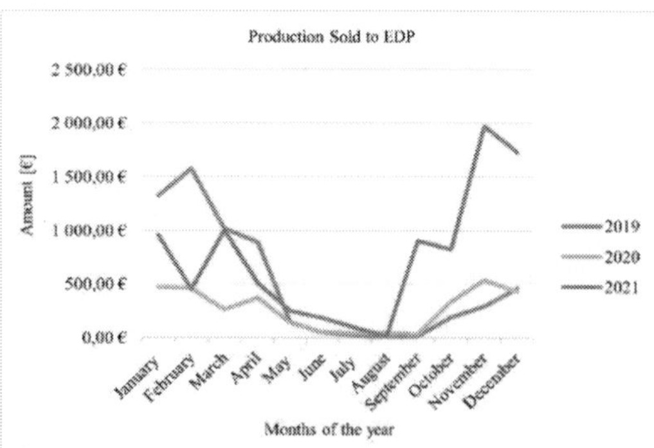

Figure 9: Monthly cashflow from surplus eletricity sold to the grid

Due to the absence of 2022 data, Figure 9 does not depict this values.

5 CONCLUSIONS

High power PV pumping provides a sustainable solution for irrigators and farmers. As shown in this work, investments in this technology should increase, providing a lower cost to pressurized water and contributing to the overall CO_2 emissions reduction. The use case presented is the first large scale system in Portugal, in a hybrid configuration with the pre-existing electricity grid. The results presented showed that depict existing some challenges regarding performing proper O&M tasks, the operation data points to good PR results, and significative savings in OPEX.

Regarding the self-consumption of this pumping station, presents a seasonal variation, as expected. The summer months gap between solar generation-consumption can be addressed in the future, with a upscale of the total installed solar power, reaching higher self-consumption rates. On winter months the existing surplus is injected into the grid and sold to an electricity trader. The existing Portuguese Decree-Law allows the creation of energy communities, which could further improve the project key performance indicators, trough peer-to-peer energy trading with existing neighbors.

The economic indicators point to favorable achieving of the AB Roxo business model goals and reached so far total savings of 198 365€.

ACKNOWLEDGMENTS

The authors would like to thank the support of this work, partly funded by European Union's H2020 Programme as part of the SolaQua project (Accessible, reliable, and affordable solar irrigation for Europe and beyond), Grant agreement ID 952789.

6 REFERENCES

[1] FENACORE - Federacíon Nacional de Comunidades de Regantes de España, "September Bulletin," 2013.

[2] L. F. Carlos Marques, "The energy issue in the distribution of water under pressure in the Roxo hydro-agricultural operation," em *15th Water Congress*, Lisbon, 2021.

[3] D. d. R. Eletrónico, "Decreto-Lei n.º 15/2022, de 14 de janeiro," [Online]. Available: https://dre.pt/dre/detalhe/decreto-lei/15-2022-177634016. [Accessed 5 9 2022].

[4] Maslowaten Project, "D5.1 Market Assessment and Business Plan," 2018.

[5] EDIA, "Empresa de Desenvolvimento e Infra-Estruturas do Alqueva, S.A," [Online]. Available: https://www.edia.pt/pt/. [Accessed 5 9 2022].

[6] "PHOTOVOLTAIC GEOGRAPHICAL INFORMATION SYSTEM," [Online]. Available: https://re.jrc.ec.europa.eu/pvg_tools/en/. [Accessed 5 9 2022].

[7] Astronergy, [Online]. Available: https://www.prosolarltd.com/wp-content/uploads/2019/09/Astronergy-Panels-310W-330W-Specs.pdf. [Accessed 6 9 2022].

COMPARISON OF LCOE FOR VARIOUS PV SYSTEM CONFIGURATIONS AND MODULE TECHNOLOGIES IN THE SOUTH AFRICAN CLIMATIC CONDITION

TJ Serameng[1], JL Crozier McCleland[2], EE van Dyk[2], FJ Vorster[2], RM Dix-Peek[2], M Vumbugwa[2], RP Roodt[2]
[1]ESKOM, Research Testing and Development, Private Bag X40175, Cleveland, 2022, South Africa
[2]Department of Physics, Nelson Mandela University, Gqeberha Port Elizabeth, 6019, South Africa

ABSTRACT: The purpose of this paper is to analyse the performance of photovoltaic (PV) systems and evaluate the energy output of various PV system configurations and module technologies in the South African climatic conditions. The project gathered actual operational data and analysed the monitored performance with the aim of understanding how technology, design and configuration affected the Levelised Cost of Energy (LCOE) for each technology and configuration. The study was conducted on a 400 kW research PV plant at the Eskom Research and Innovation Centre (ERIC) located in Rosherville, South Africa. The plant was designed to test the performance and efficiency of crystalline and thin film PV modules in different array configurations. Current-Voltage (IV) measurements of all arrays were taken from September 2016, until May 2021. Performance measurements at Standard Test Conditions (STC) of a sample of PV modules were taken at the Nelson Mandela University Photovoltaic Testing Laboratory over this period, to be the baseline measurement against which future degradation of the modules would be compared. The various performance parameters and final energy values being monitored show a lower than expected performance in terms of energy produced. This has a large impact on the expected final LCOE of the various arrays and illustrates the importance of optimising and maintaining plant performance.

Keywords: Performance, LCOE, Small Grid-connected PV Systems

1 INTRODUCTION

Photovoltaic (PV) technologies are playing a major role in reshaping the South African energy mix. The rapid increase in uptake is driven by the government's focus to promote and implement renewable projects [1] and the commercial incentives to provide stable electrical supply [2, 3]. Ranging from 1,700 kWh/m^2 to 2,300 kWh/m^2, making it among the top countries in the world in terms of expected solar energy production [4].

Performance monitoring of PV systems in the South African climatic conditions are rare, and there are few reports based on real field data in South Africa. This has created a gap in the knowledge of the real energy production of the PV systems installed with different module technologies and installations in South Africa. This work presents the results of analysing the energy production, the performance of different technologies and configurations at Eskom Research and Innovation Centre, South Africa monitored from January 2015.

For decisions to be made on the long-term investments of Photovoltaic (PV) plants it is necessary to do Yield Assessments (YA) and Long-term yield predictions (LTYP) [5]. Key to the Yield Assessments is the determination of the utilisation rate, performance loss rate and lifetime. These values together with the Cost data (CAPEX, OPEX and discount rate) provide the parameters necessary for the calculation of the Levelized Cost of Energy (LCOE). The LCOE is required to determine the cash flow, internal rate return (IRR) and Net Present Value (NPV).

The 400 kW PV research plant used in this study is located at the Eskom Research and Innovation Centre (ERIC) in Rosherville, Germiston, South Africa. The plant was designed to test the performance and efficiency of crystalline and thin film PV modules in different array configurations [6].

The project aims to evaluate the LCOE for the various PV technologies and system configurations deployed at the ERIC research PV plant. To minimise the uncertainties in these calculations actual measurements of module performance and degradation were recorded over the monitoring period. The ERIC research plant was developed for research and development and thus the costs associated with the plant are greater than what would be expected in a normal plant. The purpose of the study is to compare the performance, and hence, the LCOE of the different technologies and array configurations.

2 EXPERIMENAL PROCEDURE

The 400 kW PV plant was designed to test the performance and efficiency of crystalline and thin film PV modules in different array configurations. The plant utilized poly-crystalline Silicon (Si) and thin-film copper indium gallium diselenide (CIGS) modules configured as follows:

- Area 100 - Fixed Tilted North Facing - Crystalline
- Area 200 - Fixed Tilted North Facing - Thin Film
- Area 300 - Fixed Tilted East West - Crystalline
- Area 400 - Fixed Tilted East West - Thin Film
- Area 500 - North/South Tracking - Crystalline
- Area 600 - East/West Tracking - Crystalline
- Area 700 - East/West Tracking - Thin Film

Current-Voltage (IV) measurements of all arrays were taken from September 2016, until May 2021. Performance measurements at Standard Test Conditions (STC) of a sample of PV modules were taken at Nelson Mandela University Photovoltaic Testing Laboratory over this period, to be the baseline measurement against which future degradation of the modules would be compared. Reference (control) modules have been kept in a controlled laboratory environment in the dark to provide comparison and to establish the degradation rate of the deployed modules.

3 RESULTS AND DISCUSSION

3.1. LCOE Calculations

The levelised cost of energy (LCOE) calculations are based on the standard discounting of the total cost of the plant and operation over a given period, and the total discounted energy over that period [5]. The standard formula then applies, viz.:

$$\boldsymbol{LCOE} = \frac{\textit{NPV of total cost (total discounted costs)}}{\textit{NPV of total energy output (total discounted energy)}} \tag{1}$$

where the net present value (NPV) is determined using the standard formula:

$$V_p = V_n \cdot \frac{1}{(1+r)^n} \tag{2}$$

where; V_p is the present value and V_n is the value in year n, and r the discount rate (interest rate).
The factor $\frac{1}{(1+r)^n}$ is the discount factor.

The input variables used are listed below, with a discussion or comment on each.:

Initial Investment Cost (R/kW).

The initial investment costs were more than the typical PV plant costed at the time of construction. Previous studies have calculated the CAPEX costs for a "fixed-tilt" plant the with a range of R 11 2010 - R 13 115/kWp [7]. Taking into consideration that the plant is a research facility, many additional features such as meteorological monitoring and an extensive data monitoring system were installed. In addition, racking and tracking infrastructure cost was heavily skewed because of the plant and respective array sizes. In an attempt to normalise the cost/kWp values, only PV and balance-of-system (BOS) costs were considered, together with normalisation for area and mounting configuration.

Plant size (kWp)

The total plant size is 406.30 kWp and respective Area sizes were used in the estimation of R/kWp cost and the LCOE calculation.

Operations and Maintenance (O&M) Costs (R)

The value of O&M costs was taken as R 1 000 per month per Area.

O&M Growth Rate (inflation, %)

A default inflation rate of 5% was used, unless the value was known.

Annual Electricity Output (kWh)

The following sets of data were used: Original simulations and measured data for 2016 (for all Areas) and modified simulations were also applied for Areas 100, 200 and 600.

Annual Degradation Rate (%)

An annual degradation rate of 0.5 % was utilized in initial calculations. Part of this project is to establish a degradation rate for the technologies deployed.

Project Lifespan (years)

A project lifespan of 20 years was used.

Discount Rate (%)

A discount rate of 11.1 % was used.

Start Date

The project inception date of end of 2014 was used.

Table I lists the R/kWp costs for Areas 100 (Fixed Tilted Crystalline North Facing) and Area 600 (East/West Tracking Crystalline) for the various costing scenarios discussed above and the respective LCOE values. The costing values listed are research plant, published, and estimated and normalised (referred to as realistic). The table shows that the costing for the entire research plant does not reflect a realistic LCOE compared to the published and more realistic estimated values. The "realistic" values based on system costs normalised to area and mounting configuration were used in the analysis that follows. Note the Values given in the paper are 2014 South African Rands (R) (EUR/ZAR DEC 2014 = 14.5).

Table I: Costing scenarios

Area	Costing scenario	Cost per kWp [R/kWp]	LCOE [R/kWh]
100 (100.8 kW)	Research costing	32 726	2.71
	Published (Sklar-Chik et al.) [7]	13 115	1.15
	Realistic (PV+BOS)	17 352	1.49
600 (33.6 kW)	Research costing	38 237	2.83
	Realistic (PV+BOS)	20 274	1.63

Table II lists the energy yield and LCOE values for the different Areas. The energy yield values used for the LCOE calculations are the original simulations and for measured values for all areas. There is a discrepancy between the original simulations and the measured data. The yield values differ by 4-12 %, with the measured values always being less. There are various reasons for this discrepancy, including;

- uncertainty in simulation parameters, with meteorological data contributing the most to the uncertainty,
- instrument measurement error and uncertainty,
- system faults and downtime, including tracker problems and PV module string and problems.

3.2. Measured Degradation

The STC performance measurements on a sample of modules from the site show that crystalline Si modules have degraded as expected, while the thin-film modules have degraded more than the warranty values when compared to the reference module.

The on-site IV measurements have indicated that all areas show some level of degradation, more significantly in the thin-film areas. Thermal Infrared (TIR) and Visual inspections indicate broken modules and soiling issues. Regular TIR and Visual Inspection of the plant identified defects that lower performance and cause safety problems.

The progression of soiling can be seen on a new module (left) and a module that has been deployed for an extended period (right), figure 1. The soiling deposit on the new module can be removed with relative ease with a wet cloth. The soiling deposit on the older module can only be removed with heavy mechanical scraping. Although this soiling appears across the module, it appears to be more severe at the bottom edge of the modules, causing non-uniform shading. This is one factor that has contributed to lower than expected performance.

Figure 1: Visual Image indicating the progression of Soiling deposited on the module from a newer to older deployed module.

Table III lists the 2016 and 2020 energy yield and LCOE values for Areas 100, 200 and 600. Also listed in the table are the degradation rates for the different areas for the period 2016 to 2020. The LCOE values listed are based on the low performance values for 2020, assuming the normal accepted degradation for the remainder of the plant operation (min), and assuming the higher degradation for the remainder of the plant operation (max). Tracking, array and overall system problems make it difficult to make an inter-comparison between technologies and / or configurations in terms of LCOE for measured data. Referring to Table II, the measured yields do, however, show that the modelled yields reflect representative values. The modelled yields show a spread of LCOE from R1.49/kWh for fixed tilt crystalline Silicon to R 2.36/kWh for the fixed tilt East West facing thin film system. It is interesting to note that although the specific yield for tracked arrays is about 20% higher, the LCOE turns out to be higher. This is directly related to the cost of trackers which is disproportionately high for this research plant.

Table II: Energy yield and LCOE values for the different Areas for simulations and measured data.

Area	Configuration and Technology	DC [kWp]	Original Simulation		Measured	
			Yield [MWp]	LCOE Original [R/kWh]	Yield [MWh]	LCOE Measured [R/kWh]
100	Fixed Tilted Crystalline North Facing	100.80	184.150	1.49	169.411	1.62
200	Fixed Tilted Thin Film North Facing	102.66	190.570	1.67	182.241	1.75
300	Fixed West Tilted Crystalline	33.60	55.891	1.87	49.793	2.09
	Fixed East Tilted Crystalline	33.60	56.545	1.84	49.554	2.10
400	Fixed West Tilted Thin Film	17.11	29.330	2.36	27.214	2.54
	Fixed East Tilted Thin Film	17.11	29.615	2.34	27.965	2.47
500	North/South Tracking Crystalline	33.60	65.765	1.80	60.889	1.95
600	East/West Tracking Crystalline	33.60	72.876	1.63	64.454	1.84
700	East/West Tracking - Thin Film	34.22	76.624	1.78	70.325	1.93

Table III: Energy Yield and LCOE values for Areas 100, 200 and 600.

Area		Measured 2016			Measured 2020			
		DC [kWp]	Yield [MWh]	LCOE[R/kWh]	Yield [MWh]	min LCOE [R/kWh]	max LCOE [R/kWh]	2016-2020 Degradation
100	Fixed Tilted Crystalline North Facing	100.8	169.411	1.62	158.453	1.68	1.74	1.6%
200	Fixed Tilted Thin Film North Facing	102.66	182.241	1.75	152.514	1.91	2.11	4.1%
600	East/West Tracking Crystalline	33.6	64.454	1.84	57.781	1.92	2.05	2.6%

The impact of cost/kWp on LCOE, implicit in equation 1 and illustrated in Table I, to a large extent dictates the level of the LCOE value. Since the costing has been estimated, the final LCOE value will then depend on the estimate. This value should be refined, and an appropriate value used dependent on scale of the PV power plant. The other main factor in the LCOE determination is the energy yield. The various performance parameters and final energy values being monitored show a lower than expected performance in terms of energy produced. This has a large impact on the expected final LCOE of the various arrays as showing in Table III and illustrates the importance of maintaining plant performance.

Looking at Area 100 (mc-Si), should the plant continue to degrade at a rate of 1.6% per annum, the LCOE will increase by 7.4% to R1.74/kWh. For Area 200 (thin film) the degradation is much worse, and the 4.1% degradation translates to a 20.6% increase in the LCOE to R2.11/kWh.

This clearly demonstrates the value of monitoring performance and ensuring optimal performance of a PV plant.

4 CONCLUSIONS

The various performance parameters and final energy values being monitored show a lower-than-expected performance in terms of energy produced. This has a large impact on the expected final LCOE of the various arrays and illustrates the importance of maintaining plant performance.

The degradation of deployed PV modules has a significant influence on the Long-term investment and can be due to several different causes. The inspection of ERIC PV plant using Visual, Thermal and UV-F imaging techniques identifies features that require maintenance as they affect the performance and safety of the plant. Periodic cleaning of soiling prevents performance loss prevents irreversible damage to PV modules. The accumulation of sediment on module glass is particularly high in this plant.

The effects of Module degradation on the LCOE clearly demonstrates the value of monitoring performance and ensuring optimal performance of a PV plant.

5 REFERENCES

[1] Department of Energy. 2019. Integrated Resource Plan for Electricity (IRP2019) 2010-2030. South Africa: Department of Energy.

[2] Walwyn, D.R. and Brent, A.C. 2015. Renewable energy gathers steam in South Africa. Renewable and Sustainable Energy Reviews 41: 390–401.

[3] Bega, S. 'Incredible growth' in rooftop solar in South Africa. Mail and Guardian.4 Jul 2022. Available from: https://mg.co.za/environment/2022-07-04-incredible-growth-in-rooftop-solar-in-south-africa/

[4] Solargis, Solar resource maps of South Africa. Available from:https://solargis.com/maps-and-gis-data/download/south-africa

[5] Walter Short, Daniel J. Packey, and Thomas Holt, A Manual for the Economic Evaluation of Energy Efficiency and Renewable Energy Technologies, NREL/TP-462-5173, March 1995.

[6] Serameng T, Roro KT, Basappa Ayanna M, Cunden K, Myeni S, Bischof-Niemz T, et al, Performance monitoring of different module technologies and design configurations of PV systems in South Africa; EU PVSEC 2016; 2016. http://hdl.handle.net/10204/8778.

[7] Sklar-Chik, B and de Kock, Critical Review of the Levelised Cost of Energy Metric, SA Journal of Industrial Engineering, 27, 4, 124-133, 2016.

EXPERIMENTAL PERFORMANCE OF A PV-POWERED CENTER-PIVOT IRRIGATION SYSTEM

Juan I. Herraiz[1,2,*], Rita H. Almeida[1], Manuel Castillo-Cagigal[2], Luis Narvarte[1]
[1]Instituto de Energía Solar – Universidad Politécnica de Madrid, 28031 Madrid, España.
[2]Qualifying PhotoVoltaics. Centro de Empresas la Arboleda. Calle Alan Turing 1, 28031 Madrid, España.
*E-mail: ji.herraiz@upm.es

ABSTRACT: High-power diesel-based or grid-connected irrigation systems are giving way to high-power photovoltaic irrigation systems (PVIS) without batteries that reduce the energy costs by up to 80%. The operation of PVIS is affected by factors other than their quality, and there are no experimental data available about their performance. The traditional performance ratio (*PR*) needs to be factorized to better understand the losses strictly related with the PV system itself, the ones that vary with the crop and its irrigation period, what is intrinsic to the PVIS design, and what happens as a consequence of the behavior of the end-user. This article provides the PV community experimental performance data of a battery-free 160 kWp PV-powered constant pressure center-pivot irrigation system. The system has been analyzed over one year of real operation in which the performance ratio was 48.1%.
Keywords: Performance, PV irrigation system, Water pumping, Stand-alone PV system

1 INTRODUCTION

High-power photovoltaic irrigation systems (PVIS) are increasing worldwide [1][2][3][4], taking the place traditionally held by high-power diesel-based or grid-connected irrigation systems [5]. Rising energy costs in modernized agriculture [6][7][8] is being decisive in accelerating this process.

Until recently, PVIS consisted of a PV generator, a frequency converter (FC) and an alternating current centrifugal motor pump [9][10][11]. The maximum power of the systems was 40 kWp due to fluctuations in PV power that caused instabilities and sudden stops of the FC, producing water hammer in the hydraulic system and overvoltages between the FC output and pump motor [12][13][14]. This problem was solved by integrating batteries [15][16][17], but their high cost prevented the power of the systems from being increased.

In the last decade, the development of new control algorithms [18] solved control instabilities and the use of horizontal north-south axis trackers led on to quasi-constant daily PV power profiles that better adjust to irrigation needs [19][20][21]. New communication protocols and irrigation controllers improved the control of irrigation systems. "Delta" support structures (fixed structures with an east-west orientation) were also proposed for the PV generator to produce constant power profiles [22].

Regarding performance, as the operation of high-power PVIS is affected by external factors other than their quality, the *PR* calculation needs to be reconsidered. In 2018, Almeida et al. [1] proposed to factorize the traditional *PR* into various utilization ratios to quantify and better understand the impact of these external factors on the system performance. The new indices considered different types of losses: losses strictly related with the PV system itself (PR_{PV}); losses that vary with the crop and the irrigation period (UR_{IP}); those intrinsic to the design of the PV generator and the hydraulic infrastructure (UR_{PVIS}); and losses that happen as a consequence of the behavior of the irrigator (UR_{EF}) [1]. These indices where applied to hybrid PV-diesel and PV-grid systems to irrigate olive trees through drip systems (at constant pressure) [1] and, lately, to PVIS pumping to water pools (at variable pressure and water flow)[23]. Anyway, there is still a lack of information available in the literature on experimental performance data that allows us to know what the expected performance is in this type of systems.

This paper provides the PV community experimental performance data of a PV-powered center-pivot irrigation system and opens the door to the knowledge of the expected performance in this kind of systems.

The paper is structured including a methodology section that describes the PVIS under study and the performance indices to be calculated, a results section that presents the results of the performance evaluation and the discussion of these results and a conclusions section that summarizes the main conclusions of the work.

2 METHODOLOGY

2.1 Description of the PVIS

The system is located in Alejos, Spain (41°16'23.4''N, 5°16'48.8''E) and belongs to the cooperative "Estrella de San Juan", irrigating a total area of 150 ha.

The pre-existing system consisted of a diesel generator that fed two frequency converters which in turn fed two pumps: a 92 kW submersible pump (Caprari – E10S50S/6C+MAC10125DS-8V), pumping water from a 140m-deep borehole to an elevated water tank of 1000m^3, and a 30 kW surface centrifugal pump (Caprari – MEC-AS2/80A+FELM 30KW 2P), in charge of pumping water from the tank to the irrigation network at a constant pressure required by an accurate performance of the pivot sprinkles.

The PV system consists of a 160 kWp PV generator, divided into 8 North-South horizontal trackers, and three additional FCs: a 110 kW FC that feeds the 92 kW submersible pump, a 37 kW FC that feeds the 30 kW pump and another 37 kW FC used to feed the motors for the movement of the pivots and working at a constant frequency of 50 Hz to simulate an electrical grid. Figure 1 shows the configuration of the system.

By definition, the system is considered a stand-alone PV pumping system. However, since the pre-existing system was kept, the whole system can operate in three operation modes that the user can manually choose:

- PV mode: the entire irrigation system is fed by the PV generator.
- PV/Diesel mode: pivots and the irrigation pump

are fed by the PV generator while the borehole pump is fed by the diesel generator.

- Diesel mode: the only power supply is the diesel generator.

The PV system is being used in PV mode to pump the maximum volume of water during the day, punctually using the diesel generator when the PV system does not provide enough energy to keep irrigation working during low availability of radiation or at nights.

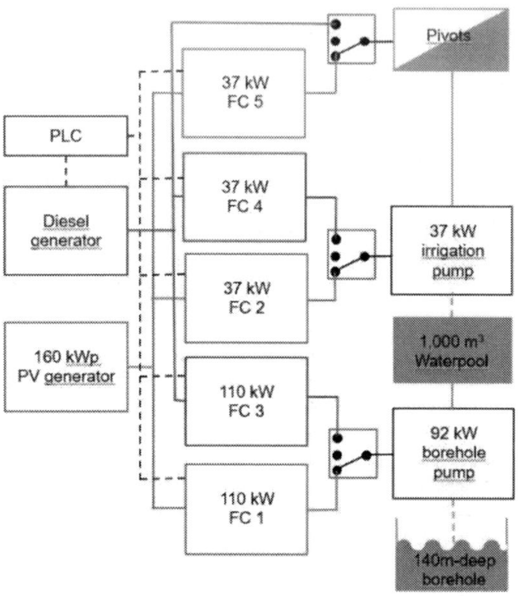

Figure 1. PVIS configuration: A PV generator, three FCs (all in green), and a PLC have been added to the pre-existing configuration (diesel generator and two FCs in red and two motor-pumps).

The system includes an external PLC that depending on the end-user irrigation needs, the selected operation mode, the estimated available PV power and the borehole and tank water levels, controls the start and stop of the FCs. The irrigation FC has priority over the borehole FC in case there is not enough PV power to make both pumps work. Otherwise, the PLC controls the start and stop of the FCs based on a set of thresholds: one start threshold and one stop threshold to operate each FC. When the estimated available power is higher than the first start threshold (35 kW), the PLC sends a start signal to the irrigation FC and the irrigation pump starts operating. Another start signal is sent to the borehole FC when the estimated PV power is greater than a second start threshold (110 kW). If the estimated PV power falls below any stop threshold (51 kW for the second FC and 30 kW for the first one) during a 60 second time interval, a signal is sent to the corresponding FC to stop. In order to avoid start/stop cycles due to tank water levels a hysteresis control based on four thresholds is used. When the tank water level is below a minimum threshold (L1), typically a 5% of the tank capacity, the tank is considered to be empty and the irrigation FC is stopped. It will only be started by the PLC when the tank water level reaches a second threshold (L2), typically a 10% of the tank capacity. Similarly, when the tank water level reaches a maximum threshold (L4), typically 100% of the tank capacity, the borehole FC is stopped and it is only restarted when the water level drops below L3 (usually 90% of the tank capacity).

Besides the PLC operating logic, all FCs have their own control programmes. The FC that feeds the borehole pump is controlled by a Maximum Power Point Tracking (MPPT) routine that maximizes the efficiency of the PV production. The irrigation FC runs a pressure control routine, to keep pressure at the irrigation network constant (there are three different pressure setpoints, 3, 4 and 5 bar, according to the field sector that is being irrigated). And the pivots FC is parameterized to work continuously at 50 Hz, simulating the grid. All FCs have passing cloud routines that allow the system not to be destabilized by fast irradiance fluctuations.

The system is monitored by means of one-minute records of: output frequency, V_{DC}, I_{DC}, P_{DC}, V_{AC}, and I_{AC} of irrigation and borehole FCs, pressure and pressure setpoint for the irrigation network, G and T_c.

2.2 Indicators used to assess the PVIS in real operating conditions

The performance ratio (PR) [24] has been traditionally used to analyse the performance of PV grid-connected plants. It is defined as $PR = E_{PV}/((P^*/G^*)\int Gdt)$, where P^* is the peak power of the PV generator, G^* is the irradiance at standard test conditions (1000W/m²), G is the irradiance on the plane of PV generator, and E_{PV} is the energy produced by the PV system,.

Due to their specificities, High-power PVIS include PR losses not only related to the quality of the PV system itself, but also to the crop and its irrigation period, the intrinsic characteristics of the PV irrigation system design and external circumstances that may affect the PR like the irrigation community decisions. Considering them, the PR can be expressed as [1]:

$$PR = \frac{E_{PV}}{P^*/G^*} \cdot \frac{1}{\int Gdt} \cdot \frac{\int_{IP} Gdt}{\int_{IP} Gdt} \cdot \frac{\int G_{useful}dt}{\int G_{useful}dt} \cdot \frac{\int G_{used}dt}{\int G_{used}dt}$$

where:

IP is the irrigation period determined by the water needs of the crop and, in case of pumping to a water pool, by the relation between water needs, pumping capacity and pumped water storage capacity.

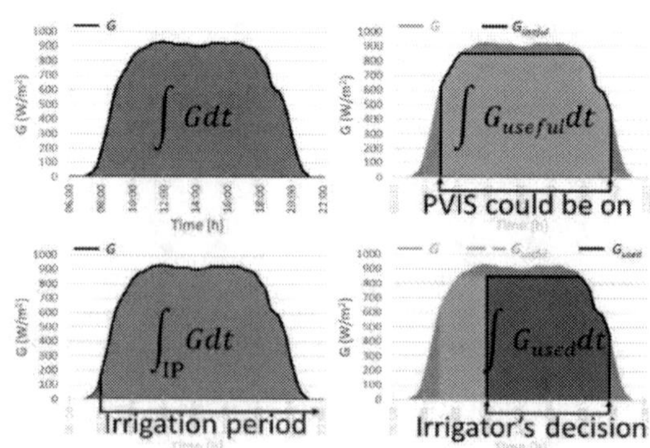

Figure 2. Graphical representation of $\int Gdt$, $\int_{IP} Gdt$, $\int G_{useful}dt$ and $\int G_{used}dt$.

G_{useful} (Figure 2) is the irradiance required to deliver the power needed to pump water, when available, during the IP. It is determined by the relationship between the PV generator nominal power (P^*), the PV generator supporting structure, and the characteristics of the irrigation system. When the available irradiance is below a minimum threshold, the system cannot pump because there is not enough power. When the available irradiance is above G_{useful}, part of the irradiance is wasted because the PV generator cannot supply more power than that consumed by the pump working at its nominal power.

G_{used} (Figure 2) is the irradiance effectively used by the system and depends on external circumstances like the availability of water or the irrigator's decisions about whether to activate the system or not.

The previous equation can be rewritten as

$$PR = PR_{PV} \times UR_{IP} \times UR_{PVIS} \times UR_{EF} \text{ [1]},$$

where:

$$PR_{PV} = \frac{E_{PV}}{P^*/G^*} \cdot \frac{1}{\int G_{used}dt}; \; UR_{IP} = \frac{\int_{IP} Gdt}{\int Gdt}$$

$$UR_{PVIS} = \frac{\int G_{useful}dt}{\int_{IP} Gdt}; \; UR_{EF} = \frac{\int G_{used}dt}{\int G_{useful}dt}$$

PR_{PV} is the PR considering losses strictly related to the PV system itself; UR_{IP} is the utilization ratio considering losses related to the irrigation period; UR_{PVIS} is the utilization ratio considering losses related to the PVIS design (type of irrigation system, the ratio PV peak power - PV power required for irrigation, the tracking geometry and the accuracy of the PLC control algorithms setup); and UR_{EF} is the utilization ratio considering losses related to the irrigator's decisions.

In addition to the *PR*, Herraiz et al. [23] describe the importance of two other indices to assess the correct functioning of PVIS: the number of abrupt stops and the passing-cloud resistance ratio. In order to calculate these indices, the PVIS requires to be monitored by means of one-second records. Since the monitoring system of the PVIS under study only collects data every minute, it has not been possible to obtain these indices in the present work.

3 RESULTS

Monthly performance indices for 2021 are shown in Table I.

Table I. Monthly real performance indices (%)

Month	PR_{PV}	UR_{IP}	UR_{PVIS}	UR_{EF}	PR
Jan	90.0	9.2	86.3	100	7.1
Feb	-	0.0	-	-	0.0
Mar	82.5	59.4	76.9	87.4	32.9
Apr	80.8	100	73.5	96.4	57.3
May	79.3	100	74.4	95.7	56.4
Jun	78.1	100	76.4	90.9	54.3
Jul	78.0	100	81.9	98.0	62.6
Aug	75.7	100	87.6	97.6	64.7
Sep	77.9	100	62.0	93.2	45.0
Oct	76.3	100	74.7	97.0	55.3
Nov	78.8	56.7	62.1	43.5	12.1
Dec	79.5	1.3	66.9	93.4	0.6
Total	78.2	85.7	76.4	93.9	48.1

PR_{PV} values (78.2% for the whole year) are similar to *PR* values expected for grid-connected PV systems.

UR_{IP} is 85.7% because the irrigation period starts on March the 17th and ends on November the 15th. For this reason UR_{IP} is 100% from April to October, close to 50% in March and November, and close to 0 in January, February and December. Values greater than cero in January and December are due to the occasional start-up of the system to fill the water tank.

UR_{PVIS} value is 76.4%. Compared to those reported by [23] for PVIS pumping to a water pool, the obtained value is similar to the 75.6 reported for Aldeanueva PVIS but lower than the 85.2 of Villena PVIS. It is reasonable to expect a lower UR_{PVIS} for a constant flow and constant pressure PVIS than that expected for a PVIS pumping to a water pool, in which the flow of pumped water can be adjusted to the available power. Similar UR_{PVIS} values for both types of systems must be explained based on the design decisions for each system (for example, an oversizing of the PV generator of the PVIS pumping to a water pool would explain the lower UR_{PVIS}).

The design factors that most affected the UR_{PVIS} value of the PVIS under study during 2021 are:

- The ratio PV peak power - PV power required for irrigation. At the beginning of the day, or whenever the irradiance is low enough, the PV generator does not generate enough power to start the motor-pumps (depending on the irradiance and the PV peak power). The PLC estimates the available power and starts the FCs when possible according to the stablished thresholds. On the other hand, when the irradiance is high the PV generator can provide more power than the maximum that can be consumed by the system. The difference between the available irradiance and the maximum irradiance needed to make the whole system work is wasted. UR_{PVIS} can be improved by adjusting the PV peak power and the start and stop thresholds for all the FCs, but there is always a part of the irradiation that is wasted.
- The time the PLC waits to start a FC after it has been stopped. To protect the pumps from continuous stops and starts, the PLC waits a certain time (8 minutes for the borehole FC) after a stop until it commands a new start. On days with many passing clouds, in which the irradiance can vary abruptly and drop below the minimum thresholds required to operate the system, the FCs remain stopped, at 8-minute intervals, part of the irrigation time.
- The water tank capacity. The tank has been empty or completely full at some points of the irrigation period. A full tank prevents de PLC to start the borehole FC, while an empty tank stops the irrigation FC. A larger tank capacity would have mitigated this effect.

UR_{EF} value (93.9%) is smaller than expected (close to 100%). The possibility of operating the system in diesel mode and the water needs of the crop have led the irrigator to activate the system in diesel mode at night and in times of low irradiance. The delay in switching to PV mode when the irradiance allowed it has caused the UR_{EF} to worsen. This effect has been increased by long stops of the system when the irrigator changed the irrigation program.

PR is below 50% (48.1%) due to the combined effect of all the indices, but specially UR_{IP} (highly dependent on the crop) and UR_{EF} (which is lower than expected). If UR_{EF} had been closer to 100%, assuming the same values for the rest of factors, the *PR* would have reached 51.2%

Data show that UR_{IP} is very dependent on the crop and that it is reasonable to expect PR_{PV} values close to or even above 80.0%, UR_{PVIS} values above 75.0%, and UR_{EF} values very close to 100% when the irrigator uses properly the PVIS (PV irrigation centred at midday, planning of maintenance tasks in cloudy days or during the night, management of water consumption,...).

PR_{PV}, UR_{IP}, UR_{PVIS} and UR_{EF} determine PR values. Their expected values make us suggest that it is reasonable to expect PR values above 50% in high-power constant pressure PVIS. But this statement must be confirmed with future measurements in other similar systems

4 CONCLUSIONS

Stand-alone high-power PV irrigation systems without batteries is a recent innovation and there are not enough experimental data about the performance of this kind of systems available. This paper is a contribution to the knowledge of the experimental performance data of a constant pressure center-pivot PV irrigation system and, thus, to the expected values for other similar systems.

To assess the systems performance, the traditional PR has been factorized into four factors: PR_{PV} (evaluates the performance quality of the PV system itself), UR_{IP} (measures the effect of the crop irrigation period), UR_{PVIS} (measures the effect of the hydraulic irrigation system) and UR_{EF} (considers losses due to the user behavior).

Relevant conclusions can be drawn from the results about the performance that can be expected from this type of system: The expected PR in good quality constant pressure PV irrigation systems is over 50%. The expected values for the different factors are: PR_{PV}>=80%, similar to what is expected in grid-connected PV systems; UR_{IP}>=85% if the needs of the crop are extended for most of the year; UR_{PVIS}>75% and $UR_{EF}\approx$100% if the end user makes appropriate use of the PV irrigation system.

Nevertheless, it is necessary to confirm these expected values with experimental measurements in other high-power constant pressure PV irrigation systems.

5 ACKNOWLEDGMENTS

This work has been possible thanks to the funding from the European Union's Horizon 2020 research and innovation programme to SOLAQUA project under grant agreement N° 952879, and thanks to the Project MADRID-PV2 (P2018/EMT-4308) funded by the Comunidad de Madrid with the support from ERDF Funds.

6 REFERENCES

[1] Almeida, R.H., Carrêlo, I.B., Lorenzo, E., Narvarte, L., Fernández-Ramos, J., Martínez-Moreno, F., Carrasco, L.M., 2018a. Development and test of solutions to enlarge the power of PV irrigation and application to a 140 kW PV-diesel representative case. Energies 11, 3538-undefined. https://doi.org/10.3390/en11123538

[2] MASLOWATEN, 2016. MASLOWATEN H2020 project. [WWW Document]. URL http://maslowaten.eu/ (accessed 4.1.21).

[3] Todde, G., Murgia, L., Deligios, P.A., Hogan, R., Carrelo, I., Moreira, M., Pazzona, A., Ledda, L., Narvarte, L., 2019. Energy and environmental performances of hybrid photovoltaic irrigation systems in Mediterranean intensive and super-intensive olive orchards. Sci. Total Environ. 651, 2514–2523. https://doi.org/10.1016/j.scitotenv.2018.10.175

[4] Todde, G., Murgia, L., Carrelo, I., Hogan, R., Pazzona, A., Ledda, L., Narvarte, L., 2018. Embodied energy and environmental impact of large-power stand-alone photovoltaic irrigation systems. Energies 11, 2110-undefined. https://doi.org/10.3390/en11082110

[5] J. Carroquino, R. Dufo-López and J. L. BarnalAugustin, "Sizing of off-grid renewable energy systems for drip irrigation in Mediterranean crops," Renewable Energy, no. 76, pp. 566-574, 2015.

[6] Abu-Aligah, M., 2011. Design of photovoltaic water pumping system and compare it with diesel powered pump. Jordan J. Mech. Ind. Eng. 5, 272–280.

[7] Langarita, R., Sánchez Chóliz, J., Sarasa, C., Duarte, R., Jiménez, S., 2016. Electricity costs in irrigated agriculture: A case study for an irrigation scheme in Spain. Renew. Sustain. Energy Rev. 68, 1008–1019. https://doi.org/10.1016/j.rser.2016.05.075

[8] Lorenzo, C., Almeida, R.H., Martínez-Núñez, M., Narvarte, L., Carrasco, L.M., 2018. Economic assessment of large power photovoltaic irrigation systems in the ECOWAS region. Energy 155, 992–1003. https://doi.org/10.1016/j.energy.2018.05.066

[9] Abella, M.A., Lorenzo, E., Chenlo, F., 2003. PV water pumping systems based on standard frequency converters. Prog. Photovoltaics Res. Appl. 11, 179–191. https://doi.org/10.1002/pip.475

[10] Posorski, R., 1996. Photovoltaic water pumps, an attractive tool for rural drinking water supply, in: Solar Energy, 58. pp. 155–163. https://doi.org/10.1016/S0038-092X(96)00060-6

[11] Sontake, V.C., Kalamkar, V.R., 2016. Solar photovoltaic water pumping system - A comprehensive review. Renew. Sustain. Energy Rev. https://doi.org/10.1016/j.rser.2016.01.021

[12] Achouri, R., Nouicer, O., Mhiri, H., Bournot, P., 2013. Probable cause analysis of cracks observed on vertical centrifugal pump. Eng. Fail. Anal. 29, 1–11. https://doi.org/10.1016/j.engfailanal.2012.11.003

[13] EIP-Water, 2013. Priority area P30: water-energy nexus. European innovation partnership on water boosting opportunities – innovating water. [WWW Document]. URL https://watereurope.eu/tag/eip-water/ (accessed 4.1.21).

[14] Ismaier, A., Schlücker, E., 2009. Fluid dynamic interaction between water hammer and centrifugal pumps. Nucl. Eng. Des. 239, 3151–3154. https://doi.org/10.1016/j.nucengdes.2009.08.028

[15] Bhayo, B.A., Al-Kayiem, H.H., Gilani, S.I., 2019. Assessment of standalone solar PV-Battery system for electricity generation and utilization of excess power for water pumping. Sol. Energy 194, 766–776. https://doi.org/10.1016/j.solener.2019.11.026

[16] Das, M., Mandal, R., 2018. A comparative performance analysis of direct, with battery,

supercapacitor, and battery-supercapacitor enabled photovoltaic water pumping systems using centrifugal pump. Sol. Energy 171, 302–309. https://doi.org/10.1016/j.solener.2018.06.069
[17] Khiareddine, A., Ben Salah, C., Mimouni, M.F., 2015. Power management of a photovoltaic/battery pumping system in agricultural experiment station. Sol. Energy 112, 319–338. https://doi.org/10.1016/j.solener.2014.11.020
[18] Fernández-Ramos, J., Narvarte-Fernández, L., Poza-Saura, F., 2010. Improvement of photovoltaic pumping systems based on standard frequency converters by means of programmable logic controllers. Sol. Energy 84, 101–109. https://doi.org/10.1016/j.solener.2009.10.013
[19] Bione, J., Vilela, O.C., Fraidenraich, N., 2004. Comparison of the performance of PV water pumping systems driven by fixed, tracking and V-trough generators. Sol. Energy 76, 703–711. https://doi.org/10.1016/j.solener.2004.01.003
[20] Narvarte, L., Fernández-Ramos, J., Martínez-Moreno, F., Carrasco, L.M., Almeida, R.H., Carrêlo, I.B., 2018. Solutions for adapting photovoltaics to large power irrigation systems for agriculture. Sustain. Energy Technol. Assessments 29, 119–130. https://doi.org/10.1016/j.seta.2018.07.004
[21] Narvarte, L., Lorenzo, E., 2008. Tracking and ground cover ratio. Prog. Photovoltaics Res. Appl. 16, 1–12. https://doi.org/10.1002/pip.847
[22] Almeida, R.H., Narvarte, L., Lorenzo, E., 2018b. PV arrays with delta structures for constant irradiance daily profiles. Sol. Energy 171, 23–30. https://doi.org/10.1016/j.solener.2018.06.066
[23] Juan Ignacio Herraiz, José Fernández-Ramos, Rita Hogan Almeida, Eva María Báguena, Manuel Castillo-Cagigal, Luis Narvarte, On the tuning and performance of Stand-Alone Large-Power PV irrigation systems, Energy Conversions and Management: X, Volume 13, 2022, 100175, ISSN 2590-1745, https://doi.org/10.1016/j.ecmx.2021.100175
[24] Luque, A., Hegedus, S., 2003. Handbook of Photovoltaic Science and Engineering, Handbook of Photovoltaic Science and Engineering. Chichester: Johm Wiley & Sons. https://doi.org/10.1002/9780470974704

MODELING INVERTERS WITH MULTIPLE INPUTS: TEST PROCEDURE FOR MEASURING EFFICIENCY

Clifford Hansen, Jay Johnson, Rachid Darbali-Zamora, Nicholas S. Gurule, Sigifredo Gonzalez, Marios Theristis
Sandia National Laboratories
Albuquerque, NM, 87185, USA

ABSTRACT: Photovoltaic (PV) inverters convert DC power to AC power. Inverters typically employ maximum power point tracking (MPPT) algorithms to maximize power production. Many modern inverters support several independent MPPT inputs to maximize energy production from arrays with different configurations or orientations. There is no consensus test procedure for evaluating the DC-to-AC conversion efficiency for multi-input inverters. Herein, we propose a test procedure based on the open-source System Validation Platform (SVP) software. We apply the procedure to a commercial inverter with six MPPT inputs to demonstrate that the resulting measurements can be used to fit a model that predicts inverter power at all conditions with reasonable accuracy.

Keywords: inverter, experimental methods, modelling

1 INTRODUCTION

Photovoltaic (PV) inverter efficiency is a critical metric for solar procurement as it directly affects power plant production and, therefore, owner or operator income. It also correlates to device lifetime because losses increase internal component temperatures that cause thermal degradation of the components [1]. In 2004, Bower et al. [2] published a procedure for measuring PV inverter efficiency over a range of test conditions. This procedure produced data that was used to fit the PV inverter model in [3]. Test results for many PV inverters are recorded in the California Energy Commission (CEC) Equipment List [4]. The Bower et al. test procedure is also being applied to multi-input PV inverters but only for conditions of equal DC voltage and DC power applied to each input [5].

In this work, we aim to create a conversion efficiency test procedure for multi-input devices that includes a representative collection of power and voltage inputs. None of the previous test protocols consider the multi-input devices, despite many products on the market with this capability. To design and test the protocol, the team used the open-source System Validation Platform (SVP) to autonomously and accurately capture and analyze hundreds of multi-input efficiency measurements. This SVP software platform has been used in the past as a distributed energy resource (DER) management system [6], a tool for PV inverter reliability measurements [7], and extensively as a platform for interoperability [8] and interconnection standard certifications [9], [10]. The advantage of using the SVP to run the experiments is that a larger range of input parameters can be measured in a short period of time.

The data collected by the test procedure can be used to fit extensions of existing PV inverter models in order to accurately model conversion efficiency for PV inverters operating with unbalanced inputs. The method for extending an PV inverter model, and demonstration of model accuracy, are reported in Hansen et al. [5].

The primary contributions of this work are (a) the establishment of the first multi-input conversion efficiency test protocol and (b) open-sourcing the evaluation environment and test script. The remainder of the paper is structured such that Section II discusses the proposed test protocol, Section III describes the test equipment and process, Section VI discusses the experimental results, and Section V presents conclusions.

2 MULTI-INPUT PHOTOVOLTAIC INVERTER TEST PROTOCOL

A PV inverter with several Maximum Power Point Trackers (MPPTs) comprises two functional stages in sequence:

- A DC-to-DC converter on each input, which holds the connected array at the array's maximum power point (MPP) and converts the input DC voltage to a DC bus at a common DC voltage.
- A DC-to-AC inverter stage which produces AC power from the DC power on the DC bus.

Typically, each DC-DC converter controls voltage (and current) from the arrays connected to its input, and the DC-AC inverter controls the DC voltage on the common bus. Fig. 1 illustrates a block diagram of a PV inverter with multiple MPPTs and assigning variables to DC voltage and power on each input and on the bus.

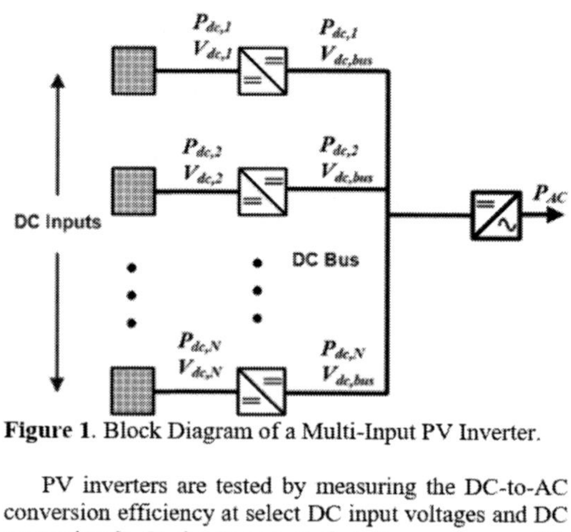

Figure 1. Block Diagram of a Multi-Input PV Inverter.

PV inverters are tested by measuring the DC-to-AC conversion efficiency at select DC input voltages and DC power levels. In the Bower et al. test procedure, multi-input PV inverters are tested by holding each input at the same DC voltage and distributing the total input DC power equally among inputs. In contrast, our test protocol measures PV inverter efficiency over a range of balanced and unbalanced input conditions.

We implement the test procedure in two test sequences. Test sequence 1 evaluates inputs at 100% or 50% of nameplate power rating with all combinations of minimum or maximum DC voltage on each input, as would be typical in installations with inputs connected to strings of different length and/or shading conditions. Test sequence 2 evaluates the PV inverter with all inputs at one of three DC voltages in combination with several inputs with DC power at 20%, 40%, 60%, or 80%, and the remaining inputs at 100% of nameplate power rating. Test 2 models a system with arrays at different tilt and/or azimuth orientations. Fig. 2 and Fig. 3 present pseudo code for the test sequences. Both test sequences are included in the python script multi_mppt_analysis.py [11].

```
for 1st n inputs derated by 50% in [1…n]
  for all input combinations of {Vmin, Vmax}
    set the I-V curves for the DC inputs
    calculate average MPPT efficiency from m
    measurements
    calculate average conversion efficiency
    from m measurements
  end
end
```

Figure. 2. Algorithm for Test Sequence 1.

```
for MPP voltage in [Vmin, Vnom, Vmax]
  for derating in [0.2, 0.4, 0.6, 0.8]
    for all string combinations in {1.0,
    derating}
      set the I-V curves for the DC inputs
      calculate average MPPT efficiency from
      m measurements
      calculate average conversion
      efficiency from m measurements
    end
  end
end
```

Figure. 3. Algorithm for Test Sequence 2.

3 TEST PROCEDURE

We applied the test protocol to a commercial SMA Tripower Core1 PV inverter, with six MPPT inputs, a power rating of 33 kVA, and an operating voltage of 480 V_{AC}. Test 1 included 448 configurations ($2^6 = 64$ DC voltage combinations with 0, 1, 2, 3, 4, 5 or 6 derated MPP inputs). Test 2 included 768 configurations (3 voltage levels with 4 derating levels and 64 input combinations). Table I summarizes the PV inverter parameters.

Table I: Description of the inverter parameters

Description	Symbol	Value	Unit
Rated Power	P_{AC}	33	kW
Nominal AC Voltage	V_{AC}	480	VAC
Nominal Frequency	f	60	Hz
Maximum DC Voltage	Vmax	800	VDC
Minimum DC Voltage	Vmin	600	VDC
Nominal DC Voltage	Vnom	720	VDC

Fig. 4 illustrates a block diagram of the laboratory configuration. DC inputs for the inverter were simulated using six channels of a ten-channel AMETEK TerraSAS PV Simulator. Each channel can source DC power that mimics a PV string's IV curve with short-circuit current up to 10 A and open-circuit voltage up to 1000 V_{DC}. AC output from the inverter was exported to the local distribution grid. An Opal-RT real-time grid simulator was used as a convenient calibrated data acquisition system (we did not use its grid simulation capability). The AC current and voltage transducers were calibrated prior to experimentation using a Fluke 5520A High Performance multi-Product Calibrator. The calibration curve for each transducer was calculated by and obtained from a LabView data acquisition system.

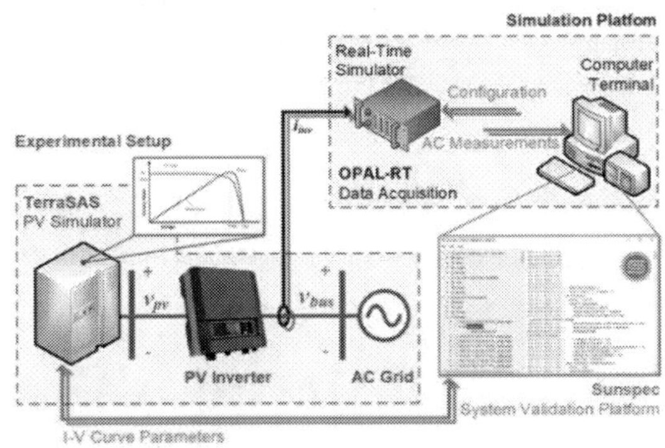

Figure 4. Block Diagram of the Laboratory Testing Configuration.

Experiments were automated by using the open-source System Validation Platform (SVP) software [12] running on a Windows 10 computer connected to the Opal-RT real-time simulator. In the configuration stage, the SVP software set the Opal-RT experiment using the OPAL-RT RT-LAB Python Application Programming Interface (API) and the TerraSAS I-V curves using IEEE 488.2 Standard Commands for Programmable Instrumentation (SCPI) commands over Ethernet.

During the measurement phase, the SVP interacted with the TerraSAS PV Simulator system to record the DC current and voltage. The SVP captured the Opal-RT AC RMS voltage, current, and power for each phase using MATLAB/Simulink.

Test 1 was completed using the SVP in 73 hours, and Test 2 was completed in 10 hours. Test 1 took significantly longer because of the changes in DC voltage. After each change, the PV simulator had to be restarted and the inverter allowed to re-establish maximum power point tracking (MPPT), a process that required several minutes. We first attempted to change the DC voltage without disconnecting the PV simulator, but the inverter would not re-establish MPPT. It may be possible to gradually change DC voltage to avoid disconnecting the PV simulator.

SVP can create a manifest of results within the script log, measurement data files, and a results summary. In each test sequence and at each configuration, 50 measurements were recorded of total AC power (sum over 3 phases), DC power at each input, and other quantities (complete files at [11]). Values of AC power, total DC power and DC power on each input were screened to discard values outside of 150% of the inter-quartile range (75th – 25th percentile). In addition, we noted three instances in Test 2 where the mean DC power on one or more inputs differed by more than 2% from the target power; we discarded these measurements. We surmise that the drift away from target occurs when the inverter MPPT algorithm steers the PV simulator away from the maximum power point on its simulated I-V curve. After filtering, the mean of remaining values is recorded and used to calculate efficiency (total AC power / total DC power) and to fit an inverter efficiency model [5]. Python scripts to perform data filtering, model fitting and analysis are published at [11].

4 RESULTS

Fig. 5 displays the standard deviation of efficiency values for the 768 configurations comprising Test 2. Efficiency is consistently measured within 0.1% (one standard deviation). Full datasets and analysis scripts are available at the GitHub repository [11].

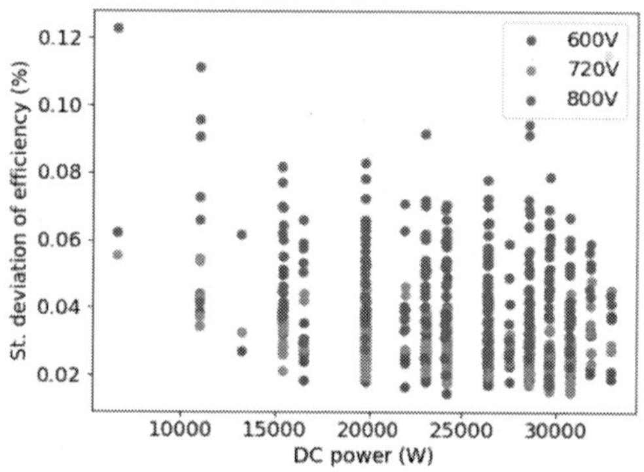

Figure 5. Variation in measured efficiency for Test 2.

Results from Test 1 and Test 2 are combined to form a data set suitable for fitting and validating models to predict inverter efficiency at all operating conditions. Hansen *et al.* [5] describe an extension of the Sandia inverter model [3] for multi-input devices. Here, we follow the current practice of the California Energy Commission (CEC) and fit the model to the "equal power" subset of test data. This subset comprises configurations where the target power is equal for each input as specified by the test procedure of Bower *et al.* [2]. The remaining data are used for validating the model predictions.

Fig. 6 illustrates measured and predicted efficiency and error in predicted AC power. The results demonstrate that the proposed model is generally unbiased with prediction accuracy between ±0.5% (Fig. 6b). Prediction error increases with AC power.

Fig. 6a compares modeled and measured efficiency for the "equal voltage" subset of data (configurations with the same V_{DC} on each input). Restricting the plot to this subset allows for comparing measured and modeled efficiency by voltage level. Fig. 6a shows variance in efficiency at each level of DC power and DC voltage that does not correlate with the DC power level nor with any specific MPPT input ([5], Fig. 4). For unknown reasons, among the 600 V_{DC} results, several configurations from Test sequence 1 show markedly lower mean efficiency than the rest of the data sample.

(a)

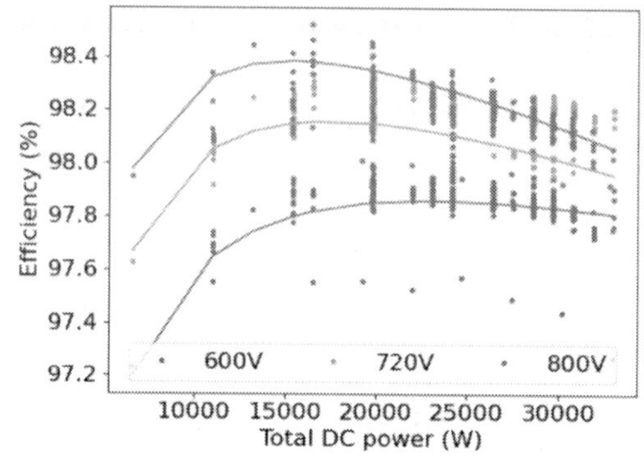

(b)

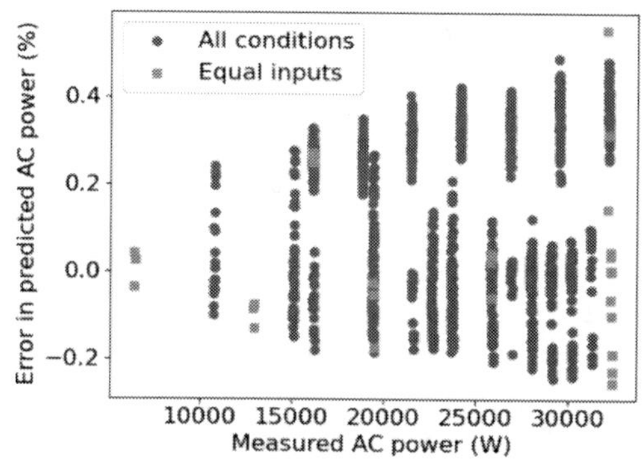

Figure 6. Experimental results: (a) measured (dots) and modeled (lines) efficiency; (b) error in modeled AC power. In (b), "Equal inputs" are points used for model fitting.

5 CONCLUSIONS

In solar installations with multiple arrays configured with different orientations, module types, or topologies, multi-input solar inverters are often used to maximize energy production. Currently a test procedure designed for single-input inverters is being used to measure efficiency of multi-input inverters, and data from these measurements are used to fit efficiency models. This work describes a procedure to evaluate multi-input inverters with different combinations of DC voltage and/or power on the PV inverter's inputs. The automated test procedure captured hundreds of efficiency measurements. Analysis shows that the results of this procedure can be used to parameterize models to predict inverter AC power. However, the measurements require careful filtering to exclude values that may be erroneous, or that indicate that the test equipment was not operating as intended.

6 ACKNOWLEDGEMENTS

Sandia National Laboratories is a multi-mission

laboratory managed and operated by National Technology and Engineering Solutions of Sandia, LLC., a wholly owned subsidiary of Honeywell International, Inc., for the U.S. Department of Energy's National Nuclear Security Administration under contract DE-NA-0003525.

7 REFERENCES

[1] J. Flicker, J. Johnson, P. Hacke, R. Thiagarajan, "Automating Component-Level Stress Measurements for Inverter Reliability Estimation," Energies, vol. 15, no. 13, p. 4828, July 2022, https://doi.org/10.3390/en15134828.

[2] W. Bower, C. Whitaker, W. Erdman, M. Behnke, M. Fitzgerald. "Performance Test Protocol for Evaluating Inverters Used in Grid-Connected Photovoltaic Systems". Available at https://www.energy.ca.gov/sites/default/files/2020-06/2004-11-22_Sandia_Test_Protocol_ada.pdf

[3] D. L. King, S. Gonzalez, G. M. Galbraith, and W. E. Boyson. "Performance Model for Grid-Connected Photovoltaic Inverters". Sandia Report SAND2007-5036, September 2007.

[4] California Energy Commission, Solar Equipment Lists, accessed 6/29/22, URL: https://www.energy.ca.gov/programs-and-topics/programs/solar-equipment-lists

[5] C. Hansen, J. Johnson, R. Darbali, N. Gurule, "Modeling Efficiency of Inverters with Multiple Inputs," 49th IEEE Photovoltaic Specialists Conference (PVSC), Philadelphia, PA, USA 2022.

[6] J. Johnson, A. Summers, R. Darbali-Zamora, J. Hernandez-Alvidrez, J. Quiroz, D. Arnold, J. Anandan, "Distribution Voltage Regulation using Extremum Seeking Control with Power Hardware-in-the-Loop," IEEE Journal of Photovoltaics, vol. 8, no. 6, pp. 1824-1832, 2018. doi: 10.1109/JPHOTOV.2018.2869758

[7] J. Flicker, J. Johnson, P. Hacke, R. Thiagarajan, "Automating Component-Level Stress Measurements for Inverter Reliability Estimation," IEEE Photovoltaic Specialists Conference (PVSC), 2021.

[8] J. Johnson, B. Fox, K. Kaur and J. Anandan, "Evaluation of Interoperable Distributed Energy Resources to IEEE 1547.1 Using SunSpec Modbus, IEEE 1815, and IEEE 2030.5," in IEEE Access, vol. 9, pp. 142129-142146, 2021, https://doi.org/10.1109/ACCESS.2021.3120304.

[9] N. Ninad, E. Apablaza-Arancibia, M. Bui, and J. Johnson, "Commercial PV Inverter IEEE 1547.1 Ride-Through Assessments Using an Automated PHIL Test Platform," Energies, vol. 14, no. 21, p. 6936, Oct. 2021, doi: 10.3390/en14216936.

[10] J. Johnson, R. Ablinger, R. Bruendlinger, B. Fox, J. Flicker, "Interconnection Standard Grid-Support Function Evaluations using an Automated Hardware-in-the-Loop Testbed," IEEE Journal of Photovoltaics, vol. 8, no. 2, pp. 565-571, Mar 2018. doi: 10.1109/JPHOTOV.2018.2794884

[11] https://github.com/jayatsandia/svp_additional_tool

[12] System Validation Platform, available at https://sunspec.org/software/

SOCIALRES: SOCIAL INNOVATIONS AS THE PATHWAY TOWARDS ENERGY DEMOCRACY AND AS SUSTAINABLE SOLUTION FOR THE MITIGATION OF THE GLOBAL WARMING

Silvia Caneva[1], Ingrid Weiss[1], Iban Lizarralde[2], Mikhail Hamwi[2], Audrey Abi Akle[2], Basma Samir[2], Volker Kromrey[3], Dimitri Vedel[3], Jung Lin[3], Andreas Schneller[4], Jakob Hoffmann[4], Kathrin Kohl[4], Lisa Richter[4], Elisabeth Schmid[5], Claudia Crippa[5], Eleanor Denny[6], Ivan Petrov[6], Sofia Mulero[7], Alejandro Hernández Serrano[7], Maria Regidor[8], Iban Lacoste[8], Rodrigo Ruiz[9], Santiago Campos[9], Eva Otero[9], Nuno Brito[10] Jorge[10], Inês Branco Lopes[10], Karl Harder[11], Tijana Šimek[12], Alexandru Costeniuc [13]

[1]WIP – RENEWABLE ENERGIES, [2]ESTIA INSTITUTE OF TECHNOLOGY, [3]BODENSEE-STIFTUNG / LAKE CONSTANCE FOUNDATION, [4]ADELPHI, [5]FONDAZIONE ICONS, [6]TRINITY COLLEGE DUBLIN, [7]CARTIF, [8]I-ENER, [9]ENERGETICA, [10]POWER PARITY, [11]ABUNDANCE, [12]REGEA, [13]TRACTEBEL

[1] WIP – Renewable Energies
Sylvenstein Strasse 2
Munich, Germany
silvia.caneva@wip-munich.de

ABSTRACT

The actual climate crisis generated by the global warming requires a strong civic participation to find a sustainable solution to overcome this critical time. The COP26, the UN climate conference, which took place last year in Glasgow, has the highest participation rate in COP´s history despite the global pandemic. The world was expecting more ambitious actions for a clean global energy transition to achieve the 1.5 degree-goal. It is now clear that this ambitious goal cannot be achieved only by political decisions. A strong civic participation is required to make sure that the global warming will not lead to an irreversible status of the climatic crisis characterized by climate events, which will make the life of human beings not more sustainable on this planet. Social innovations are the key to overcome this crisis and provide a sustainable solution to the mitigation of global warming towards the achievement of 1.5 degree-goal.

Social innovations are defined as *"New ideas (products, services and models) that simultaneously meet social needs (more effectively than alternatives) and create new social relationships or collaborations. In other words, they are innovations that are not only good for society but also enhance society's capacity to act".* Source: Empowering people, driving change, Bureau of European Advisers (BEPA), Brussels (2011).

In the renewable energy sector, social innovations are identified as the business models able to support the production of renewable energy and, at the same time, the cooperation among citizens. As common denominator they have the concept of "power of many" to highlight that, through the union of small contributions, it is possible to build projects having positive environmental, economic and social impacts. The business models implementing social innovation in the renewable energy sector are cooperatives, aggregators and crowdfunding platforms. The development of renewable energies and a greater civic participation will be the two pillars of the clean energy transition. Social innovations can combine these two fundamental aspects and lead to the energy democracy and, as consequence, to a mitigation of the global warming. This paper has been prepared in the framework of the European project SocialRES (Fig. 1), which has the aim to investigate the business models of social innovations in the renewable energy sector to provide the owners or managers of crowdfunding platforms and cooperatives and aggregators, named "the facilitators of energy democracy", with new tools to further support the production of renewable energy and the participation of citizens.

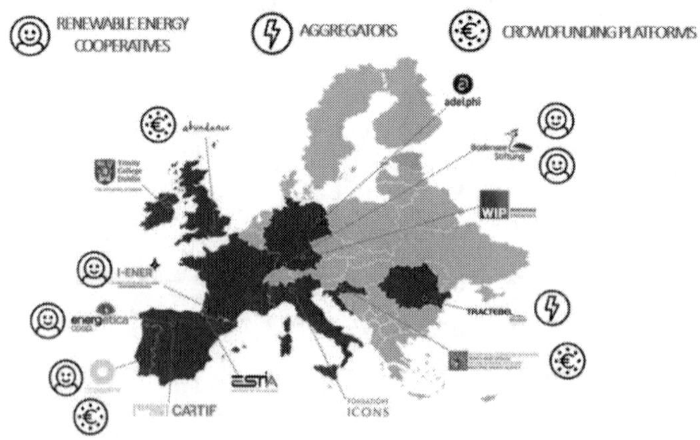

Fig. 1 SocialRES geographical and social innovation coverage

Keywords: Energy Democracy, Civic Participation, Social innovation, Business Model, Crowdfunding, Aggregation, Cooperatives, Global Warming, Climatic Crisis, Photovoltaic, Prosumer, Financing, Markets, Utilities

1 AIM AND APPROACH USED

This paper aims to provide the facilitators of energy democracy, i.e. the owners or managers of crowdfunding platforms, cooperatives and aggregators, with two new tools to further support their businesses in the production of renewable energy and the participation of citizens to the clean energy transition.

The tools presented with this paper have been developed in the frame of the H2020 European project SocialRES. The aim of SocialRES is to close non-technological research gaps that impede the widespread uptake of social innovation business and service models in the European energy sector.

The SocialRES consortium brings together thirteen partners who have been chosen specifically for their expertise and experience to ensure the successful implementation of the novel research proposed in this project and to guarantee a widespread geographical coverage (Fig.1). The SocialRES project partners are briefly described in the following paragraph.

1. WIP – Renewable Energies (WIP), based in Germany, the project coordinator.
2. ESTIA Institute of Technology (ESTIA), based in France, with extensive experience in research aspect of social innovation and innovative business models.
3. CARTIF, energy expert based in Spain.
4. Bodensee-Stiftung - Lake Constance Foundation (LCF), based in Germany, as networking and innovation expert and providers of case studies for RES cooperatives.
5. Adelphi, policy think tank based in Germany.
6. Fondazione iCons (ICONS), communication, dissemination and exploitation expert based in Italy.
7. Trinity College Dublin (TCD), University based in Ireland, with extensive experience in behavioural and energy economics, and consumer engagement.
8. I-ENER, RES cooperative based in France, involved in social innovation research projects.
9. EnergEtica, RES cooperative based in Spain.
10. Power Parity, RES cooperative and crowdfunding platform based in Portugal.
11. Abundance, RES crowdfunding platform based in United Kingdom.
12. REGEA, RES crowdfunding platform based in Croatia.
13. Tractebel, RES aggregator based in Romania.

The first tool described in this paper is the ***SOGUI framework*** developed by the SocialRES partner ESTIA in cooperation with the representatives of the SocialRES case studies of crowdfunding, cooperatives, and aggregator above-mentioned. The SOGUI framework can be seen ***as a guideline for actors that are willing to launch a social innovation in the energy sector.***

The second tool is ***toolkit for policy assessment of social innovations*** developed by adelphi with the support of the SocialRES case studies. The toolkit has been developed in the form of a ***taxonomy for measuring impacts of social innovations.*** The taxonomy provides information for each impact, including indicators, scope of the impacts, methodological approaches, as well as potential data sources for the measured impacts. The taxonomy also illustrates the social economic and environmental impacts that select case study partners from the SocialRES project have had.

2 SCIENTIFIC INNOVATION AND RELEVANCE

Social innovations for the renewable energy sector will become more and more important in the upcoming years to ensure the achievement of the ambitious target of 1.5 degree to mitigate the global warming and provide a sustainable solution for the climatic crisis. The social innovations investigated in this paper are the business models of cooperatives, aggregators and crowdfunding platforms, which have been established in the market since several years, as for the aggregators, or even decades as crowdfunding platforms and cooperatives. Nevertheless, further support is required to make sure that these forms of civic participation can further be optimized to involve more citizens in the future. The tools described in this paper provide a fundamental support to the establishment of new business models of social innovations (***SOGUI framework)*** and to assess, and therefore optimize the existing social innovations (***Toolkit for policy assessment of social innovations).***

3 PRELIMINARY RESULTS AND CONCLUSIONS

3.1 The SOGUI framework for launching a social innovation

The SOGUI framework has the aim to support stakeholders in the creation of social innovations, such as cooperatives, crowdfunding platforms, or business models for aggregation, in the renewable energy sector.

After a period of reflection on a specific topic related to energy (from production to consumption), initiatives and projects are created to face the problematics related to it. To facilitate and support the implementation of these energy-related projects, the SOGUI framework allows to define the pathway of the energy project to be developed. This will help to anticipate the organization of the different activities to be undertaken and the definition of different stakeholders depending on the needed skills and used technologies.

The SOGUI framework is divided in four parts (Fig. 2):

- The technologies that can be used for each activity in the energy field are proposed: solar PV, wind, wood, solar thermal, biogas, hydro, sea energy, biofuel, electric vehicle, batteries, hydrogen and thermal.
- Different activities in the energy field are listed, to choose the ones that will be used to develop the defined project: electricity generation, electricity retailers, energy savings, heat generation, self-consumption, storage, mobility, demand response, electricity distribution, heat distribution, energy poverty, community building, citizens meet local authorities, aggregation and funding.
- The needed competences/skills to accomplish these activities are also proposed: finance, planning and governance, energy efficiency, renewable technics, energy flexibility, digital, logistics and transport, smart grid, energy markets, communication, legal skills, teamwork and conflict management, commercial and information analysis.
- Involved stakeholders for the different activities are defined: end user, energy cooperative, crowdfunding platform, aggregator, local authority, DSO, ESCO, SME, digital service provider, social institutions.

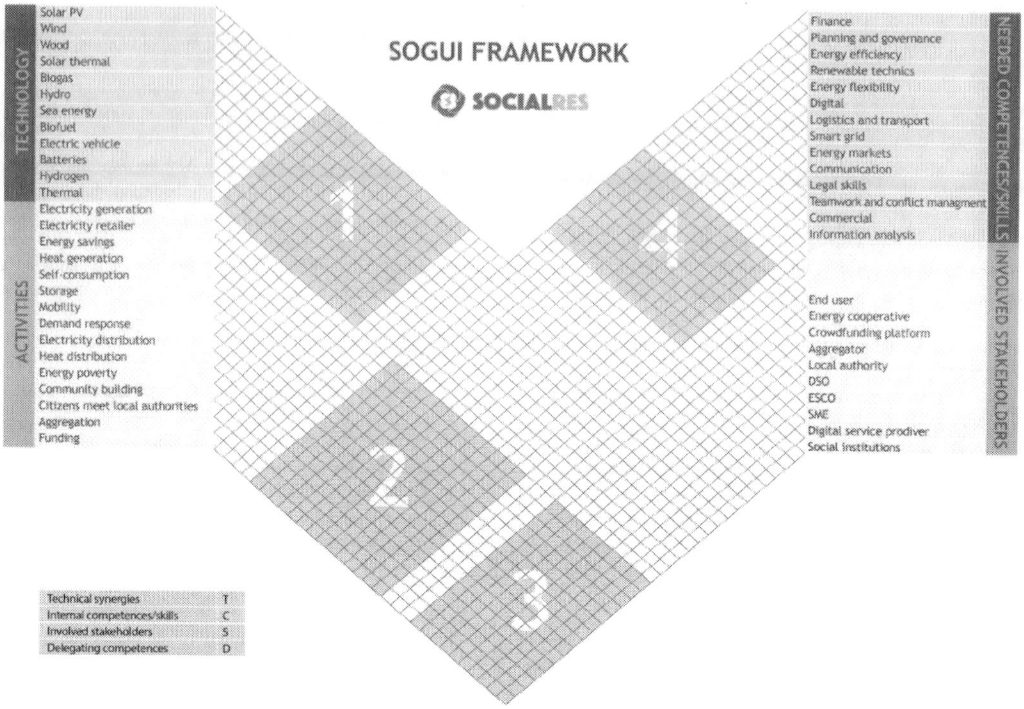

Fig. 2 The SOGUI Framework

Those four parts are cross-referenced according to four criteria:

- "Technical synergies (T)" defines the specific technologies used for each activity of the project.
- "Internal competences/skills (C)" defines which are the needed competences that should be kept internally for each activity.
- "Involved stakeholders (S)" defines the involved stakeholder for each activity.
- "Delegating competences (D)" defines which are the needed competences that are planning to be subcontracted.

The following four steps will help to define the framework of the project:

1. In the first step we are looking at technical synergies between activities and technologies that are applied.
2. The second step is to identify the competences we need internally in the energy community. There are two choices:
    - to keep the competences internally
    - these competences are subcontracted.
3. The involved stakeholders are defined for each activity. For instance, as example, planning to fund the battery system with a crowdfunding platform.
4. The 4th step defines the delegating competences for those activities that have been defined in the 2nd step. For example, the part of electric vehicles can be subcontracted to an agency for communication, legal skills and commercial. Digital skill will be managed by an external digital service provider.

All those steps will permit to define the framework of the project and will help the different initiatives/associations/companies to define the framework of their social innovation project in the energy field.

A practical example on how to implement the four steps is provided in the report "Guidelines for successful social innovations in the energy sector" developed by ESTIA and available on the SocialRES website.

3.2 The toolkit for policy assessment of social innovations

The toolkit for policy assessment developed by the SocialRES partner adelphi outlines a taxonomy to be used by stakeholders to quantify and assess the multiple impacts of social innovations.

The taxonomy has been developed based on the literature on social innovations in the energy sector (Barbiero, Blasi and Schwidtal, 2021; Bauwens and Defourny, 2017; Bauwens and Devine-Wright, 2018; Hoppe and Vries, 2019; Kirsch et al., 2015; Koltunov and Bisello, 2021; Preenen et al., 2014; RESCoop Mecise, 2019), as well as the applied impact assessments of Koltunov and Bisello (2021) and van der Waal (2020) who conducted impact assessments of RE communities in Italy and Scotland respectively using different methodologies. The methodology used for this assessment toolkit is based on the work of Koltunov and Bisello (2021), the taxonomy approach was one that can easily be translated across social innovations.

The proposed taxonomy identifies a range of indicators in the broad categories of economic, social, environmental, and health impacts. The following table gives an overview of the measured impacts, with which social innovation are measured for and the reasoning behind their inclusion.

Table 1 Description of Indicators for Social Innovation Impact Assessment

Indicator	Social Innovation Indicator is Measured for	Reasoning
Consumer Savings	RECs, Aggregators	RECs can lead to financial savings for members due to reduced energy prices compared to traditional utilities. Aggregators can improve energy efficiency at home and reduce expenditures that way.
Job Creation	RECs, Aggregators, CrowdFunding	Jobs created both during the implementation and construction phase of the projects, as well as during the operation phase which included management and maintenance.
Participant Income	RECs, CrowdFunding, Aggregators	RECs may pay out dividends to their members if profits are achieved in a fiscal year. CrowdFunding platform provide a return on investment usually in the form of an interest payment on a loan. Aggregators may pay participants based on how much energy they can produce or how much flexibility they make available.
Government Income	RECs, CrowdFunding, Aggregators	All three pay taxes. This can include land/property taxes, income taxes, corporate tax, and trade tax
Capital Investment in RE Infrastructure	RECs, CrowdFunding, Aggregators	RECs install RE infrastructure via generating units, storage capacity, charging infrastructure or other initiatives. CrowdFunding platforms raise financing to fund RE projects. Aggregators may not directly invest in RE infrastructure but may motivate participants to invest in further RE generation or storage.
Avoided GHG Emissions	RECs, CrowdFunding, Aggregators	RECs and CrowdFunding provide a RE supply which may previously have stemmed from fossil fuels. Aggregators increase energy efficiency at the household level and can contribute to peak shaving across the entire grid.
Energy Savings	RECs, Aggregators	Some RECs engage in energy efficiency projects for their members and the wider community. Aggregators save energy through optimized energy usage.
Political Mobilization	RECs, CrowdFunding, Aggregators	In all three expose citizens to RE, and active participation can increase
Energy Literacy	RECs, CrowdFunding, Aggregators	Owning or investing in an energy asset increases consumer's understanding of the energy market. It can change attitudes towards energy consumption and understanding of meter readings and bills.
Clean Energy Generated	RECs, CrowdFunding, Aggregators	RECs and CrowdFunding build generating units and thus it is important to see how much RE is produced by these units. Aggregators do not directly generate RE but the amount produced by members can be measured.

Further detailed information is available in the report available on the SoialRES website "Toolkit for Policy Assessment", which provides a detailed description on applications for use of the taxonomy, for policymakers, who may want to assess their social innovation support policies by looking beyond the outcomes at the impacts of the social innovations. Applications for use are also suggested for practitioners, as owners or managers of the social innovations in question.

The toolkit also includes the evaluation of the impacts for each SocialRES case study representing cooperatives, crowdfunding platforms and aggregators.

4 ACKNOWLEDGEMENTS AND DISCLAIMER

The SocialRES project receives funding from the European Union's Horizon 2020 research and innovation programme under grant agreement No. 837758. The sole responsibility for the content of this publication lies with the authors. It does not necessarily reflect the opinion of the European Commission. The European Commission is not responsible for any use that may be made of the information contained therein.

SocialRES (http://socialres.eu/) is on the social media at:

- Twitter: https://twitter.com/SocialRESeu
- LinkedIn: https://www.linkedin.com/company/socialres/

A SIMPLE GUIDE FOR DESIGNING A PHOTOVOLTAIC AND BATTERY SYSTEM COUPLED WITH A HEAT PUMP ACROSS EUROPE

Mattia Dallapiccola, Jennifer Adami, David Moser
Eurac Research, Institute for Renewable Energy
Via A.Volta, 13/A, 39100, Bolzano

ABSTRACT:
The aim of this work is to provide a guide for the design of a photovoltaic plus battery energy storage system coupled with a heat pump for different locations and thus different climates in Europe. Through a multi-objective optimization approach applied to the six European geoclusters and applying different solution selection criteria, it was possible to obtain the optimal photovoltaic nominal power (kWp) normalized over the thermal consumption of a reference building and the optimal capacity of the battery per kWp of photovoltaic (kWh:kWp) installed. Due to the normalization process, results can be generalized and can be used for designing similar systems in all buildings around Europe. The results of this work must be intended as a support for designers for the early-design phase of such systems or as an initial guess for an iterative process in more advanced evaluations.

1 INTRODUCTION

One of the main objectives of the European Union is the energy transition to a clean and low-carbon energy system. The building sector is one of the main areas where the decarbonization process should focus, accounting for 28% of energy-related emissions in 2019 [1]. In recent years, the adoption of heat pumps (HP) to cover the heating and cooling demand of buildings has opened a range of opportunities related to sector-coupling. At the same time, the reducing costs of rooftop photovoltaics (PV) and electric storage technologies (BESS) [2], the new national regulations for the adoption of clean energy sources in buildings and the European Climate Target Plan (55% emission reduction with respect to 1990 levels by 2030, first step to achieve carbon neutrality by 2050) make the installation of new photovoltaic system necessary [3] and the coupling of photovoltaic, electric storage and heat pumps more and more attractive. Heat pumps have a great potential to participate to the decarbonization of the heating and cooling sectors since they consume electricity to cover the thermal needs of buildings [4]. More importantly, the electricity consumed can be locally produced by a photovoltaic system. To summarize, there are multiple advantages to couple a heat pump with a photovoltaic system:

- the heat pump can consume the electricity produced by the photovoltaic system, decreasing the carbon footprint of the building
- the heat pump can improve self-consumption, since the thermal load is covered consuming electricity, rather than other sources (natural gas, etc.)
- the heat pump can be controlled by a smart controller shifting part of the electricity demand and further improving self-consumption by consuming the excess electricity that is not directly consumed or stored in the battery

On the other side, heat pumps have a strong impact on the electricity consumption of buildings (in Northern Italy can also double the electricity consumption). For this reason, the design of the coupled photovoltaic plus battery system should account of such additional consumption. The options currently available for designers are mainly two:

- Designing the photovoltaic system to produce the electricity yearly consumed (or use the same methodology on a specific season, e.g. winter). This approach can lead to oversized systems, if net-billing or net-metering tariff schemes are not available [5]
- Modelling the building, the heating system and the photovoltaic system (e.g. with TRNSYS [6]). This approach is accurate but usually in the early-stage phase of the design process the data required for a simulation are not available. Moreover, the modelling effort is generally considered too high for the early-stage phase.

The aim of this work is to answer to the need of a simple method to be used as a "Go/No-go" decision making support and to provide a simple guide for the early-design stage of photovoltaic plus battery energy storage system coupled with a heat pump for different locations and climates in Europe.

2 METHODOLOGY

In this chapter, the process used to obtain the results is explained in detail. The fundamental steps of the process are the following:

- selection of the reference locations
- selection of a reference building
- simulation of the thermal needs and the electric consumption profile
- optimization of the photovoltaic plus battery system
- solution selection
- normalization of the results

2.1 Geoclusters and reference locations

Considering that the local weather strongly impacts the heat pump consumption profile, Europe was divided in six geoclusters according to the classification reported in [7] (for the Northern geoclusters, Helsinki was used instead of Oslo, due to lack of data). For each geocluster, the typical meteorological year was used to obtain the thermal needs of the reference building (see next section) and for the simulation of the photovoltaic system. The geoclusters and the reference locations are summarized in Table I.

Table I: Geoclusters

Geocluster	Location
Northern	Helsinki
North-East	Warsaw
East	Budapest

Continental Central	Amsterdam
Mediterranean	Barcelona
Atlantic	Dublin

2.2 Reference building
A five-dwellings multi-family house was selected as the reference building for this study. It was assumed that the building is well insulated, according to the recommendation for the installation of heat pumps. It was also assumed that the roof of the reference building is optimally oriented and tilted. Thus, for each location, the optimal orientation and tilt angle of the surfaces available for PV installation were calculated with the online tool PVGIS [8]. They are reported in Table II.

Table II: roof tilt and orientation

Geocluster	Orientation [°]	Tilt [°]
Northern	2	44
North-East	-1	38
East	-2	38
Continental Central	-2	39
Mediterranean	0	37
Atlantic	-1	39

2.3 Thermal needs and electric consumption profile
The electric consumption profile is a fundamental input for the optimization of a photovoltaic system, since from the hourly demand it is possible to estimate the direct self-consumption, the electric grid, and the battery contributions. Moreover, the hourly approach unlocks the possibility to perform a detailed economic analysis. As previously introduced, the local weather has a strong impact on the optimal photovoltaic system design, since a heat pump is used to cover the thermal needs of the building. To calculate the electric consumption profile, the EDGE-HP tool introduced in [9] was used for the six geoclusters. The tool, given the building typology, the location, and the annual cumulative demands for heating, cooling and domestic hot water, calculates the hourly thermal needs of the building. Yearly cumulative thermal needs were set according to the calculations performed in the 4RinEU Horizon 2020 project [10]. The hourly electric consumption profile is then calculated using a performance map of a commercially available air-to-water heat pump. It was also assumed that the heat pump is smart enough to shift half of the night load to daytime to improve self-consumption. This assumption was done since many commercially available heat pumps are already set up to interact with a photovoltaic system.

2.4 Optimization of the photovoltaic plus battery system
The electricity load profiles obtained were used as input for an optimization process (implemented with pymoo [11]) which gives as a result the optimal photovoltaic nominal power (assuming an optimal tilt and orientation) and the optimal battery capacity of the PV+BESS system to be coupled with the heat pump. To obtain the optimal configuration of the PV+BESS+HP system, a multi-objectives optimization algorithm (NSGA-II [12]) implemented in Python was used (the tool was developed within the Horizon 2020 INFINITE project [13]). For each location, the parameters of the optimization are the photovoltaic system nominal power installed assuming an optimal orientation of the roof and the battery capacity.

The objectives used for the optimization are the maximization of the self-sufficiency (the percentage of the self-consumed energy over the total electricity consumption) and the minimization of the initial costs. The objective functions were calculated according to the following formulas:

- Minimization of the initial costs C_0 of the PV+BESS system calculated as the sum of the initial costs related to the photovoltaic system and the electric battery:

  $$C_0 = C_{0,PV} + C_{0,BESS} = P_n * c_{PV} + Q_0 * c_{BESS}$$

  where P_n is the photovoltaic system nominal power in kWp, c_{PV} is the unit cost of the photovoltaic system including inverter, cabling, mounting structure, etc. expressed in €/kWp, Q_0 is the battery capacity and c_{BESS} is the unit cost for the battery in €/kWh.

- Maximization of self-sufficiency (SS), calculated with an hourly time step according to the following formula:

  $$SS = \frac{\sum_i E_{self,i}}{\sum_i E_{cons,i}}$$

  where $E_{self,i}$ is the energy self-consumed at the time step i (directly self-consumed from the photovoltaic system or from the battery) and $E_{cons,i}$ is the energy consumption at the time step i.

During the optimization process, the tool:

- simulates with an hourly time-step the yearly power generation of the photovoltaic system using a simplified irradiance-to-power model (weather data used typical meteorological year file) represented by the following formula:

  $$P_i = PR * \eta * A * I_i * (1 - \gamma(T_{mod,i} - T_{ref}))$$

  where PR is the performance ratio of the system, η is the module efficiency, A is the module area, I_i is the plane-of-array irradiance calculated from the irradiance components included in the weather file with the Python library pvlib-python [14], γ is the temperature coefficient for power, T_{mod} is the module temperature calculated according to the formula presented in [15] and considering a Ross coefficient equal to 0.0342 K m^2/W and T_{ref} is equal to 25°C.

- Simulates the PV+BESS system calculating the energy flows between the PV system, the grid, the battery, and the heat pump to calculate the self-consumed energy. The battery model is based on an energy balance as in [16]

- Calculates the optimization objectives: the self-sufficiency index (SS) and the initial costs

- Repeats the process until the termination criteria is achieved (maximum number of generation)

The result of the optimization process is a set of optimal non-dominated solutions (Pareto front) which represent the solutions with an optimum balance between the two objectives.

2.5 Solution selection

At the end of the optimization process the tool provides a set of optimal solution (Pareto front). For selecting the final solutions, the CO_2 equivalent lifetime emissions and the Net Present Value were calculated.

The CO_2 equivalent lifetime emissions were calculated with the following formula:

$$Em_{CO2,eq} = em_{PV} * P_{n,PV} + em_{BESS} * Q_{BESS} + em_{grid} * E_{grid}$$

where em_{PV} ($kgCO_{2,eq}$/kWp) are the specific emissions per kWp for the photovoltaic system accounting of module production, installation, transport, etc., em_{BESS} ($kgCO_{2,eq}$/kWh) are the specific emissions per unit capacity related to battery production, $P_{n,PV}$ (kWp) is the photovoltaic system nominal power, Q_{BESS} (kWh) is the battery capacity, em_{grid} is the country-dependent emission factor of the grid ($kgCO_{2,eq}$/kWh) and E_{grid} (kWh) is the estimated lifetime energy purchased from the grid. Once the total emissions $Em_{CO2,eq}$ ($kgCO_{2,eq}$) are known, they are normalized over the total electricity consumption to calculate the specific emissions of the PV+BESS+HP system. The resulting value is compared to the emission factor of the country in 1990 [17]. Solutions that do not respect the target of 55% emission reduction with respect to 1990 levels were discarded (red points in Figure 1). Note that, for calculating the specific emissions, the energy consumption is considered. It differs from standard approaches, which usually take into account the overall generation. In this case, the authors decided to change perspective to suggest systems that guarantee the emission reduction requirement by their own, without relying on other possible consumers or on the grid. Considering the actual status, this assumption underestimates the real emissions reduction. In fact, in 2022 the clean energy sent to the grid will be consumed by other consumers and thus would contribute to the overall emission reduction. However, it could be that in future part of the energy produced would be curtailed to guarantee grid stability.

The Net Present Value was calculated for each Pareto-solution according to the model reported in [16]. The formula used is:

$$NPV = -C_0 + \sum_{0}^{N} \frac{el_p * \sum E_{self} - \omega_{PV} * P_{n,PV} - C_{BESS}(i) * Q_{BESS}}{(1+i)^t}$$

where el_p is the grid electricity price, E_{self} is the energy self-consumed (used to calculate the savings), ω_{PV} is the yearly cost for maintenance per kWp, $C_{BESS}(i)$ is the cost for the battery replacement at the year i assuming that it must be replaced after 2700 cycles, i is the year and t is the discount rate.

Finally, between the remaining solutions which guarantees the 55% emission reductions with respect to 1990 levels, the cheapest one (lowest initial costs, blue point in Figure 1), the one corresponding to the minimum specific emissions per kWh consumed (orange point in Figure 1) and the one which guarantees the maximum Net Present Value are selected for the normalization process discussed in the following section.

2.6 Normalization of the results

Since the aim of this work is to provide a guide/rule of thumb for the design of PV+BESS (Photovoltaic and battery energy storage system) systems around Europe to be coupled with heat pumps (HP), results were normalized to extend their applicability. They were presented in terms of optimal photovoltaic system nominal power per unit of energy consumed for covering the thermal needs for heating, cooling, and domestic hot water (DHW) preparation (kWp/MWh) and in terms of battery capacity per photovoltaic nominal power (kWh/kWp).

Thus, for each geocluster, results were normalized to calculate the optimal ratio between the PV nominal power (kWp) and the thermal needs of the building (MWh_{th}), and the battery capacity (kWh) over the PV nominal power (kWp) to generalize the output of this work. The process was repeated for all the geoclusters to obtain the results for the European area.

3 RESULTS

In this section the results obtained were reported. Assumptions and data used for the calculations were summarized in Table III.

Table III: data used for calculations

Description	Value	Unit	Reference
c_{PV}	1170	€/kWp	[2]
c_{BESS}	563.4	€/kWh	[18]
PR	82	%	[2]
η	20	%	author
A	1.6	m^2	author
γ	-0.47	%/°C	[19]
em_{PV}	1010	$kgCO_{2,eq}$/kWp	[20]
em_{BESS}	110	$kgCO_{2,eq}$/kWh	[21]
em_{GRID}	Country-dependent	$kgCO_{2,eq}$/kWh	[22]
Lifetime	25	years	author
el_p	Country-dependent	€/kWh	[23]
ω_{PV}	30	€/kWp/year	[24]
C_{BESS}	Year-dependent	€/kWh	[18]
Discount rate	3	%	author

As an example, Figure 1 represents the results obtained from the optimization process for the Mediterranean geocluster. Red points correspond to the solution discarded since they could not guarantee the 55% emissions reduction with respect to 1990 levels. The blue point corresponds to the cheapest solution that satisfies the threshold, the yellow point represents the solution with the maximum Net Present Value, while the orange point corresponds to the solution with the minimum lifetime emissions.

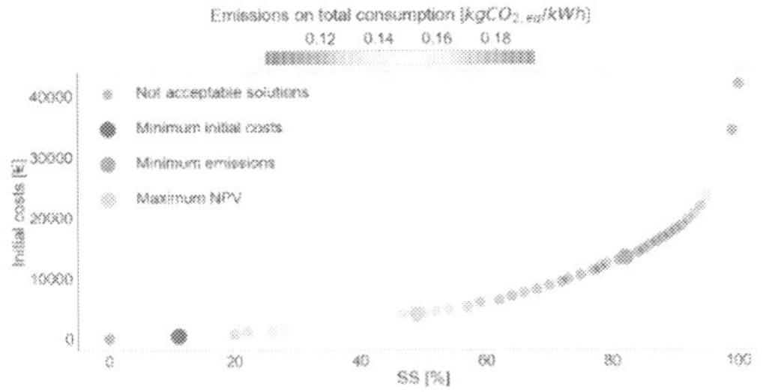

Figure 1: solutions obtained for the Mediterranean geocluster

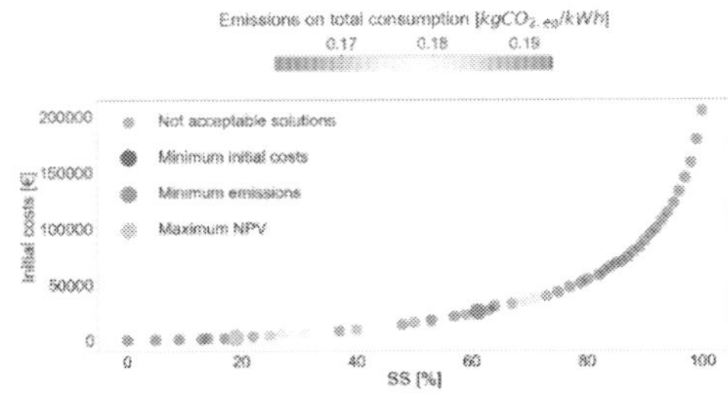

Figure 2: solutions obtained for the East geocluster

Figure 2 represents the results obtained for the East geocluster. With respect to the ones represented in Figure 1, it is possible to notice that undersized systems (low SS) are discarded since they cannot achieve the required emission reduction. On the other hand, also oversized systems are discarded, since achieving high levels of SS requires the installation of an oversized systems, connected to high emissions to manufacture the components. Moreover, mainly due to the weather conditions, the solution with minimum costs and the one with the maximum Net Present Value are the same.

The following tables reports the normalized results obtained for each geocluster. Table IV refers to the results obtained selecting the solution with the lowest initial costs, while Table V reports the solutions selected based on the minimum lifetime emissions criteria and Table IV reports solutions based on the maximum Net Present Value criteria.

Table IV: normalized results selected based on the minimum initial costs criteria

Geocluster	PV [kWp/MWh$_{th}$]	BESS [kWh/kWp]
Northern	0	0
North-East	0.23	0.93
East	0.09	0.65
Continental Central	0.28	1.47
Mediterranean	0.04	0
Atlantic	0	0

Table V: normalized results selected based on the minimum lifetime emissions criteria

Geocluster	PV [kWp/MWh$_{th}$]	BESS [kWh/kWp]
Northern	0.04	0
North-East	1.80	0.93
East	0.66	1.73
Continental Central	0.92	1.34
Mediterranean	0.64	2.55
Atlantic	0.93	1.85

Table VI: normalized results selected based on the maximum Net Present Value criteria

Geocluster	PV [kWp/MWh$_{th}$]	BESS [kWh/kWp]
Northern	0.04	0
North-East	0.03	0
East	0.04	0
Continental Central	0.06	0
Mediterranean	0.24	1.67
Atlantic	0.11	0

4 DISCUSSION

Results shown in the previous section open several points of discussion. Firstly, it is possible to notice that results may sensibly vary comparing different geoclusters, mainly due to the strong impact of the climates on the heat-pump seasonal consumptions and the photovoltaic generation. Moreover, the grid emission factor of the countries analyzed has a relevant impact on the solution selection process. For the Northern and Atlantic geoclusters, the actual grid emission factor is already 55% lower than the 1990 value. This means that the photovoltaic system is not required to achieve the decarbonization target and the cheapest solution is to power the heat pump with the national grid electricity. However, if the optimization target is to minimize the lifetime emissions, the installation of a photovoltaic system is recommended for each geocluster. This is also valid for the selection based on the maximization of the Net Present Value of the investment related to the installation of the PV+BESS system. For all geoclusters, the installation of a photovoltaic system is recommended. From Table VI, it is possible to notice that the results obtained for the Mediterranean geoclusters are sensibly different from the others. This is caused by the fact that in the Mediterranean region also in summer the heat pump consumes a relevant quantity of electric energy to cover the cooling demand. The summer consumption coincides with the peak of photovoltaic generation favoring the matching between consumption and generation. Moreover, in the Mediterranean geocluster the photovoltaic generation contributes to cover the winter electricity demand, while for the other geoclusters its impact is marginal in winter. This aspect has a strong impact also on the battery capacity suggested, that for the other geoclusters is 0. This can be explained by the fact that the modest winter production is almost directly self-consumed by the heat-pump and thus reducing the need of an energy storage, while in summer the demand is low or absent. Based on the modest systems suggested in Table VI, with the exception for the Mediterranean geocluster, it is possible to conclude that from an economic perspective the photovoltaic system should be designed accounting mainly of other consumptions (electric vehicle, domestic appliances, etc.) while the heat-pump can only marginally improve the economic outcome. On the contrary, for the Mediterranean geocluster the heat pump should be always considered since it has a strong impact on the final system configuration. According to the results reported in Table V and selected with the minimum emission criteria, it is possible to conclude that there is a wide space for improvements with respect to the use of the grid electricity to drive the heat pumps regarding lifetime emissions and that both the photovoltaic and battery systems can contribute to the environmental impact reduction.

5 CONCLUSIONS

This paper aims to provide a rule of thumb for the preliminary design of photovoltaic and electric storage systems coupled with a heat pump for different European climates. The methodology applied is made of three steps: a multi-objective optimization, a normalization process, and a filtering action. The optimization suggests optimal configurations of photovoltaic plus battery systems for a set of reference buildings, considering both economic and energy aspects since the early design phase. The normalization process, transforming the results in terms of photovoltaic nominal power per unit of thermal energy consumed (kWp/MWh_{th}) and battery capacity per photovoltaic nominal power (kWh/kWp), allows to generalize them. The filtering step, based on the environmental target, helps to make the results valid suggestions considering both the economic and environmental aspects of sustainability. The main relevance of this work is that designers can use the results to find suitable solutions easily and quickly for the design of any photovoltaic plus battery system coupled with a heat pump in the building sector around Europe once the yearly cumulative thermal demand of the building is known. Finally, it must be noted that results of this work consider only the electric consumption profile of a heat pump for suggesting the size of the photovoltaic and battery system.

6 FUTURE WORK

Results presented in this work are based on several assumptions. In future it will be interesting to analyze the impact of such assumptions on the results. In particular, the emissions factor used refers to the actual status (current energy mixes, components production in China, etc.). Such assumptions have an impact on the results as highlighted in [25] and their effect must be analyzed in detail. Possible future changes in the energy mixes and in the cost of the components should also be included in further analysis. Moreover, it will be necessary to go from geocluster to country level to improve the accuracy of the results.

7 ACKNOWLEDGEMENTS

The INFINITE project has received funding from the European Union's Horizon 2020 research and innovation programme under grant agreement No 958397. This presentation reflects only the author's view, neither the European Commission nor HADEA are responsible for any use that may be made of the information it contains.

8 REFERENCES

[1] I. E. Agency (IEA), "Tracking buildings 2020," 2020.

[2] E. Vartiainen, G. Masson, C. Breyer, and D. Moser, "Improving Competitiveness of Solar PV with Electricity Storage," in *Proceedings of the 33rd European Photovoltaic Solar Energy Conference and Exhibition, Amsterdam, The Netherlands*, 2017, pp. 25–29.

[3] IEA, "Approximately 100 million households rely on rooftop solar PV by 2030," Paris, 2022. [Online]. Available: https://www.iea.org/reports/approximately-100-million-households-rely-on-rooftop-solar-pv-by-2030

[4] IEA, "Installation of about 600 million heat pumps covering 20% of buildings heating needs required by 2030," Paris, 2022. [Online]. Available: https://www.iea.org/reports/installation-of-about-600-million-heat-pumps-covering-20-of-buildings-heating-needs-required-by-2030

[5] M. Lovati, M. Dallapiccola, J. Adami, P. Bonato, X. Zhang, and D. Moser, "Design of a residential photovoltaic system: the impact of the demand profile and the normative framework," *Renewable energy*, vol. 160, pp. 1458–1467, 2020.

[6] S. A. Klein, W. A. Beckman, J. W. Mitchell, N. A. Duffie, and et al., *TRNSYS 17, transient system simulation program*. University of Wisconsin, Madison, WI, USA, 1979. Accessed: Apr. 20, 2018. [Online]. Available: www.trnsys.com

[7] R. Pernetti, R. Pinotti, and R. Lollini, "Repository of deep renovation packages based on industrialized solutions: Definition and application," *Sustainability*, vol. 13, no. 11, p. 6412, 2021.

[8] M. Šúri, T. Huld, T. Cebecauer, and E. D. Dunlop, "Geographic aspects of photovoltaics in Europe: contribution of the PVGIS web site," *IEEE Journal of Selected Topics in Applied Earth Observations and Remote Sensing*, vol. 1, no. 1, pp. 34–41, 2008.

[9] M. Dallapiccola, G. Barchi, J. Adami, and D. Moser, "The Role of Flexibility in Photovoltaic and Battery Optimal Sizing towards a Decarbonized Residential Sector," *Energies*, vol. 14, no. 8, p. 2326, 2021.

[10] P. Roberta, P. Riccardo, L. Roberto, P. Giulia, and T. Linda, "Deep renovation packages and parametric models in different geoclusters," Eurac Research, Public D3.3, Nov. 2020. [Online]. Available: https://4rineu.eu/wp-content/uploads/2021/02/4RinEU_D3.3-4RinEU-Deep-Renovation-Packages_Annex.pdf

[11] J. Blank and K. Deb, "pymoo: Multi-objective Optimization in Python," *IEEE Access*, pp. 1–1, 2020.

[12] K. Deb, A. Pratap, S. Agarwal, and T. Meyarivan, "A fast and elitist multiobjective genetic algorithm: NSGA-II," *IEEE transactions on evolutionary computation*, vol. 6, no. 2, pp. 182–197, 2002.

[13] "INFINITE Building Renovation." https://infinitebuildingrenovation.eu/

[14] W. F. Holmgren, C. W. Hansen, and M. Mikofski, "Pvlib Python: a Python Package for Modeling Solar Energy Systems," *J. Open Source Software*, vol. 3, p. 884, 2018.

[15] N. Martín-Chivelet *et al.*, "Building-Integrated Photovoltaic (BIPV) products and systems: A review of energy-related behavior," *Energy and Buildings*, p. 111998, 2022.

[16] M. Lovati and J. Adami, "EnergyMatching (EM) Tool for optimization of RES harvesting at building and district scale," D2.2, 2020. [Online]. Available: https://www.energymatching.eu/results/?page=2

[17] European Environment Agency, "CO2-emission intensity from electricity generation." https://www.eea.europa.eu/data-and-maps/daviz/sds/co2-emission-intensity-from-

electricity-generation-5/@@view (accessed Feb. 01, 2022).

[18] M. G. Prina, M. Lionetti, G. Manzolini, W. Sparber, and D. Moser, "Transition pathways optimization methodology through EnergyPLAN software for long-term energy planning," *Applied Energy*, vol. 235, pp. 356–368, 2019.

[19] P. Dash and N. Gupta, "Effect of temperature on power output from different commercially available photovoltaic modules," *International Journal of Engineering Research and Applications*, vol. 5, no. 1, pp. 148–151, 2015.

[20] V. Fthenakis and E. Leccisi, "Updated sustainability status of crystalline silicon-based photovoltaic systems: Life-cycle energy and environmental impact reduction trends," *Progress in Photovoltaics: Research and Applications*, vol. 29, no. 10, pp. 1068–1077, 2021.

[21] J. F. Peters, M. Baumann, B. Zimmermann, J. Braun, and M. Weil, "The environmental impact of Li-Ion batteries and the role of key parameters–A review," *Renewable and Sustainable Energy Reviews*, vol. 67, pp. 491–506, 2017.

[22] European Environment Agency, "Greenhouse gas emission intensity of electricity generation," Jun. 03, 2022. https://www.eea.europa.eu/data-and-maps/daviz/co2-emission-intensity-10#tab-googlechartid_googlechartid_googlechartid_googlechartid_chart_11111

[23] Eurostat, "Electricity price statistics," 2021. https://ec.europa.eu/eurostat/statistics-explained/index.php?title=Electricity_price_statistics

[24] R. H. Wiser, M. Bolinger, and J. Seel, "Benchmarking utility-scale PV operational expenses and project lifetimes: results from a survey of US solar industry professionals," Lawrence Berkeley National Lab.(LBNL), Berkeley, CA (United States), 2020.

[25] IEA, "Solar PV Global Supply Chains," Paris, 2022. [Online]. Available: https://www.iea.org/reports/solar-pv-global-supply-chains

UPCYCLING LITHIUM-ION BATTERIES: A CASE STUDY

Ana Foles[1*], Luis Fialho[1], Joscha Winzer[2], Núria González García[2], Pedro Horta[1], Manuel Collares-Pereira[1]
[1]Renewable Energies Chair, University of Évora, Pólo da Mitra da Universidade de Évora, Edifício Ário Lobo de Azevedo, 7000-083 Nossa Senhora da Tourega, Évora, Portugal
[2]betteries AMPS GmbH, Goerzallee 299, 14167 Berlin, Germany
*Phone: +351 266 740 800; E-mail: anafoles@uevora.pt

ABSTRACT: The need for cost-effective and flexible solutions to compose the distributed energy storage makes second-life lithium-ion batteries (SLIB) a discussed option. The SLIB is a lithium-ion battery used in an energy/power-intensive first application (automotive) that still retains important performance characteristics to have a second-life application. SLIBs can represent an opportunity to meet this demand and an option for residential and off-grid applications, presenting competitive cost advantages over new batteries. However, their application within the current state of the art still needs to overcome several challenges to become technically and economically viable. In the framework of the POCITYF project (GA 864400), a prototype of a SLIB from the company *betteries AMPS GmbH* has been tested at the University of Évora, connected to the existing experimental microgrid through a commercial inverter. This work aims to contribute to the state of the art of SLIB technology, providing insights into their electrical performance and presenting technical key performance indicators in a real-operational environment. This enables the further finetuning and validation of the product, reaching a higher TRL. This work aims to improve technology acceptance by demonstrating safety and reliability under real operating conditions and promoting the ease of installation and operation of these technologies.

1 INTRODUCTION

The International Energy Agency (IEA) states that the electric automotive industry is expected to increase from 2020 to 2030 with an average annual growth rate of nearly 30% [1]. After their useful life in electric vehicles (EVs), considering the degradation of their energy retention capacity, these batteries retain capacity and power characteristics that make them suitable for applications with lower energy density requirements. The still-useful batteries can create significant value and ultimately increase the sustainability of the battery value chain. In that sense, a secondary market arises for the second application use of the EV batteries before their end of life (EOL) (recyclability, in European Union). Additionally, this second life will also contribute to the world's need for stationary or quasi-stationary energy storage.in the coming decades to meet international climate and decarbonization goals.

Current research on SLIBs still needs to overcome several challenges to become technically and economically viable, achieving the market uptake. Achieving this is currently influenced by the perceived uncertainty associated with their performance and degradation behaviour, where the major research efforts has been focused on understanding the early stages of degradation (first-life lithium-ion batteries).

In the context of the POCITYF project (GA 864400) [2], a SLIB prototype has been built and tested at the University of Évora, in cooperation with the German company *betteries AMPS GmbH* [3]. The specific SLIB, in Figure 1, presented a preliminary power and energy capacity of 2.0 kW/2.3 kWh, a voltage operating range of 45-64 V, and an approximate weight of 35 kg. With an advisable depth of discharge of 80%, the expected lifetime is estimated at 7-10 years [3]. The *betteries AMPS GmbH* solution aims to be applied to off-grid solar photovoltaic (PV) installations. In contrast to other technological approaches, *betteries AMPS GmbH* solution presents a modular design: the modules can be stacked on each other, increasing the usable energy capacity output up to 5 kW/9.2 kWh (with the stacking of 4 modules).

Figure 1: 2.0 kW/ 2.3 kWh second-life lithium-ion battery developed by *betteries AMPS GmbH* and installed in a pilot test as a result of the established cooperation between the University of Évora and *betteries AMPS GmbH*.

In this work, the SLIB and its integrated inverter are subject to performance testing at the stack level, intending to obtain a complete performance characterization of both through the development of a testing control unit. This unit is responsible for communication with the inverter, executing the data acquisition and reading. Within this set-up, the SLIB is evaluated regarding its technical feasibility and electrical performance. The results are analyzed, and key performance indicators (KPIs) are calculated to characterize this second-life technology. The KPIs obtained are relevant data for the state of the art of second-life lithium-ion batteries performance, contributing to further comparisons with other energy storage technologies and allowing further electrical modelling developments.

2 CIRCULAR BUSINESS MODEL AND INNOVATION

2.1 Overview

The global prediction of e-waste is perceived as an obstacle to the increasing future battery volumes for e-mobility. Figure 2 presents the specific global expectation market forecast of the second-life lithium-ion batteries, expected to reach 20 GWh in 2023 and 1552 GWh in 2032 [4].

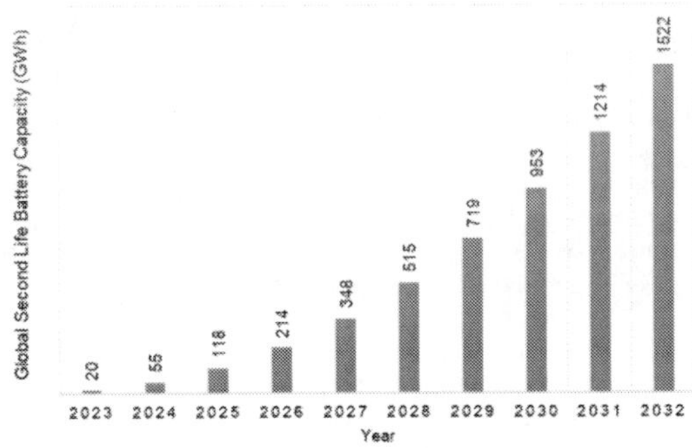

Figure 2: Expected global second life battery capacity from electric vehicle stock. Adapted from [4].

SLIB is expected to contribute to the transition from the current fossil-fuelled model to a sustainable renewable energy sources model, promoting the decentralization of generation and consumption (e.g., increasing the building's PV self-consumption rate) and reducing the carbon footprint of the overall electricity sector.

SLIBs impact the three-dimensional pillars: environmental, social, and economic. Environmentally, the carbon footprint of electric vehicles (EVs) is reduced due to the battery life extension and the second productive lifetime opportunity, minimizing natural resources usage for new batteries. It also reduces the carbon footprint of buildings, allowing a higher penetration of decentralized renewable energy generation. Socially, the SLIBs reduce households' electricity bills and can have a crucial role in energy poverty-fighting. Economically, SLIBs can be managed to provide grid-connected or off-grid competitive business models, with lower costs compared to traditional solutions (e.g., electricity grid or diesel generators). Given the current scarcity of raw materials and difficulties associated with the international logistics of goods, it can present competitive advantages due to the reuse of battery systems. The extension of the battery lifetime can contribute to stabilize high price volatility of the raw materials market. The integration of SLIBs creates a new revenue stream for the car OEMs (Original Equipment Manufacturing), creating additional revenues and job creation for installers and component suppliers.

2.2 Potential social outcomes for citizens and the community

The implementation of SLIBs should bring technology environmental awareness to end users regarding discarded EV batteries' impact on the environment. It can also promote adoption of smart energy management strategies of a household or building, empowering the prosumer. Providing similar characteristics to first life batteries, the prosumer could monitor the power management, benefiting from integration in energy communities or virtual power plants, sharing platforms and bringing awareness to new sustainable life behaviours.

2.3 Circularity and innovation strategies

The circularity of the current business model of *betteries AMPS GmbH* is associated with discarded EV batteries, which are remanufactured into smaller systems fulfilling the needs of households and creating an additional business opportunity for car OEMs. The second aspect of this circular business model is the application of the batteries in a secondary storage application after repurposing, where the batteries are designed to be easy replaced. At the end of the second productive life, optimal recycling technology is required to recover the valuable materials and reintroduce them into the battery value chain (valuable EOL batteries).

Circularity is at the core of these SLIBs. Electric mobility has a crucial contribution to meet global goals on decarbonization. However, if not properly evaluated, one could introduce the issue of how to properly deal with the expected volume of used batteries once they can no longer be applied to EVs. With 145 million EVs predicted to be on the roads by 2030, more than 12 million tons of lithium-ion batteries are expected to be retired until 2030 [5]. With at least 70% of their initial energy capacity left [6] (see Figure 3), the batteries are too valuable to be directly recycled. A productive SLIB application should be encountered so that the e-mobility sector can follow a more sustainable path.

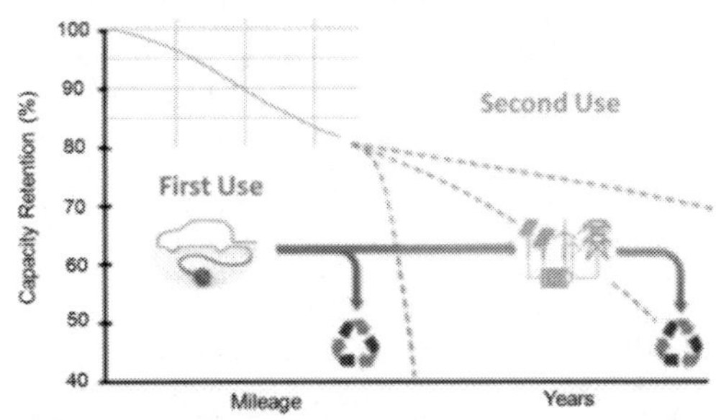

Figure 3: Energy capacity retention (%) during the lifecycle of an EV-lithium-ion battery first application, and the potential second use. Adapted from [7].

betteries AMPS GmbH is committed to supporting the implementation of a genuinely sustainable e-mobility by closing the loop in the energy sector. On the one hand, the company is using discarded EV batteries – which still have up to 70% of their initial capacity – to produce smaller energy storage systems for multiple use cases. On the other hand, deploying home storage systems supports the installation of renewable energy sources in the building sector, thus contributing to the decarbonisation of the grid.

Besides repurposing the battery packs, *betteries AMPS GmbH* reutilises several other components of the battery car pack, such as connectors, fuses, and cables. This action reinforces the business model's circularity principles and minimizes the waste generated from an electric vehicle.

2.4 Costs

The battery acquisition is the critical resource that presents the highest cost in the whole upcycling process.

Public schemes could promote using SLIBs by providing to prosumers a subsidy (similar to what some countries have promoted for other renewable technologies) associated with upcycled batteries. The end users/beneficiaries of the SLIBs will then have a better ROI (Return of Investment) for their household application, enhancing the market uptake, and reducing the higher perceived risks related to a second-life product. With a SLIB, the prosumer could then reduce costs and maximize its renewable electricity self-consumption smartly and flexibly. The primary revenue streams for the end user are the cost savings related to electricity services provided (e.g. self-consumption, ancillary services such as grid stabilization or UPS function, peak curtailment, power ramp rate control, etc.).

2.4 Environmental costs and benefits

The delay of the battery EOL translates into adding a step in the life cycle assessment (LCA). The ecological cost of the technology is related to the carbon footprint associated with the upcycling process and the transportation of storage solutions, reducing the transportation of electricity through decentralized strategies. SLIB could reduce waste generation and EVs' CO_2 footprint by >30%.

3 METHODOLOGY

The SLIB is subjected to testing at its pack level, intending to obtain a performance characterisation. The testing obeyed a protocol specifically designed to operate the batteries at their general operating conditions (presented below), through the development of a dedicated LabVIEW programming in the control unit. The batteries are fully charged and discharged at defined power limits to evaluate their response at the specified typical operating conditions. The control unit is responsible for real-time communication with the inverter to acquire data such as current, voltage or temperature from the sensors. The results are analysed, and key performance indicators (KPIs) such as energy capacity and efficiencies are calculated. This data is relevant for the state-of-the-art of second-life lithium-ion batteries, allowing further comparisons with other energy storage technologies and further modelling development.

The University of Évora tested several battery packs, although this work aims to present the results of the newest version of the product, considering a single battery module. Its preliminary specifications are gathered in Table I below.

Table I: Reference conditions and main characteristics of one pack, made available by *betteries AMPS GmbH.*

Parameter	Battery
Nominal voltage (external) (V)	52.5
Pack voltage range (V)	42.0 to 57.7
Weight (kg)	31.0
Dimensions (height x length x width) (mm)	321 x 558 x 227
Cycle lifetime (cycles)	1500
Calendar lifetime (years)	7.0
Nominal energy capacity (kWh)	2.3
Useful energy capacity (kWh)	1.8 (80% DoD)
Nominal charge power (kW)	1.4
Nominal discharge power (kW)	2.0
Maximum power output (kW)	4.0
Cooling	Passive cooling
Venting	Pressure equalization & overpressure relief
Ambient operating temperature (°C)	-10 to +40
Non-operating temperature (storage) (°C)	-20 to +50
Ingress protection	IP67

3.1 Test conditions

3.1.1. Rates

C-rates are considered reasonable limits to characterise the lithium-ion batteries, in order to normalise and enable different batteries comparison (distinct characteristic among different batteries). Lithium-ion batteries are generally tested with constant current over the voltage range (and state of charge range), resulting in a discharge and charge rate.

In the battery field, the E-rate is equivalent to C-rate. Nevertheless, it expresses the charge or discharges power in watts. In the absence of a current controller device that effectively sustains the current at a constant value, the controller presented in this work was developed by controlling the AC power setpoint (depending on the connected load), which is the controllable variable on the inverter connected to the battery. This fact conducts to relatively different outputs of these tests to the ones generally presented (if so) by the battery manufacturers. Given the absence of strict standards, the approach of testing with the E-rates is considered as similar as C-rate to describe battery discharge power [8]. Moreover, the testing on these conditions reflects the real operating conditions of the batteries.

3.1.2. Levels of charge and discharge

Lithium-ion batteries' performance and lifetime are mostly affected by charge and discharge conditions, power, and ambient temperature. In this sense, the power setpoint commands of the second-life lithium-ion battery is carefully chosen. Figure 4 (a) and (b) present the charge and discharge maximum limits chosen for current, respectively, and Table II the maximum limits for current, given the number of stacked battery modules.

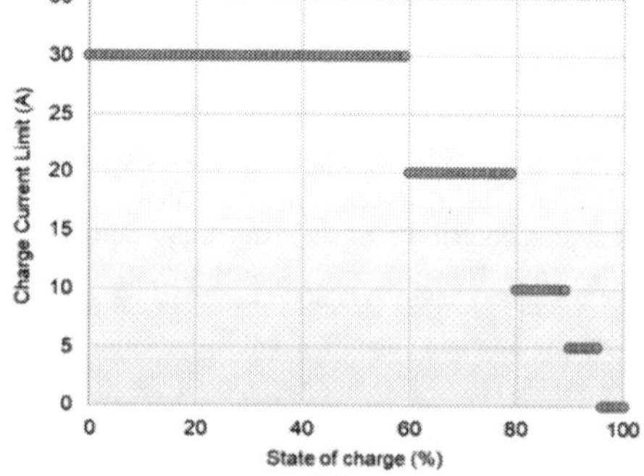

(a)

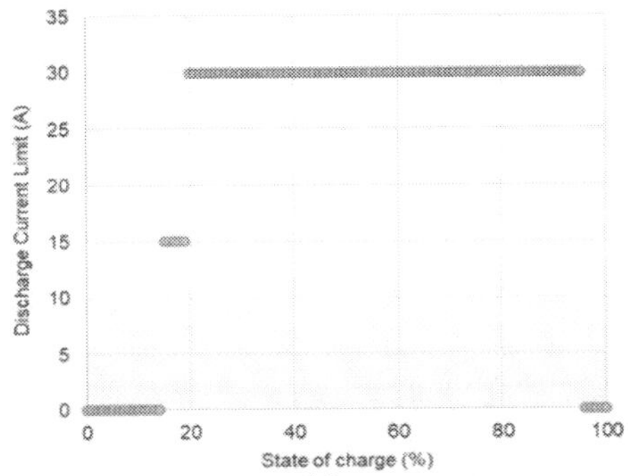

(b)

Figure 4: Charge current limits (a) and discharge current limits (b) according to the state of charge of the battery.

Table II: Maximum limits of current, given the stacked number of battery modules, associated to the same inverter.

Number of battery modules	Maximum acceptable current (A)
1	30.0
2	44.3
3	53.6
4	63.0

3.1.3. Reference test conditions

The reference conditions are the storage typical operating conditions in which the characterisation tests occur. The inverter connected to the battery limits the maximum charge and maximum discharge power limits, which in this case, is upper limited to 3.0 kW. The characterisation tests occur at a temperature range from 15 °C to 25 °C.

For the case of the tested battery, the tests occurred between the 51.1-57.68 V, representing the newer version pack voltage range (depletion and full-charge states, respectively).

3.2 LabVIEW programming control

The tests entail the development of a programming control which communicates with the inverter unit that ultimately communicates with the battery. The communication is achieved through the Modbus TCP/IP protocol, where the programming was developed within the LabVIEW environment.

The timeframe of the control application was defined as being three seconds, including the sending of requests, executing intermediate measurements, receiving answers from the battery and the inverter, and registering the acquired data (both from the inverter and the external monitoring devices, such as the precision wattmeter).

The desired E-rates were defined in real-time, adapted to the relation current-SOC, given the previously presented limits in Figure 4.

3.3 Battery performance efficiencies' calculations

A battery-suited evaluation can be achieved by determining performance indicators related to investment and energy perspectives. As the most relevant performance indicators, the analysis relies on the determination of the following indicators: voltage efficiency, coulombic efficiency, energy efficiency and power efficiency, charge and discharge energy capacities (Wh), energy densities (Wh/kg and Wh/L), power densities (W/kg) and fastest/slowest charge and discharge.

The battery performance is characterized by executing tests from the defined range of the state of charge or voltage operation limits (from depletion to full charge). The operating limits of the battery to perform these tests were chosen regarding the safety margins associated with the DOD and degradation, considering the SLIB characteristics. The evaluation KPIs are defined by:

- Voltage efficiency is the ratio of cell voltage during discharge to that during charge [9]:

$$\theta_v = \frac{U_{discharge}}{U_{charge}} \quad (1)$$

- Coulombic (charge) efficiency is the ratio of electrical charge capacity during discharge to that during charge:

$$\theta_q = \frac{q_{discharge}}{q_{charge}} \quad (2)$$

- Energy efficiency is the ratio of electrical energy during discharge to that during charge:

$$\theta_E = \frac{U_{discharge} \times q_{discharge}}{U_{charge} \times q_{charge}} \quad (3)$$

- Power efficiency is the ratio of electrical power during discharge to that during charge:

$$\theta_W = \frac{I \times U_{discharge}}{I \times U_{charge}} \quad (4)$$

The remaining calculated KPIs are defined as follows. The total energy capacity is the sum of the energy used to charge the battery. The useful energy capacity is the sum of the energy drawn from the battery (discharged from the battery). The energy densities are only calculated for discharged energy. They result from averaging the obtained discharged energy from the battery at the different power levels, rated by the weight or volume of the battery.

For the case of the power densities, it was used the averaging of the three maximum discharged power levels (rated at 1300 W AC), rated by the weight and volume of the battery. The slowest and fastest charge and discharge were calculated through the minimum and maximum values, respectively, of the sum of the period whereby the test occurred. The charge-discharge test repetition at each power level allows the averaging of the results, further reducing the associated experimental error. After that, the KPIs for each full charge-discharge of the battery were calculated.

3.4 Inverter efficiency

A dedicated inverter efficiency test was carried out to characterize the inverter input/output influence on the battery performance output. The commercial inverter connected to the SLIB is from the brand Victron, model MultiPlus-II 48/3000/32. The inverter AC-DC (battery operation in the state of charge) and DC-AC (battery operation in the state of discharge) efficiencies are calculated, for charge and discharge states, respectively:

- DC-AC and AC-DC conversion efficiency – Considers energy conversion losses from DC to AC energy and from AC to DC energy, respectively, are presented in Eqs. (5) and (6).

$$\eta_{conv(DC-AC)} = \frac{E_{AC}}{E_{DC}} \quad (5)$$

$$\eta_{conv(AC-DC)} = \frac{E_{DC}}{E_{AC}} \quad (6)$$

The inverters operates according to an efficiency profile directly related to the power level set and not always at its maximum efficiency. The inverter efficiency used in this work is calculated through Eq. (5) and Eq. (6), considering the different power levels. In this context, a LabVIEW program was developed to operate the inverter at specific power levels (from 150 up to 3000 W) to calculate the efficiency profile and evaluate its performance output. The program was chosen to operate at 15-minute intervals for each power level, regarding its nominal power.

4 RESULTS

4.1 Batteries' KPIs

Each E-rate was averaged for three repeated tests to minimize experimental-related errors. Table III presents the four batteries' efficiencies, calculated with the Equations (1) to (4). The remaining calculated KPIs are shown in Table IV.

Table III: Efficiency results of the studied battery.

	Average efficiency	Standard deviation	Variance
Coulombic efficiency	0.925	0.040	0.002
Energy efficiency	0.830	0.124	0.015
Power efficiency	0.766	0.283	0.080
Voltage efficiency	0.864	0.078	0.006

Table IV: Battery's energy performance main results.

BM / Characteristic	Battery (~50-57 V)
Total energy capacity (charge capacity) (kWh)	2.15 $\pm$ 0.11
Useful energy capacity (discharge capacity) (kWh)	2.05 $\pm$ 0.60
Energy density (discharge) (Wh/kg)	66.0 $\pm$ 10.3
Energy density (discharge) (Wh/L)	49.9 $\pm$ 7.76
Power density (discharge) (W/L)	20.4 $\pm$ 0.20
Power density (discharge) (W/kg)	27.1 $\pm$ 0.27
Slowest/ fastest charge (h)	3.13 / 12.5
Slowest/ fastest discharge (h)	2.46 / 11.8

where $\pm$ indicates the percentual error, in comparison with the average of the result.

A more extensive set of battery modules tested would improve the average results found, probably minimizing the STD values of the results. Figure 5 and Figure 6 show an example of the voltage and current-energy capacity curve in the charging operating state, while Figures 7 and 8 show an example of the voltage and current-energy capacity curve in discharging operating state.

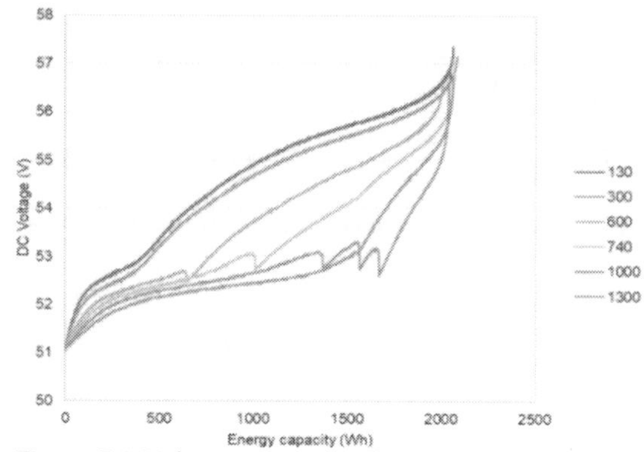

Figure 5:1 Voltage-energy capacity exemplary curve of the battery in charging state.

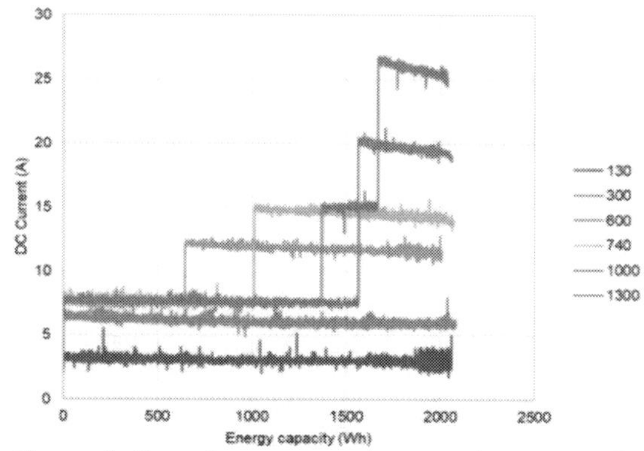

Figure 6: Current-energy capacity exemplary curve of the battery in charging state.

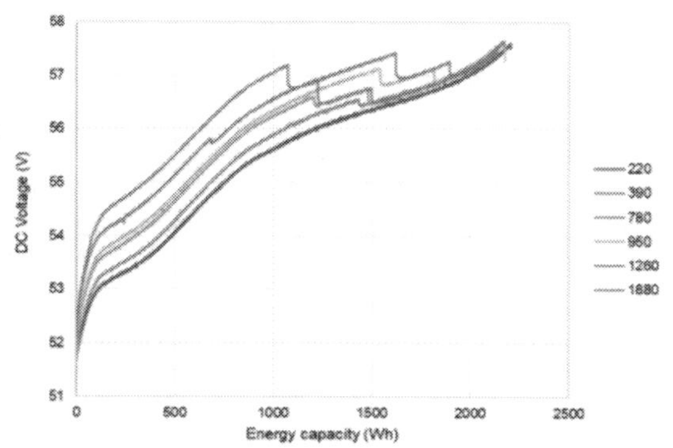

Figure 7: Voltage-energy capacity exemplary curve of the battery in discharging state.

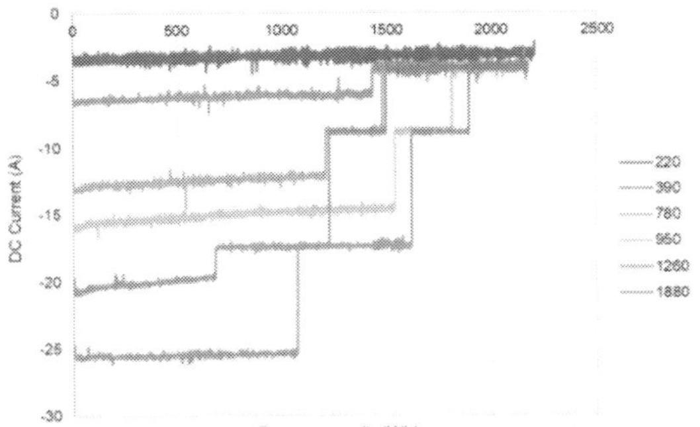

Figure 8: Current-energy capacity exemplary curve of the battery in discharging.

4.2 Inverter efficiency results

The LabVIEW developed control allowed to plot the charge and discharge efficiency curves, Figure 9 and Figure 10. This power-efficiency profile affects the battery's output and should be considered in future modelling of the energy storage system.

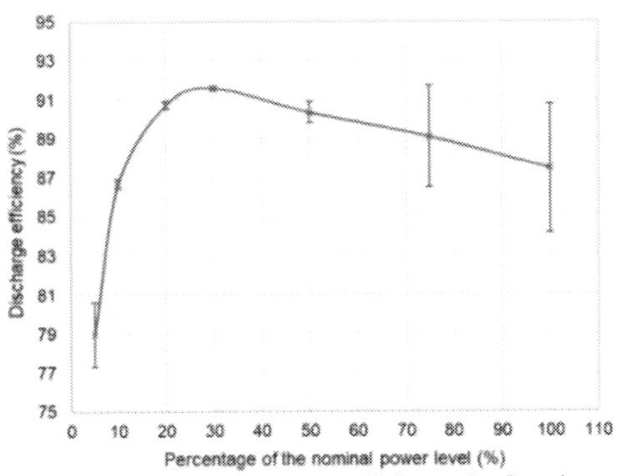

Figure 9: Inverter discharge efficiency in function of the power level.

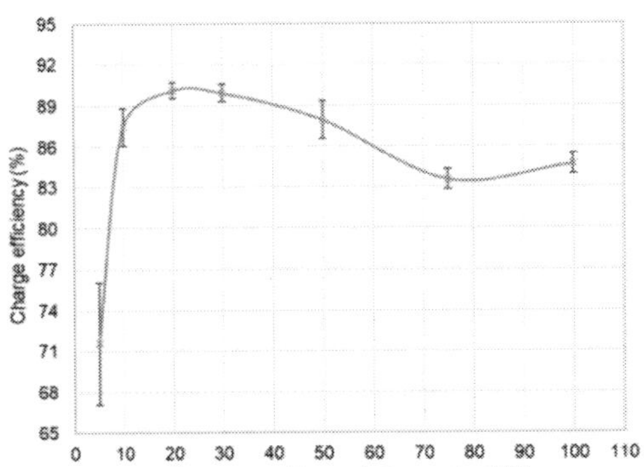

Figure 10: Inverter charge efficiency in function of the power level.

In order to better understand the results obtained of Figure 9 and Figure 10, Figure 11 is depicted below, presenting the inverter efficiency in function of its output power made available by the inverter manufacturer.

Output power (W)	Dissipation (W)	Efficiency (%)
0	11	0
10	11	47,6
20	11	64,5
30	11	73,1
50	11	81,8
100	12	89,7
200	13	93,9
400	19	95,5
800	43	94,9
1600	139	92,0
2500	324	88,5
3000	461	86,7
4000	811	83,1
5000	1261	79,9

Figure 11: Inverter efficiency in function of its output power, made available by the manufacturer [10].

5 DISCUSSION

The SLIB technology was successfully integrated into the University of Évora's microgrid, where the control with the power conditioning unit was effectively achieved, allowing to proceed with the battery and inverter characterization testing and operation evaluation.

The results in Tables III and IV are congruent to the preliminary results that *betteries AMPS GmbH*, namely the average 2.05 kWh 0.60 %. They can be presented as a reference in the SLIB technology, which facilitates the comparison of the electrical performance of the SLIB with the commercial and documented energy storage technologies. Figure 9 and Figure 10 output results can be related to the inverter available manufacturer data depicted in Figure 11. There, one can observe that for the 400-5000W power levels range, a correspondent increase in energy dissipation is observed (from the low to the highest power level), decreasing the values of efficiency.

The battery presented a controlled performance during the testing, presenting itself as a safe and robust product. In order to have an overall more representative result for this product, a higher number of battery modules are required to execute the characterization tests.

The test setup allowed the implementation of dedicated energy management strategies, producing experimental validation data and contributing to current ageing modelling approaches. In this sense, the product presented suitability for on-grid residential and off-grid applications.

6 CONCLUSION

SLIBs can be applied to improve grid performance and integrate with renewable energy generation or charging EVs. The literature lacks disclosed experiments regarding this recent technology's operation, testing and suitability.

The current work methodology allowed the quantification of the SLIB relevant KPIs, characterising the technology in real operating conditions and opening a path to its electrical modelling with experimental results validation.

The tests allowed the gathering of information that contributed to the improvement of the prior versions of the product for building a more robust solution, reaching a higher TRL. Additional testing with use-case scenarios provides technical key indicators regarding its performance in a real-operational environment under stress conditions, enabling this product's further finetuning and validation.

The integration of the SLIB with energy management strategies testing set presents itself as a suited candidate for inclusion in further project proposals and/or applications.

ACKNOWLEDGEMENTS

The authors would like to thank the support of this work, developed under the European POCITYF project, financed by 2020 Horizon under grant agreement no. 864400. This work was also supported by the PhD. Scholarship (author Ana Foles) of FCT – Fundação para a Ciência e Tecnologia –, Portugal, with the reference SFRH/BD/147087/2019.

7 REFERENCES

[1] IEA, "Global EV Outlook 2021," Paris, 2021.

[2] Pocityf, "Leading the smart evolution of historical cities," 2022. [Online]. Available: https://pocityf.eu.

[3] betteries, "betteries. Upcycle batteries. Avert climate change.," 2022. [Online]. Available: https://betteries.com.

[4] S. A. berylls, "BATTERY PRODUCTION TODAY AND TOMORROW," 2018.

[5] "Lithium batteries – 1.2m tons ready for recycling by 2030 – pv magazine International." [Online]. Available: https://www.pv-magazine.com/2019/11/04/lithium-batteries-1-2m-tons-ready-for-recycling-by-2030/. [Accessed: 19-Sep-2022].

[6] "Projected global battery demand by application | Statista." [Online]. Available: https://www.statista.com/statistics/1103218/global-battery-demand-forecast/. [Accessed: 19-Sep-2022].

[7] J. Falk, A. Nedjalkov, M. Angelmahr, and W. Schade, "Applying Lithium-Ion Second Life Batteries for Off-Grid Solar Powered System—A Socio-Economic Case Study for Rural Development," *Zeitschrift für Energiewirtschaft*, vol. 44, no. 1, pp. 47–60, 2020.

[8] MIT Electric Vehicle Team, "A Guide to Understanding Battery Specifications," 2008. [Online]. Available: https://web.mit.edu/evt/summary_battery_specifications.pdf. [Accessed: 30-May-2022].

[9] H. F. Gibbard, "Redox Flow Batteries for Energy Storage," *28th Int. Batter. Semin. Exhib. 2011*, vol. 2, pp. 394–406, Jan. 2022.

[10] "Technical notes on output rating, operating temperature and efficiency." [Online]. Available: https://www.victronenergy.com/upload/documents/Output-rating-operating-temperature-and-efficiency.pdf.

CEOTRES: A PROJECT FOR THE PHOTOVOLTAIC GENERATION OF SOLAR FUELS

A. Martí, A. González del Valle, P. García-Linares, A.B.Cristóbal,
E. Antolín, I. Durán and S. Svatek
Instituto de Energía Solar – Universidad Politécnica de Madrid
ETSI Telecomunicación, Ciudad Universitaria, 28040, Madrid, Spain.

A.L. De Lacey, M. Pita, M.V.Martínez-Huerta and J.C. Conesa
Instituto de Catálisis y Petroleoquímica – CSIC,
C. de Marie Curie, 2, 28049 Madrid, Spain.

ABSTRACT: CEOTRES is a project funded by the Regional Government of the Comunidad de Madrid that aims to explore ideas in the frontier of knowledge for the application of photovoltaics to the generation of solar fuels. In this framework, in this work we: a) describe a thermodynamic analysis that can assist to screen which solar fuel production could be economically feasible from basic principles as well as to estimate the minimum amount of energy required to capture CO_2 from air; b) we review an electrochemical process, based mainly on the use of dehydrogenases and nicotine adenine dinucleotide, that could be used to convert CO_2 into methanol and c) an electrochemical process suitable for capturing CO_2 from ambient air.
Keywords: energy options, storage, strategy, devices, solar fuels

1 INTRODUCTION

The photovoltaic production of solar fuels faces several challenges. One of these challenges lies in the development of suitable electrolyzers, a problem that becomes more relevant when the electrochemical reduction of CO_2 is considered. In contrast to water electrolysis, where only hydrogen and oxygen are produced, in the CO_2 Reduction Reaction (CO2RR) multiple products can be produced (CO, HCOOH, HCHO, CH_4...) [1] which is inconvenient. In this respect, in section 2, we shall review the operation of one of the systems proposed to reduce selectively CO_2 into methanol. The system is based on the use of dehydrogenases enzymes and nicotine adenine dinucleotides.

Another challenge relies on the obtention of CO_2 itself. In this respect CO_2, can be mainly obtained from industrial point sources or being directly captured from air (DAC) [2], [3]. In CEOTRES we focus our attention in DAC systems and in this work, in section 3, we shall review the operation of a system proposed by Wang et al. [4] based on the use of K_2SO_4 and capable of capturing CO_2 from air using electrochemistry.

Finally, in section 4, we shall describe the fundamentals of a thermodynamic analysis that allows screening what organic compounds are cost competitive from the perspective of cost associated to the energy necessary to produce them as well as to estimate the minimum amount of energy that would be required for capturing 1 metric ton (abreviated by "t" in this work and corresponding to 1000 Kg) of CO_2 from air.

2 A METHOD FOR THE ELECTROCHEMICAL PRODUCTION OF METHANOL

One of the processes being investigated in our project to achieve the reduction of CO_2 into methanol can be described by the global reaction:

$$2CO_2(g) + 4H_2O(l) \rightarrow 2CH_3OH(l) + 3O_2(g) \quad (1)$$

The idealized setup for carrying out this process is illustrated in Fig. *1*. It consists of one anode and 3 cathodes (connected in parallel). We shall start the description of the operation of this electrolyzer at the anode. Here, 6 molecules of water decompose into 12 protons and 3 oxygen molecules releasing 12 electrons to the external electric circuit:

$$6H_2O(l) \rightarrow 12H^+(aq) + 3O_2(g) + 12e^- \quad (2)$$

The PV generator drives these electrons from the anode towards the three cathodes at the time it raises their electrostatic potential energy thanks to the energy captured from the sun. Among the 12 electrons, each cathode takes 4 electrons.

Cathode 1 is composed of an electrode and an enzyme called W-FDH (tungsten-formate dehydrogenase) capable of catalysing CO_2 into formate ($COOH^-$) [5]. With 4 electrons and 2 of the protons that were also released at the anode, 2 CO_2 molecules can be catalyzed to produce 2 formate molecules:

$$2CO_2(g) + 4e^- + 2H^+ \rightarrow 2COOH^- \quad (3)$$

Cathode 2 also receives 4 electrons. Another enzyme, called diaphorase, is attached to this electrode. The solution contains also a redox cofactor called nicotine adenine dinucleotide (NAD). The following reaction occurs then at this cathode:

$$2NAD^+ + 2H^+ + 4e^- \rightarrow 2NADH \quad (4)$$

A second enzyme, called formaldehyde-dehydrogenase recycles NADH back into NAD^+ at the time it takes 4 protons to transform the formate produced in cathode 1 into formaldehide (HCOH):

$$2COOH^- + 2NADH + 4H^+ \rightarrow 2NAD^+ + 2H_2O + 2HCOH \quad (5)$$

Notice this reaction also regenerates 2 molecules of water, to be substracted from the initial six water molecules that were initally decomposed at the anode, producing as a result the net balance of 4 water molecules consumed appearing in Eq. (*1*).

Cathode 3 is also attached to the diaphorase enzyme, where the same reaction (4) than in cathode 2 takes place, and consuming the remaining 4 electrons. In this cahode, an enzyme called methanol-dehydrogenase (MDH) is attached what regenerates the NADH produced by the diaphorase at the time it finally transforms formaldehyde into methanol [6]:

$$2HCOH+2NADH+2H^{+} \rightarrow 2NAD^{+} + 2CH_3OH \quad (6)$$

Notice that Eq. (3), plus twice Eq. (4) plus Eqs. (5) and (6) give a net result of 12 protons used, which equals the 12 protons generated in the anode by means of reaction (*11*).

As for the coupling of the photovoltaic converters to the electrolyzers, we consider that indirect coupling, through DC/DC converters and maximum power point trackers is best suited than direct coupling. Although direct coupling has the advantage of avoiding losses in the DC/DC converter, we think that these losses will be largely compensated by the indirect coupling thanks to the preservation of the maximum power operation point of the solar converters, even if solar irradiance changes, and to the possibility of precisely matching in practice the operation potential required by the electrolyzers. Moreover, this is particularly important for the methanol example being illustrated here where cathode *a* and cathode *b* will require different potentials to operate optimally.

3 A METHOD FOR THE ELECTROCHEMICAL CO_2 DIRECT AIR CAPTURE.

Capturing CO_2 from industrial sources has the advantage it is highly concentrated, but it has the disadvantage that it can contain contaminants that will have to be removed before feeding the electrochemical solar fuel production systems. These contaminants are specific of the industrial process so that the design of the purifying process will likely have to be specifically adapted to the industrial process involved.

On the other hand, suppling CO_2 from direct air capture (DAC) can provide CO_2 almost free of these contaminants. Also, DAC could allow for the design of a universal process, suitable for being used at all locations of the Earth, perhaps with only minor modifications. However, this will have to be done at the cost of a higher energy consumption associated to the need of extracting the more diluted CO_2 from air.

There are also electrochemical processes allowing capturing CO_2 from air. One example has been provided by Wang et al. [4], whose results have been emphasized in the review in [3] as a process with an energy consumption lower than 400 kJ/mol, which is considered a target to be achieved for these technologies.

The process starts (Fig. 2) with an aqueous solution of K_2SO_4 (chamber 1) sandwiched by two selective membranes, one cation exchange membrane, such as Naffion 551, allowing K^+ ions passing through towards chamber 3, and an anion exchange membrane, such as AMI-7001, allowing $SO_4^{=}$ ions to pass through chamber 2, where the anode is located. At this anode, 2 water molecules are oxidized releasing 4 protons to the aqueous solution, one oxygen molecule to the air and 4 electrons to the electric circuit:

$$2H_2O \rightarrow 4H^{+} + O_2 + 4e^{-} \quad (7)$$

The $SO_4^{=}$ and H^+ ions diffuse then from chamber 2 to chamber 4.

The solar generator raises the energy of the electrons released at the anode and pumps them to the cathode located in chamber 3. At this cathode, these electrons reduce H_2O to produce hydrogen and OH^- ions:

$$4H_2O + 4e^{-} \rightarrow 4\,OH^{-} + 2H_2 \quad (8)$$

The OH^- and K^+ ions from chamber 1 combine with CO_2 to produce K_2CO_3 or $KHCO_3$.

$$2CO_2 + 2K^{+} + 2\,OH^{-} \rightarrow 2KHCO_3 \quad (9)$$
$$CO_2 + 2K^{+} + 2\,OH^{-} \rightarrow K_2CO_3 + H_2O \quad (10)$$

K_2CO_3 and $KHCO_3$ diffuse towards chamber 4 where they encounter H^+ and $SO_4^{=}$ ions that diffused from chamber 2 and the following reactions, releasing CO_2, takes place:

$$K_2CO_3 + 2H^{+} + SO_4^{=} \rightarrow CO_2 + H_2O + K_2SO_4 \quad (11)$$
$$2KHCO_3 + 2H^{+} + SO_4^{=} \rightarrow 2CO_2 + 2H_2O + K_2SO_4 \quad (12)$$

The resulting K_2SO_4 can be then recycled into chamber 1. It is interesting to note that this process, as in water electrolysis, produces, simultaneously to CO_2 capture, H_2, Eq. (8) and O_2, Eq. (7).

4 THERMODYNAMICS AND COST ANALYSIS

Thermodynamics allows to calculate the minimum amount of energy that would be required to produce a given solar fuel starting from concentrated CO_2 and H_2O only. We shall illustrate this type of calculation taking methanol (CH_3OH) as an example.

Let us assume that we initially have a system consisting of 3 mol of CO_2 and 6 mol of H_2O in thermodynamic equilibrium with a pressure and thermal reservoir at 1 atm and 298 K (that is, the atmosphere). The system is then attached to a work source (which nature does not need to be specified at this stage, but that in our case it consists of an ideal PV system providing electric work) and, together with the reservoir, it is made to evolve reversibly, and therefore quasi-statically, to an state in which 3 mol of methanol and 6 mol of O_2 are present, as illustrated by Eq. (*1*). That the system *plus* the reservoir evolve reversibly means that, considered both together, they do not increase their total entropy, although entropy can be exchanged between both. On the other hand, we recall that work sources, by definition, do not supply entropy to the system they are attached to.

In this context, thermodynamics [7] states that the minimum amount of work, ΔW, that the work source needs to supply to undertake the above process per mol of CH_3OH is given by the increment in Gibb's free energy:

$$\Delta W = \Delta_f G^0[CH_3OH(l)] + 3/2\Delta_f G^0[O_2(g)] - \Delta_f G^0[CO_2(g)] - 2\Delta_f G^0[H_2O(l)] \quad (13)$$

where $\Delta_f G^0$ is the Gibbs energy of formation of the compound indicated in square brackets. The resulting figure, taking data from [8], for methanol is ΔW=702.47 kJmol^{-1}.

Table I: Electricity cost per metric ton (t) of product assuming an electricity price of 35 €MWh^{-1} and that the minimum amount of energy allowed by thermodynamics is used to synthetize the product from CO_2 and H_2O. The extra cost assuming CO_2 is captured from air in ideal conditions is also indicated. For the calculations the following data have been assumed: $\Delta_f G^0 = -394.359$ kJmol^{-1} for CO_2(g); $\Delta_f G^0 = -237.19$ kJmol^{-1} for H_2O(l); $\Delta_f G^0 = 0$ kJmol^{-1} for O_2; [8].

Product	Reaction	Market price (€t^{-1})	$\Delta_f G^0$ (kJmol^{-1}) [8]	ΔW (kJmol^{-1})	Product molar mass, m (g) [8]	Electricity cost (€t^{-1})	CO_2 DAC electricity cost (€t^{-1})
Methanol (CH_3OH)	$2CO_2(g)+4H_2O(l) \rightarrow 2CH_3OH(l)+3O_2(g)$	555 [9]	-166.27	702.47	32.04	214	6.7
Ethylene (C_2H_4)	$2CO_2(g)+2H_2O(l) \rightarrow C_2H_4(g)+3O_2(g)$	1235 [10]	-68.15	1194.95	27.05	431	15.8
Methane (CH_4)	$CO_2(g)+2H_2O(l) \rightarrow CH_4(g)+2O_2(g)$	439 [11]	-50.72	808.02	16.04	496	13.3

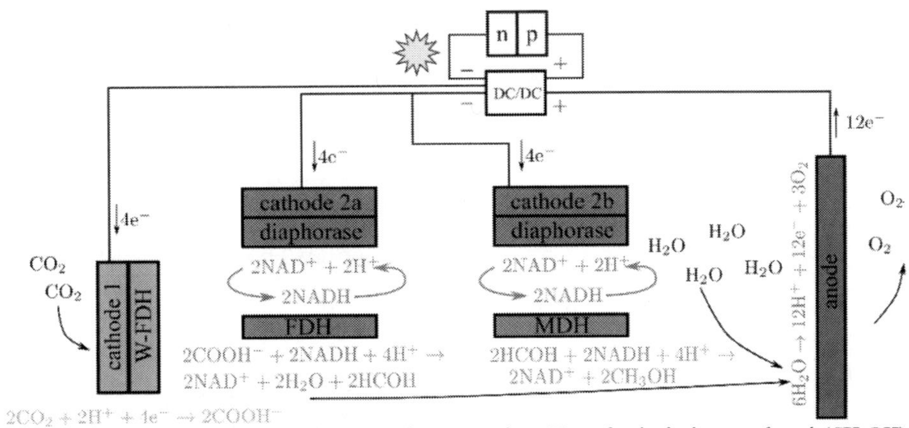

Fig. 1. Representation of an electrochemical process for converting CO_2 selectively into methanol (CH_3OH). The process involves 12 electrons in order to convert 2 molecules of CO_2 and 4 molecules of water into 2 molecules of methanol releasing 3 molecules of oxigen. W-FDH: Tungsten-formate deshydrogenase; FDH: Formaldehide dehydrogenase; MDH: methanol dehydrogenase.

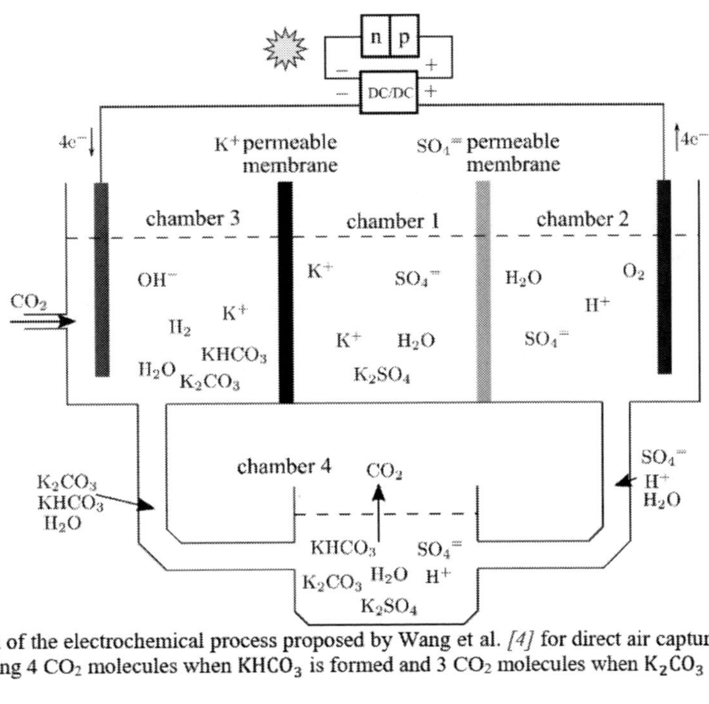

Fig. 2. Representation of the electrochemical process proposed by Wang et al. *[4]* for direct air capture. The process involves 4 electrons for capturing 4 CO_2 molecules when $KHCO_3$ is formed and 3 CO_2 molecules when K_2CO_3 is formed.

Obtaining 1 t of methanol, instead of 1 mol (32.04 g) would therefore demand 21.9 GJ or 6.1 MWh. Market photovoltaic module efficiency in the range of 21 % efficiency has allowed a levelized cost of electricity (LCOE) in the range of 35 €MWh^{-1} for crystalline silicon utility scale solar [12]. Assuming this price, 6.1 MWh of PV electricity would cost 214€. Since methanol market price is 555 €t^{-1} [9] we find out that "thermodynamics" does not forbid the production of methanol from CO_2 and H_2O below current market prices and allows a margin of 341 €t^{-1} to be spent in non-idealities of the system as well as CAPEX and OPEX costs.

Table I illustrates the results of the calculation above for the case of ethylene (resulting in a "thermodynamic" minimal cost of 431 €t^{-1} vs 1235 €t^{-1} market cost) and methane or natural gas (resulting in a "thermodynamic" minimal cost of 496 €t^{-1} vs 435 €t^{-1} market cost). As we can see, for the case of natural gas, this simple thermodynamic analysis allows to conclude that, with present costs, it would not possible to produce CH_4 as solar fuel at competitive prices.

It is also possible to use thermodynamics to calculate the minimum amount of energy required to extract 1 t of CO_2 that is initially diluted in the atmosphere. To do so, it will be necessary to clearly specify what our initial and final thermodynamic states are (Fig. 3).

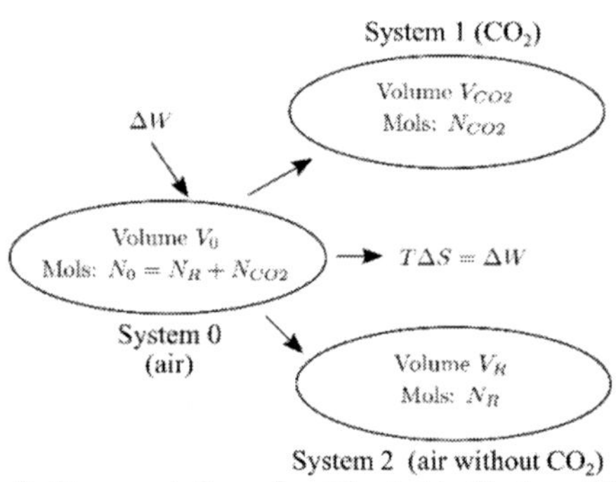

Fig. 3. Representation of a Direct Air Capture (DAC) process. All systems are at preasure p and temperature T.

We shall consider that our initial state corresponds to system 0, consisting of a volume of air V_0, in standard conditions (defined by the pressure p and temperature T) and containing 1 t of CO_2 among all its molecules. Specifically, we shall consider that the system contains N_{CO2} mol of CO_2 and a total of N_R mol corresponding to the remaining gases existing in air (oxygen, nitrogen, argon…). Along these calculations we shall assume all gases involved are ideal. Therefore, the volume V_0, must verify that:

$$V_0 = \frac{(N_R + N_{CO2})RT}{p} \tag{14}$$

where R=8.21 × 10^{-5} m^{3}atmK^{-1}mol^{-1} = 8.31 × 10^{-3} kJmol^{-1}K^{-1} is the Richardson's constant. In summary, system 0 corresponds to a volume V_0 of air before CO_2 is being extracted.

The explanation of these calculations will simplify if we calculate first the entropy S_0 of system 0, which is given by [7]:

$$S_0 = N_{CO2}\frac{S_{0,CO2}}{N_0} + N_{CO2}R\ln\left(\frac{V_{CO2,0}}{V_0}\frac{N_0}{N_{CO2}}\right) + \tag{15}$$

$$N_R\frac{S_{0,R}}{N_0} + N_R R\ln\left(\frac{V_{R,0}}{V_0}\frac{N_0}{N_R}\right)$$
$$= N_{CO2}\frac{S_{0,CO2}}{N_0} + N_{CO2}R\ln\left(\frac{N_0}{N_{CO2}}\right) +$$
$$N_R\frac{S_{0,R}}{N_0} + N_R R\ln\left(\frac{N_0}{N_R}\right)$$

where $S_{0,CO2}$ is the entropy of an amount of N_0 mol of CO_2 such that, when in standard conditions, they occupy the same volume V_0 given by Eq. (14), so that:

$$V_0 = \frac{N_0 RT}{p} = \frac{(N_R + N_{CO2})RT}{p} \tag{16}$$

Hence, we have that:

$$N_0 = N_R + N_{CO2} \tag{17}$$

as otherwise expected.

Similarly, $S_{0,R}$ is the entropy of N_0 mol of the rest of gases contained in air in standard conditions. Obviously, when considered ideal, this number of mols also occupy the volume V_0 since the pressure and temperature are the same.

The second equality in Eq. (15) comes from the fact that, in system 0, the molecules of CO_2 and the rest of molecules in air are mixed within the same volume so that they all initially share the same volume:

$$V_{CO2,0} = V_{R,0} = V_0 \tag{18}$$

The final state (Fig. 3) consists of 2 systems. System 1 contains only the N_{CO2} mol extracted from air in standard conditions. The volume of system 1, V_{CO2}, is therefore given by:

$$V_{CO2} = \frac{N_{CO2}RT}{p} \tag{19}$$

System 2 consists of the original system 0 but from which CO_2 has been extracted. It occupies a new volume, V_R, given by:

$$V_R = \frac{N_R RT}{p} \tag{20}$$

The total entropy of system 1 plus system 2, S_f, is given by:

$$S_f = N_{CO2}\frac{S_{0,CO2}}{N_0} + N_{CO2}R\ln\left(\frac{V_{CO2}}{V_0}\frac{N_0}{N_{CO2}}\right) + N_R\frac{S_{0,R}}{N_0} + N_R R\ln\left(\frac{V_R}{V_0}\frac{N_0}{N_R}\right) = S_{0,CO2} + S_{0,R} \tag{21}$$

For the last equality in Eq. (21) we have used Eqs. (16), (17), (19) and (20). By subtracting Eq. (15) from Eq. (21) we obtain the increment in entropy, ΔS, corresponding to the ideal CO_2 direct air capture process from air:

$$\Delta S = S_f - S_0 = -N_{CO2}R\ln\left(\frac{N_0}{N_{CO2}}\right) - N_R R\ln\left(\frac{N_0}{N_R}\right) \tag{22}$$

Notice that, since $N_0 > N_{CO2}$ and $N_0 > N_R$, the increase in entropy is negative since the process we are analyzing corresponds to the unmixing of two gases. However, although there has been a decrease in entropy, the total internal energy remains constant ($\Delta U = 0$) that is, the internal energy of system 0 equals the internal energy of system 1 plus system 2. This is because, for ideal gasses, the internal energy only depends on the number of mols and the temperature [7] of the gas and none of these magnitudes have changed. Therefore:

$$0 = \Delta U = \Delta W - T\Delta S \rightarrow \Delta W = T\Delta S \qquad (23)$$

where ΔW is the input work needed to decrease the entropy of the system and $T\Delta S$ the heat released from the system because of this decrease. We recall the heat released can be calculated as $T\Delta S$ because we assume the processes to be quasi-static and occur at constant temperature T.

Assuming 44.01 g CO_2 molar mass, we find that 1 t of pure CO_2 contains $N_{CO2} = 22722$ mol. Also assuming CO_2 is concentrated in the atmosphere at 400 ppm (or, equivalently, 400 μmol per mol) this leads to the fact that a total of $N_0 = 5.68 \times 10^7$ mol of air molecules are necessary to contain such amount of N_{CO2}. Taking these numbers (together with Eq. (17), that can be approximated by $N_0 \approx N_R$, if desired) into Eqs.(22) and (23) we obtain, for $T = 298$ K, that a minimum of $\Delta W = 496.7 \times 10^3$ kJ (or 137.98 kWh) are necessary to extract 1 t of CO_2 from air. This corresponds also to a figure of 22 $kJmol^{-1}$ as thermodynamic limit to be compared with 400 kJ/mol reported experimentally by Wang et al. in their actual experiment.

Assuming a photovoltaic LCOE of 35 $€MWh^{-1}$, this translates into a cost of 4.8 € per t of CO_2 captured. This cost has been included in Table I under the heading "DAC cost" and expressed as per t of solar fuel sensitized. From this result we conclude that, from a thermodynamic point view, DAC does not introduce a prohibitive energy cost into the process.

Finally, for illustrative purposes we shall mention that the volume of air, V_0, Eq. (14), required to contain 1 t of CO_2 is 1.8×10^6 m^3, corresponding to a cube with a side of 111 m; the resulting volume, V_{CO2}, Eq. (19), necessary to contain 1 t of CO_2 in standard conditions is 556 m^3, resulting in a cube of 8.2 m side. 1 t of CO_2 is approximately the emissions that corresponds to two persons flying from Madrid to this Conference, in Milan, and return [13].

5 SUMMARY

The electrochemical processes for producing solar fuels from CO_2 are more complex than the ones involved in water electrolysis. Nevertheless, they are possible, and in this work we have described in detail one possible process, based on the use of several dehydrogenases, leading to the production of methanol. Moreover, electrochemical processes for direct CO_2 capture from air are also possible and, in this respect, we have reviewed the process described by Wang et al., based on the use of K_2SO_4 and ion selective permeable membranes which is compatible with the simultaneous production of hydrogen.

We have also described the fundamental thermodynamic analysis that allows to calculate the minimum amount of energy needed to carry out the processes and translate this energy into electricity cost. Assuming photovoltaic LCOE of 35 $€MWh^{-1}$, the minimum cost for producing 1 t of methanol would be 214 €. Comparing this cost with methanol market price (555 $€t^{-1}$) this result still gives room for extra costs. The analysis of the ethylene is also favorable (431 $€t^{-1}$ vs market price of 1235 $€t^{-1}$). However, the case of methane results unfavorable (496 $€t^{-1}$ vs market price of 439 $€t^{-1}$).

The minimum cost of the energy needed for capturing 1 t of CO_2 from air has also been calculated from fundamental principles resulting in 4.8 €t that translates into a range of 6-14 € per t of solar fuel produced since each solar fuel demands a different amount of CO_2 to be produced.

6 ACKNOWLEDGMENTS

This work has been supported by the CEOTRES-CM project, funded by the Comunidad de Madrid with the Grant Number Y2020/EMT-6419.

5 REFERENCES

[1] G. Wang *et al.*, "Electrocatalysis for CO_2 conversion: from fundamentals to value-added products," *Chem. Soc. Rev.*, vol. 50, no. 8, pp. 4993–5061, Apr. 2021.

[2] N. MacDowell *et al.*, "An overview of CO_2 capture technologies," *Energy Environ. Sci.*, vol. 3, no. 11, p. 1645, 2010.

[3] E. S. Sanz-Pérez, C. R. Murdock, S. A. Didas, and C. W. Jones, "Direct Capture of CO_2 from Ambient Air," *Chem. Rev.*, vol. 116, no. 19, pp. 11840–11876, Oct. 2016.

[4] C. Wang, H. Liu, X. Li, and L. Zheng, "Importance of Ambient O_2 for Electrochemical Enrichment of Atmospheric CO_2," *Ind. Eng. Chem. Res.*, vol. 52, no. 7, pp. 2470–2476, Feb. 2013.

[5] L. B. Maia, I. Moura, and J. J. G. Moura, "Molybdenum and tungsten-containing formate dehydrogenases: Aiming to inspire a catalyst for carbon dioxide utilization," *Inorganica Chim. Acta*, vol. 455, pp. 350–363, Jan. 2017.

[6] S. Kuwabata, R. Tsuda, and H. Yoneyama, "Electrochemical conversion of carbon dioxide to methanol with the assistance of formate dehydrogenase and methanol dehydrogenase as biocatalysts," *ACS Publications*, vol. 116, no. 12, pp. 5437–5443, 1994.

[7] H. Callen, *Termodinámica*. Madrid: Editorial AC, 1981.

[8] D. D. Wagman, W. H. Evans, V. B. Parker, R. H. Schumm, and I. Halow, "The NBS Tables of Chemical Thermodynamic Properties. Selected Values for Inorganic and C1 and C2 Organic Substances in SI Units," NATIONAL STANDARD REFERENCE DATA SYSTEM, Jan. 1982. Accessed: Jul. 01, 2022. [Online]. Available: https://apps.dtic.mil/sti/citations/ADD095481

[9] "Pricing | Methanex Corporation." https://www.methanex.com/our-business/pricing (accessed Aug. 03, 2022).

[10] "Ethylene prices globally 2022," *Statista*. https://www.statista.com/statistics/1170573/price-ethylene-forecast-globally/ (accessed Aug. 03, 2022).

[11] "United States Natural Gas Industrial Price (Dollars per Thousand Cubic Feet)."

https://www.eia.gov/dnav/ng/hist/n3035us3m.htm (accessed Aug. 03, 2022).
[12] D. Ray, "Lazard's Levelized Cost of Energy Analysis—Version 15.0," 2021.
[13] "MyCLimate (www.myclimate.org/)."

ENERGY SHARING CONTROL STRATEGIES: A BENCHMARK ANALYSIS IN A CONFIGURABLE ITALIAN DEMONSTRATOR

Mattia Secchi[1,2], Grazia Barchi[1,*], David Moser[1], Stefano Nassuato[3], Daniele Pellizzari[4], Alessandro Costa[4]
[1]Eurac Research, Institute for Renewable Energy, Viale Druso/Drususallee 1, 39100 Bolzano (BZ), Italy
[2]University of Trento, Department of Industrial Engineering, Via Sommarive 9, 38123 Trento (TN), Italy
[3]Regalgrid Europe Srl, Via Manin, 73, 31100 Treviso (TV), Italy
[4]Alperia Spa, Via Dodici ville 8, 39100 Bolzano (BZ), Italy
*E-mail: grazia.barchi@eurac.edu

ABSTRACT: The EU Clean Energy for All Europeans Package introduced Renewable Energy Communities (RECs), a new legal entity that economically incentivizes energy sharing between its members. Within this context, centrally controlled battery electric storage systems (BESSs), maximizing the REC-level load-production matching, could become a viable solution. Thus, in this work, two centralized BESS controls, one of which is a commercially available solution, are benchmarked against a simulated decentralized one, which is available in the literature. The comparison is performed using real data available from a demonstrator located in Bolzano, Italy. Results highlight that the production-consumption matching increases at the REC level when centralized controls are used, and that the commercial solution performs at the same level as the simulated ones. Moreover, centralized controls make a more efficient use of the available storage capacity, thus greatly reducing the equivalent CO_2 emissions.
Keywords: *Photovoltaics, Storage Systems, Battery Control, Renewable Energy Community, Energy Sharing*

1 INTRODUCTION

The publication of the EU Clean Energy for All Europeans Package [1] and the Renewable Energy Directive II [2] emphasize the key role of citizens to collectively self-produce their energy requirement from renewable sources and develop a local energy market. The energy system is thus expected to transform from a top-down to a bottom-up one, where the prosumers and consumers are allowed to operate under collaborative schemes, as a "community". Photovoltaic (PV) systems will play a key role in this framework [3][4], as they enable for the creation of "Renewable Energy Communities" (RECs), legal entities aggregating passive consumers and PV owners to enable their participation in the electricity market, lower the electricity costs and decrease the payback period for the PV investment and increase the quota of renewables. Generally, the higher the heterogeneity of the consumption patterns, the higher are the energy benefits of joining to a REC [5]. The technical implementation of energy sharing is supported by two key elements: the existence of a smart-metering infrastructure based on information-communication technologies and the definition of a suitable control strategies for the battery energy storage systems (BESS). The first allows to gather production and consumption data, and thus interact with the BESS, while the second is responsible for the energy management process. In that regard, several examples of BESS operation in RECs are found in the literature. For example, Rodrigues et al. [6] used a centralized energy sharing BESS control algorithm to maximize the use of PV energy in a university campus system. Secchi et al. [7] instead, minimized the grid dependence of a large REC in Italy, while keeping the grid voltage levels under control. Talluri et al. [8] optimized BESS charging in a REC by simultaneously minimizing the total system operative costs and maximizing the REC members revenue. Casalicchio et al. [9] instead, found the optimal energy dispatch in presence of BESSs to minimize the total electricity costs in RECs. In that regard, the main point of interest of this work lies in the comparison between a decentralized rule-based BESSs control solution for RECs and two centralized ones, one of which is commercial available. The analysis is performed using real data with a high temporal resolution (1 minute) in a REC demonstrator located at the NOI Techpark in Bolzano (Italy). The results emphasize: i) the benefits that a centralized BESS controls enable in terms of production-demand matching at the REC-level; ii) the coherence of the results between the simulated approach and the commercial optimized solution.
The energy performance analysis is complemented by an evaluation of the equivalent emissions reduction and a sensitivity analysis on the simulation timestep.
Key aspects of the proposed work are: i) a decentralized BESS management is compared with a fully working and implemented commercially available solution and a literature based one, ii) the analyzed system is the first prototype of a fully configurable REC in the South-Tyrol (Italy) region, iii) the input profiles are collected with a high frequency from a real demonstrator, so a sensitivity analysis on the timestep can be performed.

2 METHODOLOGY

The purpose of this work is to benchmark the centralized battery controls, either the commercial Cloud-OPT[1] or the simulated peer-to-peer (P2P) one, against the business-as-usual control, called peer-to-grid (P2G). All the control strategies rely on battery management to maximize the consumption of PV produced energy, but while the P2G method considers each individual BESS and optimizes the individual-user level self-sufficiency, the P2P and Cloud-OPT strategies minimize the REC exchanges with the upstream electric network. In other words, P2G is a decentralized control, whereas P2P and Cloud-OPT are centralized ones. The P2G and P2P controls were introduced in [10] and then modified in [11]. P2G is a decentralized control where each user charges their BESS if the instantaneous production is higher than the consumption. Once the maximum storage capacity is reached, the PV system starts selling to the grid. The

[1] RegalGrid Europe - https://www.regalgrid.com/

opposite happens when the consumption is higher than the PV production, so the BESS will be discharged based on the current power mismatch. Once the battery charge is depleted, the remaining electricity is absorbed from the grid. P2P requires a preliminary step instead, where all the net-load values of the individual users are aggregated to obtain the charging/discharging signals the central coordinator sends to the BESSs. That energy requirement is then covered by discharging or charging the BESSs, the individual users state-of-charge (SOC) and depth-of-discharge (DOD) respectively. A rule-based approach is implemented, so that the BESSs with the highest SOC are discharged first, and the ones with the highest DOD are charged first. Both the P2P and Cloud-OPT algorithms will be benchmarked against the P2G algorithm over some key performance indicators (KPIs), in particular:

- *Self-consumption (SC)*: at the single-user level, the percentage of the total PV production either directly consumed by the user or stored into its batteries

$$SC_n = \frac{E_{SCdir,n} + E_{toBESS,n}}{E_{PV,n}} \cdot 100 \qquad (1)$$

  where $E_{SCdir,n}$ is the PV produced energy that is immediately consumed by the n-th user, $E_{toBESS,n}$ is the PV production stored in the BESS, and $E_{PV,n}$ is the total PV production in the analyzed period for the n-th user.
- *Self-production (SP)*: at the single-user level, the percentage of the total user consumption covered by the PV production or absorbed from the batteries

$$SP_n = \frac{E_{SCdir,n} + E_{fromBESS,n}}{E_{LOAD,n}} \cdot 100 \qquad (2)$$

  where $E_{fromBESS}$ is the total energy absorbed from the BESSs to cover the load of the n-th user, and $E_{LOAD,n}$ it's the total load consumption in the analyzed period for the n-th user.
- *Battery Usage Rate (BUR)*: represents how efficiently the control algorithm uses the installed storage capacity to cover the net-load:

$$BUR_n = \frac{E_{toBESS,n} + E_{fromBESS,n}}{E_{netLOAD,n}} \cdot 100 \qquad (3)$$

  where $E_{netLOAD,n}$ is the net-load of the n-th REC member, i.e., the residual load once the PV production is discounted. In the ideal scenario where BUR=100%, then SC=SP=100%.

At the community level the self-consumption (SC_{REC}), self-production (SP_{REC}) and battery usage rate (BUR_{REC}) are given by the sum over the all n nodes, where n∈N of the expressions in (1), (2) and (3).

- *Saved CO_2 Emissions*: relative emissions saved by using the centralized controls over P2G. After a few simple mathematical steps:

$$SE_{CO2} = \frac{SP^{CENTRALISED} - SP^{P2G}}{(100 - SP^{P2G})} \cdot 100 \qquad (4)$$

  where $SP^{CENTRALISED}$ is the centralized REC-level self-production for either the P2P or Cloud-OPT algorithms, while SP^{P2G} considers P2G.

3 CASE STUDY

The energy community has been created by virtually mapping different nodes of a large building belonging to the NOI Techpark of Bolzano (Italy). Around 60 kWp of PV and 31 kWh of storage capacity are used to feed office load such as light bulbs, printers, and computers. Table 1 shows the configuration of the seven nodes the system is composed by.

At the interface of each device (i.e., inverters, smart meters, battery energy management systems) with the electric network, a "smart node" (SN) was installed, to acquire generation and consumption real-time measurements and control the BESS with the Cloud-OPT algorithm. The period under study lasts for 15 months, from July 2020 to September 2021. The one-minute data gathered by the SN devices are used in a software-based environment to simulate the P2G and P2P controls for real-time BESSs control, while the Cloud-OPT directly run on the demonstrator.

Table 1 – Photovoltaic capacity, BESS capacity and load consumption over the whole monitoring period.

Node	Load Consumption	BESS Capacity	PV Capacity
1	9 MWh	7 kWh	10 kWp
2	5.7 MWh		
3	5.7 MWh		20 kWp
4		10 kWh	
5	1.9 MWh	7 kWh	15 kWp
6	20.1 MWh		
7	4.9 MWh	7 kWh	15 kWp

4 RESULTS

4.1 REC-Level Performance

Figure 1 shows the SC, SP and BUR values that the system can attain at the REC level, and compared to the maximum achievable result. There is a benefit in controlling the BESSs in a centralized way, since both the P2P and Cloud-OPT algorithms outperform the P2G one by up to 11% and their behavior is similar. P2P has a slight edge over Cloud-OPT because an additional 2 MWh are exchanged with the BESS at Node 4, when P2P is used. The maximum attainable SC without batteries is around 59%, and the P2P

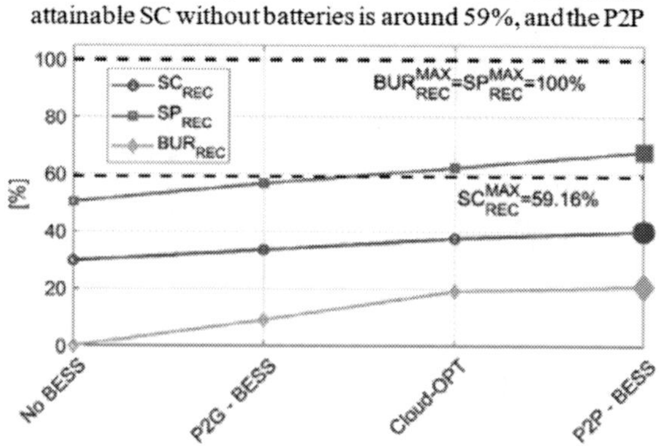

Figure 1 – Self-consumption (in blue), self-production (in red) and battery usage rate (in yellow) of the three control algorithms at community level.

control peaks at 40% SC only, which means that the overall PV generation is too high for the total load consumption. Theoretically, the PV production could be high enough to cover the entire load demand, so the maximum attainable SP is 100%, but the system only achieves around 70% SP with the P2P algorithm. It is thus possible to conclude that the system has a higher efficiency in covering the net-load than in storing the PV over-production. In that regard, a proper sizing of the PV-BESS systems and the inclusion of flexible loads, e.g., electric vehicles or heat pumps, could improve the SC values [12]. Moreover, the BUR values show that the BESSs are not used as much as they could, due to the excess of PV production sold to the grid. Larger batteries could allow for higher PV storing capacity, and thus simultaneously improve SC, SP and BUR.

4.2 Individual User-Level Performance

Figure 2 instead, represents the SC, SP and BUR results at the individual-user level, e.g., node-by-node, as defined by equations (1), (2) and (3).

In this case, P2G performs slightly better than P2P for SC and SP, since the individual user-level KPIs are better optimized by a decentralized algorithm that does not consider the other REC members. BUR instead, shows a different behavior, with P2P and Cloud-OPT exploiting the BESSs in a more efficient way. This happens because the central control unit always has access to the BESSs SOC information. It must be noted how Nodes 2, 4 and 6, which lack a PV system, are not considered in this analysis, since they have no associated PV generation, so the individual user-level KPIs cannot be calculated. Node 1 has the highest SC, and a SP still comparable to the others. This happens because the lower installed PV capacity matches the 7 kWh BESS better than the other nodes, as confirmed by the higher BUR values (30-40%). Node 3 does not have any installed storage capacity, thus

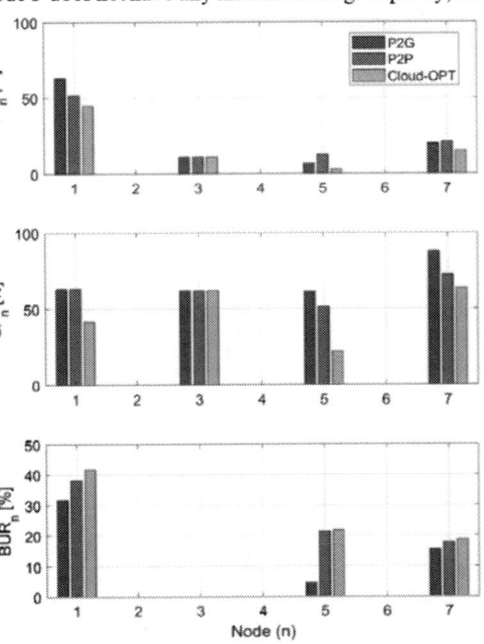

Figure 2 – Self- consumption (SC), self-production (SP) and battery energy usage rate (BUR) for the P2G (in blue), the P2P (in red) and the Cloud-Opt (in yellow) at single-user level.

implementing one strategy over the others does not produce any effect on SC and SP. The SC is quite low because there is a very high installed PV capacity, which in turns raises the SP at the same level as the other nodes. Finally, the BESSs at Nodes 5 and 7 are undersized for the installed PV capacities, as confirmed by the very low SC values. The situation at Node 5 is particularly critical, because the total energy consumption is very low compared to the other nodes (around 1.9 MWh in the analysed period).

4.3 Emissions Reduction

Table 2 shows the monthly CO_2 emissions reduction produced by either P2P or Cloud-OPT, compared to the P2G control, computed according the equation in (4). It is easy to see the seasonality effect on the additional emissions reduction in the spring/summer months, where increases of +48/+67% are obtained. Over the entire monitoring period, the application of P2P and Cloud-OPT instead of P2G leads to an additional 25% and 12% emissions reduction. As such, the introduction of PV and BESS reduces the emissions by 81% and 68%, in P2P and Cloud-OPT, which is quite an impressive result.

Table 2 – Additional relative CO_2 emissions reduction due to the application of centralised controls over P2G.

Month	Peer-to-peer(P2P)	Cloud-Opt
Jul-20	67.03 %	46.86 %
Aug-20	49.47 %	35.39 %
Sep-20	36.54 %	26.99 %
Oct-20	23.84 %	17.86 %
Nov-20	27.86 %	21.70 %
Dec-20	4.52 %	1.57 %
Jan-21	1.96 %	0.82 %
Feb-21	16.47 %	11.31 %
Mar-21	34.46 %	22.14 %
Apr-21	36.23 %	19.39 %
May-21	39.26 %	5.94 %
Jun-21	45.70 %	12.88 %
Jul-21	39.77 %	8.23 %
Aug-21	36.34 %	3.92 %
Sep-21	26.17 %	2.49 %
Total	**25.65 %**	**12.50 %**

4.4 Sensitivity Analysis on Timestep

Figure 3 instead, shows how the SC, SP and BUR at the REC-level change when using a timestep larger than the smallest available system sampling one, i.e., 1 minute. Results show that, up to 3 hours (180 minutes), the difference of the KPIs computed with higher timestep resolution compared to 1 minute is lower than 2.5%. This may seem like a very low value, but in a running phase covering 15 months, that means an error of 20 MWh on the SC calculation, and 12 MWh on SP.

Results also show that SC and SP are generally overestimated when the timestep grows, and this happens due to the combination of two contrasting effects. Firstly, the net-load power profiles, which define the BESS control signals, are smoothed out by a timestep increase, so the E_{SCdir} term in equations (1) and (2) (not considering the BESS contribution) is greatly overestimated. Secondly, if the BESSs are considered, the maximum/minimum state of charge (SOC) limits are met very easily, i.e., a very

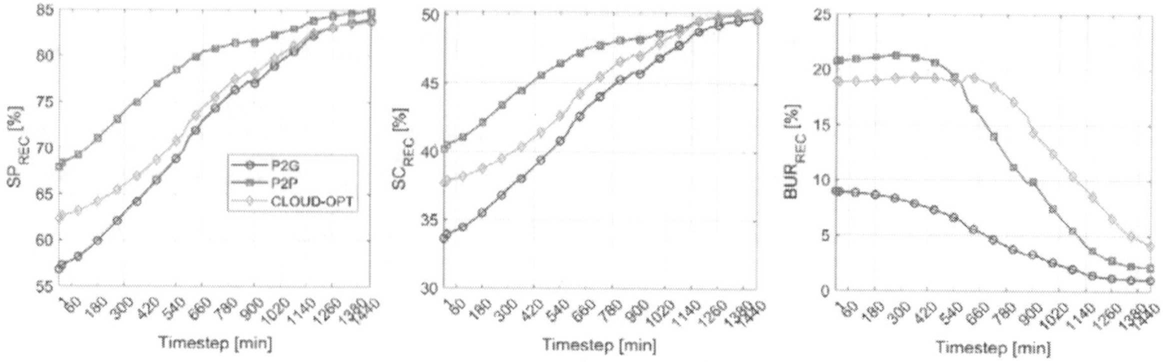

Figure 3 – Behaviour trend of self-consumption (SC), self-production (SP) and battery usage rate (BUR) resulting from the sensitivity analysis of the timestep resolution.

large timestep immediately saturates/empties the BESSs because the injected/absorbed energy will always be higher than the available storage space or stored energy. In other words, if the timestep is very high, both the $E_{fromBESS}$ and E_{toBESS} terms in equations (1) and (2) are almost null, hence the BUR being highly underestimated (up to 18%). These two contrasting effects combine in a general overestimate of SC and SP. Finally, as the timestep grows, the KPIs converge to the same value, thus a timestep increase evens out the difference between the BESS controls. Moreover, a much higher relative error is noted for shorter periods of time (e.g., monthly simulations). A good compromise for the choice of a timestep is 15 minutes, as recommended in [12], which equals the sampling period of modern smart meters and also preserve privacy issues.

5 CONCLUSIONS

The development of Renewable Energy Communities in the EU has introduced economic incentives for the energy shared and consumed by REC members. This could make centrally controlled BESSs a viable alternative to distributed storage resources. In this work, two centralized BESSs controls, one of which is a commercially available product, are benchmarked against a decentralized one by using data available from a demonstrator located in Bolzano, Italy. Results highlight that centralized controls improve the self-consumption and self-production at the REC-level, making the district more independent from the main electric grid. Additionally, the commercially available Cloud-OPT solution performs at the same level as the P2P simulated one. Both those controls, indeed, exploit the available storage capacity much better than the decentralized P2G control (as shown by the BUR values), thus greatly reducing the equivalent CO_2 emissions of the system. Finally, a sensitivity analysis on the simulation timestep proves that the errors in computing the most relevant KPIs is almost negligible up until 180 minutes for long simulated periods, thus a 15 minutes timestep is recommended as it coincides with the modern smart meters sampling time. A final remark should be made about the system itself, whose a significant quota of PV generation is injected into the grid and fails the load supply. To solve this issue, future work will consider flexible loads (e.g., electric vehicles and heat-pumps), as well as the reconfiguration of the nodes by obtaining more simultaneity between production and consumption. In conclusion, this work analyzed the performance of a demonstrator whose relevance is not limited to the proof-of-concept of what reported in the paper, but also lies in being a fully reconfigurable demonstrator to preliminary test a variety of energy sharing control strategies or advanced features to properly managed energy flows in a REC.

6 REFERENCES

[1] European Commission, "Clean energy for all Europeans," 2016.

[2] European Commission, "Renewable Energy Directive II," 2018.

[3] J. Lowitzsch, C. E. Hoicka, and F. J. van Tulder, "Renewable energy communities under the 2019 European Clean Energy Package – Governance model for the energy clusters of the future?," *Renewable and Sustainable Energy Reviews*, vol. 122, p. 109489, Apr. 2020.

[4] J. Radl, A. Fleischhacker, F. H. Revheim, G. Lettner, and H. Auer, "Comparison of Profitability of PV Electricity Sharing in Renewable Energy Communities in Selected European Countries," *Energies 2020, Vol. 13, Page 5007*, vol. 13, no. 19, p. 5007, Sep. 2020.

[5] V. Casalicchio, G. Manzolini, M. G. Prina, and D. Moser, "Optimal Allocation Method for a Fair Distribution of the Benefits in an Energy Community," *Solar RRL*, vol. 6, no. 5, p. 2100473, May 2022.

[6] D. L. Rodrigues, X. Ye, X. Xia, and B. Zhu, "Battery energy storage sizing optimisation for different ownership structures in a peer-to-peer energy sharing community," *Applied Energy*, vol. 262, no. January, p. 114498, Mar. 2020.

[7] M. Secchi, G. Barchi, D. Macii, D. Moser, and D. Petri, "Multi-objective battery sizing optimisation for renewable energy communities with distribution-level constraints: A prosumer-driven perspective," *Applied Energy*, vol. 297, p. 117171, Sep. 2021.

[8] G. Talluri, G. M. Lozito, F. Grasso, C. Iturrino Garcia, and A. Luchetta, "Optimal Battery Energy Storage System Scheduling within Renewable Energy Communities," *Energies 2021, Vol. 14, Page 8480*, vol. 14, no. 24, p. 8480, Dec. 2021.

[9] V. Casalicchio, G. Manzolini, M. G. Prina, and D. Moser, "From investment optimization to fair benefit

distribution in renewable energy community modelling," *Applied Energy*, vol. 310, p. 118447, Mar. 2022.

[10] C. Long, J. Wu, Y. Zhou, and N. Jenkins, "Peer-to-peer energy sharing through a two-stage aggregated battery control in a community Microgrid," *Applied Energy*, vol. 226, pp. 261–276, Sep. 2018.

[11] M. Secchi and G. Barchi, "Peer-to-peer electricity sharing: Maximising PV self-consumption through BESS control strategies," in *Proceedings - 2019 IEEE International Conference on Environment and Electrical Engineering and 2019 IEEE Industrial and Commercial Power Systems Europe, EEEIC/I and CPS Europe 2019*, 2019, vol. 1042, pp. 0–5.

[12] Y. Dahmane, R. Chenouard, M. Ghanes, and M. Alvarado-Ruiz, "Optimized time step for electric vehicle charging optimization considering cost and temperature," *Sustainable Energy, Grids and Networks*, vol. 26, p. 100468, Jun. 2021

HYDRO-FLOATING PV HYBRID OPERATIONAL ANALYSIS – DISPATCHABILITY EXPECTATIONS UNDER DIFFERENT OPERATIONAL MODES

Imke Meyer[1], Konstantinos Priftis[2], Christian Kaufmann[3], Brian Darling[4], Inaki Perez[5]
[1] Mott MacDonald Ltd, Victory House, Trafalgar Place, Brighton, BN1 4FY, UK, imke.meyer@mottmac.com, +44 1273365429
[2] Mott MacDonald Ltd, konstantinos.priftis@mottmac.com
[3] Mott MacDonald Ltd, christian.kaufmann@mottmac.com
[4] Mott MacDonald Ltd, brian.darling@mottmac.com
[5] Mott MacDonald Thailand, inaki.perez@mottmac.com

ABSTRACT: This analysis investigates the potential of overall dispatchability of hybrid hydro-floating PV (FPV) systems when FPV is coupled with existing hydro power facilities. The analysis examines the possible complementary nature of the FPV and hydro generation profiles in different climates for energy and profit maximization. To undertake the analysis, PLEXOS® Integrated Energy Model by Energy Exemplar has been employed. The case studies for energy dispatchability are performed in three locations (each with a different climate), namely Pakistan, Zambia and Turkey. Each location considers a hydro plant with specific operational requirements (serving peak or base load) and irrigation constraints. It is concluded that the flexibility in energy dispatch is strongly influenced by the inflow and associated available storage volume of the reservoir. Flexibility in dispatch is observed in the short term (24-72 hour window) for three of the four cases (the outlier has significant daily irrigation targets) however no seasonal storage benefits were observed. For all investigated case studies, the transmission line capacity factor increased with decrease in curtailment when compared to the standalone cases with better utilization of the infrastructure with the addition of FPV.

Keywords: hybridization, floating PV, dispatchability

1 INTRODUCTION

There has been a significant uptake of floating photovoltaic plants (FPV) over the last 5 years, with more than 2.6 GW installed worldwide in 2020 [1]. A focus area in the market is the installation of FPV on hydro dams and reservoirs (hydro) citing the inherent economic benefits given complimentary generation profiles, increased dispatchability of the combined FPV-hydro generator and better utilization of existing transmission infrastructure [2],[3]. The largest of such projects (50MW FPV- 36MW hydro) came into operation in November of 2021 in Thailand [4]. In areas where land is not a constraint, the hybridization of land-based PV plants and hydroelectric dams has also seen successful implementation, namely Longyangxia hydro-connected PV power plant in Qinghai, China.

It has been reported for the Longyangxia plant, that once PV is coupled to the hydro generator, a higher power set point was achieved and the hydro turbines are able to provide adequate response as the demand and PV varies given the inherent storage capabilities of hydro facilities [2].

The purpose of this research is to investigate to what extent the total energy produced by the PV-hydro hybrid installation becomes dispatchable when coupling FPV with hydro generation. Dispatchability of energy refers to the ability of a generator to adjust its power output to meet a given demand. PV without any storage is inherently not dispatchable given the reliance on solar meteorological conditions.

This hypothesis is challenged for different climatic cases and different modes of operation of planned or existing hydro facilities. The sizing of the FPV in relation to the hydro capacity is also investigated to determine if different ratios of FPV to hydro result in an optimum dispatchability case to help inform future feasibility.

2 METHODOLOGY

2.1 Input Parameters

To investigate the dispatchability of the hybrid operation, hydroelectric, FPV, demand and grid constraints need to be taken into consideration.

Hydroelectric parameters:

- Storage or volume limits or change thereof due to sedimentation over the lifetime of the plant;
- Water balance requirements (environmental flows, irrigation targets);
- Generator minimum stable levels and associated efficiency changes with water flow and head

FPV parameters:

- Generation is dependent on meteorological conditions and not dispatchable without storage
- PV degrades over the technology life, thus reduction in output is considered
- Water body characteristics will determine the total installed capacity of FPV on existing hydro dam

Key external parameters:

- Grid infrastructure and any constraints on power evacuation connection with transmission system. For the purpose of this paper the point of connection is limited to the size of the hydro dam with the assumption that the export limit is sized to the capacity of the existing hydro generator
- Demand (need for peaking or baseload) in the overall energy system
- Value of energy (if higher economic reward given to energy generated at specific times of the day, high demand season)

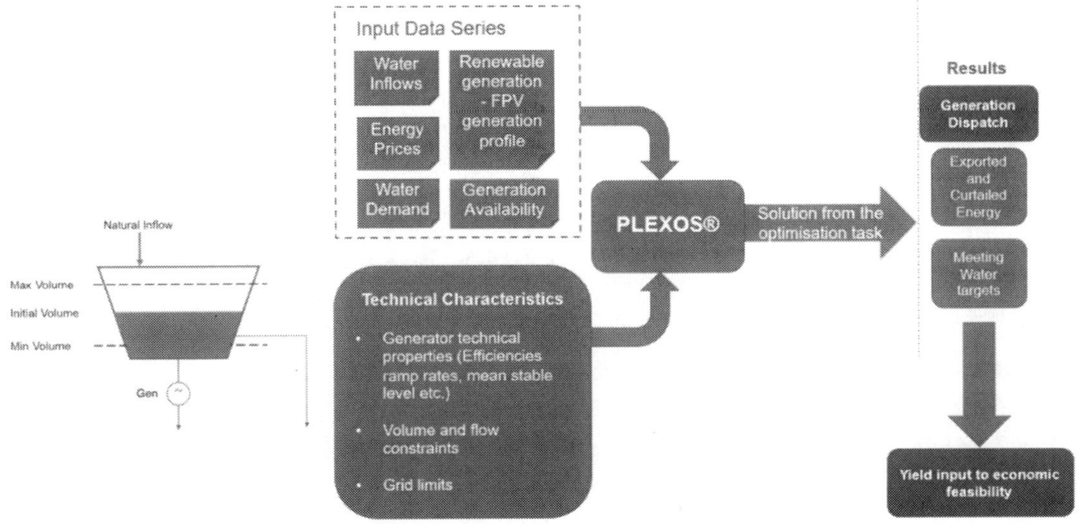

Figure 1: Modelling flow diagram

2.2 Modeling Methodology

To determine dispatchability of the hydro-FPV generator, these variables are evaluated both in the short and medium term with the use of a detailed model. The flow process is illustrated in the diagram presented in Figure 1.

PLEXOS® Integrated Energy Model (PLEXOS) version 9.1 by Energy Exemplar has been employed for the modelling and simulation of the system. PLEXOS uses mathematical optimization methods to provide an energy output solution by integrating the electric and water production over the simulated timeframes, short term (24 hours dispatch profile) to long term (25 years life of the FPV plant). The model takes the defined variables into consideration to determine economic dispatch decisions that optimize the cost or profit of power generation subject to applicable physical, environmental and operational constraints.

The system is dispatched on an hourly basis in 72-hour steps rolling daily over a horizon of a year with the aim to prioritize peak demand periods, subject to constraints including minimum stable generation, integer unit commitment, and water flow limits and targets. The FPV generation is dispatched with the occurrence of irradiation and the hydro generation scheme is limited with seasonal water flow and, if required, daily irrigation demand is respected.

The system is modelled as an energy only model, i.e. not considering any operating reserve requirements or system operational dispatch constraints imposed by the system operator.

To investigate the extent of the impact that the FPV generation has on the system and possible hydro generation displacement operation during sunlight hours, the model is run with and without the presence of the FPV. Furthermore, the hybridized FPV and hydro generator is compared to a stand alone hydro and FPV systems.

Hybrid case is defined as the FPV and hydro generators interconnected to the same substation to an evacuation line with the capacity of the hydro generator of each case, where dispatch decisions from the hydropower plant are made in consideration of the expected generation from the FPV plant, to maximise the overall generation/revenue and avoid curtailment, while respecting other constraints such as irrigation requirements. The stand alone hydro and FPV systems are modelled as two individual plants operated independently with no coordination in operation and the grid operator will determine priority for export/ curtailment. To effectively compare the hybrid system with the case of the stand alone systems it is assumed that the stand alone systems are connected separately to an evacuation line with a grid capacity limited to the size of the hydro generator.

3 CASE STUDIES

Three different climates and locations are investigated. It is noted that this is a purely theoretical exercise to stress test the hypothesis of this paper and not reflective of any actual FPV developments occurring at actual or planned hydro dams in these locations.

- **Case 1a:** Pakistan, hydro base load generator which is the main generator on the given distribution line and has daily irrigation targets which have to be met
- **Case 1b**: Pakistan, hydro peaking plant with approximately 4 hours peak operational storage in run of river and downstream from case 1a as part of a cascading scheme.

Cascade configuration of Case 1a and Case 1b entails tight correlation between generators given that the lower reservoir has inflows from the upper reservoir outflows.

- **Case 2**: Turkey, peaking generator with required daily irrigation water release and downstream from a larger storage dam
- **Case 3**: Zambia, hydro peaking generator downstream from a larger storage dam and additional feed from run of river

3.1 Meteorological conditions

Figure 2 shows the hydro inflow and long-term average GHI (global horizonal irradiation) for each site. The hydro inflow data is obtained from in-house data sources and the GHI has been sourced from SolarGIS.

The site in Turkey and Pakistan shows some correlation between the inflow peak and solar irradiation peak. For the Turkey site, the solar irradiation peak occurs after the inflow peak, and thus is it postulated that the storage volume is likely to be at its maximum when the FPV produces at its peak, limiting change in hydro operation dispatchability. Solar irradiation does not fluctuate significantly in Zambia, with a partially inverse correlation is noted for the site, thus greater flexibility is predicted.

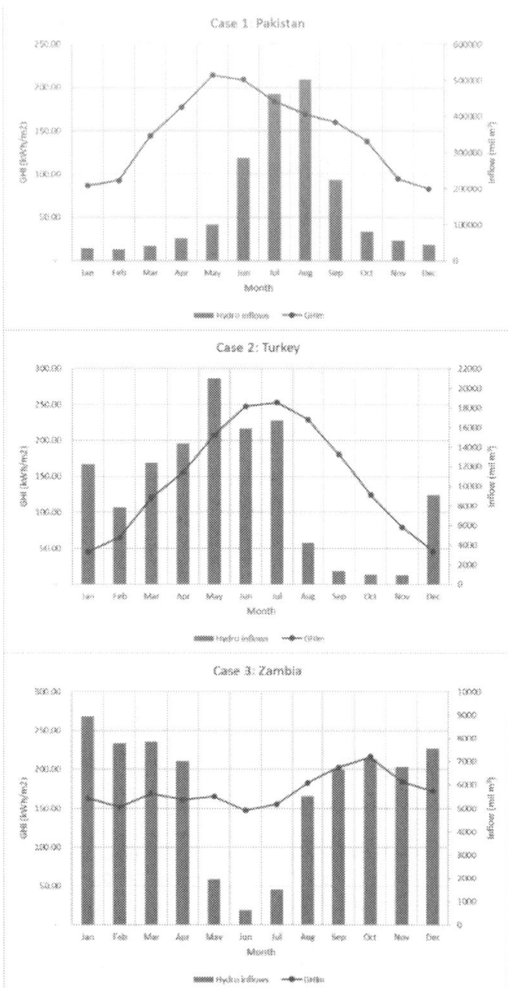

Figure 2: Hydro reservoir inflow and long term average GHI for each location examined. Top – Pakistan, middle – Turkey and bottom – Zambia

3.2 FPV - Hydro proposed configuration

Table I outlines the hydro power plant characteristics and Table II shows the general FPV properties considered. A representative FPV block is developed for each location and then replicated to evaluate different capacities of FPV coupled to the hydro generator. Water body characteristics (depth, suitable mooring, environmental aspects, etc) which may limit the technical installation of FPV were not taken into consideration when sizing the FPV on each waterbody.

Table I: Hydropower scheme properties

Object/ Property	Case 1a	Case 1b	Case 2	Case 3
Location	Pakistan	Pakistan	Turkey	Zambia
Total hydro capacity	6000 MW	1500 MW	280 MW	265 MW
Irrigation requirements	Yes	No	Yes	No
Part of a cascading hydro scheme (Y/N)	No	Yes	No	No

Table II: General FPV generator properties

Object/ Property	Value
Technology configuration	72 cell polycrystalline modules connected to string inverters with MV transformer to connect to hydro generator substation
Fixed tilt	12°
Azimuth	Due south for Turkey and Pakistan, due north for Zambia

Table III shows the FPV capacities investigated for each case. FPV capacity is presented as a percentage of the hydro capacity. Given the large size of the hydro systems in Pakistan, it is unlikely that a FPV system that matches the hydropower capacity would be feasible in practice, nonetheless these cases are modelled to test the hypothesis.

Table III: FPV capacities investigated

Case	FPV Capacity (MW)	% of hydro capacity
Case 1a	150	2%
	750	12%
	1500	25%
Case 1b	150	10%
	375	25%
	750	50%
	1500	100%
Case 2	10	4%
	50	18%
	100	36%
	200	71%
	250	89%
Case 3	10	4%
	50	19%
	100	38%
	200	76%
	250	95%

All four cases are evaluated over a one year period to assess any seasonal changes in the daily dispatch profile (short term). Given the availability of inflow data, case 1a and case 3 were also evaluated over a long term period to observe any interannual (long term) storage benefits. Case 1a was evaluated for 25 years and case 3 for 16 years given data availability.

4 RESULTS

To assess any changes in energy dispatch of the system, the duration curves, transmission line capacity factor and curtailment for each of the cases has been found. The curtailment for the fully hybridized case and the curtailment for the stand alone configuration is compared to observe the benefit of the hybridization.

4.1 Short term results

From Tables IV to VII it is observed that the hybridized scenario for each case allows for greater energy exported, or reduced curtailment when compared to the stand alone located cases. This is also seen in the increase in transmission capacity factor.

To observe the change in hourly production, yearly duration curve is presented for each case. Figure 3, for case 1a, the hydro operation without FPV (orange line) shows that the plant operates at peak for ~15% of operational time, and this does not increase when FPV is hybridized with the hydro plant (green line). This is indicative of constrained storage where the high inflow season coincides with the high irradiation, ie: once the reservoir is full, the hydro operates at maximum capacity and curtails the FPV generation. This hydro facility also has prescribed irrigation outflows from the reservoir during the high inflow season, thus minimal flexibility on changing the dispatch times of the hydro generation on a short-term basis (24 hour period) is observed.

It is also noted that there is minimal difference between the hybrid plant and standalone plants curtailment for this case and curtailment is seen when the FPV is sized at 2% of the hydro capacity.

Table IV: Case 1a

	Capacity Factor	Overall curtailment hybrid plant	Curtail-ment for stand alone sites
6000MW Hydro No FPV	42.3%		
Hydro + 150MW FPV	42.8%	0.20%	0.24%
Hydro + 750MW FPV	44.9%	0.94%	1.12%
Hydro + 1500MW FPV	47.5%	1.87%	2.15%

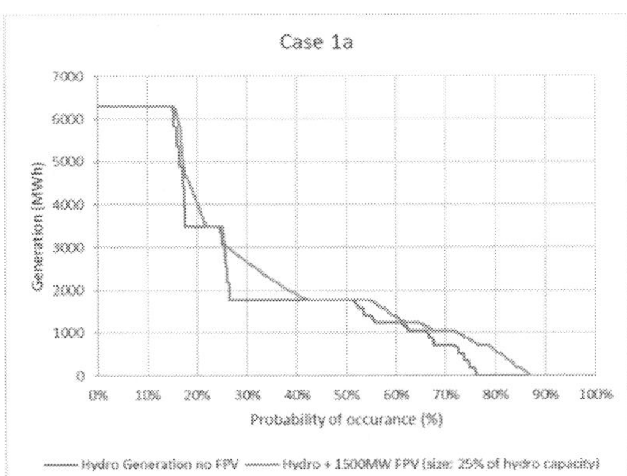

Figure 3: Duration curve for Case 1a, Hydro +1500MW FPV (FPV is sized at 25% of hydro capacity)

Case 1b is downstream from case 1a, but is a dedicated peaking hydro plant with 4 hours of storage to serve peaking demand. Table V shows that hybridizing FPV with the hydro peaking scheme increases the overall capacity factor and curtailment only occurs when the FPV system is sized at 100% of the hydro generator. Figure 4, which also illustrates the hybrid case of hydro +FPV with FPV sizing of 25% of hydro capacity, it is observed that the peaking plant with the hybridized FPV can generate at maximum capacity for an additional 4% of the year. This indicates that the hydro operation is largely unconstrained where inflows are managed upstream with flexible storage within the short term horizon, and able to accommodate the presence of FPV. With the addition of the FPV there is better use of the transmission infrastructure and sufficient flexibility in the operational profile to accommodate the FPV production with minimal grid curtailment

Table V: Case 1b

	Capacity Factor	Overall curtailment hybrid plant	Curtail-ment for stand alone sites
1500MW Hydro No FPV	55.9%		
Hydro + 150MW FPV	58.0%	0.0%	1.58%
Hydro + 375MW FPV	61.1%	0.0%	3.80%
Hydro + 750MW FPV	66.3%	0.0%	7.43%
Hydro + 1500MW FPV	75.8%	1.3%	14.51%

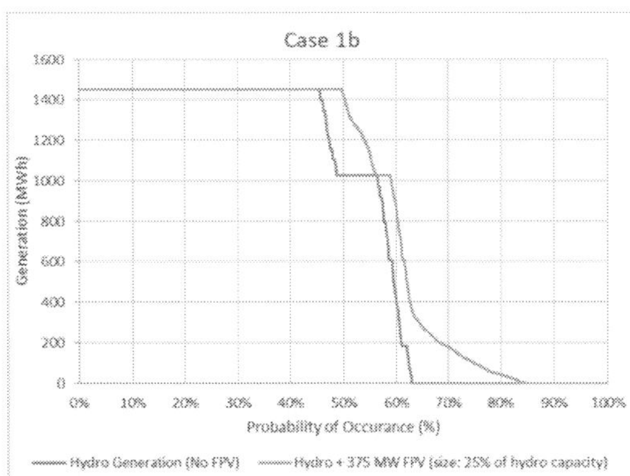

Figure 4: Duration curve for Case 1b, Hydro +375MW FPV (FPV is sized at 25% of hydro capacity)

Case 2 hydro scheme also has irrigation requirements, however there appears to be enough flexibility in the storage volume to accommodate FPV up until 71% of the hydro capacity before curtailment is seen. From the duration curve, it is observed that with the presence of FPV the peak operating time decreases but there is a longer period of served energy between 40% and 60% probability of exceedance at approximately 50% of the hybrid generator capacity considering the capacity scenario of hydro +100 MW FPV. The Plexos solver seeks the profit maximization case in dispatch, thus considering the efficiency and head profiles of the hydro generator with the presence of FPV, shifting the generation away from full capacity generation but rather running some of the turbines for a longer period results in higher energy generated. This behavior is also driven by the irrigation targets set which require a constant flow from a set of turbines as seen by the constant output at 20% generation capacity.

Table VI: Case 2

	Capacity Factor	Overall curtailment hybrid plant	Curtail-ment for stand alone sites
280 MW Hydro No FPV	42.0%		
Hydro + 10MW FPV	42.8%	0.00%	0.16%
Hydro + 50MW FPV	45.7%	0.00%	0.83%
Hydro + 100MW FPV	49.4%	0.00%	1.61%
Hydro + 200MW FPV	56.7%	0.21%	4.95%
Hydro + 250MW FPV	59.3%	1.99%	7.17%

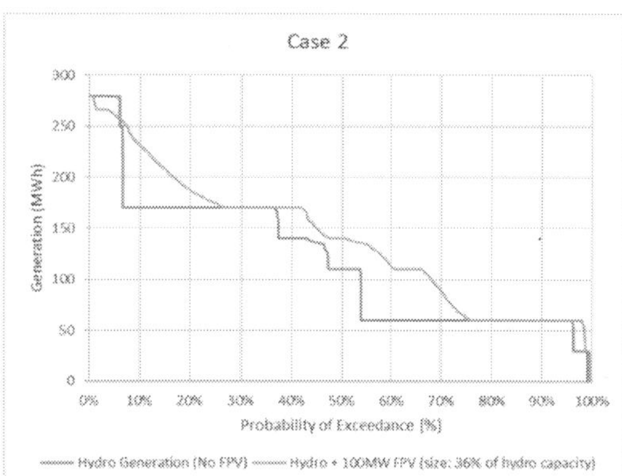

Figure 5: Duration curve for Case 2, Hydro +100MW FPV (FPV is sized at 36% of hydro capacity)

Table VII: Case 3

	Capacity Factor	Overall curtailment hybrid plant	Curtail-ment for stand alone sites
265 MW Hydro No FPV	54.1%		
Hydro + 10MW FPV	55.0%	0.31%	1.07%
Hydro + 50MW FPV	58.5%	0.49%	5.03%
Hydro + 100MW FPV	62.9%	0.52%	9.35%
Hydro + 200MW FPV	71.4%	1.18%	16.60%
Hydro + 250MW FPV	75.4%	1.66%	20.16%

Similar to case 2, with the presence of FPV, the operation of the hybridized system (as seen by the green line in Figure 6) there is a reduction in period of operation at peak capacity but increase in overall generation. Table VII shows that while curtailment occurs in the hybrid system, it is considerably lower than for the stand alone system case. The curtailment for the scenario where the FPV is sized as 95% of the hydro capacity, ie: the total hybrid generator is almost double the size of the allowed transmission line results in 1.66% curtailment where the co-located case results in a curtailment of 20.16%. This shows there is sufficient storage to allow flexibility in generation in the short-term basis (24-72 hour time scale), the absence of irrigation constraints give more flexibility in water dispatch and hydro generation.

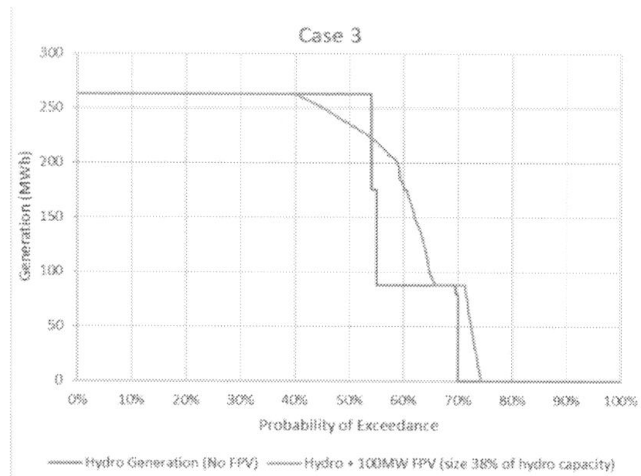

Figure 6: Duration curve for Case 3, Hydro +100MW FPV (FPV is sized at 36% of hydro capacity)

4.2 Long term results

On a medium-term basis, seasonal trends are examined for Case 1a and Case 3. Case 1a is evaluated over 25 year and case 3 over 16 years. A 5 year period is presented here for illustration.

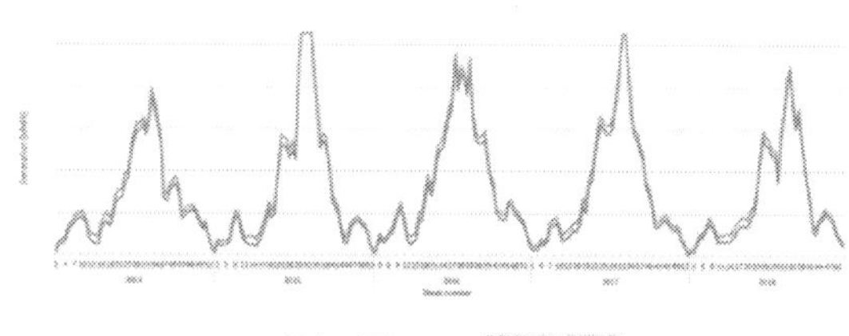

Figure 7: Time series plot on a weekly basis for 5 years of Hydro +1500MW FPV (FPV is sized at 25% of hydro capacity)

From this graph it appears that the FPV production is simply stacked on top for hydro generation. To better observe change in monthly trends, a one year period is plotted.

For Case 1a, Figure 8, no shift in the seasonal storage of the hydro facility is observed. This behaviour can be attributed to the overall system constraints, such as daily irrigation targets, storage volume constraints and thus seasonal storage is limited.

Considering the seasonal demand profile as shown in Figure 9, it is postulated that the base case hydro generation is already running at an optimum dispatch, prior to any FPV installation, when considering the demand of the system. Thus, the FPV is simply an addition to the system. For case 1a, limited flexibility in dispatch is seen in both the short and medium term.

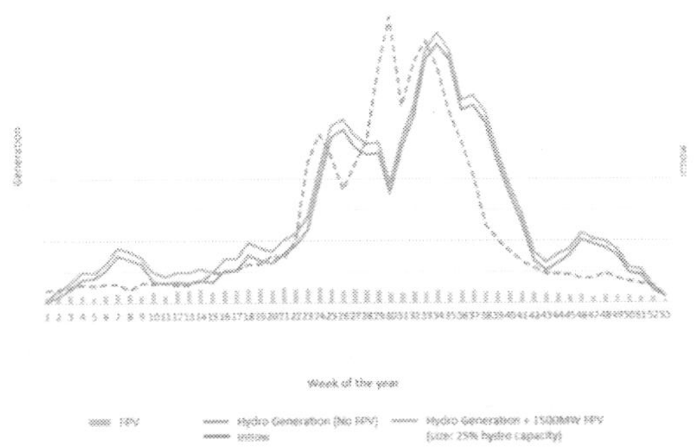

Figure 8: On year of hydro plant generation, FPV plant and hybrid hydro-FPV generation and hydro inflow showing seasonal trends

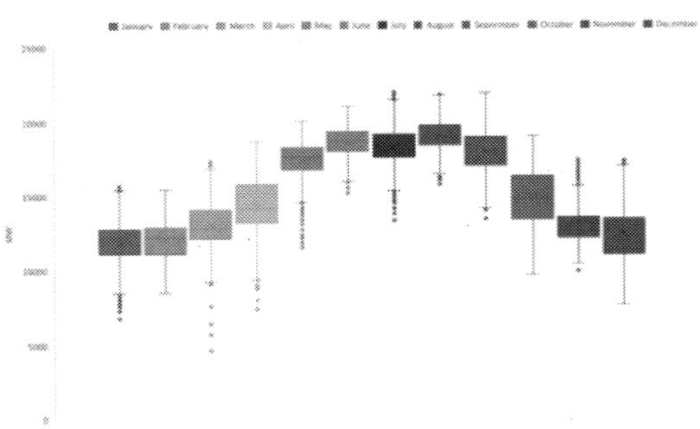

Figure 9: Case 1a, demand trend, distribution by month

The same analysis is carried out for case 3. From Figure 10, some minor changes are observed with shifting on a weekly basis but no seasonal shift is observed.

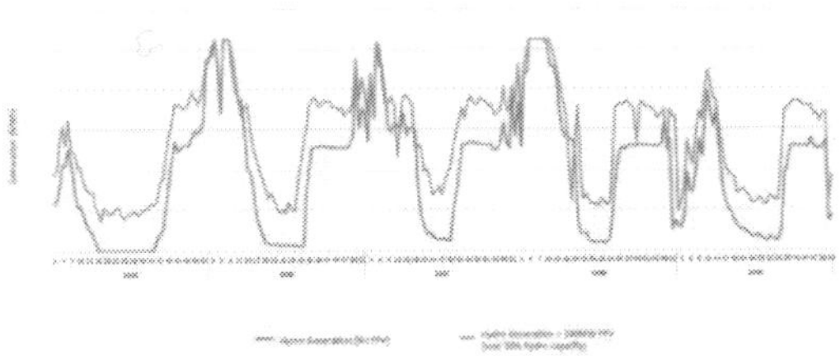

Figure 10: Time series plot on a weekly basis for 5 years of Hydro +100MW FPV (FPV is sized at 38% of hydro capacity)

A one year period is plotted, Figure 11 and no shift in the seasonal storage of the hydro facility is observed and the behaviour of the hydro generator remains dominant with the inflow dictating hydro output.

Considering Figure 12, the seasonal demand profile, the same conclusion is drawn that the base hydro case is already running at an optimum dispatch, prior to any FPV installation.

The difference however between Case 1a and Case 3, is that Case 3 exhibited dispatchable flexibility in the short-term window.

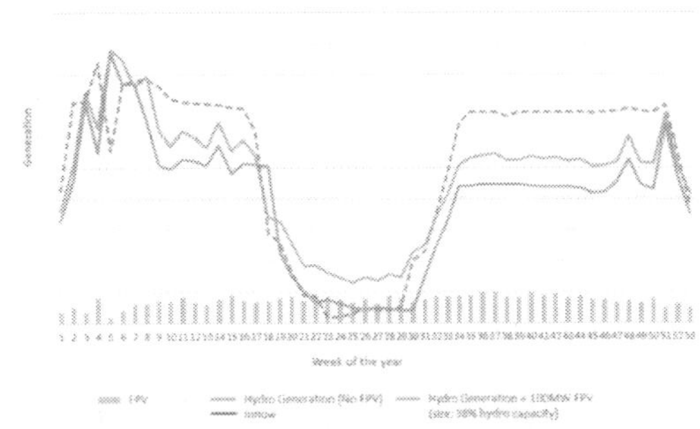

Figure 11: 1 year of hydro plant generation, FPV plant and hybrid hydro-FPV generation and hydro inflow showing seasonal trends

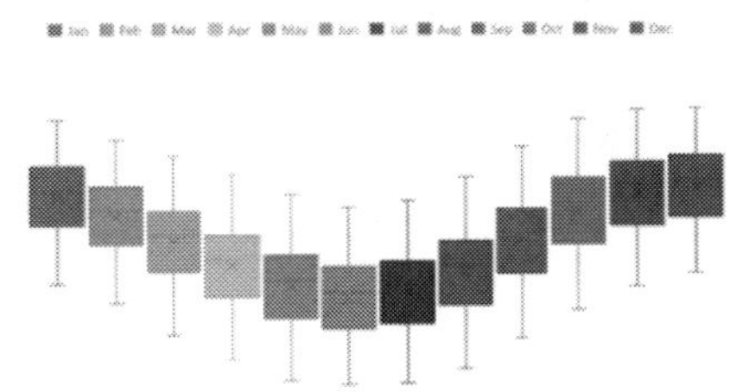

Figure 12: Case 3, demand trend, distribution by month

5 CONCLUSION

The findings show that the extent of dispatchability of a hybrid hydro – FPV plant is highly dependent on the given volume of the storage dams, inflow and outflow targets of the system, system load constraints, seasonality and grid restrictions and thus cannot be treated like a battery throughout all seasons when considering 24 hour dispatch window. Better flexibility in short term dispatch is seen when an upstream reservoir helps regulate the inflow into a downstream hydro scheme, as seen in Case 1b.

From the 2 cases analysed in the mid term, no seasonal storage benefits were observed.

For all investigated case studies, the transmission line capacity factor increased with decrease in curtailment when compared to the standalone cases with better utilization of the infrastructure with the addition of FPV.

Analysis was done on an hourly basis and thus impact on smoothing solar generation due to cloud cover intermittency was not evaluated. Further research needs to be carried out to better understand if any specific hydro-FPV hybrid operation can contribute to the firm capacity on the grid or reduce the need for spinning reserves. A techno-economic analysis will also help to better inform the optimum FPV to Hydro capacity sizing.

6 REFERENCES

[1] IRENA, 2021. "Energy from the Sea: An Action Agenda for Deploying Offshore Renewables Worldwide" September 2022. https://www.irena.org/events/2021/May/An-Action-Agenda-for-Deploying-Offshore-Renewables-Worldwide

[2] World Bank Group, ESMAP and SERIS. 2019. Where Sun Meets Water: Floating Solar Market Report. Washington, DC: World Bank.
[3] S. Gadzanku, N. Lee and A Dyreson, NREL, Michigan Technological University. 2022. Enabling Floating Solar Photovoltaic (FPV) Deployment: Exploring the Operational Benefits of Floating Solar-Hydropower Hybrids
[4] VietnamPlus. 2021. "Thailand to Put World's Largest Floating Solar Farm into Use This June." VietnamPlus (blog). January 25, 2021. https://en.vietnamplus.vn/thailand-to-put-worlds-largestfloating-solar-farm-into-use-this-june/195309.vnp

PHOTOVOLTAIC SOLAR MAPPING OF VULNERABLE AREAS AS A TOOL FOR THE DEVELOPMENT OF SOLAR ENERGY COOPERATIVES IN SLUMS

Douglas de Castro Ramos
La Salle University Center of Rio de Janeiro, UniLaSalle-RJ
Rua Gastão Gonçalves, 79 – Santa Rosa – Niteroi, Brazil

Luciane Martins Do Pilar
La Salle University Center of Rio de Janeiro, UniLaSalle-RJ
Rua Gastão Gonçalves, 79 – Santa Rosa – Niteroi, Brazil

Jhennyfer Rodrigues Parente
La Salle University Center of Rio de Janeiro, UniLaSalle-RJ
Rua Gastão Gonçalves, 79 – Santa Rosa – Niteroi, Brazil

Diego Souza Caetano
La Salle University Center of Rio de Janeiro, UniLaSalle-RJ
Rua Gastão Gonçalves, 79 – Santa Rosa – Niteroi, Brazil
Federal Fluminense University (UFF)
Rua Passo da Pátria, 156 – São Domingos – Niteroi, Brazil

Louise Land Bittencourt Lomardo
Federal Fluminense University (UFF)
Rua Passo da Pátria, 156 – São Domingos – Niteroi, Brazil

ABSTRACT: Solar energy represents an important alternative to mitigate environmental impacts related to electricity generation. Distributed generation can transform cities into power plants, as a tissue of regional self-production cells. This model fits both regional and local scales. The presence of solar cooperatives in slums in the city of Rio de Janeiro/RJ highlights the feasibility of installing photovoltaic systems on building roofs in vulnerable urban areas and shows how studies of the potential for photovoltaic energy generation can contribute to energy democracy. Through strategies for the implementation of energy resources and their infrastructure supported by solar maps, self-production of energy can contribute to more democratic and efficient cities and can reduce the impact of the cost of electric energy in an energy matrix dependent on the hydraulic base and pollutants thermoelectric plants in times of drought, which have been increasingly common, in addition to minimizing environmental and biodiversity impacts.

This paper aims to evaluate the potential of photovoltaic electricity generation in slums in the city of Niteroi/RJ, Brazil. The solar potential is being analyzed through a solar map produced from a case study of housing roofs in the "*Morro do Pé Pequeno*" community. Satellite image analysis tools are being used to classify roofs in terms of energy generation potential. Through the results, it will be possible to evaluate the best metadata structure for the development of the solar map. In addition, the potential for photovoltaic energy generation in relation to the energy consumption of community buildings will allow us to assess whether the implementation of the system is feasible. The study aims to: Disseminate information on electricity based on solar energy; democratize access to energy; Support the municipality and the community of *"Morro do Pé Pequeno"* with information on solar potential, reducing energy poverty and mitigating environmental impacts.

Keywords: Space; Developing Countries; Villages.

1 INTRODUCTION

World Bank data show that Brazil obtained 43.8% of its total final energy consumption through renewable energy sources, above the average of 18.0% of world consumption in 2015 [1]. According to the National Energy Balance, in 2019, hydropower was the main source of electricity generation in the country, with 64.9% of electricity generation. Solar Energy is in last place, currently with only 1% of the generation. However, the use of solar energy as a source has been growing relatively when compared to hydraulic energy. Between 2017 and 2019, while the production of primary energy by solar base more than tripled in the country, the hydroelectric plant kept its production at the same levels [3].

According to the Energy Research Company (EPE) [13] and Communication Brazilian Company (EBC) [5], the year 2017 was the fourth consecutive year of hard drought in several parts of the country, which considerably impacted the hydroelectric reservoirs. Dependence on hydroelectric energy has brought consequences such as the increase in the use of thermoelectric plants, generation systems with fossil and polluting sources and the increase in the cost of energy for the consumer. According to Estadão journal [6], Brazil occupies the 37th position in the most expensive energy tariffs in the world, in a ranking of 110 countries analyzed by the Global Petrol Prices 2019. According to the author, "on average, the value is behind what is charged in most European countries, but ahead of developing countries.".

On the other hand, solar energy appears as an important alternative to mitigate the impacts of problems related to energy generation. In 2019, buildings accounted for 43.5% of electricity in Brazil, with residential buildings accounting for 21.9% (142.6 TWh) of this amount [2]. According to Ministry of Mines and Energy [4], the EPE demonstrated that "currently all states would be able to fully supply their residential electrical consumption with the advent of photovoltaic energy", with residential photovoltaic technical potential estimated at

287.8 TWh according to Technical Note PR 04/18.

The development of solar potential maps has been an international strategy for the dissemination of decentralized generation of solar energy as an alternative for the adoption of photovoltaic systems in buildings of different types. The use of solar maps as a strategy to promote solar energy occurs in several cities around the world. Kanters et al [9] listed 19 cities with solar maps and categorized their information. According to Lange and Vasconcelos [11], the availability of high-resolution space satellite images allowed a great advance in urban studies. The use of geographic information systems (GIS) made it possible to link different data to cartography, configuring contents characterized as metadata. However, "there are few studies that use remote sensing techniques to map roofs for the installation of photovoltaic cells" for cities, especially in developing countries.

Lange [10] considers as important research the improvement of the studies of building roofs for the implantation of solar energy systems. The author addresses aspects such as "detailed studies of roof area by type of household based on local and regional records and classification or scanning of high resolution satellite images" and also the analysis segmented "by socioeconomic indicators" such as per capita income, for example.

According to World Economic Forum [14], the energy poverty is a dimension defined as "the absence of sufficient choice in accessing adequate, affordable, reliable, high-quality, safe, and environmentally benign energy services to support economic and human development.". In developing countries, access to energy in segregated urban areas such as slums is a challenge. According to the Brazilian Institute of Geography and Statistics (IBGE) [7], the estimate in Brazil of occupied households in slums in 2019 is 5.127.747 units. Part of these households is concentrated in the Southeast region, with São Paulo and Rio de Janeiro being the largest concentrations. In the city of Niterói, the IBGE estimates that 16% of households are in "subnormal agglomerations" or slums.

Recently, Non-Governmental Organizations, such as REVOLUSOLAR among others, have carried out projects to promote solar energy in favelas through energy generation cooperatives [8]. Moon [12] developed a study with the NGO REVOLUSOLAR in Brazil about the potential of photovoltaic systems and the perception of the value of solar energy for residents of the Morro da Babilônia community in Rio de Janeiro. The author carried out some studies to analyze the economic viability, social profile and perception of solar energy as an alternative in 120 interviews with residents. Among the results, 66% of respondents claimed to know what solar energy is for electricity generation and 94% of respondents responded that they were interested in participating in projects to implement solar energy in their homes.

Also in the PNE 2050, the EPE [4] indicates the development of "new tools, technologies and business models to predict solar generation and manage the operation of the electrical system" as recommendations to promote the advance of solar energy in Brazil. Solar mapping of cities is an alternative that allows municipalities to guide their citizens regarding the implementation of solar energy, which can be an interesting option to bring civil society closer to the topic. Internationally, solar mapping of cities already takes place in several cities around the world, mostly in Europe.

However, the solar mapping of cities still does not consider the socially vulnerable areas of the city, as in the Solar Map of Rio de Janeiro, which does not present information on the solar photovoltaic potential of the territorial areas of the slums, a sensible point of the urban network of the city. Solar maps can help initiatives such as INSOLAR and the NGO Revolusolar, which work in these areas with the installation of photovoltaic systems where local solar cooperatives have already been created. According to Grauthoff et al [15], the creation of solar maps can "create a substantial interest on solar technologies, and to encourage the organisation of local initiatives.".

2 AIM AND APPROACH

The objective of this research is to present a study of the potential of photovoltaic generation in roofs of buildings in the Morro do Pé Pequeno community in the city of Niterói/RJ, Brazil. Through the development of a solar mapping methodology for slums, the study aims to support local energy cooperatives and promote the democratization of access to clean and renewable energy and the reduction of energy poverty.

For the development of the solar map and the structure of the respective metadata, a detailed analysis of a sample of building roofs and their solar potential was carried out, defining criteria for classification of roofs. This analysis is supporting the development of the second stage of the research at this time, which aims to structure the metadata analyzed using satellite image geoprocessing tools and, with that, produce the solar map of the study area.

This research is the result of the Program for the Development of Applied Research of the municipal government of Niterói (PDPA/Niterói) in partnership with the Fluminense Federal University, and has the support of the Graduate Program in Architecture and Urbanism (PPGAU/UFF) and the La Salle University Center of Rio de Janeiro (UniLaSalle-RJ) and the Research Support Foundation for Rio de Janeiro (FAPERJ).

3 PRELIMINARY RESULTS

Preliminary results show that slums have the potential to support and promote the organization of solar energy cooperatives, and solar mapping can be an important tool for this task in defining groups of households that can form solar energy cooperatives. In the study area, it was possible to observe the possibility of defining groups of buildings for the generation of several cooperatives in the same slums, depending on the urban conditions, insolation and the roof of the buildings in the communities, mainly due to the structural needs for the implantation of the panels.

In the sample study to detail the main methodological challenges, the roof surfaces were classified according to their catchment area, considering ranges from 0.50 m^2 to 50.99 m^2 (Regular potential roof), from 51.00 m^2 to 69.99 m^2 (Good potential roof), 70.00m^2 to 99.99 m^2 (Very Good potential roof) and above 100.00 m^2 (excellent potential roof). So far, 249 roofs have been analyzed in detail, before carrying out the analysis by geoprocessing, thus making it possible to compare the results and calibrate the potential of photovoltaic energy generation to help the cooperatives of residents.

A survey of the average monthly consumption per household was carried out in a sample of 59 families, as shown in Figure 1 below:

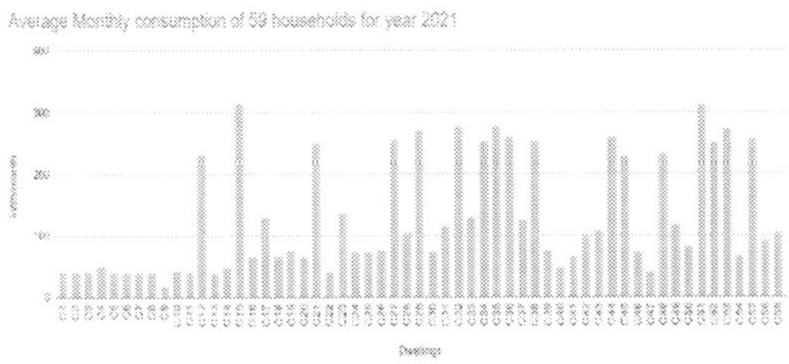

Figure 1: Average Monthly consumption of 59 households.

According to the survey, an average line between 100 to 200 kWh/month was found for energy consumption in homes. Thus, it is estimated that annual consumption is on average from 1,200 to 2,400 kWh/year per household, totaling approximately 620 MWh/year in the case of higher consumption.

Figure 2: Classification of roofs in relation to their roof area.

As shown in Figure 2, 249 roofs were classified, where 74 roofs were considered regular because they had little possible area for installing the panels. Another 46 roofs were classified as good for installing photovoltaic panels. Subsequently, 74 roofs were classified as very good for panel installation. Finally, analyzing the 249 coverages already mapped as a sample, 55 houses were found with the largest area for photovoltaic use by classification of coverage areas. It is estimated that from these 55 roofs, the energy generated by the 37 best roofs will be enough to sustain all homes, considering a generated power of 347.8 MWh/year.

Only 50% of the roofs that could be installed with panels were considered in the total area of the roofs, since the roofs have, in addition to the structural fragility found in most buildings, the presence of antennas and water tanks that make it impossible to install panels. Inverter losses and dirt characteristics on panels and shading were considered within the standards of the NREL PVWatts website, where the potential of each roof was calculated. It was not yet possible to perform the calculations automatically using GIS tools, due to the research time. However, the preliminary results are promising, since, according to the calculations developed on site, with a high rate of losses considered, only 37 roofs would provide the energy needed for the entire community. This indicates that the promotion of large-scale solar cooperatives could significantly combat energy poverty in underdeveloped countries.

It is essential to design strategies that promote renewable energy with a combination of economic development approaches for the most vulnerable areas. The results will provide the municipal government with a solar map that enables the dissemination of photovoltaic systems and promotes the democratization of electricity, reducing environmental impacts and promoting democracy in access to energy.

5 REFERENCES

[1] WORLD BANK. The World Bank | Data. Renewable energy consumption (% of total final energy consumption) - Brazil, World. Sustainable Energy for All (SE4ALL) Database of the SE4ALL Global Tracking Framework project led by the World Bank, International Energy Agency (IEA) and the Energy Sector Management Assistance Program. 2021. Available at: <https://data.worldbank.org/indicator/EG.FEC.RNEW.ZS?end=2015&locations=BR-1W&start=2015&view=bar>. Accessed on: July 1, 2021.

[2] BRAZIL. Ministry of Mines and Energy. National Energy Plan 2030 / Ministry of Mines and Energy / Energy Research Company. Brasília : MME / EPE, 2007.

[3] BRAZIL. Energy Research Company. National Energy Balance 2020: Base year 2019 / Energy Research Company. Rio de Janeiro: EPE, 2020.

[4] BRAZIL. Ministry of Mines and Energy. Potential of Energy Resources in Horizon 2050. TECHNICAL NOTE PR 04/18. Rio de Janeiro: MME / EPE. 2018

[5] EBC. Brazilian Communication Company. Emergency: entities warn of a dry period in the center-south. Organs estimate the possibility of a water crisis. Brasilia: EBC. 2021. Available at: <https://agenciabrasil.ebc.com.br/radioagencia-nacional/meio-ambiente/audio/2021-05/emergencia-entidades-alertam-para-periodo-de-seca-no-centro-south>. Accessed on: July 1, 2021.

[6] ESTADÃO. Energy from Brazil is the 37th most expensive in the world. Sao Paulo: Estadão. 2020. Available at: <https://economia.estadao.com.br/noticias/geral,energia-do-brasil-e-a-37-mais-cara-do-mundo,70003278265>. Accessed on: July 1, 2021.

[7] IBGE Brazilian Institute of Geography and Statistics. Subnormal Agglomerates 2019: preliminary classification and health information for the fight against Covid-19. Rio de Janeiro: IBGE, May 19, 2020. (Technical Note, no. 01/2020).

[8] INFOSOLAR NGO Revolusolar installs the first distributed generation cooperative in Rio de Janeiro. By Ariane Locatelli. 2021. Available at: <https://infosolar.com/mercado/consumidor/ong-revolusolar-instala-primeira-cooperativa-de-geracao-distribuida-no-rio-de-janeiro>. Accessed on: 01 Jul. 2021.

[9] KANTERS, Jouri; WALL, Maria; KJELLSSON, Elisabeth. The solar map as a knowledge base for solar energy use. Energy Procedia, v. 48, p. 1597-1606, 2014.

[10] LANGE, Wolfram Johannes. Methodology for mapping the potential area of roofs of residential buildings in Brazil for photovoltaic energy use. Rio de Janeiro: Terragis, 2012.

[11] LANGE, Wolfram Johannes; VASCONCELOS, Adriano de Oliveira. Refinement of mapping the potential area of roofs of residential buildings in Brazil. In: Deutsche Gesellschaft für Internationale Zusammenarbeit (GIZ) GmbH. 2015.

[12] MOON, Boram. A Study of Solar PV Potentials to Ensure Reliable Supply of Affordable Electricity in Favelas, Rio de Janeiro, Brazil. 2018

[13] SILVA, Renan Cabral da. The energy transition of emerging countries in the 21st century: Brazil and China in focus. Doctoral thesis. Federal University of Pernambuco. 2018. 147 p.

[14] WEA. World Economic Forum. Global Agenda Council Reports - Summaries of Global Agenda Council Discussions from the Summit on the Global Agenda 2009. WEA, 2010. Available at: < https://members.weforum.org/pdf/globalagenda2010.pdf>. Accessed on: July 1, 2021.

[15] GRAUTHOFF, Manfred et al. Identification and mobilisation of solar potentials via local strategies. Guidelines based on the experiences of the pilot actions in Lisbon, Lyon, Malmoe, Munich, Paris and Vitoria-Gasteiz. 2012.

CITIES AS PIONEERS OF THE ENERGY TRANSITION: EVIDENCE AND CHALLENGES FROM THE COVENANT OF MAYORS AND THE EU CITIES MISSION

Ulpiani G., Palermo V., Melica G., Vetters N., Bertoldi P.
European commission, Joint Research Centre, Ispra

ABSTRACT: PV-based concepts are transcending their typical applications in the stationary energy and energy generation sectors, pervading all the main sectors that are responsible for greenhouse gas emissions in cities. Cities are uniquely positioned to tackle global issues at the local level as they control and/or have the autonomy to target the highest emitting and/or energy-intensive sectors in their territory, while experimenting innovative approaches to reach climate neutrality. The potential of PV technology in future low-carbon energy scenarios is largely modulated by the complexity of cityscapes, social systems, and urban metabolic flows, with a privileged role compared to other renewable energy solutions in tackling energy poverty and fostering a just energy transition. By capitalizing on two EU initiatives that are shaping the local-scale climate dialogue, the Covenant of Mayors and the Climate-Neutral and Smart Cities Mission, this paper explores the synergies between PV systems, climate neutrality/decarbonisation, and the city/citizen dimension as pioneers of the energy transition.

Keywords: photovoltaic systems, 100 Climate Neutral and Smart Cities Mission, Covenant of Mayors, climate neutrality, greenhouse gas emissions mitigation, energy poverty

1 INTRODUCTION

In the last decade, the installed capacity of solar PV worldwide outnumbered that of any other power-generation technology with an aggregate value at the end of 2019 exceeding 600 GW [1]. Pushed by the European Green Deal's goal of 55% greenhouse gas (GHG) emissions reduction by 2030 [2], the European annual PV market could surge from approximately 16.5 GW in 2019 to 50–80 GW by 2030 [3]. Nonetheless, the real potential of this technology in future low-carbon energy scenarios is still undetermined, especially when the likely extra boost from energy poverty alleviation plans/measures and the complexity of cityscapes, social systems, and urban metabolism are contemplated. Cities have often been defined as part of the problem and the solution in the climate challenge [4, 5]. They are responsible for more than 70% of CO_2 emissions worldwide and for more than 65% of energy consumption [6]. At the same time, cities are uniquely positioned to tackle global issues at the local level as they control and/or have the autonomy to target the highest emitting and/or energy-intensive sectors in their territory, while experimenting innovation in technical actions, participation processes, and governance approaches [5]. Previous studies have highlighted that better use of the technology coupled with dual-use options would generate much more PV capacity than required to achieve a neutral economy, with no issues related to land availability [3]. Other studies have demonstrated that climate-neutral photovoltaic farms are a preferred solution to coal-fired electricity generation coupled with carbon capture and storage (CCS) [7]. However, ultimately the answers will come from the experience of cities and citizens as they take an ever more active role in the transition to climate neutrality [8]. PV-based concepts are transcending their typical applications in the stationary energy and energy generation sectors, pervading all critical high-emitting sectors (including the hard-to-eradicate transport emissions [9]) to reach climate neutrality and implying, in their technological and commercial evolution, a massive contribution from the society, at all levels. PV-systems are expected to alleviate energy poverty, given their story of seamless cost reductions [1] and to advance climate neutrality targets at district/community-scale thanks to local PV power sharing, advanced community control, and battery storage sharing [10]. As such, PV's capacity is expected to experience a significant boost in the next decade, especially while cities engage in their transformation towards climate-neutrality [3]. Against this backdrop, this study aims to track this process at EU level and to explore the synergies between PV systems, climate neutrality/decarbonisation, and the city/citizen dimension as key players in future-proofing the built environment through a capillary energy transition. By capitalizing on two EU initiatives that are shaping the local-scale climate dialogue, the Covenant of Mayors (hereinafter, CoM) [11] and the 100 Climate-Neutral and Smart Cities Mission (hereinafter, Cities Mission) [12], this paper looks for the first time at PV as one of the critical nodes in-between cities and the goal of climate neutrality, using a purely evidence-based approach, relying on direct inputs from cities.

2 THE COVENANT OF MAYORS AND THE CITIES MISSION

2.1 The Covenant of Mayors

The CoM is the first pan-European initiative having mainstreamed climate change concerns and decarbonisation into local authorities' governments and planning activities. While initially overlooked in the IPCC and officially recognised as key actors in the climate challenge at COP 21 [13], the role of local governments was acknowledged by EC already back in 2008 when the Covenant of Mayors (CoM) was launched. The initiative counts today more than 11000 cities worldwide who are taking voluntary ambitious climate commitments. Initially, the target for local authorities was to meet and exceed the European 20% greenhouse gas emissions reduction objective by 2020 established by the Climate and Energy package compared to a baseline emission inventory (BEI), by implementing a Sustainable Energy Action Plan (SEAP). In 2015, an increased 40% reduction target by 2030 was set in alignment with the EU Nationally Determined Contributions and the adaptation to climate change was integrated into the framework. Signatories have to develop a Sustainable Energy and Climate Action Plan (SECAP) to translate into action their vision for both mitigation and adaptation to climate change. With the Commission proposal to increase the EU GHG emission reduction target to 55% by 2030 as a realistic pathway to climate neutrality by 2050, the CoM has stepped up its ambition, and participating cities are setting a goal to achieve climate neutrality by 2050 while increasing resilience and tackling energy poverty.

The Covenant of Mayors initiative has brought a perspective change by enabling the contribution of lower levels of governance towards national and cross-national targets and enhancing a more integrated approach. Moreover, cities have been benefiting from exchanges with other signatories, scientific and technical support, and visibility.

The CoM framework has been allowing for dealing with a global challenge by adopting local approaches and site-specific solutions. Moreover, the structure of the initiative has ensured a strong participation of local authorities of all sizes across EU MS and worldwide, including small and medium ones in need of a structured framework and support. Following a flexible but scientifically sound approach, cities start from an inventory of GHG emissions in their territories and an assessment of climate risks and vulnerabilities (and, in future, an energy poverty assessment) to prioritise suitable strategies and actions, identify the stakeholders to involve, define methods, approaches, and timeline. All these elements have to be described in the action plan submitted through an on-line reporting platform [11]. Studies on local energy generation and related best practices of CoM signatories have been conducted [14, 15]. In this contribution, the focus is on PV-related actions and measures with the aim to understand the sectors they target, the policy instruments adopted and the geographical context.

2.2 The Cities Mission

The experience and insights gained through the CoM initiative are informing the Cities Mission, a novel EU programme that aims to step up the ambition and deliver at least 100 European climate-neutral and smart cities by 2030 to act as agents of change. Mission Cities will wade through unprecedented opportunities and challenges which will involve the entire urban energy system across three main domains: 1) urban transformation in all sectors to reduce emissions (demand side), 2) championing local renewable energy generation (supply side), and 3) better linking the supply and demand sides with system integration that also recognizes the role of cities in the broader energy system. As such, the Mission aims to unlock the cities' potential to pioneer the deepest transformative process in the energy sector (as well as in the transport, waste/wastewater, industrial, and agricultural/land use sectors), by promoting holistic and integrated approaches and by fostering Europe's twin (green and digital) transitions. On 31 January 2022, the call for Expression of Interest (EOI) [16] was closed with a total of 377 cities from all EU Member States having expressed their intention and ambition to participate in the Mission (out of which 362 passed the eligibility criteria). Through the EOI, cities were invited to reflect upon a variety of critical aspects which closely link to the role of energy technologies, policies, citizen engagement, societal and financial challenges as well as enablers and barriers towards clean energy production, renewables (RES) penetration, and energy conservation, with due attention to co-benefits and potential adverse impacts (including energy poverty). The pool of information allows insights into several aspects having a bearing/influence on the future of photovoltaics at the city level. Data analysis may help reveal how cities, the PV community, and the energy community in general can partner towards the realisation of the first climate-neutral urban environments in Europe.

3 METHODOLOGY AND RESULTS

In the SE(C)AP process, in addition to GHG emissions, signatories collect and report data on local energy production which are used to calculate local emission factors for electricity or district heat/cold. This allows understanding the energy mix behind the consumption of final energy carriers such as electricity or district heat/cold, and hence, prioritising mitigation actions. The CoM reporting framework also requires cities to report on the actions (at least three) they planned in order to reach their GHG emission reduction target by a certain year, providing information on the area of intervention and the type of policy instrument for each action. With the launch of the energy poverty pillar in the context of the CoM, cities will be required to conduct an energy poverty assessment and to plan actions specifically tackling energy poverty in their territory. However, to date reporting on energy poverty pillar is not yet mandatory and cities have been only flagging a mitigation action as also "relevant for energy poverty". The actions/measures related to PV from CoM action plans have been assessed, with the aim to understand their geographical distribution, the sectors they target and the policy instruments adopted. Table 1 summarises the elements of the reporting framework and template that may carry potential insights on PV systems, technologies, and communities.

Table I: Main aspects covered in the CoM reporting template, with potential insights on PV systems, technologies, and communities

Emission inventories
o Final energy consumption in buildings, equipment/facilities, transport and other sectors, with details by energy carrier o Emission factors, indicating the amount of GHG emissions associated with a unit of activity (e.g. energy consumed) o Certified green electricity purchases and sales o Local/distributed electricity production (by technology) o Local/distributed heat/cold production (by technology) o Non-energy related sectors o Emission inventory (calculated)
Actions
o Overview of mitigation actions per sector, with estimated impacts (energy savings, renewable energy production, GHG emission reduction) by the target year o Detailed information on selected actions (e.g. origin of the action, responsible body, implementation timeframe, stakeholders involved, implementation costs, sector, area of intervention, policy instrument, climate hazards addressed, adaptation sector, vulnerable population group(s) targeted).

Table II provides an overview of the sections of the Cities Mission questionnaire having explicit or implicit tangencies with the world of photovoltaics. The analysis of such information is used to complement the insights accrued through the CoM initiative to better estimate the potential of PV systems in a rapid acceleration towards climate neutrality by analysing how cities intend to incorporate PV or other solar systems in building renovation, district heating, and collective initiatives such as renewable energy communities and cooperatives enabling RES deployment.

Table II: Energy and social aspects addressed in the EOI carrying potential implications for PV systems, technologies, and communities

Energy
o Emissions (inventory)
o Existing plans (SECAP/SEAP/energy plans)
o % Generation from RES, % Consumption from RES, RES sources, non-RES carriers
o Current VS future policies – areas covered (incl. Nearly Zero and Positive Energy Buildings and Districts, building electrification, building integrated renewables (e.g. BIPV), citizen and renewable energy communities, on-site and nearby renewable energy generation (electricity, heat/cold), local (off-site) renewable energy generation (electricity, heat/cold), virtual power plants, RES sources to be include/expanded in the future, sector-oriented behavioural changes, digitalisation and smart city solutions.)
o Current VS future policies – measures/instruments (incl. awareness raising/training, energy management, energy certification/labelling, energy suppliers obligations, energy/carbon taxes, building standards, Public Private Partnerships, energy audits.)
o Key measures adopted in the past and scalable interventions for the future in the fields of stationary energy and energy generation.
o Co-benefits/adverse impacts (incl. energy poverty, energy security, disruption of energy, transport, water and communications networks, social inclusion, equality and justice, education and public awareness)
o City governance and staff: structure and skills.
o Data collection and reporting (frequency, sectors/sources covered), data disclosure.
o Stakeholders/Partnerships (incl. with government, academia and R&I institutions, private sector, citizen and renewable energy communities)
o Sector-specific barriers/gaps/assistance needs (incl. subsidies for competing fuels, high initial capital costs, difficulty of fuel price risk assessment, unfavourable power pricing rules, lack of enabling policy at different levels of governance, transaction costs, perceived technology performance uncertainty and risk)
Society/citizens (no sector-specific)
o Types of engagement activities (incl. deliberative practices, informative practices, participatory budgeting)
o Existing programmes/projects to engage citizens in climate change mitigation/GHG emissions reduction policies
o Actions in place targeting behavioural change (incl. infopoints, incentives/disincentives, awareness-raising campaigns)
o Social risks linked to running for climate neutrality by 2030 (e.g., related to social inequality, social inclusion, gentrification, energy poverty).

3.1 Results

In the pool of more than 6700 CoM action plans received before mid-May 2021, almost 13500 actions/measures from 5366 cities or local authorities aim at deploying photovoltaics. Over 1000 measures are indicated as key actions, meaning that signatories are considering these as central in their climate and energy strategy. The majority of the actions/measures (10493) relate to the local electricity production sector, with 8260 specifically reported under "photovoltaic" as area of intervention (Figure 1, top). The remaining are reported under other sectors, notably municipal, residential or tertiary buildings, or transport. This suggests that cities may want to put emphasis on the use of PV to decarbonise specific energy consuming sectors (e.g., buildings, electric vehicles). In terms of policy instruments, grants and subsidies represent the largest share, followed by awareness raising and training campaigns (Figure 1, bottom). Public procurement is typically used to target public buildings or public spaces. Third party financing and building standards are also common instruments.

The number of actions related to PV systems varies greatly among CoM signatories. As shown in Figure 2, the majority of cities reported less than 10 actions in their plans, with an average 2.5 actions per plan. However, there are five signatories in the sample having planned more than 25 actions related to PV, with one signatory in Italy with an exceptionally high number of actions (54). For the 89% of actions, the full implementation is estimated by 2020, meaning that it will be possible to estimate the impacts and the actual scale of the interventions in the coming years through the monitoring phase that is required within the CoM framework. The majority of signatories reporting PV actions in their plans are from Italy and Spain, in alignment with the overall distribution of signatories and climate action plan in the initiative [17]. Consistently, in Italy 90% of submitted plans included at least one PV-related action, as shown in Figure 3. Belgium, Cyprus, and Croatia show a high share as well, which may suggest that PV solutions are typically part of the decarbonisation strategies of cities in those countries. Further analysis needs to be developed to correlate these outcomes to the national legislation in force that incentivises the deployment of PV systems. In Iceland, Norway, and in the Western Balkans, CoM signatories are not developing any PV-related actions to reach their mitigation target. However, on average the overall share of SECAPs that included at least one action devoted to PV systems out of all SECAPs is 60%

To date, the number of PV-related actions and measures in the CoM dataset flagged as "energy poverty actions" is very limited (lower than 20). However, the interest demonstrated by cities in deploying PV shows the potential to contribute to address some of its causes in EU i.e improving energy efficiency, reduce the energy costs of households, and address households' energy needs.

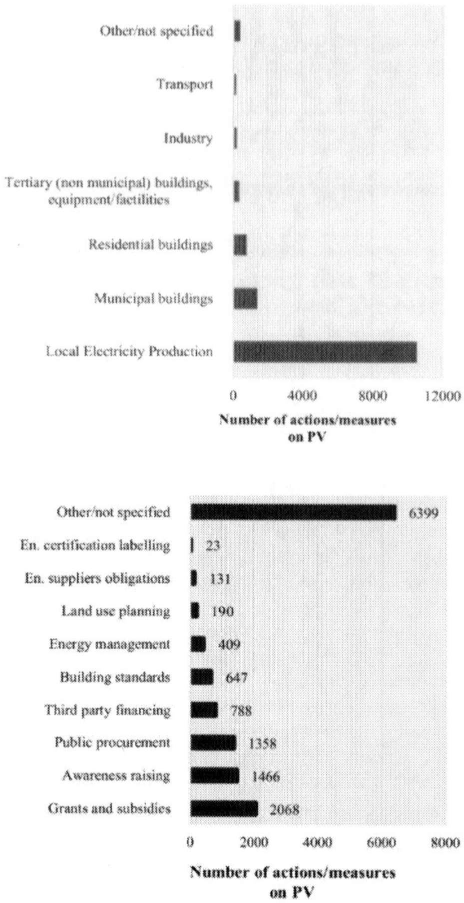

Figure 1: Actions/measures on PV by sector and by policy instrument

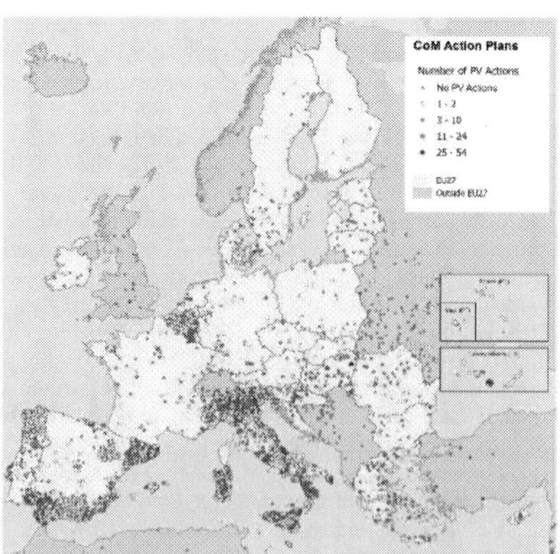

Figure 2: Geographical distribution of actions/measures on PV per SECAP

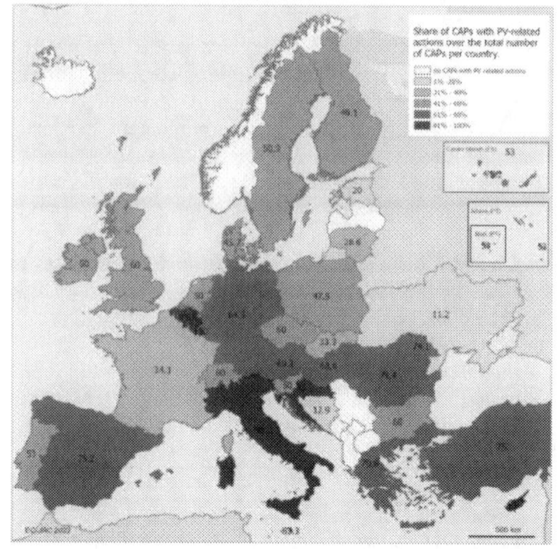

Figure 3: Share of SECAPs including at least one PV-related action over the total number of submitted SECAPs, by country

The data obtained through the Cities Mission's Expressions of Interest provide further insights on the role PV and other solar systems may play in a just transition towards climate neutrality. Notably, in describing the interventions that could be scaled up by 2030, eligible cities expressed a strong intention to disseminate photovoltaic use in a variety of forms by exploiting the full rooftop potential, as well as by turning facades, bus stops, parks, and lampposts into solar collectors. Some cities aim at boosting PV installations through group purchases and publicly available online solar registries where everyone can check whether their house is suitable for photovoltaics and get a first impression on costs, return of investment and energy generation. Cities also envision a significant substitution of fossil fuel sources used in district heating networks with renewable energy carriers (including solar, geothermal, biomass, and waste) coupled with advanced storage solutions. Finally, the creation of energy communities and renewable energy communities is regarded as a powerful instrument of transformation and aggregated action. In some contexts, the implementation of renewable energy communities is designed to create 'Solar Neighbourhoods' where consumers are families, shops and citizens, and those in condition of energy poverty may receive free energy.

4 CONCLUSIONS

Data analysis reveals that cities are emphasizing or intend to emphasize the use of PV to decarbonise specific energy consuming sectors (notably buildings, district heating) as well as, more in general, the electricity grid. This multi-sectoral interest demonstrated by cities in deploying PV systems in their territories may hint at a widespread use of the technology with potential benefits for energy poor communities.

Indeed, PV technology is broadening its range of applications and its contribution to society. Related benefits and potentials are well recognised at the local scale. More than half of the CoM signatories is relying on

PV solutions in different sectors of action to achieve decarbonisation targets, while setting the basis for counteracting the causes of energy poverty. Further, PV systems are the most popular RES solution envisioned by Mission Cities in reaching climate neutrality by 2030.

5 REFERENCES

[1] M. Victoria et al. (2021). Solar photovoltaics is ready to power a sustainable future. Joule, vol. 5, pp. 1041–1056, doi: https://doi.org/10.1016/j.joule.2021.03.005.

[2] European Commission (2019). The European Green Deal. Communication from the Commission to the European Parliament, the European council, the Council, the European Economic and Social Committee and the Committee of the Regions, Brussels, COM (2019) 640 final.

[3] A. Jäger-Waldau, I. Kougias, N. Taylor, and C. Thiel, "How photovoltaics can contribute to GHG emission reductions of 55% in the EU by 2030," Renew. Sustain. Energy Rev., vol. 126, pp. 109836, 2020, doi: https://doi.org/10.1016/j.rser.2020.109836.

[4] G. Lindseth (2004). The cities for climate protection campaign (CCPC) and the framing of local climate policy. Local Environment, 9(4), 325–336.

[5] H. Bulkeley, M. Betsill. (2003). Cities and Climate Change: Urban Sustainability and Global Environmental Governance. Routledge, ISBN 0–415–27379–X

[6] United Nations Human Settlements Programme (2016). World Cities Report 2016: Urbanization and Development - Emerging Futures. Nairobi. https://wcr.unhabitat.org/wp-content/uploads/2017/02/WCR-2016-Full-Report.pdf.

[7] J. G. Groesbeck and J. M. Pearce, "Coal with Carbon Capture and Sequestration is not as Land Use Efficient as Solar Photovoltaic Technology for Climate Neutral Electricity Production," Sci. Rep., no. August, pp. 1–17, 2018, doi: 10.1038/s41598-018-31505-3

[8] A. Jäger-Waldau, "The Untapped Area Potential for Photovoltaic Power in the European Union," Clean Technologies, vol. 2, no. 4. 2020, doi: 10.3390/cleantechnol2040027

[9] Benöhr and Alemayehu Gebremedhin, "Photovoltaic systems for road networks," Int. J. Innov. Technol. Interdiscip. Sci., vol. 4, no. 2 SE-Articles, pp. 672–684, May 2021, doi: 10.15157/IJITIS.2021.4.2.672-684.

[10] P. Huang, Y. Sun, M. Lovati, and X. Zhang, "Solar-photovoltaic-power-sharing-based design optimization of distributed energy storage systems for performance improvements," Energy, vol. 222, pp. 119931, 2021, doi: 10.1016/j.energy.2021.119931.

[11] P. Bertoldi (ed), "Guidebook 'How to develop a Sustainable Energy and Climate Action Plan (SECAP)" EUR 29412 EN, Publications Office of the European Union, Luxembourg, 2018, ISSN Online: 1831-9424.

[12] EC, "European Missions - 100 Climate-Neutral and Smart Cities by 2030 - Info Kit for Cities," 2021. [Online]. Available: https://ec.europa.eu/info/sites/default/files/research_and_innovation/funding/documents/ec_rtd_eu-mission-climate-neutral-cities-infokit.pdf.

[13] M. Hebbert, V. Jankovic (2013). Cities and Climate Change: The Precedents and Why They Matter. Urban studies, 50. https://doi.org/10.1177/0042098013480970

[14] V. Palermo, P. Bertoldi, M. Apostolou, A. Kona, and S. Rivas, "Assessment of climate change mitigation policies in 315 cities in the Covenant of Mayors initiative," Sustain. Cities Soc., vol. 60, pp. 102258, 2020, doi: https://doi.org/10.1016/j.scs.2020.102258.

[15] A. Kona, P. Bertoldi, Ş. Kilkiş (2019) Covenant of mayors: Local energy generation, methodology, policies and good practice examples. Energies, 12, doi: https://doi.org/10.3390/en12060985

[16] European Union, "Climate-Neutral and Smart Cities Mission Call for Expression of Interest," 2021. https://ec.europa.eu/info/sites/default/files/research_and_innovation/funding/documents/ec_rtd_he-missions-cities-call-questionnaire.pdf (accessed Feb. 03, 2021).

[17] G. Melica et al., "Covenant of Mayors: 2021 assessment: climate change mitigation and adaptation at local level", Publications Office of the European Union, 2022, https://data.europa.eu/doi/10.2760/58412 .

SOCIOECONOMIC AND DEMOGRAPHIC FACTORS BEHIND THE DEPLOYMENT OF DOMESTIC PHOTOVOLTAIC AND SOLAR THERMAL SYSTEMS IN THREE SWEDISH MUNICIPALITIES

Johan Lindahl[1], Sofia Ekbring[1], Robert Johansson[1], David Lingfors[2] and Joakim Munkhammar[2]
[1] Becquerel Sweden AB
[2] Uppsala University
[1]Staffansvägen 14, SE-74142 Knivsta, +46 768-511 773, johan@becquerelsweden.se

ABSTRACT: The adoption of domestic photovoltaic systems has in numerous studies been proven to be influenced by peer effects and socioeconomic factors such as income, age, gender, education etc., which has led to irregular spatial installation patterns. Only a few studies regarding domestic solar thermal systems indicate that the same effect exist for this technology. However, the interaction between photovoltaic and solar thermal deployment and the similarities or differences in socioeconomic factors have not been investigated in detail so far. This study identifies the most prominent socioeconomic factors behind both domestic photovoltaic and solar thermal adoption in three different municipalities in Sweden, based on a complete set of 452 photovoltaic and 359 solar thermal collector systems installed until 2020, which was identified and classified by a method that uses machine learning and aerial imagery. A moderate (absolute Pearson correlation, $|\rho|$, > 0.3) to intermediate ($|\rho|$ > 0.5) correlation between photovoltaic and solar thermal penetration was found on demographic statistical area level, and several of the previously reported influential socioeconomic factors for domestic photovoltaic installation were confirmed also for domestic solar thermal adoption in Sweden.
Keywords: Photovoltaic, Solar Thermal, PV Market, Solar Home System, Economic Analysis, Socioeconomic Analysis

1 INTRODUCTION

The interest in solar energy is growing rapidly in Sweden. In 2021, 26,500 grid-connected photovoltaic (PV) systems were installed with a combined installed output of 500 MW, which is an increase of 46% as compared to 2020 [1]. At the end of 2021, there were in total 92,359 grid-connected PV installations in Sweden, with a total installed capacity of 1,586 MW [1]. The positive development this year can partly be explained by the new green tax deduction for PV installations for private individuals, which was introduced in 2021 [2] and the rising electricity prices preceding the Russian invasion of Ukraine. The increasingly integrated European electricity system means that electricity prices will probably be more equalized across Europe [3,4], which means that electricity prices higher than Swedish end-consumers experienced in the last decade will become the norm the coming years.

The interest in the other major solar energy technology, solar thermal (ST), has varied the last decades in Sweden. Between 2000 and 2011, an investment subsidy for private individuals and small-scale systems was in place [5], and a majority of the current Swedish ST systems (usually with a size of less than 15 m^2) were installed in that period. In the last decade, the Swedish ST market in Sweden has declined due to low profitability [6], as there is no direct support system for ST systems at the time of writing. The shrinking market is in line with findings about the economics of unsubsidized ST in other European countries [7]. However, a majority of the previous installed systems remain in use and there is still an estimated total ST area of about 450,000 m^2 in operation in Sweden [6].

The roll out of new decentralized granular technologies, such as PV and ST, has been observed as faster than more bulky energy technologies, as they profit from shorter diffusion time scales, more attractive risk profiles for investors, and stronger potential for cost and performance improvements [8]. In addition, the diffusion pattern is usually highly influenced by social interactions, so called peer effects [9], and by socioeconomic factors. PV adoption has been found to be concentrated in certain areas or clusters [10–28]. This cluster tendency can partly be explained by peer effects, leading to an effect where the probability of additional PV installations is higher where the number of existing PV systems is high (with a certain time lag) [12–16,19,22,24–26,28–32].

In addition to peer effects, difference in socioeconomic and demographic factors of the inhabitants in different areas also affect the spatial distribution of domestic PV. Spatial, demographic and socioeconomic differences in PV installation patterns have been studied in countries such as Australia [10,33], Belgium [11], Canada [34], China [12,35], Denmark [36], Germany [13–16], Greece [37], Japan [38], the Netherlands [17], Sweden [29–31,39], UK [18–21,32] and US [22–28,40]. It has been shown that identified factors can be highly contextual and that they can vary both between and within countries [10,15,21,26,28,40]. Previous such international studies have often relied on qualitative methods based on survey studies/interviews [29–31,34,36,37,41] and/or by qualitative methods based on aggregated data over a larger area, i.e., an administrative region [14,38,42], Municipality [29], NUTS3[1] [16,19], postal or zip code [10,17,20,22,23,32,33,40], statistical sector[2] [11,13,39] or census block[3] [24,26–28,40,43].

Socioeconomic and demographic studies about the ST technology are scarce in comparison. Some literature have treated awareness, acceptance and motives [35,44,45], but when it comes to important socioeconomic factors for domestic ST adoption it has mainly been studied in Germany in the beginning of the 20th century [46–48]. An exception is [49], which through a discrete choice experiment survey investigated customer preferences of ST and PV in Boston and Atlanta and found an overall

[1] The NUTS classification (Nomenclature of territorial units for statistics) is a hierarchical system for dividing up the economic territory of the EU and the UK for the purpose of the collection, development and harmonization of European regional statistics

[2] Statistical sectors are very disaggregated local areas – typically a set of streets – grouped by socioeconomic, urban and morphological structural features.

[3] A census block is the smallest geographic unit used by the United States Census Bureau

general preference of PV over ST systems in both cities.

However, the spatial distribution of ST has so far not been investigated thoroughly, and only evidence of cluster tendencies on a high regional level has been presented in the scientific literature [42,50]. In addition, the interaction between ST and PV system distribution has not been studied.

Since the expansion of PV in Sweden is somewhat late in an international comparison, early research on the diffusion of PV in Sweden has primarily focused on qualitative methods as the statistical basis has been limited [29,30,51–54]. Through survey and interview studies, the influence of adopters on others through a so-called word-of-mouth process [29,31], and local organizations have been identified as important [29]. Especially local utility companies that bought excess electricity, offered turnkey solar systems and held information meetings about solar energy have been considered influential [29,31]. Late adopters have been found to have stronger financial, rather than environmental, motives than early adopters [51]. As the PV technology has become more common in Sweden, quantitative methods have been used in later research [31,39,55]. In one of those studies, it was investigated how socioeconomic parameters influenced the willingness to install PV systems in Uppsala municipality [39]. In this study it was found that parameters such as land ownership, average income, number of cars, age and employment rate had a large impact on the willingness to invest in PV. In a survey study, it was found that peer effects raise the likelihood of adopting PV, as does a higher subsidy from the state [31]. This article also concluded that males are marginally more likely to adopt PV, older people are marginally less likely, and that income is not a significant predictor.

The main novelty of this quantitative study is that the complete dataset contains both ST and PV systems, and therefore makes it possible to analyse the difference between the two solar energy technologies when it comes to spatial adoption and the underlaying socioeconomic and demographic factors. The motive is that an end-customer's knowledge about one technology (in Sweden ST as this was mainly built a decade before the major roll-out of PV), familiarity of a similar technology has been found to play a role in PV adoption [56]. In addition, indicative evidence of passive peer effects through spill over between PV and ST has been found in Sweden [29].

2 METHOD

The aim of this study is to analyze different socioeconomic and demographic factors' relation to and impact on solar energy systems in Sweden. This section presents the methodology of the creation of a solar energy system inventory and the choice of geographical area, data management, analysis methods and made assumptions.

2.1 Creation of solar energy systems data set

A limited geographical area of three Swedish municipalities was chosen to be able to keep a preferable high granularity [27]. Local factors have previously been shown to influence the rate of diffusion of PV in Sweden [29], and the choice of the three municipalities Falun, Knivsta and Uppvidinge was done on purpose so that three different types of municipalities in terms of size, characteristics and location were covered.

Falun is a spatially large municipality, 2,040 km^2, in central Sweden with 59,837 inhabitants in 2020 [57]. The municipality is centered around the regional center of the town Falun, but also contain ten smaller surrounding villages, and 87.3% of the inhabitants live in these eleven urban areas of the municipality.

Knivsta is a relatively spatially small municipality, 295 km^2, located close and between the two major Swedish cities Stockholm (the capital of Sweden) and Uppsala (the fourth largest city). The two urban areas in the municipality, Knivsta and Alsike, where 72.8% of the population lives, are typical commuter resorts. The average age of the 19,818 inhabitants is one of the lowest in the country [58].

Uppvidinge is a 1,178 km^2 large municipality in the southern part of Sweden with 9,449 inhabitants, which is dominated by agriculture land and forest. The municipality lacks a major town, but 74.3 % lives in urban areas including 2700 inhabitants in the regional center Åseda [59].

In this study Falun represent a large urban/rural municipality centered around a major regional city, Knivsta a suburban municipality and Uppvidinge a rural municipality. Some general socioeconomic key figures of the three municipalities are summarized in Table I.

Table I: Population density [inhabitants/km^3], average age [years], average monthly earned income for all inhabitants >20 years [SEK/month] and employment rate [%] for Falun, Knivsta and Uppvidinge municipalities in 2020 [58,60].

	Pop. density	Avg. Age	Avg. Income	Employ-ment
Falun	29.2	42.4	27,358	82.6
Knivsta	32.2	36.9	33,408	87.2
Uppvidinge	16.2	43.5	23,617	81.6

For the actual creation of an inventory of solar energy systems, a slightly modified version of the open access DeepSolar machine learning and aerial imagery PV mapping framework [61,62] was implemented and used to scan Ortophoto images of the municipalities. The identified systems were saved as polygon objects and analyzed with the QGIS3 software.

The identified PV systems were cross-checked with the local Distributed Systems Operators' (DSO) registers of grid-connected PV. In some rare cases, visual inspection through orthophoto images was not enough to remotely verify existing solar energy systems, uncertainties regarding type of solar energy system, design and/or location. For these systems, physical on-site inspections were carried out to complete the inventory. Through this procedure, a complete set of all PV and ST systems could be compiled for the three municipalities.

By using the coordinates of the identified solar energy system polygons as keys for available geodata services from the Swedish Land Survey, all the located PV and ST systems were connected to the unique Swedish property designations of the real properties (the legal division of land) within which they were installed.

The advantage of this method is that a complete inventory of PV and ST systems that include systems installed without any subsidies can be created. As a comparison, the registers of the DSO's do not contain any ST systems nor any off-grid PV systems, and the information about the location are on the level of "somewhere within the real property".

The inventory of solar energy systems contained all completed systems at the time of the aerial photos, which was (depending on spatial location in the municipality) in Falun the 2020-05-21, 2020-06-11 or 2020-06-14, in Knivsta the 2019-07-19 or 2019-07-20 and in Uppvidinge the 2020-05-31 or 2020-06-01. To enable an analysis of only domestic PV and ST, all solar energy systems on real properties owned by companies were excluded from the dataset. In addition, households that acquired the property after the aerial photos were taken, were removed from the set, as no socioeconomic information about previous owners were available. In addition, some specific uncertain ownership situations, like if the owner of a real property lived outside of the municipality, real properties with two or more owners not living at the same address, several owners all living in the municipality but not at the real property of the solar energy system, were also removed from the set.

2.2. Socioeconomic and demographic data

With the property designation as a key, socioeconomic information of the specific households with solar energy systems could be extracted from the Swedish authorities Statistic Sweden (the Swedish government agency operating responsible for producing official statistics for decision-making, debate and research) and the Swedish Land Survey, along with the company Ratsit AB (which produces a Taxation calendar that contains information about income and capital for individuals).

In addition to household level, demographic data were available at two more aggregated levels. The most aggregated level was the municipality level. The second most aggregated level was, just as in [39], the fixed demographic statistical areas (DeSO) that Statistic Sweden divide all municipalities into. The size of a DeSO is determined based on population and building concentration and can therefore have a very different spatial size, i.e., usually around 1,500 inhabitants, but can vary between 600 and 3,500 [63]. There are three different categories of DeSOs:

1. Rural – Areas mainly located outside larger population concentrations or urban areas.

2 Urban – Areas mainly located in a population concentration or urban area but not in the municipality's regional center.

3. Center – Areas mainly located in the municipality's regional urban center.

Fig. 1 shows the orthophoto of Uppvidinge municipality, with the six different DeSOs included, along with all PV and ST systems as an illustrative example.

The total number of DeSOs and distribution between the different categories is displayed in Table II for the chosen municipalities, along with the distribution of inhabitants and households.

Previous research has identified systems through different approaches and analyzed data at different levels and granularity. With the approach of this study the granularity available was on household level for some socioeconomic and demographic factors, and on either DeSO or municipality level for correlation analyses.

2.3 Dependent and independent variables

In order to evaluate different socioeconomic and demographic factors' relation to, and impact on, private residential solar energy deployment, PV or ST density in terms of number of systems per households were chosen as the dependent variables in this study.

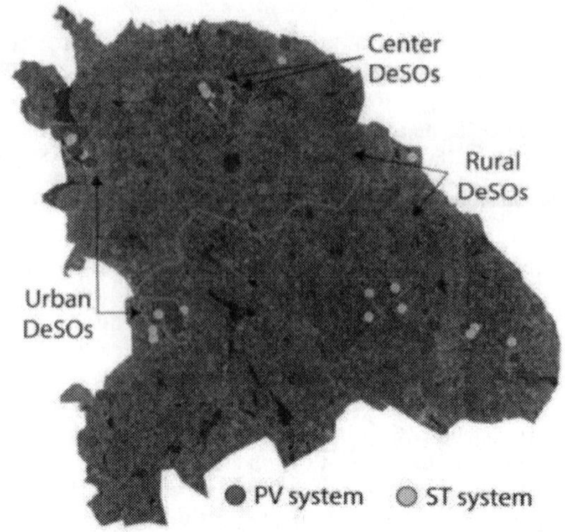

Figure 1: The orthophoto image of Uppvidinge municipality, with the different DeSO areas framed by blue boarders and all existing PV and ST systems by June 2020.

Table II: The of number of demographic statistical areas (DeSOs), inhabitants, households and average number of inhabitants per household in the different DeSOs types 'Rural', 'Urban' and 'Center' in Falun, Knivsta and Uppvidinge municipality, respectively.

	Rural	Urban	Center	Total
	Number of DeSOs			
Falun	6	5	26	37
Knivsta	3	0	7	10
Uppvidinge	2	2	2	6
	Number of inhabitants			
Falun	9,641	7,925	41,962	59,528
Knivsta	5,249	-	13,857	19,106
Uppvidinge	3,673	3,052	2,773	9,498
	Number of households			
Falun	4,119	3,648	20,222	27,989
Knivsta	2,082	-	5,463	7,545
Uppvidinge	1,724	1,448	1,219	4,391
	Average number of inhabitants per household			
Falun	2.3	2.2	2.1	2.1
Knivsta	2.5	-	2.5	2.5
Uppvidinge	2.1	2.1	2.3	2.2

The analysed independent variables were chosen based on available data and previous research (summarized in [21,64]) and were; (1) Age, (2) Sex, (3) Birth origin, (4) Education level (5) Employment rate and (6) Different economic conditions, summarized in Table III.

Age is represented through five independent variables corresponding to percentage of the population in the age groups of *0–15*, *16–24*, *25–44*, *45–64* and *>65* years.

The variable *Sex*, the distribution between the sexes, is expressed as the percentual share of males.

Birth origin as an independent variable is referred to as *BOri* and represents the percent of individuals which are born in Sweden.

Education level is denoted *Educ* and represents the percentage of the population with a post high school education.

Table III: Chosen socioeconomic factors with a detail level description, available granularity level and source of the data, where SLS is short for Swedish Land Survey.

	Detail level	Granularity Municipality	DeSO	Household	Source
Population	Total number of households	X	X	-	Statistics Sweden
	For all of age >20	-	-	X	Ratsit
Age	7 age groups	X	X	-	Statistics Sweden
	For all of age >20	-	-	X	Ratsit
Sex	Total distribution	X	X	-	Statistics Sweden
	For all of age >20	-	-	X	Ratsit
Birth origin	Sweden, Europe, RoW	X	X	-	Statistics Sweden
Education	Age 25-64	X	X	-	Statistics Sweden
Employment	Age 20-64	X	X	-	Statistics Sweden
Average income	Age ≥20	X	X	-	Statistics Sweden
	Age ≥20	-	-	X	Ratsit
Economic standard	4 groups, age 20+	X	X	-	Statistics Sweden
Property purpose	General and detailed	-	-	X	SLS
Property owner	Taxed owner	-	-	X	SLS
Tax value	Buildings, land and total	-	-	X	SLS

Employment rate as an independent variable is denoted *Emp* and expressed as the percentage of the population being employed.

The economic situation of the households is represented by several independent variables that are gathered from the different data sources listed in Table III. The first is average income, *AInc*, of the DeSO, defined as the aggregated taxable earned income, which refers to income from employment, business/entrepreneurship, pension, sick pay and other taxable transfers. The second variable is the average economic standard, *EcSt*, which is calculated as disposable income per consumption unit.

2.4 Correlation analysis

The Pearson correlation coefficient was calculated for all variables. In this study, strong correlation was defined as absolute values of the correlation coefficient of >0.7, intermediate correlation for values of 0.5–0.7, moderate for values of 0.3–0.5 and weak correlation for absolute values of 0.1–0.3.

Presenting Pearson correlation coefficients in a matrix also allow for a collinearity analysis between the independent variables. Collinearity refers to the non-independence of variables and occurs when two variables correlate due to mutual underlying factors, or when the data used is compositional. If the absolute value of the correlation coefficient between two independent variables is higher than 0.7, it indicates a severe collinearity [65].

3 RESULTS

In this section the results are presented, which includes a general description of the distribution of the 814 domestic solar energy systems (Section 3.1), a correlation analysis on the socioeconomic and demographic parameters (Section 3.2) in comparison to similar studies (Section 3.3) and finally, some characteristics on households owning a solar energy system (Section 3.4).

3.1 The domestic solar energy system data set

The final data set of domestic solar energy systems is summarized in Table IV. The table also include the number of households that have both ST and PV systems on their roof.

Table IV: Number of domestic PV and ST systems in Falun, Knivsta and Uppvidinge municipalities, along with number of households that have installed both a PV and a ST system.

	#PV	#ST	Total	*PV&ST*
Falun	276	287	563	*16*
Knivsta	103	49	152	*7*
Uppvidinge	73	23	96	*1*
Total	452	359	811	*24*

With regards to the distribution of systems in the different DeSO types, 201 of the 452 PV systems (44.5%) were found in different 'Rural' DeSOs, 54 (11.9%) in the 'Urban' DeSOs and 197 (43.6%), in the 'Center' DeSOs of the municipalities. Likewise, 161 of the total 359 ST systems (44.8%) were found in the different 'Rural' DeSOs, 66 (18.4%), in the 'Urban' DeSOs and 132 (36.8%), in the 'Center' DeSOs of the municipalities. Domestic PV systems were identified in 47 out of the 53 DeSOs and domestic ST systems were identified in 45 of the DeSOs. The DeSOs without solar energy systems were found to all be of type 'Center'. When it comes to lack of domestic PV system, all those DeSOs were in Falun, while one of the 'Center' DeSOs without a domestic ST was in Knivsta and the rest in Falun.

The excluded domestic solar energy systems, motivated by non-existing or uncertain background data, is judged to have a small effect on the results, as this number is small as compared to the final data set. 37 systems were omitted in Falun, 8 in Knivsta and 19 in Uppvidinge, as a result of the owning household living outside the municipality or acquired the real property after 2020-12-31. In addition, 12 systems were omitted in Falun due to lack of data or definition difficulties.

The number of domestic solar energy systems per 1,000 households are presented in Fig. 2. As can be seen, the concentration of both PV and ST systems per 1,000 households are higher in the 'rural' areas of the municipalities as compared to the 'urban' or 'center' areas. The reason is likely that many more households in the 'urban' and 'center' areas live in multi-family houses, which makes it impossible to own a private residential solar energy system.

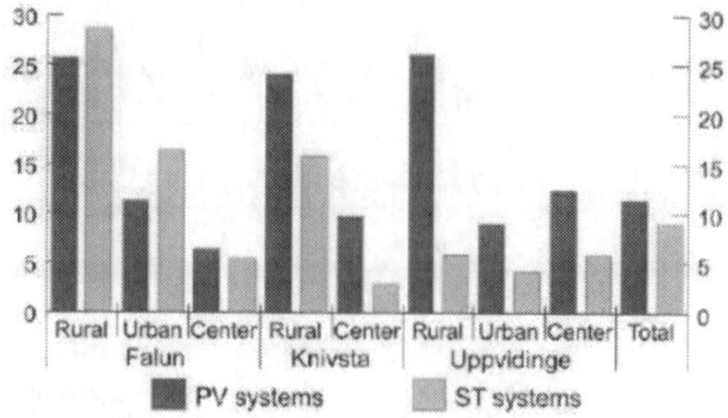

Figure 2: The average number of private residential PV or ST systems per 1,000 households in the different type of DeSOs in Falun, Knivsta and Uppvidinge.

Another observation in Fig. 2 is that the concentration of ST systems was found to be higher in the 'rural' and 'urban' areas of Falun, as compared to the concentration of PV systems, which is the opposite to the PV/ST concentration relation in the other municipalities. This indicates that local factors probably have had an important role in driving the diffusion of ST systems in Sweden as well, just as Palm found for PV systems [29].

3.2 Demographic correlation and collinearity results

The correlation analysis based on the Pearson correlation coefficient showed the relation between the dependent variables, number of PV systems per household, *#PV*, and number of ST systems per household, *#ST*, and the independent demographic variables, presented in Section 2.3. The correlation matrix presented in Fig. 3 represent the whole domestic data set. The matrix in Fig. 3 is based on installed PV systems per household, *#PV*, in each 47 DeSOs with a private residential PV system and socioeconomic and demographic statistics for each DeSO. However, in an evaluation of the result before creating the matrix, two out of the 47 DeSOs that contained PV systems were found to be outliers (one 'rural' in Falun and one 'rural' in Uppvidinge) with 0.063 and 0.046 PV systems per households, respectively, as compared to the rest of the DeSOs, which exhibit penetration levels between 0.001–0.033 PV systems per household. Omitting these two outlier DeSOs resulted in a dependent variable based on installed PV systems per household, *#PV**, in the other 45 DeSOs.

	#PV	#PV*	#ST	0–15	16–24	25–44	45–64	>65	Sex	BOri	Edu	Emp	AInc	EcSt
#PV	1	-												
#ST	0.36	0.61	1											
0–15	-0.09	0.06	-0.24	1										
16–24	-0.44	-0.31	-0.43	0.16	1									
25–44	-0.41	-0.42	-0.51	0.50	0.21	1								
45–64	0.52	0.55	0.56	-0.39	-0.35	-0.84	1							
>65	0.18	0.00	0.31	-0.93	-0.34	-0.67	0.34	1						
Sex	0.56	0.50	0.36	0.01	-0.23	-0.20	0.61	-0.07	1					
BOri	0.32	0.36	0.56	-0.07	-0.51	0.36	0.42	0.19	0.05	1				
Edu	-0.22	-0.06	-0.26	0.38	0.28	0.23	-0.22	-0.39	-0.49	0.28	1			
Emp	0.15	0.35	0.27	0.34	-0.35	-0.07	0.19	-0.21	-0.02	0.75	0.50	1		
AInc	0.00	0.21	-0.07	0.66	0.14	0.21	-0.01	-0.66	-0.07	0.33	0.79	[illegible]	1	
EcSt	0.18	0.43	0.06	0.38	-0.04	-0.05	0.27	-0.37	0.01	0.39	0.66	[illegible]	0.86	1

Figure 3: Correlation matrix with Pearson correlation coefficient between installed PV and ST systems per household and different demographic independent factors for all DeSOs with installed systems in the municipalities Falun, Knivsta and Uppvidinge combined. The colors match the strong, intermediate, moderate and weak correlations with yellow→green for the positive and yellow→red for the negative correlations.

As Fig. 3 display, negative collinearity in this matrix were found between the independent variables; (1) age groups *0–15* and *>65*, and (2) age groups *25–44* and *45–64*. Positive collinearity was uncovered for (1) employment, *Emp*, and birth origin, *BOri*, (2) education level, *Educ*, and average income, *AInc*, (3) employment, *Emp*, and average income, *AInc*, (4) employment, *Emp*, and average economic standard, *EcST*, and (5) average income, *AInc* and average economic standard, *EcST*.

As the main contribution to the literature it should be noted that a moderate positive correlation (0.36) was discovered between *#PV* and *#ST*, which increased to an intermediate positive correlation when the two outlier DeSOs were omitted (0.61). Furthermore, an intermediate positive correlation was revealed between both *#PV* and *#ST* and the age group *45–64*, while the age groups of *16–24* and *25–44* both explicit a moderate to intermediate negative correlation to solar energy system penetration on a DeSO level. The percentage of men in a demographic area also seem to have a positive effect on solar energy deployment by demonstrating intermediate positive correlation for *#PV* and moderate positive correlation for *#ST*. For the other independent demographic variables, the correlation was found to be weak or non-existing, except if the two DeSOs with exceptionally high PV penetration were omitted, which resulted in that the correlation between *#PV** and employment along with average economic standard rose to moderate positive correlation.

The presented correlation matrix in Fig. 3 does not take differences between the municipalities in general demographic terms of, e.g., overall income and education level, into account, which might have resulted in a weaker correlation between the different variables. Thus, Pearson correlation coefficients between number of PV systems per household and each independent variable are presented for the three municipalities separately in Fig. 4.

#PV	#ST	0–15	16–24	25–44	45–64	>65	Sex	BOri	Edu	Emp	AInc	EcSt
Falun	0.49	-0.08	-0.38	-0.41	0.49	0.19	0.66	0.43	-0.24	0.16	0.03	0.00
Knivsta	0.45	-0.24	-0.50	-0.55	0.59	0.37	0.74	0.50	-0.20	0.27	0.00	0.62
Uppvidinge	-0.57	-0.15	-0.81	-0.51	0.63	0.38	-0.22	0.87	0.62	0.86	0.42	0.79
#ST	**#PV**	**0–15**	**16–24**	**25–44**	**45–64**	**>65**	**Sex**	**BOri**	**Edu**	**Emp**	**AInc**	**EcSt**
Falun	0.49	-0.18	-0.43	-0.58	0.65	0.31	0.56	0.55	-0.41	0.31	-0.02	0.14
Knivsta	0.45	-0.33	-0.65	-0.45	0.73	0.33	0.81	0.55	-0.76	0.14	-0.16	0.22
Uppvidinge	-0.57	0.12	0.26	0.24	-0.25	-0.18	0.32	-0.84	0.12	-0.85	-0.70	-0.02

Figure 4: Pearson correlation coefficients between number of installed PV systems per household (top part) respective number of installed ST systems per household (bottom part) and the different demographic independent factors for all DeSOs with installed PV systems in the municipalities Falun, Knivsta and Uppvidinge.

#PV	#ST	0–15	16–24	25–44	45–64	>65	Sex	BOri	Edu	Emp	AInc	EcSt
Rural	-0.47	-0.13	-0.52	-0.28	-0.08	0.31	0.38	0.21	-0.01	-0.06	-0.13	-0.10
Urban	0.67	-0.56	0.07	-0.62	0.51	0.35	-0.37	0.28	0.34	0.57	0.56	0.59
Center	0.65	0.27	-0.11	-0.44	0.49	-0.13	0.64	0.35	0.11	0.35	0.33	0.45
#ST	**#PV**	**0–15**	**16–24**	**25–44**	**45–64**	**>65**	**Sex**	**BOri**	**Edu**	**Emp**	**AInc**	**EcSt**
Rural	-0.47	-0.07	0.26	0.04	0.06	-0.05	-0.18	0.65	0.17	0.21	0.09	-0.15
Urban	0.67	-0.38	-0.47	[illegible]	-0.09	[illegible]	[illegible]	[illegible]	[illegible]	[illegible]	[illegible]	0.81
Center	0.65	-0.03	-0.22	-0.58	0.60	0.16	0.35	0.44	0.02	0.30	0.12	0.28

Figure 5: Pearson correlation coefficients between number of installed PV systems per household (top part), respective, number of installed ST systems per household (bottom part) and the different demographic independent factors for all 'rural', 'urban' and 'center' DeSOs with installed PV systems in the municipalities Falun, Knivsta and Uppvidinge.

The result presented in Fig. 4 is relatively consistent between the municipalities when it comes to the correlation between *#PV* and the age groups and birth origin, but inconsistent regarding the sex and education, with Uppvidinge found to have an opposite correlation compared to Falun and Knivsta for these factors. Strong and intermediate correlation is found for average economic standard in Knivsta and Uppvidinge, whilst only weak in Falun.

For the number of households with ST systems the results are quite consistent with that of PV in Knivsta and Falun. Standing out is the strong negative correlation between ST system penetration and high education and that employment have changed from negative to positive correlation. The *#ST* in Uppvidinge on the other hand show quite different correlation as compared to both the two other municipalities, and the situation for *#PV*. One reason could be that the statistical base for Uppvidinge is the weakest, with only 23 ST systems and 6 DeSOs as compared to 287 ST systems in Falun and 49 in Knivsta, and 37 respective 10 DeSOs.

In Fig. 5, the Pearson's correlation coefficient between *#PV* and *#ST*, respectively, and the independent demographic variables is presented for the three types of DeSOs. For PV, there is no consistency between almost all variables, as uniformity is only shown for the variables *25–44* years and birth origin. One trend that can be observed is that the correlation coefficients are more similar between 'urban' and 'center' area DeSOs, while 'rural' DeSOs show an opposite trend as compared to the two other types.

For *#ST*, a few consistencies are found for all the DeSOs, namely birth origin, education, employment and average income. Otherwise, it is hard to distinguish any clear trends. There are similarities between 'rural' and 'urban' areas for some independent variables and between 'urban' and 'center' for some other variables. Commonly, the strongest correlations are found for both PV and ST in the 'urban' DeSOs.

3.3 Comparison with the literature

The result of the demographic correlation analysis, which resulted in non-existing or weak correlation with PV and ST penetration and income on the aggregated DeSO level, are in line with several quantitative studies that have reported statistically weak or insignificant impact of the average income for larger spatial areas [11,14,16,19,32,33,38] or even a negative effect [13,21,34]. Furthermore, just as in this study, it has also previously been shown that average economical standard, or accumulated capital, can be a demographic factor that is of higher importance as compared to the average income on an aggregated area level [26].

An intermediate positive correlation between number of, both PV and ST systems per household, and age group *45–64* years was uncovered, just as previously reported in the municipality of Uppsala [39], which indicates that a population with a higher percentage of people of this age is more likely to install solar energy systems. The opposite seems to be the case for young and middle-aged adults, as the age groups of *16–24* and *25–44* both present a moderate to intermediate negative correlation to solar energy system penetration on a DeSO level. It should be noted that the average age in Knivsta is lower (36.9 years) compared to Falun (42.4 years) and Uppvidinge (43.5 years), which might skew the result in Fig. 3 to some extent. These findings are to a large extent in line with previous research, where i.e., [22,23] in the US and [17] in the Netherlands, recognized negative correlation for age groups between 20–45 and >65 on a zip code level. Also, in Australia [33], Germany [16], Greece [37], and the UK [32], higher share of middle-aged people in an area has been found to increase PV adoption. Regarding the age group >65 years, [45] concluded that retired people in the UK are less motivated to complement their heating system with a PV or ST system due to the high upfront cost and long pay-off time. In addition, [48] found that solar heating adoption propensity declined with the age of the household head in the early market in Germany. However, the literature is not consistent as [11,40] found that household age was less important than previously shown.

The variable *sex* demonstrated intermediate positive correlation with both installed PV systems (0.56 or 0.50) and age group *45–64* years (0.51) in this study. The latter indicates a higher percentage of males in this age group in the three municipalities investigated, while the former could be an indication that a population with higher percentage males are more likely to install PV systems, which was also found in [22] on zip code level in California. However, for Flanders, Belgium, [11] found weak evidence of higher PV adoption rates for males in general, but that single male households are significantly more likely to adopt PV as compared to single female households.

The correlation between the number of inhabitants born in Sweden and solar energy system adoption was found to be moderate positive for PV (0.32 or 0.36) and intermediate positive for ST (0.56) as illustrated in Fig. 3. However, the *BOri* variable shows strong positive correlation (0.75), hence collinearity by the implemented definition, with the independent variable employment. For the municipalities evaluated separately in Fig. 4, the result is stronger for PV (0.43–0.87) and with a consistent opposite correlation of employment. The consistent strength and opposite correlation with employment also holds for ST system penetration, with a notable outlier result in Uppvidinge, where both birth origin and employment have an intermediate negative correlation of -0.64 and -0.65, respectively. The consistency in the correlation result between these two independent variables is also visible in Fig. 5, where the DeSOs are evaluated separately, but not as apparent as in Fig. 4. In summary, the consistency between these two variables could indicate that it is not Sweden as birth origin of the inhabitants per se that increases likelihood of adoption of PV. It is rather the high employment that increases the likelihood of PV adoption, and that it is the collinearity between these two social factors in Sweden that is the reason for the result. This hypothesis is to some extent strengthened by the results of [22,23,26,28,40] in the US, where racial variables where generally not statistically significant, with only weak evidence of more adoption when there is a higher percentage of white people in an area in [23,26,28,40]. On the other hand, [11] states that an increase of the number of foreigners by one percentage unit point reduces the number of PV installations by 0.38% in Belgium for the time period 2006–2012.

Regarding *Emp*, in addition to *BOri*, collinearity was obtained with the two economic conditions *AInc* and *EcSt*, which is an expected result. The correlation to installed PV and ST was found to be weak for the municipalities combined, but stronger for PV (-0.16–0.56) when analysed separately, which exemplifies the different employment rates in the municipalities. In 2020, Falun and Uppvidinge had an employment rate of 81.6% and 82.6%, respectively, while in Knivsta it was 87.2%. Knivsta, with a positive correlation of 0.27 for employment rate, is the neighbouring municipality of Uppsala, where an earlier study [39] found a positive correlation for the same dependent and independent factors of 0.3. An evaluating of the result for the different type of DeSOs show stronger positive correlation in the 'urban' and 'center' areas, while the 'rural' DeSOs show a much weaker correlation between employment and both PV and ST adoption. Compared to earlier international studies, [26] did not find any significant correlation between employment rate and installed PV systems, while [16,27] identified a positive correlation. Thus, the weak positive correlation presented in this study can be assumed to be in line with previous research. However, a difference between the different types of DeSOs can be identified. Since the average employment rates are similar in the three DeSO types within each municipality, this difference indicates that the employment rate has a larger impact on adoption of PV systems in 'urban' and 'center' DeSOs.

The independent variable education was discovered to have a weak negative correlation to both installed PV and ST systems, and an expected strong positive correlation (demonstrating collinearity) to average income (0.76). When the municipalities were evaluated separately, high education level demonstrated negative correlation with both PV and ST deployment in Falun and Knivsta (for ST in Knivsta the negative correlation was even stronger), while positive correlation in Uppvidinge. The results for different types of DeSOs show a positive correlation for 'urban' and 'center' area DeSOs, indicating that a higher education level increases the likeliness of installing PV in these areas. The results presented by [23,27,40] showed a positive impact between higher education level in an aggregated area and PV adoption in the US, and the same type of findings have been presented for Canada [34], Denmark [36], Germany [16], the UK [19], and another municipality in Sweden [39], while [11] found a negative impact by a higher share of college degrees in Belgium. The investigated areas in [21], with a negative correlation between PV deployment and education, were found to have a general lower level of education, which was assumed to be an explanation of the result. The education correlation for 'rural' DeSOs was found to be -0.01, and in Falun and Knivsta the average education levels are in general lower in the 'rural' DeSOs as compared to, for example, the 'center' DeSOs, which, if following [21], might explain the result. The average education level is more even, and in general lower, in the different DeSO areas in the rural municipality of Uppvidinge, which also might explain the positive correlation for the municipality in contrast to the other two.

3.4 Solar energy owning household characteristics

Data on the average age and average time living at the residence of each household was available through Ratsit, and a summary is presented in Table V, along with average age of all residents in the municipalities. The statistical comparison shows a slightly higher average age of the households with PV as compared to the average age of households with ST, and that owners of both PV and ST are older than the average age of all inhabitants in each municipality. This agrees with the finding that households with an average age of >40 are overrepresented as PV owners in Denmark [36], but somewhat contradictory to a Swedish survey study where age was found to have a negative effect, though not highly significant [31]. The result in Table V strengthens our correlation result on an aggregated DeSO level, where the age group *45–64* was found to have an intermediate positive correlation.

Table V: Average residence time of households with PV or ST systems in years, along with the average age of inhabitants with an age >20 years in households with PV or ST systems and the average age of all inhabitants of >20 years in Falun, Knivsta and Uppvidinge municipalities.

	Res. time PV	Res. time ST	Age PV	Age ST	Age total
Falun	22.1	22.6	57.9	57.9	51.8
Knivsta	17.1	17.8	54.4	52.3	48.3
Upp.	23.7	28.6	58.9	55.4	53.3

In addition, the average time of living at the residence for households that have installed a PV system was found to be around 20 years, with slightly longer residence times for ST owning households.

The summary of households with solar energy systems includes information about the average sex of the inhabitants, in this study expressed in terms of percentage of males. The average sex of households with solar energy system was calculated for members >20 years, while the average sex for the municipality only was available for inhabitants of all ages. Hence, there is a slight mismatch in the comparison. The results are presented in Table VI.

Table VI: Average sex in terms of share of males for households with PV and ST systems and average sex of all residents in the municipalities for Falun, Knivsta and Uppvidinge.

	Average sex PV	Average sex ST	Average sex total
Falun	49.9%	54.8%	49.6%
Knivsta	51.1%	51.4%	51.0%
Uppvidinge	54.8%	55.1%	52.3%

The conclusion is that the percentage of males in the households with PV systems was just slightly higher than the average in all three municipalities, and that in turn percentage of males in the households with ST systems was even a little bit higher. This result is in line with a previous Swedish survey study about PV adoption [31], our findings and previous research on aggregated level, and with a study from Denmark, that revealed that registered owners of PV were more often men (85.8%) and that the decision of buying a PV system was substantially more often made by men [36]. It has also been concluded that men in Germany are more prone to invest in ST systems [48].

The average income of the households with a PV or ST system was compared to the average income of the separate DeSOs. The difference in average income in percent for each DeSO is presented in Fig. 6 for households with PV systems and in Fig. 7 for households with ST systems.

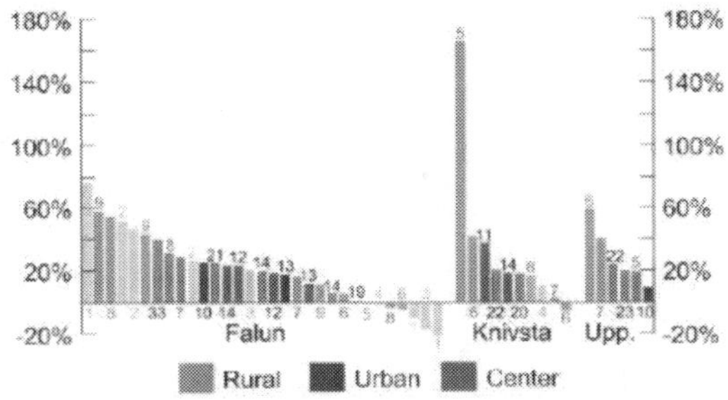

Figure 6: Difference in percent between the average income of households (average for all persons with an age of ≥20 years) with PV systems in a DeSO and the respective average in the same DeSO. The number above and below the bars represents the number of households with a PV system in the private residential set for each DeSO, and DeSOs with <10 respectively <5 households with PV are weakened in color to illustrate the less statistical strength of these results.

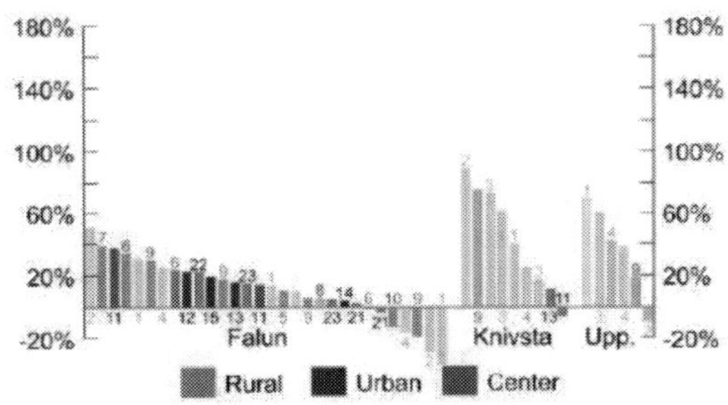

Figure 7: Difference in percent between the average income of households (average for all persons with an age of ≥20 years) with ST systems in a DeSO and the respective average in the same DeSO. The number above and below the bars represents the number of households with a ST system in the private residential set for each DeSO, and DeSOs with <10 respective <5 households with ST are weakened in color to illustrate the less statistical strength of these results.

The results show that a majority of the analyzed households were found to have a higher average income than the average income of all inhabitants in respective DeSO, and thereby confirm the importance of income as a factor for the distribution of PV, which has been recognized in many other countries [10,11,17,20,22,23,26–28,36,37,39,41,43] and in Germany for ST [46,47].

In Falun, the difference in income for all households with PV as compared to the average income in the municipality was 22%, in Knivsta 28% and in Uppvidinge 29%. Fig. 6 and Fig. 7 also show that the highest percentual differences were found in DeSOs of 'center' type for all three municipalities.

This observation is probably related to higher percental share of people living in apartments in the 'center' DeSOs, and that those that can afford a single-family house (and thereby can install a private residential PV system) in these attractive areas have a higher income than those that live in the apartments. The fact that relatively more households in the 'urban', and especially 'center' DeSOs likely live in apartments probably also influences the correlation analysis of the demographic parameters. This might explain some of the inconsistency between correlation coefficients in Fig. 5, between, above all, the 'rural' compared to the 'urban' and 'center' DeSOs.

A closer analysis of the PV owners with a lower average income than the average in the DeSO they belong to uncovered that in 53 % of these cases the average age of the owners was higher than the pension age of 65. A reasonable explanation could be that the pension income usually is lower in Sweden than the salaries of professional workers, but that the retired people living in a single-family house have had time to pay off all, or a big share, of their mortgage loans, as compared to younger people that bought the house more recently. Retired people with a solar energy system might thereby have a better economic situation than their income indicates. This hypothesis is strengthened by the long residence times for households with solar energy systems in Table V, and that the correlation coefficient for *#PV* and average economical standard is stronger than for *#PV* and average income in both Fig. 3, Fig. 4 and Fig. 5 (even if the correlation of these two factors is weak).

4 DISCUSSION

It should be made clear that in the compilation of the private residential systems set, some assumptions were made that lead to uncertainties in the result of this study. Firstly, the owner of each property was assumed to be responsible for the acquisition of the system, even though no insight in the actual financing of a solar energy system was available. The date of the acquisition of the property was either the registration date in the legal registration, or the change of address. Date of commission of PV systems was available, but not for ST systems, so there was no way of knowing if the current occupant household was the actual household that acquired the ST system. This leads to that we expect larger errors for ST systems in the data set. Some systems were justifiably removed, but those that remained was assumed to represent the owner and installer of the inventoried solar energy systems. So possible errors in the data set range from non-existing information about previous owner and lack of or imprecise data.

It should also be noted that the dataset of this study is substantially smaller compared to several of the studies the result is compared to, which can be several hundred thousand of systems [10,11,14–17,21,27,33]. However, our smaller statistical base is counterweighted by that information about the socioeconomic factors of income, age and sex could be presented on household level.

The first finding in this study was that the concentration of both domestic PV and ST systems per total number of households are higher in 'rural' areas as compared to the 'urban' or 'center' areas in the three municipalities. This conclusion is not unique, as higher penetration of PV in rural areas as compared to densely populated metropolitan areas has also been observed in several other countries, such as Australia [10], Belgium [11], Denmark [36], Germany [13], the Netherlands [17], the UK [18–21,32], US [23,26,27]. However, this is to our knowledge the first study which observes this pattern also for ST. This global pattern is at first sight somewhat incongruous, as factors with a reported positive role in adoption, such as high income and population density, generally are higher in urban and metropolitan areas. In this study, the monthly average income of all 'rural' DeSOs was 27,920 SEK, as compared to 29,459 SEK in the 'center' DeSOs. The general explanation for this contradiction in both this and other studies [10,11,13,19,26,33] is that even if there is a concurrence of high income and population density in cities, many more people live in apartment houses, and hence find it hard or even impossible to install private residential solar energy systems. Another explanation could be that a house in the city is relatively more expensive to buy compared to the income difference, which means that the purchasing power for PV or ST decreases.

In the correlation analysis, a moderate positive correlation (0.36) was found between PV penetration and ST penetration, which increased to an intermediate positive correlation (0.61) when two outlier demographic areas were omitted. This finding might be the most significant result of this study, as interaction between ST and PV system distribution has not been investigated thoroughly so far. This finding strengthens previous documentation that in addition to knowledge about a specific technology, such as PV, familiarity of a similar technology, in this study ST, facilitates higher PV adoption [56] and indicative evidence of passive spill over peer effects between PV and ST in Sweden [29].

Furthermore, it should be noted that relatively many households have installed both ST and PV system, as Table IV illustrates. With the assumption that the ST systems was installed first, which is based on the fact that the deployment of ST in Sweden in general took place a decade prior to the PV roll-out, the percentage of households with a ST system that later also have installed a PV system is 6.7% for all municipalities (5.6% in Falun, 14.3% in Knivsta and 4.3% in Uppvidinge). This could be seen as some kind of super household internal spill over peer effect between PV and ST.

In general, when our complete set of domestic PV and ST systems was divided into municipalities or 'rural', 'urban' or 'center' demographic areas, it was shown that the independent variables' effect on PV and ST deployment varied, and some substantial differences were observed. This finding is in line with [28] that found that socioeconomic profile of PV adopters in one town can be quite different from the profile in a neighbouring town and the overall profile. In addition, it confirms that socioeconomic factors can be contextual within regions and countries [10,15,21,26,28,40,49]. The statistical correlation results, the results on a household level and differences between different demographic areas in this study confirm the general image put forward by [36] that early adopters of PV tend to be men living in rural areas with a higher income and education level than others.

5. CONCLUSION

The spatial adoption of domestic PV systems has in numerous studies been proven to be influenced by socioeconomic factors. Our general conclusion, with regards to economic factors, is that the average economic situation in a Swedish demographic area seem to have a small impact on the deployment of domestic PV and ST systems in that area, but that a strong economic situation of a specific household increases the probability of PV or ST investments.

Furthermore, in line with previous studies in Sweden along with several in other countries, the result in this study shows that males, employed households and people with higher education are more prone to install PV systems. The results are basically the same for the likelihood of ST adoption, which so far only occasionally has been addressed by the scientific literature. Regarding age as a factor, the cautious conclusion is that people of age 45–64 seem to be more likely to install solar energy systems than people belonging to both younger and older age groups in Sweden.

This study is also the first of its kind to show that there seems to exist a statistical spatial correlation between solar PV and ST adoption, which can indicate that familiarity of the ST technology in an area can have a passive spill over peer effect on PV adoption.

6. REFERENCES

[1] Energimyndigheten, Nätanslutna solcellsanläggningar, (2021).

[2] J. Lindahl, A. Oller Westerberg, K. Vanky, J. Berard, National Survey Report of PV Power Applications in Sweden 2020, (2021).

[3] B. Zakeri, V. Virasjoki, S. Syri, D. Connolly, B. V. Mathiesen, M. Welsch, Impact of Germany's energy transition on the Nordic power market – A market-based multi-region energy system model, Energy 115 (2016).

[4] J. Egerer, F. Kunz, C. von Hirschhausen, Development scenarios for the North and Baltic Seas Grid — A welfare economic analysis, Utilities Policy 27 (2013) 123–134.

[5] Boverket, Utvärdering av solvärmebidraget och solvärmestödet, (2012).

[6] Energimyndigheten, Solvärme i Sverige — En studie med fokus på potential, ekonomi och bidrag till energi- och klimatmål, (2021).

[7] Y. Louvet, S. Fischer, S. Furbo, F. Giovannetti, S. Helbig, M. Köhl, D. Mugnier, D. Philippen, F. Veynandt, K. Vajen, Economic comparison of reference solar thermal systems for households in five European countries, Solar Energy 193 (2019) 85–94.

[8] C. Wilson, A. Grubler, N. Bento, S. Healey, S. de Stercke, C. Zimm, Granular technologies to accelerate decarbonization, Science 368 (2020) 36–39.

[9] H. Xiong, D. Payne, S. Kinsella, Peer effects in the diffusion of innovations: Theory and simulation, Journal of Behavioral and Experimental Economics 63 (2016) 1–13.

[10] H. Lan, Z. Gou, Y. Lu, Machine learning approach to understand regional disparity of residential solar adoption in Australia, Renewable and Sustainable Energy Reviews 136 (2021).

[11] O. de Groote, G. Pepermans, F. Verboven, Heterogeneity in the adoption of photovoltaic systems in Flanders, Energy Economics 59 (2016) 45–57.

[12] T. Zhao, Z. Zhou, Y. Zhang, P. Ling, Y. Tian, Spatio-Temporal Analysis and Forecasting of Distributed PV Systems Diffusion: A Case Study of Shanghai Using a Data-Driven Approach, IEEE Access 5 (2017) 5135–5148.

[13] S. Müller, J. Rode, The adoption of photovoltaic systems in Wiesbaden, Germany, Economics of Innovation and New Technology 22 (2013) 519–535.

[14] A.J. Schaffer, S. Brun, Beyond the sun — Socioeconomic drivers of the adoption of small-scale photovoltaic installations in Germany, Energy Research and Social Science 10 (2015) 220–227.

[15] J. Rode, A. Weber, Does localized imitation drive

technology adoption? A case study on rooftop photovoltaic systems in Germany, Journal of Environ-mental Economics and Management 78 (2016) 38–48.
[16] S. Dharshing, Household dynamics of technology adoption: A spatial econometric analysis of residential solar photovoltaic (PV) systems in Germany, Energy Research and Social Science 23 (2017) 113–124.
[17] R. Bernards, J. Morren, H. Slootweg, Development and Implementation of Statistical Models for Estimating Diversified Adoption of Energy Transition Technologies, IEEE Trans Sustain Energy 9 (2018) 1540–1554.
[18] R. Snape, C. Rynikiewicz, Peer effect and social learning in micro-generation adoption and urban smarter grids development?, Network Industries Quarterly 14 (2012) 24–27.
[19] N. Balta-Ozkan, J. Yildirim, P.M. Connor, Regional distribution of photovoltaic deployment in the UK and its determinants: A spatial econometric approach, Energy Economics 51 (2015) 417–429.
[20] J.R. Snape, Spatial and temporal characteristics of PV adoption in the UK and their implications for the smart grid, Energies 9 (2016) 1–18.
[21] N. Balta-Ozkan, J. Yildirim, P.M. Connor, I. Truckell, P. Hart, Energy transition at local level: Analyzing the role of peer effects and socio-economic factors on UK solar photovoltaic deployment, Energy Policy 148 (2021).
[22] B. Bollinger, K. Gillingham, Peer Effects in the Diffusion of Solar Photovoltaic Panels, Marketing Science 31 (2012) 900–912.
[23] C.L. Kwan, Influence of local environmental, social, economic and political variables on the spatial distribution of residential solar PV arrays across the United States, Energy Policy 47 (2012) 332–344.
[24] J. Letchford, K. Lakkaraju, Y. Vorobeychik, Individual household modeling of photovoltaic adoption, AAAI Fall Symposium - Technical Report (2014) 19–23.
[25] S.A. Robinson, V. Rai, Determinants of spatio-temporal patterns of energy technology adoption: An agent-based modeling approach, Applied Energy 151 (2015) 273–284.
[26] M. Graziano, K. Gillingham, Spatial patterns of solar photovoltaic system adoption: The influence of neighbors and the built environment, Journal of Economic Geography 15 (2015) 815–839.
[27] B. Lukanov, E. Krieger, Distributed solar and environmental justice: Exploring the demographic and socio-economic trends of residential PV adoption in California, Energy Polic. 134 (2019) 110935.
[28] M. Graziano, M. Fiaschetti, C. Atkinson-Palombo, Peer effects in the adoption of solar energy technologies in the United States: An urban case study, Energy Research and Social Science 48 (2019) 75–84.
[29] A. Palm, Local factors driving the diffusion of solar photovoltaics in Sweden: A case study of five municipalities in an early market, Energy Research and Social Science 14 (2016) 1–12.
[30] A. Palm, Peer effects in residential solar photovoltaics adoption — A mixed methods study of Swedish users, Energy Research and Social Science 26 (2017) 1–10.
[31] L. Mundaca, M. Samahita, What drives home solar PV uptake? Subsidies, peer effects and visibility in Sweden, Energy Research and Social Science 60 (2020).
[32] L.-L. Richter, Social Effects in the Diffusion of Solar Photovoltaic Technology in the UK, Cambridge Working Papers in Economics (2013).
[33] J. Sommerfeld, L. Buys, K. Mengersen, D. Vine, Influence of demographic variables on uptake of domestic solar photovoltaic technology, Renewable and Sustainable Energy Reviews 67 (2017) 315–323.
[34] T. Islam, N. Meade, The impact of attribute preferences on adoption timing: The case of photo-voltaic (PV) solar cells for household electricity generation, Energy Policy 55 (2013) 521–530.
[35] X. Yuan, J. Zuo, C. Ma, Social acceptance of solar energy technologies in China—End users' perspective, Energy Policy 39 (2011) 1031–1036.
[36] A.R. Hansen, M.H. Jacobsen, K. Gram-Hanssen, Characterizing the Danish energy prosumer: Who buys solar PV systems and why do they buy them?, Ecological Economics 193 (2022) 107333.
[37] E. Sardianou, P. Genoudi, Which factors affect the willingness of consumers to adopt renewable energies?, Renewable Energy 57 (2013) 1–4.
[38] Y. Zhang, J. Song, S. Hamori, Impact of subsidy policies on diffusion of photovoltaic power generation, Energy Policy 39 (2011) 1958–1964.
[39] U.H. Ramadhani, B. Tsegai, C. Apelryd, S. Ekbring, J. Munkhammar, J. Widén, Estimating the Socio-economic drivers of residential photovoltaic systems adoption in a Swedish city, 11th Solar & Storage Power System Integration workshop, 2021: pp. 1–6.
[40] C. Davidson, E. Drury, A. Lopez, R. Elmore, R. Margolis, Modeling photovoltaic diffusion: An analysis of geospatial datasets, Environmental Research Letters 9 (2014).
[41] C.A. Bollino, The willingness to pay for renewable energy sources: The case of Italy with socio-demographic determinants, Energy Journal 30 (2009) 81–96.
[42] G. Martinopoulos, G. Tsalikis, Diffusion and adoption of solar energy conversion systems – The case of Greece, Energy 144 (2018) 800–807.
[43] S.A. Robinson, V. Rai, Determinants of spatio-temporal patterns of energy technology adoption: An agent-based modeling approach, Applied Energy 151 (2015) 273–284.
[44] G. Guagnano, G.R. Hawkes, C. Acredolo, N. White, Innovation perception and adoption of solar heating technology, Journal of Consumer Affairs 20 (1986) 48–64.
[45] K. Willis, R. Scarpa, R. Gilroy, N. Hamza, Renewable energy adoption in an ageing population: Heterogeneity in preferences for micro-generation technology adoption, Energy Policy 39 (2011) 6021–6029.
[46] B. Mills, J. Schleich, Profits or preferences? Assessing the adoption of residential solar thermal technologies, Energy Policy 37 (2009) 4145–4154.
[47] J.S. Woersdorfer, W. Kaus, Will nonowners follow pioneer consumers in the adoption of solar thermal systems? Empirical evidence for northwestern Germany, Ecological Economics 70 (2011) 2282–2291.
[48] H. Welsch, J. Kühling, Determinants of pro-

environmental consumption: The role of reference groups and routine behavior, Ecological Economics 69 (2009) 166–176.
[49] R. Ghasemi, Y. Li, Z. Lu, J.C. Huang, W. Mo, Spatial household preferences of decentralized solar photovoltaic and thermal systems, Resources, Conservation and Recycling 185 (2022) 106487.
[50] A. Sanguinetti, S. Outcault, E. Alston-Stepnitz, M. Moezzi, A. Ingle, Residential solar water heating: California adopters and their experiences, Renewable Energy 170 (2021) 1081–1095.
[51] A. Palm, Early adopters and their motives: Differences between earlier and later adopters of residential solar photovoltaics, Renewable and Sustainable Energy Reviews 133 (2020) 110142.
[52] J. Lindahl, D. Lingfors, Å. Elmqvist, I. Mignon, Economic analysis of the early market of centralized photovoltaic parks in Sweden, Renewable Energy 185 (2022) 1192–1208.
[53] M. Warneryd, K. Karltorp, The role of values for niche expansion: The case of solar photovoltaics on large buildings in Sweden, Energy, Sustainability and Society 10 (2020).
[54] J. Palm, M. Tengvard, Motives for and barriers to household adoption of small-scale production of electricity: Examples from Sweden, Sustainability: Science, Practice, and Policy 7 (2011) 6–15.
[55] A. Palm, B. Lantz, Information dissemination and residential solar PV adoption rates: The effect of an information campaign in Sweden, Energy Policy 142 (2020).
[56] A. Kaplan, From passive to active about solar electricity: Innovation decision process and photovoltaic interest generation, Technovation 19 (1999) 467–481.
[57] Falu municipality, Falun Kommunfakta, (2022).
[58] Statistics Sweden, Average age of the population by region and sex. Year 1998–2021, (2022).
[59] Uppvidinge municipality, Uppvidinge — Kommunfakta, (2022).
[60] Statistics Sweden, Population per region by region of birth and sex. Year 2010–2021, (2022).
[61] J. Yu, Z. Wang, A. Majumdar, R. Rajagopal, DeepSolar: A Machine Learning Framework to Efficiently Construct a Solar Deployment Database in the United States, Joule 2 (2018) 2605–2617.
[62] K. Mayer, Z. Wang, M.L. Arlt, D. Neumann, R. Rajagopal, DeepSolar for Germany: A deep learning framework for PV system mapping from aerial imagery, in: Proceedings of the 3rd International Conference on Smart Energy Systems and Technologies (SEST), 2020: pp. 11–16.
[63] Statistics Sweden, Demografiska statistik-områden, en ny regional indelning under kommuner (2022).
[64] N. Nurwidiana, B.M. Sopha, A. Widyaparaga, Modelling photovoltaic system adoption for households: A systematic literature review, Evergreen 8 (2021) 69–81.
[65] C.F. Dormann, J. Elith, S. Bacher, C. Buchmann, G. Carl, G. Carré, J.R.G. Marquéz, B. Gruber, B. Lafourcade, P.J. Leitão, T. Münkemüller, C. Mcclean, P.E. Osborne, B. Reineking, B. Schröder, A.K. Skidmore, D. Zurell, S. Lautenbach, Collinearity: A review of methods to deal with it and a simulation study evaluating their performance, Ecography 36 (2013) 27–46.

This is a short version of the paper submitted to the journal EPJ Photovoltaics.

It has been selected by the Committee of the WCPEC-8 for submission to EPJ Photovoltaics.

INSIGHTS INTO CIRCULAR MATERIAL AND WASTE FLOWS FROM C-SI PV INDUSTRY

Peter Brailovsky*[1], Kerstin Baumann[2], Michael Held[3], Ann-Kathrin Briem3, Karsten Wambach[2], Estelle Gervais[1], Sina Herceg[1], Boris Mertvoy[2], Sebastian Nold[1], Jochen Rentsch[1]

[1] Fraunhofer Institute for Solar Energy Systems ISE, Heidenhofstr. 2, 79110 Freiburg i. Br., Germany
[2] bifa Umweltinstitut GmbH, Mittleren Moos 46, 86167 Augsburg, Germany
[3] Fraunhofer Institute for Building Physics IBP, Nobelstraße 12,70569 Stuttgart, Germany

*Corresponding author, e-mail to: peter.brailovsky@ise.fraunhofer.de

ABSTRACT: A material flow model for the production of Bifacial Selective Emitter 60-cell p-type Cz PERC (Passivated Emitter and Rear Contacted) glass-backsheet modules with aluminium frame was built. The selected module represents mature technologies in the PV industry and their manufacturing is considered to take place in China in a production cluster with an annual module capacity of 5 GWp. In a first step, data acquisition and validation for wafer, cell, and module fabs took place. The data were used to generate the reference system lifecycle inventories (LCI) and extended waste databases for the reference wafers, cells, and modules. A set of potential circularity actions, such as the vertical integration of the operations and waste revalorisation strategies, had been proposed and their environmental performance and cost assessed by means of a life cycle assessment (LCA) and a total cost of ownership (TCO). Our results show that 87% of the waste can be reduced and revalorised, this represents a circular flow of raw materials of 18,756 Mg per year from a 5 GWp PV module production cluster. Environmental impact reductions of 0.6 to 2.3% are estimated for different impact categories. We also estimate a cost reduction potential of 2.59% from total module costs.

Keywords: photovoltaics/ revalorized waste/ circular production/ life cycle assessment/ total cost of ownership

1 INTRODUCTION

1.1 Motivation

Photovoltaics (PV) have emerged as the backbone of the global energy transition. An expansion from the current 707 GWp PV capacity up to 63 TWp by 2050 is deemed to be both needed to comply with a 1.5°C scenario and viable from a technical and economic perspective [1]. This massive deployment and current developments on legislation of industrial sites requires from PV manufacturers to achieve exemplary sustainability. Resource consumption patterns are at the core of said sustainability, as they can affect the environmental, social, and economic performances of a product [2,3]. Hence, there is a strong need to make today's production systems more efficient. Waste flows must further be minimized and converted into streams of valuable materials, for not to be lost and to end up on landfills.

Circularity strategies in PV manufacturing are essential as they can considerably improve the sustainability of PV modules, already in the short term. Assessing the potential of such strategies requires a detailed analysis of material flows (MFA) to understand each process step, resources requirements, material losses and waste handling. Which this paper proposes to lay out through bottom-up modelling.

The generation of a broad bill of materials to derive life cycle inventories and extended waste database allows the assessment of the system circularity from many perspectives.

1.2 Methods

First, current material and solid waste flows (MFA model) for a reference production chain are identified and quantified for the production of monocrystalline silicon (c-Si) ingots and wafers, passivated emitter rear contact (PERC) solar cells and PV modules for an annual capacity of 5 GWp located in China.

By generating the MFA, potential optimizations, e.g., vertical integration, waste revalorisation, as well as reduction of wafer breakage, material loss, packaging, energy and transportation can be identified. Second, the circularity strategies cost and environmental performances are assessed by means of a total cost of ownership (TCO) and a life cycle assessment (LCA). Overall, the discussion over the advantages, disadvantages and challenges associated with each concrete optimization strategy can be of interest for researchers and industrials alike in progressing towards circular PV factories.

1.2.1 MFA and TCO

The MFA and TCO models are bottom-up calculated with the following hierarchy levels:

- Level 1: Individual equipment.
- Level 2: Production line and supply infrastructure.
- Level 3: Total factory including building requirements.

The MFA and TCO are calculated with Fraunhofer ISE technology and cost assessment model Scost [4] which follows mainly the standards SEMI E35 [5] and E10 [6] for cost of ownership and equipment utilization calculation, respectively. The costs for Process Consumables, Utilities and Waste Disposal are considered separately, instead of the aggregated Consumables cost component in SEMI E35. The Process Consumables include raw materials for the production processes. The Utilities cost considers electricity, process exhaust air, process cooling water and compressed dry air. The Waste Disposal category includes the costs of gas scrubbers and solid waste and wastewater disposal and treatment. The category Maintenance Parts includes costs of parts repair and replacement. Labour costs for maintenance and operation are included in the Labour cost category.

1.2.2 LCA

To give an estimate on the reduction potential of environmental impact of the PV module production and proposed reduction measures of production wastes, an

LCA was implemented following the ISO14040 [7] and 14044 [8] standards. The LCA is based on the described MFA model and implemented using the GaBi software and its databases (version CUP 2022_1) [9]. The Life Cycle Impact Assessment (LCIA) is based on the suggested impact method set of the EU Environmental Footprint, Version 3.0 (EF3.0) [10].

The defined functional unit and reference flow for the LCA is the production of 1 m² PV module with respective technical specifications summarized in Table I. The system boundary is cradle-to-gate, considering the impacts along the entire value chain until the PV module factory gate.

The considered reference location of the PV module production is China, which is represented in the LCA model by using country specific energy generation mixes, materials, and process auxiliaries where available. In case that no region-specific dataset is available in the GaBi database content, datasets with deviating regional references were used as a proxy (e.g., datasets referring to EU-28 productions). This is mainly the case for the chemical use in the different process steps of the PV module production.

To enable an estimation of the reduction potential of environmental impacts resulting from the analysed strategies for waste reduction and waste revalorization, an additional LCA model was created that allows the analysis of specific waste streams and waste treatment options. Finally, the LCIA results of this waste specific model are put into relation to the reference module production.

2 INVESTIGATED PV PRODUCTION CHAIN AND WASTE MANAGEMENT

2.1 Investigated PV production chain

The full production chain of the Bifacial Selective Emitter 60-cell p-type Cz PERC glass and backsheet modules with aluminium frame reference product is split into three fabs, the product specifications are shown in Table I.

Table I: Products specifications per factory

Wafers		
Crystallization Type		**Cz**
Base Doping		p-Type
Thickness	µm	170
Wafer Edge Length	mm	156.75
Wafer Mass	**g**	**9.70**
Cells		
Ø Cell Power (Pmpp)	**5.37**	**Wp**
Cell Mass	**g**	**9.19**
Modules		
Cell per Module	**60**	**Cells**
Ø Module Power (Pmpp)	**316.80**	**Wp**
Module Area	**m²**	**1.598**
Module Mass	**kg**	**18.72**

2.1.1 Ingot and wafer fab

The main flow is the one of polysilicon input to the crystallization process and its subsequent processing through the cutting and cleaning steps from ingots to rods, from rods to bricks and from bricks to wafers. The polysilicon input is composed of a virgin material feed accounting for 56% of the total flow and the polysilicon reclaim feed providing 44% of the input flow. The reclaimed polysilicon is integrated by rods sidewalls slabs, misprocessed workpieces, ingot tails and tops. In order to reclaim and reuse these workpieces for the polysilicon feed for the crystallization process, the pieces are crushed and etched in chemical baths. About 69% of the virgin polysilicon feed is transformed into wafers, the rest 31% ends up as silicon kerf loss in the industrial wastewater. With press filtration it is possible to recover 3,890 Mg of solids per year for the considered module annual production capacity of 5 GWp. From the silicon crusher we estimate a recovery potential of 237 Mg per year. The second most relevant material flow is the one of the quartz crucibles, which are used for the crystallization process; after three cycles of crystallization, they are removed with the residual silicon pot scrap from the pullers and disposed of. The silicon pot scrap and wasted crucibles sum up to 1,737 Mg per year.

2.1.2 Cell fab

In the cell factory the as-cut wafers are further processed in chemical baths to remove the sawing damages and to texturize their surface to increase the absorption of sunlight. After the diffusion process of the phosphor emitter, a phosphosilicate glass layer (PSG) is formed around the wafers, by means of chemical baths. The emitter layer is removed from one side of the wafers and the PSG is removed from both sides of the wafers. The texturized and dopped wafers are thermally oxidized and then rear passivated with two layers, one of 15 nm of aluminium oxide (Al_2O_3) and a second layer of 70 nm of silicon nitride (Si_3N_4). Afterwards, the Anti-Reflective coating is built on the front side with one layer of 85 nm of Si_3N_4. The contact points to the bulk structure are opened by lasers and then the metallic contacts are screen printed at the back and front sides, 17 g of metallic pastes are used per square meter of cells. For the annual 5 GWp module production capacity of PV modules, 193 Mg of non-metallized cell scrap and 186 Mg of metallized cell scrap are generated yearly.

2.1.3 Module fab

The main raw materials weight shares for the production of the reference PV module are 68.6% for the glass layer, 14.3% for the aluminium frame, 6.8% for the encapsulant (EVA) and 4.3% for the backsheet. Solar cells only have a weight share of 3.2% of the total raw materials required to produce a PV module. Module packaging accounts for almost 17% of the weight of packaged PV modules. From the waste perspective we estimate a generation of 0.48 kg of solid waste per square meter of produced PV module, this is mainly composed of wood, glass, and plastic waste. In terms of the annual 5 GWp module production capacity -eq. to 25.500.741 m² of PV modules- the generated mix waste in the module factory sums up to 12,268 Mg per year.

From an overall perspective, it can be stated that the main material flows occurring along the value chain from polysilicon to PV modules are the ones of glass, aluminium, module packaging materials, encapsulant and backsheet. From the waste generation perspective, we estimate a total solid mix waste flow of 21,348 Mg per year. The dominant waste partitions are composed of wood, silicon kerf, glass, and plastics. Strategies into reducing the material intensity of the industry are discussed in the section 3 of this paper.

2.1.4 Reference PV module cost

We estimate a TCO of 88.6 €/module. In terms of square meter of PV module, the TCO is 55.45 € and in terms of power units 0.28 €/Wp. This cost is just 20% higher than recent average prices shown in the PV market [11]. Our cost estimation seems plausible as we calculate our reference for the older wafer format M2.

In **Fehler! Verweisquelle konnte nicht gefunden werden.** we present the TCO distribution per cost categories. It is possible to appreciate that raw materials costs have a share of almost 70% of the total costs per PV module —polysilicon alone drives close to 30% of the TCO. These facts make very clear that scrap generation, wafer and cell breakage rates should be under control along the value chain. The category "Waste disposal" is composed of wastewater and exhaust treatment costs and solid waste disposal with a mix waste strategy -assumed as reference-. The "Yield loss" category entails the associated losses of raw materials, wafer, and cells breakage along the production chain, it is calculated for an efficient operation setting.

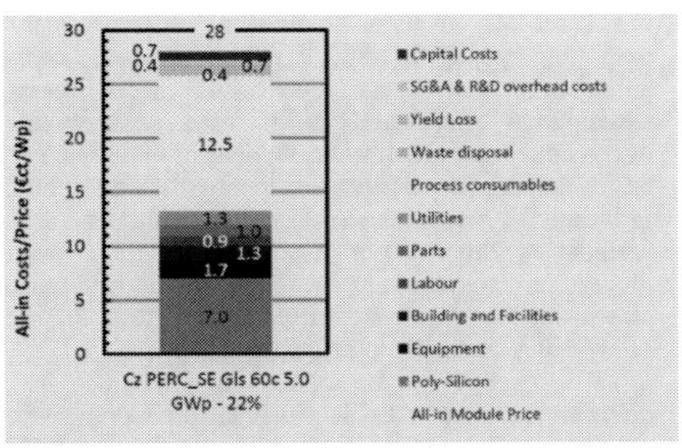

Figure 1: PV module cost per cost categories (reference)

2.2 Waste management

Based on the data of the material flow model, an extended waste database was created. The purpose of this database was to collect information about the individual solid wastes and to identify the possible hotspots for circular economy.

2.2.1 Legal framework

In addition to concrete requirements from specific waste legislation, depending on the production site, requirements from neighbouring regulations, e.g., construction, occupational health and safety, etc., have to be included in the planning of a production plant with regard to the management of waste and by-products generated. Within the scope of the present project, Germany was considered as a representative example of a European location. Since the reference model refers to the location China, the corresponding legislation was also taken into account as far as identifiable.

The most relevant European regulations are the WEEE [12], Waste Framework Directive [13] and the CLP Regulation [14], which must be transposed into national law by the member states. These are authoritative for local operations.

According to [15] the waste management of China is inspired by the organization of waste management in Europe. Ministries provide the framework (programs, laws, regulations, etc.), while the provincial and municipal governments are responsible for implementation. Nationally, the Ministry of Environmental Protection and the Department of Soil Environmental Management are responsible for electronic products and waste. Relevant laws include Environmental Protection Law, Regulations for the Collection, Transportation and Recycling of Municipal Waste, Law for the Promotion of Clean Production of the PRC, Law for the Promotion of Circular Economy and currently the 13th Five-Year Plan: Municipal Plan for the Safe Management of Municipal Waste. For the management of waste electrical and electronic equipment, a Guideline of Waste Electrical and Electronical Products Standardization Dismantling Operations and Product Management exists [15].

2.2.2 Approach to database

The database created is based on a waste register for commercial enterprises, which was expanded to include project-specific aspects. Identified wastes were grouped into waste categories (paper, metals, glass, ...). Subsequently, the wastes were described by composition (waste description) and extended by information on the waste-generating link within the production chain (waste origin). In addition, the waste register contains information on recycling and circularity. Another section provides information on hazardous and/or critical materials or conflict minerals. Beyond these aspects, the database can be extended, for example, by plant-specific key figures.

According to [15], collection and plant structure would show significant differences to the situation in Europe. The scope of documentation and definitions also differ. For example, according to [15], disposal in secured landfills is considered as treatment of waste and the main part of collected waste is seen as to be landfilled. Recyclable materials are generally collected and recycled by the so-called informal sector.

For the reference analysis, it was defined that the remaining waste from production is handed over to an approved disposal company and deposited in a secure landfill.

In order to estimate the potential, the waste volumes were divided into the disposal options "recycling", "energy recovery" and "disposal" on the basis of waste-specific quotas. The quotas are largely based on the distribution of material flows in Europe [16]. For specific wastes, in particular wastes containing silicon, the distribution was estimated based on the experience of the project participants.

3 POTENTIAL IMPROVEMENTS IN TERMS OF CIRCULARITY

3.1 Vertical integration

The occurring waste flows, including packaging and residues, were classified by type and their value assessed to suggest their treatment in terms of material and energy recovery potential.

We use the term vertical integration for stating that the manufacturing of multiple products in the production chain takes place at the same site. The boundary of our analysis starts with the reception and storage of polysilicon, glass, metallic pastes, diamond wire, polyurethane beams, quartz crucibles, backsheets, ribbons, solder, chemicals, and gases and end up with the production of the reference PV modules.

By concentrating the individual fabrications at one location, the packaging demands of the intermediate products are lower compared with external transports. The reduced number of reloading operations also reduce the risk of breakage. Therefore, the losses are minimized and fewer silicon ingots, wafers, etc., need to be produced. Furthermore, less paste is needed and there is less effluent and waste to be treated. Vertical integration can achieve a 11% reduction in the amount of waste, which accounts for 2,205 Mg per year. The demand for raw silicon is reduced by 75 Mg per year for the 5 GWp PV module cluster.

From the cost perspective, our results show that the Vertical Integration of the factories enables a total cost reduction per module of almost 1%.

3.2 Revalorized waste

Some outputs or waste fractions have a certain commercial value which should be realized. Options for recycling of kerf, quartz, graphite, defective or broken cells, polymers, glass, metal, paper, plastic, and wood are available in varying quantities and qualities depending on location of the plant. For example, Si-kerf can be reused for the production of Si ingots after processing and purification. The Si-kerf can also be used for the production of silicon nitride (Si_4N_3) crucibles or SiC. A recovery of solar cells requires a complete separation of the individual layer metallization layers. Thermoplastic polymers can be melted down, cleaned and reintroduced to the market as high-quality recyclates. For the nowadays frequently used thermoset materials no relevant recycling route (e.g., EVA) has been established yet. The glass is mainly used for the production of glass fibres or foam glass.

The metals can be automatically separated from the mixture with the cullet by metal separators. Aluminium from the frames, for example, can be reused by melting it down. Wastepaper is used to a very large extent in the manufacturing of new goods. Waste wood can be used in the furniture industry but is predominantly processed into fuel.

To estimate the potential of a consequent solid waste separation in different fractions and an available treatment and recycling plant structure like in Europe is assumed.

The occurring waste flows, including packaging and residues, were classified by type and their value assessed to suggest their treatment in terms of material recovery, energetic recovery, and waste disposal. The results also consider the implementation of a reusable wood pallet system between factories, as shown in Figure 2.

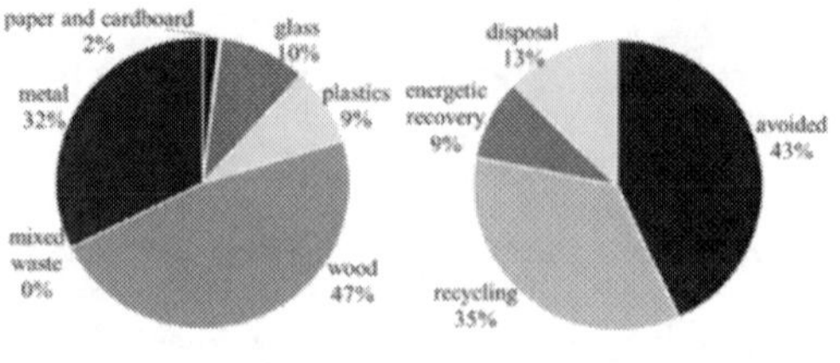

Figure 2: Poly-Si to PV modules waste flows (left side) with recycling options (right side) using reusable pallets

Out of a total cost reduction per module of 1.65%, 69% is gained due to the establishment of reusable pallet system -most of it for the glass and module transportation, 18% is contributed by the silicon waste recycling for use as ferrosilicon.

The amount of waste to be disposed of could be reduced by approx. 43% compared to the reference scenario, which represents 9,112 Mg per year. By allowing a plant for recycling and energy recovery and using reusable pallets instead of disposable pallets, 35% of the generated reference waste could be considered as recycled, 9% as energy recovered. The rest would have to be disposed of in a landfill.

3.3 LCA implementation

3.3.1 Assumptions and modelling choices for waste treatment options

Since no processes are available to represent the specific characteristics of waste treatment options for PV materials, such as silicon wastes from kerf loss, solar cells, or filtered metal dust, assumptions had to be made for the recycling, thermal recovery, and disposal of waste materials streams by using existing end-of-life treatment datasets of the GaBi database content. Due to limited availability of China (CN) specific datasets, waste treatment processes referring to the EU-28 region were chosen as a proxy. These datasets allow appropriate estimates by providing average data for the end-of-life treatments of conventional industrial materials, such as pulp and paper products, wood, plastics, copper, aluminium, and ferrous metals. As these necessary assumptions are associated with corresponding uncertainties, not all impact categories of the EF3.0 could be evaluated, especially for those that are strongly influenced by very specific process emissions or the regional conditions. For a sound evaluation, precise process and emission data are required to allow a reliable classification of the environmental impacts, e.g., in impact categories related to eco- or human toxicity.

The baseline for the evaluation, as defined in the reference scenario, considers that all production wastes streams are disposed of on landfills. For the revalorized waste scenarios, potential environmental impacts caused by landfilling and thermal treatment of wastes in waste incineration plants are taken into account. Due to the fact that some recycling options, especially for Si-wastes, are based on rough assumptions, we chose a cut-off approach for all recycled materials and energy recovery processes.

The vertical integration measures are evaluated with and without the revalorized waste scenario options, resulting in four separate LCA scenarios.

3.3.2 LCA results

Figure 3 presents the environmental impact reduction potentials of the investigated waste reduction strategies in relation to the reference PV module production on module level (represented by the 0% line).

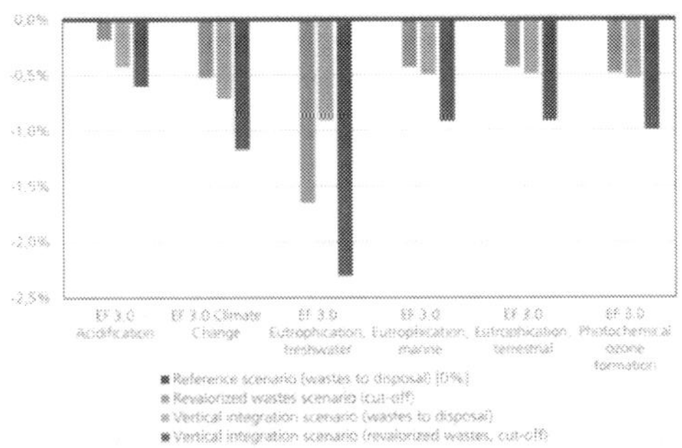

Figure 3: Environmental impact reduction potentials of investigated solid waste reduction and treatment strategies in relation to the reference PV module production (module level)

The vertically integrated production reflects savings related to the avoided production of packaging materials, such as paper and plastics as well as slightly reduced losses to wafer and cell breakage during transportation. Reductions on the production side of the revalorized waste include the change from one-way pallets to reusable wood pallets.

The environmental effect of avoiding use of packaging materials in a vertically integrated production leads to higher reductions in most of the investigated impact categories compared to the revalorized waste scenario, when potential environmental benefits for material recycling and thermal recovery are not taken into account (due to the cut-off approach). The revalorized waste scenario shows a higher reduction potential in the freshwater eutrophication potential, mainly resulting from the reduction of plastic wastes put on landfills.

Since the waste reduction strategies of the vertical integration and the revalorized wastes scenarios can be combined, the highest savings of environmental impacts are reached with this scenario in all investigated impact categories.

4 CONCLUSIONS

It can be stated that, compared to the reference scenario, the quantities of solid waste can be significantly reduced -by 48%-. Due to this fact, a reduction of the use of critical raw material flows is achieved. Most of the remaining solid waste is recyclable, which increases the industry circularity. Technical feasibility is a challenge to address, as more storage space at the production site and suitable recycling or partner installations are needed.

From the cost perspective, we estimate a TCO reduction potential per PV module of 2.59% from the reference case by applying the Vertical Integration and Revalorised Waste strategies. Additionally, we identified that raw materials cost are responsible for almost 70% of the TCO per PV module, this means that a systematic control of material flows, and waste minimization strategies should be seen as essential for manufacturing companies in this sector.

From the environmental perspective, we estimate a possible reduction of the environmental impacts of PV module production of at least 0.6 to 2.3% -depending on the impact category- compared to the defined reference case by combining both investigated circularity strategies.

5 ACKNOWLEDGMENTS

This work was partly funded by the German Ministry for Economic Affairs and Climate Action (BMWi) within the frame of the project "Green Manufacturing" (contract number 020E-100375018). The authors would like to thank the project partners for the constructive collaboration. The authors also thank Dr. Rupprecht Ackerman, for providing packaging material samples and for his supervision on laboratory; and M.Sc. Dilara Subasi, for her support on pricing research and validation.

6 AUTHOR CONTRIBUTION STATEMENT

P. Brailovsky: Conceptualization, Material flows and total cost of ownership modelling, Laboratory work, Data curation, MFA and TCO Investigation, Methodology and Validation, Writing - Original Draft, Review & Editing; K. Baumann: Conceptualization, extended waste register, Waste management Investigation and Methodology, Writing - Original Draft, Review & Editing; M. Held: Conceptualization, LCA Investigation and Methodology, Writing - Original Draft, Review & Editing; A.-K. Briem: Conceptualization, LCA Investigation and Methodology, Writing - Original Draft, Review & Editing; K. Wambach: Conceptualization, Waste management Investigation and Methodology, Writing - Original Draft, Review & Editing; E. Gervais: Conceptualization, Investigation, Methodology, Writing - Original Draft, Review & Editing; S. Herceg: Conceptualization, LCA Investigation and Methodology, Writing - Original Draft, Review & Editing; B. Mertvoy: Waste management Investigation and Methodology - Original Draft; S. Nold: Review & Editing, J. Rentsch: Supervision & Funding acquisition

7 REFERENCES

[1] D. Bogdanov, M. Ram, A. Aghahosseini, A. Gulagi, A. S. Oyewo, M. Child, U. Caldera, K. Sadovskaia, J. Farfan, L. de Souza Noel Simas Barbosa, M. Fasihi, S. Khalili, T. Traber, C. Breyer, Energy 227, 120467 (2021)

[2] E. Gervais, S. Herceg, S. Nold, K.-A. Weis, IEEE J. Photovoltaics 12, 161 (2022)

[3] P. Brailovsky, L. Friedrich, S. Nold, S. Riepe, J. Rentsch, Photovoltaics International (2021)

[4] S. Nold, Techno-ökonomische Bewertung neuer Produktionstechnologien entlang der Photovoltaik-Wertschöpfungskette: Modell zur Analyse der Total Cost of Ownership von Photovoltaik-Technologien.: Dissertation (Fraunhofer Verlag, 2019)

[5] SEMI, Guide to calculate Cost of Ownership (COO) metrics for semiconductor manufacturing equipment (2012) (2012)

[6] SEMI, Specification for Definition and Measurement of Equipment Reliability, Availability, and Maintainability (RAM) (2014) (2014)

[7] ISO, Environmental management — Life cycle assessment — Principles and framework (2006) (2006)

[8] ISO, Environmental management — Life cycle assessment — Requirements and guidelines (2006) (2006)

[9] Sphera Solutions GmbH, GaBi Software System and Database for Life Cycle Engineering 1992-2022
[10] European Commission, Product Environmental Footprint Category Rules (PEFCR). Photovoltaic modules used in photovoltaic power systems for electricity generation. (2020)
[11] PVinsights, PVinsights Weekly and Monthly Spot Price Forecast Update, 2022
[12] European Parliament, Directive 2012/19/EU of the European Parliament and of the Council of 4 July 2012 on waste electrical and electronic equipment (WEEE) (2012, Current consolidated version: 2018)
[13] European Parliament, Directive 2008/98/EC of the European Parliament and of the Council of 19 November 2008 on waste and repealing certain Directives (2008, Current consolidated version: 2018)
[14] European Parliament, Regulation (EC) No 1272/2008 of the European Parliament and of the Council of 16 December 2008 on classification, labelling and packaging of substances and mixtures, amending and repealing Directives 67/548/EEC and 1999/45/EC, and amending Regulation (EC) No 1907/2006 (2008, Current consolidated version: 2022)
[15] G. Morscheck, M. Nelles, I. Eickhoff, A. Nassour, S. Kale, H. von Ditfurth, Länderprofil zur Kreislauf- und Wasserwirtschaft in VR China, 2018
[16] B. Birnstengel, R. Simpson, R. Kölmel, M. Bijleveld, CO2 reduction potential in European waste management, 2022

This is a short version of the paper submitted to the journal Progress in Photovoltaics.

It has been selected by the Committee of the WCPEC-8 for submission to Progress in Photovoltaics.

SUSTAINABILITY OF PHOTOVOLTAIC SYSTEM'S MANUFACTURE, OPERATION AND DECOMMISSIONING IN FUTURE NET-ZERO EMISSIONS SCENARIOS-8

Antonio Urbina*
Institute for Advanced Materials and Mathematics (INAMAT2) and Department of Sciences
Public University of Navarre (UPNA), Campus de Arrosadía, 31006 Pamplona, Spain.
*email: antonio.urbina@unavarra.es

ABSTRACT: In this research work, a study about the sustainability of photovoltaic technologies has been carried out following Life Cycle Assessment (LCA) methodology. Fifteen midpoint environmental impact categories have been analyzed; the focus is put on commercial technologies comprising more than 99% of current market (crystalline silicon and thin film). Despite crystalline silicon solar cells being the dominant technology, either already commercial thin film or emerging technologies may provide the same electricity (considered as a functional unit of life cycle assessment) with lower environmental impacts. By using a combination of data from LCA inventories and the estimated capacity to be installed in the coming years, a quantitative result has been calculated at midpoint impact categories and aggregated in four endpoint categories (global warming potential, human health, resources depletion and ecosystem impacts). In the horizon of future PV system deployment for the case scenario described in the NZE2050 model of the International Energy Agency, the risk arising from mineral production is analyzed, showing for c-Si that only silver could pose some risk, while other thin film technologies depend on critical minerals. The need to increase the recyclability and material recovery for all PV technologies at the end of life is emphasized.

Keywords: LCA, CO_2 footprint, Environmental Effect, PV Materials

1 INTRODUCTION

The energy transition is accelerating the deployment of new renewable energy capacity. In particular, photovoltaic installed cumulative capacity reached 760.4GW$_{DC}$ at the end of 2020, with 139.4GW$_{DC}$ installed in a single year and the production of 821 TWh of solar electricity (3% of total electricity) despite the economic shock produced by the COVID19 pandemia [1]. On the contrary, part of the public economic expenditure approved to boost the economic recovery is being invested in building new renewable energy capacity and during year 2022, the milestone of 1TW$_P$ installed PV capacity worldwide will be achieved. The different scenarios envisaged by the International Energy Agency indicate a strong increase in the share of solar electricity in future energy production worldwide; they include a Stated Policies Scenario (STEP) which has been updated to account of COVID impact on the economy, the Sustainable Development Scenario (SDS) and in particular, the Net Zero Emissions by 2050 scenario (NZE2050) which points to reaching almost 5 TW of photovoltaic cumulative capacity in 2030 and surpassing 10 TW in 2040 [2]. This colossal growth will require the annual manufacture of millions of solar photovoltaic modules and a diversification of photovoltaic technologies in a market dominated by crystalline silicon technology. Although solar radiation is an inexhaustible renewable energy source, the future demand of raw materials and the energy required to manufacture the photovoltaic cells and the environmental impacts of production and operation of photovoltaic systems demands a sustainability analysis of solar electricity [3].

Results of LCA studies provide information to improve the yield of processing routes and to reduce its environmental impacts. Additionally, LCA results can be used to provide recommendations for more sustainable manufacture at laboratory scale and its subsequent up-scaling to industrial level, including recycling and/or landfilling at end of life of each of the components of the solar cell, especially for emerging technologies which are now starting their up-scaling towards industrial production. Design for recycling approaches should be built upon information provided by LCA inventories and impact assessment, with special attention to the pathways from inventory to impacts, which are constantly updated.

Furthermore, the end of life of photovoltaic systems will require adequate strategies at a global level when the massive amount of modules that have been deployed in recent years reaches the end of its operational life and will have to be dismantled and treated as electronic waste. According to estimations, up to 800 thousand tonnes in 2020 and more than 7 billion tonnes in 2030 of PV modules will reach its end of life. Different options can be considered: reuse of modules that still deliver enough power, recycling of modules and recovery of parts or materials, land-filling, or a combination of them. Techniques to recycle silicon and thin film modules are already available and implemented in several recycling plants, but still with low capacity and low percentage of recovered materials [6, 7]. Improved techniques, regulations and logistics for end of life of PV systems and further research and development are strongly needed. These advances may lead to a feedback from recycling strategies to original manufacture lines in order to reduce end of life impacts and to maximize materials and components recovery [8].

2 METHODOLOGY

A life cycle inventory has been compiled and used to evaluate quantitatively the impacts in 15 midpoint categories and four endpoint aggregated categories by using several life cycle impact assessment (LCIA) methods. A detailed literature review complemented with LCA calculations have been analyzed statistically to provide average maximum and minimum values. Perovskite technology has been selected as a representative for emerging technologies, and direct LCA results are included in the graphs (without statistical analysis, since there is still a very high dispersion of results for very different selection of materials and architectures for this specific technology). Results for the following categories have been obtained and presented at the conference: Climate change (Radiative forcing as Global Warming Potential,GWP100 kg CO2eq); Ozone depletion (Ozone Depletion Potential (ODP) kg CFC-11eq); Human toxicity, cancer effects (Comparative Toxic Unit for humans CTUh, c); Human toxicity, non-cancer effects (Comparative Toxic Unit for humans CTUh, n-c); Particulate matter / respiratory effects (Intake fraction for fine particles kg PM2.5eq); Ionizing radiation, human health (Human exposure efficiency relative to U235, kBq $U235_{eq}$); Photochemical ozone formation (Tropospheric ozone concentration increase, kg $NMVOC^{eq}$); Acidification (Accumulated Exceedance, mol H^{+}_{eq}); Eutrophication, terrestrial (Accumulated Exceedance, mol N_{eq}); Eutrophication, freshwater (Fraction of nutrients reaching freshwater end compartment (P)kg P_{eq}); Eutrophication, marine (Fraction of nutrients reaching marine end compartment (N)kg N_{eq}); Ecotoxicity, freshwater (Comparative Toxic Unit for ecosystems, CTU_e); Land use (Soil Organic Matter kg C deficit); Resource depletion, water (Water abstraction related to local scarcity of water, m^3 $water_{eq}$); Resource depletion, mineral, fossil (Scarcity kg Sb_{eq}).

The data used to calculate the mineral production and its trend in the past ten years have been obtained from the British Geological Survey and the United States of America Geological Survey, they are public databases that allows free access to reliable data [9]. The projected demand for minerals to be used in different PV technologies are calculated from the experimental values published for standard PV technologies (c-Si, CdTe, CIGS) and from our own inventory arising from laboratory work for the organic and hybrid perovskite technologies. The data to calculate the share of production for different countries at each step of the PV value chain for c-Si module production have been obtained from the Fraunhofer Institute for Solar Energy (ISE) annual reports and HIS Markit [10, 11]

3 RESULTS AND DISCUSSION

In order to illustrate the LCA results, two of the fifteen analyzed midpoint impact categories are presented in this communication: climate change and mineral, fossil and renewable resource depletion land occupation. For the aggregated endpoint impacts climate change (GHG emissions) and land occupation have been chosen to be presented as examples. The report of full results is included in reference [12].

The impact on climate change of solar electricity for a functional unit of $1kWh_{DC}$ is shown in Figure 1: the green-house-gases (GHG) emissions are measured in kg CO_2 equivalent. All PV technologies provide a result in the same order of magnitude, with sc-Si delivering the higher impact in this category, while the calculation for hybrid perovskite technology provides the lower values (the results for perovskite solar cells are shown in green triangle and it is a single LCA calculation and not an average of several studies which have been compiled to provide the maximum and minimum bars for the other technologies); when commercial technologies are considered, the lower GHG impact is for CdTe thin film technology, followed by CIGS technology.

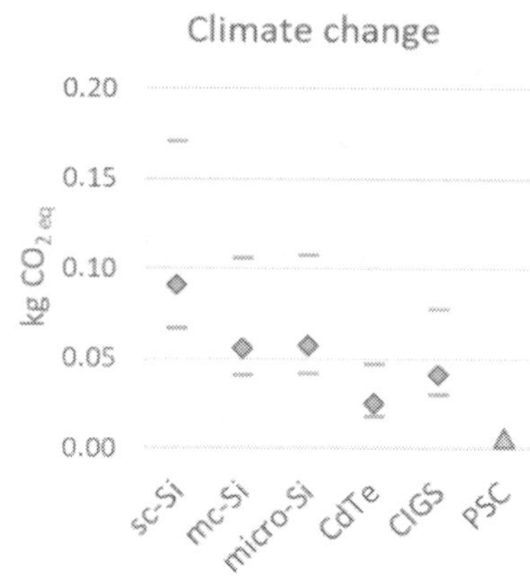

Figure 1. Average (diamond), maximum and minimum (bars) GHG emissions of 1 kWhDC electricity produced with ground mounted mono-Si, multi-Si, a-Si/micro-Si, CdTe, and CIS/CIGS PV modules. The results for perovskite solar cells are shown in green triangle.

When all the emissions are aggregated as endpoint impact categories, it enables a comparison with other means of electricity production (either renewable or fossil fuel technologies, nuclear has also been included in the graph) as shown in Figure 2 (be aware of the logarithmic scale in the axis of the figure). Even for fossil fuel technologies including carbon capture and storage, the results of PV electricity is between one and two orders of magnitude lower regarding GHG emissions. The differences between PV technologies is small compared to this quantitative huge jump for fossil fuels.

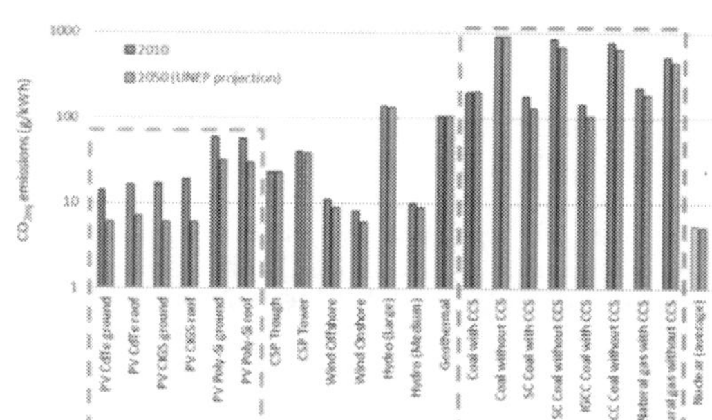

Figure 2. Life cycle carbon emissions of different electricity supply technologies, modelled for 1kWh produced in Europe (CCS: CO_2 capture and storage, IGCC: integrated gasification combined cycle, GB: gravity-based foundation). Calculations with data from references: [3,7,8].

Figure 3 is included as an example for results for other endpoint aggregated category: land occupation. Contrary to expectations, the impact on land occupation of fossil fuels is larger than utility photovoltaic plants (and much larger than roof-top photovoltaic systems); this is due to the fact that mining for fossil fuels is included in the LCA calculations, and therefore open pit mining for coal has strong impacts, also subterranean mining has land occupation impacts because the use of timber inside the tunnels create a strong impact on forests.

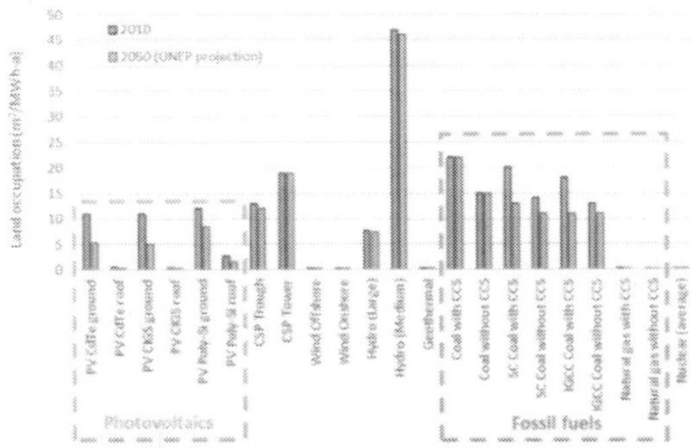

Figure 3. Land occupation required for the production of electricity (CCS: CO2 capture and storage, IGCC: integrated gasification combined cycle, SC: supercritical).

The second group of results, which in this case go beyond standard LCA methods, is the analysis of the mineral production and potential risks due to geographical location of mines or of main primary producer. The share of mineral production for seven principal materials used in commercial PV technologies is shown in Figure 4. It is clear that China holds the larger share in the seven selected minerals for the study (silver production, led by Mexico is the only exception). Confrontation will increase the risks, and therefore scientific and technological collaboration is strongly recommended.

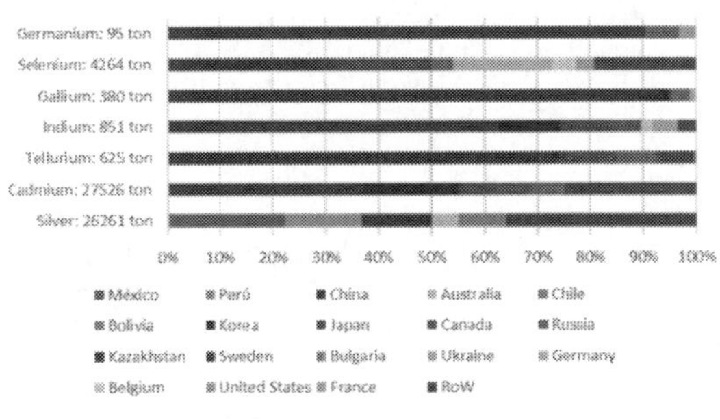

Figure 4. Share of mineral production for seven principal materials that are used to manufacture commercial PV technologies. c-Si technology is only affected by silver use which can be replaced with other metals for electrodes. CdTe and CIGS technologies are more affected by potential supply risks (tellurium and indium respectively). Data from British Geological Survey (data for 2019), Ref. [9]. Units for total production in metric tonnes.

The results shown in Figure 4 can be compared to the LCA midpoint impact category called "mineral, fossil and renewable resource depletion" for a more detailed comparative analysis between PV technologies. These results indicate that silicon based technologies have a low impact, CdTe is also a low impact technology in this category (although there is a risk arising from tellurium production), and hybrid perovkite presents the lower impacts. On the contrary, CIGS technology has the larger impact, one order of magnitude higher than the other PV technologies, thus making this technology more prone for mineral scarcity risk.

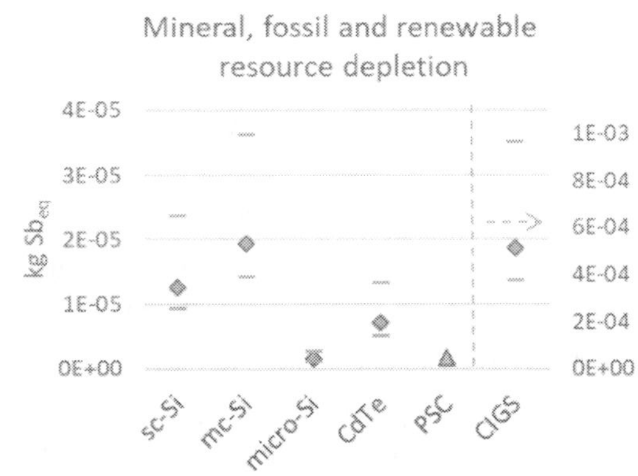

Figure 6. Average (diamond), maximum and minimum (bars) of "mineral, fossil and renewable resource depletion" midpoint impact category measured in kg Sb_{eq} of 1 kWhDC electricity produced with ground mounted mono-Si, multi-Si, a-Si/micro-Si, CdTe, and CIS/CIGS PV modules. The results for perovskite solar cells are shown in green triangle.

4 CONCLUSIONS AND RECOMMENDATIONS

The global impacts of PV future deployment have been calculated for the Net Zero Emissions by 2050 (NZE2050) scenario of the International Enegy Agency. The LCA methodology following ISOS 14040 and 14044 standards and IEA-PVPS recommendations has been used. All PV technologies under consideration (single-crystalline, multi-crystalline and micro-crystalline silicon, cadmium telluride, copper indium gallium diselenide and hybrid perovskite) have midpoint impacts of the same order of magnitude for fifteen impact categories, with crystalline silicon in the upper range of many of them, and thin films in the lower range (with the exception of CIGS in resources depletion category). Hybrid perovskite technology, chosen as a representative of emerging technologies shows a large variation of impacts, with higher impacts in human health related categories, and lower in ecosystem and green-house gas emissions categories). When aggregated endpoint categories are considered and the environmental impacts are compared with fossil fuels, photovoltaic technology shows significantly lower impacts (four times lower for human health and ecosystem categories and two orders of magnitude lower for GHG emissions), and only land occupation impacts are comparable. It is worth to mention that nuclear energy has been included in these comparative studies but in the impact analysis, the treatment of radioactive waste has not been included due to technological uncertainties and the very large time scale of the problem, additionally, the occurrence of a nuclear accident has not been included in the analysis.

Regarding the study about productions risks of the minerals required for PV manufacture, the main result is that China holds the larger share of production in six of the seven selected minerals for the study (silver production, led by Mexico is the only exception) and it also dominates the value chain for photovoltaic module manufacture, with a strong dominance in the main three stages: silicon ingot production, wafer and cell fabrication and module manufacture. Therefore, confrontation with China will increase the risks, and scientific and technological collaboration is strongly recommended. Besides, other countries, and specially Europe should consider developing (again) manufacturing capabilities to recover part of the value chain for photovoltaic module production, which may comprise all cycle from crystalline silicon ingots to module manufacture to cover, at least in part, the future capacity installation in Europe.

The already installed capacity and the estimated deployment of future capacity that will reach its end of life within three decades requires a global approach to deal with the decommissioning stage of the PV modules. Ideally, all PV modules should be recycled, but actual recycling capacity is well short of the required capacity and although there is an important research effort (and patent registrations) for PV technology recycling, the value of recovered materials still do not cover the cost of an effective recycling route. Mineral scarcity may pose some risk for CdTe and CIGS technologies, while c-Si based technology is only affected by silver dependence that can be avoided with other metals replacement for electrodes. When the risks grow higher, the investment in recycling facilities will boost the recovery ratio of minerals from PV module waste.

A final conclusion is the actual weakness of the end-of-life stage of photovoltaic systems, both at a fundamental scientific level and at an applied industrial level. On the one hand scientific knowledge progresses steadily but slowly and from relatively low recovery rates regarding elements, with recycling methods based on rough chemistry and mechanical procedures that still have a very large room for the improvement of recovery rates. On the other hand, at industrial capacity level, more facilities for dismantling modules are required; and then, either in-situ or in other specialized sites to which the dismantled parts could be sent for further treatment, facilities to increase element recovery and specially glass recycling will have to be built. A logistical approach at regional level (for example, European scale) should also be considered for the end of life of PV modules, with transport to recycling sites being a substantial part of future emissions associated to the photovoltaic systems' full life cycle.

Acknowledgements:

Funding from Agencia Estatal de Investigación (MCIN/AEI/10.13039/501100011033, Spain), grant number PID2019-104272RB-C55 including FEDER funds is acknowledged.

REFERENCES:

[1] World Energy Outlook 2020, International Energy Agency, (2020).
[2] Net Zero by 2050. A Roadmap for the Global Energy Sector, International Energy Agency, (2021).
[3] A. Urbina, Sustainable Solar Electricity, 1st ed., Springer International Publishing, (2022). ISBN: 978-3-030-91770-8 .
[4] ISO, International Organization for Standardization, ISO 14040:2006 and 14044:2006 Environmental management-Life cycle assessment (2006).
[5] EU-European Commission, Study on the EU's list of Critical Raw Materials – Final Report (2020).
[6] K. Komoto, J.-S. Lee, J. Zhang, D. Ravikumar, P. Sinha, A. Wade, G. Heath, End-of-Life Management of Photovoltaic Panels: Trends in PV Module Recycling Technologies, IEA PVPS Task12, Subtask 1, Recycling, (2018).
[7] UNEP, Green Energy Choices: The benefits, risks and trade-offs of low-carbon technologies for electricity production., United Nations Environment Programme, (2016).
[8] A. Wade, P. Stolz, R. Frischknecht, G. Heath, P. Sinha, The Product Environmental Footprint (PEF) of photovoltaic modules—Lessons learned from the environmental footprint pilot phase on the way to a single market for green products in the European Union, Prog. Photovolt. Res. Appl. 26 (2018) 553–564.
[9] BGS, World Mineral Production 2015-2019, British Geological Survey, (2021).
[10] Fraunhofer Institute for Solar Energy Systems, ISE, "Photovoltaics Report 2022" (2022).
[11] IHS Markit, "10 Cleantech Trends in 2022" (2022).
[12] A. Urbina, "Sustainability of Photovoltaic Technologies in Future Net-Zero Emissions Scenarios", Progress in Photovoltaics, special issue for WCPEC8, submitted (2022).

This is a short version of the paper submitted to the journal EPJ Photovoltaics.

It has been selected by the Committee of the WCPEC-8 for submission to EPJ Photovoltaics.

COMBINING CIRULARITY AND ENVIRONMENTAL METRICS TO GUIDE DEVELOPMENT IN PV

Aistis Rapolas Zubas, Marie Fischer, Estelle Gervais, Sina Herceg, Sebastian Nold
Fraunhofer Institute for Solar Energy Systems (ISE)
Heidenhofstr. 2, 79110 Freiburg, Germany

ABSTRACT: A product made from virgin raw materials that ends up in a landfill presents a linear supply chain model. Today's photovoltaic (PV) industry is still largely based on this model. With the increasing volume of production, and consequently the volume of waste, the application of circular economy principles in the PV sector can significantly increase its environmental efficiency. This study analyzes the impact of circularity on the supply chain of PV systems, using the example of silicon used for PV wafer production. Four scenarios based on different technological parameters and circular economy principles are defined. Their evaluation is carried out by the methodologies of Material Circularity Indicator (MCI) and Life Cycle Assessment (LCA). The State-of-art case of the PV polysilicon supply chain corresponds to the MCI score of 0.54. Closed-loop circularity solutions provide the MCI score of 0.80 presenting the potential for a circular economy approach in the field. LCA results covering 16 impact categories show the reduction of environmental impact by 12% with improved circularity and 46% by technological development. The results present the benefits of potential circularity options within the supply chain as well as the impact of technological development on the polysilicon demand.
Keywords: Circularity, Life Cycle Assessment, Polysilicon, Metrics

1 INTRODUCTION

In order to meet long-term climate change targets, intensive photovoltaic (PV) deployment is needed in the future [1]. Large amounts of materials are required to achieve this goal [2]. Possible options to meet this demand are to use virgin materials or to implement circular economy (CE) approaches to the industry. Transforming material flows to the closed loop within production steps as well as after the product's lifetime is a way to implement CE principles to the market. With the increasing production volume, the required raw materials, and the associated amount of waste, the circularity in the PV sector becomes a critical topic.

The application of circular economy approaches in photovoltaics have been analyzed in terms of environmental, technological, economic, and legislative arguments [3-5]. However, a research gap could be identified in the quantitative assessment for circularity in PV. The circular model is based on three principles: design out waste and pollution, keep products and materials in use, regenerate natural systems [6]. As these principles are represented by various parameters the need for universal metric remains. Material Circularity Indicator (MCI) includes the parameters related to the circularity and provides the single score result [6]. The metric provides a chance to estimate various CE scenarios and allows stakeholders to understand how far the products are on transitioning from a linear to a circular supply chain. Combination with additional risk indicators allows to understand the significance of circularity in environmental, economic or social contexts. More circular material flows on selected level does not necessarily lead to more environmental-friendly solutions. The interaction between circularity and life-cycle-based environmental indicators has been discussed in the literature [7,8]. Still, there is no widely-accepted universal metric combining circularity and environmental performances.

So far, the circularity in the PV sector, both for the module as a whole and its materials, has not been assessed through the herein used dedicated circularity indicator. This study aims to assess circularity of polysilicon in the PV supply chain and to measure how the implementation of circular solutions influences the performance indicators. Complementary, Life Cycle Assessment (LCA) is used for the estimation of impacts on the most relevant environmental impact categories. Therefore, this study aims to present an assessment example that can be used as an aid to decision-makers to guide technology development.

2 APPROACH

The study is carried out testing Material Circularity Indicator for circular economy approach and its combined utilization with Life Cycle Assessment for environmental impact of polysilicon under consideration of four scenarios.

The functional unit (FU) used in the study is 1 MWh of electricity produced by a PERC p-type mono-Si PV module under global average irradiation conditions.

2.1 Scenarios

Four scenarios based on technology and circularity are applied in the research. Two technological pathways are created: State-of-art PERC and 2032 ITRPV projection. The data of the IEA PVPS [9] Life Cycle Inventories provide detailed bills of materials for all production steps of PV module: metallurgical-grade silicon, Czochralski single crystal, silicon wafer, photovoltaic cell, and module production.

The technological development is carried out on expected trends for photovoltaics

- The data for 2021 is used to describe the product's specification in the State-of-art PERC scenario.
- In the case of the 2032 ITRPV projection it is assumed that improvements are implemented in the wafering process by reducing the kerf loss content and wafer thickness, as well as in the increase of the module efficiency. Values for these parameters are anticipated for the year 2032 by the International Technology Roadmap for Photovoltaic (ITRPV) [10]. A longer lifetime of the product is assumed.

Two circularity options are applied in the research. Waste treatment options are designed based on technological, economic, and legislative conditions working in the field.

- The option "Business-as-usual" represents the current situation in the industry which is based on economic reasons and legislation requirements. According to The Waste of Electrical and Electronic Equipment (WEEE) [11] silicon treatment from End-of-life (EOL) photovoltaic modules is not mandatory and therefore usually not implemented in recycling processes due to lack of economic reasons.
- The "Closed-loop" circularity option represents an example of improved Si recovery, which allows silicon to be used again in a circular way for new PV polysilicon production. Kerf loss recovery to metallurgical grade silicon is possible by methods of thermal plasma, carbothermic reduction or inductive melting (recycling efficiency – 65%) [12]. A selected EOL technology – FRELP – Full Recovery End of Life Photovoltaic [13] provides the recovery of silicon from solar cells with a purity of metallurgical grade silicon as well (recycling efficiency – 95%).

Other parameters and assumptions related to material flows and losses (production yields, collection rates, etc.) are assumed constant among the different scenarios.

2.2 Methodologies

The methodology of Material Circularity Indicator was introduced by the Ellen MacArthur Foundation. The indicator is essentially constructed from a combination of three product characteristics [6]:

1. mass V of virgin raw material used in manufacture
2. mass W of unrecoverable waste that is attributed to the product (W_0 – uncollected waste, W_C – waste generated in the recycling process; W_F – waste generated to produce any recycled content to use as a feedstock)
3. utility factor X that accounts for the length and intensity of the product's use.

Based on material flows the Linear Flow Index (*LFI*) can be computed. *LFI* measures the proportion of material flowing in a linear fashion. The index takes a value between 0 and 1, where 0 is completely restorative flow and 1 is a completely linear flow [6].

The *LFI* is calculated as follows:

$$LFI = \frac{V + W}{2M + \frac{W_F - W_C}{2}} \qquad (1)$$

The second parameter in the MCI is the utility X of a material or product. The utility X is derived from the lifetime and functional units of a product compared to an industry-average product of the same type.

$$X = \left(\frac{L}{L_{av}}\right) \cdot \left(\frac{U}{U_{av}}\right) \qquad (2)$$

The Material Circularity Indicator can be defined by considering the *LFI* of the product and a factor $F(X)$, built as a function F of X that determines the influence of the product's utility on its MCI. The equation used to calculate the MCI of a product is

$$MCI^*_P = 1 - LFI \cdot F(X) \qquad (3)$$

Where F takes the form:

$$F(X) = \frac{0.9}{X} \qquad (4)$$

MCI takes, by convention, the value 0.1 for a fully linear product (i.e., $LFI = 1$) whose utility equals the industry average (i.e., $X = 1$), while MCI value of 1 presents totally circular product [6].

In analogy to the MCI calculation, the goal of the Life Cycle Assessment is to investigate the environmental impacts that can be allocated to the silicon in a p-type mono-Si PV module. The LCA is conducted from cradle to gate, including the process steps from raw material extraction to the finished PV module. The chosen system model is cut-off. Here, recycled materials are assumed to be burden-free, as their impacts are allocated to their previous life-cycle. Any materials in the PV module other than silicon are excluded in this analysis. Auxiliaries such as electricity and heat are included on a weight-ratio-basis. Any balance-of-system components are out of scope in this study. The location of the production is assumed to be China.

As recommended in the Product Environmental Footprint Category Rules (PEFCR) the chosen impact assessment methodology is EF3.0 [14]. Additionally, the impact category *Climate Change* is investigated in more detail, as it is one of the most robust indicators and has the highest contribution to the final single score result.

3 RESULTS

3.1 Material Circularity Indicator

The resulting values for the parameters of MCI (FU = 1 MWh) are given in Table I.

The relatively high MCI result of 0.54 for the scenario State-of-art PERC/Business-as-usual case is mainly due to the reuse of the ingots' sidewall slabs, tails, and tops that are cut-off to form the polysilicon brick from an ingot. These pieces are remelted into Cz ingot and, due to high collection and internal recycling rates, have a significant impact on the MCI result. Under Closed-loop circularity, both the kerf loss and silicon from EOL waste are recovered as metallurgical grade silicon. Using it for PV polysilicon production significantly reduces the demand for virgin material.

Table I: Results of MCI and its parameters

			MATERIAL FLOW PARAMETERS			UTILITY PARAMETERS		
Scenario	Technological pathway	Circularity option	Virgin materials (g/MWh)	Unrecover. waste (g/MWh)	Linear Flow Index	Lifetime (years)	Utility	MCI
Scenario 1 (S1)	State-of-art PERC	Business-as-usual	86.6	65.1	0.52	25/25	1	0.54
Scenario 2 (S2)		Closed-loop	38.8	27.1	0.22	25/25	1	0.80
Scenario 3 (S3)	2032 ITRPV projection	Business-as-usual	48.0	37.1	0.52	35/25	1.4	0.67
Scenario 4 (S4)		Closed-loop	21.4	15.1	0.22	35/25	1.4	0.86

As mentioned above the Linear Flow Index presents how linear the material flows are. The LFI result for Business-as-usual circularity is 0.52 both for State-of-art PERC and 2032 ITRPV projection. Production efficiency remains the same in most of the processes despite the technological path. Linear Flow Index for Closed-loop option amounts to 0.22, corresponding to high circularity. Still, there is a potential for improvement in the waste treatment related to low recycling efficiency and collection rates.

The second component that influences the MCI score is product utility. The 2032 ITRPV projection scenario declares a longer lifetime (35 years) which increases the Utility (X) to 1.4. On the improved technology case LFI scores remain the same, therefore only the higher Utility ratio leads to higher MCI results: 0.54 to 0.67 and 0.80 to 0.86 on Business-as-usual and Closed-loop options, respectively. A longer lifetime satisfies one of three circular economy principles – to keep products or materials in use. Some PV manufacturers already guarantee longer lifetimes for their products than 25 years.

The methodology of the Material Circularity Indicator allows to account for the mass of virgin materials and unrecoverable waste associated with a product. The values in Table 1 are expressed by the product of polysilicon. The effect to these parameters of technology or circularity improvement can be compared. A reduction achieved in the case of the 2032 ITRPV projection is 44-45% for the aforementioned parameters in comparison to State-of-art PERC. While Closed-loop circularity lessens the mass for virgin materials by 55% and 59% for unrecoverable waste in comparison to Business-as-usual circularity.

3.2 Life Cycle Assessment

Figure 1 presents the single score results per scenario (FU = 1 MWh, Method – EF3.0).

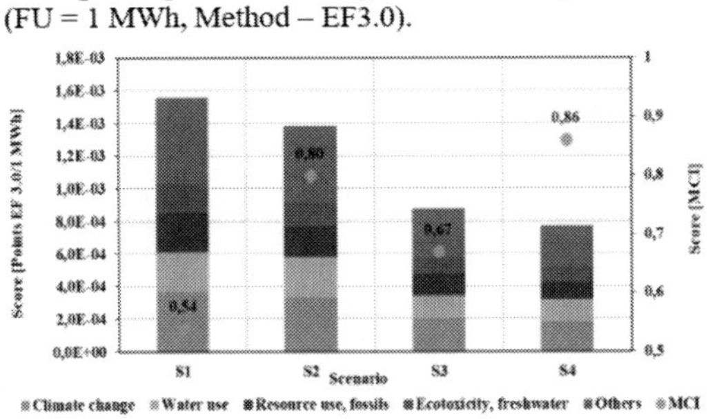

Figure 1: Normalized and weighted single score results (EF3.0) of environmental impacts by category

Further, the contributions by the different impact categories are visualized. Scenario 1 causes the largest environmental impact in this comparison, followed by scenario 2, 3 and 4, in that order. The single score for scenario 4 is about 51% lower than for scenario 1. The improved circularity decreases the environmental impacts by around 12%. At least two thirds of the single score impacts in all scenarios are due to the *Climate Change*, *Water use*, *Resource use, fossils*, and *Ecotoxicity, freshwater* categories. The results indicate, that the technological development has a more significant influence on the end result, reducing the impacts by around 46% regardless of the circularity option. The 2032 ITRPV Projection scenarios ensure higher built-in material efficiencies, which are directly related to lower production volumes. This has a larger effect on the single score result than the reduction of virgin material processing.

The cause of the specific impact can be further investigated in the category of *Climate Change* (midpoint) for all scenarios. The carbon footprint of the silicon in the PV module for 1 MWh of produced electricity is 14.25 kg CO_2 eq., 13.06 kg CO_2 eq., 8.02 kg CO_2 eq., and 7.28 kg CO_2 eq. for scenarios 1, 2, 3, and 4, respectively. Half of these impacts originate from the silicon production mix. This includes all impacts from the raw material extraction to the single crystal production in the Cz ingot pulling. The production location is China. Most of the impacts caused in the silicon production mix can be traced back to the electricity consumption in the supply chain, as it is an energy intensive process and the Chinese electricity mix relies heavily on coal-fired power plants.

The LCA results indicate that the State-of-art PERC/Business-as-usual scenario (S1) causes the most environmental impacts in this comparison. The silicon production and the electricity use along the production line are the main sources of impacts in all scenarios. The reduction in the energy and silicon consumption across the scenarios has therefore a substantial influence on the respective LCA results. Scenarios 2 and 3 cause less environmental impacts per MWh than scenario 1. When comparing scenarios 2 and 3, it becomes apparent, that technological improvement with Business-as-usual circularity outperforms the State-of-art PERC with improved circularity. The increase in electricity production due to the elongated lifetime has a greater impact on the impacts per MWh than the Closed-loop circularity. In scenario 4, where both improvements are combined, the environmental impacts are reduced by about 51% compared to the original design in scenario 1.

The improvements in the *Climate change* impact category among the scenarios (from 1 to 4) is mostly due to the reduced electricity consumption as well as the lower demand for silicon. As shown in Figure 1, the *Water use* category has also a high impact on the overall single score results. Also, the impacts in this category are greatly influenced by the reduction in silicon demand, across the four investigated scenarios. The Cz ingot pulling in the silicon production is the main cause of water consumption in the production of a PV module. While in scenario 1, the water use amounted to 32.8 m^3 depriv./MWh, this is reduced by the scenario 4 to 18 m^3 depriv./MWh.

4 DISCUSSION

The methodologies of MCI and LCA can be discussed on some critical arguments. One of the main CE goals is to lessen inputs. However, the MCI as a metric is based on a mass proportion of waste and recycled material. It means that high collection and recovery rates of wastes within production steps increase the overall MCI value of the product. In our case Poly-Si ingots are cut-off to form the shape of the wafers. However, cut pieces are remelted again to Cz crystal. Following the methodology, these steps significantly increase the MCI score. Nevertheless, it contradicts the idea of using fewer resources [15]. The chance to increase circularity by creating not relevant waste and reusing it could be an example of so-called "circular washing". The strategy used by companies to present goods as produced in a circular economy style.

The MCI analysis includes a cradle to grave approach, while the LCA is based on a cradle to gate analysis. In the LCA the recycled content is included, however the emissions of possible recycling processes are not, due to a lack of data. Further, the LCI has been adjusted to fit the MCI scope of analysis more closely, so that only the impacts allocated to the silicon within the PV module are included. Any other materials (for example the frame, or metallization pastes) that are part of a PV module have not been investigated. Auxiliaries, such as electricity, heat and transport have been included based on a weight-ratio, meaning that the weight fraction that the silicon has in a PV module, cell or wafer was applied to calculate the allocated auxiliary inputs. This entails uncertainty and can influence the end result. Finally, the normalization and weighting of the characterized results introduces additional uncertainty. However, to achieve a single score for the respective scenarios and make them more easily comparable, this step was taken.

5 CONCLUSION

The results showed that the State-of-art case of the PV polysilicon supply chain corresponds to the MCI score of 0.54. The higher-than-expected result is achieved mainly by internal recycling. Improved circularity with kerf loss and EOL waste recovery on the closed-loop model significantly increase the overall MCI value up to 0.80. The implementation of Closed-loop circularity reduces the need for virgin material by up to 55% and up to 59% for unrecoverable waste in comparison to Business-as-usual circularity. The effect of improved utility on the aforementioned parameters was not so significant. However, the evaluation by Life Cycle Assessment methodology showed that improved utility causes lower environmental impact, since it significantly reduces the volume of production. Extended lifetime contributed to decreased environmental impacts, too.

This paper contributes to research evaluating circular economy approaches in the PV industry. The results highlighted the benefit of joining metrics for circularity and environmental assessment. Interaction between them can be interpreted in the context of sustainability as guidance for development in photovoltaics.

ACKNOWLEDGMENT

Aistis Rapolas Zubas would like to thank Deutsche Bundesstiftung Umwelt (DBU) for funding the research at Fraunhofer ISE through the programme "Fellowships for university graduates from Central and Eastern Europe".

REFERENCES

[1] J. C. Goldschmidt, L. Wagner, R. Pietzcker, and L. Friedrich, *Energy Environ. Sci.*, vol. 14, no. 10, pp. 5147–5160, 2021

[2] Gervais, Estelle; Shammugam, Shivenes; Friedrich, Lorenz; Schlegl, Thomas. *Renewable and Sustainable Energy Reviews* 137, p. 110589, 2021

[3] J. A. Tsanakas *et al., Prog Photovolt Res Appl*, vol. 28, no. 6, pp. 454–464, 2020

[4] R. Deng, N. L. Chang, Z. Ouyang, and C. M. Chong, *Renewable and Sustainable Energy Reviews*, vol. 109, pp. 532–550, 2019

[5] M. A. Franco and S. N. Groesser, *Sustainability*, vol. 13, no. 17, p. 9615, 2021

[6] Ellen MacArthur Foundation, "Circularity Indicators. An Approach to Measuring Circularity. Methodology.," 2019

[7] E. Glogic, G. Sonnemann, and S. B. Young, *Sustainability*, vol. 13, no. 3, p. 1040, 2021

[8] M. Niero and P. P. Kalbar, *Resources, Conservation and Recycling*, vol. 140, pp. 305–312, 2019

[9] Frischknecht, R., Stolz, P., Krebs, L., Wild-Scholten, M. J. de, Sinha, P., Fthenakis, V., Kim, H. C., Raugei, M., & Stucki, M. International Energy Agency Photovoltaic Power Systems Program (IEA PVPS), 2020

[10] Verband Deutscher Maschinen- und Anlagenbau (VDMA). International Technology Roadmap for Photovoltaic (2021 results), 2022.

[11] *Directive 2012/19/EU on Waste Electrical and Electronic Equipment*, European Union, 2012

[12] J. Li, Y. Lin, F. Wang, L. Shi, J. Sun, B. Ban, G. Liu, and J. Chen. *Separation and Purification Technology*, vol. 254, 117581, 2021

[13] C. E. Latunussa, F. Ardente, G. A. Blengini, and L. Mancini, *Solar Energy Materials and Solar Cells*, vol. 156, pp. 101–111, 2016

[14] European Commission: Product Environmental Footprint Category Rules (PEFCR): Photovoltaic Modules used in Photovoltaic Power Systems for Electricity Generation, 2020

[15] European Environmental Agency. Circular Economy in Europe - Developing the Knowledge Base, 2016

A SNAPSHOT OF GLOBAL PV MARKETS - THE LATEST SURVEY RESULTS ON PV MARKETS AND POLICIES FROM THE IEA PVPS PROGRAMME IN 2021

Gaëtan Masson[1], Arnulf Jäger-Waldau[2], Izumi Kaizuka[3], Johan Lindahl[4], José Donoso[5]
[1]IEA PVPS Task 1, Belgium – g.masson@iea-pvps.org, +32 2 514 78 05
[2]European Commission* , Joint Research Centre (JRC), Via E. Fermi 2749, 21027 Ispra, Italy – Arnulf.jaeger-waldau@ec.europa.eu
[3]RTS Corporation, Japan – kaizuka@rts-pv.com
[4]Becquerel Sweden, Sweden – johan@becquerelsweden.se
[5]UNEF, Spain – j.donoso@unef.es

ABSTRACT: The objective of this paper is to propose a reliable and accurate perspective on key markets and policies related to PV development in 2021 and previously. It aims at offering a clear analysis of how PV markets have developed in 2021, with updated numbers, along with an analysis of the policies behind the development. In this paper are displayed and analyzed survey results for the calendar year 2021 concerning PV markets and policies, as well as other key issues. An increasing number of national markets experienced notable growth in 2021 with impacts on policy development. Figures show that over 173 GW of PV systems have been installed in the world last year. Consequently, cumulative capacity crossed the 945 GW mark in 2021.
Keywords: Photovoltaic (PV), Market, IEA PVPS

1 INTRODUCTION

This snapshot paper gives information on the development of PV power applications in the PVPS member and non-member countries and is largely based on the information provided by IEA PVPS countries plus the European Union through its European Commission. The report includes information on national market developments at the end of the former year, in this case, 2021. The International Energy Agency – Photovoltaic Power System Programme (IEA PVPS)'s Task 1 is responsible for strategy and outreach within the IEA PVPS program. This includes market and industry analysis. A key deliverable of Task 1 is the annual Snapshot of Global PV Markets publication, together with the annual flagship report Trends in PV Application.

The objective of the series of annual Snapshot and Trends reports — which have been published since 1992 (Trends) [1] and 2013 (Snapshot) [2] — is to present and interpret developments in both the PV systems and components being used in the PV power systems market and the changing applications for these products within that market. These trends are analyzed in the context of the business, policy and non-technical environment in the reporting countries.

2 THE GLOBAL PV INSTALLED CAPACITY

This paper presents the latest survey results for the calendar year 2021 concerning PV markets and policies, as well as other key issues. An increasing number of national markets experienced notable growth in 2021, and that impacted the policy development. The capacity figures given are nominal DC peak power (Wp) under standard test conditions (i.e. 1 000 W irradiance, air mass 1.5 light spectrum and 25 °C device temperature) for consistency reasons. As not all countries report DC peak power (Wp) for solar PV systems, but report AC power, these capacity figures are converted to Wp DC.

While the final figures for 2021 could slightly continue to be refined in the coming months, nearly final figures show that around 173.5 GW of PV systems have been installed in the world last year. Some important trends observed are as follows.

The global PV market grew to around 173.5 GW in 2021, compared to around 145.5 GW in 2020. This represents a year-on-year growth of close to 20%.

Asia scored the very first place again in 2021 with China (54.9 GW), followed by the USA (26.9 GW), India (13.7 GW) or Japan (6.6 GW) and more appearing in the top markets.

The market in the EU grew significantly to around 28.7 GW exceeding the 23.2 GW recorded in 2011. Germany (5.8 GW) and Spain (4.9 GW) were the key markets this year followed by Poland (3.7 GW), the Netherlands (3.6 GW) and France (3.4 GW) and several others.

Numbers show that installations rose in the Middle East and North Africa region. The PV market is particularly dynamic in Israel, Morocco, Egypt, Saudi Arabia and the United Arab Emirates.

Some development was also observed in Africa and Latin America. Brazil ranked sixth in 2021 with 5.7 GW installed.

Drivers for PV development include record-breaking competitiveness levels and development of distributed PV.

The impact of rising component prices was not clearly observed from global PV systems installations' point of view.

The total cumulative installed capacity reached around 945 GW in 2021.

3 MARKET DEVELOPMENT

In 2021, the PV market saw an important growth for the second year in a row after 2019's limited growth. Globally, the trend is upwards by exceeding the 173 GW annual installed capacity as expressed in Fig. 1.

* The views expressed are purely those of the author and may not in any circumstances be regarded as stating an official position of the European Commission

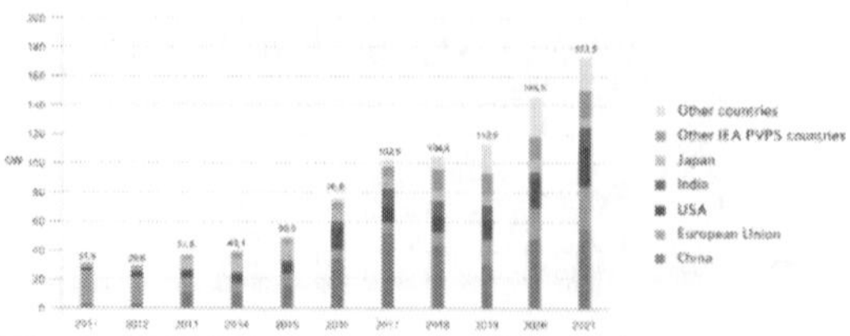

Figure 1: Evolution of annual PV installations (GW-DC)

Asia remains the leader of the global PV market. Next to China (54.9 GW), India and Japan remain a relevant presence in the global market with respectively 13.7 GW and 6.6 GW installed.

In other parts of Asia, the market observed steady growth in Taiwan (1.9 GW), stabilized in Korea (4.2 GW) and significantly decreased in Vietnam (2 GW) after the massive growth observed in 2020. Australia installed 4.9 GW in 2021, a stable level since 2018

In the Americas, the USA market continued to grow (26.9 GW), similarly to Brazil that installed at least 5.7 GW in 2021, an important increase compared to 2019 and 2020. PV installations in Chile grew tremendously in 2021 reaching a cumulative installed capacity of 6.2 GW while installations in Mexico were at a similar level (1.6 GW) compared to 2020. The market in Canada grew again in 2021 with 421 MW installed.

In the European Union, Germany kept the leading position by installing over 5.7 GW. In 2021, Spain (4.9 GW), Poland (3.7 GW), the Netherlands (3.6 GW) and France (3.4 GW) can be mentioned as leading countries as well. Greece added over 1 GW while eight countries added more than 500 MW namely Hungary, Austria, Denmark, Belgium, Italy, Portugal, Switzerland and Sweden.

New developments occurred in Africa (Egypt, South Africa) and in the Middle East (UAE) which led to GW-scale cumulative installation levels: 4.6 GW in South Africa, 3.5 GW in the UAE and 3.4 GW in Egypt for instance. Israel installed 935 MW in 2021 only.

In 2021, eighteen countries passed the GW mark concerning annual installed PV capacity. Forty-two countries reached at least 1 GW cumulatively in 2021. China alone represented 308.5 GW. Germany, which used to lead the rankings for years, lost its leading position in 2015 and now ranks fourth (59.7 GW). The USA are second (123 GW) and Japan is third (78.4 GW). With close to 198 GW of total capacity, Europe is now significantly behind the Asian leader that runs at least 540 GW, and much more to come in the coming years.

4 GRID-CONNECTED CENTRALIZED AND DISTRIBUTED

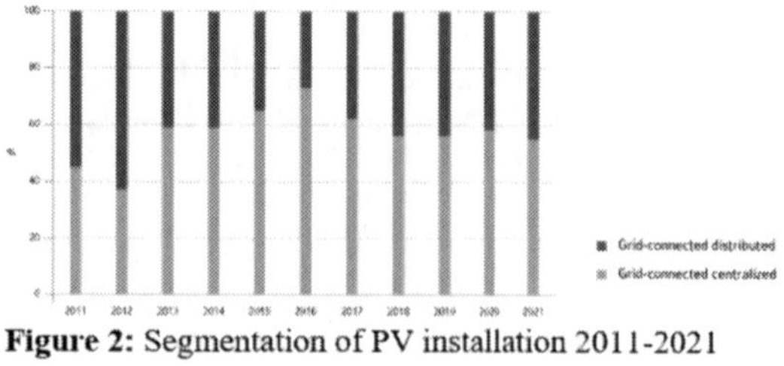

Figure 2: Segmentation of PV installation 2011-2021

Regarding the share of the grid connected and centralized installations at the global level, the trend changed a lot over the years.

Before 2013, most PV installations used to be distributed systems, i.e. mainly installed on rooftops.

In the last years, the market has seen a strong development of grid connected centralized installations. This changing trend is because centralized PV has evolved faster, e.g. in terms of cost, and most of the major PV development in emerging PV markets are coming from utility-scale PV. The success of utility-scale installations is mainly coming from the fact that the installation time and cost per Wp for utility-scale are lower than for distributed PV plants.

In the last years, tenders have driven PV development and continued to be granted in several places in the world with extremely competitive prices, well below 20 USD/MWh in the sunniest places. One of the key trends of 2021 is the wider development of utility-scale plants without financial incentives. Such development is mostly independent from financial incentives and therefore policy decisions, which makes its potential virtually unlimited.

The distributed PV, after suffering from a small decline in 2018, is exhibiting a stable level since 2019, reaching an even higher level than in 2017.

The market has also continued to diversify in 2021. While floating PV (FPV) adds to utility-scale (installed capacity of FPV worldwide has surpassed 3 GW in 2021) BIPV starts to complement BAPV in the built environment (installed capacity of BIPV globally is estimated to have been around 1 GW in 2021), although it remains a niche. Other emerging segments such as agricultural PV (APV) or PV integrated in vehicles (VIPV) are showing the potential for further diversification of PV components, but their current levels of development remain limited. Overall, dual use of land and surfaces is an option deeply investigated around the world.

At the local level, the region with the highest share of grid-connected centralized installations was Europe until 2011 which was holding 80% of the utility-scale system installed globally. Starting from 2012, the share of American and Asian countries started to grow until the end of 2013, when Asia became the main region for utility-scale projects.

5 ELECTRIC VEHICLES

The electrification of transport accelerated in many countries and almost all of which are active in the IEA PVPS program.

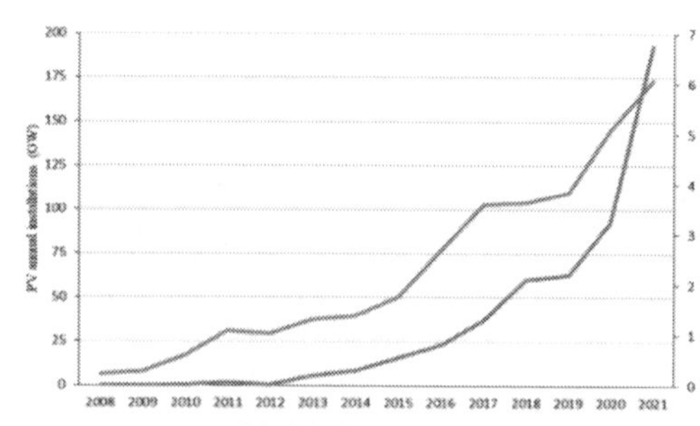

Figure 3: Comparison between PV installation and EV sales

The link between PV development and EVs is not

straightforwardly understood yet, but it is simply becoming a reality. Charging electric vehicles during peak load hours implies to rethink power generation, while concepts such as virtual and collective self-consumption could rapidly provide a framework for rapid PV development.

The accelerated development of the EV market could be compared to the development of the PV market, as shown in Fig. 3. Close to 7 million electric vehicles sold in 2021 (more than half of which were sold in China), representing 8.3% of global car sales. The sales of EVs are following a trend that is even faster than the installation of PV [3].

6 ELECTRICITY PRODUCTION

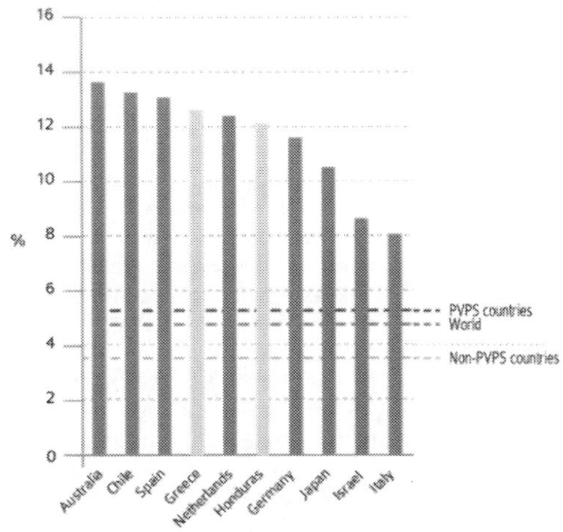

Figure 4: National PV penetration in % of the final electricity demand based on 2021 capacities

The electricity production from PV per country as shown in Fig. 4 estimates what the PV production could be, based on the cumulative PV capacity at the end of 2021 (close to optimum siting, orientation and average weather conditions). These numbers, which are not based on actual measurements, should therefore be considered as indicative, aiming at comparing different situations in different countries rather than official data. In several countries, the PV contribution to the final electricity demand has passed the 5% mark with Australia in the first place with almost 14%. Chile is second with above 13%, and Spain, Greece and the Netherlands follow. In total, PV contribution amounts to almost 5% of the electricity demand in the world.

7 CONCLUSIONS

This brief overview of the situation of the PV market and its deployment shows that in 2021 the annual PV market reached at least 173 GW worldwide and the cumulative installed capacity represented over 945 GW. While the world was facing the second year of a pandemic and despite the end-of-year disruptions in Asia, the photovoltaic market continued growing. The global PV market is, more than ever, dominated by a few leading countries even if many new PV markets are developing on all continents, at different paces. Developing photovoltaic systems with limited or no financial incentives is now an observable reality. Long-term private contracts (PPA) and the sale of electricity on wholesale markets have been observed in an increasing number of countries in 2021. Nevertheless, the market remains strongly influenced and led by incentives.

8 REFERENCES

[1] IEA PVPS, TRENDS 2022 In Photovoltaic application, ISBN 978-3-907281-35-2, https://iea-pvps.org/trends_reports/trends/

[2] IEA PVPS, 2022 Snapshot of Global PV Markets, ISBN 978-3-907281-31-4, https://iea-pvps.org/snapshot-reports/snapshot-2022/

[3] IEA, Global EV Outlook 2022, https://www.iea.org/reports/global-ev-outlook-2022

This is a short version of the paper submitted to the journal Solar RRL.

It has been selected by the Committee of the WCPEC-8 for submission to Solar RRL.

THE IMPACT OF PERFORMANCE DEGRADATION CURVES ON THE PROFITABILITY OF SOLAR PV PROJECTS

Alessandro Virtuani[1] and Luca Morganti[1]
[1] Officina del Sole (O'Sole), Milan, Italy
* Corresponding author: alessandro.virtuani@o-sole.eu

ABSTRACT: Recent literature highlights the fact that degradation curves are frequently not linear. This technical knowledge is, however, neglected by the financial world, which continues to use simplified linear degradation rates to assess the profitability of PV projects. In this work we compute the project-IRR (or unleveraged equity IRR) of utility-scale (50 MWp in size) PV projects deployed in different parts of Europe and carry out a sensitivity analysis of how the profitability of the plant is impacted when non-linear performance degradation curves are used in place of constant degradation rates. In general, a 2-step performance loss curve leads to the highest project-IRRs at equal total loss, whereas the lowest IRRs are achieved if module degradation were undergoing a negative logarithmic or exponential degradation trajectory. Further, the sensitivity of profitability to the different degradation trajectories increases at higher latitudes (i.e. in sites with lower solar resources), for which the absolute project-IRR is understandably lower. Finally, based on this analysis, we provide some indications to project developers and investors on how to make more sound and "reliable" BPs. These include, the application of stress test to BPs to assess the risk that non-linear degradation curves may have on the project profitability.

Keywords: Solar, Photovoltaics, Profitability, IRR, Reliability

1 INTRODUCTION

Solar photovoltaic (PV) electricity has presently reached the market parity in several countries worldwide and is nowadays perceived by investors and financial institutions as a low risk investment. Over the years, however, evidence has shown that the focus on quality all over the PV value chain (i.e. component selection, design, installation, operation and maintenance) is critical in safeguarding the profitability and the economical return of investments. Several aspects, in fact, may endanger the profitability of business plans (BP), particularly when considering unsubsidized solar PV projects. These include: (1) An extended duration of the investments, targeting service lifetimes of 30 years compared to the 20 to 25 years used in the past; (2) The deployment of PV in non-temperate climates, for which the technology track record is still missing; (3) A massive technology transition in the last 3 to 4 years, particularly with respect to solar cells and modules, again with a limited field experience and track record. This includes, as an example, the introduction of solar cells with novel high-efficiency architectures (PERC, SHJ, TopCons,[1] ....), the use of larger cells and modules sizes, the adoption of novel encapsulant materials (such as polyolefins) and polymeric materials (e.g. back-sheets), and novel interconnection technologies. Several recent scientific literature reports about performance degradation rates of PV modules and systems [i, ii, iii, iv, v, vi]. According to Jordan and coworkers, analyzing a large collection of data (i.e. 10'000+) from solar plants, historical median and average annual degradation rates for crystalline silicon (c-Si) cell based solar modules correspond, respectively, to -0.5%/y and -0.7%/y [vii, viii]. Performance loss rates (PLR) for thin-film technologies and for c-Si modules manufactured before year 2000 are larger, as well as PLR reported for installations in harsher climates, like desert and tropical climates [ix, x]. Other publications highlight that degradation curves are frequently not linear [xi, xii], as their temporal evolution may strongly be impacted by different failure and degradation mechanisms occurring in solar panels in different phases of their operational lifetime (i.e. early-life, mid-life, wear-out). Some degradation mechanisms may in fact evolve linearly with time (e.g. yellowing), whereas others – generally occurring in infant or wear-out phases - may lead to strong non-linear behaviors, as potential-induced degradation (PID) or the occurrence of hot-spots. Understandably, these curves are often the superposition of multiple degradation mechanism occurring in a single module or in an array of solar modules. Additional recent works, reporting about the first grid-connected PV plant in Europe (i.e. the TISO solar plant connected to the grid in 1982) [xiii, xiv], have demonstrated that service lifetimes of 35 years are a reasonable target for PV modules deployed in temperate climates. In these works, the authors could clearly correlate the long-term performance of the modules to the aging processes of the encapsulants used in module manufacturing. Further, for several of the modules in operation in the TISO plant the performance has been fairly stable over the first 20 years of operation followed by an acceleration of the degradation rates later. A performance trend consistent with a 2-step (or even a 3-step) degradation curve. This technical knowledge is, however, neglected by the financial world, which continues to use simplified linear degradation rates to assess the profitability of PV projects.

In this work we compute the **unleveraged equity-IRR** (Interest Rate of Return) – or **project IRR** - of utility-scale PV projects deployed in different parts of Europe (in sites with different availabilities of solar resources) and carry out a sensitivity analysis of how the profitability of the plant is impacted when non-linear performance degradation curves are used in place of constant degradation rates, which is the assumption usually adopted in BPs. The **project IRR** (under the assumption of a full equity project) is in fact generally used as a first discriminant to assess the profitability of solar projects,

[1] PERC: Passivated Rear-Emitter Solar Cell; SHJ: Silicon Hetero-Junction, TopCon: Tunnel Oxide Passivated Contact.

even if projects are later realized resorting to financial leverage to increase the return on equity.

2 METHOD AND APPROACH

In this work we use state-of-the-art business plans (BP) with updated figures and compute the profitability (i.e. project-IRR) of unsubsidized (i.e. merchant) utility-scale PV systems (with a size of 50 MWp) developed in Europe, considering three different locations covering a majority of the insolation levels encountered in continental Europe: Siracusa (37°N), Milan (45°N), Hamburg (53°N). This plant size allows to fully benefit from economies of scales and minimize costs that are sometimes difficult to predict (such as grid-connection costs). The linear and non-linear curves used in our sensitivity analysis are shown in Sect.3 (see Fig.1).

2.1 Assumptions and caveats

BPs rely generally on more than 20 input parameters. We provide here the key numbers used in our analysis - including numbers for CAPEX, OPEX, wholesale electricity prices, etc. - without disclosing, for confidentiality reasons, all figures (see Table I). In particular:

a. Geographical scope: the figures (cost structure and selling price of electricity) used in the BP are relative to large-scale projects realized in Italy. The adoption of the same cost structures and electricity selling prices to other countries (e.g. Germany) is understandable questionable, but allows to mitigate complexity and directly compare the different results while keeping focus on the primary research question (i.e. the impact of degradation curves on project's profitability). Additionally, at least in key European markets, the cost structure of PV projects has been notably converging over the years (see Ref. [xv]).

b. Temporal scope: the figures used in the BPs refers to numbers that were commonly used until mid to end 2021: Recent events (including shortages in component supply following the post-pandemic recovery in 2021, the outbreak of the war in Ukraine in February 2022 and a raising global inflation, unwitnessed for over 20 years) makes it extremely difficult today to make sound and resilient business plans. At the time of writing (i.e. summer 2022), CAPEX costs are 20% (or more) higher than the ones used in this work, whereas wholesale electricity prices of electricity and inflation rates are at least four-fold higher than the number used here (see Table I). Again, for the sake of simplicity, to make results directly comparable and to avoid being distract by our primary research question, in our analysis we use figures that were regularly used in the development of merchant projects in Italy until 2021. While being fully aware that some key BP parameters (e.g. inflation, whole sale electricity price, CAPEX, etc.) are presently fully outdated by the global economic situation, we are reasonably confident that we are facing a transition period that will bring us back sooner or later (months or years) to these original values. Arguably, this might not apply to all parameters.

With respect to the long-term plant's performance, again for simplicity, BPs generally make use of two major assumptions:

a. The **degradation of the** overall **plant's performance** is equated to the degradation of the solar modules only. This is understandably not true. Other components – particularly the inverters – may experience a considerable degradation in performance, but inverters and other BOS components can be more easily serviced or substituted at much lower costs (with adequate provisions generally allocated in BPs as future costs), whereas no replacement is foreseen over the lifespan of the project for the solar modules. Beside this, as of today solar modules still account for 30% to 40% of overall project costs. An unplanned substitution of the modules may therefore have a considerable impact on project economics, rather than for example an unplaned substitution of the inverters.

b. **A linear degradation curve** with a constant performance loss rate rate (PLR) of 0.5%/y is generally assumed, interchangeably, for the modules or the full system. Other, PLR found in BP vary between 0.3 and 0.7%/y depending on how conservative the BP is. Nevertheless, non-linear degradation curves are more likely to occur and have been frequently reported in the literature, as discussed in Section 1.

2.2 Key business plan figures

The key parameters used in the BPs are summarized in Table I. As discussed in Sect. 2.1, these are industry standard parameters used in BP for projects developed in Italy in 2021.

Table I: key parameters used in the business plan: general, CAPEX, COGS (cost of goods sold) -including OPEX-, and price assumptions. These can be considered industry standard figures used by the industry for the development of multi-MW_p (20 to 30+ MW_p) solar PV projects in Italy in 2021.

	General			CAPEX	
a	Years in operation [y]	30	a	HW, permitting, EPC [k€/MW_p]	550
b	Installed capacity [MWp]	50	b	Of which module costs (included in previous line) [k€/MWp]	(200)
c	**Module degradation rate [%/y]**	**-0.5**	c	Project acquisition & development [k€/MW_p]	40
d	Inflation rate [%]	1	d	Grid-connection costs [k€/MW_p]	40
			e	**Total CAPEX [k€/MW_p]**	**630**
	COGS (including Opex) & tax			**Pricing assumptions**	
a	O&M [k€/MW_p·y]	6	a	Spot electricity price [€/MWh]	50
b	Insurance [k€/MW_p·y] (*)	3	b	Wholesale electricity price [€/MWh]	47
c	Lease [k€/MWp·y]	0.6			
d	Corporate tax rate [%]	28.8			

HW includes: modules, inverters, transformers, cabling, mounting structures;
O&M includes: repair and service; EPC = Engineering, Procurement and Construction
(*) the insurance covers extreme weather events and vandalism, not underperforming solar modules covered by manufacturer warranties.

3. RESULTS

3.1 Degradation curves

There is plenty of evidence of PV system's underperformance, but only a handful of data-sets are available in the literature describing degradation trends or curves extending over multiple decades.

Figure 1 depicts different degradation curves consistent with a module performance warranty set at 80% of the

This is a short version of the paper submitted to the journal Solar RRL.

It has been selected by the Committee of the WCPEC-8 for submission to Solar RRL.

initial nominal power after 30 years of operation, together with a linear degradation curve of -0.5%/y, the value generally used in BPs (reference, REF), and a linear PLR of -0.25%/y, a hypothetical target for properly designed and maintained solar modules. A sensitivity analysis of project-*IRR* using these performance loss (PL) curves is presented in Figure 2. The *neg exp* PL curve is consistent with a degradation mode rapidly activated during the initial exposure of the modules to sunlight, which – in our assumption - is then saturating at a -20% performance loss. In the case of a *log* PL curve the evolution is progressing less rapidly, but still much faster than a linear degradation trajectory. As mentioned in Section 1, a *2-step* degradation curve is consistent with some of the works published in the literature for well-designed and performing modules [see Ref XIII, XIV]: modules that undergo only modest performance losses for most of their operational life, until a specific wear-our mechanism (e.g. cell-breakage, hot-spots, delamination, etc.) – or a combination of failure mechanisms - is initiated. At that point the loss in performance may evolve in some cases very rapidly. In the 2-step curve, a $PLR_1 = 0$ %/y is set for the first 20 years of operation, followed by a $PLR_2 = -1.8$%/y for years 20 to 30. Failure modes consistent with a negative exponential or logarithmic curve are mentioned in the next Section.

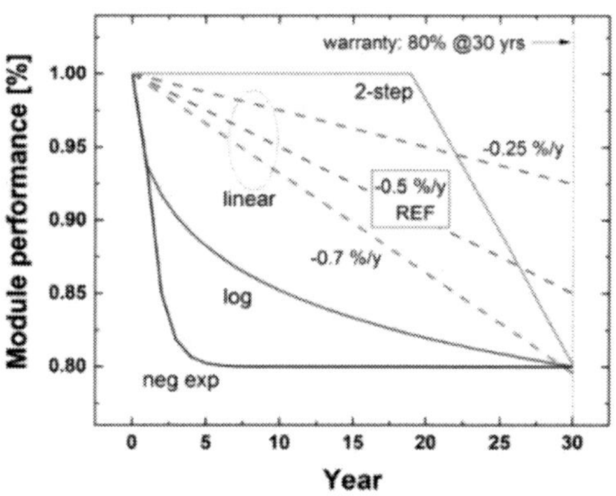

Fig. 1: **(a)** degradation curves consistent with a module performance warranty set at 80% of the initial nominal power after 30 years of operation: (i) linear degradation with a 0.7%/y PLR; (ii) a negative exponential (*neg exp*); (iii) a logarithmic (*log*) degradation; (iv) a *2-step* degradation rate. This degradation curves are compared to the one conventionally used in BPs: a linear degradation with a constant 0.5%/y PLR (reference, REF); and to a linear PLR of -0.25%/y.

All else unchanged, project-IRRs are understandably much higher at lower latitudes (corresponding to higher annual insolation levels). The impact of the different degradation curves on IRR is shown in Fig. 2 (a) and (b); the latter plotting IRR values normalized over the reference case (i.e. linear PLR of 0.5%/y).

The threshold for project selection on this parameter may vary considerably, but it is generally set at a value of 5 to 8%. The 5% treshold set in Fig. 2 (a) should be considered as a minimum value for project qualification.

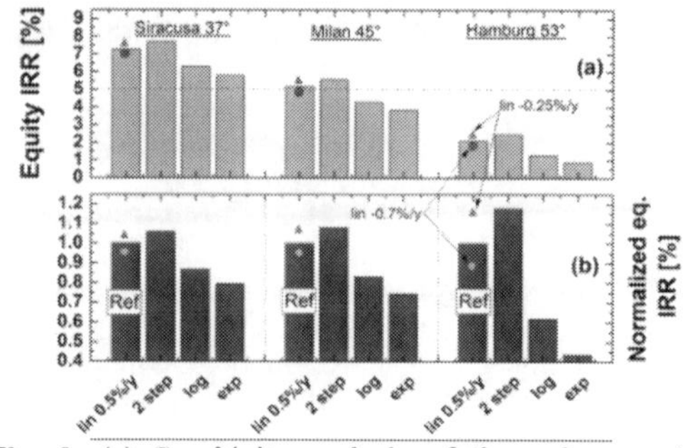

Fig. 2: (a) Sensitivity analysis of the unleveraged Equity-*IRR* computed for utility-scale projects deployed in three sites in Europe using the different degradation curves of Fig. 1; and **(b)** the corresponding IRR values normalized over the reference linear PL curve (with a PLR of -0.5%). The 5% treshold should be considered as a minimum threshold for project qualification. In reality, this threshold is generally higher (6 to 8%).

In general, a 2-step PL curve leads to the highest project-IRRs at equal total loss, whereas the lowest IRRs are achieved if module degradation were undergoing a logarithmic or negative exponential degradation trajectory. In the case of Milan, the project-IRR is just above threshold for the reference scenario (i.e. PLR .0.5%), but would be lower than the threshold using all other more severe degradation curves.

Fig. 2 (b), displaying normalized IRR values, indicates that the sensitivity of profitability to the different degradation trajectories increases with lower solar resources, as in Hamburg in our example, for which the absolute unleveraged IRR is understandably lower.

4. DISCUSSION

To assess the energy yield potential of a site, and the impact of solar resources, generally two metrics are used: P50 and P90. P50 is a statistical level of confidence suggesting that the predicted solar resource (and the corresponding energy yield) may be exceeded (or simultaneously not achieved) with a 50% probability. A P50 confidence level represents a high risk for some lenders or investors. Therefore, other probability indicators such as P90 (for which the estimate is exceeded with a 90% probability) are often used by the industry. To minimize risk, in fact, investors frequently use P90 figures to be confident that the project is bankable. Similarly to what is routinely done in the origination phase, we suggest that an analogous stress-test is applied to BP to assess the impact (and associated risks) that non-linear degradation curves may have on the profitability of BPs, particularly when the projects are deployed in non-temperate climates (with harsher climatic conditions) or when novel technologies (without adequate track-record) are adopted. For this purpose and to avoid excessive complexity, we suggest to use three different performance curves to run this sensitivity analysis and to verify if the minimum return threshold is met: (1) the conventional linear degradation rate with a -0.5%/y PLR (reference); (2) a 2-step curve (optimistic assumption); and (3) the neg log curve (pessimistic assumption). The impact of these curves on projects with different return rates is presented in absolute and relative terms in Fig. 2 (a) and (b), respectively.

5 CONCLUSIONS

Recent literature highlights the fact that degradation curves are frequently not linear, as their temporal evolution may strongly be impacted by different failure and degradation mechanisms occurring in solar panels in different phases of their operational lifetime. This technical knowledge is, however, neglected by the financial world, which continues to use simplified linear degradation rates to assess the profitability of PV projects.

In this work we compute the **project-IRR** (or unleveraged equity IRR) of utility-scale (50 MWp in size) PV projects deployed in different parts of Europe and carry out a sensitivity analysis of how the profitability of the plant is impacted when non-linear performance degradation curves are used in place of constant degradation rates (the usual assumption in BPs).

In general, a 2-step PL (performance loss) curve leads to the highest project-IRRs at equal total loss, whereas the lowest IRRs are achieved if module degradation were undergoing a negative logarithmic or exponential degradation trajectory. Further, the sensitivity of profitability to the different degradation trajectories increases at higher latitudes (i.e. in sites with lower solar resources), for which the absolute project-IRR is understandably lower. Finally, and similarly to what is routinely done by computing P50 and P90 confidence levels for site selection, we suggest that a similar stress test is applied to BPs to assess the impact (and associated risks) that non-linear degradation curves may have on the profitability of BPs, particularly when the projects are deployed in non-temperate climates (with harsher climatic conditions) or when novel technologies (without adequate track-record) are adopted.

6 ACKNOWLEDGEMENTS

We gratefully acknowledge all friends and colleagues at *Officina del Sole* (particularly Luca Parini) and elsewhere who have inspired us and given us motivation for this analysis.

6 REFERENCES

[i] Skoczek A, Sample T, Dunlop ED, Ossenbrink HA. The results of performance measurement of field-aged crystalline silicon photovoltaic modules. In: Progress in Photovoltaics: Research and Applications, *17, 4, Pages*; 2009:227-240

[ii] Polverini D, Field M, Dunlop E, Zaaiman W. Polycrystalline silicon PV modules performance and degradation over 20 years. *Prog Photovolt.Res Appl.* 2013;21(5):1004-1015

[iii] Sánchez-Friera P, Piliougine M, Peláez J, Carretero J, Sidrach de Cardona M. Analysis of degradation mechanisms of crystalline silicon PV modules after 12 years of operation in southern Europe. *Prog Photovolt Res Appl.* 2011;19(6):658-666

[iv] Han H, Dong X, Lai H, et al. Analysis of the degradation of monocrystalline silicon photovoltaic modules after long-term exposure for 18 years in a hot–humid climate in China. *IEEE J Photovoltaics* 2018;8(3):806-812

[v] Rajput P, Tiwari GN, Sastry OS, Bora B, Sharma V. Degradation of mono-crystalline photovoltaic modules after 22 years of outdoorexposure in the composite climate of India. *Sol Energy*. 2016; 135:786-795

[vi] Pozza A, Sample T. Crystalline silicon PV module degradation after 20 years of field exposure studied by electrical tests, electroluminescence, and LBIC. *Prog Photovolt Res Appl*. 2016;24(3):368-378.7

[vii] Jordan DC, Kurtz SR , Photovoltaic Degradation Rates—an Analytical Review, Prog. Photovolt: Res. Appl. 2013; 21:12–29

[viii] Jordan DC, Kurtz SR, VanSant K, Newmiller J. Compendium of photovoltaic degradation rates. *Prog Photovolt Res Appl*. 2016;24(7):978-989

[ix] L. Wei, C.E. Clement, Y. S. Khoo, Y. Wang, A.M. Khaing, T. Reindl, A. Kumar, M. Pravettoni, *Photovoltaic Module Failures After 10 Years Of Operation In The Tropics*, Renewable Energy 177 (2021) 327-335

[x] 3. Olakonu K, Belmont J, Tatapudi S, Kuitche J, TamizhMani G. Degradation and failure modes of 26-year-old 200 kW power plant in a hot-dry desert climate. In: *Conf. Proc. of the 40th ieee* Photovoltaic Specialist Conference (PVSC), Denver; 2014

[xi] DC. Jordan, TJ. Silverman, B. Sekulic, SR. Kurtz, *PV degradation curves: non-linearities and failure modes*, Prog.Photovol; Res. Appl. (2016)

[xii] Jordan DC, Silverman TJ, Wohlgemuth JH, Kurtz SR, VanSant KT. Photovoltaic failure and degradation modes. *Prog Photovolt Res Appl*. 2017;25(4):318-326

[xiii] A.Virtuani, M. Caccivio, E. Annigoni, G. Friesen, D. Chianese, C. Ballif, T. Sample, *35 Years Of Photovoltaics: Analysis Of The Tiso-10-Kw Solar Plant, Lessons Learnt In Safety And Performance—Part 1,* Prog Photovolt Res Appl. 2019;27:328–339, DOI: 10.1002/pip.3104

[xiv] E. Annigoni, A.Virtuani, M. Caccivio, G. Friesen, D. Chianese, C. Ballif, *35 Years Of Photovoltaics: Analysis Of The Tiso-10-Kw Solar Plant, Lessons Learnt In Safety And Performance—Part 2,* Prog Photovolt Res Appl. 2019;27:760–778, DOI: 10.1002/pip.3146

[xv] *Renewable Power Generation Costs in 2020*, International Renewable Energy Agency, Abu Dhabi, ISBN 978-92-9260-348-9 IRENA (2021)

A SOLAR PV PROFITABILITY STUDY AS A FUNCTION OF WHOLESALE MARKET PRICE VARIATION PATTERNS

Philippe Macé[1], Elina Bosch[1], Monica Aleman[1], André Penas[1], Adrien Van Rechem[1]
[1]Becquerel Institute
Rue Royale 146, 1000 Brussels Belgium, +32 2 514 78 04

ABSTRACT: Non-incentivized utility-scale PV systems (typically relying on selling the produced electricity on the wholesale spot day-ahead market) have been appearing on the market in the last years. In 2021, 17% of the global annual market was deployed under such business models and it is expected that this share will continue to grow [1][2]. In parallel, key components of the energy transition (higher PV penetration, electrification of heating, cooling and mobility needs, storage, power-to-X, ...) may impact wholesale electricity price patterns by altering electricity supply demand imbalance in one direction or the other. Therefore, this study is a prospective analysis which assesses how different variations of electricity price patterns could impact PV profitability. Based on the developed methodology, the impacts of different wholesale electricity price patterns (daily patterns, seasonal patterns and occurrence of negative prices) on the profitability of a 50 MWp PV system under two business models have been assessed. Results take the form of charts displaying the NPV (Net Present Value).
Keywords: Economic Analysis, Price Modelling, Electricity Markets, Large Grid-connected PV systems

1 CONTEXT

Non-incentivized utility-scale PV systems have been increasingly appearing on the market in the last years. In 2021, 17% of the global annual market was developed under such business model. Most of this share is attributable to the annual Chinese non-incentivized centralized PV market, but merchant PV development can also be seen in Europe (e.g., Germany, Spain). Moreover, this trend is anticipated to reinforce in the next decade, especially considering the phasing out of support mechanisms on mature PV markets. A recently published report suggests that between 2021 and 2030 as much as 60GW of subsidy-free solar PV could be installed in north-western Europe (UK, Germany, France, Ireland, the Netherlands and Belgium) [1][2]. For non-incentivized solar PV, the applied business models often rely on selling the produced electricity on the wholesale spot day-ahead market. This is also the case for installations with corporate Power Purchase Agreements (PPAs), where the agreements rarely cover the whole production, and the remaining electricity will also be sold on the wholesale market.

In parallel, key components of the energy transition (higher PV penetration; electrification of heating, cooling and mobility needs; battery storage deployment, ...) will impact wholesale electricity prices' patterns as they will alter the supply-demand balance. Depending on the relative pace at which these different elements develop, opposite impacts on wholesale electricity prices' variations may be observed. Therefore, electricity prices and their variations, will be a decisive variable to take into account in the future when estimating the profitability of utility-scale photovoltaic systems. However, regardless of the timeframe considered, the evaluation of this impact is complex since wholesale market prices are influenced by a great variety of factors arising from the demand side (electrification, demand-side flexibility, ...), from the supply side (renewable energy systems' penetration, fossil fuels' prices, carbon prices...) or from other independent factors such as the weather [3].

2 AIM AND APPROACH

Previous studies have focused on determining electricity wholesale market prices' patterns at different time scales, in the frame of a fully decarbonized power systems on the long term (2050 horizon) [4]. Other studies focused on determining the average electricity wholesale market price in the frame of different scenarios on the medium term or on highlighting the correlation between variable renewable energy (penetration or daily combined generation) and average wholesale electricity prices (daily or annual averages) [5][3][6]. These assessments are based on a precise quantitative description in terms of electricity generation mix, electricity demand of the industry and transport sectors, power-to-X and electrification of the demand.

However, these studies do not assess the various direct impacts of these variations of electricity wholesale market prices, for instance on the profitability of renewable energy assets. Moreover, the trajectory towards the assumed fully decarbonized power systems' scenarios is highly uncertain. Therefore, it remains very difficult to determine with a satisfactory level of certainty how electricity wholesale market prices will evolve in terms of time-dependent patterns (and not averages), on the short to medium term. Hence, it is important to consider multiple possible pathways of evolution when assessing the impacts of the variation of these electricity wholesale market prices. Otherwise, the analysis and its conclusions might reveal to be incomplete, thus irrelevant.

Such analysis is of primary importance for conventional actors of the PV sector, as it can help them foresee the evolution of their revenue streams and estimate the level of risk associated with solar PV assets. Even for those who already anticipated the paradigm shift from subsidized to unsubsidized business models, it is crucial to understand to what extent different possible patterns of electricity wholesale market prices on an hourly/daily/seasonal/annual basis as well as their variations could impact the profitability of solar PV assets. Indeed, one of the rising questions for stakeholders active in developing and operating large-scale PV power plants is related to the profitability of their installations in the context of rapidly evolving electricity markets.

This paper will address some of their key concerns by studying how two major business models could potentially be affected by changes in wholesale electricity pricing patterns. For this purpose, this paper deals with the economic and financial impact of wholesale electricity market prices' variation patterns on PV profitability, for various business models. The impact will be quantified through net present value calculations based on a Monte-Carlo analysis (given the inherent stochastic behavior of the selected modelling process). It is also strongly tied to the evolution of the renewable energy market at large (including electricity generating and storing units).

2.1 Simulation of electricity wholesale market prices

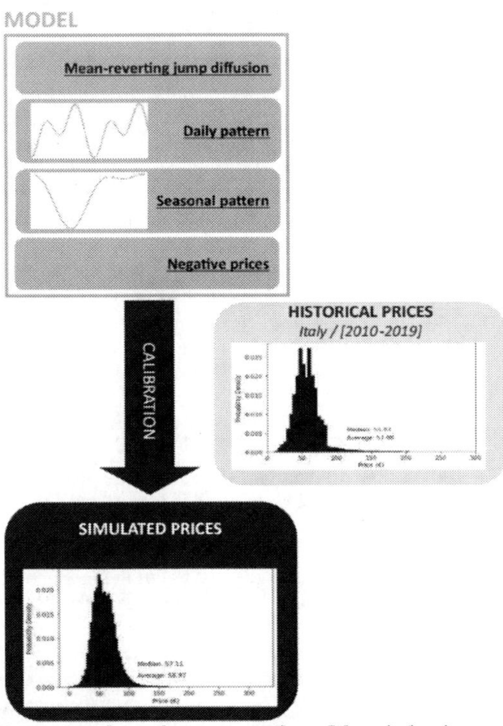

Figure 1: Schematic representation of the wholesale electricity price model developed and used.

Wholesale electricity market prices constitute the basis of this study. They have been simulated on an hourly basis, using Python thanks to a stochastic model (see formula below) based on four key characteristics of wholesale electricity price patterns (See Figure1).

a. Daily pattern: this component reflects the average evolution of electricity's supply-demand imbalance throughout the hours of the day. Typically, the highest prices are found during the morning and evening consumption peaks, while the lowest prices are found in the middle of the day and during the night when consumption is low.

b. Seasonal pattern: this component reflects the average evolution of electricity's supply-demand imbalance throughout the months of the year. Typically, the highest prices are found during the winter (respectively in the summer) when electricity consumption is increased to meet heating or lighting needs (respectively cooling needs), depending on the concerned region.

c. Mean-reverting and jump-diffusion process: this has been presented in several studies and papers as one of the relevant functions to model commodities such as electricity prices [7][8]. This process allows for volatility in prices and the occurrence of prices jumps which however rapidly revert to an average value.

d. Probability of occurrence of negative prices: this component reflects the occurrence of extreme electricity supply-demand imbalances throughout the year, on an hourly basis. Typically, the highest occurrence of negative prices is found during the night and at around mid-day when electricity consumption is particularly low while production (typically limitedly- or non-controllable) of electricity (e.g., nuclear or renewables) is high.

The following equations were used:

$$\log(Pt) = s(t_s) + d(t_d) + Xt$$

With:

Pt: the wholesale price of electricity

$s(t_s)$: a seasonal component

$$s(t_s) = a0 + a1*\sin(2*pi*t_s) + a2*\cos(2*pi*t_s) + a3*\sin(4*pi*t_s) + a4*\cos(4*pi*t_s)$$

t_s is the period of the sine and cosine functions used for the seasonal pattern (i.e., one year)

$d(t_d)$: a seasonal component

$$d(t_d) = b0 + b1*\sin(2*pi*t_d) + b2*\cos(2*pi*t_d) + b3*\sin(4*pi*t_d) + b4*\cos(4*pi*t_d)$$

t_d is the period of the sine and cosine functions used for the daily pattern (i.e., one day)

$$dXt = (\alpha - \kappa Xt)dt + \sigma dWt + J(\mu J, \sigma J)d\Pi(\lambda)$$

With:

α: mean reversion parameter

κ: mean reversion parameter

σ: volatility

Wt: standard Brownian motion

$J(\mu J, \sigma J)$: jump size with μJ a normally distributed mean and σJ a standard deviation

$\Pi(\lambda)$: Poisson process with a jump intensity of λ

In addition, a long-term decreasing trend is applied to the modelled electricity prices to reflect how the increased PV penetration will lead to an enhanced merit order effect. Considered PV penetration evolution scenario is based on the net-zero emission by 2050 in Europe scenario of researchers from Lappeeranta University of Technology (LUT) in Finland [9].

To set the different parameters used for the wholesale electricity market prices' simulation model, calibration based on real historical price data was conducted. Used historical dataset is a 10-years dataset (2010-2019) of hourly prices on the Italian market retrieved from Terna (Italian TSO) website [10]. The used historical dataset does not include 2020, 2021 and 2022, as prices during these years have been affected by highly conjunctural events (Covid-19, Ukraine war) and would have biased the wholesale electricity market prices modelling.

2.2 Studied deviations

Deviations from the current base case (based on real historical data) for three components of the wholesale electricity market price patterns have been studied. These are: daily patterns, seasonal patterns and occurrence of negative prices. These deviations can be supported by one or the other expected change(s) in terms of electricity usage, consumption and production in the energy transition context as non-exhaustively described below.

a. Daily pattern: Exacerbated daily price gaps (i.e., purple arrows) could be induced by higher PV penetration, higher electrification of energy consumption (e.g., heating, cooling and mobility needs). Attenuated daily price gaps (i.e., yellow arrows) could be induced by increased penetration of battery storage units, smart charging or stimulated electricity demand at times of the day when prices are the lowest.

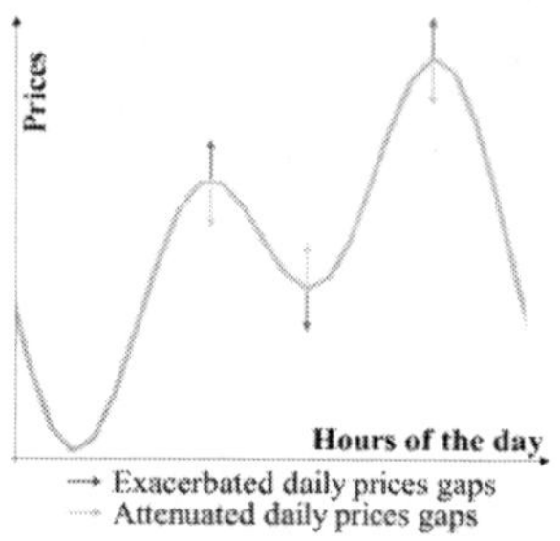

Figure 2: Studied deviations for wholesale electricity market prices' daily patterns

b. Seasonal pattern: Exacerbated seasonal price gaps (i.e., purple arrows) could be induced by higher PV penetration, higher electrification of season-dependent energy consumption (e.g., heating and cooling needs). Attenuated seasonal price gaps (i.e., yellow arrows) could be induced by increased penetration of seasonal storage units based on pumped hydro or green hydrogen, by power to X in general and by increasing interconnexion degrees of the national or regional European grids.

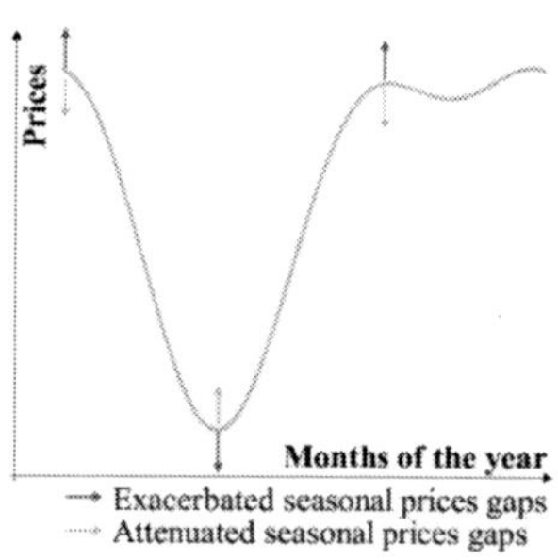

Figure 3: Studied deviations for wholesale electricity market prices' seasonal patterns

c. Occurrence of negative prices: the combination of an increased penetration of electricity generating units with low or inexistent marginal costs of production and with variable generation (such as renewables) as well as the preservation of electricity generating units with higher marginal costs (coal and gas-based power plants and to a lesser extent nuclear power plants), could lead to an increase of negative prices periods. This trend can already be observed on European wholesale electricity markets, with a twofold increase of negative prices hours in summer and spring between 2016 and 2017 and a four to fivefold increase between 2016 and 2019. [11]

As presented, given the high uncertainty related to the way and extent to which these wholesale electricity prices' patterns will be impacted in the short and medium term, they have been studied as part of a sensitivity analysis.

2.3 Reference case and business models for the analysis

The impact on PV profitability has been studied for a utility-scale ground-mounted PV system whose technical and economic characteristics are found in **Table I**.

Table I: Technical and economic characteristics of the studied utility-scale PV system

Variable	Value	Unit
Type	Ground-mounted	-
Capacity	50	MWp
Orientation	South	-
Tilt	30	°
AYI*	1,800	kWh/m².a
System lifetime	30	years
CAPEX	0,55	€/Wp
OPEX	9	€/kWp.a
WACC	6	%

* Average yearly irradiation

The impact on PV profitability has been studied for two different business models whose characteristics are found in **Table II**. The choice of these analyzed business models is supported by their increasing relevance for utility-scale PV and high dependency on wholesale market prices [2]. Note that many types of PPAs (Power Purchase Agreements) exist on the market. In order to facilitate the readability of the results and to focus on the analysis on the impact of wholesale electricity prices' pattern on PV profitability, only the case of a fixed price PPA has been studied.

Table II: Characteristics of the studied business models

Business model 1	Business model 2
Years 1 to 30: 100% of the production is merchant-PV-based	Years 1 to 15: 50% of the production is valued through a fixed price PPA* *and* 50% of the production is merchant-PV-based Years 16 to 30: 100% of the production is merchant-PV-based

* 50 €/MWh

2.4 Indicator(s) used

The two indicators used are economic indicators namely the net present value and the internal return rate. Used formula can be found below.

$$\mathbf{NPV} = -\mathbf{I}\sum_{\mathbf{i=0}}^{\mathbf{T}} \frac{\mathbf{Free\ Cash\ Flow_i}}{(\mathbf{1+d})^{\mathbf{i}}}$$

Where:

NPV: Net Present Value calculated for a theoretical system lifetime of N years

Free Cash Flow$_i$: free cash flow (positive and negative) in year i

I: the initial investment

d: the discount rate

T: theoretical system lifetime

i: a year in the theoretical system lifetime

Based on this formula, the internal return rate is then the discount rate for which the NPV equals to zero.

3 RESULTS AND SENSITIVITY ANALYSIS

3.1 Impact of different daily patterns

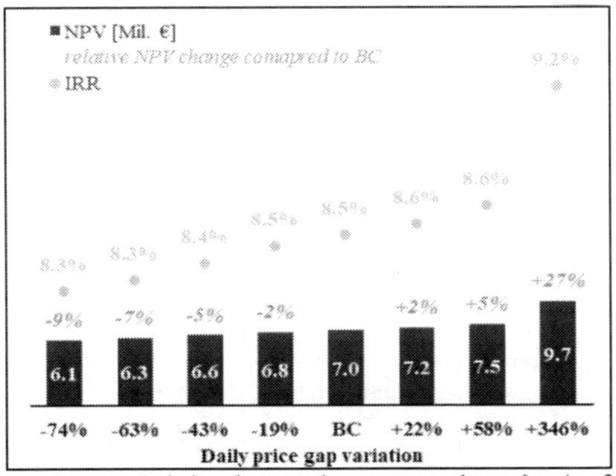

Figure 4a: Relative impact (percentage above bars) of different daily price gap variations (represented on the X axis) on the PV profitability assessed through the NPV (label on the bars) and on the IRR (label above the dots) for BM1.

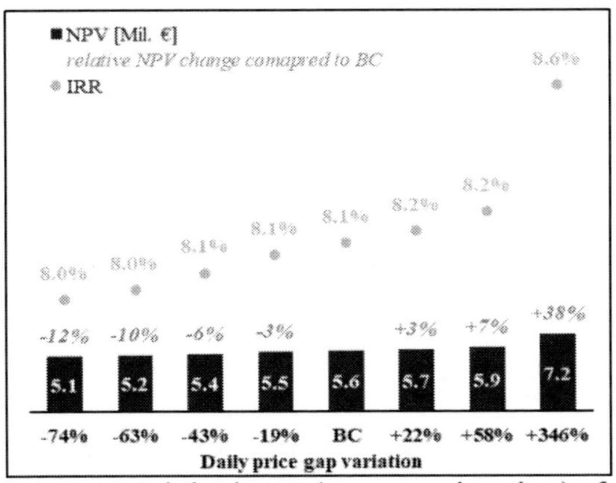

Figure 4b: Relative impact (percentage above bars) of different daily price gap variations (represented on the X axis) on the PV profitability assessed through the NPV (label on the bars) and on the IRR (label above the dots) for BM2.

An exacerbation of daily price gaps (measured between the lowest prices in the early afternoon and the highest prices during the evening) leads to an increase of PV profitability.

This can be explained by the fact that by exacerbating the daily price gaps (purple arrows in Figure 2), PV production benefits more from the higher prices during the morning consumption peak than it suffers from the lowest prices during the low consumption in the early afternoon.

The higher the exposure to hourly market prices (i.e., the higher the share of PV production which is merchant-PV-based), the higher the impact of the daily pattern variations on PV profitability.

3.2 Impact of different seasonal patterns

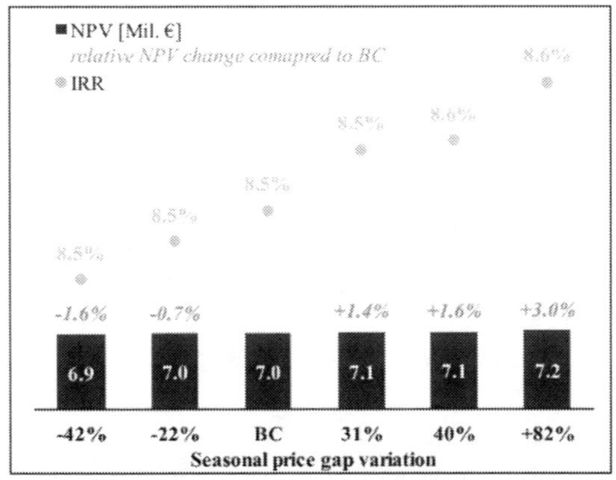

Figure 5a: Relative impact (percentage above bars) of different seasonal price gap variations (represented on the X axis) on the PV profitability assessed through the NPV (label on the bars) and on the IRR (label next to the dots) for BM1.

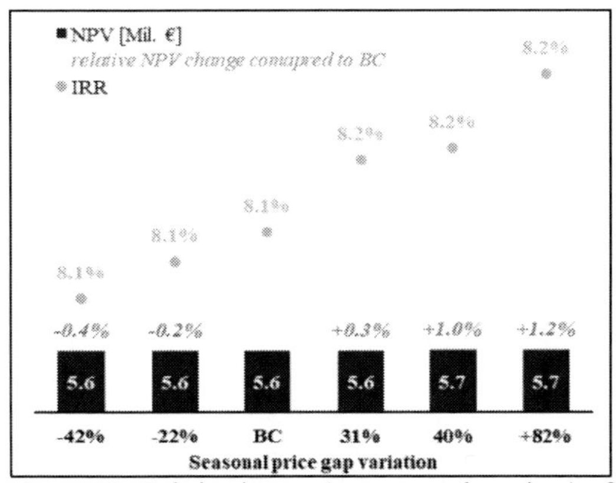

Figure 5b: Relative impact (percentage above bars) of different seasonal price gap variations (represented on the X axis) on the PV profitability assessed through the NPV (label on the bars) and on the IRR (label next to the dots) for BM2.

An exacerbation of seasonal price gaps (measured between the lowest prices in the early spring and the highest prices during the winter) leads to an increase of PV profitability.

This can be explained by the fact that by exacerbating the seasonal price gaps (purple arrows in Figure 2), PV production benefits more from the higher prices during the summer which are observed in Italy than it suffers from the lowest prices during the early spring.

The higher the exposure to hourly market prices (i.e., the higher the share of PV production which is merchant-PV-based), the higher the impact of the daily pattern variations on PV profitability.

3.3 Impact of different negative prices occurrences

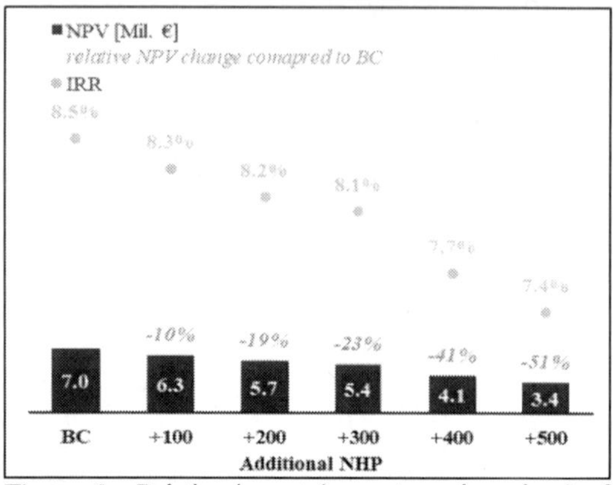

Figure 6a: Relative impact (percentage above bars) of different negative hourly prices (NHP) occurrence (represented on the X axis) on the PV profitability assessed through the NPV (label on the bars) and on the IRR (label next to the dots) for BM1.

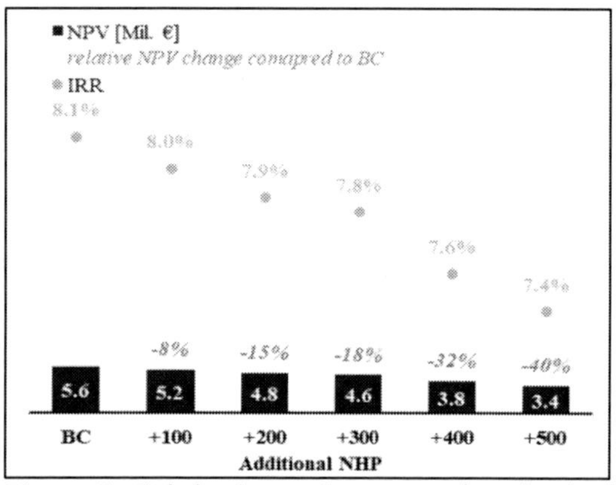

Figure 6b: Relative impact (percentage above bars) of different negative hourly prices (NHP) occurrence (represented on the X axis) on the PV profitability assessed through the NPV (label on the bars) and on the IRR (label next to the dots) for BM2.

Unsurprisingly, an increase of the occurrence of negative hourly prices leads to a deterioration of PV profitability.

Under the used model to simulate hourly wholesale market price, negative prices take place predominantly during the night but not only. Under a configuration of limited wind energy and nuclear penetration combined with a high penetration of PV energy, the impact of the same number of negative hourly prices would be more detrimental to PV profitability as episodes of negative prices would concentrate at times when PV generation is important.

The higher the exposure to hourly market prices (i.e., the higher the share of PV production which is merchant-PV-based), the higher the impact of the daily pattern variations on PV profitability.

When the number of negative hourly prices increases by 500 compared to the base case (approximately. 50 hours per year), PV profitability under the studied cases is approximately halved. As a reference, in 2020, in Germany about 300 hours with negative hours took place in a context of exceptionally low demand (COVID19-related lockdowns during spring) [6].

4 CONCLUSIONS

In the energy transition context (more solar PV, e-mobility, heat pumps, battery storage, …), wholesale electricity prices patterns are bound to vary. As the shift towards non-incentivized utility-scale PV occurs, wholesale electricity prices patterns will directly impact PV profitability.

Nonetheless, our research shows that while the profitability of solar PV plants exposed to wholesale market prices' volatility can be damaged, it is robust enough to remain attractive. Moreover, this is true even in extreme cases.

Indeed, our results show that the exacerbation of daily price gaps (e.g., through higher electrification of demand and higher PV penetration) increases the PV profitability by 5% when the gap is increased by 50%. On the contrary, the PV profitability decreases by 5% when the gap is decreased by 40%.

The exacerbation of seasonal price gaps (e.g., through higher electrification of demand and higher PV penetration) leads to a similar outcome, with an increase of the profitability of solar PV by 2% when the gap is increased by 50%. On the contrary, the PV profitability decreases by 1,5% when the gap is decreased by 40%.

The last trend tested, i.e. the occurrence of negative prices, shows a decrease of the profitability of PV of 20% when the number of negative prices is increased by 4 (from 50 h/a to 200 h/a).

Overall, this demonstrates that although the shift from subsidized-based business models to market-based ones come with an increase of exposure to market prices' shocks, profitability of solar PV under such business model remains attractive. The risk of "cannibalization" due to an increase of solar and other renewables that has been for long highlighted by multiple stakeholders thus seems limited.

Other work should be conducted to evaluate the impact of a combination of PV with various energy storage systems (batteries or electrolyzers).

5 FUNDING AND ACKNOWLEDGMENT

The work described in this publication has received funding as part of the SERENDI-PV project from the European Union's Horizon 2020 research and innovation programme under grant agreement N° 953016.

6 REFERENCES

[1] IEA PVPS, Task 1, Trends in Photovoltaic Applications 2022.

[2] Aurora Energy Research, 'Managing Market Risk in Renewables', 2021.

[3] Impacts of Variable Renewable Energy on Bulk Power System Assets, Pricing, and Costs'. https://www.researchgate.net/publication/324759287_Im

pacts_of_Variable_Renewable_Energy_on_Bulk_Power_System_Assets_Pricing_and_Costs (accessed Feb. 03, 2021).

[4] T. Boßmann, L. Fournié, and G. Verrier, METIS Study S14: Wholesale market prices, revenues and risks for producers in power system with high shares of variable RES. 2018.

[5] December 2018 – Europe's Electricity Market Design: 2030 and beyond, p. 73, 2018.

[6] Quarterly report on European electricity markets. Available : https://energy.ec.europa.eu/data-and-analysis/market-analysis_en

[7] Mayer, Klaus; Schmid, Thomas; Weber, Florian (2011): Modelling electricity spot prices - Combining mean-reversion, spikes and stochastic volatility, Working Paper, No.2011-02, Technische Universität München, Centre for Entrepreneurial and Financial Studies (CEFS), München

[8] Michel Culot, Valérie Goffin, Steve Lawford, Sébastien de Meten, Yves Smeers. Practical stochastic modelling of electricity prices. The journal of energy markets, Infopro Digital, 2013, 6 (1), pp 1-40. hal-01021603

[9] Ram., M.; Bogdanov, D.; Aghahosseini, A.; Gulagi, A.; Oyewo, A.S.; Child, M.; Caldera, U.; Sadovskaia, K.; Farfan, J.; Barbosa, LSNS.; Fasihi, M.; Khalili, S.; Dalheimer, B.; Gruber, G.; Traber, T.; De Caluwe, F.; Fell, H.-.; Breyer, C. Global Energy System based on 100% Renewable Energy – Power, Heat, Transport and Desalination. Lappeenranta University of Technology and Energy Watch Group. March 2019

[10] Terna, "Terna Driving Energy," [Online]. Available: https://www.terna.it/en. [Accessed 2021].

[11] 20200131 quarterly report on European electricity markets q3 2019.pdf'. Accessed: Feb. 03, 2021. [Online]. Available: https://www.euneighbours.eu/sites/default/files/publications/2020-01/20200131%20quarterly_report_on_european_electricity_markets_q_3_2019.pdf.

SECTOR COUPLED ENERGY MODEL FOR THE
EUROPEAN ELECTRICITY, HEAT, AND TRANSPORT SECTORS - RESULTS

Alexander Blinn, Henrik te Heesen
Trier University of Applied Sciences, Environmental Campus Birkenfeld,
Institute for Operations and Technology Management, Campusallee, 55768 Hoppstädten-Weiersbach, Germany,
Mail: h.teheesen@umwelt-campus.de

ABSTRACT: Europe can ensure a reliable, economical, and environmentally compatible energy supply by moving away from finite fossil fuels to regenerative and sustainable energy production. The extensive substitution of fossil fuels by regenerative energy sources and the associated decoupling of energy production and resource consumption takes the limitation of irreversible climatic changes as its goal. This can be achieved in particular by using photovoltaic systems and wind energy. The transition from a centralized, unidirectional energy system to a decentralized, bidirectional system can be accomplished with photovoltaics and wind power. Due to the characteristics of fluctuating power generation plants, the need for storage increases but can be reduced to a minimum through energy demand reduction measures, such as general technical efficiency improvements and changes in consumer behavior, as well as through load shifting and the expansion of domestic and cross-border transmission grid capacities. In this context, sector coupling combines the electricity, heat, and transport sectors into a comprehensive system. Energy system modeling contributes to understanding and building the future energy supply system. This publication aims to develop an energy system that is 100 percent based on renewable energy by 2050. For this purpose, a complete sector-coupled energy model (UCB SEnMod) of all energy sectors is modeled, where the interaction of energy supply and energy consumption can be investigated with an hourly resolution.
Keywords: Modelling, Photovoltaic, Software, System, Strategy

1 INTRODUCTION AND APPROACH

It is uncontested that, to achieve net-zero carbon dioxide emissions, a fundamental change from demand-driven energy supply by centralized large-scale power plants to a more complex decentralized and bidirectional energy system is necessary. In that system, there is always a balance between electricity generation from renewable sources that can only be controlled and predicted to a limited extent and energy use that is adapted to the time. Sector coupling, the temporary use of flexibly deployable generation plants, and the need for electricity storage technologies in an overall system mainly based on fluctuating energies represent the core of the future energy system. This work's central question is the feasibility of all sectors' energy supply with 100 percent renewable energy sources. In the first part of this work, the methodological approach of the modeling was presented [1, 2]. This work will show and discuss the central results of the calculations for Europe in 2050. The overall system's optimization strategy is the installed plant and storage capacity.

Figure **1** shows the basic structure of the energy model UCB-SEnMod. The essential elements of the energy model are supply and load profiles, which are projected into a possible future energy system in 2050. The power generation plants and consumption data installed in the countries of Europe are used to determine the energy profiles of today's power and heat generation and final energy consumption. It is assumed that renewable energies will cover the total energy demand in all sectors. The energy required for this is generated within Europe.

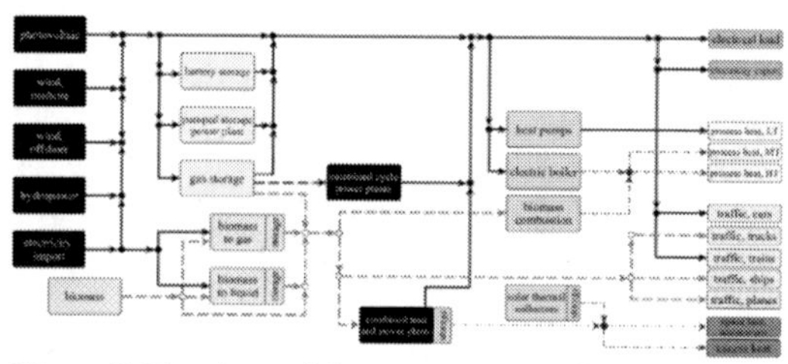

Figure 1: Topology of the system's constellation using the energy model UCB-SEnMod.

2 RESULTS

The hour-by-hour comparison of energy generation and demand shows that a reliable and environmentally compatible energy supply is possible based on 100 percent renewable energy in 2050. In the scenario considered, it is assumed that battery-electric drives will dominate in private transport. Furthermore, it is assumed that hydrogen fuel cells will prevail for trucks. *E-fuels* power ships and aircraft. Heat pumps and solar thermal energy are used mainly in the building sector. Electric boilers and green-generated methane provide process heat in industry and commerce. This results in a total electricity demand of around 6,050 TWh. With more than 4,000 TWh or two-thirds, the electricity sector accounts for the largest share. Just under 1,300 TWh, or one-fifth of the net electricity demand, is responsible for the industry's generation of process heat. The remainder is divided equally between the low-temperature sector (space heat and hot water) and transport. The capacity of photovoltaic and wind power plants required to meet the growing electricity demand is estimated at nearly 4,200 GW. An additional 2,500 GW must be considered if the fuels demanded in the transport sector for air and inland waterway transport were produced synthetically. That would correspond to 13 to 21 times the current system capacity. The transformation from today's demand-driven energy supply (based on a few large power plants) to a highly fluctuating and decentralized energy system requires a renewal of the supply model. The storage of electrical and thermal energy and the use of flexible power generation systems, preferably with heat extraction, will play a key role, accompanied by further suitable measures for flexibilization and load management. In addition, a suitable power ratio of wind and solar plants – adapted to the different climatic regions – can maximize total load hours.

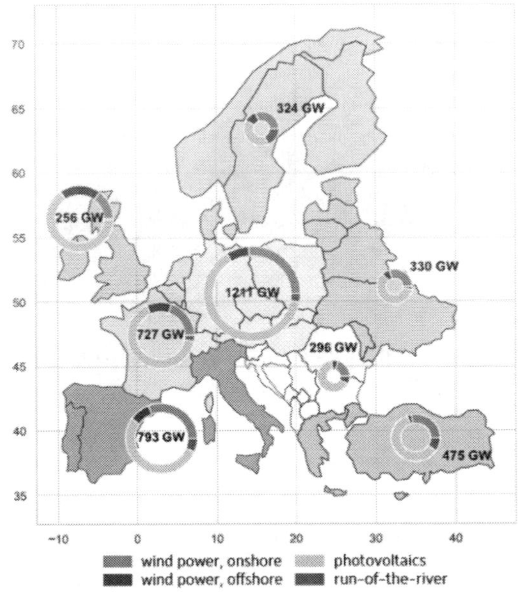

Figure 2: Assumed installed capacity of renewables in the regions of Europe in 2050.

Furthermore, the dispersion of the plant orientation of photovoltaics, adapted to the sun's course (from east to west), leads to the fact that the yield profile can be flattened, and extreme feed-in peaks can be avoided [3]. In particular, the extraordinary importance of the European interconnected grid system should not go unmentioned at this point, as it enables a large-scale and cross-border exchange of electricity, especially on windless and cloudy days, and thus increases the stability of the energy supply enormously. In Europe, the feed-in power of installed wind and photovoltaic systems over a continuous period of 48 hours was below 10 percent only once – when storage facilities were almost entirely filled. Furthermore, the reinforcement and expansion of the grids, both regionally and across country borders, reduce the need for storage [4]. This again illustrates the value of the storage technologies already discussed, and a large-scale, cross-border supply insecurity, presumed when viewed in isolation from the individual countries, is significantly reduced by the European interconnected system. The range of values of the frequencies of occurring dark periods of different periods is visualized in Figure 4 for selected countries. The gray-shaded area, whose boundaries are defined by the frequency distribution of the dark lulls of Luxembourg (cobalt blue) and Norway (turquoise blue), marks the area within which the frequency distributions of the dark recesses of all countries are located. This is highest for Luxembourg and zero for Norway - due to its well above average total load hours of onshore wind turbines of 4,471 hours, additionally supported by offshore turbines. The countries between the borders move around the frequency distribution of the European interconnected system shown in black, which integrates all European countries. Although for a dark period of one hour, the number with which this occurs is lower in eleven countries than in the interconnected system, for a sustained 12-hour dark period, the frequency already drops to eight, much faster than in most countries disconnected from the connected system.

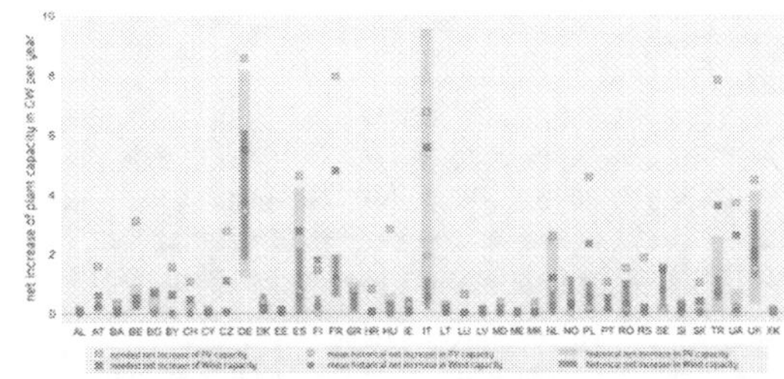

Figure 3: Net additions of photovoltaic (PV) and wind power (wind) plants between 2011 and 2019 in GW per year. The circular markers indicate the average addition rates in the respective countries during the period mentioned. To achieve the targeted plant capacities for 2050, the average annual net addition rates (as of 2018), indicated by the crosses, are required.

Only four countries, Denmark, Northern Ireland, Norway, and the United Kingdom, have lower frequencies. The installed capacity calculated in this model, which is necessary to meet energy demand in all sectors, is visualized in Figure 2. For a more straightforward presentation, the European countries have been divided into a total of eight regions, distinguished by color. In Central Europe, an average annual net addition of 7.6 GW of onshore wind, 2.6 GW of offshore wind, 22.4 GW of photovoltaics, and about 0.6 GW of run-of-river plants is required to achieve the assumed installed plant capacity in 2050. In southwestern Europe (Portugal, Spain, and Italy), an average of 6.4 GW of onshore wind, 2.4 GW of offshore wind, 12.4 GW of photovoltaics, and 1.2 GW of run-of-river hydropower plants will need to be added per year starting in 2018. In Western Europe (France, Belgium, the Netherlands, and Luxembourg), 3.9 GW of onshore wind energy, 2.7 GW of offshore wind energy, 14.3 GW of photovoltaics, and 0.7 GW of run-of-river power plants are required. In Turkey, Greece, and Cyprus, an average of 3.8 GW of onshore and 0.4 GW of offshore wind energy must be added annually, as well as 8.7 GW of photovoltaics and 1.2 GW of run-of-river power plants. In Eastern European (Belarus, Estonia, Latvia, Lithuania, Moldova, Ukraine), an average of 3 GW of onshore and 0.8 GW of offshore wind power plants, 6.2 GW of photovoltaics, and 0.2 GW of run-of-river power plants must be added per year to achieve the expansion targets. In the Scandinavian (Denmark, Norway, Sweden, and Finland), annual additions of 2.6 GW of onshore wind, 1.1 GW of offshore wind, 3.9 GW of photovoltaics, and 1.9 GW of run-of-river power plants are required. In Southeastern Europe, 1.8 GW/a of onshore and 0.6 GW/a of offshore wind, 6.3 GW/a of photovoltaics, and 0.8 GW/a of run-of-river power plants must be installed to achieve the assumed expansion target. Moreover, in Northwestern Europe (Northern Ireland and the United Kingdom), an annual addition of 0.6 GW of onshore and 1.3 GW of offshore wind, 4.8 GW of photovoltaic systems, and 37 MW of run-of-river power plants are needed to reach the installed capacity in 2050 shown in the figure. A detailed look at individual countries shows that current trends in net additions of wind power and photovoltaic systems are insufficient to achieve the required installed capacity for 2050. Figure 3 shows that the average net addition between 2011 and 2019 was far below the required addition rate in each country. The record year of 2011 shows that things can be different: Italy achieved the world's largest solar addition to date in that year, with over 9 GW of plant capacity – Germany managed to increase its plant capacity by around 8 GW in the same year. However, this solar boom in Italy and Germany, which was achieved through government subsidy programs, did not last long: in the following year, Italy's installation rate already fell to 3.6 GW, while Germany added another 8 GW in 2012, but in 2013 and the following years until 2019, it fell to well below 4 GW per year in some cases.

The cross-border exchange of electricity takes place in two steps. Our methods paper describes the applied method in detail [1]. The upstream power transit will be referred to as IE1 and the downstream storage-integrating process as IE2 in the following. Figure *5* shows the chord diagram of the summed IE1 and IE2 power flows between the countries and their neighboring countries in 2050. Energy transfer in the IE1 process is much larger than in the downstream IE2 process. Germany exports about 39.6 TWh of electrical energy to its neighboring countries. More than half of this, around 56 percent, is transferred to France, Belgium, and Sweden. A further 32 percent goes to Norway, the Czech Republic, and Austria. The remaining 12 percent is split between Switzerland, Poland, the Netherlands, Denmark, and Luxembourg. Germany obtains around 70 percent of its import requirement of 30.5 TWh from Norway, Austria, the Netherlands, and Sweden. Only for Norway, Austria, the Netherlands, Denmark, and Poland does the import of electrical energy exceed the export. The maximum interconnection capacity required to transmit the electrical power is around 58 GW. By comparison, the final report for an integrated energy concept published by the German Federal Ministry of Transport and Digital Infrastructure puts the cross-border interconnection capacity at 40 GW in its scenario for the year 2050. However, import and export capacities on a larger scale are generally possible. However, import and export capacities are generally possible on a larger scale [5] but can also be significantly reduced in some cases by slightly decreasing the transmission benefit [6].

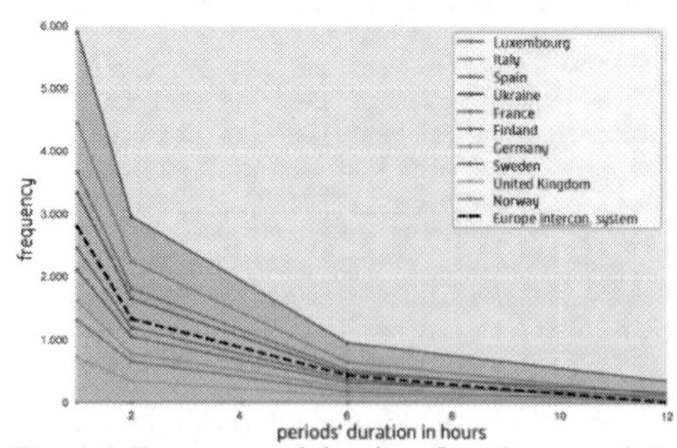

Figure 4: Frequency and duration of continuous periods of time in which the average feed-in power from photovoltaics and wind power is below 10 percent of the installed nominal power. The area in which the frequencies of all countries' periods lie is gray.

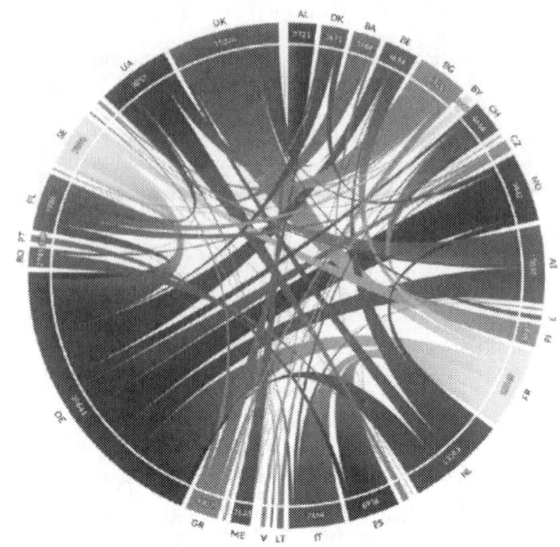

Figure 5: Chord diagram of simulated electricity flows (IE1 + IE2) between countries and their neighboring countries in 2050 in GWh.

To the greatest possible extent, the electrification of consumption areas by substituting combustion-based technologies with more efficient electricity-based technologies leads to a significant reduction in primary and final energy consumption by reducing conversion losses and shortening the conversion chains - despite the assumed increase in the demand for net energy. Most primary energy is provided by photovoltaics, wind, and run-of-river power. In addition, agriculturally produced biomass and the direct use of environmental heat make a not inconsiderable contribution to the primary energy supply. For example, the direct use of electricity and environmental heat to provide space heating and domestic hot water can significantly reduce consumption in the building sector. In addition, the energy refurbishment of buildings allows for lower heating levels, which benefits the use of heat pumps and solar thermal systems [3]. The simulation results show that direct electricity use in many areas significantly reduces the final energy demanded. In particular, battery-electric cars offer great potential for reducing consumption due to their much more efficient energy use. In addition, electromobility would serve the energy system in providing an energy counterbalance at midday when photovoltaic feed-in is high or at night when wind power reaches its maximum.

On the other hand, the use of hydrogen seems more likely in road-based freight transport. However, its increased use in other areas is also problematic since hydrogen must be produced electrolytically, liquefied, and compressed by high electrical energy input and then converted back into electricity when needed, with enormous energy loss in a fuel cell or gas turbine. The same applies to the use of synthetically or biogenically produced fluids – whose use, however, offers decisive advantages in aviation and shipping.

3 DISCUSSION AND OUTLOOK

In 2016, fossil and nuclear power generation technologies represented more than 58 percent of Europe's 834 GW of installed plant capacity. Conventional plants contributed over 72 percent of the total electrical power generation of 3,275 TWh [7]. The previous chapter showed that the share of renewables in total installed plant capacity and electrical energy generation will increase in the coming years and decades, reaching 4,432 GW and 7,190 TWh in 2050.

A fundamental change in the energy system is necessary to achieve zero carbon dioxide emissions. Sector coupling, the temporary use of flexible generation plants, and the need for electricity storage technologies are at the core of the future energy system. The choice of technologies in the heat and transport sectors is decisive for future electricity demand. The increased use of hydrogen in power generation will inevitably lead to increased generation capacity. About three kilowatt-hours of electricity generated from hydrogen are required for one kilowatt-hour of electricity. Assuming equal shares of electricity from wind power, photovoltaics and hydrogen, this would lead to a doubling of electricity demand and costs. Added to this are the immense costs of the technology and infrastructure required to convert and transport hydrogen. The use of hydrogen makes sense if it is provided with surplus capacity from renewable generation plants. However, this requires a massive expansion of renewable energies. Although some wind power plants in northern Germany are already being shut down, this is more due to grid bottlenecks and less because the electricity production would exceed its demand. In addition, local overproduction is significantly reduced in a coherent interconnected system [8].

The advantages of synchronizing the increase in renewable energy plants with grid expansion are apparent. However, the timing of restructuring within the sectors is also crucial. For instance, too early development of a hydrogen economy leads to increased use of fossil fuel power plants to provide the additional energy needed for hydrogen electrolysis and thus increase greenhouse gas emissions. The technologies used in the future heating and transport sectors will also significantly impact. On the one hand, heat can be provided by directly converting electrical energy using electric boilers. The efficiency here is almost 100 percent, and the plant technology is comparatively inexpensive [9]. However, the use of heat pumps is far more efficient. By using environmental energy, a multiple of heat can be provided relative to the electrical work used. A third option is the use of hydrogen. The gas produced using electrolyzers and renewable electricity can either be used directly to provide heat or converted into methane by a subsequent methanation process. The synthesizing of methane would allow continuing to use of the existing gas infrastructure in Europe (the existing gas infrastructure cannot be used directly for hydrogen due to its composition and combustion properties for pure hydrogen [10]). However, considering hydrogen's conversion and transport losses, its use in the low-temperature heating sector leads to an increase in electricity demand of 500 to 600 percent compared to heat pumps [11]. When methane is used, the electricity required for this purpose increases significantly again. Due to this immense electricity demand and the technology and conversion of the gas network needed for production and transport, the use of hydrogen in this segment is therefore not recommended. Since heat pumps are not suitable for the high-temperature level required for industrial processes, using synthetically produced methane is conceivable here. Compared to the direct conversion of electricity to heat, synthetic gases - from generation to transport to combustion – leads to a doubling of the electricity requirement. This shows that the increased use of hydrogen leads to significantly higher costs; therefore, the use of hydrogen should be limited to sectors where synthetic fuels are unavoidable. The heat demand for space heating and hot water in the building sector and, to a certain extent, in the industry is, therefore, not in competition with the use of hydrogen in other sectors [11].

Therefore, to limit the required renewable generation plants, heat pumps were used in this work, to a large extent in the building sector and entirely in the low-temperature segment of the industry. However, since it can be assumed that, for various reasons, not every user will choose to operate a heat pump, solar thermal systems and CHP systems were provided for a particular portion of the low-temperature heat demand in the building sector. Electric boilers are used for the vast majority of the medium- and high-temperature industrial segment, and biogenic fuels, which can be produced both synthetically and agriculturally, are kept low. Currently (year 2018), almost the entire vehicle fleet is based on combustion engines. If in the future, personal motorized transport were to be powered by synthetically generated liquid fuels (liquid e-fuels), this would lead to a significant increase in the required renewable energy capacity due to the associated conversion losses and, consequently, the additional demand for electrical energy. The amount of biodiesel or other biogenic fuels to be produced for this purpose must also be viewed critically since the significant expansion of energy crop acreage required for this purpose may compete with food cultivation, given future population growth [12] and, as a monoculture, endanger biological diversity [13]. In addition, rising demand for agrofuels may lead to an increase in food prices (cf. Food Price Crisis 2007-2008), and the climatic balance may further deteriorate due to the clearing of virgin forests for the cultivation of plantations. A report by the Swiss Federal Laboratories for Materials Testing and Research, which examined the environmental and health costs of greenhouse gas emissions from various biofuels, concluded that biogenic fuels cannot fully replace fossil fuels but that they can contribute to sustainable resource conservation if used in a supporting role [14]. Furthermore, the fundamental question arises as to how sensible the energetic use of land for the photosynthetic production of biomass is if a more than 20-fold higher yield density could be realized by photovoltaics. Nevertheless, the fundamental question arises regarding which drive systems should be recommended for which vehicle types.

For example, an internal combustion engine powered by liquid fuel generated from green electricity requires five to six times as much electrical energy as a battery-electric vehicle over the entire efficiency path. Fuel cell electric operation is also loss-intensive. A car powered by green hydrogen needs about three times the electrical energy as the exact vehicle, which would be battery electric. Helmers and Marx [15] discuss in a study of electric vehicle characteristics that the well-to-wheel efficiencies already achieved today by battery electric vehicles are 59 to 80 percent, mainly due to the low conversion losses in the powertrain. Compared to today's fossil fuel-powered vehicles, battery electric cars consume only a quarter of the energy. Therefore, according to the authors, this efficiency leap makes them the only alternative technology in personal motorized transport. The most energy-efficient drivetrain, however, is rail-based electromobility, which has already been used for many decades in trains. In principle, this form of vehicle energy supply via current-carrying contact wires in the form of an overhead line or a conductor rail is also possible for road-bound vehicles, which is currently being tested in the eHighway-Schleswig-Holstein (FESH) field trial for electrically powered overhead line trucks near Lübeck [16]. However, it remains to be seen whether this will find broad acceptance among the population for the construction of the overhead line systems and reconstruction of the road infrastructure required. Since a high vehicle weight significantly reduces the driving range with battery drives and, in addition, larger batteries are required due to the higher vehicle weight, the fuel cell offers a sensible alternative for heavier vehicles. In contrast to batteries, the energy density of compressed and liquefied hydrogen is similar to that of fossil fuels. However, the situation is different in other transportation segments. Due to the much higher energy density of liquid e-fuels, they have a decisive advantage in aviation and, to some extent, in shipping. The gravimetric energy density of liquid fuels is 60 to 100 times that of lithium-ion batteries [4]. Since the fuel tank of a passenger aircraft accounts for about 20 to 30 percent of the total weight, the use of liquid fuels seems most likely here. Thus, it is also evident here that the choice of propulsion concepts in all segments of the transportation sector strongly influences the future electricity demand and, therefore, the necessary capacities of wind and photovoltaic systems to be installed.

5 SUMMARY AND OUTLOOK

This paper showed that a reliable and environmentally compatible energy supply based on 100 percent renewable energies is possible. Energy supply from solar and wind power form the pillars of the future energy system. The capacity of photovoltaic and wind power plants required to meet the growing electricity demand is estimated at almost 4,200 GW - based on the assumptions made - and a further 2,500 GW would be required if the fuels demanded in the transport sector were produced synthetically. This would correspond to 13 to 21 times the current plant capacity. The shift from today's demand-driven energy supply - based on a few large power plants - to a highly fluctuating and decentralized energy system requires a renewal of the supply model. The storage of electrical and thermal energy and the use of flexible power generation systems, preferably with heat extraction, will play a key role here, accompanied by further suitable measures for flexibilization and load management. In addition, a suitable power ratio of wind and solar plants - adapted to the different climatic regions - can maximize total load hours.

Furthermore, the dispersion of the plant orientation of photovoltaics adapted to the sun's course (from east to west) leads to the fact that the yield profile can be flattened, and extreme feed-in peaks can be avoided [3]. In particular, the high importance of the European interconnected system should not go unmentioned at this point, which enables a large-scale and cross-border exchange of electricity, especially on windless and cloudy days, and thus increases the security of supply enormously. In Europe, for example, the feed-in power of installed wind and photovoltaic systems for 48 hours was below 10 percent only once - when storage facilities were almost full. Furthermore, the reinforcement and expansion of the grids, both regionally and across borders, reduce the need for storage. The electrification of the consumption areas to the greatest possible extent by substituting combustion-based technologies with more efficient electricity-based technologies leads to a significant reduction in primary and final energy demand by reducing conversion losses and shortening the conversion chains - despite the assumed increase in the demanded helpful energy. This is illustrated once again in Figure *6*. Most of the primary energy is provided by photovoltaics, wind, and run-of-river power. In addition, biomass produced by agriculture and the direct use of environmental heat makes a not inconsiderable contribution to the primary energy supply. For example, the direct use of electricity and environmental heat to provide space heating and domestic hot water can significantly reduce consumption in the building sector. In addition, the energy refurbishment of buildings allows for lower heating levels, which benefits the use of heat pumps and solar thermal systems. The figure illustrates that the direct use of electricity leads to a significant reduction in the final energy demand in many consumption areas. It can be seen, for example, that battery-electric cars offer great potential for reducing consumption due to their much more efficient energy use. A 100 percent electrified individual transport system would increase electricity demand by only 7 percent. In addition, electromobility would serve the energy system by providing a welcome energy counterbalance at midday when photovoltaic feed-in is high or at night when wind power reaches its maximum.

On the other hand, the use of hydrogen seems more likely in road-based freight transport. However, the increased use of hydrogen in other areas proves to make little sense since hydrogen must be produced electrolytically, liquefied, and compressed using a large amount of electrical energy and then converted back into electricity in a fuel cell if required, with further loss of energy. The same applies to the use of synthetically or biogenically produced fluids - whose use, however, offers decisive advantages in aviation and shipping.

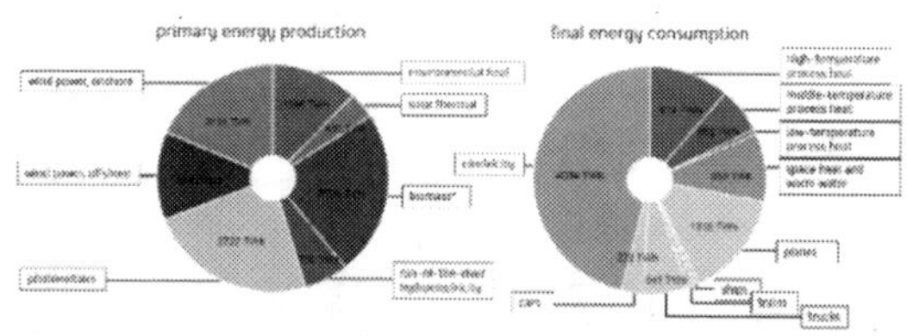

Figure 6: Energy production and consumption in Europe in 2050. In addition to photovoltaics, wind, and hydropower, primary energy is also provided by biomass, solar thermal, and environmental heat (used by heat pumps). The figure for biomass corresponds to the final energy used.

REFERENCES

[1] A. Blinn und H. te Heesen, "UCB-SEnMod: A Model for Analyzing Future Energy Systems with 100% Renewable Energy Technologies—Methodology, "*Energies,* Bd. 15, Nr. 12, p. 4383, 2022.

[2] M.-C. Leonhard und H. te Heesen, "Energy Model for a Rural Region in Germany - Methodology, "in *37th European Photovoltaic Solar Energy Conference and Exhibition*, S. 1850–1854, 2020.

[3] P. Sterchele, J. Brandes und J. Heilig, „Wege zu einem klimaneutralen Energiesystem: Die deutsche Energiewene im Kontext gesellschaftlicher

Verhaltensweisen," Fraunhofer-Institut für solare Energiesysteme ISE, 2020.

[4] M. Sterner und I. Stadler, Energiespeicher - Bedarf, Technologien, Integration, Heidelberg: Springer Berlin, 2017.

[5] B. Büttner Held, M. Altrock, B. v. Gayling-Westphal und C. Kliem, „Rechtliche Rahmenbedingungen für ein integriertes Energiekonzept 2050 und die Einbindung von EE-Kraftstoffen," 2019.

[6] R. A. Rodriguez, S. Becker, G. B. Andresen, D. Heide und M. Greiner, "Transmission needs across a fully renewable European power system, "*Renewable Energy,* Bd. 63, pp. 467-476, 2014.

[7] G. Pleßmann und P. Blechinger, "How to meet EU GHG emission reduction targets? A model based decarbonization pathway for Europe's electricity supply system until 2050, "*Energy Strategy Reviews,* pp. 19-32, 2017.

[8] S. Lehnhoff, „Dezentrales vernetztes Energiemanagement: Das europäische Energieversorgungssystem," 1994, pp. 571-622.

[9] D. Böttger und T. Bruckner, Kosten-und CO2-Effekte von Power-to-Heat im Markt für negative Sekundärregelleistung, 2014.

[10] Bundesnetzagentur, „Regulierung von Wasserstoffnetzen: Eine Bestandsaufnahme der Bundesnetzagentur," 2020.

[11] N. Gerhardt, J. Bard und R. Schmitz, „Studie zum Einsatz von Wasserstoff im zukünftigen Energiesystemmit dem besonderen Fokus auf die Gebäudewärmeversorgung," 2020.

[12] UN-Energy, "Sustainable Bioenergy. A Framework for Decision Makers, "2009.

[13] Bundesregierung, „Globale Umweltveränderungen Wissenschaftlicher Beirat der Bundesregierung, Zukunftsfähige Bioenergie und nachhaltige Landnutzung," 2008.

[14] „Ökobilanz von Energieprodukten. Ökologische Bewertung von Biotreibstoffen. Schlussbericht," 2007.

[15] E. Helmers und P. Marx, "Electric cars: technical characteristics and environmental impacts, "*Environmental Sciences Europe,* 2012.

[16] „Projektteam eHighway, FESH – Feldversuch eHighway an der BAB A1 in Schleswig-Holstein," [Online]. Available: https://www.ehighway-sh.de/de/. [Accessed 27 September 2022].

THE STATUS OF PHOTOVOLTAICS IN ITALY AND THE PROSPECTS FOR CONTRIBUTING TO THE GOAL OF ENERGY DECARBONISATION

Salvatore Guastella[1], Ezio Terzini[2]
[1]RSE, Via R. Rubattino 54, 20134 Milano, Italy
salvatore.guastella@rse-web.it
[2]ENEA - C.R. Portici, Piazzale E. Fermi 1
Località Granatello - 80055 Portici, NA, Italy

ABSTRACT: This work aims at describing the status of PV in Italy in correlation with the objectives of the Integrated National Plan for Energy and Climate (PNIEC) to contribute to energy decarbonization. The PV installations' trend in Italy (0.9 GW installed in 2021) is shown highlighting an encouraging recovery signal after several years of market stagnation. The need for a significant acceleration of the market is underlined comparing the current trend to PNIEC's targets and to the more ambitious goals of the new European climate law's package "fit for 55" (for Italy, 64 GW of cumulative PV power by 2030). Government measures and reforms for an Italian PV market boost are illustrated together with the already consolidated assets of PV industrial sector, covering the entire PV value chain, and of Research, Development and Demonstration activities of research institutes and universities supporting the knowledge advancement and industry innovation. The reported information shows the current Italian policies for the widest penetration of PV in the process of decarbonisation of energy production, but also the need of a further effort by the governmental institutions and all the sector stakeholder to reach the national PV target by 2030. The main work needed in very short term is to put in place and to valorise the objectives of the National Recovery and Resilience Plan (PNRR), the stronger source of strategic projects for the energy transition which, in Italy, relies largely on the development and diffusion of photovoltaics.
Keywords: Photovoltaic, R&D and Demonstration Programmes, National Programme, Funding and Incentives, Energy Options

1 INTRODUCTION

This work aims to describe the status of PV in Italy in correlation with the objective of the *Integrated National Plan for Energy and Climate* **(PNIEC)** [1], supported by several actions of the National Recovery and Resilience Plan **(PNRR)** [2], to contribute to energy decarbonization.

The collection and dissemination of the information contained in this paper is attributable to the continuous monitoring activities on Italian photovoltaics conducted by the authors as Italian members (on behalf of MiTE - Ministry of Ecological Transition) of the Executive Committee of the IEA TCP PVPS (PhotoVoltaic Power Systems).

2 TRENDS OF PHOTOVOLTAICS AT INTERNATIONAL AND NATIONAL LEVEL

The growth of Renewable Energy Sources (RES) worldwide is constantly increasing, supported above all by the need to decrease the dependence on energy sources based on fossil fuels, since RES are widely perceived as the most viable solution to reduce gas emissions pollutants and global warming. However, this constant increase occurs to an unequal extent in the various geographic areas of the world (Figure 1) and sees Europe in a more backward position than the Americas and, above all, compared to the Asian area.

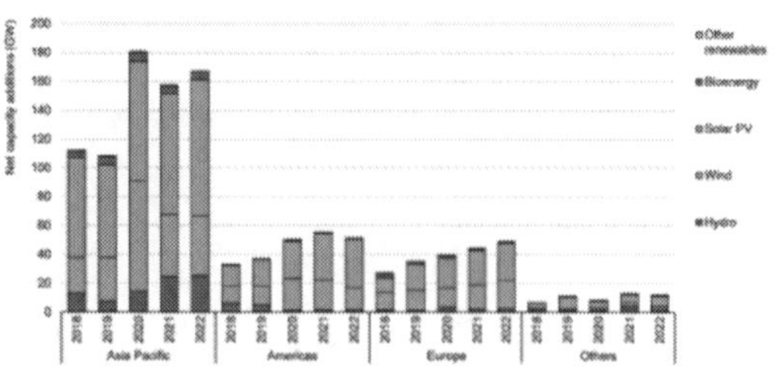

Figure 1: Net annual renewable capacity additions by region and technology (Source: based on "*IEA Renewable Energy Market Update 2021*")

The worldwide photovoltaic situation is characterized by a strong increase in the cumulative and annual installed power which at the end of 2021 was about + 21% compared to the previous year, with specific numbers showing the consolidation of the most advanced position of Asian countries). (Figure 2).

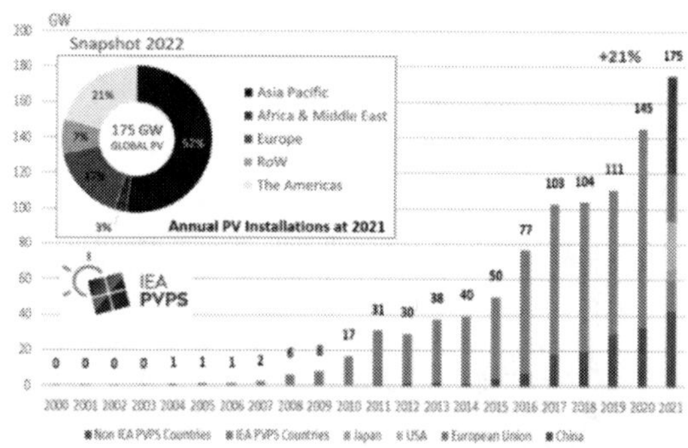

Figure 2: Trends of annual PV worldwide installations (Source: IEA PVPS Snapshot 2022) [note: It should be borne in mind that the power of photovoltaic systems is commonly accounted in W or Wp or Wdc, as it is the sum of the direct current power, therefore before the conversion equipment converts it into alternating current with an inevitable decrease in its value; for this reason, the power of the photovoltaic systems in the electric grid **Pac** is lower than the value presented in the common statistical reports **Pdc**].

The installations of photovoltaic plants in the Italian territory, on the other hand, are growing at a rather limited rate (Figure 3), although the **PV installation costs have reached market parity** in various regions and despite the fact that the objectives indicated by the *PNIEC* by 2030 (recently further boosted by the *Green Deal* and by the *REpowerEU*) are still too much far away, as described in par. 3.

The trend of PV installations in Italy remains below

the potential, even if a growth can be seen after several years of market stagnation: the installed power in 2021 was just over 0,9 GW and the cumulative PV capacity was around 23 GW.

As regards to PV energy production in Italy, the value of 2021 was at 25,1 TWh; it will have to reach a value of 73,1 TWh by 2030 (Figure 3). This means that in the coming years about 3 GW/year will have to be installed (the average in the last 5 years was about 660 MW/year); assuming that the average power of photovoltaic systems remains constant, equal to 23,7 kW (in 2019 the average power was 12,9 kW), more than one million new systems will have to be built.

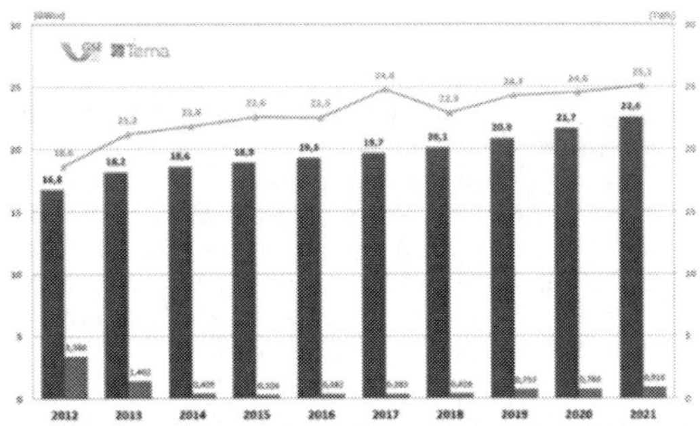

Figure 3 – Cumulative installed power and photovoltaic production in Italy (source: RSE processing on GSE and TERNA data)

The installed power with respect to the location of the plant (on the ground or not on the ground), shown in Table 1, confirms a trend of the last 10 years that sees a constantly growing ratio between systems not located on the ground and those on the ground. In 2021, the cumulative power of ground-mounted plants represents the 39% of the overall figure and it is distributed in the various regions (Figure 4) in a somewhat different percentage [5].

Table 1 – Grid connected PV installation [MWdc] in 2021 in Italy (*source data* [4] *and* [5])

Year installed power	**944**
Centralized*	52
Distributed**	892
Cumulative installed power	**22.594**
Centralized*	8.883
Distributed**	13.711

*Ground mounted, **Rooftop or façade building mounted

3 TARGET FOR PV IN EUROPE AND ITALY

The difficulties of European countries to play an important role in the PV sector, to progress towards the decarbonisation of the energy sector, entails the need for Europe to further promote and pursue the use of RES, enhancing the potential of the entire "value chain" that it possesses on RES technologies and consistently increasing production capacities in Europe.

These objectives are reflected in various important initiatives of the European Commission, such as the **Climate Target Plan** which proposes to increase the already ambitious EU goal by reducing greenhouse gas emissions by at least 55% below 1990 levels by 2030 (**Fit for 55%**) by the new EU climate law, approved on 24 June 2021, as an intermediate step towards 1) the **European Green Deal**, a set of political initiatives proposed by the European Commission with the aim of achieving climate neutrality in Europe by 2050 and 2) the **REPowerEU**, a plan to save energy, produce clean energy and diversify our energy supply.

Figure 4 – Regional distribution of PV power installations at the end of 2021 (*Source GSE data* [5])

In the above European context, in Italy the **PNIEC** has been issued to ensure compliance with the EU targets in setting important national objectives for decarbonization and penetration of RES by 2030. In particular, the electricity sector is called to play a crucial role, moving towards an ever-greater production of energy from RES, which will have to record a significant growth by 2030.

To reach the 30% share of energy to be RES in gross final consumption of energy (55% of national electricity consumption) by the end of this decade, the PNIEC established objectives and growth trajectories for each individual renewable source (Table 2). Photovoltaics, thanks to the availability of the solar resource throughout the country, is the renewable energy source on which Italy is firmly betting, so much so that in 2030 it will have to provide about 40% of the total electricity produced by RES.

Table 2 - Targets and growth trajectories of installed capacity and energy production from Renewable Sources by 2030 (*RSE data processing from PNIEC*)

	Year 2017		Year 2025		Year 2030	
Energy source	Power MW	Energy TWh	Power MW	Energy TWh	Power MW	Energy TWh
Solar	19.682	24,4	28.550	40,1	52.000	73,1
Hydro	18.863	46,0	19.140	49,0	19.200	49,3
Wind	9.766	17,2	15.950	31,0	19.300	41,5
Geothermal	813	6,2	920	6,9	950	7,1
Bioenergy	4.135	19,3	3.570	16,0	3.760	15,7
Total	53.259	113	68.130	143	95.210	187,0
Solar %	*37,0*	*21,6*	*41,9%*	*28,0%*	*54,6*	*39,2*

Moreover, the above PNIEC objectives have been revised after the "Fit for 55" EU law indicating a further acceleration of decarbonization policies to an estimate of about 64 GW of cumulative photovoltaic power, producing nearly 90 TWh out of the 340 TWh of national electricity consumption foreseen by 2030, with the coverage by RES passing from 55% to 72%.

4 NEEDS AND ACTIONS FOR AN IMPULSE TO PV INSTALLATIONS

In support of the ambitious goal mentioned in par. 3, it is worth to mention the launch in Italy of the National Recovery and Resilience Plan (PNRR), which supplies substantial support to many sectors.

In particular, PNRR allocated 2,6 billion euro for the development of photovoltaics in the agricultural sector. An investment of 1,5 billion euro, out of total above, is for the "Agrisolare Park" initiative aiming at promoting PV installation on agricultural infrastructure's roofs. The further 1,1 billion is dedicated to the Agrivoltaics for supporting the installation of at least 2 GW of PV on agricultural land, to demonstrate the synergic approach and a reduction of 1,5 million tons of CO_2 emission. On this issue it must be reported the intense activities to draw up regulations by MiTE and, at the same time, the efforts of standardization organizations, CEI (www.ceinorme.it) by means of the committee CT82 *PV systems*, and UNI (www.uni.com) to identify suitable rules for classifying agrivoltaic plant typologies and the way to evaluate their qualities; all this with the aim of providing operational tools for the application of ministerial incentives to this sector which are expected in the short term.

Another selected action is to develop an international industrial and knowledge leadership in the main energy transition's value chains, for which PNRR allocated resources for a total of 1 billion euro, to be used in next 3 years, to promote the development in Italy of competitive supply chains in the fastest growing sectors, which make it possible to reduce dependence on imported technologies and strengthen research and development in the most innovative areas: photovoltaics, electrolysers, batteries. By this measure, the PNRR gives a relevant support to the Research, Development & Demonstration activities on new PV technologies, in which various Research Institutes and PV operators are active (see par. 5).

Part of the same strategy is also the PNRR support to the realization of high-tech PV factory of gigawatts level. In this frame, it must be highlighted the action for the development of high efficiency PV able to increase the productive capacity to at least 2 GW/year by December 2025. To this scope, FuturaSun is setting up a module line of 1 GW, whereas EGP is setting up a cell and module line of 3 GW, both of which start-up are expected by 2023.

The PNRR has also given great priority at the implementation of a series of reforms useful for the elimination of non-technical obstacles to PV penetration. The highest priority has been given to the simplification and rationalization of the environmental regulations that intervene in the permitting procedures. This was done by means of several government decrees also stating by law suitable special areas for PV installation and the reference criteria to be utilized by regional governments for the autonomous localization of the "suitable areas" needed for the regional contribution to the national PV targets. For Italy this means to release a large part of 30 GW of projects presented in the last 4 years blocked in the permitting process.

It must be taken into account that some incentives are currently active for PV installations. Interesting measures to sustain PV installations are linked primarily to building renovations and energy efficiency. A tax break mechanism in favour of these projects allows the installation of residential plants (up to 20 kWdc) paying back 50% of PV installation costs. An even more advantageous measure, recently introduced, is the so called "Superbonus 110%" which is a tax relief measure providing a tax credit of 110% of the cost of the plant if it is comprised within interventions for energy improvement of buildings stepping up two levels of the building energy class. The value of tax credit can be assigned directly to the supplier or to a third party, including banks, with their financial compensation allowing the user to build the PV plant with a storage system almost without the initial investment burden.

Other incentives and facilitations currently in place for RES must be mentioned as they have a positive impact on photovoltaic installations:

- Ministerial Decree "FER1" 2019-2021

In 2019-2021, new, refurbished and upgraded plants from "mature" RES have been supported: PV, onshore wind, hydro and sewage gas for a total capacity of about 8 GW (of which 7,4 new GW), through 7 semestral rounds of competitive AUCTIONS for groups of technologies and REGISTRIES for smaller plants [4]. Unfortunately, the percentage of supported PV plant resulted low (1,9 GW), due to mainly the permitting obstacles.

- Energy communities, in which citizens, local entities and industrial subjects active in the area have the opportunity to participate directly in the energy market, not only as "consumers" but as "prosumers" through the promotion of self-consumption and the installation of RES plants;
- Decree by MiPAAF (Minister of agriculture) for PV in agricultural infrastructures
- Power Purchase Agreement (PPA) that is, contracts for the direct sale of energy between producer and consumer. This tool is being improved since in Italy there is still little diffusion of this mechanism, perhaps also due to the lack of a standard model; currently the PPAs are not able to be stipulate on a sufficiently long period of time to remunerate the PV investment; useful measures to promote its diffusion, as suggested by the EU, are to allow access also to SMEs and the obligation for public administrations to use this means to procure a share of renewable energy.

Despite all this, it is evident that the challenge by 2030 for the photovoltaic sector stated by PNIEC, reinforced by **Fit for 55% law and by the REPowerEU** perspectives, is very ambitious if we consider the values reached in 2021 (22,6 GW cumulative installed power and 25,1 TWh/year produced PV energy), described in par. 2.

Therefore, some more actions are needed to overcome the obstacle for a huge introduction of PV in Italian energy sector, essential to reduce energy costs for consumers and to achieve the objectives of the energy transition; like as also indicated by associations of PV producers, it is first of all necessary to:

- strengthen the measures for the massive use of building roofs for PV installation
- promote the maximum diffusion of self-consumption and renewable energy communities
- facilitate and incentivize the agrivoltaic plants
- allow large and medium-sized plants on the ground in suitable areas.

However, in the next future, it is expected that an improvement will be possible considering recent government innovations and these expected in the short term. These include:

- PV on the roofs; no authorization is required and unique simplified model (*Modello Unico*) use up to 200 kW;
- current and new upcoming decree on RES, now introducing incentives for plants > 1MW by descending price auctions, and <1MW by standard generation costs, direct access;
- decree defining "suitable areas" for RES installations in the Italian regions, some of which have already been introduced with the simplifying ministerial decrees (by default, suitable areas are those surrounding industrial or affected by PV plant repowering or non-protected in which a PV plants already exists);
- implementing decrees for energy communities, extending the participation limits to different subjects, increasing the single plant power limit to 1 MW and enlarging perimeter to primary electric station;
- decree on agrivoltaic plants (figure 5 and 6), providing incentives for "advanced agrivoltaic plants" on the produced energy and on the capital costs;
- decree on floating PV plants, providing incentives;
- revision of the PNIEC, to consider the new EU energy targets;
- project calls foreseen by PNRR mission 2 - *Green revolution and ecological transition*.

Figure 5 – Agrivoltaic plant in Po valley (*image by Remtec s.r.l.)*

Figure 6 - Cultivation of high-quality cedars and production of energy by PV plant at Scalea (CS) (*image by LeGreenHouse an EF Solare*)

5 R&D&D IN ITALY AND VALUE CHAIN

In the context of Research, Development & Demonstration on new PV technologies, various Research Institutes and PV operators are active in Italy.

Among them, ENEA, a public research body, is focused on high efficiency solar cells based on perovskite/c-Si tandem structure, having achieved an efficiency of 20,08% on single cells based on perovskite, on the development of module ecodesign, on the advancement PV systems digitalization and on the diffusion of a "Sustainable Agrivoltaic" giving technical and scientific support to the related National Network. Furthermore, **RSE**, a private research company on the energy system, is carrying out activities on 1) the Concentrating multi-junction solar cells, 2) the advanced PV plant O&M strategies (through improved diagnostic and predictive techniques) to contribute to their optimal production and 3) the Life Cycle Assessment (LCA) of most promising innovative PV technologies. Finally, **CNR**, a public research institution, is active in the evaluation of innovative low-cost processes for thin film cells and **EURAC Research**, a private research organization, works on the improvement of PV plant performance and reliability and on BIPV (Building Integrated PhotoVoltaic). The research framework is completed with the contribution of numerous university laboratories among which the CHOSE Centre of the University of Rome Tor Vergata, the University of Milano Bicocca and the University of Turin. All the aforementioned research institutes and labs pursue the main purpose of facilitating the development of high-efficiency and low-cost photovoltaic generation systems, to contribute to the achievement of the PNIEC objectives by 2030, but also to provide technological support to companies in the sector.

In the development of solar cell technologies and demonstration sector, many operators of the Italian photovoltaic market, growing year by year, are active and are covering the entire PV value chain.

In particular, several high-tech operators are increasing their production features related to cells/modules (mainly Enel Green Power with the initiative mentioned in par. 4), to inverters and to sun tracking systems, mostly for utility-scale plants.

Moreover, several Italian manufacturers are present on the PV market, with very different industrial capacity; among them the following are mentioned on the basis of the authors monitoring and of other indications from Italian members of the IEA PVPS' Tasks:

Module Manufactures:

Eclipse Italia, EGP, EXE Solution, FuturaSun, GruppoSTG, PEIMAR, Solbian, Trienergia, Verditek Solar Italy

Inverter Manufactures:

Fimer/PowerOne, Enertronica Santerno, Friem, Aros, Riello, SIEL

Sun Tracker Manufactures:

Convert, Comal, RemTec, Soltigua, Sunracker

The presence of a complete "PV value chain" is essential to enable the achievement of the energy decarbonization objective, leveraging the availability of PV technology at national and European level.

To strengthen this presence and reinforce the collaboration between the various Italian operators, also the *Italian PV Network for Researchers Institutes and Operators*, established in Italy with the aim of sharing PV project initiatives and research infrastructures, is supporting the enhancement of Italian PV trend, also providing technical contribution to the activities of governmental bodies.

6 CONCLUSIONS

The mix of activities from RD&D institutes, PV operators and Governmental institutions are pushing the Italian development of photovoltaics towards the achievement of the national and European goal of energy decarbonization.

However, further important efforts are needed to achieve these results, such as a greater and complete presence of the industry in the photovoltaic value chain, the availability of suitable technologies (with related infrastructures) and, last but not least, a clear planning of PV plants installations (including the identification of all suitable areas throughout the national territory and clear and simplified permitting rules).

But the adequate support for the extra efforts needed in this sector will only be seen in the achievement of the objectives of the PNRR, currently in the initial implementation phase, whose strategic projects have the challenging task of accelerating an ecological transition that cannot do without photovoltaics.

7 REFERENCES

[1] Ministero dello Sviluppo Economico, Ministero dell'Ambiente e della Tutela del Territorio e del Mare, Ministero delle Infrastrutture e dei Trasporti, *Piano Nazionale Integrato Energia e Clima*, 2019 (https://www.mise.gov.it/images/stories/documenti/PNIEC_finale_17012020.pdf)

[2] Piano Nazionale di Ripresa e Resilienza (PNRR) (https://www.mise.gov.it/images/stories/documenti/PNRR_Aggiornato.pdf)

[3] C. Fetting, *The European Green Deal*, ESDN Report, Vienna, 2020.

[4] F. Tilli, A. Scipioni, G. Maugeri, *Italy Country Presentation*, IEA PVPS Task 1

[5] Rapporto statistico GSE, *Il solare fotovoltatico in Italia – Stato di sviluppo e trend del settore*, May 2022

8 ACKNOWLEDGEMENTS

This work has been done by RSE and ENEA on funding from the Research Fund for the Italian Electrical System in compliance with the Decree of April 16, 2018.

The authors thank the experts of the IEA TCP PVPS for the data on the trend of photovoltaics in the international context, and in particular the Italian experts (Francesca Tilli - GSE, Giosuè Maugeri - RSE, Franco Roca – ENEA and Alessandro Scipioni - Elettricità Futura), for the data and information on the status of the PV in Italy collected as part of the activities of Task1.

TRANSIENT RESPONSE AND NON-LINEAR CAPACITY VARIATION AWARE UNIFIED EQUIVALENT CIRCUIT BATTERY MODEL

Dafang Zhao[1], Ittetsu Taniguchi[1], Francky Catthoor[2,3], Takao Onoye[1]
[1] Osaka University, 1-5 Yamadaoka, Suita, Osaka 565-0871, Japan
[2] imec, Kapeldreef 75, 3001 Heverlee, Belgium
[3] ESAT, KULeuven, Kasteelpark Arenberg 10, 3001 Heverlee, Belgium

ABSTRACT: Battery energy storage plays a vital role in the renewable energy system. Battery as the only dynamic element in a PV system, the current, voltage, resistivity, etc. are varying during charging and discharging. With an accurate and efficient battery model, the designer can predict and optimize battery runtime and performance during different working scenarios. However, low-complexity state-of-the-art models are too weak on the transient behavior. The primary purpose of this study is to provide an efficient dynamic battery model for the simulation of stand-alone PV applications. In this study, an enhanced and comprehensive electrical battery model has been proposed and implemented in the SPICE environment. The proposed model accounts for battery non-linear open-circuit voltage and non-linear capacity effect, which are essential for renewable energy storage applications. The proposed model consists of kinetic based runtime estimation model and transient response model. The proposed model was validated with experimental data on Li-ion batteries. The results show that the proposed model can accurately predict the battery runtime and I–V performance with acceptable execution time overhead. Moreover, the proposed model can also be easily extended to other batteries.
Keywords: Batteries, Battery Storage and Control, Hybrid, Modelling

1 INTRODUCTION

The energy crisis and power shortages have become serious problems facing humanity. Take Japan for example, the third largest economy has been running on a thinner supply of electricity since the Great East Japan Earthquake in 2011. In March 2022, a strong earthquake stretched the power grid and the situation was made worse by a surprise blast of frigid weather across Tokyo. Solar output decreased and there were not enough gas or coal-fired power plants to compensate for the difference. Under such circumstances, the introduction of an energy management system promotes the integrated use of renewable/traditional energy, and a power peak shift has been implemented. Typical renewable energy, such as solar and wind power, is highly dependent on weather conditions which result supply power change dynamically. However, when this renewable energy is integrated into the power grid, the energy storage system (ESS) is required to ensure a stable power supply. Meanwhile, the peak power shift can be achieved by charging the energy storage system with electricity generated during the night, in which excess power generation can be stored, and discharging the energy storage system during the day when then the grid reaches its peak.

With the development of chemical and material science, electrochemical energy storage devices, including lead-acid batteries, NAS (Sodium-sulfur) batteries , redox flow batteries, and lithium-ion batteries (LiBs) have become widely used energy storage technologies, such as laptops and mobile phones, because of their advantages, including high energy density, high voltage, high volumetric, and low self-discharge rate. Over the last three decades, with the popularity of the electric vehicle (EV) and consumer electronics, LiBs with its high energy density characteristic have become the mainstream energy storage component and have gained significant development. Battery energy storage systems require high power and high capacity for lithium-ion battery, thus demanding a management system to ensure a reliable and safe operation. Additionally, a battery model is very important in states estimation of the model-based battery management system.

Many battery models have been proposed to capture battery performance for various purposes. The physical-based electrochemical model is the most accurate but complex model, which is usually used to benchmark other models. Since this model is described by partial differential equations (PDE), it is not suitable for online applications [1]. In contrast, the electrical model uses an equivalent circuit to capture the current-voltage (I-V) characteristics of the battery, making it suitable for simulation with other electrical components, such as photovoltaic cells. However, this model cannot describe the nonlinear capacity variation, which is essential for battery management. The above-mentioned models are either too complex to apply in online management or only partially simulate battery characteristics. Therefore, a simple but comprehensive battery model is required for BESS applications.

In this study, a hybrid unified equivalent circuit model including the capacity and the I-V characteristic part is proposed to simultaneously simulate the nonlinear capacity effect (including the rate-capacity effect and the recovery effect) of the battery, which can more accurately reflect the transition of battery capacity and the corresponding change in SOC. In this study, a detailed parameter identification of the proposed model is presented. Finally, the correctness and validity of the proposed model are verified by simulation and experimental analysis.

2 OVERVIEW OF BATTERY MODELS

A typical LiB consists of an anode, a cathode, and liquid electrolytes where electrochemical reactions occur at the electrode-electrolyte interface and generate the current as shown in Figure 1. The battery capacity depends on two nonlinear effects: the rate capacity

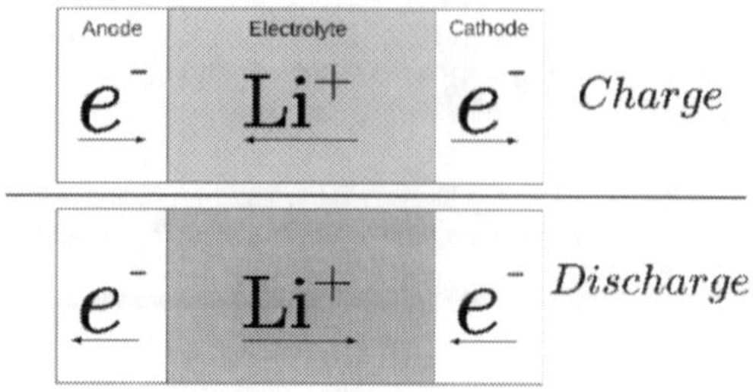

Figure 1: Schematic of lithium-ion battery

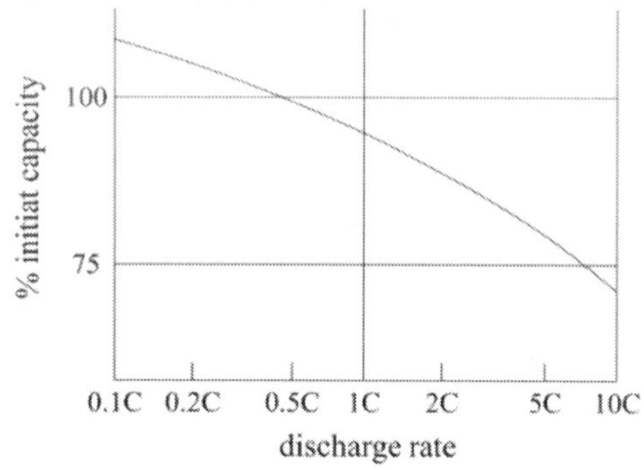

Figure 2: Rate-capacity effect

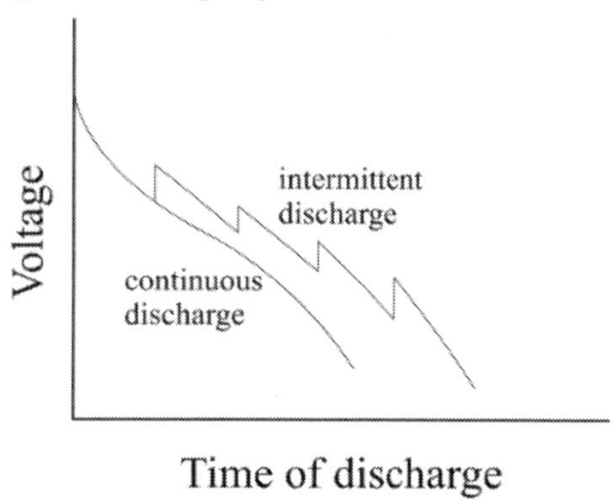

Figure 3: Recovery effect

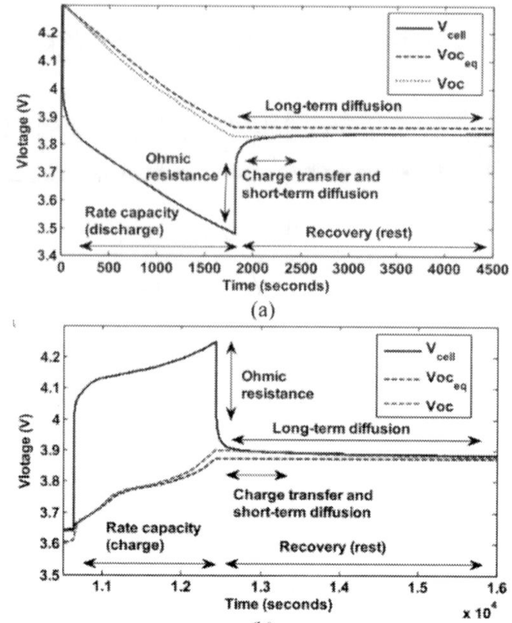

Figure 4: Time-domain response of battery terminal voltage (Vcell), open-circuit voltage (VOC), and equilibrium open-circuit voltage (V_{OCeq}), (a) discharge mode, (b) charge mode [2].

effect and the recovery effect. The ideal battery will have a constant capacity under different discharge currents. However, in reality, the dischargeable capacity of the battery will decrease as the discharge current rate (C-rate) increases because of the higher concentration gradient caused by the slow diffusion process at the active reaction site of the battery. This effect is termed the rate-capacity effect, which is expressed as a lower capacity compared to the battery discharged with a low C rate, as shown in Figure 2. Another recovery effect is that a reduction in battery load can recover the "lost" capacity over a period, for example, the battery voltage in intermittent discharge will recover after each pulse, as shown in Figure 3. This is because the diffusion process caused by a high concentration gradient will decrease after a low or no current period and make "unavailable" capacity become "available".

The effects of nonlinear capacity also cause the nonlinear voltage response of the battery. The time-domain response of a battery in discharge and charge using pulsed current profile is shown in Figure 4, due to rate-capacity effect, the battery open-circuit voltage (VOC) is higher/lower than the ready-state battery terminal voltage V_{OCeq} during charging/discharging process. Due to the recovery effect, the VOC rises/falls down to converge to V_{OCeq} in the rest period.

Various models have been developed over the years to describe battery characteristics, and these models have also been evaluated in various review papers [3–5]. Due to the nonlinear behavior of batteries, their lifetime depends on not only the material but also the consumption pattern. To improve the batteries efficiency and extend the batteries operation time, the batteries model in BEMS should be able to simulate both capacity variation and transient response in detail. Many types of battery models have been developed to estimate the state of the battery and simulate the charging and discharging behavior. Based on their modeling characteristics, modeling methods are divided into three categories: physical-based mathematics, electrical, and data-driven.

2.1 Physical-based Mathematical Model

The physical-based mathematical model includes the batteries' essential physical and electrochemical characteristics that help researchers understand and improve the performance of the battery. Although the physics-based mathematical model is the most accurate model, it can be used as a "reference" to check the performance of other models. To obtain and identify parameters of the mathematical model, a deep understanding of the battery was required. Furthermore, the mathematical model uses partial differential equations (PDEs) to describe battery processes [1]. However, these descriptions are too detailed and time

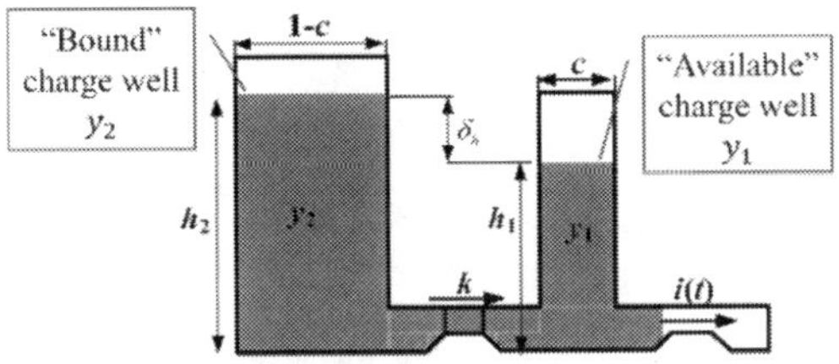

Figure 5: Kinetic battery model

consuming to assess what BESS operations and management needs. By and large, it is too complex to apply to the BESS.

The kinetic battery model was proposed by Manwell and McGowan for capturing batteries' behavior in 1993 [6] and uses a chemical kinetics process as its basis. The KiBaM is a very intuitive battery model, it describes the chemistry processes of a battery by a kinetic process. This model describes the battery as a two-well system, where the battery charge is distributed in the "available charge" y_1 well and the "bound charge" well y_2 with a capacity ratio c (0<c<1), as shown in Figure 5. During discharge, the charge of y_1 flows out to the load, and simultaneously the charge in y_2 flows to y_1 through the value *k*. Since the flow rate of y_1 is faster than y_2 , the height difference δ_h between h_1 and h_2 will increase until all of the charge in y_1 flows out. When battery stop discharge, the charge of y_2 remained flowing into y_1 due to δ_h and this charge exchange will continue until the height of y_1 and y_2 is equal. The height changes of y_1 reflects the battery rate-capacity effect and recovery effect [7].

In KiBaM, the total capacity in real time C_{total} , available capacity C_{avail} , and unavailable capacity $C_{unavail}$ can be expressed as follows:

$$\begin{cases} C_{total} = y_1(t) + y_2(t) \\ C_{unavail}(t) = (1-c)\cdot\delta_h(t) \\ C_{avail} = C_{total}(t) - \int I(t)dt - C_{unavail}(t) \end{cases} \quad (1)$$

The variation in unavailable capacity over time can be expressed by the following equation,

$$C_{unavail} = C_{unavail}(t_0)e^{-k'(t-t_0)} + (1-c)\frac{I}{c} \cdot \frac{1-e^{-k'(t-t_0)}}{k'} \quad (2)$$

where k is defined as:

$$k' = \frac{k'}{c(1-c)}. \quad (3)$$

Based on Equation (1) and (2), the real-time SOC, SOC(t) can be calculated by the initial SOC, SOCinit as:

$$SOC(t) = SOC_{init} - \frac{\int I(t)dt - C_{unavail}(t)}{C_{total}} \quad (4)$$

Regardless of whether the KiBaM can accurately estimate batteries' nonlinear capacity variation, this model cannot describe the batteries' I-V characteristics, which is essential for energy storage design and cosimulation with other electrical components.

2.2 Electrical Model

The electrical models have been developed to give an equivalent representation of different types of batteries [8-10]. The principle of electrical models is to use changes in internal parameters (e.g., resistance, capacitance, etc.) as changes in the efficiency of batteries to find a relationship between parameters' values and battery voltage [9]. In general, the electrical model can be divided into two categories: Thevenin-based models and impedance-based models. The difference between those two models is that parameters of the Thevenin-based model are extracted from time-domain measurements, while the impedance-based model' s parameters are extracted from frequency-domain measurements like electrochemical impedance spectroscopy (EIS). Moreover, the Thevenin-based models mainly describe the voltage behavior of the battery, whereas the impedance-based models mainly describe the electrochemical process.

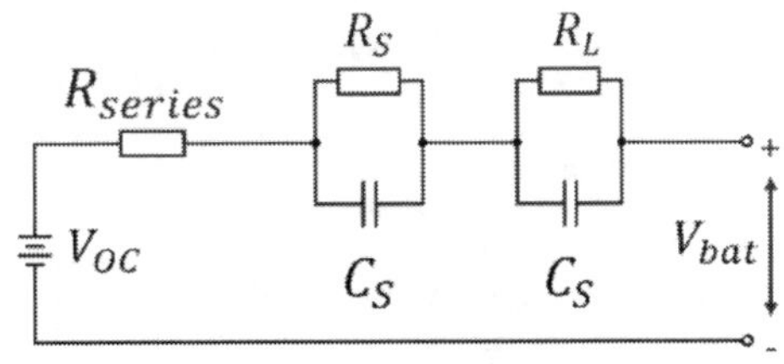

0

Figure 6: Thevenin battery model

The typical Thevenin model is illustrated in Figure 6. The ideal voltage source V_{OC} describes the battery OCV that remains constant depending on the SOC. Internal resistance includes ohmic resistance R_{series} and polarization resistances, which include electrochemical polarization equivalent resistance R_S and concentration polarization equivalent resistance R_L. The equivalent capacitance includes C_S and C_L that represent the transient response and describe the electrochemical polarization and concentration polarization, respectively. The two RC ladders are used to model batteries' fast and slow transient response of batteries during charging and discharging. The relationship between terminal voltage V_{bat}, open circuit voltage V_{OC}, voltage across RC ladders U_S and U_L, and outflow current I can be expressed in Equation (5).

$$\begin{cases} V_{bat} = V_{oc} - U_S - U_L - IR_{series} \\ \dot{U_S} = \frac{I}{C_S} - \frac{U_S}{R_S C_S} \\ \dot{U_L} = \frac{I}{C_L} - \frac{U_L}{R_L C_L} \end{cases} \quad (5)$$

2.3 Data-driven Model

With the rapid development of data science, the relationship between ambient variables and battery parameters can be easily estimated without a deep understanding of the characteristics of the battery. The main advantage of the data-driven model is that there is no need to know the complex physical and chemical processes that govern the state changes in the battery. In contrast, the data-driven model uses a learning algorithm to reveal the relationship between the states and performance characteristics by analyzing historical, real-time data.

The main objective of the data-driven battery model is the estimation of the SOC. Numerous studies have indicated that there is a potential non-linear relationship between SOC variation and battery characteristics. With battery degradation and environmental variation, this non-linear relationship will change over time. Compared to other battery models, the data-driven model that learns using operational data to meet those changes is promising, especially in battery energy storage systems. However, data-driven models are difficult to cope with uncertain battery and environment changes, such as rapid discharge in a short period or sudden temperature drops. The accuracy of those models to a great extent on the measurement accuracy. Therefore, further studies should also consider developing robust estimation technologies, which should explicitly consider the measurement error and uncertainty of the initial conditions. The estimation results show high estimation variance, probably due to over or under-fitting, which is a common problem for data-driven methods. Therefore, researchers have been developing hybrid methods to improve the accuracy of data-driven SOC models.

2.4 Comparison of Battery Modeling Methods

In this subsection, the pros, and cons of three different modeling methods are discussed, and the performance, complexity, and suitability of each modeling method are compared and evaluated.

As for the pros and cons of each model listed in Table I, the physics-based mathematical model is the most accurate model, which can be used as "reference" to check the performance of other models. However, to be able to set all parameters of the model, it is necessary to have a deep understanding of the battery, and the descriptions of this model are too detailed and time consuming to evaluate, which does not meet the requirements of the energy storage system (BESS). By and large, it is too complex to apply to the BESS. The electrical model as a "relatively" accurate and simple model is more suitable for BESS. There are two main issues for electrical models: one, how to extract and identify model parameters efficiently; two, how to choose model structure properly. The structure of the electrical model is usually "empirical" based, which cannot well reflect the electrochemical process during charging and discharging. Hence, electrical models can be further improved, potentially also by being extended and calibrated further based on mathematical model inputs. The data-driven model is a very intermediate model. Although its accuracy closely relies on the training dataset and it takes a long time to train the model, with the development of data science, the training time will gradually decrease while the accuracy remains unchanged.

Through comparisons of different battery models, none of the above modeling methods is universally suitable for all types of applications. In fact, the specific application dictates the choice of a particular battery model. Therefore, for online applications which require an optimization-friendly model, it is possible to combine the advantages of the above-mentioned models to reduce computational cost while maintaining accuracy.

Table I: Pros and cons of different modeling methods

Methods	Pros	Cons
Physocal-based	High accur.	Complex
Electrical	Simple, widely use	Transient only
Data-driven	Accurate	Need model training

3 PROPOSED HYBRID BATTERY MODEL

To overcome the shortage of inaccurate runtime estimation, this study proposed a hybrid battery model as shown in Figure 7(a), the left part is inherited from KiBaM, modeling battery runtime and capacity variation, and the right part is based on the Thevenin model to describe the transient response of the battery. In the hybrid model, both parts of the model represent all aspects of the battery, making it a more comprehensive model.

However, from the perspective of unified modeling, a unified equivalent circuit model can be achieved rather than patching different models together, such as the same concept hybrid battery model in [2]. As shown in Figure 7(b), the proposed hybrid model is equivalent to the unified equivalent circuit model in a sense. The KiBaM-based runtime estimation model was described by two capacitors and resistors, where the capacitor Cavail stores a portion of the available charge and the capacitor Ctemp stores the unavailable and parts of the available charge. The resistor Rk acts as a valve and controls the current flow between two

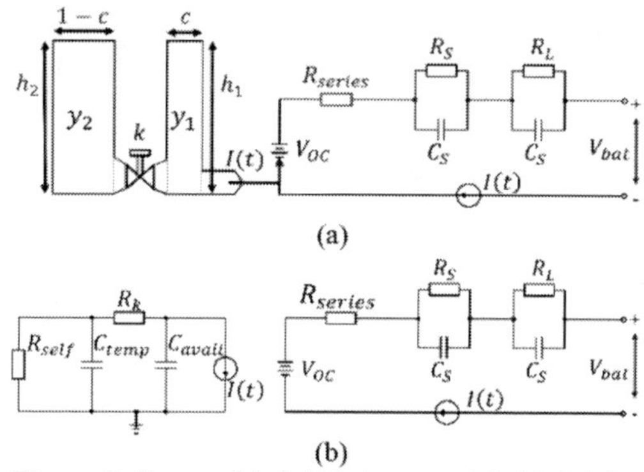

Figure 7: Proposed hybrid battery model, (a) Original schematic, (b) Unified schematic

capacitor together with the voltage difference between VCtemp and VCavail . The voltage control voltage source VOC (SOC) acts as a bridge to connect the runtime estimation and transient response parts. The voltage across Cavail and resistor Rsel f represent the SOC and the self-discharge resistance, respectively. With two different purpose submodel, the proposed hybird unified battery model can be used to capture the following battery characteristics.

- I-V characteristic: Simulate the transient response through the Thevenin submodel, where the RC ladders represent the polarization characteristics.
- Self-discharge: Simulate the battery irreversible reaction caused energy losses through parallel connected self-discharge resistances.
- Non-linear capacity characteristic: Simulate the rate-capacity effect through available capacitance variation during charge/discharge, and simulate the recovery effect through the current flow between available capacitance and unavailable capacitance during rest.

4 PARAMETER EXTRACTION

The unified equivalent circuit parameters of the proposed hybrid model can be extracted by fitting a least-squares curve from experimental data using pulse discharge and constant current discharge tests. The procedure to extract the Thevenin-based model part's parameters is similar to that in [10]. In the normal operation scenario, the battery OCV has a relatively stable relationship with SOC. Thus, the OCV can be seen as a function of the SOC. From the pulse-discharge test, as shown in Figure 8(a), the equilibrium open-circuit voltage curve (UEq) can be used to extract the OCV-SOC relationship as shown in Figure 8(b). Here UBatt − obs, UBatt (td), UBatt (td + i), and UEq are the observed terminal voltage, the terminal voltage at discharge end time td, the terminal voltage at idle time i after td, and equilibrium open-circuit voltage,

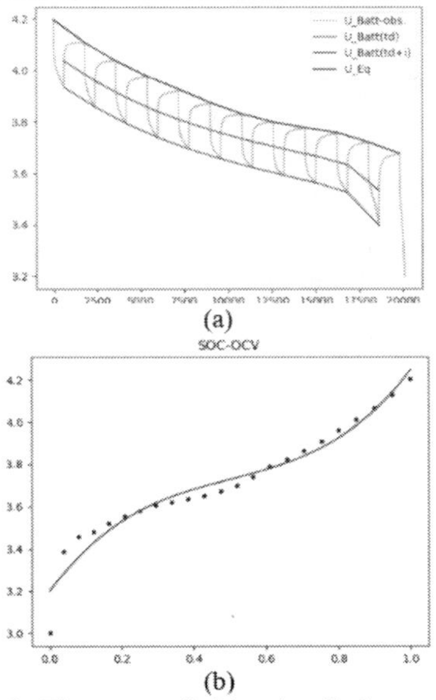

Figure 8: The curve of, (a) pulse-discharge test, (b) SOC-OCV fitting

respectively. The curve of OCV-SOC can be expressed as a polynomial function, as in Equation (6) , where a are coefficients.

$$OCV(SOC) = a_1 \cdot \exp(a_2 \cdot SOC) + a_3 + a_4 \cdot SOC + a_5 \cdot SOC^2 \quad (6)$$

Furthermore, the instantaneous voltage that increases when the discharge is completed at td, as shown in Figure 8(a), is related to series resistance Rseries , which can be calculated using the following equation.

$$R_{series} = \frac{U_{batt}(td + i) - U_{batt}(td)}{I}, \quad (7)$$

Based on Equation (5), the following equation is obtained to estimate the parametersof the RC ladders, where X is the terminal voltage U_{Batt} (td) at time t.

$$V_{bat} = a \cdot (1 - X^{-b \cdot t}) + c \cdot (1 - X^{-d \cdot t}) \quad (8)$$

The parameter a, b, c, and d can be determined by fitting the least squares curve. The RC ladder parameters are derived from

$$\begin{cases} R_S = \dfrac{a}{I} \\ C_S = \dfrac{1}{R_S \cdot b} \\ R_L = \dfrac{c}{I} \\ C_L = \dfrac{1}{R_L \cdot d} \end{cases} \quad (9)$$

The parameters c and C_{total} of the KiBaM-based runtime estimation model can be determined from the high- and low-current constant discharge test. Resistance R_k is related to the effect of nonlinear capacity. For the unified model, R_k can be calculated by the unavailable capacity $C_{unavail}$ of C_{temp}. When the battery is fully discharged, Cunavail will converge to the limitation $C_{unavail}$ (max), which can be calculated by the following equation,

$$C_{unavail}(\max) = C_{temp} \cdot \delta_h = \frac{C_{temp}}{C_{avail}} \cdot I \cdot \tau_0 \quad (10)$$

where τ0 is defined as:

$$\tau_0 = \frac{R_k C_{temp} C_{avail}}{C_{temp} + C_{avail}} \quad (11)$$

and the resistance of R_k can be calculated.

4 MODEL VALIDATION

To evaluate the performance of capturing batteries' transient response, two SPICE simulations were carried out with different discharge profiles and compared them with the experimental measured data. The proposed hybrid unified equivalent circuit model is implemented in LTSpice for a 2100 mAh LG Chem. 18650 Li-ion battery. Experimental measurements were taken from the first test cycle in NASA RW9 battery data set and the battery showed no degradation during the experiment.

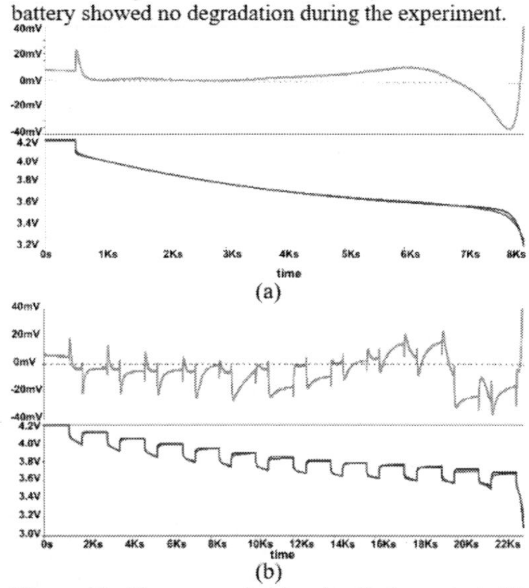

Figure 10: The curve of, (a) pulse-discharge test, (b) SOC-OCV fitting

Figure 10(a) compares the battery voltage obtained from the SPICE simulation (blue line) and the experimental results (red line) with the constant current discharge, where the fully charged battery is discharged with current at 0.5C (1A). The difference between simulation and observation as indicated in the green line. Moreover, Figure 10(b) compares the battery voltage response obtained from the SPICE simulation (blue line) and the experimental results (red line) with the pulse-current discharge. The fully charged battery will be discharged for 600 s at 1A and rest for 600 s until fully discharged. The green curve shows the voltage difference between simulation and experiment. The average voltage difference for constant current discharge and pulse-current discharge in the working SOC range is 3.01 and 7.2 mV, demonstrating that the proposed model can accurately capture battery dynamics and predict battery runtime under various discharge conditions. Furthermore, the execution time of the proposed model is 0.434-s and 2.442-s for constant current and pulse-current discharge, which is significantly lower than the physical-based electrochemical model.

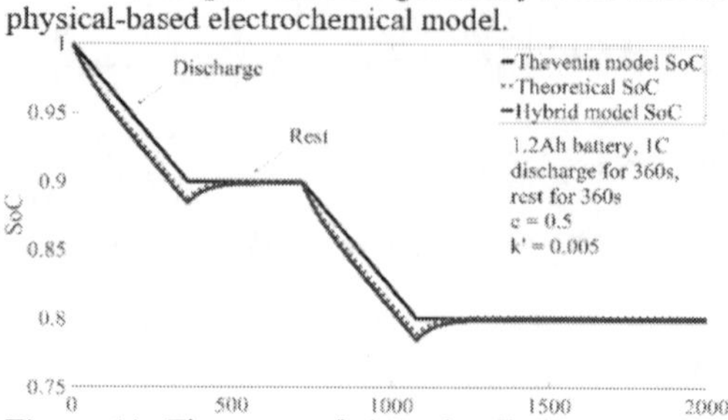

Figure 11: The curve of, (a) pulse-discharge test, (b) SOC-OCV fitting

To further validate the proposed model' s performance of capturing batteries' non-linear effect, a SPICE simulation of 1.2Ah Li-ion battery was carried out. Figure 11 compares the state of charge responses obtained from simulations using the Thevenin battery model and the proposed model for pulse-current discharge scenario, where each current pulse has 360 s of ON time and 360 s of OFF time. Compared to the Thevenin battery model, the proposed hybird model is capable of accurately capturing SOC variation and suitable for BEMS.

5 SUMMARIES

In this study, different battery modeling methods have been reviewed in detail. To accurately simulate batteries' nonlinear effects, a hybrid battery model has been proposed which is capable of estimating battery capacity and I-V characteristics. Both parts of the proposed model represent all aspects of the battery, making it a more comprehensive model. A novel equivalent circuit format of the KiBaM was carried out to unify the model. Moreover, this study also presents an efficient parameter identification method. Simulation and experiments validate the performance and effectiveness of the proposed model. The results show that the proposed hybrid unified model can accurately capture the variation of battery capacity and transient behaviors. In addition, the proposed model provides a basic battery model for future research on the control of battery energy storage systems.

REFERENCE

[1] Z. Deng, X. Hu, X. Lin, L. Xu, J. Li, and W. Guo, "A Reduced-Order Electrochemical Model for All-Solid-State Batteries," IEEE Transactions on Transportation Electrification, vol. 7, no. 2, pp. 464–473, 2021.5

[2] T. Kim, W. Qiao, and L. Qu, "An enhanced hybrid battery model," IEEE Transactions on Energy Conversion, vol. 34, no. 4, pp. 1848–1858, 2019.

[3] A. Fotouhi, D. J. Auger, K. Propp, S. Longo, and M. Wild, "A review on electric vehicle battery modelling: From Lithium-ion toward Lithium-Sulphur," Renewable and Sustainable Energy Reviews, vol. 56, pp. 1008–1021, 2016.

[4] J. Meng, G. Luo, M. Ricco, M. Swierczynski, D. I. Stroe, and R. Teodorescu, "Overview of Lithium-Ion battery modeling methods for state-of-charge estimation in electrical vehicles," Applied Sciences (Switzerland), vol. 8, no. 5, 2018.

[5] Y. Zhao, P. Stein, Y. Bai, M. Al-Siraj, Y. Yang, and B.-X. Xu, "A review onmodeling of electro-chemo-mechanics in lithium-ion batteries," Journal of Power Sources, vol. 413, pp. 259–283, 2019.

[6] J. F. Manwell and J. G. McGowan, "Lead acid battery storage model for hybrid energy systems," Solar Energy, vol. 50, no. 5, pp. 399–405, 1993.

[7] Q. Zhang, Y. Li, Y. Shang, B. Duan, N. Cui, and C. Zhang, "A fractional-order kinetic battery model of lithium-ion batteries considering a nonlinear capacity," Electronics (Switzerland), vol. 8, no. 4, 2019.

[8] [65] E. Locorotondo, L. Pugi, L. Berzi, M. Pierini, S. Scavuzzo, A. Ferraris, A. G. Airale, and M. Carello, "Modeling and simulation of Constant Phase Element for battery Electrochemical Impedance Spectroscopy," in Proc. of 5th International Forum on Research and Technologies for Society and Industry: Innovation to Shape the Future, RTSI 2019. IEEE, 2019, pp. 225–230.

[9] N. El Ghossein, J. P. Salameh, N. Karami, M. El Hassan, and M. B. Najjar, "Survey on electrical modeling methods applied on different battery types," in Proc. of 2015 3rd International Conference on Technological Advances in Electrical, Electronics and Computer Engineering, TAEECE 2015, no. January 2017. Institute of Electrical and Electronics Engineers Inc., 2015, pp. 39–44.

[10] M. Chen and G. A. Rincón-Mora, "Accurate electrical battery model capable of predicting runtime and I-V performance," IEEE Transactions on Energy Conversion, vol. 21, no. 2, pp. 504–511, 2006.

MATCHING PHOTOVOLTAICS AND HEAT PUMPS - AGAINST THE PEAK-8

Paul Grunow
Trinity Solarbeteiligungen GmbH.
Marienburger Str. 31a, 10405 Berlin, Germany, Phone: +49 173 614 3992, email: p@ulgrunow.de

ABSTRACT: This work motivates the combination of rooftop photovoltaics with electric heat pumps and decentral hydrogen storage. First of all, the efficiency is improved by adding the heat flows from the hydrogen back and forth conversion to those of the heat pump in a single device, the so-called Hybrid Storage CHP. Secondly, the power peaks from any heat pump flattens out completely. Since the electric heat pumps are synchronized by the ambient temperature, this will require countermeasures in the power grid capacities. The more heat pumps, the more pronounced the peaks. In addition, they tend to be unpredictable in winter and periods of frost can last several weeks in Central Europe. This will need additional capacities for residual power. While hydrogen systems for single-family houses are already commercially available, this work addresses multi-family houses in cities with only limited roof space per resident as well. Beyond the area constraint, energy sharing among tenants is complicated, while heat sharing is not. Smart heating with Hybrid Storage CHP is still costly, but covers all system requirements for grid-scale residual power management. These long-term storage devices consume or generate power on site and thus call the expansion of the power grid into question. The cost analysis starts with a one-sector model covering only space heating and is then extended to a full sector coupling system for Germany in 2030. The latter is compared to current energy system models with gas turbines for residual load coverage.
Keywords: Hydrogen, Storage, Heat Pump, Prosumer

1 Introduction

Decentral hydrogen [1] lowers the total system costs in a 100% renewable and sector-coupled energy system by utilizing the conversion heats on-site in buildings. It smooths out heat pump peaks on cold, windless nights in Central European winters. The automated switching between electrolysis and fuel cell operation using the existing feed-in intelligence ("ENS") of power inverters enables detailed power balancing at the nearest grid node and thus avoids bottlenecks in the power grid.

In single-family homes, these hybrid storage CHP (Combined Heat & Power) can already provide heat and power entirely from rooftop photovoltaics alone [2[. In apartment buildings with their limited roof area, this system can be operated as a fuel-free space heater, i.e. low operational costs for the tenants. But what value does this device provide for grid balancing? And could this become a business model for the owner or contractor? This paper proposes hydrogen-based space heaters as a less expensive alternative to current models [3][4] and fully meet the storage needs of an all-electric Germany in 2030.

2 APPROACH

2.1 Smart heating with solar hydrogen

Hybrid Storage CHP form the ideal link between electric heat pumps and photovoltaic rooftop systems: While the waste heat from the electrolyser is used for hot water in the summer, the waste heat from the fuel cell operation is used together with the heat pump in the heating period. In German, there are currently 21 million space heaters in operation with a total consumption of 790 TWh/a (23% of the total primary energy). Of these, 50% are operated with gas and 25% with oil and only 2.5% with heat pumps.

2.2 One-sector model vs. full sector-coupling

In the first phase, the Hybrid Storage CHP are probed in one-sector model with pure space heating, to meet the practical interest of prosumers to replace fossil heaters. Then all sectors join in for full sector coupling, i.e. 552 TWh original electric power, 790 TWh space heat or 155 TWh for electric heat pumps incl. renovations, 265 TWh all-electric industry and 212 TWh e-mobility in a second phase.

Ongoing replacement of fossil space heaters with heat pumps is targeting 6 million for Germany in 2030, where intermittent electricity from wind and photovoltaics is to be supplemented by gas turbines. In this counter-proposal, Hybrid Storage CHP will take over this task, but in the hands of prosumers

2.3 The modelled device

Hybrid storage CHP stands for an intelligent heating and storage device that combines electrolysis, fuel cell, hydrogen gas storage and heat pump, see Fig.1. It serves to replace gas heating and at the same time forms a decentral energy system without gas turbines or other central power plants and associated infrastructure installations, e.g. district heating. At a certain degree of market penetration, the existing gas might be reassigned to hydrogen via adapted compressor stations - region by region. The capacity of Germany's gas net is sufficient for hydrogen storage in full sector coupling with 100% renewable energy, see [1].

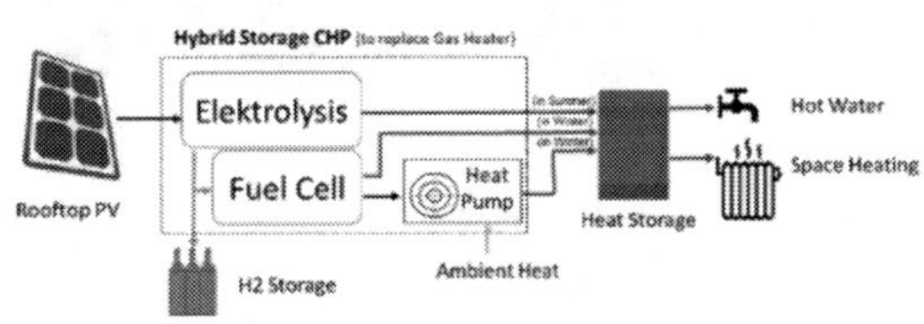

Figure 1: Hybrid Storage CHP for space heating with photovoltaics, hydrogen storage and heat pump

The heat storage tank in Fig. 1 remains an extra device. Local hydrogen bundles with internal compressors are installed outside the buildings.

3 INPUT DATA AND METHOD

3.1 Component costs

Table I lists the costs of the LUT team [3]. Since fuel cells are missing therein, their costs were taken from [5] by averaging the CAPEX ratios of electrolysis and fuel cell systems. As well, the variable OPEX for natural gas has been raised to €0.06/kWh to reflect recent increases in 2021 and 2022.

Cost optimization was performed on an hourly basis over a reference year for a snapshot in 2030. Transport costs are not included, see below.

Table I: Cost parameter for 2030 adopted from [3]

	CAPEX €/kW	OPEX fix €/kW/a	OPEX var €/kWh/a	Lifetime a
PV	390	10.6	0	35
Wind (onshore)	1000	20	0	25
Electrolysis (70%)	363	12.7	0.001	30
Fuel cell (60%)	471.9*	12.7	0.001	30
Biomass	2500	175	0.001	30
Hydro	1650	49.5	0.003	60
Battery (94%)	150/h	10/h	0.0002	10
Gas turbine (42%)	475	14.25	0.0327	30

* capex from [5]

The cost parameters from LUT were chosen to allow comparisons with competing approaches and recent results from other groups. In any case, it reflects industry costs and not consumer market prices.

3.2 Heat load

The hourly heat load profile was generated from standardized isolation characteristics of actual buildings in Germany with an additional share of 10% in renovation until 2030 and as well lower average input temperatures of 45°C instead of 55°C. The energy consumption lowers from 790 to 497 TWh/a. The heat profile is translates to electric power consumption through the efficiency curve of a standard heat pump summing up to 155 TWh/a for 21 million heat pumps in total. Both characteristics are functions of the ambient temperatures, see details in [1].

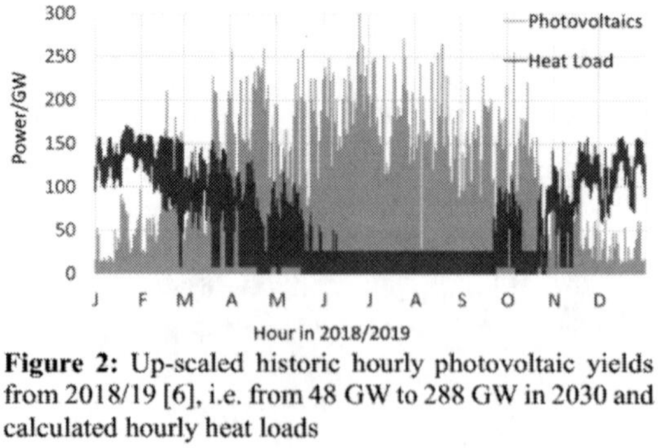

Figure 2: Up-scaled historic hourly photovoltaic yields from 2018/19 [6], i.e. from 48 GW to 288 GW in 2030 and calculated hourly heat loads

Fig. 2 shows, that there a clear seasonal mismatch between photovoltaic power generation and heat demand.

This gets even worse at the grid level, since the heat pumps are not operated independently of one another, but relate to the same or at least similar ambient temperatures.

The hourly yields in 2030 for photovoltaics, wind, hydroelectric power and biomass are extrapolations on the historical yields of the Energy-Charts [6] for 2018/2019. The capacities in GW are variable parameters in order to cover the load profiles at lowest system costs as searched by EXCEL solver. Other parameters are the capacities for electrolysis, fuel cells in GW and batteries in GWh.

3.3 Results

Three scenarios were considered in Table II. The "Heat only" and "All Sectors" scenario fully cover either heat or total residual load through 21 million Hybrid Storage CHP, respectively. In the "Reference" scenario of Fraunhofer ISE [4] the residual load is provided by gas turbines at a share of 15% using imported fuels. Hydrogen goes into industry and transport. All three scenarios are calculated at costs from Table I. While the installed capacities are free parameters in the first two scenario, they were kept fix for the cost calculation of the "Reference" scenario.

Table II: One-sector, All-sectors and ISE Fraunhofer's scenario [4] as snapshots for 2030 with costs from [3]

	Heat only	All Sectors [1]	"Reference" ISE FhG
Photovoltaics GW	299	463	415
Wind onshore GW	-	230	188
Wind offshore GW -		80	78
Electrolysis (70%) GW	133	179	75
Fuel cell (60%) GW	51	133	-
Batteries (94%) GW	0	0	150
Gas turbine (42%) GW	-	-	163
RE Generation TWh/a	287	1420*	1374*
Imported Fuels TWh/a	-	-	150
Curtailment	9.5%	3.2%	10.9%
Cost €/kWh	0.053	0.045	0.057

* incl. 7 GW Hydro and 20 GW Bio mass

The "Heat only" scenario transitions into the "All sectors" scenario by adding capacities, mostly wind. The "Heat only" scenario offers a cost advantage for the prosumers, since no fuel costs are incurred after replacing fossil heaters by Hybrid Storage CHP units. In the "All Sectors" scenario, these units draw additional power from wind, bio mass and hydro via the power grid. This reduces curtailments down to 3.2% and thus the costs per kWh. The "Reference" scenario requires less wind and photovoltaic capacities, but results in higher costs due to the higher curtailment rate of 10.9% and the high costs of natural gas since 2022. In [4], the variable OPEX for gas is at €0.023/kWh and expected to remain flat between 2020 and 2050, while in [3] it increases from €0.0222 to €0.041/kWh until 2040. The change to €0.06/kWh results in +11% or +8% in system costs. With its low original cost, gas has become the generally accepted transitional energy with the conclusion that it only needs to be replaced by imported, synthetic gas in the future. For Germany, the gas turbine capacities were scheduled to increase from 32 GW to 163 GW until 2050 in the Fraunhofer scenario, despite its high curtailment rate.

However, Hybrid Storage CHP increase the efficiency of the energy system and frees it from import dependencies. As shown in [1], the decentralization of hydrogen alone saves 7.2% of the system costs by on-site heat-coupling. Heat coupling with central power plants is feasible, but requires additional infrastructure, i.e. heat nets for district heating. This implicates rollout intervals of several decades in municipal heat planning ("Kommunale Wärmeplanung") causing additional costs on one hand and high thermal transport losses on the other. A paradigm shift from central gas turbine to fuel cell technology is very likely due to its high efficiency even on smallest system sizes. This might be the key to decentralized, distributed power and heat generation.

3.4 The role of stationary batteries

In the cost optimization, batteries do not appear in the scenarios with Hybrid Storage CHP. But after lowering the battery cost parameters by -85%, a minimum emerges in the system cost at 150GW/150GWh. This corresponds well to the "Reference" scenario.

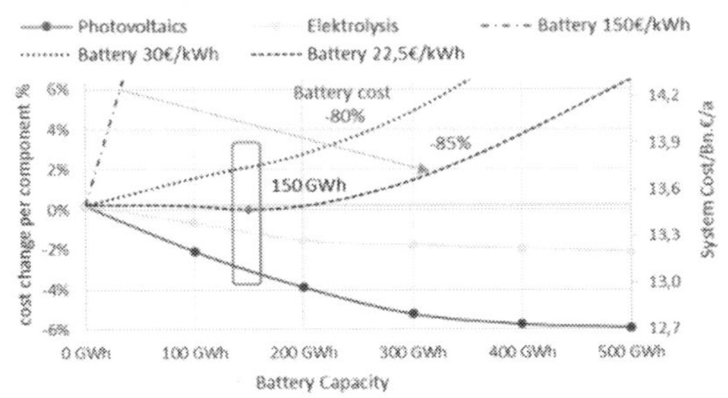

Figure 3: With lower battery costs the system costs (right axis) show a minimum at 150 GW/150 GWh due to a decrease in photovoltaic by -3% and electrolysis costs by -1% (left axis).

Batteries help to reduce photovoltaic and electrolysis capacities, but this only starts to be cost-effective at a battery capex of down to €22.5 kWh in Fig. 3. In the hourly calculations, the minute reserve was not considered, where batteries obviously make sense, as well as in battery cars.

3.5 Matching PV to heat pumps with Hybrid Storage CHP

Fig. 4 shows the effect of Hybrid Storage CHP in the one-sector scenario "Heat only": photovoltaic yields shift from summer to winter through hydrogen storage.

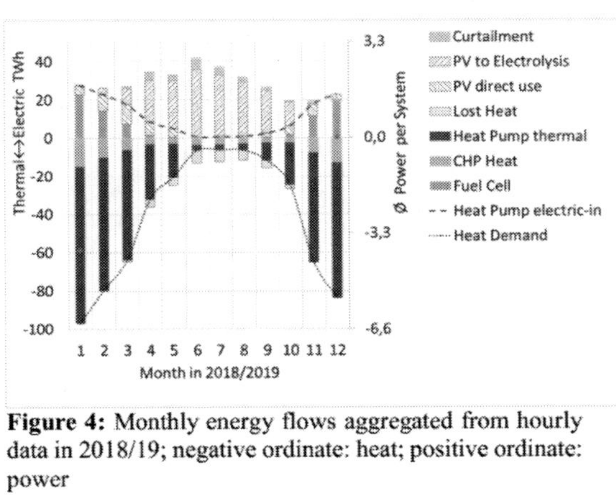

Figure 4: Monthly energy flows aggregated from hourly data in 2018/19; negative ordinate: heat; positive ordinate: power

The waste heat from the fuel cell supports the heat pump in winter. In summer, the waste heat from the electrolysis covers the heat loads for hot water. The effect of heat storage was not taken into account, but would save "Lost Heat" in summer.

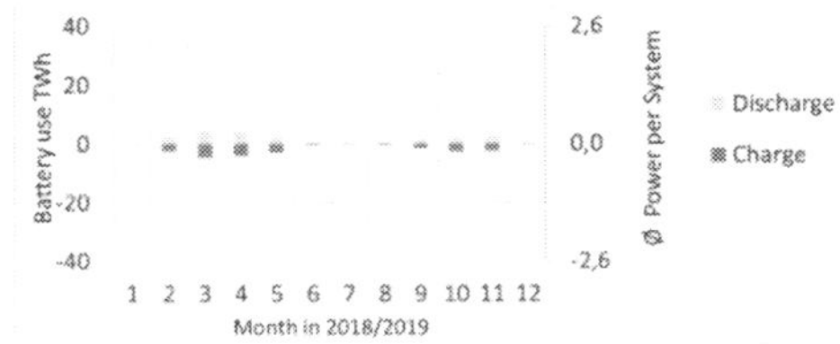

Figure 5: The impact of stationary batteries

Batteries compensate for day-night changes in spring and autumn (share 8%) whenever PV and heat pump are at the same power level, see Fig. 5, but this contribution is apparently not economical, when long term storage from Hybrid storage CHP are in place, see above.

4 GAS NET AS COLLECTIVE STORAGE

Actual market prices for the components in hybrid storage CHP are at least one order of magnitude higher than those in Table I. And the hydrogen bundles alone account for around 40% of the total cost of the device.

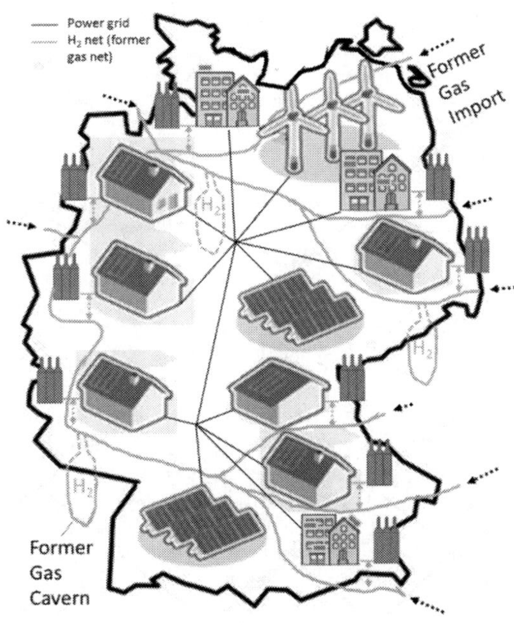

Figure 6: All electric Germany with extended hydrogen storage by means of a reassigned gas net in 2030

At first, the hydrogen components have to go through their learning curves, and at second, the hydrogen storage capacity of the existing gas network is needed as cheaper option for higher volumes. Assuming this as realized, hydrogen would be fed in and out the net instead of expensive bundles on-site, see Fig. 6. Based on the current network tariffs for power and gas, a scenario with pure hydrogen transport would already undercut the currently pursued power transmission scenario, despite the lower caloric value of hydrogen, see [1]. Hydrogen nets are state of the art in Germany and reassignment strategies are described in [7].

5 CONCLUSION

The hydrogen strategy presented here is based on Hybrid Storage CHP replacing all fossil heating systems with the following specifications: 14.2 kW rooftop photovoltaic, 6.3 kW electrolyser, 2.4 kW fuel cell and 2.5 kW electric heat pump. The electricity costs will be €0.053 per kWh or €0.027 thermal in 2030. This heating device runs without external gas and power and addresses private investors, building developers and owners. Extending the approach to full sector coupling including an electrified industry, battery cars with wind power and additional ground-mounted photovoltaics brings costs down to €0.045/kWh and would require +35% more electrolysis and +164% fuel cell capacity per device. The total system costs for this scenario amount to 56 billion €/a for Germany, which is below the German fuel imports at 65 billion €/a in 2018/19. Because Hybrid Storage CHP addresses prosumers, the corresponding market design will be a feed-in tariff. As for other feed-in tariffs, certain rules needs to be set: The tariff applies exclusively to electricity from self-generated hydrogen and only in the event of local power shortages. The latter can be detected by the inverter's feed-in intelligence.

However, the current prices for complete Hybrid Storage CHP devices including hydrogen bundles would require significantly higher feed-in tariffs than the generation costs per kWh concluded here. Photovoltaics started in Germany in 2000 at €0.46/kWh.

Conversely, if there is a power surplus in the grid, at least the power for electrolysis should be cheaper to buy.

In summary, this inherently decentral approach does not require power plants or fuel imports. Its market penetration starts with the heating sector and ends with the full energy demand across all sectors in Germany at every hour of the year. The power is generated exclusively from renewable energy sources and is only partly stored in hydrogen.

Surplus hydrogen can be marketed to heavy industry and long-distance vehicles – potentially via a reassigned gas net.

References

[1] Grunow, P., Decentral Hydrogen, Energies 2022, 15, 2820; https://doi.org/10.3390/en15082820
[2] https://www.homepowersolutions.de/en/
[3] Ram, M. et al. Global Energy System Based on 100% Renewable Energy—Power, Heat, Transport and Desalination Sectors, Lappeenranta: LUT & EWG
[4] Sterchle, P. et al., Wege Zum Klimaneutralen Energiesystem, FhG ISE 2020 incl. Annex
[5] Petkov, I.; Gabrielli, P. Power-to-hydrogen as seasonal energy storage: An uncertainty analysis for optimal design of low-carbon multi-energy systems. Appl. Energy 2020, 274,115197
[6] https://energy-charts.info
[7] Cemiauskas, S. et al. Options of natural gas pipeline reassignment for hydrogen: Cost assessment for a Germany case study. Int. J. Hydrogen Energy 2020, 45,12095.

ASSESSMENT OF NON-ISOLATED MICROGRID POTENTIAL FOR INCREASED RELIABILITY IN MEDIUM VOLTAGE FEEDERS

Pedro F. Torres[1], Roberto Zilles[1], João T. Pinho[1,2]
[1]Institute of Energy and Environment, Universidade de São Paulo
[2]Group of Studies and Development of Energy Alternatives, Universidade Federal do Pará
Avenida Professor Luciano Gualberto, 1289. São Paulo – SP, Brazil.

ABSTRACT: This paper presents a methodology for distributed energy resources (DER) valuation on distribution planning, considering the formation of non-isolated microgrids in medium voltage feeders. An algorithm is developed to allocate the microgrid-forming DER and the respective islanding switch to maximize the reliability of a feeder. The proposed DER valuation and allocation approach makes it possible to assess the feasibility of utility-led DER power and energy contracts as non-wires planning alternatives, focusing on reliability improvement, and accounting for other key aspects, such as investment deferral and system services. A set of real medium voltage feeders with low reliability is used to test the approach, considering photovoltaic plants with battery storage as microgrid-forming DER.
Keywords: Microgrid, Non-Wires Alternatives, Distribution Planning, Reliability, Islanding

LIST OF ABBREVIATIONS AND SYMBOLS

BESS – Battery Energy Storage System
CID – Customer Interruption Duration
CIF – Customer Interruption Frequency
DER – Distributed Energy Resources
DG – Distributed Generation
GUTC – Grid-use Tariff Component
LV – Low-Voltage
MFR – Microgrid-forming Resource
MG – Microgrid(s)
MV – Medium Voltage (> 2.3 kV)
NWA – Non-wires Alternatives
PVG – Photovoltaic Generator
PR – Performance Ratio
SAIDI – System Average Interruption Duration Index
SAIFI – System Average Interruption Frequency Index
b_i – i^{th} block of the system
C_{EC} – annualized saved energy contracted cost
C_{GIS} – annualized saved grid investment cost
C_{MFR} – MFR annualized energy cost
C_{MG} – Microgrid annualized value
$C_{O\&M}$ – annualized net O&M cost
$C_{O\&M,RFM}$ –annualized O&M costs due to the microgrid
$C_{O\&M,UD}$ –annualized O&M costs due to grid upgrades
C_R – annualized saved reliability cost
C_u – investment cost of a grid upgrade u
E_c – annual energy consumption of a consumer c
E_{PVG} – Expected annual energy generated by the PVG
h_p – planning horizon, in years
I_{CM} – annualized income from the capacity market
NC – number of consumers
NC – number of grid upgrades
NC_v – number of consumers with violated reliability limits
P_{MFR} – MFR power capacity during islanded operation
$P_{MFR,CM}$ – MFR power capacity available to participate in the capacity market
P_{PVG} – Installed power of PVG
$Q_{E,MFR}$ – MFR contracted energy for connected operation
$Q_{E,REF}$ – saved energy from large-scale reference supplier
$Q_{I,MFR}$ – MFR contracted energy for islanded operation
r_c – interest rate
r_{wacc} – weighted averaged capital cost rate
SDi,j – sectionalizing device between blocks i and j
T_I – MFR autonomy time at rated power
Y_r – annual reference yield for the PVG, in hours
α_d – valuation coefficient for interruption duration
α_f – valuation coefficient for interruption frequency
β_d – ordered ascending list of blocks in terms of contribution to SAIDI
γ_c – volumetric grid-use tariff component for a consumer c
ρ_{BESS} – BESS energy price
ρ_{CM} – power price of participation in the capacity market
$\rho_{E,MFR}$ – MFR energy price during connected operation
$\rho_{I,MFR}$ – MFR energy price during islanded operation
ρ_{PVG} – PVG energy price
ρ_{REF} – energy price from reference large-scale supplier
Ω_c – set of consumers
Ω_{cv} – set of consumers with violated reliability limits
Ω_{gu} – set of grid upgrades
Ω_{MFR} – set of possible blocks for MFR placement

1 INTRODUCTION

Reliability in power distribution systems is a growing concern for utilities, regulators, and consumers worldwide, given the ever-increasing dependence on electricity for daily activities. The distribution planning and operation task is challenging as technical and economic constraints involve several factors that must be considered to ensure the system's reliable and efficient operation. On top of that, the expansion of distributed energy resources (DER) requires adequate planning and operation of distribution grids to efficiently harness the benefits of distributed generation (DG), energy storage, demand response, and other DER-related technologies.

A relatively recent concept being addressed in the literature and in some power distribution industries worldwide is the Non-Wires Alternatives planning approach (NWA). In the distribution context, NWA planning aims to take advantage of the DER benefits as a cost-effective alternative to the conventional "wire-based approach", for example, using distributed generation and demand-response programs to postpone or reduce the need for substation repowering. In this context, non-isolated distribution microgrids (MG) could be a potential NWA for utilities to increase reliability, DG accommodation, and DER controllability, given the capacity of islanded operation, clearly defined electrical boundaries, and local control of resources. However, such potential is still unclear for utilities and is an open research question in the field.

Ten case studies of NWA projects implemented by

U.S. utilities are presented in [1]. As key findings, the authors list successful delays and deferrals of infrastructure upgrades, flexibility to approach load growth uncertainty, and avoidance of high up-front costs. Among the areas for further research, the need to develop NWA benefit-to-cost analysis and new incentive models for utilities and utility contracting benchmarks with technology providers and third-party owners is highlighted.

In [2], it is evaluated how a residential-level energy management system can be used for peak shaving as an alternative to grid reinforcement, considering a 20-year distribution planning study. The authors show it is a competitive alternative in some technical and economic scenarios. Similarly, a distribution capacity mechanism to fit within a distribution grid planning methodology is proposed and discussed in [3]. The result of the methodology is a volume-based capacity auction for a capacity-constrained system oriented to both active and reactive power provision.

A feeder investment model for distribution system planning, including battery energy storage, is presented in [4], where the authors formulate an optimization problem based on a mixed-integer non-linear programming algorithm to size and allocate a distribution level battery as an NWA. However, the reliability assessment considers that the average interruption duration index of the system is a function only of the distribution grid loading. In [5] a distributed power generation investment planning is proposed in an approach that considers both reliability and emissions costs.

The study shown in [6] focuses on planning microgrids to improve system resilience against severe faults. Three solution approaches are proposed aiming to determine the optimal nodes for connecting microgrids and the capacity of the internal dispatchable generation units. A similar approach is shown in [7], where a fuel-based distributed generator allocation strategy is proposed to enhance the distribution system's resilience against extreme weather. The long-term planning problem is formulated as a two-stage stochastic mixed-integer programming.

This paper investigates whether non-isolated MG can be a feasible alternative in distribution system planning in regions with low-reliability performance. The MG formation is based on the utility-led approach, where power and energy contracts are signed with local DER providers. A methodology to size, allocate and quantify the benefits of an MG is proposed in a reliability-oriented planning approach.

The remainder of the paper is organized as follows: section 2 details the utility-led DER contracts that are considered the core business model for microgrid formation; in section 3 a methodology to quantify the benefits associated to the implementation of microgrids when compared to a "wire-based" reference case is presented; in section 4 a method to size and allocate the microgrid-forming resource is discussed, as well as the allocation of the islanding switch; the methodology proposed in sections 3 and 4 is applied to a real distribution feeder, and the results are presented in section 5, followed by the conclusions in section 6.

2 UTILITY-LED DER CONTRACTS

A recent business model to consider NWA resources in distribution planning is the adoption of utility-led contracts with third-party DER providers. This approach was explored in [8], where a "contract for deferral scheme" was proposed to integrate DER as an alternative for demand-driven grid investment, where such contracts are procured in a three-step auction process. Another possibility for third-party DER procurement is to open public calls addressing specific planning objectives in the utility's areas of interest. One example is presented in [9], which procured DG with islanded operation capacity to form microgrids on a preselected set of substations and feeders with poor reliability.

The present paper considers an approach similar to the one in [9], where the utility procures third-party-owned Microgrid-Forming Resources (MFR) via power and energy contracts as an NWA focused on improving reliability. During normal operation, an MFR operates as a regular DG, selling energy to the utility, while in contingency, the MFR must be able to operate islanded, hence forming a microgrid to service a predefined set of consumers. Therefore, the MFR contract must specify the contracted energy price, $\rho_{E,MFR}$, (*e.g.* in \$/MWh); the amount of contracted energy from the MFR over the planned horizon, $Q_{E,MFR}$; the rated power capacity during islanded operation, P_{MFR}; and the islanded autonomy time, T_I. The MFR annualized cost, C_{MFR}, is given by Equation (1):

$$C_{MFR} = f_{ann}(r_c, h_p)\,(\rho_{E,MFR}Q_{E,MFR} + \rho_{I,MFR}Q_{I,MFR}) \quad (1)$$

$$f_{ann}(r,h) = \frac{r\cdot(1+r)^h}{(1+r)^h-1} \quad (2)$$

where $f_{ann}(r,h)$, given by Equation (2), is the annualization function of costs, which is defined for a given interest rate, r_c, and a planned horizon, h_p. $Q_{I,MFR}$ and $\rho_{I,MFR}$ are the MFR variables associated with the required amount of energy during islanded operation and its price, respectively.

Given the operational requirements of an MFR and its small-scale characteristics, it is reasonable to consider that its energy costs will hardly be competitive against large-scale, grid-connected only, centralized power plants. Therefore, a cost-benefit analysis must be performed, considering different aspects to verify the feasibility of an MFR contract.

3 VALUATION OF MICROGRID BENEFITS

To quantify the benefits associated with the development of a microgrid, the following aspects are considered: (1) reliability improvement; (2) operation and maintenance costs; (3) energy costs; (4) investment deferral in feeder and substation capacity expansion; (5) system-wide services. Equation (3) is used to calculate the value of the microgrid development, and each aspect is analyzed and quantified in the following items.

$$C_{MG} = C_{MFR} - C_R + C_{O\&M} - C_{GIS} - I_{CM} - C_{EC} \quad (3)$$

3.1 Reliability improvement

The customer interruption costs are used to quantify the benefits of reliability improvement due to microgrid formation. Regulated utilities are subject to penalties when interruption levels exceed maximum reference values. Such penalties are usually financial compensations to customers whose limits were violated and imposed restrictions in tariff increase in annual tariff revisions, thus reducing the utility revenue. Therefore, utilities are incentivized to improve reliability levels, which can be done in terms of assets (equipment ratings, installation methods) and managerial aspects (tree pruning, equipment maintenance, emergency team readiness).

Both collective and individual reliability indexes are considered in this work to assess the performance of grid reliability. The most commonly used reliability indexes for collective assessment are the System Average Interruption Duration Index (SAIDI) and the System Average Interruption Frequency Index (SAIFI), given by Equations (4) and (5), respectively. Furthermore, Customer Interruption Duration (CID) and Customer Interruption Frequency (CIF) are used as individual reliability metrics and relate to the annual duration, in hours, and the number of interruptions experienced by the consumer, respectively.

$$SAIDI = \frac{\sum CID_c}{NC} \ \forall\, c \in \Omega_c \quad (4)$$

$$SAIFI = \frac{\sum CIF_c}{NC} \ \forall\, c \in \Omega_c \quad (5)$$

where NC is the total number of consumers of a given system area, and Ω_c is the set of consumers of that area.

The interruption cost can be divided into two categories:

(1) The individual compensation paid to consumers whose limits were violated (CID and CIF) and is given by Equation (6);

(2) The utility performance on a system level, that is subject to fines or penalties applied by regulators. For the sake of simplicity, this cost category will not be addressed in this paper, as it varies widely with each regulation agency.

$$C_{IC} = \sum\left(CID_c \frac{\gamma_c \cdot E_c}{8760} \alpha_d + \frac{CIF_c}{CIF_l} CID_l \frac{\gamma_c \cdot E_c}{8760} \alpha_f\right) \forall\, c \in \Omega_{cv} \quad (6)$$

$$C_R = f_{ann}(r_c, h) \cdot \Delta C_{IC} \quad (7)$$

where C_{IC} is the compensation cost due to interruptions that must be paid to consumers. The sub-index l indicates the regulated limit for CID and CIF, above which the utility must compensate the consumer c. Ω_{cv} is the subset of Ω_c that contains only consumers c whose CID or CIF limits were violated. E_c is the annual energy consumption of consumer c, with a volumetric grid-use tariff component, γ_c. α_d and α_f are interruption valuation coefficients associated to the type of consumer, where the first is related to interruption duration and the latter to interruption frequency (*e.g.* an interruption to a commercial customer might be more costly than to a residential one). C_R , in Equation (7), is the annualized saved reliability cost, given by the difference between the C_{IC} before and after microgrid implementation. Therefore, higher values of C_R indicate higher savings in reliability-related compensations.

3.2 O&M costs

The costs of operation and maintenance, $C_{O\&M}$, can be calculated considering the costs associated to the microgrid implementation, $C_{O\&M,MFR}$ (*e.g.* dedicated remote monitoring and control) minus the costs avoided due to grid upgrade deferral, $C_{O\&M,UD}$, as expressed in Equation (8). Therefore, negative values of $C_{O\&M}$ indicate that the RFM implementation incurs savings in O&M.

$$C_{O\&M} = f_{ann}(r_c, h) \cdot (C_{O\&M,MFR} - C_{O\&M,UD}) \quad (8)$$

3.3 Energy contracted costs

In the considered business model, the MFR is contracted as a generator by the utility to supply part of its consumers, which large-scale centralized plants would otherwise supply. Equation (9) is used to quantify the savings incurred by reducing energy costs from contracts with a large-scale reference supplier.

$$C_{EC} = f_{ann}(r_c, h) \cdot Q_{E,REF} \cdot \rho_{REF} \quad (9)$$

where $Q_{E,REF}$ is the amount of energy that would be contracted from a conventional reference supplier over the planned horizon, with a respective contracted energy cost given by ρ_{REF}. Note that $Q_{E,REF}$ and $Q_{E,MFR}$ could be different as energy losses change depending on the source location.

3.4 Postponement of investments in feeder and substations

By upgrade deferral, it is understood the investment in assets to secure load growth and DG hosting capacity in the planned period. Therefore, such an aspect comprises investments in substations and feeder circuits, including transformers, conductors, and reactive power support. Equation (10) is used to calculate the savings in grid investment.

$$C_{GIS} = f_{ann}(r_{wacc}, h) \cdot \sum C_u \ \forall\, u \in \Omega_{gu} \quad (10)$$

where C_u is the investment cost of a given grid upgrade u, Ω_{gu} is the set of grid upgrades deferred by the MFR operation, and r_{wacc} corresponds to the interest rate given by the weighted average capital cost, which is usually the reference value that regulated utilities use when considering grid investments. Therefore, higher C_{GIS} indicates higher savings in grid upgrades from MFR installation.

3.5 System services

Apart from the energetic services, the contracted MFR could provide system services such as reserve capacity and reactive support. In this paper, it is considered that de MRF can participate in a regional capacity market, with income given by Equation (11).

$$I_{CM} = f_{ann}(r_c, h) \cdot P_{MFR,CM} \cdot \rho_{CM} \quad (11)$$

where $P_{MFR,CM}$ is the available power from the MFR to provide reserve capacity (note that $P_{MFR,CM}$ is a fraction of the rated MFR capacity, P_{MFR}). ρ_{CM} is the price due to the MFR per capacity power available.

4 MFR SIZING AND PLACEMENT

It is important to place the MFR considering the valuation indicated in the previous section to maximize its benefits. For each possible MFR location, its power and energy ratings must be determined to guarantee the reliability requirements during islanded operation and to maximize the benefits while grid-connected.

4.1 Reliability Evaluation

An analytical method is used to assess the reliability improvements after the installation of the MFR by calculating the expected SAIDI and SAIFI. For a given substation under assessment, the MV circuits are divided into blocks. Each block is a feeder section that can be completely isolated from the remainder of the circuit by triggering a switching device (SD). For each block under a sustained fault, the other blocks are classified according to their state with respect to the faulty block, which can be one of the following:

(1) not affected: block remains in service independently of the faulty block;
(2) permanently interrupted: block is out of service as long as the faulty block remains faulty;
(3) recoverable: block that can be put in service after the opening of an SD upstream of the faulty block;
(4) transferable and connected: block that can be put in service after a change in feeder configuration that leads to an alternative path to the substation;
(5) transferable and islanded: block that can be put in service after a change in feeder configuration, and the block is supplied by the MFR under islanded operation.

Based on these possibilities, utility reference values of failure rate and recovery time can be used to calculate the reliability indexes [10, 11]. Figure 1 illustrates the concept of block classification and MFR islanded operation. In Figure 1(a), a sustained fault in block 2 would cause a permanent interruption in blocks 3 to 5. Block 1 can be either unaffected if $SD_{1,2}$ is a protection device (*e.g.* fuse) or recoverable if $SD_{1,2}$ is a sectionalizing-only device (*e.g.* normally-closed switch). Blocks 6 and 7 are unaffected, given that $SD_{2,7}$ is a normally-open switch.

In Figure 1(b), blocks 3, 4, and 5 could be put in service after a sustained fault in block 2, given the alternative supply path from the MFR in islanded operation by the opening of $SD_{2,3}$. It can be noted that, for sustained faults in block 4, the MFR is useless. Also, if the microgrid islanding switch is placed in $SD_{3,4}$, block 3 would not benefit from the MFR in islanded operation. Therefore, the placement of the MFR and its respective islanding switch highly influences the effectiveness of microgrid operation concerning reliability improvement.

From the analytical reliability methodology based on block classification, it is possible to rank the blocks in terms of importance to the overall SAIDI or SAIFI: $\beta_d = [b_{d1}, b_{d2}, \ldots, b_{dN}]$, where β_d is the ordered list of blocks in terms of contribution to SAIDI, from highest to lowest contribution, in a substation area with N blocks.

4.2 Islanding Switch placement and MFR sizing

For each block in a given set of blocks where the MFR can be placed, Ω_{MFR}, there is a location for the islanding switch that will maximize the reach of the MFR considering the block failure probability associated to the block's contribution to SAIDI (β_d). For the example in Figure 1(a), assuming that $\beta_d = [6, 2, 1, 7, 3, 4, 5]$, for an MFR installed in block 5, an islanding switch in $SD_{2,3}$ will make the microgrid supply to more consumers during islanded operation (consumers in blocks 3, 4 and 5), while for an MFR is installed in block 7, the optimal location for the islanding switch is in $SD_{6,7}$, since block 6 is a major contributor to SAIDI.

After the islanding switch is selected, the MFR can be sized in terms of power and energy capacities. Minimal P_{MFR} is determined by the cumulative peak load of the blocks supplied by the MFR during islanded operation, while $Q_{I,MFR}$, is calculated by the maximum value of the moving average of the cumulative daily load demand with interval given by the average service restoration time of the utility in the substation area.

Algorithm 1 determines the islanding switch location for a given MFR location and the MFR minimal power and energy requirements.

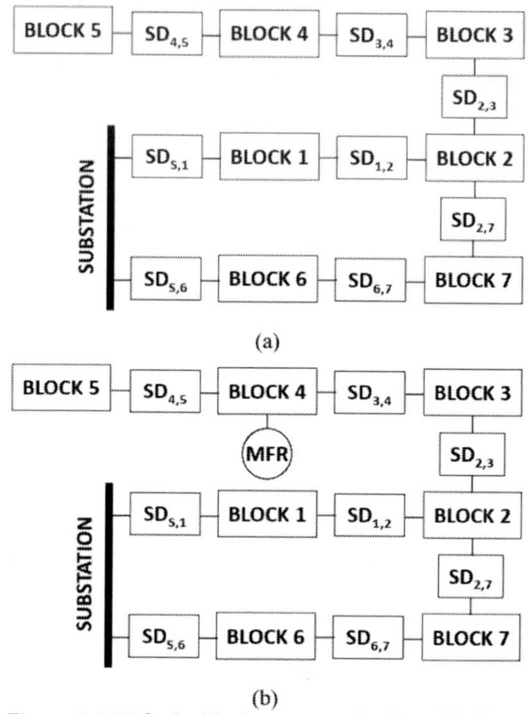

Figure 1: MV feeder block representation for reliability analysis. (a) without MFR and (b) with MFR.

Algorithm 1: Placement of the Islanding Switch.

1. For each block b_{MFR} in Ω_{MFR}:
   a. find the highest ranked block, b_d, in β_d, whose permanent fault has a transferable and islanded block;
   b. Place the islanding switch downstream from b_d;
   c. Run a daily power flow considering an MFR at b_{MFR} as the reference node and the islanding switch open (microgrid in islanded operation);
   d. Find the peak power demanded from the reference node. This will be the minimum rated MFR power capacity (P_{MFR});
   e. Calculate the moving average of the power demanded from the reference node for an interval T_I;
   f. Multiply T_I by the maximum value of the moving average. This will be $Q_{I,MFR}$.

5 SIMULATION RESULTS

The following study case illustrates the application of the proposed methodology for the valuation, sizing, and placement of an MFR. It is based on a remote system isolated from the Brazilian national grid and supplies over 4,400 consumers using a set of diesel-based generators.

The database used to carry out this study is freely available from the Brazilian Electrical Energy Regulatory Agency (ANEEL) and contains georeferenced data for both medium and low voltage circuits for each utility in Brazil. All data is given for the year 2020 unless otherwise noted. Daily power flow was performed for sequential quarter-hourly data available for the loads, and no distributed generation other than the MFR was considered. The electrical model and power flow were implemented using the open-source PandaPower [12] library for Python environment and run on the cloud via Google Colab Pro.

5.1 Reliability analysis of MFR application

This system corresponds to the area serviced by the electrical substation of the municipality of São Sebastião da Boa Vista (SSB), in the northern region of Brazil. It is located on a geographical island, isolated from the national power grid, and supplied by diesel-based generators. Access to the island is complex, affecting the response time to occurrences and therefore contributing to high SAIDI values. Furthermore, the non-urban area of the island is covered by dense vegetation, thus increasing the frequency of interruptions of the medium voltage grid related to vegetation and making it difficult for the emergency team to identify the problem and repair the grid. Table I summarizes the key characteristics of the SSB substation, and Figure 2 illustrates the two MV circuits that cover the island.

Table I: Characteristics of SSB substation service area.

MV Circuit	SSBSB-01	SSBSB-02
MV Circuit Length	3.68 km	83.94 km
Number of Blocks	14	29
Number of LV consumers	3,456	979
Number of MV consumers	3	1
Substation Transf. Capacity	5.0 MVA	
Substation Gen. Max. Power	6.06 MW	
SAIDI	33.87 h	
SAIFI	18.71	
SAIDI Regulatory Limit	85.0 h	
SAIFI Regulatory Limit	70.0	
Avg. Occurrence Clearing Time	7 h and 30 min	

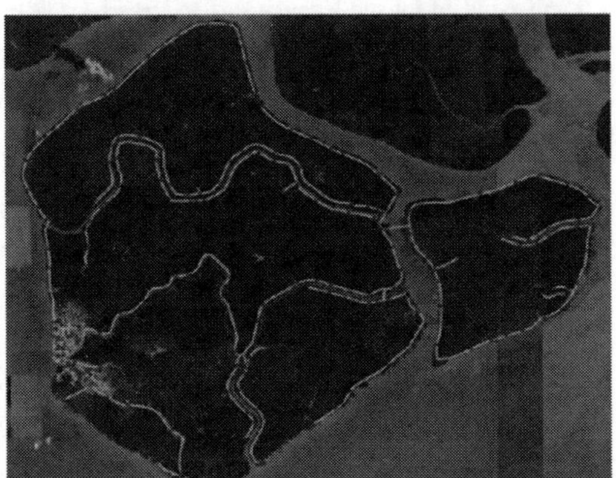

Figure 2: SSB substation MV circuits coverage area. In green is the urban-area circuit (SSBSB-01), and in orange the non-urban area circuit (SSBSB-02).

Figure 3 presents the monthly energy and peak power registered at the point of connection of the diesel generators to the substation. It can be noted that the power and energy densities of this distribution system are small when compared to reference values in more urbanized areas in the country. In terms of geographical density, the substation supplies around 51 consumers per kilometer of MV grid. An average value of 0.56 kW per consumer unit was found for peak demand. These values alone are good indicators of the potential use of small-scale distributed generation to improve system performance.

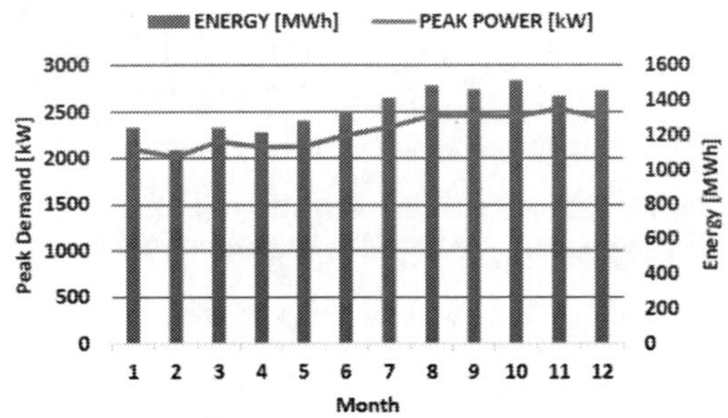

Figure 3: Monthly energy and peak power registered at the point of connection of the diesel generator to the substation.

Table II summarizes the results obtained from the application of Algorithm 1 to this system. Figure 4 shows the 16 blocks with higher SAIDI reduction per P_{MFR}.

Table II: Expected SAIDI and SAIFI for an MFR at each possible block. The block highlighted in green results in the best SAIDI and SAIFI. The block highlighted in blue results in the best SAIDI reduction per P_{MFR}. In orange are the blocks that result in virtually no improvement in SAIDI.

Block	SAIDI	SAIFI	Block	SAIDI	SAIFI
0	29.58	17.22	**23**	**33.87**	**18.71**
1	32.78	18.45	24	31.73	18.24
2	30.26	18.16	25	29.49	17.24
3	28.88	16.20	26	32.15	18.32
4	28.75	15.97	27	32.25	18.32
5	32.87	18.50	28	30.41	18.43
6	29.08	16.20	29	30.04	17.77
7	30.00	17.74	30	29.22	16.32
8	33.78	18.69	31	33.38	18.65
9	29.11	16.26	**32**	**33.87**	**18.71**
10	33.36	18.62	33	32.18	18.31
11	32.33	18.34	34	33.06	18.55
12	29.81	17.37	35	32.48	18.45
14	29.60	17.25	36	33.52	18.69
15	**30.36**	**18.35**	37	33.41	18.66
16	32.06	18.27	38	29.42	17.19
17	30.29	18.22	39	28.64	15.95
18	33.11	18.62	40	33.09	18.61
20	33.03	18.53	41	33.64	18.70
21	**28.63**	**15.94**	42	33.52	18.69
22	32.46	18.42			

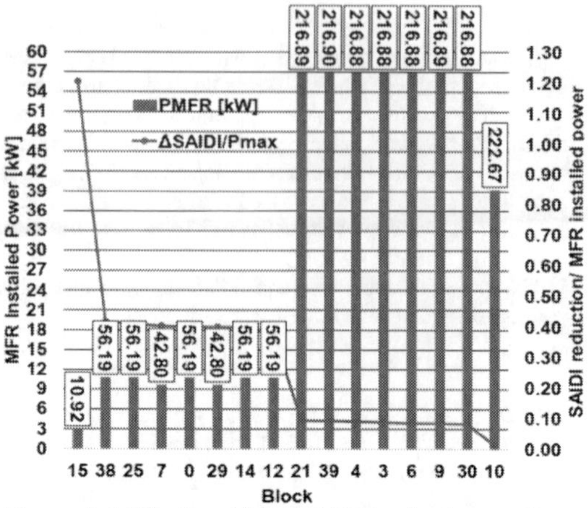

Figure 4: 16 blocks with best SAIDI reduction per P_{MFR}. The boxes above the bars indicate $Q_{I,MFR}$, in kWh.

The results indicate how the location of the MFR influences its capacity to improve overall system reliability. For instance, an MFR installed in block 21 achieves the best results regarding SAIDI and SAIFI improvements, with reductions of 15.47 % and 14.80 %, respectively. On the other hand, there are blocks where an MFR represents virtually no improvement in reliability, as in blocks 23 and 32. Apart from absolute values, from a planning perspective, it is important to note SAIDI improvements with respect to MFR power and energy required capacities: block 15 is the one with the best value for each unit of MFR installed capacity.

From the results in Table II and Figure 4, it is interesting to note that an MFR as small as 3 kW/ 11 kWh (in block 15) could improve the SAIDI by 3 hours and, together with an MFR rated at 57 kW / 217 kWh (in block 21) the SAIDI could be enhanced by more than 8 hours. The first fifteen blocks presented in Figure 4 (from block 15 to block 30) correspond to four different possible microgrids since some blocks share the same optimal location for the islanding switch. A fifth microgrid location, associated with block 10, would present only a slight reduction in SAIDI, as shown in Table II; therefore, it is not considered. Figure 5 highlights the four possible microgrids (hereafter referred to as MG1, MG2, MG3, and MG4) and their respective islanding switch location. Table III indicates the characteristics of the preselected four microgrid locations.

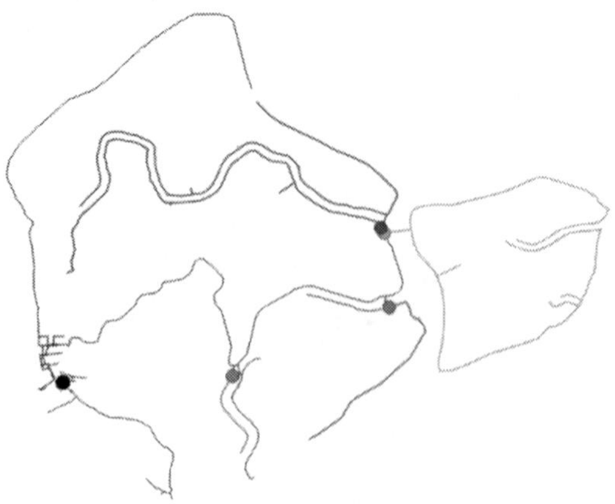

Figure 5: Four microgrid locations and their respective MFR and islanding switches (colored circles) that present the best value for SAIDI reduction: MG1 in red, MG2 in green, MG3 in magenta, and MG4 in blue. The black dot indicates the MV substation.

Table III: Summary of MG characteristics.

	Blocks	P_{MFR} [kW]	$Q_{I,MFR}$ [kWh]	Length [km]
MG1	38, 25, 0, 14, 12	10.7	56.2	23.88
MG2	21, 39, 4, 3, 6, 9, 30	56.8	216.9	18.41
MG3	7, 29	9.6	42.8	4.28
MG4	15	2.9	10.9	1.79

5.2 MFR Cost Analysis

In this paper it is considered that the MFR is composed of a PV Generator (PVG) and a Battery Energy Storage System (BESS). The BESS is sized according to the MFR power and energy sizing presented in the previous section, since it is responsible for servicing the loads during islanded operation, which cannot rely solely on the PVG.

Since the SSB system operates isolated from the national grid, the MFR cannot participate in the regional capacity market. The simulated scenario considers the actual load profiles from the dataset in the first year of simulation, followed by an expected load growth of 1 % per year. Since the lines and transformers' ratings are much oversized compared to the actual system loading, no grid upgrades are needed on the planning horizon. Therefore, the cost/benefit analysis considers the contracted energy costs, O&M costs, and reliability costs. Table IV presents the values considered in the MFR cost assessment. ρ_{PVG} is the energy price from the PVG, which is equivalent to $\rho_{E,MFR}$ in Equation (1), while ρ_{BESS} is the energy price from the BESS and is equivalent to $\rho_{I,MFR}$ in Equation (1).

Table IV: Reference values for cost assessment.

α_d	34	α_f	34
r_c	1.5 %	r_{wacc}	7 %
$C_{O\&M,UD}$	2 % C_{GIS}	$C_{O\&M,MFR}$	2 % C_{MFR}
ρ_{REF}	R$ 1,228 /MWh	γ_c	R$ 236 /MWh
ρ_{PVG} ($\rho_{E,MFR}$)	R$ 2.74 /kWh	ρ_{BESS} ($\rho_{I,MFR}$)	R$ 1,000 /kWh

The PVG yearly production, E_{PVG}, is calculated according to Equation (12), with Y_r = 1,834 h as the yearly reference yield for the location of São Sebastião da Boa Vista (Lat: -1.7005; Lon: -49,549), and PR = 0.8, the global performance ratio of the installation. P_{PVG} is the installed power of the PVG in kWp.

$$E_{PVG} = P_{PVG} \cdot Y_r \cdot PR \tag{12}$$

Table V presents the cost assessment results for each of the four microgrid locations indicated in Figure 5, considering three different sizes of P_{PVG}.

Table V: Cost assessment results (all values in 10^3 R$).

P_{PVG}		MG1	MG2	MG3	MG4
0.5 x	C_R	33.06	61.33	27.25	18.57
P_{MFR}	C_{MG}	36.54	225.21	27.42	-4.08
1 x	C_R	33.06	61.33	27.25	18.57
P_{MFR}	C_{MG}	48.84	290.53	38.45	-0,72
2 x	C_R	33.06	61.33	27.25	18.57
P_{MFR}	C_{MG}	73.43	421.17	60.49	5.99

The results in Table V for C_{MG} indicate that only MG4

is feasible from a financial point of view, considering the small size of the PVG. As expected, reliability-related savings (C_R) do not change with P_{PVG}, and are consistent with the MFR capacity to supply consumers (MG2 is the one that presents higher C_R); however, current MFR costs, especially those related to the BESS, are prohibitive and make the higher capacity microgrids infeasible in this specific case. To understand the cost/benefit composition and how BESS and PVG costs influence the overall MG feasibility, Figure 6 presents the different cost components for MG1 in three different price scenarios and $P_{PVG} = 0.5 \times P_{PVG}$.

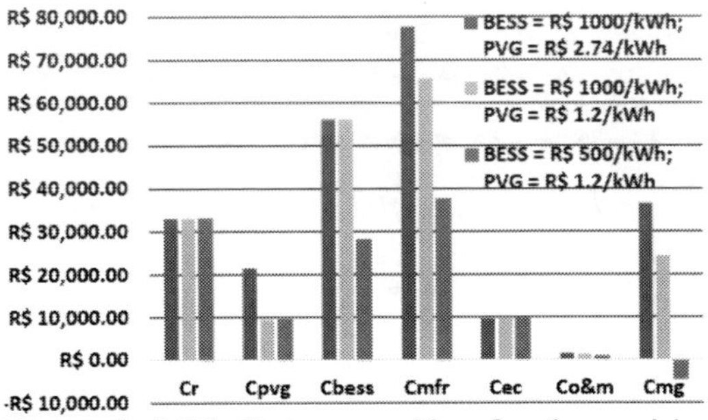

Figure 6: MG1 Cost composition for three pricing scenarios and $P_{PVG} = 0.5 \times P_{PVG}$.

The blue scenario indicated in Figure 6 is the base case ($\rho_{I,MFR}$ = R$ 1,000 /kWh and $\rho_{E,MFR}$ = R$ 2.74 /kWh), and it is possible to note that the BESS cost (C_{BESS}) has a high contribution to C_{MFR}, surpassing the benefits from C_R. In the base case, C_{MG} = R$ 36,547 (not feasible). In the orange scenario, PVG costs are reduced to $\rho_{E,MFR}$ = R$ 1.2 /kWh, and there is a noticeable reduction in C_{MFR}. However, the overall MG cost is still infeasible. In the green scenario, BESS costs are reduced by half, and the overall benefits of the MG surpass its costs, making its implementation feasible.

It is important to emphasize that in this case study no benefits other than in reliability and in energy cost (in the case where $\rho_{E,MFR}$ = R$ 1.2 /kWh) were achievable. Furthermore, a battery storage solution is quite expensive when considering distribution-scale systems. Therefore, other technologies, for example biomass-based generators, may fit better in the proposed case.

6 CONCLUSIONS

The proposed MFR sizing and allocation methodology could successfully select sections of the MV circuits that would incur the highest reliability improvement and determine MFR power and energy requirements to service the loads during contingencies. Such a methodology makes it possible to evaluate the best microgrid sizing and allocation from an economic perspective and can help support decisions from utility planners.

The case study results showed that even though MG are feasible from a technical standpoint, the costs associated with MFR implementation could be prohibitive, especially those associated with energy storage. Therefore, it is crucial to balance MFR influence area (*i.e.*, potential to improve reliability) and energy requirements.

For future works, a system that could benefit from the postponement of grid investments and the participation in the capacity market on a regional level will be analyzed. Furthermore, an economic analysis considering other technologies for the MFR could be carried out, as utility-led contracts should be technology-independent as long as power and energy requirements are guaranteed.

ACKNOWLEDGEMENTS

The authors would like to acknowledge the Brazilian Coordination for the Improvement of Higher Education Personnel – CAPES for the financial support under process n. 88887.609816/2021-00.

REFERENCES

[1] SEPA. Non-Wires Alternatives – Case Studies from Leading U.S. Projetcs. Smart Electric Power Alliance Report (2018).

[2] B. Martin, B. Feron, E. Jaeger, F. Glineur, A. Monti. Peak shaving: a planning alternative to reduce investment costs in distribution systems? In: Springer Energy Systems, v. 10, pp. 871-887 (2019).

[3] M. Alvarez, S. Ronnberg, M. Bollen, P. Farias, R. Cossent, R. Jabr, J. Zhong. A capacity mechanism design for distribution network expansion planning. In: 2018 IEEE International Conf. on Env. and Electrical Eng. and 2018 IEEE Ind. and Comm. Power Syst. Eur. (EEEIC / I&CPS Europe. DOI: 10.1109/EEEIC.2018.8493874 (2018).

[4] K. Masteri, B. Venkatesh, W. Freitas. A Feeder Investment Model for Distribution System Planning Including Battery Energy Storage. In: Canadian Jour. of Electrical and Comp. Eng., v. 41, n. 4 (2018).

[5] K. Niu, Z. Lyu, B. Zheng. Analysis of Renewable Integrated DG Investment Planning Based on Emission Cost, in IOP Conf. Series: Earth and Environmental Science, v. 632 (2021).

[6] M. Borghei, M. Ghassemi. Optimal planning of microgrids for resilient distribution networks. International Journal of Electrical Power and Energy Systems, v. 128 (2021).

[7] Q. Shi, F. Li, T. Kuruganti, M. Olama, J. Dong, X. Wang, C. Winstead. Resilience-Oriented DG Siting and Sizing Considering Stochastic Scenario Reduction. IEEE Transactions on Power Systems, v. 36, n. 4 (2021).

[8] R. Poudineh, T. Jamasb. Distributed generation, storage, demand response and energy efficiency as alternatives to grid capacity enhancement. Energy Policy, v. 67, pp. 222-231 (2014).

[9] COPEL. Chamada Pública Copel DIS GD001/2020 – Compra de Energia Elétrica de Geração Distribuída. Companhia Paranaense de Energia (2020).

[10] R. Brown. Electric Power Distribution Reliability. 2nd ed. CRC Press, ISBN: 978-0-8493-7567-5, (2009).

[11] E. Dias. Avaliação de Indicadores de Continuidade e seu Impacto no Planejamento de Sistemas de Distribuição. Master's Thesis, Escola Politécnica da Universidade de São Paulo, São Paulo (2002).

[12] L. Thurner, A. Scheidler, F. Schäfer, J. Menke, J. Dollichon, F. Meier, S. Meinecke, M. Braun. Pandapower - an Open Source Python Tool for Convenient Modeling, Analysis and Optimization of Electric Power Systems, in IEEE Trans. on Power Systems, vol. 33, no. 6, pp. 6510-6521 (2018).

ONE REALIZATION OF AN AUTONOMOUS MEASUREMENT SYSTEM FOR VERIFICATION OF THE DECLARED EFFICIENCY AND REAL-TIME MONITORING OF THE PHOTOVOLTAIC PLANT PRODUCTION BASED ON IOT PLATFORM

Ilija Batas Bjelic[1], Petar Atanasijevic[2], Pavle Dragisic[3], Goran Dragisic[3], Milos Tomic[1]

[1] Institute of Technical Sciences of SASA, Knez Mihailova 35/IV, Belgrade, Serbia ilija.batas-bjelic@itn.sanu.ac.rs

[2] University of Belgrade, School of Electrical Engineering, Bulevar kralja Aleksandra 73, 11120, Belgrade, Serbia

[3] DECODE doo, Bulevar Nikole Tesle 30A, 11080 Belgrade, Serbia

ABSTRACT: Photovoltaic technology becomes interesting in Serbia from household prosumers, to small business sector entities up to large utility scale plants have been planned and built due to real economic reasoning in the time of disruption of typical fossil fuel energy agendas. In order to meet the expectations of investors, which is increasingly important after the feed-in period is over, the more measurement, monitoring and validation of onsite production are needed. Therefore, a solution based on real-time monitoring of the photovoltaic plant has been designed and realized based on existing Internet of Things platform. The device allows those real-time measurements of each photovoltaic panel string current, voltage and temperature, together with onsite environmental and weather data to be routed by the gateway to the distant cloud server for the advanced processing. Advanced processing is important to diagnose the possible underperformance of photovoltaic plant by calculating indicators such as performance ratio, and weather corrected performance ratio. Such underperformance from temperature factor has been confirmed at the rate of ±2%. Further research will be directed towards inclusion of broader reasoning behind underperformance as a tool to maximize investors assets.

Keywords: Device, Energy Performance, Information and Communication Technologies

1 INTRODUCTION

Delaying the energy transition in Serbia led to exposure to the multiple uncertainties regarding the fossil fuel supply chains, increasing energy prices and interest in photovoltaic project up to 500 kW, self-financed by small medium enterprises. Many parts of these photovoltaic systems came from different vendor across the globe, being not compliant to interoperability, making it more difficult to verify the declared efficiencies, partially and overall. However, if the efficiency losses are caused by poor maintenance, technical failure or installer error, they can be noticed and corrected. Using the on-site weather data measurement, consumption profile and knowing the photovoltaic system conditions, the theoretical energy production can be calculated. This value can further be compared to the measured production, to detect potential possible mismatches and under-operation in real time. Since this efficiency mismatch can be in the range of 5-10%, or even larger [1] using monitoring and verification techniques, at least 1-2% of the total Serbian commercial sector electricity consumption can be saved on the yearly basis, which is equivalent to several million tons of lignite.

Several PV plant monitoring realizations [2] are developed according to standard [3], with near panel temperature measurements using low-cost analog devices [4], or open-source platform [5]. Other focused on PV plant data acquisition [6], analysis and simulation [7], fault detection [8], and further diagnostics [9].

In this paper, a realization of an autonomous measurement system for real-time monitoring and verification of the declared efficiency of the photovoltaic energy system production based on Internet of Things (IoT) platform, is proposed as a solution. The realization is based on IoT include fully interoperable sensing, efficiency verification at panel level and expertise on the *cloud* server, are elaborated in this article. Our system includes advanced string sensor unit, environmental multi-sensor units, photovoltaic facility edge gateway for data acquisition and forwarding, *cloud* server for efficiency calculation for various users.

2 METHODS

System design of autonomous measurement system for verification of the performance of photovoltaic plant production has been presented.

2.1 Designs

An own IoT platform-based design include four elements:

1. **Advanced String Sensor Unit (ASSU)**

Sensor unit that provides economic and simple solution for measuring a string current and voltage with a unique possibility to perform diagnostic function of each panel in a string.

2. **Environmental Multi-Sensor Unit (EMSU)**

Sensor unit that provides economic and simple solution for measuring several environmental values needed for modeling photovoltaic facilities performance. A technology provides easy installation, including a logger function together with reading ability from mobile platforms.

3. **Solar Facility Edge Gateway (SFEG)**

Powerful low power IoT edge gateway unit, compatible with ASSU, EMSU and variety of devices typically found in photovoltaic system installations like inverters, electricity meters, batteries or even intruder alarms. Capable to retrieve large amount of data in real-time and transfer them to cloud server using appropriate IoT protocols over wired, wireless mobile or satellite networks. Embedded processing power enable possibility for edge processing which greatly reduces data flow.

4. **Solar Facility Insight Cloud Server (SFICS)**

Besides storing data from panels, inverters, weather stations, irradiation sensors, revenue meters and batteries, SFICS is a place for applications for predictive maintenance, degradation monitoring and generation forecasting. It provides the most convenience and ensures the highest degree of protection since client monitoring specialist quickly and efficiently handles all of the data and status messages over mobile, tablet or desktop computer devices.

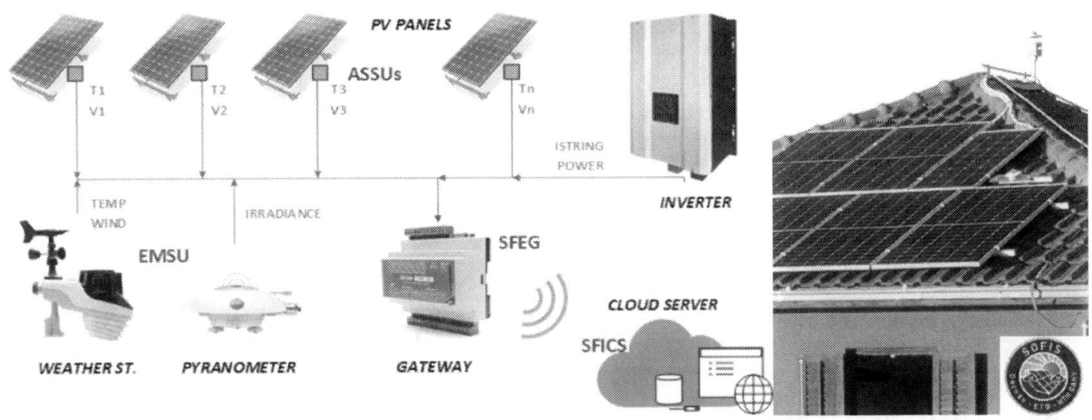

Figure 1: A system design block scheme (left) and realization outdoor of measurement site (right).

2.2 Performance and efficiency calculations

An estimate of the photovoltaic panel performance can be calculated using one of the well-established performance metrics, such as performance ratio (PR) [3], [10], or normalized efficiency η [11].

PR is usually defined by equation

$$PR = \frac{\sum_i P_i / P_{STC}}{\sum_i G_{POA_i} / G_{STC}}$$

where $\sum_i P_i$ and $\sum_i G_{POA_i}$ are the total measured panel power and plane of array (POA) irradiance, while P_{STC} and G_{STC} are the panel power and POA irradiance for the panel in standard test conditions (STC). PR as a metric is influenced not only by the panel quality and degradation, but by the weather (mainly temperature) as well. To suppress this influence, weather corrected PR (WC-PR) [12] is introduced through equation

$$PR = \frac{\sum_i P_i / P_{STC}}{\frac{\sum_i G_{POA_i}(1 - \delta(T_{typ} - T_{cell\_i}))}{G_{STC}}}$$

where δ is the panel's power temperature coefficient, T_{cell_i} is the panel's cell temperature, while T_{typ} is the average cell temperature computed from a one year of measured data. Both PR and WC-PR are usually calculated on a scale of days, months or a year.

Contrary to the averaging nature of the PR metric (over days, months and years), normalized efficiency η, as defined by the equation

$$\eta(t) = \frac{\frac{P(t)}{P_{STC}}}{\frac{G_{POA}(t)}{G_{STC}}},$$

preserves the ratio of the panel power $P(t)$ and POA irradiance $G_{POA}(t)$ on a real-time basis (even at millisecond samples), while it can later be averaged to provide insight over a larger time scale [11]. Similar to PR, η too is temperature dependent, and can be corrected, resulting in a flat line during the mid-day irradiance levels. Deviations from a constant value will thus signal a problem such as shading, or system malfunction. The η value can be calculated for individual modules or arrays, both before and after the inverter. This allows for performance of different panels and systems to be easily compared. Establishing a relationship between the panel's cell temperature and η can further provide insight into system's power temperature coefficient, thus going beyond the values found in datasheets.

To track the performance of our proposed system, PR, WC-PR and η are calculated and analysed.

3 REALIZATION AND RESULTS FROM OPERATION: MONITORED DATA AND CALCULATIONS

Performing a real test condition for photovoltaic plant in a laboratory requires significant equipment and strict procedures. Keeping the irradiance and temperature under realistic outdoor conditions requires increasing and decreasing it according to previously measured ones. The heat convection of the panel itself is different with real outdoor wind conditions. Without good simulation of real outdoor conditions in laboratory, experimental results may differ. Even if there are near outdoor conditions achieved in the lab, for new technologies equations are not fully representing the real behaviour [13].

Therefore, a representative system for outdoor measurement of the variables of interest has been realized as shown in Fig.1. The realization of outdoor measurement site has been located in southwest Serbia (near Pecka). System is compromised from 8 monocrystalline photovoltaic passivated emitter and rear contact modules each with 60 cells, max power rated to 375 watts each, 3 kW in total. The temperature effects over 1048x1758mm panels are modelled assuming temperature coefficient of maximal power (γ) to -0.36%/°C The invertor is single phase low voltage hybrid 3-5 kW GW5048-EM (Godwee) with ability to control energy flows bidirectionally. It is designed to work with Li-ion battery storage at 48 V with maximum 50A current. It is equipped with two maximum power point trackers at maximal declared efficiency of 97.6%. It has several integrated protection features: PV Insulation Resistance Detection, Residual Current Monitoring, PV Reverse Polarity Protection, Anti-islanding Protection, AC Overcurrent Protection, AC Short Circuit Protection, AC Overvoltage Protection. Also, inverter capable of communicating with battery management system, smart meter and producer monitoring portal via RS485, CAN and Wi-Fi. The weather station is low cost suitable for grid distribution SEM5000 (Semtec) micro weather station based on ultrasonic wind speed and direction and micro-electro-mechanical systems for temperature, humidity and atmospheric pressure real time monitoring. The station is equipped with digital output for serial communication (RS485). The pyranometer model is Class A SMP10 (Kipp & Zonen B.V.) with flat spectral range 285 to 2800 nm and low temperature sensitivity (< 0.3 %). Communication has been obtained using 2-wire RS-485.

The representative system operation results are outlined in the following section. The acquired monitoring data are plotted, and the standardized performance monitoring metrics are calculated to assess the system efficiency and compare it to the rated values.

3.1 Monitored data

Fig.2. (a) shows the measured real-time evolution of irradiance and panel current for three days from 26th July to 28th July 2022, while Fig.2. (b) shows the measured panel voltage, temperature, and air temperature during the same time period.

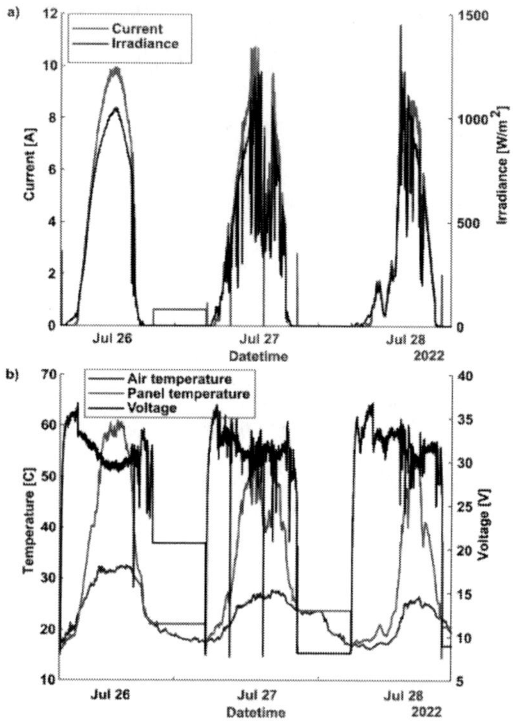

Figure 2: Time dependence of the monitored (a) panel current, irradiance, (b) air temperature, panel temperature, and panel voltage, from 26th July to 28th July 2022.

The anticipated correlation between panel current and irradiance is evident, while the increased panel temperature in high irradiances evidently decreases the voltage, as shown during the 26th July. The presence of clouds during the second (27th) and third (28th) day can be observed from fluctuations in irradiance and panel current, temperature and voltage in Fig.2. The unexpected constant line segments in panel measurands are a consequence of the setup of measurement system readings.

3.2 Performance ratio calculation, verification and normalized efficiency

From the data acquired by the proposed system, PR and η are calculated. Measuring the panel temperature allows us to further perform the weather influence correction, without the modelling of convective cooling of a panel. A linear model relating the measured panel temperature, cell temperature and irradiance is assumed [14].

Table I shows the PR and WC-PR parameters, calculated during five representative consecutive days in mid-April, late May, and late July, during which the system was fully operational.

Table I: Difference in PR

	April	June	July
Performance ratio (PR)	0.82	0.78	0.75
Weather corrected PR (WC-PR)	0.80	0.79	0.77
Difference	-0.02	0.01	0.02

As a result of a lower panel temperature in mid-April, the corresponding PR value is higher compared to the other two months. However, the WC-PR values differ less between months, as the temperature effects are suppressed, thus providing a better insight into other sources of possible system malfunction. Fig.3. (a), (b) and (c) simultaneously show the irradiance and η of a solar panel for three representative days in mid-April, late May, and late July 2022.

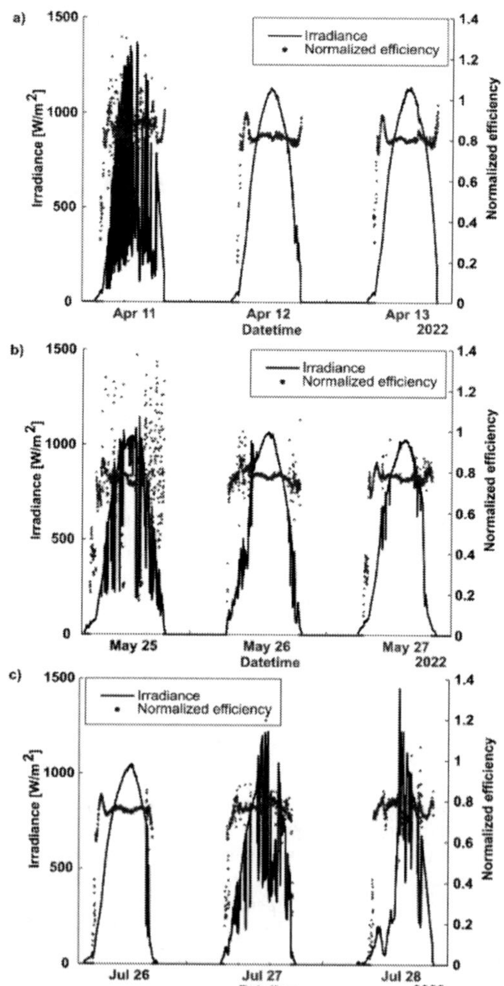

Figure 3 Time dependence of irradiance, with calculated normalized efficiency, for selected days in (a) April, (b) May, and (c) July 2022.

The normalized efficiency η is calculated on a minute level, thus providing a real-time measure of system performance. As previously discussed in literature, low irradiance levels correspond to higher deviations of η [11]. Because of this, a threshold 50 W/m^2 was introduced to η calculations. The high dispersion of η values on April 11th, or May 25th can be attributed to very high irradiance variations due to clouds.

4 CONCLUSIONS

Declared efficiency is important for commission of the photovoltaic power plant based on the data sheets.

In order to indicate the system performance loss, we used real time measurement in outdoor conditions, instead of a laboratory.

Above results show efficiency values that are consistent with the expected performance. Direct panel temperature measurements allow for more reliable temperature compensation calculations, as shown in the deviations of WC-PR to PR values.

The underperformance of the photovoltaic power plant, caused by the inverter, cables, panels and quality of works, could be observed through decrease in the WC-PR and η values.

Efficiency might be smaller due to maximum power point tracking losses, which will be analysed in future research.

ACKNOWLEDGEMENT

Authors are thankful to Innovation Fund from the budget of the Republic of Serbia from the division of the Ministry of Education, Science and Technological Development, through the Serbia Competitiveness and Jobs Project (loan agreement with the World Bank) Nr. 50257, for financing research withing project **SOFIS** (**SO**lar **F**acility **I**nsight **S**ystem). Funds for I.B.B. and M.T. are provided by the Ministry of Education, Science and Technological Development of the Republic of Serbia, the Agreement on the implementation and financing of scientific research of the Institute of Technical Sciences of SANU in 2020 (Registration number: 451-03-68 / 2020-14 / 200175).

REFERENCES

[1] F. J. Sánchez-Pacheco, P. J. Sotorrío-Ruiz, J. R. Heredia-Larrubia, F. Pérez-Hidalgo, and M. Sidrach De Cardona, "PLC-Based PV Plants Smart Monitoring System: Field Measurements and Uncertainty Estimation," *IEEE Trans. Instrum. Meas.*, vol. 63, no. 9, 2014, doi: 10.1109/TIM.2014.2308972.

[2] S. R. Madeti and S. N. Singh, "Monitoring system for photovoltaic plants: A review," *Renewable and Sustainable Energy Reviews*, vol. 67. Elsevier Ltd, pp. 1180–1207, Jan. 01, 2017. doi: 10.1016/j.rser.2016.09.088.

[3] B. of I. Standards, *IS/IEC 61724: Photovoltaic System Performance Monitoring - Guidelines for Measurement, Data exchange and Analysis.* 1998. [Online]. Available: http://archive.org/details/gov.in.is.iec.61724.1998

[4] J. Paredes-Parra, A. Mateo-Aroca, G. Silvente-Niñirola, M. Bueso, and Á. Molina-García, "PV Module Monitoring System Based on Low-Cost Solutions: Wireless Raspberry Application and Assessment," *Energies*, vol. 11, no. 11, p. 3051, Nov. 2018, doi: 10.3390/en11113051.

[5] Y. Li *et al.*, "On-line Monitoring System Based on Open Source Platform for Photovoltaic Array," in *Energy Procedia*, Jul. 2018, vol. 145, pp. 427–433. doi: 10.1016/j.egypro.2018.04.070.

[6] B. Soumia, M. K. Nallapaneni, and T. Ali, "Data acquisition system: On the solar photovoltaic module and weather parameters monitoring," *Procedia Comput. Sci.*, vol. 132, pp. 873–879, 2018, doi: 10.1016/j.procs.2018.05.099.

[7] A. Reinders *et al.*, "Development of a big data bank for PV monitoring data, analysis and simulation in COST Action 'PEARL PV,'" *2019 IEEE 46TH PHOTOVOLTAIC SPECIALISTS CONFERENCE (PVSC)*, no. IEEE 46th Photovoltaic Specialists Conference (PVSC). pp. 1388–1392, 2019.

[8] M. H. Ali, A. Rabhi, A. El Hajjaji, and G. M. Tina, "Real Time Fault Detection in Photovoltaic Systems," *Energy Procedia*, vol. 111, pp. 914–923, 2017, doi: 10.1016/j.egypro.2017.03.254.

[9] A. Triki-Lahiani, A. Bennani-Ben Abdelghani, and I. Slama-Belkhodja, "Fault detection and monitoring systems for photovoltaic installations: A review," *Renew. Sustain. Energy Rev.*, vol. 82, pp. 2680–2692, 2018, doi: 10.1016/j.rser.2017.09.101.

[10] "SRPS EN 61724:2011." https://www.iss.rs/rs/standard/?natstandard_document_id=37012

[11] B. Herteleer, B. Huyck, F. Catthoor, J. Driesen, and J. Cappelle, "Normalised efficiency of photovoltaic systems: Going beyond the performance ratio," *Sol. Energy*, vol. 157, pp. 408–418, Nov. 2017, doi: 10.1016/J.SOLENER.2017.08.037.

[12] T. Dierauf, A. Growitz, S. Kurtz, J. L. B. Cruz, E. Riley, and C. Hansen, "Weather-Corrected Performance Ratio," Apr. 2013, doi: 10.2172/1078057.

[13] J. Kettle, "Ready cells for large-scale systems," *Nat. Energy*, p. 1, 2019, doi: 10.1038/s41560-019-0422-2.

[14] D. L. King, J. A. Kratochvil, and W. E. Boyson, *Photovoltaic array performance model.* United States. Department of Energy, 2004.

LEVERAGING INDUSTRY ENERGY FLEXIBILITY TO ENABLE HIGHER SHARES OF RENEWABLE ENERGY IN THE POWER SYSTEM – AN EXPERIMENTAL CASE STUDY

Amine Aamoume, Jesus da Costa Fernandes, Rainer Gasper, Niklas Hartmann, Yahya Kenoussi, Michael Schmidt, Edgar Schmitt

Offenburg University of Applied Sciences / Institute of Sustainable Energy Systems
Badstr. 24, 77652 Offenburg, Germany

ABSTRACT: One of the major challenges impeding the energy transition is the intermittency of solar and wind electricity generation due to their dependency on weather changes. The demand-side energy flexibility contributes considerably to mitigate the energy supply/demand imbalances resulting from external influences such as the weather. As one of the largest electricity consumers, the industrial enterprises present a high demand-side flexibility potential from their production processes and on-site energy assets. In this direction, methods are needed with a focus on enabling the energy flexibility and ensure an active participation of such enterprises in the electricity markets especially with variable prices of electricity. This paper presents a generic model library for an industrial enterprise implemented with optimal control for energy flexibility purposes. The components in the model library represent the typical technical units of an industrial enterprise on material, media, and energy flow levels with their operative constraints. A case study of a plastic manufacturing plant using the generic model library is also presented, in which the results of two simulation with different electricity prices are compared and the behavior of the model can be assessed. The results show that the model provides an optimal scheduling of the manufacturing system according to the variations in the electricity prices, and ensures an optimal control for utilities and energy systems needed for the production.
Keywords: Energy systems modeling, Demand side flexibility, Digitalization, Energy management, Optimization and control

1 INTRODUCTION

With the convergence of renewable energy resources, and the reduction of the need to build or expand conventional power plants, power systems have an increasing challenge matching supply and demand at all times during their operation. The gap between supply and demand is called the flexibility gap [1]. A viable solution to this gap is to implement demand-side energy flexibility, which consists of electricity consumers adapting (increasing, decreasing or shifting) their energy consumption, also known as demand-response [2].

The implementation and operation of a local process control system that responds to dynamic price signals is also a challenge faced by industrial enterprises. To tackle this challenge and leverage energy flexibility, increased use of "digitalization" could be a key driver.

The purpose of this work is to develop models and methods that enable industrial consumers to evaluate their demand-side energy flexibility potential and participate in demand-response by developing a generic model library that enable the mapping of energy flows and energy flexibility of multiple industrial companies. Model-based control will then be used to optimize and control these flows. To achieve this, the models presented capture energy, materials, and media flows within industrial companies, considering process dependencies and boundary conditions with sufficient accuracy, while also ensuring that the mathematical optimization does not become too complex. Keeping the models generic and modular ensures their applicability to wide range of industrial enterprises.

Without a solid validation step, production processes are exposed to a high risk of interruption while trying to harness the flexibility potential. "Safe" and adequate decisions can be optimized and experimented independently from the actual ongoing production. In this context, Experimental twin as a Small and Medium Enterprise (SME) emulator for testing and developing digital twin-based energy management is an approach and suitable solution for leveraging industrial energy flexibility.

This paper investigated in detail the generic industrial enterprise model library concept, including generic modeling of manufacturing systems (MFS), technical building services (TBS), and energy systems (ES). As a case study, the model library is used to analyze and model the flexibility of a plastic manufacturing plant.

2 GENERIC INDUSTRIAL ENTERPRISE MODEL LIBRARY

Industrial companies operate as a group of industrial systems that interact to execute the intended production through core and auxiliary processes. Those systems are interconnected by material, energy, media, and data flow [3]. This research approach is taken from existing research works of [4] and [5], where they represent the industrial enterprise in many sub-units for their representation of energy flexible industrial systems.

Focusing on modelling of energy flexibility, the generic model library considers an industrial enterprise to be composed of three main technical units: Manufacturing system (MFS), Technical Building Services (TBS) and Energy System (ES) [6].

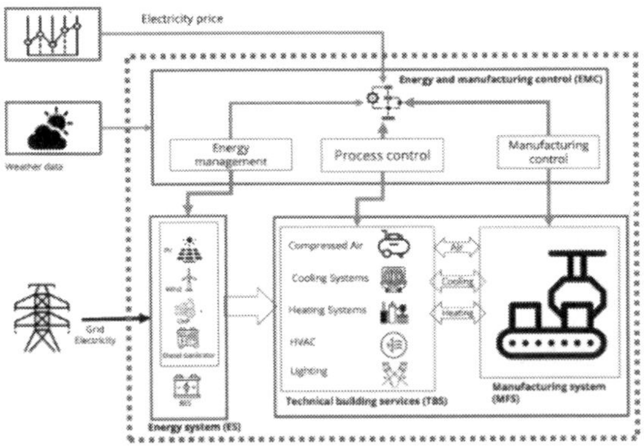

Figure 1: A generic representation of industrial enterprise for energy flexibility modelling

2.1 Manufacturing systems - MFS

The manufacturing system is the central part of every industrial enterprise and consumes the main electricity of the production process. This is the system that uses energy, raw material, media, and information to make final products [3]. Modelling this system was done in the form of a resource-task network (RTN) [7] to make it generic as possible.

A Description of manufacturing system and its constraints are listed in the table.

Table I: Description of MFS and constraints in the model library

Resources	
MFS Components	**Required Inputs**
Raw Material	- Identifier - Initial quantity (units) - Minimum quantity (units) - Storage facility id
Product	- Identifier - Initial quantity (units) - Required production quantity (units) - Type: Intermediate or final
Storage Facility	- Identifier - Capacity (units)
Machine	- Identifier - Power Rating (kW) - Processing rate (units/hr) - Air demand (m3/time) - Heat demand (kW) - Cooling demand (kW)
Tasks	
- Identifier - Products task produces - Material task consumes - Machines task use - Type: Interruptible, Uninterruptible, Constant - Minimum ON time - Minimum OFF time - Maximum allowed interruptions - Validity	
Constraints	
- Material quantities in unit shall remain between given minimum and maximum limits. - At the end of planning horizon, products of type "Final" shall be produced for the required quantities. - If two or more tasks use same machine, they cannot be executed at same time. - While operation of tasks, their run times and interruption shall be accounted. - Maximum capacities of storage facilities shall be respected.	

2.2 Technical Building Services - TBS

The technical building system provides media for the manufacturing system in the exact requested quantities and at the right time. For this purpose, the TBS system includes appropriate units for the generation and intermediate storage of these media. For the sake of simplicity, it is assumed that media doesn't flow back to the TBS system as they might in a real manufacturing company. The subunits of the TBS that are currently modelled in the library include a compressed air system (CAS), process heating system (PHS), process cooling system (PCS), and a simplified version of heating, ventilation, and air conditioning (HVAC). Only CAS and PCS are used in the present work.

Modelling parameters and constraints for compressed air systems and process cooling systems are shown in tables.

Table II: Modeling parameters and constraints for CAS

CAS components and input parameters	
Compressor	- Identifier - Electrical power rating (Kw) - Maximum air supply (m3/unit time) - Control type: Binary or Continuous - Minimum run time (MR) - Minimum off time (MO) - Number of switches allowed (Ns)
Storage tank	- Identifier - Initial pressure (bar) - Capacity (m3) - Set pressure (bar) - Maximum pressure (bar) - Minimum pressure (bar)
Constraints	
- Operational pressure shall remain in the given tolerance. - Input air supply shall be under the maximum capacity of the compressors. - If a compressor is of type binary control, then dwell time constraints (MO, MR, and Ns) shall be respected throughout the optimization horizon.	

Table III: Modeling parameters and constraints for PCS

CAS components and input parameters	
Chillers	- Identifier - Electrical power rating (kW) - Coefficient of performance (COP) - Control type: Binary or continuous - Minimum run time (MR) - Minimum off time (MO) - Number of allowed switches (Ns)
Constraints	
- The operational temperature of cooling storage shall remain in, given tolerance. - Input cooling power of chillers shall remain under the provided ratings. - If a chiller or heat-pump is of type Binary controlled, then the dwell time constraints (MO, MR, and Ns) shall be respected.	

2.3 Energy System – ES

The energy system of the industrial enterprise in the developed model library consists of on-site electricity generation devices and storage. This includes electrical power supply from locally installed photovoltaics (PV), combined heat and power (CHP) units which can also supply heat energy to the thermal storage units inside the PHS, and battery energy storage (BES). Other systems such as wind turbines or diesel generators can also be added. For simplicity, Only PV plant and BES are used in the present work.

Table IV: Modeling parameters and constraints for ES

ES components and input parameters	
PV Plant	- Identifier - Rated power (KW) - Generation factor
Battery Energy Storage (BES)	- Identifier - Energy capacity (kWh) - Maximum charging power (kW) - Maximum discharging power (kW) - Minimum SOC - Maximum SOC - Charging efficiency Discharging efficiency - Ramping rate (kW/unit time)
Constraints	
- State of Charge (SOC) of BES shall remain under defined limits. - Charging and discharging power shall not exceed maximum limits. - Ramping rates shall be respected between two consecutive charging and discharging events. - Both charging and discharging of the BES cannot happen at the same time.	

2.4 Energy and Manufacturing Control – EMC

The EMC module acts as a central control entity with model-based optimization. The primary goal is to adapt the industrial enterprise's overall load demand at the grid's connection point and therefore provide flexibility. It includes generic mixed-integer linear programming (MILP)-based optimization formulations for MFS, TBS, and ES that consider the stated input parameters and constraints (Tables 1,2,3,4). The power balance of the industrial enterprise reads for each time step t in optimization horizon T:

$$P_{Grid,t} = P_{MFS,t} + P_{PCS,t} - P_{PV,t} - P_{BES,t} \quad \forall t \in T = [0, 1, ..., T_f] \qquad \text{(Eq. 1)}$$

Based on the process constraints, each of the specified variables in Eq. 1 is ideally determined. A positive grid power $P_{Grid,t}$ represents grid consumption, while a positive electric storage power $P_{BES,t}$ represents discharge. The objective function of the optimizer is to minimize the daily electricity procurement costs. The objective function is written as:

$$J = \sum_{t=0}^{t=Tf} P_{Grid,t} * \lambda_t^{elec} * \Delta t \quad \forall t \in T = [0, 1, ..., Tf] \qquad \text{(Eq. 2)}$$

- λ_t^{elec}: The electricity price in €/Kwh at time step "t".
- Δt: the sampling time step of the optimization horizon.

Electricity price signals are taken from the Market block in the model library. The optimizer solves for all the time steps in optimization horizon T and finds a global solution. However, during this present work, a cascaded optimization with a moving horizon approach is developed. This will help to tackle and simulate the short-term disturbances in process parameters and fluctuation in price signals.

3 THE EXPERIMENTAL TWIN

The Experimental Twin concept, figure 2, consists of SME core hardware components (motors, compressors, ...) and a PLC-based SCADA system in combination with software models of those components that could not be emulated as hardware.

The experimental twin, figure 3, serves as an "SME emulator" as a test environment and showcase for the employment of Digital Twin and intelligent energy management. Thus, it provides a laboratory environment in which the industrial energy flexibility can be tested based on the optimization methods [8].

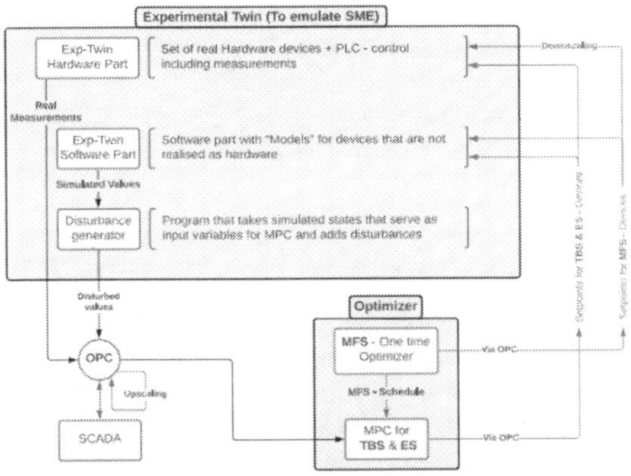

Figure 2: The Experimental Twin concept

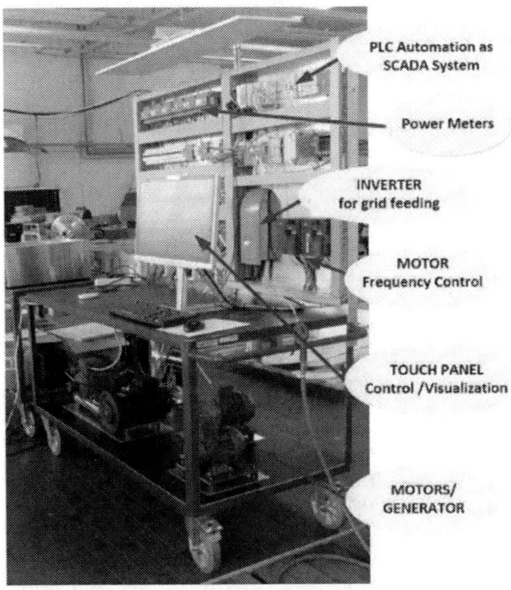

Figure 3: Setup of the experimental twin on heavy duty tables with motors, control panel, converters and switching devices [8].

4 EXPERIMENTAL CASE STUDY

4.1 Reduced SME model

An industrial company should be defined to make use of the generic optimization formulation written into the EMC module. The core components of a plastic production company as a reduced SME model is created. For simplicity, CHP, PHS, and HVAC will not be considered. The proposed model was built by three major units: MFS, TBS, and ES. (Figure 4)

The MFS unit is composed of manufacturing tasks which consist of the extrusion of raw plastic material into an intermediate product. This product is stored in an intermediate storage unit before being sent to the printer, which is capable of printing four different products. The process was kept straightforward and simplified.

The MFS unit of the considered plastic manufacturing company is composed by:

- Raw material Storage.
- Four Extruders of different power.
- Two intermediate storages (before printing).
- One printer machine.
- One final storage of final products.

The TBS unit is composed by:

- A cooling system (PCS) is composed of 2 Chillers that fill a central water storage tank.
- A compressed air system (CAS) is composed of 2 compressors that fill an air storage tank.

The ES unit is composed by:

- Battery Energy Storage.
- PV plant.

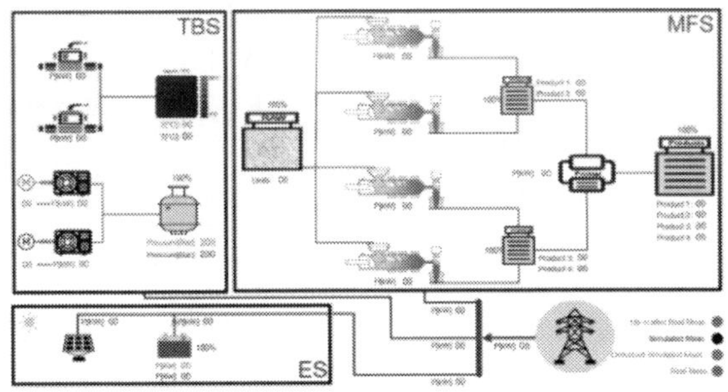

Figure 4: Representation of the plastic manufacturing company as a case study [9]

5 RESULTS

The behavior of MPC control in closed control loop was examined in different disturbances scenarios:

- Random change of electricity prices during one day.
- Two selected changes in electricity prices
- Pressure disturbance in the CAS air receiver tank.

Two simulations are run in each case: One for the expected behavior in the absence of disturbances and another in the presence of disturbances.

Figure 5 shows the impact of changing electricity prices on the MFS load profiles of the components and corresponding states. The desired optimal operation with variable prices dictates a shift of the load demands of MFS devices to intervals where prices are lower. This schedule is decided by the first stage of the optimizer.

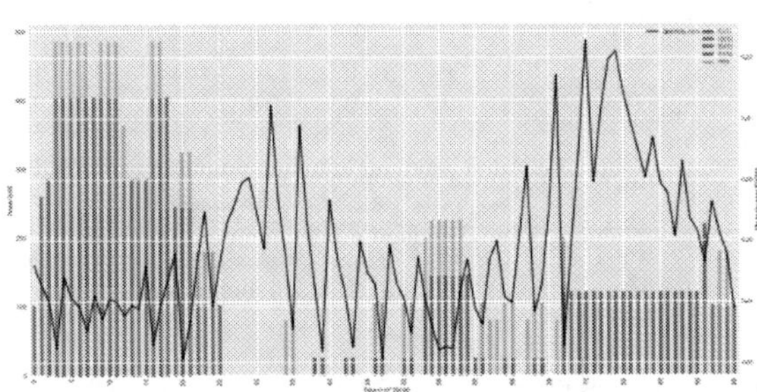

Figure 5: Load profile of MFS with variable prices

The MFS load schedule (Figure 5), which has been decided by the MILP optimizer, serves as an input of the second stage of optimization. The effect of change of electricity prices on different units is shown in figure 6 below.

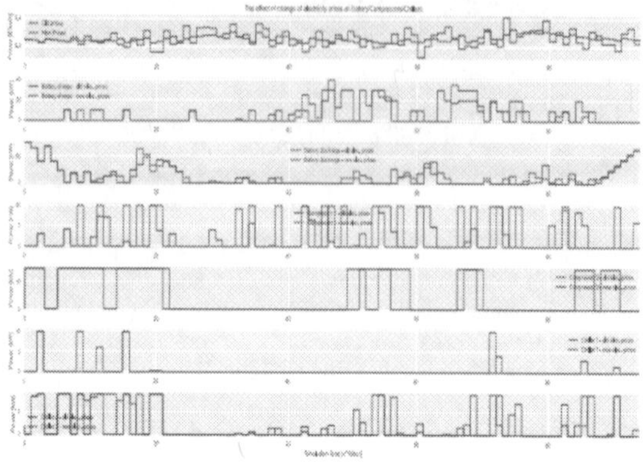

Figure 6: The effect of the variable electricity prices on Battery, Compressors and Chillers behavior

Two simulations were run separately. A day-ahead electricity price was used in the first simulation (in green). A random variation was applied to electricity prices in each step (in red).

When the prices are high, we can see that the industrial enterprise shifts its TBS loads to intervals where costs are lower. This is the case, for example, in steps 60 and 79. Also, the battery charging and discharging operate flexibly with respect to variable prices.

Due to the randomness in each step in the last results (Figure 6), we cannot interpret the results accurately. For this reason, another price disturbance was selected, as shown in figure 7. This time, Electricity prices rise following a triangular signal shape. A first disturbance was introduced to the optimizer at step 5 and lasted for 13 steps (Prices go up at step 5 and then down at step 13). Afterwards, an inversed triangular change of prices was also considered at step 60 and lasted for 13 steps as well. The results from the last evaluation were confirmed clearly during these introduced price changes.

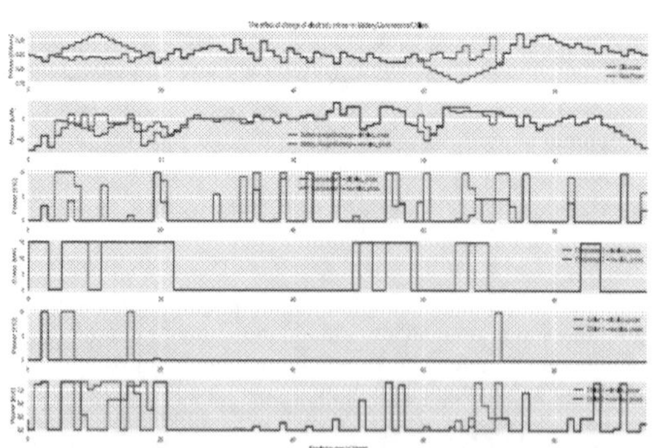

Figure 7: The effect of the variable electricity prices on Battery, Compressors and Chillers behavior – Selected prices

6 SUMMARY

This work presents an optimal energy management and control framework for industrial enterprises to define their manufacturing processes, utility processes, and energy systems as modular subunits.

Based on the framework developed, optimization formulations have been implemented to address the production constraints in the Manufacturing system, process constraints in the Technical Building Services system, and physical constraints in the Energy System, while providing flexibility for the defined industrial enterprise. A company's interface to electricity markets is provided by electricity prices, which serve as external inputs to its optimization algorithm.

For validation and for assessing energy flexibility in terms of variable price signals, the developed model framework was applied to a case study of plastic manufacturing company. Using the flexibility potential, the optimizer shifts load profiles of defined technical units towards the lower price intervals. Most of the flexibility was provided by the MFS, TBS and the BES in the modelled case.

ACKNOWLEDGEMENTS

The results presented here were partly funded by the Federal Ministry of Economics and Climate Protection of the Federal Republic as part of the research projects: GaIN (Förderkennzeichen 03EI6019E) and FlexGUIde (Förderkennzeichen 03EI6065C). We would like to thank Mr. Michael Wirwitzki for his support in setting up the Experimental Twin.

REFERENCES

[1] Papaefthymiou G, Haesen E, Sach T "Power system flexibility tracker: Indicators to track flexibility progress towards high-res systems", Renew Energy 127:1026–1035, 2018

[2] Roesch M, Bauer D, Haupt L, Keller R, Bauernhansl T, Fridgen G, Reinhart G, Sauer A *"Harnessing the full potential of industrial demand-side flexibility: An end-to-end approach connecting machines with markets through service-oriented IT platforms"* Appl Sci (Switzerland) 9(18):3796, 2019.

[3] Beier J, *"Simulation Approach Towards Energy Flexible Manufacturing Systems Introduction"* Springer International Publishing AG 2017.

[4] J. Beier, J., Thiede, S., Herrmann, *"Increasing energy exibility of manufacturing systems through exible compressed air generation"*, Procedia CIRP 37 (December):18–23, 2015.

[5] A. Tristán, F. Heuberger, and A. Sauer, *"A Methodology to Systematically Identify and Characterize Energy Flexibility Measures in Industrial Systems,"* Energies, Vol. 13, Page 5887, vol. 13, no. 22, p. 5887, 2020.

[6] R. Khatri, M. Schmidt, and R. Gasper, *"Energy Informatics Active participation of industrial enterprises in electricity markets-a generic modeling approach"* Energy Informatics, 2021.

[7] Castro PM, Harjunkoski I, Grossmann IE *"New continuous-time scheduling formulation for continuous plants under variable electricity cost"*, Ind Eng Chem Res 48(14):6701–6714, 2009.

[8] J.S. da Costa Fernandes, M. Schmidt, R. Khatri, R. Gasper, N. Hartmann, *"Smart Energy Management for SME Using Digital Twins."*, EU PVSEC, 2021.

[9] Kenoussi, Yahya, *"Contributions to the Development of a Test Environment for Leveraging Industrial Energy Flexibility"*, Master-Thesis, Hochschule Offenburg, 2021.

ANALYSIS OF THE SPATIAL AND TEMPORAL VARIABILITY OF SOLAR RADIATION AND ITS USE IN THE DESIGN OF ENERGY SYSTEMS

Jannik Langenecker, Edgar Schmitt, Amine Aamoume, Michael Schmidt
Offenburg University of Applied Sciences / Institute of Sustainable Energy Systems
Badstr. 24, 77652 Offenburg, Germany

ABSTRACT: Solar energy plays a central role in the energy transition. Clouds generate locally large fluctuations in the generation output of photovoltaic systems, which is a major problem for energy systems such as microgrids, among others. For an optimal design of a power system, this work analyzed the variability using a spatially distributed sensor network at Stuttgart Airport. It has been shown that the spatial distribution partially reduces the variability of solar radiation. A tool was also developed to estimate the output power of photovoltaic systems using irradiation time series and assumptions about the photovoltaic sites. For days with high fluctuations of the estimated photovoltaic power, different energy system scenarios were investigated. It was found the approach can be used to have a more realistic representation of aggregated PV power taking spatial smoothing into account and that the resulting PV power generation profiles provide a good basis for energy system design considerations like battery sizing.

1 INTRODUCTION

Clouds that shade individual large photovoltaic systems or a large number of smaller photovoltaic systems, for example in a local grid, lead to fluctuations in the solar power feed-in. For the efficient design and operation of power systems with a high PV share, it is therefore important to characterize and understand these fluctuations well.

At the Institute for Sustainable Energy Systems (INES) of the Offenburg University of Applied Sciences, the modular measuring system "Irradiance and Sky Imager sensor network" (ISInet) was developed, which collects measured values of solar radiation and temperature as well as cloud images. Through the data acquisition of this sensor network it is possible to analyze the variability of solar radiation at its location [1].

Comparable approaches have been followed by other groups and in other locations and have provided very valuable results for understanding cloud trains [2,3].

The synchronized irradiance time series allow a quantitative analysis, while the cloud images additionally allow a visual assessment of the cloud situations, which can be very helpful for the interpretation of the time series. Furthermore, the camera images allow additional evaluations such as estimating the direction and speed of cloud trains. Such cloud images can also be used as a basis for short-term forecasts for solar irradiance [4,5,6].

This work investigated the specific use case of an airport that wants to meet a significant portion of its electricity needs through locally distributed PV systems and, to do so, wants to understand the power fluctuations that can be expected in cloudy conditions and the impact on certain energy system design decisions [7].

2 METHODS

2.1 Sensor network ISInet

The sensor network ISInet consists of up to 10 sensor measurement modules, which can be distributed locally and which each record solar irradiance and cloud images with GPS-synchronized time stamps with high temporal resolution and send them via mobile radio to a central server (see Figure 1 and Figure 2).

The sensor measurement systems are completely self-sufficient and require no data or supply interface at the site. An important requirement was the lowest possible installation and maintenance effort, so that locations that are not easily accessible are also possible. A PV module with battery unit is available for year-round energy supply. The measuring task is performed by a small computer, which was designed for resource-saving operation. This makes ISInet particularly suitable for temporary measurement campaigns without the need for extensive advance planning and installation work.

Figure 1: ISInet module of the second generation. The left image shows the solar panel that powers the ISInet module and the right image above shows the cloud camera, reference cell and pyranometer. Source: Own images

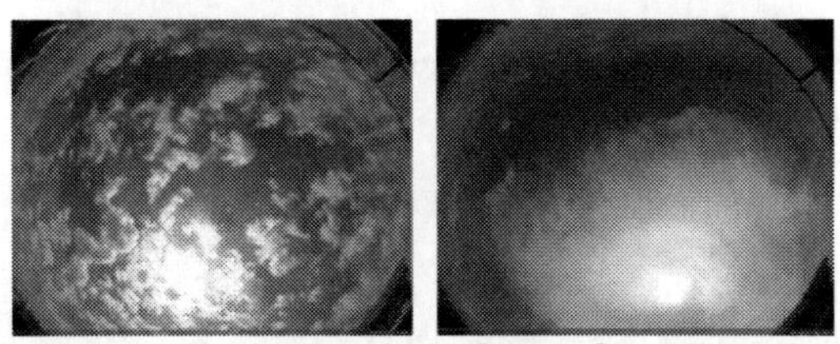

Figure 2: The images show photos of an ISInet cloud camera with cluster layer clouds (left) and veil clouds (right), respectively, which cause high-frequency and low-frequency fluctuations, respectively, and also allow a visual interpretation of the measurement data.

The data collected with ISInet can be used in the following use cases, for example:

a. Validation and disturbance identification for existing solar power projections and forecasts for microgrids or distribution grids.
b. Characterization of irradiance fluctuations and resulting solar power fluctuations in microgrids and distribution grids caused by clouds. These results can then be used for the planning of back-up generators or batteries and the design of

energy management systems.

c. Time-synchronized cloud images from spatially distributed locations as a basis for local area-based short-term solar power forecasts.

In this work, the focus was on application (b).

2.2 Exemplary use case airport energy supply

The ISInet sensor network was deployed at Stuttgart airport with the focus currently on use case (b). The use of ISInet is connected with plans for climate-neutral operation of the airport by the year 2050. Application case (c) for the automated management of an energy system with a high proportion of solar power is also of interest for Stuttgart Airport.

The goal of this work was to place the sensor modules spatially distributed at the Stuttgart airport where – in the future - the photovoltaic systems are to be installed. By measuring the irradiation, these photovoltaic systems can be simulated in advance, which also allows a better design of the battery storage of a microgrid and prepare for the airport energy system for the resulting fluctuations of cloud-induced PV power fluctuations.

Figure 3: Map section of Stuttgart Airport of the first generation ISInet

2.3 Data validation

Data was collected with 4 ISInet modules in the period of July 2020 to March 2021. An extensive data quality assessment was carried out. For instance, clear-sky days were used to compare and correct the data from the different sensors of the different ISInet modules and to reconcile them with a theoretical clear-sky model. Here the model of Ineichen and Perez was used [8]. A comparison with external weather stations was also performed in order to ensure reliable sensor data. The clear-sky model also allows to calculate the clear-sky index that is needed in later steps.

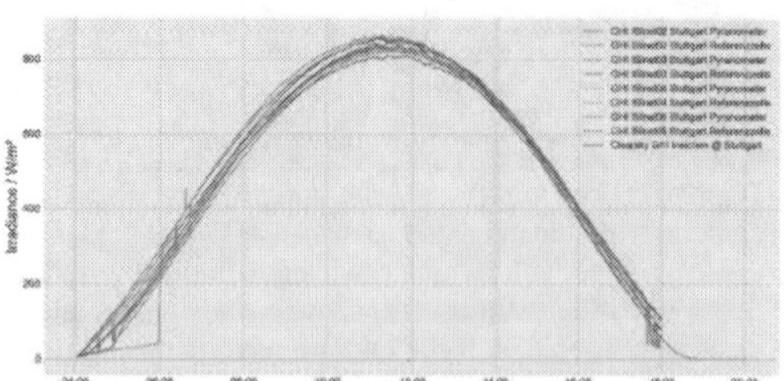

Figure 4: Comparison of irradiance measurements from different ISInet measurement modules on a clear-sky day and comparison with Ineichen-Perez clear-sky model

2.4 PV power estimation

To estimate the power of an imaginary or real PV system at the location of an irradiance measurement, the irradiance in the respective inclined module plane of the PV system must first be inferred from the measured horizontal irradiance. Since the diffuse and direct radiation components are not measured separately for the ISInet modules, a decomposition model according to Erbs et al. was applied, which in turn requires the clear-sky index as input variable [9].

Otherwise, we follow the three-component model and the consideration of heat losses as described in numerous textbooks, such as [10], in converting global horizontal irradiance to tilted global irradiance and then estimating the resulting PV power.

3 RESULTS

3.1 Variability characterization

Short-term fluctuations - caused by clouds - are the variability that is of greatest interest to grid operators [3]. In this case, affected PV systems are shaded by these clouds, which can lead to rapid changes in irradiance or the generated PV power. In extreme cases, the irradiance can change by more than 60% of the maximum irradiance within a few seconds [2]. The speed of the clouds as well as the type of clouds have an influence on how fast the irradiance or the output power of a PV system changes. Small individual clouds from cloud fields, for example, are responsible for high-frequency fluctuations and weather fronts for the low-frequency fluctuations [3].

This could also be seen in the data from Stuttgart airport. Figure 5 shows the time series of the global irradiance - measured by the pyranometers of the ISInet modules - and the rate of change of the irradiance (RR) on July 22, 2020 at different time intervals. In addition, the average RR - which represents how the irradiance of all ISInet modules changes on average - is also shown in red and the RR of the individual modules in gray. The rate of change is related to an interval of 15 s, 120 s and 30 min which corresponds to integer multiples of the measurement interval. It is clearly visible that the average rate of change of irradiance (red) is much lower than that of the individual ISInet (gray), especially for the small and medium interval. For the large interval, the average rate of change is not much smaller than that of the individual modules.

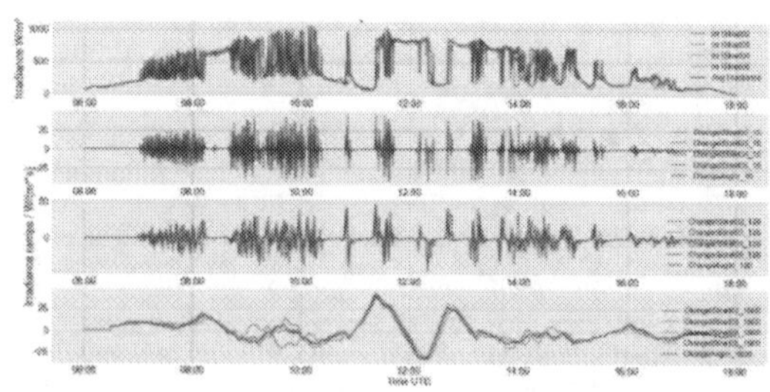

Figure 5: The irradiance and its rates of change were investigated for all ISInet modules (gray lines). The rates of change were related to 15 s, 120 s, and 30 min. In red, the averaging over all ISInet modules was calculated and a strong attenuation of the steep gradients can be observed.

3.2 Cloud velocity estimation

Based on the time offset of similarities in irradiance patterns recorded by the different ISInet modules at different locations, cloud directions and cloud velocities could be inferred to some degree. This information could be useful, for example, for short-term forecasts of cloud-based irradiance variations.

Figure 6 shows the four irradiance time series of the ISInet modules for a partly cloudy situation on 20.08.2020.

It can be seen that the irradiance increases steeply at different times in each case. Figure 7 shows these time differences on a map with the four ISInet modules. Here, the irradiance of ISInet03 changes before that of ISInet05 (by Δt2) and the irradiance of ISInet04 changes before that of ISInet02 (by Δt1). Thus, the cloud must have shadowed ISInet03 and ISInet04 first. It can be concluded from the geographical position that the direction of the clouds should be to the northeast (north is at the top of the image).

It is complementary possible to determine cloud direction and speed from consecutive cloud images, see Figure 8.

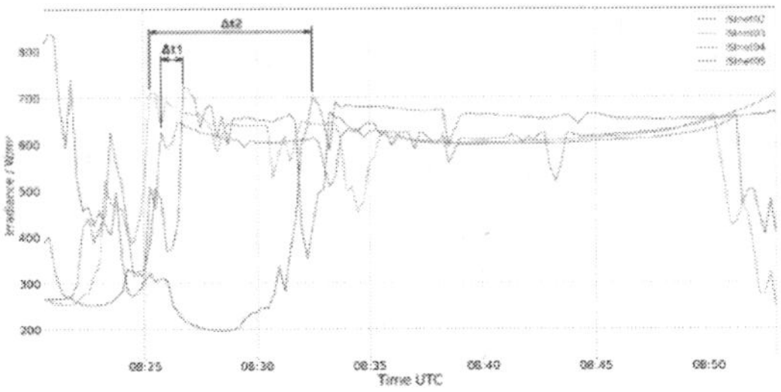

Figure 6: Temporal offset in the irradiation time series of the four ISInet modules.

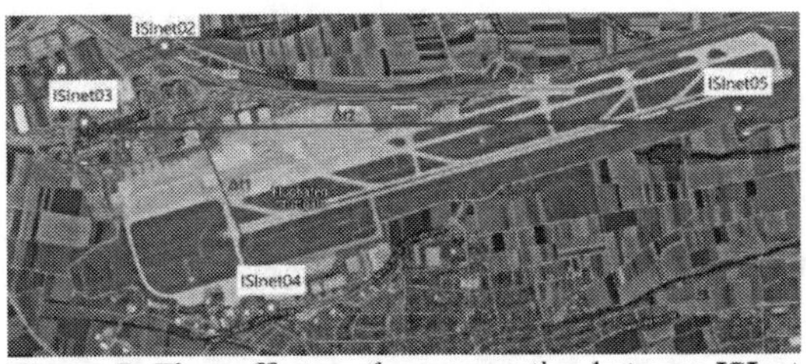

Figure 7: Time offset on the map section between ISInet modules at Stuttgart airport. Source: www.maps.google.com

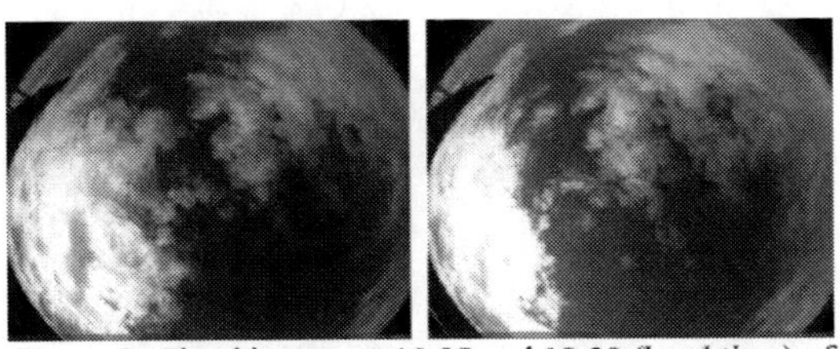

Figure 8: Cloud images at 10:28 and 10:30 (local time) of the ISInet05 module with marked direction of cloud drift.

3.3 Estimation of PV power generation

It was now assumed that larger PV systems are or will be installed on the airport in the positions of the 4 ISInet modules, with module areas of 3000 m², 2900 m², 2300 m² as well as 3500 m² and different tilt angles of 30°, 32°, 25° and 20° and also slightly different azimuth angles, see Table I. Temperature coefficients and system efficiencies were assumed to be the same in this case.

Based on the method described in Section 2, the PV power of each system was then estimated for each site. In each case, the inclined irradiance was scaled up accordingly. The sum of the outputs of the 4 PV systems at the 4 sites then shows spatial smoothing, especially of the fast ramps, see Figure 9. Spatial smoothing within the individual PV plants was not considered.

Table I: Parameters of the considered PV installations at Stuttgart airport used in this study

ISInet	Latitude / °	Longitude / °	Module tilt / °	Azimuth / °	System eff	Module are /m²	Albedo factor	Temp coeff. %/K
ISInet02	48.694890	9.195004	30	160	0.14	3000	0.25	-0.43
ISInet03	48.690903	9.188644	32	164	0.14	2900	0.25	-0.43
ISInet04	48.681030	9.203267	25	165	0.14	2300	0.25	-0.43
ISInet05	48.691497	9.240114	20	150	0.14	3500	0.25	-0.43

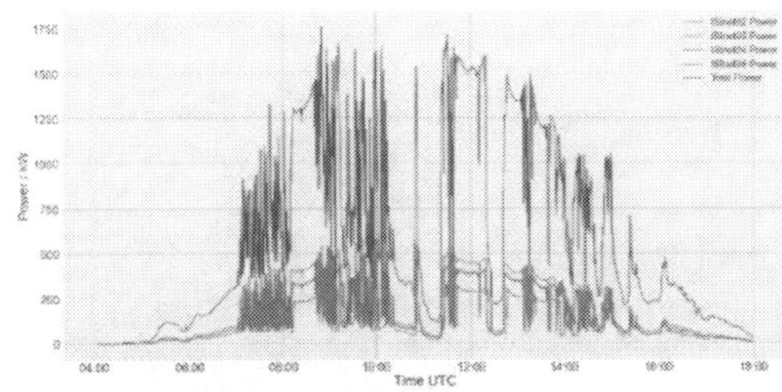

Figure 9: Aggregated PV power (in blue) and PV power of the individual simulated PV systems (in red) at the ISInet sites

3.4 Use for energy system planning

The estimation of aggregated PV power production in a microgrid that realistically covers the effect of spatial smoothing as shown in Figure 9 allows for better decisions on system design, especially if the compensation of fast fluctuations is a design issue.

In this case study, we investigated how batteries can be used to smooth the fast fluctuations of aggregated PV power and how different battery sizes affect the smoothing behavior here, see Table 2. For this purpose, a simple battery model with a simple charging and discharging algorithm was implemented.

Table 2: Different large batteries for fluctuation compensation

	Scenario 1	Scenario 2
Battery capacity in MWh	3.0	1.2
Max. charge power in MW	1.7	0.9
Max. discharge power in MW	1.7	0.9

In a first approach, it was assumed that no load is present in order to exclusively investigate the smoothing of the PV power with the help of a battery storage. The designed algorithm significantly reduced the power change rates. However, the capacity of the battery storage was 3 MWh in Scenario 1 and the charge and discharge power limits were very large at 1.7 MW. This battery size is grossly over-dimensioned for a practical use in a microgrid of this size to buffer short-term fluctuations in power. Despite the significant reduction to 1.2 MWh in Scenario 2, the capacity was still significantly too large. Due to the additional reduction of the charging and discharging power limits, there were also jumps in the smoothed power in some cases, see Figure 10 and Figure 11.

Numerous other investigations with different battery sizes and also taking measured load profiles into account have been carried out, although the results cannot be shown for reasons of brevity.

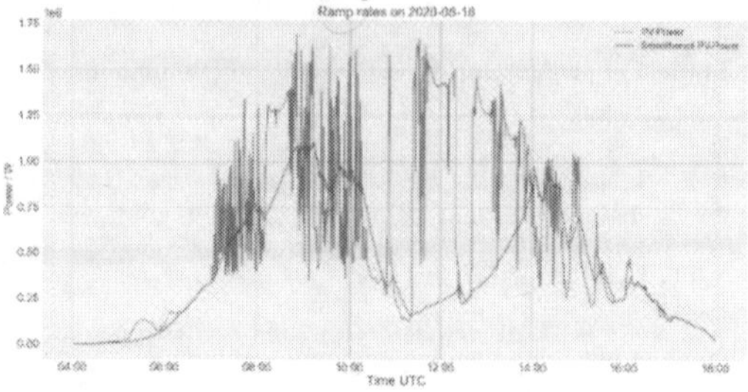

Figure 10: Aggregated PV power smoothed with battery from scenario 1

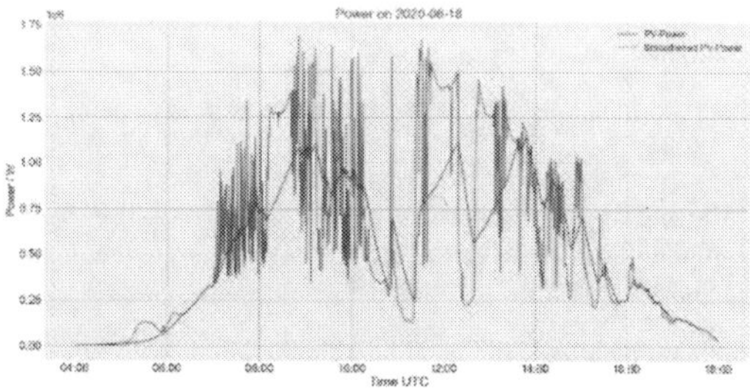

Figure 11: Aggregated PV power smoothed with battery from scenario 2

4 SUMMARY

If industrial areas (like the example of an airport here) want to secure more and more of their energy in the future also with their own large PV systems, it plays a greater role to understand the cloud-based fluctuations of the total power in order to be able to respond to these fluctuations through batteries, load management or fast controllable generators like CHP. The measurement system and analysis methods presented here in this paper help to better design these future power systems.

In the study presented here, it has been shown that the ISInet sensor network is very well suited to characterize the variability of solar irradiance under cloud cover for a larger area without much effort and to use this information to estimate the generation of future distributed solar arrays as well as their variability. This information can be used in many ways for system design, e.g. to design batteries to compensate for fluctuations.

ACKNOWLEDGEMENTS

The results presented here were partly funded by the Federal Ministry of Economics and Climate Protection of the Federal Republic as part of the research program "SINTEG-Schaufenster: C/sells; Teilprojekt: Ökonomische Analysen, Prädiktive Steuerung und Musterliegenschaft Offenburg" (Förderkennzeichen 03SIN103). We would like to thank Stuttgart Airport for its support.

REFERENCES

[1] Langenecker, Jannik: *Analyse der räumlichen und zeitlichen Variabilität der Sonneneinstrahlung und ihre Verwendung zur Auslegung von Energiesystemen*, Bachelor-Thesis, Hochschule Offenburg, 2021.

[2] Mills, Andrew, et al. *Understanding variability and uncertainty of photovoltaics for integration with the electric power system*. No. LBNL-2855E. Lawrence Berkeley National Lab.(LBNL), Berkeley, CA (United States), 2009.

[3] Kleissl, Jan. *Solar energy forecasting and resource assessment*. Academic Press, 2013.

[4] Caldas, M., and R. Alonso-Suárez. "Very short-term solar irradiance forecast using all-sky imaging and real-time irradiance measurements." *Renewable energy* 143 (2019): 1643-1658.

[5] Dittmann, Anna, Nicolas Holland, and Elke Lorenz. "*A new sky imager based global irradiance forecasting model with analyses of cirrus situations.*" Meteorologische Zeitschrift (2021): 101-113.

[6] Dittmann, Anna, et al. "Optimierung der PV-Speisung von Wärmepumpen Durch Kurzfristprognosen mit Wolkenkameras." *34. PV-Symposium 2019: 19. bis 21. März 2019, Kloster Banz, Bad Staffelstein Pforzheim, Conexio 2019.* Conexio GMBH, 2019.

[7] Stuttgart Airport Energie GmbH: Photovoltaik-anlagen. (https://www.flughafen-stuttgart-energie.de/photovoltaikanlage/), (14. Sep. 2022)

[8] Perez, Richard, et al. "Modeling daylight availability and irradiance components from direct and global irradiance." *Solar energy* 44.5 (1990): 271-289.

[9] Erbs, D. G., S. A. Klein, and J. A. Duffie. "Estimation of the diffuse radiation fraction for hourly, daily and monthly-average global radiation." *Solar energy* 28.4 (1982): 293-302.

[10] Häberlin, H.."Photovoltaik: Strom aus Sonnenlicht für Verbundnetz und Inselanlagen, Fehraltorf: Electrosuisse, 2." (2010).

EFFECT OF LOW-TEMPERATURE THERMAL ENERGY STORAGE ON HYBRID PV- COMPRESSED AIR ENERGY STORAGE OPERATION

Bazdar Elaheh, Nasiri Fuzhan, Haghighat Fariborz
Concordia University, Department of Building, Civil and Environmental Engineering
Montreal, Quebec, Canada

ABSTRACT: Unlike conventional energy storage, compressed air energy storage (CAES) is a complex system mix of mechanical, electrical, and thermal engineering that causes relatively low efficiency due to multi-step energy conversion and losses. Energy in CAES is partly stored in the form of high-pressure air and partially converted into thermal energy. One way to improve the CAES cycle efficiency and reduce its carbon footprint is recycling the thermal energy by integrating regenerative heat units into CAES. Introducing decentralized CAES for building applications means that the compressor and turbine operation must be adjusted by intermittent solar power and fluctuated energy demand, respectively, to increase solar power penetration and building autonomy. In this study, a novel operation strategy is proposed to investigate the effect of limitation and dynamic behavior of thermal energy storage (TES) on the thermodynamic and techno-environmental performance of CAES and the whole hybrid power system, including PV/CAES/grid. This operation strategy enables the investigation of CAES and hybrid systems' time-dependent operation under several CAES structures characterized by heat management and fuel consumption concept. The results show a strong relationship between TES states with the turbine/compressor operation and the air tank state.
Keywords: compressed air energy storage, photovoltaic, hybrid, operation strategy, performance

1 INTRODUCTION

Implementing on-site distributed hybrid renewable energy projects to meet the energy demand of building as one of the main energy consumers is one way toward intelligent grids and smart cities. But intermittency nature of renewables adds uncertainty to the hybrid energy system (HES). Nowadays, the battery system is the preferable solution to address the renewable intermittency for building applications. However, the limited lifetime of battery storage, along with their waste disposal and recycling issues, raised attention paid to other bulk energy storage systems [1,2]. Compressed air energy storage (CAES) as a mechanical system is promising bulk-energy storage with a longer lifespan, less environmental impact, low self-discharge, high durability, and almost low cost per unit of stored energy [2]. However, unlike conventional energy storages, the CAES system is a complex system mix of mechanical, electrical, and thermal engineering that causes relatively low efficiency due to multi-step energy conversion and losses. Furthermore, employing CAES as decentralized energy storages cause it to operate at the off-design condition, especially when it is integrated with intermittent renewables and fluctuated building energy demand, affecting all components' power dispatch [3,4].

Although several studies have been conducted on the technical feasibility of incorporating CAES for decentralized applications, none considered the environmental aspect of CAES under different structures and operational conditions [1,5–7]. On the other hand, the effects of TES dynamic behavior on the operation and performance of CAES and the whole system were overlooked in previous studies [8–12]. This paper aims to advance the knowledge for a more precise investigation of TES integration, its impact on CAES's thermodynamic and techno-environmental performance, and the decentralized hybrid system that could contribute to better CAES heat and emission management along with building renewable integration. The novelty of this study is developing a flexible operation strategy for CAES integrated with TES for building renewable integration considering the limitations and characteristics of all interdependent components of CAES (mechanical parts, thermal part, and air storage).

The main contribution of this research, according to the literature summarily, is presented as follows:

- Presenting a time-dependent and dynamic model for CAES components and TES unit to investigate the cycle to cycle and long-term operation and performance of decentralized CAES according to power generation units and building load demand.
- Developing a novel operation strategy for decentralized CAES by adopting TES characteristics along with the limitation of compressor/motor, turbine/generator, and air tank energy storage.
- Proposing a flexible approach applicable for thermodynamic and techno-environmental investigation of decentralized CAES under different CAES structures concerning heat management and fuel consumption.

2 HYBRID SYSTEM DESCRIPTION

The proposed HES comprises five significant parts: photovoltaic panels (PV), CAES, grid, control system, and building. Besides, the CAES system consists of different components such as a motor/ compressor train, generator/ turbine train, heating/cooling heat exchangers, double cold/hot tank as TES, and a combustion chamber in the presence of axillary heat. Fig.1 illustrates the schematics of the proposed HES.

3 OPERATION STRATEGY

Adopting an appropriate operation strategy to manage the energy flow in HES is essential to ensure the system's reliable operation. Fig. 2 represents the flowchart of the novel operation strategy developed in this study with the role of solar energy accommodation. According to fig.2, PV and grid are considered primary and backup power sources, respectively.

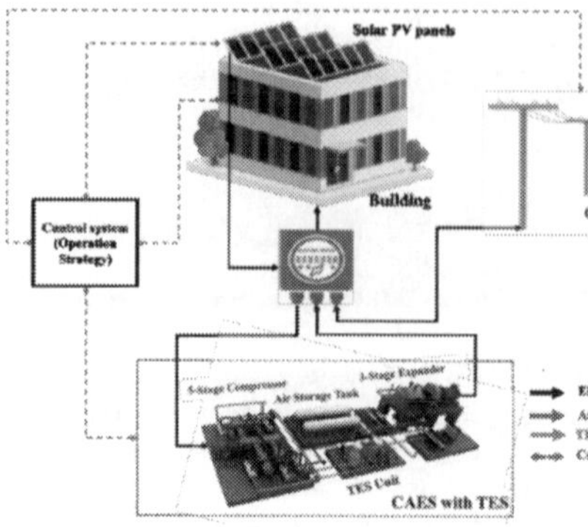

Figure 1: Schematic diagram of the utilized hybrid energy system.

During the times when PV output power exceeds the building electricity demand (W_{Gap}<0), the excess electricity can be stored in CAES considering the rated capacity of motor/ compressors and depending on the state of the air storage tank. When the air tank is full, or the excess power violates the operating range of compressors' power, the extra electricity is sold back to the grid. Meanwhile, the thermal energy generated by compressors can be stored in TES depending on its state of charge (SOC_{TES}). Otherwise, thermal energy is either dissipated or used for other purposes, such as domestic hot water, etc.

During the time when electricity generation is less than building electricity demand (W_{Gap}>0), at first, CAES compensates for the electricity shortage by discharging the energy via expanding the pressurized air depending on the SOC of both air tank and TES while considering the operating range of turbines' capacity. If the air tank has enough compressed air to be used by turbines, the air mass flow rate is released from the air tank and passed through turbines to supply electricity considering the turbines' rated operating range. Otherwise, the grid is used to compensate for the rest of the electricity shortage and during times of non-availability of solar energy. Meanwhile, the thermal energy required for expansion should be supplied by a heating source, affecting CAES's operation and performance. In such a case, two scenarios are defined for the operation condition of the CAES and consequently HES as follows:

Scenario 1 (Scen.1): The required heat is provided only by TES. Therefore, when TES does not have enough energy (less than the amount needed for expansion), the mass flowrate of high-pressure air released from the air tank is regulated accordingly. If compressed air is less than the minimum mass flowrate required by turbines, CAES stops discharging, and further power deficit is compensated by purchasing electricity from the grid (TES scenario).

Scenario 2 (Scen.2): For the first, the required heat is supplied by TES. If the thermal energy is lower than that needed by turbines, the axillary heat unit compensates for the further thermal energy, which is the combustion chamber in this study (TES and axillary heat scenario).

4 MODEL DEVELOPMENT

In this section, the formulation of different components that are applied in the calculation are expressed: 1) PV panel, 2) motor/compressor train, 3) generator/turbine train, 4) air storage tank, 5) heat exchanger, 6) TES, 7) combustion chamber.

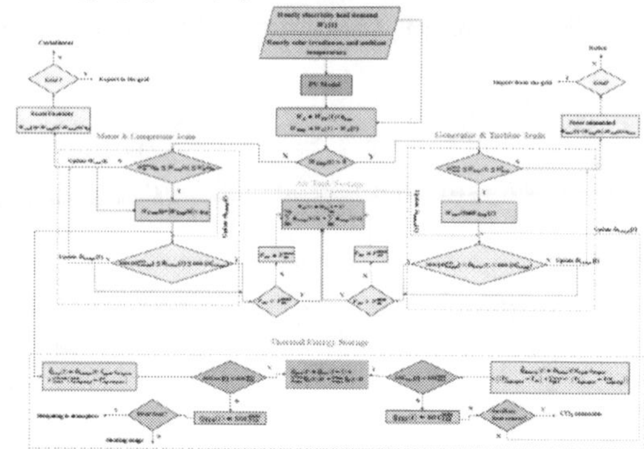

Figure 2: The flowchart of the proposed operation strategy.

4.1 PV module

Based on the hourly solar irradiance and considering the number of PV (N_{PV}) and the efficiency of the inverter (η_{inv}), the total power of PV panels can be calculated by:

$$P_{PV-AC}(t) = N_{PV} \times P_{rated-PV} \times \left(\frac{I(t)}{I_{ref}}\right) \times [1 + N_T\,(T_c\,(t) - T_{ref})] \times \eta_{inv} \quad (1)$$

$$T_c(t) = T_a(t) - \left(\frac{(NOCT-20)}{800} \times I(t)\right) \quad (2)$$

Where $P_{rated-PV}$ refers to the single PV rated output power under standard test conditions in watts (W), I shows the solar irradiance at time t in W/m^2, $I_{ref}$is solar irradiance at reference conditions and set usually as 1000 W/m^2, N_T is the temperature coefficient of the photovoltaic panel, and it equals -3.7×10 $^{(-3)}$ (1/°C) for monocrystalline and polycrystalline silicon [16], T_{ref} is the cell temperature at the reference conditions and set usually as 25 °C, T_c is the cell temperature °C at time t, and can be calculated by equation 2 in which T_a is the ambient air temperature in °C, $NOCT$ is the normal operating cell temperature in °C (approximately 44-48 °C).

4.2 Compressed air energy storage (CAES)

4.2.1 Compressor train (charge phase)

The real charging power of the compressor train at time t can be calculated by [6,13]:

$$W_{CAES,crg}(t) = K_c \times \dot{m}_{com,a}(t) \quad (3)$$

Where $\dot{m}_{com,a}$ is the air mass flowrate of the compressor train, K_c is the power used by the compressor train per unit of air mass flowrate in the charging phase, and it is written as [14] :

$$K_c = \frac{1}{\eta_M}\sum_{i=1}^{N_{com}} \frac{C_{p,air}\, T^{air}_{in,com,i}}{\eta_{com,ist,i}}\left(\pi_{com,i}^{\frac{\gamma-1}{\gamma}} - 1\right) \quad (4)$$

Where η_M presents the motor's energy conversion efficiency, N_{com} represents the number of compressor stages, $C_{p,air}$ is the specific heat capacity of air at constant pressure, $\eta_{com,ist,i}$ refers isentropic efficiency of each compressor (i), $T^{air}_{in,com,i}$ represents the intake air temperature of the ith compressor, γ is the specific heat ratio and $\pi_{com,i}$ shows compression ratio for the ith compressor that is expressed by:

$$\pi_{com,i} = \frac{p^{air}_{out,com,i}}{p^{air}_{in,com,i}} \quad (5)$$

Where $p^{air}_{out,com,i}$ and $p^{air}_{in,com,i}$ refers to the inlet and outlet air pressure of the ith compressor.

4.2.2 Expansion Train

The real discharging power of turbine train at time t can be calculated by:

$$W_{CAES,dis}(t) = K_t \times \dot{m}_{tur,a}(t) \quad (6)$$

Where $\dot{m}_{tur,a}$ is air mass flowrate of the turbine train and K_t is the power delivered by turbine train per unit of air mass flowrate in the discharging phase as expressed below:

$$K_t = \eta_{Gen} \sum_{j=1}^{N_{tur}} \eta_{tur,ist,j} C_{p,air}\, T^{air}_{in,tur,j} \left(1 - \pi_{tur,j}^{\frac{\gamma-1}{\gamma}}\right) \quad (7)$$

Where η_{Gen} presents the generator's energy conversion efficiency, N_{tur} shows the number of turbine stages, $\eta_{tur,ist,j}$ refers isentropic efficiency of each turbine (jth), $T^{air}_{in,tur,j}$ and $\pi_{tur,j}$ are intake air temperature and expansion ratio of the jth turbine, respectively, with below formulation:

$$\frac{T^{air}_{out,tur,j}}{T^{air}_{in,tur,j}} = \left(\pi_{tur,j}\right)^{\frac{\gamma-1}{\gamma}} \quad (8)$$

$$\pi_{tur,j} = \frac{p^{air}_{out,tur,j}}{p^{air}_{in,tur,j}} \quad (9)$$

Where $p^{air}_{out,tur,i}$ and $p^{air}_{in,tur,i}$ refers to the inlet and outlet air pressure of the jth turbine.

4.2.3 Air storage tank

Based on the ideal gas law (pV=mRT) and assuming the isochoric air reservoir (ΔV_{ac}=0), the air pressure (p_{ac}) in the air storage tank at the time t can be obtained by:

$$p_{ac}(t) = \frac{M_{ac}(t).R_a.T_{ac}}{V_{ac}} \quad (10)$$

Where Tac presents the air container temperature, $M_{ac}(t)$ is accumulated air mass in the tank at time t and expressed as [6]:

$$M_{ac}(t) = M_{ac}(t-1) + \sum_{t=1}^{t_{crg}} \dot{m}_{com,a}(t)\,\Delta t - \sum_{t=1}^{t_{dis}} \dot{m}_{tur,a}(t)\,\Delta t \quad (11)$$

Where Δt stands for time resolution and t_{crg} and t_{dis} represent the charge and discharge time of the air tank, respectively.

4.2.4 Heat exchangers

The overall heat transfer of the heat exchanger train during the charge and discharge phase can be determined as follows:

Charge phase:

$$\dot{Q}_{crg}(t) = \dot{m}_{com,a}(t)\, C_{p,air}\, \varepsilon_{HX,com} \sum_{i=1}^{Ncom}\left(T^{air}_{out,com,i} - T^{w}_{in,cold,com}\right) \quad (12)$$

Discharging phase:

$$\dot{Q}_{discrg}(t) = \dot{m}_{tur,a}(t) C_{P,air}\, \varepsilon_{HX,tur} \left(\left(T^{w}_{in,hot,tur} - T_{ac}\right) + \sum_{j=1}^{Ntur-1}\left(T^{w}_{in,hot,tur} - T^{air}_{out,tur,j}\right)\right) \quad (13)$$

Where $\varepsilon_{HX,tur}$ shows the effectiveness of the heat exchanger in preheating the air before each expansion stage, $\varepsilon_{HX,com}$ shows the effectiveness of the cooling heat exchanger after each compression stage, $T^{air}_{out,com,i}$ is the temperature of the hot air outlet from the ith compressor, $T^{air}_{out,tur,j}$ is the temperature of cold air outlet from the jth turbine, $T^{w}_{in,cold,com}$ and $T^{w}_{in,hot,tur}$ are the temperature of cold and hot water coming from TES, respectively.

4.2.5 TES model

The dynamic behavior of TES can be calculated by [15,16]:

$$Q_{TES}(t) = Q_{TES}(t-1) + \sum_{t=1}^{t_{crg}} \dot{Q}_{Com}(t)\,\Delta t - \sum_{t=1}^{t_{dis}} \dot{Q}_{tur}(t)\,\Delta t \quad (14)$$

Where $Q_{TES}(t)$ presents the state of the thermal energy (kWh) in the hot tank with a volume of:

$$V_{Tank} = Q^{Max}_{TES} / (C_w \rho_W (T_{hot} - T_{cold})) \quad (15)$$

In which $C_w\rho_W$ equals 1.163 $\left(\frac{kwh}{m3,k}\right)$.

4.2.6 Combustion chamber

When using axillary heat, the compressed air should be burned with natural gas (NG) in the combustion chamber. Therefore the required heat can be calculated by [17]:

$$\dot{Q}_t(t) = \dot{m}_{tur,air,ax}(t) \times C_{P,air} \times \left(\sum_{i=1}^{Nt} T^{air}_{in,tur,j} - T_{ac} - \sum_{j=1}^{Nt-1} T^{air}_{in,tur,j}\left(\pi_{tur,j}\right)^{\frac{\gamma-1}{\gamma}}\right) \quad (16)$$

Where $\dot{m}_{tur,air,ax}$ is part of the compressed air mass flowrate burned in the combustion chamber. The equivalent fuel consumption in m3 can be estimated as follow:

$$FC = \frac{\dot{Q}_t(t)}{H_f \times \eta_f} \quad (17)$$

Where H_f shows the heat value of gas (38.8 MJ/m3) and η_f is gas combustion efficiency (85% in this study).

5 ASSESSMENT CRITERIA

Several indicators are defined to analyze the performance of the proposed HES as presented below:

Cycle efficiency of CAES: which is ratio of CAES output and input energies during the one complete charging and discharging cycle.

$$\eta_{CAES\_cycle} = \frac{\sum_{t=1}^{t_{dis}} W_{CAES,dis}(t)}{\sum_{t=1}^{t_{crg}} W_{CAES,crg}(t) + \frac{\dot{Q}_t(t)}{\eta_f}} \quad (18)$$

Integrated efficiency of CAES: which is ratio of CAES output and input energies during the year.

$$\eta_{CAES\_int} = \frac{\sum_{t=1}^{8760} W_{CAES,dis}(t)}{\sum_{t=1}^{8760} W_{CAES,crg}(t) + \frac{\dot{Q}_t(t)}{\eta_f}} \quad (19)$$

Coverage ratio of CAES: which implies the share of CAES in satisfying the building energy demand.

$$C.R. = \frac{\sum_{t=1}^{8760} W_{CAES,dis}(t)}{\sum_{t=1}^{8760} W_L(t)} \quad (20)$$

Absorption ratio of CAES: which refers the share of CAES in absorbing excess PV output power.

$$A.R. = \frac{\sum_{t=1}^{8760} W_{CAES,crg}(t)}{\sum_{t=1}^{8760} W_G(t)} \quad (21)$$

Building autonomy: which is the share of hybrid PV-CAES in the meeting the building energy demand.

$$B.A. = \frac{\sum_{t=1}^{8760} min(W_G(t), W_L(t)) + W_{CAES,dis}(t)}{\sum_{t=1}^{8760} W_L(t)} \quad (22)$$

6 CASE STUDY

6.1 Load profile

The Concordia University's electricity consumption data for 2019 was provided by Gina Cody School of Engineering and Computer Science (ENCS) located in Montreal, Canada, which is shown on an hourly basis in fig.3. The annual average hourly electricity demand is 513.5 kWh.

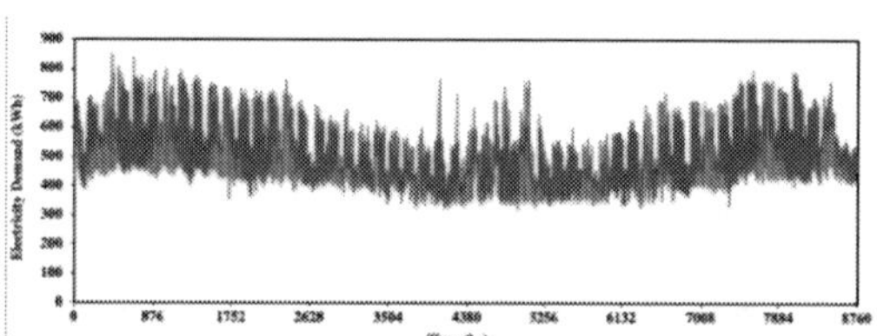

Figure 3: Hourly electrical load profile of the educational building over a year.

6.2 Weather conditions

Solar radiation and ambient temperature in the city of Montreal for the year 2019 is extracted from the NASA surface meteorology data set [18] based on the location of the case study (45°50′ N of latitude and 73°56′W of longitude.). Fig.4 presents the hourly solar radiation and ambient temperature. The annual average daily radiation and ambient temperatures are 3.58 kWh/m^2/day and 5.67°C, respectively.

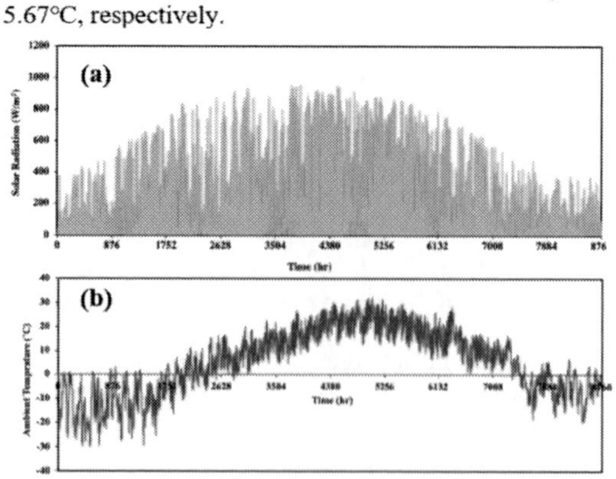

Figure 4: Hourly profile of (a) solar radiation, (b) ambient temperature for one-year horizon.

6.3 System configuration and design

The same PV module used in [19] is selected. Based on the available roof area of Concordia University (e.g., 24,487 m^2) and the surface area of chosen PV module, around 11000 PV panel is considered to be installed.

The specifications of the TICC-500 adiabatic-CAES (A-CAES) pilot plant [20] are employed as a reference for CAES configuration to validate and verify the developed operation strategy. More detail can be found in [20]. In this facility, a double hot/cold tank is used as a TES with a temperature of 120 / 45 °C and a volume of 12 m^3 each, equaling 1040 kWh energy capacity according to equation (15).

7 RESULT AND DISCUSSION

As mentioned before, the design and behavior of TES as a sub-system of CAES influence the whole system's performance. In this section, the aforementioned issue is investigated considering two defined scenarios (section 3) to evaluate and verify the developed operation strategy for every possible condition that might be happened in the CAES system, especially when it is adopted for decentralized application. MATLAB software is used to implement the code according to the mathematical model of components and developed operation strategies. Table I shows the results of simulation during the full charge and discharge process of CAES compared with the TICC-500 tested result. It visibly can be seen the results of the proposed model and the TICC-500 facility are well-matched. It should be noted that the results are obtained under steady-state conditions.

7.1 Thermal energy flow in the system

This section investigates the heat generated during the compression process and the heat required for expansion under the proposed operation strategy [19]. The behavior of the compressor and turbine in generating/ consuming heat during charging/discharging is displayed in fig.5.

As evident from fig.5, adopting rated power limitation of energy conversion units (compressors/turbines) in operation strategy cause CAES works in its operating range to avoid any failure and surge /choke issues. For example, the compressor train generates maximum thermal energy of 178 kWh in each hour compared to 716 kWh in case of neglecting the limitation of power conversion units in operation strategy as proposed in the literature [19]. This fact is also true for the turbine side.

7.2 Effect of TES volume

CAES is a complex system with interdependent components. Therefore, each element's design and operation influence one another's performance and, consequently, the whole hybrid system's function. This section analyzes the effect of TES capacity on the proposed system's operation under two defined scenarios (section 3).

Table I: Results of the developed model at design condition compared to the TICC-500 plant.

Parameter	Current model	TICC-500 Plant
Cycle efficiency (%)	37.08 %	37 %
CAES charging energy (kWh)	1447	---
CAES discharging energy (kWh)	563	---
CAES charging time (hr)	5.46	5.59
CAES discharging time (hr)	1.08	1.03

Charging thermal energy (kWh)	973	747
Discharging thermal energy (kWh)	574	600
TES charging time (hr)	5.84	---
TES discharging time (hr)	1.73	---

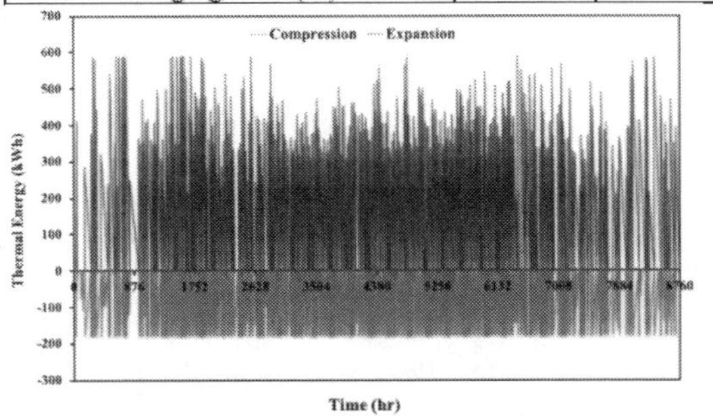

Figure 5: Hourly heat flow in A-CAES for one year horizon.

7.2.1 Performance analysis

Figure 6 displays the impact of TES capacity on the charging and discharging times of the air tank and TES. From figure 6a, it is observed that TES's charging/discharging times increase monotonously by raising the TES volume because it can capture and release more thermal energy. However, the increase in charging time is sharper than discharging time because the heat generated during the compression phase is more than the required heat for expansion according to the characteristics of considered CAES in the current study. It should be noted that this behavior is the same for both scenarios because the charging and discharging time of TES depends on its capacity and the rated heat of the compressor/turbine. Therefore, it is not affected by adopting other axillary heat sources.

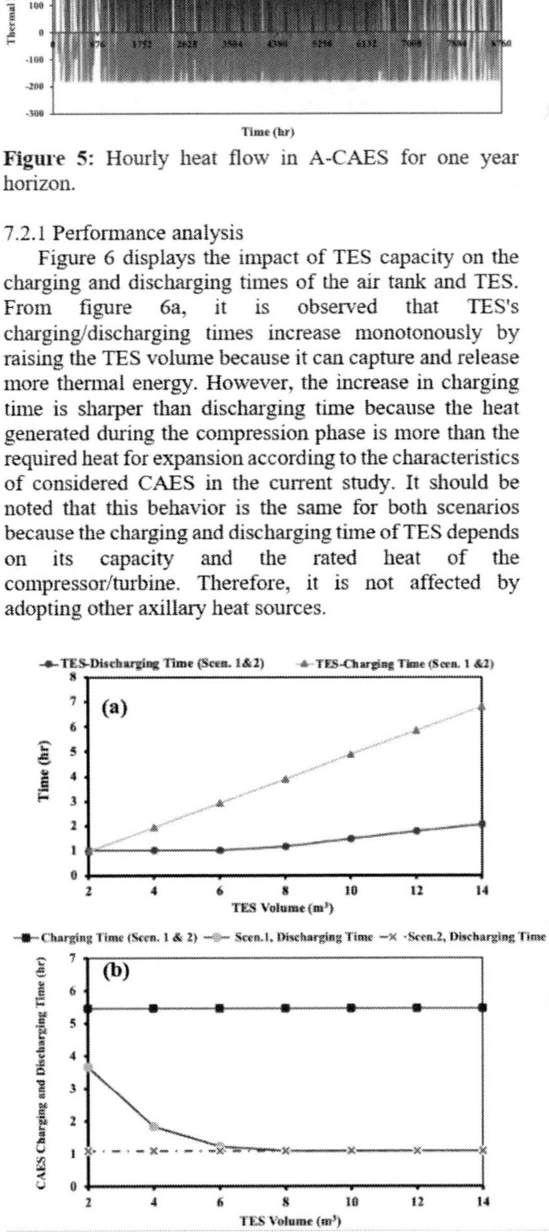

Figure 6: The effect of TES capacity on the charging and discharging time of a) TES and b) CAES.

In contrast, with raised TES capacity, the discharging time of the air tank increases in scen.1 while remaining constant in scen.2 as shown in fig.6b. This phenomenon is due to the strong dependence between TES capacity and mass flowrate of the turbine, which is regulated according to the available thermal energy when TES is the only source of heat (scen.1). In such a case, the turbine operation depends on the available heat capacity in the system. However, discharging time does not change when there is another axillary heat source because it can be compensated for the capacity shortage of TES. On the other hand, the full charging time of the air tank under both scenarios is constant because there is no regulation on the rated mass flowrate of the compressors at steady-state conditions based on the TES capacity.

Fig.7 shows the impact of TES capacity on the performance of the proposed system. From fig.7a illustrates that with increased TES capacity, CAES's integrated, and cycle efficiencies remain constant in scen.1 while they increase in scen.2. That is because in scen.1, raised TES capacity affects the turbine's rated mass flowrate positively and discharging time negatively. Therefore, the turbine and compressor work ratio remains constant considering one complete cycle. However, building autonomy (B.A.) has the exact opposite trend under both scenarios, as demonstrated in fig 7b.

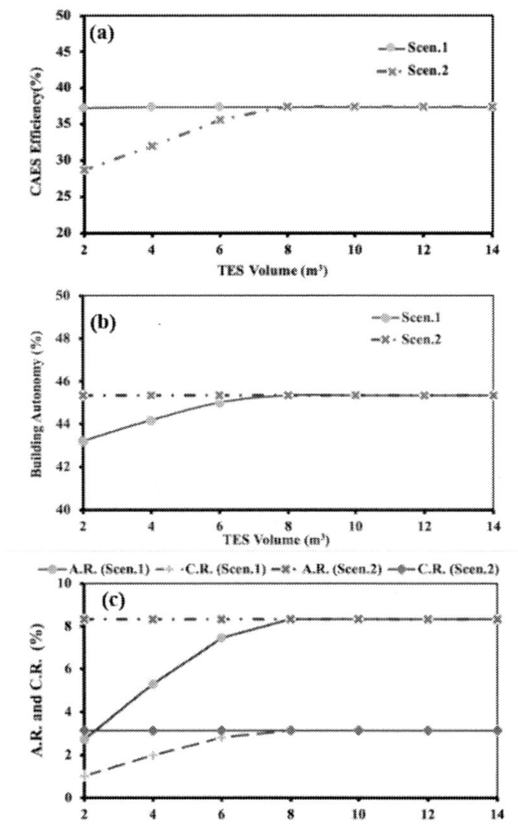

Figure 7: The effect of TES on the a) A-CAES efficiency, b) building autonomy, and c) A.R. and C.R.

Since the TES affects the turbine operation and subsequently the air tank state and the compressor train operation in the following steps, the integrated efficiency of CAES is also constant. However, both C.R. and A.R. are improved first by raising TES volume in Scen.1, as shown in fig.7c. However, for TES volume higher than 8

m^3, the variation of C.R. and A.R. is not considerable. In contrast, in scen.2, increased TES volume positively impacts both integrated and cycle efficiencies (fig.7a). At the same time, A.R. and C.R. remain constant, as can be seen in fig.7c. The reason is the reduction of the NG usage share in preheating the compressed air.

7.2.2 Heat management in the system

As mentioned before, using TES to recover the heat generated during the compression phase mitigates the system energy loss and emission while improving efficiency. However, finding the optimal size for TES to reach the above-mentioned goal is essential. Fig.8 shows the system's heat loss during a year. Obviously, with increased TES capacity, the system's heat loss reduces in scen.2 because, along with the raised TES volume, more thermal energy generated during the charging process can be captured by TES. These phenomena can be proved by fig.9, which displays the share of heat loss and thermal energy managed by TES in the total thermal energy generated during the compression phase.

On the other hand, in scen.1, changing the TES volume alters the operation hour of turbine and compressor trains. Hence, with raised TES capacity, the heat released from the compression phase becomes larger, resulting in a higher heat loss, as demonstrated in fig.9. Thus, an increase of TES volume from 2 m^3 to 8 m^3 increases and decreases the yearly heat loss by about 3 and 2.3 times in scen.1 and scen.2, respectively. However, the further increase in the TES volume (more than 8 m^3) does not impact heat loss for both scenarios.

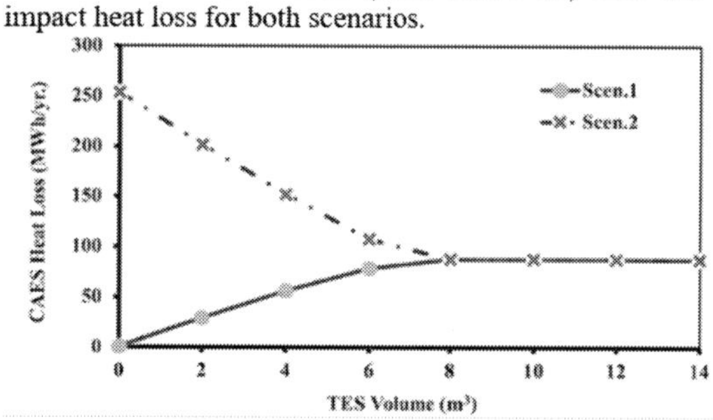

Figure 8: The effect of TES capacity on yearly heat loss from CAES.

Fig.10 also illustrates the TES impact on the share of TES and NG in providing the heat required by the turbine train during the discharging process. As can be seen, TES volume positively impacts TES share in supplying heat for both scenarios. However, in scen.2 the rest of desired thermal energy is compensated by burning NG, resulting in carbon emission.

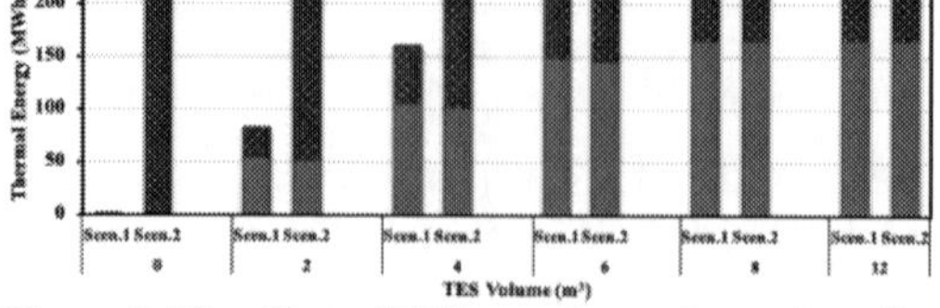

Figure 9: The effect of TES capacity on the portion of captured heat by TES and heat loss during the charging phase.

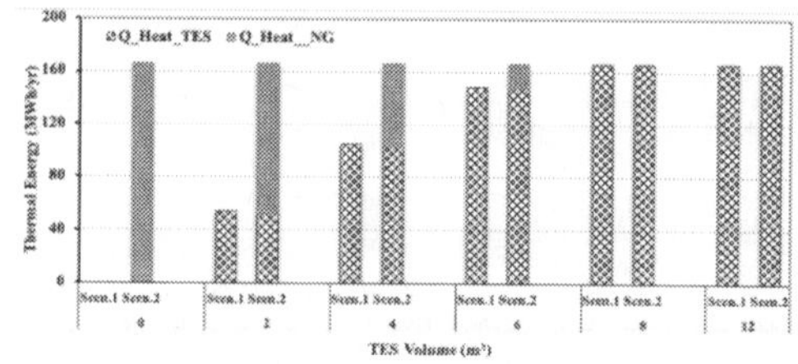

Figure 10: The effect of TES capacity on the portion of TES and NG in supplying thermal energy during discharging phase.

7.2.3 Operational condition of the system

To better understand the system's dynamic operation, three days of January (from 10th to 12th) are selected as a representative to evaluate the different possible operational conditions of HES components. Fig. 11 displays the PV panel's output power and electric demand of the building for three aforementioned days.

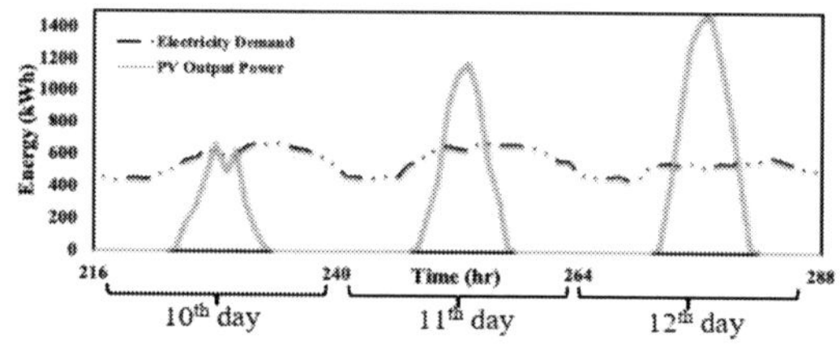

Figure 11: Profile of the PV output power and building electric load demand during the three days of January.

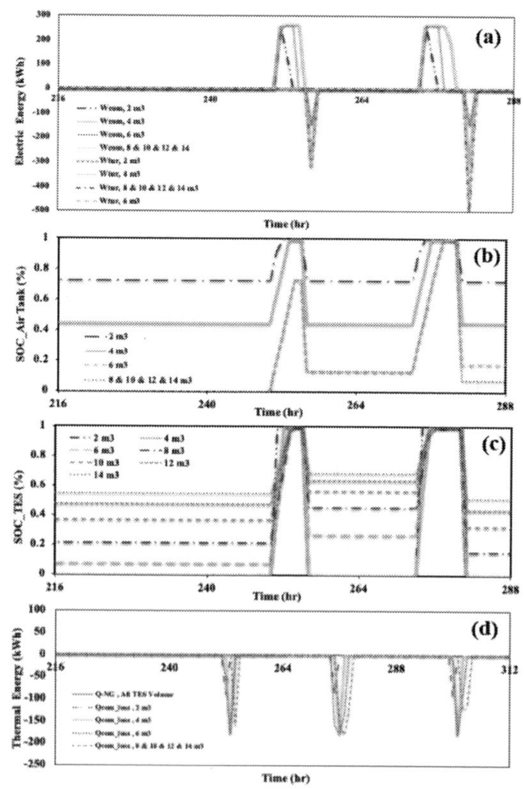

Figure 12: The effect of TES capacity on the dynamic operation of a) compressors and turbines, b) air tank SOC, c) SOC of TES, and d) compressor heat loss and NG usage under Scen.1.

As can be seen, a part of building demand is fulfilled by solar energy in these three days. However, there is an excess PV power output on days 11th and 12th that can be stored in CAES. While on the day 9th, since the excess electricity is less than the compressor start-up power, extra energy is sold back to the grid. This also can be seen in fig.12a and fig.13a, which show the hourly work of the compressor (W_{com}) and turbine (W_{tur}).

Furthermore, the effect of TES capacity on the operational condition of compressor/turbine, air storage tank, TES, and hourly heat loss /required by turbine train are shown in fig. 12 and fig. 13 for scen.1 and scen.2 respectively.

Obviously, the TES has the same operational behavior in two scenarios (fig 12c and 13c), and the change of TES volume will alter the amount of stored thermal energy and thus TES behavior. However, unlike scen.2, in scen.1, since the operation hour of the turbine train and, subsequently, the compressor train are affected by TES volume variation (fig 12a), the SOC of the air tank is changed (fig.12b). Therefore, the increased TES capacity implies a larger thermal energy discharge. Thus, more air mass flow is discharged from the air storage tank resulting in higher electric power generation by turbine train. The results also indicate that for the TES volume higher than 8 m^3, the system's operational condition variation is the same (except SOC of TES).

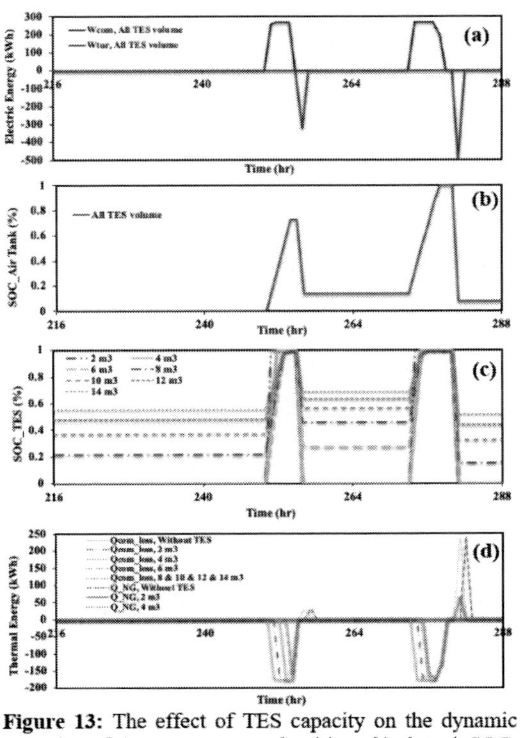

Figure 13: The effect of TES capacity on the dynamic operation of a) compressors and turbines, b) air tank SOC, c) SOC of TES, and d) compressor heat loss and NG usage under Scen.2.

As highlighted before, in scen.2 the variation of TES capacity does not affect the operation of the turbine/compressor train and thus air tank SOC. Such behavior is due to the presence of axillary heat in the system that can compensate for the TES shortage to provide the desired heat of the turbine train.

7.2.4 Emission

It is assumed that the combustion chamber of CAES considered in scen.2 is the only source of carbon emission. The result shows that increasing TES volume from 0 to 8 m^3 mitigates the amount of CO_2 emission of the system by around 100% (from 32.5 to 0 m^3) while converting it to zero-emission energy storage.

8 CONCLUSIONS

The present paper investigates the influence of thermal energy storage (TES) design and operation on the CAES system's techno-environmental performance in a hybrid system. To achieve this goal, a comprehensive operation strategy is developed in which the limitations of all components are considered. Besides, two scenarios were defined considering heat management and fuel utilization in CAES to investigate the different possible operational conditions. The proposed hybrid system includes PV, grid, CAES, and building electricity demand. The overall characteristics of the TICC-500 A-CAES pilot plant [20] are considered as a reference for CAES units to validate the developed approach. Some of the main results are highlighted as follows:

- The result shows that the proposed operation strategy and TICC-500 tested results are well-matched.
- The optimum volume for the TES is about 8 m^3 which equals an energy capacity of 698 kWh.
- Changing the TES rated capacity influences the turbine operation resulting in variation of CAES full discharging time at the steady-state condition when TES is the only heat source (scenario 1). Besides, four times increase in TES capacity leads to a 7.4% improvement in the building autonomy under the same scenario.
- By increasing TES volume from 2 m^3 to 8 m^3, the efficiency of CAES improves by around 48% while converting to zero-emission storage with 100% emission reduction (scenario 2).
- With raised TES capacity, the yearly heat loss (exergy destruction) from CAES is increased and decreased under scenarios 1 and 2 due to the compressor's operation hour augmenting and capturing more thermal energy generated by compressors, respectively.
- The result also indicated that dynamic operational conditions of compressor/turbine and air tank stats are all affected by altering the TES rated capacity when TES is the only source of heat.

Furthermore, this operation strategy allows studying other types of CAES with different characteristics for the sake of adopting TES and fuel utilization by adjusting the design parameter range of the TES unit. In addition, it provides a framework for finding the optimum TES volume for further development of the CAES system's practical application, either grid-connected or stand-alone considering techno-economic and environmental aspects.

ACKNOWLEDGMENT

The authors would like to express appreciation to the Concordia university's facility management for providing the electric energy demand data of Gina Cody School of Engineering and Computer Science (ENCS) building. This work was funded by the India-Canada Centre for Innovative Multidisciplinary Partnerships to Accelerate Community Transformation and Sustainability (IC-IMPACTS), dedicated to the development of scientific collaborations between Canada and India.

REFERENCES

[1] P. Zhao, W. Xu, S. Zhang, J. Wang, Y. Dai, Technical feasibility assessment of a standalone photovoltaic/wind/adiabatic compressed air energy storage based hybrid energy supply system for rural mobile base station, Energy Convers. Manag. 206 (2020) 112486. https://doi.org/10.1016/j.enconman.2020.112486

[2] E. Bazdar, M. Sameti, F. Nasiri, F. Haghighat, Compressed air energy storage in integrated energy systems : A review Organic Rankin cycle, Renew. Sustain. Energy Rev. 167 (2022) 112701. https://doi.org/10.1016/j.rser.2022.112701.

[3] R. Jiang, X. Yang, Y. Xu, M. Yang, Design/off-design performance analysis and comparison of two different storage modes for trigenerative compressed air energy storage system, Appl. Therm. Eng. 175 (2020) 115335. https://doi.org/10.1016/j.applthermaleng.2020.115335.

[4] S.W. Mei, J.J. Wang, F. Tian, L.J. Chen, X.D. Xue, Q. Lu, Y. Zhou, X.X. Zhou, Design and engineering implementation of non-supplementary fired compressed air energy storage system: TICC-500, Sci. China Technol. Sci. 58 (2015) 600–611. https://doi.org/10.1007/s11431-015-5789-0.

[5] X. Zhang, C. (Chris) C. Qin, Y. Xu, W. Li, X. Zhou, R. Li, Y. Huang, H. Chen, Integration of small-scale compressed air energy storage with wind generation for flexible household power supply, J. Energy Storage. 37 (2021) 102430. https://doi.org/10.1016/j.est.2021.102430.

[6] X. Xu, W. Hu, D. Cao, Q. Huang, W. Liu, Z. Liu, Z. Chen, H. Lund, Designing a standalone wind-diesel-CAES hybrid energy system by using a scenario-based bi-level programming method, Energy Convers. Manag. 211 (2020) 112759. https://doi.org/10.1016/j.enconman.2020.112759

[7] P.M. Congedo, C. Baglivo, L. Carrieri, Application of an unconventional thermal and mechanical energy storage coupled with the air conditioning and domestic hot water systems of a residential building, Energy Build. 224 (2020) 110234. https://doi.org/10.1016/j.enbuild.2020.110234.

[8] G. Dib, P. Haberschill, R. Rullière, Q. Perroit, S. Davies, R. Revellin, Thermodynamic simulation of a micro advanced adiabatic compressed air energy storage for building application, Appl. Energy. 260 (2020) 114248. https://doi.org/10.1016/j.apenergy.2019.114248.

[9] E. Jannelli, M. Minutillo, A. Lubrano Lavadera, G. Falcucci, A small-scale CAES (compressed air energy storage) system for stand-alone renewable energy power plant for a radio base station: A sizing-design methodology, Energy. 78 (2014) 313–322. https://doi.org/10.1016/j.energy.2014.10.016.

[10] Y. Huang, A. Rolfe, I. Vorushylo, P. Keatley, R. Byrne, P. MacArtain, D. Flynn, and N. Hewitt, Integration of compressed air energy storage with wind generation into the electricity grid, (2018).

[11] S. Sadeghi, I.B. Askari, Prefeasibility techno-economic assessment of a hybrid power plant with photovoltaic, fuel cell and Compressed Air Energy Storage (CAES), Energy. 168 (2019) 409–424. https://doi.org/10.1016/j.energy.2018.11.108.

[12] S. Simpore, F. Garde, M. David, O.M. Marc, J. Castaing-Lasvignottes, Design and Dynamic Simulation of a Compressed Air Energy Storage System (CAES) Coupled with a Building, an Electric Grid and Photovoltaic Power Plant, CLIMA 2016, 12th REHVA World Congr. (2016). https://hal.archives-ouvertes.fr/hal-01467204.

[13] Y. Yan, C. Zhang, K. Li, Z. Wang, An integrated design for hybrid combined cooling, heating and power system with compressed air energy storage, Appl. Energy. 210 (2018) 1151–1166. https://doi.org/10.1016/j.apenergy.2017.07.005.

[14] X. Xu, W. Hu, D. Cao, Q. Huang, W. Liu, Z. Liu, Z. Chen, H. Lund, Designing a standalone wind–diesel–CAES hybrid energy system by using a scenario–based bi–level programming method, Elsevier, n.d. https://doi.org/10.1016/j.enconman.2020.112759

[15] Y. Li, S. Miao, S. Zhang, B. Yin, X. Luo, M. Dooner, J. Wang, A reserve capacity model of AA-CAES for power system optimal joint energy and reserve scheduling, Int. J. Electr. Power Energy Syst. 104 (2019) 279–290. https://doi.org/10.1016/j.ijepes.2018.07.012.

[16] R. Li, L. Chen, T. Yuan, C. Li, Optimal dispatch of zero-carbon-emission micro Energy Internet integrated with non-supplementary fired compressed air energy storage system, J. Mod. Power Syst. Clean Energy. 4 (2016) 566–580. https://doi.org/10.1007/s40565-016-0241-4.

[17] Y. Li, S. Miao, S. Zhang, B. Yin, X. Luo, M. Dooner, J. Wang, A reserve capacity model of AA-CAES for power system optimal joint energy and reserve scheduling, Int. J. Electr. Power Energy Syst. 104 (2019) 279–290. https://doi.org/10.1016/j.ijepes.2018.07.012.

[18] NASA Langley - POWER, (n.d.). https://power.larc.nasa.gov/common/php/SSE_ExSummary.php (accessed May 21, 2019).

[19] P. Zhao, W. Xu, S. Zhang, J. Wang, Y. Dai, Technical feasibility assessment of a standalone photovoltaic/wind/adiabatic compressed air energy storage based hybrid energy supply system for rural mobile base station, Energy Convers. Manag. 206 (2020) 112486. https://doi.org/10.1016/j.enconman.2020.112486

[20] X. Luo, M. Dooner, W. He, J. Wang, Y. Li, D. Li, O. Kiselychnyk, Feasibility study of a simulation software tool development for dynamic modelling and transient control of adiabatic compressed air energy storage with its electrical power system applications, Appl. Energy. 228 (2018) 1198–1219. https://doi.org/10.1016/j.apenergy.2018.06.068.

CO_2 EMISSIONS OF SILICON PHOTOVOLTAIC MODULES – IMPACT OF MODULE DESIGN AND PRODUCTION LOCATION

Christian Reichel, Amelie Müller, Lorenz Friedrich, Sina Herceg, Max Mittag, Alexander Protti, Dirk Holger Neuhaus

Fraunhofer Institute for Solar Energy Systems ISE, Heidenhofstrasse 2, 79102 Freiburg, Germany

ABSTRACT: CO_2 emissions of two different module designs (conventional glass-backsheet or novel frameless glass-glass modules) produced at three different locations (China, Germany or the European Union, EU) are determined and compared in a life cycle assessment (LCA), using current inventory data and differentiated electricity yields of the modules. The results show that lower environmental impacts are obtained for glass-glass compared to glass-backsheet modules and for a production in the EU and Germany compared to China. Glass-backsheet (glass-glass) modules produced in China, Germany or the EU are linked to 810 (750), 580 (520) and 480 (420) kg CO_2-eq/kW$_p$, respectively. This corresponds to CO_2 emission reductions of 30% for Germany and 40% for the EU compared to Chinese production, and 8% to 12.5% reduction in glass-glass compared to glass-backsheet modules. The CO_2 emissions of the produced electricity, excluding balance of system (BoS), amount to 13-30 g CO_2-eq/kWh, depending on the production location and electricity yield of the modules which is based on warranty yield calculations with longer lifetimes for glass-glass than glass-backsheet modules. It is shown that module efficiency, energy requirements, silicon consumption and electricity mix used at the production location are significant levers for future reductions of environmental impacts. Furthermore, it is emphasized that up-to-date inventories and differentiated electricity yield calculations as well as current modelling of electricity mixes are important to incentivize the development of sustainable module designs.

Keywords: life cycle assessment, crystalline silicon, glass-backsheet module, glass-glass module

1 INTRODUCTION

Modules based on silicon solar cells are dominating the photovoltaic (PV) market and are considered as a green technology for the supply of renewable and emission-free energy. However, the production of the solar cells, the encapsulation and interconnection of them in the modules, as well as the transport and disposal or recycling, have a non-negligible impact on the environment and resource consumption. Therefore, an overall assessment of the modules is essential for the success of the envisaged energy transition with its expected growth towards a multi-terawatt scale in the next decades.

In general, the CO_2 emissions of PV systems are significantly lower than those of conventional energy generation using, for instance, coal, but there are considerable differences in modules due to the module design (classic glass-backsheet or novel frameless glass-glass modules) and, in particular due to the production location (China, Germany or the European Union, EU).

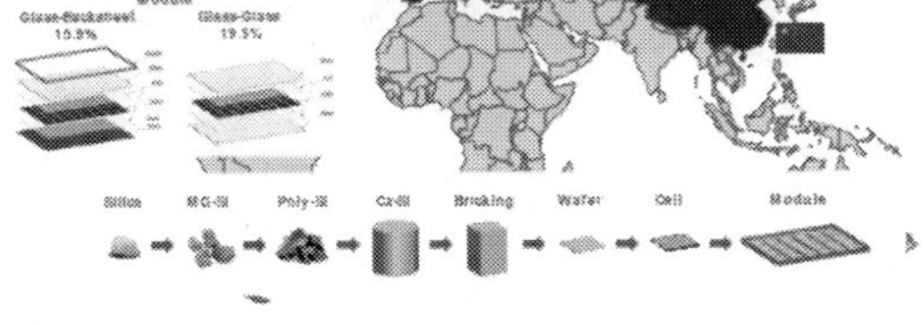

Figure 1: World map depicting the production location China (red), Germany (blue with green stripes) and the EU as of 2020 (blue) and the installation site in Freiburg, Germany, including the representation of the entire value chain (bottom) and the investigated glass-backsheet and glass-glass module designs with the corresponding efficiency (left).

In order to identify these differences, the CO_2 emissions of the PV modules were determined and compared in a life cycle assessment (LCA), using current inventory data and differentiated electricity yields of the modules.

In comparison to previous studies, more recent developments such as diamond wire saws for wafer production and current data in silicon, solar cell and module production of the years 2019/2020 were used as foreground data [1,2]. The existing literature also pays little attention to new developments in the module designs of PV systems and alternatives to the conventional glass-backsheet designs, such as frameless glass-glass designs with a higher expected market share [3], longer lifetimes [4,5] and reduced costs [6]. Furthermore, the recent shift in production to China is not always accounted for in PV LCAs and, therefore, the historic focus on inventory data from European producers prevails [7].

This study addresses the aforementioned aspects and reveals climate change impacts of the different module designs produced at different locations to identify benefits for the development of sustainable module designs.

It is shown that module efficiency, energy requirements, silicon consumption and electricity mix used at the production location are significant levers for future reductions of environmental impacts. Furthermore, it is emphasized that up-to-date inventories and differentiated electricity yield calculations as well as current modelling of electricity mixes are important to incentivize the development of sustainable module designs.

2 METHODOLOGY

The environmental impact of the modules was investigated over the entire life cycle from manufacturing to recycling in various categories, using the environmental footprint (EF 3.0) by the European Commission as the methodological approach, especially in terms of climate change, whereby the LCA performed is based on the guidelines of the International Energy Agency for PV systems (IEA PVPS) [8]. For the assessment of the impact on climate change, an indicator developed by the Intergovernmental Panel on Climate Change (IPCC) 2013 was used to determine the CO_2 equivalent emissions of the modules for a period of 100 years [9]. For this purpose, the material and energy consumptions to produce the modules along the entire value chain were implemented in SimaPro 9.0 [11] with background data from ecoinvent 3.7 [11] In ecoinvent 3.7 the electricity mix for China is from 2014 and for Germany and EU from 2017. An update of the current electricity mix of China, Germany and the EU is in preparation.

3 RESULTS AND DISCUSSION

The comparison of the entire value chain for a complete production of the PV modules in China, Germany and the EU shows clearly that a complete glass-backsheet module production in the EU or Germany causes about 41% and 28% less CO_2 emissions, respectively, compared to China (480 and 580 kg versus 810 kg CO_2-eq/kW$_p$), see Figure 2 (top).

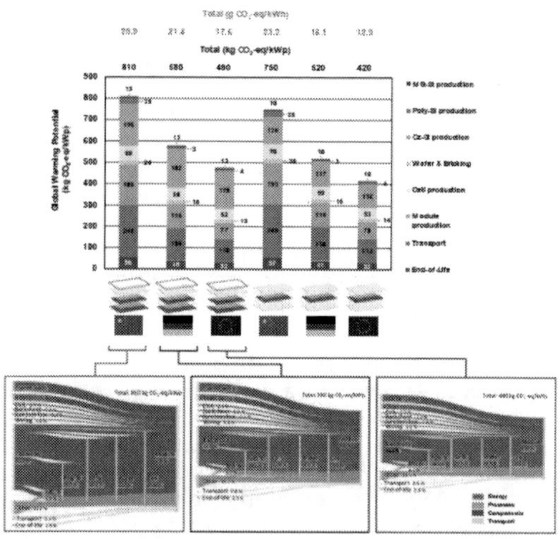

Figure 2: CO_2 emissions of glass-backsheet and glass-glass module designs for the different production processes (top) and share of energy, processes, components and transport for the production locations China, Germany and the EU (bottom). The total global warming potential (GWP) in shown as kg CO_2-eq/kW$_p$ (top, black) and g CO_2-eq/kWh (top, green).

The main reason for this difference can be found in the specific electricity mix at the production location. While in China the energy is predominantly generated from fossil fuels using coal and renewable energy represents only a small share of the electricity mix, the total energy provided from renewable sources in Germany and the EU is much higher. In addition, it is also shown that frameless glass-glass modules have even lower CO_2 emissions compared to glass-backsheet modules. For a glass-backsheet module produced in Germany, the shares of energy, processes, components, and transport were also broken down in more detail, see Figure 2 (bottom). It can be seen that the energy required for the production of the module is the main cause of the GWP with up to 62.7%, 53.7% and 44.2% for China, Germany and the EU, respectively.

The CO_2 emissions of the produced electricity, excluding BoS, amounts to 23.2-29.9 g CO_2-eq/kWh for China, 16.1-21.4 g CO_2-eq/kWh for Germany and 12.9-17.6 g CO_2-eq/kWh for the EU, assuming an electricity yield calculation based on different module performance warranties given by module producers with almost 30 years lifetime for glass-glass modules and 25 years for glass-backsheet modules and at a central Europan installation location (Freiburg, Germany, with 1391 kWh/(m^{2}a) solar irradiation).

The results underline that powerful and durable modules are indeed a promising pillar for a climate-neutral energy supply in the future. The production generates fewer CO_2 emissions than assumed in previous studies and migrating the complete production to the EU or Germany instead of China reduces CO_2 emissions, thus mitigating climate change. Therefore, a paradigm shift must take place in the future to ensure a sustainable and cost-effective development of the PV industry by re-evaluating the production locations, especially since the cost of CO_2 emissions will increase from the currently 25 €/t to up to 55 €/t by 2025 [12] in Germany and even up to 75 €/t worldwide, which is discussed for 2030 [13].

5 CONCLUSION AND OUTLOOK

The results show that lower environmental impacts are obtained for glass-glass compared to glass-backsheet modules and for a production in the EU and Germany compared to China. Glass-backsheet (glass-glass) modules produced in China, Germany or the EU are linked to CO_2 emissions of 810 (750), 580 (520) and 480 (420) kg CO_2-eq/kW$_p$, respectively. This corresponds to CO_2 emission reductions of 30% for Germany and 40% for the EU compared to Chinese production, and 8% to 12.5% reduction in glass-glass compared to glass-backsheet modules. The CO_2 emissions of the produced electricity, excluding balance of system (BoS), amounts to 13-30 g CO_2-eq/kWh, depending on the production location and electricity yield of the modules which is based on warranty yield calculations with longer lifetimes for glass-glass than glass-backsheet modules.

It is shown that module efficiency, energy requirements, silicon consumption and electricity mix used at the production location are significant levers for future reductions of environmental impacts. Furthermore, it is emphasized that up-to-date inventories and differentiated electricity yield calculations as well as current modelling of electricity mixes are important to incentivize the development of sustainable module designs.

ACKNOWLEDGEMENTS

The authors acknowledge the funding of the work by the German Ministry of Economic Affairs and Climate Action under Grant No. 03EE1095B "KoMoGER".

6 REFERENCES

[1] L. Friedrich, S. Nold, A. Müller, J. Rentsch, R. Preu, "Global Warming Potential and Energy Payback Time Analysis of Photovoltaic Electricity by Passivated Emitter

and Rear Cell (PERC) Solar Modules," submitted: IEEE Journal of Photovoltaics, 2021.
[2] A. Müller, L. Friedrich, C. Reichel, S. Herceg, M. Mittag, D. H. Neuhaus, "A Comparative Life Cylce Assemssment of Silicon PV Modules: Impact of Module Design, Manufacturing Location and Inventory," Solar Energy Materials and Solar Cells, vol. 230, 2021
[3] I. Miller, E. Gencer, H.S. Vogelbaum, P.R. Brown, S. Torkamani, F.M. O'Sullivan, "Parametric modeling of life cycle greenhouse gas emissions from photovoltaic power, Applied Energy 238 (2019) 760–774.
[4] G. Cattaneo, A. Faes, H.-Y. Li, Y. Yao, Lamination process and encapsulation materials for glass–glass PV module design, Photovoltaics International 82 (2015).
[5] Y. Zhang, J. Xu, J. Mao, J. Tao, H. Shen, Y. Chen, Z. Feng, P.J. Verlinden, P. Yang, J. Chu, Long-term reliability of silicon wafer-based traditional backsheet modules and double glass modules, RSC Advances 5 (2015) 65768–65774.
[6] M. Fischer, M. Woodhouse, S. Herritsch, J. Trube, International Technology Roadmap for Photovoltaic (ITRPV): 2020 Results, twelfth ed., 2021. Frankfurt, Germany.
[7] J. Hong, W. Chen, C. Qi, L. Ye, C. Xu, Life cycle assessment of multicrystalline silicon photovoltaic cell production in China, Solar Energy 133 (2016) 283–293.
[8] R. Frischknecht, P. Stolz, G. Heath, M. Raugei, P. Sinha, M. de Wild-Scholten, Methodology Guidelines on Life Cycle Assessment of Photovoltaic 2020, fourth ed., Report IEA-PVPS T12-18: 2020
[9] S. Fazio, F. Biganzioli, V. de Laurentiis, L. Zampori, S. Sala, E. Diaconu, Supporting Information to the Characterisation Factors of Recommended EF Life Cycle Impact Assessment Methods: Version 2 from ILCD to EF 3.0, Ispra, Italy, 2018
[10] SimaPro Analyst v9.0: PRé Sustainability. [Online]. Available: https://simapro.com/
[11] Ecoinvent database v.3.7. [Online]. Available: https://www.ecoinvent.org/
[12] Entwurf eines Ersten Gesetzes zur Änderung des Brennstoffemissionshandelsgesetzes. Deutscher Bundestag. Accessed: October, 8 2020.
[13] "High greenhouse gas emitters should pay for carbon they produce, says IMF". The Guardian, accessed: June, 19 2021.

Technical, business and regulatory approaches to enhance renewable energy capabilities to take part actively in the electricity and ancillary services markets.

Hanna Bartoszewicz-Burczy
Institute of Power Engineering, Warsaw, Poland
+48602681704 fax: +48228366363
Hanna.burczy@ien.com.pl
Przemysław Starzyński
ENEA Operator, Poznań, Poland
Filip Schraube
ENEA Operator, Poznań, Poland

ABSTRACT: This paper is based on the work carried out in the EU project: "DRES2Market: Technical, business and regulatory approaches to enhance renewable energy capabilities to take part actively in the electricity and ancillary services markets" is a European Union's Horizon 2020, granted research program project, grant number: 952851.
Main goal of the DRES2Market project is developing a comprehensive and affordable approach to facilitate the effective participation in distributed generation, based upon wind and solar PV energies in the electricity markets to enable provision of balancing and reserve services according to market criteria.
One of the main objective of the project is to evaluate and get the results of the impact of large penetration of renewable energies (RES) on the electricity markets and the system operation according to different time scopes: the electricity demand coverage using an hourly simulation dispatch taking into account the volatility of solar photovoltaic and wind energy, the participation of renewable energies and distributed generation based on solar photovoltaic in the European Continuous Intraday Electricity Markets and the Local Electricity Markets, as well as the impact of the System Operation and the Distribution Network.
Keywords: photovoltaic, wind energy, barriers of PV and wind energy, energy market

1 BARRIERS FOR THE SECURE AND RELIABLE CONNECTION OF DISTRIBUTED SOLAR PV AND ONSHORE WIND

The European Union has adopted ambitious and wide-ranging binding targets for reduction of greenhouse gases emissions and transformation of energy systems. The EU goals are to become the world's first climate-neutral continent by 2050 and climate resilient society.
In December 2020, the EU and the Member States submitted an enhanced Nationally Determined Contribution (NDC), which aims to reduce GHG emissions by at least 55% by 2030 from 1990 levels. This updated NDCs are part of the EU climate action and the European Green Deal. The goals are to become the world's first climate-neutral continent by 2050 and to be a climate resilient society[1] . It includes:

- European Climate Law to enshrine the 2050 climate-neutrality objective into EU law;
- European Climate Pact to engage citizens and all parts of society in climate action;
- 2030 Climate Target Plan to further reduce net greenhouse gas emissions by at least 55% by 2030;
- New EU Strategy on Climate Adaptation to make Europe a climate-resilient society by 2050, fully adapted to the unavoidable impacts of climate change.

Solar PV and wind energy have become key climate component of NDCs and climate solutions, by reducing CO_2, SO_2, NO_x and other emissions pollutants from power generation and one of the most important sources of electricity generation in transition towards to low-carbon energy system. Among other effects all these legal frameworks increase the use of renewable energy, especially solar PV and wind power in European countries. In order to successfully meet the EU's ambitious 2050 greenhouse gas emission reduction targets and significantly increase installed renewable energy capacity, additional work is needed to overcome existing RES barriers.

The EU countries made significant progress in the field of RES regulations, and in particular implementation of the RED II Directive. However, several barriers still remain, and development of photovoltaics and wind is being slowed by regulatory, administrative, economic and social obstacles.
Some hurdles are intended to ensure that the system functions properly and customers benefit from high standards of service, they may:

- be excessively complex and time consuming,
- require payments or financial guarantees,
- unduly increase costs or make it harder to do business,
- favour certain market participants or business models over others.[2]

The development of an internal electricity market in the European Union is an ongoing process. The ranking of European retail electricity and markets based on the

[1] https://ec.europa.eu/clima/policies/eu-climate-action_en
[2] Lewis, P., Granroth-Wilding, H., Napolitano, L., Zabala, C., Vékony, A., Felsmann, B., & Hirschbichler, F. (2021). European Barriers in Retail Energy Markets Project: Final Report. Luxembourg: Publications Office of the European Union. ISBN 978-92-76-30269-8, doi:10.2833/5217.

Barriers Index (details in the "European Barriers in Retail Energy Markets")[3] indicate that the Norwegian market faces the fewest barriers, (but there is high public resistance against grid extension) and the electricity markets of Slovenia, Sweden, Netherlands, and Finland are also outstandingly entrant friendly.
On the other side, Cyprus is the market presenting the most difficulties for a supplier, while Bulgaria, Poland, Lithuania and Romania are also amongst the high-barrier countries.

2.1 Regulatory and administrative barriers

Regulatory, legal stability and market structure are of key importance for all RES investments. The regulatory and administrative requirements for solar PV and wind energy installations can limit possibilities of expansion of distributed generation in European countries.
Transparent, unambiguous and stable regulatory systems favour investments in solar PV and wind, especially in the initial stages.

The administrative and legal procedures that delay development of the PV and wind installations are as follows:

- Duration of the grid connection permit process;
- Insufficient spatial planning or RES insufficiently taken into account;
- Lack of coordination between involved authorities;
- Lack of stable legal and financial regulations;
- Administrative hurdles such as planning delays;
- Large amount of administrative processing and long deadlines in time;
- Frequent changes in law, as well as interpretative uncertainties related to newly introduced regulations;
- Long lasting environmental impact assessment and zoning plan.

Regulatory instability causes investors uncertainty regarding the provision of support in the coming years, and the level of the current costs of their operations. Frequent legal amendments cause difficulties in interpretation of entries and increase risk in implementation of long-term development strategies.
Bureaucracy, prolonging of administrative procedures by both authorities and grid operators, still pose barriers to the rapid deployment of PV and wind in many European countries.

2.2. Technical barriers

Technically PV and wind systems are mature, so there are no technical barriers regarding the systems themselves, but more regarding integration of these systems in the local grid. Some questions of technical grid management with very high penetration of renewables (new flexibility needs) are still open along next to market challenges, but up to now renewables' penetration rates are still too low in most of European countries.

The technical condition of distribution networks is a basic factor determining development potential of photovoltaics and wind energy in many European countries.
The distribution system in many EU countries needs to be reinforced by creating new lines, or by increasing their cross-sections, which requires financial expenditure for development of smart grids, network management technologies, and interconnections.

Modernization of the existing network in all EU countries allows for optimal use of solar PV systems and wind power, reduction of network load, minimization of blackout and emergency threats.

2.3 Economic barriers

Capital constraints are an important economic barrier to solar PV and wind energy development, occurring primarily in case of households and small and medium-sized enterprises.
Relatively high initial costs of the solar PV, and wind energy insufficient economic mechanisms, in particular fiscal mechanisms, which would enable obtaining appropriate financial benefits in relation to the amount of investment outlays on facilities, installation costs, equipment intended for the production of energy from PV and wind, may be the most important barriers to their rapid development.

The policy of the European Union and the member states supports projects in the field of RES installations. To increase the number of consumers willing to engage in prosumer activity various types of incentives, subsidies, low-interest loans were introduced, which shorten payback periods and increase the attractiveness of the investment opportunity. In 2015, more than 1300 support measures (economic, financial, regulatory, administrative) for the development of renewables were in place in the EU countries.[4] At the same time, more and more PV projects are being implemented without support. In many EU countries solar PV investments are competitive, and can be justified based upon market revenues alone without the need for additional subsidies.

2.4. Social and environmental barriers

The level of investment in photovoltaics and wind energy is influenced also by social factors, environmental and energy consciousness of society, approach to energy saving, and the development of renewable energies.
Solar PV investments are generally accepted by local communities, and social protests associated with planned investments, are rare. The approach of societies to photovoltaics is shaped by the ecological and renewable nature of this form of energy, i.e., clean energy, whose use in local and global dimensions, brings several social and environmental benefits.

Wind energy projects have a wider landscape impact than PV installation. Wind turbines may result in effects, such as: shadow flicker, audible noise, low-frequency noise, electromagnetic fields, and many others.

Social acceptance can have a significant impact in respect of implementing wind energy, in many cases, lack of local political – decision-makers, stakeholders, populations i.e., local citizens, household organisations,

[3] Lewis, P., Granroth-Wilding, H., Napolitano, L., Zabala, C., Vékony, A., Felsmann, B., & Hirschbichler, F. (2021). European Barriers in Retail Energy Markets Project: Final Report. Luxembourg: Publications Office of the European Union. ISBN 978-92-76-30269-8, doi:10.2833/5217.

[4] Banja M., Jégard M. Monforti-Ferrario F., Dallemand J.-F.Taylor N., Motola V., Sikkema R.Renewables in the EU: the support framework. towards a single energy market.EU countries reporting under Article 22(1) b, e and f of Renewable Energy Directive. JRC 2017.

and individual end-users' acceptance, can limit the overall wind energy resource. Social acceptance of wind energy is a potential barrier to faster expansion of onshore wind energy, and appears to be more challenging, than in respect of solar PV energy.

Social opposition to new energy infrastructure development is present in many countries. The opposition, ranging from spontaneous neighbourhood protests to professional campaigns, and legal suits at the national level, is perceived as a barrier to wind energy growth.

It should be noted that in recent years, significant technological progress has been made in the development of wind energy production. As a result of this, the negative impact of wind farms on the environment has been significantly reduced. This progress relates to the implementation of modern construction technologies of wind farms and methods of their operation, as well as the development of software, in order to optimise the location and estimate their potential impact on the environment, etc.

Photovoltaic systems are more environmentally friendly, compared to wind farms. PV installations do not emit greenhouse gases into the atmosphere, do not cause noise or vibrations. They do not cause any negative influence on landscapes, and a large part of the local residents and local government institutions accept photovoltaics.

2. ANALYZING THE IMPACT OF RES LARGE PENETRATION FOR GRID CONGESTIONS

The objective of this Test Case was to analyse the technical and economic impacts of massive penetration of PV self-consumption. In particular, this Test Case evaluated PV installations in Poland that has seen a huge increase of micro PV installations– from 50 000 at the end of 2018 to 845 505 at the end of 2021 with further growth expected.

The most promising distributed solar PV solutions identified in Poland (like homeowner, company and municipal organisation) with assumed range of PV sizes and energy storages were evaluated by analysing proposed criteria:

Economic:

- The estimation of the cost of the solutions (investments, operation and maintenance costs, etc.);
- Impact of net – metering;
- The identification of the economic flows;
- The revenue and profitability model, (IRR, NPV, payback);
- The calculation of the LCOE&LCOS of the solution;

Technical:

- Solar generation;
- Energy flows (energy bought from the grid, consumed from net metering, consumed from storage systems, consumed from panels, fed-in energy lost);
- Reduction of carbon emissions;
- Cycles of the storage system consumed;
- Degree of self-sufficient;

Test conditions used were the following:

- Radiation profiles, PV generation and consumption data, specific characteristics of the demand profile;
- Real historical and current data from Polish markets (electricity tariffs and prices, economic and technical data, storage system);
- Real net-metering scheme;
- Real prices of PV panels, PV installations, energy storage;
- Inclusion of replacement costs, corporate taxes, inflation rate, charge and discharge rate;
- Assumed ranges of PV installations in selected scenarios;
- Number of analysed solutions, kind of prosumers;
- Range of PV sizes;
- Net – metering scheme;
- Investment and maintenance costs ("what – if" scenarios possible).

A European CIGRE Benchmark Network was used to determine any technical issues (line congestions, voltage excursions etc.) in running the scenario based on the use of real demand and generation data. European configuration of CIGRE MV Network used in simulation was described by the following parameters:

- Structure – three phase feeders (mesh or radial structure), nominal voltage 20kV, frequency 50Hz;
- Symmetry – loads and feeders are phase balanced;
- Line types – overhead lines are bare conductors of aluminium with or without steel reinforcements.

A software tool OrigAMI developed by Enea Operator was used for aggregation, analyses and reporting of data collected by AMI smart meters. The data was obtain using advanced metering infrastructures data. OrigAMI system allows to export 15 minutes profiles reports of active and reactive power for each smart meter.

For the purpose of the analysis, following profiles were generated:

- Generation profiles (2 x photovoltaics farms, 4 x wind farms);
- Loads - different types, described in the table 1.

Table 1 Selected profile parameters

		Active Power				Reactive Power			
type	object	max E [kWh]	max P [kW]	percentile 95% E[kWh]	percentile 95% P[kW]	max Q [kvarh]	max Q [kvar]	percentile 95% Q[kvarh]	percentile 95% Q[kvar]
wind	Wind 1	1530	6131	1170	4675	136	554	88	354
	Wind 2	1472	5886	1313	5252	117	468	1	4
	Wind 3	1406	5625	1390	5558	69	277	1	4
	Wind 4	1110	4438	932	3730	187	746	124	497
solar	PV 1	231	538	150	600	14	57	6	25
	PV 2	195	782	141	562	15	59	9	36
load	Bus charging station	131	524	74	296	1	4	0	2
	Industrial 1	20	81	8	34	5	15	1	4
	Industrial 2	15	61	11	44	6	25	3	12
	Substation MV/LV Farm	39	157	8	33	29	116	1	3
	Substation MV/LV Estate Modern	10	42	7	29	-2	-6	-2	-7
	Substation MV/LV Estate 2	51	205	30	121	-18	-73	-17	-67
	Substation MV/LV Mall	100	400	81	322	29	114	18	74
	Substation MV/LV Railway	3	14	1	6	2	7	1	3
	Substation MV/LV1	15	61	11	44	6	25	3	12
	Substation MV/LV2	23	93	13	52	8	31	3	10
	Substation MV/LV3 rural	18	72	11	42	4	16	2	7
	Substation MV/LV4 rural	3	10	1	5	1	5	0	1
	Substation MV/LV5 rural	9	36	5	21	3	14	1	3
	Substation MV/LV6 rural	32	128	23	94	11	45	4	17
	Substation MV/LV7 city	51	205	30	121	18	71	17	67
	Substation MV/LV8 rural	23	95	13	52	7	26	2	10

Example profiles in 15 minutes intervals were provided for one year for:

- Wind Farms;
- Photovoltaics farms;
- Industrial loads;
- Building loads;
- Railway loads;
- Shopping mall loads;

- Rural demand loads.

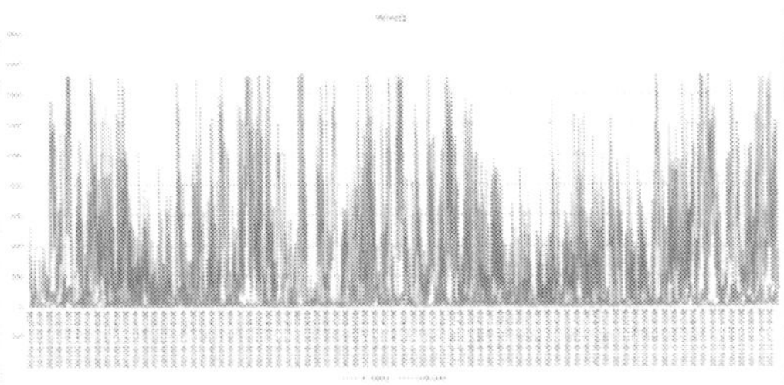

Fig. 1 Wind 1 – visualisation of year profile

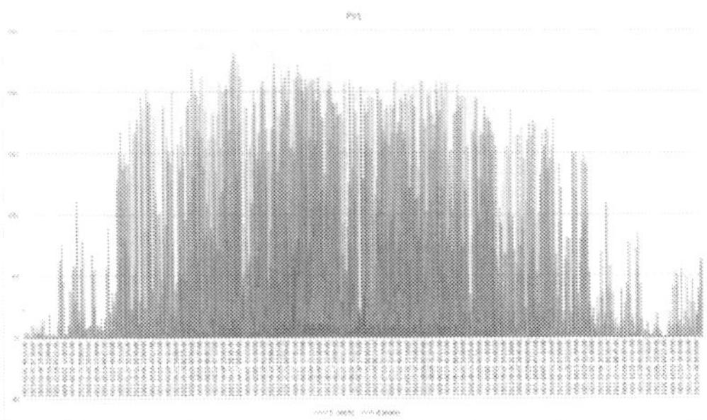

Fig. 2 PV 1 – visualisation of year profile

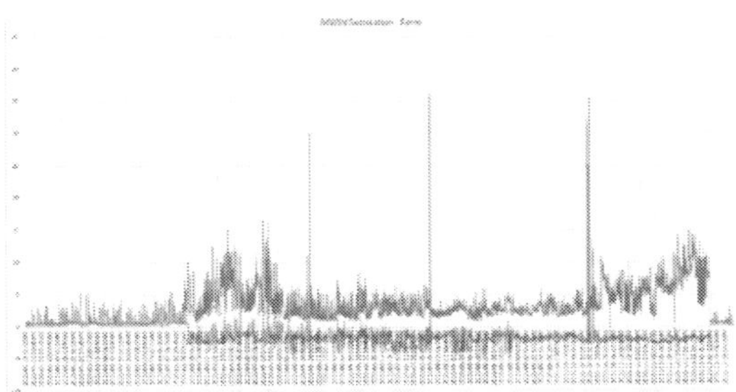

Fig. 3 Load – MV/LV Substation - Farm - visualisation of year profile

A multiple scenario concept has been proposed to study the impact of RES large penetration for grid congestions. Each scenario consisted of different configuration of power generation which was reflecting different ratios of scaling factor parameter which acts directly on energy profiles by scaling the values (expressed in individual values of kWp of generation and load). Scaling factor equals 1 reflects base input values and another specific scaling factors have been chosen to effectively determine the impact of RES on the grid.

The main result from conducting analyses in Power Factory for each scenario was line loading in %. Level of 100% should be considered as moment in time when the network protections are triggered causing the power supply to be cut off and thus saving the equipment and infrastructure elements from irreversible damage.

Topology of CIGRE network with proposed specific and permanent location of loads and generation sources caused maximum load of Line 1-2 in each scenario, which is considered as "critical line".

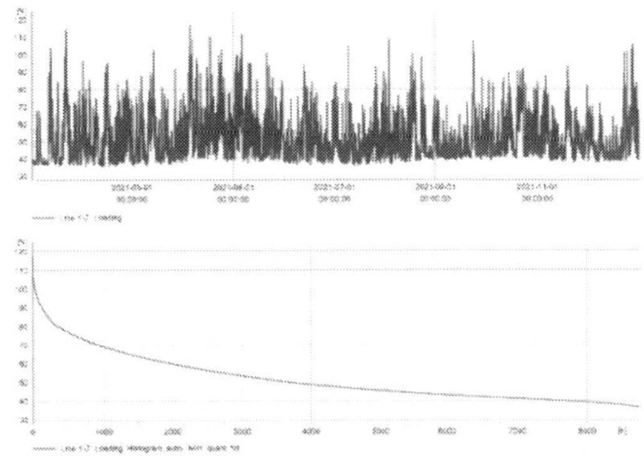

Fig. 4. Line 1-2 year's loading and histogram chart

Conclusions from Test Case:

- For each scenario maximum load is for the line 1-2, directly behind the transformer;
- Loads of particular line differs due to localisation of renewable generation and loads entities and its profiles;
- Comparing different scenarios it's visible that the more RES generation the higher maximum load value in year period (for specific 15 minute value);
- The dates in one particular scenario are different for analysed lines (maximum values occurs in different moments in time for different "places" of network). It's because specific RES and loads localisations in the network topology and their profiles of work.

3. CONCLUDING REMARKS

The transformation of the energy sector towards more sustainable electricity production increases the importance of distributed generation from renewable sources, such as solar, photovoltaics, and wind energy.

Achieving EU energy goals and climate targets requires overcoming existing barriers and regulatory frameworks. Several barriers have been identified that hamper the development of PV and wind energy implementation, including regulatory, technical, administrative, financial and social.

For faster development of wind energy and photovoltaics, regulatory changes are recommended, increasing flexibility of the distribution system and adapting it to a higher share of energy from renewable sources.

Distribution network operators must prepare for further rapid development of micro-installations and facilitation of the process of connecting new sources. Network management must be planned carefully, with accurate prediction of RES generation.

The technical condition of distribution networks – stable, secure, and high capacity is a basic factor for high levels of electricity to be generated from PV and onshore wind.

In general, increased levels of RES have positive effects on the local decarbonisation of distribution networks and local markets.

ACKNOWLEDGEMENTS

This paper is based on the work carried out in the EU project :DRES2Market: Technical, business, and regulatory approaches to enhance renewable energy capabilities to take part actively in the electricity and ancillary services markets", a European Union's Horizon 2020, grant number 952851.

The authors want to thank the project members for valuable contributions and fruitful discussions, in this to: L. Dolera from Asociacion de Empresas de Energias Renovables; Y-M. Bourien from Commissariat À L'énergie Atomique et Aux Énergies Alternatives; M. Garcia Mora, G. B. Alcubilla and M. I. Gonzalez Barreiro from Acciona; R. Gogou from National Technical University of Athens; E. Doheijo and S. De Palacios from Deloitte SL.; and M. Prodanovic from IMDEA.

REFERENCES:

[1] DRES2Market: Technical, business, and regulatory approaches to enhance renewable energy capabilities to take part actively in the electricity and ancillary services markets", a European Union's Horizon 2020 - LC – SC3 – 2020 - RES-IA-CSA, granted research program project, grant number 952851.

[2] Barriers for large integration of renewable energies in electricity and ancillary services markets. Deliverable D2.3, Technical, business, and regulatory approaches to enhance renewable energy capabilities to take part actively in the electricity and ancillary services markets. Horizon 2020 - LC – SC3 – 2020 - RES-IA-CSA. July, 2021.

[3] M. Banja, M. Jégard, F. Monforti-Ferrario, J.-F Dallemand,N. Taylor, V. Motola and R. Sikkema. 2017. Renewables in the EU: the support framework. towards a single energy market.EU countries reporting under Article 22(1) b, e and f of Renewable Energy Directive, EUR 29100 EN, Publication Office of the European Union, Luxembourg, 2017, ISBN 978-92-79-79361-5, doi10.2760/521847, JRC110415.

[4] H. Bartoszewicz-Burczy, P. Starzyński, and M. Wachowiak. 2019. "Solar PV on the Distribution Grid: Smart Integrated Solutions of Distributed Generation based on Solar PV, Energy Storage Devices and Active Demand Management." In 36th European Photovoltaic Solar Energy Conference and Exhibition, edited by F. Lambert, R. Kenny, and P. Helm.1994 – 1999.

[5] European Commission. 2018. A Clean Planet for All a European Long-Term Strategic Vision for a Prosperous, Modern, Competitive and Climate Neutral Economy. Communication COM(2018) 773 final. Brussels: European Commission 28.11.2018. Accessed August 12, 2022. https://eurlex.europa.eu/legalcontent/EN/TXT/PDF/?uriE LEX:52018DC0773&from=EN

[6] European Commission. 2022 European Green Deal. . Accessed August 12, 2022. https://ec.europa.eu/clima/eu action/european-green-deal_en. Accessed February, 1 2022.

[7] European Commission. 2022b. *A European Green* Deal. Striving to be the first climate-neutral continent. https://ec.europa.eu/info/strategy/priorities-2019 2024/european-green-deal_en. Accessed August 12, 2022.

[8] IEA: High penetration of PV in local distribution grids: Subtasks 2 case study collection. Photovoltaic Power Systems Programme. Report IEA PVPS T14-02:2014. ISBN: 978-3-9060421-23-7.

[9] P. Lewis, H. Granroth-Wilding, L. Napolitano, C. Zabala, A. Vékony, B. Felsmann, & F. Hirschbichler, (2021). European Barriers in Retail Energy Markets Project: Final Report. Luxembourg: Publications Office of the European Union. ISBN 978-92-76-30269-8, doi:10.2833/5217.

A COMPARATIVE LIFE-CYCLE ASSESSMENT OF RENEWABLE ENERGY FROM HIGH EFFICIENCY SOLAR PHOTOVOLTAIC TECHNOLOGIES

Andrea Danelli, Elisabetta Brivio
RSE SpA - Research on Energy System, Sustainable Development and Energy Sources
Via R. Rubattino 54, 20134 Milano, Italy

ABSTRACT: This paper presents a comparative life-cycle assessment of photovoltaic (PV) electricity generation by various mono-crystalline silicon (mono-Si) PV technologies; in particular, three ground mounted PV systems using different high-efficiency technologies (cell and module) are explored: (1) bifacial HJT (Heterojunction Technology) modules; (2) PERC (Passivated Emitter and Rear Cell) modules; and (3) IBC (Interdigitated Back Contact) modules. Furthermore, the LCA study – undertaken in accordance with the ISO 14040 and 14044 standards, and IEA-PVPS guidelines on the matter – concerned a hypothetical 85 MW power system in two possible plant configurations: i) ground mounted on a mono-axial tracker; ii) ground mounted on a fixed structure. The goal is to compare the different configurations and to identify, if any, the optimal solution. To assess the influence of the incident solar radiation value on the results obtained from the LCA analysis, two possible locations were considered, one in the north and the other in southern Italy. Results show that the greenhouse gas emissions of a PV system based on high-efficiency technologies, ranging from 10.5 to 18 g CO_2eq/kWh, are significantly lower over the conventional Al-BSF technology. As regards plant configurations, the installation on a mono-axial solar tracker generates the least environmental impact for all the impact categories taken into consideration. As a matter of fact, the increase in energy production due to the mono-axial tracker generates environmental advantages that exceed those connected with the tracker production. The key finding of this study is that the improvements in the modules' efficiency result in lower environmental impacts when compared to traditional silicon PV systems. The analysis relies on primary data for cells, modules, inverter and trackers production.
Keywords: LCA, carbon footprint, HJT, PERC, IBC

1 INTRODUCTION

Solar Photovoltaic (PV) has seen stable growth in recent years and will dominate the renewable energy market in the coming future [1]. This exceptional growth is due to the fact that solar energy is seen as key factor for meeting world growing electricity demand, still reducing greenhouse gas emissions from the electricity sector [2]. The PV growth has been pushed by strong policy support; in Italy, for example, the National Energy and Climate Plan (NECP) foresees for the national photovoltaic sector a cumulative installed capacity of 52 GW in 2030 with an annual electricity production of 73 TWh [3]; this means that at least 3 GWh/year should be installed between now and 2030 (see Fig. 1).

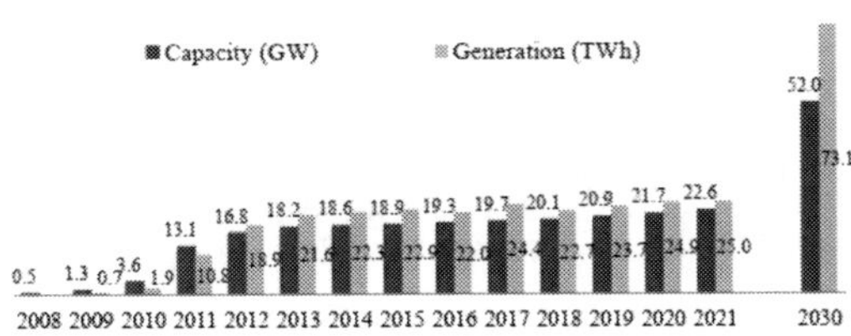

Figure 1: Trend of growth over the years for the PV sector in Italy and target for 2030 (source GSE)

However, in order to reduce costs and to further decrease the Greenhouse Gasses (GHG) emission, PV systems have to increase their efficiency and their lifetime [4]. In this framework, the emerging PV technologies (i.e bifacial HJT, PERC and IBC cells) characterized by higher efficiency compared to conventional crystalline silicon solar cells, are occupying ever larger market shares [5]. Of course, for an effective transition toward a sustainable energy sector, the GHGs directly emitted during the generation of electricity should be tracked, but also all the indirect emissions associated with construction, upstream processes and decommissioning. Moreover, the gamut of possible environmental life cycle impacts should be considered [6]. A good, standardized method to reach this goal is the Life Cycle Assessment (LCA) [4] in accordance with ISO 14040 [7] and ISO 14044 [8].

In this framework, this paper presents the LCA of innovative photovoltaic solutions based on bifacial HJT, PERC and IBC modules. Two possible plant configurations are analyzed: ground mounted on a mono-axial tracker; ground mounted on a fixed structure.

To assess the influence of the incident solar radiation value on the LCA results, two possible locations are considered, in the north (Piacenza) and in south of Italy (Catania). In comparison to other studies based on secondary data from previous analysis, the authors have been able to work closely with the manufacturers of cells, modules, inverters, trackers and with photovoltaic systems operators. The resulting LCA is therefore based on primary data rather than just literature information.

2 GOAL AND SCOPE DEFINITION

The main goal of this study is the assessment of the potential environmental impacts related to the electricity generation by high efficiency PV technologies. The study aims to evaluate the environmental impacts at PV plant level, considering its entire life cycle (from realization to dismantling), and above all, it wants to highlight the components and processes mainly involved in the generation of environmental impacts, such as greenhouse gas emissions and raw material consumption.

2.1 Method

The main rules and principles that should be carefully followed in an LCA study are specified and detailed in the ISO 14040 and 14044 standards; in the present study, besides having complied with the forementioned standards, we have followed the recommendations from the "Methodology Guidelines on Life Cycle Assessment of Photovoltaic" by IEA-PVPS [9]. The ISO standards, in fact, provide the general framework for LCA, leaving the individual practitioner with a range of choices that can affect the results and thus the conclusions of an LCA study. An LCA study on photovoltaic (PV) electricity generation systems in compliance with IEA-PVPS guidelines assures consistency, quality, credibility and reliability to the obtained results.

The potential environmental impacts have been evaluated by resorting to the midpoint indicators of the Product Environmental Footprint Category Rules (PEFCR). The only exception is the ionizing radiation category which is omitted as it is not relevant for the system investigated. The indicators are listed in Table I.

The attributional approach [10], which is the most prevalent type used in the published LCAs [6], has been adopted.

Table I: List of impact categories addressed in this PV LCA study, together with their indicator, unit and source

Impact category	Indicator	Unit	LCIA method	Robustness
Climate change	Radiative forcing as Global Warming Potential (GWP100)	Kg CO_2 eq	Baseline model of 100 years of the IPCC [11]	I
Ozone depletion	Ozone Depletion Potential (ODP)	Kg CFC-11 eq	Steady-state ODPs 1999 as in WMO assessment [12]	I
Photochemical ozone formation	Tropospheric ozone concentration increase	Kg NMVOC eq	[13] as applied in ReCiPe	II
Respiratory inorganics	Impact on human health	disease incidence	PM method recommended by UNEP [14]	I
Acidification	Accumulated Exceedance (AE)	mol H+ eq	Accumulated Exceedance [15], [16]	II
Freshwater eutrophication	Fraction of nutrients reaching freshwater end compartment (P)	kg P eq	EUTREND model [17] as implemented in ReCiPe [14]	II
Marine eutrophication	Fraction of nutrients reaching marine end compartment (N)	kg N eq	EUTREND model [17] as implemented in ReCiPe [14]	II
Terrestrial eutrophication	Accumulated Exceedance (AE)	mol N eq	Accumulated Exceedance [15], [16]	II
Land use	Soil quality index[i]	Dimensionless (Pt)	Soil quality index based on LANCA model	III
Resource use, energy carriers	Abiotic resource depletion – fossil fuels (ADP-fossil)	MJ	CML 2002 [18] and [19]	III
Resource use, minerals and metals	Abiotic resource depletion (ADP ultimate reserves)	kg Sb eq	CML 2002 [18] and [19]	III

2.2 Functional Unit

The functional unit (FU) is the quantified performance of a product system, to be used as a reference unit (e.g., the FU of paint could be described as providing protection of 1 m^2 of substrate for 50 years with a minimum 98% opacity) [7]. In synthesis, the FU represents the unit to which all inputs and outputs of the studied system refer to.

In accordance with the methodology guidelines by IEA-PVPS [9], we assume 1 kWh of AC electricity produced by the plant as functional unit. This choice allows to make comparison among PV technologies, module technologies, and electricity-generating technologies in general.

2.2 System boundary

As regards the boundaries of the analyzed system, the modules, the tracker and the inverter have been considered as components of the PV plant, while the connections to the network are not included in the analysis. The construction phases, the operation of the plant and its dismantling (only the end-of-life of modules) are comprised in the assessment. This means that inputs of raw materials, energy supply, manufacturing of panels, mounting system, cables, inverters, and all other components needed to produce electricity [2] have been considered. To deal with the operating phase, we have considered the replacement of the inverters after 17 years and the substitution of tracker actuators (the latter based on the probability of failure). Only the installation of the tracker and of the mounting system (for which primary data from the manufacturer were available) was included for the in-situ construction phase. The system boundaries adopted in this LCA study are outlined in Fig. 2.

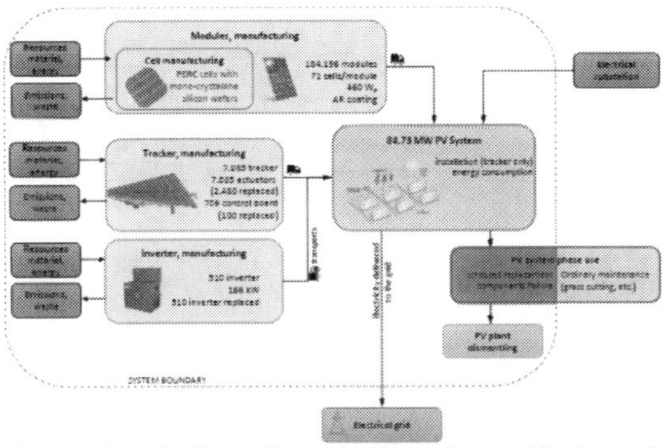

Figure 2 : Outline of system boundary with the unit processes included in the LCA study (PERC technology)

3 LIFE CYCLE INVENTORY

The LCA study considered a hypothetical photovoltaic plant, equipped with innovative technologies and solutions.

The photovoltaic system has a peak power of 84.73 MW and is based on high efficiency technologies: bifacial heterojunction (HJT), PERC and IBC modules among them as an alternative. High efficiency technologies aim at reducing the costs of construction per power unit installed and at optimizing the management of PV plants, improving their energy efficiency, and extending their useful life. All these improvements are expected to also have a positive impact in terms of environmental sustainability [6]. Furthermore, this LCA study considered the use of:

- latest generation of bifacial heterojunction / PERC / IBC modules with the use of thin cells;
- mono-axial tracker manufactured with long-lasting and low-cost materials like weathering steel;
- high efficiency inverter (> 99%);
- new modules treatment, including anti-reflective coating, to ensure a useful life of 35 years.

The quantity and the main characteristics of the PV components taken into account in this study are shown in Table II. The general data regarding the cells and modules for the three different scenarios are reported in Table III, while the characteristics of trackers and inverters are shown in Table IV.

In this LCA study, two different PV system configurations are analysed, with or without tracker, and two sites in Italy. Trackers allow to increase the energy production from PV plant. However, the tracker production and assembling requires increased raw materials and energy compared to fixed mounting systems.

Furthermore, the influence of the irradiation levels, derived from the PVGIS database [20], on the LCA was explored by selecting two different sites (Table V): one in the north of Italy (Piacenza) and one in the south (Catania). The lifetime of the system, which is relevant parameter in PV LCAs [4] is set to 35 years, as indicated by the manufacturer of the photovoltaic modules. Table V shows the different configurations of the photovoltaic system analysed in this study.

In the inventory phase, both primary data (provided by some manufacturers) and secondary data from Ecoinvent database v3.3, cut-off system model [21] have been used. It is worth mentioning that for the silicon wafer production [22] used in the PC cells, we referred to the global market [21]. The manufacture of silicon wafer is a highly energy demanding process [23], [24] due to the Czochralski method of growing silicon ingots.

Table II : Number of components for the realization of a 84.73 MW PV plant

	Bifacial HJT	PERC	IBC
Module Peak Power	400 W	460 W	430 W
N. of cells per module	72	72	66
N. of modules	211.825	184.196	197.047
Type of supporting system	fixed tilt or tracker realized in weathering steel (26 modules each, 13x2 portrait)		
N. supporting system	8148	7085	7579
Typo of inverter	nominal power of 166 kW		
N. of inverter	510	425	425

Table III : general data about the modules considered in this LCA study

	HJT	PERC	IBC
Cells	72 bifacial cells	144 PERC half-cut cells	132 IBC half-cut cells
Dimension (mm)	1983 x 998 x 7	2108 x 1048 x 35	1895 x 1039 x 35
Area (m^2)	1.979	2.209	1.969
Life (years)	35	35	35
Weight (kg)	31	24.3	21
Nominal power (W)	400	460	430
Efficiency	20.21%	20.82%	21.84%
Additional note	frameless, glass-glass module		

Table IV : general data about tracker and inverter

	Tracker	Inverter
Typology	Mono-axial tracker for 26 modules in portrait configuration supported by 3 poles. Made of Corten steel	166 kVA Inverter (two stage)
Dimension	14 x 4.2 m	
Life (years)	35	17
Weight (kg)	622	100
Producer	Convert Italia SPA	REFU Elektronik

Table V: List of the PV solutions considered in the LCA study, with the related global horizontal irradiance (GHI) and the amount of electricity production in entire useful life

Module	PV plant config.	Location	GHI kWh/m²/y	Production GWh
HJT	single axis tracker	Catania	1.819	6.895
HJT	fixed tilt	Catania	1.819	5.561
HJT	single axis tracker	Piacenza	1.368	4.992
HJT	fixed tilt	Piacenza	1.368	4.341
PERC	single axis tracker	Catania	1.819	6 273
PERC	fixed tilt	Catania	1.819	5 063
PERC	single axis tracker	Piacenza	1.368	4 615
PERC	fixed tilt	Piacenza	1.368	4 008
IBC	single axis tracker	Catania	1.819	6 259
IBC	fixed tilt	Catania	1.819	5 055
IBC	single axis tracker	Piacenza	1.368	4 615
IBC	fixed tilt	Piacenza	1.368	4 011

The amount of electricity production over the lifetime of the analysed PV plant configurations considerably affects the LCA results. Core data for the assessment of this crucial parameter are both the annual irradiation (i.e., GHI) at the installation site and the technical features of the PV system such as: inverter and module efficiency rate, mounting and supporting structures of modules [25]. The lifetime electricity production has been determined by coupling the PVGIS tool developed by JRC [20] with an Excel spreadsheet; this latter has been prepared by amending the original version proposed in GOPV project[1]. Through this method, starting from the data related to the geographic coordinates of installation, PV technology and the mounting system adopted, it is possible to determine the irradiation incident on the PV modules surface first, and then the electricity production in the first year of operation of the plant. Assuming a linear degradation of 0.4% per year, as declared by the module producers, we calculated the electricity production over the lifetime of the PV plant. Table V shows lifetime production for the analysed configurations. It is worth to underline that the degradation rate used in this LCA study is a bit lower than the figures reported in the IEA-PVPS guidelines (linear degradation of 0.7 %-points per years) [26] and in a NREL report [27].

4 RESULTS

To better understand the advantages introduced thanks to the use of innovative photovoltaic technologies instead of conventional energy technologies, the results obtained are compared with the impact generated by the Italian energy mix (Fig. 4 and Table VI). The energy mix reflects the energy situation in 2019 (Fig. 3). In Fig. 4 the impacts were normalized to the highest score in each category. The photovoltaic plants are in Catania and are characterized by a lifetime of 35 years. The modules are placed on a mono-axis solar tracker.

The outcomes reveal that innovative photovoltaic technologies allow a reduction of more than 70% in all the impact categories except for freshwater eutrophication and resource use, mineral, and metals. In the latter one, the use of noble metals, e.g. silver, in cell production negatively affects the results. Among the three technologies analysed, the HJT plant appears the most beneficial, from the environmental point of view thanks to the lower impact generated by module production.

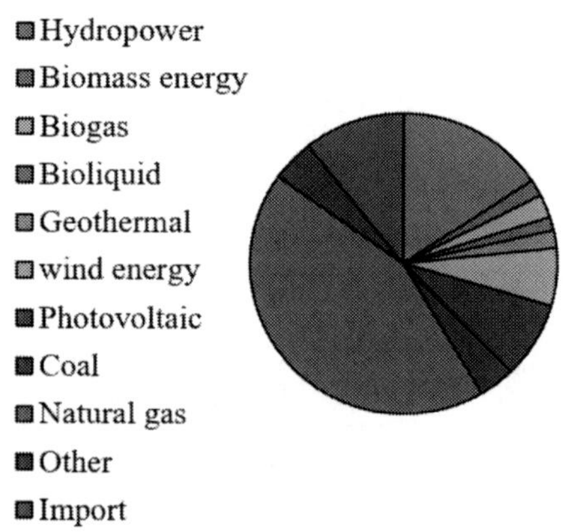

Figure 3: Italian energy mix 2019 [25]

[1] The GoPV (Global Optimization of Integrated Photovoltaic System for Low Electricity Cost) research project has received funding from the European Union's Horizon 2020 research and innovation programme under grant agreement No 792059. https://www.gopvproject.eu/

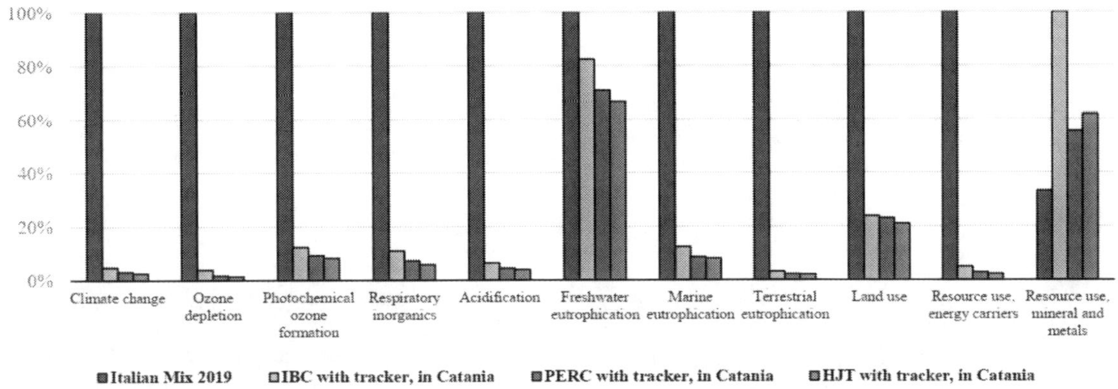

Figure 4: Comparison of the potential environmental impact associated with the three configurations of photovoltaic plants and the Italian energy mix 2019. The best solution is set equal to 100%

Table VI: Environmental impact relating to the three photovoltaic plants analyzed (IBC plant; PERC plant and HJT plant) located in Catania and to the Italian energy mix. The results are referred to the FU

Impact category	Mix 2019	IBC CT tracker	PERC CT tracker	HJT CT tracker
Climate change (kg CO_2 eq)	3.84E-01	1.82E-02	1.21E-02	9.88E-03
Ozone depletion (kg CFC11eq)	4.39E-08	1.71E-09	7.20E-10	6.02E-10
Photochemical ozone formation (kg NMVOCeq)	4.52E-04	5.60E-05	4.20E-05	3.77E-05
Respiratory inorganics (disease inc.)	1.10E-08	1.21E-09	8.13E-10	6.54E-10
Acidification (mol H+ eq)	1.86E-03	1.21E-04	8.31E-05	7.31E-05
Freshwater eutrophication (kg P eq)	1.02E-05	8.44E-06	7.26E-06	6.83E-06
Marine eutrophication (kg N eq)	2.03E-04	2.53E-05	1.76E-05	1.64E-05
Terrestrial eutrophication (mol N eq)	6.13E-03	1.98E-04	1.40E-04	1.28E-04
Land use (Pt)	3.79E+00	9.03E-01	8.75E-01	7.95E-01
Resource use, energy carriers (MJ)	4.69E+00	2.31E-01	1.33E-01	1.06E-01
Resource use, mineral and metals (kg Sb eq)	2.46E-07	7.40E-07	4.11E-07	4.58E-07

In Fig. 5 the contribution analysis regarding the Climate Change, Freshwater eutrophication and Resource use, mineral and metals is shown. The three categories selected are considered the most representative for the objective of this study.

Focusing on the contribution analysis, it can be observed that, in all the technologies, the main contribution is associate with module production (from the raw material extraction to the module production) which ranges from 34% (PERC module – Resource use, mineral and metals) to 66% (IBC module – Climate Change). More specifically the energy intensive process and chemicals involved in wafer manufacturing increase the environmental burden of this component (Fig. 5). The wafer is produced through the Czochralski process which represents the main energy demand and the main responsible for CO_2-eq emissions (covering about 50% of the overall emissions). It's worth noting that most of the production of silicon wafer is realized in China, where the share of non-renewable resources is still significant in the energy mix. Producing silicon wafer in areas characterized by higher penetration of renewables in the electric mix will further reduce the GHG emissions per kWh of PV plants.

In the IBC cell production, N_2O is also considered, the chemical could contribute to reduce the LID phenomenon, however, the real advantages are still under investigations. Beside module manufacturing also inverter adversely influences the freshwater eutrophication and resource use, mineral and metals categories due to the use of noble metals in the production process. The eutrophication category is also affected by the metals involved in the tracker manufacturing and more specifically the materials used to produce tracker electronics. Finally, the impacts generated by the components transport and land use and the dismantling and treatment of the plant (End of Life) are negligible.

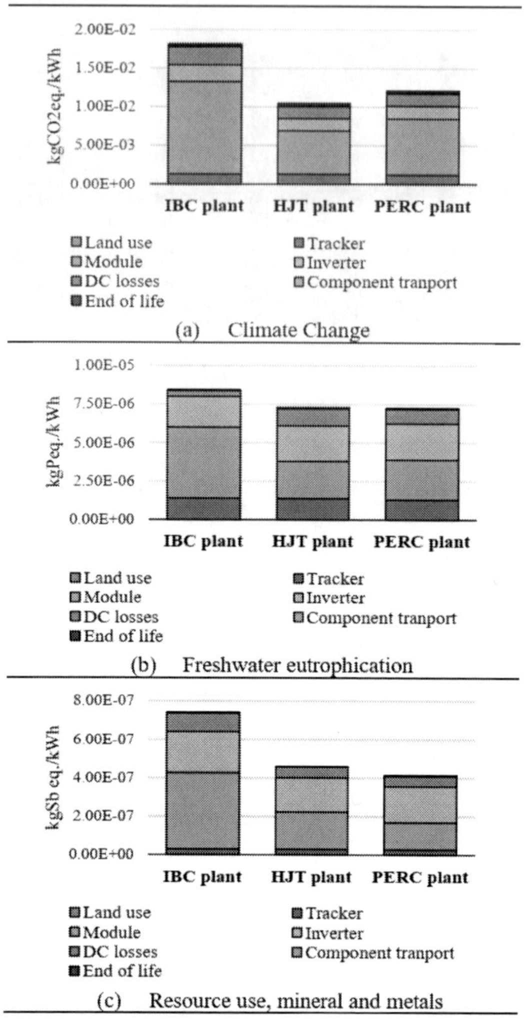

Figure 5: Component contribution for the (A) Climate change, (B) freshwater eutrophication and (C) Resource use, mineral and metals categories.

5 CONCLUSIONS

The innovative photovoltaic technologies represent a good solution to deal with environmental problems and achieve the ambitious European target. The analysis of the environmental impact provides a theoretical overview of the advantages and drawbacks introduced by the use of the photovoltaic modules: IBC, PERC, and HJT. The comparison between the three PV plants and the Italian energy mix highlights the significant benefits arising (a reduction of more than 70% in most of the categories) from the use of solar energy instead of other technologies. However, module production and more specifically the wafer production turn out a critical element causing an increase in the environmental burden. The analysis also reveals that for inverters and trackers (whose contributions are however far less than that of modules) use phase energy losses and steel production are respectively the main relevant processes.

6 ACKNOWLEDGMENTS

This work has been financed by the Research Fund for the Italian Electrical System under the Contract Agreement between RSE S.p.A. and the Ministry of Economic Development – General Directorate for the Electricity Market, Renewable Energy and Energy Efficiency, Nuclear Energy in compliance with the Decree of April 16th, 2018.

7 REFERENCES

[1] IEA, «Solar - Fuels & Technologies - IEA», 24 novembre 2020. https://www.iea.org/fuels-and-technologies/solar (consultato 23 febbraio 2021).

[2] N. A. Ludin et al., «Prospects of life cycle assessment of renewable energy from solar photovoltaic technologies: A review», Renewable and Sustainable Energy Reviews, vol. 96, pagg. 11–28, nov. 2018, doi: 10.1016/j.rser.2018.07.048.

[3] Ministry of Economic Development, «Integrated National Energy and Climate Plan», Rome, 2019.

[4] V. Muteri et al., «Review on life cycle assessment of solar photovoltaic panels», Energies, vol. 13, n. 1, 2020, doi: 10.3390/en13010252.

[5] J. Zhao et al., «>24% Silicon Heterojunction Solar Cells on Meyer Burger's on Mass Production Tools and How Wafer Material Impacts Cell Parameters», in 2018 IEEE 7th World Conference on Photovoltaic Energy Conversion (WCPEC) (A Joint Conference of 45th IEEE PVSC, 28th PVSEC & 34th EU PVSEC), 2018, pagg. 1514–1519. doi: 10.1109/PVSC.2018.8547908.

[6] D. D. Hsu et al., «Life Cycle Greenhouse Gas Emissions of Crystalline Silicon Photovoltaic Electricity Generation: Systematic Review and Harmonization», Journal of Industrial Ecology, vol. 16, n. SUPPL.1, 2012, doi: 10.1111/j.1530-9290.2011.00439.x.

[7] International Organization for Standardization, «ISO 14040: Environmental management–life cycle assessment—Principles and framework», International organization for standardization, vol. 2006, 2006.

[8] International Organization for Standardization, «ISO 14044: Environmental management — Life cycle assessment — Requirements and guidelines», International organization for standardization, vol. 2006, n. 7555, pag. 1418.1, 2006, doi: 10.1136/bmj.332.7555.1418.

[9] G. Heath et al., Methodology Guidelines on Life Cycle Assessment of Photovoltaic Electricity, 3rd edition, IEA PVPS Task 12, 3rd ed. IEA PVPS Task 12, International Energy Agency Photovoltaic Power Systems Programme, 2016.

[10] T. Ekvall, A. Azapagic, G. Finnveden, T. Rydberg, B. P. Weidema, e A. Zamagni, «Attributional and consequential LCA in the ILCD handbook», International Journal of Life Cycle Assessment, vol. 21, n. 3, pagg. 293–296, mar. 2016, doi: 10.1007/s11367-015-1026-0.

[11] IPCC, Climate Change 2013: The Physical Science Basis. Contribution of Working Group I to the Fifth Assessment Report of the Intergovernmental Panel on Climate Change, vol. AR5. Cambridge, United Kingdom and New York, NY, USA: Cambridge

University Press, 2013. doi: 10.1017/CBO9781107415324.
[12] WMO, «Scientific Assessment of Ozone Depletion», Geneva, Switzerland, 1998.
[13] R. van Zelm et al., «European characterization factors for human health damage of PM10 and ozone in life cycle impact assessment», Atmospheric Environment, vol. 42, n. 3, pagg. 441–453, gen. 2008, doi: 10.1016/j.atmosenv.2007.09.072.
[14] M. Goedkoop, R. Heijungs, M. Huijbregts, A. De Schryver, J. Struijs, e R. Van Zelm, «ReCiPe 2008», Potentials, n. May 2014, pagg. 1–44, 2009.
[15] M. Posch, J. Seppälä, J.-P. Hettelingh, M. Johansson, M. Margni, e O. Jolliet, «The role of atmospheric dispersion models and ecosystem sensitivity in the determination of characterisation factors for acidifying and eutrophying emissions in LCIA», The International Journal of Life Cycle Assessment, vol. 13, n. 6, pag. 477, 2008, doi: 10.1007/s11367-008-0025-9.
[16] J. Seppälä, M. Posch, M. Johansson, e J.-P. Hettelingh, «Country-dependent Characterisation Factors for Acidification and Terrestrial Eutrophication Based on Accumulated Exceedance as an Impact Category Indicator (14 pp)», The International Journal of Life Cycle Assessment, vol. 11, n. 6, pagg. 403–416, 2006, doi: 10.1065/lca2005.06.215.
[17] J. Struijs, A. Beusen, D. De Zwart, e M. Huijbregts, «Characterization factors for inland water eutrophication at the damage level in life cycle impact assessment», International Journal of Life Cycle Assessment, vol. 16, n. 1, pagg. 59–64, 2011, doi: 10.1007/s11367-010-0232-z.
[18] J. B. Guinée, «Handbook on life cycle assessment - Operational guide to the ISO standards», International Journal of Life Cycle Assessment, vol. 6, n. 5. Springer Verlag, pag. 255, 2001. doi: 10.1007/BF02978784.
[19] L. van Oers, A. de Koning, J. B. Guinée, e G. Huppes, «Abiotic Resource Depletion in LCA: Improving characterisation factors for abiotic resource depletion as recommended in the new Dutch LCA Handbook», Road and Hydraulic Engineering Institute, n. June, pagg. 1–75, 2002.
[20] T. Huld, R. Müller, e A. Gambardella, «A new solar radiation database for estimating PV performance in Europe and Africa», Solar Energy, vol. 86, n. 6, pagg. 1803–1815, giu. 2012, doi: 10.1016/j.solener.2012.03.006.
[21] G. Wernet, C. Bauer, B. Steubing, J. Reinhard, E. Moreno-Ruiz, e B. Weidema, «The ecoinvent database version 3 (part I): overview and methodology», International Journal of Life Cycle Assessment, vol. 21, n. 9, pagg. 1218–1230, set. 2016, doi: 10.1007/s11367-016-1087-8.
[22] N. Jungbluth e M. Stucki, «Life cycle inventories of photovoltaics», ESU-services Ltd., …, n. 17, pag. 250, 2012.
[23] J. Friedrich, W. von Ammon, e G. Müller, «Czochralski Growth of Silicon Crystals», in Handbook of Crystal Growth: Bulk Crystal Growth: Second Edition, vol. 2, Elsevier Inc., 2015, pagg. 45–104. doi: 10.1016/B978-0-444-63303-3.00002-X.
[24] J. H. Wong, M. Royapoor, e C. W. Chan, «Review of life cycle analyses and embodied energy requirements of single-crystalline and multi-crystalline silicon photovoltaic systems», Renewable and Sustainable Energy Reviews, vol. 58. Elsevier Ltd, pagg. 608–618, 1 maggio 2016. doi: 10.1016/j.rser.2015.12.241.
[25] A. Agrillo, M. del Verme, P. Liberatore, D. Lipari, G. Lucido, V. Maio, V. Surace, «RAPPORTO STATISTICO 2020 - Energia da fonti rinnovabili in Italia». Gestione Servizi Energetici (GSE), marzo 2022.

[i] This index is the result of the aggregation, performed by JRC, of the 4 indicators provided by LANCA model as indicators for land use: Biotic production, Erosion resistance, Mechanical filtration, Groundwater replenishment.

SCIENCE DISSEMINATION IN PHOTOVOLTAICS? ... A "GAME" FOR YOUNGSTERS

M. Canino[1], M. Seri[2], A. Le Donne[4], S. Binetti[4], A. Torreggiani[3]
[1] CNR-IMM - Consiglio Nazionale delle Ricerche, Istituto per la Microelettronica e Microsistemi, Bologna (Italy)
[2] CNR-ISMN- Consiglio Nazionale delle Ricerche, Istituto per lo Studio dei Materiali Nanostrutturati, Italy
[3] CNR-ISOF- Consiglio Nazionale delle Ricerche, Istituto per la Sintesi Organica e la Fotoreattività, Italy
[4] Dipartimento di Scienza dei Materiali and Centro MIBSolar, Università di Milano Bicocca, Italy

ABSTRACT: In this work we present and discuss the learning pathways related to photovoltaic (PV) technology developed by the European project Raw Matters Ambassadors @Schools addressed to pupils aged 10-19. The project has received funding from the European Institute for Innovation and Technology (EIT) since 2016, with the aim of catching the interest of youngsters towards STEM subject and making careers in raw materials sector attractive for youngsters.
The topics treated in the pathways cover the whole raw material value chain. Research topics are presented with a practical slant and connected with some of the most urgent problems felt by youngsters. In this work we focus on 3 different pathways which are of interest for the PV community: the rising price and subsequent need for replacement of Indium Tin Oxide, lifecycle-related issues answered by solar cell recycling, and evolving technologies, such as organic PV. The pathways include hands-on activities in the form of experiment: "ITO vs graphene", showing alternative materials and the issues related to their implementation in industrial processes; "Silicon solar cells recycling", where a solar cell is dismantled; "Solar cells from cranberry juice", dealing with the principle of working and technology of dye sensitized solar cells (DSSCs).
It is apparent that the topics presented in these learning pathways are strictly connected with research and innovation in the field of materials. As these are not part of the background knowledge of pupils of secondary school, the issues in finding a connection with the traditional school curricula and the advantages of including communication actions in the teaching strategy are discussed.
Keywords: Education and Training, ITO, Recycling, Sustainable, Raw Materials

1 INTRODUCTION

The energy transition relies on the secure supply of the materials needed in photovoltaic and battery technologies. The strategies are varied and include intensive research in recycling and substitution of critical raw materials as well as the availability of highly educated professionals trained for solar jobs. Besides, citizen awareness and acceptance are mandatory elements for a successful accomplishment. This paper gives an overview of the methodology developed within the European project "Raw Matters Ambassadors at Schools" (RM@Schools) [1] to raise awareness about RM-related challenges and promote careers in science & technology. The project has been funded since 2016 by the EIT Raw Materials; in 2018 RM@Schools became the Flagship project in the Wider Society Learning segment of the EIT Raw Materials counting, at present, 24 partners from 19 European Countries. The idea on which the project is based, is to propose to schools active learning pathways, involving students in experiments and in science communication activities.

The learning pathways proposed by RM@Schools have a modular structure as described in Figure 1. In this paper we describe the outcome from learning pathways made up of a lesson and an experiment dealing with materials used in PV technologies and sustainable issues. These are presented in classrooms by researchers who act as RM Ambassadors. The experiments are carried out basing on hands-on educational kits developed within the project. The pupils are then asked to become themselves ambassadors by creating a dissemination product inspired by the lesson and experiment. The dissemination products can be multimedia tools, such as videos, comics, websites, as well as interactive tools such as games or experimental kits. They can be used by the students within dissemination activities in which students act as protagonists. Examples can be the European Researchers Night, science fairs, open days in their school, and peer to peer education. The best dissemination products are also presented to the project community (students, teachers, researchers, professionals, and officers) during a European Conference held at the end of each year [1]. The project methodology is intended to address the needs of improving students' understanding of scientific phenomena and to and to reinforce the link between society and research.

Table I: scheme of a learning pathway proposed by CNR, Italy. The columns represent the modules and each module includes independent optional activities. The main actors in each activity are indicated.

	Learning pathway modules			
	Introduction to the RM-related topic	Strengthening/ widening of knowledge	Creation	Communication /dissemination
	Researcher	Researcher/ Teacher	Student	Student
Activity	Lesson	Experiment Visit to the labs Company visits Second lesson	Dissemi nation product Experim ental kit	Society Peer education Science fair EU conference

Although the contents of this dissemination project are focused on the RM value chain, some of the pathways cover topics that are strictly related to the deployment of PV. These include the rising price and subsequent need for replacement of Indium Tin Oxide [2], used in thin film solar cells as well as flat panel displays, and lifecycle-related issues answered by some technologies such as Organic PV [3] and solar cell recycling [4].

2 METHODOLOGY

The didactic material developed within the RM@Schools project is released through Virtual Center, that groups the lessons by categories related to the RM value chain. The material available for each lesson includes: information for teachers; presentation for the classroom, and instructions to carry out laboratory activities.

2.1 ITO vs Graphene

The pathway starts with a lesson named "Don't throw away your mobile", about raw materials in electronic devices. In order to illustrate an example of substitution of critical raw materials the chosen topic is the search for an alternative to the expensive and rare Indium Tin Oxide (ITO) as a transparent electrode in optoelectronic devices, for example in flat screens and solar cells [2]. The proposed experiment consists in switching on a light emitting diode inserted in a circuit made up of the diode itself, a battery, and a resistor; this latter is presented in the form of a glass platelet either uncoated, or coated with transparent conductive layers in different materials. One coating is 100-nm thick ITO; the other coating is single-layer graphene. It turns out that graphene is more resistive than ITO due to the difficulties in manipulating a 2-dimensional material, which can eventually cause rips in it. The experiment setup is shown in Figure 1.

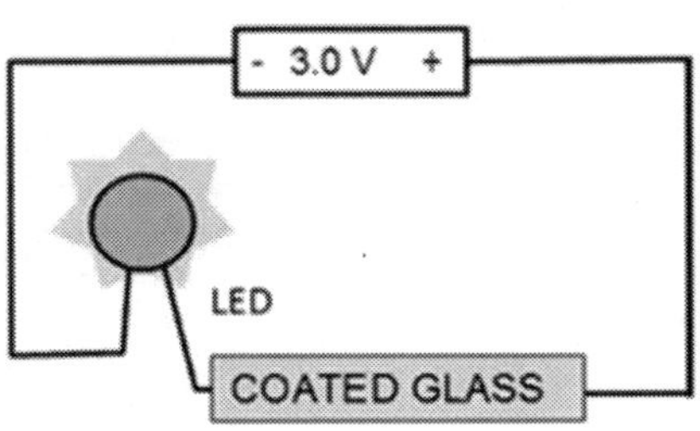

Figure 1. Circuit scheme: battery, LED and resistor (coated glass).

This experiment allows to speak about materials in electronics with Ohm law as the only prerequisite knowledge. Dealing with transparent electrodes allows to avoid spending time on extracurricular concepts like semiconductor theory [5]. Indeed, in Italy, semiconductor science is an optional topic in the background of the last year of secondary school in Lyceum; on the contrary it is a basic knowledge in Technical and Professional Schools, where the main topic is digital electronics. Besides making students get in contact with graphene, a two-dimensional material whose discoverers were awarded the Nobel Prize in Physics in 2010, the experiment modular structure allows for hosting more materials if desired.

2.2 Solar Cells from cranberry juice

The pathway starts with a frontal lesson titled "Solar cells from cranberry juice", introducing the potentialities and specific properties of organic semiconductors (natural or artificial) in the PV field. The aim of the lesson is to introduce these innovative materials that are able to offer new and complementary opportunities/application areas compared to conventional PV systems. The lesson is followed by an experimental part in which, under the supervision of a researcher, the students can fabricate (in classroom or in a school lab) a working DSSC (Figure 2). At the end, it is possible to test the quality of the work by illuminating the resulting device with an artificial light, such as the light of a smartphone, and simply measuring (or not, in case of failure) the generated photovoltage. Note that, cranberry juice has been used as dye (photoactive material), because it is safe, edible and contains natural photoactive (organic) pigments suitable for our purpose.

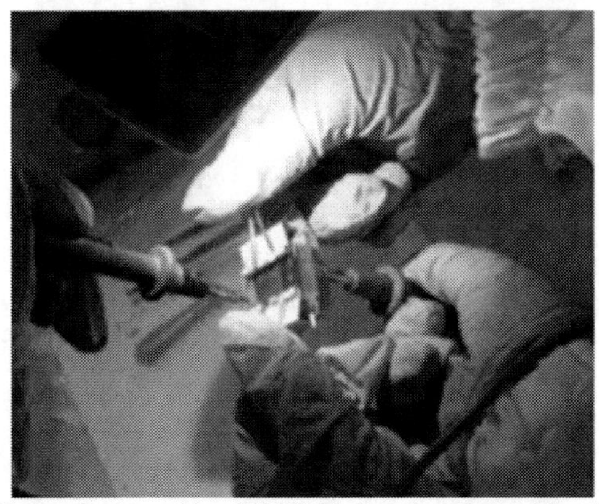

Figure 2. A working DSSC fabricated and tested by the students.

The proposed learning pathways offer to the students the possibility to learn more about emerging photovoltaic technologies and their remarkable merits and specific properties such as simple manufacturing, semi-transparency and lightness. In addition, they can also learn that organic materials can be manipulated and chemically modified in a relatively simple way, thus giving the possibility to tune and modulate the resulting properties as a function of the final application.

2.3 Solar cell recycling

The pathway starts with a lesson related to the global energy demand and to the need of a sustainable clean energy supply. The silicon based solar technology and its outstanding development in the last years in terms of cost and efficiency are also stressed in the introductive lessons. One of the main outcomes of this learning pathway is to inform students and teachers and make them aware that PV panels are included into the new Waste of Electrical and Electronic Equipment (WEEE) directive. Basing on this principle, PV panel manufacturers and distributors have to ensure the right collection and recovery of End of life products within the European borders. This a very important achievement which is not very well known by the society.

All this considering, the industrial processes involved in

both the assembling and disassembling of a PV panel cannot take place in an ordinary classroom or school lab, first of all due to the large size and weight of a PV module, the pupils will firstly assemble and then disassemble a photovoltaic (PV) mini- panel in order to understand the basic principles behind the important recycling process of PV wastes (see Figure 3).

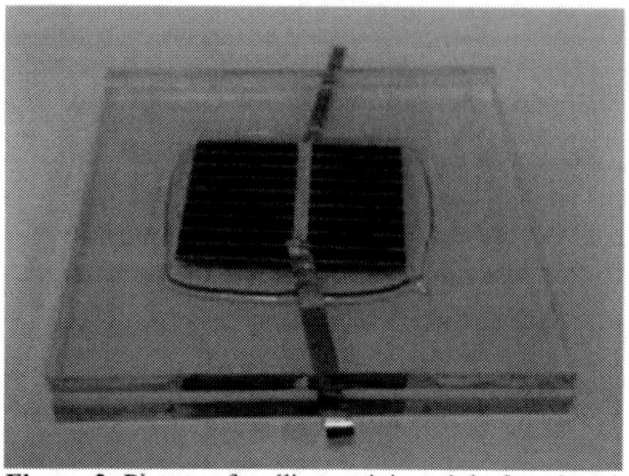

Figure 3. Picture of a silicon mini-module for education purposes (including 2 glass slices 5x5 cm2, 2 EVA sheets 3x3 cm^2, 1 silicon PV cell 2.5x2 cm^2).

After this learning path, the pupils know that large scale silicon solar cells recycling can ensure both a long term sustainability of the supply chain and a shortening of both the energy supply and the energy pay back time of PV technology and make the silicon based PV technology absolutely sustainable. Last but not least, they learn that the use of recycled materials in the PV production may improve economics by providing a less expensive feedstock supply.

3 RESULTS AND DISCUSSION

These pathways can be declined in different ways depending on the background knowledge of the students and on the learning outcome expected by the teacher and the researcher. Professional training, reinforcement of the curricular knowledge, and science dissemination are no exhaustive examples of goals that can be achieved. In all the classes a special focus was given to the re-elaboration of the scientific knowledge related to the experience in order to present it to different kinds of audience: to a generic public, during the open-access events; to younger students through peer tutoring experiences; to the commissioners in the final examination for the diploma and in a future CV drawing.

3.1 Professional training

The realization of a printed board circuit (PCB) that allows for a visual estimate of the voltage drop in different coatings was the outcome of a Technical School. The students outlined the circuit design and realized the circuit by bromographic printing [6].

Figure 4. "Real world" view of the circuit design produced by using the software PCB Wizard.

Figure 4 shows the design of the PCB in "real world" mode. It is constituted by a parallel of 4 branches each constituted by a series of one LED (D1, D2, D3, and D4) and one variable resistor (VR1, VR2, VR3 and VR4); it is powered by a battery and the coated glass acting as resistor is connected between two free terminals. The display DS1 represents a voltmeter measuring the voltage drop in the resistor. This circuit offers also the possibility of visualizing the voltage drop in different coatings. Indeed, the variable resistors connected to the LEDs are set in order to increase in resistance going from 1 to 4. Each LED can light up if the bias is enough to compensate the voltage drop in the trimmer, thus creating a scale of LEDs lighting up as the coating resistance decreases. This work is suitable for Technical Schools where students are familiar with circuit fabrication and electronic devices. In this case the novelty consists in analyzing materials for electronics other than silicon, such as graphene, which represents a promising candidate for a new generation of electronic devices. In addition, since the branches measuring and visualizing the voltage drop can work independently, a simplified version of this circuit can be used as a demonstration material to visualize the comparison ITO/graphene in science exhibitions.

3.2 Strengthening of knowledge

The re-elaboration of the acquired knowledge allows for better understanding and long-term retention of the concepts. Many classes realized games such as quizzes or table games to accompany their science dissemination activities. Indeed, the use of games is a powerful tool to strengthen the knowledge acquired in classroom [7], and can be even more powerful by having the students realize their game by themselves.

A class realized a table game in the form of quiz in order to convey messages related to raw material issues. The students wrote ten questions with two possible answers, a right one and a wrong one. Each question is printed on two labels: on one label it is followed by the correct answer and on the other by the wrong answer. The labels are glued on the two sides of an ITO coated glass platelet (coating on one side only) by paying attention to glue the right answer on the ITO- coated side.

The circuit, composed of a battery and a red LED, is inserted in a wooden structure able to lodge a coated glass. The role of ITO is to close the circuit allowing for the LED lighting up if the player chooses the correct answer; otherwise, the LED will not light up. For didactic purposes, the topic of each question is treated in more detail on separate small cards, in order to make the player more aware of the matter in hand.

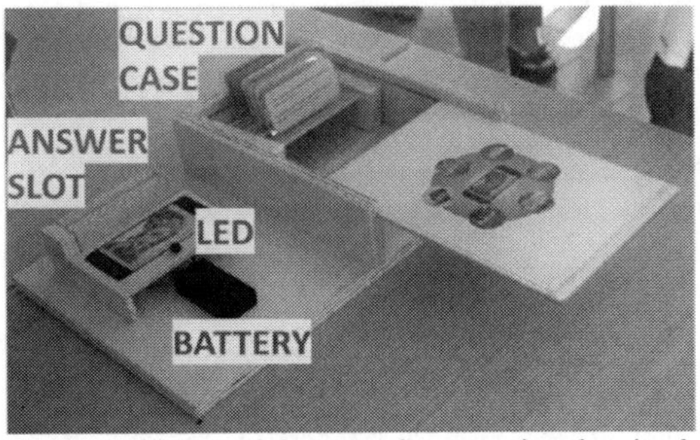

Figure 5. Picture of the game incorporating the circuit based on the experiment "ITO vs Graphene".

Figure 5 shows a picture of the game: the circuit is on the left while the box containing the glass platelets where the questions are printed is on the right.

3.3 Science dissemination

The ultimate goal of the RM@Schools pathways is to involve students in science communication events. -Also for these 3 learning paths some of th pupils involved with their teachers created presentations and video to be presented in science communication events. We observed this activity enhances the results obtained by creating dissemination material, the depth of the scientific insight, and the satisfaction towards the pathway [8].

For instance, the table games described in 3.2 were used in the event of a "White Night" hosted by the School [9] and in peer education activities for classes of 13-year old students. The students gave seminars about different aspects of Raw Materials used for energy production and storage devices. The allowed for a more dynamic lesson and for the interaction with the public.

4 CONCLUSIONS

This work shows an experience of science communication activity which combines training and science dissemination on PV technology . Applying this approach to the field of Raw Materials in energy production and storage devices allows to address some of the requirements for the energy transition: student awareness about solar jobs, student training about research topics, and citizen information about some of the issues related to Energy and Raw Materials.

We also observed that energy is a topic that is well suited for the application of this methodology because secure energy supply is a theme very close to the citizen experience. For this reason, the RM@Schools Methodology could also be applied to aspects of energy production and storage other than Raw Materials.

This activity was carried out within the project RM@Schools 3.0 - Raw Matters Ambassadors at Schools (project agreement No. 17146) under the framework partnership agreement No. [FPA 2016/EIT/EIT Raw Materials], Specific Grant Agreement No. [EIT/RAW MATERIALS/SGA2019/1], KIC RawMaterials Internal Agreement of 04 December 2015.

Dr. Alessandra Degli Esposti is kindly acknowledged for the RM@Schools website and Virtual Center setup. Professors Alessandra Sia, Cristina Erba, Ofelia Molteni, and Silvia Castoldi are kindly acknowledged for tutoring the involved classes.

4.1 References

[1] http://www.rmschools.eu
[2] https://www.youtube.com/watch?v=YPSkjjXcIMs&list=PLjD8NWQ60GCWlIVmGNw7vgNY1gy6D9pO0&index=9
[3] https://rmschools.isof.cnr.it/moodle/course/index.php?categoryid=4
[4] https://rmschools.isof.cnr.it/moodle/course/index.php?categoryid=3
[5] A. García-Carmona, A.M. Criado, International Journal of Science Education, 31 (16) (2009) 2205.
[6] https://padlet.com/alessandrasia/alternanza5a
[7] B. Bahar Taspinar *et al.*, Procedia Computer Science 99 (2016) 101.
[8] M. Canino *et al.*, Frontiers in education (2021). DOI: 10.3389/feduc.2021.690294
[9] https://rmschools.isof.cnr.it/gallery/Lissone_2019/

INVENTORY OF WASTE MATERIALS FROM SILICON INGOT AND WAFER MANUFACTURING FOR A CIRCULAR PV VALUE CHAIN

Yijiang Xu[a], Gaute Stokkan[a], Anne-Karin Søiland[b], Birgit Ryningen[a], Irene Bragstad[a], Martin Bellmann[a]
a SINTEF Industry, b ReSiTec AS
a Richard Birkelands vei 3, 7034 Trondheim, Norway. b Setesdalsveien 110, 4617 Kristiansand, Norway.

ABSTRACT: Photovoltaics (PV) is one of the most important renewable energy. During the production of Si-based PV panels, large resource flows are generated, and many waste materials are generated. The challenge for recycling is connected to the material properties: The composition, purity and morphology will determine how the current waste should be processed in order to realize the potential value. We have therefore performed a waste inventory for three important material flows occurring from monocrystalline silicon ingot and wafer manufacturing in the PV production value chain: Si kerf from sawing of wafers, silica from crucibles used in CZ-Si ingot pulling and graphite from CZ crystal growth furnaces. A series of characterization and analysis methods, have been used to investigate the chemical composition, phases, size and morphology etc. of the three waste materials. Possible contamination sources were discussed. The waste inventory results give an overview of the materials quality and property, and their potential for recycling and re-use in the PV section and other application fields.
Keywords: Si kerf, Recycling, Materials characterization

1 INTRODUCTION

The PV module market in 2021 [1] showed a significant growth to 183 GW, where 95% is based on the c-Si technology. Among c-Si module, mono-Si is the dominant in the market while mc-Si products reduced to 13%. The upstream silicon ingot and water manufacturing process is very energy intensive, and consume large quantity of raw material and parts, and generate large waste. In the silicon wafer slicing processes, diamond wire sawing, nearly 40% of the high purity silicon become kerf loss waste. The amount of Si kerf waste is estimated to be at around 160, 000 metric ton in 2021. Besides, in the silicon ingot production process, ancillary materials, such as graphite used as crystal growth furnace parts, and silica from fused quartz crucibles turn to waste. According to the recent report [2], the global isostatic graphite consumed in PV filed is estimated around 50, 000 metric tons in 2021. And the silica waste volume reached 205,400 mt in 2019.

The diamond wire sawing (DWS) process has replaced the traditional slurry process for silicon wafer manufacturing. The steel wire are covered with diamond grits by electroplated nickel coating. Silicon bricks which are mounted to a beam are fed into the moving wire web while being bathed in water-based cutting fluid. Depending on the wire thickness, around 35% of silicon is lost as Si-kerf which is dispersed in the water-based cutting fluid. The cutting fluid and the Si-kerf are later separated through a filter press providing a liquid and a solid fraction (Si-kerf filter cake). The investigation and research on recycling these Si kerf for different application has started since 2013 [3], but it is still at its infant stage. It is short of a systematic characterization and inventory of these materials, to clarify the composition, phases, impurities, morphology etc.

In silicon ingot growth furnace, such as Czochralski single crystal furnace, there are about 30 kinds of high purity isostatic graphite parts [4], including vortex, heater, heat shields and insulation components, susceptors, bases for rotation. Graphite used in the hottest sections of the furnace has a lifetime of only about a few months, either due to the structure and property degradation or due to the impurity deposition and formation. This material is currently down-cycled as low-value carbon for example as a reducing agent in smelting operations. In order to recycle it as a high-value material, such as reuse in existing graphite production plants and use for anode materials in lithium-ion batteries, a detailed analysis of these graphite is demanded.

In addition, in Czochralski process for silicon ingot growth, high purity fused quartz crucible is another key component [5]. It is used to hold the silicon melt during ingot production. The crucible has two layers, a bubble containing layer with certain mechanical strength and a bubble-free inner layer. During the CZ-process, the silica goes through a phase change, forming cristobalite on the surfaces. At the same time, bubbles could form and grow in the bubble-free layer after several runs, which is detrimental for the growing ingot. Therefore, the life time of these crucibles is very limited. After cooled down, the remaining silicon is sticked to the bottom part of the crucible, and the crucibles cannot be reused. This waste materials is being landfilled. Some research has been reported [6] to study the silica crucible waste from the multicrystalline silicon ingot growth, and the method for recycling. However, those crucibles are different from the high purity fused quartz crucibles for CZ mono-crystalline silicon regarding the original state, purity and the using condition etc.. Until now, there is no investigation on the used crucibles from CZ Si growth process, regarding the phases, impurity and its potential for recycling etc.

In this work, the materials' quality of three important waste material from silicon-ingot and wafer manufacturing in the PV production value chain: Si kerf from sawing of wafers, silica from crucibles used in crystal pulling and graphite from crystal growth furnaces, has been investigated. The quality is mainly consisted of chemical composition and impurity content, phases, size and morphology of particles in powder form etc.

2 EXPERIMENTAL

2.1 Materials

The materials investigated are summarized in Table I. Si kerf samples with different source and condition are labled with different ID number. While graphite and silica crucible scrap are only from Norway.

Table I: Type, origin, ID and condition of waste materials investigated in this study.

Waste type	Origin	ID	Condition
Si-kerf dry (1x)	Norway	RST-ODIN-0921	Dried
Si-kerf dry (3x)	Asia	RST-HAN-0120/-MING-020/-SONG1-0721	Dried
Hot-zone graphite	Norway	RST-ODIN-C	crushed
Silica crucible scrap	Norway	RST-ODIN-Q-0921	crushed

For the crushed graphite samples, they have been sorted according to their visual appearance: 1: "clean" samples, 2: yellow surface, 3: silver-grey surface, 4: silver with droplet. Here in this work, sample type 1&type 4 will be studied.

For the silica crucible samples, they were categorized as well by 4 different types, as shown in Fig. 1. 1: Clean samples with a white layer on the crucible inner and outer surface, 2: Inner surface with orange and brown dots, while outer surface is white, 3: Outer surface with orange dots and areal spreading, while inner surface is white, 4: Non-flat inner surface with silver-like particles.

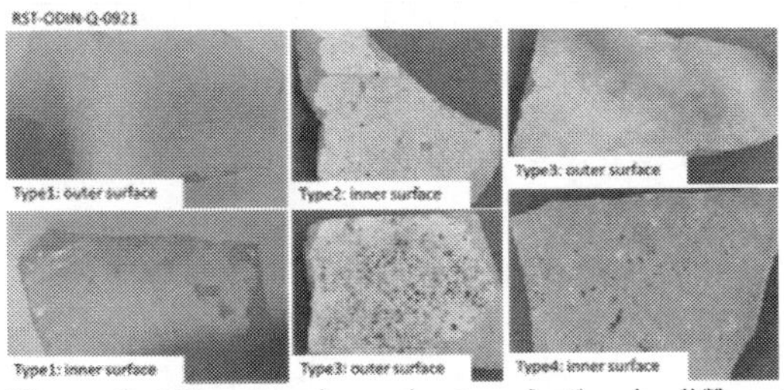

Figure 1: Silica samples and categorization in different types: type 1: white inner and outer surface, type 2: small yellow and black dots at the inner surface, type 3: yellow dots and spread on the outer surface, type 4: silver-like particles on the non-flat inner surface.

2.2 Analysis and Characterization methods

A series of chemical analysis and materials characterization, such as Inductively coupled plasma mass spectrometry (ICP-MS), LECO, X-ray photoelectron spectroscopy (XPS), X-ray diffraction analysis (XRD), light microscopy and scanning electron microscopy, and laser diffraction has been performed.

3 RESULTS

3.1 Silicon kerf

Fig. 2 shows the concentration (in ppmw) of common dopant elements (B, Ga and P) in different Si-kerf samples measured by ICP-MS. The B content in ODIN source samples and most other source samples is around 0.35 to 0.4 ppm. Ming-0120 has slightly higher B, which reaches 0.6 ppm. The Ga content is in the range of 0.05 to 0.65 ppm depending on the source. Generally, ODIN source has higher Ga concentration. For P, the highest content is 15 ppm while lowest is 4 ppm, which is much higher than B and Ga. Comparing the dry samples from different sources, it can be seen that for Asian sources the P content is higher than for the ODIN sources.

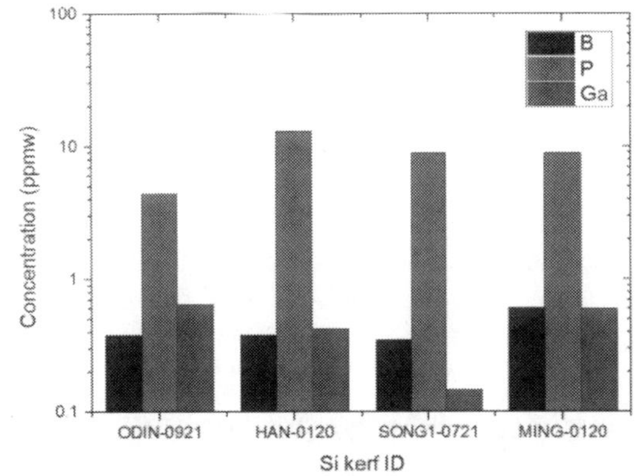

Figure 2: Content of common dopants (B, Ga and P) in different Si-kerf samples measured by ICP-MS.

The metal contaminations, such as Al, Fe, and Ni, determined by ICP-MS are shown in Fig. 3. ODIN source sample has the highest Al (~0.72 wt.%), while the three Asian samples have much lower Al (<1000 ppm), especially SONG1 only has 10 ppm. Al contamination in ODIN source should be mainly from aluminium tri-hydroxide used as beam material in the DWS process. Ming has a very high Fe concentration (~1000 ppm), while other samples have less than 100 ppm Fe. The Fe content in ODIN source samples is lower than that in Asian samples. Fe contamination could come from the steel wire or other steel-made equipment/tools in the filtration and drying process. In addition, Ni is a common contamination element, due to the use of electroplated nickel coating on the steel wire. Typical Ni content is 100-200 ppm.

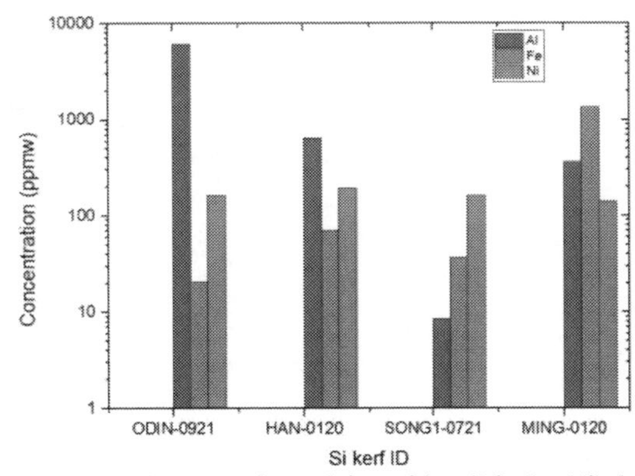

Figure 3: Content of metal impurities (Al, Fe, Ni) in different Si-kerf samples measured by ICP-MS.

Table II shows the carbon and oxygen content measured by LECO. The typical carbon content for most of the samples is around 1-2 wt.%, while SONG1 is more than double this. Carbon may come from the organic compound in the cutting fluid. The oxygen content is about 4-5 wt.%.

Table II Carbon and oxygen content in different Si kerf samples measured by LECO.

Sample-ID (wt.%)	RST-ODIN-0921	RST-HAN-0120	RST-MING-0120	RST-SONG1-0721
Carbon	2.0	2.0	1.0	4.4
Oxygen	4.2	4.3	4.8	5.6

Fig. 4 shows an SEM image of RST-ODIN-0921-dry Si-kerf sample. It is seen that the Si-kerf consists of ribbon-like, irregular shaped angular particles, agglomerates, and nano-sized particles. The ribbon shaped particles are up to around 4 μm long.

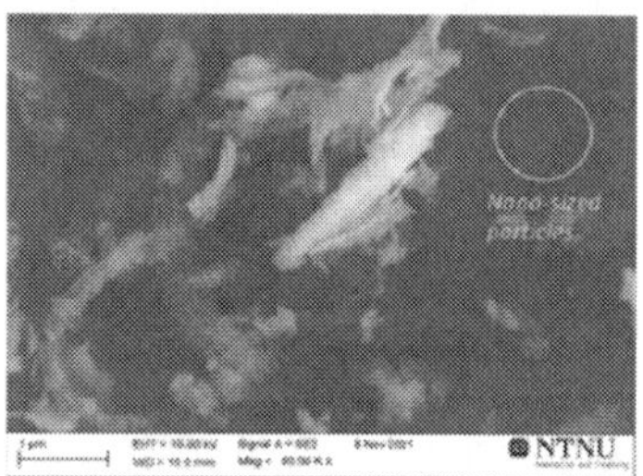

Figure 4: SEM images of RST-ODIN-0921-dry Si-kerf sample.

The particle size distribution (spherical equivalent) of two representative Si kerf samples from Norway and Asia measured by laser diffraction analysis is shown in Fig. 5. As can be seen, ODIN-0921 has more small particles with size below 0.34 μm, and more particles in the range of 0.66-1.8 μm, but less big particles larger than 2 μm.

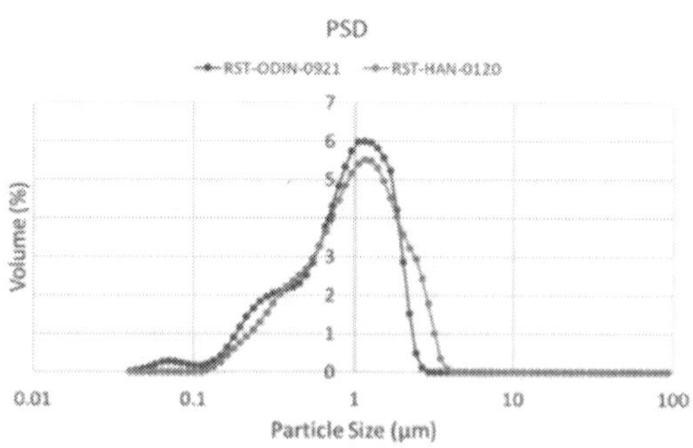

Figure 5: Particle size distribution of 2 Si-kerf samples (RST-ODIN-0921 and RST-HAN-0120) measured by laser diffraction method.

3.2 Graphite

The ash content of graphite sample is 0.16% and 4.1% respectively for sample 1 and sampel4. The metal impurity content in the graphite was calculated from the impurity content measured in the ash by ICP-MS, and the results are listed in Table III. It can be seen that the highest impurity is Si, which is at 0.39 wt.% and 10.9 wt.% in graphite sample 1 and sample 4. Other impurities are in ppm level. Ca and Co are slightly higher in sample 4.

Table III: Impurity content in the graphite measured by ICP-MS of ash samples.

	Si wt%	Ca ppmw	Al ppmw	Fe ppmw	Co ppmw	Ni ppmw	Ti	B ppmw	P ppmw
#1	0.39	4.7	1.3	2.0	5.5	0.2	0.2	0.1	6.7
#4	10.9	21.8	0.6	2.1	41.1	0.7	0.3	0.2	10.2

Fig. 6 shows the XRD results of graphite sample 4. Both Si and SiC are identified on the surface. The surface is further characterized by SEM. Two representative Secondary electron images are shown in Fig. 7. In Fig. 7a, we see a region with flat surface and a region with rough surface. The flat region is consisted of Si grains. While in the rough region, there are several phases with different morphologies. From magnified images of Fig. 7b, it shows that some facetted particles are imbedded inside some matrix, while only exposing some surfaces, which is SiC. At the same time, there are some small spherical shaped phases and irregular particles on the surface.

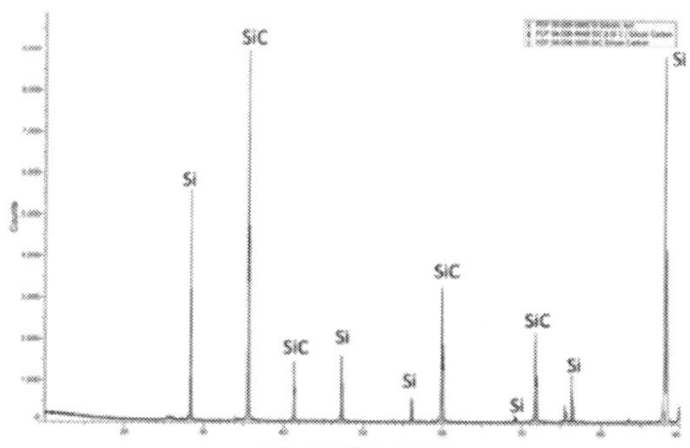

Figure 6: XRD results of the surface of graphite type 4.

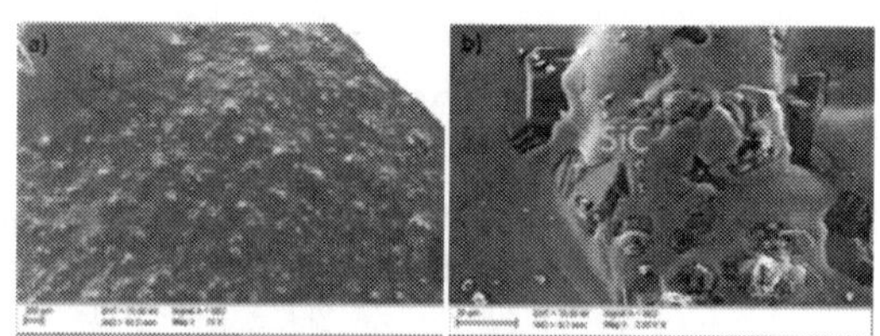

Figure 7: SEM secondary electron image of graphite type 4 sample surface.

3.3 Silica

Fig. 8 shows the XRD results of powder sample of silica type 4, which is similar to the pattern obtained for the other three types. A sharp cristobalite peak is observed, which is the only crystalline phase been detected. Si phase was not detected by current XRD, probably due to the small volume fraction. There is a bump at 2θ=23°, which corresponds to the amorphous silica phase.

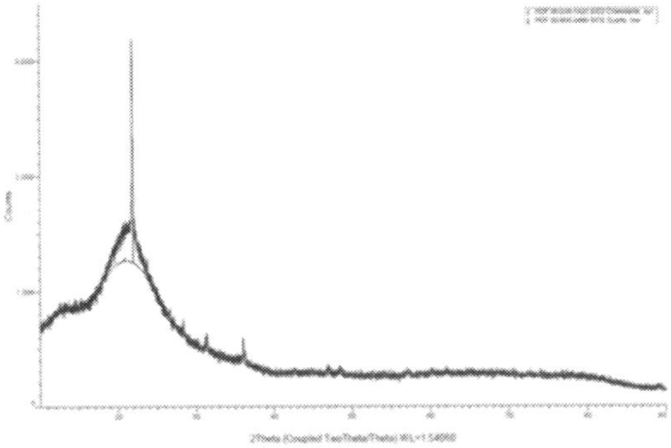

Figure 8: XRD results of milled powder sample of silica sample type 4. Same pattern was obtained for the other 3 sample types as well.

The impurity content determined by ICP-MS is shown in Table IV. The highest impurity is Ca, which is much higher than the typical amount in the new crucibles for CZ process. Co is around 21.4 ppm, which is also higher than the typical value (<0.01 ppm), which could be a result from contamination during the milling process. Similar for Fe, the samples could have been contaminated during crushing process if steel tools have been used. Comparing the 4 samples, it can be seen that type 3 has much higher Al content than the other 3, which are almost the same as in the original quartz sand and new crucible.

Table IV Impurity content (ICP-MS) in the 4 categorized silica samples. All values in ppmw.

#	Na	Al	P	Ca	Ti	Fe	Co	Ba	B
1	11.8	16.7	1.4	518.4	6.7	10.1	25.2	2.1	0.2
2	13.4	10.6	0.6	433.7	6.2	2.4	24.0	1.4	0.6
3	30.1	135.4	1.5	403.5	2.7	37.0	17.4	1.1	<0.2
4	25.0	7.5	1.1	431.2	2.0	26.6	19.0	1.1	<0.2

4 DISCUSSION

For recycling of Si kerf back into the PV value chain as new feedstock, the contamination level is an important parameter. A further refining process is needed to reduce the impurity level to the required amount. Besides, the original wafer sawing process could be optimized to reduce and avoid the contamination of some impurity such as Al and Fe.

Use of Si-kerf in other application or products can be another way to recycle. For example, as feedstock for metallurgical Si production, alloying into Al-Si alloys, and Li-ion battery. The processability will also depend on morphology and surface conditions. It is seen that some differences exist between materials in terms of particle size and surface area.

The graphite is commonly contaminated with Si, SiC etc. The ash content with relatively clean surface graphite is acceptable, while the one with contaminated surface is relatively high. The generally low levels of Ni and Co are promising for typical graphite uses, such as isostatic graphite products and Li-ion battery application.

The differences in impurity concentration between the types are not significant, except type 3 which has an order of magnitude higher Al concentration. The original of high Al in type 3 needs to be further investigated. Ca is also very high in all samples. The recycling and reuse of those waste materials are less explored.

5 CONCLUSIONS

A detailed characterization and analysis of 3 waste materials from solar cell CZ-Si ingot and wafter production process have been performed, including Si kerf, graphite and silica crucibles.

Si-kerf is consisted of ribbon like, irregular shaped angular particles, agglomerates, and nano-sized particles. In Si kerf, the Si concentration is about 92 wt.%, while the main impurities are C (~1-2 wt.%), O (~4-5 wt.%), metallic elements (Al, Ca, Ni, Fe etc.) and doping elements (P, B, Ga).

Graphite contains mainly Si and surface infiltrations of SiC originating from the chemical reactions inside the furnace. The ash content for type 1 graphite is 0.16 %, which is quite low. Other detected impurities are at much lower levels, Ca and Co could reach 20-40 ppm in type 4 and are ~5ppm in type 1. P is about 10 ppm. Other element like Al, Fe Ni, Ti B are below 1ppm.

Silica is mainly present in the form of amorphous silica and crystalline cristobalite. In type 4, it can contain silicon. Ca has the highest level (~500 ppmw) among other impurities. Other impurities like Na, Al, P, Ti, Fe, Co, Ba and B are also present but at much lower levels (1-40 ppmw).

6 ACKNOWLEDGMENT

This project has received funding from the European Union's Horizon 2020 research and innovation programme under grant agreement number 958365.

7 REFERENCES

[1] M. Fisher, M. Woodhouse, S. Herritsch, J. Trube, International technology road map for photovoltaic (ITRPV), 2022.

[2] SPECIAL GRAPHITE MARKET - GROWTH, TRENDS, COVID-19 IMPACT, AND FORECASTS (2022 - 2027) Industry Report, mordor intelligence 2021.

[3] K. Tomono, Separation and Purification Technology 120 (2013) 304.

[4] O. Anttila, Handbook of Silicon Based MEMS Materials and Technologies (Third Edition), 2020

[5] A. Hirsh, Journal of Crystal Growth 570 (2021) 126231.

[6] Y. Zhou, Refractories and Industrial Ceramics 54 (2014) 489.

PROPOSAL OF AN ALTERNATIVE FOR DEVELOPING A GLOBAL GRID ABLE TO PROVIDE 100% RENEWABLE POWER TO THE WORLD

Ralf Leutz and Ana Rosa Lagunas
TOMATO:GSL
ralf.leutz@tomato-gsl.org, ana.lagunas@tomato-gsl.org

ABSTRACT: A global electrical grid can support 100% renewable electricity generation, that could be used at distant places at any time. We started simulating the Global Solar Link in stochastic fashion, showing that daily and seasonal storage/backup can be reduced in favour of a grid connection. That is exciting, and a paradigm change in the energy world, showing the way forward in the climate crisis.
Keywords: 100% solar, adequacy, capacity factor, global solution, grid stability, hybrid PV (PV/storage)

1 INTRODUCTION

As the earth rotates around the sun, its surface, at any given time, receives 100 times the amount of primary energy demand, on only 1% of its surface. Well spread out, and efficiently connected by high-voltage direct current (HVDC) power lines, solar power plants can provide all energy required for human well-being, at the same time slowing global warming by allowing us to keep all carbon in the ground. A global solar grid offers the potential to satisfy the world's energy demand by 100% renewable energy. Technologies of renewable generation of electricity (wind, PV) and intercontinental HVDC transmission lines are mature.

Strengthening and modifying earlier approaches [1-8], TOMATO:GSL (an acronym for The Only Meaningful Alternative To Oblivion – Global Solar Link) aims to solve the challenge of seasonal storage caused by the inclination of earth on its orbit, i.e. the variation of solar irradiance in higher latitudes between day and night, as well as between summer and winter. The transmission capacity of the global grid allows all energy to be locally intermittent, providing a constant global level of electric power.

In this paper, we detail a stochastic and modular approach towards a model allowing to oversee the transition of the world to 100% renewable power.

2 APPROACH

2.1 Methodology

Energy modelling is based on statistics, on data selection, data cleaning, and selection of scope. These processes require input data to be of higher quality than it usually is. Should the modelling be deterministic, in the sense that only actual data is to be processed in the model, data gaps will appear, and assumptions need to be made.

Rather than making assumptions after the gaps appear, we suggest a methodology that is based on arbitrarily chosen data, which we call a stochastic model; that way the model is kept simple and modular. Granularity in time and area, as well as accuracy in input data and output model results can be increased when data becomes available; all the while learning about the model's behavior improves. The metric used throughout our work are capacity factors for power generators, storage, grid, and energy demand.

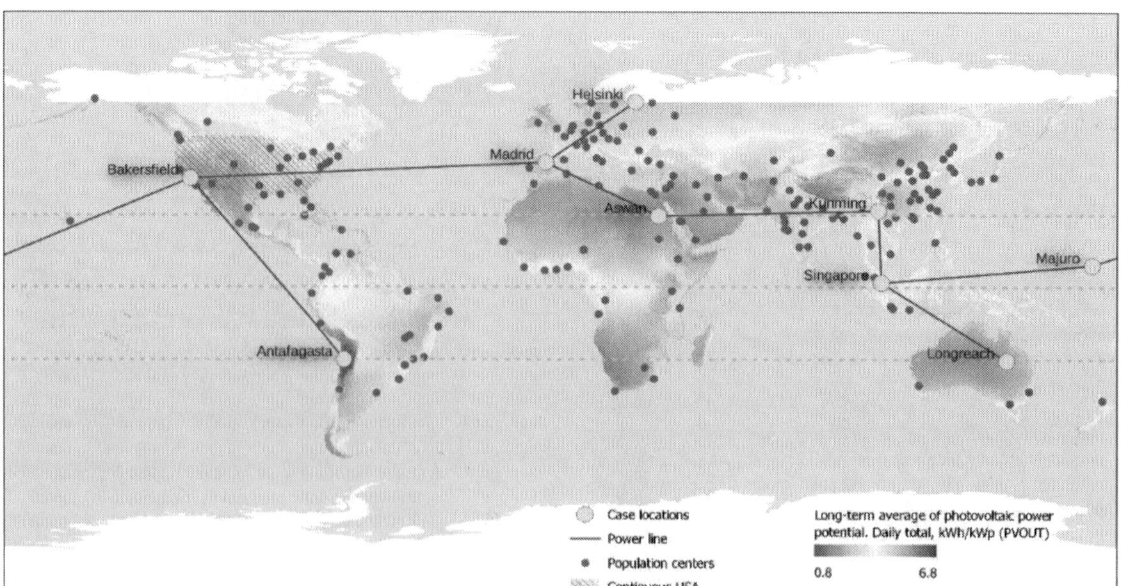

Figure 1: Global Solar Link. Exemplary Locations of solar energy generation (Cases), connected by a high-voltage power line spanning the globe, connecting continental grids. The Case locations have been chosen arbitrarily

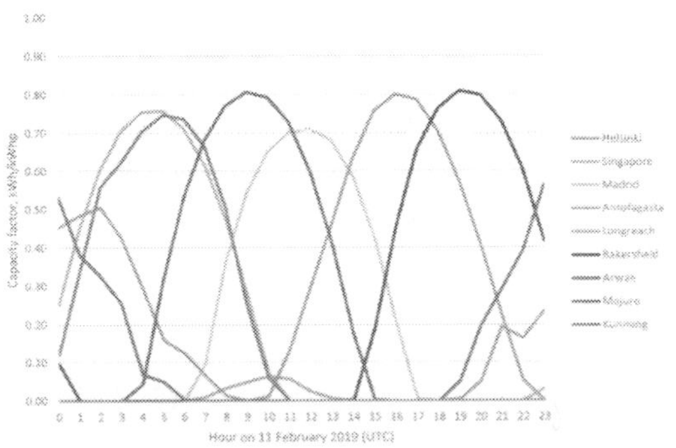

Figure 2: Capacity factors of PV power plants at Case locations, for the 11th Feb 2019 (UTC)

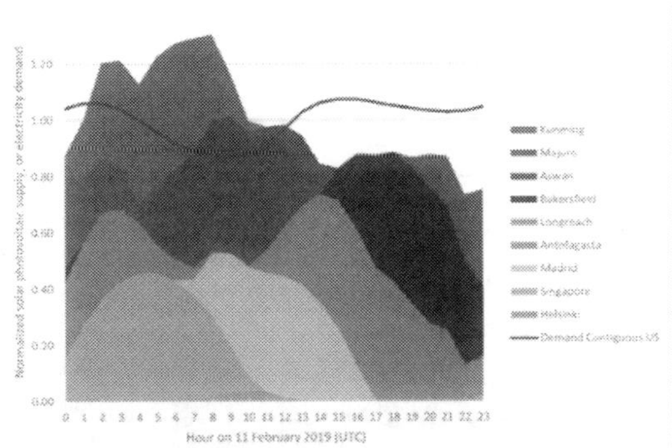

Figure 3: Normalized cumulated photovoltaic power generation for the Cases on 11th Feb 2019 (UTC); normalized demand of the contiguous United States

Table 1: Announced intercontinental transmission lines in high-voltage direct current (HVDC) technology for solar electricity

Project (status)	Source - Destination	Description	Link
Xlinks (announced 2021, commissioning planned 2027)	Morocco – Great Britain	2x1.8 GW HVDC undersea cable. Losses 10-12%. Invest EUR21.3 bn	xlinks.co/morocco-uk-power-project
Sun Cable Australia ASEAN Power Link (AAPL) (announced 2019, commissioning planned 2027)	Australia – ASEAN	3,800 km submarine + 750 km overhead 3.2 GW HVDC cable. Investment EUR23 bn	suncable.sg
Antipodas (announced 2021)	Chile – Asia	15,000 km submarine 0.6 GW(?) HVDC cable. Investment 25 bn EUR(?)	www.pv-magazine.com/2021/11/15/chile-wants-to-export-solar-energy-to-asia-via-15000km-submarine-cable
Saudi-Egypt Interconnector (to be commissioned 2025)	Egypt – Saudi Arabia	1,350 km, 3,000 MW at 500 kV HVDC, 1.8 bn USD	www.hitachienergy.com/de/de/case-studies/saudi-arabia-egypt
EuroAsia Interconnector (ongoing)	Greece – Cyprus – Israel	1,518 km, 1,000 MW at 500 kV HVDC, subsea cable	www.euroasia-interconnector.com
Morocco-Spain Interconnector (commissioned 1997 and 2006, one cable successfully repaired)	Morocco – Spain	2x400 kV, 700 MW, third cable proposed at 800 MW, $169 m	northafricapost.com/43565-morocco-spain-interconnection-submarine-power-cable-repaired.html

2.2 Date

The methodology shall be demonstrated for the 11th February 2019, and the 11th August 2019 (with the demand data for the same day one year earlier, due to unavailability). Dates have been selected at random, all times are universal (UTC). Time slice is one hour.

2.3 Cases and solar resources

We select Case locations arbitrarily, knowing that solar resources are freely available for the whole world in hourly resolution [10]. The locations are to cover most regions of the world, shown in Figure 1. Photovoltaic generators in the locations should produce with hourly capacity factors given in Figure 2, energies shown in Figure 3. The locations shall be connected by a high-voltage grid, examples of existing and planned power lines are given in Table 1.

2.4 Demand

Data of hourly electricity demand data is available for the contiguous United States [9]. The Global Solar Link is used to connect existing grids.

2.5 Capacity factor

The capacity factor describes the ratio of the actual energy generated to generation potential (or, nominal generation capacity), both for the same period of time. The definition allows matching generation and demand as both happen at the same time and works for all energy generation technologies. As fossil fuels, hydropower, or nuclear are the basis of the existing electrical system, same criteria should be applied to renewables, which can have inherently intermittent resource (wind or sun). Thinking about electricity generation by PV, the demand of the day (24h) cannot be matched, as the sun shines only during certain hours in every location. The same applies to wind. Other technologies for generation have resources available 24 hours per day.

A photovoltaic generator may have a capacity factor near unity for any peak hour, depending on losses calculated. Since half the year is dark, cloud cover and temperature are influencing performance, the yearly capacity factor of PV is around 20%.

Storage can have a capacity factor of its own, illustrated by the way a solar field can be used to directly feed electricity into a grid, and charge a battery. The latter may independently of the solar field discharge into the grid. A grid may be understood as a large storage unit balancing electricity supply and demand, as we all know from our daily lives. Thus, by analogy to storage, a grid can be assigned a capacity factor. This makes sense, as grid losses reduce the capacity factors of generators.

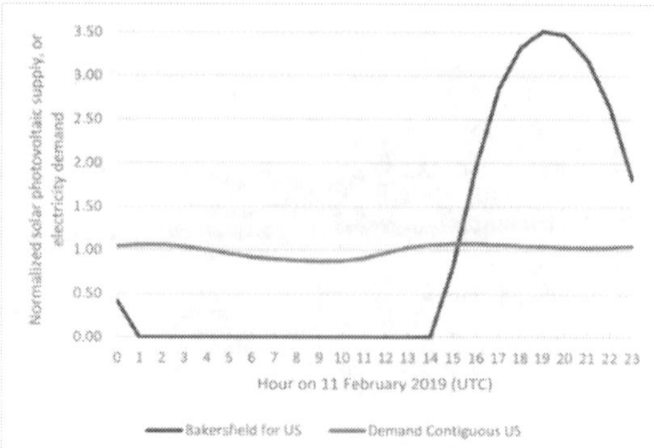

Figure 4: Normalized solar photovoltaic generation in Bakersfield, and normalized electricity demand of the Contingous United States, for 11th Feb 2019

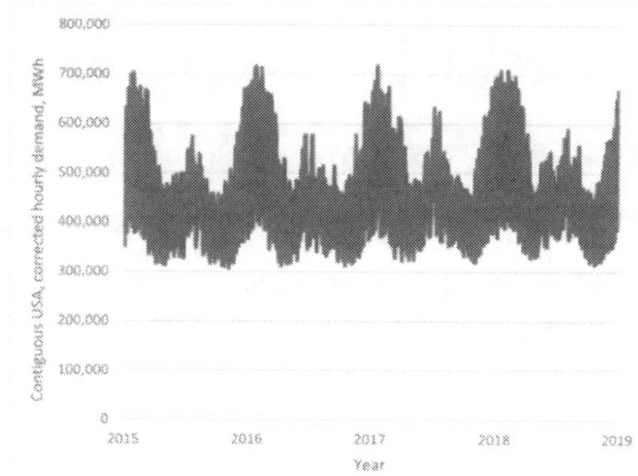

Figure 5: Corrected long-term hourly electricity demand of the Contiguous United States. Source [9]

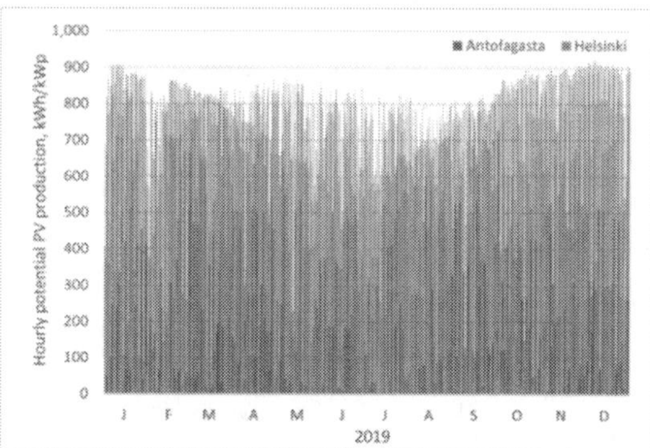

Figure 6: Capacity factors of photovoltaic power plants in Antofagasta, Chile, and Helsinki, Finland, hourly values for 2019

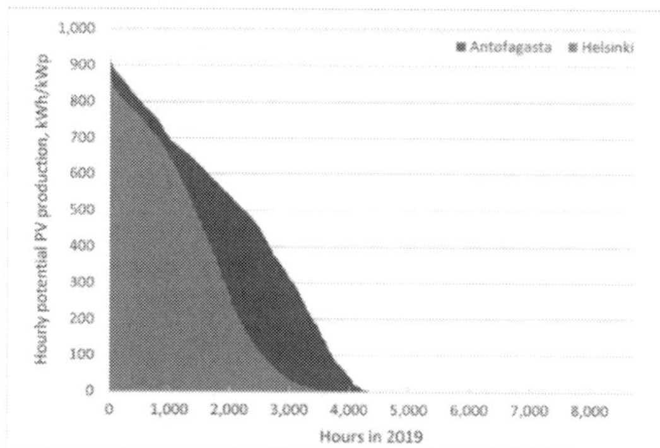

Figure 7: Histogram of capacity factors of photovoltaic power plants in Antofagasta, Chile, and Helsinki, Finland, hourly values for 2019

(At least as seen from the energy consumer). On the other hand, the grid increases the area under consideration by the model.

The capacity factor of interlinked distant solar power plants can be greater than the sum of the individual capacity factors from these power plants, in the event of local curtailing due to oversupply.

3 DIURNAL AND SEASONAL SETTLEMENTS

3.1 Hybrid PV/storage mode

Assume that any one local grid, or continental grid, needs to supply 100% solar energy to satisfy demand. In this example, Bakersfield, CA should satisfy the whole contiguous United States with clean electricity. As the night settles, storage needs to be employed. This storage must have been locally charged during the preceeding daytime. The solar field required to charge the batteries must be several times larger than the solar field used to satisfy daytime demand; this ratio is called the Solar Multiple. Plotting the normalized photovoltaic energy supply and the demand (Figure 4), we confirm that the two areas enclosed between the curves, one for the time when demand exceeds supply, and the other vice versa, are of equal area. Figure 5 puts the hourly demand into a long-term perspective.

It can be seen that the solar field must be 3.5 times larger for 24 hour operation in hybrid PV/storage mode than for solar-only operations.

The need for seasonal balancing is implicitly shown in the comparison of the hourly potential PV production on opposite sides of the equator, for Antofagasta, Chile, and Helsinki, Finland, in Figure 6 and Figure 7.

The need for seasonal balancing can be deducted from the two final graphs (Figure 8 and Figure 9), showing surprisingly similar demand curves for 11th Feb and 11th. Aug. The associated Solar Multiples shown are 10% reduction in the North American summer. However this result should be reviewed in the case of 100% electrified economy. For example, the actual demand used in the model considers mostly electrified air conditioning while heating is mainly produced through coal or oil.

3.2 Inter-continental grid connection

By connecting Bakersfield, as supply hub, to the grid of the contiguous United States and by the Global Solar Link to the rest of the world, storage interventions become small. Electricity corresponding to the area circumscribed by the yellow lines in Figure 8 and Figure 9 needs to be imported by the grid of the contiguous US.

Electricity transmitted through the Global Solar Link to the contiguous US grid amounts to less than 20% of the electricity to be stored for the satisfaction of diurnal variations in local solar supply demand.

The seasonal differences are small, in Feb the ratio of electricity to be imported to electricity to be stored (and produced by a solar field of Solar Multiple 3.5) is 0.17; in Aug this ratio is 0.19, for this simple example.

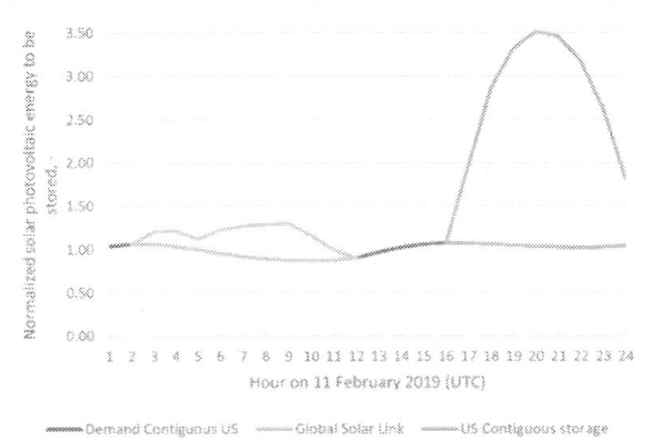

Figure 8: Normalized solar photovoltaic energy to be stored, or transmitted for the cases of world transmission (Global Solar Link), and generation/storage in the Contiguous United States, for 11th Feb 2019

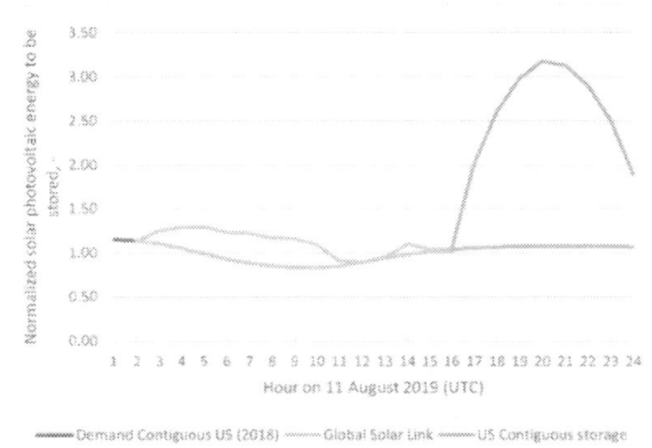

Figure 9: Normalized solar photovoltaic energy to be stored, or transmitted for the cases of world transmission (Global Solar Link), and generation/storage in the Contiguous United States, for 11th Aug 2019 (demand data 11th Aug 2018)

4 RESULTS

4.1 Battery storage versus grid connection

Global grid and large storage units can be compared in their capability to balance generated energy and demand. On a global scale, solar energy can supply all energy needs through the nights and through seasons.

As shown in the example of one location providing the electricity needs of the contiguous United States, diurnal balancing is possible with local storage units. Once storage is allowed anywhere on the globe, within the reach of the Global Solar Link, the size of the storage unit to balance diurnal differences in solar supply and demand reduces to about 17% of the local storage size, 19% for the cases of 11th Feb, and 11th Aug, respectively, illustrated by the comparison of areas under the curves in Figure 8 and Figure 9.

The global interconnector, without storage, reduces energy handling (by means of transmission, or storage) requirements to one fifth, for the United States, in this simple model. Our results indicate that there might be little differences between the seasons, if the world is grid-connected.

4.2 Grid dimensioning

The transmission capacity of the global grid needs to be calculated in a next step of the modelling. Absolute values should be assessed to dimension the power line. The existence of local short-term storage might be an advantage in absorbing supply and demand peaks.

The Global Solar Link is a transmission network connecting grids with renewable energy generators. Diurnal or seasonal balancing may be taken care of by the strength of the existing grid, without the need to improve the local grids' capacities.

4.3 Conclusions

Through a simple stochastic model, it has been demonstrated that international cooperation by means of a Global Solar Link might be reducing the need of local storage capacity.

The simple model uses seven, arbitrarily chosen locations for photovoltaic energy generation. Results regarding aggregated uniformity might be improved by adding locations. For a fully grid-connected world, uniformity should approach unity. This reduces the need for storage capacity further.

Connecting the whole earth would be an appropriate answer to the global challenge of global warming, at the same time opening chances for international development and harmonization of living standards.

A high-voltage grid is an established technology available immediately. Electricity is the most versatile carrier of energy, transformable into all other forms of energy, again with standard technologies. Electricity is safe and clean, if produced from the sun and is available through PV technology all around the world. It is available in most places, and if connected in a mesh by a Global Solar Link, can be safe and reliable.

5 REFERENCES

[1] S. Chatzivasileiadis, D. Ernst, and G. Andersson, The global grid, Renewable Energy 57, 372-383, 2013.

[2] Z. Liu, Global energy interconnection, China Electric Power Press, Academic Press, 2015.

[3] M. Brinkerink, P. Deane, S. Collins, and B. O'Gallachóir, Developing a global interconnected power system model, Journal of Global Energy Interconnection 1:3, Aug 2018, doi 10.14171/j.2096-5117.gei.2018.03.004.

[4] C. Wu, X. Zhang, and M.J.H. Sterling, Economic Analysis of Power Grid Interconnections among Europe, North-East Asia, and North America with 100% renewable energy generation', IEEE Open Access Journal of Power and Energy, 8:268-280, 2021, doi.org/10.1109/OAJPE.2021.3085776.

[5] L. Reichenberg, F. Hedenus, N. Mattson, and V. Verendel, Deep decarbonization and the supergrid – prospects for electricity transmission between Europe and China, Energy 239 (2022) 122335.

[6] T.W. Brown, T. Bischof-Niemz, K. Blok, C. Breyer, H. Lund, and B.V. Mathiesen, Response to 'Burden of proof: A comprehensive review of the feasibility of 100% renewable-electricity systems', Renewable and Sustainable Energy Reviews, 92:834-847, 2018, doi.org/10.1016/j.rser.2018.04.113.

[7] M. Ram, D. Bogdanow, A. Aghahosseini, A.S. Oyewo, A. Gulagi, M. Child, H.-J. Fell, and C. Breyer, Global energy system based on 100% renewable energy - power sector. Study by Lappeenranta University of Technology and Energy Watch Group, Nov 2017.

[8] M. Parzen, H. Abdel-Khalek, E. Fedorova, M.

Mahmood, M.M. Frysztacki, J.Hampp, L. Franken, L. Schumm, F. Neumann, D. Poli, A. Kiprakis, and D. Fioriti, PyPSA-Earth. A New Global Open Energy System Optimization Model Demonstrated in Africa, under review, arXiv:2209.04663v1 [physics.soc-ph] 10 Sep 2022.

[9] T.H. Ruggles, D.J Farnham, D. Tong, and K. Caldeira, Developing reliable hourly electricity demand data through screening and imputation. Sci Data 7, 155 (2020). doi.org/10.1038/s41597-020-0483-x.

[10] Photovoltaic Geographical Information System (PVGIS), 2022. re.jrc.ec.europa.eu/pvg_tools/en /.

UPDATES FROM THE AGROFOSSILFREE PROJECT - STRATEGIES AND TECHNOLOGIES TO ACHIEVE A EUROPEAN FOSSIL-ENERGY-FREE AGRICULTURE

Chuan Ma[1], Dominik Rutz[1], Veronika Hofmeier[1], Rainer Janssen[1],
Thanos Balafoutis[2], Konstantinos Vaiopoulos[2], Claus Aage Grøn Sørensen [3], Dimitris Manolakos[4], Alex Koutsouris[4], George Papadakis[4], Magdalena Borzecka[5], Vanja Bisevac[6], Daan Creupelandt[7], Julio Román[8], Frank Oudshoorn[9], Daniele Rossi[10], Martyna Próchniak[11], Zisis Tsiropoulos[12], Harm Brinks[13], Barry Caslin[14], Jorge Sneij[15], Maite Zarranz[16]

[2]Centre for Research and Technology Hellas (coordinator), [3]Aarhus University, [4]Agricultural University of Athens, [5]Institute of Soil Science and Plant Cultivation, [6]European Agricultural Machinery Association, [7]REScoop, [8]European Conservation Agriculture Federation, [9]Landbrug & Fodevarer, [10]Confagricoltura, [11]Lublin Agricultural Advisory Center in Konskowola, [12]AGENSO, [13]DELPHY BV, [14]TEAGASC, [15]Trama TecnoAmbiental, [16]Initiativas Innovadoras

[1]WIP – Renewable Energies, Sylvensteinstr. 2, D - 81369 Munich, Germany
Tel. +49 89 720 12732, Fax +49 89 720 12791
E-Mail: chuan.ma@wip-munich.de
Internet: www.wip-munich.de, www.agrofossilfree.eu

ABSTRACT: The AgroFossilFree project promotes fossil-energy-free technologies and strategies (FEFTS) in the agricultural sector by integrating various stakeholder initiatives and disseminating knowledge on FEFTS in agriculture. Applications boosting RES integration in important agricultural sectors, notably solar technology, are given considerable attention. The project's main goal is to create a platform where farmers, advisors, academics, business partners, and other stakeholders may discover pertinent information about FEFTS in agriculture. The AgroFossilFree initiative works to reduce the use of fossil fuels in all aspects of farming, which supports the EU Green Deal and Farm to Fork strategies. By encouraging the productive sharing of novel technologies, it helps to bridge the gap between the available FEFTS and common agricultural practices throughout the EU. By facilitating the effective flow of innovative information and ideas amongst stakeholders, it contributes to close the gap between the available FEFTS and common EU agricultural practices, thus European agriculture could become more environmentally, socioeconomically, and energetically sustainable. The European agricultural sector can support and abide by policies to reduce the use of fossil energy, the relevant legal and regulatory frameworks, and sustainable food production practices with the help of the knowledge that is gathered, classified, and presented to actors in the agricultural field. This article provides an update on the European Commission founded AgroFossilFree project.
Keywords: Photovoltaics, Agri-PV, RES integration, Agriculture, Strategy, Policies implementation, Sustainability, Defossilisation

1 INTRODUCTION

Compared to other sectors, global agriculture emissions constitute a major contributor to climate change. About 10% of the EU's total GHG emissions in 2015 were attributable to the agricultural sector, excluding land use, land use change, and net forestry removals [1]. These emissions are dominantly associated with livestock production (both enteric fermentation and manure management) and the application of fertilisers to agricultural soils. A recent study [2] found that the emissions related to the production of energy in agriculture (except for the horticultural crop type) are of minor importance compared to other emission sources.

Nevertheless, in absolute numbers, large amounts of fossil energy are used in agriculture for fertilizer and pesticide production as well as for heating, cooling, transport and electricity. The AgroFossilFree project aims to support any fossil fuel saving measures and replacements by renewable energies, also named as Fossil-Energy-Free Technologies and Strategies (FEFTS).

FEFTS refer to the tools that are required to address cleaner and more efficient energy production and use in agriculture. FEFTS are categorised based on the associated agricultural technology application. The two main categories are Clean Energy Production and Energy Efficiency Improvement, while an auxiliary category refers to Carbon Sequestration.

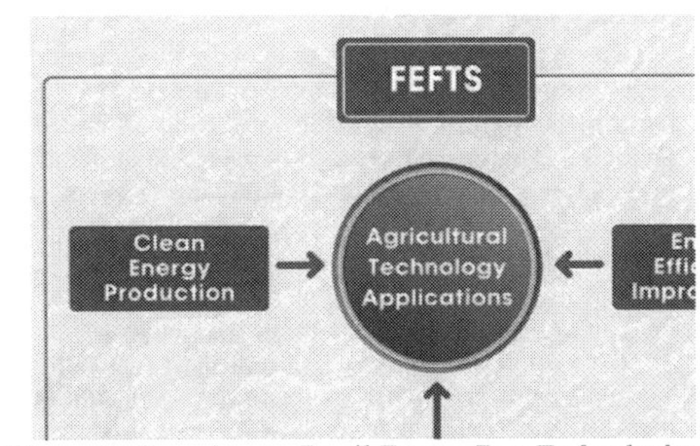

Figure 1: Overview on Fossil-Energy-Free Technologies and Strategies (FEFTS)

2 THE AGROFOSSILFREE PROJECT

The AgroFossilFree project (Figure 1) is supporting integration of FEFTS in agriculture by coordinating several stakeholder actions and by providing information on FEFTS in agriculture. Special emphasis is placed on applications promoting RES integration in major agricultural sectors and especially solar energy e.g., photovoltaics, Agri-PV and solar thermal. As it is evident from our research that the focus of the research and the industry sectors is focused on solar energy and biomass applications.

The core objective of the project is to set-up an online platform (AgEnergy Platform) for farmers, advisors, researchers, industrial partners and other actors on which they can find useful information on FEFTS in agriculture.

The AgroFossilFree project contributes to the High Level EU Strategies (i.e. EU Green Deal and Farm to Fork strategy) as it aims to decrease the use of fossil energy in any farming process. It also contributes to close the gap between the available FEFTS and the everyday EU agricultural practices by promoting effective exchange of novel ideas and by exchanging information between research, industry, extension services and the farming community. If this gap closes, European agriculture can become more sustainable with minimum energy, environmental and socioeconomic impacts. With the knowledge that is collected, classified and presented to actors in the agricultural field, the European agricultural sector is able to promote and comply with the fossil energy use reduction policies, the related legal and regulatory frameworks and sustainable food production practices.

The expected results of the project are:

- Assessment and evaluation of the EU's agricultural sector's present energy usage and the necessary steps to establish a farming sector free of fossil fuels.
- Identifying and presenting readily available and immediately usable FEFTS from market solutions and examination of the current finance tools for de-fossilising efforts.
- Producing community-based ideas for FEFTS integration in agricultural systems at the regional and EU levels requires collaboration with all pertinent stakeholders in thematic groups utilizing interactive physical and online approaches.
- Creation of an online platform with all available FEFTS for stakeholders and a Decision Support Toolkit to propose interventions and financing tools.
- The development of policy directives at the national, regional, and EU levels that will be disseminated, as well as the promotion of FEFTS in upcoming agricultural operations.

The AgroFossilFree project is funded by the Horizon 2020 Programme of the European Commission (Grant Agreement Number 101000496), started in October 2020 and has a duration of 3 years. An overview on the status of the project is presented in the following chapters.

3 RESULTS

3.1 Current energy use status in EU agriculture

After an evaluation of the current energy situation in EU agriculture, the major conclusions show that fossil fuel dependence and high energy use are present throughout the EU. Among the findings of the AgroFossilFree Project, 3 Scientific Papers were produced, with 2 out of 3 already being published in prestigious Scientific Journals. The 1st paper called "Energy use in open-field agriculture in the EU: A critical review recommending energy efficiency measures and renewable energy sources adoption" is published in Renewable and Sustainable Energy Reviews. The second paper called "Energy Use in the EU Livestock Sector: A Review Recommending Energy Efficiency Measures and Renewable Energy Sources Adoption" is published in Applied Sciences. Whereas, the 3rd paper, which is about the Energy usage in the Greenhouse Sector, is expected to be published soon. These papers are categorized based on the agricultural sector (Open-field, Livestock and Greenhouse) in which the search is being conducted. The most important findings of these papers are presented in the following section.

According to Eurostat, agriculture accounts for 3.2% of total energy consumption, 56% of which is derived directly from crude oil and petroleum products, 17% from electricity, 14% from gas and 9% from renewables and biofuels. However, AgroFossilFree results suggest that if indirect energy use associated with the production and transport of fertilizers and pesticides is included, the proportion of energy use in the EU-27 would be 62% higher overall.

The results also show that energy use, its concentrations and breakdown, vary significantly per production system (open-field, livestock, greenhouses). According to AgroFossilFree estimates of the crops and production systems, the annual energy inputs for arable agriculture are 1227 PJ, for orchards and vineyards 208 PJ, for meat production systems 501 PJ, and for dairy production systems 543 PJ.

For open-field agriculture, the use of fertilizer is the largest energy consuming activity in EU agriculture, accounting for around 50% of all energy inputs [3]. On-farm diesel use accounts for 30%, while other uses are mainly dedicated to irrigation, storage and drying which accounts for 8%. Pesticides and seeds each account for 5% of total energy inputs. In all livestock systems, except for beef production, animal feed is the main energy consumer accounting for around three quarters of all energy requirements. The production of animal feed consumes around 60% of the cereal production in the EU and requires significant high-protein imports. On-farm electricity use, which currently mainly comes from fossil sources, is also significant but varies depending on the production system [4]. In high yielding and high-energy intensive greenhouses, energy demand is dominated by heating and cooling. Lower yielding and less energy-intensive systems, on the other hand, require little to no heating/cooling and instead mainly rely on fertilizers, diesel use for machinery, irrigation and other activities.

The aforementioned shows that in order for the EU to meet the objectives outlined in the Green Deal and Farm to Fork strategy, it will likely be necessary to adopt cutting-edge technologies, increase energy efficiency, and continue the development and deployment of non-fossil energy sources for agriculture. In addition, a changeover to farming systems and agricultural techniques that are more sustainable is required. For instance, conservation agriculture can significantly reduce on-farm energy use and carbon emissions as well as sequester considerable amounts of carbon (up to 190M tonnes per year) and improve the overall climate resilience of the agricultural sector.

Providing an overview of energy use in EU agriculture is a challenging topic due to the plethora of the available, and in most cases inconsistent data. Further study may be useful in a number of areas, however. Additional research on the energy usage of the three primary production systems, particularly greenhouse agriculture, would help to better understand and estimate the amount of energy used in EU agriculture. It is also necessary to conduct additional study on the relationships between farm indicators (size, location, and specialty) and energy usage, as well as the energy use of non-conventional systems and crops that are grown on a small scale.

In addition to these findings, significant preliminary effort has already been done to identify and promote creative case studies on FEFTS. Evidently, both adopters and non-adopters took part in a farmer's survey. The acquired data will be used to produce multimedia content demonstrating novel FEFTS, highlighting the shift towards a more sustainable EU agriculture.

3.2 Inventories of FEFTS

A methodology for a data gathering inventory was developed to feed the AgEnergy Platform, which was released in early 2022. The methodology includes a number of categories that will help with the web platform's search function. Information about FEFTS was included by the project consortium and fed into the web platform:

- Scientific papers
- Research projects
- Commercial technologies
- Training and advising material
- Financial support mechanisms

A database was created and in total 1556 FEFTS entries registered (838 of which were published on the platform as more intriguing and comprehensive content provision). The remaining entries are enlisted in the AgEnergy Repository and will be made publicly available after a thorough review. It is important to note that many of the FEFTS entries submitted so far—more than 240—concern solar energy solutions, implementation strategies, policy recommendations, funding schemes, and other topics. Hence, emphasizing the transition to solar-powered solutions. The public AgEnergy platform now allows external users to register new entries. As a result, new cutting-edge technologies are added daily and the number of FEFTS on the platform keeps increasing.

3.3 AgEnergy Platform

As previously mentioned, AgroFossilFree created the AgEnergy Platform (https://platform.agrofossilfree.eu/en) in which all the available FEFTS are included in the form of easily accessible and comprehensive end-user material (Figure 2).

Figure 2: Homepage of the AgEnergy platform

In January 2022, the official platform launch was conducted. All the submitted FEFTS can be found under 3 basic categories: Clean Energy Supply, Energy Efficiency Improvement and Soil Carbon Sequestration. The platform (Figure 3) and the FEFTS included in it, are currently being translated into the consortium's spoken languages. These languages are: English, Greek, Polish, Spanish, German, Dutch and Danish. This is done so that all interested stakeholders from different countries can access the available information in their native language.

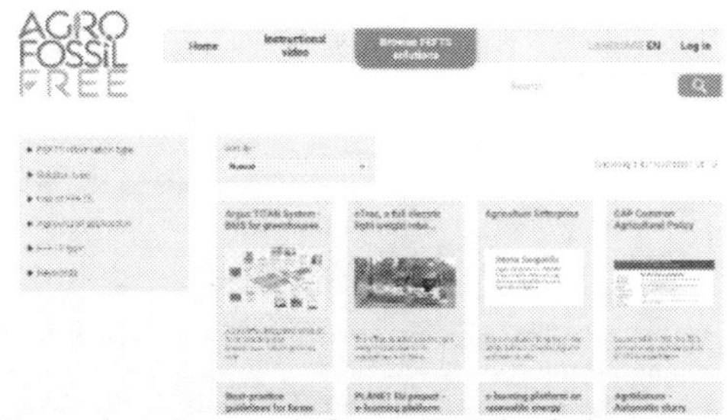

Figure 3: AgEnergy platform's FEFTS solutions

The platform provides detailed filters so that end users could find any targeting FEFTS by using a variety of methods, such as FEFTS information type, use of FEFTS, FEFTS type and keywords.

Furthermore, all registered users can evaluate the available FEFTS solutions by answering a 10-question survey on general info of the FEFTS as well as the environmental and socioeconomic impacts.

In addition, a decision support toolkit (DST) is currently under construction. This tool serves as an advisory mechanism to provide interested stakeholders with the most appropriate solution according to their needs. This tool, when finished, will be integrated into the AgEnergy Platform.

3.4 Interactive and multi-actor regional workshops

Twenty-four multi-actor workshops were organized by AFF partners in 8 European countries (3 per country) based on (i) Open-field agriculture; (ii) Greenhouses; and (iii) Livestock buildings. These workshops enable the direct distribution of innovative FEFTS solutions and the collection of grassroots innovations and requirements at the regional level by bringing stakeholders from research, extension services, industry, and farmers inside and outside the project together. More particularly, from the regional workshops, useful information is acquired about the requirements and issues that the stakeholders are facing. To achieve a sustainable EU agriculture, it is also necessary to consider what additional policy proposals, incentives, and new partnerships should be made.

After the completion of the regional workshops, three transitional workshops are scheduled. The first transitional workshop was held in Greece on Greenhouse, the second transitional workshop took place in Poland on open-field agriculture, and the last one will be held in Denmark on livestock on December 2022. These workshops will incorporate all the information gathered from the national workshops in order to produce a comprehensive outcome at the EU level. New policy recommendations will be made using these findings, and they will also serve to emphasize the direction that future research should focus.

4 CONCLUSIONS

Along with other important industries, agriculture needs to actively pursue a variety of mitigation strategies in order to help Europe reach its goal of climate neutrality as soon as possible.

Innovations in animal production, the application of fertilizers to agricultural soils, the use of pesticides, and the overall energy use in agriculture are some examples of these strategies. The AgroFossilFree project contributes to the dissemination of knowledge and facilitates easy access to data on fossil-free agriculture technology, such as biogas production, precision agriculture, and soil covering, through the development of the AgEnergy platform.

Additionally, the regional and international workshops being held all around Europe help to spread the word about how important it is to reduce the use of fossil fuels while also offering fantastic chances for knowledge sharing between stakeholders.

With the completion of the workshops and the AgEnergy platform, extensive feedback will be accumulated regarding new policy proposals, initiatives, and collaborations that should be carried out to achieve a sustainable EU agriculture.

5 REFERENCES

[1] European Parliament (2020) Briefing: EU agricultural policy and climate change. - https://www.europarl.europa.eu/RegData/etudes/BRIE/2020/651922/EPRS_BRI(2020)651922_EN.pdf

[2] Gołasa P. et al. (2021) Sources of Greenhouse Gas Emissions in Agriculture, with Particular Emphasis on Emissions from Energy Used. - Energies 2021, 14, 3784. https://doi.org/10.3390/en14133784

[3] Paris, B.; Vandorou, F.; Balafoutis, A.T.; Vaiopoulos, K.; Kyriakarakos, G.; Manolakos, D.; Papadakis, G. Energy use in open-field agriculture in the EU: A critical review recommending energy efficiency measures and renewable energy sources adoption. Renew. Sustain. Energy Rev. 2022, 158, 112098.

[4] Paris, B.; Vandorou, F.; Tyris, D.; Balafoutis, A.T.; Vaiopoulos, K.; Kyriakarakos, G.; Manolakos, D.; Papadakis, G. Energy Use in the EU Livestock Sector: A Review Recommending Energy Efficiency Measures and Renewable Energy Sources Adoption. Appl. Sci. 2022, 12, 2142. https://doi.org/10.3390/app12042142

6 ACKNOWLEDGEMENT & DISCLAIMER

This project has received funding from the European Union's Horizon 2020 research and innovation programme under grant agreement No 101000496. The contents of this publication are the sole responsibility of the authors and do not necessarily reflect the opinion of the European Union.

PV WILL SHINE IN THE NEW CLEAN ENERGY TRANSITION PARTNERSHIP

Stefan Nowak and Marcel Gutschner, SOLAR-ERA.NET Coordination Office, c/o NET Nowak Energy & Technology Ltd., Waldweg 8, 1717 St. Ursen, Switzerland, stefan.nowak@netenergy.ch and marcel.gutschner@netenergy.ch, +41 (0)26 494 00 30, https://www.solar-era.net
Michael Hübner, Coordinator Clean Energy Transition Partnership, Federal Ministry for Climate Action, Environment, Energy, Mobility, Innovation and Technology, Radetzkystraße 2, 1030 Vienna, Austria, michael.huebner@bmk.gv.at, +43 1 711 62-652922, https://cetpartnership.eu
Marko Topič, Chairman of European Technology and Innovation Platform Photovoltaics, University of Ljubljana, Trzaska 25, 1000 Ljubljana, Slovenia, marko.topic@fe.uni-lj.si, +386 1 4768 470, https://etip-pv.eu
Further co-authors from the networks involved:
https://www.solar-era.net, https://etip-pv.eu and https://cetpartnership.eu

ABSTRACT: SOLAR-ERA.NET – a network that brings together more than 20 national and regional RTD and innovation programmes in the field of solar electricity technologies in the European Research Area (ERA) – has been working together for more than a decade, providing funding to highly innovative transnational RDTI projects involving some 500 organisations, and further improving transnational cooperation both in the solar power sector and between national funding agencies [1]. Eight Joint Calls have been launched promoting priority topics identified by the European Technology and Innovation Platform Photovoltaics (ETIP-PV). Similar to other relevant energy ERA-Net's, the SOLAR-ERA.NET network is about to be integrated into a thematically much larger and energy system-oriented network called Clean Energy Transition Partnership (CETPartnership) that can seamlessly build on the momentum the ERA-Net's have created over the last 15 years. CETPartnership is a new transnational initiative on joint research, technological development and innovation (RTDI) programming to boost and accelerate the energy transition in Europe, building upon regional and national RTDI funding programmes. CETPartnership enables some 50 national and regional RTDI programmes from 30 countries to align their priorities, pool national budgets, as well as to implement annual joint calls each with an average budget of some 100 million euros from 2022 to 2027. The first joint call is launched in September 2022 and is based on the CETPartnership Strategic Research and Innovation Agenda (SRIA) [2] with respect to the topics and challenges addressed in this joint call. Via the European Technology and Innovation Platform Photovoltaics and its Strategic Research and Innovation Agenda" (SRIA) for Photovoltaics [3]. This contribution shall present i) key figures and findings from the long cooperation of national programmes in SOLAR-ERA.NET and ii) the new CETPartnership and opportunities for the PV industry and research community CETPartnership offers as a network and in the upcoming joint calls.
Keywords: Funding and Incentives, National Programmes

1 INTRODUCTION AND CONTEXT

The European Union, member and associated states have been supporting networks of national and regional funding organisations and research and innovation programmes for some twenty years in different formats. This transnational cooperation has become a key instrument and success factor for Europe, both for programmes as well as the industry and research community. This transnational cooperation allows for strengthening the competitiveness of European stakeholders in research and industry and more efficiently tackling major challenges related to energy and climate issues.

The European Strategic Energy Technology Plan (SET Plan) is a key stepping-stone to boost the transition towards a climate neutral energy system through the development of low-carbon technologies in a fast and cost-competitive way. By improving new technologies and bringing down costs through coordinated national research efforts, the SET Plan helps promote cooperation among EU countries and associated states, companies and research institutions, and in so doing also deliver on the key objectives in the energy and climate field. The European Technology and Innovation Platforms (ETIPs) were created to support the implementation of the SET Plan by bringing together EU countries and associated states, industry, and researchers in key areas. They promote the market uptake of key energy technologies by pooling funding, experts, skills, and research facilities.

In the PV area, the European Technology and Innovation Platforms for Photovoltaics has published the regularly updated European Strategic Research and Innovation Agenda for Photovoltaics [3, for the most recent one] and related Implementation Plans. They serve(d) as a reference for SOLAR-ERA.NET, especially for its eight joint calls carried out between 2012 and 2022. The first section shall present key figures and findings from the long cooperation of national programmes in SOLAR-ERA.NET.

With the new EU's Framework Programme for research and innovation, Horizon Europe, larger networks (called partnerships) of national and regional funding organisations and research and innovation programmes with a larger scope are supported. For the PV sector, the Clean Energy Transition Partnership (CETPartnership) provides the most interesting opportunities. As for the more single technology oriented ERA-Net's, joint calls to be carried out by the CETPartnership will refer to the Strategic Research and Innovation Agenda (SRIA) specifically defined, including PV. The CETPartnership SRIA [2] will serve as a guidance and compass for the multilateral collaboration in Europe and beyond for the next years. The second section shall present the new CETPartnership and opportunities for the PV industry and research community CETPartnership offers as a network and in the upcoming joint calls.

2 SOLAR-ERA.NET

2.1 Network

SOLAR-ERA.NET started in 2012 and has been supported by the EU in three projects (see section on acknowledgement). The network has been composed of some 20 organisations being programme owners and managers. In total, SOLAR-ERA.NET has organised eight joint calls.

2.2 Participation in joint calls – some key figures

The joint calls found good interest in the solar power industry sector and research community. 326 preproposals were submitted with a total budget of around 423 MEUR (of which 296 MEUR funding requested) involving 1'546 organisations (multiple counting – some organisations have been involved in several preproposals). About half of the preproposals were selected for the second stage of the call procedure, resulting in 177 full proposals.

In total, 93 transnational projects have been supported. The total public funding amounts to approx. 79 MEUR for a total project volume of around 117 MEUR. 488 organisations are involved in these transnational projects funded (multiple counting). A slight majority of the organisations involved belong to the categories of small and medium sized enterprises and large enterprises.

63 projects are finalized by the time of writing this paper. Results of the transnational projects can be consulted on the solar-era.net website. Another 30 projects are running.

Figure 1 displays the thematic split of the 93 projects.

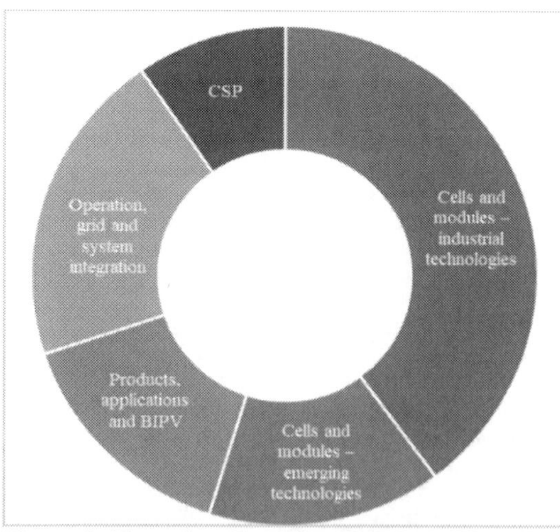

Figure 1: Thematic split of the 93 SOLAR-ERA.NET projects.

Figures 2 and 3 show the number of projects and the funding share by country / region resulting from the eight SOLAR-ERA.NET calls comprising 93 projects.

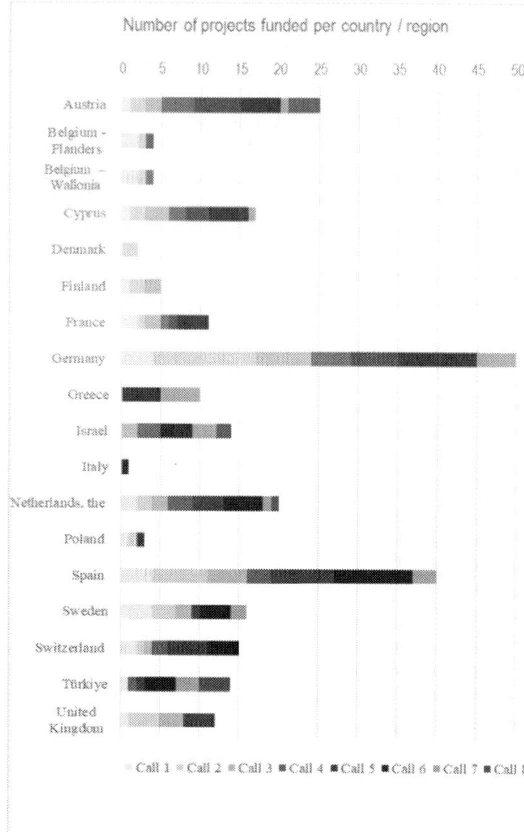

Figure 2: Number of projects funded - by country / region as well as by call resulting from the eight SOLAR-ERA.NET joint calls.

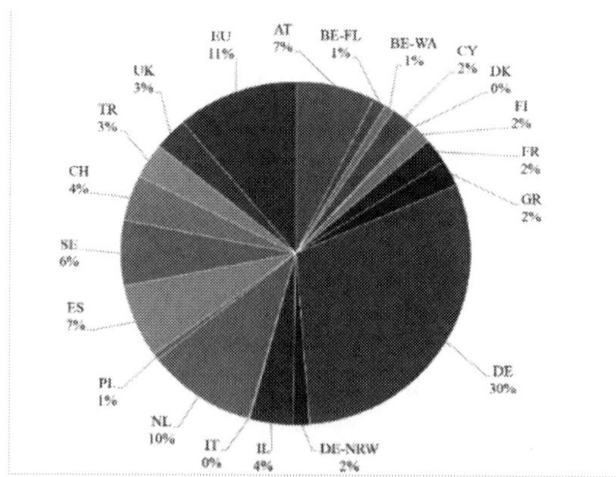

Figure 3: Funding share per country / region for the transnational projects supported through the eight joint calls carried out by SOLAR-ERA.NET.

2.3 Outcomes of the transnational cooperation – some key findings

The successful projects demonstrate i) potential commercial impact / relevance to industrial and market needs / contribution to the SET-Plan and added transnational value, ii) scientific and technological excellence and iii) high quality and efficiency of the implementation and the management.

SOLAR-ERA.NET has provided added-value and opportunities for the research and industry sector, namely:

- Complementarity between national and EU funded programmes
- Building a more robust and consistent research landscape
- Increasing strengths and reducing weaknesses
- Stepping stone for international cooperation
- Flexibility and bottom-up nature, no "one size fits all" approach
- New routes and innovative transnational research for medium-sized projects

Considering the experience made in eight joint calls, the following success factors and recommendations can be identified for potential applications also valid for the upcoming joint calls in the CETPartnership:

- Start with a good idea and have a vision how to realize it
- Build strong, complementary and trustful partnerships
- Communicate with partners and funding agencies
- Respect formal requirements / conditions
- Avoid redundancies with existing projects / proposals
- Identify innovation – excellence
- Head for industrial orientation – impact
- Provide convincing project plan – implementation
- Secure a high transnational value
- SMART Goals (Specific, Measurable, Ambitious, Realistic, Time Bound) / Key Performance Indicators

SOLAR-ERA.NET has not only allowed for supporting innovative research projects on a transnational level but also provided added value in terms of networking and programming, namely:

- New industry-led innovation projects and partnerships in a strongly competitive environment
- More transnational cooperation bottom-up and "cut to measure"
- Larger and more diverse portfolio
- Common initiatives and implementation activities
- Networking and collaboration between countries across Europe
- Overview on European research and insight into trends
- Best practice for supporting projects on the transnational level

Yet, there is a number of challenges respectively opportunities for further improvements related to public-public-partnerships in the future Horizon Europe programme.

- Adequacy and adaptability of European, national and regional funding budgets (Funding needs identified in the SET-Plan are considerably higher than funding budgets actually made available.)
- Flexibility to overcome the "funding gap" by following the ranking list with individual funding budgets running out already further up in the ranking list
- Consideration and inclusion of less strong ERA countries
- Visibility of impact through transnational projects and related projects
- Targeted and fair redistribution / use of Top-Up (Ideally, all funding agencies and transnational projects get the same share of the Top-Up.)
- Minimum extra administrative workload for agencies and project partners on transnational level
- Synchronisation of the timeline with all funding agencies (especially in the negotiation and contractual phase)
- Pragmatic handling (e.g. making reserve list possible)

SOLAR-ERA.NET's transnational activities have had a strong focus on innovation in new concepts, materials, processes, components and applications, by joining complementary scientific and industrial competences across the European RTDI landscape in view of bringing new solutions to the growing solar electricity market. Industrial relevance, cost effectiveness and scalability of innovative approaches are key criteria to be fulfilled. Increasingly, systemic aspects and integration have become more important and will gain a new dimension within CETPartnership.

This dimension is also aimed at by the European Technology Innovation Platform for Photovoltaics (ETIP-PV). In its recently developed Strategic Research and Innovation Agenda for photovoltaic science, technology and engineering in Europe [3], two overarching challenges to PV's large potential are identified: i) making the energy transition a European success and ii) supporting economic recovery and building the value chains for renewables. These challenges as well as the topics of highest scientific innovation and relevance in the PV field are also integrated in the SRIA developed by and for the CETPartnership.

3 Clean Energy Transition Partnership

3.1 Set-up and mission

With the creation of the CETPartnership, the network and thematic scope will grow considerably, capitalising on SOLAR-ERA.NET and other ERA-Net's networks, RDTI portfolio and know-how. CETPartnership brings together some 50 participating RDTI programme owners / managers of 30 countries and covers virtually any renewable and efficient energy technology.

The CETPartnership follows a common vision reflected in its Strategic Research and Innovation Agenda (SRIA) [2] and a common ambition all CETPartnership members are committed to:

1) Build a transnational transformative RTDI joint programming platform by coordinating, pooling and strengthening regional, national and international RTDI funding programmes for Europe's pathway towards the clean energy transition.

2) Accelerate clean energy technology development and transition to a widely decarbonised energy systems through demonstration and innovation in technology development and integration as well as system change.

3) Build an innovation ecosystem that fosters capacity building at all governance and actor levels, faster market diffusion, upscaling and replication and enabling of the clean energy transition.

Concretely, CETPartnership will implement a series of annual joint calls for RTDI projects from 2022 to 2027. Accompanying activities to build a Knowledge Community will ensure a maximum quality, dissemination and consolidation of the findings from the funded projects.

By establishing an Impact Network on programme level, the outputs of the funded projects will be taken to a higher level and the "wheel grip" of the initiative will be enabled by regularly listening to the voice of the need owners, potential buyers and implementers of solutions.

3.2 First joint call

CETPartnership translates the challenges identified for the energy technologies into so-called Transition Initiatives (TRIs) which will allow to systematically articulate the upcoming joint calls and activities. The seven TRIs are shortly described below. Out of the seven TRIs, TRI 2 and 7 are of particular interest for the PV sector.

- TRI 1 „Integrated Net-zero-emissions Energy System": The main objective is to develop the optimised, integrated European net-zero emissions energy system, where electricity distribution and transmission grids are seen as the "backbone" of the future low-carbon energy systems with a high level of integration among all energy carrier networks, by e.g. coupling electricity networks with gas, heating and cooling networks, supported by energy storage and power conversion processes.
- TRI 2 „Enhanced zero emission Power Technologies": The goal is to develop a pool of zero-emission power technologies and solutions based on Renewable Energy Sources as the backbone of the future energy system, being able to deliver carbon-neutral electricity accessible to all and to contribute to the resilience of the system. TRI 2 is articulated in two submodules. „Advancing RE technologies for power production through cost reduction" targets demonstration projects aiming at reducing the LCoE and/or CAPEX through technology development of (primarily) components or at system level. „Breakthrough R&D to increase RE power technologies efficiency" supports breakthrough research (TRL 4 or above) leading to increase the conversion/transfer efficiency and reliability of RE technologies and sustainability.
- TRI 3 „Enabling Climate Neutrality with Storage Technologies, Renewable Fuels and CCU/CCS": The goal is to provide technological cleaner solutions for storage technologies, hydrogen and renewable fuels, CCS (Carbon Capture and Storage) and CCU (Carbon Capture and Utilisation). In particular, projects are sought that have a significant bearing on accelerating the technologies and provide results showing significant CO_2 reduction by 2030 and demonstrate a contribution to the climate neutrality by 2050.
- TRI 4 „Efficient zero emission Heating and Cooling Solutions": The goals are to provide enhanced and improved heating and cooling technologies and systems for all major parts of Europe by 2030 and to enable 100% climate-neutral heating and cooling by 2050.
- TRI 5 „Integrated Regional Energy Systems": The goal is to develop and validate integrated regional and local energy systems, that make it possible to efficiently provide, host and utilize high shares of renewables, up to and beyond 100% in the dynamic local or regional supply by 2030. Such systems shall provide tailor-made solutions that meet the individual regional and local requirements and demand.
- TRI 6 „Integrated Industrial Energy Systems": The goal is to develop and demonstrate a set of technical solutions for integrated industrial energy systems that enables efficient carbon-neutral industrial production sites and takes industrial energy systems into development as part of the entire energy system. It focuses specifically on integrated solutions across industries, across energy sectors and across public and private sectors.
- TRI 7 „Integration in the Built Environment": The goal is to provide solutions and technologies for existing and new buildings to become an active element in the energy system, with enhanced capability to produce, store and efficiently use energy in the residential and non-residential sector, comprising public and commercial buildings, service and mobility infrastructure buildings, etc. Two modules are defined according to the TRL targeted, i.e. „Solutions to energy transition in the built environment" as an Innovation and Demonstration Action (IDA), targeting TRL 5-9, and „R&I in clean energy integration in the built environment" as a Research and Innovation Action (RIA), targeting TRL 3-6.

The first joint call opened on 14 September 2022; preproposals can be submitted latest by 23 November 2022. Applications selected in January 2023 for the full proposal phase have then a submission deadline running till 20 March 2023. Projects selected for funding in June 2023 are expected to start in September 2023.

4 OUTLOOK

As CETPartnership is in the starting blocks, results will, obviously, still have to be materialised. The set-up of CETPartnership is however already promising in terms of large programmers' network and a series of annual joint calls with substantial funding budgets.

CETPartnership is based on and anticipating the following results and conclusions: Our energy system is not just an "infrastructure" or a "supply system". It is - to a growing extent - a complex and dynamic system of systems, with not only technical but also social, economic, environmental, and political dimensions. In order to enable the clean energy transition, innovation is needed on different levels.

A remarkable portfolio of enabling zero emission technologies for the energy system has become available in recent years, with a significant industrial and economic potential, which is especially true for PV. The ongoing deployment is driving a transformation of energy systems around the world while exponential learning curves of these technologies are leading to a decrease of production cost per unit as well as gains in efficiency and performance.

On the next level, these energy system components have to interact in user groups' specific energy systems. In the context of the built environment, they can help transform existing or new buildings, blocks and districts into active elements in the energy system, providing high-level services to their inhabitants or commercial users. The integrated energy system will meet the users' demands throughout the seasons, with the aid of storage technologies. Integrated industrial power, heating and cooling systems as well as enhanced processes enable efficient carbon-neutral industrial sites and production that at the same time provide flexibility to the power system or make excess heat available for local networks. The efficient utilisation of high shares of renewables in the regional supply becomes possible with a smart management of available energy sources and infrastructures as well as users from different sectors like communities, commerce and industry or the transportation system. Such integrated regional systems can increase acceptance and uptake of new solutions, by ensuring that citizens, companies, communities and other stakeholders take part in the related exchange of different kinds of values on different levels.

In the end, all these elements are requested to seamlessly link to an overarching European energy system compatible with climate neutrality, with a high level of integration among all energy carriers, infrastructures and networks, and based on the "energy efficiency first" and "Do-No-Significant-Harm" principles. This will ensure an appropriate level of reliability, resilience and economic efficiency.

Before and for getting there, highly innovative RDTI projects shall continue to and will make their contribution. PV is becoming part of the backbone of the clean energy transition and the future energy system. New bold initiatives are needed, taking advantage of the wide experience and good practice gained over the past decade in SOLAR-ERA.NET, IEA-PVPS and other networks, keeping the momentum and accelerating the efforts. CETPartnership is one of these initiatives.

5 REFERENCES AND ACKNOWLEDGEMENT

References:

[1] SOLAR-ERA.NET, European Cooperation in PV RTDI and Beyond – Highlights, Lessons Learned and Future Perspectives, paper and presentation at 38th EU PV SEC, September 2021

[2] Clean Energy Transition Partnership (CETPartnership), Strategic Research and Innovation Agenda (SRIA), November 2020

[3] European Technology Innovation Platform for Photovoltaics (ETIP-PV), The European Strategic Research and Innovation Agenda for Photovoltaics, May 2022

Acknowledgement:

SOLAR-ERA.NET joint activities have been supported by the EU: SOLAR-ERA.NET under grant agreement N° 321571 within the 7th Framework Programme, SOLAR-ERA.NET Cofund under grant agreement N° 691664 and SOLAR-ERA.NET Cofund 2 under grant agreement N° 786483 within Horizon 2020. CETPartnership is supported by the European Union within Horizon Europe under grant agreement N° 101069750. This support is gratefully acknowledged.

SAVINGS BY NET-METERING PV IN A SPANISH HOUSEHOLD

Rodríguez San Segundo HJ*[1] and López N[1,2]
[1]Edhuna Consulting
Alhaja 5
28054 Madrid (Spain)
[2]Universidad Autónoma de Madrid
Ciudad Universitaria de Cantoblanco
28049 Madrid (Spain)
*Author for correspondence. E-mail: edhunas@gmail.com

ABSTRACT: The purpose of this work is to evaluate the real impact of a solar PV installation in net-metering scheme on the household economics of a Spanish home located in Madrid, central Spain, with over 1,600 average equivalent sun-hours per year. To that purpose, a 1.5 kW_p PV system has been monitored from its installation date in February, 2020. The impact of PV production on consumption, compared to the time before PV, has been assessed, both in terms of energy (kWh) and savings in the monthly electricity bill (€). The work has also taken into account the kWh retail price variations. Year-on-year energy (kWh) savings have reached an average of 48%. Economic (€) savings are in the range of 41%. These significant savings are, however, hindered by the fact that the Spanish net-metering regulation for PV power plants includes aspects that are detrimental for the PV plant owners. On one hand, solar kWh fed by the prosumers into the grid are paid by the utility at an average of 32% lower than the price the prosumer pays for the consumed retail kWh bought to the utility. On the other hand, there is a limit to the amount of solar kWh that can be paid. Every kWh over this limit is given for free. This amounts to 14% of the paid solar kWh.
Keywords: economic analysis, performance, prosumer, residential systems

1 INTRODUCTION

When profiting from the advantages of a grid-connected solar photovoltaic (PV) system, a residential owner is usually legally bound to one of basically three schemes: a.) feed-in-tariff, where the producer is paid per kWh fed into the grid at a state-fixed tariff; b.) self-consumption, where the prosumer (producer + consumer) optimizes consumption in order to couple it to PV production, or c.) net-metering, a mix of the previous two, where the prosumer sells the surplus (i.e., non-consumed) energy to the grid at an agreed price (equal or not to the price paid for consumption), therefore avoiding the need to optimize consumption.

Even though self-consumption has been the main driver of stand-alone PV systems for decades, when grid-connected PV irrupted the market in the mid-2000's, the preferred scheme was not self-consumption, but feed-in-tariff, where the state fixed a tariff at which each PV-produced kWh was paid. Quite slowly, grid-connected self-consumption schemes entered, first in Germany and then in many other countries, which very quickly passed to net-metering. This latter scheme has proven successful in many countries around the world, and many studies show[1] that it is profitable already, even in the absence of any subsidy[2]. This profitability can grow even higher with the "prosumer" behavior that couples both consumption and production[3].

The case of Spain, however, is quite different. While being one of the PV leaders and promoting the most favorable feed-in-tariff in the late 2000's, the Spanish PV market stagnated with the arrival of the financial crisis, with legislation being worse every year and prognoses poor[4,5]. Only very recently, it has begun to slowly recover with self-consumption, being Europe's biggest market in 2019[6], and having finally introduced net-metering with effect from the second quarter of 2020[7,8]. This recent legislation has, however, controversial and potentially discouraging corners. For instance, PV-produced kWh are not paid at the same price than PV consumed from the grid, but, in average, at 32% less (own data from electricity bills along the whole measuring period). More importantly, PV production is paid in the monthly bill, but only up to a limit related to the consumption from the grid made during that month. This means that every excess kWh over this limit is given for free to the utility, which in turn sells it at market prices. In times of high-energy prices such as the present times, this not only means big earnings for the utilities, but also loss of opportunities for the PV plant owners. Therefore, careful design and dimensioning of the PV plant is now more important than ever. Unfortunately, this aspect is usually not taken into account by plant designers, who tend to couple consumption with production when designing, and therefore offer PV plants with a usual average power ranging between 3.5 and 5 kW_p. Their customers, the future plant owners, often ignore the real savings and potential losses when giving their produced energy for free, and trust their PV plant providers to take care of that.

In this paper, we intend to demonstrate the importance of taking into account all aspects of the country's legislation (in this case, Spain) when designing a residential PV power plant. We will not only see how much money and energy is saved when compared to having no PV, but also how much is lost by giving the excess solar energy for free.

We present the performance of one mono-Si 1.5 kW_p PV system (i.e., much less than the average residential power plant currently offered by providers). It is located in Madrid, central Spain. Data have been taken from the monthly electricity bills in terms of retail electricity prices, consumption from the grid and PV-produced kWh paid by the utility. PV-production data have been extracted from the plant monitoring system.

Therefore, this works presents one of the few studies of real impact of solar photovoltaics on the plant owners' economies.

2 SAVINGS IN kWh AND EUROS

2.1 Savings in kWh (consumption)

To demonstrate first that it is indeed profitable to install a PV power plant, a comparison between "PV" and "No PV" has been carried out. Figure 1 shows the household's monthly energy consumption in kWh from December 2017 to July 2022, covering over 4.5 years. The 1.5 kWp PV system was installed in February 2020 (red bar in the graphs), approximately in the middle of the study period. Figure 1 clearly shows how the average consumption drops from 270 kWh/month before PV to 78 kWh/month after PV, i.e., a 71% drop in energy consumption. Note: in all Figures, the red bar represents the month of installation of the PV plant, the orange bar indicates the month from which PV production is paid by the utility (the delay is due to administrative issues), and the green bar refers to a change in electricity provider (utility) to decrease electricity prices (this happened before prices skyrocketed with the current energy prices). Therefore, despite the current energy crisis situation, it is clear that PV drastically decreases consumption.

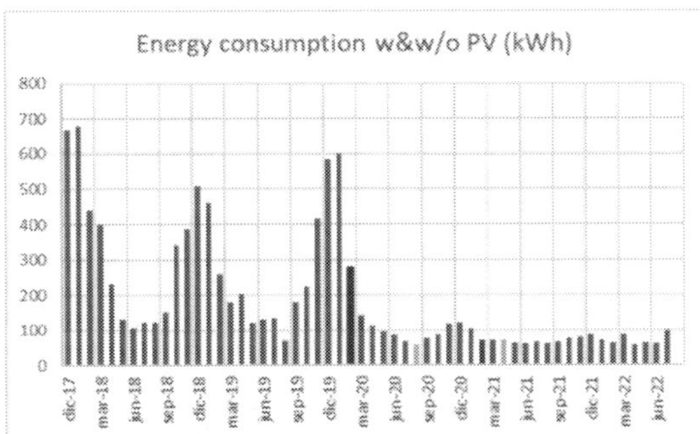

Figure 1: Energy consumption from December 2017 to July 2022. Red, orange and green bars are explained in the text (see Section 2.1).

To reduce the effect of the energy crisis, a one-year period has also been considered. In this case, the year-on-year (YoY) savings can be more accurately estimated, i.e., the comparison between the period from February 2019 to January 2020 and the period from February 2020 to January 2021. In this case, the energy savings still show the dramatic savings, adding up to 48%.

2.2 Savings in euros (monetary)

Figure 2 shows the electricity bills that were paid by the PV system owner in the same period, from March 2017 to July 2022. Again, a drop from an average of 89 €/month before PV to 34 €/month after PV is observed, which corresponds to 62% savings.

If the figures Year-on-Year (YoY) are taken, i.e., from February 2019 to January 2020 and from February 2020 to January 2021, the numbers reach 42% savings at the end of the month in the electricity bill, in euros.

Figure 3 shows the kWh retail price variation in the electricity bill (retail electricity price) in the same period. Although the owner change the electricity provider in April 2021 to decrease their costs (green bar), it is clearly seen how the prices start to escalate by the Fall of 2021, when the energy crisis started. Figure 3 shows that the retail price increased from an average of 0.157 €/kWh from December 2017 to July 2022, to an average of 0.201 €/kWh from June 2018 to May 2019. This implies an increase of retail electricity prices of 28% in average.

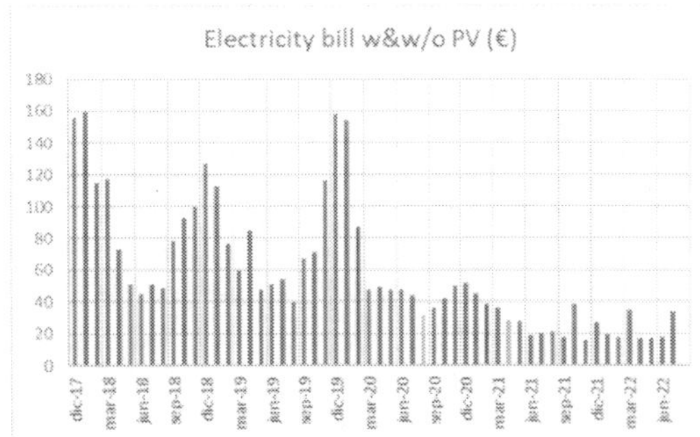

Figure 2: Electricity bill in euros, from December 2017 to July 2022. Red, orange and green bars are explained in the text (see Section 2.1).

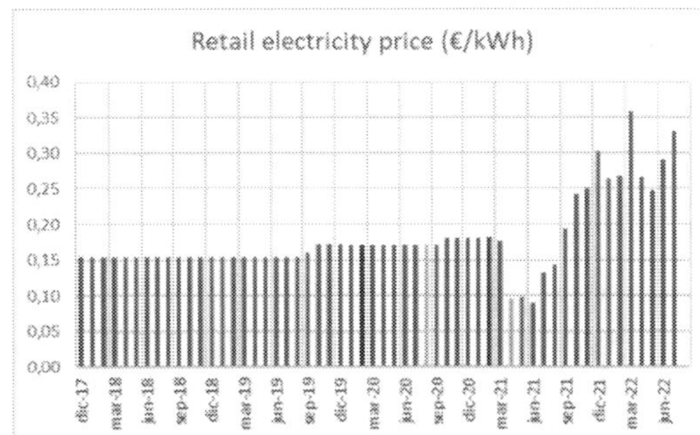

Figure 3: kWh retail prices in euros, extracted from the electricity bill, from December 2017 to July 2022. Red, orange and green bars are explained in the text (see Section 2.1).

Figure 4 summarizes all previous results in the form of YoY savings, both in energy consumption (kWh, orange line) and money (€, blue line), together with the YoY kWh retail price variations (grey line) in the electricity bill. The mentioned savings are clearly seen in the 12-month period right after the PV system was installed, i.e., from February 2020 to January 2021.

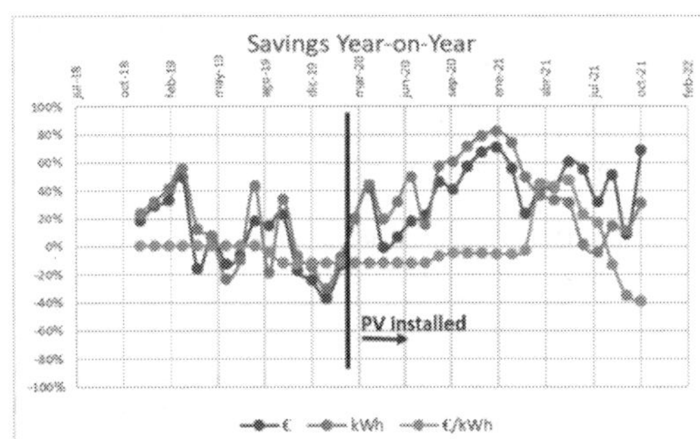

Figure 4: Year-on-Year savings in the electricity bill, in kWh (orange line), euros (blue line), and kWh retail price (grey line) from December 2017 to July 2022. The red vertical line marks the PV system start (February 2020).

In other works[9], payback time of the PV power plant was also calculated. However, in the current circumstances, where market electricity prices are affected by external factors and are extremely high, it is assumed that they would disturb the results.

3 LOSSES DUE TO REGULATION

3.1 Explanation of the regulation

The electricity bill in Spain is composed of two main items, each of them contributing around half of the total amount. One is a so-called "fixed" term, which is paid for the availability of a certain wattage. This term has not been taken into account for the calculations included in this work, as have not been either the general taxes, such as VAT, among others. The other term relates to the actually consumed kWh each month, with a final price per kWh ("retail electricity price"), which is the one used in this work for comparison.

Furthermore, as mentioned above, the regulation for net-metering PV power plants in Spain pays the prosumer for the solar energy that the latter feeds into the grid ("PV electricity price"). This price, which is set by the regulation, is as commented an average 32% lower than the retail price (see Figure 5).

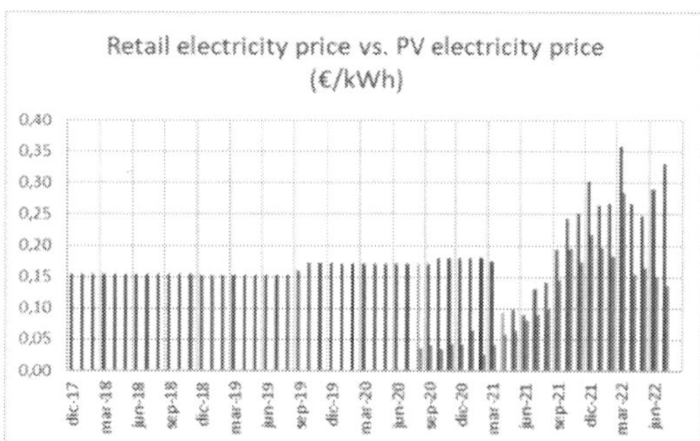

Figure 5: Retail electricity price (blue bars) and PV electricity price (orange bars, starting August 2020). The red vertical line marks the PV system start (February 2020).

On top of that, there is a limitation as to how many solar kWh are paid each month. This limit is related (not equal) to the amount of kWh consumed by the prosumer from the grid in the same monthly period. This is an important aspect of the Spanish regulation, and therefore an explanatory example will be given to explain more clearly how it works. As the regulation is complicated, these numbers must not be taken as real, but as an example of what happens:

Case A. The prosumer consumes 100 kWh from the grid, and feeds 80 kWh from the PV plant into the grid. Then all 80 solar kWh are paid at the PV electricity price.

Case B. The prosumer consumes 100 kWh from the grid, and feeds 100 kWh from the PV plant into the grid. Then all solar 100 kWh are paid at the PV electricity price.

Case C. The prosumer consumes 100 kWh from the grid, and feeds 120 kWh from the PV plant into the grid. Then, 100 kWh are paid at the PV electricity price, while the remaining 20 kWh are given to the utility for free. i.e., at the price of zero euros per kWh.

As will be seen in the results, case C is detrimental for the PV plant owner. The simplest solution to this problem would be to optimize the "prosumption" and self-consume as much solar kWh as possible, ideally the whole monthly production. However, in real-life situations this is nearly impossible. The next step would be to produce always below the regulated limit for payment. This is where solar plant design plays an important role: solar plants in Spain, given the current situation, should be designed below the yearly average payment limit that would balance the production, which is barely considered by PV plant providers. Even if this were the case, there would be extreme situations where a significant amount of solar kWh would be given for free to the utility. For example, whenever the owner goes on vacation and consumes zero kWh from the grid. In that period, the whole solar production would be simply given out at zero euros.

3.2 Implications for the prosumer (PV plant owner)

Figure 6 shows the monthly losses due to those solar kWh that remain unpaid per regulation, since the start of solar kWh payments in August 2020. As seen in Figure 6, even in such a low-dimensioned PV plant as the one in study, i.e., 1.5 kW_p as compared to the usual 3.5 - 5 kW_p, unpaid solar kWh happen in 18 out of the total 24 months in study (75% of the time). The losses can be over 60% and as high as over 90% over the paid solar kWh, such as the case of June and August 2021, where the owner had vacation and did not consume for significant periods of time. In average, the losses amount to a significant 14% of the paid solar kWh, which implies that the prosumer gives away to the utility 14% of their paid solar production for free. This number escalates when the usual sizes of power PV plants in Spain (3.5 – 5 kW_p) are taken into account.

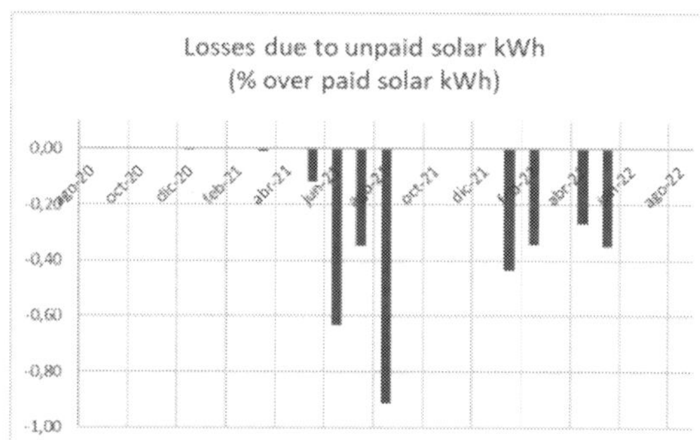

Figure 6: Retail electricity price (blue bars) and PV electricity price (orange bars, starting August 2020). The red vertical line marks the PV system start (February 2020).

4 CONCLUSIONS

A study of savings in energy (kWh) and money (€) of a 1.5 kW_p self-consumption PV system located in Madrid, central Spain, has been presented, with respect to the previous time with no PV. This PV power plant size is two to three times smaller than the usual PV plants designed for residential homes in Spain, which range from 3.5 kW_p to 5 kW_p. The household consumption in kWh and the money spent on the electricity bill have been monitored for a period of over four-and-a-half years, at around the middle of which this PV system was installed. An average of 71% energy savings and 62% monetary savings have been recorded. These numbers are, however, affected by the current energy crisis where anomaly high electricity prices are in place. To have a more accurate understanding of the savings, Year-on-Year data have been analyzed, and still 48% energy and 42% monetary savings have been observed, clearly showing the high direct impact of PV on the prosumer's monthly electricity bill.

The Spanish net-metering regulation nevertheless presents some controversial aspects. On one hand, during the period of study, the solar kWh (i.e., the kWh produced by the PV power plant and fed into the grid) have been paid by the utility at a price (PV electricity price) which is in average 32% lower than the price at which the consumer buys the kWh from the utility (retail price). On the other hand, the amount of kWh that the prosumer is allowed to sell to the utility is limited by the amount of kWh the prosumer buys from the utility. Every excess solar kWh fed into the grid by the prosumer will be given for free to the utility. In the period of study, this amount has been, in average, 14% of the paid solar kWh. These losses correspond to a size of only 1.5 kW_p. For bigger installations, such as the sizes of 3.5 – 5 kW_p usually designed in Spain for residential homes, the losses are expected to increase accordingly.

Given the significance of the losses, it is therefore important to take into account these aspects at the time of designing the power plant.

5 ACKNOWLEDGEMENTS

The work of N. López was partially supported by the Grant ERC-Stg-2017-758885 from the European Commission and by the Ramon y Cajal Program from the Spanish Ministry of Education.

6 LITERATURE

[1] Luthander, R, et. al., Applied Energy, Vol.142, pp. 80-94, 2015
[2] Lang, T, et al., Ren. Energy, Vol. 87, Part I, pp. 77-87, 2016
[3] Widén, J, Applied Energy, Vol. 126, pp. 199-212, 2014
[4] Talavera, DL, et al., Ren. and Sust. Energy Reviews, Vol. 33, pp. 752-771, 2014
[5] López Prol, J, and Steininger, KW, Energy Policy, Vol. 108, pp. 742-754, 2017
[6] https://www.pveurope.eu/News/Markets-Money/EU-Over-100-PV-market-increase-in-2019, consulted on 07/30/2022
[7] https://www.boe.es/buscar/doc.php?id=BOE-A-2019-5089, consulted on 07/30/2022
[8] https://www.boe.es/buscar/act.php?id=BOE-A-2014-3376, consulted on 08/10/2022
[9] Rodriguez San Segundo, HJ and López, N, in Proceedings of the 37th EUPVSEC, 2020, 1865-1869

ECONOMIC FEASIBILITY STUDY OF DIFFERENT CONFIGURATIONS OF A HIGH-POWER PHOTOVOLTAIC IRRIGATION SYSTEM

Rita Hogan Almeida[1], Celena Lorenzo[1], Luis Narvarte[1], Juan Ignacio Herraiz[1,2]
[1]Instituto de Energía Solar – Universidad Politécnica de Madrid. Campus Sur UPM. Ctra. Valencia km. 7. ETSIS Telecomunicación, 28031 Madrid, España.
[2]Qualifying PhotoVoltaics. Centro de Empresas la Arboleda. Calle Alan Turing 1, 28031 Madrid, España.

ABSTRACT: This paper presents an economic feasibility study of different configurations of a PV-grid high-power PV irrigation system in Alto Aragón, Spain. The 1574 kWp PV generator is installed in a previously only-grid pumping station with five motorpumps and, in the new configurations, PV can feed 3 of these 5 motorpumps. Conf.1 is a stand-alone solution since 3 pumps are fed by PV and the energy that is not consumed is wasted, Conf.2 is a stand-alone plus grid-connected solution in which the motorpumps are fed by PV during the irrigation period and PV energy is sold to the grid outside this period, and Conf.3 considers that PV energy can be directly used or exported throughout the year according to the pumping station needs. The PV energy consumed by the system is 1241.7 kWh/kWp, while the total energy used achieves 2016.8 kWh/kWp for Conf.3. Economic results show Profitability Index values in the 2.0-2.4 €/€ range, Internal Rate of Return values in the 5.52-7.23 % range, and Payback Period values in the 12-15 years range. The Levelized Cost of Electricity are in the 4.82-6.40 c€/kWh range, which means that savings can achieve up to 53% when compared to the only-grid system.
Keywords: Economic Analysis, Energy Performance, PV Pumping

1 INTRODUCTION

High-power photovoltaic irrigation systems (PVIS) are growing worldwide due to the increasing price of electricity from the electric grid and diesel generators [1], as well as the trend to phase-out fossil fuels. Currently, there is a lack of experience about this kind of high-power systems, which limits the feasibility assessment of its investment.

A 1574 kWp PVIS is currently under construction in Alto Aragón, Spain. The aim of this article is to provide the PV community an economic assessment of a high-power PVIS, evaluating the economic feasibility of partially substituting grid-powered irrigation systems with PVIS. Three different possible configurations will be studied and a discussion about the advantages and disadvantages of each one is presented.

The relevance of this paper lies on the study of the economic feasibility of a PVIS with more than 1 MWp. Up to now, this was only performed for small power systems [2-6] and for systems up to 360 kWp [7]. For this reason, there is the need to explore this solution for professional agriculture exploitations with high power systems (large irrigation communities, large farms, and agro-industries). For example, the system under analysis belongs to one of the irrigation communities of Riegos del Alto Aragón (which manages the largest irrigation system in Europe that consumes more than 50 GWh of energy per year) [8]. Accordingly, this study and this installation can be used as a reference for other similar systems.

To evaluate the economic feasibility, firstly, the energy and water productivity of the system is obtained through a simulation exercise done with SISIFO (a free software that allows the simulation of PVIS) [9].

Then, the economic analysis is performed. The economic profitability of the systems is evaluated in terms of the Net Present Cost (*NPC*), the Profitability Index (*PI*), the Internal Rate of Return (*IRR*) and the Payback Period (*PBP*). The Levelized Cost of Electricity (*LCOE*) is also calculated.

2 METHODOLOGY

2.1 Description of the system

A pumping station made up of five motor-pumps of 450 kW, pumping water from a channel to a water pool, is located in Alto Aragón, Spain (latitude 41.85°N, longitude 0.387°W). The irrigation period (IP) runs from February to October, which means that energy is only necessary in this period, and the annual energy demand is around 3 GWh. Figure 1 shows the monthly energy consumption of the system, in which can be seen that the maximum energy demand is in August, followed by July and September.

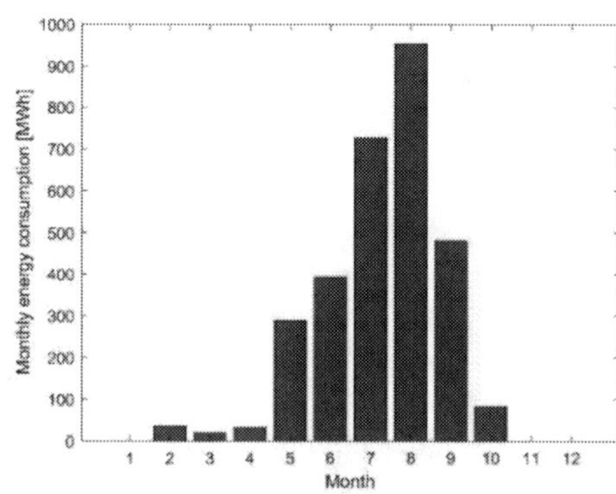

Figure 1: Monthly energy consumption.

In the next months, this system will be fed by a 1574 kWp PV generator. A PV-only solution is not enough to feed the irrigation system because, in some months of the year, the number of irrigation hours per day is higher than the number of hours of sunshine. Accordingly, in the configurations under analysis, a hybrid PV-grid solution is needed [10] – three motorpumps can be fed by PV during the day and all the five motorpumps can be fed by the national grid during the night. Considering this constrain, three different configurations are analyzed:

- A stand-alone solution, in which the PV generator will feed three of the five pumps. The PV energy not used by the pumping station is wasted (Conf.1, Figure 2).

- A stand-alone plus grid-connection solution in which the system will be a stand-alone configuration during the IP, and a grid-connected configuration outside the IP. This way, the PV generator can inject energy into the grid (Conf.2, Figure 3). The switch will be like in Figure 3 during the IP and will be in its other position outside the IP.
- A stand-alone plus grid-connection solution in which the PV system will be able to simultaneously feed the pumping station and export the energy surplus to the grid during the whole year (Conf.3, Figure 4);

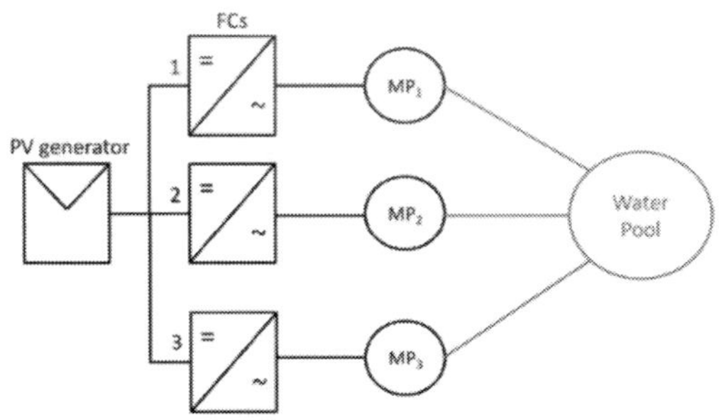

Figure 2: Conf.1: stand-alone solution.

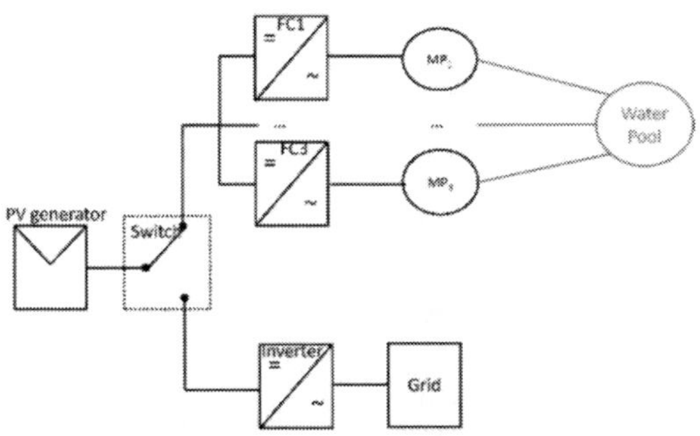

Figure 3: Conf.2: the PV generator fed the motorpumps during the irrigation period with the switch in the position of this figure, while outside the irrigation period the PV energy is sold to the national grid.

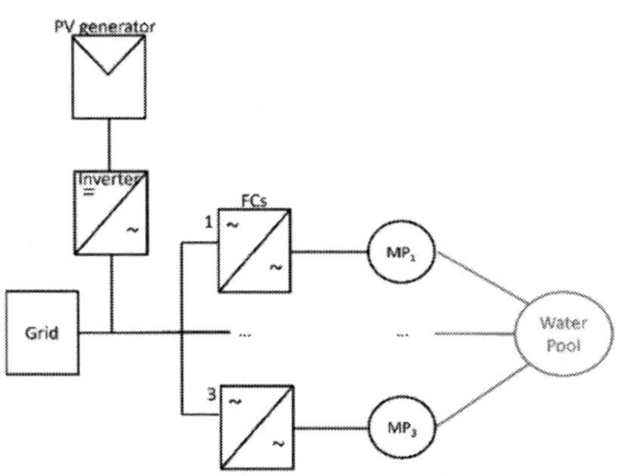

Figure 4: Conf.3: the energy from the PV generator can be directly used to feed the motorpumps or sold to the national grid along the year.

2.2 Energy performance

Firstly, the energy and water productivity of the system is obtained through a simulation exercise done with SISIFO (a free software that allows the simulation of PVIS) [9].

The difference between the energy obtained with PV and the real consumption of the system will be supplied by the national grid during night hours.

2.3 Economic study

For the estimations of the economic indicators, the annual cash flows were calculated for the whole lifetime of the system (25 years), considering the initial investment cost and the annual profits obtained. The initial investment cost was 0.8 €/Wp for Conf.1 and 0.9 €/Wp for Conf.2 and Conf.3 (this happens because in the last configurations a grid inverter needs to be installed). The annual profits are given by the economic savings obtained by substituting the national grid for the PV energy.

The installation of a PVIS, instead of a traditional grid-connected system, will be economically feasible when the resulting *PI* is bigger than 1, the *IRR* is higher than the discount rate, and the *PBP* is significantly lower than the lifetime of the system.

In this study, the irrigator community must have 20% of the investment and the remaining 80% is obtained as a loan and returned in 10 years.

The energy is sold to the national grid at 4.69c€/kWh – the average value in Spain in the last 10 years [11].
The full methodology to calculate the economic indicators can be found at [12].

3 RESULTS

3.1 Energy productivity

Table I shows the PV energy productivity (the total energy used, the energy consumed by the pumping station, and the one exported to the national grid for the different configurations under analysis). The energy consumed by the pumping station is 1241.7 kWh/kWp for the three configurations under analysis, while the energy coming from the national grid to fulfill the remaining consumption is 673,82 kWh/kWp. In what concerns PV used, as expected, it increased from Conf.1 to Conf.3 because in the first configuration energy not used in the pumping station is wasted, in Conf.2 the energy not used in the IP can be sold to the national grid and only in Conf.3 all the PV energy potentially generated is used.

It is important to highlight that Conf.1 is a stand-alone solution and, in general, permissions and authorizations are easier to obtain. For Conf.2 and Conf.3 there is the need to require permissions and authorizations that can be denied by the energy distribution companies.

Table I: PV energy for the three configurations under analysis

	PV Energy Used [kWh/kWp]	PV Energy Consumed [kWh/kWp]	PV Energy Exported [kWh/kWp]
Conf.1	1241.70	1241.70	0.00
Conf.2	1553.57	1241.70	311.87
Conf.3	2016.78	1241.70	775.08

3.2 Economic feasibility

Figure 5 shows the *LCOE* of the previous system and of the three different configurations under analysis. As can be seen, the use of PV will decrease the *LCOE* of the system when compared to the only-grid solution. The new *LCOE* will range from 4.82 c€/kWh to 6.40 c€/kWh, which represents *LCOE* savings from around 38% to 53%. As expected, the higher the use of PV, the lower the *LCOE*.

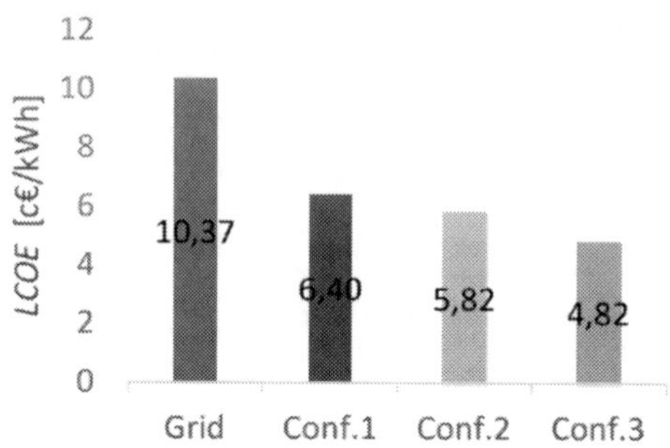

Figure 5: *LCOE* of the previously system fed by the national grid and of the three different configurations under analysis.

Figure 6 shows the *NPC* of the different configurations under analysis. As expected, the highest the energy used the highest the *NPC*.

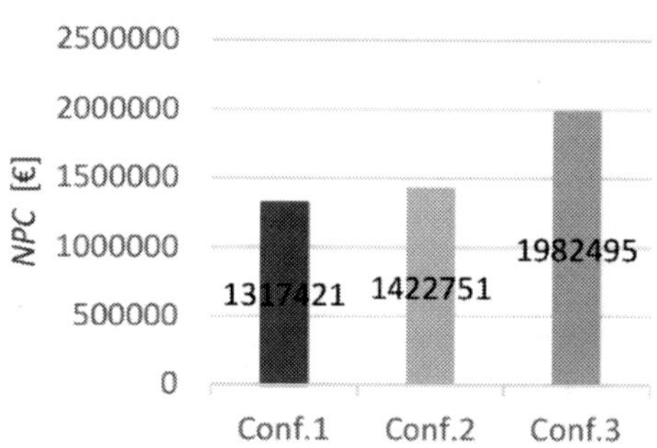

Figure 6: *NPC* of the three different configurations under analysis.

In terms of *PI* (Figure 7), values are in the 2.00-2.40 €/€ range, and in terms of *IRR* (Figure 8), values are in the 5.52-7.23%. The highest values are obtained for Conf.3, the one that maximizes the energy used. Results for Conf.1 and Conf.2 are simular since Conf.1 has a lower use of PV but also a lower *ICC* than Conf.2.

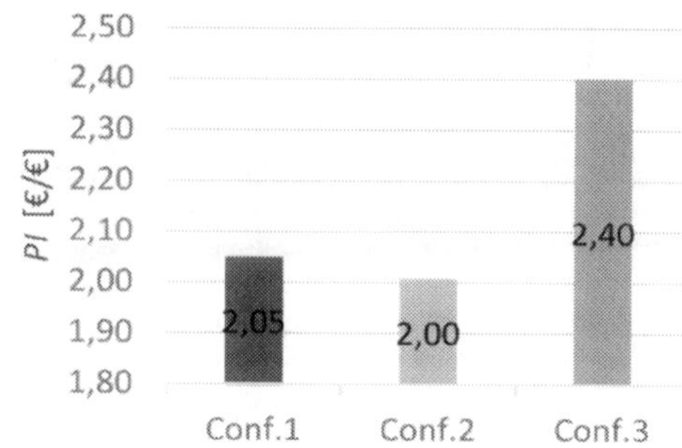

Figure 7: *PI* of the three different configurations under analysis.

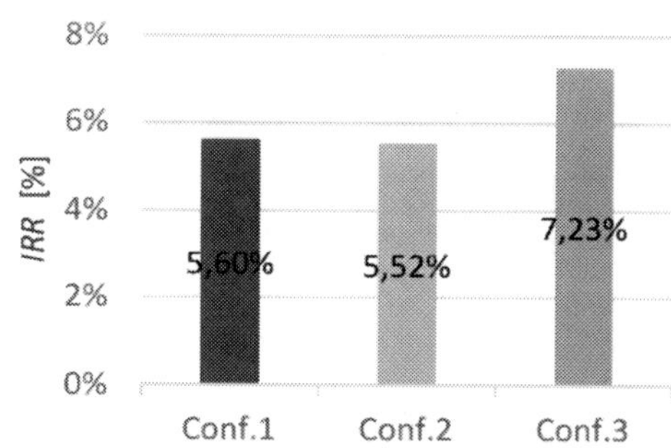

Figure 8: *IRR* of the three different configurations under analysis.

The *PBPs* (Figure 9) are lower than the lifetime of the system: ranging from 13 years in Conf. 3 to 15 years in Conf.1 and Conf.2.

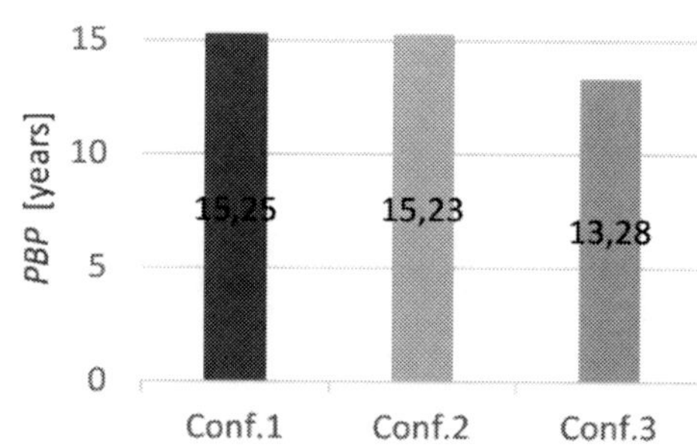

Figure 9: *PBP* of the three different configurations under analysis.

4 CONCLUSIONS

Results show the economic feasibility of partially substituting the energy source of a pumping station in Alto Aragón, Spain. Previously, the station was only fed by the national grid and this study considers the introduction of a 1574 kWp PV generator to decrease grid dependence and, as a consequence, energy costs. Three different configurations of PV systems are studied: a stand-alone configuration (Conf.1), and two stand-alone plus grid-connection solution (Conf.2 in which PV energy can be sold to the national grid outside the irrigation period, and Conf.3 in which all the PV energy not used by the PV generator can be sold to the national grid).

The *LCOE* of the only-grid system was 10.37 c€/kWh and savings up to 53% can be seen with the use of PV. For the different configurations under analysis, *LCOE* ranges from 4.82 c€/kWh to 6.40 c€/kWh, *NPC* values are all positive, *PI* values values are all higher than 1 €/€, *IRR* are higher than 5% and the maximum *PBP* is 15 years.

As expected, the possibility to export PV energy to the national grid increases the economic profitability of the systems and must be considered, if possible, to increase the economic feasibility of high-power PVIS.

Authors believe that this configuration is better than the standard self-consumption solution since it allows costs reductions in terms of both energy used from the grid and contracted power, but a future research line is the comparision of the results presented in this work with the self-consumption solution.

5 AKNOWLEDGEMENT

This work has been possible thanks to the funding from the European Union's Horizon 2020 research and innovation programme to SOLAQUA project under grant agreement No 952879, and thanks to the Project MADRID-PV2 (P2018/EMT-4308) funded by the Comunidad de Madrid with the support from ERDF Funds.

6 REFERENCES

[1] R. Langarita, J. Sánchez Chóliz, C. Sarasa, R. Duarte, S. Jiménez, Electricity costs in irrigated agriculture: a case study for an irrigation scheme in Spain, Renew. Sustain. Energy Rev. 68 (2) (February 2017) 1008-1019.

[2] I. Odeh, Y. Yohanis, B. Norton, Economic viability of photovoltaic water pumping systems, Sol. Energy 80 (2006).

[3] P. Purohit, Financial evaluation of renewable energy technologies for irrigation water pumping in India, Energy Policy 35 (2007).

[4] R. Foster, A. Cota, Solar water pumping advances and comparative economics, Energy Procedia 57 (2014).[5] M.A. Hossain, M.S. Hassan, M.A. Mottalib, Technical and economic feasibility of solar pump irrigations for eco-friendly environment, Procedia Eng. 105 (2015).

[6] A. Allouhi, M. Buker, H. El-houari, A. Boharb, M. Benzakour Amine, T. Kousksou, A. Jamil, PV water pumping systems for domestic uses in remote areas: sizing process, simulation and economic evaluation, Renew. Energy (2019).

[7] I. B. Carrêlo, R. H. Almeida, L. Narvarte, F. Martínez-Moreno, L. M. Carrasco, Comparative analysis of the economic feasibility of five large-power photovoltaic irrigation systems in the Mediterranean region, Renewable Energy 145 (2020) 2671-2682.

[8] http://riegosaltoaragon.es/

[9] https://www.sisifo.info/es/default

[10] R. H. Almeida, I. B. Carrêlo, E. Lorenzo, L. Narvarte, J. Fernández-Ramos, F. Martínez-Moreno, L. M. Carrasco, Development and Test of Solutions to Enlarge the Power of PV Irrigation and Application to a 140 kW PV-Diesel Representative Case, Energies, 11 (2018)

[11] https://www.omie.es/es/market-results/interannual/average-final-prices/components-spanish-demand?scope=interannual

[12] C. Lorenzo, L. Narvarte, and A. B. Cristóbal, A Comparative Economic Feasibility Study of Photovoltaic Heat Pump Systems for Industrial Space Heating and Cooling, Energies, 13 (2020)

TOWARDS A NEW APPROACH IN EVALUATING THE PRACTICAL ECONOMIC PERFORMANCE OF OFF-GRID STORAGE-LESS HYBRID ENERGY SYSTEM WITH INTEGRATED SOLAR FORECAST

Authors: Fuqiang Zhuang[1,2], Rodrigo Amaro e Silva[1], Thomas Carriere[3],
Benoit Gschwind[1], Yves-Marie Saint-Drenan[1], Philippe Blanc[1]
Institute: [1]Centre for Observation, Impacts, Energy, MINES Paris, PSL university
[2]Spie Industrie & Tertiaire
[3]SOLAÏS
Address: [1]1 Rue Claude Daunesse, CS 10207, 06904 Sophia Antipolis, France
[2]125 rue de Bielle, 64121 Serres Castet
[3]55 allée Pierre Ziller Atlantis, 06560, Sophia Antipolis, France

ABSTRACT: Photovoltaic (PV) energy plays an important role in the energy transition, especially for off-grid contexts, through a format of Hybrid Energy System (HES) with storage system. Yet, using a storage system comes with additional investment costs, extra environmental impact and maintenance issues, etc. For an off-grid PV/Diesel HES, a precisely designed Power Management System (PMS) with short-term solar forecast could be used to reduce or even avoid the use of energy storage system. Hence, this work aims to find out the feasibility of this concept, and introduce a new approach to evaluate the HES performance rather than using classic statistical metrics like Root Mean Square Error (RMSE) and MAE (Mean Absolute Error). Besides, several scenarios are tested to explore the influence of different elements on HES performance.

Keywords: Stand-alone PV Systems, Plant Control, Economic Analysis

1 CONTEXT & OBJECTIVE

In an off-grid context, the energy systems are mostly relied on fossil fuel energy [1]. During the energy transition, Photovoltaic (PV) energy is largely used through a format of Hybrid Energy System (HES) with storage system [2]. Yet, using a storage system comes with some extra adverse effects, such as high additional investment costs [1], extra environmental impacts [3] and maintenance issues, etc. A precisely designed Power Management System (PMS) with short-term solar forecast could be used as a transitory solution to reduce or even avoid the use of energy storage systems.

This work aims to find out the feasibility of this concept through evaluating the storage-less HES performance using PMS and solar forecast techniques. Yet, the classic metrics like Root Mean Square Error (RMSE) and MAE (Mean Absolute Error) could not always correctly describe the HES performance, especially when using different forecast methods [4]. For example, an over-estimated and under-estimated forecast with same level of RMSE could have quite different practical results. Hence, in this work, a new cost-based approach is introduced to properly evaluate the performance of stand-alone storage-less HES.

2 CASE STUDY & APPROACH

2.1 Case study

The case study of this work is an energy system for a mining site located in a tropical area in West Africa, with a rather 50 MW energy demand 24/7. This energy system is initially a diesel Genset-based energy system with 70MW capacity. Due to the more and more important environmental issues and the increase of the fossil fuel price, the PV energy is integrated into the energy system to reduce the environmental impacts and the Levelized Cost of Energy (LCOE) of electricity production.

When using PV energy, its intrinsic temporal variability has to be considered to ensure grid stability [5]. The common choice to tackle this issue is to use an energy storage system [6], [7], such as the pumped-hydroelectricity station, batteries, etc. However, while accomplishing this issue with the use of energy storage system, it also brings side effects like high investment costs, extra environmental impacts and maintenance issues, etc. Hence, the energy system of case study uses an innovative format of HES with PMS and short-term solar forecast rather than energy storage system.

2.2 Approach

The main idea of using a PMS with solar forecast is to predict the future residual demand, which is the difference between the total demand and the PV production, then committed by smartly dispatching the diesel generators.

To be able to simulate the behavior of HES using solar forecast, a python-based simulation platform using ***oemof*** is designed and implemented for the simulation [8], [9].

In this simulation work, probabilistic climatological and perfect forecast are used to define the full range of performance gain. The former forecast is used as a simple forecast method to provide the lower bound of economic gain, the latter forecast is the prognostic forecast which aims to provide the upper bound of possible economic gain. When using this probabilistic forecast information, the P_{50} quantile level is used to generate the Genset dispatch as the common choice [10], the uncertainty range derived from other inter-quantile levels like $P_0 - P_{50}$ is used to size the spinning reserve, which is the energy buffer of the HES to mitigate the solar variability. This solution aims to cover almost all the over-estimated forecast cases since the over-estimated forecast could be dangerous if the spinning reserve is not correctly sized. Yet, the under-estimated forecast case could be dealt with PV curtailment, which is relatively easier to handle.

As mentioned, the over-estimated and under-estimated forecast could bring different results in energy system performance. Thus, the classic metrics like RMSE and MAE could not always correctly describe the HES performance. For example, an over-estimated and underestimated forecast with same level of RMSE could have quite different practical results.

In this work, a new cost-based approach is therefore used to evaluate both system stability, economic and solar forecast performance. To explore the influence of different elements on HES performance, several scenarios are tested.

3 RESULTS & DISCUSSIONS

3.1 Scenario with different PV penetration at clear day

In the first work, different PV penetration rates from 0% to 200%, with a step of 25% are tested. The two forecast methods mentioned before are used in the simulation. Different sky states like clear day and variable day are also tested to provide a relatively complete evaluation of different elements.

For a given clear day and given energy system setup, the summary of energy system result is shown in Figure 1. When using perfect prognostic forecast, the optimal penetration rate is 175%, then the system cost increases due to the PV curtailment. When using the more basic climatological forecast, the optimal penetration rate is 50%, after that, the system cost increases for the same reason as PV curtailment. The system cost in blue square decreases significantly since gensets are stopped when PV energy is sufficient for all energy demand without energy excess, the associated fuel cost is hence avoided.

Unlike the perfect forecast has no uncertainty to consider, the climatological forecast has a conservative uncertainty, which means that to cover the uncertainty range between $P_0 - P_{50}$ would need a huge Genset capacity for the spinning reserve sizing. Due to this reason, the energy system becomes unfeasible when the PV penetration rate becomes too important.

The detail of system costs using different forecast

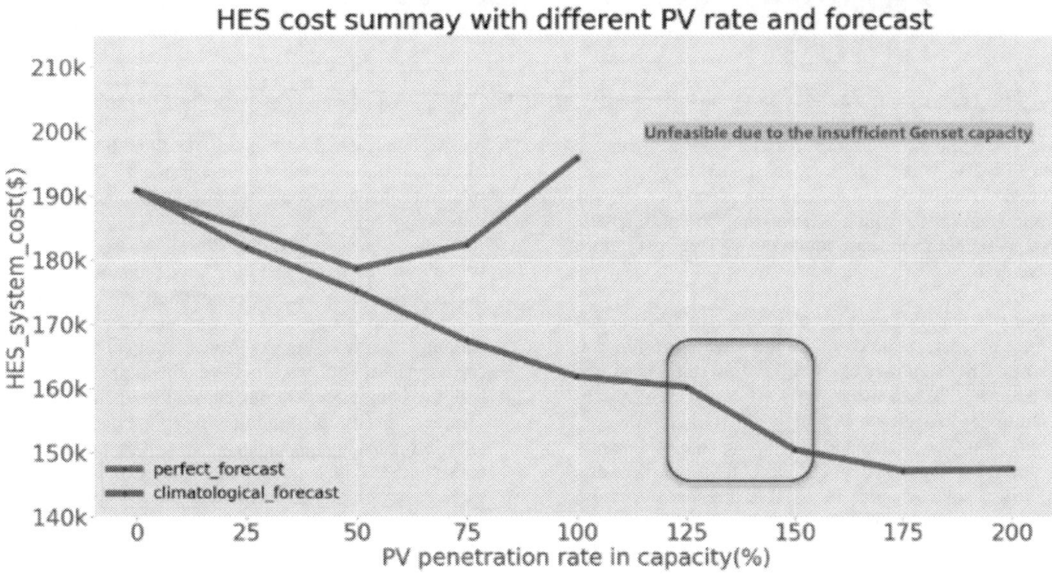

Figure 1: Summary of system performance with different PV penetration rates at clear day situation

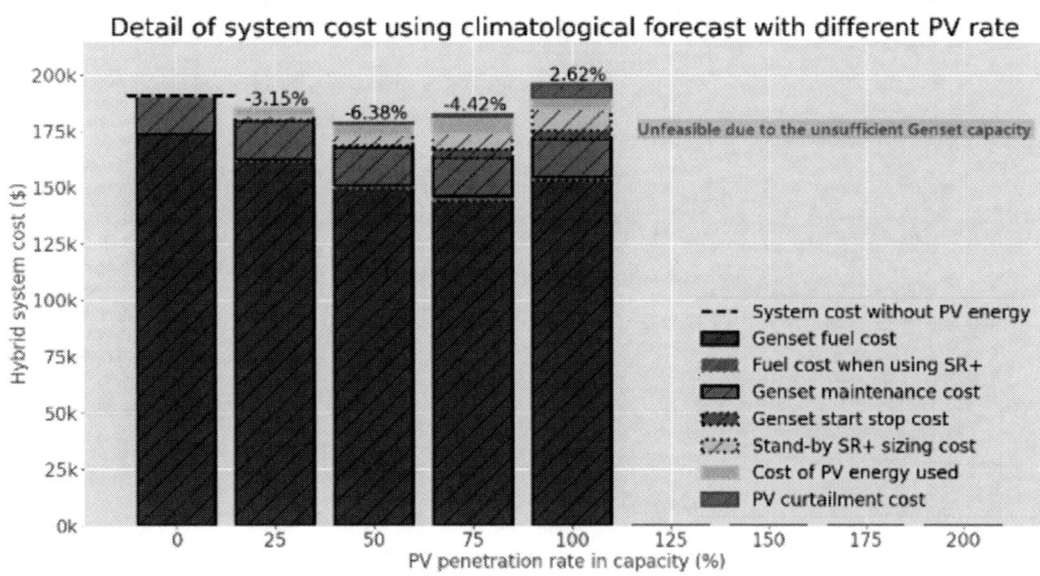

Figure 2: Detail of system cost with different PV penetration rates using climatological forecast at clear day situation

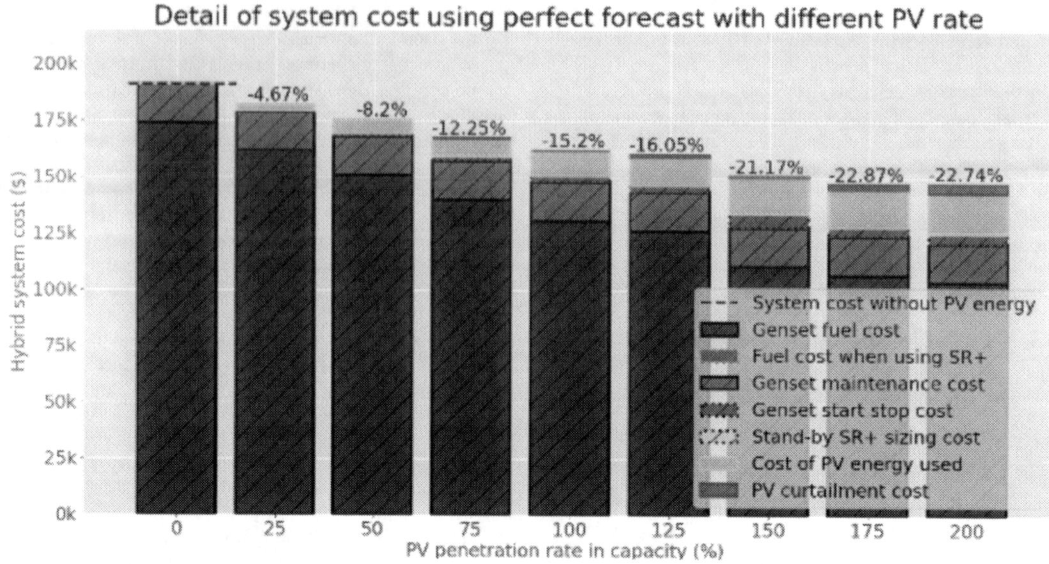

Figure 3: Detail of system cost with different PV penetration rates using perfect forecast at clear day situation

methods are shown in Figure 2 and Figure 3, with the increase in PV penetration rate, the general system cost and fuel consumption continue to decrease till its saturation moment, which means that Integrating PV energy in HES can indeed reduce the system cost before its saturation.

3.2 Scenario with different PV penetration at variable day

For a given variable day with the same configuration at clear day, the summary of energy system result is shown as in Figure 3. When using perfect forecast, the system cost continues to decrease with higher PV penetration rate, even at 200%. The situation in blue square like at clear day is postponed at variable day. In fact, both of them are due to the lower solar resource in variable day, the saturation of PV energy integration is therefore later.

When using climatological forecast, the optimal penetration rate is still at 50%, then the system cost increases due to the PV curtailment. When the PV penetration rate becomes too high, the energy system is unfeasible.

Both in variable and clear sky states, the energy system becomes unfeasible when the PV penetration rate becomes too important. A higher PV penetration rate with same level of uncertainty range asks for a bigger Genset capacity for spinning reserve sizing. As shown in Figure 5, the available capacity is the difference between the total Genset capacity and the committed Genset capacity. When the need for Genset capacity for spinning reserve is larger than the available capacity, the energy system is unfeasible. The maximum PV penetration rate is hence limited by the forecast uncertainty and available genset capacity. In other words, less uncertain models allow higher PV capacity integration.

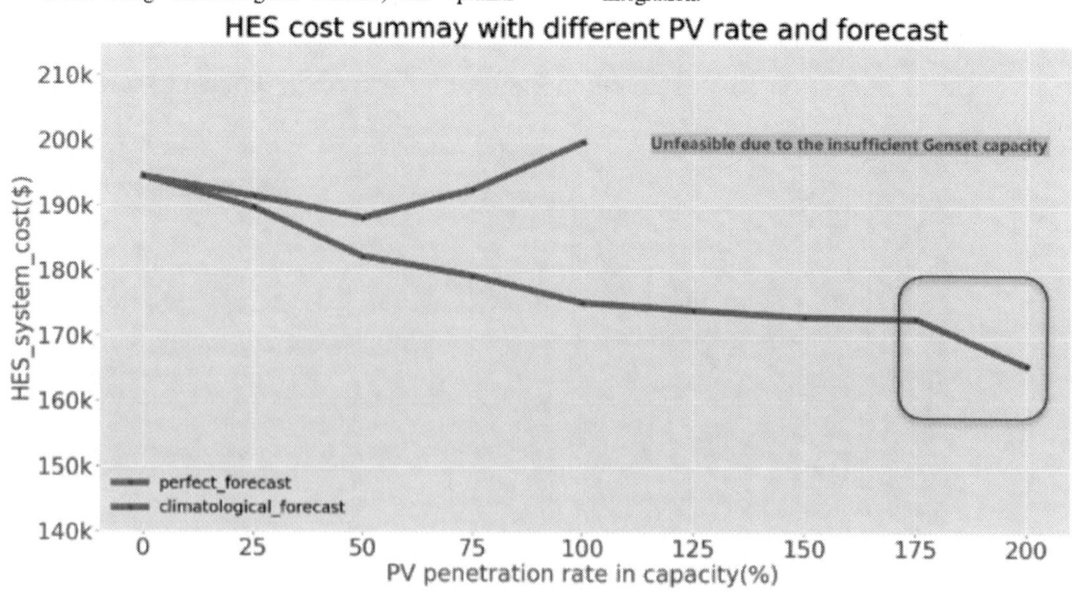

Figure 4: Summary of system performance with different PV penetration rates at variable day situation

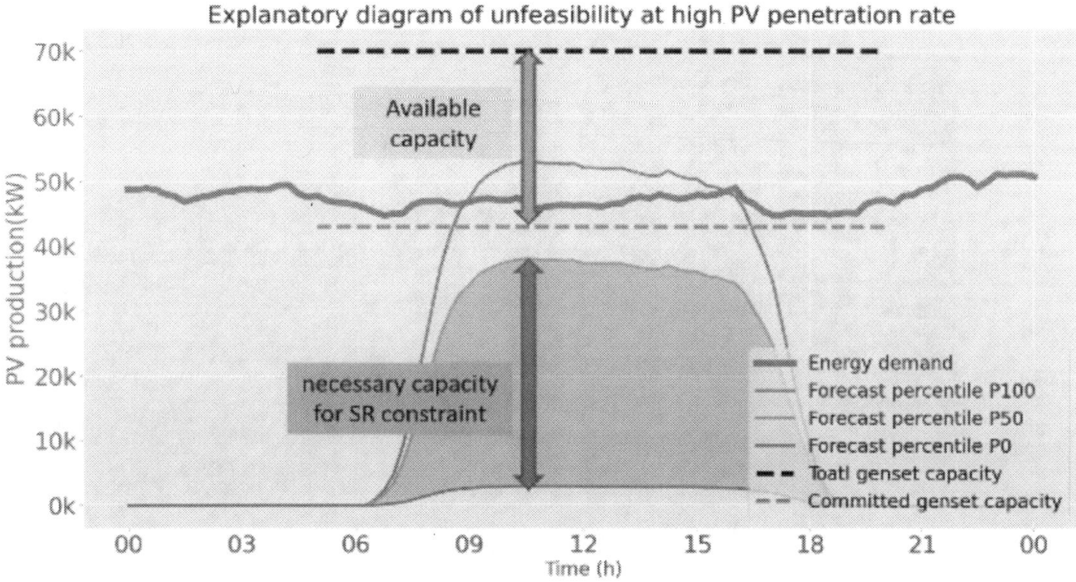

Figure 5: Explanation diagram of system unfeasibility cases

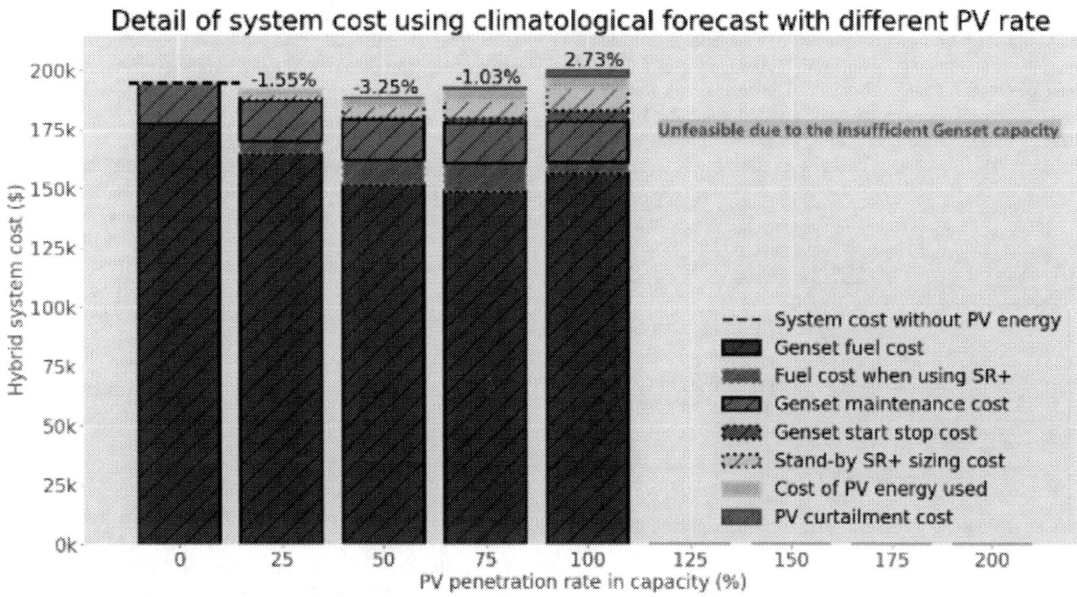

Figure 6: Detail of system cost with different PV penetration rates at variable day situation

The detail of system costs using different forecast methods at variable day are shown in Figure 6 and Figure 7. Due to the lower solar resource, the PV production is naturally lower, the fuel consumption is hence higher in variable day. The gain of using PV energy with a same level of capacity is therefore lower than clear day situation.

By comparing Figure 2 and Figure 6, an interesting phenomenon is noticeable. For climatological, no matter at clear day or variable day, the uncertainty range for a given period remains the same. Hence, the cost of preparing a steady spinning reserve is the same. Yet, the solar variability at clear day is much less than at clear day, which results in high spinning reserve preparation cost and low use rate of spinning reserve at clear day. Hence, for basic forecast model like climatological forecast, the strategy of sizing spinning reserve should be adapted to the sky state.

3.4 Scenario with different forecast horizons

In the second scenario, two forecast horizons (60-min and 120-min) are tested to evaluate the influence of forecast horizons on system performance. The forecast methods used are still probabilistic climatological and perfect prognostic forecast. The PV penetration rate used here is 50%, which is the actual penetration rate of the case study. The reference case used is the case without PV energy.

The result is shown in Figure 8, which shows that a relatively longer forecast horizon could help in reducing the system cost.

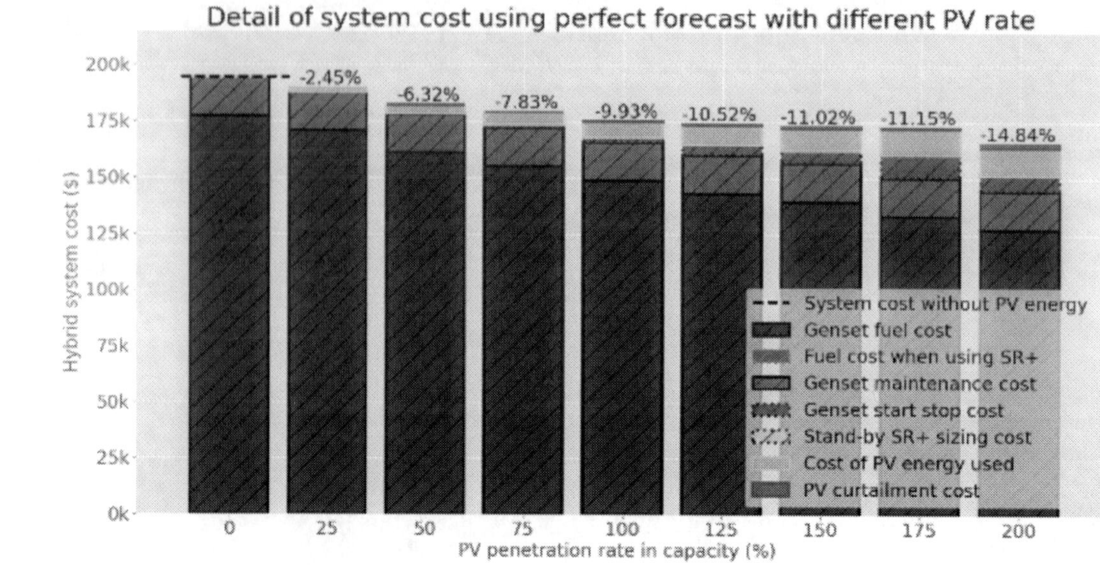

Figure 7: Detail of system cost with different PV penetration rates at variable day situation

When using perfect forecast, 120-min forecast horizon could reduce more than 10% system cost compared to a no PV case, and 2% better than the case with 60-min horizon. When using climatological, the gain of 120-min is around 7.6% lower than the no PV case, and around 1% better than the case of 60-min forecast horizon. It means that a longer forecast horizon has a more positive influence for high-quality forecast methods. For a basic model like persistence, the impact is relatively smaller.

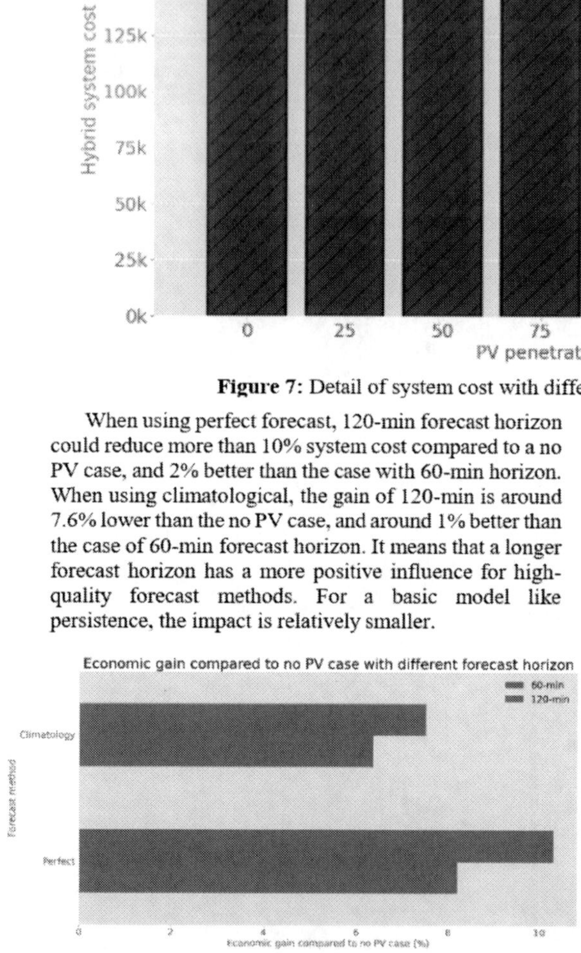

Figure 8: Influence of forecast horizon in economic gain

4 CONCLUSION & PERSPECTIVE

4.1 Feasibility of Diesel-PV concept

The concept of using PMS with solar forecast for a storage-less HES is shown to be feasible. In this concept, the main enablers of this concept are either an automated dispatch or a strict manual operation of Gensets, cooperated with the smart PMS and accurate solar forecasts.

4.2 Gain of using PMS with solar forecast

The gain of using solar forecast depends on many elements, such as the weather condition, PV penetration rate, forecast model, the Genset characteristic, etc. For the given data, the simulation results show a range up to 22.9% of decrease in daily system costs with perfect forecast and 6.4% with climatological forecast.

Yet, this result has its limit due to several elements, such as the dataset, the forecast methods used, the strategy of using forecast uncertainty to size the spinning reserve, the forecast horizon used and the Genset order update frequency. To have a more complete evaluation, a further study with above parameters would be achieved.

4.3 Limit of storage-less design

Due to the storage-less design, the ability to integrate PV energy is limited. Hence, when the PV penetration rate is too high and the strategy of spinning reserve sizing is conservative, the maximum PV penetration would not be higher than 100% in PV installation capacity over energy demand. To be able to accommodate more PV energy, a less uncertain forecast model, a less conservative spinning reserve sizing strategy, or installing an energy storage system are the possible solution to consider. The former two solutions aim to reduce the need for available Genset capacity and the latter solution is to add the possibility of storing energy.

4.4 Perspective

This work shows the feasibility of this concept, as a transitory solution towards 100% renewable energy, the first result seems interesting when building a storage-less Diesel-PV HES. Yet, there is no study about the complete evaluation of environmental impact and economic gain from a life-cycle perspective, for the comparison between the classic PV-storage system and this storage-less Diesel-PV system. A further associated study should be therefore carried out.

5 REFERENCE

[1] J. J. S. Guilbaud, "Hybrid Renewable Power Systems for the Mining Industry: System Costs, Reliability Costs, and Portfolio Cost Risks," Doctoral, UCL (University College London), 2016. Accessed: Jul. 01, 2019. [Online]. Available: http://discovery.ucl.ac.uk/1528681/

[2] S. Baurzhan and G. P. Jenkins, "Off-grid solar PV: Is it an affordable or appropriate solution for rural electrification in Sub-Saharan African countries?," *Renewable and Sustainable Energy Reviews*, vol. 60, pp. 1405–1418, Jul. 2016, doi: 10.1016/j.rser.2016.03.016.

[3] R. Besseau, "Analyse de cycle de vie de scénarios énergétiques intégrant la contrainte d'adéquation temporelle production-consommation," 2019.

[4] L. Vallance, "Synergie des mesures pyranométriques et des images hémisphériques in situ avec des images satellites météorologiques pour la prévision photovoltaïque," 2018. Accessed: Jun. 28, 2019. [Online]. Available: https://tel.archives-ouvertes.fr/tel-02078908

[5] J. Antonanzas, N. Osorio, R. Escobar, R. Urraca, F. J. Martinez-de-Pison, and F. Antonanzas-Torres, "Review of photovoltaic power forecasting," *Solar Energy*, vol. 136, pp. 78–111, Oct. 2016, doi: 10.1016/j.solener.2016.06.069.

[6] D. Lew *et al.*, "The Western Wind and Solar Integration Study Phase 2," NREL/TP--5500-55588, 1220243, Sep. 2013. doi: 10.2172/1220243.

[7] A. Urtasun, P. Sanchis, D. Barricarte, and L. Marroyo, "Energy management strategy for a battery-diesel stand-alone system with distributed PV generation based on grid frequency modulation," *Renewable Energy*, vol. 66, pp. 325–336, Jun. 2014, doi: 10.1016/j.renene.2013.12.020.

[8] S. Hilpert, C. Kaldemeyer, U. Krien, S. Günther, C. Wingenbach, and G. Plessmann, "The Open Energy Modelling Framework (oemof) - A new approach to facilitate open science in energy system modelling," *Energy Strategy Reviews*, vol. 22, pp. 16–25, Nov. 2018, doi: 10.1016/j.esr.2018.07.001.

[9] U. Krien, P. Schönfeldt, J. Launer, S. Hilpert, C. Kaldemeyer, and G. Pleßmann, "oemof.solph—A model generator for linear and mixed-integer linear optimisation of energy systems," *Software Impacts*, vol. 6, p. 100028, Nov. 2020, doi: 10.1016/j.simpa.2020.100028.

[10] A. Dobos and P. Gilman, "P50/P90 Analysis forSolar Energy Systems Using the System Advisor Model: Preprint," p. 8.

Competitiveness Perspectives for Vertically Integrated Local PV Manufacturing Until 2025

Philippe Macé, Elina Bosch, Monica Aleman, Gaëtan Masson, Adrien Van Rechem, Damien Gautier
Becquerel Institute, Rue Royale 146, 1000 Brussels (Belgium), p.mace@becquerelinstitute.org

ABSTRACT: The pandemic, combined with additional conjunctural events, led to logistics prices hike as well as increases of raw materials prices. As many other industries, the solar PV sector was impacted, with substantial inflation of PV module prices, which approached the 0,30 US$/Wp. But such turmoil across the supply chain can also create opportunities, e.g. by enabling the competitiveness of local manufacturing in comparison with centralized, large scale manufacturing bases located in Asia. To assess these opportunities, this paper provides estimations of the factory gate cost of manufacturing PV modules using mono c-Si p-type PERC cells, in 2022 and 2025, through vertically integrated factories located in France, Morocco and South Africa of 1GW of production capacity
Results show that it remains challenging to be cost competitive with China-based vertically integrated manufacturing. The gap remains higher than 20%, even in the best-case situation, with 0,37 US$/Wp in France, 0,36 US$/Wp in South Africa and 0,34 US$/Wp in Morocco achievable in 2022. The most influential factors are economies of scale, polysilicon price, cell efficiency or the cost of electricity. Competitiveness can be reached in the medium-term, only by focusing on these factors simultaneously, and with the support of public authorities, direct and indirect.
Keywords: Manufacturing, Techno-economic assessment, Factory, Cost competitiveness

1 CONTEXT

The pandemic, combined with additional conjunctural events, led to logistics prices hike as well as increases of raw materials prices. As many other industries, the solar PV sector was impacted, with substantial inflation of PV module prices, which approached the 0,30 US$/Wp while they stood once below 0,20 US$/Wp.

These price shocks can have severe negative effects, for instance for PV project developers, and endanger PV market development. It is therefore of primary importance for many stakeholders in the sector to be able to anticipate, at least partially, the likely evolution of PV module prices, in order to adapt their risk management strategy accordingly. At the same time, such turmoil across the supply chain can also create opportunities, for example by enabling the competitiveness of local manufacturing in comparison with centralized, large scale manufacturing bases located in Asia.

To assess these opportunities, this paper provides estimations of the cost of manufacturing PV modules (factory gate cost) using mono c-Si p-type PERC cells, in 2022 and 2025, through vertically integrated factories located outside of Asia, namely France, Morocco and South Africa. This paper differs from previous research by focusing on original locations, in Europe, but also in other regions which no or limited history as solar PV manufacturing hubs.

2 AIM AND APPROACH

2.1 Processes' simulation

In order to estimate the cost competitiveness of local manufacturing of solar PV modules with a vertically integrated factory, i.e. starting from ingots up to modules, a multi-step approach has been followed, as described on Figure 1. First, a "bottom-up" simulation model is built, which is a detailed modelling of manufacturing processes along the value chain. This step follows a deterministic approach, with assumptions on various key parameters, for instance uptime, efficiency or energy consumption of the production equipment as well as labour intensity.

2.2 Technical assumptions

As the developed techno-economic model follows a

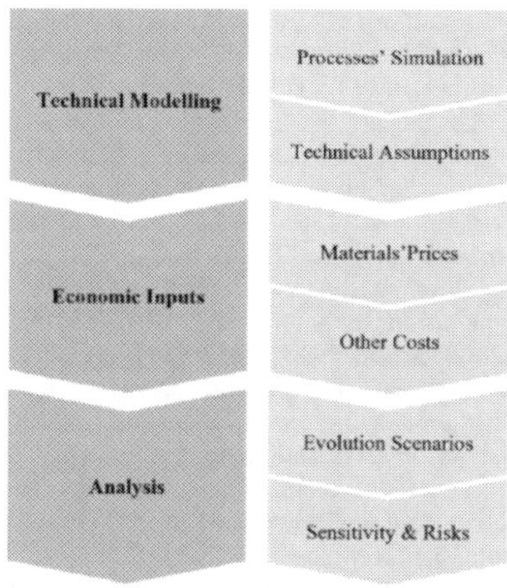

Figure 1: Methodolological approach scheme

deterministic approach, parameters and assumptions are manually defined for the various time-horizons, based on a thorough desk research and interviews with researchers as well as production equipment suppliers or manufacturers, allowing to estimate what is currently achievable in the industry and how it should most likely evolve in the short term.

The technical assumptions feeding the model concern, among others, the following points:

- Ingots & wafers:
  - Consumption of dopants
  - Consumption of polysilicon
  - Lifetime of diamond wires
- Cells:
  - Consumption of silver paste
  - Consumption of aluminum paste
- Modules:
  - Consumption of ribbon
  - Lifetime of laminators' joints

2.3 Materials' prices

In addition to this technical modelling of PV module manufacturing, research on input materials' and consumables' prices has been conducted. This allows to evaluate their impact on module prices.

Among these prices used as inputs to the techno-economic model are the following key materials:

- Ingots & wafers:
  - Polysilicon
  - Diamond wire
- Cells:
  - Gazes & chemicals
  - Silver paste
- Modules:
  - Module BOM components (glass, ...)
  - Laminators' joints

2.4 Other costs

Then, as part of the definition of economic inputs' value required to feed the techno-economic model, some cost constraints had to be defined. These mostly relate to CAPEX, although it also concerns some OPEX. The same methodology as for the prices of materials was used (desk research and interviews). These costs mainly are:

- Ingots to modules:
  - Equipment cost
  - Building & facility cost
  - Financing cost (debt & equity)
  - Labour cost
  - Utility cost
- Modules:
  - Packaging cost

2.5 Evolution scenarios

As part of the analysis of the results of the techno-economic modelling, it is important to be able to compare different years for the start of the production, or different situations. For this purpose, various evolution scenarios are defined, considering possible improvements such as:

- Economies of scale
- Increasing efficiencies
- Process optimization
- Prices' variations
- Consumption of input materials and consumables
- Equipment cost
- Labour cost

Note that in our case, the main parameters evolving between a manufacturing plant starting today and 2025 are the polysilicon cost, the cell efficiency, the wafer thickness, the silver consumption, the labour cost as well as the building and facility cost. Due to a too high uncertainty of the future cost of equipment or on the prices of input materials, they have been maintained constant in the analysis.

2.6 Sensitivity & risks

The last step is dedicated to a sensitivity and uncertainty analysis, allowing to identify main influencing factors as well as possible deviations of results, and associated risks. Note that all aforementioned parameters can be tested, although only most meaningful ones are.

2.7 Regions analyzed

This paper focuses on the potential competitiveness of local manufacturing outside of the large-scale, centralized manufacturing bases located in Asia. The three countries targeted here are France, Morocco and South Africa.

3 MAIN ASSUMPTIONS

A vertically integrated factory with 4 main stages has been simulated: ingots' pulling (mono c-Si p-type), wafers' slicing (M10), cells' manufacturing (PERC) and modules' assembling (144 half-cells glass-backsheet). A production capacity of 1GW at each step is considered.

The computed cost is the factory gate cost, i.e. without transportation cost.

Two different years for the start of the production (2022 & 2025) have been tested. The evolving factors between these two years are the following:

- Labour, facility and building costs, whose evolutions are due to the inflation considered for these three countries.
- The polysilicon price, which is expected to significantly decrease in the next few years. This decrease is foreseen as many production capacities' expansions have been announced.
- Technical assumptions, like cell efficiency, CTM ratio, wafer thickness or silver consumption. These are made according to the latest ITRPV report [1].

The values of these evolving factors as well as other assumptions are presented in Table I on the next page. The value of annual inflation is from the year 2021i [2]. It is considered that this value will remain study until 2025. Also, this inflation has not been considered for cost items such as input materials (glass, frames, ...) or energy, whose prices are currently very high and are not expected to remain so by 2025. We thus anticipate a stagnation of their nominal value but a decrease of their real value.

Finally, values of other assumptions such as the price of consumables, labour intensity or the performances of production lines (uptime, rate of defects) are not presented here. These are indeed sensitive data and displaying them all would make the present paper too long.

Table I Main assumptions used in the simulations for 2022 and 2025

2022 / 2025	France	Morocco	South Africa
Cost of capital			
Cost of debt	6%	7%	8%
Cost of equity	10%	12%	12%
Debt to equity ratio	70 / 30	70 / 30	70 / 30
Labour (k$ per year)			
Operator	35 / 37	6 / 7	25 / 29
Technician	46 / 49	10 / 11	41 / 47
Engineer	73 / 78	14 / 16	75 / 86
Energy			
Electricity cost (US$/kWh)	0.06	0.101	0.065
Inflation			
Annual average inflation	2%	1.4%	4.6%
Market			
Polysilicon price (US$/kg)	40 / 10	40 / 10	40 / 10
Equipment			
Ingots & Wafers (MUS$/GW)		80.2	
Cells (MUS$/GW)		35.8	
Modules (MUS$/GW)		18	
Other cost			
Facility (MUS$)	19.5 / 20.7	13.6 / 14.2	17.5 / 20.1
Building (MUS$)	11.7 / 12.4	8.2 / 8.5	10.5 / 12.0
Technical assumptions			
Cell efficiency (M10)		23% / 23.8%	
CTM ratio		1.009 / 1.0105	
Wafer thickness (µm)		155 / 145	
Silver consumption (mg/cell)		94 / 85	

4 RESULTS

The following bar charts represent the factory gate cost of solar PV modules manufactured in a vertically integrated plant that would be launched in 2022 or 2025, in each country studied.

These costs are divided into the four main stages of the production line: ingots' pulling (mono c-Si p-type), wafers' slicing (M10), cells' manufacturing (PERC) and modules' assembling (144 half-cells glass-backsheet). The production capacity at each stage of the production line is estimated at 1GW. The result is given in US dollar per Wp for each stage.

4.1 Manufacturing activity starting in 2022

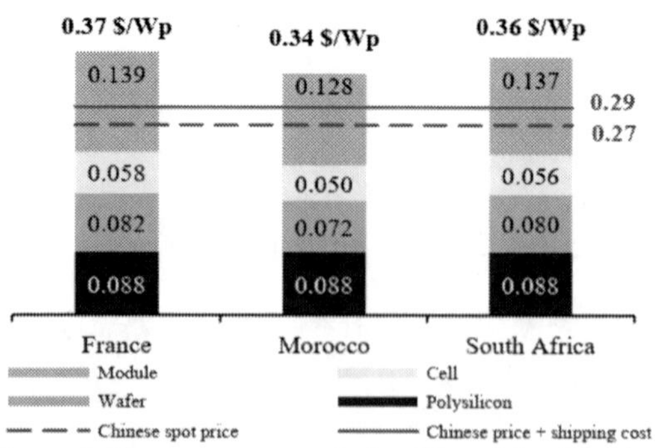

Figure 2: Factory gate cost of solar PV modules (launch in 2022)

The results for the tested countries range between 0,34 and 0,37 US$/Wp in 2022, as shown on Figure 2 above. This is >20% more than the average spot prices for the same types of modules, produced in China, around 0,27 $/Wp currently [3]. Even adding the costs of shipping Chinese modules (0,02-0,03 US$/Wp) to Europe or Africa, the gap remains substantial. Especially as these costs have been decreasing for a few months [4].

This difference can be explained by various factors. First, the very small scale of the manufacturing activity simulated here, compared to the scale of factories in China, where 5 GW is a minimum and 10 to 20 GW the norm, especially at upstream steps.

Then, one can also mention further key advantages, such as grants, tax exemptions, access to cheap financing, public support for the construction of buildings, facilities, infrastructures or networks (electricity, gas, water), public contribution to workforce's training or transportation, special conditions in terms of labour taxation, …

4.2 Manufacturing activity starting in 2025

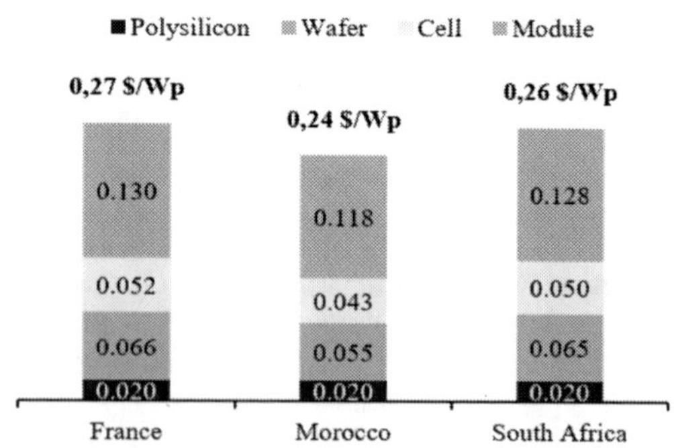

Figure 3: Cost of manufacturing PV modules in 2025

Then, in the case where the manufacturing activity starts in 2025, the situation is expected to improve. A decrease of 0,10 US/Wp of the total factory gate cost of modules has been estimated, as shown above on Figure 3.

But this decrease is mainly achieved thanks to external

factors, as further detailed on the graph below.

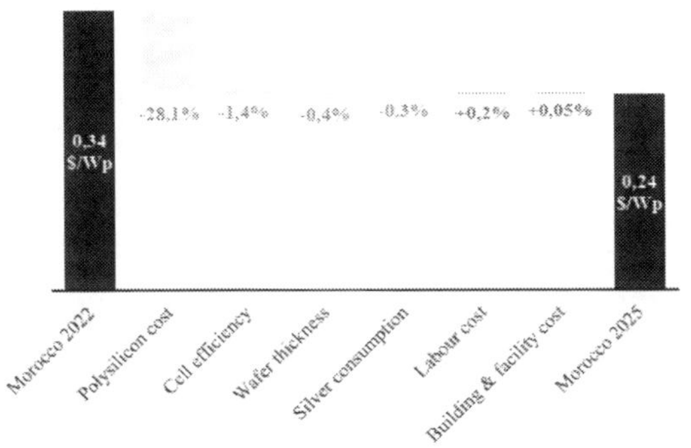

Figure 4: Main influencing factors of the decrease between 2022 and 2025

This graph shows the sources of the cost differences between starting a manufacturing activity in Morocco in 2022 or in 2025. The main decrease in module manufacturing cost by 2025 should come from the expected decrease in the price of polysilicon. Improvements in cell efficiency, wafer thickness or silver consumption will also contribute to the cost reduction, but to a much lesser extent. On the other hand, the increase in labour and construction costs, assumed to be caused by inflation, is expected to have a limited negative effect. Note that to simplify calculations, this inflation has not been considered for cost items such as input materials (glass, frames, …) or energy, whose prices are currently very high and are not expected to remain so by 2025.

5 ANALYSIS OF INFLUENCING FACTORS

In this section, the main influencing factors on the cost of manufacturing c-Si solar PV modules will be studied and their impact quantified. First, the effect of economies of scale, with a shift towards a factory with 5GW of production capacity instead of 1GW, as in our base case. Here, we focus on the situation of Morocco.

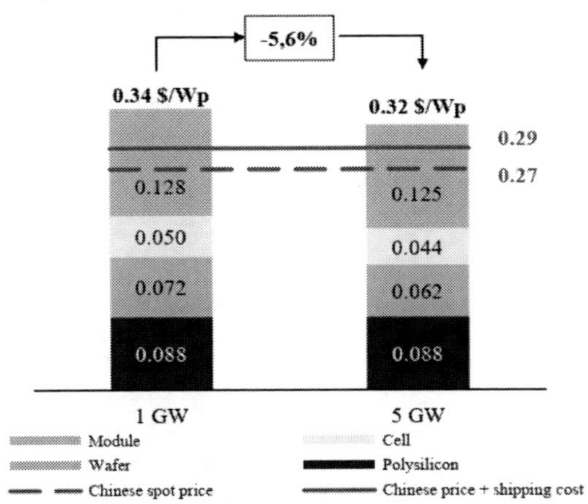

Figure 5: Effect of economies of scale on the factory gate cost of solar PV modules made in Morocco, in 2022

As shown on the graph above, the increase in scale could permit to decrease the factory gate cost of locally manufactured solar PV module by 0,02 US$/Wp, or approximately 6%. While limited in scale, this is still a significant improvement. It is also worth highlighting that this scale effect is particularly important at upstream steps, i.e. ingots and wafers as well as cells manufacturing. Focusing on the assembling of modules, the effect is negligible.

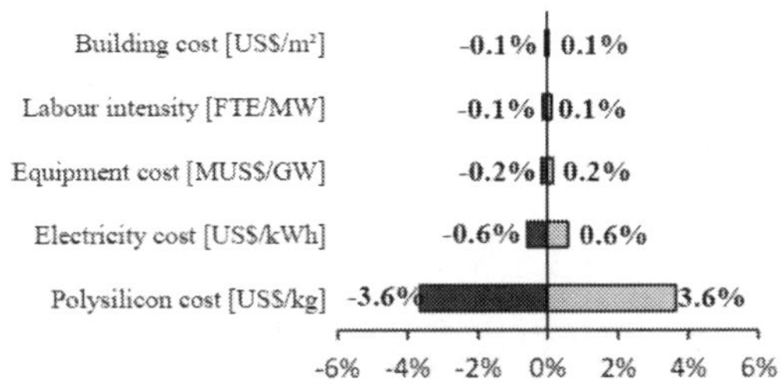

Figure 6: Variation of PV modules' factory gate cost in case of 10% variation of input parameters' values

Looking at other parameters' influence, the conclusions from Figure 4 focusing on the evolution between 2022 and 2025 are confirmed. Indeed, as shown by the graph above with results of the sensitivity analysis, polysilicon pricing is by far the most influential factor of the manufacturing cost of PV modules.

Electricity cost being a far second, with only 0,6% module cost variation when electricity cost varies by 10%. As shown by the Figure 4, a parameter like cell efficiency is more than twice as impactful.

While not studied in this paper, other consumables' price such as silver paste also have a significant influence on the module price. Although, it is not as high as polysilicon's

Then, CAPEX (e.g., equipment or building cost) and OPEX (e.g., labour intensity and cost) elements are much less influential.

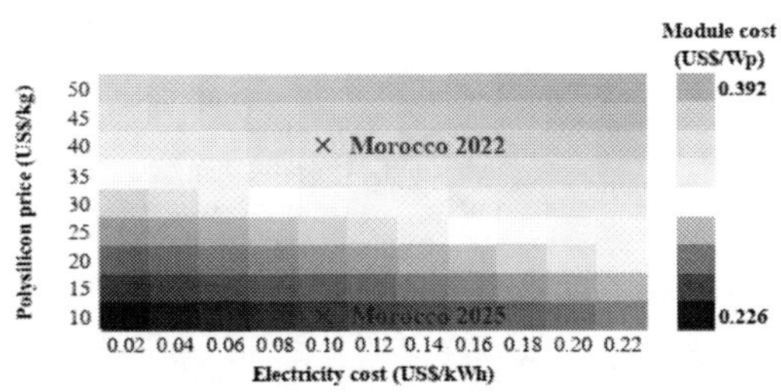

Figure 7: Sensitivity analysis based on the prices of poly-Si & electricity

Focusing specifically on these two influential cost factors, i.e. electricity cost as well as polysilicon pricing, it appears that by 2022 already, in a very favorable scenario (poly-Si at 10 US$/kg and electricity at 0,04 US$/kWh), the factory gate cost of PV modules could approach 0,20 US$/Wp in Morocco.

Although, such decrease of polysilicon is an external factor and would benefit all actors in the sector, including Chinese ones. Thus, the gap between Chinese-made modules and local ones would persist.

6 CONCLUSIONS

In this paper, we demonstrated that it remains challenging to be cost competitive with China-based vertically integrated manufacturing in different regions of the world. Indeed, the gap in terms of factory gate cost remains higher than 20%, even in the best-case situation.

Another conclusion is that vertical integration only makes sense if sufficient scale can be reached at all steps, otherwise specialization is better. The main steps where scale is crucial are upstream, in our case from ingots to cells, as shown by the results on Figure 5.

Although challenging, competitiveness can be reached in the medium-term, by playing on different factors simultaneously. First, by leveraging the currently high cost of shipping (0,02-0,03 \$/Wp), which can advantage local manufacturing. Nonetheless, these cast are decreasing. Hence, this card should not be overplayed.

Then, our calculations showed that even the most influential factor, i.e. polysilicon price, only influences the factory cost of modules by ~3,5% when changing by 10%. Thus, even if no factor will be single-handedly impactful enough to reach the competitiveness of Chinese manufacturing hubs, a combination of the following factors could allow to approach it.

- Technical innovations (cell efficiency, wafer thickness, material intensity, module-level improvements, …).
- Reaching sufficient scale, which is crucial to enable cost savings, especially at upstream steps. 5GW appears as a minimum, while 10GW or even 20GW are the optimum (1GW only was considered here).
- Public authorities' support through:
    - Access to low-cost loans to reduce the cost of capital
    - Direct (e.g., grants) or indirect reductions of CAPEX
    - Guaranteed off-taking to reduce risk, thereby increasing the confidence of potential investors and decreasing the average cost of capital. This can be achieved via mechanisms such as public procurement of locally manufactured PV modules, or tenders where feed-in tariffs are conditioned by the development of local factories.

Finally, it is crucial to keep in mind that local manufacturing is much more than just an economic question. It is a matter of energy independency, local job creations, social acceptance of renewables, supply risk as well as geopolitical risk management.

7 REFERENCES

[1] Verband Deutscher Maschinen- und Anlagenbau (VDMA), "International Technology Roadmap for Photovoltaic (ITRPV)," 2022.

[2] World Bank Group, "Inflation, consumer prices (annual %)," 2022. [Online]. Available: https://data.worldbank.org/indicator/FP.CPI.TOTL.ZG.

[3] InfoLink, "Upstream prices sustain before holiday, cell price hikes continue," 28 September 2022. [Online]. Available: https://www.infolink-group.com/spot-price/.

[4] Drewry, "World Container Index," Drewry Supply Chain Advisors, 2022. [Online]. Available: https://www.drewry.co.uk/supply-chain-advisors/supply-chain-expertise/world-container-index-assessed-by-drewry.

QUALITY INFRASTRUCTURE FOR SOLAR PHOTOVOLTAIC IN INDIA

Saurabh Kumar[1]
[1]Physikalisch-Technische Bundesanstalt
Bundesallee 100, 38116 Braunschweig, Germany, india@ptb.de

ABSTRACT: The objective of this paper is to make the stakeholders aware about steps to be taken to strengthen the national quality infrastructure of solar PV industryin India to build the trust of financial institutions to finance the solar PV plants also enabling the solar PV component manufacturers to get their products certified against international standards at lower prices enabling them to sell their products in global market and compete with international players. The need for national certification body for PV products, calibration laboratory, quick adoption/adaption of international standards by the national standard body is also underlined in this paper. The current status of national quality infrastructure in India has been described in this paper ie. national & international standards, testing labs, metrology institutes and general recommendations on systematic development of PV sector and the relevant quality infrastructure. The paper highlights the need for certification of PV power plants, PV installers as at present Government only focusing the certification of solar components but installation of PV power plant has also a major contribution in the performance of PV power plants.
Keywords: Quality Infrastructure, QI, accreditation, metrology, refrence module, certification body

1 INTRODUCTION

India has enhanced its ambition to install 500GW of renewable energy capacity by 2030 during COP-21.Earlier, Government of India was aiming at 100 GW of Solar Energy installations by the end of 2022 and 59.30 GW has been already achieved in August'22. In order to facilitate the expansion of renewable energy sector the Govt of India on 20th Nov' 20 has announced Production Linked Incentive Scheme 'National Programme on High Efficiency Solar PV Modules' for enhancing India's manufacturing capabilities and exports under Atmanirbhar Bharat initiative. Currently quality issues are common in installed PV power plants and the solar PV components. Additionally, more than 85% of solar PV modules being installed in India are being imported as the Solar Module manufacturing capacity of India is around 10 GW so the quality of imported solar PV components can't be always be assured. Quality assurance needs to be maintained during all the phases of project execution including manufacturing, installation and O&M. The national Quality Infrastructure plays a significant role ensuring the quality of PV power plants. At present, Ministry of New & Renewable Energy (MNRE) has issued a new lab policy to improve the quality and reliability of solar PV projects where all the solar components being imported or manufactured locally need to be tested by a laboratory recognized by National Standard Body of India-BIS. Additionally, to control the quality of Solar PV modules and solar PV cells, Ministry of New & Renewable Energy (MNRE) has issued Approved Models & Manufacturers of Solar PV Modules (ALMM) order for enlistment of module and cell manufacturers with MNRE for the sale of their products in India. This paper provides recommendations to strengthen the quality infrastructure and address quality gaps and suggestion to policy makers to adopt international quality standards and fasten the ALMM process for quick achievement of renewable energy goals.

2 QUALITY INFRASTRUCTURE AND ITS COMPONENTS

The national quality infrastructure(QI) of a country is comprised of internationally recognized metrology, standards, conformity assessment and accreditation[1] as shown in Figure 1.

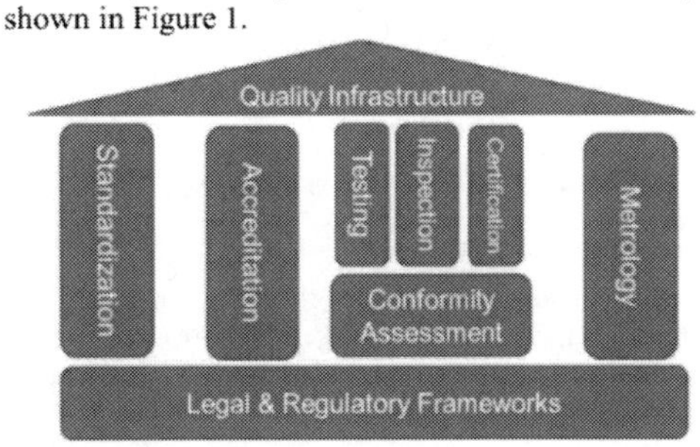

Figure 1: Quality Infrastructure

The life cycle of a solar plant does not start with installation, but begins much earlier and comprises all the stages shown in Figure 2.
QI services across all stages ensure that the full energy generation potential of the PV plant is exploited while at the same time not compromising safety and sustainability aspects. These services include all QI components, namely metrology, standardisation, accreditation,and conformity assessment (testing, certification, and inspection)[2].

Figure 2: Components of Solar PV QI System

3 NATIONAL SOLAR PV QUALITY INFRASTRUCTURE IN INDIA

In India, well established organizations comprise the national QI:

Standardization
Standardisation in India is largely government-driven and led by a national standards body, the Bureau of Indian Standards (BIS), which is exclusively authorised to publish Indian Standards. The BIS develops standards through several hundred sectional committees, made-up of different stakeholder groups, and draft standards are shared for public consultation. ETD 28 sectional

committee in BIS is responsible for solar PV standards. The solar PV modules are tested against Indian standards IS 14286 and IS 61730 for product certification purpose.

Metrology

The National Physical Laboratory (NPL), an institution of the autonomous government body Council for Scientific & Industrial Research, is India's National Metrology Institute (NMI). It provides traceability for calibration of measurements. The primary refrence cell calibration facility is under development in scientific cooperation with PTB Germany.

Conformity Assessment

There any few public sector labs ie National Institute of Solar Energy (NISE), CPRI (Central Power Research Institute) and ERTL(Electronics Regional Test Laboratory) which are testing against Indian/International standards. The product certification of solar modules against Indian/IEC standards is mainly provided by the private conformity assessment bodies.

Accreditation

NABL (National Accreditation Board for Testing & Calibration labs) is the accreditation body of India with its accreditation system established in accordance with ISO/ IEC 17011.

4 METROLOGICAL, TESTING &CERTIFICATION SERVICES

National Physical Laboratory (NPL) is setting up primary refrence cell calibration lab and secondary refrence module lab in under development at NISE. The status of metrological services in India is shown in figure 3.

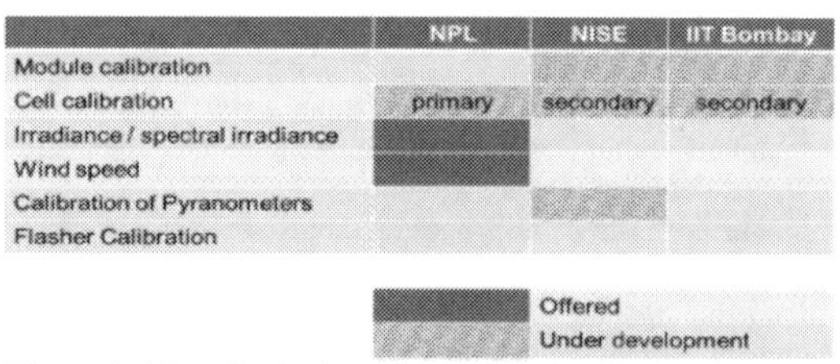

Figure 3: Metrological services for Solar PV Sector

Testing laboratories in India are testing the PV modules against the international standards ie. IEC 61215, IEC 61730 etc and their equivalent standards IS 14286, IS 61730.Test facilities available in India are shown in figure 4.

IEC 61215/IS 14286	√
IEC 61730/IS 61730	√
PID: IEC TS62804	√
Salt Mist: IEC 61701	√
NH3 Corrosion: IEC 62716	X
Dust & Sand: IEC 60068-2-68	X

Figure:4 PV Module test facilities in India

At present, there is no certification body for international standards IEC 61215 and IEC 61730. Few Testing labs like TUV Rheinland and UL recognised by foreign certification bodies ie Germany or USA are providing the product certificates against these standards.

5 RESULTS & RECOMMENDATIONS

India added around 15.4 gigawatts (GW) of renewable power capacity in 2021, the third highest after China (136 GW) and the United States (43 GW). By strengthning of quality infrastructure for solar PV industry the quality and performance of solar PV plants cqn be inproved and it can set an example for the world to deploy clean and green energy. On the basis of analysys of current quality infrastructure for Solar PV, following recommendations have been made.

5.1 Standardization

a) Standards according to Indian specific climatic conditions to be developed
b) Indian Standard as per latest IEC standard to be developed/adapted

5.2 Metrology

a) Primary and secondary cell and module calibration facilities to be established very soon.
b) Pyranometer calibration lab still to be developed
c) Flasher calibration in testing labs as well as in module manufacturing units is required

5.3 Accreditation

a) As solar PV is still an emerging domain in India, there is a need for capacity development of NABL staff & Assessors.
b) Inspection bodies ie organizations involved in ALMM/factory inspections need to be accredited.

5.4 Conformity assessment

a) Public sector labs to develop their Infrastructure is for full testing ie IEC 61215 & IEC 61730.
b) Test facilities for Salt mist, Ammonia corrosion needed
c) Transportation testing of PV modules ie, IEC 62759 to be established
d) National Certification body to be established for international Solar PV standards.
e) In India, the quality is predominantly focused on the components certification and less on the design and engineering of the solar energy plants.So, the regulatory authorities should aslo ask for PV power plant certification according to the relevant international standards.

Indian can become the manufacturing hub for solar PV panels but there is a great need to strengthen its quality infrastructure for solar photovoltaic

6 REFRENCES

[1] D.K.Aswal "Quality Infrastructure of India and Its Importance for Inclusive National Growth" MAPAN-Journal of Metrology Society of India (June 2020) 35(2):139–150.Online available at https://doi.org/10.1007/s12647-020-00376-3123

[2] "Technical Requirements in public auctions to make solar plant shine" IASS Policy Brief 1/2022 online available at https://publications.iass-potsdam.de/rest/items/item_6001518_4/component/file_6001522/content

SMART STRATEGIES FOR THE TRANSITION IN COAL INTENSIVE REGIONS

Rita Mergner[1], Rainer Janssen[1]

[2]Charalampos Malamatenios, [2]Georgia Veziryianni, [3]Dirk Knoche, [3]Anne Rademacher, [3]Rainer Schlepphorst, [4]Liliana Fonseca, [4]Rona Michie, [5]Angel Nikolaev, [6]Christian Doczekal, [7]Greg Arrowsmith, [7]Nicolas de la Vega, [8]Gloria Popescu, [8]Marian Dobrin, [9]Jasmina Mandic Lukic, [10]Igor Volchyn, [10]Dmytro Bondzyk, [11]Marcin Pietrzykowski, [11]Justyna Likus-Cieslik, [12]Lee Guilfoyle, [12]Kenn Palmer, [13]Jan Frouz, [14]Marketa Hendrychova, [15]Sabina Irimie

[1]WIP – Renewable Energies, Sylvensteinstr. 2, D - 81369 Munich, Germany
Tel. +49 89 720 12743, Fax +49 89 720 12791
E-Mail: rita.mergner@wip-munich.de; rainer.janssen@wip-munich.de
Internet: www.wip-munich.de, www.tracer-h2020.eu

[2]Centre for Renewable Energy Sources and Saving (CRES), Greece, [3]Research Institute for Post-Mining Landscapes (FIB), Germany, [4]University of Strathclyde, UK,[5] Black Sea Energy Research Centre (BSERC), Bulgaria, [6]GET, Austria, [7]EUREC, Belgium, [8]ISPE, Romania, [9]Energoprojekt ENTEL, Serbia, [10]Coal Energy Technology Institute, Ukraine, [11]University of Agriculture in Krakow, Poland, [12]Welsh Government, UK, [13]Charles University, Czech Republic, [14]Czech University of Life Sciences Prague, Czech Republic, [15]Jiu Valley Social Institute Association, Romania

ABSTRACT: The TRACER project supports a number of coal-intensive regions around Europe to design (or re-design) their Research and Innovation (R&I) strategies in order to facilitate their transition towards a sustainable energy system. The TRACER consortium consists of different target regions: South East Bulgaria, North West Bohemia - Czech Republic, Lusatian Lignite District - Germany, Western Macedonia - Greece, Upper Silesian Coalfield - Poland, West Region, Jiu Valley - Romania, Wales - UK, Kolubara - Serbia, Donetsk - Ukraine. Core activities of TRACER include the implementation of an EDP (Entrepreneurial Discovery Process) to mobilise a wide range of stakeholders in each target region to develop an appropriate governance structure and to bring regional stakeholders together to discuss and agree on a shared vision and priorities for coal transition. R&I strategies, industrial roadmaps and decision support tools will be developed jointly with key stakeholders of the TRACER target regions. Further TRACER activities include the identification and analysis of best practice examples of successful and ambitious transition processes in coal intensive regions, a detailed assessment of social, environmental and technological challenges, the elaboration of guidelines on how to mobilise investment as well as dedicated activities to stimulate R&I cooperation among coal intensive regions in Europe and beyond.

Keywords: energy transition, coal intensive regions, R&I strategies, industrial roadmaps, entrepreneurial discovery process, social challenges, re-skilling needs, SET Plan, mobilising investment.

1 INTRODUCTION

The EU aims to be climate-neutral by 2050 – an economy with net-zero greenhouse gas emissions. This objective is at the heart of the European Green Deal and in line with the EU's commitment to global climate action under the Paris Agreement. In view of these objectives, some Member States are in disadvantageous position due to the current state of the technologies used and the available local energy resources, therefore, these specificities must be taken into account when planning and implementing the national and/or regional energy strategies.

A number of EU Member States are actively mining coal. This provides direct employment to about 185,000 people across the EU, with additional indirect jobs relying on coal production. However, the EU's decision to move to emission-free technology is not subject to discussion and it is therefore necessary to take into account the difficulties of coal regions, to anticipate and propose the energy policy that is in line with the tasks. These regions need an effective roadmap to make the necessary transition to a more diversified economic base and a more sustainable energy system, while safeguarding the social cohesion. At the same time, the production and consumption of coal in the EU has been in steady decline over the past few decades, partly due to relatively high extraction costs, which have turned Europe's coal industry less competitive, but also due to the commitment by a number of Member States to phase out coal use for power generation.

In a changing world, the EU needs to become a smart, sustainable and inclusive economy. These three mutually reinforcing priorities should help the EU and the Member States to deliver high levels of employment, productivity and social cohesion.

Lately, the European Commission proposed an outline of a plan to make Europe independent from Russian fossil fuels well before 2030, starting with gas, in light of Russia's invasion of Ukraine. Europe has been facing increased energy prices for several months, but now uncertainty on supply is becomming a challenge. REPowerEU will seek to diversify gas supplies, speed up the roll-out of renewable gases and replace gas in heating and power generation. This can reduce EU demand for Russian gas by two thirds by the end of the year 2022.[1]

2 THE TRACER PROJECT

The overall objective of TRACER is to support a number of coal-intensive regions around Europe to design (or re-design) their Research and Innovation (R&I) strategies in order to facilitate their transition

[1] REPowerEU: Joint European action for more affordable, secure and sustainable energy, https://ec.europa.eu/commission/presscorner/detail/en/ip_22_1511

towards a sustainable energy system. Nine (9) European regions are involved in this procedure, 6 of them in EU Members States (Bulgaria, Czech Republic, Germany, Greece, Poland, Romania), and 3 in countries outside the EU (UK, Serbia and Ukraine).

For this purpose, regional policy makers and other key stakeholders are assisted by the TRACER partners through the provision of information, expertise and advice, as well as by decision support tools. Concrete roadmaps and blueprints on the energy transition paths will be developed in the framework of TRACER project.

In order to achieve this a number of activities were implemented. In the first year of the project an investigation of the current status in the target regions and the analysis of the best practices were made. This included an analysis of renewable energy technologies and the extent of their application in partner countries, the identification of best practices on social issues and tourism as well as the exploration of best practices of smart specialisation strategies and on financing.

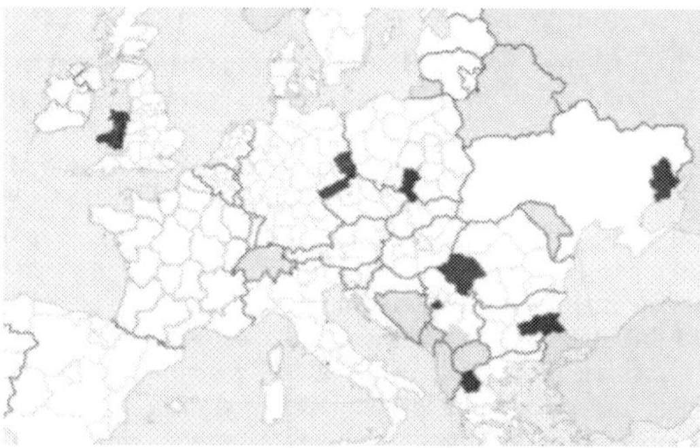

Figure 1: TRACER target regions (in red)

TRACER project is based on the cooperative approach. One of the important actions of the project is the establishment of close collaboration with other sections of Horizon 2020 and other EU funded programs, as well as the R&I Cooperation in the framework of the Implementation Plans jointly developed by European countries, as part of the EU's SET Plan and with the EU Coal Regions in Transition Platform.

3 PROJECTIONS FOR THE TRANSITION TO 2030/2050

While coal remains a central fuel in the European energy mix, the transition to clean forms of energy (renewables) and innovative technologies is an imperative to meet the EU's commitment to reduce CO_2 emissions by at least 55% by 2030 and to become the world's first climate-neutral bloc by 2050. The decline of coal-based energy production is an ongoing reality in Europe.

Since 2012, total coal power generation has dropped by almost a third in the EU. The declining use of coal has led to mines closing down and power plants being decommissioned in a number of regions across Europe. By 2030, approximately two thirds of the current coal-fired power generation capacity could retire, posing a key challenge on energy sufficiency. Moreover, around half of the relevant direct jobs could be lost by 2030, representing a major challenge in the transformation of coal regions.

By examining the projected energy mix (by source/fuel) in each target region, the energy technologies that are going to play a significant role in the area (by replacing the already existing and operating coal fired power plants, both for the region's power/energy coverage, but also for producing energy for the entire country) were identified.

With the cutting-edge technologies for the target regions known, the corresponding R&I activities in this region in the field of energy will have to focus on the improvement of the efficiency and/or functioning of these specific technologies. Thus, the stakeholders will be aware in which direction R&I (on national or regional level) will have to concentrate.

4 RESEARCH AND INNOVATION STRATEGIES

4.1 Southeast Bulgaria (BG34)

The Southeast Region is a strategic region not only for Bulgaria but also from an EU point of view, being the host of the biggest energy complex in South-eastern Europe. In the next decade, the coal region will need to undergo substantial transformation of its energy system towards decarbonisation.

R&I has a vital role to play in the region's planned energy transition. Despite its current state of underdevelopment, Bulgaria's R&I sector has large untapped potential for becoming a driving force behind the country's forthcoming energy transition and earning a competitive advantage in the decarbonisation process.

The national R&I legislative framework, which is crucial for delivering working solutions and system reforms, is being aligned with the European goal for the sector's growth, which is a positive indicator for the sector's future development. Furthermore, the current legislative framework in the energy sector for research and development is developing and is geared to promote R&I in the country.

The presence of a number of R&I entities in the region and the availability of funding opportunities are not sufficient to drive R&I excellence. Long-term effort needs to be put on the improvement of R&I human capacity, including better education and attracting leading researchers.

Additionally, the national planning, funding, and monitoring systems need to ensure that research activities effectively support energy decarbonisation. Despite the lack of regional decentralisation in Bulgaria and the limited empowerment of local and regional governments in energy matters, there are optimistic signs that the region would manage to enhance its economy and energy through R&I. The evident political intent, when combined with the considerable funding resources that are expected to be invested in the region, provides a promising potential to make the Just Transition into a beneficial, well-functioning, and desired phenomenon in the region and nationally.

The current Strategy identified several priority areas for energy-related R&I in the region. In the field of renewables, it is suggested that efforts are directed towards energy scenario modelling and socio-economic research focused on consumers, cost-benefit analyses, and policies. Another key priority is hydrogen, where R&I need to address both the technological development and socio-economic aspects ensuring positive attitude and supporting legal environment. The next priority is dedicated to smart grids and energy system digitalisation – these will support the market liberalisation and the flexibility of market players. The last priority is energy storage systems, where research efforts are necessary to assess the role, they could play in the energy system and to support the planned batteries manufacturing in the region.

The R&I and the promotion of its results may be particularly essential for the preparation of societal attitudes and behaviour change towards energy decarbonisation. R&I would assist the society to better understand the benefits and support the European Green Deal.

4.2 North West Bohemia (CZ049), Czech Republic

Considering the large share of coal in the energy production of the Czech Republic, replacement energy provided by coal will represent the complex issue. So far state energy policy counts on partial replacement of coal by other fossil fuels namely by natural gas. In the sector of smaller energy sources that use cogeneration to produce heat and electricity, there is already use of other fuel namely biomass and household waste and it is expected that share of these fuels in the mix will increase over time.

Traditionally, biofuel has a dominant position among renewables in the Czech Republic and as already explained above post-mining sites provide opportunities to increase biofuel production, which may even bring benefits in terms of ecosystem restoration and carbon sequestration in the soil at the same time. Among other renewables, potential for solar energy in terms of natural conditions is similar as in the rest of the country of slightly below average. However, a large area of post-mining heaps may offer possibilities for large scale installations. There are even plants to install floating power plants in lakes which will be formed by the flooding of residual mining pit.

The Fair Transformation Operational Program is completely new in the period 2021–2027 aimed at addressing the negative effects of coal diversion in the most affected regions. In the Czech Republic, this applies to the Karlovy Vary, Moravian-Silesian and Ústí regions. The support aims to enable regions and people to address the social, economic and environmental impact of a transformation aimed at achieving the Union's 2030 energy and climate goals and the Union's climate-neutral economy by 2050. A wide range of topics can be supported in this programme: small and medium enterprises, research and innovation, digitisation, clean energy and energy savings, circular economy, reclamation and new land use, retraining and job search assistance. The programme aims to direct resources from the Fair Transformation Fund to areas not covered by other operational programmes. These are extra resources for the regions concerned compared to other regions. The exact wording of the programme is under preparation, and the draft program will be published in the first half of 2022. The programme will be based on the Plan of Fair Territorial Transformation, which is prepared by the Ministry of Regional Development in cooperation with members of the Transformation Platform.

In addition to these programmes supported by individual ministries, the government established program RE:START to support coordination of funding from various sources and to give governments and regions a transparent, systemic and long-term tool to address the specific problems of coal intensive regions.

The RE:START program is an open and publicly discussed process of restructuring the coal intensive regions concerned with the involvement of many stakeholders.

4.3 Lusatia region, Brandenburg (DE40) and Dresden (DED2), Germany

How is Lusatia positioned with regard to its medium-to long-term future strategy for research and development in the energy sector? To answer this question, it should be noted that Lusatia is not a separate administrative unit, but a cross-border region between the federal states of Brandenburg and Saxony. With the Lusatia 2050 development strategy (Wirtschaftsregion Lausitz GmbH, 2020), a broadly coordinated future concept, that affects all relevant areas of public life, already exists since autumn 2020. Aspects of research and development in the energy sector are an integral part. Therefore, there is neither need nor a basis to develop a new regional R&I strategy within the TRACER project.

The current research landscape in the region is well positioned basically. In the core area there are the Technical University Cottbus-Senftenberg, the University of Applied sciences Zittau Görlitz and a number of non-university research institutes, like FIB Finsterwalde, that deal more or less with energy topics and renewables. Financing is currently provided, and will be in the foreseeable future, through institutional funding from the federal states Brandenburg and Saxony or funding programmes, particularly from the German federal government.

The region´s research institutions are basically well-connected. They have many references to the topic of energy, although in Saxony many institutes and activities are concentrated in Dresden, which is not part of Lusatia, but adjacent. One challenge, however, is that young and well-educated people still too often leave the region to work and live elsewhere. The ageing of the population has already a detrimental effect on the labour market in Lusatia, not only in the area of research and development. Accordingly, an important task is to get more attractive to young people from the region and highly qualified people from outside Lusatia.

Even after the phase-out of coal-fired power generation, Lusatia should retain its character as an energy region and industrial location (Agora Energiewende, 2017). In the Development Strategy Lusatia 2050, reference was made to this and corresponding goals and fields of action were formulated.

The Berlin-Brandenburg Energy Technology Cluster of Wirtschaftsförderung Land Brandenburg GmbH (WFBB) is also worth mentioning in this context. Since 2011, the cluster has served to network energy technology companies in the capital region with research institutions. The core topics are renewable energies, energy efficiency, energy grids and storage as well as turbomachinery and power plant technology.

In order to achieve the climate goals, the state government of Brandenburg supports the further expansion of photovoltaic systems. An area potential analysis is currently being carried out for usable areas, both open areas and sealed, but also roof and building areas. However, the preliminary recommendation for action by the MLUK indicates the use of mining areas for ground-mounted photovoltaic systems and/or agri-photovoltaics, so that an interesting further use for the energy generation of the mining-impaired areas is foreseeable here for the mining region Lusatia. In this context, the FIB carried out a workshop about photovoltaic systems on arable land in the post-mining landscape as a part of the future energy region Lusatia with stakeholders from Lusatia and external experts on 17th November 2021. It was an exciting exchange for finding strategies to merge agricultural and energy production with the needs of developing the reclaimed land.

4.4 Western Macedonia (EL53), Greece

The key role of R&I in contributing to and accelerating the attainment of regional ambitions is acknowledged at all levels (EU, national and regional). Bolstering the support for and guiding the direction of R&I so that it is tailored to local specificities and national priorities is thus considered an integral component for development and progress within the energy, and in a second step, the environmental sectors. In addition, driving targeted R&I for both excellence and impact will feed into boosting industrial competitiveness and economic development through instruments of knowledge.

The "Research & Innovation strategy in the field of energy for the Region of Western Macedonia", aims to ensure complementarity and synergy between the regional, national and EU-level R&I related policies and programmes, as well as alignment with the expected developments in the energy sector and the long-term drive for sustainability, climate change resilience and environmental protection. Working in line with national targets and policy documents, this Regional Strategy aims to encourage investment across multiple sectors throughout value chains and a wide array of scientific disciplines, as well as different actors and stakeholders.

The R&I priority activities related to biomass are the following:

1. Development of high efficiency heat and electricity cogeneration using biomass as fuel on a large scale.
2. Development, demonstration and scaling up of solid, liquid and gaseous bio-energy intermediates through biochemical/thermochemical/chemical conversion from sustainable biomass.

To assess the available biomass capacity in Western Macedonia and especially in the areas of interest for the DH networks of Ptolemaida, Amynteo and Kozani, several studies have been conducted and the available data have been presented in publications. According to the Public Enterprise for District Heating in Ptolemaida (DETIP), the available biomass capacity in Western Macedonia mainly originates from forest biomass (125,000 tons/year) and agricultural residues (201,000 tons/year) with a respective total thermal content of 1,630 GWh/year.

To ensure an easier transition to a clean energy economy, some challenges need to be overcome. They are related not only to the technology, but also to society. The region has been supplying lignite to the energy industry for more than seven decades, enough to shape a specific coal-oriented mentality of the local population. It seems to be very difficult to persuade employees, miners, engineers that this specific activity will stop. If the support consists in an early retirement bonus for the generations close to retirement age, then the younger generations will need to learn a new profession. Development of RES could be one of the options in the overall package.

4.5 Upper Silesia coal field (PL22), Poland

The Silesia Voivodeship authorities have elaborated a development strategy for the region until 2030. The vision behind the elaborated strategy is "Green Silesia 2030". According to this vision, by 2030, Silesia Voivodeship will be a modern European region with a competitive economy as the result of a responsible transformation, providing its inhabitants with development opportunities and offering a high quality of life in a clean environment.

In summarising the analysis of data on the research and innovation (R&I) strategy in the field of energy for Upper Silesia, it can be concluded that in 2021–2030, the demand for solutions supporting energy efficiency in industry, energy-efficient construction, energy generation from renewable sources and integration of electric transport infrastructure with building infrastructure will continue. This will contribute to the development of new technologies and their implementation by IT companies, manufacturers of RES equipment and systems, service providers for the utility power industry, and companies in the construction sector. At this stage, no official vision or development strategy has been prepared for the Silesian Province for the year 2050. Therefore, it should be assumed that the visions and strategies set for 2030 will be continued.

The development of the energy sector in Poland is described in the document "Energy Policy of Poland until 2040" (Ministry of Energy, 2019). According to this strategy, hard coal will continue to play an important role in the energy mix. A complete departure from hard coal as a source of electricity is not currently planned, although it is planned to increase the importance of renewable energy sources in the overall energy mix. The directions of actions indicated in "The Energy Policy of Poland until 2040" emphasise, above all, the necessity of improving energy efficiency, increasing the security of fuel and energy supplies, diversification of the structure of electricity generation through the introduction of nuclear power, development of the use of renewable energy sources, development of competitive fuel and energy markets, as well as limiting the impact of the power generation industry on the environment. To meet the above challenges, it is necessary to build new power plants and thermal power plants, as well as modernise and expand existing plants throughout the country.

It should be remembered, however, that this relatively new document is already somewhat outdated, as the European Union's energy policy is changing dynamically and, consequently, Polish legal regulations and strategies are changing. For example, the recent agreement between the Polish government and mining unions assumes that all hard coal mines will be closed by 2049. This means that in 2050 there will no longer be any active hard coal mining in Upper Silesia. In the consultations carried out with the inter-assembly groups, via email, telephone contact and face to face at the meeting on 15 December 2021, additional attention was paid to the following aspects:

- the horizon of the energy transition and hard deadlines given the current state of the

developing economy seem too sharp;
- energy security is crucial for the average citizen, but the existing trend and direction of the energy transition is not fully understood. This is a challenge in the perspective of innovation and energy generation from alternative sources;
- consideration of the post-mining landscape as a unique feature of Upper Silesia in the national context, and even in the European context, with an aim to preserve it during the process of revitalisation of post-industrial areas;
- putting more emphasis on research and innovation in the field of alternative energy sources, and the need for better recognition of the region's potential in RES; and
- the need for research and innovation in the use of methane.

Biomass is an alternative energy source with the largest theoretical potential in the Śląskie Voivodeship. The average area of an individual farm in the voivodship is 3.7 ha, and there are than 110,000 farms in the voivodship. However, only those with area over 10 ha have really good conditions for biomass production. The largest technical potentials have the areas of Częstochowa, Gliwice, Kłobuck and Racibórz.

The biogas plant market is one of the most underrated and at the same time promising markets for the development of energy production from renewable sources. The slow development of the Polish biogas market is mainly due to high investment costs and low public interest in this technology. This sector needs an active participation of the state for development.

It is important to mention that experiments confirmed the usefulness of black locust on post-mining areas. This could open potential for additional biomass production that goes hand in hand with land reclamation and ecological goals.

4.6 Jiu Valley / West region (RO42), Romania

According to the EC Country Report Romania's weak research and innovation performance hampers the transition towards a knowledge-based economy. The country continues to have one of the lowest levels of public and private expenditure on research and development in the EU, negatively affecting scientific quality and the diffusion of technology amongst firms.

Increasing R&D investment and quality and supporting innovative firms remain important challenges. Otherwise, regional initiatives to enhance growth exist, but risk is being hampered by a lack of a robust national innovation and entrepreneurship ecosystem; and, without a sufficient level of public R&D funding, the economic competitiveness, research and innovation and smart specialisation strategies cannot achieve their stated objectives. Unfortunately, without a significant increase in the public R&D budget plus related regulatory measures to increase R&D quality and innovation, little progress is expected (EC, 2020).

Even if the national and regional figures and EC assessment are not optimistic in terms of R&I, the Jiu Valley micro-region takes over the regional model and is keen to contribute to and synchronise with ADR Vest effort to boost the public and private R&I potential, currently underexploited in the region.

Innovation depends on cooperation, which can allow underused knowledge and innovation capacities to be identified and used more effectively. Willingness, constant and committed engagement of key local stakeholders, together with good governance will be the main drivers for effectively developing and implementing the R&I solutions and projects which will be further identified and promoted.

According to Jiu Valley common vision, investing in human's potential, education will trigger the development of the local economy by implementing innovative ideas. In the future, supporting the involvement of young people, by openly promoting R&I activities and increasing their attractiveness, by avoiding migration and encouraging the return, will be essential for the effective implementation of Jiu Valley R&I Strategy and Roadmap in the field of energy.

The transformation into a RES-based sustainable energy system with low carbon emissions, will be made gradually, with natural gas as a transition fuel, in order to provide a time buffer for upgrading and retrofitting the electricity distribution networks, as many of them are obsolete. The gradual transition of Jiu Valley micro-region to a RES-based energy system, by switching power/cogeneration units from hard-coal to natural gas is also necessary, knowing that wind and solar energy are variable resources, unable to ensure the entire energy demand, especially in extreme weather conditions generated by the increasing negative impact of climate change.

Currently, there are 3 projects of biomass or biogas-based cogeneration plants:

- Apa Serv Valea Jiului, 2 biogas-based micro-CHP summing up 0.38 MWel, 0.436 MWth by revaluating the Danutoni WTP sewage sludge in Petrosani municipality, Jiu Valley;
- energy Serv (0.160 MW), Brad locality – not yet in operation (Transelectrica, 2020);
- biomass-based cogeneration unit in Aninoasa city, Jiu Valley – under development with ESIF 2014-2020 support.

EU Just Transition Fund will provide other investment opportunities in areas such as site regeneration and decontamination, land reclamation, implementing LCT and infrastructure for clean and affordable energy supply, reducing GHG emissions, increasing energy efficiency and RES use, SMEs, including start-ups, which support economic diversification and transition, creation of new companies, including consulting services and "business incubators", R&I, including advanced technologies transfer etc.

RES potential (hydro, wind, solar, biomass, geothermal) is defined in specialised studies performed by prestigious R&I institutions. Estimated biomass potential is app. 2,370 TJ in Hunedoara (78% agriculture, 22% forestry), however there is no data available for Jiu Valley.

The existence of a centralised DHS in Jiu Valley, including a network of approx. 185 km (primary and secondary circuit), can be the starting point for the transition to a semi-decentralised DHS, by implementing a retrofitting & upgrading program including also RES.

It is recommended that until 2035-2040, population supply with firewood for heating purposes should be switched to green biomass - pellets and wheat/straw briquettes, through the development of an integrated collecting and processing mechanism of wood wastes, resulting from wood industrial processing, and agricultural wastes, for the entire Jiu Valley.

4.7 Kolubara region (RS11 and RS21), Serbia

Energy transition towards the non-carbon energy technologies is not an easy effort for Serbia due to the fact that its major primary energy source is the indigenous coal-lignite, which is used to generate three quarters of the country's electricity. With an aim to ensure security of energy supply, especially of electricity and heat, the transition away from coal will mean closure of the coal fired thermal power plants before the expiration of their (and of coal mines) lifetime. For an expected country's economic growth and continuous raise of power demand, adequate measures and activities should be undertaken to ensure that a substitute with the low-carbon generation could be realised in the meantime.

In terms of biomass, EPS is piloting mixing biomass with coal in its thermal power plants. The project secures the reclamation of unused and degraded land in the former open pit mines areas in Kolubara basin on one side and production of energy from biomass on the other. The fast-growing willows are planted on the edge of the Field G of the open-pit mines in mining basin Kolubara. It is a pilot phase of the SLLES Project (Sustainable Land, Livelihoods and Energy Initiative Serbia) in EPS. Domestic white willow and Scandinavian Inger willow were planted on 3.17 hectares. These trees could give up to 112 tons of biomass per hectare.

It has been explored that there are 1.6 million hectares of unused and degraded land in Serbia, including mines and landfills. It should be examined if the implementation of the mentioned pilot project should be suitable in these areas as well, to secure its reclamation, but also the production of energy from biomass. It can also create new jobs and enable an easier energy transition.

The willows, planted near an open pit mine, can be cut 12 times within 25 years in order to be used in thermal power plants together with lignite for electricity production. Joint use of coal and biomass in power plants will result for cleaner air. It is estimated that wood biomass from 75,000 ha of SRPs can substitute 10% of Serbia's coal-based energy through co-firing in coal-powered plants.

The inevitable transitional process of the energy sector in Serbia would result in a significant shift of the current energy development strategy of Serbia towards the low-carbon development strategy. This would imply a shift towards climate adaptable economy in the Kolubara target region. According to the current position of the stakeholders in the region, stepwise energy transition towards an efficient carbon-free energy sector should be provided. In order to implement the energy transition properly, to take care of all employees, as well as taking into account the energy security of the system, a lot of knowledge and experience is needed. All available domestic experts and institutions should be involved in the transition process, trying to implement available research and innovation procedures in an aim to support the transition process.

The new energy-related research and innovation (R&I) activities need to be developed, focused on the application of innovative technologies in line with the resources present in the Kolubara target region. For the R&I on that and other hot subjects related to the energy future, more trust was gained by local experts and R&I institutes than by foreign consultants. The idea of the need to form a specialised multidisciplinary institute for national energy matters was proposed. Nevertheless, it was stated that there is a lack of continued education of energy professionals, resulting in a lack of ability to participate in the multidisciplinary research required to support or direct and monitor transitional processes.

4.8 Donetsk region, Ukraine

The use of biomass for home heating is only beginning in the Donetsk region. The number of households that have installed biomass boilers is not significant yet. At the end of 2012 Sloviansk Local Education Department and LLC "Promshlennoe Oborudovanie" established boilers on biofuel in two schools of the city. Thanks to the support of the United Nations Development Programme in Ukraine within the framework of the project "Development and commercialisation of bioenergy technologies in the municipal sector in Ukraine", implemented with the support of the Global Environment Fund, in 2017 the installation of new boilers working on agricultural biomass at the objects of social infrastructure in the village of Cherkasy of Slavic district of Donetsk region has started.

In general, the potential of Donetsk region allows efficient use of renewable energy sources. Thus, according to experts, the technically achievable potential of the region is 25.4 billion kWh per year. One of the largest shares in its structure belongs to biomass (32.9%).

Energy strategy of Ukraine until 2035 is aimed at innovative development of the energy sector and investment in new generation capacity to replace decommissioning one. The choice of the type of generation will depend on the forecast price situation for fuel and the intensity of development of each type of generation, which will increase the level of competition between them, from the introduction of smart technologies to equalise consumption peaks.

It is envisaged to constantly expand the use of all types of renewable energy, which will become one of the tools to ensure the energy security of the state. In the short and medium term (until 2025), the share of renewable energy in Ukraine is projected to increase to 12% of primary energy and at least to 25% by 2035. At the same time, the share of energy produced from coal is planned to be reduced to 12% by 2035.

In Ukraine, the cooperation of industrial enterprises in production and innovation stays at a low level. The most pressing issues are the low level of cooperation of innovative industrial enterprises with the science sector; insufficient level of integration of small and medium enterprises (SMEs) into added value chains; low level of internationalisation of value chains; insufficient use of digital technologies by industrial enterprises.

The solution to these problems will be facilitated by the use of public policy tools devised by the developed countries aimed at building cross-sectoral cooperation. At the local level, attracting the potential of the scientific and educational sector to the commercialisation of innovations is due to the introduction of the concept of entrepreneurial universities, which activates the university's own business initiatives, projects with external investors, development of cooperation with business and other organisations.

The development of cooperation within innovation

ecosystems is stimulated through the introduction of the concept of open innovation platforms, which expands the possibilities of partnership based on research projects, stimulates the attraction of investors and accelerates the commercialisation of scientific developments.

At the regional and national/supranational levels, ensuring the inclusiveness of economic development is due to the concept of smart specialisation, which allows realizing the unique potential of economic development of regions and countries based on specific competitive advantages. An important component of the concept of smart specialisation is the concept of cluster development, the focus of which is to create world-class clusters with the active involvement of SMEs.

Ukraine's steps towards implementing these concepts in state economic policy include integrating their principles into strategic documents - the Concept of Digital Economy and Society of Ukraine for 2018-2020, the Strategy of Ukraine's Industrial Complex until 2025, the Government's Medium-Term Priority Action Plan until 2020 and Government Priority Action Plan for 2018; entry of certain regions of Ukraine into the EU Smart Specialisation Platform; implementation of pilot projects of smart specialisation of industry in the regions.

4.9 Wales (UKL1, UKL2), United Kingdom

The governance support framework and R&I strategies and plans for the transition have been characterised by OECD as fragmented. Horizontal and vertical collaboration and cross departmental/sectoral efforts have been recommended, as well as the continued development of an integrated R&I strategy for Wales. This is particularly important in the context of relative uncertainty (post-Brexit), especially with regard to future funding. Efforts to increase coordination are already underway. Since publication of the OECD report, and following an election in 2021, a new climate change ministry was announced within Welsh Government, bringing together the portfolios for environment and energy (along with housing, planning and transport).

Further, a new integrated innovation strategy is currently under development. The European Union's Cohesion Policy current research and innovation policy framework is based on the concept of Smart Specialisation. It involves the process of identification of a region's areas of uniqueness and strength and the design and implementation of a strategy (RIS3 or Research and Innovation Smart Specialisation Strategy) that can materialise an ideal regional development pathway. The Smart Specialisation policy framework was announced as an ex-ante conditionality for the access to European Structural and Investment Funds (ESIF), meaning that in its near-mandatory adoption it spurred a widespread innovation governance experiment across EU regions.

The Welsh context benefits from multiple strengths and advantages related to energy and environment. Wales possesses both the ample natural resources and the institutions and collaborative networks that can allow it to gain competitive advantage in the field of decarbonisation. This is evidenced in the rapid developments in the field of renewable energies in Wales, namely offshore and onshore wind and marine and tidal. Similarly, higher education and research institutions in Wales are developing expertise in both the technological and the behavioural aspects of the green transition, and there is promotion of collaborative links with industry to allow for the development of innovative solutions and processes that can allow for the scale-up of projects.

However, current plans for the energy and environment sectors in Wales lack integration with broader research and innovation strategies and goals. This is potentially related to the existence of several different government silos that generate fragmentation in efforts. At the same time, the post-Brexit context brings with it uncertainty regarding changes in priorities, funding opportunities and the governance of research and innovation in these sectors.

In terms of biomass, it is important to mention that bioenergy and waste use are expected to grow modestly by 30% to 2050 at UK level. R&I in biomass should focus on the following aspects:

- utilisation of biomass as source of renewable heat and electricity;
- research on converting biomass and bio-industry wastes into biobased products with commercial applications, via biorefining and bio processing (e.g., Beacon project44)

Biomass generating over 6% of Welsh energy is stagnating as an energy generation technology. According to Welsh Government, the pathway for decarbonising heat by using biomass as fuel is unclear. Major concerns are the emissions associated with long-distance transport and the air quality impacts of burning biomass in power stations. Although the UK-wide government identified its value as a transition technology, subsidy incentives are ending in 2021 and tariff cuts have curtailed the number of new projects.

An important aspect of renewable energy projects, particularly in comparison with a fossil fuel-based system, is their degree of local ownership. This covers projects owned by Welsh households, communities, local authorities, housing associations, charities (including religious organisations), education establishments, and Wales-based businesses, farms and estates. Data sources usually underestimate their numbers, but locally owned renewable energy project accumulated to an amount of over 64,000 in 2018, with a total capacity of 540 MWel and 243 MWth. This is respectively 17% and 37% of the total renewable electrical and heat energy capacity. In comparison, 19% of gas and coal electricity generation capacity is locally owned. One of the key principles of the Welsh energy strategy is the support of local ownership, in order to reduce costs, climate change effects and reliance on import. In addition, community ownership delivers key opportunities for changing local attitudes towards new renewable energy developments, such as wind farms. Concrete local ownership targets include:

- all new energy projects in Wales should have an element of local ownership from 2020;
- at least 1 GW of renewable energy capacity should be locally owned by 2030;
- concrete investment in public sector organisations, using interest-free loans, to finance renewable energy installations (in particular solar PV).

5 NEXT STEPS

TRACER will continue working on a key challenge to mobilise a wide range of stakeholders in 9 coal regions across Europe and discuss further steps for just energy transition. The project will be finalised in September 2022.

ACKNOWLEDGEMENTS

The authors would like to thank the European Commission and the CINEA for the support of the TRACER project. TRACER has received funding from the European Union's Horizon 2020 research and innovation programme under grant agreement No. 836819.

Disclaimer: The sole responsibility for the content of this paper lies with the authors. It does not necessarily reflect the opinion of the European Union. Neither the CINEA nor the European Commission are responsible for any use that may be made of the information contained therein.

Simple and Smart Communities for All

Rita Mergner[1], Dominik Rutz[1]
[2]Angela Holzmann, [2]Kerstin Schilcher, [2]Altan Sahin, [3]Natascha Fenz, [3]Ulfert Höhne, [4]Angel Nikolaev, [4]Mariya Trifonova, [5]Thekla Heinel, [6]Benjamin Dannemann, [7]Eszter Süle, [7]Nikoloz Sumbadze, [8]Slavica Robić, [8]Tijana Simek, [9]Bence Kovács, [10]Bea Lugosi

[1]WIP – Renewable Energies, Sylvensteinstr. 2, D - 81369 Munich, Germany
Tel. +49 89 720 12743, Fax +49 89 720 12791
E-Mail: rita.mergner@wip-munich.de; dominik.rutz@wip-munich.de
Internet: www.wip-munich.de

[2]Austrian Energy Agency (AEA), Austria, [3]OurPower Energiegenossenschaft SCE (OurPower), Austria, [4]Black Sea Energy Research Centre (BSERC), Bulgaria, [5]B.&S.U. Beratungs- und Service-Gesellschaft Umwelt mbH (BSU) Germany, [6]Deutscher Genossenschafts- und Raiffeisenverband (DGRV), Germany, [7]Association of Young Professionals in Energy of Georgia (AYPEG), Georgia, [8]North-West Croatia Regional Energy Agency (REGEA), Croatia, [9]National Society of Conservationists – Friends of the Earth Hungary (MTVSZ), Hungary, [10]Reflex Environmental Association (Reflex), Hungary

ABSTRACT: Increased energy efficiency and/or optimised energy management and/or integrate a higher share of renewables. Thus, the term energy communities in SHAREs refers to all forms of collective actions by and for consumers such as cooperatives, collective purchase groups or other consumer-driven collective actions. Six countries are involved in the project implementation: Austria, Germany, Croatia, Bulgaria, Georgia and Hungary. In order to achieve a clean energy transition, all consumer groups need to be involved and convinced to contribute to the energy market. Founders of energy communities are by definition very motivated and "on-board" but often lack the tools to engage with other consumer groups. The concept of the SHAREs project is simple and straightforward - to create a Gateway for those interested in setting up collective actions to help them initiate an action and grow. The Gateway will consist of: (1) A country-specific implementation toolkit to equip local heroes with the technical and logistic capacity to set up their energy community (legal framework, such as master contracts, technical and IT solutions, business models, etc.); (2) The building blocks of a tailored "pick-and-mix" communication campaign to enable local heroes to promote their idea effectively to their most relevant stakeholder groups. The Gateway will combine the insights and tools from existing initiatives as well as input of established energy communities (pioneers) and policy developers. After having been put to test and refined in at least 20 aspiring energy communities in 6 European countries, the Gateway will be broadly disseminated to relevant stakeholders all over Europe.

Keywords: prosumers, local heroes, energy community, communication campaign.

1 INTRODUCTION

Energy communities bear the great potential to enable European citizens to produce, consume, store and sell their own energy. They are member-based organisations whose primary purpose is to provide environmental, social and economic benefits to community members or to the local areas in which they operate. A key aspect of energy communities is the collective ownership and management of the respective facility(ies).

Since the main purpose of an energy community is to generate benefits for the community, energy communities usually reinvest a large part or all of the profit in their operations before distributing any kind of dividend to their members.

Today, there are an estimated 3,500 energy cooperatives in Europe, representing over one million EU citizens. As shown in the map below, only a minority of them can be found in Eastern Europe because of several factors, including financial constraints, unsupportive legal frameworks and non-liberalised energy markets[1].

In countries where energy communities have a wide presence, dedicated cooperative networks have been established to monitor the yearly progress in the country (i.e., DGRV in Germany[2], Hier Opgewekt in the Netherlands[3], while in Austria a country-wide advisory network is currently being created with the backing of existing regional energy agencies and institutes to support the uptake of energy communities[4]).

Research conducted by CE Delft in 2016[5] revealed that over 264 million European citizens (which amounts to half of the European population) could be producing their own energy by 2050 at home or together in energy communities. These energy citizens could be producing 611 TWh of electricity by 2030 and 1,557 TWh by 2050. This means that by 2030, energy citizens could be delivering 19% of Europe's electricity demand, and 45% by 2050. This is a significant contribution to achieving the EU's 2030 renewable energy target and moving towards a 100% renewable future.

2 THE SHARES PROJECT

The Clean Energy for all Europeans package paves the way for an energy revolution by putting the consumer in the driving seat of the green transition process. Energy communities offer the unique opportunity for consumers to form critical mass and become renewable energy producers and offer their demand flexibility to the

[1]Stanislas d'Herbemont and Compile project, 2021, "D4.1 The Energy Community Toolkit"
[2]DGRV

[3]Hier Opgewekt
[4]Österreichische Koordinationsstelle für Energiegemeinschaften
[5] B. Kampman, 2016, "The potential of energy citizens in the European Union"

market. However, the current requirements of becoming a prosumer bear a crucial risk that ultimately there will be a two-speed energy world: those who have the knowledge, access and opportunity to become a prosumer, and those who lack the resources, know-how, education or time to participate in and benefit from the new opportunities. Therefore, there is a clear need to support local heroes (i.e. those interested in setting up collective actions) in establishing their energy community and enable them to motivate and target consumers directly. This is particularly important for countries where energy communities are yet to be established.

The SHAREs project supports the set-up of new energy communities as well as other collective actions and helps existing energy communities to grow on a large scale. It aims to equip local heroes – organisations, public authorities or individuals who push the initiation of energy communities – with a country-specific implementation toolkit and a communication campaign that allows them to launch and expand a large variety of collective actions in a simple and smart way.

Figure 1: SHAREs target countries (in green)

3 ENERGY LIFESTYLES

Research on energy-related lifestyles tries to find out whether and to what extent different lifestyles explain the energy-related behaviour of different individuals. In recent years, interest in the role of citizens or consumers as central agents in systemic change has increased in studies of sustainability transitions. As the "energy lifestyle" of a person we mean activities that are part of the person's private sphere of influence and have an impact on this person's energy-related behaviour and climate impact.[6]

Analysis of energy lifestyles enables a target group-specific presentation of the attitudes towards the energy transition and the personal contribution in the form of CO_2 emissions as well as the willingness to change the behaviour. A closer look at energy lifestyles helps to uncover the problems in the implementation of the energy transition. In addition, points of criticism can also be better understood. Finally, they provide a differentiated social context of attitudes towards the energy transition and energy conservation.

There are certain driving factors behind different patterns of energy and climate relevant behaviour, therefore it is important to understand them and develop different approaches that help to change lifestyles in a sustainable direction. In the SHAREs project we aim to build on available research results and develop targeted communication strategies to reach different consumer groups and motivate them to join community energy and collective actions. Energy lifestyles are practical, so that target group-specific recommendations for action can be made and obstacles to action can be addressed. This will help to improve communication with different citizens' groups leading to a higher engagement in energy communities and collective actions.

Different energy lifestyles show that messaging around environmental friendliness and climate change will resonate rather with "green advocates" but will not engage other groups that consider other aspects more important (e.g., saving money, increasing their comfort of life etc.). There are different competing priorities in each energy lifestyle. Therefore, for a broader communication campaign many different messages should be communicated to cover all different energy lifestyles (environment, costs, attractiveness of technology, home improvement). Figure 1 shows five different customer clusters based on their energy lifestyle.[7]

Community based initiatives (collective actions, community energy) could lead to long-term behaviour change because they facilitate the introduction of new, pro-environmental social norms. The literature research shows that for these types of initiatives to be successful, it is important that they are part of a wider programme that has clear objectives. These could include reducing the environmental footprint or delivering energy savings. A pre-existing relationship between the participants and whether participants share pro-environmental views also contribute to the success of such measures. Such initiatives are likely to be more successful if they involve a role model respected by the selected group. Many successful community-based activities have a local "champion" driving the initiative, often a trusted and well-known figure in the local community.[8]

It is important to mention that there is a certain percentage of the human population who could be regarded as "hard-to-reach energy users". These are the people that policymakers and research experts are often struggling to engage with. This is despite many efforts and interventions to change behaviours around energy-efficient technology uptake and management, as well as energy consumption. This audience's barriers are often related to trust, which is why it is important to identify

[6]Schwarzinger, S.; Bird, D.N.; Skjølsvold, T.M. Identifying Consumer Lifestyles through Their Energy Impacts: Transforming Social Science Data into Policy-Relevant Group-Level Knowledge. Sustainability 2019, 11, 6162. https://doi.org/10.3390/su11216162

[7]Using a consumer-segmentation approach to make energy-efficiency gains in the residential market https://www.mckinsey.com/~/media/mckinsey/dotcom/client_service/epng/pdfs/using_a_consumer_segmentation_approach_to_make_ee_gains.pdf

[8]Achieving energy efficiency through behaviour change: what does it take? (EEA Technical report No 5/2013) (europa.eu)

and engage trusted messengers (often from the community or not-for-profit sector).[9]

4 BENEFITS OF ENERGY COMMUNITIES

There are numerous reasons for citizens and households to join a community energy project, as listed by REScoop, the European federation of citizen energy cooperatives[10] and reported below. Energy communities initiators should leverage on raising awareness about such benefits in order to attract more members.

You'll be building the system needed to stop the climate crisis. Community-owned renewable energy projects significantly cut carbon emissions by replacing fossil fuels. They also put renewable energy into the hands of people, making possible an energy system that is 100% renewable, democratically owned and doesn't compromise the well-being of future generations.

Community renewables redirect money supporting the fossil fuel system. By joining a renewable energy community, you will stop paying bills to energy companies and utilities that might burn fossil fuels, and the benefits of the electricity that you consume will go to a community that you are part of, and you will have a say in how the money will be used.

It can reduce energy poverty in your area (depending on legal conditions). Many community-owned energy projects provide an allowance of electricity at low cost to the people involved. When communities own the means to produce their own energy, they have more control over the costs, don't have an incentive to overcharge people, and don't demand higher prices like big energy companies.

You'll get to know your neighbours and strengthen your community. Communities that have worked together on a clean energy project are more likely to set in motion other projects which also benefit their community.

You will produce your own renewable energy. When you participate in community energy project, you own the energy production resources along with the other members of the community. This means that you will have freedom to choose the type of technology you want to use (solar panels, windmills, biomass, etc.), where to install it, the price at which you sell the electricity, etc. The profits of producing renewable energy will also belong to the community.

It creates spaces to educate people on the issues of energy, climate and democracy. Many energy communities provide education and training for their members and the general public. The sessions are usually around the electricity bill, energy saving, energy efficiency, self-production of renewable energy, climate action, sustainable life, democratic forms of organisation, soft skills, etc. These actions raise awareness about the energy transition in the community and empower citizens to take action.

It keeps the money locally in your community. Research shows that community energy projects generate 2 to 8 times more local revenue than a project carried out by an external actor.

You will show other communities what is possible. A successful project increases the confidence in community energy and it is an inspiring and guiding example for other citizens to launch their own initiative or to join an existing one.

It will contribute to creating a more local, more circular economy. Because they are local, communities are more likely to support local or regional manufacturers applying high social and environmental standards and reinvest revenues in local projects.

You will be building the kind of world you want to see. Community energy can help find a new balance between local economies and the global economy. It can help overcome the urban and rural divide, and close the gap between north and south, between rich and poor - because it empowers local people.

5 SHARES COMMUNICATION STRATEGY FOR LOCAL HEROES

A socially fair energy transition means putting renewable energy into the hands of communities and citizens. Through the Clean energy for all Europeans package, the EU has introduced the concept of energy communities in its legislation, notably as citizen energy communities and renewable energy communities. Across Europe, energy communities are gaining momentum. A first step forward, still there's a lot to be done. Not only implementing EU directives into national law, but also raising public awareness, giving information, and doing convincing work. An energy transition throughout Europe can only succeed if people know about their own power in this process. This is where the SHAREs project picks up.

The SHAREs project turns people on the spot, the so-called local heroes, into multipliers who proactively promote the energy transition. Those "ambassadors of the energy transition" are well-informed and in position to encourage the establishment of energy communities in their municipalities. The local heroes know about contact persons, how they can reach them and, most importantly, oversee the relevant communication materials.

Based on literature analysis and discussions with consortium partners, 14 target groups were defined and dedicated communication profiles were elaborated.

ACKNOWLEDGEMENTS

The authors would like to thank the European Commission for the support of the SHAREs project.

Disclaimer: This project has received funding from the European Union's Horizon 2020 Research and innovation program under grant agreement No 101033722.

Neither the European Commission nor any person acting on behalf of the Commission is responsible for the use which might be made of the following information. The views expressed in this publication are the sole

[9]Hard-to-Reach Energy Users – UsersTCP

[10]Güell Oscar, 2021, "Ten reasons to start or join a community energy project"

responsibility of the author and do not necessarily reflect the views of the European Commission.

Reproduction and translation for non-commercial purposes are authorised, provided the source is acknowledged.

Website: https://shares-project.eu/

SHAREs
empowering communities

HOW TO MAKE A SOLAR PV-BASED ENERGY COMMUNITY ECONOMICALLY ATTRACTIVE IN BELGIUM?

Philippe Macé[1], Elina Bosch[1], Monica Aleman[1]
[1] Becquerel Institute
Rue Royale 146, 1000 Brussels Belgium

ABSTRACT: Over the last years, the concept of energy communities has gained popularity. They are increasingly perceived as a citizen empowerment opportunity, and a mean to involve them in renewable energy deployment. Yet, deployment of energy communities remains limited. Among the explanatory factors of this observation, one can cite the lack of appropriate legislation, but also the lack of knowledge (of citizen but also professionals) around new forms of business models, with an ensured added value for all stakeholders.
The research results presented here have been obtained through the work conducted in R2EC (Regional Renewable Energy Cells) ERANet project, which intends to evaluate revenue model(s) for an energy community (EC) created in a neighborhood of the municipality of Flobecq (Wallonia, Belgium). The objective is also to evaluate to which extent this model could be replicated in similar settings, elsewhere in Wallonia.
Results show that the local regulation defining renewable energy communities should be adapted, as it is currently too restrictive and prevents sufficient economic value creation to make shared energy generation concepts viable. Peer-to-peer exchange, currently excluded from regulatory frameworks, was identified by our research as a non-optimal business model given the current context in Wallonia (i.e., net-metering).
Keywords: Economic Analysis, Modelling, PV System

1 AIM AND APPROACH

The purpose of this paper is to identify the most promising revenue models for the investigated case study, a neighborhood in the municipality of Flobecq (Wallonia, Belgium), that could be replicated in similar settings. To conduct the analysis of the selected business cases, various elements were defined. Key ones are presented in the rest of this section.

1.1 The participants of the energy community

The participating households have been defined based on two main parameters. Namely:

- Presence (and size if any) of individual PV system on their roof
- Annual electricity consumption, including the consumption for space heating and domestic hot water if covered by electricity (e.g., electric boiler or heat pumps) as well as for electricity car charging if any.

These different parameters allowed to define categories of households. Each category being defined by a certain value or value range for the two abovementioned parameters.

This categorization permits to make conclusions with regards to the interest in participating in an energy community for a large, yet manageable, number of households, based on their production and consumption profile

Note that the profiles of the participants and the household categories have been defined based on actual historical data received from the local DSO (ORES) with the consent of inhabitants of the municipality.

1.2 The composition of the energy community

The created techno-economic model allows to model energy communities in the frame of different evolution scenarios for the studied municipality. These scenarios refer to the level of PV penetration, and the electrification rate.

The PV penetration level is directly linked to the share of households inside the studied energy community equipped with an individual PV system. The electrification rate level is directly linked to the share of households inside the studied energy community having a heat pump and/or one or more electric vehicle(s).

For each considered scenario of PV penetration (Base Case (BC), Accelerated (Acc.) and Massive (Mass.)) and electrification rate (Base Case (BC), Accelerated (Acc.) and Massive (Mass.)), the energy communities pools are constituted accordingly (See Table I).

Table I: Deployment of PV, EV and HP in the different considered scenarios.

PV penetration scenario	Share of households equipped with PV
BC	≈ 35
Acc.	≈ 55
Mass.	≈ 70

Electrification scenario	Share of households equipped with an EV
BC	≈ 1
Acc.	≈ 10
Mass.	≈ 20

Electrification scenario	Share of households equipped with a HP
BC	≈ 2
Acc.	≈ 13
Mass.	≈ 20

Regardless of the considered PV penetration and electrification scenario, a sport center which hosts the 200 kWp common PV system is included in the theoretically energy community.

1.3 Distribution keys

To determine the order of priority by which the energy community members can have access to excess PV production, a distribution key must be defined. A dynamic distribution key was tested here. In other words, at each given point in time, residual production is distributed proportionally to each member's residual demand.

1.4 Applied business models

Two different main business models have been considered.

The first one is based on peer-to-peer exchange. Electricity produced by the different individual PV systems of the community is exchanged inside the community. Any individual PV system owner can sell electricity they did not self-consume at a price higher than the wholesale market price, but lower than the variable retail price so that neighbors buying this electricity have an economic incentive to do so. Based on current observed conditions for EC, the assumption is made that individual PV system owner can no longer be eligible to the local support scheme (net-metering) if they participate in the community. This concept is currently not allowed in the regulation in Wallonia. Nevertheless, it is still relevant to mention as with the end of the net-metering, such concept could become possible as part of the regulation.

The second business model is based on shared generation. Electricity produced by a collectively owned, larger, rooftop PV system is used inside the community. Any member of the community can buy electricity from this PV installation at a price slightly lower than the variable retail price. A community manager (external or internal to the community) bears the investment cost in the PV system, the operation and maintenance costs and the energy community management costs. Each kWh sold to an EC member generates a revenue for the EC manager, amounting to a value higher than the wholesale price. Individual PV system owners behave as they would without the community according to local support schemes.

1.5 Energy flows within the community

For each member of the energy community, annual 15-min time series for their individual PV production profile (if any) and their consumption profile are simulated for 20 years, based on historical real data.

In addition, different "energy" business models related to the energy flows can be studied. These "energy" business models define whether some energy exchanges between EC members are allowed or not, i.e. peer-to-peer.

Combining these elements, the analysis allows to determine for each member of the community: (1) How the coverage of its annual consumption is distributed between individual self-consumption, self-consumption from the common PV system and the grid; and (2) How its individual production (if any) is distributed between individual self-consumption, distribution to other energy community members (if possible) and injection to the grid.

1.6 Cash flows

The quantification of the cash flows relies on the association of a value to each of the energy flows determined above. The value associated with each energy flow is function of the considered business model. The latter defines how value is created in the energy community and by whom this value is captured. In particular, the business model defines the price at which the electricity produced inside the community is sold and bought.

1.7 Final KPIs

There are several aspects and indicators which can allow to identify whether creating an EC is attractive or not, and what are the most influencing factors. The first aspect is economic. It consists in evaluating, for each EC participant (i.e., in the studied case: the participating households, the sport center and the community manager) the NPV (Net Present Value) for all energy community-related expenses and revenues. A positive NPV would indicate that the participant's situation has improved with the EC compared to the case without EC.

In addition, or as an alternative, this economic assessment can be conducted at the scale of the EC. In this case the assessment is conducted on the sum of all energy community-related expenses and revenues (i.e., in the studied case: for the participating households, the sport center and the community manager combined). The drawback of this collective assessment when conducted alone, is that it does not allow to identify if the created value is corned by a few participants only or if it is distributed relatively evenly.

Other indicators such as the total environmental footprint for all energy-related uses, the average individual or total collective self-consumption and self-sufficiency can also allow to assess the attractiveness of the EC.

2 MAIN ASSUMPTIONS

2.1 Energy flows

Theoretical ECs of 100 households and 1 sport center are formed. Different proportions (Base case (BC), Accelerated (Acc.), Massive (Mass.)) of individual PV systems, electric vehicles (EVs) and heat pumps (HPs) are considered (See Table I). Electricity exchanged within the EC follows a one-to-many scheme and comes from a 200 kWp PV system ((S) in Figure 1) installed on the sport center's roof when an EC is considered.

In addition, a fully proportional distribution key is considered. [1]

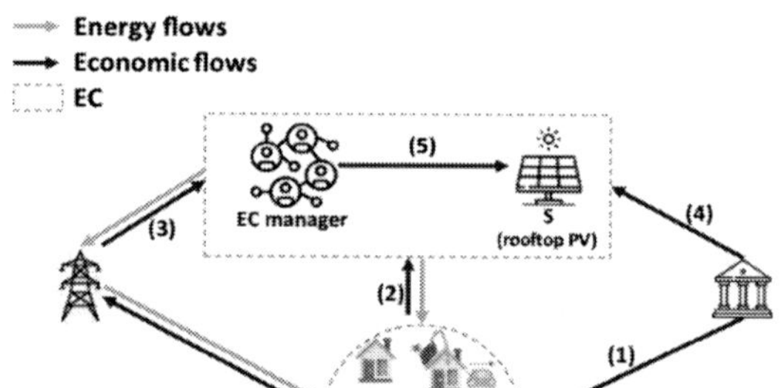

Figure 1: Schematic overview of energy flows between the different entities and economic flows between the different stakeholders in the energy community under the shared generation business model.

2.2 Economic flows*

There are multiple revenues and cost streams in the modelled energy community. Consuming electricity from S enables savings on the electricity bill amounting to the GEC (2,5 c€/kWh) ((1) in Figure 1) [2]. Injecting electricity from S to the community enables revenues amounting to the energy component (7c€/kWh) ((2) in Figure 1). Injecting electricity from S to the grid enables revenues equal to the wholesale electricity price (4,5 c€/kWh) ((3) in Figure 1) [3]. Producing electricity from S enables revenues amounting to green certificates value for 10 years (4,2 c€/kWh) ((4) in Figure 1) [3]. Covering investment in/maintenance of the 200 kWp PV system (0,8 €/Wp – 8 €/kWp.a) ((5) in Figure 1).

Energy community-related costs such as the investment in the control devices and the annual management costs are treated separately as they are associated with a high level of uncertainty because of the limited EC deployment observed yet.

*Early 2021 data

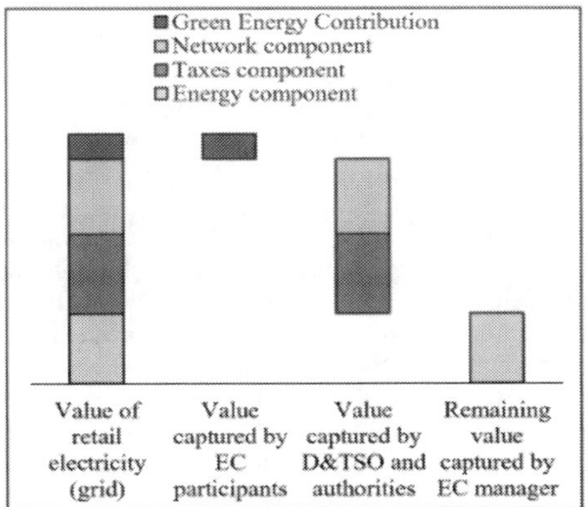

Figure 2: Visual representation of the different revenue streams for the EC participants and for the EC manager from PV electricity produced and exchanged within the community.

3 RESULTS & SENSITIVITY ANALYSIS

First of all, results concerning the peer-to-peer business model are not presented extensively here. Given the existence of net-metering in Wallonia, and the very high economic attractiveness of this scheme for excess electricity (which enables the same savings as self-consumed electricity) and despite the existence of a prosumer tax, the peer-to-peer is currently an unattractive business model from an economic point of view. Moreover, this concept is currently not included in the proposition of regulatory framework for energy communities in Wallonia.

3.1 Main results for the shared generation business model

Presented results are expressed in terms of Net Present Value (NPV) calculated over 20 years and accounting for all revenues and costs associated with the formation of an EC compared to the case without EC (i.e., without 200 kWp system on the sport center).

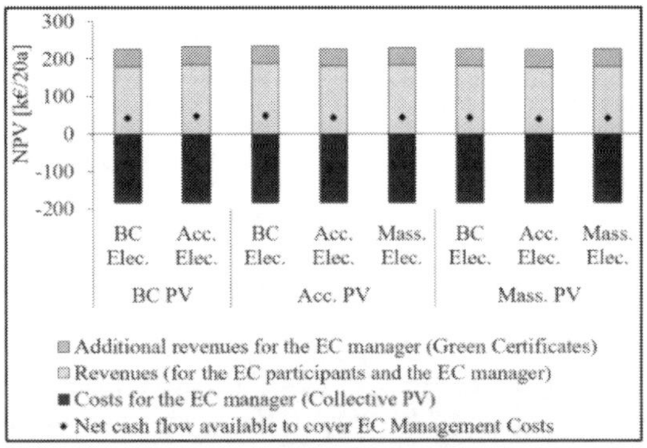

Figure 3: Revenues and costs associated with the formation of an EC and Net Cash Flow available to cover EC Management costs for the different studied EC pools (Note: Discount rate used to discount the households cashflows (respectively the EC manager's cash flows) is 2% (respectively 4%)

Assuming that the energy community manager only targets break-even, the yielded NPV is positive in all cases. With these revenues, some energy community-related costs still need to be covered such as the investment in the control devices and the annual management costs. Therefore, a sensitivity analysis is conducted in Table II. This shows that under the three most optimistic assumptions, all cases reach break-even when factoring in EC-related costs. Under 3 additional moderate assumptions, some cases become profitable, while the remaining assumption do not allow to reach break-even. On the 18 EC management costs combinations, only a third are sufficiently low so that break-even at the scale of the whole EC can be reached.

Table II: Sensitivity analysis on total EC Management costs (monthly management costs and investment in a control device for each participant)
(Note: A: assumptions which allow All studied cases to be competitive, S: assumptions which allow Some studied cases to be competitive, N: assumptions which allow None of the studied cased to be competitive.)

		Control device (CAPEX) [€/device]					
		Changed after 10 years			no change		
		1000	500	250	1000	500	250
Management	50	N	N	N	N	N	A
Costs	25	N	N	S	N	N	A
[€/month]	0	N	N	S	N	S	A

3.2 Influence of consumers' profiles on revenues:

Figure 4 shows that there is strong correlation between higher electricity consumption and higher revenues. This is highly explained by the chosen distribution key (i.e., fully proportional to electricity consumption). With a hybrid distribution key[1], this would be altered and would reduce the possible rebound effect of encouraging higher electricity consumption among participants. [1] [2]

Furthermore, EC members with individual PV have more limited benefits from the EC because part of their

[1] "A hybrid distribution key is a mix between a static and dynamic distribution key. With a hybrid distribution key, a certain share of the total electricity production at a time t is distributed according to a static distribution key (e.g. proportional to each participant's initial contribution to the collective investment) and the remaining share of electricity production is distributed dynamically (i.e., purely based on consumption at this time t).

consumption is already covered by individual self-consumption. Eventually, EC members with flexibilities (e.g., EVs) can gain more revenues if associated consumption takes place during the middle of the day (e.g., top two grey bubbles), when the solar PV system produces the most.

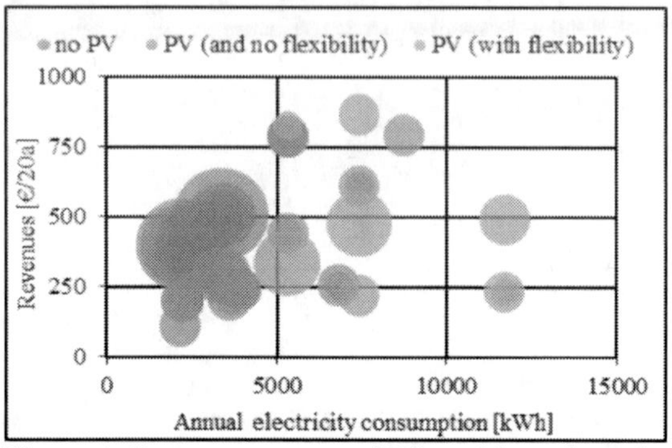

Figure 4: Revenues per type of EC participant (annual electricity consumption, ownership of an individual PV system, ownership of a flexibility (e.g., EV or HP). (Note: Bubble sizes represent the number of participants with this profile)

3.3 Influence of policy framework and electricity prices on profitability

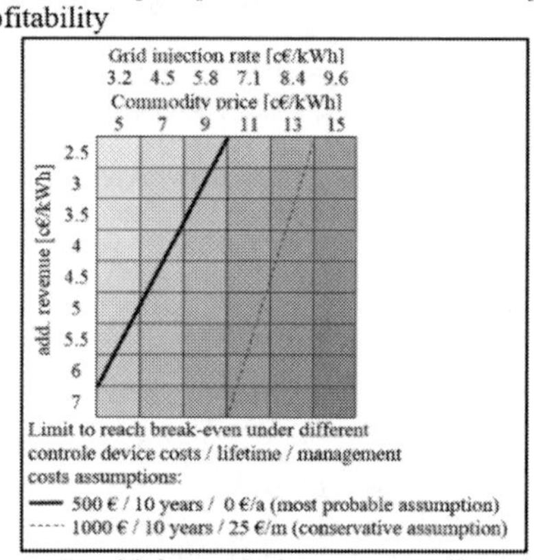

Figure 5: Sensitivity analysis on commodity prices and additional savings.
(Notes: Required combinations to reach break-even under two different EC Management cost assumptions are highlighted with blue lines. Net Cash Flow available to cover EC Management costs range from 10 k€ (top left) to 290 k€ (bottom right)).

From the results of the sensitivity analysis conducted on commodity prices, grid injection rates and additional saving presented in Figure 5, the following remarks can be made:

Even if all combinations to the left of the blue bold line in Figure 5 yield a positive NPV, its value is not sufficient to cover most probable EC management costs, thus not allowing the EC manager to reach break-even.

Unsurprisingly, higher commodity prices and feed-in value increase the profitability of EC. In particular, the higher the gap between these two values, the higher the incentive to self-consume electricity within the EC rather than injecting it to the grid.

Additional revenues which include possible exemptions on the electricity bill (e.g., existing exemption on the GEC or potential exemption on network tariffs) but also EC-specific incentives and can significantly enhance the economic attractiveness of ECs.

4 CONCLUSIONS

The local regulation defining renewable energy communities in Wallonia is currently too restrictive and prevents sufficient economic value creation as it only foresees the exemption of the Green Energy Contribution (GEC) for electricity exchanged within the EC.

Additional revenue streams such as further exemptions (e.g., partial exemption on network tariffs), or EC-specific incentives (e.g., higher green certificate allowance) would greatly enhance the economic attractiveness of ECs in Wallonia.

Investigating ECs' profitability in Wallonia under alternative configurations (e.g., with the inclusion of individual or shared storage, with a hybrid distribution key, with a more diversified consumer profile pool (e.g., commercial consumers) or under a different regulatory framework (e.g., analyzing the attractiveness of P2P after net-metering phase-out in 2024) would be of interest and provides perspectives for future work on this topic.

5 REFERENCES

[1] Apere, "Liste d'eNemples de clés de répartition," 2020.

[2] C. W. P. L. Cwape, interviewee, [interview]. 2021.

[3] Fraunhofer ISE, "Energy Charts," [Online]. Available: https://energy-charts.info/charts/price_spot_market/chart.htm?L=en&c=BE&stacking=stacked_absolute_area&year=2021&interval=year. [Accessed 2021].

[4] Wallonie Energie SPW, "Energie Wallonie," [Online]. Available: https://energie.wallonie.be/fr/certificats-verts.html?IDC=8609. [Accessed 2021].

[5] SIBELGA, "Autoconsommation collective - NOTE ENPLICATIVE DES MÉTHODES DE RÉPARTITION," 2020.

6 FUNDING SOURCES AND ACKOWLEDGMENT

This project (R2EC (Regional Renewable Energy Cells) ERA Net project) has received funding from the SPW (Wallonia) under the ERA-NET SES RegSys programme

We thank the R2EC project partners for their expertise, assistance and work in the research project which were of great help in writing this paper.

AUTHOR INDEX

WIP – Renewable Energies
Sylvensteinstr. 2
81369 Munchen
Germany

ISBN 978-1-7138-7180-4
9 781713 871804

CURRAN ASSOCIATES INC.
proceedings.com